# These in-book study tools help you succeed in the course

INVESTIGATING LIFE

**44.10 The Dorsal Lip Induces Embryonic Organization**
In a classic experiment, Hans Spemann and Hilde Mangold transplanted the dorsal blastopore lip mesoderm of an early gastrula stage salamander embryo. The results showed that the cells of this embryonic region, which they dubbed "the organizer," could direct the formation of an entire embryo.

**HYPOTHESIS** Cytoplasmic factors in the early dorsal blastopore lip organize cell differentiation in amphibian embryos.

**METHOD**
1. Excise a patch of mesoderm tissue from above the dorsal blastopore lip of an early gastrula stage salamander embryo (the donor).
2. Transplant the donor tissue onto a recipient embryo at the same stage. The donor tissue is transplanted onto a region of ectoderm that should become epidermis (skin).

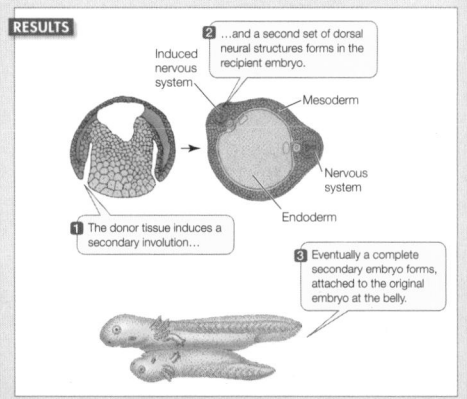



Presumptive mesoderm
Blastocoel

**RESULTS**



Nervous system
Endoderm
1 The donor tissue induces a secondary involution...
3 Eventually a complete secondary embryo forms, attached to the original embryo at the belly.

**CONCLUSION** The cells of the dorsal blastopore lip can induce other cells to change their developmental fates.

Go to **yourBioPortal.com** for original citations, discussions, and relevant links for all INVESTIGATING LIFE figures.

- **INVESTIGATING LIFE** and **TOOLS FOR INVESTIGATING LIFE** figures emphasize the process of scientific inquiry and give you a realistic sense of how science is done.

- **BALLOON CAPTIONS** guide you step by step through experiments and biological processes with clear explanations.

- A **RECAP** at the end of each numbered section offers a very brief summary and 2–3 questions so you can check your mastery of the material you just read.

38.2 RECAP

**Flowering of some angiosperms is controlled by night length, a phenomenon called photoperiodism. Gibberellins can induce flowering in some species, as can exposure to low temperatures (vernalization). Some species flower when their stems have grown by a certain amount, independent of environmental cues. All pathways to flowering converge on the meristem identity genes.**

- What are the differences between apical meristems, inflorescence meristems, and floral meristems? What genes control the transitions between them? See p. 803 and Figure 38.10
- Explain why "short-day plant" is a misleading term. See p. 805 and Figure 38.13
- What is the evidence for florigen? What is its molecular mechanism of action? See p. 807 and Figures 38.15 and 38.16

CHAPTER SUMMARY

**25.1 How Do Scientists Date Ancient Events?**
- The relative ages of organisms can be determined by the dating of fossils and the **strata** of **sedimentary rocks** in which they are found.
- Paleontologists use a variety of radioisotopes with different **half-lives** to date events at different times in the remote past. Review Figure 25.1
- Geologists divide the history of life into eras and periods, based on major differences in the fossil assemblages found in successive layers of rocks. Review Table 25.1

**25.2 How Have Earth's Continents and Climates Changed over Time?**
- Earth's crust consists of solid lithospheric plates that float on fluid magma. **Continental drift** caused by convection currents in the magma moves these plates and the continents that lie on top of them. Review Figure 25.2, ANIMATED TUTORIAL 25.1
- Conditions on E[illegible] ...d dramatically over time...
- Major physical events on Earth, such as the collision of continents that formed the supercontinent **Pangaea**, have affected Earth's surface, climate, and atmosphere. In addition, extraterrestrial events such as meteorite strikes created sudden and dramatic environmental shifts. All of these changes have affected the history of life.

**25.3 What Are The Major Events in Life's History?**
- Paleontologists use fossils and evidence of geological changes to determine what Earth and its **biota** may have looked like at different times.
- During most of its history, life was confined to the oceans. Multicellular life diversified extensively during the **Cambrian explosion**. Review Figure 25.11
- The periods of the Paleozoic era were each characterized by the diversification of specific groups of organisms. **Amniotes**—vertebrates whose eggs can be laid in dry places—first appeared during the Carboniferous period.

- The **CHAPTER SUMMARY** provides a thorough review of the chapter's concepts and key terms, suggests key figures to review, and reminds you of Web Activities and Animated Tutorials that support the chapter.

ADDITIONAL INVESTIGATION

1. The interpretation of Pasteur's experiment (see Figure 4.7) depended on the inactivation of microorganisms by heat. We now know of microorganisms that can survive extremely high temperatures (see Chapter 26). Does this change the interpretation of Pasteur's experiment? What experiments would you do to inactivate such microbes?
2. The Miller–Urey experiment (see Figure 4.9) showed that it was possible for amino acids to be formed from gases that were hypothesized to have been in Earth's early atmosphere. These amino acids were dissolved in water. Knowing what you do about the polymerization of amino acids into proteins (see Figure 3.6), how would you set up experiments to show that proteins can form under the conditions of early Earth? What properties would you expect of those proteins?

- **ADDITIONAL INVESTIGATION** offers thought-provoking questions based on material in the chapter, and asks you to formulate hypotheses and design experiments to test them.

## ALSO IN EACH CHAPTER

Presumptive epidermis
Primary involution (recipient of dorsal lip)

Induced nervous system
2 ...and a second set of dorsal neural structures forms in the recipient embryo.

- **IN THIS CHAPTER** provides you with a concise preview and roadmap of the chapter.
- A **CHAPTER OUTLINE** introduces the major section headings, all of which are posed as questions to get you started in seeing the study of biology as an inquiry-based experience.
- Each chapter opens with an **INTRODUCTORY STORY** that provides a historical, medical, or social context for the chapter subject. Chapters revisit these stories to drive home their relevance.
- **SELF-QUIZZES** at the end of each chapter (with all answers in the back of the book) offer an opportunity to check your mastery of the material.

# *Life*

The Science of Biology

NINTH EDITION

Sinauer Associates, Inc.

W. H. Freeman and Company

NINTH EDITION

# Life
## The Science of Biology

DAVID **SADAVA**
The Claremont Colleges
Claremont, California

DAVID M. **HILLIS**
University of Texas
Austin, Texas

H. CRAIG **HELLER**
Stanford University
Stanford, California

MAY R. **BERENBAUM**
University of Illinois
Urbana-Champaign, Illinois

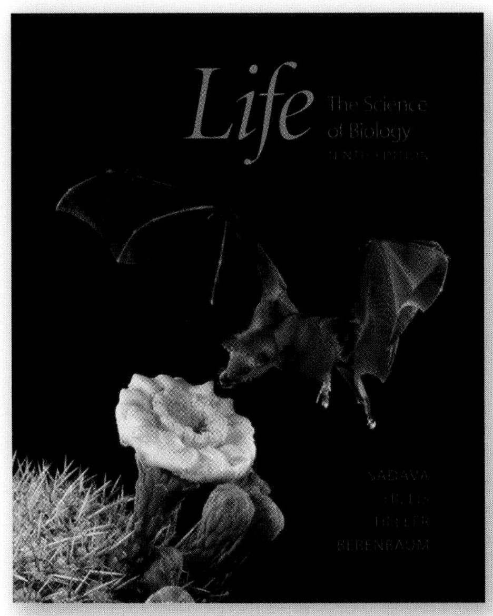


## About the Cover

The cover of *Life* captures many themes that echo throughout the book. The photograph shows a lesser long-nosed bat pollinating a saguaro cactus. This cactus has evolved large flowers that produce copious quantities of nectar. The nectar attracts many species that pollinate the cactus, including bats. The ability of bats to hover as they feed on the nectar of the cactus is an excellent example of adaptation of body form and physiology. These themes of adaptation, evolution, nutrition, reproduction, species interactions, and integrated form and function are ideas that are repeated throughout the chapters of *Life*. Photograph copyright © Dr. Merlin D. Tuttle/Photo Researchers, Inc.

## The Frontispiece

Blue wildebeest and Burchell's zebra migrate together through Serengeti National Park, Tanzania. Copyright © Art Wolfe, www.artwolfe.com.

**LIFE: The Science of Biology,** Ninth Edition



Address editorial correspondence to:
Sinauer Associates Inc., 23 Plumtree Road, Sunderland, MA 01375 U.S.A.
www.sinauer.com
publish@sinauer.com

Address orders to:
MPS / W. H. Freeman & Co., Order Dept., 16365 James Madison Highway,
U.S. Route 15, Gordonsville, VA 22942 U.S.A.
Examination copy information: 1-800-446-8923
Orders: 1-888-330-8477

**Library of Congress Cataloging-in-Publication Data**

Life, the science of biology / David Sadava .. [et al.]. — 9th ed.
p. cm.
Includes index.
ISBN 978-1-4292-1962-4 (hardcover) — 978-1-4292-4645-3 (pbk. : v. 1 ) —
ISBN 978-1-4292-4644-6 (pbk. : v. 2 ) — ISBN 978-1-4292-4647-7 (pbk. : v. 3)
1. Biology. I. Sadava, David E.
QH308.2.L565 2011
570—dc22 2009036693

Printed in U.S.A.
First Printing October 2009
The Courier Companies, Inc.

*To Bill Purves and Gordon Orians,*
*extraordinary colleagues, biologists, and teachers,*
*and the original authors of LIFE*

# The Authors

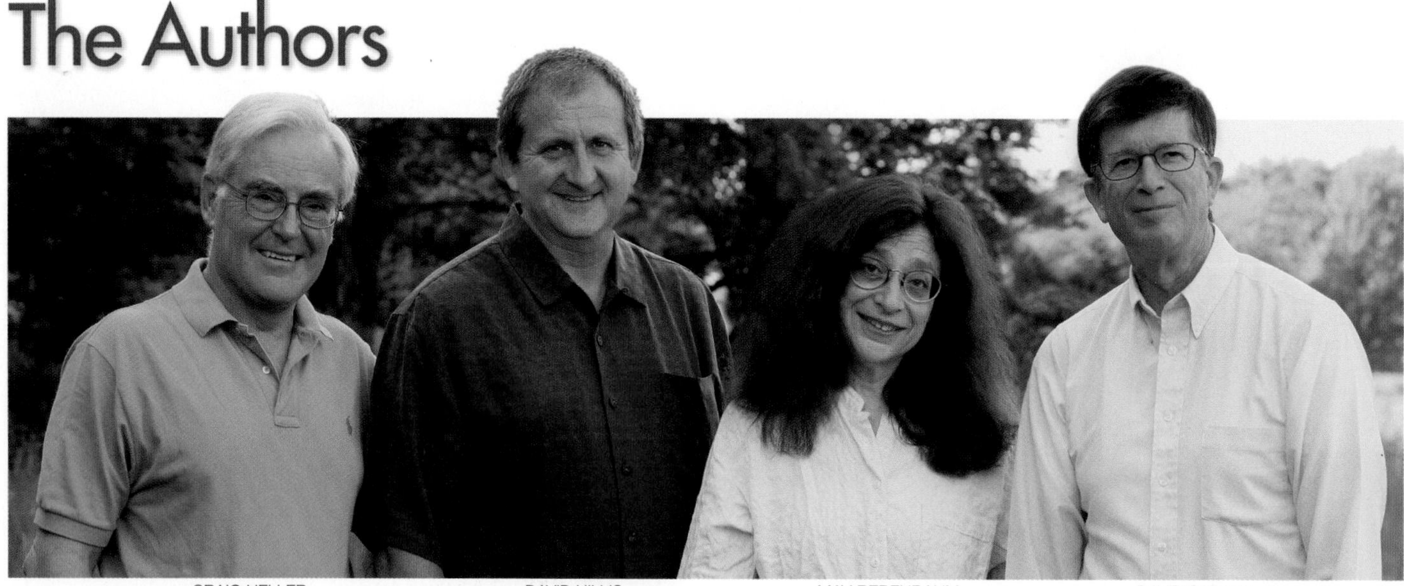

CRAIG HELLER DAVID HILLIS MAY BERENBAUM DAVID SADAVA

**DAVID SADAVA** is the Pritzker Family Foundation Professor of Biology, Emeritus, at the Keck Science Center of Claremont McKenna, Pitzer, and Scripps, three of The Claremont Colleges. In addition, he is Adjunct Professor of Cancer Cell Biology at the City of Hope Medical Center. Twice winner of the Huntoon Award for superior teaching, Dr. Sadava taught courses on introductory biology, biotechnology, biochemistry, cell biology, molecular biology, plant biology, and cancer biology. In addition to *Life: The Science of Biology*, he is the author or coauthor of books on cell biology and on plants, genes, and crop biotechnology. His research has resulted in many papers coauthored with his students, on topics ranging from plant biochemistry to pharmacology of narcotic analgesics to human genetic diseases. For the past 15 years, he has investigated multi-drug resistance in human small-cell lung carcinoma cells with a view to understanding and overcoming this clinical challenge. At the City of Hope, his current work focuses on new anti-cancer agents from plants.

**DAVID HILLIS** is the Alfred W. Roark Centennial Professor in Integrative Biology and the Director of the Center for Computational Biology and Bioinformatics at the University of Texas at Austin, where he also has directed the School of Biological Sciences. Dr. Hillis has taught courses in introductory biology, genetics, evolution, systematics, and biodiversity. He has been elected into the membership of the National Academy of Sciences and the American Academy of Arts and Sciences, awarded a John D. and Catherine T. MacArthur Fellowship, and has served as President of the Society for the Study of Evolution and of the Society of Systematic Biologists. His research interests span much of evolutionary biology, including experimental studies of evolving viruses, empirical studies of natural molecular evolution, applications of phylogenetics, analyses of biodiversity, and evolutionary modeling. He is particularly interested in teaching and research about the practical applications of evolutionary biology.

**CRAIG HELLER** is the Lorry I. Lokey/BusinessWire Professor in Biological Sciences and Human Biology at Stanford University. He earned his Ph.D. from the Department of Biology at Yale University in 1970. Dr. Heller has taught in the core biology courses at Stanford since 1972 and served as Director of the Program in Human Biology, Chairman of the Biological Sciences Department, and Associate Dean of Research. Dr. Heller is a fellow of the American Association for the Advancement of Science and a recipient of the Walter J. Gores Award for excellence in teaching. His research is on the neurobiology of sleep and circadian rhythms, mammalian hibernation, the regulation of body temperature, the physiology of human performance, and the neurobiology of learning. Dr. Heller has done research on a huge variety of animals and physiological problems ranging from sleeping kangaroo rats, diving seals, hibernating bears, photoperiodic hamsters, and exercising athletes. Some of his recent work on the effects of temperature on human performance is featured in the opener to Chapter 40, "Physiology, Homeostasis, and Temperature Regulation."

**MAY BERENBAUM** is the Swanlund Professor and Head of the Department of Entomology at the University of Illinois at Urbana-Champaign. She has taught courses in introductory animal biology, entomology, insect ecology, and chemical ecology, and has received awards at the regional and national level for distinguished teaching from the Entomological Society of America. A fellow of the National Academy of Sciences, the American Academy of Arts and Sciences, and the American Philosophical Society, she served as President of the American Institute for Biological Sciences in 2009. Her research addresses insect–plant coevolution, from molecular mechanisms of detoxification to impacts of herbivory on community structure. Concerned with the practical application of ecological and evolutionary principles, she has examined impacts of genetic engineering, global climate change, and invasive species on natural and agricultural ecosystems. Devoted to fostering science literacy, she has published numerous articles and five books on insects for the general public.

# Contents in Brief

# Investigating Life/Tools for Investigating Life

## INVESTIGATING LIFE

## TOOLS FOR INVESTIGATING LIFE

# Preface

Biology is a dynamic, exciting, and important subject. It is dynamic because it is constantly changing, with new discoveries about the living world being made every day. (Although it is impossible to pinpoint an exact number, approximately 1 million new research articles in biology are published each year.) The subject is exciting because life in all of its forms has always fascinated people. As active scientists who have spent our careers teaching and doing research in a wide variety of fields, we know this first hand.

Biology has always been important in peoples' daily lives, if only through the effects of achievements in medicine and agriculture. Today more than ever the science of biology is at the forefront of human concerns as we face challenges raised both by recent advances in genome science and by the rapidly changing environment.

*Life's* new edition brings a fresh approach to the study of biology while retaining the features that have made the book successful in the past. A new coauthor, the distinguished entomologist May R. Berenbaum (University of Illinois at Urbana-Champaign) has joined our team, and the role of evolutionary biologist David Hillis (University of Texas at Austin) is greatly expanded in this edition. The authors hail from large, medium-sized, and small institutions. Our multiple perspectives and areas of expertise, as well as input from many colleagues and students who used previous editions, have informed our approach to this new edition.

## Enduring Features

We remain committed to blending the presentation of core ideas with an emphasis on introducing students to the *process of scientific inquiry*. Having pioneered the idea of depicting seminal experiments in specially designed figures, we continue to develop this here, with 79 **INVESTIGATING LIFE** figures. Each of these figures sets the experiment in perspective and relates it to the accompanying text. As in previous editions, these figures employ a structure: Hypothesis, Method, Results, and Conclusion. They often include questions for further research that ask students to conceive an experiment that would explore a related question. Each *Investigating Life* figure has a reference to BioPortal (*yourBioPortal.com*), where citations to the original work as well as additional discussion and references to follow-up research can be found.

A related feature is the **TOOLS FOR INVESTIGATING LIFE** figures, which depict laboratory and field methods used in biology. These, too, have been expanded to provide more useful context for their importance.

Over a decade ago—in *Life's* Fifth Edition—the authors and publishers pioneered the much-praised use of **BALLOON CAPTIONS** in our figures. We recognized then, and it is even truer today, that many students are visual learners. The balloon captions bring explanations of intricate, complex processes directly into the illustration, allowing students to integrate information without repeatedly going back and forth between the figure, its legend, and the text.

*Life* is the only introductory textbook for biology majors to begin each chapter with a story. These **OPENING STORIES** provide historical, medical, or social context and are intended to intrigue students while helping them see how the chapter's biological subject relates to the world around them. In the new edition, all of the opening stories (some 70 percent of which are new) are revisited in the body of the chapter to drive home their relevance.

We continue to refine our well-received *chapter organization.* The chapter-opening story ends with a brief **IN THIS CHAPTER** preview of the major subjects to follow. A **CHAPTER OUTLINE** asks questions to emphasize scientific inquiry, each of which is answered in a major section of the chapter. A **RECAP** at the end of each section asks the student to pause and answer questions to review and test their mastery of the previous material. The end-of-chapter summary continues this inquiry framework and highlights key figures, bolded terms, and activities and animated tutorials available in BioPortal.

## New Features

Probably the most important new feature of this edition is *new authorship.* Like the biological world, the authorship team of *Life* continues to evolve. While two of us (Craig Heller and David Sadava) continue as coauthors, David Hillis has a greatly expanded role, with full responsibility for the units on evolution and diversity. New coauthor May Berenbaum has rewritten the chapters on ecology. The perspectives of these two acclaimed experts have invigorated the entire book (as well as their coauthors).

Even with the enduring features (see above), this edition has a different look and feel from its predecessor. A fresh *new design* is more open and, we hope, more accessible to students. The extensively *revised art program* has a contemporary style and color palette. The information flow of the figures is easier to follow, with numbered balloons as a guide for students. There are new conceptual figures, including a striking visual timeline for the evolution of life on Earth (Figure 25.12) and a single overview figure that summarizes the information in the genome (Figure 17.4).

In response to instructors who asked for more real-world data, we have incorporated a feature introduced online in the Eighth Edition, **WORKING WITH DATA**. There are now 36 of these exercises, most of which relate to an *Investigating Life* figure. Each is referenced at the end of the relevant chapter and is available online via BioPortal (*yourBioPortal.com*). In these exercises, we describe in detail the context and approach of the research paper that forms the basis of the figure. We then ask the student to examine the data, to make calculations, and to draw conclusions.

We are proud that this edition is a *greener Life*, with the goal of reducing our environmental impact. This is the first introductory biology text to be printed on paper earning the Forest Stewardship Council label, the "gold standard" in green paper products, and it is manufactured from wood harvested from sustainable forests. And, of course, we also offer *Life* as an eBook.

## The Ten Parts

We have reorganized the book into ten parts. **Part One, The Science of Life and Its Chemical Basis**, sets the stage for the book: the opening chapter focuses on biology as an exciting science. We begin with a startling observation: the recent, dramatic decline of amphibian species throughout the world. We then show how biologists have formed hypotheses for the causes of this environmental problem and are testing them by carefully designed experiments, with a view not only to understanding the decline, but reversing it. This leads to an outline of the basic principles of biology that are the foundation for the rest of the book: the unity of life at the cellular level and how evolution unites the living world. This is followed by chapters on the basic chemical building blocks that underlie life. We have added a new chapter on nucleic acids and the origin of life, introducing the concepts of genes and gene expression early and expanding our coverage of the major ideas on how life began and evolved at its earliest stages.

In **Part Two, Cells**, we describe the view of life as seen through cells, its structural units. In response to comments by users of our previous edition, we have moved the chapter on cell signaling and communication from the genetics section to this part of the book, with a change in emphasis from genes to cells. There is an updated discussion of ideas on the origin of cells and organelles, as well as expanded treatment of water transport across membranes.

**Part Three, Cells and Energy**, presents an integrated view of biochemistry. For this edition, we have worked to clarify such challenging concepts as energy transfer, allosteric enzymes, and biochemical pathways. There is extensive revision of the discussions of alternate pathways of photosynthetic carbon fixation, as well as a greater emphasis on applications throughout these chapters.

**Part Four, Genes and Heredity**, is extensively revised and reorganized to improve clarity, link related concepts, and provide updates from recent research results. Separate chapters on prokaryotic genetics and molecular medicine have been removed and their material woven into relevant chapters. For example, our chapter on cell reproduction now includes a discussion of how the basic mechanisms of cell division are altered in cancer cells. The chapter on transmission genetics now includes coverage of this phenomenon in prokaryotes. New chapters on gene expression and gene regulation compare prokaryotic and eukaryotic mechanisms and include a discussion of epigenetics. A new chapter on mutation describes updated applications of medical genetics.

In **Part Five, Genomes**, we reinforce the concepts of the previous part, beginning with a new chapter on genomes—how they are analyzed and what they tell us about the biology of prokaryotes and eukaryotes, including humans. This leads to a chapter describing how our knowledge of molecular biology and genetics underpins biotechnology (the application of this knowledge to practical problems). We discuss some of the latest uses of biotechnology, including environmental cleanup. Part Five finishes with two chapters on development that explore the themes of molecular biology and evolution, linking these two parts of the book.

**Part Six, The Patterns and Processes of Evolution**, emphasizes the importance of evolutionary biology as a basis for comparing and understanding all aspects of biology. These chapters have been extensively reorganized and revised, as well as updated with the latest thinking of biologists in this rapidly changing field. This part now begins with the evidence and mechanisms of evolution, moves into a discussion of phylogenetic trees, then covers speciation and molecular evolution, and concludes with the evolutionary history of life on Earth. An integrated timeline of evolutionary history shows the timing of major events of biological evolution, the movements of the continents, floral and faunal reconstructions of major time periods, and depicts some of the fossils that form the basis of the reconstructions.

In **Part Seven, The Evolution of Diversity**, we describe the latest views on biodiversity and evolutionary relationships. Each chapter has been revised to make it easier for the reader to appreciate the major changes that have evolved within the various groups of organisms. We emphasize understanding the big picture of organismal diversity, as opposed to memorizing a taxonomic hierarchy and names (although these are certainly important). Throughout the book, the tree of life is emphasized as a way of understanding and organizing biological information. A *Tree of Life Appendix* allows students to place any group of organisms mentioned in the text of our book into the context of the rest of life. The web-based version of this appendix provides links to photos, keys, species lists, distribution maps, and other information to help students explore biodiversity of specific groups in greater detail.

After modest revisions in the past two editions, **Part Eight, Flowering Plants: Form and Function**, has been extensively reorganized and updated with the help of Sue Wessler, to include both classical and more recent approaches to plant physiology. Our emphasis is not only on the basic findings that led to the elucidation of mechanisms for plant growth and reproduction, but also on the use of genetics of model organisms. There is expanded coverage of the cell signaling events that regulate gene expression in plants, integrating concepts introduced earlier in the book. New material on how plants respond to their environment is included, along with links to both the book's earlier descriptions of plant diversity and later discussions of ecology.

**Part Nine, Animals: Form and Function,** continues to provide a solid foundation in physiology through comprehensive coverage of basic principles of function of each organ system and then emphasis on mechanisms of control and integration. An important reorganization has been moving the chapter on immunology from earlier in the book, where its emphasis was on molecular genetics, to this part, where it is more closely allied to the information systems of the body. In addition, we have added a number of new experiments and made considerable effort to clarify the sometimes complex phenomena shown in the illustrations.

**Part Ten, Ecology,** has been significantly revised by our new coauthor, May Berenbaum. A new chapter of biological interactions has been added (a topic formerly covered in the community ecology chapter). Full of interesting anecdotes and discussions of field studies not previously described in biology texts, this new ecology unit offers practical insights into how ecologists acquire, interpret, and apply real data. This brings the book full circle, drawing upon and reinforcing prior topics of energy, evolution, phylogenetics, Earth history, and animal and plant physiology.

## Exceptional Value Formats

We again provide *Life* both as the full book and as a cluster of *paperbacks*. Thus, instructors who want to use less than the whole book can choose from these split volumes, each with the book's front matter, appendices, glossary, and index.

Volume I, The Cell and Heredity, includes: Part One, The Science of Life and Its Chemical Basis (Chapters 1–4); Part Two, Cells (Chapters 5–7); Part Three, Cells and Energy (Chapters 8–10); Part Four, Genes and Heredity (Chapters 11–16); and Part Five, Genomes (Chapters 17–20).

Volume II, Evolution, Diversity, and Ecology, includes: Chapter 1, Studying Life; Part Six, The Patterns and Processes of Evolution (Chapters 21–25); Part Seven, The Evolution of Diversity (Chapters 26–33); and Part Ten, Ecology (Chapters 54–59).

Volume III, Plants and Animals, includes: Chapter 1, Studying Life; Part Eight, Flowering Plants: Form and Function (Chapters 34–39); and Part Nine, Animals: Form and Function (Chapters 40–53).

Responding to student concerns, we offer two options of the entire book at a *significantly reduced cost.* After it was so well received in the previous edition, we again provide *Life* as a *loose-leaf version*. This shrink-wrapped, unbound, 3-hole punched version fits into a 3-ring binder. Students take only what they need to class and can easily integrate any instructor handouts or other resources.

*Life* was the first comprehensive biology text to offer the entire book as a truly robust *eBook*. For this edition, we continue to offer a flexible, interactive ebook that gives students a new way to read the text and learn the material. The ebook integrates the student media resources (animations, quizzes, activities, etc.) and offers instructors a powerful way to customize the textbook with their own text, images, Web links, documents, and more.

## Media and Supplements for the Ninth Edition

The wide range of media and supplements that accompany *Life*, Ninth Edition have all been created with the dual goal of helping students learn the material presented in the textbook more efficiently and helping instructors teach their courses more effectively. Students in majors introductory biology are faced with learning a tremendous number of new concepts, facts, and terms, and the more different ways they can study this material, the more efficiently they can master it.

All of the *Life* media and supplemental resources have been developed specifically for this textbook. This provides strong consistency between text and media, which in turn helps students learn more efficiently. For example, the animated tutorials and activities found in BioPortal were built using textbook art, so that the manner in which structures are illustrated, the colors used to identify objects, and the terms and abbreviations used are all consistent.

For the Ninth Edition, a new set of Interactive Tutorials gives students a new way to explore many key topics across the textbook. These new modules allow the student to learn by doing, including solving problem scenarios, working with experimental techniques, and exploring model systems. All new copies of the Ninth Edition include access to the robust new version of BioPortal, which brings together all of *Life's* student and instructor resources, powerful assessment tools, and new integration with Prep-U adaptive quizzing.

The rich collection of visual resources in the Instructor's Media Library provides instructors with a wide range of options for enhancing lectures, course websites, and assignments. Highlights include: layered art PowerPoint® presentations that break down complex figures into detailed, step-by-step presentations; a collection of approximately 200 video segments that can help capture the attention and imagination of students; and PowerPoint slides of textbook art with editable labels and leaders that allow easy customization of the figures.

For a detailed description of all the media and supplements available for the Ninth Edition, please turn to "*Life's* Media and Supplements," on page xvii.

## Many People to Thank

"If I have seen farther, it is by standing on the shoulders of giants." The great scientist Isaac Newton wrote these words over 330 years ago and, while we certainly don't put ourselves in his lofty place in science, the words apply to us as coauthors of this text. This is the first edition that does not bear the names of Bill Purves and Gordon Orians. As they enjoy their "retirements," we are humbled by their examples as biologists, educators, and writers.

One of the wisest pieces of advice ever given to a textbook author is to "be passionate about your subject, but don't put your ego on the page." Considering all the people who looked over our shoulders throughout the process of creating this book, this advice could not be more apt. We are indebted to many people who gave invaluable help to make this book what it is. First and foremost are our colleagues, biologists from over 100 institutions. Some were users of the previous edition, who suggested many improvements. Others reviewed our chapter drafts in detail, including advice on how to improve the illustrations. Still others acted as accuracy reviewers when the book was almost completed. All of these biologists are listed in the Reviewer credits.

Of special note is Sue Wessler, a distinguished plant biologist and textbook author from the University of Georgia. Sue looked critically at Part Eight, Flowering Plants: Form and Function, wrote three of the chapters (34–36), and was important in the revision of the other three (37–39). The new approach to plant biology in this edition owes a lot to her.

The pace of change in biology and the complexities of preparing a book as broad as this one necessitated having two developmental editors. James Funston coordinated Parts 1–5, and Carol Pritchard-Martinez coordinated Parts 6–10. We benefitted from the wide experience, knowledge, and wisdom of both of them. As the chapter drafts progressed, we were fortunate to have experienced biologist Laura Green lending her critical eye as in-house editor. Elizabeth Morales, our artist, was on her third edition with us. As we have noted, she extensively revised almost all of the prior art and translated our crude sketches into beautiful new art. We hope you agree that our art program remains superbly clear and elegant. Our copy editors, Norma Roche, Liz Pierson, and Jane Murfett, went far beyond what such people usually do. Their knowledge and encyclopedic recall of our book's chapters made our prose sharper and more accurate. Diane Kelly, Susan McGlew, and Shannon Howard effectively coordinated the hundreds of reviews that we described above. David McIntyre was a terrific photo editor, finding over 550 new photographs, including many new ones of his own, that enrich the book's content and visual statement. Jefferson Johnson is responsible for the design elements that make this edition of *Life* not just clear and easy to learn from, but beautiful as well. Christopher Small headed the production department—Joanne Delphia, Joan Gemme, Janice Holabird, and Jefferson Johnson—who contributed in innumerable ways to bringing *Life* to its final form. Jason Dirks once again coordinated the creation of our array of media and supplements, including our superb new Web resources. Carol Wigg, for the ninth time in nine editions, oversaw the editorial process; her influence pervades the entire book.

W. H. Freeman continues to bring *Life* to a wider audience. Associate Director of Marketing Debbie Clare, the Regional Specialists, Regional Managers, and experienced sales force are effective ambassadors and skillful transmitters of the features and unique strengths of our book. We depend on their expertise and energy to keep us in touch with how *Life* is perceived by its users. And thanks also to the Freeman media group for eBook and BioPortal production.

Finally, we are indebted to Andy Sinauer. Like ours, his name is on the cover of the book, and he truly cares deeply about what goes into it. Combining decades of professionalism, high standards, and kindness to all who work with him, he is truly our mentor and friend.

DAVID SADAVA

DAVID HILLIS

CRAIG HELLER

MAY BERENBAUM

# Reviewers for the Ninth Edition

### Between-Edition Reviewers

David D. Ackerly, University of California, Berkeley
Amy Bickham Baird, University of Leiden
Jeremy Brown, University of California, Berkeley
John M. Burke, University of Georgia
Ruth E. Buskirk, University of Texas, Austin
Richard E. Duhrkopf, Baylor University
Casey W. Dunn, Brown University
Erika J. Edwards, Brown University
Kevin Folta, University of Florida
Lynda J. Goff, University of California, Santa Cruz
Tracy A. Heath, University of Kansas
Shannon Hedtke, University of Texas, Austin
Richard H. Heineman, University of Texas, Austin
Albert Herrera, University of Southern California
David S. Hibbett, Clark University
Norman A. Johnson, University of Massachusetts
Walter S. Judd, University of Florida
Laura A. Katz, Smith College
Emily Moriarty Lemmon, Florida State University
Sheila McCormick, University of California, Berkeley
Robert McCurdy, Independence Creek Nature Preserve
Jacalyn Newman, University of Pittsburgh
Juliet F. Noor, Duke University
Theresa O'Halloran, University of Texas, Austin
K. Sata Sathasivan, University of Texas, Austin
H. Bradley Shaffer, University California, Davis
Rebecca Symula, Yale University
Christopher D. Todd, University of Saskatchewan
Elizabeth Willott, University of Arizona
Kenneth Wilson, University of Saskatchewan

### Manuscript Reviewers

Tamarah Adair, Baylor University
William Adams, University of Colorado, Boulder
Gladys Alexandre, University of Tennessee, Knoxville
Shivanthi Anandan, Drexel University
Brian Bagatto, University of Akron
Lisa Baird, University of San Diego
Stewart H. Berlocher, University of Illinois, Urbana-Champaign
William Bischoff, University of Toledo
Meredith M. Blackwell, Louisiana State University
David Bos, Purdue University
Jonathan Bossenbroek, University of Toledo
Nicole Bournias-Vardiabasis, California State University, San Bernardino
Nancy Boury, Iowa State University
Sunny K. Boyd, University of Notre Dame
Judith L. Bronstein, University of Arizona
W. Randy Brooks, Florida Atlantic University
James J. Bull, University of Texas, Austin
Darlene Campbell, Cornell University
Domenic Castignetti, Loyola University, Chicago
David T. Champlin, University of Southern Maine
Shu-Mei Chang, University of Georgia
Samantha K. Chapman, Villanova University
Patricia Christie, MIT
Wes Colgan, Pikes Peak Community College
John Cooper, Washington University
Ronald Cooper, University of California, Los Angeles
Elizabeth Cowles, Eastern Connecticut State University
Jerry Coyne, University of Chicago
William Crampton, University of Central Florida
Michael Dalbey, University of California, Santa Cruz
Anne Danielson-Francois, University of Michigan, Dearborn
Grayson S. Davis, Valparaiso University
Kevin Dixon, Florida State University
Zaldy Doyungan, Texas A&M University, Corpus Christi
Ernest F. Dubrul, University of Toledo
Roland Dute, Auburn University
Scott Edwards, Harvard University
William Eldred, Boston University
David Eldridge, Baylor University
Joanne Ellzey, University of Texas, El Paso
Susan H. Erster, State University of New York, Stony Book
Brent E. Ewers, University of Wyoming
Kevin Folta, University of Florida
Brandon Foster, Wake Technical Community College
Richard B. Gardiner, University of Western Ontario
Douglas Gayou, University of Missouri, Columbia
John R. Geiser, Western Michigan University
Arundhati Ghosh, University of Pittsburgh
Alice Gibb, Northern Arizona University
Scott Gilbert, Swarthmore College
Matthew R. Gilg, University of North Florida
Elizabeth Godrick, Boston University
Lynda J. Goff, University of California, Santa Cruz
Elizabeth Blinstrup Good, University of Illinois, Urbana-Champaign
John Nicholas Griffis, University of Southern Mississippi
Cameron Gundersen, University of California, Los Angeles
Kenneth Halanych, Auburn University
E. William Hamilton, Washington and Lee University
Monika Havelka, University of Toronto at Mississauga
Tyson Hedrick, University of North Carolina, Chapel Hill
Susan Hengeveld, Indiana University, Bloomington
Albert Herrera, University of Southern California

Kendra Hill, South Dakota State University
Richard W. Hill, Michigan State University
Erec B. Hillis, University of California, Berkeley
Jonathan D. Hillis, Carleton College
William Huddleston, University of Calgary
Dianne B. Jennings, Virginia Commonwealth University
Norman A. Johnson, University of Massachusetts, Amherst
William H. Karasov, University of Wisconsin, Madison
Susan Keen, University of California, Davis
Cornelis Klok, Arizona State University, Tempe
Olga Ruiz Kopp, Utah Valley University
William Kroll, Loyola University, Chicago
Allen Kurta, Eastern Michigan University
Rebecca Lamb, Ohio State University
Brenda Leady, University of Toledo
Hugh Lefcort, Gonzaga University
Sean C. Lema, University of North Carolina, Wilmington
Nathan Lents, John Jay College, City University of New York
Rachel A. Levin, Amherst College
Donald Levin, University of Texas, Austin
Bernard Lohr, University of Maryland, Baltimore County
Barbara Lom, Davidson College
David J. Longstreth, Louisiana State University
Catherine Loudon, University of California, Irvine
Francois Lutzoni, Duke University
Charles H. Mallery, University of Miami
Kathi Malueg, University of Colorado, Colorado Springs
Richard McCarty, Johns Hopkins University
Sheila McCormick, University of California, Berkeley
Francis Monette, Boston University
Leonie Moyle, Indiana University, Bloomington
Jennifer C. Nauen, University of Delaware
Jacalyn Newman, University of Pittsburgh
Alexey Nikitin, Grand Valley State University
Shawn E. Nordell, Saint Louis University
Tricia Paramore, Hutchinson Community College
Nancy J. Pelaez, Purdue University
Robert T. Pennock, Michigan State University
Roger Persell, Hunter College
Debra Pires, University of California, Los Angeles
Crima Pogge, City College of San Francisco
Jaimie S. Powell, Portland State University
Susan Richardson, Florida Atlantic University
David M. Rizzo, University of California, Davis
Benjamin Rowley, University of Central Arkansas
Brian Rude, Mississippi State University
Ann Rushing, Baylor University
Christina Russin, Northwestern University
Udo Savalli, Arizona State University, West
Frieder Schoeck, McGill University
Paul J. Schulte, University of Nevada, Las Vegas
Stephen Secor, University of Alabama
Vijayasaradhi Setaluri, University of Wisconsin, Madison
H. Bradley Shaffer, University of California, Davis
Robin Sherman, Nova Southeastern University
Richard Shingles, Johns Hopkins University
James Shinkle, Trinity University
Richard M. Showman, University of South Carolina
Felisa A. Smith, University of New Mexico
Ann Berry Somers, University of North Carolina, Greensboro
Ursula Stochaj, McGill University
Ken Sweat, Arizona State University, West
Robin Taylor, Ohio State University
William Taylor, University of Toledo
Mark Thogerson, Grand Valley State University
Sharon Thoma, University of Wisconsin, Madison
Lars Tomanek, California Polytechnic State University
James Traniello, Boston University
Jeffrey Travis, State University of New York, Albany
Terry Trier, Grand Valley State University
John True, State University of New York, Stony Brook
Elizabeth Van Volkenburgh, University of Washington
John Vaughan, St. Petersburg College
Sara Via, University of Maryland
Suzanne Wakim, Butte College (Glenn Community College District)
Randall Walikonis, University of Connecticut
Cindy White, University of Northern Colorado
Elizabeth Willott, University of Arizona
Mark Wilson, Humboldt State University
Stuart Wooley, California State University, Stanislaus
Lan Xu, South Dakota State University
Heping Zhou, Seton Hall University

### Accuracy Reviewers

John Alcock, Arizona State University
Gladys Alexandre, University of Tennessee, Knoxville
Lawrence A. Alice, Western Kentucky University
David R. Angelini, American University
Fabia U. Battistuzzi, Arizona State University
Arlene Billock, University of Louisiana, Lafayette
Mary A. Bisson, State University of New York, Buffalo
Meredith M. Blackwell, Louisiana State University
Nancy Boury, Iowa State University
Eldon J. Braun, University of Arizona
Daniel R. Brooks, University of Toronto
Jennifer L. Campbell, North Carolina State University
Peter C. Chabora, Queens College, CUNY
Patricia Christie, MIT
Ethan Clotfelter, Amherst College
Robert Connour, Owens Community College

Peter C. Daniel, Hofstra University
D. Michael Denbow, Virginia Polytechnic Institute
Laura DiCaprio, Ohio State University
Zaldy Doyungan, Texas A&M University, Corpus Christi
Moon Draper, University of Texas, Austin
Richard E. Duhrkopf, Baylor University
Susan A. Dunford, University of Cincinnati
Brent E. Ewers, University of Wyoming
James S. Ferraro, Southern Illinois University
Rachel D. Fink, Mount Holyoke College
John R. Geiser, Western Michigan University
Elizabeth Blinstrup Good, University of Illinois, Urbana-Champaign
Melina E. Hale, University of Chicago
Patricia M. Halpin, University of California, Los Angeles
Jean C. Hardwick, Ithaca College
Monika Havelka, University of Toronto at Mississauga
Frank Healy, Trinity University
Marshal Hedin, San Diego State University
Albert Herrera, University of Southern California
David S. Hibbett, Clark University
James F. Holden, University of Massachusetts, Amherst
Margaret L. Horton, University of North Carolina, Greensboro
Helen Hull-Sanders, Canisius College
C. Darrin Hulsey, University of Tennessee, Knoxville
Timothy Y. James, University of Michigan
Dianne B. Jennings, Virginia Commonwealth University
Norman A. Johnson, University of Massachusetts, Amherst
Susan Jorstad, University of Arizona
Ellen S. Lamb, University of North Carolina, Greensboro
Dennis V. Lavrov, Iowa State University
Hugh Lefort, Gonzaga University
Rachel A. Levin, Amherst College
Bernard Lohr, University of Maryland, Baltimore County
Sharon E. Lynn, College of Wooster
Sarah Mathews, Harvard University
Susan L. Meacham, University of Nevada, Las Vegas
Mona C. Mehdy, University of Texas, Austin
Bradley G. Mehrtens, University of Illinois, Urbana-Champaign
James D. Metzger, Ohio State University
Thomas W. Moon, University of Ottowa
Thomas M. Niesen, San Francisco State University
Theresa O'Halloran, University of Texas, Austin
Thomas L. Pannabecker, University of Arizona
Nancy J. Pelaez, Purdue University
Nicola J. R. Plowes, Arizona State University
Gregory S. Pryor, Francis Marion University
Laurel B. Roberts, University of Pittsburgh
Anjana Sharma, Western Carolina University
Richard M. Showman, University of South Carolina
John B. Skillman, California State University, San Bernadino
John J. Stachowicz, University of California, Davis
Brook O. Swanson, Gonzaga University
Robin A. J. Taylor, Ohio State University
William Taylor, University of Toledo
Steven M. Theg, University of California, Davis
Mark Thogerson, Grand Valley State University
Christopher D. Todd, University of Saskatchewan
Jeffrey Travis, State University of New York, Albany
Joseph S. Walsh, Northwestern University
Andrea Ward, Adelphi University
Barry Williams, Michigan State University
Kenneth Wilson, University of Saskatchewan
Carol L. Wymer, Morehead State University

# *LIFE's* Media and Supplements

## BIOPORTAL featuring Prep-U

**yourBioPortal.com**

BioPortal is the new gateway to all of *Life's* state-of-the-art online resources for students and instructors. BioPortal includes the breakthrough quizzing engine, Prep-U; a fully interactive eBook; and additional premium learning media. The textbook is tightly integrated with BioPortal via in-text references that connect the printed text and media resources. The result is a powerful, easily-managed online course environment. BioPortal includes the following features and resources:

### *Life*, Ninth Edition eBook

- Integration of all activities, animated tutorials, and other media resources.
- Quick, intuitive navigation to any section or subsection, as well as any printed book page number.
- In-text links to all glossary entries.
- Easy text highlighting.
- A bookmarking feature that allows for quick reference to any page.
- A powerful Notes feature that allows students to add notes to any page.
- A full glossary and index.
- Full-text search, including an additional option to search the glossary and index.
- Automatic saving of all notes, highlighting, and bookmarks.

***Additional eBook features for instructors:***

- Content Customization: Instructors can easily add pages of their own content and/or hide chapters or sections that they do not cover in their course.
- Instructor Notes: Instructors can choose to create an annotated version of the eBook with their own notes on any page. When students in the course log in, they see the instructor's personalized version of the eBook. Instructor notes can include text, Web links, images, links to all BioPortal content, and more.

Built by educators, Prep-U focuses student study time exactly where it should be, through the use of personalized, adaptive quizzes that move students toward a better grasp of the material—and better grades. For *Life,* Ninth Edition, Prep-U is fully integrated into BioPortal, making it easy for instructors to take advantage of this powerful quizzing engine in their course. Features include:

- Adaptive quizzing
- Automatic results reporting into the BioPortal gradebook
- Misconception index
- Comparison to national data

### Student Resources

***Diagnostic Quizzing.*** The diagnostic quiz for each chapter of *Life* assesses student understanding of that chapter, and generates a Personalized Study Plan to effectively focus student study time. The plan includes links to specific textbook sections, animated tutorials, and activities.

***Interactive Summaries.*** For each chapter, these dynamic summaries combine a review of important concepts with links to all of the key figures from the chapter as well as all of the relevant animated tutorials, activities, and key terms.

***Animated Tutorials.*** Over 100 in-depth animated tutorials, in a new format for the Ninth Edition, present complex topics in a clear, easy-to-follow format that combines a detailed animation with an introduction, conclusion, and quiz.

***Activities.*** Over 120 interactive activities help students learn important facts and concepts through a wide range of exercises, such as labeling steps in processes or parts of structures, building diagrams, and identifying different types of organisms.

***NEW! Interactive Tutorials.*** New for the Ninth Edition, these tutorial modules help students master key concepts through hands-on activities that allow them to learn through action. With these tutorials, students can solve problem scenarios by applying concepts from the text, by working with experimental techniques, and by using interactive models to discover how biological mechanisms work. Each tutorial includes a self-assessment quiz that can be assigned.

***Interactive Quizzes.*** Each question includes an image from the textbook, thorough feedback on both correct and incorrect answer choices, references to textbook pages, and links to eBook pages, for quick review.

***BioNews from* Scientific American.** BioNews makes it easy for instructors to bring the dynamic nature of the biological sciences and up-to-the minute currency into their course. Accessible from within BioPortal, BioNews is a continuously updated feed of current news, podcasts, magazine articles, science blog entries, "strange but true" stories, and more.

***NEW! BioNavigator.*** This unique visual resource is an innovative way to access the wide variety of *Life* media resources. Starting from the whole-Earth view, instructors and students can zoom to any level of biological inquiry, encountering links to a wealth of animations, activities, and tutorials on the full range of topics along the way.

***Working with Data.*** Built around some of the original experiments depicted in the Investigating Life figures, these exercises help build quantitative skills and encourage student interest in how scientists do research, by looking at real experimental data and answering questions based on those data.

***Flashcards.*** For each chapter of the book, there is a set of flashcards that allows the student to review all the key terminology from the chapter. Students can review the terms in study mode, and then quiz themselves on a list of terms.

***Experiment Links.*** For each Investigating Life figure in the textbook, BioPortal includes an overview of the experiment featured in the figure and related research or applications that followed, a link to the original paper, and links to additional information related to the experiment.

***Key Terms.*** The key terminology introduced in each chapter is listed, with definitions and audio pronunciations from the glossary.

***Suggested Readings.*** For each chapter of the book, a list of suggested readings is provided as a resource for further study.

***Glossary.*** The language of biology is often difficult for students taking introductory biology to master, so BioPortal includes a full glossary that features audio pronunciations of all terms.

***Statistics Primer.*** This brief introduction to the use of statistics in biological research explains why statistics are integral to biology, and how some of the most common statistical methods and techniques are used by biologists in their work.

***Math for Life.*** A collection of mathematical shortcuts and references to help students with the quantitative skills they need in the laboratory.

***Survival Skills.*** A guide to more effective study habits. Topics include time management, note-taking, effective highlighting, and exam preparation.

### Instructor Resources

#### *Assessment*

- Diagnostic Quizzing provides instant class comprehension feedback to instructors, along with targeted lecture resources for those areas requiring the most attention.
- Question banks include questions ranked according to Bloom's taxonomy.
- Question filtering: Allows instructors to select questions based on Bloom's category and/or textbook section.
- Easy-to-use customized assessment tools allow instructors to quickly create quizzes and many other types of assignments using any combination of the questions and resources provided along with their own materials.
- Comprehensive question banks include questions from the test bank, study guide, textbook self-quizzes, and diagnostic quizzes.

#### *Media Resources* *(see Instructor's Media Library below for details)*

- Videos
- PowerPoint® Presentations (Textbook Figures, Lectures, Layered Art)
- Supplemental Photos
- Clicker Questions
- Instructor's Manual
- Lecture Notes

#### *Course Management*

- Complete course customization capabilities
- Custom resources/document posting
- Robust Gradebook
- Communication Tools: Announcements, Calendar, Course Email, Discussion Boards

*Note:* The printed textbook, the eBook, BioPortal, and Prep-U can all be purchased individually as stand-alone items, in addition to being available in a package with the printed textbook.

## Student Supplements

### Study Guide (ISBN 978-1-4292-3569-3)

Jacalyn Newman, *University of Pittsburgh;* Edward M. Dzialowski, *University of North Texas;* Betty McGuire, *Cornell University;* Lindsay Goodloe, *Cornell University;* and Nancy Guild, *University of Colorado*

For each chapter of the textbook, the *Life* Study Guide offers a variety of study and review tools. The contents of each chapter are broken down into both a detailed review of the Important Concepts covered and a boiled-down Big Picture snapshot. New for the Ninth Edition, Diagram Exercises help students synthesize what they have learned in the chapter through exercises such as ordering concepts, drawing graphs, linking steps in processes, and labeling diagrams. In addition, Common Problem Areas and Study Strategies are highlighted. A set of study questions (both multiple-choice and short-answer) allows students to test their comprehension. All questions include answers and explanations.

### Lecture Notebook (ISBN 978-1-4292-3583-9)

This invaluable printed resource consists of all the artwork from the textbook (more than 1,000 images with labels) presented in the order in which they appear in the text, with ample space for note-taking. Because the Notebook has already done the drawing, students can focus more of their attention on the concepts. They will absorb the material more efficiently during class, and their notes will be clearer, more accurate, and more useful when they study from them later.

### Companion Website www.thelifewire.com

(Also available as a CD, which can be optionally packaged with the textbook.)

For those students who do not have access to BioPortal, the *Life,* Ninth Edition Companion Website is available free of charge (no access code required). The site features a variety of resources, including animations, flashcards, activities, study ideas, help with math and statistics, and more.

### CatchUp Math & Stats

Michael Harris, Gordon Taylor, and Jacquelyn Taylor (ISBN 978-1-4292-0557-3)

This primer will help your students quickly brush up on the quantitative skills they need to succeed in biology. Presented in brief, accessible units, the book covers topics such as working with powers, logarithms, using and understanding graphs, calculating standard deviation, preparing a dilution series, choosing the right statistical test, analyzing enzyme kinetics, and many more.

### Student Handbook for Writing in Biology, Third Edition

Karen Knisely, *Bucknell University* (ISBN 978-1-4292-3491-7)

This book provides practical advice to students who are learning to write according to the conventions in biology. Using the standards of journal publication as a model, the author provides, in a user-friendly format, specific instructions on: using biology databases to locate references; paraphrasing for improved comprehension; preparing lab reports, scientific papers, posters; preparing oral presentations in PowerPoint®, and more.

### Bioethics and the New Embryology: Springboards for Debate

Scott F. Gilbert, Anna Tyler, and Emily Zackin (ISBN 978-0-7167-7345-0)

Our ability to alter the course of human development ranks among the most significant changes in modern science and has brought embryology into the public domain. The question that must be asked is: Even if we can do such things, should we?

### BioStats Basics: A Student Handbook

James L. Gould and Grant F. Gould (ISBN 978-0-7167-3416-1)

*BioStats Basics* provides introductory-level biology students with a practical, accessible introduction to statistical research. Engaging and informal, the book avoids excessive theoretical and mathematical detail, and instead focuses on how core statistical methods are put to work in biology.

## Instructor Media & Supplements

### Instructor's Media Library

The *Life*, Ninth Edition Instructor's Media Library (available both online via BioPortal and on disc) includes a wide range of electronic resources to help instructors plan their course, present engaging lectures, and effectively assess student comprehension. The Media Library includes the following resources:

***Textbook Figures and Tables.*** Every image and table from the textbook is provided in both JPEG (high- and low-resolution) and PDF formats. Each figure is provided both with and without balloon captions, and large, complex figures are provided in both a whole and split version.

***Unlabeled Figures.*** Every figure is provided in an unlabeled format, useful for student quizzing and custom presentation development.

***Supplemental Photos.*** The supplemental photograph collection contains over 1,500 photographs (in addition to those in the text), giving instructors a wealth of additional imagery to draw upon.

***Animations.*** Over 100 detailed animations, revised and enlarged for the Ninth Edition, all created from the textbook's art program, and viewable in either narrated or step-through mode.

***Videos.*** A collection of over 200 video segments that covers topics across the entire textbook and helps demonstrate the complexity and beauty of life. Includes the Cell Visualization Videos.

***PowerPoint® Resources.*** For each chapter of the textbook, several different PowerPoint presentations are available. These give instructors the flexibility to build presentations in the manner that best suits their needs. Included are:

- Textbook Figures and Tables
- Lecture Presentation
- Figures with Editable Labels
- Layered Art Figures
- Supplemental Photos
- Videos
- Animations

***Clicker Questions.*** A set of questions written specifically to be used with classroom personal response systems, such as the iClicker system, is provided for each chapter. These questions are designed to reinforce concepts, gauge student comprehension, and engage students in active participation.

***Chapter Outlines, Lecture Notes,*** and the complete ***Test File*** are all available in Microsoft Word® format for easy use in lecture and exam preparation.

***Intuitive Browser Interface*** provides a quick and easy way to preview and access all of the content on the Instructor's Media Library.

### Instructor's Resource Kit

The *Life*, Ninth Edition Instructor's Resource Kit includes a wealth of information to help instructors in the planning and teaching of their course. The Kit includes:

***Instructor's Manual,*** featuring (by chapter):

- A "What's New" guide to the Ninth Edition
- Brief chapter overview
- Chapter outline
- Key terms section with all of the boldface terms from the text

***Lecture Notes.*** Detailed notes for each chapter, which can serve as the basis for lectures, including references to figures and media resources.

***Media Guide.*** A visual guide to the extensive media resources available with the Ninth Edition of *Life*. The guide includes thumbnails and descriptions of every video, animation, lecture PowerPoint®, and supplemental photo in the Media Library, all organized by chapter.

### Overhead Transparencies

This set includes over 1,000 transparencies—including all of the four-color line art and all of the tables from the text—along with convenient binders. All figures have been formatted and color-enhanced for clear projection in a wide range of conditions. Labels and images have been resized for improved readability.

### Test File

Catherine Ueckert, *Northern Arizona University;* Norman Johnson, *University of Massachusetts;* Paul Nolan, *The Citadel;* Nicola Plowes, *Arizona State University*

The Test File offers more than 5,000 questions, covering the full range of topics presented in the textbook. All questions are referenced to textbook sections and page numbers, and are ranked according to Bloom's taxonomy. Each chapter includes a wide range of multiple choice and fill-in-the-blank questions. In addition, each chapter features a set of diagram questions that involve the student in working with illustrations of structures, graphs, steps in processes, and more. The electronic versions of the Test File (within BioPortal, the Instructor's Media Library, and the Computerized Test Bank CD) also include all of the textbook end-of-chapter Self-Quiz questions, all of the BioPortal Diagnostic Quiz questions, and all of the Study Guide multiple-choice questions.

### Computerized Test Bank

The entire printed Test File, plus the textbook end-of-chapter Self-Quizzes, the BioPortal Diagnostic Quizzes, and the Study Guide multiple-choice questions are all included in Wimba's easy-to-use Diploma® software. Designed for both novice and advanced users, Diploma enables instructors to quickly and easily create or edit questions, create quizzes or exams with a "drag-and-drop" feature, publish to online courses, and print paper-based assignments.

### Course Management System Support

As a service for *Life* adopters using WebCT, Blackboard, or ANGEL for their courses, full electronic course packs are available.

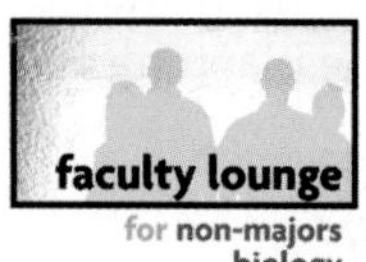


**www.whfreeman.com/facultylounge/majorsbio**

***NEW!*** The new Faculty Lounge for Majors Biology is the first publisher-provided website for the majors biology community that lets instructors freely communicate and share peer-reviewed lecture and teaching resources. It is continually updated and vetted by majors biology instructors—there is always something new to see. The Faculty Lounge offers convenient access to peer-recommended and vetted resources, including the following categories: Images, News, Videos, Labs, Lecture Resources, and Educational Research.

In addition, the site includes special areas for resources for lab coordinators, resources and updates from the *Scientific Teaching* series of books, and information on biology teaching workshops.

iclicker

Developed for educators by educators, iclicker is a hassle-free radio-frequency classroom response system that makes it easy for instructors to ask questions, record responses, take attendance, and direct students through lectures as active participants. For more information, visit www.iclicker.com.

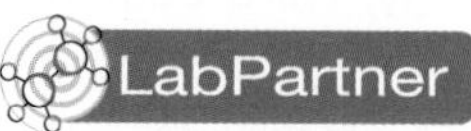


**www.whfreeman.com/labpartner**

***NEW!*** LabPartner is a site designed to facilitate the creation of customized lab manuals. Its database contains a wide selection of experiments published by W. H. Freeman and Hayden-McNeil Publishing. Instructors can preview, choose, and re-order labs, interleave their original experiments, add carbonless graph paper and a pocket folder, and customize the cover both inside and out. LabPartner offers a variety of binding types: paperback, spiral, or loose-leaf. Manuals are printed on-demand once W. H. Freeman receives an order from a campus bookstore or school.

The Scientific Teaching Book Series is a collection of practical guides, intended for all science, technology, engineering and mathematics (STEM) faculty who teach undergraduate and graduate students in these disciplines. The purpose of these books is to help faculty become more successful in all aspects of teaching and learning science, including classroom instruction, mentoring students, and professional development. Authored by well-known science educators, the Series provides concise descriptions of best practices and how to implement them in the classroom, the laboratory, or the department. For readers interested in the research results on which these best practices are based, the books also provide a gateway to the key educational literature.

### Scientific Teaching

Jo Handelsman, Sarah Miller, and Christine Pfund, *University of Wisconsin-Madison* (ISBN 978-1-4292-0188-9)

### *NEW!* Transformations: Approaches to College Science Teaching

***A Collection of Articles from CBE Life Sciences Education***
Deborah Allen, *University of Delaware;* Kimberly Tanner, *San Francisco State University* (ISBN 978-1-4292-5335-2)

# Contents

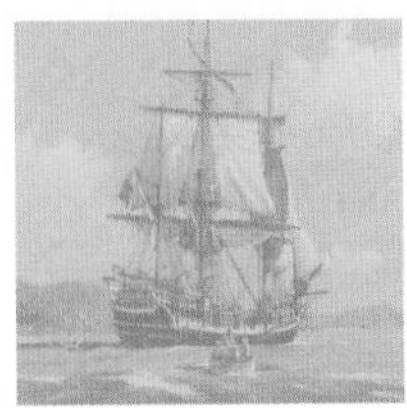

## PART ONE
## THE SCIENCE OF LIFE AND ITS CHEMICAL BASIS



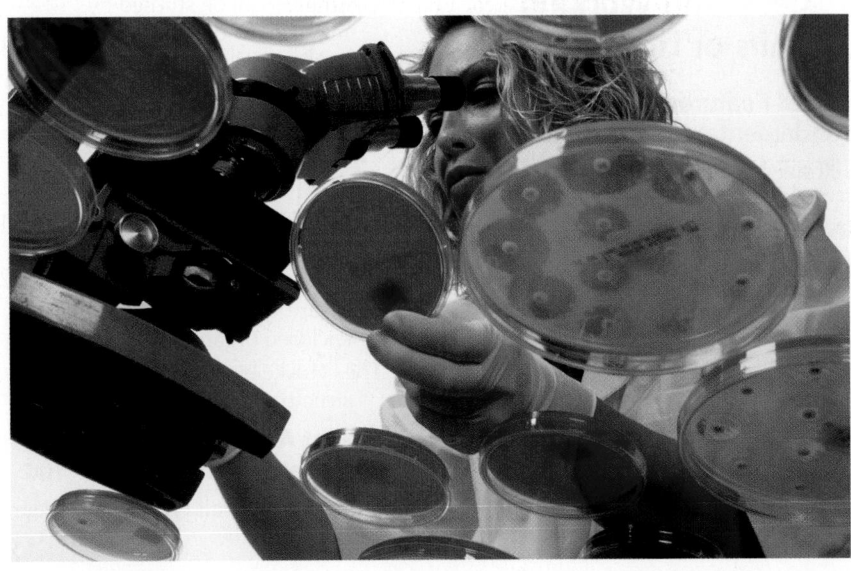

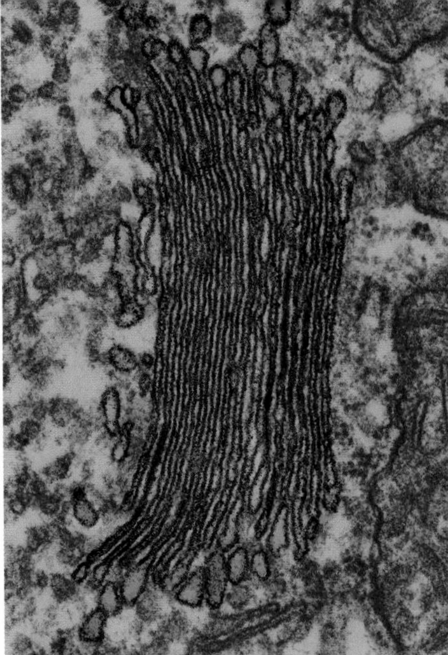



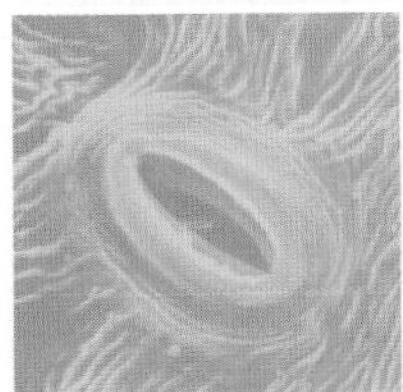

PART THREE

CELLS AND ENERGY

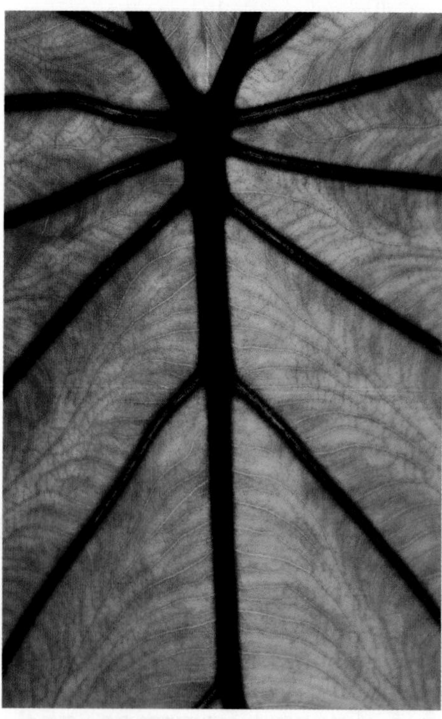

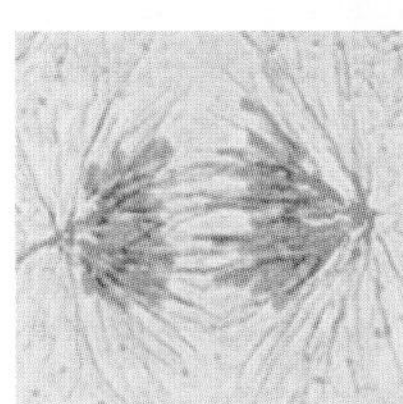

# PART FOUR
# GENES AND HEREDITY

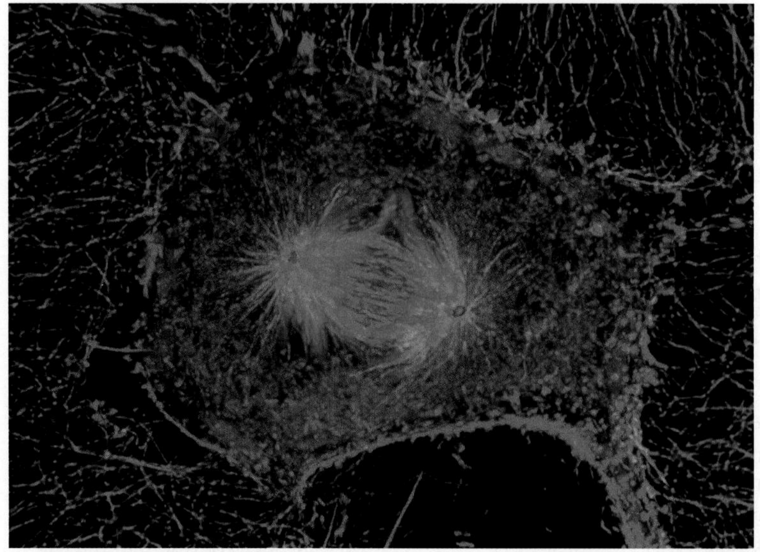

# PART FIVE GENOMES



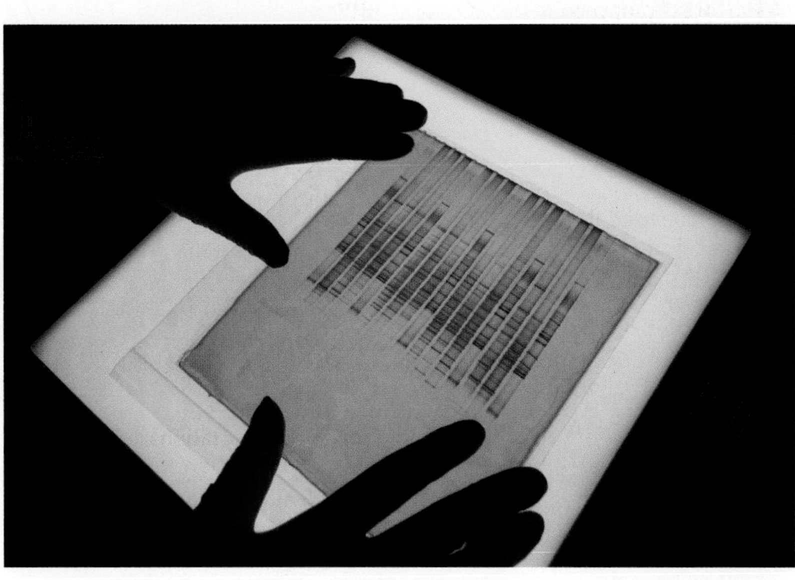

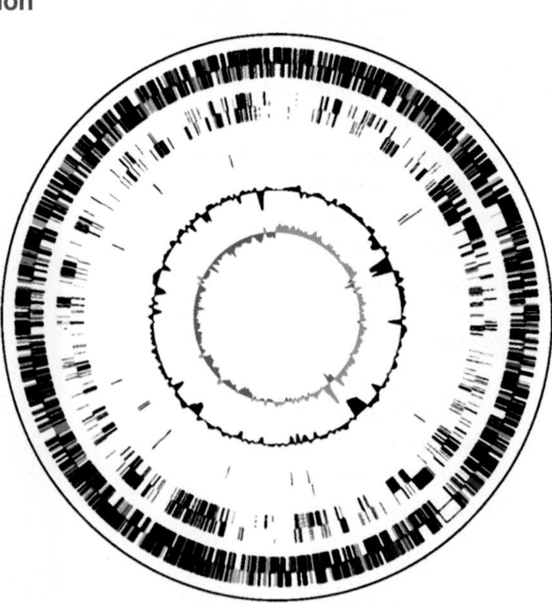

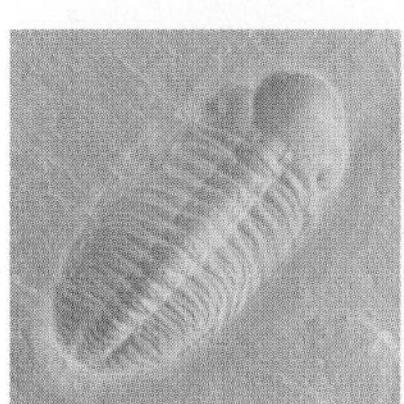

# PART SIX
# THE PATTERNS AND PROCESSES OF EVOLUTION

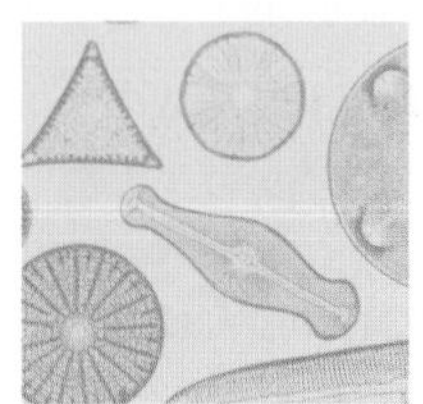

## PART SEVEN
## THE EVOLUTION OF DIVERSITY

# PART EIGHT
# FLOWERING PLANTS: FORM AND FUNCTION

# PART NINE
# ANIMALS: FORM AND FUNCTION

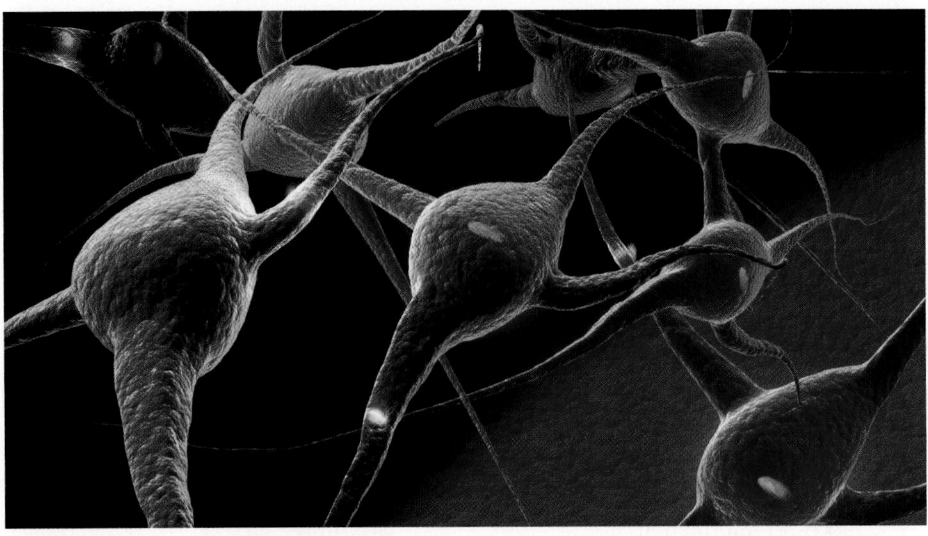

# PART TEN ECOLOGY

# Life

## The Science of Biology

NINTH EDITION

PART ONE THE SCIENCE OF LIFE AND ITS CHEMICAL BASIS

# 1 Studying Life

## Why are frogs croaking?

Amphibians—frogs, toads, and salamanders—have been around for a long time. They watched the dinosaurs come and go. But today amphibian populations around the world are in dramatic decline, with more than a third of the world's amphibian species threatened with extinction. Why?

Biologists work to answer this question by making observations and doing experiments. A number of factors may be involved, and one possible cause may be the effects of agricultural pesticides and herbicides. Several studies have shown that many of these chemicals tested at realistic concentrations do not kill amphibians. But Tyrone Hayes, a biologist at the University of California at Berkeley, probed deeper.

Hayes focused on atrazine, the most widely used herbicide in the world and a common contaminant in fresh water. More than 70 million pounds of atrazine are applied to farmland in the United States every year, and it is used in at least 20 countries. Atrazine is usually applied in the spring, when many amphibians are breeding and thousands of tadpoles swim in the ditches, ponds, and streams that receive runoff from farms.

In his laboratory, Hayes and his associates raised frog tadpoles in water containing no atrazine and in water with concentrations ranging from 0.01 parts per billion (ppb) up to 25 ppb. The U.S. Environmental Protection Agency considers environmental levels of atrazine of 10 to 20 ppb of no concern; the level it considers safe in drinking water is 3 ppb. Rainwater in Iowa has been measured to contain 40 ppb. In Switzerland, where the use of atrazine is illegal, the chemical has been measured at approximately 1 ppb in rainwater.

In the Hayes laboratory, concentrations as low as 0.1 ppb had a dramatic effect on tadpole development: it feminized the males. In some of the adult males that developed from these larvae, the vocal structures used in mating calls were smaller than normal, female sex organs developed, and eggs were found growing in the testes. In other studies, normal adult male frogs exposed to 25 ppb had a tenfold reduction in testosterone levels and did not produce sperm. You can imagine the disastrous effects these developmental and hormonal changes could have on the capacity of frogs to breed and reproduce.

But Hayes's experiments were performed in the laboratory, with a species of frog bred for laboratory use. Would his results be the same in nature? To find out, he and his students traveled from Utah to Iowa, sampling water and collecting frogs. They analyzed the water

**Frogs Are Having Serious Problems** An alarming number of species of frogs, such as this tiny leaf frog (*Agalychnis calcarifer*) from Ecuador, are in danger of becoming extinct. The numerous possible reasons for the decline in global amphibian populations have been a subject of widespread scientific investigation.

**A Biologist at Work** Tyrone Hayes grew up near the great Congaree Swamp in South Carolina collecting turtles, snakes, frogs, and toads. Now a professor of biology at the University of California at Berkeley, he has more than 3,000 frogs in his laboratory and studies hormonal control of their development.

for atrazine and examined the frogs. In the only site where atrazine was undetectable in the water, the frogs were normal; in all the other sites, male frogs had abnormalities of the sex organs.

Like other biologists, Hayes made observations. He then made predictions based on those observations, and designed and carried out experiments to test his predictions. Some of the conclusions from his experiments, described at the end of this chapter, could have profound implications not only for amphibians but also for other animals, including humans.

**IN THIS CHAPTER** we identify and examine the most common features of living organisms and put those features into the context of the major principles that underlie all biology. Next we offer a brief outline of how life evolved and how the different organisms on Earth are related. We then turn to the subjects of biological inquiry and the scientific method. Finally we consider how knowledge discovered by biologists influences public policy.

## CHAPTER OUTLINE



## 1.1 What Is Biology?

**Biology** is the scientific study of living things. Biologists define "living things" as all the diverse organisms descended from a single-celled ancestor that evolved almost 4 billion years ago. Because of their common ancestry, living organisms share many characteristics that are not found in the nonliving world. Living organisms:

- consist of one or more cells
- contain genetic information
- use genetic information to reproduce themselves
- are genetically related and have evolved
- can convert molecules obtained from their environment into new biological molecules
- can extract energy from the environment and use it to do biological work
- can regulate their internal environment

This simple list, however, belies the incredible complexity and diversity of life. Some forms of life may not display all of these characteristics all of the time. For example, the seed of a desert plant may go for many years without extracting energy from the environment, converting molecules, regulating its internal environment, or reproducing; yet the seed is alive.

And what about viruses? Viruses do not consist of cells, and they cannot carry out physiological functions on their own; they must parasitize host cells to do those jobs for them. Yet viruses contain genetic information, and they certainly mutate and evolve (as we know, because evolving flu viruses require constant changes in the vaccines we create to combat them). The existence of viruses depends on cells, and it is highly probable that viruses evolved from cellular life forms. So, are viruses alive? What do you think?

This book explores the characteristics of life, how these characteristics vary among organisms, how they evolved, and how they work together to enable organisms to survive and reproduce. *Evolution* is a central theme of biology and therefore of this book. Through differential survival and reproduction, living systems evolve and become adapted to Earth's many environments. The processes of evolution have generated the enormous diversity that we see today as life on Earth.

### Cells are the basic unit of life

We lay the chemical foundation for our study of life in the next three chapters, after which we will turn to cells and the processes by which they live, reproduce, age, and die. Some organisms are *unicellular,* consisting of a single cell that carries out

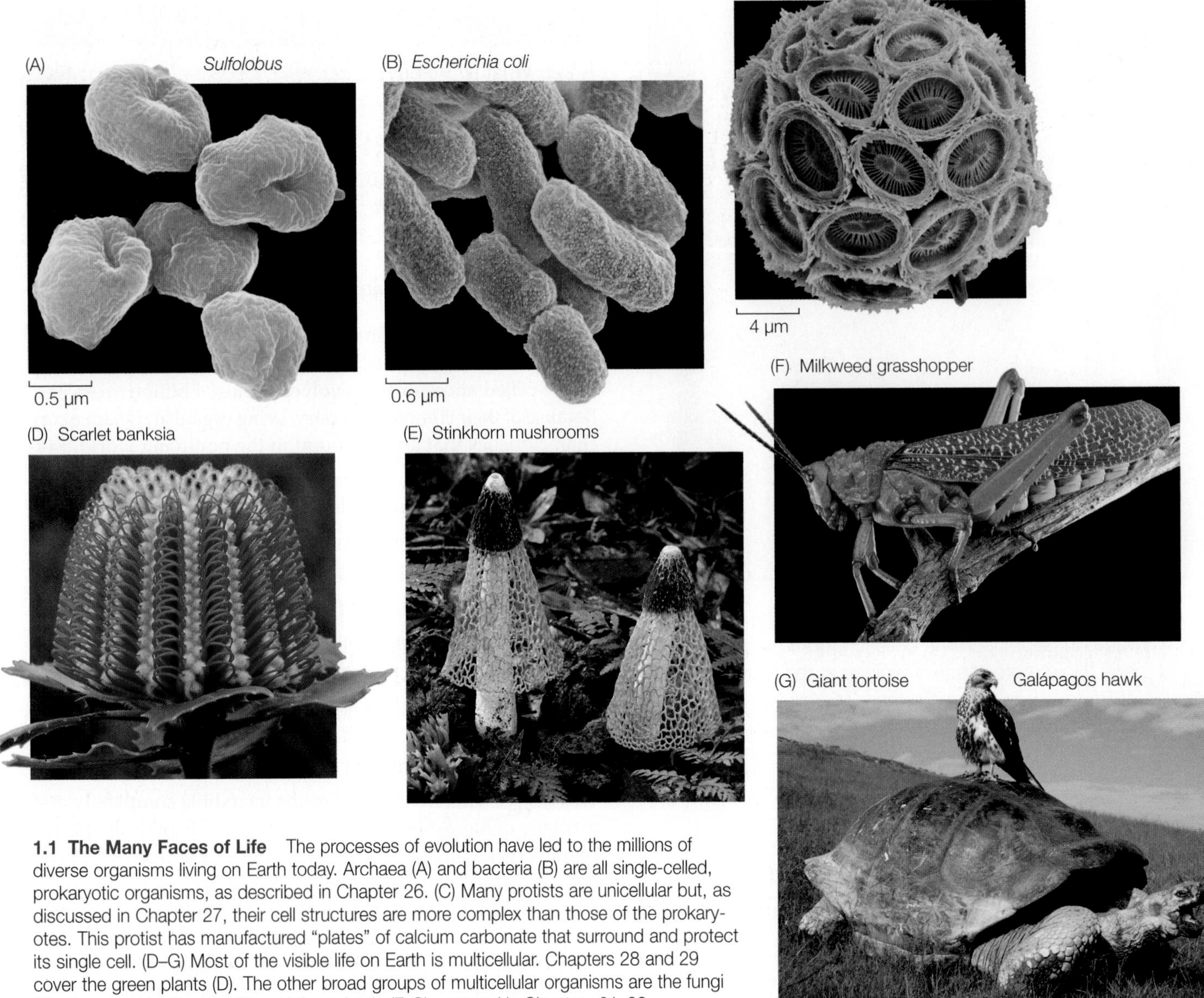


**1.1 The Many Faces of Life** The processes of evolution have led to the millions of diverse organisms living on Earth today. Archaea (A) and bacteria (B) are all single-celled, prokaryotic organisms, as described in Chapter 26. (C) Many protists are unicellular but, as discussed in Chapter 27, their cell structures are more complex than those of the prokaryotes. This protist has manufactured "plates" of calcium carbonate that surround and protect its single cell. (D–G) Most of the visible life on Earth is multicellular. Chapters 28 and 29 cover the green plants (D). The other broad groups of multicellular organisms are the fungi (E), discussed in Chapter 30, and the animals (F, G), covered in Chapters 31–33.

all the functions of life (**Figure 1.1A–C**). Others are *multicellular*, made up of many cells that are specialized for different functions (**Figure 1.1D–G**). Viruses are *acellular*, although they depend on cellular organisms.

The discovery of cells was made possible by the invention of the microscope in the 1590s by the Dutch spectacle makers Hans and Zaccharias Janssen (father and son). In the mid- to late 1600s, Antony van Leeuwenhoek of Holland and Robert Hooke of England both made improvements on the Janssens' technology and used it to study living organisms. Van Leeuwenhoek discovered that drops of pond water teemed with single-celled organisms, and he made many other discoveries as he progressively improved his microscopes over a long lifetime of research. Hooke put pieces of plants under his microscope and observed that they were made up of repeated units he called *cells* (**Figure 1.2**). In 1676, Hooke wrote that van Leeuwenhoek had observed "a vast number of small animals in his Excrements which were most abounding when he was troubled with a Loosenesse and very few or none when he was well." This simple observation represents the discovery of bacteria—and makes one wonder why scientists do some of the things they do.

More than a hundred years passed before studies of cells advanced significantly. As they were dining together one evening in 1838, Matthias Schleiden, a German biologist, and Theodor Schwann, from Belgium, discussed their work on plant and animal tissues, respectively. They were struck by the similarities in their observations and came to the conclusion that the basic structural elements of plants and animals were essentially the same. They formulated their conclusion as the **cell theory**, which states that:

- Cells are the basic structural and physiological units of all living organisms.
- Cells are both distinct entities and building blocks of more complex organisms.

But Schleiden and Schwann also believed (wrongly) that cells emerged by the self-assembly of nonliving materials, much as crystals form in a solution of salt. This conclusion was in ac-

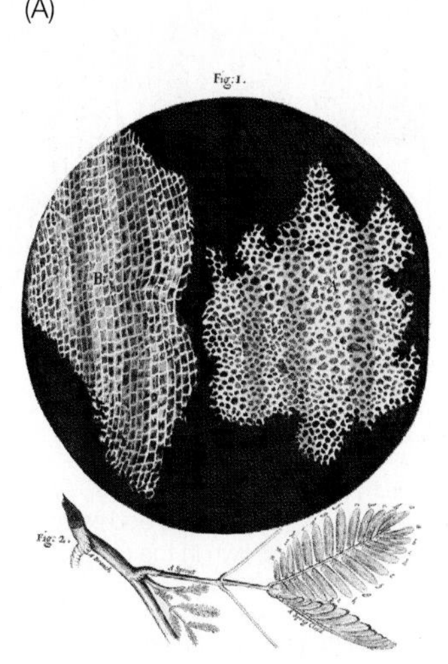


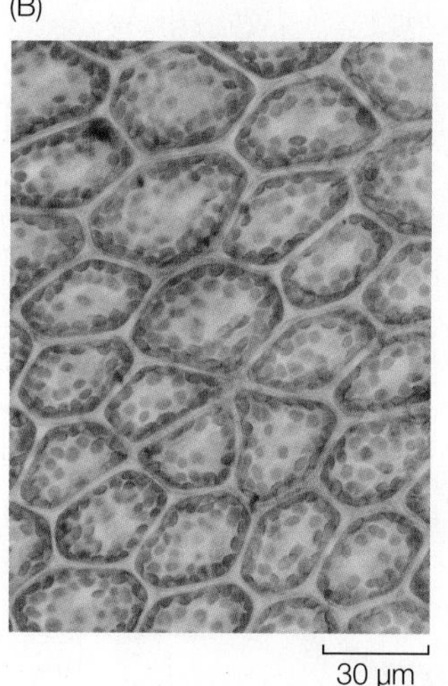


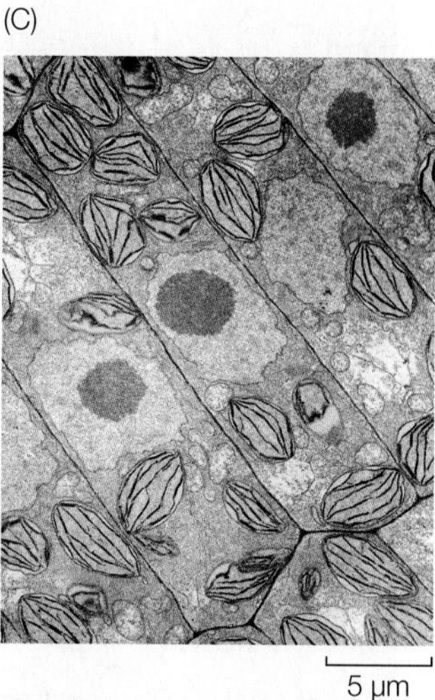


**1.2 Cells Are the Building Blocks of Life** The development of microscopes revealed the microbial world to seventeenth-century scientists such as Robert Hooke, who proposed the concept of cells based on his observations. (A) Hooke drew the cells of a slice of plant tissue (cork) as he saw them under his optical microscope. (B) A modern optical, or "light," microscope reveals the intricacies of cells in a leaf. (C) Transmission electron microscopes (TEMs) allow scientists to see even smaller objects. TEMs do not visualize color; here color has been added to a black-and-white micrograph of cells in a duckweed stem.

cordance with the prevailing view of the day, which was that life can arise from non-life by spontaneous generation—mice from dirty clothes, maggots from dead meat, or insects from pond water.

The debate continued until 1859, when the French Academy of Sciences sponsored a contest for the best experiment to prove or disprove spontaneous generation. The prize was won by the great French scientist Louis Pasteur, who demonstrated that sterile broth directly exposed to the dirt and dust in air developed a culture of microorganisms, but a similar container of broth not directly exposed to air remained sterile (see Figure 4.7). Pasteur's experiment did not prove that it was microorganisms in the air that caused the broth to become infected, but it did uphold the conclusion that life must be present in order for new life to be generated.

Today scientists accept the fact that all cells come from preexisting cells and that the functional properties of organisms derive from the properties of their cells. Since cells of all kinds share both essential mechanisms and a common ancestry that goes back billions of years, modern cell theory has additional elements:

- All cells come from preexisting cells.
- All cells are similar in chemical composition.
- Most of the chemical reactions of life occur in aqueous solution within cells.
- Complete sets of genetic information are replicated and passed on during cell division.
- Viruses lack cellular structure but remain dependent on cellular organisms.

At the same time Schleiden and Schwann were building the foundation for the cell theory, Charles Darwin was beginning to understand how organisms undergo evolutionary change.

### All of life shares a common evolutionary history

**Evolution**—change in the genetic makeup of biological populations through time—is the major unifying principle of biology. Charles Darwin compiled factual evidence for evolution in his 1859 book *On the Origin of Species*. Since then, biologists have gathered massive amounts of data supporting Darwin's theory that all living organisms are descended from a common ancestor. Darwin also proposed one of the most important processes that produce evolutionary change. He argued that differential survival and reproduction among individuals in a population, which he termed **natural selection**, could account for much of the evolution of life.

Although Darwin proposed that living organisms are descended from common ancestors and are therefore related to one another, he did not have the advantage of understanding the mechanisms of genetic inheritance. Even so, he observed that offspring resembled their parents; therefore, he surmised, such mechanisms had to exist. That simple fact is the basis for the concept of a **species**. Although the precise definition of a species is complicated, in its most widespread usage it refers to a group of organisms that can produce viable and fertile offspring with one another.

But offspring do differ from their parents. Any population of a plant or animal species displays variation, and if you select breeding pairs on the basis of some particular trait, that trait is more likely to be present in their offspring than in the general population. Darwin himself bred pigeons, and was well aware of how pigeon fanciers selected breeding pairs to produce offspring with unusual feather patterns, beak shapes, or body sizes (see Figure 21.2). He realized that if humans could select for specific traits in domesticated animals, the same process could operate in nature; hence the term *natural selection* as opposed to artificial (human-imposed) selection.

How would natural selection function? Darwin postulated that different probabilities of survival and reproductive success would do the job. He reasoned that the reproductive capacity of plants and animals, if unchecked, would result in unlimited growth of populations, but we do not observe such growth in nature; in most species, only a small percentage of offspring survive to reproduce. Thus any trait that confers even a small increase in the probability that its possessor will survive and reproduce would be spread in the population.

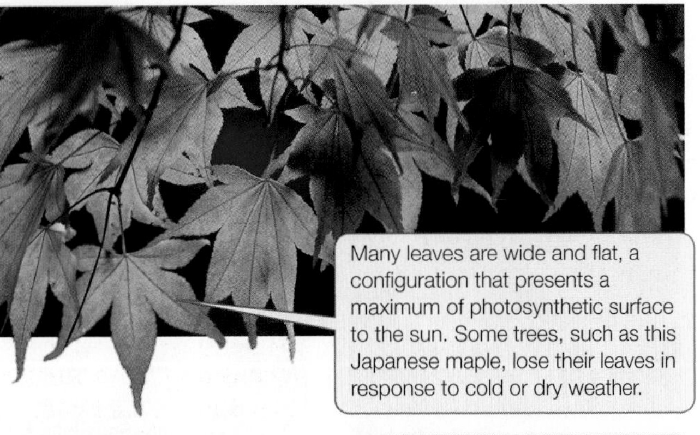

Many leaves are wide and flat, a configuration that presents a maximum of photosynthetic surface to the sun. Some trees, such as this Japanese maple, lose their leaves in response to cold or dry weather.

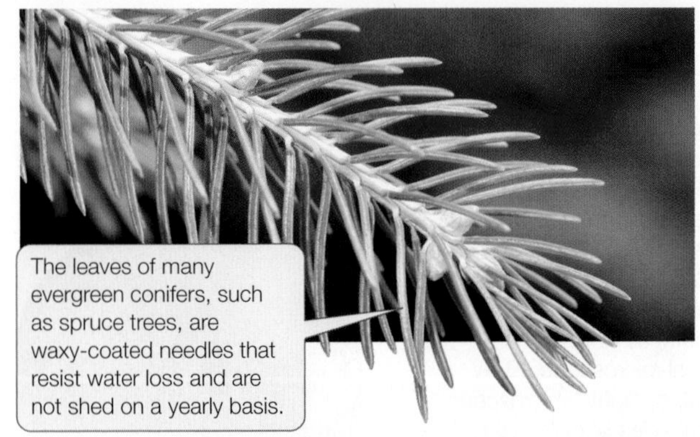

The leaves of many evergreen conifers, such as spruce trees, are waxy-coated needles that resist water loss and are not shed on a yearly basis.

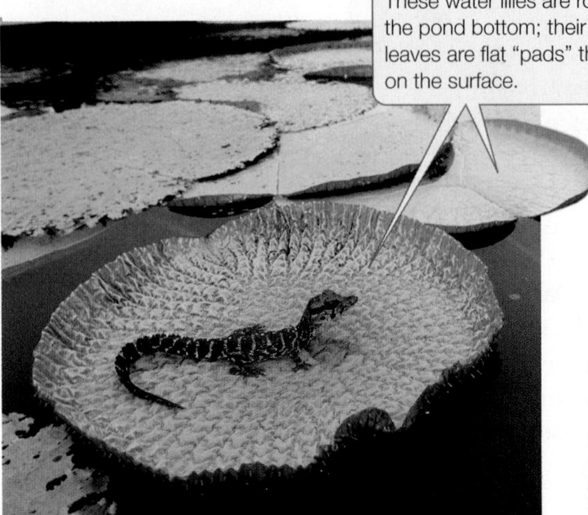

These water lilies are rooted in the pond bottom; their large leaves are flat "pads" that float on the surface.

The ability to climb can be advantageous to a plant, enabling it to reach above other plants to obtain more sunlight. Some of the leaves of this climbing cucumber are tightly furled tendrils that wrap around a stake.

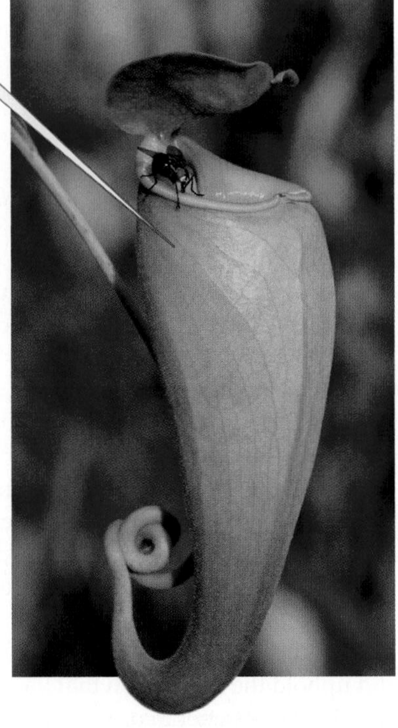

The leaves of pitcher plants form a vessel that holds water. The plant receives extra nutrients from the decomposing bodies of insects that drown in the pitcher.

**1.3 Adaptations to the Environment** The leaves of all plants are specialized for photosynthesis—the sunlight-powered transformation of water and carbon dioxide into larger structural molecules called carbohydrates. The leaves of different plants, however, display many different adaptations to their individual environments.

Because organisms with certain traits survive and reproduce best under specific sets of conditions, natural selection leads to **adaptations**: structural, physiological, or behavioral traits that enhance an organism's chances of survival and reproduction in its environment (**Figure 1.3**). In addition to natural selection, evolutionary processes such as sexual selection (selection due to mate choice) and genetic drift (the random fluctuation of gene frequencies in a population due to chance events) contribute to the rise of diverse adaptations. These processes operating over evolutionary history have led to the remarkable array of life on Earth.

If all cells come from preexisting cells, and if all the diverse species of organisms on Earth are related by descent with modification from a common ancestor, then what is the source of information that is passed from parent to daughter cells and from parental organisms to their offspring?

### Biological information is contained in a genetic language common to all organisms

Cells are the basic building blocks of organisms, but even a single cell is complex, with many internal structures and many functions that depend on information. The information required for a cell to function and interact with other cells—the "blueprint" for existence—is contained in the cell's **genome**, the sum total of all the DNA molecules it contains. **DNA** (deoxyribonucleic acid) molecules are long sequences of four different subunits called **nucleotides**. The sequence of the nucleotides contains genetic information. **Genes** are specific segments of DNA encoding the information the cell uses to make **proteins** (**Figure 1.4**). Protein molecules govern the chemical reactions within cells and form much of an organism's structure.

By analogy with a book, the nucleotides of DNA are like the letters of an alphabet. Protein molecules are the sentences. Combinations of proteins that form structures and control biochemical processes are the paragraphs. The structures and processes that are organized into different systems with specific tasks (such as digestion or transport) are the chapters of the book, and the complete book is the organism. If you were to write out your own genome using four letters to represent the four nucleotides, you would write more than 3 billion letters. Using the size type you are reading now, your genome would fill about a thousand books the size of this one. The mechanisms of evolution, including natural selection, are the authors and editors of all the books in the library of life.

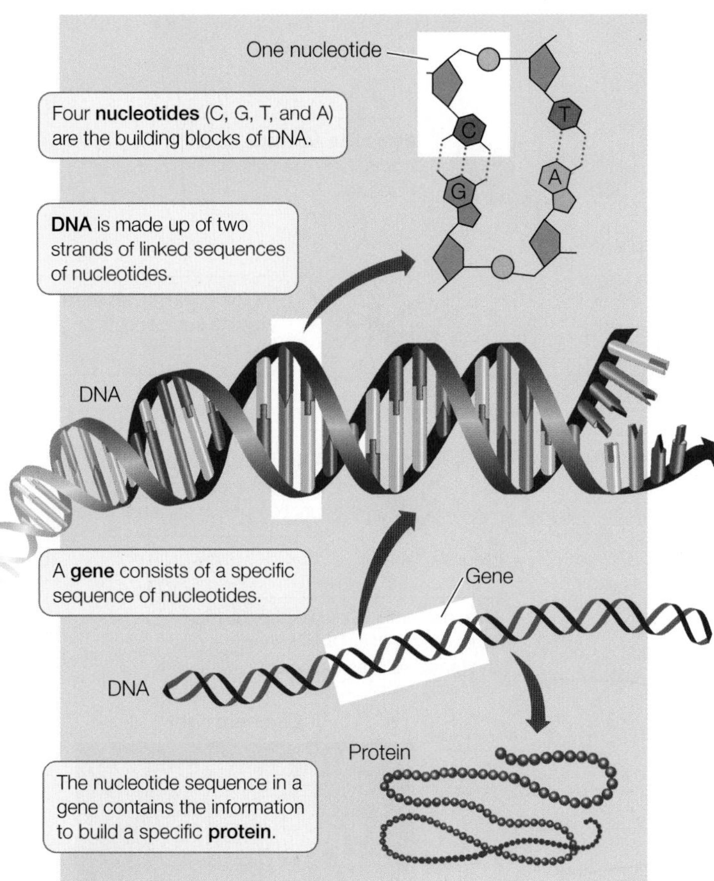

One nucleotide
Four **nucleotides** (C, G, T, and A) are the building blocks of DNA.
C
G
T
A
**DNA** is made up of two strands of linked sequences of nucleotides.
DNA
A **gene** consists of a specific sequence of nucleotides.
Gene
DNA
Protein
The nucleotide sequence in a gene contains the information to build a specific **protein**.


**1.4 DNA Is Life's Blueprint** The instructions for life are contained in the sequences of nucleotides in DNA molecules. Specific DNA nucleotide sequences comprise genes. The average length of a single human gene is 16,000 nucleotides. The information in each gene provides the cell with the information it needs to manufacture molecules of a specific protein.

All the cells of a multicellular organism contain the same genome, yet different cells have different functions and form different structures—contractile proteins form in muscle cells, hemoglobin in red blood cells, digestive enzymes in gut cells, and so on. Therefore, different types of cells in an organism must express different parts of the genome. How cells control gene expression in ways that enable a complex organism to develop and function is a major focus of current biological research.

The genome of an organism consists of thousands of genes. If the nucleotide sequence of a gene is altered, it is likely that the protein that gene encodes will be altered. Alterations of the genome are called *mutations*. Mutations occur spontaneously; they can also be induced by outside factors, including chemicals and radiation. Most mutations are either harmful or have no effect, but occasionally a mutation improves the functioning of the organism under the environmental conditions it encounters. Such beneficial mutations are the raw material of evolution and lead to adaptations.

### Cells use nutrients to supply energy and to build new structures

Living organisms acquire *nutrients* from the environment. Nutrients supply the organism with energy and raw materials for carrying out biochemical reactions. Life depends on thousands of biochemical reactions that occur inside cells. Some of these reactions break down nutrient molecules into smaller chemical units, and in the process some of the energy contained in the chemical bonds of the nutrients is captured by high-energy molecules that can be used to do different kinds of cellular work.

One obvious kind of work cells do is mechanical—moving molecules from one cellular location to another, moving whole cells or tissues, or even moving the organism itself, as muscles do (**Figure 1.5A**). The most basic cellular work is the building, or *synthesis*, of new complex molecules and structures from smaller chemical units. For example, we are all familiar with the fact that carbohydrates eaten today may be deposited in the body as fat tomorrow (**Figure 1.5B**). Still another kind of work is the electrical work that is the essence of information processing in nervous systems. The sum total of all the chemical transformations and other work done in all the cells of an organism is its **metabolism**, or **metabolic rate**.

The myriad of biochemical reactions that go on in cells are integrally linked in that the products of one are the raw materials of the next. These complex networks of reactions must be integrated and precisely controlled; when they are not, the result is disease.

### Living organisms regulate their internal environment

Multicellular organisms have an *internal environment* that is not cellular. That is, their individual cells are bathed in extracellular fluids, from which they receive nutrients and into which they excrete waste products of metabolism. The cells of multicellu-

**1.5 Energy Can Be Used Immediately or Stored** (A) Animal cells break down and release the energy contained in the chemical bonds of food molecules to do mechanical work—in this kangaroo's case, to jump. (B) The cells of this Arctic ground squirrel have broken down the complex carbohydrates in plants and converted their molecules into fats, which are stored in the animal's body to provide an energy supply for the cold months.

lar organisms are specialized, or *differentiated*, to contribute in some way to the maintenance of the internal environment. With the evolution of specialization, differentiated cells lost many of the functions carried out by single-celled organisms, and must depend on the internal environment for essential services.

To accomplish their specialized tasks, assemblages of differentiated cells are organized into *tissues.* For example, a single muscle cell cannot generate much force, but when many cells combine to form the tissue of a working muscle, considerable force and movement can be generated (see Figure 1.5B). Different tissue types are organized to form *organs* that accomplish specific functions. For example, the heart, brain, and stomach are each constructed of several types of tissues. Organs whose functions are interrelated can be grouped into *organ systems;* the stomach, intestine, and esophagus, for example, are parts of the digestive system. The functions of cells, tissues, organs, and organ systems are all integral to the multicellular *organism.* We cover the biology of organisms in Parts Eight and Nine of this book.

### Living organisms interact with one another

The internal hierarchy of the individual organism is matched by the external hierarchy of the biological world (**Figure 1.6**). Organisms do not live in isolation. A group of individuals of the same species that interact with one another is a *population,* and populations of all the species that live and interact in the same area are called a *community*. Communities together with their abiotic environment constitute an *ecosystem.*

Individuals in a population interact in many different ways. Animals eat plants and other animals (usually members of another species) and compete with other species for food and other resources. Some animals will prevent other individuals of their own species from exploiting a resource, whether it be food, nesting sites, or mates. Animals may also *cooperate* with members of their species, forming social units such as a termite colony or a flock of birds. Such interactions have resulted in the evolution of social behaviors such as communication.

Plants also interact with their external environment, which includes other plants, animals, and microorganisms. All terrestrial plants depend on complex partnerships with fungi, bacteria, and animals. Some of these partnerships are necessary to obtain nutrients, some to produce fertile seeds, and still others to disperse seeds. Plants compete with each other

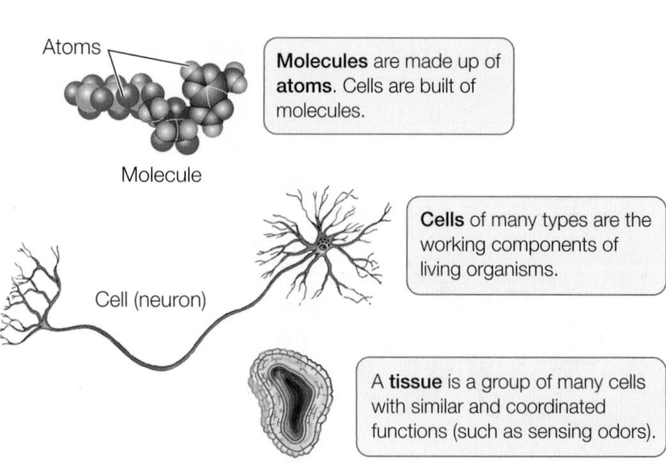
Atoms
Molecule
**Molecules** are made up of **atoms**. Cells are built of molecules.
Cell (neuron)
**Cells** of many types are the working components of living organisms.
Neural tissue
A **tissue** is a group of many cells with similar and coordinated functions (such as sensing odors).

Organ (brain)
**Organs** combine several tissues that function together. Organs form **systems**, such as the nervous system.
Organism (fish)
An **organism** is a recognizable, self-contained individual. Complex multicellular organisms are made up of organs and organ systems.
Population (school of fish)
A **population** is a group of many organisms of the same species.

**Communities** consist of populations of many different species.
Community (coral reef)
Biological communities in the same geographical location form **ecosystems**. Ecosystems exchange energy and create Earth's **biosphere**.
Biosphere

**1.6 Biology Is Studied at Many Levels of Organization** Life's properties emerge when DNA and other molecules are organized in cells. Energy flows through all the biological levels shown here.

**yourBioPortal.com**
GO TO Web Activity 1.1 • The Hierarchy of Life

for light and water, and they have ongoing evolutionary interactions with the animals that eat them, evolving anti-predation adaptations or ways to attract the animals that assist in their reproduction. The interactions of populations of different plant and animal species in a community are major evolutionary forces that produce specialized adaptations.

Communities interacting over a broad geographic area with distinguishing physical features form ecosystems; examples might include an Arctic tundra, a coral reef, or a tropical rainforest. The ways in which species interact with one another and with their environment in communities and in ecosystems is the subject of *ecology* and of Part Ten of this book.

### Discoveries in biology can be generalized

Because all life is related by descent from a common ancestor, shares a genetic code, and consists of similar building blocks—cells—knowledge gained from investigations of one type of organism can, with care, be generalized to other organisms. Biologists use **model systems** for research, knowing that they can extend their findings to other organisms, including humans. For example, our basic understanding of the chemical reactions in cells came from research on bacteria but is applicable to all cells, including those of humans. Similarly, the biochemistry of photosynthesis—the process by which plants use sunlight to produce biological molecules—was largely worked out from experiments on *Chlorella*, a unicellular green alga (see Figure 10.13). Much of what we know about the genes that control plant development is the result of work on *Arabidopsis thaliana*, a relative of the mustard plant. Knowledge about how animals develop has come from work on sea urchins, frogs, chickens, roundworms, and fruit flies. And recently, the discovery of a major gene controlling human skin color came from work on zebrafish. Being able to generalize from model systems is a powerful tool in biology.

### 1.1 RECAP

**Living organisms are made of (or depend on) cells, are related by common descent and evolve, contain genetic information and use it to reproduce, extract energy from their environment and use it to do biological work, synthesize complex molecules to construct biological structures, regulate their internal environment, and interact with one another.**

- Describe the relationship between evolution by natural selection and the genetic code. See pp. 6–7
- Why can the results of biological research on one species often be generalized to very different species? See p. 9

Now that you have an overview of the major features of life that you will explore in depth in this book, you can ask how and when life first emerged. In the next section we will summarize briefly the history of life from the earliest simple life forms to the complex and diverse organisms that inhabit our planet today.

## 1.2 How Is All Life on Earth Related?

What do biologists mean when they say that all organisms are *genetically related*? They mean that species on Earth share a *common ancestor*. If two species are similar, as dogs and wolves are, then they probably have a common ancestor in the fairly recent past. The common ancestor of two species that are more different—say, a dog and a deer—probably lived in the more distant past. And if two organisms are very different—such as a dog and a clam—then we must go back to the *very* distant past to find their common ancestor. How can we tell how far back in time the common ancestor of any two organisms lived? In other words, how do we discover the evolutionary relationships among organisms?

For many years, biologists have investigated the history of life by studying the *fossil record*—the preserved remains of organisms that lived in the distant past (**Figure 1.7**). Geologists supplied knowledge about the ages of fossils and the nature of the environments in which they lived. Biologists then inferred the evolutionary relationships among living and fossil organisms by comparing their anatomical similarities and differences. Frequently big gaps existed in the fossil record, forcing biologists to predict the nature of the "missing links" between two lineages of organisms. As the fossil record became more complete, those missing links were filled in.

Molecular methods for comparing genomes, described in Chapter 24, are enabling biologists to more accurately establish the degrees of relationship between living organisms and to use that information to interpret the fossil record. Molecular information can occasionally be gleaned from fossil specimens, such as recently deciphered genetic material from fossil bones of Ne-

**1.7 Fossils Give Us a View of Past Life** This fossil, formed some 150 million years ago, is that of an *Archaeopteryx*, the earliest known representative of the birds. Birds evolved from the same group of reptiles as the modern crocodiles.

anderthals that led to the conclusion that even though Neanderthals and modern humans coexisted, they did not interbreed.

In general, the greater the differences between the genomes of two species, the more distant their common ancestor. Using molecular techniques, biologists are exploring fundamental questions about life. What were the earliest forms of life? How did simple organisms give rise to the great diversity of organisms alive today? Can we reconstruct a family tree of life?

### Life arose from non-life via chemical evolution

Geologists estimate that Earth formed between 4.6 and 4.5 billion years ago. At first, the planet was not a very hospitable place. It was some 600 million years or more before the earliest life evolved. If we picture the history of Earth as a 30-day month, life first appeared somewhere toward the end of the first week (**Figure 1.8**).

When we consider how life might have arisen from nonliving matter, we must take into account the properties of the young Earth's atmosphere, oceans, and climate, all of which were very different than they are today. Biologists postulate that complex biological molecules first arose through the random physical association of chemicals in that environment. Experiments simulating the conditions on early Earth have confirmed that the generation of complex molecules under such conditions is possible, even probable. The critical step for the evolution of life, however, had to be the appearance of molecules that could reproduce themselves and also serve as templates for the synthesis of large molecules with complex but stable shapes. The variation of the shapes of these large, stable molecules (described in Chapters 3 and 4) enabled them to participate in increasing numbers and kinds of chemical reactions with other molecules.

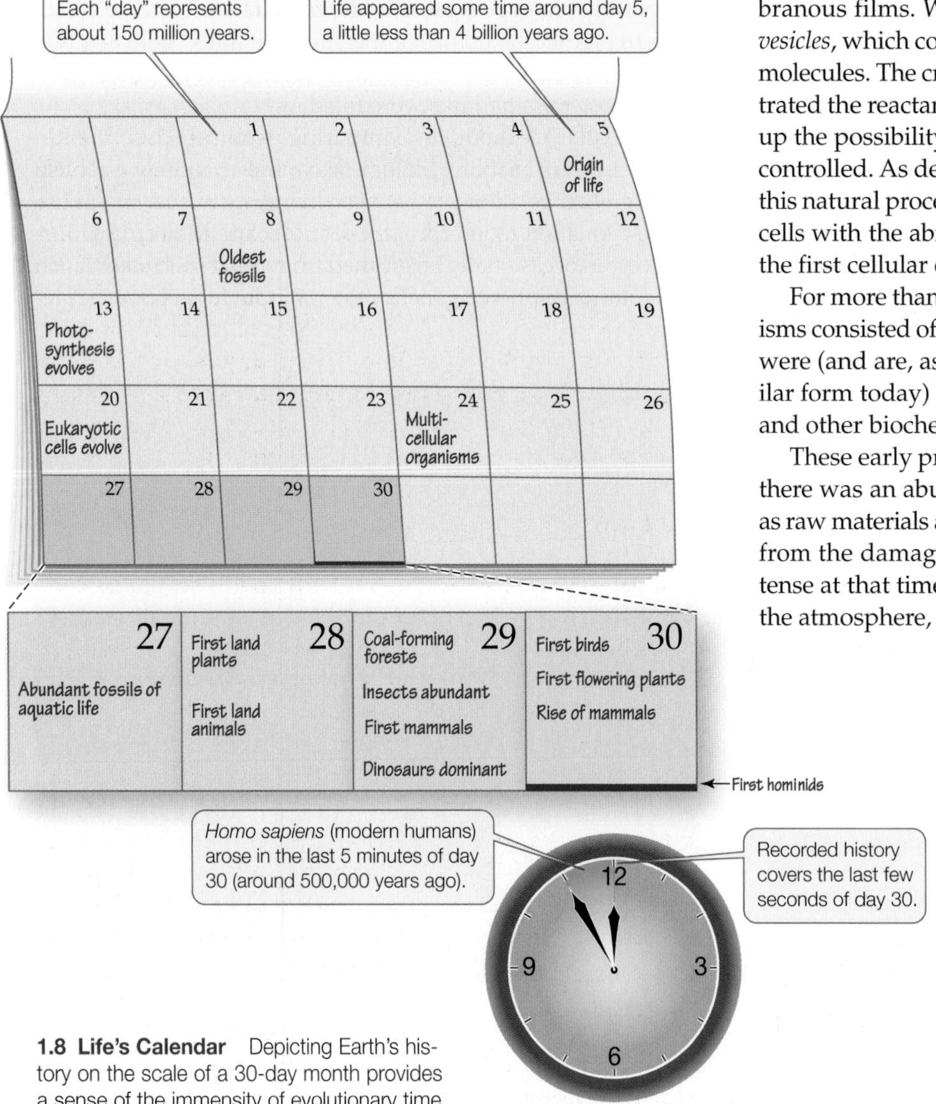


**1.8 Life's Calendar** Depicting Earth's history on the scale of a 30-day month provides a sense of the immensity of evolutionary time.

### Cellular structure evolved in the common ancestor of life

The second critical step in the origin of life was the enclosure of complex biological molecules by *membranes* that contained them in a compact internal environment separate from the surrounding external environment. Fatlike molecules played a critical role because they are not soluble in water and they form membranous films. When agitated, these films can form spherical *vesicles*, which could have enveloped assemblages of biological molecules. The creation of an internal environment that concentrated the reactants and products of chemical reactions opened up the possibility that those reactions could be integrated and controlled. As described in Section 4.4, scientists postulate that this natural process of membrane formation resulted in the first cells with the ability to replicate themselves—the evolution of the first cellular organisms.

For more than 2 billion years after cells originated, all organisms consisted of only one cell. These first unicellular organisms were (and are, as multitudes of their descendants exist in similar form today) **prokaryotes**. Prokaryotic cells consist of DNA and other biochemicals enclosed in a membrane.

These early prokaryotes were confined to the oceans, where there was an abundance of complex molecules they could use as raw materials and sources of energy. The ocean shielded them from the damaging effects of ultraviolet light, which was intense at that time because there was little or no oxygen ($O_2$) in the atmosphere, and hence no protective ozone ($O_3$) layer.

### Photosynthesis changed the course of evolution

To fuel their cellular metabolism, the earliest prokaryotes took in molecules directly from their environment and broke these small molecules down to release and use the energy contained in their chemical bonds. Many modern species of prokaryotes still function this way, and very successfully. During the early eons of life on Earth, there was no oxygen in the atmosphere. In fact, oxygen was toxic to the life forms that existed then.

**1.9 Photosynthetic Organisms Changed Earth's Atmosphere** These strands are composed of many cells of cyanobacteria. This modern species (*Oscillatoria tenuis*) may be very similar to the early photosynthetic prokaryotes responsible for the buildup of oxygen in Earth's atmosphere.

About 2.7 billion years ago, the evolution of **photosynthesis** changed the nature of life on Earth. The chemical reactions of photosynthesis transform the energy of sunlight into a form of biological energy that can power the synthesis of large molecules (see Chapter 10). These large molecules are the building blocks of cells, and they can be broken down to provide metabolic energy. Photosynthesis is the basis of much of life on Earth today because its energy-capturing processes provide food for other organisms.

Early photosynthetic cells were probably similar to present-day prokaryotes called *cyanobacteria* (**Figure 1.9**). Over time, photosynthetic prokaryotes became so abundant that vast quantities of $O_2$, which is a by-product of photosynthesis, slowly began to accumulate in the atmosphere. Oxygen was poisonous to many of the prokaryotes that lived at that time. Those organisms that did tolerate oxygen, however, were able to proliferate as the presence of oxygen opened up vast new avenues of evolution. *Aerobic metabolism* (energy production based on the conversion of $O_2$) is more efficient than *anaerobic* (non-$O_2$-using) *metabolism*, and today it is used by the majority of Earth's organisms. Aerobic metabolism allowed cells to grow larger.

Oxygen in the atmosphere also made it possible for life to move onto land. For most of life's history, ultraviolet (UV) radiation falling on Earth's surface was too intense to allow life to exist outside the shielding water. But the accumulation of photosynthetically generated oxygen in the atmosphere for more than 2 billion years gradually produced a layer of ozone in the upper atmosphere. By about 500 million years ago, the ozone layer was sufficiently dense and absorbed enough UV radiation to make it possible for organisms to leave the protection of the water and live on land.

### Eukaryotic cells evolved from prokaryotes

Another important step in the history of life was the evolution of cells with discrete intracellular compartments, called **organelles**, which were capable of taking on specialized cellular functions. This event happened about 3 weeks into our calendar of Earth's history (see Figure 1.8). One of these organelles, the dense-appearing *nucleus* (Latin *nux*, "nut" or "core"), came to contain the cell's genetic information and gives these cells their name: **eukaryotes** (Greek *eu*, "true"; *karyon*, "kernel" or "core"). The eukaryotic cell is completely distinct from the cells of prokaryotes (*pro*, "before"), which lack nuclei and other internal compartments.

Some organelles are hypothesized to have originated by **endosymbiosis** when cells ingested smaller cells. The *mitochondria* that generate a cell's energy probably evolved from engulfed prokaryotic organisms. And *chloroplasts*—organelles specialized to conduct photosynthesis—could have originated when photosynthetic prokaryotes were ingested by larger eukaryotes. If the larger cell failed to break down this intended food object, a partnership could have evolved in which the ingested prokaryote provided the products of photosynthesis and the host cell provided a good environment for its smaller partner.

### Multicellularity arose and cells became specialized

Until just over a billion years ago, all the organisms that existed—whether prokaryotic or eukaryotic—were unicellular. An important evolutionary step occurred when some eukaryotes failed to separate after cell division, remaining attached to each other. The permanent association of cells made it possible for some cells to specialize in certain functions, such as reproduction, while other cells specialized in other functions, such as absorbing nutrients and distributing them to neighboring cells. This **cellular specialization** enabled multicellular eukaryotes to increase in size and become more efficient at gathering resources and adapting to specific environments.

### Biologists can trace the evolutionary tree of life

If all the species of organisms on Earth today are the descendants of a single kind of unicellular organism that lived almost 4 billion years ago, how have they become so different? A simplified answer is that as long as individuals within a population mate with one another, structural and functional changes can evolve within that population, but the population will remain one species. However, if something happens to isolate some members of a population from the others, the structural and functional differences between the two groups may accumulate over time. The two groups may diverge to the point where their members can no longer reproduce with each other and are thus distinct species. We discuss this evolutionary process, called *speciation*, in Chapter 23.

Biologists give each species a distinctive scientific name formed from two Latinized names (a **binomial**). The first name identifies the species' *genus*—a group of species that share a recent common ancestor. The second is the name of the species. For example, the scientific name for the human species is *Homo sapiens*: *Homo* is our genus and *sapiens* our species. *Homo* is Latin for "man"; *sapiens* is from the Latin for word for "wise" or "rational."

Tens of millions of species exist on Earth today. Many times that number lived in the past but are now extinct. Many millions of speciation events created this vast diversity, and the unfolding of these events can be diagrammed as an evolutionary "tree" whose branches describe the order in which populations split and eventually evolved into new species, as described in Chapter 22. Much of biology is based on comparisons among species, and these comparisons are useful precisely because we can place species in an evolutionary context relative to one another. Our ability to do this has been greatly enhanced in recent decades by our ability to sequence and compare the genomes of different species.

Genome sequencing and other molecular techniques have allowed *systematists*—scientists who study the evolution and classification of life's diverse organisms—to augment evolutionary knowledge based on the fossil record with a vast array of molecular evidence. The result is the ongoing compilation of *phylogenetic trees* that document and diagram evolutionary relationships as part of an overarching tree of life, the broadest categories of which are shown in **Figure 1.10**. (The tree is expanded in this book's Appendix; you can also explore the tree interactively at http://tolweb.org/tree.)

Although many details remain to be clarified, the broad outlines of the tree of life have been determined. Its branching patterns are based on a rich array of evidence from fossils, structures, metabolic processes, behavior, and molecular analyses of genomes. Molecular data in particular have been used to separate the tree into three major **domains**: Archaea, Bacteria, and Eukarya. The organisms of each domain have been evolving separately from those in the other domains for more than a billion years.

Organisms in the domains **Archaea** and **Bacteria** are single-celled prokaryotes. However, members of these two groups differ so fundamentally in their metabolic processes that they are believed to have separated into distinct evolutionary lineages very early. Species belonging to the third domain—**Eukarya**—have eukaryotic cells whose mitochondria and chloroplasts may have originated from the ingestion of prokaryotic cells, as described on page 11.

The three major groups of multicellular eukaryotes—plants, fungi, and animals—each evolved from a different group of the eukaryotes generally referred to as *protists*. The chloroplast-containing, photosynthetic protist that gave rise to plants was completely distinct from the protist that was ancestral to both animals and fungi, as can be seen from the branching pattern of Figure 1.10. Although most protists are unicellular (and thus sometimes called *microbial eukaryotes*), multicellularity has evolved in several protist lineages.

### The tree of life is predictive

There are far more species alive on Earth than biologists have discovered and described to date. In fact, most species on Earth

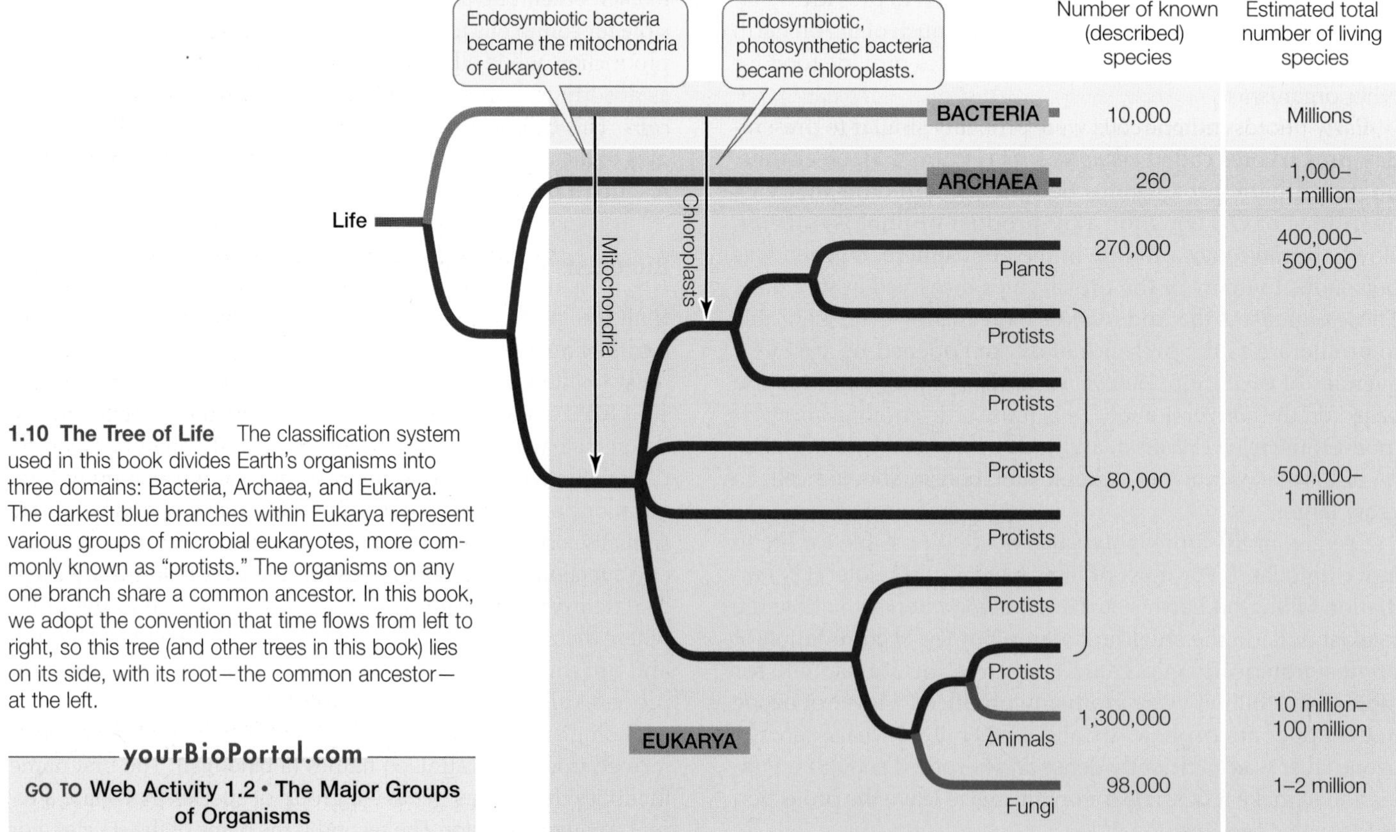


**1.10 The Tree of Life** The classification system used in this book divides Earth's organisms into three domains: Bacteria, Archaea, and Eukarya. The darkest blue branches within Eukarya represent various groups of microbial eukaryotes, more commonly known as "protists." The organisms on any one branch share a common ancestor. In this book, we adopt the convention that time flows from left to right, so this tree (and other trees in this book) lies on its side, with its root—the common ancestor—at the left.

**yourBioPortal.com**
GO TO **Web Activity 1.2 • The Major Groups of Organisms**

have yet to be discovered by humans (see Section 32.4 for a discussion of how we know this). When we encounter a new species, its placement on the tree of life immediately tells us a great deal about its biology. In addition, understanding relationships among species allows biologists to make predictions about species that have not yet been studied, based on our knowledge of those that have.

For example, until phylogenetic methods were developed, it took years of investigation to isolate and identify most newly encountered human pathogens, and even longer to discover how these pathogens moved into human populations. Today, pathogens that cause diseases such as the flu are identified quickly on the basis of their evolutionary relationships. Placement in an evolutionary tree also gives us clues about the disease's biology, possible effective treatments, and the origin of the pathogen (see Chapters 21 and 22).

### 1.2 RECAP

**The first cellular life on Earth was prokaryotic and arose about 4 billion years ago. The complexity of the organisms that exist today is the result of several important evolutionary events, including the evolution of photosynthesis, eukaryotic cells, and multicellularity. The genetic relationships of all organisms can be shown as a branching tree of life.**

- Discuss the evolutionary significance of photosynthesis. **See pp. 10–11**
- What do the domains of life represent? What are the major groups of eukaryotes? **See p. 12 and Figure 1.10**

In February of 1676, Robert Hooke received a letter from the physicist Sir Isaac Newton in which Newton famously remarked, "If I have seen a little further, it is by standing on the shoulders of giants." We all stand on the shoulders of giants, building on the research of earlier scientists. By the end of this course, you will know more about evolution than Darwin ever could have, and you will know infinitely more about cells than Schleiden and Schwann did. Let's look at the methods biologists use to expand our knowledge of life.

## 1.3 How Do Biologists Investigate Life?

Regardless of the many different tools and methods used in research, all scientific investigations are based on *observation* and *experimentation*. In both, scientists are guided by the *scientific method,* one of the most powerful tools of modern science.

### Observation is an important skill

Biologists have always observed the world around them, but today our ability to observe is greatly enhanced by technologies such as electron microscopes, DNA chips, magnetic resonance imaging, and global positioning satellites. These technologies have improved our ability to observe at all levels, from the distribution of molecules in the body to the distribution of fish in the oceans. For example, not too long ago marine biologists were only able to observe the movement of fish in the ocean by putting physical tags on the fish, releasing them, and hoping that a fisherman would catch that fish and send back the tag—and even that would reveal only where the fish ended up. Today we can attach electronic recording devices to fish that continuously record not only where the fish is, but also how deep it swims and the temperature and salinity of the water around it (**Figure 1.11**). The tags download this information to a satellite, which relays it back to researchers. Suddenly we are acquiring a great deal of knowledge about the distribution of life in the oceans—information that is relevant to studies of climate change.

Technologies that enable us to *quantify* observations are very important in science. For example, for hundreds of years species were classified by generally qualitative descriptions of the physical differences between them. There was no way of objectively calculating evolutionary distances between organisms, and biologists had to depend on the fossil record for insight. Today our ability to rapidly analyze DNA sequences enables quantitative estimates of evolutionary distances, as described in Parts Five and Six of this book. The ability to gather quantitative observations adds greatly to the biologist's ability to make strong conclusions.

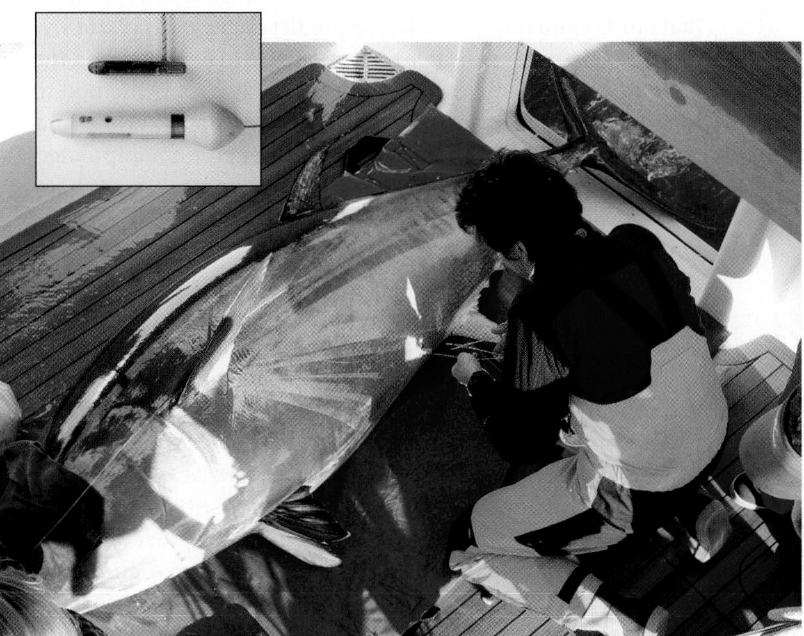

**1.11 Tuna Tracking** Marine biologist Barbara Block attaches computerized data recording tags (inset) to a live bluefin tuna before returning it to the ocean. Such tags make it possible to track an individual tuna wherever it travels in the world's oceans.

## The scientific method combines observation and logic

Observations lead to questions, and scientists make additional observations and do experiments to answer those questions. The conceptual approach that underlies most modern scientific investigations is the **scientific method**. This powerful tool, also called the *hypothesis–prediction* (*H–P*) *method*, has five steps: (1) making *observations*; (2) asking *questions*; (3) forming *hypotheses*, or tentative answers to the questions; (4) making *predictions* based on the hypotheses; and (5) *testing* the predictions by making additional observations or conducting experiments (**Figure 1.12**).

After posing a question, a scientist uses *inductive logic* to propose a tentative answer. Inductive logic involves taking observations or facts and creating a new proposition that is compatible with those observations or facts. Such a tentative proposition is called a **hypothesis**. In formulating a hypothesis, scientists put together the facts they already know to formulate one or more possible answers to the question. For example, at the opening of this chapter you learned that scientists have observed the rapid decline of amphibian populations worldwide and are asking why. Some scientists have hypothesized that a fungal disease is a cause; other scientists have hypothesized that increased exposure to ultraviolet radiation is a cause. Tyrone Hayes hypothesized that exposure to agricultural chemicals could be a cause. He knew that the most widely used chemical herbicide is atrazine; that it is mostly applied in the spring, when amphibians are breeding; and that atrazine is a common contaminant in the waters in which amphibians live as they develop into adults.

The next step in the scientific method is to apply a different form of logic—*deductive logic*—to make predictions based on the hypothesis. Deductive logic starts with a statement believed to be true and then goes on to predict what facts would also have to be true to be compatible with that statement. Based on his hypothesis, Tyrone Hayes predicted that frog tadpoles exposed to atrazine would show adverse effects of the chemical once they reached adulthood.

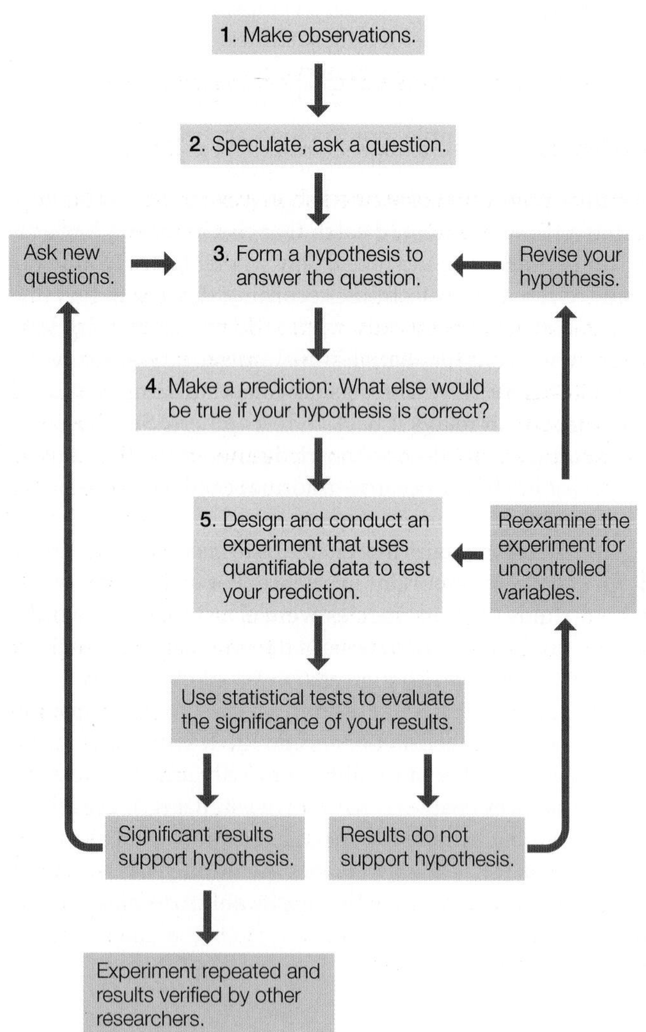


**1.12 The Scientific Method** The process of observation, speculation, hypothesis, prediction, and experimentation is the cornerstone of modern science. Answers gleaned through experimentation lead to new questions, more hypotheses, further experiments, and expanding knowledge.

## Good experiments have the potential to falsify hypotheses

Once predictions are made from a hypothesis, experiments can be designed to test those predictions. The most informative experiments are those that have the ability to show that the prediction is wrong. If the prediction is wrong, the hypothesis must be questioned, modified, or rejected.

There are two general types of experiments, both of which compare data from different groups or samples. A *controlled* experiment manipulates one or more of the factors being tested; *comparative* experiments compare unmanipulated data gathered from different sources. As described at the opening of this chapter, Tyrone Hayes and his colleagues conducted both types of experiment to test the prediction that the herbicide atrazine, a contaminant in freshwater ponds and streams throughout the world, affects the development of frogs.

In a **controlled experiment**, we start with groups or samples that are as similar as possible. We predict on the basis of our hypothesis that some critical factor, or **variable**, has an effect on the phenomenon we are investigating. We devise some method to manipulate *only that variable* in an "experimental" group and compare the resulting data with data from an unmanipulated "control" group. If the predicted difference occurs, we then apply statistical tests to ascertain the probability that the manipulation created the difference (as opposed to the difference being the result of random chance). **Figure 1.13** describes one of the many controlled experiments performed by the Hayes laboratory to quantify the effects of atrazine on male frogs.

The basis of controlled experiments is that one variable is manipulated while all others are held constant. The variable that is manipulated is called the *independent variable*, and the response that is measured is the *dependent variable*. A good controlled experiment is not easy to design because biological variables are so interrelated that it is difficult to alter just one.

A **comparative experiment** starts with the prediction that there will be a difference between samples or groups based on the hypothesis. In comparative experiments, however, we can-

INVESTIGATING LIFE

**1.13 Controlled Experiments Manipulate a Variable** The Hayes laboratory created controlled environments that differed only in the concentrations of atrazine in the water. Eggs from leopard frogs (*Rana pipiens*) raised specifically for laboratory use were allowed to hatch and the tadpoles were separated into experimental tanks containing water with different concentrations of atrazine.

**HYPOTHESIS** **Exposure to atrazine during larval development causes abnormalities in the reproductive system of male frogs.**

**METHOD**

1. Establish 9 tanks in which all attributes are held constant except the water's atrazine concentrations. Establish 3 atrazine conditions (3 replicate tanks per condition): 0 ppb (control condition), 0.1 ppb, and 25 ppb.
2. Place *Rana pipiens* tadpoles from laboratory-reared eggs in the 9 tanks (30 tadpoles per replicate).
3. When tadpoles have transitioned into adults, sacrifice the animals and evaluate their reproductive tissues.
4. Test for correlation of degree of atrazine exposure with the presence of abnormalities in the reproductive systems of male frogs.

**RESULTS**

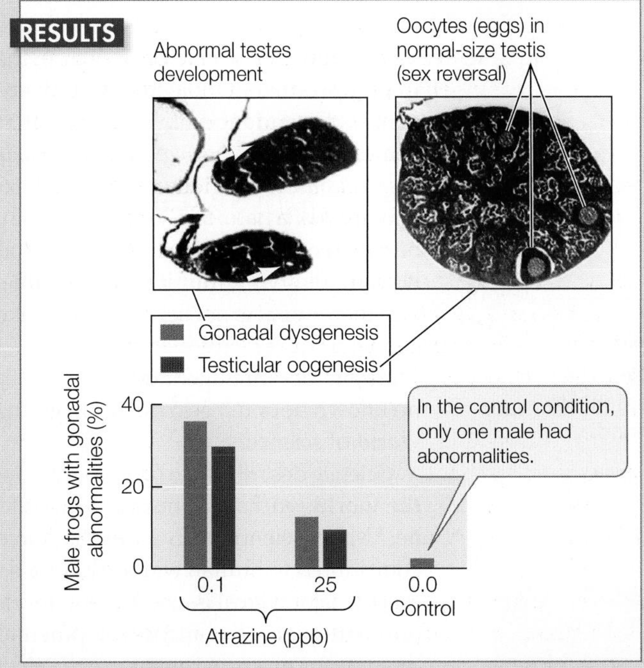


**CONCLUSION** **Exposure to atrazine at concentrations as low as 0.1 ppb induces abnormalities in the male reproductive systems of frogs. The effect is not proportional to the level of exposure.**

Go to **yourBioPortal.com** for original citations, discussions, and relevant links for all INVESTIGATING LIFE figures.

not control the variables; often we cannot even identify all the variables that are present. We are simply gathering and comparing data from different sample groups.

When his controlled experiments indicated that atrazine indeed affects reproductive development in frogs, Hayes and his colleagues performed a comparative experiment. They collected frogs and water samples from eight widely separated sites across the United States and compared the incidence of abnormal frogs from environments with very different levels of atrazine (**Figure 1.14**). Of course, the sample sites differed in many ways besides the level of atrazine present.

The results of experiments frequently reveal that the situation is more complex than the hypothesis anticipated, thus raising new questions. In the Hayes experiments, for example, there was no clear direct relationship between the *amount* of atrazine present and the percentage of abnormal frogs: there were fewer abnormal frogs at the highest concentrations of atrazine than at lower concentrations. There are no "final answers" in science. Investigations consistently reveal more complexity than we expect. The scientific method is a tool to identify, assess, and understand that complexity.

**yourBioPortal.com**

GO TO Animated Tutorial 1.1 • The Scientific Method

### Statistical methods are essential scientific tools

Whether we do comparative or controlled experiments, at the end we have to decide whether there is a difference between the samples, individuals, groups, or populations in the study. How do we decide whether a measured difference is enough to support or falsify a hypothesis? In other words, how do we decide in an unbiased, objective way that the measured difference is significant?

Significance can be measured with statistical methods. Scientists use statistics because they recognize that variation is always present in any set of measurements. Statistical tests calculate the probability that the differences observed in an experiment could be due to random variation. The results of statistical tests are therefore probabilities. A statistical test starts with a **null hypothesis**—the premise that no difference exists. When quantified observations, or **data**, are collected, statistical methods are applied to those data to calculate the likelihood that the null hypothesis is correct.

More specifically, statistical methods tell us the probability of obtaining the same results by chance even if the null hypothesis were true. We need to eliminate, insofar as possible, the chance that any differences showing up in the data are merely the result of random variation in the samples tested. Scientists generally conclude that the differences they measure are significant if statistical tests show that the *probability of error* (that is, the probability that the same results can be obtained by mere chance) is 5 percent or lower.

### Not all forms of inquiry are scientific

Science is a unique human endeavor that is bounded by certain standards of practice. Other areas of scholarship share with science the practice of making observations and asking ques-

# INVESTIGATING LIFE

## 1.14 Comparative Experiments Look for Differences among Groups

To see whether the presence of atrazine correlates with reproductive system abnormalities in male frogs, the Hayes lab collected frogs and water samples from different locations around the U.S. The analysis that followed was "blind," meaning that the frogs and water samples were coded so that experimenters working with each specimen did not know which site the specimen came from.

**HYPOTHESIS** **Presence of the herbicide atrazine in environmental water correlates with reproductive system abnormalities in frog populations.**

**METHOD**

1. Based on commercial sales of atrazine, select 4 sites (sites 1–4) less likely and 4 sites (sites 5–8) more likely to be contaminated with atrazine.
2. Visit all sites in the spring (i.e., when frogs have transitioned from tadpoles into adults); collect frogs and water samples.
3. In the laboratory, sacrifice frogs and examine their reproductive tissues, documenting abnormalities.
4. Analyze the water samples for atrazine concentration (the sample for site 7 was not tested).
5. Quantify and correlate the incidence of reproductive abnormalities with environmental atrazine concentrations.

**RESULTS**

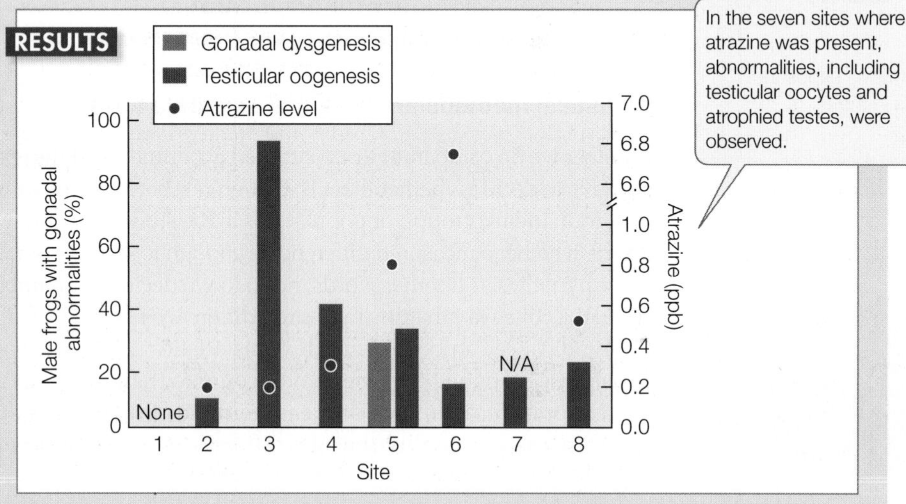


**CONCLUSION** **Reproductive abnormalities exist in frogs from environments in which aqueous atrazine concentration is 0.2 ppb or above. The incidence of abnormalities does not appear to be proportional to atrazine concentration at the time of transition to adulthood.**

**FURTHER INVESTIGATION:** The highest proportion of abnormal frogs was found at site 3, located on a wildlife reserve in Wyoming. What kind of data and observations would you need to suggest possible explanations for this extremely high incidence?

Go to **yourBioPortal.com** for original citations, discussions, and relevant links for all INVESTIGATING LIFE figures.

tions, but scientists are distinguished by what they do with their observations and how they answer their questions. Data, subjected to appropriate statistical analysis, are critical in the testing of hypotheses. The scientific method is the most powerful way humans have devised for learning about the world and how it works.

Scientific explanations for natural processes are objective and reliable because the hypotheses proposed *must be testable* and *must have the potential of being rejected* by direct observations and experiments. Scientists must clearly describe the methods they use to test hypotheses so that other scientists can repeat their results. Not all experiments are repeated, but surprising or controversial results are always subjected to independent verification. Scientists worldwide share this process of testing and rejecting hypotheses, contributing to a common body of scientific knowledge.

If you understand the methods of science, you can distinguish science from non-science. Art, music, and literature all contribute to the quality of human life, but they are not science. They do not use the scientific method to establish what is fact. Religion is not science, although religions have historically purported to explain natural events ranging from unusual weather patterns to crop failures to human diseases. Most such phenomena that at one time were mysterious can now be explained in terms of scientific principles.

The power of science derives from the uncompromising objectivity and absolute dependence on evidence that comes from *reproducible and quantifiable observations*. A religious or spiritual explanation of a natural phenomenon may be coherent and satisfying for the person holding that view, but it is not testable, and therefore it is not science. To invoke a supernatural explanation (such as a "creator" or "intelligent designer" with no known bounds) is to depart from the world of science.

Science describes the facts about how the world works, not how it "ought to be." Many scientific advances that have contributed to human welfare have also raised major ethical issues. Recent developments in genetics and developmental biology, for example, enable us to select the sex of our children, to use stem cells to repair our bodies, and to modify the human genome. Although scientific knowledge allows us to do these things, science cannot tell us whether or not we should do them, or, if we choose to do so, how we should regulate them.

To make wise decisions about public policy, we need to employ the best possible ethical reasoning in deciding which outcomes we should strive for.

## 1.3 RECAP

**The scientific method of inquiry starts with the formulation of hypotheses based on observations and data. Comparative and controlled experiments are carried out to test hypotheses.**

- Explain the relationship between a hypothesis and an experiment. **See p. 14 and Figure 1.12**
- What is controlled in a controlled experiment? **See p. 14 and Figure 1.13**
- What features characterize questions that can be answered only by using a comparative approach? **See pp. 14–15 and Figure 1.14**
- Do you understand why arguments must be supported by quantifiable and reproducible data in order to be considered scientific? **See pp. 15–16**

The vast scientific knowledge accumulated over centuries of human civilization allows us to understand and manipulate aspects of the natural world in ways that no other species can. These abilities present us with challenges, opportunities, and above all, responsibilities.

## 1.4 How Does Biology Influence Public Policy?

Agriculture and medicine are two important human activities that depend on biological knowledge. Our ancestors unknowingly applied the principles of evolutionary biology when they domesticated plants and animals, and people have speculated about the causes of diseases and searched for methods to combat them since ancient times. Long before the microbial causes of diseases were known, people recognized that infections could be passed from one person to another, and the isolation of infected persons has been practiced as long as written records have been available.

Today, thanks to the deciphering of genomes and our newfound ability to manipulate them, vast new possibilities exist for controlling human diseases and increasing agricultural productivity, but these capabilities raise ethical and policy issues. How much and in what ways should we tinker with the genes of humans and other species? Does it matter whether the genomes of our crop plants and domesticated animals are changed by traditional methods of controlled breeding and crossbreeding or by the biotechnology of gene transfer? What rules should govern the release of genetically modified organisms into the environment? Science alone cannot provide all the answers, but wise policy decisions must be based on accurate scientific information.

Biologists are increasingly called on to advise government agencies concerning the laws, rules, and regulations by which society deals with the increasing number of challenges that have at least a partial biological basis. As an example of the value of scientific knowledge for the assessment and formulation of public policy, let's return to the tracking study of bluefin tuna introduced in Section 1.3. Prior to this study, both scientists and fishermen knew that bluefins had a western breeding ground in the Gulf of Mexico and an eastern breeding ground in the Mediterranean Sea (**Figure 1.15**). Overfishing had led to declining numbers of fish in the western-breeding populations, to the point of these populations being endangered.

**1.15 Bluefin Tuna Do Not Recognize Boundaries** It was assumed that tuna from western-breeding populations and those from eastern-breeding populations also fed on their respective sides of the Atlantic, so separate fishing quotas were established on either side of 45° W longitude (dashed line) to allow the endangered western population to recover. However, tracking data shows that the two populations *do not* remain separate after spawning, so in fact the established policy does not protect the western population.

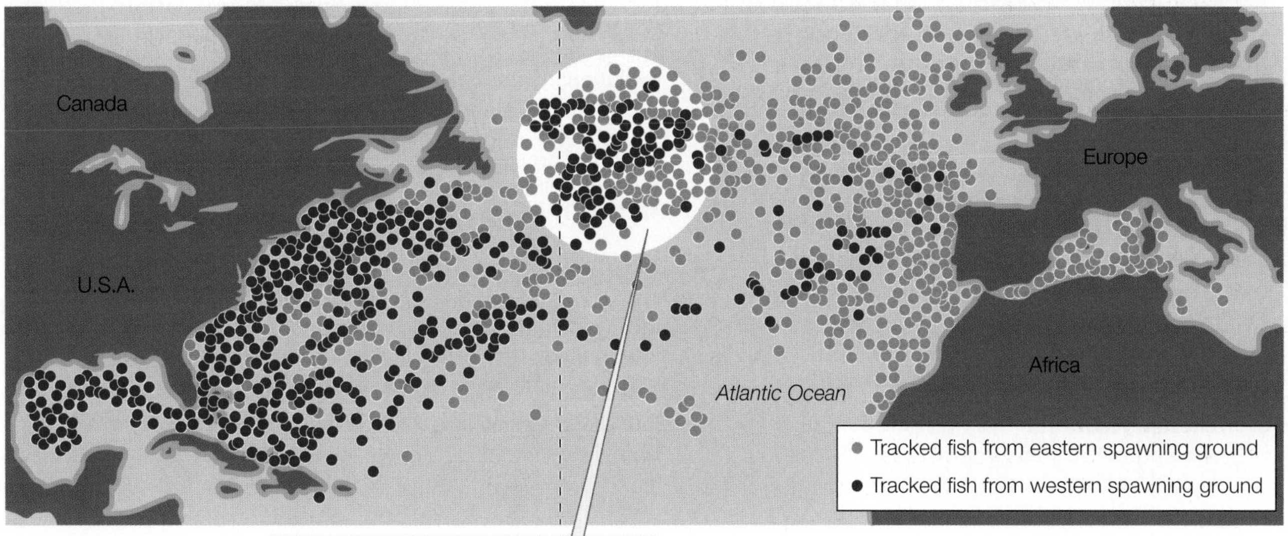

Initially it was assumed by scientists, fishermen, and policy makers alike that the eastern and western populations had geographically separate feeding grounds as well as separate breeding grounds. Acting on this assumption, an international commission drew a line down the middle of the Atlantic Ocean and established strict fishing quotas on the western side of the line, with the intent of allowing the western population to recover. New tracking data, however, revealed that in fact the eastern and western bluefin populations mix freely on their feeding grounds across the entire North Atlantic—a swath of ocean that includes the most heavily fished waters in the world. Tuna caught on the eastern side of the line could just as likely be from the western breeding population as the eastern; thus the established policy was not achieving its intended goal.

Policy makers take more things into consideration than scientific knowledge and recommendations. For example, studies on the effects of atrazine on amphibians have led one U.S. group, the Natural Resources Defense Council, to take legal action to have atrazine banned on the basis of the Endangered Species Act. The U.S. Environmental Protection Agency, however, must also consider the potential loss to agriculture that such a ban would create and has continued to approve atrazine's use as long as environmental levels do not exceed 30 to 40 ppb—which is 300 to 400 times the levels shown to induce abnormalities in the Hayes studies. Scientific conclusions do not always prevail in the political world.

Another reason for studying biology is to understand the effects of the vastly increased human population on its environment. Our use of natural resources is putting stress on the ability of Earth's ecosystems to continue to produce the goods and services on which our society depends. Human activities are changing global climates, causing the extinctions of a large number of species like the amphibians featured in this chapter, and spreading new diseases while facilitating the resurgence of old ones. The rapid spread of flu viruses has been facilitated by modern modes of transportation, and the recent resurgence of tuberculosis is the result of the evolution of bacteria that are resistant to antibiotics. Biological knowledge is vital for determining the causes of these changes and for devising wise policies to deal with them.

Beyond issues of policy and pragmatism lies the human "need to know." Human beings are fascinated by the richness and diversity of life, and most people want to know more about organisms and how they interact. Human curiosity might even be seen as an adaptive trait—it is possible that such a trait could have been selected for if individuals who were motivated to learn about their surroundings were likely to have survived and reproduced better, on average, than their less curious relatives. Far from ending the process, new discoveries and greater knowledge typically engender questions no one thought to ask before. There are vast numbers of questions for which we do not yet have answers, and the most important motivator of most scientists is curiosity.

## CHAPTER SUMMARY

### 1.1 What Is Biology?

- **Biology** is the scientific study of living organisms, including their characteristics, functions, and interactions. Cells are the basic structural and physiological units of life. The **cell theory** states that all life consists of cells and that all cells come from preexisting cells.
- All living organisms are related to one another through descent with modification. **Evolution** by **natural selection** is responsible for the diversity of **adaptations** found in living organisms.
- The instructions for a cell are contained in its **genome**, which consists of **DNA** molecules made up of sequences of **nucleotides**. Specific segments of DNA called **genes** contain the information the cell uses to make **proteins**. Review Figure 1.4
- Living organisms regulate their internal environment. They also interact with other organisms of the same and different species. Biologists study life at all these levels of organization. Review Figure 1.6, **WEB ACTIVITY 1.1**
- Biological knowledge obtained from a **model system** may be generalized to other species.

### 1.2 How Is All Life on Earth Related?

- Biologists use fossils, anatomical similarities and differences, and molecular comparisons of genomes to reconstruct the history of life. Review Figure 1.8
- Life first arose by chemical evolution. Cells arose early in the evolution of life.
- **Photosynthesis** was an important evolutionary step because it changed Earth's atmosphere and provided a means of capturing energy from sunlight.
- The earliest organisms were **prokaryotes**. Organisms called **eukaryotes**, with more complex cells, arose later. Eukaryotic cells have discrete intracellular compartments, called **organelles**, including a nucleus that contains the cell's genetic material.
- The genetic relationships of **species** can be represented as an evolutionary tree. Species are grouped into three **domains**: **Archaea**, **Bacteria**, and **Eukarya**. Archaea and Bacteria are domains of unicellular prokaryotes. Eukarya contains diverse groups of protists (most but not all of which are unicellular) and the multicellular plants, fungi, and animals. Review Figure 1.10, **WEB ACTIVITY 1.2**

### 1.3 How Do Biologists Investigate Life?

- The **scientific method** used in most biological investigations involves five steps: making observations, asking questions, forming hypotheses, making predictions, and testing those predictions. Review Figure 1.12
- **Hypotheses** are tentative answers to questions. Predictions made on the basis of a hypothesis are tested with additional

observations and two kinds of **experiments**: **comparative** and **controlled experiments**. Review Figures 1.13 and 1.14, **ANIMATED TUTORIAL 1.1**

- Statistical methods are applied to **data** to establish whether or not the differences observed are significant or whether they could be the result of chance. These methods start with the **null hypothesis** that there are no differences.
- Science can tell us how the world works, but it cannot tell us what we should or should not do.

### 1.4 How Does Biology Influence Public Policy?

- Biologists are often called on to advise government agencies on the solution of important problems that have a biological component.

## FOR DISCUSSION

1. Even if we knew the sequences of all of the genes of a single-celled organism and could cause those genes to be expressed in a test tube, it would still be incredibly difficult to create a functioning organism. Why do you think this is so? In light of this fact, what do you think of the statement that the genome contains all of the information for a species?
2. Why is it so important in science that we design and perform tests capable of falsifying a hypothesis?
3. What features characterize questions that can be answered only by using a comparative approach?
4. Cite an example of how you apply aspects of the scientific method to solve problems in your daily life.

## ADDITIONAL INVESTIGATION

1. The abnormalities of frogs in Tyrone Hayes's studies were associated with the presence of a herbicide in the environment. That herbicide did not kill the frogs, but it feminized the males. How would you investigate whether this effect could lead to decreased reproductive capacity for the frog populations in nature?
2. Just as all cells come from preexisting cells, all mitochondria—the cell organelles that convert energy in food to a form of energy that can do biological work—come from preexisting mitochondria. Cells do not synthesize mitochondria from the genetic information in their nuclei. What investigations would you carry out to understand the nature of mitochondria?

## WORKING WITH DATA (GO TO yourBioPortal.com)

**Feminization of Frogs** Analogous to the experiment shown in Figure 1.13, this exercise asks you to graph data about the size of the laryngeal (throat) muscles required to produce male mating calls in the frog *Xenopus laevis*. After plotting data from frogs exposed to different levels of the herbicide atrazine during their development, you will formulate conclusions about the effects of the herbicide on this physical attribute and speculate about what these effects might mean.

# 2 Small Molecules and the Chemistry of Life

## A hairy story

"You are what you eat—and that is recorded in your hair." Two scientists at the University of Utah are responsible for adding the last phrase to this famous saying about body chemistry. Ecologist Jim Ehleringer and chemist Thure Cerling showed that the composition of human hair reflects the region where a person lives.

As we pointed out in Chapter 1, living things are made up of the same kinds of atoms that make up the inanimate universe. Two of those atoms are hydrogen (H) and oxygen (O), which combine to form water ($H_2O$). Both atoms have naturally occurring variants called *isotopes*, which have the same chemical properties but different weights because their nuclei have different numbers of particles called *neutrons*.

When water evaporates from the ocean, it forms clouds that move inland and release rain. Water made up of the heavier H and O isotopes is heavier and tends to fall more readily than water containing the lighter isotopes. Warm rains tend to be heavier than cooler precipitation. People living on the coast or in regions where there are frequent warm rains consume heavier water and foods made from water than people living in cooler, inland areas (assuming, of course, that their beverages and produce come from the same area they live in). And, since you are what you eat, the heavy H and O atoms become part of their bodies.

Our hair contains abundant H and O atoms, many obtained from local water. Ehleringer and Cerling wondered whether the ratios of heavy-to-light H and O in hair reflected the ratio of heavy-to-light $H_2O$ in the local water. To address this question, Ehleringer's wife and Cerling's children and their friends went on a hair-collecting trip across the United States, collecting hair trimmings from barbershop floors while at the same time filling test tubes with local water. Back at the lab, scientists tested the samples and found that the ratios of heavy to light isotopes in the hair did indeed reflect these same ratios in the local water.

While this information is intrinsically fascinating, it is also potentially useful. For example, police could use hair analysis to evaluate a suspect's alibi: "You say you've been in Montana for the past month? Your hair sample indicates that you were in a warm coastal area." Such conflicting evidence could form the basis of further investigation.

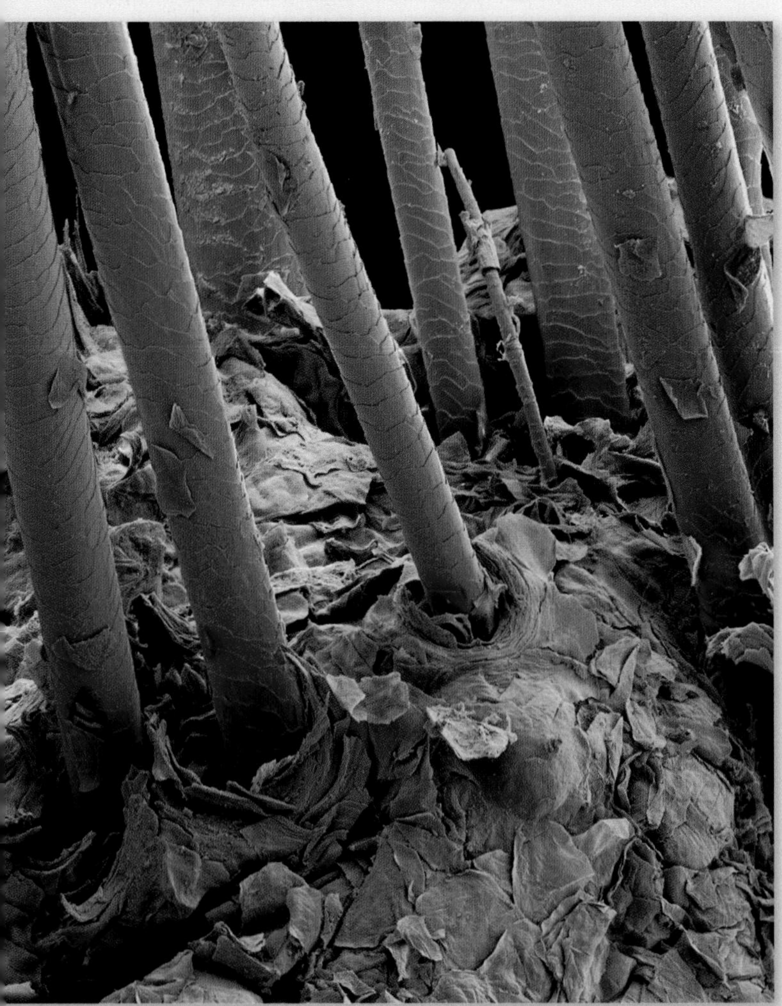

**Hair Tells a Tale** The ratio in hair protein of the heavy isotope $^{18}O$ to its lighter counterpart $^{16}O$ reflects the ratios in local water.

**Free Samples** Need hair samples for a research project? Try the local barber shop.

Or anthropologists might analyze hair samples from graves to work out migration patterns of human groups.

The understanding that life is based on chemistry and obeys universal laws of chemistry and physics is relatively new in human history. Until the nineteenth century, a "vital force" (from the Latin *vitalis*, "of life") was presumed be responsible for life. This vital force was seen as distinct from the mechanistic forces governing physics and chemistry. Many people still assume that a vital force exists, but the physical–chemical view of life has led to great advances in biological science and is the cornerstone of modern medicine and agriculture.

**IN THIS CHAPTER** we will introduce the constituents of matter: atoms, their variety, their properties, and their capacity to combine with other atoms. We will consider how matter changes, including changes in state (solid to liquid to gas), and changes caused by chemical reactions. We will examine the structure and properties of water and its relationship to chemical acids and bases.

## CHAPTER OUTLINE



## 2.1 How Does Atomic Structure Explain the Properties of Matter?

All matter is composed of **atoms**. Atoms are tiny—more than a trillion ($10^{12}$) of them could fit on top of the period at the end of this sentence. Each atom consists of a dense, positively charged **nucleus**, around which one or more negatively charged **electrons** move (**Figure 2.1**). The nucleus contains one or more positively charged **protons** and may contain one or more **neutrons** with no electrical charge. Atoms and their component particles have volume and mass, which are characteristics of all matter. *Mass* is a measure of the quantity of matter present; the greater the mass, the greater the quantity of matter.

The mass of a proton serves as a standard unit of measure called the *dalton* (named after the English chemist John Dalton) or **atomic mass unit** (**amu**). A single proton or neutron has a mass of about 1 dalton (Da), which is $1.7 \times 10^{-24}$ grams (0.0000000000000000000000017 g). That's tiny, but an electron is even tinier at $9 \times 10^{-28}$ g (0.0005 Da). Because the mass of an electron is negligible compared with the mass of a proton or a neutron, the contribution of electrons to the mass of an atom can usually be ignored when measurements and calculations are made. It is electrons, however, that determine how atoms will combine with other atoms to form stable associations.

Each proton has a positive electric charge, defined as +1 unit of charge. An electron has a negative charge equal and opposite to that of a proton (–1). The neutron, as its name suggests, is electrically neutral, so its charge is 0. Charges that are different (+/–) attract each other, whereas charges that are alike (+/+, –/–) repel each other. Atoms are electrically neutral because the number of electrons in an atom equals the number of protons.

### An element consists of only one kind of atom

An **element** is a pure substance that contains only one kind of atom. The element hydrogen consists only of hydrogen atoms; the element iron consists only of iron atoms. The atoms of each element have certain characteristics or properties that distinguish them from the atoms of other elements. These properties include their mass and how they interact and associate with other atoms.

The more than 100 elements found in the universe are arranged in the *periodic table* (**Figure 2.2**). Each element has its own one- or two-letter chemical symbol. For example, H stands for hydrogen, C for carbon, and O for oxygen. Some symbols come from other languages: Fe (from the Latin, *ferrum*) stands for iron, Na (Latin, *natrium*) for sodium, and W (German, *wolfram*) for tungsten.

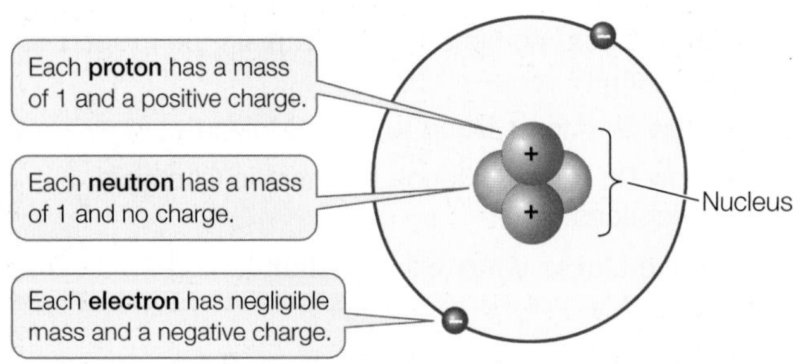

Each **proton** has a mass of 1 and a positive charge.
Each **neutron** has a mass of 1 and no charge.
Each **electron** has negligible mass and a negative charge.
Nucleus


**2.1 The Helium Atom** This representation of a helium atom is called a Bohr model. It exaggerates the space occupied by the nucleus. In reality, although the nucleus accounts for virtually all of the atomic mass, it occupies only about 1/10,000 of the atom's volume. The Bohr model is also inaccurate in that it represents the electron as a discrete particle in a defined orbit around the nucleus.

The elements of the periodic table are not found in equal amounts. Stars have abundant amounts of hydrogen and helium. Earth's crust, and the surfaces of the neighboring planets, are almost half oxygen, 28 percent silicon, 8 percent aluminum, and between 2 and 5 percent each of sodium, magnesium, potassium, calcium, and iron. They contain much smaller amounts of the other elements.

About 98 percent of the mass of every living organism (bacterium, turnip, or human) is composed of just six elements: carbon, hydrogen, nitrogen, oxygen, phosphorus, and sulfur. The chemistry of these six elements will be our primary concern in this chapter, but other elements found in living organisms are important as well. Sodium and potassium, for example, are essential for nerve function; calcium can act as a biological signal; iodine is a component of a vital hormone; and magnesium is bound to chlorophyll in plants. The physical and chemical (reactive) properties of atoms depend on the number of subatomic particles they contain.

### Each element has a different number of protons

An element differs from other elements by the number of protons in the nucleus of each of its atoms; the number of protons is designated the **atomic number**. This atomic number is unique

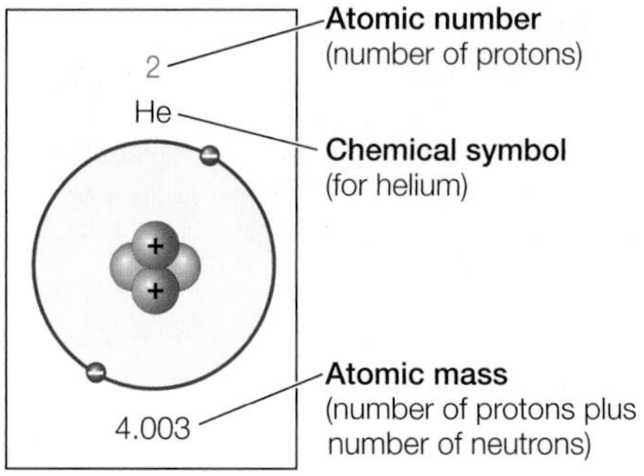

2
He
4.003

**Atomic number** (number of protons)
**Chemical symbol** (for helium)
**Atomic mass** (number of protons plus number of neutrons)

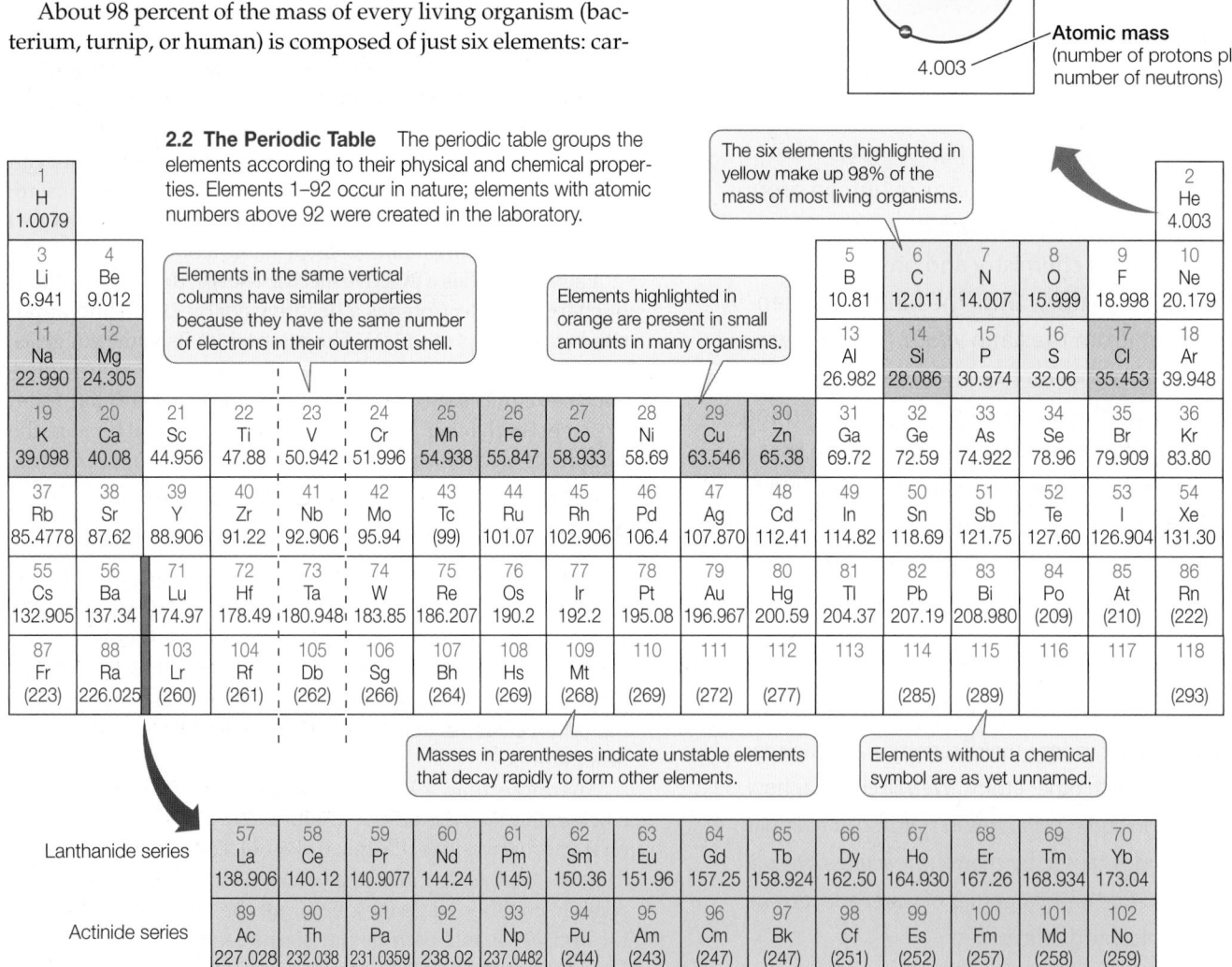

**2.2 The Periodic Table** The periodic table groups the elements according to their physical and chemical properties. Elements 1–92 occur in nature; elements with atomic numbers above 92 were created in the laboratory.

The six elements highlighted in yellow make up 98% of the mass of most living organisms.

Elements in the same vertical columns have similar properties because they have the same number of electrons in their outermost shell.

Elements highlighted in orange are present in small amounts in many organisms.

| | | | | | | | | | | | | | | | | | |
|---|---|---|---|---|---|---|---|---|---|---|---|---|---|---|---|---|---|
| 1<br>H<br>1.0079 | | | | | | | | | | | | | | | | | 2<br>He<br>4.003 |
| 3<br>Li<br>6.941 | 4<br>Be<br>9.012 | | | | | | | | | | | 5<br>B<br>10.81 | 6<br>C<br>12.011 | 7<br>N<br>14.007 | 8<br>O<br>15.999 | 9<br>F<br>18.998 | 10<br>Ne<br>20.179 |
| 11<br>Na<br>22.990 | 12<br>Mg<br>24.305 | | | | | | | | | | | 13<br>Al<br>26.982 | 14<br>Si<br>28.086 | 15<br>P<br>30.974 | 16<br>S<br>32.06 | 17<br>Cl<br>35.453 | 18<br>Ar<br>39.948 |
| 19<br>K<br>39.098 | 20<br>Ca<br>40.08 | 21<br>Sc<br>44.956 | 22<br>Ti<br>47.88 | 23<br>V<br>50.942 | 24<br>Cr<br>51.996 | 25<br>Mn<br>54.938 | 26<br>Fe<br>55.847 | 27<br>Co<br>58.933 | 28<br>Ni<br>58.69 | 29<br>Cu<br>63.546 | 30<br>Zn<br>65.38 | 31<br>Ga<br>69.72 | 32<br>Ge<br>72.59 | 33<br>As<br>74.922 | 34<br>Se<br>78.96 | 35<br>Br<br>79.909 | 36<br>Kr<br>83.80 |
| 37<br>Rb<br>85.4778 | 38<br>Sr<br>87.62 | 39<br>Y<br>88.906 | 40<br>Zr<br>91.22 | 41<br>Nb<br>92.906 | 42<br>Mo<br>95.94 | 43<br>Tc<br>(99) | 44<br>Ru<br>101.07 | 45<br>Rh<br>102.906 | 46<br>Pd<br>106.4 | 47<br>Ag<br>107.870 | 48<br>Cd<br>112.41 | 49<br>In<br>114.82 | 50<br>Sn<br>118.69 | 51<br>Sb<br>121.75 | 52<br>Te<br>127.60 | 53<br>I<br>126.904 | 54<br>Xe<br>131.30 |
| 55<br>Cs<br>132.905 | 56<br>Ba<br>137.34 | 71<br>Lu<br>174.97 | 72<br>Hf<br>178.49 | 73<br>Ta<br>180.948 | 74<br>W<br>183.85 | 75<br>Re<br>186.207 | 76<br>Os<br>190.2 | 77<br>Ir<br>192.2 | 78<br>Pt<br>195.08 | 79<br>Au<br>196.967 | 80<br>Hg<br>200.59 | 81<br>Tl<br>204.37 | 82<br>Pb<br>207.19 | 83<br>Bi<br>208.980 | 84<br>Po<br>(209) | 85<br>At<br>(210) | 86<br>Rn<br>(222) |
| 87<br>Fr<br>(223) | 88<br>Ra<br>226.025 | 103<br>Lr<br>(260) | 104<br>Rf<br>(261) | 105<br>Db<br>(262) | 106<br>Sg<br>(266) | 107<br>Bh<br>(264) | 108<br>Hs<br>(269) | 109<br>Mt<br>(268) | 110<br><br>(269) | 111<br><br>(272) | 112<br><br>(277) | 113 | 114<br><br>(285) | 115<br><br>(289) | 116 | 117 | 118<br><br>(293) |

Masses in parentheses indicate unstable elements that decay rapidly to form other elements.

Elements without a chemical symbol are as yet unnamed.

| | | | | | | | | | | | | | | |
|---|---|---|---|---|---|---|---|---|---|---|---|---|---|---|
| Lanthanide series | 57<br>La<br>138.906 | 58<br>Ce<br>140.12 | 59<br>Pr<br>140.9077 | 60<br>Nd<br>144.24 | 61<br>Pm<br>(145) | 62<br>Sm<br>150.36 | 63<br>Eu<br>151.96 | 64<br>Gd<br>157.25 | 65<br>Tb<br>158.924 | 66<br>Dy<br>162.50 | 67<br>Ho<br>164.930 | 68<br>Er<br>167.26 | 69<br>Tm<br>168.934 | 70<br>Yb<br>173.04 |
| Actinide series | 89<br>Ac<br>227.028 | 90<br>Th<br>232.038 | 91<br>Pa<br>231.0359 | 92<br>U<br>238.02 | 93<br>Np<br>237.0482 | 94<br>Pu<br>(244) | 95<br>Am<br>(243) | 96<br>Cm<br>(247) | 97<br>Bk<br>(247) | 98<br>Cf<br>(251) | 99<br>Es<br>(252) | 100<br>Fm<br>(257) | 101<br>Md<br>(258) | 102<br>No<br>(259) |

to each element and does not change. The atomic number of helium is 2, and an atom of helium always has two protons; the atomic number of oxygen is 8, and an atom of oxygen always has eight protons.

Along with a definitive number of protons, every element except hydrogen has one or more neutrons in its nucleus. The **mass number** of an atom is the total number of protons and neutrons in its nucleus. The nucleus of a carbon atom contains six protons and six neutrons, and has a mass number of 12. Oxygen has eight protons and eight neutrons, and has a mass number of 16. The mass number is essentially the mass of the atom in daltons (see below).

By convention, we often print the symbol for an element with the atomic number at the lower left and the mass number at the upper left, both immediately preceding the symbol. Thus hydrogen, carbon, and oxygen can be written as ${}^{1}_{1}H$, ${}^{12}_{6}C$, and ${}^{16}_{8}O$, respectively.

## The number of neutrons differs among isotopes

In some elements, the number of neutrons in the atomic nucleus is not always the same. Different **isotopes** of the same element have the same number of protons, but different numbers of neutrons. Many elements have several isotopes. The isotopes of hydrogen shown below have special names, but the isotopes of most elements do not have distinct names.

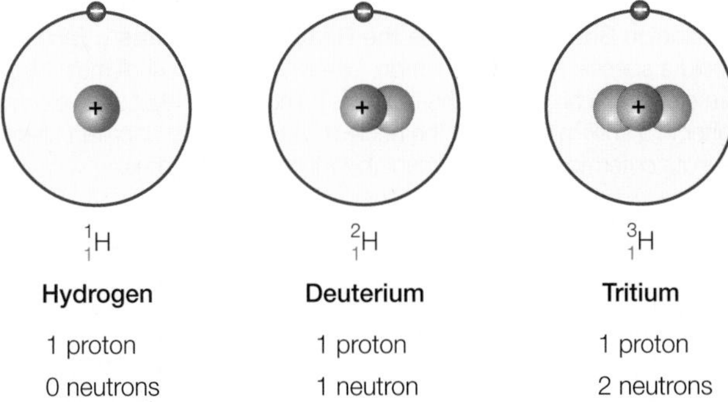


The natural isotopes of carbon, for example, are ${}^{12}C$ (six neutrons in the nucleus), ${}^{13}C$ (seven neutrons), and ${}^{14}C$ (eight neutrons). Note that all three (called "carbon-12," "carbon-13," and "carbon-14") have six protons, so they are all carbon. Most carbon atoms are ${}^{12}C$, about 1.1 percent are ${}^{13}C$, and a tiny fraction are ${}^{14}C$. But all have virtually the same chemical reactivity, which is an important property for their use in experimental biology and medicine. An element's **atomic weight** (or atomic mass) is the average of the mass numbers of a representative sample of atoms of that element, with all the isotopes in their normally occurring proportions. The atomic weight of carbon, taking into account all of its isotopes and their abundances, is thus 12.011. The fractional atomic weight results from averaging the contributing weights of all of the isotopes.

Most isotopes are stable. But some, called **radioisotopes**, are unstable and spontaneously give off energy in the form of $\alpha$ (alpha), $\beta$ (beta), or $\gamma$ (gamma) radiation from the atomic nucleus. Known as *radioactive decay*, this release of energy transforms the original atom. The type of transformation varies depending on the radioisotope, but some can change the number of protons, so that the original atom becomes a different element.

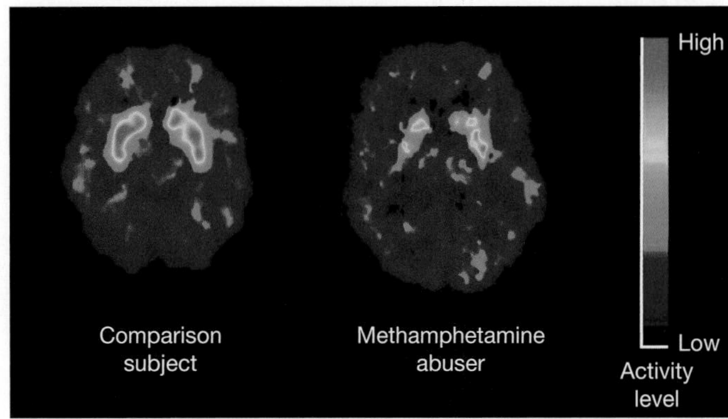


**2.3 Tagging the Brain** In these images from live people, a radioactively labeled sugar is used to detect differences between the brain activity of a healthy person and that of a person who abuses methamphetamines. The more active a brain region is, the more sugar it takes up. The healthy brain (left) shows more activity in the region involved in memory (the red area) than the drug abuser's brain does.

With sensitive instruments, scientists can use the released radiation to detect the presence of radioisotopes. For instance, if an earthworm is given food containing a radioisotope, its path through the soil can be followed using a simple detector called a Geiger counter. Most atoms in living organisms are organized into stable associations called **molecules**. If a radioisotope is incorporated into a molecule, it acts as a tag or label, allowing researchers or physicians to trace the molecule in an experiment or in the body (**Figure 2.3**). Radioisotopes are also used to date fossils, an application described in Section 25.1.

Although radioisotopes are useful in research and in medicine, even a low dose of the radiation they emit has the potential to damage molecules and cells. However, these damaging effects are sometimes used to our advantage; for example, the radiation from ${}^{60}Co$ (cobalt-60) is used in medicine to kill cancer cells.

## The behavior of electrons determines chemical bonding and geometry

The characteristic number of electrons in an atom determines how it will combine with other atoms. Biologists are interested in how chemical changes take place in living cells. When considering atoms, they are concerned primarily with electrons because the behavior of electrons explains how chemical *reactions* occur. Chemical reactions alter the atomic compositions of substances and thus alter their properties. Reactions usually involve changes in the distribution of electrons between atoms.

The location of a given electron in an atom at any given time is impossible to determine. We can only describe a volume of space within the atom where the electron is likely to be. The region of space where the electron is found at least 90 percent of the time is the electron's **orbital**. Orbitals have characteristic shapes and orientations, and a given orbital can be occupied by

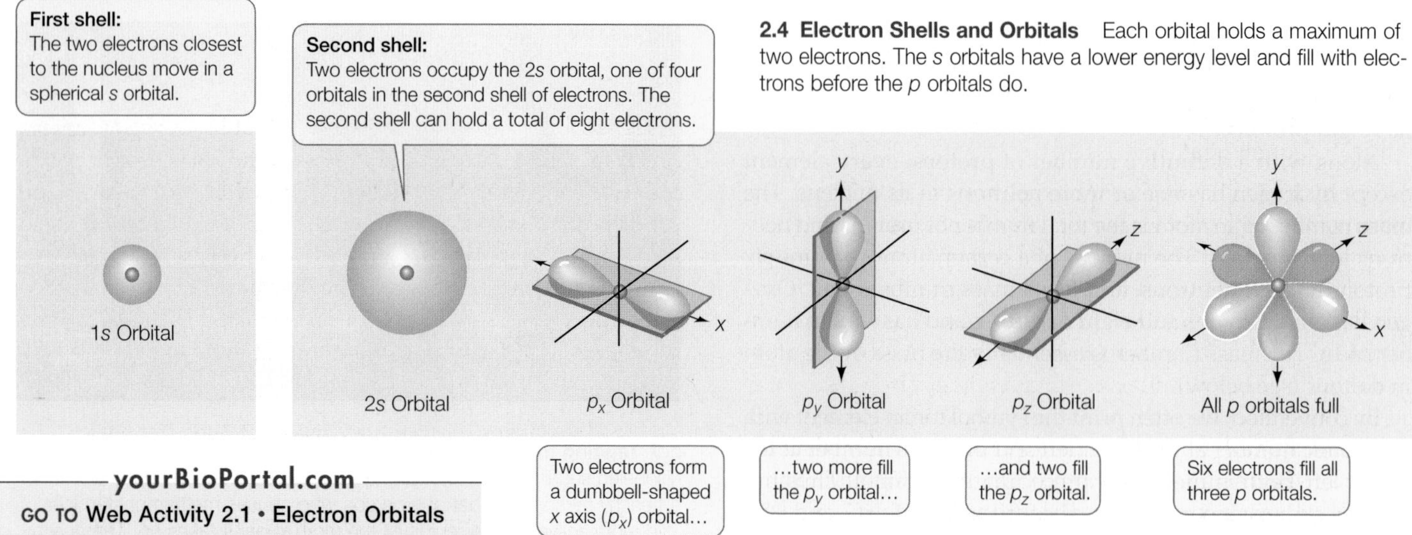


**2.4 Electron Shells and Orbitals** Each orbital holds a maximum of two electrons. The *s* orbitals have a lower energy level and fill with electrons before the *p* orbitals do.

yourBioPortal.com
GO TO Web Activity 2.1 • Electron Orbitals

a maximum of two electrons (**Figure 2.4**). Thus any atom larger than helium (atomic number 2) must have electrons in two or more orbitals. As we move from lighter to heavier atoms in the periodic chart, the orbitals are filled in a specific sequence, in a series of what are known as **electron shells**, or *energy levels*, around the nucleus.

- *First shell*: The innermost electron shell consists of just one orbital, called an *s* orbital. A hydrogen atom ($_1$H) has one electron in its first shell; helium ($_2$He) has two. Atoms of all other elements have two or more shells to accommodate orbitals for additional electrons.
- *Second shell*: The second shell contains four orbitals (an *s* orbital and three *p* orbitals), and hence holds up to eight electrons. As depicted in Figure 2.4, the *s* orbitals have the shape of a sphere, while the *p* orbitals are directed at right angles to one another. The orientations of these orbitals in space contribute to the three-dimensional shapes of molecules when atoms link to other atoms.

**2.5 Electron Shells Determine the Reactivity of Atoms** Each shell can hold a specific maximum number of electrons. Each shell must be filled before electrons can occupy the next shell. The energy level of an electron is higher in a shell farther from the nucleus. An atom with unpaired electrons in its outermost shell can react (bond) with other atoms.

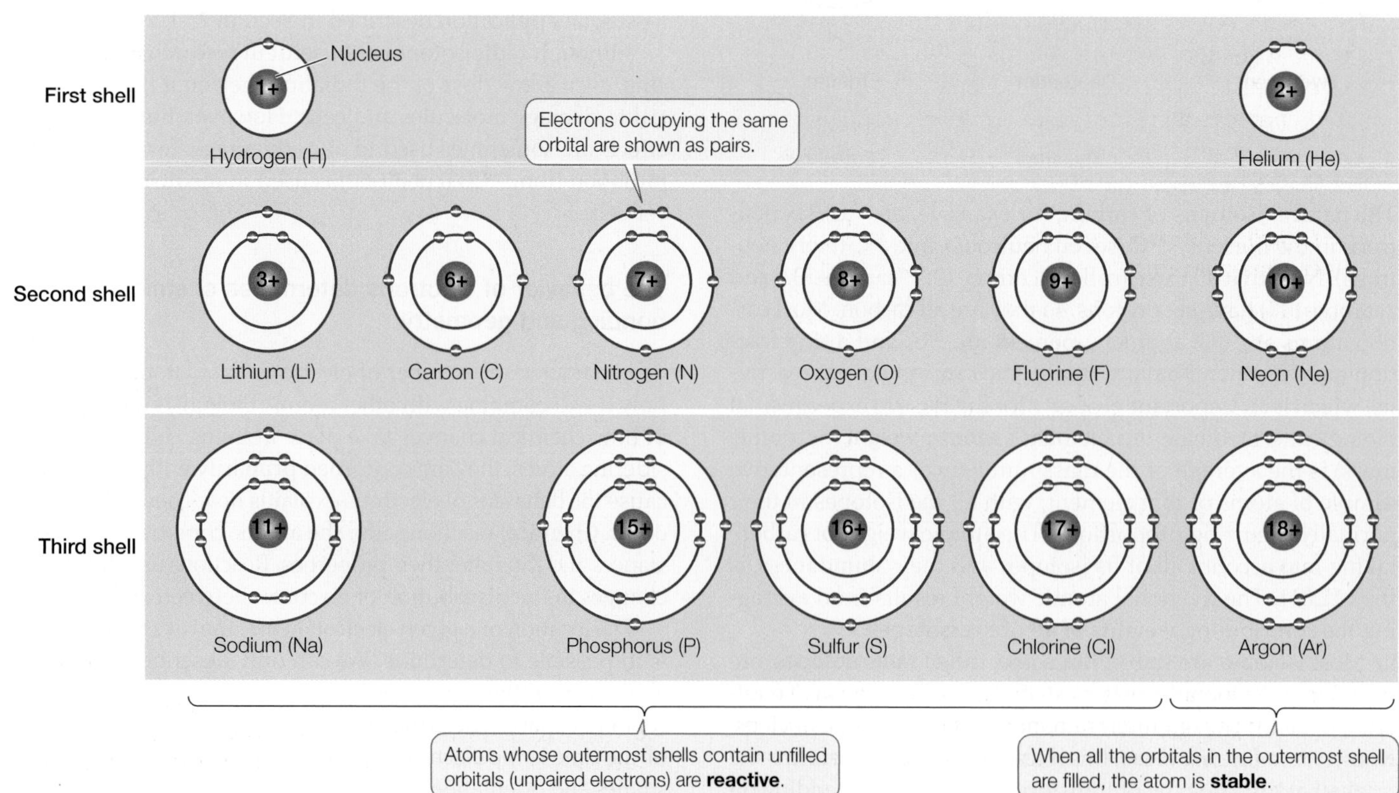

- *Additional shells*: Elements with more than ten electrons have three or more electron shells. The farther a shell is from the nucleus, the higher the energy level is for an electron occupying that shell.

The *s* orbitals fill with electrons first, and their electrons have the lowest energy level. Subsequent shells have different numbers of orbitals, but the outermost shells usually hold only eight electrons. In any atom, the outermost electron shell (the *valence shell*) determines how the atom combines with other atoms—that is, how the atom behaves chemically. When a valence shell with four orbitals contains eight electrons, there are no unpaired electrons, and the atom is *stable*—it will not react with other atoms (**Figure 2.5**). Examples of chemically stable elements are helium, neon, and argon. On the other hand, atoms that have one or more unpaired electrons in their outer shells are capable of reacting with other atoms.

Atoms with unpaired electrons (i.e., partially filled orbitals) in their outermost electron shells are unstable, and will undergo reactions in order to fill their outermost shells. Reactive atoms can attain stability either by sharing electrons with other atoms or by losing or gaining one or more electrons. In either case, the atoms involved are *bonded* together into stable associations called molecules. The tendency of atoms to form stable molecules so that they have eight electrons in their outermost shells is known as the *octet rule*. Many atoms in biologically important molecules—for example, carbon (C) and nitrogen (N)—follow this rule. An important exception is hydrogen (H), which attains stability when two electrons occupy its single shell (consisting of just one *s* orbital).

## 2.1 RECAP

**The living world is composed of the same set of chemical elements as the rest of the universe. An atom consists of a nucleus of protons and neutrons, and a characteristic configuration of electrons in orbitals around the nucleus. This structure determines the atom's chemical properties.**

- Describe the arrangement of protons, neutrons, and electrons in an atom. **See Figure 2.1**
- Use the periodic table to identify some of the similarities and differences in atomic structure among different elements (for example, oxygen, carbon, and helium). How does the configuration of the valence shell influence the placement of an element in the periodic table? **See p. 25 and Figures 2.2 and 2.5**
- How does bonding help a reactive atom achieve stability? **See p. 25 and Figure 2.5**

We have introduced the individual players on the biochemical stage—the atoms. We have shown how the energy levels of electrons drive an atomic "quest for stability." Next we will describe the different types of chemical bonds that can lead to stability, joining atoms together into molecular structures with hosts of different properties.

# 2.2 How Do Atoms Bond to Form Molecules?

A **chemical bond** is an attractive force that links two atoms together in a molecule. There are several kinds of chemical bonds (**Table 2.1**). In this section we will begin with *covalent bonds*, the strong bonds that result from the sharing of electrons. Next we will examine *ionic bonds*, which form when an atom gains or loses one or more electrons to achieve stability. We will then consider other, weaker, kinds of interactions, including hydrogen bonds, which are enormously important to biology.

**yourBioPortal.com**
GO TO Animated Tutorial 2.1 • Chemical Bond Formation

### Covalent bonds consist of shared pairs of electrons

A **covalent bond** forms when two atoms attain stable electron numbers in their outermost shells by *sharing* one or more pairs of electrons. Consider two hydrogen atoms coming into close proximity, each with an unpaired electron in its single shell (**Figure 2.6**). When the electrons pair up, a stable association is formed, and this links the two hydrogen atoms in a covalent bond, resulting in $H_2$.

A **compound** is a substance made up of molecules with two or more elements bonded together in a fixed ratio. Methane gas ($CH_4$), water ($H_2O$), and table sugar (sucrose, $C_{12}H_{22}O_{11}$) are examples of compounds. The chemical symbols identify the different elements in a compound, and the subscript numbers indicate how many atoms of each element are present. Every compound has a **molecular weight** (molecular mass) that is the

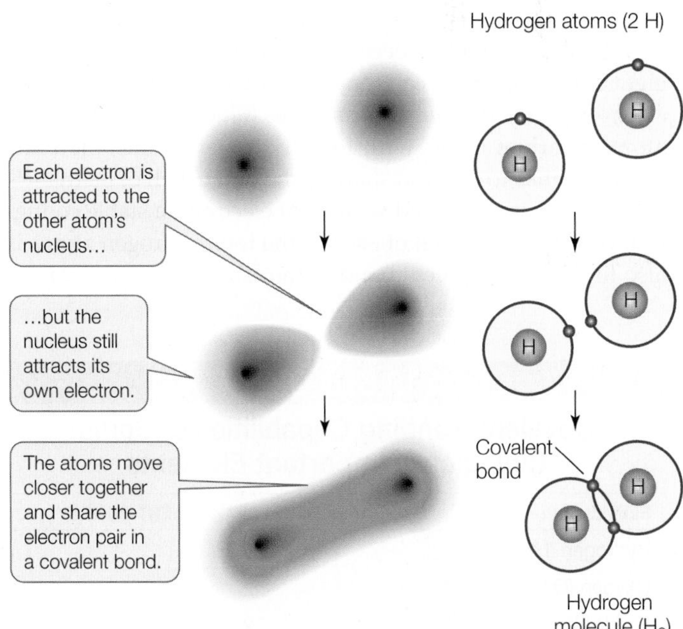

Hydrogen atoms (2 H)
H
H
Each electron is attracted to the other atom's nucleus...
...but the nucleus still attracts its own electron.
H
H
The atoms move closer together and share the electron pair in a covalent bond.
Covalent bond
H
H
Hydrogen molecule ($H_2$)

**2.6 Electrons Are Shared in Covalent Bonds** Two hydrogen atoms can combine to form a hydrogen molecule. A covalent bond forms when the electron orbitals of the two atoms overlap in an energetically stable manner.

## TABLE 2.1
### Chemical Bonds and Interactions

| NAME | BASIS OF INTERACTION | STRUCTURE | BOND ENERGY[a] (KCAL/MOL) |
|---|---|---|---|
| Covalent bond | Sharing of electron pairs | 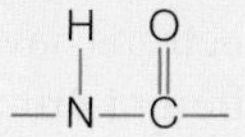 H O<br>—N—C— | 50–110 |
| Ionic bond | Attraction of opposite charges | 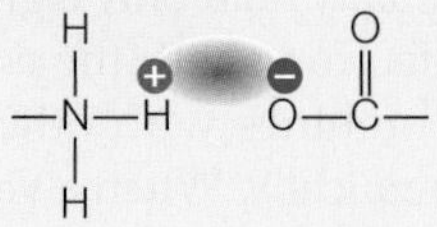 H O<br>—N—H ⊕ ⊖ O—C—<br>H | 3–7 |
| Hydrogen bond | Sharing of H atom | 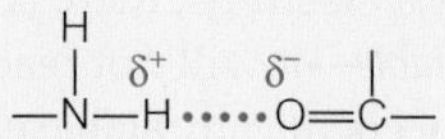 H<br>—N—H•••••O=C—<br>δ+ δ− | 3–7 |
| Hydrophobic interaction | Interaction of nonpolar substances in the presence of polar substances (especially water) | 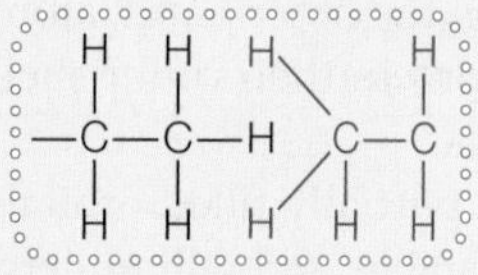 H H H H<br>—C—C—H C—C<br>H H H H H | 1–2 |
| van der Waals interaction | Interaction of electrons of nonpolar substances | 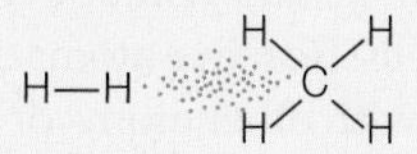 H—H· H H<br>C<br>H H | 1 |

[a]Bond energy is the amount of energy needed to separate two bonded or interacting atoms under physiological conditions.

sum of the atomic weights of all atoms in the molecule. Looking at the periodic table in Figure 2.2, you can calculate the molecular weights of the three compounds listed above to be 16.04, 18.01, and 342.29, respectively. Molecules that make up living organisms range in molecular weight from two to half a billion, and covalent bonds are common to all.

How are covalent bonds formed in a molecule of methane gas ($CH_4$)? The carbon atom in this compound has six electrons: two electrons fill its inner shell, and four unpaired electrons travel in its outer shell. Because its outer shell can hold up to eight electrons, carbon can share electrons with up to four other atoms—*it can form four covalent bonds* (**Figure 2.7A**). When an atom of carbon reacts with four hydrogen atoms, methane forms. Thanks to electron sharing, the outer shell of methane's carbon atom is now filled with eight electrons, a stable configuration. The outer shell of each of the four hydrogen atoms is also filled. Four covalent bonds—four shared electron pairs—hold methane together. **Figure 2.7B** shows several different ways to represent the molecular structure of methane. **Table 2.2** shows the covalent bonding capacities of some biologically significant elements.

## TABLE 2.2
### Covalent Bonding Capabilities of Some Biologically Important Elements

| ELEMENT | USUAL NUMBER OF COVALENT BONDS |
|---|---|
| Hydrogen (H) | 1 |
| Oxygen (O) | 2 |
| Sulfur (S) | 2 |
| Nitrogen (N) | 3 |
| Carbon (C) | 4 |
| Phosphorus (P) | 5 |

**STRENGTH AND STABILITY** Covalent bonds are very strong, meaning that it takes a lot of energy to break them. At temperatures in which life exists, the covalent bonds of biological molecules are quite stable, as are their three-dimensional structures. However, this stability does not preclude change, as we will discover.

**ORIENTATION** For a given pair of elements—for example, carbon bonded to hydrogen—the length of the covalent bond is always the same. And for a given atom within a molecule, the angle of each covalent bond, with respect to the other bonds, is generally the same. This is true regardless of the type of larger molecule that contains the atom. For example, the four filled orbitals around the carbon atom in methane are always distributed in space so that the bonded hydrogens point to the corners of a regular tetrahedron, with carbon in the center (see Figure 2.7B). Even when carbon is bonded to four atoms other than hydrogen, this three-dimensional orientation is more or less maintained. The orientation of covalent bonds in space gives the molecules their three-dimensional geometry, and the shapes of molecules contribute to their biological functions, as we will see in Section 3.1.

**MULTIPLE COVALENT BONDS** A covalent bond can be represented by a line between the chemical symbols for the linked atoms:

- A *single bond* involves the sharing of a single pair of electrons (for example, H—H or C—H).

(A) 1 C and 4 H

Methane ($CH_4$)

Covalent bond

Bohr models

Carbon can complete its outer shell by sharing the electrons of four hydrogen atoms, forming methane.

(B)

Each line or pair of dots represents a shared pair of electrons.

The hydrogen atoms form corners of a regular tetrahedron.

This model shows the shape methane presents to its environment.

Structural formulas

Ball-and-stick model

Space-filling model

**2.7 Covalent Bonding Can Form Compounds** (A) Bohr models showing the formation of covalent bonds in methane, whose molecular formula is $CH_4$. Electrons are shown in shells around the nucleus. (B) Three additional ways of representing the structure of methane. The ball-and-stick model and the space-filling model show the spatial orientations of the bonds. The space-filling model indicates the overall shape and surface of the molecule. In the chapters that follow, different conventions will be used to depict molecules. Bear in mind that these are models to illustrate certain properties, and not the most accurate portrayal of reality.

- A *double bond* involves the sharing of four electrons (two pairs) (C=C).
- *Triple bonds*—six shared electrons—are rare, but there is one in nitrogen gas (N≡N), which is the major component of the air we breathe.

**UNEQUAL SHARING OF ELECTRONS** If two atoms of the same element are covalently bonded, there is an equal sharing of the pair(s) of electrons in their outermost shells. However, when the two atoms are of different elements, the sharing is not necessarily equal. One nucleus may exert a greater attractive force on the electron pair than the other nucleus, so that the pair tends to be closer to that atom.

The attractive force that an atomic nucleus exerts on electrons in a covalent bond is called its **electronegativity**. The electronegativity of a nucleus depends on how many positive charges it has (nuclei with more protons are more positive and thus more attractive to electrons) and on the distances between the electrons in the bond and the nucleus (the closer the electrons, the greater the electronegative pull). **Table 2.3** shows the electronegativities (which are calculated to produce dimensionless quantities) of some elements important in biological systems.

If two atoms are close to each other in electronegativity, they will share electrons equally in what is called a *nonpolar covalent bond*. Two oxygen atoms, for example, each with an electronegativity of 3.5, will share electrons equally. So will two hydrogen atoms (each with an electronegativity of 2.1). But when hydrogen bonds with oxygen to form water, the electrons involved are unequally shared: they tend to be nearer to the oxygen nucleus because it is the more electronegative of the two. When electrons are drawn to one nucleus more than to the other, the result is a *polar covalent bond* (**Figure 2.8**).

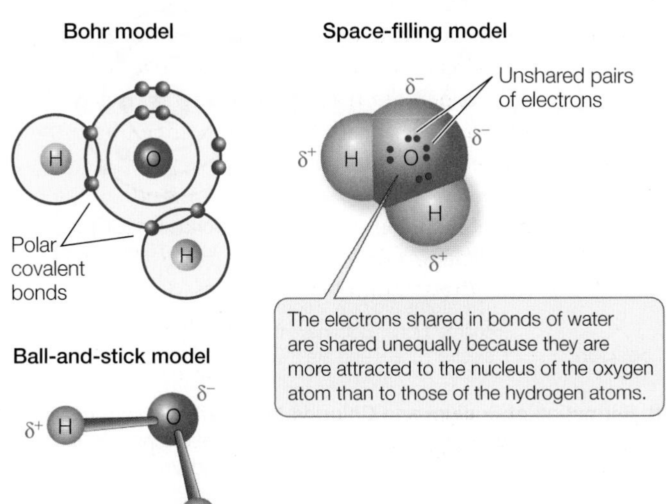

Bohr model
Space-filling model
Unshared pairs of electrons
Polar covalent bonds
Ball-and-stick model
The electrons shared in bonds of water are shared unequally because they are more attracted to the nucleus of the oxygen atom than to those of the hydrogen atoms.


**2.8 Water's Covalent Bonds Are Polar** These three representations all illustrate polar covalent bonding in water ($H_2O$). When atoms with different electronegativities, such as oxygen and hydrogen, form a covalent bond, the electrons are drawn to one nucleus more than to the other. A molecule held together by such a polar covalent bond has partial ($\delta^+$ and $\delta^-$) charges at different surfaces. In water, the shared electrons are displaced toward the oxygen atom's nucleus.

TABLE 2.3
**Some Electronegativities**

| ELEMENT | ELECTRONEGATIVITY |
|---|---|
| Oxygen (O) | 3.5 |
| Chlorine (Cl) | 3.1 |
| Nitrogen (N) | 3.0 |
| Carbon (C) | 2.5 |
| Phosphorus (P) | 2.1 |
| Hydrogen (H) | 2.1 |
| Sodium (Na) | 0.9 |
| Potassium (K) | 0.8 |

Because of this unequal sharing of electrons, the oxygen end of the hydrogen–oxygen bond has a slightly negative charge (symbolized by $\delta^-$ and spoken of as "delta negative," meaning a partial unit of charge), and the hydrogen end has a slightly positive charge ($\delta^+$). The bond is **polar** because these opposite charges are separated at the two ends, or poles, of the bond. The partial charges that result from polar covalent bonds produce polar molecules or polar regions of large molecules. Polar bonds within molecules greatly influence the interactions that they have with other polar molecules. Water ($H_2O$) is a polar compound, and this polarity has significant effects on its physical properties and chemical reactivity, as we will see in later chapters.

### Ionic bonds form by electrical attraction

When one interacting atom is much more electronegative than the other, a complete transfer of one or more electrons may take place. Consider sodium (electronegativity 0.9) and chlorine (3.1). A sodium atom has only one electron in its outermost shell; this condition is unstable. A chlorine atom has seven electrons in its outermost shell—another unstable condition. Since the electronegativity of chlorine is so much greater than that of sodium, any electrons involved in bonding will tend to transfer completely from sodium's outermost shell to that of chlorine (**Figure 2.9**). This reaction between sodium and chlorine makes the resulting atoms more stable because they both have eight fully paired electrons in their outer shells. The result is two *ions*.

**Ions** are electrically charged particles that form when atoms gain or lose one or more electrons:

- The sodium ion ($Na^+$) in our example has a +1 unit of charge because it has one less electron than it has protons. The outermost electron shell of the sodium ion is full, with eight electrons, so the ion is stable. Positively charged ions are called **cations**.
- The chloride ion ($Cl^-$) has a –1 unit of charge because it has one more electron than it has protons. This additional electron gives $Cl^-$ a stable outermost shell with eight electrons. Negatively charged ions are called **anions**.

Some elements can form ions with multiple charges by losing or gaining *more than one* electron. Examples are $Ca^{2+}$ (the calcium ion, a calcium atom that has lost two electrons) and $Mg^{2+}$ (the magnesium ion). Two biologically important elements can each yield more than one stable ion. Iron yields $Fe^{2+}$ (the ferrous ion) and $Fe^{3+}$ (the ferric ion), and copper yields $Cu^+$ (the cuprous ion) and $Cu^{2+}$ (the cupric ion). Groups of covalently bonded atoms that carry an electric charge are called *complex ions;* examples include $NH_4^+$ (the ammonium ion), $SO_4^{2-}$ (the sulfate ion), and $PO_4^{3-}$ (the phosphate ion). Once formed, ions are usually stable and no more electrons are lost or gained.

**Ionic bonds** are bonds formed as a result of the electrical attraction between ions bearing opposite charges. Ions can form bonds that result in stable solid compounds, which are referred to by the general term *salts*. Examples are sodium chloride (NaCl) and potassium phosphate ($K_3PO_4$). In sodium chloride—familiar to us as table salt—cations and anions are held together by ionic bonds. In solids, the ionic bonds are strong because the ions are close together. However, when ions are dispersed in water, the distance between them can be large; the strength of their attraction is thus greatly reduced. Under the conditions in living cells, an ionic attraction is less strong than a nonpolar covalent bond (see Table 2.1).

Not surprisingly, ions can interact with polar molecules, since they both carry electric charges. Such an interaction results when a solid salt such as NaCl dissolves in water. Water molecules surround the individual ions, separating them (**Figure**

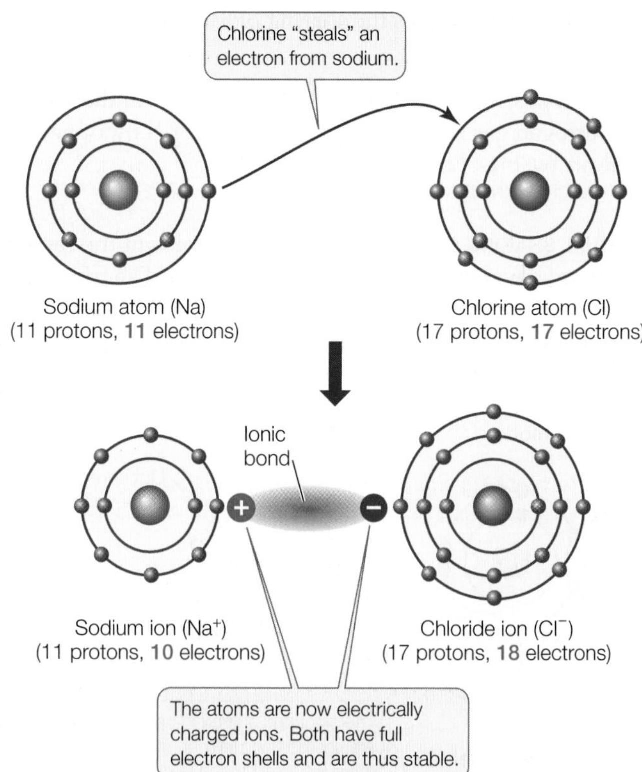


**2.9 Formation of Sodium and Chloride Ions** When a sodium atom reacts with a chlorine atom, the more electronegative chlorine fills its outermost shell by "stealing" an electron from the sodium. In so doing, the chlorine atom becomes a negatively charged chloride ion ($Cl^-$). With one less electron, the sodium atom becomes a positively charged sodium ion ($Na^+$).

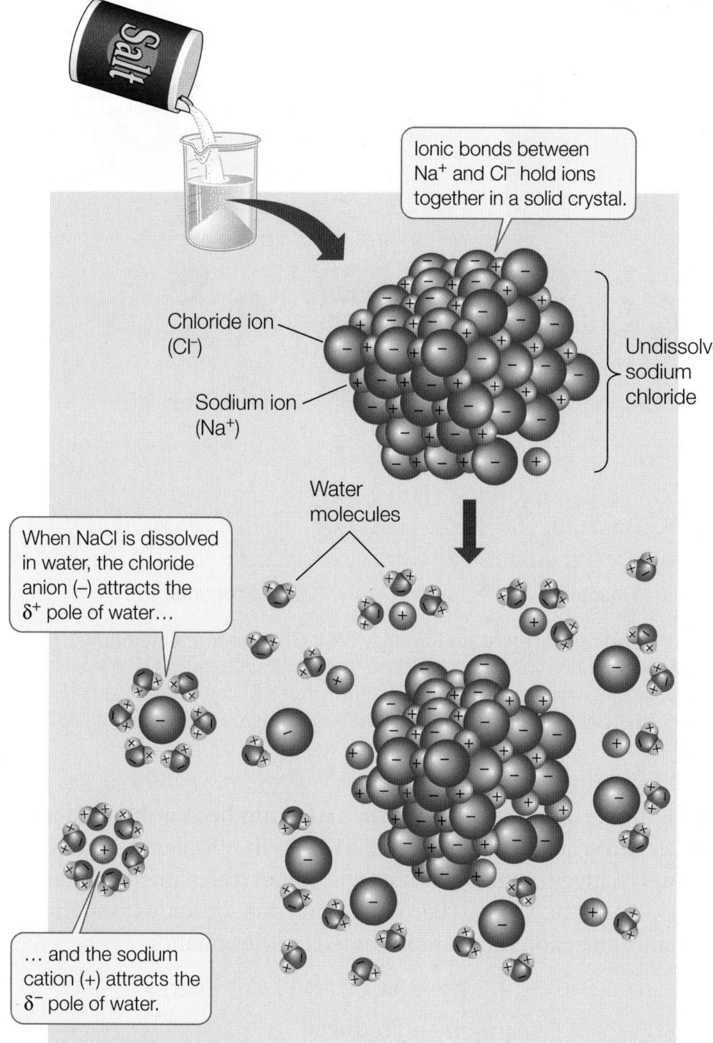
Salt
Ionic bonds between $Na^+$ and $Cl^-$ hold ions together in a solid crystal.
Chloride ion ($Cl^-$)
Sodium ion ($Na^+$)
Undissolved sodium chloride
Water molecules
When NaCl is dissolved in water, the chloride anion (–) attracts the $\delta^+$ pole of water…
… and the sodium cation (+) attracts the $\delta^-$ pole of water.

**2.10 Water Molecules Surround Ions** When an ionic solid dissolves in water, polar water molecules cluster around the cations and anions, preventing them from re-associating.

2.10). The negatively charged chloride ions attract the positive poles of the water molecules, while the positively charged sodium ions attract the negative poles of the water molecules. This is one of the special properties of water molecules, due to their polarity.

### Hydrogen bonds may form within or between molecules with polar covalent bonds

In liquid water, the negatively charged oxygen ($\delta^-$) atom of one water molecule is attracted to the positively charged hydrogen ($\delta^+$) atoms of another water molecule (**Figure 2.11A**). The bond resulting from this attraction is called a **hydrogen bond**. Hydrogen bonds are not restricted to water molecules; they may also form between a strongly electronegative atom and a hydrogen atom that is covalently bonded to a different electronegative atom, as shown in **Figure 2.11B**.

A hydrogen bond is weaker than most ionic bonds because its formation is due to partial charges ($\delta^+$ and $\delta^-$). It is much weaker than a covalent bond between a hydrogen atom and an oxygen atom (see Table 2.1). Although individual hydrogen bonds are weak, many of them can form within one molecule or

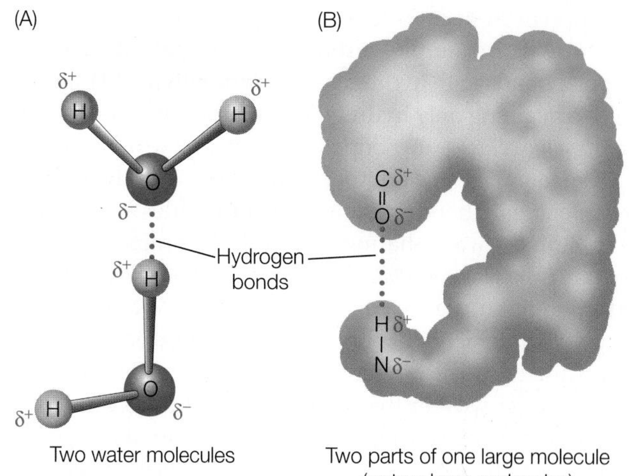
(A)
(B)
δ+
H
O
δ–
Hydrogen bonds
C δ+
O δ–
H δ+
N δ–
Two water molecules
Two parts of one large molecule (or two large molecules)

**2.11 Hydrogen Bonds Can Form Between or Within Molecules** (A) A hydrogen bond between two molecules is an attraction between a negative charge on one molecule and the positive charge on a hydrogen atom of the second molecule. (B) Hydrogen bonds can form between different parts of the same large molecule.

between two molecules. In these cases, the hydrogen bonds together have considerable strength, and greatly influence the structure and properties of substances. Later in this chapter we'll see how hydrogen bonding between water molecules contributes to many of the properties that make water so significant for living systems. Hydrogen bonds also play important roles in determining and maintaining the three-dimensional shapes of giant molecules such as DNA and proteins (see Section 3.2).

### Polar and nonpolar substances: Each interacts best with its own kind

Just as water molecules can interact with one another through hydrogen bonds, any molecule that is polar can interact with other polar molecules through the weak ($\delta^+$ to $\delta^-$) attractions of hydrogen bonds. If a polar molecule interacts with water in this way, it is called **hydrophilic** ("water-loving") (**Figure 2.12A**).

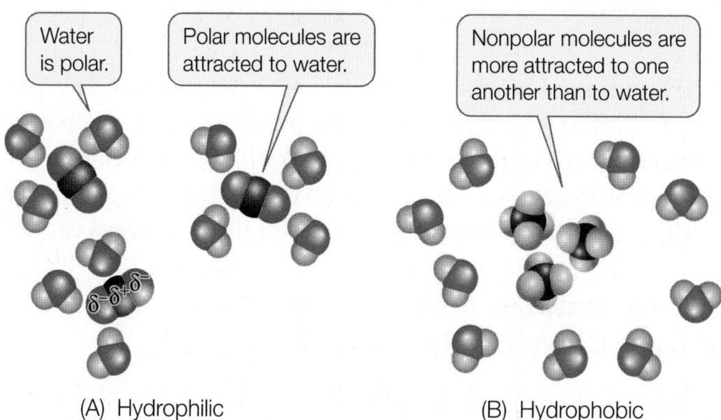
Water is polar.
Polar molecules are attracted to water.
Nonpolar molecules are more attracted to one another than to water.
(A) Hydrophilic
(B) Hydrophobic

**2.12 Hydrophilic and Hydrophobic** (A) Molecules with polar covalent bonds are attracted to polar water (they are hydrophilic). (B) Molecules with nonpolar covalent bonds show greater attraction to one another than to water (they are hydrophobic).

Nonpolar molecules tend to interact with other nonpolar molecules. For example, carbon (electronegativity 2.5) forms nonpolar bonds with hydrogen (electronegativity 2.1), and molecules containing only hydrogen and carbon atoms—called *hydrocarbon molecules*—are nonpolar. In water these molecules tend to aggregate with one another rather than with the polar water molecules. Therefore, nonpolar molecules are known as **hydrophobic** ("water-hating"), and the interactions between them are called *hydrophobic interactions* (**Figure 2.12B**). Of course, hydrophobic substances do not really "hate" water; they can form weak interactions with it, since the electronegativities of carbon and hydrogen are not exactly the same. But these interactions are far weaker than the hydrogen bonds between the water molecules, so the nonpolar substances tend to aggregate.

The interactions between nonpolar substances are enhanced by **van der Waals forces**, which occur when the atoms of two nonpolar molecules are in close proximity. These brief interactions result from random variations in the electron distribution in one molecule, which create opposite charge distributions in the adjacent molecule. Although a single van der Waals interaction is brief and weak, the sum of many such interactions over the entire span of a large nonpolar molecule can result in substantial attraction. This makes nonpolar molecules stick together in the polar (aqueous) environment inside organisms. We will see this many times, for example in the structure of biological membranes.

## 2.2 RECAP

**Some atoms form strong covalent bonds with other atoms by sharing one or more pairs of electrons. Unequal sharing of electrons produces polarity. Other atoms become ions by losing or gaining electrons, and they interact with other ions or polar molecules.**

- Why is a covalent bond stronger than an ionic bond? See pp. 26–28 and Table 2.1
- How do variations in electronegativity result in the unequal sharing of electrons in polar molecules? See pp. 27–28 and Figure 2.8
- What is a hydrogen bond and how is it important in biological systems? See p. 29 and Figure 2.11

The bonding of atoms into molecules is not necessarily a permanent affair. The dynamic of life involves constant change, even at the molecular level. Let's look at how molecules interact with one another—how they break up, how they find new partners, and what the consequences of those changes can be.

## 2.3 How Do Atoms Change Partners in Chemical Reactions?

A **chemical reaction** occurs when moving atoms collide with sufficient energy to combine or change their bonding partners. Consider the combustion reaction that takes place in the flame of a propane stove. When propane ($C_3H_8$) reacts with oxygen gas ($O_2$), the carbon atoms become bonded to oxygen atoms instead

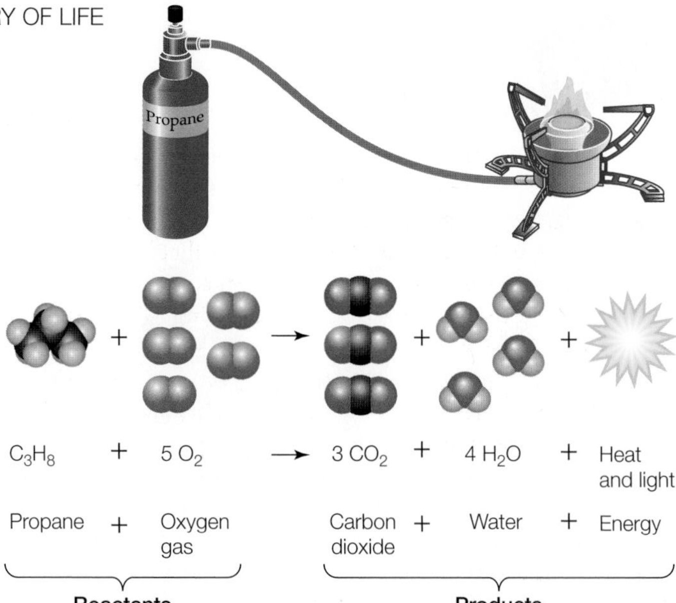


**2.13 Bonding Partners and Energy May Change in a Chemical Reaction** One molecule of propane from this burner reacts with five molecules of oxygen gas to give three molecules of carbon dioxide and four molecules of water. This reaction releases energy in the form of heat and light.

of hydrogen atoms, and the hydrogen atoms become bonded to oxygen instead of carbon (**Figure 2.13**). As the covalently bonded atoms change partners, the composition of the matter changes; propane and oxygen gas become carbon dioxide and water. This chemical reaction can be represented by the equation

$$C_3H_8 + 5\,O_2 \rightarrow 3\,CO_2 + 4\,H_2O + \text{Energy}$$
$$\text{Reactants} \rightarrow \text{Products}$$

In this equation, the propane and oxygen are the **reactants**, and the carbon dioxide and water are the **products**. In fact, this is a special type of reaction called an oxidation–reduction reaction. Electrons and protons are transferred from propane (the reducing agent) to oxygen (the oxidizing agent) to form water. You will see this kind of reaction involving electron/proton transfer many times in later chapters.

The products of a chemical reaction have very different properties from the reactants. In the case shown in Figure 2.13, the reaction is *complete*: all the propane and oxygen are used up in forming the two products. The arrow symbolizes the direction of the chemical reaction. The numbers preceding the molecular formulas indicate how many molecules are used or produced.

Note that in this and all other chemical reactions, *matter is neither created nor destroyed*. The total number of carbon atoms on the left (3) equals the total number of carbon atoms on the right (3). In other words, the equation is *balanced*. However, there is another aspect of this reaction: the heat and light of the stove's flame reveal that the reaction between propane and oxygen releases a great deal of energy.

**Energy** is defined as the capacity to do work, but in the context of chemical reactions, it can be thought of as the capacity for change. Chemical reactions do not create or destroy energy, but *changes in the form of energy* usually accompany chemical reactions.

In the reaction between propane and oxygen, a large amount of heat energy is released. This energy was present in another form, called *potential chemical energy*, in the covalent bonds within the propane and oxygen gas molecules. Not all reactions release energy; indeed, many chemical reactions require that energy be supplied from the environment. Some of this energy is then stored as potential chemical energy in the bonds formed in the products. We will see in future chapters how reactions that release energy and reactions that require energy can be linked together.

Many chemical reactions take place in living cells, and some of these have a lot in common with the oxidation–reduction reaction that happens in the combustion of propane. In cells, the reactants are different (they may be sugars or fats), and the reactions proceed by many intermediate steps that permit the released energy to be harvested and put to use by the cells. But the products are the same: carbon dioxide and water. We will discuss energy changes, oxidation–reduction reactions, and several other types of chemical reactions that are prevalent in living systems in Part Three of this book.

## 2.3 RECAP

**In a chemical reaction, a set of reactants is converted to a set of products with different chemical compositions. This is accomplished by breaking and making bonds. Reactions may release energy or require its input.**

- Explain how a chemical equation is balanced. See p. 30 and Figure 2.13
- How can the form of energy change during a chemical reaction? See p. 30

We will present and discuss energy changes, oxidation–reduction reactions, and several other types of chemical reactions that are prevalent in living systems in Part Two of this book. First, however, we must understand the unique properties of the substance in which most biochemical reactions take place: water.

# 2.4 What Makes Water So Important for Life?

Water is an unusual substance with unusual properties. Under conditions on Earth, water exists in solid, liquid, and gas forms, all of which have relevance to living systems. Water allows chemical reactions to occur inside living organisms, and it is necessary for the formation of certain biological structures. In this section we will explore how the structure and interactions of water molecules make water essential to life.

### Water has a unique structure and special properties

The molecule $H_2O$ has unique chemical features. As we have already learned, water is a polar molecule that can form hydrogen bonds. The four pairs of electrons in the outer shell of the oxygen atom repel one another, giving the water molecule a tetrahedral shape:

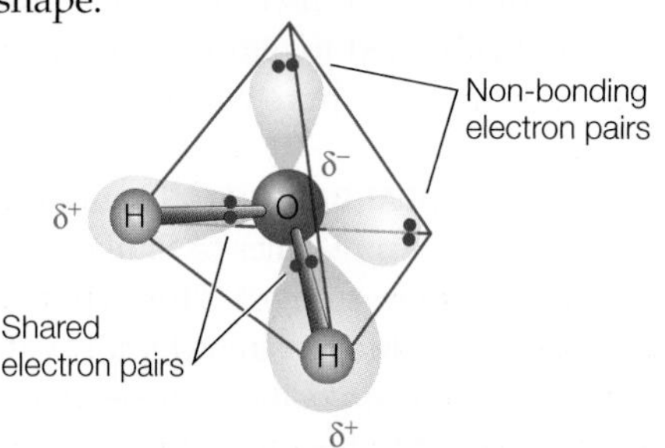


These chemical features explain some of the interesting properties of water, such as the ability of ice to float, the melting and freezing temperatures of water, the ability of water to store heat, the formation of water droplets, and water's ability to dissolve—and not dissolve—many substances.

**ICE FLOATS** In water's solid state (ice), individual water molecules are held in place by hydrogen bonds. Each molecule is bonded to four other molecules in a rigid, crystalline structure (**Figure 2.14**). Although the molecules are held firmly in place, they are not as tightly packed as they are in liquid water. In other words, *solid water is less dense than liquid water*, which is why ice floats.

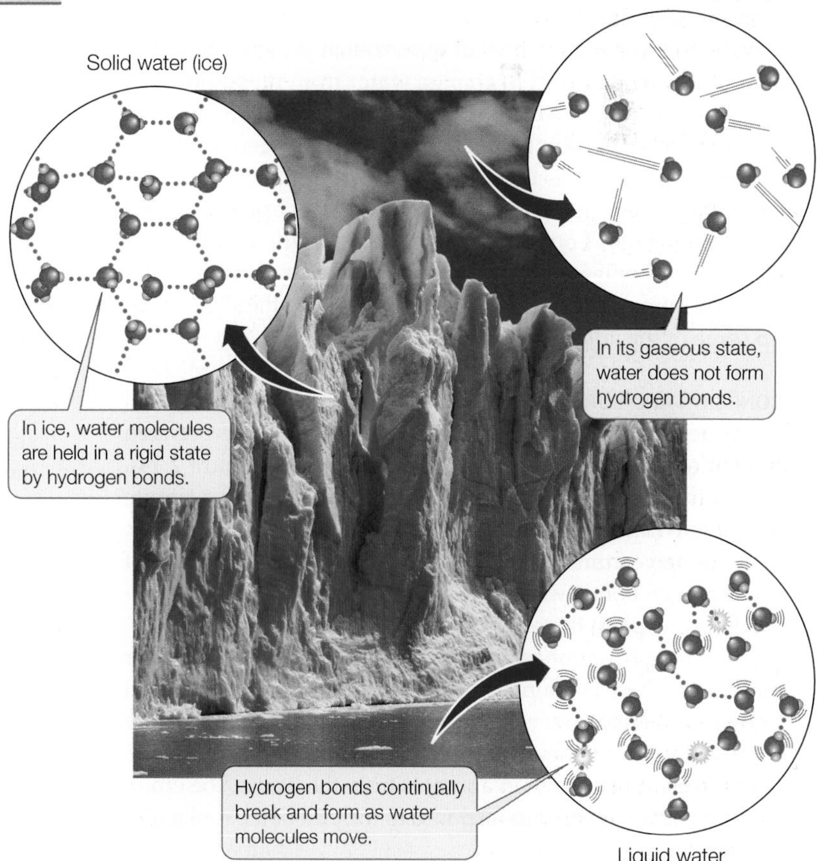


**2.14 Hydrogen Bonding and the Properties of Water** Hydrogen bonding exists between the molecules of water in both its liquid and solid states. Ice is more structured but less dense than liquid water, which is why ice floats. Water forms a gas when its hydrogen bonds are broken and the molecules move farther apart.

Think of the biological consequences if ice were to sink in water. A pond would freeze from the bottom up, becoming a solid block of ice in winter and killing most of the organisms living there. Once the whole pond is frozen, its temperature could drop well below the freezing point of water. But in fact ice floats, forming an insulating layer on the top of the pond, and reducing heat flow to the cold air above. Thus fish, plants, and other organisms in the pond are not subjected to temperatures lower than 0°C, which is the freezing point of pure water.

**MELTING, FREEZING, AND HEAT CAPACITY** Compared with many other substances that have molecules of similar size, ice requires a great deal of heat energy to melt. This is because so many hydrogen bonds must be broken in order for water to change from solid to liquid. In the opposite process—freezing—a great deal of energy is released to the environment.

This property of water contributes to the surprising constancy of the temperatures found in oceans and other large bodies of water throughout the year. The temperature changes of coastal land masses are also moderated by large bodies of water. Indeed, water helps minimize variations in atmospheric temperature across the planet. This moderating ability is a result of the high *heat capacity* of liquid water, which is in turn a result of its high specific heat.

The **specific heat** of a substance is the amount of heat energy required to raise the temperature of 1 gram of that substance by 1°C. Raising the temperature of liquid water takes a relatively large amount of heat because much of the heat energy is used to break the hydrogen bonds that hold the liquid together. Compared with other small molecules that are liquids, water has a high specific heat.

Water also has a high **heat of vaporization**, which means that a lot of heat is required to change water from its liquid to its gaseous state (the process of *evaporation*). Once again, much of the heat energy is used to break the many hydrogen bonds between the water molecules. This heat must be absorbed from the environment in contact with the water. Evaporation thus has a cooling effect on the environment—whether a leaf, a forest, or an entire land mass. This effect explains why sweating cools the human body: as sweat evaporates from the skin, it uses up some of the adjacent body heat.

**COHESION AND SURFACE TENSION** In liquid water, individual molecules are able to move about. The hydrogen bonds between the molecules continually form and break (see Figure 2.14). Chemists estimate that this occurs about a trillion times a minute for a single water molecule, making it a truly dynamic structure.

At any given time, a water molecule will form an average of 3.4 hydrogen bonds with other water molecules. These hydrogen bonds explain the *cohesive strength* of liquid water. This cohesive strength, or **cohesion**, is defined as the capacity of water molecules to resist coming apart from one another when placed under tension. Water's cohesive strength permits narrow columns of liquid water to move from the roots to the leaves of tall trees. When water evaporates from the leaves, the entire column moves upward in response to the pull of the molecules at the top.

**2.15 Surface Tension** Water droplets form "beads" on the surface of a leaf because hydrogen bonds keep the water molecules together. The leaf is coated in a nonpolar wax that does not interact with the water molecules.

The surface of liquid water exposed to the air is difficult to puncture because the water molecules at the surface are hydrogen-bonded to other water molecules below them (**Figure 2.15**). This *surface tension* of water permits a container to be filled slightly above its rim without overflowing, and it permits insects to walk on the surface of a pond.

## Water is an excellent solvent—the medium of life

A human body is over 70 percent water by weight, excluding the minerals contained in bones. Water is the dominant component of virtually all living organisms, and most biochemical reactions take place in this watery, or aqueous, environment.

A **solution** is produced when a substance (the **solute**) is dissolved in a liquid (the **solvent**). If the solvent is water, then the solution is an *aqueous solution*. Many of the important molecules in biological systems are polar, and therefore soluble in water. Many important biochemical reactions occur in aqueous solutions within cells. Biologists study these reactions in order to identify the reactants and products and to determine their amounts:

- *Qualitative analyses* deal with the identification of substances involved in chemical reactions. For example, a qualitative analysis would be used to investigate the steps involved, and the products formed, during the combustion of glucose in living tissues.
- *Quantitative analyses* measure concentrations or amounts of substances. For example, a biochemist would seek to describe *how much* of a certain product is formed during the combustion of a given amount of glucose using a quantitative analysis. What follows is a brief introduction to some of the quantitative chemical terms you will see in this book.

Fundamental to quantitative thinking in chemistry and biology is the concept of the mole. A **mole** is the amount of a substance (in grams) that is numerically equal to its molecular weight.

So a mole of table sugar ($C_{12}H_{22}O_{11}$) weighs about 342 grams; a mole of sodium ion ($Na^+$) weighs 23 grams; and a mole of hydrogen gas ($H_2$) weighs 2 grams.

Quantitative analyses do not yield direct counts of molecules. Because the amount of a substance in 1 mole is directly related to its molecular weight, it follows that the number of molecules in 1 mole is constant for all substances. So 1 mole of salt contains the same number of molecules as 1 mole of table sugar. This constant number of molecules in a mole is called **Avogadro's number**, and it is $6.02 \times 10^{23}$ molecules per mole. Chemists work with moles of substances (which can be weighed out in the laboratory) instead of actual molecules (which are too numerous to be counted). Consider 34.2 grams (just over 1 ounce) of table sugar, $C_{12}H_{22}O_{11}$. This is one-tenth of a mole, or as Avogadro puts it, $6.02 \times 10^{22}$ molecules.

If you have trouble grasping the concept of a mole, compare it with the concept of a dozen. We buy a dozen eggs or a dozen doughnuts, knowing that we will get 12 of whichever we buy, even though they don't weigh the same or take up the same amount of space.

A chemist can dissolve a mole of sugar (342 g) in water to make 1 liter of solution, knowing that the mole contains $6.02 \times 10^{23}$ individual sugar molecules. This solution—1 mole of a substance dissolved in water to make 1 liter—is called a 1 molar (1 *M*) solution. When a physician injects a certain molar concentration of a drug into the bloodstream of a patient, a rough calculation can be made of the actual number of drug molecules that will interact with the patient's cells.

The many molecules dissolved in the water of living tissues are not present at concentrations anywhere near 1 molar. Most are in the micromolar (millionths of a mole per liter of solution; μ*M*) to millimolar (thousandths of a mole per liter; m*M*) range. Some, such as hormone molecules, are even less concentrated than that. While these molarities seem to indicate very low concentrations, remember that even a 1 μ*M* solution has $6.02 \times 10^{17}$ molecules of the solute per liter.

## Aqueous solutions may be acidic or basic

When some substances dissolve in water, they release *hydrogen ions* ($H^+$), which are actually single, positively charged protons. Hydrogen ions can attach to other molecules and change their properties. For example, the protons in "acid rain" can damage plants, and you probably have experienced the excess of hydrogen ions that we call "acid indigestion."

Here we will examine the properties of **acids** (defined as substances that release $H^+$) and **bases** (defined as substances which accept $H^+$). We will distinguish between strong and weak acids and bases and provide a quantitative means for stating the concentration of $H^+$ in solutions: the pH scale.

**ACIDS RELEASE H+** When hydrochloric acid (HCl) is added to water, it dissolves, releasing the ions $H^+$ and $Cl^-$:

$$HCl \rightarrow H^+ + Cl^-$$

Because its $H^+$ concentration has increased, such a solution is *acidic*.

Acids are substances that *release* $H^+$ ions in solution. HCl is an acid, as is $H_2SO_4$ (sulfuric acid). One molecule of sulfuric acid will ionize to yield two $H^+$ and one $SO_4^{2-}$. Biological compounds that contain —COOH (the carboxyl group) are also acids because

$$—COOH \rightarrow —COO^- + H^+$$

Acids that fully ionize in solution, such as HCl and $H_2SO_4$ are called *strong acids*. However, not all acids ionize fully in water. For example, if acetic acid ($CH_3COOH$) is added to water, some will dissociate into two ions ($CH_3COO^-$ and $H^+$), but some of the original acetic acid remains as well. Because the reaction is *not complete*, acetic acid is a *weak acid*.

**BASES ACCEPT H+** Bases are substances that *accept* $H^+$ in solution. Just as with acids, there are strong and weak bases. If NaOH (sodium hydroxide) is added to water, it dissolves and ionizes, releasing $OH^-$ and $Na^+$ ions:

$$NaOH \rightarrow Na^+ + OH^-$$

Because the concentration of $OH^-$ increases and $OH^-$ absorbs $H^+$ to form water ($OH^- + H^+ \rightarrow H_2O$), such a solution is *basic*. Because this reaction is complete, NaOH is a *strong base*.

Weak bases include the bicarbonate ion ($HCO_3^-$), which can accept a $H^+$ ion and become carbonic acid ($H_2CO_3$), and ammonia ($NH_3$), which can accept a $H^+$ and become an ammonium ion ($NH_4^+$). Biological compounds that contain $—NH_2$ (the amino group) are also bases because

$$—NH_2 + H^+ \rightarrow —NH_3^+$$

**ACID–BASE REACTIONS MAY BE REVERSIBLE** When acetic acid is dissolved in water, two reactions happen. First, the acetic acid forms its ions:

$$CH_3COOH \rightarrow CH_3COO^- + H^+$$

Then, once the ions are formed, some of them re-form acetic acid:

$$CH_3COO^- + H^+ \rightarrow CH_3COOH$$

This pair of reactions is reversible. A **reversible reaction** can proceed in either direction—left to right or right to left—depending on the relative starting concentrations of the reactants and products. The formula for a reversible reaction can be written using a double arrow:

$$CH_3COOH \rightleftharpoons CH_3COO^- + H^+$$

In terms of acids and bases, there are two types of reactions, depending on the extent of the reversibility:

- The ionization of strong acids and bases in water is virtually irreversible.
- The ionization of weak acids and bases in water is somewhat reversible.

**WATER IS A WEAK ACID AND A WEAK BASE** The water molecule has a slight but significant tendency to ionize into a hydroxide ion ($OH^-$) and a hydrogen ion ($H^+$). Actually, two water molecules participate in this reaction. One of the two molecules "captures" a hydrogen ion from the other, forming a hydroxide ion and a hydronium ion:

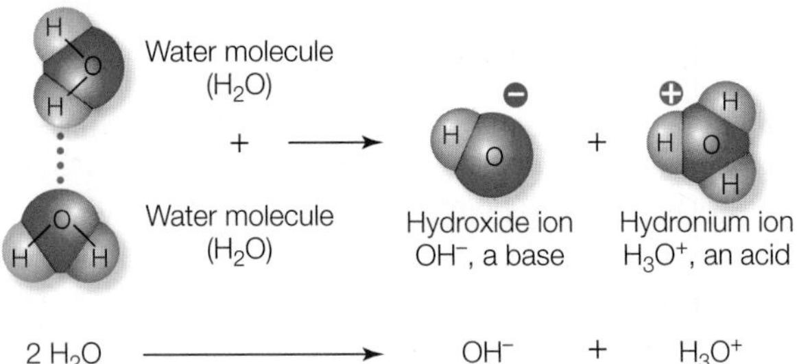


The hydronium ion is, in effect, a hydrogen ion bound to a water molecule. For simplicity, biochemists tend to use a modified representation of the ionization of water:

$$H_2O \rightarrow H^+ + OH^-$$

The ionization of water is important to all living creatures. This fact may seem surprising, since only about one water molecule in 500 million is ionized at any given time. But this is less surprising if we focus on the abundance of water in living systems, and the reactive nature of the $H^+$ ions produced by ionization.

**pH: HYDROGEN ION CONCENTRATION** Compounds or ions can be acids or bases, and thus, solutions can be acidic or basic. We can measure how acidic or basic a solution is by measuring its concentration of $H^+$ in moles per liter (its *molarity*; see page 33). Here are some examples:

- Pure water has a $H^+$ concentration of $10^{-7}$ *M*.
- A 1 *M* HCl solution has a $H^+$ concentration of 1 *M* (recall that all the HCl dissociates into its ions).
- A 1 *M* NaOH solution has a $H^+$ concentration of $10^{-14}$ *M*.

This is a very wide range of numbers to work with—think about the decimals! It is easier to work with the *logarithm* of the $H^+$ concentration, because logarithms compress this range: the $\log_{10}$ of 100, for example is 2, and the $\log_{10}$ of 0.01 is –2. Because most $H^+$ concentrations in living systems are less than 1M, their $\log_{10}$ values are negative. For convenience, we convert these negative numbers into positive ones, by using the *negative* of the logarithm of the $H^+$ molar concentration (the molar concentration is designated by square brackets: $[H^+]$). This number is called the **pH** of the solution.

Since the $H^+$ concentration of pure water is $10^{-7}$ *M*, its pH is $-\log(10^{-7}) = -(-7)$, or 7. A smaller negative logarithm means a larger number. In practical terms, a lower pH means a higher $H^+$ concentration, or greater acidity. In 1 *M* HCl, the $H^+$ concentration is 1 *M*, so the pH is the negative logarithm of 1 ($-\log 10^0$), or 0. The pH of 1 *M* NaOH is the negative logarithm of $10^{-14}$, or 14.

A solution with a pH of less than 7 is acidic—it contains more $H^+$ ions than $OH^-$ ions. A solution with a pH of 7 is *neutral* (without net charge), and a solution with a pH value greater than 7 is basic. **Figure 2.16** shows the pH values of some common substances.

Why is this discussion of pH so important in biology? Many biologically important molecules contain charged groups

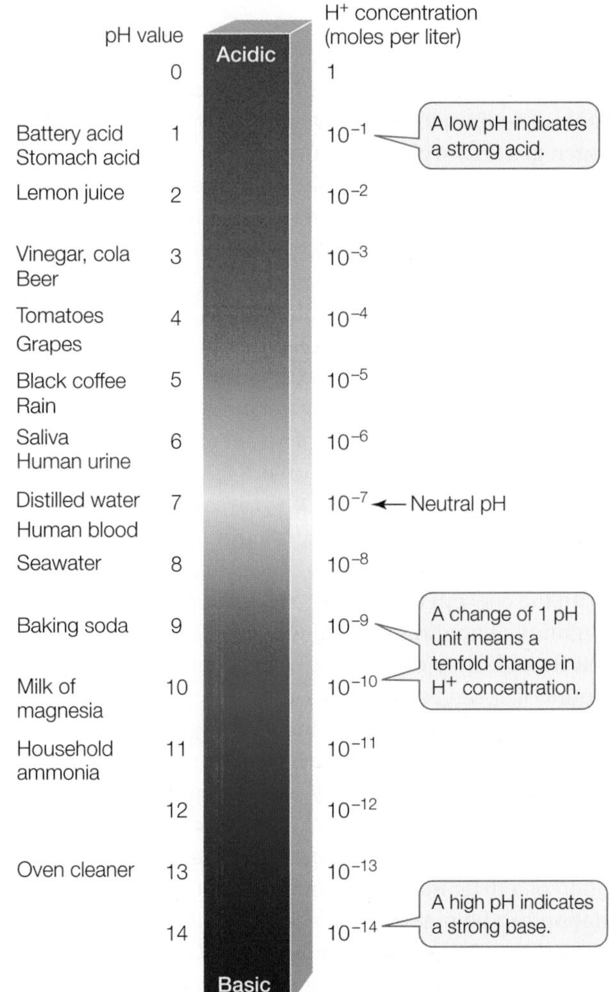


**2.16 pH Values of Some Familiar Substances**

(e.g., —$COO^-$) that can interact with the polar regions of water to form their structures. But these groups can combine with $H^+$ or other ions in their environment to form uncharged groups (e.g., —COOH, see above). These uncharged groups have much less tendency to interact with water. If such a group is part of a larger molecule, it might now induce the molecule to fold in such a way that it stays away from water because it is hydrophobic. In a more acidic environment, a negatively charged group such as —$COO^-$ is more likely to combine with $H^+$. So the pH of a biological tissue is a key to the three-dimensional structures of many of its constituent molecules. Organisms do all they can to minimize changes in the pH of their watery medium. An important way to do this is with buffers.

**BUFFERS** The maintenance of internal constancy—*homeostasis*—is a hallmark of all living things and extends to pH. As we mentioned earlier, if biological molecules lose or gain $H^+$ ions their properties can change, thus upsetting homeostasis. Internal constancy is achieved with buffers: solutions that maintain a relatively constant pH even when substantial amounts of acid or base are added. How does this work?

A **buffer** is a solution of a weak acid and its corresponding base—for example, carbonic acid ($H_2CO_3$) and bicarbonate ions ($HCO_3^-$). If an acid is added to a solution containing this buffer,

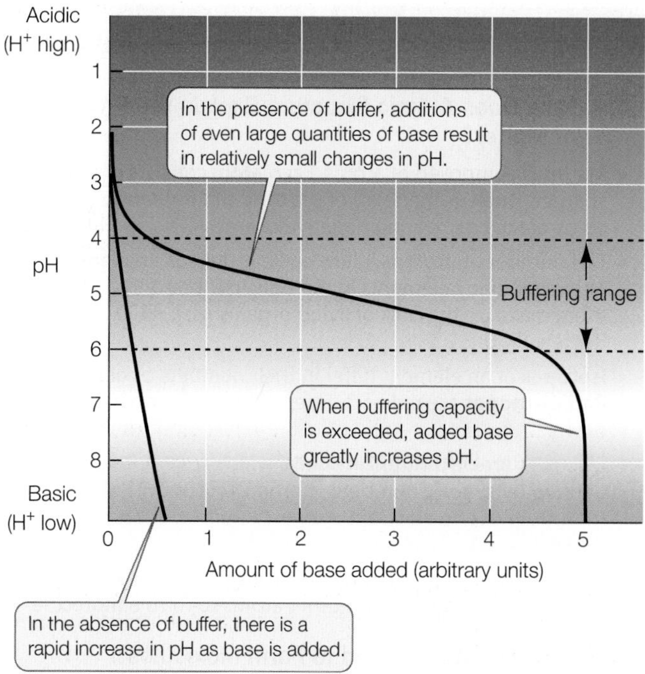


**2.17 Buffers Minimize Changes in pH** With increasing amounts of added base, the overall slope of a graph of pH is downward. Without a buffer, the slope is steep. Inside the buffering range of an added buffer, however, the slope is shallow. At very high and very low values of pH, where the buffer is ineffective, the slopes are much steeper.

not all the $H^+$ ions from the acid stay in solution. Instead, many of them combine with the bicarbonate ions to produce more carbonic acid:

$$HCO_3^- + H^+ \rightarrow H_2CO_3$$

This reaction uses up some of the $H^+$ ions in the solution and decreases the acidifying effect of the added acid. If a base is added, the reaction essentially reverses. Some of the carbonic acid ionizes to produce bicarbonate ions and more $H^+$, which counteracts some of the added base. In this way, the buffer minimizes the effect that an added acid or base has on pH. This buffering system is present in the blood, where it is important for preventing significant changes in pH that could disrupt the ability of the blood to carry vital oxygen to tissues. A given amount of acid or base causes a smaller pH change in a buffered solution than in a non-buffered one (**Figure 2.17**).

Buffers illustrate an important chemical principle of reversible reactions, called the *law of mass action*. Addition of a reactant on one side of a reversible system drives the reaction in the direction that uses up that compound. In the case of buffers, addition of an acid drives the reaction in one direction; addition of a base drives the reaction in the other direction.

We use a buffer to relieve the common problem of indigestion. The lining of the stomach constantly secretes hydrochloric acid, making the stomach contents acidic. Excessive stomach acid inhibits digestion and causes discomfort. We can relieve this discomfort by ingesting a salt such as $NaHCO_3$ ("bicarbonate of soda"), which acts as a buffer.

## 2.4 RECAP

**Most of the chemistry of life occurs in water, which has molecular properties that make it suitable for its important biochemical roles. A special property of water is its ability to ionize (release hydrogen ions). The presence of hydrogen ions in solution can change the properties of biological molecules.**

- Describe some of the biologically important properties of water arising from its molecular structure. **See pp. 31–32 and Figure 2.14**
- What is a solution, and why do we call water "the medium of life"? **See pp. 32–33**
- What is the relationship between hydrogen ions, acids, and bases? Explain what the pH scale measures. **See pp. 33–34 and Figure 2.16**
- How does a buffer work, and why is buffering important to living systems? **See pp. 34–35 and Figure 2.17**

## An Overview and a Preview

Now that we have covered the major properties of atoms and molecules, let's review them and see how these properties relate to the major molecules of biological systems.

- *Molecules vary in size.* Some are small, such as those of hydrogen gas ($H_2$) and methane ($CH_4$). Others are larger, such as a molecule of table sugar ($C_{12}H_{22}O_{11}$), which has 45 atoms. Still others, especially proteins and nucleic acids, are gigantic, containing tens of thousands or even millions of atoms.
- *All molecules have a specific three-dimensional shape.* For example, the orientations of the bonding orbitals around the carbon atom give the methane molecule ($CH_4$) the shape of a regular tetrahedron (see Figure 2.7B). Larger molecules have complex shapes that result from the numbers and kinds of atoms present, and the ways in which they are linked together. Some large molecules, such as the protein hemoglobin (the oxygen carrier in red blood cells), have compact, ball-like shapes. Others, such as the protein called keratin that makes up your hair, have long, thin, ropelike structures. Their shapes relate to the roles these molecules play in living cells.
- *Molecules are characterized by certain chemical properties* that determine their biological roles. Chemists use the characteristics of composition, structure (three-dimensional shape), reactivity, and solubility to distinguish a pure sample of one molecule from a sample of a different molecule. The presence of certain groups of atoms can impart distinctive chemical properties to a molecule.

Between the small molecules discussed in this chapter and the world of the living cell are the macromolecules. These larger molecules—proteins, lipids, carbohydrates, and nucleic acids—will be discussed in the next two chapters.

## CHAPTER SUMMARY

### 2.1 How Does Atomic Structure Explain the Properties of Matter?

- Matter is composed of atoms. Each **atom** consists of a positively charged **nucleus** made up of **protons** and **neutrons**, surrounded by **electrons** bearing negative charges. Review Figure 2.1
- The number of protons in the nucleus defines an **element**. There are many elements in the universe, but only a few of them make up the bulk of living organisms: C, H, O, P, N, and S. Review Figure 2.2
- **Isotopes** of an element differ in their numbers of neutrons. **Radioisotopes** are radioactive, emitting radiation as they break down.
- Electrons are distributed in **shells**, which are volumes of space defined by specific numbers of orbitals. Each **orbital** contains a maximum of two electrons. Review Figures 2.4 and 2.5, WEB ACTIVITY 2.1
- In losing, gaining, or sharing electrons to become more stable, an atom can combine with other atoms to form a **molecule**.

### 2.2 How Do Atoms Bond to Form Molecules?

SEE ANIMATED TUTORIAL 2.1

- A **chemical bond** is an attractive force that links two atoms together in a molecule. Review Table 2.1
- A **compound** is a substance made up of molecules with two or more elements bonded together in a fixed ratio, such as water ($H_2O$) or table sugar ($C_6H_{12}O_6$).
- **Covalent bonds** are strong bonds formed when two atoms share one or more pairs of electrons. Review Figure 2.6
- When two atoms of unequal electronegativity bond with each other, a **polar** covalent bond is formed. The two ends, or poles, of the bond have partial charges ($\delta^+$ or $\delta^-$). Review Figure 2.8
- **Ions** are electrically charged bodies that form when an atom gains or loses one or more electrons in order to form more stable electron configurations. **Anions** and **cations** are negatively and positively charged ions, respectively. Different charges attract, and like charges repel each other.
- **Ionic bonds** are electrical attractions between oppositely charged ions. Ionic bonds are strong in solids (salts), but weaken when the ions are separated from one another in solution. Review Figure 2.9
- A **hydrogen bond** is a weak electrical attraction that forms between a $\delta^+$ hydrogen atom in one molecule and a $\delta^-$ atom in another molecule (or in another part of a large molecule). Hydrogen bonds are abundant in water.
- Nonpolar molecules interact very little with polar molecules, including water. Nonpolar molecules are attracted to one another by very weak bonds called **van der Waals forces**.

### 2.3 How Do Atoms Change Partners in Chemical Reactions?

- In **chemical reactions**, atoms combine or change their bonding partners. **Reactants** are converted into **products**.
- Some chemical reactions release **energy** as one of their products; other reactions can occur only if energy is provided to the reactants.
- Neither matter nor energy is created or destroyed in a chemical reaction, but both change form. Review Figure 2.13
- Some chemical reactions, especially in biology, are reversible. That is, the products formed may be converted back to the reactants.
- In living cells, chemical reactions take place in multiple steps so that the released energy can be harvested for cellular activities.

### 2.4 What Makes Water So Important for Life?

- Water's molecular structure and its capacity to form hydrogen bonds give it unique properties that are significant for life. Review Figure 2.14
- The high **specific heat** of water means that water gains or loses a great deal of heat when it changes state. Water's high **heat of vaporization** ensures effective cooling when water evaporates.
- The **cohesion** of water molecules refers to their capacity to resist coming apart from one another. Hydrogen bonds between water molecules play an essential role in these properties.
- A **solution** is produced when a solid substance (the **solute**) dissolves in a liquid (the **solvent**). Water is the critically important solvent for life.
- **Acids** are solutes that release hydrogen ions in aqueous solutions. **Bases** accept hydrogen ions.
- The **pH** of a solution is the negative logarithm of its hydrogen ion concentration. Values lower than pH 7 indicate that a solution is acidic; values above pH 7 indicate a basic solution. Review Figure 2.16
- A **buffer** is a mixture of a weak acid and a base that limits changes in the pH of a solution when acids or bases are added.

## SELF-QUIZ

1. The atomic number of an element
   *a.* equals the number of neutrons in an atom.
   *b.* equals the number of protons in an atom.
   *c.* equals the number of protons minus the number of neutrons.
   *d.* equals the number of neutrons plus the number of protons.
   *e.* depends on the isotope.
2. The atomic weight (atomic mass) of an element
   *a.* equals the number of neutrons in an atom.
   *b.* equals the number of protons in an atom.
   *c.* equals the number of electrons in an atom.
   *d.* equals the number of neutrons plus the number of protons.
   *e.* depends on the relative abundances of its electrons and neutrons.
3. Which of the following statements about the isotopes of an element is *not* true?
   *a.* They all have the same atomic number.
   *b.* They all have the same number of protons.
   *c.* They all have the same number of neutrons.
   *d.* They all have the same number of electrons.
   *e.* They all have identical chemical properties.

4. Which of the following statements about covalent bonds is *not* true?
   *a.* A covalent bond is stronger than a hydrogen bond.
   *b.* A covalent bond can form between atoms of the same element.
   *c.* Only a single covalent bond can form between two atoms.
   *d.* A covalent bond results from the sharing of electrons by two atoms.
   *e.* A covalent bond can form between atoms of different elements.
5. Hydrophobic interactions
   *a.* are stronger than hydrogen bonds.
   *b.* are stronger than covalent bonds.
   *c.* can hold two ions together.
   *d.* can hold two nonpolar molecules together.
   *e.* are responsible for the surface tension of water.
6. Which of the following statements about water is *not* true?
   *a.* It releases a large amount of heat when changing from liquid into vapor.
   *b.* Its solid form is less dense than its liquid form.
   *c.* It is the most effective solvent for polar molecules.
   *d.* It is typically the most abundant substance in a living organism.
   *e.* It takes part in some important chemical reactions.
7. The reaction $HCl \rightarrow H^+ + Cl^-$ in the human stomach is an example of the
   *a.* cleavage of a hydrophobic bond.
   *b.* formation of a hydrogen bond.
   *c.* elevation of the pH of the stomach.
   *d.* formation of ions by dissolving an acid.
   *e.* formation of polar covalent bonds.
8. The hydrogen bond between two water molecules arises because water is
   *a.* polar.
   *b.* nonpolar.
   *c.* a liquid.
   *d.* small.
   *e.* hydrophobic.
9. When table salt (NaCl) is added to water,
   *a.* a covalent bond is broken.
   *b.* an acidic solution is formed.
   *c.* the $Na^+$ and $Cl^-$ ions are separated.
   *d.* the $Na^+$ ions are attracted to the hydrogen atoms of water.
   *e.* water molecules surround the $Na^+$ (but not $Cl^-$) ions.
10. The three most abundant elements in a human skin cell are
   *a.* calcium, carbon, and oxygen.
   *b.* carbon, hydrogen, and oxygen.
   *c.* carbon, hydrogen, and sodium.
   *d.* carbon, nitrogen, and potassium.
   *e.* nitrogen, hydrogen, and argon.

## FOR DISCUSSION

1. Using the information in the periodic table (Figure 2.2), draw a Bohr model (see Figures 2.5 and 2.7) of silicon dioxide, showing electrons shared in covalent bonds.
2. Compare a covalent bond between two hydrogen atoms with a hydrogen bond between a hydrogen and an oxygen atom, with regard to the electrons involved, the role of polarity, and the strength of the bond.
3. Write an equation describing the combustion of glucose ($C_6H_{12}O_6$) to produce carbon dioxide and water.
4. The pH of the human stomach is about 2.0, while the pH of the small intestine is about 10.0. What are the hydrogen ion concentrations [$H^+$] inside these two organs?

## ADDITIONAL INVESTIGATION

Would you expect the elemental composition of Earth's crust to be the same as that of the human body? How could you find out?

# 3 Proteins, Carbohydrates, and Lipids

## Molecular fossils

About 68 million years ago, a *Tyrannosaurus rex*, the fearsome dinosaur of movie stardom, died in what is now Wyoming in the United States. Over time, the giant carcass became buried 60 feet below the surface of what geologists call the Hell Creek Formation. In 2003, a thigh bone from the long-dead beast was found by the famous dinosaur hunter/biologist, John Horner from the Museum of the Rockies. Mary Schweitzer, a molecular paleontologist, was visiting Horner's Montana lab from North Carolina State University. She cut into the bone and found that it contained the remnants of soft tissues (such as bone marrow). This discovery was remarkable, because up until then scientists had thought that after about a million years, all the soft tissues in bone were replaced with minerals.

Back on the east coast, Lewis Cantley, a biochemist at Harvard University, read about Schweitzer's find in a newspaper and saw the possibility for a unique opportunity: for the first time, a scientist would be able to isolate and study the complex molecules of soft tissues from an extinct organism. He asked Schweitzer to send him a sample, and when he and his colleagues analyzed the dinosaur material, they found fragments of protein molecules.

Protein molecules are composed of long chains of individual molecules called amino acids. The protein fragments extracted from the *T. rex* bone were identified as collagen, a substance found in many modern animals. Moreover, the identity and specific order of the amino acids in the dinosaur collagen fragments closely matched that of collagen from chickens, and the dinosaur collagen folded into shapes very similar to those of bird collagen. This similarity to birds is not surprising, because, based on other evidence, scientists believe that birds are evolutionarily closely related to dinosaurs. Cantley's molecular analysis further confirmed this belief.

Proteins are one of the four major kinds of large molecules that characterize living systems. These *macromolecules*, which also include *carbohydrates*, *lipids*, and *nucleic acids*, differ in several significant ways from the small molecules and ions described in Chapter 2. First—no surprise—they are larger; the molecular weights of some

**Molecular Clues** A thigh bone from a *Tyrannosaurus rex* that died 68 million years ago contained fragments of the protein collagen.

**Molecular Evolution** The sequence of amino acids in collagen dictates the shape the protein folds into. Collagen's amino acid sequence is similar in *T. rex* and in chickens, indicating that the two species share a common evolutionary ancestor.

nucleic acids reach billions of daltons. Second, these molecules all contain carbon atoms, and so belong to a group of what are known as *organic* chemicals. Third, the atoms of individual macromolecules are held together mostly by covalent bonds, which gives them important structural stability and distinctive three-dimensional geometries. These distinctive shapes are the basis of many of the functions of macromolecules, particularly the proteins.

Finally, carbohydrates, proteins, lipids, and nucleic acids are all unique to the living world. None of these molecular classes occurs in inanimate nature. You aren't likely to find protein in a rock—but if you do, you can be sure it came from a living organism.

**IN THIS CHAPTER** we will describe the chemical and biological properties of proteins, carbohydrates, and lipids. We will identify the components that make up these larger molecules, describe their assembly and geometries, as well as the roles they play in living organisms.

## CHAPTER OUTLINE



## 3.1 What Kinds of Molecules Characterize Living Things?

Four kinds of molecules are characteristic of living things: proteins, carbohydrates, lipids, and nucleic acids. With the exception of the lipids, these *biological molecules* are **polymers** (*poly*, "many"; *mer*, "unit") constructed by the covalent bonding of smaller molecules called **monomers**. The monomers that make up each kind of biological molecule have similar chemical structures:

- *Proteins* are formed from different combinations of 20 *amino acids*, all of which share chemical similarities.
- *Carbohydrates* can form giant molecules by linking together chemically similar sugar monomers (*monosaccharides*) to form polysaccharides.
- *Nucleic acids* are formed from four kinds of nucleotide monomers linked together in long chains.
- *Lipids* also form large structures from a limited set of smaller molecules, but in this case noncovalent forces maintain the interactions between the lipid monomers.

Polymers with molecular weights exceeding 1,000 grams per mole are considered to be **macromolecules**. The proteins, carbohydrates, and nucleic acids of living systems certainly fall into this category. Although large lipid structures are not polymers in the strictest sense, it is convenient to treat them as a special type of macromolecule (see Section 3.4).

How the macromolecules function and interact with other molecules depends on the properties of certain chemical groups in their monomers, the *functional groups*.

**yourBioPortal.com**

GO TO Animated Tutorial 3.1 • Macromolecules

### Functional groups give specific properties to biological molecules

Certain small groups of atoms, called **functional groups**, are consistently found together in very different biological molecules. You will encounter several functional groups repeatedly in your study of biology (**Figure 3.1**). Each functional group has specific chemical properties and, when it is attached to a larger molecule, it confers those properties on the larger molecule. One of these properties is polarity. Looking at the structures in Figure 3.1, can you determine which functional groups are the most

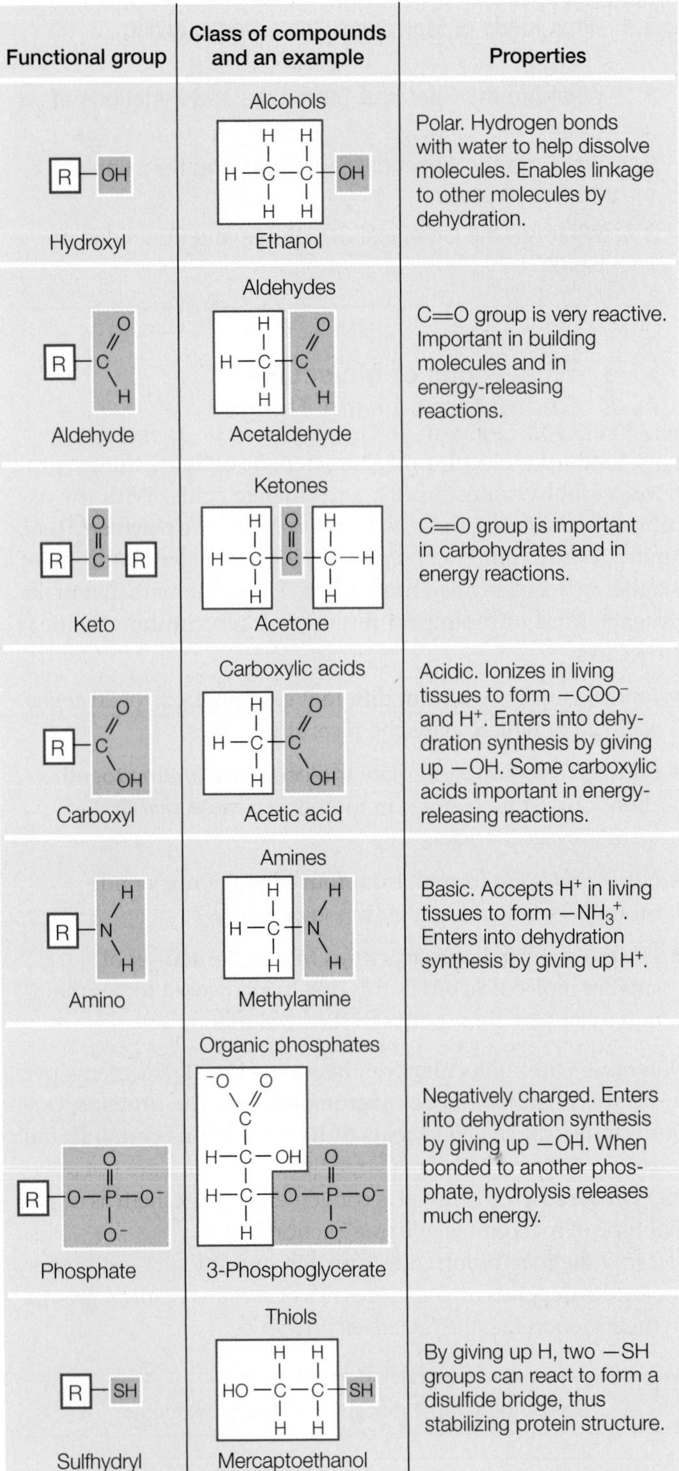

| Functional group | Class of compounds and an example | Properties |
|---|---|---|
| Hydroxyl | Alcohols — Ethanol | Polar. Hydrogen bonds with water to help dissolve molecules. Enables linkage to other molecules by dehydration. |
| Aldehyde | Aldehydes — Acetaldehyde | C═O group is very reactive. Important in building molecules and in energy-releasing reactions. |
| Keto | Ketones — Acetone | C═O group is important in carbohydrates and in energy reactions. |
| Carboxyl | Carboxylic acids — Acetic acid | Acidic. Ionizes in living tissues to form $-COO^-$ and $H^+$. Enters into dehydration synthesis by giving up $-OH$. Some carboxylic acids important in energy-releasing reactions. |
| Amino | Amines — Methylamine | Basic. Accepts $H^+$ in living tissues to form $-NH_3^+$. Enters into dehydration synthesis by giving up $H^+$. |
| Phosphate | Organic phosphates — 3-Phosphoglycerate | Negatively charged. Enters into dehydration synthesis by giving up $-OH$. When bonded to another phosphate, hydrolysis releases much energy. |
| Sulfhydryl | Thiols — Mercaptoethanol | By giving up H, two $-SH$ groups can react to form a disulfide bridge, thus stabilizing protein structure. |

**3.1 Some Functional Groups Important to Living Systems** Highlighted here are the seven functional groups most commonly found in biologically important molecules. "R" is a variable chemical grouping.

yourBioPortal.com
GO TO Web Activity 3.1 • Functional Groups

polar? (Hint: Look for C—O, N—H, and P—O bonds.) The consistent chemical behavior of functional groups helps us understand the properties of the molecules that contain them.

Because macromolecules are so large, they contain many different functional groups (see Figure 3.1). A single large protein may contain hydrophobic, polar, and charged functional groups, each of which gives different specific properties to local sites on the macromolecule. As we will see, sometimes these different groups interact on the same macromolecule. These diverse groups and their properties help determine the shapes of macromolecules as well as how they interact with other macromolecules and with smaller molecules.

### Isomers have different arrangements of the same atoms

**Isomers** are molecules that have the same chemical formula—the same kinds and numbers of atoms—but the atoms are arranged differently. (The prefix *iso-*, meaning "same," is encountered in many biological terms.) Of the different kinds of isomers, we will consider two: structural isomers and optical isomers.

**Structural isomers** differ in how their atoms are joined together. Consider two simple molecules, each composed of four carbon and ten hydrogen atoms bonded covalently, both with the formula $C_4H_{10}$. These atoms can be linked in two different ways, resulting in different molecules:

Butane    Isobutane

The different bonding relationships in butane and isobutane are distinguished by their structural formulas, and the two molecules have different chemical properties.

**Optical isomers** occur when a carbon atom has four different atoms or groups of atoms attached to it. This pattern allows two different ways of making the attachments, each the mirror image of the other (**Figure 3.2**). Such a carbon atom is called an *asymmetrical carbon*, and the two resulting molecules are optical isomers of each other. You can envision your right and left hands as optical isomers. Just as a glove is specific for a particular hand, some biochemical molecules that can interact with one optical isomer of a carbon compound are unable to "fit" the other.

### The structures of macromolecules reflect their functions

The four kinds of biological macromolecules are present in roughly the same proportions in all living organisms (**Figure 3.3**). Furthermore, a protein that has a certain function in an apple tree probably has a similar function in a human being because its chemistry is the same wherever it is found. Such *biochemical unity* reflects the evolution of all life from a common ancestor, by descent with modification. An important advantage of biochemical unity is that some organisms can acquire

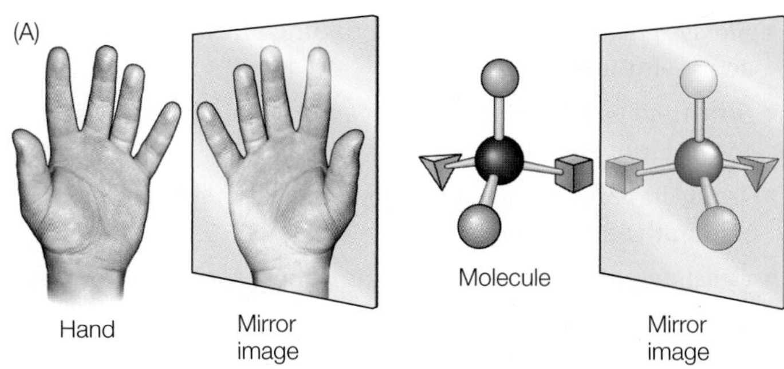


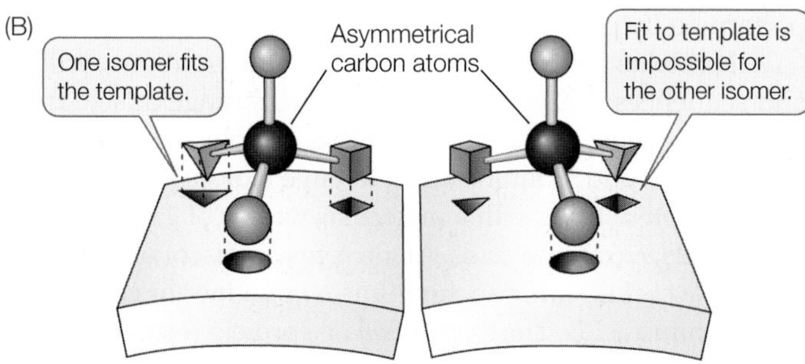


**3.2 Optical Isomers** (A) Optical isomers are mirror images of each other. (B) Molecular optical isomers result when four different atoms or groups are attached to a single carbon atom. If a template (representing a larger biological molecule in a living system) is laid out to match the groups on one carbon atom, the groups on that carbon's optical isomer cannot be rotated to fit the same template. This is a source of specificity in biological structure and biochemical transformations.

needed raw materials by eating other organisms. When you eat an apple, the molecules you take in include carbohydrates, lipids, and proteins that can be broken down and rebuilt into the varieties of those molecules needed by humans.

Each type of macromolecule performs some combination of functions, such as energy storage, structural support, protection, catalysis (speeding up a chemical reaction), transport, defense, regulation, movement, and information storage. These roles are not necessarily exclusive; for example, both carbohydrates and proteins can play structural roles, supporting and protecting tissues and organs. However, only the nucleic acids specialize in information storage and transmission. These macromolecules function as hereditary material, carrying the traits of both species and individuals from generation to generation.

The functions of macromolecules are directly related to their three-dimensional shapes and to the sequences and chemical properties of their monomers. Some macromolecules fold into compact spherical forms with surface features that make them water-soluble and capable of intimate interaction with other molecules. Some proteins and carbohydrates form long, fibrous systems (such as those found in hair) that provide strength and rigidity to cells and tissues. The long, thin assemblies of proteins such as those in muscles can contract, resulting in movement.

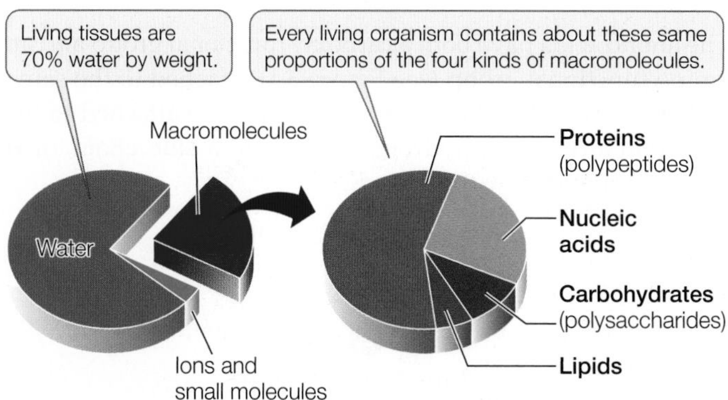


**3.3 Substances Found in Living Tissues** The substances shown here make up the nonmineral components of living tissues (bone would be an example of a mineral component).

### Most macromolecules are formed by condensation and broken down by hydrolysis

Polymers are constructed from monomers by a series of reactions called **condensation reactions** (sometimes called *dehydration* reactions; both terms refer to the loss of water). Condensation reactions result in covalent bonds between monomers. A molecule of water is released with each covalent bond formed (**Figure 3.4A**). The condensation reactions that produce the different kinds of polymers differ in detail, but in

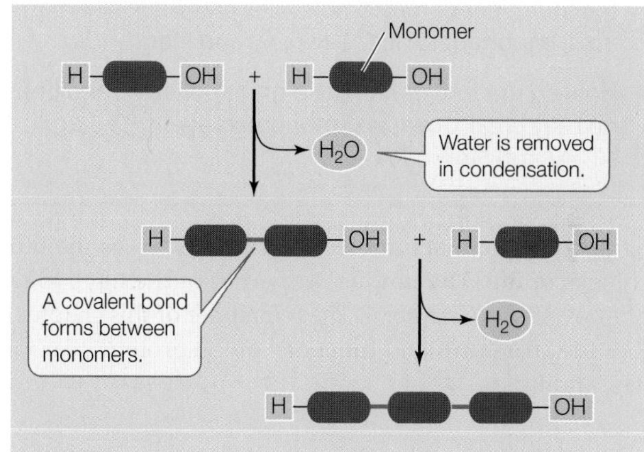


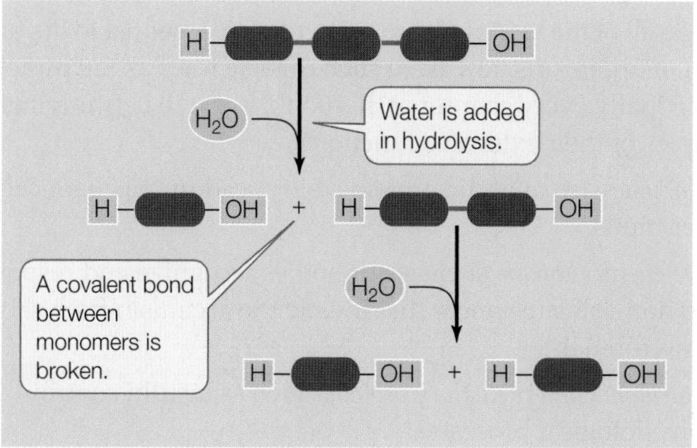


**3.4 Condensation and Hydrolysis of Polymers** (A) Condensation reactions link monomers into polymers and produce water. (B) Hydrolysis reactions break polymers into individual monomers and consume water.

all cases, polymers form only if water molecules are removed and energy is added to the system. In living systems, specific energy-rich molecules supply the necessary energy.

The reverse of a condensation reaction is a **hydrolysis reaction** (*hydro*, "water"; *lysis*, "break"). Hydrolysis reactions result in the breakdown of polymers into their component monomers. Water reacts with the covalent bonds that link the polymer together. For each covalent bond that is broken, a water molecule splits into two ions ($H^+$ and $OH^-$), which each become part of one of the products (**Figure 3.4B**). The linkages between monomers can thus be formed and broken inside living tissues.

## 3.1 RECAP

**The four kinds of large molecules that distinguish living tissues are proteins, lipids, carbohydrates, and nucleic acids. These biological molecules carry out a wide range of life-sustaining functions. Most of them are polymers, made up of linked monomeric subunits. Very large polymers are called macromolecules.**

- How do functional groups affect the structure and function of macromolecules? (Keep this question in mind as you read the rest of this chapter.) **See pp. 39–40 and Figure 3.1**
- Why is biochemical unity, as seen in the proportions of the four types of macromolecules present in all organisms, important for life? **See p. 40 and Figure 3.3**
- How do monomers link up to make polymers and how do they break down into monomers again? **See pp. 41–42 and Figure 3.4**

The four types of macromolecules can be seen as the building blocks of life. The unique properties of the nucleic acids will be covered in Chapter 4. The remainder of this chapter describes the structures and functions of the proteins, carbohydrates, and lipids.

## 3.2 What Are the Chemical Structures and Functions of Proteins?

While all of the kinds of large molecules are essential to the function of organisms, few have such diverse roles as the proteins. In virtually every chapter of this book, you will be studying examples of their extensive functions:

- *Enzymes* are catalytic proteins that speed up biochemical reactions.
- *Defensive proteins* such as antibodies recognize and respond to non-self substances that invade the organism from the environment.
- *Hormonal and regulatory proteins* such as insulin control physiological processes.
- *Receptor proteins* receive and respond to molecular signals from inside and outside the organism.
- *Storage proteins* store chemical building blocks—amino acids—for later use.
- *Structural proteins* such as collagen provide physical stability and movement.
- *Transport proteins* such as hemoglobin carry substances within the organism.
- *Genetic regulatory proteins* regulate when, how, and to what extent a gene is expressed.

Among the functions of macromolecules listed earlier, only two—energy storage and information storage—are not usually performed by proteins.

All **proteins** are polymers made up of different proportions and sequences of 20 amino acids. Proteins range in size from small ones such as insulin, which has a molecular weight of 5,733 daltons and 51 amino acids, to huge molecules such as the muscle protein titin, with a molecular weight of 2,993,451 daltons and 26,926 amino acids. All proteins consist of one or more *polpeptide chains*—unbranched (linear) polymer of covalently linked amino acids. The *composition* of a protein refers to the relative amounts of the different amino acids present in its polypeptide chains. Variation in the *sequence* of the amino acids in polypeptide chains is the source of the diversity in protein structure and function, because each chain folds into specific three-dimensional shape that is defined by the precise sequence of the amino acids present in the chain.

Many proteins are made up of more than one polypeptide chain. For example, the oxygen-carrying protein hemoglobin has four chains that are folded separately and come together to make up the functional protein. Proteins can also associate with one another, forming multi-protein complexes that carry out intricate tasks such as DNA synthesis.

To understand the many functions of proteins, we must first explore protein structure. We begin by examining the properties of amino acids and how they link together to form polypeptide chains. Then we will describe how a linear chain of amino acids is consistently folded into a specific, compact, three-dimensional shape. Finally, we will see how this three-dimensional structure provides a definitive physical and chemical environment that influences how other molecules can interact with the protein.

### Amino acids are the building blocks of proteins

The amino acids have both a carboxyl functional group and an amino functional group (see Figure 3.1) attached to the same carbon atom, called the α (alpha) carbon. Also attached to the α carbon atom are a hydrogen atom and a **side chain**, or **R group**, designated by the letter R.

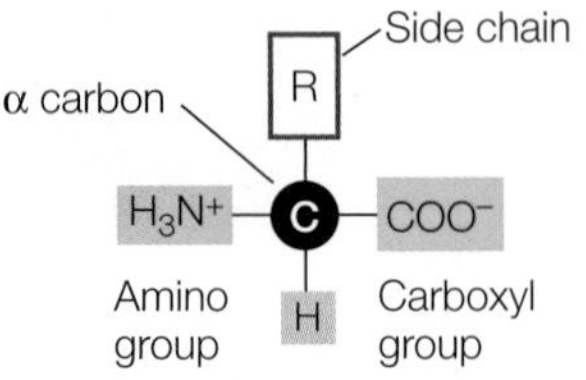


The α carbon is asymmetrical because it is bonded to four different atoms or groups of atoms. Therefore, amino acids exist in two isomeric forms, called D-amino acids and L-amino acids. D and L are abbreviations of the Latin terms for right (*dextro*) and left (*levo*). Only L-amino acids are commonly found in proteins in most organisms, and their presence is an important chemical "signature" of life.

At the pH values commonly found in cells, both the carboxyl and amino groups of amino acids are ionized: the carboxyl group has lost a hydrogen ion, and the amino group has gained one. Thus *amino acids are simultaneously acids and bases.*

The side chains of amino acids contain functional groups that are important in determining the three-dimensional structure and thus the function of the protein. As **Table 3.1** shows, the 20 amino acids found in living organisms are grouped and distinguished by their side chains:

- The five amino acids that have electrically charged side chains (+1, –1) attract water (are hydrophilic) and attract oppositely charged ions of all sorts.
- The five amino acids that have polar side chains ($\delta^+$, $\delta^-$) tend to form hydrogen bonds with water and with other polar or charged substances. These amino acids are also hydrophilic.
- Seven amino acids have side chains that are nonpolar hydrocarbons or very slightly modified hydrocarbons. In the watery environment of the cell, these hydrophobic side chains may cluster together in the interior of the protein. These amino acids are hydrophobic.

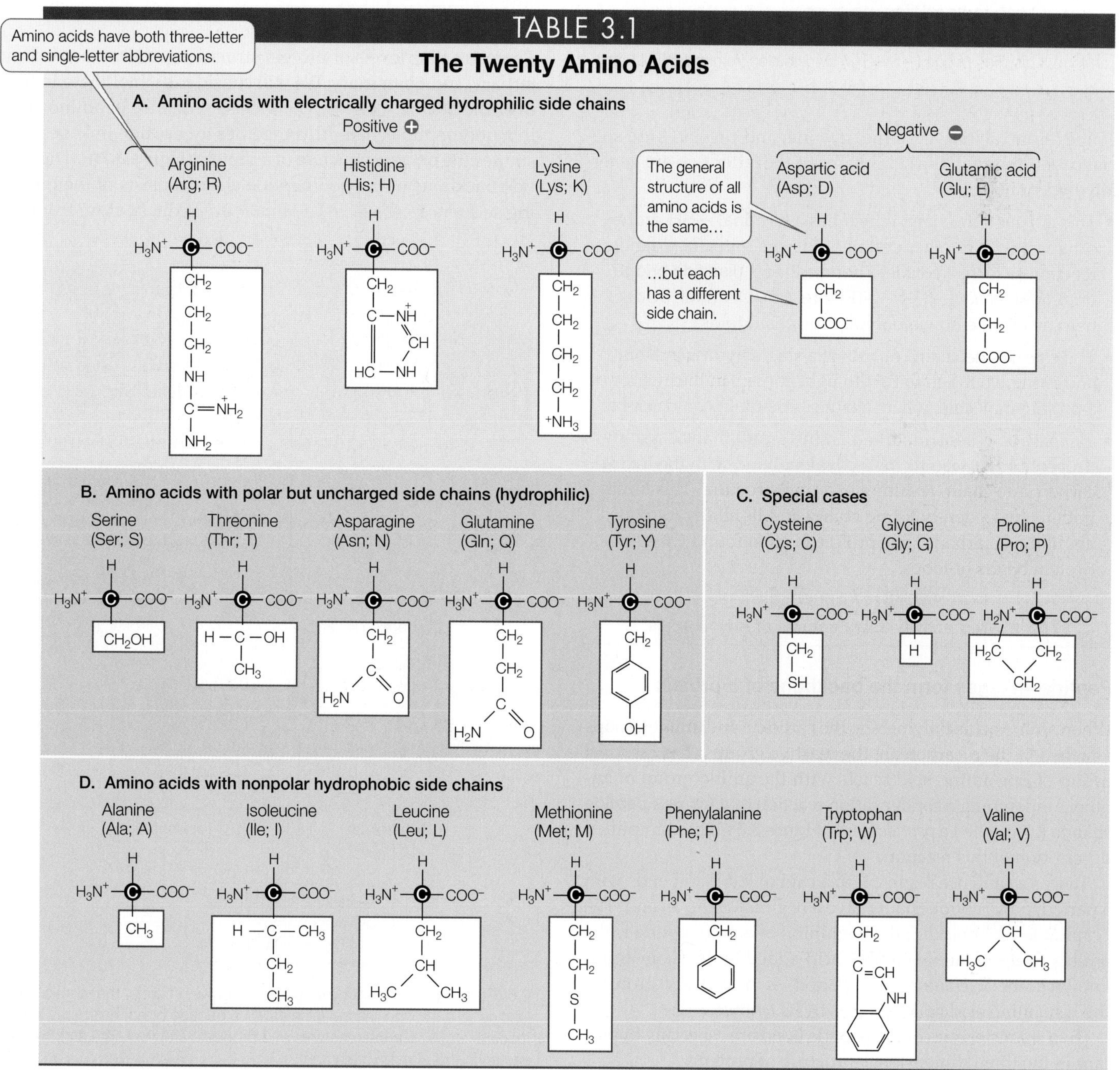

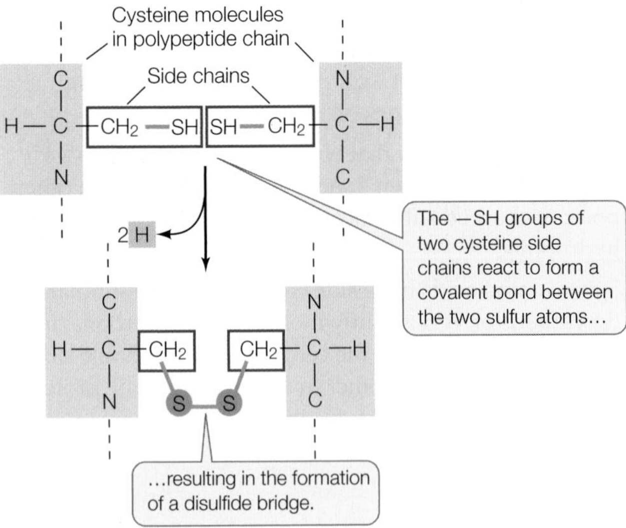


**3.5 A Disulfide Bridge** Two cysteine molecules in a polypeptide chain can form a disulfide bridge (–S–S–) by oxidation (removal of H atoms).

Three amino acids—cysteine, glycine, and proline—are special cases, although the side chains of the latter two are generally hydrophobic.

- The *cysteine* side chain, which has a terminal —SH group, can react with another cysteine side chain in an oxidation reaction to form a covalent bond called a **disulfide bridge**, or *disulfide bond* (—S—S—) (**Figure 3.5**). Disulfide bridges help determine how a polypeptide chain folds.
- The *glycine* side chain consists of a single hydrogen atom and is small enough to fit into tight corners in the interior of a protein molecule, where a larger side chain could not fit.
- *Proline* possesses a modified amino group that lacks a hydrogen and instead forms a covalent bond with the hydrocarbon side chain, resulting in a ring structure. This limits both its hydrogen-bonding ability and its ability to rotate about the α carbon. Thus proline is often found where a protein bends or loops.

**yourBioPortal.com**

GO TO Web Activity 3.2 • Features of Amino Acids

### Peptide linkages form the backbone of a protein

When amino acids polymerize, the carboxyl and amino groups attached to the α carbon are the reactive groups. The carboxyl group of one amino acid reacts with the amino group of another, undergoing a condensation reaction that forms a **peptide linkage** (also called a *peptide bond*). **Figure 3.6** gives a simplified description of this reaction.

Just as a sentence begins with a capital letter and ends with a period, polypeptide chains have a beginning and an end. The "capital letter" marking the beginning of a polypeptide is the amino group of the first amino acid added to the chain and is known as the *N terminus*. The "period" is the carboxyl group of the last amino acid added; this is the *C terminus*.

Two characteristics of the peptide bond are especially important in the three-dimensional structure of proteins:

- In the C—N linkage, the adjacent α carbons (αC—C—N—αC) are not free to rotate fully, which limits the folding of the polypeptide chain.
- The oxygen bound to the carbon (C═O) in the carboxyl group carries a slight negative charge ($\delta^-$), whereas the hydrogen bound to the nitrogen (N—H) in the amino group is slightly positive ($\delta^+$). This asymmetry of charge favors hydrogen bonding within the protein molecule itself and with other molecules, contributing to both the structure and the function of many proteins.

Before we explore the significance of these characteristics of the peptide linkage, however, we will describe the significance of the sequence of amino acids in determining a protein's structure.

### The primary structure of a protein is its amino acid sequence

There are four levels of protein structure: primary, secondary, tertiary, and quaternary. We will consider each of these in turn over the next few pages. The precise sequence of amino acids in a polypeptide chain held together by peptide linkages constitutes the **primary structure** of a protein (**Figure 3.7A**). The peptide backbone of the polypeptide chain consists of the repeating sequence —N—C—C—made up of the N atom from the

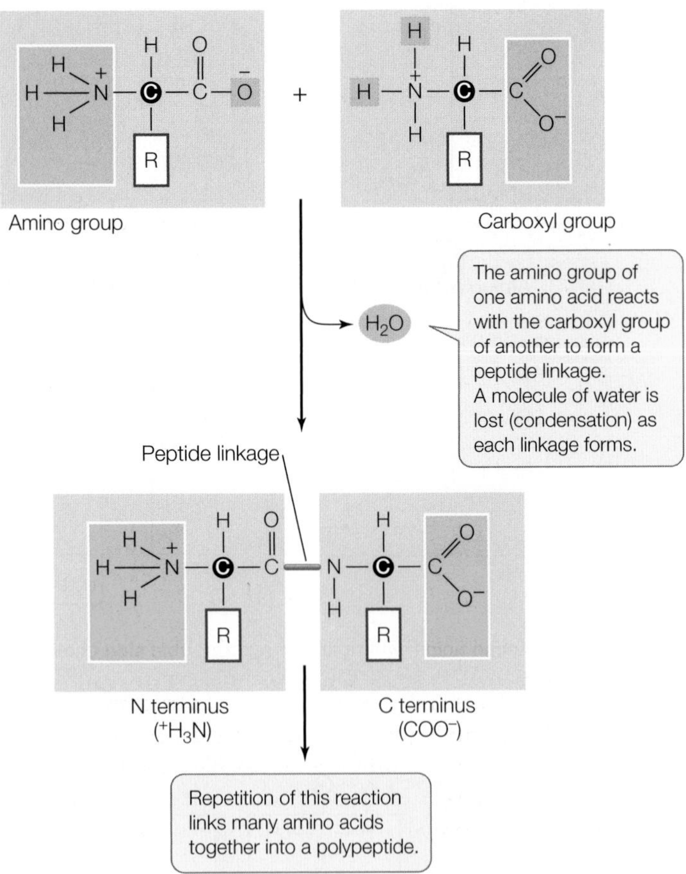

Amino group
Carboxyl group
The amino group of one amino acid reacts with the carboxyl group of another to form a peptide linkage. A molecule of water is lost (condensation) as each linkage forms.
$H_2O$
Peptide linkage
N terminus ($^+H_3N$)
C terminus ($COO^-$)
Repetition of this reaction links many amino acids together into a polypeptide.


**3.6 Formation of Peptide Linkages** In living things, the reaction leading to a peptide linkage (also called a peptide bond) has many intermediate steps, but the reactants and products are the same as those shown in this simplified diagram.

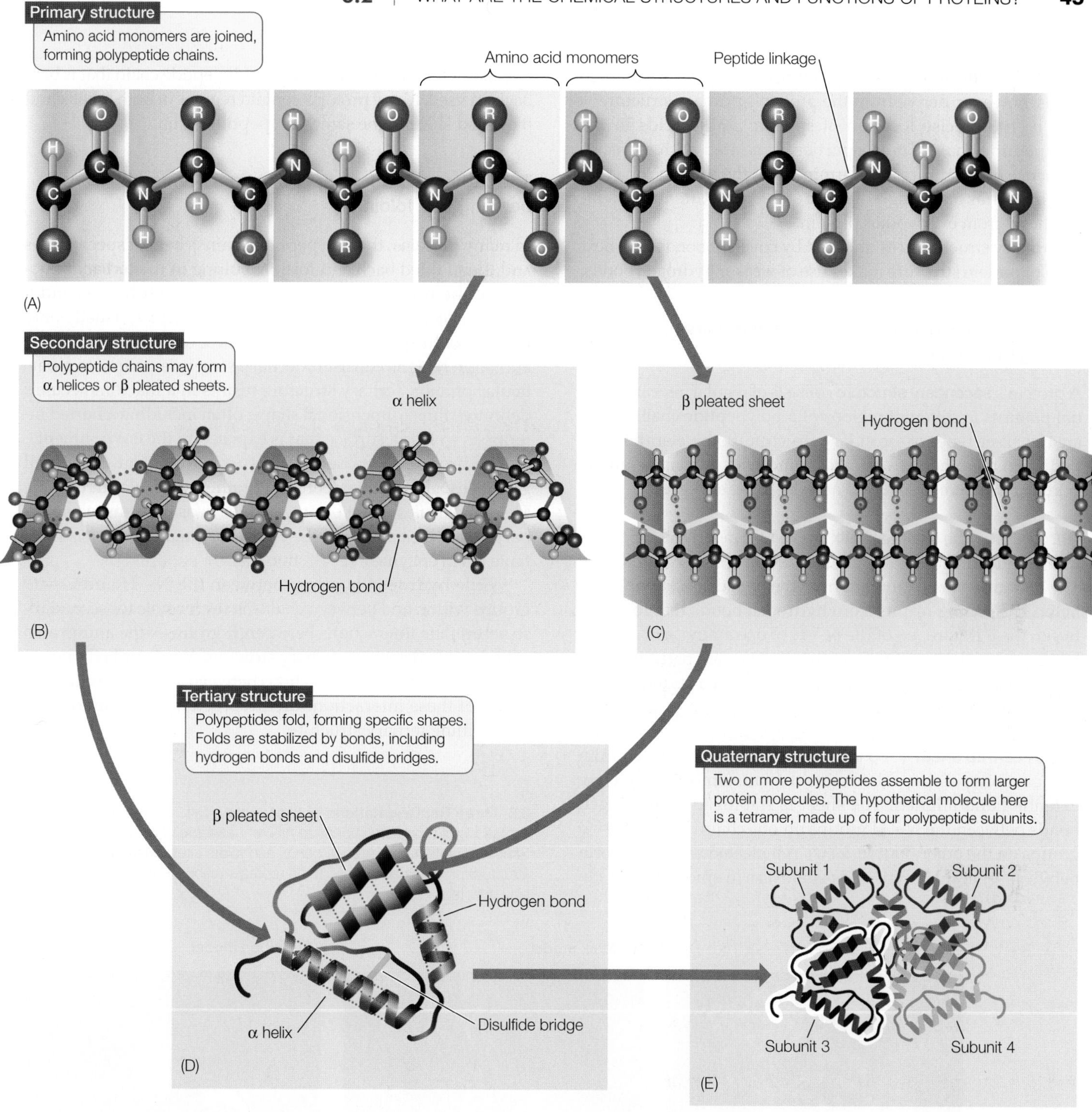
Primary structure
Amino acid monomers are joined, forming polypeptide chains.
Amino acid monomers
Peptide linkage
(A)
Secondary structure
Polypeptide chains may form α helices or β pleated sheets.
α helix
Hydrogen bond
(B)
β pleated sheet
Hydrogen bond
(C)
Tertiary structure
Polypeptides fold, forming specific shapes. Folds are stabilized by bonds, including hydrogen bonds and disulfide bridges.
β pleated sheet
Hydrogen bond
α helix
Disulfide bridge
(D)
Quaternary structure
Two or more polypeptides assemble to form larger protein molecules. The hypothetical molecule here is a tetramer, made up of four polypeptide subunits.
Subunit 1
Subunit 2
Subunit 3
Subunit 4
(E)

**3.7 The Four Levels of Protein Structure** Secondary, tertiary, and quaternary structure all arise from the primary structure of the protein.

amino group, the α carbon atom, and the C atom from the carboxyl group of each amino acid.

Scientists have determined the primary structure of many proteins. The single-letter abbreviations for amino acids (see Table 3.1) are used to record the amino acid sequence of a protein. Here, for example, are the first 20 amino acids (out of a total of 124) in the protein ribonuclease from a cow:

KETAAAKFERQHMDSSTSAA

The theoretical number of different proteins is enormous. Since there are 20 different amino acids, there could be $20 \times 20 = 400$ distinct dipeptides (two linked amino acids), and $20 \times 20 \times 20 =$ 8,000 different tripeptides (three linked amino acids). Imagine this process of multiplying by 20 extended to a protein made up of 100 amino acids (which would be considered a small protein). There could be $20^{100}$ (that's approximately $10^{130}$) such small proteins, each with its own distinctive primary structure. How large is the number $20^{100}$? Physicists tell us that there aren't that many electrons in the entire universe.

At the higher levels of protein structure (secondary, tertiary and quaternary), local coiling and folding of the polypeptide chain(s) give the molecule its final functional shape. All of these levels, however, derive from the protein's primary structure—that is, the precise location of specific amino acids in the polypeptide chain. The properties associated with a precise sequence of amino acids determine how the protein can twist and fold, thus adopting a specific stable structure that distinguishes it from every other protein.

Primary structure is established by covalent bonds. The next level of protein structure makes use of weaker hydrogen bonds.

### The secondary structure of a protein requires hydrogen bonding

A protein's **secondary structure** consists of regular, repeated spatial patterns in different regions of a polypeptide chain. There are two basic types of secondary structure, both determined by hydrogen bonding between the amino acids that make up the primary structure, the α helix and the β pleated sheet.

**THE α HELIX** The **α (alpha) helix** is a right-handed coil that turns in the same direction as a standard wood screw (**Figure 3.7B**). The R groups extend outward from the peptide backbone of the helix. The coiling results from hydrogen bonds that form between the $\delta^+$ hydrogen of the N—H of one amino acid and the $\delta^-$ oxygen of the C=O of another. When this pattern of hydrogen bonding is established repeatedly over a segment of the protein, it stabilizes the coil.

**THE β PLEATED SHEET** A **β (beta) pleated sheet** is formed from two or more polypeptide chains that are almost completely extended and aligned. The sheet is stabilized by hydrogen bonds between the N—H groups on one chain and the C=O groups on the other (**Figure 3.7C**). A β pleated sheet may form between separate polypeptide chains, as in spider silk, or between different regions of a single polypeptide chain that is bent back on itself. Many proteins contain regions of both α helix and β pleated sheet in the same polypeptide chain.

### The tertiary structure of a protein is formed by bending and folding

In many proteins, the polypeptide chain is bent at specific sites and then folded back and forth, resulting in the **tertiary structure** of the protein (**Figure 3.7D**). Although α helices and β pleated sheets contribute to the tertiary structure, usually only portions of the macromolecule have these secondary structures, and large regions consist of tertiary structure unique to a particular protein. Tertiary structure results in a macromolecule's definitive three-dimensional shape, often including a buried interior as well as a surface that is exposed to the environment.

The protein's exposed outer surfaces present functional groups capable of interacting with other molecules in the cell. These molecules might be other proteins (as happens in quaternary structure, as we will see below) or smaller chemical reactants (as in enzymes; see Section 7.4).

While hydrogen bonding between the N—H and C=O groups within and between chains is responsible for secondary structure, the interactions between R groups—the amino acid side chains—determine tertiary structure. We described the various strong and weak interactions between atoms in Section 2.2. Many of these interactions are involved in determining and maintaining tertiary structure.

**3.8 Three Representations of Lysozyme** Different molecular representations of a protein emphasize different aspects of its tertiary structure: surface features, sites of bends and folds, sites where alpha or beta structure predominate. These three representations of lysozyme are similarly oriented.

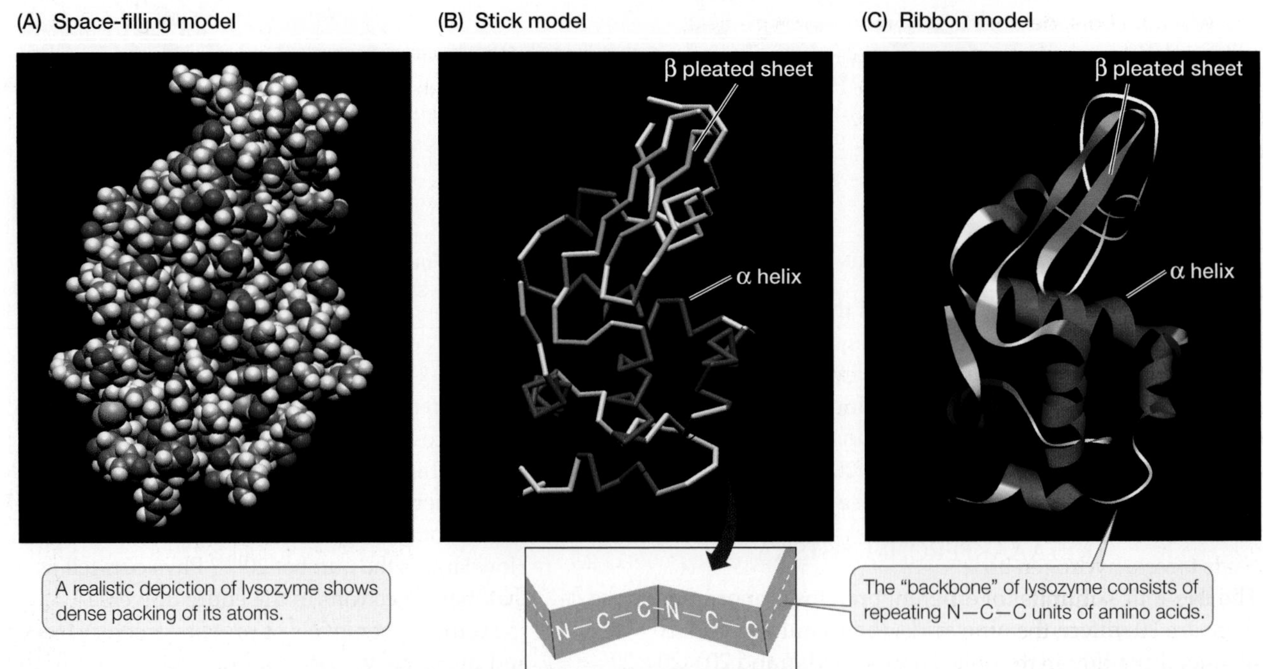
(A) Space-filling model
(B) Stick model
(C) Ribbon model
β pleated sheet
α helix
β pleated sheet
α helix
A realistic depiction of lysozyme shows dense packing of its atoms.
N–C–C–N–C–C
The "backbone" of lysozyme consists of repeating N–C–C units of amino acids.

- Covalent *disulfide bridges* can form between specific cysteine side chains (see Figure 3.5), holding a folded polypeptide in place.
- *Hydrogen bonds* between side chains also stabilize folds in proteins.
- *Hydrophobic* side chains can aggregate together in the interior of the protein, away from water, folding the polypeptide in the process.
- *van der Waals forces* can stabilize the close interactions between hydrophobic side chains.
- *Ionic bonds* can form between positively and negatively charged side chains, forming *salt bridges* between amino acids. Ionic bonds can also be buried deep within a protein, away from water.

A complete description of a protein's tertiary structure would specify the location of every atom in the molecule in three-dimensional space relative to all the other atoms. Such a description is available for the protein lysozyme (**Figure 3.8**).

The different ways of depicting the molecule have their uses. The space-filling model might be used to study how other molecules interact with specific sites and R groups on a protein's surface. The stick model emphasizes the sites where bends occur in order to make the folds of the polypeptide chain. The ribbon model, perhaps the most widely used, shows the different types of secondary structure and how they fold into the tertiary structure.

Remember that both secondary and tertiary structure derive from primary structure. If a protein is heated slowly, the heat energy will disrupt only the weak interactions, causing the secondary and tertiary structure to break down. The protein is then said to be **denatured**. But the protein can return to its normal tertiary structure when it cools, demonstrating that all the information needed to specify the unique shape of a protein is contained in its primary structure. This was first shown (using chemicals instead of heat to denature the protein) by biochemist Christian Anfinsen for the protein ribonuclease (**Figure 3.9**).

### The quaternary structure of a protein consists of subunits

Many functional proteins contain two or more polypeptide chains, called *subunits*, each of them folded into its own unique tertiary structure. The protein's **quaternary structure** results from the ways in which these subunits bind together and interact (**Figure 3.7E**).

The models of hemoglobin in **Figure 3.10** illustrate quaternary structure. Hydrophobic interactions, van der Waals forces, hydrogen bonds, and ionic bonds all help hold the four subunits together to form a hemoglobin molecule. However, the weak nature of these forces permits small changes in the quaternary structure to aid the

## INVESTIGATING LIFE

**3.9 Primary Structure Specifies Tertiary Structure**

Using the protein ribonuclease, Christian Anfinsen showed that proteins spontaneously fold into a functionally correct three-dimensional configuration. As long as the primary structure is not disrupted, the information for correct folding under the right conditions is retained.

**HYPOTHESIS** **Under controlled conditions that simulate normal cellular environment in the laboratory, the primary structure of a denatured protein can reestablish the protein's three-dimensional structure.**

**METHOD** Chemically denature functional ribonuclease, disrupting disulfide bridges and other intramolecular interactions that maintain the protein's shape, so that only primary structure (i.e., the amino acid sequence) remains. Once denaturation is complete, remove the disruptive chemicals.

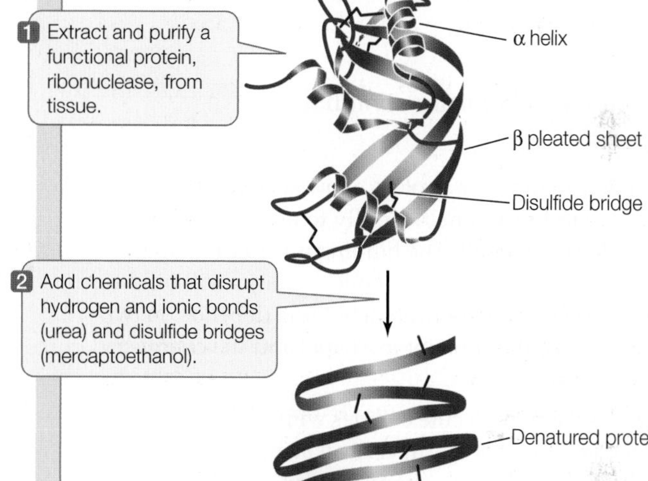

1 Extract and purify a functional protein, ribonuclease, from tissue.
α helix
β pleated sheet
Disulfide bridge
2 Add chemicals that disrupt hydrogen and ionic bonds (urea) and disulfide bridges (mercaptoethanol).
Denatured protein

3 Slowly remove the chemical agents.

**RESULTS** When the disruptive agents are removed, three-dimensional structure is restored and the protein once again is functional.

**CONCLUSION** **In normal cellular conditions, the primary structure of a protein specifies how it folds into a functional, three-dimensional structure.**

Go to **yourBioPortal.com** for original citations, discussions, and relevant links for all INVESTIGATING LIFE figures.

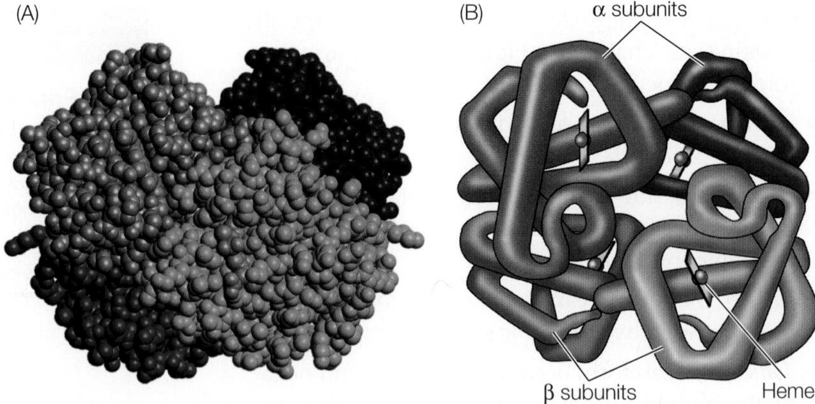


**3.10 Quaternary Structure of a Protein** Hemoglobin consists of four folded polypeptide subunits that assemble themselves into the quaternary structure shown here. In these two graphic representations, each type of subunit is a different color. The heme groups contain iron and are the oxygen-carrying sites.

protein's function—which is to carry oxygen in red blood cells. As hemoglobin binds one $O_2$ molecule, the four subunits shift their relative positions slightly, changing the quaternary structure. Ionic bonds are broken, exposing buried side chains that enhance the binding of additional $O_2$ molecules. The quaternary structure changes again when hemoglobin releases its $O_2$ molecules to the cells of the body.

### Shape and surface chemistry contribute to protein function

The shape and structure of a protein allow specific sites on its exposed surface to bind noncovalently to another molecule, which may be large or small. The binding is said to be specific because only certain compatible chemical groups will bind to one another. The specificity of protein binding depends on two general properties of the protein: its shape, and the chemistry of its exposed surface groups.

- *Shape*. When a small molecule collides with and binds to a much larger protein, it is like a baseball being caught by a catcher's mitt: the mitt has a shape that binds to the ball and fits around it. Just as a hockey puck or a ping-pong ball does not fit a baseball catcher's mitt, a given molecule will not bind to a protein unless there is a general "fit" between their two three-dimensional shapes.
- *Chemistry*. The exposed amino acid R groups on the surface of a protein permit chemical interactions with other substances (**Figure 3.11**). Three types of interactions may be involved: ionic, hydrophobic, and hydrogen bonding. Many important functions of proteins involve interactions between exposed-surface R groups and other molecules.

### Environmental conditions affect protein structure

Because it is determined by weak forces, the three-dimensional structure of proteins is influenced by environmental conditions. Conditions that would not break covalent bonds can disrupt the weaker, noncovalent interactions that determine secondary and tertiary structure. Such alterations may affect a protein's shape and thus its function. Various conditions can alter the weak, noncovalent interactions:

- *Increases in temperature* cause more rapid molecular movements and thus can break hydrogen bonds and hydrophobic interactions.
- *Alterations in pH* can change the pattern of ionization of exposed carboxyl and amino groups in the R groups of amino acids, thus disrupting the pattern of ionic attractions and repulsions.
- *High concentrations of polar substances* such as urea can disrupt the hydrogen bonding that is crucial to protein structure. This was used in the experiment on reversible protein denaturation shown in Figure 3.9.
- *Nonpolar substances* may also disrupt normal protein structure in cases where hydrophobic groups are essential to maintain the structure.

Denaturation can be irreversible when amino acids that were buried in the interior of the protein become exposed at the surface, and vice versa, causing a new structure to form or different molecules to bind to the protein. Boiling an egg denatures its proteins and is, as you know, not reversible.

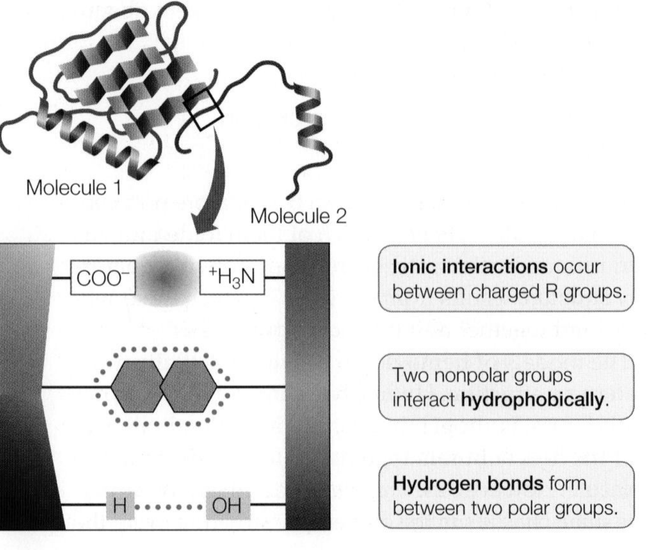

Molecule 1
Molecule 2
$COO^-$
$^+H_3N$
**Ionic interactions** occur between charged R groups.
Two nonpolar groups interact **hydrophobically**.
H
OH
**Hydrogen bonds** form between two polar groups.


**3.11 Noncovalent Interactions Between Proteins and Other Molecules** Noncovalent interactions allow a protein (brown) to bind tightly to another molecule (green) with specific properties. Noncovalent interactions also allow regions within the same protein to interact with one another.

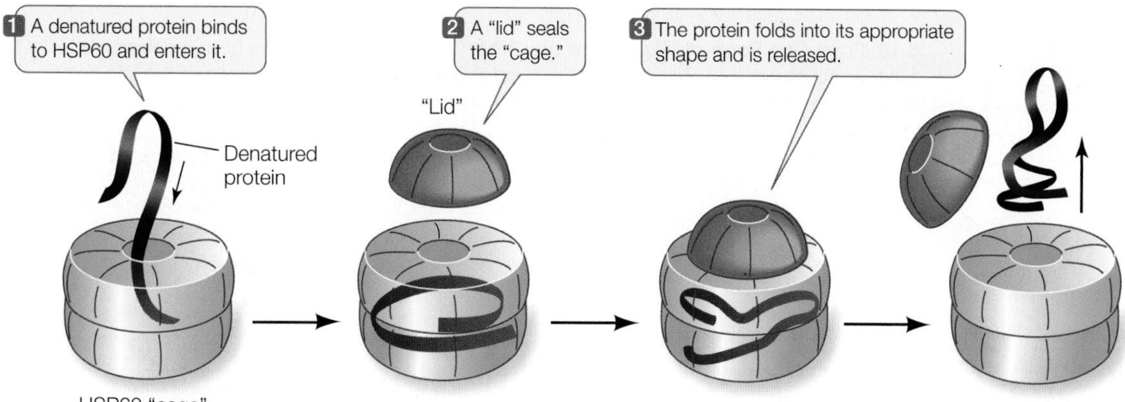

1 A denatured protein binds to HSP60 and enters it.
2 A "lid" seals the "cage."
3 The protein folds into its appropriate shape and is released.
"Lid"
Denatured protein
HSP60 "cage"

**3.12 Chaperones Protect Proteins from Inappropriate Binding** Chaperone proteins surround new or denatured proteins and prevent them from binding to the wrong substance. Heat shock proteins such as HSP60, whose actions are illustrated here, are one class of chaperone proteins.

### Molecular chaperones help shape proteins

Because of their specific shapes and the exposure of chemical groups on their surfaces, proteins can bind specific substances. Within a living cell, a polypeptide chain is sometimes in danger of binding the wrong substance. Two important examples of such a situation are:

- Following denaturation: Inappropriate environmental conditions in a cell, such as elevated temperature, can cause the denatured protein to re-fold incorrectly.
- Just after a protein is made: When a protein has not yet folded completely, it can present a surface that binds the wrong molecule.

In these cases, change may be irreversible. Eukaryotic cells have a special class of proteins that act to counteract threats to three-dimensional structure. Proteins in this class, called **chaperones**, act as molecular caretakers for other proteins. Like the chaperones at a high school dance, they prevent inappropriate interactions and enhance the appropriate ones.

Molecular chaperones were discovered by accident in 1962, when the temperature of an incubator holding fruit flies was accidentally turned up. Italian geneticist Ferruccio Ritossa noticed that this "heat shock" did not kill the flies. Instead, there was enhanced synthesis of a set of proteins that were later described as chaperones. They bound to many target proteins in the fruit fly cells and kept them from being denatured, and in some cases facilitated the correct refolding of proteins.

The general class of stress-induced chaperone proteins is called the **heat shock proteins** (**HSPs**), after this discovery. HSPs are made by most eukaryotic cells, and many enhance protein folding in addition to their protective role during periods of stress. As an example, HSP60 forms a cage that sucks a protein in, causes it to fold into the correct shape, and then releases it (**Figure 3.12**). Tumors make abundant HSPs, possibly to stabilize proteins important in the cancer process, and so HSP-inhibiting drugs are being designed. In some clinical situations, treatment with these inhibitors results in the inappropriate folding of tumor-cell proteins, causing the tumors to stop growing and even disappear.

### 3.2 RECAP

**Proteins are polymers of amino acids. The sequence of amino acids in a protein determines its primary structure. Secondary, tertiary, and quaternary structures arise through interactions between the amino acids. A protein's three-dimensional shape and exposed chemical groups establish binding specificity for other substances.**

- What are the attributes of an amino acid's R group that would make it hydrophobic? Hydrophilic? **See pp. 42–43 and Table 3.1**
- Sketch and explain how two amino acids link together to form a peptide linkage. **See p. 44 and Figure 3.6**
- What are the four levels of protein structure and how are they all ultimately determined by the protein's primary structure (i.e., its amino acid sequence)? **See pp. 44–48 and Figure 3.7**
- How do environmental factors such as temperature and pH affect the weak interactions that give a protein its specific shape and function? **See p. 48**

The seemingly infinite number of protein configurations made possible by the biochemical properties of the 20 amino acids has driven the evolution of life's diversity. The linkage configurations of sugar monomers (monosaccharides) drives the structure of the next group of macromolecules, the carbohydrates that provide energy for life.

## 3.3 What Are the Chemical Structures and Functions of Carbohydrates?

**Carbohydrates** are a large group of molecules that all have a similar atomic composition but differ greatly in size, chemical properties, and biological functions. Carbohydrates have the general formula $C_n(H_2O)_n$, which makes them appear as hydrates of carbon (association between water molecules and carbon in the ratio $C_1H_2O_1$), hence their name. When their molecular structures are examined, the linked carbon atoms are seen to be bonded with hydrogen atoms (—H) and hydroxyl groups

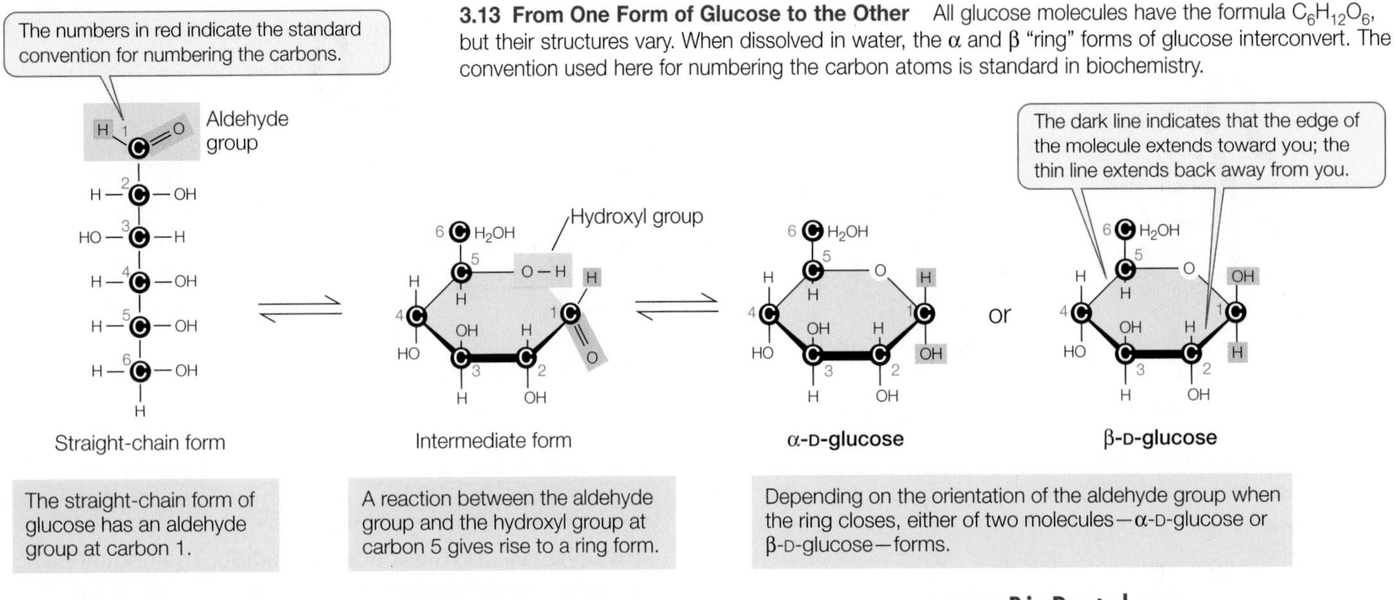

The numbers in red indicate the standard convention for numbering the carbons.
Aldehyde group
Hydroxyl group
The dark line indicates that the edge of the molecule extends toward you; the thin line extends back away from you.
or
Straight-chain form
Intermediate form
α-D-glucose
β-D-glucose
The straight-chain form of glucose has an aldehyde group at carbon 1.
A reaction between the aldehyde group and the hydroxyl group at carbon 5 gives rise to a ring form.
Depending on the orientation of the aldehyde group when the ring closes, either of two molecules—α-D-glucose or β-D-glucose—forms.

**3.13 From One Form of Glucose to the Other** All glucose molecules have the formula $C_6H_{12}O_6$, but their structures vary. When dissolved in water, the α and β "ring" forms of glucose interconvert. The convention used here for numbering the carbon atoms is standard in biochemistry.

yourBioPortal.com
GO TO Web Activity 3.3 • Forms of Glucose

(—OH), the components of water. Carbohydrates have three major biochemical roles:

- They are a source of stored energy that can be released in a form usable by organisms.
- They are used to transport stored energy within complex organisms.
- They serve as *carbon skeletons* that can be rearranged to form new molecules.

Some carbohydrates are relatively small, with molecular weights of less than 100 Da. Others are true macromolecules, with molecular weights in the hundreds of thousands.

There are four categories of biologically important carbohydrates:

- **Monosaccharides** (*mono*, "one"; *saccharide*, "sugar"), such as glucose, ribose, and fructose, are *simple sugars*. They are the monomers from which the larger carbohydrates are constructed.
- **Disaccharides** (*di*, "two") consist of two monosaccharides linked together by covalent bonds. The most familiar is sucrose, which is made up of covalently bonded glucose and fructose molecules.
- **Oligosaccharides** (*oligo*, "several") are made up of several (3–20) monosaccharides.
- **Polysaccharides** (*poly*, "many"), such as starch, glycogen, and cellulose, are polymers made up of hundreds or thousands of monosaccharides.

### Monosaccharides are simple sugars

All living cells contain the monosaccharide **glucose**; it is the familiar "blood sugar," used to transport energy in humans. Cells use glucose as an energy source, breaking it down through a series of reactions that release stored energy and produce water and carbon dioxide; this is a cellular form of the combustion reaction described in Chapter 2.

Glucose exists in straight chains and in ring forms. The ring forms predominate in virtually all biological circumstances because they are more stable under physiological conditions. There are two versions of glucose ring, called α- and β-glucose, which differ only in the orientation of the —H and —OH attached to carbon 1 (**Figure 3.13**). The α and β forms interconvert and exist in equilibrium when dissolved in water.

Different monosaccharides contain different numbers of carbons. Some monosaccharides are structural isomers, with the same kinds and numbers of atoms, but in different arrangements (**Figure 3.14**). Such seemingly small structural changes can significantly alter properties. Most of the monosaccharides in living systems belong to the D (right-handed) series of isomers.

**Pentoses** (*pente*, "five") are five-carbon sugars. Two pentoses are of particular biological importance: the backbones of the nucleic acids RNA and DNA contain ribose and deoxyribose, respectively (see Section 4.1). These two pentoses are not isomers of each other; rather, one oxygen atom is missing from carbon 2 in deoxyribose (*de-*, "absent"). The absence of this oxygen atom is an important distinction between RNA and DNA.

The **hexoses** (*hex*, "six"), a group of structural isomers, all have the formula $C_6H_{12}O_6$. Included among the hexoses are glucose, fructose (so named because it was first found in fruits), mannose, and galactose.

### Glycosidic linkages bond monosaccharides

The disaccharides, oligosaccharides, and polysaccharides are all constructed from monosaccharides that are covalently bonded together by condensation reactions that form **glycosidic linkages**. A single glycosidic linkage between two monosaccharides forms a disaccharide. For example, sucrose—common table sugar in the human diet and a major disaccharide in plants—is a disaccharide formed from a glucose and a fructose molecule.

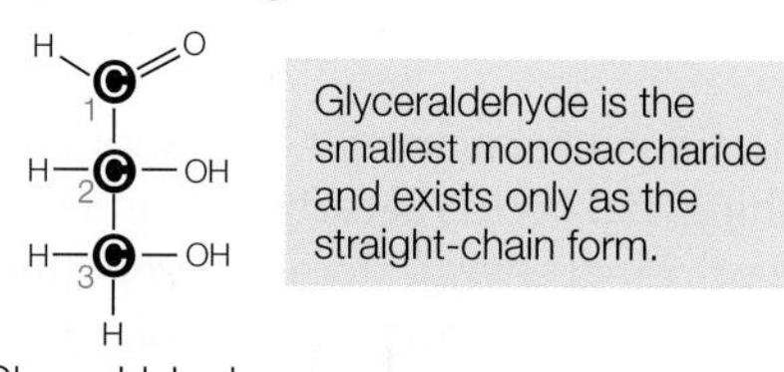

Three-carbon sugar
Glyceraldehyde is the smallest monosaccharide and exists only as the straight-chain form.
Glyceraldehyde


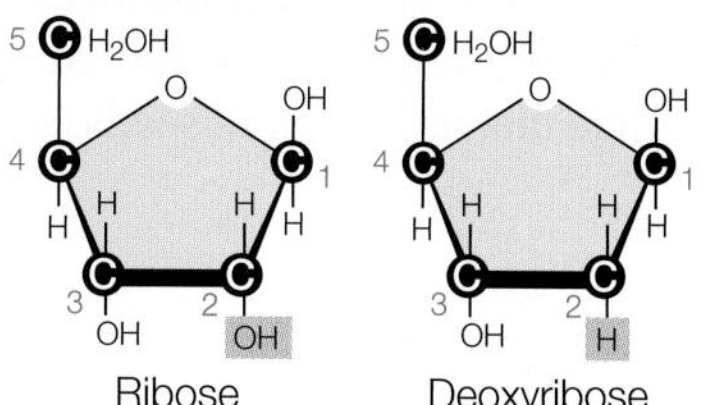


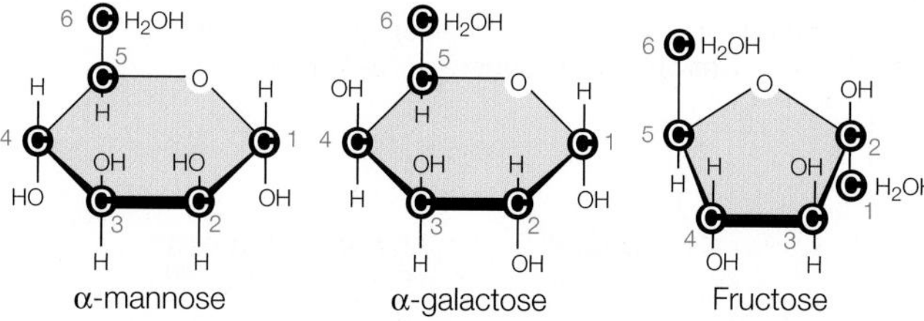


**3.14 Monosaccharides Are Simple Sugars** Monosaccharides are made up of varying numbers of carbons. Some hexoses are structural isomers that have the same kind and number of atoms, but the atoms are arranged differently. Fructose, for example, is a hexose, but forms a five-membered ring like the pentoses.

The disaccharides maltose and cellobiose are made from two glucose molecules (**Figure 3.15**). Maltose and cellobiose are structural isomers, both having the formula $C_{12}H_{22}O_{11}$. However, they have different chemical properties and are recognized by different enzymes in biological tissues. For example, maltose can be hydrolyzed into its monosaccharides in the human body, whereas cellobiose cannot.

Oligosaccharides contain several monosaccharides bound by glycosidic linkages at various sites. Many oligosaccharides have additional functional groups, which give them special properties. Oligosaccharides are often covalently bonded to proteins and lipids on the outer cell surface, where they serve as recognition signals. The different human blood groups (for example, the ABO blood types) get their specificity from oligosaccharide chains.

**3.15 Disaccharides Form by Glycosidic Linkages** Glycosidic linkages between two monosaccharides can create many different disaccharides. Which disaccharide is formed depends on which monosaccharides are linked; on the site of linkage (i.e., which carbon atoms are involved); and on the form (α or β) of the linkage.

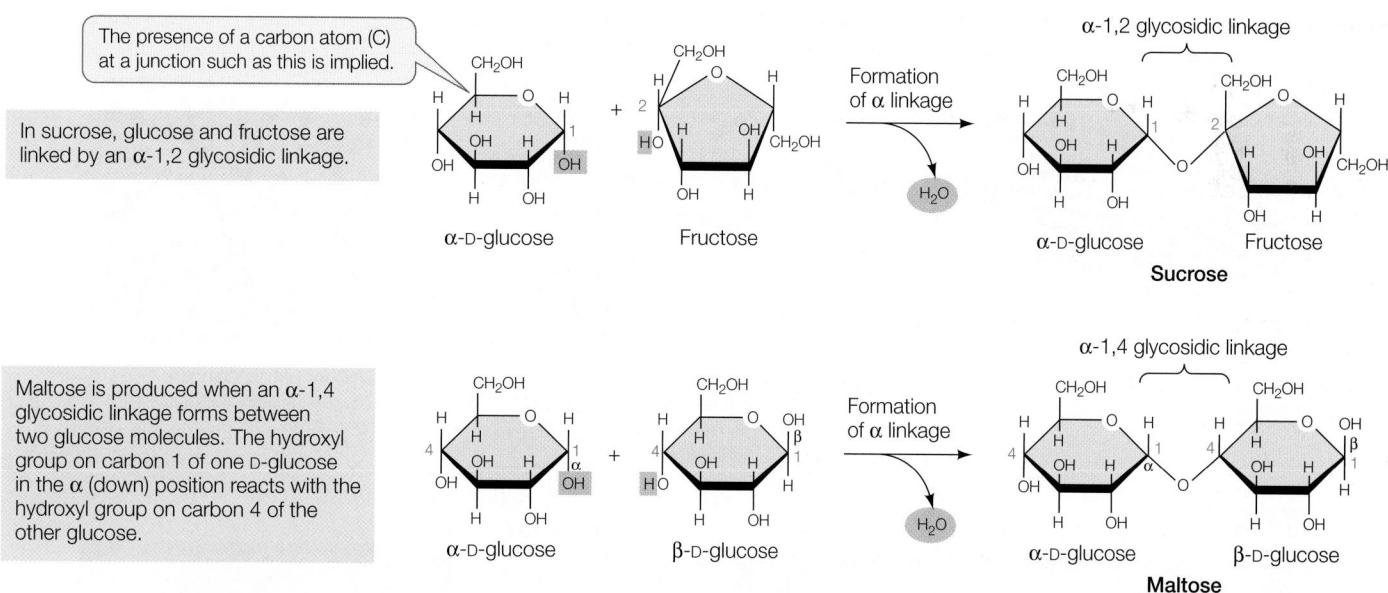

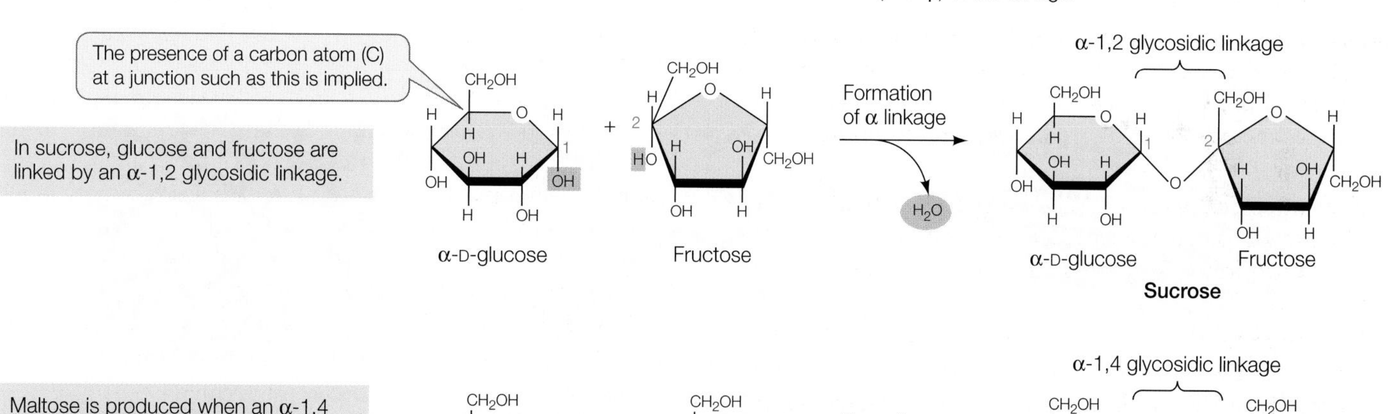

### Polysaccharides store energy and provide structural materials

Polysaccharides are large (sometimes gigantic) polymers of monosaccharides connected by glycosidic linkages (**Figure 3.16**). In contrast to proteins, polysaccharides are not necessarily linear chains of monomers. Each monomer unit has several sites that may be capable of forming glycosidic linkages, and thus branched molecules are possible.

**STARCH** Starches comprise a family of giant molecules of broadly similar structure. While all starches are polysaccharides of glucose with α-glycosidic linkages (α–1,4 and α–1,6 glycosidic bonds; Figure 3.16A), the different starches can be distinguished by the amount of branching that occurs at carbons 1 and 6 (Figure 3.16B). Starch is the principal energy storage compound of plants. Some plant starches, such as amylose, are unbranched; others are moderately branched (amylopectin, for example). Starch readily binds water. When that water is removed, however, hydrogen bonds tend to form between the unbranched polysaccharide chains, which then aggregate, as in the large starch grains observed in the storage material of plant seeds (see Figure 3.16C).

**3.16 Representative Polysaccharides** Cellulose, starch, and glycogen have different levels of branching and compaction of the polysaccharides.

(A) Molecular structure

Cellulose

Hydrogen bonding to other cellulose molecules can occur at these points.

Cellulose is an unbranched polymer of glucose with β-1,4 glycosidic linkages that are chemically very stable.

Starch and glycogen

Branching occurs here.

Glycogen and starch are polymers of glucose with α-1,4 glycosidic linkages. α-1,6 glycosidic linkages produce branching at carbon 6.

(B) Macromolecular structure

Linear (cellulose)

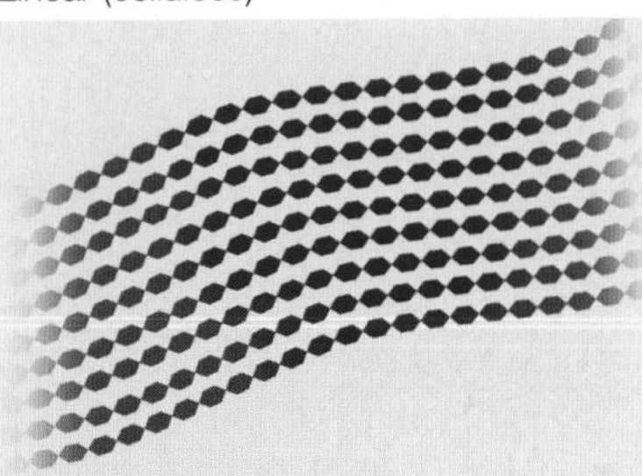

Parallel cellulose molecules form hydrogen bonds, resulting in thin fibrils.

Branched (starch)

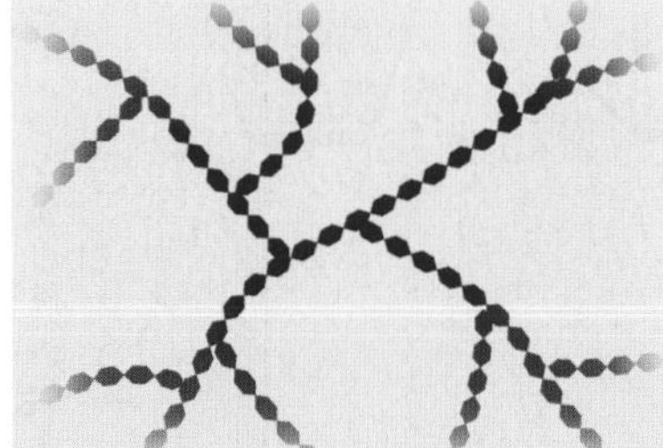

Branching limits the number of hydrogen bonds that can form in starch molecules, making starch less compact than cellulose.

Highly branched (glycogen)

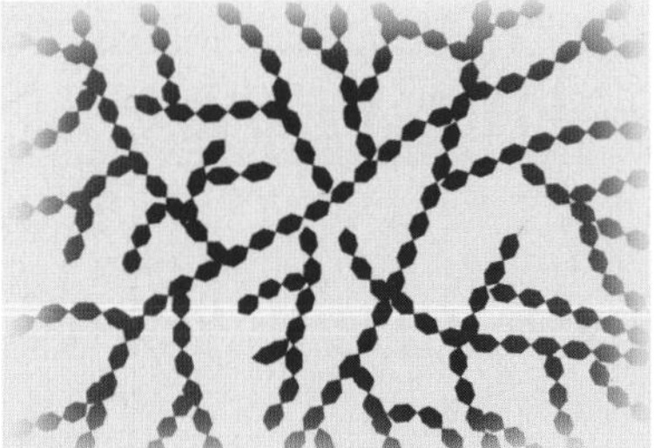

The high amount of branching in glycogen makes its solid deposits more compact than starch.

(C) Polysaccharides in cells

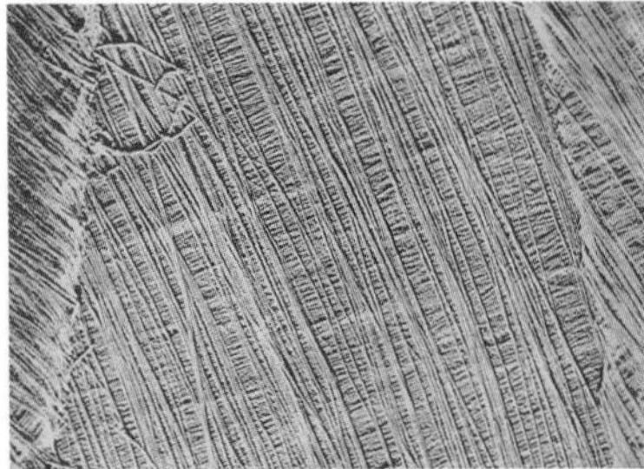

Layers of cellulose fibrils, as seen in this scanning electron micrograph, give plant cell walls great strength.

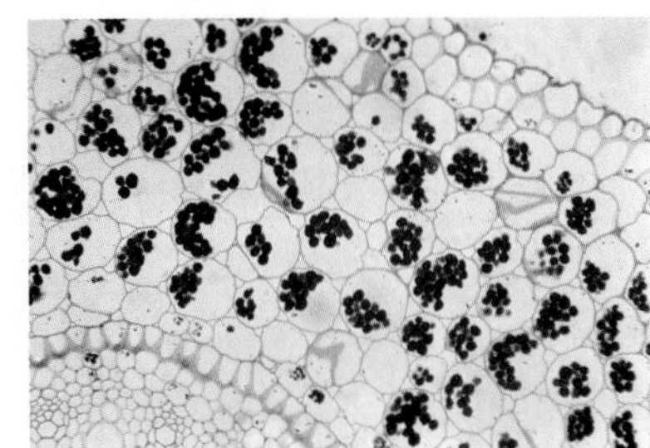

Within these plant cells, starch deposits (dyed purple in this micrograph) have a granular shape.

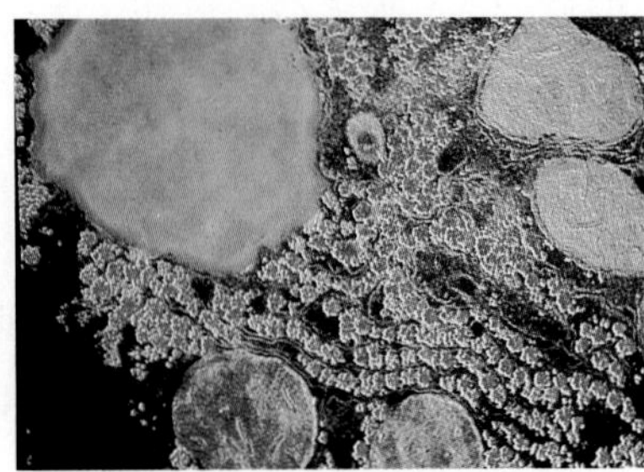

The pink-stained granules in this electron micrograph are glycogen deposits in the human liver.

**GLYCOGEN** Glycogen is a water-insoluble, highly branched polymer of glucose. It stores glucose in liver and muscle, serving as an energy storage compound for animals as starch does for plants. Both glycogen and starch are readily hydrolyzed into glucose monomers, which in turn can be broken down to liberate their stored energy.

But if it is glucose that is needed for fuel, why store it in the form of glycogen? The reason is that 1,000 glucose molecules would exert 1,000 times the *osmotic pressure* of a single glycogen molecule, causing water to enter the cells (see Section 6.3). If it were not for polysaccharides, many organisms would expend a lot of energy expelling excess water from their cells.

**CELLULOSE** As the predominant component of plant cell walls, cellulose is by far the most abundant organic compound on Earth. Like starch and glycogen, cellulose is a polysaccharide of glucose, but its individual monosaccharides are connected by β- rather than by α-glycosidic linkages. Starch is easily degraded by the actions of chemicals or enzymes. Cellulose, however, is chemically more stable because of its β-glycosidic linkages. Thus, whereas starch is easily broken down to supply glucose for energy-producing reactions, cellulose is an excellent structural material that can withstand harsh environmental conditions without substantial change.

### Chemically modified carbohydrates contain additional functional groups

Some carbohydrates are chemically modified by the addition of functional groups, such as phosphate and amino groups (**Figure 3.17**). For example, carbon 6 in glucose may be oxidized from —$CH_2OH$ to a carboxyl group (—COOH), producing glucuronic acid. Or a phosphate group may be added to one or more of the —OH sites. Some of the resulting *sugar phosphates*, such as fructose 1,6-bisphosphate, are important intermediates in cellular energy reactions, which will be discussed in Chapter 9.

When an amino group is substituted for an —OH group, *amino sugars*, such as glucosamine and galactosamine, are produced. These compounds are important in the extracellular matrix (see Section 5.4), where they form parts of glycoproteins, which are molecules involved in keeping tissues together. Galactosamine is a major component of cartilage, the material that forms caps on the ends of bones and stiffens the ears and nose. A derivative of glucosamine is present in the polymer *chitin*, the principal structural polysaccharide in the external skeletons of insects and many crustaceans (e.g., crabs and lobsters) and a component of the cell walls of fungi. Because these organisms are among the most abundant eukaryotes on Earth, chitin rivals cellulose as one of the most abundant substances in the living world.

**(A) Sugar phosphate**

Fructose 1,6 bisphosphate is involved in the reactions that liberate energy from glucose. (The numbers in its name refer to the carbon sites of phosphate bonding; *bis-* indicates that two phosphates are present.)

Phosphate groups

Fructose

**Fructose 1,6 bisphosphate**

**(B) Amino sugars**

The monosaccharides glucosamine and galactosamine are amino sugars with an amino group in place of a hydroxyl group.

Amino group

**Glucosamine** **Galactosamine**

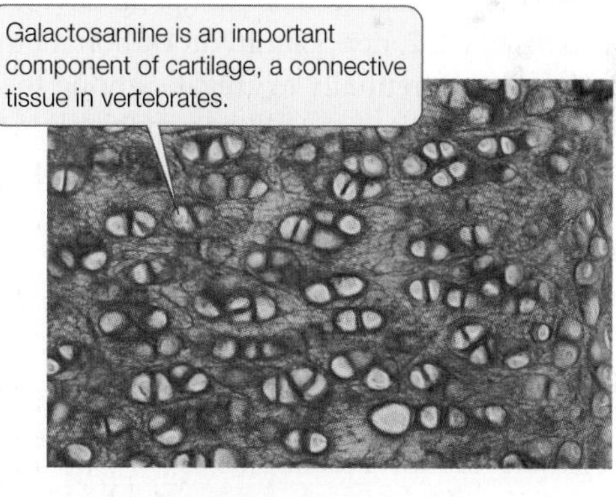

Galactosamine is an important component of cartilage, a connective tissue in vertebrates.

**(C) Chitin**

Chitin is a polymer of *N*-acetylglucosamine; *N*-acetyl groups provide additional sites for hydrogen bonding between the polymers.

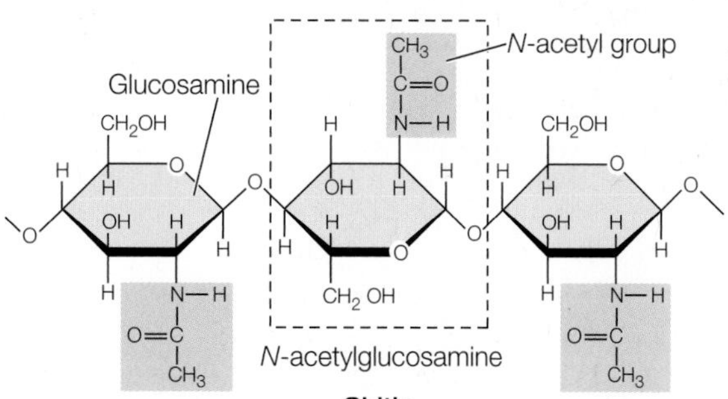


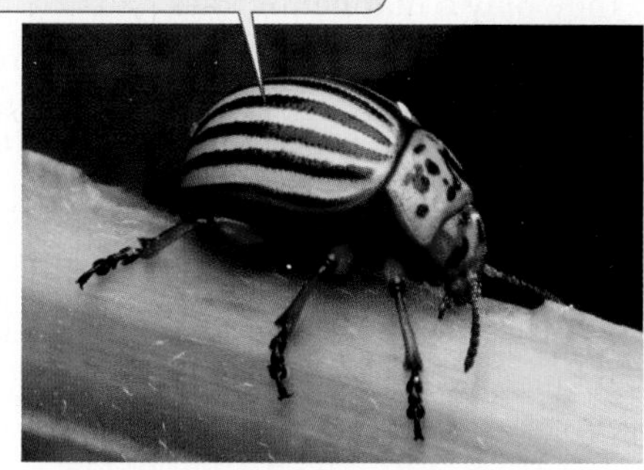


**3.17 Chemically Modified Carbohydrates** Added functional groups can modify the form and properties of a carbohydrate.

## 3.3 RECAP

**Carbohydrates are composed of carbon, hydrogen, and oxygen in the general ratio of 1:2:1. They provide energy and structure to cells and are precursors of numerous important biological molecules. Monosaccharide monomers can be connected by glycosidic linkages to form disaccharides, oligosaccharides, and polysaccharides.**

- Draw the chemical structure of a disaccharide formed by two monosaccharides. **See Figure 3.15**
- What qualities of the polysaccharides starch and glycogen make them useful for energy storage? **See pp. 52–53 and Figure 3.16**
- From looking at the cellulose molecules in Figure 3.16A, can you see where a large number of hydrogen bonds are present in the linear structure of cellulose shown in Figure 3.16B? Why is this structure so strong?

We have seen how amino acid monomers form protein polymers and how sugar monomers form the polymers of carbohydrates. Now we will look at the lipids, which are unique among the four classes of large biological molecules in that they are not, strictly speaking, polymers.

## 3.4 What Are the Chemical Structures and Functions of Lipids?

**Lipids**—colloquially called *fats*—are hydrocarbons that are insoluble in water because of their many nonpolar covalent bonds. As we saw in Section 2.2, nonpolar hydrocarbon molecules are hydrophobic and preferentially aggregate among themselves, away from water (which is polar). When nonpolar hydrocarbons are sufficiently close together, weak but additive van der Waals forces hold them together. The huge macromolecular aggregations that can form are not polymers in a strict chemical sense, because the individual lipid molecules are not covalently bonded. With this understanding, it is still useful to consider aggregations of individual lipids as a different sort of polymer.

There are several different types of lipids, and they play a number of roles in living organisms:

- Fats and oils store energy.
- Phospholipids play important structural roles in cell membranes.
- Carotenoids and chlorophylls help plants capture light energy.
- Steroids and modified fatty acids play regulatory roles as hormones and vitamins.
- Fat in animal bodies serves as thermal insulation.
- A lipid coating around nerves provides electrical insulation.
- Oil or wax on the surfaces of skin, fur, and feathers repels water.

### Fats and oils are hydrophobic

Chemically, fats and oils are *triglycerides*, also known as *simple lipids*. Triglycerides that are solid at room temperature (around 20°C) are called **fats**; those that are liquid at room temperature are called **oils**. Triglycerides are composed of two types of building blocks: *fatty acids* and *glycerol*. **Glycerol** is a small molecule with three hydroxyl (—OH) groups (thus it is an alcohol). A **fatty acid** is made up of a long nonpolar hydrocarbon chain and a polar carboxyl group (—COOH). These chains are very hydrophobic, with their abundant C—H and C—C bonds, which have low electronegativity and are nonpolar (see Section 2.2).

A **triglyceride** contains three fatty acid molecules and one molecule of glycerol. Synthesis of a triglyceride involves three condensation (dehydration) reactions. In each reaction, the carboxyl group of a fatty acid bonds with a hydroxyl group of glycerol, resulting in a covalent bond called an **ester linkage** and the release of a water molecule (**Figure 3.18**). The three fatty acids in a triglyceride molecule need not all have the same hydrocarbon chain length or structure; some may be saturated fatty acids, while others may be unsaturated:

- In **saturated fatty acids**, all the bonds between the carbon atoms in the hydrocarbon chain are single bonds—there are no double bonds. That is, all the bonds are saturated with

**3.18 Synthesis of a Triglyceride** In living things, the reaction that forms a triglyceride is more complex, but the end result is the same as shown here.

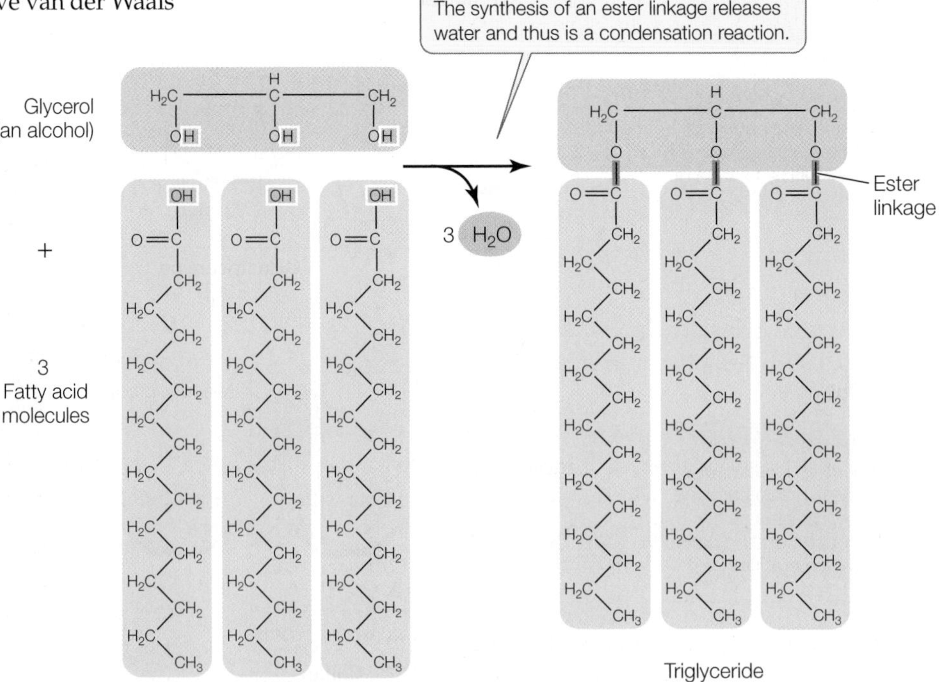

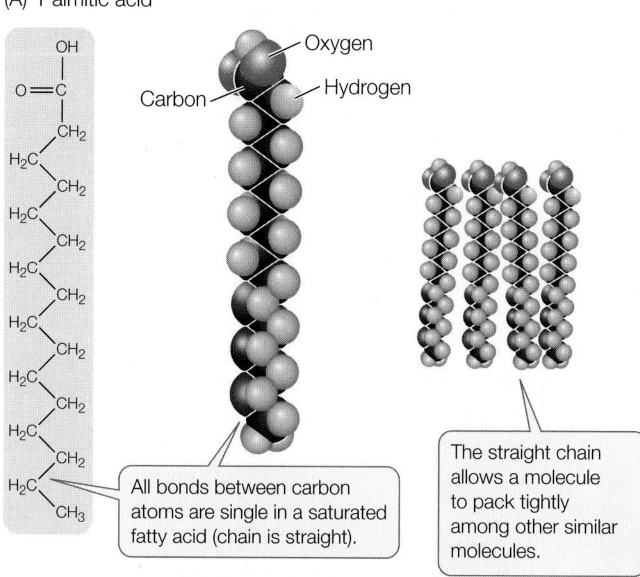


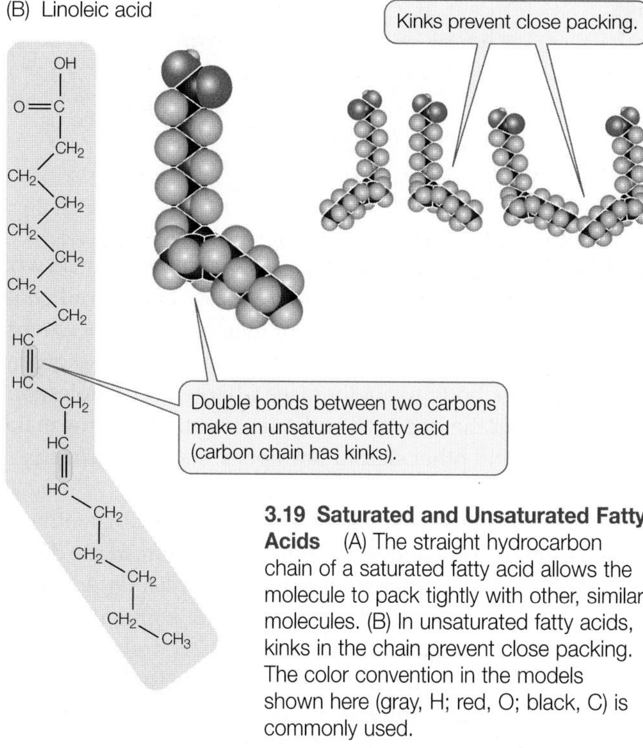


**3.19 Saturated and Unsaturated Fatty Acids** (A) The straight hydrocarbon chain of a saturated fatty acid allows the molecule to pack tightly with other, similar molecules. (B) In unsaturated fatty acids, kinks in the chain prevent close packing. The color convention in the models shown here (gray, H; red, O; black, C) is commonly used.

hydrogen atoms (**Figure 3.19A**). These fatty acid molecules are relatively rigid and straight, and they pack together tightly, like pencils in a box.

- In **unsaturated fatty acids**, the hydrocarbon chain contains one or more double bonds. Linoleic acid is an example of a *polyunsaturated* fatty acid that has two double bonds near the middle of the hydrocarbon chain, which causes kinks in the molecule (**Figure 3.19B**). Such kinks prevent the unsaturated fat molecules from packing together tightly.

The kinks in fatty acid molecules are important in determining the fluidity and melting point of a lipid. The triglycerides of animal fats tend to have many long-chain saturated fatty acids, packed tightly together; these fats are usually solids at room temperature and have a high melting point. The triglycerides of plants, such as corn oil, tend to have short or unsaturated fatty acids. Because of their kinks, these fatty acids pack together poorly and have a low melting point, and these triglycerides are usually liquids at room temperature.

Fats are excellent storehouses for chemical energy. As you will see in Chapter 9, when the C—H bond is broken, it releases significant energy that an organism can use for its own purposes, such as movement or building up complex molecules. On a per weight basis, broken-down fats yield more than twice as much energy as do degraded carbohydrates.

### Phospholipids form biological membranes

We have mentioned the hydrophobic nature of the many C—C and C—H bonds in fatty acids. But what about the carboxyl functional group at the end of the molecule? When it ionizes and forms $COO^-$, it is strongly hydrophilic. So a fatty acid is a molecule with a hydrophilic end and a long hydrophobic tail. It has two opposing chemical properties; the technical term for this is **amphipathic**. This explains what happens when oil (fatty acid) and water mix: the fatty acids orient themselves so that their polar ends face outward (i.e., toward the water) and their nonpolar tails face inward (away from water). Although no covalent bonds link individual lipids in large aggregations, such stable aggregations form readily in aqueous conditions. So these large lipid structures can be considered a different kind of macromolecule.

Like triglycerides, **phospholipids** contain fatty acids bound to glycerol by ester linkages. In phospholipids, however, any one of several phosphate-containing compounds replaces one of the fatty acids, giving these molecules amphipathic properties—that is properties of both water soluble and water insoluble molecules (**Figure 3.20A**). The phosphate functional group has a negative electric charge, so this portion of the molecule is hydrophilic, attracting polar water molecules. But the two fatty acids are hydrophobic, so they tend to avoid water and aggregate together or with other hydrophobic substances.

In an aqueous environment, phospholipids line up in such a way that the nonpolar, hydrophobic "tails" pack tightly together and the phosphate-containing "heads" face outward, where they interact with water. The phospholipids thus form a **bilayer**: a sheet two molecules thick, with water excluded from the core (**Figure 3.20B**). Biological membranes have this kind of **phospholipid bilayer** structure, and we will devote Chapter 6 to their biological functions.

### Lipids have roles in energy conversion, regulation, and protection

In the previous section, we focused on lipids involved in energy storage and cell structure, whose molecular structures are variations on the glycerol–fatty acid structure. However, there are other nonpolar and amphipathic lipids that are not based on this structure.

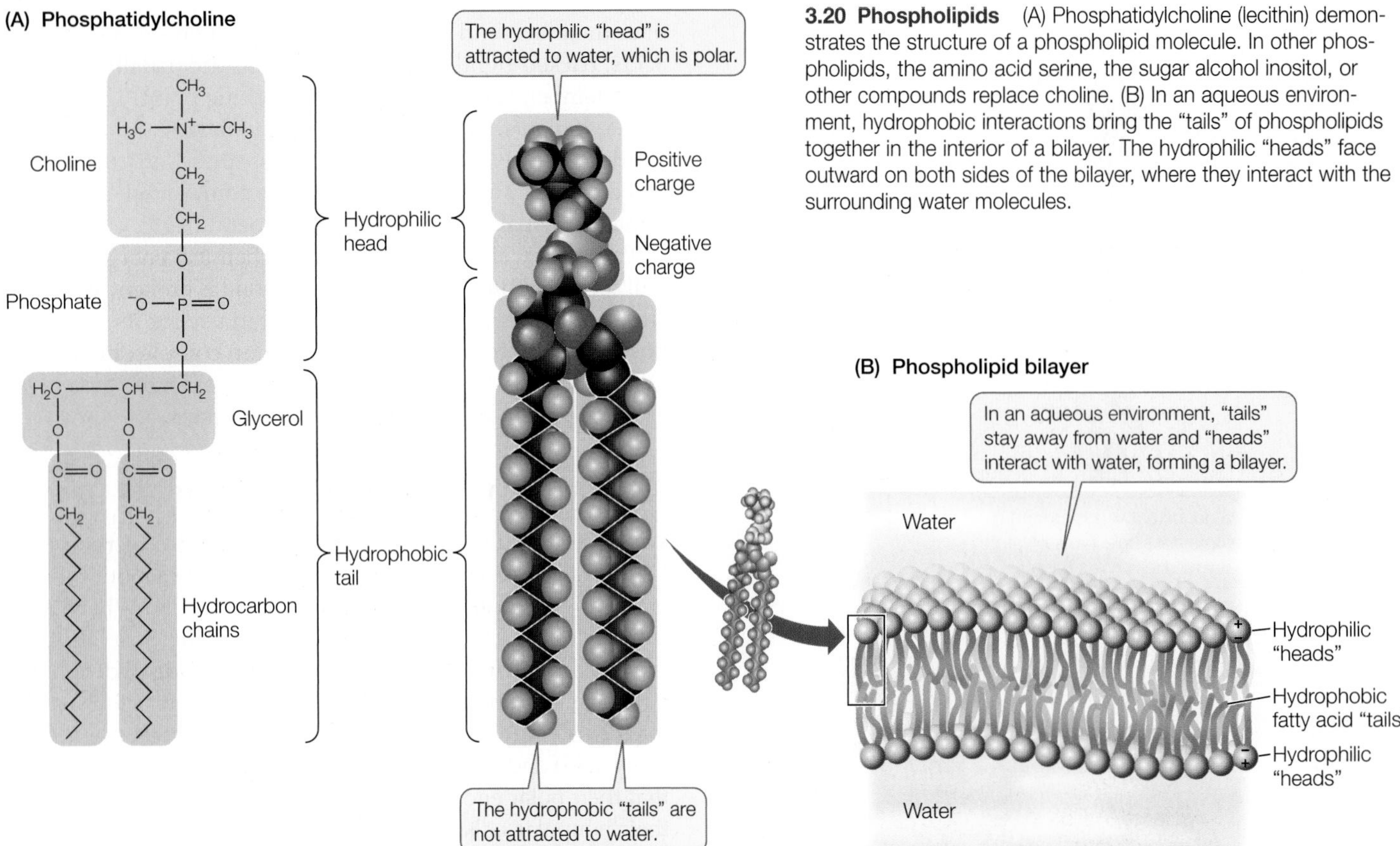
(A) Phosphatidylcholine
Choline
Phosphate
Glycerol
Hydrocarbon chains
Hydrophilic head
Hydrophobic tail
The hydrophilic "head" is attracted to water, which is polar.
Positive charge
Negative charge
The hydrophobic "tails" are not attracted to water.
(B) Phospholipid bilayer
In an aqueous environment, "tails" stay away from water and "heads" interact with water, forming a bilayer.
Water
Hydrophilic "heads"
Hydrophobic fatty acid "tails"
Hydrophilic "heads"
Water

**3.20 Phospholipids** (A) Phosphatidylcholine (lecithin) demonstrates the structure of a phospholipid molecule. In other phospholipids, the amino acid serine, the sugar alcohol inositol, or other compounds replace choline. (B) In an aqueous environment, hydrophobic interactions bring the "tails" of phospholipids together in the interior of a bilayer. The hydrophilic "heads" face outward on both sides of the bilayer, where they interact with the surrounding water molecules.

**CAROTENOIDS** The carotenoids are a family of light-absorbing pigments found in plants and animals. Beta-carotene (β-carotene) is one of the pigments that traps light energy in leaves during photosynthesis. In humans, a molecule of β-carotene can be broken down into two vitamin A molecules (**Figure 3.21**), from which we make the pigment *cis*-retinal, which is required for vision. Carotenoids are responsible for the colors of carrots, tomatoes, pumpkins, egg yolks, and butter.

**STEROIDS** The steroids are a family of organic compounds whose multiple rings share carbons (**Figure 3.22**). The steroid cholesterol is an important constituent of membranes. Other steroids function as hormones, chemical signals that carry messages from one part of the body to another (see Chapter 41). Cholesterol is synthesized in the liver and is the starting material for making testosterone and other steroid hormones, such as estrogen.

**VITAMINS** *Vitamins* are small molecules that are not synthesized by the human body and so must be acquired from the diet (see Chapter 50). For example, *vitamin A* is formed from the β-carotene found in green and yellow vegetables (see Figure 3.21). In humans, a deficiency of vitamin A leads to dry skin, eyes, and internal body surfaces, retarded growth and development, and night blindness, which is a diagnostic symptom for the deficiency. Vitamins D, E, and K are also lipids.

**3.21 β-Carotene is the Source of Vitamin A** The carotenoid β-carotene is symmetrical around its central double bond. When that bond is broken, two molecules of vitamin A are formed. The structural formula presented here is standard chemical shorthand for large organic molecules with many carbon atoms; it is simplified by omitting the C (indicating a carbon atom) at the intersections representing covalent bonds. The presence of hydrogen atoms (H) to fill all the available bonding sites on each C is assumed.

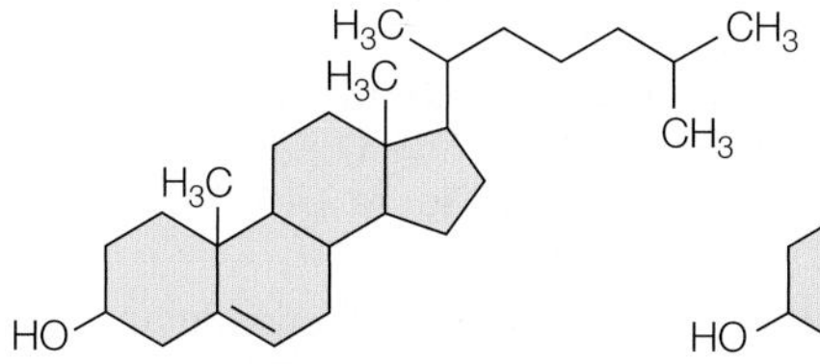

**Cholesterol** is a constituent of membranes and is the source of steroid hormones.

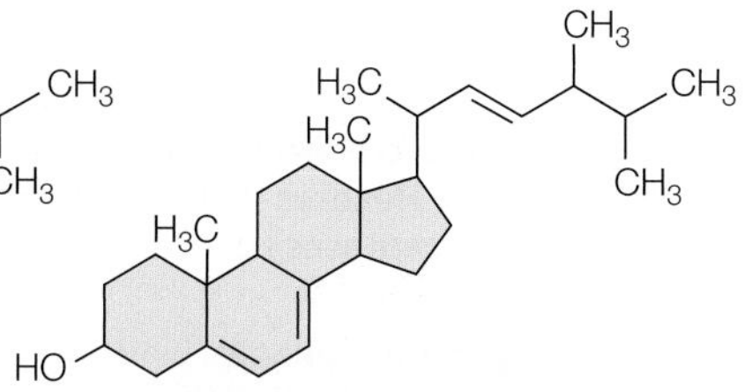

**Vitamin $D_2$** can be produced in the skin by the action of light on a cholesterol derivative.

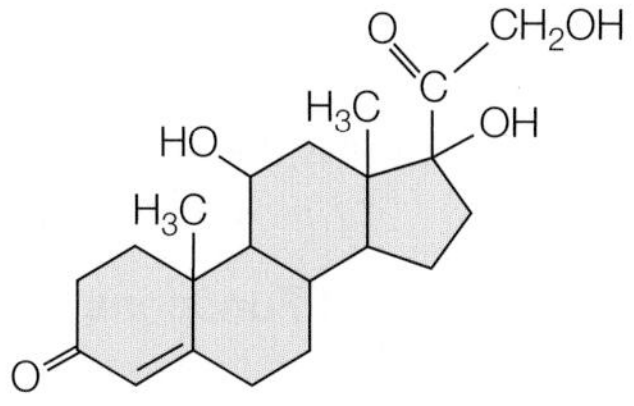

**Cortisol** is a hormone secreted by the adrenal glands.

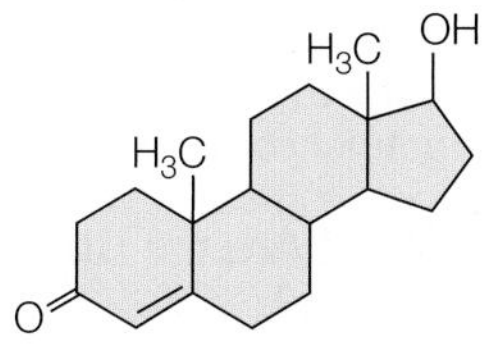

**Testosterone** is a male sex hormone.

**3.22 All Steroids Have the Same Ring Structure** The steroids shown here, all important in vertebrates, are composed of carbon and hydrogen and are highly hydrophobic. However, small chemical variations, such as the presence or absence of a hydroxyl group, can produce enormous functional differences among these molecules.

**WAXES** The sheen on human hair is more than cosmetic. Glands in the skin secrete a waxy coating that repels water and keeps the hair pliable. Birds that live near water have a similar waxy coating on their feathers. The shiny leaves of plants such as holly, familiar during winter holidays, also have a waxy coating. Finally, bees make their honeycombs out of wax. All waxes have the same basic structure: they are formed by an ester linkage between a saturated, long-chain fatty acid and a saturated, long-chain alcohol. The result is a very long molecule, with 40–60 $CH_2$ groups. For example, here is the structure of beeswax:

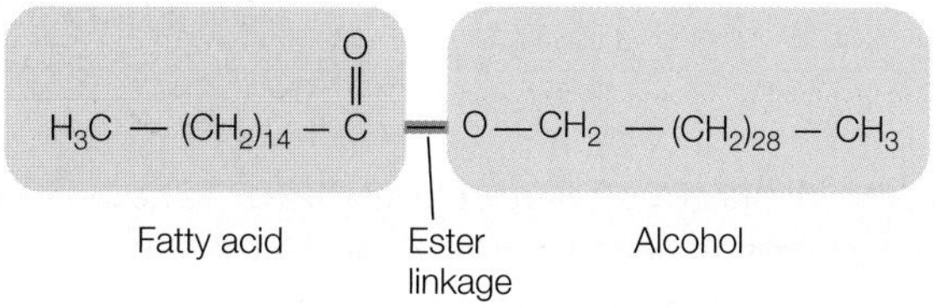

This highly nonpolar structure accounts for the impermeability of wax to water.

## 3.4 RECAP

**Lipids include both nonpolar and amphipathic molecules that are largely composed of carbon and hydrogen. They are important in energy storage, light absorption, regulation and biological structures. Cell membranes contain phospholipids, which are composed of hydrophobic fatty acids linked to glycerol and a hydrophilic phosphate group.**

- Draw the molecular structures of fatty acids and glycerol and show how they are linked to form a triglyceride. See p. 54 and Figure 3.18
- What is the difference between fats and oils? See p. 54
- How does the polar nature of phospholipids result in their forming a bilayer? See p. 55 and Figure 3.20
- Why are steroids and some vitamins classified as lipids? See p. 56

All the types of molecules we have discussed in this chapter are found only in living organisms, but a final class of biological macromolecules has special importance to the living world. The function of the nucleic acids is nothing less than the transmission of life's "blueprint" to each new organism. This chapter showed the wonderful biochemical unity of life, a unity that implies all life has a common origin. Essential to this origin were the monomeric nucleotides and their polymers, nucleic acids. In the next chapter, we turn to the related topics of nucleic acids and the origin of life.

# CHAPTER SUMMARY

## 3.1 What Kinds of Molecules Characterize Living Things?

**SEE ANIMATED TUTORIAL 3.1**

- **Macromolecules** are **polymers** constructed by the formation of covalent bonds between smaller molecules called **monomers**. Macromolecules in living organisms include polysaccharides, proteins, and nucleic acids. Large lipid structures may also be considered macromolecules.
- **Functional groups** are small groups of atoms that are consistently found together in a variety of different macromolecules. Functional groups have particular chemical properties that they confer on any larger molecule of which they are a part. **Review Figure 3.1, WEB ACTIVITY 3.1**
- Structural and optical **isomers** have the same kinds and numbers of atoms, but differ in their structures and properties. **Review Figure 3.2**
- The many functions of macromolecules are directly related to their three-dimensional shapes, which in turn result from the sequences and chemical properties of their monomers.
- Monomers are joined by **condensation reactions**, which release a molecule of water for each bond formed. **Hydrolysis reactions** use water to break polymers into monomers. **Review Figure 3.4**

## 3.2 What Are the Chemical Structures and Functions of Proteins?

- The functions of proteins include support, protection, catalysis, transport, defense, regulation, and movement.
- **Amino acids** are the monomers from which proteins are constructed. Four groups are attached to a central carbon atom: a hydrogen atom, an amino group, a carboxyl group, and a variable R group. The particular properties of each amino acids depend on its **side chain**, or **R group**, which may be charged, polar, or hydrophobic. Review Table 3.1, **WEB ACTIVITY 3.2**
- **Peptide linkages**, also called peptide bonds, covalently link amino acids into polypeptide chains. These bonds form by condensation reactions between the carboxyl and amino groups. Review Figure 3.6
- The **primary structure** of a protein is the sequence of amino acids in the chain. This chain is folded into a **secondary structure**, which in different parts of the protein may form an **α helix** or a **β pleated sheet**. Review Figure 3.7A–C
- **Disulfide bridges** and noncovalent interactions between amino acids cause polypeptide chains to fold into three-dimensional **tertiary structures** and allow multiple chains to interact in a **quaternary structure**. Review Figure 3.7D,E
- Heat, alterations in pH, or certain chemicals can all result in protein **denaturation**, which involves the loss of tertiary and/or secondary structure as well as biological function. Review Figure 3.9
- The specific shape and structure of a protein allows it to bind noncovalently to other molecules. Review Figure 3.11
- **Chaperone proteins** enhance correct protein folding and prevent binding to inappropriate ligands. Review Figure 3.12

## 3.3 What Are the Chemical Structures and Functions of Carbohydrates?

- **Carbohydrates** contain carbon bonded to hydrogen and oxygen atoms in a ratio of 1:2:1, or $(CH_2O)_n$.
- **Monosaccharides** are the monomers that make up carbohydrates. **Hexoses** such as **glucose** are six-carbon monosaccharides; **pentoses** have five carbons. Review Figure 3.14, **WEB ACTIVITY 3.3**
- **Glycosidic linkages**, which have either an α or a β orientation in space, covalently link monosaccharides into larger units such as **disaccharides**, **oligosaccharides**, and **polysaccharides**. Review Figure 3.15
- **Starch** stores energy in plants. Starch and **glycogen** are formed by α-glycosidic linkages between glucose monomers and are distinguished by the amount of branching they exhibit. They can be easily broken down to release stored energy. Review Figure 3.16
- **Cellulose** is a very stable glucose polymer and is the principal structural component of plant cell walls.

## 3.4 What Are the Chemical Structures and Functions of Lipids?

- Fats and oils are **triglycerides**, composed of three fatty acids covalently bonded to a molecule of glycerol by ester linkages. Review Figure 3.18
- **Saturated** fatty acids have a hydrocarbon chain with no double bonds. The hydrocarbon chains of **unsaturated** fatty acids have one or more double bonds that bend the chain, making close packing less possible. Review Figure 3.19
- **Phospholipids** have a hydrophobic hydrocarbon "tail" and a hydrophilic phosphate "head"; that is, they are **amphipathic**. In water, the interactions of the tails and heads of phospholipids generate a **phospholipid bilayer**. The heads are directed outward, where they interact with the surrounding water. The tails are packed together in the interior of the bilayer, away from water. Review Figure 3.20
- Other lipids include vitamins A and D, steroids and plant pigments such as carotenoids.

## SELF-QUIZ

1. The most abundant molecule in the cell is
   *a.* a carbohydrate.
   *b.* a lipid.
   *c.* a nucleic acid.
   *d.* a protein.
   *e.* water.
2. All lipids are
   *a.* triglycerides.
   *b.* polar.
   *c.* hydrophilic.
   *d.* polymers of fatty acids.
   *e.* more soluble in nonpolar solvents than in water.
3. All carbohydrates
   *a.* are polymers.
   *b.* are simple sugars.
   *c.* consist of one or more simple sugars.
   *d.* are found in biological membranes.
   *e.* are more soluble in nonpolar solvents than in water.
4. Which of the following is *not* a carbohydrate?
   *a.* Glucose
   *b.* Starch
   *c.* Cellulose
   *d.* Hemoglobin
   *e.* Deoxyribose
5. All proteins
   *a.* are enzymes.
   *b.* consist of one or more polypeptide chains.
   *c.* are amino acids.
   *d.* have quaternary structures.
   *e.* are more soluble in nonpolar solvents than in water.
6. Which of the following statements about the primary structure of a protein is *not* true?
   *a.* It may be branched.
   *b.* It is held together by covalent bonds.
   *c.* It is unique to that protein.
   *d.* It determines the tertiary structure of the protein.
   *e.* It is the sequence of amino acids in the protein.

7. The amino acid leucine
   *a.* is found in all proteins.
   *b.* cannot form peptide linkages.
   *c.* has a hydrophobic side chain.
   *d.* has a hydrophilic side chain.
   *e.* is identical to the amino acid lysine.
8. The quaternary structure of a protein
   *a.* consists of four subunits—hence the name quaternary.
   *b.* is unrelated to the function of the protein.
   *c.* may be either alpha or beta.
   *d.* depends on covalent bonding among the subunits.
   *e.* depends on the primary structures of the subunits.
9. The amphipathic nature of phospholipids is
   *a.* determined by the fatty acid composition.
   *b.* important in membrane structure.
   *c.* polar but not nonpolar.
   *d.* shown only if the lipid is in a nonpolar solvent.
   *e.* important in energy storage by lipids.
10. Which of the following statements about condensation reactions is *not* true?
   *a.* Protein synthesis results from them.
   *b.* Polysaccharide synthesis results from them.
   *c.* They involve covalent bonds.
   *d.* They consume water as a reactant.
   *e.* Different condensation reactions produce different kinds of macromolecules.

## FOR DISCUSSION

1. Suppose that, in a given protein, one lysine is replaced by aspartic acid (see Table 3.1). Does this change occur in the primary structure or in the secondary structure? How might it result in a change in tertiary structure? In quaternary structure?
2. If there are 20 different amino acids commonly found in proteins, how many different dipeptides are there? How many different tripeptides?

## ADDITIONAL INVESTIGATION

Human hair is composed of a protein, keratin. At the hair salon, two techniques are used to modify the three-dimensional shape of hair. Styling involves heat, and a perm involves cleaving and reforming disulfide bonds. How would you investigate these phenomena in terms of protein structure?

## WORKING WITH DATA (GO TO yourBioPortal.com)

**Primary Structure Specifies Tertiary Structure** In this hands-on exercise based on Figure 3.9, you will learn about the methods used to disrupt the chemical interactions that determine the tertiary structure of proteins. You will examine the original data that led Anfinsen to conclude that denaturation of ribonuclease is reversible.

# 4 Nucleic Acids and the Origin of Life

## Looking for life

The trip had lasted a long and anxious ten months when, in the summer of 1976, the first of two visitors from Earth landed on a plain on the Martian surface. A second spacecraft arrived in September. The task of these robotic laboratories, part of NASA's Viking project, was to search for life.

On Earth, life has existed for several billion years and has spread over most of the planet's surface. Determining life's origins is difficult, however, because (with few exceptions) simple organisms leave no fossils. On Mars, scientists thought, things might be different. A primitive form of life might exist there now, or might have left chemical signatures that remain in place, untouched by other organisms.

The two Viking spacecraft that landed on Mars in 1976 analyzed soil samples for the small molecules of life, including simple sugars and amino acids. None were found. The robotic laboratories immersed soil samples in an aqueous solution of sugars, amino acids, and minerals. Living organisms take in and break down such substances from their environment, releasing gases such as $CO_2$. A small amount of $CO_2$ was detected in one experiment, but, frustratingly, no gases were released in further experiments.

The results from the Viking landers remain controversial. Why did that one experiment detect a sign of life? The 1976 robotic landers are still on Mars but have long since stopped working. In 2008, more probes were sent from Earth, carrying more sophisticated instruments. One of them, the Phoenix lander, is in a northern region of Mars, at a latitude corresponding to that of Alaska on Earth. Phoenix has a robotic arm like the backhoes used in a construction site. When the arm dug a small trench into the Martian soil, shiny dice-sized beads of what turned out to be ice were exposed, although the beads disappeared in a few days as exposure to the atmosphere caused them to vaporize. Dissolved ions such as sodium, magnesium, potassium and chloride were all present in the frozen water, indicating that at least those requirements for life are present on Mars. Once again, the soil was analyzed for traces of current or past organisms; once again, the results were negative. But even if there probably is no life on Mars today, there might have been in the past.

**Lab Seeking Life** Landers such as the robotic space laboratory Phoenix, shown here on Earth, have been sent to look for traces of life on Mars.

**Ice on Mars** The Phoenix landing site (blue dot) is near the Martian north pole, where chemical traces of life might be preserved in the hypercold environment. When the lander scooped up a patch of soil for analysis, it also took photos that revealed ice crystals just below the surface of the Red Planet.

As we saw in Chapter 2, water is a key requirement for life. Remote measurements from orbiting spacecraft and chemical measurements using special telescopes have shown that water is present on Mars and, indeed, on some of the moons of other planets in our solar system.

Scientists are using their knowledge of the small and large molecules that are present in living organisms to search for the chemical signatures of life on other planets. Chapters 2 and 3 described molecules that are important for biological structure and function. In Chapter 4, we turn to certain molecules involved in the origin and perpetuation of life itself.

**IN THIS CHAPTER** we first describe the structure of nucleic acids, the informational macromolecules needed for the perpetuation of life. We then turn to biologists' speculations on the origin of life and describe early experimental evidence that life on Earth today comes from pre-existing life. We present some ideas on the formation of the building blocks of life, including the monomers and polymers that characterize biological systems. Finally, we describe some proposals for the origin of cells.

## CHAPTER OUTLINE



## 4.1 What Are the Chemical Structures and Functions of Nucleic Acids?

From medicine to evolution, from agriculture to forensics, the properties of nucleic acids impact our lives every day. It is with nucleic acids that the concept of "information" entered the biological vocabulary. Nucleic acids are uniquely capable of coding for and transmitting biological information.

The **nucleic acids** are polymers specialized for the storage, transmission between generations, and use of genetic information. There are two types of nucleic acids: **DNA** (*d*eoxyribo*n*ucleic *a*cid) and **RNA** (*r*ibo*n*ucleic *a*cid). DNA is a macromolecule that encodes hereditary information and passes it from generation to generation. Through an RNA intermediate, the information encoded in DNA is used to specify the amino acid sequences of proteins. Information flows from DNA to DNA during reproduction. In the non-reproductive activities of the cell, information flows from DNA to RNA to proteins. It is the proteins that ultimately carry out life's functions.

### Nucleotides are the building blocks of nucleic acids

Nucleic acids are composed of monomers called **nucleotides**, each of which consists of a pentose sugar, a phosphate group, and a nitrogen-containing **base**. (Molecules consisting of a pentose sugar and a nitrogenous base—but no phosphate group—are called *nucleosides*.) The bases of the nucleic acids take one of two chemical forms: a six-membered single-ring structure called a **pyrimidine**, or a fused double-ring structure called a **purine** (**Figure 4.1**). In DNA, the pentose sugar is **deoxyribose**, which differs from the **ribose** found in RNA by the absence of one oxygen atom (see Figure 3.14).

In both RNA and DNA, the backbone of the macromolecule consists of a chain of alternating pentose sugars and phosphate groups (sugar–phosphate–sugar–phosphate). The bases are attached to the sugars and project from the polynucleotide chain (**Figure 4.2**). The nucleotides are joined by **phosphodiester linkages** between the sugar of one nucleotide and the phosphate of the next (*diester* refers to the two covalent bonds formed by —OH groups reacting with acidic phosphate groups). The phosphate groups link carbon 3 in one pentose sugar to carbon 5 in the adjacent sugar.

Most RNA molecules consist of only one polynucleotide chain. DNA, however, is usually double-stranded; its two polynucleotide chains are held together by hydrogen bonding between their nitrogenous bases. The two strands of DNA run in opposite directions. You can see what this means by drawing an arrow through a phosphate group from carbon 5 to

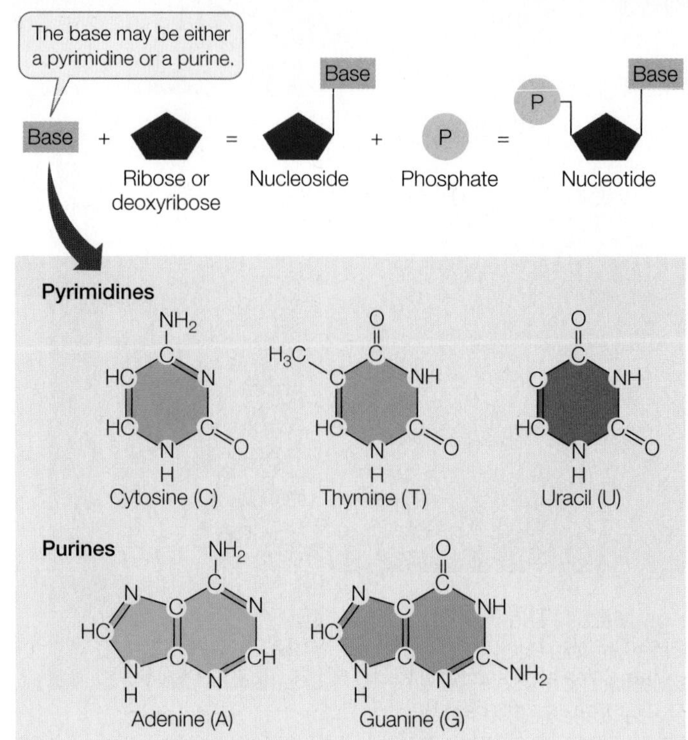


**4.1 Nucleotides Have Three Components** Nucleotide monomers are the building blocks of DNA and RNA polymers.

**yourBioPortal.com**

GO TO **Web Activity 4.1 • Nucleic Acid Building Blocks**

carbon 3 in the next ribose. If you do this for both strands of the DNA in Figure 4.2, the arrows will point in opposite directions. This *antiparallel* orientation allows the strands to fit together in three-dimensional space.

## Base pairing occurs in both DNA and RNA

Only four nitrogenous bases—and thus only four nucleotides—are found in DNA. The DNA bases and their abbreviations are **adenine (A)**, **cytosine (C)**, **guanine (G)**, and **thymine (T)**. Adenine and guanine are purines; thymine and cytosine are pyrimidines. RNA is also made up of four different monomers, but its nucleotides differ from those of DNA. In RNA the nucleotides are termed *ribonucleotides* (the ones in DNA are *deoxyribonucleotides*). They contain ribose rather than deoxyribose, and in-

**4.2 Distinguishing Characteristics of DNA and RNA Polymers** RNA is usually a single strand. DNA usually consists of two strands running in opposite directions (antiparallel).

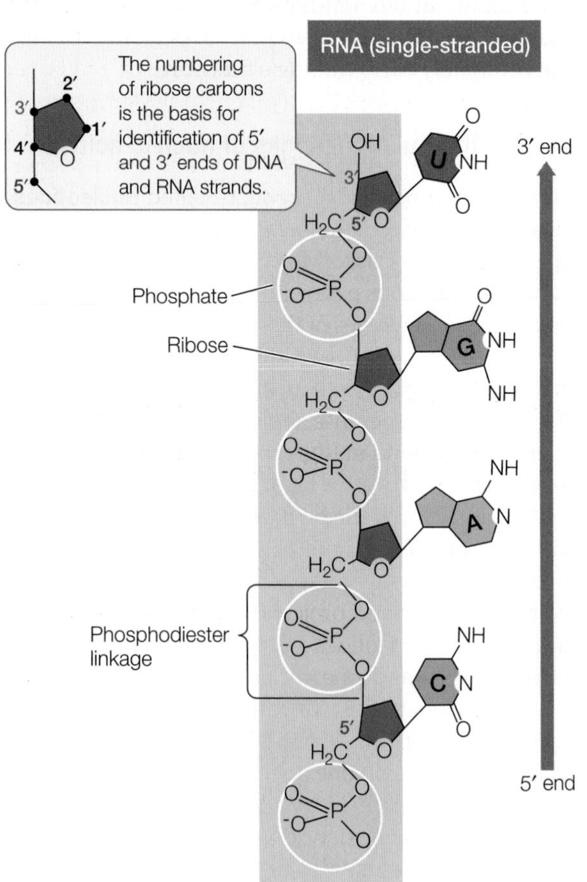


In RNA, the bases are attached to ribose. The bases in RNA are the purines adenine (A) and guanine (G) and the pyrimidines cytosine (C) and uracil (U).

In DNA, the bases are attached to deoxyribose, and the base thymine (T) is found instead of uracil. Hydrogen bonds between purines and pyrimidines hold the two strands of DNA together.

**TABLE 4.1**
**Distinguishing RNA from DNA**

| NUCLEIC ACID | SUGAR | BASES | STRANDS |
|---|---|---|---|
| RNA | Ribose | Adenine<br>Cytosine<br>Guanine<br>Uracil | Single |
| DNA | Deoxyribose | Adenine<br>Cytosine<br>Guanine<br>Thymine | Double |

stead of the base thymine, RNA uses the base **uracil (U)**. The other three bases are the same in RNA and DNA (**Table 4.1**).

The key to understanding the structure and function of nucleic acids is the principle of **complementary base pairing**. In double-stranded DNA, adenine and thymine always pair (A-T), and cytosine and guanine always pair (C-G).

Three factors make base pairing complementary:

- The sites for hydrogen bonding on each base
- The geometry of the sugar–phosphate backbone, which brings complementary bases near each other
- The molecular sizes of the paired bases; the pairing of a larger purine with a smaller pyrimidine ensures stability and uniformity in the double-stranded molecule of DNA

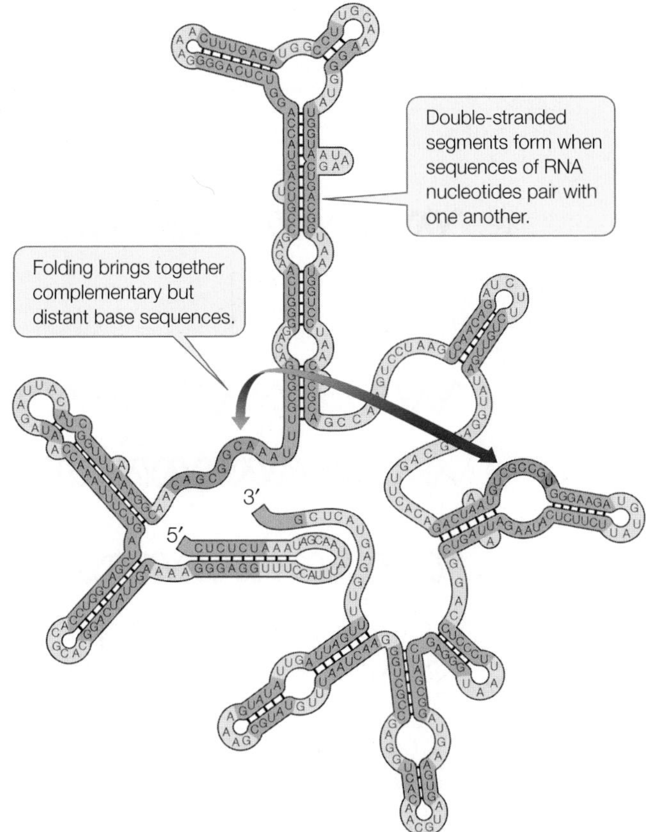

Double-stranded segments form when sequences of RNA nucleotides pair with one another.
Folding brings together complementary but distant base sequences.
3′
5′


**4.3 Hydrogen Bonding in RNA** When a single-stranded RNA folds in on itself, hydrogen bonds between complementary sequences can stabilize it into a three-dimensional shape with complicated surface characteristics.

Although RNA is generally single-stranded, complementary hydrogen bonding between ribonucleotides plays important roles in determining the three-dimensional shapes of some types of RNA molecules, since portions of the single-stranded RNA can fold back and pair with each other (**Figure 4.3**). Complementary base pairing can also take place between ribonucleotides and deoxyribonucleotides. In RNA, guanine and cytosine pair (G-C), as in DNA, but adenine pairs with uracil (A-U). Adenine in an RNA strand can pair either with uracil (in another RNA strand) or with thymine (in a DNA strand).

The three-dimensional physical appearance of DNA is strikingly uniform. The segment shown in **Figure 4.4** could be from any DNA molecule. The variations in DNA—the different sequences of bases—are strictly internal. Through hydrogen bonding, the two complementary polynucleotide strands pair and twist to form a **double helix**. When compared with the complex and varied tertiary structures of proteins, this uniformity is surprising. But this structural contrast makes sense in terms of the functions of these two classes of macromolecules. As we describe in Section 3.2, the different and unique shapes of proteins permit these macromolecules to recognize specific "target" molecules. The area on the surface of a protein that interacts with the target molecule must match the shape of at least part of the target molecule. In other words, structural diversity in the target molecules requires corresponding diversity in the structures of the proteins themselves. Structural diversity is necessary in DNA as well. However, the diversity of DNA is found in its base sequence rather than in the physical shape of the molecule. Different DNA base sequences encode specific information.

**yourBioPortal.com**
GO TO **Web Activity 4.2 • DNA Structure**

### DNA carries information and is expressed through RNA

DNA is a purely *informational* molecule. The information is encoded in the sequence of bases carried in its strands—the infor-

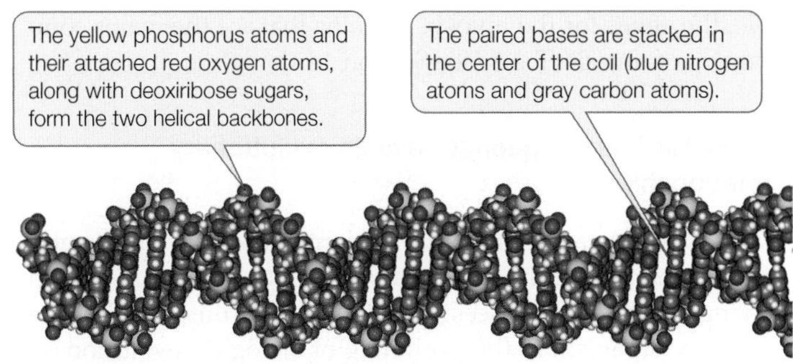

The yellow phosphorus atoms and their attached red oxygen atoms, along with deoxiribose sugars, form the two helical backbones.
The paired bases are stacked in the center of the coil (blue nitrogen atoms and gray carbon atoms).


**4.4 The Double Helix of DNA** The backbones of the two strands in a DNA molecule are coiled in a double helix that is held together by hydrogen bonds between the purines and pyrimidines in the interior of the structure. In this model, the small white atoms represent hydrogen.

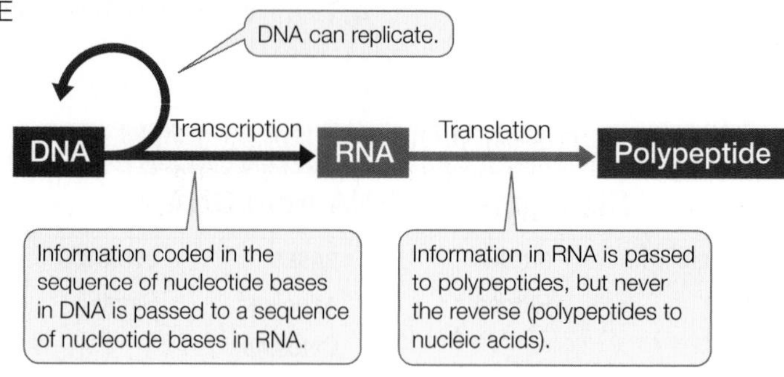


**4.5 DNA Stores Information** The DNA macromolecule stores information that can either be copied (replicated) or transcribed into RNA. RNA can then be translated into protein.

mation encoded in the sequence TCAGCA is different from the information in the sequence CCAGCA. DNA has two functions in terms of information. Taken together, they comprise the *central dogma of molecular biology* (**Figure 4.5**).

- DNA can reproduce itself exactly. This is called *DNA replication*. It is done by polymerization on a template.
- DNA can copy its information into RNA, in a process called *transcription*. The nucleotide sequence in RNA can specify a sequence of amino acids in a polypeptide. This is called *translation*.

While the details of these important processes are described in later chapters, it is important to realize two things at this point:

1. *DNA replication and transcription depend on the base pairing properties of nucleic acids.* The hydrogen-bonded base pairs are A-T and G-C in DNA and A-U and G-C in RNA (see Figure 4.2). Consider this double-stranded DNA region:

TCAGCA
AGTCGT

Transcription of the lower strand will result in a single strand of RNA with the sequence UCAGCA. Can you figure out what the top strand would produce?

2. *DNA replication usually involves the entire DNA molecule, but only relatively small sections of the DNA are transcribed into RNA molecules.* Since DNA holds essential information, it must be replicated completely so that each new cell or new organism receives a complete set of DNA from its parent. The complete set of DNA in a living organism is called its **genome**. However, not all of the information in the genome is needed at all times (**Figure 4.6A**).

The sequences of DNA that encode specific proteins are transcribed into RNA and are called **genes** (**Figure 4.6B**). In humans, the genes that encode the subunits of the protein hemoglobin (see Figure 3.10) are expressed only in the precursors of red blood cells. The genetic information in each globin gene is transcribed into RNA and then translated into a globin polypeptide. In other tissues, such as the muscles, the genes that encode the globin subunits are not transcribed, but others are—for example, the genes for the myosin proteins that are the major component of muscle fibers (see Section 48.1).

### The DNA base sequence reveals evolutionary relationships

Because DNA carries hereditary information from one generation to the next, a theoretical series of DNA molecules, with changes in base sequences, stretches back through the lineage of every organism to the beginning of biological evolution on Earth, about 4 billion years ago. Therefore, closely related living species should have more similar base sequences than species that are more distantly related. The details of how scientists use this information are covered in Chapter 24.

The elucidation and examination of DNA base sequences has confirmed many of the evolutionary relationships that were inferred from more traditional comparisons of body structures, biochemistry, and physiology. Many studies of anatomy, physiology, and behavior have concluded that the closest living relative of humans (*Homo sapiens*) is the chimpanzee (genus *Pan*). In fact, the chimpanzee genome shares more than 98 percent of its DNA base sequence with the human genome. Increasingly, scientists turn to DNA analyses to elucidate evolutionary relationships when other comparisons are not possible or are not conclusive. For example, DNA studies revealed a close relationship between starlings and mockingbirds that was not expected on the basis of their anatomy or behavior.

### Nucleotides have other important roles

Nucleotides are more than just the building blocks of nucleic acids. As we will describe in later chapters, there are several nucleotides with other functions:

- ATP (adenosine triphosphate) acts as an energy transducer in many biochemical reactions (see Section 8.2).

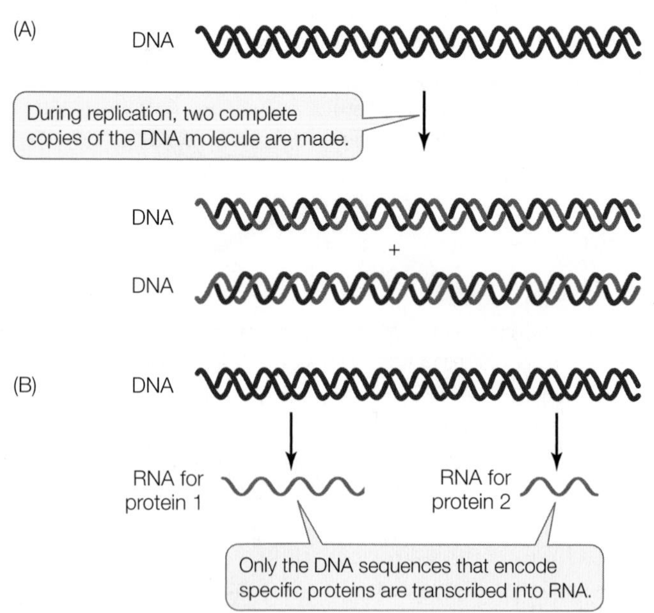


**4.6 DNA Replication and Transcription** DNA is usually completely replicated (A) but only partially transcribed (B). RNA transcripts encode the genes for specific proteins. Transcription of the many different proteins is activated at different times and, in multicellular organisms, in different cells of the body.

- GTP (guanosine triphosphate) serves as an energy source, especially in protein synthesis. It also plays a role in the transfer of information from the environment to cells (see Section 7.2).
- cAMP (cyclic adenosine monophosphate) is a special nucleotide in which an additional bond forms between the sugar and phosphate group. It is essential in many processes, including the actions of hormones and the transmission of information by the nervous system (see Section 7.3).

## 4.1 RECAP

**The nucleic acids DNA and RNA are polymers made up of nucleotide monomers. The sequence of nucleotides in DNA carries the information that is used by RNA to specify primary protein structure. The genetic information in DNA is passed from generation to generation and can be used to understand evolutionary relationships.**

- List the key differences between DNA and RNA. Between purines and pyrimidines. **See p. 61, Table 4.1, and Figure 4.1**
- How do purines and pyrimidines pair up in complementary base pairing? **See pp. 62–63 and Figure 4.2**
- What are the differences between DNA replication and transcription? **See pp. 63–64 and Figures 4.5 and 4.6**
- How can DNA molecules be very diverse, even though they appear to be structurally similar? **See p. 64**

We have seen that the nucleic acids RNA and DNA carry the blueprint of life, and that the inheritance of these macromolecules reaches back to the beginning of evolutionary time. But when, where, and how did nucleic acids arise on Earth? How did the building blocks of life such as amino acids and sugars originally arise?

# 4.2 How and Where Did the Small Molecules of Life Originate?

Chapter 2 points out that living things are composed of the same atomic elements as the inanimate universe—the 92 naturally occurring elements of the periodic table (see Figure 2.2). But the arrangements of these atoms into molecules are unique in biological systems. You will not find biological molecules in inanimate matter unless they came from a once-living organism.

It is impossible to know for certain how life on Earth began. But one thing is sure: life (or at least life as we know it) is not constantly being re-started. That is, *spontaneous generation* of life from inanimate nature is not happening before our eyes. Now and for many millenia past, all life has come from life that existed before. But people, including scientists, did not always believe this.

### Experiments disproved spontaneous generation of life

The idea that life could have originated from nonliving matter is common in many cultures and religions. During the European Renaissance (from about 1450 to 1700, a period that witnessed the birth of modern science), most people thought that at least some forms of life arose repeatedly and directly from inanimate or decaying matter by *spontaneous generation*. Many thought that mice arose from sweaty clothes placed in dim light; that frogs sprang directly from moist soil; and that rotting meat produced flies. Scientists such as the Italian physician and poet Francesco Redi, however, doubted these assumptions. Redi proposed that flies arose not by some mysterious transformation of decaying meat, but from other flies that laid their eggs on the meat. In 1668, Redi performed a scientific experiment—a relatively new concept at the time—to test his hypothesis. He set out several jars containing chunks of meat.

- One jar contained meat exposed to both air and flies.
- A second jar was covered with a fine cloth so that the meat was exposed to air, but not to flies.
- The third jar was sealed so the meat was exposed to neither air nor flies.

As he had hypothesized, Redi found maggots, which then hatched into flies, only in the first jar. This finding demonstrated that maggots could occur only where flies were present. The idea that a complex organism like a fly could appear *de novo* from a nonliving substance in the meat, or from "something in the air," was laid to rest. Well, perhaps not quite to rest.

In the 1660s, newly developed microscopes revealed a vast new biological world. Under microscopic observation, virtually every environment on Earth was found to be teeming with tiny organisms. Some scientists believed these organisms arose spontaneously from their rich chemical environment, by the action of a "life force." But experiments by the great French scientist Louis Pasteur showed that microorganisms can arise only from other microorganisms, and that an environment without life remains lifeless (**Figure 4.7**).

**yourBioPortal.com**
GO TO **Animated Tutorial 4.1 • Pasteur's Experiment**

Pasteur's and Redi's experiments showed that living organisms cannot arise from nonliving materials *under the conditions that existed on Earth during their lifetimes*. But their experiments did not prove that spontaneous generation never occured. Eons ago, conditions on Earth and in the atmosphere above it were vastly different. Indeed, conditions similar to those found on primitive Earth may have existed, or may exist now, on other bodies in our solar system and elsewhere. This has led scientists to ask whether life has originated on other bodies in space, as it did on Earth.

### Life began in water

As we emphasize in Chapter 2 and in the opening story of this chapter, the presence of water on a planet or moon is a necessary prerequisite for life as we know it. Astronomers believe our solar system began forming about 4.6 billion years ago, when a

INVESTIGATING LIFE

**4.7 Disproving the Spontaneous Generation of Life**

Previous experiments disproving spontaneous generation were called into question in regard to microorganisms, whose abundance and diversity were appreciated but whose living processes were not understood. Louis Pasteur's classic experiments disproved the spontaneous generation of microorganisms.

**HYPOTHESIS** **Microorganisms come only from other microorganisms and cannot arise by spontaneous generation.**

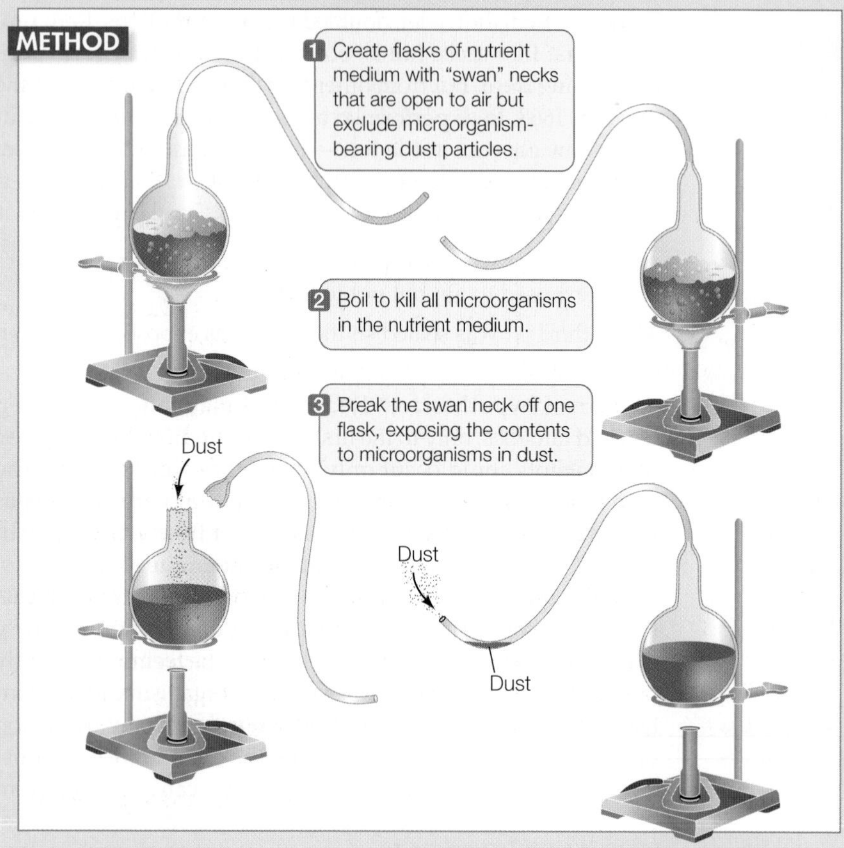

METHOD
1 Create flasks of nutrient medium with "swan" necks that are open to air but exclude microorganism-bearing dust particles.
2 Boil to kill all microorganisms in the nutrient medium.
3 Break the swan neck off one flask, exposing the contents to microorganisms in dust.
Dust
Dust
Dust

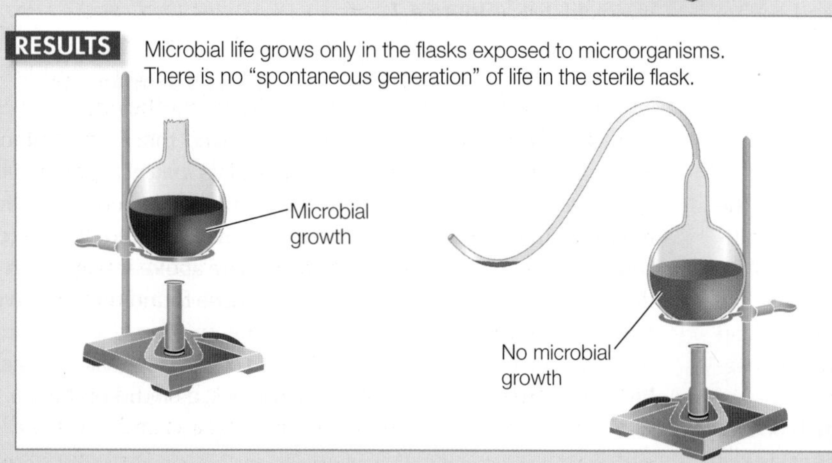

RESULTS
Microbial life grows only in the flasks exposed to microorganisms. There is no "spontaneous generation" of life in the sterile flask.
Microbial growth
No microbial growth

**CONCLUSION** **All life comes from pre-existing life. An environment without life remains lifeless.**

Go to **yourBioPortal.com** for original citations, discussions, and relevant links for all INVESTIGATING LIFE figures.

star exploded and collapsed to form the sun and 500 or so bodies, called planetesimals. These planetesimals collided with one another to form the inner planets, including Earth and Mars. The first chemical signatures indicating the presence of life on Earth appear to be about 4 billion years old. So it took 600 million years, during a geological time frame called the Hadean, for the chemical conditions on Earth to become just right for life. Key among those conditions was the presence of water.

Ancient Earth probably had a lot of water high in its atmosphere. But the new planet was hot, and the water remained in vapor form and dissipated into space. As Earth cooled, it became possible for water to condense on the planet's surface—but where did that water come from? One current view is that comets (loose agglomerations of dust and ice that have orbited the sun since the planets formed) struck Earth and Mars repeatedly, bringing to those planets not only water but other chemical components of life, such as nitrogen.

As the planets cooled and chemicals from their crusts dissolved in the water, simple chemical reactions would have taken place. Some of these reactions might have led to life, but impacts by large comets and rocky meteorites released enough energy to heat the developing oceans almost to boiling, thus destroying any early life. On Earth, these large impacts eventually subsided, and some time around 3.8 to 4 billion years ago life gained a foothold. There has been life on Earth ever since. Because Mars and some other celestial bodies have a similar geological history, the possibility exists that life exists or has existed on them. This possibility was an impetus for sending the Viking and Phoenix landers to Mars.

Several models have been proposed to explain the origin of life on Earth. The next sections discuss two alternative theories: that life came from outside of Earth, or that life arose on Earth through chemical evolution.

### Life may have come from outside Earth

In 1969, a remarkable event led to the discovery that a meteorite from space carried molecules that were characteristic of life on Earth. On September 28 of that year, fragments of a meteorite fell around the town of Murchison, Australia. Using gloves to avoid Earth-derived contamination, scientists immediately shaved off tiny pieces of the rock, put them in

**4.8 The Murchison Meteorite** Pieces from a fragment of the meteorite that landed in Australia in 1969 were put into test tubes with water. Soluble molecules present in the rock, including amino acids, nucleotide bases, and sugars, dissolved in the water. Plastic gloves and sterile instruments were used to reduce the possibility of contamination with substances from Earth.

test tubes and extracted them in water (**Figure 4.8**). They found a number of the molecules that are unique to life, including purines, pyrimidines, sugars, and ten amino acids.

Were these molecules truly brought from space as part of the meteorite, or did they get there after the rock landed on Earth? There were a number of reasons to believe the molecules were not Earthly contaminants:

- The scientists took great care to avoid contamination. They used gloves and sterile instruments, took pieces from below the rock's surface, and did their work very soon after it landed (hopefully before Earth organisms could contaminate the samples).
- Amino acids found in living organisms on Earth are left-handed (see Figure 3.2). The amino acids in the meteorite were a mixture of right- and left-handed forms, with a slight preponderance of the left-handed. Thus the amino acids in the meteorite were not likely to have come from a living organism on Earth.
- In the story that opens Chapter 2, we describe how the ratio of isotopes in a living organism reflects that isotope ratio in the environment where the organism lives. The isotope ratios for carbon and hydrogen in the sugars from the meteorite were different from the ratios of those elements found on Earth.

In 1984, another informative meteorite, this one the size of a softball, was found in Antarctica. We know that the meteorite, ALH 84001, came from Mars because the composition of the gases trapped within the rock was identical to the composition found in the Martian atmosphere, which is quite different from Earth's atmosphere. Radioactive dating and mineral analyses determined that ALH 84001 was 4.5 billion years old and was blasted off the Martian surface 16 million years ago. It landed on Earth fairly recently, about 13,000 years ago.

Scientists found water trapped below the Martian meteorite's surface. This discovery was not surprising, given that surface observations had already shown that water was once abundant on Mars (see the chapter-opening story). Because water is essential for life, scientists wondered whether the meteorite might contain other signs of life as well. Their analysis revealed two substances related to living systems. First, simple carbon-containing molecules called polycyclic aromatic hydrocarbons were present in small but unmistakable amounts; these substances can be formed by living organisms. Second, crystals of magnetite, an iron oxide mineral made by many living organisms on Earth, were found in the interior of the rock.

ALH 84001 and the Murchison meteorite are not the only visitors from outer space that have been shown to contain chemical signatures of life. While the presence of such molecules in rocks may suggest that those rocks once harbored life, it does not prove that there were living organisms in the rocks when they landed on Earth. Most scientists find it hard to believe that an organism could survive thousands of years of traveling through space in a meteorite, followed by intense heat as the meteorite passed through Earth's atmosphere. But there is some evidence that the heat inside some meteorites may not have been severe. When weakly magnetized rock is heated, it reorients its magnetic field to align with the magnetic field around it. In the case of ALH 84001, this would have been Earth's powerful magnetic field, which would have affected the meteorite as it approached our planet.

Careful measurements indicate that, while reorientation did occur at the surface of the rock, it did not occur on the inside. The scientists who took these measurements, Benjamin Weiss and Joseph Kirschvink at the California Institute of Technology, concluded that the inside of ALH 84001 was never heated over 40°C as it entered Earth's atmosphere. This suggests that a long interplanetary trip by living organisms could be possible.

### Prebiotic synthesis experiments model the early Earth

It is clear that other bodies in the solar system have, or once had, water and other simple molecules. Possibly, a meteorite was the source of the simple molecules that were the original building blocks for life on Earth. But a second theory for the origin of life on Earth, **chemical evolution**, holds that conditions on primitive Earth led to the formation of these simple molecules (prebiotic synthesis), and these molecules led to the formation of life forms. Scientists have sought to reconstruct those primitive conditions, both physically (hot or cold) and chemically (by re-creating the combinations and proportions of elements that may have been present).

**HOT CHEMISTRY** The amounts of trace metals such as molybdenum and rhenium in sediments under oceans and lakes is directly proportional to the amount of oxygen gas ($O_2$) present in and above the water. Measurements of dated sedimentary cores indicate that none of these rare metals was present prior to 2.5 billion years ago. This and other lines of evidence suggest that there was little oxygen gas in Earth's early atmosphere. Oxygen gas is thought to have accumulated about 2.5 billion years ago as the by-product of photosynthesis by single-celled life forms; today 21 percent of our atmosphere is $O_2$.

In the 1950s, Stanley Miller and Harold Urey at the University of Chicago set up an experimental "atmosphere" containing the gases thought to have been present in Earth's early atmosphere: hydrogen gas, ammonia, methane gas, and water vapor. They passed an electric spark through these gases, to simulate lightning as a source of energy to drive chemical reactions. Then, they cooled the system so the gases would condense and collect in a watery solution, or "ocean" (**Figure 4.9**). After a few days of continuous operation, the system contained numerous complex molecules, including amino acids, purines, and pyrimidines—some of the building blocks of life.

**yourBioPortal.com**

**GO TO Animated Tutorial 4.2 • Synthesis of Prebiotic Molecules**

The results of this experiment were profoundly important in giving weight to speculations about the chemical origin of life on Earth and elsewhere in the universe. Decades of experimental work and critical evaluation followed. The experiments showed that, under the conditions used by Miller and Urey, many small molecular building blocks of life could be formed:

- All five bases that are present in DNA and RNA (i.e., A, T, C, G and U)
- 17 of the 20 amino acids used in protein synthesis
- 3- to 6-carbon sugars

However, the 5-carbon sugar ribose was not produced in these experiments.

In science, an experiment and its results must be repeated, reinterpreted, and refined as more knowledge accumulates. The results of the Miller–Urey experiments have undergone several such refinements.

The amino acids in living things are always L-isomers (see Figure 3.2 and p. 43). But a mixture of D- and L-isomers appeared in the amino acids formed in the Miller–Urey experiments. Recent experiments show that natural processes could have selected the L-amino acids from the mixture. Some minerals, especially calcite-based rocks, have unique crystal structures that selectively bind to D- or L-amino acids, separating the two. Such rocks were abundant on early Earth. This suggests that while both kinds of amino acid structures were made, binding to certain rocks may have eliminated the D- amino acids. (Interestingly, some meteorites, such as the Murchison meteorite, also have this selectivity.)

Ideas about Earth's original atmosphere have changed since Miller and Urey did their experiments. There is abundant evidence indicating that major volcanic eruptions occurred 4 bil-

## INVESTIGATING LIFE

**4.9 Miller and Urey Synthesized Prebiotic Molecules in an Experimental Atmosphere**

With an increased understanding of the atmospheric conditions that existed on primitive Earth, the researchers devised an experiment to see if these conditions could lead to the formation of organic molecules.

**HYPOTHESIS** **Organic chemical compounds can be generated under conditions similar to those that existed in the atmosphere of primitive Earth.**

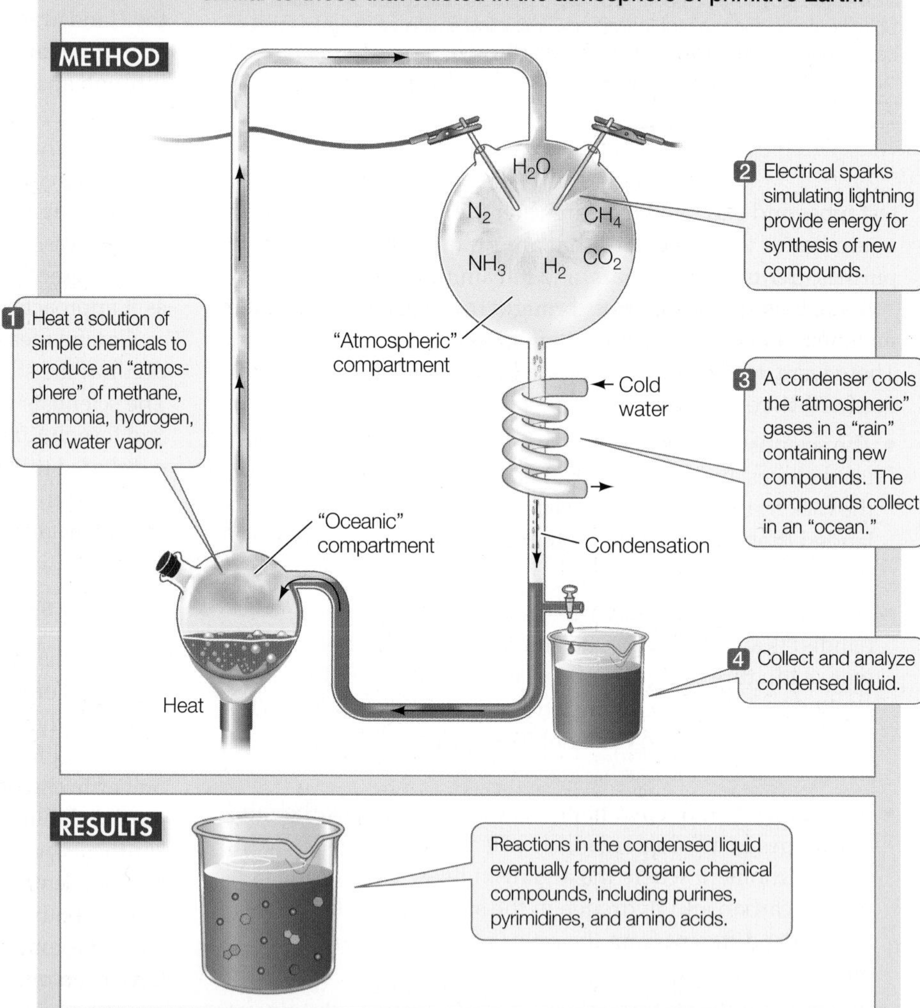

METHOD
$H_2O$
$N_2$
$CH_4$
$NH_3$
$H_2$
$CO_2$
2 Electrical sparks simulating lightning provide energy for synthesis of new compounds.
1 Heat a solution of simple chemicals to produce an "atmosphere" of methane, ammonia, hydrogen, and water vapor.
"Atmospheric" compartment
Cold water
3 A condenser cools the "atmospheric" gases in a "rain" containing new compounds. The compounds collect in an "ocean."
"Oceanic" compartment
Condensation
4 Collect and analyze condensed liquid.
Heat
RESULTS
Reactions in the condensed liquid eventually formed organic chemical compounds, including purines, pyrimidines, and amino acids.


**CONCLUSION** **The chemical building blocks of life could have been generated in the probable atmosphere of early Earth.**

**FURTHER INVESTIGATION:** What result would you predict if $O_2$ were present in the "atmosphere" in this experiment?

Go to **yourBioPortal.com** for original citations, discussions, and relevant links for all INVESTIGATING LIFE figures.

lion years ago, which would have released carbon dioxide ($CO_2$), nitrogen ($N_2$), hydrogen sulfide ($H_2S$), and sulfur dioxide ($SO_2$) into the atmosphere. Experiments using these gases in addition to the ones in the original experiment have produced more diverse molecules, including:

- Vitamin $B_6$, pantothenic acid (a component of coenzyme A), and nicotinamide (part of NAD, which is involved in energy metabolism).
- Carboxylic acids such as succinic and lactic acids (also involved in energy metabolism) and fatty acids.
- Ribose, a key component of RNA, which can be formed from formaldehyde gas (HCHO), evidence of which has been found in space.

**COLD CHEMISTRY** Stanley Miller also performed a long-term experiment in which the electric spark was not used. In 1972, he filled test tubes with ammonia gas, water vapor and cyanide (HCN), another molecule that is thought to have formed on primitive Earth. After checking that there were no contaminating substances or organisms that might confound the results, he sealed the tubes and cooled them to −100°C, the temperature of the ice that covers Europa, one of Jupiter's moons. Opening the tubes 25 years later, he found amino acids and nucleotide bases. Apparently, pockets of liquid water within the ice had allowed high concentrations of the starting materials to accumulate, thereby speeding up chemical reactions. The important conclusion is that the cold water within ice on ancient Earth, and other celestial bodies such as Mars, Europa, and Enceladus (one of Saturn's moons; satellite photos have revealed geysers of liquid water coming from its interior) may have provided environments for the prebiotic synthesis of molecules required for the subsequent formation of simple living systems.

## 4.2 RECAP

**Life does not arise repeatedly through spontaneous generation, but comes from pre-existing life. Water is an essential ingredient for the emergence of life. Meteorites that have landed on Earth provide some evidence for an extraterrestrial origin of life. Prebiotic chemical synthesis experiments provide support for the idea that life's simple molecules formed in the primitive Earth environment.**

- Explain how Redi's and Pasteur's experiments disproved spontaneous generation. See p. 65 and Figure 4.7
- What is the evidence that life on Earth came from other bodies in the solar system? See pp. 66–67
- What is the significance of the Miller–Urey experiment, what did it find, and what were its limitations? See p. 68 and Figure 4.9

Chemistry experiments using conditions modeling the ancient Earth's environment suggest an origin for the monomers (such as amino acids) that make up the polymers (such as proteins) that characterize life. How did these polymers develop on the ancient Earth?

## 4.3 How Did the Large Molecules of Life Originate?

The Miller–Urey experiment and other experiments that followed it provide a plausible scenario for the formation of the building blocks of life under conditions that prevailed on primitive Earth. The next step in forming and supporting a general theory on the origin of life on Earth would be an explanation of the formation of polymers from these monomers.

### Chemical evolution may have led to polymerization

Scientists have used a number of model systems to try to simulate conditions under which polymers might have been made. Each of these systems is based on several observations and speculations:

- *Solid mineral surfaces*, such as powder-like clays, have large surface areas. Scientists speculate that the silicates within clay may have been catalytic (speeded up the reactions) in the formation of early carbon-based molecules.
- *Hydrothermal vents* deep in the ocean, where hot water emerges from beneath Earth's crust, lack oxygen gas and contain metals such as iron and nickel. In laboratory experiments, these metals have been shown to catalyze the polymerization of amino acids to polypeptides in the absence of oxygen.
- *Hot pools* at the edges of oceans may, through evaporation, have concentrated monomers to the point where polymerization was favored (the "primordial soup" hypothesis).

In whatever ways the earliest stages of chemical evolution occurred, they resulted in the emergence of monomers and polymers that have probably remained unchanged in their general structure and function for several billion years.

### There are two theories for the emergence of nucleic acids, proteins, and complex chemistry

Earlier in this chapter, we described the key roles of nucleic acids as informational molecules that are passed on from one generation to the next. We also described how DNA is transcribed to RNA, which can then be translated into protein (see Figure 4.5). Chapter 3 describes the roles of proteins as catalysts, speeding up biochemical transformations (see Section 3.2). In existing life forms, nucleic acids and proteins require one another in order to perpetuate life. For the origin of life, this results in a chicken-or-egg problem. Which came first, the genetic material (nucleic acids) or proteins? Two ideas have emerged. One suggests that sequential catalytic changes (primitive metabolism) came first. The other suggests that replication by nucleic acids preceded metabolism (**Figure 4.10**).

**CHEMICAL CHANGES (METABOLISM) FIRST** In this model, life began in tiny droplets, or compartments, that concentrated and sepa-

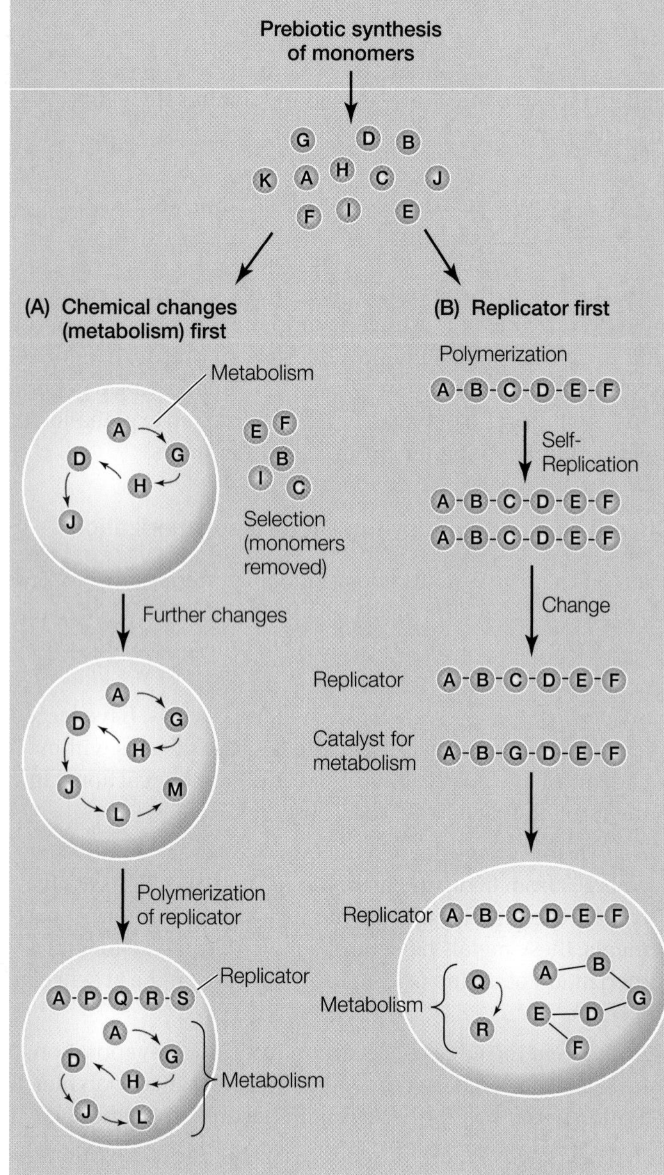


**4.10 Two Pathways to Life** Biologists have proposed two ways in which simple monomers could have become self-replicating systems capable of biological functions. (A) The chemical changes (metabolism) first pathway. (B) The replicator first pathway.

rated their contents from the external environment. Within such a chemically rich environment, some substances could occasionally and randomly undergo chemical changes. Proponents of this model speculate that those compartments where the changes were effective for survival in the environment might even have been selected for growth and some primitive form of reproduction. Could catalysis, the speeding up of reactions essential for life, occur in such an environment? The German scientist Günter Wächtershäuser proposed that catalysis and reproduction could have occurred without proteins on a mineral called pyrite (iron disulfide), which has been found at hydrothermal vents and which could serve as a source of energy for polymerization reactions. Over time, nucleic acids and eventually proteins might have formed in the concentrated droplets. Then, in some of these proteins, the ability to catalyze biochemical reactions—including the replication of nucleic acids—could have evolved.

**REPLICATOR FIRST** In this model, the genetic material—nucleic acids—came first. The nucleotide building blocks made by prebiotic chemistry came together to form polymers. Some of these polymers might have had the right shape to be catalytic so that they could reproduce themselves and catalyze other chemical transformations. Such transformations might have included the synthesis of proteins, just as RNA is translated into proteins in living organisms today (see Figure 4.5). Along the way, those molecules that were best adapted to the environment would survive and reproduce. Eventually they would have become incorporated into living cells.

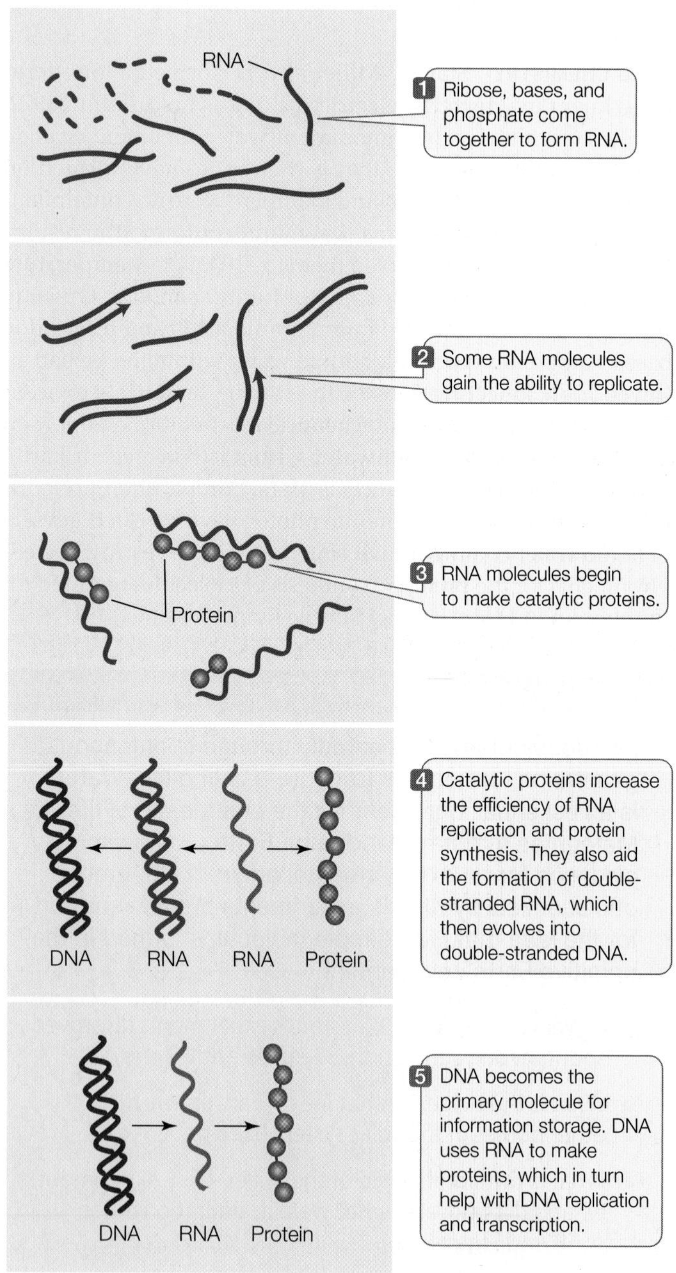

RNA
1 Ribose, bases, and phosphate come together to form RNA.
2 Some RNA molecules gain the ability to replicate.
Protein
3 RNA molecules begin to make catalytic proteins.
DNA
RNA
RNA
Protein
4 Catalytic proteins increase the efficiency of RNA replication and protein synthesis. They also aid the formation of double-stranded RNA, which then evolves into double-stranded DNA.
DNA
RNA
Protein
5 DNA becomes the primary molecule for information storage. DNA uses RNA to make proteins, which in turn help with DNA replication and transcription.

**4.11 The "RNA World" Hypothesis** In a world before DNA, this view postulates that RNA alone was both the blueprint for protein synthesis and a catalyst for its own replication. Eventually, the more compact information storage molecules of DNA could have evolved from RNA.

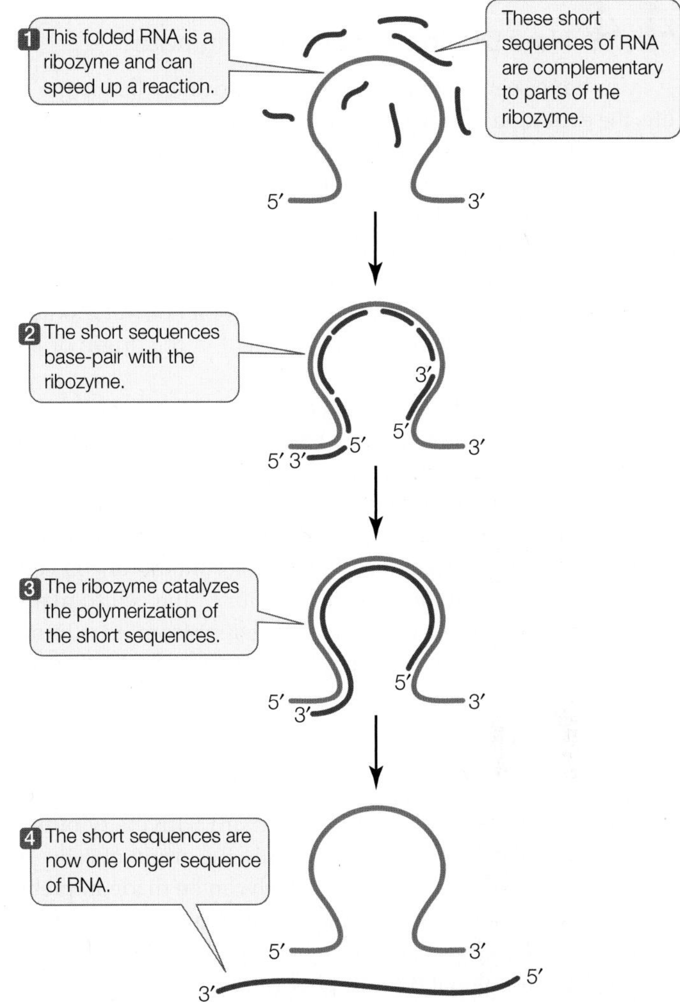
1 This folded RNA is a ribozyme and can speed up a reaction.
These short sequences of RNA are complementary to parts of the ribozyme.
5′
3′
2 The short sequences base-pair with the ribozyme.
5′ 3′
5′
5′
3′
3′
3 The ribozyme catalyzes the polymerization of the short sequences.
5′
3′
5′
3′
4 The short sequences are now one longer sequence of RNA.
5′
3′
3′
5′

**4.12 An Early Catalyst for Life?** In the laboratory, a ribozyme (a folded RNA molecule) can catalyze the polymerization of several short RNA strands into a longer molecule. Such a process could be a precursor for the copying of nucleic acids, which is essential for their replication and for gene expression.

There are two major problems with the replicator first model:

- Nucleic acid polymers have not been observed in prebiotic chemistry simulations.
- DNA, the genetic material in almost all current organisms, is not self-catalytic.

The first problem remains, but the second has a plausible solution: RNA can be a catalyst and can catalyze its own synthesis.

### RNA may have been the first biological catalyst

The three-dimensional structure of a folded RNA molecule presents a unique surface to the external environment (see Figure 4.3). The surfaces of RNA molecules can be every bit as specific as those of proteins. Just as the shapes of proteins allow them to function as catalysts, speeding up reactions that would ordinarily take place too slowly to be biologically useful, the three-dimensional shapes and other chemical properties of certain RNA molecules allow them to function as catalysts. Catalytic RNAs, called **ribozymes**, can catalyze reactions on their own nucleotides as well as in other cellular substances. Although in retrospect it is not too surprising, the discovery of catalytic RNAs was a major shock to a community of biologists who were convinced that all biological catalysts were proteins (enzymes). It took almost a decade for the work of the scientists involved, Thomas Cech and Sidney Altman, to be fully accepted by other scientists. Later, they were awarded the Nobel Prize.

Given that RNA can be both informational (in its nucleotide sequence) and catalytic (due to its ability to form unique three-dimensional shapes), it has been hypothesized that early life consisted of an "RNA world"—a world before DNA. It is thought that when RNA was first made, it could have acted as a catalyst for its own replication as well as for the synthesis of proteins. DNA could eventually have evolved from RNA (**Figure 4.11**). Some laboratory evidence supports this scenario:

- When certain short RNA sequences are added to a mixture of nucleotides, RNA polymers can be formed at a rate 7 million times greater than the formation of polymers without the added RNA. This added RNA is not a template, but a catalyst.
- In the test tube, a ribozyme can catalyze the assembly of short RNAs into a longer molecule (**Figure 4.12**). This may be how nucleic acid replication evolved.
- In living organisms today, the formation of peptide linkages (see Figure 3.6) is catalyzed by ribozymes.
- In certain viruses called retroviruses, there is an enzyme called reverse transcriptase that catalyzes the synthesis of DNA from RNA.

## 4.3 RECAP

**The emergence of the chemical reactions characteristic of life (metabolism), and the polymerization of monomers to polymers, may have occurred on the surfaces of hydrothermal vents. One theory proposes that metabolism came before polymerization; another suggests that the reverse occurred. RNA may have been the first genetic material and catalyst.**

- What are the two theories for the emergence of metabolism and polymers? **See pp. 69–71 and Figure 4.10**
- How does RNA self-replicate? **See p. 71 and Figure 4.12**

The discovery of mechanisms for the formation of small and large molecules is essential to answering questions about the origin of life on Earth. But we also need to understand how organized systems formed that include these molecules and display the characteristic properties of life, such as reproduction, energy processing, and responsiveness to the environment. These properties are present in cells, and we now turn to ideas on their origin.

## 4.4 How Did the First Cells Originate?

As you have seen from many of the theories for the origin of life, the evolution of biochemistry occurred under localized conditions. That is, the chemical reactions of metabolism, polymerization, and replication could not occur in a dilute aqueous environment. There had to be a compartment of some sort that brought together and concentrated the compounds involved in these events. Biologists have proposed that initially this compartment may have simply been a tiny droplet of water on the surface of a rock. But another major event in the origin of life was necessary.

Life as we know it is separated from the environment within structurally defined units called **cells**. The internal contents of a cell are separated from the nonbiological environment by a special barrier—a **membrane**. The membrane is not just a barrier; it regulates what goes into and out of the cell, as we describe in Chapter 6. This role of the surface membrane is very important because it permits the interior of the cell to maintain a chemical composition that is different from its external environment. How did the first cells with membranes come into existence?

### Experiments describe the origin of cells

Jack Szostak and his colleagues at Harvard University built a laboratory model that gives insights into the origin of cells. To do this, they first put fatty acids (which can be made in prebiotic experiments) into water. Recall from Chapter 3 that fatty acids are *amphipathic*: they have a hydrophilic polar end and a long, nonpolar tail that is hydrophobic (see Figure 3.20). When placed in water, fatty acids will arrange themselves in a round "huddle" much like a football team: the hydrophilic ends point outward to interact with the aqueous environment and the fatty acid tails point inward, away from the water molecules.

What if some water becomes trapped in the interior of this "huddle"? Now the layer of hydrophobic fatty acid tails is in water, which is an unstable situation. To stabilize this, a second layer of fatty acids forms. This *lipid bilayer* has the polar ends of the fatty acids facing both outward and inward, because they are attracted to the polar water molecules present on each side of the double layer. The nonpolar tails form the interior of the bilayer (**Figure 4.13**). These prebiotic, water-filled structures, defined by a lipid bilayer membrane, very much resemble living cells. Scientists refer to these compartments as **protocells**. Examining their properties revealed that

- Large molecules such as DNA or RNA could not pass through the bilayer to enter the protocells, but small molecules such as sugars and individual nucleotides could.
- Nucleic acids inside the protocells could replicate using the nucleotides from outside. When the investigators placed a short nucleic acid strand capable of self-replication inside protocells and added nucleotides to the watery environment outside, the nucleotides crossed the barrier, entered the protocells, and became incorporated into new polynucleotide chains. This may have been the first step toward cell reproduction, and it took place without protein catalysis.

Were these protocells truly cells, and was the lipid bilayer produced in these experiments a true cell membrane? Certainly not. The protocells could not fully reproduce, nor could they carry out all the metabolic reactions that take place in modern cells. The simple lipid bilayer had few of the sophisticated functions

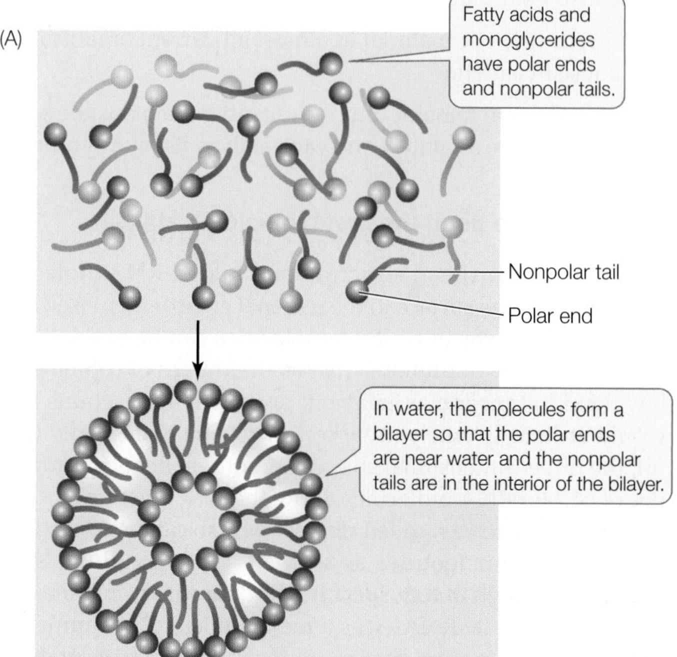

(A)
Fatty acids and monoglycerides have polar ends and nonpolar tails.
Nonpolar tail
Polar end
In water, the molecules form a bilayer so that the polar ends are near water and the nonpolar tails are in the interior of the bilayer.


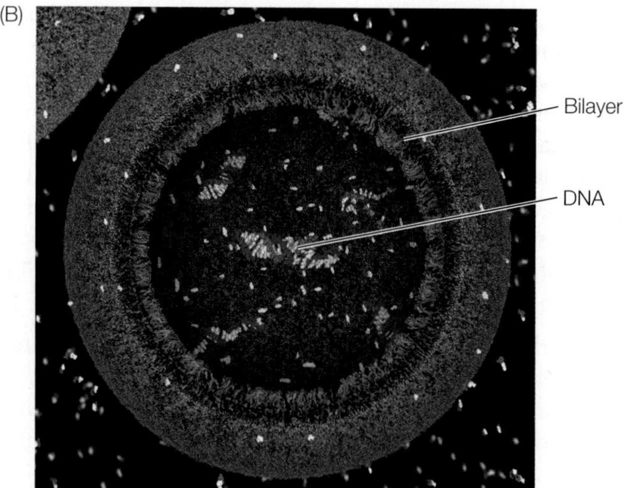


**4.13 Protocells** (A) In a series of experiments in the Szostak lab, researchers mixed fatty acid molecules in water. The molecules formed bilayers that have some of the properties of a cell membrane. The bilayers and the water "trapped" inside them are essential to form a protocell. (B) A model of the protocell. Nutrients and nucleotides (blue and white particles) pass through the "membrane" and enter the protocell, where they copy an already present DNA template. The new copies of DNA remain in the protocell.

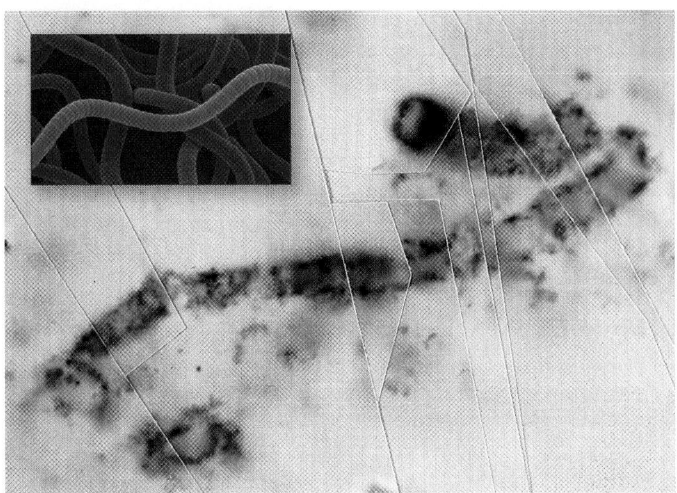

**4.14 The Earliest Cells?** This fossil from Western Australia is 3.5 billion years old. Its form is similar to that of modern filamentous cyanobacteria (inset).

of modern cell membranes. Nevertheless, the protocell may be a reasonable facsimile of a cell as it evolved billions of years ago:

- It can act as a system of interacting parts
- It is capable of organization and self-catalysis
- It includes an interior that is distinct from the exterior environment.

These are all fundamental characteristics of living cells.

### Some ancient cells left a fossil imprint

In the 1990s, scientists made an extremely rare find: a formation of ancient rocks in Australia that had remained relatively unchanged since they first formed 3.5 billion years ago. In one of these rock samples, geologist J. William Schopf of the University of California, Los Angeles, saw chains and clumps of what looked tantalizingly like contemporary cyanobacteria, or "blue-green" bacteria (**Figure 4.14**). Cyanobacteria are believed to have been among the first organisms, because they can perform photosynthesis, converting $CO_2$ from the atmosphere and water into carbohydrates. Schopf needed to prove that the chains were once alive, not just the results of simple chemical reactions. He and his colleagues looked for chemical evidence of photosynthesis in the rock samples.

The use of carbon dioxide in photosynthesis is a hallmark of life and leaves a unique chemical signature—a specific ratio of isotopes of carbon ($^{13}C$:$^{12}C$) in the resulting carbohydrates. Schopf showed that the Australian material had this isotope signature. Furthermore, microscopic examination of the chains revealed *internal* substructures that are characteristic of living systems and were not likely to be the result of simple chemical reactions. Schopf's evidence suggests that the Australian sample is indeed the remains of a truly ancient living organism.

Taking geological, chemical, and biological evidence into account, it is plausible that it took about 500 million to a billion years from the formation of the Earth until the appearance of the first cells (**Figure 4.15**). Life has been cellular ever since. In the next chapter, we begin our study of cell structure and function.

## 4.4 RECAP

**The chemical reactions that preceded living organisms probably occurred in specialized compartments, such as water droplets on the surfaces of minerals. Life as we know it did not begin until the emergence of cells. Protocells made in the laboratory have some of the properties of modern cells. Cell-like structures fossilized in ancient rocks date the first cells to about 3.5 billion years ago.**

- Explain the importance of the cell membrane to the evolution of living organisms. See p. 72
- What is the evidence that ancient rocks contain the fossils of cells? See p. 73

**4.15 The Origin of Life** This highly simplified timeline gives a sense of the major events that culminated in the origin of life more than 3.5 billion years ago.

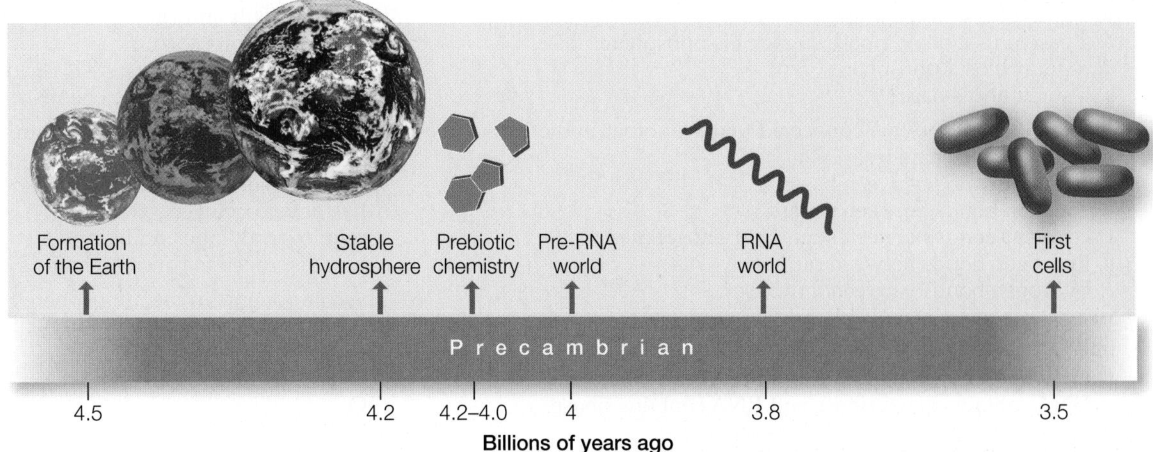

# CHAPTER SUMMARY

## 4.1 What Are the Chemical Structures and Functions of Nucleic Acids?

- The unique function of the nucleic acids—DNA and RNA—is information storage. They form the hereditary material that passes genetic information from one generation to the next.
- Nucleic acids are polymers of nucleotides. A **nucleotide** consists of a phosphate group, a pentose sugar (**ribose** in RNA and **deoxyribose** in DNA), and a nitrogen-containing **base**. Review Figure 4.1
- In DNA, the nucleotide bases are **adenine**, **guanine**, **cytosine**, and **thymine**. **Uracil** replaces thymine in RNA. The nucleotides are joined by **phosphodiester linkages** between the sugar of one and the phosphate of the next, forming a nucleic acid polymer. **WEB ACTIVITY 4.1**
- DNA is a **double helix** with two separate strands in which there is **complementary base pairing** based on hydrogen bonds between adenine and thymine (A-T) and between guanine and cytosine (G-C). The two strands of the DNA double helix run in opposite directions. RNA consists of one chain of nucleotides. Hydrogen bonding can occur within the single strand of RNA, forming double-stranded regions and giving the molecule a three-dimensional surface shape. Review Figures 4.2 and 4.3; **WEB ACTIVITY 4.2**
- The information content of DNA and RNA resides in their **base sequences**.
- DNA is expressed as RNA in **transcription**. RNA can then specify the amino acid sequence of a protein in **translation**. Review Figures 4.5 and 4.6

## 4.2 How and Where did the Small Molecules of Life Originate?

- Historically, many cultures believed that life originates repeatedly by **spontaneous generation**. This was disproven experimentally. Review Figure 4.7; **ANIMATED TUTORIAL 4.1**
- Life probably originated from chemical reactions. A prerequisite for life is the presence of water.
- The presence of chemical traces of life on meteorites that have landed on Earth suggests that life might have originated extraterrestrially.
- Chemical experiments modeling the prebiotic conditions on Earth have shown that the small molecules that characterize life could have been formed from atmospheric chemicals. Review Figure 4.9; **ANIMATED TUTORIAL 4.2**

## 4.3 How Did the Large Molecules of Life Originate?

- Polymerization of small molecules to polymers could occur in small compartments such as droplets or on surfaces. Both of these conditions concentrate molecules such that reactions are favored.
- The "metabolism first" theory of poylmerization proposes that chemical reactions involving small molecules evolved first, and some of them formed polymers that acted as genetic information and catalysts.
- The "replicator first" theory proposes that RNA formed early, and acted as both genetic material and catalyst. Then reactions involving small molecules could occur. Review Figure 4.10
- In contemporary organisms, RNA can act as both an information molecule and as a catalyst. This favors the replicator first model. The **RNA world** may have been an important step on the way to life. Review Figure 4.11

## 4.4 How Did Cells Originate?

- A key to the emergence of living cells was the prebiotic chemical generation of compartments enclosed by **membranes**. Such enclosed compartments permitted the generation and maintenance of internal chemical conditions that were different from those in the exterior environment.
- In the laboratory, fatty acids and related lipids assemble into **protocells** that have some of the characteristics of cells. Review Figure 4.13
- Ancient rocks (3.5 billion years old) have been found with imprints that are probably fossils of early cells.

# SELF-QUIZ

1. A nucleotide in DNA is made up of
   *a.* four bases.
   *b.* a base plus a ribose sugar.
   *c.* a base plus a deoxyribose sugar plus phosphate.
   *d.* a sugar plus a phosphate.
   *e.* a sugar and a base.
2. Nucleotides in RNA are connected to one another in the polynucleotide chain by
   *a.* covalent bonds between bases.
   *b.* covalent bonds between sugars.
   *c.* covalent bonds between sugar and phosphate.
   *d.* hydrogen bonds between purines.
   *e.* hydrogen bonds between any bases.
3. Which is a difference between DNA and RNA?
   *a.* DNA is single-stranded and RNA is double-stranded.
   *b.* DNA is only informational and RNA is only catalytic.
   *c.* DNA contains deoxyribose and RNA contains ribose.
   *d.* DNA is transcribed and RNA is replicated.
   *e.* DNA contains uracil (U) and RNA contains thymine (T).
4. The nucleotide sequence of DNA
   *a.* is the same in all organisms of a species.
   *b.* contains only information for translation.
   *c.* evolved before RNA.
   *d.* contains the four bases, A, T, G, and C.
   *e.* is produced by prebiotic chemistry experiments.
5. Spontaneous generation of life from nonliving materials
   *a.* can occur in dark places.
   *b.* has not been a belief of humans.
   *c.* has never occurred.
   *d.* requires only nucleotides and fatty acids.
   *e.* was disproven for microorganisms by Pasteur's experiment.
6. The components in the atmosphere for the Miller–Urey experiment on prebiotic synthesis did not include
   *a.* $H_2$.
   *b.* $H_2O$.
   *c.* $O_2$.
   *d.* $NH_3$.
   *e.* $CH_4$.

7. All of the major building blocks of macromolecules were made in Miller–Urey prebiotic synthesis experiments *except*
   *a.* amino acids.
   *b.* hexose sugars.
   *c.* bases for nucleotides.
   *d.* fatty acids.
   *e.* ribose.
8. The "RNA world" hypothesis proposes that
   *a.* RNA formed from DNA.
   *b.* RNA was both a catalyst and genetic material.
   *c.* RNA was a catalyst only.
   *d.* RNA formed after proteins.
   *e.* DNA formed after RNA was broken down.
9. Ribozymes are
   *a.* enzymes that are made up of ribose sugar.
   *b.* ancient catalysts that no longer exist.
   *c.* RNA catalysts.
   *d.* present in bacterial cells only.
   *e.* less active than protein enzymes.
10. Findings in ancient rocks indicate cells first appeared
    *a.* about 4.5 billion years ago.
    *b.* about 3.5 billion years ago.
    *c.* about 2 billion years ago.
    *d.* before rocks were formed.
    *e.* before water arrived on Earth.

## FOR DISCUSSION

1. Are the statements "all life comes from pre-existing life" and "life on Earth could have arisen from prebiotic molecules" truly paradoxical? What conditions existing on Earth today might preclude the origin of life from such molecules?
2. Why might RNA have preceded proteins in the evolution of biological macromolecules?
3. Do you consider the two alternative theories presented in this chapter as possible explanations of the origin of life on Earth (that life came from outside of Earth, or that life arose on Earth through chemical evolution) to be equally plausible? Which do you favor, and why?
4. Why was the evolution of a self-contained cell essential for life as we know it?

## ADDITIONAL INVESTIGATION

1. The interpretation of Pasteur's experiment (see Figure 4.7) depended on the inactivation of microorganisms by heat. We now know of microorganisms that can survive extremely high temperatures (see Chapter 26). Does this change the interpretation of Pasteur's experiment? What experiments would you do to inactivate such microbes?
2. The Miller–Urey experiment (see Figure 4.9) showed that it was possible for amino acids to be formed from gases that were hypothesized to have been in Earth's early atmosphere. These amino acids were dissolved in water. Knowing what you do about the polymerization of amino acids into proteins (see Figure 3.6), how would you set up experiments to show that proteins can form under the conditions of early Earth? What properties would you expect of those proteins?

## WORKING WITH DATA (GO TO yourBioPortal.com)

**Synthesis of Prebiotic Molecules in an Experimental Atmosphere** In this hands-on exercise, you will examine the original research paper of Miller and Urey to see the experimental approach they used to show that amino acids could be made in a simulation of Earth's early atmosphere (Figure 4.9). You will also analyze more recent data using the same apparatus.

**Disproving the Spontaneous Generation of Life** In this hands-on exercise, you will examine data from an experiment similar to Pasteur's famous experiments (Figure 4.7). By calculating growth rates in the different flasks, you will be able to see how Pasteur came to the conclusion he did.

# 5 Cells: The Working Units of Life

## How to mend a broken heart

It is a day in the not-too-distant future. Decades of eating fatty foods, combined with an inherited tendency to deposit cholesterol in his arteries, have finally caught up with 70-year-old Don. A blood clot has closed off blood flow to part of his heart, leading to a heart attack and severe damage to that vital organ.

If this had happened today, Don would have been faced with a long period of rehabilitation, taking medications to manage his weakened heart. Instead, his physicians take a pinch of skin tissue from his arm and bring it to a laboratory. After certain DNA sequences are added, Don's skin cells no longer look and act like skin cells: They are undifferentiated (unspecialized) and reproduce continuously in the laboratory dish. These cells are also multipotent stem cells, able to differentiate into almost any type of cell in the body if given the right environment. When they are injected directly into Don's heart, his stem cells soon become heart muscle cells, repairing the damage caused by the heart attack. Don leaves the hospital with full cardiac function and recommendations for a healthy diet.

You are probably familiar with another type of multi–potent cell, the fertilized human egg. This single cell ultimately produces the tens of trillions of cells that make up the human body. The fertilized egg is programmed to generate an entire organism—not just the heart and skin, but blood, nerves, liver, brain, and even bones—and for this reason is called totipotent ("toti" means all; "multi" means most). In contrast, the stem cells derived from Don's skin need specific external signals to differentiate into other kinds of cells, and could not develop into an entire person.

The potential uses of stem cells in medicine have generated a lot of excitement in recent years. Such widely read periodicals as *Time* have hailed advances in stem cell research as "breakthroughs of the year." Patients with the neurological disorder Parkinson's disease dream of the day when their skin cells can be turned into brain cells to fix their damaged nervous systems. People with diabetes hope for stem cells to repair their pancreases. The list is long.

Behind all of this hope and the research it inspires is a cornerstone of biological science: the cell theory. As you saw in the last

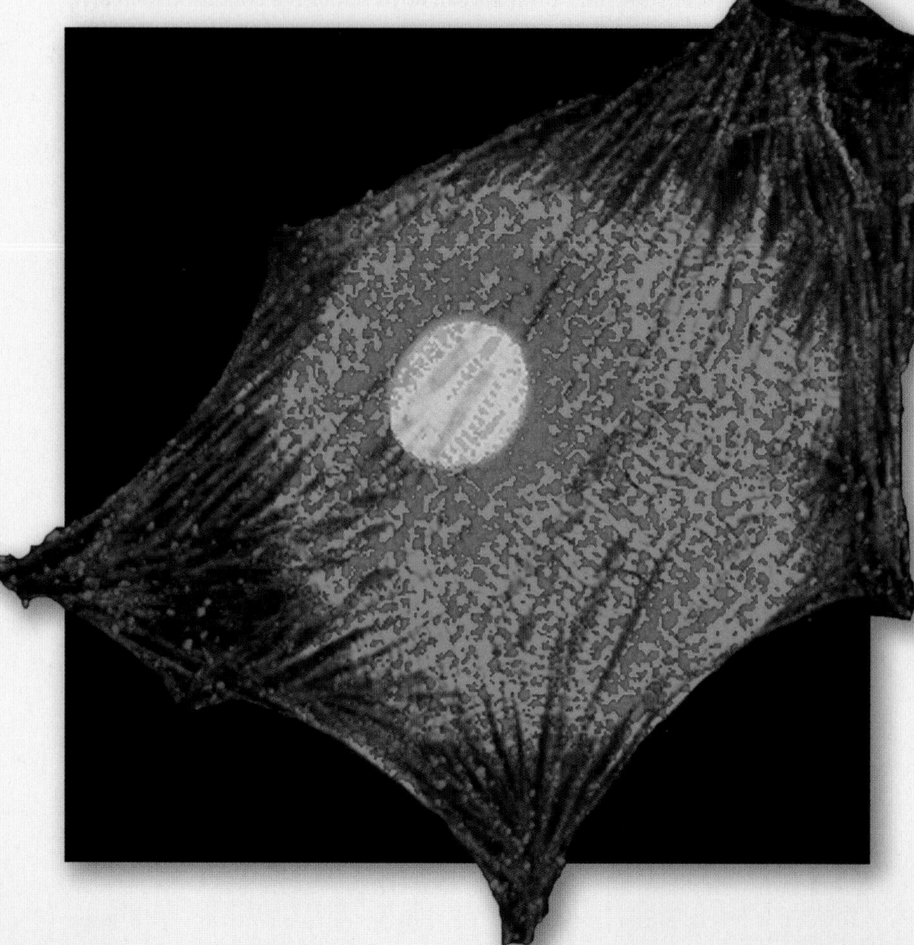

**A New Heart Cell** This cardiac stem cell is developing into a fully differentiated heart cell. The hope is to be able to coax stem cells to follow this path or to produce other cell types to repair damaged tissues.

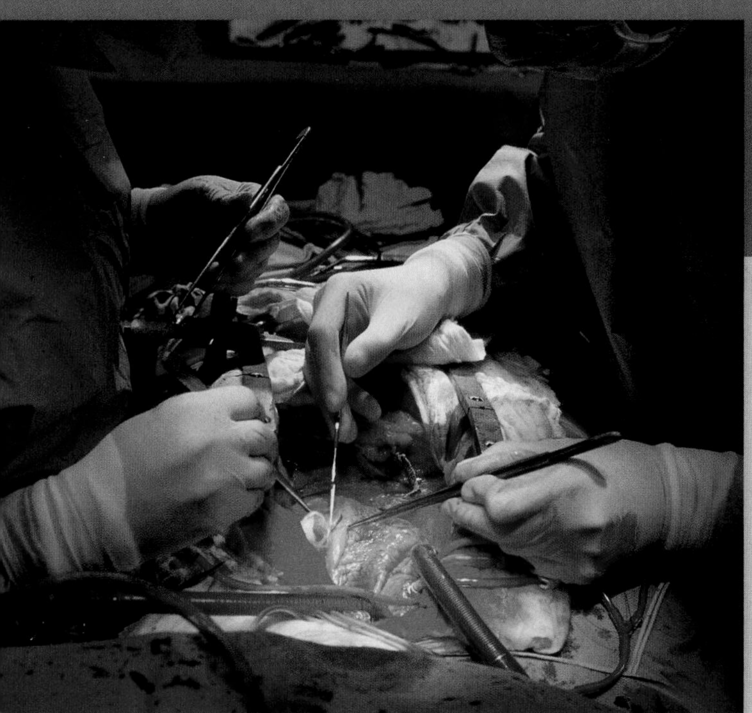

**Open Heart Surgery** Stem cell therapies may provide alternative approaches to treating heart disease in the future.

chapter, a key event in the emergence of life was the enclosure of biochemical reactions inside a cell, thus concentrating them and separating them from the external environment. These are the first two tenets of the cell theory, that the cell is the unit of life and that the activities of life either happen inside cells or are caused by them. Don's stem cells contain not just the activities of a living entity, but also the potential to change those activities in new directions. The third tenet of the cell theory—equally important—is that the cell is the unit of reproduction: all cells come from pre-existing cells. Stem cell therapy does not create new cells out of thin air; it coaxes existing ones to differentiate and reproduce along the desired path.

**IN THIS CHAPTER** we examine the structure and some of the functions of cells. We will begin with a fuller explanation of cell theory. Then, we will examine the relatively simple cells of prokaryotes. This is followed by a tour of the more complex eukaryotic cell and its various internal compartments, each of which performs specific functions. Finally, we discuss ideas on how complex cells evolved.

## CHAPTER OUTLINE



## 5.1 What Features Make Cells the Fundamental Units of Life?

In Chapter 1 we introduced some of the characteristics of life: chemical complexity, growth and reproduction, the ability to refashion substances from the environment, and the ability to determine what substances can move into and out of the organism. These characteristics are all demonstrated by cells. Just as atoms are the building blocks of chemistry, cells are the building blocks of life.

The **cell theory** is described in Section 1.1 as the first unifying principle of biology. There are three critical components of the cell theory:

- Cells are the fundamental units of life.
- All living organisms are composed of cells.
- All cells come from preexisting cells.

Cells contain water and the other small and large molecules, which we examined in Chapters 2–4. Each cell contains at least 10,000 different types of molecules, most of them present in many copies. Cells use these molecules to transform matter and energy, to respond to their environments, and to reproduce themselves.

The cell theory has three important implications:

- Studying cell biology is in some sense the same as studying life. The principles that underlie the functions of the single cell of a bacterium are similar to those governing the approximately 60 trillion cells of your body.
- Life is continuous. All those cells in your body came from a single cell, a fertilized egg. That egg came from the fusion of two cells, a sperm and an egg, from your parents. The cells of your parents' bodies were all derived from their parents, and so on back through generations and evolution to the first living cell.
- The origin of life on Earth was marked by the origin of the first cells (see Chapter 4).

Even the largest creatures on Earth are composed of cells, but the cells themselves are usually too small for the naked eye to see. Why are cells so small?

### Cell size is limited by the surface area-to-volume ratio

Most cells are tiny. In 1665, the early microscopist Robert Hooke estimated that in one square inch of cork, which he examined under his magnifying lens, there were 1,259,712,000 cells! The volumes of cells range from 1 to 1,000 cubic micrometers ($\mu m^3$). There are some exceptions: the eggs of birds are single cells that

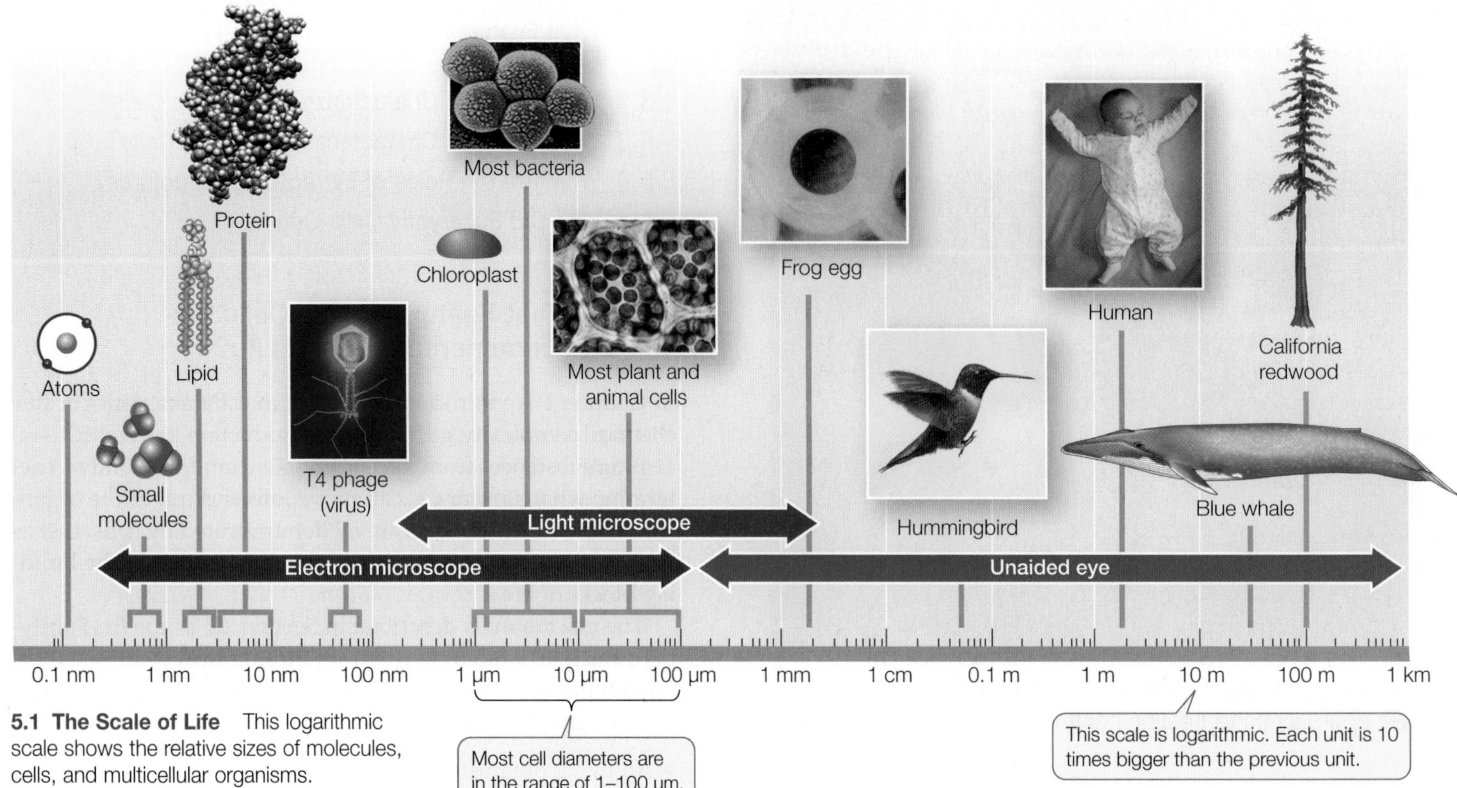


**5.1 The Scale of Life** This logarithmic scale shows the relative sizes of molecules, cells, and multicellular organisms.

**yourBioPortal.com**
GO TO **Web Activity 5.1 • The Scale of Life**

are, relatively speaking, enormous, and individual cells of several types of algae and bacteria are large enough to be viewed with the unaided eye (**Figure 5.1**). And although neurons (nerve cells) have volumes that are within the "usual" range, they often have fine projections that may extend for meters, carrying signals from one part of a large animal to another. So there is enormous diversity among cells in their dimensions and volumes, but cells are usually very small.

Small cell size is a practical necessity arising from the change in the **surface area-to-volume ratio** of any object as it increases in size. As an object increases in volume, its surface area also increases, but not at the same rate (**Figure 5.2**). This phenomenon has great biological significance for two reasons:

- The volume of a cell determines the amount of chemical activity it carries out per unit of time.
- The surface area of a cell determines the amount of substances that can enter it from the outside environment, and the amount of waste products that can exit to the environment.

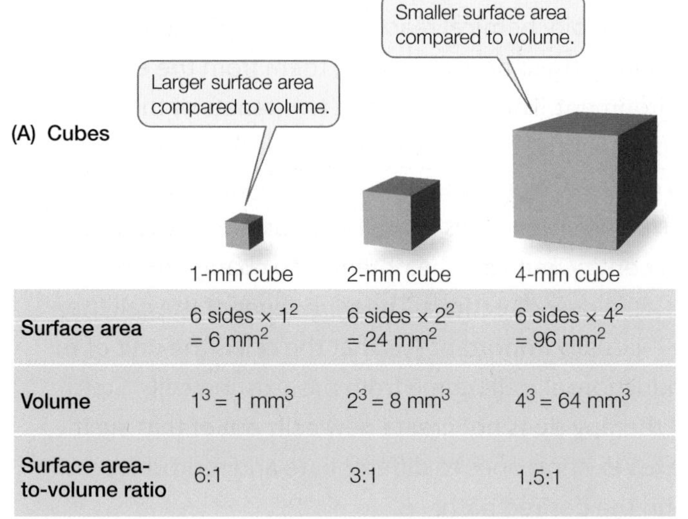


| | 1-mm cube | 2-mm cube | 4-mm cube |
|---|---|---|---|
| **Surface area** | 6 sides × $1^2$ = 6 $mm^2$ | 6 sides × $2^2$ = 24 $mm^2$ | 6 sides × $4^2$ = 96 $mm^2$ |
| **Volume** | $1^3$ = 1 $mm^3$ | $2^3$ = 8 $mm^3$ | $4^3$ = 64 $mm^3$ |
| **Surface area-to-volume ratio** | 6:1 | 3:1 | 1.5:1 |

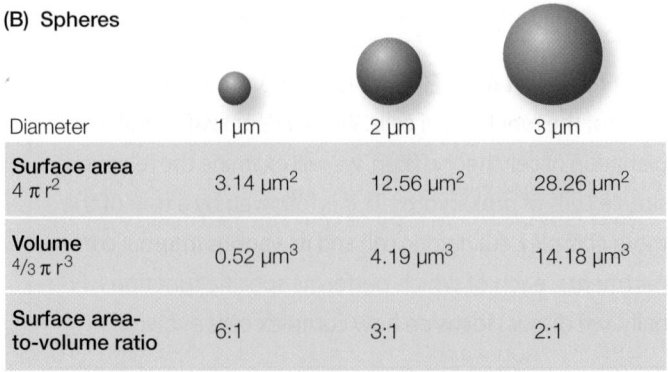

(B) Spheres

| Diameter | 1 µm | 2 µm | 3 µm |
|---|---|---|---|
| **Surface area** $4 \pi r^2$ | 3.14 $\mu m^2$ | 12.56 $\mu m^2$ | 28.26 $\mu m^2$ |
| **Volume** $4/3 \pi r^3$ | 0.52 $\mu m^3$ | 4.19 $\mu m^3$ | 14.18 $\mu m^3$ |
| **Surface area-to-volume ratio** | 6:1 | 3:1 | 2:1 |

**5.2 Why Cells Are Small** Whether it is cuboid (A) or spheroid (B), as an object grows larger its volume increases more rapidly than its surface area. Cells must maintain a large surface area-to-volume ratio in order to function. This fact explains why large organisms must be composed of many small cells rather than a few huge ones.

As a living cell grows larger, its chemical activity, and thus its need for resources and its rate of waste production, increases faster than its surface area. (The surface area increases in proportion to the square of the radius, while the volume increases much more—in proportion to the cube of the radius.) In addition, substances must move from one site to another within the cell; the smaller the cell, the more easily this is accomplished. This explains why large organisms must consist of many small cells: cells must be small in volume in order to maintain a large enough surface area-to-volume ratio and an ideal internal volume. The large surface area represented by the many small cells of a multicellular organism enables it to carry out the many different functions required for survival.

### Microscopes reveal the features of cells

Microscopes do two different things to allow cells and details within them to be seen by the human eye. First, they increase the apparent size of the object: this is called *magnification*. But just increasing the magnification does not necessarily mean that the object will be seen clearly. In addition to being larger, a magnified object must be sharp, or clear. This is a property called *resolution*. Formally defined, resolution is the minimum distance two objects can be apart and still be seen as two objects. Resolution for the human eye is about 0.2 mm (200 μm). Most cells are much smaller than 200 μm, and thus are invisible to the human eye. Microscopes magnify and increase resolution so that cells and their internal structures can be seen clearly (**Figure 5.3**).

There are two basic types of microscopes—*light microscopes* and *electron microscopes*—that use different forms of radiation (see Figure 5.3). While the resolution is better in electron microscopy, we should emphasize that because cells are prepared in a vacuum, only dead, dehydrated cells are visualized. Therefore, the preparation of cells for electron microscopy may alter them, and this must be taken into consideration when interpreting the images produced. On the other hand, light microscopes can be used to visualize living cells (for example, by phase-contrast microscopy; see Figure 5.3).

Before we delve into the details of cell structure, it is useful to consider the many uses of microscopy. An entire branch of medicine, *pathology*, makes use of many different methods of microscopy to aid in the analysis of cells and the diagnosis of diseases. For instance, a surgeon might remove from a body some tissue suspected of being cancerous. The pathologist might:

- examine the tissue quickly by phase-contrast microscopy or interference-contrast microscopy to determine the size, shape, and spread of the cells
- stain the tissue with a general dye and examine it by bright-field microscopy to bring out features such as the shape of the nucleus, or cell division characteristics
- stain the tissue with a fluorescent dye and examine it by fluorescence microscopy or confocal microscopy for the presence of specific proteins that are diagnostic of a particular cancer
- examine the tissue under the electron microscope to observe its most minute internal structures, such as the shapes of the mitochondria and the chromatin. (These structures are described in Section 5.3.)

### The plasma membrane forms the outer surface of every cell

While the structural diversity of cells can often be observed using light microscopy, the **plasma membrane** is best observed with an electron microscope. This very thin structure forms the outer surface of every cell, and it has more or less the same thickness and molecular structure in all cells. Biochemical methods have shown that membranes have great functional diversity. These methods have revealed that the thin, almost invisible plasma membrane is actively involved in many cellular functions—it is not a static structure. The plasma membrane separates the interior of the cell from its outside environment, creating a segregated (but not isolated) compartment. The presence of this outer limiting membrane is a feature of all cells. What is the composition and molecular architecture of this amazing structure?

The plasma membrane is composed of a *phospholipid bilayer* (or simply *lipid bilayer*), with the hydrophilic "heads" of the lipids facing the cell's aqueous interior on one side of the membrane and the extracellular environment on the other (see Figure 3.20). Proteins and other molecules are embedded in the lipids. The membrane is not a rigid, static structure. Rather, it is an oily fluid, in which the proteins and lipids are in constant motion. This allows the membrane to move and change the shape of the cell. A detailed description of the structure and functions of the plasma membrane is given in Chapter 6. Here is a brief summary:

- The plasma membrane acts as a *selectively permeable barrier*, preventing some substances from crossing it while permitting other substances to enter and leave the cell. For example, macromolecules such as DNA and proteins cannot normally cross the plasma membrane, but some smaller molecules such as oxygen can. In addition to size, other factors such as polarity determine a molecule's ability to cross the plasma membrane: because the membrane is composed mostly of hydrophobic fatty acids, nonpolar molecules cross it more easily than polar or charged molecules.
- The plasma membrane allows the cell to maintain a more or less *constant internal environment*. A self-maintaining, constant internal environment (known as *homeostasis*) is a key characteristic of life that will be discussed in detail in Chapter 40. One way that the membrane does this is by actively regulating the transport of substances across it. This dynamic process is distinct from the more passive process of diffusion, which is dependent on the size of a molecule.
- As the cell's boundary with the outside environment, the plasma membrane is important in *communicating* with adjacent cells and receiving signals from the environment. We will describe this function in Chapter 7.
- The plasma membrane often has proteins protruding from it that are responsible for *binding and adhering* to adjacent

## TOOLS FOR INVESTIGATING LIFE

### 5.3 Looking at Cells

The six images on this page show some techniques used in light microscopy. The three images on the following page were created using electron microscopes. All of these images are of a particular type of cultured cell known as HeLa cells. Note that the images in most cases are flat, two-dimensional views. As you look at images of cells, keep in mind that they are three-dimensional structures.

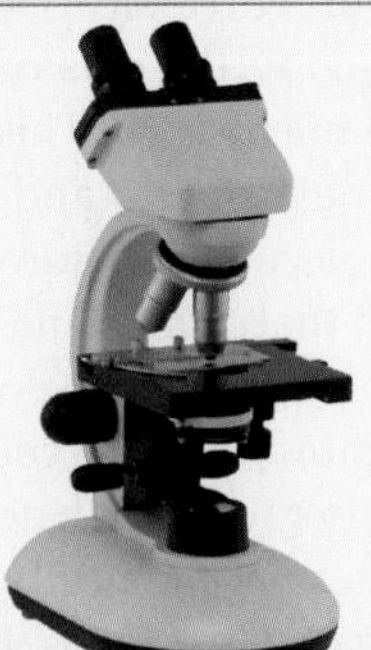

In a *light microscope*, glass lenses and visible light are used to form an image. The resolution is about 0.2 μm, which is 1,000 times greater than that of the human eye. Light microscopy allows visualization of cell sizes and shapes and some internal cell structures. Internal structures are hard to see under visible light, so cells are often chemically treated and stained with various dyes to make certain structures stand out by increasing contrast.

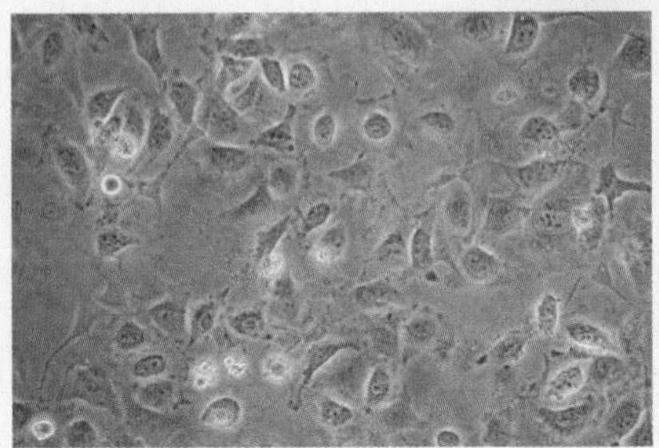


In **bright-field microscopy**, light passes directly through these human cells. Unless natural pigments are present, there is little contrast and details are not distinguished.

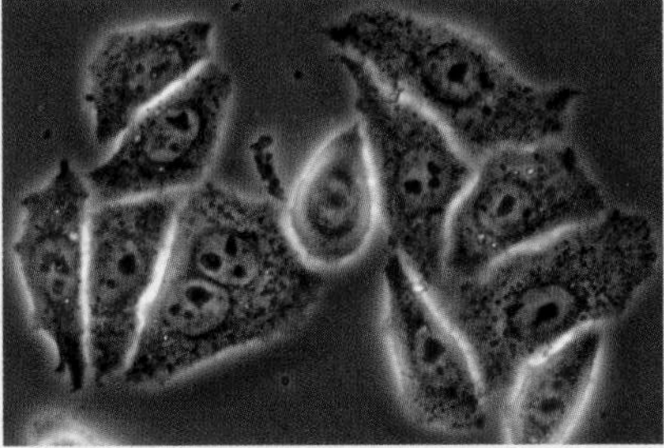


In **phase-contrast microscopy**, contrast in the image is increased by emphasizing differences in refractive index (the capacity to bend light), thereby enhancing light and dark regions in the cell.

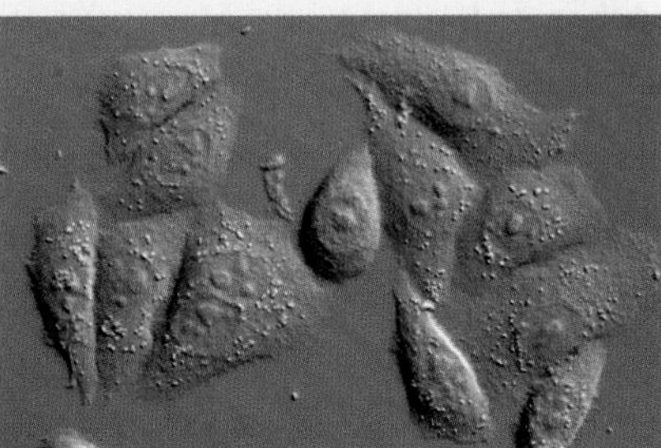


**Differential interference-contrast microscopy** uses two beams of polarized light. The combined images look as if the cell is casting a shadow on one side.

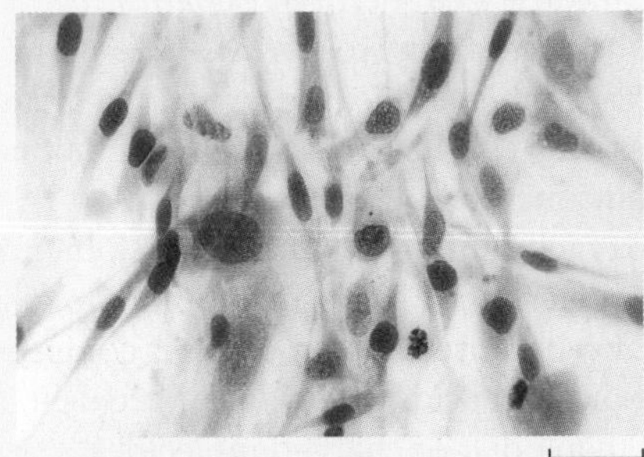


In **stained bright-field microscopy**, a stain enhances contrast and reveals details not otherwise visible. Stains differ greatly in their chemistry and their capacity to bind to cell materials, so many choices are available.

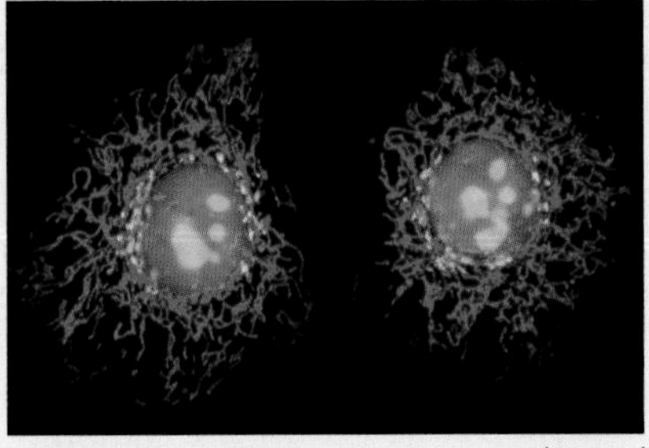


In **fluorescence microscopy**, a natural substance in the cell or a fluorescent dye that binds to a specific cell material is stimulated by a beam of light, and the longer-wavelength fluorescent light is observed coming directly from the dye.

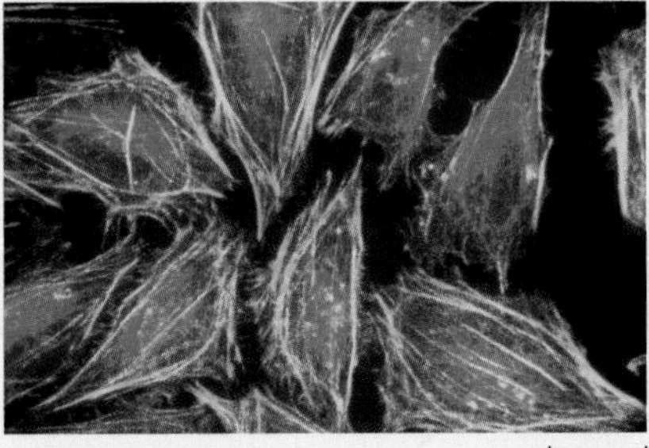


**Confocal microscopy** uses fluorescent materials but adds a system of focusing both the stimulating and emitted light so that a single plane through the cell is seen. The result is a sharper two-dimensional image than with standard fluorescence microscopy.

cells. Thus the plasma membrane plays an important structural role and contributes to cell shape.

### All cells are classified as either prokaryotic or eukaryotic

As we learned in Section 1.2, biologists classify all living things into three domains: Archaea, Bacteria, and Eukarya. The organisms in Archaea and Bacteria are collectively called **prokaryotes** because they have in common a prokaryotic cell organization. A prokaryotic cell does not typically have membrane-enclosed internal compartments; in particular, it does not have a nucleus. The first cells were probably similar in organization to those of modern prokaryotes.

## TOOLS FOR INVESTIGATING LIFE

### 5.3 Looking at Cells (*continued*)

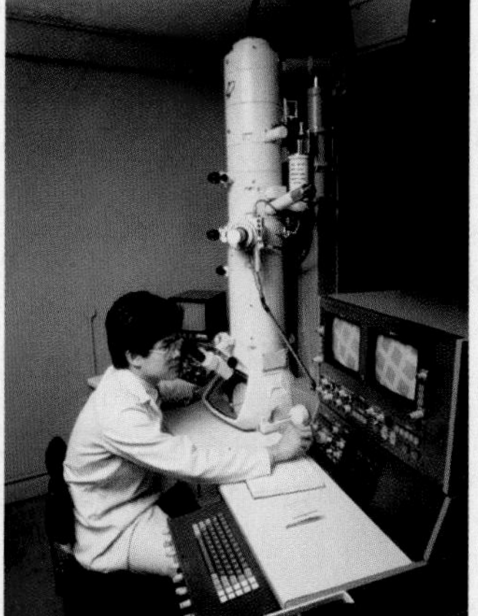

In an *electron microscope*, electromagnets are used to focus an electron beam, much as a light microscope uses glass lenses to focus a beam of light. Since we cannot see electrons, the electron microscope directs them through a vacuum at a fluorescent screen or photographic film to create a visible image. The resolution of electron microscopes is about 2 nm, which is about 100,000 times greater than that of the human eye. This resolution permits the details of many subcellular structures to be distinguished.

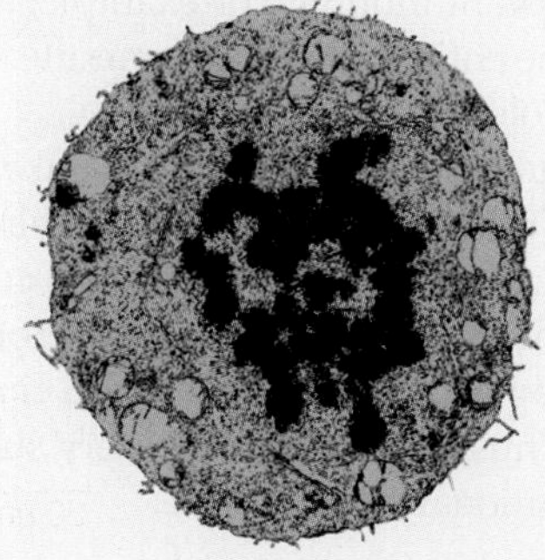


In **transmission electron microscopy** (TEM), a beam of electrons is focused on the object by magnets. Objects appear darker if they absorb the electrons. If the electrons pass through they are detected on a fluorescent screen.

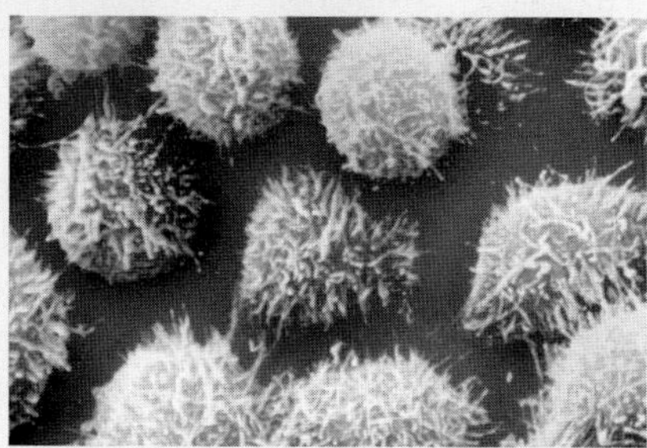


**Scanning electron microscopy** (SEM) directs electrons to the surface of the sample, where they cause other electrons to be emitted. These electrons are viewed on a screen. The three-dimensional surface of the object can be visualized.

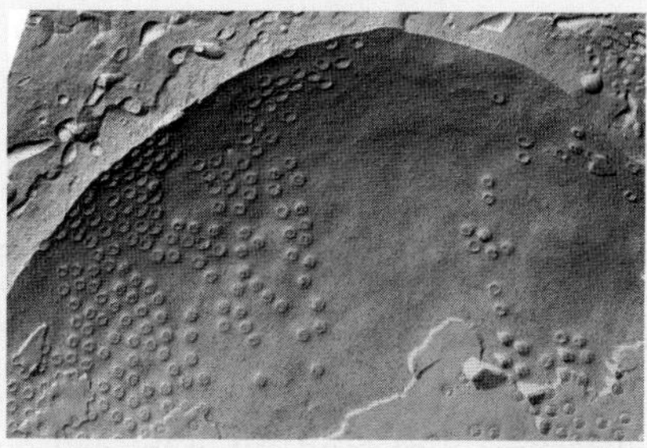


In **freeze-fracture microscopy**, cells are frozen and then a knife is used to crack them open. The crack often passes through the interior of plasma and internal membranes. The "bumps" that appear are usually large proteins or aggregates embedded in the interior of the membrane.

**yourBioPortal.com**
GO TO Web Activity 5.2 • Know Your Techniques

Eukaryotic cell organization, on the other hand, is found in members of the domain Eukarya (**eukaryotes**), which includes the protists, plants, fungi, and animals. As we will discuss later in this chapter, eukaryotic cells probably evolved from prokaryotes. In contrast to the prokaryotes, the genetic material (DNA) of eukaryotic cells is contained in a special membrane-enclosed compartment called the **nucleus**. Eukaryotic cells also contain other membrane-enclosed compartments in which specific chemical reactions occur. For example, some of the key reactions that generate usable chemical energy for cells take place in mitochondria. The internal membranes that enclose these compartments have the same basic composition, structure and properties as the plasma membrane. The efficiency afforded by these compartments has led to the impressive functions that can occur in eukaryotic cells, and their specialization into tissues as diverse as the parts of a flower, muscles, and nerves.

## 5.1 RECAP

**The cell theory is a unifying principle of biology. Surface area-to-volume ratios limit the sizes of cells. Both prokaryotic and eukaryotic cells are enclosed within a plasma membrane, but prokaryotic cells lack the membrane-enclosed internal compartments found in eukaryotes.**

- How does cell biology embody all the principles of life? See p. 77
- Why are cells small? See pp. 77–79 and Figure 5.2
- Explain the importance of the plasma membrane to cells. See pp. 79–80

As we mentioned in this section, there are two structural themes in cell architecture: prokaryotic and eukaryotic. We now turn to the organization of prokaryotic cells.

## 5.2 What Features Characterize Prokaryotic Cells?

Prokaryotes can derive energy from more diverse sources than any other living organisms. They can tolerate environmental extremes—such as very hot springs with temperatures up to 100°C (*Thermus aquaticus*) or very salty water (*Halobacterium*)—that would kill other organisms. As we examine prokaryotic cells in this section, bear in mind that there are vast numbers of prokaryotic species, and that the Bacteria and Archaea are distinguished in numerous ways. These differences, and the vast diversity of organisms in these two domains, will be the subject of Chapter 26.

The volume of a prokaryotic cell is generally about one fiftieth of the volume of a eukaryotic cell. Prokaryotic cells range from about 1 to 10 μm in length or diameter. Each individual prokaryote is a single cell, but many types of prokaryotes are usually seen in chains or small clusters, and some occur in large clusters containing hundreds of cells. In this section we will first consider the features shared by cells in the domains Bacteria and Archaea. Then we will examine structural features that are found in some, but not all, prokaryotes.

### Prokaryotic cells share certain features

All prokaryotic cells have the same basic structure (**Figure 5.4**):

- The plasma membrane encloses the cell, regulating the traffic of materials into and out of the cell, and separating its interior from the external environment.
- The **nucleoid** is a region in the cell where the DNA is located. As we described in Section 4.1, DNA is the hereditary material that controls cell growth, maintenance, and reproduction.

The rest of the material enclosed in the plasma membrane is called the **cytoplasm**. The cytoplasm has two components: the cytosol and insoluble suspended particles, including ribosomes:

- The **cytosol** consists mostly of water that contains dissolved ions, small molecules, and soluble macromolecules such as proteins.
- **Ribosomes** are complexes of RNA and proteins that are about 25 nm in diameter. They can only be visualized with the electron microscope. They are the sites of protein synthesis, where information coded for in nucleic acids directs the sequential linking of amino acids to form proteins.

The cytoplasm is not a static region. Rather, the substances in this environment are in constant motion. For example, a typical protein moves around the entire cell within a minute, and it collides with many other molecules along the way.

Although they are structurally less complex than eukaryotic cells, prokaryotic cells are functionally complex, carrying out thousands of biochemical reactions. Based on our current knowledge about the origins of the first cells (see Section 4.4), some prokaryotic cell lineages must stretch back in time for more than 3 billion years. Thus, prokaryotes are very successful organisms from an evolutionary perspective.

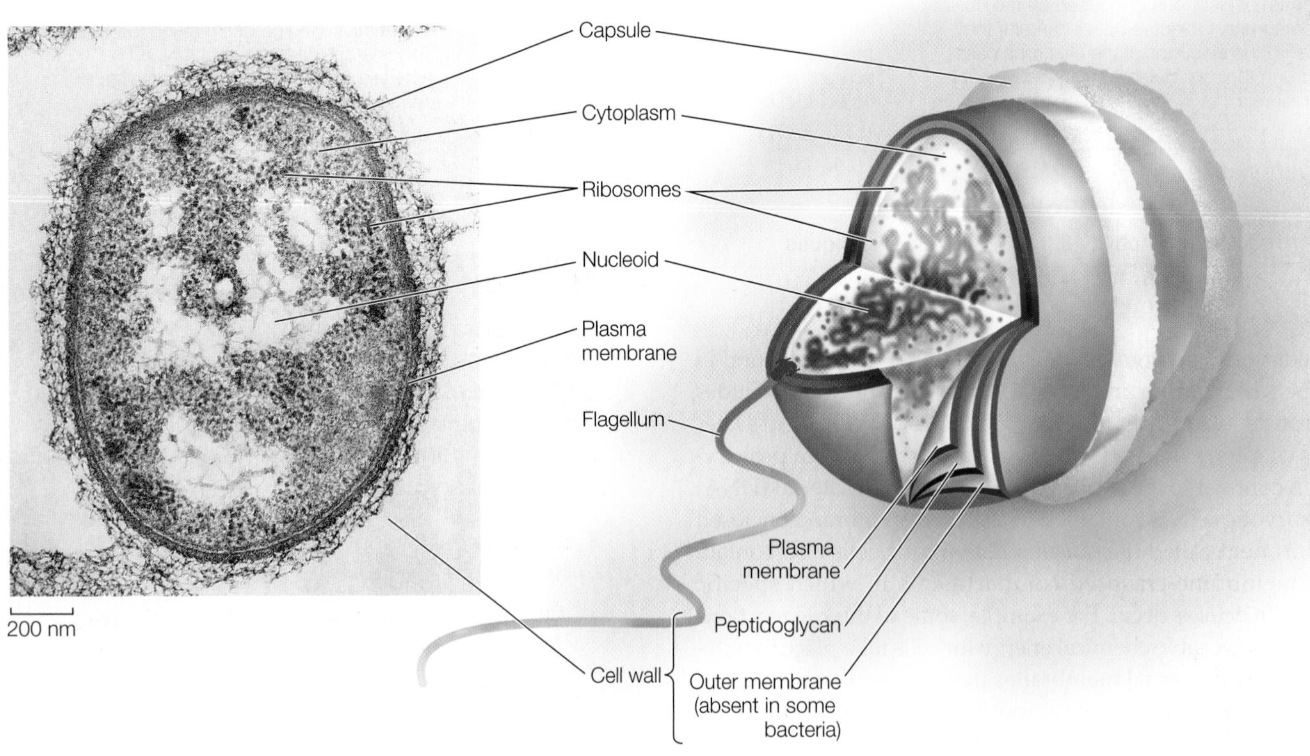


**5.4 A Prokaryotic Cell** The bacterium *Pseudomonas aeruginosa* illustrates the typical structures shared by all prokaryotic cells. This bacterium also has a protective outer membrane that not all prokaryotes have. The flagellum and capsule are also structures found in some, but not all, prokaryotic cells.

### Specialized features are found in some prokaryotes

As they evolved, some prokaryotes developed specialized structures that gave a selective advantage to those that had them: cells with these structures were better able to survive and reproduce in particular environments than cells lacking them. These structures include a protective cell wall, an internal membrane for compartmentalization of some chemical reactions, flagella for cell movement through the watery environment, and a rudimentary internal skeleton.

**CELL WALLS** Most prokaryotes have a cell wall located outside the plasma membrane. The rigidity of the cell wall supports the cell and determines its shape. The cell walls of most bacteria, but not archaea, contain peptidoglycan, a polymer of amino sugars that are cross-linked by covalent bonds to peptides, to form a single giant molecule around the entire cell. In some bacteria, another layer, the outer membrane (a polysaccharide-rich phospholipid membrane), encloses the peptidoglycan layer (see Figure 5.4). Unlike the plasma membrane, this outer membrane is not a major barrier to the movement of molecules across it.

Enclosing the cell wall in some bacteria is a slimy layer composed mostly of polysaccharides, and referred to as a capsule. In some cases these capsules protect the bacteria from attack by white blood cells in the animals they infect. Capsules also help to keep the cells from drying out, and sometimes they help bacteria attach to other cells. Many prokaryotes produce no capsule, and those that do have capsules can survive even if they lose them, so the capsule is not essential to prokaryotic life.

As you will see later in this chapter, eukaryotic plant cells also have a cell wall, but it differs in composition and structure from the cell walls of prokaryotes.

**INTERNAL MEMBRANES** Some groups of bacteria—including the cyanobacteria—carry out photosynthesis: they use energy from the sun to convert carbon dioxide and water into carbohydrates. These bacteria have an internal membrane system that contains molecules needed for photosynthesis. The development of photosynthesis, which requires membranes, was an important event in the early evolution of life on Earth. Other prokaryotes have internal membrane folds that are attached to the plasma membrane. These folds may function in cell division or in various energy-releasing reactions.

**FLAGELLA AND PILI** Some prokaryotes swim by using appendages called **flagella**, which sometimes look like tiny corkscrews (**Figure 5.5A**). In bacteria a single flagellum is made of a protein called flagellin. A complex motor protein spins the flagellum on its axis like a propeller, driving the cell along. The motor protein is anchored to the plasma membrane and, in some bacteria, to the outer membrane of the cell wall (**Figure 5.5B**). We know that the flagella cause the motion of cells because if they are removed, the cells do not move.

**Pili** are structures made of protein that project from the surfaces of some types of bacterial cells. These hairlike structures are shorter than flagella, and are used for adherence. The sex-pili help bacteria join to one another to exchange genetic material. The *fimbriae* are similar to pili but shorter, and help cells to adhere to surfaces such as animal cells, for food and protection.

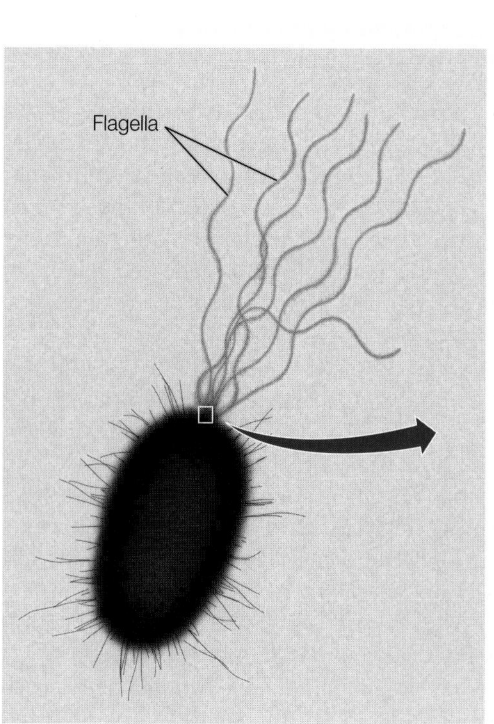


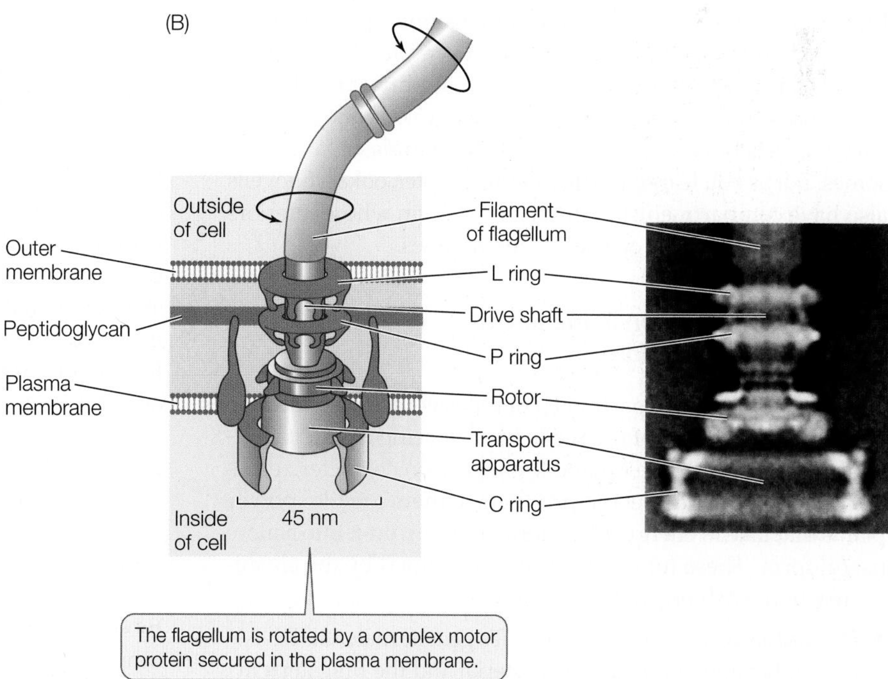


**5.5 Prokaryotic Flagella** (A) Flagella contribute to the movement and adhesion of prokaryotic cells. (B) Complex protein ring structures anchored in the plasma membrane form a motor unit that rotates the flagellum and propels the cell.

**CYTOSKELETON** Some prokaryotes, especially rod-shaped bacteria, have a helical network of filamentous structures that extend down the length of the cell just inside the plasma membrane. The proteins that make up this structure are similar in amino acid sequence to actin in eukaryotic cells. Since actin is part of the cytoskeleton in eukaryotes (see Section 5.3), it has been suggested that the helical filaments in prokaryotes play a role in maintaining the rod-like cell shape.

## 5.2 RECAP

**Prokaryotic organisms can live on diverse energy sources and in extreme environments. Unlike eukaryotic cells, prokaryotic cells do not have extensive internal compartments.**

- What structures are present in all prokaryotic cells? See p. 82 and Figure 5.4
- Describe the structure and function of a specialized prokaryotic cell feature, such as the cell wall, capsule, flagellum, or pilus. See pp. 83–84 and Figure 5.5

As we mentioned earlier, the prokaryotic cell is one of two types of cell structure recognized in cell biology. The other is the eukaryotic cell. Eukaryotic cells, and multicellular eukaryotic organisms, are more structurally and functionally complex than prokaryotic cells.

## 5.3 What Features Characterize Eukaryotic Cells?

Eukaryotic cells generally have dimensions up to 10 times greater than those of prokaryotes; for example, the spherical yeast cell has a diameter of about 8 μm, in contrast to a typical bacterium with a diameter of 1 μm. Like prokaryotic cells, eukaryotic cells have a plasma membrane, cytoplasm, and ribosomes. But as you learned earlier in this chapter, eukaryotic cells also have compartments within the cytoplasm whose interiors are separated from the cytosol by membranes.

### Compartmentalization is the key to eukaryotic cell function

The membranous compartments of eukaryotic cells are called **organelles**. Each type of organelle has a specific role in its particular cell. Some of the organelles have been characterized as factories that make specific products. Others are like power plants that take in energy in one form and convert it into a more useful form. These functional roles are defined by the chemical reactions each organelle can carry out:

- The *nucleus* contains most of the cell's genetic material (DNA). The replication of the genetic material and the first steps in expressing genetic information take place in the nucleus.
- The *mitochondrion* is a power plant and industrial park, where energy stored in the bonds of carbohydrates and fatty acids is converted into a form that is more useful to the cell (ATP; see Section 9.1).
- The *endoplasmic reticulum* and *Golgi apparatus* are compartments in which some proteins synthesized by the ribosomes are packaged and sent to appropriate locations in the cell.
- *Lysosomes* and vacuoles are cellular digestive systems in which large molecules are hydrolyzed into usable monomers.
- *Chloroplasts* (found in only some cells) perform photosynthesis.

The membrane surrounding each organelle has two essential roles. First, it keeps the organelle's molecules away from other molecules in the cell, to prevent inappropriate reactions. Second, it acts as a traffic regulator, letting important raw materials into the organelle and releasing its products to the cytoplasm. In some organelles, the membrane also has proteins that have functional roles in chemical reactions that occur at the organelle surface.

There are a number of other structures in eukaryotic cells that have specialized functions, but are not generally called organelles because they lack membranes:

- Ribosomes, where protein synthesis takes place
- The cytoskeleton, composed of several types of protein-based filaments, which has both structural and functional roles
- The extracellular matrix, which also has structural and functional roles

The evolution of compartments was an important development that enabled eukaryotic cells to specialize, forming the organs and tissues of complex multicellular organisms.

### Organelles can be studied by microscopy or isolated for chemical analysis

Cell organelles and structures were first detected by light and then by electron microscopy. The functions of the organelles could sometimes be inferred by observations and experiments, leading, for example, to the hypothesis (later confirmed) that the nucleus contained the genetic material. Later, the use of stains targeted to specific macromolecules allowed cell biologists to determine the chemical compositions of organelles (see Figure 5.17, which shows a single cell stained for three different proteins).

Another way to analyze cells is to take them apart in a process called cell fractionation. This process permits cell organelles and other cytoplasmic structures to be separated from each other and examined using chemical methods. Cell fractionation begins with the destruction of the plasma membrane, which allows the cytoplasmic components to flow out into a test tube. The various organelles can then be separated from one another on the basis of size or density (**Figure 5.6**). Biochemical analyses can then be done on the isolated organelles.

Microscopy and cell fractionation have complemented each other, giving us a more complete picture of the composition and function of each organelle and structure.

TOOLS FOR INVESTIGATING LIFE

**5.6 Cell Fractionation**

Organelles can be separated from one another after cells are broken open and their contents suspended in an aqueous medium. The medium is placed in a tube and spun in a centrifuge, which rotates about an axis at high speed. Centrifugal forces cause particles to sediment at the bottom of the tube where they may be collected for biochemical study. Heavier particles sediment at lower speeds than do lighter particles. By adjusting the speed of centrifugation, cellular organelles and even large particles like ribosomes can be separated and partially purified.

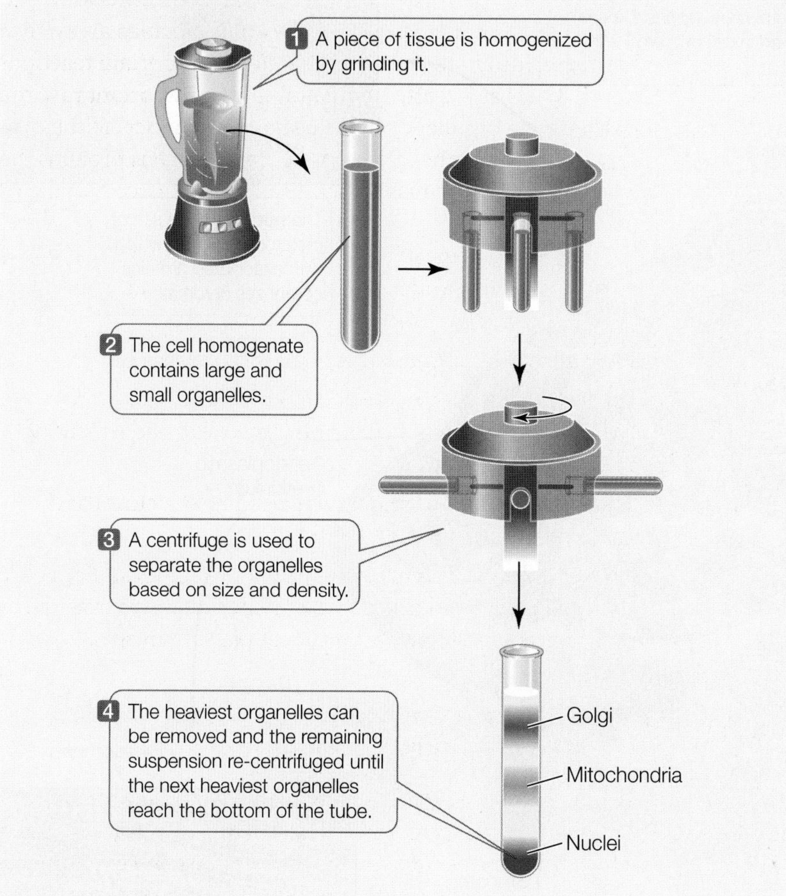

1 A piece of tissue is homogenized by grinding it.
2 The cell homogenate contains large and small organelles.
3 A centrifuge is used to separate the organelles based on size and density.
4 The heaviest organelles can be removed and the remaining suspension re-centrifuged until the next heaviest organelles reach the bottom of the tube.
Golgi
Mitochondria
Nuclei

Microscopy of plant and animal cells has revealed that many of the organelles are similar in appearance in each cell type (**Figure 5.7**). By comparing the illustrations in Figure 5.7 and Figure 5.4 you can see some of the prominent differences between eukaryotic cells and prokaryotic cells.

### Ribosomes are factories for protein synthesis

The ribosomes of prokaryotes and eukaryotes are similar in that both types consist of two different-sized subunits. Eukaryotic ribosomes are somewhat larger than those of prokaryotes, but the structure of prokaryotic ribosomes is better understood. Chemically, ribosomes consist of a special type of RNA called ribosomal RNA (rRNA). Ribosomes also contain more than 50 different protein molecules, which are noncovalently bound to the rRNA.

In prokaryotic cells, ribosomes float freely in the cytoplasm. In eukaryotic cells they are found in two places: in the cytoplasm, where they may be free or attached to the surface of the endoplasmic reticulum (a membrane-bound organelle, see below), and inside mitochondria and chloroplasts. In each of these locations, the ribosomes are molecular factories where proteins are synthesized with their amino acid sequences specified by nucleic acids. Although they seem small in comparison to the cells that contain them, by molecular standards ribosomes are huge complexes (about 25 nm in diameter), made up of several dozen different molecules.

### The nucleus contains most of the genetic information

Organisms depend on accurate information—internal signals, environmental cues, and stored instructions—in order to respond appropriately to changing conditions, to maintain a constant internal environment, and to reproduce. In the cell, hereditary information is stored in the sequence of nucleotides in DNA molecules. Most of the DNA in eukaryotic cells resides in the nucleus (see Figure 5.7). Information encoded in the DNA is *translated* into proteins at the ribosomes. This process is described in detail in Chapter 14.

Most cells have a single nucleus, which is usually the largest organelle (**Figure 5.8**). The nucleus of a typical animal cell is approximately 5 μm in diameter—substantially larger than most prokaryotic cells. The nucleus has several functions in the cell:

- It is the location of the DNA and the site of DNA replication.
- It is the site where gene transcription is turned on or off.
- A region within the nucleus, the **nucleolus**, is where ribosomes begin to be assembled from RNA and proteins.

The nucleus is surrounded by two membranes, which together form the *nuclear envelope* (see Figure 5.8). This structure separates the genetic material from the cytoplasm. Functionally, it separates DNA transcription (which occurs in the nucleus) from translation (in the cytoplasm) (see Figure 4.5). The two membranes of the nuclear envelope are perforated by thousands of nuclear pores, each measuring approximately 9 nm in diameter, which connect the interior of the nucleus with the cytoplasm (see Figure 5.8). The pores regulate the traffic between these two cellular compartments by allowing some molecules to enter the nucleus and blocking others. This allows the nucleus to regulate the information-processing functions.

At the nuclear pore, small substances, including ions and other molecules with molecular weights of less than 10,000 dal-

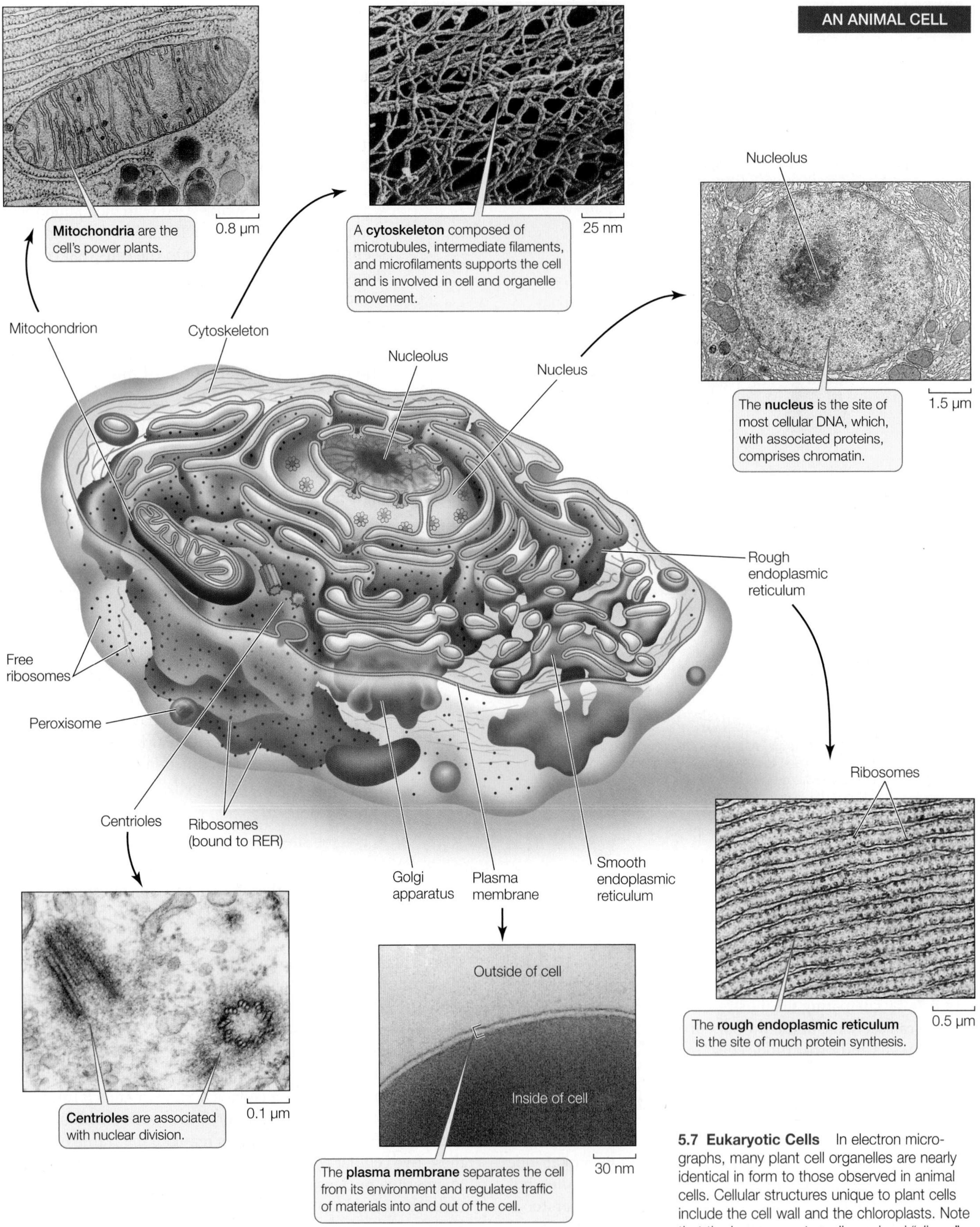

AN ANIMAL CELL
0.8 μm
**Mitochondria** are the cell's power plants.
A **cytoskeleton** composed of microtubules, intermediate filaments, and microfilaments supports the cell and is involved in cell and organelle movement.
25 nm
Nucleolus
The **nucleus** is the site of most cellular DNA, which, with associated proteins, comprises chromatin.
1.5 μm
Mitochondrion
Cytoskeleton
Nucleolus
Nucleus
Rough endoplasmic reticulum
Free ribosomes
Peroxisome
Centrioles
Ribosomes (bound to RER)
Golgi apparatus
Plasma membrane
Smooth endoplasmic reticulum
Ribosomes
Outside of cell
Inside of cell
The **rough endoplasmic reticulum** is the site of much protein synthesis.
0.5 μm
**Centrioles** are associated with nuclear division.
0.1 μm
The **plasma membrane** separates the cell from its environment and regulates traffic of materials into and out of the cell.
30 nm

**5.7 Eukaryotic Cells** In electron micrographs, many plant cell organelles are nearly identical in form to those observed in animal cells. Cellular structures unique to plant cells include the cell wall and the chloroplasts. Note that the images are two-dimensional "slices," while cells are three-dimensional structures.

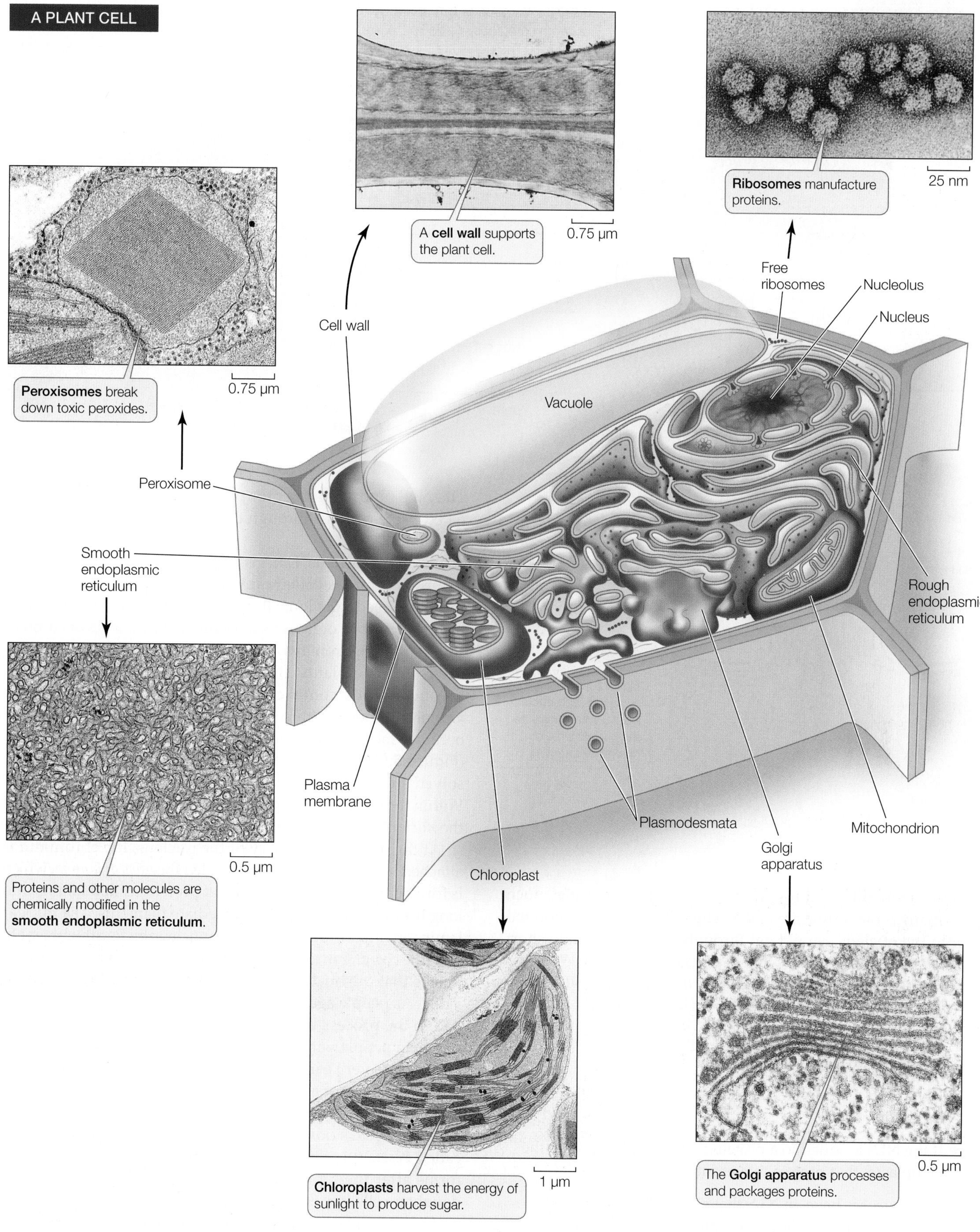
A PLANT CELL
A **cell wall** supports the plant cell.
0.75 µm
**Ribosomes** manufacture proteins.
25 nm
**Peroxisomes** break down toxic peroxides.
0.75 µm
Free ribosomes
Nucleolus
Nucleus
Cell wall
Vacuole
Peroxisome
Smooth endoplasmic reticulum
Rough endoplasmic reticulum
Plasma membrane
Plasmodesmata
Mitochondrion
Golgi apparatus
Chloroplast
0.5 µm
Proteins and other molecules are chemically modified in the **smooth endoplasmic reticulum**.
**Chloroplasts** harvest the energy of sunlight to produce sugar.
1 µm
The **Golgi apparatus** processes and packages proteins.
0.5 µm

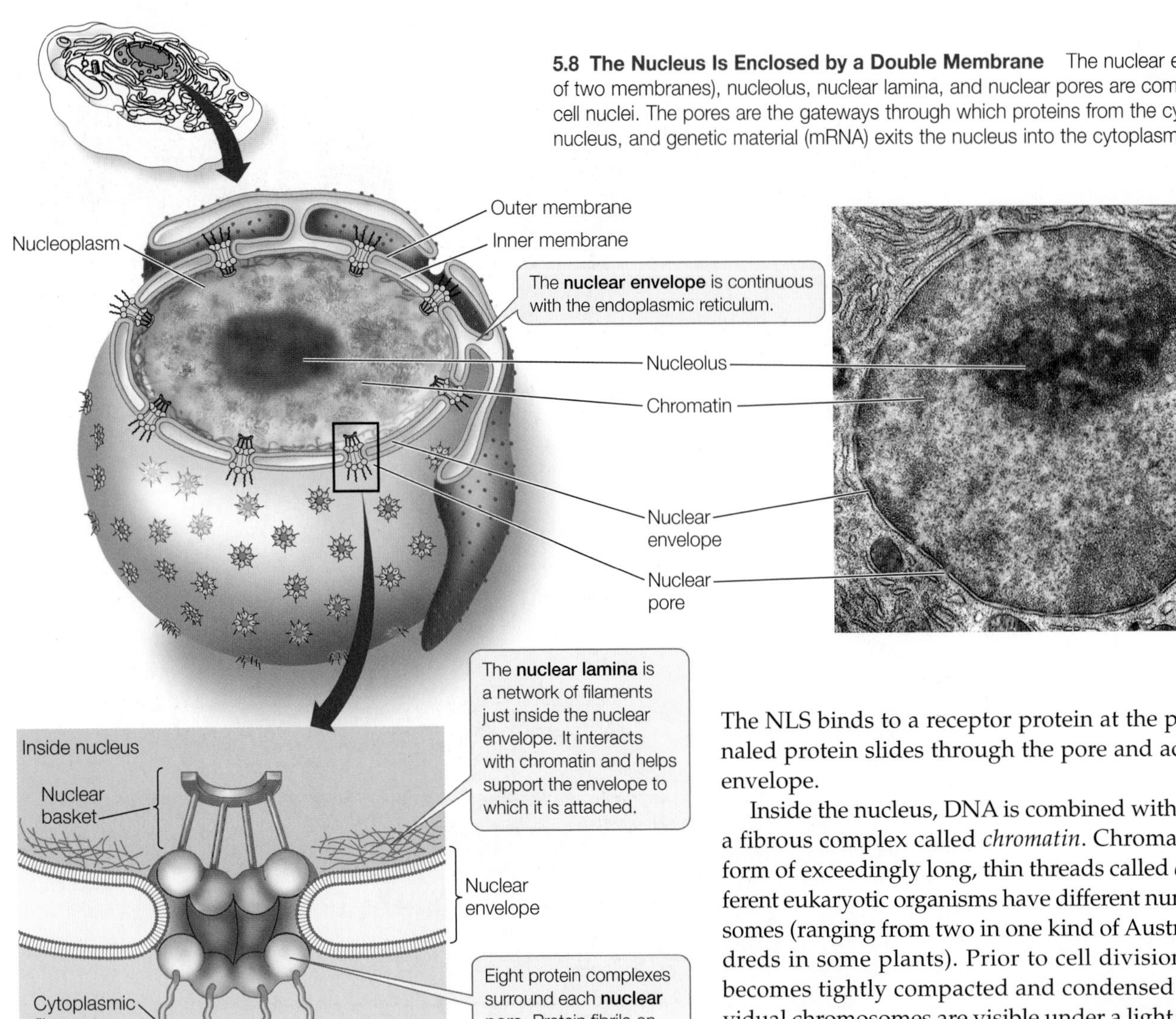
Outer membrane
Inner membrane
Nucleoplasm
The **nuclear envelope** is continuous with the endoplasmic reticulum.
Nucleolus
Chromatin
Nuclear envelope
Nuclear pore
1 μm
The **nuclear lamina** is a network of filaments just inside the nuclear envelope. It interacts with chromatin and helps support the envelope to which it is attached.
Inside nucleus
Nuclear basket
Nuclear envelope
Eight protein complexes surround each **nuclear pore**. Protein fibrils on the nuclear side form a basketlike structure.
Cytoplasmic filament
Outside nucleus (cytoplasm)

**5.8 The Nucleus Is Enclosed by a Double Membrane** The nuclear envelope (made up of two membranes), nucleolus, nuclear lamina, and nuclear pores are common features of all cell nuclei. The pores are the gateways through which proteins from the cytoplasm enter the nucleus, and genetic material (mRNA) exits the nucleus into the cytoplasm.

tons, freely diffuse through the pore. Larger molecules, such as many proteins that are made in the cytoplasm and imported into the nucleus, cannot get through without a certain short sequence of amino acids that is part of the protein. We know that this sequence is the *nuclear localization signal* (*NLS*) from several lines of evidence (see also Figure 14.20):

- The NLS occurs in most proteins targeted to the nucleus, but not in proteins that remain in the cytoplasm.
- If the NLS is removed from a protein, the protein stays in the cytoplasm.
- If the NLS is added to a protein that normally stays in the cytoplasm, that protein moves into the nucleus.
- Some viruses have an NLS that allows them to enter the nucleus; viruses without the signal sequence do not enter the nucleus as virus particles.

The NLS binds to a receptor protein at the pore, and the signaled protein slides through the pore and across the nuclear envelope.

Inside the nucleus, DNA is combined with proteins to form a fibrous complex called *chromatin*. Chromatin occurs in the form of exceedingly long, thin threads called *chromosomes*. Different eukaryotic organisms have different numbers of chromosomes (ranging from two in one kind of Australian ant to hundreds in some plants). Prior to cell division, the chromatin becomes tightly compacted and condensed so that the individual chromosomes are visible under a light microscope. This occurs to facilitate distribution of the DNA during cell division. (**Figure 5.9**). Surrounding the chromatin are water and dissolved substances collectively referred to as the *nucleoplasm*. Within the nucleoplasm, a network of structural proteins called the *nuclear matrix* helps organize the chromatin.

At the interior periphery of the nucleus, the chromatin is attached to a protein meshwork, called the *nuclear lamina*, which is formed by the polymerization of proteins called lamins into long thin structures called intermediate filaments. The nuclear lamina maintains the shape of the nucleus by its attachment to both the chromatin and the nuclear envelope. There is some evidence that the nuclear lamina may be involved with human aging. As people age, the nuclear lamina begins to disintegrate and in the process the structural integrity of the nucleus declines. In people with the rare disease called progeria, this decline begins very early in life and their aging is accelerated.

During most of a cell's life cycle, the nuclear envelope is a stable structure. When the cell reproduces, however, the nuclear envelope breaks down into small, membrane-bound droplets, called *vesicles*, containing pore complexes. The envelope reforms after the replicated DNA has been distributed to the daughter cells (see Section 11.3).

At certain sites, the outer membrane of the nuclear envelope folds outward into the cytoplasm and is continuous with the

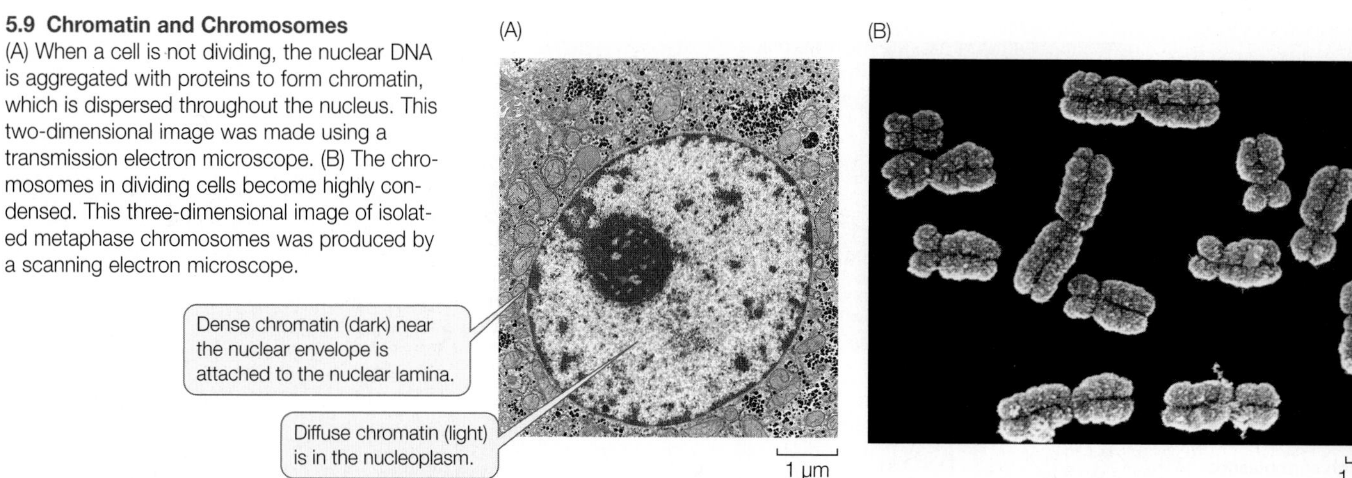
(A)
(B)
Dense chromatin (dark) near the nuclear envelope is attached to the nuclear lamina.
Diffuse chromatin (light) is in the nucleoplasm.
1 μm
1.4 μm

**5.9 Chromatin and Chromosomes**
(A) When a cell is not dividing, the nuclear DNA is aggregated with proteins to form chromatin, which is dispersed throughout the nucleus. This two-dimensional image was made using a transmission electron microscope. (B) The chromosomes in dividing cells become highly condensed. This three-dimensional image of isolated metaphase chromosomes was produced by a scanning electron microscope.

membrane of another organelle, the endoplasmic reticulum, which we will discuss next.

### The endomembrane system is a group of interrelated organelles

Much of the volume of some eukaryotic cells is taken up by an extensive **endomembrane system**. This is an interconnected system of membrane-enclosed compartments that are sometimes flattened into sheets and sometimes have other characteristic shapes (see Figure 5.7). The endomembrane system includes the plasma membrane, nuclear envelope, endoplasmic reticulum, Golgi apparatus, and lysosomes, which are derived from the Golgi. Tiny, membrane-surrounded droplets called vesicles shuttle substances between the various components of the endomembrane system (**Figure 5.10**). In drawings and electron microscope pictures this system appears static, fixed in space and time. But these depictions are just snapshots; in the living cell, membranes and the materials they contain are in constant motion. Membrane components have been observed to shift from one organelle to another within the endomembrane system. Thus, all these membranes must be functionally related.

**ENDOPLASMIC RETICULUM** Electron micrographs of eukaryotic cells reveal networks of interconnected membranes branching throughout the cytoplasm, forming tubes and flattened sacs. These membranes are collectively called the **endoplasmic reticulum**, or **ER**. The interior compartment of the ER, referred to as the lumen, is separate and distinct from the surrounding cytoplasm (see Figure 5.10). The ER can enclose up to 10 percent of the interior volume of the cell, and its foldings result in a surface area many times greater than that of the plasma membrane. There are two types of endoplasmic reticulum, the so-called rough and smooth.

**Rough endoplasmic reticulum** (**RER**) is called "rough" because of the many ribosomes attached to the outer surface of the membrane, giving it a "rough" appearance in electron microscopy (see Figure 5.7). The attached ribosomes are actively involved in protein synthesis, but that is not the entire story:

- The RER receives into its lumen certain newly synthesized proteins, segregating them away from the cytoplasm. The RER also participates in transporting these proteins to other locations in the cell.
- While inside the RER, proteins can be chemically modified to alter their functions and to chemically 'tag' them for delivery to specific cellular destinations.
- Proteins are shipped to cellular destinations enclosed within vesicles that pinch off from the ER.
- Most membrane-bound proteins are made in the RER.

A protein enters the lumen of the RER through a pore as it is synthesized. As with a protein passing through a nuclear pore, this is accomplished via a sequence of amino acids on the protein, which acts as a RER localization signal (see Section 14.6). Once in the lumen of the RER, proteins undergo several changes, including the formation of disulfide bridges and folding into their tertiary structures (see Figure 3.7).

Some proteins are covalently linked to carbohydrate groups in the RER, thus becoming glycoproteins. In the case of proteins directed to the lysosomes, the carbohydrate groups are part of an "addressing" system that ensures that the right proteins are directed to those organelles. This addressing system is very important because the enzymes within the lysosomes are some of the most destructive the cell makes. Were they not properly addressed and contained, they could destroy the cell.

The **smooth endoplasmic reticulum** (**SER**) lacks ribosomes and is more tubular (and less like flattened sacs) than the RER, but it shows continuity with portions of the RER (see Figure 5.10). Within the lumen of the SER, some proteins that have been synthesized on the RER are chemically modified. In addition, the SER has three other important roles:

- It is responsible for the chemical modification of small molecules taken in by the cell, including drugs and pesticides. These modifications make the targeted molecules more polar, so they are more water-soluble and more easily removed.
- It is the site for glycogen degradation in animal cells. We discuss this important process in Chapter 9.
- It is the site for the synthesis of lipids and steroids.

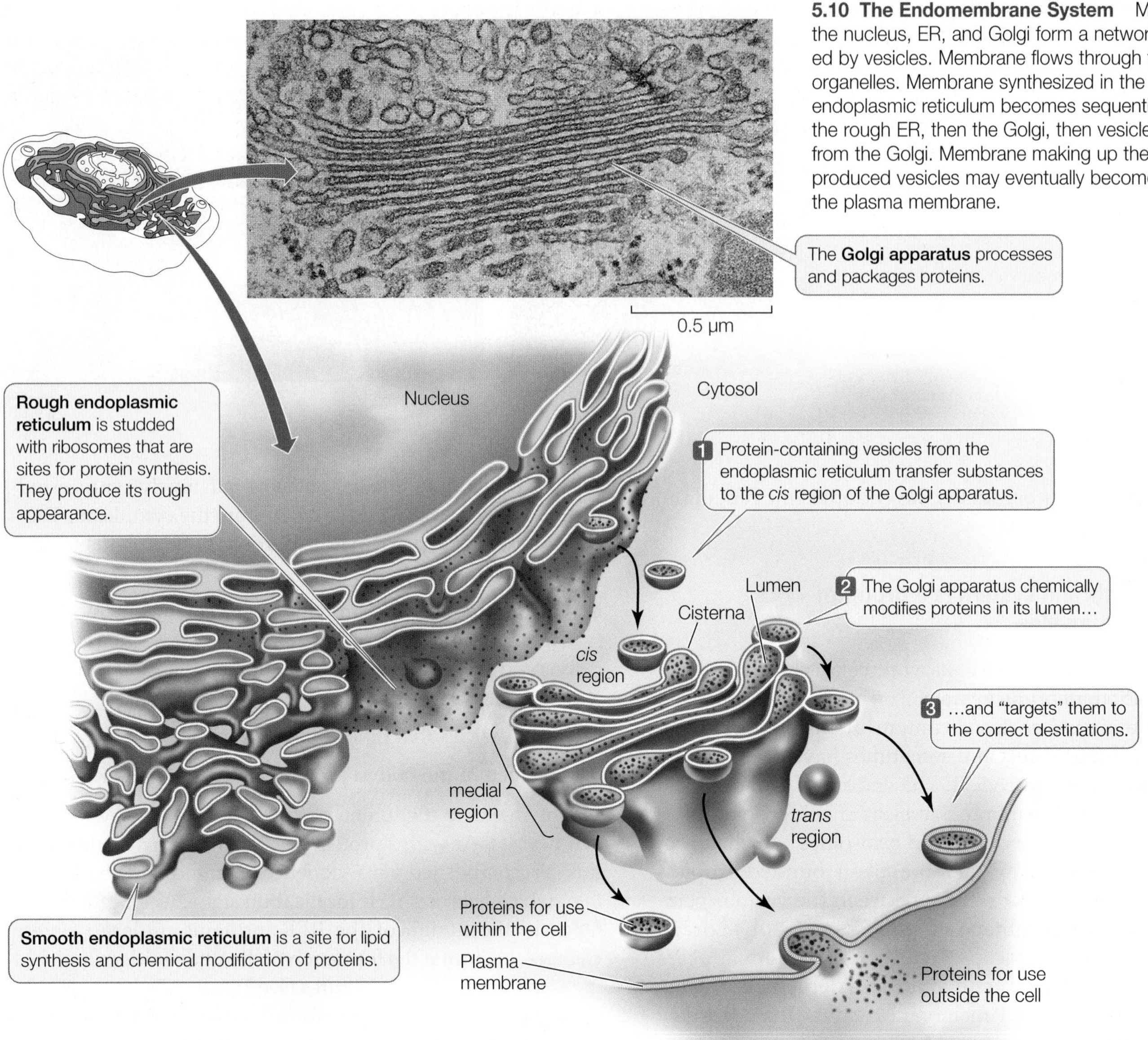

The Golgi apparatus processes and packages proteins.
0.5 µm
Rough endoplasmic reticulum is studded with ribosomes that are sites for protein synthesis. They produce its rough appearance.
Nucleus
Cytosol
1 Protein-containing vesicles from the endoplasmic reticulum transfer substances to the cis region of the Golgi apparatus.
Lumen
Cisterna
2 The Golgi apparatus chemically modifies proteins in its lumen…
cis region
3 …and "targets" them to the correct destinations.
medial region
trans region
Proteins for use within the cell
Plasma membrane
Proteins for use outside the cell
Outside of cell
Smooth endoplasmic reticulum is a site for lipid synthesis and chemical modification of proteins.

**5.10 The Endomembrane System** Membranes of the nucleus, ER, and Golgi form a network, connected by vesicles. Membrane flows through these organelles. Membrane synthesized in the smooth endoplasmic reticulum becomes sequentially part of the rough ER, then the Golgi, then vesicles formed from the Golgi. Membrane making up the Golgi-produced vesicles may eventually become part of the plasma membrane.

Cells that synthesize a lot of protein for export are usually packed with RER. Examples include glandular cells that secrete digestive enzymes and white blood cells that secrete antibodies. In contrast, cells that carry out less protein synthesis (such as storage cells) contain less RER. Liver cells, which modify molecules (including toxins) that enter the body from the digestive system, have abundant SER.

**GOLGI APPARATUS** The **Golgi apparatus** (or Golgi complex), more often referred to merely as the Golgi, is another part of the diverse, dynamic, and extensive endomembrane system (see Figure 5.10). The exact appearance of the Golgi apparatus (named for its discoverer, Camillo Golgi) varies from species to species, but it almost always consists of two components: flattened membranous sacs called *cisternae* (singular *cisterna*) that are piled up like saucers, and small membrane-enclosed vesicles. The entire apparatus is about 1 µm long.

**yourBioPortal.com**
GO TO Animated Tutorial 5.1 • The Golgi Apparatus

The Golgi has several roles:

- When protein-containing vesicles from the RER fuse with the Golgi membranes, the Golgi receives the proteins and may further modify them.
- It concentrates, packages, and sorts proteins before they are sent to their cellular or extracellular destinations.
- It adds some carbohydrates to proteins and modifies others that were attached to proteins in the ER.
- It is where some polysaccharides for the plant cell wall are synthesized.

While there is a characteristic form for all Golgi, there are also variations in its size and appearance in different cell types. In the cells of plants, protists, fungi, and many invertebrate animals, the stacks of cisternae are individual units scattered throughout the cytoplasm. In vertebrate cells, a few such stacks usually form a single, larger, more complex Golgi apparatus.

The cisternae of the Golgi apparatus appear to have three functionally distinct regions: the *cis* region lies nearest to the nucleus or a patch of RER, the *trans* region lies closest to the plasma membrane, and the *medial* region lies in between (see Figure 5.10). (The terms *cis*, *trans*, and *medial* derive from Latin words meaning, respectively, "on the same side," "on the opposite side," and "in the middle.") These three parts of the Golgi apparatus contain different enzymes and perform different functions.

The Golgi apparatus receives proteins from the ER, packages them, and sends them on their way. Since there is often no direct membrane continuity between the ER and Golgi apparatus, how does a protein get from one organelle to the other? The protein could simply leave the ER, travel across the cytoplasm, and enter the Golgi apparatus. But that would expose the protein to interactions with other molecules in the cytoplasm. On the other hand, segregation from the cytoplasm could be maintained if a piece of the ER could "bud off," forming a membranous vesicle that contains the protein—and that is exactly what happens.

Proteins make the passage from the ER to the Golgi apparatus safely enclosed in vesicles. Once it arrives, a vesicle fuses with the *cis* membrane of the Golgi apparatus, releasing its cargo into the lumen of the Golgi cisterna. Other vesicles may move between the cisternae, transporting proteins, and it appears that some proteins move from one cisterna to the next through tiny channels. Vesicles budding off from the trans region carry their contents away from the Golgi apparatus. These vesicles go to the plasma membrane, or to another organelle in the endomembrane system called the lysosome.

**LYSOSOMES** The **primary lysosomes** originate from the Golgi apparatus. They contain digestive enzymes, and they are the sites where macromolecules—proteins, polysaccharides, nucleic acids, and lipids—are hydrolyzed into their monomers (see Figure 3.4). Lysosomes are about 1 μm in diameter; they are surrounded by a single membrane and have a densely staining, featureless interior (**Figure 5.11**). There may be dozens of lysosomes in a cell, depending on its needs.

Lysosomes are sites for the breakdown of food, other cells, or foreign objects that are taken up by the cell. These materials get into the cell by a process called *phagocytosis* (*phago*, "eat"; *cytosis*, "cellular"). In this process, a pocket forms in the plasma membrane and then deepens and encloses material from outside the cell. The pocket becomes a small vesicle called a phagosome, containing food or other material, which breaks free of the plasma membrane to move into the cytoplasm. The phagosome fuses with a primary lysosome to form a **secondary lysosome**, in which digestion occurs.

The effect of this fusion is rather like releasing hungry foxes into a chicken coop: the enzymes in the secondary lysosome quickly hydrolyze the food particles. These reactions are en-

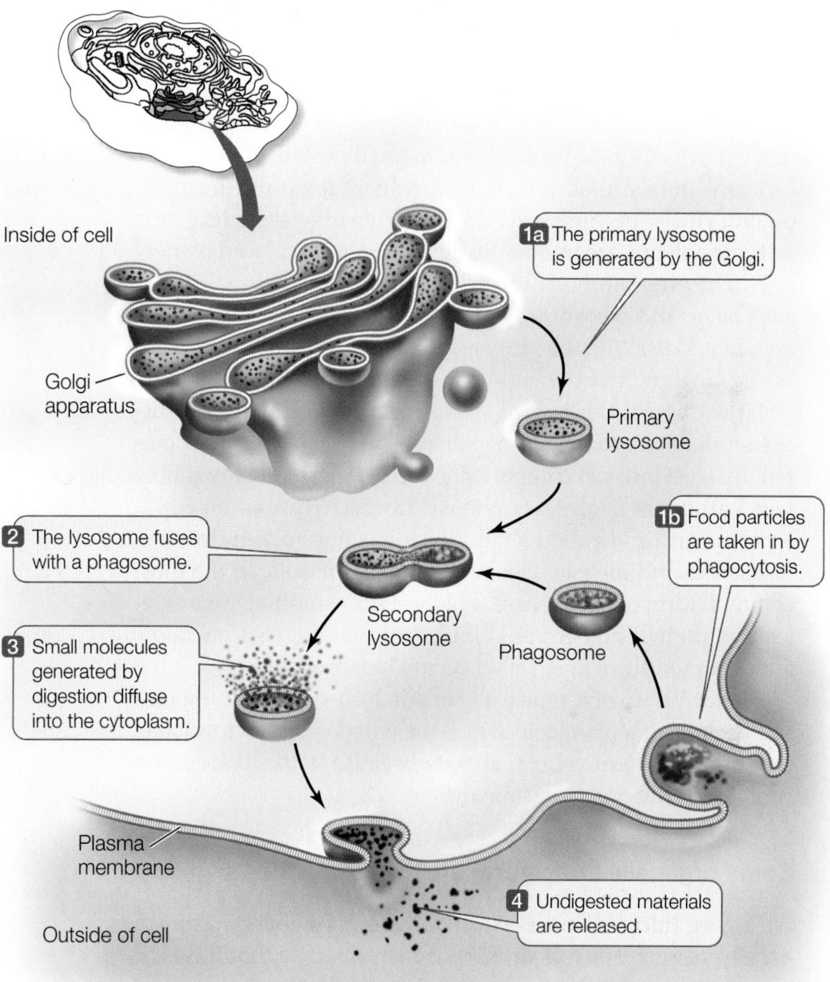
Inside of cell
1a The primary lysosome is generated by the Golgi.
Golgi apparatus
Primary lysosome
1b Food particles are taken in by phagocytosis.
2 The lysosome fuses with a phagosome.
Secondary lysosome
Phagosome
3 Small molecules generated by digestion diffuse into the cytoplasm.
Plasma membrane
4 Undigested materials are released.
Outside of cell

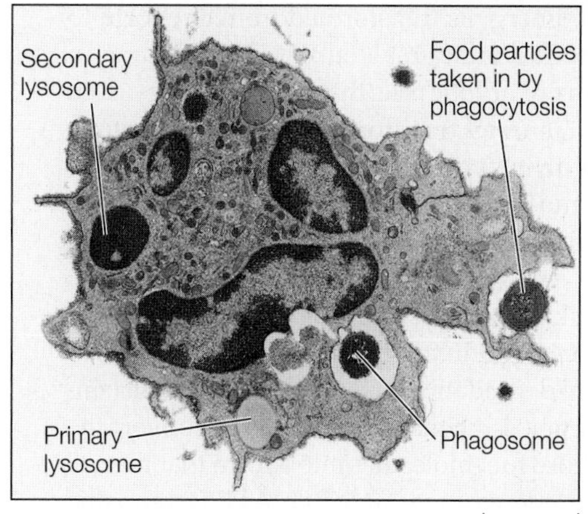
Secondary lysosome
Food particles taken in by phagocytosis
Primary lysosome
Phagosome
1 μm

**5.11 Lysosomes Isolate Digestive Enzymes from the Cytoplasm** Lysosomes are sites for the hydrolysis of material taken into the cell by phagocytosis.

**yourBioPortal.com**
GO TO **Web Activity 5.3 • Lysosomal Digestion**

hanced by the mild acidity of the lysosome's interior, where the pH is lower than in the surrounding cytoplasm. The products of digestion pass through the membrane of the lysosome, providing energy and raw materials for other cellular processes. The "used" secondary lysosome, now containing undigested particles, then moves to the plasma membrane, fuses with it, and releases the undigested contents to the environment.

Phagocytes are specialized cells that have an essential role in taking up and breaking down materials; they are found in nearly all animals and many protists. You will encounter them and their activities again at many places in this book, but at this point one example suffices: in the human liver and spleen, phagocytes digest approximately 10 billion aged or damaged blood cells each day! The digestion products are then used to make new cells to replace those that are digested.

Lysosomes are active even in cells that do not perform phagocytosis. Because cells are such dynamic systems, some cell components are frequently destroyed and replaced by new ones. The programmed destruction of cell components is called *autophagy*, and lysosomes are where the cell breaks down its own materials. With the proper signal, lysosomes can engulf entire organelles, hydrolyzing their constituents.

How important is autophagy? An entire class of human diseases called lysosomal storage diseases occur when lysosomes fail to digest internal components; these diseases are invariably very harmful or fatal. An example is Tay-Sachs disease, in which a particular lipid called a ganglioside is not broken down in lysosomes and instead accumulates in brain cells. In the most common form of this disease, a baby starts exhibiting neurological symptoms and becomes blind, deaf, and unable to swallow after six months of age. Death occurs before age 4.

Plant cells do not appear to contain lysosomes, but the central vacuole of a plant cell (which we will describe below) may function in an equivalent capacity because it, like lysosomes, contains many digestive enzymes.

### Some organelles transform energy

All living things require external sources of energy. The energy from such sources must be transformed so that it can be used by cells. A cell requires energy to make the molecules it needs for activities such as growth, reproduction, responsiveness, and movement. Energy is transformed from one form to another in mitochondria (found in all eukaryotic cells) and in chloroplasts (found in eukaryotic cells that harvest energy from sunlight). In contrast, energy transformations in prokaryotic cells are associated with enzymes attached to the inner surface of the plasma membrane or to extensions of the plasma membrane that protrude into the cytoplasm.

**MITOCHONDRIA** In eukaryotic cells, the breakdown of fuel molecules such as glucose begins in the cytosol. The molecules that result from this partial degradation enter the **mitochondria** (singular *mitochondrion*), whose primary function is to convert the chemical energy of those fuel molecules into a form that the cell can use, namely the energy-rich molecule ATP (adenosine triphosphate) (see Section 8.2). The production of ATP in the mitochondria, using fuel molecules and molecular oxygen ($O_2$), is called *cellular respiration.*

Typical mitochondria are somewhat less than 1.5 μm in diameter and 2–8 μm in length—about the size of many bacteria. They can divide independently of the central nucleus. The number of mitochondria per cell ranges from one gigantic organelle in some unicellular protists to a few hundred thousand in large egg cells. An average human liver cell contains more than a thousand mitochondria. Cells that are active in movement and growth require the most chemical energy, and these tend to have the most mitochondria per unit of volume.

Mitochondria have two membranes. The outer membrane is smooth and protective, and it offers little resistance to the movement of substances into and out of the organelle. Immediately inside the outer membrane is an inner membrane, which folds inward in many places, and thus has a surface area much greater than that of the outer membrane (**Figure 5.12**). The folds tend to be quite regular, giving rise to shelf-like structures called *cristae.*

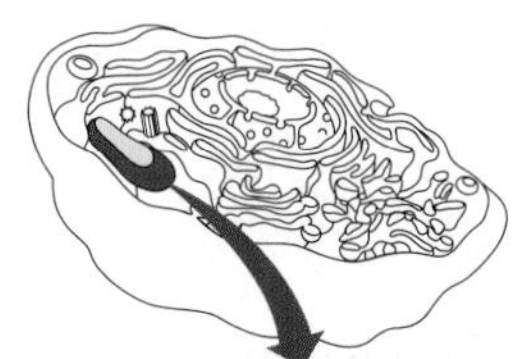

**5.12 A Mitochondrion Converts Energy from Fuel Molecules into ATP** The electron micrograph is a two-dimensional slice through a three-dimensional organelle. As the drawing emphasizes, the cristae are extensions of the inner mitochondrial membrane.

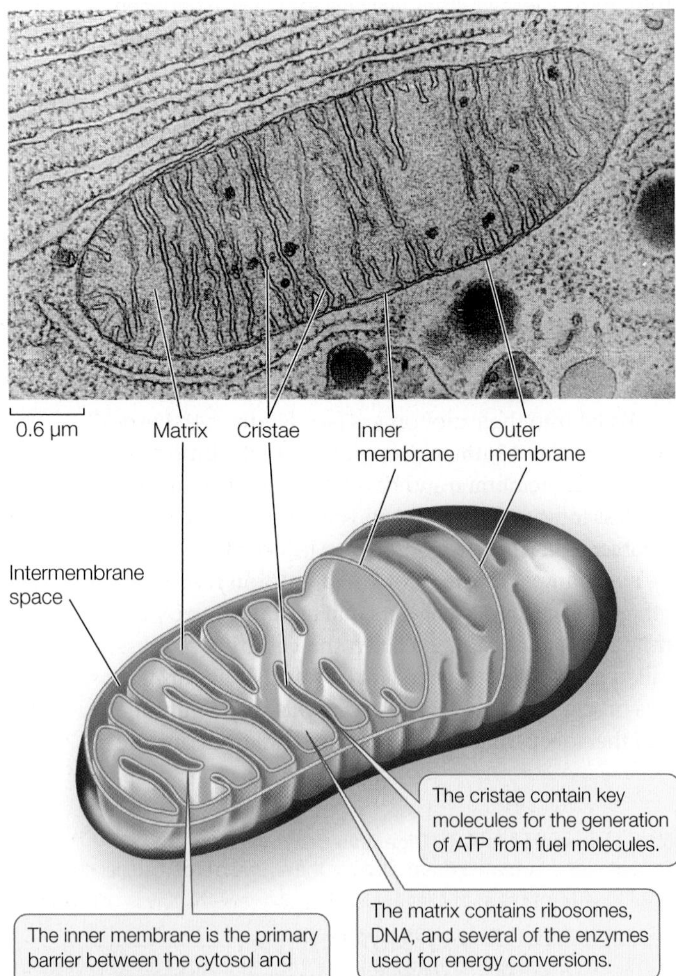

0.6 μm
Matrix
Cristae
Inner membrane
Outer membrane
Intermembrane space
The cristae contain key molecules for the generation of ATP from fuel molecules.
The matrix contains ribosomes, DNA, and several of the enzymes used for energy conversions.
The inner membrane is the primary barrier between the cytosol and mitochondrial enzymes.

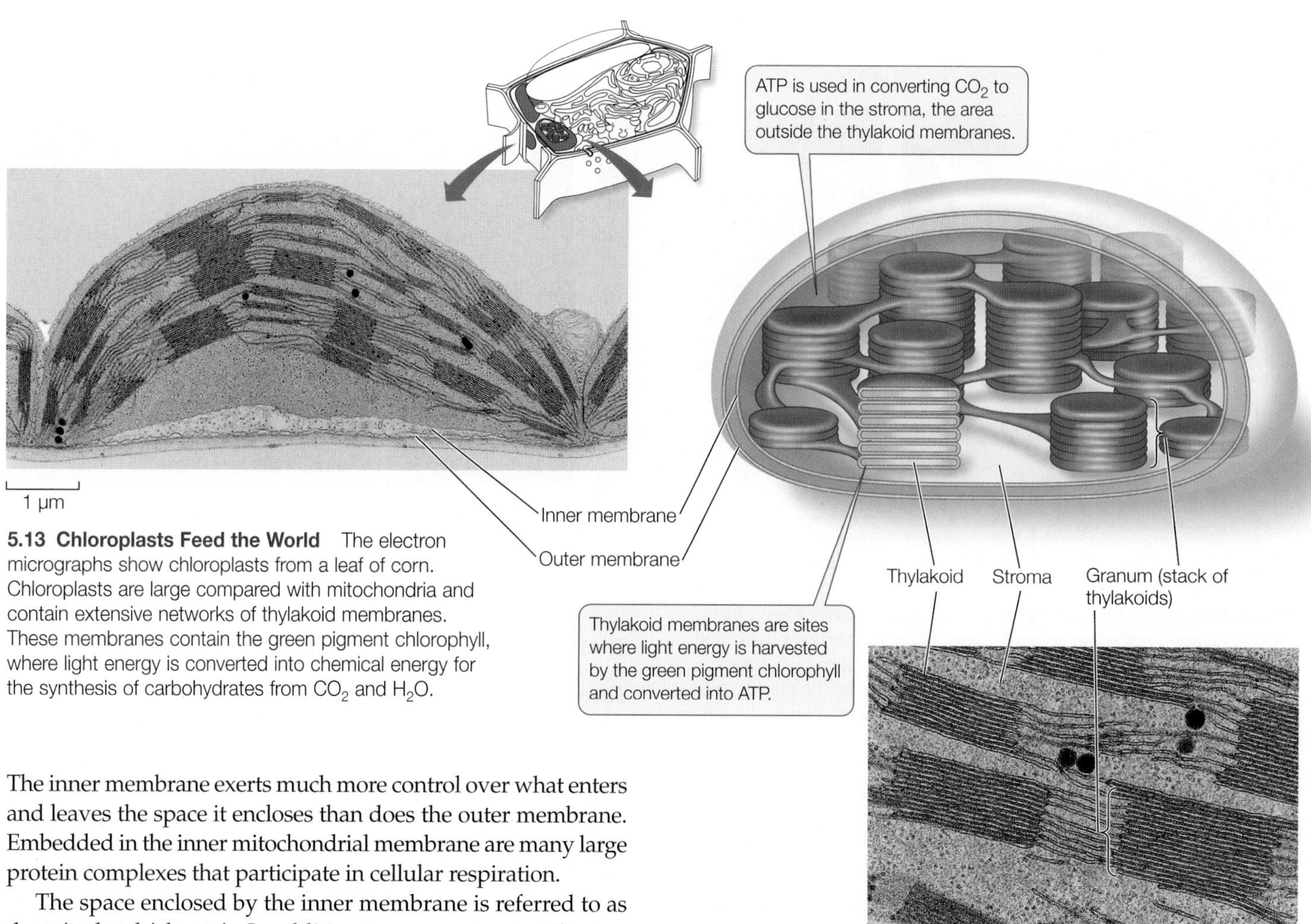
ATP is used in converting $CO_2$ to glucose in the stroma, the area outside the thylakoid membranes.
1 µm
Inner membrane
Outer membrane
Thylakoid
Stroma
Granum (stack of thylakoids)
Thylakoid membranes are sites where light energy is harvested by the green pigment chlorophyll and converted into ATP.
0.5 µm

**5.13 Chloroplasts Feed the World** The electron micrographs show chloroplasts from a leaf of corn. Chloroplasts are large compared with mitochondria and contain extensive networks of thylakoid membranes. These membranes contain the green pigment chlorophyll, where light energy is converted into chemical energy for the synthesis of carbohydrates from $CO_2$ and $H_2O$.

The inner membrane exerts much more control over what enters and leaves the space it encloses than does the outer membrane. Embedded in the inner mitochondrial membrane are many large protein complexes that participate in cellular respiration.

The space enclosed by the inner membrane is referred to as the *mitochondrial matrix*. In addition to many enzymes, the matrix contains ribosomes and DNA that are used to make some of the proteins needed for cellular respiration. As you will see later in this chapter, this DNA is the remnant of a much larger, complete chromosome of a prokaryote that may have been the mitochondrion's progenitor (see Figure 5.26). In Chapter 9 we discuss how the different parts of the mitochondrion work together in cellular respiration.

**PLASTIDS** One class of organelles—the plastids—is present only in the cells of plants and certain protists. Like mitochondria, plastids can divide autonomously. There are several types of plastids, with different functions.

**Chloroplasts** contain the green pigment chlorophyll and are the sites of photosynthesis (**Figure 5.13**). In photosynthesis, light energy is converted into the chemical energy of bonds between atoms. The molecules formed by photosynthesis provide food for the photosynthetic organism and for other organisms that eat it. Directly or indirectly, photosynthesis is the energy source for most of the living world.

Chloroplasts are variable in size and shape (**Figure 5.14**). Like a mitochondrion, a chloroplast is surrounded by two membranes. In addition, there is a series of internal membranes whose structure and arrangement vary from one group of photosynthetic organisms to another. Here we concentrate on the chloroplasts of the flowering plants.

The internal membranes of chloroplasts look like stacks of flat, hollow pita bread. Each stack, called a *granum* (plural *grana*), consists of a series of flat, closely packed, circular compartments called **thylakoids** (see Figure 5.13). Thylakoid lipids are distinctive: only 10 percent are phospholipids, while the rest are galactose-substituted diglycerides and sulfolipids. Because of the abundance of chloroplasts, these are the most abundant lipids in the biosphere.

In addition to lipids and proteins, the membranes of the thylakoids contain chlorophyll and other pigments that harvest light energy for photosynthesis (we see how they do this in Section 10.2). The thylakoids of one granum may be connected to those of other grana, making the interior of the chloroplast a highly developed network of membranes, much like the ER.

The fluid in which the grana are suspended is called the *stroma*. Like the mitochondrial matrix, the chloroplast stroma contains ribosomes and DNA, which are used to synthesize some, but not all, of the proteins that make up the chloroplast.

Animal cells typically do not contain chloroplasts, but some do contain functional photosynthetic organisms. The green color of some corals and sea anemones comes from chloroplasts in algae that live within those animals (see Figure 5.14C). The animals derive some of their nutrition from the photosynthesis that their chloroplast-containing "guests" carry out. Such an intimate relationship between two different organisms is called *symbiosis*.

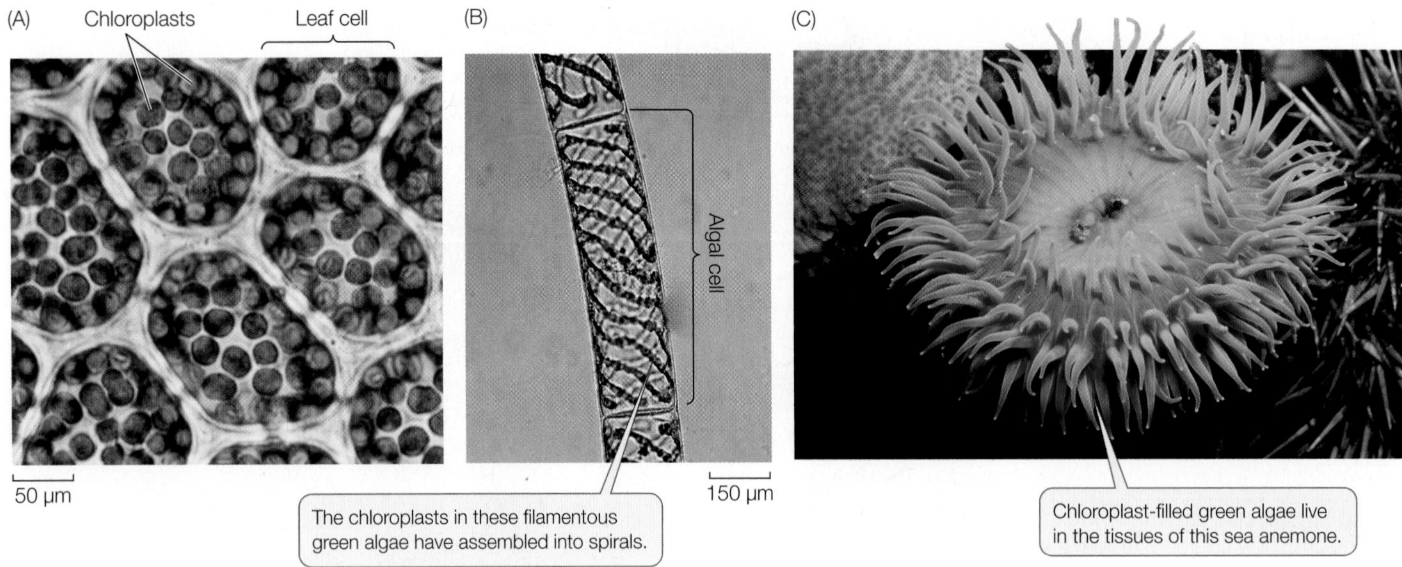


**5.14 Chloroplasts Are Everywhere** (A) In green plants, chloroplasts are concentrated in the leaf cells. (B) Green algae are photosynthetic and filled with chloroplasts. (C) No animal species produces its own chloroplasts, but this sea anemone (an animal) is nourished by the chloroplasts of unicellular green algae living within its tissues, in what is termed a symbiotic relationship.

Other types of plastids such as *chromoplasts* and *leucoplasts* have functions different from those of chloroplasts (**Figure 5.15**). Chromoplasts make and store red, yellow, and orange pigments, especially in flowers and fruits. Leucoplasts are storage organelles that do not contain pigments. An amyloplast is a leucoplast that stores starch.

### There are several other membrane-enclosed organelles

There are several other organelles whose boundary membranes separate their specialized chemical reactions and contents from the cytoplasm: peroxisomes, glyoxysomes, and vacuoles, including contractile vacuoles.

**Peroxisomes** are organelles that accumulate toxic peroxides, such as hydrogen peroxide ($H_2O_2$), that occur as byproducts of some biochemical reactions. These peroxides can be safely broken down inside the peroxisomes without mixing with other parts of the cell. Peroxisomes are small organelles, about 0.2 to 1.7 μm in diameter. They have a single membrane and a granular interior containing specialized enzymes. Peroxisomes are found in at least some of the cells of almost every eukaryotic species.

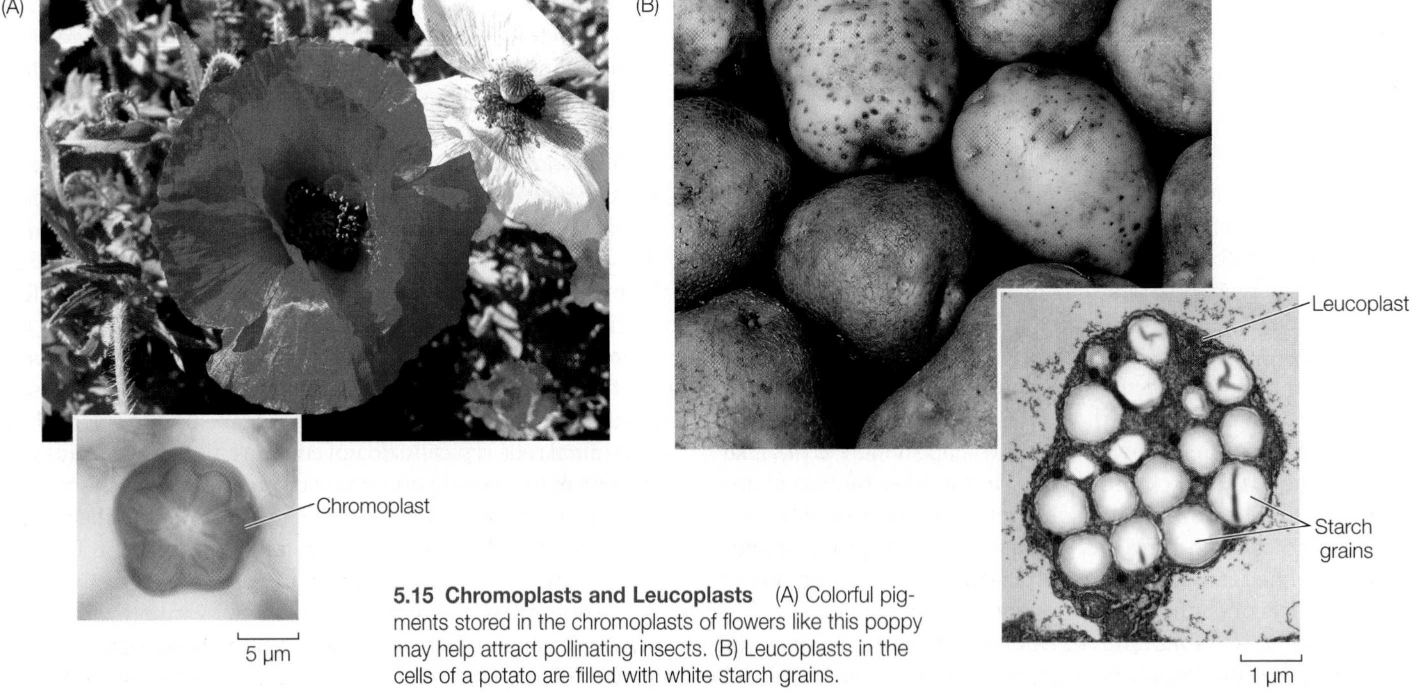


**5.15 Chromoplasts and Leucoplasts** (A) Colorful pigments stored in the chromoplasts of flowers like this poppy may help attract pollinating insects. (B) Leucoplasts in the cells of a potato are filled with white starch grains.

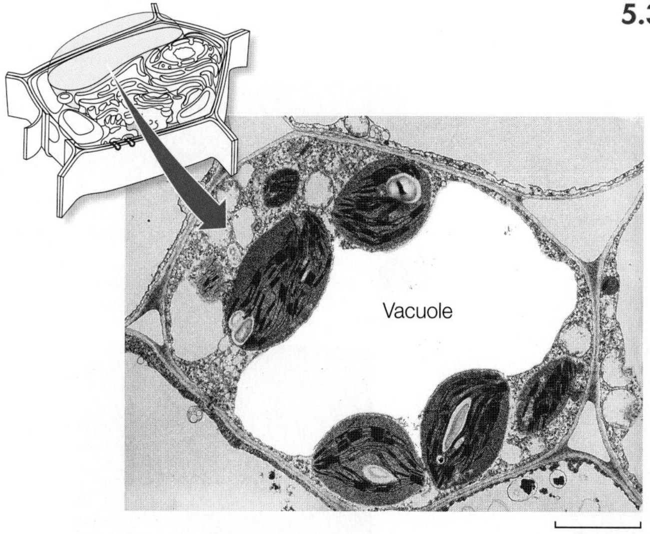


**5.16 Vacuoles in Plant Cells Are Usually Large** The large central vacuole in this cell is typical of mature plant cells. Smaller vacuoles are visible toward each end of the cell.

**Glyoxysomes** are similar to peroxisomes and are found only in plants. They are most abundant in young plants, and are the locations where stored lipids are converted into carbohydrates for transport to growing cells.

**Vacuoles** occur in many eukaryotic cells, but particularly those of plants and protists. Plant vacuoles (**Figure 5.16**) have several functions:

- *Storage*: Plant cells produce a number of toxic by-products and waste products, many of which are simply stored within vacuoles. Because they are poisonous or distasteful, these stored materials deter some animals from eating the plants, and may thus contribute to plant defenses and survival.
- *Structure*: In many plant cells, enormous vacuoles take up more than 90 percent of the cell volume and grow as the cell grows. The presence of dissolved substances in the vacuole causes water to enter it from the cytoplasm, making the vacuole swell like a balloon. The plant cell does not swell when the vacuole fills with water, since it has a rigid cell wall. Instead, it stiffens from the increase in water pressure (called turgor), which supports the plant (see Figure 6.10).
- *Reproduction*: Some pigments (especially blue and pink ones) in the petals and fruits of flowering plants are contained in vacuoles. These pigments—the anthocyanins—are visual cues that help attract the animals that assist in pollination or seed dispersal.
- *Digestion*: In some plants, vacuoles in seeds contain enzymes that hydrolyze stored seed proteins into monomers that the developing plant embryo can use as food.

*Contractile vacuoles* are found in many freshwater protists. Their function is to get rid of the excess water that rushes into the cell because of the imbalance in solute concentration between the interior of the cell and its freshwater environment. The contractile vacuole enlarges as water enters, then abruptly contracts, forcing the water out of the cell through a special pore structure.

So far, we have discussed numerous membrane-enclosed organelles. Now we turn to a group of cytoplasmic structures without membranes.

### The cytoskeleton is important in cell structure and movement

From the earliest observations, light microscopy revealed distinctive shapes of cells that would sometimes change, and within cells rapid movements were observed. With the advent of electron microscopy, a new world of cellular substructure was revealed, including a meshwork of filaments inside cells. Experimentation showed that this **cytoskeleton** fills several important roles:

- It supports the cell and maintains its shape.
- It holds cell organelles in position within the cell.
- It moves organelles within the cell.
- It is involved with movements of the cytoplasm, called cytoplasmic streaming.
- It interacts with extracellular structures, helping to anchor the cell in place.

There are three components of the cytoskeleton: microfilaments (smallest diameter), intermediate filaments, and microtubules (largest diameter). These filaments have very different functions.

**MICROFILAMENTS** **Microfilaments** can exist as single filaments, in bundles, or in networks. They are about 7 nm in diameter and up to several micrometers long. Microfilaments have two major roles:

- They help the entire cell or parts of the cell to move.
- They determine and stabilize cell shape.

Microfilaments are assembled from *actin* monomers, a protein that exists in several forms and has many functions, especially in animals. The actin found in microfilaments (which are also known as actin filaments) has distinct ends designated "plus" and "minus." These ends permit actin monomers to interact with one another to form long, double helical chains (**Figure 5.17A**). Within cells, the polymerization of actin into microfilaments is reversible, and the microfilaments can disappear from cells by breaking down into monomers of free actin. Special actin-binding proteins mediate these events.

In the muscle cells of animals, actin filaments are associated with another protein, the "motor protein" *myosin*, and the interactions of these two proteins account for the contraction of muscles (described in Section 48.1). In non-muscle cells, actin filaments are associated with localized changes in cell shape. For example, microfilaments are involved in the flowing movement of the cytoplasm called cytoplasmic streaming, in amoeboid movement, and in the "pinching" contractions that divide an animal cell into two daughter cells. Microfilaments are also involved in the formation of cellular extensions called pseudopodia (*pseudo*, "false"; *podia*, "feet") that enable some cells to move (**Figure 5.18**). As you will see in Chapter 42, cells of the immune system must move toward other cells during the immune response.

In some cell types, microfilaments form a meshwork just inside the plasma membrane. Actin-binding proteins then cross-link the microfilaments to form a rigid net-like structure that supports the cell. For example, microfilaments support the tiny

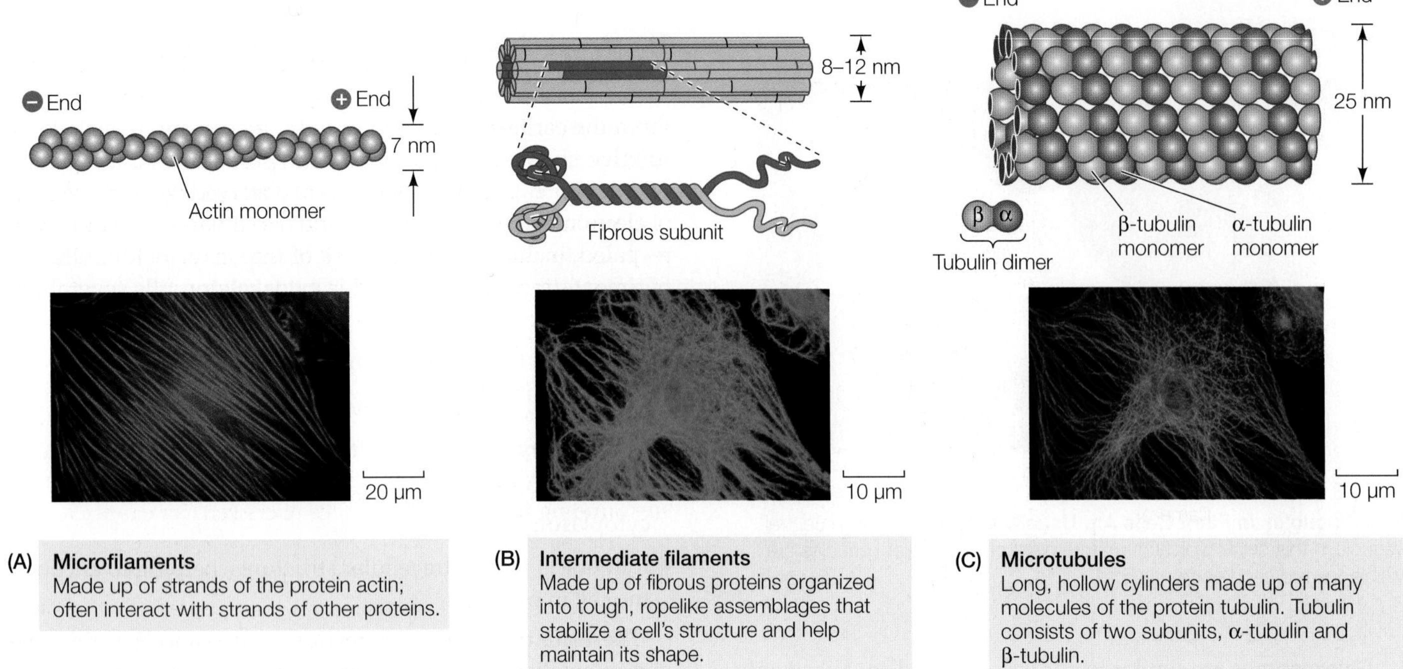


**5.17 The Cytoskeleton** Three highly visible and important structural components of the cytoskeleton are shown here in detail. These structures maintain and reinforce cell shape and contribute to cell movement.

microvilli that line the human intestine, giving it a larger surface area through which to absorb nutrients (**Figure 5.19**).

**INTERMEDIATE FILAMENTS** There are at least 50 different kinds of **intermediate filaments**, many of them specific to a few cell types. They generally fall into six molecular classes (based on amino acid sequence) that share the same general structure. One of these classes consists of fibrous proteins of the keratin family, which also includes the proteins that make up hair and fingernails. The intermediate filaments are tough, ropelike protein assemblages 8 to 12 nm in diameter (**Figure 5.17B**). Intermediate filaments are more permanent than the other two types; in cells they do not form and re-form, as the microtubules and microfilaments do.

Intermediate filaments have two major structural functions:

- They anchor cell structures in place. In some cells, intermediate filaments radiate from the nuclear envelope and help maintain the positions of the nucleus and other organelles in the cell. The lamins of the nuclear lamina are intermediate filaments (see Figure 5.8). Other kinds of intermediate filaments help hold in place the complex apparatus of microfilaments in the microvilli of intestinal cells (see Figure 5.19).
- They resist tension. For example, they maintain rigidity in body surface tissues by stretching through the cytoplasm and connecting specialized membrane structures called desmosomes (see Figure 6.7).

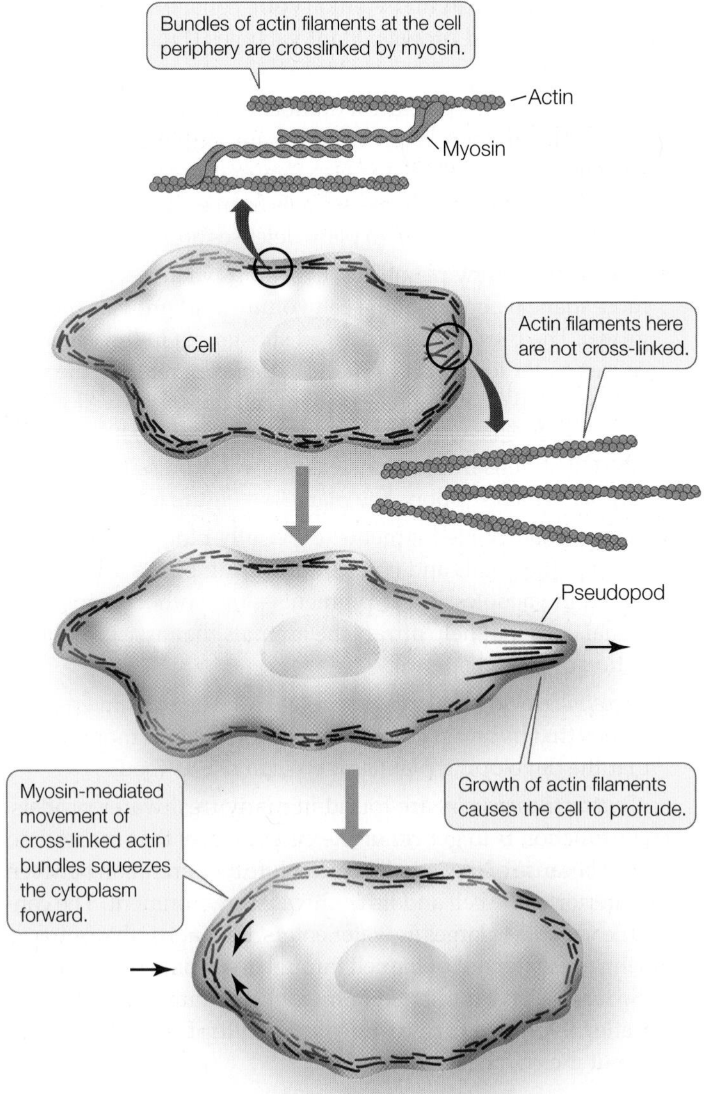

Bundles of actin filaments at the cell periphery are crosslinked by myosin.
Actin
Myosin
Cell
Actin filaments here are not cross-linked.
Pseudopod
Growth of actin filaments causes the cell to protrude.
Myosin-mediated movement of cross-linked actin bundles squeezes the cytoplasm forward.

**5.18 Microfilaments and Cell Movements** Microfilaments mediate the movement of whole cells (as illustrated here for amoebic movement), as well as the movement of cytoplasm within a cell.

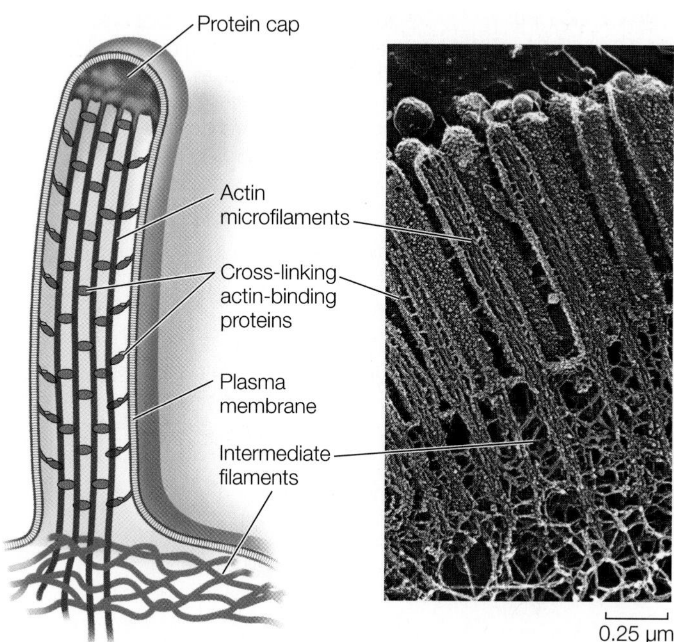


**5.19 Microfilaments for Support** Cells that line the intestine are folded into tiny projections called microvilli, which are supported by microfilaments. The microfilaments interact with intermediate filaments at the base of each microvillus. The microvilli increase the surface area of the cells, facilitating their absorption of small molecules.

**MICROTUBULES** The largest diameter components of the cytoskeletal system, **microtubules**, are long, hollow, unbranched cylinders about 25 nm in diameter and up to several micrometers long. Microtubules have two roles in the cell:

- They form a rigid internal skeleton for some cells.
- They act as a framework along which motor proteins can move structures within the cell.

Microtubules are assembled from dimers of the protein *tubulin*. A dimer is a molecule made up of two monomers. The polypeptide monomers that make up a tubulin dimer are known as α-tubulin and β-tubulin. Thirteen chains of tubulin dimers surround the central cavity of the microtubule (**Figure 5.17C**; see also Figure 5.20). The two ends of a microtubule are different: one is designated the plus (+) end, and the other the minus (–) end. Tubulin dimers can be rapidly added or subtracted, mainly at the plus end, lengthening or shortening the microtubule. This capacity to change length rapidly makes microtubules dynamic structures, permitting some animal cells to rapidly change shape.

Many microtubules radiate from a region of the cell called the microtubule organizing center. Tubulin polymerization results in a rigid structure, and tubulin depolymerization leads to its collapse.

In plants, microtubules help control the arrangement of the cellulose fibers of the cell wall. Electron micrographs of plants frequently show microtubules lying just inside the plasma membranes of cells that are forming or extending their cell walls. Experimental alteration of the orientation of these microtubules leads to a similar change in the cell wall and a new shape for the cell.

Microtubules serve as tracks for **motor proteins**, specialized molecules that use cellular energy to change their shape and move. Motor proteins bond to and move along the microtubules, carrying materials from one part of the cell to another. Microtubules are also essential in distributing chromosomes to daughter cells during cell division. Because of this, drugs such as vincristine and taxol that disrupt microtubule dynamics also disrupt cell division. These drugs are useful for treating cancer, where cell division is excessive.

**CILIA AND FLAGELLA** Microtubules are also intimately associated with movable cell appendages: the **cilia** and **flagella**. Many eukaryotic cells have one or both of these appendages. Cilia are smaller than flagella—only 0.25 μm in length. They may move surrounding fluid over the surface of the cell (for example, protists or cells lining tubes through which eggs move, the oviducts). Eukaryotic flagella are 0.25 μm in diameter and 100–200 μm in length. (The structure and operation of eukaryotic flagella are very different from those of prokaryotic flagella; see Figure 5.5.) They may push or pull the cell through its aqueous environment (for example, protists or sperm). Cilia and eukaryotic flagella are both assembled from specialized microtubules and have identical internal structures, but differ in their length and pattern of beating:

- Cilia (singular *cilium*) are usually present in great numbers (**Figure 5.20A**). They beat stiffly in one direction and recover flexibly in the other direction (like a swimmer's arm), so that the recovery stroke does not undo the work of the power stroke.
- Eukaryotic flagella are usually found singly or in pairs. Waves of bending propagate from one end of a flagellum to the other in a snakelike undulation. Forces exerted by these waves on the surrounding fluid medium move the cell.

In cross section, a typical *cilium* or eukaryotic flagellum is surrounded by the plasma membrane and contains a "9 + 2" array of microtubules. As **Figure 5.20B** shows, nine fused pairs of microtubules—called doublets—form an outer cylinder, and one pair of unfused microtubules runs up the center. A spoke radiates from one microtubule of each doublet and connects the doublet to the center of the structure. These structures are essential to the bending motions of both cilia and flagella.

In the cytoplasm at the base of every eukaryotic flagellum and cilium is an organelle called a **basal body**. The nine microtubule doublets extend into the basal body. In the basal body, each doublet is accompanied by another microtubule, making nine sets of three microtubules. The central, unfused microtubules in the cilium do not extend into the basal body.

*Centrioles* are almost identical to the basal bodies of cilia and flagella. Centrioles are found in the microtubule organizing centers (sites of tubulin storage where microtubules polymerize) of all eukaryotes except the seed plants and some protists. Under the light microscope, a centriole looks like a small, featureless particle, but the electron microscope reveals that it contains a precise bundle of microtubules arranged in nine sets of three. Centrioles are involved in the formation of the mitotic spindle, to which the chromosomes attach during cell division (see Figure 11.10).

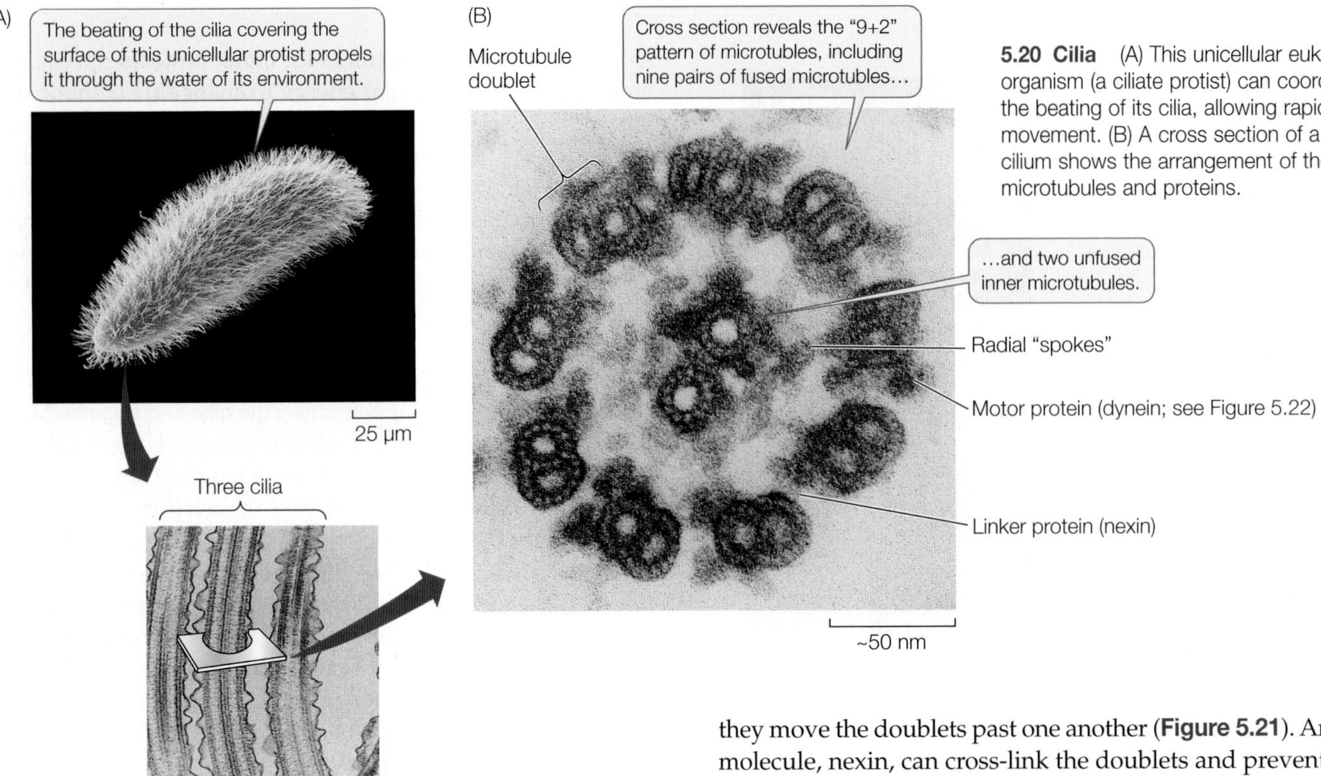
(A)
The beating of the cilia covering the surface of this unicellular protist propels it through the water of its environment.
25 μm
Three cilia
250 nm
(B)
Microtubule doublet
Cross section reveals the "9+2" pattern of microtubules, including nine pairs of fused microtubules...
...and two unfused inner microtubules.
Radial "spokes"
Motor protein (dynein; see Figure 5.22)
Linker protein (nexin)
~50 nm

**5.20 Cilia** (A) This unicellular eukaryotic organism (a ciliate protist) can coordinate the beating of its cilia, allowing rapid movement. (B) A cross section of a single cilium shows the arrangement of the microtubules and proteins.

**MOTOR PROTEINS AND MOVEMENT** The nine microtubule doublets of cilia and flagella are linked by proteins. The motion of cilia and flagella results from the sliding of the microtubule doublets past each other. This sliding is driven by a motor protein called *dynein*, which can change its three-dimensional shape. All motor proteins work by undergoing reversible shape changes powered by energy from ATP hydrolysis. Dynein molecules that are attached to one microtubule doublet bind to a neighboring doublet. As the dynein molecules change shape, they move the doublets past one another (**Figure 5.21**). Another molecule, nexin, can cross-link the doublets and prevent them from sliding past one another; in this case, the cilium bends.

Another motor protein, *kinesin*, carries protein-laden vesicles from one part of the cell to another (**Figure 5.22**). Kinesin and similar motor proteins bind to a vesicle or other organelle, then "walk" it along a microtubule by a repeated series of shape changes. Recall that microtubules are directional, with a plus end and a minus end. Dynein moves attached organelles toward the minus end, while kinesin moves them toward the plus end (see Figure 5.17).

**DEMONSTRATING CYTOSKELETON FUNCTIONS** How do we know that the structural fibers of the cytoskeleton can achieve all these dynamic functions? We can observe an individual structure under the microscope and a function in a living cell that contains that structure. These observations may suggest that the structure carries out that function, but in science mere correlation does not show cause and effect. For example, light microscopy of living cells reveals that the cytoplasm is actively streaming around the cell, and that cytoplasm flows into an extended portion of an amoeboid cell during movement. The observed presence of cytoskeletal components *suggests, but does not prove*, their role in this process. Science seeks to show the specific links that relate one process, "A," to a function, "B." In cell biology, there are two ways to show that a structure or process "A" causes function "B":

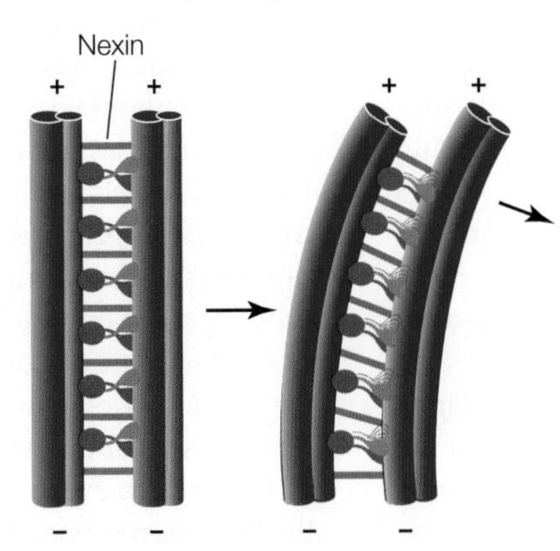
Dynein
Microtubule doublet
Nexin
In isolated cilia without nexin cross-links, movement of dynein motor proteins causes microtubule doublets to slide past one another.
When nexin is present to cross-link the doublets, they cannot slide and the force generated by dynein movement causes the cilium to bend.

**5.21 A Motor Protein Moves Microtubules in Cilia and Flagella** A motor protein, dynein, causes microtubule doublets to slide past one another. In a flagellum or cilium, anchorage of the microtubule doublets to one another results in bending.

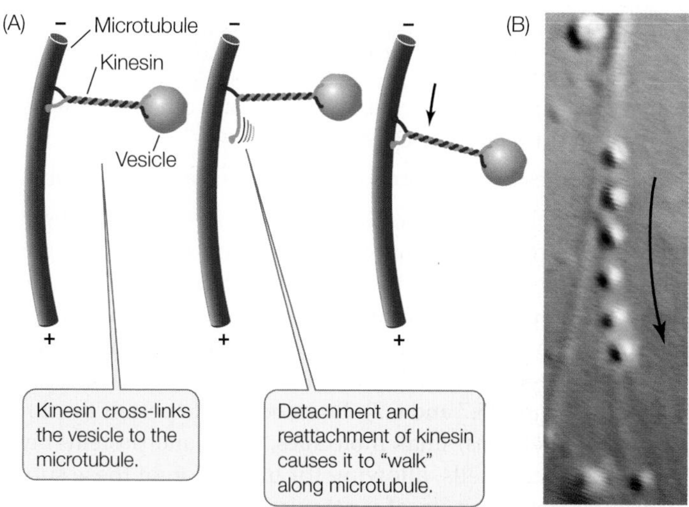


Detachment and reattachment of kinesin causes it to "walk" along microtubule.

**5.22 A Motor Protein Drives Vesicles along Microtubules**
(A) Kinesin delivers vesicles or organelles to various parts of the cell by moving along microtubule "railroad tracks." Kinesin moves things from the minus toward the plus end of a microtubule; dynein works similarly, but moves from the plus toward the minus end. (B) Powered by kinesin, a vesicle moves along a microtubule track in the protist *Dictyostelium*. The time sequence (time-lapse micrography at half-second intervals) is shown by the color gradient of purple to blue.

- *Inhibition*: use a drug that inhibits A and see if B still occurs. If it does not, then A is probably a causative factor for B. **Figure 5.23** shows an experiment with such a drug (an inhibitor) that demonstrates cause and effect in the case of the cytoskeleton and cell movement.
- *Mutation*: examine a cell that lacks the gene (or genes) for A and see if B still occurs. If it does not, then A is probably a causative factor for B. Part Four of this book describes many experiments using this genetic approach.

**yourBioPortal.com**
GO TO **Animated Tutorial 5.2 • Eukaryotic Cell Tour**

## 5.3 RECAP

**The hallmark of eukaryotic cells is compartmentalization. Membrane-enclosed organelles process information, transform energy, form internal compartments for transporting proteins, and carry out intracellular digestion. An internal cytoskeleton plays several structural roles.**

- What are some advantages of organelle compartmentalization? **See p. 84**
- Describe the structural and functional differences between rough and smooth endoplasmic reticulum. **See pp. 89–90 and Figure 5.10**
- Explain how motor proteins and microtubules move materials within the cell. **See pp. 95–98 and Figures 5.21 and 5.22**

## INVESTIGATING LIFE

**5.23 The Role of Microfilaments in Cell Movement—Showing Cause and Effect in Biology**
After a test tube demonstration that the drug cytochalasin B prevented microfilament formation from monomeric precursors, the question was asked: Will the drug work like this in living cells and inhibit cell movement in *Amoeba*? Complementary experiments showed that the drug did not poison other cellular processes.

**HYPOTHESIS** Amoeboid cell movements are caused by the cytoskeleton.

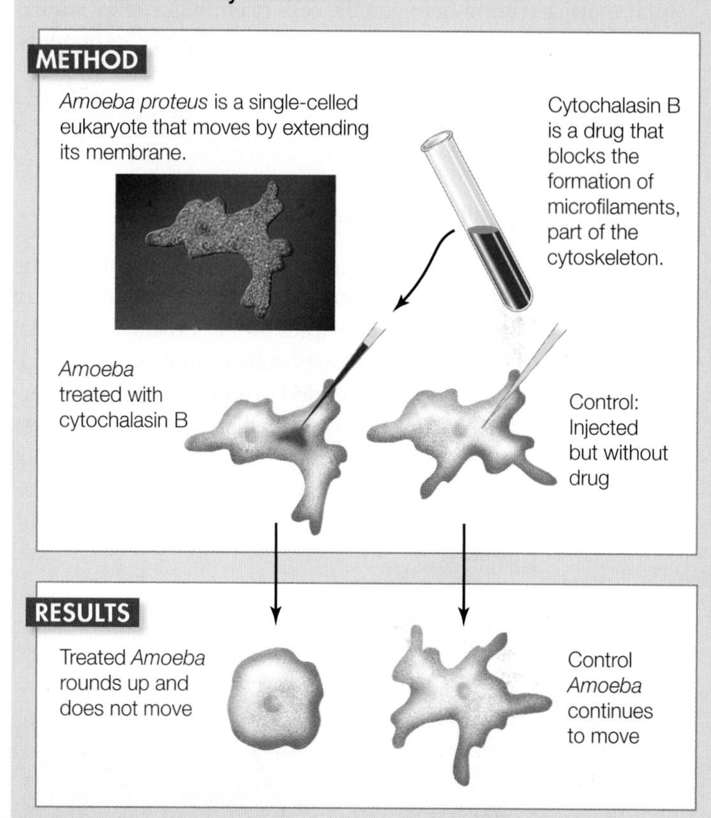

METHOD
Amoeba proteus is a single-celled eukaryote that moves by extending its membrane.
Cytochalasin B is a drug that blocks the formation of microfilaments, part of the cytoskeleton.
Amoeba treated with cytochalasin B
Control: Injected but without drug
RESULTS
Treated Amoeba rounds up and does not move
Control Amoeba continues to move


**CONCLUSION** Microfilaments of the cytoskeleton are essential for amoeboid cell movement.

**FURTHER INVESTIGATION:** The drug colchicine breaks apart microtubules. How would you show that these components of the cytoskeleton are not involved in cell movement in *Amoeba*?

Go to **yourBioPortal.com** for original citations, discussions, and relevant links for all INVESTIGATING LIFE figures.

All cells interact with their environments, and many eukaryotic cells are parts of multicellular organisms and must interact, and closely coordinate activities, with other cells. The plasma membrane plays a crucial role in these interactions, but other structures outside that membrane are involved as well.

## 5.4 What Are the Roles of Extracellular Structures?

Although the plasma membrane is the functional barrier between the inside and the outside of a cell, many structures are produced by cells and secreted to the outside of the plasma membrane, where they play essential roles in protecting, supporting, or attaching cells to each other. Because they are outside the plasma membrane, these structures are said to be *extracellular*. The peptidoglycan cell wall of bacteria is an example of an extracellular structure (see Figure 5.4). In eukaryotes, other extracellular structures—the cell walls of plants and the extracellular matrices found between the cells of animals—play similar roles. Both of these structures are made up of two components:

- a prominent fibrous macromolecule
- a gel-like medium in which the fibers are embedded

### The plant cell wall is an extracellular structure

The plant **cell wall** is a semirigid structure outside the plasma membrane (**Figure 5.24**). We consider the structure and role of the cell wall in more detail in Chapter 34. For now, we note that it is typical of a two-component extracellular matrix, with cellulose fibers (see Figure 3.16) embedded in other complex polysaccharides and proteins. The plant cell wall has three major roles:

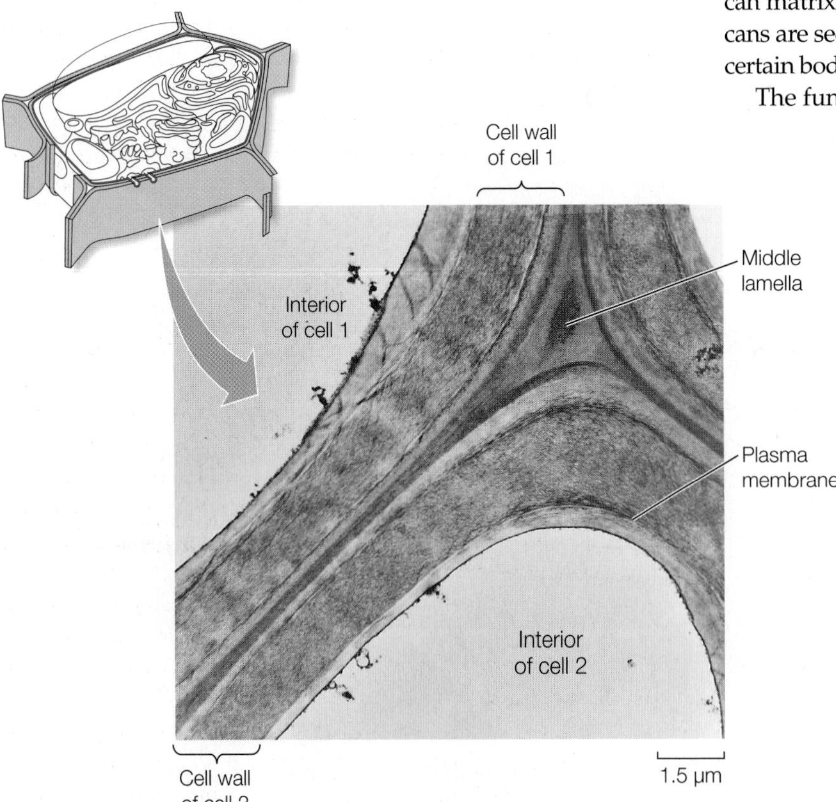


**5.24 The Plant Cell Wall** The semirigid cell wall provides support for plant cells. It is composed of cellulose fibrils embedded in a matrix of polysaccharides and proteins.

- It provides support for the cell and limits its volume by remaining rigid.
- It acts as a barrier to infection by fungi and other organisms that can cause plant diseases.
- It contributes to plant form by growing as plant cells expand.

Because of their thick cell walls, plant cells viewed under a light microscope appear to be entirely isolated from one another. But electron microscopy reveals that this is not the case. The cytoplasms of adjacent plant cells are connected by numerous plasma membrane–lined channels, called **plasmodesmata**, that are about 20–40 nm in diameter and extend through the cell walls (see Figures 5.7 and 6.7). Plasmodesmata permit the diffusion of water, ions, small molecules, RNA, and proteins between connected cells, allowing for utilization of these substances far from their site of synthesis.

### The extracellular matrix supports tissue functions in animals

Animal cells lack the semirigid wall that is characteristic of plant cells, but many animal cells are surrounded by, or in contact with, an **extracellular matrix**. This matrix is composed of three types of molecules: fibrous proteins such as **collagen** (the most abundant protein in mammals, constituting over 25 percent of the protein in the human body); a matrix of glycoproteins termed **proteoglycans**, consisting primarily of sugars; and a third group of proteins that link the fibrous proteins and the gel-like proteoglycan matrix together (**Figure 5.25**). These proteins and proteoglycans are secreted, along with other substances that are specific to certain body tissues, by cells that are present in or near the matrix.

The functions of the extracellular matrix are many:

- It holds cells together in tissues. In Chapter 6 we see how there is an intercellular "glue" that is involved in both cell recognition and adhesion.
- It contributes to the physical properties of cartilage, skin, and other tissues. For example, the mineral component of bone is laid down on an organized extracellular matrix.
- It helps filter materials passing between different tissues. This is especially important in the kidney.
- It helps orient cell movements during embryonic development and during tissue repair.
- It plays a role in chemical signaling from one cell to another. Proteins connect the cell's plasma membrane to the extracellular matrix. These proteins (for example, *integrin*) span the plasma membrane and are involved with transmitting signals to the interior of the cell. This allows communication between the extracellular matrix and the cytoplasm of the cell.

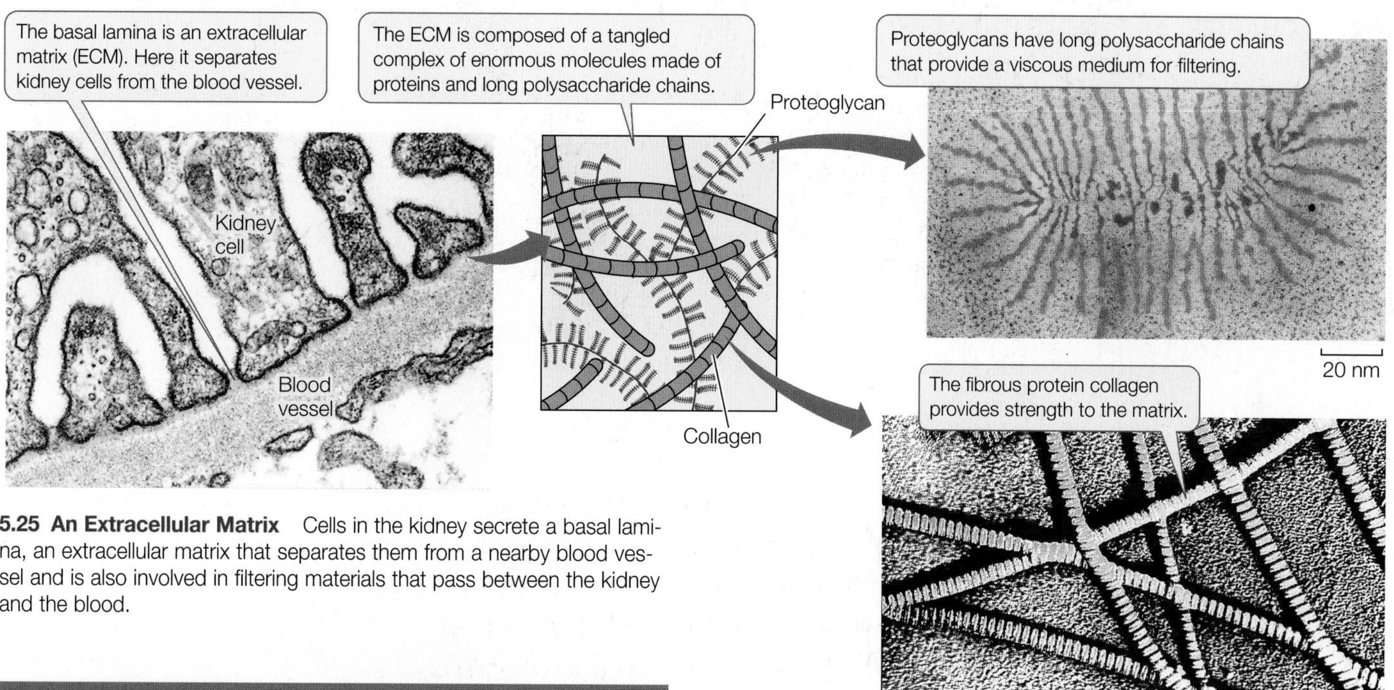

The basal lamina is an extracellular matrix (ECM). Here it separates kidney cells from the blood vessel.
The ECM is composed of a tangled complex of enormous molecules made of proteins and long polysaccharide chains.
Proteoglycans have long polysaccharide chains that provide a viscous medium for filtering.
Proteoglycan
Kidney cell
Blood vessel
Collagen
20 nm
The fibrous protein collagen provides strength to the matrix.
100 nm

**5.25 An Extracellular Matrix** Cells in the kidney secrete a basal lamina, an extracellular matrix that separates them from a nearby blood vessel and is also involved in filtering materials that pass between the kidney and the blood.

## 5.4 RECAP

**Extracellular structures are produced by cells and secreted outside the plasma membrane. Most consist of a fibrous component in a gel-like medium.**

- What are the functions of the cell wall in plants and the extracellular matrix in animals? See p. 100

We have now discussed the structures and some functions of prokaryotic and eukaryotic cells. Both exemplify the cell theory, showing that cells are the basic units of life and of biological continuity. Much of the rest of this part of the book will deal with these two aspects of cells. There is abundant evidence that the simpler prokaryotic cells are more ancient than eukaryotic cells, and that the first cells were probably prokaryotic. We now turn to the next step in cellular evolution, the origin of eukaryotic cells.

## 5.5 How Did Eukaryotic Cells Originate?

For about 2 billion years, life on Earth was entirely prokaryotic—from the time when prokaryotic cells first appeared until about 1.5 billion years ago, when eukaryotic cells arrived on the scene. The advent of compartmentalization—the hallmark of eukaryotes—was a major event in the history of life, as it permitted many more biochemical functions to coexist in the same cell than had previously been possible. Compared to the typical eukaryote, a single prokaryotic cell is often biochemically specialized, limited in the resources it can use and the functions it can perform.

What is the origin of compartmentalization? We will describe the evolution of eukaryotic organelles in more detail in Section 27.1. Here, we outline two major themes in this process.

### Internal membranes and the nuclear envelope probably came from the plasma membrane

We noted earlier that some bacteria contain internal membranes. How could these arise? In electron micrographs, the internal membranes of prokaryotes often appear to be inward folds of the plasma membrane. This has led to a theory that the endomembrane system and cell nucleus originated by a related process (**Figure 5.26A**). The close relationship between the ER and the nuclear envelope in today's eukaryotes is consistent with this theory.

A bacterium with enclosed compartments would have several evolutionary advantages. Chemicals could be concentrated within particular regions of the cell, allowing chemical reactions to proceed more efficiently. Biochemical activities could be segregated within organelles with, for example, a different pH from the rest of the cell, creating more favorable conditions for certain metabolic processes. Finally, gene transcription could be separated from translation, providing more opportunities for separate control of these steps in gene expression.

### Some organelles arose by endosymbiosis

**Symbiosis** means "living together," and often refers to two organisms that coexist, each one supplying something that the other needs. Biologists have proposed that some organelles—the mitochondria and the plastids—arose not by an infolding of the plasma membrane but by one cell ingesting another cell, giving rise to a symbiotic relationship. Eventually, the ingested cell lost its autonomy and some of its functions. In addition, many of the ingested cell's genes were transferred to the host's DNA. Mitochondria and plastids in today's eukaryotic cells are the remnants of these *symbionts*, retaining some specialized functions that benefit their host cells. This is the essence of the **endosymbiosis theory** for the origin of organelles.

Consider the case of the plastid. About 2.5 billion years ago some prokaryotes (the cyanobacteria) developed photosynthesis (see Figure 1.9). The emergence of these prokaryotes was a key event in the evolution of complex organisms, because they increased the $O_2$ concentration in Earth's atmosphere (see Section 1.2).

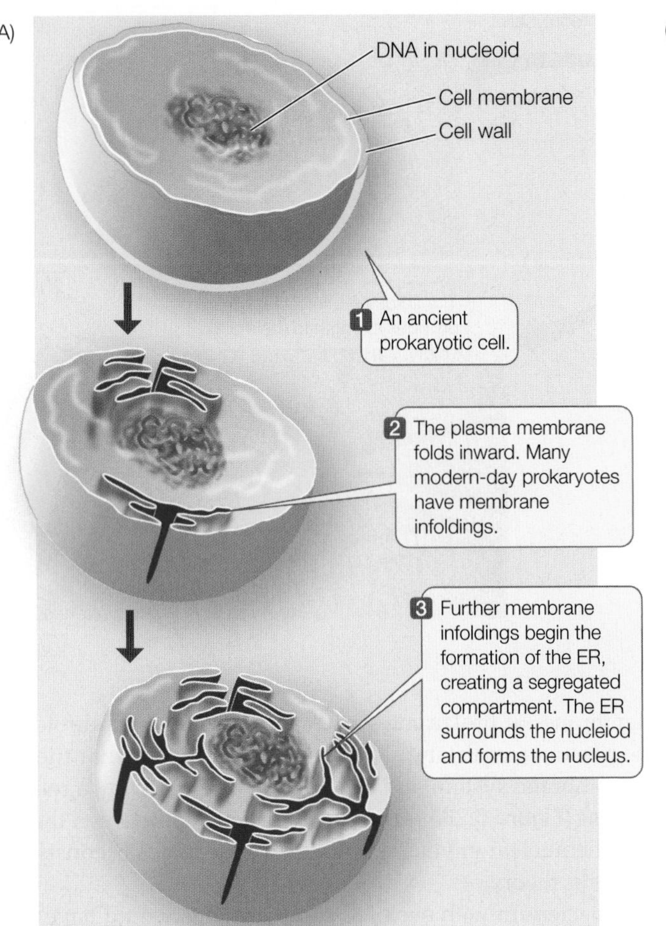


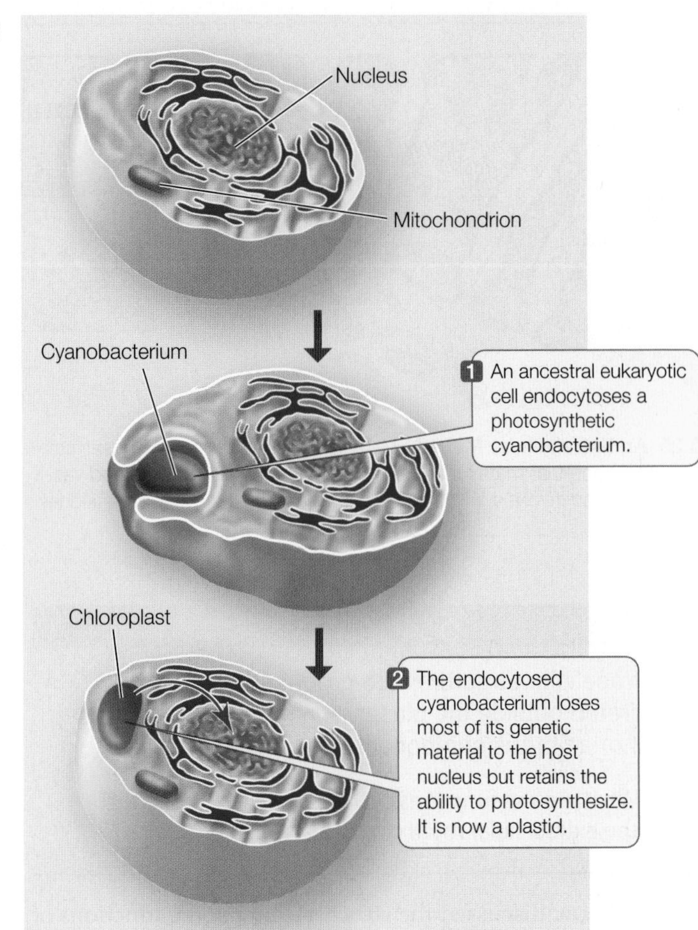


**5.26 The Origin of Organelles** (A) The endomembrane system and cell nucleus may have been formed by infolding and then fusion of the plasma membrane. (B) The endosymbiosis theory proposes that some organelles may be descended from prokaryotes that were engulfed by other, larger cells.

According to endosymbiosis theory, photosynthetic prokaryotes also provided the precursor of the modern-day plastid. Cells without cell walls can engulf relatively large particles by phagocytosis (see Figure 5.11). In some cases, such as that of phagocytes in the human immune system, the engulfed particle can be an entire cell, such as a bacterium. Plastids may have arisen by a similar event involving an ancestral eukaryote and a cyanobacterium (**Figure 5.26B**).

Among the abundant evidence supporting the endosymbiotic origin of plastids (see Section 27.1), perhaps the most remarkable comes from a sandy beach in Japan. Noriko Okamoto and Isao Inouye recently discovered a single-celled eukaryote that contains a large "chloroplast," and named it *Hatena* (**Figure 5.27**). It turns out that the "chloroplast" is the remains of a green alga, *Nephroselmis*, which lives among the *Hatena* cells. When living autonomously, this algal cell has flagella, a cytoskeleton, ER, Golgi, and mitochondria in addition to a plastid. Once ingested by *Hatena*, all of these structures, and presumably their associated functions, are lost. What remains is essentially a plastid.

When *Hatena* divides, only one of the two daughter cells ends up with the "chloroplast." The other cell finds and ingests its own *Nephroselmis* alga—almost like a "replay" of what may have occurred in the evolution of eukaryotic cells. No wonder the Japanese scientists call the host cell *Hatena*: in Japanese, it means "how odd"!

## 5.5 RECAP

**Eukaryotic cells arose long after prokaryotic cells. Some organelles may have evolved by infolding of the plasma membrane, while others evolved by endosymbiosis.**

- How could membrane infolding in a prokaryotic cell lead to the endomembrane system? **See p. 101 and Figure 5.26A**
- Explain the endosymbiosis theory for the origin of chloroplasts. **See Figure 5.26B**

In this chapter, we presented an overview of the structures of cells, with some ideas about their relationships and origins. As you now embark on the study of major cell functions, keep in

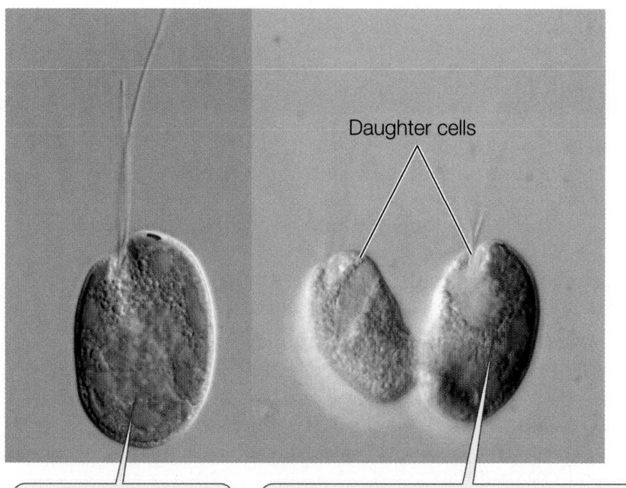

Daughter cells
Cell with ingested green photosynthetic plastid.
After cell division, only one of the daughter cells inherits the plastid; the other cell must ingest a new one from the environment.

**5.27 Endosymbiosis in Action** A *Hatena* cell engulfs an algal cell, which then loses most of its cellular functions other than photosynthesis. This re-enacts a possible event in the origin of plastids in eukaryotic cells.

mind that the structures in a cell do not exist in isolation. They are part of a dynamic, interacting cellular system. In Chapter 6 we show that the plasma membrane is far from a passive barrier, but instead is a multi-functional system that connects the inside of the cell with its extracellular environment.

## CHAPTER SUMMARY

### 5.1 What Features Make Cells the Fundamental Units of Life?

**SEE WEB ACTIVITIES 5.1 AND 5.2**

- All cells come from preexisting cells.
- Cells are small because a cell's surface area must be large compared with its volume to accommodate exchanges with its environment. **Review Figure 5.2**
- All cells are enclosed by a selectively permeable **plasma membrane** that separates their contents from the external environment.
- While certain biochemical processes, molecules, and structures are shared by all kinds of cells, two categories of cells—**prokaryotes** and **eukaryotes**—are easily distinguished.

### 5.2 What Features Characterize Prokaryotic Cells?

- Prokaryotic cells have no internal compartments, but have a **nucleoid** region containing DNA, and a **cytoplasm** containing **cytosol**, **ribosomes**, proteins, and small molecules. Some prokaryotes have additional protective structures, including a **cell wall**, an **outer membrane**, and a **capsule**. **Review Figure 5.4**
- Some prokaryotes have folded membranes that may be photosynthetic membranes, and some have **flagella** or **pili** for motility or attachment. **Review Figure 5.5**

### 5.3 What Features Characterize Eukaryotic Cells?

- Eukaryotic cells are larger than prokaryotic cells and contain many membrane-enclosed **organelles**. The membranes that envelop organelles ensure compartmentalization of their functions. **Review Figure 5.7**
- **Ribosomes** are sites of protein synthesis.
- The **nucleus** contains most of the cell's DNA and participates in the control of protein synthesis. **Review Figure 5.8**
- The **endomembrane system**—consisting of the **endoplasmic reticulum** and **Golgi apparatus**—is a series of interrelated compartments enclosed by membranes. It segregates proteins and modifies them. **Lysosomes** contain many digestive enzymes. **Review Figures 5.10 and 5.11, WEB ACTIVITY 5.3, ANIMATED TUTORIAL 5.1**
- **Mitochondria** and **chloroplasts** are semi-autonomous organelles that process energy. Mitochondria are present in most eukaryotic organisms and contain the enzymes needed for cellular respiration. The cells of photosynthetic eukaryotes contain chloroplasts that harvest light energy for photosynthesis. **Review Figures 5.12 and 5.13**
- **Vacuoles** are prominent in many plant cells and consist of a membrane-enclosed compartment full of water and dissolved substances.
- The **microfilaments**, **intermediate filaments**, and **microtubules** of the **cytoskeleton** provide the cell with shape, strength, and movement. **Review Figure 5.18**

**SEE ANIMATED TUTORIAL 5.2**

### 5.4 What Are the Roles of Extracellular Structures?

- The plant **cell wall** consists principally of **cellulose**. Cell walls are pierced by **plasmodesmata** that join the cytoplasms of adjacent cells.
- In animals, the **extracellular matrix** consists of different kinds of proteins, including collagen and proteoglycans. **Review Figure 5.25**

### 5.5 How Did Eukaryotic Cells Originate?

- Infoldings of the plasma membrane could have led to the formation of some membrane-enclosed organelles, such as the endomembrane system and the nucleus. **Review Figure 5.26A**
- The **endosymbiosis theory** states that mitochondria and chloroplasts originated when larger prokaryotes engulfed, but did not digest, smaller prokaryotes. Mutual benefits permitted this symbiotic relationship to be maintained, allowing the smaller cells to evolve into the eukaryotic organelles observed today. **Review Figure 5.26B**

## SELF-QUIZ

1. Which structure is generally present in both prokaryotic cells and eukaryotic plant cells?
   *a.* Chloroplasts
   *b.* Cell wall
   *c.* Nucleus
   *d.* Mitochondria
   *e.* Microtubules
2. The major factor limiting cell size is the
   *a.* concentration of water in the cytoplasm.
   *b.* need for energy.
   *c.* presence of membrane-enclosed organelles.
   *d.* ratio of surface area to volume.
   *e.* composition of the plasma membrane.
3. Which statement about mitochondria is *not* true?
   *a.* The inner mitochondrial membrane folds to form cristae.
   *b.* The outer membrane is relatively permeable to macromolecules.
   *c.* Mitochondria are green because they contain chlorophyll.
   *d.* Fuel molecules from the cytosol are used for respiration in mitochondria.
   *e.* ATP is synthesized in mitochondria.
4. Which statement about plastids is true?
   *a.* They are found in prokaryotes.
   *b.* They are surrounded by a single membrane.
   *c.* They are the sites of cellular respiration.
   *d.* They are found only in fungi.
   *e.* They may contain several types of pigments or polysaccharides.
5. If all the lysosomes within a cell suddenly ruptured, what would be the most likely result?
   *a.* The macromolecules in the cytosol would break down.
   *b.* More proteins would be made.
   *c.* The DNA within mitochondria would break down.
   *d.* The mitochondria and chloroplasts would divide.
   *e.* There would be no change in cell function.
6. The Golgi apparatus
   *a.* is found only in animals.
   *b.* is found in prokaryotes.
   *c.* is the appendage that moves a cell around in its environment.
   *d.* is a site of rapid ATP production.
   *e.* modifies and packages proteins.
7. Which structure is *not* surrounded by one or more membranes?
   *a.* Ribosome
   *b.* Chloroplast
   *c.* Mitochondrion
   *d.* Peroxisome
   *e.* Vacuole
8. The cytoskeleton consists of
   *a.* cilia, flagella, and microfilaments.
   *b.* cilia, microtubules, and microfilaments.
   *c.* internal cell walls.
   *d.* microtubules, intermediate filaments, and microfilaments.
   *e.* calcified microtubules.
9. Microfilaments
   *a.* are composed of polysaccharides.
   *b.* are composed of actin.
   *c.* allow cilia and flagella to move.
   *d.* make up the spindle that aids the movement of chromosomes.
   *e.* maintain the position of the chloroplast in the cell.
10. Which statement about the plant cell wall is *not* true?
   *a.* Its principal chemical components are polysaccharides.
   *b.* It lies outside the plasma membrane.
   *c.* It provides support for the cell.
   *d.* It completely isolates adjacent cells from one another.
   *e.* It is semirigid.

## FOR DISCUSSION

1. The drug vincristine is used to treat many cancers. It apparently works by causing microtubules to depolymerize. Vincristine use has many side effects, including loss of dividing cells and nerve problems. Explain why this might be so.
2. Through how many membranes would a molecule have to pass in moving from the interior (stroma) of a chloroplast to the interior (matrix) of a mitochondrion? From the interior of a lysosome to the outside of a cell? From one ribosome to another?
3. How does the possession of double membranes by chloroplasts and mitochondria relate to the endosymbiosis theory of the origins of these organelles? What other evidence supports the theory?
4. Compare the extracellular matrix of the animal cell with the plant cell wall, with respect to composition of the fibrous and nonfibrous components, rigidity, and connectivity of cells.

## ADDITIONAL INVESTIGATION

The pathway of newly synthesized proteins can be followed through the cell using a "pulse-chase" experiment. During synthesis, proteins are tagged with a radioactive isotope (the "pulse"), and then the cell is allowed to process the proteins for varying periods of time. The locations of the radioactive proteins are then determined by isolating cell organelles and quantifying their radioactivity. How would you use this method, and what results would you expect for (*a*) a lysosomal enzyme and (*b*) a protein that is released from the cell?

# 6 Cell Membranes

## Membranes and memory

James noticed the changes in his grandfather when he was home from college for the winter holiday. He and grandpa John had always joked about grandpa John's missing keys and glasses; the old man, who had lived with James' family since his wife died, was forever searching for them. Now the memory lapses had become more pronounced. When James introduced his new girlfriend to the family, he was relieved (as was she) when she was welcomed with open arms. But an hour later, grandpa John just stared at her, unable to remember who she was. By the time James came home for the summer, his grandfather had become withdrawn; he could no longer talk about current events, and often he became confused and lashed out in anger.

James' grandfather had Alzheimer's disease. This condition is most common in (but not limited to) the elderly, and as more people today are living to advanced ages, more and more Alzhiemer's cases are diagnosed. But the symptoms are not new to human experience or to medicine. The condition was first recognized as a disease in 1901. That year, the family of 51-year-old Frau Auguste D. brought her to Dr. Alois Alzheimer at the Frankfurt hospital in Germany. She had severe memory lapses, accused her husband of infidelity, and had difficulty communicating. These symptoms got worse before she died several years later. When Alzheimer autopsied her brain, he saw that the parts of the brain that are important in thought and speech were shrunken. Moreover, when he examined these areas through the microscope he saw abnormal protein deposits in and around the brain cells.

In the century since Alzheimer's original case, cell biologists have investigated the nature of these abnormal deposits, now known as *plaques*. It turns out that the key events that produce plaques take place in the plasma membrane of nerve cells in the brain. Plaques are clumps of the protein amyloid beta, which at high levels is toxic to brain cells. Amyloid beta is a small piece of a larger amyloid precursor protein (APP), which is embedded in the nerve cell plasma membrane; APP is cut twice by two other membrane proteins, β-secretase and γ-secretase, to produce amyloid beta, which is released from the membrane to fall outside of the cell. All these proteins are present in a variety of animal cells and have multiple important

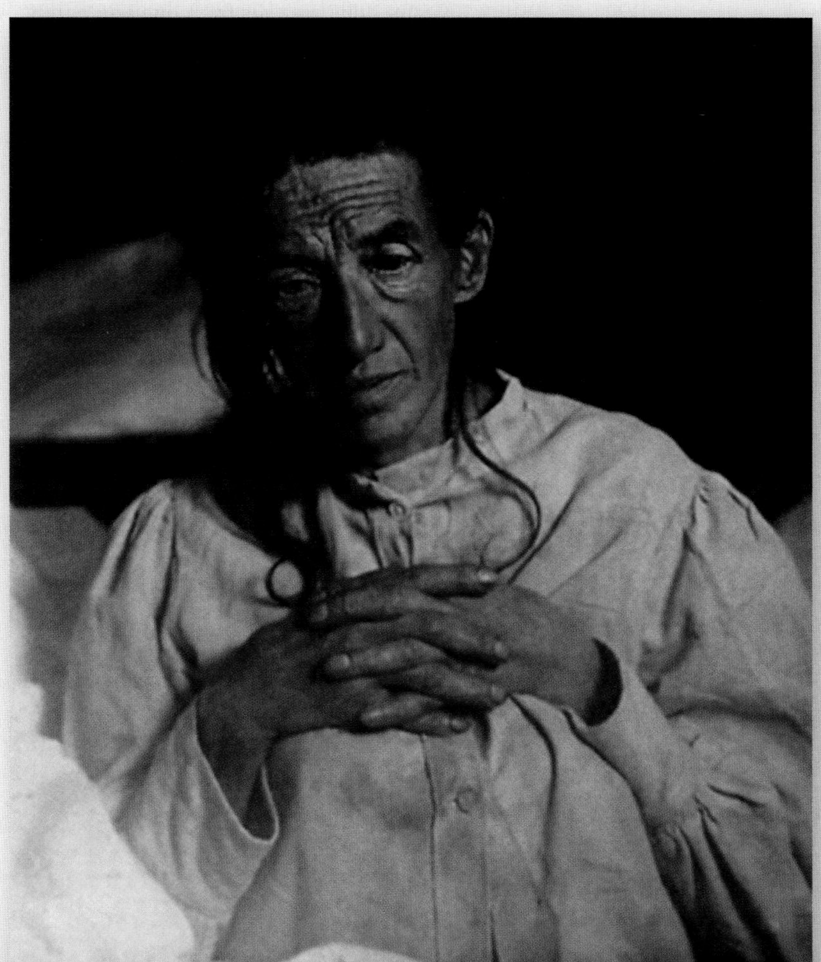

**Dr. Alzheimer's Patient** Frau Auguste D., who died in 1906, was the first patient described with progressive dementia by Dr. Alois Alzheimer.

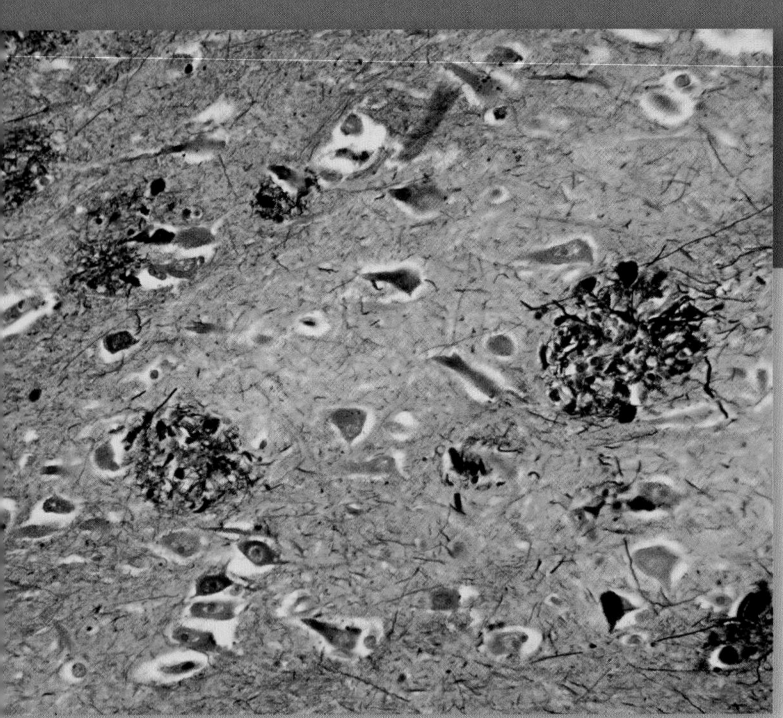

**Plaques in the Brain** At autopsy, the brain of an Alzheimer's disease patient accumulates plaques (dark fibers in this micrograph) composed of protein fragments produced by an enzyme in the nerve cell membrane.

roles in the dynamic cell membrane; they may even be essential for normal nervous system development and function.

So what goes wrong in Alzheimer's disease? Cells in the diseased brain might be producing too much amyloid beta (e.g., because $\gamma$-secretase is too active) or producing it at the wrong time (e.g., in old age instead of infancy). One form of the disease is caused by a mutant form of $\gamma$-secretase, which has a tendency to cut APP in the "wrong" place, thereby producing a particularly toxic form of amyloid beta. Because of their role in producing plaques, APP and $\gamma$-secretase are potential targets for Alzheimer's disease therapies.

Learning how membranes are made and how they work has been a key to understanding, and perhaps treating, this increasingly prevalent disease.

**IN THIS CHAPTER** we focus on the structure and functions of biological membranes. First we describe the composition and structure of biological membranes. We go on to discuss their functions—how membranes are involved in intercellular interactions, and how membranes regulate which substances enter and leave the cell.

CHAPTER OUTLINE



## 6.1 What Is the Structure of a Biological Membrane?

The physical organization and functioning of all biological membranes depend on their constituents: lipids, proteins, and carbohydrates. You are already familiar with these molecules from Chapter 3; it may be useful to review that chapter now. The lipids establish the physical integrity of the membrane and create an effective barrier to the rapid passage of hydrophilic materials such as water and ions. In addition, the phospholipid bilayer serves as a lipid "lake" in which a variety of proteins "float" (**Figure 6.1**). This general design is known as the **fluid mosaic model**.

In the fluid mosaic model for biological membranes, the proteins are noncovalently embedded in the phospholipid bilayer by their hydrophobic regions (or *domains*), but their hydrophilic domains are exposed to the watery conditions on either side of the bilayer. These membrane proteins have a number of functions, including moving materials through the membrane and receiving chemical signals from the cell's external environment. Each membrane has a set of proteins suitable for the specialized functions of the cell or organelle it surrounds.

The carbohydrates associated with membranes are attached either to the lipids or to protein molecules. In plasma membranes, carbohydrates are located on the outside of the cell, where they may interact with substances in the external environment. Like some of the membrane proteins, carbohydrates are crucial in recognizing specific molecules, such as those on the surfaces of adjacent cells.

Although the fluid mosaic model is largely valid for membrane structure, it does not say much about membrane composition. As you read about the different molecules in membranes in the next sections, keep in mind that some membranes have more protein than lipids, others are lipid-rich, others have significant amounts of cholesterol or other sterols, and still others are rich in carbohydrates.

### Lipids form the hydrophobic core of the membrane

The lipids in biological membranes are usually *phospholipids*. Recall from Section 2.2 that some compounds are hydrophilic ("water-loving") and others are hydrophobic ("water-hating"), and from Section 3.4 that a phospholipid molecule has regions of both kinds:

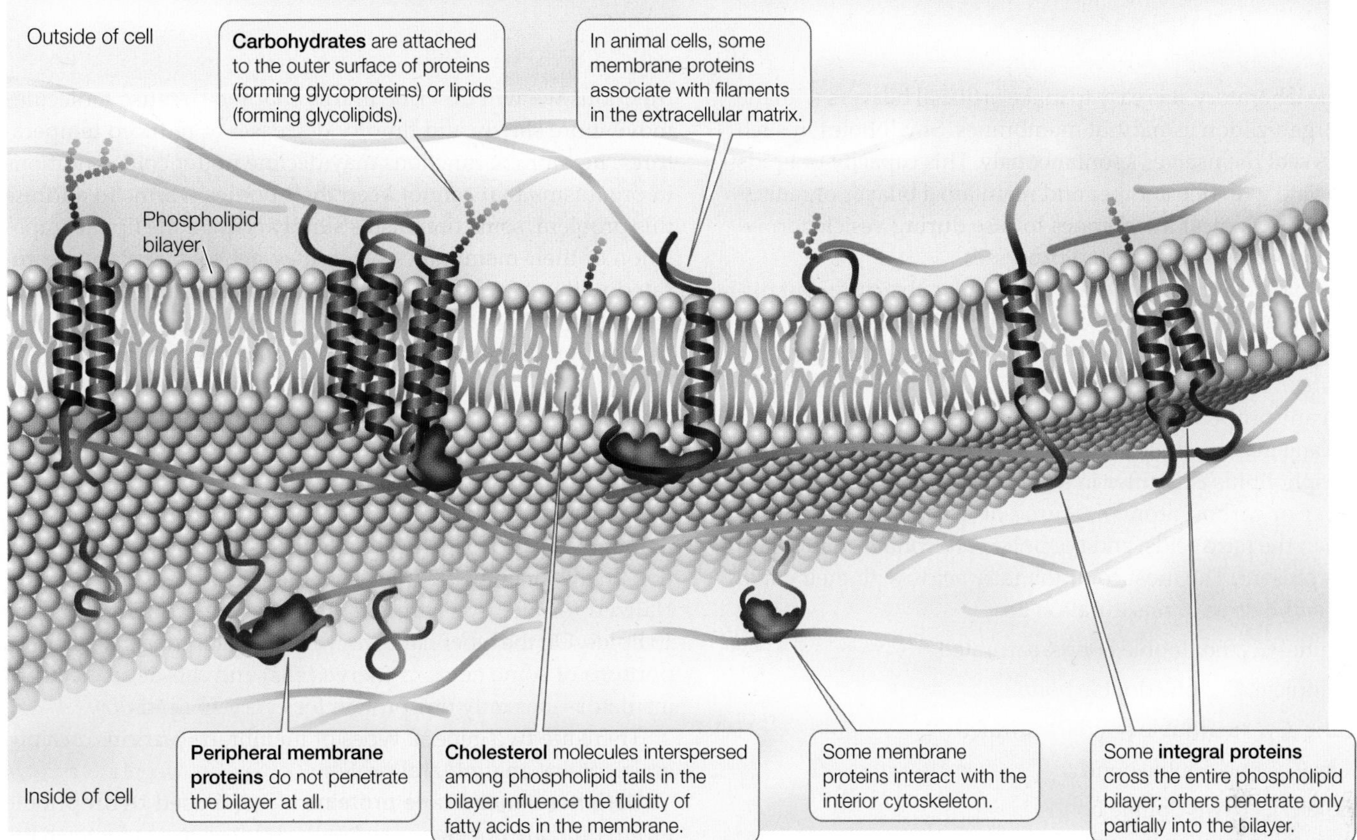


**6.1 The Fluid Mosaic Model** The general molecular structure of biological membranes is a continuous phospholipid bilayer which has proteins embedded in or associated with it.

**yourBioPortal.com**
GO TO **Web Activity 6.1 • The Fluid-Mosaic Model**

- *Hydrophilic regions*: The phosphorus-containing "head" of the phospholipid is electrically charged and therefore associates with polar water molecules.
- *Hydrophobic regions*: The long, nonpolar fatty acid "tails" of the phospholipid associate with other nonpolar materials, but they do not dissolve in water or associate with hydrophilic substances.

Because of these properties, one way in which phospholipids can coexist with water is to form a *bilayer*, with the fatty acid "tails" of the two layers interacting with each other and the polar "heads" facing the outside aqueous environment (**Figure 6.2**). The thickness of a biological membrane is about 8 nm (0.008 µm), which is twice the length of a typical phospholipid—another indication that the membrane consists of a lipid bilayer. This thickness is about 8,000 times thinner than a piece of paper.

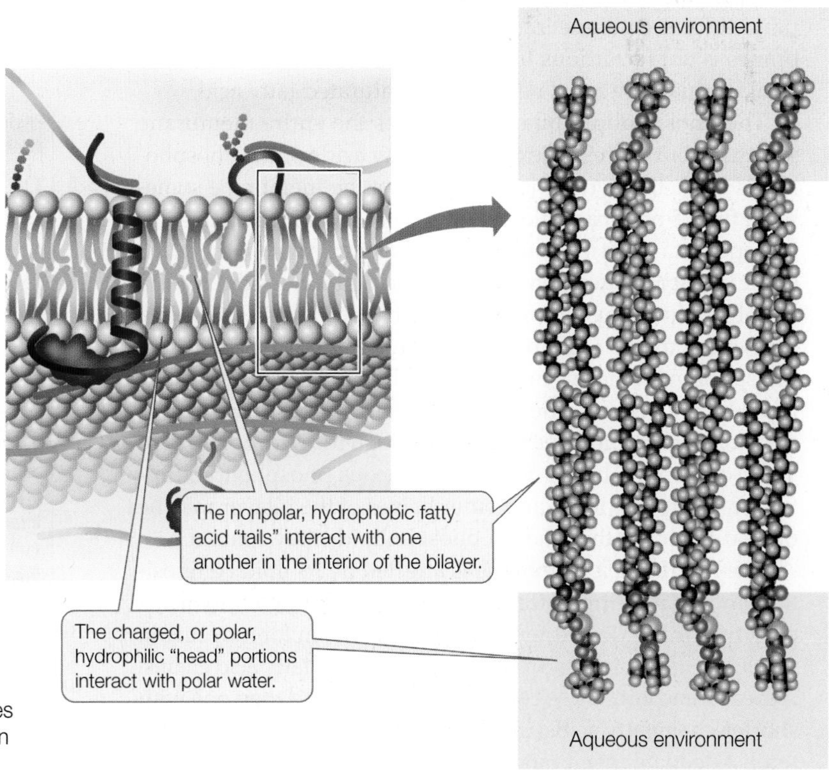

Aqueous environment
The nonpolar, hydrophobic fatty acid "tails" interact with one another in the interior of the bilayer.
The charged, or polar, hydrophilic "head" portions interact with polar water.
Aqueous environment

**6.2 A Phospholipid Bilayer** The phospholipid bilayer separates two aqueous regions. The eight phospholipid molecules shown on the right represent a small cross section of a membrane bilayer.

In the laboratory, it is easy to make artificial bilayers with the same organization as natural membranes. Small holes in such bilayers seal themselves spontaneously. This capacity of lipids to associate with one another and maintain a bilayer organization helps biological membranes to fuse during vesicle formation, phagocytosis, and related processes.

All biological membranes have a similar structure, but differ in the kinds of proteins and lipids they contain. Membranes from different cells or organelles may differ greatly in their *lipid composition*. Not only are phospholipids highly variable, but a significant proportion of the lipid content in an animal cell membrane may be cholesterol.

Phospholipids can differ in terms of fatty acid chain length (number of carbon atoms), degree of unsaturation (double bonds) in the fatty acids, and the polar (phosphate-containing) groups present. The most common fatty acids with their chain length and degree of unsaturation are:

- Palmitic: $C_{14}$, no double bonds, saturated
- Palmitoleic: $C_{16}$, one double bond
- Stearic: $C_{18}$, no double bonds, saturated
- Oleic: $C_{18}$, one double bond
- Linoleic: $C_{18}$, two double bonds
- Linolenic: $C_{18}$, three double bonds

The saturated fatty acid chains allow close packing of fatty acids in the bilayer, while the "kinks" in unsaturated fatty acids (see Figure 3.19) make for a less dense, more fluid packing. These less-dense membranes in animal cells can accommodate cholesterol molecules.

Up to 25 percent of the lipid content of an animal cell plasma membrane may be cholesterol. When present, cholesterol is important for membrane integrity; the cholesterol in your membranes is not hazardous to your health. A molecule of cholesterol is usually situated next to an unsaturated fatty acid.

The phospholipid bilayer stabilizes the entire membrane structure, but leaves it flexible. The fatty acids of the phospholipids make the hydrophobic interior of the membrane somewhat fluid—about as fluid as lightweight machine oil. This fluidity permits some molecules to move laterally within the plane of the membrane. A given phospholipid molecule in the plasma membrane can travel from one end of the cell to the other in a little more than a second! On the other hand, seldom does a phospholipid molecule in one half of the bilayer spontaneously flip over to the other side. For that to happen, the polar part of the molecule would have to move through the hydrophobic interior of the membrane. Since spontaneous phospholipid flip-flops are rare, the inner and outer halves of the bilayer may be quite different in the kinds of phospholipids they contain.

The fluidity of a membrane is affected by its lipid composition and by its temperature. Long-chain, saturated fatty acids pack tightly beside one another, with little room for movement. Cholesterol interacts hydrophobically with the fatty acid chains. A membrane with these components is less fluid than one with shorter-chain fatty acids, unsaturated fatty acids, or less cholesterol. Adequate membrane fluidity is essential for many of the functions we will describe in this chapter. Because molecules move more slowly and fluidity decreases at reduced temperatures, membrane functions may decline under cold conditions in organisms that cannot keep their bodies warm. To address this problem, some organisms simply change the lipid composition of their membranes when they get cold, replacing saturated with unsaturated fatty acids and using fatty acids with shorter tails. These changes play a role in the survival of plants, bacteria, and hibernating animals during the winter.

### Membrane proteins are asymmetrically distributed

All biological membranes contain proteins. Typically, plasma membranes have one protein molecule for every 25 phospholipid molecules. This ratio varies depending on membrane function. In the inner membrane of the mitochondrion, which is specialized for energy processing, there is one protein for every 15 lipids. On the other hand, myelin—a membrane that encloses portions of some neurons (nerve cells) and acts as an electrical insulator—has only one protein for every 70 lipids.

There are two general types of membrane proteins: peripheral proteins and integral proteins.

**Peripheral membrane proteins** lack exposed hydrophobic groups and are not embedded in the bilayer. Instead, they have polar or charged regions that interact with exposed parts of integral membrane proteins, or with the polar heads of phospholipid molecules (see Figure 6.1).

**Integral membrane proteins** are at least partly embedded in the phospholipid bilayer (see Figure 6.1). Like phospholipids, these proteins have both hydrophilic and hydrophobic regions (**Figure 6.3**).

- *Hydrophilic domains*: Stretches of amino acids with hydrophilic side chains (see Table 3.1) give certain regions of the

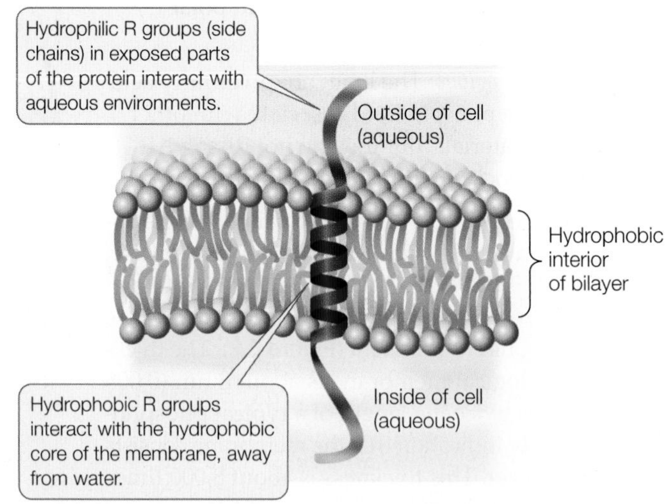
Hydrophilic R groups (side chains) in exposed parts of the protein interact with aqueous environments.
Outside of cell (aqueous)
Hydrophobic interior of bilayer
Hydrophobic R groups interact with the hydrophobic core of the membrane, away from water.
Inside of cell (aqueous)

**6.3 Interactions of Integral Membrane Proteins** An integral membrane protein is held in the membrane by the distribution of the hydrophilic and hydrophobic side chains on its amino acids. The hydrophilic parts of the protein extend into the aqueous cell exterior and the internal cytoplasm. The hydrophobic side chains interact with the hydrophobic lipid core of the membrane.

protein a polar character. These hydrophilic domains interact with water and stick out into the aqueous environment inside or outside the cell.

- *Hydrophobic domains*: Stretches of amino acids with hydrophobic side chains give other regions of the protein a nonpolar character. These domains interact with the fatty acids in the interior of the phospholipid bilayer, away from water.

A special preparation method for electron microscopy, called **freeze-fracturing**, reveals proteins that are embedded in the phospholipid bilayers of cellular membranes (**Figure 6.4**). When the two lipid *leaflets* (or layers) that make up the bilayer are separated, the proteins can be seen as bumps that protrude from the interior of each membrane. The bumps are not observed when artificial bilayers of pure lipid are freeze-fractured.

According to the fluid mosaic model, the proteins and lipids in a membrane are somewhat independent of each other and interact only noncovalently. The polar ends of proteins can interact with the polar ends of lipids, and the nonpolar regions of both molecules can interact hydrophobically.

However, some membrane proteins have fatty acids or other lipid groups covalently attached to them. Proteins in this subgroup of integral membrane proteins are referred to as *anchored membrane proteins*, because their hydrophobic lipid components allow them to insert themselves into the phospholipid bilayer.

Proteins are asymmetrically distributed on the inner and outer surfaces of membranes. An integral protein that extends all the way through the phospholipid bilayer and protrudes on both sides is known as a **transmembrane protein**. In addition to one or more *transmembrane domains* that extend through the bilayer, such a protein may have domains with other specific functions on the inner and outer sides of the membrane. Peripheral membrane proteins are localized on one side of the membrane or the other. This asymmetrical arrangement of membrane proteins gives the two surfaces of the membrane different properties. As we will soon see, these differences have great functional significance.

Like lipids, some membrane proteins move around relatively freely within the phospholipid bilayer. Experiments that involve the technique of cell fusion illustrate this migration dramatically. When two cells are fused, a single continuous membrane forms and surrounds both cells, and some proteins from each cell distribute themselves uniformly around this membrane (**Figure 6.5**).

Although some proteins are free to migrate in the membrane, others are not, but rather appear to be "anchored" to a specific region of the membrane. These membrane regions are like a corral of horses on a farm: the horses are free to move around within the fenced area, but not outside it. An example is the protein in the plasma membrane of a muscle cell that recognizes a chemical signal from a neuron. This protein is normally found only at the specific region where the neuron meets the muscle cell.

TOOLS FOR INVESTIGATING LIFE

**6.4 Membrane Proteins Revealed by the Freeze-Fracture Technique** This HeLa cell (a human cell) membrane was first frozen to immobilize the lipids and proteins, and then fractured so that the bilayer was split open.

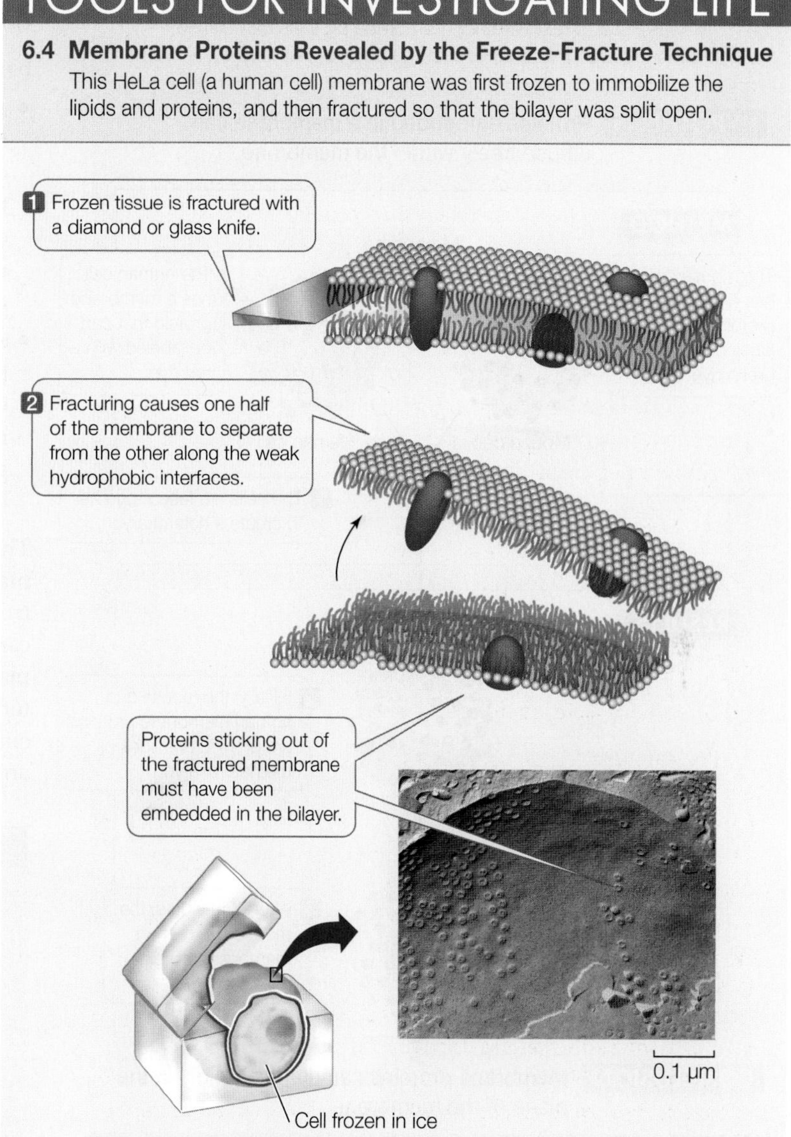

1 Frozen tissue is fractured with a diamond or glass knife.
2 Fracturing causes one half of the membrane to separate from the other along the weak hydrophobic interfaces.
Proteins sticking out of the fractured membrane must have been embedded in the bilayer.
0.1 µm
Cell frozen in ice

Proteins inside the cell can restrict the movement of proteins within a membrane. The cytoskeleton may have components just below the inner face of the membrane that are attached to membrane proteins protruding into the cytoplasm. The stability of the cytoskeletal components may thus restrict movement of attached membrane proteins.

### Membranes are constantly changing

Membranes in eukaryotic cells are constantly forming, transforming from one type to another, fusing with one another, and breaking down. As we discuss in Chapter 5, fragments of membrane move, in the form of vesicles, from the endoplasmic reticulum (ER) to the Golgi, and from the Golgi to the plasma membrane (see Figure 5.10). Secondary lysosomes form when primary lysosomes from the Golgi fuse with phagosomes from the plasma membrane (see Figure 5.11).

# INVESTIGATING LIFE

**6.5 Rapid Diffusion of Membrane Proteins**

Two animal cells can be fused together in the laboratory, forming a single large cell (heterokaryon). This phenomenon was used to test whether membrane proteins can diffuse independently in the plane of the plasma membrane.

**HYPOTHESIS** Proteins embedded in a membrane can diffuse freely within the membrane.

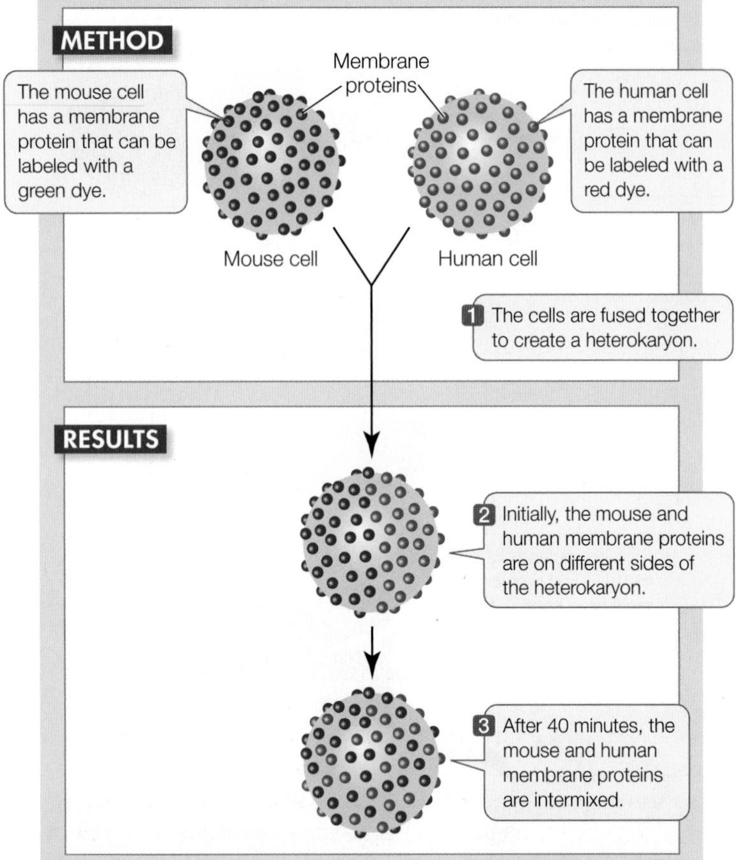

**METHOD**

Membrane proteins

The mouse cell has a membrane protein that can be labeled with a green dye.

The human cell has a membrane protein that can be labeled with a red dye.

Mouse cell

Human cell

1 The cells are fused together to create a heterokaryon.

**RESULTS**

2 Initially, the mouse and human membrane proteins are on different sides of the heterokaryon.

3 After 40 minutes, the mouse and human membrane proteins are intermixed.

**CONCLUSION** Membrane proteins can diffuse rapidly in the plane of the membrane.

Go to **yourBioPortal.com** for original citations, discussions, and relevant links for all INVESTIGATING LIFE figures.

Because all membranes appear similar under the electron microscope, and because they interconvert readily, we might expect all subcellular membranes to be chemically identical. However, that is not the case, for there are major chemical differences among the membranes of even a single cell. Membranes are changed chemically when they form parts of certain organelles. In the Golgi apparatus, for example, the membranes of the *cis* face closely resemble those of the endoplasmic reticulum in chemical composition, but those of the *trans* face are more similar to the plasma membrane.

### Plasma membrane carbohydrates are recognition sites

In addition to lipids and proteins, the plasma membrane contains carbohydrates. The carbohydrates are located on the outer surface of the plasma membrane and serve as recognition sites for other cells and molecules, as you will see in Section 6.2.

Membrane-associated carbohydrates may be covalently bonded to lipids or to proteins:

- A **glycolipid** consists of a carbohydrate covalently bonded to a lipid. Extending outside the cell surface, the carbohydrate may serve as a recognition signal for interactions between cells. For example, the carbohydrates on some glycolipids change when cells become cancerous. This change may allow white blood cells to target cancer cells for destruction.
- A **glycoprotein** consists of a carbohydrate covalently bonded to a protein. The bound carbohydrate is an oligosaccharide, usually not exceeding 15 monosaccharide units in length (see Section 3.3). The oligosaccharides of glycoproteins often function as signaling sites, as do the carbohydrates attached to glycolipids.

The "alphabet" of monosaccharides on the outer surfaces of membranes can generate a large diversity of messages. Recall from Section 3.3 that sugar molecules consist of three to seven carbons that are attached at different sites to one another. They may form linear or branched oligosaccharides with many different three-dimensional shapes. An oligosaccharide of a specific shape on one cell can bind to a complementary shape on an adjacent cell. This binding is the basis of cell–cell adhesion.

## 6.1 RECAP

**The fluid mosaic model applies to both the plasma membrane and the membranes of organelles. An integral membrane protein has both hydrophilic and hydrophobic domains, which affect its position and function in the membrane. Carbohydrates that attach to lipids and proteins on the outside of the membrane serve as recognition sites.**

- What are some of the features of the fluid mosaic model of biological membranes? See p. 106
- Explain how the hydrophobic and hydrophilic regions of phospholipids cause a membrane bilayer to form. See Figures 6.1 and 6.2
- What differentiates an integral protein from a peripheral protein? See p. 108 and Figure 6.1
- What is the experimental evidence that membrane proteins can diffuse in the plane of the membrane? See pp. 109–110 and Figure 6.5

Now that you understand the structure of biological membranes, let's see how their components function. In the next section we'll focus on the membrane that surrounds individual cells: the plasma membrane. We'll look at how the plasma membrane allows individual cells to be grouped together into multicellular systems of tissues.

## 6.2 How Is the Plasma Membrane Involved in Cell Adhesion and Recognition?

Some organisms, such as bacteria, are unicellular; that is, the entire organism is a single cell. Others, such as plants and animals, are multicellular—composed of many cells. Often these cells exist in specialized groups with similar functions, called tissues. Your body has about 60 trillion cells, arranged in different kinds of tissues (such as muscle, nerve, and epithelium).

Two processes allow cells to arrange themselves in groups:

- **Cell recognition**, in which one cell specifically binds to another cell of a certain type
- **Cell adhesion**, in which the connection between the two cells is strengthened

Both processes involve the plasma membrane. They are most easily studied if a tissue is separated into its individual cells, which are then allowed to adhere to one another again. Simple organisms provide a good model for studying processes that also occur in the complex tissues of larger species. Studies of sponges, for example, have revealed how cells associate with one another.

A sponge is a multicellular marine animal with a simple body plan that consists of only a few distinct tissues (see Section 31.5). The cells of a sponge adhere to one another, but can be separated mechanically by passing the animal several times through a fine wire screen (**Figure 6.6**). Through this process, what was a single animal becomes hundreds of individual cells suspended in seawater. Remarkably, if the cell suspension is shaken for a few hours, the cells bump into one another and stick together in the same shape and organization as the original sponge! The cells recognize and adhere to one another, and re-form the original tissues.

There are many different species of sponges. If disaggregated sponge cells from two different species are placed in the same container and shaken, individual cells will stick only to other cells of the same species. Two different sponges form, just like the ones at the start of the experiment. This demonstrates not just adhesion, but species-specific cell recognition.

Such tissue-specific and species-specific cell recognition and cell adhesion are essential to the formation and maintenance of tissues and multicellular organisms. Think of your own body. What keeps muscle cells bound to muscle cells and skin to skin? Specific cell adhesion is so obvious a characteristic of complex organisms that it is easy to overlook. You will see many examples of specific cell adhesion throughout this book; here, we describe its general principles. As you will see, cell recognition and cell adhesion depend on plasma membrane proteins.

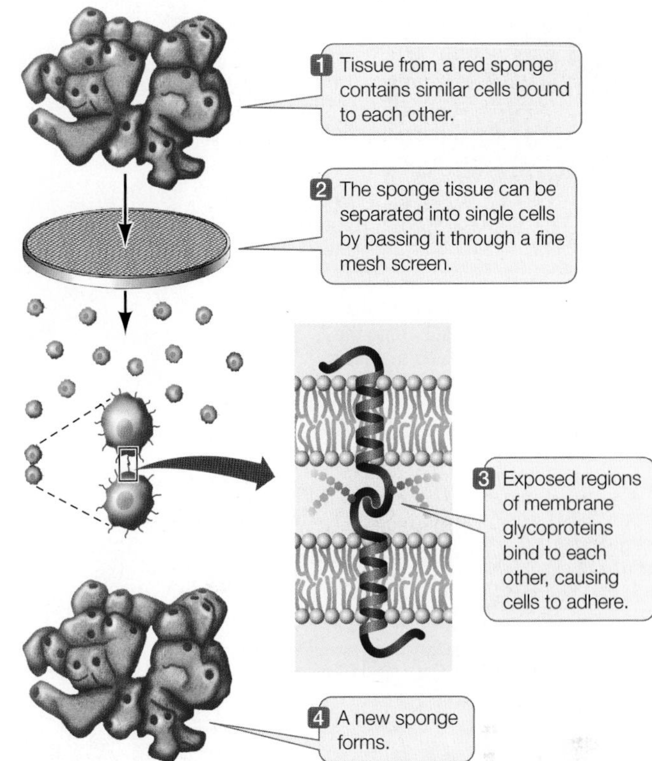

1 Tissue from a red sponge contains similar cells bound to each other.
2 The sponge tissue can be separated into single cells by passing it through a fine mesh screen.
3 Exposed regions of membrane glycoproteins bind to each other, causing cells to adhere.
4 A new sponge forms.

**6.6 Cell Recognition and Adhesion** In most cases (including the aggregation of animal cells into tissues), protein binding is homotypic.

### Cell recognition and cell adhesion involve proteins at the cell surface

The molecule responsible for cell recognition and adhesion in sponges is a huge integral membrane glycoprotein (which is 80 percent carbohydrate by molecular weight) that is partly embedded in the plasma membrane, with the carbohydrate part sticking out and exposed to the environment (and to other sponge cells). As we describe in Section 3.2, a protein not only has a specific shape, but also has specific chemical groups exposed on its surface where they can interact with other substances, including other proteins. Both of these features allow binding to other specific molecules. The cells of the disaggregated sponge in Figure 6.6 find one another again through the recognition of exposed chemical groups on their membrane glycoproteins. Adhesion proteins are not restricted to animal cells. In most plant cells, the plasma membrane is covered with a thick cell wall, but this structure also has adhesion proteins that allow cells to bind to one another.

In most cases, the binding of cells in a tissue is **homotypic**; that is, the same molecule sticks out of both cells, and the exposed surfaces bind to each other. But **heterotypic** binding (of cells with different proteins) can also occur. In this case, different chemical groups on different surface molecules have an affinity for one another. For example, when the mammalian sperm meets the egg, different proteins on the two types of cells have complementary binding surfaces. Similarly, some algae form male and female reproductive cells (analogous to sperm and eggs) that have flagella to propel them toward each other. Male and female cells can recognize each other by heterotypic proteins on their flagella.

### Three types of cell junctions connect adjacent cells

In a complex multicellular organism, cell recognition proteins allow specific types of cells to bind to one another. Often, after

(A)

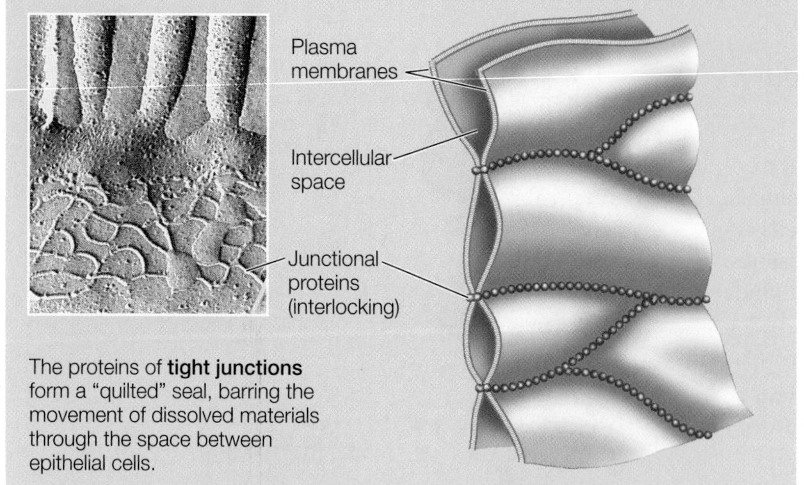


The proteins of **tight junctions** form a "quilted" seal, barring the movement of dissolved materials through the space between epithelial cells.

(B)

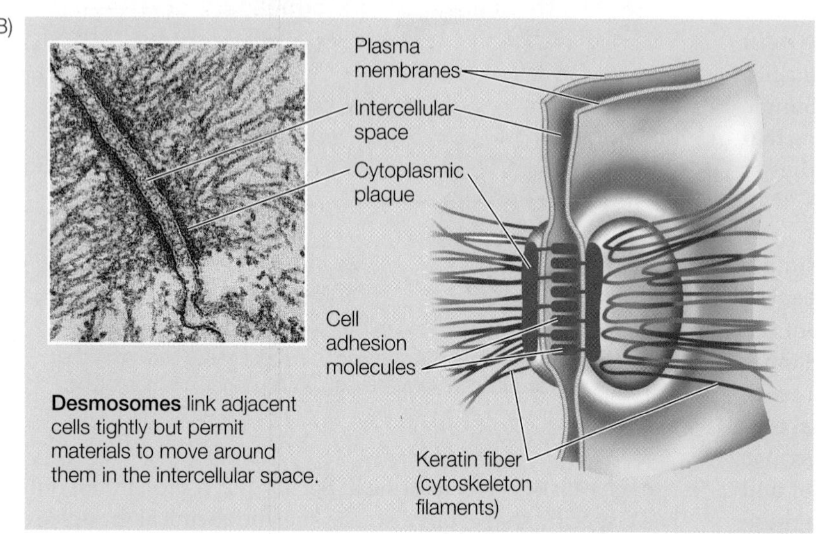


**Desmosomes** link adjacent cells tightly but permit materials to move around them in the intercellular space.

(C)

Plasma membranes

Intercellular space

Hydrophilic channel

Molecules pass between cells

Connexins (channel proteins)

**Gap junctions** let adjacent cells communicate.

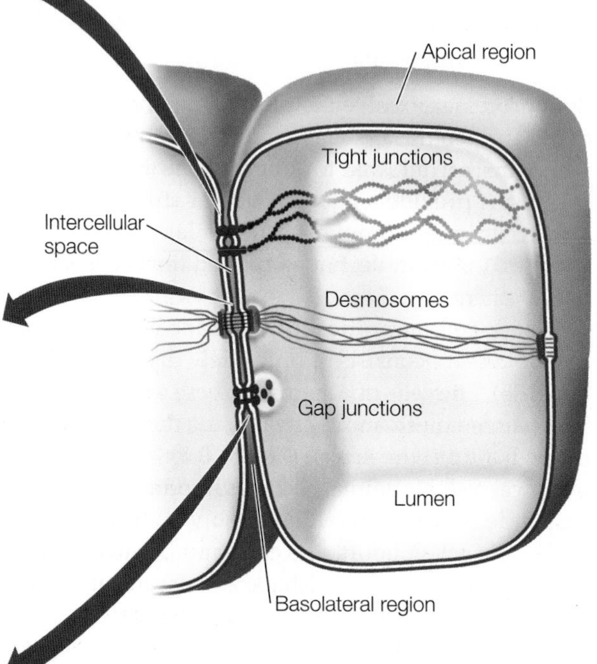


**6.7 Junctions Link Animal Cells Together** Tight junctions (A) and desmosomes (B) are abundant in epithelial tissues. Gap junctions (C) are also found in some muscle and nerve tissues, in which rapid communication between cells is important. Although all three junction types are shown in the cell at the right, all three are not necessarily seen at the same time in actual cells.


**yourBioPortal.com**

GO TO **Web Activity 6.2 • Animal Cell Junctions**


the initial binding, both cells contribute material to form additional membrane structures that connect them to one another. These specialized structures, called **cell junctions**, are most evident in electron micrographs of *epithelial* tissues, which are layers of cells that line body cavities or cover body surfaces. These surfaces often receive stresses, or must retain contents under pressure, or both, so it is particularly important that their cells adhere tightly. We will examine three types of cell junctions that enable animal cells to seal intercellular spaces, reinforce attachments to one another, and communicate with each other. Tight junctions, desmosomes, and gap junctions, respectively, perform these three functions.

**TIGHT JUNCTIONS SEAL TISSUES** **Tight junctions** are specialized structures that link adjacent epithelial cells, and they result from the mutual binding of specific proteins in the plasma membranes of the cells. These proteins are arrayed in bands so that they form a series of joints encircling each cell (**Figure 6.7A**). Tight junctions are found in the lining of lumens (cavities) in organs such as the stomach and intestine. They have two major functions:

- They prevent substances from moving from the lumen through the spaces between cells. For example, the presence of tight junctions means that substances must pass through, rather than between, the epithelial cells that form the lining of the digestive tract. In another example, the cells lining the bladder have tight junctions so urine cannot leak out into the body cavity. Thus, tight junctions help to establish cellular control over what enters and leaves the body.

- They define specific functional regions of membranes by restricting the migration of membrane proteins and phospholipids from one region of the cell to another. Thus the membrane proteins and phospholipids in the apical ("tip") region of an intestinal epithelial cell (facing the lumen) are different from those in the basolateral (*basal*, "bottom"; *lateral*, "side") regions of the cell (facing the body cavity or blood capillary outside the lumen).

By forcing materials to enter certain cells, and by allowing different areas of the same cell to have different membrane proteins with different functions, tight junctions in the digestive tract help ensure the directional movement of materials into the body.

**DESMOSOMES HOLD CELLS TOGETHER** **Desmosomes** connect adjacent plasma membranes. Desmosomes hold neighboring cells firmly together, acting like spot welds or rivets (**Figure 6.7B**). Each desmosome has a dense structure called a plaque on the cytoplasmic side of the plasma membrane. To this plaque are attached special cell adhesion molecules that stretch from the plaque through the plasma membrane of one cell, across the intercellular space, and through the plasma membrane of the adjacent cell, where they bind to the plaque proteins in that adjacent cell.

The plaque is also attached to fibers in the cytoplasm. These fibers, which are intermediate filaments of the cytoskeleton (see Figure 5.18), are made of a protein called keratin. They stretch from one cytoplasmic plaque across the cell to another plaque on the other side of the cell. Anchored thus on both sides of the cell, these extremely strong fibers provide great mechanical stability to epithelial tissues. This stability is needed for these tissues, which often receive rough wear while protecting the integrity of the organism's body surface, or the surface of an organ.

**GAP JUNCTIONS ARE A MEANS OF COMMUNICATION** Whereas tight junctions and desmosomes have mechanical roles, **gap junctions** facilitate communication between cells. Each gap junction is made up of specialized channel proteins, called *connexins*, which interact to form a structure (called a *connexon*) that spans the plasma membranes of adjacent cells and the intercellular space between them (**Figure 6.7C**). Water, dissolved small molecules, and ions can pass from cell to cell through these junctions. This allows groups of cells to coordinate their activities. In Chapter 7 we discuss cell communication and signaling, and in that chapter we describe in more detail the roles of gap junctions and plasmodesmata, which perform a similar role in plants.

### Cell membranes adhere to the extracellular matrix

In Section 5.4 we describe the extracellular matrix of animal cells, which is composed of collagen protein arranged in fibers in a gelatinous matrix of proteoglycans. The attachment of a cell to the extracellular matrix is important in maintaining the integrity of a tissue. In addition, some cells can detach from their neighbors, move, and attach to other cells; this is often mediated by interactions with the extracellular matrix.

A transmembrane protein called **integrin** often mediates the attachment of epithelial cells to the extracellular matrix (**Figure 6.8**). More than 24 different integrins have been described in human cells. All of them bind to a protein in the extracellular matrix on the outside of the cell, and to actin filaments, which are part of the cytoskeleton, on the inside of the cell. So, in addition to adhesion, integrin has a role in maintaining cell structure via its interaction with the cytoskeleton.

The binding of integrin to the extracellular matrix is noncovalent and reversible. When a cell moves its location within a tissue or organism, the first step is detachment of the cell's integrin from the matrix. The integrin protein changes its three-dimensional structure and no longer maintains its link to the matrix. These events are important for cell movement within the developing embryo, and for the spread of cancer cells.

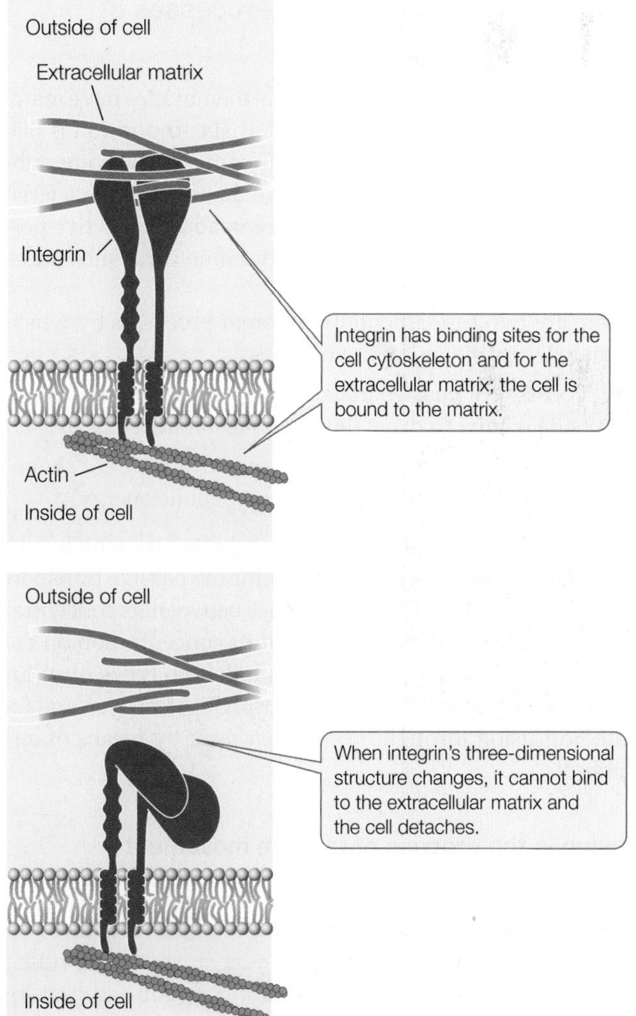

Outside of cell
Extracellular matrix
Integrin
Integrin has binding sites for the cell cytoskeleton and for the extracellular matrix; the cell is bound to the matrix.
Actin
Inside of cell
Outside of cell
When integrin's three-dimensional structure changes, it cannot bind to the extracellular matrix and the cell detaches.
Inside of cell

**6.8 Integrins Mediate the Attachment of Animal Cells to the Extracellular Matrix**

## 6.2 RECAP

**In multicellular organisms, cells arrange themselves in groups by two processes: cell recognition and cell adhesion. Both processes are mediated by integral proteins in the plasma membrane. Cell membrane proteins also interact with the extracellular matrix.**

- Describe the difference between cell recognition and cell adhesion. See p. 111
- How do the three types of cell junctions regulate the passage of materials between cells and through the intercellular space? See pp. 111–113 and Figure 6.7

We have just examined how the plasma membrane structure accommodates the binding and maintenance of cell adhesion. We turn now to another major function of membranes: regulating the substances that enter or leave a cell or organelle.

# 6.3 What Are the Passive Processes of Membrane Transport?

As you have already learned, biological membranes have many functions, and control of the cell's internal composition is one of the most significant. Biological membranes allow some substances, but not others, to pass through them. This characteristic of membranes is called **selective permeability**. Selective permeability allows the membrane to determine what substances enter or leave a cell or organelle.

There are two fundamentally different processes by which substances cross biological membranes:

- The processes of **passive transport** do not require any input of outside energy to drive them (no metabolic energy).
- The processes of **active transport** require the input of chemical energy from an outside source (metabolic energy).

This section focuses on the passive processes by which substances cross membranes. The energy for the passive transport of a substance is found in the difference between its concentration on one side of the membrane and its concentration on the other. Passive transport processes include two types of diffusion: simple diffusion through the phospholipid bilayer, and facilitated diffusion through *channel proteins* or by means of *carrier proteins*.

## Diffusion is the process of random movement toward a state of equilibrium

Nothing in this world is ever absolutely at rest. Everything is in motion, although the motions may be very small. An important consequence of all this random vibration, rotation and translocation (moving from one location to another) of molecules is that all the components of a solution tend eventually to become evenly distributed. For example, if a drop of ink is allowed to fall into a container of water, the pigment molecules of the ink are initially very concentrated. Without human intervention, such as stirring, the pigment molecules move about at random, spreading slowly through the water until eventually the concentration of pigment—and thus the intensity of color—is exactly the same in every drop of liquid in the container.

A solution in which the solute particles are uniformly distributed is said to be at *equilibrium* because there will be no future net change in their concentration. Equilibrium does not mean that the particles have stopped moving; it just means that they are moving in such a way that their overall distribution does not change.

**Diffusion** is the process of random movement toward a state of equilibrium. Although the motion of each individual particle is absolutely random, the net movement of particles is directional until equilibrium is reached. Diffusion is thus a net movement from regions of greater concentration to regions of lesser concentration (**Figure 6.9**).

In a complex solution (one with many different solutes), the diffusion of each solute is independent of those of the others. How fast a substance diffuses depends on three factors:

- The *diameter* of the molecules or ions: smaller molecules diffuse faster.
- The *temperature* of the solution: higher temperatures lead to faster diffusion because ions or molecules have more energy, and thus move more rapidly, at higher temperatures.
- The *concentration gradient* in the system—that is, the change in solute concentration with distance in a given direction: the greater the concentration gradient, the more rapidly a substance diffuses.

We'll see how these factors influence membrane transport in the detailed discussions that follow.

**DIFFUSION WITHIN CELLS AND TISSUES** Within cells, or wherever distances are very short, solutes distribute themselves rapidly by diffusion. Small molecules and ions may move from one end of an organelle to another in a millisecond ($10^{-3}$ s, or one-thousandth of a second). However, the usefulness of diffusion as a transport mechanism declines drastically as distances become greater. In the absence of mechanical stirring, diffusion across more than a centimeter may take an hour or more, and diffusion across meters may take years! Diffusion would not be adequate to distribute materials over the length of a human body, much less that of a larger organism. But within our cells or across layers of one or two cells, diffusion is rapid enough to distribute small molecules and ions almost instantaneously.

**DIFFUSION ACROSS MEMBRANES** In a solution without barriers, all the solutes diffuse at rates determined by temperature, their physical properties, and their concentration gradients. If a biological membrane divides the solution into separate compartments, then the movement of the different solutes can be affected by the properties of the membrane. The membrane is said to be *permeable* to solutes that can cross it more or less easily, but *impermeable* to substances that cannot move across it.

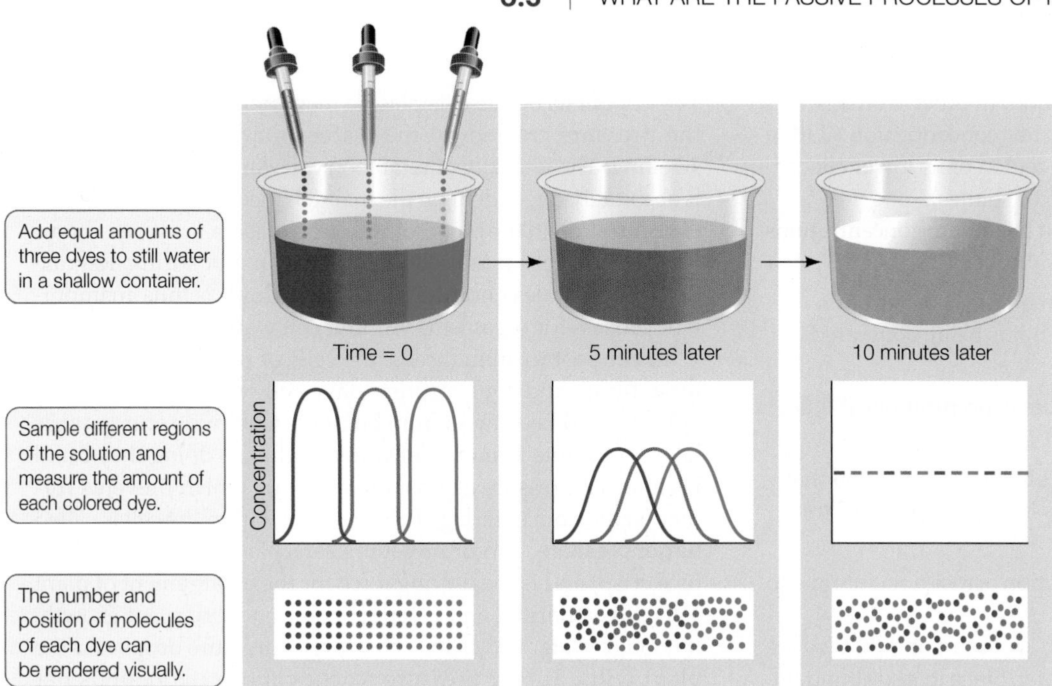

Add equal amounts of three dyes to still water in a shallow container.
Sample different regions of the solution and measure the amount of each colored dye.
The number and position of molecules of each dye can be rendered visually.
Time = 0
5 minutes later
10 minutes later
Concentration

**6.9 Diffusion Leads to Uniform Distribution of Solutes** A simple experiment demonstrates that solutes move from regions of greater concentration to regions of lesser concentration until equilibrium is reached.

Molecules to which the membrane is impermeable remain in separate compartments, and their concentrations may be different on the two sides of the membrane. Molecules to which the membrane is permeable diffuse from one compartment to the other until their concentrations are equal on both sides of the membrane. When the concentrations of a diffusing substance on the two sides of the permeable membrane are identical, equilibrium is reached. Individual molecules continue to pass through the membrane after equilibrium is established, but equal numbers of molecules move in each direction, so *at equilibrium there is no net change in concentration.*

### Simple diffusion takes place through the phospholipid bilayer

In **simple diffusion**, small molecules pass through the phospholipid bilayer of the membrane. A molecule that is itself hydrophobic, and is therefore soluble in lipids, enters the membrane readily and is able to pass through it. The more lipid-soluble the molecule is, the more rapidly it diffuses through the membrane bilayer. This statement holds true over a wide range of molecular weights.

On the other hand, electrically charged or polar molecules, such as amino acids, sugars, and ions, do not pass readily through a membrane for two reasons. First, such charged or polar molecules are not very soluble in the hydrophobic interior of the bilayer. Second, such charged and polar substances form many hydrogen bonds with water and ions in the aqueous environment, be it the cytoplasm or the cell exterior. The multiplicity of these hydrogen bonds prevent the substances from moving into the hydrophobic interior of the membrane.

Consider two molecules: a small protein made up of a few polar amino acids, and a cholesterol-based steroid of equivalent size. If a membrane separates high and low concentrations of these substances, the protein, being polar, will diffuse only very slowly through the membrane, while the nonpolar steroid will diffuse through it readily.

### Osmosis is the diffusion of water across membranes

Water molecules pass through specialized channels in membranes (see below) by a diffusion process called **osmosis**. This completely passive process uses no metabolic energy and can be understood in terms of solute concentrations. Recall that a solute dissolves in a solvent and the solute's constituents are dispersed throughout the solution. Osmosis depends on the *number* of solute particles present, not on the *kinds* of particles. We will describe osmosis using red blood cells and plant cells as examples. In these examples, the plasma membranes are considered to be permeable to water and impermeable to most solutes.

Red blood cells are normally suspended in a fluid called plasma, which contains salts, proteins, and other solutes. Examining a drop of blood under the light microscope reveals that these red cells have a characteristic flattened disk shape with a depressed center, sometimes called "biconcave." If pure water is added to the drop of blood, drastically reducing the solute concentration of the plasma, the red cells quickly swell and burst. Similarly, if slightly wilted lettuce is placed in pure water, it soon becomes crisp; by weighing it before and after, we can show that it has taken up water. If, on the other hand, red blood cells or crisp lettuce leaves are placed in a relatively concentrated solution of salt or sugar, the leaves become limp (they wilt), and the red blood cells pucker and shrink.

From such observations we know that the difference in solute concentration between a cell and its surrounding environment determines whether water will move from the environment into the cell or out of the cell into the environment. Other things being equal, if two different solutions are separated by a membrane that allows water, *but not solutes*, to pass through, water molecules will move across the membrane toward the solution with a higher solute concentration. In other words, water will diffuse from a region of its higher concentration (with a lower concentration of solutes) to a region of its lower concentration (with a higher concentration of solutes).

Three terms are used to compare the solute concentrations of two solutions separated by a membrane:

- A **hypertonic** solution has a higher solute concentration than the other solution with which it is being compared (**Figure 6.10A**).
- **Isotonic** solutions have equal solute concentrations (**Figure 6.10B**).
- A **hypotonic** solution has a lower solute concentration than the other solution with which it is being compared (**Figure 6.10C**).

Water moves from a hypotonic solution across a membrane to a hypertonic solution.

When we say that "water moves," bear in mind that we are referring to the net movement of water. Since it is so abundant, water is constantly moving through protein channels across the plasma membrane into and out of cells. What concerns us here is whether the overall movement is greater in one direction or the other.

The concentration of solutes in the environment determines the direction of osmosis in all animal cells. A red blood cell takes up water from a solution that is hypotonic to the cell's contents. The cell bursts because its plasma membrane cannot withstand the pressure created by the water entry and the resultant swelling. The integrity of red blood cells (and other blood cells) is absolutely dependent on the maintenance of a constant solute concentration in the blood plasma: the plasma must be isotonic to the blood cells if the cells are not to burst or shrink. Regulation of the solute concentration of body fluids is thus an important process for organisms without cell walls.

In contrast to animal cells, the cells of plants, archaea, bacteria, fungi, and some protists have cell walls that limit their volumes and keep them from bursting. Cells with sturdy walls take up a limited amount of water, and in so doing they build up internal pressure against the cell wall, which prevents further water from entering. This pressure within the cell is called **turgor pressure**. Turgor pressure keeps plants upright (and lettuce crisp) and is the driving force for the enlargement of plant cells. It is a normal and essential component of plant growth. If enough water leaves the cells, turgor pressure drops and the plant wilts. Turgor pressure reaches about 100 pounds per square inch (0.7 kg/cm$^2$)—several times greater than the pres-

**6.10 Osmosis Can Modify the Shapes of Cells** In a solution that is isotonic with the cytoplasm (center column), a plant or animal cell maintains a consistent, characteristic shape because there is no net movement of water into or out of the cell. In a solution that is hypotonic to the cytoplasm (right), water enters the cell. An environment that is hypertonic to the cytoplasm (left) draws water out of the cell.

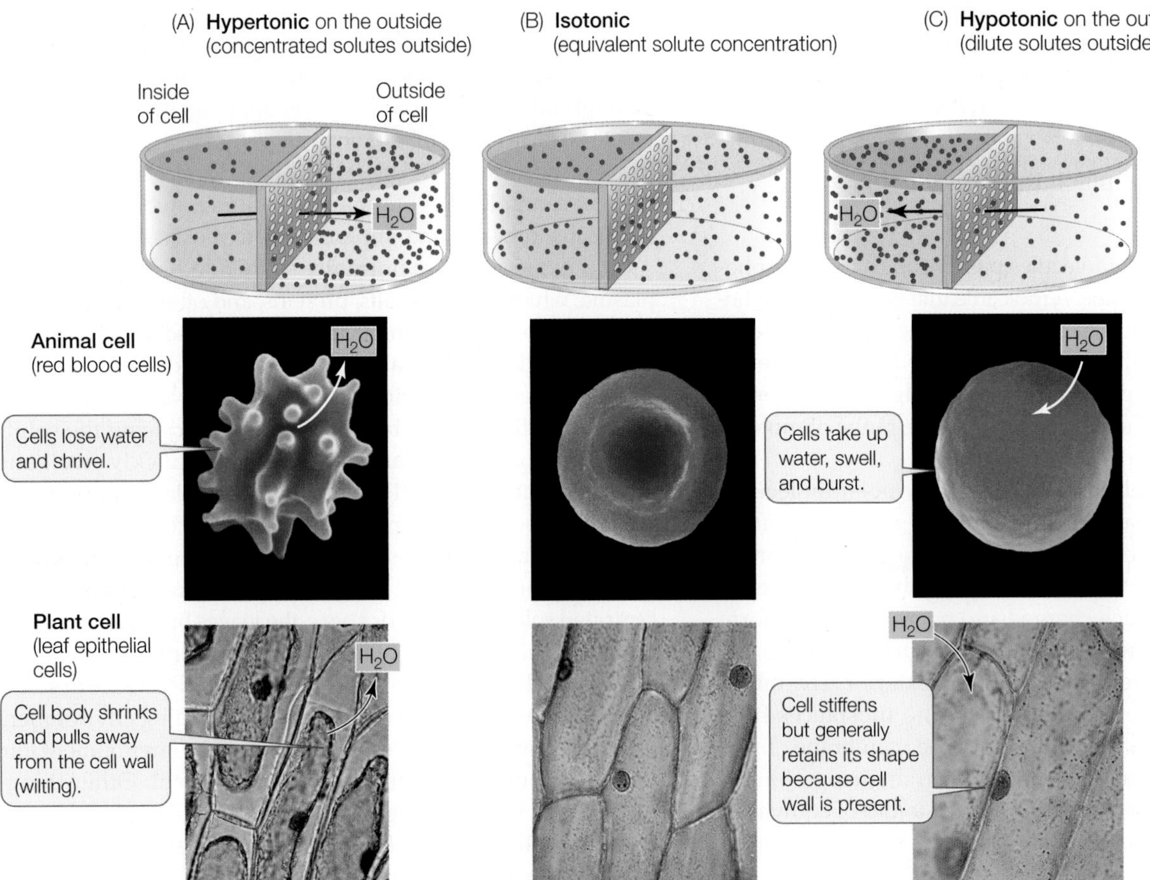
(A) Hypertonic on the outside (concentrated solutes outside)
(B) Isotonic (equivalent solute concentration)
(C) Hypotonic on the outside (dilute solutes outside)
Inside of cell
Outside of cell
H₂O
Animal cell (red blood cells)
Cells lose water and shrivel.
Cells take up water, swell, and burst.
Plant cell (leaf epithelial cells)
Cell body shrinks and pulls away from the cell wall (wilting).
Cell stiffens but generally retains its shape because cell wall is present.

sure in automobile tires. This pressure is so great that the cells would change shape and detach from one another, were it not for adhesive molecules in the plant cell wall.

### Diffusion may be aided by channel proteins

As we saw earlier, polar or charged substances such as water, amino acids, sugars and ions do not readily diffuse across membranes. But they can cross the hydrophobic phospholipid bilayer passively (that is, without the input of energy) in one of two ways, depending on the substance:

- **Channel proteins** are integral membrane proteins that form channels across the membrane through which certain substances can pass.
- Some substances can bind to membrane proteins called **carrier proteins** that speed up their diffusion through the phospholipid bilayer.

Both of these processes are forms of **facilitated diffusion**. That is, the substances diffuse according to their concentration gradients, but their diffusion is facilitated by protein channels or carriers.

**ION CHANNELS** The best-studied channel proteins are the **ion channels**. As you will see in later chapters, the movement of ions across membranes is important in many biological processes, ranging from respiration within the mitochondria, to the electrical activity of the nervous system and the opening of the pores in leaves that allow gas exchange with the environment. Several types of ion channels have been identified, each of them specific for a particular ion. All of them show the same basic structure of a hydrophilic pore that allows a particular ion to move through it (**Figure 6.11**).

Just as a fence may have a gate that can be opened or closed, most ion channels are gated: they can be opened or closed to ion passage. A **gated channel** opens when a stimulus causes a change in the three-dimensional shape of the channel. In some cases, this stimulus is the binding of a chemical signal, or **ligand** (see Figure 6.11). Channels controlled in this way are called *ligand-gated channels*. In contrast, a *voltage-gated channel* is stimulated to open or close by a change in the voltage (electrical charge difference) across the membrane.

**THE MEMBRANE POTENTIAL** All living cells maintain an imbalance of ion concentrations across the plasma membrane, and consequently a small voltage or **membrane potential** exists across that membrane. When a gated ion channel opens, millions of ions can rush through it per second. How fast the ions move, and in which direction (into or out of the cell), depends on two factors, the concentration gradient and the magnitude of the voltage. Let's consider how these factors affect the concentration of potassium ions ($K^+$) inside an animal cell:

- *The concentration gradient*: Because of active transport (discussed below), the concentration of $K^+$ is usually much higher inside the cell than outside, so $K^+$ will tend to diffuse out of the cell through an open potassium channel.
- *The distribution of electrical charge*: As $K^+$ diffuses out of the cell it leaves behind an excess of chloride ($Cl^-$) and other negatively charged ions. These negatively charged substances cannot readily diffuse through the plasma membrane to follow $K^+$ out of the cell, and this results in a charge difference (negative inside) across the membrane. $K^+$ is attracted to the negative charge inside the cell, creating a tendency for $K^+$ to stay inside the cell, even though it is more concentrated there than outside.

Now, consider what happens when the $K^+$ channel is opened. Two forces are at work: diffusion draws $K^+$ out of the cell through the channel, and electrical attraction keeps $K^+$ inside the cell. The system exists in a state of equilibrium, in which the ion's rate of diffusion out through the channel is balanced by the rate of movement in through the channel due to electrical attraction. Obviously, the concentrations of $K^+$ on each side of the membrane will not be equal, as we would expect if diffusion were the only force involved. Instead, the attraction of electrical charges keeps some extra $K^+$ inside the cell. This imbalance in $K^+$ is a major factor in generating a voltage across the plasma membrane called the *membrane potential*.

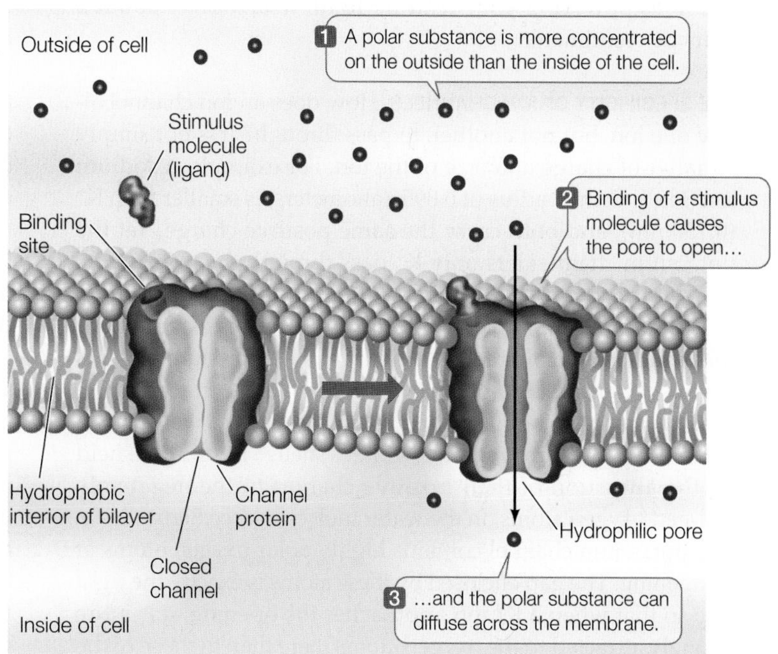
Outside of cell
1 A polar substance is more concentrated on the outside than the inside of the cell.
Stimulus molecule (ligand)
Binding site
2 Binding of a stimulus molecule causes the pore to open...
Hydrophobic interior of bilayer
Channel protein
Hydrophilic pore
Closed channel
3 ...and the polar substance can diffuse across the membrane.
Inside of cell

**6.11 A Gated Channel Protein Opens in Response to a Stimulus** The channel protein has a pore of polar amino acids and water. It is anchored in the hydrophobic bilayer interior by its outer coating of nonpolar R groups of its amino acids. The protein changes its three-dimensional shape when a stimulus molecule (ligand) binds to it, opening the pore so that hydrophilic polar substances can pass through. Other gated channels open in response to an electrical potential (voltage).

The membrane potential is related to the concentration imbalance of $K^+$ by the Nernst equation:

$$E_K = 2.3\frac{RT}{zF}\log\frac{[K]_o}{[K]_i}$$

where $R$ is the gas constant, $F$ is the Faraday constant (both familiar to chemistry students), $T$ is the temperature, and $z$ is the charge on the ion (+1). Solving for 2.3 $RT/zF$ at 20°C ("room temperature"), the equation becomes much simpler:

$$E_K = 58\log\frac{[K]_o}{[K]_i}$$

where $E_K$ is the membrane potential (in millivolts, mV) that results from the ratio of $K^+$ concentrations outside the cell $[K]_o$ and inside the cell $[K]_i$.

What does this equation tell us about cells? It shows that a small change in $K^+$ concentration, due to the opening of a ligand-gated $K^+$ channel, for example, can have a large effect on the electrical potential ($E$) across the membrane. This change in potential might be enough to cause other proteins in the membrane, such as voltage-gated channels, to change configuration. As we discuss in Chapter 45, this is exactly what happens in the nervous system. Many drugs that act on electrically sensitive tissues work as ligands that open ion channels and thereby affect membrane potential. And as you will see shortly, membrane potential drives secondary active transport.

Actual measurements from animal cells give a total membrane potential between –60 and –70 mV across the membrane, where the inside is negative with respect to the outside (see Figure 45.5). Cells have a tremendous amount of potential energy stored in their membrane potentials. In fact, the brain cells you are using to read this book have more potential energy—about 200,000 volts per centimeter—than the high-voltage electric lines powering your reading light, which carry about 2 volts per centimeter.

**THE SPECIFICITY OF ION CHANNELS** How does an ion channel allow one ion, but not another, to pass through? It is not simply a matter of charge and size of the ion. For example, a sodium ion ($Na^+$), with a radius of 0.095 nanometers, is smaller than $K^+$ (0.130 nm), and both carry the same positive charge. Yet the potassium channel lets only $K^+$ pass through the membrane, and not the smaller $Na^+$. Nobel Laureate Roderick MacKinnon at The Rockefeller University found an elegant explanation for this when he deciphered the structure of a potassium channel from a bacterium (**Figure 6.12**).

Being charged, both $Na^+$ and $K^+$ are attracted to water molecules. They are surrounded by water "shells" in solution, held by the attraction of their positive charges to the negatively charged oxygen atoms on the water molecules (see Figure 2.10). The potassium channel contains highly polar oxygen atoms at its opening. The gap enclosed by these atoms is exactly the right size so that when a $K^+$ ion approaches the opening, it is more strongly attracted to the oxygen atoms there than to those of the

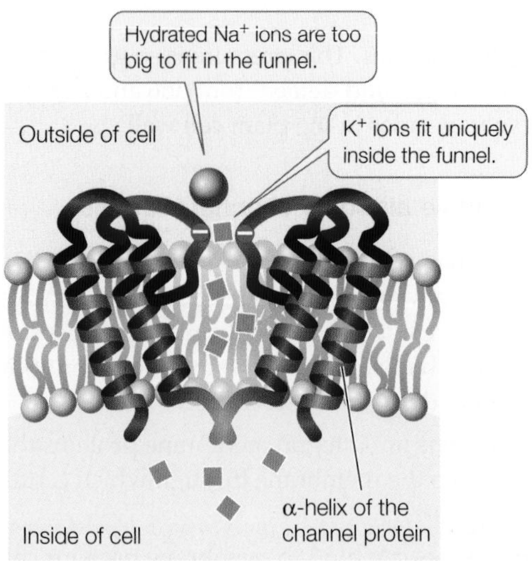


**6.12 The Potassium Channel** The positively charged potassium ions are attracted by the polar (negatively charged) oxygen atoms in the R groups (side chains) of the channel protein, and the ions funnel through the channel. This channel is a "custom fit" for $K^+$; other ions do not pass through.

water molecules in its shell. It sheds its water shell and passes through the channel. The smaller $Na^+$ ion, on the other hand, is kept a bit more distant from the oxygen atoms at the opening of the channel because extra water molecules can fit between the ion (with its shell) and the oxygen atoms at the opening. So $Na^+$ does not enter the potassium channel. The gate that opens or closes the channel appears to be an interaction between positively charged arginine residues on the protein and negative charges on membrane phospholipids. This is an example of the functional interactions between membrane proteins and lipids.

**AQUAPORINS FOR WATER** Water crosses membranes at a much faster rate than would be expected for simple diffusion through the hydrophobic phospholipid bilayer. One way that water can do this is by "hitchhiking" with some ions, such as $Na^+$, as they pass through ion channels. Up to 12 water molecules may coat an ion as it traverses a channel. But there is an even faster way to get water across membranes. Plant cells and some animal cells, such as red blood cells and kidney cells, have membrane channels called **aquaporins**. These channels function as a cellular plumbing system for moving water. Like the $K^+$ channel, the aquaporin channel is highly specific. Water molecules move in single file through the channel, which excludes ions so that the electrical properties of the cell are maintained.

Aquaporins were first identified by Peter Agre at Duke University, who shared the Nobel Prize with Rod McKinnon (see above). Agre noticed a membrane protein that was present in red blood cells, kidney cells, and plant cells but did not know its function. A colleague suggested that it might be a water channel, because these cell types show rapid diffusion of water across their membranes. Agre inserted the protein into the membrane of an oocyte, which normally does not permit much diffusion of water. He injected the oocyte with mRNA for aquaporin, from which the protein was produced and inserted into the membrane. Remarkably, the oocyte began swelling immediately after being transferred to a hypotonic solution, indicating rapid diffusion of water into the cell (**Figure 6.13**).

### Carrier proteins aid diffusion by binding substances

As we described earlier, another kind of facilitated diffusion involves not just the opening of a channel, but also the actual binding of the transported substance to a membrane protein called a carrier protein. Like channel proteins, carrier proteins allow diffusion both into and out of the cell or organelle. In other words, carrier proteins operate in both directions. Carrier proteins transport polar molecules such as sugars and amino acids.

## INVESTIGATING LIFE

**6.13 Aquaporin Increases Membrane Permeability to Water**
A protein was isolated from the membranes of cells in which water diffuses rapidly across the membranes. When the protein was inserted into oocytes, which do not normally have it, the water permeability of the oocytes was greatly increased.

**HYPOTHESIS** Aquaporin increases membrane permeability to water.

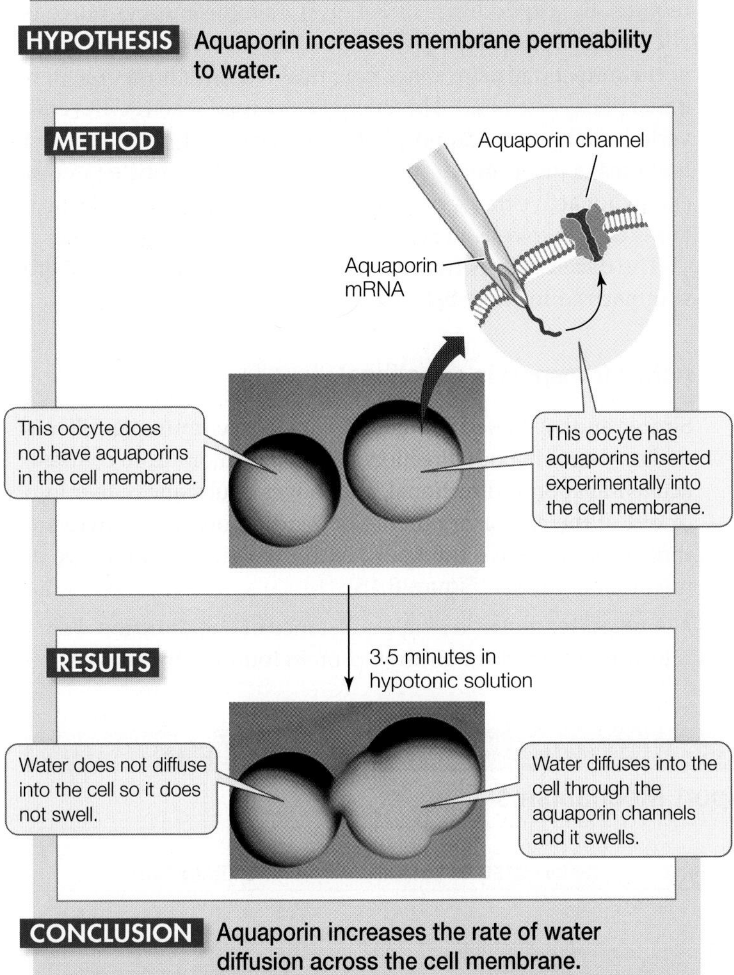

METHOD
Aquaporin channel
Aquaporin mRNA
This oocyte does not have aquaporins in the cell membrane.
This oocyte has aquaporins inserted experimentally into the cell membrane.
3.5 minutes in hypotonic solution
RESULTS
Water does not diffuse into the cell so it does not swell.
Water diffuses into the cell through the aquaporin channels and it swells.

**CONCLUSION** Aquaporin increases the rate of water diffusion across the cell membrane.

Go to **yourBioPortal.com** for original citations, discussions, and relevant links for all INVESTIGATING LIFE figures.

Glucose is the major energy source for most mammalian cells, and they require a great deal of it. Their membranes contain a carrier protein—the glucose transporter—that facilitates glucose uptake into the cell. Binding of glucose to a specific three-dimensional site on one side of the transporter protein causes the protein to change its shape and release glucose on the other side of the membrane (**Figure 6.14A**). Since glucose is broken down almost as soon as it enters a cell, there is almost always a strong concentration gradient favoring glucose entry (that is, a higher concentration outside the cell than inside). The transporter allows glucose molecules to cross the membrane and enter the cell much faster than they would by simple diffusion through the bilayer. This rapid entry is necessary to ensure that the cell receives enough glucose for its energy needs.

Transport by carrier proteins is different from simple diffusion. In both processes, the rate of movement depends on the concentration gradient across the membrane. However, in carrier-mediated transport, a point is reached at which increases in the concentration gradient are not accompanied by an increased rate of diffusion. At this point, the facilitated diffusion system is said to be *saturated* (**Figure 6.14B**). Because there are only a limited number of carrier protein molecules per unit of membrane area, the rate of diffusion reaches a maximum when all the carrier molecules are fully loaded with solute molecules. Think of waiting for the elevator on the ground floor of a hotel with 50 other people. They can't all get in the elevator (carrier) at once, so the rate of transport (say 10 people at a time) is saturated.

**yourBioPortal.com**
GO TO Animated Tutorial 6.1 • Passive Transport

## 6.3 RECAP

**Diffusion is the movement of ions or molecules from a region of greater concentration to a region of lesser concentration. Water can diffuse through cell membranes by a process called osmosis. Channel proteins, which can be open or closed, and carrier proteins facilitate diffusion of charged and polar substances, including water. The diffusion of ions across cell membranes sets up an electrochemical potential gradient across the membranes.**

- What properties of a substance determine whether, and how fast, it will diffuse across a membrane? See p. 114
- Describe osmosis and explain the terms hypertonic, hypotonic, and isotonic. See p. 116 and Figure 6.10
- How does a channel protein facilitate diffusion? See p. 118 and Figures 6.11 and 6.12

The process of diffusion tends to equalize the concentrations of substances outside and inside cells. However, one hallmark of

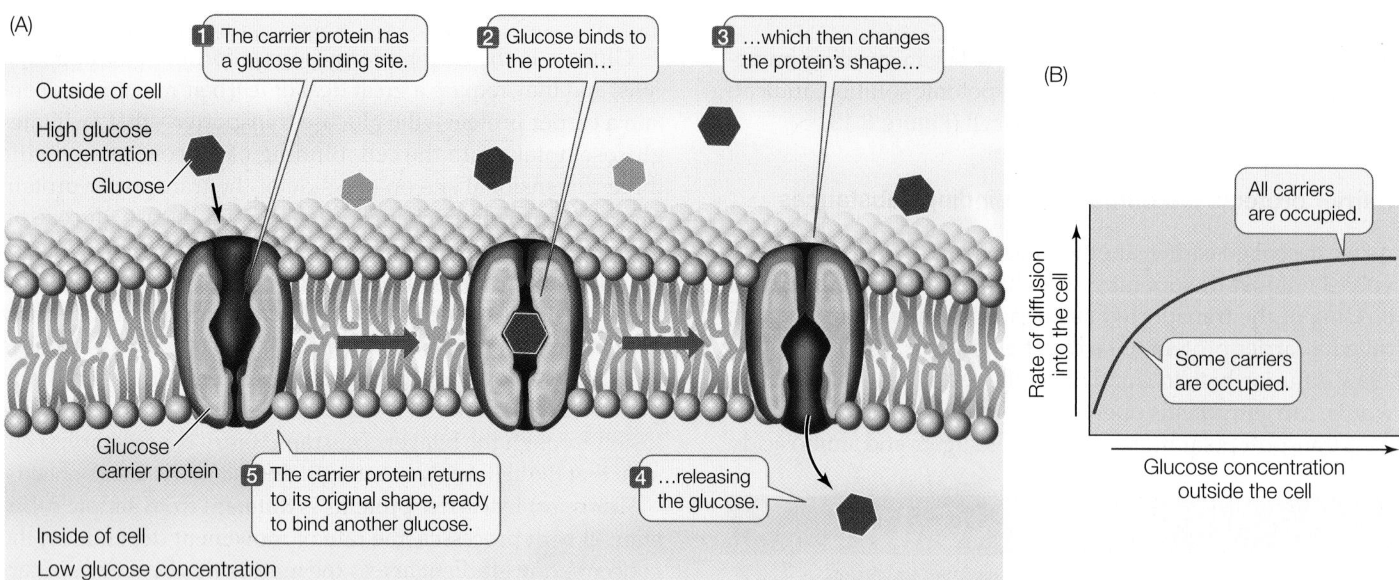


**6.14 A Carrier Protein Facilitates Diffusion** The glucose transporter is a carrier protein that allows glucose to enter the cell at a faster rate than would be possible by simple diffusion. (A) The transporter binds to glucose, brings it into the membrane interior, then changes shape, releasing glucose into the cell cytoplasm. (B) The graph shows the rate of glucose entry via a carrier versus the concentration of glucose outside the cell. As the glucose concentration increases, the rate of diffusion increases until the point at which all the available transporters are being used (the system is saturated).

a living thing is that it can have an internal composition quite different from that of its environment. To achieve this it must sometimes move substances in opposite directions from the ones in which they would naturally tend to diffuse. That is, substances must sometimes be moved against concentration gradients and/or against the cell's membrane potential (electrical gradient). This process requires work—the input of energy—and is known as *active transport*.

## 6.4 What are the Active Processes of Membrane Transport?

In many biological situations, there is a different concentration of a particular ion or small molecule inside compared with outside a cell. In these cases, the imbalance is maintained by a protein in the plasma membrane that moves the substance against its concentration and/or electrical gradient. This is called *active transport*, and because it is acting "against the normal flow," it requires the expenditure of energy. Often the energy source is adenosine triphosphate (ATP). In eukaryotes, ATP is produced in the mitochondria and has chemical energy stored in its terminal phosphate bond. This energy is released when ATP is converted to adenosine diphosphate (ADP) in a hydrolysis reaction that breaks the terminal phosphate bond. This is one source of energy for active transport. (We give the details of how ATP provides energy to cells in Section 8.2.)

The differences between diffusion and active transport are summarized in **Table 6.1**.

### Active transport is directional

Simple and facilitated diffusion follow concentration gradients and can occur in both directions across a membrane. In contrast, active transport is directional, and moves a substance either into or out of the cell or organelle, depending on need. There are three types of active transport, each involving its own type of membrane protein (**Figure 6.15**):

- A **uniporter** moves a single substance in one direction. For example, a calcium-binding protein found in the plasma

**TABLE 6.1**
**Membrane Transport Mechanisms**

| | SIMPLE DIFFUSION | DIFFUSION THROUGH CHANNEL | FACILITATED DIFFUSION | ACTIVE TRANSPORT |
|---|---|---|---|---|
| Cellular energy required? | No | No | No | Yes |
| Driving force | Concentration gradient | Concentration gradient | Concentration gradient | ATP hydrolysis (against concentration gradient) |
| Membrane protein required? | No | Yes | Yes | Yes |
| Specificity | No | Yes | Yes | Yes |

**Uniporter** transports one substance in one direction.

**Symporter** transports two different substances in the same direction.

**Antiporter** transports two different substances in opposite directions.

Outside of cell

Transported ions

Inside of cell

**6.15 Three Types of Proteins for Active Transport** Note that in each of the three cases, transport is directional. Symporters and antiporters are examples of coupled transporters. All three types of transporters are coupled to energy sources in order to move substances against their concentration gradients.

membrane and endoplasmic reticulum of many cells actively transports $Ca^{2+}$ to locations where it is more highly concentrated, either outside the cell or inside the ER.

- A **symporter** moves two substances in the same direction. For example, a symporter in the cells that line the intestine must bind $Na^+$ in addition to an amino acid in order to absorb amino acids from the intestine.
- An **antiporter** moves two substances in opposite directions, one into the cell (or organelle) and the other out of the cell (or organelle). For example, many cells have a sodium–potassium pump that moves $Na^+$ out of the cell and $K^+$ into it.

Sympporters and antiporters are also known as *coupled transporters* because they move two substances at once.

### Different energy sources distinguish different active transport systems

There are two basic types of active transport:

- **Primary active transport** involves the direct hydrolysis of ATP, which provides the energy required for transport.
- **Secondary active transport** does not use ATP directly. Instead, its energy is supplied by an ion concentration and electrical gradient established by primary active transport. This transport system uses the energy of ATP indirectly to set up the gradient.

In primary active transport, energy released by the hydrolysis of ATP drives the movement of specific ions against their concentration gradients. For example, we mentioned earlier that concentrations of potassium ions ($K^+$) inside a cell are often much higher than in the fluid bathing the cell. On the other hand, the concentration of sodium ions ($Na^+$) is often much higher outside the cell. A protein in the plasma membrane pumps $Na^+$ out of the cell and $K^+$ into the cell against these concentration and electrochemical gradients, ensuring that the gradients are maintained (**Figure 6.16**). This **sodium–potassium ($Na^+$–$K^+$) pump** is

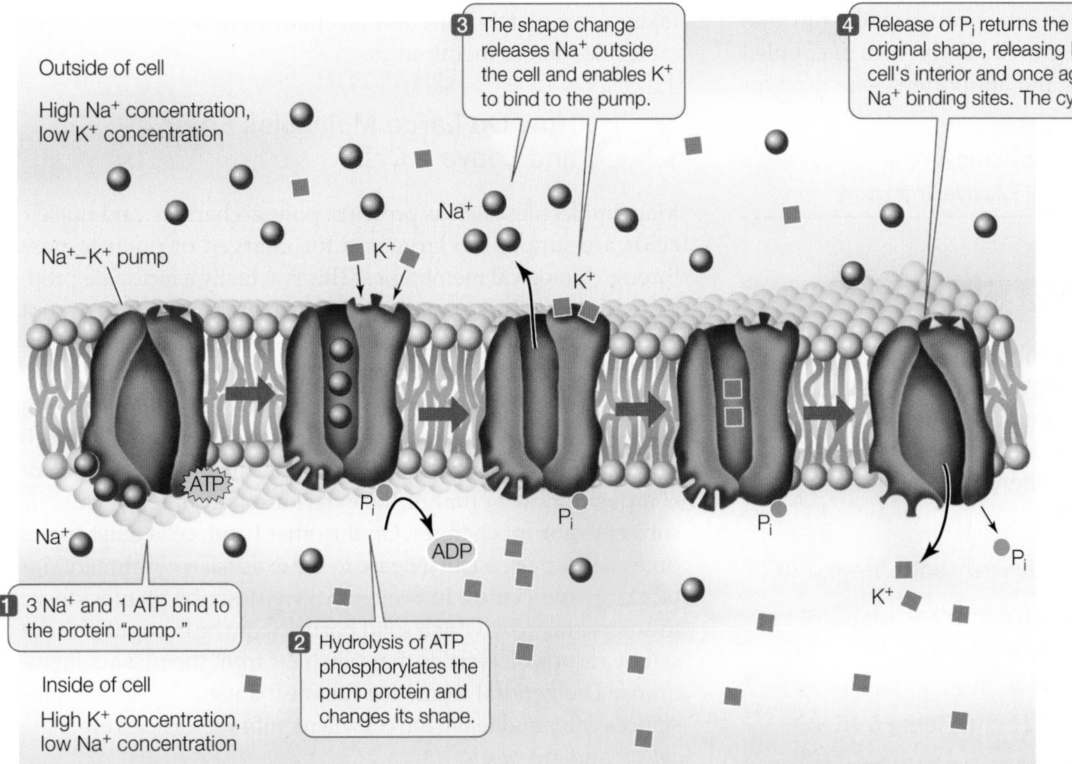
Outside of cell
High Na⁺ concentration, low K⁺ concentration
Na⁺–K⁺ pump
Na⁺
K⁺
ATP
ADP
Pᵢ
Inside of cell
High K⁺ concentration, low Na⁺ concentration
1 3 Na⁺ and 1 ATP bind to the protein "pump."
2 Hydrolysis of ATP phosphorylates the pump protein and changes its shape.
3 The shape change releases Na⁺ outside the cell and enables K⁺ to bind to the pump.
4 Release of Pᵢ returns the pump to its original shape, releasing K⁺ to the cell's interior and once again exposing Na⁺ binding sites. The cycle repeats.

**6.16 Primary Active Transport: The Sodium–Potassium Pump** In active transport, energy is used to move a solute against its concentration gradient. Here, energy from ATP is used to move $Na^+$ and $K^+$ against their concentration gradients.

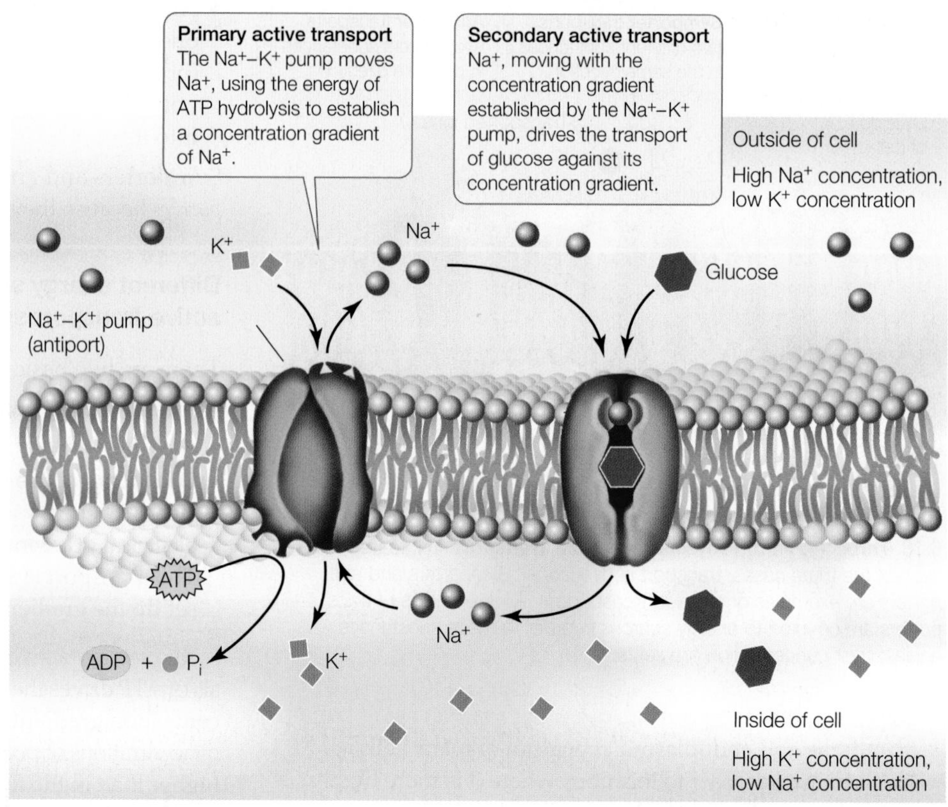

Primary active transport
The Na+–K+ pump moves Na+, using the energy of ATP hydrolysis to establish a concentration gradient of Na+.
Secondary active transport
Na+, moving with the concentration gradient established by the Na+–K+ pump, drives the transport of glucose against its concentration gradient.
Outside of cell
High Na+ concentration, low K+ concentration
K+
Na+
Glucose
Na+–K+ pump (antiport)
ATP
ADP + Pi
K+
Na+
Inside of cell
High K+ concentration, low Na+ concentration


**6.17 Secondary Active Transport** The $Na^+$ concentration gradient established by primary active transport (left) powers the secondary active transport of glucose (right). A symporter protein couples the movement of glucose across the membrane against its concentration gradient to the passive movement of $Na^+$ into the cell.

found in all animal cells. The pump is an integral membrane glycoprotein. It breaks down a molecule of ATP to ADP and a free phosphate ion ($P_i$) and uses the energy released to bring two $K^+$ ions into the cell and export three $Na^+$ ions. The $Na^+$–$K^+$ pump is thus an antiporter because it moves two substances in different directions.

In secondary active transport, the movement of a substance against its concentration gradient is accomplished using energy "regained" by letting ions move across the membrane with their electrochemical and concentration gradients. For example, once the sodium–potassium pump establishes a concentration gradient of sodium ions, the passive diffusion of some $Na^+$ back into the cell can provide energy for the secondary active transport of glucose into the cell (**Figure 6.17**). This occurs when glucose is absorbed into the bloodstream from the digestive tract. Secondary active transport aids in the uptake of amino acids and sugars, which are essential raw materials for cell maintenance and growth. Both types of coupled transport proteins—symporters and antiporters—are used for secondary active transport.

**yourBioPortal.com**
GO TO Animated Tutorial 6.2 • Active Transport

## 6.4 RECAP

**Active transport across a membrane is directional and requires an input of energy to move substances against their concentration gradients. Active transport allows a cell to maintain small molecules and ions at concentrations very different from those in the surrounding environment.**

- Why is energy required for active transport? See p. 120
- Explain the difference between primary active transport and secondary active transport. See p. 121
- Why is the sodium–potassium ($Na^+$–$K^+$) pump classified as an antiporter? See p. 122 and Figure 6.16

We have examined a number of passive and active ways in which ions and small molecules can enter and leave cells. But what about large molecules such as proteins? Many proteins are so large that they diffuse very slowly, and their bulk makes it difficult for them to pass through the phospholipid bilayer. It takes a completely different mechanism to move intact large molecules across membranes.

## 6.5 How Do Large Molecules Enter and Leave a Cell?

Macromolecules such as proteins, polysaccharides, and nucleic acids are simply too large and too charged or polar to pass through biological membranes. This is actually a fortunate property—think of the consequences if such molecules diffused out of cells. A red blood cell would not retain its hemoglobin! Indeed, as we discuss in Chapter 5, the development of a selectively permeable membrane was essential for the functioning of the first cells when life on Earth began. The interior of a cell can be maintained as a separate compartment with a different composition from that of the exterior environment, which is subject to abrupt changes. On the other hand, cells must sometimes take up or *secrete* (release to the external environment) intact large molecules. In Section 5.3 we describe phagocytosis, the mechanism by which solid particles can be brought into the cell by means of vesicles that pinch off from the plasma membrane. The general terms for the mechanisms by which substances enter and leave the cell via membrane vesicles are *endocytosis* and *exocytosis*.

### Macromolecules and particles enter the cell by endocytosis

**Endocytosis** is a general term for a group of processes that bring small molecules, macromolecules, large particles, and even small cells into the eukaryotic cell (**Figure 6.18A**). There are three types of endocytosis: phagocytosis, pinocytosis, and receptor-mediated endocytosis. In all three, the plasma membrane invaginates (folds inward), forming a small pocket around materials from the environment. The pocket deepens, forming a vesicle. This vesicle separates from the plasma membrane and migrates with its contents to the cell's interior.

- In **phagocytosis** ("cellular eating"), part of the plasma membrane engulfs large particles or even entire cells. Unicellular protists use phagocytosis for feeding, and some white blood cells use phagocytosis to defend the body by engulfing foreign cells and substances. The food vacuole or phagosome that forms usually fuses with a lysosome, where its contents are digested (see Figure 5.11).
- In **pinocytosis** ("cellular drinking"), vesicles also form. However, these vesicles are smaller, and the process operates to bring dissolved substances, including proteins or fluids, into the cell. Like phagocytosis, pinocytosis can be relatively nonspecific regarding what it brings into the cell. For example, pinocytosis goes on constantly in the endothelium, the single layer of cells that separates a tiny blood capillary from the surrounding tissue. Pinocytosis allows cells of the endothelium to rapidly acquire fluids and dissolved solutes from the blood.
- In **receptor-mediated endocytosis**, molecules at the cell surface recognize and trigger the uptake of specific materials.

Let's take a closer look at this last process.

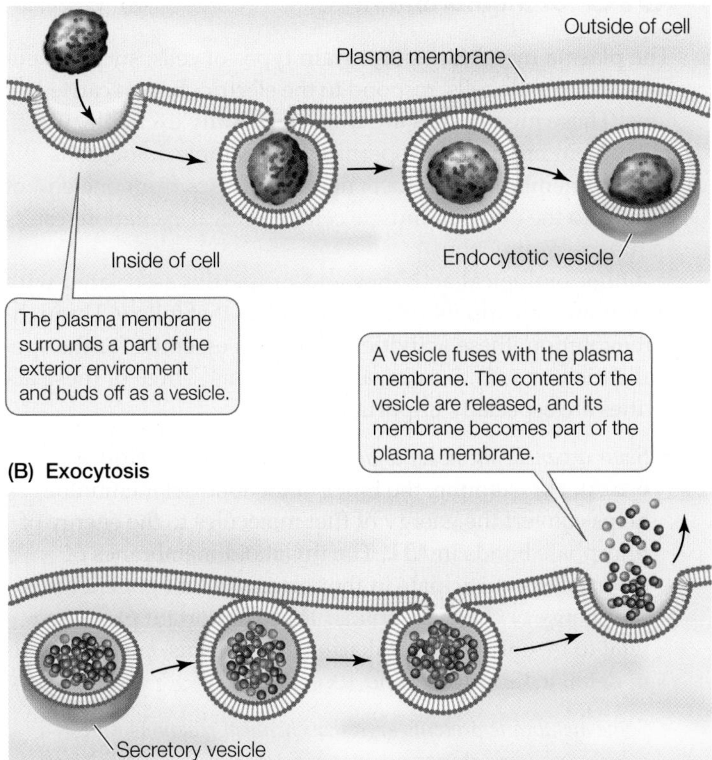
(A) Endocytosis
Outside of cell
Plasma membrane
Inside of cell
Endocytotic vesicle
The plasma membrane surrounds a part of the exterior environment and buds off as a vesicle.
A vesicle fuses with the plasma membrane. The contents of the vesicle are released, and its membrane becomes part of the plasma membrane.
(B) Exocytosis
Secretory vesicle

**6.18 Endocytosis and Exocytosis** Endocytosis (A) and exocytosis (B) are used by eukaryotic cells to take up and release large molecules and particles, and small cells.

### Receptor-mediated endocytosis is highly specific

Receptor-mediated endocytosis is used by animal cells to capture specific macromolecules from the cell's environment. This process depends on **receptor proteins**, which are proteins that can bind to specific molecules within the cell or in the cell's external environment. In receptor-mediated endocytosis, the receptors are integral membrane proteins located at particular regions on the extracellular surface of the plasma membrane. These membrane regions are called *coated pits* because they form slight depressions in the plasma membrane and their cytoplasmic surfaces are coated by other proteins, such as clathrin. The uptake process is similar to that in phagocytosis.

When a receptor protein binds to its specific ligand (in this case, the macromolecule to be taken into the cell), its coated pit invaginates and forms a coated vesicle around the bound macromolecule. The clathrin molecules strengthen and stabilize the vesicle, which carries the macromolecule away from the plasma membrane and into the cytoplasm (**Figure 6.19**). Once inside, the vesicle loses its clathrin coat and may fuse with a lysosome, where the engulfed material is digested (by the hydrolysis of polymers to monomers) and the products released into the cytoplasm. Because of its specificity for particular macromolecules, receptor-mediated endocytosis is an efficient method of taking up substances that may exist at low concentrations in the cell's environment.

Receptor-mediated endocytosis is the method by which cholesterol is taken up by most mammalian cells. Water-insoluble cholesterol and triglycerides are packaged by liver cells into lipoprotein particles. Most of the cholesterol is packaged into a type of lipoprotein particle called *low-density lipoprotein*, or LDL, which is circulated via the bloodstream. When a particular cell requires cholesterol, it produces specific LDL receptors, which are inserted into the plasma membrane in clathrin-coated pits. Binding of LDLs to the receptor proteins triggers the uptake of the LDLs via receptor-mediated endocytosis. Within the resulting vesicle, the LDL particles are freed from the receptors. The receptors segregate to a region that buds off and forms a new vesicle, which is recycled to the plasma membrane. The freed LDL particles remain in the original vesicle, which fuses with a lysosome. There, the LDLs are digested and the cholesterol made available for cell use.

In healthy individuals, the liver takes up unused LDLs for recycling. People with the inherited disease *familial hypercholesterolemia* have a deficient LDL receptor in their livers. This prevents receptor-mediated endocytosis of LDLs, resulting in

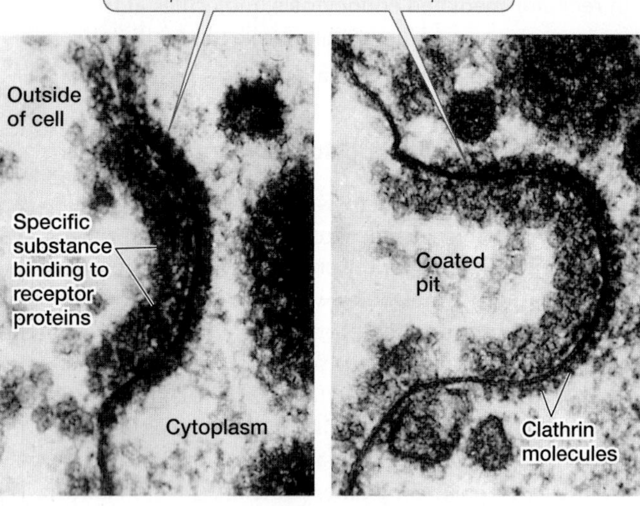


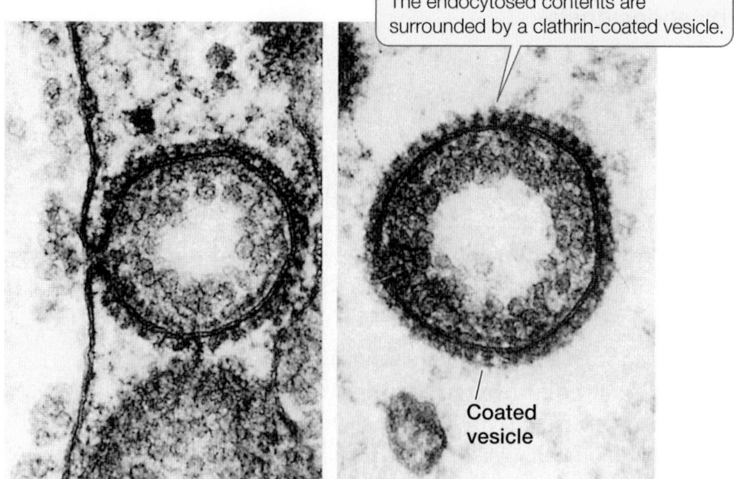


**6.19 Receptor-Mediated Endocytosis** The receptor proteins in a coated pit bind specific macromolecules, which are then carried into the cell by a coated vesicle.

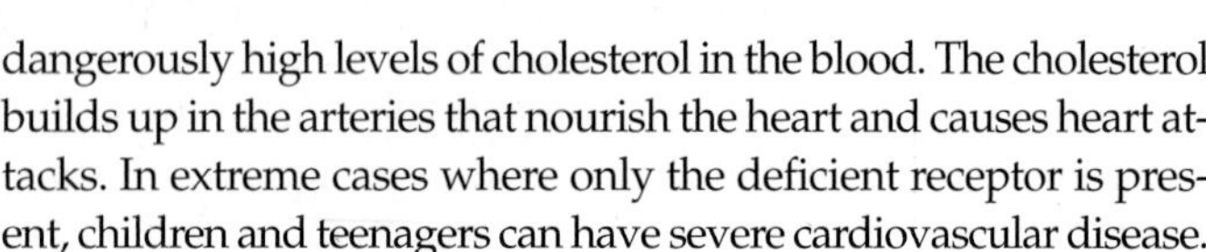

dangerously high levels of cholesterol in the blood. The cholesterol builds up in the arteries that nourish the heart and causes heart attacks. In extreme cases where only the deficient receptor is present, children and teenagers can have severe cardiovascular disease.

### Exocytosis moves materials out of the cell

**Exocytosis** is the process by which materials packaged in vesicles are secreted from a cell when the vesicle membrane fuses with the plasma membrane (see Figure 6.18B). This fusing makes an opening to the outside of the cell. The contents of the vesicle are released into the environment, and the vesicle membrane is smoothly incorporated into the plasma membrane.

In Chapter 5 we encounter exocytosis as the last step in the processing of material engulfed by phagocytosis—the release of undigested materials back to the extracellular environment. Exocytosis is also important in the secretion of many different substances, including digestive enzymes from the pancreas, neurotransmitters from neurons, and materials for the construction of the plant cell wall. You will encounter these processes in later chapters.

**yourBioPortal.com**
GO TO **Animated Tutorial 6.3 • Endocytosis and Exocytosis**

### 6.5 RECAP

**Endocytosis and exocytosis are the processes by which large particles and molecules are transported into and out of the cell. Endocytosis may be mediated by a receptor protein in the plasma membrane.**

- Explain the difference between phagocytosis and pinocytosis. See p. 123
- Describe an example of receptor-mediated endocytosis. See p. 123 and Figure 6.19

We have now examined the structures and some of the functions of biological membranes. We have seen how macromolecules on the plasma membrane surface allow cells to recognize and adhere to each other, so that tissues and organs can form. We have also seen how membranes selectively regulate the traffic of small and large molecules, and how large particles such as LDLs can be taken up by cells. These are crucial functions, but they are not the only functions of biological membranes.

## 6.6 What Are Some Other Functions of Membranes?

The plasma membranes of certain types of cells, such as neurons and muscle cells, respond to the electric charges carried by ions. These membranes are thus electrically excitable, which gives them important properties. For example, in neurons, the plasma membrane conducts nerve impulses from one end of the cell to the other. In muscle cells, electrical excitation results in muscle contraction.

Other biological activities and properties associated with membranes are discussed in the chapters that follow. Throughout evolution, these activities have been essential for the specialization of cells, tissues, and organisms. Three of these activities are especially important:

- *Some organelle membranes help transform energy* (**Figure 6.20A**). For example, the inner mitochondrial membrane helps convert the energy of fuel molecules to the energy of phosphate bonds in ATP. The thylakoid membranes of chloroplasts participate in the conversion of light energy to the energy of chemical bonds. These important processes, vital to the life of most eukaryotic organisms, are discussed in detail in Chapters 9 and 10.
- *Some membrane proteins organize chemical reactions.* Often a cellular process depends on a series of enzyme-catalyzed

(A) Energy transformation

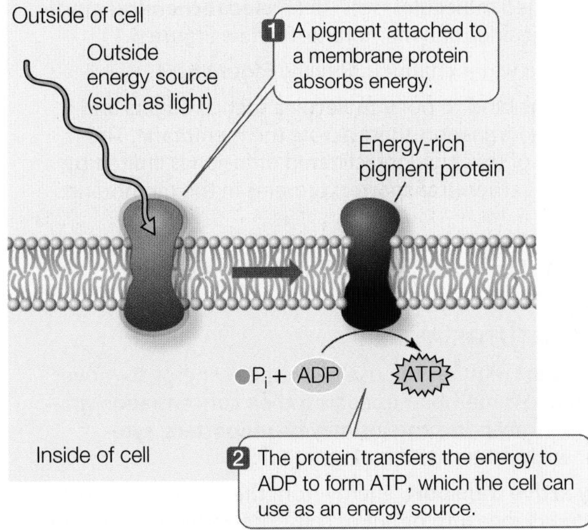


Energy-rich pigment protein

2 The protein transfers the energy to ADP to form ATP, which the cell can use as an energy source.

(C) Information processing

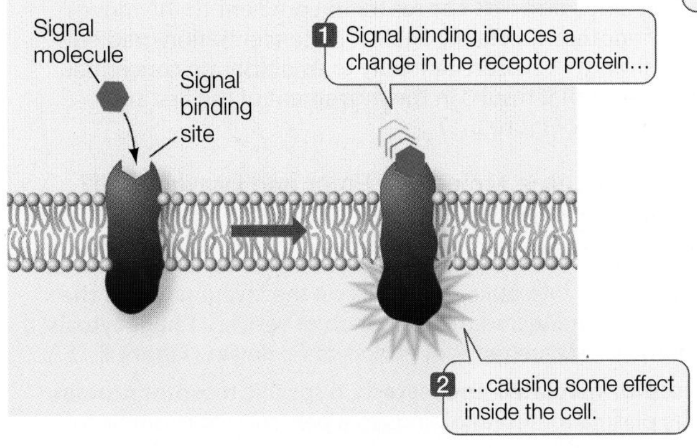


Signal binding site

2 ...causing some effect inside the cell.

**6.20 Other Membrane Functions** The compartmentation afforded by a lipid bilayer or protein membrane was a key event in the emergence of cells. Functions such as energy transformation (A), organization of chemical reactions (B), and signaling (C) probably evolved later and conferred a selective advantage on cells and organisms that had them.

(B) Organizing chemical reactions

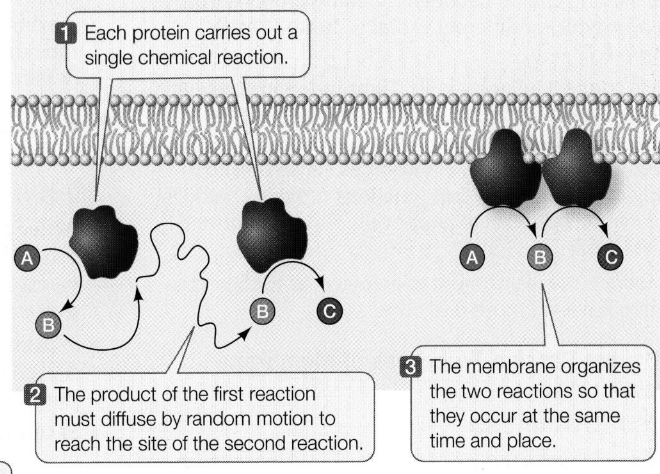

1 Each protein carries out a single chemical reaction.
A
B
B
C
A
B
C
2 The product of the first reaction must diffuse by random motion to reach the site of the second reaction.
3 The membrane organizes the two reactions so that they occur at the same time and place.

reactions, in which the products of one reaction serve as reactants in the next. For such a series of reactions to occur, all the necessary molecules must come together. In a solution, reactant and enzyme molecules are randomly distributed and collisions among them are random. Because these collisions are necessary for chemical reactions to occur, a complete series of chemical reactions may occur only very slowly in a solution. However, if the different enzymes are bound to a membrane in sequential order, the product of one reaction can be released close to the enzyme for the next reaction. Such an "assembly line" allows reactions to proceed rapidly and efficiently (**Figure 6.20B**).

- *Some membrane proteins process information.* As we have seen, biological membranes may have integral membrane proteins or attached carbohydrates that can bind to specific substances in the environment. Without entering a cell, a specific ligand can bind to a receptor and serve as a signal to initiate, modify, or turn off a cell function (**Figure 6.20C**). In this type of information processing, specificity in binding is essential.

We have seen the informational role of the LDL receptor protein in the recognition and endocytosis of LDL, with its cargo of cholesterol. Another example is the binding of a hormone such as insulin to specific receptors on a target cell. When insulin binds to receptors on a liver cell, it elicits the uptake of glucose. In Chapter 7 there are many other examples of the role of membrane proteins in information processing.

# CHAPTER SUMMARY

## 6.1 What Is the Structure of a Biological Membrane?

- Biological membranes consist of lipids, proteins, and carbohydrates. The **fluid mosaic model** of membrane structure describes a phospholipid bilayer in which proteins can move about within the plane of the membrane. **SEE WEB ACTIVITY 6.1**
- The two leaflets of a membrane may have different properties because of their different phospholipid compositions, exposed domains of **integral membrane proteins**, and **peripheral membrane proteins**. Some proteins, called **transmembrane proteins**, span the membrane. Review Figure 6.1
- Carbohydrates, attached to proteins in **glycoproteins** or to phospholipids in **glycolipids**, project from the external surface of the plasma membrane and function as recognition signals.
- Membranes are not static structures, but are constantly forming, exchanging, and breaking down.

## 6.2 How Is the Plasma Membrane Involved in Cell Adhesion and Recognition?

- In order for cells to assemble into tissues they must recognize and adhere to one another. **Cell recognition** and **cell adhesion** depend on integral membrane proteins that protrude from the cell surface. Binding can be between the same proteins from two cells (**homotypic**) or different proteins (**heterotypic**). Review Figure 6.6
- Cell junctions connect adjacent cells. **Tight junctions** prevent the passage of molecules through the intercellular spaces between cells, and they restrict the migration of membrane proteins over the cell surface. **Desmosomes** cause cells to adhere firmly to one another. **Gap junctions** provide channels for communication between adjacent cells. Review Figure 6.7, WEB ACTIVITY 6.2
- **Integrins** mediate the attachment of animal cells to the extracellular matrix. Review Figure 6.8

## 6.3 What Are the Passive Processes of Membrane Transport?

SEE ANIMATED TUTORIAL 6.1

- Membranes exhibit **selective permeability**, regulating which substances pass through them.
- A substance can diffuse passively across a membrane by one of two processes: **simple diffusion** through the phospholipid bilayer or **facilitated diffusion** either through a **channel** or by means of a **carrier protein**.
- A solute diffuses across a membrane from a region with a greater concentration of that solute to a region with a lesser concentration of that solute. Equilibrium is reached when the solute concentrations on both sides of the membrane show no net change over time. Review Figure 6.9
- In **osmosis**, water diffuses from a region of higher water concentration to a region of lower water concentration.
- Most cells are in an **isotonic** environment, where total solute concentrations on both sides of the plasma membrane are equal. If the solution surrounding a cell is **hypotonic** to the cell interior, more water enters the cell than leaves it. In plant cells, this leads to **turgor pressure**. In a **hypertonic** solution, more water leaves the cell than enters it. Review Figure 6.10
- **Ion channels** are membrane proteins that allow the rapid facilitated diffusion of ions through membranes. **Gated channels** can be opened or closed by certain conditions or chemicals. The opening or closing of channels, as well as an asymmetric distribution of charged molecules, sets up an **electrochemical gradient** on different sides of a membrane. Review Figure 6.11
- **Aquaporins** are water channels. Review Figure 6.13
- **Carrier proteins** bind to polar molecules such as sugars and amino acids and transport them across the membrane. The maximum rate of this type of facilitated diffusion is limited by the number of carrier (transporter) proteins in the membrane. Review Figure 6.14

## 6.4 What Are the Active Processes of Membrane Transport?

SEE ANIMATED TUTORIAL 6.2

- **Active transport** requires the use of chemical energy to move substances across membranes against their concentration gradients. Active transport proteins may be **uniporters**, **symporters**, or **antiporters**. Review Figure 6.15
- In **primary active transport**, energy from the hydrolysis of ATP is used to move ions into or out of cells. The **sodium-potassium pump** is an important example. Review Figure 6.16
- **Secondary active transport** couples the passive movement of one substance down its concentration gradient to the movement of another substance against its concentration gradient. Energy from ATP is used indirectly to establish the concentration gradient that results in the movement of the first substance. Review Figure 6.17

## 6.5 How Do Large Molecules Enter and Leave a Cell?

SEE ANIMATED TUTORIAL 6.3

- **Endocytosis** is the transport of macromolecules, large particles, and small cells into eukaryotic cells via the invagination of the plasma membrane and the formation of vesicles. **Phagocytosis** and **pinocytosis** are types of endocytosis. Review Figure 6.18A
- In **receptor-mediated endocytosis**, a specific **receptor protein** on the plasma membrane binds to a particular macromolecule.
- In **exocytosis**, materials in vesicles are secreted from the cell when the vesicles fuse with the plasma membrane. Review Figure 6.18B

## 6.6 What Are Some Other Functions of Membranes?

- Membranes function as sites for energy transformations, for organizing chemical reactions, and for recognition and initial processing of extracellular signals. Review Figure 6.20

# SELF-QUIZ

1. Which statement about membrane phospholipids is *not* true?
   *a.* They associate to form bilayers.
   *b.* They have hydrophobic "tails."
   *c.* They have hydrophilic "heads."
   *d.* They give the membrane fluidity.
   *e.* They flip-flop readily from one side of the membrane to the other.
2. When a hormone molecule binds to a specific protein on the plasma membrane, the protein it binds to is called a
   *a.* ligand.
   *b.* clathrin.
   *c.* receptor protein.
   *d.* hydrophobic protein.
   *e.* cell adhesion molecule.
3. Which statement about membrane proteins is *not* true?
   *a.* They all extend from one side of the membrane to the other.
   *b.* Some serve as channels for ions to cross the membrane.
   *c.* Many are free to migrate laterally within the membrane.
   *d.* Their position in the membrane is determined by their tertiary structure.
   *e.* Some play roles in photosynthesis.

4. Which statement about membrane carbohydrates is *not* true?
   *a.* Some are bound to proteins.
   *b.* Some are bound to lipids.
   *c.* They are added to proteins in the Golgi apparatus.
   *d.* They show little diversity.
   *e.* They are important in recognition reactions at the cell surface.
5. Which statement about animal cell junctions is *not* true?
   *a.* Tight junctions are barriers to the passage of molecules between cells.
   *b.* Desmosomes allow cells to adhere firmly to one another.
   *c.* Gap junctions block communication between adjacent cells.
   *d.* Connexons are made of protein.
   *e.* The fibers associated with desmosomes are made of protein.
6. You are studying how the protein transferrin enters cells. When you examine cells that have taken up transferrin, you find it inside clathrin-coated vesicles. Therefore, the most likely mechanism for uptake of transferrin is
   *a.* facilitated diffusion.
   *b.* an antiporter.
   *c.* receptor-mediated endocytosis.
   *d.* gap junctions.
   *e.* ion channels.
7. Which statement about ion channels is *not* true?
   *a.* They form pores in the membrane.
   *b.* They are proteins.
   *c.* All ions pass through the same type of channel.
   *d.* Movement through them is from regions of high concentration to regions of low concentration.
   *e.* Movement through them is by simple diffusion.
8. Facilitated diffusion and active transport both
   *a.* require ATP.
   *b.* require the use of proteins as carriers or channels.
   *c.* carry solutes in only one direction.
   *d.* increase without limit as the concentration gradient increases.
   *e.* depend on the solubility of the solute in lipids.
9. Primary and secondary active transport both
   *a.* generate ATP.
   *b.* are based on passive movement of $Na^+$ ions.
   *c.* include the passive movement of glucose molecules.
   *d.* use ATP directly.
   *e.* can move solutes against their concentration gradients.
10. Which statement about osmosis is *not* true?
    *a.* It obeys the laws of diffusion.
    *b.* In animal tissues, water moves into cells if they are hypertonic to their environment.
    *c.* Red blood cells must be kept in a plasma that is hypotonic to the cells.
    *d.* Two cells with identical solute concentrations are isotonic to each other.
    *e.* Solute concentration is the principal factor in osmosis.

## FOR DISCUSSION

1. Muscle function requires calcium ions ($Ca^{2+}$) to be pumped into a subcellular compartment against a concentration gradient. What types of molecules are required for this to happen?
2. Section 27.5 describes the diatoms, which are protists that have complex glassy structures in their cell walls (see Figure 27.7B). These structures form within the Golgi apparatus. How do these structures reach the cell wall without having to pass through a membrane?
3. Organisms that live in fresh water are almost always hypertonic to their environment. In what way is this a serious problem? How do some organisms cope with this problem?
4. Contrast nonspecific endocytosis and receptor-mediated endocytosis.
5. The emergence of the phospholipid membrane was important to the origin of cells. Describe the properties of membranes that might have allowed cells to thrive in comparison with molecular aggregates without membranes.

## ADDITIONAL INVESTIGATION

When a normal lung cell becomes a lung cancer cell, there are several important changes in plasma membrane properties. How would you investigate the following phenomena? *(a)* The cancer cell membrane is more fluid, with more rapid diffusion in the plane of the membrane of both lipids and proteins. *(b)* The cancer cell has altered cell adhesion properties, binding to other tissues in addition to lung cells.

## WORKING WITH DATA (GO TO yourBioPortal.com)

**Aquaporin Increases Membrane Permeability to Water** In this hands-on exercise based on Figure 6.13, you will investigate how Agre and colleagues used an egg cell to show that expression of aquaporin results in rapid water uptake when the cell is placed in a hypotonic medium. Analyzing their experimental design and data, you will see how this model cell system and control experiments confirmed the important role of aquaporin as a water channel.

# 7 Cell Signaling and Communication

## Love signals

Prairie voles (*Microtus ochrogaster*) are small rodents that live in temperate climates, where they dig tunnels in fields. When a male prairie vole encounters a female, mating often ensues. After mating (which can take as long as a day), the couple stays together, building a nest and raising their pups together. The two voles bond so tightly that they stay together for life. Contrast this behavior with that of the montane vole (*M. montanus*), which is closely related to the prairie vole and lives in the hills not far away. In this species, mating is quick, and afterwards the couple separates. The male looks for new mates and the female abandons her young soon after they are born.

The explanation for these dramatic behavioral differences lies in the brains of these two species. Neuroscientist Thomas Insel and his colleagues found that when prairie voles mate for all those hours, their brains release a 9-amino-acid peptide. In females, this peptide is oxytocin; in males, it is vasopressin. The peptide is circulated in the bloodstream and reaches all tissues in the body, but it binds to only a few cell types. These cells have surface proteins, called receptors, that specifically bind the peptide, like a key inserting into a lock.

The interaction of peptide and receptor causes the receptor, which extends across the plasma membrane, to change shape. Within the cytoplasm, this change sets off a series of events called a signal transduction pathway. Such a pathway can cause many different cellular responses, but in this case, the notable changes are in behavior. The receptors for oxytocin and vasopressin in prairie voles are most concentrated in the regions of the brain that are responsible for behaviors such as bonding and caring for the young. In montane voles, there are far fewer receptors and as a result, fewer postmating behaviors.

These cause-and-effect relationships between peptides, receptors, and behavior have been established through experiments. For example, a female prairie vole that is injected before mating with a molecule that blocks oxytocin does not bond with the male. Also, a female injected with oxytocin will bond with a male even without mating. Experiments with vasopressin in males give similar results. Furthermore, promiscuous vole males that were genetically manipulated to express prairie vole amounts of the vasopressin receptor grew up to behave more like prairie vole males. These experiments show that oxytocin and

**Voles** Prairie voles display extensive bonding behaviors after mating. These behaviors are mediated by peptides acting as intercellular signals.

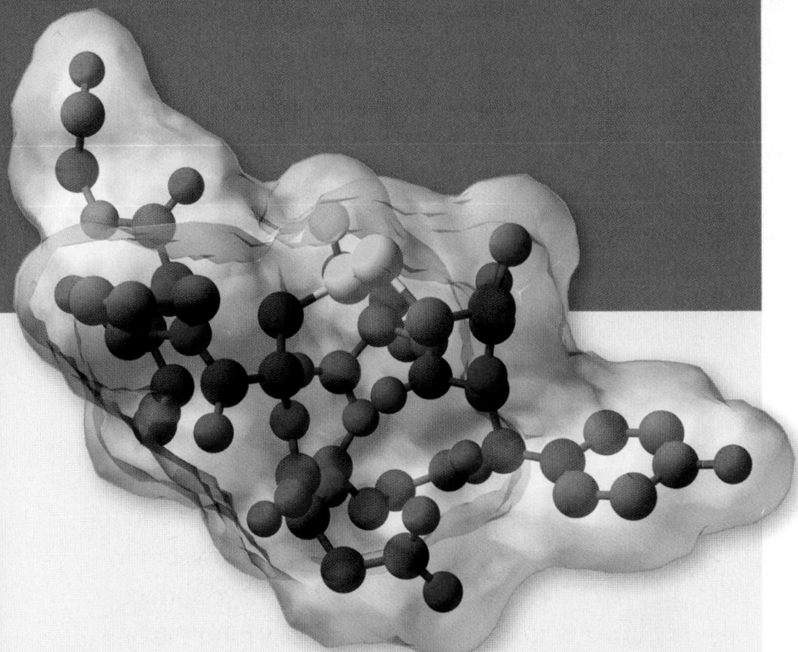

**Oxytocin** This peptide with 9 amino acids acts as a signal for postmating behaviors.

vasopressin are signals that induce bonding and caring behaviors in voles. Could this also be true of humans?

Neuroeconomist Paul Zaks thinks so. He has done experiments with human volunteers, who were asked to "invest" funds with a stranger. A group of investors that was given a nasal spray containing oxytocin was more trusting of the stranger (and invested more funds) than a group that got an inert spray. So the oxytocin signaling pathway is important in human behavior too.

A cell's response to any signal molecule takes place in three sequential steps. First, the signal binds to a receptor in the cell, often on the outside surface of the plasma membrane. Second, signal binding conveys a message to the cell. Third, the cell changes its activity in response to the signal. And in a multicellular organism, this leads to changes in that organism's functioning.

**IN THIS CHAPTER** we first describe the types of signals that affect cells. These include chemicals produced by other cells and substances from outside the body, as well as physical and environmental factors such as light. Then we show how a signal affects only those cells that have the specific receptor to recognize that signal. Next, we describe the steps of signal transduction in which the receptor communicates to the cell that a signal has been received, thus causing a change in cell function.

CHAPTER OUTLINE



## 7.1 What Are Signals, and How Do Cells Respond to Them?

Both prokaryotic and eukaryotic cells process information from their environments. This information can be in the form of a physical stimulus, such as the light reaching your eyes as you read this book, or chemicals that bathe a cell, such as lactose in a bacterial growth medium. It may come from outside the organism, such as the scent of a female moth seeking a mate in the dark, or from a neighboring cell within the organism, such as in the heart, where thousands of muscle cells contract in unison by transmitting signals to one another.

Of course, the mere presence of a signal does not mean that a cell will respond to it, just as you do not pay close attention to every image in your environment as you study. To respond to a signal, the cell must have a specific receptor that can detect it. This section provides examples of some types of cellular signals and one model of *signal transduction*. A **signal transduction pathway** is a sequence of molecular events and chemical reactions that lead to a cell's response to a signal. After discussing signals in this section, we will consider their receptors in Section 7.2.

### Cells receive signals from the physical environment and from other cells

The physical environment is full of signals. Our sense organs allow us to respond to light, odors and tastes (chemical signals), temperature, touch, and sound. Bacteria and protists can respond to minute chemical changes in their environments. Plants respond to light as a signal as well as an energy source. The amount and wavelengths of light reaching a plant's surface differ from day to night and in direct sunlight versus shade. These variations act as signals that affect plant growth and reproduction. Some plants also respond to temperature: when the weather gets cold, they may respond either by becoming tolerant to cold or by accelerating flowering.

A cell deep inside a large multicellular organism is far away from the exterior environment. Such a cell's environment consists of other cells and extracellular fluids. Cells receive their nutrients from, and pass their wastes into, extracellular fluids. Cells also receive signals—mostly chemical signals—from their extracellular fluid environment. Most of these chemical signals come from other cells, but they can also come from the environment via the digestive and respiratory systems. And cells can respond to changes in the extracellular concentrations of certain chemicals, such as $CO_2$ and $H^+$, which are affected by the metabolic activities of other cells.

Inside a large multicellular organism, chemical signals made by the body itself reach a target cell by local diffusion or by circulation within the blood. These signals are usually in tiny concentrations (as low as $10^{-10}$ *M*) (see Chapter 2 for an explanation of *molar* concentrations). **Autocrine** signals diffuse to and affect the cells that make them; for example, part of the reason many tumor cells reproduce uncontrollably is because they self-stimulate cell division by making their own division signals. **Paracrine** signals diffuse to and affect nearby cells; an example is a neurotransmitter made by one nerve cell that diffuses to an adjacent cell and stimulates it. (**Figure 7.1A**). Signals to distant cells called hormones travel through the circulatory system (**Figure 7.1B**).

## A signal transduction pathway involves a signal, a receptor, and responses

For the information from a signal to be transmitted to a cell, the target cell must be able to receive or sense the signal and respond to it, and the response must have some effect on the function of the cell. In a multicellular organism, all cells may receive chemical signals that are circulated in the blood, such as the peptides oxytocin and vasopressin that are released following mating in voles (see the opening of this chapter), but most body cells are not capable of responding to the signals. Only the cells with the necessary receptors can respond.

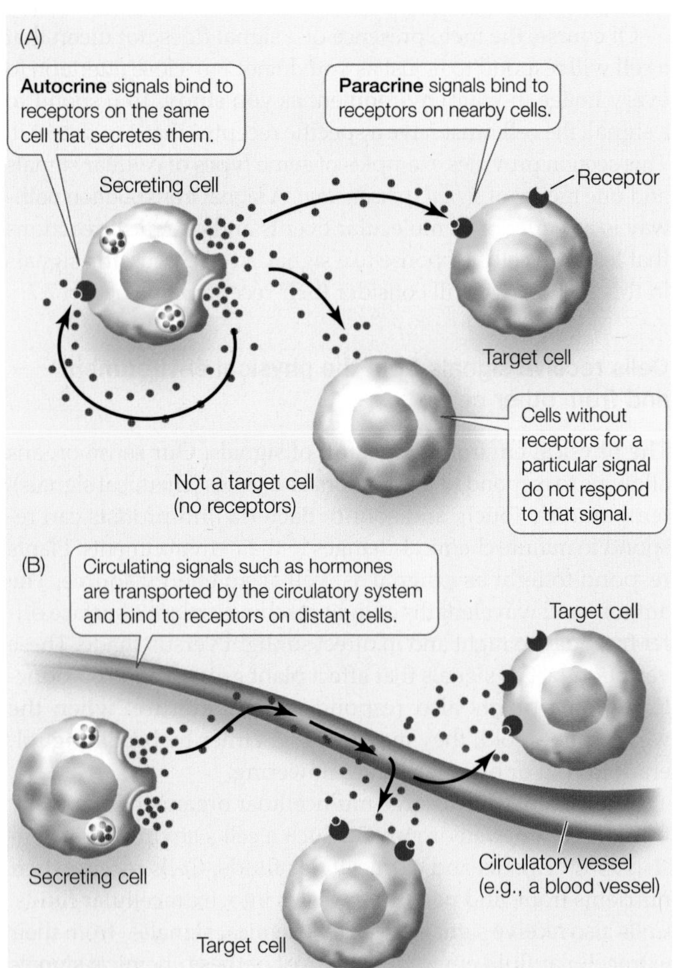


**7.1 Chemical Signaling Systems** (A) A signal molecule can diffuse to act on the cell that produces it, or on a nearby cell. (B) Many signals act on distant cells and must be transported by the organism's circulatory system.

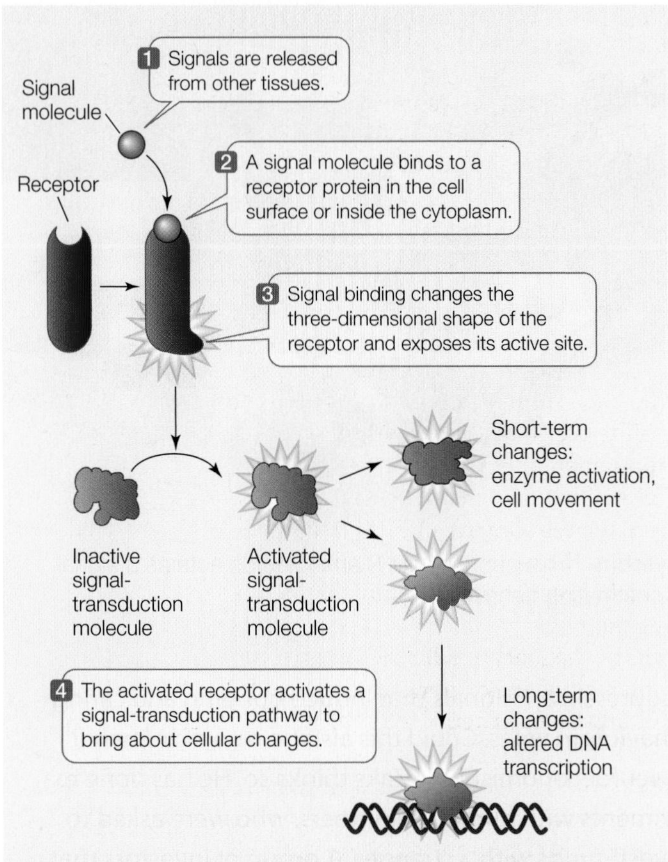


**7.2 A Signal Transduction Pathway** This general pathway is common to many cells and situations. The ultimate effects on the cell are either short-term or long-term molecular changes, or both.

The kinds of responses vary greatly depending on the signal and the target cell. Just a few examples are: a skin cell initiating cell division to heal a wound; a cell moving to a new location in the embryo to form a tissue; a cell releasing enzymes to digest food; a plant cell loosening bonds that hold its cell wall polymers together so that it can expand; and a cell in the eye sending messages to the brain about the book you are reading. A signal transduction pathway involves a signal, a receptor, and a response (**Figure 7.2**).

Let's look at an example of such a pathway in the bacterium *Escherichia coli* (*E. coli*). Follow the features of this pathway in general (see Figure 7.2) and in particular (**Figure 7.3**).

**SIGNAL** As a prokaryotic cell, a bacterium is very sensitive to changes in its environment. One thing that can change is the total solute concentration (osmotic concentration—see Section 6.3) in the environment surrounding the cell. In the mammalian intestine where *E. coli* lives, the solute concentration around

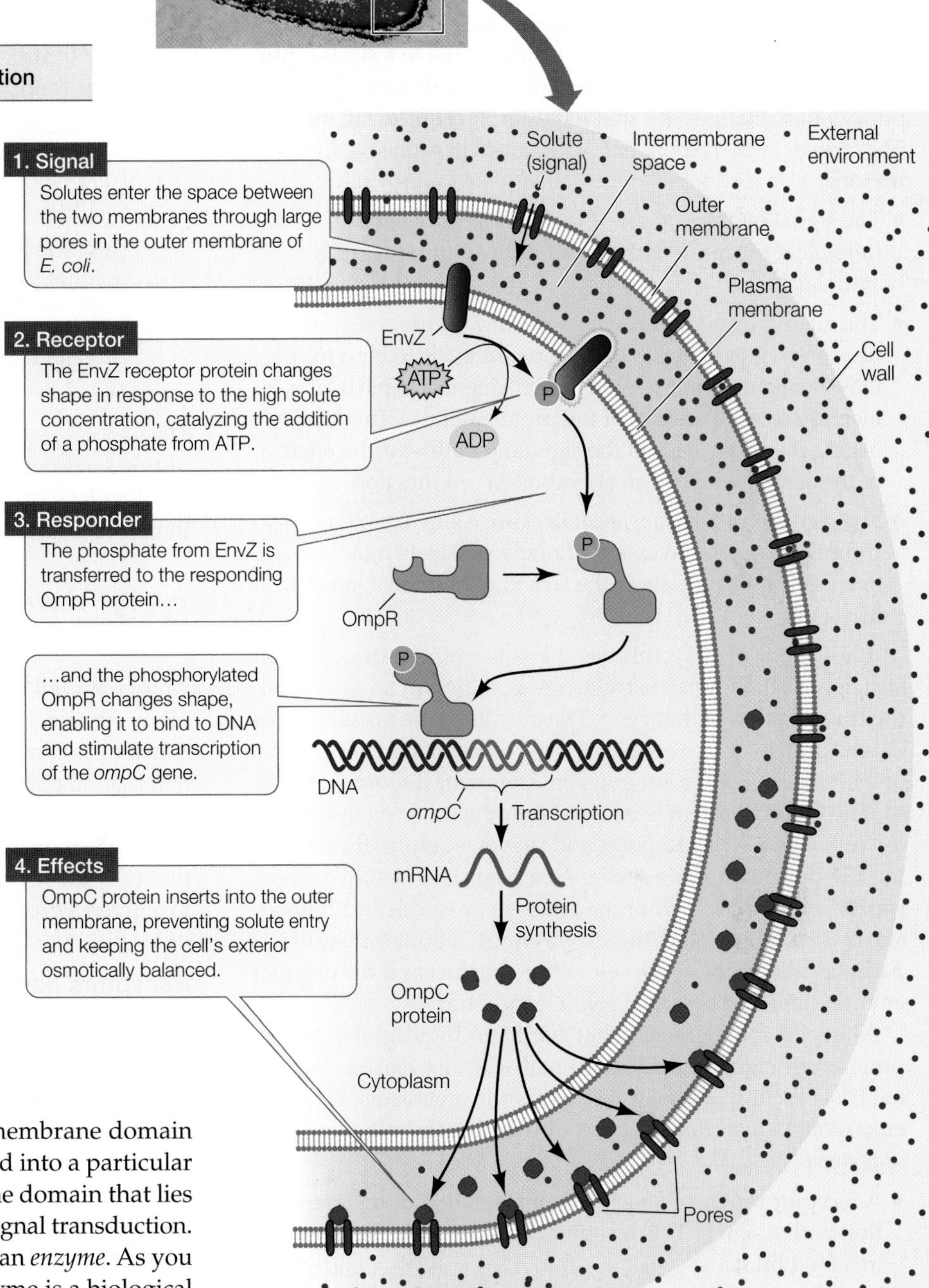

Bacterial cell
Solute (signal)
Intermembrane space
External environment
Outer membrane
Plasma membrane
Cell wall
1. Signal
Solutes enter the space between the two membranes through large pores in the outer membrane of E. coli.
2. Receptor
The EnvZ receptor protein changes shape in response to the high solute concentration, catalyzing the addition of a phosphate from ATP.
EnvZ
ATP
ADP
P
3. Responder
The phosphate from EnvZ is transferred to the responding OmpR protein...
OmpR
...and the phosphorylated OmpR changes shape, enabling it to bind to DNA and stimulate transcription of the ompC gene.
DNA
ompC
Transcription
mRNA
Protein synthesis
4. Effects
OmpC protein inserts into the outer membrane, preventing solute entry and keeping the cell's exterior osmotically balanced.
OmpC protein
Cytoplasm
Pores


**7.3 A Model Signal Transduction Pathway** *E. coli* responds to the signal of an increase in solute concentration in its environment. The basic steps of such a signal transduction pathway occur in all living organisms.

**yourBioPortal.com**
GO TO **Web Activity 7.1 • Signal Transduction**

the bacterium often rises far above the solute concentration inside the cell. A fundamental characteristic of all living cells is that they maintain a constant internal environment, or homeostasis. To do this, the bacterium must perceive and quickly respond to this environmental signal (**Figure 7.3, step 1**). The cell does this by a signal transduction pathway involving two major components: a receptor and a responder.

**RECEPTOR** The *E. coli* receptor protein for changes in solute concentration is called EnvZ. EnvZ is a transmembrane protein that extends across the bacterium's plasma membrane into the space between the plasma membrane and the highly porous outer membrane, which forms a complex with the cell wall. When the solute concentration of the extracellular environment rises, so does the solute concentration in the space between the two membranes. This change in the aqueous solution causes the part of the receptor protein that sticks out into the intermembrane space to undergo a change in conformation (its three-dimensional shape).

The conformational change in the intermembrane domain (a *domain* is a sequence of amino acids folded into a particular shape) causes a conformational change in the domain that lies in the cytoplasm and initiates the events of signal transduction. The cytoplasmic domain of EnvZ can act as an *enzyme*. As you will see in more detail in Chapter 8, an enzyme is a biological catalyst that greatly speeds up a chemical reaction, and the active site is the region where the reaction actually takes place. The conformational change in EnvZ exposes an active site that was previously buried within the protein, so that EnvZ becomes a **protein kinase**—an enzyme that catalyzes the transfer of a phosphate group from ATP to another molecule. EnvZ transfers the phosphate group to one of its own histidine amino acids. In other words, EnvZ *phosphorylates* itself (**Figure 7.3, step 2**).

$$\text{EnvZ + ATP} \xrightarrow{\text{EnvZ}} \text{EnvZ–P + ADP}$$

What does phosphorylation do to a protein? As discussed in Section 3.2, proteins can have both hydrophilic regions (which tend to interact with water on the outside of the protein macromolecule) and hydrophobic regions (which tend to interact with one another on the inside of the macromolecule). These regions are important in giving a protein its three-dimensional shape. Phosphate groups are charged, so an amino acid with such a group tends to be on the outside of the protein. Thus phosphorylation leads to a change in the shape and function of a protein by changing its charge.

**RESPONDER** A **responder** is the second component of a signal transduction pathway. The charged phosphate group added to the histidine of the EnvZ protein causes its cytoplasmic domain to change its shape again. It now binds to a second protein, OmpR, and transfers the phosphate to it. In turn, this phosphorylation changes the shape of OmpR (**Figure 7.3, step 3**). The change in the responder is a key event in signaling, for three reasons:

- The signal on the outside of the cell has now been *transduced* to a protein that lies totally within the cell's cytoplasm.
- The altered responder can *do something*. In the case of the phosphorylated OmpR, that "something" is to bind to DNA to alter the expression of many genes; in particular, it increases the expression of the protein OmpC. This binding begins the final phase of the signaling pathway: the effect of the signal, which is an alteration in cell function.
- The signal has been *amplified*. Because a single enzyme can catalyze the conversion of many substrate molecules, one EnvZ molecule alters the structure of many OmpR molecules.

Phosphorylated OmpR has the correct three-dimensional structure to bind to the *ompC* DNA, resulting in an increase in the transcription of that gene. This results in the production of OmpC protein, which enables the cell to respond to the increase in osmotic concentration in its environment (**Figure 7.3, step 4**). The OmpC protein is inserted into the outer membrane of the cell, where it blocks pores and prevents solutes from entering the intermembrane space. As a result, the solute concentration in the intermembrane space is lowered, and homeostasis is restored. Thus the EnvZ-OmpR signal transduction pathway allows the *E. coli* cell to function just as if the external environment had a normal solute concentration.

Many of the elements that we have highlighted in this prokaryotic signal transduction pathway also exist in the signal transduction pathways of eukaryotic organisms. A typical eukaryotic signal transduction pathway has the following general steps:

- A receptor protein changes its conformation upon interaction with a signal. This receptor protein may or may not be in a membrane.
- A conformational change in the receptor protein activates its protein kinase activity, resulting in the transfer of a phosphate group from ATP to a target protein.
- This phosphorylation alters the function of a responder protein.
- The signal is amplified.
- A protein that binds to DNA is activated.
- The expression of one or more specific genes is turned on or off.
- Cell activity is altered.

## 7.1 RECAP

**Cells are constantly exposed to molecular signals that can come from the external environment or from within the body of a multicellular organism. To respond to a signal, the cell must have a specific receptor that detects the signal and activates some cellular response.**

- What are the differences between an autocrine signal, a paracrine signal, and a hormone? See p. 130 and Figure 7.1
- Describe the three components in a cell's response to a signal. See pp. 130–132 and Figure 7.2
- What are the elements of signal transduction that are described at the close of this section?

The general features of signal transduction pathways described in this section will recur in more detail throughout the chapter. First let's consider more closely the nature of the receptors that bind signal molecules.

## 7.2 How Do Signal Receptors Initiate a Cellular Response?

Any given cell in a multicellular organism is bombarded with many signals. However, it responds to only some of them, because no cell makes receptors for all signals. A receptor protein that binds to a chemical signal does so very specifically, in much the same way that a membrane transport protein binds to the substance it transports. This *specificity* of binding ensures that only those cells that make a specific receptor will respond to a given signal.

### Receptors have specific binding sites for their signals

A specific chemical signal molecule fits into a three-dimensional site on its protein receptor (**Figure 7.4A**). A molecule that binds to a receptor site on another molecule in this way is called a **ligand**. Binding of the signaling ligand causes the receptor protein to change its three-dimensional shape, and that conformational change initiates a cellular response. The ligand does not contribute further to this response. In fact, the ligand is usually not metabolized into a useful product; its role is purely to "knock on the door." (This is in sharp contrast to the enzyme–substrate interaction, which is described in Chapter 8. The whole purpose of that interaction is to change the substrate into a useful product.)

Receptors bind to their ligands according to chemistry's *law of mass action*:

$$R + L \rightleftharpoons RL$$

This means that the binding is reversible, although for most ligand–receptor complexes, the equilibrium point is far to the right—that is, binding is favored. Reversibility is important, however, because if the ligand were never released, the receptor would be continuously stimulated.

**7.4 A Signal and Its Receptor** (A) The adenosine 2A receptor occurs in the human brain, where it is involved in inhibiting arousal. (B) Adenosine is the normal ligand for the receptor. Caffeine has a similar structure to that of adenosine and can act as an antagonist that binds the receptor and prevents its normal functioning.

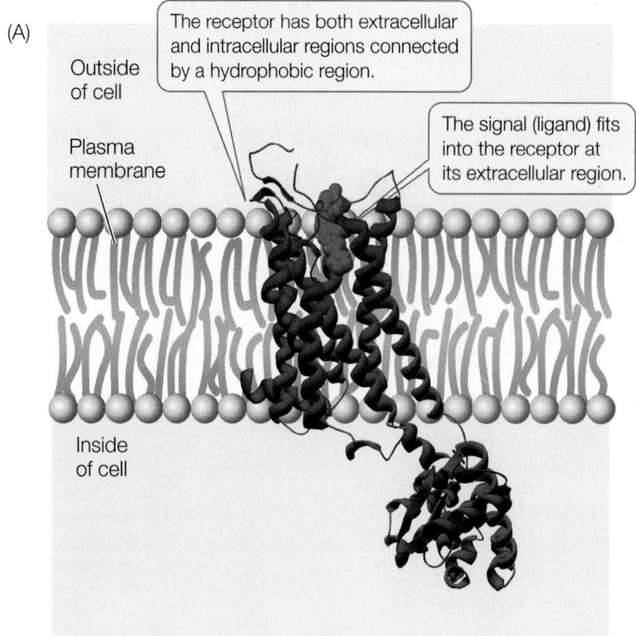


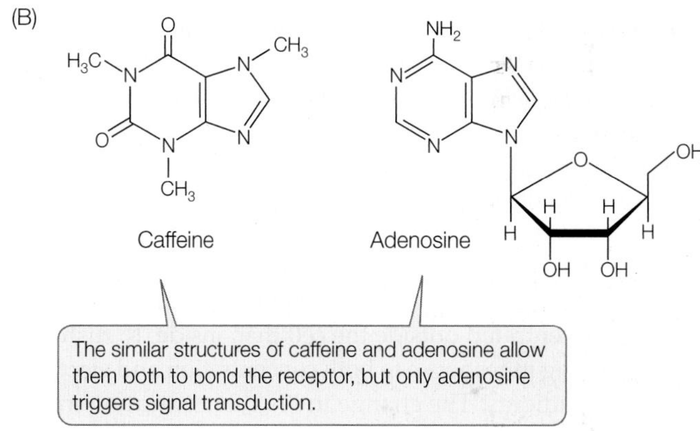


An inhibitor (or *antagonist*) can also bind to a receptor protein, instead of the normal ligand. There are both natural and artificial antagonists of receptor binding. For example, many substances that alter human behavior bind to specific receptors in the brain, and prevent the binding of the receptors' specific ligands. An example is caffeine, which is probably the world's most widely consumed stimulant. In the brain, the nucleoside adenosine acts as a ligand that binds to a receptor on nerve cells, initiating a signal transduction pathway that reduces brain activity, especially arousal. Because caffeine has a similar molecular structure to that of adenosine, it also binds to the adenosine receptor (**Figure 7.4B**). But in this case binding does not initiate a signal transduction pathway. Rather, it "ties up" the receptor, preventing adenosine binding and thereby allowing nerve cell activity and arousal.

## Receptors can be classified by location and function

The chemistry of ligand signals is quite variable, but they can be divided into two groups, based on whether or not they can diffuse through membranes. Correspondingly, a receptor can be classified by its location in the cell, which largely depends on the nature of its ligand (**Figure 7.5**):

- *Cytoplasmic receptors*: Small or nonpolar ligands can diffuse across the nonpolar phospholipid bilayer of the plasma membrane and enter the cell. Estrogen, for example, is a lipid-soluble steroid hormone that can easily diffuse across the plasma membrane; it binds to a receptor in the cytoplasm.
- *Membrane receptors*: Large or polar ligands cannot cross the lipid bilayer. Insulin, for example, is a protein hormone that cannot diffuse through the plasma membrane; instead, it binds to a transmembrane receptor with an extracellular binding domain.

In complex eukaryotes such as mammals and higher plants, there are three well-studied categories of plasma membrane receptors that are grouped according to their functions: ion channels, protein kinase receptors, and G protein-linked receptors.

**ION CHANNEL RECEPTORS** As described in Section 6.3, the plasma membranes of many types of cells contain gated **ion channels** for ions such as $Na^+$, $K^+$, $Ca^{2+}$, or $Cl^-$ to enter or leave the cell (see Figure 6.11). The gate-opening mechanism is an alteration in the three-dimensional shape of the channel protein upon ligand binding; thus these proteins function as receptors. Each type of ion channel has its own signal, and these include sensory stimuli such as light, sound, and electric charge

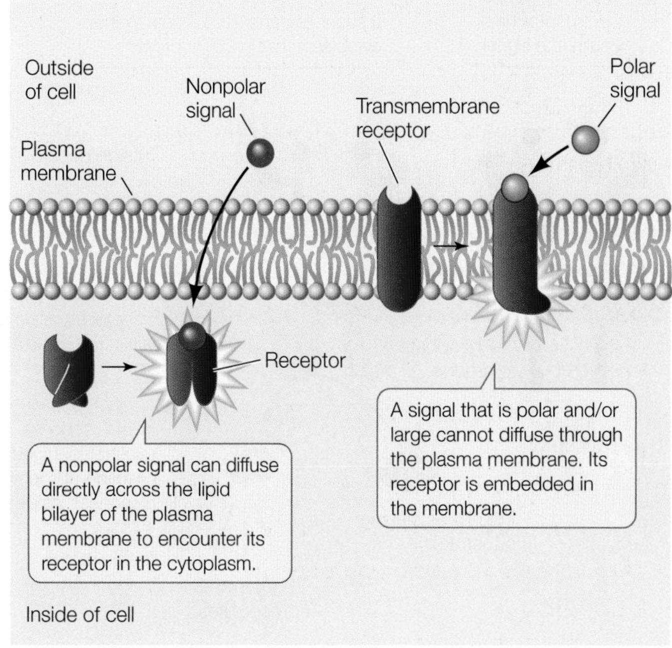


**7.5 Two Locations for Receptors** Receptors can be located in the cytoplasm or in the plasma membrane of the cell.

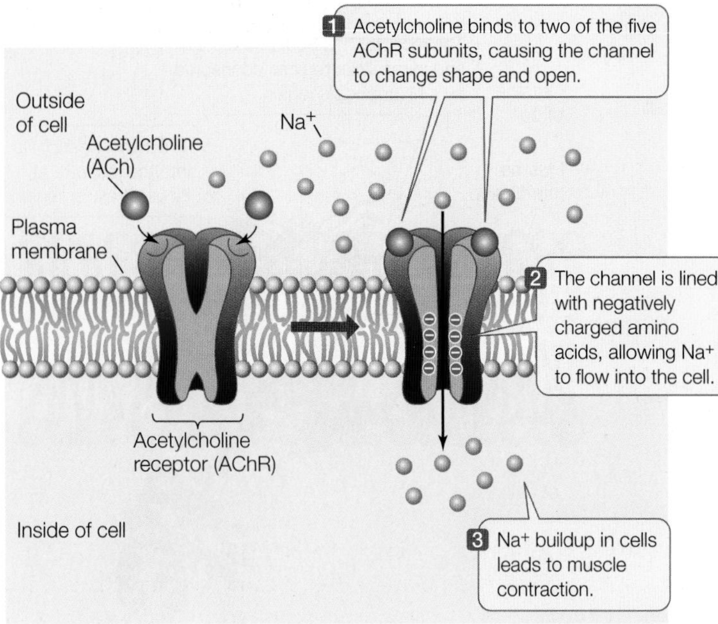
1 Acetylcholine binds to two of the five AChR subunits, causing the channel to change shape and open.
Outside of cell
Na+
Acetylcholine (ACh)
Plasma membrane
2 The channel is lined with negatively charged amino acids, allowing Na+ to flow into the cell.
Acetylcholine receptor (AChR)
Inside of cell
3 Na+ buildup in cells leads to muscle contraction.

**7.6 A Gated Ion Channel** The acetylcholine receptor (AChR) is a ligand-gated ion channel for sodium ions. It is made up of five polypeptide subunits. When acetylcholine molecules (ACh) bind to two of the subunits, the gate opens and $Na^+$ flows into the cell. This channel helps regulate membrane polarity (see Chapter 6).

differences across the plasma membrane, as well as chemical ligands such as hormones and neurotransmitters.

The *acetylcholine receptor*, which is located in the plasma membrane of skeletal muscle cells, is an example of a gated ion channel. This receptor protein is a sodium channel that binds the ligand acetylcholine, which is a neurotransmitter—a chemical signal released from neurons (nerve cells) (**Figure 7.6**). When two molecules of acetylcholine bind to the receptor, it opens for about a thousandth of a second. That is enough time for $Na^+$, which is more concentrated outside the cell than inside, to rush into the cell, moving in response to both concentration and electrical potential gradients. The change in $Na^+$ concentration in the cell initiates a series of events that result in muscle contraction.

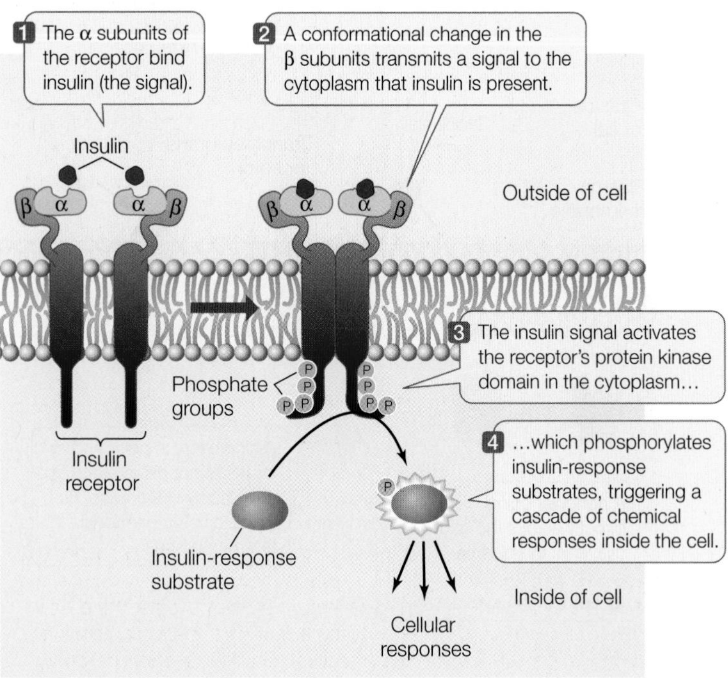
1 The α subunits of the receptor bind insulin (the signal).
2 A conformational change in the β subunits transmits a signal to the cytoplasm that insulin is present.
Insulin
Outside of cell
3 The insulin signal activates the receptor's protein kinase domain in the cytoplasm...
Phosphate groups
Insulin receptor
4 ...which phosphorylates insulin-response substrates, triggering a cascade of chemical responses inside the cell.
Insulin-response substrate
Cellular responses
Inside of cell

**PROTEIN KINASE RECEPTORS** Like the EnvZ receptor of *E. coli*, some eukaryotic receptor proteins become protein kinases when they are activated. They catalyze the phosphorylation of themselves and/or other proteins, thus changing their shapes and therefore their functions.

The receptor for insulin is an example of a protein kinase receptor. Insulin is a protein hormone made by the mammalian pancreas. Its receptor has two copies each of two different polypeptide subunits (**Figure 7.7**). When insulin binds to the receptor, the receptor becomes activated and able to phosphorylate itself and certain cytoplasmic proteins that are appropriately called *insulin response substrates*. These proteins then initiate many cellular responses, including the insertion of glucose transporters (see Figure 6.14) into the plasma membrane.

**G PROTEIN-LINKED RECEPTORS** A third category of eukaryotic plasma membrane receptors is the G protein-linked receptors, also referred to as the seven transmembrane domain receptors. This descriptive name identifies a fascinating group of receptors, each of which is composed of a single protein with seven transmembrane domains. These seven domains pass through the phospholipid bilayer and are separated by short loops that extend either outside or inside the cell. Ligand binding on the extracellular side of the receptor changes the shape of its cytoplasmic region, exposing a site that binds to a mobile membrane protein called a **G protein**. The G protein is partially inserted into the lipid bilayer and partially exposed on the cytoplasmic surface of the membrane.

Many G proteins have three polypeptide subunits and can bind three different molecules (**Figure 7.8A**):

- The receptor
- GDP and GTP (guanosine diphosphate and triphosphate, respectively; these are nucleoside phosphates like ADP and ATP)
- An effector protein

When the G protein binds to an activated receptor protein, GDP is exchanged for GTP (**Figure 7.8B**). At the same time, the ligand is usually released from the extracellular side of the receptor. GTP binding causes a conformational change in the G protein. The GTP-bound subunit then separates from the rest of the protein, diffusing in the plane of the phospholipid bilayer until it encounters an **effector protein** to which it can bind. An effector protein is just what its name implies: it causes an effect in the cell. The binding of the GTP-bearing G protein

**7.7 A Protein Kinase Receptor** The mammalian hormone insulin binds to a receptor on the outside surface of the cell and initiates a response.

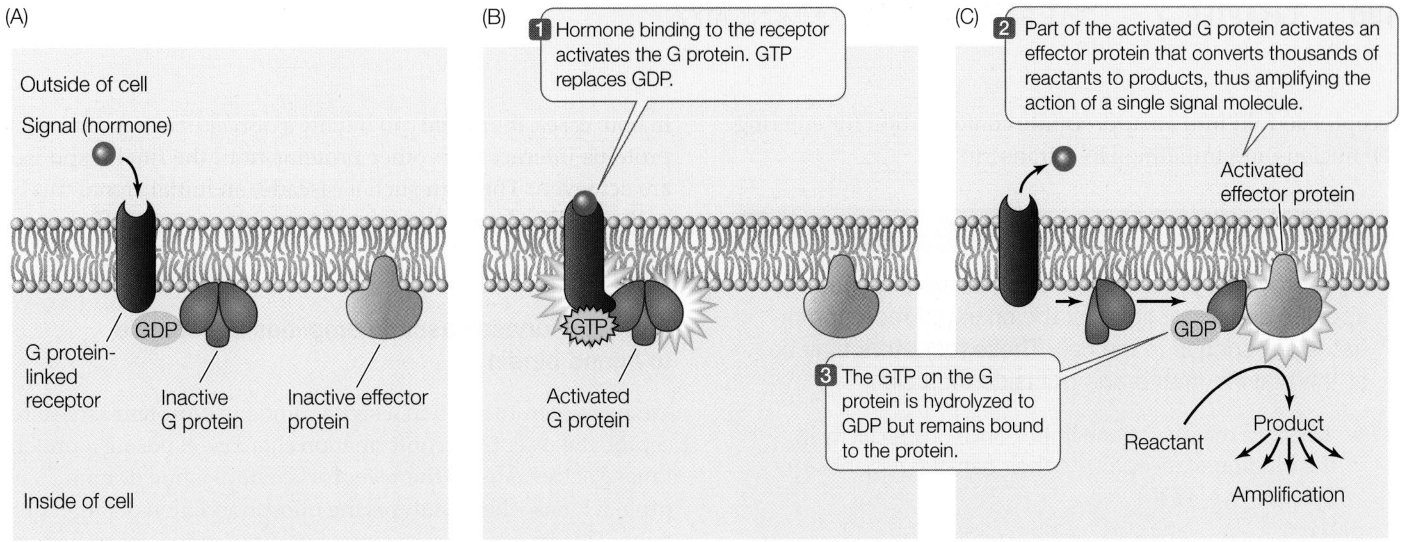

(A)

Outside of cell
Signal (hormone)
GDP
G protein-linked receptor
Inactive G protein
Inactive effector protein
Inside of cell

(B)

1 Hormone binding to the receptor activates the G protein. GTP replaces GDP.
GTP
Activated G protein

(C)

2 Part of the activated G protein activates an effector protein that converts thousands of reactants to products, thus amplifying the action of a single signal molecule.
Activated effector protein
GDP
3 The GTP on the G protein is hydrolyzed to GDP but remains bound to the protein.
Reactant
Product
Amplification

**7.8 A G Protein-Linked Receptor** The G protein is an intermediary between the receptor and an effector.

**yourBioPortal.com**

GO TO **Animated Tutorial 7.1 • Signal Transduction Pathway**

subunit activates the effector—which may be an enzyme or an ion channel—thereby causing changes in cell function (**Figure 7.8C**).

After activation of the effector protein, the GTP on the G protein is hydrolyzed to GDP. The now inactive G protein subunit separates from the effector protein and diffuses in the membrane to collide with and bind to the other two G protein subunits. When the three components of the G protein are reassembled, the protein is capable of binding again to an activated receptor. After binding, the activated receptor exchanges the GDP on the G protein for a GTP, and the cycle begins again.

There are variations in all three G protein subunits, giving different G protein complexes different functions. A G protein can either activate or inhibit an effector protein. An example in humans of an *activating* response involves the receptor for epinephrine (adrenaline), which is a hormone made by the adrenal gland in response to stress or heavy exercise. In heart muscle, this hormone binds to its G protein-linked receptor, activating a G protein. The GTP-bound subunit then activates a membrane-bound enzyme to produce a small molecule, cyclic adenosine monophosphate (cAMP). This molecule, in turn, has many effects on the cell (as we will see below), including the mobilization of glucose for energy and muscle contraction.

G protein-mediated *inhibition* occurs when the same hormone, epinephrine, binds to its receptor in the smooth muscle cells surrounding blood vessels lining the digestive tract. Again, the epinephrine-bound receptor changes its shape and activates a G protein, and the GTP-bound subunit binds to a target enzyme. But in this case, the enzyme is inhibited instead of being activated. As a result, the muscles relax and the blood vessel diameter increases, allowing more nutrients to be carried away from the digestive system to the rest of the body. Thus the same signal and signaling mechanism can have different consequences in different cells, depending on the presence of specific receptor and effector molecules.

**CYTOPLASMIC RECEPTORS** **Cytoplasmic receptors** are located inside the cell and bind to signals that can diffuse across the plasma membrane. Binding to the signaling ligand causes the receptor to change its shape so that it can enter the cell nucleus, where it affects expression of specific genes. But this general view is somewhat simplified. The receptor for the steroid hormone cortisol, for example, is normally bound to a chaperone protein, which blocks it from entering the nucleus. Binding of the hormone causes the receptor to change its shape so that the chaperone is released (**Figure 7.9**). This release allows the

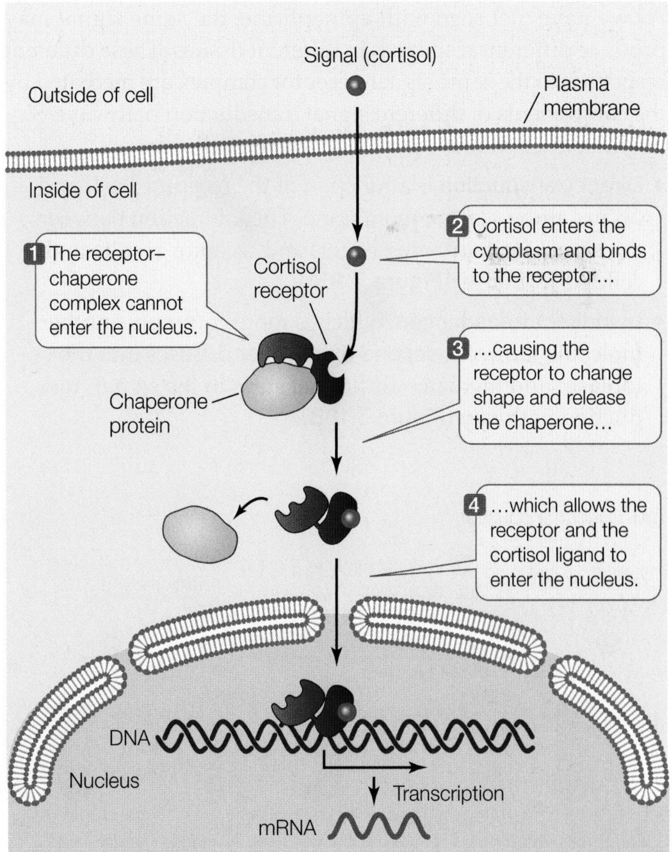

Signal (cortisol)
Outside of cell
Plasma membrane
Inside of cell
1 The receptor–chaperone complex cannot enter the nucleus.
Cortisol receptor
2 Cortisol enters the cytoplasm and binds to the receptor...
Chaperone protein
3 ...causing the receptor to change shape and release the chaperone...
4 ...which allows the receptor and the cortisol ligand to enter the nucleus.
DNA
Nucleus
Transcription
mRNA

**7.9 A Cytoplasmic Receptor** The receptor for cortisol is bound to a chaperone protein. Binding of the signal to the receptor releases the chaperone and allows the ligand–receptor complex to enter the cell's nucleus, where it binds to DNA. Changes in DNA transcription are long-term in comparison to the more immediate changes in enzyme activity observed in other pathways (see Figure 7.20).

receptor to fold into an appropriate conformation for entering the nucleus and initiating DNA transcription.

## 7.2 RECAP

**Receptors are proteins that bind, or are changed by, specific signals or ligands; the changed receptor initiates a response in the cell. These receptors may be at the plasma membrane or inside the cell.**

- What are the nature and importance of specificity in the binding of receptors to their particular ligands? See pp. 132–133
- What are three important categories of plasma membrane receptors seen in complex eukaryotes? See pp. 133–134 and Figures 7.6, 7.7, and 7.8

Now that we have discussed signals and receptors, let's examine the characteristics of the molecules (*transducers*) that mediate between the receptor and the cellular response.

# 7.3 How Is the Response to a Signal Transduced through the Cell?

As we have just seen with epinephrine, the same signal may produce different responses in different tissues. These different responses to the same signal–receptor complex are mediated by the components of different signal transduction pathways. Signal transduction may be either direct or indirect:

- **Direct transduction** is a function of the receptor itself and occurs at the plasma membrane. The interaction between the signal (primary messenger) and receptor results in the cellular response. (**Figure 7.10A**).
- In **indirect transduction**, which is more common, another molecule termed a **second messenger** diffuses into the cytoplasm and mediates additional steps in the signal transduction pathway (**Figure 7.10B**).

In both cases, the signal can initiate a *cascade* of events, in which proteins interact with other proteins until the final responses are achieved. Through such a cascade, an initial signal can be both amplified and distributed to cause several different responses in the target cell.

### A protein kinase cascade amplifies a response to ligand binding

We have seen that when a signal binds to a protein kinase receptor, the receptor's conformation changes, exposing a protein kinase active site on the receptor's cytoplasmic domain. The protein kinase then catalyzes the phosphorylation of target proteins. This process is an example of direct signal transduction, because the amplifying enzyme is the receptor itself. Protein kinase receptors are important in binding signals called growth factors that stimulate cell division in both plants and animals.

A complete signal transduction pathway that occurs after a protein kinase receptor binds a growth factor was discovered in studies on a cell that went wrong. Many human bladder cancers contain an abnormal form of a protein called Ras (so named because a similar protein was previously isolated from a *rat* sarcoma tumor). Investigations of these bladder cancers showed that Ras was a G protein, and the abnormal form was always active because it was permanently bound to GTP, and thus caused continuous cell division (**Figure 7.11**). If this abnormal form of Ras was inhibited, the cells stopped dividing. This discovery has led to a major effort to develop specific Ras inhibitors for cancer treatment.

Other cancers have abnormalities in different aspects of signal transduction. Biologists have compared the defects in these cells with the normal signaling process in non-cancerous cells, and thus worked out the entire signaling pathway. It is an ex-

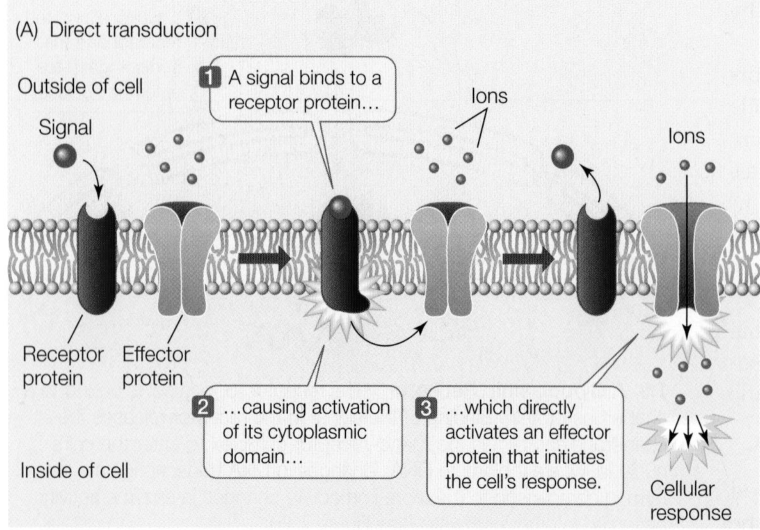
(A) Direct transduction
Outside of cell
Signal
1 A signal binds to a receptor protein...
Ions
Ions
Receptor protein
Effector protein
2 ...causing activation of its cytoplasmic domain...
3 ...which directly activates an effector protein that initiates the cell's response.
Inside of cell
Cellular response

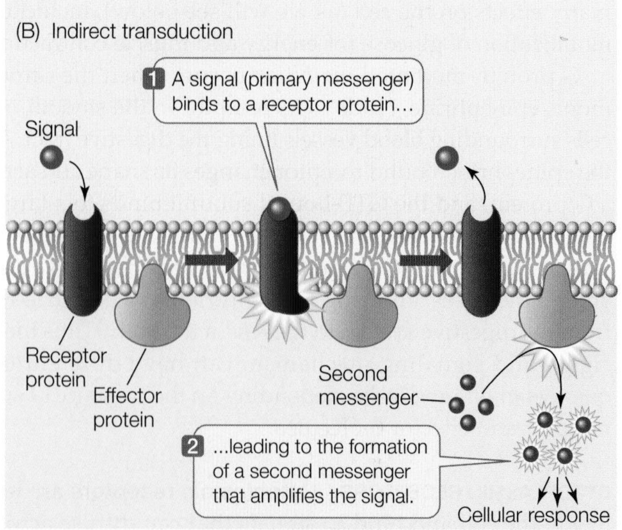
(B) Indirect transduction
Signal
1 A signal (primary messenger) binds to a receptor protein...
Receptor protein
Effector protein
Second messenger
2 ...leading to the formation of a second messenger that amplifies the signal.
Cellular response

**7.10 Direct and Indirect Signal Transduction** (A) All the events of direct transduction occur at or near the receptor (in this case, at the plasma membrane). (B) In indirect transduction, a second messenger mediates the events inside the cell. The signal is considered to be the first messenger.

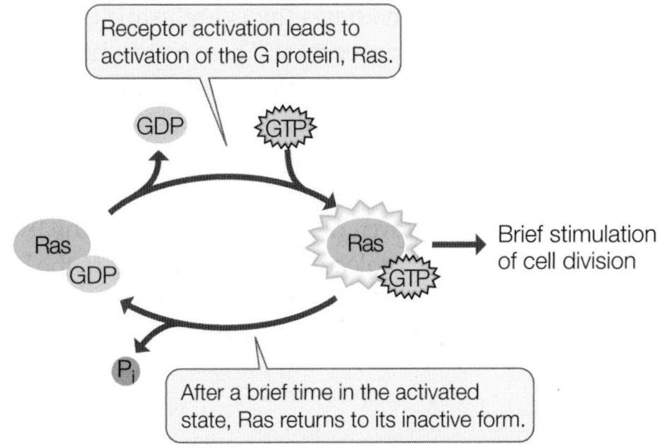


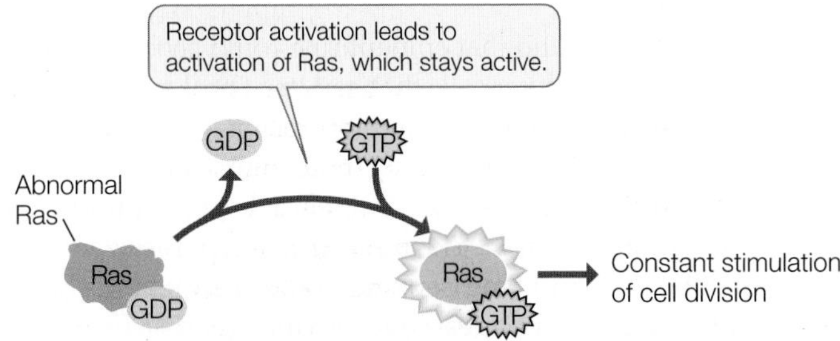


**7.11 Signal Transduction and Cancer** (A) Ras is a G protein that regulates cell division. (B) In some tumors, the Ras protein is permanently active, resulting in uncontrolled cell division.

ample of a more general phenomenon, called a **protein kinase cascade**, where one protein kinase activates the next, and so on (**Figure 7.12**). Such cascades are key to the external regulation of many cellular activities. Indeed, the eukaryotic genome codes for hundreds, even thousands, of such kinases.

Protein kinase cascades are useful signal transducers for four reasons:

- At each step in the cascade of events, the signal is *amplified*, because each newly activated protein kinase is an enzyme that can catalyze the phosphorylation of many target proteins.
- The information from a signal that originally arrived at the plasma membrane is *communicated* to the nucleus.
- The multitude of steps provides some *specificity* to the process.
- Different target proteins at each step in the cascade can provide *variation* in the response.

**yourBioPortal.com**
GO TO **Animated Tutorial 7.2 • Signal Transduction and Cancer**

## Second messengers can stimulate protein kinase cascades

As we have just seen, protein kinase receptors initiate protein kinase cascades right at the plasma membrane. However, the stimulation of events in the cell is more often indirect. In a series of clever experiments, Earl Sutherland and his colleagues at Case Western Reserve University discovered that a small water-solu-

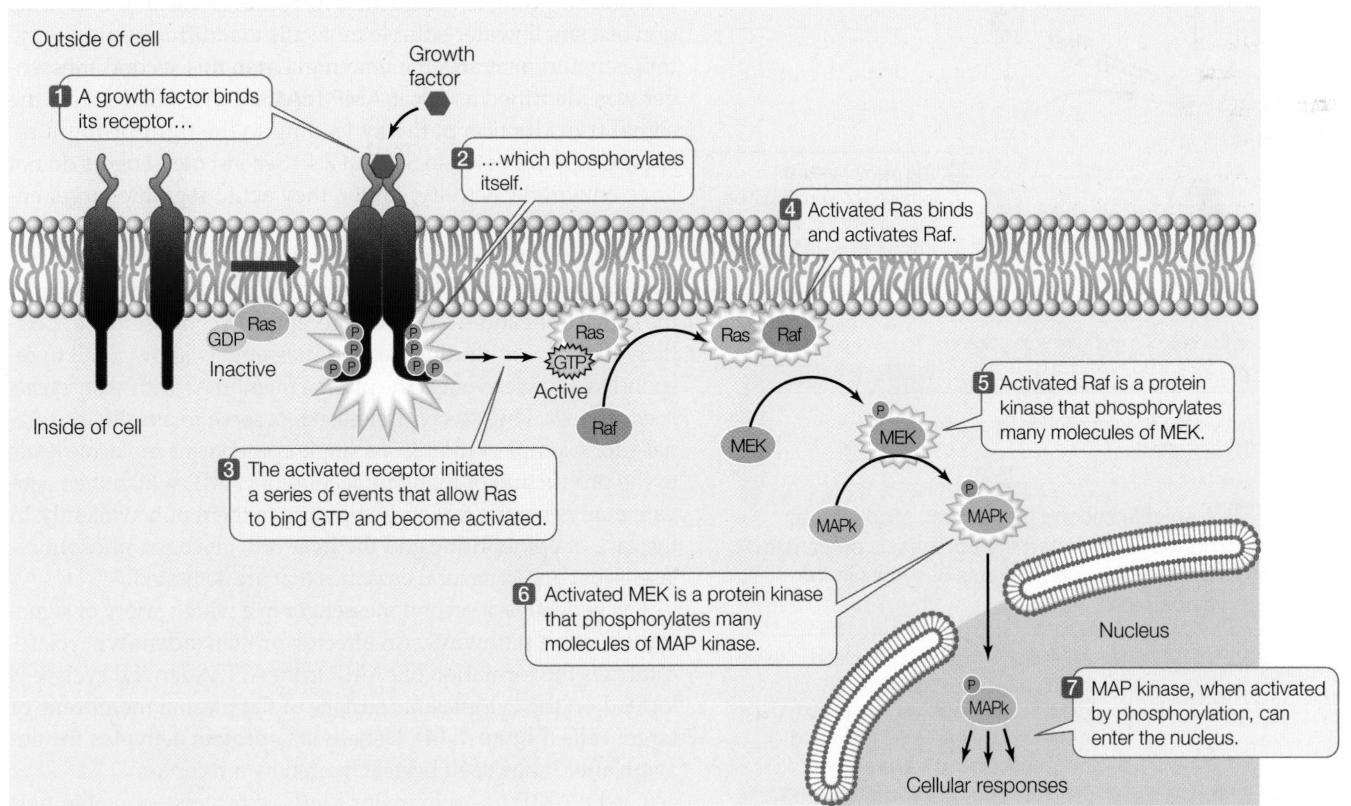


**7.12 A Protein Kinase Cascade** In a protein kinase cascade, a series of proteins are sequentially activated.

## INVESTIGATING LIFE

### 7.13 The Discovery of a Second Messenger

Glycogen phosphorylase is activated in liver cells after epinephrine binds to a membrane receptor. Sutherland and his colleagues observed that this activation could occur in vivo only if fragments of the plasma membrane were present. They designed experiments to show that a second messenger caused the activation of glycogen phosphorylase.

**HYPOTHESIS** **A second messenger mediates between receptor activation at the plasma membrane and enzyme activation in the cytoplasm.**

**METHOD**

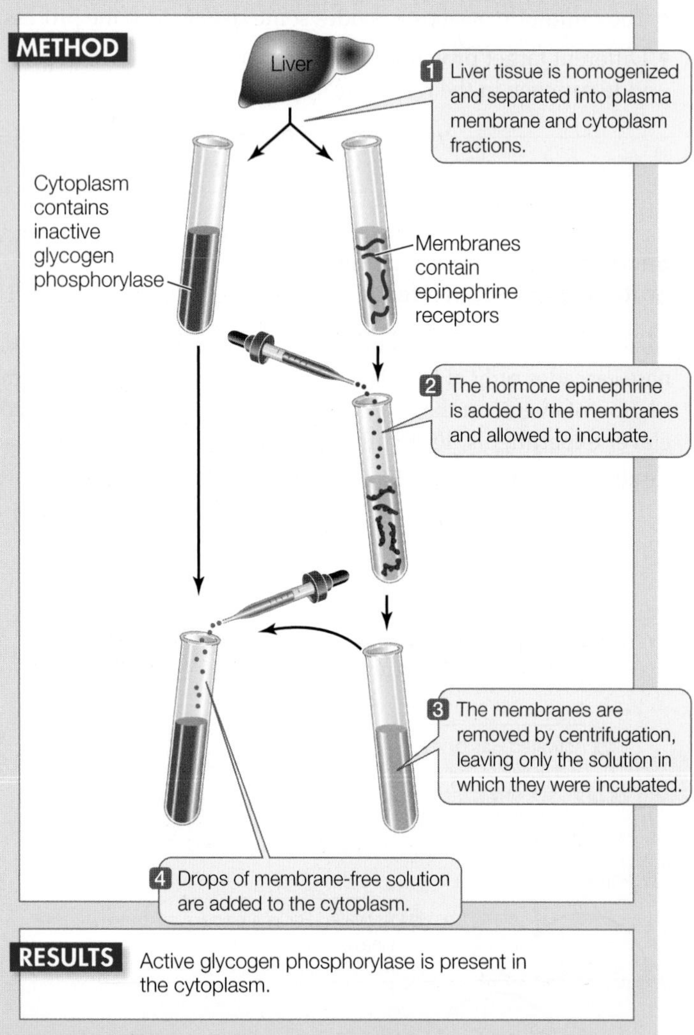

Liver
1 Liver tissue is homogenized and separated into plasma membrane and cytoplasm fractions.
Cytoplasm contains inactive glycogen phosphorylase
Membranes contain epinephrine receptors
2 The hormone epinephrine is added to the membranes and allowed to incubate.
3 The membranes are removed by centrifugation, leaving only the solution in which they were incubated.
4 Drops of membrane-free solution are added to the cytoplasm.

**RESULTS** Active glycogen phosphorylase is present in the cytoplasm.

**CONCLUSION** **A soluble second messenger, produced by hormone-activated membranes, is present in the solution and activates enzymes in the cytoplasm.**

**FURTHER INVESTIGATION:** The soluble molecule produced in this experiment was later identified as cAMP. How would you show that cAMP, and not ATP, is the second messenger in this system?

Go to **yourBioPortal.com** for original citations, discussions, and relevant links for all INVESTIGATING LIFE figures.

ble chemical messenger mediates the cytoplasmic events initiated by a plasma membrane receptor. These researchers were investigating the activation of the liver enzyme glycogen phosphorylase by the hormone epinephrine. The enzyme is released when an animal faces life-threatening conditions and needs energy fast for the fight-or-flight response. Glycogen phosphorylase catalyzes the breakdown of glycogen stored in the liver so that the resulting glucose molecules can be released to the blood. The enzyme is present in the liver cell cytoplasm, but is inactive except in the presence of epinephrine.

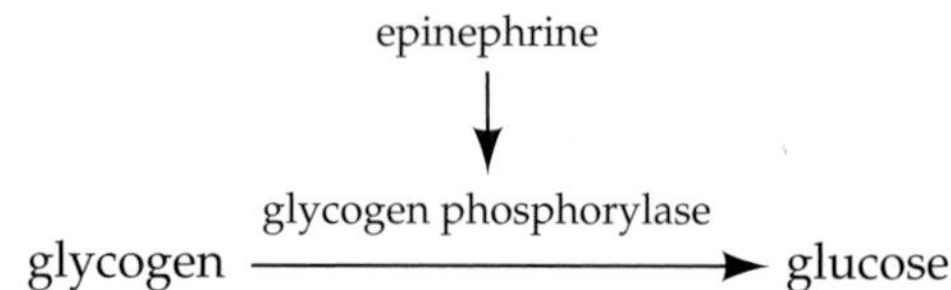


The researchers found that epinephrine could activate glycogen phosphorylase in liver cells that had been broken open, but only if the entire cell contents, including plasma membrane fragments, were present. Under these circumstances epinephrine bound to the plasma membranes, but the active phosphorylase was present in the solution. The researchers hypothesized that there must be a second "messenger" that transmits the signal of epinephrine (the "first messenger," which binds to a receptor at the plasma membrane) to the phosphorylase (in the cytoplasm). To investigate the production of this messenger, they separated plasma membrane fragments from the cytoplasms of broken liver cells and followed the sequence of steps described in **Figure 7.13**. This experiment confirmed their hypothesis that hormone binding to the membrane receptor causes the production of a small, water-soluble molecule that diffuses into the cytoplasm and activates the enzyme. Later, this second messenger was identified as **cyclic AMP** (**cAMP**). (We will describe the signal transduction pathway leading to the fight-or-flight response in more detail in Section 7.4.) Second messengers do not have enzymatic activity; rather, they act to regulate target enzymes (see Chapter 8).

A second messenger is a small molecule that mediates later steps in a signal transduction pathway after the first messenger—the signal or ligand—binds to its receptor. In contrast to the specificity of receptor binding, second messengers allow a cell to respond to a single event at the plasma membrane with *many events inside the cell*. Thus, second messengers serve to amplify the signal—for example, binding of a single epinephrine molecule leads to the production of many molecules of cAMP, which then activate many enzyme targets by binding to them noncovalently. In the case of epinephrine and the liver cell, glycogen phosphorylase is just one of several enzymes that are activated.

Cyclic AMP is a second messenger in a wide variety of signal transduction pathways. An effector protein, adenylyl cyclase, catalyzes the formation of cAMP from ATP. Adenylyl cyclase is located on the cytoplasmic surface of the plasma membrane of target cells (**Figure 7.14**). Usually a G protein activates the enzyme after it has itself been activated by a receptor.

Cyclic AMP has two major kinds of targets—ion channels and protein kinases. In many sensory cells, cAMP binds to ion channels and thus opens them. Cyclic AMP may also bind to a

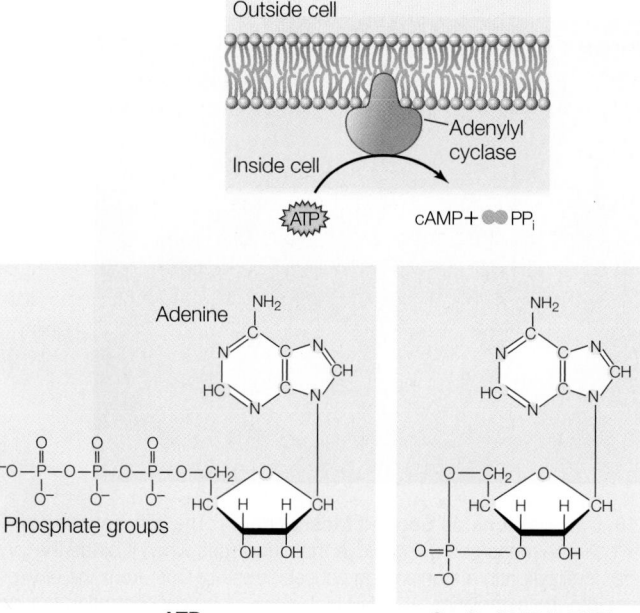

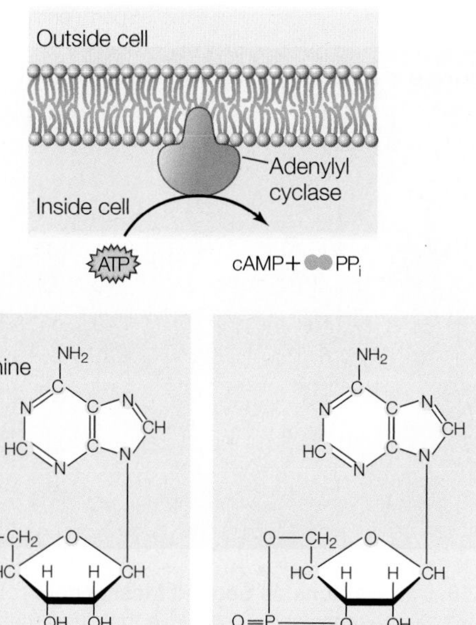


**7.14 The Formation of Cyclic AMP** The formation of cAMP from ATP is catalyzed by adenylyl cyclase, an enzyme that is activated by G proteins.

protein kinase in the cytoplasm, activating its catalytic function. A protein kinase cascade (see Figure 7.12) ensues, leading to the final effects in the cell.

### Second messengers can be derived from lipids

In addition to their role as structural components of the plasma membrane, phospholipids are also involved in signal transduction. When certain phospholipids are hydrolyzed into their component parts by enzymes called **phospholipases**, second messengers are formed.

The best-studied examples of lipid-derived second messengers come from the hydrolysis of the phospholipid **phosphatidyl inositol-bisphosphate (PIP2)**. Like all phospholipids, PIP2 has a hydrophobic portion embedded in the plasma membrane: two fatty acid tails attached to a molecule of glycerol, which together form **diacylglycerol**, or **DAG**. The hydrophilic portion of PIP2 is **inositol trisphosphate**, or **$IP_3$**, which projects into the cytoplasm.

As with cAMP, the receptors involved in this second-messenger system are often G protein-linked receptors. A G protein subunit is activated by the receptor, then diffuses within the plasma membrane and activates phospholipase C, an enzyme that is also located in the membrane. This enzyme cleaves off the $IP_3$ from PIP2, leaving the diacylglycerol (DAG) in the phospholipid bilayer:

$$\underset{\text{in membrane}}{\text{PIP2}} \xrightarrow{\text{phospholipase C}} \underset{\substack{\text{released to}\\ \text{cytoplasm}}}{IP_3} + \underset{\text{in membrane}}{\text{DAG}}$$

$IP_3$ and DAG, both second messengers, have different modes of action that build on each other, activating protein kinase C (PKC) (**Figure 7.15**). PKC refers to a family of protein kinases that can phosphorylate a wide variety of target proteins, leading to a multiplicity of cellular responses that vary depending on the tissue or cell type.

The $IP_3$/DAG pathway is apparently a target for the ion lithium ($Li^+$), which was used for many years as a psychoactive drug to treat bipolar (manic-depressive) disorder. This serious illness occurs in about 1 in every 100 people. In these patients, an overactive $IP_3$/DAG signal transduction pathway in the

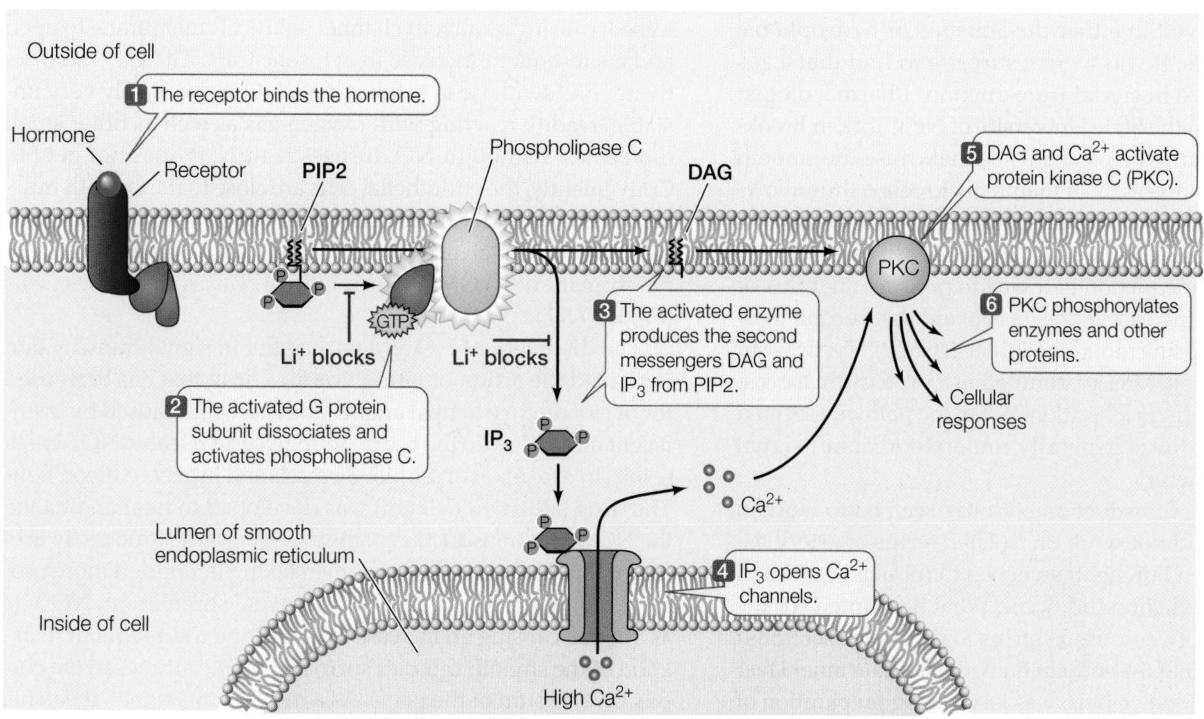


**7.15 The $IP_3$/DAG Second-Messenger System** Phospholipase C hydrolyzes the phospholipid PIP2 into its components, $IP_3$ and DAG, both of which are second messengers. Lithium ions ($Li^+$) block this pathway and are used to treat bipolar disorder (red type).

brain leads to excessive brain activity in certain regions. Lithium "tones down" this pathway in two ways, as indicated by the red notations in Figure 7.15. It inhibits G protein activation of phospholipase C, and also inhibits the synthesis of $IP_3$. The overall result is that brain activity returns to normal.

### Calcium ions are involved in many signal transduction pathways

Calcium ions ($Ca^{2+}$) are scarce inside most cells, which have cytosolic $Ca^{2+}$ concentrations of only about 0.1 mM. $Ca^{2+}$ concentrations outside cells and within the endoplasmic reticulum are usually much higher. Active transport proteins in the plasma and ER membranes maintain this concentration difference by pumping $Ca^{2+}$ out of the cytosol. In contrast to cAMP and the lipid-derived second messengers, $Ca^{2+}$ cannot be made in order to increase the intracellular $Ca^{2+}$ concentration. Instead, $Ca^{2+}$ ion levels are regulated via the opening and closing of ion channels, and the action of membrane pumps.

There are many signals that can cause calcium channels to open, including $IP_3$ (see Figure 7.15). The entry of a sperm into an egg is a very important signal that causes a massive opening of calcium channels, resulting in numerous and dramatic changes that prepare the now fertilized egg for cell divisions and development (**Figure 7.16**). Whatever the initial signal that causes the calcium channels to open, their opening results in a dramatic increase in cytosolic $Ca^{2+}$ concentration, which can increase up to one hundredfold within a fraction of a second. As we saw earlier, this increase activates protein kinase C. In addition, $Ca^{2+}$ controls other ion channels and stimulates secretion by exocytosis in many cell types.

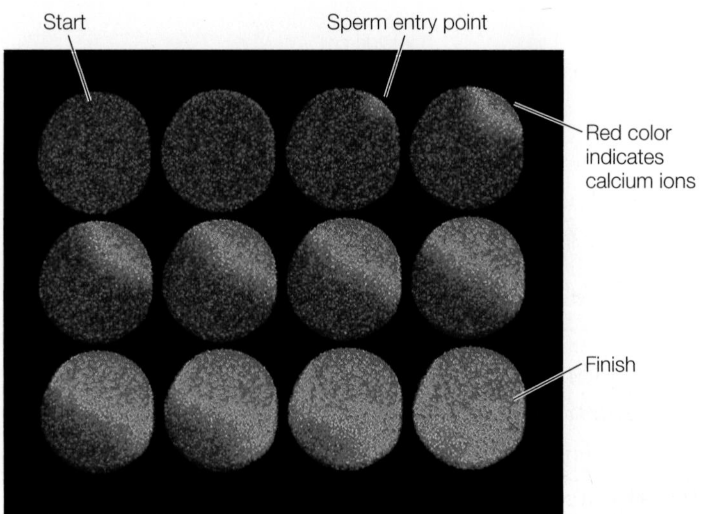


**7.16 Calcium Ions as Second Messengers** The concentration of $Ca^{2+}$ can be measured using a dye that fluoresces when it binds the ion. Here, fertilization in a starfish egg causes a rush of $Ca^{2+}$ from the environment into the cytoplasm. Areas of high calcium ion concentration are indicated by the red color and the events are photographed at 5-second intervals. Calcium signaling occurs in virtually all animal groups and triggers cell division in fertilized eggs, initiating the development of new individuals.

### Nitric oxide can act in signal transduction

Most signaling molecules and second messengers are solutes that remain dissolved in either the aqueous or hydrophobic components of cells. It was a great surprise to find that a gas could also be active in signal transduction. Pharmacologist Robert Furchgott, at the State University of New York in Brooklyn, was investigating the mechanisms that cause the smooth muscles lining blood vessels in mammals to relax, thus allowing more blood to flow to certain organs. The neurotransmitter acetylcholine (see Section 7.2) appeared to stimulate the $IP_3$/DAG signal transduction pathway to produce an influx of $Ca^{2+}$, leading to an increase in the level of another second messenger, cyclic guanosine monophosphate (cGMP). Cyclic GMP then binds to a protein kinase, stimulating a protein kinase cascade that leads to muscle relaxation. So far, the pathway seemed to conform to what was generally understood about signal transduction in general.

While this signal transduction pathway seemed to work in intact animals, it did not work on isolated strips of artery tissue. However, when Furchgott switched to tubular sections of artery, signal transduction did occur. What accounted for the different results between tissue strips and tubular sections? Furchgott realized that the endothelium, the delicate inner layer of cells lining the blood vessels, was lost during preparation of the tissue strips. He hypothesized that the endothelium was producing some chemical that diffused into the smooth muscle cells and was needed for their response to acetylcholine. However, the substance was not easy to isolate. It seemed to break down quickly, with a half-life (the time in which half of it disappeared) of 5 seconds in living tissue.

Furchgott's elusive substance turned out to be a gas, **nitric oxide** (**NO**), which formerly had been recognized only as a toxic air pollutant! In the body, NO is made from the amino acid arginine by the enzyme NO synthase. When the acetylcholine receptor on the surface of an endothelial cell is activated, $IP_3$ is released, causing a calcium channel on the ER membrane to open and a subsequent increase in cytosolic $Ca^{2+}$. The $Ca^{2+}$ then activates NO synthase to produce NO. NO is chemically very unstable, readily reacting with oxygen gas as well as other small molecules. Although NO diffuses readily, it does not get far. Conveniently, the endothelial cells are close to the smooth muscle cells, where NO acts as a paracrine signal. In smooth muscle, NO activates an enzyme called guanylyl cyclase, catalyzing the formation of cGMP, which in turn relaxes the muscle cells (**Figure 7.17**).

The discovery of NO as a participant in signal transduction explained the action of nitroglycerin, a drug that has been used for over a century to treat angina, the chest pain caused by insufficient blood flow to the heart. Nitroglycerin releases NO, which results in relaxation of the blood vessels and increased blood flow. The drug sildenafil (Viagra) was developed to treat angina via the NO signal transduction pathway, but was only modestly useful for that purpose. However, men taking it reported more pronounced penile erections. During sexual stimulation, NO acts as a signal causing an increase in cGMP and a subsequent relaxation of the smooth muscles surrounding the arteries in the corpus cavernosum of the penis. As a result of this signal, the penis

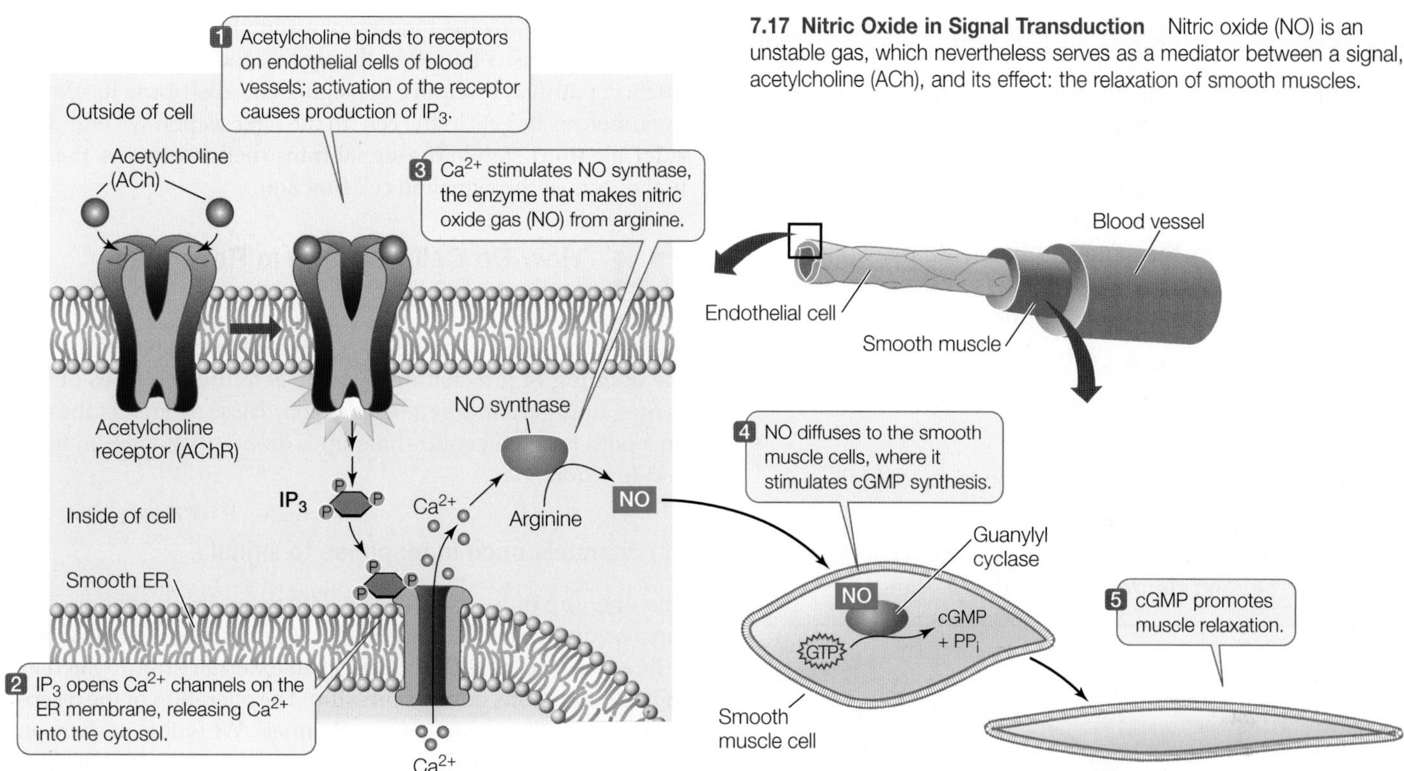


**7.17 Nitric Oxide in Signal Transduction** Nitric oxide (NO) is an unstable gas, which nevertheless serves as a mediator between a signal, acetylcholine (ACh), and its effect: the relaxation of smooth muscles.

fills with blood, producing an erection. Sildenafil acts by inhibiting an enzyme (a phosphodiesterase) that breaks down cGMP—resulting in more cGMP and better erections.

## Signal transduction is highly regulated

There are several ways in which cells can regulate the activity of a transducer. The concentration of NO, which breaks down quickly, can be regulated only by how much of it is made. On the other hand, membrane pumps and ion channels regulate the concentration of $Ca^{2+}$, as we have seen. To regulate protein kinase cascades, G proteins, and cAMP, there are enzymes that convert the activated transducer back to its inactive precursor (**Figure 7.18**).

The balance between the activities of enzymes that activate transducers (for example, protein kinase) and enzymes that inactivate them (for example, protein phosphatase) is what determines the ultimate cellular response to a signal. Cells can alter this balance in several ways:

- *Synthesis or breakdown of the enzymes involved.* For example, synthesis of adenylyl cyclase and breakdown of phosphodiesterase (which breaks down cAMP) would tilt the balance in favor of more cAMP in the cell.
- *Activation or inhibition of the enzymes by other molecules.* Examples include the activation of a G protein-linked receptor by ligand binding, and inhibition of phosphodiesterase (which also breaks down cGMP) by sildenafil.

Because cell signaling is so important in diseases such as cancer, a search is under way for new drugs that can modulate the activities of enzymes that participate signal transduction pathways.

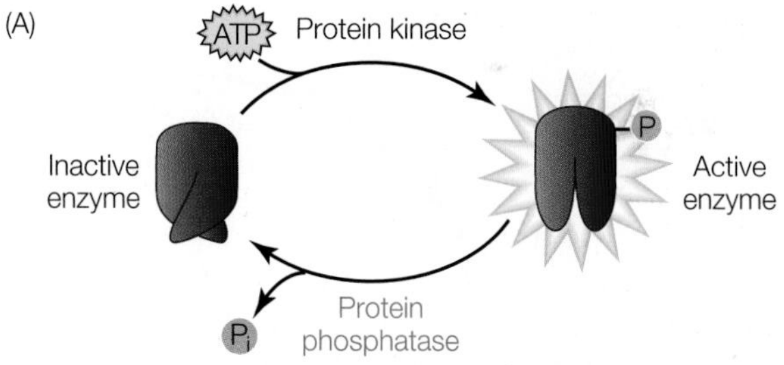


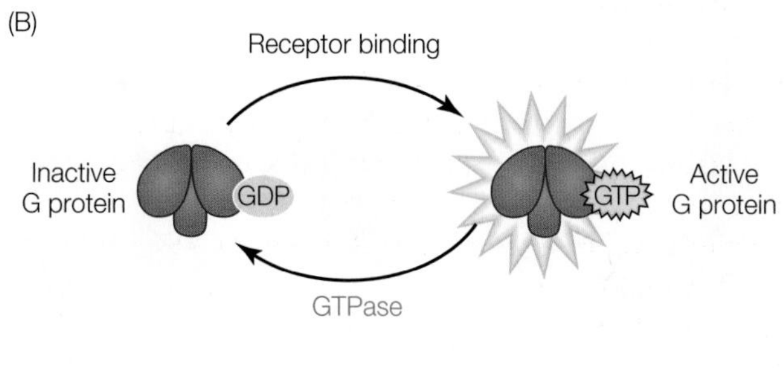


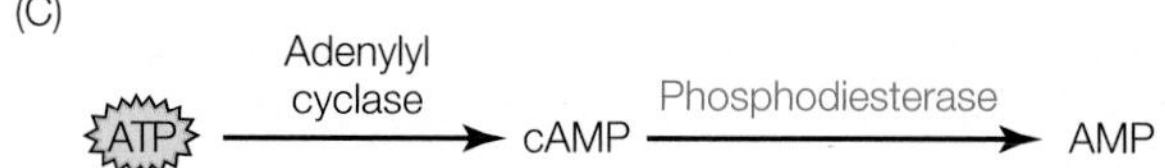


**7.18 Regulation of Signal Transduction** Some signals lead to the production of active transducers such as (A) protein kinases, (B) G proteins, and (C) cAMP. Other enzymes (red type) inactivate or remove these transducers.

## 7.3 RECAP

**Signal transduction is the series of steps between the binding of a signal to a receptor and the ultimate cellular response. A receptor can activate a signal transduction pathway, such as a protein kinase cascade, directly. In many cases, a second messenger serves to amplify the signal and activate the signaling pathway indirectly. Protein kinase cascades amplify, distribute, and regulate signaling.**

- How does a protein kinase cascade amplify a signal's message inside the cell? **See pp. 136–137 and Figure 7.12**
- What is the role of cAMP as a second messenger? **See p. 138**
- How are signal transduction cascades regulated? **See p. 141 and Figure 7.18**

We have seen how the binding of a signal to its receptor initiates the response of a cell to the signal, and how signal transduction pathways amplify the signal and distribute its effects to numerous targets in the cell. In the next section we will consider the third step in the signal transduction process, the actual effects of the signal on cell function.

## 7.4 How Do Cells Change in Response to Signals?

The effects of a signal on cell function take three primary forms: the opening of ion channels, changes in the activities of enzymes, or differential gene expression. These events set the cell on a path for further and sometimes dramatic changes in form and function.

### Ion channels open in response to signals

The opening of ion channels is a key step in the response of the nervous system to signals. In the sense organs, specialized cells have receptors that respond to external stimuli such as light, sound, taste, odor, or pressure. The alteration of the receptor results in the opening of ion channels. We will focus here on one such signal transduction pathway, that for the sense of smell, which responds to gaseous molecules in the environment.

The sense of smell is well developed in mammals. Each of the thousands of neurons in the nose expresses one of many different odorant receptors. The identification of which chemical signal, or odorant, activates which receptor is just getting under way. Humans have the genetic capacity to make about 950 different odorant receptor proteins, but very few people express more than 400 of them. Some express far fewer, which may explain why you are able to smell certain things that your roommate cannot, or vice versa.

Odorant receptors are G-protein linked, and signal transduction leads to the opening of ion channels for sodium and calcium ions, which have higher concentrations outside the cell than in the cytosol (**Figure 7.19**). The resulting influx of $Na^+$ and

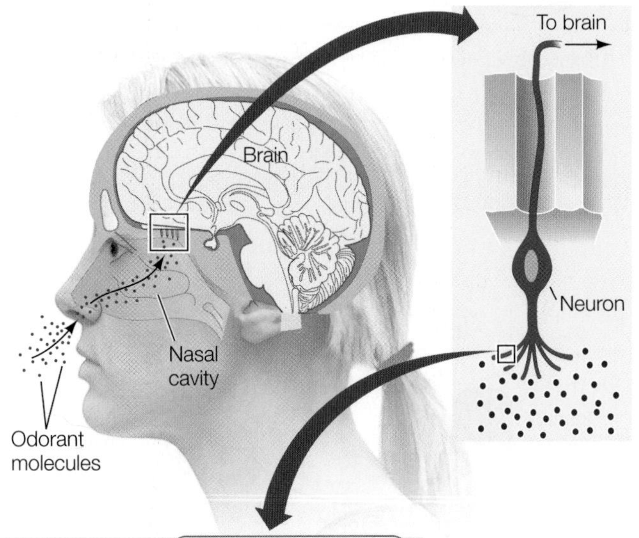


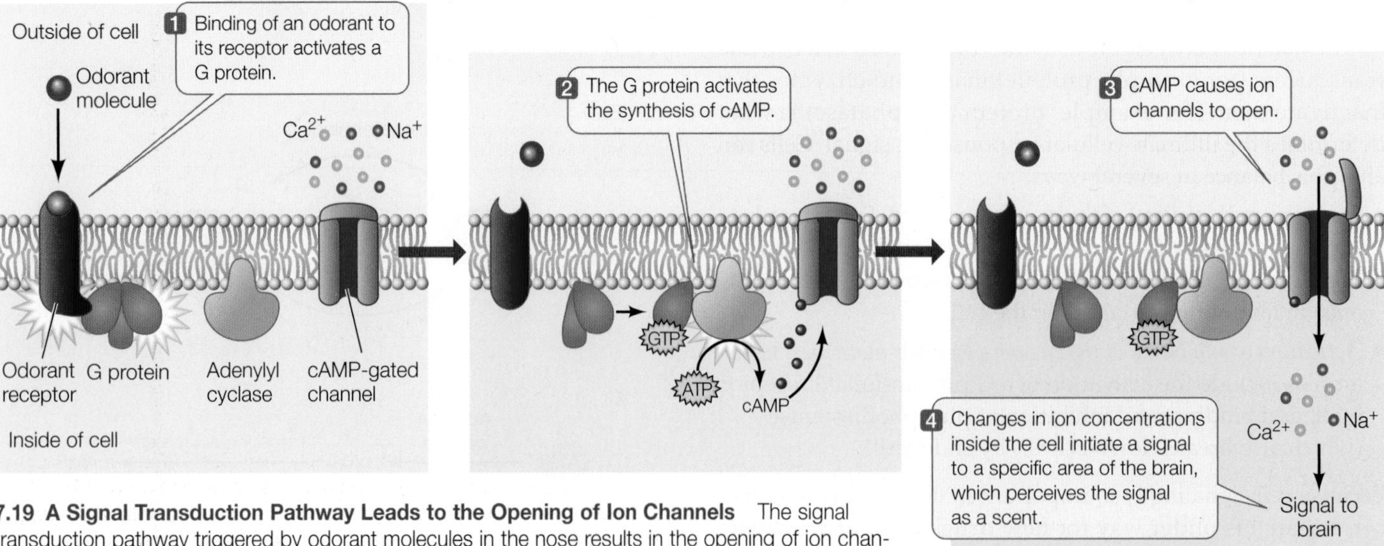


**7.19 A Signal Transduction Pathway Leads to the Opening of Ion Channels** The signal transduction pathway triggered by odorant molecules in the nose results in the opening of ion channels. The resulting influx of $Na^+$ and $Ca^{2+}$ into the neuron cells of the nose stimulates the transmission of a scent message to a specific region of the brain.

$Ca^{2+}$ causes the neuron to become stimulated so that it sends a signal to the brain that a particular odor is present.

### Enzyme activities change in response to signals

Proteins will change their shapes if they are modified either covalently or noncovalently. We have seen examples of both types of modification in our description of signal transduction. A protein kinase adds a phosphate group to a target protein, and this covalent change alters the protein's conformation and activates or inhibits a function. Cyclic AMP binds noncovalently to a target protein, and this changes the protein's shape, activating or inhibiting its function. In the case of activation, a previously inaccessible active site is exposed, and the target protein goes on to perform a new cellular role.

The G protein-mediated protein kinase cascade that is stimulated by epinephrine in liver cells results in the activation by cAMP of a key signaling molecule, protein kinase A. In turn, protein kinase A phosphorylates two other enzymes, with opposite effects:

- *Inhibition*. Glycogen synthase, which catalyzes the joining of glucose molecules to synthesize the energy-storing molecule glycogen, is inactivated when a phosphate group is added to it by protein kinase A. Thus the epinephrine signal *prevents glucose from being stored* in glycogen (**Figure 7.20, step 1**).
- *Activation*. Phosphorylase kinase is activated when a phosphate group is added to it. It is part of a protein kinase cascade that ultimately leads to the activation of glycogen phosphorylase, another key enzyme in glucose metabolism. This enzyme results in the *liberation of glucose molecules* from glycogen (**Figure 7.20, steps 2 and 3**).

The amplification of the signal in this pathway is impressive; as detailed in Figure 7.20, each molecule of epinephrine that arrives at the plasma membrane ultimately results in 10,000 molecules of blood glucose:

**1** molecule of epinephrine bound to the membrane activates
**20** molecules of cAMP, which activate
**20** molecules of protein kinase A, which activate
**100** molecules of phosphorylase kinase, which activate
**1,000** molecules of glycogen phosphorylase, which produce
**10,000** molecules of glucose 1-phosphate, which produce
**10,000** molecules of blood glucose

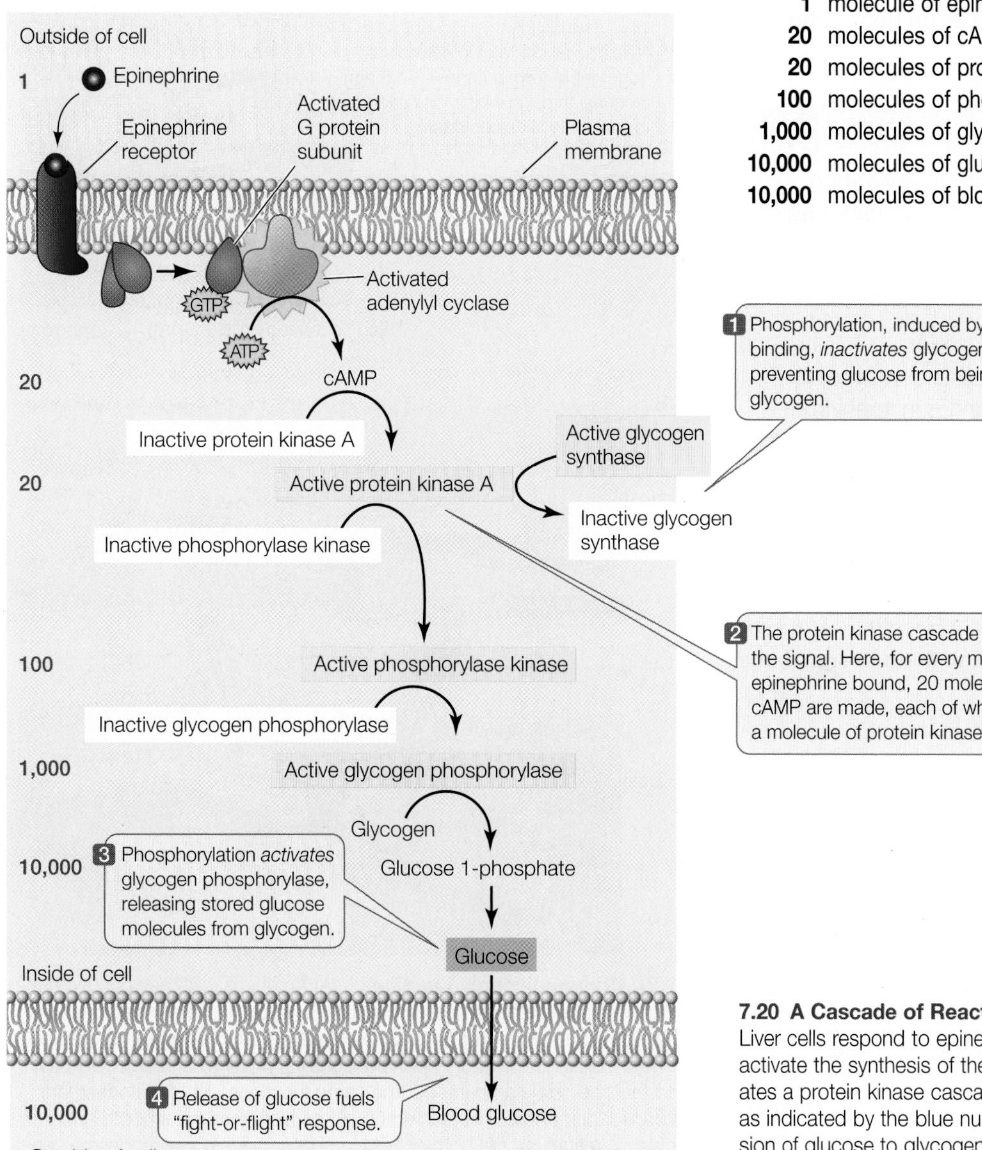


**7.20 A Cascade of Reactions Leads to Altered Enzyme Activity** Liver cells respond to epinephrine by activating G proteins, which in turn activate the synthesis of the second messenger cAMP. Cyclic AMP initiates a protein kinase cascade, greatly amplifying the epinephrine signal, as indicated by the blue numbers. The cascade both inhibits the conversion of glucose to glycogen and stimulates the release of previously stored glucose.

### Signals can initiate DNA transcription

As we introduce in Section 4.1, the genetic material, DNA, is expressed by transcription as RNA, which is then translated into a protein whose amino acid sequence is specified by the original DNA sequence. Proteins are important in all cellular functions, so a key way to regulate specific functions in a cell is to regulate which proteins are made, and therefore, which DNA sequences are transcribed.

Signal transduction plays an important role in determining which DNA sequences are transcribed. Common targets of signal transduction are proteins called transcription factors, which bind to specific DNA sequences in the cell nucleus and activate or inactivate transcription of the adjacent DNA regions. For example, the Ras signaling pathway ends in the nucleus (see Figure 7.12). The final protein kinase in the Ras signaling cascade, MAPk, enters the nucleus and phosphorylates a protein which stimulates the expression of a number of genes involved in cell proliferation.

In this chapter we have concentrated on signaling pathways that occur in animal cells. However, as you will see in Part Eight of this book, plants also have signal transduction pathways, with equally important roles.

## 7.4 RECAP

**Cells respond to signal transduction by activating enzymes, opening membrane channels, or initiating gene transcription.**

- What role does cAMP play in the sense of smell? **See pp. 142–143 and Figure 7.19**
- How does amplification of a signal occur and why is it important in a cell's response to changes in its environment? **See p. 143 and Figure 7.20**

We have described how signals from a cell's environment can influence the cell. But the environment of a cell in a multicellular organism is more than the extracellular medium—it includes neighboring cells as well. In the next section we'll look at specialized junctions between cells that allow them to signal one another directly.

## 7.5 How Do Cells Communicate Directly?

Most cells are in contact with their neighbors. Section 6.2 describes various ways in which cells adhere to one another, such as via recognition proteins that protrude from the cell surface, or via tight junctions and desmosomes. But as we know from our own experience with our neighbors (and roommates), just being in proximity does not necessarily mean that there is functional communication. Neither tight junctions nor desmosomes are specialized for intercellular communication. However, many multicellular organisms have specialized cell junctions that allow their cells to communicate directly. In animals, these structures are gap junctions; in plants, they are plasmodesmata.

### Animal cells communicate by gap junctions

**Gap junctions** are channels between adjacent cells that occur in many animals, occupying up to 25 percent of the area of the plasma membrane (**Figure 7.21A**). Gap junctions traverse the narrow space between the plasma membranes of two cells (the "gap") by means of channel structures called **connexons**. The walls of a connexon are composed of six subunits of the integral membrane protein connexin. In adjacent cells, two connexons come together to form a gap junction that links the cytoplasms of the two cells. There may be hundreds of these channels between a cell and its neighbors. The channel pores are about 1.5 nm in diameter—far too narrow for the passage of large molecules such as proteins. But they are wide enough to allow small mol-

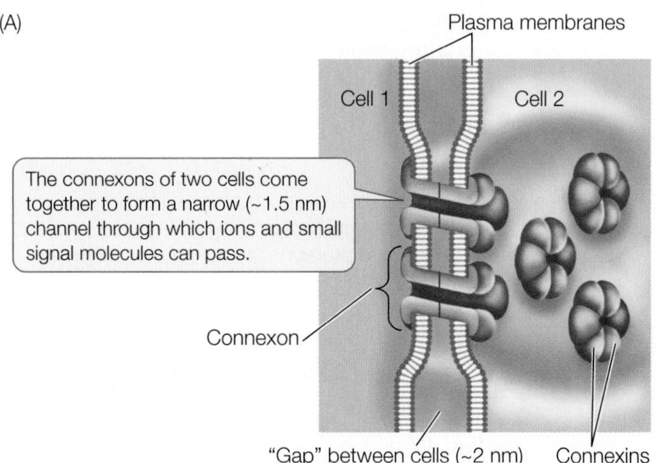


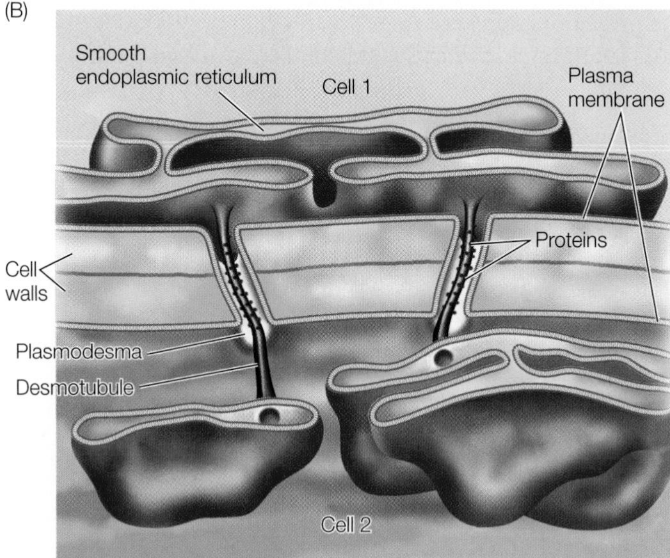


**7.21 Communicating Junctions** (A) An animal cell may contain hundreds of gap junctions connecting it to neighboring cells. The pores of gap junctions allow small molecules to pass from cell to cell, assuring similar concentrations of important signaling molecules in adjacent cells so that the cells can carry out the same functions. (B) Plasmodesmata connect plant cells. The desmotubule, derived from the smooth endoplasmic reticulum, fills up most of the space inside a plasmodesma, leaving a tiny gap through which small metabolites and ions can pass.

ecules to pass between the cells. Experiments in which labeled signal molecules or ions are injected into one cell show that they can readily pass into adjacent cells if the cells are connected by gap junctions. Why is it necessary to have these linkages between the cytoplasms of adjacent cells?

Gap junctions permit *metabolic cooperation* between the linked cells. Such cooperation ensures the sharing between cells of important small molecules such as ATP, metabolic intermediates, amino acids, and coenzymes (see Section 8.4). In some tissues, metabolic cooperation is needed so that signals and metabolic products can be passed from cells at the edges of tissues to cells in the interior and vice versa. It is not clear how important this function is in many tissues, but it is known to be vital in some. For example, in the lens of the mammalian eye only the cells at the periphery are close enough to the blood supply to allow diffusion of nutrients and wastes. But because lens cells are connected by large numbers of gap junctions, material can diffuse between them rapidly and efficiently.

As mentioned above, there is evidence that signal molecules such as hormones and second messengers such as cAMP can move through gap junctions. If this is true, then only a few cells would need receptors for a signal in order for the signal to be transduced throughout the tissue. In this way, a tissue can have a coordinated response to the signal.

### Plant cells communicate by plasmodesmata

Instead of gap junctions, plants have **plasmodesmata** (singular *plasmodesma*), which are membrane-lined tunnels that traverse the thick cell walls separating plant cells from one another. A typical plant cell has several thousand plasmodesmata.

Plasmodesmata differ from gap junctions in one fundamental way: unlike gap junctions, in which the wall of the channel is made of integral proteins from the adjacent plasma membranes, plasmodesmata are lined by the fused plasma membranes themselves. Plant biologists are so familiar with the notion of a tissue as cells interconnected in this way that they refer to these continuous cytoplasms as a *symplast* (see Figure 35.6).

The diameter of a plasmodesma is about 6 nm, far larger than a gap junction channel. But the actual space available for diffusion is about the same—1.5 nm. Examination of the interior of the plasmodesma by transmission electron microscopy reveals that a tubule called the **desmotubule**, apparently derived from the endoplasmic reticulum, fills up most of the opening of the plasmodesma (**Figure 7.21B**). Typically, only small metabolites and ions can move between plant cells. This fact is important in plant physiology because the bulk transport system in plants, the vascular system, lacks the tiny circulatory vessels (capillaries) that many animals have for bringing gases and nutrients to every cell. Diffusion from cell to cell across plasma membranes is probably inadequate to account for the movement of a plant hormone from the site of production to the site of action. Instead, plants rely on more rapid diffusion through plasmodesmata to ensure that all cells of a tissue respond to a signal at the same time. There are cases in which larger molecules or particles can pass between cells via plasmodesmata. For example, some viruses can move through plasmodesmata by using "movement proteins" to assist their passage.

### 7.5 RECAP

**Cells can communicate with their neighbors through specialized cell junctions. In animals, these structures are gap junctions; in plants, they are plasmodesmata.**

- What are the roles that gap junctions and plasmodesmata play in cell signaling?

## CHAPTER SUMMARY

### 7.1 What Are Signals, and How Do Cells Respond to Them?

- Cells receive many signals from the physical environment and from other cells. Chemical signals are often at very low concentrations. **Autocrine** signals affect the cells that make them; **paracrine** signals diffuse to and affect nearby cells. Review Figure 7.1, **WEB ACTIVITY 7.1**
- A **signal transduction pathway** involves the interaction of a signal molecule with a **receptor**; the transduction and amplification of the signal via a series of steps within the cell; and effects on the function of the cell. Review Figure 7.2

### 7.2 How Do Signal Receptors Initiate a Cellular Response?

- Cells respond to signals only if they have specific receptor proteins that can bind those signals. Depending on the nature of its signal or **ligand**, a receptor may be located in the plasma membrane or in the cytoplasm of the target cell. Review Figure 7.5
- Receptors located in the plasma membrane include **ion channels**, **protein kinases**, and **G protein-linked receptors**.
- Ion channel receptors are "gated": the gate "opens" when the three-dimensional structure of the channel protein is altered by ligand binding. Review Figure 7.6
- A **G protein** has three important binding sites, which bind a G protein-linked receptor, GDP or GTP, and an **effector protein**. A G protein can either activate or inhibit an effector protein. Review Figure 7.8, **ANIMATED TUTORIAL 7.1**
- Lipid-soluble signals, such as steroid hormones, can diffuse through the plasma membrane and meet their receptors in the cytoplasm; the ligand–receptor complex may then enter the nucleus to affect gene expression. Review Figure 7.9

### 7.3 How Is the Response to a Signal Transduced through the Cell?

- **Direct signal transduction** is a function of the receptor itself and occurs at the plasma membrane. **Indirect transduction** involves a soluble **second messenger**. Review Figure 7.10
- A **protein kinase cascade** amplifies the response to receptor binding. Review Figure 7.12, **ANIMATED TUTORIAL 7.2**

- Second messengers include **cyclic AMP (cAMP)**, **inositol trisphosphate ($IP_3$)**, **diacylglycerol (DAG)**, and **calcium** ions. $IP_3$ and DAG are derived from the phospholipid **phosphatidyl inositol-bisphosphate (PIP2)**.
- The gas **nitric oxide (NO)** is involved in signal transduction in human smooth muscle cells. Review Figure 7.17
- Signal transduction can be regulated in several ways. The balance between activating and inactivating the molecules involved determines the ultimate cellular response to a signal. Review Figure 7.18

### 7.4 How Do Cells Change in Response to Signals?

- The cellular responses to signals may be the opening of ion channels, the alteration of enzyme activities, or changes in gene expression. Review Figure 7.19
- Protein kinases covalently add phosphate groups to target proteins; cAMP binds target proteins noncovalently. Both kinds of binding change the target protein's conformation to expose or hide its active site.
- Activated enzymes may activate other enzymes in a signal transduction pathway, leading to impressive amplification of a signal. Review Figure 7.20

### 7.5 How Do Cells Communicate Directly?

- Many adjacent animal cells can communicate with one another directly through small pores in their plasma membranes called **gap junctions**. Protein structures called **connexons** form thin channels between two adjacent cells through which small signal molecules and ions can pass. Review Figure 7.21A
- Plant cells are connected by somewhat larger pores called **plasmodesmata**, which traverse both plasma membranes and cell walls. The **desmotubule** narrows the opening of the plasmodesma. Review Figure 7.21B

**SEE WEB ACTIVITY 7.2 for a concept review of this chapter.**

## SELF-QUIZ

1. What is the correct order for the following events in the interaction of a cell with a signal? (1) Alteration of cell function; (2) signal binds to receptor; (3) signal released from source; (4) signal transduction.
   *a.* 1234
   *b.* 2314
   *c.* 3214
   *d.* 3241
   *e.* 3421
2. Why do some signals ("first messengers") trigger "second messengers" to activate target cells?
   *a.* The first messenger requires activation by ATP.
   *b.* The first messenger is not water soluble.
   *c.* The first messenger binds to many types of cells.
   *d.* The first messenger cannot cross the plasma membrane.
   *e.* There are no receptors for the first messenger.
3. Steroid hormones such as estrogen act on target cells by
   *a.* initiating second messenger activity.
   *b.* binding to membrane proteins.
   *c.* initiating gene expression.
   *d.* activating enzymes.
   *e.* binding to membrane lipids.
4. The major difference between a cell that responds to a signal and one that does not is the presence of a
   *a.* DNA sequence that binds to the signal.
   *b.* nearby blood vessel.
   *c.* receptor.
   *d.* second messenger.
   *e.* transduction pathway.
5. Which of the following is *not* a consequence of a signal binding to a receptor?
   *a.* Activation of receptor enzyme activity
   *b.* Diffusion of the receptor in the plasma membrane
   *c.* Change in conformation of the receptor protein
   *d.* Breakdown of the receptor to amino acids
   *e.* Release of the signal from the receptor
6. A nonpolar molecule such as a steroid hormone usually binds to a
   *a.* cytoplasmic receptor.
   *b.* protein kinase.
   *c.* ion channel.
   *d.* phospholipid.
   *e.* second messenger.
7. Which of the following is *not* a common type of receptor?
   *a.* Ion channel
   *b.* Protein kinase
   *c.* G protein–linked receptor
   *d.* Cytoplasmic receptor
   *e.* Adenylyl cyclase
8. Which of the following is *not* true of a protein kinase cascade?
   *a.* The signal is amplified.
   *b.* A second messenger is formed.
   *c.* Target proteins are phosphorylated.
   *d.* The cascade ends up at the mitochondrion.
   *e.* The cascade begins at the plasma membrane.
9. Which of the following is *not* a second messenger?
   *a.* Calcium ion
   *b.* Inositol trisphosphate
   *c.* ATP
   *d.* Cyclic AMP
   *e.* Diacylglycerol
10. Plasmodesmata and gap junctions
   *a.* allow small molecules and ions to pass rapidly between cells.
   *b.* are both membrane-lined channels.
   *c.* are channels about 1 mm in diameter.
   *d.* are present only once per cell.
   *e.* are involved in cell recognition.

## FOR DISCUSSION

1. Like the Ras protein itself, the various components of the Ras signaling pathway were discovered when cancer cells showed changes (mutations) in the genes encoding one or another of the components. What might be the biochemical consequences of mutations in the genes coding for (*a*) Raf and (*b*) MAP kinase that resulted in rapid cell division?
2. Cyclic AMP is a second messenger in many different responses. How can the same messenger act in different ways in different cells?
3. Compare direct communication via plasmodesmata or gap junctions with receptor-mediated communication between cells. What are the advantages of one method over the other?
4. The tiny invertebrate Hydra has an apical region with tentacles and a long, slender body. Hydra can reproduce asexually when cells on the body wall differentiate and form a bud, which then breaks off as a new organism. Buds form only at certain distances from the apex, leading to the idea that the apex releases a signal molecule that diffuses down the body and, at high concentrations (i.e., near the apex), inhibits bud formation. Hydra lacks a circulatory system, so this inhibitor must diffuse from cell to cell. If you had an antibody that binds to connexons and plugs up the gap junctions, how would you test the hypothesis that Hydra's inhibitory factor passes through these junctions?

## ADDITIONAL INVESTIGATION

Endosymbiotic bacteria in the marine invertebrate *Begula neritina* synthesize bryostatins, a name derived from the invertebrate's animal group Ectoprocta, once known as *bryo*zoans ("moss animals"), and *stat* (stop). When used as drugs, bryostatins curtail cell division in many cell types, including several cancers. It has been proposed that bryostatins inhibit protein kinase C (see Figure 7.15). How would you investigate this hypothesis, and how would you relate this inhibition to cell division?

## WORKING WITH DATA (GO TO yourBioPortal.com)

**The Discovery of a Second Messenger** In this hands-on exercise, you will examine the experiments that Sutherland and his colleagues performed (Figure 7.13) using liver tissue to demonstrate that there can be a second, soluble chemical messenger between a hormone binding to a receptor and its eventual effects in the cell. By analyzing their data, you will see how controls were important in their reasoning.

# 8 Energy, Enzymes, and Metabolism

## Lactase deficiency

United Nations officials first noticed the problem during the 1950s when massive food relief efforts were made to alleviate famines in Asia and Africa. The conventional wisdom was that donated food should provide a balanced diet, and that an important component of the diet (one of the "four major food groups") was dairy products. Reports started coming in of people developing bloating, nausea, and diarrhea after consuming donated dairy products.

At first, this problem was attributed to contamination by bacteria during shipping, or to errors in the preparation of powdered milk products by the recipients. It never occurred to the donors that the scientific principles of nutrition that they had so carefully developed did not apply to people everywhere. But it soon became apparent that the donors in Europe and other wealthy countries, who were usually of European descent, were atypical of humanity in their ability to hydrolyze the disaccharide lactose, the major milk sugar, to its constituent monosaccharides, glucose and galactose. Their small intestines make a protein called lactase (β-galactosidase) that acts to speed up the hydrolysis reaction millions-fold. Such catalytic proteins are called enzymes, and their names often end with the suffix "ase." Most people around the world are born with the ability to make the enzyme lactase, but soon after infancy they lose it. People of European descent are unusual in that they do not lose their lactase production after infancy.

When many non-European adults consume lactose it does not get hydrolyzed in their small intestine, because they do not produce lactase. Disaccharides such as lactose are not absorbed into the blood stream by cells lining the small intestine. So the lactose remains intact and travels onward to the colon (large intestine). Among the billions of bacteria in the colon, there are species that make lactase. But as a side product, these bacteria produce the gases that cause all the discomfort. The condition of discomfort after eating lactose is called lactose intolerance.

Why does lactase production go down after infancy in most humans? The explanation lies with diet: an infant first consumes mother's milk, which contains abundant lactose. This stimulates the intestinal cells to make lactase. But many humans—and other mammals—consume little or no milk after weaning, and the ability to make lactase in the small intestine is not needed. So most mammals have evolved to produce lactase only during infancy. Lactose intolerance is not a problem in many human societies because the people simply don't consume dairy products—

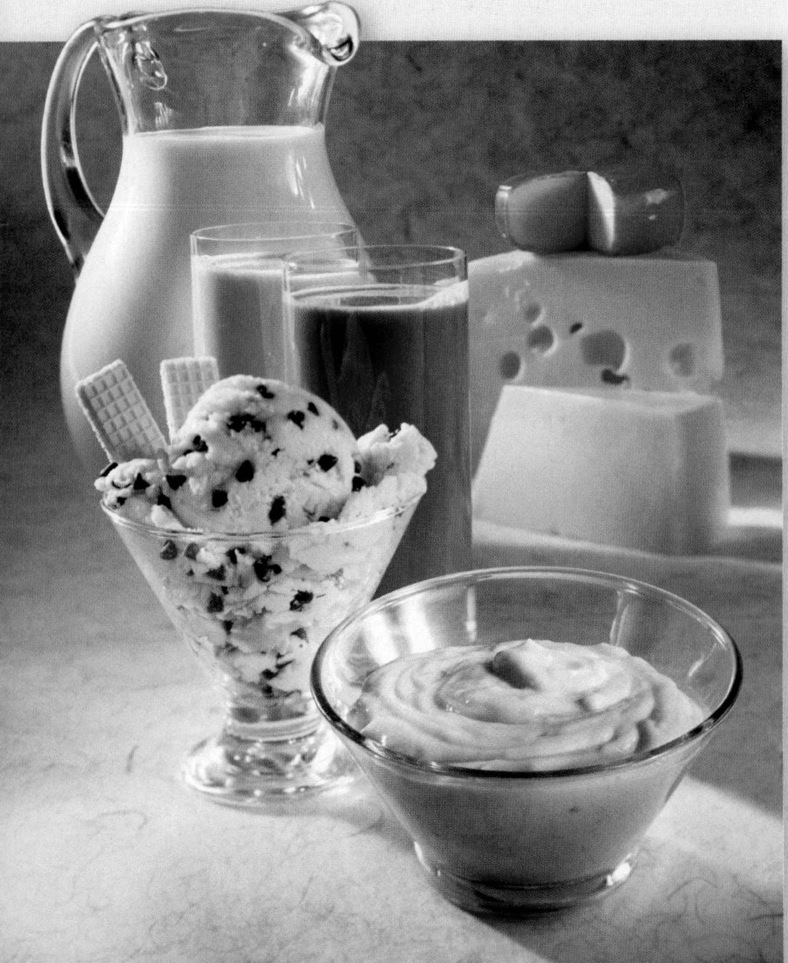

**A Precursor to Trouble** Many adults do not produce the enzyme lactase in their small intestines. When they consume dairy products, these people have ill effects.

**Maasai Herders** The Maasai are unusual among Africans in that they consume milk throughout their lives. They can do this because they produce lactase after weaning.

unless they are given them by well-meaning donors! They get their carbohydrates from other sources.

Then why are many people of European descent still able to make lactase as adults? It turns out that they carry a mutation (a change in their DNA sequence) that eliminates the shutdown in lactase production after weaning. This mutation became predominant in European (and some east African) populations after those people began to keep grazing animals and to use their milk.

Lactase activity is an example of an enzyme-catalyzed biochemical transformation. The hydrolysis of lactose is the beginning of its transformation to simpler molecules—ultimately $CO_2$—and this transformation releases energy.

**IN THIS CHAPTER** we begin our study of biochemical transformations, focusing on the role of energy. We first describe the physical principles that underlie energy transformations and how these principles apply to biology. Then, we go on to show how the energy carrier ATP plays an important role in the cell. Finally, we follow up on the lactase story by describing the nature, activities, and regulation of enzymes, which speed up biochemical transformations and are essential for life.

CHAPTER OUTLINE



## 8.1 What Physical Principles Underlie Biological Energy Transformations?

Metabolic reactions and catalysts are essential to the biochemical transformation of energy by living things. Whether it is a plant using light energy to produce carbohydrates or a cat transforming food energy so it can leap to a countertop (where it hopes to find food so it can obtain more energy), the transformation of energy is a hallmark of life.

Physicists define energy as the capacity to do work. Work occurs when a force operates on an object over a distance. In biochemistry, it is more useful to consider energy as *the capacity for change*. In biochemical reactions these energy changes are usually associated with changes in the chemical composition and properties of molecules. No cell creates energy; all living things must obtain energy from the environment. Indeed one of the fundamental laws of physics is that energy can neither be created nor destroyed. However, energy can be transformed from one form into another, and living cells carry out many such transformations. For example, green plant cells convert light energy into chemical energy; the jumping cat transforms chemical energy into movement. Energy transformations are linked to the chemical transformations that occur in cells—the breaking and creating of chemical bonds, the movement of substances across membranes, cell reproduction, and so forth.

### There are two basic types of energy and of metabolism

Energy comes in many forms: chemical, electrical, heat, light, and mechanical. But all forms of energy can be considered as one of two basic types:

- *Potential energy* is the energy of state or position—that is, stored energy. It can be stored in many forms: in chemical bonds, as a concentration gradient, or even as an electric charge imbalance (as in the membrane potential; see Section 6.3). Think of a crouching cat, holding still as it prepares to pounce.
- *Kinetic energy* is the energy of movement—that is, the type of energy that does work, that makes things change. Think of the cat leaping as some of the potential energy stored in its muscles is converted into the kinetic energy of muscle contractions.

Potential energy can be converted into kinetic energy and vice versa, and the form that the energy takes can also be converted. The potential energy in the cat's muscles is in covalent bonds (chemical energy), while the kinetic energy of the pouncing cat is mechanical (**Figure 8.1**). You can think of many other such

Potential chemical energy is converted into kinetic mechanical energy when the cat leaps.
Potential chemical energy is stored in the muscles of the cat.

**8.1 Energy Conversions and Work** A leaping cat illustrates both the conversion between potential and kinetic energy and the conversion of energy from one form (chemical) to another (mechanical).

conversions: while reading this book, for example, light energy is converted into chemical energy in your eyes, and then is converted into electric energy in the nerve cells that carry messages to your brain. When you decide to turn a page, the electrical and chemical energy of nerve and muscle are converted into kinetic energy.

In any living organism, chemical reactions are occurring continuously. **Metabolism** is defined as the totality of these reactions. While particular cells carry out many reactions at any given instant, scientists usually focus on a few reactions at a time. Two broad categories of metabolic reactions occur in all cells of all organisms:

- **Anabolic reactions** (anabolism) link simple molecules to form more complex molecules (for example, the synthesis of a protein from amino acids). Anabolic reactions require an input of energy and capture it in the chemical bonds that are formed.
- **Catabolic reactions** (catabolism) break down complex molecules into simpler ones and release the energy stored in chemical bonds. For example, when the polysaccharide starch is hydrolyzed to simpler molecules, energy is released.

*Catabolic and anabolic reactions are often linked.* The energy released in catabolic reactions is often used to drive anabolic reactions—that is, to do biological work. For example, the energy released by the breakdown of glucose (catabolism) is used to drive anabolic reactions such as the synthesis of nucleic acids and proteins.

Catabolic reactions also provide energy for movement: muscle contraction is driven by the catabolism (hydrolysis) of ATP (see Section 8.2). In this case, the potential energy released by catabolism is converted to kinetic energy.

The **laws of thermodynamics** (thermo, "energy"; dynamics, "change") were derived from studies of the fundamental physical properties of energy, and the ways it interacts with matter. The laws apply to all matter and all energy transformations in the universe. Their application to living systems helps us to understand how organisms and cells harvest and transform energy to sustain life.

### The first law of thermodynamics: Energy is neither created nor destroyed

The first law of thermodynamics states that in any conversion of energy, it is neither created nor destroyed. Another way of saying this is: in any conversion of energy, the total energy before and after the conversion is the same (**Figure 8.2A**). As you will see in the next two chapters, the potential energy present in the chemical bonds of carbohydrates and lipids can be converted to potential energy in the form of ATP. This can then be converted into kinetic energy to do mechanical work (such as in muscle contractions), or used to do biochemical work (such as protein synthesis).

### The second law of thermodynamics: Disorder tends to increase

Although energy cannot be created or destroyed, the second law of thermodynamics states that when energy is converted from one form to another, some of that energy becomes unavailable for doing work (**Figure 8.2B**). In other words, no physical process or chemical reaction is 100 percent efficient; some of the released energy is lost to a form associated with disorder. Think of disorder as a kind of randomness due to the thermal motion of particles; this energy is of such a low value and so dispersed that it is unusable. *Entropy* is a measure of the disorder in a system.

It takes energy to impose order on a system. Unless energy is applied to a system, it will be randomly arranged or disordered. The second law applies to all energy transformations, but we will focus here on chemical reactions in living systems.

**NOT ALL ENERGY CAN BE USED** In any system, the total energy includes the usable energy that can do work and the unusable energy that is lost to disorder:

$$\text{total energy} = \text{usable energy} + \text{unusable energy}$$

In biological systems, the total energy is called **enthalpy (*H*)**. The usable energy that can do work is called **free energy (*G*)**. Free energy is what cells require for all the chemical reactions needed for growth, cell division, and maintenance. The unusable en-

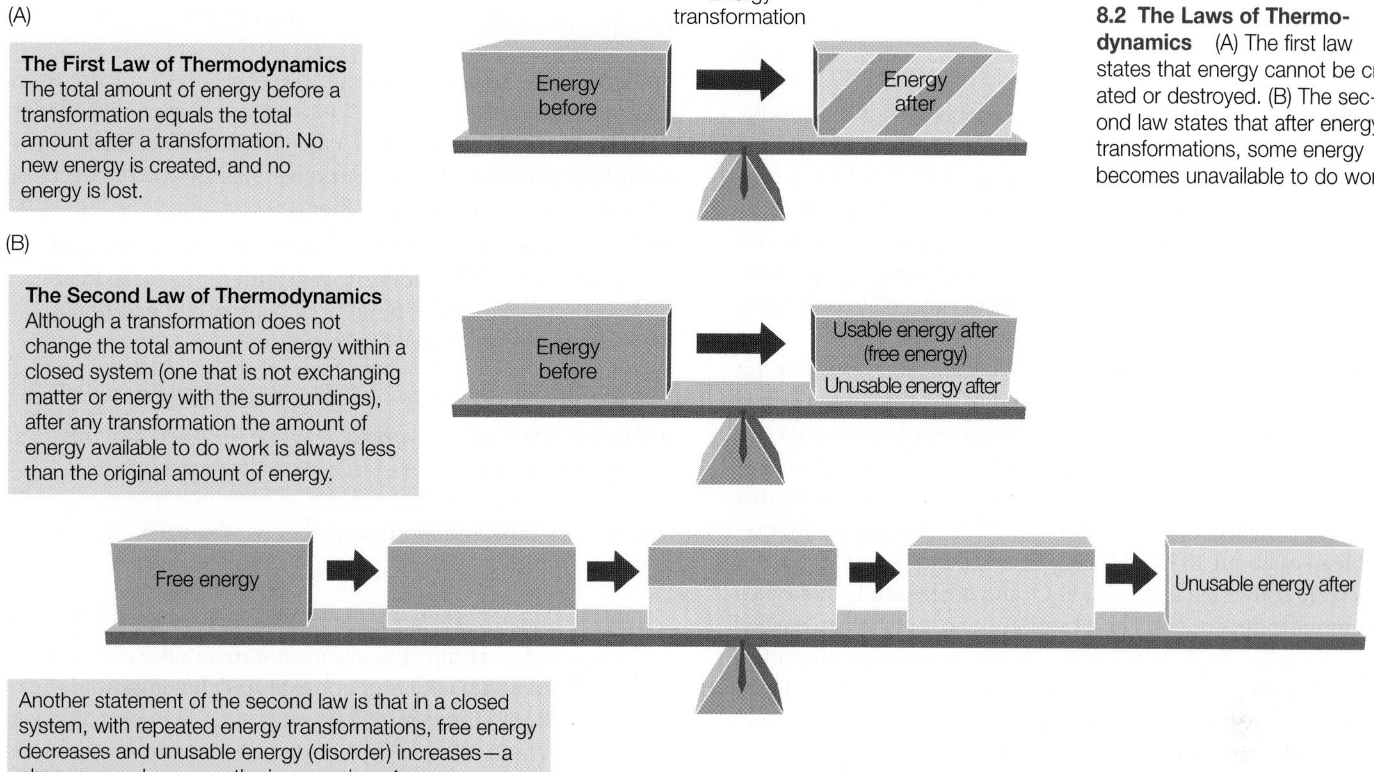

(A)
The First Law of Thermodynamics
The total amount of energy before a transformation equals the total amount after a transformation. No new energy is created, and no energy is lost.
Energy transformation
Energy before
Energy after
(B)
The Second Law of Thermodynamics
Although a transformation does not change the total amount of energy within a closed system (one that is not exchanging matter or energy with the surroundings), after any transformation the amount of energy available to do work is always less than the original amount of energy.
Energy before
Usable energy after (free energy)
Unusable energy after
Free energy
Unusable energy after
Another statement of the second law is that in a closed system, with repeated energy transformations, free energy decreases and unusable energy (disorder) increases—a phenomenon known as the increase in entropy.


**8.2 The Laws of Thermodynamics** (A) The first law states that energy cannot be created or destroyed. (B) The second law states that after energy transformations, some energy becomes unavailable to do work.

ergy is represented by **entropy (*S*)** multiplied by the absolute temperature (*T*). Thus we can rewrite the word equation above more precisely as:

$$H = G + TS$$

Because we are interested in usable energy, we rearrange this expression:

$$G = H - TS$$

Although we cannot measure *G*, *H*, or *S* absolutely, we can determine the change in each at a constant temperature. Such energy changes are measured in calories (cal) or joules (J).* A change in energy is represented by the Greek letter delta (Δ). The change in free energy (Δ*G*) of any chemical reaction is equal to the difference in free energy between the products and the reactants:

$$\Delta G_{\text{reaction}} = G_{\text{products}} - G_{\text{reactants}}$$

Such a change can be either positive or negative; that is, the free energy of the products can be more or less than the free energy of the reactants. If the products have more free energy than the reactants, then there must have been some input of energy into the reaction. (Remember that energy cannot be created, so some energy must have been added from an external source.) At a constant temperature, Δ*G* is defined in terms of the change in total energy (Δ*H*) and the change in entropy (Δ*S*):

$$\Delta G = \Delta H - T\Delta S$$

This equation tells us whether free energy is released or consumed by a chemical reaction:

- If Δ*G* is negative ($\Delta G < 0$), free energy is released.
- If Δ*G* is positive ($\Delta G > 0$), free energy is required (consumed).

If the necessary free energy is not available, the reaction does not occur. The sign and magnitude of Δ*G* depend on the two factors on the right of the equation:

- Δ*H*: In a chemical reaction, Δ*H* is the total amount of energy added to the system ($\Delta H > 0$) or released ($\Delta H < 0$).
- Δ*S*: Depending on the sign and magnitude of Δ*S*, the entire term, *T*Δ*S*, may be negative or positive, large or small. In other words, in living systems at a constant temperature (no change in *T*), the magnitude and sign of Δ*G* can depend a lot on changes in entropy.

If a chemical reaction increases entropy, its products are more disordered or random than its reactants. If there are more products than reactants, as in the hydrolysis of a protein to its amino acids, the products have considerable freedom to move around. The disorder in a solution of amino acids will be large compared with that in the protein, in which peptide bonds and other forces prevent free movement. So in hydrolysis, the change in entropy (Δ*S*) will be positive. Conversely, if there are fewer products and they are more restrained in their movements than the reac-

*A calorie is the amount of heat energy needed to raise the temperature of 1 gram of pure water from 14.5°C to 15.5°C. In the SI system, energy is measured in joules. 1 J = 0.239 cal; conversely, 1 cal = 4.184 J. Thus, for example, 486 cal = 2,033 J, or 2.033 kJ. Although defined here in terms of heat, the calorie and the joule are measures of any form of energy—mechanical, electrical, or chemical. When you compare data on energy, always compare joules with joules and calories with calories.

tants (as for amino acids being joined in a protein), $\Delta S$ will be negative.

**DISORDER TENDS TO INCREASE** The second law of thermodynamics also predicts that, as a result of energy transformations, disorder tends to increase; some energy is always lost to random thermal motion (entropy). Chemical changes, physical changes, and biological processes all tend to increase entropy (see Figure 8.2B), and this tendency gives direction to these processes. It explains why some reactions proceed in one direction rather than another.

How does the second law apply to organisms? Consider the human body, with its highly organized tissues and organs composed of large, complex molecules. This level of complexity appears to be in conflict with the second law but is not for two reasons. First, the construction of complexity also generates disorder. Constructing 1 kg of a human body requires the metabolism of about 10 kg of highly ordered biological materials, which are converted into $CO_2$, $H_2O$, and other simple molecules that move independently and randomly. So metabolism creates far more disorder (more energy is lost to entropy) than the amount of order (total energy; enthalpy) stored in 1 kg of flesh. Second, life requires a constant input of energy to maintain order. Without this energy, the complex structures of living systems would break down. Because energy is used to generate and maintain order, there is no conflict with the second law of thermodynamics.

Having seen that the laws of thermodynamics apply to living things, we will now turn to a consideration of how these laws apply to biochemical reactions.

### Chemical reactions release or consume energy

Since anabolic reactions link simple molecules to form more complex molecules, they tend to increase complexity (order) in the cell. On the other hand, catabolic reactions break down complex molecules into simpler ones, so they tend to decrease complexity (generate disorder).

- *Catabolic* reactions may break down an ordered reactant into smaller, more randomly distributed products. Reactions that release free energy ($-\Delta G$) are called **exergonic** reactions (**Figure 8.3A**). For example:

  complex molecules → free energy + small molecules

- *Anabolic* reactions may make a single product (a highly ordered substance) out of many smaller reactants (less ordered). Reactions that require or consume free energy ($+\Delta G$) are called **endergonic** reactions (**Figure 8.3B**). For example:

  free energy + small molecules → complex molecules

In principle, chemical reactions are reversible and can run both forward and backward. For example, if compound A can be converted into compound B ($A \rightarrow B$), then B, in principle, can be converted into A ($B \rightarrow A$), although *the concentrations of A and B determine which of these directions will be favored*. Think of the overall reaction as resulting from competition between forward and reverse reactions ($A \rightleftharpoons B$). Increasing the concentration of A speeds up the forward reaction, and increasing the concentration of B favors the reverse reaction.

At some concentration of A and B, the forward and reverse reactions take place at the same rate. At this concentration, no further net change in the system is observable, although individual molecules are still forming and breaking apart. This balance between forward and reverse reactions is known as **chemical equilibrium**. Chemical equilibrium is a state of no net change, and a state in which $\Delta G = 0$.

**(A) Exergonic reaction**

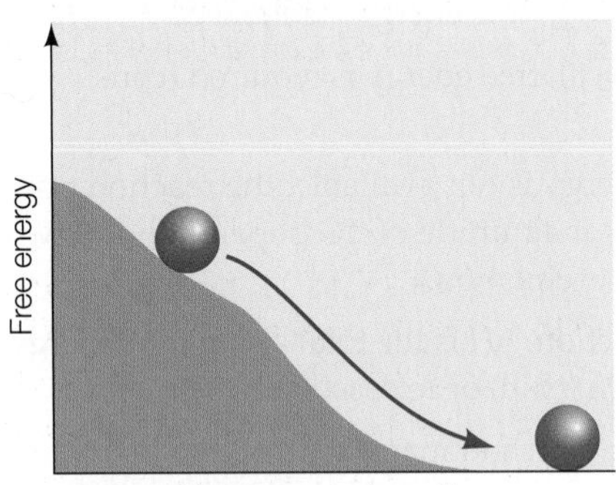
Free energy
Time course of reaction

Reactants
Amount of energy released
Products
Time course of reaction
In an exergonic reaction, *energy is released* as the reactants form lower-energy products. ΔG is negative.

**(B) Endergonic reaction**

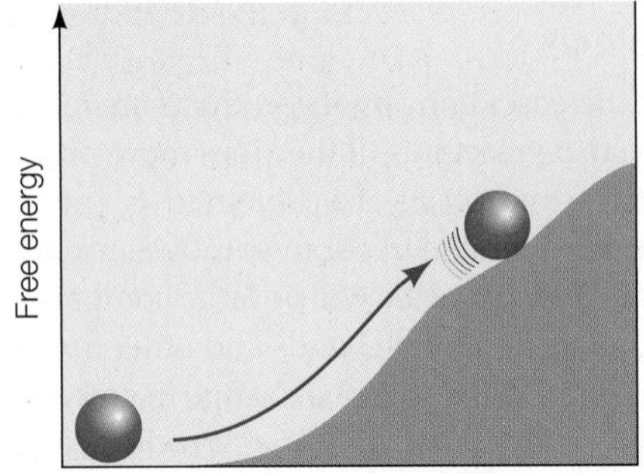
Free energy
Time course of reaction

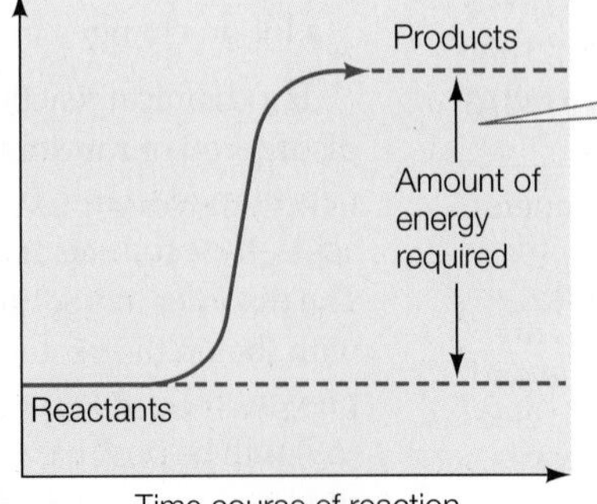
Products
Amount of energy required
Reactants
Time course of reaction

*Energy must be added* for an endergonic reaction, in which reactants are converted to products with a higher energy level. ΔG is positive.

**8.3 Exergonic and Endergonic Reactions** (A) In an exergonic reaction, the reactants behave like a ball rolling down a hill, and energy is released. (B) A ball will not roll uphill by itself. Driving an endergonic reaction, like moving a ball uphill, requires the addition of free energy.

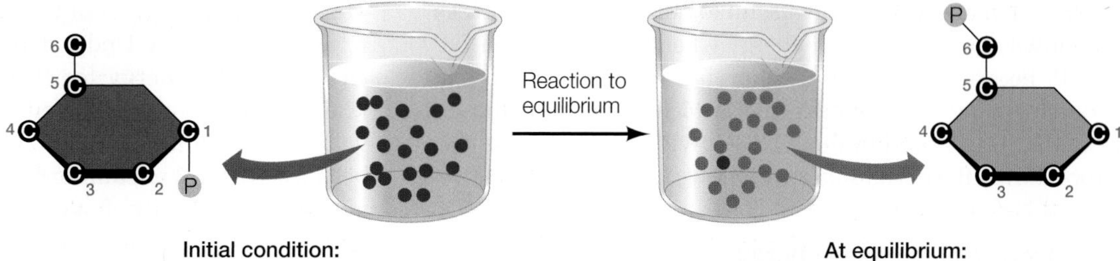


**8.4 Chemical Reactions Run to Equilibrium** No matter what quantities of glucose 1-phosphate and glucose 6-phosphate are dissolved in water, when equilibrium is attained, there will always be 95 percent glucose 6-phosphate and 5 percent glucose 1-phosphate.

### Chemical equilibrium and free energy are related

Every chemical reaction proceeds to a certain extent, but not necessarily to completion (all reactants converted into products). Each reaction has a specific equilibrium point, which is related to the free energy released by the reaction under specified conditions. To understand the principle of equilibrium, consider the following example.

Most cells contain glucose 1-phosphate, which is converted into glucose 6-phosphate.

$$\text{glucose 1-phosphate} \rightleftharpoons \text{glucose 6-phosphate}$$

Imagine that we start out with an aqueous solution of glucose 1-phosphate that has a concentration of 0.02 *M*. (*M* stands for molar concentration; see Section 2.4). The solution is maintained under constant environmental conditions (25°C and pH 7). As the reaction proceeds to equilibrium, the concentration of the product, glucose 6-phosphate, rises from 0 to 0.019 *M*, while the concentration of the reactant, glucose 1-phosphate, falls to 0.001 *M*. At this point, equilibrium is reached (**Figure 8.4**). At equilibrium, the reverse reaction, from glucose 6-phosphate to glucose 1-phosphate, progresses at the same rate as the forward reaction.

At equilibrium, then, this reaction has a product-to-reactant ratio of 19:1 (0.019/0.001), so the forward reaction has gone 95 percent of the way to completion ("to the right," as written above). This result is obtained every time the experiment is run under the same conditions.

The change in free energy ($\Delta G$) for any reaction is related directly to its point of equilibrium. The further toward completion the point of equilibrium lies, the more free energy is released. In an exergonic reaction, such as the conversion of glucose 1-phosphate to glucose 6-phosphate, $\Delta G$ is a negative number (in this example, $\Delta G = -1.7$ kcal/mol, or $-7.1$ kJ/mol).

A large, positive $\Delta G$ for a reaction means that it proceeds hardly at all to the right ($A \rightarrow B$). If the concentration of B is initially high relative to that of A, such a reaction runs "to the left" ($A \leftarrow B$), and nearly all B is converted into A. A $\Delta G$ value near zero is characteristic of a readily reversible reaction: reactants and products have almost the same free energies.

In Chapters 9 and 10 we examine the metabolic reactions that harvest energy from food and light. In turn, this energy is used to synthesize carbohydrates, lipids and proteins. All of the chemical reactions carried out by living organisms are governed by the principles of thermodynamics and equilibrium.

### 8.1 RECAP

**Two laws of thermodynamics govern energy transformations in biological systems. A biochemical reaction can release or consume energy, and it may not run to completion, but instead end up at a point of equilibrium.**

- What is the difference between potential energy and kinetic energy? Between anabolism and catabolism? **See pp. 149–150**
- What are the laws of thermodynamics? How do they relate to biology? **See pp. 150–152 and Figure 8.2**
- What is the difference between endergonic and exergonic reactions and what is the importance of $\Delta G$? **See p. 152 and Figure 8.3**

The principles of thermodynamics that we have been discussing apply to all energy transformations in the universe, so they are very powerful and useful. Next, we'll apply them to reactions in cells that involve the currency of biological energy, ATP.

## 8.2 What Is the Role of ATP in Biochemical Energetics?

Cells rely on adenosine triphosphate (ATP) for the capture and transfer of the free energy they need to do chemical work. ATP operates as a kind of "energy currency." Just as it is more effective, efficient, and convenient for you to trade money for a lunch than to trade your actual labor, it is useful for cells to have a single currency for transferring energy between different reactions and cell processes. So some of the free energy that is released by exergonic reactions is captured in the formation of ATP from adenosine diphosphate (ADP) and inorganic phosphate ($HPO_4^{2-}$, which is commonly abbreviated to $P_i$). The ATP can then be hydrolyzed at other sites in the cell to release free energy to drive endergonic reactions. (In some reactions, guanosine triphosphate [GTP] is used as the energy transfer molecule instead of ATP, but we will focus on ATP here.)

ATP has another important role in the cell beyond its use as an energy currency: it is a nucleotide that can be converted into a building block for nucleic acids (see Chapter 4). The structure of ATP is similar to those of other nucleotides, but two things about ATP make it especially useful to cells. First, ATP releases a relatively large amount of energy when hydrolyzed to ADP and $P_i$. Second, ATP can phosphorylate (donate a phosphate group to) many different molecules, which gain some of the energy that was stored in the ATP. We will examine these two properties in the discussion that follows.

### ATP hydrolysis releases energy

An ATP molecule consists of the nitrogenous base adenine bonded to ribose (a sugar), which is attached to a sequence of three phosphate groups (**Figure 8.5A**). The hydrolysis of a molecule of ATP yields free energy, as well as ADP and the inorganic phosphate ion ($P_i$). Thus:

$$ATP + H_2O \rightarrow ADP + P_i + \text{free energy}$$

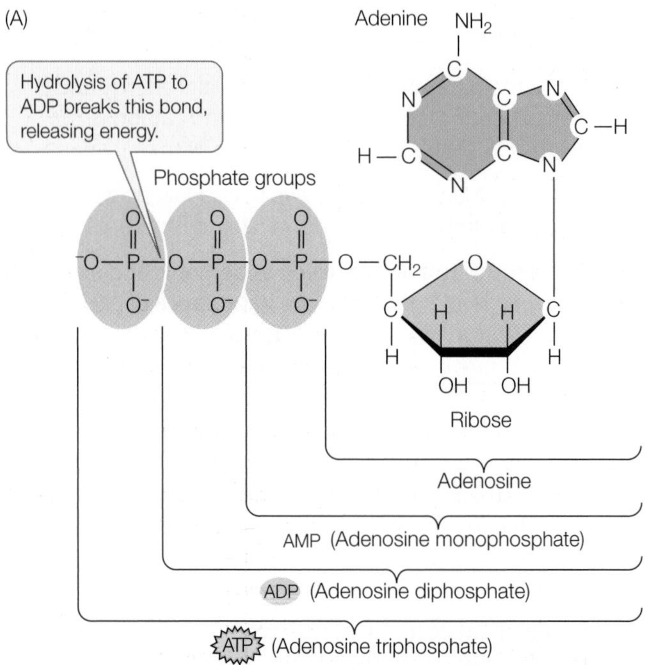


(B) *Luciola cruciata*

**8.5 ATP** (A) ATP is richer in energy than its relatives ADP and AMP. (B) Fireflies use ATP to initiate the oxidation of luciferin. This process converts chemical energy into light energy, emitting rhythmic flashes that signal the insect's readiness to mate.

The important property of this reaction is that it is exergonic, releasing free energy. Under standard laboratory conditions, the change in free energy for this reaction ($\Delta G$) is about –7.3 kcal/mol (–30 kJ/mol). However, under cellular conditions, the value can be as much as –14 kcal/mol. We give both values here because you will encounter both values, and you should be aware of their origins. Both are correct, but in different conditions.

Two characteristics of ATP account for the free energy released by the loss of one or two of its phosphate groups:

- The free energy of the P—O bond between phosphate groups (called a phosphoric acid anhydride bond) is much higher than the energy of the O—H bond that forms after hydrolysis. So some usable energy is released by hydrolysis.

$$\text{Adenine—ribose—O—}\underset{\underset{O}{\|}}{\overset{\overset{O^-}{|}}{P}}\text{—O}\sim\underset{\underset{O}{\|}}{\overset{\overset{O^-}{|}}{P}}\text{—O}\sim\underset{\underset{O}{\|}}{\overset{\overset{O^-}{|}}{P}}\text{—O}^- + H_2O \quad (\textbf{ATP})$$

$$\downarrow$$

$$\text{Adenine—ribose—O—}\underset{\underset{O}{\|}}{\overset{\overset{O^-}{|}}{P}}\text{—O}\sim\underset{\underset{O}{\|}}{\overset{\overset{O^-}{|}}{P}}\text{—OH} \; (\textbf{ADP}) + \text{HO—}\underset{\underset{O}{\|}}{\overset{\overset{O^-}{|}}{P}}\text{—O}^- \; (\mathbf{P_i})$$

- Because phosphate groups are negatively charged and so repel each other, it takes energy to get phosphates near enough to each other to make the covalent bond that links them together (e.g., to add a phosphate to ADP to make ATP). Some of this energy is conserved when the third phosphate is attached.

A molecule of ATP can be hydrolyzed either to ADP and $P_i$, or to adenosine monophosphate (AMP) and a pyrophosphate ion ($P_2O_7^{4-}$; commonly abbreviated to $PP_i$). Cells use the energy released by ATP hydrolysis to fuel endergonic reactions (such as the biosynthesis of complex molecules), for active transport, and for movement. Another interesting example of the use of ATP involves converting its chemical energy into light energy.

**BIOLUMINESCENCE** The production of light by living organisms is referred to as **bioluminesence** (**Figure 8.5B**). It is an example of an endergonic reaction driven by ATP hydrolysis that involves an interconversion of energy forms (chemical to light). The chemical that becomes luminescent is called luciferin (after the light-bearing fallen angel, Lucifer):

$$\text{luciferin} + O_2 + \text{ATP} \xrightarrow{\text{luciferase}} \text{oxyluciferin} + \text{AMP} + PP_i + \text{light}$$

This reaction and the enzyme that catalyzes it (luciferase) occur in a wide variety of organisms in addition to the familiar firefly. These include a variety of marine organisms, microorganisms, worms, and mushrooms. The light is generally used to avoid predators or for signaling to mates.

Soft-drink companies use the firefly proteins luciferin and luciferase to detect bacterial contamination. Where there are

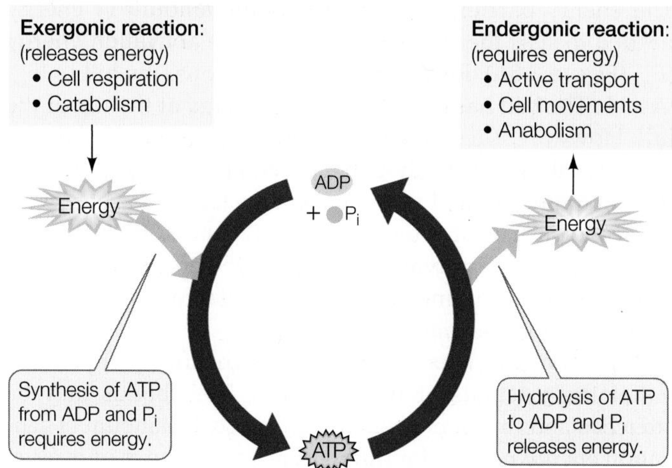


**8.6 Coupling of Reactions** Exergonic cellular reactions release the energy needed to make ATP from ADP. The energy released from the conversion of ATP back to ADP can be used to fuel endergonic reactions.

**yourBioPortal.com**

GO TO **Web Activity 8.1 • ATP and Coupled Reactions**

living cells there is ATP, and when the firefly proteins encounter ATP and oxygen, they give off light. Thus, a sample of soda that lights up in the test is contaminated with bacteria and is discarded.

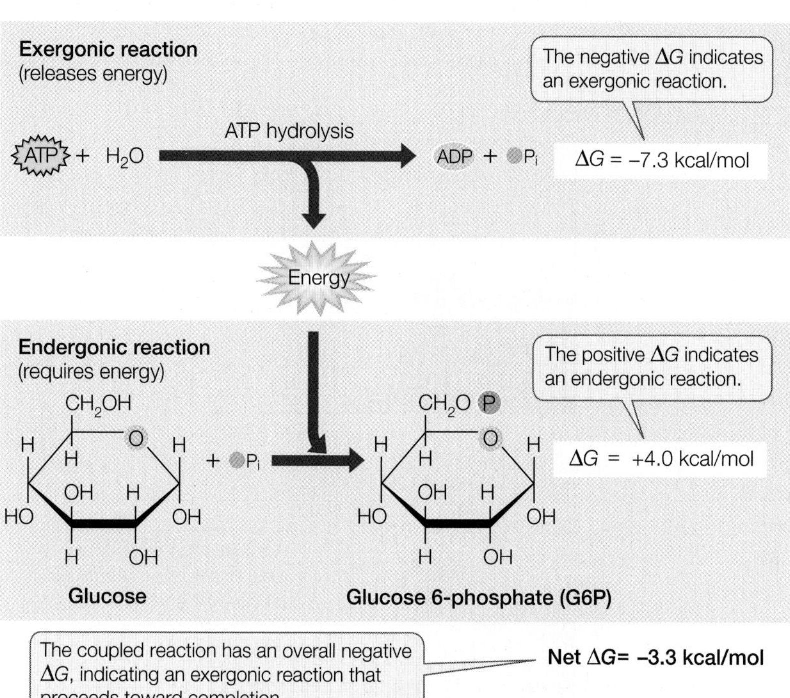


**8.7 Coupling of ATP Hydrolysis to an Endergonic Reaction** The addition of phosphate derived from the hydrolysis of ATP to glucose forms the molecule glucose 6-phosphate (in a reaction catalyzed by hexokinase). ATP hydrolysis is exergonic and the energy released drives the second reaction, which is endergonic.

## ATP couples exergonic and endergonic reactions

As we have just seen, the hydrolysis of ATP is exergonic and yields ADP, $P_i$, and free energy. The reverse reaction, the formation of ATP from ADP and $P_i$, is endergonic and consumes as much free energy as is released by the hydrolysis of ATP:

$$ADP + P_i + \text{free energy} \rightarrow ATP + H_2O$$

Many different exergonic reactions in the cell can provide the energy to convert ADP into ATP. For eukaryotes and many prokaryotes, the most important of these reactions is cellular respiration, in which some of the energy released from fuel molecules is captured in ATP. The formation and hydrolysis of ATP constitute what might be called an "energy-coupling cycle," in which ADP picks up energy from exergonic reactions to become ATP, which then donates energy to endergonic reactions. ATP is the common component of these reactions and is the agent of coupling, as illustrated in **Figure 8.6**.

Coupling of exergonic and endergonic reactions is very common in metabolism. Free energy is captured and retained in the P—O bonds of ATP. ATP then diffuses to another site in the cell, where its hydrolysis releases the free energy to drive an endergonic reaction. For example, the formation of glucose 6-phosphate from glucose (**Figure 8.7**), which has a positive $\Delta G$ (is endergonic), will not proceed without the input of free energy from ATP hydrolysis, which has a negative $\Delta G$ (is exergonic). The overall $\Delta G$ for the coupled reactions (when the two $\Delta G$s are added together) is negative. Hence the reactions proceed exergonically when they are coupled, and glucose 6-phosphate is synthesized. As you will see in Chapter 9, this is the initial reaction in the catabolism of glucose.

An active cell requires the production of millions of molecules of ATP per second to drive its biochemical machinery. An ATP molecule is typically consumed within a second of its formation. At rest, an average person produces and hydrolyzes about 40 kg of ATP per day—as much as some people weigh. This means that each ATP molecule undergoes about 10,000 cycles of synthesis and hydrolysis every day!

## 8.2 RECAP

**ATP is the "energy currency" of cells. Some of the free energy released by exergonic reactions can be captured in the form of ATP. This energy can then be released by ATP hydrolysis and used to drive endergonic reactions.**

- How does ATP store energy? See p. 153
- What are coupled reactions? See p. 155 and Figure 8.7

ATP is synthesized and used up very rapidly. But these biochemical reactions could not proceed so rapidly without the help of *enzymes*.

# 8.3 What Are Enzymes?

When we know the change in free energy ($\Delta G$) of a reaction, we know where the equilibrium point of the reaction lies: the more negative $\Delta G$ is, the further the reaction proceeds toward completion. However, $\Delta G$ tells us nothing about the *rate* of a reaction—the speed at which it moves toward equilibrium. The reactions that cells depend on have spontaneous rates that are so slow that the cells would not survive without a way to speed up the reactions. That is the role of catalysts: substances that speed up reactions without themselves being permanently altered. A catalyst does not cause a reaction to occur that would not proceed without it, *but merely increases the rate of the reaction*, allowing equilibrium to be approached more rapidly. This is an important point: *no catalyst makes a reaction occur that cannot otherwise occur.*

Most biological catalysts are proteins called *enzymes*. Although we will focus here on proteins, some catalysts—perhaps the earliest ones in the origin of life—are RNA molecules called ribozymes (see Section 4.3). A biological catalyst, whether protein or RNA, is a framework or scaffold within which chemical catalysis takes place. This molecular framework binds the reactants and can participate in the reaction itself; however, such participation does not permanently change the enzyme. The catalyst ends up in exactly the same chemical condition after a reaction as before it. Over time, cells have evolved to utilize proteins rather than RNA as catalysts in most biochemical reactions, probably because of the great diversity in the three-dimensional structures of proteins, and because of the variety of chemical functions provided by their functional groups (see Figure 3.1).

In this section we will discuss the energy barrier that controls the rate of a chemical reaction. Then we will focus on the roles of enzymes: how they interact with specific reactants, how they lower the energy barrier, and how they permit reactions to proceed more quickly.

### To speed up a reaction, an energy barrier must be overcome

An exergonic reaction may release a great deal of free energy, but take place very slowly. Such reactions are slow because there is an energy barrier between reactants and products. Think about the propane stove we describe in Section 2.3.

The burning of propane ($C_3H_8 + 5\ O_2 \rightarrow 3\ CO_2 + 4\ H_2O$ + energy) is an exergonic reaction—energy is released in the form of heat and light. Once started, the reaction goes to completion: all of the propane reacts with oxygen to form carbon dioxide and water vapor.

Because burning propane liberates so much energy, you might expect this reaction to proceed rapidly whenever propane is exposed to oxygen. But this does not happen; propane will start burning only if a spark, an input of energy such as a burning match, is provided. A spark is needed because there is an energy barrier between the reactants and the products.

In general, exergonic reactions proceed only after the reactants are pushed over the energy barrier by some added energy. The energy barrier thus represents the amount of energy needed to start the reaction, known as the **activation energy** (**$E_a$**) (**Figure 8.8A**). Recall the ball rolling down the hill in Figure 8.3. The ball has a lot of potential energy at the top of the hill. However, if it is stuck in a small depression, it will not roll down the hill, even though that action is exergonic. To start the ball rolling, a small amount of energy (activation energy) is needed to push it out of the depression (**Figure 8.8B**). In a chemical reaction, the activation energy is the energy needed to change the reactants into unstable molecular forms called transition-state intermediates.

**Transition-state intermediates** have higher free energies than either the reactants or the products. Their bonds may be stretched and therefore unstable. Although the amount of activation energy needed for different reactions varies, it is often small compared with the change in free energy of the reaction. The activation energy put in to start a reaction is recovered during the ensuing "downhill" phase of the reaction, so it is not a part of the net free energy change, $\Delta G$ (see Figure 8.8A).

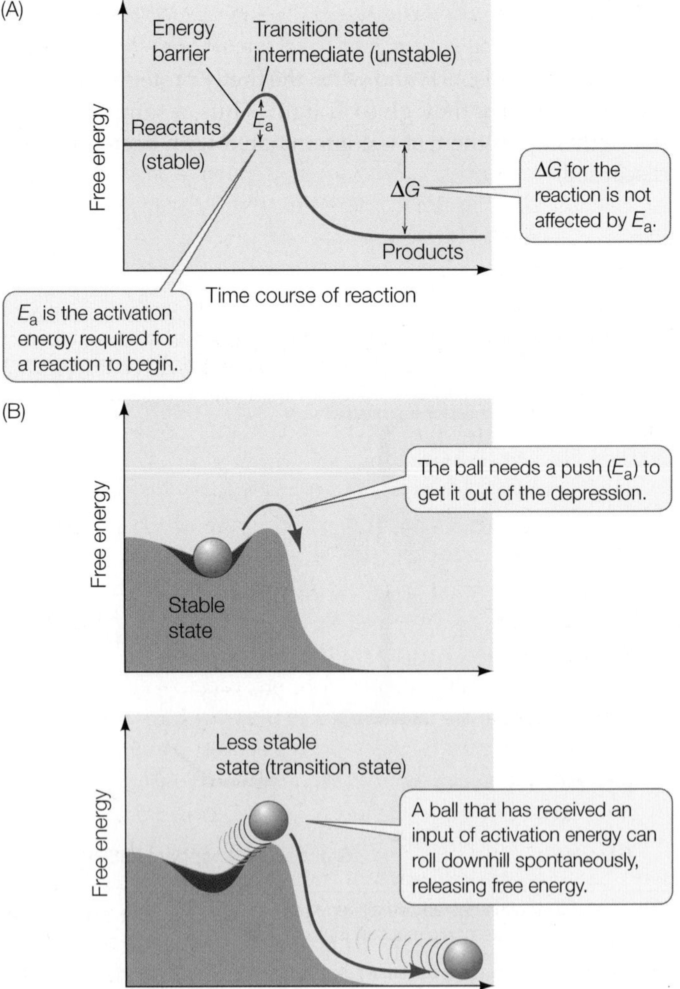


**8.8 Activation Energy Initiates Reactions** (A) In any chemical reaction, an initial stable state must become less stable before change is possible. (B) A ball on a hillside provides a physical analogy to the biochemical principle graphed in (A).

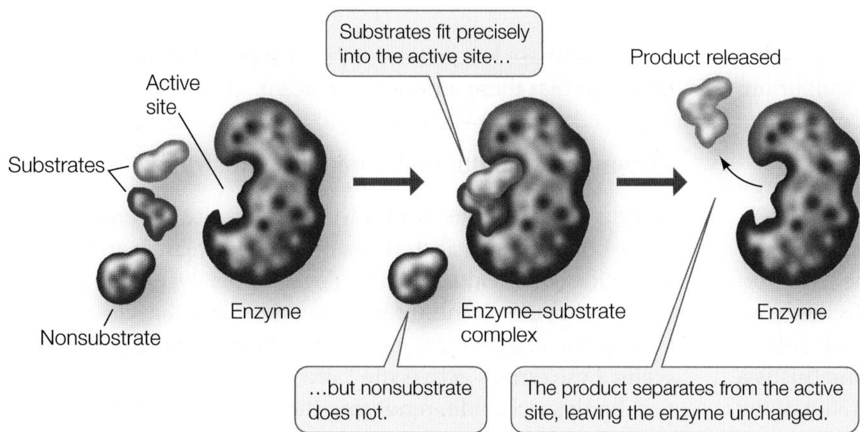
Substrates fit precisely into the active site...
Product released
Active site
Substrates
Nonsubstrate
Enzyme
Enzyme–substrate complex
Enzyme
...but nonsubstrate does not.
The product separates from the active site, leaving the enzyme unchanged.

**8.9 Enzyme and Substrate** An enzyme is a protein catalyst with an active site capable of binding one or more substrate molecules.

Where does the activation energy come from? In any collection of reactants at room or body temperatures, the molecules are moving around. A few are moving fast enough that their kinetic energy can overcome the energy barrier, enter the transition state, and react. However, the reaction takes place very slowly at room or body temperatures. If the system were heated, all the reactant molecules would move faster and have more kinetic energy, and the reaction would speed up. You have probably used this technique in the chemistry laboratory.

However, adding enough heat to increase the average kinetic energy of the molecules would not work in living systems. Such a nonspecific approach would accelerate all reactions, including destructive ones such as the denaturation of proteins (see Figure 3.9). A more effective way to speed up a reaction in a living system is to lower the energy barrier by bringing the reactants close together. In living cells, enzymes and ribozymes accomplish this task.

## Enzymes bind specific reactants at their active sites

Catalysts increase the rates of chemical reactions. Most nonbiological catalysts are nonspecific. For example, powdered platinum catalyzes virtually any reaction in which molecular hydrogen ($H_2$) is a reactant. In contrast, most biological catalysts are highly specific. An enzyme or ribozyme usually recognizes and binds to only one or a few closely related reactants, and it catalyzes only a single chemical reaction. In the discussion that follows, we focus on enzymes, but remember that similar rules of chemical behavior apply to ribozymes as well.

In an enzyme-catalyzed reaction, the reactants are called **substrates**. Substrate molecules bind to a particular site on the enzyme, called the active site, where catalysis takes place (**Figure 8.9**). The specificity of an enzyme results from the exact three-dimensional shape and structure of its active site, into which only a narrow range of substrates can fit. Other molecules—with different shapes, different functional groups, and different properties—cannot fit properly and bind to the active site. This specificity is comparable to the specific binding of a membrane transport protein or receptor protein to its specific ligand, as described in Chapters 6 and 7.

The names of enzymes reflect their functions and often end with the suffix "ase." For example the enzyme lactase, which you encountered in the opening story for this chapter, catalyzes the hydrolysis of lactose but not another disaccharide, sucrose. The enzyme hexokinase accelerates the phosphorylation of glucose, but not ribose, to make glucose 6-phosphate (see Figure 8.7).

The binding of a substrate to the active site of an enzyme produces an **enzyme–substrate complex** (**ES**) that is held together by one or more means, such as hydrogen bonding, electrical attraction, or temporary covalent bonding. The enzyme–substrate complex gives rise to product and free enzyme:

$$E + S \rightarrow ES \rightarrow E + P$$

where E is the enzyme, S is the substrate, P is the product, and ES is the enzyme–substrate complex. The free enzyme (E) is in the same chemical form at the end of the reaction as at the beginning. While bound to the substrate, it may change chemically, but by the end of the reaction it has been restored to its initial form and is ready to bind more substrate.

## Enzymes lower the energy barrier but do not affect equilibrium

When reactants are bound to the enzyme, forming an enzyme–substrate complex, they require less activation energy than the transition-state species of the corresponding uncatalyzed reaction (**Figure 8.10**). Thus the enzyme lowers the energy barrier for the reaction—it offers the reaction an easier path, speeding it up. When an enzyme lowers the energy bar-

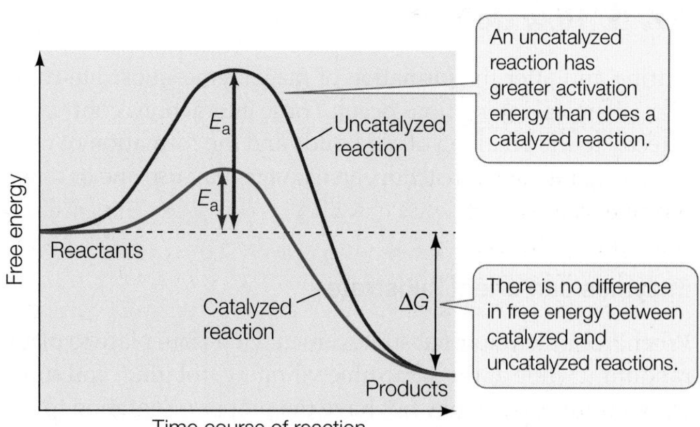
An uncatalyzed reaction has greater activation energy than does a catalyzed reaction.
$E_a$
Uncatalyzed reaction
$E_a$
Free energy
Reactants
Catalyzed reaction
$\Delta G$
There is no difference in free energy between catalyzed and uncatalyzed reactions.
Products
Time course of reaction

**8.10 Enzymes Lower the Energy Barrier** Although the activation energy is lower in an enzyme-catalyzed reaction than in an uncatalyzed reaction, the energy released is the same with or without catalysis. In other words, $E_a$ is lower, but $\Delta G$ is unchanged. A lower activation energy means the reaction will take place at a faster rate.

**yourBioPortal.com**

GO TO Web Activity 8.2 • Free Energy Changes

rier, both the forward and the reverse reactions speed up, so the enzyme-catalyzed overall reaction proceeds toward equilibrium more rapidly than the uncatalyzed reaction. *The final equilibrium is the same with or without the enzyme.* Similarly, adding an enzyme to a reaction does not change the difference in free energy ($\Delta G$) between the reactants and the products (see Figure 8.10).

Enzymes can change the rate of a reaction substantially. For example, if 600 molecules of a protein with arginine as its terminal amino acid just sit in solution, the protein molecules tend toward disorder and the terminal peptide bonds break, releasing the arginines ($\Delta S$ increases). Without an enzyme this is a very slow reaction—it takes about 7 years for half (300) of the proteins to undergo this reaction. However, with the enzyme carboxypeptidase A catalyzing the reaction, the 300 arginines are released in half a second! The important consequence of this for living cells is not difficult to imagine. Such speeds make new realities possible.

### 8.3 RECAP

**A chemical reaction requires a "push" over the energy barrier to get started. Enzymes provide this activation energy by binding specific reactants (substrates).**

- Explain how the structure of an enzyme makes that enzyme specific. **See p. 157 and Figure 8.9**
- What is the relationship between an enzyme and the equilibrium point of a reaction? **See pp. 157–158**

Now that you have a general understanding of the structures, functions, and specificities of enzymes, let's see how they work to speed up chemical reactions between the substrate molecules.

## 8.4 How Do Enzymes Work?

During and after the formation of the enzyme–substrate complex, chemical interactions occur. These interactions contribute directly to the breaking of old bonds and the formation of new ones. In catalyzing a reaction, an enzyme may use one or more mechanisms.

### Enzymes can orient substrates

When free in solution, substrates are moving from place to place randomly while at the same time vibrating, rotating, and tumbling around. They may not have the proper orientation to interact when they collide. Part of the activation energy needed to start a reaction is used to bring together specific atoms so that bonds can form (**Figure 8.11A**). For example, if acetyl coenzyme A (acetyl CoA) and oxaloacetate are to form citrate (a step in the metabolism of glucose; see Section 9.2), the two substrates must be oriented so that the carbon atom of the methyl group of acetyl CoA can form a covalent bond with the carbon atom of the carbonyl group of oxaloacetate. The active site of the enzyme citrate synthase has just the right shape to bind these two molecules so that these atoms are adjacent.

### Enzymes can induce strain in the substrate

Once a substrate has bound to its active site, an enzyme can cause bonds in the substrate to stretch, putting it in an unstable transition state (**Figure 8.11B**). For example, lysozyme is a protective enzyme abundant in tears and saliva that destroys invading bacteria by cleaving polysaccharide chains in their cell walls. Lysozyme's active site "stretches" the bonds of the bacterial polysaccharide, rendering the bonds unstable and more reactive to lysozyme's other substrate, water.

### Enzymes can temporarily add chemical groups to substrates

The side chains (R groups) of an enzyme's amino acids may be direct participants in making its substrates more chemically reactive (**Figure 8.11C**).

- In *acid–base catalysis*, the acidic or basic side chains of the amino acids in the active site transfer $H^+$ to or from the substrate, destabilizing a covalent bond in the substrate, and permitting it to break.
- In *covalent catalysis*, a functional group in a side chain forms a temporary covalent bond with a portion of the substrate.
- In *metal ion catalysis,* metal ions such as copper, iron, and manganese, which are often firmly bound to side chains of enzymes, can lose or gain electrons without detaching from the enzymes. This ability makes them important participants in oxidation–reduction reactions, which involve the loss or gain of electrons.

### Molecular structure determines enzyme function

Most enzymes are much larger than their substrates. An enzyme is typically a protein containing hundreds of amino acids and may consist of a single folded polypeptide chain or of several subunits (see Section 3.2). Its substrate is generally a small molecule or a small part of a large molecule. The active site of the enzyme is usually quite small, not more than 6–12 amino acids. Two questions arise from these observations:

- What features of the active site allow it to recognize and bind the substrate?
- What is the role of the rest of the huge protein?

**THE ACTIVE SITE IS SPECIFIC TO THE SUBSTRATE** The remarkable ability of an enzyme to select exactly the right substrate depends on a precise interlocking of molecular shapes and interactions of chemical groups at the active site. The binding of the substrate to the active site depends on the same kinds of forces that maintain the tertiary structure of the enzyme: hydrogen bonds, the attraction and repulsion of electrically charged groups, and hydrophobic interactions.

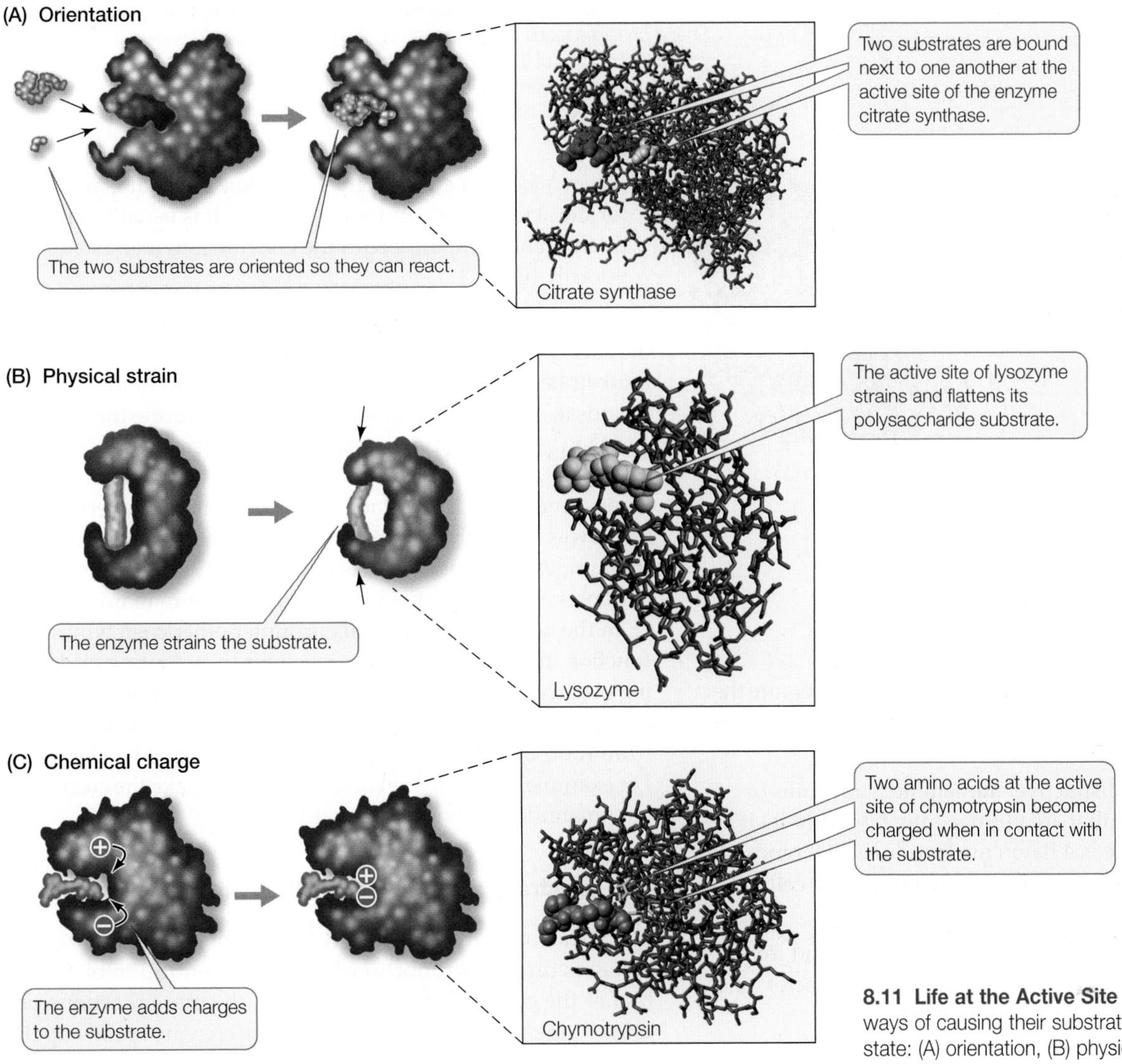

(A) Orientation

The two substrates are oriented so they can react.

Citrate synthase

Two substrates are bound next to one another at the active site of the enzyme citrate synthase.

(B) Physical strain

The enzyme strains the substrate.

Lysozyme

The active site of lysozyme strains and flattens its polysaccharide substrate.

(C) Chemical charge

The enzyme adds charges to the substrate.

Chymotrypsin

Two amino acids at the active site of chymotrypsin become charged when in contact with the substrate.

**8.11 Life at the Active Site** Enzymes have several ways of causing their substrates to enter the transition state: (A) orientation, (B) physical strain, and (C) chemical charge.

In 1894, the German chemist Emil Fischer compared the fit between an enzyme and its substrate to that of a lock and key. Fischer's model persisted for more than half a century with only indirect evidence to support it. The first direct evidence came in 1965, when David Phillips and his colleagues at the Royal Institution in London crystallized the enzyme lysozyme and determined its tertiary structure using the technique of X-ray crystallography (described in Section 13.2). They observed a pocket in lysozyme that neatly fits its substrate (see Figure 8.11B).

**AN ENZYME CHANGES SHAPE WHEN IT BINDS A SUBSTRATE** Just as a membrane receptor protein may undergo precise changes in conformation upon binding to its ligand (see Chapter 7), some enzymes change their shapes when they bind their substrate(s). These shape changes, which are called **induced fit**, expose the active site (or sites) of the enzyme.

An example of induced fit can be seen in the enzyme hexokinase (see Figure 8.7), which catalyzes the reaction

$$\text{glucose} + \text{ATP} \rightarrow \text{glucose 6-phosphate} + \text{ADP}$$

Induced fit brings reactive side chains from the hexokinase active site into alignment with the substrates (**Figure 8.12**), facilitating its catalytic mechanisms. Equally important, the folding of hexokinase to fit around the substrates (glucose and ATP) excludes water from the active site. This is essential, because if water were present, the ATP could be hydrolyzed to ADP and $P_i$. But since water is absent, the transfer of a phosphate from ATP to glucose is favored.

Induced fit at least partly explains why enzymes are so large. The rest of the macromolecule may have three roles:

- It provides a framework so that the amino acids of the active site are properly positioned in relation to the substrate(s).
- It participates in significant changes in protein shape and structure that result in induced fit.
- It provides binding sites for regulatory molecules (see Section 8.5).

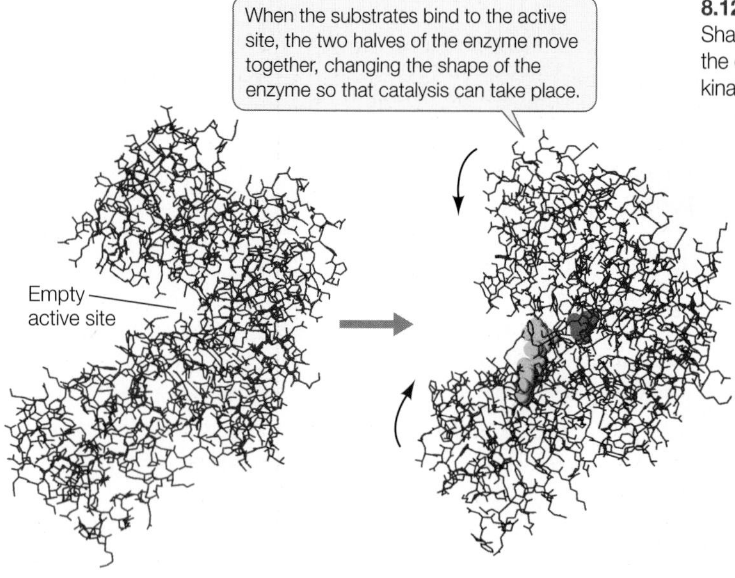

When the substrates bind to the active site, the two halves of the enzyme move together, changing the shape of the enzyme so that catalysis can take place.
Empty active site

**8.12 Some Enzymes Change Shape When Substrate Binds to Them** Shape changes result in an induced fit between enzyme and substrate, improving the catalytic ability of the enzyme. Induced fit can be observed in the enzyme hexokinase, seen here with and without its substrates, glucose (red) and ATP (yellow).

### Some enzymes require other molecules in order to function

As large and complex as enzymes are, many of them require the presence of nonprotein chemical "partners" in order to function (**Table 8.1**):

- *Prosthetic groups* are distinctive, non-amino acid atoms or molecular groupings that are permanently bound to their enzymes. An example is a flavin nucleotide, which binds to succinate dehydrogenase, an important enzyme in cellular respiration (see Section 9.2).
- *Cofactors* are inorganic ions such as copper, zinc, and iron that bind to certain enzymes. For example, the cofactor zinc binds to the enzyme alcohol dehydrogenase.

TABLE 8.1

**Some Examples of Nonprotein "Partners" of Enzymes**

| TYPE OF MOLECULE | ROLE IN CATALYZED REACTIONS |
|---|---|
| **COFACTORS** | |
| Iron ($Fe^{2+}$ or $Fe^{3+}$) | Oxidation/reduction |
| Copper ($Cu^{+}$ or $Cu^{2+}$) | Oxidation/reduction |
| Zinc ($Zn^{2+}$) | Helps bind NAD |
| **COENZYMES** | |
| Biotin | Carries $-COO^{-}$ |
| Coenzyme A | Carries $-CO-CH_3$ |
| NAD | Carries electrons |
| FAD | Carries electrons |
| ATP | Provides/extracts energy |
| **PROSTHETIC GROUPS** | |
| Heme | Binds ions, $O_2$, and electrons; contains iron cofactor |
| Flavin | Binds electrons |
| Retinal | Converts light energy |

- A *coenzyme* is a carbon-containing molecule that is required for the action of one or more enzymes. It is usually relatively small compared with the enzyme to which it temporarily binds.

A coenzyme moves from enzyme to enzyme, adding or removing chemical groups from the substrate. A coenzyme is like a substrate in that it does not permanently bind to the enzyme; it binds to the active site, changes chemically during the reaction, and then separates from the enzyme to participate in other reactions. ATP and ADP, as energy carriers, can be considered coenzymes, even though they are really substrates. The term coenzyme was coined before the functions of these molecules were fully understood. Biochemists continue to use the term, and to be consistent with the field, we will use the term in this book.

In the next chapter we will encounter other coenzymes that function in energy-harvesting reactions by accepting or donating electrons or hydrogen atoms. In animals, some coenzymes are produced from vitamins—substances that must be obtained from food because they cannot be synthesized by the body. For example, the B vitamin niacin is used to make the coenzyme nicotinamide adenine dinucleotide (NAD).

### The substrate concentration affects the reaction rate

For a reaction of the type A → B, the rate of the uncatalyzed reaction is directly proportional to the concentration of A. The higher the concentration of substrate, the more reactions per unit of time. Addition of the appropriate enzyme speeds up the reaction, of course, but it also changes the shape of a plot of rate versus substrate concentration (**Figure 8.13**). For a given concentration of enzyme, the rate of the enzyme-catalyzed reaction initially increases as the substrate concentration increases from zero, but then it levels off. At some point, further increases in the substrate concentration do not significantly increase the reaction rate—the maximum rate has been reached.

Since the concentration of an enzyme is usually much lower than that of its substrate and does not change as substrate concentration changes, what we see is a saturation phenomenon like the one that occurs in facilitated diffusion (see Figure 6.14). When all the enzyme molecules are bound to substrate molecules, the enzyme is working as fast as it can—at its maximum rate. Nothing is gained by adding more substrate, because no free enzyme molecules are left to act as catalysts. Under these conditions the active sites are said to be saturated.

The maximum rate of a catalyzed reaction can be used to measure how efficient the enzyme is—that is, how many molecules of substrate are converted into product per unit of time when there is an excess of substrate present. This *turnover number* ranges from one molecule every two seconds for lysozyme to an amazing 40 million molecules per second for the liver enzyme catalase.

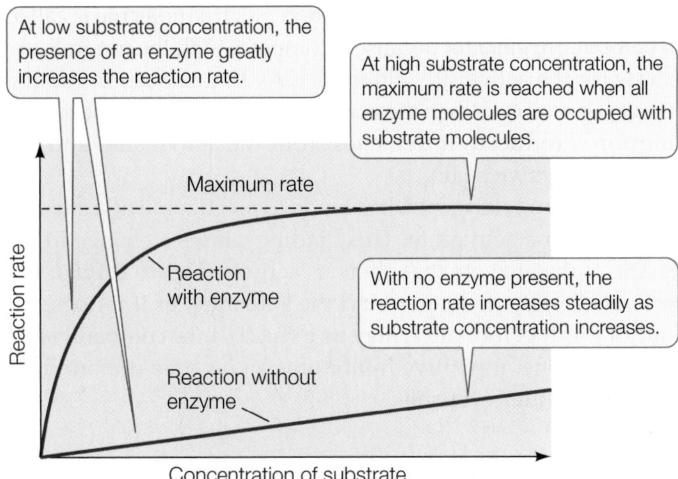

At low substrate concentration, the presence of an enzyme greatly increases the reaction rate.
At high substrate concentration, the maximum rate is reached when all enzyme molecules are occupied with substrate molecules.
Maximum rate
Reaction with enzyme
With no enzyme present, the reaction rate increases steadily as substrate concentration increases.
Reaction without enzyme
Reaction rate
Concentration of substrate

**8.13 Catalyzed Reactions Reach a Maximum Rate** Because there is usually less enzyme than substrate present, the reaction rate levels off when the enzyme becomes saturated.

## 8.4 RECAP

**Enzymes orient their substrates to bring together specific atoms so that bonds can form. An enzyme can participate in the reaction it catalyzes by temporarily changing shape or destabilizing the enzyme-substrate complex. Some enzymes require cofactors, coenzymes, or prosthetic groups in order to function.**

- What are three mechanisms of enzyme catalysis? See p. 158 and Figure 8.11
- What are the chemical roles of coenzymes in enzymatic reactions? See p. 160

Now that you understand more about how enzymes function, let's see how different enzymes work together in a complex organism.

# 8.5 How Are Enzyme Activities Regulated?

A major characteristic of life is homeostasis—the maintenance of stable internal conditions (see Chapter 40). How does a cell maintain a relatively constant internal environment while thousands of chemical reactions are going on? These chemical reactions operate within *metabolic pathways* in which the product of one reaction is a reactant for the next. The pathway for the metabolism of lactose begins with lactase (as we described in the chapter's opening story), and is just one of many pathways that regulate the internal environment of the cell. These pathways have such diverse functions as the catabolism of glucose to yield energy, $CO_2$, and $H_2O$, and the anabolism of amino acids to yield proteins. Metabolic pathways do not exist in isolation, but interact extensively, and each reaction in each pathway is catalyzed by a specific enzyme.

Within a cell or organism, the presence and activity of enzymes determine the "flow" of chemicals through different metabolic pathways. The amount of enzyme activity, in turn, is controlled in part via the regulation of gene expression. Many signal transduction pathways (described in Chapter 7) end with changes in gene expression, and often the genes that are switched on or off encode enzymes. But the simple presence of an enzyme does not ensure that it is functioning. Another means by which cells can control which pathways are active at a particular time is by the activation or inactivation of enzymes. If one enzyme in the pathway is inactive, that step and all subsequent steps shut down. Thus, enzymes are target points for the regulation of entire sequences of chemical reactions.

Regulation of the rates at which thousands of different enzymes operate contributes to homeostasis within an organism. Such control permits cells to make orderly changes in their functions in response to changes in the external environment. In Chapter 7 we describe a number of enzymes that become activated in signal transduction pathways, illustrating how enzyme activation can dramatically alter cell functions. (For example, see the activation of glycogen phosphorylase in Figure 7.20.)

The flow of chemicals such as carbon atoms through interacting metabolic pathways can be studied, but this process becomes complicated quickly, because each pathway influences the others. Computer algorithms are used to model these pathways and show how they mesh in an interdependent system (**Figure 8.14**). Such models can help predict what will happen if the concentration of one molecule or another is altered. This new field of biology is called **systems biology**, and it has numerous applications.

In this section we will investigate the roles of enzymes in organizing and regulating metabolic pathways. In living cells, enzymes can be activated or inhibited in various ways, and there are also mechanisms for controlling the rates at which some enzymes catalyze reactions. We will also examine how the environment—particularly temperature and pH—affects enzyme activity.

### Enzymes can be regulated by inhibitors

Various chemical inhibitors can bind to enzymes, slowing down the rates of the reactions they catalyze. Some inhibitors occur naturally in cells; others are artificial. Naturally occurring inhibitors regulate metabolism; artificial ones can be used to treat disease, to kill pests, or to study how enzymes work. In some cases the inhibitor binds the enzyme irreversibly, and the enzyme becomes permanently inactivated. In other cases the inhibitor has reversible effects; it can separate from the enzyme, allowing the enzyme to function fully as before. The removal of a natural reversible inhibitor increases an enzyme's rate of catalysis.

**IRREVERSIBLE INHIBITION** If an inhibitor covalently binds to certain side chains at the active site of an enzyme, it will permanently inactivate the enzyme by destroying its capacity to interact with its normal substrate. An example of an irreversible inhibitor is DIPF (diisopropyl phosphorofluoridate), which

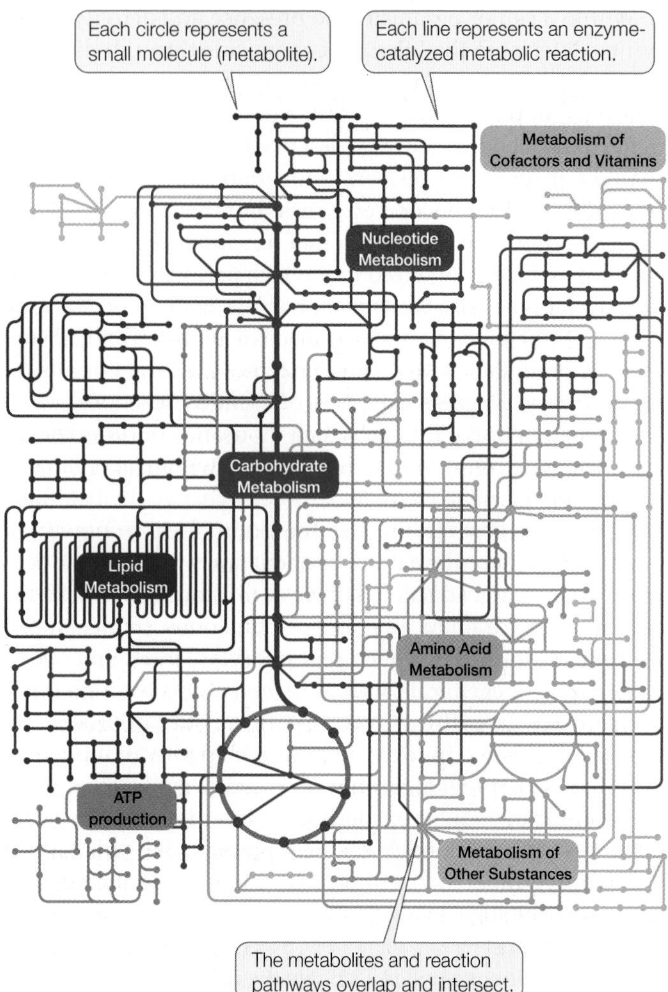


**8.14 Metabolic Pathways** The complex interactions of metabolic pathways can be modeled by the tools of systems biology. In cells, the main elements controlling these pathways are enzymes.

reacts with serine (**Figure 8.15**). DIPF is an irreversible inhibitor of acetylcholinesterase, whose operation is essential for the normal functioning of the nervous system. Because of their effect on acetylcholinesterase, DIPF and other similar compounds are classified as nerve gases, and were developed for biological warfare. One of these compounds, Sarin, was used in an attack on the Tokyo subway in 1995, resulting in a dozen deaths and the hospitalization of hundreds more. The widely used insecticide malathion is a derivative of DIPF that inhibits only insect acetylcholinesterase, not the mammalian enzyme. The irreversible inhibition of enzymes is of practical use to humans, but this form of regulation is not common in the cell, because the enzyme is permanently inactivated and cannot be recycled. Instead, cells use reversible inhibition.

**REVERSIBLE INHIBITION** In some cases an inhibitor is similar enough to a particular enzyme's natural substrate to bind noncovalently to its active site, yet different enough that the enzyme catalyzes no chemical reaction. While such a molecule is bound to the enzyme, the natural substrate cannot enter the active site and the enzyme is unable to function. Such a molecule is called a **competitive inhibitor** because it competes with the natural substrate for the active site (**Figure 8.16A**). In this case, the inhibition is reversible. When the concentration of the competitive inhibitor is reduced, it detaches from the active site, and the enzyme is active again.

A **noncompetitive inhibitor** binds to an enzyme at a site distinct from the active site. This binding causes a change in the shape of the enzyme that alters its activity (**Figure 8.16B**). The active site may no longer bind the substrate, or if it does, the rate of product formation may be reduced. Like competitive inhibitors, noncompetitive inhibitors can become unbound, so their effects are reversible.

### Allosteric enzymes control their activity by changing shape

The change in enzyme shape due to noncompetitive inhibitor binding is an example of allostery (*allo*, "different"; *stereos*, "shape"). **Allosteric regulation** occurs when an effector molecule binds to a site other than the active site of an enzyme, *inducing the enzyme to change its shape*. The change in shape alters the affinity of the active site for the substrate, and so the rate of the reaction is changed.

Often, an enzyme will exist in the cell in more than one possible shape (**Figure 8.17**):

- The *active form* of the enzyme has the proper shape for substrate binding.
- The *inactive form* of the enzyme has a shape that cannot bind the substrate.

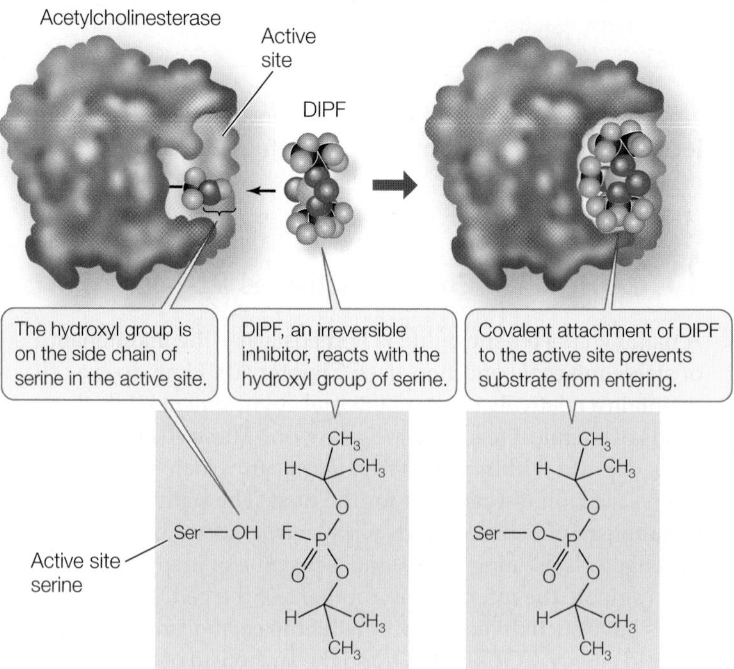

Acetylcholinesterase
Active site
DIPF
The hydroxyl group is on the side chain of serine in the active site.
DIPF, an irreversible inhibitor, reacts with the hydroxyl group of serine.
Covalent attachment of DIPF to the active site prevents substrate from entering.
Active site serine
Ser — OH
Ser — O
F
P
O
H
$CH_3$


**8.15 Irreversible Inhibition** DIPF forms a stable covalent bond with the side chain of the amino acid serine at the active site of the enzyme acetylcholinesterase, thus irreversibly disabling the enzyme.

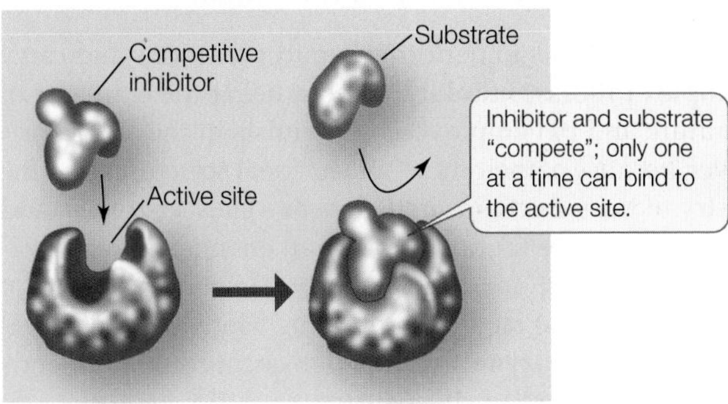
(A) Competitive inhibition
Competitive inhibitor
Substrate
Inhibitor and substrate "compete"; only one at a time can bind to the active site.
Active site

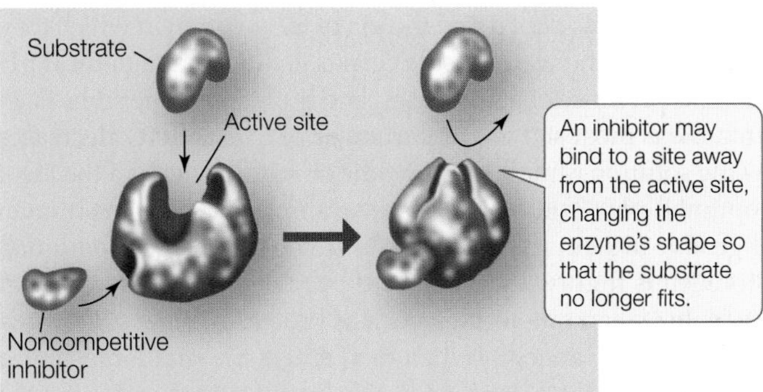
(B) Noncompetitive inhibition
Substrate
Active site
An inhibitor may bind to a site away from the active site, changing the enzyme's shape so that the substrate no longer fits.
Noncompetitive inhibitor

**8.16 Reversible Inhibition** (A) A competitive inhibitor binds temporarily to the active site of an enzyme. (B) A noncompetitive inhibitor binds temporarily to the enzyme at a site away from the active site. In both cases, the enzyme's function is disabled for only as long as the inhibitor remains bound.

**yourBioPortal.com**
GO TO **Animated Tutorial 8.1 • Enzyme Catalysis**

Other molecules, collectively referred to as effectors, can influence which form the enzyme takes:

- Binding of an inhibitor to a site separate from the active site can stabilize the inactive form of the enzyme, making it less likely to convert to the active form.
- The active form can be stabilized by the binding of an activator to another site on the enzyme.

Like substrate binding, the binding of inhibitors and activators to their regulatory sites (also called allosteric sites) is highly specific. Most (but not all) enzymes that are allosterically regulated are proteins with quaternary structure; that is, they are made up of multiple polypeptide subunits. The polypeptide that has the active site is called the catalytic subunit. The allosteric sites are often on different polypeptides, called the regulatory subunits.

Some enzymes have multiple subunits containing active sites, and the binding of substrate to one of the active sites causes allosteric effects. When substrate binds to one subunit, there is a slight change in protein structure that influences the adjacent subunit. The slight change to the second subunit makes its active site more likely to bind to the substrate. So the reaction speeds up as the sites become sequentially activated.

As a result, an allosteric enzyme with multiple active sites and a nonallosteric enzyme with a single active site differ greatly in their reaction rates when the substrate concentration is low. Graphs of reaction rates plotted against substrate concentrations show this relationship. For a nonallosteric enzyme, the plot looks like that in **Figure 8.18A**. The reaction rate first increases sharply with increasing substrate concentration, then tapers off to a constant maximum rate as the supply of enzyme becomes saturated.

The plot for a multisubunit allosteric enzyme is radically different, having a sigmoid (S-shaped) appearance (**Figure 8.18B**). At low substrate concentrations, the reaction rate increases gradually as substrate concentration increases. But within a certain range, the reaction rate is extremely sensitive to relatively small changes in substrate concentration. In addition, allosteric enzymes are very sensitive to low concentrations of inhibitors. Because of this sensitivity, allosteric enzymes are important in regulating entire metabolic pathways.

## Allosteric effects regulate metabolism

Metabolic pathways typically involve a starting material, various intermediate products, and an end product that is used for some purpose by the cell. In each pathway there are a number of reactions, each

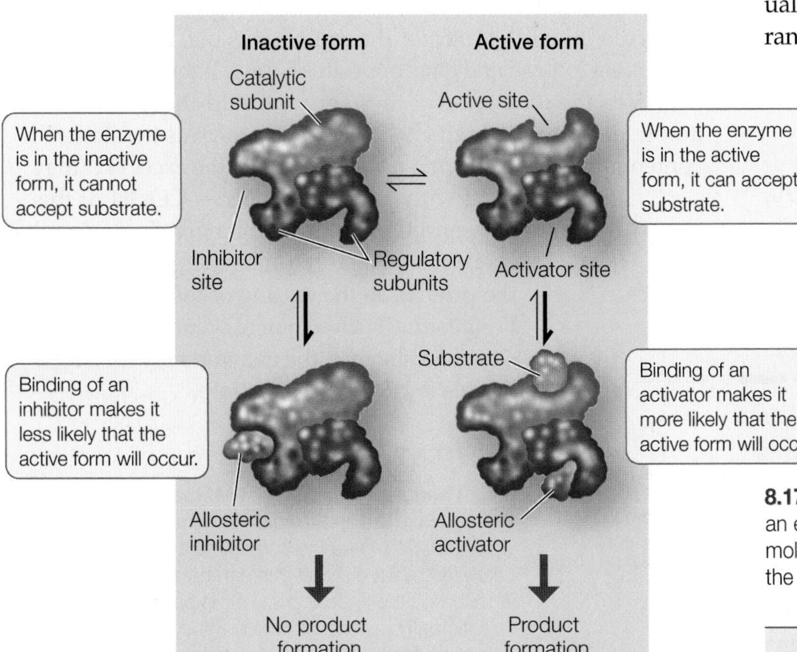
Inactive form
Active form
Catalytic subunit
Active site
When the enzyme is in the inactive form, it cannot accept substrate.
When the enzyme is in the active form, it can accept substrate.
Inhibitor site
Regulatory subunits
Activator site
Substrate
Binding of an inhibitor makes it less likely that the active form will occur.
Binding of an activator makes it more likely that the active form will occur.
Allosteric inhibitor
Allosteric activator
No product formation
Product formation

**8.17 Allosteric Regulation of Enzymes** Active and inactive forms of an enzyme can be interconverted, depending on the binding of effector molecules at sites other than the active site. Binding an inhibitor stabilizes the inactive form and binding an activator stabilizes the active form.

**yourBioPortal.com**
GO TO **Animated Tutorial 8.2 • Allosteric Regulation of Enzymes**

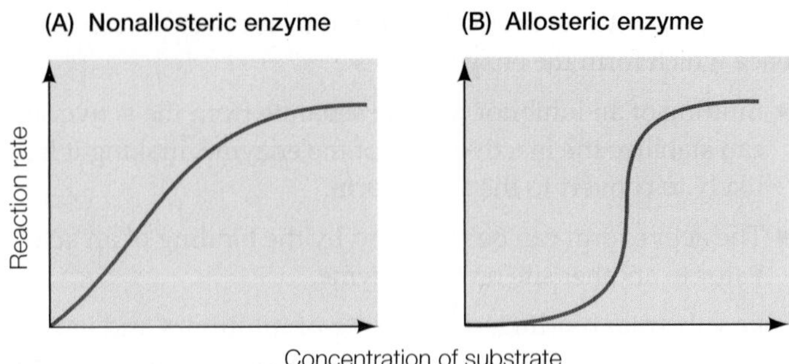


**8.18 Allostery and Reaction Rate** The number of active sites on an enzyme determines how the rate of the enzyme-catalyzed reaction changes as substrate concentration increases. A sigmoid curve (B) is typical for an enzyme with multiple subunits, each with an active site. After one subunit binds the substrate, changes in structure make it more likely that the next subunit will also bind substrate. So the reaction speeds up more rapidly than in the case of an enzyme with a single active site (A).

forming an intermediate product and each catalyzed by a different enzyme. The first step in a pathway is called the commitment step, meaning that once this enzyme-catalyzed reaction occurs, the "ball is rolling," and the other reactions happen in sequence, leading to the end product. But what if the cell has no need for that product—for example, if that product is available from its environment in adequate amounts? It would be energetically wasteful for the cell to continue making something it does not need.

One way to avoid this problem is to shut down the metabolic pathway by having the final product inhibit the enzyme that catalyzes the commitment step (**Figure 8.19**). Often this inhibition occurs allosterically. When the end product is present at a high concentration, some of it binds to an allosteric site on the commitment step enzyme, thereby causing it to become inactive. Thus, the final product acts as a *noncompetitive inhibitor* (described earlier in this section) of the first enzyme in the pathway. This mechanism is known as feedback inhibition or end-product inhibition. We will describe many other examples of such inhibition in later chapters.

## Enzymes are affected by their environment

Enzymes enable cells to perform chemical reactions and carry out complex processes rapidly without using the extremes of temperature and pH employed by chemists in the laboratory. However, because of their three-dimensional structures and the chemistry of the side chains in their active sites, enzymes (and their substrates) are highly sensitive to changes in temperature and pH. In Section 3.2 we describe the general effects of these environmental factors on proteins. Here we will examine their effects on enzyme function (which, of course, depends on enzyme structure and chemistry).

**pH AFFECTS ENZYME ACTIVITY** The rates of most enzyme-catalyzed reactions depend on the pH of the solution in which they occur. While the water inside cells is generally at a neutral pH of 7, the presence of acids, bases, and buffers can alter this. Each enzyme is most active at a particular pH; its activity decreases as the solution is made more acidic or more basic than the ideal (optimal) pH (**Figure 8.20**). As an example, consider the human digestive system (see Section 51.3). The pH inside the human stomach is highly acidic, around pH 1.5. Many enzymes that hydrolyze macromolecules, such as proteases, have pH optima in the neutral range. So when food enters the small intestine, a buffer (bicarbonate) is secreted into the intestine to raise the pH to 6.5. This allows the hydrolytic enzymes to be active and digest the food.

Several factors contribute to this effect. One factor is ionization of the carboxyl, amino, and other groups on either the substrate or the enzyme. In neutral or basic solutions, carboxyl groups (—COOH) release $H^+$ to become negatively charged carboxylate groups (—$COO^-$). On the other hand, in neutral or acidic solutions, amino groups (—$NH_2$) accept $H^+$ to become positively charged —$NH_3^+$ groups (see the discussion of acids and bases in Section 2.4). Thus, in a neutral solution, an amino group is electrically attracted to a carboxyl group on another molecule or another part of the same molecule, because both groups are ionized and have opposite charges. If the pH changes, however, the ionization of these groups may change. For example, at a low pH (high $H^+$ concentration, such as the stomach contents where the enzyme pepsin is active), the excess $H^+$ may react with —$COO^-$ to form —COOH. If this happens, the group is no longer charged and cannot interact with other charged groups in the protein, so the folding of the protein may be altered. If such a change occurs at the active site of an enzyme, the enzyme may no longer be able to bind to its substrate.

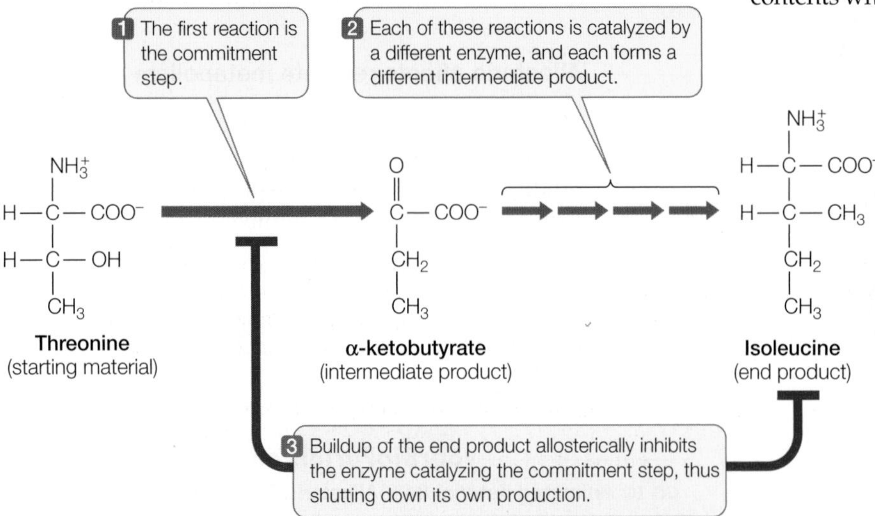

1 The first reaction is the commitment step.
2 Each of these reactions is catalyzed by a different enzyme, and each forms a different intermediate product.
Threonine
(starting material)
α-ketobutyrate
(intermediate product)
Isoleucine
(end product)
3 Buildup of the end product allosterically inhibits the enzyme catalyzing the commitment step, thus shutting down its own production.


**8.19 Feedback Inhibition of Metabolic Pathways** The first reaction in a metabolic pathway is referred to as the commitment step. It is often catalyzed by an enzyme that can be allosterically inhibited by the end product of the pathway. The specific pathway shown here is the synthesis of isoleucine from threonine in bacteria. It is typical of many enzyme-catalyzed biosynthetic pathways.

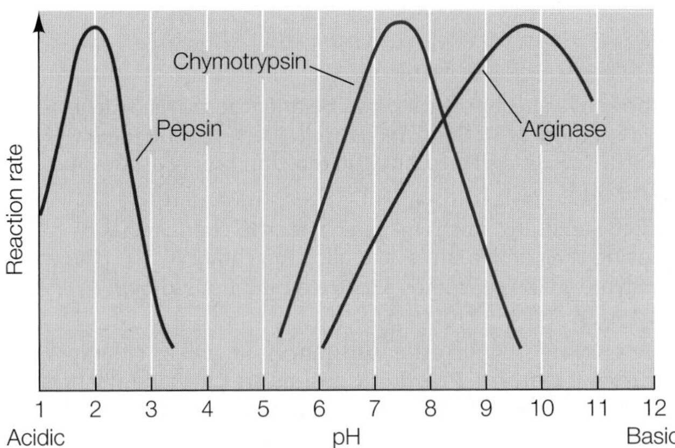


**8.20 pH Affects Enzyme Activity** An enzyme catalyzes its reaction at a maximum rate. The activity curve for each enzyme peaks at its optimal pH. For example, pepsin is active in the acidic environment of the stomach, while chymotrypsin is active in the small intestine.

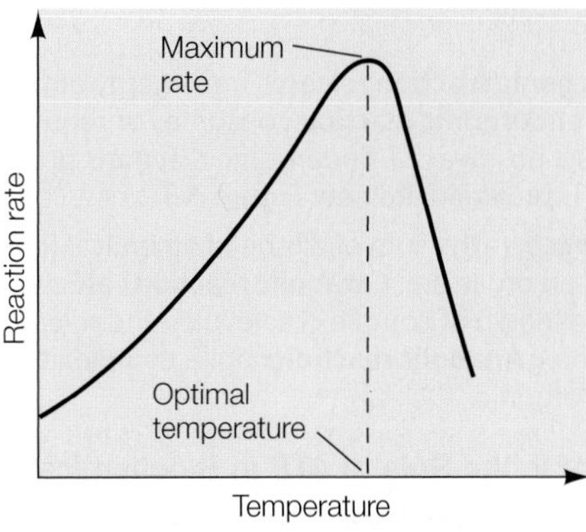


**8.21 Temperature Affects Enzyme Activity** Each enzyme is most active at a particular optimal temperature. At higher temperatures the enzyme becomes denatured and inactive; this explains why the activity curve falls off abruptly at temperatures above the optimal.

**TEMPERATURE AFFECTS ENZYME ACTIVITY** In general, warming increases the rate of a chemical reaction because a greater proportion of the reactant molecules have enough kinetic energy to provide the activation energy for the reaction. Enzyme-catalyzed reactions are no different (**Figure 8.21**). However, temperatures that are too high inactivate enzymes, because at high temperatures enzyme molecules vibrate and twist so rapidly that some of their noncovalent bonds break. When an enzyme's tertiary structure is changed by heat it loses its function. Some enzymes denature at temperatures only slightly above that of the human body, but a few are stable even at the boiling point (or freezing point) of water. All enzymes, however, have an optimal temperature for activity.

Individual organisms adapt to changes in the environment in many ways, one of which is based on groups of enzymes, called *isozymes*, that catalyze the same reaction but have different chemical compositions and physical properties. Different isozymes within a given group may have different optimal temperatures. The rainbow trout, for example, has several isozymes of the enzyme acetylcholinesterase. If a rainbow trout is transferred from warm water to near-freezing water (2°C), the fish produces an isozyme of acetylcholinesterase that is different from the one it produces at the higher temperature. The new isozyme has a lower optimal temperature, allowing the fish's nervous system to perform normally in the colder water.

In general, enzymes adapted to warm temperatures do not denature at those temperatures because their tertiary structures are held together largely by covalent bonds, such as charge interactions or disulfide bridges, instead of the more heat-sensitive weak chemical interactions. Most enzymes in humans are more stable at high temperatures than those of the bacteria that infect us, so that a moderate fever tends to denature bacterial enzymes, but not our own.

## 8.5 RECAP

**The rates of most enzyme-catalyzed reactions are affected by interacting molecules (such as inhibitors and activators) and by environmental factors (such as temperature and pH).**

- What is the difference between reversible and irreversible enzyme inhibition? **See pp. 161–162**
- How are allosteric enzymes regulated? **See pp. 162–163 and Figure 8.17**
- Explain the concept of feedback inhibition. How might the reactions shown in Figure 8.19 fit into a systems diagram such as the one shown in Figure 8.14?

## CHAPTER SUMMARY

### 8.1 What Physical Principles Underlie Biological Energy Transformations?

- Energy is the capacity to do work. In a biological system, the usable energy is called **free energy** (*G*). The unusable energy is **entropy**, a measure of the disorder in the system.
- **Potential energy** is the energy of state or position; it includes the energy stored in chemical bonds. **Kinetic energy** is the energy of motion; it is the type of energy that can do work.
- The **laws of thermodynamics** apply to living organisms. The first law states that energy cannot be created or destroyed. The second law states that energy transformations decrease the amount of energy available to do work (free energy) and increase disorder. **Review Figure 8.2**
- The **change in free energy** ($\Delta G$) of a reaction determines its point of **chemical equilibrium**, at which the forward and reverse reactions proceed at the same rate.

- An **exergonic reaction** releases free energy and has a negative $\Delta G$. An **endergonic reaction** consumes or requires free energy and has a positive $\Delta G$. Endergonic reactions proceed only if free energy is provided. **Review Figure 8.3**
- **Metabolism** is the sum of all the biochemical (metabolic) reactions in an organism. **Catabolic reactions** are associated with the breakdown of complex molecules and release energy (are exergonic). **Anabolic reactions** build complexity in the cell and are endergonic.

## 8.2 What Is the Role of ATP in Biochemical Energetics?

- **Adenosine triphosphate (ATP)** serves as an energy currency in cells. Hydrolysis of ATP releases a relatively large amount of free energy.
- The **ATP cycle** couples exergonic and endergonic reactions, harvesting free energy from exergonic reactions, and providing free energy for endergonic reactions. **Review Figure 8.6, WEB ACTIVITY 8.1**

## 8.3 What Are Enzymes?

- The rate of a chemical reaction is independent of $\Delta G$, but is determined by the **energy barrier**. **Enzymes** are protein catalysts that affect the rates of biological reactions by lowering the energy barrier, supplying the **activation energy** ($E_a$) needed to initiate reactions. **Review Figure 8.10, WEB ACTIVITY 8.2**
- A **substrate** binds to the enzyme's **active site**—the site of catalysis—forming an **enzyme–substrate complex**. Enzymes are highly specific for their substrates.

## 8.4 How Do Enzymes Work?

- At the active site, a substrate can be oriented correctly, chemically modified, or strained. As a result, the substrate readily forms its **transition state**, and the reaction proceeds. **Review Figure 8.11**
- Binding substrate causes many enzymes to change shape, exposing their active site(s) and allowing catalysis. The change in enzyme shape caused by substrate binding is known as **induced fit**. **Review Figure 8.12**
- Some enzymes require other substances, known as **cofactors**, to carry out catalysis. **Prosthetic groups** are permanently bound to enzymes; **coenzymes** are not. A coenzyme can be considered a substrate, as it is changed by the reaction and then released from the enzyme.
- Substrate concentration affects the rate of an enzyme-catalyzed reaction.

## 8.5 How Are Enzyme Activities Regulated?

- Metabolism is organized into pathways in which the product of one reaction is a reactant for the next reaction. Each reaction in the pathway is catalyzed by an enzyme.
- Enzyme activity is subject to regulation. Some inhibitors bind irreversibly to enzymes. Others bind reversibly. **Review Figures 8.15 and 8.16, ANIMATED TUTORIAL 8.1**
- An **allosteric effector** binds to a site other than the active site and stabilizes the active or inactive form of an enzyme. **Review Figure 8.17, ANIMATED TUTORIAL 8.2**
- The end product of a metabolic pathway may inhibit an enzyme that catalyzes the **commitment step** of that pathway. **Review Figure 8.19**
- Enzymes are sensitive to their environments. Both pH and temperature affect enzyme activity. **Review Figures 8.20 and 8.21**

# SELF-QUIZ

1. Coenzymes differ from enzymes in that coenzymes are
   *a.* only active outside the cell.
   *b.* polymers of amino acids.
   *c.* smaller molecules, such as vitamins.
   *d.* specific for one reaction.
   *e.* always carriers of high-energy phosphate.
2. Which statement about thermodynamics is true?
   *a.* Free energy is used up in an exergonic reaction.
   *b.* Free energy cannot be used to do work.
   *c.* The total amount of energy can change after a chemical transformation.
   *d.* Free energy can be kinetic but not potential energy.
   *e.* Entropy has a tendency to increase.
3. In a chemical reaction,
   *a.* the rate depends on the value of $\Delta G$.
   *b.* the rate depends on the activation energy.
   *c.* the entropy change depends on the activation energy.
   *d.* the activation energy depends on the value of $\Delta G$.
   *e.* the change in free energy depends on the activation energy.
4. Which statement about enzymes is *not* true?
   *a.* They usually consist of proteins.
   *b.* They change the rate of the catalyzed reaction.
   *c.* They change the $\Delta G$ of the reaction.
   *d.* They are sensitive to heat.
   *e.* They are sensitive to pH.
5. The active site of an enzyme
   *a.* never changes shape.
   *b.* forms no chemical bonds with substrates.
   *c.* determines, by its structure, the specificity of the enzyme.
   *d.* looks like a lump projecting from the surface of the enzyme.
   *e.* changes the $\Delta G$ of the reaction.
6. The molecule ATP is
   *a.* a component of most proteins.
   *b.* high in energy because of the presence of adenine.
   *c.* required for many energy-transforming biochemical reactions.
   *d.* a catalyst.
   *e.* used in some exergonic reactions to provide energy.
7. In an enzyme-catalyzed reaction,
   *a.* a substrate does not change.
   *b.* the rate decreases as substrate concentration increases.
   *c.* the enzyme can be permanently changed.
   *d.* strain may be added to a substrate.
   *e.* the rate is not affected by substrate concentration.

8. Which statement about enzyme inhibitors is *not* true?
   *a.* A competitive inhibitor binds the active site of the enzyme.
   *b.* An allosteric inhibitor binds a site on the active form of the enzyme.
   *c.* A noncompetitive inhibitor binds a site other than the active site.
   *d.* Noncompetitive inhibition cannot be completely overcome by the addition of more substrate.
   *e.* Competitive inhibition can be completely overcome by the addition of more substrate.
9. Which statement about the feedback inhibition of enzymes is *not* true?
   *a.* It is usually exerted through allosteric effects.
   *b.* It is directed at the enzyme that catalyzes the commitment step in a metabolic pathway.
   *c.* It affects the rate of reaction, not the concentration of enzyme.
   *d.* It acts by permanently modifying the active site.
   *e.* It is an example of reversible inhibition.
10. Which statement about temperature effects is *not* true?
   *a.* Raising the temperature may reduce the activity of an enzyme.
   *b.* Raising the temperature may increase the activity of an enzyme.
   *c.* Raising the temperature may denature an enzyme.
   *d.* Some enzymes are stable at the boiling point of water.
   *e.* All enzymes have the same optimal temperature.

## FOR DISCUSSION

1. What makes it possible for endergonic reactions to proceed in organisms?
2. Consider two proteins: one is an enzyme dissolved in the cytosol of a cell, the other is an ion channel in its plasma membrane. Contrast the structures of the two proteins, indicating at least two important differences.
3. Plot free energy versus the time course of an endergonic reaction, and the same for an exergonic reaction. Include the activation energy on both plots. Label $E_a$ and $\Delta G$ on both graphs.
4. Consider an enzyme that is subject to allosteric regulation. If a competitive inhibitor (not an allosteric inhibitor) is added to a solution containing such an enzyme, the ratio of enzyme molecules in the active form to those in the inactive form increases. Explain this observation.

## ADDITIONAL INVESTIGATION

In humans, hydrogen peroxide ($H_2O_2$) is a dangerous toxin produced as a by-product of several metabolic pathways. The accumulation of $H_2O_2$ is prevented by its conversion to harmless $H_2O$, a reaction catalyzed by the appropriately named enzyme catalase. Air pollutants can inhibit this enzyme and leave individuals susceptible to tissue damage by $H_2O_2$. How would you investigate whether catalase has an allosteric or a nonallosteric mechanism, and whether the pollutants are acting as competitive or noncompetitive inhibitors?

# 9 Pathways that Harvest Chemical Energy

## Of mice and marathons

Like success in your biology course, winning a prestigious marathon comes only after a lot of hard work. Distance runners have more mitochondria in the leg muscles than most of us. The chemical energy stored in the bonds of ATP in those mitochondria is converted into mechanical energy to move the muscles.

There are two types of muscle fibers. Most people have about equal proportions of each type. But in a top marathon racer, 90 percent of the body's muscle is made up of so-called *slow-twitch* fibers. Cells of these fibers have lots of mitochondria and use oxygen to break down fats and carbohydrates, forming ATP. In contrast, the muscles of sprinters are about 80 percent *fast-twitch* fibers, which have fewer mitochondria. Fast-twitch fibers generate short bursts of ATP in the absence of $O_2$, but the ATP is soon used up. Extensive research with athletes has shown that training can improve the efficiency of blood circulation to the muscle fibers, providing more oxygen, and can even change the ratio of fast-twitch to slow-twitch fibers.

Now enter Marathon Mouse. No, this is not a cartoon character or a computer game, but a very real mouse that was genetically programmed by Ron Evans at the Salk Institute to express high levels of the protein PPARδ in its muscles. This protein is a receptor located inside cell nuclei, where it regulates the transcription of genes involved with the breakdown of fat to yield ATP. Evans's mouse was supposed to break down fats better, and thus be leaner—but there was an unexpected bonus. With high levels of PPARδ came an increase in slow-twitch fibers and a decrease in fast-twitch ones. It was as if the mouse had been in marathon training for a long time!

Marathon mice are leaner and meaner than ordinary mice. Leaner, because they are good at burning fat; and meaner in terms of their ability to run long distances. On an exercise wheel, a normal mouse can run for 90 minutes and about a half-mile (900 meters) before it gets tired. PPARδ-enhanced mice can run almost twice as long and twice as far—marks of true distance runners. Could we also manipulate genes to enhance performance (and fat burning) in humans?

The genetic engineering of people, if it is feasible, is probably far in the future. But implanting genetically altered muscle tissue is actually not such a far-fetched idea, and has already raised concerns over improper athletic enhancement. More likely in the near term is the use of an experimental drug called Aicar, which activates the PPARδ

**Marathon Men** It takes a lot of training to run a marathon. One of the results of all that training is that the leg muscles become packed with slow-twitch muscle fibers, containing cells rich in energy-metabolizing mitochondria.

**Marathon Mouse** This mouse can run for much longer than a normal mouse because its energy metabolism has been genetically altered.

protein. When Evans and colleagues gave the drug to normal mice, they achieved the same results as with the genetically modified mice. A test for Aicar in blood and urine has been developed to prevent its use by human athletes to gain a competitive advantage. Of more importance is the drug's potential in the treatment of obesity and diabetes, since the drug stimulates fat breakdown. Obesity is a key part of a disorder called metabolic syndrome, which also includes high blood pressure, heart disease, and diabetes.

The free energy trapped in ATP is the energy you use all the time to fuel both conscious actions, like running a marathon or turning the pages of a book, and your body's automatic actions, such as breathing or contracting your heart muscles.

**IN THIS CHAPTER** we will describe how cells extract usable energy from food, usually in the form of ATP. We describe the general principles of energy transformations in cells, and illustrate these principles by describing the pathways for the catabolism of glucose in the presence and absence of $O_2$. Finally, we describe the relationships between the metabolic pathways that use and produce the four biologically important classes of molecules—carbohydrates, fats, proteins, and nucleic acids.

CHAPTER OUTLINE



## 9.1 How Does Glucose Oxidation Release Chemical Energy?

Energy is stored in the covalent bonds of fuels, and it can be released and transformed. Wood burning in a campfire releases energy as heat and light. In cells, fuel molecules release chemical energy that is used to make ATP, which in turn drives endergonic reactions. ATP is central to the energy transformations of all living organisms. Photosynthetic organisms use energy from sunlight to synthesize their own fuels, as we describe in Chapter 10. In nonphotosynthetic organisms, the most common chemical fuel is the sugar glucose ($C_6H_{12}O_6$). Other molecules, including other carbohydrates, fats, and proteins, can also supply energy. However, to release their energy they must be converted into glucose or intermediate compounds that can enter into the various pathways of glucose metabolism.

In this section we explore how cells obtain energy from glucose by the chemical process of oxidation, which is carried out through a series of metabolic pathways. Five principles govern metabolic pathways:

- A complex chemical transformation occurs in a series of separate reactions that form a metabolic pathway.
- Each reaction is catalyzed by a specific enzyme.
- Most metabolic pathways are similar in all organisms, from bacteria to humans.
- In eukaryotes, many metabolic pathways are compartmentalized, with certain reactions occurring inside specific organelles.
- Each metabolic pathway is regulated by key enzymes that can be inhibited or activated, thereby determining how fast the reactions will go.

### Cells trap free energy while metabolizing glucose

As we saw in Section 2.3, the familiar process of combustion (burning) is very similar to the chemical processes that release energy in cells. If glucose is burned in a flame, it reacts with oxygen gas ($O_2$), forming carbon dioxide and water and releasing energy in the form of heat. The balanced equation for the complete combustion reaction is

$$C_6H_{12}O_6 + 6\ O_2 \rightarrow 6\ CO_2 + 6\ H_2O + \text{free energy} \quad (\Delta G = -686 \text{ Kcal/mol})$$

This is an oxidation-reduction reaction. Glucose ($C_6H_{12}O_6$) becomes completely oxidized and six molecules of $O_2$ are reduced to six molecules of water. The energy that is released can be used to do work. The same equation applies to the overall metabolism of glucose in cells. However, in contrast to combustion, the metabolism of glucose is a multistep pathway—each step is catalyzed by an enzyme, and the process is compartmentalized. Unlike combustion, glucose metabolism is tightly regulated and occurs at temperatures compatible with life.

The glucose metabolism pathway "traps" the energy stored in the covalent bonds of glucose and stores it instead in ATP molecules, via the phosphorylation reaction:

$$\text{ADP} + \text{P}_\text{i} + \text{free energy} \rightarrow \text{ATP}$$

As we introduce in Chapter 8, ATP is the energy currency of cells. The energy trapped in ATP can be used to do cellular work—such as movement of muscles or active transport across membranes—just as the energy captured from combustion can be used to do work.

The change in free energy ($\Delta G$) resulting from the complete conversion of glucose and $O_2$ to $CO_2$ and water, whether by combustion or by metabolism, is –686 kcal/mol (–2,870 kJ/mol). Thus the overall reaction is highly exergonic and can drive the endergonic formation of a great deal of ATP from ADP and phosphate. Note that in the discussion that follows, "energy" means free energy.

Three metabolic processes harvest the energy in the chemical bonds of glucose: glycolysis, cellular respiration, and fermentation (**Figure 9.1**). All three processes involve pathways made up of many distinct chemical reactions.

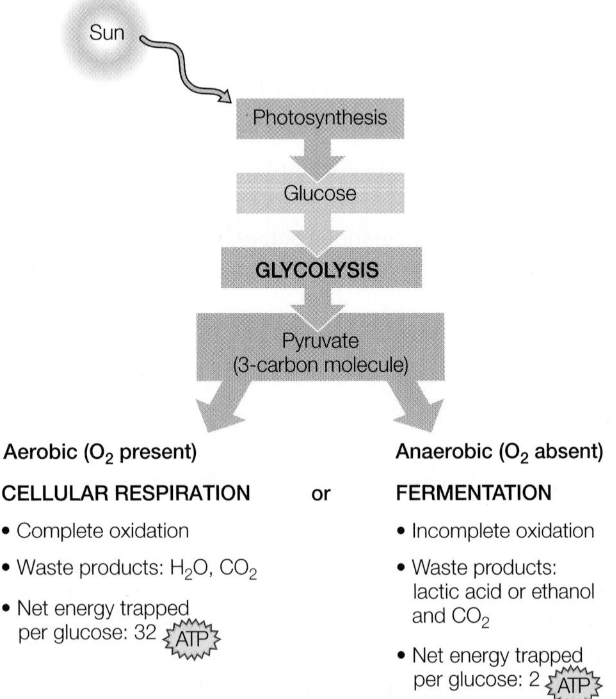


**9.1 Energy for Life** Living organisms obtain their energy from the food compounds produced by photosynthesis. They convert these compounds into glucose, which they metabolize to trap energy in ATP.

- **Glycolysis** begins glucose metabolism in all cells. Through a series of chemical rearrangements, glucose is converted to two molecules of the three-carbon product **pyruvate**, and a small amount of energy is captured in usable forms. Glycolysis is an **anaerobic** process because it does not require $O_2$.
- **Cellular respiration** uses $O_2$ from the environment, and thus it is **aerobic**. Each pyruvate molecule is completely converted into three molecules of $CO_2$ through a set of metabolic pathways including pyruvate oxidation, the citric acid cycle, and an electron transport system (the respiratory chain). In the process, a great deal of the energy stored in the covalent bonds of pyruvate is captured to form ATP.
- **Fermentation** does not involve $O_2$ (it is anaerobic). Fermentation converts pyruvate into lactic acid or ethyl alcohol (ethanol), which are still relatively energy-rich molecules. Because the breakdown of glucose is incomplete, much less energy is released by fermentation than by cellular respiration.

### Redox reactions transfer electrons and energy

As is illustrated in Figure 8.6, the addition of a phosphate group to ADP to make ATP is an endergonic reaction that can extract and transfer energy from exergonic to endergonic reactions. Another way of transferring energy is to transfer electrons. A reaction in which one substance transfers one or more electrons to another substance is called an oxidation–reduction reaction, or **redox** reaction.

- **Reduction** is the gain of one or more electrons by an atom, ion, or molecule.
- **Oxidation** is the loss of one or more electrons.

Oxidation and reduction *always occur together*: as one chemical is oxidized, the electrons it loses are transferred to another chemical, reducing it. In a redox reaction, we call the reactant that becomes reduced an oxidizing agent and the one that becomes oxidized a reducing agent:

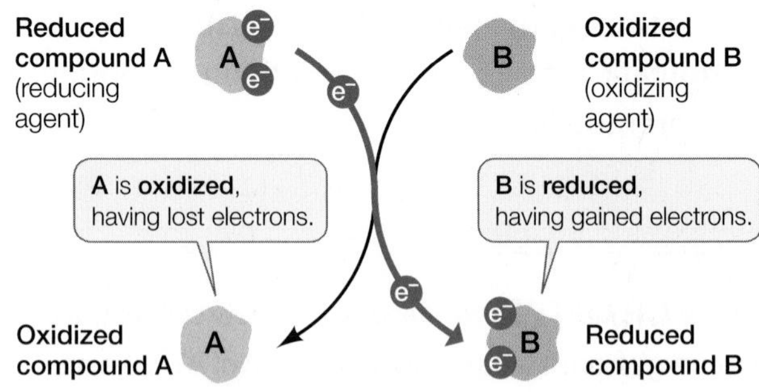


In both the combustion and the metabolism of glucose, glucose is the reducing agent (electron donor) and $O_2$ is the oxidizing agent (electron acceptor).

Although oxidation and reduction are always defined in terms of traffic in electrons, it is often helpful to think in terms of the gain or loss of hydrogen atoms. Transfers of hydrogen atoms involve transfers of electrons ($H = H^+ + e^-$). So when a molecule loses hydrogen atoms, it becomes oxidized.

Methane ($CH_4$) Methanol ($CH_3OH$) Formaldehyde ($CH_2O$) Formic acid (HCOOH) Carbon dioxide ($CO_2$)

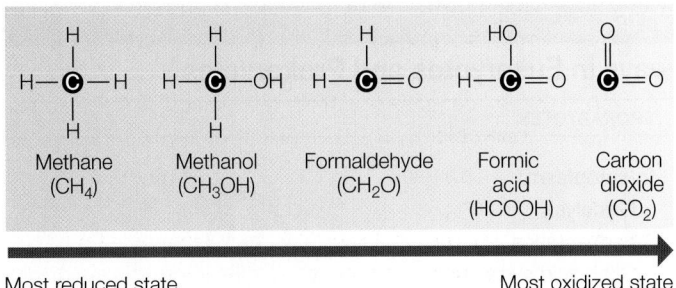

Most reduced state
Highest free energy

Most oxidized state
Lowest free energy

**9.2 Oxidation, Reduction, and Energy** The more oxidized a carbon atom in a molecule is, the less its free energy.

In general, the more reduced a molecule is, the more energy is stored in its covalent bonds (**Figure 9.2**). In a redox reaction, some energy is transferred from the reducing agent to the reduced product. The rest remains in the reducing agent or is lost to entropy. As we will see, some of the key reactions of glycolysis and cellular respiration are highly exergonic redox reactions.

### The coenzyme NAD+ is a key electron carrier in redox reactions

Section 8.4 describes the role of coenzymes, small molecules that assist in enzyme-catalyzed reactions. ADP acts as a coenzyme when it picks up energy released in an exergonic reaction and packages it to form ATP. On the other hand, the coenzyme nicotinamide adenine dinucleotide ($NAD^+$) acts as an electron carrier in redox reactions:

AH NAD+ BH
Oxidation Reduction Oxidation Reduction
A NADH B

As you can see, $NAD^+$ exists in two chemically distinct forms, one oxidized ($NAD^+$) and the other reduced (NADH) (**Figure 9.3**). Both forms participate in redox reactions. The reduction reaction

$$NAD^+ + H^+ + 2\,e^- \rightarrow NADH$$

is actually the transfer of a proton (the hydrogen ion, $H^+$) and two electrons, which are released by the accompanying oxidization reaction.

The electrons do not remain with the coenzyme. Oxygen is highly electronegative and readily accepts electrons from NADH. The oxidation of NADH by $O_2$ (which occurs in several steps)

$$NADH + H^+ + \frac{1}{2}\,O_2 \rightarrow NAD^+ + H_2O$$

is highly exergonic, with a $\Delta G$ of –52.4 kcal/mol (–219 kJ/mol). Note that the oxidizing agent appears here as "$\frac{1}{2}\,O_2$" instead of "O." This notation emphasizes that it is molecular oxygen, $O_2$, that acts as the oxidizing agent.

Just as a molecule of ATP can be thought of as a package of about 12 kcal/mol (50 kJ/mol) of free energy, NADH can be thought of as a larger package of free energy (approximately 50 kcal/mol, or 200 kJ/mol). $NAD^+$ is a common electron carrier in cells, but not the only one. Another carrier, flavin adenine dinucleotide (FAD), also transfers electrons during glucose metabolism.

### An overview: Harvesting energy from glucose

The energy-harvesting processes in cells use different combinations of metabolic pathways depending on the presence or absence of $O_2$:

- Under aerobic conditions, when $O_2$ is available as the final electron acceptor, four pathways operate (**Figure 9.4A**). Glycolysis is followed by the three pathways of cellular respiration: pyruvate oxidation, the citric acid cycle (also called the Krebs cycle or the tricarboxylic acid cycle), and electron transport/ATP synthesis (also called the respiratory chain).
- Under anaerobic conditions when $O_2$ is unavailable, pyruvate oxidation, the citric acid cycle, and the respiratory chain do not function, and the pyruvate produced by glycolysis is further metabolized by fermentation (**Figure 9.4B**).

These five metabolic pathways occur in different locations in the cell (**Table 9.1**).

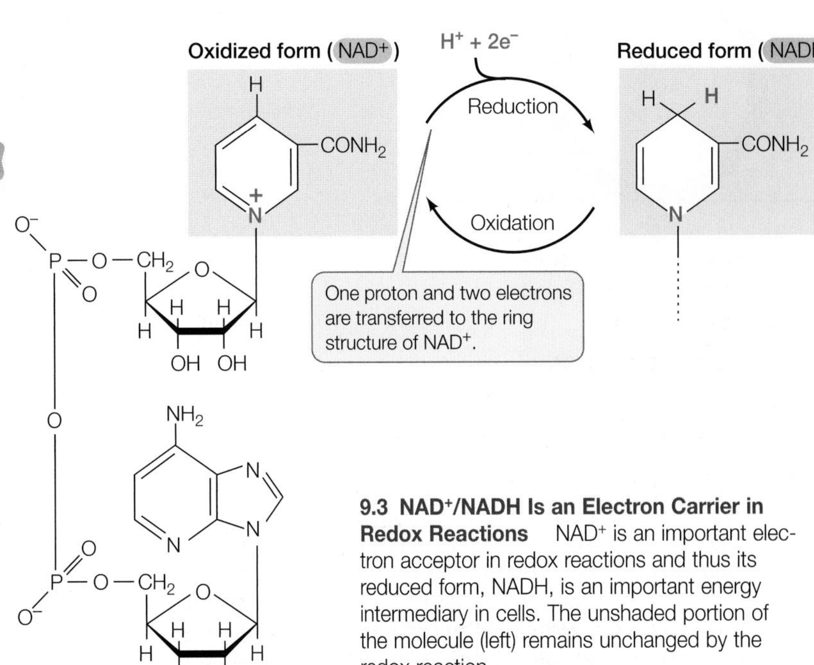


**9.3 NAD+/NADH Is an Electron Carrier in Redox Reactions** $NAD^+$ is an important electron acceptor in redox reactions and thus its reduced form, NADH, is an important energy intermediary in cells. The unshaded portion of the molecule (left) remains unchanged by the redox reaction.

**TABLE 9.1**

**Cellular Locations for Energy Pathways in Eukaryotes and Prokaryotes**

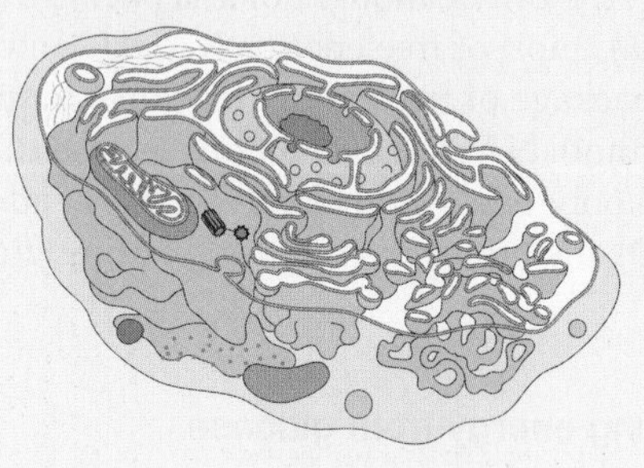

| EUKARYOTES | PROKARYOTES |
|---|---|
| **External to mitochondrion** | **In cytoplasm** |
| Glycolysis | Glycolysis |
| Fermentation | Fermentation |
| | Citric acid cycle |
| **Inside mitochondrion** | **On plasma membrane** |
| Inner membrane | Pyruvate oxidation |
| Respiratory chain | Respiratory chain |
| Matrix | |
| Citric acid cycle | |
| Pyruvate oxidation | |

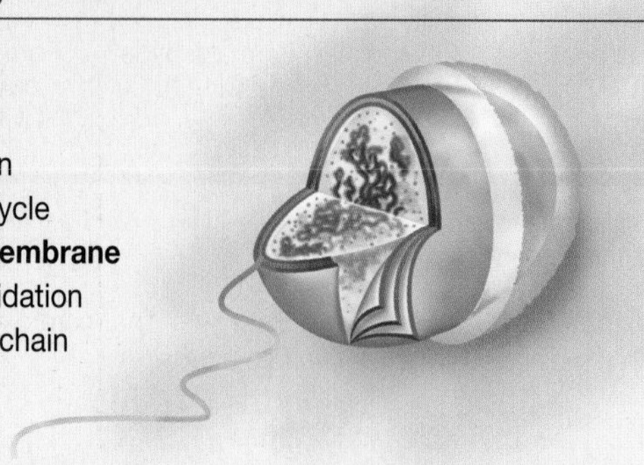

**yourBioPortal.com**

GO TO Web Activity 9.1 • Energy Pathways in Cells

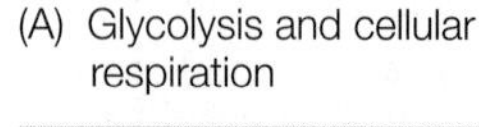
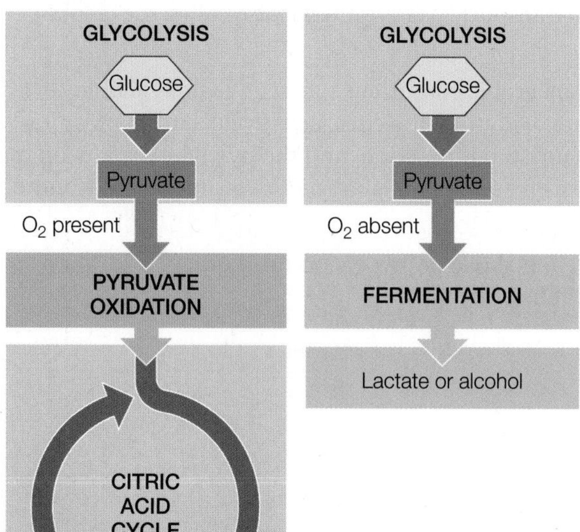

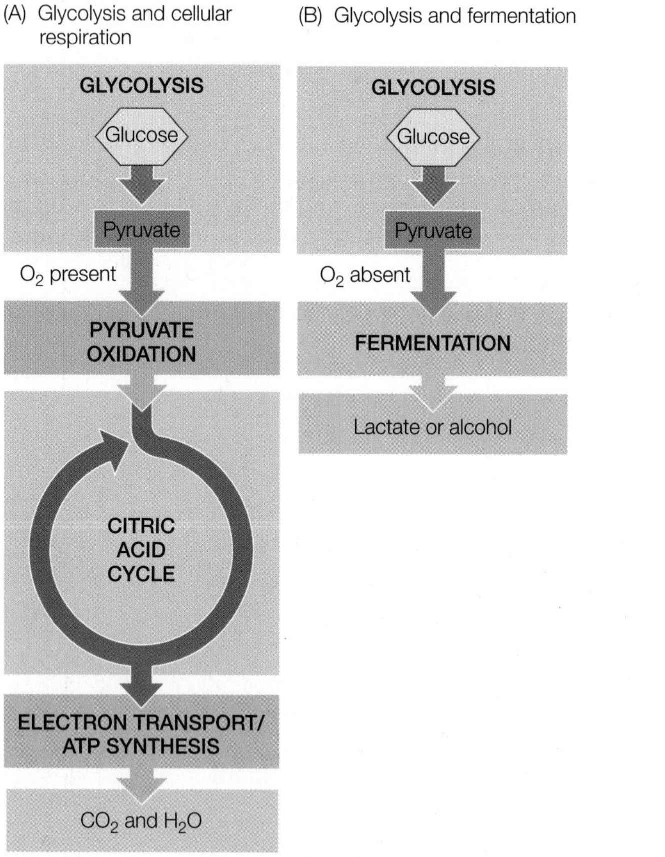

**9.4 Energy-Producing Metabolic Pathways** Energy-producing reactions can be grouped into five metabolic pathways: glycolysis, pyruvate oxidation, the citric acid cycle, the respiratory chain/ATP synthesis, and fermentation. (A) The three lower pathways occur only in the presence of $O_2$ and are collectively referred to as cellular respiration. (B) When $O_2$ is unavailable, glycolysis is followed by fermentation.

**yourBioPortal.com**

GO TO Web Activity 9.2 • Glycolysis and Fermentation

**9.1 RECAP**

**The free energy released from the oxidation of glucose is trapped in the form of ATP. Five metabolic pathways combine in different ways to produce ATP, which supplies the energy for myriad other reactions in living cells.**

- What principles govern metabolic pathways in cells? **See p. 169**
- Describe how the coupling of oxidation and reduction transfers energy from one molecule to another. **See pp. 170–171**
- Explain the roles of $NAD^+$ and $O_2$ with respect to electrons in a redox reaction. **See p. 171 and Figure 9.3**

Now that you have an overview of the metabolic pathways that harvest energy from glucose, let's take a closer look at the three pathways involved in aerobic catabolism: glycolysis, pyruvate oxidation, and the citric acid cycle.

## 9.2 What Are the Aerobic Pathways of Glucose Metabolism?

The aerobic pathways of glucose metabolism oxidize glucose completely to $CO_2$ and $H_2O$. Initially, the glycolysis reactions convert the six-carbon glucose molecule to two 3-carbon pyruvate molecules (**Figure 9.5**). Pyruvate is then converted to $CO_2$ in a second series of reactions beginning with pyruvate oxidation and followed by the citric acid cycle. In addition to generating $CO_2$, the oxidation events are coupled with the reduction of electron carriers, mostly $NAD^+$. So much of the chemical energy in the C—C and C—H bonds of glucose is transferred to NADH. Ultimately, this energy will be transferred to ATP, but this comes in a separate series of reactions involving electron transport, called the respiratory chain. In the respiratory chain, redox reactions result in the oxidative phosphorylation of ADP by ATP synthase. We will begin our consideration of the metabolism of glucose with a closer look at glycolysis.

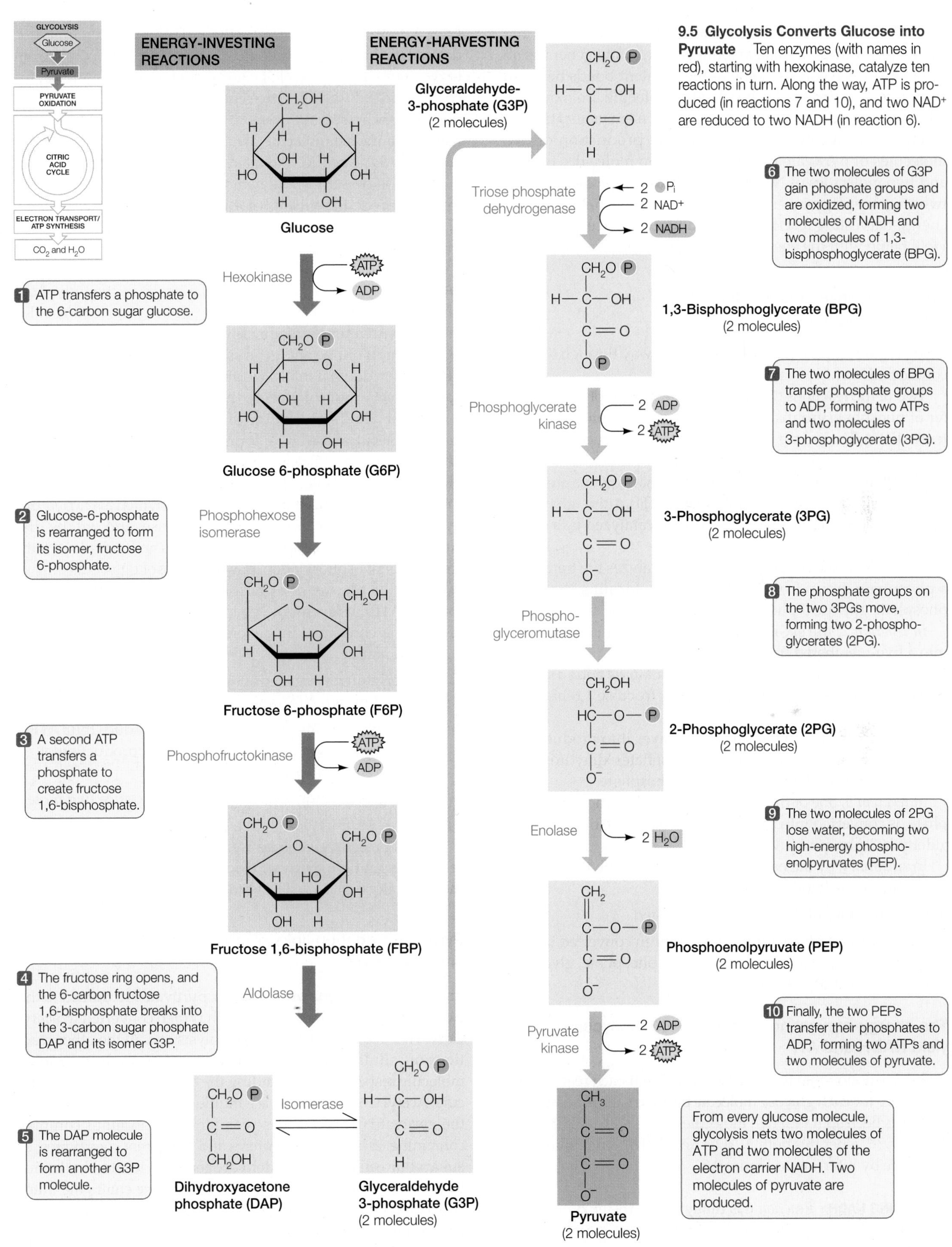

GLYCOLYSIS
Glucose
Pyruvate
PYRUVATE OXIDATION
CITRIC ACID CYCLE
ELECTRON TRANSPORT/ ATP SYNTHESIS
$CO_2$ and $H_2O$
ENERGY-INVESTING REACTIONS
ENERGY-HARVESTING REACTIONS
Glucose
Hexokinase
ATP
ADP
1 ATP transfers a phosphate to the 6-carbon sugar glucose.
Glucose 6-phosphate (G6P)
2 Glucose-6-phosphate is rearranged to form its isomer, fructose 6-phosphate.
Phosphohexose isomerase
Fructose 6-phosphate (F6P)
3 A second ATP transfers a phosphate to create fructose 1,6-bisphosphate.
Phosphofructokinase
ATP
ADP
Fructose 1,6-bisphosphate (FBP)
4 The fructose ring opens, and the 6-carbon fructose 1,6-bisphosphate breaks into the 3-carbon sugar phosphate DAP and its isomer G3P.
Aldolase
5 The DAP molecule is rearranged to form another G3P molecule.
Dihydroxyacetone phosphate (DAP)
Isomerase
Glyceraldehyde 3-phosphate (G3P) (2 molecules)
Glyceraldehyde-3-phosphate (G3P) (2 molecules)
Triose phosphate dehydrogenase
2 $P_i$
2 $NAD^+$
2 NADH
6 The two molecules of G3P gain phosphate groups and are oxidized, forming two molecules of NADH and two molecules of 1,3-bisphosphoglycerate (BPG).
1,3-Bisphosphoglycerate (BPG) (2 molecules)
Phosphoglycerate kinase
2 ADP
2 ATP
7 The two molecules of BPG transfer phosphate groups to ADP, forming two ATPs and two molecules of 3-phosphoglycerate (3PG).
3-Phosphoglycerate (3PG) (2 molecules)
Phospho-glyceromutase
8 The phosphate groups on the two 3PGs move, forming two 2-phospho-glycerates (2PG).
2-Phosphoglycerate (2PG) (2 molecules)
Enolase
2 $H_2O$
9 The two molecules of 2PG lose water, becoming two high-energy phospho-enolpyruvates (PEP).
Phosphoenolpyruvate (PEP) (2 molecules)
Pyruvate kinase
2 ADP
2 ATP
10 Finally, the two PEPs transfer their phosphates to ADP, forming two ATPs and two molecules of pyruvate.
Pyruvate (2 molecules)
From every glucose molecule, glycolysis nets two molecules of ATP and two molecules of the electron carrier NADH. Two molecules of pyruvate are produced.


**9.5 Glycolysis Converts Glucose into Pyruvate** Ten enzymes (with names in red), starting with hexokinase, catalyze ten reactions in turn. Along the way, ATP is produced (in reactions 7 and 10), and two $NAD^+$ are reduced to two NADH (in reaction 6).

Glycolysis takes place in the cytosol. It converts glucose into pyruvate, produces a small amount of energy, and generates no $CO_2$. During glycolysis, some of the covalent bonds between carbon and hydrogen in the glucose molecule are oxidized, releasing some of the stored energy. The ten enzyme-catalyzed reactions of glycolysis result in the net production of two molecules of pyruvate (pyruvic acid), two molecules of ATP, and two molecules of NADH. Glycolysis can be divided into two stages: energy-investing reactions that consume ATP, and energy-harvesting reactions that produce ATP (see Figure 9.5). We'll begin with the energy-investing reactions.

### The energy-investing reactions 1–5 of glycolysis require ATP

Using Figure 9.5 as a guide, let's work our way through the glycolytic pathway.

Two of the reactions (1 and 3 in Figure 9.5), involve the transfer of phosphate groups from ATP to form phosphorylated intermediates. The second of these intermediates, fructose 1,6-bisphosphate, has a free energy substantially higher than that of glucose. Later in the pathway, these phosphate groups are transferred to ADP to make new molecules of ATP. Although both of these steps use ATP as a substrate, each is catalyzed by a different, specific enzyme.

*In reaction 1*, the enzyme hexokinase catalyzes the transfer of a phosphate group from ATP to glucose, forming the sugar phosphate glucose 6-phosphate.

*In reaction 2*, the six-membered glucose ring is rearranged into a five-membered fructose ring.

*In reaction 3*, the enzyme phosphofructokinase adds a second phosphate to the fructose ring, forming fructose 1,6-bisphosphate.

*Reaction 4* opens up the ring and cleaves it to produce two different three-carbon sugar (triose) phosphates: dihydroxyacetone phosphate and glyceraldehyde 3-phosphate.

*In reaction 5*, one of those products, dihydroxyacetone phosphate, is converted into a second molecule of the other, glyceraldehyde 3-phosphate (G3P).

In summary, by the halfway point of the glycolytic pathway, two things have happened:

- Two molecules of ATP have been invested.
- The six-carbon glucose molecule has been converted into two molecules of a three-carbon sugar phosphate, glyceraldehyde 3-phosphate (G3P).

### The energy-harvesting reactions 6–10 of glycolysis yield NADH and ATP

In the discussion that follows, remember that each reaction occurs twice for each glucose molecule because each glucose molecule has been split into two molecules of G3P. The transformation of G3P generates both NADH and ATP. Again, follow the sequence by referring to Figure 9.5.

**PRODUCING NADH** *Reaction 6* is catalyzed by the enzyme triose phosphate dehydrogenase, and its end product is a phosphate ester, 1,3-bisphosphoglycerate (BPG). This is an exergonic oxidation reaction, and it is accompanied by a large drop in free energy—more than 100 kcal of energy is released per mole of glucose (**Figure 9.6, left**). The free energy released in this reaction is not lost to heat, but is captured by the accompanying reduction reaction. For each molecule of G3P that is oxidized, one molecule of $NAD^+$ is reduced to make a molecule of NADH.

$NAD^+$ is present in only small amounts in the cell, and it must be recycled to allow glycolysis to continue. As we will see, NADH is oxidized back to $NAD^+$ in the metabolic pathways that follow glycolysis.

**PRODUCING ATP** *In reactions 7–10* of glycolysis, the two phosphate groups of BPG are transferred one at a time to molecules of ADP, with a rearrangement in between. More than 20 kcal (83.6 kJ/mol) of free energy is stored in ATP for every mole of BPG broken down. Finally, we are left with two moles of pyruvate for every mole of glucose that entered glycolysis.

The enzyme-catalyzed transfer of phosphate groups from donor molecules to ADP to form ATP is called **substrate-level phosphorylation**. (Phosphorylation is the addition of a phosphate group to a molecule.) Substrate-level phosphorylation is distinct from oxidative phosphorylation, which is carried out by the respiratory chain and ATP synthase, and will be discussed later in this chapter. Reaction 7 is an example of substrate-level phosphorylation, in which phosphoglycerate kinase catalyzes the transfer of a phosphate group from BPG to ADP, forming ATP. It is exergonic, even though a substantial amount of energy is consumed in the formation of ATP.

To summarize:

- The energy-investing steps of glycolysis use the energy of hydrolysis of two ATP molecules per glucose molecule.
- The energy-releasing steps of glycolysis produce four ATP molecules per glucose molecule, so the net production of ATP is two molecules.
- The energy-releasing steps of glycolysis produce two molecules of NADH.

If $O_2$ is present, glycolysis is followed by the three stages of cellular respiration: pyruvate oxidation, the citric acid cycle, and the respiratory chain/ATP synthesis.

### Pyruvate oxidation links glycolysis and the citric acid cycle

In the process of **pyruvate oxidation**, pyruvate is oxidized to the two-carbon acetate molecule, which is then converted to acetyl CoA. This is the link between glycolysis and all the other reactions of cellular respiration. **Coenzyme A** (**CoA**) is a complex molecule responsible for binding the two-carbon acetate molecule. Acetyl CoA formation is a multi-step reaction catalyzed by the pyruvate dehydrogenase complex, an enormous complex containing 60 individual proteins and 5 different coenzymes. In eukaryotic cells, pyruvate dehydrogenase is located in the mitochondrial matrix (see Figure 5.12). Pyruvate enters the mitochondrion by active transport, and then a series of coupled reactions takes place:

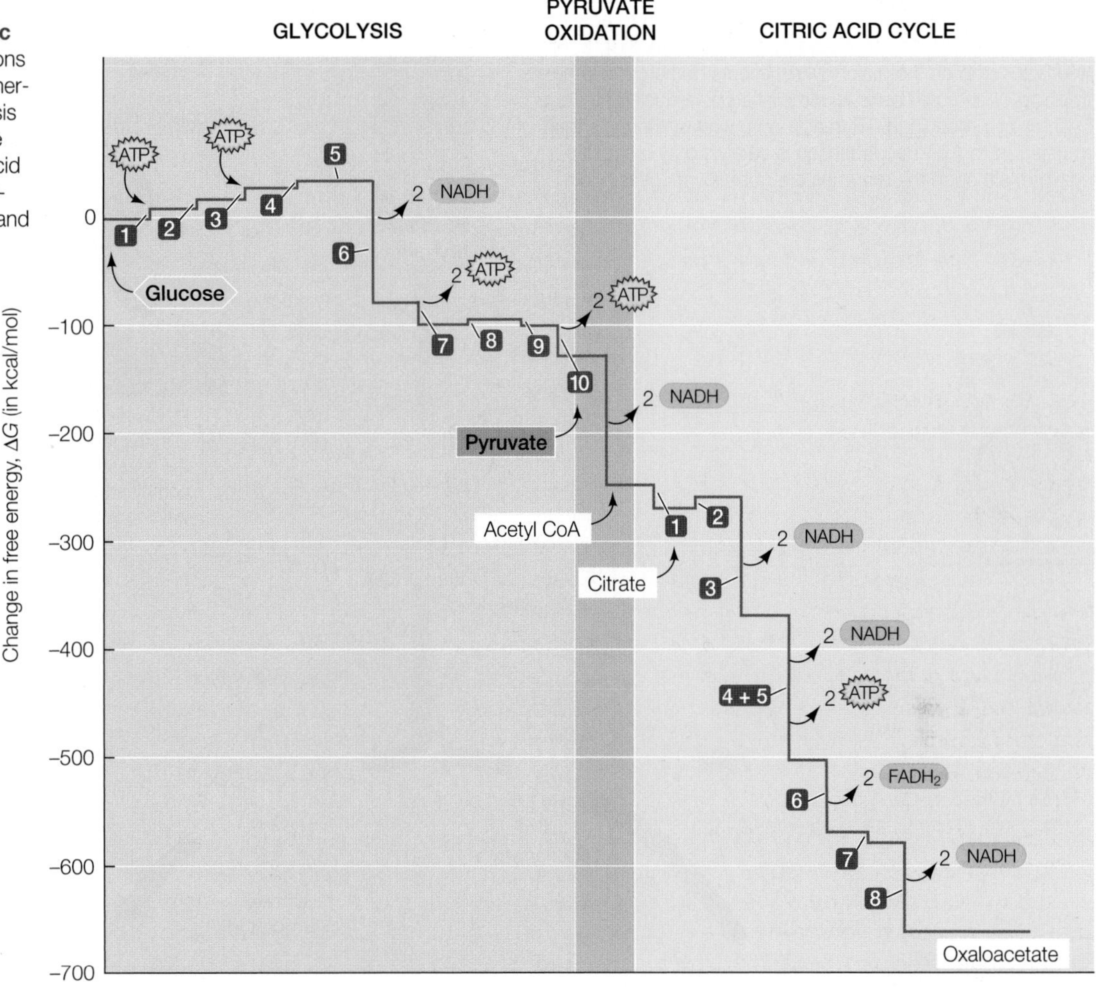


**9.6 Changes in Free Energy During Glycolysis and the Citric Acid Cycle** The first five reactions of glycolysis (left) consume free energy, and the remaining five glycolysis reactions release energy. Pyruvate oxidation (middle) and the citric acid cycle (right) both release considerable energy. Refer to Figures 9.5 and 9.7 for the reaction numbers.

1. Pyruvate is oxidized to a two-carbon acetyl group (acetate), and $CO_2$ is released (decarboxylation).
2. Part of the energy from this oxidation is captured by the reduction of $NAD^+$ to NADH.
3. Some of the remaining energy is stored temporarily by combining the acetyl group with CoA, forming acetyl CoA:

$$\text{pyruvate} + NAD^+ + \text{CoA} + H^+ \rightarrow \text{acetyl CoA} + \text{NADH} + CO_2$$

(In this reaction, the proton and electrons used to reduce $NAD^+$ are derived from the oxidation of both pyruvate and CoA.) Acetyl CoA has 7.5 kcal/mol (31.4 kJ/mol) more energy than simple acetate. Acetyl CoA can donate its acetyl group to various acceptor molecules, much as ATP can donate phosphate groups to various acceptors. But the main role of acetyl CoA is to donate its acetyl group to the four-carbon compound oxaloacetate, forming the six-carbon molecule citrate. This initiates the citric acid cycle, one of life's most important energy-harvesting pathways.

Arsenic, the classic poison of rodent exterminators and murder mysteries, acts by inhibiting pyruvate dehydrogenase, thus decreasing acetyl CoA production. The lack of acetyl CoA stops the citric acid cycle and all the subsequent reactions that depend on it. Consequently, cells eventually run out of ATP and cannot perform essential processes that are powered by ATP hydrolysis.

### The citric acid cycle completes the oxidation of glucose to $CO_2$

Acetyl CoA is the starting point for the **citric acid cycle**. This pathway of eight reactions completely oxidizes the two-carbon acetyl group to two molecules of carbon dioxide. The free energy released from these reactions is captured by ADP and the electron carriers $NAD^+$ and FAD. **Figure 9.6 right** shows the free energy changes during each step of the pathway.

The citric acid cycle is maintained in a steady state—that is, although the intermediate compounds in the cycle enter and leave it, the concentrations of those intermediates do not change much. Refer to the numbered reactions in **Figure 9.7** as you read the description of each reaction.

**9.7 Pyruvate Oxidation and the Citric Acid Cycle** Pyruvate enters the mitochondrion and is oxidized to acetyl CoA, which enters the citric acid cycle. Reactions 3, 4, 5, 6, and 8 accomplish the major overall effects of the cycle—the trapping of energy. This is accomplished by reducing NAD+ or FAD, or by producing GTP (reaction 5), whose energy is then transferred to ATP. Each reaction is catalyzed by a specific enzyme, although the enzymes are not shown in this figure.

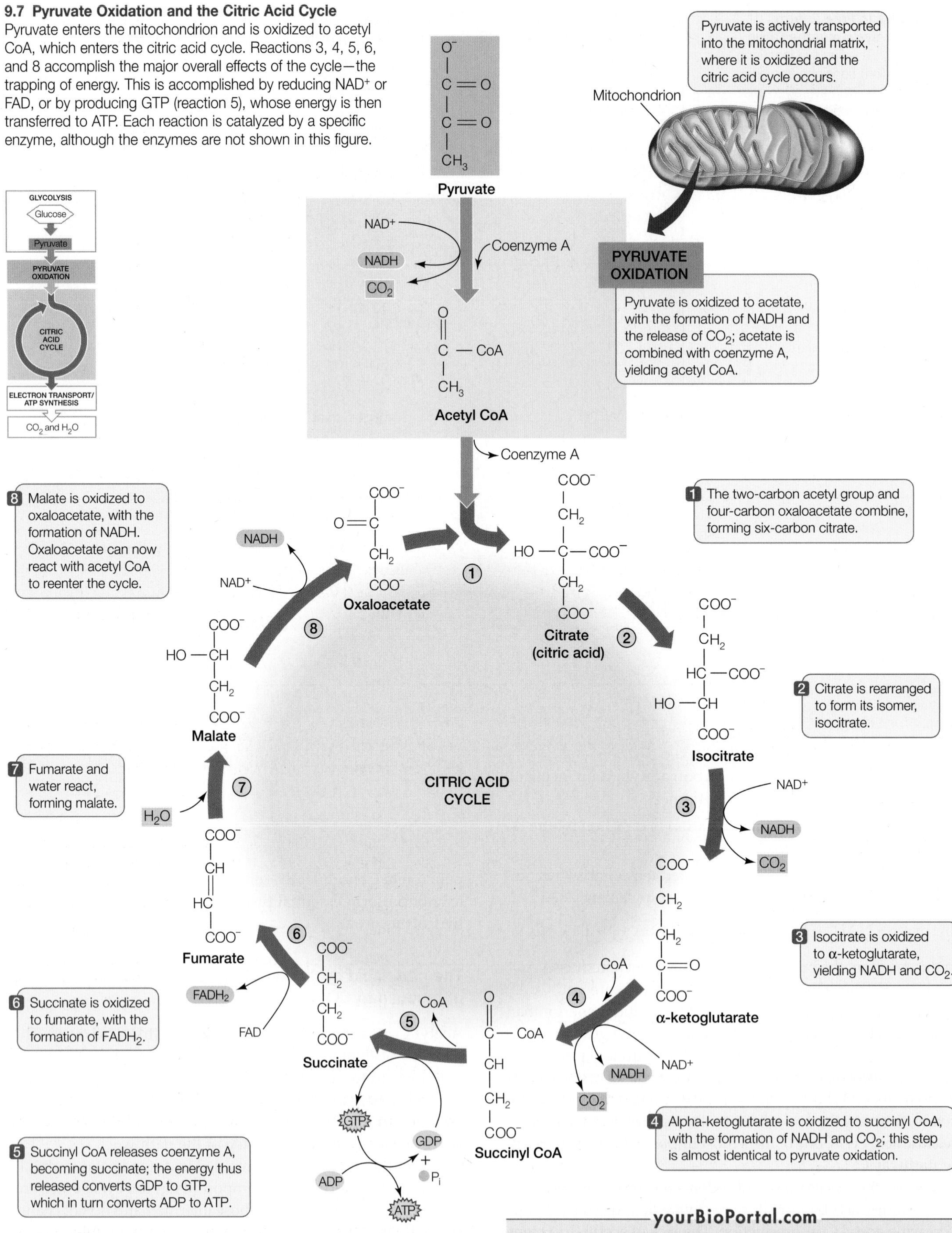

Pyruvate is actively transported into the mitochondrial matrix, where it is oxidized and the citric acid cycle occurs.
Mitochondrion
Pyruvate
GLYCOLYSIS
Glucose
Pyruvate
PYRUVATE OXIDATION
CITRIC ACID CYCLE
ELECTRON TRANSPORT/ ATP SYNTHESIS
CO2 and H2O
NAD+
NADH
CO2
Coenzyme A
Acetyl CoA
PYRUVATE OXIDATION
Pyruvate is oxidized to acetate, with the formation of NADH and the release of CO2; acetate is combined with coenzyme A, yielding acetyl CoA.
Coenzyme A
1 The two-carbon acetyl group and four-carbon oxaloacetate combine, forming six-carbon citrate.
8 Malate is oxidized to oxaloacetate, with the formation of NADH. Oxaloacetate can now react with acetyl CoA to reenter the cycle.
Oxaloacetate
Citrate (citric acid)
Malate
Isocitrate
2 Citrate is rearranged to form its isomer, isocitrate.
CITRIC ACID CYCLE
7 Fumarate and water react, forming malate.
H2O
Fumarate
α-ketoglutarate
3 Isocitrate is oxidized to α-ketoglutarate, yielding NADH and CO2.
FADH2
FAD
Succinate
CoA
Succinyl CoA
6 Succinate is oxidized to fumarate, with the formation of FADH2.
GTP
GDP
+ Pi
ADP
ATP
5 Succinyl CoA releases coenzyme A, becoming succinate; the energy thus released converts GDP to GTP, which in turn converts ADP to ATP.
4 Alpha-ketoglutarate is oxidized to succinyl CoA, with the formation of NADH and CO2; this step is almost identical to pyruvate oxidation.

yourBioPortal.com
GO TO Web Activity 9.3 • The Citric Acid Cycle

*In reaction 1,* the energy temporarily stored in acetyl CoA drives the formation of citrate from oxaloacetate. During this reaction, the CoA molecule is removed and can be reused by pyruvate dehydrogenase.

*In reaction 2,* the citrate molecule is rearranged to form isocitrate.

*In reaction 3,* a $CO_2$ molecule, a proton, and two electrons are removed, converting isocitrate into α-ketoglutarate. This reaction releases a large amount of free energy, some of which is stored in NADH.

*In reaction 4,* α-ketoglutarate is oxidized to succinyl CoA. This reaction is similar to the oxidation of pyruvate to form acetyl CoA. Like that reaction, it is catalyzed by a multi-enzyme complex and produces $CO_2$ and NADH.

*In reaction 5,* some of the energy in succinyl CoA is harvested to make GTP (guanosine triphosphate) from GDP and $P_i$. This is another example of substrate-level phosphorylation. GTP is then used to make ATP from ADP and $P_i$.

*In reaction 6,* the succinate released from succinyl CoA in reaction 5 is oxidized to fumarate. In the process, free energy is released and two hydrogens are transferred to the electron carrier FAD, forming $FADH_2$.

*Reaction 7* is a molecular rearrangement in which water is added to fumarate, forming malate.

In *reaction 8,* one more $NAD^+$ reduction occurs, producing oxaloacetate from malate. Reactions 7 and 8 illustrate a common biochemical mechanism: in reaction 7, water ($H_2O$) is added to form a hydroxyl (—OH) group, and then in reaction 8 the H from the hydroxyl group is removed, generating a carbonyl group and reducing $NAD^+$ to NADH.

The final product, oxaloacetate, is ready to combine with another acetyl group from acetyl CoA and go around the cycle again. The citric acid cycle operates twice for each glucose molecule that enters glycolysis (once for each pyruvate that enters the mitochondrion).

To summarize:

- The *inputs* to the citric acid cycle are acetate (in the form of acetyl CoA), water, and the oxidized electron carriers $NAD^+$, FAD, and GDP.
- The *outputs* are carbon dioxide, reduced electron carriers (NADH and $FADH_2$), and a small amount of GTP. Overall, the citric acid cycle releases two carbons as $CO_2$ and produces four reduced electron carrier molecules.

### The citric acid cycle is regulated by the concentrations of starting materials

We have seen how pyruvate, a three-carbon molecule, is completely oxidized to $CO_2$ by pyruvate dehydrogenase and the citric acid cycle. For the cycle to continue, the starting molecules—acetyl CoA and oxidized electron carriers—must all be replenished. The electron carriers are reduced during the cycle and in reaction 6 of glycolysis (see Figure 9.5), and they must be reoxidized:

$$NADH \rightarrow NAD^+ + H^+ + 2\,e^-$$

$$FADH_2 \rightarrow FAD + 2\,H^+ + 2\,e^-$$

The oxidation of these electron carriers take place in coupled redox reactions, in which other molecules get reduced. When it is present, $O_2$ is the molecule that eventually accepts these electrons and gets reduced to form $H_2O$.

### 9.2 RECAP

**The oxidation of glucose in the presence of $O_2$ involves glycolysis, pyruvate oxidation, and the citric acid cycle. In glycolysis, glucose is converted to pyruvate with some energy capture. Following the initial oxidation of pyruvate, the citric acid cycle completes its oxidation to $CO_2$ and more energy is captured in the form of reduced electron carriers.**

- What is the net energy yield of glycolysis in terms of energy invested and energy harvested? **See p. 174 and Figure 9.6**
- What role does pyruvate oxidation play in the citric acid cycle? **See pp. 174–175 and Figure 9.7**
- Explain why reoxidation of NADH is crucial for the continuation of the citric acid cycle. **See p. 177**

Pyruvate oxidation and the citric acid cycle cannot continue operating unless $O_2$ is available to receive electrons during the reoxidation of reduced electron carriers. However, these electrons are not passed directly to $O_2$, as you will learn next.

## 9.3 How Does Oxidative Phosphorylation Form ATP?

The overall process of ATP synthesis resulting from the reoxidation of electron carriers in the presence of $O_2$ is called **oxidative phosphorylation**. Two components of the process can be distinguished:

1. *Electron transport.* The electrons from NADH and $FADH_2$ pass through the **respiratory chain**, a series of membrane-associated electron carriers. The flow of electrons along this pathway results in the active transport of protons out of the mitochondrial matrix and across the inner mitochondrial membrane, creating a proton concentration gradient.
2. *Chemiosmosis.* The protons diffuse back into the mitochondrial matrix through a channel protein, **ATP synthase**, which couples this diffusion to the synthesis of ATP. The inner mitochondrial membrane is otherwise impermeable to protons, so the only way for them to follow their concentration gradient is through the channel.

Before we proceed with the details of these pathways, let's consider an important question: Why should the respiratory chain be such a complex process? Why don't cells use the following single step?

$$NADH + H^+ + \tfrac{1}{2}\,O_2 \rightarrow NAD^+ + H_2O$$

The answer is that this reaction would be untamable. It is extremely exergonic—and would be rather like setting off a stick of dynamite in the cell. There is no biochemical way to harvest that burst of energy efficiently and put it to physiological use (that is, no single metabolic reaction is so endergonic as to consume a significant fraction of that energy in a single step). To control the release of energy during the oxidation of glucose, cells have evolved a lengthy respiratory chain: a series of reactions, each of which releases a small, manageable amount of energy, one step at a time.

### The respiratory chain transfers electrons and releases energy

The respiratory chain is located in the inner mitochondrial membrane and contains several interactive components, including large integral proteins, smaller mobile proteins, and a small lipid molecule. **Figure 9.8** shows a plot of the free energy released as electrons are passed between the carriers.

- Four large protein complexes (I, II, III, and IV) contain electron carriers and associated enzymes. In eukaryotes they are integral proteins of the inner mitochondrial membrane (see Figure 5.12), and three are transmembrane proteins.
- Cytochrome *c* is a small peripheral protein that lies in the intermembrane space. It is loosely attached to the outer surface of the inner mitochondrial membrane.
- Ubiquinone (abbreviated Q) is a small, nonpolar, lipid molecule that moves freely within the hydrophobic interior of the phospholipid bilayer of the inner mitochondrial membrane.

As illustrated in Figure 9.8, NADH passes electrons to protein complex I (called NADH-Q reductase), which in turn passes the electrons to Q. This electron transfer is accompanied by a large drop in free energy. Complex II (succinate dehydrogenase) passes electrons to Q from $FADH_2$, which was generated in reaction 6 of the citric acid cycle (see Figure 9.7). These electrons enter the chain later than those from NADH and will ultimately produce less ATP.

Complex III (cytochrome *c* reductase) receives electrons from Q and passes them to cytochrome *c*. Complex IV (cytochrome *c* oxidase) receives electrons from cytochrome *c* and passes them to oxygen. Finally the reduction of oxygen to $H_2O$ occurs:

$$\tfrac{1}{2}\,O_2 + 2\,H^+ + 2\,e^- \rightarrow H_2O$$

Notice that two protons ($H^+$) are also consumed in this reaction. This contributes to the proton gradient across the inner mitochondrial membrane.

During electron transport, protons are also actively transported across the membrane—electron transport within each of the three transmembrane complexes (I, III, and IV) results in the transfer of protons from the matrix to the intermembrane space (**Figure 9.9**). So an imbalance of protons is set up, with the impermeable inner mitochondrial membrane as a barrier. The concentration of $H^+$ in the intermembrane space is higher than in the matrix, and this gradient represents a source of potential energy. The diffusion of those protons across the membrane is coupled with the formation of ATP. Thus the energy originally contained in glucose and other fuel molecules is finally captured in the cellular energy currency, ATP. For each pair of electrons passed along the chain from NADH to oxygen, about 2.5 molecules of ATP are formed. $FADH_2$ oxidation produces about 1.5 ATP molecules.

### Proton diffusion is coupled to ATP synthesis

All the electron carriers and enzymes of the respiratory chain, except cytochrome *c*, are embedded in the inner mitochondrial membrane. As we have just seen, the operation of the respiratory chain results in the active transport of protons from the mitochon-

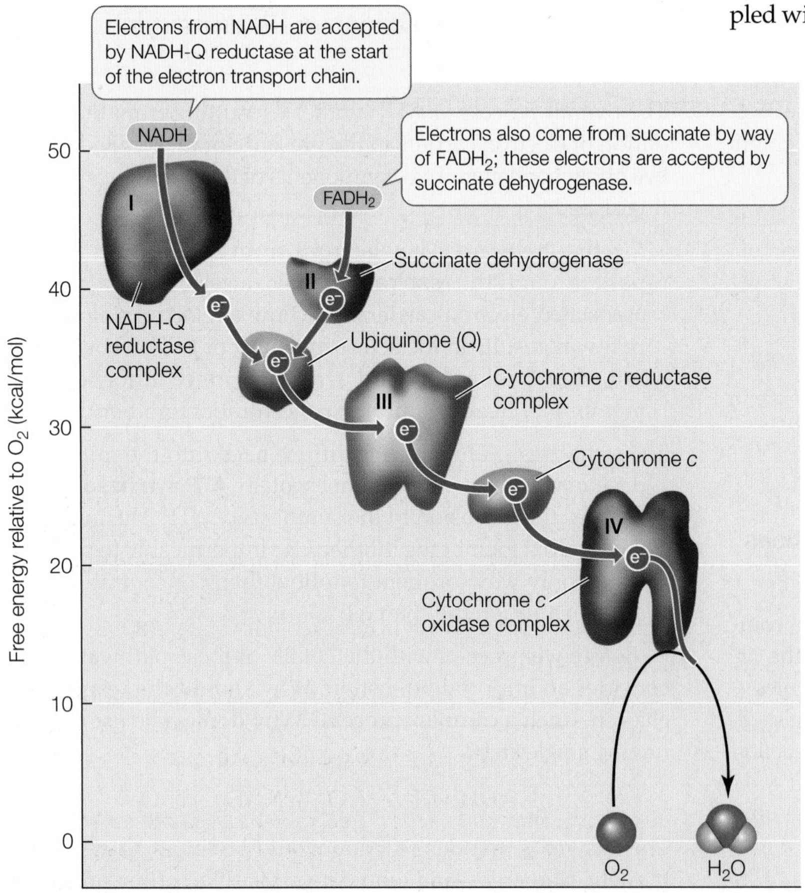

**9.8 The Oxidation of NADH and $FADH_2$ in the Respiratory Chain** Electrons from NADH and $FADH_2$ are passed along the respiratory chain, a series of protein complexes in the inner mitochondrial membrane containing electron carriers and enzymes. The carriers gain free energy when they become reduced and release free energy when they are oxidized.

yourBioPortal.com
GO TO Web Activity 9.4 • Respiratory Chain

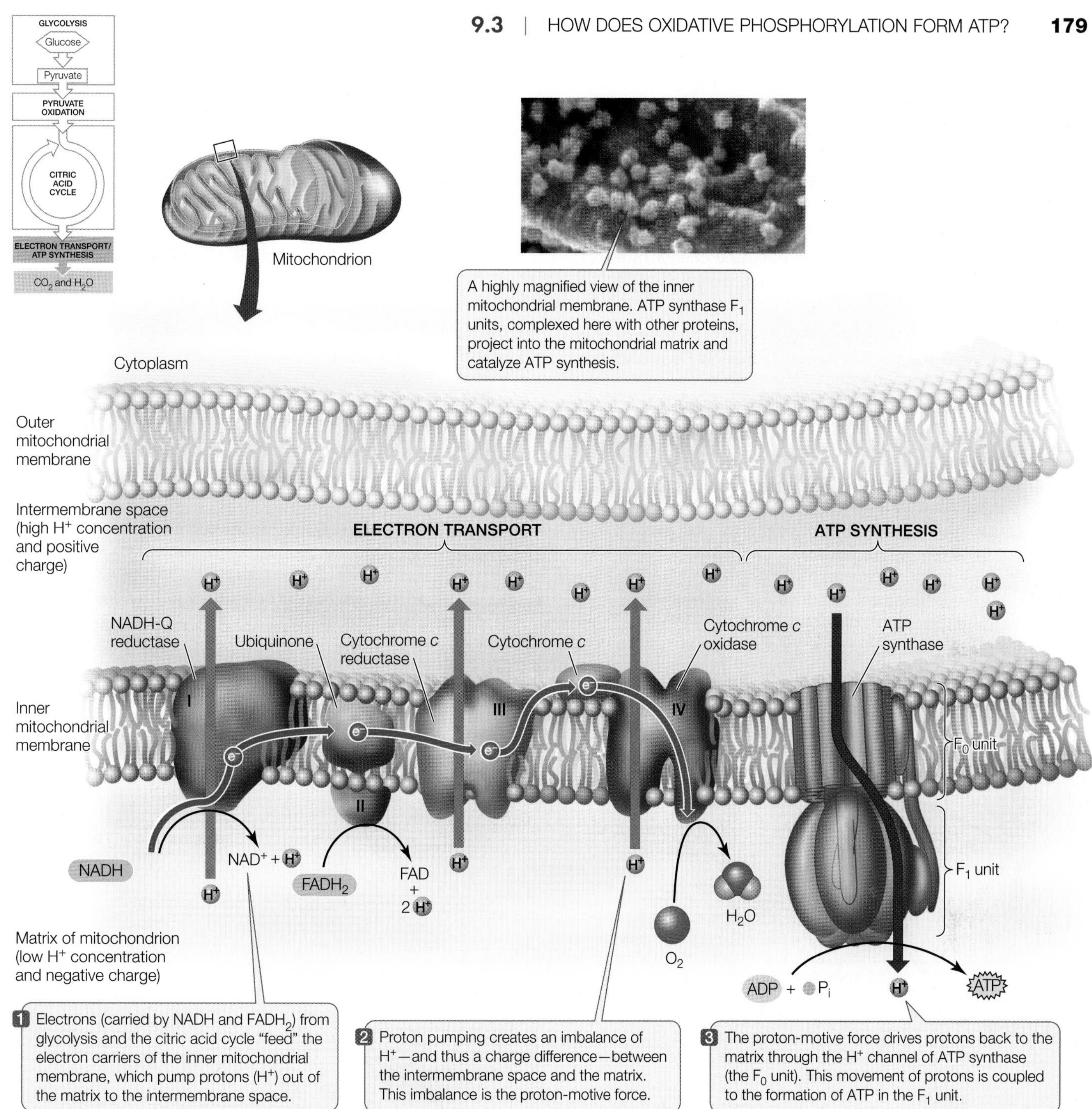


**9.9 The Respiratory Chain and ATP Synthase Produce ATP by a Chemiosmotic Mechanism** As electrons pass through the transmembrane protein complexes in the respiratory chain, protons are pumped from the mitochondrial matrix into the intermembrane space. As the protons return to the matrix through ATP synthase, ATP is formed.

yourBioPortal.com

GO TO Animated Tutorial 9.1 • Electron Transport and ATP Synthesis

drial matrix to the intermembrane space. The transmembrane protein complexes (I, III, and IV) act as proton pumps, and as a result, the intermembrane space is more acidic than the matrix.

Because of the positive charge carried by a proton ($H^+$), this pumping creates not only a concentration gradient but also a difference in electric charge across the inner mitochondrial membrane, making the mitochondrial matrix more negative than the intermembrane space. Together, the proton concentration gradient and the electrical charge difference constitute a source of potential energy called the **proton-motive force**. This force tends to drive the protons back across the membrane, just as the charge on a battery drives the flow of electrons to discharge the battery.

The hydrophobic lipid bilayer is essentially impermeable to protons, so the potential energy of the proton-motive force cannot be discharged by simple diffusion of protons across the membrane. However, protons can diffuse across the membrane by passing through a specific proton channel, called ATP synthase, which couples proton movement to the synthesis of ATP. This coupling of proton-motive force and ATP synthesis is called the chemiosmotic mechanism (or **chemiosmosis**) and is found in all respiring cells.

**THE CHEMIOSMOTIC MECHANISM FOR ATP SYNTHESIS** The chemiosmotic mechanism involves transmembrane proteins, including a proton channel and the enzyme ATP synthase, that couple proton diffusion to ATP synthesis. The potential energy of the $H^+$ gradient, or the proton-motive force (described above), is harnessed by ATP synthase. This protein complex has two roles: it acts as a channel allowing $H^+$ to diffuse back into the matrix, and it uses the energy of that diffusion to make ATP from ADP and $P_i$.

ATP synthesis is a reversible reaction, and ATP synthase can also act as an ATPase, hydrolyzing ATP to ADP and $P_i$:

$$ATP \rightleftharpoons ADP + P_i + \text{free energy}$$

If the reaction goes to the right, free energy is released and is used to pump $H^+$ out of the mitochondrial matrix—not the usual mode

yourBioPortal.com
GO TO Animated Tutorial 9.2 • Two Experiments Demonstrate the Chemiosmotic Mechanism

## INVESTIGATING LIFE

**9.10 Two Experiments Demonstrate the Chemiosmotic Mechanism**
The chemiosmosis hypothesis was a bold departure for the conventional scientific thinking of the time. It required an intact compartment separated by a membrane. Could a proton gradient drive the synthesis of ATP? And was this capacity entirely due to the ATP synthase enzyme?

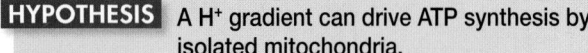

**HYPOTHESIS** A $H^+$ gradient can drive ATP synthesis by isolated mitochondria.

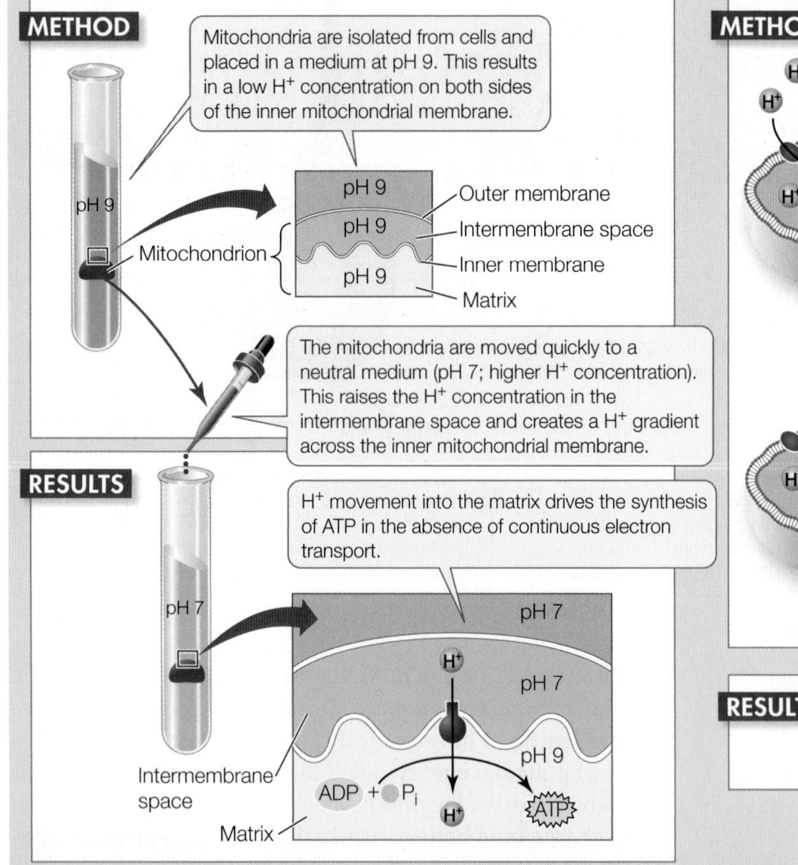


**CONCLUSION** In the absence of electron transport, an artificial $H^+$ gradient is sufficient for ATP synthesis by mitochondria.

**HYPOTHESIS** ATP synthase is needed for ATP synthesis.

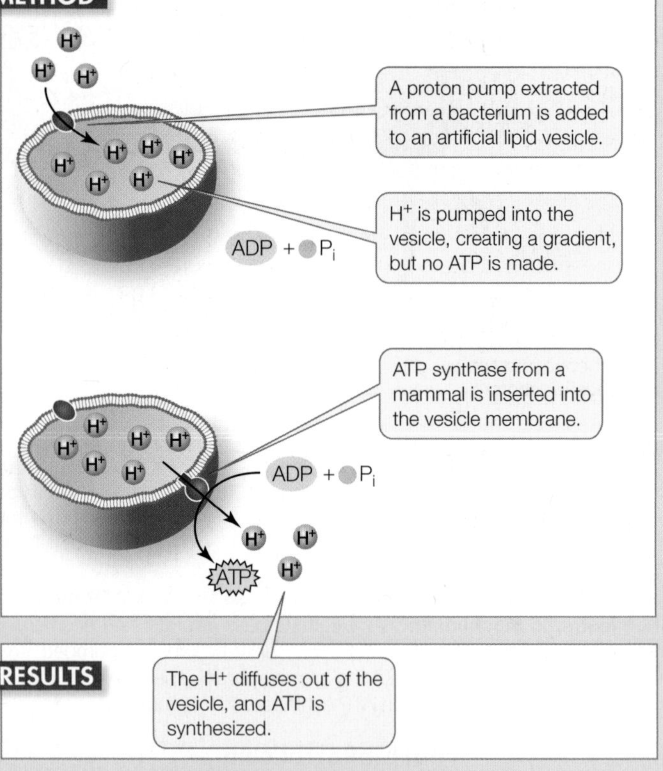

METHOD
H+
A proton pump extracted from a bacterium is added to an artificial lipid vesicle.
H+ is pumped into the vesicle, creating a gradient, but no ATP is made.
ADP + Pi
ATP synthase from a mammal is inserted into the vesicle membrane.
ADP + Pi
ATP
RESULTS
The H+ diffuses out of the vesicle, and ATP is synthesized.

**CONCLUSION** ATP synthase, acting as a $H^+$ channel, is necessary for ATP synthesis.

**FURTHER INVESTIGATION:** What would happen in the experiment on the right if a second ATP synthase, oriented in the opposite way to the one originally inserted in the membrane, were added?

Go to **yourBioPortal.com** for original citations, discussions, and relevant links for all INVESTIGATING LIFE figures.

of operation. If the reaction goes to the left, it uses the free energy from $H^+$ diffusion into the matrix to make ATP. What makes it prefer ATP synthesis? There are two answers to this question:

- ATP leaves the mitochondrial matrix for use elsewhere in the cell as soon as it is made, keeping the ATP concentration in the matrix low, and driving the reaction toward the left.
- The $H^+$ gradient is maintained by electron transport and proton pumping.

Every day a person hydrolyzes about $10^{25}$ ATP molecules to ADP. This amounts to 9 kg, a significant fraction of almost everyone's entire body weight! The vast majority of this ADP is "recycled"—converted back to ATP—using free energy from the oxidation of glucose.

**EXPERIMENTS DEMONSTRATE CHEMIOSMOSIS** When it was first proposed almost a half-century ago, the idea that a proton gradient was the energy intermediate linking electron transport to ATP synthesis was a departure from the current conventional thinking. Scientists had been searching for a mitochondrial intermediate that they believed would carry energy in much the same way as the ATP produced by substrate level phosphorylation. The search for this intermediate was not successful, and this led to the idea that chemiosmosis was the mechanism of oxidative phosphorylation. Experimental evidence was needed to support this hypothesis. Two key experiments demonstrated (1) that a proton ($H^+$) gradient across a membrane can drive ATP synthesis; and (2) that the enzyme ATP synthase is the catalyst for this reaction (**Figure 9.10**).

In the first experiment, mitochondria without a food source were "fooled" into making ATP by raising the $H^+$ concentration in their environment. In the second experiment, a light-driven proton pump isolated from bacteria was inserted into artificial lipid vesicles. This generated a proton gradient, but since ATP synthase was absent, ATP was not made. Then, ATP synthase was inserted into the vesicles and ATP was generated.

**UNCOUPLING PROTON DIFFUSION FROM ATP PRODUCTION** The tight coupling between $H^+$ diffusion and the formation of ATP provides further evidence for the chemiosmotic mechanism. If a second type of $H^+$ diffusion channel (that does not synthesize ATP) is present in the mitochondrial membrane, the energy of the $H^+$ gradient is released as heat rather than being coupled to ATP synthesis. Such uncoupling molecules actually exist in the mitochondria of some organisms to generate heat instead of ATP. For example, the natural uncoupling protein thermogenin plays an important role in regulating the temperatures of newborn human infants, who lack hair to keep warm, and in hibernating animals.

A popular weight loss drug in the 1930s was the uncoupler molecule, dinitrophenol. There were claims of dramatic weight loss when the drug was administered to obese patients. Unfortunately, the heat that was released caused fatally high fevers, and the effective dose and fatal dose were quite close. So the use of this drug was discontinued in 1938. However, the general strategy of using an uncoupler for weight loss remains a subject of research.

**HOW ATP SYNTHASE WORKS: A MOLECULAR MOTOR** Now that we have established that the $H^+$ gradient is needed for ATP synthesis, a question remains: how does the enzyme actually make ATP from ADP and $P_i$? This is certainly a fundamental question in biology, as it underlies energy harvesting in most cells. Look at the structure of ATP synthase in Figure 9.9. It is a molecular motor composed of two parts: the $F_0$ unit, a transmembrane region that is the $H^+$ channel, and the $F_1$ unit, which contains the active sites for ATP synthesis. $F_1$ consists of six subunits (three each of two polypeptide chains), arranged like the segments of an orange around a central polypeptide. ATP synthesis is coupled with conformational changes in the ATP synthase enzyme, which are induced by proton movement through the complex. The potential energy set up by the proton gradient across the inner membrane drives the passage of protons through the ring of polypeptides that make up the $F_0$ component. This ring rotates as the protons pass through the membrane, causing the $F_1$ unit to rotate as well. ADP and $P_i$ bind to active sites that become exposed on the $F_1$ unit as it rotates, and ATP is made. The structure and function of ATP synthase are shared by living organisms as diverse as bacteria and humans. These molecular motors make ATP at rates up to 100 molecules per second.

### 9.3 RECAP

**The oxidation of reduced electron carriers in the respiratory chain drives the active transport of protons across the inner mitochondrial membrane, generating a proton-motive force. Diffusion of protons down their electrochemical gradient through ATP synthase is coupled to the synthesis of ATP.**

- What are the roles of oxidation and reduction in the respiratory chain? See Figures 9.8 and 9.9
- What is the proton motive force and how does it drive chemiosmosis? See pp. 179–180
- Explain how the two experiments described in Figure 9.10 demonstrate the chemiosmotic mechanism. See p. 181

Oxidative phosphorylation captures a great deal of energy in ATP. But it does not occur if $O_2$ is absent. We turn now to the metabolism of glucose in anaerobic conditions.

## 9.4 How Is Energy Harvested from Glucose in the Absence of Oxygen?

In the absence of $O_2$ (anaerobic conditions), a small amount of ATP can be produced by glycolysis and fermentation. Like glycolysis, fermentation pathways occur in the cytoplasm. There are many different types of fermentation, but they all operate to regenerate $NAD^+$ so that the NAD-requiring reaction of glycolysis can continue (see reaction 6 in Figure 9.5). Of course, if a necessary reactant such as $NAD^+$ is not present, the reaction will not take place. How do fermentation reactions regenerate $NAD^+$ and permit ATP formation to continue?

Prokaryotic organisms often live in $O_2$-deficient environments and are known to use many different fermentation pathways. But the two best understood fermentation pathways are found in a wide variety of organisms including eukaryotes. These two short pathways are lactic acid fermentation, whose end product is lactic acid (lactate); and alcoholic fermentation, whose end product is ethyl alcohol (ethanol).

In lactic acid fermentation, pyruvate serves as the electron acceptor and lactate is the product (**Figure 9.11**). This process takes place in many microorganisms and complex organisms, including higher plants and vertebrates. A notable example of lactic acid fermentation occurs in vertebrate muscle tissue. Usually, vertebrates get their energy for muscle contraction aerobically, with the circulatory system supplying $O_2$ to muscles. In small vertebrates, this is almost always adequate: for example, birds can fly long distances without resting. But in larger vertebrates such as humans, the circulatory system is not up to the task of delivering enough $O_2$ when the need is great, such as during high activity. At this point, the muscle cells break down glycogen (a stored polysaccharide) and undergo lactic acid fermentation.

Lactic acid buildup becomes a problem after prolonged periods because the acid ionizes, forming $H^+$ and lowering the pH of the cell. This affects cellular activities and causes muscle cramps, resulting in muscle pain, which abates upon resting. Lactate dehydrogenase, the enzyme that catalyzes the fermentation reaction, works in both directions. That is, it can catalyze the oxidation of lactate as well as the reduction of pyruvate. When lactate levels are decreased, muscle activity can resume.

Alcoholic fermentation takes place in certain yeasts (eukaryotic microbes) and some plant cells under anaerobic conditions. This process requires two enzymes, pyruvate dehydrogenase and alcohol dehydrogenase, which metabolize pyruvate to ethanol (**Figure 9.12**). As with lactic acid fermentation, the reactions are essentially reversible. For thousands of years, humans have used anaerobic fermentation by yeast cells to produce alcoholic beverages. The cells use sugars from plant sources (glucose from grapes or maltose from barley) to produce the end product, ethanol, in wine and beer.

By recycling $NAD^+$, fermentation allows glycolysis to continue, thus producing small amounts of ATP through substrate-level phosphorylation. The net yield of two ATPs per glucose

GLYCOLYSIS
Glucose ($C_6H_{12}O_6$)
2 ADP + 2 $P_i$
2 ATP
2 $NAD^+$
2 NADH
$COO^-$ | C=O | $CH_3$
2 Pyruvate
Lactate dehydrogenase reaction
2 NADH
2 $NAD^+$
FERMENTATION
$COO^-$ | H—C—OH | $CH_3$
2 Lactic acid (lactate)

Summary of reactants and products:
$C_6H_{12}O_6$ + 2 ADP + 2 $P_i$ ⟶ 2 lactic acid + 2 ATP

**9.11 Lactic Acid Fermentation** Glycolysis produces pyruvate, ATP, and NADH from glucose. Lactic acid fermentation uses NADH as a reducing agent to reduce pyruvate to lactic acid (lactate), thus regenerating $NAD^+$ to keep glycolysis operating.

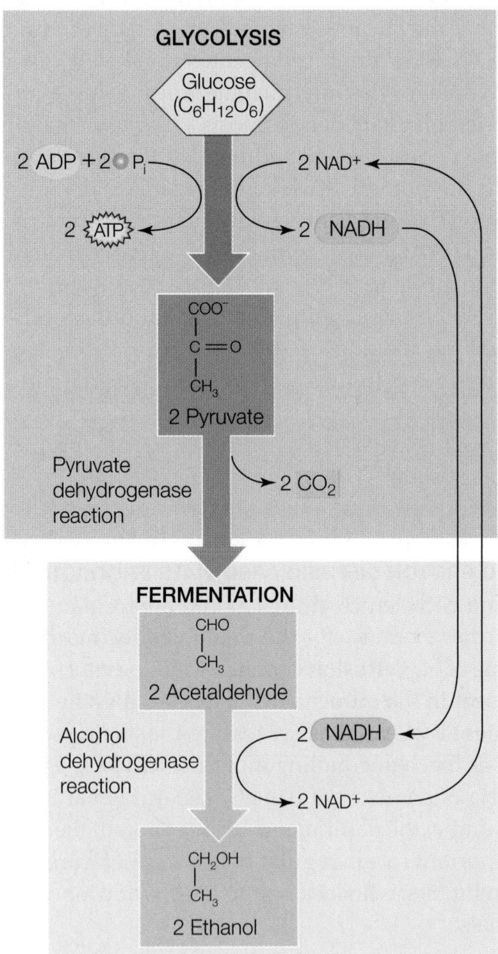


Summary of reactants and products:
$C_6H_{12}O_6$ + 2 ADP + 2 $P_i$ ⟶ 2 ethanol + 2 $CO_2$ + 2 ATP

**9.12 Alcoholic Fermentation** In alcoholic fermentation, pyruvate from glycolysis is converted into acetaldehyde, and $CO_2$ is released. NADH from glycolysis is used to reduce acetaldehyde to ethanol, thus regenerating $NAD^+$ to keep glycolysis operating.

molecule is much lower than the energy yield from cellular respiration. For this reason, most organisms existing in anaerobic environments are small microbes that grow relatively slowly.

### Cellular respiration yields much more energy than fermentation

The total net energy yield from glycolysis plus fermentation is two molecules of ATP per molecule of glucose oxidized. The maximum yield of ATP that can be harvested from a molecule of glucose through glycolysis followed by cellular respiration is much greater—about 32 molecules of ATP (**Figure 9.13**). (Review Figures 9.5, 9.7, and 9.9 to see where all the ATP molecules come from.)

Why do the metabolic pathways that operate in aerobic environments produce so much more ATP? Glycolysis and fermentation only partially oxidize glucose, as does fermentation. Much more energy remains in the end products of fermentation (lactic acid and ethanol) than in $CO_2$, the end product of cellular respiration. In cellular respiration, carriers (mostly $NAD^+$) are reduced in pyruvate oxidation and the citric acid cycle. Then the reduced carriers are oxidized by the respiratory chain, with the accompanying production of ATP by chemiosmosis (2.5 ATP for each NADH and 1.5 ATP for each $FADH_2$). In an aerobic environment, a cell or organism capable of aerobic metabolism will have the advantage over one that is limited to fermentation, in terms of its ability to harvest chemical energy. Two key events in the evolution of multicellular organisms were the rise in atmospheric $O_2$ levels (see Chapter 1) and the development of metabolic pathways to utilize that $O_2$.

### The yield of ATP is reduced by the impermeability of some mitochondria to NADH

The total gross yield of ATP from the oxidation of one molecule of glucose to $CO_2$ is 32. However, in some animal cells the inner mitochondrial membrane is impermeable to NADH, and a "toll" of one ATP must be paid for each NADH molecule that is produced in glycolysis and must be "shuttled" into the mitochondrial matrix. So in these animals, the net yield of ATP is 30.

NADH *shuttle systems* transfer the electrons captured by glycolysis onto substrates that are capable of movement across the mitochondrial membranes. In muscle and liver tissues, an important shuttle involves glycerol 3-phosphate. In the cytosol,

$$\text{NADH (from glycolysis)} + \text{dihydroxyacetone phosphate (DHAP)} \rightarrow NAD^+ + \text{glycerol 3-phosphate}$$

Glycerol 3-phosphate crosses both mitochondrial membranes. In the mitochondrial matrix,

$$\text{FAD} + \text{glycerol 3-phosphate} \rightarrow FADH_2 + \text{DHAP}$$

DHAP is able to move back to the cytosol, where it is available to repeat the process. Note that the reducing electrons are transferred from NADH outside the mitochondrion to $FADH_2$ inside the mitochondrion. As you know from Figures 9.8 and 9.9, the energy yield in terms of ATP from $FADH_2$ is lower than that from NADH. This lowers the overall energy yield.

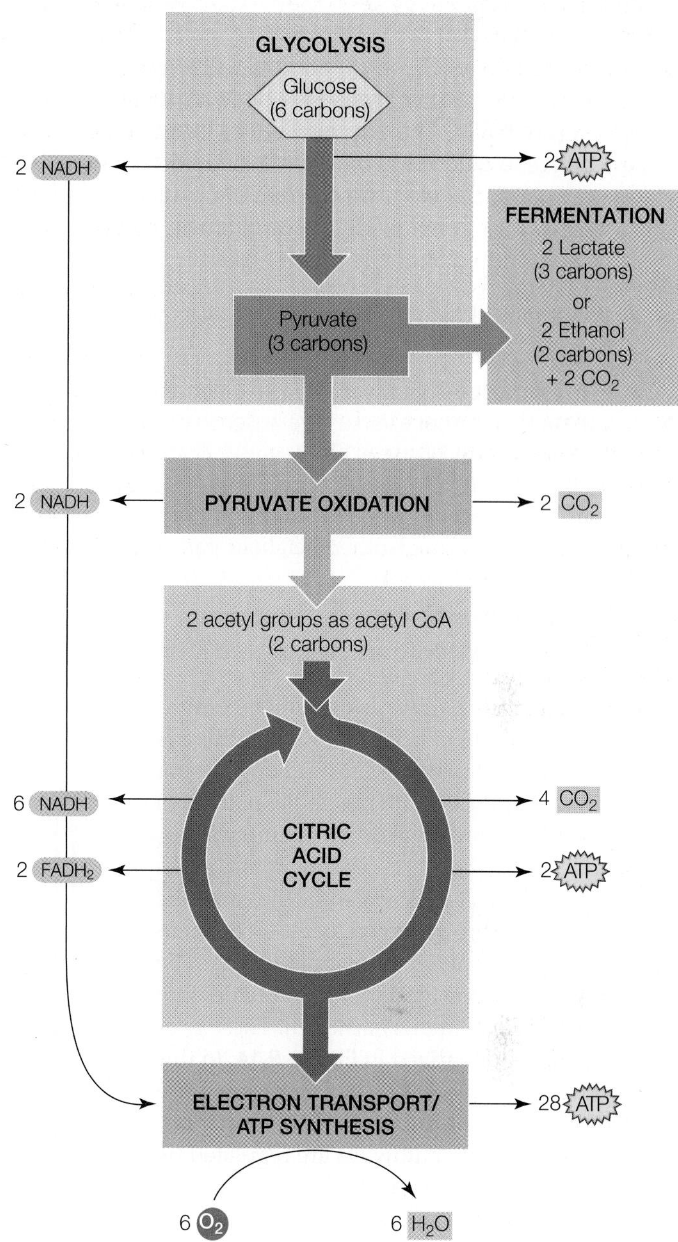


**9.13 Cellular Respiration Yields More Energy Than Fermentation** Electron carriers are reduced in pyruvate oxidation and the citric acid cycle, then oxidized by the respiratory chain. These reactions produce ATP via chemiosmosis.

**yourBioPortal.com**
GO TO Web Activity 9.5 • Energy Levels

## 9.4 RECAP

**In the absence of $O_2$, fermentation pathways use NADH formed by glycolysis to reduce pyruvate and regenerate $NAD^+$. The energy yield of fermentation is low because glucose is only partially oxidized. When $O_2$ is present, the electron carriers of cellular respiration allow for the full oxidation of glucose, so the energy yield from glucose is much higher.**

- Why is replenishing $NAD^+$ crucial to cellular metabolism? **See pp. 182–183**
- What is the total energy yield from glucose in human cells in the presence versus the absence of $O_2$? **See p. 183 and Figure 9.13**

Now that you've seen how cells harvest energy, let's see how that energy moves through other metabolic pathways in the cell.

## 9.5 How Are Metabolic Pathways Interrelated and Regulated?

Glycolysis and the pathways of cellular respiration do not operate in isolation. Rather, there is an interchange of molecules into and out of these pathways, to and from the metabolic pathways for the synthesis and breakdown of amino acids, nucleotides, fatty acids, and other building blocks of life. Carbon skeletons can enter the catabolic pathways and be broken down to release their energy, or they can enter anabolic pathways to be used in the formation of the macromolecules that are the major constituents of the cell. These relationships are summarized in **Figure 9.14**. In this section we will explore how pathways are interrelated by the sharing of intermediate substances, and we will see how pathways are regulated by the inhibitors of key enzymes.

### Catabolism and anabolism are linked

A hamburger or veggie burger on a bun contains three major sources of carbon skeletons: carbohydrates, mostly in the form of starch (a polysaccharide); lipids, mostly as triglycerides (three fatty acids attached to glycerol); and proteins (polymers of amino acids). Look at Figure 9.14 to see how each of these three types of macromolecules can be hydrolyzed and used in catabolism or anabolism.

**CATABOLIC INTERCONVERSIONS** Polysaccharides, lipids, and proteins can all be broken down to provide energy:

- *Polysaccharides* are hydrolyzed to glucose. Glucose then passes through glycolysis and cellular respiration, where its energy is captured in ATP.
- *Lipids* are broken down into their constituents, glycerol and fatty acids. Glycerol is converted into dihydroxyacetone phosphate (DHAP), an intermediate in glycolysis. Fatty acids are highly reduced molecules that are converted to acetyl CoA inside the mitochondrion by a series of oxidation enzymes, in a process known as β-oxidation. For example, the β-oxidation of a $C_{16}$ fatty acid occurs in several steps:

$$C_{16} \text{ fatty acid} + \text{CoA} \rightarrow C_{16} \text{ fatty acyl CoA}$$

$$C_{16} \text{ fatty acyl CoA} + \text{CoA} \rightarrow C_{14} \text{ fatty acyl CoA} + \text{acetyl CoA}$$

$$\text{repeat 6 times} \rightarrow 8 \text{ acetyl CoA}$$

The acetyl CoA can then enter the citric acid cycle and be catabolized to $CO_2$.

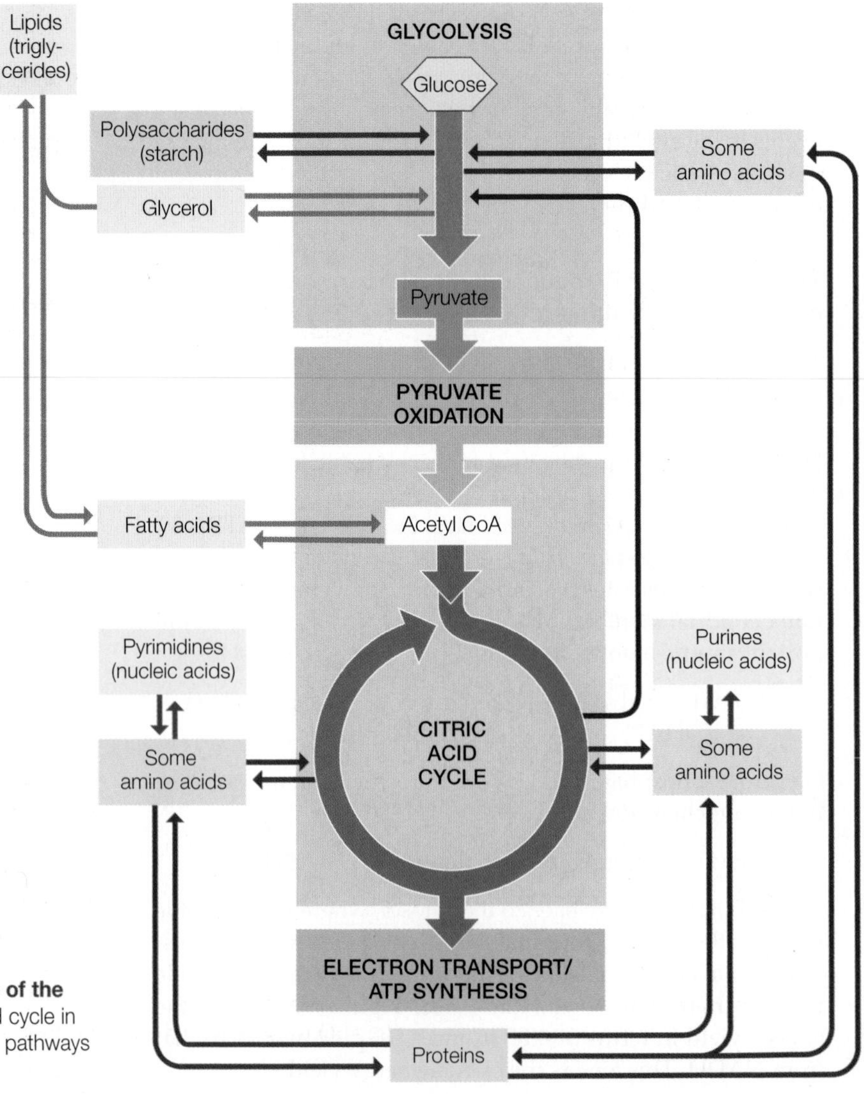


**9.14 Relationships among the Major Metabolic Pathways of the Cell** Note the central positions of glycolysis and the citric acid cycle in this network of metabolic pathways. Also note that many of the pathways can operate essentially in reverse.

- *Proteins* are hydrolyzed to their amino acid building blocks. The 20 different amino acids feed into glycolysis or the citric acid cycle at different points. For example, the amino acid glutamate is converted into α-ketoglutarate, an intermediate in the citric acid cycle.

Glutamate is an amino acid.

Alpha-ketoglutarate is an intermediate in the citric acid cycle.

$$\underset{\text{Glutamate}}{{}^{-}OOC{-}CH_2{-}CH_2{-}CH(NH_3^+){-}COO^-} \;\underset{NAD^+ \;\to\; NADH}{\rightleftharpoons}\; \underset{\alpha\text{-Ketoglutarate}}{{}^{-}OOC{-}CH_2{-}CH_2{-}C({=}O){-}COO^-} + NH_4^+$$

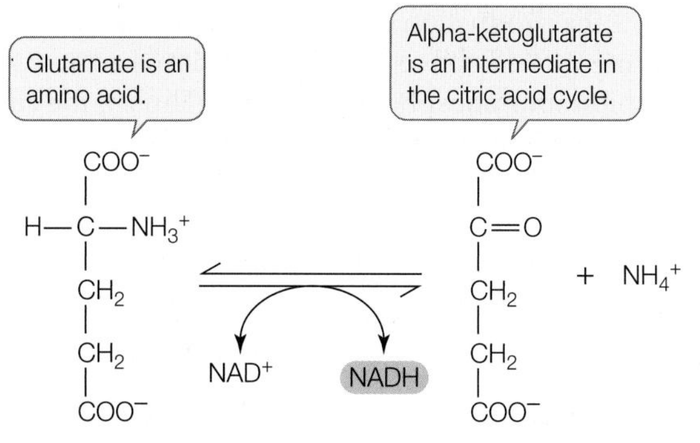

**ANABOLIC INTERCONVERSIONS** Many catabolic pathways can operate essentially in reverse, with some modifications. Glycolytic and citric acid cycle intermediates, instead of being oxidized to form $CO_2$, can be reduced and used to form glucose in a process called **gluconeogenesis** (which means "new formation of glucose"). Likewise, acetyl CoA can be used to form fatty acids. The most common fatty acids have even numbers of carbons: 14, 16, or 18. These are formed by the addition of two-carbon acetyl CoA "units" one at a time until the appropriate chain length is reached. Acetyl CoA is also a building block for various pigments, plant growth substances, rubber, steroid hormones, and other molecules.

Some intermediates in the citric acid cycle are reactants in pathways that synthesize important components of nucleic acids. For example, α-ketoglutarate is a starting point for purines, and oxaloacetate for pyrimidines. In addition, α-ketoglutarate is a starting point for the synthesis of chlorophyll (used in photosynthesis; see Chapter 10) and the amino acid glutamate (used in protein synthesis).

## Catabolism and anabolism are integrated

A carbon atom from a protein in your burger can end up in DNA, fat, or $CO_2$, among other fates. How does the organism "decide" which metabolic pathways to follow, in which cells? With all of the possible interconversions, you might expect that cellular concentrations of various biochemical molecules would vary widely. Remarkably, the levels of these substances in what is called the metabolic pool—the sum total of all the biochemical molecules in a cell—are quite constant. Organisms regulate the enzymes of catabolism and anabolism in various cells in order to maintain a balance. This metabolic homeostasis gets upset only in unusual circumstances. Let's look one such unusual circumstance: undernutrition.

Glucose is an excellent source of energy, but lipids and proteins can also be broken down and their constituents used as energy sources. Any one or all three of these types of molecules could be used to provide the energy your body needs. But normally these substances are not equally available for energy metabolism and ATP formation. Proteins, for example, have essential roles as enzymes and as structural elements, providing support and movement; they are not stored for energy, and using them for energy might deprive the body of other vital functions.

Fats (trigylcerides) do not have catalytic roles. Because they are nonpolar, fats do not bind water, and they are therefore less dense than polysaccharides in aqueous environments. In addition, fats are more reduced than carbohydrates (have more C—H bonds and fewer C—OH bonds) and thus have more energy stored in their bonds (see Figure 9.2). So it is not surprising that fats are the preferred energy store in many organisms. The human body stores fats and carbohydrates; fats are stored in adipose tissue, and glucose is stored as the polysaccharide glycogen in muscles and the liver. A typical person has about one day's worth of food energy stored as glycogen (a polysaccharide) and over a month's food energy stored as fats.

What happens if a person does not eat enough to produce sufficient ATP and NADH for anabolism and biological activities? This situation can be deliberate (to lose weight), but for too many people, it is forced upon them because not enough food is available, resulting in undernutrition and starvation. Initially, homeostasis can be maintained. The first energy stores to be used are the glycogen stores in muscle and liver cells. These stores do not last long, and next come the fats.

In cells that have access to fatty acids, their breakdown produces acetyl CoA for cellular respiration. However, a problem remains: because fatty acids cannot cross from the blood to the brain, the brain can use only glucose as its energy source. With glycogen already depleted, the body must convert something else to make glucose for the brain. This is accomplished by the breakdown of proteins and the conversion of their amino acids to glucose by gluconeogenesis. Without sufficient food intake, proteins and fats are used up. After several weeks of starvation, fat stores become depleted, and the only energy source left is protein. At this point, essential structural proteins, enzymes, and antibodies get broken down. The loss of such proteins can lead to severe illness and eventual death.

## Metabolic pathways are regulated systems

We have described the relationships between metabolic pathways and noted that these pathways work together to provide homeostasis in the cell and organism. But how does the cell regulate the interconversions between pathways to maintain constant metabolic pools? This is a problem of systems biology, which seeks to understand how biochemical pathways interact (see Figure 8.15). It is a bit like trying to predict traffic patterns in a city: if an accident blocks traffic on a major road, drivers take alternate routes, where the traffic volume consequently changes.

Consider what happens to the starch in your burger bun. In the digestive system, starch is hydrolyzed to glucose, which enters the blood for distribution to the rest of the body. But before the glucose is distributed, a regulatory check must be made: if there is already enough glucose in the blood to supply the body's needs, the excess glucose is converted into glycogen and

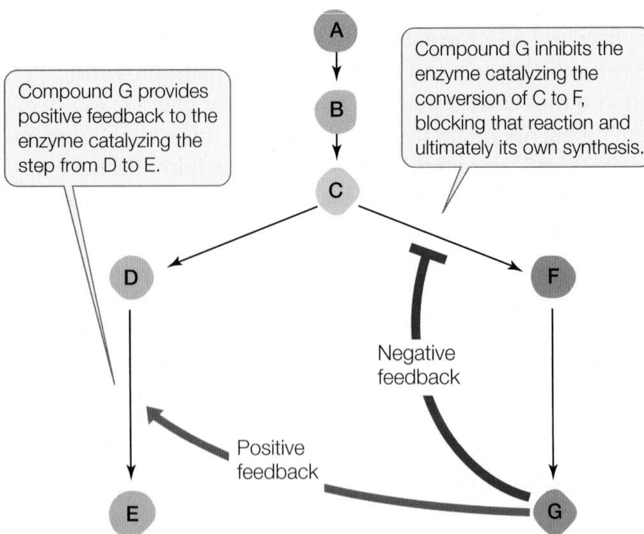


**9.15 Regulation by Negative and Positive Feedback**
Allosteric feedback regulation plays an important role in metabolic pathways. The accumulation of some products can shut down their synthesis, or can stimulate other pathways that require the same raw materials.

stored in the liver. If not enough glucose is supplied by food, glycogen is broken down, or other molecules are used to make glucose by gluconeogenesis.

The end result is that the level of glucose in the blood is remarkably constant. How does the body accomplish this?

Glycolysis, the citric acid cycle, and the respiratory chain are subject to *allosteric regulation* (see Section 8.5) of the enzymes involved. An example of allosteric regulation is feedback inhibition, illustrated in Figure 8.19. In a metabolic pathway, a high concentration of the final product can inhibit the action of an enzyme that catalyzes an earlier reaction. On the other hand, an excess of the product of one pathway can speed up reactions in another pathway, diverting raw materials away from synthesis of the first product (**Figure 9.15**). These negative and positive feedback mechanisms are used at many points in the energy-harvesting pathways, and are summarized in **Figure 9.16**.

- The main control point in glycolysis is the enzyme *phosphofructokinase* (reaction 3 in Figure 9.5). This enzyme is allosterically inhibited by ATP or citrate, and activated by ADP or AMP. Under anaerobic conditions, fermentation yields a relatively small amount of ATP, and phosphofructokinase operates at a high rate. However when conditions are aerobic, respiration makes 16 times more ATP than fermentation does, and the abundant ATP allosterically inhibits phosphofructokinase. Consequently, the conversion of fructose 6-phosphate to fructose 1,6-bisphosphate declines, and so does the rate of glucose utilization.
- The main control point in the citric acid cycle is the enzyme *isocitrate dehydrogenase*, which converts isocitrate to α-ketoglutarate (reaction 3 in Figure 9.7). NADH and ATP are feedback inhibitors of this reaction, while ADP and $NAD^+$ are activators. If too much ATP or NADH accumulates, the conversion of isocitrate is slowed, and the citric acid cycle shuts down. A shutdown of the citric acid cycle would cause large amounts of isocitrate and citrate to accumulate if the production of citrate were not also slowed. But, as mentioned above, an excess of citrate acts as a feedback inhibitor of phosphofructokinase. Thus, if the citric acid cycle has been slowed or shut down because of abun-

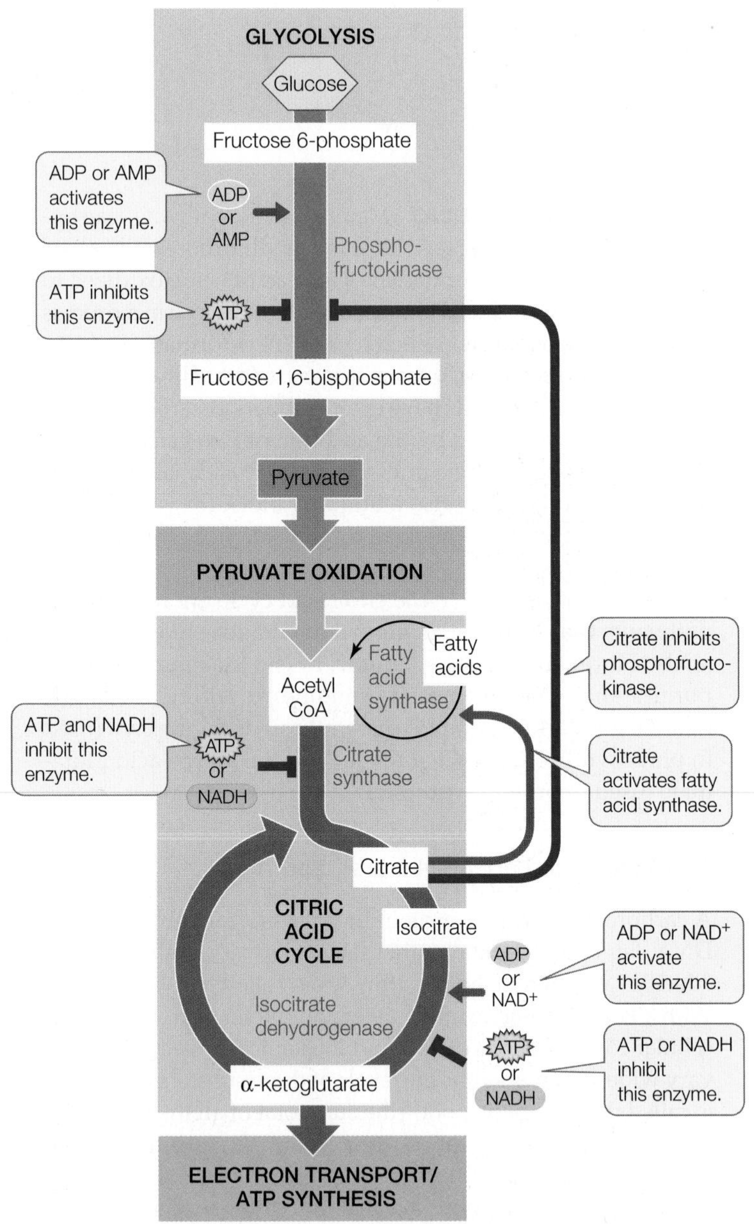


**9.16 Allosteric Regulation of Glycolysis and the Citric Acid Cycle**
Allosteric regulation controls glycolysis and the citric acid cycle at crucial early steps, increasing their efficiency and preventing the excessive buildup of intermediates.

**yourBioPortal.com**
GO TO Web Activity 9.6 • Regulation of Energy Pathways

dant ATP (and not because of a lack of oxygen), glycolysis is slowed as well. Both processes resume when the ATP level falls and they are needed again. Allosteric regulation keeps these processes in balance.

- Another control point involves *acetyl CoA*. If the level of ATP is high and the citric acid cycle shuts down, the accumulation of citrate activates fatty acid synthase, diverting acetyl CoA to the synthesis of fatty acids for storage. That is one reason why people who eat too much accumulate fat. These fatty acids may be metabolized later to produce more acetyl CoA.

## 9.5 RECAP

**Glucose can be made from intermediates in glycolysis and the citric acid cycle by a process called gluconeogenesis. The metabolic pathways for the production and breakdown of lipids and amino acids are tied to those of glucose metabolism. Reaction products regulate key enzymes in the various pathways.**

- Give examples of a catabolic interconversion of a lipid and of an anabolic interconversion of a protein. See pp. 184–185 and Figure 9.14
- How does phosphofructokinase serve as a control point for glycolysis? See p. 186 and Figure 9.16
- Describe what would happen if there was no allosteric mechanism for modulating the level of acetyl CoA.

# CHAPTER SUMMARY

### 9.1 How Does Glucose Oxidation Release Chemical Energy?

- As a material is **oxidized**, the electrons it loses are transferred to another material, which is thereby **reduced**. Such **redox reactions** transfer large amounts of energy. Review Figure 9.2, **WEB ACTIVITIES 9.1 and 9.2**
- The coenzyme **$NAD^+$** is a key electron carrier in biological redox reactions. It exists in two forms, one oxidized ($NAD^+$) and the other reduced (NADH).
- **Glycolysis** operates in the presence or absence of $O_2$. Under **aerobic** conditions, **cellular respiration** continues the process of breaking down glucose. Under **anaerobic** conditions, **fermentation** occurs. Review Figure 9.4
- The pathways of cellular respiration after glycolysis are **pyruvate oxidation**, the **citric acid cycle**, and the **electron transport/ATP synthesis**.

### 9.2 What Are the Aerobic Pathways of Glucose Metabolism?

- Glycolysis consists of 10 enzyme-catalyzed reactions that occur in the cell cytoplasm. Two **pyruvate** molecules are produced for each partially oxidized molecule of glucose, providing the starting material for both cellular respiration and fermentation. Review Figure 9.5
- The first five reactions of glycolysis require an investment of energy; the last five produce energy. The net gain is two molecules of ATP. Review Figure 9.6
- The enzyme-catalyzed transfer of phosphate groups to ADP by enzymes other than ATPase is called **substrate-level phosphorylation** and produces ATP.
- Pyruvate oxidation follows glycolysis and links glycolysis to the citric acid cycle. This pathway converts pyruvate into **acetyl CoA**.
- Acetyl CoA is the starting point of the citric acid cycle. It reacts with oxaloacetate to produce citrate. A series of eight enzyme-catalyzed reactions oxidize citrate and regenerate oxaloacetate, continuing the cycle. Review Figure 9.7, **WEB ACTIVITY 9.3**

### 9.3 How Does Oxidative Phosphorylation Form ATP?

- Oxidation of electron carriers in the presence of $O_2$ releases energy that can be used to form ATP in a process called **oxidative phosphorylation**.
- The NADH and $FADH_2$ produced in glycolysis, pyruvate oxidation, and the citric acid cycle are oxidized by the respiratory chain, regenerating $NAD^+$ and FAD. Oxygen ($O_2$) is the final acceptor of electrons and protons, forming water ($H_2O$). Review Figure 9.8, **WEB ACTIVITY 9.4**
- The respiratory chain not only transports electrons, but also pumps protons across the inner mitochondrial membrane, creating the **proton-motive force**.
- Protons driven by the proton-motive force can return to the mitochondrial matrix via **ATP synthase**, a molecular motor that couples this movement of protons to the synthesis of ATP. This process is called **chemiosmosis**. Review Figure 9.9, **ANIMATED TUTORIALS 9.1 and 9.2**

### 9.4 How Is Energy Harvested from Glucose in the Absence of Oxygen?

- In the absence of $O_2$, glycolysis is followed by fermentation. Together, these pathways partially oxidize pyruvate and generate end products such as **lactic acid** or **ethanol**. In the process, $NAD^+$ is regenerated from NADH so that glycolysis can continue, thus generating a small amount of ATP. Review Figures 9.11 and 9. 12
- For each molecule of glucose used, fermentation yields 2 molecules of ATP. In contrast, glycolysis operating with pyruvate oxidation, the citric acid cycle, and the respiratory chain/ATP synthase yields up to 32 molecules of ATP per molecule of glucose. Review Figure 9.13, **WEB ACTIVITY 9.5**

### 9.5 How Are Metabolic Pathways Interrelated and Regulated?

- The **catabolic pathways** for the breakdown of carbohydrates, fats, and proteins feed into the energy-harvesting metabolic pathways. Review Figure 9.14
- **Anabolic pathways** use intermediate components of the energy-harvesting pathways to synthesize fats, amino acids, and other essential building blocks.
- The formation of glucose from intermediates of glycolysis and the citric acid cycle is called **gluconeogenesis**.
- The rates of glycolysis and the citric acid cycle are controlled by **allosteric regulation** and by the diversion of excess acetyl CoA into fatty acid synthesis. Key regulated enzymes include phosphofructokinase, citrate synthase, isocitrate dehydrogenase, and fatty acid synthase. See Figure 9.16, **WEB ACTIVITY 9.6**

## SELF-QUIZ

1. The role of oxygen gas in our cells is to
   *a.* catalyze reactions in glycolysis.
   *b.* produce $CO_2$.
   *c.* form ATP.
   *d.* accept electrons from the respiratory chain.
   *e.* react with glucose to split water.
2. Oxidation and reduction
   *a.* entail the gain or loss of proteins.
   *b.* are defined as the loss of electrons.
   *c.* are both endergonic reactions.
   *d.* always occur together.
   *e.* proceed only under aerobic conditions.
3. $NAD^+$ is
   *a.* a type of organelle.
   *b.* a protein.
   *c.* present only in mitochondria.
   *d.* a part of ATP.
   *e.* formed in the reaction that produces ethanol.
4. Glycolysis
   *a.* takes place in the mitochondrion.
   *b.* produces no ATP.
   *c.* has no connection with the respiratory chain.
   *d.* is the same thing as fermentation.
   *e.* reduces two molecules of $NAD^+$ for every glucose molecule processed.
5. Fermentation
   *a.* takes place in the mitochondrion.
   *b.* takes place in all animal cells.
   *c.* does not require $O_2$.
   *d.* requires lactic acid.
   *e.* prevents glycolysis.
6. Which statement about pyruvate is *not* true?
   *a.* It is the end product of glycolysis.
   *b.* It becomes reduced during fermentation.
   *c.* It is a precursor of acetyl CoA.
   *d.* It is a protein.
   *e.* It contains three carbon atoms.
7. The citric acid cycle
   *a.* has no connection with the respiratory chain.
   *b.* is the same thing as fermentation.
   *c.* reduces two $NAD^+$ for every glucose processed.
   *d.* produces no ATP.
   *e.* takes place in the mitochondrion.
8. The respiratory chain
   *a.* is located in the mitochondrial matrix.
   *b.* includes only peripheral membrane proteins.
   *c.* always produces ATP.
   *d.* reoxidizes reduced coenzymes.
   *e.* operates simultaneously with fermentation.
9. Compared with fermentation, the aerobic pathways of glucose metabolism produce
   *a.* more ATP.
   *b.* pyruvate.
   *c.* fewer protons for pumping in the mitochondria.
   *d.* less $CO_2$.
   *e.* more oxidized coenzymes.
10. Which statement about oxidative phosphorylation is *not* true?
    *a.* It forms ATP by the respiratory chain/ATP synthesis.
    *b.* It is brought about by chemiosmosis.
    *c.* It requires aerobic conditions.
    *d.* It takes place in mitochondria.
    *e.* Its functions can be served equally well by fermentation.

## FOR DISCUSSION

1. Trace the sequence of chemical changes that occurs in mammalian tissue when the oxygen supply is cut off. The first change is that the cytochrome *c* oxidase system becomes totally reduced, because electrons can still flow from cytochrome *c*, but there is no oxygen to accept electrons from cytochrome *c* oxidase. What are the remaining steps?
2. Some cells that use the aerobic pathways of glucose metabolism can also thrive by using fermentation under anaerobic conditions. Given the lower yield of ATP (per molecule of glucose) in fermentation, how can these cells function so efficiently under anaerobic conditions?
3. The drug antimycin A blocks electron transport in mitochondria. Explain what would happen if the experiment on the left in Figure 9.10 were repeated in the presence of this drug.
4. You eat a burger that contains polysaccharides, proteins, and lipids. Using what you know of the integration of biochemical pathways, explain how the amino acids in the proteins and the glucose in the polysaccharides can end up as fats.

## ADDITIONAL INVESTIGATION

A protein in the fat of newborns uncouples the synthesis of ATP from electron transport and instead generates heat. How would you investigate the hypothesis that this uncoupling protein adds a second proton channel to the mitochondrial membrane?

## WORKING WITH DATA (GO TO yourBioPortal.com)

**Two Experiments Demonstrate the Chemiosmotic Mechanism**
In this real-life exercise, you will examine the background and data from the original research paper by Jagendorf and Uribe in which they showed that an artificially induced $H^+$ gradient could drive ATP synthesis (Figure 9.10). You will see how they measured ATP by two different methods, and what control experiments they performed to confirm their interpretation.

# 10 Photosynthesis: Energy from Sunlight

## Photosynthesis and global climate change

If all the carbohydrates produced by photosynthesis in a year were in the form of sugar cubes, there would be 300 quadrillion of them. Lined up, these cubes would extend from Earth to Pluto—a lot of photosynthesis! As you may have learned from previous courses, photosynthetic organisms use atmospheric carbon dioxide ($CO_2$) to produce carbohydrates. The simplified equation says it all:

$$CO_2 + H_2O \rightarrow O_2 + \text{carbohydrates}$$

Given the role of $CO_2$, how will photosynthesis change with increasing levels of atmospheric $CO_2$? Over the past 200 years, the concentration of atmospheric $CO_2$ has increased—from 280 parts per million (ppm) in 1800 to 386 ppm in 2008. This increase is correlated with industrialization and the accompanying use of fossil fuels such as coal and oil, which release $CO_2$ into the atmosphere when they are burned. The Intergovernmental Panel of Climate Change, sponsored by the United Nations, estimates that atmospheric $CO_2$ will continue to rise over the next century.

Carbon dioxide is a "greenhouse gas" that traps heat in the atmosphere, and the rising $CO_2$ level is predicted to result in global climate change. Policy makers concerned about climate change are asking plant biologists to answer two questions about the rise in $CO_2$: will it lead to increased photosynthesis, and if so, will it lead to increased plant growth? To answer these questions, scientists initially measured the rate of photosynthesis of plants grown in greenhouses with elevated concentrations of $CO_2$. The results were surprising: at first, the rate of photosynthesis went up, but then it returned to near normal as the plants adapted to the higher $CO_2$ levels.

To determine how plants might respond under more realistic conditions, scientists developed a way to expose plants to high levels of $CO_2$ in the field. *F*ree-*a*ir *c*oncentration *e*nrichment (FACE) involves the use of rings of pipes that release $CO_2$ to the air surrounding plants in fields or forests. Wind speed and direction are monitored by a computer, which constantly controls which pipes release $CO_2$. Data from these experiments confirm that photosynthetic rates increase as the concentration of $CO_2$ rises—although generally the increase is not as high as that seen initially in the greenhouse experiments. Nevertheless, these measurements indicate that as atmospheric $CO_2$ rises globally, there will be an increase in photosynthesis.

Will this increase in photosynthesis result in an increase in plant growth? Keep in mind that plants, like all organisms, use carbohydrates as an energy source. They perform cellular respiration with the general equation:

$$\text{carbohydrates} + O_2 \rightarrow CO_2 + H_2O$$

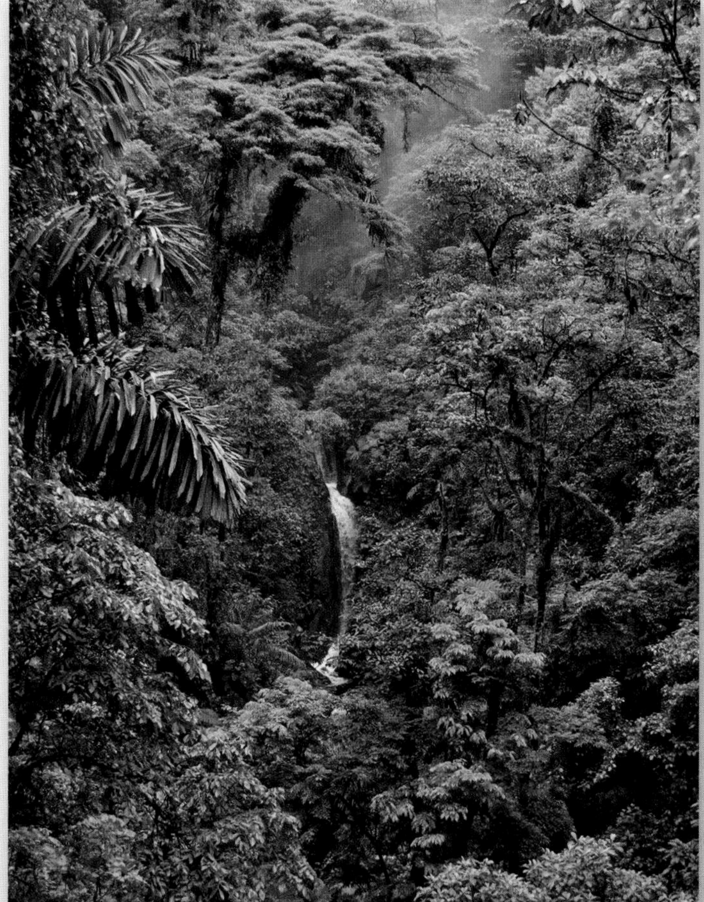

**Primary Producers** Covering less than 2 percent of Earth's surface, rainforests are photosynthetic dynamos. They may act as a "sink" for increasing atmospheric $CO_2$.

**FACE** Free-air carbon dioxide enrichment uses pipes to release $CO_2$ around plants in the field, to estimate the effects of rising atmospheric $CO_2$ on photosynthesis and plant growth.

The challenge facing plant biologists is to determine the balance between photosynthesis and respiration and how this affects the rate of plant growth. The FACE experiments indicate that crop yields increase under higher $CO_2$ concentrations, suggesting that the overall increase in photosynthesis is greater than the increase in respiration. But climate change alters rainfall patterns as well as temperatures. These changes affect where plants grow, and could shift the balance between plant growth and cellular respiration.

As with much in science, the initial questions at first appeared amenable to simple answers. Instead, they led to more questions, and more data are needed. An understanding of the processes of photosynthesis, described in this chapter, provides us with a foundation for asking and answering these urgent questions about climate change and its effects on our world.

**IN THIS CHAPTER** we begin with a consideration of light energy, and move on to describe how photosynthesis converts light energy into chemical energy, in the form of reduced electron carriers and ATP. Then, we show how these two sources of chemical energy are used to drive the synthesis of carbohydrates from $CO_2$. Finally, we describe how these processes relate to plant metabolism and growth.

CHAPTER OUTLINE



## 10.1 What Is Photosynthesis?

Photosynthesis (literally, "synthesis from light") is a metabolic process by which the energy of sunlight is captured and used to convert carbon dioxide ($CO_2$) and water ($H_2O$) into carbohydrates (which we will represent as a six-carbon sugar, $C_6H_{12}O_6$) and oxygen gas ($O_2$). By early in the nineteenth century, scientists had grasped these broad outlines of photosynthesis and had established several facts about the way the process works:

- The water for photosynthesis in land plants comes primarily from the soil, and must travel from the roots to the leaves.
- Plants take in carbon dioxide, producing carbohydrates (sugars) for growth, and plants release $O_2$ (**Figure 10.1**).
- Light is absolutely necessary for the production of oxygen and sugars.

By 1804, scientists had summarized photosynthesis as follows:

$$\text{carbon dioxide} + \text{water} + \text{light energy} \rightarrow \text{sugar} + \text{oxygen}$$

In molecular terms, this equation seems to be the reverse of the overall equation for cellular respiration (see Section 9.1). More precisely, photosynthesis can be written as:

$$6\,CO_2 + 6\,H_2O \rightarrow C_6H_{12}O_6 + 6\,O_2$$

While this equation and the one for cellular respiration (given in the chapter opening story) are essentially correct, they are too general for a real understanding of the processes involved. A number of questions arise: What are the precise chemical reactions of photosynthesis? What role does light play in these reactions? How do carbons become linked to form carbohydrates? What carbohydrates are formed? And where does the oxygen gas come from: $CO_2$ or $H_2O$?

### Experiments with isotopes show that in photosynthesis $O_2$ comes from $H_2O$

In 1941 Samuel Ruben and Martin Kamen, at the University of California, Berkeley, performed experiments using the isotopes $^{18}O$ and $^{16}O$ to identify the source of the $O_2$ produced during photosynthesis (**Figure 10.2**). Their results showed that all the oxygen gas produced during photosynthesis comes from water, as is reflected in the revised balanced equation:

$$6\,CO_2 + 12\,H_2O \rightarrow C_6H_{12}O_6 + 6\,O_2 + 6\,H_2O$$

**10.1 The Ingredients for Photosynthesis** A typical terrestrial plant uses light from the sun, water from the soil, and carbon dioxide from the atmosphere to form organic compounds by photosynthesis.

Water appears on both sides of the equation because it is both used as a reactant (the twelve molecules on the left) and released as a product (the six new ones on the right). This revised equation accounts for all the water molecules needed for all the oxygen gas produced.

The realization that water was the source of photosynthetic $O_2$ led to an understanding of photosynthesis in terms of *oxidation and reduction*. As we describe in Chapter 9, oxidation–reduction (redox) reactions are coupled: when one molecule becomes oxidized in a reaction, another gets reduced. In this case, oxygen atoms in the reduced state in $H_2O$ get oxidized to $O_2$:

$$12\ H_2O \rightarrow 24\ H^+ + 24\ e^- + 6\ O_2$$

while carbon atoms in the oxidized state in $CO_2$ get reduced to carbohydrate, with the simultaneous production of water:

$$6\ CO_2 + 24\ H^+ + 24\ e^- \rightarrow C_6H_{12}O_6 + 6\ H_2O$$

Adding these two equations (chemistry students will recognize them as *half-cell reactions*) gives the overall equation shown above. As you will see, there is an intermediary carrier of the $H^+$ and electrons between these two processes—the redox coenzyme, nicotinamide adenine dinucleotide phosphate ($NADP^+$).

### Photosynthesis involves two pathways

The equations above summarize the overall process of photosynthesis, but not the stages by which it is completed. Like glycolysis and the other metabolic pathways that harvest energy in cells, photosynthesis is a process consisting of many reactions. These reactions are commonly divided into two main pathways:

- The **light reactions** convert light energy into chemical energy in the form of ATP and the reduced electron carrier NADPH. This molecule is similar to NADH (see Section 9.1) but with an additional phosphate group attached to the sugar of its adenosine. In general, NADPH acts as a reducing agent in photosynthesis and other anabolic reactions.
- The **light-independent reactions** (carbon-fixation reactions) do not use light directly, but instead use ATP, NADPH (*made by the light reactions*), and $CO_2$ to produce carbohydrate.

## INVESTIGATING LIFE

**10.2 The Source of the Oxygen Produced by Photosynthesis**

Although it was clear that $O_2$ was made during photosynthesis, its molecular source was not known. Two possibilities were the reactants, $CO_2$ and $H_2O$. In two separate experiments, Samuel Ruben and Martin Kamen labeled the oxygen in these molecules with the isotope $^{18}O$, then tested the $O_2$ produced by a green plant to find out which molecule contributed the oxygen.

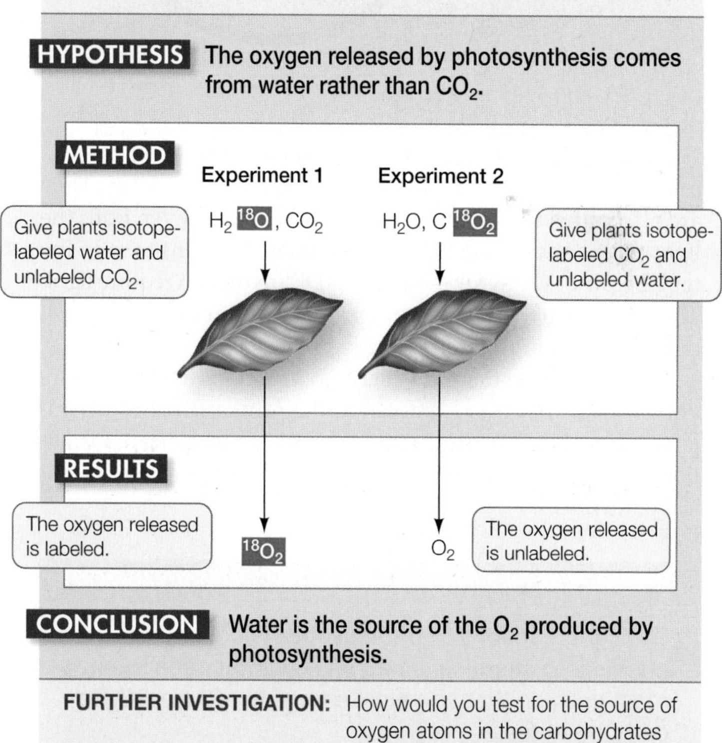


**FURTHER INVESTIGATION:** How would you test for the source of oxygen atoms in the carbohydrates made by photosynthesis?

Go to **yourBioPortal.com** for original citations, discussions, and relevant links for all INVESTIGATING LIFE figures.

**yourBioPortal.com**

GO TO Animated Tutorial 10.1 • The Source of the Oxygen Produced by Photosynthesis

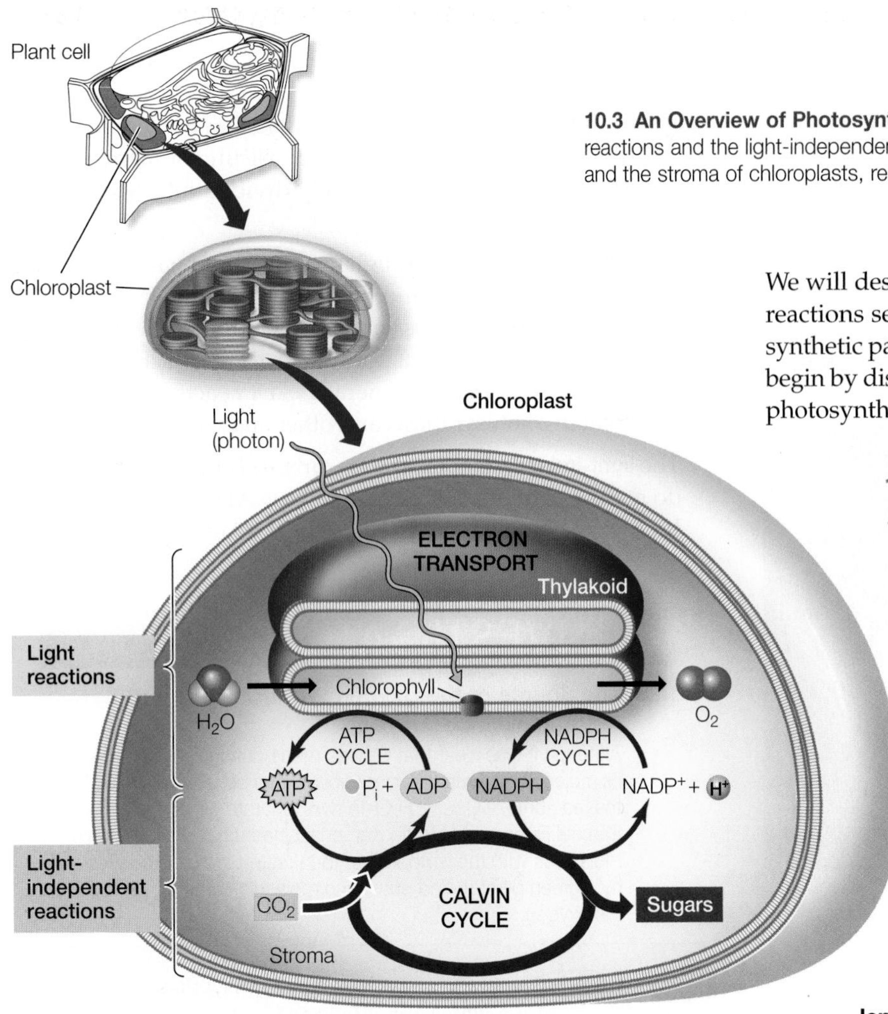


**10.3 An Overview of Photosynthesis** Photosynthesis consists of two pathways: the light reactions and the light-independent reactions. These reactions take place in the thylakoids and the stroma of chloroplasts, respectively.

The light-independent reactions are sometimes called the *dark reactions* because they do not directly require light energy. They are also called the *carbon-fixation reactions*. However, both the light reactions and the light-independent reactions stop in the dark because ATP synthesis and $NADP^+$ reduction require light. The reactions of both pathways proceed within the chloroplast, but they occur in different parts of that organelle (**Figure 10.3**).

As we describe these two series of reactions in more detail, you will see that they conform to the principles of biochemistry that we discuss in Chapters 8 and 9: energy transformations, oxidation-reduction, and the stepwise nature of biochemical pathways.

## 10.1 RECAP

**The light reactions of photosynthesis convert light energy into chemical energy. The light-independent reactions use that chemical energy to reduce $CO_2$ to carbohydrates.**

- What is the experimental evidence that water is the source of the $O_2$ produced during photosynthesis? **See pp. 190–191 and Figure 10.2**
- What is the relationship between the light reactions and the light-independent reactions of photosynthesis? **See pp. 191–192 and Figure 10.3**

We will describe the light reactions and the light-independent reactions separately and in detail. But since these two photosynthetic pathways are powered by the energy of sunlight, let's begin by discussing the physical nature of light and the specific photosynthetic molecules that capture its energy.

## 10.2 How Does Photosynthesis Convert Light Energy into Chemical Energy?

Light is a form of energy, and it can be converted to other forms of energy such as heat or chemical energy. Our focus here will be on light as the source of energy to drive the formation of ATP (from ADP and $P_i$) and NADPH (from $NADP^+$ and $H^+$).

### Light is a form of energy with dual properties

Light is a form of **electromagnetic radiation**. It is propagated in waves, and the amount of energy in light is inversely proportional to its **wavelength**—the shorter the wavelength, the greater the energy. The visible portion of the electromagnetic spectrum (**Figure 10.4**) encompasses a wide range of wavelengths and energy levels. In addition to traveling in waves, light also behaves as particles, called **photons**, which have no mass. In plants and other photosynthetic organisms, receptive molecules absorb photons in order to harvest their energy for biological processes. Because these receptive molecules absorb only specific wavelengths of light, the photons must have the correct amount of energy—they must be of the appropriate wavelength.

### Molecules become excited when they absorb photons

When a photon meets a molecule, one of three things can happen:

- The photon may bounce off the molecule—it may be scattered or reflected.
- The photon may pass through the molecule—it may be transmitted.
- The photon may be absorbed by the molecule, adding energy to the molecule.

Neither of the first two outcomes causes any change in the molecule. However, in the case of absorption, the photon disappears and its energy is absorbed by the molecule. The photon's energy cannot disappear, because according to the first law of thermodynamics, energy is neither created nor destroyed. When the molecule acquires the energy of the photon it is raised from a ground state (with lower energy) to an excited state (with higher energy) (**Figure 10.5A**).

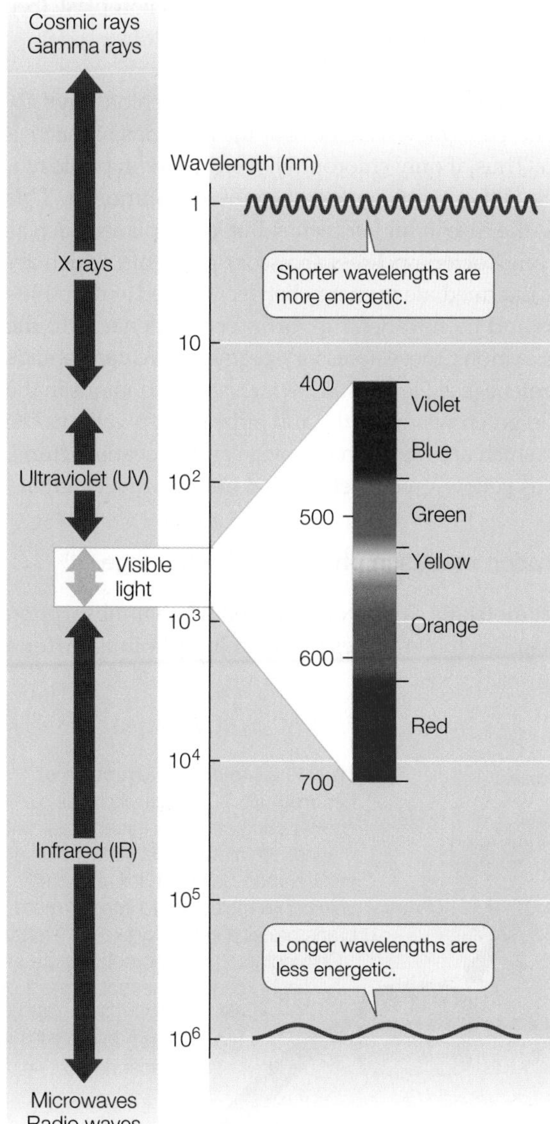


**10.4 The Electromagnetic Spectrum** The portion of the electromagnetic spectrum that is visible to humans as light is shown in detail at the right.

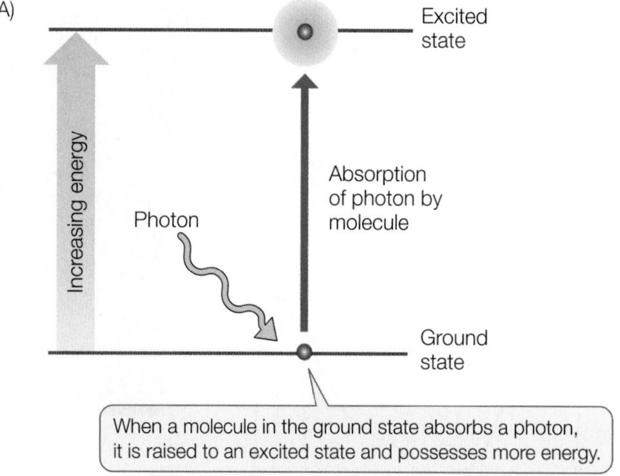


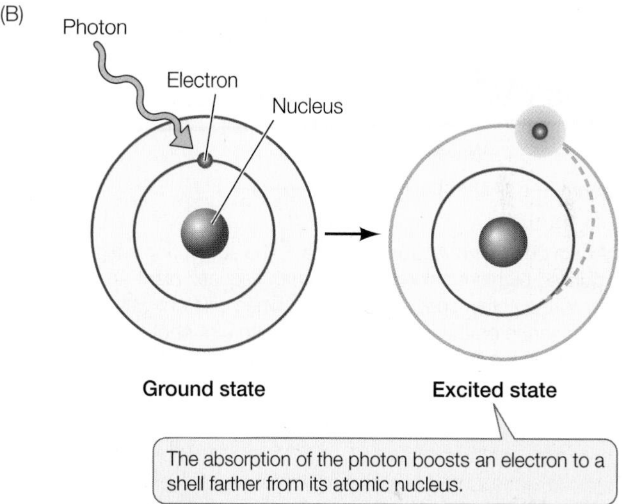


**10.5 Exciting a Molecule** (A) When a molecule absorbs the energy of a photon, it is raised from a ground state to an excited state. (B) In the excited state, an electron is boosted to a shell more distant from the atomic nucleus, where it is held less firmly.

The difference in free energy between the molecule's excited state and its ground state is approximately equal to the free energy of the absorbed photon (a small amount of energy is lost to entropy). The increase in energy boosts one of the electrons within the molecule into a shell farther from its nucleus; this electron is now held less firmly (**Figure 10.5B**), making the molecule unstable and more chemically reactive.

## Absorbed wavelengths correlate with biological activity

The specific wavelengths absorbed by a particular molecule are characteristic of that type of molecule. Molecules that absorb wavelengths in the visible spectrum are called **pigments**.

When a beam of white light (containing all the wavelengths of visible light) falls on a pigment, certain wavelengths are absorbed. The remaining wavelengths, which are scattered or transmitted, make the pigment appear to us as colored. For example, if a pigment absorbs both blue and red light (as does chlorophyll) what we see is the remaining light, which is primarily green. If we plot light absorbed by a purified pigment against wavelength, the result is an **absorption spectrum** for that pigment.

In contrast to the absorption spectrum, an **action spectrum** is a plot of the *biological activity* of an organism as a function of the wavelengths of light to which it is exposed. The experimental determination of an action spectrum might be performed as follows:

1. Place a plant (a water plant with thin leaves is convenient) in a closed container.
2. Expose the plant to light of a certain wavelength for a period of time.
3. Measure photosynthesis by the amount of $O_2$ released.
4. Repeat with light of other wavelengths.

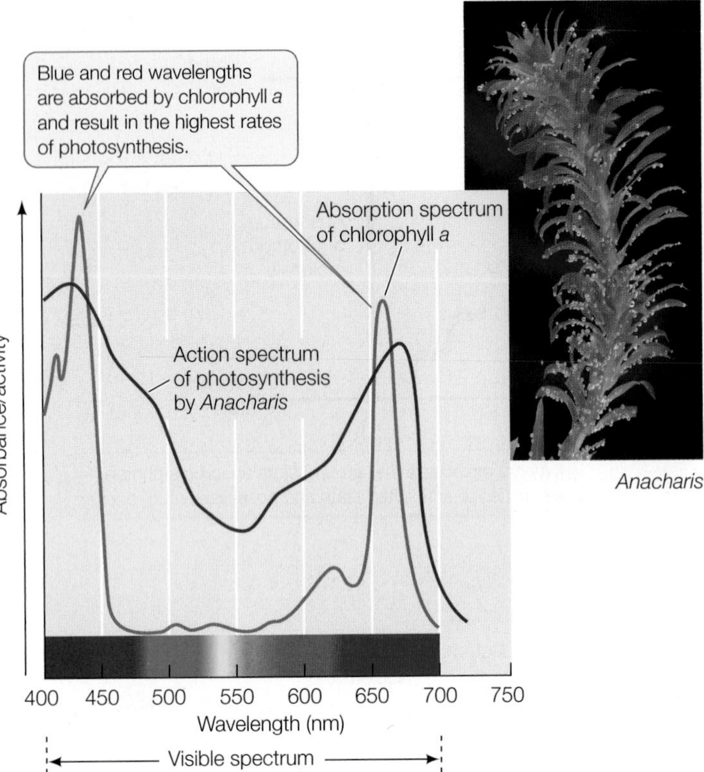


**10.6 Absorption and Action Spectra** The absorption spectrum of the purified pigment chlorophyll *a* from the aquatic plant *Anacharis* is similar to the action spectrum obtained when different wavelengths of light are shone on the intact plant and the rate of photosynthesis is measured. In the thicker leaves of land plants, the action spectra show less of a dip in the green region (500–650 nm).

**Figure 10.6** shows the absorption spectrum of the pigment chlorophyll *a*, which was isolated from the leaves of *Anacharis*, a common aquarium plant. Also shown is the action spectrum for photosynthetic activity by the same plant. A comparison of the two spectra shows that the wavelengths at which photosynthesis is highest are the same wavelengths at which chlorophyll *a* absorbs light.

## Several pigments absorb energy for photosynthesis

The light energy used for photosynthesis is not absorbed by just one type of pigment. Instead, several different pigments with different absorption spectra collect the energy that is eventually used for photosynthesis. In photosynthetic organisms as diverse as green algae, protists, and bacteria, these pigments include chlorophylls, carotenoids, and phycobilins.

**CHLOROPHYLLS** In plants, two chlorophylls are responsible for absorbing the light energy that is used to drive the light reactions: **chlorophyll *a*** and **chlorophyll *b***. These two molecules differ only slightly in their molecular structures. Both have a complex ring structure similar to that of the heme group of hemoglobin (**Figure 10.7**). In the center of the chlorophyll ring is a magnesium atom. Attached at a peripheral location on the ring is a long hydrocarbon "tail," which anchors the chlorophyll molecule to integral proteins in the thylakoid membrane of the chloroplast. (See Figure 5.13 to review the anatomy of a chloroplast.)

**ACCESSORY PIGMENTS** We saw in Figure 10.6 that chlorophyll absorbs blue and red light, which are near the two ends of the visible spectrum. Thus, if only chlorophyll were active in photosynthesis, much of the visible spectrum would go unused. This appears to be the case in higher plants. But lower plants (such as algae) and cyanobacteria possess accessory pigments, which absorb photons intermediate in energy between the red and the blue wavelengths and then transfer a portion of that energy to the chlorophylls. Among these accessory pigments are **carotenoids** such as β-carotene (see Figure 3.21), which absorb photons in the blue and blue-green wavelengths and appear deep yellow. The **phycobilins**, which are found in red algae and in cyanobacteria, absorb various yellow-green, yellow, and orange wavelengths.

## Light absorption results in photochemical change

Any pigment molecule can become excited when its absorption spectrum matches the energies of incoming photons. After a

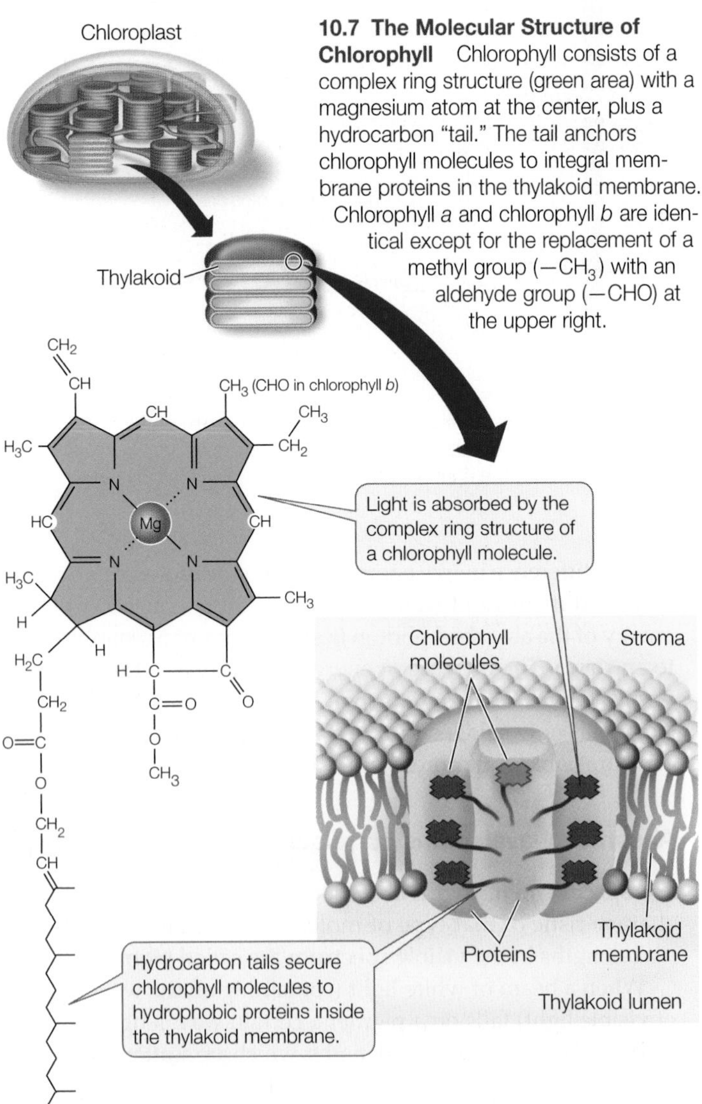


**10.7 The Molecular Structure of Chlorophyll** Chlorophyll consists of a complex ring structure (green area) with a magnesium atom at the center, plus a hydrocarbon "tail." The tail anchors chlorophyll molecules to integral membrane proteins in the thylakoid membrane. Chlorophyll *a* and chlorophyll *b* are identical except for the replacement of a methyl group ($-CH_3$) with an aldehyde group ($-CHO$) at the upper right.

pigment molecule absorbs a photon and enters an excited state (see Figure 10.5), there are several alternative fates for the absorbed energy:

- It can be released as heat and/or light.
- It may be rapidly transferred to a neighboring pigment molecule.
- It can be used as free energy to drive a chemical reaction.

When the excited molecule gives up the absorbed energy it returns to the ground state.

Sometimes the absorbed energy is given off as heat and light, in a process called *fluorescence*. Because some of the energy of the original absorbed photon is lost as heat, the photon that is released as fluorescence has less energy and a longer wavelength than the absorbed light. When there is fluorescence, there are no permanent chemical changes made or biological functions performed—no chemical work is done.

On the other hand, the excited pigment molecule may pass the absorbed energy along to another molecule—provided that the target molecule is very near, has the right orientation, and has the appropriate structure to receive the energy. This is what happens in photosynthesis.

The pigments in photosynthetic organisms are arranged into energy-absorbing **antenna systems**, also called *light-harvesting complexes*. These form part of a large multi-protein complex called a **photosystem**. The photosystem spans the thylakoid membrane, and consists of multiple antenna systems, with their associated pigment molecules, all surrounding a **reaction center**. The pigment molecules in the antenna systems are packed together in such a way that the excitation energy from an absorbed photon can be passed along from one pigment molecule to another (**Figure 10.8**). Excitation energy moves from pigments that absorb shorter wavelengths (higher energy) to pigments that absorb longer wavelengths (lower energy). Thus the excitation ends up in the pigment molecules that absorb the longest wavelengths. These pigment molecules are in the reaction center of the photosystem, and form special associations with the photosystem proteins (see Figure 10.8). The ratio of antenna pigments to reaction center pigments can be quite high (over 300:1).

The reaction center converts the absorbed light energy into chemical energy. A pigment molecule in the reaction center absorbs sufficient energy that it actually gives up its excited electron (is chemically oxidized) and becomes positively charged. In plants, the reaction center contains a pair of chlorophyll *a* molecules. There are many other chlorophyll *a* molecules in the antenna systems, but because of their interactions with antenna proteins, all of them absorb light at shorter wavelengths than the pair in the reaction center.

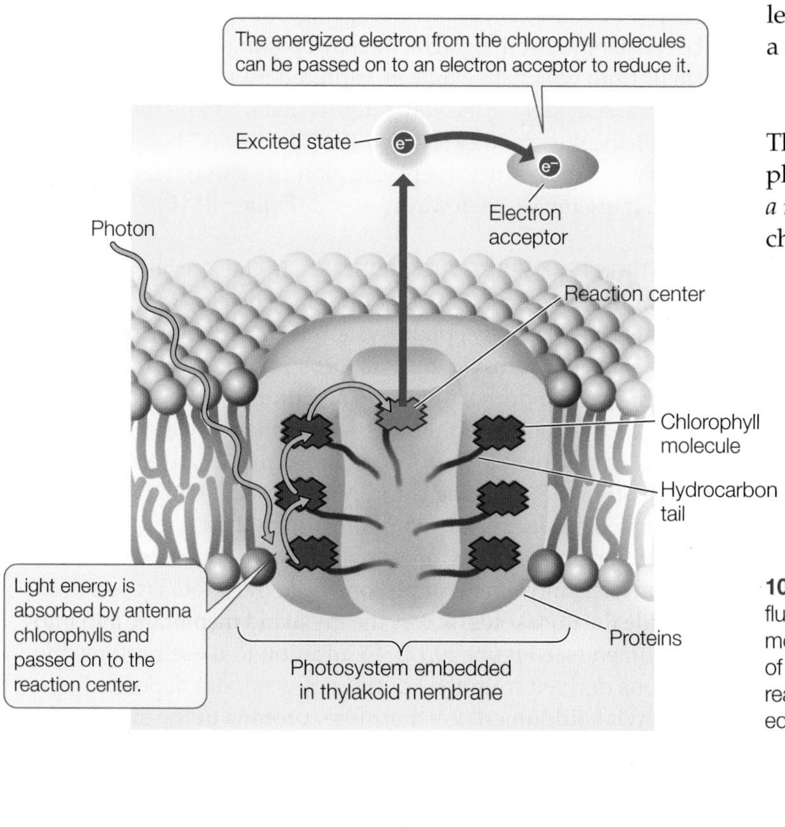

### Excited chlorophylls in the reaction center act as electron donors

Chlorophyll has two vital roles in photosynthesis:

- It absorbs light energy and transforms it into excited electrons.
- It transfers those electrons to other molecules, initiating chemical changes.

We have dealt with the first role; now we turn to the second.

Photosynthesis harvests chemical energy by using the excited chlorophyll molecules in the reaction center as electron donors (reducing agents) to reduce a stable electron acceptor (see Figure 10.8). Ground-state chlorophyll (symbolized by Chl) is not much of a reducing agent, but excited chlorophyll (Chl*) is a good one. This is because in the excited molecule, one of the electrons has moved to a shell that is farther away from the nucleus than the shell it normally occupies. This electron is held less tightly than in the normal state, and it can be transferred in a redox reaction to an electron acceptor (an oxidizing agent):

$$\text{Chl}^* + \text{acceptor} \rightarrow \text{Chl}^+ + \text{acceptor}^-$$

This, then, is the first consequence of light absorption by chlorophyll: *a reaction center chlorophyll (Chl*) loses its excited electron in a redox reaction and becomes* $Chl^+$ (because it gives up a negative charge—it gets oxidized).

### Reduction leads to electron transport

The electron acceptor that is reduced by Chl* is the first in a chain of electron carriers in the thylakoid membrane that participate in a process termed *electron transport*. This energetically "downhill" series of reductions and oxidations is similar to what occurs in the respiratory chain of

**10.8 Energy Transfer and Electron Transport** Rather than being lost as fluorescence, energy from a photon may be transferred from one pigment molecule to another. In a photosystem, energy is transferred through a series of molecules to one or more pigment molecules in the reaction center. If a reaction center molecule becomes sufficiently excited, it will give up its excited electron to an electron acceptor.

mitochondria (see Section 9.3). The final electron acceptor is $NADP^+$ (nicotinamide adenine dinucleotide phosphate), which gets reduced:

$$NADP^+ + H^+ + 2\ e^- \rightarrow NADPH$$

The energy-rich NADPH is a stable, reduced coenzyme.

There are two different systems of electron transport in photosynthesis:

- **Noncyclic electron transport** produces NADPH and ATP. Essentially, the excited electron is "lost" from chlorophyll and the transport process ends up with a reduced coenzyme.
- **Cyclic electron transport** produces only ATP. Essentially, the transport process ends up with the excited electron returning to chlorophyll, after giving up energy to make ATP.

We'll consider these two systems before describing the production of ATP from ADP and $P_i$.

### Noncyclic electron transport produces ATP and NADPH

In noncyclic electron transport, light energy is used to oxidize water, forming $O_2$, $H^+$, and electrons. In quantitative terms this would be

$$H_2O \rightarrow 2\ H^+ + 1/2\ O_2 + 2\ e^-$$

We saw above that a key reaction in photosynthesis occurs when chlorophyll that is excited by absorbing light (Chl*) gives up its excited electron, becoming oxidized:

$$Chl^* \rightarrow Chl^+ + e^-$$

Because it lacks an electron, $Chl^+$ is very unstable; it has a very strong tendency to "grab" an electron from another molecule to replenish the one it lost.

$$Chl^+ + e^- \rightarrow Chl$$

So in chemical terms, $Chl^+$ is a strong oxidizing agent. The replenishing electrons come from water, splitting the H–O–H bonds.

$$H_2O \rightarrow 1/2\ O_2 + 2\ H^+ + 2\ e^-$$
$$2\ e^- + 2\ Chl^+ \rightarrow 2\ Chl$$
$$\text{Overall: } 2\ Chl^* + H_2O \rightarrow 2\ Chl + 2\ H^+ + 1/2\ O_2$$

Notice that this is a more precise description of what Ruben and Kamen had found, namely that the source of $O_2$ in photosynthesis is $H_2O$ (see Figure 10.2).

The electrons are passed from chlorophyll to $NADP^+$ through a chain of electron carriers in the thylakoid membrane. These redox reactions are exergonic, and some of the released free energy is ultimately used to form ATP by *chemiosmosis* (see p. 180).

**TWO PHOTOSYSTEMS ARE REQUIRED** Noncyclic electron transport requires the participation of two different photosystems in the thylakoid membrane. What is the evidence of the existence of these two photosystems? In 1957, Robert Emerson at the University of Illinois shone light of various wavelengths onto cells of *Chlorella*, a freshwater protist. Both red light (wavelength 680 nm) and far-red light (700 nm) resulted in modest rates of photosynthesis, as measured by $O_2$ production. But when the two lights were combined, the rate of photosynthesis was much greater than the rates under either red light or far red light. In fact it was greater than the two rates added together. This phenomenon was termed photo enhancement (**Figure 10.9**). A few years later, photo enhancement was explained by the existence of *not one but two reaction centers*, which act together to enhance photosynthesis.

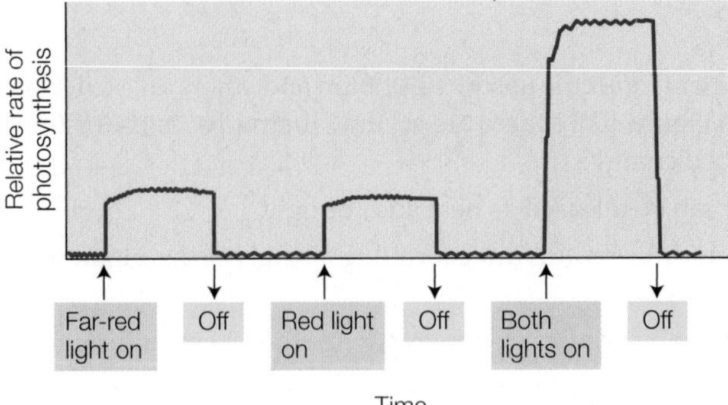


**10.9 Two Photosystems** The absorption and action spectra for chlorophyll and photosynthesis indicated that the rate of photosynthesis would increase in red light. Robert Emerson shone red (660 nm) and far-red (>700 nm) light both separately and together on algal cells to look for cooperative effects.

- **Photosystem I** uses light energy to pass an excited electron to $NADP^+$, reducing it to NADPH.
- **Photosystem II** uses light energy to oxidize water molecules, producing electrons, protons ($H^+$), and $O_2$.

The reaction center for photosystem I contains a pair of chlorophyll *a* molecules called $P_{700}$ because it can best absorb light with a wavelength of 700 nm. Similarly, the pair of chlorophyll *a* molecules in the photosystem II reaction center is called $P_{680}$ because it absorbs light maximally at 680 nm. Thus photosystem II requires photons that are somewhat more energetic (i.e., have shorter wavelengths) than those required by photosystem I. To keep noncyclic electron transport going, both photosystems must be constantly absorbing light, thereby boosting electrons to higher shells from which they may be captured by specific electron acceptors. A model for the way photosystems I and II interact and complement each other is called the "Z scheme," because when the path of the electrons is placed along an axis of rising energy level, it resembles a sideways letter Z (**Figure 10.10**).

**ELECTRON TRANSPORT: THE Z SCHEME** In the Z scheme model, which describes the reactions of noncyclic electron transport from water to $NADP^+$, photosystem II comes before photosystem I. When photosystem II absorbs photons, electrons pass from $P_{680}$ to the primary electron acceptor and $P_{680}{}^*$ is oxidized to $P_{680}{}^+$. Then an electron from the oxidation of water is passed to $P_{680}{}^+$, reducing it to $P_{680}$ once again, so that it can receive more energy from neighboring chlorophyll molecules in the antenna systems. The electrons from photosystem II pass through a series of transfer reactions, one of which is directly responsible for the physical movement of protons from the stroma (the matrix outside the thylakoids) across the thylakoid membrane and into the lumen (see Figure 10.12). In addition to these protons, the protons derived from the splitting of water are deposited into the thylakoid lumen. Furthermore, protons in the stroma are

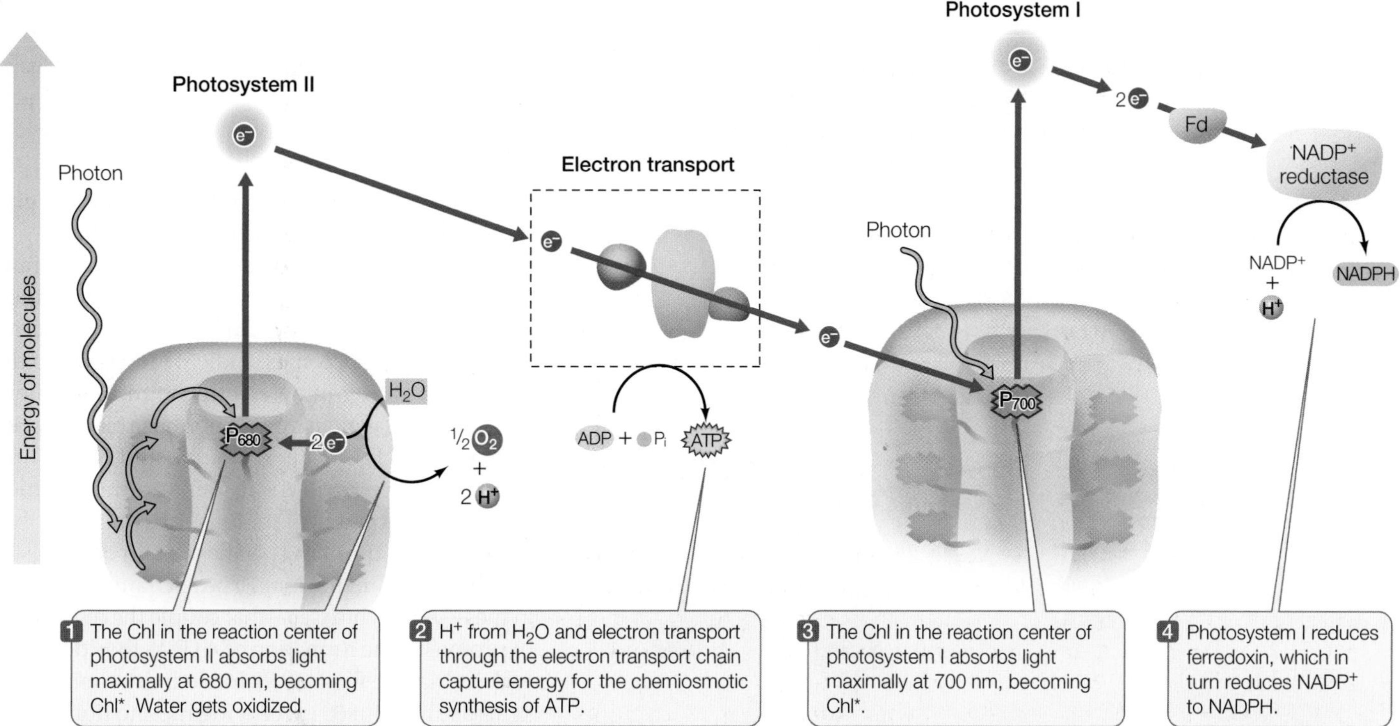

Photosystem I
Photosystem II
Photon
Electron transport
Photon
Fd
NADP+ reductase
NADP+ + H+
NADPH
Energy of molecules
P680
2e−
H2O
½O2 + 2H+
ADP + Pi
ATP
P700
1 The Chl in the reaction center of photosystem II absorbs light maximally at 680 nm, becoming Chl*. Water gets oxidized.
2 H+ from H2O and electron transport through the electron transport chain capture energy for the chemiosmotic synthesis of ATP.
3 The Chl in the reaction center of photosystem I absorbs light maximally at 700 nm, becoming Chl*.
4 Photosystem I reduces ferredoxin, which in turn reduces NADP+ to NADPH.


**10.10 Noncyclic Electron Transport Uses Two Photosystems** Absorption of light energy by chlorophyll molecules in the reaction centers of photosystems I and II allows them to pass electrons into a series of redox reactions. The term "Z scheme" describes the path (blue arrows) of electrons as they travel through the two photosystems. On this scheme the vertical positions represent the energy levels of the molecules in the electron transport system.

consumed during the reduction of $NADP^+$, and together these reactions create a proton gradient across the thylakoid membrane, which provides the energy for ATP synthesis.

In photosystem I, the $P_{700}$ molecules in the reaction center become excited to $P_{700}{}^*$, leading to the reduction of an electron carrier called ferredoxin (Fd) and the production of $P_{700}{}^+$. $P_{700}{}^+$ returns to the reduced state by accepting electrons passed through the electron transport system from photosystem II. Having identified the role of the electrons produced by photosystem II, we can now ask, "What is the role of the electrons transferred to Fd from photosystem I?" These electrons are used in the last step of noncyclic electron transport, in which two electrons and a proton are used to reduce a molecule of $NADP^+$ to NADPH.

In summary:

- Noncyclic electron transport extracts electrons from water and passes them ultimately to NADPH, utilizing energy absorbed by photosystems I and II, and resulting in ATP synthesis.
- Noncyclic electron transport yields NADPH, ATP, and $O_2$.

### Cyclic electron transport produces ATP but no NADPH

Noncyclic electron transport results in the production of ATP and NADPH. However, as we will see, the light-independent reactions of photosynthesis require more ATP than NADPH + $H^+$. If only the noncyclic pathway is operating, there is the possibility that there will not be enough ATP formed. **Cyclic electron transport** makes up for the imbalance. This pathway, which produces only ATP, is called *cyclic* because an electron passed from an excited chlorophyll molecule at the outset cycles back to the same chlorophyll molecule at the end of the chain of reactions (**Figure 10.11**).

Cyclic electron transport begins and ends in photosystem I. A $P_{700}$ chlorophyll molecule in the reaction center absorbs a photon and enters the excited state, $P_{700}{}^*$. The excited electron is passed from $P_{700}{}^*$ to a primary acceptor, and then to oxidized ferredoxin ($Fd_{ox}$), producing reduced ferredoxin ($Fd_{red}$). $Fd_{red}$ passes its added electron to a different oxidizing agent, plastoquinone (PQ, a small organic molecule), resulting in the transfer of two $H^+$ from the stroma to the thylakoid lumen. The electron passes from reduced PQ through the electron transport system until it completes its cycle by returning to $P_{700}{}^+$, restoring it to its uncharged form, $P_{700}$. This electron transport is carried out by plastocyanin (PC) and cytochromes that are similar to those of the mitochondrial respiratory chain.

By the time the electron from $P_{700}{}^*$ travels through the electron transport system and comes back to reduce $P_{700}{}^+$, all the energy from the original photon has been released. The released energy is stored in the form of a proton gradient that can be used to produce ATP.

### Chemiosmosis is the source of the ATP produced in photophosphorylation

In Chapter 9 we describe the chemiosmotic mechanism for ATP formation in the mitochondrion. A similar mechanism, called **photophosphorylation**, operates in the chloroplast, where electron transport is coupled to the transport of protons ($H^+$) across

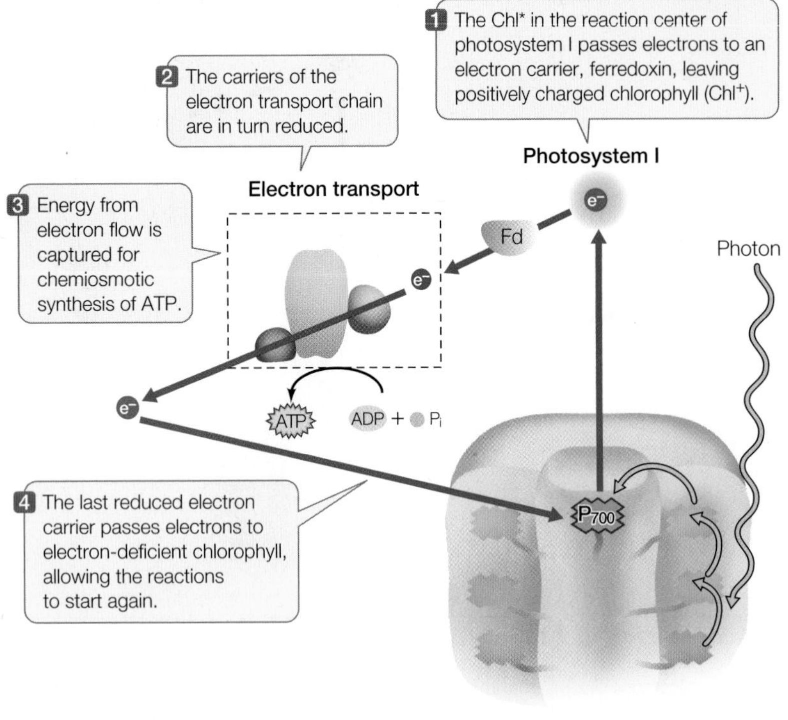


**10.11 Cyclic Electron Transport Traps Light Energy as ATP** Cyclic electron transport produces ATP, but no NADPH.

the thylakoid membrane, resulting in a proton gradient across the membrane (**Figure 10.12**).

The electron carriers in the thylakoid membrane are oriented so that protons are actively pumped from the stroma into the lumen of the thylakoid. Thus the lumen becomes acidic with respect to the stroma, resulting in an electrochemical gradient across the thylakoid membrane, whose bilayer is not permeable to $H^+$. Water oxidation and $NADP^+$ reduction also contribute to this gradient, which drives the movement of protons back out of the thylakoid lumen through specific protein channels in the thylakoid membrane. These channels are enzymes—ATP synthases—that couple the movement of protons to the formation of ATP, as they do in mitochondria (see Figure 9.9). Indeed, chloroplast ATP synthase is about 60 percent identical to human mitochondrial ATP synthase—a remarkable similarity, given that plants and animals had their most recent

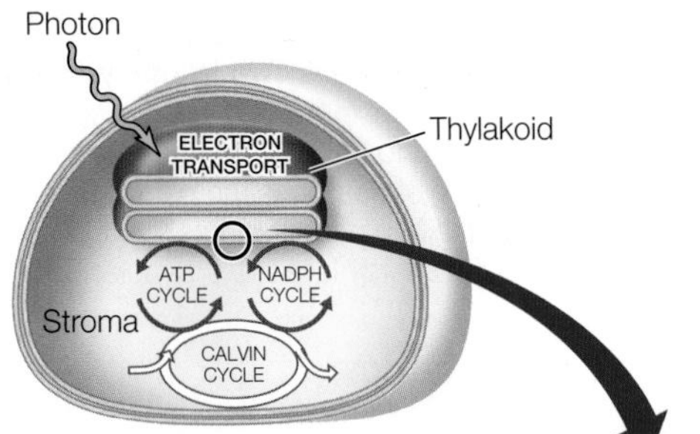


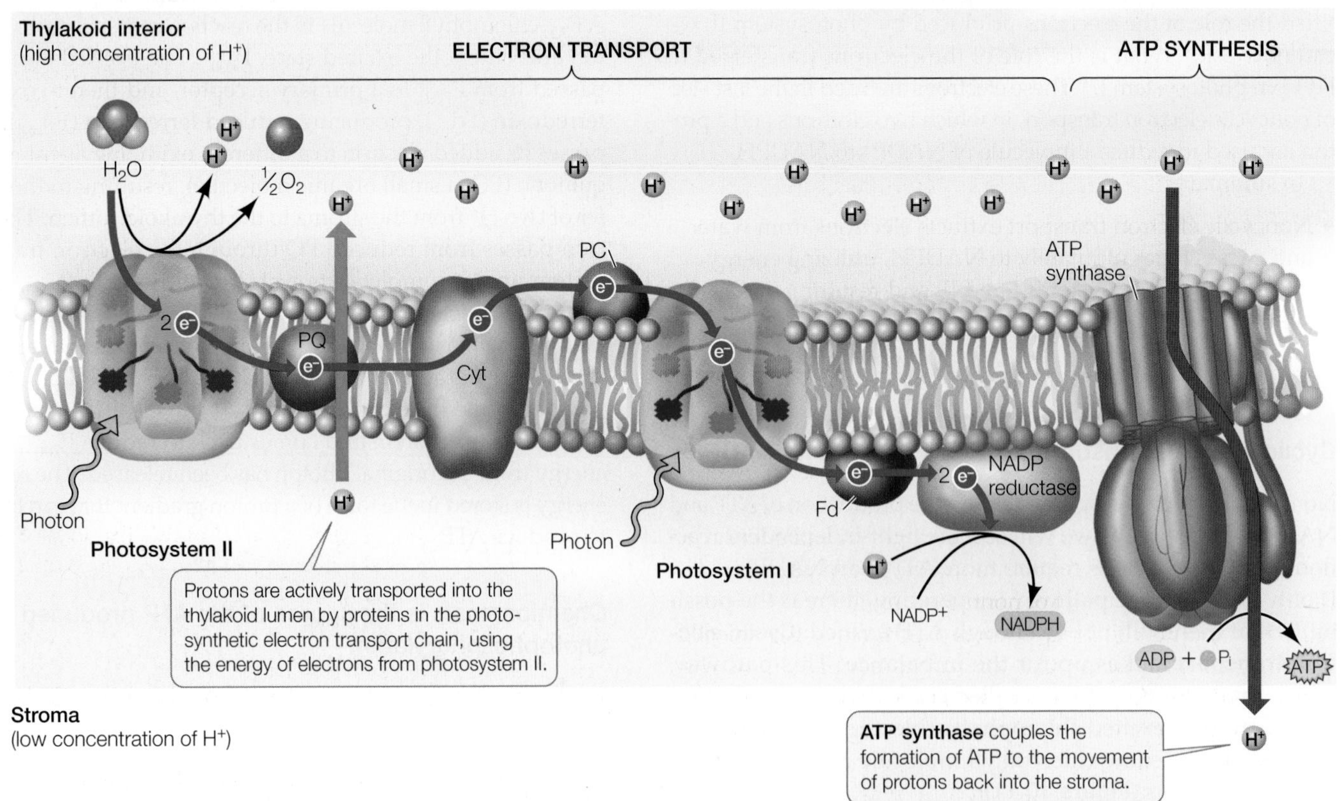


**10.12 Chloroplasts Form ATP Chemiosmotically** Compare this illustration with Figure 9.9, where a similar process is depicted in mitochondria.

yourBioPortal.com
GO TO **Animated Tutorial 10.2 • Photophosphorylation**

common ancestor more than a billion years ago. This is testimony to the evolutionary unity of life.

The mechanisms of the two enzymes are similar, but their orientations differ. In chloroplasts, protons flow through the ATP synthase out of the thylakoid lumen into the stroma (where the ATP is synthesized) but in mitochondria they flow out of the cytosol into the mitochondrial matrix.

## 10.2 RECAP

**Conversion of light energy into chemical energy occurs when pigments absorb photons. Light energy is used to drive a series of protein-associated redox reactions in the thylakoid membranes of the chloroplast.**

- How does chlorophyll absorb and transfer light energy? **See pp. 194–195 and Figure 10.8**
- How are electrons produced in photosystem II and how do they flow to photosystem I? **See pp. 196–197 and Figure 10.10**
- How does cyclic electron transport in photosystem I result in the production of ATP? **See p. 197 and Figure 10.11**

We have seen how light energy drives the synthesis of ATP and NADPH in the stroma of chloroplasts. We now turn to the light-independent reactions of photosynthesis, which use energy-rich ATP and NADPH to reduce $CO_2$ and form carbohydrates.

# 10.3 How Is Chemical Energy Used to Synthesize Carbohydrates?

Most of the enzymes that catalyze the reactions of $CO_2$ fixation are dissolved in the stroma of the chloroplast, where those reactions take place. These enzymes use the energy in ATP and NADPH to reduce $CO_2$ to carbohydrates. Therefore, with some exceptions, $CO_2$ fixation occurs only in the light, when ATP and NADPH are being generated.

### Radioisotope labeling experiments revealed the steps of the Calvin cycle

To identify the reactions by which the carbon from $CO_2$ ends up in carbohydrates, scientists found a way to label $CO_2$ so that they could isolate and identify the compounds formed from it during photosynthesis. In the 1950s, Melvin Calvin, Andrew Benson, and their colleagues used radioactively labeled $CO_2$ in which some of the carbon atoms were the radioisotope $^{14}C$ rather than the normal $^{12}C$. Although $^{14}C$ emits radiation, its chemical behavior is virtually identical to that of nonradioactive $^{12}C$.

Calvin and his colleagues exposed cultures of the unicellular green alga *Chlorella* to $^{14}CO_2$ for various lengths of time. Then they rapidly killed the cells and extracted the organic compounds. They separated the different compounds from one another by paper chromatography and exposed the paper to X-ray film (**Figure 10.13**). When the film was developed, dark

## INVESTIGATING LIFE

**10.13 Tracing the Pathway of $CO_2$**

How is $CO_2$ incorporated into carbohydrate during photosynthesis? What is the first stable covalent linkage that forms with the carbon of $CO_2$? Short exposures to $^{14}CO_2$ were used to identify the first compound formed from $CO_2$.

**HYPOTHESIS** **The first product of $CO_2$ fixation is a 3-carbon molecule.**

**METHOD**

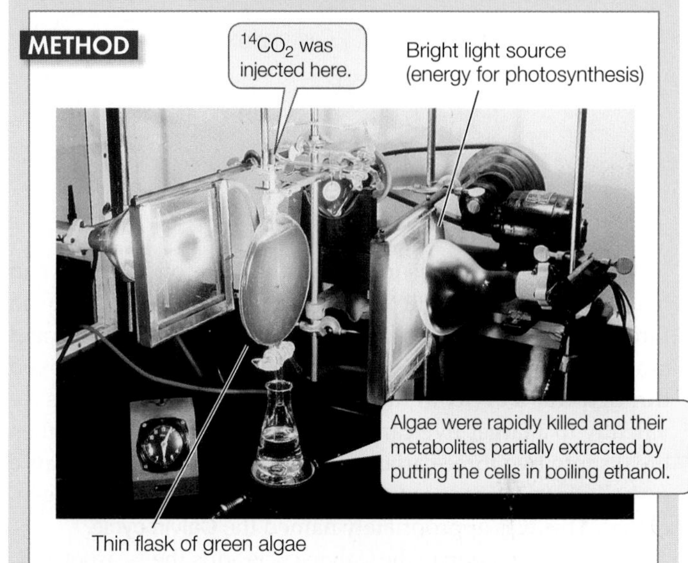
$^{14}CO_2$ was injected here.
Bright light source (energy for photosynthesis)
Algae were rapidly killed and their metabolites partially extracted by putting the cells in boiling ethanol.
Thin flask of green algae

The algal extract was spotted here and run in two directions to separate compounds from one another.

First run

Second run

After separation of the compounds, the chromatogram was overlaid with X-ray film.

Paper chromatogram

**RESULTS**

3PG

GLUT, ALA, GLY, SER, ASP, CIT, SUC, G3P, 3PG, HEXOSE-P

A chromatogram made after 3 seconds of exposure to $^{14}CO_2$ shows $^{14}C$ only in 3PG (3-phosphoglycerate).

A chromatogram made after 30 seconds of exposure to $^{14}CO_2$ shows $^{14}C$ in many molecules.

**CONCLUSION** **The initial product of $CO_2$ fixation is 3PG. Later, the carbon from $CO_2$ ends up in many molecules.**

Go to **yourBioPortal.com** for original citations, discussions, and relevant links for all INVESTIGATING LIFE figures.

**yourBioPortal.com**

GO TO Animated Tutorial 10.3 • Tracing the Pathway of $CO_2$

spots indicated the locations of compounds containing $^{14}C$ in the paper.

To discover the first compound in the pathway of $CO_2$ fixation, Calvin and his team exposed the algae to $^{14}CO_2$ for shorter and shorter periods of time. The 3-second exposure revealed that only one compound was labeled—a 3-carbon sugar phosphate called 3-phosphoglycerate (3PG) (the $^{14}C$ is shown in red):

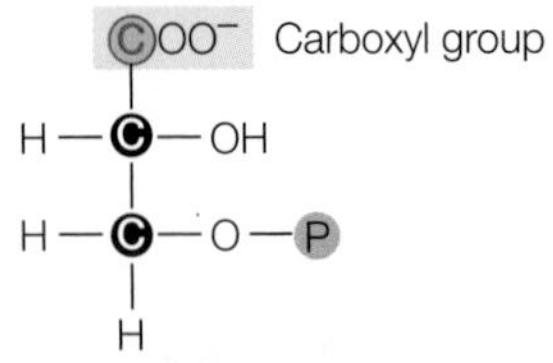

COO⁻ Carboxyl group

H—C—OH

H—C—O—P

H

3-Phosphoglycerate (3PG)

With successive exposures longer than 3 seconds, Calvin and his colleagues were able to trace the route of $^{14}C$ as it moved through a series of compounds, including monosaccharides and amino acids. It turned out that the pathway the $^{14}C$ moved through was a cycle. In this cycle, the $CO_2$ initially bonds covalently to a larger five-carbon acceptor molecule, which then breaks into two three-carbon molecules. As the cycle repeats a carbohydrate is produced and the initial $CO_2$ acceptor is regenerated. This was appropriately named the **Calvin cycle**.

The initial reaction in the Calvin cycle adds the 1-carbon $CO_2$ to an acceptor molecule, the 5-carbon compound ribulose 1,5-bisphosphate (RuBP). The product is an intermediate 6-carbon compound, which quickly breaks down and forms two molecules of 3PG (**Figure 10.14**). The intermediate compound is broken down so rapidly that Calvin did not observe radioactive label appearing in it first. But the enzyme that catalyzes its formation, **ribulose bisphosphate carboxylase/oxygenase (rubisco)**, is the most abundant protein in the world! It constitutes up to 50 percent of all the protein in every plant leaf.

### The Calvin cycle is made up of three processes

The Calvin cycle uses the ATP and NADPH made in the light to reduce $CO_2$ in the stroma to a carbohydrate. Like all biochemical pathways, each reaction is catalyzed by a specific enzyme. The cycle is composed of three distinct processes (**Figure 10.15**):

- *Fixation* of $CO_2$. As we have seen, this reaction is catalyzed by rubisco, and its stable product is 3PG.
- *Reduction* of 3PG to form glyceraldehyde 3-phosphate (G3P). This series of reactions involves a phosphorylation (using the ATP made in the light reactions) and a reduction (using the NADPH made in the light reactions).
- *Regeneration* of the $CO_2$ acceptor, RuBP. Most of the G3P ends up as ribulose monophosphate (RuMP), and ATP is used to convert this compound into RuBP. So for every "turn" of the cycle, with one $CO_2$ fixed, the $CO_2$ acceptor is regenerated.

The product of this cycle is **glyceraldehyde 3-phosphate (G3P)**, which is a 3-carbon sugar phosphate, also called triose phosphate:

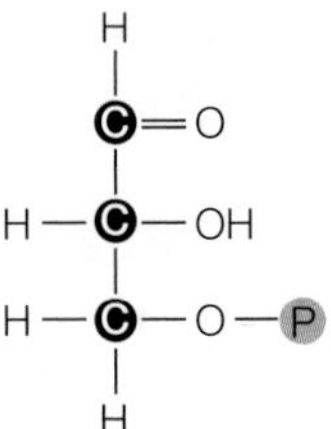

Glyceraldehyde 3-phosphate (G3P)

In a typical leaf, five-sixths of the G3P is recycled into RuBP. There are two fates for the remaining G3P, depending on the time of day and the needs of different parts of the plant:

- Some of it is exported out of the chloroplast to the cytosol, where it is converted to hexoses (glucose and fructose). These molecules may be used in glycolysis and mitochondrial respiration to power the activities of photosynthetic cells (see Chapter 9) or they may be converted into the disaccharide sucrose, which is transported out of the leaf to other organs in the plant. There it is hydrolyzed to its constituent monosaccharides, which can be used as sources of energy or as building blocks for other molecules.

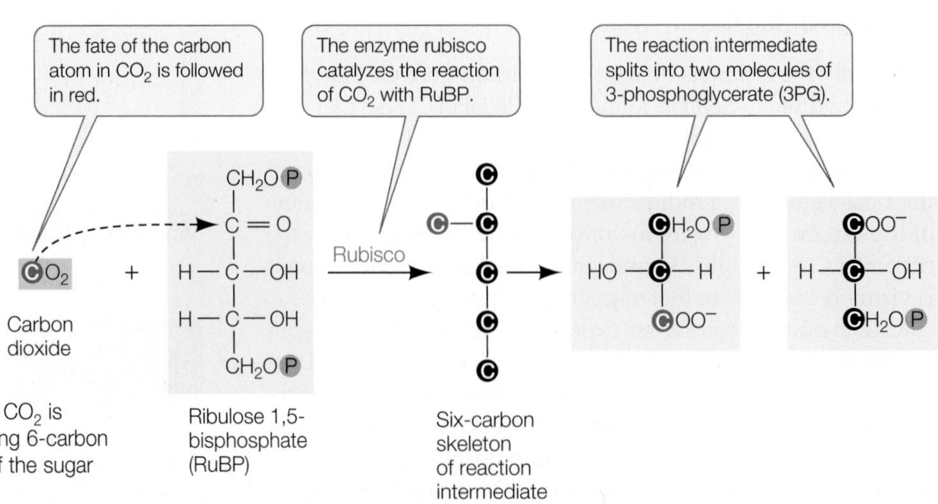


**10.14 RuBP Is the Carbon Dioxide Acceptor** $CO_2$ is added to a 5-carbon compound, RuBP. The resulting 6-carbon compound immediately splits into two molecules of the sugar phosphate 3PG.

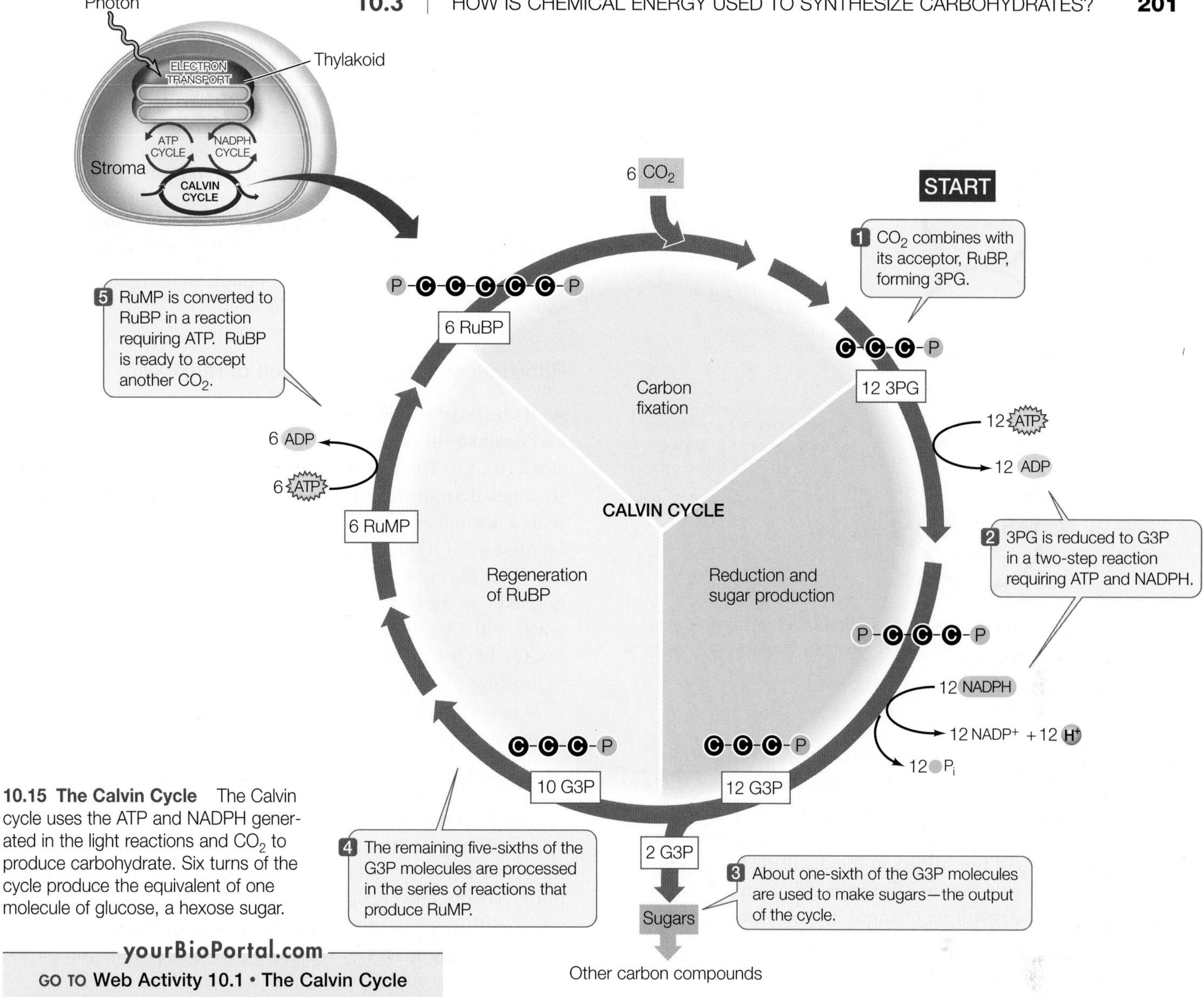


**10.15 The Calvin Cycle** The Calvin cycle uses the ATP and NADPH generated in the light reactions and $CO_2$ to produce carbohydrate. Six turns of the cycle produce the equivalent of one molecule of glucose, a hexose sugar.

**yourBioPortal.com**
GO TO **Web Activity 10.1 • The Calvin Cycle**

- As the day wears on, glucose accumulates inside of the chloroplast, and these glucose units are linked to form the polysaccharide starch. This stored carbohydrate can then be drawn upon during the night so that the photosynthetic tissues can continue to export sucrose to the rest of the plant, even when photosynthesis is not taking place. In addition, starch is abundant in nonphotosynthetic organs such as roots, underground stems and seeds, where it provides a ready supply of glucose to fuel cellular activities, including plant growth.

The carbohydrates produced in photosynthesis are used by the plant to make other compounds. The carbon molecules are incorporated into amino acids, lipids, and the building blocks of nucleic acids—in fact all the organic molecules in the plant.

The products of the Calvin cycle are of crucial importance to the Earth's entire biosphere. For the majority of living organisms on Earth, the C—H covalent bonds generated by the cycle provide almost all of the energy for life. Photosynthetic organisms, which are also called **autotrophs** ("self-feeders"), release most of this energy by glycolysis and cellular respiration, and use it to support their own growth, development, and reproduction. But plants are also the source of energy for other organisms. Much plant matter ends up being consumed by **heterotrophs** ("other-feeders"), such as animals, which cannot photosynthesize. Heterotrophs depend on autotrophs for both raw materials and energy. Free energy is released from food by glycolysis and cellular respiration in heterotroph cells.

### Light stimulates the Calvin cycle

As we have seen, the Calvin cycle uses NADPH and ATP, which are generated using energy from light. Two other processes connect the light reactions with this $CO_2$ fixation pathway. Both connections are indirect but significant:

- Light-induced pH changes in the stroma activate some Calvin cycle enzymes. Proton pumping from the stroma into the thylakoid lumen causes an increase in the pH of the stroma from 7 to 8 (a tenfold decrease in $H^+$ concentration). This favors the activation of rubisco.
- The light-induced electron transport reduces disulfide bonds in four of the Calvin cycle enzymes, thereby activat-

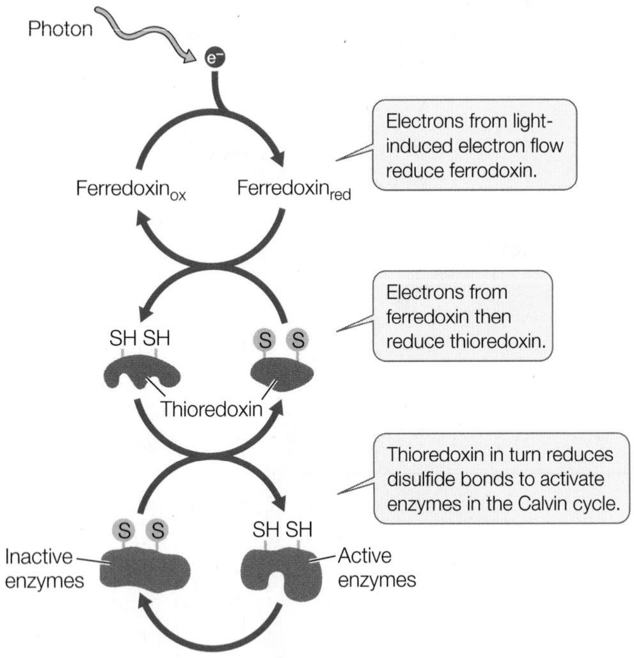

**10.16 The Photochemical Reactions Stimulate the Calvin Cycle** By reducing (breaking) disulfide bridges, electrons from the light reactions activate enzymes in $CO_2$ fixation.

ing them (**Figure 10.16**). When ferredoxin is reduced in photosystem I (see Figure 10.10), it passes some electrons to a small, soluble protein called thioredoxin, and this protein passes electrons to four enzymes in the $CO_2$ fixation pathway. Reduction of the sulfurs in the disulfide bridges of these enzymes (see Figure 3.5) forms SH groups and breaks the bridges. The resulting changes in their three-dimensional shapes activate the enzymes and increase the rate at which the Calvin cycle operates.

## 10.3 RECAP

**ATP and NADPH produced in the light reactions power the synthesis of carbohydrates by the Calvin cycle. This cycle fixes $CO_2$, reduces it, and regenerates the acceptor, RuBP, for further fixation.**

- Describe the experiments that led to the identification of RuBP as the initial $CO_2$ acceptor in photosynthesis. See pp. 199–200 and Figure 10.13
- What are the three processes of the Calvin cycle? See pp. 200–201 and Figure 10.15
- In what ways does light stimulate the Calvin cycle? See pp. 201–202 and Figure 10.16

Although all green plants carry out the Calvin cycle, some plants have evolved variations on, or additional steps in, the light-independent reactions. These variations and additions have permitted plants to adapt to and thrive in certain environmental conditions. Let's look at these environmental limitations and the metabolic bypasses that have evolved to circumvent them.

## 10.4 How Do Plants Adapt to the Inefficiencies of Photosynthesis?

In addition to fixing $CO_2$ during photosynthesis, rubisco can react with $O_2$. This reaction leads to a process called photorespiration, which lowers the overall rate of $CO_2$ fixation in some plants. After examining this problem, we'll look at some biochemical pathways and features of plant anatomy that compensate for the limitations of rubisco.

### Rubisco catalyzes the reaction of RuBP with $O_2$ or $CO_2$

As its full name indicates, rubisco is an **oxygenase** as well as a **carboxylase**—it can add $O_2$ to the acceptor molecule RuBP instead of $CO_2$. The affinity of rubisco for $CO_2$ is about ten times stronger than its affinity for $O_2$. This means that inside a leaf with a normal exchange of air with the outside, $CO_2$ fixation is favored even though the concentration of $CO_2$ in the air is far less than that of $O_2$. But if there is an even higher concentration of $O_2$ in the leaf, it acts as a competitive inhibitor, and RuBP reacts with $O_2$ rather than $CO_2$. This reduces the overall amount of $CO_2$ that is converted into carbohydrates, and may play a role in limiting plant growth.

When $O_2$ is added to RuBP, one of the products is a 2-carbon compound, phosphoglycolate:

$$\text{RuBP} + O_2 \rightarrow \text{phosphoglycolate} + \text{3-phosphoglycerate (3PG)}$$

The 3PG formed by oxygenase activity enters the Calvin cycle but the phosphoglycolate does not. Plants have evolved a metabolic pathway that can partially recover the carbon in phosphoglycolate. The phosphoglycolate is hydrolyzed to glycolate, which diffuses into membrane-enclosed organelles called peroxisomes (**Figure 10.17**). There, a series of reactions converts it into the amino acid glycine:

$$\text{glycolate} + O_2 \rightarrow \text{glycine}$$

The glycine then diffuses into a mitochondrion, where two glycine molecules are converted in a series of reactions into the amino acid serine, which goes back to the peroxisome and is converted into glycerate (a 3-carbon molecule) and $CO_2$:

$$\text{2 glycine} \rightarrow\rightarrow \text{glycerate} + CO_2$$

The glycerate moves into the chloroplast, where it is phosphorylated to make 3PG, which enters the Calvin cycle. So overall:

$$\text{phosphoglycolate (4 carbons)} + O_2 \rightarrow \text{3PG (3 carbons)} + CO_2$$

This pathway thus reclaims 75 percent of the carbons from phosphoglycolate for the Calvin cycle. In other words, the reaction of RuBP with $O_2$ instead of $CO_2$ reduces the net carbon fixed by the Calvin cycle by 25 percent. The pathway is called **photorespiration** because it consumes $O_2$ and releases $CO_2$ and because it occurs only in the light (due to the same enzyme activation processes that were mentioned above with regard to the Calvin cycle).

Why does rubisco act as an oxygenase as well as a carboxylase? Several factors are involved: active site affinities, concentrations of $CO_2$ and $O_2$, and temperature.

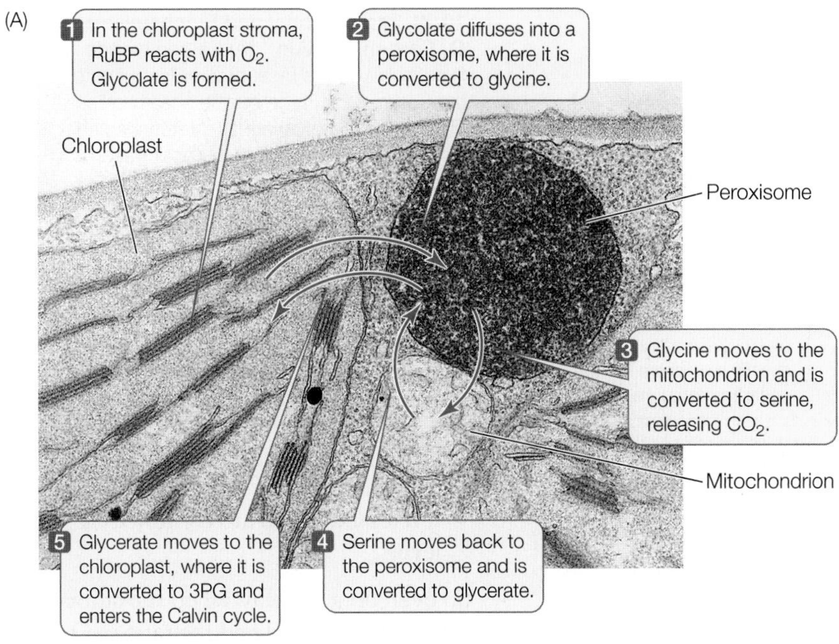

(A)
1 In the chloroplast stroma, RuBP reacts with $O_2$. Glycolate is formed.
2 Glycolate diffuses into a peroxisome, where it is converted to glycine.
Chloroplast
Peroxisome
3 Glycine moves to the mitochondrion and is converted to serine, releasing $CO_2$.
Mitochondrion
5 Glycerate moves to the chloroplast, where it is converted to 3PG and enters the Calvin cycle.
4 Serine moves back to the peroxisome and is converted to glycerate.


(B)
Carbon gained
Carbon lost
2-PG
$O_2$
$CO_2$
Carboxylase reaction
Photorespiration
Rubisco
Calvin cycle
Oxygenase reaction
$CO_2$
$O_2$
3-PGA

**10.17 Organelles of Photorespiration** (A) The reactions of photorespiration take place in the chloroplasts, peroxisomes, and mitochondria. (B) Overall, photorespiration consumes $O_2$ and releases $CO_2$.

- As noted above, rubisco has a ten times higher affinity for $CO_2$ than for $O_2$, and this favors $CO_2$ fixation.
- In the leaf, the relative concentrations of $CO_2$ and $O_2$ vary. If $O_2$ is relatively abundant, rubisco acts as an oxygenase and photorespiration ensues. If $CO_2$ predominates, rubisco fixes it for the Calvin cycle.
- Photorespiration is more likely at high temperatures. On a hot, dry day, small pores in the leaf surface called **stomata** close to prevent water from evaporating from the leaf (see Figure 10.1). But this also prevents gases from entering and leaving the leaf. The $CO_2$ concentration in the leaf falls as $CO_2$ is used up in photosynthetic reactions, and the $O_2$ concentration rises because of these same reactions. As the ratio of $CO_2$ to $O_2$ falls, the oxygenase activity of rubisco is favored, and photorespiration proceeds.

### $C_3$ plants undergo photorespiration but $C_4$ plants do not

Plants differ in how they fix $CO_2$, and can be distinguished as $C_3$ or $C_4$ plants, based on whether the first product of $CO_2$ fixation is a 3- or 4-carbon molecule. In **$C_3$ plants** such as roses, wheat, and rice, the first product is the 3-carbon molecule 3PG—as we have just described for the Calvin cycle. In these plants the cells of the mesophyll, which makes up the main body of the leaf, are full of chloroplasts containing rubisco (**Figure 10.18A**). On a hot day, these leaves close their stomata to conserve water, and as a result, rubisco acts as an oxygenase as well as a carboxylase, and photorespiration occurs.

**$C_4$ plants**, which include corn, sugarcane, and tropical grasses, make the 4-carbon molecule **oxaloacetate** as the first product of $CO_2$ fixation (**Figure 10.18B**). On a hot day, they partially close their stomata to conserve water, but their rate of photosynthesis does not fall, nor does photorespiration occur. What do they do differently?

(A) Arrangement of cells in a $C_3$ leaf

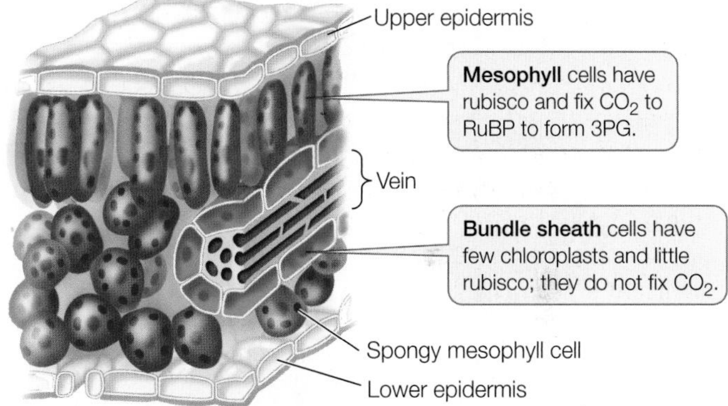


(B) Arrangement of cells in a $C_4$ leaf

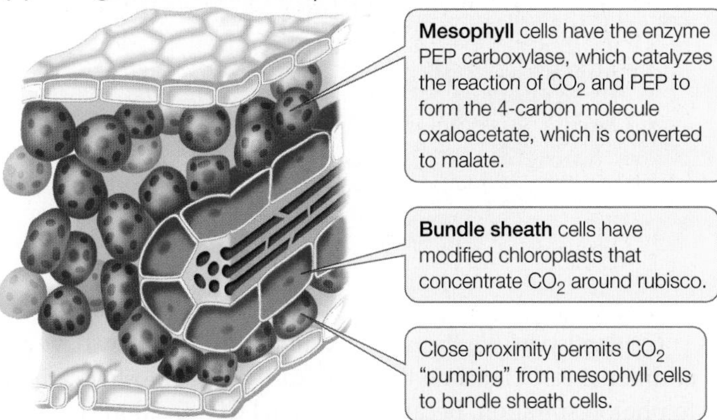

**Mesophyll** cells have the enzyme PEP carboxylase, which catalyzes the reaction of $CO_2$ and PEP to form the 4-carbon molecule oxaloacetate, which is converted to malate.
**Bundle sheath** cells have modified chloroplasts that concentrate $CO_2$ around rubisco.
Close proximity permits $CO_2$ "pumping" from mesophyll cells to bundle sheath cells.


**10.18 Leaf Anatomy of $C_3$ and $C_4$ Plants** Carbon dioxide fixation occurs in different organelles and cells of the leaves in (A) $C_3$ plants and (B) $C_4$ plants. Cells that are tinted blue have rubisco.

yourBioPortal.com
GO TO Web Activity 10.2 • $C_3$ and $C_4$ Leaf Anatomy

(A)

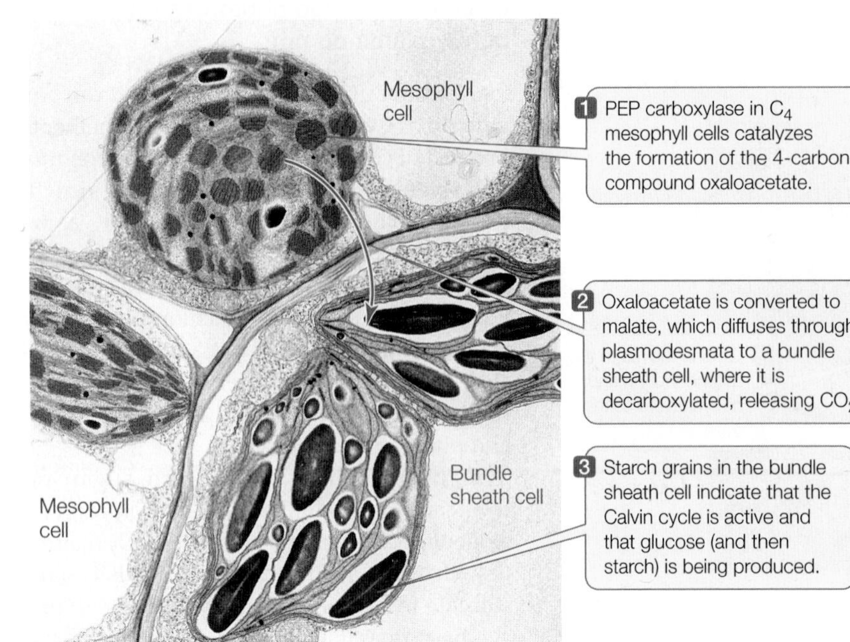

Mesophyll cell
1 PEP carboxylase in $C_4$ mesophyll cells catalyzes the formation of the 4-carbon compound oxaloacetate.
2 Oxaloacetate is converted to malate, which diffuses through plasmodesmata to a bundle sheath cell, where it is decarboxylated, releasing $CO_2$.
3 Starch grains in the bundle sheath cell indicate that the Calvin cycle is active and that glucose (and then starch) is being produced.
Bundle sheath cell
Mesophyll cell


(B)



**10.19 The Anatomy and Biochemistry of $C_4$ Carbon Fixation** (A) Carbon dioxide is fixed initially in the mesophyll cells, but enters the Calvin cycle in the bundle sheath cells. (B) The two cell types share an interconnected biochemical pathway for $CO_2$ assimilation.

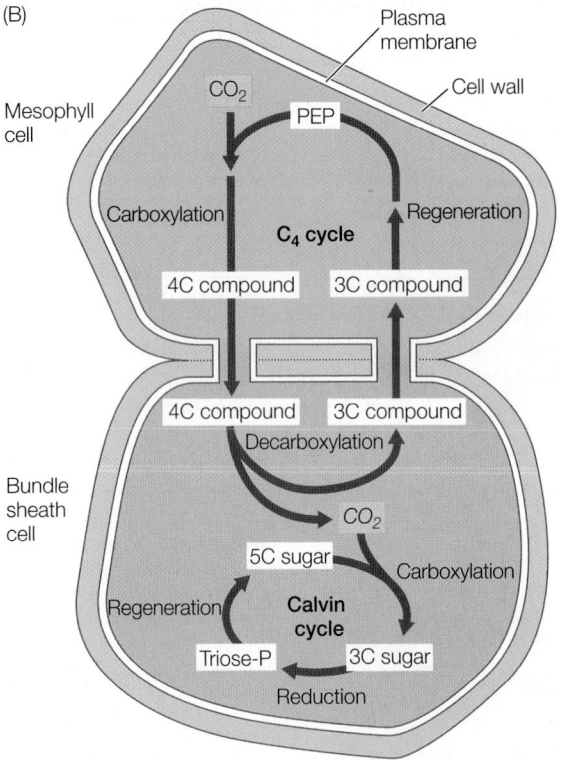

$C_4$ plants perform the normal Calvin cycle, but they have an additional early reaction that fixes $CO_2$ without losing carbon to photorespiration. Because this initial $CO_2$ fixation step can function even at low levels of $CO_2$ and high temperatures, $C_4$ plants very effectively optimize photosynthesis under conditions that inhibit it in $C_3$ plants. $C_4$ plants have *two separate enzymes for $CO_2$ fixation, located in different parts of the leaf* (**Figure 10.19**; see also Figure 10.18B). The first enzyme, called **PEP carboxylase**, is present in the cytosols of mesophyll cells near the surface of the leaf. It fixes $CO_2$ to a 3-carbon acceptor compound, **phosphoenolpyruvate** (**PEP**), to produce the 4-carbon fixation product, oxaloacetate. PEP carboxylase has two advantages over rubisco:

- It does not have oxygenase activity.
- It fixes $CO_2$ even at very low $CO_2$ levels.

So even on a hot day when the stomata are partially closed and the ratio of $O_2$ to $CO_2$ rises, PEP carboxylase just keeps on fixing $CO_2$.

Oxaloacetate is converted to malate, which diffuses out of the mesophyll cells and into the **bundle sheath cells** (see Figure 10.18B), located in the interior of the leaf. (Some $C_4$ plants convert the oxaloacetate to aspartate instead of malate, but we will only discuss the malate pathway here.) The bundle sheath cells contain modified chloroplasts that are designed to concentrate $CO_2$ around the rubisco. There, the 4-carbon malate loses one carbon (is decarboxylated), forming $CO_2$ and pyruvate. The latter moves back to the mesophyll cells where the 3-carbon acceptor compound, PEP, is regenerated at the expense of ATP. Thus the role of PEP is to bind $CO_2$ from the air in the leaf so that it can be transferred to the bundle sheath cells, where it is delivered to rubisco. This process essentially "pumps up" the $CO_2$ concentration around rubisco, so that it acts as a carboxylase and begins the Calvin cycle.

$C_3$ plants have an advantage over $C_4$ plants in that they don't expend extra ATP to "pump up" the concentration of $CO_2$ near rubisco. But this advantage begins to be outweighed under conditions that favor photorespiration, such as warmer seasons and climates. Under these conditions $C_4$ plants have the advantage. For example, Kentucky bluegrass is a $C_3$ plant that thrives on lawns in April and May. But in the heat of summer it does not do as well and Bermuda grass, a $C_4$ plant, takes over the lawn. The same is true on a global scale for crops: $C_3$ plants such as soybean, wheat, and barley have been adapted for human food production in temperate climates, while $C_4$ plants such as corn and sugarcane originated and are grown in the tropics.

**THE EVOLUTION OF $CO_2$ FIXATION PATHWAYS** $C_3$ plants are certainly more ancient than $C_4$ plants. While $C_3$ photosynthesis appears to have begun about 3.5 billion years ago, $C_4$ plants appeared about 12 million years ago. A possible factor in the emergence of the $C_4$ pathway is the decline in atmospheric $CO_2$. When dinosaurs dominated Earth 100 million years ago, the concentration of $CO_2$ in the atmosphere was four times what it is now. As $CO_2$ levels declined thereafter, the more efficient $C_4$

TABLE 10.1

**Comparison of Photosynthesis in $C_3$, $C_4$, and CAM Plants**

| | $C_3$ PLANTS | $C_4$ PLANTS | CAM PLANTS |
|---|---|---|---|
| Calvin cycle used? | Yes | Yes | Yes |
| Primary $CO_2$ acceptor | RuBP | PEP | PEP |
| $CO_2$-fixing enzyme | Rubisco | PEP carboxylase | PEP carboxylase |
| First product of $CO_2$ fixation | 3PG (3-carbon) | Oxaloacetate (4-carbon) | Oxaloacetate (4-carbon) |
| Affinity of carboxylase for $CO_2$ | Moderate | High | High |
| Photosynthetic cells of leaf | Mesophyll | Mesophyll and bundle sheath | Mesophyll with large vacuoles |
| Photorespiration | Extensive | Minimal | Minimal |

plants would have gained an advantage over their $C_3$ counterparts.

As we described in the opening essay of this chapter, $CO_2$ levels have been increasing over the past 200 years. Currently, the level of $CO_2$ is not enough for maximal $CO_2$ fixation by rubisco, so photorespiration occurs, reducing the growth rates of $C_3$ plants. Under hot conditions, $C_4$ plants are favored. But if $CO_2$ levels in the atmosphere continue to rise, the reverse will occur and $C_3$ plants will have a comparative advantage. The overall growth rates of crops such as rice and wheat should increase. This may or may not translate into more food, given that other effects of the human-spurred $CO_2$ increase (such as global warming) will also alter Earth's ecosystems.

### CAM plants also use PEP carboxylase

Other plants besides the $C_4$ plants use PEP carboxylase to fix and accumulate $CO_2$. They include some water-storing plants (called succulents) of the family Crassulaceae, many cacti, pineapples, and several other kinds of flowering plants. The $CO_2$ metabolism of these plants is called **crassulacean acid metabolism**, or **CAM**, after the family of succulents in which it was discovered. CAM is much like the metabolism of $C_4$ plants in that $CO_2$ is initially fixed into a 4-carbon compound. But in CAM plants the initial $CO_2$ fixation and the Calvin cycle are separated in time rather than space.

- At night, when it is cooler and water loss is minimized, the stomata open. $CO_2$ is fixed in mesophyll cells to form the 4-carbon compound oxaloacetate, which is converted into malate and stored in the vacuole.
- During the day, when the stomata close to reduce water loss, the accumulated malate is shipped to the chloroplasts, where its decarboxylation supplies the $CO_2$ for the Calvin cycle and the light reactions supply the necessary ATP and NADPH.

CAM benefits the plant by allowing it to close its stomata during the day. As you will learn in Chapter 35, plants lose most of the water that they take up in their roots by evaporation through the leaves (transpiration). In dry climates, closing stomata is a key to water conservation and survival.

**Table 10.1** compares photosynthesis in $C_3$, $C_4$, and CAM plants.

## 10.4 RECAP

**Rubisco catalyzes the carboxylation of RuBP to form two 3PG, and the oxygenation of RuBP to form one 3PG and one phosphoglycolate. The diversion of rubisco to its oxygenase function decreases net $CO_2$ fixation. $C_4$ photosynthesis and CAM allow plants to adapt to environmental conditions that result in a limited availability of $CO_2$ inside the leaf.**

- Explain how photorespiration recovers some of the carbon that is channeled away from the Calvin cycle. **See pp. 202–203 and Figure 10.17**
- What do $C_4$ plants do to keep the concentration of $CO_2$ around rubisco high, and why? **See pp. 203–204 and Figure 10.19**
- What is the pathway for $CO_2$ fixation in CAM plants? **See p. 205**

Now that we understand how photosynthesis produces carbohydrates, let's see how the pathways of photosynthesis are connected to other metabolic pathways.

## 10.5 How Does Photosynthesis Interact with Other Pathways?

Green plants are autotrophs and can synthesize all the molecules they need from simple starting materials: $CO_2$, $H_2O$, phosphate, sulfate, ammonium ions ($NH_4^+$), and small quantities of other mineral nutrients. The $NH_4^+$ is needed to synthesize amino acids and nucleotides, and it comes from either the conversion of nitrogen-containing molecules in soil water or the conversion of $N_2$ gas from the atmosphere by bacteria, as we will see in Chapter 36.

Plants use the carbohydrates generated in photosynthesis to provide energy for processes such as active transport and anabolism. Both cellular respiration and fermentation can occur in plants, although the former is far more common. Unlike photosynthesis, plant cellular respiration takes place both in the light and in the dark.

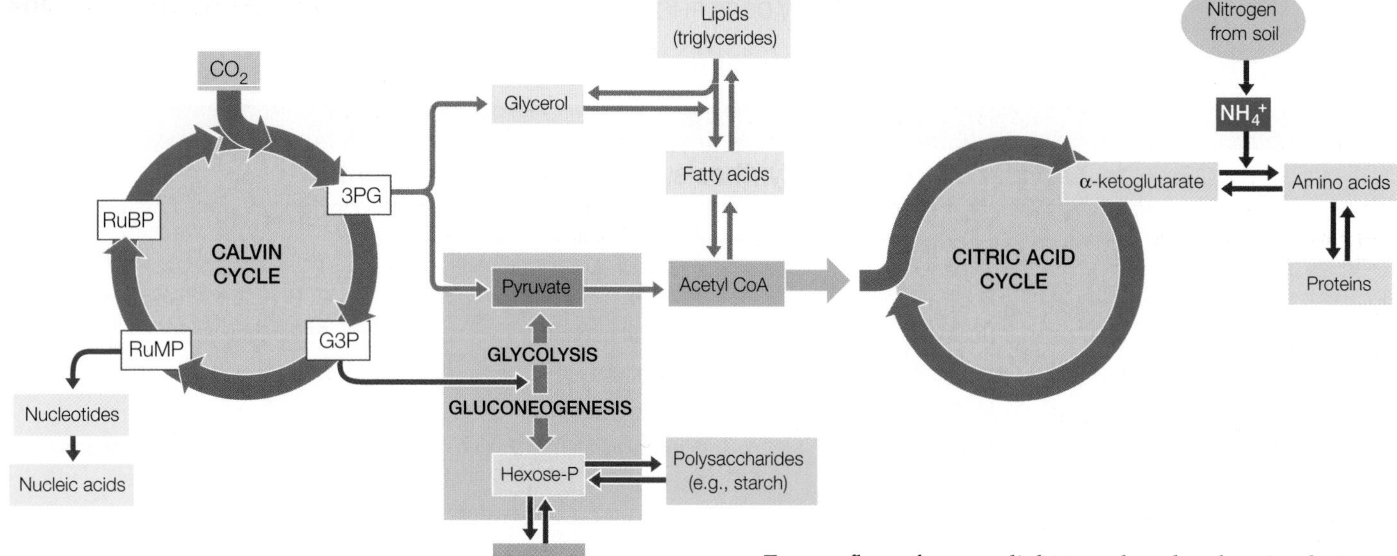


**10.20 Metabolic Interactions in a Plant Cell** The products of the Calvin cycle are used in the reactions of cellular respiration (glycolysis and the citric acid cycle).

Photosynthesis and respiration are closely linked through the Calvin cycle (**Figure 10.20**). The partitioning of G3P is particularly important:

- Some G3P from the Calvin cycle takes part in the glycolysis pathway and is converted into pyruvate in the cytosol. This pyruvate can be used in cellular respiration for energy, or its carbon skeletons can be used in anabolic reactions to make lipids, proteins, and other carbohydrates (see Figure 9.14).
- Some G3P can enter a pathway that is the reverse of glycolysis (gluconeogenesis; see Section 9.5). In this case, hexose-phosphates and then sucrose are formed and transported to the nonphotosynthetic tissues of the plant (such as the root).

Energy flows from sunlight to reduced carbon in photosynthesis, then to ATP in respiration. Energy can also be stored in the bonds of macromolecules such as polysaccharides, lipids, and proteins. For a plant to grow, energy storage (as body structures) must exceed energy release; that is, overall carbon fixation by photosynthesis must exceed respiration. This principle is the basis of the ecological food chain, as we will see in later chapters.

Photosynthesis provides most of the energy that we need for life. Given the uncertainties about the future of photosynthesis (due to changes in $CO_2$ levels and climate change), it would be wise to seek ways to improve photosynthetic efficiency. **Figure 10.21** shows the various ways in which solar energy is utilized by plants or lost. In essence, only about 5 percent of the sunlight that reaches Earth is converted into plant growth. The inefficiencies of photosynthesis involve basic chemistry and physics (some light energy is not absorbed by photosynthetic pigments) as well as biology (plant anatomy and leaf exposure, photorespiration, and inefficiencies in metabolic pathways). While it is hard to change chemistry and physics, biologists might be able to use their knowledge of plants to improve on the basic biology of photosynthesis. This could result in a more efficient use of resources and better food production.

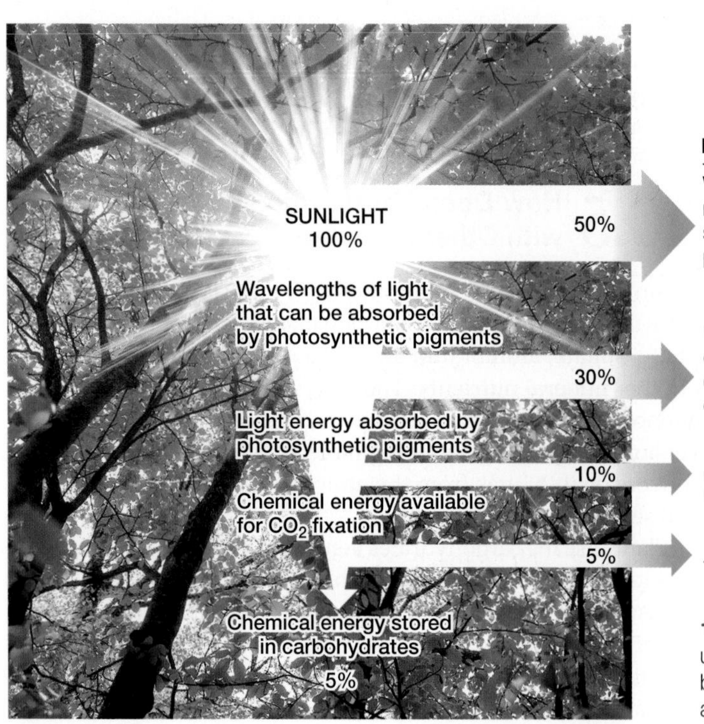


**10.21 Energy Losses During Photosynthesis** As we face an increasingly uncertain future for photosynthesis on Earth, understanding its inefficiencies becomes increasingly important. Photosynthetic pathways preserve at most about 5 percent of the sun's energy input as chemical energy in carbohydrates.

## 10.5 RECAP

**The products of photosynthesis are utilized in glycolysis and the citric acid cycle, as well as in the synthesis of lipids, proteins, and other large molecules.**

- How do common intermediates link the pathways of glycolysis, the citric acid cycle, and photosynthesis? **See p. 206 and Figure 10.20**

## CHAPTER SUMMARY

### 10.1 What Is Photosynthesis?

- In the process of **photosynthesis**, plants and other organisms take in $CO_2$, water, and light energy, producing $O_2$ and carbohydrates. **SEE ANIMATED TUTORIAL 10.1**
- The **light reactions** of photosynthesis convert light energy into chemical energy. They produce ATP and reduce $NADP^+$ to NADPH. Review Figure 10.3
- The **light-independent reactions** do not use light directly but instead use ATP and NADPH to reduce $CO_2$, forming carbohydrates.

### 10.2 How Does Photosynthesis Convert Light Energy into Chemical Energy?

- Light is a form of electromagnetic radiation. It is emitted in particle-like packets called **photons** but has wavelike properties.
- Molecules that absorb light in the visible spectrum are called **pigments**. Photosynthetic organisms have several pigments, most notably **chlorophylls**, but also accessory pigments such as carotenoids and phycobilins.
- Absorption of a photon puts a pigment molecule in an excited state that has more energy than its ground state. Review Figure 10.5
- Each compound has a characteristic **absorption spectrum**. An **action spectrum** reflects the biological activity of a photosynthetic organism for a given wavelength of light. Review Figure 10.6
- The pigments in photosynthetic organisms are arranged into **antenna systems** that absorb energy from light and funnel this energy to a pair of chlorophyll *a* molecules in the reaction center of the **photosystem**. Chlorophyll can act as a reducing agent, transferring excited electrons to other molecules. Review Figure 10.8
- **Noncyclic electron transport** uses photosystems I and II to produce ATP, NADPH and $O_2$. **Cyclic electron transport** uses only photosystem I and produces only ATP. Review Figures 10.10 and 10.11
- Chemiosmosis is the mechanism of ATP production in **photophosphorylation**. Review Figure 10.12, **ANIMATED TUTORIAL 10.2**

### 10.3 How Is Chemical Energy Used to Synthesize Carbohydrates?

- The **Calvin cycle** makes carbohydrates from $CO_2$. The cycle consists of three processes: fixation of $CO_2$, reduction and carbohydrate production, and regeneration of RuBP. **SEE ANIMATED TUTORIAL 10.3**
- **RuBP** is the initial $CO_2$ acceptor, and **3PG** is the first stable product of $CO_2$ fixation. The enzyme **rubisco** catalyzes the reaction of $CO_2$ and RuBP to form 3PG. Review Figure 10.14, **WEB ACTIVITY 10.1**
- ATP and NADPH formed by the light reactions are used in the reduction of 3PG to form **G3P**. Review Figure 10.15
- Light stimulates enzymes in the Calvin cycle, further integrating the light-dependent and light-independent pathways.

### 10.4 How Do Plants Adapt to the Inefficiencies of Photosynthesis?

- Rubisco can catalyze a reaction between $O_2$ and RuBP in addition to the reaction between $CO_2$ and RuBP. At high temperatures and low $CO_2$ concentrations, the **oxygenase** function of rubisco is favored over its **carboxylase** function.
- When rubisco functions as an oxygenase, the result is **photorespiration**, which significantly reduces the efficiency of photosynthesis.
- In **$C_4$ plants**, $CO_2$ reacts with **PEP** to form a 4-carbon intermediate in mesophyll cells. The 4-carbon product releases its $CO_2$ to rubisco in the **bundle sheath** cells in the interior of the leaf. Review Figure 10.18, **WEB ACTIVITY 10.2**
- **CAM** plants operate much like $C_4$ plants, but their initial $CO_2$ fixation by PEP carboxylase is temporally separated from the Calvin cycle, rather than spatially separated as in $C_4$ plants.

### 10.5 How Does Photosynthesis Interact with Other Pathways?

- Photosynthesis and cellular respiration are linked through the **Calvin cycle**, the **citric acid cycle**, and **glycolysis**. Review Figure 10.20
- To survive, a plant must photosynthesize more than it respires.
- Photosynthesis utilizes only a small portion of the energy of sunlight. Review Figure 10.21

## SELF-QUIZ

1. In noncyclic photosynthetic electron transport, water is used to
   *a.* excite chlorophyll.
   *b.* hydrolyze ATP.
   *c.* reduce $P_i$.
   *d.* oxidize NADPH.
   *e.* reduce chlorophyll.
2. Which statement about light is true?
   *a.* An absorption spectrum is a plot of biological effectiveness versus wavelength.
   *b.* An absorption spectrum may be a good means of identifying a pigment.
   *c.* Light need not be absorbed to produce a biological effect.
   *d.* A given kind of molecule can occupy any energy level.
   *e.* A pigment loses energy as it absorbs a photon.
3. Which statement about chlorophylls is *not* true?
   *a.* Chlorophylls absorb light near both ends of the visible spectrum.
   *b.* Chlorophylls can accept energy from other pigments, such as carotenoids.
   *c.* Excited chlorophyll can either reduce another substance or release light energy.
   *d.* Excited chlorophyll cannot be an oxidizing agent.
   *e.* Chlorophylls contain magnesium.

4. In cyclic electron transport,
   *a.* oxygen gas is released.
   *b.* ATP is formed.
   *c.* water donates electrons and protons.
   *d.* NADPH forms.
   *e.* $CO_2$ reacts with RuBP.
5. Which of the following does *not* happen in noncyclic electron transport?
   *a.* Oxygen gas is released.
   *b.* ATP forms.
   *c.* Water donates electrons and protons.
   *d.* NADPH forms.
   *e.* $CO_2$ reacts with RuBP.
6. In chloroplasts,
   *a.* light leads to the flow of protons out of the thylakoids.
   *b.* ATP is formed when protons flow into the thylakoid lumen.
   *c.* light causes the thylakoid lumen to become less acidic than the stroma.
   *d.* protons return passively to the stroma through protein channels.
   *e.* proton pumping requires ATP.
7. Which statement about the Calvin cycle is *not* true?
   *a.* $CO_2$ reacts with RuBP to form 3PG.
   *b.* RuBP forms by the metabolism of 3PG.
   *c.* ATP and NADPH form when 3PG is oxidized.
   *d.* The concentration of 3PG rises if the light is switched off.
   *e.* Rubisco catalyzes the reaction of $CO_2$ and RuBP.
8. In $C_4$ photosynthesis,
   *a.* 3PG is the first product of $CO_2$ fixation.
   *b.* rubisco catalyzes the first step in the pathway.
   *c.* 4-carbon acids are formed by PEP carboxylase in bundle sheath cells.
   *d.* photosynthesis continues at lower $CO_2$ levels than in $C_3$ plants.
   *e.* $CO_2$ released from RuBP is transferred to PEP.
9. Photosynthesis in green plants occurs only during the day. Respiration in plants occurs
   *a.* only at night.
   *b.* only when there is enough ATP.
   *c.* only during the day.
   *d.* all the time.
   *e.* in the chloroplast after photosynthesis.
10. Photorespiration
    *a.* takes place only in $C_4$ plants.
    *b.* includes reactions carried out in peroxisomes.
    *c.* increases the yield of photosynthesis.
    *d.* is catalyzed by PEP carboxylase.
    *e.* is independent of light intensity.

## FOR DISCUSSION

1. Both photosynthetic electron transport and the Calvin cycle stop in the dark. Which specific reaction stops first? Which stops next? Continue answering the question "Which stops next?" until you have explained why both pathways have stopped.
2. In what principal ways are the reactions of electron transport in photosynthesis similar to the reactions of oxidative phosphorylation discussed in Section 9.3?
3. Differentiate between cyclic and noncyclic electron transport in terms of (1) the products and (2) the source of electrons for the reduction of oxidized chlorophyll.
4. If water labeled with $^{18}O$ is added to a suspension of photosynthesizing chloroplasts, which of the following compounds will first become labeled with $^{18}O$: ATP, NADPH, $O_2$, or 3PG? If water labeled with $^3H$ is added, which of the same compounds will first become radioactive? Which will be first if $CO_2$ labeled with $^{14}C$ is added?
5. The Viking lander was sent to Mars in 1976 to detect signs of life. Explain the rationale behind the following experiments this unmanned probe performed:
   *a.* A scoop of dirt was inserted into a container and $^{14}CO_2$ was added. After a while during the Martian day, the $^{14}CO_2$ was removed and the dirt was heated to a high temperature. Scientists monitoring the experiment back on Earth looked for the release of $^{14}CO_2$ as a sign of life.
   *b.* The same experiment was performed, except that the dirt was heated to a high temperature for 30 minutes and then allowed to cool to Martian temperature right after scooping, and before the $^{14}CO_2$ was added. If living things were present, then $^{14}CO_2$ would be released in experiment (*a*), but not this one.

## ADDITIONAL INVESTIGATION

Calvin's experiment (see Figure 10.13) laid the foundations for a full description of the pathway of $CO_2$ fixation. Given the interrelationships between metabolic pathways in plants, how would you do an experiment to follow the pathway of fixed carbon from photosynthesis to proteins?

## WORKING WITH DATA (GO TO yourBioPortal.com)

**Water is the Source of the Oxygen Produced by Photosynthesis** The proposal that the source of $O_2$ in photosynthesis was $H_2O$ rather than $CO_2$ was first made in 1932. But it took the invention of isotope tracing a decade later to prove this. In this exercise, you will examine the methods that Ruben and Kamen used (Figure 10.2) to identify the isotopes of oxygen and the data they obtained.

**Tracing the Pathway of $CO_2$** Studies of radioactive isotopes were intensified during World War II as an offshoot of the development of nuclear weapons. This led Calvin and his colleagues to perform the experiments designed to trace the path of carbon in photosynthesis (Figure 10.13). In this hands-on exercise, you will examine their data and see the reasoning that led to the $CO_2$ fixation pathway.

PART FOUR GENES AND HEREDITY

# 11 The Cell Cycle and Cell Division

## An enemy of the cell reproduction cycle

Ruth felt healthy and was surprised when she was called back to her physician's office a week after her annual checkup. "Your lab report indicates you have early cervical cancer," said the doctor. "I ordered a follow-up test, and it came back positive—at some point, you were infected with HPV."

Ruth felt numb as soon as she heard the word "cancer." Her mother had died of breast cancer in the previous year. The doctor's statement about HPV (human papillomavirus) did not register in her consciousness. Sensing Ruth's discomfort, the doctor quickly reassured her that the cancer was caught at an early stage, and that a simple surgical procedure would remove it. Two weeks later, the cancer was removed and Ruth remains cancer-free. She was fortunate that her annual medical exam included a Papanicolau (Pap) test, in which the cells lining the cervix are examined for abnormalities. Since they were begun almost 50 years ago in Europe, Pap tests have resulted in the early detection and removal of millions of early cervical cancers, and the death rate from this potentially lethal disease has plummeted.

Only recently was HPV found to be the cause of most cervical cancers. The German physician Harald Zur-Hansen was awarded the Nobel Prize in 2008 for this discovery and it has led to a vaccine to prevent future infections. There are many different types of HPV and many of the ones that infect humans cause warts, which are small, rough growths on the skin. The types of HPV that infect tissues at the cervix get there by sexual transmission, and this is a common infection in Western societies.

When HPV arrives at the tissues lining the cervix, it has one of two fates. Most of the time, it gets into the cells, turning them into HPV factories, producing a lot of HPV in the mucus outside the uterus. These viruses can infect another person during a sexual encounter. In some cases, however, the virus follows a different and—for the host cells—more sinister path. It infects the cervical cells and causes them to make a viral protein called E7, a protein that can deregulate human cell reproduction.

Cell reproduction in healthy humans is tightly controlled, and one of the strongest regulators that prevent a cell from dividing is the

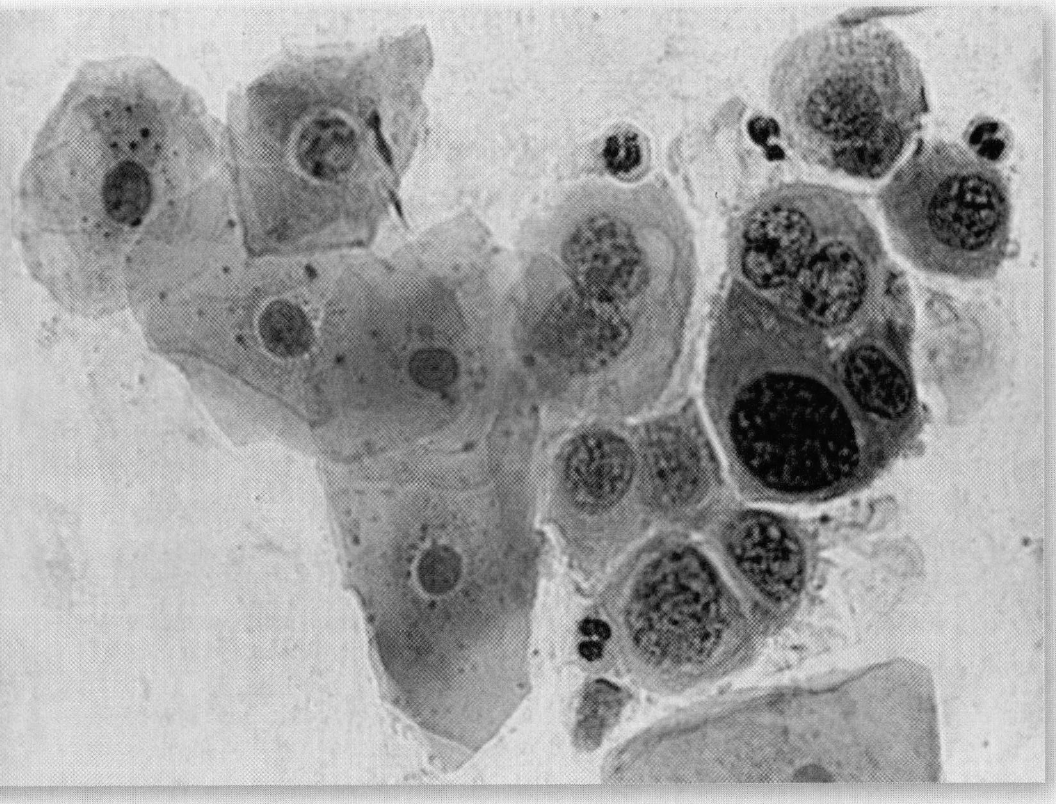

**Abnormal Cells** In this Pap test, cervical cancer cells at right differ from the normal cells at left. The cancer cells have larger nuclei.

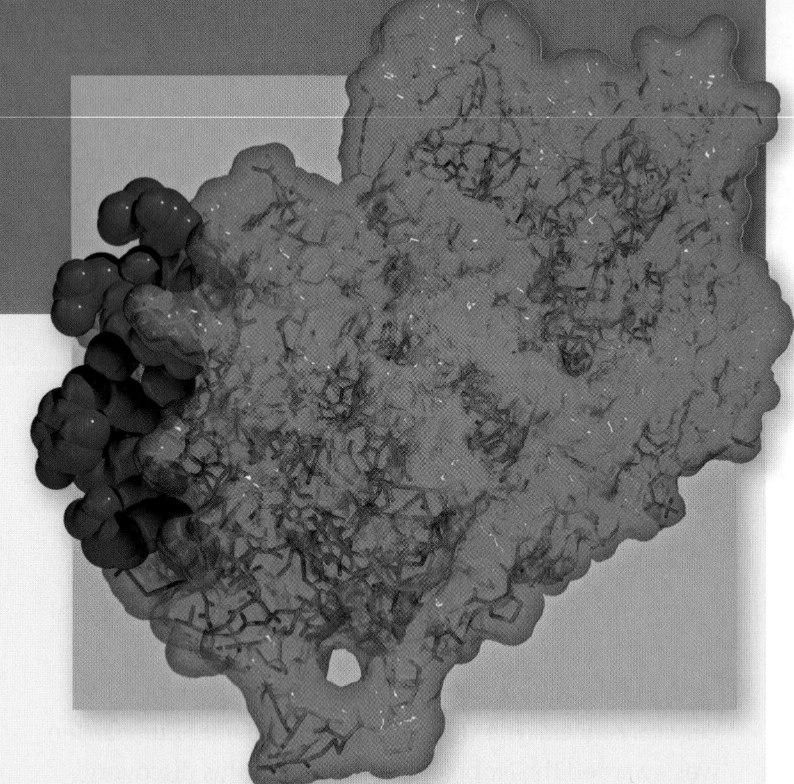

**E7, RB, and Cell Reproduction** The E7 protein (blue) from human papillomavirus binds to the RB protein (red) to inhibit RB's ability to block cell division. This results in cancer.

retinoblastoma protein (RB), which you will encounter later in the chapter. One of the viral gene products is the protein E7, which has a three-dimensional shape that just fits into the protein-binding site of RB, thereby inactivating it. With no active RB to put the brakes on, cell division proceeds. As you know, uncontrolled cell reproduction is a hallmark of cancer—and so cervical cancer begins.

Understanding the cell division cycle and its control is clearly an important subject for understanding cancer. But cell division is not just important in medicine. It underlies the growth, development and reproduction of all organisms.

**IN THIS CHAPTER** we will see how cells give rise to more cells. We first describe how prokaryotic cells divide to produce new, single-celled organisms. Then we turn to the two types of nuclear division in eukaryotes—mitosis and meiosis—and relate them to asexual and sexual reproduction. Cell reproduction is linked to cell death, so we then consider the process of programmed cell death, also known as apoptosis. Finally, we relate these processes to the loss of cell reproduction control in cancer cells.

## CHAPTER OUTLINE



## 11.1 How Do Prokaryotic and Eukaryotic Cells Divide?

The life cycle of an organism, from birth to death, is intimately linked to cell division. Cell division plays important roles in the growth and repair of tissues in multicellular organisms, as well as in the reproduction of all organisms (**Figure 11.1**).

In order for any cell to divide, four events must occur:

- There must be a reproductive signal. This signal initiates cell division and may originate from either inside or outside the cell.
- **Replication** of DNA (the genetic material) must occur so that each of the two new cells will have identical genes and complete cell functions.
- The cell must distribute the replicated DNA to each of the two new cells. This process is called **segregation**.
- In addition to synthesizing needed enzymes and organelles, new material must be added to the plasma membrane (and the cell wall, in organisms that have one), in order to separate the two new cells by a process called **cytokinesis**.

These four events proceed somewhat differently in prokaryotes and eukaryotes.

### Prokaryotes divide by binary fission

In prokaryotes, cell division results in the reproduction of the entire single-celled organism. The cell grows in size, replicates its DNA, and then separates the cytoplasm and DNA into two new cells by a process called **binary fission**.

**REPRODUCTIVE SIGNALS** The reproductive rates of many prokaryotes respond to conditions in the environment. The bacterium *Escherichia coli*, a species commonly used in genetic studies, is a "cell division machine"; if abundant sources of carbohydrates and mineral nutrients are available, it can divide as often as every 20 minutes. Another bacterium, *Bacillus subtilis*, does not just slow its growth when nutrients are low but stops dividing and then resumes dividing when conditions improve. Clearly, external factors such as environmental conditions and nutrient concentrations are signals for the initiation of cell division in prokaryotes.

**REPLICATION OF DNA** As we saw in Section 5.3, a **chromosome** can be defined in molecular terms as a DNA molecule containing genetic information. When a cell divides, all of its chromo-

(A) Reproduction
(B) Growth
(C) Regeneration
These yeast cells divide by budding.
Cell division contributes to the growth of this root tissue.
Cell division contributes to the regeneration of a lizard's tail.

**11.1 Important Consequences of Cell Division** Cell division is the basis for (A) reproduction, (B) growth, and (C) repair and regeneration of tissues.

somes must be replicated, and one copy of each chromosome must find its way into one of the two new cells.

Most prokaryotes have only one chromosome—a single long DNA molecule with proteins bound to it. In *E. coli*, the ends of the DNA molecule are joined to create a circular chromosome. Circular chromosomes are characteristic of most prokaryotes as well as some viruses, and are also found in the chloroplasts and mitochondria of eukaryotic cells.

If the *E. coli* DNA were spread out into an actual circle, it would be about 500 μm in diameter. The bacterium itself is only about 2 μm long and 1μm in diameter. Thus if the bacterial DNA were fully extended, it would form a circle over 200 times larger than the cell! To fit into the cell, bacterial DNA must be compacted. The DNA folds in on itself, and positively charged (basic) proteins bound to the negatively charged (acidic) DNA contribute to this folding.

Two regions of the prokaryotic chromosome play functional roles in cell reproduction:

- *ori*: the site where replication of the circular chromosome starts (the *ori*gin of replication)
- *ter*: the site where replication ends (the *ter*minus of replication)

Chromosome replication takes place as the DNA is threaded through a "replication complex" of proteins near the center of the cell. (These proteins include the enzyme DNA polymerase, whose important role in replication is discussed further in Section 13.3.) Replication begins at the *ori* site and moves toward the *ter* site. While the DNA replicates, anabolic metabolism is active and the cell grows. When replication is complete, the two daughter DNA molecules separate and segregate from one another at opposite ends of the cell. In rapidly dividing prokaryotes, DNA replication occupies the entire time between cell divisions.

**SEGREGATION OF DNAS** Replication begins near the center of the cell, and as it proceeds, the *ori* regions move toward opposite ends of the cell (**Figure 11.2A**). DNA sequences adjacent to the *ori* region bind proteins that are essential for this segregation. This is an active process, since the binding proteins hydrolyze ATP. The prokaryotic cytoskeleton (see Section 5.2) may be involved in DNA segregation, either actively moving the DNA along, or passively acting as a "railroad track" along which DNA moves.

**CYTOKINESIS** The actual division of a single cell and its contents into two cells is called cytokinesis and begins immediately after chromosome replication is finished in rapidly growing cells. The first event of cytokinesis is a pinching in of the plasma membrane to form a ring of fibers similar to a purse string. The major component of these fibers is a protein that is related to eukaryotic tubulin (which makes up microtubules). As the membrane pinches in, new cell wall materials are deposited, which finally separate the two cells (**Figure 11.2B**).

### Eukaryotic cells divide by mitosis or meiosis followed by cytokinesis

As in prokaryotes, cell reproduction in eukaryotes entails reproductive signals, DNA replication, segregation, and cytokinesis. The details, however, are quite different:

- *Signal*. Unlike prokaryotes, eukaryotic cells do not constantly divide whenever environmental conditions are adequate. In fact, most eukaryotic cells that are part of a multicellular organism and have become specialized seldom divide. In a eukaryotic organism, the signals for cell division are related not to the environment of a single cell, but to the needs of the entire organism.
- *Replication*. While most prokaryotes have a single main chromosome, eukaryotes usually have many (humans have 46). Consequently the processes of replication and segregation are more intricate in eukaryotes than in prokaryotes. In eukaryotes, DNA replication is usually limited to a portion of the period between cell divisions.

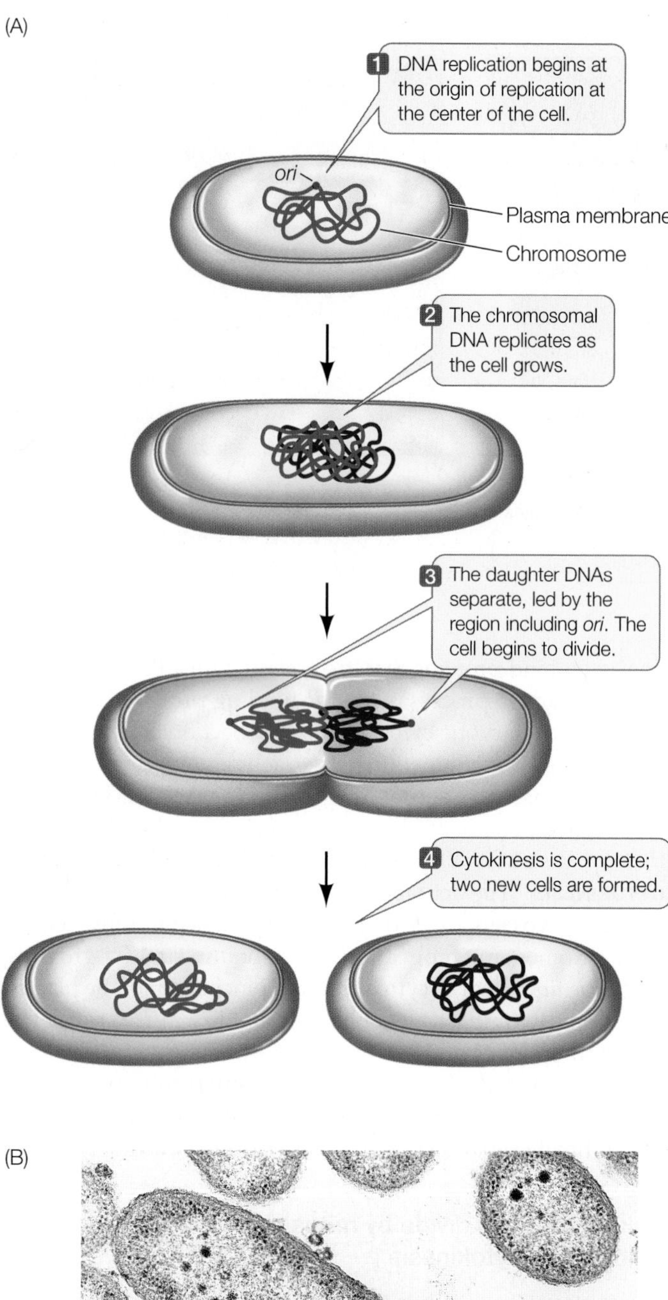
(A)
1 DNA replication begins at the origin of replication at the center of the cell.
ori
Plasma membrane
Chromosome
2 The chromosomal DNA replicates as the cell grows.
3 The daughter DNAs separate, led by the region including ori. The cell begins to divide.
4 Cytokinesis is complete; two new cells are formed.

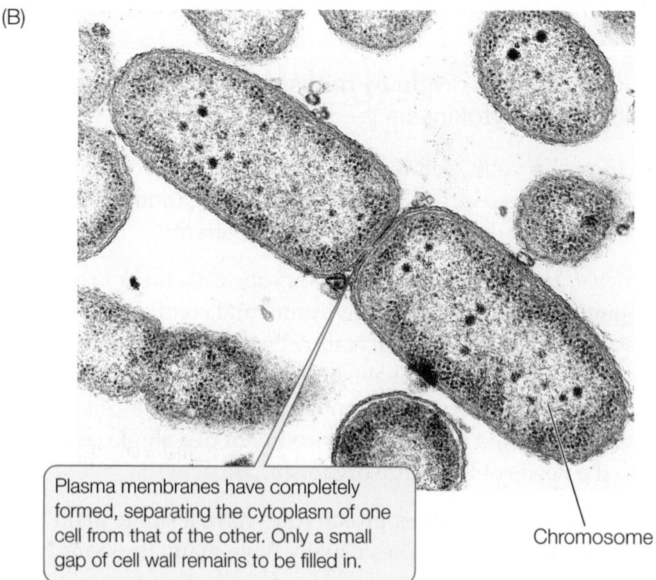
(B)
Plasma membranes have completely formed, separating the cytoplasm of one cell from that of the other. Only a small gap of cell wall remains to be filled in.
Chromosome

**11.2 Prokaryotic Cell Division** (A) The process of cell division in a bacterium. (B) These two cells of the bacterium *Pseudomonas aeruginosa* have almost completed cytokinesis.

- *Segregation*. In eukaryotes, the newly replicated chromosomes are closely associated with each other (thus they are known as **sister chromatids**), and a mechanism called **mitosis** segregates them into two new nuclei.
- *Cytokinesis*. Cytokinesis proceeds differently in plant cells (which have a cell wall) than in animal cells (which do not).

The cells resulting from mitosis are identical to the parent cell in the amount and kind of DNA that they contain. This contrasts with the second mechanism of nuclear division, meiosis.

**Meiosis** is the process of nuclear division that occurs in cells involved with sexual reproduction. While the two products of mitosis are genetically identical to the cell that produced them—they both have the same DNA—the products of meiosis are not. As we will see in Section 11.5, meiosis generates diversity by shuffling the genetic material, resulting in new gene combinations. Meiosis plays a key role in the sexual life cycle.

## 11.1 RECAP

**Four events are required for cell division: a reproductive signal, replication of the genetic material (DNA), segregation of replicated DNA, and separation of the two daughter cells (cytokinesis). In prokaryotes, cell division can be rapid; in eukaryotes, the process is more intricate, and the chromosomes must be duplicated before cell division can occur.**

- What is the reproductive signal that leads the bacterium *Bacillus subtilis* to divide? **See p. 210**
- Explain why DNA must be replicated and segregated *before* a cell can divide. **See p. 210**
- What are the differences between cell division in prokaryotes (binary fission) and mitosis in eukaryotes? **See pp. 211–212**

What determines whether a cell will divide? How does mitosis lead to identical cells, and meiosis to diversity? Why do most eukaryotic organisms reproduce sexually? In the sections that follow, we will describe the details of mitosis and meiosis, and discuss their roles in development and evolution.

## 11.2 How Is Eukaryotic Cell Division Controlled?

As you will see throughout the book, different cells have different rates of cell division. Some cells, such as those in an early embryo, stem cells in bone marrow, or cells in the growing tip of a plant root, divide rapidly and continuously. Others, such as neurons in the brain or phloem cells in a plant stem, don't divide at all. Clearly, the signaling pathways for cells to divide are highly controlled.

The period between cell divisions is referred to as the **cell cycle**. The cell cycle can be divided into mitosis/cytokinesis and interphase. During **interphase**, the cell nucleus is visible and typical cell functions occur, including DNA replication. This phase of

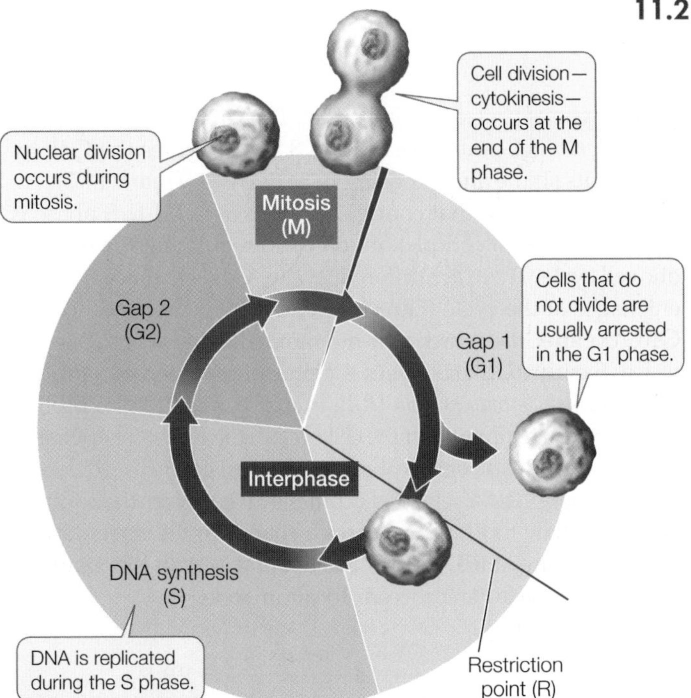
Cell division—cytokinesis—occurs at the end of the M phase.
Nuclear division occurs during mitosis.
Mitosis (M)
Cells that do not divide are usually arrested in the G1 phase.
Gap 2 (G2)
Gap 1 (G1)
Interphase
DNA synthesis (S)
DNA is replicated during the S phase.
Restriction point (R)

**11.3 The Eukaryotic Cell Cycle** The cell cycle consists of a mitotic (M) phase, during which mitosis and cytokinesis take place, and a long period of growth known as interphase. Interphase has three subphases (G1, S, and G2) in cells that divide.

the cell cycle begins when cytokinesis is completed and ends when mitosis begins (**Figure 11.3**). In this section we will describe the events of interphase, especially those that trigger mitosis.

Cells, even when rapidly dividing, spend most of their time in interphase. So if we take a snapshot through the microscope of a cell population, most of the cells will be in interphase; only a small percentage will be in mitosis or cytokinesis at any given moment.

Interphase has three subphases, called G1, S, and G2. The cell's DNA replicates during **S phase** (the *S* stands for synthesis) (see Figure 11.3). The period between the end of cytokinesis and the onset of S phase is called **G1**, or Gap 1. Another gap phase—**G2**—separates the end of S phase and the beginning of mitosis. Mitosis and cytokinesis are referred to as the **M phase** of the cell cycle.

Let's look at the events of interphase in more detail:

- *G1 phase.* During G1, a cell is preparing for S phase, so at this stage each chromosome is a single, unreplicated structure. G1 is quite variable in length in different cell types. Some rapidly dividing embryonic cells dispense with it entirely, while other cells may remain in G1 for weeks or even years. In many cases these cells enter a resting phase called G0. Special internal and external signals are needed to prompt a cell to leave G0 and reenter the cell cycle at G1.
- *The G1-to-S transition.* At the G1-to-S transition, called the **restriction point** (**R**), the commitment is made to DNA replication and subsequent cell division (and thus another cell cycle).
- *S phase.* DNA replication occurs during S phase (see Section 13.3 for a detailed description). Each chromosome is duplicated and thereafter consists of two sister chromatids joined together and awaiting segregation into two new cells.
- *G2 phase.* During G2, the cell makes preparations for mitosis—for example, by synthesizing components of the microtubules that will move the chromatids to opposite ends of the dividing cell.

The initiation, termination, and operations of these phases are regulated by specific signals.

### Specific signals trigger events in the cell cycle

What events cause a cell to enter the S or M phases? A first indication that there were substances that control these transitions came from experiments involving *cell fusion.* Polyethylene glycol can be used to make different cells fuse together. Membrane lipids tend to partially dissolve in this nonpolar solvent, so that when it is present, cells will fuse their plasma membranes. Experiments involving the fusion of mammalian cells at different phases of the cell cycle showed that a cell in S phase produces a substance that activates DNA replication (**Figure 11.4**).

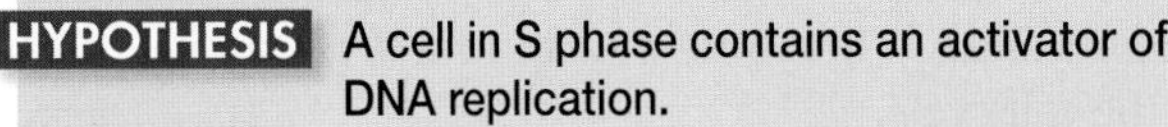
INVESTIGATING LIFE

**11.4 Regulation of the Cell Cycle**
Nuclei in G1 do not undergo DNA replication, but nuclei in S phase do. To determine if there is some signal in the S cells that stimulates G1 cells to replicate their DNA, cells in G1 and S phases were fused together, creating cells with both G1 and S properties.

**HYPOTHESIS** A cell in S phase contains an activator of DNA replication.

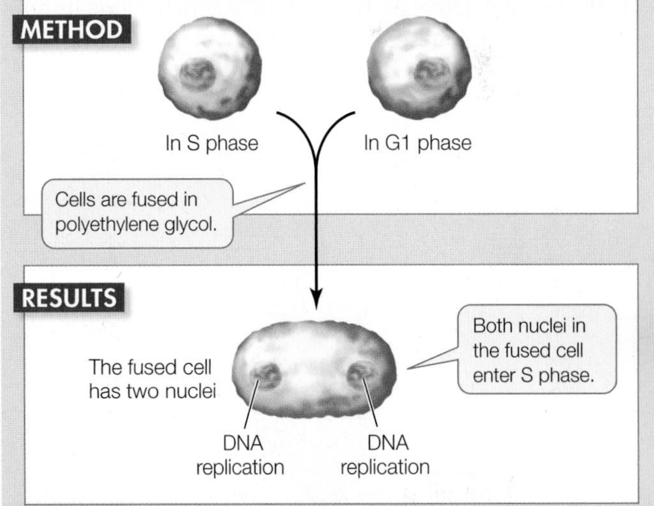
METHOD
In S phase
In G1 phase
Cells are fused in polyethylene glycol.
RESULTS
The fused cell has two nuclei
Both nuclei in the fused cell enter S phase.
DNA replication
DNA replication

**CONCLUSION** The S phase cell produces a substance that diffuses to the G1 nucleus and activates DNA replication.

**FURTHER INVESTIGATION:** How would you use this method to show that a cell in M phase produces an activator of mitosis?

Go to **yourBioPortal.com** for original citations, discussions, and relevant links for all INVESTIGATING LIFE figures.

Similar experiments point to a molecular activator for entry into M phase. As you will see, the signals that control progress through the cell cycle act through protein kinases.

Progress through the cell cycle depends on the activities of **cyclin-dependent kinases**, or **Cdk's**. Recall from Section 7.1 that a *protein kinase* is an enzyme that catalyzes the transfer of a phosphate group from ATP to a target protein; this phosphate transfer is called *phosphorylation*.

$$\text{protein} + \text{ATP} \xrightarrow{\text{protein kinase}} \text{protein-P} + \text{ADP}$$

By catalyzing the phosphorylation of certain target proteins, Cdk's play important roles at various points in the cell cycle. The discovery that Cdk's induce cell division is a beautiful example of how research on different organisms and different cell types can converge on a single mechanism. One group of scientists, led by James Maller at the University of Colorado, was studying immature sea urchin eggs, trying to find out how they are stimulated to divide and eventually form a mature egg. A protein called *maturation promoting factor* was purified from maturing eggs, which by itself prodded immature egg cells to divide.

Meanwhile, Leland Hartwell at the University of Washington was studying the cell cycle in yeast (a single-celled eukaryote, see Figure 11.1A), and found a strain that was stalled at the G1–S boundary because it lacked a Cdk. It turned out that this yeast Cdk and the sea urchin maturation promoting factor had similar properties, and further work confirmed that the sea urchin protein was indeed a Cdk. Similar Cdk's were soon found to control the G1-to-S transition in many other organisms, including humans. Then others were found to control other parts of the cell cycle.

Cdk's are not active by themselves. As their name implies, cyclin-dependent kinases need to be activated by binding to a second type of protein, called **cyclin**. This binding—an example of *allosteric regulation* (see Section 8.5)—activates the Cdk by altering its shape and exposing its active site (**Figure 11.5**).

The cyclin–Cdk that controls passage from G1 to S phase is not the only such complex involved in regulating the eukaryotic cell cycle. There are different cyclin–Cdk's that act at different stages of the cycle (**Figure 11.6**). Let's take a closer look at G1–S cyclin–Cdk, which was the first to be discovered.

G1–S cyclin–Cdk catalyzes the phosphorylation of a protein called *retinoblastoma protein* (*RB*). In many cells, RB or a protein like it acts as an inhibitor of the cell cycle at the R (for "restriction") point in late G1. To begin S phase, a cell must get by the RB block. Here is where G1–S cyclin–Cdk comes in: it catalyzes the addition of a phosphate to RB. This causes a change in the three-dimensional structure of RB, thereby inactivating it. With RB out of the way, the cell cycle can proceed. To summarize:

$$\underset{\text{(active—blocks cell cycle)}}{\text{RB}} \xrightarrow{\text{G1–S cyclin–Cdk}} \underset{\text{(inactive—allows cell cycle)}}{\text{RB-P}}$$

Progress through the cell cycle is regulated by the activities of Cdk's, and so regulating *them* is a key to regulating cell division. An effective way to regulate Cdk's is to regulate the presence or absence of cyclins (**Figure 11.7**). Simply put, if a cyclin is not present, its partner Cdk is not active. As their name suggests, the presence of cyclins is cyclic: they are made only at certain times of in the cell cycle.

The different cyclin–Cdk's act at **cell cycle checkpoints**, points at which a cell cycle's progress is regulated. For example, if a cell's DNA is substantially damaged by radiation or

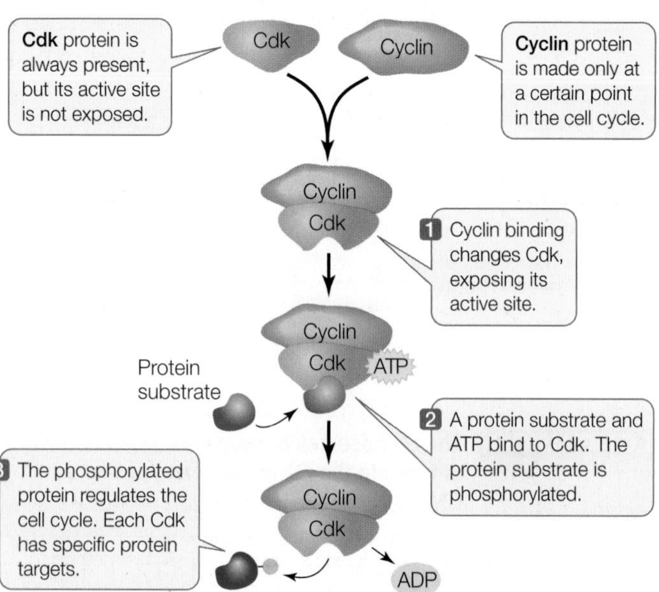


**11.5 Cyclin Binding Activates Cdk** Binding of a cyclin changes the three-dimensional structure of an inactive Cdk, making it an active protein kinase. Each cyclin–Cdk complex phosphorylates a specific target protein in the cell cycle.

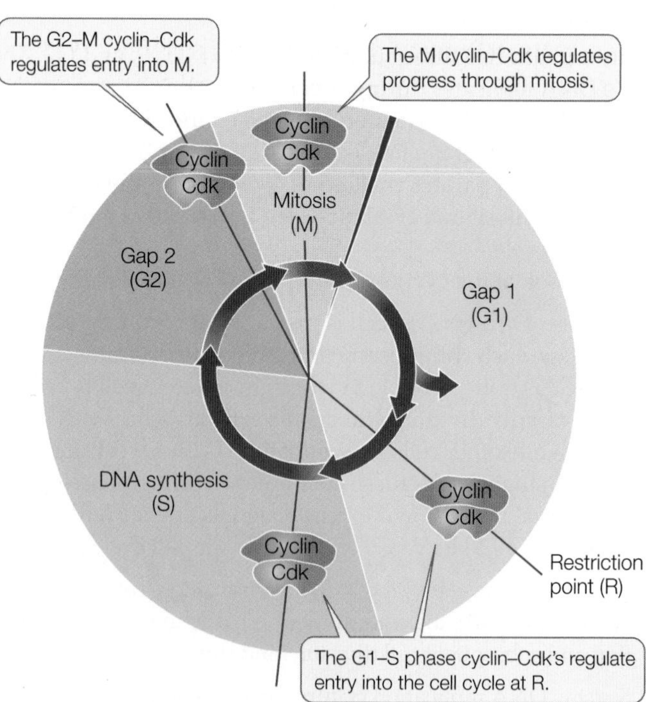


**11.6 Cyclin-Dependent Kinases Regulate Progress Through the Cell Cycle** By acting at checkpoints (red lines), different cyclin–Cdk complexes regulate the orderly sequence of events in the cell cycle.

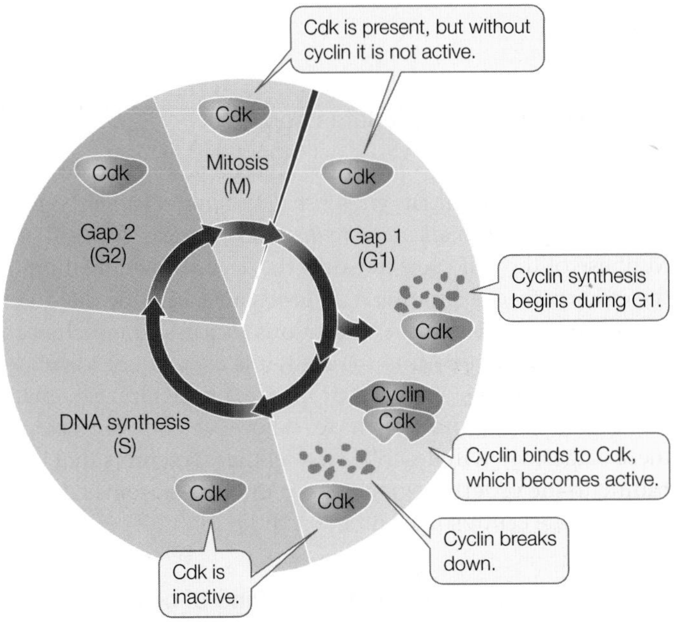

**11.7 Cyclins Are Transient in the Cell Cycle** Cyclins are made at a particular time and then break down. In this case, the cyclin is present during $G_1$ and activates a Cdk at that time.

toxic chemicals, it may be prevented from successfully completing a cell cycle. For DNA damage, there are three checkpoints:

- During G1, before the cell enters S phase (restriction point)
- During S phase
- After S phase, during G2

Let's consider the G1 checkpoint. If DNA is damaged by radiation during G1, a protein called p21 is made. (The *p* stands for "protein" and the 21 stands for its molecular weight—about 21,000 daltons.) The p21 protein can bind to the G1–S Cdk, preventing its activation by cyclin. So the cell cycle stops while repairs are made to the DNA (you will learn more about DNA repair in Section 13.4). The p21 protein breaks down after the DNA is repaired, allowing cyclin to bind to the Cdk so that the cell cycle can proceed. If DNA damage is severe and it cannot be repaired, the cell will undergo programmed cell death (apoptosis, which we will discuss in Section 11.6).

In addition to these internal signals, the cell cycle is influenced by signals from the extracellular environment.

### Growth factors can stimulate cells to divide

Cyclin–Cdk's provide cells with internal controls of their progress through the cell cycle. Not all cells in an organism go through the cell cycle on a regular basis. Some cells either no longer go through the cell cycle and enter G0, or go through it slowly and divide infrequently. If such cells are to divide, they must be stimulated by external chemical signals called **growth factors**. These proteins activate a signal transduction pathway that often ends up with the activation of Cdk's (signal transduction is discussed in Chapter 7):

- If you cut yourself and bleed, specialized cell fragments called *platelets* gather at the wound to initiate blood clotting. The platelets produce and release a protein called *platelet-derived growth factor* that diffuses to the adjacent cells in the skin and stimulates them to divide and heal the wound.
- Red and white blood cells have limited lifetimes and must be replaced through the division of immature, unspecialized blood cell precursors in the bone marrow. Two types of growth factors, *interleukins* and *erythropoietin*, stimulate the division and specialization, respectively, of precursor cells.

In these and other examples, growth factors bind to specific receptors on target cells, and activate signal transduction pathways the end with cyclin synthesis, thereby activating Cdk's and the cell cycle. As you can see from the examples, growth factors are important in maintaining homeostasis.

## 11.2 RECAP

**The eukaryotic cell cycle is under both external and internal control. Cdk's control the eukaryotic cell cycle and are themselves controlled by cyclins. External signals such as growth factors can initiate the cell cycle.**

- Draw a cell cycle diagram showing the various stages of interphase. **See pp. 212–213 and Figure 11.3**
- How do cyclin–Cdk's control the progress of the cell cycle? **See pp. 214–215 and Figure 11.6**
- What are the differences between external and internal controls of the cell cycle? **See p. 215**

## 11.3 What Happens during Mitosis?

The third essential step in the process of cell division—segregation of the replicated DNA—occurs during mitosis. Prior to segregation, the huge DNA molecules and their associated proteins in each chromosome become condensed into more compact structures. After segregation by mitosis, cytokinesis separates the two cells. Let's now look at these steps more closely.

### Prior to mitosis, eukaryotic DNA is packed into very compact chromosomes

A eukaryotic chromosome consists of one or two gigantic, linear, double-stranded DNA molecules complexed with many proteins (the complex of DNA and proteins is referred to as **chromatin**). Before S phase, each chromosome contains only one double-stranded DNA molecule. After it replicates during S phase, however, there are two double-stranded DNA molecules, known as sister chromatids. The sister chromatids are held together along most of their length by a protein complex called *cohesin*. They stay this way throughout interphase G2 until mitosis, when most of the cohesin is removed, except in a region called the **centromere** at which the chromatids remain held together. At the end of G2, a second group of proteins called *condensins* coat the DNA molecules and makes them more compact (**Figure 11.8**).

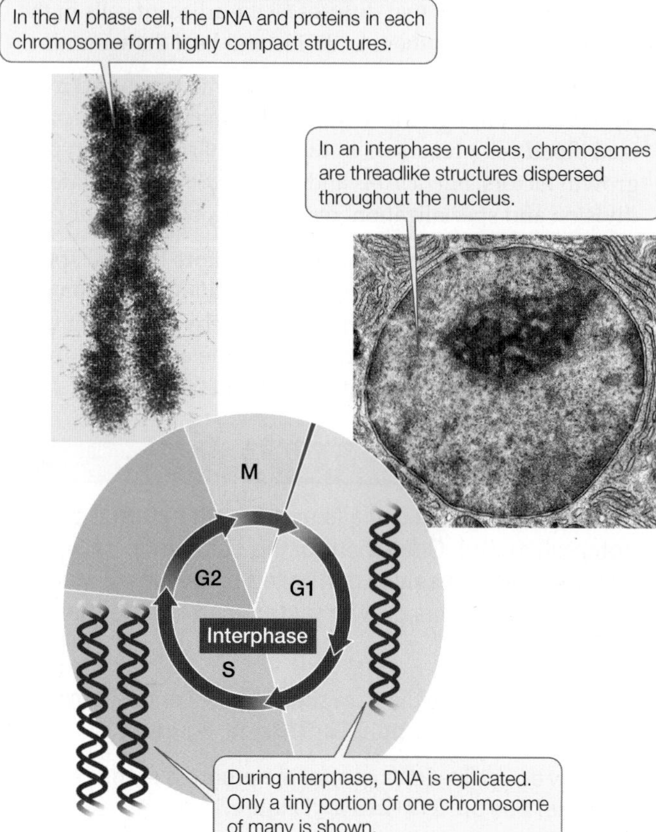

In the M phase cell, the DNA and proteins in each chromosome form highly compact structures.
In an interphase nucleus, chromosomes are threadlike structures dispersed throughout the nucleus.
M
G2
G1
Interphase
S
During interphase, DNA is replicated. Only a tiny portion of one chromosome of many is shown.

**11.8 Chromosomes, Chromatids, and Chromatin** DNA in the interphase nucleus is diffuse and becomes compacted as mitosis begins.

If all the DNA in a typical human cell were put end to end, it would be nearly 2 meters long. Yet the nucleus is only 5 μm (0.000005 meters) in diameter. So eukaryotic DNA, like that in prokaryotes, is extensively packaged in a highly organized way (**Figure 11.9**). This packing is achieved largely by proteins that are closely associated with the DNA; these proteins are called **histones** (*histos*, "web" or "loom"). They are positively charged at cellular pH levels because of their high content of the basic amino acids lysine and arginine. These positive charges attract the negative phosphate groups on DNA. These DNA–histone interactions, as well as histone–histone interactions, result in the formation of beadlike units called **nucleosomes**.

During interphase, the chromatin that makes up each chromosome is much less densely packaged, and consists of single DNA molecules running around vast numbers of nucleosomes like beads on a string. During this phase of the cell cycle, the DNA is accessible to proteins involved in replication and transcription. Once a mitotic chromosome is formed its compact nature makes it inaccessible to replication and transcription factors, and so these processes cannot occur.

During the early stages of both mitosis and meiosis, the chromatin becomes ever more tightly coiled and condensed as the nucleosomes pack together. Further coiling of the chromatin continues up to the time at which the chromatids begin to move apart.

### Overview: Mitosis segregates copies of genetic information

*In mitosis, a single nucleus gives rise to two nuclei that are genetically identical to each other and to the parent nucleus.* Mitosis (the M phase of the cell cycle) ensures the accurate segregation of the eukaryotic cell's multiple chromosomes into the daughter nuclei. While mitosis is a continuous process in which each event flows smoothly into the next, it is convenient to subdivide it into a series of stages: prophase, prometaphase, metaphase, anaphase, and telophase. Before we consider each of these stages, we will describe two cellular structures that contribute to the orderly segregation of the chromosomes during mitosis—the centrosome and the spindle.

#### The centrosomes determine the plane of cell division

Before the spindle apparatus for chromosome segregation forms, the orientation of this spindle is determined. This is accomplished by the **centrosome** ("central body"), an organelle in the cytoplasm near the nucleus. In many organisms, each centrosome consists of a pair of **centrioles**, each one a hollow tube formed by nine triplets of microtubules. The two tubes are at right angles to each other.

During S phase the centrosome doubles to form a pair of centrosomes. At the G2-to-M transition, the two centrosomes separate from one another, moving to opposite ends of the nuclear envelope. Eventually these will identify "poles" toward which chromosomes will move during segregation. The positions of the centrosomes determine the plane at which the cell will divide; therefore they determine the spatial relationship between the two new cells. This relationship may be of little consequence to single free-living cells such as yeasts, but it is important for cells in a multicellular organism. For example, during development from a fertilized egg to an embryo, the daughter cells from some divisions must be positioned correctly to receive signals to form new tissues.

The centrioles are surrounded by high concentrations of tubulin dimers, and these proteins aggregate to form the microtubules that orchestrate chromosomal movement. (Plant cells lack centrosomes, but distinct microtubule organizing centers at each end of the cell play the same role.) These microtubules are the major part of the spindle structure, which is required for the orderly segregation of the chromosomes.

#### The spindle begins to form during prophase

During interphase, only the nuclear envelope, the nucleoli (see Section 5.3), and a barely discernible tangle of chromatin are visible under the light microscope. The appearance of the nucleus changes as the cell enters **prophase**—the beginning of mitosis. Most of the cohesin that has held the two products of DNA replication together since S phase is removed, so the individual chromatids become visible. They are still held together by a small amount of cohesin at the centromere. Late in prophase, specialized three-layered structures called **kinetochores** develop in the centromere region, one on each chro-

DNA double helix
2 nm
Core of eight histone molecules
"Tails" protrude from histones and allow them to interact with other molecules in the nucleus.
Nucleosomes form "beads" on a DNA "string."
Histone
"Linker" DNA
Nucleosome
DNA wraps around histones, forming a vast number of nucleosomes.
Nucleosomes pack into a coil that twists into another larger coil, and so forth, producing condensed, supercoiled chromatin fibers.
30 nm
Chromatin
Scaffold-associated chromatin
300 nm
700 nm
The fibers fold to form loops.
The loops coil even further, forming a chromosome.
1400 nm
700 nm
Metaphase chromosome

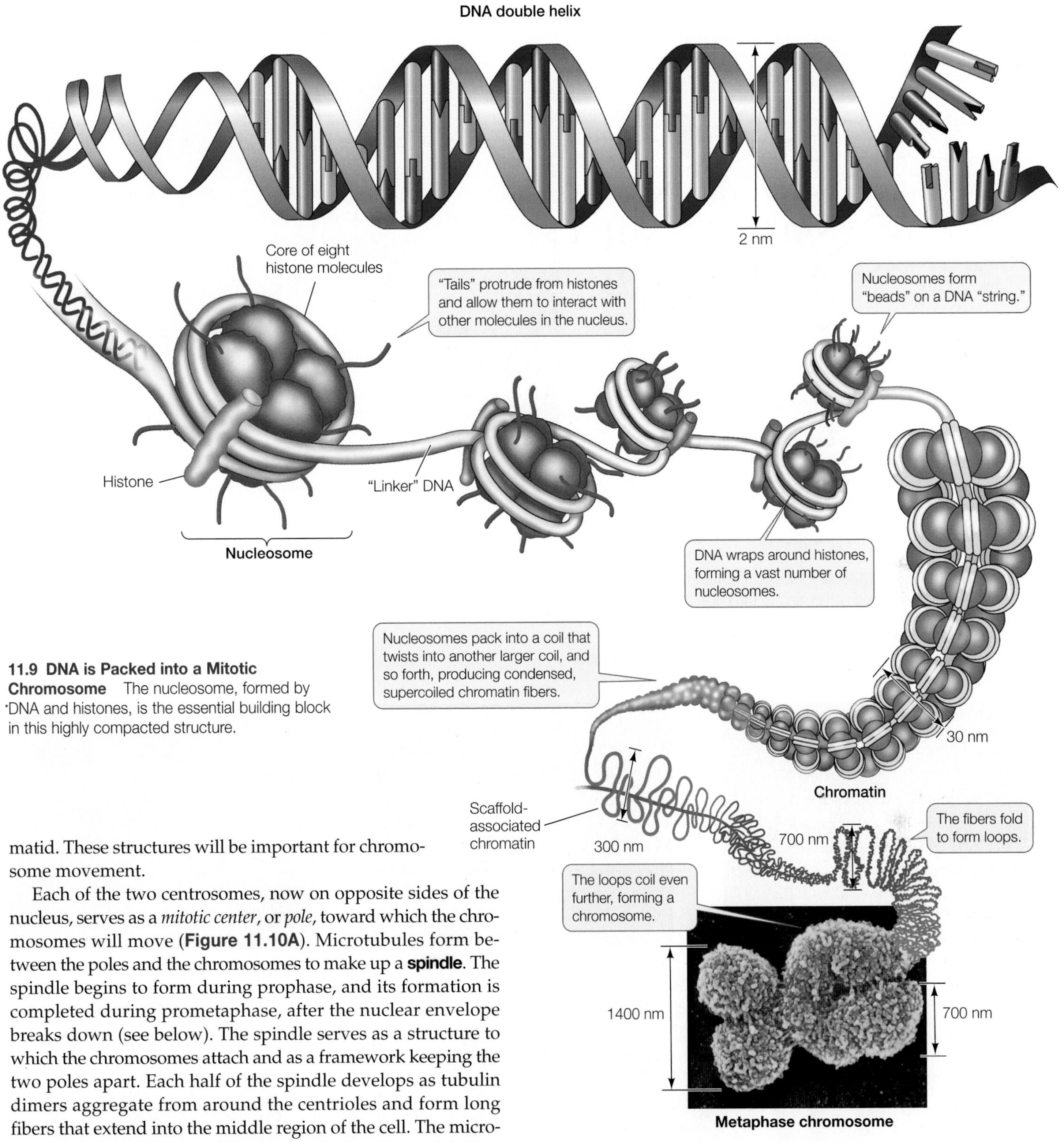

**11.9 DNA is Packed into a Mitotic Chromosome** The nucleosome, formed by DNA and histones, is the essential building block in this highly compacted structure.

matid. These structures will be important for chromosome movement.

Each of the two centrosomes, now on opposite sides of the nucleus, serves as a *mitotic center*, or *pole*, toward which the chromosomes will move (**Figure 11.10A**). Microtubules form between the poles and the chromosomes to make up a **spindle**. The spindle begins to form during prophase, and its formation is completed during prometaphase, after the nuclear envelope breaks down (see below). The spindle serves as a structure to which the chromosomes attach and as a framework keeping the two poles apart. Each half of the spindle develops as tubulin dimers aggregate from around the centrioles and form long fibers that extend into the middle region of the cell. The microtubules are initially unstable, constantly forming and falling apart, until they contact kinetochores or microtubules from the other half-spindle and become more stable.

There are two types of microtubule in the spindle:

- *Polar microtubules* form the framework of the spindle, and run from one pole to the other.
- *Kinetochore microtubules*, which form later, attach to the kinetochores on the chromosomes. The two sister chromatids in each chromosome pair become attached to kinetochore microtubules in opposite halves of the spindle (**Figure 11.10B**). This ensures that the two chromatids will eventually move to opposite poles.

Movement of the chromatids is the central feature of mitosis. It accomplishes the segregation that is needed for cell division and completion of the cell cycle. Prophase prepares for this movement, and the actual segregation takes place in the next three phases of mitosis.

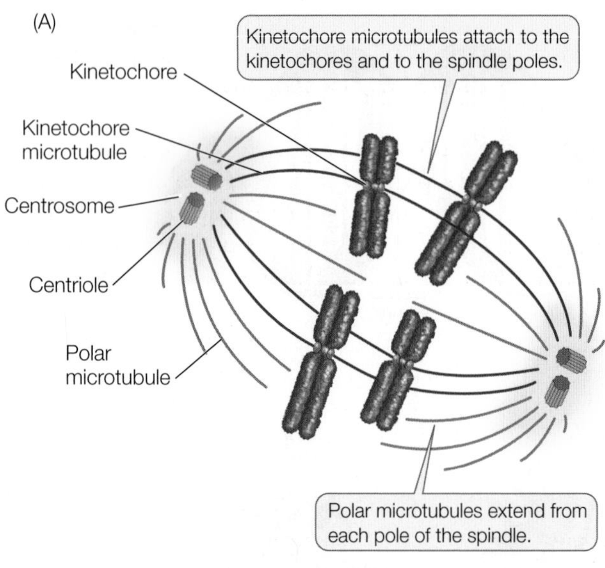
(A)
Kinetochore microtubules attach to the kinetochores and to the spindle poles.
Kinetochore
Kinetochore microtubule
Centrosome
Centriole
Polar microtubule
Polar microtubules extend from each pole of the spindle.

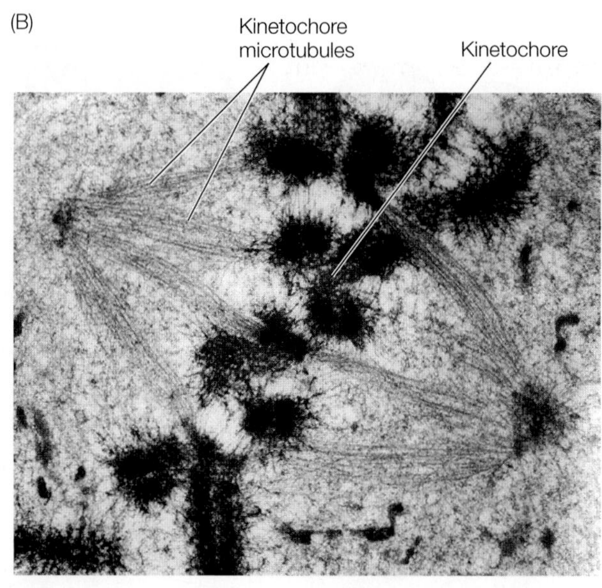
(B)
Kinetochore microtubules
Kinetochore

**11.10 The Mitotic Spindle Consists of Microtubules**
(A) The spindle apparatus in an animal cell at metaphase. In plant cells, centrioles are not present. (B) An electron micrograph of metaphase emphasizing the kinetochore microtubules.

yourBioPortal.com
GO TO Web Activity 11.1 • The Mitotic Spindle

**11.11 The Phases of Mitosis** Mitosis results in two new nuclei that are genetically identical to each other and to the nucleus from which they were formed. In the micrographs, the green dye stains microtubules (and thus the spindle); the red dye stains the chromosomes. The chromosomes in the diagrams are stylized to emphasize the fates of the individual chromatids.

yourBioPortal.com
GO TO Web Activity 11.2 • Images of Mitosis

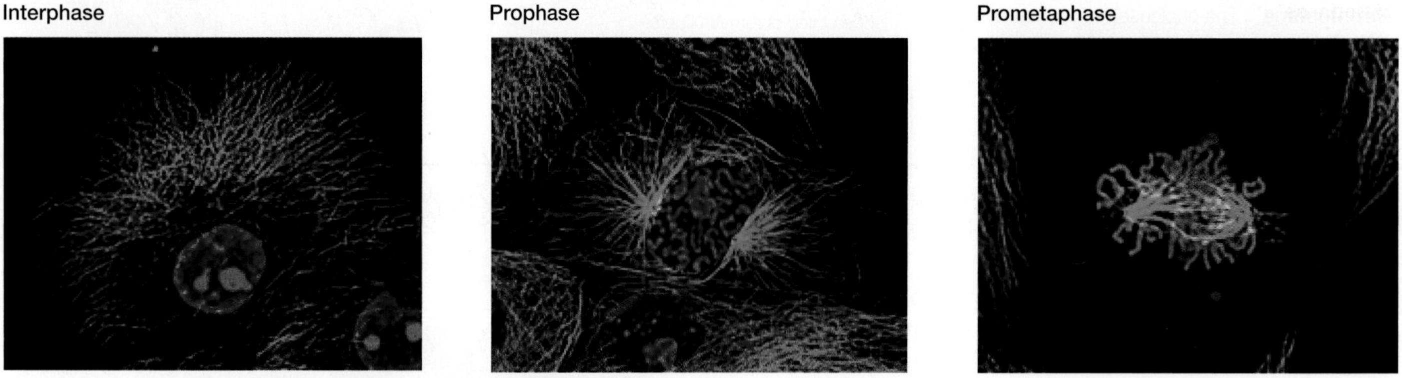

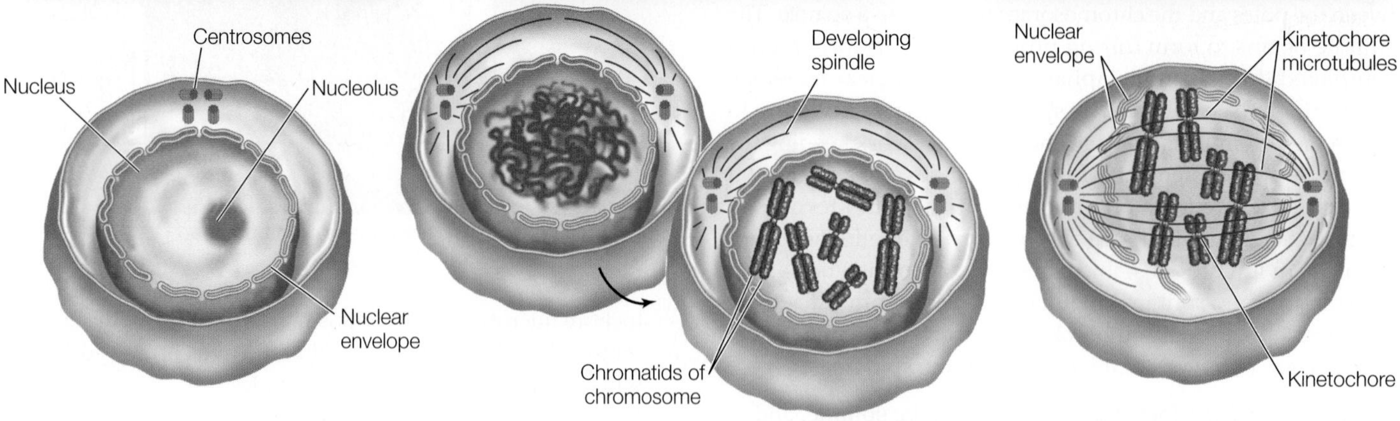

**Interphase**

Centrosomes
Nucleus
Nucleolus
Nuclear envelope

1 During the S phase of interphase, the nucleus replicates its DNA and centrosomes.

**Prophase**

Developing spindle
Chromatids of chromosome

2 The chromatin coils and supercoils, becoming more and more compact and condensing into visible chromosomes. The chromosomes consist of identical, paired sister chromatids. Centrosomes move to opposite poles.

**Prometaphase**

Nuclear envelope
Kinetochore microtubules
Kinetochore

3 The nuclear envelope breaks down. Kinetochore microtubules appear and connect the kinetochores to the poles.

## Chromosome separation and movement are highly organized

During the next three phases, prometaphase, metaphase, and anaphase, dramatic changes take place in the cell and the chromosomes.

- **Prometaphase**. The nuclear envelope breaks down and the compacted chromosomes consisting of two chromatids attach to the kinetochore microtubules.
- **Metaphase**. The chromosomes line up at the midline of the cell (equatorial position).
- **Anaphase**. The chromatids separate and move away from each other toward the poles.

You will find these events depicted and described in **Figure 11.11**. Here, we will consider two key processes: separation of the chromatids, and the mechanism of their actual movement toward the poles.

**CHROMATID SEPARATION** The separation of chromatids occurs at the beginning of anaphase. It is controlled by M phase cyclin–Cdk (see Figure 11.6), which activates another protein complex called the *anaphase-promoting complex* (APC). Separation occurs because one subunit of the cohesin protein holding the sister chromatids together is hydrolyzed by a specific protease, appropriately called *separase* (**Figure 11.12**). After they separate, the chromatids are called **daughter chromosomes**.

**CHROMOSOME MOVEMENT** The migration of the two sets of daughter chromosomes to the poles of the cell is a highly organized, active process. Two mechanisms operate to move the chromosomes along. First, the kinetochores contain a protein called *cytoplasmic dynein* that acts as a "molecular motor." It hydrolyzes ATP to ADP and phosphate, thus releasing energy that may move the chromosomes along the microtubules toward the poles. This accounts for about 75 percent of the force of motion. Second, the kinetochore microtubules shorten from the poles, drawing the chromosomes toward them, accounting for about 25 percent of the force of motion.

- **Telophase** occurs after the chromosomes have separated and is the last phase of mitosis. During this period, a nuclear envelope forms around each set of chromosomes, nucleoli appear, and the chromosomes become less compact. The spindle also disappears at this time. As a result, there are two new nuclei in a single cell.

## Cytokinesis is the division of the cytoplasm

Mitosis refers only to the division of the nucleus. The division of the cell's cytoplasm, which follows mitosis, is called cytoki-

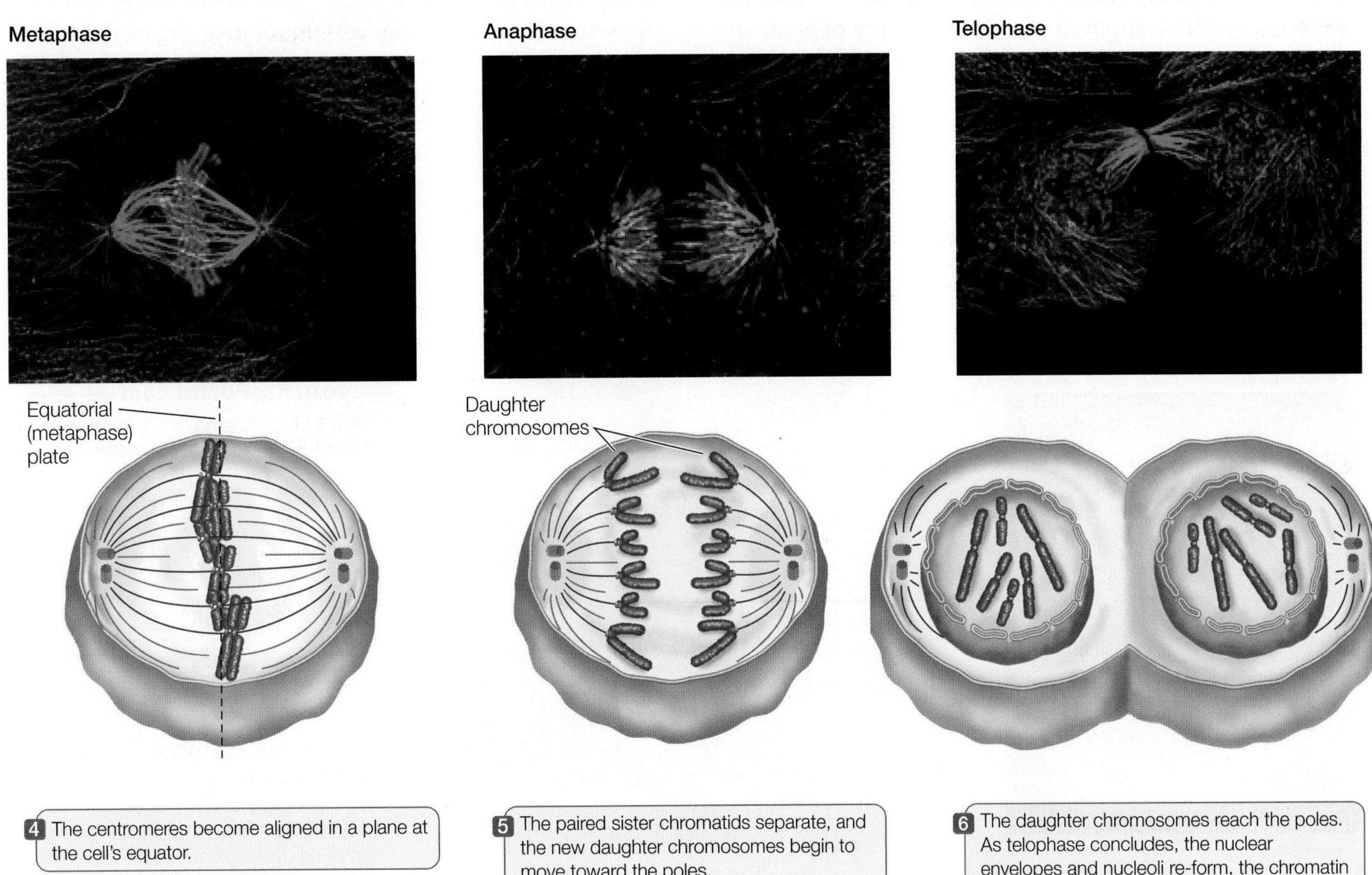


4 The centromeres become aligned in a plane at the cell's equator.

5 The paired sister chromatids separate, and the new daughter chromosomes begin to move toward the poles.

6 The daughter chromosomes reach the poles. As telophase concludes, the nuclear envelopes and nucleoli re-form, the chromatin decondenses, and, after cytokinesis, the daughter cells enter interphase once again.

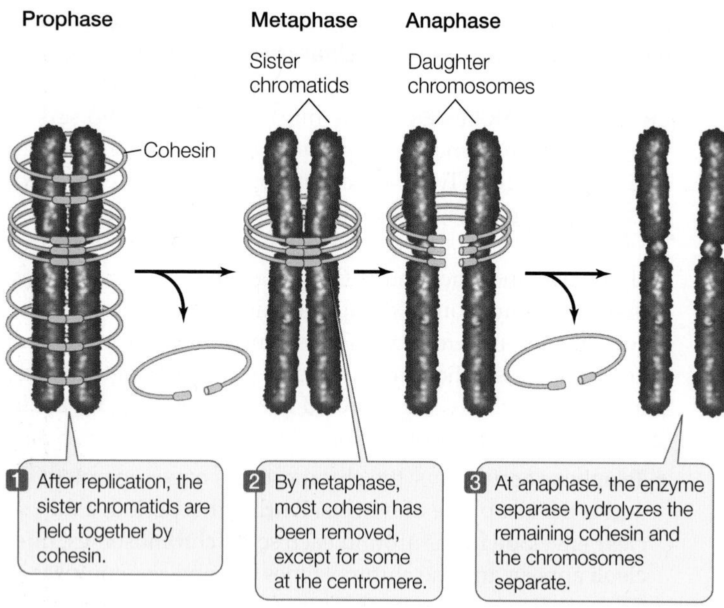

Prophase
Metaphase
Anaphase
Sister chromatids
Daughter chromosomes
Cohesin
1 After replication, the sister chromatids are held together by cohesin.
2 By metaphase, most cohesin has been removed, except for some at the centromere.
3 At anaphase, the enzyme separase hydrolyzes the remaining cohesin and the chromosomes separate.

**11.12 Chromatid Attachment and Separation** The cohesin protein complex holds sister chromatids together at the centromere. The enzyme separase hydrolyzes cohesin at the onset of anaphase, allowing the chromatids to separate into daughter chromosomes.

nesis. Cytokinesis occurs in different ways, depending on the type of organism. In particular there are substantial differences between the process in plants and in animals.

In animal cells, cytokinesis usually begins with a furrowing of the plasma membrane, as if an invisible thread were cinching the cytoplasm between the two nuclei (**Figure 11.13A**). This *contractile ring* is composed of microfilaments of actin and myosin (see Figure 5.18), which form a ring on the cytoplasmic surface of the plasma membrane. These two proteins interact to produce a contraction, just as they do in muscles, thus pinching the cell in two. The microfilaments assemble rapidly from actin monomers that are present in the interphase cytoskeleton. Their assembly is under the control of calcium ions that are released from storage sites in the center of the cell.

The plant cell cytoplasm divides differently because plants have cell walls that are rigid. In plant cells, as the spindle breaks down after mitosis, membranous vesicles derived from the Golgi apparatus appear along the plane of cell division, roughly midway between the two daughter nuclei. The vesicles are propelled along microtubules by the motor protein kinesin, and fuse to form a new plasma membrane. At the same time they contribute their contents to a *cell plate*, which is the beginning of a new cell wall (**Figure 11.13B**).

Following cytokinesis, each daughter cell contains all the components of a complete cell. A precise distribution of chromosomes is ensured by mitosis. In contrast, organelles such as ribosomes, mitochondria, and chloroplasts need not be distributed equally between daughter cells as long as some of each are present in each cell. Accordingly, there is no mechanism with a precision comparable to that of mitosis to provide for their equal allocation to daughter cells. As we will see in Chapter 19, the unequal distribution of cytoplasmic components during development can have functional significance for the two new cells.

yourBioPortal.com
GO TO Animated Tutorial 11.1 • Mitosis

(A)

Contractile ring

The contractile ring has completely separated the cytoplasms of these two daughter cells, although their surfaces remain in contact.

(B)

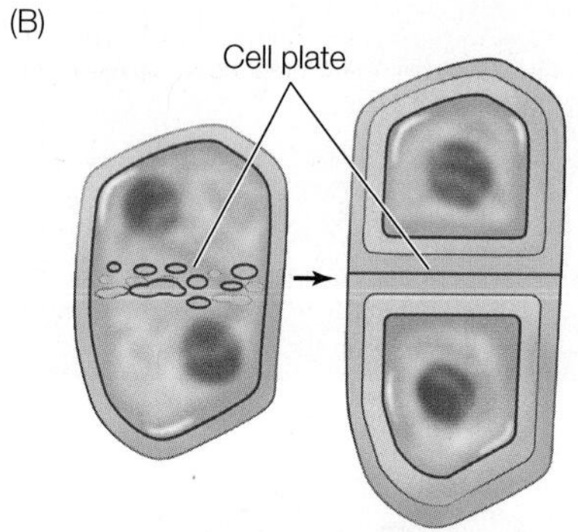


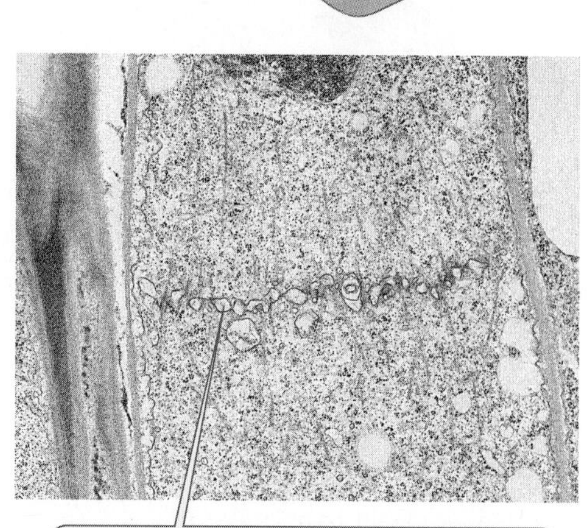

This row of vesicles will fuse to form a cell plate between the cell above and the cell below.

**11.13 Cytokinesis Differs in Animal and Plant Cells** (A) A sea urchin zygote (fertilized egg) that has just completed cytokinesis at the end of the first cell division of its development into an embryo. (B) A dividing plant cell in late telophase. Plant cells divide differently from animal cells because plant cells have cell walls.

## 11.3 RECAP

**Mitosis is the division of the nucleus of a eukaryotic cell into two nuclei identical to each other and to the parent nucleus. The process of mitosis, while continuous, can be viewed as a series of events (prophase, prometaphase, metaphase, anaphase, and telophase). Once two identical nuclei have formed, the cell divides into two cells by cytokinesis.**

- What is the difference between a chromosome, a chromatid, and a daughter chromosome? **See Figures 11.8 and 11.11**
- What are the various levels of "packing" by which the genetic information contained in linear DNA is condensed during prophase? **See p. 216 and Figure 11.9**
- Describe how chromosomes move during mitosis. **See p. 219 and Figure 11.11**
- What are the differences in cytokinesis between plant and animal cells? **See p. 220 and Figure 11.13**

The intricate process of mitosis results in two cells that are genetically identical. But, as mentioned earlier, there is another eukaryotic cell division process, called meiosis, that results in genetic diversity. What is the role of that process?

# 11.4 What Role Does Cell Division Play in a Sexual Life Cycle?

The mitotic cell cycle repeats itself and by this process, a single cell can give rise to a vast number of cells with identical nuclear DNA. Meiosis, on the other hand, produces just four daughter cells. Mitosis and meiosis are both involved in reproduction, but they have different roles: asexual reproduction involves only mitosis, while sexual reproduction involves both mitosis and meiosis.

### Asexual reproduction by mitosis results in genetic constancy

**Asexual reproduction**, sometimes called *vegetative reproduction*, is based on the mitotic division of the nucleus. An organism that reproduces asexually may be single-celled like yeast, reproducing itself with each cell cycle, or it may be multicellular like the cholla cactus, that breaks off a piece to produce a new multicellular organism (**Figure 11.14**). Asexual reproduction is a rapid and effective means of making new individuals, and it is common in nature. In asexual reproduction, the offspring are **clones** of the parent organism; that is, the offspring are *genetically identical* to the parent. Any genetic variation among the offspring is most likely due to small environmentally caused changes in the DNA, called *mutations*. As you will see, this small amount of variation contrasts with the extensive variation possible in sexually reproducing organisms.

### Sexual reproduction by meiosis results in genetic diversity

Unlike asexual reproduction, **sexual reproduction** results in an organism that is not identical to its parents. Sexual reproduction requires **gametes** created by meiosis; two parents each contribute one gamete to each of their offspring. Meiosis can produce gametes—and thus offspring—that differ genetically from each other and from the parents. Because of this genetic variation, some offspring may be better adapted than others to sur-

(A)

(B)

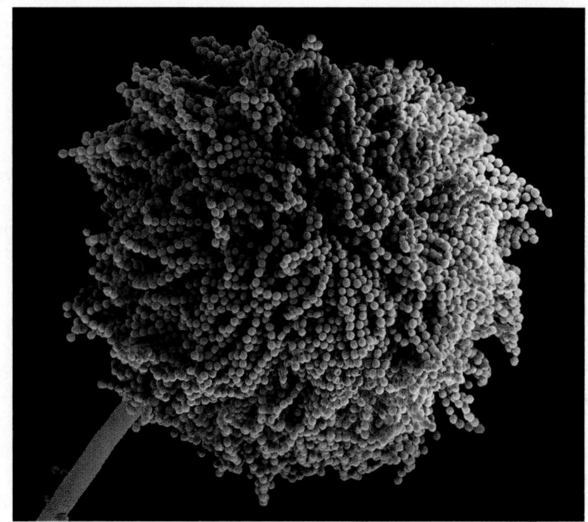

**11.14 Asexual Reproduction in the Large and the Small** (A) Some cacti like this cholla have brittle stems that break off easily. Fragments on the ground set down roots and develop by mitotic cell divisions into new plants that are genetically identical to the plant they came from. (B) These strings of cells are asexual spores formed by a fungus. Each spore contains a nucleus produced by a mitotic division and is genetically identical to the parent that produced it. It can divide to form a new fungus.

vive and reproduce in a particular environment. Meiosis thus generates the genetic diversity that is the raw material for natural selection and evolution.

In most multicellular organisms, the body cells that are *not* specialized for reproduction, called **somatic cells**, each contain two sets of chromosomes, which are found in pairs. One chromosome of each pair comes from each of the organism's two parents; for example, in humans with 46 chromosomes, 23 come from the mother and 23 from the father. The members of such a **homologous pair** are similar in size and appearance, except for the sex chromosomes found in some species (see Section 12.4). The two chromosomes in a homologous pair (called **homologs**) bear corresponding, though often not identical, genetic information. For example, a homologous pair of chromosomes in a plant may carry different versions of a gene that controls seed shape. One homolog may carry the version for wrinkled seeds while the other may carry the version for smooth seeds.

**Gametes**, on the other hand, contain only a single set of chromosomes—that is, one homolog from each pair. The number of chromosomes in a gamete is denoted by $n$, and the cell is said to be **haploid**. Two haploid gametes fuse to form a **zygote**, in a process called **fertilization**. The zygote thus has two sets of chromosomes, just as somatic cells do. Its chromosome number is denoted by $2n$, and the zygote is said to be **diploid**. Depending on the organism, the zygote may divide by either meiosis or mitosis. Either way, a new mature organism develops that is capable of sexual reproduction.

All sexual life cycles involve meiosis to produce gametes or cells that are haploid. Eventually, the haploid cells or gametes fuse to produce a zygote, beginning the diploid stage of the lifecycle. Since the origin of sexual reproduction, evolution has generated many different versions of the sexual life cycle. **Figure 11.15** presents three examples.

- In *haplontic* organisms, including most protists, fungi, and some green algae, the tiny zygote is the only diploid cell in the life cycle. After it is formed it immediately undergoes meiosis to produce more haploid cells. These are usually *spores*, which are the dispersal units for the organism, like the seeds of a plant. A spore germinates to form a new haploid organism, which may be single-celled or multicellular. Cells of the mature haploid organism fuse to form the diploid zygote.
- Most plants and some fungi display **alternation of generations**. As for many haplontic organisms, meiosis gives rise

**11.15 Fertilization and Meiosis Alternate in Sexual Reproduction** In sexual reproduction, haploid ($n$) cells or organisms alternate with diploid ($2n$) cells or organisms.

yourBioPortal.com
GO TO Web Activity 11.3 • Sexual Life Cycle

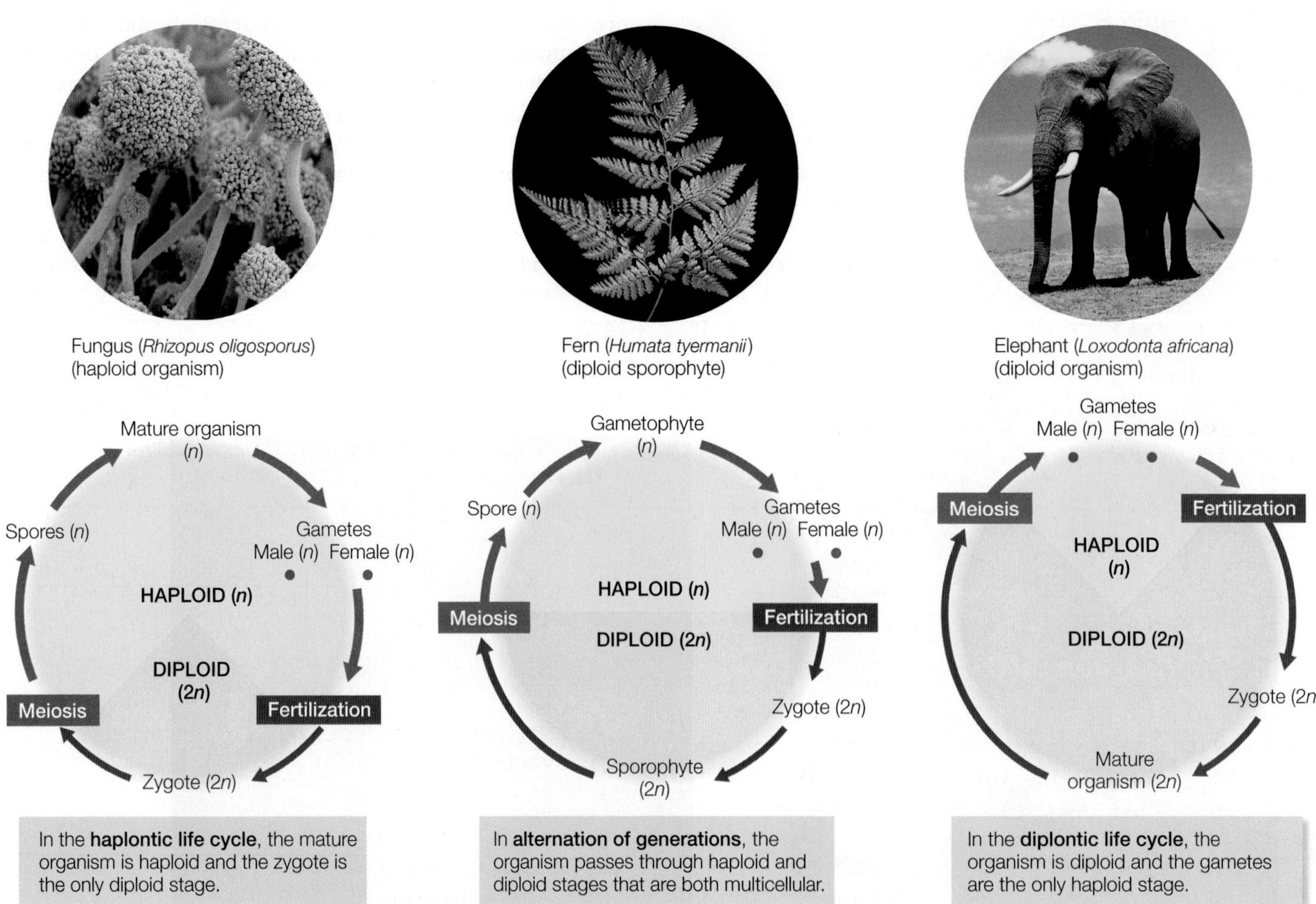

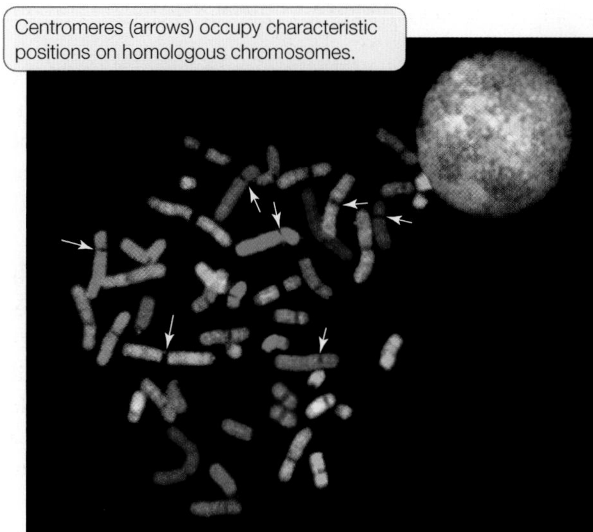

(A)
Centromeres (arrows) occupy characteristic positions on homologous chromosomes.


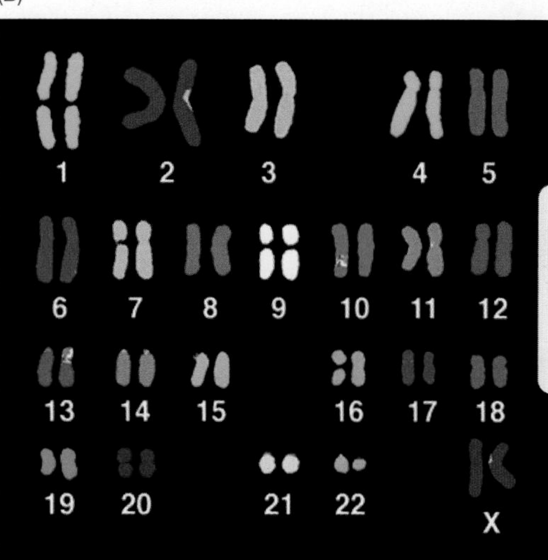


**11.16 The Human Karyotype** (A) Chromosomes from a human cell in metaphase. The DNA of each chromosome pair has a specific nucleotide sequence that is stained by a particular colored dye, so that the chromosomes in a homologous pair share a distinctive color. Each chromosome at this stage is composed of two chromatids, but they cannot be distinguished. At the upper right is an interphase nucleus. (B) This karyogram, produced by computerized analysis of the image on the left, shows homologous pairs lined up together and numbered, clearly revealing the individual's karyotype.

to haploid spores, which divide by mitosis to form a haploid life stage called the *gametophyte*. The gametophyte forms gametes by mitosis, which fuse to form a diploid zygote. The zygote divides by mitosis to become the diploid *sporophyte*, which in turn produces the gametes by meiosis.

- In *diplontic* organisms, which include animals, brown algae and some fungi, the gametes are the only haploid cells in the life cycle, and the mature organism is diploid.

These life cycles are described in greater detail in Part Seven. For now we will focus on the role of sexual reproduction in generating diversity among individual organisms.

The essence of sexual reproduction is the *random selection of half of the diploid chromosome set* to make a haploid gamete, followed by fusion of two haploid gametes to produce a diploid cell. Both of these steps contribute to a shuffling of genetic information in the population, so that no two individuals have exactly the same genetic constitution. The diversity provided by sexual reproduction opens up enormous opportunities for evolution.

### The number, shapes, and sizes of the metaphase chromosomes constitute the karyotype

When cells are in metaphase of mitosis, it is often possible to count and characterize their individual chromosomes. If a photomicrograph of the entire set of chromosomes is made, the images of the individual chromosomes can be manipulated, pairing and placing them in an orderly arrangement. Such a rearranged photomicrograph reveals the number, shapes, and sizes of the chromosomes in a cell, which together constitute its **karyotype** (**Figure 11.16**). In humans, karyotypes can aid in the diagnosis of certain diseases, and this has led to an entire branch of medicine called *cytogenetics*. However, as you will see in Chapter 15, chromosome analysis with the microscope is being replaced by direct analysis of DNA.

Individual chromosomes can be recognized by their lengths, the positions of their centromeres, and characteristic banding patterns that are visible when the chromosomes are stained and observed at high magnification. In diploid cells, the karyotype consists of homologous pairs of chromosomes—for example, there are 23 pairs and a total of 46 chromosomes in humans. There is no simple relationship between the size of an organism and its chromosome number. A housefly has 5 chromosome pairs and a horse has 32, but the smaller carp (a fish) has 52 pairs. Probably the highest number of chromosomes in any organism is in the fern *Ophioglossum reticulatum*, which has 1,260 (630 pairs)!

## 11.4 RECAP

**Meiosis is necessary for sexual reproduction, in which haploid gametes fuse to produce a diploid zygote. Sexual reproduction results in genetic diversity, the foundation of evolution.**

- What is the difference, in terms of genetics, between asexual and sexual reproduction? See p. 221
- How does fertilization produce a diploid organism? See p. 222
- What general features do all sexual life cycles have in common? See p. 222 and Figure 11.15

Meiosis, unlike mitosis, results in daughter cells that have half as many chromosomes as the parent cell. Next we will look at the processes of meiosis.

MEIOSIS I

Early prophase I

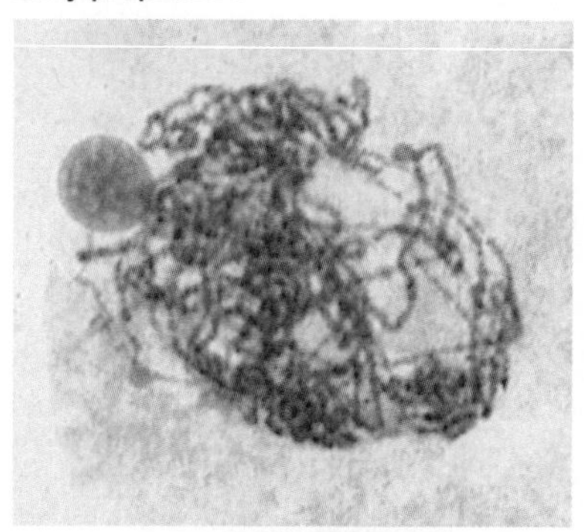



1 The chromatin begins to condense following interphase.

Mid-prophase I

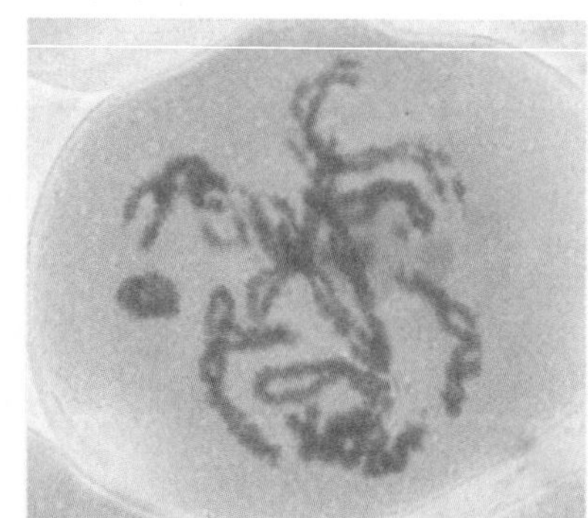

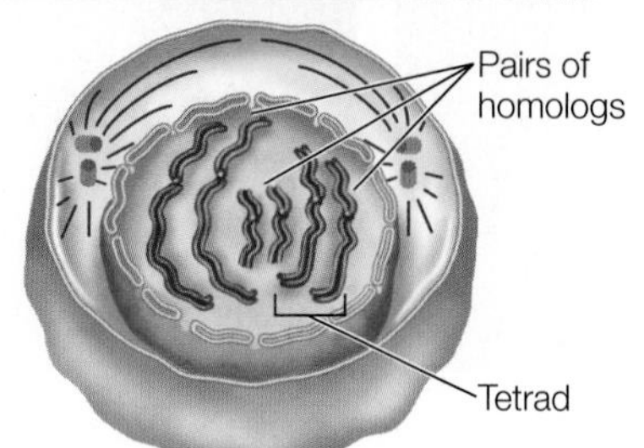



2 Synapsis aligns homologs, and chromosomes condense further.

Late prophase I–Prometaphase

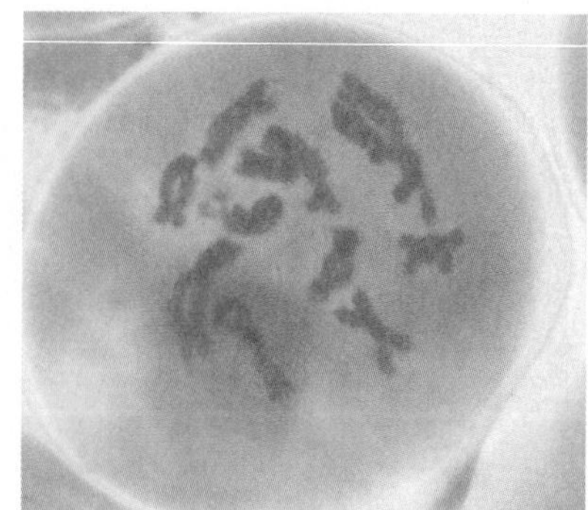

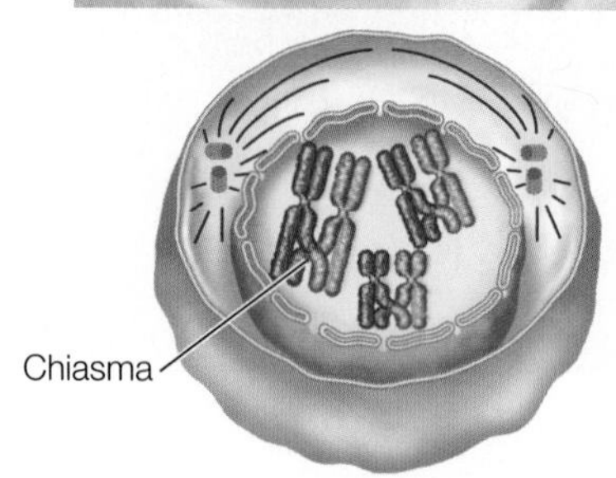



3 The chromosomes continue to coil and shorten. The chiasmata reflect crossing over, the exchange of genetic material between nonsister chromatids in a homologous pair. In prometaphase the nuclear envelope breaks down.

MEIOSIS II

Prophase II

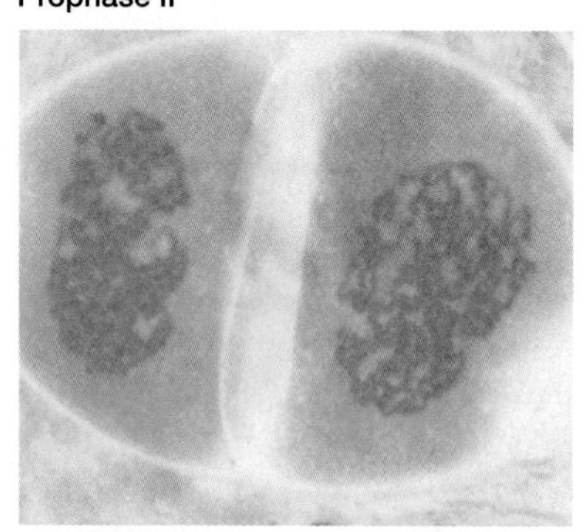

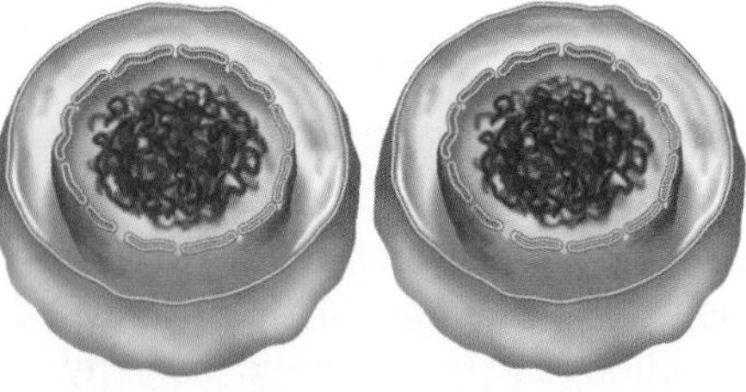

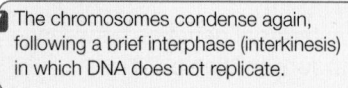

7 The chromosomes condense again, following a brief interphase (interkinesis) in which DNA does not replicate.

Metaphase II

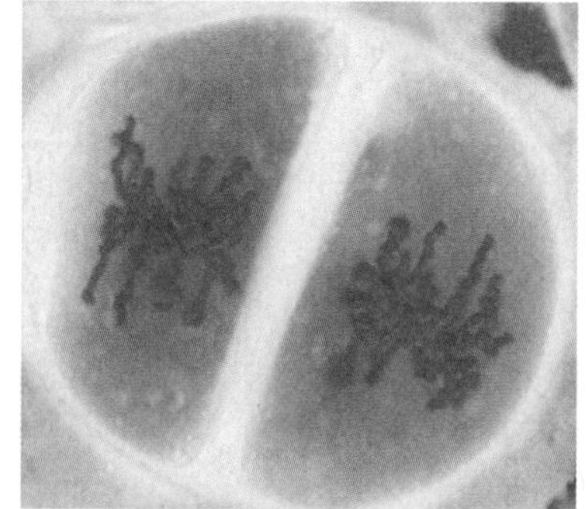

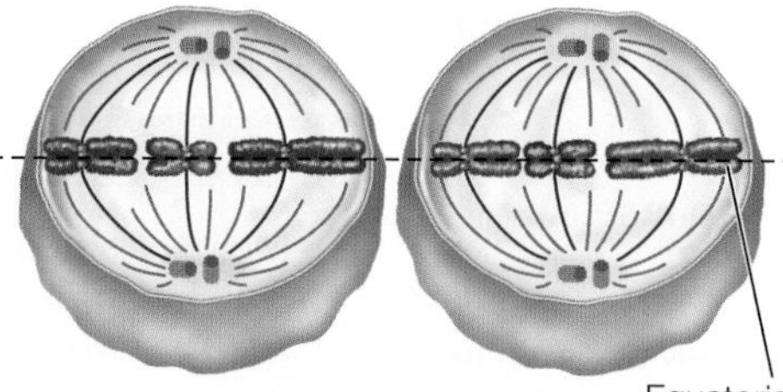



8 The centromeres of the paired chromatids line up across the equatorial plates of each cell.

Anaphase II

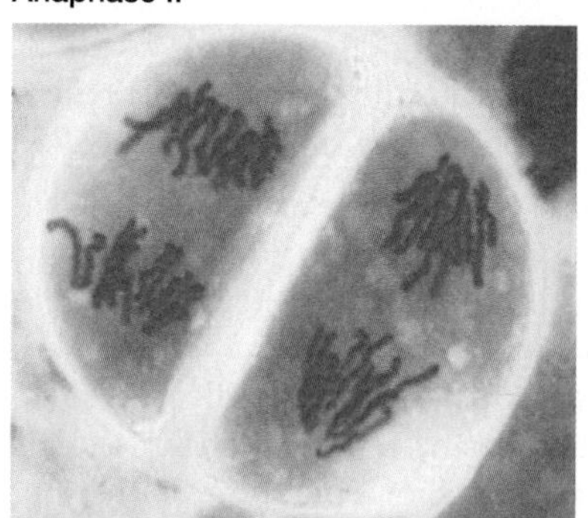

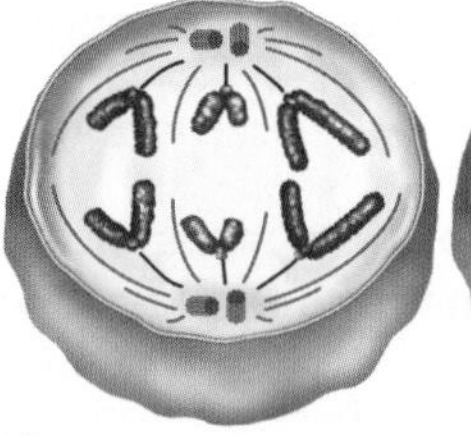

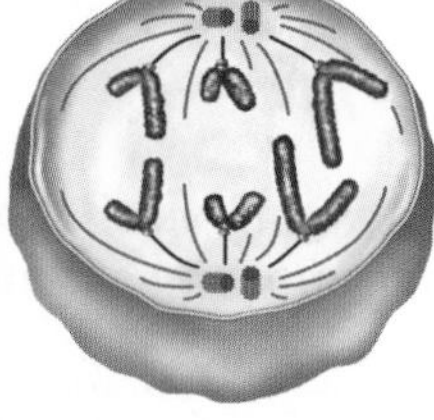

9 The chromatids finally separate, becoming chromosomes in their own right, and are pulled to opposite poles. Because of crossing over and independent assortment, each new cell will have a different genetic makeup.

# 11.5 What Happens during Meiosis?

In the last section we described the role and importance of meiosis in sexual reproduction. Now we will see how meiosis accomplishes the orderly and precise generation of haploid cells.

Meiosis consists of *two* nuclear divisions that reduce the number of chromosomes to the haploid number, in preparation for sexual reproduction. Although the *nucleus divides twice* during meiosis, the *DNA is replicated only once*. Unlike the products of mitosis, the products of meiosis are genetically different from one another and from the parent cell. To understand the process of meiosis and its specific details, it is useful to keep in mind the overall functions of meiosis:

Metaphase I

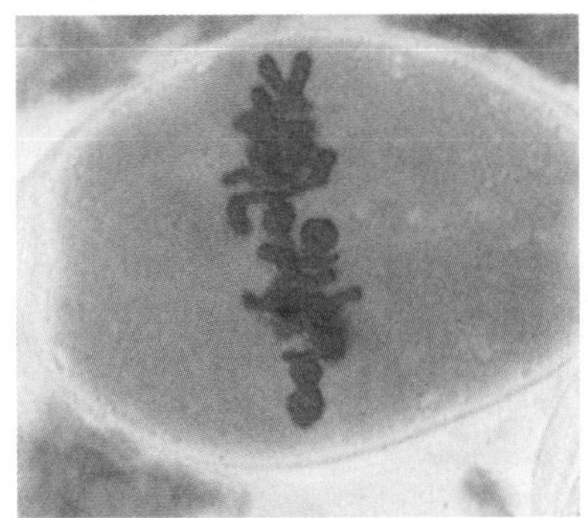

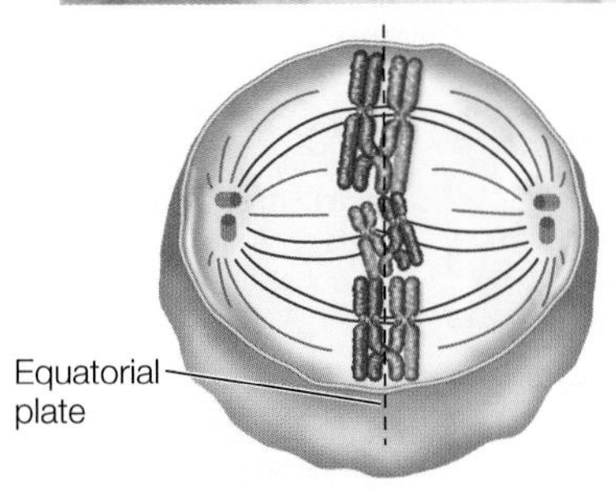


4 The homologous pairs line up on the equatorial (metaphase) plate.

Anaphase I

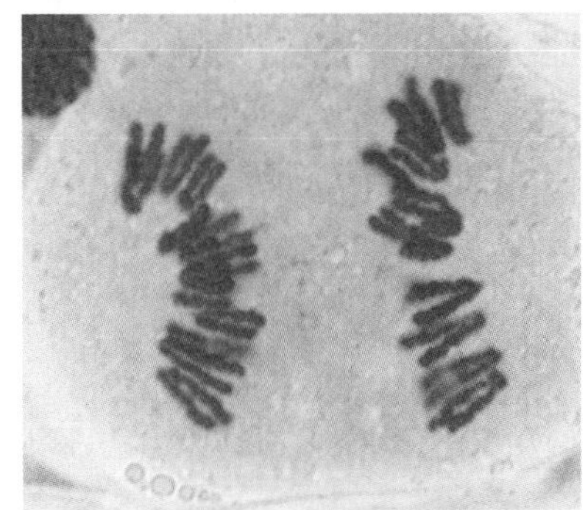

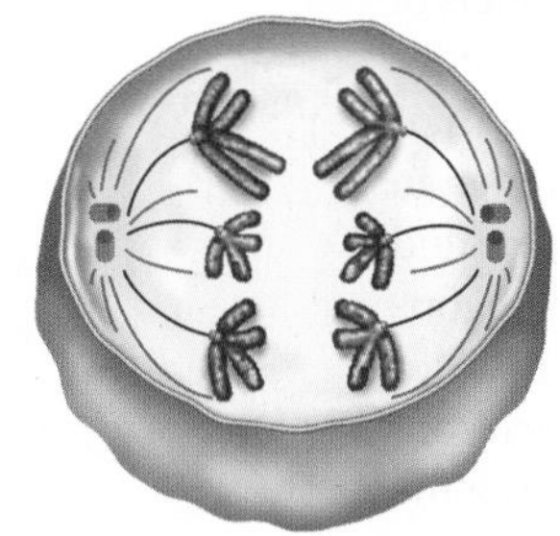

5 The homologous chromosomes (each with two chromatids) move to opposite poles of the cell.

Telophase I

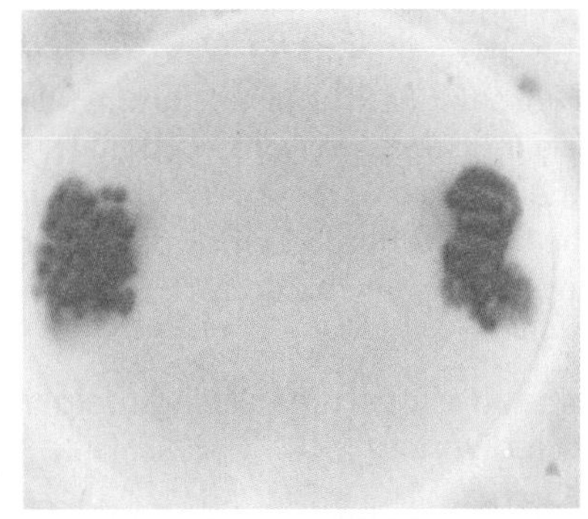

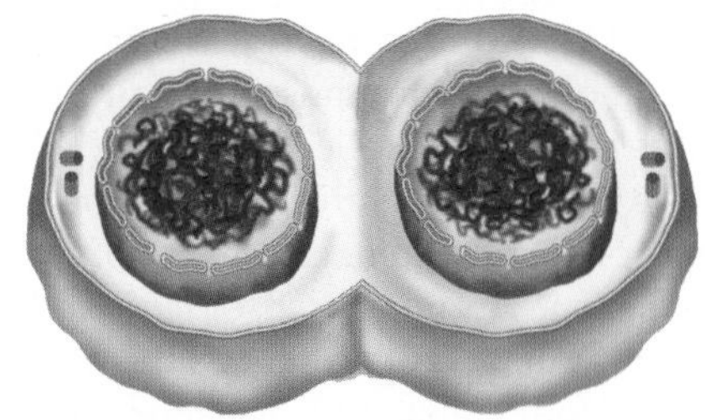

6 The chromosomes gather into nuclei, and the original cell divides.

Telophase II

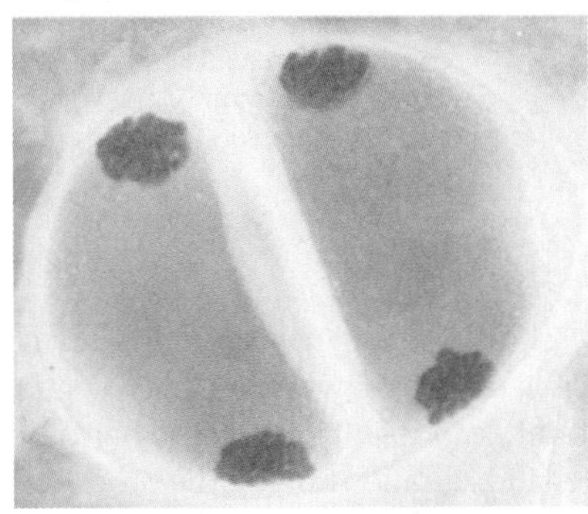

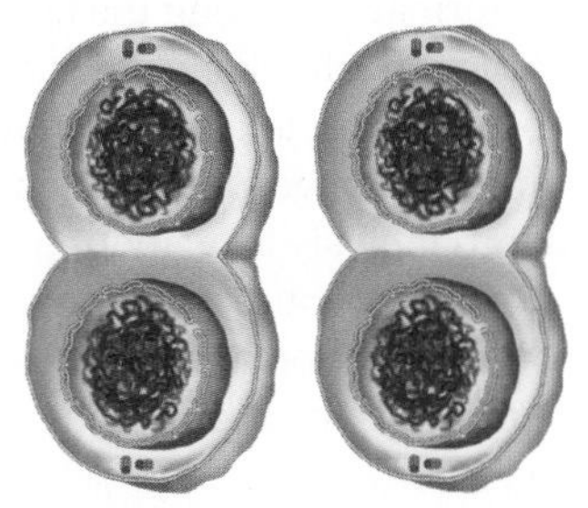

10 The chromosomes gather into nuclei, and the cells divide.

Products

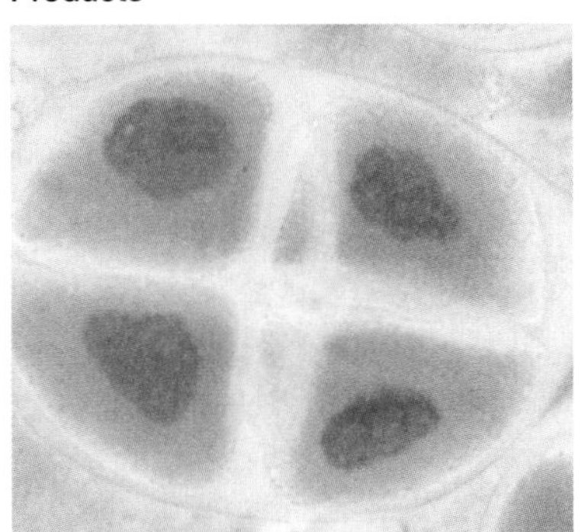

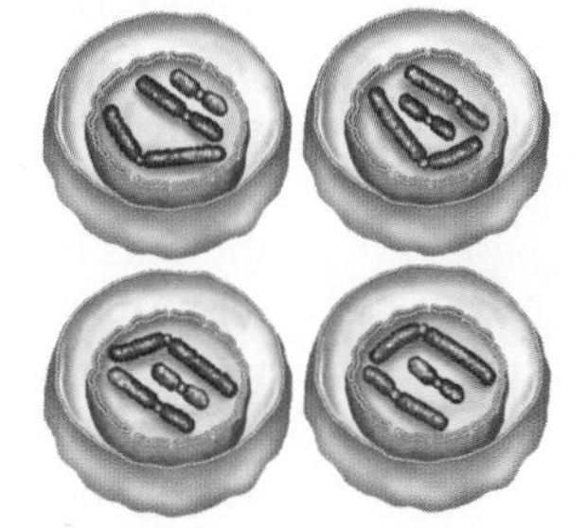

11 Each of the four cells has a nucleus with a haploid number of chromosomes.

**11.17 Meiosis: Generating Haploid Cells** In meiosis, two sets of chromosomes are divided among four daughter nuclei, each of which has half as many chromosomes as the original cell. Four haploid cells are the result of two successive nuclear divisions. The micrographs show meiosis in the male reproductive organ of a lily; the diagrams show the corresponding phases in an animal cell. (For instructional purposes, the chromosomes from one parent are colored blue and those from the other parent are red.)

**yourBioPortal.com**
GO TO **Web Activity 11.4 • Images of Meiosis**

- To reduce the chromosome number from diploid to haploid
- To ensure that each of the haploid products has a complete set of chromosomes
- To generate genetic diversity among the products

The events of meiosis are illustrated in **Figure 11.17**. In this section, we discuss some of the key features that distinguish meiosis from mitosis.

### Meiotic division reduces the chromosome number

As noted above, meiosis consists of two nuclear divisions, *meiosis I* and *meiosis II*. Two unique features characterize **meiosis I**.

- *Homologous chromosomes come together to pair* along their entire lengths. No such pairing occurs in mitosis.
- *The homologous chromosome pairs separate*, but the individual chromosomes, each consisting of two sister chromatids, remain intact. (The chromatids will separate during meiosis II.)

Like mitosis, meiosis I is preceded by an interphase with an S phase, during which each chromosome is replicated. As a result, each chromosome consists of two sister chromatids, held together by cohesin proteins. At the end of meiosis I, two nuclei form, each with half of the original chromosomes (one member of each homologous pair). Since the centromeres did not separate, these chromosomes are still double—composed of two sister chromatids. The sister chromatids are separated during **meiosis II**, which is *not* preceded by DNA replication. As a result, the products of meiosis I and II are four cells, each containing the haploid number of chromosomes. But these four cells are not genetically identical.

### Chromatid exchanges during meiosis I generate genetic diversity

Meiosis I begins with a long prophase I (the first three panels of Figure 11.17), during which the chromosomes change markedly. The homologous chromosomes pair by adhering along their lengths in a process called **synapsis**. (This does not happen in mitosis.) This pairing process lasts from prophase I to the end of metaphase I. The four chromatids of each pair of homologous chromosomes form a **tetrad**, or *bivalent*. For example, in a human cell at the end of prophase I there are 23 tetrads, each consisting of four chromatids. The four chromatids come from the two partners in each homologous pair of chromosomes.

Throughout prophase I and metaphase I, the chromatin continues to coil and compact, so that the chromosomes appear ever thicker. At a certain point, the homologous chromosomes appear to repel each other, especially near the centromeres, but they remain held together by physical attachments mediated by cohesins. Later in prophase, regions having these attachments take on an X-shaped appearance (**Figure 11.18**) and are called **chiasmata** (singular *chiasma*, "cross").

A chiasma reflects an *exchange of genetic material* between nonsister chromatids on homologous chromosomes—what geneticists call **crossing over** (**Figure 11.19**). The chromosomes usually begin exchanging material shortly after synapsis begins, but chiasmata do not become visible until later, when the homologs are repelling each other. Crossing over results in **recombinant** chromatids, and it increases genetic variation among the products of meiosis by reshuffling genetic information among the homologous pairs. In Chapter 12 we explore further the genetic consequences of crossing over. Mitosis seldom takes more than an hour or two, but meiosis can take *much* longer. In human males, the cells in the testis that undergo meiosis take about a week for prophase I and about a month for the entire meiotic cycle. In females, prophase I begins long before a woman's birth, during her early fetal development, and ends as much as decades later, during the monthly ovarian cycle.

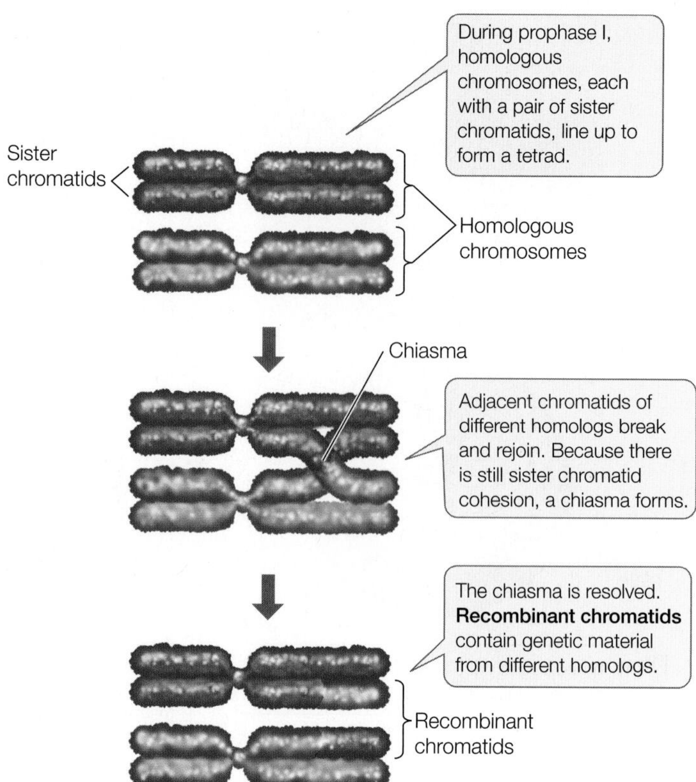

During prophase I, homologous chromosomes, each with a pair of sister chromatids, line up to form a tetrad.
Sister chromatids
Homologous chromosomes
Chiasma
Adjacent chromatids of different homologs break and rejoin. Because there is still sister chromatid cohesion, a chiasma forms.
The chiasma is resolved. **Recombinant chromatids** contain genetic material from different homologs.
Recombinant chromatids


**11.19 Crossing Over Forms Genetically Diverse Chromosomes** The exchange of genetic material by crossing over results in new combinations of genetic information on the recombinant chromosomes. The two different colors distinguish the chromosomes contributed by the male and female parents.

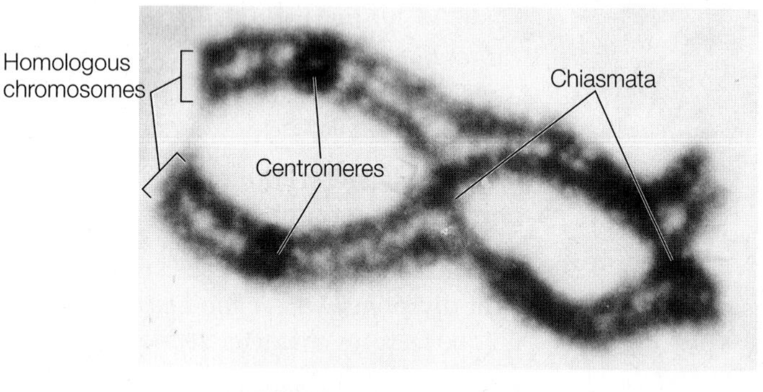


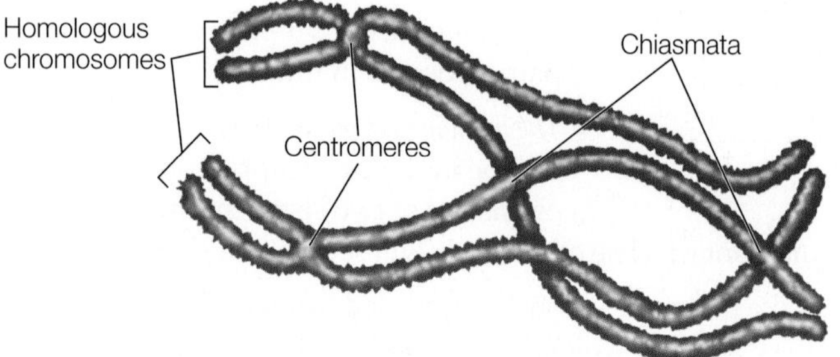


**11.18 Chiasmata: Evidence of Genetic Exchange between Chromatids** This micrograph shows a pair of homologous chromosomes, each with two chromatids, during prophase I of meiosis in a salamander. Two chiasmata are visible.

### During meiosis homologous chromosomes separate by independent assortment

A diploid organism has two sets of chromosomes ($2n$); one set derived from its male parent, and the other from its female parent. As the organism grows and develops, its cells undergo mitotic divisions. In mitosis, each chromosome behaves independently of its homolog, and its two chromatids are sent to opposite poles during anaphase. Each daughter nucleus ends up with $2n$ chromosomes. In meiosis, things are very different. **Figure 11.20** compares the two processes.

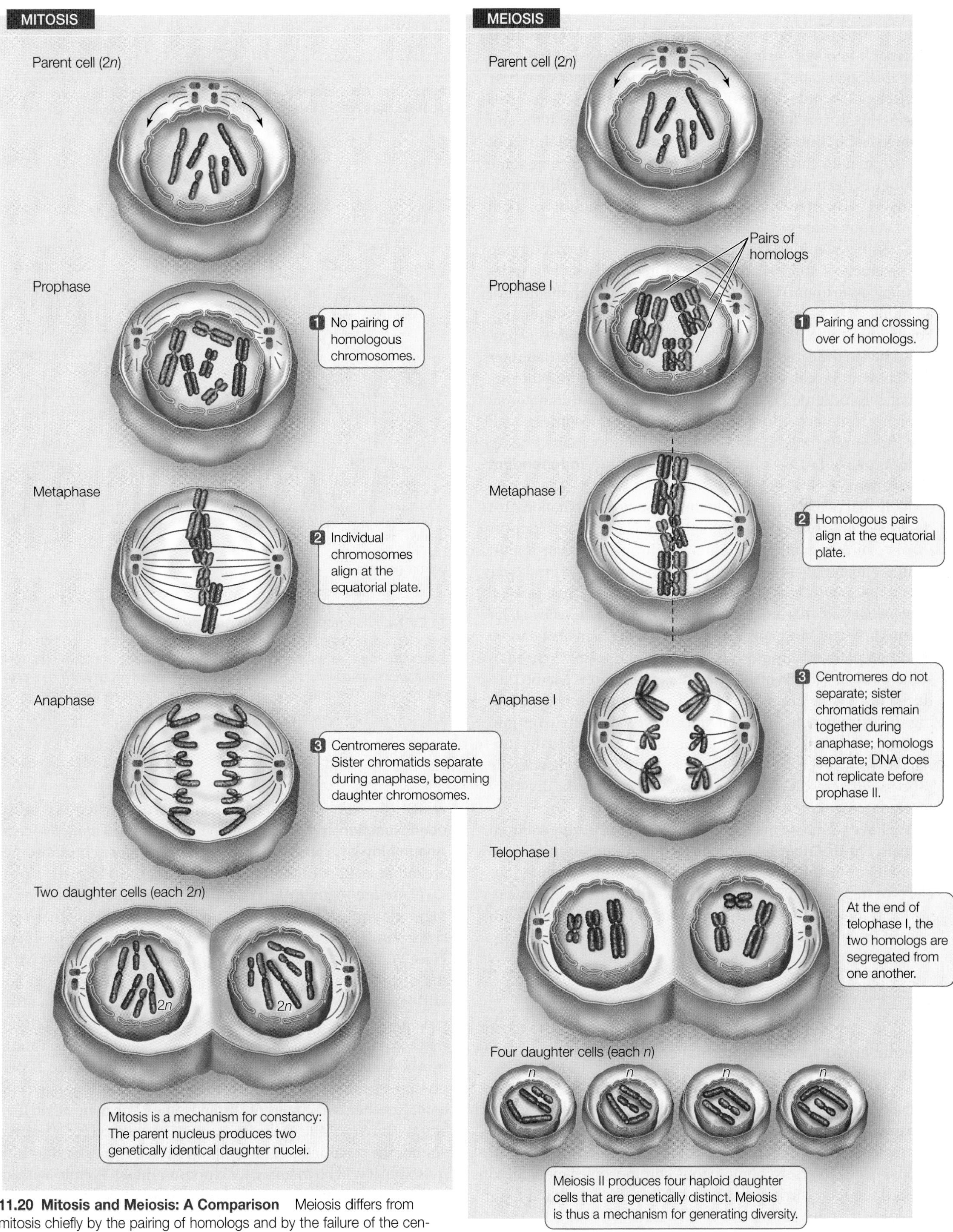
MITOSIS
Parent cell (2n)
Prophase
1 No pairing of homologous chromosomes.
Metaphase
2 Individual chromosomes align at the equatorial plate.
Anaphase
3 Centromeres separate. Sister chromatids separate during anaphase, becoming daughter chromosomes.
Two daughter cells (each 2n)
2n
2n
Mitosis is a mechanism for constancy: The parent nucleus produces two genetically identical daughter nuclei.
MEIOSIS
Parent cell (2n)
Pairs of homologs
Prophase I
1 Pairing and crossing over of homologs.
Metaphase I
2 Homologous pairs align at the equatorial plate.
Anaphase I
3 Centromeres do not separate; sister chromatids remain together during anaphase; homologs separate; DNA does not replicate before prophase II.
Telophase I
At the end of telophase I, the two homologs are segregated from one another.
Four daughter cells (each n)
n
n
n
n
Meiosis II produces four haploid daughter cells that are genetically distinct. Meiosis is thus a mechanism for generating diversity.

**11.20 Mitosis and Meiosis: A Comparison** Meiosis differs from mitosis chiefly by the pairing of homologs and by the failure of the centromeres to separate at the end of metaphase I.

In meiosis I, chromosomes of maternal origin pair with their paternal homologs during synapsis. *This pairing does not occur in mitosis.* Segregation of the homologs during meiotic anaphase I (see steps 4–6 of Figure 11.17) ensures that each pole receives one member of each homologous pair. For example, at the end of meiosis I in humans, each daughter nucleus contains 23 of the original 46 chromosomes. In this way, the chromosome number is decreased from diploid to haploid. Furthermore, meiosis I guarantees that each daughter nucleus gets one full set of chromosomes.

Crossing over is one reason for the genetic diversity among the products of meiosis. The other source of diversity is independent assortment. It is a matter of chance which member of a homologous pair goes to which daughter cell at anaphase I. For example, imagine there are two homologous pairs of chromosomes in the diploid parent nucleus. A particular daughter nucleus could receive the paternal chromosome 1 and the maternal chromosome 2. Or it could get paternal 2 and maternal 1, or both maternal, or both paternal chromosomes. It all depends on the way in which the homologous pairs line up at metaphase I. This phenomenon is termed **independent assortment**.

Note that of the four possible chromosome combinations just described, only two produce daughter nuclei with full complements of either maternal or paternal chromosome sets (apart from the material exchanged by crossing over). *The greater the number of chromosomes, the less probable that the original parental combinations will be reestablished, and the greater the potential for genetic diversity.* Most species of diploid organisms have more than two pairs of chromosomes. In humans, with 23 chromosome pairs, $2^{23}$ (8,388,608) different combinations can be produced just by the mechanism of independent assortment. Taking the extra genetic shuffling afforded by crossing over into account, the number of possible combinations is virtually infinite. Crossing over and independent assortment, along with the processes that result in mutations, provide the genetic diversity needed for evolution by natural selection.

We have seen how meiosis I is fundamentally different from mitosis. On the other hand, meiosis II is similar to mitosis, in that it involves the separation of chromatids into daughter nuclei (see steps 7–11 in Figure 11.17). The final products of meiosis I and meiosis II are four haploid daughter cells, each with one set ($n$) of chromosomes.

yourBioPortal.com
GO TO Animated Tutorial 11.2 • Meiosis

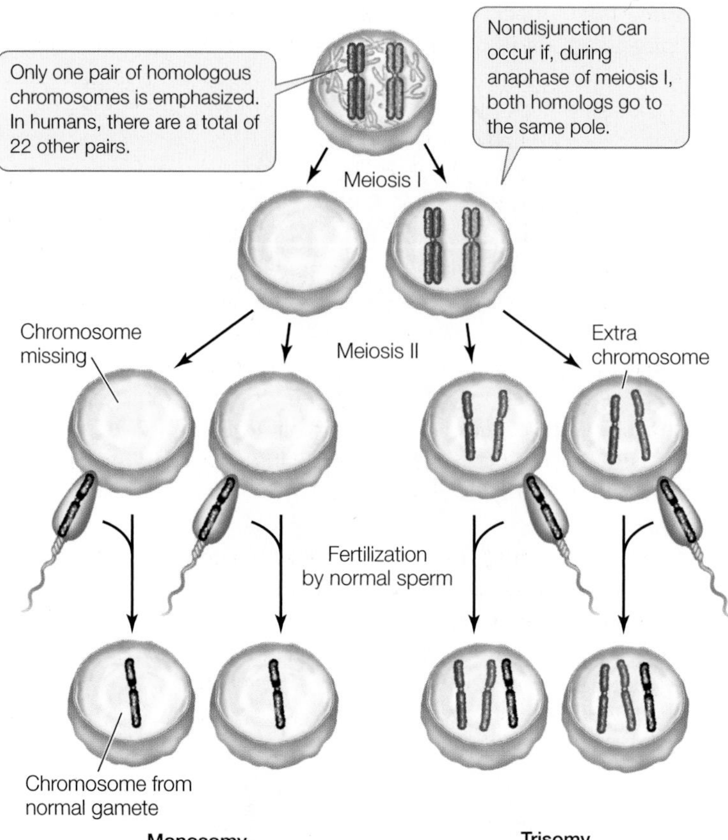


**11.21 Nondisjunction Leads to Aneuploidy** Nondisjunction occurs if homologous chromosomes fail to separate during meiosis I, or if chromatids fail to separate during meiosis II. The first case is shown here. The result is aneuploidy: one or more chromosomes are either lacking or present in excess. Generally, aneuploidy is lethal to the developing embryo.

### Meiotic errors lead to abnormal chromosome structures and numbers

In the complex process of meiosis, things occasionally go wrong. A pair of homologous chromosomes may fail to separate during meiosis I, or sister chromatids may fail to separate during meiosis II. Conversely, homologous chromosomes may fail to remain together during metaphase I, and then both may migrate to the same pole in anaphase I. This phenomenon is called **nondisjunction** and it results in the production of *aneuploid* cells. **Aneuploidy** is a condition in which one or more chromosomes are either lacking or present in excess (**Figure 11.21**).

There are many different causes of aneuploidy, but one of them may result from a breakdown in the cohesins that keep sister chromatids and tetrads joined together during prophase I (see Figure 11.17). These and other proteins ensure that when the chromosomes line up at the equatorial plate, one homolog will face one pole and the other homolog will face the other pole. If the cohesins break down at the wrong time, both homologs may go to one pole. If, for example, during the formation of a human egg, both members of the chromosome 21 pair go to the same pole during anaphase I, the resulting eggs will contain either two copies of chromosome 21 or none at all. If an egg with two of these chromosomes is fertilized by a normal sperm, the resulting zygote will have three copies of the chromosome: it will be **trisomic** for chromosome 21. A child with an extra chromosome 21 has the symptoms of *Down syndrome*: impaired intelligence; characteristic abnormalities of the hands, tongue, and eyelids; and an increased susceptibility to cardiac abnormalities and diseases such as leukemia. If an egg that did not receive chromosome 21 is fertilized by a normal sperm, the zygote will have only one copy: it will be **monosomic** for chromosome 21, and this is lethal.

Trisomies and the corresponding monosomies are surprisingly common in human zygotes, with 10–30 percent of all conceptions showing aneuploidy. But most of the embryos that develop from such zygotes do not survive to birth, and those that do often die before the age of 1 year (trisomies for chromosome 21 are the viable exception). At least one-fifth of all recognized pregnancies are spontaneously terminated (miscarried) during the first 2 months, largely because of trisomies and monosomies. The actual proportion of spontaneously terminated pregnancies is certainly higher, because the earliest ones often go unrecognized.

Other abnormal chromosomal events can also occur. In a process called **translocation**, a piece of a chromosome may break away and become attached to another chromosome. For example, a particular large part of one chromosome 21 may be translocated to another chromosome. Individuals who inherit this translocated piece along with two normal chromosomes 21 will have Down syndrome.

### Polyploids have more than two complete sets of chromosomes

As mentioned in Section 11.4, mature organisms are often either diploid (for example, most animals) or haploid (for example, most fungi). Under some circumstances, triploid ($3n$), tetraploid ($4n$), or higher-order **polyploid** nuclei may form. Each of these *ploidy levels* represents an increase in the number of complete chromosome sets present. Organisms with complete extra sets of chromosomes may sometimes be produced by artificial breeding or by natural accidents. Polyploidy occurs naturally in some animals and many plants, and it has probably led to speciation (the evolution of a new species) in some cases.

A diploid nucleus can undergo normal meiosis because there are two sets of chromosomes to make up homologous pairs, which separate during anaphase I. Similarly, a tetraploid nucleus has an even number of each kind of chromosome, so each chromosome can pair with its homolog. However, a triploid nucleus cannot undergo normal meiosis because one-third of the chromosomes would lack partners. Polyploidy has implications for agriculture, particularly in the production of hybrid plants. For example, ploidy must be taken into account in wheat breeding because there are diploid, tetraploid, and hexaploid wheat varieties. Polyploidy can be a desirable trait in crops and ornamental plants because it often leads to more robust plants with larger flowers, fruits, and seeds. Triploidy can be useful in some circumstances. For example, rivers and lakes can be stocked with triploid trout, which are sterile and will not escape to reproduce in waters where they might upset the natural ecology.

## 11.5 RECAP

**Meiosis produces four daughter cells in which the chromosome number is reduced from diploid to haploid. Because of the independent assortment of chromosomes and the crossing over of homologous chromatids, the four products of meiosis are not genetically identical. Meiotic errors, such as the failure of a homologous chromosome pair to separate, can lead to abnormal numbers of chromosomes.**

- How do crossing over and independent assortment result in unique daughter nuclei? See p. 226 and Figure 11.19
- What are the differences between meiosis and mitosis? See pp. 224–228 and Figure 11.20
- What is aneuploidy, and how can it arise from nondisjunction during meiosis? See p. 228 and Figure 11.21

An essential role of cell division in complex eukaryotes is to replace cells that die. What happens to those cells?

## 11.6 In a Living Organism, How Do Cells Die?

Cells die in one of two ways. The first type of cell death, **necrosis**, occurs when cells are damaged by mechanical means or toxins, or are starved of oxygen or nutrients. These cells usually swell up and burst, releasing their contents into the extracellular environment. This process often results in inflammation (see Section 42.2).

More typically, cell death is due to **apoptosis** (Greek, "falling apart"). Apoptosis is a programmed series of events that result in cell death. Why would a cell initiate apoptosis, which is essentially cell suicide? In animals, there are two possible reasons:

- *The cell is no longer needed by the organism.* For example, before birth, a human fetus has weblike hands, with connective tissue between the fingers. As development proceeds, this unneeded tissue disappears as its cells undergo apoptosis in response to specific signals.
- *The longer cells live, the more prone they are to genetic damage that could lead to cancer.* This is especially true of epithelial cells of the surface of an organism, which may be exposed to radiation or toxic substances. Such cells normally die after only days or weeks and are replaced by new cells.

The outward events of apoptosis are similar in many organisms. The cell becomes detached from its neighbors, cuts up its chromatin into nucleosome-sized pieces, and forms membranous lobes, or "blebs," that break up into cell fragments (**Figure 11.22A**). In a remarkable example of the economy of nature, the surrounding living cells usually ingest the remains of the dead cell by phagocytosis. Neighboring cells digest the apop-

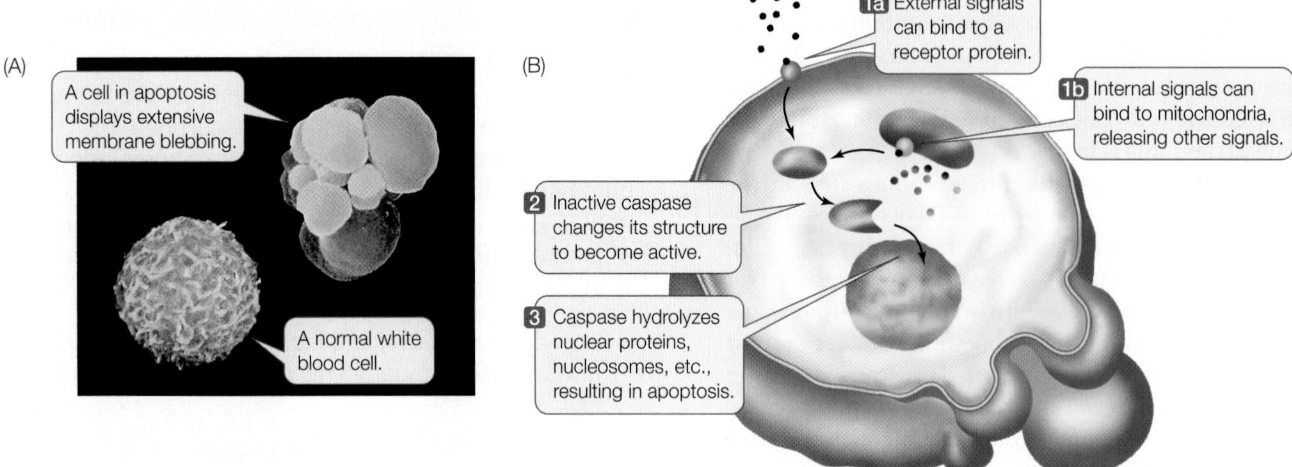
(A)
A cell in apoptosis displays extensive membrane blebbing.
A normal white blood cell.
(B)
1a External signals can bind to a receptor protein.
1b Internal signals can bind to mitochondria, releasing other signals.
2 Inactive caspase changes its structure to become active.
3 Caspase hydrolyzes nuclear proteins, nucleosomes, etc., resulting in apoptosis.

**11.22 Apoptosis: Programmed Cell Death** (A) Many cells are programmed to "self-destruct" when they are no longer needed, or when they have lived long enough to accumulate a burden of DNA damage that might harm the organism. (B) Both external and internal signals stimulate caspases, the enzymes that break down specific cell constituents, resulting in apoptosis.

totic cell contents in their lysosomes and the digested components are recycled.

Apoptosis is also used by plant cells, in an important defense mechanism called the *hypersensitive response*. Plants can protect themselves from disease by undergoing apoptosis at the site of infection by a fungus or bacterium. With no living tissue to grow in, the invading organism is not able to spread to other parts of the plant. Because of their rigid cell wall, plant cells do not form blebs the way that animal cells do. Instead, they digest their own cell contents in the vacuole and then release the digested components into the vascular system.

Despite these differences between plant and animal cells, they share many of the signal transduction pathways that lead to apoptosis. Like the cell division cycle, programmed cell death is controlled by signals, which may come from inside or outside the cell (**Figure 11.22B**). Internal signals may be linked to the absence of mitosis or the recognition of damaged DNA. External signals (or a lack of them) can cause a receptor protein in the plasma membrane to change its shape, and in turn activate a signal transduction pathway. Both internal and external signals can lead to the activation of a class of enzymes called **caspases**. These enzymes are proteases that hydrolyze target molecules in a cascade of events. As a result, the cell dies as the caspases hydrolyze proteins of the nuclear envelope, nucleosomes, and plasma membrane.

## 11.6 RECAP

**Cell death can occur either by necrosis or by apoptosis. Apoptosis is governed by precise molecular controls.**

- What are some differences between apoptosis and necrosis? See p. 229
- In what situation is apoptosis necessary? See p. 229
- How is apoptosis regulated? See Figure 11.22

# 11.7 How Does Unregulated Cell Division Lead to Cancer?

Perhaps no malady affecting people in the industrialized world instills more fear than cancer, and most people realize that it involves an inappropriate increase in cell numbers. One in three Americans will have some form of cancer in their lifetimes, and at present, one in four will die of it. With 1.5 million new cases and half a million deaths in the United States annually, cancer ranks second only to heart disease as a killer.

### Cancer cells differ from normal cells

Cancer cells differ from the normal cells from which they originate in two ways:

- Cancer cells lose control over cell division.
- Cancer cells can migrate to other locations in the body.

Most cells in the body divide only if they are exposed to extracellular signals such as growth factors. Cancer cells do not respond to these controls, and instead divide more or less continuously, ultimately forming **tumors** (large masses of cells). By the time a physician can feel a tumor or see one on an X-ray film or CAT scan, it already contains millions of cells. Tumors can be benign or malignant.

**Benign** tumors resemble the tissue they came from, grow slowly, and remain localized where they develop. For example, a lipoma is a benign tumor of fat cells that may arise in the armpit and remain there. Benign tumors are not cancers, but they must be removed if they impinge on an organ, obstructing its function.

**Malignant** tumors do not look like their parent tissue at all. A flat, specialized epithelial cell in the lung wall may turn into a relatively featureless, round, malignant lung cancer cell (**Figure 11.23**). Malignant cells often have irregular structures, such as variable nucleus sizes and shapes. Recall the opening story of this chapter, in which cervical cancer was diagnosed by cell structure.

The second and most fearsome characteristic of cancer cells is their ability to invade surrounding tissues and spread to other parts of the body by traveling through the bloodstream or lymphatic ducts. When malignant cells become lodged in some distant part of the body they go on dividing and growing, establishing a tumor at that new site. This spreading, called **metastasis**, results in organ failures and makes the cancer very hard to treat.

### Cancer cells lose control over the cell cycle and apoptosis

Earlier in this chapter you learned about proteins that regulate the progress of a eukaryotic cell through the cell cycle:

- Positive regulators such as growth factors stimulate the cell cycle: they are like "gas pedals."
- Negative regulators such as RB inhibit the cell cycle: they are like "brakes."

Just as driving a car requires stepping on the gas pedal *and* releasing the brakes, a cell will go through a division cycle only if the positive regulators are active and the negative regulators are inactive.

In most cells, the two regulatory systems ensure that cells divide only when needed. In cancer cells, these two processes are abnormal.

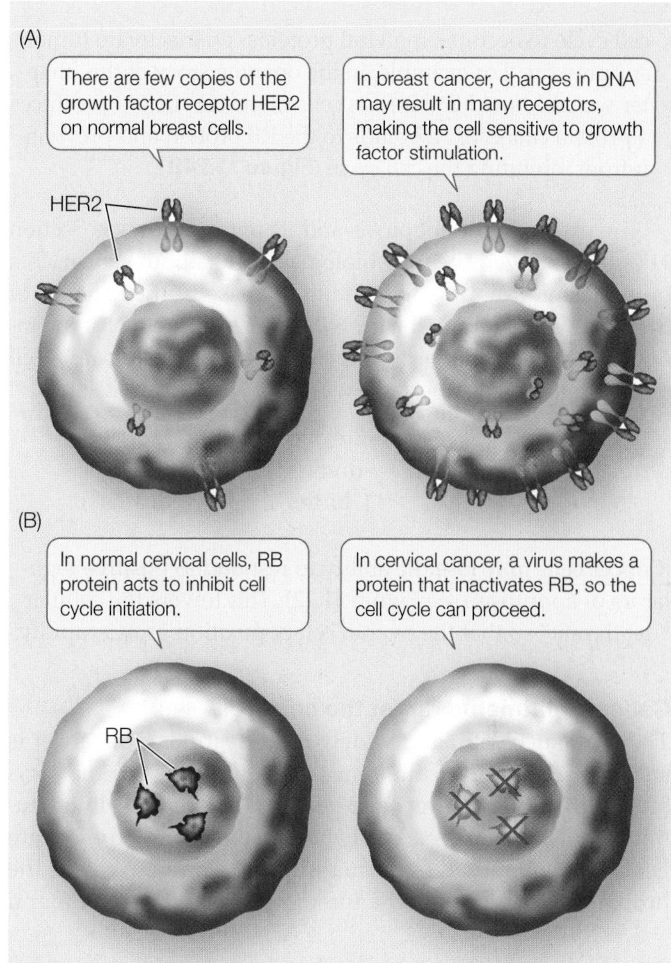


**11.24 Molecular Changes in Cancer Cells** In cancer, oncogene proteins become active (A) and tumor suppressor proteins become inactive (B).

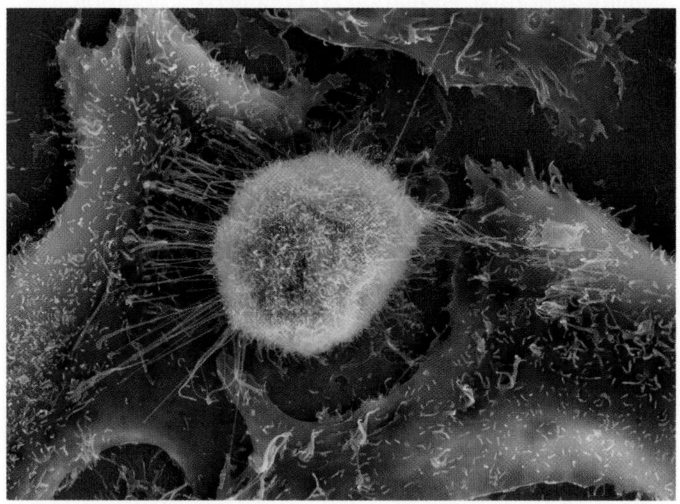

**11.23 A Cancer Cell with its Normal Neighbors** This lung cancer cell (yellow-green) is quite different from the normal lung cells surrounding it. The cancer cell can divide more rapidly than its normal counterparts, and it can spread to other organs. This form of small-cell cancer is lethal, with a 5-year survival rate of 10 percent. Most cases are caused by tobacco smoking.

- **Oncogene** proteins are positive regulators in cancer cells. They are derived from normal positive regulators that have become mutated to be overly active or that are present in excess, and they stimulate the cancer cells to divide more often. Oncogene products could be growth factors, their receptors, or other components in the signal transduction pathway. An example of an oncogene protein is the growth factor receptor in a breast cancer cell (**Figure 11.24A**). Normal breast cells have relatively low numbers of the growth factor receptor HER2. So when this growth factor is made, it doesn't find many breast cell receptors with which to bind and initiate cell division. In about 25 percent of breast cancers, a DNA change results in the increased production of the HER2 receptor. This results in positive stimulation of the cell cycle, and a rapid proliferation of cells with the altered DNA.
- **Tumor suppressors** are negative regulators in both cancer and normal cells, but in cancer cells they are inactive. An example is the RB protein that acts at R (the restriction point) in G1 (see Figure 11.6). When RB is active the cell cycle does not proceed, but it is inactive in cancer cells, allowing the

cell cycle to occur. Some viral proteins can inactivate tumor suppressors. For example, in the opening story of this chapter we saw how HPV infects cells of the cervix and produces a protein called E7. E7 binds to the RB protein and prevents it from inhibiting the cell cycle (**Figure 11.24B**).

The discovery of apoptosis and its importance (see Section 11.6) has changed the way biologists think about cancer. In a population of organisms, the net increase in the number of individuals over time (the growth rate) is a function of the individuals added (the birth rate) and lost (the death rate). Cell populations behave the same way:

$$\text{growth rate of cell population} = \text{rate of cell division ("births")} - \text{rate of apoptosis ("deaths")}$$

Cancer cells may lose the ability to respond to positive regulators of apoptosis (see Figure 11.22). This lowers the cellular "death rate" so that the overall cell population grows rapidly.

### Cancer treatments target the cell cycle

The most successful and widely used treatment for cancer is surgery. While physically removing a tumor is optimal, it is often difficult for a surgeon to get all of the tumor cells. (A tumor about 1 cm in size already has a billion cells!) Tumors are generally embedded in normal tissues. Added to this is the probability that cells of the tumor may have broken off and spread to other organs. This makes it unlikely that localized surgery will be curative. So other approaches are taken to treat or cure cancer, and these generally target the cell cycle (**Figure 11.25**).

An example of a cancer drug that targets the cell cycle is *5-fluorouracil*, which blocks the synthesis of thymine, one of the four bases in DNA. The drug *taxol* prevents the functioning of microtubules in the mitotic spindle. Both drugs inhibit the cell cycle, and apoptosis causes tumor shrinkage. More dramatic is radiation treatment, in which a beam of high-energy radiation is focused on the tumor. DNA damage is extensive, and the cell cycle checkpoint for DNA repair is overwhelmed. As a result, the cell undergoes apoptosis. A major problem with these treatments is that they target normal cells as well as the tumor cells. These treatments are toxic to tissues with large populations of normal dividing cells such as those in the intestine, skin and bone marrow (producing blood cells).

A major effort in cancer research is to find treatments that target only cancer cells. A promising recent example is *Herceptin*, which targets the HER2 growth factor receptor that is expressed at high levels on the surfaces of some breast cancer cells (see Figure 11.24A). Herceptin binds specifically to the HER2 receptor but does not stimulate it. This prevents the natural growth factor from binding, and so the cells are not stimulated to divide. As a result, the tumor shrinks because the apoptosis rate remains the same. More such treatments are on the way.

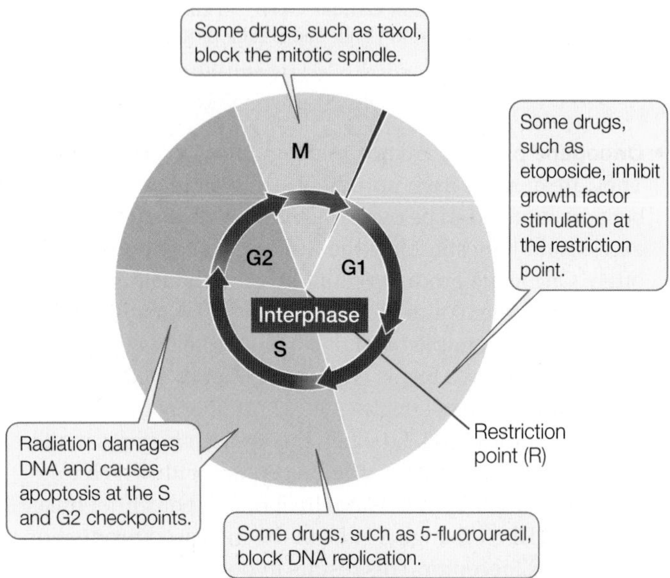


**11.25 Cancer Treatment and the Cell Cycle** To prevent cancer cells from dividing, physicians use combinations of therapies that attack the cell cycle at different points.

## 11.7 RECAP

**Cancer cells differ from normal cells in terms of their rapid cell division and their ability to spread (metastasis). Many proteins regulate the cell cycle, either positively or negatively. In cancer, one or another of these proteins is altered in some way, making its activity abnormal. Radiation and many cancer drugs target proteins involved in the cell cycle.**

- How are oncogene proteins and tumor suppressor proteins involved in cell cycle control in normal and cancer cells? **Review p. 231 and Figure 11.24**
- How does cancer treatment target the cell cycle? **Review p. 232 and Figure 11.25**

We have now looked at the cell cycle and at cell division by binary fission, mitosis, and meiosis. We have described the normal cell cycle and how it is upset in cancer. We have seen how meiosis produces haploid cells in sexual life cycles. In the coming chapters we examine heredity, genes, and DNA. In Chapter 12 we see how Gregor Mendel studied heredity in the nineteenth century and how the enormous power of his discoveries founded the science of genetics, and changed forever the science of biology.

# CHAPTER SUMMARY

## 11.1 How Do Prokaryotic and Eukaryotic Cells Divide?

- Cell division is necessary for the reproduction, growth, and repair of organisms.
- Cell division must be initiated by a reproductive signal. Before a cell can divide, the genetic material (DNA) must be **replicated** and **segregated** to separate portions of the cell. **Cytokinesis** then divides the cytoplasm into two cells.
- In prokaryotes, most cellular DNA is a single molecule, usually in the form of a circular **chromosome**. Prokaryotes reproduce by **binary fission.** Review Figure 11.2
- In eukaryotes, cells divide by either **mitosis** or **meiosis.** Eukaryotic cell division follows the same general pattern as binary fission, but with significant differences. For example, a eukaryotic cell has a distinct nucleus that must be replicated prior to separating the two daughter cells.
- Cells that produce gametes undergo a special kind of nuclear division called meiosis; the four daughter cells produced by meiosis are not genetically identical.

## 11.2 How Is Eukaryotic Cell Division Controlled?

- The eukaryotic cell cycle has two main phases: **interphase**, during which cells are not dividing and the DNA is replicating, and mitosis or **M phase**, when the cells are dividing.
- During most of the eukaryotic cell cycle, the cell is in interphase, which is divided into three subphases: **S, G1**, and **G2**. DNA is replicated during the S phase. Mitosis (M phase) and cytokinesis follow. Review Figure 11.3
- **Cyclin–Cdk complexes** regulate the passage of cells through checkpoints in the cell cycle. The suppressor protein **RB** inhibits the cell cycle. The G1–S cyclin–Cdk functions by inactivating RB and allows the cell cycle to progress beyond the **restriction point**. Review Figures 11.5 and 11.6
- External controls such as **growth factors** can also stimulate the cell to begin a division cycle.

## 11.3 What Happens during Mitosis?

SEE ANIMATED TUTORIAL 11.1

- In mitosis, a single nucleus gives rise to two nuclei that are genetically identical to each other and to the parent nucleus.
- DNA is wrapped around proteins called **histones**, forming beadlike units called **nucleosomes**. A eukaryotic chromosome contains strings of nucleosomes bound to proteins in a complex called **chromatin**. Review Figure 11.9
- At mitosis, the replicated chromosomes, called **sister chromatids**, are held together at the **centromere**. Each chromatid consists of one double-stranded DNA molecule. Review Figure 11.10, **WEB ACTIVITY 11.1**
- Mitosis can be divided into several phases called **prophase**, **prometaphase**, **metaphase**, **anaphase**, and **telophase**.
- During mitosis sister chromatids, attached by **cohesin**, line up at the equatorial plate and attach to the **spindle**. The chromatids separate (becoming **daughter chromosomes**) and migrate to opposite ends of the cell. Review Figure 11.11, **WEB ACTIVITY 11.2**
- Nuclear division is usually followed by cytokinesis. Animal cell cytoplasms divide via a contractile ring made up of actin microfilaments. In plant cells, cytokinesis is accomplished by vesicles that fuse to form a cell plate. Review Figure 11.13

## 11.4 What Role Does Cell Division Play in a Sexual Life Cycle?

- **Asexual reproduction** produces clones, new organisms that are genetically identical to the parent. Any genetic variation is the result of mutations.
- In **sexual reproduction**, two **haploid** gametes—one from each parent—unite in **fertilization** to form a genetically unique, **diploid zygote**. There are many different sexual life cycles that can be **haplontic**, **diplontic**, or involve **alternation of generations**. Review Figure 11.15, **WEB ACTIVITY 11.3**
- In sexually reproducing organisms, certain cells in the adult undergo meiosis, a process by which a diploid cell produces haploid gametes.
- Each gamete contains of one of each pair of **homologous chromosomes** from the parent.
- The numbers, shapes, and sizes of the chromosomes constitute the **karyotype** of an organism.

## 11.5 What Happens during Meiosis?

SEE ANIMATED TUTORIAL 11.2

- Meiosis consists of two nuclear divisions, **meiosis I** and **meiosis II**, that collectively reduce the chromosome number from diploid to haploid. It ensures that each haploid cell contains one member of each chromosome pair, and results in four genetically diverse haploid cells, usually gametes. Review Figure 11.17, **WEB ACTIVITY 11.4**
- In meiosis I, entire chromosomes, each with two chromatids, migrate to the poles. In meiosis II, the sister chromatids separate.
- During prophase I, homologous chromosomes undergo **synapsis** to form pairs in a **tetrad**. Chromatids can form junctions called **chiasmata** and genetic material may be exchanged between the two homologs by **crossing over**. Review Figure 11.18
- Both crossing over during prophase I and **independent assortment** of the homologs as they separate during anaphase I ensure that the gametes are genetically diverse.
- In **nondisjunction**, two members of a homologous pair of chromosomes go to the same pole during meiosis I, or two chromatids go to the same pole during meiosis II. This leads to one gamete having an extra chromosome and another lacking that chromosome. Review Figure 11.21
- The union between a gamete with an abnormal chromosome number and a normal haploid gamete results in **aneuploidy**. Such genetic abnormalities are harmful or lethal to the organism.

## 11.6 In a Living Organism, How Do Cells Die?

- A cell may die by **necrosis**, or it may self-destruct by **apoptosis**, a genetically programmed series of events that includes the fragmentation of its nuclear DNA.

- Apoptosis is regulated by external and internal signals. These signals result in activation of a class of enzymes called **caspases** that hydrolyze proteins in the cell. Review Figure 11.22

## 11.7 How Does Unregulated Cell Division Lead to Cancer?

- Cancer cells divide more rapidly than normal cells and can be **metastatic**, spreading to distant organs in the body.
- Cancer can result from changes in either of two types of proteins that regulate the cell cycle. **Oncogene** proteins stimulate cell division and are activated in cancer. **Tumor suppressor** proteins normally inhibit the cell cycle but in cancer they are inactive. Review Figure 11.24
- Cancer treatment often targets the cell cycle in tumor cells. Review Figure 11.25

# SELF-QUIZ

1. Which statement about eukaryotic chromosomes is *not* true?
   *a.* They sometimes consist of two chromatids.
   *b.* They sometimes consist only of a single chromatid.
   *c.* They normally possess a single centromere.
   *d.* They consist only of proteins.
   *e.* During metaphase they are visible under the light microscope.
2. Nucleosomes
   *a.* are made of chromosomes.
   *b.* consist entirely of DNA.
   *c.* consist of DNA wound around a histone core.
   *d.* are present only during mitosis.
   *e.* are present only during prophase.
3. Which statement about the cell cycle is *not* true?
   *a.* It consists of interphase, mitosis, and cytokinesis.
   *b.* The cell's DNA replicates during G1.
   *c.* A cell can remain in G1 for weeks or much longer.
   *d.* DNA is not replicated during G2.
   *e.* Cells enter the cell cycle as a result of internal or external signals.
4. Which statement about mitosis is *not* true?
   *a.* A single nucleus gives rise to two identical daughter nuclei.
   *b.* The daughter nuclei are genetically identical to the parent nucleus.
   *c.* The centromeres separate at the onset of anaphase.
   *d.* Homologous chromosomes synapse in prophase.
   *e.* The centrosomes organize the microtubules of the spindle fibers.
5. Which statement about cytokinesis is true?
   *a.* In animals, a cell plate forms.
   *b.* In plants, it is initiated by furrowing of the membrane.
   *c.* It follows mitosis.
   *d.* In plant cells, actin and myosin play an important part.
   *e.* It is the division of the nucleus.
6. Apoptosis
   *a.* occurs in all cells.
   *b.* involves the formation of the plasma membrane.
   *c.* does not occur in an embryo.
   *d.* is a series of programmed events resulting in cell death.
   *e.* is the same as necrosis.
7. In meiosis,
   *a.* meiosis II reduces the chromosome number from diploid to haploid.
   *b.* DNA replicates between meiosis I and meiosis II.
   *c.* the chromatids that make up a chromosome in meiosis II are identical.
   *d.* each chromosome in prophase I consists of four chromatids.
   *e.* homologous chromosomes separate from one another in anaphase I.
8. In meiosis,
   *a.* a single nucleus gives rise to two daughter nuclei.
   *b.* the daughter nuclei are genetically identical to the parent nucleus.
   *c.* the centromeres separate at the onset of anaphase I.
   *d.* homologous chromosomes synapse in prophase I.
   *e.* no spindle forms.
9. An animal has a diploid chromosome number of 12. An egg cell of that animal has 5 chromosomes. The most probable explanation is
   *a.* normal mitosis.
   *b.* normal meiosis.
   *c.* nondisjunction in meiosis I.
   *d.* nondisjunction in meiosis I or II.
   *e.* nondisjunction in mitosis.
10. The number of daughter chromosomes in a human cell (diploid number 46) in anaphase II of meiosis is
    *a.* 2.
    *b.* 23.
    *c.* 46.
    *d.* 69.
    *e.* 92.

# FOR DISCUSSION

1. Compare the roles of cohesins in mitosis, meiosis I, and meiosis II.
2. Compare and contrast cell division in animals and plants.
3. Contrast mitotic prophase and prophase I of meiosis. Contrast mitotic anaphase and anaphase I of meiosis.
4. Compare the sequence of events in the mitotic cell cycle with the sequence of events in apoptosis.
5. Cancer-fighting drugs are rarely used alone. Usually, there are several drugs given in combination that target different stages of the cell cycle. Why might this be a better approach than single drugs?

## ADDITIONAL INVESTIGATION

1. Suggest two ways in which one might use a microscope to determine the relative durations of the various phases of mitosis.
2. Studying the events and controls of the cell cycle is much easier if the cells under investigation are synchronous; that is, if they are all in the same stage of the cell cycle. This can be accomplished with various chemicals. But some populations of cells are naturally synchronous. The anther (male sex organ) of a lily plant contains cells that become pollen grains (male gametes). As anthers develop in the flower, their lengths correlate precisely with the stage of the meiotic cycle in those cells. These stages each take many days, so an anther that is 1.5 millimeters long, for example, contains cells in early prophase I. How would you use lily anthers to investigate the roles of cyclins and Cdk's in the meiotic cell cycle?

# 12 Inheritance, Genes, and Chromosomes

## Genetic piracy

The Nazis were infamous for plundering the art collections of Europe during World War II, but why did they also spirit away a collection of seeds? The answer lies in the power and promise offered by the scientific understanding of genetics, the science of *heredity*.

A key step in the rise of human civilizations was the development of agriculture—the cultivation of plants and animals for food and other human needs. Some 10,000 years ago, early farmers began preferentially cultivating plants with certain traits (e.g., that survived drought better). Over time, the cultivated varieties (*cultivars*) became quite different from their wild relatives, an example of evolution by selection. In this case, it was not the result of natural selection (see Chapter 21), but of artificial selection by the practices of ancient farmers.

Early in the twentieth century, Russian scientist Nicolai Vavilov began systematically collecting seeds from thousands of cultivars and their wild relatives. He convinced Lenin, the leader of the new Communist regime, that his seed collection would be useful in breeding crops that would be more productive in the difficult Russian climate. Lenin put Vavilov in charge of a large research institute. But when Lenin died in 1929, his successor, Josef Stalin, had little interest in science.

A politically ambitious student of Vavilov's, Trofim Lysenko, proposed to Stalin that favorable characteristics in plants could be rendered heritable by manipulating the parent plant's *phenotype* (physical state). This idea was at odds with what scientists knew about heredity and evolution, but it appealed to Stalin's political ideology. Stalin put Lysenko in charge of Vavilov's institute and sent Vavilov to a prison camp, where he died in 1943. Vavilov's unique seed collection—a *gene bank*—was ignored.

Meanwhile, in Germany, the Nazi leader Heinrich Himmler learned of the collection and was convinced that Vavilov's seeds could be a valuable key to providing better crops for the expanding German empire. Himmler put Heinz Brücher, a young SS officer with a doctorate in botany, in charge of obtaining the seeds. When the German army invaded Russia, Brücher's team removed thousands of seeds to a castle in Austria that already housed a seed collection Brücher had brought from Tibet.

Brücher's aim was to cross-breed plants from Tibet with plants from Russia to develop new crops that would grow well at high elevations and in cold climates; these plans came to a halt with the end of World War II. However,

**Genetics Pioneer** Collecting thousands of crop plant varieties from all over the world, Nikolai Vavilov laid the foundations for theories about the genetic origins of modern crops.

**Hardy Grain** Early geneticists hoped to increase food production by breeding crop varieties adapted to harsh climates (such as those in Tibet) with varieties with other desirable traits.

Brücher ignored a superior's order to blow up the castle, thus preserving most of Vavilov's seed bank. The collection was returned to Russia, where it continued to be used in breeding programs.

The ideas of Vavilov and the breeding plans of Brücher depended on the principles of *genetics*, a science born in an Austrian monastery in the 1860s, where Gregor Mendel performed—and, importantly, correctly interpreted—experiments on pea plants. It was almost 50 years before the scientific community recognized the significance of Mendel's work, but once that recognition was achieved, science and medicine sprang forward at a rapid pace.

**IN THIS CHAPTER** we will discuss how the units of inheritance—genes—are transmitted from generation to generation. We will show that many of the rules that govern inheritance can be explained by the behavior of chromosomes during meiosis. We will describe the interactions of genes with one another and with the environment, and we will see how the specific positions of genes on chromosomes affect diversity.

CHAPTER OUTLINE



## 12.1 What Are the Mendelian Laws of Inheritance?

Much of the early study of biological inheritance was done with plants and animals of economic importance. Records show that people were deliberately cross-breeding date palm trees and horses as early as 5,000 years ago. By the early nineteenth century plant breeding was widespread, especially for ornamental flowers such as tulips. Plant breeders of that time were operating under two key assumptions about how inheritance worked. Only one of those assumptions turned out to be supported by experimental evidence.

- *Each parent contributes equally to offspring* (supported by experiments). In the 1770s, the German botanist Josef Gottlieb Kölreuter studied the offspring of **reciprocal crosses**, in which plants were crossed (mated with each other) in both directions. For example, in one cross, plants with white flowers were used as males to pollinate related plants with red flowers. In the complementary crosses, the red-flowered plants were used as males in crosses with the white flowered plants. In Kölreuter's studies, such reciprocal crosses always gave identical results, showing that both parents contributed equally to the offspring.
- *Hereditary determinants blend in offspring* (not supported by experiments). Kölreuter and others proposed that there were hereditary determinants in the egg and sperm cells. When these determinants came together in a single cell after mating, they were believed to blend together. If a plant with one form of a character (say, red flowers) was crossed with a plant with a different form of that character (blue flowers), the offspring would have a blended combination of the two parents' characteristics (purple flowers). According to the blending theory, once heritable elements were combined, they could not be separated again (like inks of different colors mixed together). The red and blue hereditary determinants were thought to be forever blended into the new purple one.

In his experiments in the 1860s, Gregor Mendel confirmed the first of these two assumptions but refuted the second.

### Mendel brought new methods to experiments on inheritance

Gregor Mendel was an Austrian monk, not an academic scientist (**Figure 12.1**). He was well qualified, however, to under-

**12.1 Gregor Mendel and His Garden** The Austrian monk Gregor Mendel (left) did his groundbreaking genetics experiments in a garden at the monastery at Brno, in what is now the Czech Republic.

take scientific investigations. In 1850 he failed an examination for a teaching certificate in natural science, so he undertook intensive studies in physics, chemistry, mathematics, and various aspects of biology at the University of Vienna. His studies in physics and mathematics under the famous physicist Christian Doppler strongly influenced his use of experimental and quantitative methods in his studies of heredity, and it was those quantitative experiments that were key to his successful deductions.

Over the seven years he spent working out some principles of inheritance in plants, Mendel made crosses with hundreds of plants and noted the resulting characteristics of 24,034 progeny. Analysis of his meticulously gathered data suggested to him a new theory of how inheritance might work. He presented this theory in a public lecture in 1865 and a detailed written publication in 1866. Mendel's paper appeared in a journal that was received by 120 libraries, and he sent reprinted copies (of which he had obtained 40) to several distinguished scholars, including Charles Darwin. However, his theory was not readily accepted. In fact, it was mostly ignored.

One reason Mendel's paper received so little attention was that most prominent biologists of his time were not in the habit of thinking in mathematical terms, even the simple terms Mendel used. Even Charles Darwin, whose theory of evolution by natural selection was predicated on heritable variations among individuals, failed to understand the significance of Mendel's findings. In fact, Darwin performed breeding experiments on snapdragons that were similar to Mendel's work with peas. Although Darwin's data were similar to Mendel's, he failed to question the assumption that parental contributions blend in the offspring.

By 1900, the events of meiosis had been observed and described, and Mendel's discoveries burst into sudden prominence as a result of independent experiments by three plant geneticists: Hugo DeVries, Carl Correns, and Erich von Tschermak. Each carried out crossing experiments, each published his principal findings in 1900, and each cited Mendel's 1866 paper. These three men realized that chromosomes and meiosis provided a physical explanation for the theory that Mendel had proposed to explain the data from his crosses.

That Mendel was able to achieve his remarkable insights before the discovery of genes and meiosis was largely due to his experimental methods. His work is a definitive example of extensive preparation, meticulous execution, and imaginative yet logical interpretation. He was also fortunate in his choice of experimental subjects. Let's take a closer look at these experiments and the conclusions and hypotheses that emerged.

### Mendel devised a careful research plan

Mendel chose to study the common garden pea because of its ease of cultivation, the feasibility of controlled pollination, and

TOOLS FOR INVESTIGATING LIFE

**12.2 A Controlled Cross between Two Plants**
Plants were widely used in early genetic studies because it is easy to control which individuals mate with which. Mendel used the garden pea (*Pisum sativum*) in many of his experiments.

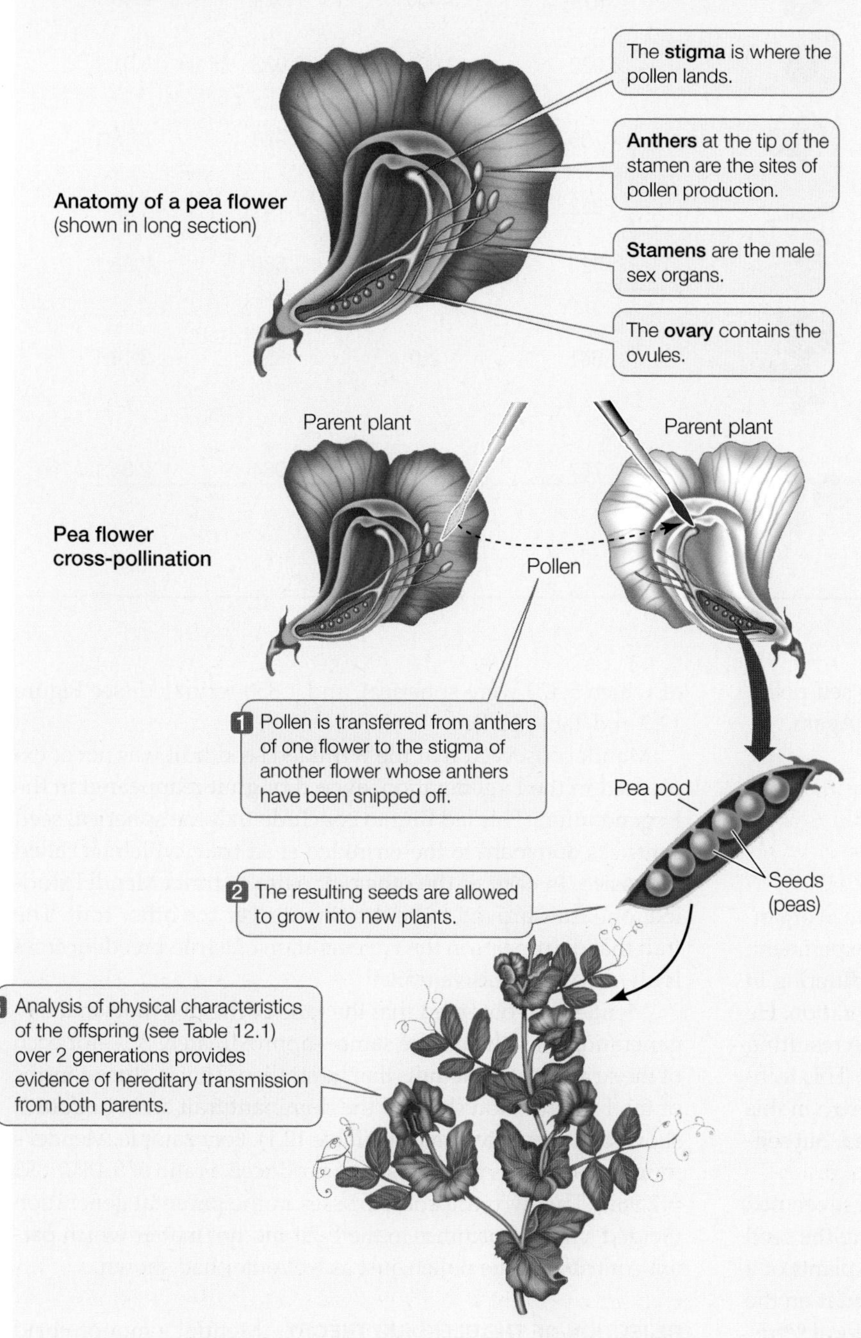

The stigma is where the pollen lands.
Anthers at the tip of the stamen are the sites of pollen production.
Anatomy of a pea flower (shown in long section)
Stamens are the male sex organs.
The ovary contains the ovules.
Parent plant
Parent plant
Pea flower cross-pollination
Pollen
1 Pollen is transferred from anthers of one flower to the stigma of another flower whose anthers have been snipped off.
Pea pod
Seeds (peas)
2 The resulting seeds are allowed to grow into new plants.
3 Analysis of physical characteristics of the offspring (see Table 12.1) over 2 generations provides evidence of hereditary transmission from both parents.

the availability of varieties with contrasting traits. He controlled pollination, and thus fertilization, of his parent plants by manually moving pollen from one plant to another (**Figure 12.2**). Thus he knew the parentage of the offspring in his experiments. The pea plants Mendel studied produce male and female sex organs and gametes—sex cells such as eggs and sperm—in the same flower. If untouched, they naturally self-pollinate—that is, the female organ of each flower receives pollen from the male organs of the same flower. Mendel made use of this natural phenomenon in some of his experiments.

Mendel began by examining different varieties of peas in a search for heritable characters and traits suitable for study:

- A **character** is an observable physical feature, such as flower color.
- A **trait** is a particular form of a character, such as purple flowers or white flowers.
- A **heritable trait** is one that is passed from parent to offspring.

Mendel looked for characters with well-defined, contrasting alternative traits, such as purple flowers versus white flowers. Furthermore, these traits had to be **true-breeding**, meaning that the observed trait was the only form present for many generations. In other words, if they were true-breeding, peas with white flowers would give rise only to progeny with white flowers when self-pollinated or crossed with one another for repeated generations. Similarly, tall plants bred with other tall plants would produce only tall progeny.

Mendel isolated each of his true-breeding strains by repeated inbreeding (done by crossing sibling plants that were seemingly identical or by allowing individuals to self-pollinate) and selection. In most of his work, Mendel concentrated on the seven pairs of contrasting traits shown in **Table 12.1** (left side). His use of true-breeding strains for experimental crosses was an essential feature of his work.

Mendel then performed his crosses in the following manner:

- He removed the anthers from the flowers of one parental strain so that it couldn't self-pollinate. Then he collected pollen from the other parental strain and placed it on the stigmas of flowers of the strain whose anthers had been removed. The plants providing and receiving the pollen were the **parental generation**, designated **P**.
- In due course, seeds formed and were planted. The seeds and the resulting new plants constituted the **first filial generation**, or $\mathbf{F_1}$. (The word "filial" refers to the relationship between offspring and parents, from the Latin, *filius*, "son.") Mendel and his assistants examined each $F_1$ plant to see which traits it bore and then recorded the number of $F_1$ plants expressing each trait.

TABLE 12.1
**Mendel's Results from Monohybrid Crosses**

| PARENTAL GENERATION PHENOTYPES | | $F_2$ GENERATION PHENOTYPES | | | |
|---|---|---|---|---|---|
| DOMINANT | RECESSIVE | DOMINANT | RECESSIVE | TOTAL | RATIO |
| Spherical seeds × | Wrinkled seeds | 5,474 | 1,850 | 7,324 | 2.96:1 |
| Yellow seeds × | Green seeds | 6,022 | 2,001 | 8,023 | 3.01:1 |
| Purple flowers × | White flowers | 705 | 224 | 929 | 3.15:1 |
| Inflated pods × | Constricted pods | 882 | 299 | 1,181 | 2.95:1 |
| Green pods × | Yellow pods | 428 | 152 | 580 | 2.82:1 |
| Axial flowers × | Terminal flowers | 651 | 207 | 858 | 3.14:1 |
| Tall stems (1 m) × | Dwarf stems (0.3 m) | 787 | 277 | 1,064 | 2.84:1 |

- In some experiments the $F_1$ plants were allowed to self-pollinate and produce a **second filial generation**, the **$F_2$**. Again, each $F_2$ plant was characterized and counted.

## Mendel's first experiments involved monohybrid crosses

The term *hybrid* refers to the offspring of crosses between organisms differing in one or more traits. In Mendel's first experiment, he crossed two true-breeding parental (P) lineages differing in just one trait, producing monohybrids in the $F_1$ generation. He subsequently planted the $F_1$ seeds and allowed the resulting plants to self-pollinate to produce the $F_2$ generation. This technique is referred to as a **monohybrid cross**, even though in this case, the monohybrid plants were not literally crossed, but self-pollinated.

Mendel performed the same experiment for all seven pea traits. His method is illustrated in **Figure 12.3**, using the seed shape trait as an example. He took pollen from pea plants of a true-breeding strain with wrinkled seeds and placed it on the stigmas of flowers of a true-breeding strain with spherical seeds. He also performed the reciprocal cross, in which the parental source of each trait is reversed: he placed pollen from the spherical-seeded strain on the stigmas of flowers of the strain with wrinkled seeds. In all cases, the $F_1$ seeds were spherical—it was as if the wrinkled seed trait had disappeared completely.

The following spring, Mendel grew 253 $F_1$ plants from these spherical seeds. Each of the $F_1$ plants was allowed to self-pollinate to produce $F_2$ seeds. In all, 7,324 $F_2$ seeds were produced, of which 5,474 were spherical, and 1,850 wrinkled (see Figure 12.3 and Table 12.1).

Mendel observed that the wrinkled seed trait was never expressed in the $F_1$ generation, even though it reappeared in the $F_2$ generation. This led him to conclude that the spherical seed trait was **dominant** to the wrinkled seed trait, which he called **recessive**. In each of the other six pairs of traits Mendel studied, one trait proved to be dominant over the other trait. The trait that disappears in the $F_1$ generation of a true-breeding cross is always the recessive trait.

Mendel also observed that the ratio of the two traits in the $F_2$ generation was always the same—approximately 3:1—for each of the seven pea-plant traits that he studied. That is, three-fourths of the $F_2$ generation showed the dominant trait and one-fourth showed the recessive trait (see Table 12.1). For example, Mendel's monohybrid cross for seed shape produced a ratio of 5,474:1,850 = 2.96:1. The two reciprocal crosses in the parental generation yielded similar outcomes in the $F_2$; it did not matter which parent contributed the pollen, just as Kölreuter had shown.

**REJECTION OF THE BLENDING THEORY** Mendel's monohybrid cross experiments showed that inheritance cannot be the result of a blending phenomenon. According to the blending theory, Mendel's $F_1$ seeds should have had an appearance that was intermediate between those of the two parents—for example, the $F_1$ seeds from the cross between strains with wrinkled and spherical seeds should have been slightly wrinkled. Furthermore, the blending theory offered no explanation for the reappearance of the recessive trait in the $F_2$ seeds after its absence in the $F_1$ seeds.

INVESTIGATING LIFE

**12.3 Mendel's Monohybrid Experiments**

Mendel performed crosses with pea plants and carefully analyzed the outcomes to show that genetic determinants are particulate.

**HYPOTHESIS** When two strains of peas with contrasting traits are bred, their characteristics are irreversibly blended in succeeding generations.

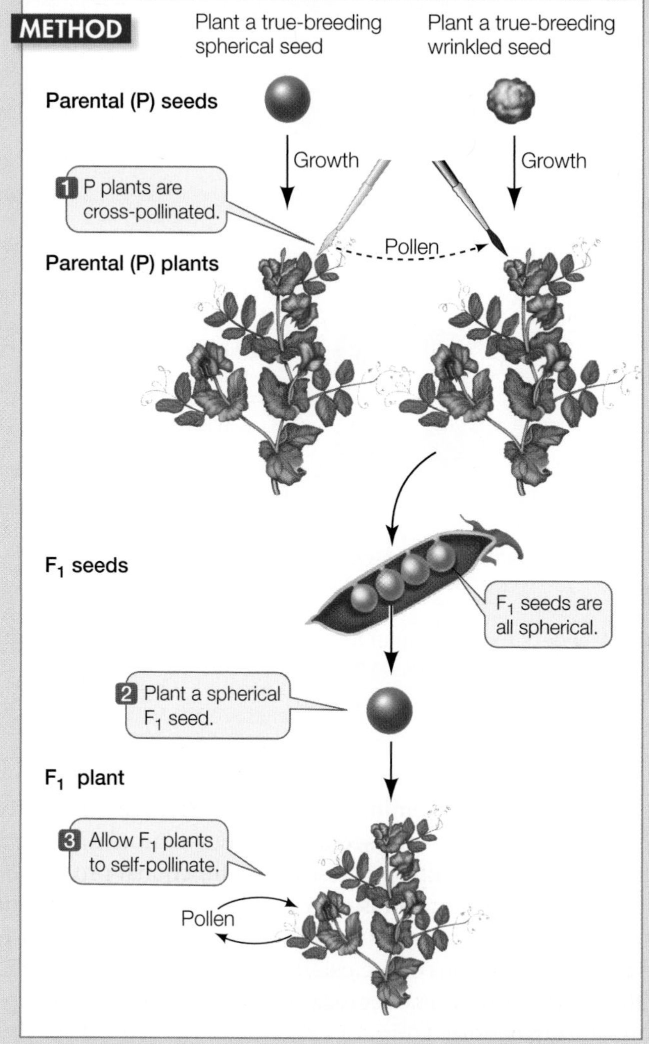


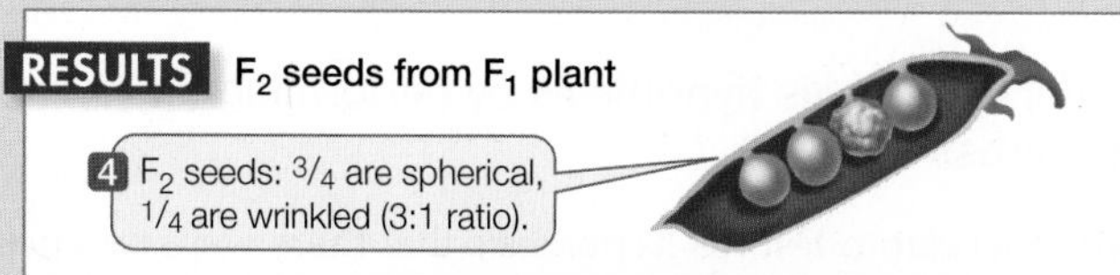

RESULTS $F_2$ seeds from $F_1$ plant
4 $F_2$ seeds: 3/4 are spherical, 1/4 are wrinkled (3:1 ratio).


**CONCLUSION** The hypothesis is rejected. There is no irreversible blending of characteristics, and a recessive trait can reappear in succeeding generations.

Go to **yourBioPortal.com** for original citations, discussions, and relevant links for all INVESTIGATING LIFE figures.

**SUPPORT FOR THE PARTICULATE THEORY** Given the absence of blending and the reappearance of the recessive seed traits in the $F_2$ generations of his monohybrid cross experiments, Mendel proposed that the units responsible for the inheritance of specific traits are present as discrete particles that occur in pairs and segregate (separate) from one another during the formation of gametes. According to his **particulate theory**, the units of inheritance retain their integrity in the presence of other units. Mendel concluded that each pea plant has two units (particles) of inheritance for each character, one from each parent. We now use the term **diploid** to refer to the two copies of each heritable unit in an organism. Mendel proposed that during the production of gametes, only one of these paired units is given to a gamete. We now use the term **haploid** to refer to the single set of heritable units. Mendel concluded that while each gamete contains one unit, the resulting zygote contains two, because it is produced by the fusion of two gametes. This conclusion is the core of Mendel's model of inheritance. Mendel's unit of inheritance is now called a **gene**. The totality of all the genes of an organism is that organism's **genome**.

Mendel reasoned that in his experiments, the two true-breeding parent plants had different forms of the gene affecting a particular character, such as seed shape (although he did not use the term "gene"). The true-breeding spherical-seeded parent had two genes of the same form, which we will call *S*, and the parent with wrinkled seeds had two copies of an alternative form of the gene, which we will call *s*. The *SS* parent would produce gametes having a single *S* gene, and the *ss* parent would produce gametes having a single *s* gene. The cross producing the $F_1$ generation would donate an *S* from one parent and an *s* from the other to each seed; the $F_1$ offspring would thus be *Ss*. We say that *S* is dominant over *s* because the trait specified by *s* is not evident—is not expressed—when both forms of the gene are present.

## Alleles are different forms of a gene

The different forms of a gene (*S* and *s* in this case) are called **alleles**. Individuals that are true-breeding for a trait contain two copies of the same allele. For example, all the individuals in a population of true-breeding peas with wrinkled seeds must have the allele pair *ss*; if the dominant *S* allele were present, some of the plants would produce spherical seeds.

We say that the individuals that produce wrinkled seeds are **homozygous** for the allele *s*, meaning that they have two copies of the same allele (*ss*). Some peas with spherical seeds—the ones with the genotype *SS*—are also homozygous. However, not all plants with spherical seeds have the *SS* genotype. Some spherical-seeded plants, like Mendel's $F_1$, are **heterozygous**: they have two different alleles of the gene in question (in this case, *Ss*). An individual that is homozygous for a character is sometimes called a **homozygote**; an individual that is heterozygous for a character is termed a **heterozygote**.

As a somewhat more complex example of inheritance, let's consider three gene pairs. An individual with the three genes and alleles *AABbcc* is homozygous for the *A* and *C* genes, because it has two *A* alleles and two *c* alleles, but heterozygous for the *B* gene, because it contains the *B* and *b* alleles.

The physical appearance of an organism is its **phenotype**. Mendel correctly supposed the phenotype to be the result of the **genotype**, or genetic constitution, of the organism showing the phenotype. Spherical seeds and wrinkled seeds are two phenotypes, which are the result of three genotypes: the wrinkled seed phenotype is produced by the genotype *ss*, whereas the spherical seed phenotype is produced by either of the genotypes *SS* or *Ss*.

### Mendel's first law says that the two copies of a gene segregate

How does Mendel's model of inheritance explain the ratios of traits seen in the $F_1$ and $F_2$ generations? Consider first the $F_1$, in which all progeny have the spherical seed phenotype. According to Mendel's model, *when any individual produces gametes, the two copies of a gene separate, so that each gamete receives only one copy*. This is Mendel's first law, the **law of segregation**. Thus, every individual in the offspring from a cross between the P generation parents inherits one gene copy from each parent, and has the genotype *Ss* (**Figure 12.4**).

Now let's consider the composition of the $F_2$ generation. Half of the gametes produced by the $F_1$ generation have the *S* allele and the other half the *s* allele. Since both *SS* and *Ss* plants produce spherical seeds while *ss* plants produce wrinkled seeds, in the $F_2$ generation there are three ways to get a spherical-seeded plant (*SS*, *Ss*, or *sS*), but only one way to get a plant with wrinkled seeds (*ss*). Therefore, we predict a 3:1 ratio, remarkably close to the values Mendel found experimentally for all seven of the traits he compared (see Table 12.1).

The allele combinations that will result from a cross can be predicted using a **Punnett square**, a method devised in 1905 by the British geneticist Reginald Crundall Punnett. This device ensures that we consider all possible combinations of gametes when calculating expected genotype frequencies. A Punnett square looks like this:

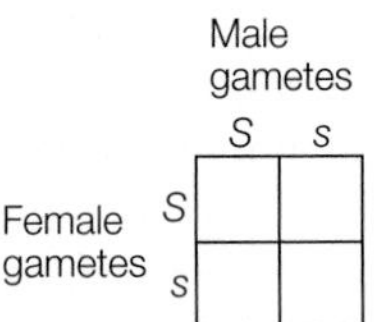


It is a simple grid with all possible male gamete (haploid sperm) genotypes shown along the top and all possible female gamete (haploid egg) genotypes along the left side. The grid is completed by filling in each square with the diploid genotype that can be generated from each combination of gametes (see Figure 12.4). In this example, to fill in the top right square, we put in the *S* from the egg (female gamete) and the *s* from the pollen (male gamete), yielding *Ss*.

Mendel did not live to see his theory placed on a sound physical footing with the discoveries of chromosomes and DNA. Genes are now known to be regions of the DNA molecules in chromosomes. More specifically, a gene is a sequence of DNA that resides at a particular site on a chromosome, called a **locus** (plural **loci**). Genes are expressed in the phenotype mostly as proteins with particular functions, such as enzymes.

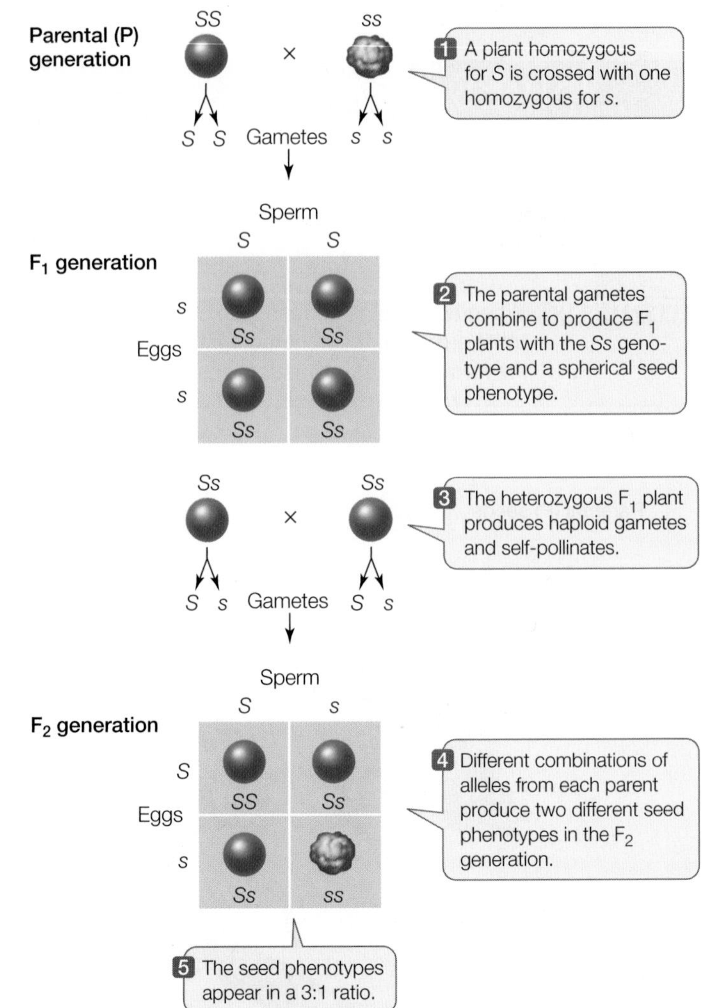


**12.4 Mendel's Explanation of Inheritance** Mendel concluded that inheritance depends on discrete factors from each parent that do not blend in the offspring.

So, in many cases, a dominant gene can be thought of as a region of DNA that is expressed as a functional protein, while a recessive gene typically expresses a nonfunctional protein, or a protein whose function is overshadowed by the dominant form. Mendel arrived at his law of segregation with no knowledge of chromosomes or meiosis, but today we can picture the different alleles of a gene segregating as the chromosomes separate during meiosis I (**Figure 12.5**).

### Mendel verified his hypothesis by performing a test cross

Mendel set out to test his hypothesis that there were two possible allele combinations (*SS* and *Ss*) in the spherical-seeded $F_1$ generation. He did so by performing a **test cross**, which is a way of finding out whether an individual showing a dominant trait is homozygous or heterozygous. In a test cross, the individual in question is crossed with an individual that is known to be homozygous for the recessive trait—an easy individual to identify, because all individuals with the recessive phenotype are homozygous for the recessive trait.

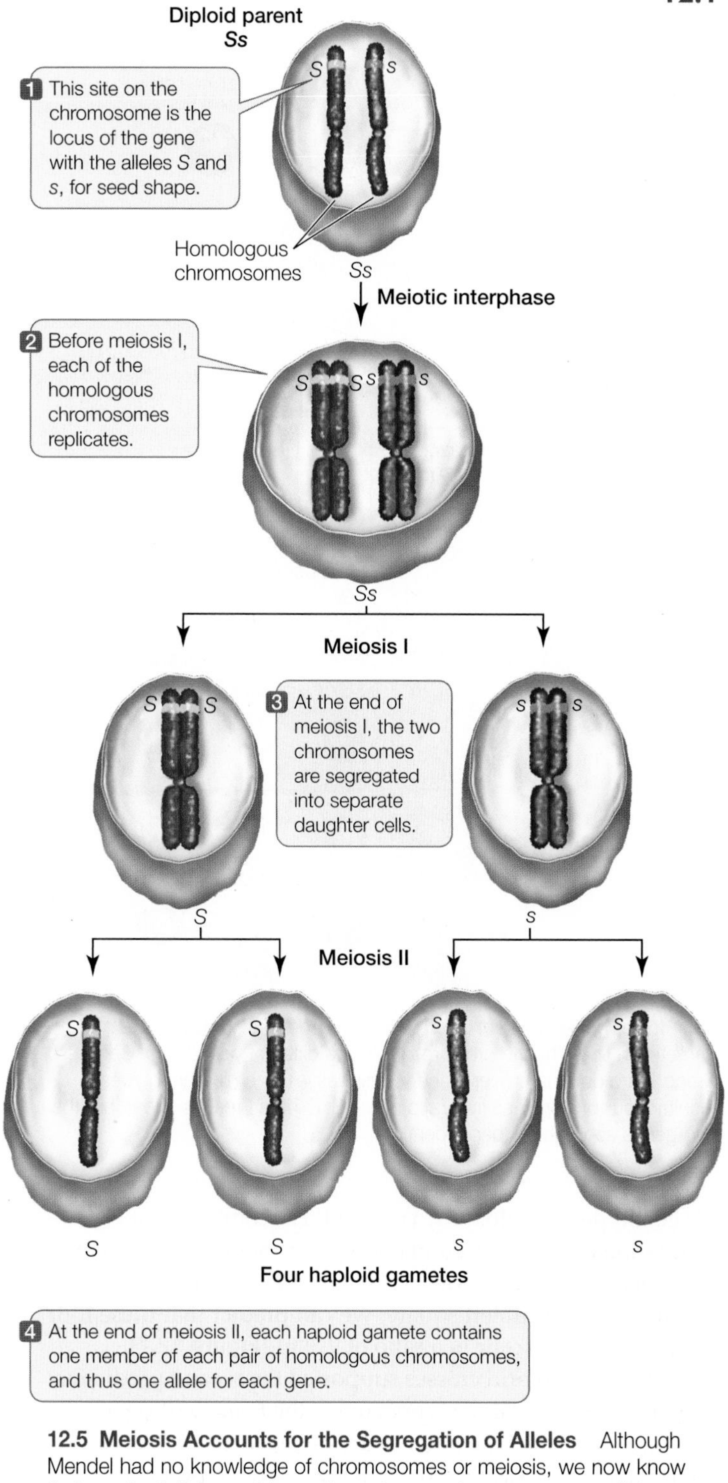

Diploid parent
Ss
1 This site on the chromosome is the locus of the gene with the alleles S and s, for seed shape.
S
s
Homologous chromosomes
Ss
Meiotic interphase
2 Before meiosis I, each of the homologous chromosomes replicates.
S S s s
Ss
Meiosis I
S S
3 At the end of meiosis I, the two chromosomes are segregated into separate daughter cells.
s s
S
s
Meiosis II
S S s s
S S s s
Four haploid gametes
4 At the end of meiosis II, each haploid gamete contains one member of each pair of homologous chromosomes, and thus one allele for each gene.

**12.5 Meiosis Accounts for the Segregation of Alleles** Although Mendel had no knowledge of chromosomes or meiosis, we now know that a pair of alleles resides on homologous chromosomes, and that those alleles segregate during meiosis.

## INVESTIGATING LIFE

**12.6 Homozygous or Heterozygous?**

An individual with a dominant phenotype may have either a homozygous or a heterozygous genotype. The test cross determines which.

**HYPOTHESIS** The progeny of a test cross can reveal whether an organism is homozygous or heterozygous.

**METHOD**

1a Test spherical peas of undetermined genotype...

S_ × ss

1b ...by crossing them with wrinkled peas with a known genotype (homozygous recessive).

2a If the plant being tested is homozygous...

SS × ss

Gametes: S S s s

2b If the plant being tested is heterozygous...

Ss × ss

Gametes: S s s s

**RESULTS**

| Eggs \ Sperm | s | s |
|---|---|---|
| S | Ss | Ss |
| S | Ss | Ss |

| Eggs \ Sperm | s | s |
|---|---|---|
| S | Ss | Ss |
| s | ss | ss |

3a ...then all progeny will show the dominant phenotype (spherical).

3b ...then half the seeds from the cross will be wrinkled, and half will be spherical.

**CONCLUSION** The plant being tested is homozygous.

**CONCLUSION** The plant being tested is heterozygous.

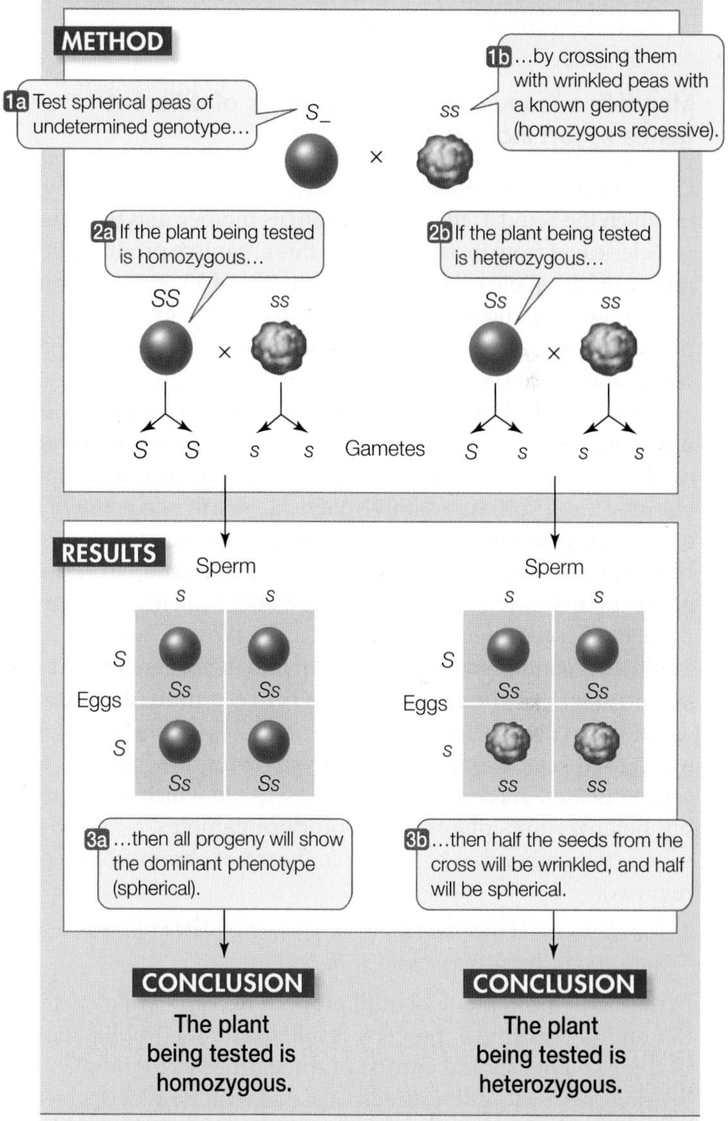

**FURTHER INVESTIGATIONS:** What would be the result if the "tester" plant was homozygous for spherical instead of wrinkled seeds?

Go to **yourBioPortal.com** for original citations, discussions, and relevant links for all INVESTIGATING LIFE figures.

For the seed shape gene that we have been considering, the recessive homozygote used for the test cross is *ss*. The individual being tested may be described initially as *S*_ because we do not yet know the identity of the second allele. We can predict two possible results:

- If the individual being tested is homozygous dominant (*SS*), all offspring of the test cross will be *Ss* and show the dominant trait (spherical seeds) (**Figure 12.6, left**).

**yourBioPortal.com**

GO TO **Web Activity 12.1 • Homozygous or Heterozygous?**

- If the individual being tested is heterozygous (*Ss*), then approximately half of the offspring of the test cross will be heterozygous and show the dominant trait (*Ss*), but the other half will be homozygous for, and will show, the recessive trait (*ss*) (**Figure 12.6, right**).

Mendel obtained results consistent with both of these predictions; thus Mendel's hypothesis accurately predicted the results of his test crosses.

With his first hypothesis confirmed, Mendel went on to ask another question: How do different pairs of genes behave in crosses when considered together?

### Mendel's second law says that copies of different genes assort independently

Consider an organism that is heterozygous for two genes (*SsYy*), in which the *S* and *Y* alleles came from its mother, and the *s* and *y* alleles came from its father. When this organism produces gametes, do the alleles of maternal origin (*S* and *Y*) go together in one gamete and those of paternal origin (*s* and *y*) in another gamete? Or can a single gamete receive one maternal and one paternal allele, *S* and *y* (or *s* and *Y*)?

To answer these questions, Mendel performed another series of experiments. He began with peas that differed in two seed characters: seed shape and seed color. One true-breeding parental strain produced only spherical, yellow seeds (*SSYY*), and the other produced only wrinkled, green ones (*ssyy*). A cross between these two strains produced an $F_1$ generation in which all the plants were *SsYy*. Because the *S* and *Y* alleles are dominant, the $F_1$ seeds were all spherical and yellow.

Mendel continued this experiment into the $F_2$ generation by performing a **dihybrid cross** (a cross between individuals that are identical double heterozygotes) with the $F_1$ plants (although again, in this case, this was done by allowing the $F_1$ plants to self-pollinate). There are two possible ways in which such doubly heterozygous plants might produce gametes, as Mendel saw it (remember that he had never heard of chromosomes or meiosis):

1. *The alleles could maintain the associations they had in the parental generation (that is, they could be linked).*

In this case, the $F_1$ plants should produce two types of gametes (*SY* and *sy*), and the $F_2$ progeny resulting from self-pollination of the $F_1$ plants should consist of three times as many plants bearing spherical, yellow seeds as plants with wrinkled, green seeds. If such results were obtained, there might be no reason to suppose that two different genes regulated seed shape and seed color, because spherical seeds would always be yellow and wrinkled ones always green.

2. *The segregation of* S *from* s *could be independent of the segregation of* Y *from* y *(that is, the two genes could be unlinked).*

In this case, four kinds of gametes should be produced by the $F_1$ in equal numbers: *SY*, *Sy*, *sY*, and *sy*. When these gametes combine at random, they should produce an $F_2$ having nine different genotypes. The $F_2$ progeny could have any of three possible genotypes for shape (*SS*, *Ss*, or *ss*) and any of three possible genotypes for color (*YY*, *Yy*, or *yy*). The combined nine genotypes should produce four phenotypes (spherical yellow, spherical green, wrinkled yellow, wrinkled green). Putting these possibilities into a Punnett square, we can predict that these four phenotypes will occur in a ratio of 9:3:3:1 (**Figure 12.7**).

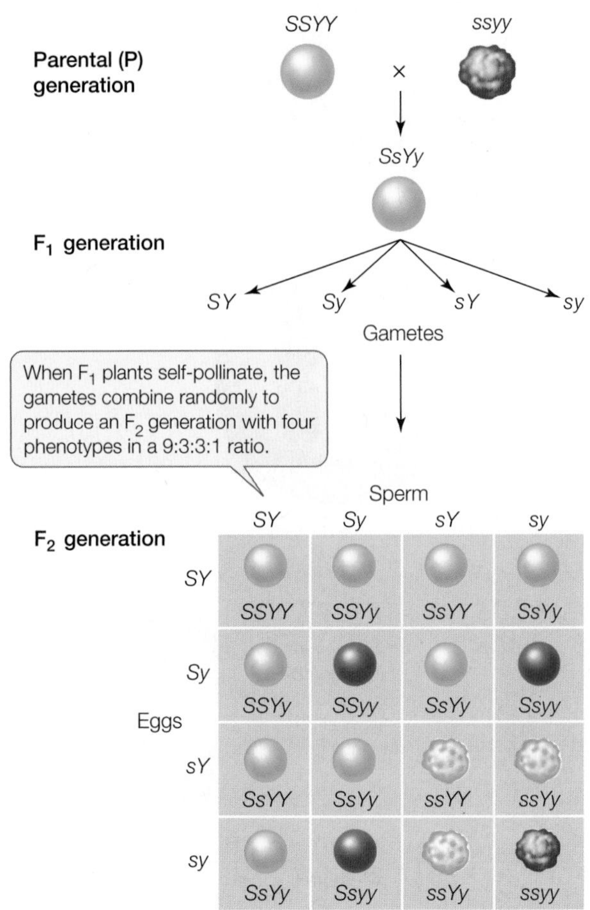


**12.7 Independent Assortment** The 16 possible combinations of gametes in this dihybrid cross result in nine different genotypes. Because *S* and *Y* are dominant over *s* and *y*, respectively, the nine genotypes result in four phenotypes in a ratio of 9:3:3:1. These results show that the two genes segregate independently.

Mendel's dihybrid crosses supported the second prediction: four different phenotypes appeared in the $F_2$ generation in a ratio of about 9:3:3:1. The parental traits appeared in new combinations (spherical green and wrinkled yellow) in some progeny. Such new combinations are called **recombinant** phenotypes.

These results led Mendel to the formulation of what is now known as Mendel's second law: *alleles of different genes assort independently of one another during gamete formation.* That is, the segregation of gene A alleles is independent of the segregation of gene B alleles. We now know that this **law of independent assortment** is not as universal as the law of segregation, because it applies to genes located on separate chromosomes, but not to those located near one another on the same chromosome, as we will see in Section 12.4. However, it is correct to say that chromosomes segregate independently during the formation of gametes, and so do any two genes on separate homologous chromosome pairs (**Figure 12.8**).

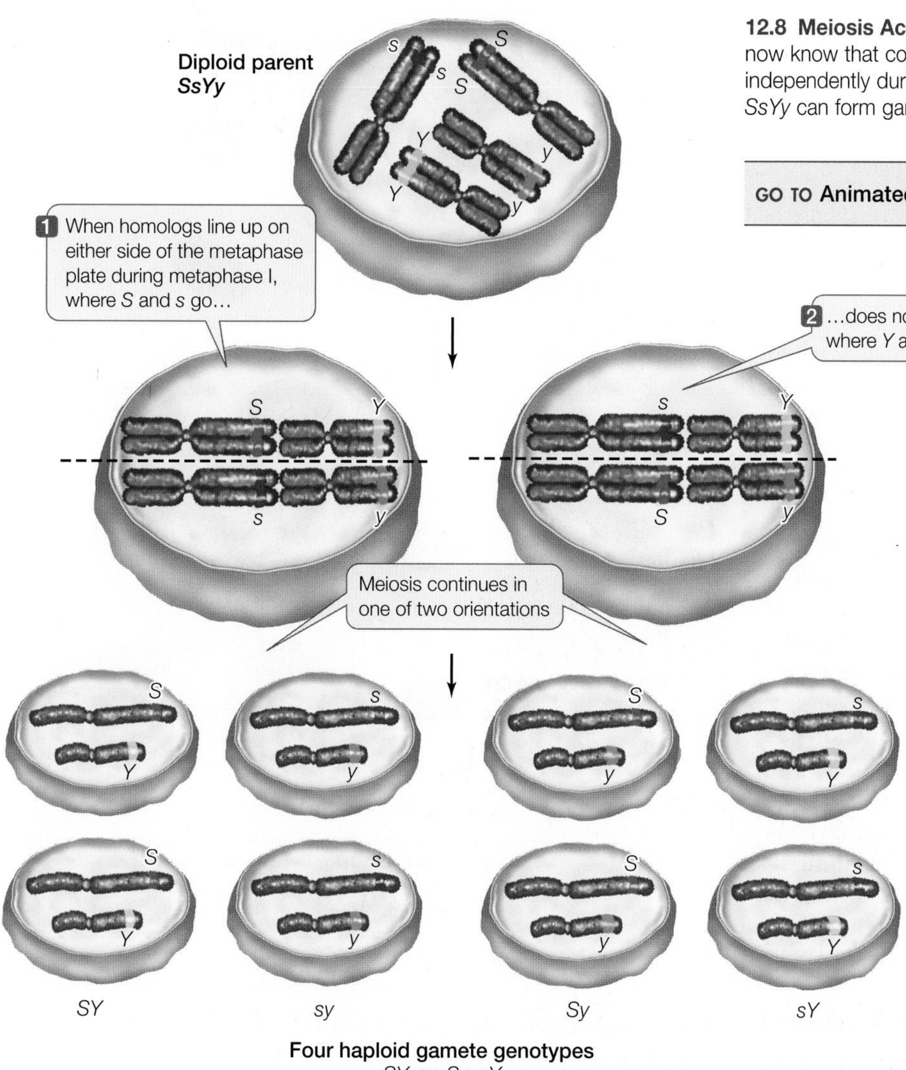
Diploid parent
SsYy
1 When homologs line up on either side of the metaphase plate during metaphase I, where S and s go…
2 …does not determine where Y and y go.
Meiosis continues in one of two orientations
SY
sy
Sy
sY
Four haploid gamete genotypes
SY, sy, Sy, sY

**12.8 Meiosis Accounts for Independent Assortment of Alleles** We now know that copies of genes on different chromosomes are segregated independently during metaphase I of meiosis. Thus a parent of genotype *SsYy* can form gametes with four different genotypes.

**yourBioPortal.com**
GO TO **Animated Tutorial 12.1 • Independent Assortment of Alleles**

One of Mendel's major contributions to the science of genetics was his use of the rules of statistics and probability to analyze his masses of data from hundreds of crosses resulting in thousands of progeny plants. His mathematical analyses revealed clear patterns in the data that allowed him to formulate his hypotheses. Ever since his work became widely recognized, geneticists have used simple mathematics in the same ways that Mendel did.

## Punnett squares or probability calculations: A choice of methods

Punnett squares provide one way of solving problems in genetics, and probability calculations provide another. Many people find it easier to use the principles of probability, some of which are intuitive and familiar. For example, when we flip a coin, the law of probability states that it has an equal probability of landing "heads" or "tails." For any given toss of a fair coin, the probability of heads is independent of what happened in all the previous tosses. A run of ten straight heads implies nothing about the next toss. No "law of averages" increases the likelihood that the next toss will come up tails, and no "momentum" makes an eleventh occurrence of heads any more likely. On the eleventh toss, the odds of getting heads are still 50-50.

The basic conventions of probability are simple:

- If an event is absolutely certain to happen, its probability is 1.
- If it cannot possibly happen, its probability is 0.
- All other events have a probability between 0 and 1.

A coin toss results in heads approximately half the time, so the probability of heads is ½—as is the probability of tails.

**MULTIPLYING PROBABILITIES** How can we determine the probability of two independent events happening together? If two coins (say a penny and a dime) are tossed, each acts independently of the other. What is the probability of both coins coming up heads? In half of the tosses, the penny comes up heads; of that fraction, the dime also comes up heads half of the time. Therefore, the joint probability of both coins coming up heads is half of one-half, or $\frac{1}{2} \times \frac{1}{2} = \frac{1}{4}$. So, to find the joint probability of independent events, we multiply the probabilities of the individual events (**Figure 12.9**). How does this method apply to genetics?

To see how joint probability is calculated in genetics problems, let's consider the monohybrid cross. The probabilities of two events are involved: gamete formation and random fertilization.

Calculating the probabilities involved in gamete formation is straightforward. A homozygote can produce only one type of gamete, so, for example, the probability of an *SS* individual producing gametes with the genotype *S* is 1. The heterozygote *Ss* produces *S* gametes with a probability of ½ and *s* gametes with a probability of ½.

Now let's see how the rules of probability might predict the ratio of the $F_2$ progeny of the cross shown in Figure 12.4. These plants are obtained by the self-pollination of $F_1$ plants of genotype *Ss*. The probability that an $F_2$ plant will have the genotype *SS* must be $\frac{1}{2} \times \frac{1}{2} = \frac{1}{4}$, because there is a 50-50 chance that the

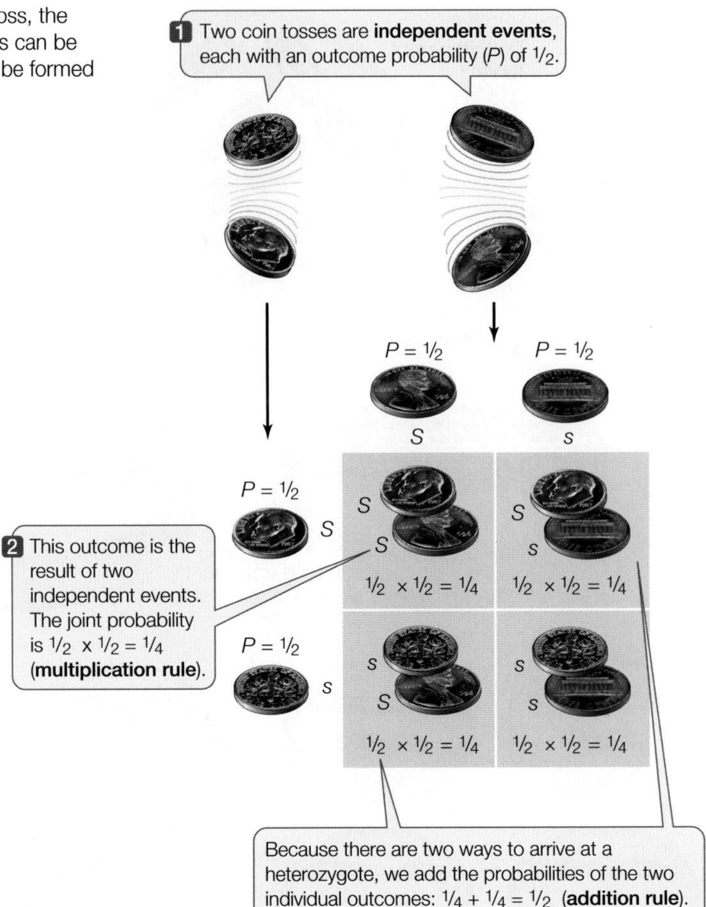
1 Two coin tosses are **independent events**, each with an outcome probability ($P$) of $1/2$.
$P = 1/2$
$P = 1/2$
$S$
$s$
$P = 1/2$
$S$
$P = 1/2$
$s$
$S$ $S$
$S$ $s$
$1/2 \times 1/2 = 1/4$
$1/2 \times 1/2 = 1/4$
$s$ $S$
$s$ $s$
$1/2 \times 1/2 = 1/4$
$1/2 \times 1/2 = 1/4$
2 This outcome is the result of two independent events. The joint probability is $1/2 \times 1/2 = 1/4$ (**multiplication rule**).
Because there are two ways to arrive at a heterozygote, we add the probabilities of the two individual outcomes: $1/4 + 1/4 = 1/2$ (**addition rule**).

**12.9 Using Probability Calculations in Genetics** Like the results of a coin toss, the probability of any given combination of alleles appearing in the offspring of a cross can be obtained by multiplying the probabilities of each event. Since a heterozygote can be formed in two ways, these two probabilities are added together.

sperm will have the genotype $S$, and an independent chance of 50-50 that the egg will have the genotype $S$. Similarly, the probability of $ss$ offspring is also $\frac{1}{2} \times \frac{1}{2} = \frac{1}{4}$.

**ADDING PROBABILITIES** How are probabilities calculated when an event can happen in different ways? The probability of an $F_2$ plant getting an $S$ allele from the sperm and an $s$ allele from the egg is $\frac{1}{4}$. In addition, there is a probability of $\frac{1}{4}$ that the $F_2$ plant will get an $s$ from the sperm and an $S$ from the egg, resulting in the same genotype of $Ss$. The probability of an event that can occur in two or more different ways is the sum of the individual probabilities of those ways. Thus the probability that an $F_2$ plant will be a heterozygote is equal to the sum of the probabilities of the two ways of forming a heterozygote: $\frac{1}{4} + \frac{1}{4} = \frac{1}{2}$ (see Figure 12.9). The three genotypes are therefore expected to occur in the ratio $\frac{1}{4}\ SS : \frac{1}{2}\ Ss : \frac{1}{4}\ ss$, resulting in the 1:2:1 ratio of genotypes and the 3:1 ratio of phenotypes seen in Figure 12.4.

**PROBABILITY AND THE DIHYBRID CROSS** If $F_1$ plants heterozygous for two independent characters self-pollinate, the resulting $F_2$ plants express four different phenotypes. The proportions of these phenotypes are easily determined by probability calculations. Let's see how this works for the experiment shown in Figure 12.7.

Using the principles described above, we can calculate that the probability that an $F_2$ seed will be spherical is $\frac{3}{4}$. This is found by adding the probability of an $Ss$ heterozygote ($\frac{1}{2}$) and the probability of an $SS$ homozygote ($\frac{1}{4}$) = a total of $\frac{3}{4}$. By the same reasoning, the probability that a seed will be yellow is also $\frac{3}{4}$. The two characters are determined by separate genes and are independent of each other, so the joint probability that a seed will be both spherical and yellow is $\frac{3}{4} \times \frac{3}{4} = \frac{9}{16}$. What is the probability of $F_2$ seeds being both wrinkled and yellow? The probability of being yellow is again $\frac{3}{4}$; the probability of being wrinkled is $\frac{1}{2} \times \frac{1}{2} = \frac{1}{4}$. The joint probability that a seed will be *both* wrinkled and yellow is $\frac{1}{4} \times \frac{3}{4} = \frac{3}{16}$.

The same probability applies, for similar reasons, to spherical, green $F_2$ seeds. Finally, the probability that $F_2$ seeds will be both wrinkled and green is $\frac{1}{4} \times \frac{1}{4} = \frac{1}{16}$. Looking at all four phenotypes, we see that they are expected to occur in the ratio of 9:3:3:1.

Probability calculations and Punnett squares give the same results. Learn to do genetics problems both ways, and then decide which method you prefer.

## Mendel's laws can be observed in human pedigrees

How are Mendel's laws of inheritance applied to humans? Mendel worked out his laws by performing many planned crosses and counting many offspring. Neither of these approaches is possible with humans, so human geneticists rely on **pedigrees**: family trees that show the occurrence of phenotypes (and alleles) in several generations of related individuals.

Because humans have such small numbers of offspring, human pedigrees do not show the clear proportions of offspring phenotypes that Mendel saw in his pea plants. For example, when a man and a woman who are both heterozygous for a recessive allele (say, *Aa*) have children together, each child has a 25 percent probability of being a recessive homozygote (*aa*). Thus if this couple were to have dozens of children, one-fourth of them would be recessive homozygotes. But the offspring of a single couple are likely to be too few to show the exact one-fourth proportion. In a family with only two children, for example, both could easily be *aa* (or *Aa*, or *AA*).

What if we want to know whether a recessive allele is carried by both the mother and the father? Human geneticists assume that any allele that causes an abnormal phenotype (such as a genetic disease) is rare in the human population. This means that if some members of a given family have a rare allele, it is highly unlikely that an outsider marrying into that family will have that same rare allele.

Human geneticists may wish to know whether a particular rare allele that causes an abnormal phenotype is dominant or recessive. **Figure 12.10A** is a pedigree showing the pattern of

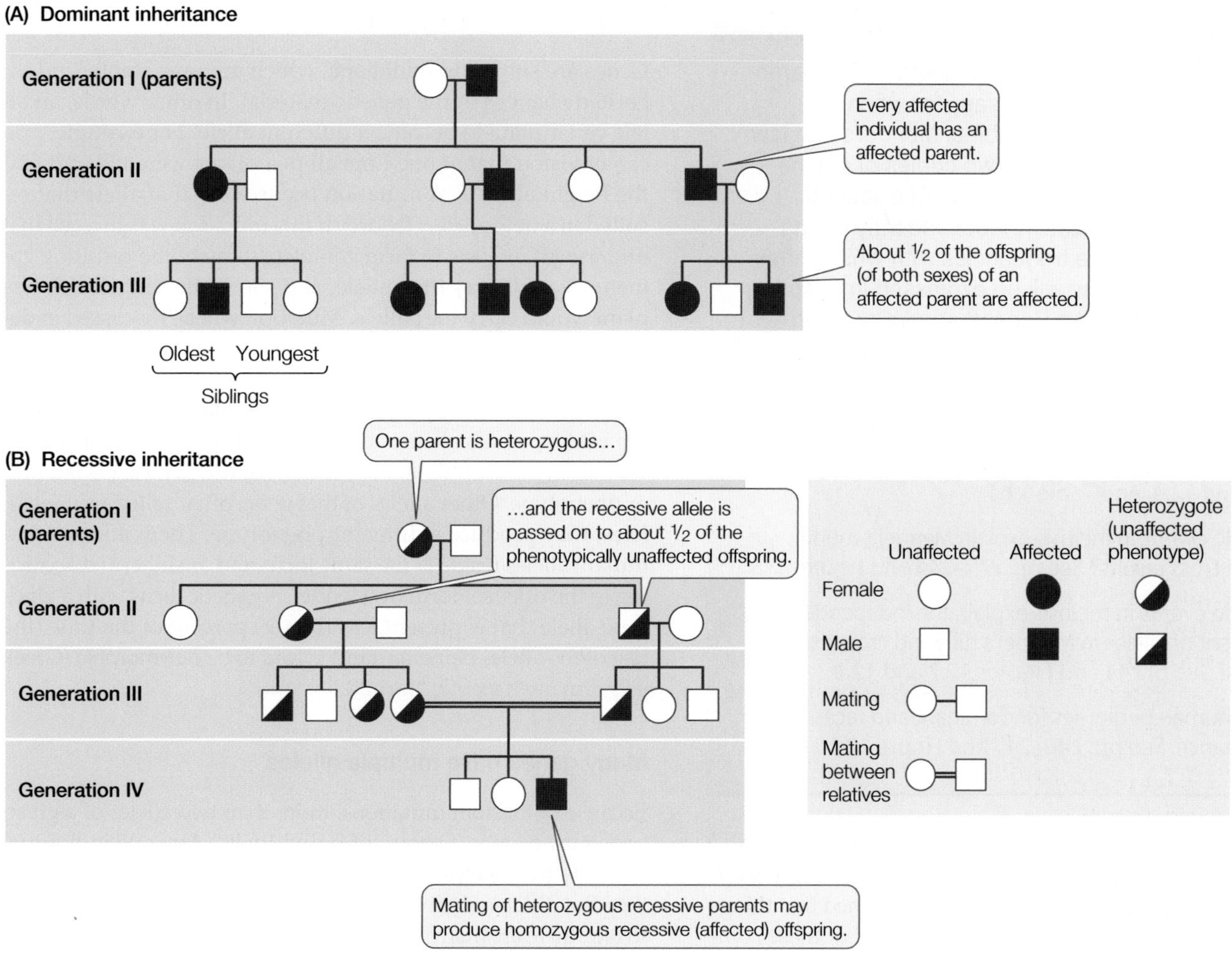

**12.10 Pedigree Analysis and Inheritance** (A) This pedigree represents a family affected by Huntington's disease, which results from a rare dominant allele. Everyone who inherits this allele is affected. (B) The family in this pedigree carries the allele for albinism, a recessive trait. Because the trait is recessive, heterozygotes do not have the albino phenotype, but they can pass the allele on to their offspring. Affected persons must inherit the allele from two heterozygous parents, or (rarely) from one homozygous recessive and one heterozygous parent, or (very rarely) two homozygous recessive parents. In this family, in generation III the heterozygous parents are cousins; however, the same result could occur if the parents were unrelated but heterozygous.

inheritance of a rare dominant allele. The following are the key features to look for in such a pedigree:

- Every affected person has an affected parent.
- About half of the offspring of an affected parent are also affected.
- The phenotype occurs equally in both sexes.

Compare this pattern with the one shown in **Figure 12.10B**, which is typical for the inheritance of a rare recessive allele:

- Affected people usually have two parents who are not affected.
- In affected families, about one-fourth of the children of unaffected parents are affected.
- The phenotype occurs equally in both sexes.

In pedigrees showing inheritance of a recessive phenotype, it is not uncommon to find a marriage of two relatives. This observation is a result of the rarity of recessive alleles that give rise to abnormal phenotypes. For two phenotypically normal parents to have an affected child (*aa*), the parents must both be heterozygous (*Aa*). If a particular recessive allele is rare in the general population, the chance of two people marrying who are both carrying that allele is quite low. On the other hand, if that allele is present in a family, two cousins might share it (see Figure 12.10B). For this reason, studies on populations that are isolated either culturally (by religion, as with the Amish in the United States) or geographically (as on islands) have been extremely valuable to human geneticists. People in these groups are more likely to marry relatives who may carry the same rare recessive alleles.

Because the major use of pedigree analysis is in the clinical evaluation and counseling of patients with inherited abnormalities in their families, a single pair of alleles is usually followed. However, pedigree analysis can also show independent assortment if two different allele pairs are considered.

## 12.1 RECAP

**Mendel showed that genetic determinants are particulate and do not "blend" or disappear when the genes from two gametes combine. Mendel's first law of inheritance states that the two copies of a gene segregate during gamete formation. His second law states that genes assort independently during gamete formation. The frequencies with which different allele combinations will be expressed in offspring can be calculated with a Punnett square or using probability theory.**

- What results seen in the $F_1$ and $F_2$ generations of Mendel's monohybrid cross experiments refuted the blending theory of inheritance? See p. 240, Figures 12.3 and 12.4, and Table 12.1
- How do events in meiosis explain Mendel's monohybrid cross results? See pp. 242–244 and Figure 12.5
- How do events in meiosis explain the independent assortment of alleles in Mendel's dihybrid cross experiments? See p. 244 and Figures 12.7 and 12.8
- Draw human pedigrees for dominant and recessive inheritance. See pp. 246–247 and Figure 12.10

The laws of inheritance as articulated by Mendel remain valid today; his discoveries laid the groundwork for all future studies of genetics. Inevitably, however, we have learned that things are more complicated. Let's take a look at some of these complications, beginning with the interactions between alleles at different loci.

## 12.2 How Do Alleles Interact?

Existing alleles are subject to change, and thus may give rise to new alleles, so there can be many alleles for a single character. In addition, alleles do not always show simple dominant-recessive relationships. Furthermore, a single allele may have multiple phenotypic effects.

### New alleles arise by mutation

Genes are subject to **mutations**, which are rare, stable, and inherited changes in the genetic material. In other words, an allele can mutate to become a different allele. For example, you can envision that at one time all pea plants were tall and had the height allele *T*. A mutation occurred in that allele that resulted in a new allele, *t* (short). If this mutation was in a cell that underwent meiosis to form gametes, some of the resulting gametes would carry the *t* allele, and some offspring of this pea plant would carry the *t* allele. Mutation will be discussed in detail in Chapter 15. By creating variety, mutations are the raw material for evolution.

Geneticists usually define one particular allele of a gene as the **wild type**; this allele is the one that is present in most individuals in nature ("the wild") and gives rise to an expected trait or phenotype. Other alleles of that gene, often called mutant alleles, may produce a different phenotype. The wild-type and mutant alleles reside at the same locus and are inherited according to the rules set forth by Mendel. A genetic locus with a wild-type allele that is present less than 99 percent of the time (the rest of the alleles being mutant) is said to be **polymorphic** (Greek *poly*, "many"; *morph*, "form").

### Many genes have multiple alleles

Because of random mutations, more than two alleles of a given gene may exist in a group of individuals. (Any one individual has only two alleles—one from its mother and one from its father. But different individuals may carry several different alleles.) In fact, there are many examples of such multiple alleles, and they often show a hierarchy of dominance.

Coat color in rabbits, for example, is determined by one gene with four alleles:

- $C$ determines dark gray
- $c$ determines albino
- $c^{ch}$ determines chinchilla
- $c^h$ determines light gray

**12.11 Inheritance of Coat Color in Rabbits** There are four alleles of the gene for coat color in these Netherlands dwarf rabbits. Different combinations of two alleles give different coat colors. The dominance hierarchy is $C > c^{ch} > c^h > c$.

| Possible genotypes | $CC$, $Cc^{ch}$, $Cc^h$, $Cc$ | $c^{ch}c^{ch}$ | $c^{ch}c^h$, $c^{ch}c$ | $c^hc^h$, $c^hc$ | $cc$ |
|---|---|---|---|---|---|
| Phenotype | Dark gray | Chinchilla | Light gray | Point restricted | Albino |

Any rabbit with the *C* allele (paired with any of the four) is dark gray, and a rabbit with *cc* is albino. The intermediate colors result from the different allele combinations shown in **Figure 12.11**.

Multiple alleles increase the number of possible phenotypes. Each of Mendel's monohybrid crosses involved just one pair of alleles (for example, *S* and *s*) and two possible phenotypes (resulting from *SS* or *Ss* and *ss*). The four alleles of the rabbit coat color gene produce five different phenotypes.

### Dominance is not always complete

In the pairs of alleles studied by Mendel, dominance is complete when an individual is heterozygous. That is, an *Ss* individual always expresses the *S* phenotype. However, many genes have alleles that are not dominant or recessive to one another. Instead, the heterozygotes show an intermediate phenotype—at first glance, like that predicted by the old blending theory of inheritance. For example, if a true-breeding red snapdragon is crossed with a true-breeding white one, all the $F_1$ flowers are pink. However, further crosses indicate that this apparent blending phenomenon can still be explained in terms of Mendelian genetics (**Figure 12.12**). The red and white alleles have not disappeared, as those colors reappear when the $F_1$ plants are interbred.

We can understand these results in terms of the Mendelian laws of inheritance. When heterozygotes show a phenotype that is intermediate between those of the two homozygotes, the gene is said to be governed by **incomplete dominance**. In other words, neither of the two alleles is dominant. Incomplete dominance is common in nature, and at the biochemical level, most examples of incomplete dominance are actually codominance (see below). In fact, Mendel's study of seven pea-plant traits is unusual in that all seven traits happened to be characterized by complete dominance.

### In codominance, both alleles at a locus are expressed

Sometimes the two alleles at a locus produce two different phenotypes that *both* appear in heterozygotes, a phenomenon called **codominance**. Note that this is different from incomplete dominance, where the phenotype of a heterozygote is a blend of the phenotypes of the parents. A good example of codominance is seen in the ABO blood group system in humans.

There are numerous glycoproteins on the surfaces of red blood cells and they are all encoded by genes. One genetic locus is called the ABO locus, with three alleles, $I^A$, $I^B$ and $I^O$, that encode variants of a surface glycoprotein designated A, B, and O (the "ABO system"). Since people inherit one allele from each parent, they may have any combination of these alleles: $I^AI^B$, $I^AI^O$, $I^AI^A$, and so on. In terms of gene expression, it is important to note that in a codominant system, all alleles are expressed in a heterozygote. So people with $I^AI^B$ express both $I^A$ and $I^B$ alleles on their red blood cell surfaces.

Early attempts at blood transfusion frequently killed the patient. Around 1900, the Austrian scientist Karl Landsteiner mixed blood cells and serum (blood from which cells have been removed) from different individuals. He found that only certain combinations of blood and serum are compatible. In other combinations, the red blood cells from one individual form clumps in the presence of serum from the other individual. This discovery led to our ability to administer compatible blood transfusions that do not kill the recipient.

Incompatible transfusions result in the formation of clumps because of genetic systems like the ABO locus. People make specific proteins in the serum, called antibodies, that react with foreign, or "nonself," molecules called antigens. The A and B glycoproteins can act as antigens if present on the surfaces of red

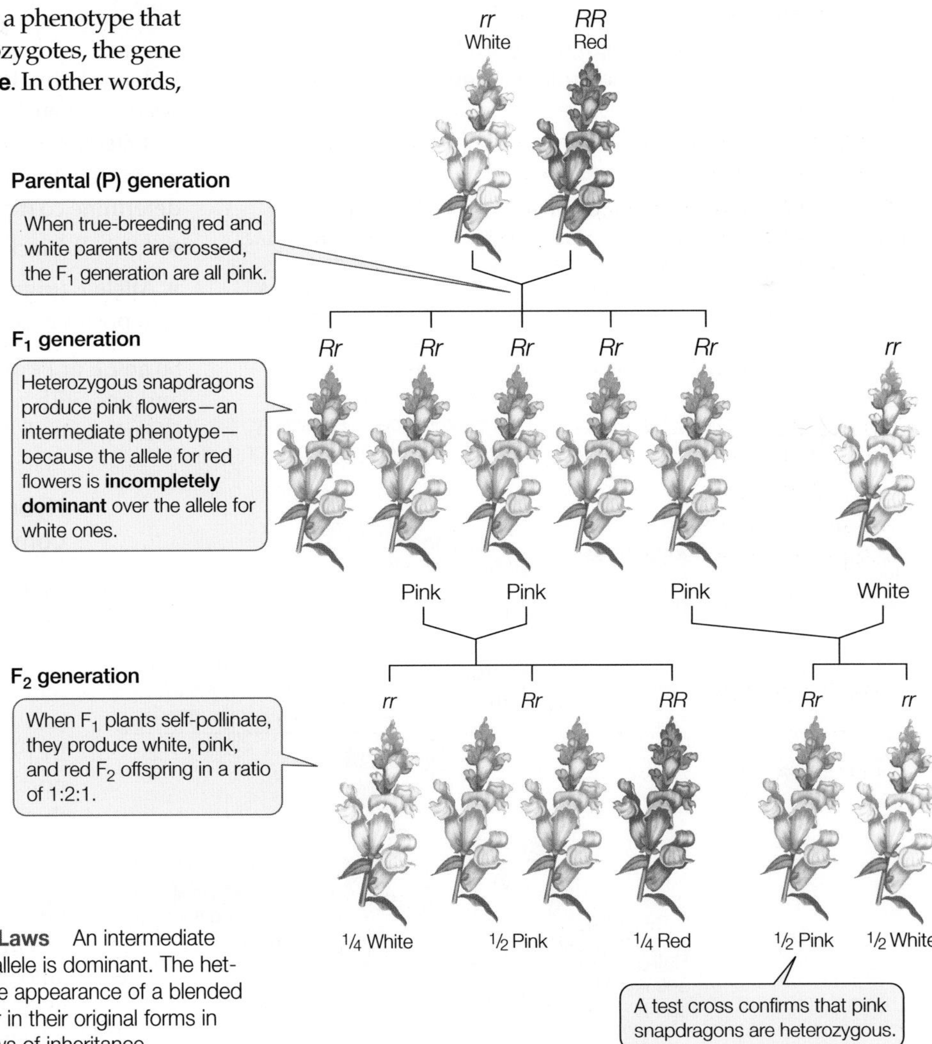

rr
White
RR
Red
Parental (P) generation
When true-breeding red and white parents are crossed, the F1 generation are all pink.
F1 generation
Rr
Rr
Rr
Rr
Rr
rr
Heterozygous snapdragons produce pink flowers—an intermediate phenotype—because the allele for red flowers is incompletely dominant over the allele for white ones.
Pink
Pink
Pink
White
F2 generation
rr
Rr
RR
Rr
rr
When F1 plants self-pollinate, they produce white, pink, and red F2 offspring in a ratio of 1:2:1.
1/4 White
1/2 Pink
1/4 Red
1/2 Pink
1/2 White
A test cross confirms that pink snapdragons are heterozygous.

**12.12 Incomplete Dominance Follows Mendel's Laws** An intermediate phenotype can occur in heterozygotes when neither allele is dominant. The heterozygous phenotype (here, pink flowers) may give the appearance of a blended trait, but the traits of the parental generation reappear in their original forms in succeeding generations, as predicted by Mendel's laws of inheritance.

blood cells in donated blood. If the person receiving the blood does not carry the $I^A$ or $I^B$ alleles, their antibodies will react with the nonself glycoproteins and the red blood cells will form clumps. The O glycoprotein does not act as an antigen. You can see these relationships in **Figure 12.13**. We will learn much more about the functions of antibodies and antigens in Chapter 42.

Interestingly, a recent development may make it possible to circumvent the ABO system of blood incompatibility. Enzymes have been isolated from bacteria that can convert the A and B glycoproteins into O glycoprotein. So blood from any genotype in the ABO system could be treated with these enzymes to make O-type blood, which is not antigenic. Since $I^O$ is not a common allele in most human populations, this technology may be important in overcoming shortages of genetically suitable blood for transfusions.

### Some alleles have multiple phenotypic effects

Mendel's principles were further extended when it was discovered that a single allele can influence more than one phenotype. When a single allele has more than one distinguishable phenotypic effect, we say that the allele is **pleiotropic**. A familiar example of pleiotropy involves the allele responsible for the coloration pattern (light body, darker extremities) of Siamese cats. The same allele is also responsible for the characteristic crossed eyes of Siamese cats. Although these effects appear to be unrelated, both are caused by the protein encoded by this allele.

**12.13 ABO Blood Reactions Are Important in Transfusions** This table shows the results of mixing red blood cells of types A, B, AB, and O with serum containing anti-A or anti-B antibodies. As you look down the columns, note that each of the types, when mixed separately with anti-A and with anti-B, gives a unique pair of results; this is the basic method by which blood is typed. People with type O blood are good blood donors because O cells do not react with either anti-A or anti-B antibodies. People with type AB blood are good recipients, since they make neither type of antibody. When blood transfusions are incompatible, the reaction (clumping of red blood cells) can have severely adverse consequences for the recipient.

| Blood type of cells | Genotype | Antibodies made by body | Reaction to added antibodies: Anti-A | Reaction to added antibodies: Anti-B |
|---|---|---|---|---|
| A | $I^AI^A$ or $I^Ai^O$ | Anti-B | | |
| B | $I^BI^B$ or $I^Bi^O$ | Anti-A | | |
| AB | $I^AI^B$ | Neither anti-A nor anti-B | | |
| O | $i^Oi^O$ | Both anti-A and anti-B | | |

Red blood cells that do not react with antibody remain evenly dispersed.

Red blood cells that react with antibody clump together (speckled appearance).

> ## 12.2 RECAP
>
> **Genes are subject to random mutations that give rise to new alleles; thus many genes have more than two alleles within a population. Dominance is not necessarily an all-or-nothing phenomenon.**
>
> - How does the experiment in Figure 12.12 demonstrate incomplete dominance? See p. 249
> - Explain how blood type AB results from codominance. See pp. 249–250 and Figure 12.13

Thus far we have treated the phenotype of an organism, with respect to a given character, as a simple result of the alleles of a single gene. In many cases, however, several genes interact to determine a phenotype. To complicate things further, the physical environment may interact with the genetic constitution of an individual in determining the phenotype.

## 12.3 How Do Genes Interact?

We have just seen how two alleles of the same gene can interact to produce a phenotype. If you consider most complex phenotypes, such as human height, you will realize that they are influenced by the products of many genes. We now turn to the genetics of such gene interactions.

**Epistasis** occurs when the phenotypic expression of one gene is affected by another gene. For example, two genes (*B* and *E*) determine coat color in Labrador retrievers:

- Allele *B* (black pigment) is dominant to *b* (brown)
- Allele *E* (pigment deposition in hair) is dominant to *e* (no deposition, so hair is yellow)

So an *EE* or *Ee* dog with *BB* or *Bb* is black; one with *bb* is brown; and one with *ee* is yellow regardless of the *Bb* alleles present. Clearly, gene *E* determines the expression of *Bb* (**Figure 12.14**).

### Hybrid vigor results from new gene combinations and interactions

In 1876, Charles Darwin reported that when he crossed two different true-breeding, homozygous genetic strains of corn, the offspring were 25 percent taller than either of the parent strains. Darwin's observation was largely ignored for the next 30 years. In 1908, George Shull "rediscovered" this idea, reporting that not just plant height but the weight of the corn grain produced was dramatically higher in the offspring. Agricultural scientists took note, and Shull's paper had a lasting impact on the field of applied genetics (**Figure 12.15**).

Farmers have known for centuries that matings among close relatives (known as **inbreeding**) can result in offspring of lower quality than matings between unrelated individuals. Agricultural scientists call this *in-*

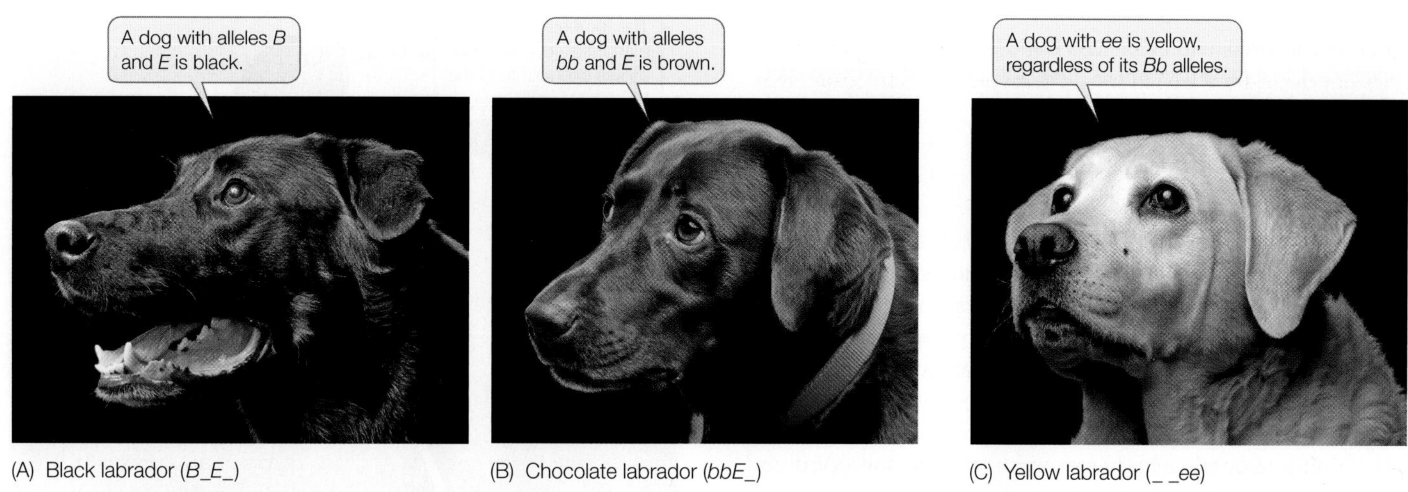

A dog with alleles B and E is black.
A dog with alleles bb and E is brown.
A dog with ee is yellow, regardless of its Bb alleles.
(A) Black labrador (B_E_)
(B) Chocolate labrador (bbE_)
(C) Yellow labrador (_ _ee)


BbEe × BbEe

| Eggs \ Sperm | *BE* | *Be* | *bE* | *be* |
|---|---|---|---|---|
| *BE* | Black *BBEE* | Black *BBEe* | Black *BbEE* | Black *BbEe* |
| *Be* | Black *BBEe* | Yellow *BBee* | Black *BbEe* | Yellow *Bbee* |
| *bE* | Black *BbEE* | Black *BbEe* | Brown *bbEE* | Brown *bbEe* |
| *be* | Black *BbEe* | Yellow *Bbee* | Brown *bbEe* | Yellow *bbee* |

**12.14 Genes May Interact Epistatically** Epistasis occurs when one gene alters the phenotypic effect of another gene. In Labrador retrievers, the *Ee* gene determines the expression of the *Bb* gene.

*breeding depression.* The problems with inbreeding arise because close relatives tend to have the same recessive alleles, some of which may be harmful. The "hybrid vigor" after crossing inbred lines is called **heterosis** (short for heterozygosis). The cultivation of hybrid corn spread rapidly in the United States and all over the world, quadrupling grain production. Unfortunately, as we saw in the opening story, this scientific advance was not universally adopted, and regions such as the Russian empire fell far behind in corn production. The practice of hybridization has spread to many other crops and animals used in agriculture. For example, beef cattle that are crossbred are larger and live longer than cattle bred within their own genetic strain.

The mechanism by which heterosis works is not known. A widely accepted hypothesis is overdominance, in which the heterozygous condition in certain important genes whose products interact is superior to the homozygous condition in either or both genes. Another hypothesis is that the homozygotes have alleles that inhibit growth, and these are less active or absent in the heterozygote.

## The environment affects gene action

The phenotype of an individual does not result from its genotype alone. *Genotype and environment interact to determine the phenotype of an organism.* This is especially important to remember in the era of genome sequencing (see Chapter 17). When the sequence of the human genome was completed in 2003, it was hailed as the "book of life," and public expectations of the benefits gained from this knowledge were (and are) high. But this kind of "genetic determinism" is wrong. Common knowledge tells us that environmental variables such as light, temperature, and nutrition can affect the phenotypic expression of a genotype.

*B73* Hybrid *Mo17*

**12.15 Hybrid Vigor in Corn** Two homozygous parent lines of corn (cobs shown), B73 (left) and Mo17 (right), were crossed to produce the more vigorous hybrid line (center).

The temperature of the extremities is lower and allows expression of the black coat color gene.
The temperature of most of the body is too high for the expression of the black coat color gene.

**12.16 The Environment Influences Gene Expression** This rabbit expresses a coat pattern known as "chocolate point." Its genotype specifies dark fur, but the enzyme for dark fur is inactive at normal body temperature, so only the rabbit's extremities—the coolest regions of the body—express this phenotype.

A familiar example of this phenomenon involves "point restriction" coat patterns found in Siamese cats and certain rabbit breeds (**Figure 12.16**). These animals carry a mutant allele of a gene that controls the growth of black fur all over the body. As a result of this mutation, the enzyme encoded by the gene is inactive at temperatures above a certain point (usually around 35°C). The animals maintain a body temperature above this point, and so their fur is mostly light. However, the extremities—feet, ears, nose, and tail—are cooler, about 25°C, so the fur on these regions is dark. These animals are all white when they are born, because the extremities were kept warm in the mother's womb.

A simple experiment shows that the dark fur is temperature-dependent. If a patch of white fur on a point-restricted rabbit's back is removed and an ice pack is placed on the skin where the patch was, the fur that grows back will be dark. This indicates that although the gene for dark fur was expressed all along, the environment inhibited the activity of the mutant enzyme.

Two parameters describe the effects of genes and environment on the phenotype:

- **Penetrance** is the proportion of individuals in a group with a given genotype that actually show the expected phenotype.
- **Expressivity** is the degree to which a genotype is expressed in an individual.

Penetrance affects, for example, the incidence of Huntington's disease in humans. The disease results from the presence of a dominant allele, but 5 percent of people with the allele do not express the disease. So this allele is said to be 95 percent penetrant. For an example of environmental effects on expressivity, consider how Siamese cats kept indoors or outdoors in different climates might look.

### Most complex phenotypes are determined by multiple genes and the environment

The differences between individual organisms in simple characters, such as those that Mendel studied in pea plants, are discrete and **qualitative**. For example, the individuals in a population of pea plants are either short or tall. For most complex characters, however, such as height in humans, the phenotype varies more or less continuously over a range. Some people are short, others are tall, and many are in between the two extremes. Such variation within a population is called **quantitative**, or continuous, variation (**Figure 12.17**).

**12.17 Quantitative Variation** Quantitative variation is produced by the interaction of genes at multiple loci and the environment. These students (women in white on the left are shorter; men in blue on the right are taller) show continuous variation in height that is the result of interactions between many genes and the environment.

Sometimes this variation is largely genetic. For instance, much of human eye color is the result of a number of genes controlling the synthesis and distribution of dark melanin pigment. Dark eyes have a lot of it, brown eyes less, and green, gray, and blue eyes even less. In the latter cases, the distribution of other pigments in the eye is what determines light reflection and color.

In most cases, however, quantitative variation is due to *both genes and environment*. Height in humans certainly falls into this category. If you look at families, you often see that parents and their offspring all tend to be tall or short. However, nutrition also plays a role in height: American 18-year-olds today are about 20 percent taller than their great-grandparents were at the same age. Three generations are not enough time for mutations that would exert such a dramatic effect to occur, so the height difference must not be due to genetics.

Geneticists call the genes that together determine such complex characters **quantitative trait loci**. Identifying these loci is a major challenge, and an important one. For example, the amount of grain that a variety of rice produces in a growing season is determined by many interacting genetic factors. Crop plant breeders have worked hard to decipher these factors in order to breed higher-yielding rice strains. In a similar way, human characteristics such as disease susceptibility and behavior are caused in part by quantitative trait loci. Recently, one of the many genes involved with human height was identified. The gene, *HMGA2*, has an allele that apparently has the potential to add 4 mm to human height.

## 12.3 RECAP

**In epistasis, one gene affects the expression of another. Perhaps the most challenging problem for genetics is the explanation of complex phenotypes that are caused by many interacting genes and the environment.**

- Explain the difference between penetrance and expressivity. See p. 252
- How is quantitative variation different from qualitative variation? See pp. 252–253

In the next section we'll see how the discovery that genes occupy specific positions on chromosomes enabled Mendel's successors to provide a physical explanation for his model of inheritance, and to provide an explanation for those cases where Mendel's second law does not apply.

# 12.4 What Is the Relationship between Genes and Chromosomes?

There are far more genes than chromosomes. Studies of different genes that are physically linked on the same chromosome reveal inheritance patterns that are not Mendelian. These patterns have been useful not only in detecting linkage of genes, but also in determining how far apart they are from one another on the chromosome.

The organism that revealed genetic linkage is the fruit fly *Drosophila melanogaster*. Its small size, the ease with which it can be bred, and its short generation time make this animal an attractive experimental subject. Beginning in 1909, Thomas Hunt Morgan and his students at Columbia University pioneered the study of *Drosophila*, and it remains a very important organism in studies of genetics.

**yourBioPortal.com**

GO TO **Animated Tutorial 12.2 • Alleles That Do Not Sort Independently**

### Genes on the same chromosome are linked

Some of the crosses Morgan performed with fruit flies yielded phenotypic ratios that were not in accordance with those predicted by Mendel's law of independent assortment. Morgan crossed *Drosophila* with two known genotypes, *BbVgvg* × *bbvgvg*,* for two different characters, body color and wing shape:

- *B* (wild-type gray body), is dominant over *b* (black body)
- *Vg* (wild-type wing) is dominant over *vg* (vestigial, a very small wing)

Morgan expected to see four phenotypes in a ratio of 1:1:1:1, but that is not what he observed. The body color gene and the wing size gene were not assorting independently; rather, they were, for the most part, inherited together (**Figure 12.18**).

These results became understandable to Morgan when he considered the possibility that the two loci are on the same chromosome—that is, that they might be linked. Suppose that the *B* and *Vg* loci are indeed located on the same chromosome. Why didn't all of Morgan's $F_1$ flies have the parental phenotypes—that is, why didn't his cross result in gray flies with normal wings (wild type) and black flies with vestigial wings, in a 1:1 ratio? If linkage were absolute—that is, if chromosomes always remained intact and unchanged—we would expect to see just those two types of progeny. However, this does not always happen.

### Genes can be exchanged between chromatids

**ABSOLUTE LINKAGE IS RARE** If linkage were absolute, Mendel's law of independent assortment would apply only to loci on different chromosomes. What actually happens is more complex, and therefore more interesting. Genes at different loci on the same chromosome *do* sometimes separate from one another during meiosis. Genes may recombine when two homologous chromosomes physically exchange corresponding segments during prophase I of meiosis—that is, by crossing over (**Figure 12.19**; see also Figures 11.18 and 11.19). As described in Section 11.2, DNA is replicated during the S phase, so that by prophase I, when homologous chromosome pairs come together to form tetrads, each chromosome consists of two chromatids.

*Do you recognize this type of cross? It is a test cross for the two gene pairs; see Figure 12.6.

## INVESTIGATING LIFE

**12.18 Some Alleles Do Not Assort Independently**
Morgan's studies showed that the genes for body color and wing size in *Drosophila* are linked, so that their alleles do not assort independently.

**HYPOTHESIS** Alleles for different characteristics always assort independently.

**METHOD**

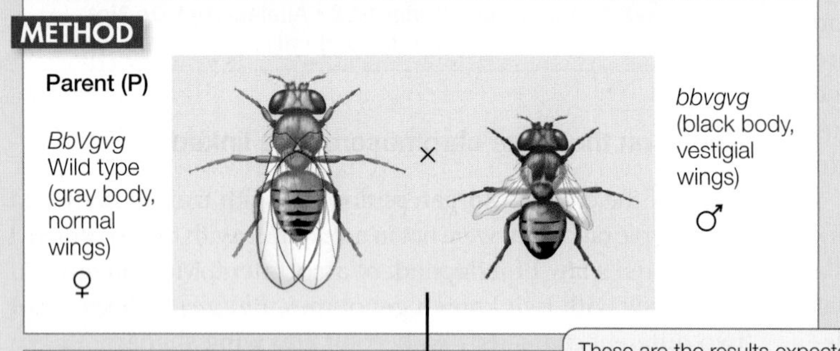


**RESULTS**

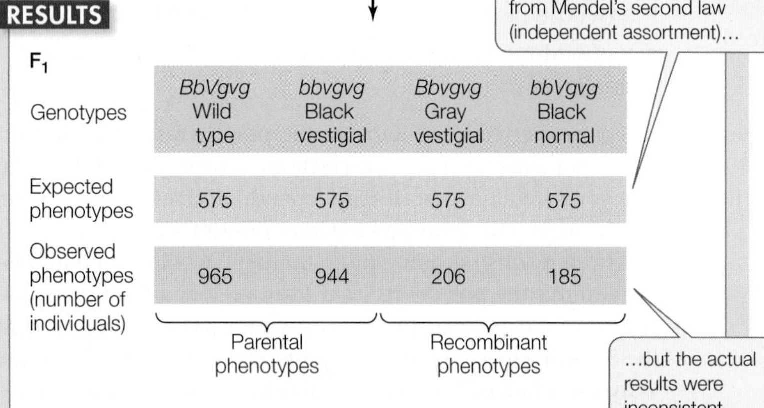

$F_1$

| | | | | |
|---|---|---|---|---|
| Genotypes | *BbVgvg* Wild type | *bbvgvg* Black vestigial | *Bbvgvg* Gray vestigial | *bbVgvg* Black normal |
| Expected phenotypes | 575 | 575 | 575 | 575 |
| Observed phenotypes (number of individuals) | 965 | 944 | 206 | 185 |
| | Parental phenotypes | | Recombinant phenotypes | |



**CONCLUSION** The hypothesis is rejected. These two genes do not assort independently, but are linked (on the same chromosome).

**FURTHER INVESTIGATIONS:** Look again at Mendel's dihybrid cross (see Figure 12.7). If the genes for seed shape and seed color were linked, what would the results be?

Go to **yourBioPortal.com** for original citations, discussions, and relevant links for all INVESTIGATING LIFE figures.

Note that the exchange event involves *only two of the four chromatids* in a tetrad, one from each member of the homologous pair, and can occur at any point along the length of the chromosome. The chromosome segments involved are exchanged reciprocally, so both chromatids involved in crossing over become recombinant (that is, each chromatid ends up with genes from both of the organism's parents). Usually several exchange events occur along the length of each homologous pair.

When crossing over takes place between two linked genes, not all the progeny of a cross have the parental phenotypes. Instead, recombinant offspring appear as well, as they did in Morgan's cross. They appear in proportions called **recombinant frequencies**, which are calculated by dividing the number of recombinant progeny by the total number of progeny (**Figure 12.20**). Recombinant frequencies will be *greater for loci that are farther apart* on the chromosome than for loci that are closer together because an exchange event is more likely to occur between genes that are far apart. Genetic recombination is another way to generate the diversity that is the raw material for natural selection and evolution.

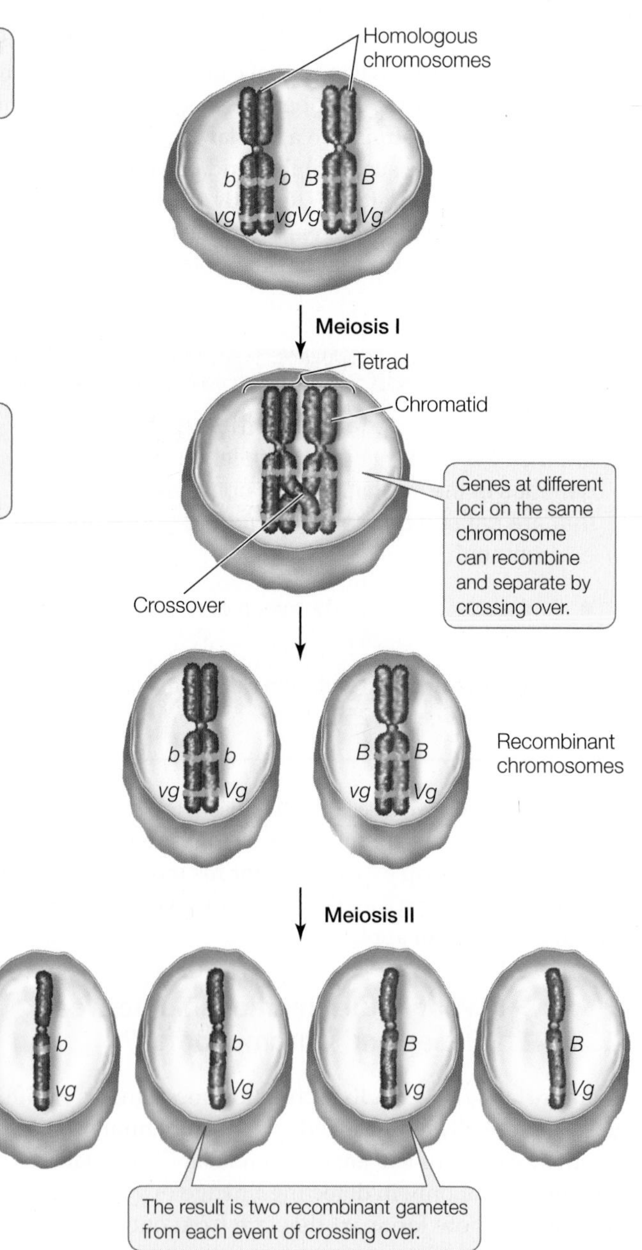


**12.19 Crossing Over Results in Genetic Recombination**
Recombination accounts for why linked alleles are not always inherited together. Alleles at different loci on the same chromosome can be recombined by crossing over, and separated from one another. Such recombination occurs during prophase I of meiosis.

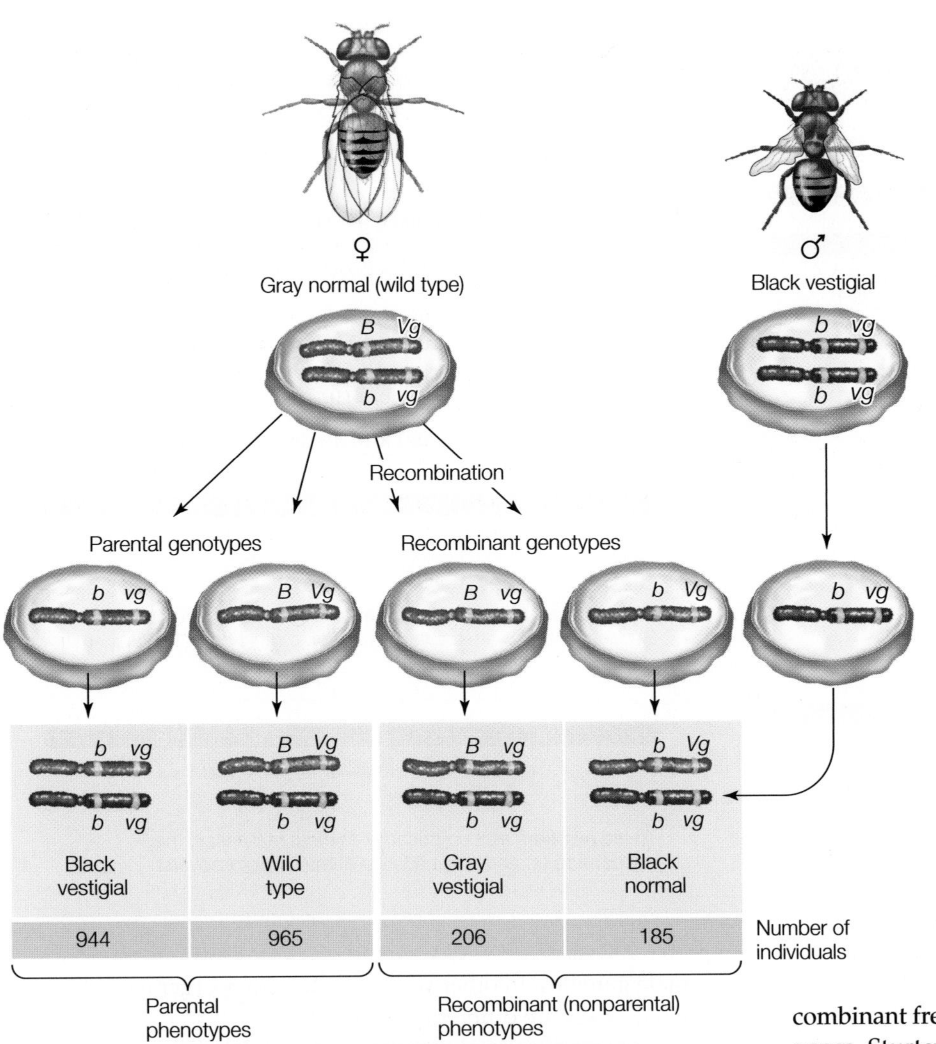


**12.20 Recombinant Frequencies** The frequency of recombinant offspring (those with a phenotype different from either parent) can be calculated.

## Geneticists can make maps of chromosomes

If two loci are very close together on a chromosome, the odds of a crossover occurring between them are small. In contrast, if two loci are far apart, crossing over could occur between them at many points. This pattern is a consequence of the mechanism of crossing over: the farther apart two genes are, the more places there are in the chromosome for breakage and reunion of chromatids to occur. In a population of cells undergoing meiosis, a greater proportion of the cells will undergo recombination between two loci that are far apart than between two loci that are close together. In 1911, Alfred Sturtevant, then an undergraduate student in T. H. Morgan's fly room, realized how this simple insight could be used to show where different genes lie on a chromosome in relation to one another.

The Morgan group had determined recombinant frequencies for many pairs of linked *Drosophila* genes. Sturtevant used those recombinant frequencies to create **genetic maps** that showed the arrangements of genes along the chromosomes (**Figure 12.21**). Ever since Sturtevant demonstrated this method, geneticists have mapped the chromosomes of eukaryotes, prokaryotes, and viruses, assigning distances between genes in **map units**. A map unit corresponds to a recombinant frequency of 0.01; it is also referred to as a **centimorgan (cM)**, in honor of the founder of the fly room. You, too, can work out a genetic map (**Figure 12.22**).

**12.21 Steps toward a Genetic Map** The chance of a crossing over between two loci on a chromosome increases with the distance between the loci. Thus, Sturtevant was able to derive this partial map of a *Drosophila* chromosome using the Morgan group's data on the recombinant frequencies of five recessive traits. He used an arbitrary unit of distance—the map unit, or centimorgan (cM)—equivalent to a recombinant frequency of 0.01.

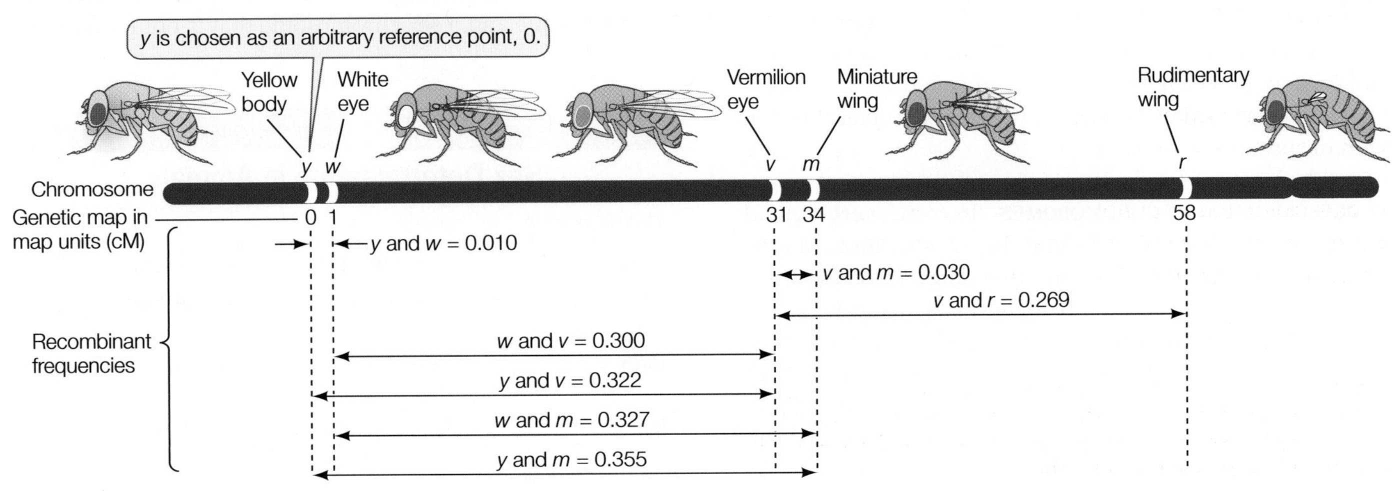

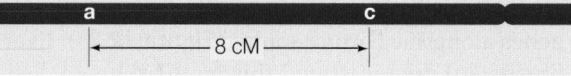

**12.22 Map These Genes** The object of this exercise is to determine the order of three loci (*a*, *b*, and *c*) on a chromosome, as well as the map distances (in cM) between them.

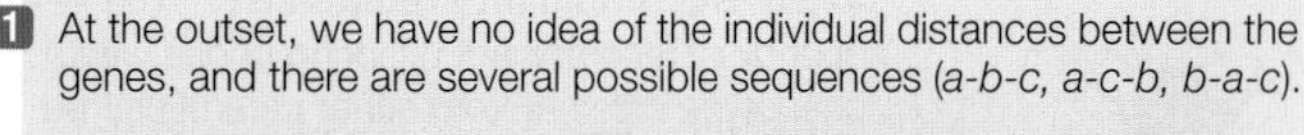

**1** At the outset, we have no idea of the individual distances between the genes, and there are several possible sequences (*a-b-c*, *a-c-b*, *b-a-c*).

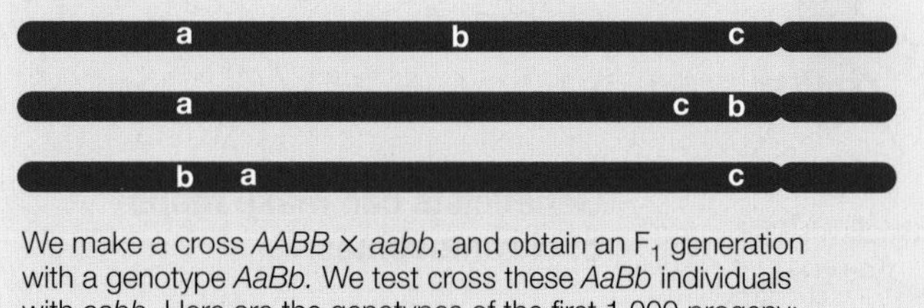


We make a cross *AABB* × *aabb*, and obtain an $F_1$ generation with a genotype *AaBb*. We test cross these *AaBb* individuals with *aabb*. Here are the genotypes of the first 1,000 progeny:

450 *AaBb*, 450 *aabb*, 50 *Aabb*, and 50 *aaBb*.
(parental types) (recombinant types)

**2 How far apart are the *a* and *b* genes?**

What is the recombinant frequency? Which are the recombinant types, and which are the parental types?

Recombinant frequency (*a* to *b*) = (50 + 50)/1,000 = 0.1
So the map distance is

Map distance = 100 × recombinant frequency =
100 × 0.1 = 10 cM

a b
10 cM

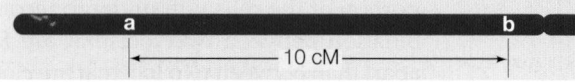

**3 How far apart are the *a* and *c* genes?**

Now we make a cross *AACC* × *aacc*, obtain an $F_1$ generation, and test cross it, obtaining

460 *AaCc*, 460 *aacc*, 40 *Aacc*, and 40 *aaCc*

Recombinant frequency (*a* to *c*) = (40 + 40)/1,000 = 0.08

Map distance = 100 × recombinant frequency =
100 × 0.08 = 8 cM

a c
8 cM

**4 How far apart are the *b* and *c* genes?**

We make a cross *BBCC* × *bbcc*, obtain an $F_1$ generation, and test cross it, obtaining

490 *BbCc*, 490 *bbcc*, 10 *Bbcc*, and 10 *bbCc*

Recombinant frequency (*b* to *c*) = (10 + 10)/1,000 = 0.02

Map distance = 100 × recombinant frequency =
100 × 0.02 = 2 cM

b c
2 cM

**5 Which of the three genes is between the other two?**
Because *a* and *b* are the farthest apart, *c* must be between them.

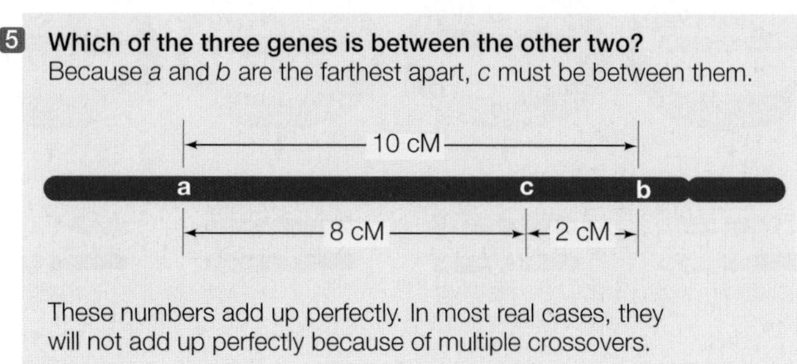


These numbers add up perfectly. In most real cases, they will not add up perfectly because of multiple crossovers.

### Linkage is revealed by studies of the sex chromosomes

In Mendel's work, reciprocal crosses always gave identical results; it did not matter whether a dominant allele was contributed by the mother or by the father. But in some cases, the parental origin of a chromosome does matter. For example, human males inherit a bleeding disorder called hemophilia from their mothers, not from their fathers. To understand the types of inheritance in which the parental origin of an allele is important, we must consider the ways in which sex is determined in different species.

**SEX DETERMINATION BY CHROMOSOMES** In corn, every diploid adult has both male and female reproductive structures. The tissues in these two types of structure are genetically identical, just as roots and leaves are genetically identical. Plants such as corn, in which the same individual produces both male and female gametes, are said to be *monoecious* (Greek, "one house"). Other plants, such as date palms and oak trees, and most animals are *dioecious* ("two houses"), meaning that some individuals can produce only male gametes and the others can produce only female gametes. In other words, in dioecious organisms the different sexes are different individuals.

In most dioecious organisms, sex is determined by differences in the chromosomes, but such determination operates in different ways in different groups of organisms. For example, in many animals including mammals, sex is determined by a single **sex chromosome**, or by a pair of them. Both males and females have two copies of each of the rest of the chromosomes, which are called **autosomes**. In other animals, the chromosomal basis of sex determination is different from that of mammals (**Table 12.2**).

The sex chromosomes of female mammals consist of a pair of X chromosomes. Male mammals, on the other hand, have one X chromosome and a sex chromosome that is not found in fe-

TABLE 12.2

**Sex Determination in Animals**

| ANIMAL GROUP | MECHANISM |
|---|---|
| Bees | Males are diploid, females are haploid |
| Fruit Flies | Fly is female if ratio of sex chromosomes to autosomes is ≥ 1 |
| Birds | Males WW (homogametic), females WZ (heterogametic) |
| Mammals | Males XY (heterogametic), females XX (homogametic) |

males, the Y chromosome. Females may be represented as XX and males as XY.

**MALE MAMMALS PRODUCE TWO KINDS OF GAMETES** Each gamete produced by a male mammal has a complete set of autosomes, but half the gametes carry an X chromosome, and the other half carry a Y. When an X-bearing sperm fertilizes an egg, the resulting XX zygote is female; when a Y-bearing sperm fertilizes an egg, the resulting XY zygote is male.

**SEX CHROMOSOME ABNORMALITIES REVEALED THE GENE THAT DETERMINES SEX** Can we determine which chromosome, X or Y, carries the sex-determining gene, and can the gene be identified? One way to determine cause (e.g., the presence of a gene on the Y chromosome) and effect (e.g., maleness) is to look at cases of biological error, in which the expected outcome does not happen.

Abnormal sex chromosome arrangements resulting from nondisjunction during meiosis (see Section 11.5) tell us something about the functions of the X and Y chromosomes. As you will recall, nondisjunction occurs when a pair of homologous chromosomes (in meiosis I) or sister chromatids (in meiosis II) fail to separate. As a result, a gamete may have one too few or one too many chromosomes. If this gamete fuses with another gamete that has the full haploid chromosome set, the resulting offspring will be aneuploid, with fewer or more chromosomes than normal.

In humans, XO individuals sometimes appear. (The O implies that a chromosome is missing—that is, individuals that are XO have only one sex chromosome.) Human XO individuals are females who are moderately abnormal physically but normal mentally; usually they are also sterile. The XO condition in humans is called Turner syndrome. It is the only known case in which a person can survive with only one member of a chromosome pair (here, the XY pair), although most XO conceptions are spontaneously terminated early in development. XXY individuals also occur; this condition, which affects males, is called Klinefelter syndrome, and results in overlong limbs and sterility.

These observations suggest that the gene controlling maleness is located on the Y chromosome. Observations of people with other types of chromosomal abnormalities helped researchers to pinpoint the location of that gene:

- Some women are genetically XY but lack a small portion of the Y chromosome.
- Some men are genetically XX but have a small piece of the Y chromosome attached to another chromosome.

The Y fragments that are respectively missing and present in these two cases are the same and contain the maleness-determining gene, which was named *SRY* (sex-determining region on the *Y* chromosome).

The *SRY* gene encodes a protein involved in **primary sex determination**—that is, the determination of the kinds of gametes that an individual will produce and the organs that will make them. In the presence of the functional SRY protein, an embryo develops sperm-producing testes. (Notice that *italic type* is used for the name of a gene, but roman type is used for the name of a protein.) If the embryo has no Y chromosome, the *SRY* gene is absent, and thus the SRY protein is not made. In the absence of the SRY protein, the embryo develops egg-producing ovaries. In this case, a gene on the X chromosome called *DAX1* produces an anti-testis factor. So the role of SRY in a male is to inhibit the maleness inhibitor encoded by *DAX1*. The SRY protein does this in male cells, but since it is not present in females, DAX1 can act to inhibit maleness.

Primary sex determination is not the same as **secondary sex determination**, which results in the outward manifestations of maleness and femaleness (such as body type, breast development, body hair, and voice). These outward characteristics are not determined directly by the presence or absence of the Y chromosome. Instead, they are determined by genes that are scattered on the autosomes and the X chromosome. These genes control the actions of hormones, such as testosterone and estrogen.

## Genes on sex chromosomes are inherited in special ways

Genes on sex chromosomes do not show the Mendelian patterns of inheritance. In *Drosophila* and in humans, the Y chromosome carries few known genes, but the X chromosome carries a substantial number of genes that affect a great variety of characters. These genes are present in two copies in females but only one copy in males. Therefore, males are always **hemizygous** for genes on the X chromosome—they have only one copy of each, and it is expressed. So reciprocal crosses do not give identical results for characters whose genes are carried on the sex chromosomes, and these characters do not show the usual Mendelian inheritance ratios.

Eye color in *Drosophila* is a good example of inheritance of a character that is governed by a locus on a sex chromosome (**sex-linked** inheritance). The wild-type eye color of these flies is red. In 1910, Morgan discovered a mutation that causes white eyes. He crossed flies of the wild-type and mutant phenotypes, and demonstrated that the eye color locus is on the X chromosome. If we abbreviate the eye color alleles as *R* (red eyes) and *r* (white eyes), the presence of the alleles on the X chromosome is designated by $X^R$ and $X^r$.

When a homozygous red-eyed female ($X^RX^R$) was crossed with a (hemizygous) white-eyed male ($X^rY$), all the sons and daughters had red eyes, because red (*R*) is dominant over white (*r*) and all the progeny had inherited a wild-type X chromosome ($X^R$) from their mothers (**Figure 12.23A**).

In the reciprocal cross, in which a white-eyed female ($X^rX^r$) was mated with a red-eyed male ($X^RY$), all the sons were white-eyed and all the daughters were red-eyed (**Figure 12.23B**). The sons from the reciprocal cross inherited their only X chromosome from their white-eyed mother; the Y chromosome they inherited from their father did not carry the eye color locus. On the other hand, the daughters got an X chromosome bearing the white allele from their mother and an X chromosome bearing the red allele from their father; therefore they were red-eyed heterozygotes.

When heterozygous females were mated with red-eyed males, half their sons had white eyes, but all their daughters

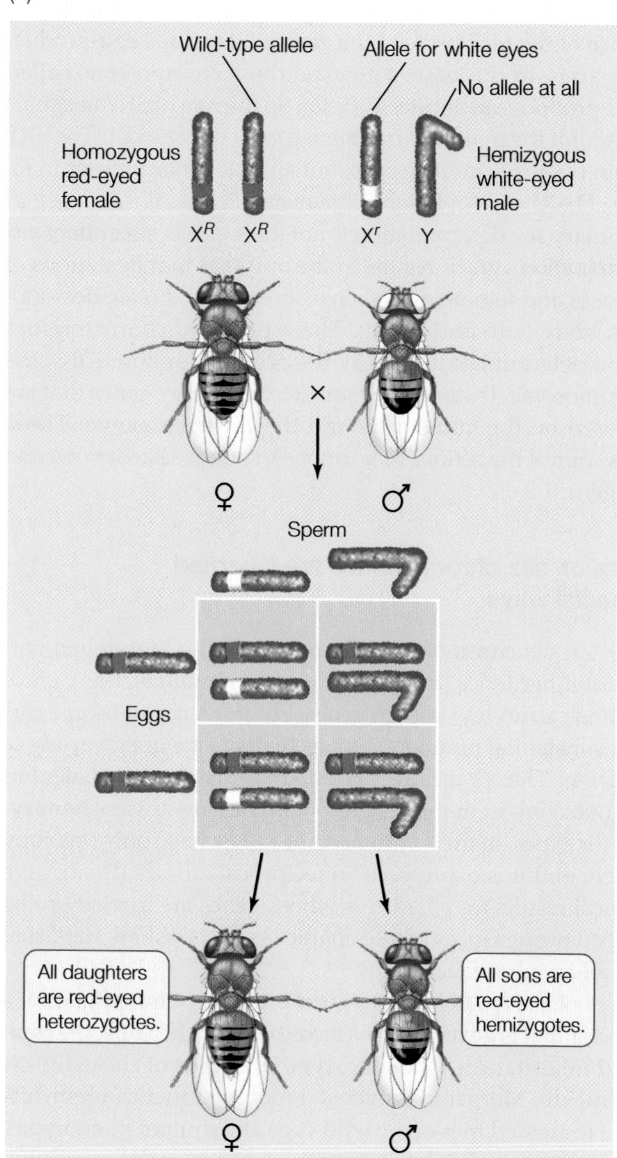


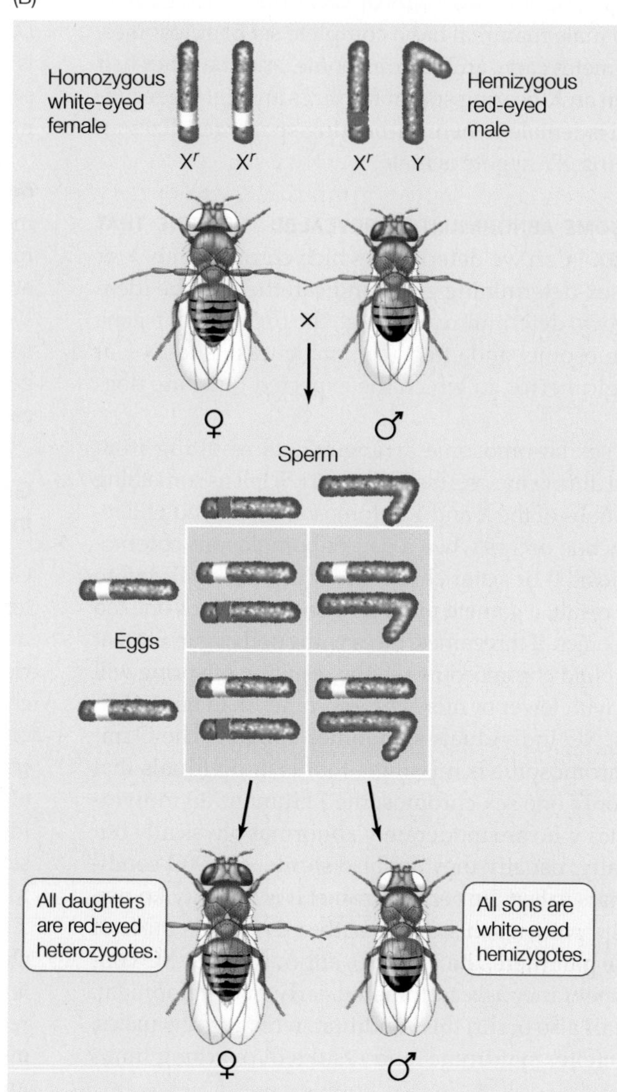


**12.23 Eye Color Is a Sex-Linked Trait in *Drosophila*** Morgan demonstrated that a mutant allele that causes white eyes in *Drosophila* is carried on the X chromosome. Note that in this case, the reciprocal crosses do not have the same results.

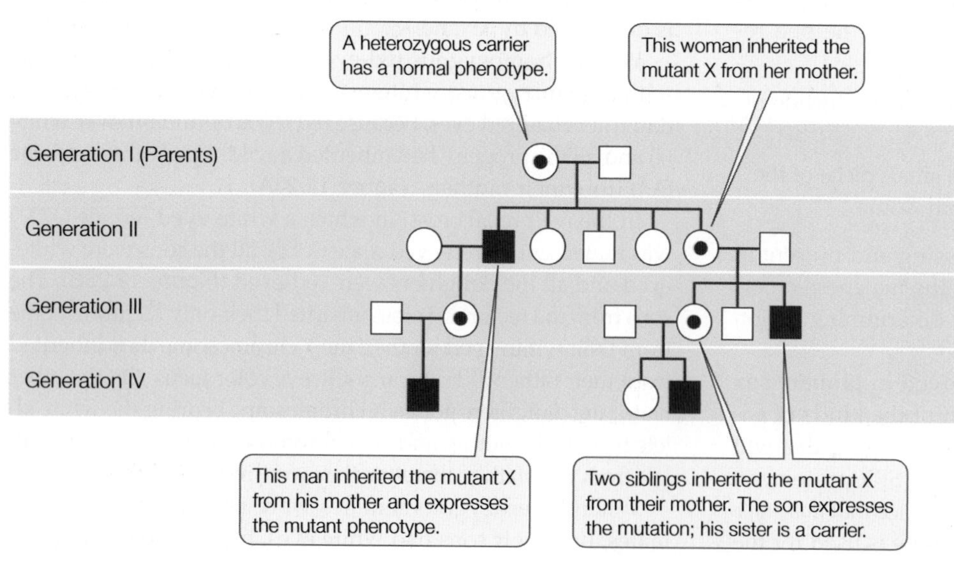


**12.24 Red-Green Color Blindness Is a Sex-Linked Trait in Humans** The mutant allele for red-green color blindness is expressed as an X-linked recessive trait, and therefore is always expressed in males when they carry that allele.

had red eyes. Together, these three results showed that eye color was carried on the X chromosome and not on the Y.

### Humans display many sex-linked characters

The human X chromosome carries about 2,000 known genes. The alleles at these loci follow the same pattern of inheritance as those for eye color in *Drosophila*. For example, one gene on the human X chromosome has a mutant recessive allele that leads to red–green color blindness, and it appears in individuals who are homozygous or hemizygous for the recessive mutant allele.

Pedigree analyses of X-linked recessive phenotypes (like the one in **Figure 12.24**) reveal the following patterns:

- The phenotype appears much more often in males than in females, because only one copy of the rare allele is needed for its expression in males, while two copies must be present in females.
- A male with the mutation can pass it on only to his daughters; all his sons get his Y chromosome.
- Daughters who receive one X-linked mutation are heterozygous **carriers**. They are phenotypically normal, but they can pass the mutant allele to either sons or daughters. (On average only half their children inherit the mutant allele, since half of their X chromosomes carry the normal allele.)
- The mutant phenotype can skip a generation if the mutation passes from a male to his daughter (who will be phenotypically normal) and thus to her son.

The small human Y chromosome carries several dozen genes. Among them is the maleness determinant, *SRY*. Interestingly, for some genes on the Y chromosome there are similar, but not identical, genes on the X chromosome. For example, one of the proteins that make up ribosomes is encoded by a gene on the Y chromosome that is expressed only in male cells, while the X-linked counterpart is expressed in both sexes. This means that there are "male" and "female" ribosomes; the significance of this phenomenon is unknown. Y-linked alleles are passed only from father to son. (Verify this with a Punnett square.)

## 12.4 RECAP

**Simple Mendelian ratios are not observed when genes are linked on the same chromosome. Linkage is indicated by atypical frequencies of phenotypes in the offspring from a test cross. Sex linkage in humans refers to genes on the X chromosome that have no counterpart on the Y chromosome.**

- What is the concept of linkage and what are its implications for the results of genetic crosses? See pp. 253–254 and Figures 12.19 and 12.20
- How does a sex-linked gene behave differently in genetic crosses than a gene on an autosome? See pp. 257–259 and Figure 12.23

The genes we've discussed so far in this chapter are all in the cell nucleus. But other organelles, including mitochondria and plastids, also carry genes. What are they, and how are they inherited?

## 12.5 What Are the Effects of Genes Outside the Nucleus?

The nucleus is not the only organelle in a eukaryotic cell that carries genetic material. As described in Section 5.5, mitochondria and plastids contain small numbers of genes, which are remnants of the entire genomes of colonizing prokaryotes that eventually gave rise to these organelles. For example, in humans, there are about 24,000 genes in the nuclear genome and 37 in the mitochondrial genome. Plastid genomes are about five times larger than those of mitochondria. In any case, several of the genes carried by cytoplasmic organelles are important for organelle assembly and function, so it is not surprising that mutations of these genes can have profound effects on the organism.

The inheritance of organelle genes differs from that of nuclear genes for several reasons:

- In most organisms, mitochondria and plastids are inherited only from the mother. As you will learn in Chapter 43, eggs contain abundant cytoplasm and organelles, but the only part of the sperm that survives to take part in the union of haploid gametes is the nucleus. So you have inherited your mother's mitochondria (with their genes), but not your father's.
- There may be hundreds of mitochondria or plastids in a cell. So a cell is not diploid for organelle genes.
- Organelle genes tend to mutate at much faster rates than nuclear genes, so there are multiple alleles of organelle genes.

The phenotypes resulting from mutations in organelle genes reflect the organelles' roles. For example, in plants and some photosynthetic protists, certain plastid gene mutations affect the proteins that assemble chlorophyll molecules into photosystems. These mutations result in a phenotype that is essentially white instead of green. The inheritance of this phenotype follows a non-Mendelian, maternal pattern (**Figure 12.25**).

Mitochondrial gene mutations that affect one of the complexes in the respiratory chain result in less ATP production. These mutations have particularly noticeable effects in tissues with high energy requirements, such as the nervous system, muscles, and kidneys. In 1995, Greg LeMond, a professional cyclist who had won the famous Tour de France three times, was forced to retire because of muscle weakness caused by a mitochondrial mutation.

## 12.5 RECAP

**Genes in the genomes of organelles, specifically plastids and mitochondria, do not behave in a Mendelian fashion.**

- Why are genes carried in the organelle genomes usually inherited only from the mother?

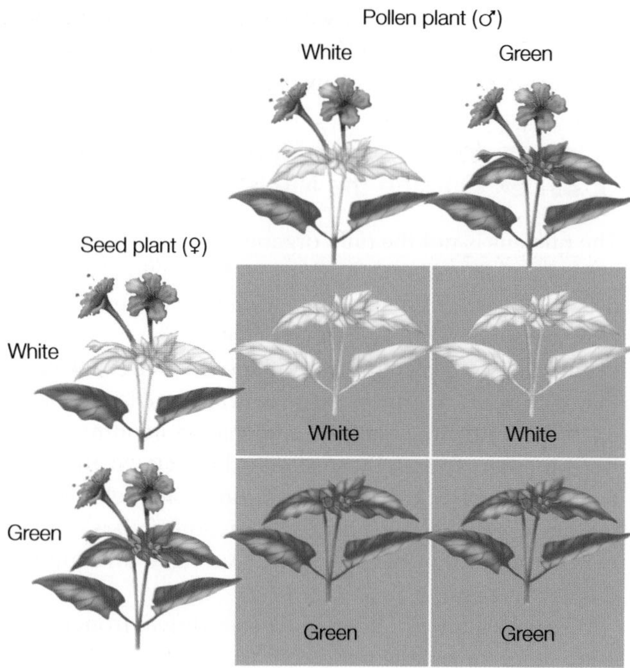


**12.25 Cytoplasmic Inheritance** In four o'clock plants, leaf color is inherited through the female plant only. The white leaf color is caused by a chloroplast mutation that occurs during the life of the parent plant; the leaves that form before the mutation occurs are green. The mutation is passed on to the germ cells, and the offspring that inherit the mutation are entirely white.

Mendel and those who followed him scientifically focused on eukaryotes, with diploid organisms and haploid gametes. A half-century after the rediscovery of Mendel's work, a sexual process that allows genetic recombination was discovered in prokaryotes as well. We now turn to that process.

## 12.6 How Do Prokaryotes Transmit Genes?

As you saw in Chapter 5, prokaryotic cells lack a nucleus but contain their genetic material as mostly a single chromosome in a central region of the cell. In Chapter 11, you saw that bacteria reproduce asexually by cell division, a process that gives rise to virtually genetically identical products. That is, the offspring of cell reproduction in bacteria constitute a clone. However, mutations occur in bacteria just as they do in eukaryotes; the resulting new alleles increase genetic diversity.

You might expect, therefore, that there is no way for individuals of these species to exchange genes, as in sexual reproduction. It turns out, though, that prokaryotes do have a sexual process.

### Bacteria exchange genes by conjugation

The bacterial chromosome, like the bacterial cell, is considerably smaller than its eukaryotic counterpart. In humans, each of the 23 chromosomes in a haploid set may have thousands of linked genes and be a highly compacted linear strand several centimeters in length. In contrast, *E. coli* has a single, circular chromosome that carries a few thousand genes and is only about 1 μm in circumference. Genetic recombination in bacteria occurs after a chromosome is transferred from one cell to another, which brings the chromosomes of two cells into close proximity within a single cell.

Joshua Lederberg and Edward Tatum discovered this recombination process in 1946. They worked with two genetic strains of *E. coli* that had different alleles for each of six genes (each of the genes coded for the synthesis of certain small molecules). Simply put, the two strains had the following genotypes (remember that bacteria are haploid):

*ABCdef* and *abcDEF*

where capital letters stand for wild-type alleles and lower case letters stand for mutant alleles.

When the two strains were grown in the same environment in the laboratory, most of the cells produced clones. That is, almost all of the cells that grew had the original genotypes:

*ABCdef* and *abcDEF*

However, very rarely, Lederberg and Tatum detected bacteria that had the genotype

*ABCDEF*

How could these completely wild-type bacteria have arisen? One possibility was mutation: the *d* allele could have mutated to D, and so on for *e* and *f*. The problem with this explanation was that the probabilities of mutation from *d* to *D*, *e* to *E*, and *f* to *F* were each very low. The probability of all three events occurring in the same cell would be the product of the three individual probabilities—an extremely low number and millions of times lower than the actual rate of appearance of cells with the genotype *ABCDEF*.

Electron microscopy showed how sexual transmission in bacteria might happen, via physical contact between the cells (**Figure 12.26A**). Physical contact is initiated by a thin projection called a **sex pilus** (plural *pili*). Once sex pili bring the two cells together, the actual transfer of the chromosome occurs through a thin cytoplasmic bridge called a **conjugation tube** that forms between the cells.

The chromosome moves in a linear fashion from a donor cell to a recipient cell. Since the bacterial chromosome is circular, it must be made linear (cut) before it can pass through the tube. Contact between the cells is brief—only rarely long enough for the entire donor genome to enter the recipient cell. Therefore, the recipient cell usually receives only a portion of the donor chromosome. There is no reciprocal transfer of a chromosome from the recipient to the donor.

Once the donor chromosome fragment is inside the recipient cell, it can recombine with the recipient cell's chromosome. In much the same way that chromosomes pair up, gene for gene, in prophase I of meiosis, the donor chromosome can line up beside its homologous genes in the recipient, and crossing over can occur. Gene(s) from the donor can become integrated into the genome of the recipient, thus changing the recipient's genetic constitution (**Figure 12.26B**), although only about half the transferred genes become integrated in this way. When the recipient cells proliferate, the donor genes are passed on to all progeny cells.

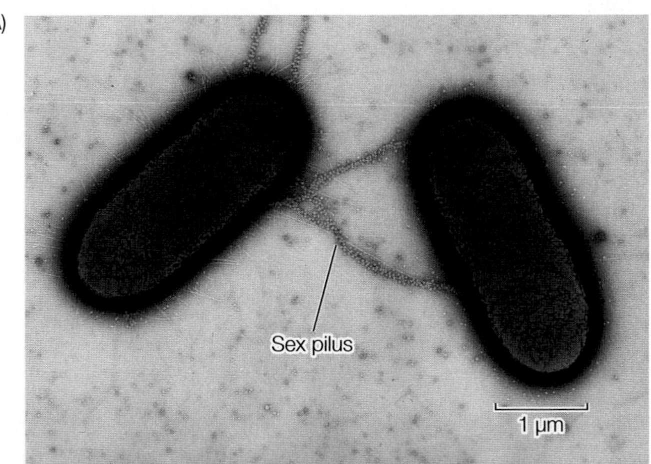


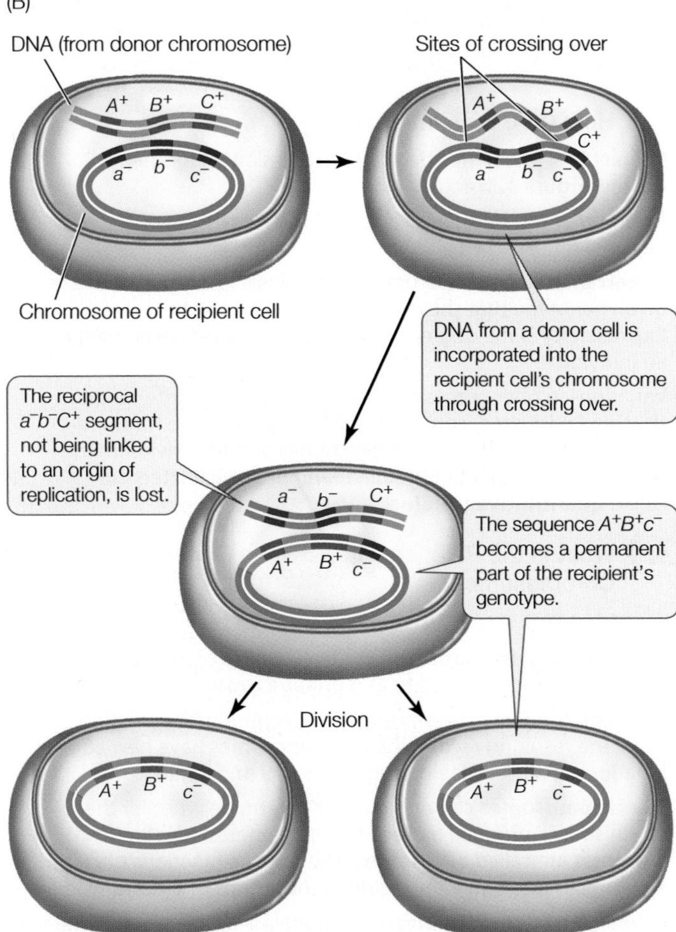


The reciprocal $a^-b^-C^+$ segment, not being linked to an origin of replication, is lost.

**12.26 Bacterial Conjugation and Recombination** (A) Sex pili draw two bacteria into close contact, so that a cytoplasmic conjugation tube can form. DNA is transferred from one cell to the other via the conjugation tube. (B) DNA from a donor cell can become incorporated into a recipient cell's chromosome through crossing over.

## Plasmids transfer genes between bacteria

In addition to their main chromosome, many bacteria harbor additional smaller, circular chromosomes called **plasmids**. They typically contain at most a few dozen genes, which, depending on the particular plasmid, may fall into one of several categories:

- Genes for unusual metabolic capacities, such as the ability to break down hydrocarbons; bacteria carrying these plasmids can be used to clean up oil spills.
- Genes for conjugation, including the ability to make a sex pilus; bacteria carrying this type of plasmid, called fertility factor, are designated $F^+$ and conjugate with bacteria that lack the plasmid ($F^-$).
- Genes for antibiotic resistance; bacteria carrying such gene(s)—the plasmids are called R factors—are a major threat to human health.

Plasmids can move between cells during conjugation, thereby transferring new genes to the recipient bacterium (**Figure 12.27**). Because plasmids can replicate independently of the main chromosome, they do not need to recombine with the main chromosome to add their genes to the recipient cell's genome.

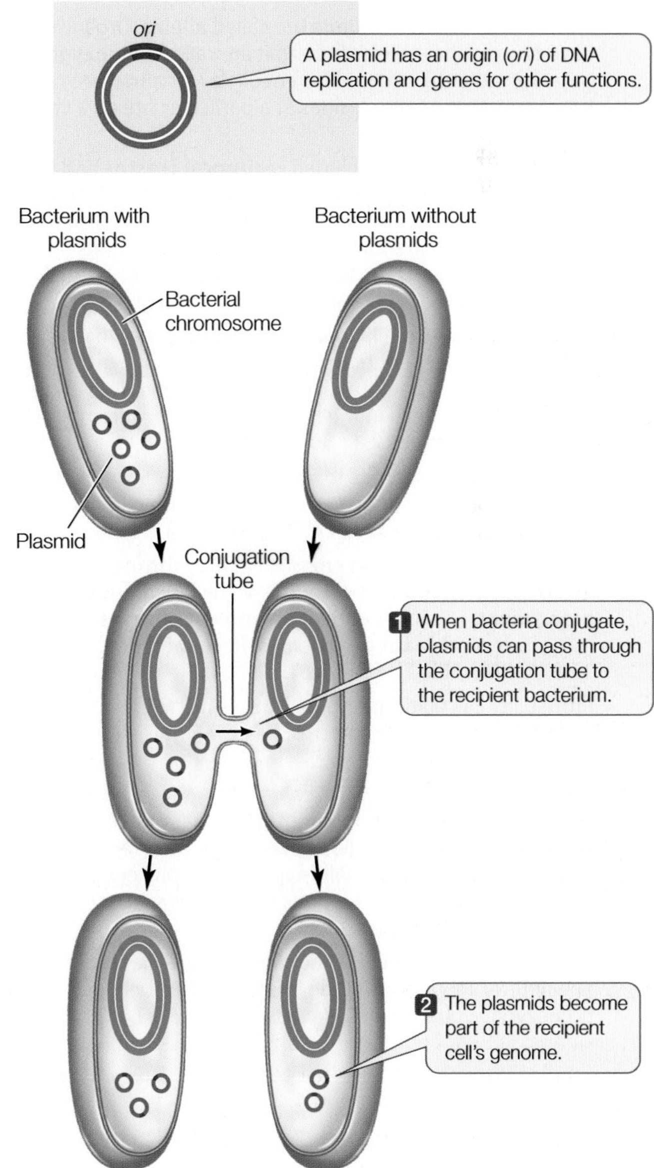

ori
A plasmid has an origin (*ori*) of DNA replication and genes for other functions.
Bacterium with plasmids
Bacterium without plasmids
Bacterial chromosome
Plasmid
Conjugation tube
1 When bacteria conjugate, plasmids can pass through the conjugation tube to the recipient bacterium.
2 The plasmids become part of the recipient cell's genome.

**12.27 Gene Transfer by Plasmids** When plasmids enter a cell via conjugation, their genes can be expressed in the recipient cell.

## 12.6 RECAP

Although they are haploid and reproduce primarily asexually, prokaryotes have the ability to transfer genes from one cell to another. These genes can be part of the main single chromosome or on a small chromosome called a plasmid.

- How were prokaryotic gene transfer and recombination discovered? See p. 260
- What are the differences between recombination after conjugation in prokaryotes and recombination during meiosis in eukaryotes?

## CHAPTER SUMMARY

### 12.1 What Are the Mendelian Laws of Inheritance?

- Physical features of organisms, or **characters**, can exist in different forms, or **traits**. A **heritable trait** is one that can be passed from parent to offspring. A **phenotype** is the physical appearance of an organism; a **genotype** is the genetic constitution of the organism.
- The different forms of a **gene** are called **alleles**. Organisms that have two identical alleles for a trait are called **homozygous**; organisms that have two different alleles for a trait are called **heterozygous**. A gene resides at a particular site on a chromosome called a **locus**.
- Mendel's experiments included **reciprocal crosses** and **monohybrid crosses** between **true-breeding** pea plants. Analysis of his meticulously tabulated data led Mendel to propose a **particulate theory** of inheritance stating that discrete units (now called genes) are responsible for the inheritance of specific traits, to which both parents contribute equally.
- Mendel's first law, the **law of segregation**, states that when any individual produces gametes, the two copies of a gene separate, so that each gamete receives only one member of the pair. Thus every individual in the $F_1$ inherits one copy from each parent. Review Figures 12.4 and 12.5
- Mendel used a **test cross** to find out whether an individual showing a dominant phenotype was homozygous or heterozygous. Review Figure 12.6, **WEB ACTIVITY 12.1**
- Mendel's use of **dihybrid crosses** to study the inheritance of two characters led to his second law: the **law of independent assortment**. The independent assortment of genes in meiosis leads to **recombinant** phenotypes. Review Figures 12.7 and 12.8, **ANIMATED TUTORIAL 12.1**
- Probability calculations and **pedigrees** help geneticists trace Mendelian inheritance patterns. Review Figures 12.9 and 12.10

### 12.2 How Do Alleles Interact?

- New alleles arise by random **mutation**. Many genes have multiple alleles. A **wild-type** allele gives rise to the predominant form of a trait. When the wild-type allele is present at a locus less than 99 percent of the time, the locus is said to be **polymorphic**. Review Figure 12.11
- In **incomplete dominance**, neither of two alleles is dominant. The heterozygous phenotype is intermediate between the homozygous phenotypes. Review Figure 12.12
- **Codominance** exists when two alleles at a locus produce two different phenotypes that both appear in heterozygotes.
- An allele that affects more than one trait is said to be **pleiotropic**.

### 12.3 How Do Genes Interact?

- In **epistasis**, one gene affects the expression of another. Review Figure 12.14
- Environmental conditions can affect the expression of a genotype.
- **Penetrance** is the proportion of individuals in a group with a given genotype that show the expected phenotype. **Expressivity** is the degree to which a genotype is expressed in an individual.
- Variations in phenotypes can be **qualitative** (discrete) or **quantitative** (graduated, continuous). Most quantitative traits are the result of the effects of several genes and the environment. Genes that together determine quantitative characters are called **quantitative trait loci**.

### 12.4 What Is the Relationship between Genes and Chromosomes?

**SEE ANIMATED TUTORIAL 12.2**

- Each chromosome carries many genes. Genes on the same chromosome are referred to as a **linkage group**.
- Genes on the same chromosome can recombine by crossing over. The resulting recombinant chromosomes have new combinations of alleles. Review Figures 12.19 and 12.20
- **Sex chromosomes** carry genes that determine whether the organism will produce male or female gametes. All other chromosomes are called **autosomes**. The specific functions of X and Y chromosomes differ among different groups of organisms.
- **Primary sex determination** in mammals is usually a function of the presence or absence of the *SRY* gene. **Secondary sex determination** results in the outward manifestations of maleness or femaleness.
- In fruit flies and mammals, the X chromosome carries many genes, but the Y chromosome has only a few. Males have only one allele (are **hemizygous**) for X-linked genes, so recessive **sex-linked** mutations are expressed phenotypically more often in males than in females. Females may be unaffected **carriers** of such alleles.

## 12.5 What Are the Effects of Genes Outside the Nucleus?

- Cytoplasmic organelles such as plastids and mitochondria contain small numbers of genes. In many organisms, cytoplasmic genes are inherited only from the mother because the male gamete contributes only its nucleus (i.e., no cytoplasm) to the zygote at fertilization. **Review Figure 12.25**

## 12.6 How Do Prokaryotes Transmit genes?

- Prokaryotes reproduce primarily asexually but can exchange genes in a sexual process called conjugation. **Review Figure 12.26**
- Plasmids are small, extra chromosomes in bacteria that carry genes involved in important metabolic processes and that can be transmitted from one cell to another. **Review Figure 12.27**

**SEE WEB ACTIVITIES 12.2 and 12.3 for a concept review of this chapter.**

# SELF-QUIZ

1. In a simple Mendelian monohybrid cross, true-breeding tall plants are crossed with short plants, and the $F_1$ plants, which are all tall, are allowed to self-pollinate. What fraction of the $F_2$ generation are both tall and heterozygous?
   *a.* 1/8
   *b.* 1/4
   *c.* 1/3
   *d.* 2/3
   *e.* 1/2
2. The phenotype of an individual
   *a.* depends at least in part on the genotype.
   *b.* is either homozygous or heterozygous.
   *c.* determines the genotype.
   *d.* is the genetic constitution of the organism.
   *e.* is either monohybrid or dihybrid.
3. The ABO blood groups in humans are determined by a multiple-allele system in which $I^A$ and $I^B$ are codominant and are both dominant to $I^O$. A newborn infant is type A. The mother is type O. Possible phenotypes of the father are
   *a.* A, B, or AB.
   *b.* A, B, or O.
   *c.* O only.
   *d.* A or AB.
   *e.* A or O.
4. Which statement about an individual that is homozygous for an allele is *not* true?
   *a.* Each of its cells possesses two copies of that allele.
   *b.* Each of its gametes contains one copy of that allele.
   *c.* It is true-breeding with respect to that allele.
   *d.* Its parents were necessarily homozygous for that allele.
   *e.* It can pass that allele to its offspring.
5. Which statement about a test cross is *not* true?
   *a.* It tests whether an unknown individual is homozygous or heterozygous.
   *b.* The test individual is crossed with a homozygous recessive individual.
   *c.* If the test individual is heterozygous, the progeny will have a 1:1 ratio.
   *d.* If the test individual is homozygous, the progeny will have a 3:1 ratio.
   *e.* Test cross results are consistent with Mendel's model of inheritance for unlinked genes.
6. Linked genes
   *a.* must be immediately adjacent to one another on a chromosome.
   *b.* have alleles that assort independently of one another.
   *c.* never show crossing over.
   *d.* are on the same chromosome.
   *e.* always have multiple alleles.
7. In the $F_2$ generation of a dihybrid cross
   *a.* four phenotypes appear in the ratio 9:3:3:1 if the loci are linked.
   *b.* four phenotypes appear in the ratio 9:3:3:1 if the loci are unlinked.
   *c.* two phenotypes appear in the ratio 3:1 if the loci are unlinked.
   *d.* three phenotypes appear in the ratio 1:2:1 if the loci are unlinked.
   *e.* two phenotypes appear in the ratio 1:1 whether or not the loci are linked.
8. The genetic sex of a human is determined by
   *a.* ploidy, with the male being haploid.
   *b.* the Y chromosome.
   *c.* X and Y chromosomes, the male being XX.
   *d.* the number of X chromosomes, the male being XO.
   *e.* Z and W chromosomes, the male being ZZ.
9. In epistasis
   *a.* nothing changes from generation to generation.
   *b.* one gene alters the effect of another.
   *c.* a portion of a chromosome is deleted.
   *d.* a portion of a chromosome is inverted.
   *e.* the behavior of two genes is entirely independent.
10. In humans, spotted teeth are caused by a dominant sex-linked gene. A man with spotted teeth whose father had normal teeth marries a woman with normal teeth. Therefore,
    *a.* all of their daughters will have normal teeth.
    *b.* all of their daughters will have spotted teeth.
    *c.* all of their children will have spotted teeth.
    *d.* half of their sons will have spotted teeth.
    *e.* all of their sons will have spotted teeth.

## GENETICS PROBLEMS

1. In guinea pigs, black body color ($B$) is completely dominant over albino ($b$). For the crosses below, give the genotypes of the parents:

| *Parental phenotypes* | *Black offspring* | *Albino offspring* | *Parental genotypes?* |
|---|---|---|---|
| Black × albino | 12 | 0 | |
| Albino × albino | 0 | 12 | |
| Black × albino | 5 | 7 | |
| Black × black | 9 | 3 | |

2. In the genetic cross, $AaBbCcDdEE \times AaBBCcDdEe$, what fraction of the offspring will be heterozygous for all of these genes ($AaBbCcDdEe$)? Assume all genes are unlinked and the alleles show simple dominance.
3. The pedigree below shows the inheritance of a rare mutant phenotype in humans, congenital cataracts (black symbols).

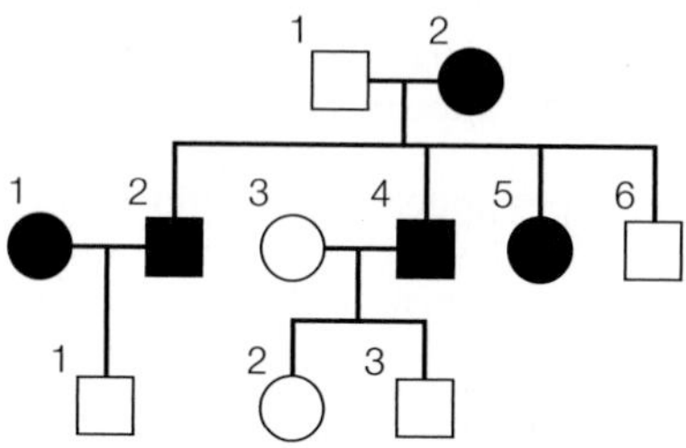

*a.* Are cataracts inherited as an autosomal dominant trait? Autosomal recessive? Sex-linked dominant? Sex-linked recessive?

*b.* Person #5 in the second generation marries a man who does not have cataracts. Two of their four children, a boy and a girl, develop cataracts. What is the chance that their next child will be a girl with cataracts?

4. In cats, black coat ($B$) is codominant with yellow ($b$). The coat color gene is on the X chromosome. Calico cats, which have coats with black and yellow patches, are heterozygous for the coat color alleles.

*a.* Why are most calico cats females?

*b.* A calico female, Pickle, had a litter with one yellow male, two black males, two yellow females and three calico females. What were the genotype and phenotype of the father?

5. In *Drosophila,* three autosomal genes have alleles as follows:
Gray body color ($G$) is dominant over black ($g$)
Full wings ($A$) is dominant over vestigial ($a$)
Red eye ($R$) is dominant over sepia ($r$)

Two crosses were performed, with the following results:

Cross I: Parents: heterozygous red, full × sepia, vestigial
Offspring: 131 red, full
120 sepia, vestigial
122 red, vestigial
127 sepia, full

Cross II: Parents: heterozygous gray, full × black, vestigial
Offspring: 236 gray, full
253 black, vestigial
50 gray, vestigial
61 black, full

Are any of the three genes linked on the same chromosome? If so, what is the map distance between the linked genes?

6. In a particular plant species, two alleles control flower color, which can be yellow, blue, or white. Crosses of these plants produce the following offspring:

| *Parental phenotypes* | *Offspring phenotypes (ratio)* |
|---|---|
| Yellow × yellow | All yellow |
| Blue × yellow | Blue or yellow (1:1) |
| Blue × white | Blue or white (1:1) |
| White × white | All white |

What will be the phenotype, and ratio, of the offspring of a cross of blue × blue?

7. In *Drosophila melanogaster*, the recessive allele $p$, when homozygous, determines pink eyes. $Pp$ or $PP$ results in wild-type eye color. Another gene on a different chromosome has a recessive allele, $sw$, that produces short wings when homozygous. Consider a cross between females of genotype $PPSwSw$ and males of genotype $ppswsw$. Describe the phenotypes and genotypes of the $F_1$ generation and of the $F_2$ generation, produced by allowing the $F_1$ progeny to mate with one another.
8. On the same chromosome of *Drosophila melanogaster* that carries the $p$ (pink eyes) locus, there is another locus that affects the wings. Homozygous recessives, $byby$, have blistery wings, while the dominant allele $By$ produces wild-type wings. The $P$ and $By$ loci are very close together on the chromosome; that is, the two loci are tightly linked. In answering Questions 8*a* and 8*b*, assume that no crossing over occurs, and that the $F_2$ generation is produced by interbreeding the $F_1$ progeny.

*a.* For the cross $PPByBy \times ppbyby$, give the phenotypes and genotypes of the $F_1$ and $F_2$ generations.

*b.* For the cross $PPbyby \times ppByBy$, give the phenotypes and genotypes of the $F_1$ and $F_2$ generations.

*c.* For the cross of Question 8*b*, what further phenotype(s) would appear in the $F_2$ generation if crossing over occurred?

*d.* Draw a nucleus undergoing meiosis at the stage in which the crossing over (Question 8*c*) occurred. In which generation (P, $F_1$, or $F_2$) did this crossing over take place?

9. In chickens, when the dominant alleles of the genes for rose comb ($R$) and pea comb ($A$) are present together ($R\_A\_$), the result is a bird with a walnut comb. Chickens that are homozygous recessive for both genes produce a single comb. A rose-combed bird mated with a walnut-combed bird and produced offspring in the proportion:
3/8 walnut:3/8 rose:1/8 pea:1/8 single
What were the genotypes of the parents?
10. In *Drosophila melanogaster*, white ($w$), eosin ($w^e$), and wild-type red ($w^+$) are multiple alleles at a single locus for eye color. This locus is on the X chromosome. A female that has eosin (pale orange) eyes is crossed with a male that has wild-type eyes. All the female progeny are red-eyed; half the male progeny have eosin eyes, and half have white eyes.

*a.* What is the order of dominance of these alleles?

*b.* What are the genotypes of the parents and progeny?

11. In humans, red–green color blindness is determined by an X-linked recessive allele (*a*), while eye color is determined by an autosomal gene, where brown (*B*) is dominant over blue (*b*).
    *a.* What gametes can be formed with respect to these genes by a heterozygous, brown-eyed, color-blind male?
    *b.* If a blue-eyed mother with normal vision has a brown eyed, color-blind son and a blue-eyed, color-blind daughter, what are the genotypes of both parents and children?
12. If the dominant allele *A* is necessary for hearing in humans, and another allele, *B*, located on a different chromosome, results in deafness no matter what other genes are present, what percentage of the offspring of the marriage of *aaBb* × *Aabb* will be deaf?
13. The disease Leber's optic neuropathy is caused by a mutation in a gene carried on mitochondrial DNA. What would be the phenotype of their first child if a man with this disease married a woman who did not have the disease? What would be the result if the wife had the disease and the husband did not?

## ADDITIONAL INVESTIGATION

Sometimes scientists get lucky. Consider Mendel's dihybrid cross shown in Figure 12.7. Peas have a haploid number of seven chromosomes, so many of their genes are linked. What would Mendel's results have been if the genes for seed color and seed shape were linked with a map distance of ten units? Now, consider Morgan's fruit flies (see Figure 12.21). Suppose that the genes for body color and wing shape were not linked? What results would Morgan have obtained?

## WORKING WITH DATA (GO TO yourBioPortal.com)

**Mendel's Monohybrid Experiments** Mendel's experiments with pea plants (Figure 12.3) laid the foundations of genetics. In this real-world exercise, you will analyze Mendel's data from his published paper and see how he came to his conclusions about the nature of genes.

# 13 DNA and Its Role in Heredity

## A structure for our times

*Jurassic Park*, in both its literary and film incarnations, features a fictional theme park populated with live dinosaurs. In the story, scientists isolate DNA from dinosaur blood extracted from the digestive tracts of fossil insects. The insects supposedly sucked the reptiles' blood right before being preserved in amber (fossilized tree resin). This DNA, according to the novel, could be manipulated to produce living individuals of long-extinct organisms such as velociraptors and the ever-memorable *Tyrannosaurus rex*.

The late Michael Crichton got the idea for his novel from an actual scientific paper in which the authors claimed to have detected reptilian DNA sequences in a fossil insect. Unfortunately, upon additional study, the "preserved" DNA turned out to be a contaminant from modern organisms.

Despite the facts that (1) the preservation of intact DNA over millions of years is highly improbable, and (2) DNA alone cannot generate a new organism, the huge success of Crichton's book brought DNA to the attention of millions. But even before *Jurassic Park*, the DNA double helix was a familiar secular icon.

The double helix first appeared in 1953, in a short paper by James Watson and Francis Crick in the journal *Nature*. An illustration of the molecule's structure drawn by Crick's wife, Odile, accompanied the article, and its simplicity and elegance caught the imagination of the general public as well as the intellect of scientists. As Watson later put it, "A structure this pretty just had to exist."

The double-helical structure of *deoxyribonucleic acid* is perhaps the most widely recognized symbol of modern science, and "DNA" has become part of everyday speech. One sees advertisements for a company whose customers get "into the DNA of business." A digital media software system is called the "DNA Server." A perfume called DNA bills itself as "the essence of life."

Salvador Dali was the first well-known artist to use the DNA double helix in his whimsical creations in 1958. Today, sculptures representing the DNA double helix abound, and it is not only DNA's appearance that stirs our imagination. The DNA nucleotide sequence itself, the "code for life," has inspired unique works of art that incorporate real DNA molecules. A portrait of Sir John Sulston, a Nobel prize-winning geneticist, is made of tiny bacterial colonies, each containing a piece of Sulston's DNA. The Brazilian artist Eduardo Kac translated a sentence from the Bible into Morse code, and from Morse

**Reviving the Velociraptor** Scientists and artists have been creating inanimate reconstructions of dinosaurs for more than 100 years. Michael Crichton's novel *Jurassic Park* was based on the fictional premise that DNA retrieved from fossils could produce living dinosaurs, such as this velociraptor.

**In the Nature of Things** The double helix of DNA has become an iconic symbol of modern science and culture. Artists and designers make use of the widely recognized shape in many ways.

code into a DNA sequence. The sequence was synthesized and incorporated into bacteria. Viewers could turn on an ultraviolet lamp to create mutations in the DNA (and thus in the biblical verse it encoded).

For many people, DNA has come to symbolize the promise and perils of our rapidly expanding knowledge of genetics. Although DNA sequences alone cannot generate a new organism, *biotechnologies* using DNA can modify existing organisms into essentially new organisms. As we will see in Chapter 18, such use of this iconic molecule has generated both excitement and concern about potential risks.

**IN THIS CHAPTER** we will describe the key experiments that led to the identification of DNA as the genetic material. We will then describe the structure of the DNA molecule and how this structure determines its function. We will describe the processes by which DNA is replicated, repaired, and maintained. Finally, we present an important practical application arising from our knowledge of DNA replication: the polymerase chain reaction.

## CHAPTER OUTLINE



## 13.1 What Is the Evidence that the Gene Is DNA?

By the early twentieth century, geneticists had associated the presence of genes with chromosomes. Research began to focus on exactly which chemical component of chromosomes comprised this genetic material.

By the 1920s, scientists knew that chromosomes were made up of DNA and proteins. At this time a new dye was developed by Robert Feulgen that could bind specifically to DNA and that stained cell nuclei red in direct proportion to the amount of DNA present in the cell. This technique provided circumstantial evidence that DNA was the genetic material:

- *It was in the right place.* DNA was confirmed to be an important component of the nucleus and the chromosomes, which were known to carry genes.
- *It varied among species.* When cells from different species were stained with the dye and their color intensity measured, each species appeared to have its own specific amount of nuclear DNA.
- *It was present in the right amounts.* The amount of DNA in somatic cells (body cells not specialized for reproduction) was twice that in reproductive cells (eggs or sperm)—as might be expected for diploid and haploid cells, respectively.

But circumstantial evidence is *not* a scientific demonstration of cause and effect. After all, proteins are also present in cell nuclei. Science relies on experiments to test hypotheses. The convincing demonstration that DNA is the genetic material came from two sets of experiments, one on bacteria and the other on viruses.

### DNA from one type of bacterium genetically transforms another type

The history of biology is filled with incidents in which research on one specific topic has—with or without answering the question originally asked—contributed richly to another, apparently unrelated area. Such a case of serendipity is seen in the work of Frederick Griffith, an English physician.

In the 1920s, Griffith was studying the bacterium *Streptococcus pneumoniae*, or pneumococcus, one of the agents that cause pneumonia in humans. He was trying to develop a vaccine against this devastating illness (antibiotics had not yet been discovered). Griffith was working with two strains of pneumococcus:

- Cells of the S strain produced colonies that looked smooth (S). Covered by a polysaccharide capsule, these cells were

## INVESTIGATING LIFE

**13.1 Genetic Transformation**

Griffith's experiments demonstrated that something in the virulent S strain of pneumococcus could transform nonvirulent R strain bacteria into a lethal form, even when the S strain bacteria had been killed by high temperatures.

**HYPOTHESIS** Material in dead bacterial cells can genetically transform living bacterial cells.

**METHOD**

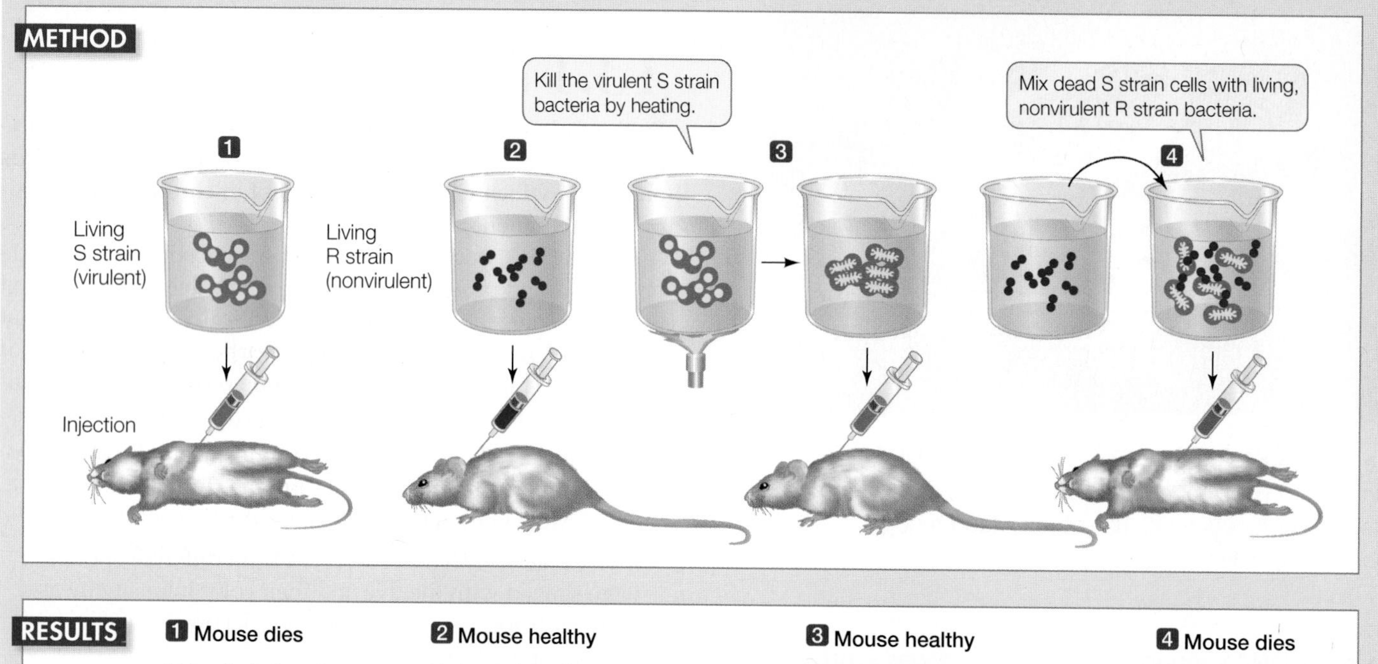
Kill the virulent S strain bacteria by heating.
Mix dead S strain cells with living, nonvirulent R strain bacteria.
1
2
3
4
Living S strain (virulent)
Living R strain (nonvirulent)
Injection

**RESULTS**

**1 Mouse dies**
Living S strain cells found in heart

**2 Mouse healthy**
No bacterial cells found in heart

**3 Mouse healthy**
No bacterial cells found in heart

**4 Mouse dies**
Living S strain cells found in heart

**CONCLUSION** A chemical substance from one cell is capable of genetically transforming another cell.

**FURTHER INVESTIGATION:** How would you show that heat-killed R strain bacteria can transform living S strain bacteria?

Go to **yourBioPortal.com** for original citations, discussions, and relevant links for all INVESTIGATING LIFE figures.

protected from attack by a host's immune system. When S cells were injected into mice, they reproduced and caused pneumonia (the strain was *virulent*).

- Cells of the R strain produced colonies that looked rough (R), lacked the protective capsule, and were not virulent.

Griffith inoculated some mice with heat-killed S-type pneumococcus cells. These heat-killed bacteria did not produce infection. However, when Griffith inoculated other mice with a mixture of living R bacteria and heat-killed S bacteria, to his astonishment, the mice died of pneumonia (**Figure 13.1**). When he examined blood from the hearts of these mice, he found it full of living bacteria—many of them with characteristics of the virulent S strain! Griffith concluded that in the presence of the dead S-type pneumococcus cells, some of the living R-type cells had been transformed into virulent S cells. The fact that these S-type cells reproduced to make more S-type cells showed that the change from R-type to S-type was genetic.

Did this transformation of the bacteria depend on something that happened in the mouse's body? No. It was shown that simply incubating living R and heat-killed S bacteria together in a test tube yielded the same transformation. Years later, another group of scientists discovered that a cell-free extract of heat-killed S cells could also transform R cells. (A cell-free extract contains all the contents of ruptured cells, but no intact cells.) This result demonstrated that some substance—called at the time a chemical **transforming principle**—from the dead S pneumococcus cells could cause a heritable change in the affected R cells. This was an extraordinary discovery: treatment with a chemical substance could permanently change an inherited characteristic. Now it remained to identify the chemical structure of this substance.

INVESTIGATING LIFE

**13.2 Genetic Transformation by DNA**
Experiments by Avery, MacLeod, and McCarty showed that DNA from the virulent S strain of pneumococcus was responsible for the transformation in Griffith's experiments (see Figure 13.1).

**HYPOTHESIS** The chemical nature of the transforming substance from pneumococcus is DNA.

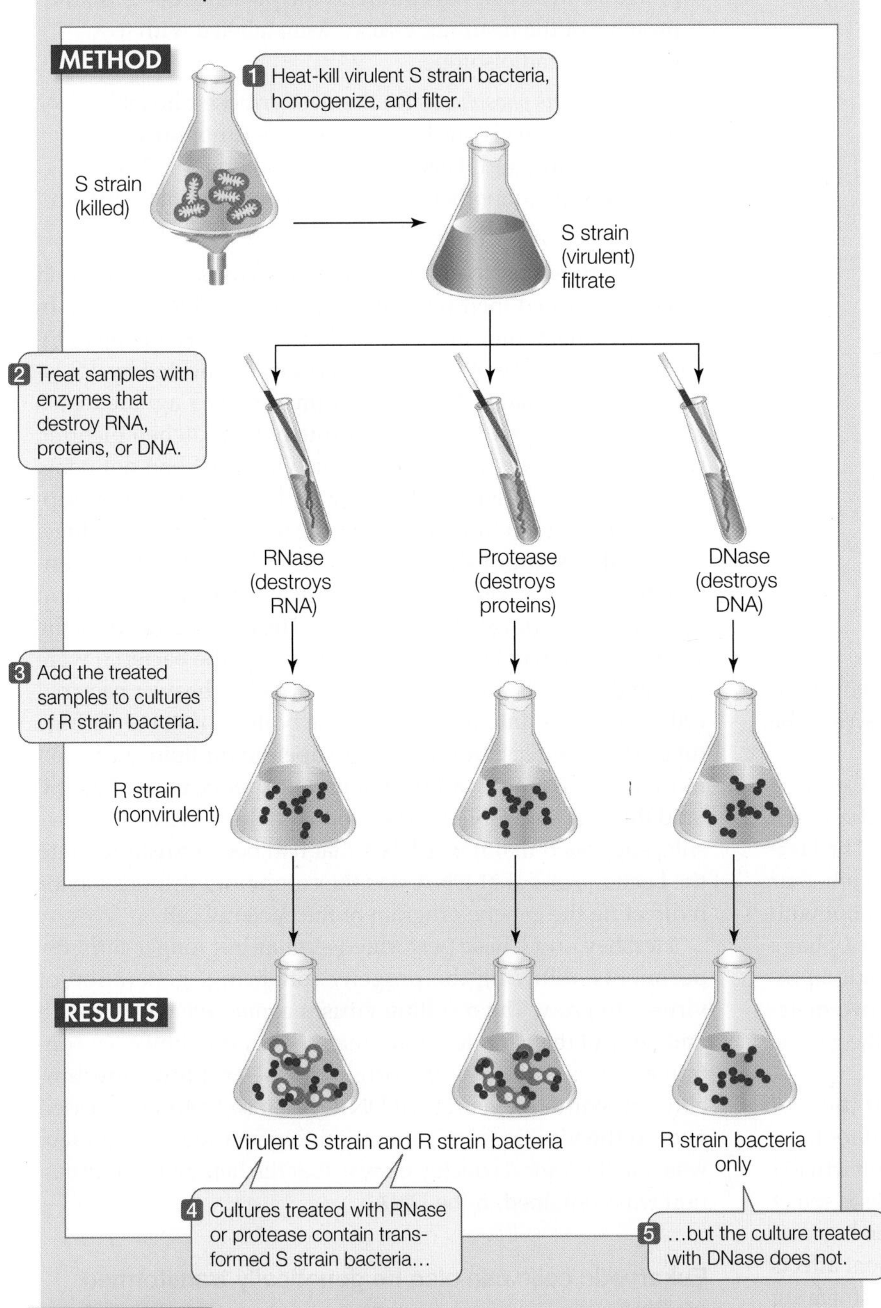
METHOD
1 Heat-kill virulent S strain bacteria, homogenize, and filter.
S strain (killed)
S strain (virulent) filtrate
2 Treat samples with enzymes that destroy RNA, proteins, or DNA.
RNase (destroys RNA)
Protease (destroys proteins)
DNase (destroys DNA)
3 Add the treated samples to cultures of R strain bacteria.
R strain (nonvirulent)
RESULTS
Virulent S strain and R strain bacteria
R strain bacteria only
4 Cultures treated with RNase or protease contain transformed S strain bacteria...
5 ...but the culture treated with DNase does not.

**CONCLUSION** Because only DNase destroyed the transforming substance, the transforming substance is DNA.

Go to **yourBioPortal.com** for original citations, discussions, and relevant links for all INVESTIGATING LIFE figures.

### The transforming principle is DNA

Identifying the transforming principle was a crucial step in the history of biology. Work on identifying the transforming principle was completed by Oswald Avery and his colleagues at what is now The Rockefeller University.

They treated samples known to contain the pneumococcal transforming principle in a variety of ways to destroy different types of molecules—proteins, nucleic acids, carbohydrates, and lipids—and tested the treated samples to see if they had retained the transforming activity. The answer was always the same: if the DNA in the sample was destroyed, transforming activity was lost, but there was no loss of activity when proteins, carbohydrates, or lipids were destroyed (**Figure 13.2**). As a final step, Avery and his colleagues Colin MacLeod and Maclyn McCarty isolated virtually pure DNA from a sample containing pneumococcal-transforming principle, and showed that it caused bacterial transformation. We now know that the gene for the enzyme that catalyzes the synthesis of the polysaccharide capsule, which makes the bacteria look "smooth," was transferred during transformation.

Genetic transformation occurs in nature, although only in certain species of bacteria such as *Pneumococcus*. It does not occur, for example, in *E. coli*. Cells can pick up DNA fragments released into the environment by dead and ruptured cells. Only a small part of the genome is taken up by the transformed cells. Once the new DNA enters the cell, a transforming event very similar to recombination occurs (see Figure 12.26), and new genes can be incorporated into the host chromosome.

The work of Avery's group was a milestone in establishing that DNA is the genetic material in bacterial cells. However, when it was first published in 1944, it had little impact, for two reasons. First, most scientists did not believe that DNA was chemically complex enough to be the genetic material, especially given the much greater chemical complexity of proteins. Second, and perhaps more important, bacterial genetics was a new field of study—it was not yet clear that bacteria even had genes.

### Viral replication experiments confirmed that DNA is the genetic material

The questions about bacteria and other simple organisms were soon resolved, as researchers identified genes and mutations. Bacteria and viruses seemed to undergo genetic processes

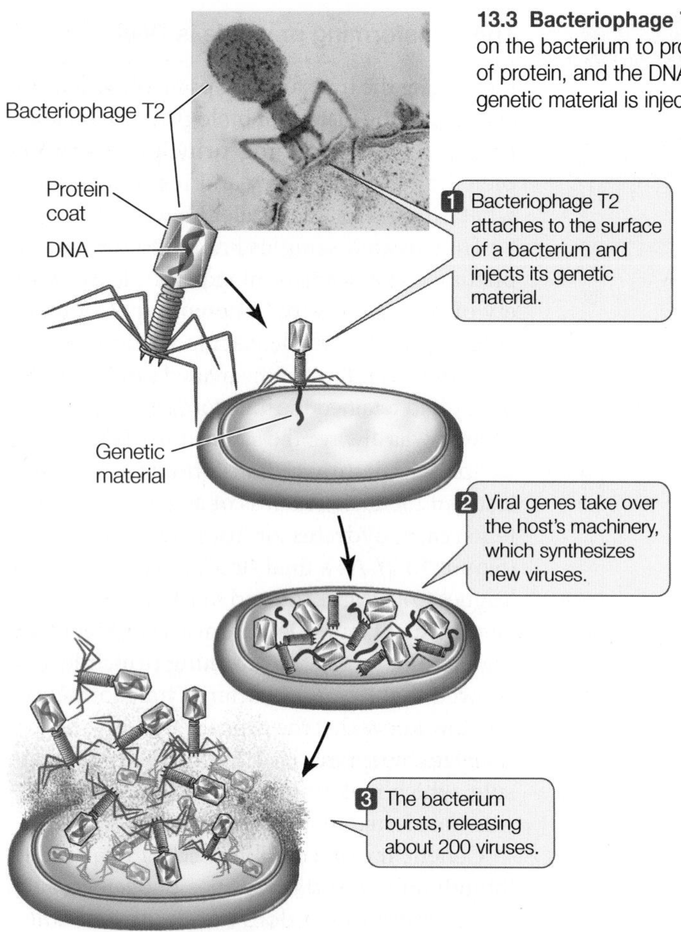

Bacteriophage T2
Protein coat
DNA
Genetic material
1 Bacteriophage T2 attaches to the surface of a bacterium and injects its genetic material.
2 Viral genes take over the host's machinery, which synthesizes new viruses.
3 The bacterium bursts, releasing about 200 viruses.

**13.3 Bacteriophage T2: Reproduction Cycle** Bacteriophage T2 is parasitic on *E. coli*, depending on the bacterium to produce new viruses. The external structures of bacteriophage T2 consist entirely of protein, and the DNA is contained within the protein coat. When the virus infects an *E. coli* cell, its genetic material is injected into the host bacterium.

similar to those in fruit flies and pea plants. Experiments with these relatively simple organisms were designed to discover the nature of the genetic material.

In 1952, Alfred Hershey and Martha Chase of the Carnegie Laboratory of Genetics published a paper that had a much greater immediate impact than Avery's 1944 paper. The Hershey–Chase experiment, which sought to determine whether DNA or protein was the genetic material, was carried out with a virus that infects bacteria. This virus, called bacteriophage T2, consists of little more than a DNA core packed inside a protein coat (**Figure 13.3**). Thus the virus is made of the two materials that were, at the time, the leading candidates for the genetic material.

When bacteriophage T2 attacks a bacterium, part (but not all) of the virus enters the bacterial cell. About 20 minutes later, the cell bursts, releasing dozens of particles that are virtually identical to the infecting virus particle. Clearly the virus is somehow able to replicate itself inside the bacterium. Hershey and Chase deduced that the entry of some viral component affects the genetic program of the host bacterial cell, transforming it into a bacteriophage factory. They set out to determine which part of the virus—protein or DNA—enters the bacterial cell. To trace the two components of the virus over its life cycle, Hershey and Chase labeled each component with a specific radioisotope:

- *Proteins contain some sulfur* (in the amino acids cysteine and methionine). Sulfur is not present in DNA, and has a radioactive isotope, $^{35}S$. Hershey and Chase grew bacteriophage T2 in a bacterial culture in the presence of $^{35}S$, so the proteins of the resulting viruses were labeled with (contained) the radioisotope.
- *DNA contains phosphorus* (in the deoxyribose-phosphate backbone—see Figure 4.2). Phosphorus is not present in most proteins, and it also has a radioisotope, $^{32}P$. The researchers grew another batch of T2 in a bacterial culture in the presence of $^{32}P$, thus labeling the viral DNA with $^{32}P$.

Using these radioactively labeled viruses, Hershey and Chase performed their revealing experiments (**Figure 13.4**). In one experiment, they allowed $^{32}P$-labeled bacteriophage to infect bacteria; in the other, the bacteria were infected by $^{35}S$-labeled bacteriophage. After a few minutes, they agitated each mixture of infected bacteria vigorously in a kitchen blender, which stripped away the parts of the virus that had not penetrated the bacteria, without bursting the bacteria. Then they separated the bacteria from the rest of the material in a centrifuge.

Spinning solutions or suspensions at high speed in a centrifuge causes the solutes and/or particles to separate and form a gradient according to their densities. The lighter remains of the viruses (those parts that had not penetrated the bacteria) were captured in the "supernatant" fluid, while the heavier bacterial cells segregated into a "pellet" in the bottom of the centrifuge tube. The scientists found that the supernatant fluid contained most of the $^{35}S$ (and thus the viral protein), while most of the $^{32}P$ (and thus the viral DNA) had stayed with the bacteria. These results suggested that it was DNA that had been transferred into the bacteria, and that DNA was the compound responsible for redirecting the genetic program of the bacterial cell.

Hershey and Chase performed similar but longer-term experiments, allowing the progeny (offspring) generation of viruses to grow. The resulting viruses contained almost no $^{35}S$ and none of the parental viral protein. They did, however, contain about one-third of the original $^{32}P$—and thus, presumably, one-third of the original DNA. Because DNA was carried over in the viruses from generation to generation but protein was not, the logical conclusion was that the hereditary information was contained in the DNA.

### Eukaryotic cells can also be genetically transformed by DNA

With the publication of the evidence for DNA as the genetic material in bacteria and viruses, the question arose as to whether DNA was also the genetic material in complex eukaryotes. Some dubious experimental results were reported. For example, a

## INVESTIGATING LIFE

### 13.4 The Hershey–Chase Experiment

When bacterial cells were infected with radioactively labeled T2 bacteriophage, only labeled DNA was found in the bacteria. After centrifuging the culture to make the bacteria form a pellet, the labeled protein remained in the supernatant. This showed that DNA, not protein, is the genetic material.

**HYPOTHESIS** **Either component of a bacteriophage—DNA or protein—might be the hereditary material that enters a bacterial cell to direct the assembly of new viruses.**

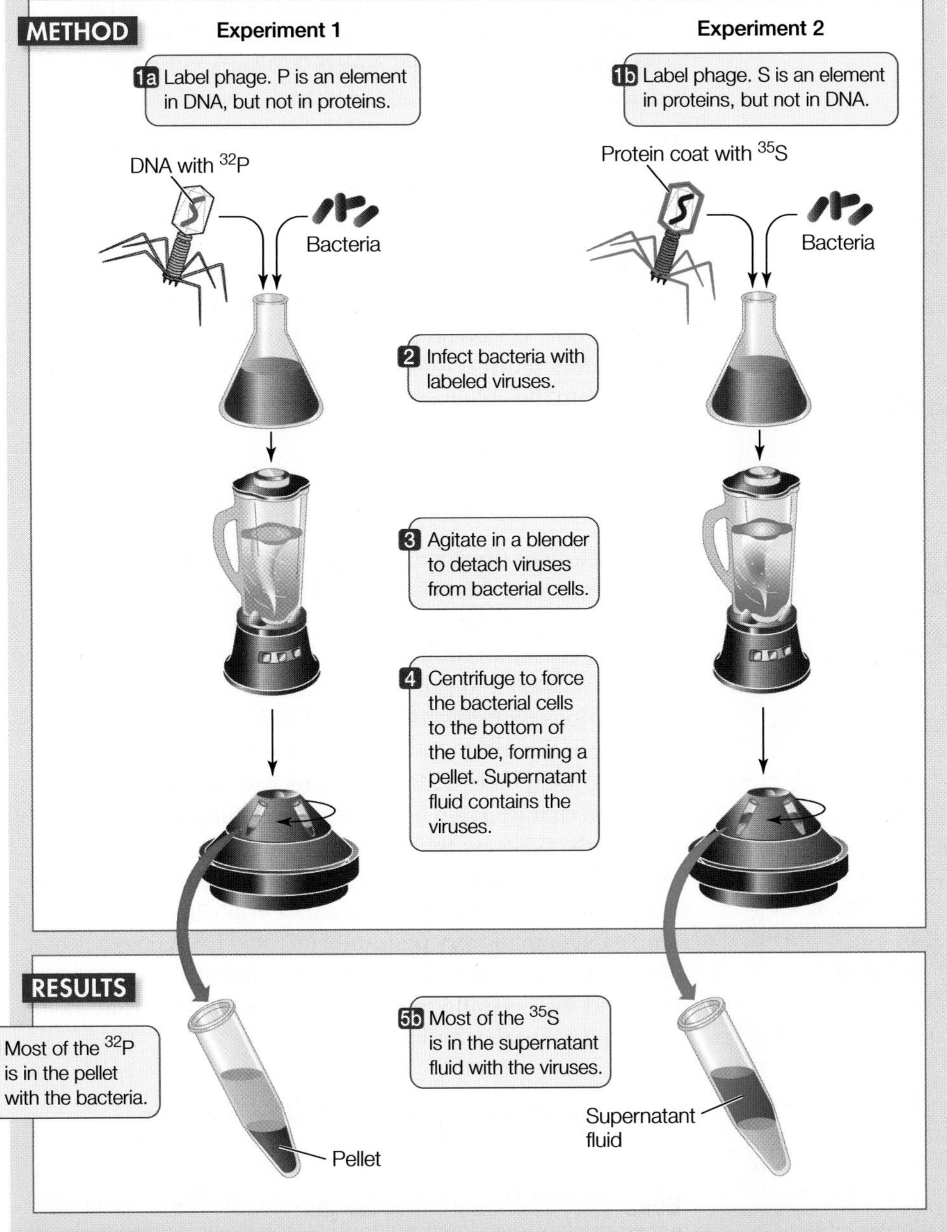

METHOD
Experiment 1
Experiment 2
1a Label phage. P is an element in DNA, but not in proteins.
1b Label phage. S is an element in proteins, but not in DNA.
DNA with 32P
Protein coat with 35S
Bacteria
Bacteria
2 Infect bacteria with labeled viruses.
3 Agitate in a blender to detach viruses from bacterial cells.
4 Centrifuge to force the bacterial cells to the bottom of the tube, forming a pellet. Supernatant fluid contains the viruses.
RESULTS
5a Most of the 32P is in the pellet with the bacteria.
5b Most of the 35S is in the supernatant fluid with the viruses.
Pellet
Supernatant fluid

**CONCLUSION** **DNA, not protein, enters bacterial cells and directs the assembly of new viruses.**

Go to **yourBioPortal.com** for original citations, discussions, and relevant links for all INVESTIGATING LIFE figures.

white duck was injected with DNA from a brown duck and the recipient was reported to turn brown. In another example, flatworms were fed DNA from worms that had learned a simple task, and the recipient worms were reported to immediately get smarter. However, no one could duplicate these results. This episode underscores a central aspect of experimental biology: that published research should be repeated with the same results before the conclusions can be considered valid.

It would be impossible for a large molecule such as DNA to avoid hydrolysis into nucleotides in the digestive system, let alone get into all the cells of the body, after being ingested by an animal. However, genetic transformation of eukaryotic cells by DNA (called **transfection**) can be demonstrated. The key is to use a **genetic marker**, a gene whose presence in the recipient cell confers an observable phenotype. In the experiments with pneumococcus, these phenotypes were the smooth polysaccharide capsule and virulence. In eukaryotes, researchers usually use a nutritional or antibiotic resistance marker gene that permits the growth of transformed recipient cells but not of nontransformed cells. For example, thymidine kinase is an enzyme needed to make use of thymidine in the synthesis of deoxythymidine triphosphate (dTTP), one of the four deoxyribonucleoside triphosphates used in the synthesis of DNA. Mammalian cells that lack the gene for thymidine kinase cannot grow in a medium that contains thymidine as the only source for dTTP synthesis. When DNA containing the marker gene encoding thymidine kinase is added to a culture of mammalian cells lacking this gene, some cells will grow in the thymidine medium, demonstrating that they have been transfected with the gene (**Figure 13.5**). Any cell can be transfected in this way, even an egg cell. In this case, a whole new genetically transformed organism can result; such an organism is referred to as *transgenic*. Transformation in eukaryotes is the final line of evidence for DNA as the genetic material.

INVESTIGATING LIFE

**13.5 Transfection in Eukaryotic Cells**
The use of a marker gene shows that mammalian cells can be genetically transformed by DNA. Usually, the marker gene is carried by a larger molecule (a virus or a small chromosome).

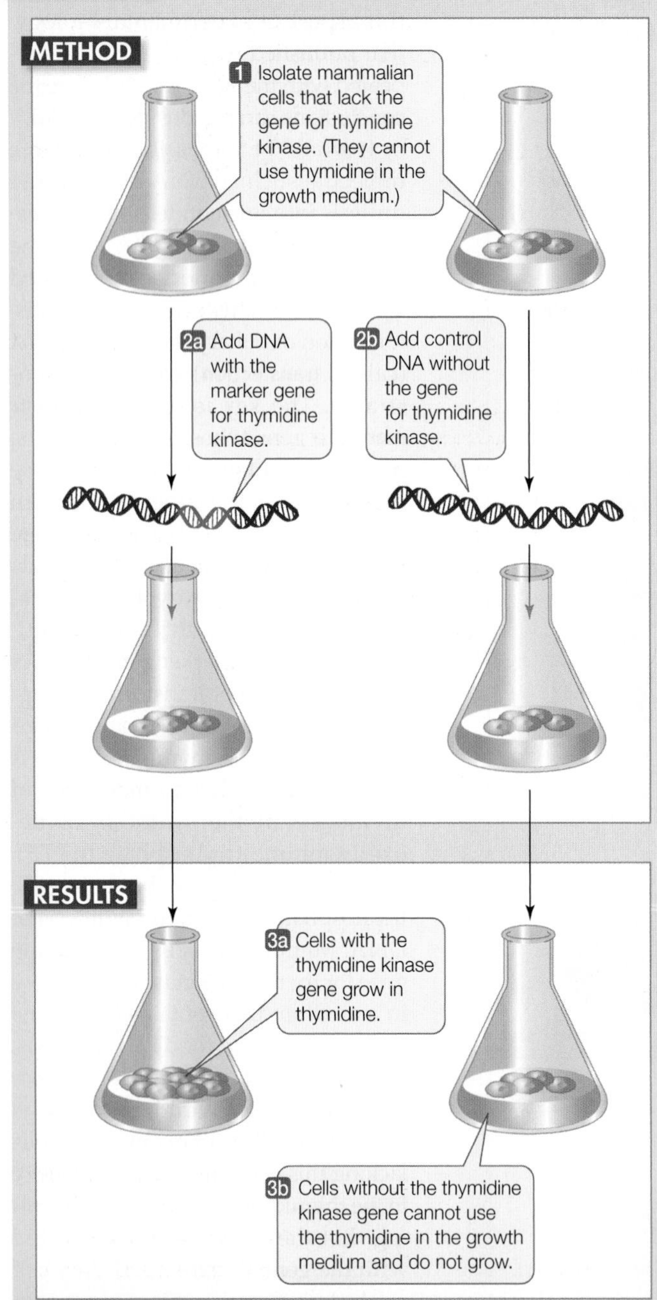
HYPOTHESIS DNA can transform eukaryotic cells.
METHOD
1 Isolate mammalian cells that lack the gene for thymidine kinase. (They cannot use thymidine in the growth medium.)
2a Add DNA with the marker gene for thymidine kinase.
2b Add control DNA without the gene for thymidine kinase.
RESULTS
3a Cells with the thymidine kinase gene grow in thymidine.
3b Cells without the thymidine kinase gene cannot use the thymidine in the growth medium and do not grow.
CONCLUSION The cells were transformed by DNA.

Go to **yourBioPortal.com** for original citations, discussions, and relevant links for all INVESTIGATING LIFE figures.

## 13.1 RECAP

**Experiments on bacteria and on viruses demonstrated that DNA is the genetic material.**

- At the time of Griffith's experiments in the 1920s, what circumstantial evidence suggested to scientists that DNA might be the genetic material? See p. 267
- Why were the experiments of Avery, MacLeod, and McCarty definitive evidence that DNA was the genetic material? See p. 269 and Figure 13.2
- What attributes of bacteriophage T2 were key to the Hershey–Chase experiments demonstrating that DNA is the genetic material? See p. 270 and Figure 13.4

As soon as scientists became convinced that the genetic material was DNA, they began efforts to learn its precise three-dimensional chemical structure. The chemical makeup of DNA, as a polymer made up of nucleotide monomers, had been known for several decades. In determining the structure of DNA, scientists hoped to find the answers to two questions: (1) how is DNA replicated between cell divisions, and (2) how does it direct the synthesis of specific proteins? They were eventually able to answer both questions.

# 13.2 What Is the Structure of DNA?

The structure of DNA was deciphered only after many types of experimental evidence were considered together in a theoretical framework. The most crucial evidence was obtained using X-ray crystallography. Some chemical substances, when they are isolated and purified, can be made to form crystals. The positions of atoms in a crystallized substance can be inferred from the diffraction pattern of X rays passing through the substance (**Figure 13.6A**). The structure of DNA would not have been characterized without the crystallographs prepared in the early 1950s by the English chemist Rosalind Franklin (**Figure 13.6B**). Franklin's work, in turn, depended on the success of the English biophysicist Maurice Wilkins, who prepared samples containing very uniformly oriented DNA fibers. These DNA samples were far better for diffraction than previous ones, and the crystallographs Franklin prepared from them suggested a spiral or helical molecule.

### The chemical composition of DNA was known

The chemical composition of DNA also provided important clues to its structure. Biochemists knew that DNA was a polymer of nucleotides. Each nucleotide consists of a molecule of the sugar deoxyribose, a phosphate group, and a nitrogen-containing base (see Figures 4.1 and 4.2). The only differences among the four nucleotides of DNA are their nitrogenous bases: the purines **adenine** (**A**) and **guanine** (**G**), and the pyrimidines **cytosine** (**C**) and **thymine** (**T**).

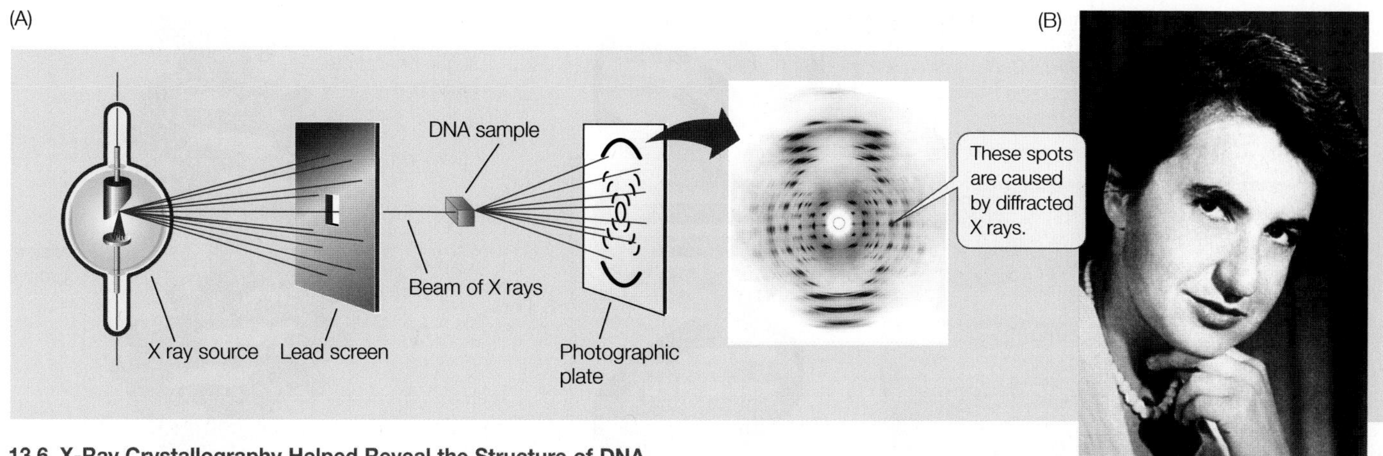


**13.6 X-Ray Crystallography Helped Reveal the Structure of DNA** (A) The positions of atoms in a crystallized chemical substance can be inferred by the pattern of diffraction of X rays passed through it. The pattern of DNA is both highly regular and repetitive. (B) Rosalind Franklin's crystallography helped scientists to visualize the helical structure of the DNA molecule.

In 1950, biochemist Erwin Chargaff at Columbia University reported some observations of major importance. He and his colleagues found that DNA from many different species—and from different sources within a single organism—exhibits certain regularities. In almost all DNA, the following rule holds: The amount of adenine equals the amount of thymine (A = T), and the amount of guanine equals the amount of cytosine (G = C) (**Figure 13.7**). As a result, the total abundance of purines (A + G) equals the total abundance of pyrimidines (T + C). The structure of DNA could not have been worked out without this observation, now known as Chargaff's rule, yet its significance was overlooked for at least three years.

## Watson and Crick described the double helix

The solution to the structure of DNA was finally achieved through model building: the assembly of three-dimensional representations of possible molecular structures using known relative molecular dimensions and known bond angles. This technique was originally applied to molecular structural studies by the American biochemist Linus Pauling. The English physicist Francis Crick and the American geneticist James D. Watson (**Figure 13.8A**), who were both then at the Cavendish Laboratory of Cambridge University, used model building to solve the structure of DNA.

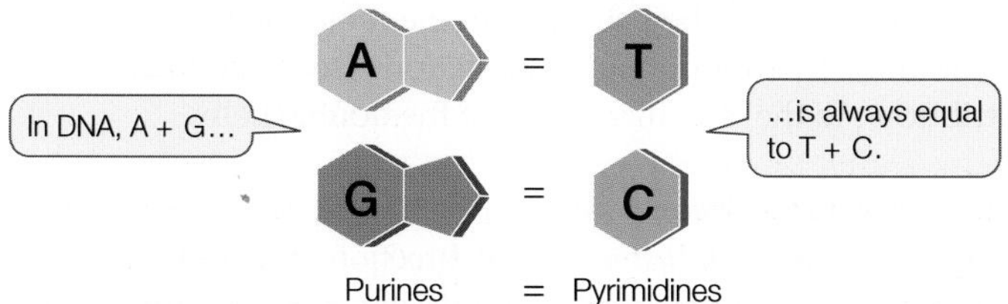


**13.7 Chargaff's Rule** In DNA, the total abundance of purines is equal to the total abundance of pyrimidines.

Watson and Crick attempted to combine all that had been learned so far about DNA structure into a single coherent model. Rosalind Franklin's crystallography results (see Figure 13.6) convinced Watson and Crick that the DNA molecule must be **helical** (cylindrically spiral). Density measurements and previous model building results suggested that there are two polynucleotide chains in the molecule. Modeling studies also showed that the strands run in opposite directions, that is, they are **antiparallel**; that two strands would not fit together in the model if they were parallel.

How are the nucleotides oriented in these chains? Watson and Crick suggested that:

- The nucleotide bases are on the interior of the two strands, with a sugar-phosphate backbone on the outside:

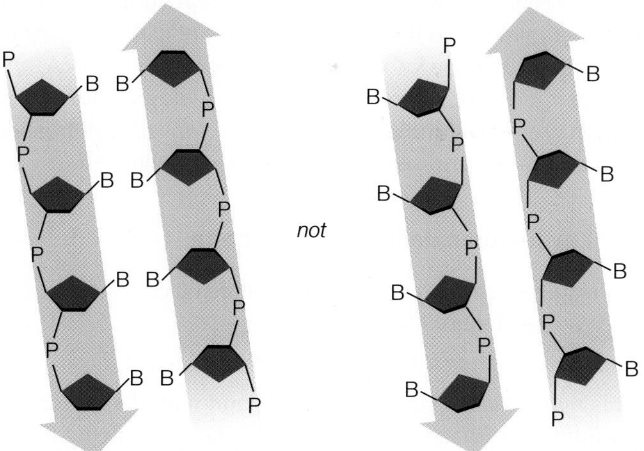


- To satisfy Chargaff's rule (purines = pyrimidines), a purine on one strand is always paired with a pyrimidine on the opposite strand. These **base pairs** (A-T and G-C) have the same width down the double helix, a uniformity shown by x-ray diffraction.

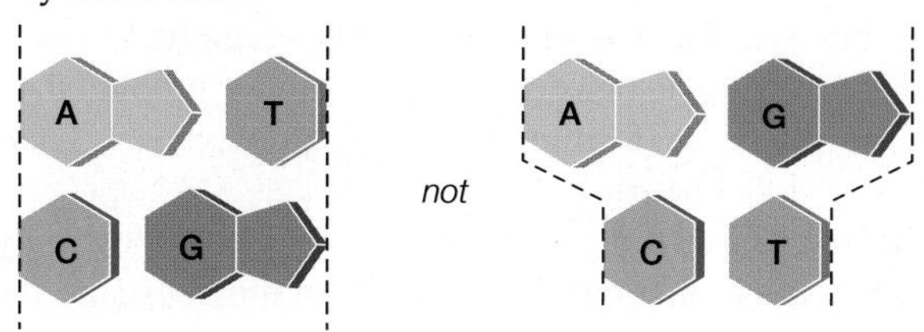

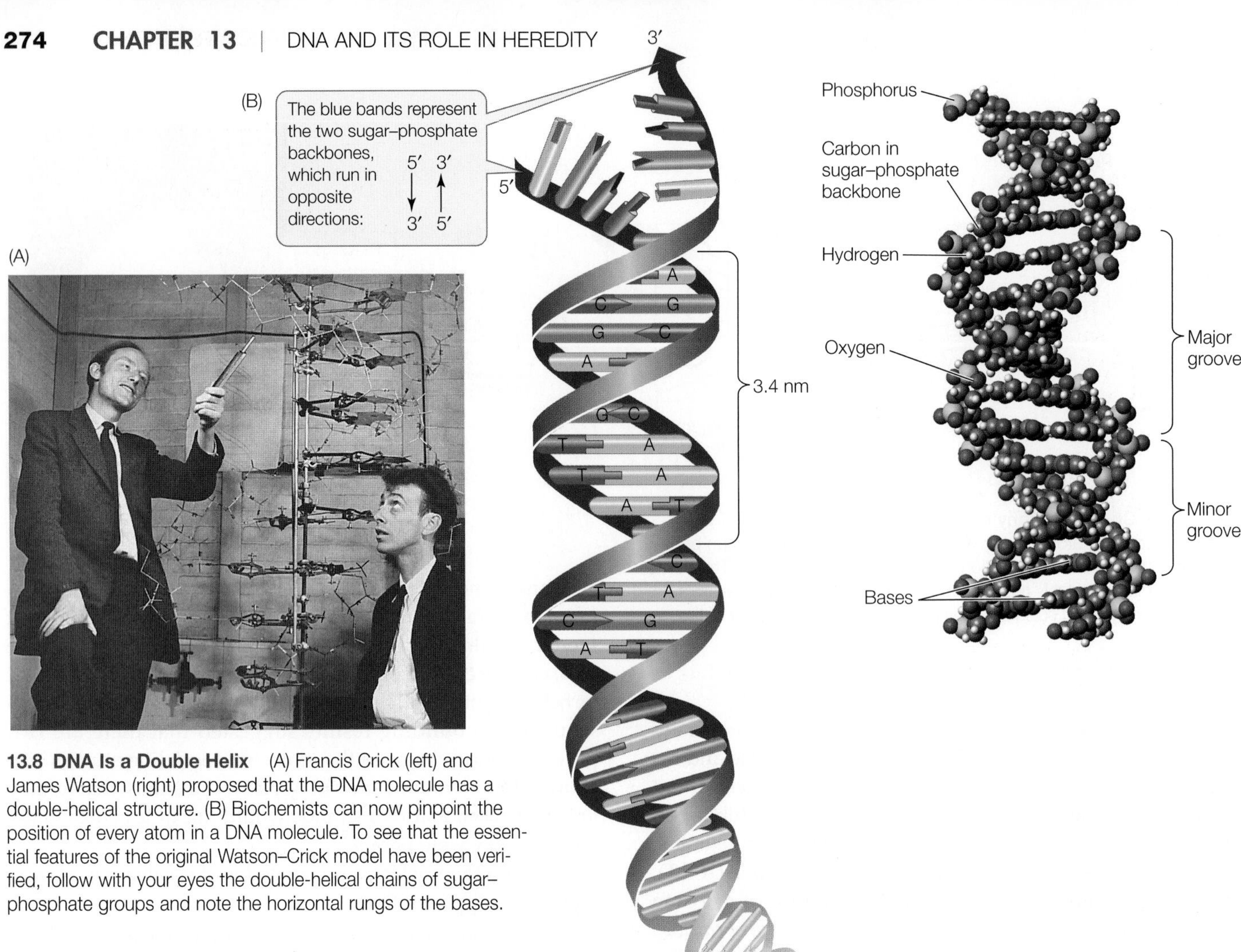


**13.8 DNA Is a Double Helix** (A) Francis Crick (left) and James Watson (right) proposed that the DNA molecule has a double-helical structure. (B) Biochemists can now pinpoint the position of every atom in a DNA molecule. To see that the essential features of the original Watson–Crick model have been verified, follow with your eyes the double-helical chains of sugar–phosphate groups and note the horizontal rungs of the bases.

In late February of 1953, Crick and Watson built a model out of tin that established the general structure of DNA. This structure explained all the known chemical properties of DNA, and it opened the door to understanding its biological functions. There have been minor amendments to that first published structure, but its principal features remain unchanged.

### Four key features define DNA structure

Four features summarize the molecular architecture of the DNA molecule (see **Figure 13.8B**):

- It is a *double-stranded helix* of uniform diameter.
- It is *right-handed*. (Hold your right hand with the thumb pointing up. Imagine the curve of the helix following the direction of your fingers as it winds upward and you have the idea.)
- It is *antiparallel* (the two strands run in opposite directions).
- The outer edges of the nitrogenous bases are *exposed* in the major and minor grooves. These grooves exist because the backbones of the two strands are closer together on one side of the double helix (forming the minor groove) than on the other side (forming the major groove).

**THE HELIX** The sugar–phosphate "backbones" of the polynucleotide chains coil around the outside of the helix, and the nitrogenous bases point toward the center. The two chains are held together by hydrogen bonding between specifically paired bases (**Figure 13.9**). Consistent with Chargaff's rule,

- Adenine (A) pairs with thymine (T) by forming two hydrogen bonds.
- Guanine (G) pairs with cytosine (C) by forming three hydrogen bonds.

Every base pair consists of one purine (A or G) and one pyrimidine (T or C). This pattern is known as **complementary base pairing**.

Because the A-T and G-C pairs are of equal length, they fit into a fixed distance between the two chains (like rungs on a ladder), and the diameter of the helix is thus uniform. The base pairs are flat, and their stacking in the center of the molecule is stabilized by hydrophobic interactions (see Section 2.2), contributing to the overall stability of the double helix.

**ANTIPARALLEL STRANDS** What does it mean to say that the two DNA strands are *antiparallel*? The direction of each strand is determined by examining the bonds between the alternating phosphate and sugar groups that make up the backbone of each strand. Look closely at the five-carbon sugar (deoxyribose)

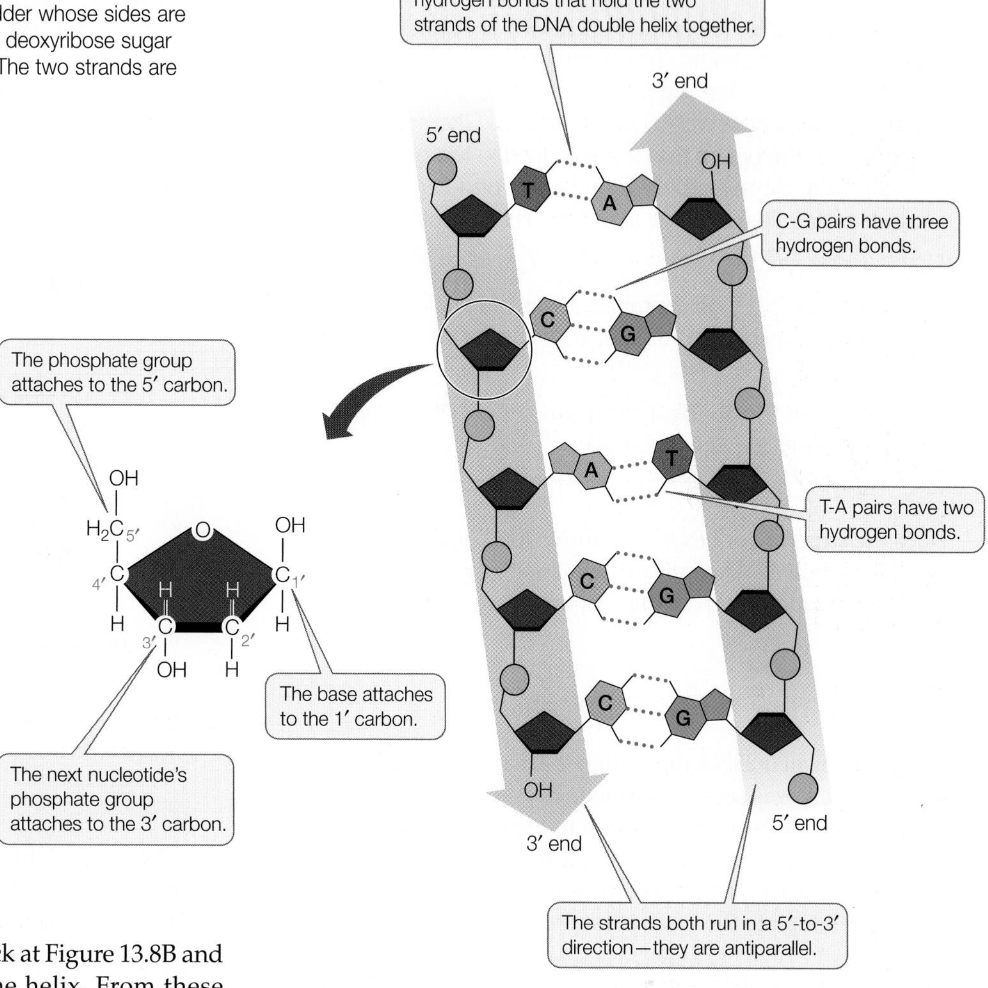
Pairs of complementary bases form hydrogen bonds that hold the two strands of the DNA double helix together.
3′ end
5′ end
OH
T
A
C-G pairs have three hydrogen bonds.
C
G
The phosphate group attaches to the 5′ carbon.
OH
$H_2C_{5'}$
O
OH
C
$C_{1'}$
4′
H
H
H
C
C
H
3′
2′
OH
H
The base attaches to the 1′ carbon.
The next nucleotide's phosphate group attaches to the 3′ carbon.
T-A pairs have two hydrogen bonds.
OH
3′ end
5′ end
The strands both run in a 5′-to-3′ direction—they are antiparallel.

**13.9 Base Pairing in DNA Is Complementary** The purines (A and G) pair with the pyrimidines (T and C, respectively) to form base pairs that are equal in size and resemble the rungs on a ladder whose sides are formed by the sugar–phosphate backbones. The deoxyribose sugar (left) is where the 3′ and 5′ carbons are located. The two strands are antiparallel.

molecule in Figure 13.9. The number followed by a prime (′) designates the position of a carbon atom in the sugar. In the sugar–phosphate backbone of DNA, the phosphate groups are connected to the 3′ carbon of one deoxyribose molecule and the 5′ carbon of the next, linking successive sugars together.

Thus the two ends of a polynucleotide chain differ. At one end of a chain is a free (not connected to another nucleotide) 5′ phosphate group (—$OPO_3^-$); this is called the **5′ end.** At the other end is a free 3′ hydroxyl group (—OH); this is called the **3′ end.** In a DNA double helix, the 5′ end of one strand is paired with the 3′ end of the other strand, and vice versa. In other words, if you drew an arrow for each strand running from 5′ to 3′, the arrows would point in opposite directions.

**BASE EXPOSURE IN THE GROOVES** Look back at Figure 13.8B and note the major and minor grooves in the helix. From these grooves, the exposed outer edges of the flat, hydrogen-bonded base pairs are accessible for additional hydrogen bonding. As seen in Figure 13.9, two hydrogen bonds join each A-T base pair, while three hydrogen bonds join each G-C base pair. Hydrogen-bonding opportunities also exist at an unpaired C=O group in T and an "N" in A. The G-C base pair offers additional hydrogen bonding possibilities as well. Thus the surfaces of the A-T and G-C base pairs are chemically distinct, allowing other molecules, such as proteins, to recognize specific base pair sequences and bind to them. Access to the exposed base-pair sequences in the major and minor grooves is the key to protein–DNA interactions, which are necessary for the replication and expression of the genetic information in DNA.

## The double-helical structure of DNA is essential to its function

The genetic material performs four important functions, and the DNA structure proposed by Watson and Crick was elegantly suited to three of them.

- *The genetic material stores an organism's genetic information.* With its millions of nucleotides, the base sequence of a DNA molecule can encode and store an enormous amount of information. Variations in DNA sequences can account for species and individual differences. DNA fits this role nicely.
- *The genetic material is susceptible to mutations* (permanent changes) *in the information it encodes.* For DNA, mutations might be simple changes in the linear sequence of base pairs.
- *The genetic material is precisely replicated in the cell division cycle.* Replication could be accomplished by complementary base pairing, A with T and G with C. In the original publication of their findings in 1953, Watson and Crick coyly pointed out, "It has not escaped our notice that the specific pairing we have postulated immediately suggests a possible copying mechanism for the genetic material."
- *The genetic material (the coded information in DNA) is expressed as the phenotype.* This function is not obvious in the structure of DNA. However, as we will see in the next chapter, the nucleotide sequence of DNA is copied into RNA, which uses the coded information to specify a linear sequence of amino acids—a protein. The folded forms of proteins determine many of the phenotypes of an organism.

## 13.2 RECAP

**DNA is a double helix made up of two antiparallel polynucleotide chains. The two chains are joined by hydrogen bonds between the nucleotide bases, which pair specifically: A with T, and G with C. Chemical groups on the bases that are exposed in the grooves of the helix are available for hydrogen bonding with other molecules, such as proteins. These molecules can recognize specific sequences of nucleotide bases.**

- Describe the evidence that Watson and Crick used to come up with the double helix model for DNA. See p. 273
- How does the double-helical structure of DNA relate to its function? See p. 275

Once the structure of DNA was understood, it was possible to investigate how DNA replicates itself. Let's examine the experiments that taught us how this elegant process works.

# 13.3 How Is DNA Replicated?

The mechanism of DNA replication that suggested itself to Watson and Crick was soon confirmed. First, experiments showed that DNA could be replicated in a test tube containing simple substrates and an enzyme. Then a truly classic experiment showed that each of the two strands of the double helix can serve as a template for a new strand of DNA.

**yourBioPortal.com**

GO TO **Animated Tutorial 13.1 • DNA Replication, Part 1: Replication of a Chromosome and DNA Polymerization**

### Three modes of DNA replication appeared possible

The prediction that the DNA molecule contains the information needed for its own replication was confirmed by the work of Arthur Kornberg, then at Washington University in St. Louis. He showed that new DNA molecules with the same base composition as the original molecules could be synthesized in a test tube containing the following substances:

- The substrates were the deoxyribonucleoside triphosphates dATP, dCTP, dGTP, and dTTP.
- A **DNA polymerase** enzyme catalyzed the reaction.
- DNA served as a **template** to guide the incoming nucleotides.
- The reaction also contained salts and a pH buffer, to create an appropriate chemical environment for the DNA polymerase to function.

Recall that a nucleoside is a nitrogen base attached to a sugar. The four deoxyribonucleoside triphosphates (dNTPs) each consist of a nitrogen base attached to deoxyribose, which in turn is attached to three phosphate groups. When a dNTP is added

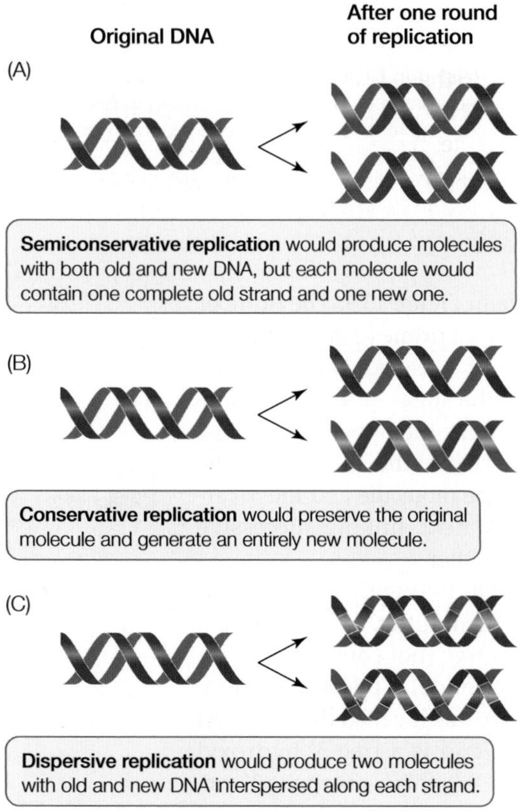

Original DNA

After one round of replication

(A)

Semiconservative replication would produce molecules with both old and new DNA, but each molecule would contain one complete old strand and one new one.

(B)

Conservative replication would preserve the original molecule and generate an entirely new molecule.

(C)

Dispersive replication would produce two molecules with old and new DNA interspersed along each strand.

**13.10 Three Models for DNA Replication** In each model, the original DNA is shown in blue and the newly synthesized DNA is in red.

to a DNA strand during DNA synthesis, the two terminal phosphates are removed, resulting in a monophosphate nucleotide.

The next challenge was to determine which of three possible replication patterns occurs during DNA replication:

- *Semiconservative replication*, in which each parent strand serves as a template for a new strand, and the two new DNA molecules each have one old and one new strand (**Figure 13.10A**)
- *Conservative replication*, in which the original double helix serves as a template for, but does not contribute to, a new double helix (**Figure 13.10B**)
- *Dispersive replication*, in which fragments of the original DNA molecule serve as templates for assembling two new molecules, each containing old and new parts, perhaps at random (**Figure 13.10C**)

Watson and Crick's original paper suggested that DNA replication was semiconservative, but Kornberg's experiment did not provide a basis for choosing among these three models.

### An elegant experiment demonstrated that DNA replication is semiconservative

The work of Matthew Meselson and Franklin Stahl convinced the scientific community that DNA is reproduced by **semiconservative replication**. Working at the California Institute of Technology, Meselson and Stahl devised a simple way to distinguish between old parent strands of DNA and newly copied ones: *density labeling*.

The key to their experiment was the use of a "heavy" isotope of nitrogen. Heavy nitrogen ($^{15}N$) is a rare, nonradioactive isotope that makes molecules containing it denser than chemically identical molecules containing the common isotope, $^{14}N$. Meselson, Stahl, and Jerome Vinograd grew two cultures of the bacterium *Escherichia coli* for many generations:

- One culture was grown in a medium whose nitrogen source (ammonium chloride, $NH_4Cl$) was made with $^{15}N$ instead of $^{14}N$. As a result, all the DNA in the bacteria was "heavy."
- Another culture was grown in a medium containing $^{14}N$, and all the DNA in these bacteria was "light."

When DNA extracts from the two cultures were combined and centrifuged, two separate bands formed, showing that this method could be used to distinguish between DNA samples of slightly different densities.

Next, the researchers grew another *E. coli* culture on $^{15}N$ medium, then transferred it to normal $^{14}N$ medium and allowed the bacteria to continue growing (**Figure 13.11**). Under the conditions they used, *E. coli* cells replicate their DNA and divide

**yourBioPortal.com**
GO TO **Animated Tutorial 13.2 • The Meselson–Stahl Experiment**

## INVESTIGATING LIFE

### 13.11 The Meselson–Stahl Experiment

A centrifuge was used to separate DNA molecules labeled with isotopes of different densities. This experiment revealed a pattern that supports the semiconservative model of DNA replication.

**HYPOTHESIS** **DNA replicates semiconservatively.**

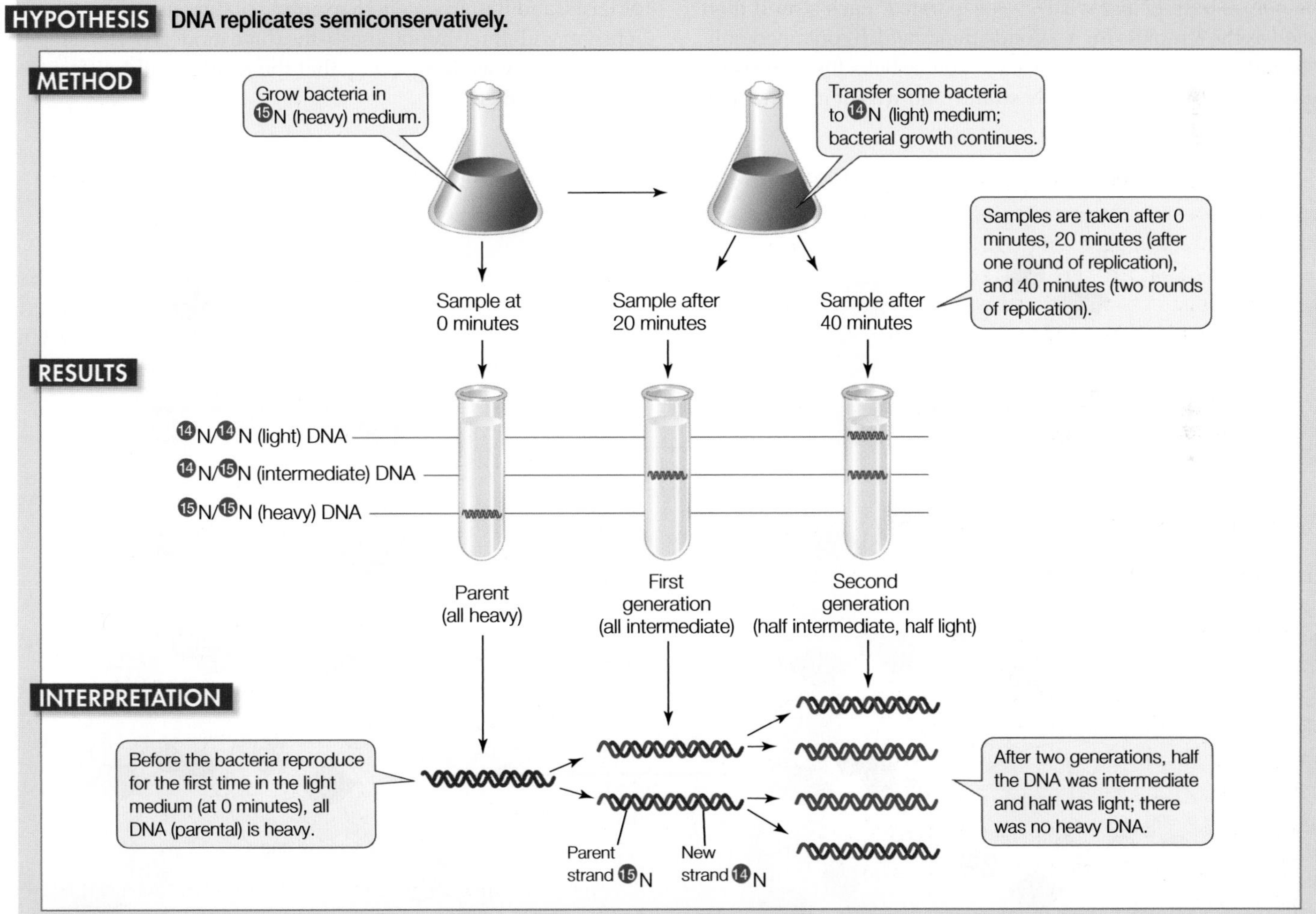
METHOD
Grow bacteria in 15N (heavy) medium.
Transfer some bacteria to 14N (light) medium; bacterial growth continues.
Samples are taken after 0 minutes, 20 minutes (after one round of replication), and 40 minutes (two rounds of replication).
Sample at 0 minutes
Sample after 20 minutes
Sample after 40 minutes
RESULTS
14N/14N (light) DNA
14N/15N (intermediate) DNA
15N/15N (heavy) DNA
Parent (all heavy)
First generation (all intermediate)
Second generation (half intermediate, half light)
INTERPRETATION
Before the bacteria reproduce for the first time in the light medium (at 0 minutes), all DNA (parental) is heavy.
Parent strand 15N
New strand 14N
After two generations, half the DNA was intermediate and half was light; there was no heavy DNA.

**CONCLUSION** **This pattern could only have been observed if each DNA molecule contains a template strand from the parental DNA; thus DNA replication is semiconservative.**

**FURTHER INVESTIGATION:** If you continued this experiment for two more generations (as Meselson and Stahl actually did), what would be the composition (in terms of density) of the fourth generation DNA?

Go to **yourBioPortal.com** for original citations, discussions, and relevant links for all INVESTIGATING LIFE figures.

every 20 minutes. Meselson and Stahl collected some of the bacteria after each division and extracted DNA from the samples. They found that the density gradient was different in each bacterial generation:

- At the time of the transfer to the $^{14}N$ medium, the DNA was uniformly labeled with $^{15}N$, and hence formed a single band corresponding with dense DNA.
- After one generation in the $^{14}N$ medium, when the DNA had been duplicated once, all the DNA was of intermediate density.
- After two generations, there were two equally large DNA bands: one of low density and one of intermediate density.
- In samples from subsequent generations, the proportion of low-density DNA increased steadily.

The results of this experiment can be explained only by the semiconservative model of DNA replication. In the first round of DNA replication in the $^{14}N$ medium, the strands of the double helix—both heavy with $^{15}N$—separated. Each strand then acted as the template for a second strand, which contained only $^{14}N$ and hence was less dense. Each double helix then consisted of one $^{15}N$ strand and one $^{14}N$ strand, and was of intermediate density. In the second replication, the $^{14}N$-containing strands directed the synthesis of partners with $^{14}N$, creating low-density DNA, and the $^{15}N$ strands formed new $^{14}N$ partners.

The crucial observation demonstrating the semiconservative model was that intermediate-density DNA ($^{15}N$–$^{14}N$) appeared in the first generation and continued to appear in subsequent generations. With the other models, the results would have been quite different (see Figure 13.10):

- If conservative replication had occurred, the first generation would have had both high-density DNA ($^{15}N$–$^{15}N$) and low-density DNA ($^{14}N$–$^{14}N$), but no intermediate-density DNA.
- If dispersive replication had occurred, the density of the new DNA would have been intermediate, but DNA of this density would not continue to appear in subsequent generations.

Some scientists consider the Meselson–Stahl experiment to be one of the most elegant experiments ever performed by biologists, and it is an excellent example of the scientific method. It began with three hypotheses—the three models of DNA replication—and was designed so that the results could differentiate between them.

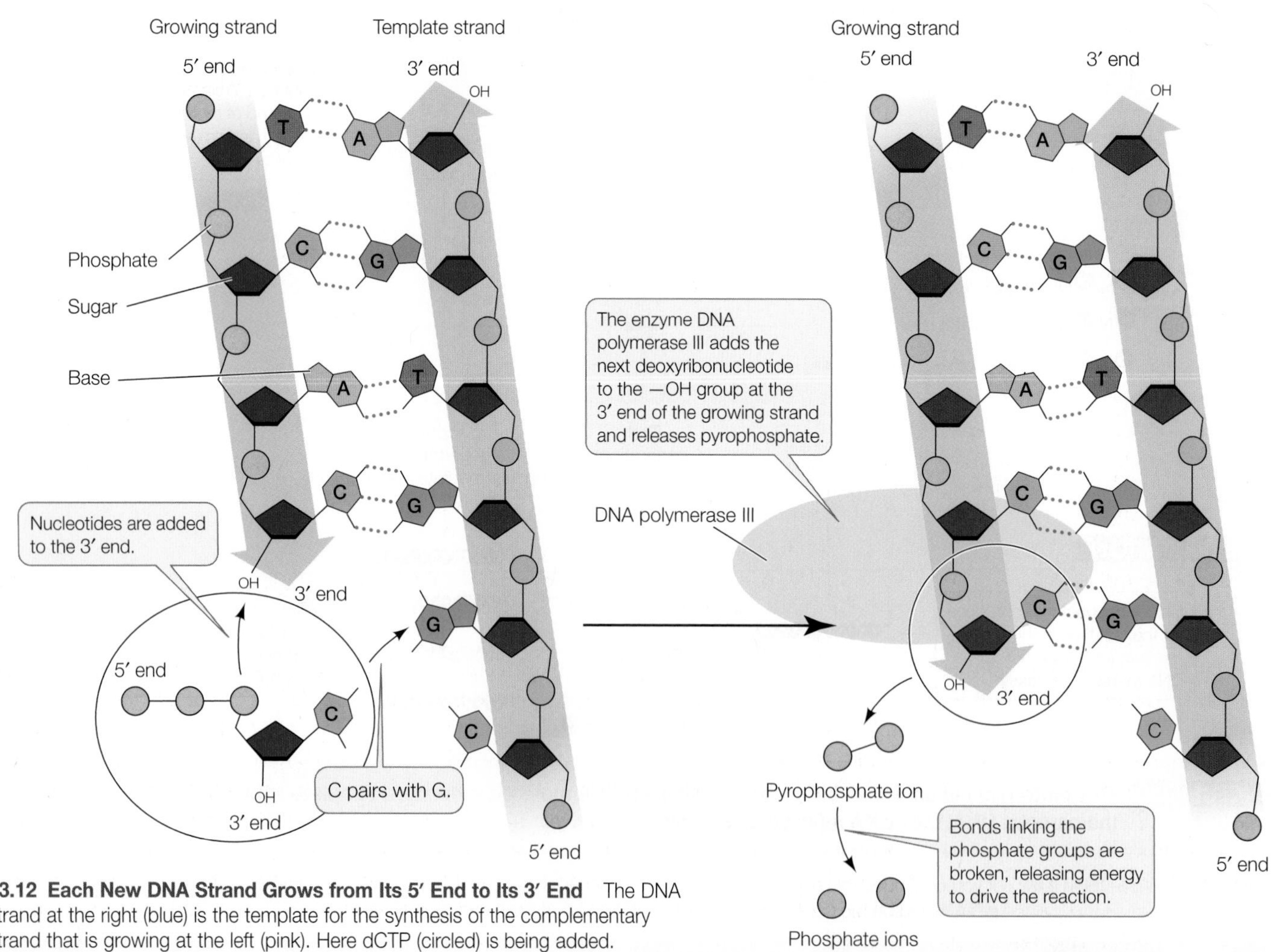


**13.12 Each New DNA Strand Grows from Its 5′ End to Its 3′ End** The DNA strand at the right (blue) is the template for the synthesis of the complementary strand that is growing at the left (pink). Here dCTP (circled) is being added.

## There are two steps in DNA replication

Semiconservative DNA replication in the cell involves a number of different enzymes and other proteins. It takes place in two general steps:

- The DNA double helix is unwound to separate the two template strands and make them available for new base pairing.
- As new nucleotides form complementary base pairs with template DNA, they are covalently linked together by phosphodiester bonds, forming a polymer whose base sequence is complementary to the bases in the template strand.

A key observation is that *nucleotides are added to the growing new strand at the 3′ end*—the end at which the DNA strand has a free hydroxyl (—OH) group on the 3′ carbon of its terminal deoxyribose (**Figure 13.12**). One of the three phosphate groups in a dNTP is attached to the 5′ position of the sugar. The bonds linking the other two phosphate groups to the dNTP are broken, resulting in a monophosphate nucleotide, and releasing energy for the reaction.

## DNA polymerases add nucleotides to the growing chain

DNA is replicated through the interaction of the template strand with a huge protein complex called the **replication complex**, which contains at least four proteins, including DNA polymerase. All chromosomes have at least one region called the **origin of replication (*ori*)**, to which the replication complex binds. Binding occurs when proteins in the complex recognize a specific DNA sequence within the origin of replication.

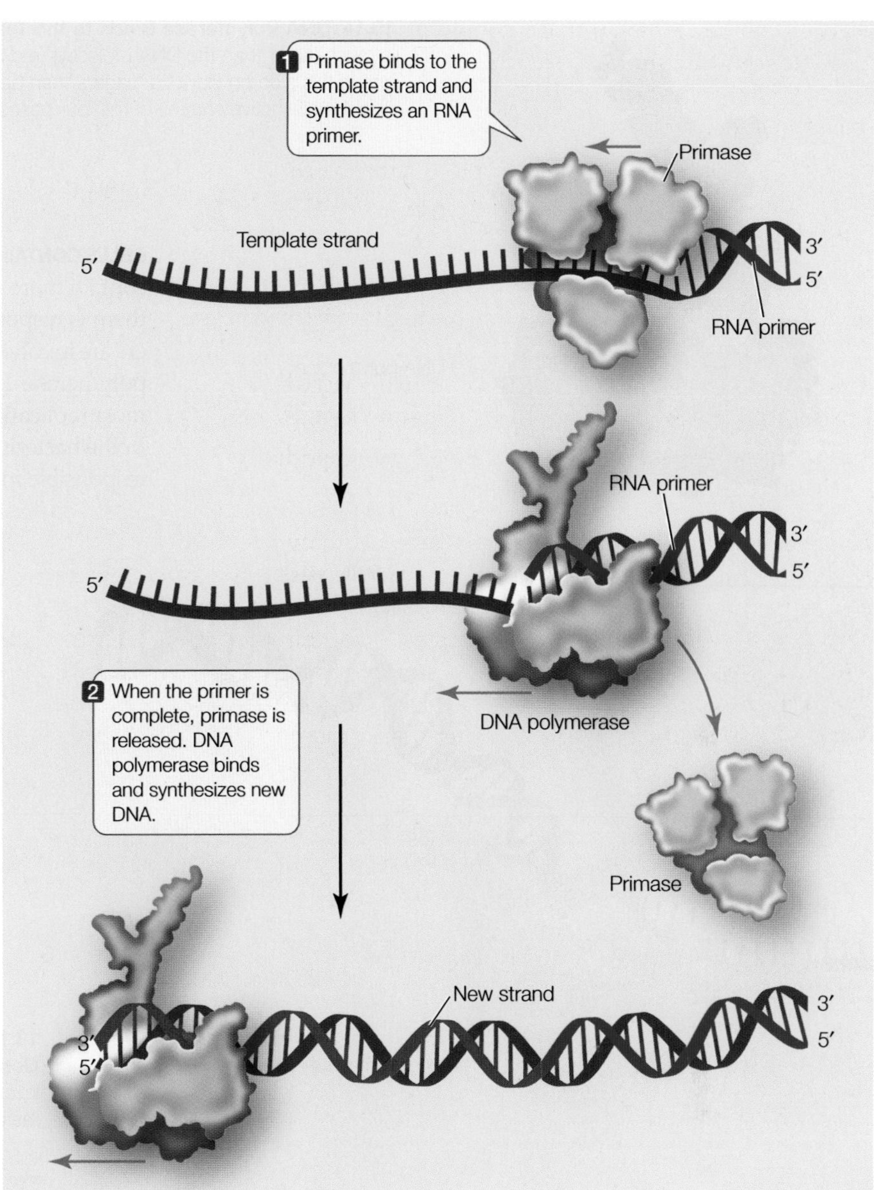

1 Primase binds to the template strand and synthesizes an RNA primer.
Primase
Template strand
5′
3′
5′
RNA primer
RNA primer
5′
3′
5′
2 When the primer is complete, primase is released. DNA polymerase binds and synthesizes new DNA.
DNA polymerase
Primase
New strand
3′
5′
3′
5′

**13.13 DNA Forms with a Primer** DNA polymerases require a primer—a "starter" strand of DNA or RNA to which they can add new nucleotides.

**DNA REPLICATION BEGINS WITH A PRIMER** A DNA polymerase elongates a polynucleotide strand by covalently linking new nucleotides to a previously existing strand. However, it cannot start this process without a short "starter" strand, called a **primer**. In DNA replication, the primer is usually a short single strand of RNA (**Figure 13.13**) but in some organisms it is DNA. This RNA primer strand is complementary to the DNA template, and is synthesized one nucleotide at a time by an enzyme called a **primase**. The DNA polymerase then adds nucleotides to the 3′ end of the primer and continues until the replication of that section of DNA has been completed. Then the RNA primer is degraded, DNA is added in its place, and the resulting DNA fragments are connected by the action of other enzymes. When DNA replication is complete, each new strand consists only of DNA.

**DNA POLYMERASES ARE LARGE** DNA polymerases are much larger than their substrates, the dNTPs, and the template DNA, which is very thin (**Figure 13.14A**). Molecular models of the enzyme–substrate–template complex from bacteria show that the enzyme is shaped like an open right hand with a palm, a thumb, and fingers (**Figure 13.14B**). The palm holds the active site of the enzyme and brings together each substrate and the template. The finger regions rotate inward and have precise shapes that can recognize the different shapes of the four nucleotide bases.

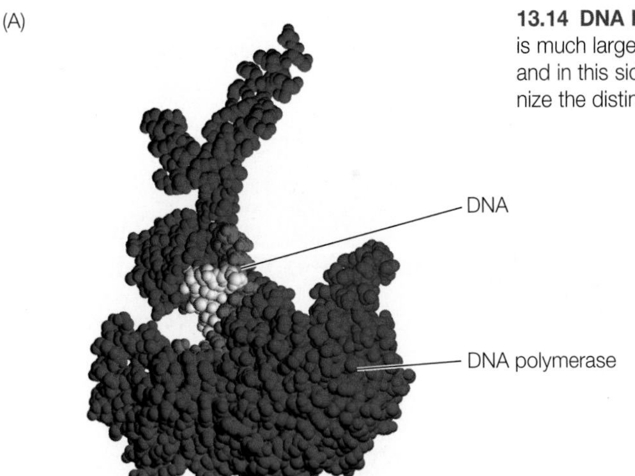


**13.14 DNA Polymerase Binds to the Template Strand** (A) The DNA polymerase enzyme (blue) is much larger than the DNA molecule (red and white). (B) DNA polymerase is shaped like a hand, and in this side-on view, its "fingers" can be seen curling around the DNA. These "fingers" can recognize the distinctive shapes of the four bases.

**CELLS CONTAIN SEVERAL DIFFERENT DNA POLYMERASES** Most cells contain more than one kind of DNA polymerase, but only one of them is responsible for chromosomal DNA replication. The others are involved in primer removal and DNA repair. Fifteen DNA polymerases have been identified in humans; the ones catalyzing most replication are DNA polymerases δ (delta) and ε (epsilon). In the bacterium *E. coli* there are five DNA polymerases; the one responsible for replication is DNA polymerase III.

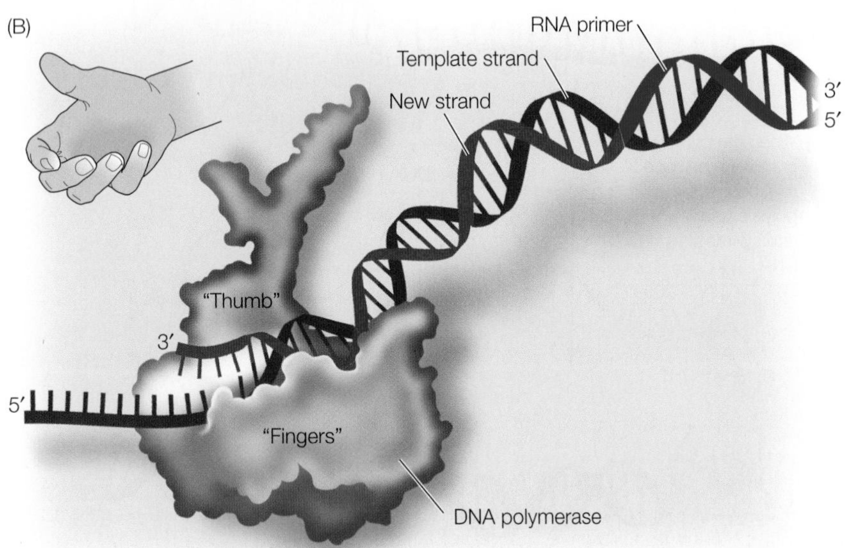


## Many other proteins assist with DNA polymerization

Various other proteins play roles in other replication tasks; some of these are shown in **Figure 13.15**. The first event at the origin of replication is the localized unwinding and separation (denaturation) of the DNA strands. As we saw in Chapter 4, there are several forces that hold the two strands together, including hydrogen bonding and the hydrophobic interactions of the bases. An enzyme called **DNA helicase** uses energy from ATP hydrolysis to unwind and separate the strands, and spe-

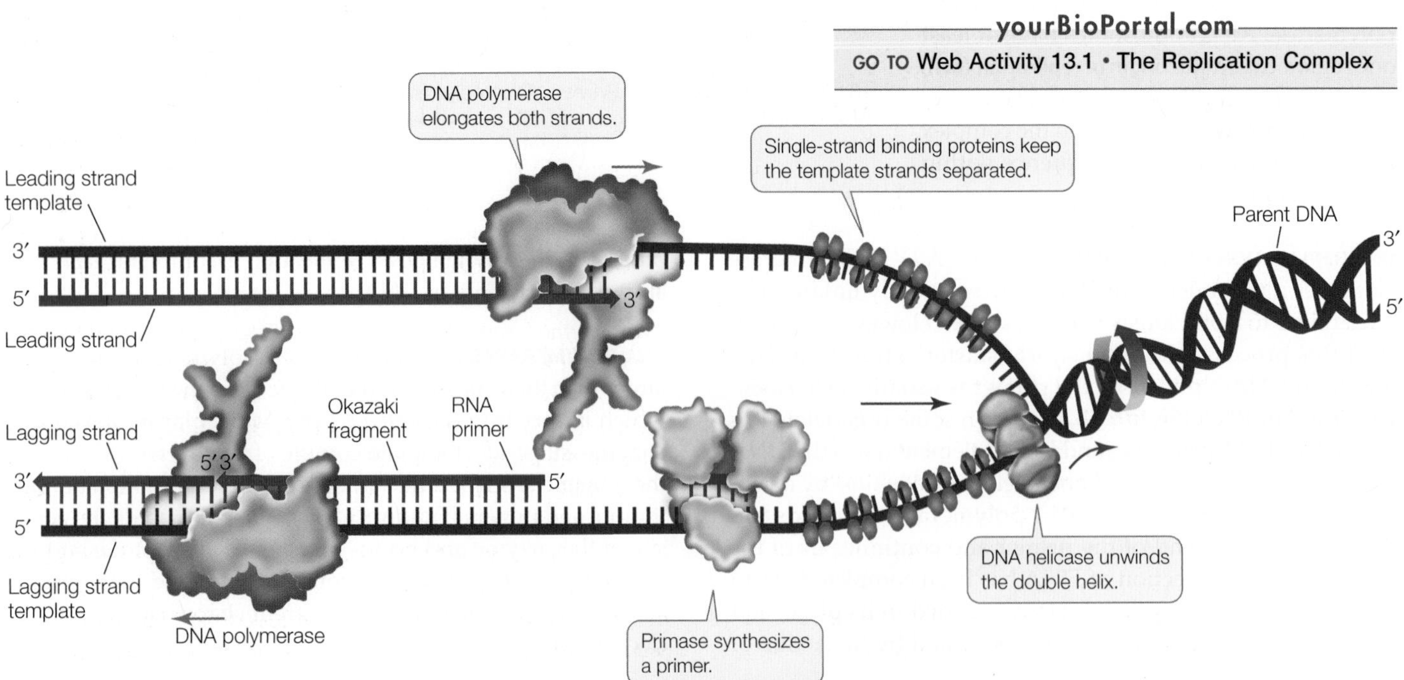


**13.15 Many Proteins Collaborate in the Replication Complex** Several proteins in addition to DNA polymerase are involved in DNA replication. The two molecules of DNA polymerase shown here are actually part of the same complex.

yourBioPortal.com
GO TO Web Activity 13.1 • The Replication Complex

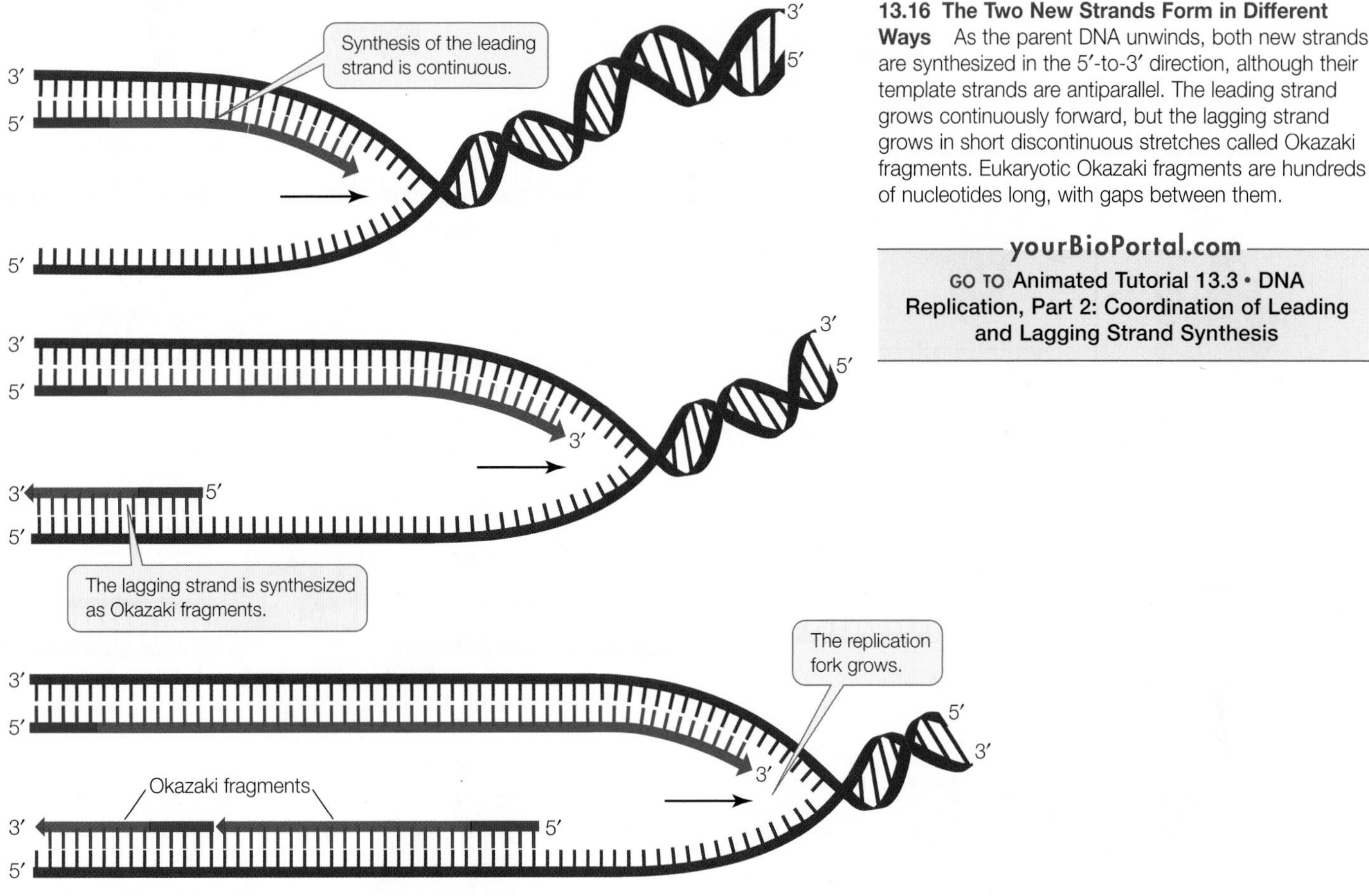


**13.16 The Two New Strands Form in Different Ways** As the parent DNA unwinds, both new strands are synthesized in the 5′-to-3′ direction, although their template strands are antiparallel. The leading strand grows continuously forward, but the lagging strand grows in short discontinuous stretches called Okazaki fragments. Eukaryotic Okazaki fragments are hundreds of nucleotides long, with gaps between them.

yourBioPortal.com

GO TO Animated Tutorial 13.3 • DNA Replication, Part 2: Coordination of Leading and Lagging Strand Synthesis

cial proteins called **single-strand binding proteins** bind to the unwound strands to keep them from reassociating into a double helix. This process makes each of the two template strands available for complementary base pairing.

**THE TWO DNA STRANDS GROW DIFFERENTLY** As Figure 13.15 shows, the DNA at the **replication fork**—the site(s) where DNA unwinds to expose the bases so that they can act as templates—opens up like a zipper in one direction. Study **Figure 13.16** and try to imagine what is happening over a short period of time. Remember that the two DNA strands are antiparallel; that is, the 3′ end of one strand is paired with the 5′ end of the other.

- One newly replicating strand (the **leading strand**) is oriented so that it can grow continuously at its 3′ end as the fork opens up.
- The other new strand (the **lagging strand**) is oriented so that as the fork opens up, its exposed 3′ end gets farther and farther away from the fork, and an unreplicated gap is formed. This gap would get bigger and bigger if there were not a special mechanism to overcome this problem.

Synthesis of the lagging strand requires the synthesis of relatively small, discontinuous stretches of sequence (100 to 200 nucleotides in eukaryotes; 1,000 to 2,000 nucleotides in prokaryotes). These discontinuous stretches are synthesized just as the leading strand is, by the addition of new nucleotides one at a time to the 3′ end of the new strand, but the synthesis of this new strand moves in the direction opposite to that in which the replication fork is moving. These stretches of new DNA are called **Okazaki fragments** (after their discoverer, the Japanese biochemist Reiji Okazaki). While the leading strand grows continuously "forward," the lagging strand grows in shorter, "backward" stretches with gaps between them.

A single primer is needed for synthesis of the leading strand, but each Okazaki fragment requires its own primer to be synthesized by the primase. In bacteria, DNA polymerase III then synthesizes an Okazaki fragment by adding nucleotides to one primer until it reaches the primer of the previous fragment. At this point, DNA polymerase I (discovered by Arthur Kornberg) removes the old primer and replaces it with DNA. Left behind is a tiny nick—the final phosphodiester linkage between the adjacent Okazaki fragments is missing. The enzyme **DNA ligase** catalyzes the formation of that bond, linking the fragments and making the lagging strand whole (**Figure 13.17**).

Working together, DNA helicase, the two DNA polymerases, primase, DNA ligase, and the other proteins of the replication complex do the job of DNA synthesis with a speed and accuracy that are almost unimaginable. In *E. coli*, the replication complex makes new DNA at a rate in excess of 1,000 base pairs per second, committing errors in fewer than one base in a million.

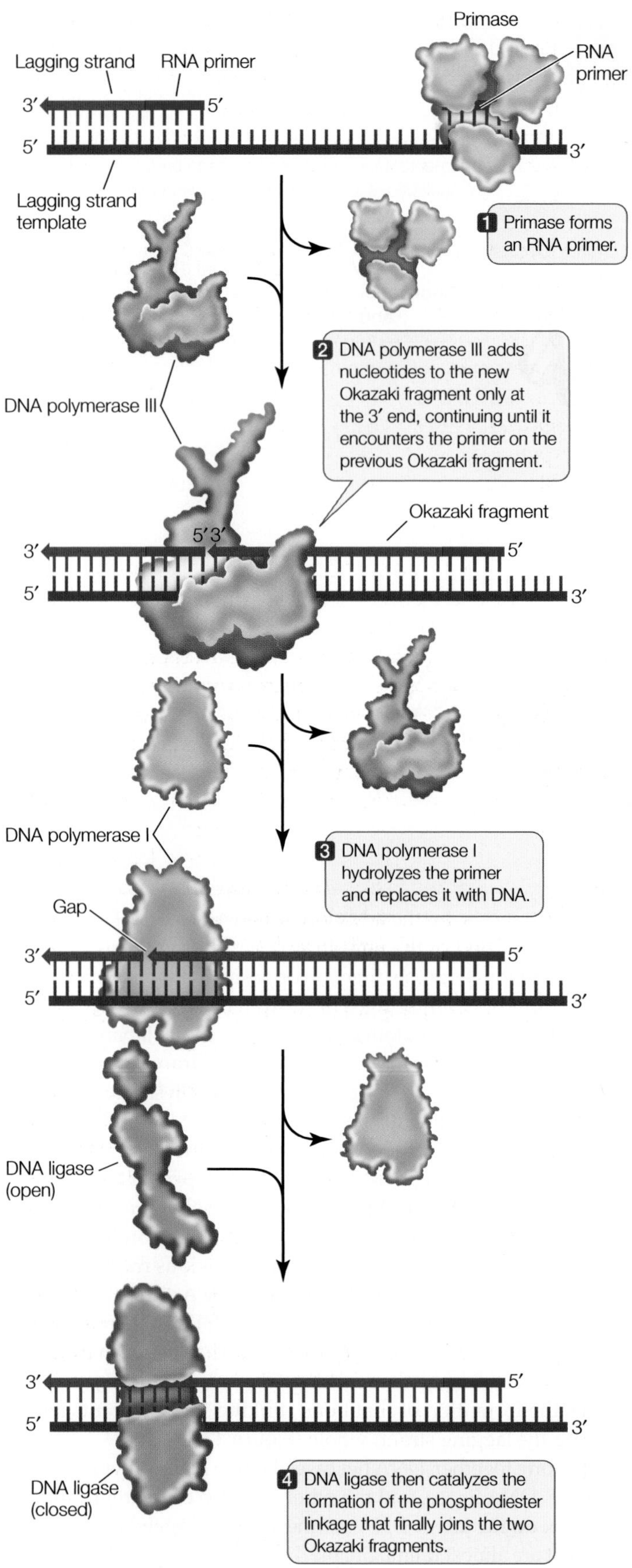

Primase
Lagging strand
RNA primer
RNA primer
Lagging strand template
1 Primase forms an RNA primer.
DNA polymerase III
2 DNA polymerase III adds nucleotides to the new Okazaki fragment only at the 3′ end, continuing until it encounters the primer on the previous Okazaki fragment.
Okazaki fragment
DNA polymerase I
3 DNA polymerase I hydrolyzes the primer and replaces it with DNA.
Gap
DNA ligase (open)
DNA ligase (closed)
4 DNA ligase then catalyzes the formation of the phosphodiester linkage that finally joins the two Okazaki fragments.

**13.17 The Lagging Strand Story** In bacteria, DNA polymerase I and DNA ligase cooperate with DNA polymerase III to complete the complex task of synthesizing the lagging strand.

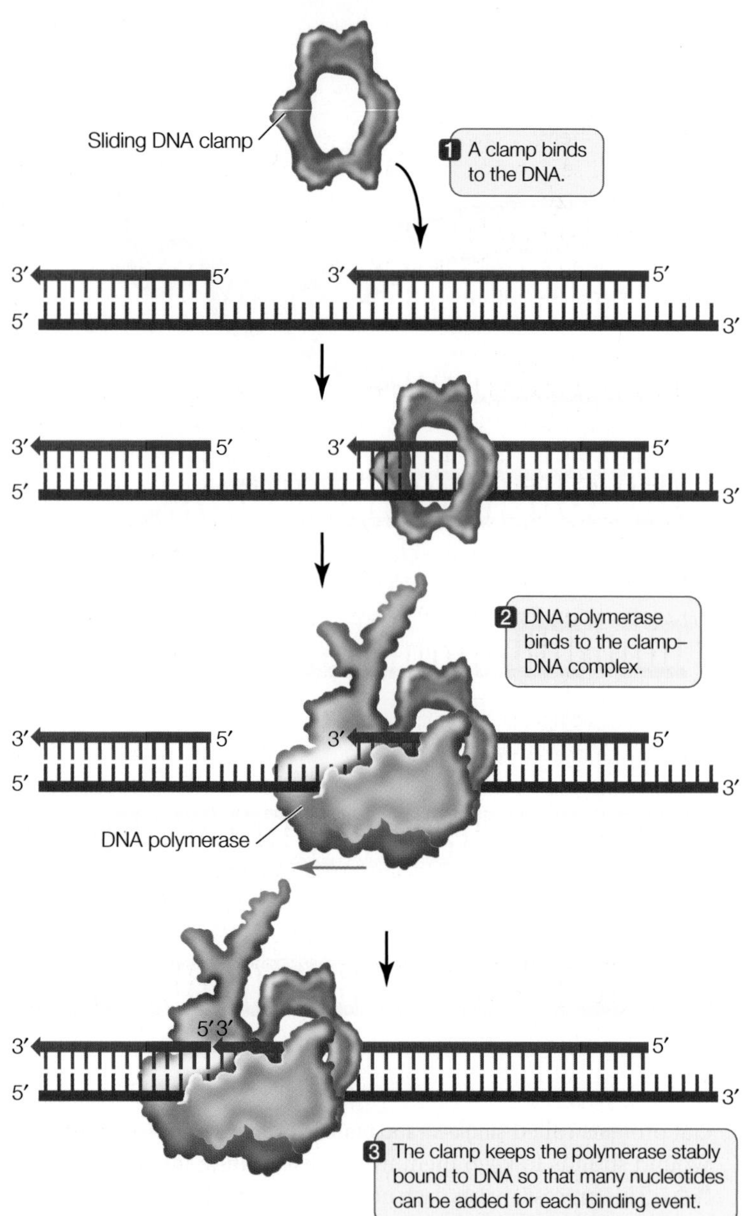

Sliding DNA clamp
1 A clamp binds to the DNA.
2 DNA polymerase binds to the clamp–DNA complex.
DNA polymerase
3 The clamp keeps the polymerase stably bound to DNA so that many nucleotides can be added for each binding event.

**13.18 A Sliding DNA Clamp Increases the Efficiency of DNA Polymerization** The clamp increases the efficiency of polymerization by keeping the enzyme bound to the substrate, so the enzyme does not have to repeatedly bind to template and substrate.

**A SLIDING CLAMP INCREASES THE RATE OF DNA REPLICATION** How do DNA polymerases work so fast? We saw in Section 8.3 that an enzyme catalyzes a chemical reaction:

$$\text{substrate binds to enzyme} \rightarrow \text{one product is formed} \rightarrow \text{enzyme is released} \rightarrow \text{cycle repeats}$$

DNA replication would not proceed as rapidly as it does if it went through such a cycle for each nucleotide. Instead, DNA polymerases are **processive**—that is, they catalyze many polymerizations each time they bind to a DNA molecule:

$$\text{substrates bind to one enzyme} \rightarrow \text{many products are formed} \rightarrow \text{enzyme is released} \rightarrow \text{cycle repeats}$$

The newly replicated strand is stabilized by a **sliding DNA clamp**, which is shaped like a screw cap on a bottle (**Figure 13.18**). This protein has multiple identical subunits assembled into a doughnut shape. The doughnut's "hole" is just large enough to en-le the DNA double helix, along with a single layer of water molles for lubrication. The clamp binds to the DNA polymerase–DNA nplex, keeping the enzyme and the DNA associated tightly with h other. If the clamp is absent, DNA polymerase dissociates from A after 20–100 polymerizations. With the clamp, it can polymerup to 50,000 nucleotides before it detaches.

**A IS THE MAESTRO OF THE REPLICATION FORK** In mammals, the sliding clamp was first recognized in rapidly dividing cells and is called **liferating cell nuclear antigen (PCNA)**. PCNA does more than just keep DNA polymerase bound to the DNA; it also helps to orient the polyrase for binding to the substrates. Furthermore, PCNA has binding s for many other proteins, including chromosome structural proteins, A ligase, DNA methylation enzymes (see Section 16.4) and enzymes olved in DNA repair (see below). It also removes the prereplication nplex from *ori*, ensuring that replication only happens once per cell le. For all that it does, PCNA has been called the "maestro of the replion fork."

**DNA IS THREADED THROUGH A REPLICATION COMPLEX** Until recently, DNA replication was always depicted to look like a locomotive (the replication complex) moving along a railroad track (the DNA). While this does occur in some organisms, most commonly in eukaryotes the replication complex seems to be stationary, attached to chromatin structures, and it is the DNA that moves, essentially threading through the complex as single strands and emerging as double strands).

**SMALL CIRCULAR CHROMOSOMES REPLICATE FROM A SINGLE ORIGIN** Small circular chromosomes, such as those of bacteria (consisting of 1–4 million base pairs), have a single origin of replication. Two replication forks form at this *ori*, and as the DNA moves through the replication complex, the replication forks extend around the circle (**Figure 13.19A**). Two interlocking circular DNA molecules are formed, and they are separated by an enzyme called **DNA topoisomerase**. As we mentioned above, DNA polymerases are very fast. In *E. coli*, replication can be as fast as 1,000 bases per second, and it takes 20–40 minutes to replicate the bacterium's 4.7 million base pairs.

**LARGE LINEAR CHROMOSOMES HAVE MANY ORIGINS** Human DNA polymerases are slower than those of *E. coli*, and can replicate DNA at a rate of about 50 bases per second. Human chromosomes are much larger than those of bacteria (about 80 million base pairs) and linear. Large linear chromosomes such as those of humans contain hundreds of origins of replication. Numerous replication complexes bind to these sites at the same time and catalyze simultaneous replication. Thus there are many replication forks in eukaryotic DNA (**Figure 13.19B**).

## Telomeres are not fully replicated and are prone to repair

As we have just seen, replication of the lagging strand occurs by the addition of Okazaki fragments to RNA primers. When the terminal RNA primer is removed, no DNA can be synthesized to replace it because there is no 3′ end to extend (**Figure 13.20A**). So the new chromosome has a bit of single-stranded DNA at each end. This situation activates a mechanism for cutting off the single-stranded region, along with some of the

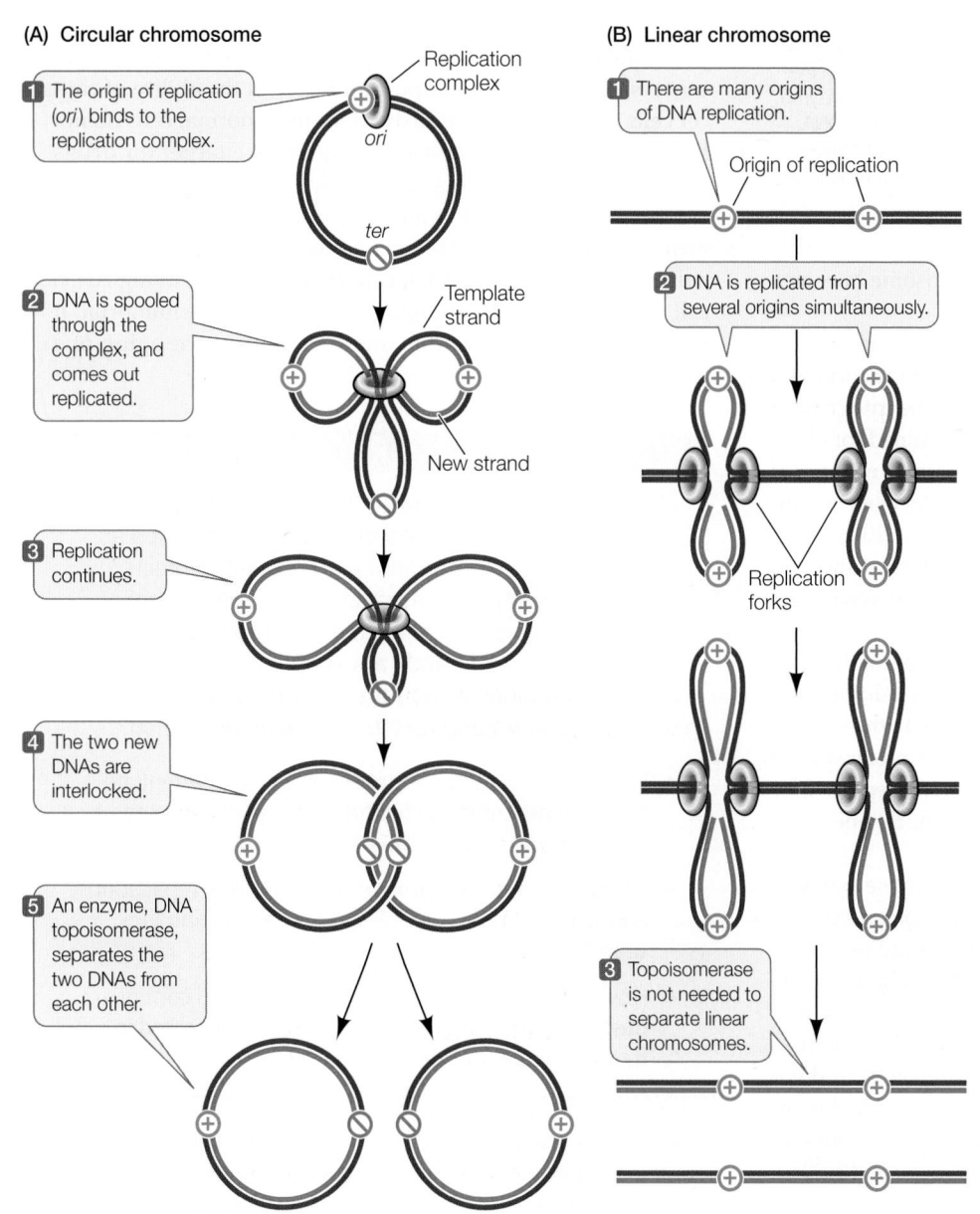



4 The two new DNAs are interlocked.

2 DNA is replicated from several origins simultaneously.

**13.19 Replication of Small Circular and Large Linear Chromosomes**
(A) Small circular chromosomes, typical of prokaryotes, have a single origin (*ori*) and terminus (*ter*) of replication. (B) Larger linear chromosomes, typical of nuclear DNA in eukaryotes, have many origins of replication.

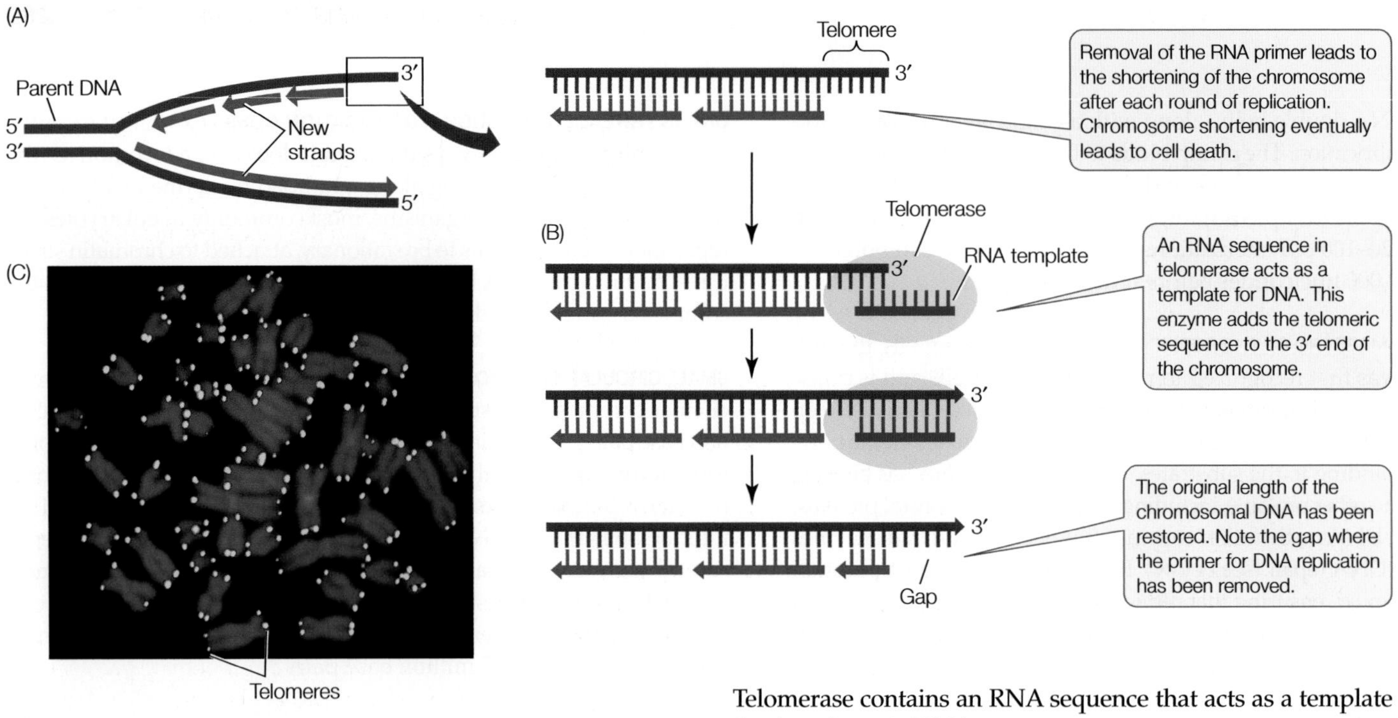

**13.20 Telomeres and Telomerase** (A) Removal of the RNA primer at the 3′ end of the template for the lagging strand leaves a region of DNA—the telomere—unreplicated. (B) In continuously dividing cells, the enzyme telomerase binds to the 3′ end and extends the lagging strand of DNA, so the chromosome does not get shorter. (C) Bright fluorescent staining marks the telomeric regions on these blue-stained human chromosomes.

intact double-stranded DNA. Thus the chromosome becomes slightly shorter with each cell division.

There is another, more serious problem at the ends of chromosomes, and that is simply that they are ends! In Section 11.2 we described checkpoints in the cell cycle for the integrity of DNA. At one of the checkpoints, the DNA is examined for DNA breaks (due to radiation, etc.) and DNA repair is initiated if breaks are found. This involves joining the breaks via a combination of DNA synthesis and DNA ligase activity. This system might recognize the ends of chromosomes as breaks and join two chromosomes together. This would create havoc with genomic integrity.

In many eukaryotes, there are repetitive sequences at the ends of chromosomes called **telomeres**. In humans, the telomere sequence is TTAGGG, and it is repeated about 2,500 times. These repeats bind special proteins that prevent the DNA repair system from recognizing the ends as breaks. In addition, the repeats may form loops that have a similar protective role. So the telomere acts like the plastic tip of shoelaces to prevent fraying.

Each human chromosome can lose 50–200 base pairs of telomeric DNA after each round of DNA replication and cell division. After 20–30 divisions, the chromosomes are unable to participate in cell division, and the cell dies. This phenomenon explains, in part, why many cell lineages do not last the entire lifetime of the organism: their telomeres are lost. Yet continuously dividing cells, such as bone marrow stem cells and gamete-producing cells, maintain their telomeric DNA. An enzyme, appropriately called **telomerase**, catalyzes the addition of any lost telomeric sequences in these cells (**Figure 13.20B**). Telomerase contains an RNA sequence that acts as a template for the telomeric DNA repeat sequence.

Telomerase is expressed in more than 90 percent of human cancers, and may be an important factor in the ability of cancer cells to divide continuously. Since most normal cells do not have this ability, telomerase is an attractive target for drugs designed to attack tumors specifically.

There is also interest in telomerase and aging. When a gene expressing high levels of telomerase is added to human cells in culture, their telomeres do not shorten. Instead of living 20–30 cell generations and then dying, the cells become immortal. It remains to be seen how this finding relates to the aging of a whole organism.

## 13.3 RECAP

**Meselson and Stahl showed that DNA replication is semiconservative: each parent DNA strand serves as a template for a new strand. A complex of proteins, most notably DNA polymerases, is involved in replication. New DNA is polymerized in one direction only, and since the two strands are antiparallel, one strand is made continuously and the other is synthesized in short Okazaki fragments that are eventually joined.**

- How did the Meselson–Stahl experiment differentiate between the three models for DNA replication? See pp. 276–278 and Figures 13.10 and 13.11
- What are the five enzymes needed for DNA replication and what are their roles? See pp. 279–283 and Figures 13.13–13.17
- How is the leading strand of DNA replicated continuously while the lagging strand must be replicated in fragments? See p. 281and Figure 13.16

The complex process of DNA replication is amazingly accurate, but it is not perfect. What happens when things go wrong?

## 13.4 How Are Errors in DNA Repaired?

DNA must be accurately replicated and faithfully maintained. The price of failure can be great; the accurate transmission of genetic information is essential for the functioning and even the life of a single cell or multicellular organism. Yet the replication of DNA is not perfectly accurate, and the DNA of nondividing cells is subject to damage by natural chemical alterations and by environmental agents. In the face of these threats, how has life gone on so long?

DNA repair mechanisms help to preserve life. DNA polymerases initially make significant numbers of mistakes in assembling polynucleotide strands. Without DNA repair, the observed error rate of one for every $10^5$ bases replicated would result in about 60,000 mutations every time a human cell divided. Fortunately, our cells can repair damaged nucleotides and DNA replication errors, so that very few errors end up in the replicated DNA. Cells have at least three DNA repair mechanisms at their disposal:

- A **proofreading** mechanism corrects errors in replication as DNA polymerase makes them.
- A **mismatch repair** mechanism scans DNA immediately after it has been replicated and corrects any base-pairing mismatches.
- An **excision repair** mechanism removes abnormal bases that have formed because of chemical damage and replaces them with functional bases.

Most DNA polymerases perform a **proofreading** function each time they introduce a new nucleotide into a growing DNA strand (**Figure 13.21A**). When a DNA polymerase recognizes a mispairing of bases, it removes the improperly introduced nucleotide and tries again. (Other proteins in the replication complex also play roles in proofreading.) The error rate for this process is only about 1 in 10,000 repaired base pairs, and it lowers the overall error rate for replication to about one error in every $10^{10}$ bases replicated.

After the DNA has been replicated, a second set of proteins surveys the newly replicated molecule and looks for mismatched base pairs that were missed in proofreading (**Figure**

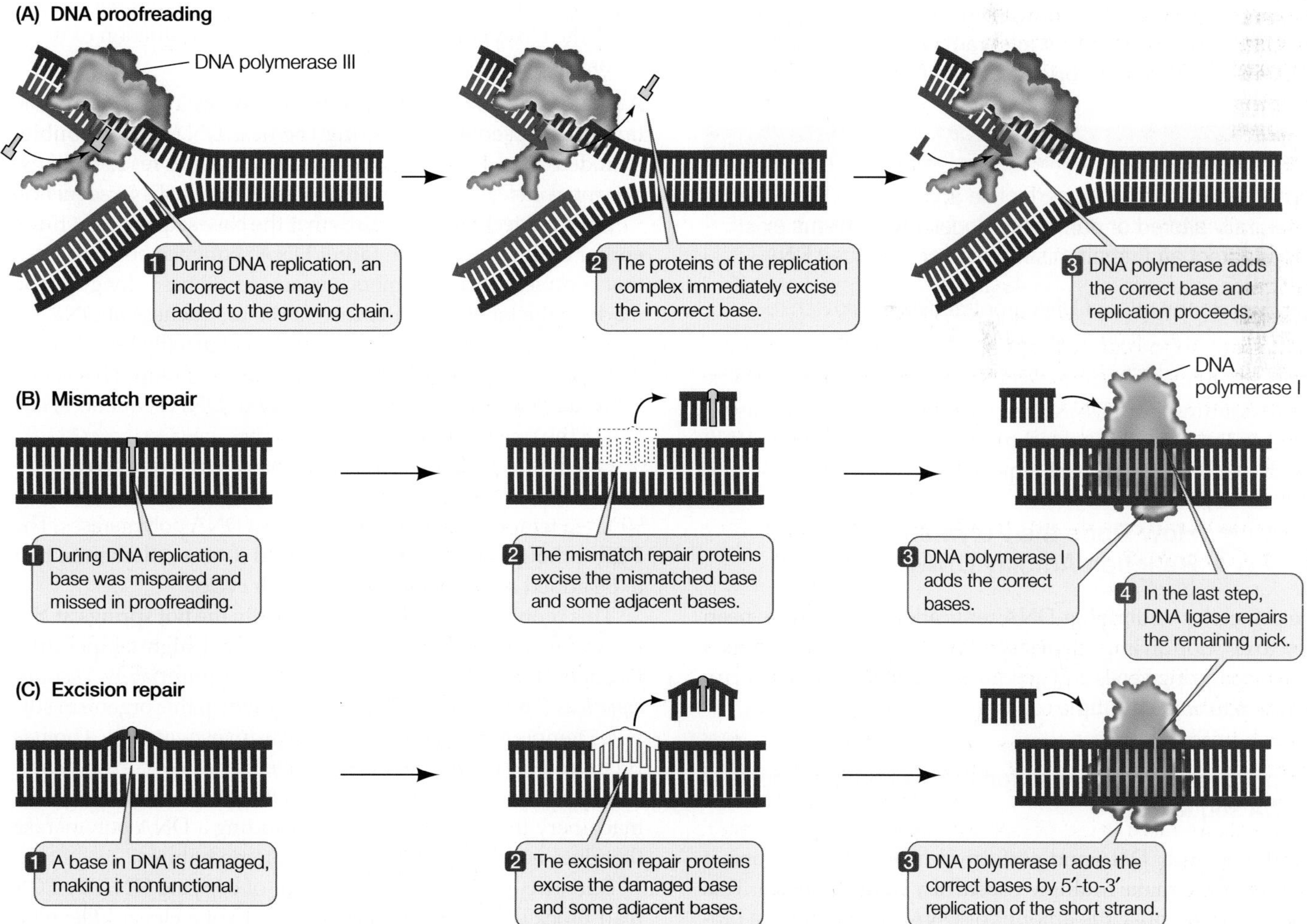

(A) DNA proofreading
DNA polymerase III
1 During DNA replication, an incorrect base may be added to the growing chain.
2 The proteins of the replication complex immediately excise the incorrect base.
3 DNA polymerase adds the correct base and replication proceeds.
(B) Mismatch repair
DNA polymerase I
1 During DNA replication, a base was mispaired and missed in proofreading.
2 The mismatch repair proteins excise the mismatched base and some adjacent bases.
3 DNA polymerase I adds the correct bases.
4 In the last step, DNA ligase repairs the remaining nick.
(C) Excision repair
1 A base in DNA is damaged, making it nonfunctional.
2 The excision repair proteins excise the damaged base and some adjacent bases.
3 DNA polymerase I adds the correct bases by 5′-to-3′ replication of the short strand.

**13.21 DNA Repair Mechanisms** The proteins of the replication complex function in DNA repair mechanisms, reducing the rate of errors in the replicated DNA. Another mechanism (excision repair) repairs damage to existing DNA molecules.

**13.21B**). For example, this **mismatch repair** mechanism might detect an A-C base pair instead of an A-T pair. But how does the repair mechanism "know" whether the A-C pair should be repaired by removing the C and replacing it with T or by removing the A and replacing it with G?

The mismatch repair mechanism can detect the "wrong" base because a DNA strand is chemically modified some time after replication. In prokaryotes, methyl groups (—$CH_3$) are added to some adenines. In eukaryotes, cytosine bases are methylated. Immediately after replication, methylation has not yet occurred on the newly replicated strand, so the new strand is "marked" (distinguished by being unmethylated) as the one in which errors should be corrected.

When mismatch repair fails, DNA sequences are altered. One form of colon cancer arises in part from a failure of mismatch repair.

DNA molecules can also be damaged during the life of a cell (for example, when it is in $G_1$). High-energy radiation, chemicals from the environment, and random spontaneous chemical reactions can all damage DNA. **Excision repair** mechanisms deal with these kinds of damage (**Figure 13.21C**). Individuals who suffer from a condition known as xeroderma pigmentosum lack an excision repair mechanism that normally corrects the damage caused by ultraviolet radiation. They can develop skin cancers after even a brief exposure to sunlight.

## 13.4 RECAP

**DNA replication is not perfect; in addition, DNA may be naturally altered or damaged. Repair mechanisms exist that detect and repair mismatched or damaged DNA.**

- Explain the roles of DNA proofreading, mismatch repair, and excision repair. See **Figure 13.21**

Understanding how DNA is replicated and repaired has allowed scientists to develop techniques for studying genes. We'll look at just one of those techniques next.

# 13.5 How Does the Polymerase Chain Reaction Amplify DNA?

The principles underlying DNA replication in cells have been used to develop an important laboratory technique that has been vital in analyzing genes and genomes. This technique allows researchers to make multiple copies of short DNA sequences.

### The polymerase chain reaction makes multiple copies of DNA sequences

In order to study DNA and perform genetic manipulations, it is necessary to make multiple copies of a DNA sequence. This is necessary because the amount of DNA isolated from a biological sample is often too small to work with. The **polymerase chain reaction** (**PCR**) technique essentially automates this replication process by copying a short region of DNA many times in a test tube. This process is referred to as *DNA amplification*.

The PCR reaction mixture contains:

- a sample of double stranded DNA from a biological sample, to act as the template,
- two short, artificially synthesized primers that are complementary to the ends of the sequence to be amplified,
- the four dNTPs (dATP, dTTP, dCTP and dGTP),
- a DNA polymerase that can tolerate high temperatures without becoming degraded, and
- salts and a buffer to maintain a near-neutral pH.

The PCR amplification is a cyclic process in which a sequence of steps is repeated over and over again (**Figure 13.22**):

- The first step involves heating the reaction to near boiling point, to separate (*denature*) the two strands of the DNA template.
- The reaction is then cooled to allow the primers to bind (or *anneal*) to the template strands.
- Next, the reaction is warmed to an optimum temperature for the DNA polymerase to catalyze the production of the complementary new strands.

A single cycle takes a few minutes to produce two copies of the target DNA sequence, leaving the new DNA in the double-stranded state. Repeating the cycle many times leads to an exponential increase in the number of copies of the DNA sequence.

The PCR technique requires that the base sequences at the 3′ end of each strand of the target DNA sequence be known, so that complementary primers, usually 15–30 bases long, can be made in the laboratory. Because of the uniqueness of DNA sequences, a pair of primers this length will usually bind to only a single region of DNA in an organism's genome. This specificity, despite the incredible diversity of DNA sequences, is a key to the power of PCR.

One initial problem with PCR was its temperature requirements. To denature the DNA, it must be heated to more than 90°C—a temperature that destroys most DNA polymerases. The PCR technique would not be practical if new polymerase had to be added after denaturation in each cycle.

This problem was solved by nature: in the hot springs at Yellowstone National Park, as well as in other high-temperature locations, there lives a bacterium called, appropriately, *Thermus aquaticus* ("hot water"). The means by which this organism survives temperatures of up to 95°C was investigated by Thomas Brock and his colleagues at the University of Wisconsin, Madison. They discovered that *T. aquaticus* has an entire metabolic machinery that is heat-resistant, including a DNA polymerase that does not denature at these high temperatures.

Scientists pondering the problem of copying DNA by PCR read Brock's basic research articles and got a clever idea: why not use *T. aquaticus* DNA polymerase in the PCR technique? It

TOOLS FOR INVESTIGATING LIFE

**13.22 The Polymerase Chain Reaction**

The steps in this cyclic process are repeated many times to produce multiple identical copies of a DNA fragment. This makes enough DNA for chemical analysis and genetic manipulations.

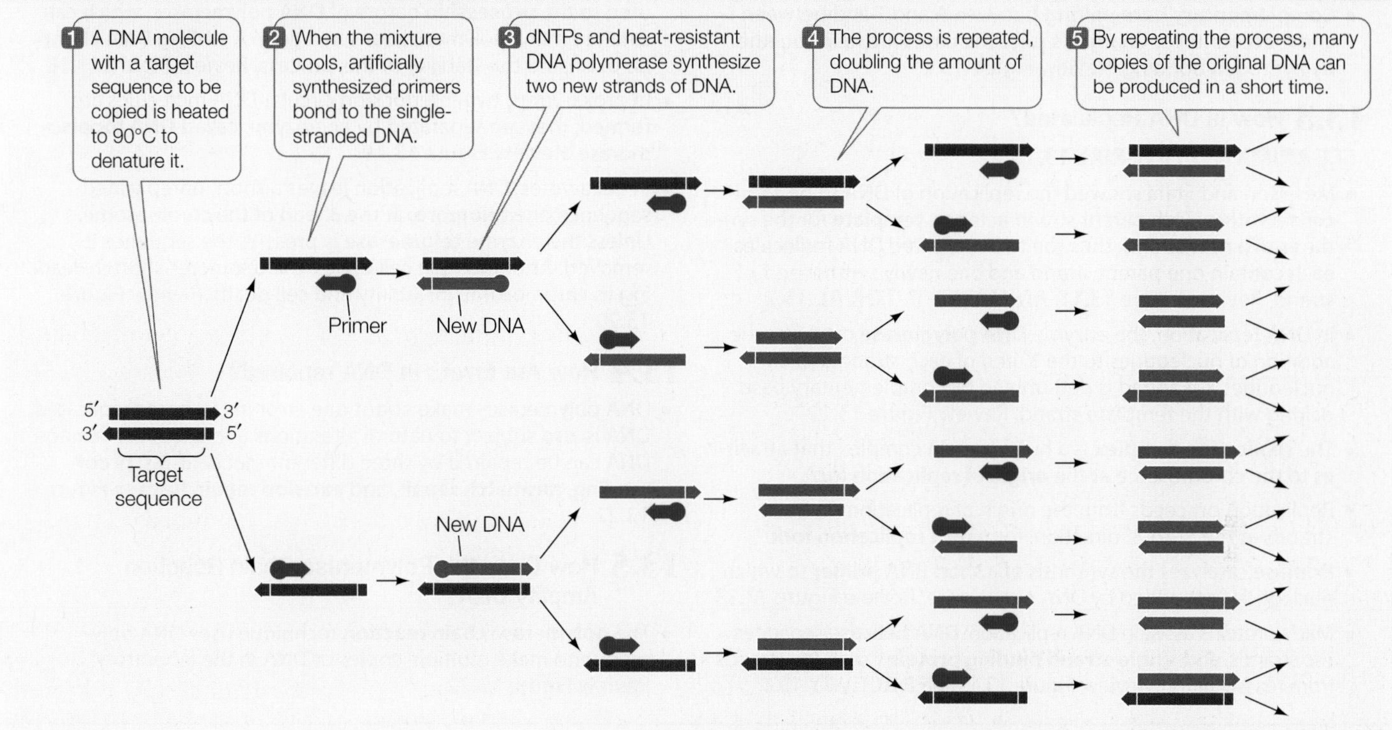


could withstand the 90°C denaturation temperature and would not have to be added during each cycle. The idea worked, and it earned biochemist Kary Mullis a Nobel prize. PCR has had an enormous impact on genetic research. Some of its most striking applications will be described in Chapters 15–18. These applications range from amplifying DNA in order to identify an individual person or organism, to detection of diseases.

1 A DNA molecule with a target sequence to be copied is heated to 90°C to denature it.
2 When the mixture cools, artificially synthesized primers bond to the single-stranded DNA.
3 dNTPs and heat-resistant DNA polymerase synthesize two new strands of DNA.
4 The process is repeated, doubling the amount of DNA.
5 By repeating the process, many copies of the original DNA can be produced in a short time.
5′
3′
3′
5′
Target sequence
Primer
New DNA
New DNA

## 13.5 RECAP

**Knowledge of the mechanisms of DNA replication led to the development of a technique for making multiple copies of DNA sequences.**

- What is the role of primers in PCR? **See pp. 286 and Figure 13.22**

# CHAPTER SUMMARY

## 13.1 What Is the Evidence that the Gene Is DNA?

- Griffith's experiments in the 1920s demonstrated that some substance in cells—then called a **transforming principle**—can cause heritable changes in other cells. **Review Figure 13.1**
- The location and quantity of DNA in the cell suggested that DNA might be the genetic material. Avery, MacLeod, and McCarty isolated the transforming principle from bacteria and identified it as DNA. **Review Figure 13.2**
- The Hershey–Chase experiment established conclusively that DNA (and not protein) is the genetic material, by tracing the DNA of radioactively labeled viruses, with which they infected bacterial cells. **Review Figure 13.4**
- Genetic transformation of eukaryotic cells is called **transfection**. Transformation and transfection can be studied with the aid of a **marker** gene that confers a known and observable phenotype. **Review Figure 13.5**

## 13.2 What Is the Structure of DNA?

- Chargaff's rule states that the amount of **adenine** in DNA is equal to the amount of **thymine**, and that the amount of **guanine** is equal to the amount of **cytosine**; thus the total abundance of purines (A + G) equals the total abundance of pyrimidines (T + C).

- X-ray crystallography showed that the DNA molecule is **helical**. Watson and Crick proposed that DNA is a double-stranded helix in which the strands are **antiparallel**. Review Figure 13.8
- **Complementary base pairing** between A and T and between G and C accounts for Chargaff's rule. The bases are held together by hydrogen bonding. Review Figure 13.9

## 13.3 How Is DNA Replicated?

SEE ANIMATED TUTORIAL 13.1

- Meselson and Stahl showed the replication of DNA to be **semiconservative**. Each parent strand acts as a **template** for the synthesis of a new strand; thus the two replicated DNA molecules each contain one parent strand and one newly synthesized strand. Review Figure 13.11, ANIMATED TUTORIAL 13.2
- In DNA replication, the enzyme **DNA polymerase** catalyzes the addition of nucleotides to the 3′ end of each strand. Which nucleotides are added is determined by complementary base pairing with the template strand. Review Figure 13.12
- The **replication complex** is a huge protein complex that attaches to the chromosome at the **origin of replication** (*ori*).
- Replication proceeds from the origin of replication on both strands in the 5′-to-3′ direction, forming a **replication fork**.
- **Primase** catalyzes the synthesis of a short RNA **primer** to which nucleotides are added by DNA polymerase. Review Figure 13.13
- Many proteins assist in DNA replication. **DNA helicase** separates the strands, and **single-strand binding proteins** keep the strands from reassociating. Review Figure 13.13, WEB ACTIVITY 13.1
- The **leading strand** is synthesized continuously and the **lagging strand** in pieces called **Okazaki fragments**. The fragments are joined together by **DNA ligase**. Review Figures 13.16 and 13.17, ANIMATED TUTORIAL 13.3
- The speed with which DNA polymerization proceeds is attributed to the **processive** nature of DNA polymerases, which can catalyze many polymerizations at a time. A **sliding DNA clamp** helps ensure the stability of this process. Review Figure 13.18
- In prokaryotes, two interlocking circular DNA molecules are formed; they are separated by an enzyme called **DNA topoisomerase**. Review Figure 13.19
- In eukaryotes, DNA replication leaves a short, unreplicated sequence, the **telomere**, at the 3′ end of the chromosome. Unless the enzyme **telomerase** is present, the sequence is removed. After multiple cell cycles, the telomeres shorten, leading to chromosome instability and cell death. Review Figure 13.20

## 13.4 How Are Errors in DNA repaired?

- DNA polymerases make about one error in $10^5$ bases replicated. DNA is also subject to natural alterations and chemical damage. DNA can be repaired by three different mechanisms: **proofreading**, **mismatch repair**, and **excision repair**. Review Figure 13.21

## 13.5 How Does the Polymerase Chain Reaction Amplify DNA?

- The **polymerase chain reaction** technique uses DNA polymerase to make multiple copies of DNA in the laboratory. Review Figure 13.22

## SELF-QUIZ

1. Griffith's studies of *Streptococcus pneumoniae*
   *a.* showed that DNA is the genetic material of bacteria.
   *b.* showed that DNA is the genetic material of bacteriophage.
   *c.* demonstrated the phenomenon of bacterial transformation.
   *d.* proved that prokaryotes reproduce sexually.
   *e.* proved that protein is not the genetic material.
2. In the Hershey–Chase experiment,
   *a.* DNA from parent bacteriophage appeared in progeny bacteriophage.
   *b.* most of the phage DNA never entered the bacteria.
   *c.* more than three-fourths of the phage protein appeared in progeny phage.
   *d.* DNA was labeled with radioactive sulfur.
   *e.* DNA formed the coat of the bacteriophage.
3. Which statement about complementary base pairing is *not* true?
   *a.* Complementary base pairing plays a role in DNA replication.
   *b.* In DNA, T pairs with A.
   *c.* Purines pair with purines, and pyrimidines pair with pyrimidines.
   *d.* In DNA, C pairs with G.
   *e.* The base pairs are of equal length.
4. In semiconservative replication of DNA,
   *a.* the original double helix remains intact and a new double helix forms.
   *b.* the strands of the double helix separate and act as templates for new strands.
   *c.* polymerization is catalyzed by RNA polymerase.
   *d.* polymerization is catalyzed by a double-helical enzyme.
   *e.* DNA is synthesized from amino acids.
5. Which of the following does *not* occur during DNA replication?
   *a.* Unwinding of the parent double helix
   *b.* Formation of short pieces that are connected by DNA ligase
   *c.* Complementary base pairing
   *d.* Use of a primer
   *e.* Polymerization in the 3′-to-5′ direction
6. The primer used for DNA replication
   *a.* is a short strand of RNA added to the 3′ end.
   *b.* is needed only once on a leading strand.
   *c.* remains on the DNA after replication.
   *d.* ensures that there will be a free 5′ end to which nucleotides can be added.
   *e.* is added to only one of the two template strands.

7. One strand of DNA has the sequence 5′-ATTCCG-3′ The complementary strand for this is
   *a.* 5′-TAAGGC-3′
   *b.* 5′-ATTCCG-3′
   *c.* 5′-ACCTTA-3′
   *d.* 5′-CGGAAT-3′
   *e.* 5′-GCCTTA-3′
8. The role of DNA ligase in DNA replication is to
   *a.* add more nucleotides to the growing strand one at a time.
   *b.* open up the two DNA strands to expose template strands.
   *c.* ligate base to sugar to phosphate in a nucleotide.
   *d.* bond Okazaki fragments to one another.
   *e.* remove incorrectly paired bases.
9. The polymerase chain reaction
   *a.* is a method for sequencing DNA.
   *b.* is used to transcribe specific genes.
   *c.* amplifies specific DNA sequences.
   *d.* does not require DNA replication primers.
   *e.* uses a DNA polymerase that denatures at 55°C.
10. What is the correct order for the following events in excision repair of DNA? (1) DNA polymerase I adds correct bases by 5′ to 3′ replication; (2) damaged bases are recognized; (3) DNA ligase seals the new strand to existing DNA; (4) part of a single strand is excised.
    *a.* 1, 2, 3, 4
    *b.* 2, 1, 3, 4
    *c.* 2, 4, 1, 3
    *d.* 3, 4, 2, 1
    *e.* 4, 2, 3, 1

## FOR DISCUSSION

1. Suppose that Meselson and Stahl had continued their experiment on DNA replication for another ten bacterial generations. Would there still have been any $^{14}N$–$^{15}N$ hybrid DNA present? Would it still have appeared in the centrifuge tube? Explain.
2. If DNA replication were conservative rather than semiconservative, what results would Meselson and Stahl have observed? Draw a diagram of the results using the conventions of Figure 13.10.
3. Using the following information, calculate the number of origins of DNA replication on a human chromosome: DNA polymerase adds nucleotides at 3,000 base pairs per minute in one direction; replication is bidirectional; S phase lasts 300 minutes; there are 120 million base pairs per chromosome. In a typical chromosome 3 μm long, how many origins are there per μm?
4. The drug dideoxycytidine, used to treat certain viral infections, is a nucleotide made with 2′,3′-dideoxyribose. This sugar lacks —OH groups at both the 2′ and the 3′ positions. Explain why this drug stops the growth of a DNA chain when added to DNA.

## ADDITIONAL INVESTIGATION

Outline a series of experiments using radioactive isotopes (such as $^{32}P$ and $^{35}S$) to show that it is bacterial DNA and not bacterial protein that enters the host cell and is responsible for bacterial transformation.

## WORKING WITH DATA (GO TO yourBioPortal.com)

**The Hershey-Chase Experiment** The experiments in which labeled bacteriophage were used to infect host *E. coli* cells were key evidence for the identification of DNA as the gene (Figure 13.4). In this exercise, you will analyze the data that Hershey and Chase obtained, as well as important control experiments that ruled out protein and pointed to DNA as the gene.

**The Meselson-Stahl Experiment** Because of its elegant simplicity, this experiment has been called one of the most beautiful in the history of biology (Figure 13.11). In this real-world exercise, you will examine the experimental protocol and make calculations based on the actual centrifuge photographs that the experimenters obtained.

# 14 From DNA to Protein: Gene Expression

## An unexpected wedding gift

The wedding and honeymoon began spectacularly. Andrew Speaker, an Atlanta lawyer, and law student Sarah Cooksey began their honeymoon in Rome. Days later, they got shocking news from the U.S. Centers for Disease Control and Prevention: Andrew had drug-resistant tuberculosis (TB) and would have to be quarantined to prevent him from spreading the disease to others.

Several months before, Speaker had gone to see his physician, complaining of a sore rib. The doctor ordered an X-ray and saw some fluid in Speaker's lungs. Suspicious, the physician sent samples of the fluid to a lab, which confirmed the diagnosis of TB. Moreover, the TB appeared to be resistant to several drugs.

Before the nineteenth-century German microbiologist Robert Koch identified the bacterium *Mycobacterium tuberculosis* as its causative agent, TB was known as consumption. What started as a bloody cough with fever, chills, and night sweats would usually progress to other organs, including the nervous system. Death was almost inevitable. Today tuberculosis cases are fairly rare in the United States and Europe. Worldwide, however, there are more than 8 million cases and 1.6 million deaths annually, and TB remains the scourge it was in the nineteenth century. Speaker probably picked up the bacterium on his travels and it hid in his tissues (possibly for years) before flaring up at the time of his wedding.

Two drugs are used as the first approach to treating TB. One of them, isoniazid, is activated inside the bacterial cell, and the activated form blocks an enzyme essential for the assembly of the bacterial cell walls. Without functional cell walls, new bacterial cells cannot survive. The second drug, rifampin, binds to a part of the enzyme RNA polymerase that is necessary for gene expression. Without the appropriate expression of its genes, the bacterium soon dies.

In both cases, the targets of the antibiotics are proteins, each encoded by a gene (a sequence of DNA). Mutations in these genes can lead to altered amino acid sequences in the proteins, so that they no longer have three-dimensional structures that bind to the antibiotics. Unfortunately, these altered genes can be transferred from one bacterium to another, so a single *M. tuberculosis* strain can evolve to have both mutations and be resistant to both antibiotics. That is what happened in the case of Andrew Speaker.

He made his way to Denver for treatment, this time with a third antibiotic, kanamycin, which binds to the bacterial ribosome. The ribosome is the cell's protein synthesis factory, and is also essential for gene expression. Finally, the treatment was successful.

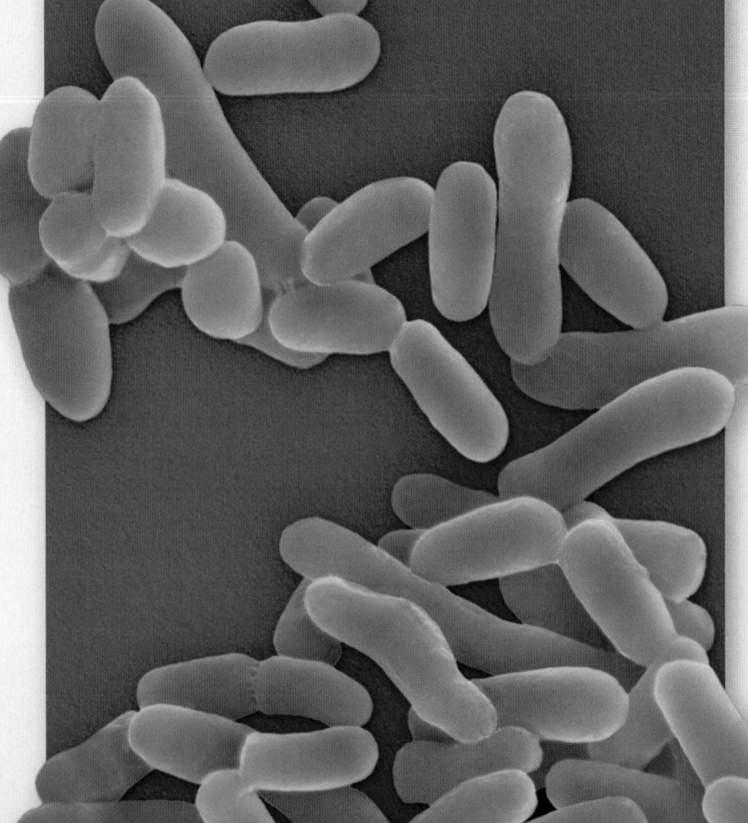

***Mycobacterium tuberculosis*** The causative agent of TB can be killed with antibiotics, but resistance sometimes occurs.

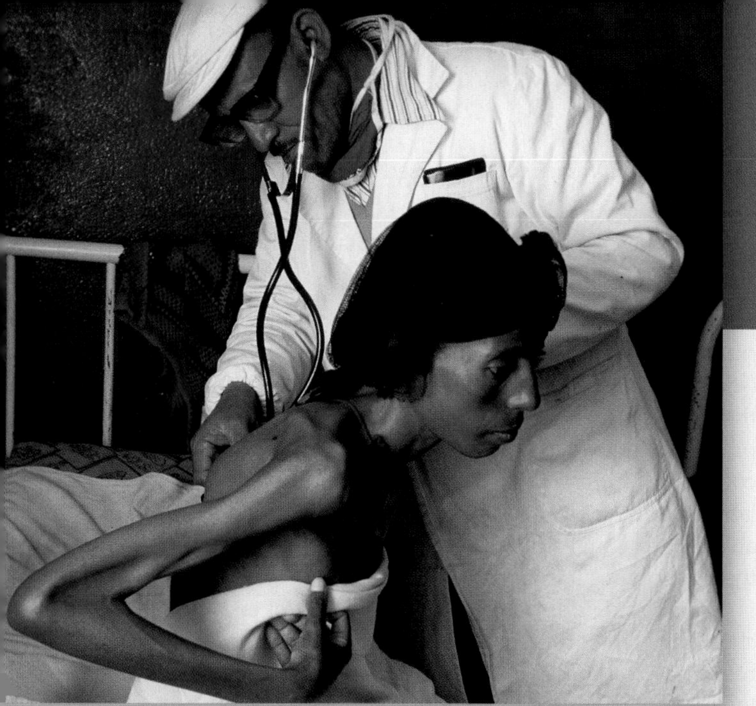

**Tuberculosis Is a World Health Issue** Drug-resistant TB has become a major medical problem throughout the world. Here a doctor examines a patient in Ethiopia, which ranks high among the world's nations in number of TB cases.

Proteins are the major products of gene expression. Some proteins play vital structural roles in cells, and others act as enzymes, which are essential for most aspects of phenotypic expression. So, when protein synthesis is inhibited, cells cannot survive. This is what happened to the TB-causing bacteria when Speaker was treated with kanamycin. But genes can mutate, and the alleles that result may encode proteins that have altered surfaces. The mutant alleles in resistant TB encoded proteins that would no longer bind antibiotics.

**IN THIS CHAPTER** we will describe how genes are expressed as proteins, first discussing the evidence for the relationship between genes and proteins. We will then describe how the DNA sequence of a gene is copied (transcribed) into a sequence of RNA, and how the RNA sequence is translated to make a polypeptide with a defined sequence of amino acids. We will discuss some of the modifications to proteins that occur after they are made by the ribosomes. Following Mendel's definition of the gene as a physical entity, scientists characterized the gene as DNA (see Chapter 13). In this chapter we see how the gene gets expressed as a phenotype at the molecular level.

CHAPTER OUTLINE



## 14.1 What Is the Evidence that Genes Code for Proteins?

In Chapter 12, we defined genes as sequences of DNA and learned that genes are expressed as physical characteristics known as the phenotype. Here, we show that in most cases, genes code for proteins, and it is the proteins that determine the phenotype. What is the evidence for this?

The molecular basis of phenotypes was actually discovered before it was known that DNA was the genetic material. Scientists had studied the chemical differences between individuals carrying wild-type and mutant alleles in organisms as diverse as humans and bread molds. They found that the major phenotypic differences resulted from differences in specific proteins.

### Observations in humans led to the proposal that genes determine enzymes

The identification of a gene product as a protein began with a mutation. In the early twentieth century, the English physician Archibald Garrod saw a number of children with a rare disease. One symptom of the disease was that the urine turned dark brown in air. This was especially noticeable on the infants' diapers. The disease was given the descriptive name alkaptonuria ("black urine").

Garrod noticed that the disease was most common in children whose parents were first cousins. Mendelian genetics had just been "rediscovered," and Garrod realized that because first cousins share alleles (can you calculate what fraction?), the children of first cousins might inherit a rare mutant allele from both parents. He proposed that alkaptonuria was a phenotype caused by a recessive, mutant allele.

Garrod took the analysis one step further. He identified the biochemical abnormality in the affected children. He isolated from them an unusual substance, homogentisic acid, which accumulated in blood, joints (where it crystallized and caused severe pain), and urine (where it turned black). The chemical structure of homogentisic acid is similar to that of the amino acid tyrosine:

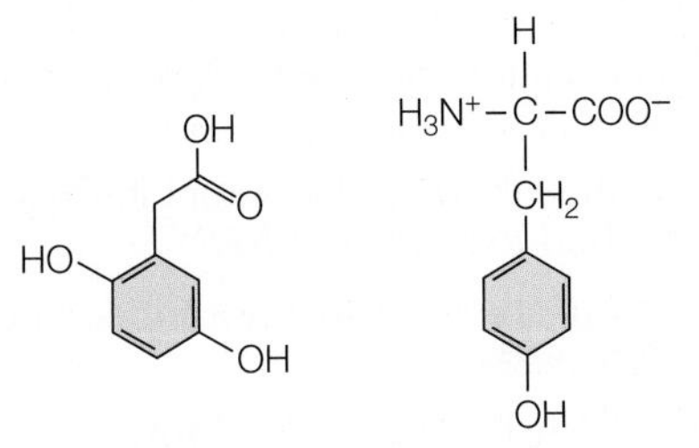

Homogentisic acid    Tyrosine

Enzymes as biological catalysts had just been discovered. Garrod proposed that homogentisic acid was a breakdown product of tyrosine. Normally, homogentisic acid is converted to a harmless product. According to Garrod, there is a normal (wild-type) human allele that determines the synthesis of an enzyme that catalyzes this conversion:

normal allele
↓
active enzyme
↓
tyrosine → homogentisic acid → harmless product

When the allele has been mutated, the enzyme is inactive and homogentisic acid accumulates instead:

mutant allele
↓
inactive enzyme
↓
tyrosine → homogentisic acid ⊣ (HA accumulates)

Therefore, Garrod correlated *one gene to one enzyme* and coined the term "inborn error of metabolism" to describe this genetically determined biochemical disease. But his hypothesis needed direct confirmation by the identification of the specific enzyme and specific gene mutation involved. This did not occur until the enzyme, homogentisic acid oxidase, was described as active in healthy people and inactive in alkaptonuria patients in 1958, and the specific DNA mutation was described in 1996.

To directly relate genes and enzymes, biologists first turned to simpler organisms that could be manipulated in the laboratory.

### Experiments on bread mold established that genes determine enzymes

As they work to explain the principles that underlie life, biologists often turn to organisms that they can manipulate experimentally. It wasn't possible to follow up on Garrod's hypothesis relating genes to enzymes in humans, because it is unethical to perform genetics experiments on people. Instead, biologists use *model organisms* that are easy to grow in the laboratory or greenhouse, and use them to develop principles of genetics that can then be applied more generally to other organisms. You have seen some of these model organisms in previous chapters:

- Pea plants (*Pisum sativum*) were used by Mendel in his genetics experiments.
- Fruit flies (*Drosophila*) were used by Morgan in his genetics experiments.
- *E. coli* was used by Meselson and Stahl to study DNA replication.

To this list we now add the common bread mold, *Neurospora crassa*. *Neurospora* is a type of fungus known as an ascomycete (see Chapter 30). This mold is haploid for most of its life, so that there are no dominant or recessive alleles: all alleles are expressed phenotypically and are not masked by a heterozygous condition. *Neurospora* is simple to culture and grows well in the laboratory. In the 1940s, George W. Beadle and Edward L. Tatum at Stanford University undertook studies to chemically define the phenotypes in *Neurospora*.

Beadle and Tatum knew about the roles of enzymes in biochemistry when they began their work, and like Garrod, they hypothesized that the expression of a specific gene results in the activity of a specific enzyme. Now, they set out to *test this hypothesis directly*. They grew *Neurospora* on a minimal nutritional medium containing sucrose, minerals, and a vitamin. Using this medium, the enzymes of wild-type *Neurospora* could catalyze all the metabolic reactions needed to make all the chemical constituents of their cells, including amino acids and proteins. These wild-type strains are called *prototrophs* ("original eaters"). From these wild type strains, they were able to produce and isolate distinct mutant strains that showed specific biochemical deficiencies.

Mutations provide a powerful way to determine cause and effect in biology. Nowhere has this been so evident as in the elucidation of biochemical pathways. Such pathways consist of sequential events (chemical reactions) in which each event is dependent on the occurrence of the preceding event. The general reasoning is as follows:

- *Observation.* Condition (1) occurs and condition (2) occurs; that is, (1) and (2) are *correlated.*
- *Hypothesis.* Condition (1) results in condition (2); that is, (1) *causes* (2).

In biochemical genetics, this can be stated as follows:

- *Observation.* A particular gene (*a*) is present and a particular reaction catalyzed by a particular enzyme (A) occurs; the two are correlated.
- *Hypothesis.* The gene (*a*) encodes (causes the synthesis of) the specific enzyme (A).
- *Test of hypothesis.* A mutant gene (*a*′) encodes a nonfunctional enzyme (A′) and the reaction does not occur.

Beadle and Tatum treated wild-type *Neurospora* with X rays, which act as a **mutagen** (something known to cause mutations—inherited genotypic changes). When they tested the treated molds, they found that some mutant strains could no longer grow on the minimal medium, but grew only if they were supplied with additional nutrients, such as amino acids. The scientists hypothesized that these *auxotrophs* ("increased eaters") must have suffered mutations in genes that encoded the enzymes used to synthesize the nutrients that they now needed to obtain from their environment.

For each auxotrophic mutant strain, Beadle and Tatum were able to find a single compound that, when added to the minimal medium, supported the growth of that strain. These results suggested that mutations have simple effects, and that each mutation causes a defect in only one enzyme in a metabolic pathway. These conclusions confirmed Garrod's **one-gene, one-enzyme hypothesis** (**Figure 14.1**).

One group of auxotrophs, for example, could grow only if the minimal medium was supplemented with the amino acid arginine. These strains were designated *arg* mutants. Beadle and Tatum found several different *arg* mutant strains. They proposed two alternative hypotheses to explain why these different genetic strains had the same phenotype:

- The different *arg* mutants could have mutations in the same gene, as is the case for some eye color mutations in fruit flies. In this case, the gene might code for an enzyme involved in arginine synthesis.
- The different *arg* mutants could have mutations in different genes, each coding for a separate function that leads to arginine production. These independent functions might be different enzymes along the same biochemical pathway.

Some of the *arg* mutant strains fell into one of these two categories, and some into the other. Genetic crosses showed that some of the mutations were at the same chromosomal locus, and were different alleles of the same gene. Other mutations were at different loci, or on different chromosomes, and so were not alleles of the same gene. Beadle and Tatum concluded that these different genes participated in governing a single biosynthetic pathway—in this case, the pathway leading to arginine synthesis. Next, they set out to elucidate each step in this pathway (see the Interpretation in Figure 14.1).

By growing different *arg* mutants in the presence of various compounds suspected to be intermediates in the biosynthetic pathway for arginine, Beadle and Tatum were able to classify each mutation as affecting one enzyme or another, and to order the compounds along the pathway. Then they broke open the wild-type and mutant cells and examined them for enzyme activities. The results confirmed their hypothesis: each mutant strain was indeed missing a single enzyme activity in the pathway. In general, gene expression controls metabolism.

## INVESTIGATING LIFE

**14.1 One Gene, One Enzyme**

Beadle and Tatum had several mutant strains of *Neurospora* that could not make arginine (*arg*). Several compounds are needed for arginine synthesis. By testing these compounds in the growth media for the mutant strains, the researchers deduced that each mutant strain was deficient in one enzyme along a biochemical pathway.

**HYPOTHESIS** **Each gene determines an enzyme in a biochemical pathway.**

**METHOD** Place spores (single cells that divide to produce mold colonies) of each *arg* mutant strain on a minimal nutritional medium with and without supplements.

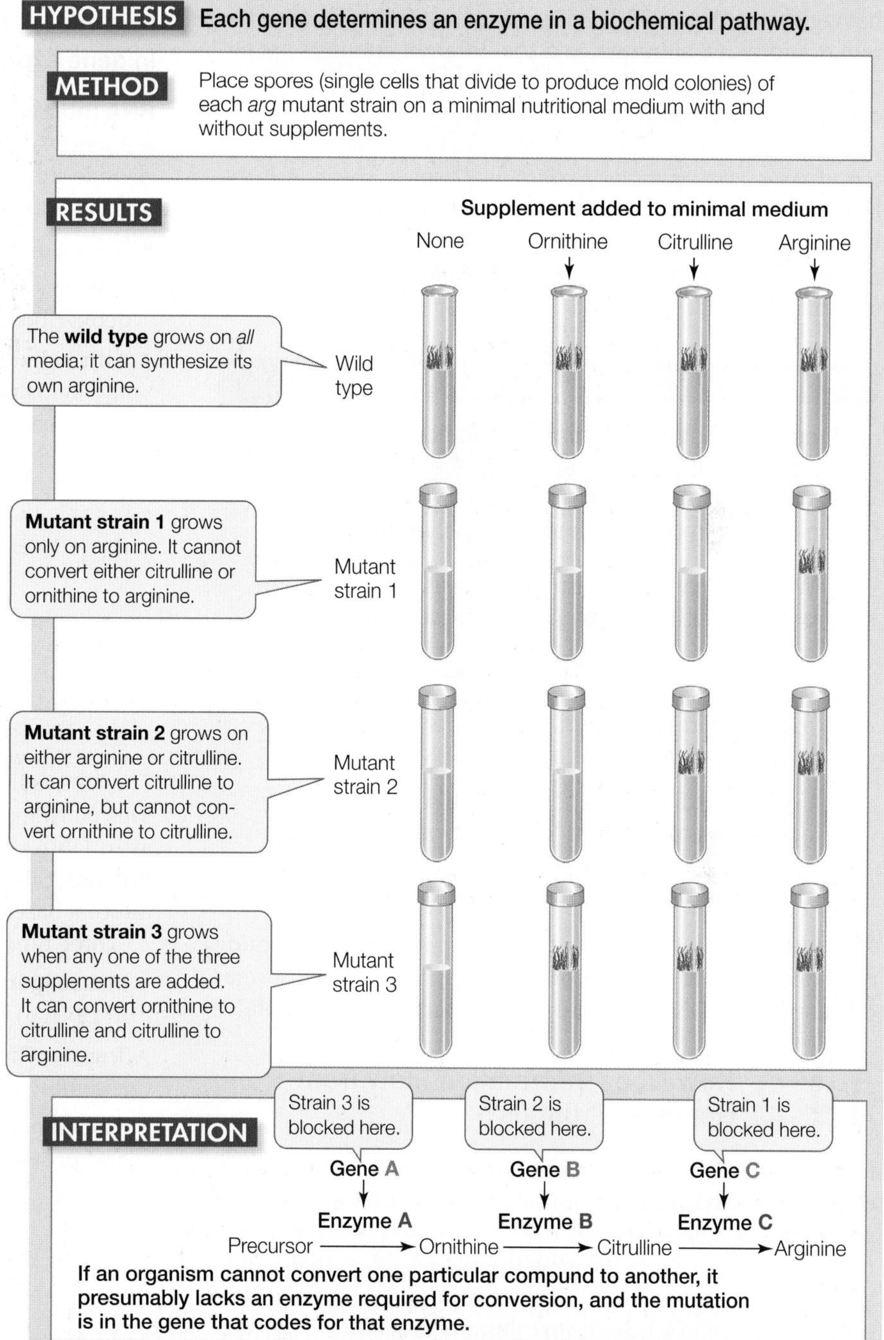

RESULTS
Supplement added to minimal medium
None
Ornithine
Citrulline
Arginine
The **wild type** grows on *all* media; it can synthesize its own arginine.
Wild type
**Mutant strain 1** grows only on arginine. It cannot convert either citrulline or ornithine to arginine.
Mutant strain 1
**Mutant strain 2** grows on either arginine or citrulline. It can convert citrulline to arginine, but cannot convert ornithine to citrulline.
Mutant strain 2
**Mutant strain 3** grows when any one of the three supplements are added. It can convert ornithine to citrulline and citrulline to arginine.
Mutant strain 3
INTERPRETATION
Strain 3 is blocked here.
Strain 2 is blocked here.
Strain 1 is blocked here.
Gene A
Gene B
Gene C
Enzyme A
Enzyme B
Enzyme C
Precursor → Ornithine → Citrulline → Arginine
**If an organism cannot convert one particular compund to another, it presumably lacks an enzyme required for conversion, and the mutation is in the gene that codes for that enzyme.**


**CONCLUSION** **Each gene specifies a particular enzyme.**

**FURTHER INVESTIGATION:** If a diploid *Neurospora* spore were made from two haploid cells, one with mutant 3 and the other with mutant 2, what would be its phenotype?

Go to **yourBioPortal.com** for original citations, discussions, and relevant links for all INVESTIGATING LIFE figures.

### One gene determines one polypeptide

The gene–enzyme relationship has undergone several modifications in light of our current knowledge of molecular biology. Many proteins, including many enzymes, are composed of more than one polypeptide chain, or subunit (that is, they have a quaternary structure; see Section 3.2). Look at the illustration of hemoglobin in Figure 3.10. This protein has four polypeptides—two α and two β subunits, and the different subunits are encoded by separate genes. Thus it is more correct to speak of a **one-gene, one-polypeptide relationship**.

So far we have seen that in terms of protein synthesis, the *function of a gene is to inform the production of a single, specific polypeptide*. But not all genes code for polypeptides. As we will see below and in Chapter 16, there are many DNA sequences that code for RNA molecules that are not translated into polypeptides but instead have other functions.

### 14.1 RECAP

**Beadle and Tatum's studies of mutations in bread molds led to our understanding of the one-gene, one-polypeptide relationship. In most cases, the function of a gene is to code for a specific polypeptide.**

- What is a model organism, and why is *Neurospora* a good model for studying biochemical genetics? See p. 292
- How were Beadle and Tatum's experiments on *Neurospora* set up to determine the order of steps in a biochemical pathway? See pp. 292–293 and Figure 14.1
- Explain the distinction between the phrases "one-gene, one-enzyme" and "one-gene, one-polypeptide." See pp. 293–294

Now that we have established the one-gene, one-polypeptide relationship, how does it work? That is, how is the information encoded in DNA used to produce a particular polypeptide?

## 14.2 How Does Information Flow from Genes to Proteins?

Much of the biochemical genetics in the middle of the twentieth century was directed at revealing the relationship between genes and protein synthesis. As we discussed in Section 14.1, the expression of a specific gene usually results in the synthesis of a specific polypeptide. The process of gene expression was outlined in Section 4.1. To review, this process occurs in two major steps:

- During **transcription**, the information in a DNA sequence (a gene) is copied into a complementary RNA sequence.
- During **translation**, this RNA sequence is used to create the amino acid sequence of a polypeptide.

In 1958, Francis Crick described this process as "the **central dogma** of molecular biology."

### RNA differs from DNA and plays a vital role in gene expression

**RNA (ribonucleic acid)** is a key intermediary between a DNA sequence and a polypeptide. RNA is an informational polynucleotide similar to DNA (see Figure 4.2), but it differs from DNA in three ways:

- RNA generally consists of only one polynucleotide strand.
- The sugar molecule found in RNA is ribose, rather than the deoxyribose found in DNA.
- Although three of the nitrogenous bases (adenine, guanine, and cytosine) in RNA are identical to those in DNA, the fourth base in RNA is **uracil (U)**, which is similar to thymine but lacks the methyl (—$CH_3$) group.

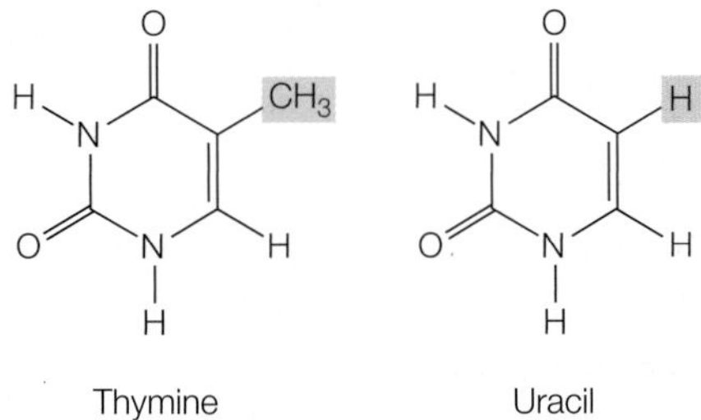


The bases in RNA can pair with those in a single strand of DNA. This pairing obeys the same complementary base-pairing rules as in DNA, except that adenine pairs with uracil instead of thymine. Single-stranded RNA can fold into complex shapes by internal base pairing, as seen below.

Three types of RNA participate in protein synthesis:

- **Messenger RNA (mRNA)** carries a copy of a gene sequence in DNA to the site of protein synthesis at the ribosome.
- **Transfer RNA (tRNA)** carries amino acids to the ribosome for assembly into polypeptides.
- **Ribosomal RNA (rRNA)** catalyzes peptide bond formation and provides a structural framework for the ribosome.

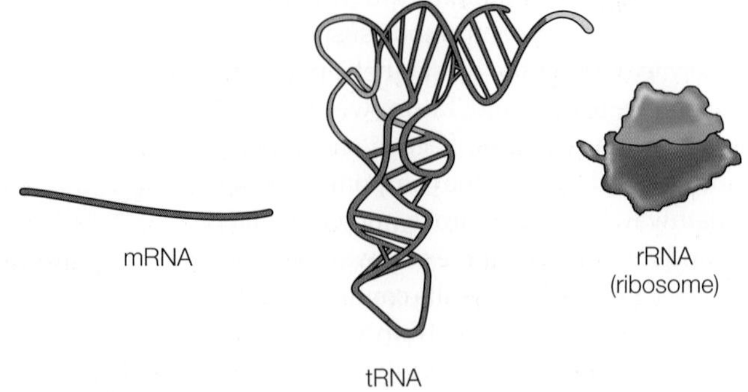

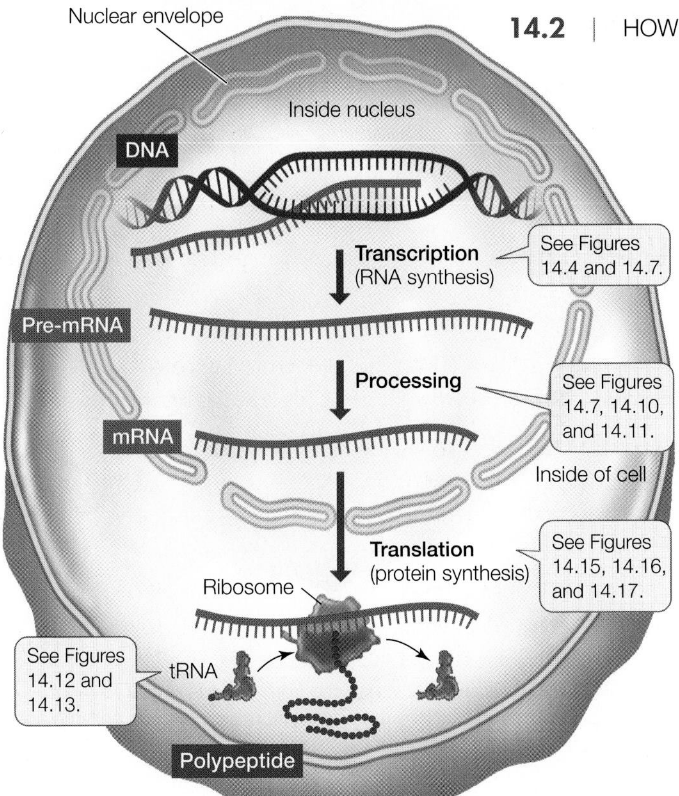


**14.2 From Gene to Protein** This diagram summarizes the processes of gene expression in eukaryotes.

yourBioPortal.com
GO TO Web Activity 14.1 • Eukaryotic Gene Expression

### Two hypotheses were proposed to explain information flow from DNA to protein

The central dogma suggested that information flows from DNA to RNA to protein, and not in the reverse direction. It raised two questions:

- How does genetic information get from the nucleus of a eukaryotic cell to the cytoplasm? (As Section 5.3 explains, most of the DNA of a eukaryotic cell is confined to the nucleus, but proteins are synthesized in the cytoplasm.)
- What is the relationship between a specific nucleotide sequence in DNA and a specific amino acid sequence in a protein?

To answer these questions, Crick proposed two hypotheses, the messenger hypothesis and the adapter hypothesis.

**THE MESSENGER HYPOTHESIS AND TRANSCRIPTION** Crick and his colleagues proposed that an RNA molecule forms as a complementary copy of one DNA strand in a gene. This messenger RNA, or mRNA, then travels from the nucleus to the cytoplasm, where it serves as an informational sequence of **codons**. Each codon consists of three consecutive nucleotides, and different codons encode particular amino acids. Thus the mRNA sequence determines the ordered sequence of amino acids in a polypeptide chain, which is built by the ribosome. The process by which RNA forms is called transcription (**Figure 14.2**).

This hypothesis has been tested repeatedly, and the result is always the same: each DNA sequence that encodes a protein is transcribed as a sequence of mRNA. Today it is routine in thousands of laboratories around the world to test for gene expression by examining the mRNA copy of the gene, which is often called the **transcript**. There is no longer any question that Crick's model was correct.

**THE ADAPTER HYPOTHESIS AND TRANSLATION** To answer the question of how a DNA sequence gets transformed into the specific amino acid sequence of a polypeptide, Crick proposed the adapter hypothesis: that there must be an adapter molecule that can both bind a specific amino acid and recognize a specific sequence of nucleotides. He proposed that this recognition function occurs because the adapter molecule contains an **anticodon** complementary to the codon in the mRNA. He envisioned such adapters as molecules with two regions, one serving the binding function and the other serving the recognition function.

In due course, such adapter molecules were found: they are known as transfer RNA, or tRNA. Each tRNA recognizes a specific codon in the mRNA and simultaneously carries the specific amino acid corresponding to that codon. Thus, the tRNAs together can translate the language of DNA into the language of proteins. The tRNA adapters, carrying bound amino acids, line up on the mRNA sequence so that the amino acids are in the proper sequence for a growing polypeptide chain—in the process of *translation* (see Figure 14.2). Once again, actual observations of the expression of thousands of genes in all types of organisms have confirmed the hypothesis that tRNA acts as the intermediary between the nucleotide sequence information in mRNA and the amino acid sequence in a protein.

We can summarize the main features of the central dogma, the messenger hypothesis, and the adapter hypothesis as follows: a given gene is transcribed to produce an mRNA molecule that is complementary to one of the DNA strands, and then the tRNA molecules translate the sequence of codons in the mRNA into a sequence of linked amino acids, to form a polypeptide.

### RNA viruses are exceptions to the central dogma

Certain viruses present exceptions to the central dogma. As we saw in Section 13.1, a virus is a non-cellular infectious particle that reproduces inside cells. Many viruses, such as the tobacco mosaic virus, influenza viruses, and poliovirus, have *RNA* rather than DNA as their genetic material. With its nucleotide sequence, RNA could potentially act as an information carrier and be expressed as a protein. But if RNA is usually single-stranded, how do these viruses replicate? They generally solve this problem by transcribing from RNA to RNA, making an RNA strand that is complementary to their genomes. This "opposite" strand is then used to make multiple copies of the viral genome by transcription:

Human immunodeficiency viruses (HIV) and certain rare tumor viruses also have RNA as their genomes, but do not replicate by transcribing from RNA to RNA. Instead, after infecting a host cell, such a virus makes a DNA copy of its genome, which becomes incorporated into the host's genome. The virus relies on the host cell's transcription machinery to make more RNA. This RNA can be either translated to produce viral proteins, or incorporated as the viral genome into new viral particles. Synthesis of DNA from RNA is called **reverse transcription**, and not surprisingly, such viruses are called **retroviruses**.

## 14.2 RECAP

**The central dogma of molecular biology states that the DNA code is used to produce RNA and the RNA sequence determines the sequence of amino acids in a polypeptide. Transcription is the process by which a DNA sequence is copied into mRNA. Translation is the process by which this information is converted into polypeptide chain. Transfer RNAs recognize the genetic information in messenger RNA and bring the appropriate amino acids into position in a growing polypeptide chain.**

- What is the central dogma of molecular biology? See p. 294
- What are the roles of mRNA and tRNA in gene expression? See p. 294 and Figure 14.2

The central dogma is indeed central to gene expression in all organisms. Understanding its details is essential for understanding how organisms function at the molecular level, and this understanding is key to the application of biology to human welfare, in areas such as agriculture and medicine. Much of the remainder of this book will in one way or another involve DNA and proteins. Let's begin by describing how the information in DNA is transcribed to produce RNA.

## 14.3 How Is the Information Content in DNA Transcribed to Produce RNA?

In normal prokaryotic and eukaryotic cells, RNA synthesis is directed by DNA. Transcription—the formation of a specific RNA sequence from a specific DNA sequence—requires several components:

- A DNA template for complementary base pairing; one of the two strands of DNA
- The appropriate nucleoside triphosphates (ATP, GTP, CTP, and UTP) to act as substrates
- An RNA polymerase enzyme

Not only mRNA is produced by transcription. The same process is responsible for the synthesis of tRNA and ribosomal RNA (rRNA), whose important roles in protein synthesis will be described below. Like polypeptides, these RNAs are encoded by specific genes. In addition, as we will see in Chapter 16, many small RNAs, called microRNAs, are transcribed. These molecules stay in the nucleus, where they play roles in stimulating or inhibiting gene expression.

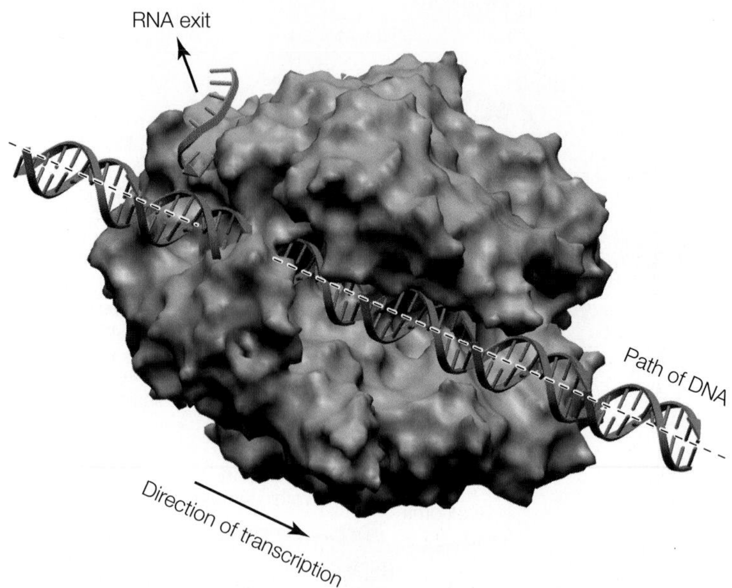


**14.3 RNA Polymerase** This enzyme from yeast is similar to most other RNA polymerases. Note the size relationship between enzyme and DNA. See Figure 14.4 for details.

### RNA polymerases share common features

**RNA polymerases** from both prokaryotes and eukaryotes catalyze the synthesis of RNA from the DNA template. There is only one kind of RNA polymerase in bacteria, while there are several kinds in eukaryotes; however, they all share a common structure (**Figure 14.3**). Like DNA polymerases, RNA polymerases are *processive*; that is, a single enzyme–template binding event results in the polymerization of hundreds of RNA bases. But unlike DNA polymerases, RNA polymerases *do not require a primer* and *do not have a proofreading function.*

### Transcription occurs in three steps

Transcription can be divided into three distinct processes: initiation, elongation, and termination. You can follow these processes in **Figure 14.4**.

**INITIATION** Transcription begins with initiation, which requires a **promoter**, a special sequence of DNA to which the RNA polymerase binds very tightly (see Figure 14.4A). Eukaryotic genes generally have one promoter each, while in prokaryotes and viruses, several genes often share one promoter. Promoters are important control sequences that "tell" the RNA polymerase two things:

- Where to start transcription
- Which strand of DNA to transcribe

A promoter, which is a specific sequence in the DNA that reads in a particular direction, orients the RNA polymerase and thus "aims" it at the correct strand to use as a template. Promoters

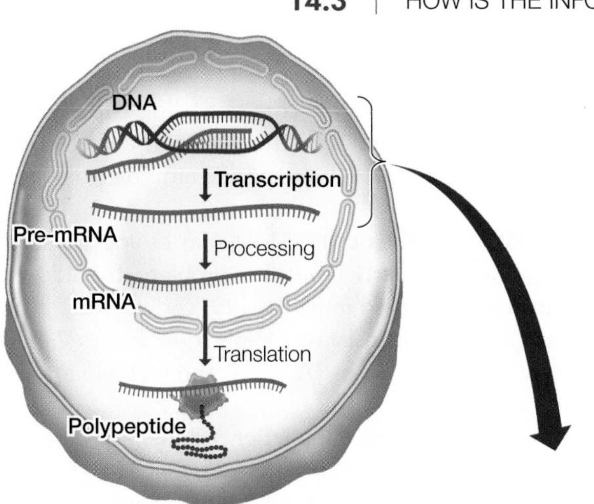


**14.4 DNA Is Transcribed to Form RNA** DNA is partially unwound by RNA polymerase to serve as a template for RNA synthesis. The RNA transcript is formed and then peels away, allowing the DNA that has already been transcribed to rewind into a double helix. Three distinct processes—initiation, elongation, and termination—constitute DNA transcription. RNA polymerase is much larger in reality than indicated here, covering about 50 base pairs.

**yourBioPortal.com**
GO TO **Animated Tutorial 14.1 • Transcription**

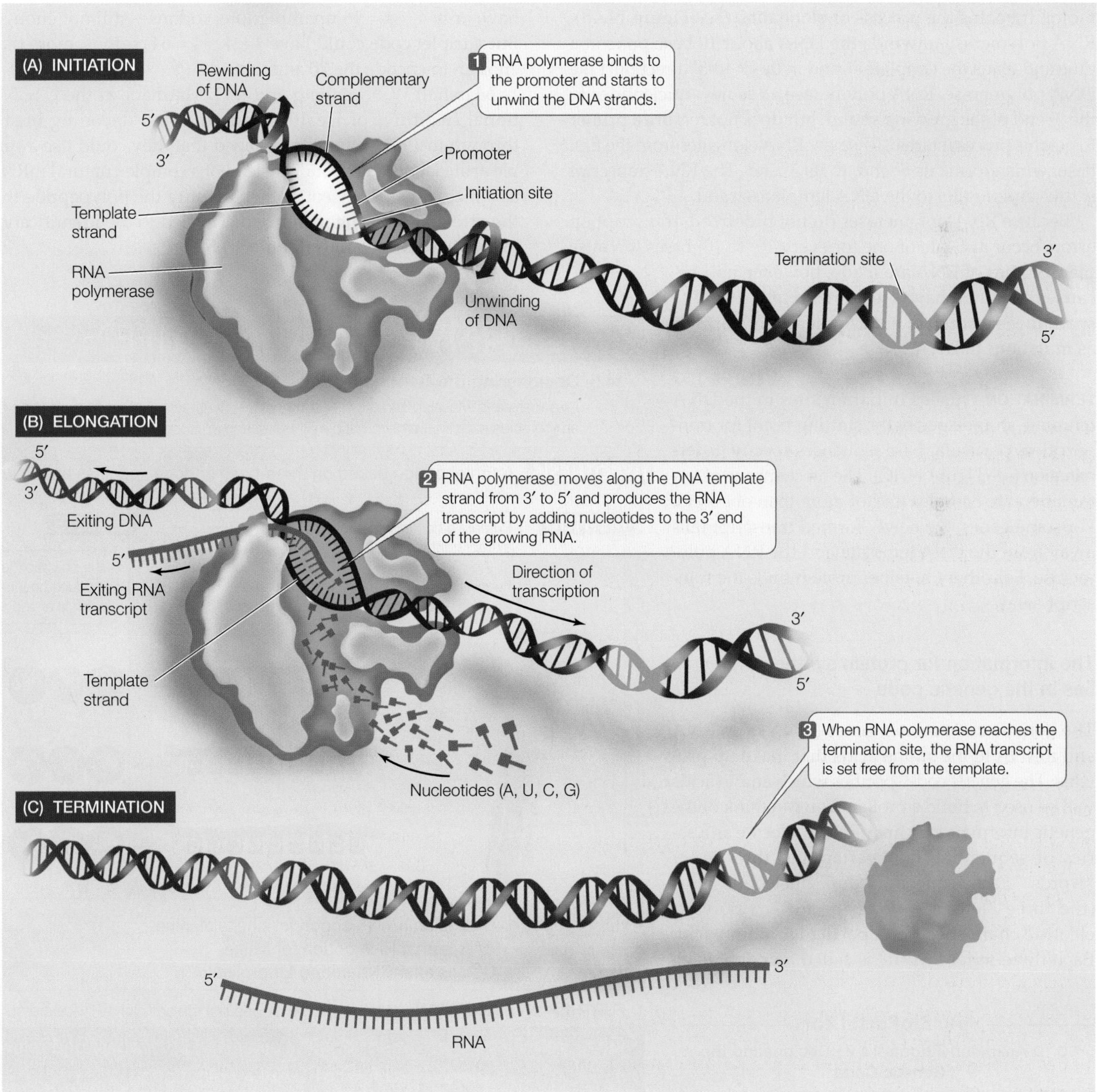

(A) INITIATION
Rewinding of DNA
Complementary strand
1 RNA polymerase binds to the promoter and starts to unwind the DNA strands.
5′
3′
Promoter
Initiation site
Template strand
RNA polymerase
Termination site
3′
5′
Unwinding of DNA

(B) ELONGATION
5′
3′
Exiting DNA
2 RNA polymerase moves along the DNA template strand from 3′ to 5′ and produces the RNA transcript by adding nucleotides to the 3′ end of the growing RNA.
5′
Exiting RNA transcript
Direction of transcription
3′
5′
Template strand
Nucleotides (A, U, C, G)

(C) TERMINATION
3 When RNA polymerase reaches the termination site, the RNA transcript is set free from the template.
5′
3′
RNA

function somewhat like the capital letter at the beginning of a sentence, indicating how the sequence of words should be read. Part of each promoter is the **initiation site**, where transcription begins. Groups of nucleotides lying "upstream" from the initiation site (5′ on the non-template strand, and 3′ on the template strand) help the RNA polymerase bind.

Although every gene has a promoter, not all promoters are identical. Some promoters are more effective at transcription initiation than others. Furthermore, there are differences between transcription initiation in prokaryotes and in eukaryotes. But despite these variations, the basic mechanisms of initiation are the same throughout the living world.

**ELONGATION** Once RNA polymerase has bound to the promoter, it begins the process of **elongation** (see Figure 14.4B). RNA polymerase unwinds the DNA about 10 base pairs at a time and reads the template strand in the 3′-to-5′ direction. Like DNA polymerase, RNA polymerase adds new nucleotides to the 3′ end of the growing strand, but does not require a primer to get this process started. The new RNA elongates from the first base, which forms its 5′ end, to its 3′ end. The RNA transcript is thus antiparallel to the DNA template strand.

Because RNA polymerases do not proofread, transcription errors occur at a rate of one for every $10^4$ to $10^5$ bases. Because many copies of RNA are made, however, and because they often have only a relatively short life span, these errors are not as potentially harmful as mutations in DNA.

**TERMINATION** Just as initiation sites in the DNA template strand specify the starting point for transcription, particular base sequences specify its **termination** (see Figure 14.4C). The mechanisms of termination are complex and of more than one kind. For some genes, the newly formed transcript falls away from the DNA template and the RNA polymerase. For others, a helper protein pulls the transcript away.

### The information for protein synthesis lies in the genetic code

The **genetic code** relates genes (DNA) to mRNA and mRNA to the amino acids that make up proteins. The genetic code specifies which amino acids will be used to build a protein. You can think of the genetic information in an mRNA molecule as a series of sequential, nonoverlapping three-letter "words." Each sequence of three nucleotide bases (the three "letters") along the mRNA polynucleotide chain specifies a particular amino acid. Each three-letter "word" is called a **codon**. Each codon is complementary to the corresponding triplet of bases in the DNA molecule from which it was transcribed. The genetic code relates codons to their specific amino acids.

**CHARACTERISTICS OF THE CODE** Molecular biologists "broke" the genetic code in the early 1960s. The problem they addressed was perplexing: how could more than 20 "code words" be written with an "alphabet" consisting of only four "letters"? In other words, how could four bases (A, U, G, and C) code for 20 different amino acids?

A triplet code, based on three-letter codons, was considered likely. Since there are only four letters (A, G, C, and U), a one-letter code clearly could not unambiguously encode 20 amino acids; it could encode only four of them. A two-letter code could have only $4 \times 4 = 16$ unambiguous codons—still not enough. But a triplet code could have $4 \times 4 \times 4 = 64$ codons, more than enough to encode the 20 amino acids.

Marshall W. Nirenberg and J. H. Matthaei, at the U.S. National Institutes of Health, made the first decoding breakthrough in 1961 when they realized that they could use a simple artificial polynucleotide instead of a complex natural mRNA as a messenger. They could then identify the polypeptide that the artificial messenger encoded. This led to the identification of the first three codons (**Figure 14.5**).

INVESTIGATING LIFE

**14.5 Deciphering the Genetic Code**
Nirenberg and Matthaei used a test-tube protein synthesis system to determine the amino acids specified by synthetic mRNAs of known codon compositions.

**HYPOTHESIS** A triplet codon based on three-base codons specifies amino acids.

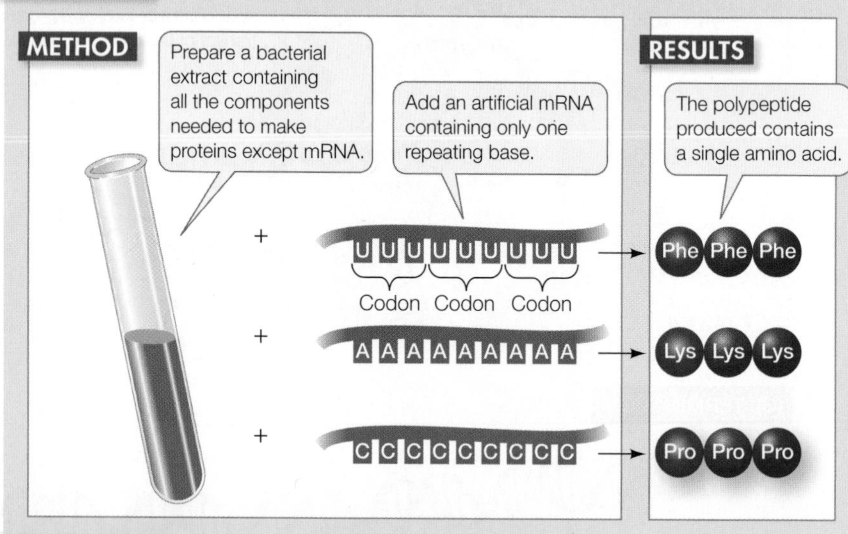
METHOD
Prepare a bacterial extract containing all the components needed to make proteins except mRNA.
Add an artificial mRNA containing only one repeating base.
RESULTS
The polypeptide produced contains a single amino acid.
+
UUUUUUUUU
Codon Codon Codon
Phe Phe Phe
+
AAAAAAAAA
Lys Lys Lys
+
CCCCCCCCC
Pro Pro Pro

**CONCLUSION** UUU is an mRNA codon for phenylalanine.
AAA is an mRNA codon for lysine.
CCC is an mRNA codon for proline.

**FURTHER INVESTIGATION:** What would be the result if the artificial mRNA were poly-G?

Go to **yourBioPortal.com** for original citations, discussions, and relevant links for all INVESTIGATING LIFE figures.

**yourBioPortal.com**

GO TO Animated Tutorial 14.2 • Deciphering the Genetic Code

Other scientists later found that simple artificial mRNAs only three nucleotides long—each amounting to one codon—could bind to a ribosome, and that the resulting complex could then bind to the corresponding tRNA with its specific amino acid. Thus, for example, a simple UUU mRNA caused the tRNA carrying phenylalanine to bind to the ribosome. After this discovery, the complete deciphering of the genetic code was relatively simple. To discover which amino acid a codon represented, the scientists simply repeated the experiment using a sample of artificial mRNA for that codon, and observed which amino acid became bound to it.

The complete genetic code is shown in **Figure 14.6**. Notice that there are many more codons than there are different amino acids in proteins. All possible combinations of the four available "letters" (the bases) give 64 ($4^3$) different three-letter codons, yet these codons determine only 20 amino acids. AUG, which codes for methionine, is also the **start codon**, the initiation signal for translation. Three of the codons (UAA, UAG, UGA) are **stop codons**, or termination signals for translation. When the translation machinery reaches one of these codons, translation stops, and the polypeptide is released from the translation complex.

What happens if a stop codon isn't there? In humans a severe anemic condition, α-thalassemia, results from various mutations in either of the two genes that encode the α-polypeptide chain of hemoglobin. In one of these mutant alleles, the stop codon UAA has been converted to GAA, a codon for glutamine. The next stop codon doesn't occur until much further along the mRNA, resulting in a protein molecule with larger, defective α subunits.

**THE GENETIC CODE IS REDUNDANT BUT NOT AMBIGUOUS** The 60 codons that are not start or stop codons are far more than enough to code for the other 19 amino acids—and indeed, for almost all amino acids, there is more than one codon. Thus we say that the genetic code is redundant (or degenerate). For example, leucine is represented by six different codons (see Figure 14.6). Only methionine and tryptophan are represented by just one codon each.

A *redundant* code should not be confused with an *ambiguous* code. If the code were ambiguous, a single codon could specify either of two (or more) different amino acids, and there would be doubt about which amino acid should be incorporated into a growing polypeptide chain. Redundancy in the code simply means that there is more than one clear way to say, "Put leucine here." The genetic code is not ambiguous: a given amino acid may be encoded by more than one codon, but a codon can code for only one amino acid.

**THE GENETIC CODE IS (NEARLY) UNIVERSAL** The same genetic code is used by all the species on our planet. Thus the code must be an ancient one that has been maintained intact throughout the evolution of living organisms. Exceptions are known: within mitochondria and chloroplasts, the code differs slightly from that in prokaryotes and in the nuclei of eukaryotic cells; and in one group of protists, UAA and UAG code for glutamine rather than functioning as stop codons. The significance of these differences is not yet clear. What is clear is that the exceptions are few.

The common genetic code means that there is also a common language for evolution. Natural selection acts on phenotypic variations that result from genetic variation. The genetic code probably originated early in the evolution of life. As we saw in Chapter 4, simulation experiments indicate the plausibility of individual nucleotides and nucleotide polymers arising spontaneously on the primeval Earth. The common code also has profound implications for genetic engineering, as we will see in Chapter 18, since it means that the code for a human gene is the same as that for a bacterial gene. It is therefore impressive, but not surprising, that a human gene can be expressed in *E. coli* via laboratory manipulations, since these cells speak the same "molecular language."

The codons in Figure 14.6 are mRNA codons. The base sequence of the DNA strand that is transcribed to produce the mRNA is complementary and antiparallel to these codons. Thus, for example,

- 3′-AAA-5′ in the template DNA strand corresponds to phenylalanine (which is encoded by the mRNA codon 5′-UUU-3′)

| First letter | Second letter: U | Second letter: C | Second letter: A | Second letter: G | Third letter |
|---|---|---|---|---|---|
| U | UUU Phenylalanine | UCU Serine | UAU Tyrosine | UGU Cysteine | U |
| | UUC | UCC | UAC | UGC | C |
| | UUA Leucine | UCA | UAA Stop codon | UGA Stop codon | A |
| | UUG | UCG | UAG Stop codon | UGG Tryptophan | G |
| C | CUU Leucine | CCU Proline | CAU Histidine | CGU Arginine | U |
| | CUC | CCC | CAC | CGC | C |
| | CUA | CCA | CAA Glutamine | CGA | A |
| | CUG | CCG | CAG | CGG | G |
| A | AUU Isoleucine | ACU Threonine | AAU Asparagine | AGU Serine | U |
| | AUC | ACC | AAC | AGC | C |
| | AUA | ACA | AAA Lysine | AGA Arginine | A |
| | AUG Methionine; start codon | ACG | AAG | AGG | G |
| G | GUU Valine | GCU Alanine | GAU Aspartic acid | GGU Glycine | U |
| | GUC | GCC | GAC | GGC | C |
| | GUA | GCA | GAA Glutamic acid | GGA | A |
| | GUG | GCG | GAG | GGG | G |

**14.6 The Genetic Code** Genetic information is encoded in mRNA in three-letter units—codons—made up of nucleoside monophosphates with the bases uracil (U), cytosine (C), adenine (A), and guanine (G) and is read in a 5′ to 3′ direction on mRNA. To decode a codon, find its first letter in the left column, then read across the top to its second letter, then read down the right column to its third letter. The amino acid the codon specifies is given in the corresponding row. For example, AUG codes for methionine, and GUA codes for valine.

**yourBioPortal.com**
GO TO **Web Activity 14.2 • The Genetic Code**

- 3′-ACC-5′ in the template DNA corresponds to tryptophan (which is encoded by the mRNA codon 5′-UGG-3′)

The non-template strand has the same sequence as the mRNA (but with T′s instead of U′s), and is often referred to as the "coding strand." By convention, DNA sequences are usually shown beginning with the 5′ end of the coding sequence.

## 14.3 RECAP

**Transcription, which is catalyzed by an RNA polymerase, proceeds in three steps: initiation, elongation, and termination. The genetic code relates the information in mRNA (as a linear sequence of codons) to protein (a linear sequence of amino acids).**

- What are the steps of gene transcription that produce mRNA? See pp. 296–298 and Figure 14.4
- How do RNA polymerases work? See pp. 296–298
- How was the genetic code elucidated? See pp. 298–299 and Figure 14.5

The features of transcription that we have described were first elucidated in model prokaryotes, such as *E. coli*. Biologists then used the same methods to analyze this process in eukaryotes, and, although the basics are the same, there are some notable (and important) differences. We now turn to eukaryotic gene expression.

## 14.4 How Is Eukaryotic DNA Transcribed and the RNA Processed?

Since the genetic code is the same, you might expect the process of gene expression to be the same in eukaryotes as it is in prokaryotes. And basically it is. However, there are significant differences in gene structure between prokaryotes and eukaryotes, that is, there are differences in the organization of the nucleotide sequences in the genes. In addition, in eukaryotes but not prokaryotes, a nucleus separates transcription and translation (**Table 14.1**). Let's look at the distinctive *eukaryotic* process of transcription.

### Eukaryotic genes have noncoding sequences

A diagram of the structure and transcription of a typical eukaryotic gene is shown in **Figure 14.7**. In prokaryotes, several adjacent genes sometimes share one promoter; however, in eukaryotes, each gene has its own promoter, which usually precedes the coding region. Unlike the prokaryotic RNA polymerase, a eukaryotic RNA polymerase does not recognize the promoter sequence by itself, but requires help from other molecules, as we'll see in more detail in Chapter 16. At the other end of the gene, downstream from the coding region, is a DNA sequence appropriately called the **terminator**, which signals the end of transcription.

Eukaryotic genes may also contain noncoding base sequences, called **introns** (*int*ervening *regions*). One or more introns may be interspersed with the coding sequences, which are called **exons** (*ex*pressed regi*ons*). Both introns and exons appear in the primary mRNA transcript, called **pre-mRNA**, but the introns are removed by the time the mature mRNA—the mRNA that will be translated—leaves the nucleus. Pre-mRNA processing involves cutting introns out of the pre-mRNA transcript and splicing together the remaining exon transcripts (see Figure 14.7). If this seems surprising, you are in good company. For scientists who were familiar with prokaryotic genes and gene expression, the discovery of introns in eukaryotic genes was entirely unexpected.

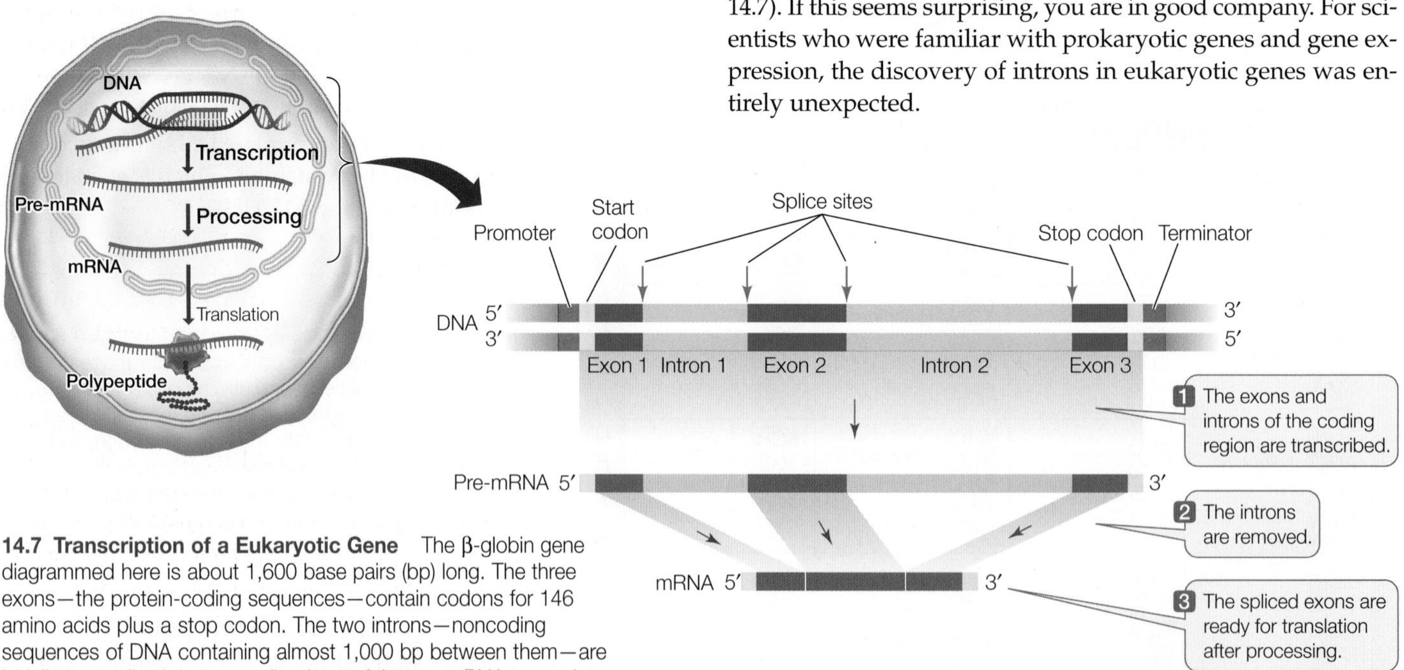


**14.7 Transcription of a Eukaryotic Gene** The β-globin gene diagrammed here is about 1,600 base pairs (bp) long. The three exons—the protein-coding sequences—contain codons for 146 amino acids plus a stop codon. The two introns—noncoding sequences of DNA containing almost 1,000 bp between them—are initially transcribed, but are spliced out of the pre-mRNA transcript.

TABLE 14.1

**Differences between Prokaryotic and Eukaryotic Gene Expression**

| CHARACTERISTIC | PROKARYOTES | EUKARYOTES |
|---|---|---|
| Transcription and translation occurrence | At the same time in the cytoplasm | Transcription in the nucleus, then translation in the cytoplasm |
| Gene structure | DNA sequence is read in the same order as the amino acid sequence | Noncoding introns within coding sequence |
| Modification of mRNA after initial transcription but before translation | None | Introns spliced out; 5′ cap and 3′ poly A added |

How can we locate introns within a eukaryotic gene? One way is by **nucleic acid hybridization**, the method that originally revealed the existence of introns. This method, outlined in **Figure 14.8**, has been crucial for studying the relationship between eukaryotic genes and their transcripts. It involves two steps:

- The target DNA is denatured by heat to break the hydrogen bonds between the base pairs and separate the two strands.
- A single-stranded nucleic acid from another source (called a **probe**) is incubated with the denatured DNA. If the probe has a base sequence complementary to the target DNA, a probe–target double helix forms by hydrogen bonding between the bases. Because the two strands are from different sources, the resulting double-stranded molecule is called a hybrid.

TOOLS FOR INVESTIGATING LIFE

**14.8 Nucleic Acid Hybridization**

Base pairing permits the detection of a sequence that is complementary to the probe.

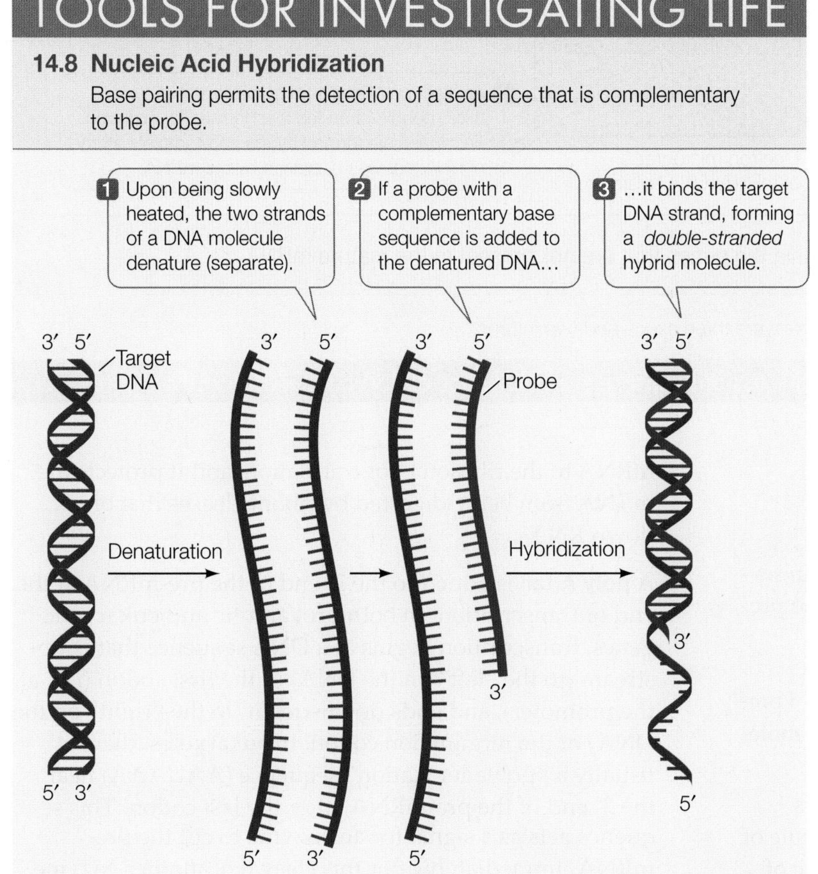

1 Upon being slowly heated, the two strands of a DNA molecule denature (separate).
2 If a probe with a complementary base sequence is added to the denatured DNA...
3 ...it binds the target DNA strand, forming a *double-stranded* hybrid molecule.
Target DNA
Probe
Denaturation
Hybridization
3′ 5′
5′ 3′


Biologists used nucleic acid hybridization to examine the β-globin gene, which encodes one of the globin polypeptides that make up hemoglobin. Follow the experiment in **Figure 14.9** carefully as we describe what they did and what happened.

The researchers first denatured DNA containing the β-globin gene by heating it slowly, then added previously isolated, mature β-globin mRNA. They were able to view the hybridized molecules using electron microscopy. As expected, the mRNA bound to the DNA by complementary base pairing. The researchers expected to obtain a linear (1:1) matchup of the mRNA to the coding DNA. That expectation was only partly met: there were indeed stretches of RNA–DNA hybrid, but some looped structures were also visible. These loops were not expected, and initially the scientists thought that something must be wrong with the experimental procedure. However, when they repeated the experiment, they got the same results, and when they did it with other genes and mRNAs, loops again appeared. The loops turned out to be the introns, stretches of DNA that did not have complementary base sequences on the mature mRNA.

When pre-mRNA was used instead of mature mRNA to hybridize to the DNA, there was complete hybridization, revealing that the introns were indeed part of the pre-mRNA transcript. Somewhere on the path from primary transcript (pre-mRNA) to mature mRNA, the introns had been removed, and the exons had been spliced together. We will examine this splicing process in the next section.

Introns *interrupt, but do not scramble*, the DNA sequence of a gene. The base sequences of the exons in the template strand, if joined and taken in order, form a continuous sequence that is complementary to that of the mature mRNA. In some cases, the separated exons encode different functional regions, or **domains**, of the protein. For example, the globin polypeptides that make up hemoglobin each have two domains: one for binding to a nonprotein pigment called heme, and another for binding to the other globin subunits. These two domains are encoded by different exons in the globin genes. Most (but not all) eukaryotic genes contain introns, and in rare cases, introns are also found in prokaryotes. The largest human gene encodes a muscle protein called titin; it has 363 exons, which together code for 38,138 amino acids.

INVESTIGATING LIFE

**14.9 Demonstrating the Existence of Introns**

When an mRNA transcript of the β-globin gene was hybridized with the double-stranded DNA of that gene, the introns in the DNA "looped out." This demonstrated that the coding region of a eukaryotic gene can contain noncoding DNA that is not present in the mature mRNA transcript.

**HYPOTHESIS** Some regions within the coding sequence of a gene do not end up in its mRNA.

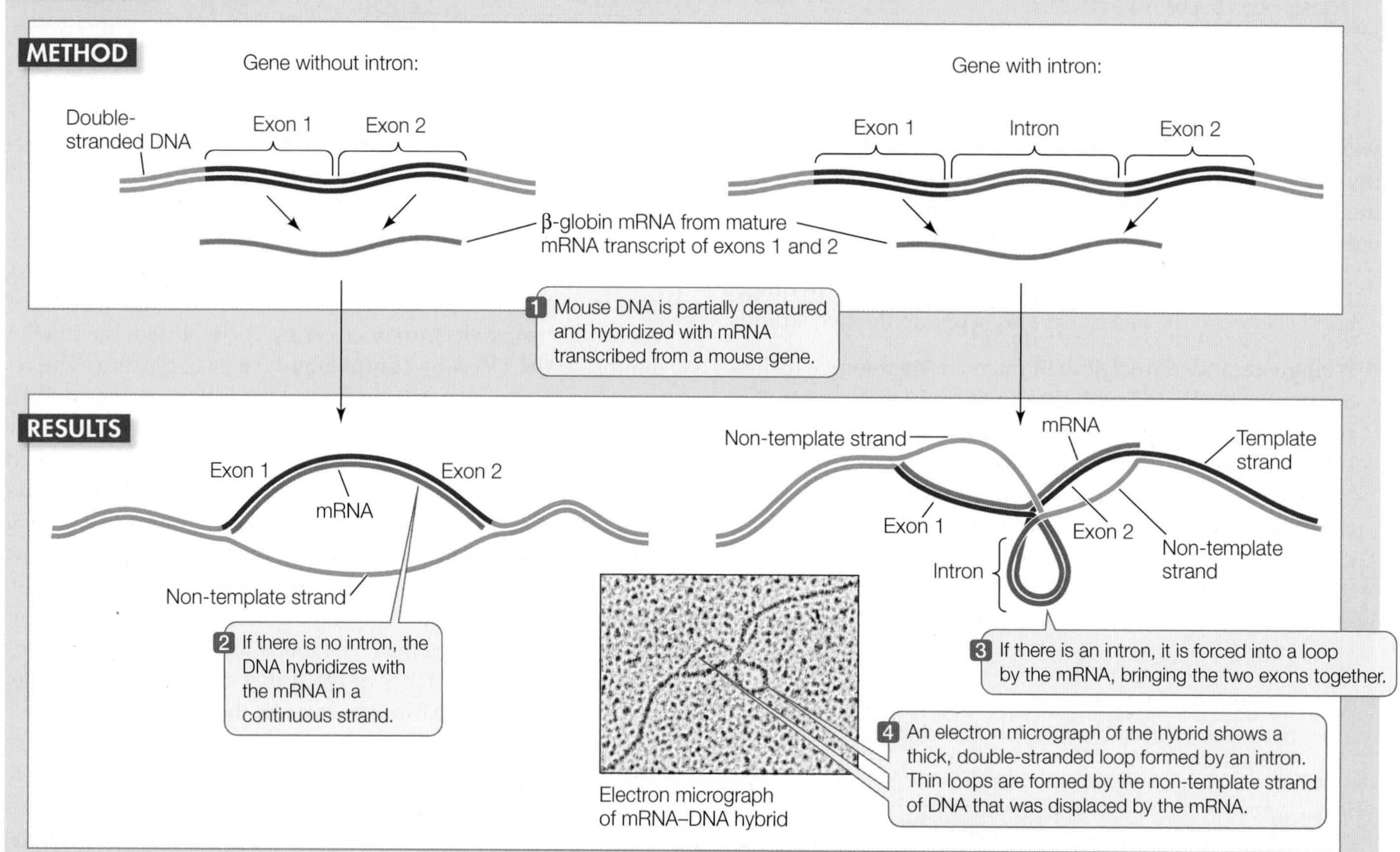

METHOD
Gene without intron:
Gene with intron:
Double-stranded DNA
Exon 1
Exon 2
Exon 1
Intron
Exon 2
β-globin mRNA from mature mRNA transcript of exons 1 and 2
1 Mouse DNA is partially denatured and hybridized with mRNA transcribed from a mouse gene.
RESULTS
Exon 1
Exon 2
mRNA
Non-template strand
2 If there is no intron, the DNA hybridizes with the mRNA in a continuous strand.
Non-template strand
mRNA
Template strand
Exon 1
Exon 2
Non-template strand
Intron
3 If there is an intron, it is forced into a loop by the mRNA, bringing the two exons together.
Electron micrograph of mRNA–DNA hybrid
4 An electron micrograph of the hybrid shows a thick, double-stranded loop formed by an intron. Thin loops are formed by the non-template strand of DNA that was displaced by the mRNA.

**CONCLUSION** The DNA contains noncoding regions within the genes that are not present in the mature mRNA.

**FURTHER INVESTIGATION:** Draw the result assuming that there were three exons and two introns.

Go to **yourBioPortal.com** for original citations, discussions, and relevant links for all INVESTIGATING LIFE figures.

### Eukaryotic gene transcripts are processed before translation

The primary transcript of a eukaryotic gene is modified in several ways before it leaves the nucleus: both ends of the pre-mRNA are modified, and the introns are removed.

**MODIFICATION AT BOTH ENDS** Two steps in the processing of pre-mRNA take place in the nucleus, one at each end of the molecule (**Figure 14.10**):

- A **G cap** is added to the 5′ end of the pre-mRNA as it is transcribed. The G cap is a chemically modified molecule of guanosine triphosphate (GTP). It facilitates the binding of mRNA to the ribosome for translation, and it protects the mRNA from being digested by ribonucleases that break down RNAs.
- A **poly A tail** is added to the 3′ end of the pre-mRNA at the end of transcription. In both prokaryotic and eukaryotic genes, transcription begins at a DNA sequence that is upstream (to the "left" on the DNA) of the first codon (i.e., at the promoter), and ends downstream (to the "right" on the DNA) of the termination codon. In eukaryotes, there is usually a "polyadenylation" sequence (AAUAAA) near the 3′ end of the pre-mRNA, after the last codon. This sequence acts as a signal for an enzyme to cut the pre-mRNA. Immediately after this cleavage, another enzyme

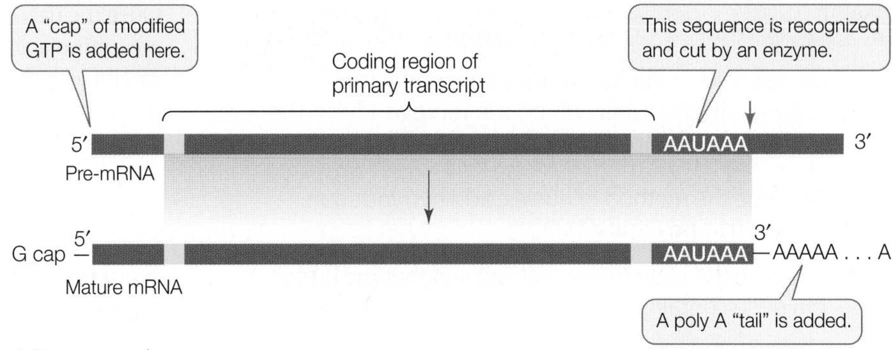


A "cap" of modified GTP is added here.

**14.10 Processing the Ends of Eukaryotic Pre-mRNA** Modifications at each end of the pre-mRNA transcript—the G cap and the poly A tail—are important for mRNA function.

adds 100 to 300 adenine nucleotides (a "poly A" sequence) to the 3′ end of the pre-mRNA. This "tail" may assist in the export of the mRNA from the nucleus and is important for mRNA stability.

**SPLICING TO REMOVE INTRONS** The next step in the processing of eukaryotic pre-mRNA within the nucleus is removal of the introns. If these RNA sequences were not removed, a very different amino acid sequence, and possibly a nonfunctional protein, would result. A process called **RNA splicing** removes the introns and splices the exons together.

As soon as the pre-mRNA is transcribed, several **small nuclear ribonucleoprotein particles** (**snRNPs**) bind at each end. There are several types of these RNA–protein particles in the nucleus.

At the boundaries between introns and exons are **consensus sequences**—short stretches of DNA that appear, with little variation ("consensus"), in many different genes. The RNA in one of the snRNPs has a stretch of bases complementary to the consensus sequence at the 5′ exon–intron boundary, and it binds to the pre-mRNA by complementary base pairing. Another snRNP binds to the pre-mRNA near the 3′ intron–exon boundary (**Figure 14.11**).

Next, using energy from adenosine triphosphate (ATP), proteins are added to form a large RNA–protein complex called a **spliceosome**. This complex cuts the pre-mRNA, releases the introns, and joins the ends of the exons together to produce mature mRNA.

Molecular studies of human genetic diseases have provided insights into intron consensus sequences and splicing machinery. For example, people with the genetic disease β-thalassemia, like those with α-thalassemia discussed earlier in the chapter, have a defect in the production of one of the hemoglobin subunits. These people suffer from severe anemia because they have an inadequate supply of red blood cells. In some cases, the genetic mutation that causes the disease occurs at an intron consensus sequence in the β-globin gene. Consequently, β-globin pre-mRNA cannot be spliced correctly, and nonfunctional β-globin mRNA is made. This finding offers another example of how biologists can use mutations to elucidate cause-and-effect relationships.

After processing is completed in the nucleus, the mature mRNA moves out into the cytoplasm through the nuclear pores. In the nucleus, a protein called TAP binds to the 5′ end of processed mRNA. This protein in turn binds to others, which are recognized by a receptor at the nuclear pore. Together, these proteins lead the mRNA through the pore. Unprocessed or incompletely processed pre-mRNAs remain in the nucleus.

**yourBioPortal.com**
**GO TO Animated Tutorial 14.3 • RNA Splicing**

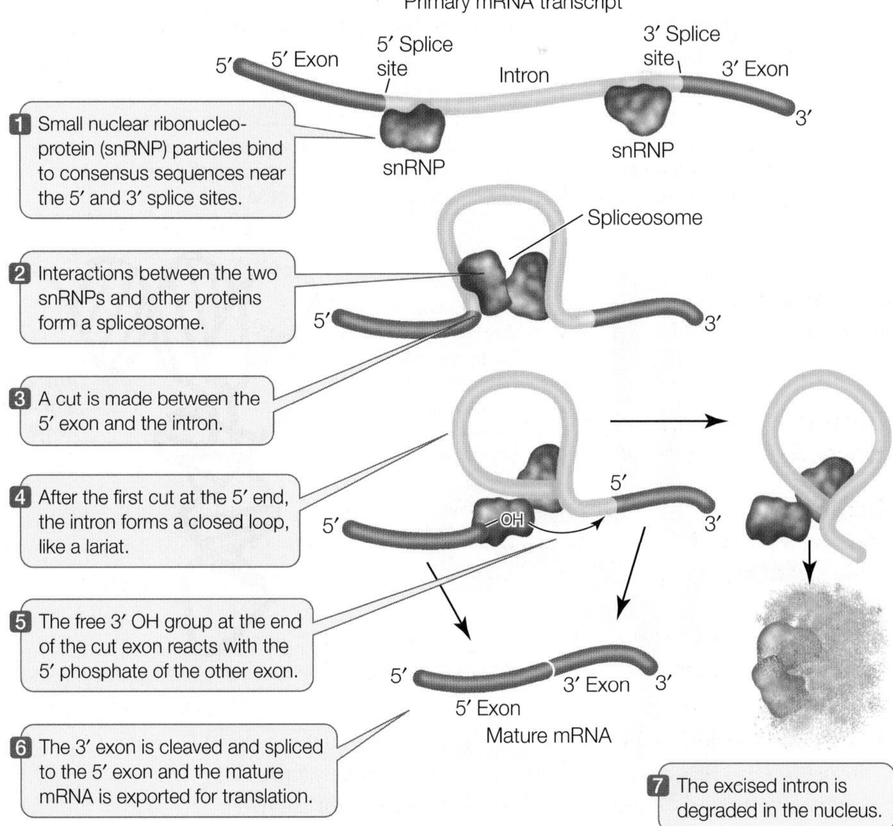

Primary mRNA transcript
5′
5′ Exon
5′ Splice site
Intron
3′ Splice site
3′ Exon
3′
1 Small nuclear ribonucleoprotein (snRNP) particles bind to consensus sequences near the 5′ and 3′ splice sites.
snRNP
snRNP
Spliceosome
2 Interactions between the two snRNPs and other proteins form a spliceosome.
3 A cut is made between the 5′ exon and the intron.
4 After the first cut at the 5′ end, the intron forms a closed loop, like a lariat.
OH
5 The free 3′ OH group at the end of the cut exon reacts with the 5′ phosphate of the other exon.
5′ Exon
3′ Exon
Mature mRNA
6 The 3′ exon is cleaved and spliced to the 5′ exon and the mature mRNA is exported for translation.
7 The excised intron is degraded in the nucleus.

**14.11 The Spliceosome: An RNA Splicing Machine** The binding of snRNPs to consensus sequences bordering the introns on the pre-mRNA results in a series of proteins binding and forming a large complex called a spliceosome. This structure determines the exact position of each cut in the pre-mRNA with great precision.

## 14.4 RECAP

**Most eukaryotic genes contain noncoding sequences called introns, which are removed from the pre-mRNA transcript.**

- Describe the method of nucleic acid hybridization. **See p. 301 and Figure 14.8**
- Describe the experiment that showed that the β-globin gene contains introns. **See p. 301 and Figure 14.9**
- How is the pre-mRNA transcript modified at the 5′ and 3′ ends? **See p. 302 and Figure 14.10**
- How does RNA splicing happen? What are the consequences if it does not happen correctly? **See p. 303 and Figure 14.11**

Transcription and post-transcriptional events produce an mRNA that is ready to be translated into a sequence of amino acids in a polypeptide. We turn now to the events of translation.

# 14.5 How Is RNA Translated into Proteins?

As Crick's adapter hypothesis proposed, the translation of mRNA into proteins requires a molecule that links the information contained in mRNA codons with specific amino acids in proteins. That function is performed by transfer RNA (tRNA). Two key events must take place to ensure that the protein made is the one specified by the mRNA:

- The tRNAs must read mRNA codons correctly.
- The tRNAs must deliver the amino acids that correspond to each mRNA codon.

Once the tRNAs "decode" the mRNA and deliver the appropriate amino acids, components of the ribosome catalyze the formation of peptide bonds between amino acids. We now turn to these two steps.

**yourBioPortal.com**
GO TO **Animated Tutorial 14.4 • Protein Synthesis**

### Transfer RNAs carry specific amino acids and bind to specific codons

A codon in mRNA and an amino acid in a protein are related by way of an adapter—a specific tRNA with an attached amino acid. For each of the 20 amino acids, there is at least one specific type ("species") of tRNA molecule. The tRNA molecule has three functions:

- It binds to a particular amino acid. When it is carrying an amino acid, the tRNA is said to be "charged."
- It associates with mRNA.
- It interacts with ribosomes.

The tRNA molecular structure relates clearly to all of these functions. The molecule has about 75 to 80 nucleotides. It has a conformation (a three-dimensional shape) that is maintained by complementary base pairing (hydrogen bonding) between bases within its own sequence (**Figure 14.12**).

The conformation of a tRNA molecule is exquisitely suited for its interaction with specific binding sites on ribosomes. In addition, at the 3′ end of every tRNA molecule is its amino acid attachment site: a site to which its specific amino acid binds co-

**14.12 Transfer RNA** The stem and loop structure of a tRNA molecule is well suited to its functions: binding to amino acids, associating with mRNA molecules, and interacting with ribosomes.

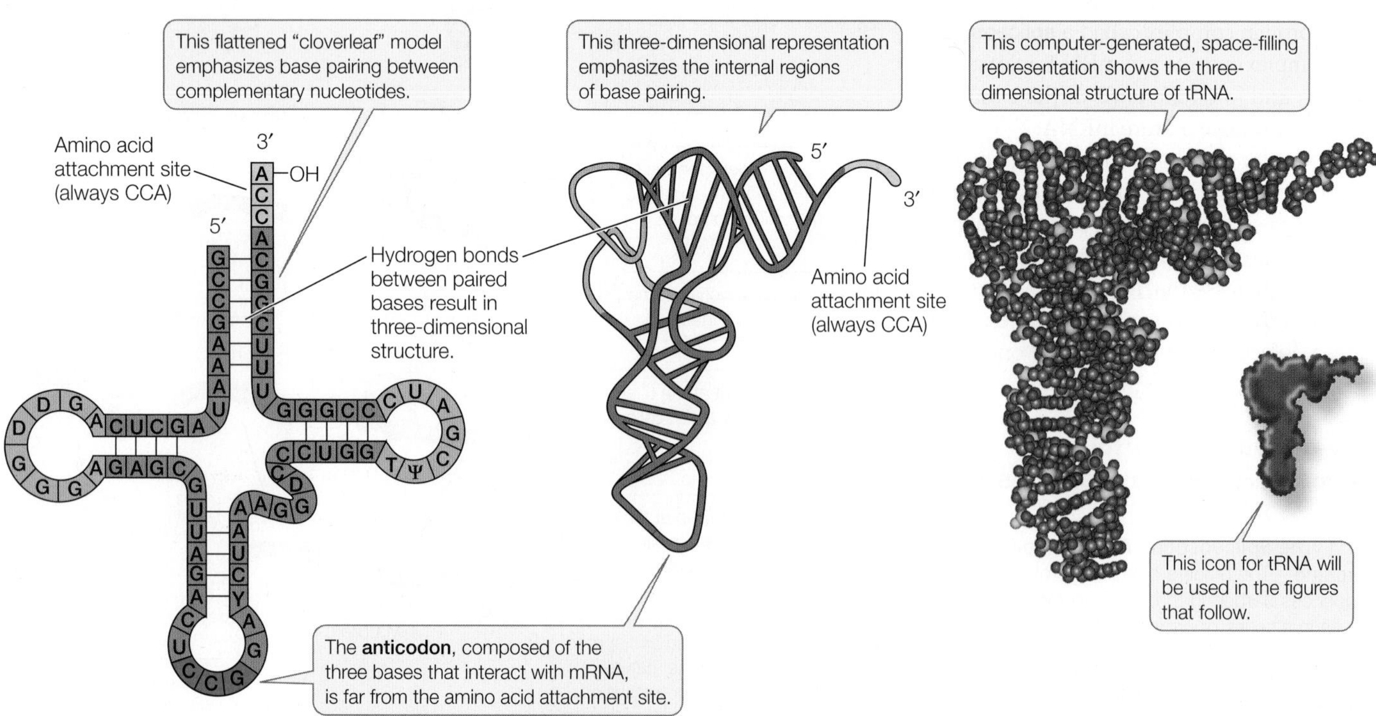
This flattened "cloverleaf" model emphasizes base pairing between complementary nucleotides.
Amino acid attachment site (always CCA)
3′
5′
—OH
Hydrogen bonds between paired bases result in three-dimensional structure.
This three-dimensional representation emphasizes the internal regions of base pairing.
5′
3′
Amino acid attachment site (always CCA)
This computer-generated, space-filling representation shows the three-dimensional structure of tRNA.
This icon for tRNA will be used in the figures that follow.
The **anticodon**, composed of the three bases that interact with mRNA, is far from the amino acid attachment site.

valently. At about the midpoint of the tRNA sequence is a group of three bases, called the anticodon, which is the site of complementary base pairing (via hydrogen bonding) with the codon on the mRNA. Thus, each tRNA species has a unique anticodon that corresponds to the amino acid it carries. When the tRNA and the mRNA come into contact on the surface of the ribosome, the codon and anticodon are antiparallel, permitting hydrogen bonding to occur between the complementary bases. As an example of this process, consider the amino acid arginine:

- The template strand DNA sequence that codes for arginine is 3′-GCC-5′, which is transcribed, by complementary base pairing, to produce the mRNA codon 5′-CGG-3′
- That mRNA codon binds by complementary base pairing to a tRNA with the anticodon 3′-GCC-5′, which is charged with arginine.

Recall that 61 different codons encode the 20 amino acids in proteins (see Figure 14.6). Does this mean that the cell must produce 61 different tRNA species, each with a different anticodon? No. The cell gets by with about two-thirds of that number of tRNA species because the specificity for the base at the 3′ end of the codon (and the 5′ end of the anticodon) is not always strictly observed. This phenomenon, called *wobble*, allows the alanine codons GCA, GCC, and GCU, for example, all to be recognized by the same tRNA. Wobble is allowed in some matches but not in others; of most importance, it does not allow the genetic code to be ambiguous. That is, each mRNA codon binds to just one tRNA species, carrying a specific amino acid.

### Activating enzymes link the right tRNAs and amino acids

The charging of each tRNA with its correct amino acid is achieved by a family of activating enzymes, known more formally as aminoacyl-tRNA synthases (**Figure 14.13**). Each activating enzyme is specific for one amino acid and for its corresponding tRNA. The enzyme has a three-part active site that recognizes three molecules: a specific amino acid, ATP, and a specific tRNA. Since tRNA has a complex three-dimensional structure, the activating enzyme recognizes a specific tRNA with a very low error rate. Remarkably, the error rate for amino acid recognition is also low, on the order of one in 1,000. Because the activating enzymes are so highly specific, the process of tRNA charging is sometimes called the *second genetic code*. Follow the events of activation in Figure 14.13.

A clever experiment by Seymour Benzer and his colleagues at Purdue University demonstrated the importance of specificity in the attachment of tRNA to its amino acid. In their laboratory, the amino acid cysteine, already properly attached to its tRNA, was chemically modified to become a different amino acid, alanine. Which component—the amino acid or the tRNA—would be recognized when this hybrid charged tRNA was put

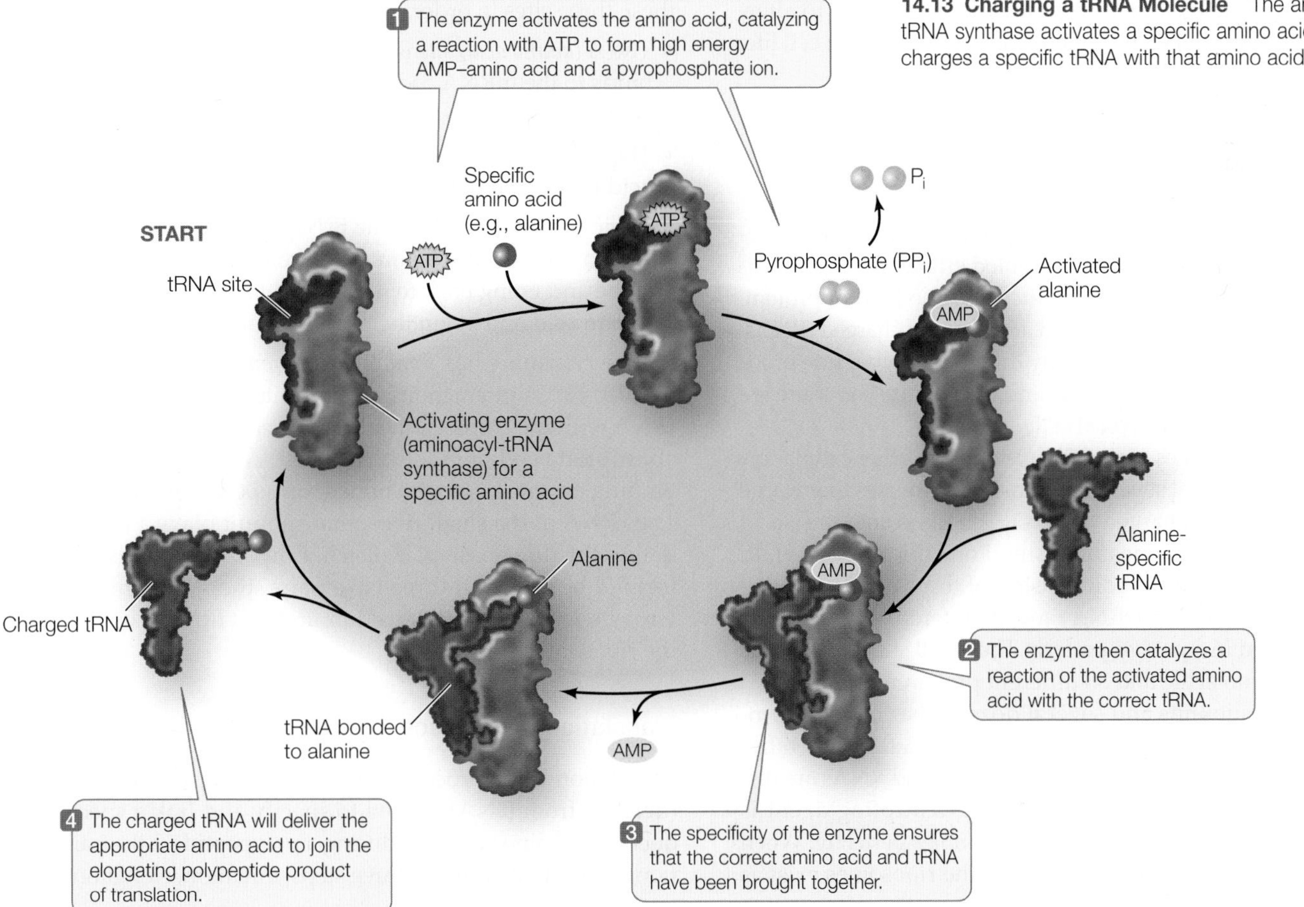
1 The enzyme activates the amino acid, catalyzing a reaction with ATP to form high energy AMP–amino acid and a pyrophosphate ion.
START
tRNA site
ATP
Specific amino acid (e.g., alanine)
ATP
Pyrophosphate ($PP_i$)
$P_i$
Activated alanine
AMP
Activating enzyme (aminoacyl-tRNA synthase) for a specific amino acid
Alanine-specific tRNA
AMP
Alanine
Charged tRNA
2 The enzyme then catalyzes a reaction of the activated amino acid with the correct tRNA.
tRNA bonded to alanine
AMP
4 The charged tRNA will deliver the appropriate amino acid to join the elongating polypeptide product of translation.
3 The specificity of the enzyme ensures that the correct amino acid and tRNA have been brought together.

**14.13 Charging a tRNA Molecule** The aminoacyl-tRNA synthase activates a specific amino acid and charges a specific tRNA with that amino acid.

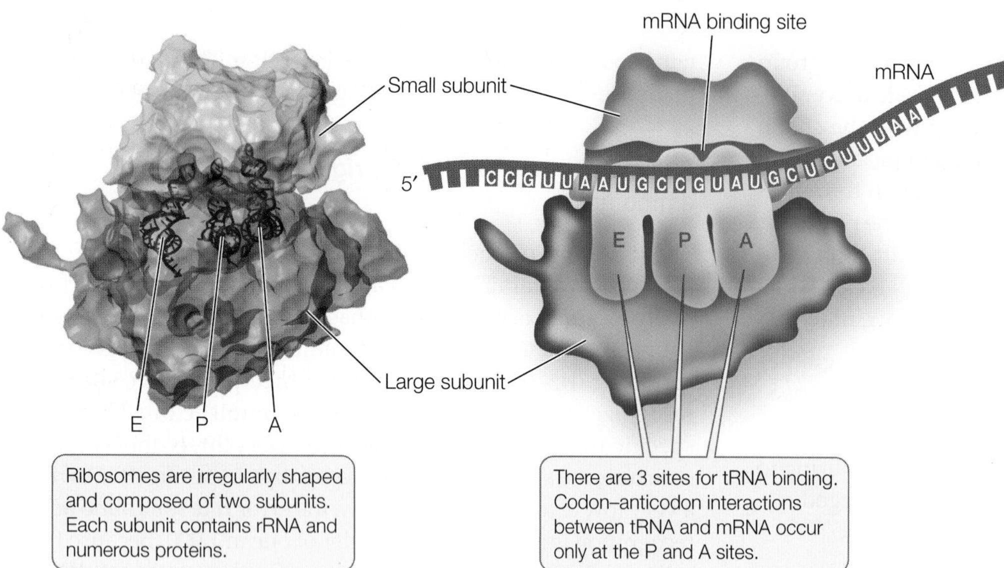

mRNA binding site
Small subunit
mRNA
3′
5′
CCGUUAAUGCCGUAUGCUCUUUAA
E
P
A
Large subunit
E
P
A
Ribosomes are irregularly shaped and composed of two subunits. Each subunit contains rRNA and numerous proteins.
There are 3 sites for tRNA binding. Codon–anticodon interactions between tRNA and mRNA occur only at the P and A sites.

**14.14 Ribosome Structure** Each ribosome consists of a large and a small subunit. The subunits remain separate when they are not in use for protein synthesis.

into a protein-synthesizing system? The answer was the tRNA. Everywhere in the synthesized protein where cysteine was supposed to be, alanine appeared instead. The cysteine-specific tRNA had delivered its cargo (alanine) to every mRNA codon for cysteine. This experiment showed that the protein synthesis machinery recognizes the anticodon of the charged tRNA, not the amino acid attached to it. If activating enzymes in nature did what Benzer did in the laboratory and charged tRNAs with the wrong amino acids, those amino acids would be inserted into proteins at inappropriate places, leading to alterations in protein shape and function and endangering cell life.

### The ribosome is the workbench for translation

The **ribosome** is the molecular workbench where the task of translation is accomplished. Its structure enables it to hold mRNA and charged tRNAs in the right positions, thus allowing a polypeptide chain to be assembled efficiently. A given ribosome does not specifically produce just one kind of protein. A ribosome can use any mRNA and all species of charged tRNAs, and thus can be used to make many different polypeptide products. Ribosomes can be used over and over again, and there are thousands of them in a typical cell.

Although ribosomes are small in contrast to other cellular organelles, their mass of several million daltons makes them large in comparison with charged tRNAs. Each ribosome consists of two subunits, a large one and a small one (**Figure 14.14**). In eukaryotes, the large subunit consists of three different molecules of ribosomal RNA (rRNA) and 49 different protein molecules, arranged in a precise pattern. The small subunit consists of one rRNA molecule and 33 different protein molecules.

These two subunits and several dozen other molecules interact non-covalently, like a jigsaw puzzle. In fact, when hydrophobic interactions between the proteins and RNAs are disrupted, the ribosome falls apart. If the disrupting agent is removed, the complex structure self-assembles perfectly! When not active in the translation of mRNA, the ribosomes exist as two separate subunits.

The ribosomes of prokaryotes are somewhat smaller than those of eukaryotes, and their ribosomal proteins and RNAs are different. Mitochondria and chloroplasts also contain ribosomes, some of which are similar to those of prokaryotes (see Chapter 5).

On the large subunit of the ribosome there are three sites to which a tRNA can bind, and these are designated A, P, and E (see Figure 14.14). The mRNA and ribosome move in relation to one another, and as they do so, a charged tRNA traverses these three sites in order:

- The *A (amino acid) site* is where the charged tRNA anticodon binds to the mRNA codon, thus lining up the correct amino acid to be added to the growing polypeptide chain.
- The *P (polypeptide) site* is where the tRNA adds its amino acid to the polypeptide chain.
- The *E (exit) site* is where the tRNA, having given up its amino acid, resides before being released from the ribosome and going back to the cytosol to pick up another amino acid and begin the process again.

The ribosome has a *fidelity function* that ensures that the mRNA–tRNA interactions are accurate; that is, that a charged tRNA with the correct anticodon (e.g., 3′-UAC-5′) binds to the appropriate codon in mRNA (e.g., 5′-AUG-3′). When proper binding occurs, hydrogen bonds form between the base pairs. The rRNA of the small ribosomal subunit plays a role in validating the three-base-pair match. If hydrogen bonds have not formed between all three base pairs, the tRNA must be the wrong one for that mRNA codon, and that tRNA is ejected from the ribosome.

### Translation takes place in three steps

Translation is the process by which the information in mRNA (derived from DNA) is used to specify and link a specific sequence of amino acids, producing a polypeptide. Like transcription, translation occurs in three steps: initiation, elongation, and termination.

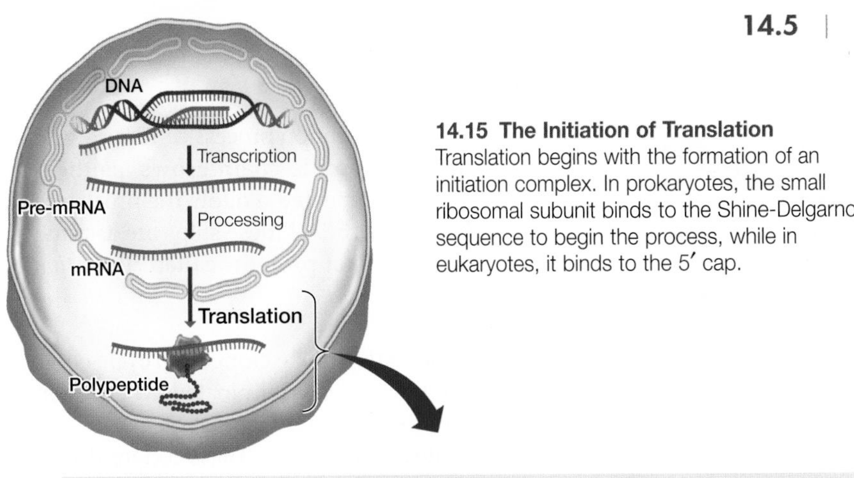


**14.15 The Initiation of Translation** Translation begins with the formation of an initiation complex. In prokaryotes, the small ribosomal subunit binds to the Shine-Delgarno sequence to begin the process, while in eukaryotes, it binds to the 5′ cap.

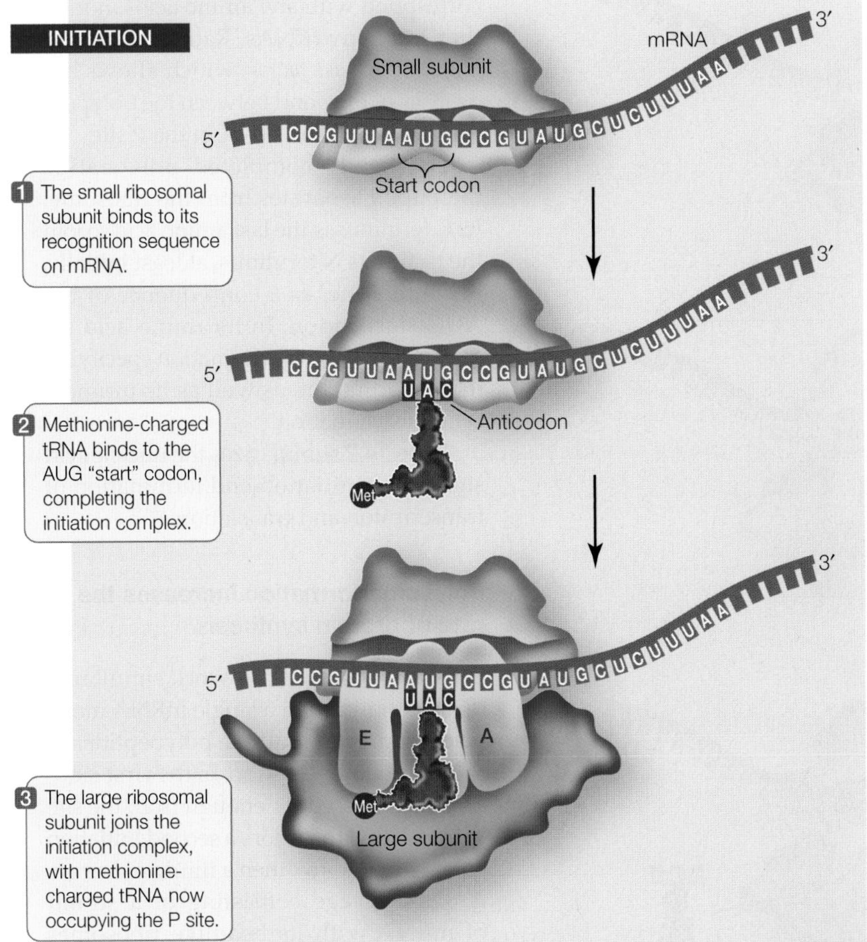

INITIATION
Small subunit
mRNA
5′
3′
Start codon
1 The small ribosomal subunit binds to its recognition sequence on mRNA.
Anticodon
Met
2 Methionine-charged tRNA binds to the AUG "start" codon, completing the initiation complex.
E
A
Large subunit
3 The large ribosomal subunit joins the initiation complex, with methionine-charged tRNA now occupying the P site.


**INITIATION** The translation of mRNA begins with the formation of an **initiation complex**, which consists of a charged tRNA and a small ribosomal subunit, both bound to the mRNA (**Figure 14.15**).

In prokaryotes, the rRNA of the small ribosomal subunit first binds to a complementary ribosome binding site (known as the Shine–Dalgarno sequence) on the mRNA. This sequence is upstream of the actual start codon, but lines up the start codon so that it is will be adjacent to the P site of the large subunit.

Eukaryotes do this somewhat differently: the small ribosomal subunit binds to the 5′cap on the mRNA. After binding, the small subunit moves along the mRNA until it reaches the start codon.

Recall that the mRNA start codon in the genetic code is AUG (see Figure 14.6). The anticodon of a methionine-charged tRNA binds to this start codon by complementary base pairing to complete the initiation complex. Thus the first amino acid in a polypeptide chain is always methionine. However, not all mature proteins have methionine as their N-terminal amino acid. In many cases, the initiator methionine is removed by an enzyme after translation.

After the methionine-charged tRNA has bound to the mRNA, the large subunit of the ribosome joins the complex. The methionine-charged tRNA now lies in the P site of the ribosome, and the A site is aligned with the second mRNA codon. These ingredients—mRNA, two ribosomal subunits, and methionine-charged tRNA—are put together properly by a group of proteins called *initiation factors*.

**ELONGATION** A charged tRNA whose anticodon is complementary to the second codon of the mRNA now enters the open A site of the large ribosomal subunit. The large subunit then catalyzes two reactions:

- It breaks the bond between the tRNA and its amino acid in the P site.
- It catalyzes the formation of a peptide bond between that amino acid and the one attached to the tRNA in the A site.

Because the large ribosomal subunit performs these two actions, it is said to have **peptidyl transferase** activity. In this way, methionine (the amino acid in the P site) becomes the N terminus of the new protein. The second amino acid is now bound to methionine, but remains attached to its tRNA at the A site.

How does the large ribosomal subunit catalyze this binding? Harry Noller and his colleagues at the University of California at Santa Cruz did a series of experiments and found that:

- If they removed almost all of the proteins from the large subunit, it still catalyzed peptide bond formation.
- If the rRNA was destroyed, so was peptidyl transferase activity.

Thus *rRNA is the catalyst*. The purification and crystallization of ribosomes has allowed scientists to examine their structure in detail, and the catalytic role of rRNA in peptidyl transferase activity has been confirmed. This supports the hypothesis that RNA, and catalytic RNA in particular, evolved before DNA (see Section 4.3).

After the first tRNA releases its methionine, it moves to the E site and is then dissociated from the ribosome, returning to the cytosol to become charged with another methionine. The second tRNA, now bearing a dipeptide (a two-amino-acid chain), is shifted to the P site as the ribosome moves one codon along the mRNA in the 5′-to-3′ direction.

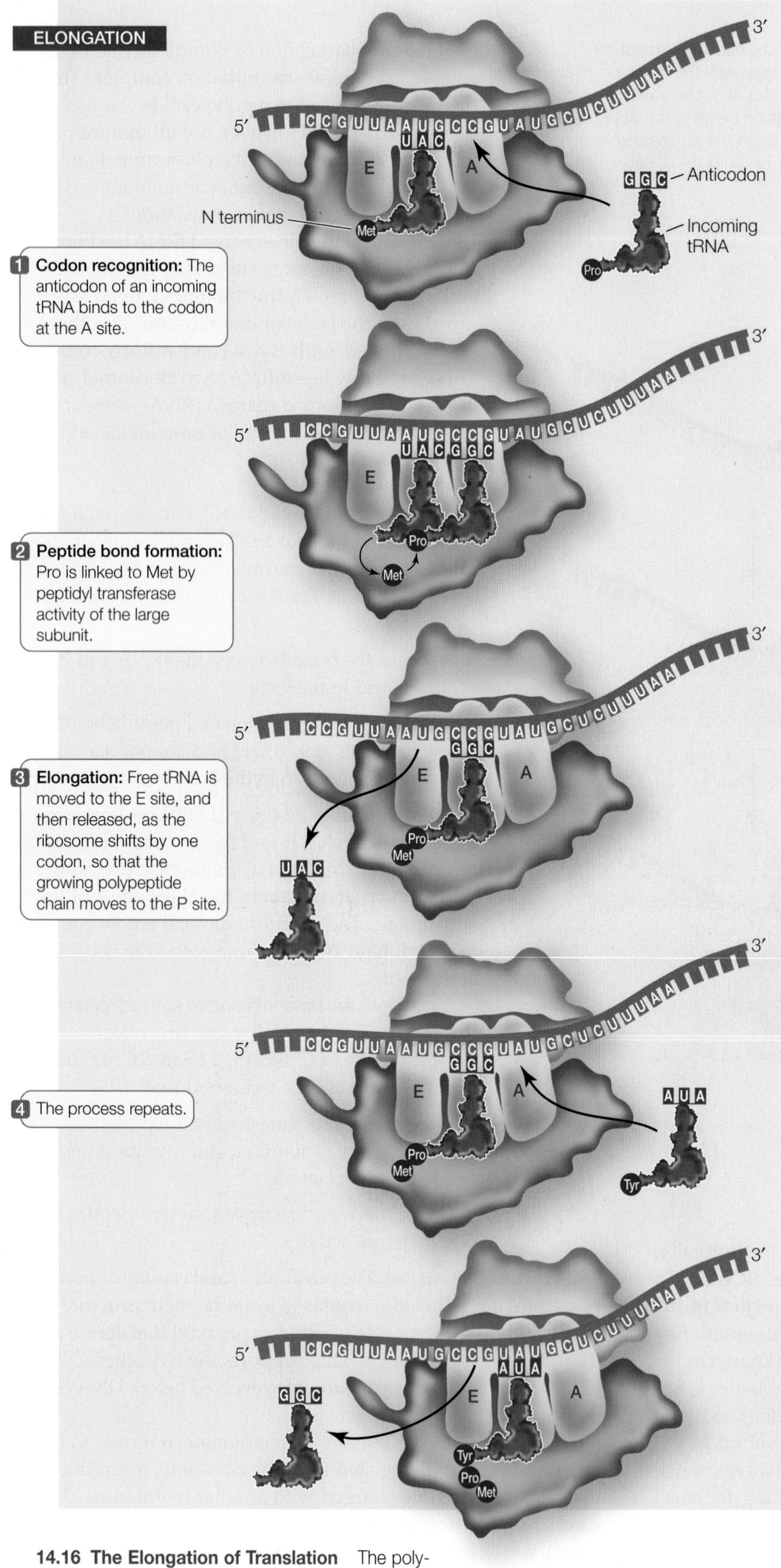
ELONGATION
3′
5′
CCGUUAAUGCCGUAUGCUCUUUAA
UAC
E
A
GGC
Anticodon
N terminus
Met
Incoming tRNA
Pro
1 Codon recognition: The anticodon of an incoming tRNA binds to the codon at the A site.
UACGGC
2 Peptide bond formation: Pro is linked to Met by peptidyl transferase activity of the large subunit.
3 Elongation: Free tRNA is moved to the E site, and then released, as the ribosome shifts by one codon, so that the growing polypeptide chain moves to the P site.
UAC
4 The process repeats.
AUA
Tyr
GGC
AUA

**14.16 The Elongation of Translation** The polypeptide chain elongates as the mRNA is translated.

The elongation process continues, and the polypeptide chain grows, as these steps are repeated. Follow the process in **Figure 14.16**. All these steps are assisted by ribosomal proteins called *elongation factors*.

**TERMINATION** The elongation cycle ends, and translation is terminated, when a stop codon—UAA, UAG, or UGA—enters the A site (**Figure 14.17**). These codons do not correspond with any amino acids, nor do they bind any tRNAs. Rather, they bind a *protein release factor*, which allows hydrolysis of the bond between the polypeptide chain and the tRNA in the P site.

The newly completed polypeptide thereupon separates from the ribosome. Its C terminus is the last amino acid to join the chain. Its N terminus, at least initially, is methionine, as a consequence of the AUG start codon. In its amino acid sequence, it contains information specifying its conformation, as well as its ultimate cellular destination.

**Table 14.2** summarizes the nucleic acid signals for initiation and termination of transcription and translation.

### Polysome formation increases the rate of protein synthesis

Several ribosomes can work simultaneously at translating a single mRNA molecule, producing multiple polypeptides at the same time. As soon as the first ribosome has moved far enough from the site of translation initiation, a second initiation complex can form, then a third, and so on. An assemblage consisting of a strand of mRNA with its beadlike ribosomes and their growing polypeptide chains is called a **polyribosome**, or **polysome** (**Figure 14.18**). Cells that are actively synthesizing proteins contain large numbers of polysomes and few free ribosomes or ribosomal subunits.

A polysome is like a cafeteria line in which patrons follow one another, adding items to their trays. At any moment, the person at the start has a little food (a newly initiated protein); the person at the end has a complete meal (a completed protein). However, in the polysome cafeteria, everyone gets the same meal: many copies of the same protein are made from a single mRNA.

**14.17 The Termination of Translation** Translation terminates when the A site of the ribosome encounters a stop codon on the mRNA.

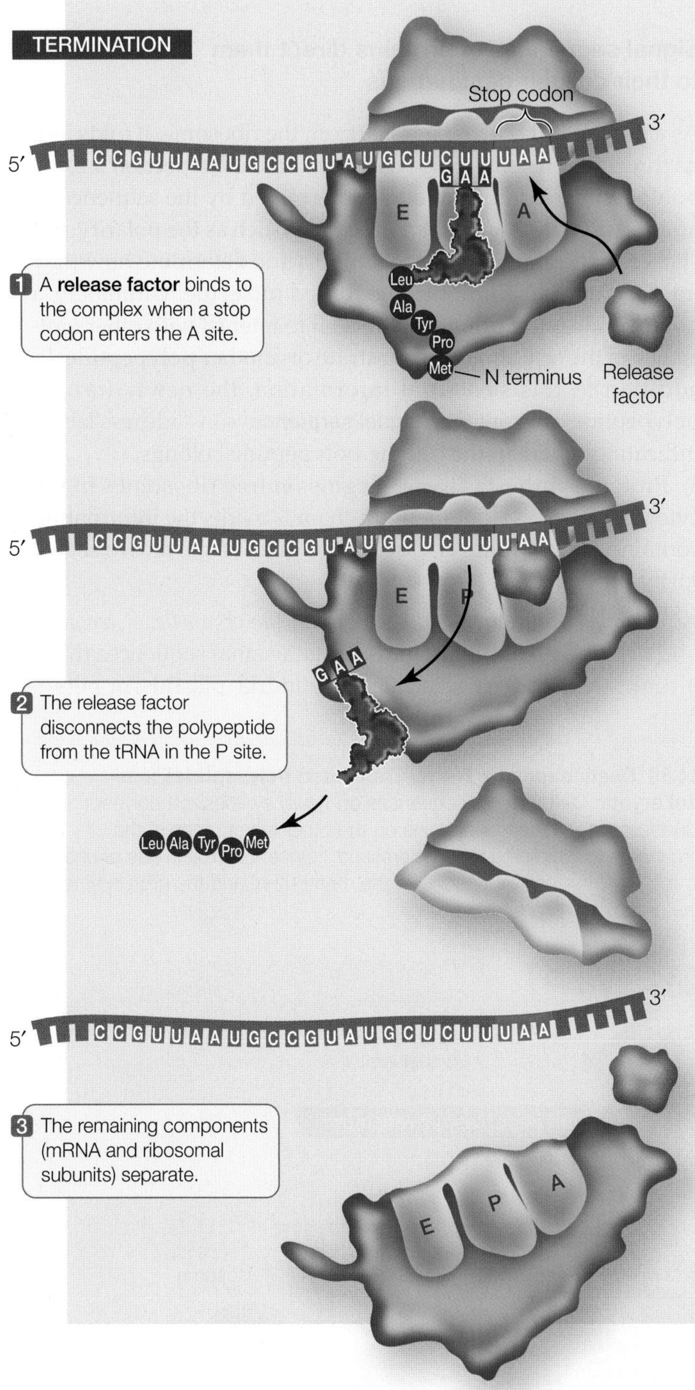


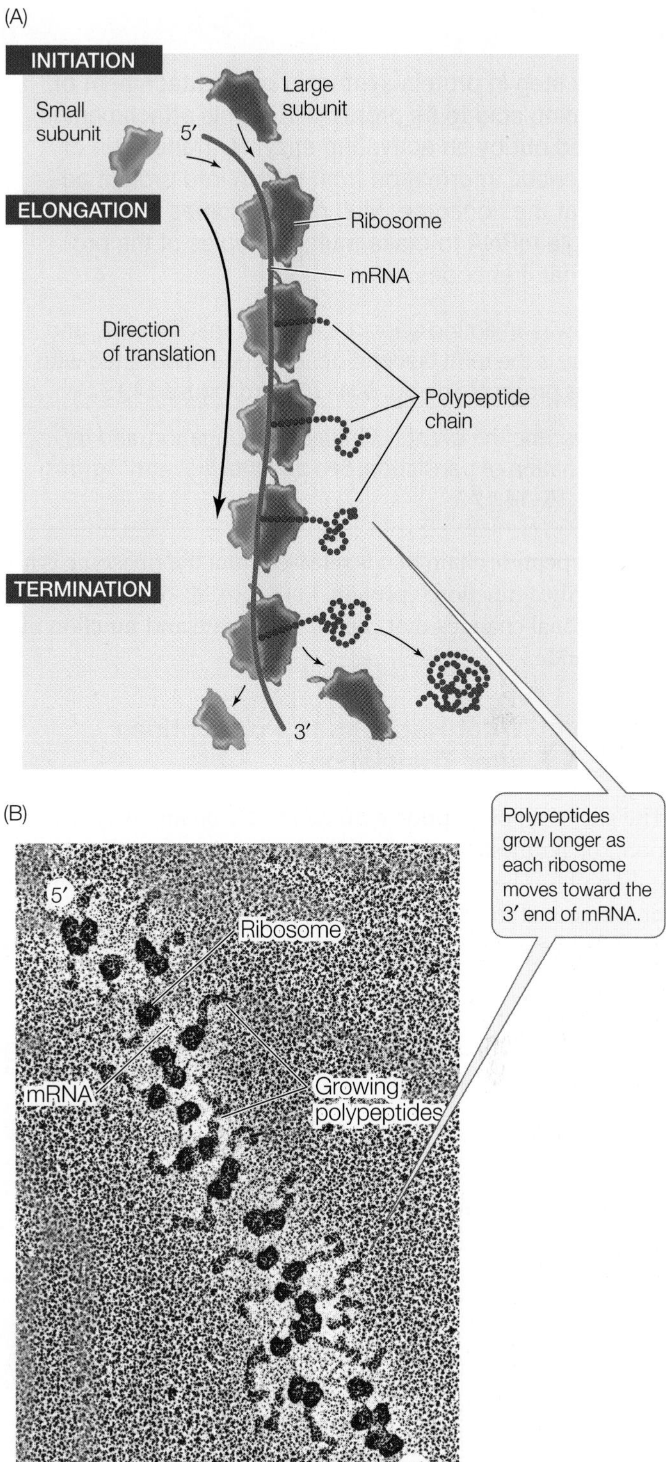


**14.18 A Polysome** (A) A polysome consists of multiple ribosomes and their growing polypeptide chains moving along an mRNA molecule. (B) An electron micrograph of a polysome.

TABLE 14.2

**Signals that Start and Stop Transcription and Translation**

| | TRANSCRIPTION | TRANSLATION |
|---|---|---|
| Initiation | Promoter DNA | AUG start codon in the mRNA |
| Termination | Terminator DNA | UAA, UAG, or UGA in the mRNA |

## 14.5 RECAP

**A key step in protein synthesis is the attachment of an amino acid to its proper tRNA. This attachment is carried out by an activating enzyme. Translation of the genetic information from mRNA into protein occurs at the ribosome. Multiple ribosomes may act on a single mRNA to make multiple copies of the protein that it encodes.**

- How is an amino acid attached to a specific tRNA, and why is the term "second genetic code" associated with this process? **See pp. 304–305 and Figure 14.13**
- Describe the events of initiation, elongation, and termination of translation. **See pp. 306–308 and Figures 14.15–14.17**

The polypeptide chain that is released from the ribosome is not necessarily a functional protein. Let's look at some of the post-translational changes that can affect the fate and function of a polypeptide.

## 14.6 What Happens to Polypeptides after Translation?

The site of a polypeptide's function may be far away from its point of synthesis in the cytoplasm. This is especially true for eukaryotes. The polypeptide may be moved into an organelle, or even out of the cell. In addition, polypeptides are often modified by the addition of new chemical groups that have functional significance. In this section we examine these *posttranslational* aspects of protein synthesis.

### Signal sequences in proteins direct them to their cellular destinations

As a polypeptide chain emerges from the ribosome, it folds into its three-dimensional shape. As we described in Section 3.2, the polypeptide's conformation is determined by the sequence of amino acids that make it up. Properties such as the polarity and charge of the R groups in the amino acids determine how they interact with each other in the folded molecule. Ultimately, a polypeptide's conformation allows it to interact with other molecules in the cell, such as a substrate or another polypeptide. In addition to this structural information, the newly formed polypeptide can contain a **signal sequence**—an "address label" indicating where in the cell the polypeptide belongs.

Protein synthesis always begins on free ribosomes in the cytoplasm. But as a polypeptide chain is made, the information contained in its amino acid sequence gives it one of two sets of further instructions (**Figure 14.19**):

- *"Complete translation and be released to an organelle, or remain in the cytosol."* Some proteins contain signal sequences that direct them to the nucleus, mitochondria, plastids, or per-

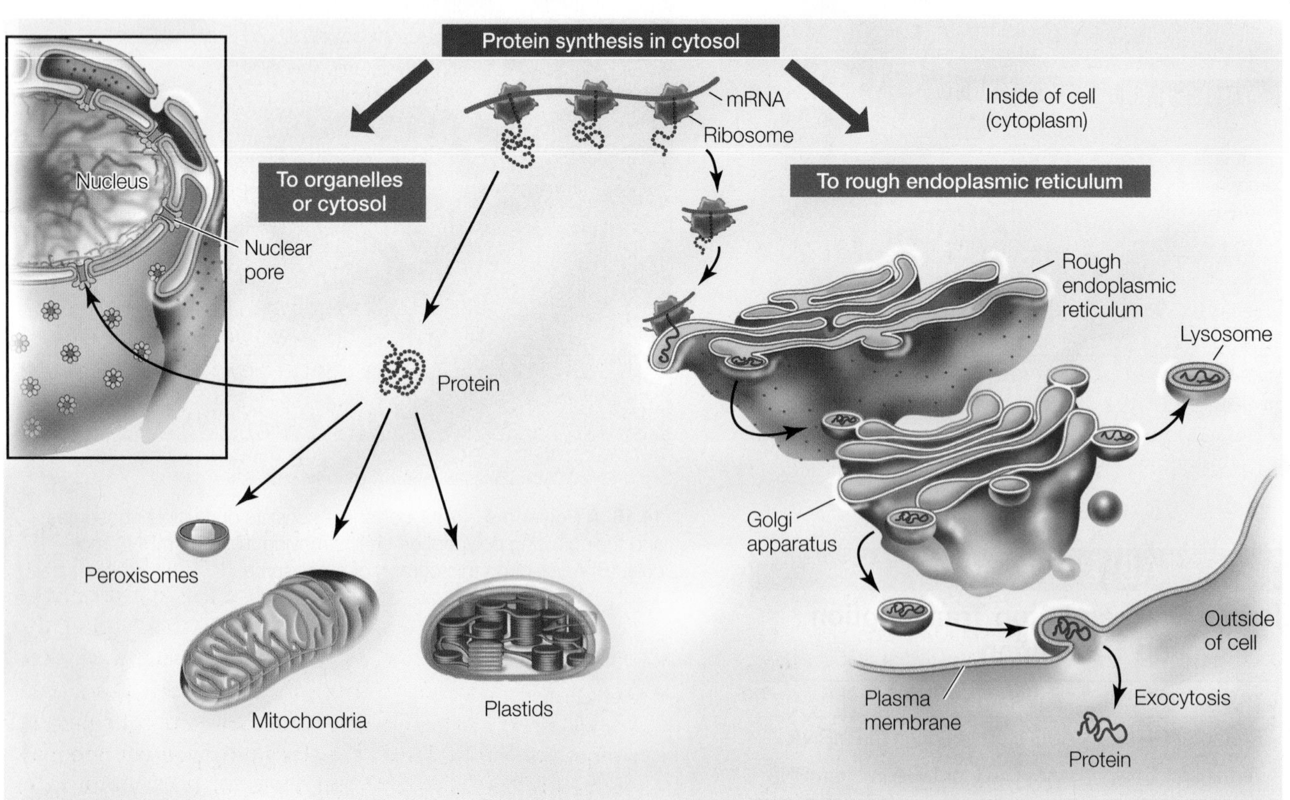


**14.19 Destinations for Newly Translated Polypeptides in a Eukaryotic Cell** Signal sequences on newly synthesized polypeptides bind to specific receptor proteins on the outer membranes of the organelles to which they are "addressed." Once the protein has bound to it, the receptor forms a channel in the membrane, and the protein enters the organelle.

oxisomes. If they lack a signal sequence, they remain in the cytosol by default.

- *"Stop translation, go to the endoplasmic reticulum, and finish synthesis there."* Other proteins contain a signal sequence that directs them to the endoplasmic reticulum (ER) before translation is complete. Such proteins may be retained in the lumen (the inside) of the ER, or be sent to the Golgi apparatus. From there, they may be sent to the lysosomes or the plasma membrane. Alternatively if they lack such specific instructions, they may be secreted from the cell via vesicles that fuse with the plasma membrane.

**DESTINATION: NUCLEUS, MITOCHONDRION, OR CHLOROPLAST** After translation, some folded polypeptides have a short exposed sequence of amino acids that acts like a postal "zip code," directing them to an organelle. These signal (or localization) sequences are either at the N terminus or in the interior of the amino acid chain. For example, the following sequence directs a protein to the nucleus:

—Pro—Pro—Lys—Lys—Lys—Arg—Lys—Val—

A nuclear localization sequence would occur in histone proteins associated with nuclear DNA, but not in citric acid cycle enzymes, which are addressed to the mitochondria. Signal sequences for a particular organelle vary, so not all the polypeptides destined for the nucleus have the same signal sequence.

How do we know that the amino acid sequence shown above is the signal? To investigate this question, Stephen Dilworth and colleagues at the University of Cambridge injected cells with nuclear and cytoplasmic proteins (**Figure 14.20**). The experiments involved the nuclear protein nucleoplasmin, the cytoplasmic protein pyruvate kinase, the nuclear localization signal (see above) and a "mix-and-match" procedure. For example, the putative nuclear signal was removed from nucleoplasmin, which normally carries it, or attached to pyruvate kinase, which does not normally carry it. The result was that it did not matter where in the cell the protein normally resided. If it had the signal, it went to the nucleus and if it did not have the signal, it stayed in the cytoplasm.

A signal sequence binds to a specific receptor protein, appropriately called a **docking protein**, on the outer membrane of the appropriate organelle. Once the signal sequence has bound to it, the docking protein forms a channel in the membrane, allowing the signal-bearing protein to pass through the membrane and enter the organelle. In this process, the protein is usually unfolded by a chaperonin protein (see Figure 3.12) so that it can pass through the channel; then it refolds into its normal conformation.

**DESTINATION: ENDOPLASMIC RETICULUM** If a specific hydrophobic sequence of 15–30 amino acids occurs at the N terminus of an elongated polypeptide chain, the polypeptide is sent initially to the ER. Some proteins are retained in the ER, but most move on to the Golgi, where they can be modified for eventual transport to the lysosomes, the plasma membrane, or out of the cell. In the cytoplasm, before translation is finished and while the polypeptide is still attached to a ribosome, this signal sequence binds to a **signal recognition particle** composed of protein and RNA (**Figure 14.21**). This binding blocks further protein synthesis until the ribosome becomes attached to a specific receptor protein in the membrane of the rough ER. Once again, the receptor protein is converted into a channel, through which the growing polypeptide passes. After the formation of the channel, protein synthesis resumes, and the chain grows longer until its sequence is completed. The elongating polypeptide may be retained in the ER membrane itself, or it may enter the

## INVESTIGATING LIFE

**14.20 Testing the Signal**

A series of experiments were used to test whether the nuclear localization signal (NLS) sequence is all that is needed to direct a protein to the nucleus.

**HYPOTHESIS** A nuclear localization signal (NLS) is necessary for import of a protein into the nucleus.

**METHOD**

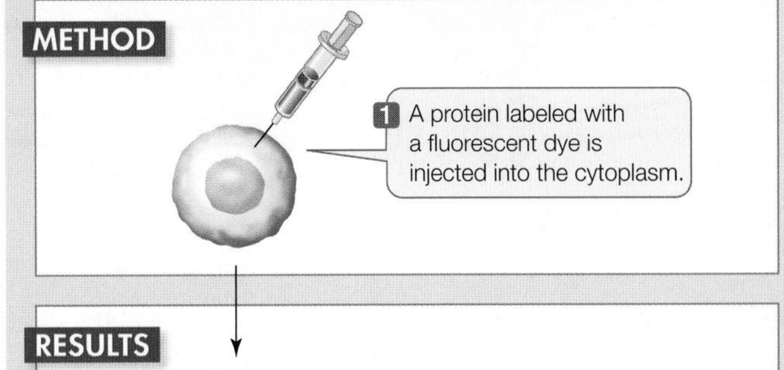

1 A protein labeled with a fluorescent dye is injected into the cytoplasm.


**RESULTS**

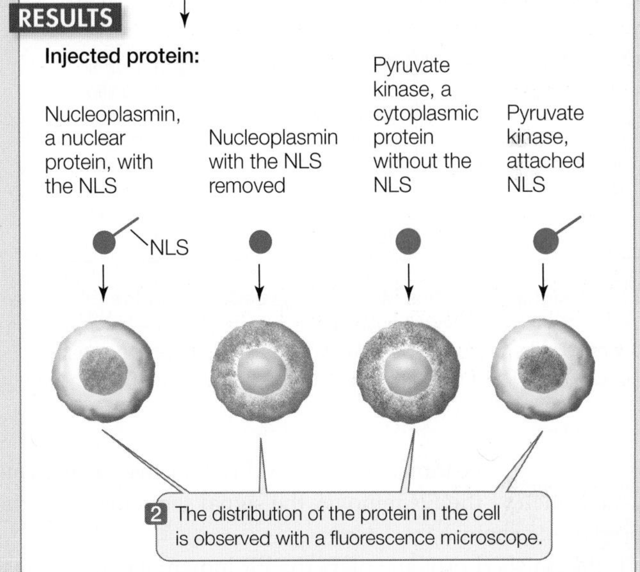


**CONCLUSION** The NLS is essential for nuclear protein import and will direct a normally cytoplasmic protein to the nucleus.

**FURTHER INVESTIGATION:** How would you test for a chloroplast signal sequence?

Go to **yourBioPortal.com** for original citations, discussions, and relevant links for all INVESTIGATING LIFE figures.

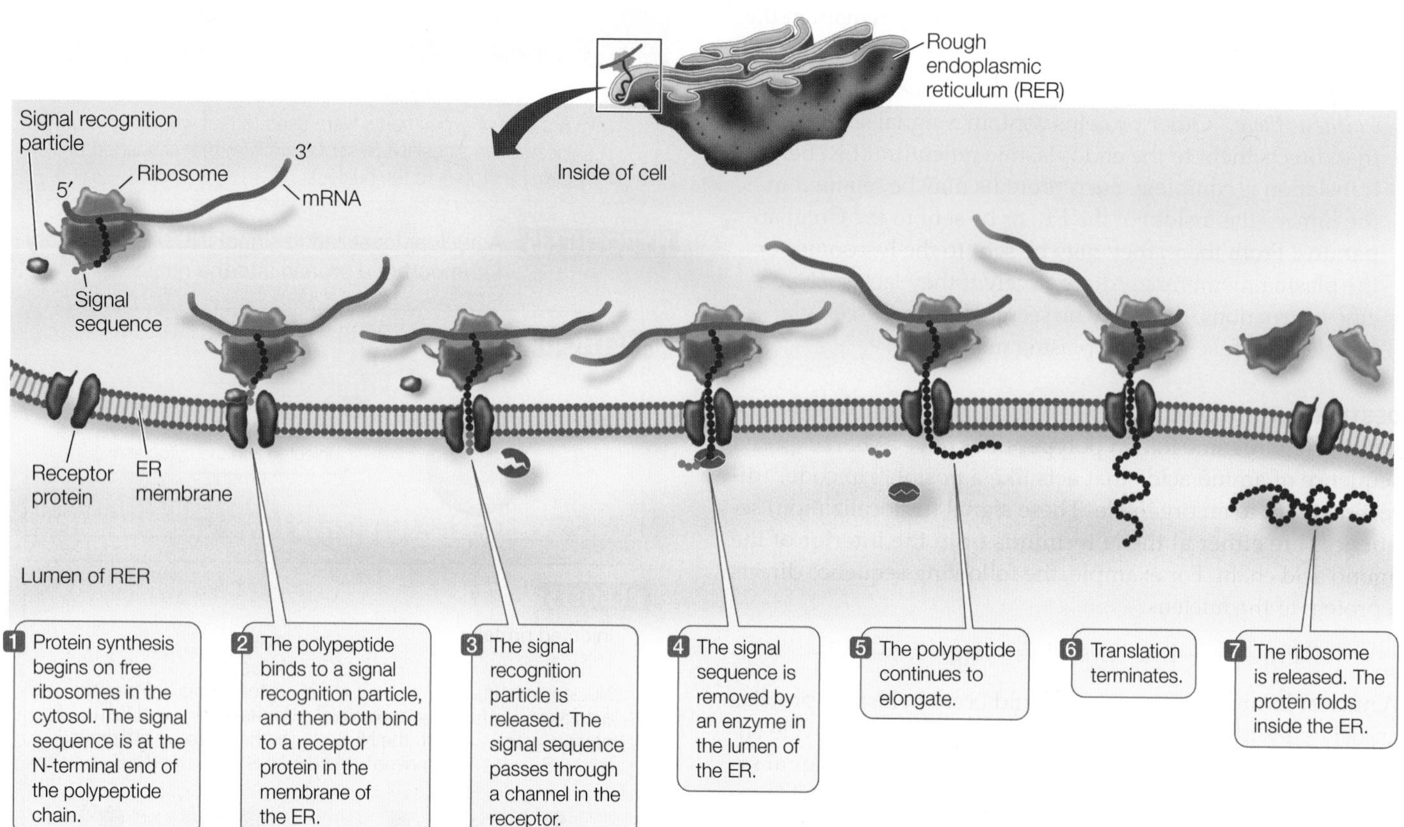

Rough endoplasmic reticulum (RER)
Inside of cell
Signal recognition particle
Ribosome
5′
3′
mRNA
Signal sequence
Receptor protein
ER membrane
Lumen of RER
1 Protein synthesis begins on free ribosomes in the cytosol. The signal sequence is at the N-terminal end of the polypeptide chain.
2 The polypeptide binds to a signal recognition particle, and then both bind to a receptor protein in the membrane of the ER.
3 The signal recognition particle is released. The signal sequence passes through a channel in the receptor.
4 The signal sequence is removed by an enzyme in the lumen of the ER.
5 The polypeptide continues to elongate.
6 Translation terminates.
7 The ribosome is released. The protein folds inside the ER.


**14.21 A Signal Sequence Moves a Polypeptide into the ER** When a certain signal sequence of amino acids is present at the beginning of a polypeptide chain, the polypeptide will be taken into the endoplasmic reticulum (ER). The finished protein is thus segregated from the cytosol.

interior space—the lumen—of the ER. In either case, an enzyme in the lumen of the ER removes the signal sequence from the polypeptide chain.

If the finished protein enters the ER lumen, it can be transported to its appropriate location—to other cellular compartments or to the outside of the cell—via the ER and the Golgi apparatus, without mixing with other molecules in the cytoplasm.

After removal of the terminal signal sequence in the lumen of the ER, additional signals are needed to direct the protein to its destination. These signals are of two kinds:

- Some are sequences of amino acids that allow the protein's retention within the ER.
- Others are sugars, which are added in the Golgi apparatus. The resulting *glycoproteins* end up either at the plasma membrane or in a lysosome (or plant vacuole), or are secreted, depending on which sugars are added.

Proteins with no additional signals pass from the ER through the Golgi apparatus and are secreted from the cell.

The importance of signals is shown by Inclusion-cell (I-cell) disease, an inherited disease that causes death in early childhood. People with this disease have a mutation in the gene encoding a Golgi enzyme that adds targeting sugars to proteins destined for the lysosomes. As a result, enzymes that are essential for the hydrolysis of various macromolecules cannot reach the lysosomes, where they are normally active. The macromolecules accumulate in the lysosomes, and this lack of cellular recycling has drastic effects, resulting in early death.

### Many proteins are modified after translation

Most mature proteins are not identical to the polypeptide chains that are translated from mRNA on the ribosomes. Instead, most polypeptides are modified in any of a number of ways after translation (**Figure 14.22**). These modifications are essential to the final functioning of the protein.

- **Proteolysis** is the cutting of a polypeptide chain. Cleavage of the signal sequence from the growing polypeptide chain in the ER is an example of proteolysis; the protein might move back out of the ER through the membrane channel if the signal sequence were not cut off. Some proteins are actually made from *polyproteins* (long polypeptides) that are cut into final products by enzymes called *proteases*. These protease enzymes are essential to some viruses, including human immunodeficiency virus (HIV), because the large viral polyprotein cannot fold properly unless it is cut. Certain drugs used to treat acquired immune deficiency syndrome (AIDS) work by inhibiting the HIV protease, thereby preventing the formation of proteins needed for viral reproduction.
- **Glycosylation** is the addition of sugars to proteins to form glycoproteins. In both the ER and the Golgi apparatus, resident enzymes catalyze the addition of various sugars or short sugar chains to certain amino acid R groups on pro-

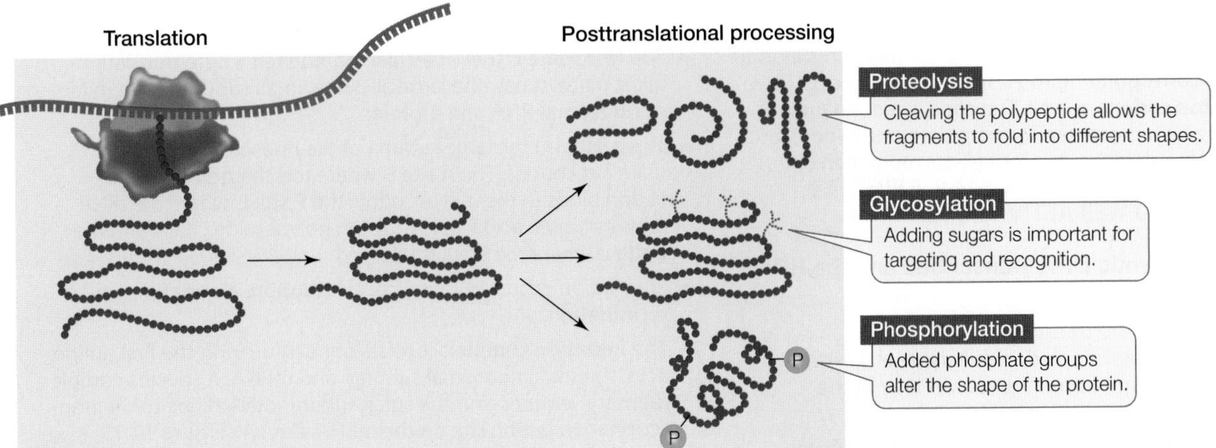


**14.22 Posttranslational Modifications of Proteins** Most polypeptides must be modified after translation in order to become functional proteins.

teins. One such type of "sugar coating" is essential for directing proteins to lysosomes, as mentioned above. Other types are important in the conformation of proteins and their recognition functions at the cell surface. Other attached sugars help to stabilize extracellular proteins, or proteins stored in vacuoles in plant seeds.

- **Phosphorylation** is the addition of phosphate groups to proteins, and is catalyzed by *protein kinases*. The charged phosphate groups change the conformation of a protein, often exposing the active site of an enzyme or the binding site for another protein. We have seen the role of phosphorylation in cell signaling (see Chapter 7).

## 14.6 RECAP

**Signal sequences in polypeptides direct them to their appropriate destinations inside or outside the cell. Many polypeptides are modified after translation.**

- How do signal sequences determine where a protein will go after it is made? See pp. 310–312 and Figure 14.21
- What are some ways in which posttranslational modifications alter protein structure and function? See pp. 312–313 and Figure 14.22

All of the processes we have just described result in a functional protein, but only if the amino acid sequence of that protein is correct. If the sequence is not correct, cellular dysfunction may result. Changes in the DNA—mutations—are a major source of errors in amino acid sequences. This is the subject of the next chapter.

## CHAPTER SUMMARY

### 14.1 What Is the Evidence that Genes Code for Proteins?

- Beadle and Tatum's experiments on metabolic enzymes in the bread mold *Neurospora* led to the **one-gene, one-enzyme hypothesis**. We now know that there is a **one-gene, one-polypeptide relationship**. Review Figure 14.1

### 14.2 How Does Information Flow from Genes to Proteins?

- The **central dogma** of molecular biology states that DNA encodes RNA, and RNA encodes proteins. Proteins do not encode proteins, RNA, or DNA.
- The process by which the information in DNA is copied to RNA is called **transcription**. The process by which a protein is built from the information in RNA is called **translation**. Review Figure 14.2, **WEB ACTIVITY 14.1**
- Certain RNA viruses are exceptions to the central dogma. These **retroviruses** synthesize DNA from RNA in a process called **reverse transcription**.
- The product of transcription is **messenger RNA** (**mRNA**). **Transfer RNA** (**tRNA**) molecules are adapters that translate the genetic information in the mRNA into a corresponding sequence of amino acids to produce a polypeptide.

### 14.3 How Is the Information Content in DNA Transcribed to Produce RNA?

- In a given gene, only one of the two strands of DNA (the **template strand**) acts as a template for transcription. **RNA polymerase** is the catalyst for transcription.
- RNA transcription from DNA proceeds in three steps: **initiation**, **elongation**, and **termination**. Review Figure 14.4, **ANIMATED TUTORIAL 14.1**
- Initiation requires a **promoter**, to which RNA polymerase binds. Part of each promoter is the **initiation site**, where transcription begins.
- Elongation of the RNA molecule proceeds from the 5′ to 3′ end.
- Particular base sequences specify termination, at which point transcription ends and the RNA transcript separates from the DNA template.

- The **genetic code** is a "language" of triplets of mRNA nucleotide bases (**codons**) corresponding to 20 specific amino acids; there are **start** and **stop codons** as well. The code is redundant (an amino acid may be represented by more than one codon), but not ambiguous (no single codon represents more than one amino acid). Review Figures 14.5 and 14.6, **ANIMATED TUTORIAL 14.2 AND WEB ACTIVITY 14.2**

## 14.4 How Is Eukaryotic DNA Transcribed and the RNA Processed?

- Unlike prokaryotes, where transcription and translation occur in the cytoplasm and are coupled, in eukaryotes transcription occurs in the nucleus and translation occurs later in the cytoplasm.
- The initial transcript of a eukaryotic protein-coding gene is modified with a **5′ cap** and a **3′ poly A sequence**. Review Figure 14.10
- Eukaryotic genes contain **introns**, which are noncoding sequences within the transcribed regions of genes.
- Pre-mRNA contains the introns. They are removed in the nucleus via **mRNA splicing** by the **small nuclear ribonucleoprotein particles**. Then the mRNA passes through the nuclear pore into the cytoplasm, where it is translated on the surfaces of **ribosomes**. Review Figure 14.11, **ANIMATED TUTORIAL 14.3**

## 14.5 How Is RNA Translated into Proteins?

**SEE ANIMATED TUTORIAL 14.4**

- During translation, amino acids are linked together in the order specified by the codons in the mRNA. This task is achieved by tRNAs, which bind to (are charged with) specific amino acids.
- Each tRNA species has an amino acid attachment site as well as an **anticodon** complementary to a specific mRNA codon. A specific activating enzyme charges each tRNA with its specific amino acid. Review Figures 14.12 and 14.13
- The **ribosome** is the molecular workbench where translation takes place. It has one large and one small subunit, both made of **ribosomal RNA** and proteins.
- Three sites on the large subunit of the ribosome interact with tRNA anticodons. The **A site** is where the charged tRNA anticodon binds to the mRNA codon; the **P site** is where the tRNA adds its amino acid to the growing polypeptide chain; and the **E site** is where the tRNA is released.
- Translation occurs in three steps: **initiation**, **elongation**, and **termination**.
- The **initiation complex** consists of tRNA bearing the first amino acid, the small ribosomal subunit, and mRNA. A specific complementary sequence on the small subunit rRNA binds to the transcription initiation site on the mRNA. Review Figure 14.15
- The growing polypeptide chain is elongated by the formation of peptide bonds between amino acids, catalyzed by the rRNA. Review Figure 14.16
- When a stop codon reaches the A site, it terminates translation by binding a release factor. Review Figure 14.17
- In a **polysome**, more than one ribosome moves along a strand of mRNA at one time. Review Figure 14.18

## 14.6 What Happens to Polypeptides after Translation?

- **Signal sequences** of amino acids direct polypeptides to their cellular destinations. Review Figure 14.19
- Destinations in the cytoplasm include organelles, which proteins enter after being recognized and bound by surface receptors called **docking proteins**.
- Proteins "addressed" to the ER bind to a **signal recognition particle**. Review Figure 14.21
- Posttranslational modifications of polypeptides include **proteolysis**, in which a polypeptide is cut into smaller fragments; **glycosylation**, in which sugars are added; and **phosphorylation**, in which phosphate groups are added. Review Figure 14.22

## SELF-QUIZ

1. Which of the following is *not* a difference between RNA and DNA?
   *a.* RNA has uracil; DNA has thymine.
   *b.* RNA has ribose; DNA has deoxyribose.
   *c.* RNA has five bases; DNA has four.
   *d.* RNA is a single polynucleotide strand; DNA is a double strand.
   *e.* RNA molecules are smaller than human chromosomal DNA molecules.
2. Normally, *Neurospora* can synthesize all 20 amino acids. A certain strain of this mold cannot grow in minimal nutritional medium, but grows only when the amino acid leucine is added to the medium. This strain
   *a.* is dependent on leucine for energy.
   *b.* has a mutation affecting a biochemical pathway leading to the synthesis of carbohydrates.
   *c.* has a mutation affecting the biochemical pathways leading to the synthesis of all 20 amino acids.
   *d.* has a mutation affecting the biochemical pathway leading to the synthesis of leucine.
   *e.* has a mutation affecting the biochemical pathways leading to the syntheses of 19 of the 20 amino acids.
3. An mRNA has the sequence 5′-AUGAAAUCCUAG-3′. What is the template DNA strand for this sequence?
   *a.* 5′-TACTTTAGGATC-3′
   *b.* 5′-ATGAAATCCTAG-3′
   *c.* 5′-GATCCTAAAGTA-3′
   *d.* 5′-TACAAATCCTAG-3′
   *e.* 5′-CTAGGATTTCAT-3′
4. The adapters that allow translation of the four-letter nucleic acid language into the 20-letter protein language are called
   *a.* aminoacyl-tRNA synthetases.
   *b.* transfer RNAs.
   *c.* ribosomal RNAs.
   *d.* messenger RNAs.
   *e.* ribosomes.
5. Which of the following does *not* occur after eukaryotic mRNA is transcribed?
   *a.* Binding of RNA polymerase to the promoter
   *b.* Capping of the 5′ end
   *c.* Addition of a poly A tail to the 3′ end
   *d.* Splicing out of the introns
   *e.* Transport to the cytosol

6. Transcription
   *a.* produces only mRNA.
   *b.* requires ribosomes.
   *c.* requires tRNAs.
   *d.* produces RNA growing from the 5′ end to the 3′ end.
   *e.* takes place only in eukaryotes.
7. Which statement about translation is *not* true?
   *a.* Translation is RNA-directed polypeptide synthesis.
   *b.* An mRNA molecule can be translated by only one ribosome at a time.
   *c.* The same genetic code operates in almost all organisms and organelles.
   *d.* Any ribosome can be used in the translation of any mRNA.
   *e.* There are both start and stop codons.
8. Which statement about RNA is *not* true?
   *a.* Transfer RNA functions in translation.
   *b.* Ribosomal RNA functions in translation.
   *c.* RNAs are produced by transcription.
   *d.* Messenger RNAs are produced on ribosomes.
   *e.* DNA codes for mRNA, tRNA, and rRNA.
9. The genetic code
   *a.* is different for prokaryotes and eukaryotes.
   *b.* has changed during the course of recent evolution.
   *c.* has 64 codons that code for amino acids.
   *d.* has more than one codon for many amino acids
   *e.* is ambiguous.
10. Which statement about RNA splicing is *not* true?
   *a.* It removes introns.
   *b.* It is performed by small nuclear ribonucleoprotein particles (snRNPs).
   *c.* It removes the introns at the ribosome.
   *d.* It is usually directed by consensus sequences.
   *e.* It shortens the RNA molecule.

## FOR DISCUSSION

1. In rats, a gene 1,440 base pairs (bp) long codes for an enzyme made up of 192 amino acids. Discuss this apparent discrepancy. How long would the initial and final mRNA transcripts be?
2. Har Gobind Khorana at the University of Wisconsin synthesized artificial mRNAs such as poly CA (CACA … ) and poly CAA (CAACAACAA … ). He found that poly CA codes for a polypeptide consisting of alternating threonine (Thr) and histidine (His) residues. There are two possible codons in poly CA, CAC and ACA. One of these must encode histidine and the other threonine—but which is which? The answer comes from results with poly CAA, which produces three different polypeptides: poly Thr, poly Gln (glutamine), and poly Asn (asparagine). (An artificial mRNA can be read, inefficiently, beginning at any point in the chain; there is no specific initiation signal. Thus poly CAA can be read as a polymer of CAA, of ACA, or of AAC.) Compare the results of the poly CA and poly CAA experiments, and determine which codon corresponds with threonine and which with histidine.
3. Look back at Question 2. Using the genetic code in Figure 14.6 as a guide, deduce what results Khorana would have obtained had he used poly UG and poly UGG as artificial mRNAs. In fact, very few such artificial mRNAs would have given useful results. For an example of what could happen, consider poly CG and poly CGG. If poly CG were the mRNA, a mixed polypeptide of arginine and alanine (Arg–Ala–Ala–Arg … ) would be obtained; poly CGG would give three polypeptides: poly Arg, poly Ala, and poly Gly (glycine). Can any codons be determined from only these data? Explain.
4. Errors in transcription occur about 100,000 times as often as errors in DNA replication. Why can this high rate be tolerated in RNA synthesis but not in DNA synthesis?

## ADDITIONAL INVESTIGATION

Beadle and Tatum's experiments showed that a biochemical pathway could be deduced from mutant strains. In bacteria, the biosynthesis of the amino acid tryptophan (T) from the precursor chorismate (C) involves four intermediate chemical compounds, which we will call D, E, F, and G. Here are the phenotypes of various mutant strains. Each strain has a mutation in a gene for a different enzyme; + means growth with the indicated compound added to the medium, and 0 means no growth. Based on these data, order the compounds (C, D, E, F, G, and T) and enzymes (1, 2, 3, 4, and 5) in a biochemical pathway.

| | Addition to medium | | | | | |
|---|---|---|---|---|---|---|
| Mutant strain | C | D | E | F | G | T |
| 1 | 0 | 0 | 0 | 0 | + | + |
| 2 | 0 | + | + | 0 | + | + |
| 3 | 0 | + | 0 | 0 | + | + |
| 4 | 0 | + | + | + | + | + |
| 5 | 0 | 0 | 0 | 0 | 0 | + |

## WORKING WITH DATA (GO TO yourBioPortal.com)

**Deciphering the Genetic Code** The identification of the first mRNA codons associated with specific amino acids was a landmark in molecular biology (Figure 14.5). In this hands-on exercise, you will learn about the experimental protocol that Nirenberg and Matthei followed, using artificial mRNA, and analyze the results they obtained.

# 15 Gene Mutation and Molecular Medicine

## Baby 81

The tsunami of December 26, 2004, struck the coastal town of Kalmunai, Sri Lanka, with such force that 4-month-old Abilass Jeyarajah was torn from his mother's arms and swept away. Hours later, while his parents desperately searched the devastated town, their tiny son washed up on the beach a kilometer away, alive. A local schoolteacher found him and brought him to the hospital—the eighty-first patient admitted that day. The hospital was overwhelmed with 1,000 dead bodies, many of them children. Since Abilass was alive and healthy, he was dubbed "Baby 81, the miracle baby" and became an instant celebrity among the staff as they went about their grim duties of caring for the injured and dying.

Meanwhile, the parents kept looking. Two days later, they met the schoolteacher, who told them about the baby he had found. Rushing to the hospital, the Jeyarajahs were elated to find their son, but were in for a rude shock. Eight other couples who had also lost infants were claiming Baby 81 as theirs. The baby remained in the hospital while the case went to court.

Judge M. P. Mohaideen faced a situation not unlike one faced by King Solomon 3,000 years ago, who was asked to decide which of two women was the mother of an infant. Solomon's method of determining parentage is told in a famous biblical passage—he ordered the baby cut in two, and the real mother indicated that she would rather give the baby away than have the baby killed. The Sri Lankan judge had a different method: he called in molecular biologists.

With 6 billion base pairs of DNA packaged in 46 chromosomes, each one of us is unique. Although our protein-coding sequences are similar (after all, our phenotypes are similar), only 1.5 percent of the DNA in the human genome actually codes for proteins. The eukaryotic genome contains many repeated sequences, and the repeat frequencies may differ between individuals, offering one way to differentiate one individual from another. A base pair at a particular site may also vary between individuals, due to DNA replication errors or random muta-

**After the Tsunami** In December of 2004, a tsunami originating in the Indian Ocean struck a broad region that encompassed many nations in Southeast Asia. The result was an unprecedented humanitarian disaster that left almost a quarter of a million people dead and many more homeless.

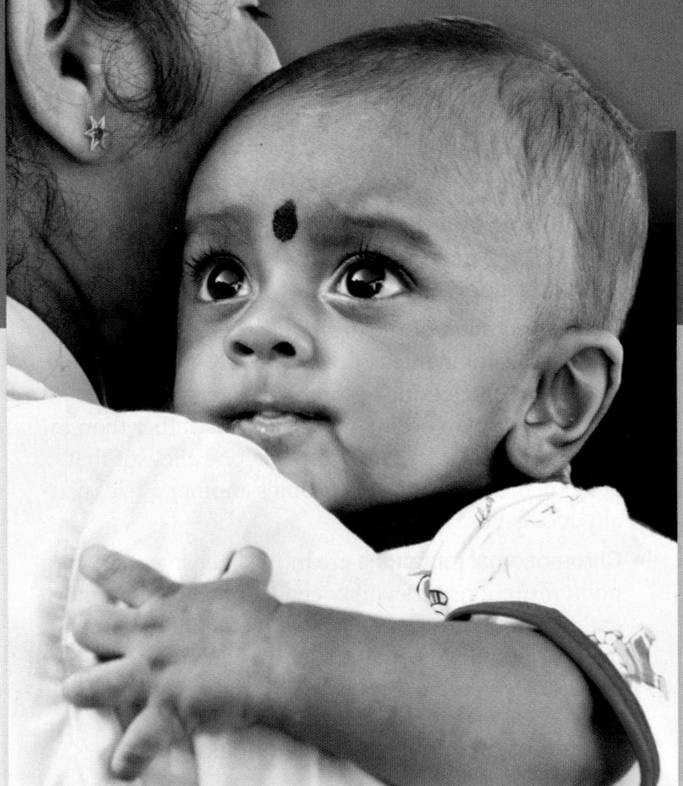

**Baby 81** Abilass Jeyarajah survived the tsunami and was reunited with his parents by court order after DNA testing proved that he is indeed their son.

tions. Both of these types of differences are mutations, defined as inherited changes in DNA.

It is now possible to analyze these differences in DNA sequences (amplified by PCR) to identify people, in a process called DNA fingerprinting. The most common DNA fingerprinting technique used today involves the detection of variations in repeat sequences at different loci throughout the genome. When DNA samples from the nine sets of contesting parents were analyzed and compared with a sample from Baby 81, only one pair of parents carried sequences that were the same as those of the baby. On February 14, 2005, the judge ruled that the Jeyarajahs were the biological parents, and Baby 81 got his real name and parents back.

**IN THIS CHAPTER** we will discuss the nature and detection of mutations at the molecular and chromosomal levels. We will describe how abnormal proteins can cause human genetic diseases, and how these diseases and the alleles that produce them can be detected. Finally, we'll see how this knowledge of mutations has been applied in the development of new treatments.

CHAPTER OUTLINE



## 15.1 What Are Mutations?

In Chapter 12, we described mutations as inherited changes in genes, and we saw that different alleles may produce different phenotypes (short pea plants versus tall, for example). Now that we understand the chemical nature of genes and how they are expressed as phenotypes (in particular, proteins) we can return to the concept of mutations for a more specific definition. We can now state that mutations are changes in the nucleotide sequence of DNA that are passed on from one cell, or organism, to another.

As an example of just one cause of mutations, recall from Chapter 13 that DNA polymerases make errors. Repair systems such as proofreading are in place to correct them. But some errors escape being corrected and are passed on to the daughter cells.

Mutations in multicellular organisms can be divided into two types:

- **Somatic mutations** are those that occur in somatic (body) cells. These mutations are passed on to the daughter cells during mitosis, and to the offspring of those cells in turn, but are not passed on to sexually produced offspring. For example, a mutation in a single human skin cell could result in a patch of skin cells that all have the same mutation, but it would not be passed on to the person's children.
- **Germ line mutations** are those that occur in the cells of the germ line—the specialized cells that give rise to gametes. A gamete with the mutation passes it on to a new organism at fertilization.

In either case, the mutations may or may not have phenotypic effects.

### Mutations have different phenotypic effects

Phenotypically, we can understand mutations in terms of their effects on proteins and their function (**Figure 15.1**).

- **Silent mutations** do not affect protein function. They can be mutations in noncoding DNA, such as the repeat sequences that were used to identify Baby 81 in the opening story of this chapter. Or they can be in the coding portion of DNA but not have any effect on the protein.
- **Loss of function mutations** affect protein function. These mutations may lead to nonfunctional proteins that no longer work as structural proteins or enzymes. They almost

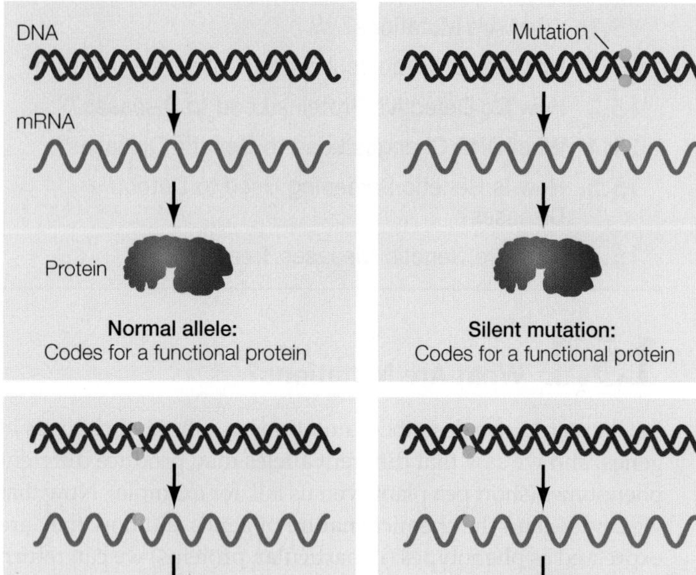

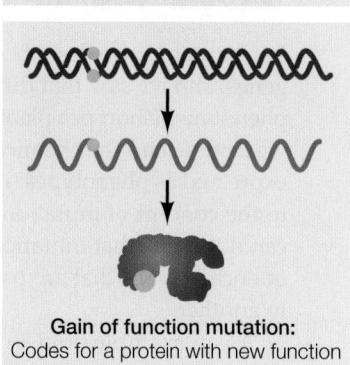


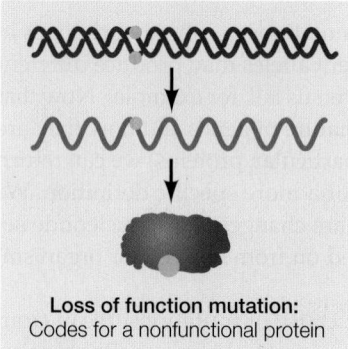

**15.1 Mutation and Phenotype** Mutations may or may not affect the protein phenotype.

always show recessive inheritance in a diploid organism, because the presence of one wild-type allele will usually result in sufficient functional protein for the cell. For example, the familiar wrinkled-seed allele in pea plants, originally studied by Mendel (see Figure 12.3), is due to a mutation in the gene *SBE1* (*s*tarch *b*ranching *e*nzyme). Normally the protein made by this gene catalyzes the branching of starch as seeds develop. In the mutant, the SBE1 protein is not functional and that leads to osmotic changes, causing the wrinkled appearance.

- A **gain of function mutation** leads to a protein with an altered function. This kind of mutation usually shows dominant inheritance, because the presence of the wild-type allele does not prevent the mutant allele from functioning. This is common in cancer. For example, a receptor for a growth factor normally requires binding of the growth factor (the ligand) to activate the cell division cycle. Some cancers are caused by mutations in genes coding for these receptors such that they no longer require stimulation by their particular ligands. The mutant receptors are "always on," leading to the unrestrained cell proliferation that is characteristic of cancer cells.
- **Conditional mutations** cause their phenotypes only under certain *restrictive* conditions. They are not detectable under other, *permissive* conditions. Many conditional mutants are temperature-sensitive; that is, they show the altered phenotype only at a certain temperature (recall the rabbit in Figure 12.11). The mutant allele in such an organism may code for an enzyme with an unstable tertiary structure that is altered at the restrictive temperature.

All mutations are alterations in the nucleotide sequence of DNA. At the molecular level, we can divide mutations into two categories:

- A **point mutation** results from the gain, loss, or substitution of a single nucleotide. After DNA replication, the altered nucleotide becomes a mutant base pair. If a point mutation occurs within a gene (rather than in a noncoding DNA sequence), then one allele of that gene (usually dominant) becomes another allele (usually recessive).
- **Chromosomal mutations** are more extensive than point mutations. They may change the position or orientation of a DNA segment without actually removing any genetic information, or they may cause a segment of DNA to be duplicated or irretrievably lost.

### Point mutations change single nucleotides

Point mutations result from the addition or subtraction of a nucleotide base, or the substitution of one base for another. Point mutations can arise due to errors in DNA replication that are not corrected during proofreading, or they may be caused by environmental **mutagens** (substances that cause mutations, such as radiation or certain chemicals).

Point mutations in the coding regions of DNA usually result in changes in the mRNA, but changes in the mRNA may or may not result in changes in the protein. Silent mutations by definition have no effect on the protein. Missense and nonsense mutations result in changes in the protein, some of them drastic (**Figure 15.2**).

**SILENT MUTATIONS** Silent mutations have no effect on amino acid sequences. This is because they are often found in noncoding DNA. Also, because of the redundancy of the genetic code, a base substitution in a coding region will not always cause a change in the amino acid sequence when the altered mRNA is translated. Silent mutations are quite common, and they result in genetic diversity that is not expressed as phenotypic differences.

**MISSENSE MUTATIONS** Some base substitutions change the genetic code such that one amino acid substitutes for another in a protein. These changes are called **missense mutations**. A specific example of a missense mutation is the one that causes sickle-cell disease, a serious heritable blood disorder. The disease occurs in people who carry two copies of the sickle allele of the gene for human β-globin (a subunit of hemoglobin, the protein in human blood that carries oxygen). The sickle allele differs from the normal allele by one base pair, resulting in a polypeptide that differs by one amino acid from the normal protein. Individuals who are homozygous for this recessive allele have defective, sickle-shaped red blood cells (**Figure 15.3**).

A missense mutation may result in a defective protein, but often it has no effect on the protein's function. For example, a

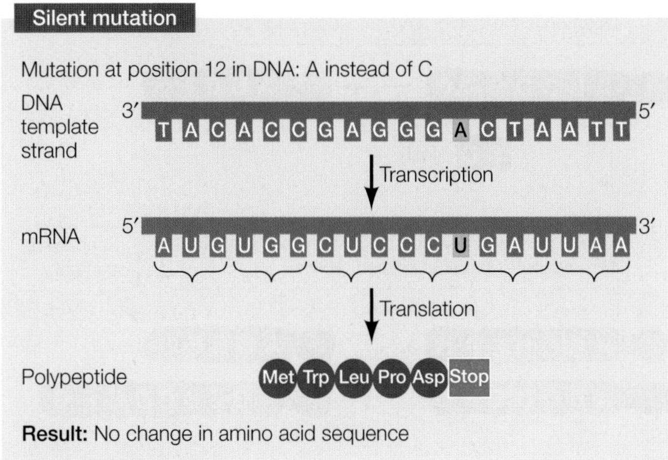


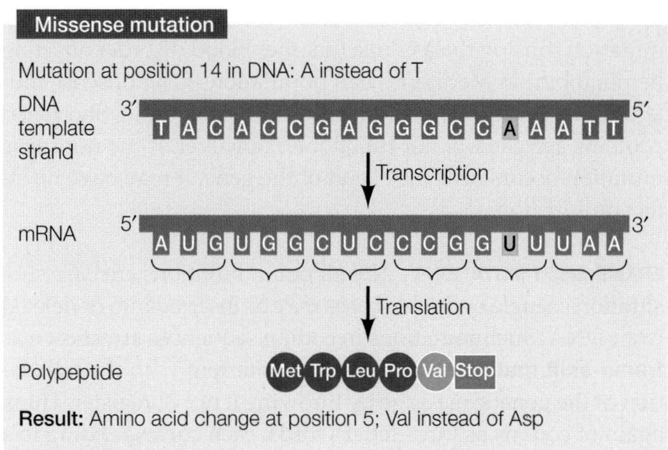


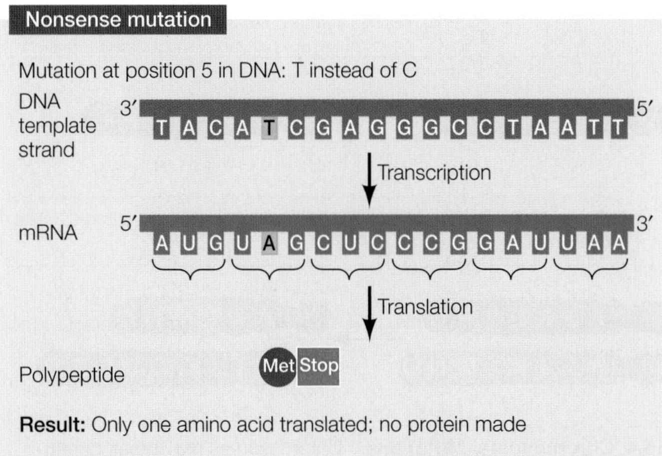


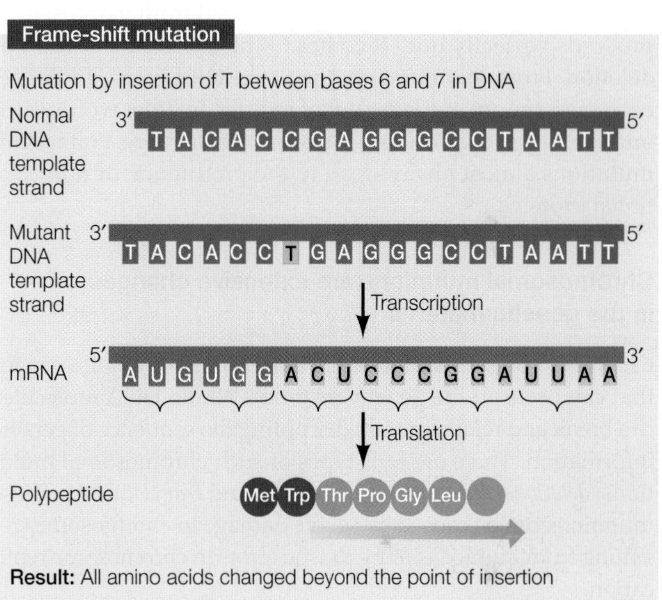


**15.2 Point Mutations** When they occur in the coding regions of proteins, single-base pair changes can cause missense, nonsense, or frame-shift mutations. Some of these mutations are silent, while others affect the protein's amino acid sequence.

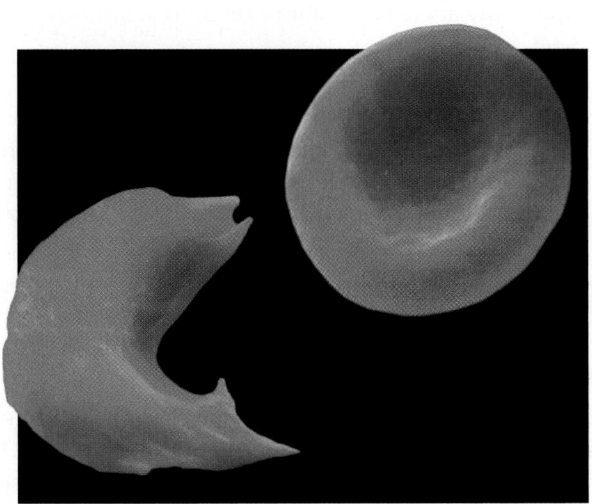

**15.3 Sickled and Normal Red Blood Cells** The misshapen red blood cell on the left is caused by a missense mutation and an incorrect amino acid in one of the two polypeptides of hemoglobin.

hydrophilic amino acid may be substituted for another hydrophilic amino acid, so that the shape of the protein is unchanged. Or a missense mutation might reduce the functional efficiency of a protein rather than completely inactivating it. Therefore, individuals homozygous for a missense mutation in a protein essential for life may survive if enough of the protein's function is retained.

In some cases, a gain of function missense mutation occurs. An example is a mutation in the human *TP53* gene, which codes for a tumor suppressor; that is, the TP53 protein normally functions to inhibit the cell cycle. Certain mutations of the *TP53* gene cause this protein to no longer inhibit cell division, but to promote it and prevent programmed cell death. So a TP53 protein mutated in this way has a gain of oncogenic (cancer-causing) function.

**NONSENSE MUTATIONS** A **nonsense mutation** involves a base substitution that causes a stop codon (for translation) to form somewhere in the mRNA. A nonsense mutation results in a shortened protein, since translation does not proceed beyond the point where the mutation occurred. For example, a common mutation causing thalassemia (another blood disorder affecting hemoglobin) in Mediterranean populations is a nonsense mutation that drastically shortens the α-globin subunit. Shortened proteins are usually not functional; however, if the nonsense mutation occurs near the 3′ end of the gene, it may have no effect on function.

**FRAME-SHIFT MUTATIONS** Not all point mutations are base substitutions. Single or double bases may be inserted into or deleted from DNA. Such mutations in coding sequences are known as **frame-shift mutations** because they interfere with the translation of the genetic message by throwing it out of register. Think again of codons as three-letter words, each corresponding to a particular amino acid. Translation proceeds codon by codon; if a base is added to the mRNA or subtracted from it, translation proceeds perfectly until it comes to the one-base insertion or deletion. From that point on, the three-letter words in the genetic message are one letter out of register. In other words, such mutations shift the "reading frame" of the message. Frame-shift mutations almost always lead to the production of nonfunctional proteins.

### Chromosomal mutations are extensive changes in the genetic material

Changes in single nucleotides are not the most dramatic changes that can occur in the genetic material. Whole DNA molecules can break and rejoin, grossly disrupting the sequence of genetic information. There are four types of such chromosomal mutations: *deletions*, *duplications*, *inversions*, and *translocations*. These mutations can be caused by severe damage to chromosomes resulting from mutagens or by drastic errors in chromosome replication.

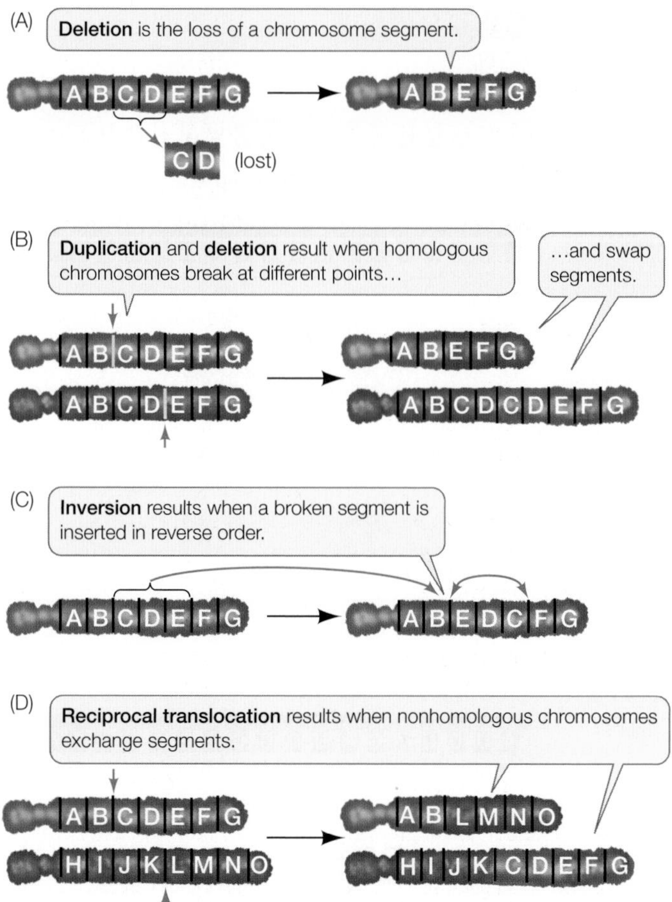


**15.4 Chromosomal Mutations** Chromosomes may break during replication, and parts of chromosomes may then rejoin incorrectly. The letters on these chromosome illustrations represent large segments of the chromosomes. Each segment may include anywhere from zero to hundreds or thousands of genes.

- **Deletions** result from the removal of part of the genetic material (**Figure 15.4A**). Like frame-shift point mutations, their consequences can be severe unless they affect noncoding DNA or unnecessary genes, or are masked by the presence of normal alleles of the deleted genes in the same cell. It is easy to imagine one mechanism that could produce deletions: a DNA molecule might break at two points and the two end pieces might rejoin, leaving out the DNA between the breaks.
- **Duplications** can be produced at the same time as deletions (**Figure 15.4B**). A duplication would arise if homologous chromosomes broke at different positions and then reconnected to the wrong partners. One of the two chromosomes produced by this mechanism would lack a segment of DNA (it would have a deletion), and the other would have two copies (a duplication) of the segment that was deleted from the first chromosome.
- **Inversions** can also result from breaking and rejoining of chromosomes. A segment of DNA may be removed and reinserted into the same location in the chromosome, but "flipped" end over end so that it runs in the opposite direction (**Figure 15.4C**). If the break site includes part of a DNA segment that codes for a protein, the resulting protein will be drastically altered and almost certainly nonfunctional.
- **Translocations** result when a segment of a chromosome breaks off and is inserted into a different chromosome. Translocations may involve reciprocal exchanges of chromosome segments, as in **Figure 15.4D**. Translocations often lead to duplications and deletions and may result in sterility if normal chromosome pairing in meiosis cannot occur.

### Mutations can be spontaneous or induced

It is useful to distinguish two types of mutations in terms of their causes:

- **Spontaneous mutations** are permanent changes in the genetic material that occur without any outside influence. In other words, they occur simply because cellular processes are imperfect.
- **Induced mutations** occur when some agent from outside the cell—a mutagen—causes a permanent change in DNA.

*Spontaneous* mutations may occur by several mechanisms:

- *The four nucleotide bases of DNA can have different structures.* Each can exist in two different forms (called *tautomers*), one of which is common and one rare. When a base temporarily forms its rare tautomer, it can pair with the wrong base. For example, C normally pairs with G, but if C is in its rare tautomer at the time of DNA replication, it pairs with (and DNA polymerase will insert) an A. The result is a point mutation: G $\rightarrow$ A (**Figure 15.5A and C**).
- *Bases in DNA may change because of a chemical reaction*—for example, loss of an amino group in cytosine (a reaction called *deamination*). If this occurs in a DNA molecule, the error will usually be repaired. However, since the repair mechanism is not perfect, the altered nucleotide will sometimes remain during replication. Then, DNA polymerase will add an A (which base-pairs with U) instead of G (which normally pairs with C).
- *DNA polymerase can make errors in replication* (see Section 13.4)—for example, inserting a T opposite a G. Most of these errors are repaired by the proofreading function of the replication complex, but some errors escape detection and become permanent.
- *Meiosis is not perfect.* Nondisjunction—failure of homologous chromosomes to separate during meiosis—can occur, leading to one too many or one too few chromosomes (aneuploidy; see Figure 11.21). Random chromosome breakage and rejoining can produce deletions, duplications, inversions, or translocations.

*Induced* mutations result from alterations of DNA by mutagens:

- *Some chemicals can alter the nucleotide bases.* For example, nitrous acid ($HNO_2$) and similar molecules can react with cytosine and convert it to uracil by deamination. More specifically, they convert an amino group on the cytosine (—$NH_2$) into a keto group (—C=O). This alteration has the same result as spontaneous deamination: instead of a G, DNA polymerase inserts an A (**Figure 15.5B and C**).
- *Some chemicals add groups to the bases.* For instance, benzopyrene, a component of cigarette smoke, adds a large chemical group to guanine, making it unavailable for base pairing. When DNA polymerase reaches such a modified guanine, it inserts any one of the four bases; of course, three-fourths of the time the inserted base will not be cytosine, and a mutation results.
- *Radiation damages the genetic material.* Radiation can damage DNA in two ways. First, ionizing radiation (including X rays, gamma rays, and radiation from unstable isotopes) produces highly reactive chemicals called *free radicals*. Free radicals can change bases in DNA to forms that are not recognized by DNA polymerase. Ionizing radiation can also

**15.5 Spontaneous and Induced Mutations** (A) All four nitrogenous bases in DNA exist in both a prevalent (common) form and a rare form. When a base spontaneously forms its rare tautomer, it can pair with a different base. (B) Mutagens such as nitrous acid can induce changes in the bases. (C) The results of both spontaneous and induced mutations are permanent changes in the DNA sequence following replication.

**(A) A spontaneous mutation**

Cytosine (common tautomer)
Cytosine (rare tautomer)
This C cannot hydrogen-bond with G but instead pairs with A.

**(B) An induced mutation**

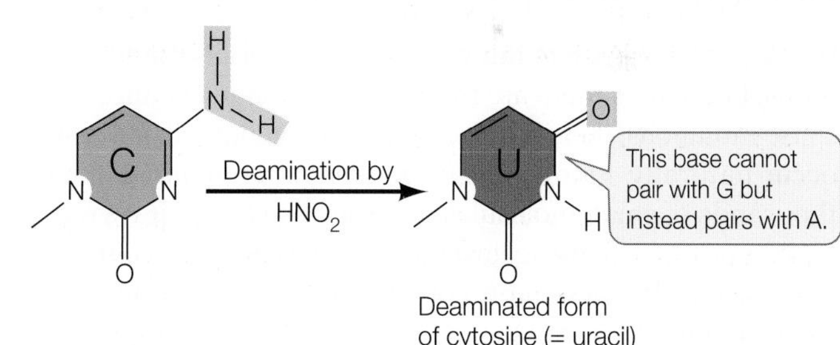
Deamination by $HNO_2$
This base cannot pair with G but instead pairs with A.
Deaminated form of cytosine (= uracil)

**(C) The consequences of either mutation**

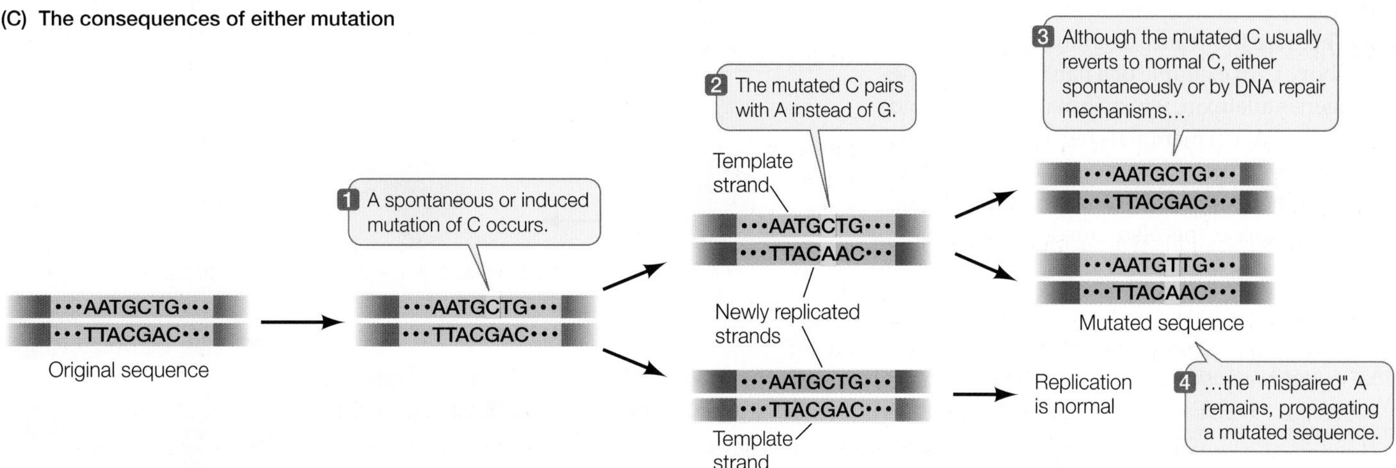
1 A spontaneous or induced mutation of C occurs.
2 The mutated C pairs with A instead of G.
3 Although the mutated C usually reverts to normal C, either spontaneously or by DNA repair mechanisms...
4 ...the "mispaired" A remains, propagating a mutated sequence.
•••AATGCTG•••
•••TTACGAC•••
Original sequence
Template strand
•••TTACAAC•••
Newly replicated strands
•••AATGTTG•••
Mutated sequence
Replication is normal

break the sugar–phosphate backbone of DNA, causing chromosomal abnormalities. Second, ultraviolet radiation (from the sun or a tanning lamp) can damage DNA in another way. It is absorbed by thymine, causing it to form covalent bonds with adjacent bases. This, too, plays havoc with DNA replication by distorting the double helix.

### Some base pairs are more vulnerable than others to mutation

DNA sequencing has revealed that mutations occur most often at certain base pairs. These "hot spots" are often located where cytosine has been methylated to 5-methylcytosine.

As we discussed above, unmethylated cytosine can lose its amino group, either spontaneously or because of a chemical mutagen, to form uracil (see Figure 15.5B). This type of error is usually detected by the cell and repaired. The DNA repair mechanism recognizes uracil as inappropriate for DNA (since uracil occurs only in RNA) and replaces it with cytosine.

However, when 5-methylcytosine loses its amino group, the product is thymine, a natural base for DNA. The DNA repair mechanism ignores this thymine (**Figure 15.6**). During replication, however, the mismatch repair mechanism recognizes that G-T is a mismatched pair, although it cannot tell which base was incorrectly inserted into the sequence. So half of the time it matches a new C to the G, but the other half of the time it matches a new A to the T, resulting in a mutation. It is not surprising that 5-methylcytosine residues are hot spots for mutation.

### Mutagens can be natural or artificial

Many people associate mutagens with materials made by humans, but just as there are many human-made chemicals that cause mutations, there are also many mutagenic substances that occur naturally. Plants (and to a lesser extent animals) make thousands of small molecules that they use for their own purposes, such as defense against pathogens (see Chapter 39). Some of these are mutagenic and potentially carcinogenic. Examples of human-made mutagens are nitrites, which are used to preserve meats. Once in mammals, nitrites get converted by the smooth endoplasmic reticulum (ER) to nitrosamines, which are strongly mutagenic because they cause deamination of cytosine (see above). An example of a naturally occurring mutagen is aflatoxin, which is made by the mold *Aspergillus*. When mammals ingest the mold, the aflatoxin is converted by the ER into a product that, like benzopyrene from cigarette smoke, binds to guanine; this also causes mutations.

Radiation can also be human-made or natural. Some of the isotopes made in nuclear reactors and nuclear bomb explosions are certainly harmful. For example, extensive studies have shown increased mutations in the survivors of the atom bombs dropped on Japan in 1945. You probably know that natural ultraviolet radiation in sunlight also causes mutations, in this case by affecting thymine and, to a lesser extent, other bases in DNA.

Biochemists have estimated how much DNA damage occurs in the human genome under normal circumstances: among the genome's 2.3 billion base pairs there are about 16,000 DNA-damaging events per cell per day, of which 80 percent are repaired.

### Mutations have both benefits and costs

What is the overall effect of mutation? For an organism, there are benefits and costs.

- *Mutations are the raw material of evolution.* Without mutation, there would be no evolution. As we will see in Part Seven of this book, mutation alone does not drive evolution, but it provides the genetic diversity that makes natural selection possible. This diversity can be beneficial in two ways. First, a mutation in somatic cells may benefit the organism immediately. Second, a mutation in germ line cells may have no immediate selective advantage to the organism but may cause a phenotypic change in offspring. If the environment changes in a later generation, that mutation may be advantageous and thus selected for under these conditions.
- *Germ line and somatic mutations can be harmful.* Mutations in germ line cells that get carried to the next generation are often deleterious, especially if the offspring are homozygous for a harmful recessive allele. In their extreme form, such mutations produce phenotypes that are lethal. Lethal mutations can kill an organism during early development, or the organism may die before maturity and reproduction.

In Chapter 11 we described how genetic changes in somatic cells can lead to cancer. Typically these are mutations in oncogenes (the "gas pedal") that result in the stimulation of cell division, or mutations in tumor suppressor genes (the "brakes") that result in a lack of inhibition of cell division. These muta-

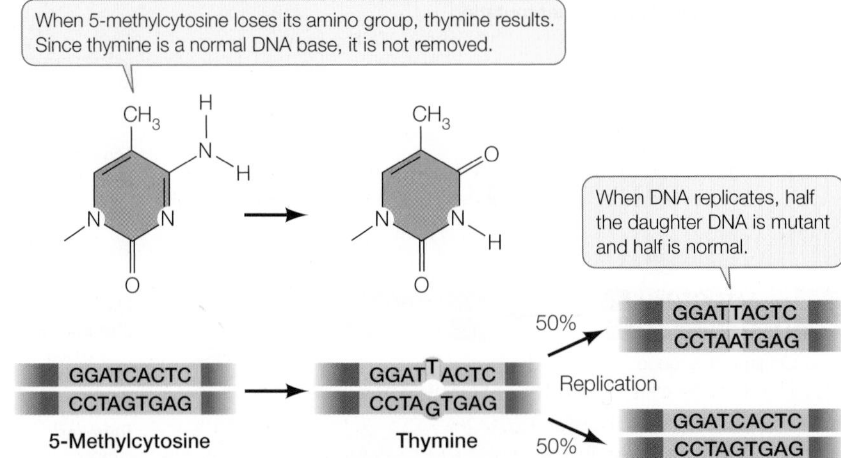


**15.6 5-Methylcytosine in DNA Is a "Hot Spot" for Mutations** If cytosine has been methylated to 5-methylcytosine, the mutation is unlikely to be repaired and a C-G base pair is replaced with a T-A pair.

tions can occur by either spontaneous or induced mutagenesis. While spontaneous mutagenesis is not in our control, we can certainly try to avoid mutagenic substances and radiation. Not surprisingly, many things that cause cancer (carcinogens) are also mutagens. A good example is benzopyrene (discussed above), which is found in coal tar, car exhaust fumes, and charbroiled foods, as well as in cigarette smoke.

A major environmental issue is the effect of both human-made and natural mutagens on public health. Identifying mutagens to which people are exposed, and estimating their risk for both mutagenesis and carcinogenesis, is a major public policy goal. Here are two recent examples:

- The Montreal Protocol is the only international environmental agreement signed and adhered to by all nations. It bans chlorofluorocarbons and other substances that cause depletion of the ozone layer in the upper atmosphere of Earth. Such depletion can result in increased ultraviolet radiation reaching Earth's surface. This would cause more somatic mutations which lead to skin cancer.
- Bans on cigarette smoking have rapidly spread throughout the world. Cigarette smoking causes cancer due to increased exposure of somatic cells in the lungs and throat to benzopyrene and other carcinogens.

## 15.1 RECAP

**Mutations are alterations in the nucleotide sequence of DNA. They may be changes in single nucleotides or extensive rearrangements of chromosomes. If they occur in somatic cells, they will be passed on to daughter cells; if they occur in germ line cells, they will be passed on to offspring.**

- What are the various kinds of point mutations? See pp. 318–320 and Figure 15.2
- What distinguishes the various kinds of chromosomal mutations: deletions, duplications, inversions, and translocations? See p. 320 and Figure 15.4
- Explain the difference between spontaneous and induced mutagenesis. Give an example of each. See pp. 320–322 and Figure 15.5
- Why do many mutations involve G-C base pairs? See p. 322 and Figure 15.6

We have seen that there are many different ways in which DNA can be altered, in terms of both the types of changes and the mechanisms by which they occur. We turn now to the ways that biologists detect mutations in DNA.

# 15.2 How Are DNA Molecules and Mutations Analyzed?

Once biologists understood the connections between phenotype and proteins, and between genes and DNA, they were faced with the important challenge of precisely describing the specific DNA changes that lead to specific protein changes—an area of research called *molecular genetics*. To begin this work, biologists needed tools to analyze DNA molecules for mutations. In this section we will see how some of the numerous naturally occurring enzymes that cleave DNA have now become one of the most important tools used in molecular genetics laboratories.

### Restriction enzymes cleave DNA at specific sequences

All organisms, including bacteria, must have ways of dealing with their enemies. As we saw in Section 13.1, bacteria are attacked by viruses called bacteriophage. These viruses inject their genetic material into the host cell and turn it into a virus-producing factory, eventually killing the cell. Some bacteria defend themselves against such invasions by producing **restriction enzymes** (also known as *restriction endonucleases*), which cut double-stranded DNA molecules—such as those injected by bacteriophage—into smaller, noninfectious fragments (**Figure 15.7**). These enzymes break the bonds of the DNA backbone between the 3′ hydroxyl group of one nucleotide and the 5′ phosphate group of the next nucleotide. This cutting process is called **restriction digestion**.

There are many such restriction enzymes, each of which cleaves DNA at a specific sequence of bases called a **recognition sequence** or a **restriction site**. Most recognition sequences are 4–6 base pairs long. The sequence is recognized through the principles of protein–DNA interactions (see Section 13.2). That is, the base pairs inside the DNA double helix vary slightly in shape, so that a particular short sequence of base pairs will fit a specific three-dimensional structure on an enzyme.

Why doesn't a restriction enzyme cut the DNA of the bacterial cell that makes it? One way that the cell protects itself is by modifying the restriction sites on its own DNA. Specific modifying enzymes called *methylases* add methyl (—$CH_3$) groups to certain bases at the restriction sites on the host's DNA after it has been replicated. The methylation of the host's bases makes

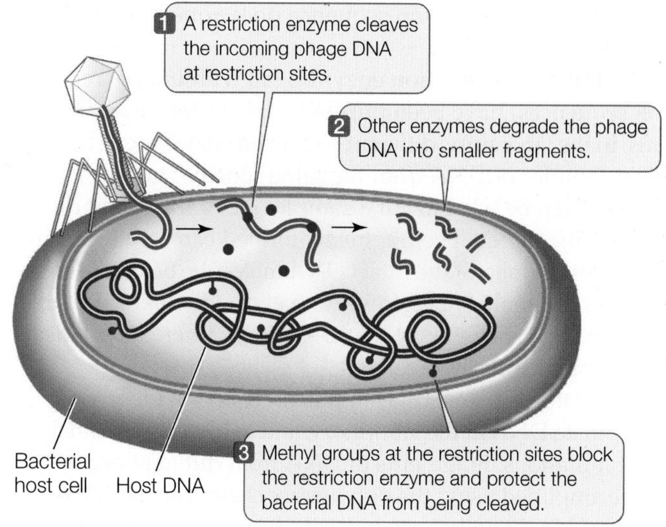

1 A restriction enzyme cleaves the incoming phage DNA at restriction sites.
2 Other enzymes degrade the phage DNA into smaller fragments.
Bacterial host cell
Host DNA
3 Methyl groups at the restriction sites block the restriction enzyme and protect the bacterial DNA from being cleaved.

**15.7 Bacteria Fight Invading Viruses by Making Restriction Enzymes**

the recognition sequence unrecognizable to the restriction enzyme. But unmethylated phage DNA is efficiently recognized and cleaved.

Bacterial restriction enzymes can be isolated from the cells that make them and used as biochemical reagents in the laboratory to give information about the nucleotide sequences of DNA molecules from other organisms. If DNA from any organism is incubated in a test tube with a restriction enzyme (along with buffers and salts that help the enzyme to function), that DNA will be cut wherever the restriction site occurs. A specific sequence of bases defines each restriction site. For example, the enzyme *Eco*RI (named after its source, a strain of the bacterium *E. coli*) cuts DNA only where it encounters the following paired sequence in the DNA double helix:

$$\begin{array}{c} \downarrow \\ 5'\ldots\mathrm{GAATTC}\ldots 3' \\ 3'\ldots\mathrm{CTTAAG}\ldots 5' \\ \uparrow \end{array}$$

Note that this sequence is palindromic, like the word "mom," in that the opposite strands have the same sequences when they are read from their 5′ ends. The *Eco*RI enzyme has two identical active sites on its two subunits, which cleave the two strands simultaneously between the G and the A of each strand.

The *Eco*RI recognition sequence occurs, on average, about once in every 4,000 base pairs in a typical prokaryotic genome, or about once per four prokaryotic genes. So *Eco*RI can chop a large piece of DNA into smaller pieces containing, on average, just a few genes. Using *Eco*RI in the laboratory to cut small genomes, such as those of viruses that have tens of thousands of base pairs, may result in a few fragments. For a huge eukaryotic chromosome with tens of millions of base pairs, a very large number of fragments will be created.

Of course, "on average" does not mean that the enzyme cuts all stretches of DNA at regular intervals. For example, the *Eco*RI recognition sequence does not occur even once in the 40,000 base pairs of the T7 phage genome—a fact that is crucial to the survival of this virus, since its host is *E. coli*. Fortunately for *E. coli*, the *Eco*RI recognition sequence does appear in the DNA of other bacteriophage.

Hundreds of restriction enzymes (all with unique recognition sequences) have been purified from various microorganisms. In the laboratory, different restriction enzymes can be used to cut samples of DNA from the same source. Thus restriction enzymes can be used to cut a sample of DNA in many different, specific places. The fragments formed can be used to create a physical map of the intact DNA molecule. Before DNA sequencing technology became automated and widely available, this was the principal way that DNA from different organisms was mapped and characterized.

Restriction enzyme digestion is used very widely today to manipulate DNA in the laboratory, and to identify and analyze point mutations. Fragments of DNA from different organisms can be amplified using the polymerase chain reaction (PCR; see Section 13.5). Even between closely related individuals, these amplified fragments often contain variations in DNA sequences (due in most cases to silent mutations). If these variations affect restriction sites, then digestion of the fragments with restriction enzymes can be used to distinguish between the samples. Restriction enzymes are also used to cut DNA for use in genetic engineering experiments, and in many other types of experiments aimed at understanding how organisms function at the molecular level.

### Gel electrophoresis separates DNA fragments

After a laboratory sample of DNA has been cut with a restriction enzyme, the DNA is in fragments, which must be separated to identify (map) where the cuts were made. Because the recognition sequence does not occur at regular intervals, the fragments are not all the same size, and this property provides a way to separate them from one another. Separating the fragments is necessary to determine the number and molecular sizes (in base pairs) of the fragments produced, or to identify and purify an individual fragment for further analysis or for use in an experiment.

A convenient way to separate or purify DNA fragments is by **gel electrophoresis**. Samples containing the fragments are placed in wells at one end of a semisolid gel (usually made of agarose or polyacrylamide polymers) and an electric field is applied to the gel (**Figure 15.8**). Because of its phosphate groups, DNA is negatively charged at neutral pH; therefore, because opposite charges attract, the DNA fragments move through the gel toward the positive end of the field. Because the spaces between the polymers of the gel are small, small DNA molecules can move through the gel faster than larger ones. Thus, DNA fragments of different sizes separate from one another and can be detected with a dye. This gives us three types of information:

- *The number of fragments.* The number of fragments produced by digestion of a DNA sample with a given restriction enzyme depends on how many times that enzyme's recognition sequence occurs in the sample. Thus gel electrophoresis can provide some information about the presence of specific DNA sequences in the DNA sample.
- *The sizes of the fragments.* DNA fragments of known size are often placed in one well of the gel to provide a standard for comparison. This tells us how large the DNA fragments in the other wells are. By comparing the fragment sizes obtained with two or more restriction enzymes, the locations of their recognition sites relative to one another can be worked out (mapped).
- *The relative abundance of a fragment.* In many experiments, the investigator is interested in how much DNA is present. The relative *intensity* of a band produced by a specific fragment can indicate the amount of that fragment.

After separation on a gel, a fragment with a specific DNA sequence can be revealed with a single-stranded DNA probe (as we will see later in this chapter; see Figure 15.16). The gel region containing the desired fragment (in size or sequence) can be cut out as a lump of gel, and the pure DNA fragment can then be removed from the gel by diffusion into a small volume of water. This fragment can then be analyzed in terms of sequence or amplified and used experimentally.

TOOLS FOR INVESTIGATING LIFE

**15.8 Separating Fragments of DNA by Gel Electrophoresis** A mixture of DNA fragments is placed in a gel and an electric field is applied across the gel. The negatively charged DNA moves toward the positive end of the field, with smaller molecules moving faster than larger ones. After minutes to hours for separation, the electric power is shut off and the separated fragments can be analyzed.

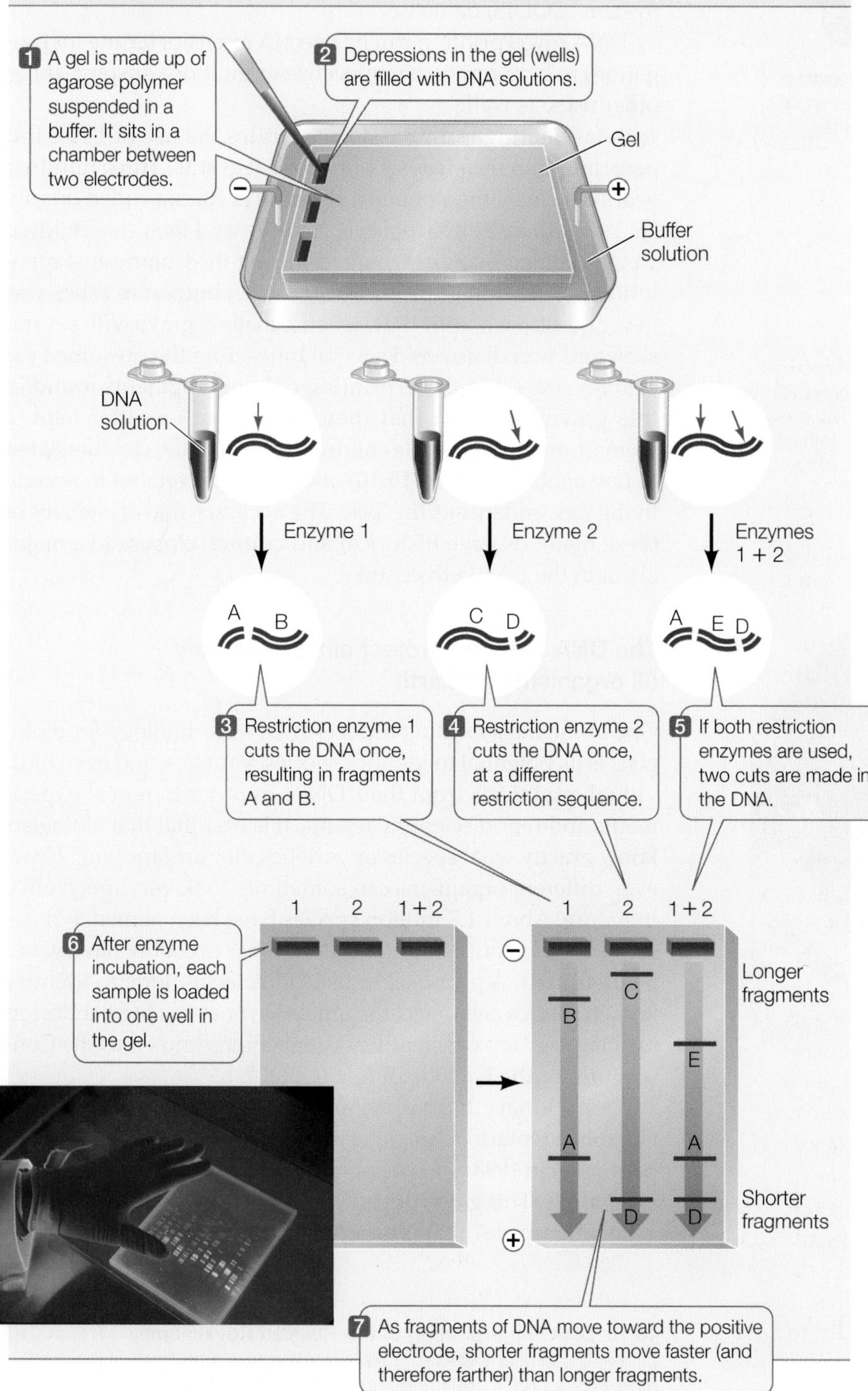
1 A gel is made up of agarose polymer suspended in a buffer. It sits in a chamber between two electrodes.
2 Depressions in the gel (wells) are filled with DNA solutions.
Gel
Buffer solution
DNA solution
Enzyme 1
Enzyme 2
Enzymes 1 + 2
A B
C D
A E D
3 Restriction enzyme 1 cuts the DNA once, resulting in fragments A and B.
4 Restriction enzyme 2 cuts the DNA once, at a different restriction sequence.
5 If both restriction enzymes are used, two cuts are made in the DNA.
1 2 1+2
1 2 1+2
6 After enzyme incubation, each sample is loaded into one well in the gel.
Longer fragments
Shorter fragments
B
C
E
A
A
D
D
7 As fragments of DNA move toward the positive electrode, shorter fragments move faster (and therefore farther) than longer fragments.

**yourBioPortal.com**
GO TO **Animated Tutorial 15.1 • Separating Fragments of DNA by Gel Electrophoresis**

### DNA fingerprinting uses restriction analysis and electrophoresis

The two methods we have just described—restriction digestion to cut DNA into fragments and gel electrophoresis to separate them by size—are techniques used in **DNA fingerprinting**, which identifies individuals based on their DNA profiles. DNA fingerprinting works best with sequences that are highly polymorphic—that is, sequences that have multiple alleles (due to many point mutations during the evolution of the organism) and are therefore likely to be different in different individuals. Two types of polymorphisms are especially informative:

- **Single nucleotide polymorphisms** (**SNPs**; pronounced "snips") are inherited variations involving a single nucleotide base (so SNPs are point mutations). These polymorphisms have been mapped for many organisms. If one parent is homozygous for A at a certain point on the genome, and the other parent has a G at that point, the offspring will be heterozygous: one chromosome will have A at that point and the other will have G. If a SNP occurs in a restriction enzyme recognition site, such that one variant is recognized by the enzyme and the other isn't, then individuals can be distinguished from one another by amplifying a DNA fragment containing that site from a sample of total DNA isolated from each individual. The fragments are then cut with the restriction enzyme and analyzed by gel electrophoresis.
- **Short tandem repeats** (**STRs**) are short, repetitive DNA sequences that occur side by side on the chromosomes, usually in the noncoding regions. These repeat patterns, which contain 1–5 base pairs, are also inherited. For example, at a particular locus on chromosome 15 there may be an STR of "AGG." An individual may inherit an allele with six copies of the repeat (AGGAGGAGGAGGAGGAGG) from her mother and an allele with two copies (AGGAGG) from her father. Again, PCR is used to amplify DNA fragments containing these repeat sequences, and the fragments are distinguished by gel electrophoresis (**Figure 15.9**).

The method of DNA fingerprinting used most commonly today involves STR analysis. When several different STR loci, each with numerous alleles, are analyzed, an individual's unique pat-

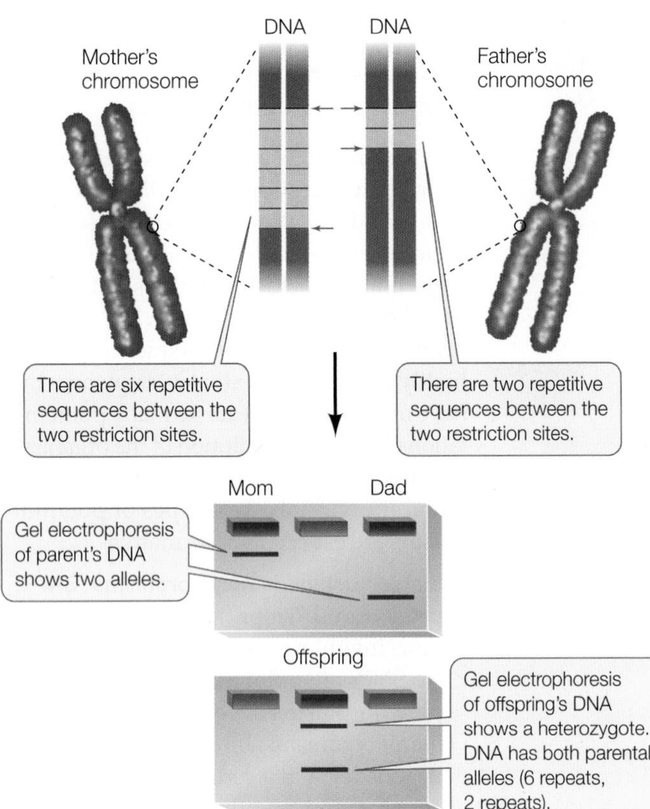


**15.9 DNA Fingerprinting with Short Tandem Repeats** A particular STR locus can be analyzed to determine the number of repeat sequences that were inherited by an individual from each parent. The two alleles can be identified in an electrophoresis gel on the basis of their sizes. When several STR loci are analyzed, the pattern can constitute a definitive identification of an individual.

tern becomes apparent. The Federal Bureau of Investigation in the United States uses 13 STR loci in its Combined DNA Index System (CODIS) database.

DNA fingerprinting can be used in forensics (crime investigation) to help prove the innocence or guilt of a suspect. It has other uses, as well.

A fascinating example demonstrates the use of DNA fingerprinting in the analysis of historical events. Three hundred years of rule by the Romanov dynasty in Russia ended on July 16, 1918, when Tsar Nicholas II, his wife, and their five children were executed by a firing squad during the Communist revolution. A report that the bodies had been burned to ashes was never questioned until 1991, when a shallow grave with several skeletons was discovered several miles from the presumed execution site. DNA fingerprinting of bone fragments found in this grave indicated that they came from an older man, a woman, and three female children, who all were clearly related to one another (**Figure 15.10**) and were also related to several living descendants of the Tsar. The accuracy and specificity of these methods gave historical and cultural closure to a major event in the twentieth century.

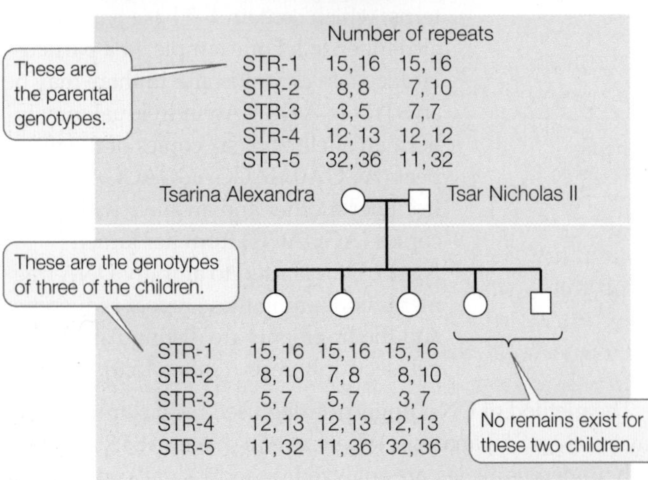


**15.10 DNA Fingerprinting of the Russian Royal Family** The skeletal remains of Tsar Nicholas II, his wife Alexandra, and three of their children were found in 1991 and subjected to DNA fingerprinting. Five STRs were tested. The results can be interpreted by looking at the inheritance of alleles from each parent in the children. In STR-2, for example, the parents had genotypes 8,8 (homozygous) and 7,10 (heterozygous). The three children all inherited type 8 from Alexandra and either type 7 or type 10 from Nicholas.

### The DNA barcode project aims to identify all organisms on Earth

One of the most exciting aspects of DNA technology for biologists is its potential to identify species, varieties, and even individual organisms from their DNA. In order to repeat experiments and report scientific results, it is essential that biologists know exactly what species or varieties they are studying. However, different organisms can sometimes look very much alike in nature. About 1.7 million species have been named and described, but about ten times that number probably have yet to be identified. A proposal to use DNA technology to identify known species and detect the unknown ones has been endorsed by a large group of scientific organizations known as the Consortium for the Barcode of Life (CBOL).

Evolutionary biologist Paul Hebert at the University of Guelph in Ontario, Canada was walking down the aisle of a supermarket in 1998 when he noticed the barcodes on all the packaged foods. This gave him an idea to identify each species with a "DNA barcode" that is based on a short sequence from a sin-

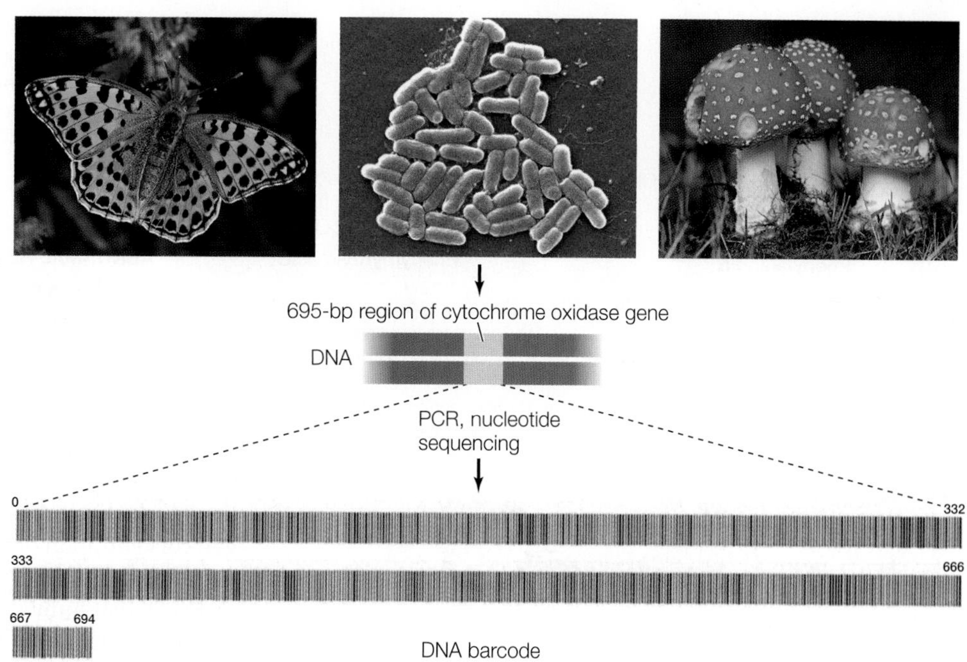


**15.11 A DNA Barcode** A 650- to 750-base-pair region of the cytochrome oxidase gene can be amplified by PCR from any organism and then sequenced. This knowledge is used to make a bar code in which each of the four DNA bases is represented as a different color. Such a species barcode permits accurate and rapid identification of a particular species for experimental, ecological or evolutionary studies.

gle gene. The gene he chose is the cytochrome oxidase gene, a component of the respiratory chain that is present in most organisms. Because this gene mutates readily, there are many allelic differences between species. A fragment of 650–750 base pairs in this gene is being sequenced for all organisms, and so far sufficient variation has been detected to make it diagnostic for each species (**Figure 15.11**).

Once the DNA of the targeted gene fragment has been sequenced for all known species, a simple device for conducting field analyses can be developed. The barcode project has the potential to advance biological research on evolution, to track species diversity in ecologically significant areas, to help identify new species, and even to detect undesirable microbes or bioterrorism agents.

### 15.2 RECAP

**Large DNA molecules can be cut into smaller pieces by restriction digestion and then sorted by gel electrophoresis. PCR is used to amplify sequences of interest from complex samples. These techniques are used in DNA fingerprinting to analyze DNA polymorphisms for the purpose of identifying individuals. Scientists hope to identify all species using DNA analysis.**

- How does a restriction enzyme recognize a restriction site on DNA? See p. 323
- How does gel electrophoresis separate DNA fragments? See p. 324 and Figure 15.8
- What are STRs and how are they used to identify individuals? See pp. 325–326

We have seen that molecular methods can be used to identify individuals because of mutations in their DNA. Many of the STRs and SNPs used in these analyses do not occur in protein-coding regions, and so probably do not affect the phenotype. Nevertheless, they are mutations—inherited changes in the DNA. We now turn to mutations that affect phenotype, using humans as our model organism.

## 15.3 How Do Defective Proteins Lead to Diseases?

The biochemistry that relates genotype (DNA) and phenotype (proteins) has been most completely described for model organisms, such as the prokaryote *E. coli* and the eukaryotes yeast and *Drosophila*. While the details vary, there is great similarity in the fundamental processes among these forms of life. These similarities have permitted the application of knowledge and methods discovered using these model organisms to the study of human biochemical genetics. Of particular interest are the effects of mutations on human phenotypes, sometimes leading to diseases.

### Genetic mutations may make proteins dysfunctional

Genetic mutations are often expressed phenotypically as proteins that differ from normal (wild-type) proteins. Abnormalities in enzymes, receptor proteins, transport proteins, structural proteins, and most of the other functional classes of proteins have all been implicated in genetic diseases.

**DYSFUNCTIONAL ENZYMES** In 1934, the urine of two mentally retarded young siblings was found to contain phenylpyruvic acid, an unusual by-product of the metabolism of the amino acid phenylalanine. It was not until two decades later, however, that the complex clinical phenotype of the disease that afflicted these children, called *phenylketonuria* (*PKU*), was traced back to its molecular cause. The disease resulted from an abnormality in a single enzyme, phenylalanine hydroxylase (**Figure 15.12**). This enzyme normally catalyzes the conversion of dietary phenylalanine to tyrosine, but it was not active in the livers of PKU patients. Lack of this conversion led to excess phenylalanine in the blood and explained the accumulation of phenylpyruvic acid. Later, the amino acid sequence of phenylalanine hydroxylase (PAH) in normal people was compared with the amino acid sequences

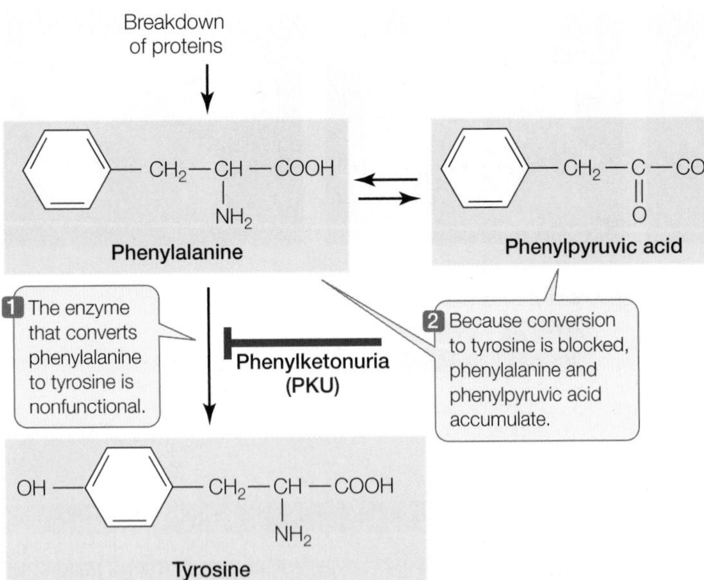


**15.12 One Gene, One Enzyme** Phenylketonuria is caused by an abnormality in a specific enzyme that metabolizes the amino acid phenylalanine. Knowing the molecular causes of such single-gene, single-enzyme metabolic diseases can aid researchers in developing screening tests as well as treatments.

from individuals with PKU. Many people with PKU had tryptophan instead of arginine at position 408 of this long polypeptide chain of 452 amino acids (**Table 15.1**).

The exact cause of mental retardation in PKU remains elusive, although, as we will see later in this chapter, it can be prevented. We can, however, understand why most people with PKU have light skin and hair color. The pigment melanin, which is responsible for dark skin and hair, is made from tyrosine, which people with PKU cannot synthesize adequately.

Hundreds of human genetic diseases that result from enzyme abnormalities have been discovered, many of which lead to mental retardation and premature death. Most of these diseases are rare; PKU, for example, shows up in one out of every 12,000 newborns. But these diseases are just the tip of the mutation iceberg. Some mutations result in amino acid changes that have no obvious clinical effects. In fact, amino acid differences among individuals have been detected in at least 30 percent of all human proteins whose sequences are known. Thus polymorphism does not necessarily mean disease. There can be numerous alleles of a gene, some producing proteins that function normally, while others produce variants that cause disease—as we will now see for hemoglobin.

**ABNORMAL HEMOGLOBIN** The first human genetic disease known to be caused by an amino acid sequence abnormality was sickle-cell disease. This blood disorder most often afflicts people whose ancestors came from the tropics or from the Mediterranean. About 1 in 655 African-Americans are homozygous for the sickle allele and have the disease. The abnormal allele produces abnormal hemoglobin that results in sickle-shaped red blood cells (see Figure 15.3). These cells tend to block narrow blood capillaries, especially when the oxygen concentration of the blood is low. The result is tissue damage and eventually death by organ failure.

Recall that human hemoglobin is a protein with quaternary structure, containing four globin subunits—two α chains and two β chains—as well as the pigment heme (see Figure 3.10). In sickle-cell disease, one of the 146 amino acids in the β-globin chain is abnormal: at position 6, the normal glutamic acid has been replaced by valine. This replacement changes the charge of the protein (glutamic acid is negatively charged and valine is neutral), causing it to form long, needle-like aggregates in the red blood cells. The phenotypic result is anemia, a deficiency of normal red blood cells and an impaired ability of the blood to carry oxygen.

Because hemoglobin is easy to isolate and study, its variations in the human population have been extensively documented (**Figure 15.13**). Hundreds of single amino acid alterations in β-globin have been reported. For example, at the same position that is mutated in sickle-cell disease (resulting in hemoglobin S), the normal glutamic acid may be replaced by lysine, causing hemoglobin C disease. In this case, the resulting anemia is usually not severe. Many alterations of hemoglobin do not affect the protein's function. That is fortunate, because about 5 percent of all humans are carriers for one of these variants.

There are hundreds of inherited diseases in humans in which the primary phenotypes are caused by specific mutations leading to protein abnormalities. Some of the more common examples are listed in **Table 15.2**. These mutations can be domi-

TABLE 15.1

**Two Common Mutations That Cause Phenylketonuria**

| | NORMAL CODON 408 | MUTANT CODON 408 (20% OF PKU CASES) | NORMAL CODON 280 | MUTANT CODON 280 (2% OF PKU CASES) |
|---|---|---|---|---|
| Length of PAH protein | 452 amino acids | 452 amino acids | 452 amino acids | 452 amino acids |
| DNA at codon | xxCGGxx<br>xxGCCxx | xxTGGxx<br>xxACCxx | xxGAAxx<br>xxCTTxx | xxAAAxx<br>xxTTTxx |
| mRNA at codon | xxCGGxx | xxUGGxx | xxGAAxx | xxAAAxx |
| Amino acid at codon | Arginine | Tryptophan | Glutamic acid | Lysine |
| Active PAH enzyme? | Yes | No | Yes | No |

## TABLE 15.2
### Some Human Genetic Diseases

| DISEASE NAME | INHERITANCE PATTERN; FREQUENCY | GENE MUTATED; PROTEIN PRODUCT | CLINICAL PHENOTYPE |
|---|---|---|---|
| Familial hypercholesterolemia | Autosomal codominant; 1 in 500 heterozygous | *LDLR*; low-density lipoprotein receptor | High blood cholesterol, heart disease |
| Cystic fibrosis | Autosomal recessive; 1 in 4000 | *CFTR*; chloride ion channel in membrane | Immune, digestive, and respiratory illness |
| Duchenne muscular dystrophy | Sex-linked recessive; 1 in 3500 males | *DMD*; the muscle membrane protein dystrophin | Muscle weakness |
| Hemophilia A | Sex-linked recessive; 1 in 5000 males | *HEMA*; factor VIII blood clotting protein | Inability to clot blood after injury, hemorrhage |

nant, codominant, or recessive, and some are sex-linked. Before we examine how these diseases can be analyzed at the molecular level, we turn briefly to a fascinating exception to the association between genes and proteins.

### Prion diseases are disorders of protein conformation

*Transmissible spongiform encephalopathies* (TSEs) are degenerative brain diseases that occur in many mammals, including humans. The brain gradually develops holes, making it look like a sponge. Scrapie is a TSE that has been known for 250 years. It causes affected sheep and goats to show the abnormal behavior of rubbing ("scraping") the wool off their bodies (as well as causing more severe neurological problems). In the 1980s, a TSE that appeared in cows in Britain was traced to the cows having eaten products from sheep that had scrapie. These cows would shake and rub their bodies against fences, and their staggering led farmers to dub them "mad cows." In the 1990s, some people who ate beef from these cows got a human version of the disease, dubbed "mad cow disease" by the media. Those with the disease eventually died.

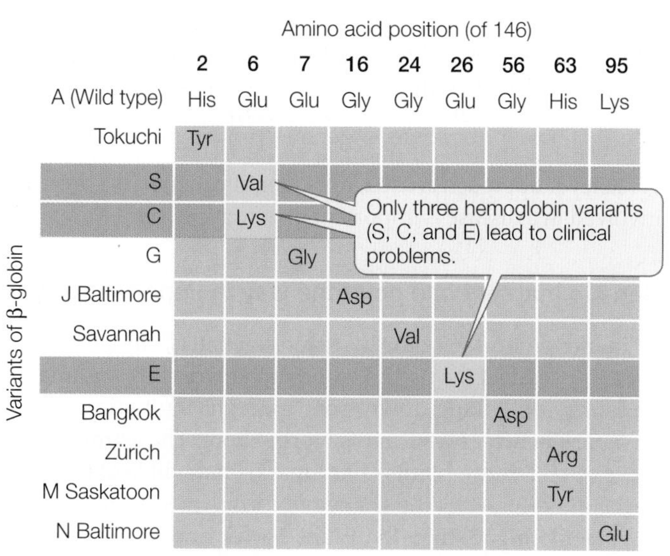


**15.13 Hemoglobin Polymorphism** Each of these mutant alleles codes for a protein with a single amino acid change in the 146-amino acid chain of β-globin. Only three of the hundreds of known variants of β-globin, shown on the left, are known to lead to clinical abnormalities. "S" is the sickle-cell anemia allele.

At first, viruses were suspected to cause TSEs. But when Tikva Alper at Hammersmith Hospital, London, treated infectious extracts with high doses of ultraviolet light to inactivate nucleic acids, they still caused TSEs. She proposed that the causative agent was a protein, not a virus. Later, Stanley Prusiner at the University of California purified the protein responsible and showed it to be free of DNA or RNA. He called it a *proteinaceous infective particle*, or **prion**. This is a violation of the central dogma of molecular biology (DNA → RNA → protein; see Chapter 14), because in this case the protein was "doing it all." There was no genetic material involved.

This is a rare case of a mutant phenotype without a mutant gene. Normal brain cells contain a membrane protein called $PrP^{c}$. A protein with the *same amino acid sequence* is present in TSE-affected brain tissues, but that protein, called $PrP^{sc}$, has a different three-dimensional shape (**Figure 15.14**). Thus TSEs are not caused by a mutated gene (the primary structures of the two proteins are the same), but are somehow caused by an alteration in protein conformation. The altered three-dimensional structure of the protein has profound effects on its function in the cell. $PrP^{sc}$ is insoluble, and it piles up as fibers in brain tissue, causing cell death.

How can the exposure of a normal cell to material containing $PrP^{sc}$ result in a TSE? The abnormal $PrP^{sc}$ protein seems to induce a conformational change in the normal $PrP^{c}$ protein so that it, too, becomes abnormal. Just how this conversion occurs, and how it causes a TSE, is unclear.

To try to understand how TSEs develop, scientists are asking "What is the *normal* role of the prion protein?" Recently, it was shown that in the brain the prion protein blocks a key enzyme in the synthesis of a protein called β-amyloid. This is the protein that accumulates in the brains of patients with Alzheimer's disease. People with early-onset Alzheimer's (age 40) have less $PrP^{c}$ in their brains than people who age normally. So the $PrP^{c}$ pro-

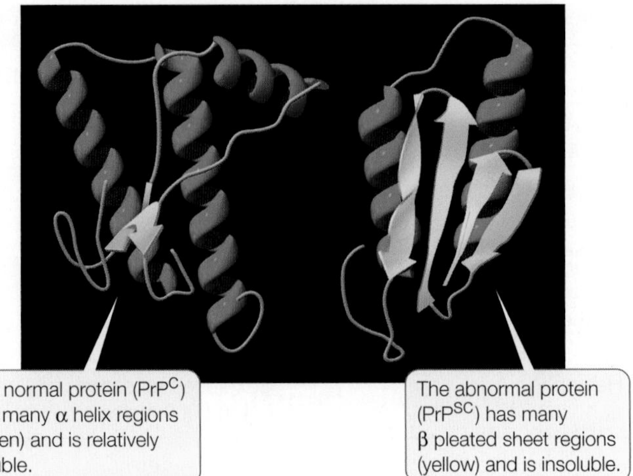

The normal protein ($PrP^C$) has many α helix regions (green) and is relatively soluble.
The abnormal protein ($PrP^{SC}$) has many β pleated sheet regions (yellow) and is insoluble.


**15.14 Prion Diseases are Disorders of Protein Conformation** A normal membrane protein in brain cells ($PrP^c$, left) can be converted to the disease-causing form ($PrP^{sc}$, right), which has a different three-dimensional structure.

tein appears to play a role in protecting against Alzheimer's disease. Other functions for this protein are also being discovered, but the mechanism by which it appears to spread TSE disease is not yet understood.

Prions are an unusual phenomenon in human disease. The vast majority of inherited diseases are caused by mutations in genes that reduce the levels of their protein products, or make the proteins dysfunctional. But the expression of these genes, like that of all genes, is influenced by the environment.

### Most diseases are caused by multiple genes and environment

The human diseases for which clinical phenotypes can be traced to a single altered protein and its altered gene may number in the thousands. Taken together, these diseases have a frequency of about one percent in the human population.

Far more common, however, are diseases that are **multifactorial**; that is, diseases that are caused by the interactions of many genes and proteins with one or more factors in the environment. When studying genetics, we tend to call individuals either normal (wild type) or abnormal (mutant); however, in reality every individual contains thousands or millions of genetic variations that arose through mutations. Our susceptibility to disease is often determined by complex interactions between these genotypes and factors in the environment, such as the foods we eat or the pathogens we encounter. For example, a complex set of genotypes determine who among us can eat a high-fat diet and not experience a heart attack, or will succumb to disease when exposed to infectious bacteria. Estimates suggest that up to 60 percent of all people are affected by diseases that are *genetically influenced*. Identifying these genetic influences is a major task of molecular medicine and human genome sequencing.

## 15.3 RECAP

**Many genetic mutations are expressed as nonfunctional enzymes, structural proteins, or membrane proteins. Human genetic diseases may be inherited in dominant, codominant, or recessive patterns, and they may be sex-linked.**

- Describe an example of an abnormal protein in humans that results from a genetic mutation and causes a disease. See pp. 327–328
- Describe an example of an abnormal protein in humans that results from a genetic mutation and does *not* cause a disease. See p. 328
- How is the brain cell membrane protein $PrP^c$ related to diseases caused by prions? See p. 329

The abnormal proteins that cause disease result (with the exception of TSEs) from genetic mutations. We now turn to the identification of such mutations, an important task for molecular medicine.

## 15.4 What DNA Changes Lead to Genetic Diseases?

We have seen for diseases such as PKU and sickle-cell anemia that the clinical phenotype of inherited diseases could be traced to individual proteins, and that the genes could then be identified. With the advent of new ways to identify DNA variations, a new pattern of human genetic analysis has emerged. In these cases, the clinical phenotype is first related to a DNA variation, and then the protein involved is identified. This pattern of discovery is called **reverse genetics**, because it proceeds in the opposite direction to genetic analyses done before the mid-1980s. For example, in sickle-cell anemia, the protein abnormality in hemoglobin was described first (a single amino acid change), and then the gene for β-globin was isolated and the DNA mutation was pinpointed.

clinical phenotype → protein phenotype → gene

On the other hand, for cystic fibrosis (see Table 15.2), a mutant version of the gene *CFTR* was isolated first, and then the protein was characterized:

clinical phenotype → gene → protein phenotype

Whichever approach is used, final identification of the protein(s) involved in a disease is important in designing specific therapies.

### Genetic markers can point the way to important genes

To identify a mutant gene by reverse genetics, close linkage to a marker sequence is used. To understand this linkage, imagine an astronaut looking down from space, trying to find her son on a park bench on Chicago's North Shore. The astronaut first picks out reference points—landmarks that will lead her to the park. She recognizes the shape of North America, then moves to Lake Michigan, then the Willis Tower, and so on. Once she has zeroed in on the North Shore Park, she can use advanced optical instruments to find her son. The reference points for gene isolation are the **genetic markers**.

- Knowledge of at least *two mutations* is needed. One mutation determines the disease phenotype and the other mutation is a closely linked "marker mutation" that does not affect the disease phenotype but is easy to identify. In early genetic studies, markers that produced visible phenotypes were used to follow the inheritance of important traits. Today, single nucleotide polymorphisms (SNPs) or STRs are usually used.
- *Genetic linkage* is the co-inheritance of the marker and the disease-causing allele. If they are always together, they must be close together on the chromosome.

A key requirement for a genetic marker is that it has allelic polymorphisms (differences in sequence) that are identifiable by current methods of rapid DNA analysis. As we saw in Section 15.2, an STR can have varying numbers of a short repeat sequence, and thus there are multiple alleles of these markers. We also saw in Section 15.2 that restriction enzymes can be used to identify SNPs, provided the SNP occurs within a restriction site. Restriction enzymes can also be used to identify mutations such as insertions or deletions if the affected sequences contain restriction sites. We will examine in more detail the use of restriction enzymes to identify genetic polymorphisms, and other SNP markers, before returning to the discussion of human genes and their abnormalities.

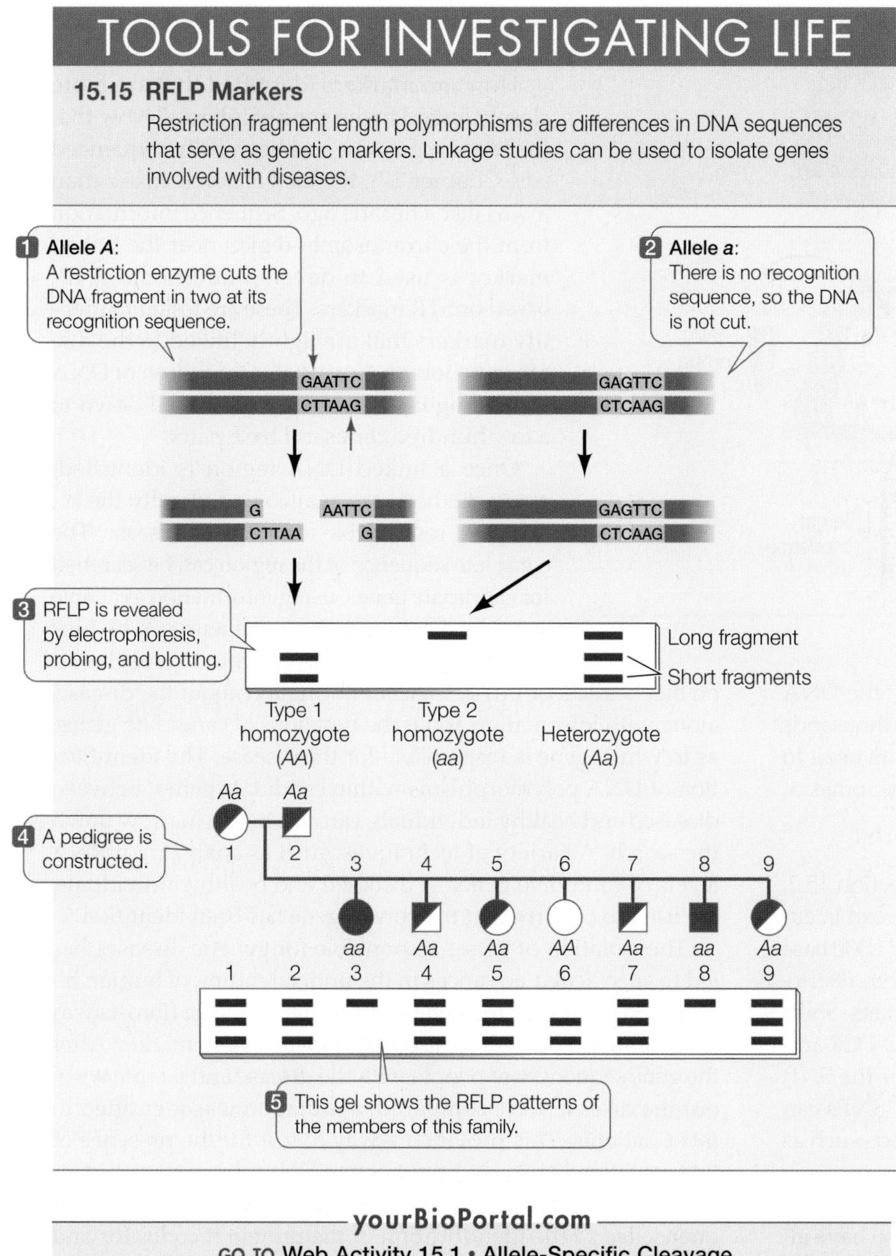

TOOLS FOR INVESTIGATING LIFE
1 Allele A: A restriction enzyme cuts the DNA fragment in two at its recognition sequence.
2 Allele a: There is no recognition sequence, so the DNA is not cut.
GAATTC
CTTAAG
GAGTTC
CTCAAG
G
CTTAA
AATTC
G
GAGTTC
CTCAAG
3 RFLP is revealed by electrophoresis, probing, and blotting.
Long fragment
Short fragments
Type 1 homozygote (AA)
Type 2 homozygote (aa)
Heterozygote (Aa)
4 A pedigree is constructed.
Aa
Aa
1
2
3
4
5
6
7
8
9
aa
Aa
Aa
AA
Aa
aa
Aa
1
2
3
4
5
6
7
8
9
5 This gel shows the RFLP patterns of the members of this family.

**15.15 RFLP Markers** Restriction fragment length polymorphisms are differences in DNA sequences that serve as genetic markers. Linkage studies can be used to isolate genes involved with diseases.

yourBioPortal.com

GO TO Web Activity 15.1 • Allele-Specific Cleavage

**RESTRICTION FRAGMENT LENGTH POLYMORPHISMS** As Section 15.2 describes, restriction enzymes cut DNA molecules at specific recognition sequences. On a particular human chromosome, a given restriction enzyme may make thousands of cuts, producing many DNA fragments. The enzyme *Eco*RI, for example, cuts DNA at

5′… GAATTC … 3′

Suppose this recognition sequence exists in a certain stretch of human chromosome 7. The restriction enzyme will cut this stretch once and make two fragments of DNA. Now suppose that, in some people, this sequence contains a SNP and is mutated as follows:

5′… GAGTTC … 3′

This sequence will not be recognized by the restriction enzyme; thus it will remain intact and yield one larger fragment of DNA.

Differences in DNA sequences due to mutations in restriction sites are called **restriction fragment length polymorphisms** (**RFLPs**) (**Figure 15.15**). They can be easily visualized as bands on an electrophoresis gel. An RFLP band pattern is inherited in a Mendelian fashion and can be followed through a pedigree. Thousands of such markers have been described for humans and many other organisms.

Before the advent of PCR technology, the only way to analyze RFLPs was by digesting total genomic DNA samples with restriction enzymes. These samples contain thousands of DNA fragments of various sizes. In order to visualize a particular fragment, the DNA from the gel is transferred (blotted) onto a nylon membrane, denatured to separate the double-stranded molecules, and mixed with a single-stranded DNA fragment (a *probe*) containing at least part of the sequence within the RFLP fragment of interest (**Figure 15.16**). The probe hybridizes (by base pairing) with the DNA band containing the RFLP. Because the probe is "la-

TOOLS FOR INVESTIGATING LIFE

**15.16 Analyzing DNA Fragments by DNA Gel Blotting**
A probe can be used to locate a specific DNA fragment on an electrophoresis gel.

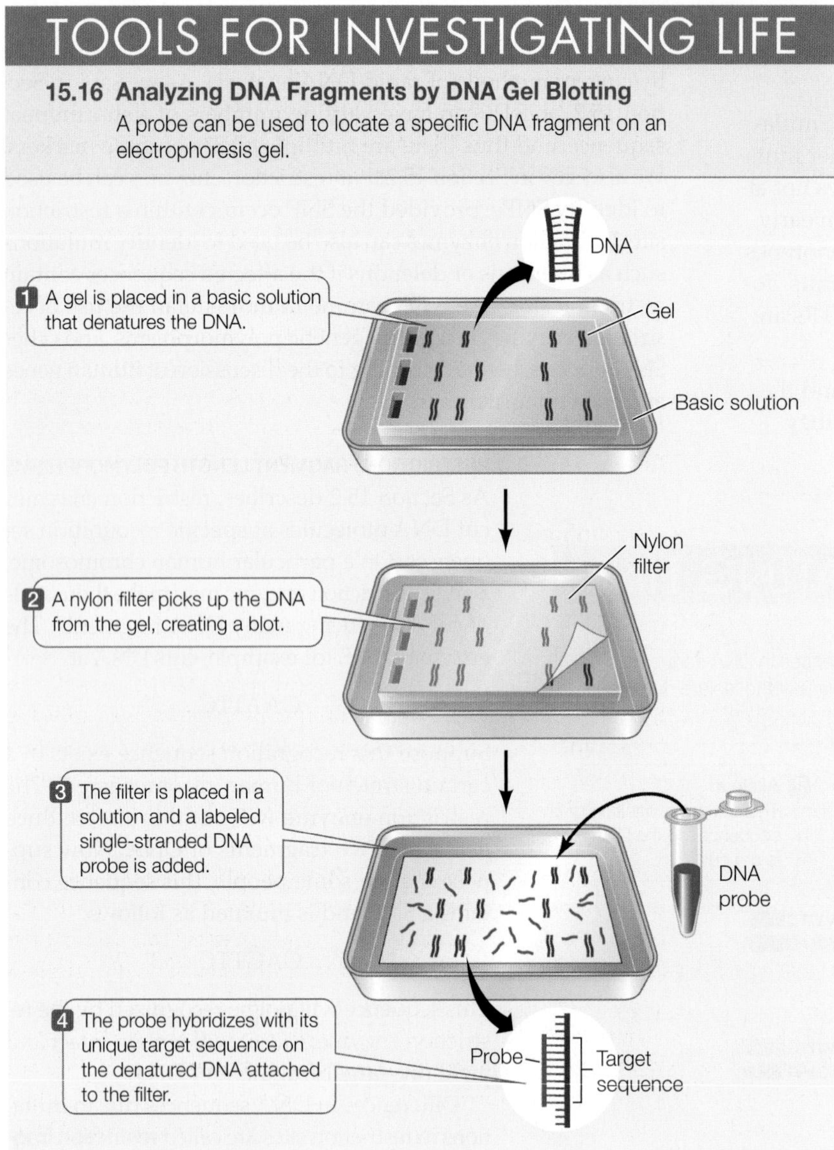

DNA
1 A gel is placed in a basic solution that denatures the DNA.
Gel
Basic solution
Nylon filter
2 A nylon filter picks up the DNA from the gel, creating a blot.
3 The filter is placed in a solution and a labeled single-stranded DNA probe is added.
DNA probe
4 The probe hybridizes with its unique target sequence on the denatured DNA attached to the filter.
Probe
Target sequence

beled" with a radioactive isotope or a chemical tag, the DNA fragment containing the RFLP can be seen among the thousands of other fragments on the blot. This technology was used to create "RFLP maps" of the human genome and of genomes of many other organisms.

**SINGLE NUCLEOTIDE POLYMORPHISMS** As noted in Section 15.2, single nucleotide polymorphisms (SNPs) are widespread in eukaryotic genomes. There is roughly one SNP for every 1,330 base pairs in the human genome. Not all SNPs occur within restriction sites, but those that don't can still be used as markers. SNPs can be detected by direct sequence comparisons, or by PCR amplification using primers that contain one version of the SNP, so that only one allele will be amplified efficiently. SNPs can also be detected using sophisticated chemical methods such as mass spectrometry (see Section 17.5).

Genetic markers such as STRs, RFLPs, and SNPs can be used as landmarks to find genes of interest if the genes also have alleles and are polymorphic. The key to this method is the well-known observation that if two genes are located near each other on the same chromosome, they are usually passed on together from parent to offspring. The same holds true for any pair of genetic markers. The idea is to find markers that are close by on the chromosome to get to the gene of interest.

To narrow down the location of a gene, a scientist must find a marker and a gene that are *always inherited together*. To do this, family medical histories are taken and pedigrees are constructed. If a genetic marker and a genetic disease are inherited together, then they must be near each other on the same chromosome. Unfortunately, "near each other" still might be as much as several million base pairs apart. The process of locating a gene is thus similar to that of the astronaut looking for her son: the first landmarks lead to only an approximate location.

How are markers identified that are more closely linked to the gene of interest? Now that the human genome has been fully sequenced (see Chapter 17), the task is much easier than it was just a decade ago. Sequence information from the chromosomal region near the linked marker is used to develop additional SNP-based or STR markers. These are tested to identify markers that are tightly linked to the disease phenotype. Eventually, the region of DNA containing the gene can be narrowed down to a few hundred thousand base pairs.

Once a linked DNA region is identified, many methods are available to identify the actual gene responsible for a genetic disease. The complete sequence of the region can be searched for candidate genes, using information available from databases of genome sequences. With luck a scientist can make an educated guess, based on biochemical or physiological information about the disease, along with information about the functions of candidate genes, as to which gene is responsible for the disease. The identification of DNA polymorphisms within candidate genes, between diseased and healthy individuals, can also help to narrow down the search. A variety of techniques, such as analyzing mRNA levels of candidate genes in diseased and healthy individuals, are used to confirm that the correct gene has been identified.

The isolation of genes responsible for genetic diseases has led to spectacular advances in the understanding of human biology. For example, the gene responsible for cystic fibrosis was first identified by its close association with a SNP marker. After the gene sequences of people with the disease and people without the disease were compared, a mutation was identified in most patients. This provided a way to test for the presence of that mutation in people, by extracting DNA from cells or tissues that could be easily sampled. Moreover, knowing the gene sequence led to the identification of the protein it codes for and a characterization of the abnormal protein. Treatments that specifically target this protein are now being devised. Research on the protein in normal people has led to an understanding of its role in the body. So reverse genetics can lead to diagnosis, treatment, and biological understanding.

### Disease-causing mutations may involve any number of base pairs

Disease-causing mutations may involve a single base pair (as we saw in the case of hemophilia), a long stretch of DNA (as in cases of Duchenne muscular dystrophy, which we will discuss shortly), multiple segments of DNA (as in fragile-X syndrome), or even entire chromosomes (as we saw with Down syndrome in Section 11.5).

**POINT MUTATIONS** There are many examples of point mutations in human genetic diseases. In some cases, all of the people with the disease have the same genetic mutation. This is the case with sickle-cell anemia, where a single base pair change in the β-globin gene causes a single amino acid change, which leads to the abnormal protein and phenotype. This is not the situation with most other genetic diseases. For example, over 500 different mutations in the *PAH* (phenylalanine hydroxylase) gene have been discovered in different patients with phenylketonuria (PKU; see Table 15.1). This makes sense if you think about the three-dimensional structure of an enzyme protein and the many amino acid changes that could affect its activity.

**LARGE DELETIONS** Larger mutations may involve many base pairs of DNA. For example, deletions in the X chromosome that include the gene for the protein dystrophin result in Duchenne muscular dystrophy. Dystrophin is important in organizing the structure of muscles, and people who have only the abnormal form have severe muscle weakness. Sometimes only part of the dystrophin gene is missing, leading to an incomplete but partly functional protein and a mild form of the disease. In other cases, however, deletions span the entire sequence of the gene, so that the protein is missing entirely, resulting in a severe form of the disease. In yet other cases, deletions involve millions of base pairs and cover not only the dystrophin gene but adjacent genes as well; the result may be several diseases simultaneously.

**CHROMOSOMAL ABNORMALITIES** Chromosomal abnormalities also cause human diseases. Such abnormalities include the gain or loss of one or more chromosomes (aneuploidy) (see Figure 11.21), loss of a piece of chromosome (deletion), and the transfer of a piece of one chromosome to another chromosome (translocation) (see Figure 15.4). About one newborn in 200 has a chromosomal abnormality. While some of these abnormalities are inherited as preexisting aberrations from one or both parents, others are the result of meiotic events, such as nondisjunction, that occurred during the formation of gametes in one of the parents.

One common cause of mental retardation is *fragile-X syndrome* (**Figure 15.17**). About one male in 1,500 and one female in 2,000 are affected. These people have a constriction near the tip of the X chromosome. Although the basic pattern of inheritance is that of an X-linked recessive trait, there are departures from this pattern. Not all people with the fragile-X chromosomal abnormality are mentally retarded, as we will see.

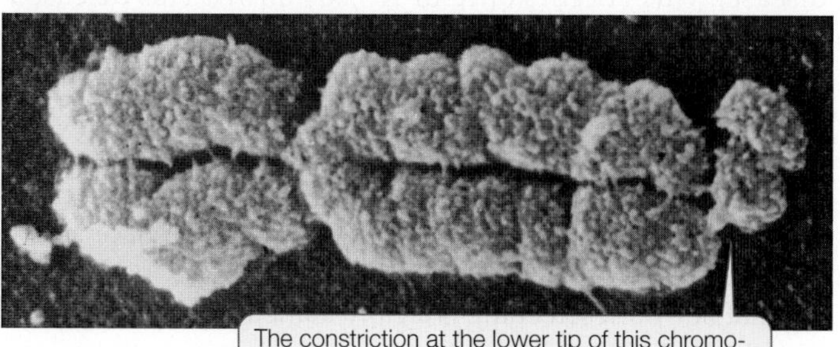

The constriction at the lower tip of this chromosome is the location of the fragile-X abnormality.

**15.17 A Fragile-X Chromosome at Metaphase** The chromosomal abnormality associated with fragile-X syndrome shows up under the microscope as a constriction in the chromosome. This occurs during preparation of the chromosome for microscopy.

### Expanding triplet repeats demonstrate the fragility of some human genes

About one-fifth of all males that have the fragile-X chromosomal abnormality are phenotypically normal, as are most of their daughters. But many of those daughters' sons are mentally retarded. In a family in which the fragile-X syndrome appears, later generations tend to show earlier onset and more severe symptoms of the disease. It is almost as if the abnormal allele itself is changing—and getting worse. And that's exactly what is happening.

The gene responsible for fragile-X syndrome (*FMR1*) contains a repeated triplet, CGG, at a certain point in the promoter region. In normal people, this triplet is repeated 6 to 54 times (the average is 29). In mentally retarded people with fragile-X syndrome, the CGG sequence is repeated 200 to 2,000 times.

Males carrying a moderate number of repeats (55–200) show no symptoms and are called premutated. These repeats become more numerous as the daughters of these men pass the chromosome on to their children (**Figure 15.18**). With more than 200 repeats, increased methylation of the cytosines in the CGG triplets is likely, accompanied by transcriptional inactivation of the *FMR1* gene. The normal role of the protein product of this gene is to bind to mRNAs involved in neuron function and regulate their translation at the ribosome. When the FMR1 protein is not made in adequate amounts, these mRNAs are not properly translated, and nerve cells die. Their loss often results in mental retardation.

This phenomenon of **expanding triplet repeats** has been found in over a dozen other diseases, such as myotonic dystrophy (involving repeated CTG triplets) and Huntington's disease (in which CAG is repeated). Such repeats, which may be found within a protein-coding region or outside it, appear to be present in many other genes without causing harm. How the repeats expand is not known; one theory is that DNA polymerase may slip after copying a repeat and then fall back to copy it again.

**15.18 The CGG Repeats in the FMR1 Gene Expand with Each Generation** The genetic defect in fragile-X syndrome is caused by 200 or more repeats of the CGG triplet.

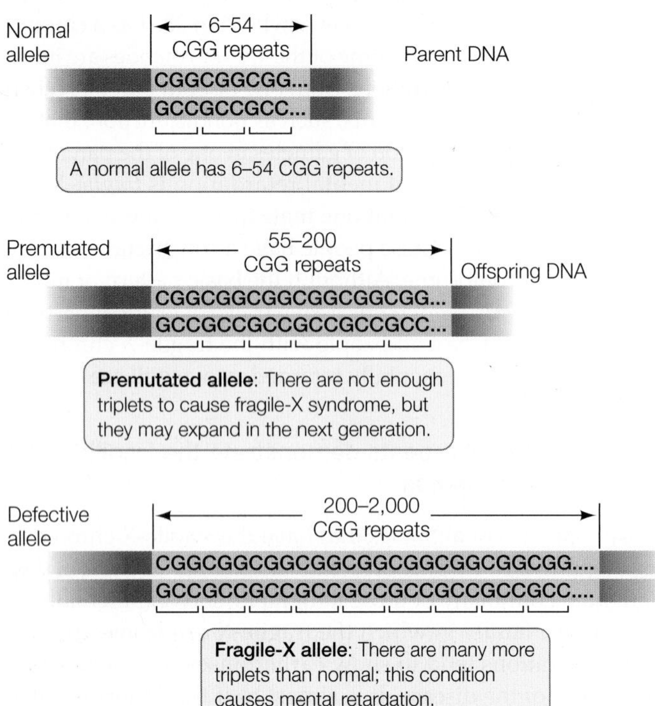
Normal allele
6–54 CGG repeats
Parent DNA
CGGCGGCGG...
GCCGCCGCC...
A normal allele has 6–54 CGG repeats.
Premutated allele
55–200 CGG repeats
Offspring DNA
CGGCGGCGGCGGCGGCGG...
GCCGCCGCCGCCGCCGCC...
Premutated allele: There are not enough triplets to cause fragile-X syndrome, but they may expand in the next generation.
Defective allele
200–2,000 CGG repeats
CGGCGGCGGCGGCGGCGGCGGCGGCGG....
GCCGCCGCCGCCGCCGCCGCCGCCGCC....
Fragile-X allele: There are many more triplets than normal; this condition causes mental retardation.

## 15.4 RECAP

**Genes involved in disease can be identified by first detecting the abnormal DNA sequence and then the protein that the wild-type allele encodes. Unusual features such as expanding triplet repeats have been detected in the human genome.**

- How can a gene be identified before its protein product is known? See pp. 330–331 and Figure 15.15
- How do expanding repeats cause genetic diseases? See p. 333 and Fig. 15.18

The determination of the precise molecular phenotypes and genotypes of various human genetic diseases has made it possible to diagnose these diseases even before symptoms first appear. Let's take a detailed look at some of these genetic screening techniques.

# 15.5 How Is Genetic Screening Used to Detect Diseases?

**Genetic screening** is the use of a test to identify people who have, are predisposed to, or are carriers of a genetic disease. It can be done at many times of life and used for many purposes.

- *Prenatal screening* can be used to identify an embryo or fetus with a disease so that medical intervention can be applied or decisions can be made about whether or not to continue the pregnancy.
- *Newborn babies* can be screened so that proper medical intervention can be initiated quickly for those babies who need it.
- *Asymptomatic people* who have a relative with a genetic disease can be screened to determine whether they are carriers of the disease or are likely to develop the disease themselves.

Genetic screening can be done at the level of either the phenotype or the genotype.

## Screening for disease phenotypes involves analysis of proteins

At the level of the phenotype, genetic screening involves examining a protein relevant to the phenotype for abnormal structure or function. Since many proteins are enzymes, low enzyme activity is strongly suggestive of a mutation, as we saw in Section 15.1. Perhaps the best example of this kind of protein screening is a test for phenylketonuria (PKU), which has made it possible to identify the disease in newborns, so that treatment of the disease can be started. It is very likely that you were screened for PKU.

Initially, babies born with PKU have a normal phenotype because excess phenylalanine in their blood before birth diffuses across the placenta to the mother's circulatory system. Since the mother is almost always heterozygous, and therefore has adequate phenylalanine hydroxylase activity, her body metabolizes the excess phenylalanine from the fetus. After birth, however, the baby begins to consume protein-rich food (milk) and to break down some of his or her own proteins. Phenylalanine enters the baby's blood and accumulates. After a few days, the phenylalanine level in the baby's blood may be ten times higher than normal. Within days, the developing brain is damaged, and untreated children with PKU become severely mentally retarded. If detected early, PKU can be treated with a special diet low in phenylalanine to avoid the brain damage that would otherwise result. Thus, early detection is imperative.

Newborn screening for PKU and other diseases began in 1963 with the development of a simple, rapid test for the presence of excess phenylalanine in blood serum (**Figure 15.19**). This method uses dried blood spots from newborn babies and can be automated so that a screening laboratory can process many samples in a day.

Screening using newborn babies' blood is now done for up to 25 genetic diseases. Some are rare, such as maple syrup urine disease, which occurs once in 185,000 births. This disease is caused by a defect in an enzyme that metabolizes certain amino acids, and results in sweet-smelling urine and severe brain damage. Other genetic diseases are more common, such as congenital hypothyroidism, which occurs about once in 4,000 births, and causes reduced growth and mental retardation due to low levels of thyroid hormone. With early intervention, many of these infants can be successfully treated. So it is not surprising that newborn screening is legally mandatory in many countries, including the United States and Canada.

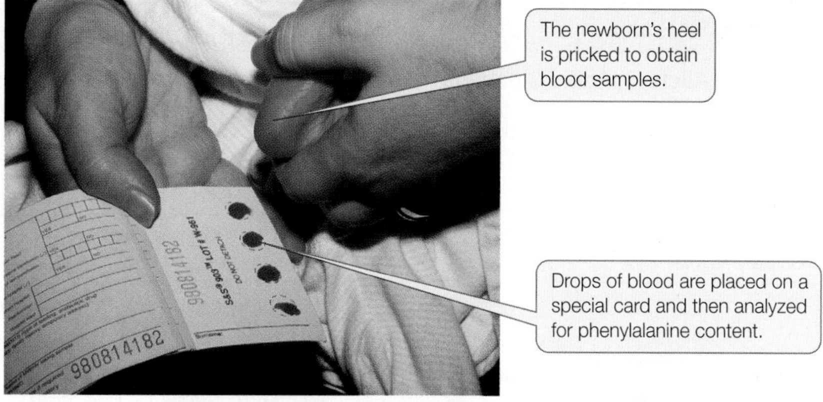

The newborn's heel is pricked to obtain blood samples.
Drops of blood are placed on a special card and then analyzed for phenylalanine content.
980814182

**15.19 Genetic Screening of Newborns for Phenylketonuria** A blood test is used to screen newborns for phenylketonuria. Small samples of blood are taken from a newborn's heel. The samples are placed in a machine that measures the phenylalanine concentration in the blood. Early detection means that the symptoms of the condition can be prevented by putting the baby on a therapeutic diet.

## DNA testing is the most accurate way to detect abnormal genes

The level of phenylalanine in the blood is an indirect measure of phenylalanine hydroxylase activity in the liver. But how can we screen for genetic diseases that are not detectable by blood tests? What if blood is difficult to obtain, as it is in a fetus? How are genetic abnormalities in heterozygotes, who express the normal protein at some level, identified?

**DNA testing** is the direct analysis of DNA for a mutation, and it offers the most direct and accurate way of detecting an abnormal allele. Now that the mutations responsible for many human diseases have been identified, any cell in the body can be examined at any time of life for mutations. With the amplification power of PCR, only one or a few cells are needed for testing. These methods work best for diseases caused by only one or a few different mutations.

Consider, for example, two parents who are both heterozygous for the cystic fibrosis allele, who have had a child with the disease, and want a normal child. If treated with the appropriate hormones, the mother can be induced to "superovulate," releasing several eggs. An egg can be injected with a single sperm from her husband and the resulting zygote allowed to divide to the 8-cell stage. If one of these embryonic cells is removed, it can be tested for the presence of the cystic fibrosis allele. If the test is negative, the remaining 7-cell embryo can be implanted in the mother's womb where with luck, it will develop normally.

Such *preimplantation screening* is performed only rarely. More typical are analyses of fetal cells after normal fertilization and implantation in the womb. Fetal cells can be analyzed at about the tenth week of pregnancy by chorionic villus sampling, or during the thirteenth to seventeenth weeks by amniocentesis. In either case, only a few fetal cells are necessary to perform DNA testing.

DNA testing can also be performed with newborns. The blood samples used for screening for PKU and other disorders contain enough of the baby's blood cells to permit DNA analysis using PCR-based techniques. Screening tests using DNA analysis are now being used for sickle-cell disease and cystic fibrosis; similar tests for other diseases will surely follow.

Of the numerous methods of DNA testing available, two are the most widespread. We will describe their use to detect the mutation in the β-globin gene that results in sickle-cell disease.

**SCREENING FOR ALLELE-SPECIFIC CLEAVAGE DIFFERENCES** The first method uses RFLP analysis, as we described earlier. There is a difference between the normal and the sickle allele of the β-globin gene, with respect to a restriction enzyme recognition sequence. Around the sixth codon in the normal gene is the sequence

5′... CCTGAGGAG... 3′

This sequence is recognized by the restriction enzyme *Mst*II, which will cleave DNA at

5′... CCTNAGGAG... 3′

where N is any base. In the sickle allele, the DNA sequence is

5′... CCTGTGGAG... 3′

The point mutation at codon 6 makes this sequence unrecognizable by *Mst*II. The sequence surrounding the mutant site can be amplified by PCR and digested with *Mst*II. Gel electrophoresis is used to distinguish between PCR products derived from the normal allele, which are cut by the enzyme, and products from the sickle allele, which are not cut (**Figure 15.20**).

This *allele-specific cleavage* method of DNA testing works only if a restriction enzyme exists that can recognize the sequence of either the normal or the mutant allele.

**SCREENING BY ALLELE-SPECIFIC OLIGONUCLEOTIDE HYBRIDIZATION** The *allele-specific oligonucleotide hybridization* method uses short synthetic DNA strands called *oligonucleotide probes* that will hybridize with denatured PCR products from either the normal or the mutant allele. Usually, an oligonucleotide probe of at least a dozen bases is needed to form a stable double helix with the target DNA. If the probe is radioactively or fluorescently labeled, hybridization can be readily detected (**Figure 15.21**).

Detection of a mutation by either DNA screening method can be used for diagnosis of a genetic disease, so that appropriate treatment can begin. In addition, identification provides a person with important information about his or her genome.

## TOOLS FOR INVESTIGATING LIFE

### 15.20 DNA Testing by Allele-Specific Cleavage

Allele-specific cleavage can be used to detect mutations such as the one that causes sickle-cell disease.

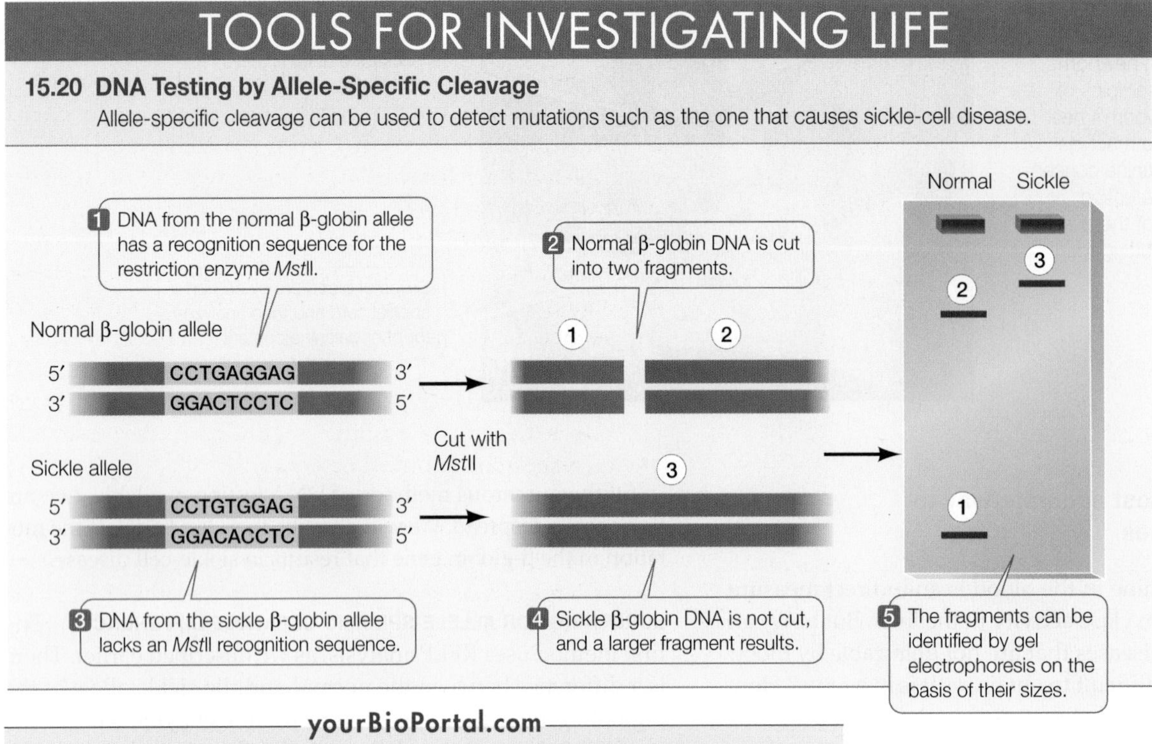
1 DNA from the normal β-globin allele has a recognition sequence for the restriction enzyme *Mst*II.
Normal β-globin allele
5′ CCTGAGGAG 3′
3′ GGACTCCTC 5′
Sickle allele
5′ CCTGTGGAG 3′
3′ GGACACCTC 5′
3 DNA from the sickle β-globin allele lacks an *Mst*II recognition sequence.
Cut with *Mst*II
2 Normal β-globin DNA is cut into two fragments.
1
2
3
4 Sickle β-globin DNA is not cut, and a larger fragment results.
Normal
Sickle
5 The fragments can be identified by gel electrophoresis on the basis of their sizes.

**yourBioPortal.com**

GO TO **Animated Tutorial 15.2 • DNA Testing by Allele-Specific Cleavage**

## TOOLS FOR INVESTIGATING LIFE

### 15.21 DNA Testing by Allele-Specific Oligonucleotide Hybridization

Testing of this family reveals that three of them are heterozygous carriers of the sickle allele. The first child, however, has inherited two normal alleles and is neither affected by the disease nor a carrier.

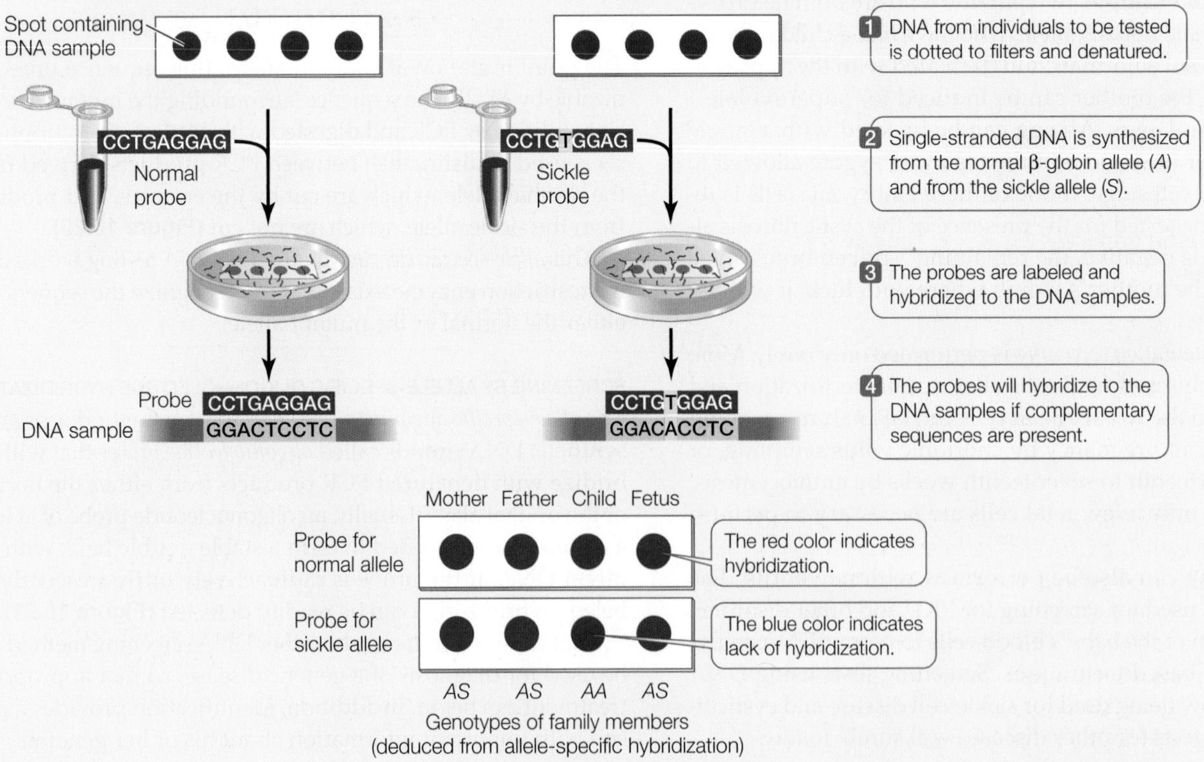

Spot containing DNA sample

1 DNA from individuals to be tested is dotted to filters and denatured.

CCTGAGGAG Normal probe

CCTGTGGAG Sickle probe

2 Single-stranded DNA is synthesized from the normal β-globin allele (*A*) and from the sickle allele (*S*).

3 The probes are labeled and hybridized to the DNA samples.

Probe CCTGAGGAG
DNA sample GGACTCCTC

CCTGTGGAG
GGACACCTC

4 The probes will hybridize to the DNA samples if complementary sequences are present.

Mother Father Child Fetus

Probe for normal allele

Probe for sickle allele

The red color indicates hybridization.

The blue color indicates lack of hybridization.

*AS* *AS* *AA* *AS*

Genotypes of family members (deduced from allele-specific hybridization)

## 15.5 RECAP

**Genetic screening can be used to identify people who have, are predisposed to, or are carriers of, genetic diseases. Screening can be done at the phenotype level by identifying an abnormal protein such as an enzyme with altered activity. It can also be done at the genotype level by direct testing of DNA.**

- How are newborn babies screened for PKU? See p. 334 and Figure 15.19
- What is the advantage of screening for genetic mutations by allele-specific oligonucleotide hybridization relative to screening for allele-specific cleavage differences? See p. 335 and Figures 15.20 and 15.21

Ongoing research has resulted in the development of increasingly accurate diagnostic tests and a better understanding of various genetic diseases at the molecular level. This knowledge is now being applied to the development of new treatments for genetic diseases. In the next section we will survey various approaches to treatment, including modifications of the mutant phenotype and gene therapy, in which the normal version of a mutant gene is supplied.

# 15.6 How Are Genetic Diseases Treated?

Most treatments for genetic diseases simply try to alleviate the patient's symptoms. But to effectively treat these diseases—whether they affect all cells, as in inherited disorders such as PKU, or only somatic cells, as in cancer—physicians must be able to diagnose the disease accurately, understand how the disease works at the molecular level, and intervene early, before the disease ravages or kills the individual. There are two main approaches to treating genetic diseases: modifying the disease phenotype, or replacing the defective gene.

### Genetic diseases can be treated by modifying the phenotype

Altering the phenotype of a genetic disease so that it no longer harms an individual is commonly done in one of three ways: by restricting the substrate of a deficient enzyme, by inhibiting a harmful metabolic reaction, or by supplying a missing protein product (**Figure 15.22**).

**RESTRICTING THE SUBSTRATE** Restricting the substrate of a deficient enzyme is the approach taken when a newborn is diagnosed with PKU. In this case, the deficient enzyme is phenylalanine hydroxylase, and the substrate is phenylalanine. The infant's inability to break down phenylalanine in food leads to a buildup of the substrate, which causes the clinical symptoms. So the infant is immediately put on a special diet that contains only enough phenylalanine for immediate use. Lofenelac, a milk-based product that is low in phenylalanine, is fed to these infants just like formula. Later, certain fruits, vegetables, cereals, and noodles low in phenylalanine can be added to the diet. Meat, fish, eggs, dairy products, and bread, which contain high amounts of phenylalanine, must be avoided, especially during childhood, when brain development is most rapid. The artificial sweetener aspartame must also be avoided because it is made of two amino acids, one of which is phenylalanine.

People with PKU are generally advised to stay on a low-phenylalanine diet for life. Although maintaining these dietary restrictions may be difficult, it is effective. Numerous follow-up studies since newborn screening was initiated have shown that people with PKU who stay on the diet are no different from the rest of the population in terms of mental ability. This is an impressive achievement in public health, given the severity of mental retardation in untreated patients.

**METABOLIC INHIBITORS** In Section 11.7, we described how drugs that are inhibitors of various cell cycle processes are used to treat cancer. Drugs are also used to treat the symptoms of many genetic diseases. As biologists have gained insight into the molec-

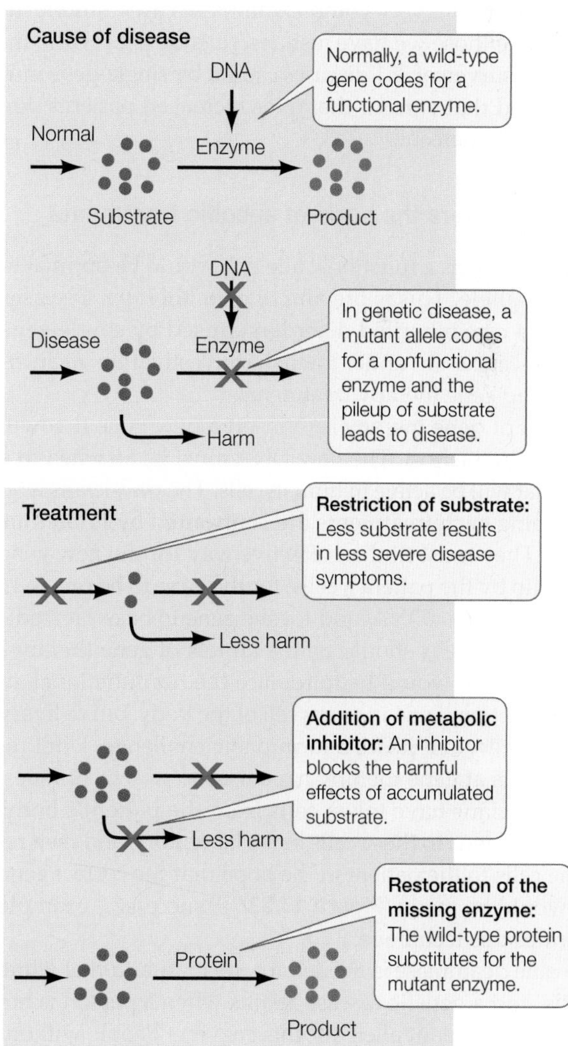


**15.22 Strategies for Treating Genetic Diseases**

ular characteristics of these diseases and the specific proteins involved, a more specific approach to treatment is taking shape. This is called *molecular medicine.*

An example of this approach is the treatment of chronic myelogenous leukemia. In this cancer, certain white blood cells undergo a gain-of-function mutation, making a totally new protein that is not made in any other cells. This new protein was isolated and a drug was made that specifically targets and inactivates the protein, thereby preventing the proliferation of the cancerous cells. The result has been greatly improved survival in these patients.

**SUPPLYING THE MISSING PROTEIN** An obvious way to treat a disease phenotype in which a functional protein is missing is to supply that protein. This approach is the basis of treatment for hemophilia, in which the missing blood factor VIII is supplied to the patient. At first this protein was obtained from blood and was sometimes contaminated with viruses or other pathogens. Now, however, the production of human clotting proteins by recombinant DNA technology (see Chapter 18) has made it possible to provide the protein in a much purer form.

Unfortunately, the phenotypes of many diseases caused by genetic mutations are very complex. In these cases, simple interventions like those we have just described do not work. Indeed, a recent survey of 351 diseases caused by single-gene mutations showed that current therapies increased patients' life spans by only 15 percent.

### Gene therapy offers the hope of specific treatments

Clearly, if a cell lacks a functional allele, it would be optimal to provide that allele. This is the aim of gene therapy. Diseases ranging from rare inherited disorders caused by single-gene mutations to cancer are under intensive investigation, in an effort to develop gene therapy treatments.

The object of **gene therapy** is to insert a new gene that will be expressed in the host. The new DNA must be attached to a promoter that will be active in human cells. The physicians who are developing such treatments are confronted by numerous challenges. They must find an effective way for the new gene to be taken up by the patient's cells, for the gene to be precisely inserted into the host DNA, and for the gene to be expressed.

Which human cells should be the targets of gene therapy? The best approach would be to replace the nonfunctional allele with a functional one in every cell of the body. But delivery of a gene to every cell poses a formidable challenge. Until recently, attempts at gene therapy have used *ex vivo* techniques. That is, physicians have taken cells from the patient's body, added the new gene to those cells in the laboratory, and then returned the cells to the patient in the hope that the correct gene product would be made (**Figure 15.23**). A successful example demonstrates this technique.

Adenosine deaminase is needed for the maturation of white blood cells, and a genetic disease results when a person is homozygous for a mutant allele for this enzyme. People without this enzyme have severe immune system deficiencies. The wild-type gene for adenosine deaminase has been isolated and in-

INVESTIGATING LIFE

**15.23 Gene Therapy: The Ex Vivo Approach** New genes are added to somatic cells taken from a patient's body. These cells are then returned to the body to make the missing gene product.

**HYPOTHESIS** **The introduction and expression of a normal allele can help a patient with two defective alleles for an important gene.**

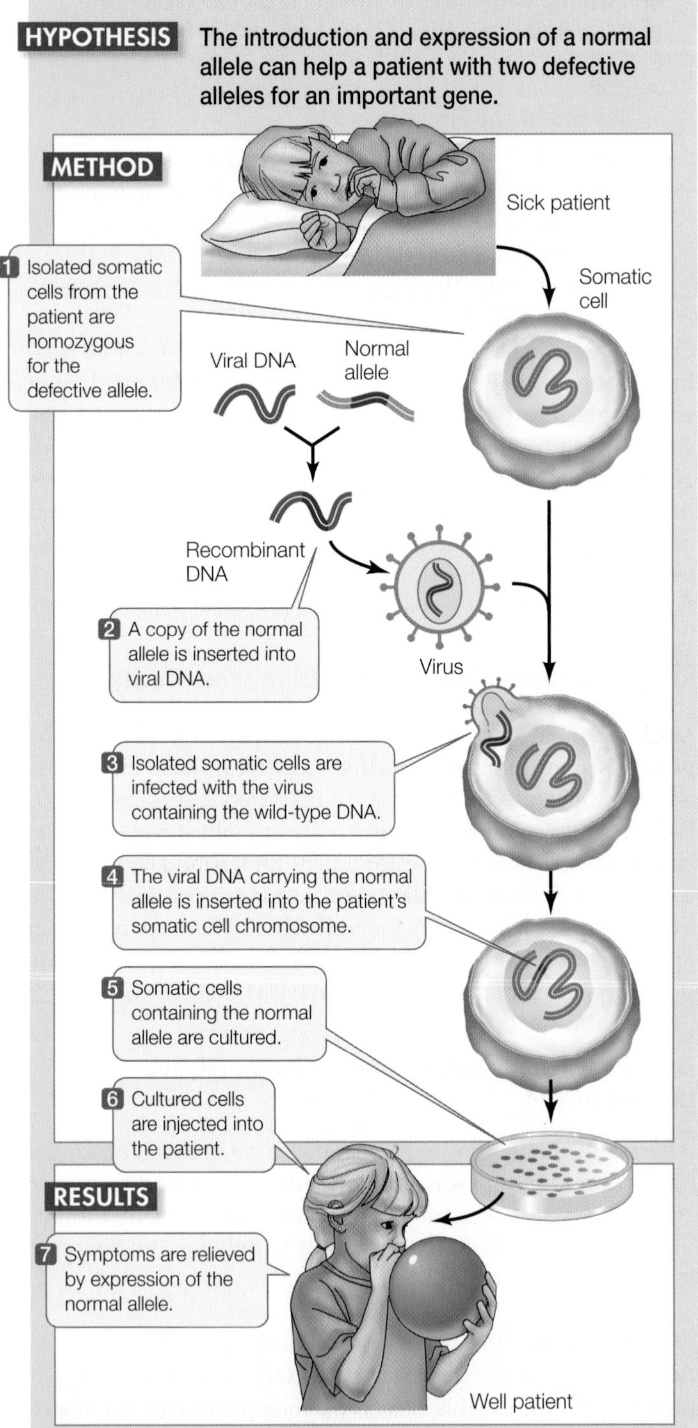

METHOD
Sick patient
Somatic cell
1 Isolated somatic cells from the patient are homozygous for the defective allele.
Viral DNA
Normal allele
Recombinant DNA
2 A copy of the normal allele is inserted into viral DNA.
Virus
3 Isolated somatic cells are infected with the virus containing the wild-type DNA.
4 The viral DNA carrying the normal allele is inserted into the patient's somatic cell chromosome.
5 Somatic cells containing the normal allele are cultured.
6 Cultured cells are injected into the patient.
RESULTS
7 Symptoms are relieved by expression of the normal allele.
Well patient


**CONCLUSION** **Gene therapy can be effective in relieving symptoms caused by a genetic disease.**

Go to **yourBioPortal.com** for original citations, discussions, and relevant links for all INVESTIGATING LIFE figures.

serted into a virus that can carry the gene into white blood cells of a patient lacking the enzyme. The recombinant virus lacks the genes for reproduction inside cells, but retains the genes coding for cell uptake and insertion into the host DNA. The recombinant virus was added to white blood cells from a patient that had inherited the mutant form of adenosine deaminase. The wild-type adenosine deaminase gene became inserted into the cells' chromosomes, along with viral DNA. When these transformed white blood cells with the wild-type gene were put back into the patient, the cells made adenosine deaminase and the patient's condition improved.

The other approach to gene therapy is to insert the gene directly into cells in the body of the patient. This in vivo approach is being attempted for various types of cancer. Lung cancer cells, for example, are accessible to such treatment if the DNA is given as an aerosol through the respiratory system. Several thousand patients, over half of them with cancer, have undergone this treatment. In preliminary clinical trials, people are given the therapy to see whether it has any toxicity and whether the new gene is actually incorporated into the patients' genomes. In more ambitious trials, larger numbers of patients receive the therapy with the hope that their disease will disappear, or at least improve.

## 15.6 RECAP

**Treatment of a human genetic disease may involve an attempt to modify the abnormal phenotype by restricting the substrate of a deficient enzyme, inhibiting a harmful metabolic reaction, or supplying a missing protein. On the other hand, gene therapy aims to address a genetic defect by inserting a normal allele into a patient's cells.**

- How do metabolic inhibitors used in chemotherapy function in treating cancer? See pp. 337–338 and Figure 5.22
- How does ex vivo gene therapy work? Can you give an example? See p. 338 and Figure 15.23

In this chapter, we dealt with mutations in general, focusing on DNA changes that affect phenotypes through specific protein products. But there is much more to molecular genetics than genes and proteins. Determining which genes will be expressed when and where is a major function of the genome. In Chapter 16 we turn to gene regulation.

## CHAPTER SUMMARY

### 15.1 What Are Mutations?

- **Mutations** are heritable changes in DNA. **Somatic mutations** are passed on to daughter cells, but only **germ line mutations** are passed on to sexually produced offspring.
- **Point mutations** result from alterations in single base pairs of DNA. **Silent mutations** can occur in noncoding DNA or in coding regions of genes and do not affect the amino acid sequences of proteins. **Missense**, **nonsense**, and **frame-shift** mutations all cause changes in protein sequences. Review Figure 15.2
- Chromosomal mutations (**deletions**, **duplications**, **inversions**, or **translocations**) involve large regions of chromosomes. Review Figure 15.4
- **Spontaneous mutations** occur because of instabilities in DNA or chromosomes. **Induced mutations** occur when a mutagen damages DNA. Review Figure 15.5
- Mutations can occur in hot spots where cytosine has been methylated to 5-methylcytosine. Review Figure 15.6
- Mutations, although often detrimental to an individual organism, are the raw material of evolution.

### 15.2 How Are DNA Molecules and Mutations Analyzed?

- **Restriction enzymes**, which are made by microorganisms as a defense against viruses, bind to and cut DNA at specific **recognition sequences** (also called **restriction sites**). These enzymes can be used to produce small fragments of DNA for study, a technique known as restriction digestion. Review Figure 15.7
- DNA fragments can be separated by size using **gel electrophoresis**. Review Figure 15.8, **ANIMATED TUTORIAL 15.1**
- **DNA fingerprinting** is used to distinguish between specific individuals, or to reveal which individuals are most closely related to one another. It involves the detection of DNA polymorphisms, including **single nucleotide polymorphisms** (**SNPs**) and **short tandem repeats** (**STRs**). Review Figure 15.9
- The goal of the DNA barcoding project is to sequence a single region of DNA in all species for identification purposes.

### 15.3 How Do Defective Proteins Lead to Diseases?

- Abnormalities in nearly all classes of proteins, including enzymes, transport proteins, receptor proteins, and structural proteins, have been implicated in genetic diseases.
- While a single amino acid difference can be the cause of disease, amino acid variations have been detected in many functional proteins. Review Figure 15.13
- Transmissible spongiform encephalopathies (TSEs) are degenerative brain diseases that can be transmitted from one animal to another by consumption of infected tissues. The infective agent is a **prion**, a protein with an abnormal conformation.
- **Multifactorial** diseases are caused by the interactions of many genes and proteins with the environment. They are much more common than diseases caused by mutations in a single gene.
- Predictable patterns of inheritance are associated with some human genetic diseases. Autosomal recessive, autosomal dominant, and sex-linked patterns are common.

### 15.4 What DNA Changes Lead to Genetic Diseases?

- It is possible to isolate both the mutant genes and the abnormal proteins responsible for human diseases. Review Figure 15.15, **WEB ACTIVITY 15.1**
- The effects of fragile-X syndrome worsen with each generation. This pattern is the result of an **expanding triplet repeat**. Review Figure 15.18

## 15.5 How Is Genetic Screening Used to Detect Human Diseases?

- **Genetic screening** is used to detect human genetic diseases, alleles predisposing people to those diseases, or carriers of those diseases.
- Genetic screening can be done by looking for abnormal protein expression. **Review Figure 15.19**
- **DNA testing** is the direct identification of mutant alleles. Any cell can be tested at any time in the life cycle.
- The two predominant methods of DNA testing are the allele-specific cleavage method and allele-specific oligonucleotide hybridization method. **Review Figures 15.20 and 15.21, ANIMATED TUTORIAL 15.2**

## 15.6 How Are Genetic Diseases Treated?

- There are three ways to modify the phenotype of a genetic disease: restrict the substrate of a deficient enzyme, inhibit a harmful metabolic reaction, or supply a missing protein. **Review Figure 15.22**
- Cancer is treated with metabolic inhibitors.
- In **gene therapy**, a mutant gene is replaced with a normal gene. Both ex vivo and in vivo therapies are being developed. **Review Figure 15.23**

## SELF-QUIZ

1. Phenylketonuria is an example of a genetic disease in which
   *a.* a single enzyme is not functional.
   *b.* inheritance is sex-linked.
   *c.* two parents without the disease cannot have a child with the disease.
   *d.* mental retardation always occurs, regardless of treatment.
   *e.* a transport protein does not work properly.
2. Mutations of the gene for β-globin
   *a.* are usually lethal.
   *b.* occur only at amino acid position 6.
   *c.* number in the hundreds.
   *d.* always result in sickling of red blood cells.
   *e.* can always be detected by gel electrophoresis.
3. Multifactorial (complex) diseases
   *a.* are less common than single-gene diseases.
   *b.* involve the interaction of many genes with the environment.
   *c.* affect less than 1 percent of humans.
   *d.* involve the interactions of several mRNAs.
   *e.* are exemplified by sickle-cell disease.
4. In fragile-X syndrome,
   *a.* females are affected more severely than males.
   *b.* a short sequence of DNA is repeated many times to create the fragile site.
   *c.* both the X and Y chromosomes tend to break when prepared for microscopy.
   *d.* all people who carry the gene that causes the syndrome are mentally retarded.
   *e.* the basic pattern of inheritance is autosomal dominant.
5. Most genetic diseases are rare because
   *a.* each person is unlikely to be a carrier for harmful alleles.
   *b.* genetic diseases are usually sex-linked and so uncommon in females.
   *c.* genetic diseases are always dominant.
   *d.* two parents probably do not carry the same recessive alleles.
   *e.* mutation rates in humans are low.
6. Mutational "hot spots" in human DNA
   *a.* always occur in genes that are transcribed.
   *b.* are common at cytosines that have been modified to 5-methylcytosine.
   *c.* involve long stretches of nucleotides.
   *d.* occur only where there are long repeats.
   *e.* are very rare in genes that code for proteins.
7. Newborn genetic screening for PKU
   *a.* is very expensive.
   *b.* detects phenylketones in urine.
   *c.* has not led to the prevention of mental retardation resulting from this disorder.
   *d.* should be done during the second or third day of an infant's life.
   *e.* uses bacterial growth to detect excess phenylketones in blood.
8. Genetic diagnosis by DNA testing
   *a.* detects only mutant and not normal alleles.
   *b.* can be done only on eggs or sperm.
   *c.* involves hybridization to rRNA.
   *d.* often utilizes restriction enzymes and a polymorphic site.
   *e.* cannot be done with PCR.
9. Which of the following is *not* a way to treat a genetic disease?
   *a.* Inhibiting a harmful biochemical reaction
   *b.* Adding the wild-type allele to cells expressing the mutation
   *c.* Restricting the substrate of a harmful biochemical reaction
   *d.* Replacing a mutant allele with the wild-type allele in the fertilized egg
   *e.* Supplying a wild-type protein that is missing due to mutation
10. Current treatments for genetic diseases include all of the following *except*
   *a.* restricting a dietary substrate.
   *b.* replacing the mutant gene in all cells.
   *c.* alleviating the patient's symptoms.
   *d.* inhibiting a harmful metabolic reaction.
   *e.* supplying a protein that is missing.

## FOR DISCUSSION

1. In the past, it was common for people with phenylketonuria (PKU) who were placed on a low-phenylalanine diet after birth to be allowed to return to a normal diet during their teenage years. Although the levels of phenylalanine in their blood were high, their brains were thought to be beyond the stage when they could be harmed. If a woman with PKU becomes pregnant, however, a problem arises. Typically, the fetus is heterozygous, but is unable, at early stages of development, to metabolize the high levels of phenylalanine that arrive from the mother's blood. Why is the fetus likely to be heterozygous? What do you think would happen to the fetus during this "maternal PKU" situation? What would be your advice to a woman with PKU who wants to have a child?
2. Cystic fibrosis is an autosomal recessive disease in which thick mucus is produced in the lungs and airways. The gene responsible for this disease encodes a protein composed of 1,480 amino acids. In most patients with cystic fibrosis, the protein has 1,479 amino acids: a phenylalanine is missing at position 508. A baby is born with cystic fibrosis. He has an older brother who is not affected. How would you test the DNA of the older brother to determine whether he is a carrier for cystic fibrosis? How would you design a gene therapy protocol to "cure" the cells in the younger brother's lungs and airways?
3. A number of efforts are under way to identify human genetic polymorphisms that correlate with multifactorial diseases such as diabetes, heart disease, and cancer. What would be the uses of such information? What concerns do you think are being raised about this kind of genetic testing?

## ADDITIONAL INVESTIGATION

Tay-Sachs disease is caused by a recessively inherited mutation in the gene coding for the enzyme hexosaminidase A (HexA), which normally breaks down a lipid called GM2 ganglioside. Accumulation of this lipid in the brain leads to progressive deterioration of the nervous system and death, usually by age 4. HexA activity in blood serum is 0–6 percent in homozygous recessives and 7–35 percent in heterozygous carriers, compared to non-carriers (100 percent). The most common mutation in the *HexA* gene is an insertion of four base pairs, which presumably leads to a premature stop codon. How would you do genetic screening for carriers of this disease by enzyme testing and by DNA testing? What are the advantages of DNA testing? How would you investigate the premature stop codon hypothesis?

## WORKING WITH DATA (GO TO yourBioPortal.com)

**Gene Therapy: The Ex Vivo Approach** In this exercise, you use the original research paper to examine the protocol used to treat two patients with gene therapy for adenosine deaminase deficiency (Figure 15.23). You will examine the kinds of evidence used to detect the wild-type gene in the cells of these patients, and will analyze the results in terms of immune system cell function.

# 16 Regulation of Gene Expression

## Alcoholism and the control of gene expression

Many people drink alcoholic beverages but relatively few of them become addicted (alcoholic). When they do, the results are often disastrous, both socially and physiologically. Alcoholism often disrupts relationships with family, friends, and colleagues. Lost productivity leads to economic costs estimated at over $100 billion per year in the U.S. alone. Physiologically, alcoholism is characterized by a compulsion to consume alcohol, tolerance (increasing doses are needed for the same effect), and dependence (abrupt cessation of consumption leads to severe withdrawal symptoms). In most of these people, alcohol acts not just to provide pleasant sensations (positive reinforcement) but also to alleviate unpleasant ones such as anxiety (negative reinforcement).

Why do only some people become alcoholic? Alcoholism is a complex behavioral disease. Psychologists sometimes speak of "addictive personalities," and genetic studies indicate there may be inherited factors. It would help both alcoholics and those who treat them if we understood the differences in brain chemistry between alcoholic and nonalcoholic individuals. But we can't do the necessary experiments on humans; instead, animal models are used to study alcoholism at the molecular level. James Murphy at Indiana University has bred a strain of rats, called P rats, that prefer alcohol when given the choice of alcohol-containing or alcohol-free water. These rats show many of the symptoms of true addiction, including compulsive drinking, tolerance, and withdrawal. In effect, they are a genetic strain of alcoholic animals.

People often drink alcoholic beverages to relieve anxiety, and there are clear links between anxiety disorders and alcoholism. Like many of their human counterparts, the P rats appear more anxious than wild-type rats, spending more time in a closed rather than an open environment. Drinking alcohol alters this behavior and seems to relieve their anxiety.

There may also be a link between the transcription factor CREB and alcohol consumption. CREB (or *c*yclic AMP *r*esponse *e*lement *b*inding protein) is especially abundant in the brain and regulates the expression of hundreds of genes that are important in metabolism. CREB becomes activated when it is phosphorylated by the enzyme protein kinase A, which in turn is activated by the second messenger cyclic AMP. In an effort to understand the molecular basis of alcoholism and anxiety, neuroscientist Subhash Pandey and his colleagues at the University of Illinois compared CREB levels in the brains of P rats and wild-type rats.

**Alcoholism** Huge social and economic costs are associated with alcohol abuse. Scientists are trying to understand its molecular basis.

**An Explanation for Alcoholism?** The transcription factor, CREB, binds to DNA and activates promoters of genes involved in addictive behaviors in alcoholism.

They found that P rats have inherently lower levels of CREB in certain parts of the brain. When these rats consumed alcohol, the total levels of CREB did not increase, but the levels of phosphorylated CREB did. It is the phosphorylated version of CREB that binds to DNA and regulates gene transcription.

The prospect that CREB, a transcription factor that regulates gene expression, is a key element in the genetic propensity for alcoholism is important because it begins to explain the molecular nature of a complex behavioral disease. Such understanding may permit more effective treatment of alcohol abuse or its prevention. Equally important to our purpose here, it underscores the importance of the regulation of gene expression in biological processes.

**IN THIS CHAPTER** we will focus on the control of gene expression in many types of organisms. We begin with the simplest systems, viruses, which undertake an ordered series of molecular events when they infect a host cell. Then we turn to prokaryotes, which respond to changes in their environment with coordinated changes in gene expression. In eukaryotes, similar principles are used to regulate gene expression, but with added levels of complexity. Finally, we turn to the regulation of gene expression by modification of the genome—the field of epigenetics.

## CHAPTER OUTLINE



# 16.1 How Do Viruses Regulate Their Gene Expression?

"A virus is a piece of bad news wrapped in protein." This quote from immunologist Sir Peter Medawar is certainly true for the cells that viruses infect. As we describe in Chapter 13, a virus injects its genetic material into a host cell and turns that cell into a virus factory (see Figure 13.3). Viral life cycles are very efficient. Perhaps the record is held by poliovirus: a single poliovirus infecting a mammalian cell can produce over 100,000 new virus particles!

Unlike organisms, **viruses** are *acellular*; that is, they are not cells, do not consist of cells, and do not carry out many of the processes characteristic of life. Most virus particles, called **virions**, are composed of only nucleic acid and a few proteins. Viruses do not carry out two of the basic functions of cellular life: they do not regulate the transport of substances into and out of themselves by membranes, and they do not perform metabolic functions involved with taking in nutrients, refashioning them, and expelling wastes. But they can reproduce in systems that do perform these metabolic functions—namely, living cells. By studying the relatively simple viral reproductive cycle, biologists have discovered principles of gene expression and its regulation that apply to cellular systems that may be much more complex.

As we describe in Chapter 14, gene expression begins at the *promoter*, where RNA polymerase binds to initiate transcription. In a genome with many genes, not all promoters are active at a given time—there is *selective gene transcription*. The "decision" regarding which genes to activate involves two types of regulatory proteins that bind to DNA: repressor proteins and activator proteins. In both cases, these proteins bind to the promoter to regulate the gene (**Figure 16.1**):

- In **negative regulation**, the gene is normally transcribed. Binding of a repressor protein prevents transcription.
- In **positive regulation**, the gene is normally not transcribed. An activator protein binds to stimulate transcription.

You will see these mechanisms, or combinations of them, as we examine regulation in viruses, prokaryotes, and eukarytotes.

### Bacteriophage undergo a lytic cycle

The Hershey–Chase experiment (see Figure 13.4) involved the typical viral reproductive cycle, the **lytic cycle**, so named because the infected host cell lyses (bursts), releasing progeny viruses. Once a virus has injected its nucleic acid into a cell, that

(A) Negative regulation
DNA
Repressor binding site
5′
3′
3′
5′
Transcription
DNA
5′
3′
3′
5′
No transcription
Binding of repressor protein blocks transcription.
(B) Positive regulation
DNA
Activator binding site
5′
3′
3′
5′
No transcription
DNA
5′
3′
3′
5′
Transcription
Binding of activator protein stimulates transcription.

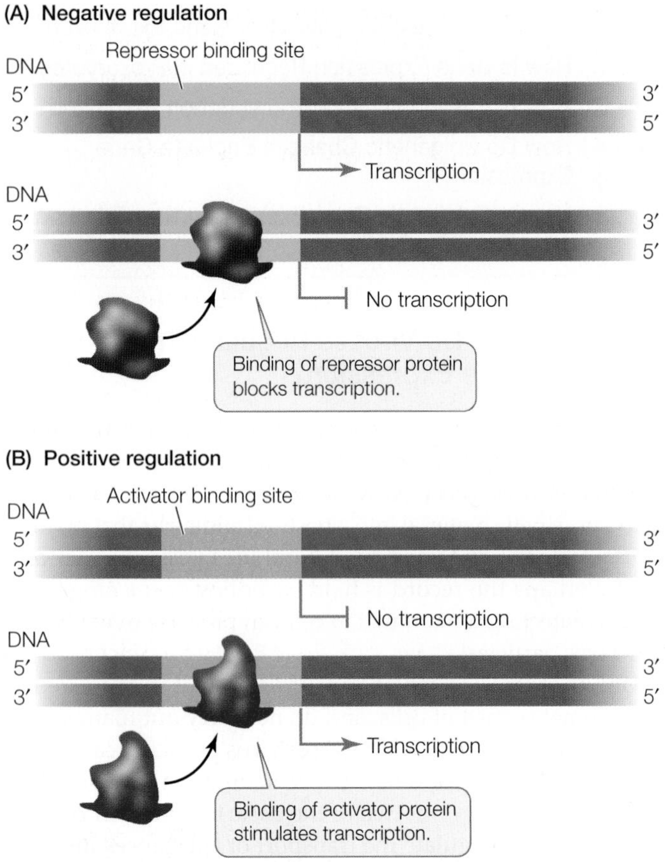

**16.1 Positive and Negative Regulation** Proteins regulate gene expression by binding to DNA and preventing or allowing RNA polymerase to bind DNA at the promotor region to control transcription.

(A)
0.50 μm
(B)

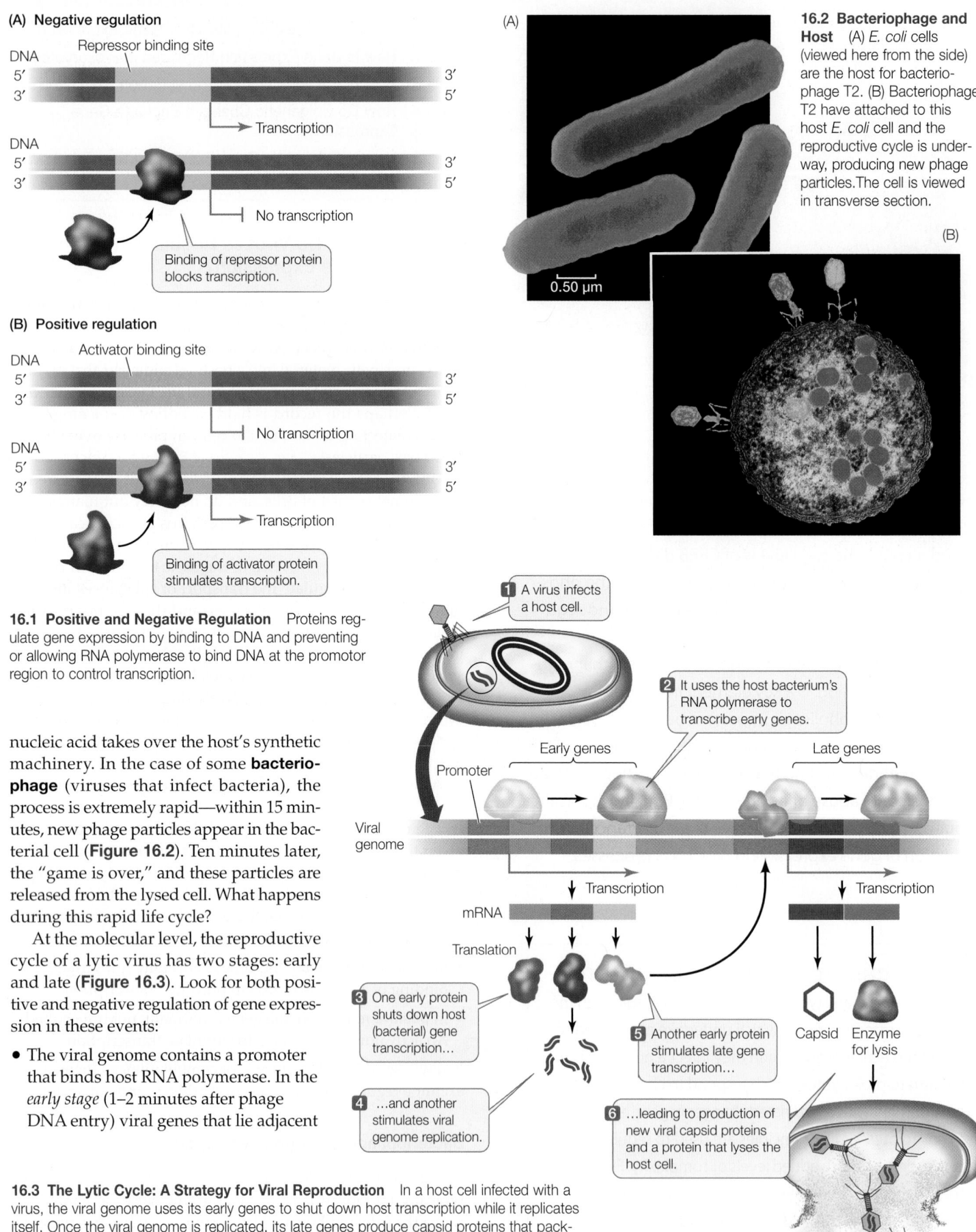

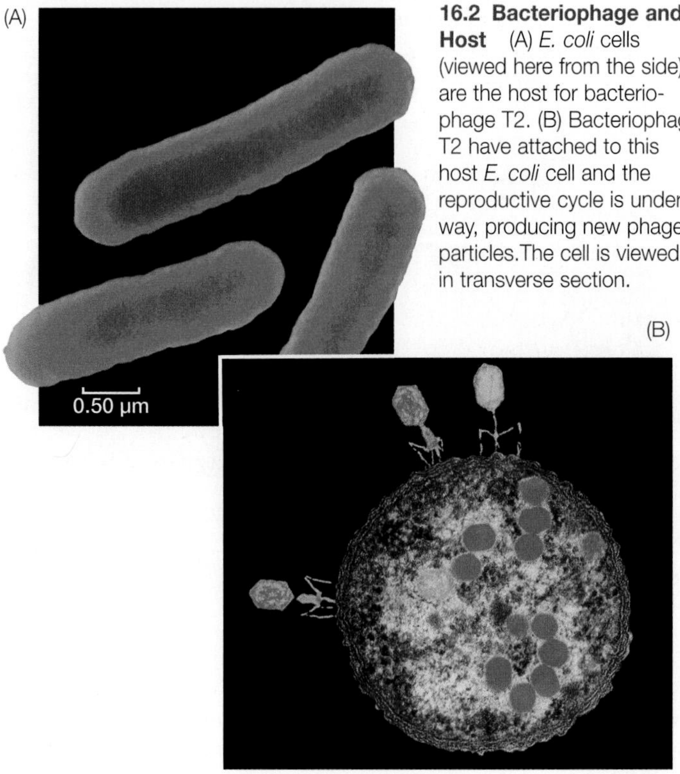

**16.2 Bacteriophage and Host** (A) *E. coli* cells (viewed here from the side) are the host for bacteriophage T2. (B) Bacteriophage T2 have attached to this host *E. coli* cell and the reproductive cycle is underway, producing new phage particles.The cell is viewed in transverse section.

nucleic acid takes over the host's synthetic machinery. In the case of some **bacteriophage** (viruses that infect bacteria), the process is extremely rapid—within 15 minutes, new phage particles appear in the bacterial cell (**Figure 16.2**). Ten minutes later, the "game is over," and these particles are released from the lysed cell. What happens during this rapid life cycle?

At the molecular level, the reproductive cycle of a lytic virus has two stages: early and late (**Figure 16.3**). Look for both positive and negative regulation of gene expression in these events:

- The viral genome contains a promoter that binds host RNA polymerase. In the *early stage* (1–2 minutes after phage DNA entry) viral genes that lie adjacent

1 A virus infects a host cell.
2 It uses the host bacterium's RNA polymerase to transcribe early genes.
Early genes
Late genes
Promoter
Viral genome
Transcription
Transcription
mRNA
Translation
3 One early protein shuts down host (bacterial) gene transcription...
4 ...and another stimulates viral genome replication.
5 Another early protein stimulates late gene transcription...
Capsid
Enzyme for lysis
6 ...leading to production of new viral capsid proteins and a protein that lyses the host cell.

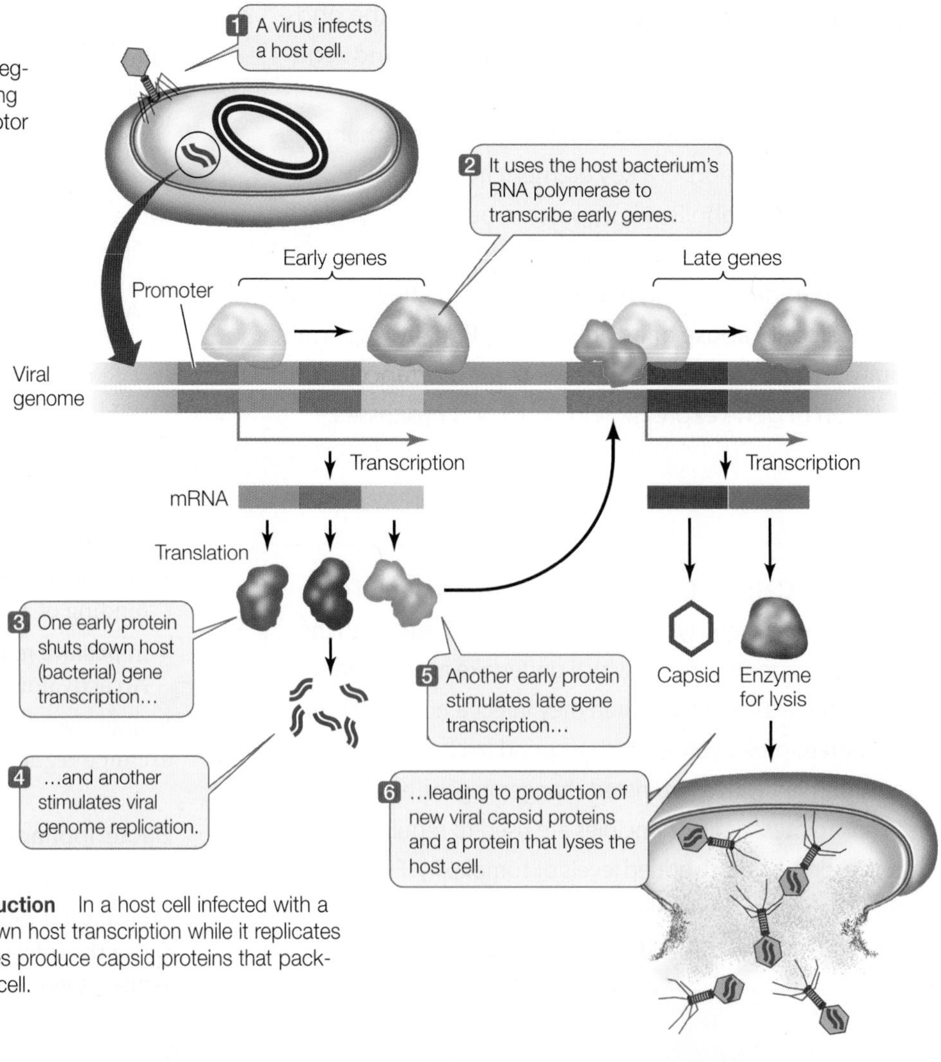

**16.3 The Lytic Cycle: A Strategy for Viral Reproduction** In a host cell infected with a virus, the viral genome uses its early genes to shut down host transcription while it replicates itself. Once the viral genome is replicated, its late genes produce capsid proteins that package the genome and other proteins that lyse the host cell.

to this promoter are transcribed. These early genes often encode proteins that shut down host transcription and stimulate viral genome replication and transcription of viral late genes. Three minutes after DNA entry, viral nuclease enzymes digest the host's chromosome, providing nucleotides for the synthesis of viral genomes.

- In the *late stage*, viral late genes are transcribed; they encode the viral capsid proteins and enzymes that lyse the host cell to release the new virions. This begins 9 minutes after DNA entry and 6 minutes before the first new phage particles appear.

The whole process—from binding and infection to release of new phage—takes about half an hour. During this period, the sequence of transcriptional events is carefully controlled to produce complete, infective virons.

### Some bacteriophage can carry bacterial genes from one cell to another

During the lytic cycle some bacteriophage package their DNA in **capsids** (outer shells). In rare cases, a bacterial DNA fragment is inserted into a capsid instead of, or along with, the phage DNA. When such a virion infects another bacterium, the bacterial DNA is injected into the new host cell, a mechanism of gene transfer called **transduction**. The viral infection does not produce new viruses. Instead, the incoming DNA fragment can recombine with the host chromosome, replacing host genes with genes from the virus's former host. The recipeint cell survives under these conditions because there is no virus replication.

### Some bacteriophage can undergo a lysogenic cycle

Like all nucleic acid genomes, those of viruses can mutate and evolve by natural selection. Some viruses have evolved an advantageous process called **lysogeny** that postpones the lytic cycle. In lysogeny, the viral DNA becomes integrated into the host DNA and becomes a **prophage** (**Figure 16.4**). As the host cell divides, the viral DNA gets replicated along with that of the host. The prophage can remain inactive within the bacterial genome for thousands of generations, producing many copies of the original viral DNA.

However, if the host cell is not growing well, the virus "cuts its losses." It immediately switches to a lytic cycle, in which the prophage excises itself from the host chromosome and reproduces. In other words, the virus is able to enhance its chances of multiplication and survival by inserting its DNA into the host chromosome, where it sits as a silent partner until conditions are right for lysis.

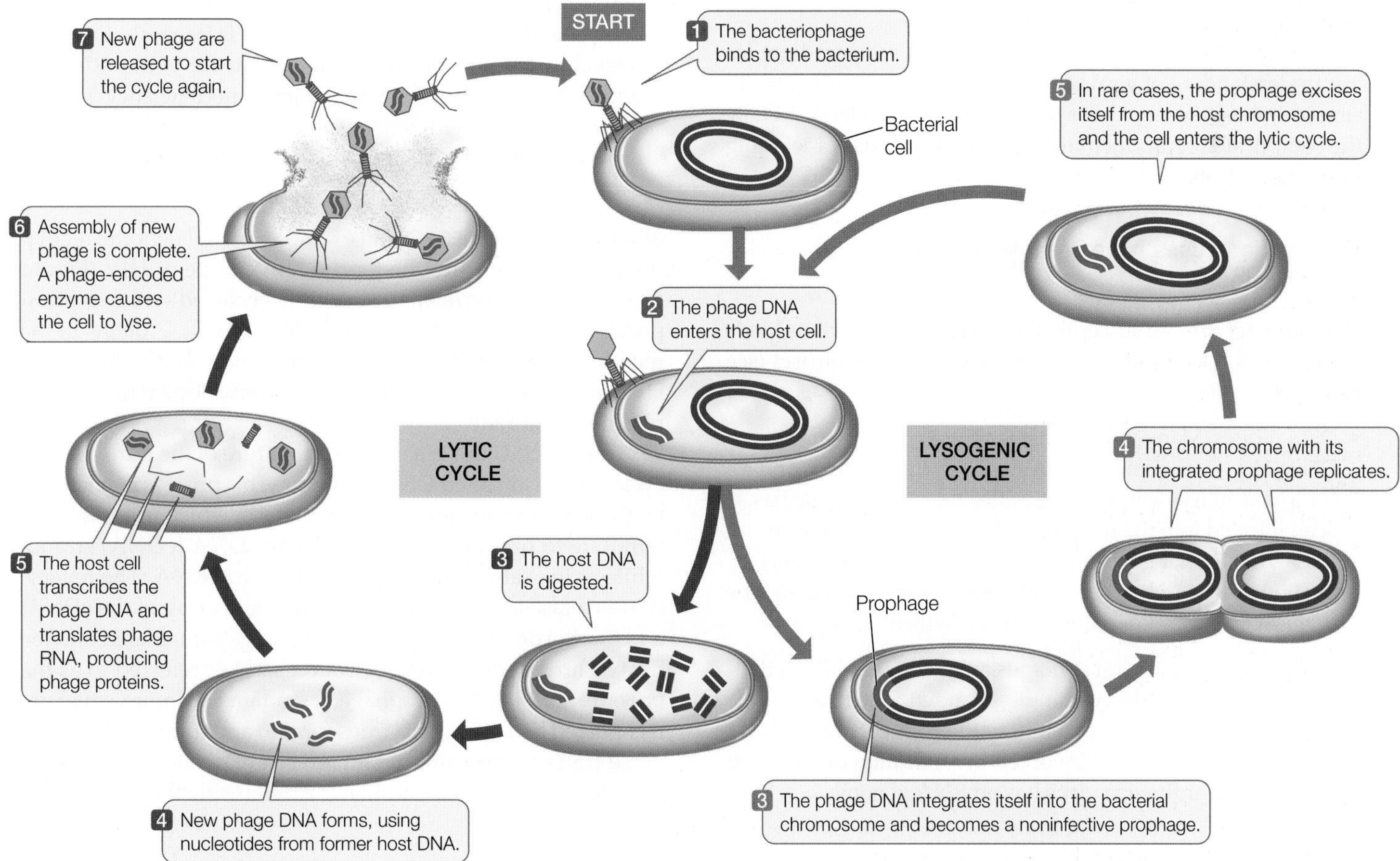

START
1 The bacteriophage binds to the bacterium.
Bacterial cell
2 The phage DNA enters the host cell.
3 The host DNA is digested.
4 New phage DNA forms, using nucleotides from former host DNA.
5 The host cell transcribes the phage DNA and translates phage RNA, producing phage proteins.
6 Assembly of new phage is complete. A phage-encoded enzyme causes the cell to lyse.
7 New phage are released to start the cycle again.
LYTIC CYCLE
LYSOGENIC CYCLE
Prophage
3 The phage DNA integrates itself into the bacterial chromosome and becomes a noninfective prophage.
4 The chromosome with its integrated prophage replicates.
5 In rare cases, the prophage excises itself from the host chromosome and the cell enters the lytic cycle.

**16.4 The Lytic and Lysogenic Cycles of Bacteriophage** In the lytic cycle, infection of a bacterium by viral DNA leads directly to the multiplication of the virus and lysis of the host cell. In the lysogenic cycle, an inactive prophage is integrated into the host DNA where it is replicated during the bacterial life cycle.

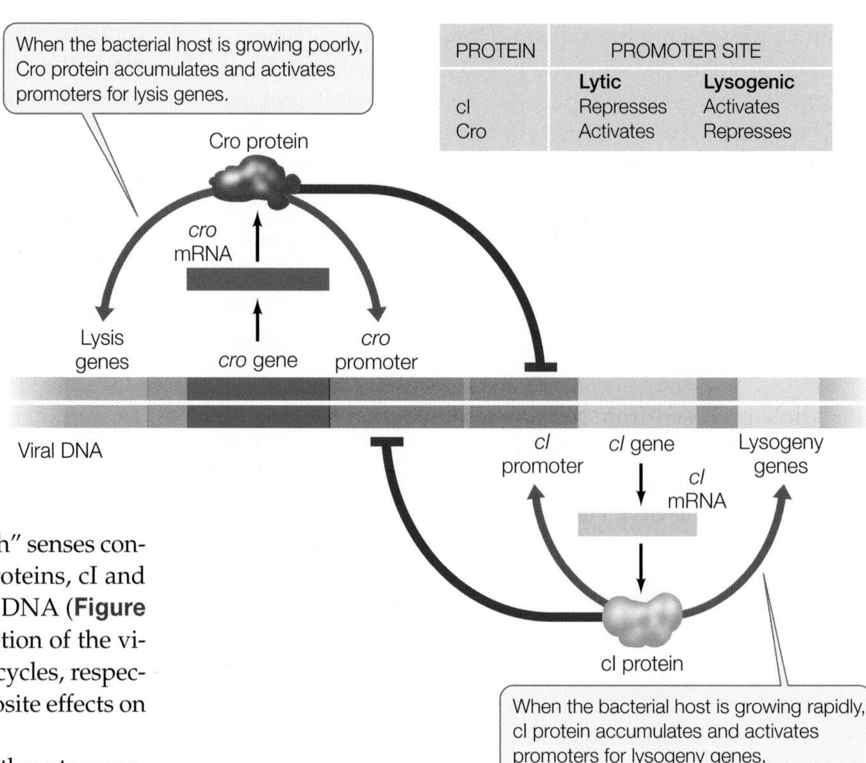
When the bacterial host is growing poorly, Cro protein accumulates and activates promoters for lysis genes.
Cro protein
cro mRNA
Lysis genes
cro gene
cro promoter
Viral DNA
cI promoter
cI gene
cI mRNA
Lysogeny genes
cI protein
When the bacterial host is growing rapidly, cI protein accumulates and activates promoters for lysogeny genes.

| PROTEIN | PROMOTER SITE | |
|---|---|---|
| | **Lytic** | **Lysogenic** |
| cI | Represses | Activates |
| Cro | Activates | Represses |

**16.5 Control of Bacteriophage λ Lysis and Lysogeny** Two regulatory proteins, Cro and cI, compete to control expression of one another and genes for viral lysis and lysogeny.

Uncovering the regulation of gene expression that underlies the lysis/lysogeny switch was a major achievement of molecular biologists. Here we present just an outline of the process to give you an idea of the positive and negative regulatory mechanisms involved (**Figure 16.5**). The model virus *bacteriophage λ* (lambda) has been used extensively to study the lysogenic mechanism.

How does the phage "know" when to switch to the lytic cycle? A kind of "genetic switch" senses conditions within the host. Two viral regulatory proteins, cI and Cro, compete for two promoters on the phage DNA (**Figure 16.5**). These two promoters control the transcription of the viral genes involved in the lytic and the lysogenic cycles, respectively, and the two regulatory proteins have opposite effects on each promoter.

Phage infection is essentially a "race" between these two regulatory proteins. In a rapidly growing *E. coli* host cell, Cro synthesis is low, so cI "wins," and the phage enters a lysogenic cycle. If the host cell is growing slowly, Cro synthesis is higher, and the genes involved in lysis are activated. The two regulatory proteins are made very early in phage infection, and each binds to a specific DNA sequence.

The reproductive cycle of bacteriophage λ is a paradigm for our understanding of viral life cycles in general. This relatively simple system has served as a model to help us understand how the complicated reproductive cycles of other viruses, including HIV, are controlled.

### Eukaryotic viruses have complex regulatory mechanisms

Many eukaryotes are susceptible to infections by various kinds of viruses: RNA and DNA viruses, as well as retroviruses (see also Section 26.6).

- *DNA viruses*. Many viral particles contain double-stranded DNA. However, some contain single-stranded DNA, and a complementary strand is made after the viral genome has been injected into the host cell. Like some bacteriophage, DNA viruses that infect eukaryotes are capable of undergoing both lytic and lysogenic life cycles. Examples include the herpes viruses and papillomaviruses (which cause warts).
- *RNA viruses*. Some viral genomes are made up of RNA that is usually, but not always, single-stranded. The RNA is translated by the host's machinery to produce viral proteins, some of which are involved in replication of the RNA genome. The influenza virus has an RNA genome.
- *Retroviruses*. The retroviral genome is RNA, and the **retrovirus** encodes a protein that makes a DNA strand that is complementary to the RNA. The DNA is integrated into the host chromosome and acts as a template for both mRNA and new viral genomes. Human immunodeficiency virus (HIV) is the retrovirus that causes acquired immune deficiency syndrome (AIDS).

**REGULATING HIV GENES** As an example of viral genome regulation, we will consider the reproductive cycle of HIV (**Figure 16.6**). HIV is an **enveloped virus**; it is enclosed within a phospholipid membrane derived from its host cell. Proteins in the membrane are involved in infection of new host cells, which HIV enters by direct fusion of the viral envelope with the host plasma membrane.

As indicated above, a distinctive feature of the retroviral life cycle is RNA-directed DNA synthesis. This process is catalyzed by the viral enzyme **reverse transcriptase**, which uses the RNA template to produce a complementary DNA (cDNA) strand, while at the same time degrading the viral RNA. The reverse transcriptase also makes a complementary copy of the cDNA, and it is the double-stranded cDNA that gets integrated into the host's chromosome. The integrated DNA is referred to as the **provirus** and, like the prophage, it contains promoters that are recognized by the host cell transcription apparatus. Both the reverse transcriptase and the integrase are needed for the very early stages of infection and are carried inside the HIV virion.

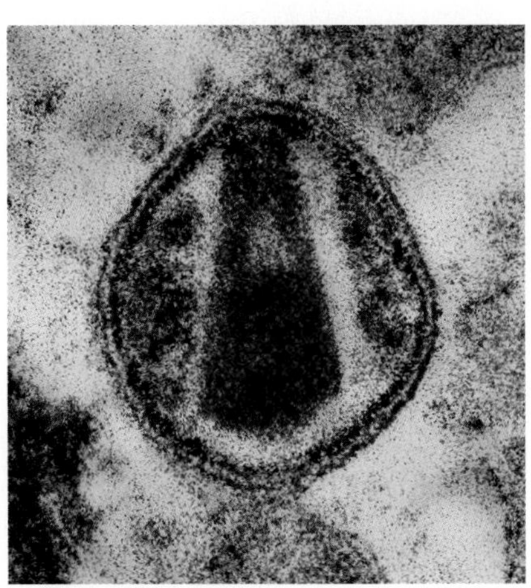

**16.6 The Reproductive Cycle of HIV** This retrovirus enters a host cell via fusion of its envelope with the host's plasma membrane. Reverse transcription of retroviral RNA then produces a DNA provirus—a molecule of complementary DNA that inserts itself into the host's genome.

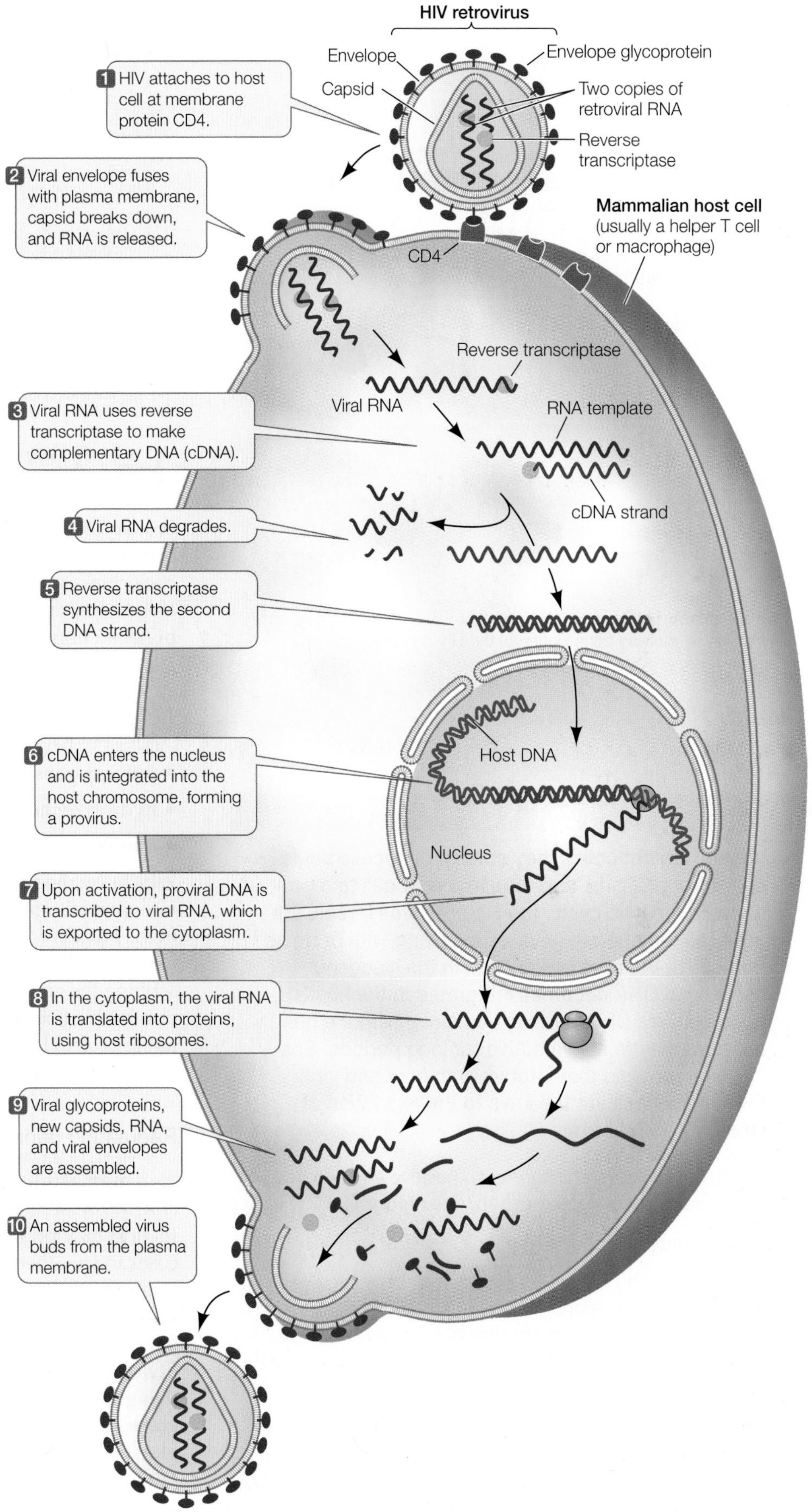

HIV retrovirus
Envelope
Envelope glycoprotein
Capsid
Two copies of retroviral RNA
Reverse transcriptase
Mammalian host cell (usually a helper T cell or macrophage)
CD4
1 HIV attaches to host cell at membrane protein CD4.
2 Viral envelope fuses with plasma membrane, capsid breaks down, and RNA is released.
Reverse transcriptase
Viral RNA
RNA template
3 Viral RNA uses reverse transcriptase to make complementary DNA (cDNA).
cDNA strand
4 Viral RNA degrades.
5 Reverse transcriptase synthesizes the second DNA strand.
Host DNA
6 cDNA enters the nucleus and is integrated into the host chromosome, forming a provirus.
Nucleus
7 Upon activation, proviral DNA is transcribed to viral RNA, which is exported to the cytoplasm.
8 In the cytoplasm, the viral RNA is translated into proteins, using host ribosomes.
9 Viral glycoproteins, new capsids, RNA, and viral envelopes are assembled.
10 An assembled virus buds from the plasma membrane.

The provirus resides permanently in the host chromosome and is occasionally activated to produce new virions. When this happens, the provirus is transcribed as mRNA, which is then translated by the host cell's protein-synthesizing machinery.

Under normal circumstances, the host cell regulates viral gene expression using proteins that may have originated as a defense mechanism against invaders. Host proteins bind to viral mRNA as it is being made and causes RNA polymerase to fall off the viral DNA, thereby terminating transcription. However, HIV can counteract this regulation with a virus-encoded protein called tat (*trans*activator of *t*ranscription), which binds to the terminator proteins and blocks their action. This *antitermination* allows viral gene transcription and the rest of the viral reproductive cycle to proceed (**Figure 16.7**).

Almost every step in the complex reproductive cycle of HIV is, in principle, a potential target for drugs to treat AIDS.

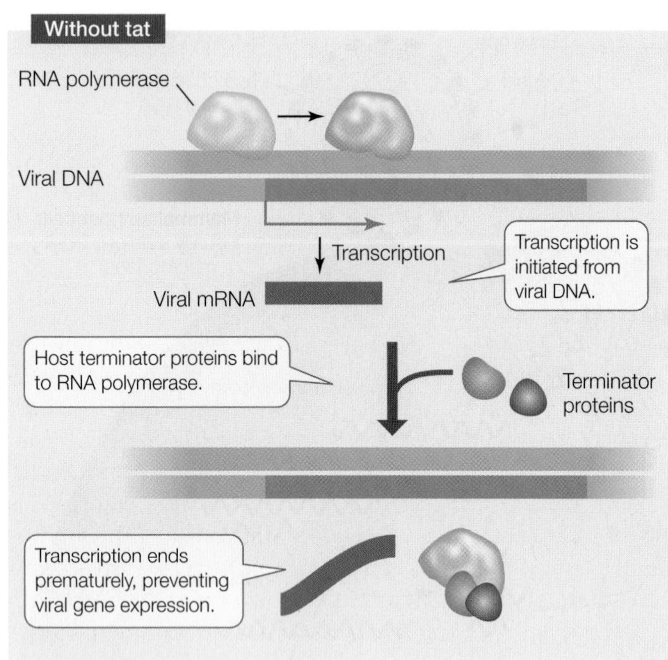


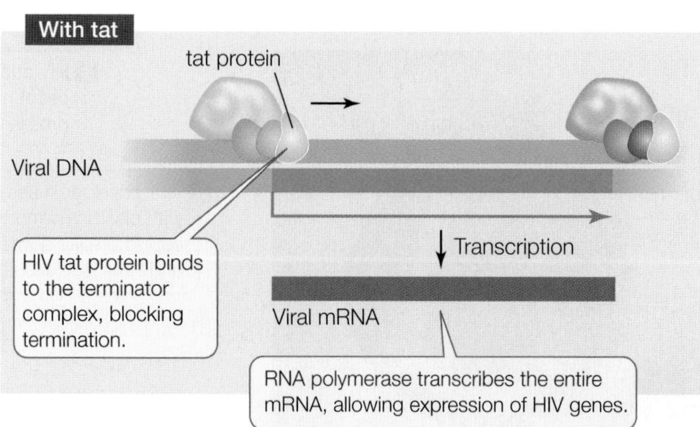


**16.7 Regulation of Transcription by HIV** The tat protein acts as an antiterminator, allowing transcription of the HIV genome.

## 16.1 RECAP

**Viruses are not cells. They consist of nucleic acids and a few proteins, and require a host cell to reproduce. In the lytic cycle, the viral genome directs the host cell to generate new virions along with proteins that cause the host cell to lyse. In the lysogenic cycle, viral DNA becomes integrated in the host's genome. This DNA is multiplied along with the host cells but may remain inactive for long periods. Special viral proteins that interact with host and viral DNA sequences are the keys to the regulation of viral gene expression.**

- What is the difference between positive and negative regulation of gene expression? See Figure 16.1
- What are the lytic and lysogenic cycles of bacteriophage? See p. 345 and Figure 16.4
- Describe positive and negative regulation of gene expression in bacteriophage and HIV life cycles. See pp. 346–347 and Figures 16.5 and 16.7

The environment surrounding prokaryotic cells can change abruptly, requiring rapid responses by the cell. We now turn to these responses, which often involve, as in viruses, the positive and negative regulation of gene expression by proteins binding to DNA.

# 16.2 How Is Gene Expression Regulated in Prokaryotes?

Prokaryotes conserve energy and resources by making certain proteins only when they are needed. The protein content of a bacterium can change rapidly when conditions warrant. There are several ways in which a prokaryotic cell can shut off the supply of an unneeded protein. The cell can:

- downregulate the transcription of mRNA for that protein;
- hydrolyze the mRNA after it is made, thereby preventing translation;
- prevent translation of the mRNA at the ribosome;
- hydrolyze the protein after it is made; or
- inhibit the function of the protein.

Whichever mechanism is used, it must be both responsive to environmental signals and efficient. The earlier the cell intervenes in the process of protein synthesis, the less energy it wastes. Selective blocking of transcription is far more efficient than transcribing the gene, translating the message, and then degrading or inhibiting the protein. While all five mechanisms for regulating protein levels are found in nature, prokaryotes generally use the most efficient one: transcriptional regulation.

### Regulating gene transcription conserves energy

As a normal inhabitant of the human intestine, *E. coli* must be able to adjust to sudden changes in its chemical environment. Its host may present it with one foodstuff one hour (e.g., glucose) and another the next (e.g., lactose). Such changes in nutrients present the bacterium with a metabolic challenge. Glucose is its preferred energy source, and is the easiest sugar to metabolize, but not all of its host's foods contain an abundant supply of glucose. For example, the bacterium may suddenly be deluged with milk, whose main sugar is lactose. Lactose is a β-galactoside—a disaccharide containing galactose β-linked to glucose (see Section 3.3). Three proteins are involved in the initial uptake and metabolism of lactose by *E. coli*:

- *β-galactoside permease* is a carrier protein in the bacterial plasma membrane that moves the sugar into the cell.

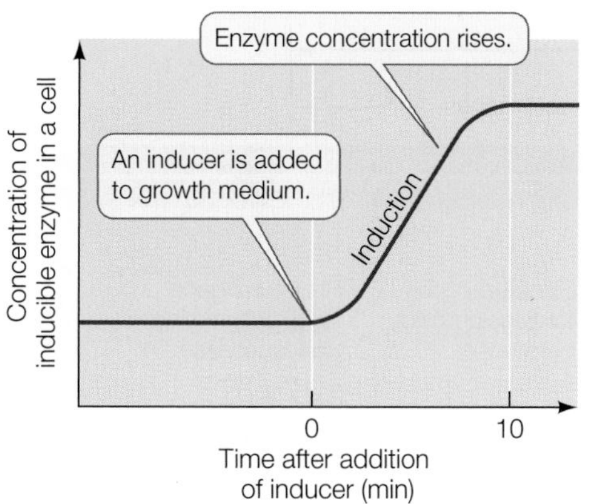


**16.8 An Inducer Stimulates the Expression of a Gene for an Enzyme** It is most efficient for a cell to produce an enzyme only when it is needed. Some enzymes are induced by the presence of the substance they act upon (for example, β-galactosidase is induced by the presence of lactose).

- *β-galactosidase* is an enzyme that hydrolyses lactose to glucose and galactose.
- *β-galactoside transacetylase* transfers acetyl groups from acetyl CoA to certain β-galactosides. Its role in the metabolism of lactose is not clear.

When *E. coli* is grown on a medium that contains glucose but no lactose or other β-galactosides, the levels of these three proteins are extremely low—the cell does not waste energy and materials making the unneeded enzymes. But if the environment changes such that lactose is the predominant sugar available and very little glucose is present, the bacterium promptly begins making all three enzymes. There are only two molecules of β-galactosidase present in an *E. coli* cell when glucose is present in the medium. But when glucose is absent, the presence of lactose can induce the synthesis of 3,000 molecules of β-galactosidase per cell!

If lactose is removed from *E. coli's* environment, synthesis of the three enzymes stops almost immediately. The enzyme molecules already present do not disappear; they are merely diluted during subsequent cell divisions until their concentration falls to the original low level within each bacterium.

Compounds that, like lactose, stimulate the synthesis of a protein are called **inducers** (**Figure 16.8**). The proteins that are produced are called **inducible proteins**, whereas proteins that are made all the time at a constant rate are called **constitutive proteins**. (Think of the constitution of a country, a document that does not change under normal circumstances.)

We have now seen two basic ways of regulating the rate of a metabolic pathway. In Section 8.5 we described allosteric regulation of enzyme activity (the rate of enzyme-catalyzed reactions); this mechanism allows rapid fine-tuning of metabolism. Regulation of protein synthesis—that is, regulation of the concentration of enzymes—is slower, but results in greater savings of energy and resources. Protein synthesis is a highly endergonic process, since assembling mRNA, charging tRNA, and moving the ribosomes along mRNA all require the hydrolysis of ATP. **Figure 16.9** compares these two modes of regulation.

### Operons are units of transcriptional regulation in prokaryotes

The genes that encode the three enzymes for processing lactose in *E. coli* are **structural genes**; they specify the primary structure (the amino acid sequence) of a protein molecule. Structural genes are genes that can be transcribed into mRNA.

The three structural genes involved in the metabolism of lactose lie adjacent to one another on the *E. coli* chromosome. This arrangement is no coincidence: the genes share a single promoter, and their DNA is transcribed into a single, continuous molecule of mRNA. Because this particular mRNA governs the synthesis of all three lactose-metabolizing enzymes, either all or none of these enzymes are made, depending on whether their common message—their mRNA—is present in the cell.

A cluster of genes with a single promoter is called an **operon**, and the operon that encodes the three lactose-metabolizing enzymes in *E. coli* is called the *lac operon*. The *lac* operon promoter can be very efficient (the maximum rate of mRNA synthesis can be high) but mRNA synthesis can be shut down when the enzymes are not needed. This example of negative regulation was elegantly worked out by Nobel Prize winners François Jacob and Jacques Monod.

In addition to the promoter, an operon has other regulatory sequences that are not transcribed. A typical operon consists of a promoter, an operator, and two or more

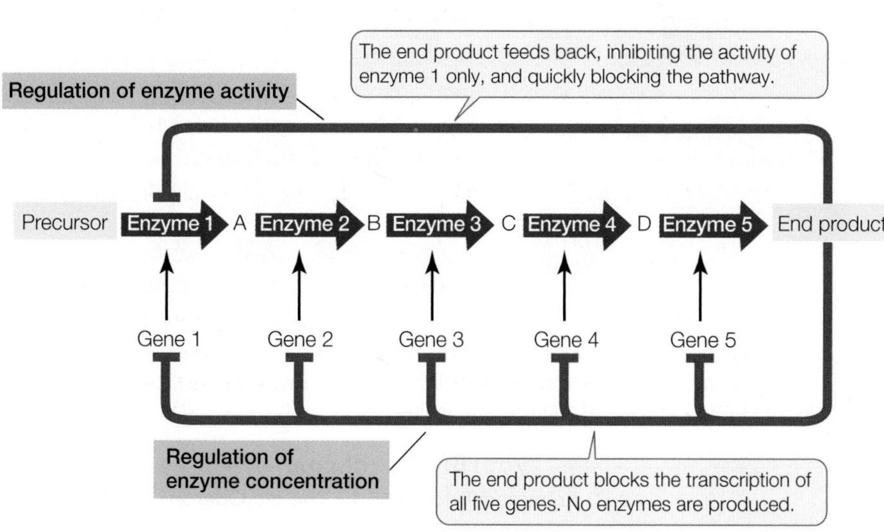


**16.9 Two Ways to Regulate a Metabolic Pathway** Feedback from the end product of a metabolic pathway can block enzyme activity (allosteric regulation), or it can stop the transcription of genes that code for the enzymes in the pathway (transcriptional regulation).

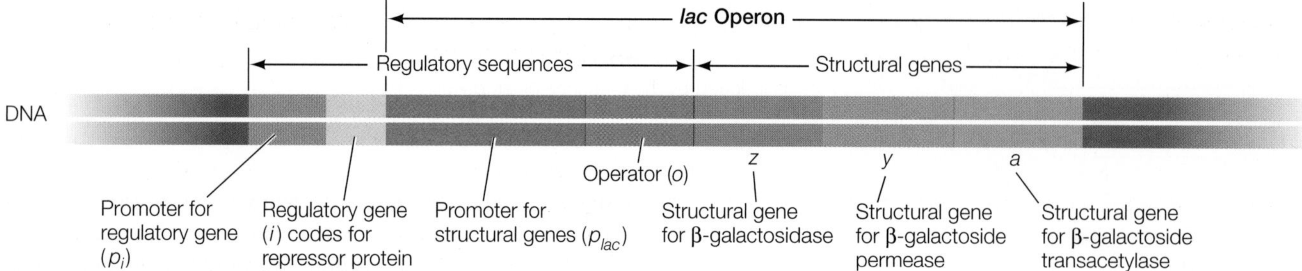


**16.10 The *lac* Operon of *E. coli*** The *lac* operon of *E. coli* is a segment of DNA that includes a promoter, an operator, and the three structural genes that code for lactose-metabolizing enzymes.

structural genes (**Figure 16.10**). The **operator** is a short stretch of DNA that lies between the promoter and the structural genes. It can bind very tightly with regulatory proteins that either activate or repress transcription. There are numerous mechanisms to control the transcription of operons; here we will focus on three examples:

- An inducible operon regulated by a repressor protein
- A repressible operon regulated by a repressor protein
- An operon regulated by an activator protein

### Operator–repressor interactions control transcription in the *lac* and *trp* operons

The *lac* operon contains a promoter, to which RNA polymerase binds to initiate transcription, and an operator, to which a **repressor** protein can bind. When the repressor is bound, transcription of the operon is blocked.

The repressor protein has two binding sites: one for the operator and the other for the inducer, lactose. Binding with the inducer changes the shape of the repressor protein. This change in three-dimensional structure (conformation) prevents the repressor from binding to the operator (**Figure 16.11**). As a result, RNA polymerase can bind to the promoter and start transcribing the structural genes of the *lac* operon.

Study Figure 16.11 for the features of this negative control. You will notice that:

- in the absence of inducer, the operon is turned off;
- control is exerted by a regulatory protein—the repressor—that turns the operon off;
- the inducer, when present, binds to and changes the shape of the repressor so that it no longer binds to the operator, turning the operon on;

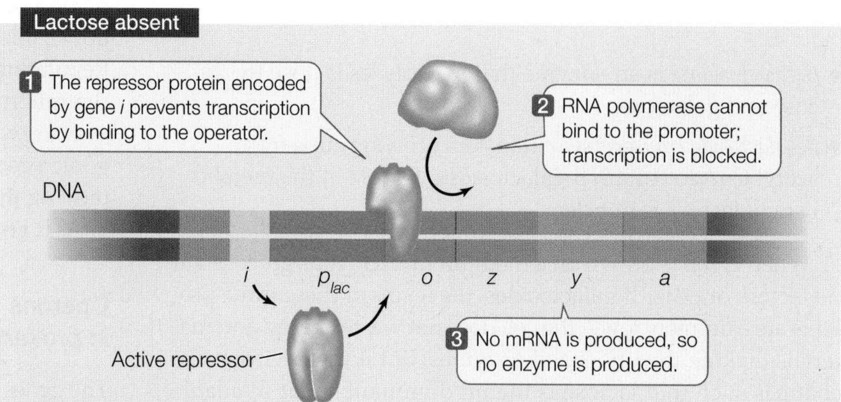


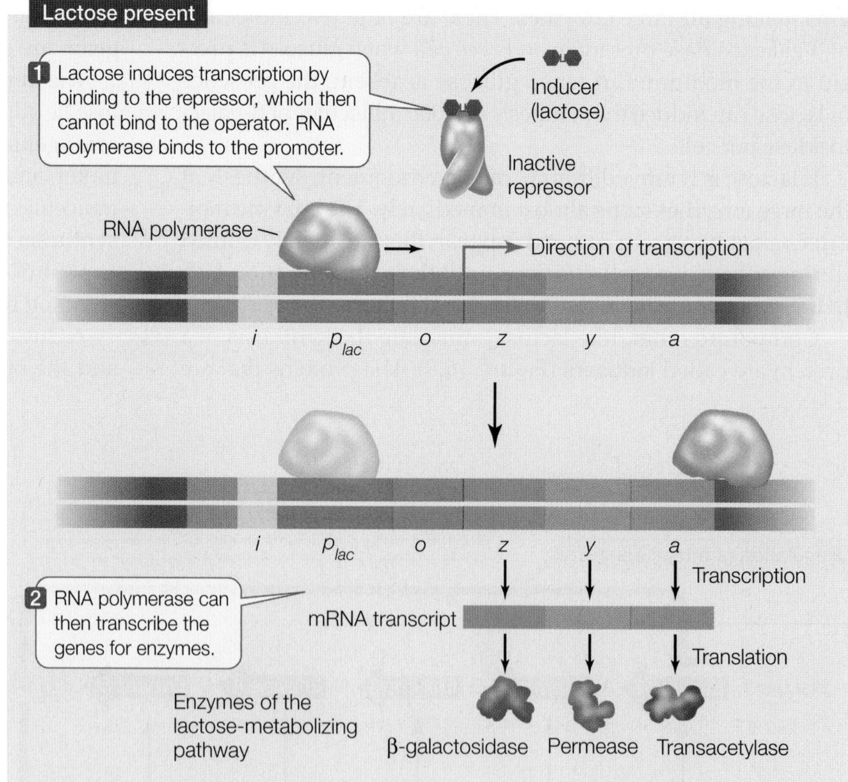


**16.11 The *lac* Operon: An Inducible System** Lactose (the inducer) leads to synthesis of the enzymes in the lactose-metabolizing pathway by binding to the repressor protein and preventing its binding to the operator.

yourBioPortal.com
GO TO Animated Tutorial 16.1 • The *lac* Operon

- the **regulatory gene** produces a protein whose sole function is to regulate expression of the other genes; and
- certain DNA sequences (operators and promoters) do not code for proteins, but are binding sites for regulatory or other proteins.

In contrast to the inducible system of the *lac* operon, other operons in *E. coli* are repressible; that is, they are repressed when molecules called **co-repressors** bind to their repressors. This binding causes the repressor to change shape and bind to the operator, thereby inhibiting transcription.

An example is the operon whose structural genes catalyze the synthesis of the amino acid tryptophan:

$$\text{precursor molecules} \xrightarrow{\text{5 enzyme-catalyzed reactions}} \rightarrow \rightarrow \rightarrow \rightarrow \text{tryptophan}$$

When tryptophan is present in the cell in adequate concentrations, it is advantageous to stop making the enzymes for tryptophan synthesis. To do this, the cell uses a repressor that binds to an operator upstream of the genes of the *trp* operon. But the repressor of the *trp* operon is not normally bound to the operator; it only binds when its shape is changed by binding to tryptophan, the co-repressor. To summarize the differences between these two types of operons:

- In *inducible* systems, the substrate of a metabolic pathway (the inducer) interacts with a regulatory protein (the repressor), rendering the repressor incapable of binding to the operator and thus allowing transcription.
- In *repressible* systems, the product of a metabolic pathway (the co-repressor) binds to a regulatory protein, which is then able to bind to the operator and block transcription.

**yourBioPortal.com**
GO TO Animated Tutorial 16.2 • The *trp* Operon

In general, inducible systems control catabolic pathways (which are turned on only when the substrate is available), whereas repressible systems control anabolic pathways (which are turned on until the concentration of the product becomes excessive). In both of the systems described here, the regulatory protein is a repressor that functions by binding to the operator. Next we will consider an example of positive control involving an activator.

## Protein synthesis can be controlled by increasing promoter efficiency

The examples described in the previous section are termed negative control because transcription is *decreased* in the presence of a repressor protein. *E. coli* can also use positive control to *increase* transcription through the presence of an **activator** protein. For an example we return to the *lac* operon, where the relative levels of glucose and lactose determine the amount of transcription. When lactose is present and glucose is low, the *lac* operon is activated by binding of a protein called *c*AMP *r*eceptor *p*rotein (CRP) to the *lac* operon promoter. CRP is an activator of transcription, because its binding results in more efficient binding of RNA polymerase to the promoter, and thus increased transcription of the structural genes (**Figure 16.12**).

In the presence of abundant glucose, CRP does not bind to the promoter and so the efficiency of transcription of the *lac* operon is reduced. This is an example of **catabolite repression**, a system of gene regulation in which the presence of the preferred energy source represses other catabolic pathways. The signaling pathway that controls catabolite repression of the *lac* operon involves the second messenger cAMP (see Section 7.3). The mechanisms controlling positive and negative regulation of the *lac* operon are summarized in **Table 16.1**.

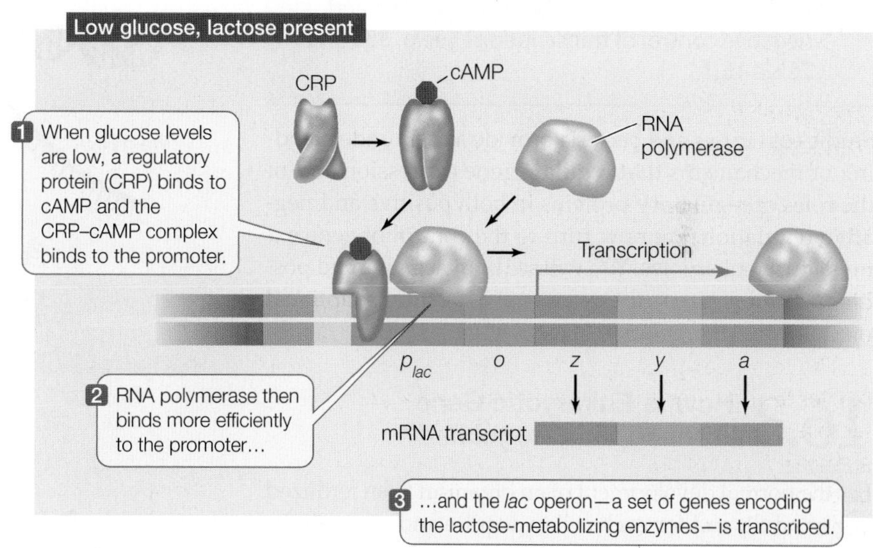


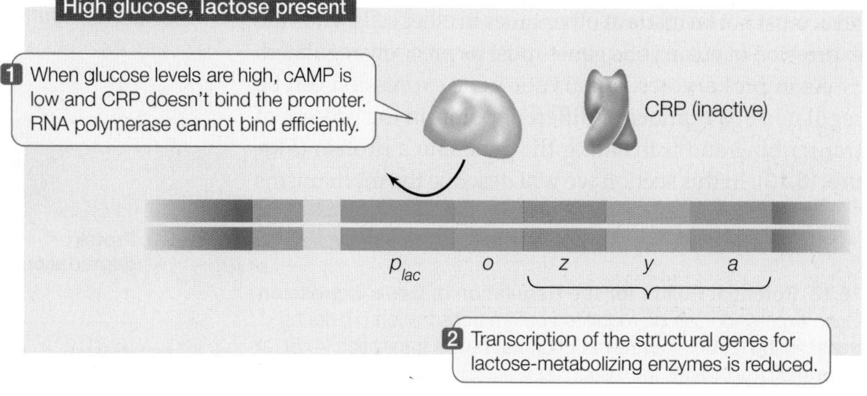


**16.12 Catabolite Repression Regulates the *lac* Operon** The promoter for the *lac* operon does not function efficiently in the absence of cAMP, as occurs when glucose levels are high. High glucose levels thus repress the enzymes that metabolize lactose.

## TABLE 16.1

### Positive and Negative Regulation in the *lac* Operon[a]

| GLUCOSE | cAMP LEVELS | RNA POLYMERASE BINDING TO PROMOTER | LACTOSE | LAC REPRESSOR | TRANSCRIPTION OF *lac* GENES? | LACTOSE USED BY CELLS? |
|---|---|---|---|---|---|---|
| Present | Low | Absent | Absent | Active and bound to operator | No | No |
| Present | Low | Present, not efficient | Present | Inactive and not bound to operator | Low level | No |
| Absent | High | Present, very efficient | Present | Inactive and not bound to operator | High level | Yes |
| Absent | High | Absent | Absent | Active and bound to operator | No | No |

[a]Negative regulators are in red type.

## 16.2 RECAP

**Gene expression in prokaryotes is most commonly regulated through control of transcription. An operon consists of a set of closely linked structural genes and the DNA sequences (promoter and operator) that control their transcription. Operons can be regulated by both negative and positive controls.**

- Describe the molecular conditions at the *lac* operon promoter in the presence versus absence of lactose. **See Figure 16.11**
- What are the key differences between an inducible system and a repressible system? **See p. 351**
- What are the differences between positive and negative control of transcription? **See p. 351 and Table 16.1**

Studies of viruses and bacteria provide a basic understanding of mechanisms that regulate gene expression and of the roles of regulatory proteins in both positive and negative regulation. We now turn to the control of gene expression in eukaryotes. You will see both negative and positive control of transcription, as well as posttranscriptional mechanisms of regulation.

# 16.3 How Is Eukaryotic Gene Transcription Regulated?

For the normal development of an organism from fertilized egg to adult, and for each cell to acquire and maintain its proper specialized function, certain proteins must be made at just the right times and in just the right cells; these proteins must not be made at other times in other cells. Thus the expression of eukaryotic genes must be precisely regulated.

As in prokaryotes, eukaryotic gene expression can be regulated at a number of different points in the process of transcribing and translating the gene into a protein (**Figure 16.13**). In this section we will describe the mechanisms

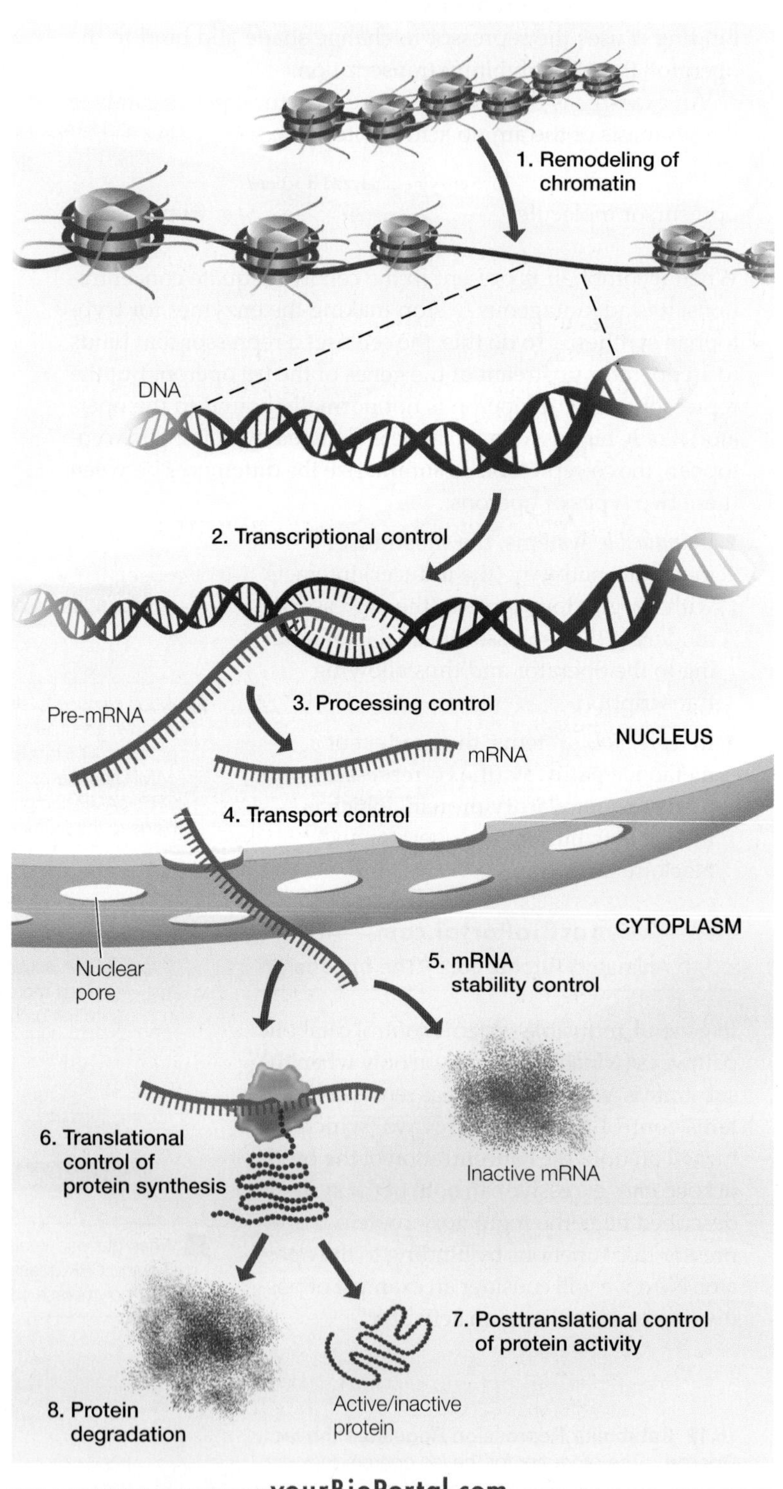


**16.13 Potential Points for the Regulation of Gene Expression** Gene expression can be regulated before transcription (1), during transcription (2, 3), after transcription but before translation (4, 5), at translation (6), or after translation (7).

yourBioPortal.com
GO TO Web Activity 16.1 • Eukaryotic Gene Expression Control Points

TABLE 16.2

**Transcription in Prokaryotes and Eukaryotes**

| | PROKARYOTES | EUKARYOTES |
|---|---|---|
| Locations of functionally related genes | Often clustered in operons | Often distant from one another with separate promoters |
| RNA polymerases | One | Three:<br>I transcribes rRNA<br>II transcribes mRNA<br>III transcribes tRNA and small RNAs |
| Promoters and other regulatory sequences | Few | Many |
| Initiation of transcription | Binding of RNA polymerase to promoter | Binding of many proteins, including RNA polymerase |

that result in the selective transcription of specific genes. The mechanisms for regulating gene expression in eukaryotes have similar themes to those of prokaryotes. Both types of cells use DNA–protein interactions and negative and positive control. However, there are many differences, some of them dictated by the presence of a nucleus, which physically separates transcription and translation (**Table 16.2**).

### Transcription factors act at eukaryotic promoters

As in prokaryotes, a promoter in eukaryotes is a sequence of DNA near the 5′ end of the coding region of a gene where RNA polymerase binds and initiates transcription. There are typically two important sequences in a promoter: One is the **recognition sequence**—the sequence recognized by RNA polymerase. The second, closer to the transcription initiation site, is the **TATA box** (so called because it is rich in AT base pairs), where DNA begins to denature so that the template strand can be exposed.

Eukaryotic RNA polymerase II cannot simply bind to the promoter and initiate transcription. Rather, it does so only after various regulatory proteins, called **transcription factors**, have assembled on the chromosome (**Figure 16.14**). First, the protein TFIID ("TF" stands for transcription factor) binds to the TATA box. Binding of TFIID changes both its own shape and that of the DNA, presenting a new surface that attracts the binding of other transcription factors to form a transcription complex. RNA polymerase II binds only after several other proteins have bound to this complex.

Some regulatory DNA sequences, such as the TATA box, are common to the promoters of many eukaryotic genes and are recognized by transcription factors that are found in all the cells of an organism. Other sequences found in promoters are specific

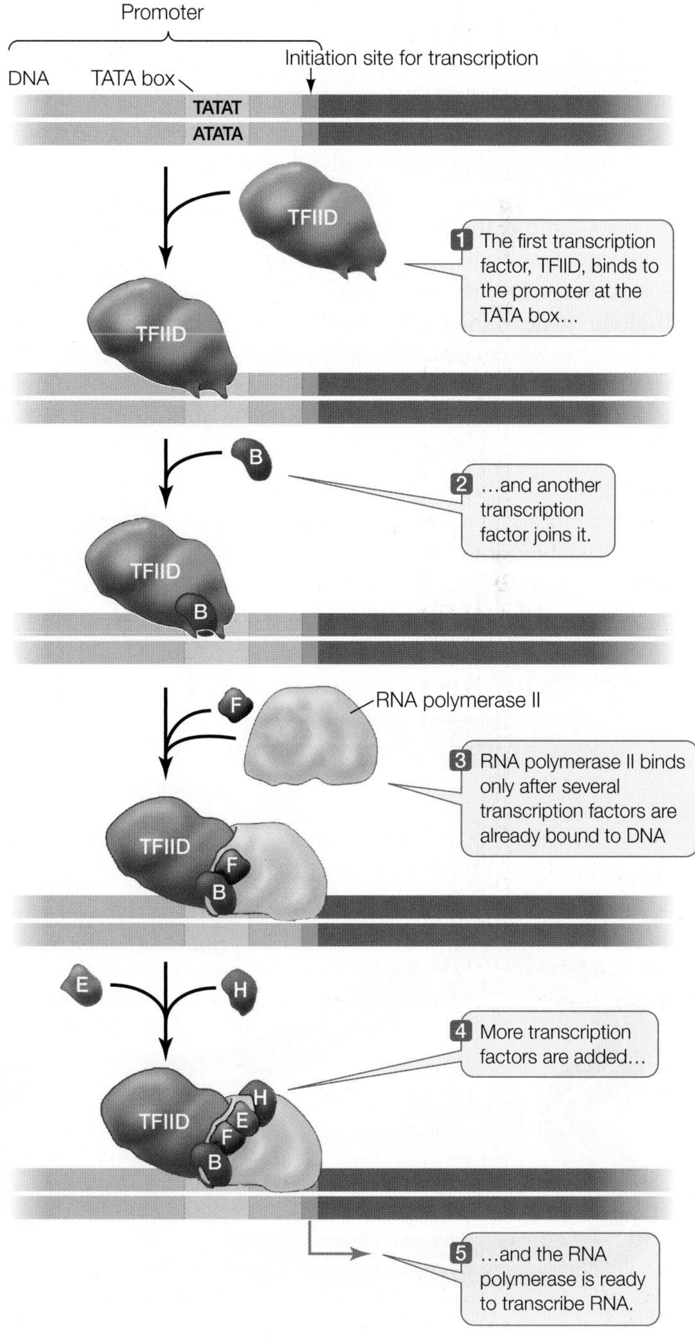
Promoter
Initiation site for transcription
DNA
TATA box
TATAT
ATATA
TFIID
1 The first transcription factor, TFIID, binds to the promoter at the TATA box…
B
2 …and another transcription factor joins it.
F
RNA polymerase II
3 RNA polymerase II binds only after several transcription factors are already bound to DNA
E
H
4 More transcription factors are added…
5 …and the RNA polymerase is ready to transcribe RNA.

**16.14 The Initiation of Transcription in Eukaryotes** Apart from TFIID, which binds to the TATA box, each transcription factor in this transcription complex has binding sites only for the other proteins in the complex, and does not bind directly to DNA. B, E, F, and H are transcription factors.

**yourBioPortal.com**

GO TO **Animated Tutorial 16.3 • Initiation of Transcription**

to only a few genes and are recognized by transcription factors found only in certain types of cells. These specific transcription factors play an important role in cell differentiation, the structural and functional specialization of cells during development.

### Other proteins can recognize and bind to DNA sequences and regulate transcription

In addition to the promoter, there are other short sequences (elements) of DNA that bind regulatory proteins, which in turn interact with RNA polymerase to regulate the rate of transcription (**Figure 16.15**). Some of these DNA elements are positive regulators (termed enhancers, which bind activator proteins) and others are negative (silencers, which bind repressor proteins). Some occur near the promoter and others as far as 20,000 base pairs away. One example of a transcription factor is CREB, which you read about in the opening essay of this chapter. When the activators and/or repressors (collectively termed transcription factors) bind to these elements, they interact with the RNA polymerase complex, causing DNA to bend. Often many such binding proteins are involved, and *the combination of factors present determines the rate of transcription.*

For example, the immature red blood cells in bone marrow make large amounts of β-globin. At least thirteen different transcription factors are involved in regulating transcription of the β-globin gene in these cells. Not all of these factors are present or active in other cells, such as the immature white blood cells produced by the same bone marrow. As a result the β-globin gene is not transcribed in those cells. So although the same genes are present in all cells, the fate of the cell is determined by which of its genes are expressed. How do transcription factors recognize specific DNA sequences?

### Specific protein–DNA interactions underlie binding

As we have seen, transcription factors with specific DNA binding domains are involved in the activation and inactivation of specific genes. There are four common structural themes in the protein domains that bind to DNA. These themes, or **structural motifs**, consist of different combinations of structural elements (protein conformations) and may include special components such as zinc. The four common structural motifs in DNA binding domains are: helix-turn-helix, leucine zipper, zinc finger, and helix-loop-helix (**Figure 16.16**).

Let's look at how one of these motifs works. As pointed out in Section 13.2, the complementary bases in DNA not only form hydrogen bonds with each other, but also can form additional hydrogen bonds with proteins, particularly at points exposed in the major and minor grooves. In this way, an intact DNA double helix can be recognized by a protein motif whose structure:

- fits into the major or minor groove;
- has amino acids that can project into the interior of the double helix; and
- has amino acids that can form hydrogen bonds with the interior bases.

The helix-turn-helix motif, in which two α-helices are connected via a non-helical turn, fits these three criteria. The interior-facing "recognition" helix is the one whose amino acids interact with the bases inside the DNA. The exterior-facing helix sits on the sugar–phosphate backbone, ensuring that the interior helix is presented to the bases in the correct configuration. Many repressor proteins have this helix-turn-helix motif in their structure.

Repressors can inhibit transcription in several different ways. They can prevent the binding of transcriptional activators to DNA, or they can interact with other DNA binding proteins to decrease the rate of transcription.

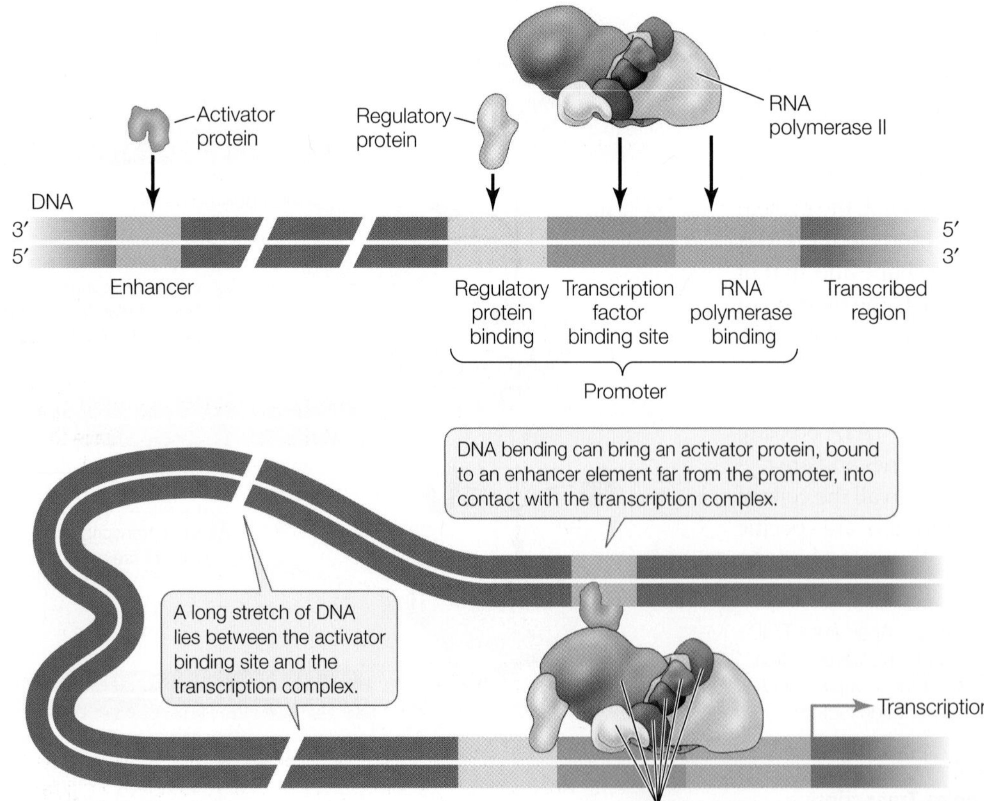


**16.15 Transcription Factors, Repressors, and Activators** The actions of many proteins determine whether and where RNA polymerase II will transcribe DNA.

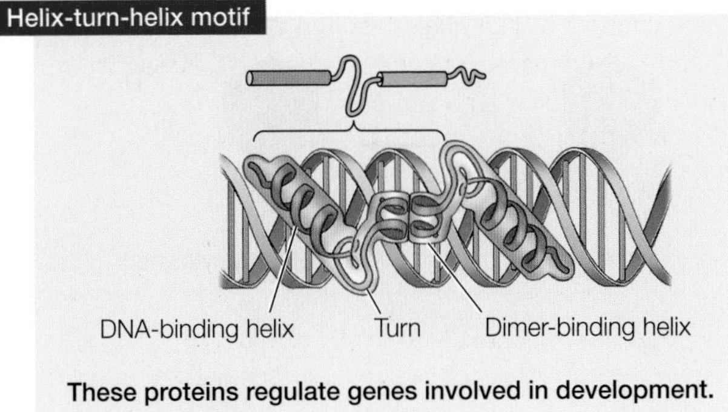
Helix-turn-helix motif
DNA-binding helix
Turn
Dimer-binding helix
These proteins regulate genes involved in development.

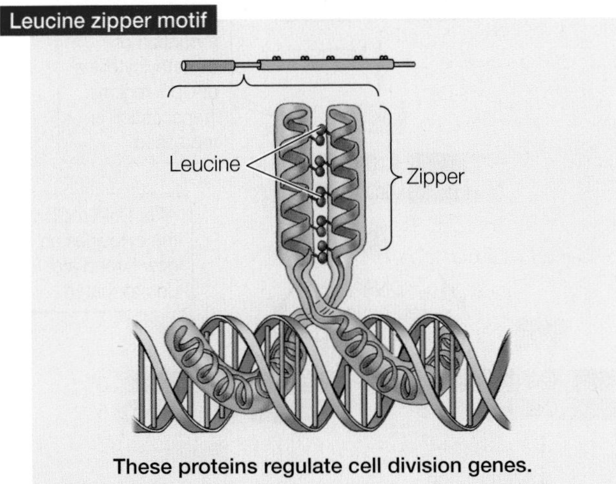
Leucine zipper motif
Leucine
Zipper
These proteins regulate cell division genes.

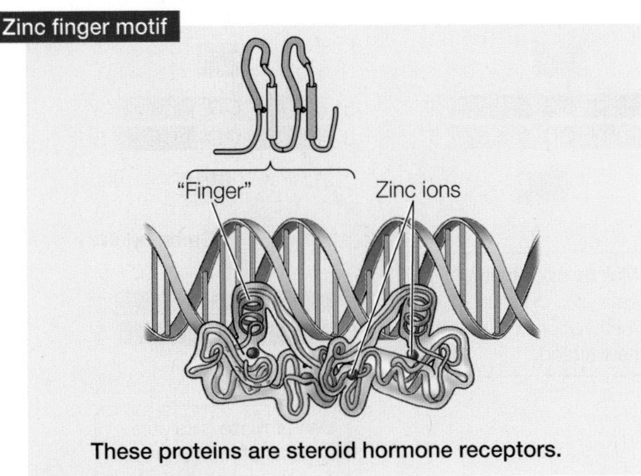
Zinc finger motif
"Finger"
Zinc ions
These proteins are steroid hormone receptors.

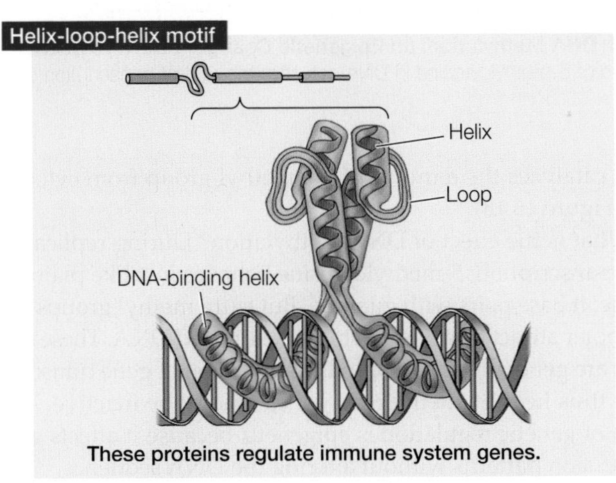
Helix-loop-helix motif
Helix
Loop
DNA-binding helix
These proteins regulate immune system genes.

**16.16 Protein–DNA Interactions** The DNA-binding domains of most regulatory proteins contain one of four structural motifs.

### The expression of sets of genes can be coordinately regulated by transcription factors

How do eukaryotic cells coordinate the regulation of several genes whose transcription must be turned on at the same time? Prokaryotes solve this problem by arranging multiple genes in an operon that is controlled by a single promoter. But most eukaryotic genes have their own separate promoters, and genes that are coordinately regulated may be far apart. In these cases, the expression of genes can be coordinated if they share regulatory sequences that bind the same transcription factors.

This type of coordination is used by organisms to respond to stress—for example, by plants in response to drought. Under conditions of drought stress, a plant must simultaneously synthesize a number of proteins whose genes are scattered throughout the genome. The synthesis of these proteins comprises the stress response. To coordinate expression, each of these genes has a specific regulatory sequence near its promoter called the *stress response element* (*SRE*). A transcription factor binds to this element and stimulates mRNA synthesis (**Figure 16.17**). The stress re-

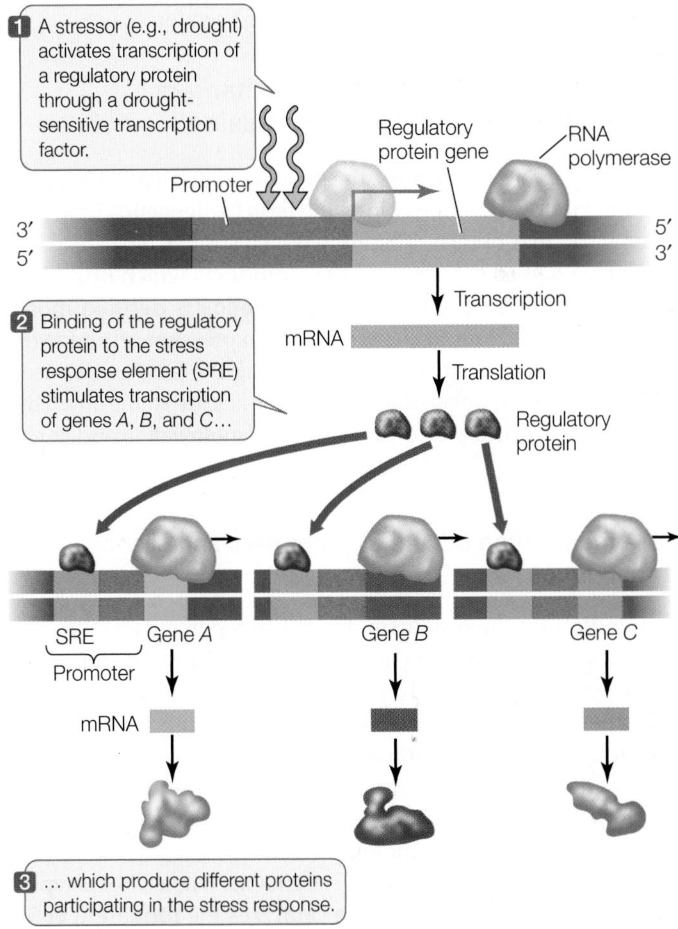
1 A stressor (e.g., drought) activates transcription of a regulatory protein through a drought-sensitive transcription factor.
Promoter
Regulatory protein gene
RNA polymerase
3′
5′
5′
3′
Transcription
2 Binding of the regulatory protein to the stress response element (SRE) stimulates transcription of genes A, B, and C…
mRNA
Translation
Regulatory protein
SRE
Promoter
Gene A
Gene B
Gene C
mRNA
3 … which produce different proteins participating in the stress response.

**16.17 Coordinating Gene Expression** A single environmental signal, such as drought stress, causes the synthesis of a transcriptional regulatory protein that acts on many genes.

sponse proteins not only help the plant conserve water, but also protect the plant against excess salt in the soil and freezing. This finding has considerable importance for agriculture because crops are often grown under less than optimal conditions.

## 16.3 RECAP

**A number of transcription factors must bind to a eukaryotic promoter before RNA polymerase will bind to it and begin transcription. This provides a number of ways to increase or decrease transcription.**

- Describe some of the different ways in which transcription factors regulate gene transcription. See pp. 353–354 and Figure 16.15
- How can more than one gene be regulated at the same time? See p. 355 and Figure 16.17

The mechanisms for control of gene expression that we have discussed so far involve direct interactions between proteins and specific DNA elements. If the sequences of the DNA elements are altered, then transcription of the gene will be affected. However, there are other mechanisms for controlling gene expression that do not depend on specific DNA sequences. We will discuss these mechanisms in the next section.

## 16.4 How Do Epigenetic Changes Regulate Gene Expression?

In the mid-twentieth century, the great developmental biologist Conrad Hal Waddington coined the term "epigenetics" and defined it as "that branch of biology which studies the causal interactions between genes and their products which bring the phenotype into being." Today **epigenetics** is defined more specifically, referring to changes in the expression of a gene or set of genes that occur without changing the DNA sequence. These changes are reversible, but sometimes are stable and heritable. They include two processes: DNA **methylation** and chromosomal protein alterations.

### DNA methylation occurs at promoters and silences transcription

Depending on the organism, from 1 to 5 percent of cytosine residues in the DNA are chemically modified by the addition of a methyl group (—$CH_3$) to the 5′-carbon, to form 5-methylcytosine (**Figure 16.18**). This covalent addition is catalyzed by the enzyme **DNA methyltransferase** and, in mammals, usually occurs in C residues that are adjacent to G residues. DNA regions rich in these doublets are called **CpG islands**, and are especially abundant in promoters.

This covalent change in DNA is heritable: when DNA is replicated, a **maintenance methylase** catalyzes the formation of 5-methylcytosine in the new DNA strand. However, the pattern of cytosine methylation can also be altered, because methylation is reversible: a third enzyme, appropriately called **demethylase**, catalyzes the removal of the methyl group from cytosine (see Figure 16.18).

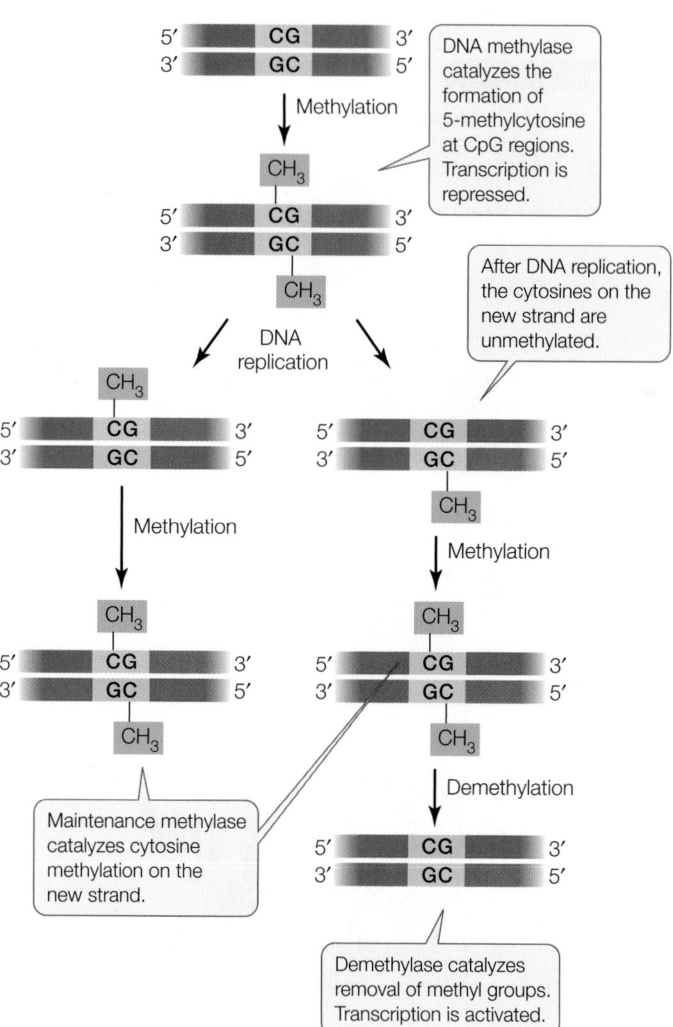


**16.18 DNA Methylation: an Epigenetic Change** The reversible formation of 5-methylcytosine in DNA can alter the rate of transcription.

What is the effect of DNA methylation? During replication and transcription, 5-methylcytosine behaves just like plain cytosine: it base pairs with guanine. But extra methyl groups in a promoter attract proteins that bind methylated DNA. These proteins are generally involved in the repression of gene transcription; thus heavily methylated genes tend to be inactive. This form of genetic regulation is epigenetic because it affects gene expression patterns without altering the DNA sequence.

DNA methylation is important in development from egg to embryo. For example, when a mammalian sperm enters an egg, many genes in first the male and then the female genome become demethylated. Thus many genes that are usually inactive are expressed during early development. As the embryo develops and its cells become more specialized, genes whose products are not needed in particular cell types become methylated. These methylated genes are "silenced"; their transcription is repressed. However, unusual or abnormal events can sometimes turn silent genes back on.

For example, DNA methylation may play roles in the genesis of some cancers. In cancer cells, oncogenes get activated and promote cell division, and tumor suppressor genes (that normally inhibit cell division) are turned off (see Chapter 11). This misregulation can occur when the promoters of oncogenes become demethylated while those of tumor suppressor genes become methylated. This is the case in colorectal cancer.

### Histone protein modifications affect transcription

Another mechanism for epigenetic gene regulation is the alteration of chromatin structure, or *chromatin remodeling*. DNA is packaged with histone proteins into nucleosomes, which can make DNA physically inaccessible to RNA polymerase and the rest of the transcription apparatus. Each histone protein has a "tail" of approximately 20 amino acids at its N terminus that sticks out of the compact structure and contains certain positively charged amino acids (notably lysine). Enzymes called histone acetyltransferases can add acetyl groups to these positively charged amino acids, thus changing their charges:

$$\cdots\mathrm{NH{-}CH[(CH_2)_3NH_3^+]{-}C(=O)}\cdots + \mathrm{CoA{-}S{-}C(=O){-}CH_3} \longrightarrow \cdots\mathrm{NH{-}CH[(CH_2)_3NH{-}C(=O){-}CH_3]{-}C(=O)}\cdots + \mathrm{CoA{-}SH}$$

Lysine in histone — Acetyl-CoA — Acetyl-lysine

Ordinarily, there is strong electrostatic attraction between the positively charged histone proteins and DNA, which is negatively charged because of its phosphate groups. Reducing the positive charges of the histone tails reduces the affinity of the histones for DNA, opening up the compact nucleosome. Additional chromatin remodeling proteins can bind to the loosened nucleosome–DNA complex, opening up the DNA for gene expression (**Figure 16.19**). Histone acetyltransferases can thus activate transcription.

Another kind of chromatin remodeling protein, histone deacetylase, can remove the acetyl groups from histones and

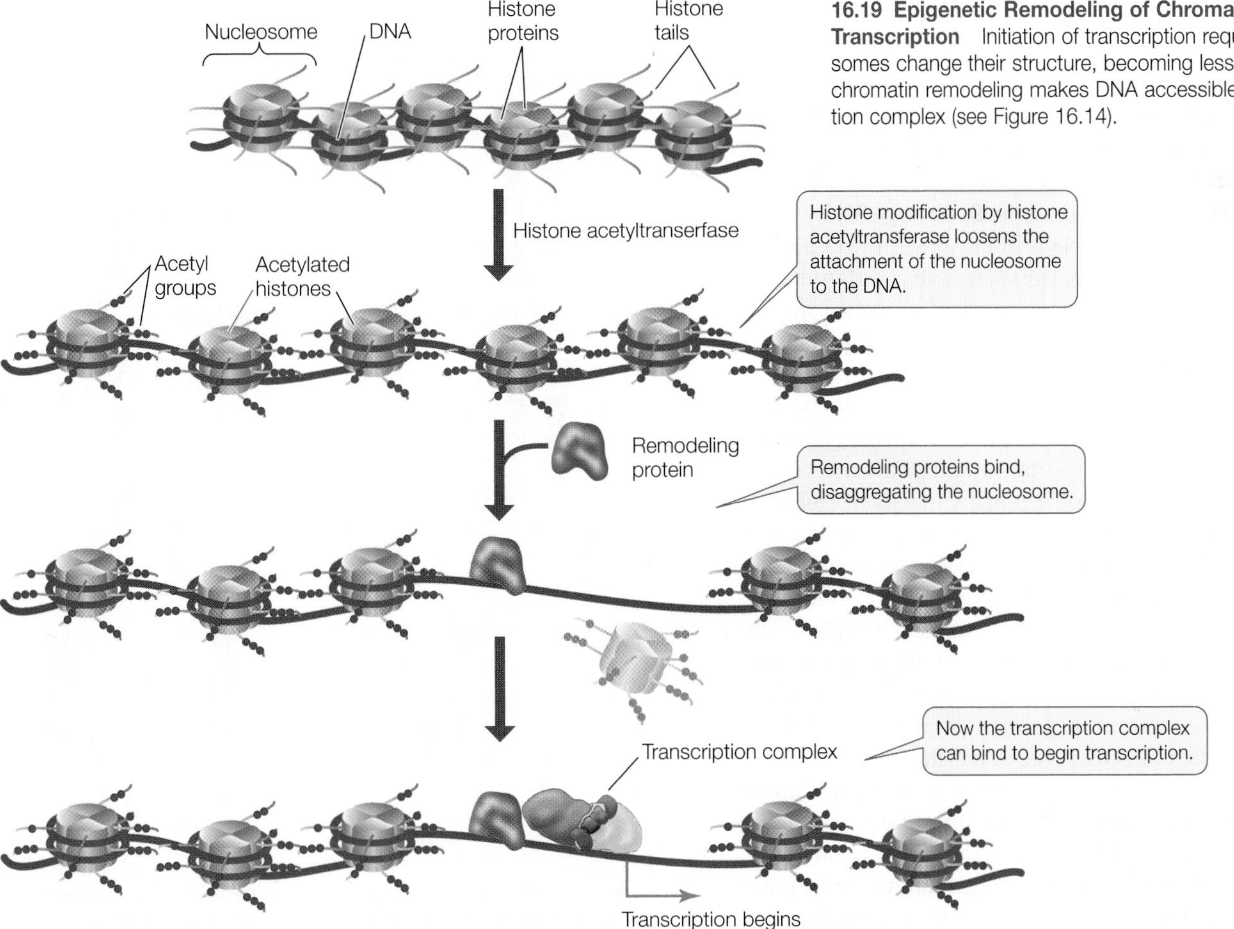
Nucleosome
DNA
Histone proteins
Histone tails
Histone acetyltranserfase
Acetyl groups
Acetylated histones
Histone modification by histone acetyltransferase loosens the attachment of the nucleosome to the DNA.
Remodeling protein
Remodeling proteins bind, disaggregating the nucleosome.
Transcription complex
Now the transcription complex can bind to begin transcription.
Transcription begins

**16.19 Epigenetic Remodeling of Chromatin for Transcription** Initiation of transcription requires that nucleosomes change their structure, becoming less compact. This chromatin remodeling makes DNA accessible to the transcription complex (see Figure 16.14).

thereby repress transcription. Histone deacetylases are targets for drug development to treat some forms of cancer. As noted above, certain genes block cell division in normal specialized tissues. In some cancers these genes are less active than normal, and the histones near them show excessive levels of deacetylation. Theoretically, a drug acting as a histone deacetylase inhibitor could tip the balance toward acetylation and this might activate genes that normally inhibit cell division.

Other types of histone modification can affect gene activation and repression. For example, histone methylation is associated with gene inactivation and histone phosphorylation also affects gene expression, the specific effect depending on which amino acid is modified. All of these effects are reversible and so the activity of a eukaryotic gene may be determined by very complex patterns of histone modification. David Allis of the Rockefeller University in New York City has dubbed this epigenetic system the "histone code."

### Epigenetic changes induced by the environment can be inherited

Despite that fact that they are reversible, many epigenetic changes such as DNA methylation and histone modification can permanently alter gene expression patterns in a cell. If the cell is a germ line cell that forms gametes, the epigenetic changes can be passed on to the next generation. But what determines these epigenetic changes? A clue comes from a recent study of monozygotic twins.

Monozygotic twins come from a single fertilized egg that divides to produce two separate cells; each of these goes on to develop a separate individual. Twin brothers or sisters thus have identical genomes. But are they identical in their *epigenomes*? A comparison of DNA in hundreds of such twin pairs shows that in tissues of three-year-olds, the DNA methylation patterns are virtually the same. But by age 50, by which time the twins have usually been living apart for decades, in different environments, the patterns are quite different. This indicates that the *environment plays an important role in epigenetic modifications* and, therefore, in the regulation of genes that these modifications affect.

What factors in the environment lead to epigenetic changes? One might be stress: when mice are put in a stressful situation, genes that are involved in important brain pathways become heavily methylated (and transcriptionally inactive). Treatment of the stressed mice with an antidepressant drug "hits the undo button," reversing these changes. Transcription factors such as CREB that mediate addiction (see the opening story of this chapter) are involved with histone acetylation, which leads to subsequent gene activation. The sperm of men with psychosis have different methylation patterns than sperm from nonpsychotic men. This last observation is especially provocative, as it suggests that epigenetic patterns, some of which may have formed during life, can be passed on to the next generation. This means that some phenotypic characteristics acquired during the lifetime of an organism might be heritable, contrary to biologists' long-held views. The idea that epigenetic changes can be inherited remains controversial.

### DNA methylation can result in genomic imprinting

In mammals specific patterns of methylation develop for each sex during gamete formation. This happens in two stages: first, the existing methyl groups are removed from the 5-methylcytosines by a demethylase, and then a DNA methylase adds methyl groups to the appropriate cytosines. When the gametes form they carry this new pattern of methylation (epigenetic information).

The DNA methylation pattern in male gametes (sperm) differs from that in female gametes (eggs) at about 200 genes in the mammalian genome. That is, a given gene in this group may be methylated in eggs but unmethylated in sperm (**Figure 6.20**). In this case the offspring would inherit a maternal gene that is transcriptionally inactive (methylated) and a paternal gene that is transcriptionally active (demethylated). This is called **genomic imprinting**.

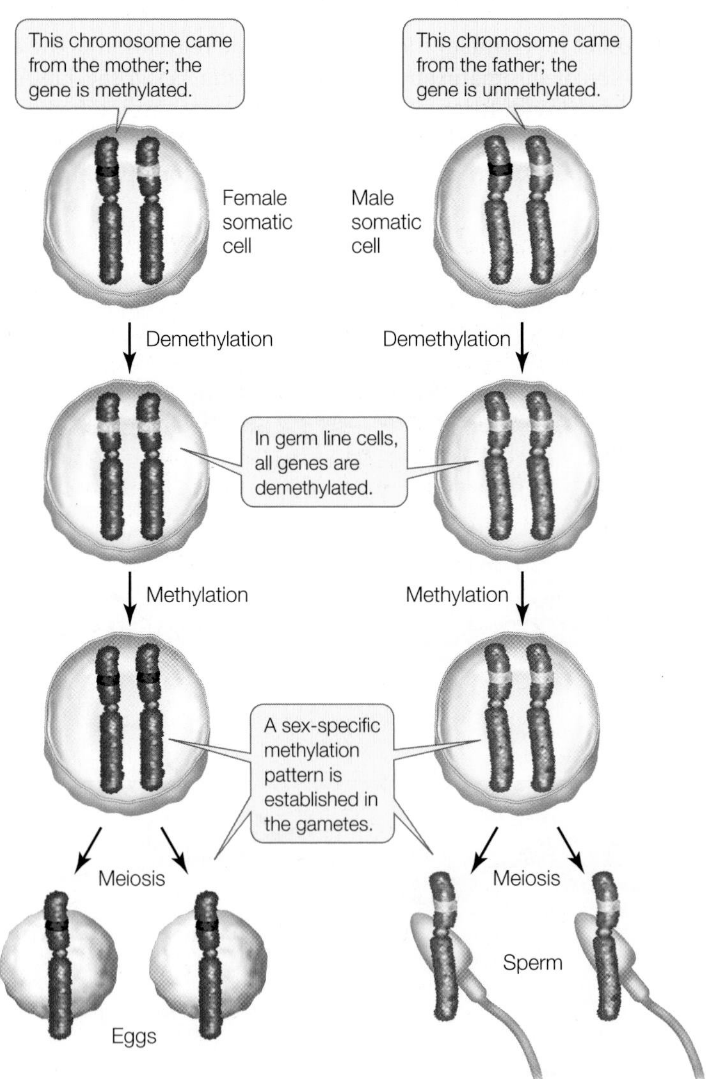


**16.20 Genomic Imprinting** For some genes, epigenetic DNA methylation differs in male and female gametes. As a result, an individual might inherit an allele from the female parent that is transcriptionally silenced; but the same allele from the male parent would be expressed.

An example of imprinting is found in a region on human chromosome 15 called 15q11. This region is imprinted differently during the formation of male and female gametes, and offspring normally inherit both the paternally and maternally derived patterns. In rare cases, there is a chromosome deletion in one of the gametes, and the newborn baby inherits just the male or the female imprinting pattern in this particular chromosome region. If the male pattern is the only one present (female region deleted), the baby develops Angelman Syndrome, characterized by epilepsy, tremors, and constant smiling. If the female pattern is the only one present (male region deleted), the baby develops a quite different phenotype called Prader-Willi syndrome, marked by muscle weakness and obesity. Note that the *gene sequences are the same* in both cases: it is *the epigenetic patterns that are different.*

Imprinting of specific genes occurs primarily in mammals and flowering plants. Most imprinted genes are involved with embryonic development. An embryo must have both the paternally and maternally imprinted gene patterns to develop properly. In fact, attempts to make an embryo that has chromosomes from only one sex (for example, by chemically treating an egg cell to double its chromosomes) usually fail. So imprinting has an important lesson for genetics: *males and females may be the same genetically (except for the X and Y chromosomes), but they differ epigenetically.*

### Global chromosome changes involve DNA methylation

Like single genes, large regions of chromosomes or even entire chromosomes can have distinct patterns of DNA methylation. Under a microscope, two kinds of chromatin can be distinguished in the stained interphase nucleus: *euchromatin* and *heterochromatin.* The euchromatin appears diffuse and stains lightly; it contains the DNA that is transcribed into mRNA. Heterochromatin is condensed and stains darkly; any genes it contains are generally not transcribed.

Perhaps the most dramatic example of heterochromatin is the inactive X chromosome of mammals. A normal female mammal has two X chromosomes; a normal male has an X and a Y (see Section 12.4). The X and Y chromosomes probably arose from a pair of autosomes (non–sex chromosomes) about 300 million years ago. Over time, mutations in the Y chromosome resulted in maleness-determining genes, and the Y chromosome gradually lost most of the genes it once shared with its X homolog. As a result, females and males differ greatly in the "dosage" of X-linked genes. Each female cell has two copies of each gene on the X chromosome, and therefore has the potential to produce twice as much of each protein product. Nevertheless, for 75 percent of the genes on the X chromosome, transcription is generally the same in males and in females. How does this happen?

Mary Lyon, Liane Russell, and Ernest Beutler independently hypothesized in 1961 that one of the X chromosomes in each cell of a female is, to a significant extent, transcriptionally inactivated early in embryonic development. They proposed that one copy of X becomes inactive in each embryonic cell, and the same X remains inactive in all that cell's descendants. Several lines of evidence have since confirmed this hypothesis.

In a given embryonic cell, the "choice" of which X in the pair to inactivate is random. Recall that one X in a female comes from her father and one from her mother. Thus, in one embryonic cell the paternal X might be the one remaining transcriptionally active, but in a neighboring cell the maternal X might be active.

The inactivated X chromosome does not vanish, but is identifiable within the nucleus. During interphase a single, stainable nuclear body called a Barr body (after its discoverer, Murray Barr) can be seen in cells of human females under the light microscope (**Figure 16.21A**). This clump of heterochromatin, which is not present in normal males, is the inactivated X chromosome, and it consists of heavily methylated DNA. A female with the normal two X chromosomes will have one Barr body, while a rare female with three Xs will have two, and an XXXX female will have three. Males that are XXY will have one. These observations suggest that the interphase cells of each person, male

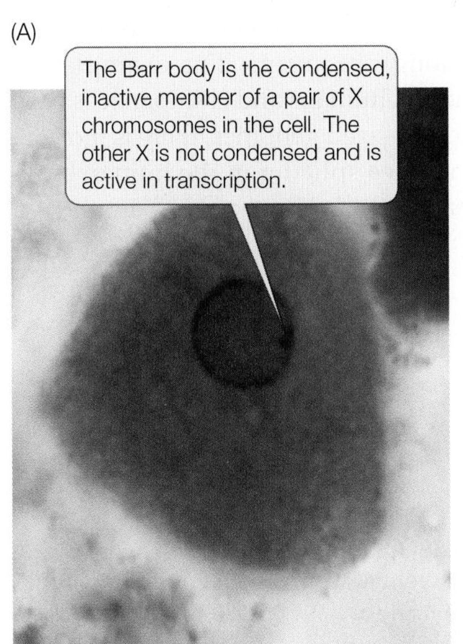

(A)
The Barr body is the condensed, inactive member of a pair of X chromosomes in the cell. The other X is not condensed and is active in transcription.


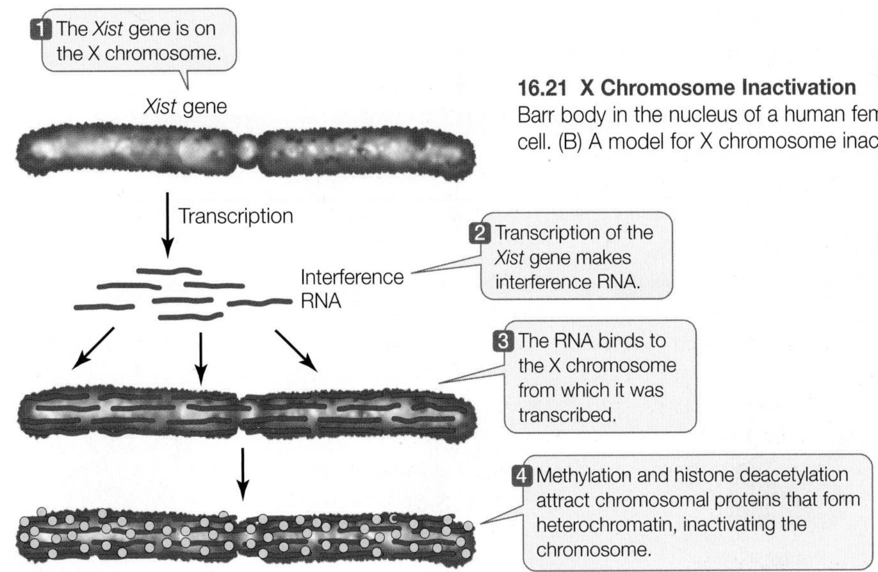


**16.21 X Chromosome Inactivation** (A) A Barr body in the nucleus of a human female cell. (B) A model for X chromosome inactivation.

or female, have a single active X chromosome, and thus a constant dosage of expressed X chromosome genes.

Condensation of the inactive X chromosome makes its DNA sequences physically unavailable to the transcriptional machinery. Most of the genes of the inactive X are heavily methylated. However, one gene, *Xist* (for *X inactivation-specific transcript*), is only lightly methylated and is transcriptionally active. On the active X chromosome, *Xist* is heavily methylated and not transcribed. The RNA transcribed from *Xist* binds to the X chromosome from which it is transcribed, and this binding leads to a spreading of inactivation along the chromosome. The *Xist* RNA transcript is an example of **interference RNA** (**Figure 16.21B**).

## 16.4 RECAP

**Epigenetics describes stable changes in gene expression that do not involve changes in DNA sequences. These changes involve modifications of DNA (cytosine methylation) or of histone proteins bound to DNA. Epigenetic changes can be affected by the environment, and can also result in genome imprinting, in which expression of some genes depends on their parental origin.**

- How are DNA methylation patterns established and how do they affect gene expression? See p. 356 and Figure 16.18
- Explain how histone modifications affect transcription. See pp. 357–358 and Figure 16.19
- Why and how does X chromosome inactivation occur? See p. 359

Gene expression involves transcription and then translation. So far we have described how eukaryotic gene expression is regulated at the transcriptional level. But as Figure 16.13 shows, there are many points at which regulation can occur after the initial gene transcript is made.

## 16.5 How Is Eukaryotic Gene Expression Regulated After Transcription?

Eukaryotic gene expression can be regulated both in the nucleus prior to mRNA export, and after the mRNA leaves the nucleus. Posttranscriptional control mechanisms can involve alternative splicing of pre-mRNA, microRNAs, repressors of translation, or regulation of protein breakdown in the proteasome.

### Different mRNAs can be made from the same gene by alternative splicing

Most primary mRNA transcripts contain several introns (see Figure 14.7). We have seen how the splicing mechanism recognizes the boundaries between exons and introns. What would happen if the β-globin pre-mRNA, which has two introns, were spliced from the start of the first intron to the end of the second? The middle exon would be spliced out along with the two introns. An entirely new protein (certainly not a β-globin) would be made, and the functions of normal β-globin would be lost. Such **alternative splicing** can be a deliberate mechanism for generating a family of different proteins with different activities and functions from a single gene (**Figure 16.22**).

Before the human genome was sequenced, most scientists estimated that they would find between 80,000 and 150,000 protein-coding genes. You can imagine their surprise when the actual sequence revealed only about 24,000 genes! In fact, there are many more human mRNAs than there are human genes, and most of this variation comes from alternative splicing. Indeed, recent surveys show that about half of all human genes are alternatively spliced. Alternative splicing may be a key to the differences in levels of complexity among organisms. For example, although humans and chimpanzees have similar-sized genomes, there is more alternative splicing in the human brain than in the brain of a chimpanzee.

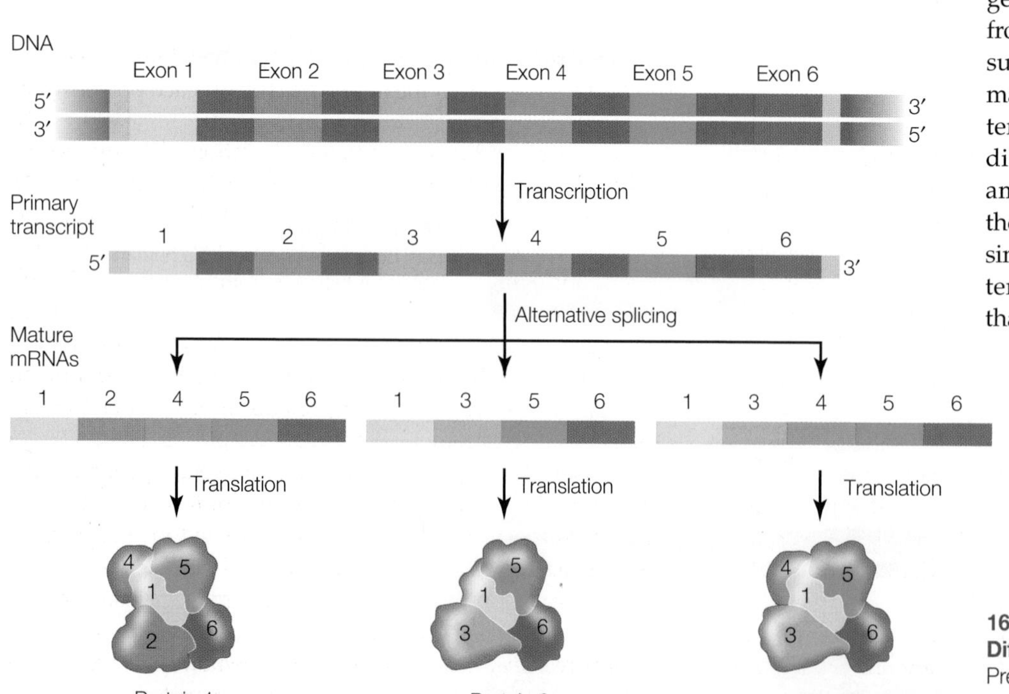


**16.22 Alternative Splicing Results in Different Mature mRNAs and Proteins** Pre-mRNA can be spliced differently in different tissues, resulting in different proteins.

## MicroRNAs are important regulators of gene expression

As we discuss in the next chapter, less than 5 percent of the genome in most plants and animals codes for proteins. Some of the genome encodes ribosomal RNA and transfer RNAs, but until recently biologists thought that the rest of the genome was not transcribed; some even called it "junk." Recent investigations, however, have shown that some of these noncoding regions are transcribed. The noncoding RNAs are often very small and therefore difficult to detect. These tiny RNA molecules are called **microRNA** (miRNA).

The first miRNA sequences were found in the worm *Caenorhabditis elegans.* This model organism, which has been studied extensively by developmental biologists, goes through several larval stages. Victor Ambros at the University of Massachusetts found mutations in two genes that had different effects on progress through these stages:

- *lin-14* mutations (named for abnormal cell *lin*eage) caused the larvae to skip the first stage and go straight to the second stage. Thus the gene's normal role is to facilitate events of the first larval stage.
- *lin-4* mutations caused certain cells in later larval stages to repeat a pattern of development normally shown in the first larval stage. It was as if the cells were stuck in that stage. So the normal role of this gene is to *negatively regulate lin-14*, turning its expression off so the cells can progress to the next stage.

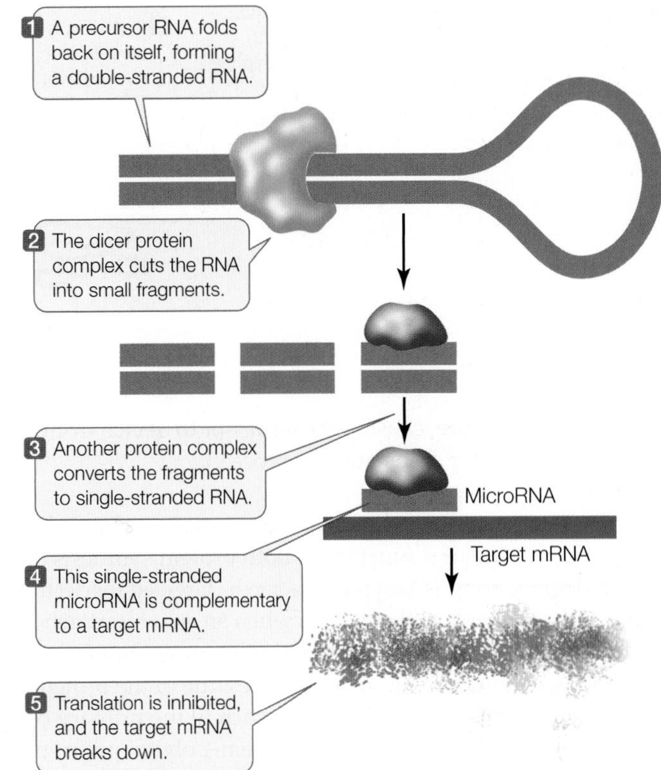

1 A precursor RNA folds back on itself, forming a double-stranded RNA.
2 The dicer protein complex cuts the RNA into small fragments.
3 Another protein complex converts the fragments to single-stranded RNA.
MicroRNA
Target mRNA
4 This single-stranded microRNA is complementary to a target mRNA.
5 Translation is inhibited, and the target mRNA breaks down.


**16.23 mRNA Inhibition by MicroRNAs** MicroRNAs result in inhibition of translation and in breakdown of the target mRNA.

Not surprisingly, further investigation showed that *lin-14* encodes a transcription factor that affects the transcription of genes involved in larval cell progression. It was originally expected that *lin-4,* the negative regulator, would encode a protein that downregulates genes activated by the lin-14 protein. But this turned out to be incorrect. Instead, *lin-4* encodes a 22-base miRNA that inhibits *lin-14* expression posttranscriptionally by binding to its mRNA.

Several hundred miRNAs have now been described in many eukaryotes. Each one is about 22 bases long and usually has dozens of mRNA targets. These miRNAs are transcribed as longer precursors that are then cleaved through a series of steps to double-stranded miRNAs. A protein complex guides the miRNA to its target mRNA, where translation is inhibited and the mRNA is degraded (**Figure 16.23**). The remarkable conservation of the miRNA gene silencing mechanism in eukaryotes indicates that it is evolutionarily ancient and biologically important.

## Translation of mRNA can be regulated

Is the amount of a protein in a cell determined by the amount of its mRNA? Recently, scientists examined the relationship between mRNA abundance and protein abundance in yeast cells. For about a third of the many genes surveyed, there was a clear correlation between mRNA and protein: more of one led to more of the other. But for two-thirds of the proteins, there was no apparent relationship between the two: sometimes there was lots of mRNA and little or no protein, or lots of protein and little mRNA. The concentrations of these proteins must therefore be determined by factors acting after the mRNA is made. Cells do this in two major ways: by blocking the translation of mRNA, or by altering how long newly synthesized proteins persist in the cell (protein longevity).

**REGULATION OF TRANSLATION** There are three known ways in which the translation of mRNA can be regulated. One way, as we saw in the previous section, is to inhibit translation with miRNAs. A second way involves modification of the guanosine triphosphate cap on the 5′ end of the mRNA (see Section 14.4). An mRNA that is capped with an unmodified GTP molecule is not translated. For example, stored mRNAs in the egg cells of the tobacco hornworm moth are capped with unmodified GTP molecules and are not translated. After the egg is fertilized, however, the caps are modified, allowing the mRNA to be translated to produce the proteins needed for early embryonic development.

In another system, repressor proteins directly block translation. For example, in mammalian cells the protein ferritin binds free iron ions ($Fe^{2+}$). When iron is present in excess, ferritin synthesis rises dramatically, but the amount of ferritin mRNA remains constant, indicating that the increase in ferritin synthesis is due to an increased rate of mRNA translation. Indeed, when the iron level in the cell is low, a translational repressor protein binds to ferritin mRNA and prevents its translation by blocking its attachment to a ribosome. When the iron level rises, some of the excess $Fe^{2+}$ ions bind to the repressor and alter its three-

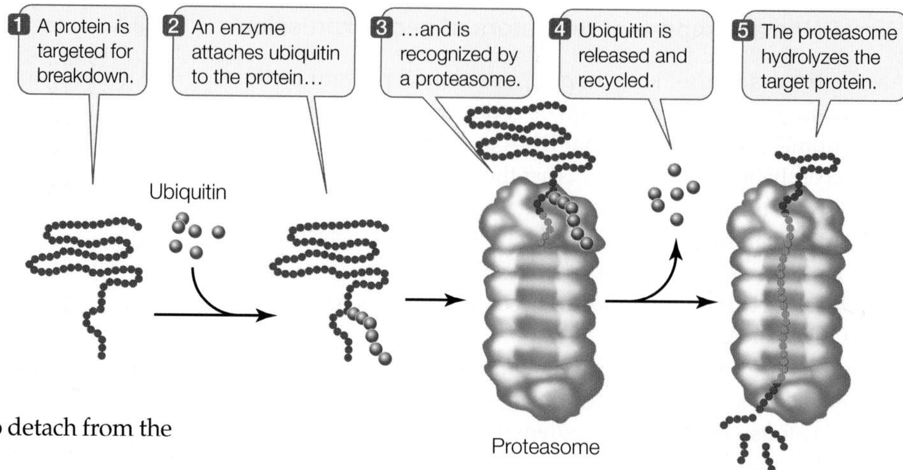

1 A protein is targeted for breakdown.
2 An enzyme attaches ubiquitin to the protein…
3 …and is recognized by a proteasome.
4 Ubiquitin is released and recycled.
5 The proteasome hydrolyzes the target protein.
Ubiquitin
Proteasome

**16.24 A Proteasome Breaks Down Proteins** Proteins targeted for degradation are bound to ubiquitin, which then binds the targeted protein to a proteasome. The proteasome is a complex structure where proteins are digested by several powerful proteases.

dimensional structure, causing the repressor to detach from the mRNA and allowing translation to proceed.

**REGULATION OF PROTEIN LONGEVITY** The protein content of any cell at a given time is a function of both protein synthesis and protein degradation. Certain proteins can be targeted for destruction in a chain of events that begins when an enzyme attaches a 76-amino acid protein called **ubiquitin** (so named because it is ubiquitous, or widespread) to a lysine residue of the protein to be destroyed. Other ubiquitins then attach to the primary one, forming a polyubiquitin chain. The protein–polyubiquitin complex then binds to a huge protein complex called a **proteasome** (from *protease* and *soma*, body) (**Figure 16.24**). Upon entering the proteasome, the polyubiquitin is removed and ATP energy is used to unfold the target protein. Three different proteases then digest the protein into small peptides and amino acids.

You may recall from Section 11.2 that cyclins are proteins that regulate the activities of key enzymes at specific points in the cell cycle. Cyclins must be broken down at just the right time, and this is done by proteasomes. Viruses can hijack this system. For example, some strains of the human papillomavirus target p53 protein and retinoblastoma protein, which normally inhibit the cell cycle, for proteasomal degradation, resulting in unregulated cell division (cancer).

## 16.5 RECAP

**One of the most important means of posttranscriptional regulation is alternative RNA splicing, which allows more than one protein to be made from a single gene. The stability of mRNA in the cytoplasm can also be regulated. MicroRNAs, mRNA modifications, and translational repressors can prevent mRNA translation. Proteins in the cell can be targeted for breakdown by ubiquitin and then hydrolyzed in proteasomes.**

- How can a single pre-mRNA sequence encode several different proteins? See p. 360 and Figure 16.22
- How do miRNAs regulate gene expression? See p. 361 and Figure 16.23
- Explain the role of the proteasome. See p. 362 and Figure 16.24

## CHAPTER SUMMARY

### 16.1 How Do Viruses Regulate Their Gene Expression?

- **Viruses** are not cells, and rely on host cells to reproduce.
- The basic unit of a virus is a **virion**, which consists of a nucleic acid genome (DNA or RNA) and a protein coat, called a **capsid**.
- **Bacteriophage** are viruses that infect bacteria.
- Viruses undergo a **lytic cycle**, which causes the host cell to burst, releasing new virions.
- Some viruses have promoters that bind host RNA polymerase, which they use to transcribe their own genes and proteins. Review Figure 16.3
- Rarely, a phage will transfer bacterial genes to a new host in the process of **transduction**.
- Some viruses can also undergo a **lysogenic cycle**, in which a molecule of their DNA, called a **prophage**, is inserted into the host chromosome, where it replicates for generations. Review Figure 16.4
- The cellular environment determines whether a phage undergoes a lytic or a lysogenic cycle. Regulatory proteins that compete for promoters on phage DNA control the switch between the two life cycles. Review Figure 16.5
- A **retrovirus** uses reverse transcriptase to generate a cDNA **provirus** from its RNA genome. The provirus is incorporated into the host's DNA and can be activated to produce new virions. Review Figure 16.6

### 16.2 How Is Gene Expression Regulated in Prokaryotes?

- Some proteins are synthesized only when they are needed. Proteins that are made only in the presence of a particular compound—an **inducer**—are **inducible** proteins. Proteins that are made at a constant rate regardless of conditions are **constitutive** proteins.
- An **operon** consists of a promoter, an **operator**, and two or more **structural genes**. Promoters and operators do not code

for proteins, but serve as binding sites for regulatory proteins. Review Figure 16.10

- **Regulatory genes** code for regulatory proteins, such as **repressors**. When a repressor binds to an operator, transcription of the structural gene is inhibited. Review Figure 16.11, **ANIMATED TUTORIALS 16.1 AND 16.2**
- The *lac* operon is an example of an inducible system, in which the presence of an inducer (lactose) keeps the repressor from binding the operator, allowing the transcription of structural genes for lactose metabolism.
- Transcription can be enhanced by the binding of an **activator** protein to the promoter. Review Figure 16.12
- **Catabolite repression** is the inhibition of a catabolic pathway for one energy source by a different, preferred energy source.

## 16.3 How Is Eukaryotic Gene Transcription Regulated?

- Eukaryotic gene expression is regulated both during and after transcription. Review Figure 16.13, **WEB ACTIVITY 16.1**
- **Transcription factors** and other proteins bind to DNA and affect the rate of initiation of transcription at the promoter. Review Figure 16.14 and 16.15, **ANIMATED TUTORIAL 16.3**
- The interactions of these proteins with DNA are highly specific and depend on protein domains and DNA sequences.
- Genes at distant locations from one another can be coordinately regulated by common transcription factors and promoter elements. Review Figure 16.17

## 16.4 How Do Epigenetic Changes Regulate Gene Expression?

- **Epigenetics** refers to changes in gene expression that do not involve changes in DNA sequences.
- **Methylation** of cytosine residues generally inhibits transcription. Review Figure 16.18
- Modifications of histone proteins in nucleosomes make transcription either easier or more difficult. Review Figure 16.19
- Epigenetic changes can occur because of the environment.
- DNA methylation can explain **genome imprinting**, where the expression of a gene depends on its parental origin. Review Figure 16.20

## 16.5 How Is Eukaryotic Gene Expression Regulated After Transcription?

- **Alternative splicing** of pre-mRNA can produce different proteins. Review Figure 16.22
- **MicroRNAs** are small RNAs that do not code for proteins, but regulate the translation and longevity of mRNA. Review Figure 16.23
- The translation of mRNA to proteins can be regulated by translational repressors.
- The **proteasome** can break down proteins, thus affecting protein longevity. Review Figure 16.24

**SEE WEB ACTIVITY 16.2 for a concept review of this chapter.**

# SELF-QUIZ

1. Which of the following statements about the *lac* operon is *not* true?
   *a.* When lactose binds to the repressor, the repressor can no longer bind to the operator.
   *b.* When lactose binds to the operator, transcription is stimulated.
   *c.* When the repressor binds to the operator, transcription is inhibited.
   *d.* When lactose binds to the repressor, the shape of the repressor is changed.
   *e.* The repressor has binding sites for both DNA and lactose.
2. Which of the following is *not* a type of viral reproduction?
   *a.* DNA virus in a lytic cycle
   *b.* DNA virus in a lysogenic cycle
   *c.* DNA virus (single-stranded) with a double-stranded DNA intermediate
   *d.* RNA virus with reverse transcription to make cDNA
   *e.* RNA virus acting as tRNA
3. In the lysogenic cycle of bacteriophage λ,
   *a.* a repressor, cI, blocks the lytic cycle.
   *b.* the bacteriophage carries DNA between bacterial cells.
   *c.* both early and late phage genes are transcribed.
   *d.* the viral genome is made into RNA, which stays in the host cell.
   *e.* many new viruses are made immediately, regardless of host health.
4. An operon is
   *a.* a molecule that can turn genes on and off.
   *b.* an inducer bound to a repressor.
   *c.* a series of regulatory sequences controlling transcription of protein-coding genes.
   *d.* any long sequence of DNA.
   *e.* a promoter, an operator, and a group of linked structural genes.
5. Which of the following is true of both positive and negative gene regulation?
   *a.* They reduce the rate of transcription of certain genes.
   *b.* They involve regulatory proteins (or RNA) binding to DNA.
   *c.* They involve transcription of all genes in the genome.
   *d.* They are not both active in the same organism or virus.
   *e.* They act away from the promoter.
6. In DNA, 5-methylcytosine
   *a.* forms a base pair with adenine.
   *b.* is not recognized by DNA polymerase.
   *c.* is related to transcriptional silencing of genes.
   *d.* does not occur at promoters.
   *e.* is an irreversible modification of cytosine.
7. Which statement about selective gene transcription in eukaryotes is *not* true?
   *a.* Regulatory proteins can bind at a site on DNA distant from the promoter.
   *b.* Transcription requires transcription factors.
   *c.* Genes are usually transcribed as groups called operons.
   *d.* Both positive and negative regulation occur.
   *e.* Many proteins bind at the promoter.
8. Control of gene expression in eukaryotes includes all of the following except
   *a.* alternative RNA splicing.
   *b.* binding of proteins to DNA.
   *c.* transcription factors.
   *d.* stabilization of mRNA by miRNA.
   *e.* DNA methylation.

9. The promoter in the *lac* operon is
   *a.* the region that binds the repressor.
   *b.* the region that binds RNA polymerase.
   *c.* the gene that codes for the repressor.
   *d.* a structural gene.
   *e.* an operon.

10. Epigenetic changes
    *a.* can involve DNA methylation.
    *b.* are due to nonhistone protein acetylation.
    *c.* are due to changes in the genetic code.
    *d.* are an example of positive control of translation.
    *e.* are never reversible.

## FOR DISCUSSION

1. Compare the life cycles of a lysogenic bacteriophage and HIV (Figures 16.4 and 16.6) with respect to:
   *a.* how the virus enters the cell.
   *b.* how the virion is released in the cell.
   *c.* how the viral genome is replicated.
   *d.* how new viruses are produced.
2. Compare promoters adjacent to early and late genes in the bacteriophage lytic cycle.
3. The repressor protein that acts on the *lac* operon of *E. coli* is encoded by a regulatory gene. The repressor is made in small quantities and at a constant rate. Would you surmise that the promoter for this repressor protein is efficient or inefficient? Is synthesis of the repressor constitutive, or is it under environmental control?
4. A protein-coding gene in a eukaryote has three introns. How many different proteins could be made by alternative splicing of the pre-mRNA from this gene?

## ADDITIONAL INVESTIGATION

In colorectal cancer, tumor suppressor genes are not active. This is an important factor resulting in uncontrolled cell division. Two possible explanations for the inactive genes are: a mutation in the coding region, resulting in an inactive protein, or epigenetic silencing at the promoter of the gene, resulting in reduced transcription. How would you investigate these two possibilities?

# 17 Genomes

## The dog genome

*Canis lupus familiaris*, the dog, was domesticated by humans from the gray wolf thousands of years ago. While there are many kinds of wolves, they all look more or less the same. Not so with "man's best friend." The American Kennel Club recognizes about 155 different breeds. Dog breeds not only look different, they vary greatly in size. For example, an adult Chihuahua weighs just 1.5 kg, while a Scottish deerhound weighs 70 kg. No other mammal shows such large phenotypic variation, and biologists are curious about how this occurs. Also, there are hundreds of genetic diseases in dogs, and many of these diseases have counterparts in humans. To find out about the genes behind the phenotypic variation, and to elucidate the relationships between genes and diseases, the Dog Genome Project began in the late 1990s. Since then the sequences of several dog genomes have been published.

Two dogs—a boxer and a poodle—were the first to have their entire genomes sequenced. The dog genome contains 2.8 billion base pairs of DNA in 39 pairs of chromosomes. There are 19,000 protein-coding genes, most of them with close counterparts in other mammals, including humans. The whole genome sequence made it easy to create a map of genetic markers—specific nucleotides or short sequences of DNA at particular locations on the genome that differ between individual dogs and/or breeds.

Genetic markers are used to map the locations of (and thus identify) genes that control particular traits. For example, Dr. Elaine Ostrander and her colleagues at the National Institutes of Health studied Portuguese water dogs to identify genes that control size. Taking samples of cells for DNA isolation was relatively easy: a cotton swab was swept over the inside of the cheek. As Dr. Ostrander said, the dogs "didn't care, especially if they thought they were going to get a treat or if there was a tennis ball in our other hand." It turned out that the gene for *i*nsulin-like *g*rowth *f*actor 1 (IGF-1) is important in determining size: large breeds have an allele that codes for an active IGF-1 and small breeds have a different allele that codes for a less active IGF-1.

Another gene important to phenotypic variation was found in whippets, sleek dogs that run fast and are often raced. A mutation in the gene for myostatin, a protein that inhibits overdevelopment of muscles, results in a

**Variation in Dogs** The Chihuahua (bottom) and the Brazilian mastiff (top) are the same species, *Canis lupus familiaris*, and yet show great variation in size. Genome sequencing has revealed insights into how size is controlled by genes.

**Genetic Bully** These dogs are both whippets, but the muscle-bound dog (right) has a mutation in a gene that limits muscle buildup.

whippet that is more muscular and runs faster. Myostatin is important in human muscles as well.

Inevitably, some scientists have set up companies to test dogs for genetic variations, using DNA supplied by anxious owners and breeders. Some traditional breeders frown on this practice, but others say it will improve the breeds and give more joy (and prestige) to owners. So the issues surrounding the Dog Genome Project are not very different from ones arising from the Human Genome Project.

Powerful methods have been developed to analyze DNA sequences, and the resulting information is accumulating at a rapid rate. Comparisons of sequenced genomes are providing new insights into evolutionary relationships and confirming old ones. We are in a new era of biology.

**IN THIS CHAPTER** we look at genomes. First we look at how large molecules of DNA are cut and sequenced, and what kinds of information these genome sequences provide. Then we turn to the results of ongoing sequencing efforts in both prokaryotes and eukaryotes. We next consider the human genome and some of the real and potential uses of human genome information. Finally, we will describe the emerging fields of proteomics and metabolomics, which attempt to give a complete inventory of a cell's proteins and metabolic activity.

## CHAPTER OUTLINE



# 17.1 How Are Genomes Sequenced?

As you saw in the opening story on dogs, one reason for sequencing genomes is to compare different organisms. Another is to identify changes in the genome that result in disease. In 1986, the Nobel laureate Renato Dulbecco and others proposed that the world scientific community be mobilized to undertake the sequencing of the entire human genome. One challenge discussed at the time was to detect DNA damage in people who had survived the atomic bomb attacks and been exposed to radiation in Japan during World War II. But in order to detect changes in the human genome, scientists first needed to know its normal sequence.

The result was the publicly funded **Human Genome Project**, an enormous undertaking that was successfully completed in 2003. This effort was aided and complemented by privately funded groups. The project benefited from the development of many new methods that were first used in the sequencing of smaller genomes—those of prokaryotes and simple eukaryotes.

### Two approaches were used to sequence the human genome

Many prokaryotes have a single chromosome, while eukaryotes have several to many. Because of their differing sizes, chromosomes can be separated from one another, identified, and experimentally manipulated. It might seem that the most straightforward approach to sequencing a chromosome would be to start at one end and simply sequence the entire DNA molecule. However, this approach is not practical since only about 700 base pairs can be sequenced at a time using current methods. Prokaryotic chromosomes contain 1–4 million base pairs and human chromosome 1 contains 246 million base pairs.

To sequence an entire genome, chromosomal DNA must be cut into short fragments about 500 base pairs long, which are separated and sequenced. For the haploid human genome, which has about 3.3 billion base pairs, there are more than 6 million such fragments. When all of the fragments have been sequenced, the problem becomes how to put these millions of sequences together. This task can be accomplished using larger, *overlapping fragments*.

Let's illustrate this process using a single, 10 base-pair (bp) DNA molecule. (This is a double-stranded molecule, but for convenience we show only the sequence of the noncoding strand.) The molecule is cut three ways. The first cut generates the fragments:

TG, ATG, and CCTAC

The second cut of the same molecule generates the fragments:

AT, GCC, and TACTG

The third cut results in:

CTG, CTA, and ATGC

Can you put the fragments into the correct order? (The answer is ATGCCTACTG.) Of course, the problem of ordering 6 million fragments, each about 500 bp long, is more of a challenge! The field of **bioinformatics** was developed to analyze DNA sequences using complex mathematics and computer programs.

Until recently, two broad approaches were used to analyze DNA fragments for alignment: hierarchical sequencing and shotgun sequencing. These were developed for the Human Genome Project, but have been applied to other organisms as well.

**HIERARCHICAL SEQUENCING** The publicly funded human genome sequencing team developed a method known as **hierarchical sequencing**. The first step was to systematically identify short marker sequences along the chromosomes, ensuring that every fragment of DNA to be sequenced would contain a marker (**Figure 17.1A**). Genetic markers can be short tandem repeats (STRs), single nucleotide polymorphisms (SNPs), or the recognition sites for *restriction enzymes*, which recognize and cut DNA at specific sequences (see Chapter 15).

Some restriction enzymes recognize sequences of 4–6 base pairs and generate many fragments from a large DNA molecule. For example, the enzyme *Sau*3A cuts DNA every time it encounters GATC. Other restriction enzymes recognize sequences of 8–12 base pairs (*Not*I cuts at GCGGCCGC, for example) and generate far fewer, but much larger, fragments.

In hierarchical sequencing, genomic DNA is cut up into a set of relatively large (55,000 to 2 million bp) fragments. If different enzymes are used in separate digests, the fragments will overlap so that some fragments share particular markers. Each fragment is inserted into a bacterial plasmid to create a **bacterial artificial chromosome** (**BAC**), which is then inserted into bacteria. Each bacterium gets just one plasmid with its fragment of (for example) the human genome and is allowed to grow into a colony containing millions of genetically identical bacteria (called a *clone*). Clones differ from one another in that each has a different fragment from the human genome. A collection of clones, containing many different fragments of a genome, is called a **genomic library**.

The DNA from each clone is then extracted and cut into smaller overlapping pieces, which in turn are cloned, purified, and sequenced. The overlapping parts of the sequences allow researchers (with the aid of computers) to align them to create the complete sequence of the BAC clone. The genetic markers on each BAC clone are used to arrange the larger fragments in the proper order along the chromosome map. This method works, but it is slow. An alternative approach, shotgun sequencing, makes far greater use of use of computers to align the sequences.

yourBioPortal.com
GO TO Animated Tutorial 17.1 • Sequencing the Genome

TOOLS FOR INVESTIGATING LIFE

**17.1 Sequencing Genomes Involves Fragment Overlaps**

Short fragments of the whole genome can be sequenced, but then the fragments must be correctly aligned. Historically two approaches were used. Both involved the use of bacterial clones to separate and amplify individual DNA fragments.

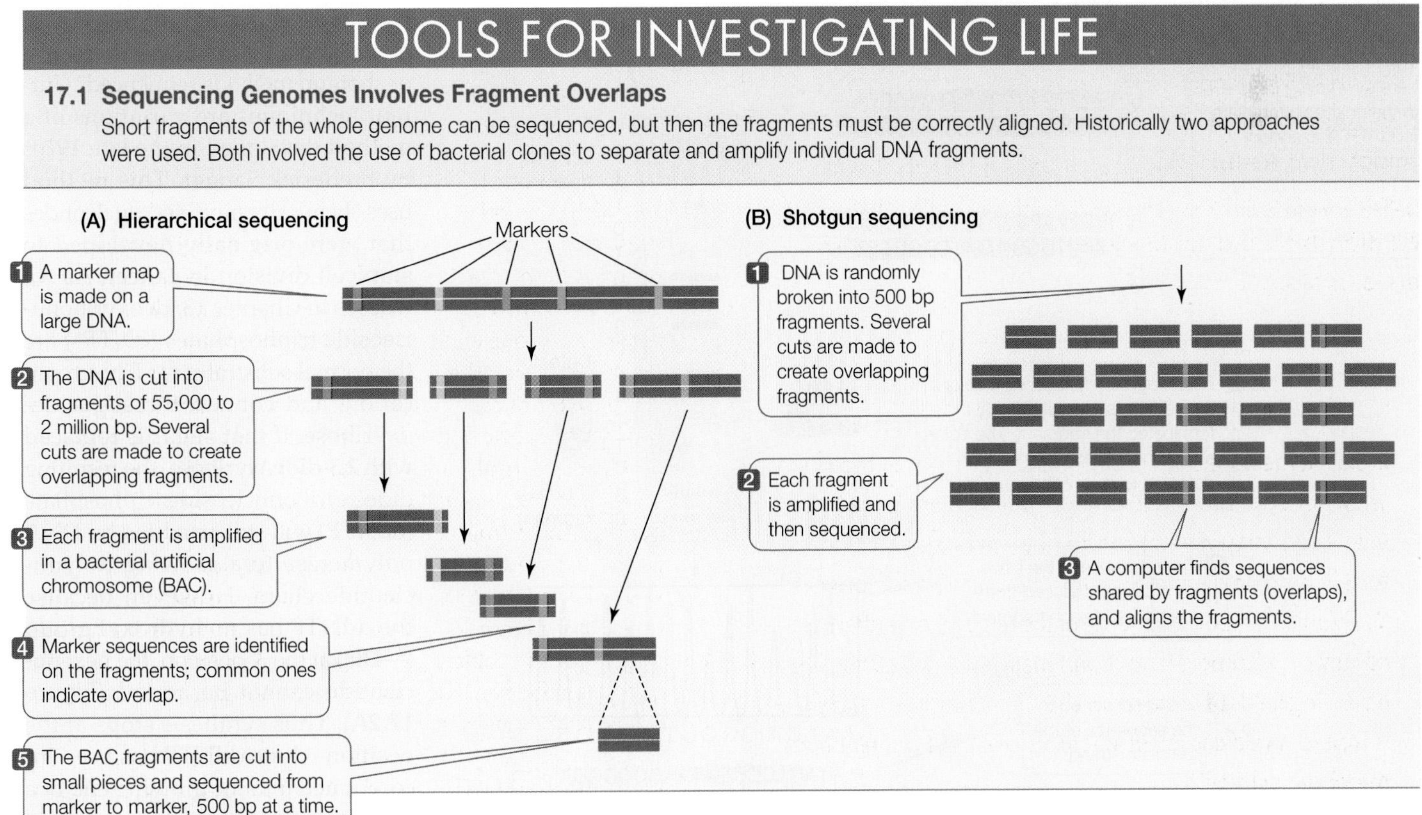

(A) Hierarchical sequencing
Markers
1 A marker map is made on a large DNA.
2 The DNA is cut into fragments of 55,000 to 2 million bp. Several cuts are made to create overlapping fragments.
3 Each fragment is amplified in a bacterial artificial chromosome (BAC).
4 Marker sequences are identified on the fragments; common ones indicate overlap.
5 The BAC fragments are cut into small pieces and sequenced from marker to marker, 500 bp at a time.
(B) Shotgun sequencing
1 DNA is randomly broken into 500 bp fragments. Several cuts are made to create overlapping fragments.
2 Each fragment is amplified and then sequenced.
3 A computer finds sequences shared by fragments (overlaps), and aligns the fragments.

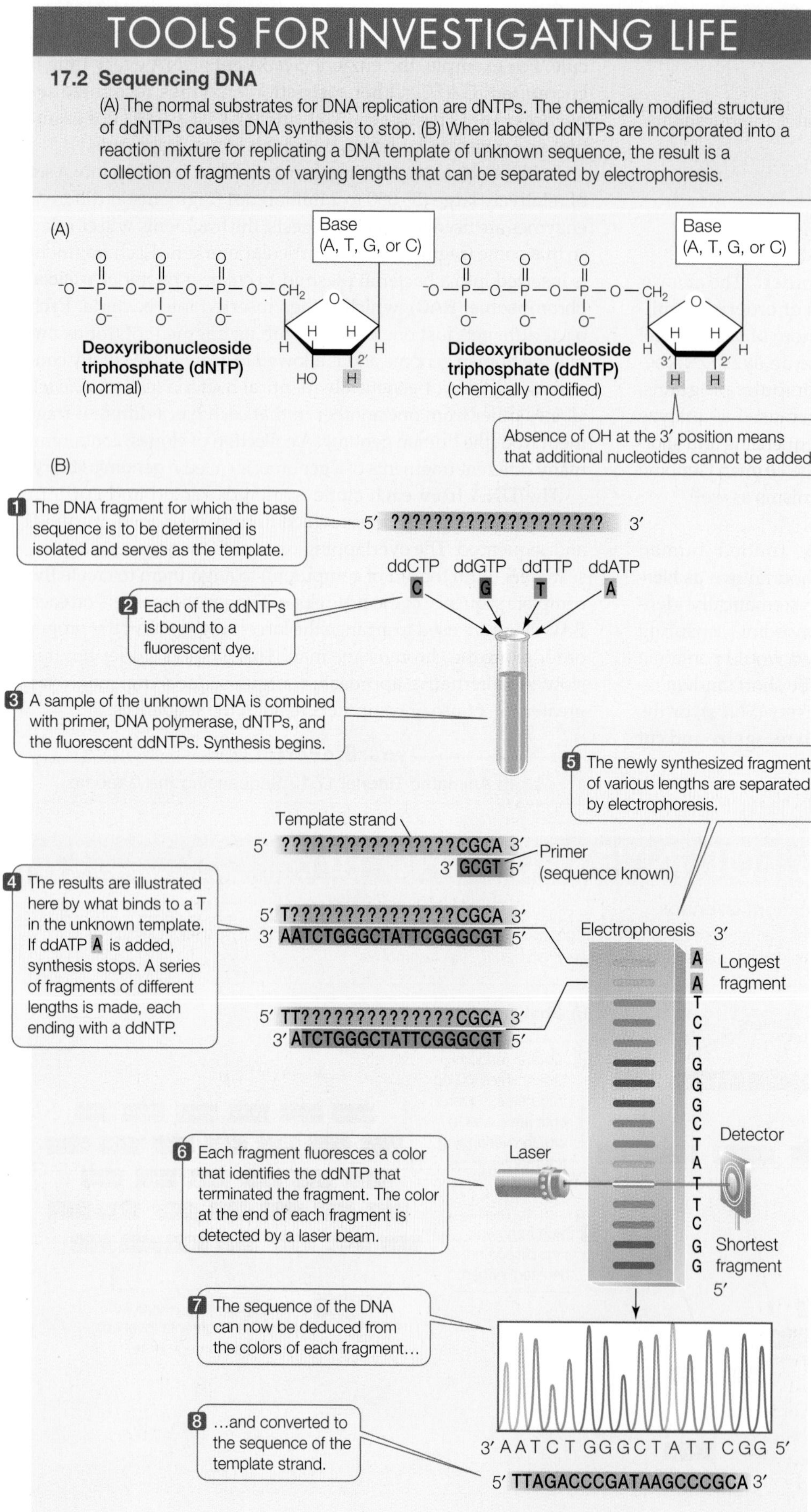

TOOLS FOR INVESTIGATING LIFE
(A)
Base
(A, T, G, or C)
Deoxyribonucleoside triphosphate (dNTP)
(normal)
Dideoxyribonucleoside triphosphate (ddNTP)
(chemically modified)
Absence of OH at the 3′ position means that additional nucleotides cannot be added.
(B)
1 The DNA fragment for which the base sequence is to be determined is isolated and serves as the template.
5′ ???????????????????? 3′
ddCTP ddGTP ddTTP ddATP
2 Each of the ddNTPs is bound to a fluorescent dye.
3 A sample of the unknown DNA is combined with primer, DNA polymerase, dNTPs, and the fluorescent ddNTPs. Synthesis begins.
5 The newly synthesized fragments of various lengths are separated by electrophoresis.
Template strand
5′ ????????????????CGCA 3′
3′ GCGT 5′
Primer
(sequence known)
4 The results are illustrated here by what binds to a T in the unknown template. If ddATP A is added, synthesis stops. A series of fragments of different lengths is made, each ending with a ddNTP.
5′ T???????????????CGCA 3′
3′ AATCTGGGCTATTCGGGCGT 5′
5′ TT??????????????CGCA 3′
3′ ATCTGGGCTATTCGGGCGT 5′
Electrophoresis
Longest fragment
Detector
Laser
Shortest fragment
6 Each fragment fluoresces a color that identifies the ddNTP that terminated the fragment. The color at the end of each fragment is detected by a laser beam.
7 The sequence of the DNA can now be deduced from the colors of each fragment…
8 …and converted to the sequence of the template strand.
3′ AATCTGGGCTATTCGG 5′
5′ TTAGACCCGATAAGCCCGCA 3′


**17.2 Sequencing DNA** (A) The normal substrates for DNA replication are dNTPs. The chemically modified structure of ddNTPs causes DNA synthesis to stop. (B) When labeled ddNTPs are incorporated into a reaction mixture for replicating a DNA template of unknown sequence, the result is a collection of fragments of varying lengths that can be separated by electrophoresis.

**SHOTGUN SEQUENCING** Instead of mapping the genome and creating a BAC library, the **shotgun sequencing** method involves directly cutting genomic DNA into smaller, overlapping fragments that are cloned and sequenced. Powerful computers align the fragments by finding sequence homologies in the overlapping regions (**Figure 17.1B**). As sequencing technologies and computers have improved, the shotgun approach has become much faster and cheaper than the hierarchical approach.

As a demonstration, researchers used this method to sequence a 1.8 million-base-pair prokaryotic genome in just a few months. Next came larger genomes. The entire 180 million-base-pair fruit fly genome was sequenced by the shotgun method in little over a year. This success proved that the shotgun method might work for the much larger human genome, and in fact it was used to sequence the human genome rapidly relative to the hierarchical method.

## The nucleotide sequence of DNA can be determined

How are the individual DNA fragments generated by the hierarchical or shotgun methods sequenced? Current techniques are variations of a method developed in the late 1970s by Frederick Sanger. This method uses chemically modified nucleotides that were originally developed to stop cell division in cancer. As we discuss in Chapter 13, deoxyribonucleoside triphosphates (dNTPs) are the normal substrates for DNA replication, and contain the sugar deoxyribose. If that sugar is replaced with 2,3-dideoxyribose, the resulting dideoxyribonucleoside triphosphate (ddNTP) will still be added by DNA polymerase to a growing polynucleotide chain. However, because the ddNTP has no hydroxyl group (—OH) at the 3′ position, the next nucleotide cannot be added (**Figure 17.2A**). Thus synthesis stops at the position where ddNTP has been incorporated into the growing end of a DNA strand.

To determine the sequence of a DNA fragment (usually no more than 700 base pairs long), it is isolated and mixed with

- DNA polymerase
- A short primer appropriate for the DNA sequence
- The four dNTPs (dATP, dGTP, dCTP, and dTTP)
- Small amounts of the four ddNTPs, each bonded to a differently colored fluorescent "tag"

In the first step of the reaction, the DNA is heated to denature it (separate it into single strands). Only one of these strands will act as a template for sequencing—the one to which the primer binds. DNA replication proceeds, and the test tube soon contains a mixture of the original DNA strands and shorter, new complementary strands. The new strands, each ending with a fluorescent ddNTP, are of varying lengths. For example, each time a T is reached on the template strand, DNA polymerase adds either a dATP or a ddATP to the growing complementary strand. If dATP is added, the strand continues to grow. If ddATP is added, growth stops (**Figure 17.2B**).

After DNA replication has been allowed to proceed for a while, the new DNA fragments are denatured and the single-stranded fragments separated by electrophoresis (see Figure 15.8), which sorts the DNA fragments by length. During the electrophoresis run, the fragments pass through a laser beam that excites the fluorescent tags, and the distinctive color of light emitted by each ddNTP is detected. The color indicates which ddNTP is at the end of each strand. A computer processes this information and prints out the DNA sequence of the fragment (see Figure 17.2B).

The delivery of chemical reagents by automated machines, coupled with automated analysis, has made DNA sequencing faster than ever. Huge laboratories often have 80 sequencing machines operating at once, each of which can sequence and analyze up to 70,000 bp in a typical 4-hour run. This may be fast enough for a prokaryotic genome with 1.5 million base pairs (20 runs), but when it comes to routine sequencing of larger genomes (like the 3.3 billion-base-pair human genome), even more speed is needed.

## High-throughput sequencing has been developed for large genomes

The first decade of the new millennium has seen rapid development of **high-throughput sequencing** methods—fast, cheap ways to sequence and analyze large genomes. A variety of different approaches are being used. They generally involve the amplification of DNA templates by the polymerase chain reaction (PCR; see Section 13.5), and the physical binding of template DNA to a solid surface or to tiny beads called microbeads. These techniques are often referred to as *massively parallel DNA sequencing*, because thousands or millions of sequencing reactions are run at once to greatly speed up the process. One such high-throughput method is illustrated in **Figure 17.3**. In one 7-hour run, these machines can sequence 50,000,000 base pairs of DNA! How does it work?

yourBioPortal.com

GO TO Animated Tutorial 17.2 • High-Throughput Sequencing

TOOLS FOR INVESTIGATING LIFE

**17.3 High-Throughput Sequencing**

High-speed sequencing is faster and cheaper than traditional methods, and involves the chemical amplification of DNA fragments. One example of high-throughput sequencing is shown here.

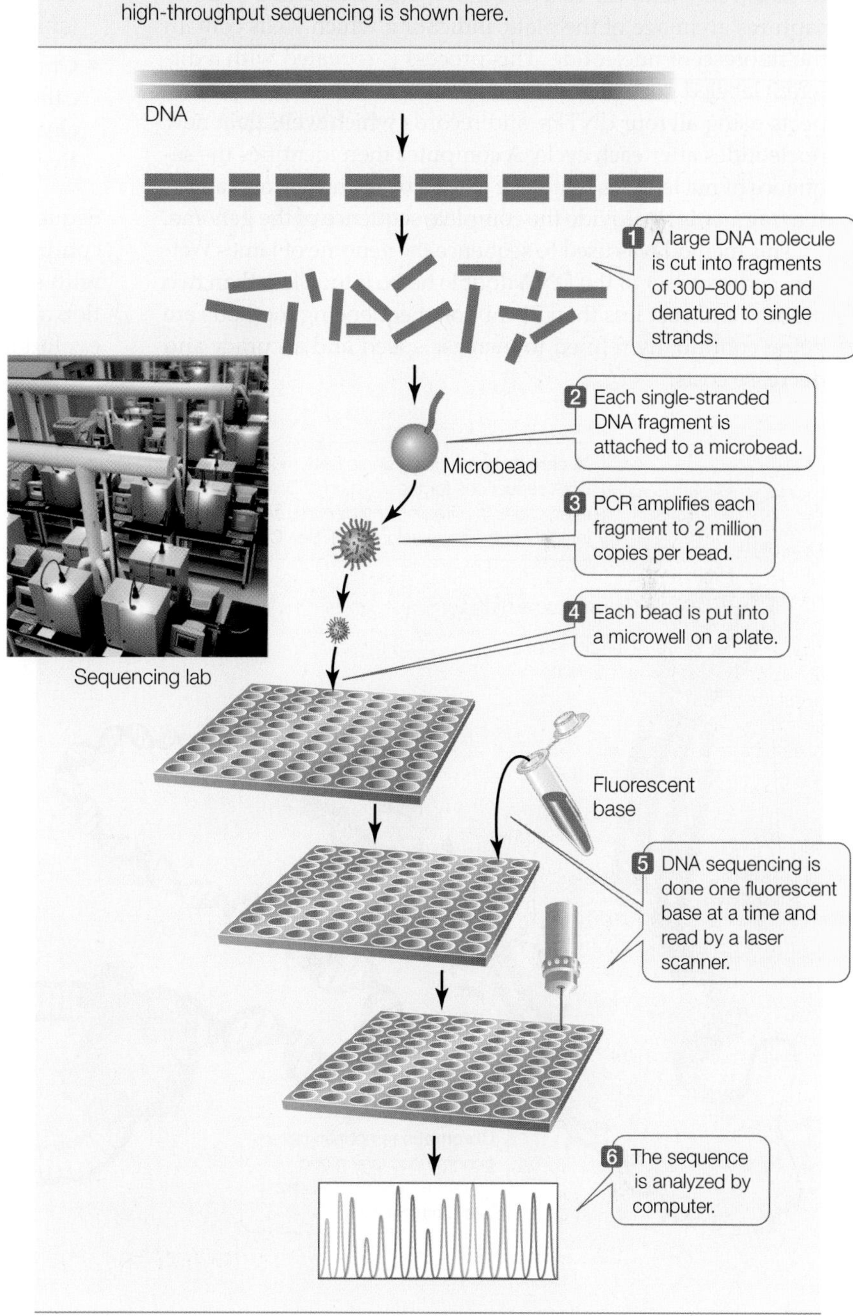

DNA
1 A large DNA molecule is cut into fragments of 300–800 bp and denatured to single strands.
2 Each single-stranded DNA fragment is attached to a microbead.
Microbead
3 PCR amplifies each fragment to 2 million copies per bead.
4 Each bead is put into a microwell on a plate.
Sequencing lab
Fluorescent base
5 DNA sequencing is done one fluorescent base at a time and read by a laser scanner.
6 The sequence is analyzed by computer.

For massively parallel sequencing using microbeads, the genomic DNA is first cut into 300- to 800-base-pair fragments. The fragments are denatured to single strands and attached to tiny beads that are less than 20 μm in diameter, one DNA fragment (template) per bead. PCR is used to create several million identical copies of the fragment on each bead. Then each bead is loaded into a tiny (40 μm diameter) well in a multi-well plate, and the sequencing begins.

The automated sequencer adds a reaction mix like the one described above, but containing only one of four fluorescently labeled dNTPs. That nucleotide will become incorporated as the first nucleotide in a complementary strand only in wells where the first nucleotide in the template strand can base-pair with it. For example, if the first nucleotide on the template in well #1 has base T, then a fluorescent nucleotide with base A will bind to that well. Next, the reaction mix is removed and a scanner captures an image of the plate, indicating which wells contain the fluorescent nucleotide. This process is repeated with a different labeled nucleotide. The machine cycles through many repeats using all four dNTPs, and records which wells gain new nucleotides after each cycle. A computer then identifies the sequence of nucleotides that were gained by each well, and aligns the fragments to provide the complete sequence of the genome.

This method was used to sequence the genome of James Watson, codiscoverer of the DNA double helix. It took less than two months and cost less than $1 million. Sequencing methods are being continually refined to increase speed and accuracy and decrease costs.

### Genome sequences yield several kinds of information

New genome sequences are published more and more frequently, creating a torrent of biological information (**Figure 17.4**). In general, biologists use sequence information to identify:

- *Open reading frames*, the coding regions of genes. For protein-coding genes, these regions can be recognized by the start and stop codons for translation, and by intron consensus sequences that indicate the locations of introns.
- *Amino acid sequences* of proteins, which can be deduced from the DNA sequences of open reading frames by applying the genetic code (see Figure 14.6).
- *Regulatory sequences*, such as promoters and terminators for transcription.
- *RNA genes*, including rRNA, tRNA, and small nuclear RNA (snRNA) genes.
- *Other noncoding sequences* that can be classified into various categories including centromeric and telomeric regions, nuclear matrix attachment regions, transposons, and repetitive sequences such as short tandem repeats.

Sequence information is also used for *comparative genomics*, the comparison of a newly sequenced genome (or parts thereof) with sequences from other organisms. This can give information about the functions of sequences, and can be used to trace evolutionary relationships among different organisms.

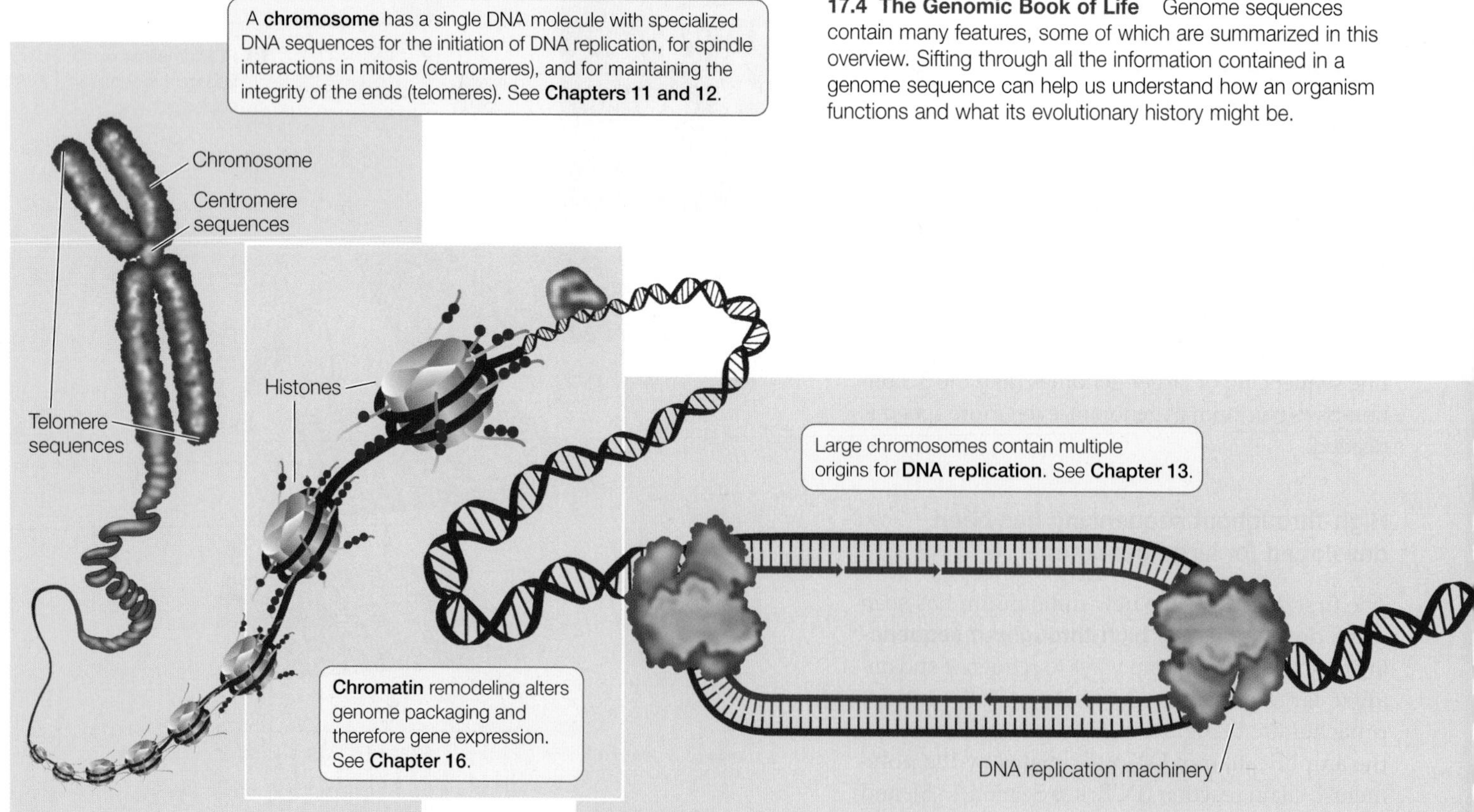

A **chromosome** has a single DNA molecule with specialized DNA sequences for the initiation of DNA replication, for spindle interactions in mitosis (centromeres), and for maintaining the integrity of the ends (telomeres). See **Chapters 11 and 12**.
Chromosome
Centromere sequences
Telomere sequences
Histones
Large chromosomes contain multiple origins for **DNA replication**. See **Chapter 13**.
**Chromatin** remodeling alters genome packaging and therefore gene expression. See **Chapter 16**.
DNA replication machinery


**17.4 The Genomic Book of Life** Genome sequences contain many features, some of which are summarized in this overview. Sifting through all the information contained in a genome sequence can help us understand how an organism functions and what its evolutionary history might be.

## 17.1 RECAP

**The sequencing of genomes required the development of ways to cut large chromosomes into fragments, sequence the fragments, and then line them up on the chromosome. Two ways to do this are hierarchical sequencing and shotgun sequencing. Today new procedures are being developed that require automation and powerful computers. Actual DNA sequencing involves labeled nucleotides that are detected at the ends of growing polynucleotide chains.**

- What are the hierarchical and shotgun approaches to genome analysis? **See pp. 367–368 and Figure 17.1**
- What is the dideoxy method for DNA sequencing? **See pp. 368–369 and Figure 17.2**
- Explain how high-throughput sequencing methods work. **See pp. 369–370 and Figure 17.3**
- How are open reading frames recognized in a genomic sequence? What kind of information can be derived from an open reading frame? **See p. 370**

We now turn to the first organisms whose sequences were determined, prokaryotes, and the information these sequences provided.

# 17.2 What Have We Learned from Sequencing Prokaryotic Genomes?

When DNA sequencing became possible in the late 1970s, the first life forms to be sequenced were the simplest viruses with their relatively small genomes. The sequences quickly provided new information on how these viruses infect their hosts and reproduce. But the manual sequencing techniques used on viruses were not up to the task of studying the genomes of prokaryotes and eukaryotes. The newer, automated sequencing techniques we just described made such studies possible. We now have genome sequences for many prokaryotes, to the great benefit of microbiology and medicine.

### The sequencing of prokaryotic genomes led to new genomics disciplines

In 1995 a team led by Craig Venter and Hamilton Smith determined the first complete genomic sequence of a free-living cellular organism, the bacterium *Haemophilus influenzae*. Many more prokaryotic sequences have followed, revealing not only how prokaryotes apportion their genes to perform different cellular functions, but also how their specialized functions are carried out. Soon we may even be able to ask the provocative question of what the minimal requirements of a living cell might be.

**FUNCTIONAL GENOMICS** **Functional genomics** is the biological discipline that assigns functions to the products of genes. This

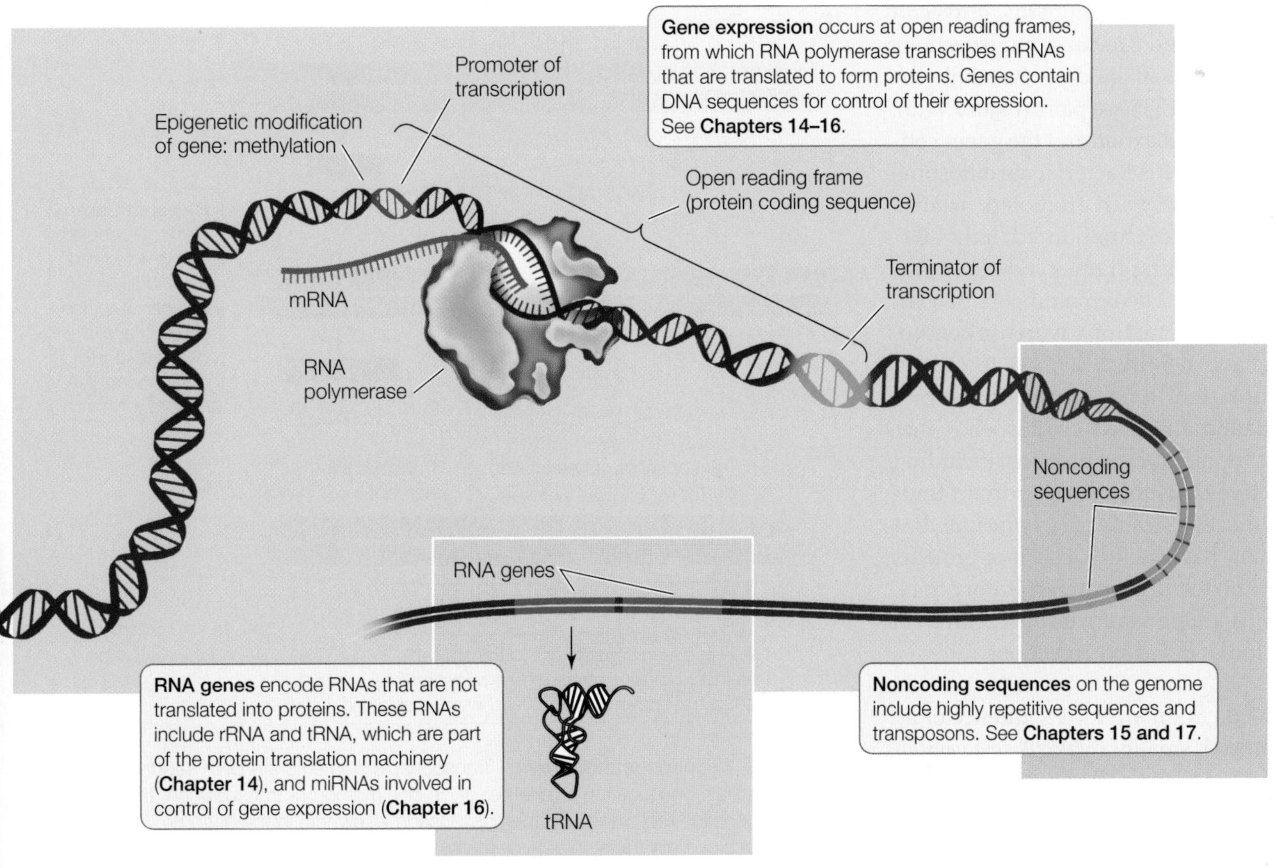
**Gene expression** occurs at open reading frames, from which RNA polymerase transcribes mRNAs that are translated to form proteins. Genes contain DNA sequences for control of their expression. See **Chapters 14–16**.
Promoter of transcription
Epigenetic modification of gene: methylation
Open reading frame (protein coding sequence)
Terminator of transcription
mRNA
RNA polymerase
Noncoding sequences
RNA genes
**RNA genes** encode RNAs that are not translated into proteins. These RNAs include rRNA and tRNA, which are part of the protein translation machinery (**Chapter 14**), and miRNAs involved in control of gene expression (**Chapter 16**).
tRNA
**Noncoding sequences** on the genome include highly repetitive sequences and transposons. See **Chapters 15 and 17**.

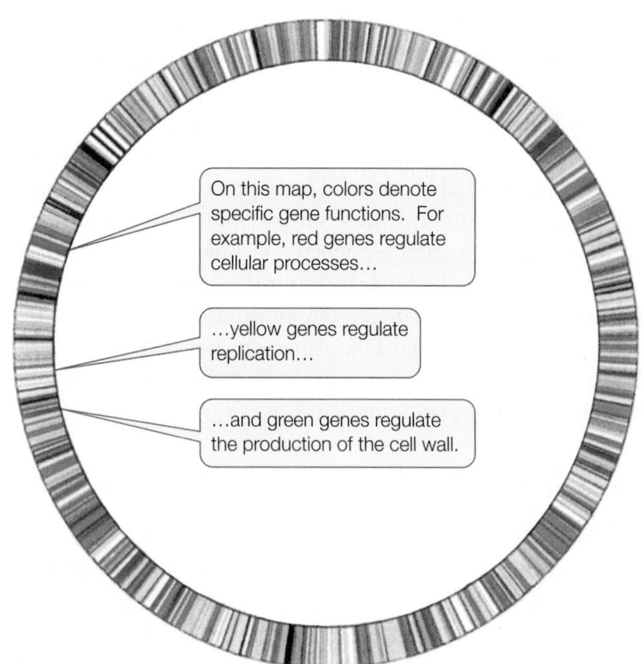


**17.5 Functional Organization of the Genome of *H. influenzae*** The entire DNA sequence has 1,830,137 base pairs. Different colors reflect different classes of gene function.

field, less than 15 years old, is now a major occupation of biologists. Let's see how funtional genomics methods were applied to the bacterium *H. influenzae* once its sequence was known.

The only host for *H. influenzae* is humans. It lives in the upper respiratory tract and can cause ear infections or, more seriously, meningitis in children. Its single circular chromosome has 1,830,138 base pairs (**Figure 17.5**). In addition to its origin of replication and the genes coding for rRNAs and tRNAs, this bacterial chromosome has 1,738 open reading frames with promoters nearby.

When this sequence was first announced, only 1,007 (58 percent) of the open reading frames coded for proteins with known functions. The remaining 42 percent coded for proteins whose functions were unknown. Since then scientists have identified many of these proteins' roles. For example, they found genes for enzymes of glycolysis, fermentation, and electron transport. Other gene sequences code for membrane proteins, including those involved in active transport. An important finding was that highly infective strains of *H. influenzae*, but not noninfective strains, have genes for surface proteins that attach the bacterium to the human respiratory tract. These surface proteins are now a focus of research on possible treatments for *H. influenzae* infections.

**COMPARATIVE GENOMICS** Soon after the sequence of *H. influenzae* was announced, smaller (*Mycoplasma genitalium*; 580,073 base pairs) and larger (*E. coli*; 4,639,221 base pairs) prokaryotic sequences were completed. Thus began a new era in biology, that of **comparative genomics**, which compares genome sequences from different organisms. Scientists can identify genes that are present in one bacterium and missing in another, allowing them to relate these genes to bacterial function.

*M. genitalium*, for example, lacks the enzymes needed to synthesize amino acids, which *E. coli* and *H. influenzae* both possess. This finding reveals that *M. genitalium* must obtain all its amino acids from its environment (usually the human urogenital tract). Furthermore, *E. coli* has 55 regulatory genes coding for transcriptional activators and 58 for repressors; *M. genitalium* only has 3 genes for activators. What do such findings tell us about an organism's lifestyle? For example, is the biochemical flexibility of *M. genitalium* limited by its relative lack of control over gene expression?

### Some sequences of DNA can move about the genome

Genome sequencing allowed scientists to study more broadly a class of DNA sequences that had been discovered by geneticists decades earlier. Segments of DNA called **transposable elements** can move from place to place in the genome and can even be inserted into another piece of DNA in the same cell (e.g., a plasmid). A transposable element might be at one location in the genome of one *E. coli* strain, and at a different location in another strain. The insertion of this movable DNA sequence from elsewhere in the genome into the middle of a protein-coding gene disrupts that gene (**Figure 17.6A**). Any mRNA expressed from the disrupted gene will have the extra sequence and the

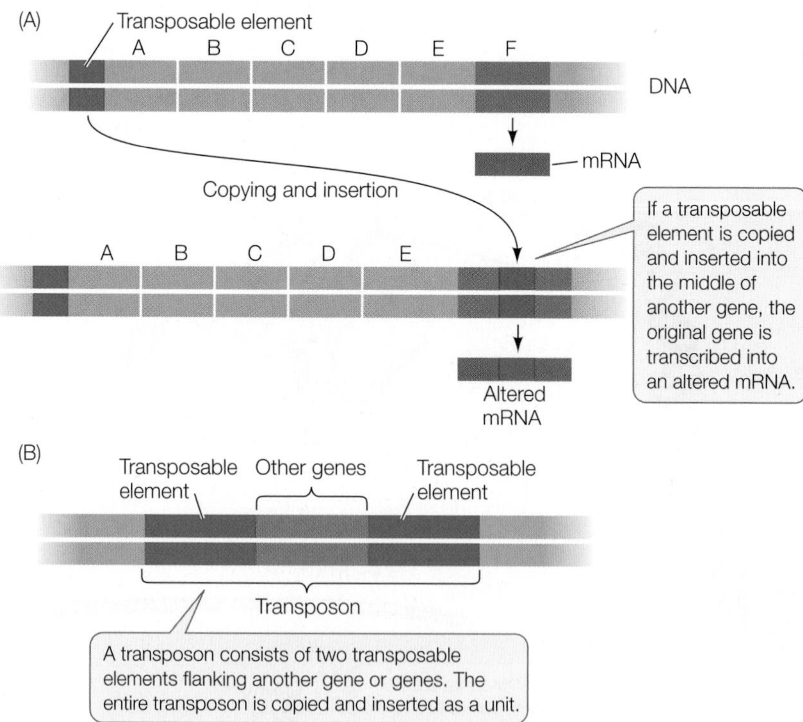


**17.6 DNA Sequences that Move** Transposable elements are DNA sequences that move from one location to another. (A) In one method of transposition, the DNA sequence is replicated and the copy inserts elsewhere in the genome. (B) Transposons contain transposable elements and other genes.

protein will be abnormal. So transposable elements can produce significant phenotypic effects by inactivating genes.

Transposable elements are often short sequences of 1,000–2,000 base pairs, and are found at many sites in prokaryotic genomes. The mechanisms that allow them to move vary. For example, a transposable element may be replicated, and then the copy inserted into another site in the genome. Or the element might splice out of one location and move to another location.

Longer transposable elements (up to 5,000 bp) carry additional genes and are called **transposons** (**Figure 17.6B**). Sometimes these DNA regions contain a gene for antibiotic resistance.

### The sequencing of prokaryotic and viral genomes has many potential benefits

Prokaryotic genome sequencing promises to provide insights into microorganisms that cause human diseases. Genome sequencing has revealed unknown genes and proteins that can be targeted for isolation and functional study. Such studies are revealing new methods to combat pathogens and their infections. Sequencing has also revealed surprising relationships between some pathogenic organisms, suggesting that genes may be transferred between different strains.

- *Chlamydia trachomatis* causes the most common sexually transmitted disease in the United States. Because it is an intracellular parasite, it has been very hard to study. Among its 900 genes are several for ATP synthesis—something scientists used to think this bacterium could not accomplish on its own.
- *Rickettsia prowazekii* causes typhus; it is carried by lice and infects people bitten by the lice. Of its 634 genes, 6 encode proteins that are essential for virulence. These virulence proteins are being used to develop vaccines.
- *Mycobacterium tuberculosis* causes tuberculosis. It has a relatively large genome, coding for 4,000 proteins. Over 250 of these are used to metabolize lipids, so this may be the main way that this bacterium gets its energy. Some of its genes code for previously unidentified cell surface proteins; these proteins are targets for potential vaccines.
- *Streptomyces coelicolor* and its close relatives are the source for the genes for two-thirds of all naturally occuring antibiotics currently in clinical use. These antibiotics include streptomycin, tetracycline, and erythromycin. The genome sequence of *S. coelicolor* reveals 22 clusters of genes responsible for antibiotic production, of which only four were previously known. This finding may lead to new antibiotics to combat pathogens that have evolved resistance to conventional antibiotics.
- *E. coli* strain O157:H7 causes illness (sometimes severe) in at least 70,000 people a year in the United States. Its genome has 5,416 genes, of which 1,387 are different from those in the familiar (and harmless) laboratory strains of this bacterium. Many of these unique genes are also present in other pathogenic bacteria, such as *Salmonella* and *Shigella*. This finding suggests that there is extensive genetic exchange among these species, and that "superbugs" that share genes for antibiotic resistance may be on the horizon.
- *Severe acute respiratory syndrome* (*SARS*) was first detected in southern China in 2002 and rapidly spread in 2003. There is no effective treatment and 10 percent of infected people die. Isolation of the causative agent, a virus, and the rapid sequencing of its genome revealed several novel proteins that are possible targets for antiviral drugs or vaccines. Research is underway on both fronts, since another outbreak is anticipated.

Genome sequencing also provides insights into organisms involved in global ecological cycles (see Chapter 58). In addition to the well-known carbon dioxide, another important gas contributing to the atmospheric "greenhouse effect" and global warming is methane ($CH_4$; see Figure 2.7). Some bacteria, such as *Methanococcus*, produce methane in the stomachs of cows. Others, such as *Methylococcus*, remove methane from the air and use it as an energy source. The genomes of both of these bacteria have been sequenced. Understanding the genes involved in methane production and oxidation may help us to slow the progress of global warming.

### Metagenomics allows us to describe new organisms and ecosystems

If you take a microbiology laboratory course you will learn how to identify various prokaryotes on the basis of their growth in lab cultures. For example, staphylococci are a group of bacteria that infect skin and nasal passages. When grown on a special medium called blood agar they form round, raised colonies. Microorganisms can also be identified by their nutritional requirements or the conditions under which they will grow (for example, aerobic versus anaerobic). Such culture methods have been the mainstay of microbial identification for over a century and are still useful and important. However, scientists can now use PCR and modern DNA analysis techniques to analyze microbes *without* culturing them in the laboratory.

In 1985, Norman Pace, then at Indiana University, came up with the idea of isolating DNA directly from environmental samples. He used PCR to amplify specific sequences from the samples to determine whether particular microbes were present. The PCR products were sequenced to explore their diversity. The term **metagenomics** was coined to describe this approach of analyzing genes without isolating the intact organism. It is now possible to perform shotgun sequencing with samples from almost any environment. The sequences can be used to detect the presence of known microbes and pathogens, and perhaps even the presence of heretofore unidentified organisms (**Figure 17.7**). For example:

- Shotgun sequencing of DNA from 200 liters of seawater indicated that it contained 5,000 different viruses and 2,000 different bacteria, many of which had not been described previously.
- One kilogram of marine sediment contained a million different viruses, most of them new.

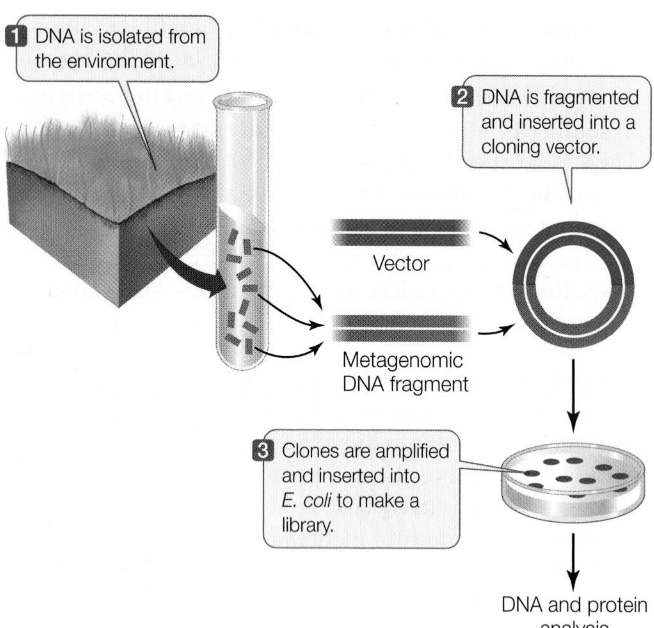

1 DNA is isolated from the environment.
2 DNA is fragmented and inserted into a cloning vector.
Vector
Metagenomic DNA fragment
3 Clones are amplified and inserted into E. coli to make a library.
DNA and protein analysis

**17.7 Metagenomics** Microbial DNA extracted from the environment can be amplified and analyzed. This has led to the description of many new genes and species.

- Water runoff from a mine contained many new species of prokaryotes thriving in this apparently inhospitable environment. Some of these organisms exhibited metabolic pathways that were previously unknown to biologists. These organisms and their capabilities may be useful in cleaning up pollutants from the water.

These and other discoveries are truly extraordinary and potentially very important. It is estimated that 90 percent of the microbial world has been invisible to biologists and is only now being revealed by metagenomics. Entirely new ecosystems of bacteria and viruses are being discovered in which, for example, one species produces a molecule that another metabolizes. It is hard to overemphasize the importance of such an increase in our knowledge of the hidden world of microbes. This knowledge will help us to understand natural ecological processes, and has the potential to help us find better ways to manage environmental catastrophes such as oil spills, or remove toxic heavy metals from soil.

### Will defining the genes required for cellular life lead to artificial life?

When the genomes of prokaryotes and eukaryotes are compared, a striking conclusion arises: certain genes are present in all organisms (universal genes). There are also some (nearly) universal gene segments that are present in many genes in many organisms; for example, the sequence that codes for an ATP binding site. These findings suggest that there is some ancient, minimal set of DNA sequences common to all cells. One way to identify these sequences is to look for them in computer analyses of sequenced genomes.

Another way to define the minimal genome is to take an organism with a simple genome and deliberately mutate one gene at a time to see what happens. *M. genitalium* has one of the smallest known genomes—only 482 protein-coding genes. Even so, some of its genes are dispensable under some circumstances. For example, it has genes for metabolizing both glucose and fructose, but it can survive in the laboratory on a medium containing only one of these sugars.

What about other genes? Researchers have addressed this question with experiments involving the use of transposons as mutagens. When transposons in the bacterium are activated, they insert themselves into genes at random, mutating and inactivating them (**Figure 17.8**). The mutated bacteria are tested for growth and survival, and DNA from interesting mutants is sequenced to find out which genes contain transposons.

The astonishing result of these studies is that *M. genitalium* can survive in the laboratory with a minimal genome of only 382 functional genes! Is this really all it takes to make a viable organism? Experiments are underway to make a synthetic genome based on that of *M. genitalium*, and then insert it into an empty bacterial cell. If the cell starts transcribing mRNA and making proteins—is in fact viable—it may turn out to be the first life created by humans.

In addition to the technical feat of creating artificial life, this technique could have important applications. New microbes could be made with entirely new abilities, such as degrading oil spills, making synthetic fibers, reducing tooth decay, or converting cellulose to ethanol for use as fuel. On the other hand, fears of the misuse or mishandling of this knowledge are not unfounded. For example, it might also be possible to develop synthetic bacteria harmful to people, animals or plants, and use them as agents of biological warfare or bioterrorism. The "genomics genie" is, for better or worse, already out of the bottle. Hopefully human societies will use it to their benefit.

## 17.2 RECAP

**DNA sequencing is used to study the genomes of prokaryotes that are important to humans and to ecosystems. Functional genomics uses gene sequences to determine the functions of the gene products. Comparative genomics compares gene sequences from different organisms to help identify their functions and evolutionary relationships. Transposable elements and transposons move from one place to another in the genome.**

- Give some examples of prokaryotic genomes that have been sequenced. What have the sequences shown? **See pp. 371–373**
- What is metagenomics and how is it used? **See pp. 373–374 and Figure 17.7**
- How are selective inactivation studies being used to determine the minimal genome? **See p. 374 and Figure 17.8**

INVESTIGATING LIFE

**17.8 Using Transposon Mutagenesis to Determine the Minimal Genome** *Mycoplasma genitalium* has the smallest number of genes of any prokaryote. But are all of its genes essential to life? By inactivating the genes one by one, scientists determined which of them are essential for the cell's survival. This research may lead to the construction of artificial cells with customized genomes, designed to perform functions such as degrading oil and making plastics.

**HYPOTHESIS** Only some of the genes in a bacterial genome are essential for cell survival.

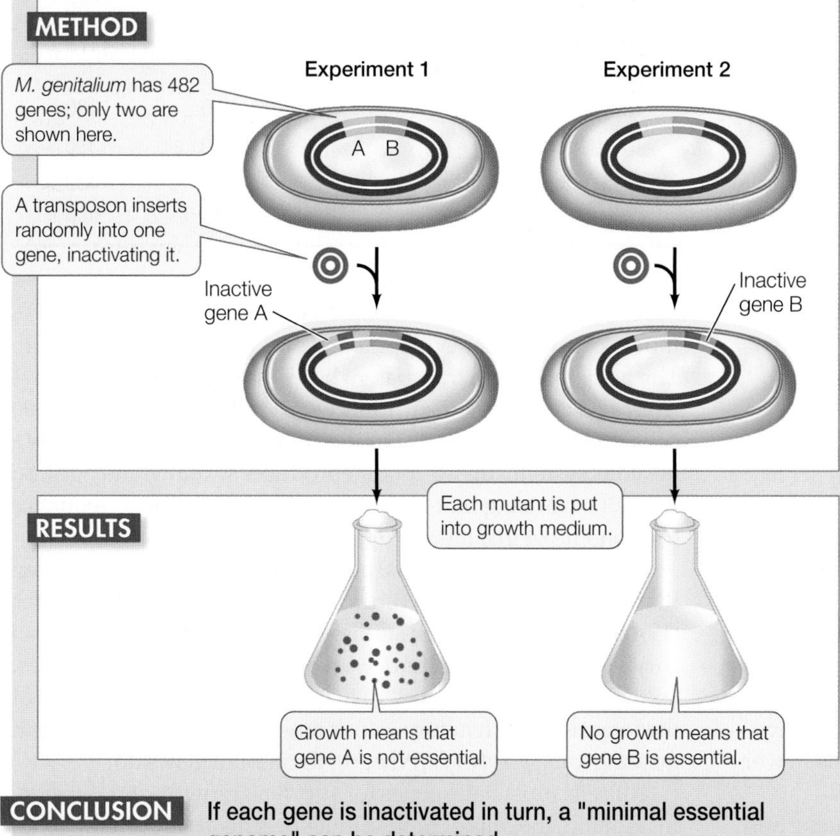

METHOD
Experiment 1
Experiment 2
M. genitalium has 482 genes; only two are shown here.
A B
A transposon inserts randomly into one gene, inactivating it.
Inactive gene A
Inactive gene B
RESULTS
Each mutant is put into growth medium.
Growth means that gene A is not essential.
No growth means that gene B is essential.

**CONCLUSION** If each gene is inactivated in turn, a "minimal essential genome" can be determined.

Go to **yourBioPortal.com** for original citations, discussions, and relevant links for all INVESTIGATING LIFE figures.

Advances in DNA sequencing and analysis have led to the rapid sequencing of eukaryotic genomes. We now turn to the results of these analyses.

## 17.3 What Have We Learned from Sequencing Eukaryotic Genomes?

As genomes have been sequenced and described, a number of major differences have emerged between eukaryotic and prokaryotic genomes (**Table 17.1**). Key differences include:

- *Eukaryotic genomes are larger than those of prokaryotes*, and they have more protein-coding genes. This difference is not surprising, given that multicellular organisms have many cell types with specific functions. Many proteins are needed to do those specialized jobs. A typical virus contains enough DNA to code for only a few proteins—about 10,000 base pairs (bp). As we saw above, the simplest prokaryote, *Mycoplasma*, has several hundred protein-coding genes in a genome of 0.5 million bp. A rice plant, in contrast, has 37,544 genes.
- *Eukaryotic genomes have more regulatory sequences*—and many more regulatory proteins—than prokaryotic genomes. The greater complexity of eukaryotes requires much more regulation, which is evident in the many points of control associated with the expression of eukaryotic genes (see Figure 16.13).
- *Much of eukaryotic DNA is noncoding*. Distributed throughout many eukaryotic genomes are various kinds of DNA sequences that are not transcribed into mRNA, most notably introns and gene control sequences. As we discuss in Chapter 16, some noncoding sequences are transcribed into microRNAs. In addition, eukaryotic genomes contain various kinds of repeated sequences. These features are rare in prokaryotes.
- *Eukaryotes have multiple chromosomes*. The genomic "encyclopedia" of a eukaryote is separated into multiple "volumes." Each chromosome must have, at a minimum, three defining DNA sequences that we have described in previous chapters: an origin of replication (*ori*) that is recognized by the DNA replication machinery; a centromere region that holds the replicated chromosomes together before mitosis; and a telomeric sequence at each end of the chromosome that maintains chromosome integrity.

### Model organisms reveal many characteristics of eukaryotic genomes

Most of the lessons learned from eukaryotic genomes have come from several simple model organisms that have been studied extensively: the yeast *Saccharomyces cerevisiae*, the nematode (roundworm) *Caenorhabditis elegans*, the fruit fly *Drosophila melanogaster*, and—representing plants—the thale cress, *Arabidopsis thaliana*. Model organisms have been chosen because they are relatively easy to grow and study in a laboratory, their genetics are well studied, and they exhibit characteristics that represent a larger group of organisms.

**YEAST: THE BASIC EUKARYOTIC MODEL** Yeasts are single-celled eukaryotes. Like most eukaryotes, they have membrane-enclosed organelles, such as the nucleus and endoplasmic reticulum, and a life cycle that alternates between haploid and diploid generations (see Figure 11.15).

## TABLE 17.1
### Representative Sequenced Genomes

| ORGANISM | HAPLOID GENOME SIZE (Mb) | NUMBER OF GENES | PROTEIN-CODING SEQUENCE |
|---|---|---|---|
| Bacteria | | | |
| *M. genitalium* | 0.58 | 485 | 88% |
| *H. influenzae* | 1.8 | 1,738 | 89% |
| *E. coli* | 4.6 | 4,377 | 88% |
| Yeasts | | | |
| *S. cerevisiae* | 12.5 | 5,770 | 70% |
| *S. pombe* | 12.5 | 4,929 | 60% |
| Plants | | | |
| *A. thaliana* | 115 | 28,000 | 25% |
| Rice | 390 | 37,544 | 12% |
| Animals | | | |
| *C. elegans* | 100 | 19,427 | 25% |
| *D. melanogaster* | 123 | 13,379 | 13% |
| Pufferfish | 342 | 27,918 | 10% |
| Chicken | 1,130 | 25,000 | 3% |
| Human | 3,300 | 24,000 | 1.2% |

Mb = millions of base pairs

While the prokaryote *E. coli* has a single circular chromosome with about 4.6 million bp and 4,290 protein-coding genes, budding yeast (*Saccharomyces cerevisiae*) has 16 linear chromosomes and a haploid content of more than 12.5 million bp, with 5,770 protein-coding genes. Gene inactivation studies similar to those carried out for *M. genitalium* (see Figure 17.7) indicate that fewer than 20 percent of these genes are essential to survival.

The most striking difference between the yeast genome and that of *E. coli* is in the number of genes for targeting proteins to organelles (**Table 17.2**). Both of these single-celled organisms appear to use about the same numbers of genes to perform the basic functions of cell survival. It is the compartmentalization of the eukaryotic yeast cell into organelles that requires it to have many more genes. This finding is direct, quantitative confirmation of something we have known for a century: the eukaryotic cell is structurally more complex than the prokaryotic cell.

**THE NEMATODE: UNDERSTANDING EUKARYOTIC DEVELOPMENT** In 1965 Sydney Brenner, fresh from being part of the team that first isolated mRNA, looked for a simple organism in which to study multicellularity. He settled on *Caenorhabditis elegans*, a millimeter-long nematode (roundworm) that normally lives in the soil. It can also live in the laboratory, where it has become a favorite model organism of developmental biologists (see Section 19.4). The nematode has a transparent body that develops over 3 days from a fertilized egg to an adult worm made up of nearly 1,000 cells. In spite of its small number of cells, the nematode has a nervous system, digests food, reproduces sexually, and ages. So it is not surprising that an intense effort was made to sequence the genome of this model organism.

## TABLE 17.2
### Comparison of the Genomes of *E. coli* and Yeast

| | *E. COLI* | YEAST |
|---|---|---|
| Genome length (base pairs) | 4,640,000 | 12,068,000 |
| Number of protein-coding genes | 4,290 | 5,770 |
| Proteins with roles in: | | |
| Metabolism | 650 | 650 |
| Energy production/storage | 240 | 175 |
| Membrane transport | 280 | 250 |
| DNA replication/repair/recombination | 120 | 175 |
| Transcription | 230 | 400 |
| Translation | 180 | 350 |
| Protein targeting/secretion | 35 | 430 |
| Cell structure | 180 | 250 |

The *C. elegans* genome (100 million bp) is eight times larger than that of yeast and has 3.5 times as many protein-coding genes (19,427). Gene inactivation studies have shown that the worm can survive in laboratory cultures with only 10 percent of these genes. So the "minimum genome" of a worm is about twice the size of that of yeast, which in turn is four times the size of the minimum genome for *Mycoplasma*. What do these extra genes do?

All cells must have genes for survival, growth, and division. In addition, the cells of multicellular organisms must have genes for holding cells together to form tissues, for cell differentiation, and for intercellular communication. Looking at **Table 17.3**, you will recognize functions that we discussed in earlier chapters,

## TABLE 17.3
### *C. elegans* Genes Essential to Multicellularity

| FUNCTION | PROTEIN/DOMAIN | NUMBER OF GENES |
|---|---|---|
| Transcription control | Zinc finger; homeobox | 540 |
| RNA processing | RNA binding domains | 100 |
| Nerve impulse transmission | Gated ion channels | 80 |
| Tissue formation | Collagens | 170 |
| Cell interactions | Extracellular domains; glycotransferases | 330 |
| Cell–cell signaling | G protein-linked receptors; protein kinases; protein phosphatases | 1,290 |

including gene regulation (see Chapter 16) and cell communication (see Chapter 7).

***DROSOPHILA MELANOGASTER*: RELATING GENETICS TO GENOMICS** The fruit fly *Drosophila melanogaster* is a famous model organism. Studies of fruit fly genetics resulted in the formulation of many basic principles of genetics (see Section 12.4). Over 2,500 mutations of *D. melanogaster* had been described by the 1990s when genome sequencing began, and this fact alone was a good reason for sequencing the fruit fly's DNA. The fruit fly is a much larger organism than *C. elegans*, both in size (it has 10 times more cells) and complexity, and it undergoes complicated developmental transformations from egg to larva to pupa to adult.

Not surprisingly, the fly's genome (about 123 million bp) is larger than that of *C. elegans*. But as we mentioned earlier, genome size does not necessarily correlate with the number of genes encoded. In this case, the larger fruit fly genome contains fewer genes (13,379) than the smaller nematode genome. **Figure 17.9** summarizes the functions of the *Drosophila* genes that have been characterized so far; this distribution is typical of complex eukaryotes.

***ARABIDOPSIS*: STUDYING THE GENOMES OF PLANTS** About 250,000 species of flowering plants dominate the land and fresh water. But in the context of the history of life, the flowering plants are fairly young, having evolved only about 200 million years ago. The genomes of some plants are huge—for example, the genome of corn is about 3 billion bp, and that of wheat is 16 billion bp. So although we are naturally most interested in the genomes of plants we use as food and fiber, it is not surprising that scientists first chose to sequence a simpler flowering plant.

*Arabidopsis thaliana*, thale cress, is a member of the mustard family and has long been a favorite model organism of plant biologists. It is small (hundreds could grow and reproduce in the space occupied by this page) and easy to manipulate, and has a relatively small (115 million bp) genome.

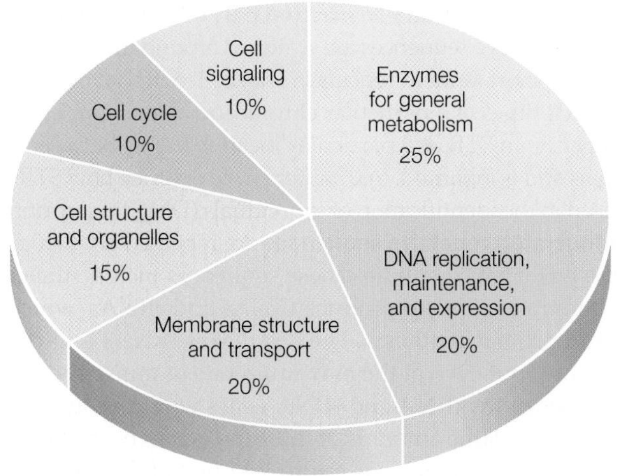


**17.9 Functions of the Eukaryotic Genome** The distribution of gene functions in *Drosophila melanogaster* shows a pattern that is typical of many complex organisms.

**TABLE 17.4**
***Arabidopsis* Genes Unique to Plants**

| FUNCTION | NUMBER OF GENES |
|---|---|
| Cell wall and growth | 42 |
| Water channels | 300 |
| Photosynthesis | 139 |
| Defense and metabolism | 94 |

The *Arabidopsis* genome has about 28,000 protein-coding genes but, remarkably, many of these genes are duplicates and probably originated by chromosomal rearrangements. When these duplicate genes are subtracted from the total, about 15,000 unique genes are left—similar to the gene numbers found in fruit flies and nematodes. Indeed, many of the genes found in these animals have homologs (genes with very similar sequences) in *Arabidopsis* and other plants, suggesting that plants and animals have a common ancestor.

But *Arabidopsis* has some genes that distinguish it as a plant (**Table 17.4**). These include genes involved in photosynthesis, in the transport of water into the root and throughout the plant, in the assembly of the cell wall, in the uptake and metabolism of inorganic substances from the environment, and in the synthesis of specific molecules used for defense against microbes and herbivores (organisms that eat plants). These plant defense molecules may be a major reason why the number of protein-coding genes in plants is higher than in animals. Plants cannot escape their enemies or other adverse conditions as animals can, and so they must cope with the situation where they are. So they make tens of thousands of molecules to fight their enemies and adapt to the environment (see Chapter 39).

These plant-specific genes are also found in the genomes of other plants, including rice, the first major crop plant whose sequence has been determined. Rice (*Oryza sativa*) is the world's most important crop; it is a staple in the diet of 3 billion people. The larger genome in rice has a set of genes remarkably similar to that of *Arabidopsis*. More recently the genome of the poplar tree, *Populus trichocarpa*, was sequenced to gain insight into the potential for this rapidly growing tree to be used as a source of fixed carbon for making fuel. A comparison of the three genomes shows many genes in common, comprising the *basic plant genome* (**Figure 17.10**).

## Eukaryotes have gene families

About half of all eukaryotic protein-coding genes exist as only one copy in the haploid genome (two copies in somatic cells). The rest are present in multiple copies, which arose from gene duplications. Over evolutionary time, different copies of genes have undergone separate mutations, giving rise to groups of closely related genes called **gene families**. Some gene families, such as those encoding the globin proteins that make up hemoglobin, contain only a few members; other families, such as the genes encoding the immunoglobulins that make up antibodies, have hundreds of members. In the human genome,

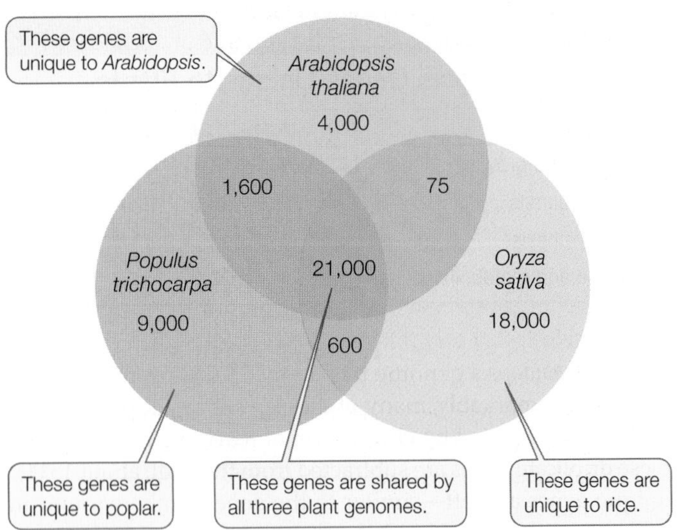


**17.10 Plant Genomes** Three plant genomes share a common set of approximately 21,000 genes that appear to comprise the "minimal" plant genome.

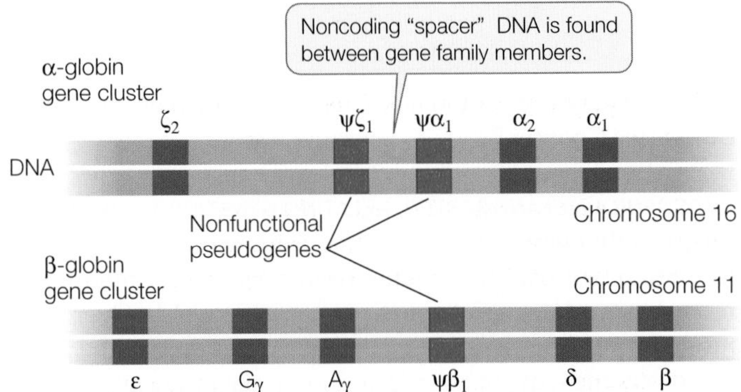


**17.11 The Globin Gene Family** The α-globin and β-globin clusters of the human globin gene family are located on different chromosomes. The genes of each cluster are separated by noncoding "spacer" DNA. The nonfunctional pseudogenes are indicated by the Greek letter psi (ψ). The γ gene has two variants, $A_\gamma$ and $G_\gamma$.

there are 24,000 protein-coding genes, but 16,000 distinct gene families. So only one-third of the human genes are unique.

The DNA sequences in a gene family are usually different from one another. As long as at least one member encodes a functional protein, the other members may mutate in ways that change the functions of the proteins they encode. For evolution, the availability of multiple copies of a gene allows for selection of mutations that provide advantages under certain circumstances. If a mutated gene is useful, it may be selected for in succeeding generations. If the mutated gene is a total loss, the functional copy is still there to carry out its role.

The gene family encoding the globins is a good example of the gene families found in vertebrates. These proteins are found in hemoglobin and myoglobin (an oxygen-binding protein present in muscle). The globin genes all arose long ago from a single common ancestral gene. In humans, there are three functional members of the α-globin cluster and five in the β-globin cluster (**Figure 17.11**). In adults, each hemoglobin molecule is a tetramer containing two identical α-globin subunits, two identical β-globin subunits, and four heme pigments (see Figure 3.10).

During human development, different members of the globin gene cluster are expressed at different times and in different tissues. This differential gene expression has great physiological significance. For example, hemoglobin containing γ-globin, a subunit found in the hemoglobin of the human fetus, binds $O_2$ more tightly than adult hemoglobin does. This specialized form of hemoglobin ensures that in the placenta, $O_2$ will be transferred from the mother's blood to the developing fetus's blood. Just before birth the liver stops synthesizing fetal hemoglobin and the bone marrow cells take over, making the adult forms (2α and 2β). Thus hemoglobins with different binding affinities for $O_2$ are provided at different stages of human development.

In addition to genes that encode proteins, many gene families include nonfunctional **pseudogenes**, which are designated with the Greek letter psi (Ψ) (see Figure 17.11). These pseudogenes result from mutations that cause a loss of function rather than an enhanced or new function. The DNA sequence of a pseudogene may not differ greatly from that of other family members. It may simply lack a promoter, for example, and thus fail to be transcribed. Or it may lack a recognition site needed for the removal of an intron, so that the transcript it makes is not correctly processed into a useful mature mRNA. In some gene families pseudogenes outnumber functional genes. Because some members of the family are functional, there appears to be little selection pressure for the deletion of pseudogenes.

### Eukaryotic genomes contain many repetitive sequences

Eukaryotic genomes contain numerous repetitive DNA sequences that do not code for polypeptides. These include highly repetitive sequences, moderately repetitive sequences, and transposons.

**Highly repetitive sequences** are short (less than 100 bp) sequences that are repeated thousands of times in tandem (side-by-side) arrangements in the genome. They are not transcribed. Their proportion in eukaryotic genomes varies, from 10 percent in humans to about half the genome in some species of fruit flies. Often they are associated with heterochromatin, the densely packed, transcriptionally inactive part of the genome. Other highly repetitive sequences are scattered around the genome. For example, short tandem repeats (STRs) of 1–5 bp can be repeated up to 100 times at a particular chromosomal location. The copy number of an STR at a particular location varies between individuals and is inherited. In Chapter 15 we describe how STRs can be used in the identification of individuals (DNA fingerprinting).

**Moderately repetitive sequences** are repeated 10–1000 times in the eukaryotic genome. These sequences include the genes that are transcribed to produce tRNAs and rRNAs, which are used in protein synthesis. The cell makes tRNAs and rRNAs constantly, but even at the maximum rate of transcription, single copies of the tRNA and rRNA genes would be inadequate to supply the large amounts of these molecules needed by most cells. Thus the genome has multiple copies of these genes.

In mammals, four different rRNA molecules make up the ribosome: the 18S, 5.8S, 28S, and 5S rRNAs. (The S stands for Svedberg unit, which is a measure of size.) The 18S, 5.8S, and

28S rRNAs are transcribed together as a single precursor RNA molecule (**Figure 17.12**). As a result of several posttranscriptional steps, the precursor is cut into the final three rRNA products, and the noncoding "spacer" RNA is discarded. The sequence encoding these RNAs is moderately repetitive in humans: a total of 280 copies of the sequence are located in clusters on five different chromosomes.

**TRANSPOSONS** Apart from the RNA genes, most moderately repetitive sequences are not stably integrated into the genome. Instead, these sequences can move from place to place, and are thus called transposable elements or transposons. Prokaryotes also have transposons (see Figure 17.6). Transposons make up over 40 percent of the human genome and about 50 percent of the maize genome, although the percentage is smaller (3–10 percent) in many other eukaryotes.

There are four main types of transposons in eukaryotes:

1. *SINEs* (short *in*terspersed *e*lements) are up to 500 bp long and are transcribed but not translated. There are about 1.5 million of them scattered over the human genome, making up about 15 percent of the total DNA content. A single type, the 300-bp Alu element, accounts for 11 percent of the human genome; it is present in a million copies.
2. *LINEs* (long *in*terspersed *e*lements) are up to 7,000 bp long, and some are transcribed and translated into proteins. They constitute about 17 percent of the human genome.

   SINEs and LINEs move about the genome in a distinctive way: they are transcribed into RNA, which then acts as a template for new DNA. The new DNA becomes inserted at a new location in the genome. This "copy and paste" mechanism results in two copies of the transposon: one at the original location and the other at a new location.
3. *Retrotransposons* also make RNA copies of themselves when they move about the genome. Some of them encode proteins needed for their own transposition, and others do not. SINEs and LINEs are types of retrotransposons. Non-SINE, non-LINE retrotransposons constitute about 8 percent of the human genome.
4. *DNA transposons* do not use RNA intermediates. Like some prokaryotic transposable elements, they are excised from the original location and become inserted at a new location without being replicated.

What role do these moving sequences play in the cell? The best answer so far seems to be that transposons are simply cellular parasites that can be replicated. The insertion of a transposon at a new location can have important consequences. For example, the insertion of a transposon into the coding region of a gene results in a mutation (see Figure 17.8). This phenomenon accounts for a few rare forms of several genetic diseases in humans, including hemophilia and muscular dystrophy. If the insertion of a transposon takes place in the germ line, a gamete with a new mutation results. If the insertion takes place in a somatic cell, cancer may result.

Sometimes an adjacent gene can be replicated along with a transposon, resulting in a gene duplication. A transposon can carry a gene, or a part of it, to a new location in the genome, shuffling the genetic material and creating new genes. Clearly, transposition stirs the genetic pot in the eukaryotic genome and thus contributes to genetic variation.

Section 5.5 describes the theory of *endosymbiosis*, which proposes that chloroplasts and mitochondria are the descendants of once free-living prokaryotes. Transposons may have played a role in endosymbiosis. In living eukaryotes the chloroplasts and mitochondria contain some DNA, but the nucleus contains most of the genes

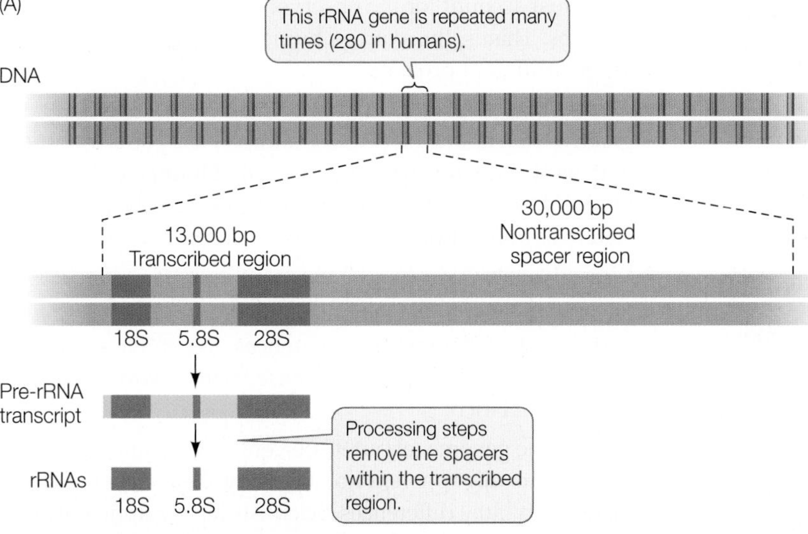
(A)
This rRNA gene is repeated many times (280 in humans).
DNA
13,000 bp
Transcribed region
30,000 bp
Nontranscribed spacer region
18S
5.8S
28S
Pre-rRNA transcript
Processing steps remove the spacers within the transcribed region.
rRNAs
18S
5.8S
28S

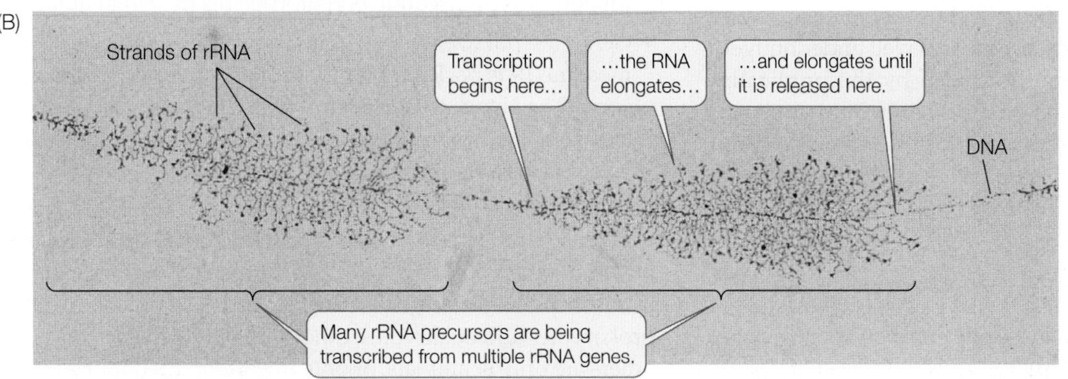
(B)
Strands of rRNA
Transcription begins here...
...the RNA elongates...
...and elongates until it is released here.
DNA
Many rRNA precursors are being transcribed from multiple rRNA genes.

**17.12 A Moderately Repetitive Sequence Codes for rRNA** (A) This rRNA gene, along with its nontranscribed spacer region, is repeated 280 times in the human genome, with clusters on five chromosomes. (B) This electron micrograph shows transcription of multiple rRNA genes.

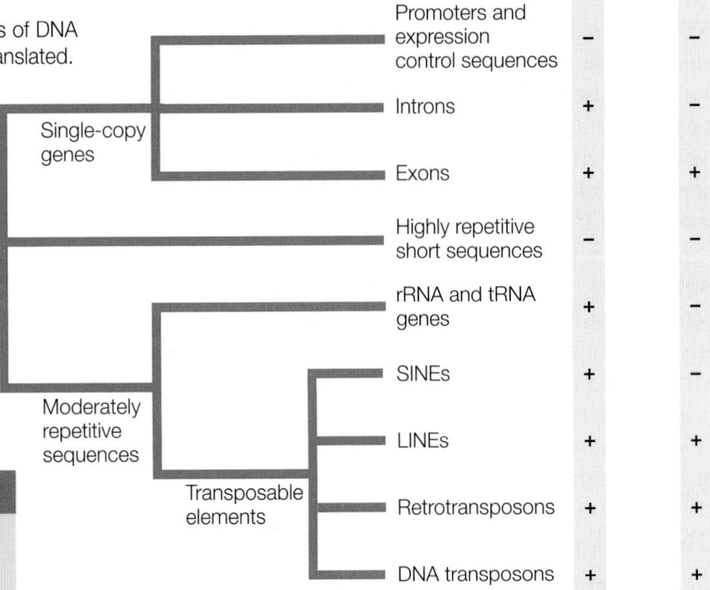


**17.13 Sequences in the Eukaryotic Genome** There are many types of DNA sequences. Some are transcribed, and some of those sequences are translated.

that encode the organelles' proteins. If the organelles were once independent, they must originally have contained all of those genes. How did the genes move to the nucleus? They may have done so by DNA transpositions between organelles and the nucleus, which still occur today. The DNA that remains in the organelles may be the remnants of more complete prokaryotic genomes.

See **Figure 17.13** for a summary of the various types of sequences in the human genome.

## 17.3 RECAP

**The sequencing of the genomes of model organisms demonstrated common features of the eukaryotic genome, including the presence of repetitive sequences and transposons. Some eukaryotic genes are in families, which may include members that are mutated and nonfunctional. Some sequences are transcribed, but others are not.**

- What are the major differences between prokaryotic and eukaryotic genomes? See p. 375
- Describe one function of genes found in *C. elegans* that has no counterpart in the genome of yeast. See p. 376 and Table 17.3
- What is the evolutionary role of eukaryotic gene families? See p. 377
- Why are there multiple copies of sequences coding for rRNA in the mammalian genome? See p. 378
- What effects can transposons have on a genome? See p. 379

The analysis of eukaryotic genomes has resulted in an enormous amount of useful information, as we have seen. In the next section we look more closely at the human genome.

# 17.4 What Are the Characteristics of the Human Genome?

By the start of 2005 the first human genome sequences were completed, two years ahead of schedule and well under budget. The published sequences, one produced by the publically funded Human Genome Project, and the other by a private company, were haploid genomes that were composites of several people. Since 2005, the diploid genomes of several individuals have been sequenced and published.

### The human genome sequence held some surprises

The following are just some of the interesting facts that we have learned about the human genome:

- Of the 3.3 billion base pairs in the haploid human genome, fewer than 2 percent (about 24,000 genes) make up protein-coding regions. This was a surprise. Before sequencing began, humans were estimated, based on the diversity of their proteins, to have 80,000–150,000 genes. The actual number of genes—not many more than in a fruit fly—means that posttranscriptional mechanisms (such as alternative splicing) must account for the observed number of proteins in humans. That is, the average human gene must code for several different proteins.
- The average gene has 27,000 base pairs. Gene sizes vary greatly, from about 1,000 to 2.4 million base pairs. Variation in gene size is to be expected given that human proteins (and RNAs) vary in size, from 100 to about 5,000 amino acids per polypeptide chain.
- Virtually all human genes have many introns.
- Over 50 percent of the genome is made up of transposons and other highly repetitive sequences. Repetitive sequences near genes are GC-rich, while those farther away from genes are AT-rich.
- Most of the genome (about 97 percent) is the same in all people. Despite this apparent homogeneity, there are, of course, many individual differences. Scientists have mapped over 7 million single nucleotide polymorphisms (SNPs) in humans.
- Genes are not evenly distributed over the genome. Chromosome 19 is packed densely with genes, while chromosome 8 has long stretches without coding regions. The Y chromosome has the fewest genes (231), while chromosome 1 has the most (2,968).

Comparisons between sequenced genomes from prokaryotes and eukaryotes have revealed some of the evolutionary relationships between genes. Some genes are present in both prokaryotes and eukaryotes; others are only in eukaryotes; still others are only in animals, or only in vertebrates (**Figure 17.14**).

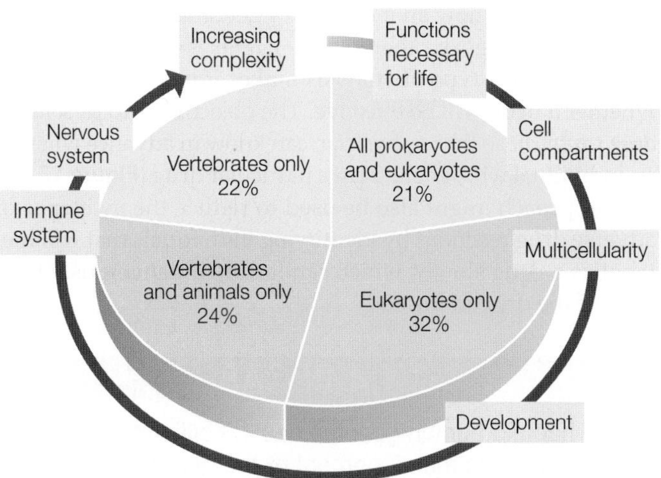


**17.14 Evolution of the Genome** A comparison of the human and other genomes has revealed how genes with new functions have been added over the course of evolution. Each percentage number refers to genes in the human genome. Thus, 21 percent of human genes have homologs in prokaryotes and other eukaryotes, 32 percent of human genes occur only in other eukaryotes, and so on.

More comparative genomics is possible now that the genomes of two other primates, the chimpanzee and the rhesus macaque, have been sequenced. The chimpanzee is evolutionarily close to humans, and shares 95 percent of the human genome sequence. The more distantly related rhesus macaque shares 91 percent of the human sequence. The search is on for a set of human genes that differ from the other primates and "make humans human."

## Human genomics has potential benefits in medicine

Complex phenotypes are determined not by single genes, but by multiple genes interacting with the environment. The single-allele models of phenylketonuria and sickle-cell anemia (see Chapter 15) do not apply to such common disorders as diabetes, heart disease, and Alzheimer's disease. To understand the genetic bases of these diseases, biologists are now using rapid genotyping technologies to create "haplotype maps," which are used to identify SNPs (pronounced "snips") that are linked to genes involved in disease.

**HAPLOTYPE MAPPING** The SNPs that differ between individuals are not inherited as independent alleles. Rather, *a set of SNPs that are present on a segment of chromosome are usually inherited as a unit.* This linked piece of a chromosome is called a **haplotype**. You can think of the haplotype as a sentence and the SNP as a word in the sentence. Analyses of haplotypes in humans from all over the world have shown that there are at most 500,000 common variations.

**GENOTYPING TECHNOLOGY AND PERSONAL GENOMICS** New technologies are continually being developed to analyze thousands or millions of SNPs in the genomes of individuals. Such technologies include rapid sequencing methods and DNA microarrays that depend on DNA hybridization to identify specific SNPs. For example, a microarray of 500,000 SNPs has been used to analyze thousands of people to find out which SNPs are associated with specific diseases. The aim is to *correlate the SNP-defined haplotype with a disease state*. The amount of data is prodigious: 500,000 SNPs, thousands of people, thousands of medical records. With so much natural variation, statistical measures of association between a haplotype and a disease need to be very rigorous.

These association tests have revealed particular haplotypes or alleles that are associated with modestly increased risks for such diseases as breast cancer, diabetes, arthritis, obesity, and coronary heart disease (**Figure 17.15** and **Table 17.5**). Private companies will now scan a human genome for these variants—and the price for this service keeps getting lower. However, at this point it is unclear what a person without symptoms should do with the information, since multiple genes, environmental influences, and epigenetic effects all contribute to the development of these diseases.

Of course, the best way to analyze a person's genome is by actually sequencing it. Until recently, this was prohibitively expensive. As we mentioned earlier, DNA pioneer James Watson's genome cost over $1 million, certainly too much for a typical person or insurance company to afford in the context of health care. But with advances in sequencing technologies the cost is decreasing rapidly. One new method automatically sequences protein-coding exons only, for example. Once the cost of genome sequencing is within an affordable range, SNP testing will be superseded.

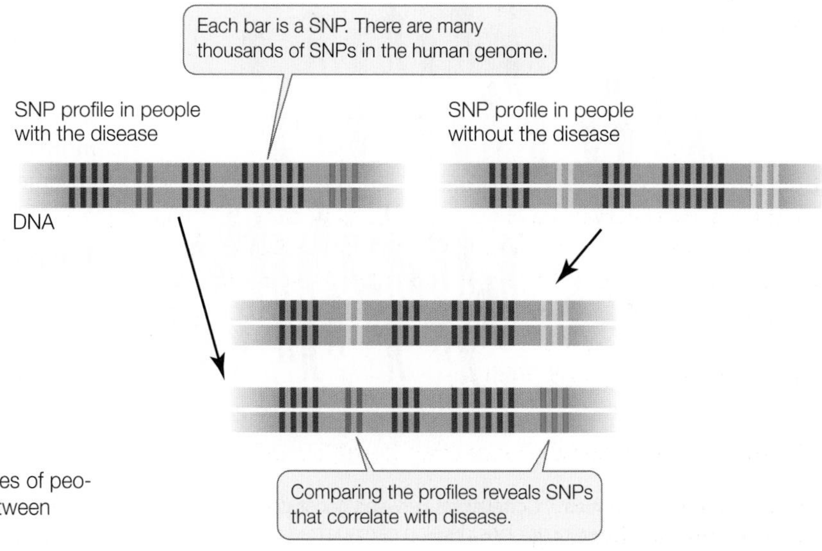


**17.15 SNP Genotyping and Disease** Scanning the genomes of people with and without particular diseases reveals correlations between SNPs and complex diseases.

TABLE 17.5

**SNP Human Genome Scans and Diseases**

| DISEASE | LOCATION OF SNP (CHROMOSOME NUMBER) | % INCREASED RISK | |
|---|---|---|---|
| | | HETEROZYGOTES | HOMOZYGOTES |
| Breast cancer | 8 | 20 | 63 |
| Coronary heart disease | 9 | 20 | 56 |
| Heart attack | 9 | 25 | 64 |
| Obesity | 16 | 32 | 67 |
| Diabetes | 10 | 65 | 277 |
| Prostate cancer | 8 | 26 | 58 |

**PHARMACOGENOMICS** Genetic variation can affect how an individual responds to a particular drug. For example, a drug may be chemically modified in the liver to make it more or less active. Consider an enzyme that catalyzes the following reaction:

active drug → less active drug

A mutation in the gene that encodes this enzyme may make the enzyme less active. For a given dose of the drug, a person with the mutation would have more active drug in the bloodstream than a person without the mutation. So the effective dose of the drug would be lower in these people.

Now consider a different case, in which the liver enzyme is needed to make the drug active:

inactive drug → active drug

A person carrying a mutation in the gene encoding this liver enzyme would not be affected by the drug, since the activating enzyme is not present.

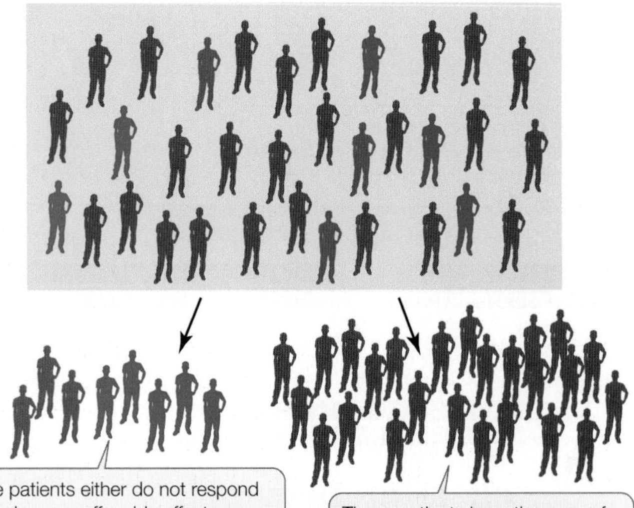

All patients with the same diagnosis
These patients either do not respond to the drug or suffer side effects. They need an alternative drug or dose.
These patients have the genes for an effective response to the drug.

**17.16 Pharmacogenomics** Correlations between genotypes and responses to drugs will help physicians develop personalized medical care.

The study of how an individual's genome affects his or her response to drugs or other outside agents is called **pharmacogenomics**. This type of analysis makes it possible to predict whether a drug will be effective. The objective is to *personalize drug treatment* so that a physician can know in advance whether an individual will benefit from a particular drug (**Figure 17.16**). This approach might also be used to reduce the incidence of adverse drug reactions by identifying individuals that will metabolize a drug slowly, which can lead to a dangerously high level of the drug in the body.

### 17.4 RECAP

**The haploid human genome has 3.3 billion base pairs, but less than 2 percent of the genome codes for proteins. Most human genes are subject to alternative splicing; this may account for the fact that there are more proteins than genes. SNP mapping to find correlations with disease and drug susceptibility holds promise for personalized medicine.**

- What are some of the major characteristics of the human genome? See p. 380
- How does SNP mapping work in personalized medicine? See pp. 381–382 and Figures 17.15 and 17.16

Genome sequencing has had great success in advancing biological understanding. High-throughput technologies are now being applied to other components of the cell: proteins and metabolites. We now turn to the results of these studies.

## 17.5 What Do the New Disciplines Proteomics and Metabolomics Reveal?

"The human genome is the book of life." Statements like this were common at the time the human genome sequence was first revealed. They reflect "genetic determinism," that a person's phenotype is determined by his or her genotype. But is an organism just a product of gene expression? We know that it is not. The proteins and small molecules present in any cell at a given point in time reflect not just gene expression but modifications by the intracellular and extracellular environment. Two new fields have emerged to complement genomics and take a more complete snapshot of a cell and organism—proteomics and metabolomics.

### The proteome is more complex than the genome

As mentioned above, many genes encode more than a single protein (**Figure 17.17A**). Alternative splicing leads to different combinations of exons in the mature mRNAs transcribed from a single gene (see Figure 16.22). Posttranslational modifications also increase the number of proteins that can be derived from one gene (see Figure 14.22). The **proteome** is the sum total of the proteins produced by an organism, and it is more complex than its genome.

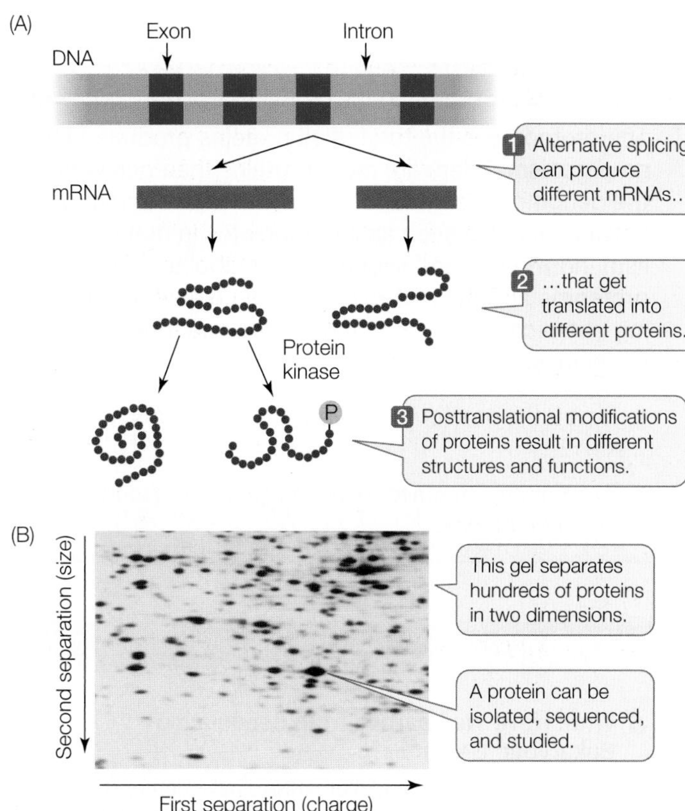


**17.17 Proteomics** (A) A single gene can code for multiple proteins. (B) A cell's proteins can be separated on the basis of charge and size by two-dimensional gel electrophoresis. The two separations can distinguish most proteins from one another.

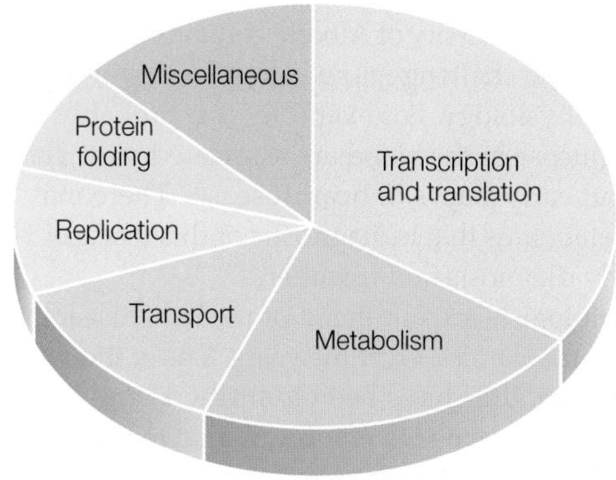


**17.18 Proteins of the Eukaryotic Proteome** About 1,300 proteins are common to all eukaryotes and fall into these categories. Although their amino acid sequences may differ to a limited extent, they perform the same essential functions in all eukaryotes.

Two methods are commonly used to analyze the proteome:

- Because of their unique amino acid compositions (primary structures), most proteins have unique combinations of electric charge and size. On the basis of these two properties, they can be separated by two-dimensional gel electrophoresis. Thus isolated, individual proteins can be analyzed, sequenced, and studied (**Figure 17.17B**).
- Mass spectrometry uses electromagnets to identify proteins by the masses of their atoms and displays them as peaks on a graph.

The ultimate aim of proteomics is just as ambitious as that of genomics. While genomics seeks to describe the genome and its expression, proteomics seeks to identify and characterize all of the expressed proteins.

Comparisons of the proteomes of humans and other eukaryotic organisms have revealed a common set of proteins that can be categorized into groups with similar amino acid sequences and similar functions. Forty-six percent of the yeast proteome, 43 percent of the worm proteome, and 61 percent of the fly proteome are shared by the human proteome. Functional analyses indicate that this set of 1,300 proteins provide the basic metabolic functions of a eukaryotic cell, such as glycolysis, the citric acid cycle, membrane transport, protein synthesis, DNA replication, and so on. (**Figure 17.18**).

Of course, these are not the only human proteins. There are many more, which presumably distinguish us as *human* eukaryotic organisms. As we have mentioned before, proteins have different functional regions called domains (for example, a domain for binding a substrate, or a domain for spanning a membrane). While a particular organism may have many unique proteins, those proteins are often just unique combinations of domains that exist in other organisms. *This reshuffling of the genetic deck is a key to evolution.*

### Metabolomics is the study of chemical phenotype

Studying genes and proteins gives a limited picture of what is going on in a cell. But as we have seen, both gene function and protein function are affected by the internal and external environments of the cell. Many proteins are enzymes and their activities affect the concentrations of their substrates and products. So as the proteome changes, so will the abundances of these often-small molecules, called metabolites. The **metabolome** is the quantitative description of all of the *small molecules* in a cell or organism. These include:

- *Primary metabolites* involved in normal processes, such as intermediates in pathways like glycolysis. This category also includes hormones and other signaling molecules.
- *Secondary metabolites*, which are often unique to particular organisms or groups of organisms. They are often involved in special responses to the environment. Examples are antibiotics made by microbes, and the many chemicals made by plants that are used in defense against pathogens and herbivores.

Not surprisingly, measuring metabolites involves sophisticated analytical instruments. If you have studied organic or analytical chemistry, you may be familiar with gas chromatography and high-performance liquid chromatography, which separate molecules, and mass spectrometry and nuclear magnetic resonance spectroscopy, which are used to identify them. These measurements result in "chemical snapshots" of cells or organisms, which can be related to physiological states.

There has been some progress in defining the human metabolome. A database created by David Wishart and colleagues at the University of Alberta contains over 6,500 metabolite entries. The challenge now is to relate levels of these substances to physiology. For example, you probably know high levels of glucose in the blood are associated with diabetes. But what about early stages of heart disease? There may be a pattern of metabolites that is diagnostic of this disease. This could aid in early diagnosis and treatment.

Plant biologists are far ahead of medical researchers in the study of metabolomics. Over the years, tens of thousands of secondary metabolites have been identified in plants, many of them made in response to environmental challenges. Some of these are discussed in Chapter 39. The metabolome of the model organism *Arabidopsis thaliana* is being described, and will give insight into how a plant copes with stresses such as drought or pathogen attack. This knowledge could be helpful in optimizing plant growth for agriculture.

## 17.5 RECAP

**The proteome is the total of all proteins produced by an organism. There are more proteins than genes in the genome. The metabolome is the total content of small molecules such as intermediates in metabolism, hormones, and secondary metabolites. The proteome and the metabolome can be analyzed using chemical methods that separate and identify molecules.**

- How is the proteome analyzed? See p. 383 and Figure 17.17
- Explain the differences between genome, protoeome, and metabolome.

## CHAPTER SUMMARY

### 17.1 How Are Genomes Sequenced?

- The sequencing of genomes required the development of ways to cut large chromosomes into fragments, sequence each of the fragments, and then line them up on the chromosome. Review Figure 17.1, **ANIMATED TUTORIAL 17.1**
- **Hierarchical sequencing** involves mapping the genome with genetic markers, cutting the genome into smaller pieces and sequencing them, then lining up the sequences using the markers.
- **Shotgun sequencing** involves directly cutting the genome into overlapping fragments, sequencing them, and using a computer to line up the sequences.
- DNA sequencing technologies involve labeled nucleotides that terminate the growing polynucleotide chain. Review Figure 17.2
- Rapid, automated methods for **high-throughput sequencing** are being developed. Review Figure 17.3, **ANIMATED TUTORIAL 17.2**

### 17.2 What Have We Learned from Sequencing Prokaryotic Genomes?

- DNA sequencing is used to study the genomes of prokaryotes that are important to humans and ecosystems.
- **Functional genomics** aims to determine the functions of gene products. **Comparative genomics** involves comparisons of genes and genomes from different organisms to identify common features and functions.
- **Transposable elements** and **transposons** can move about the genome. Review Figure 17.6
- **Metagenomics** is the identification of DNA sequences without first isolating, growing and identifying the organisms present in an environmental sample. Many of these sequences are from prokaryotes that were heretofore unknown to biologists. Review Figure 17.7
- Transposon mutagenesis can be used to inactivate genes one by one. Then the organism can be tested for survival. In this way, a minimal genome of less than 350 genes was identified for the bacterium *Mycoplasma genitalium*. Review Figure 17.8

### 17.3 What Have We Learned from Sequencing Eukaryotic Genomes?

- Genome sequences from model organisms have demonstrated some common features of the eukaryotic genome. In addition, there are specialized genes for cellular compartmentation, development, and features unique to plants. Review Tables 17.1–17.4 and Figures 17.9 and 17.10
- Some eukaryotic genes exist as members of **gene families.** Proteins may be made from these closely related genes at different times and in different tissues. Some members of gene families may be nonfunctional **pseudogenes.**
- Repeated sequences are present in the eukaryotic genome.
- **Moderately repeated sequences** include those coding for rRNA. Review Figure 17.12

### 17.4 What Are the Characteristics of the Human Genome?

- The haploid human genome has 3.3 billion base pairs.
- Only 2 percent of the genome codes for proteins; the rest consists of repeated sequences and noncoding DNA.
- Virtually all human genes have introns, and alternative splicing leads to the production of more than one protein per gene.
- SNP genotyping correlates variations in the genome with diseases or drug sensitivity. It may lead to personalized medicine. Review Figure 17.15
- **Pharmacogenomics** is the analysis of genetics as applied to drug metabolism.

### 17.5 What Do the New Disciplines of Proteomics and Metabolomics Reveal?

- The **proteome** is the total protein content of an organism.
- There are more proteins than protein-coding genes in the genome.
- The proteome can be analyzed using chemical methods that separate and identify proteins. These include two-dimensional electrophoresis and mass spectrometry. See Figure 17.17
- The **metabolome** is the total content of small molecules, such as intermediates in metabolism, hormones, and secondary metabolites.

**SEE WEB ACTIVITY 17.1 for a concept review of this chapter.**

## SELF-QUIZ

1. Eukaryotic protein-coding genes differ from their prokaryotic counterparts in that eukaryotic genes
   *a.* are double-stranded.
   *b.* are present in only a single copy.
   *c.* contain introns.
   *d.* have promoters.
   *e.* are transcribed into mRNA.
2. A comparison of the genomes of yeast and bacteria shows that only yeast has many genes for
   *a.* energy metabolism.
   *b.* cell wall synthesis.
   *c.* intracellular protein targeting.
   *d.* DNA-binding proteins.
   *e.* RNA polymerase.
3. The genomes of the fruit fly and the nematode are similar to that of yeast, except that the former organisms have many genes for
   *a.* intercellular signaling.
   *b.* synthesis of polysaccharides.
   *c.* cell cycle regulation.
   *d.* intracellular protein targeting.
   *e.* transposable elements.
4. The minimum genome of *Mycoplasma genitalium*
   *a.* has 100 genes.
   *b.* has been used to create new species.
   *c.* has an RNA genome.
   *d.* is larger than the genome of *E. coli*.
   *e.* was derived by transposon mutagenesis.
5. Which is *not* true of metagenomics?
   *a.* It has been done with bacteria.
   *b.* It is done on rRNA sequences.
   *c.* It has revealed many new metabolic capacities.
   *d.* It involves extracting DNA from the environment.
   *e.* It cannot be done on seawater.
6. Transposons
   *a.* always use RNA for replication.
   *b.* are approximately 50 bp long.
   *c.* are made up of either DNA or RNA.
   *d.* do not contain genes coding for proteins.
   *e.* make up about 40 percent of the human genome.
7. Vertebrate gene families
   *a.* have mostly inactive genes.
   *b.* include the globins.
   *c.* are not produced by gene duplications.
   *d.* increase the number of unique genes in the genome.
   *e.* are not transcribed.
8. The DNA sequences that code for eukaryotic rRNA
   *a.* are transcribed only at the ribosome.
   *b.* are repeated hundreds of times.
   *c.* contain all the genes clustered directly beside one another.
   *d.* are on only one human chromosome.
   *e.* are identical to the sequences that code for miRNA.
9. The human genome
   *a.* contains very few repeated sequences.
   *b.* has 3.3 billion base pairs.
   *c.* was sequenced by hierarchical sequencing only.
   *d.* has genes evenly distributed along chromosomes.
   *e.* has few genes with introns.
10. Which of the following about genome sequencing is true?
    *a.* In hierarchical sequencing, but not high-throughput sequencing, DNA is amplified in BAC vectors.
    *b.* In hierarchical sequencing, a genetic map is made after the DNA is sequenced.
    *c.* Shotgun sequencing is considerably slower than hierarchical sequencing.
    *d.* The human genome was first sequenced by high-throughput methods.
    *e.* DNA sequence determination by chain termination is the basis of shotgun sequencing only.

## FOR DISCUSSION

1. In rats, a protein-coding gene 1,440 bp long codes for an enzyme made up of 192 amino acids. Discuss this apparent discrepancy. How long would the initial and final mRNA transcripts be?
2. The genomes of rice, wheat, and corn are similar to one another and to that of *Arabidopsis* in many ways. Discuss how these plants might nevertheless have very different proteins.
3. Why are the proteome and the metabolome more complex than the genome?

## ADDITIONAL INVESTIGATION

It is the year 2025. You are taking care of a patient who is concerned about having an early stage of kidney cancer. His mother died from this disease.

*a.* Assume that the SNPs linked to genes involved in the development of this type of cancer have been identified. How would you determine if this man has a genetic predisposition for developing kidney cancer? Explain how you would do the analysis.

*b.* How might you develop a metabolomic profile for kidney cancer and then use it to determine whether your patient has kidney cancer?

*c.* If the patient was diagnosed with cancer by the methods in (*a*) and (*b*), how would you use pharmacogenomics to choose the right medications to treat the tumor in this patient?

# 18 Recombinant DNA and Biotechnology

## Pollution fighters

In the summer of 1990, soldiers from Iraq invaded neighboring Kuwait. The reason was oil: the Iraqis were angry because Kuwait was pumping too much of it, keeping prices low. Six months later, a United Nations–sponsored coalition army from more than 30 countries drove the Iraqis out of Kuwait and back to their homeland. For Kuwait, the Gulf War was a success, but it left an environmental disaster. As they fled, the Iraqi soldiers set fire to more than 700 oil wells. It took over six months to put the fires out, and in the meantime an astounding 250 million gallons of crude oil were released into the desert. Twenty years later, much of the oil remains as a gooey coating, severely affecting the organisms that live there.

The government of Kuwait is using a variety of processes to get rid of the contaminating oil. Among them is the addition of bacteria that break down and consume the oil, utilizing the hydrocarbons in it as an energy source for growth. This process—using an organism to remove a pollutant—is called bioremediation. The Kuwait episode is not the first major use of bacteria for bioremediation. In 1989, the oil tanker *Exxon Valdez* ran aground near the Alaskan shore, releasing 11 million gallons of crude oil along 500 miles of shoreline. Physical methods such as skimming the water were used to remove more than half of the oil. Nitrogen and phosphorus salts were then sprayed on the oily rocks to stimulate the growth of oil-consuming bacteria already there, and other bacteria were added as part of the recovery effort. The oil gradually disappeared.

Some species of bacteria, because of their genetic capacity to produce unusual enzymes and biochemical pathways, thrive on all sorts of nutrients besides the usual glucose, including pollutants. Scientists have discovered these organisms simply by mixing polluted soil with water and seeing what grows. Many of the genes coding for enzymes involved in breaking down crude oil are carried on small chromosomes called plasmids. In 1971, Ananda Chakrabarty at the General Electric Research Center in New York used genetic crosses to develop a single strain of the bacterium *Pseudomonas* with multiple plasmids carrying genes for the breakdown of various hydrocarbons in oil. He and his company applied for a patent to legally protect their discovery and profit from it. In a landmark case, the U.S. Supreme Court ruled in 1980 that "a live, human-made microorganism is

**The Spoils of War** Massive oil spills occurred in Kuwait during the 1991 Gulf War.

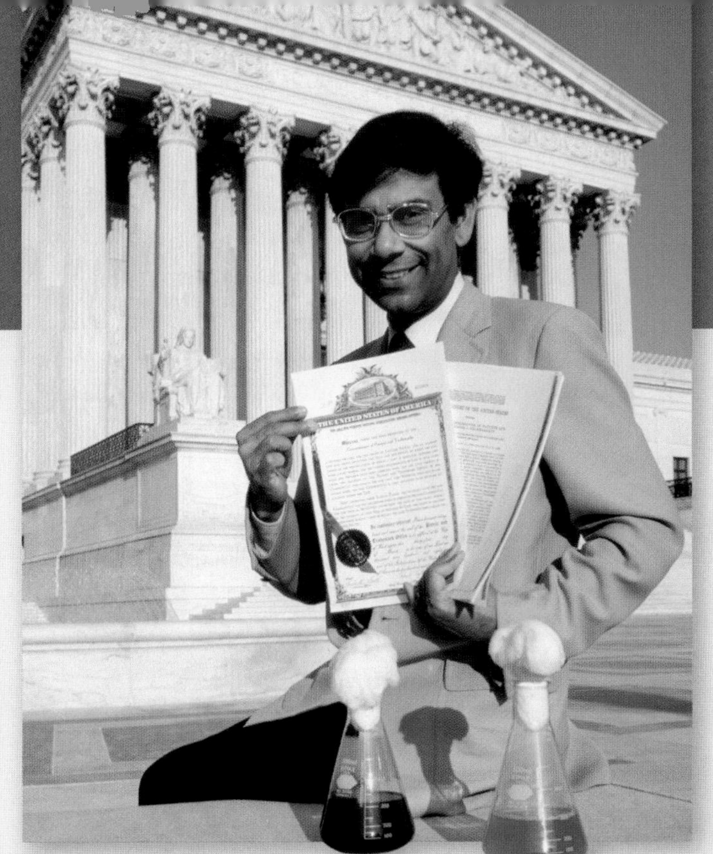

**Using Biotechnology to Clean Up the Environment** Ananda Chakrabarty received the first patent for a genetically modified organism, a bacterium that breaks down crude oil.

patentable" under the U.S. Constitution. Since then other bacteria have been patented that remove toxic metals such as mercury and copper from soils. In these cases, the bacteria use metabolic pathways to convert the metals to biologically inert forms.

The Supreme Court ruling came at a time when new laboratory methods were being developed to insert specific DNA sequences into organisms by recombinant DNA technology. Since then, an entirely new biotechnology industry has sprung up, its activities legally protected. The resulting flood of patents for DNA sequences and genetically modified organisms continues to this day.

**IN THIS CHAPTER** we will describe some of the techniques that are used to manipulate DNA. First, we will describe how DNA molecules are cut into smaller fragments and how these fragments are spliced together to create recombinant DNA. This will lead to a discussion of how recombinant DNA is introduced into suitable host cells. After describing some other ways to manipulate DNA, we will show how scientists have applied these methods to create a new biotechnology industry.

## CHAPTER OUTLINE



## 18.1 What Is Recombinant DNA?

You are familiar with *restriction endonucleases* (restriction enzymes), which occur naturally in bacteria and are used in the laboratory to cut DNA into fragments (see Chapter 15). Our focus in Chapter 15 was on the use of these enzymes for detecting mutations. In this chapter we examine how they are used, along with other enzymes, to construct recombinant DNA.

During the late 1960s, scientists discovered other enzymes that act on DNA. One of these is **DNA ligase**, which catalyzes the joining of DNA fragments. This is the enzyme that joins Okazaki fragments during DNA replication (see Section 13.3). Once they had isolated restriction enzymes and DNA ligase, scientists could use these enzymes to cut DNA into fragments and then splice them together in new combinations. Stanley Cohen and Herbert Boyer did just that in 1973. They used restriction enzymes to cut sequences from two *E. coli* plasmids (small chromosomal DNAs—see Figure 12.27) containing different antibiotic resistance genes. Then they used DNA ligase to join the fragments together. The resulting plasmid, when inserted into new *E. coli* cells, gave those cells resistance to both antibiotics (**Figure 18.1**). The era of **recombinant DNA**—a DNA molecule made in the laboratory that is derived from at least two genetic sources—was born.

Hundreds of different restriction enzymes are now available. They recognize *palindromic* DNA sequences—sequences that read the same way in both directions. For example, you can read the DNA recognition sequence for the restriction enzyme *Eco*R1 from 5′ to 3′ as GAATTC on both strands:

5′.......GAATTC......3′
3′.......CTTAAG......5′

Some restriction enzymes cut the DNA straight through the middle of the palindrome, generating "blunt-ended" fragments. Others, such as *Eco*RI, make staggered cuts—they cut one strand of the double helix several bases away from where they cut the other (**Figure 18.2**). After *Eco*RI makes its two cuts in the complementary strands, the ends of the strands are held together only by the hydrogen bonds between four base pairs. These hydrogen bonds are too weak to persist at warm temperatures (above room temperature), so the DNA separates into fragments when it is warmed. As a result, each fragment carries a single-stranded "overhang" at the location of each cut. These overhangs are called **sticky ends** because they have specific base sequences that can bind by base pairing with complementary sticky ends.

## INVESTIGATING LIFE

### 18.1 Recombinant DNA

With the discovery of restriction enzymes and DNA ligase, it became possible to combine DNA fragments from different sources in the laboratory. But would such "recombinant DNA" be functional when inserted into a living cell? The results of this experiment completely changed the scope of genetic research, increasing our knowledge of gene structure and function, and ushered in the new field of biotechnology.

**HYPOTHESIS** **Biologically functional recombinant chromosomes can be made in the laboratory.**

**METHOD** *E. coli* plasmids carrying a gene for resistance to either the antibiotic kanamycin (K) or tetracycline (T) are cut with a restriction enzyme.

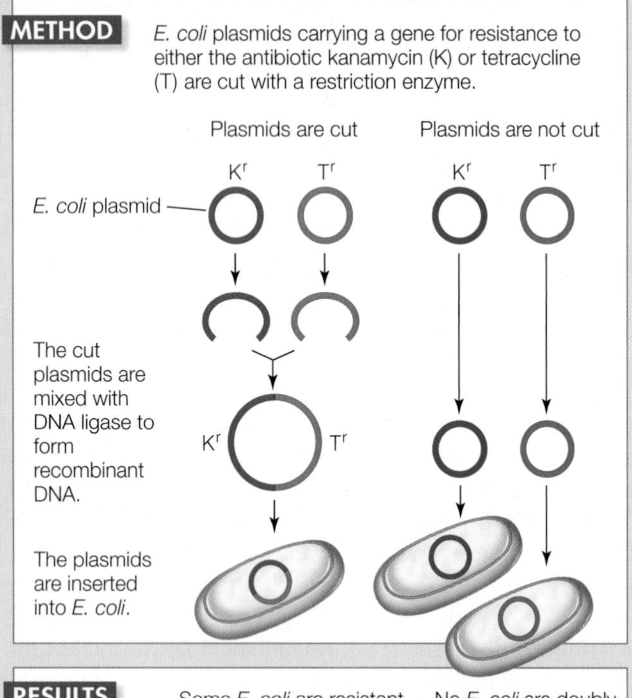


**RESULTS** Some *E. coli* are resistant to both antibiotics. No *E. coli* are doubly resistant.

**CONCLUSION** **Two DNA fragments with different genes can be joined to make a recombinant DNA molecule, and the resulting DNA is functional.**

**FURTHER INVESTIGATION:** Only one cell in 10,000 took up the plasmid in the experiment. The spontaneous mutation rate to $T^r$ or $K^r$ is one cell in $10^6$. How would you distinguish between genetic transformation and spontaneous mutation in this experiment?

Go to **yourBioPortal.com** for original citations, discussions, and relevant links for all INVESTIGATING LIFE figures.

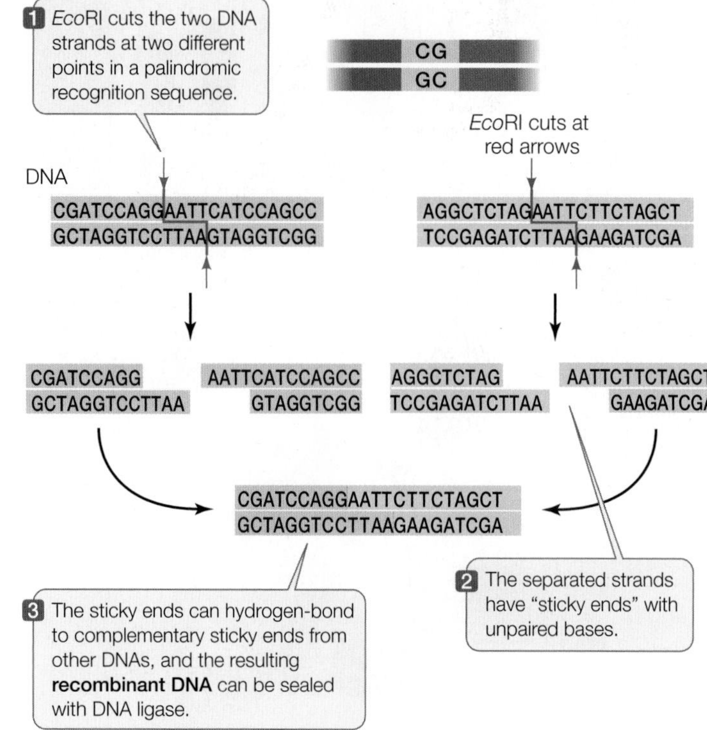


**18.2 Cutting, Splicing, and Joining DNA** Some restriction enzymes (*Eco*RI is shown here) make staggered cuts in DNA. *Eco*RI can be used to cut two different DNA molecules (blue and orange). The exposed bases can hydrogen bond with complementary exposed bases on other DNA fragments, forming recombinant DNA. DNA ligase stabilizes the recombinant molecule by forming covalent bonds in the DNA backbone.

After a DNA molecule has been cut with a restriction enzyme, complementary sticky ends can form hydrogen bonds with one another. The original ends may rejoin, or two different fragments with complementary sticky ends may join. Indeed, a fragment from one source, such as a human, can be joined to a fragment from another source, such as a bacterium. Initially the fragments are held together by weak hydrogen bonds, but then the enzyme ligase catalyzes the formation of covalent bonds between adjacent nucleotides at the ends of the fragments, joining them to form a single, larger molecule.

With these tools—restriction enzymes and DNA ligase—scientists can cut and rejoin different DNA molecules from any and all sources, including artificially synthesized DNA sequences.

## 18.1 RECAP

**DNA fragments from different sources can be linked together to make recombinant DNA.**

- How did Cohen and Boyer make the first recombinant DNA? See Figure 18.1
- How does a staggered cut in DNA create a "sticky end"? See p. 387 and Figure 18.2

Recombinant DNA has no biological significance until it is inserted inside a living cell, which can replicate and transcribe the transplanted genetic information. How can recombinant DNA made in the laboratory be inserted and expressed in living cells?

## 18.2 How Are New Genes Inserted into Cells?

One goal of recombinant DNA technology is to **clone**—that is, to produce many identical copies of—a particular gene. Cloning might be done for analysis, to produce a protein product in quantity, or as a step toward creating an organism with a new phenotype. Recombinant DNA is cloned by inserting it into host cells in a process known as **transformation** (or **transfection** if the host cells are derived from an animal). A host cell or organism that contains recombinant DNA is referred to as a **transgenic** cell or organism. Later in this chapter we will encounter many examples of transgenic cells and organisms, including yeast, mice, wheat plants, and even cows.

Various methods are used to create transgenic cells. Generally, these methods are inefficient in that only a few of the cells that are exposed to the recombinant DNA actually become transformed with it. In order to grow only the transgenic cells, **selectable marker** genes, such as genes that confer resistance to antibiotics, are often included as part of the recombinant DNA molecule. Antibiotic resistance genes were the markers used in Cohen and Boyer's experiment (see Figure 18.1).

### Genes can be inserted into prokaryotic or eukaryotic cells

The initial successes with recombinant DNA technology were achieved using bacteria as hosts. As we have seen in preceding chapters, bacterial cells are easily grown and manipulated in the laboratory. Much of their molecular biology is known, especially for certain well-studied bacteria such as *E. coli*. Furthermore, bacteria contain plasmids, which are easily manipulated to carry recombinant DNA into the cell.

In some important ways, however, bacteria are not ideal organisms for studying and expressing eukaryotic genes. Consider how differently the processes of transcription and translation proceed in prokaryotes and eukaryotes, and recall that DNA often contains the signals for these specific functions (see Chapter 14). Furthermore, scientists often want to study how genes function in multicellular eukaryotic organisms rather than in cells grown in cultures. Or they might want to create a crop plant or farm animal with a new phenotype for use in agriculture. For these reasons, scientists have developed methods to transform or transfect eukaryotic cells.

Yeasts such as *Saccharomyces* are commonly used as eukaryotic hosts for recombinant DNA studies. The advantages of using yeasts include rapid cell division (a life cycle completed in 2–8 hours), ease of growth in the laboratory, and a relatively small genome size (about 12 million base pairs and 6,000 genes). In addition, yeasts have most of the characteristics of other eukaryotes, except for those characteristics involved in multicellularity.

Plant cells can also be used as hosts. One property that makes plant cells good hosts is the ability to make stem cells (unspecialized, totipotent cells; see Chapter 5 opener) from mature plant tissues. When these unspecialized plant cells are isolated and grown in culture, they can be transformed with recombinant DNA. These transgenic cells can be studied in culture, or manipulated to form entire new plants. There are also methods for making whole transgenic plants without going through the cell culture step. These methods result in plants that carry the recombinant DNA in all their cells, including the germ line cells.

If biologists want to study expression of human or animal genes, for example for medical purposes, they use cultured animal cells as hosts. Whole transgenic animals can also be created.

### Recombinant DNA enters host cells in a variety of ways

Methods for inserting DNA into host cells vary. The cells may be chemically treated to make their outer membranes more permeable, and then mixed with the DNA so that it can diffuse into the cells. Another approach is called *electroporation*; a short electric shock is used to create temporary pores in the membranes, through which the DNA can enter. Viruses can be altered so that they carry recombinant DNA into cells. Plants are often transformed using a bacterium that has evolved mechanisms to transfer its DNA into cells and then insert the DNA into a plant chromosome. Transgenic animals can be produced by injecting recombinant DNA into the nuclei of fertilized eggs. There are even "gene guns," which "shoot" the host cells with tiny particles carrying the DNA.

The challenge of inserting new DNA into a cell lies not just in getting it into the host cell, but in getting it to replicate as the host cell divides. DNA polymerase does not bind to just any sequence. If the new DNA is to be replicated, it must become part of a segment of DNA that contains an origin of replication. Such a DNA molecule is called a **replicon**, or replication unit.

There are two general ways in which the newly introduced DNA can become part of a replicon:

- It may be inserted into a host chromosome. Although the site of insertion is usually random, this is nevertheless a common method of integrating new genes into host cells.
- It can enter the host cell as part of a carrier DNA sequence, called a **vector**, that already has an origin of replication.

Several types of vectors are used to get DNA into cells. Once inside the cells, some vectors replicate independently, while others incorporate all or part of their DNA into the host chromosomes.

**PLASMIDS AS VECTORS** As you learned in Chapter 12, plasmids are small chromosomes that exist in prokaryotic cells in addition to the main chromosomes. Yeast cells can also harbor plasmids. A number of characteristics make plasmids useful as transformation vectors:

- They are relatively small (an *E. coli* plasmid has 2,000–6,000 base pairs) and therefore easy to manipulate in the laboratory.
- A plasmid will usually have one or more restriction enzyme recognition sequences that each occur only once in the plasmid sequence. These sites make it easy to insert additional DNA into the plasmid before it is used to transform host cells.
- Many plasmids contain genes that confer resistance to antibiotics, which can serve as selectable markers.

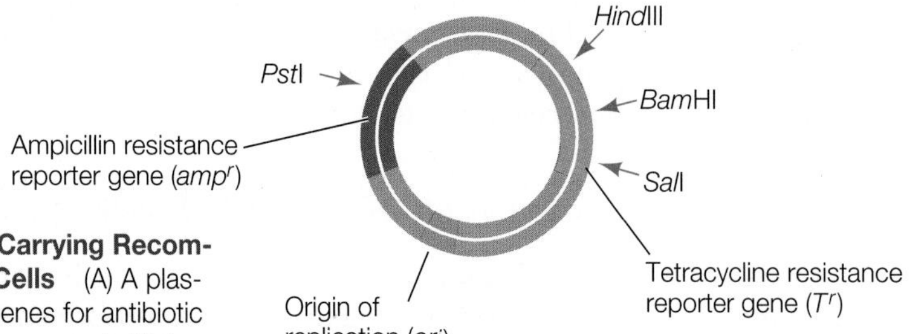


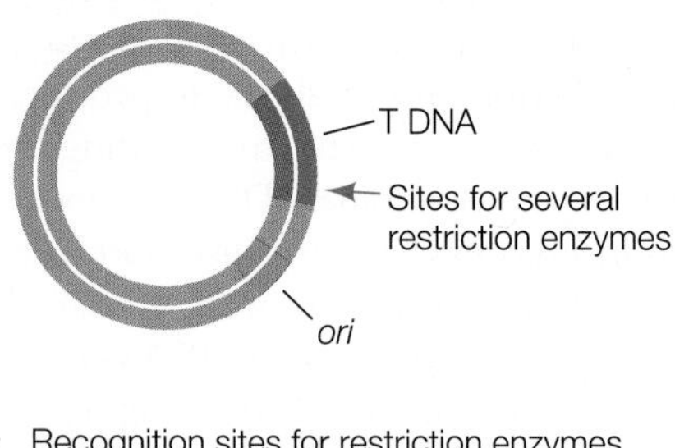


**18.3 Vectors for Carrying Recombinant DNA into Cells** (A) A plasmid with reporter genes for antibiotic resistance can be incorporated into an *E. coli* cell. (B) The Ti plasmid, isolated from the bacterium *Agrobacterium tumefaciens*, is used to insert DNA into many types of plants.

- Plasmids have a bacterial origin of replication (*ori*) and can replicate independently of the host chromosome. It is not uncommon for a bacterial cell to contain hundreds of copies of a recombinant plasmid. For this reason, the power of bacterial transformation to amplify a gene is extraordinary. A one-liter culture of bacteria harboring the human β-globin gene in a typical plasmid has as many copies of that gene as the sum total of all the cells in a typical adult human being ($10^{14}$). A typical bacterial plasmid is shown in **Figure 18.3A**.

The plasmids used as vectors in the laboratory have been extensively altered by recombinant DNA technology to include convenient features: multiple cloning sites with 20 or more unique restriction enzyme sites for cloning purposes; origins of replication for a variety of host cells; and various kinds of reporter genes and selectable marker genes.

**VIRUSES AS VECTORS** Constraints on plasmid replication limit the size of the new DNA that can be inserted into a plasmid to about 10,000 base pairs. Although many prokaryotic genes may be smaller than this, most eukaryotic genes—with their introns and extensive flanking sequences—are bigger. A vector that accommodates larger DNA inserts is needed for these genes.

Both prokaryotic and eukaryotic viruses are often used as vectors for eukaryotic DNA. Bacteriophage λ, which infects *E. coli*, has a DNA genome of about 45,000 base pairs. If the genes that cause the host cell to die and lyse—about 20,000 base pairs—are eliminated, the virus can still attach to a host cell and inject its DNA. The deleted 20,000 base pairs can be replaced with DNA from another organism. Because viruses infect cells naturally, they offer a great advantage over plasmids, which often require artificial means to coax them to enter host cells. As we saw in Section 15.6, viruses are important vectors in human gene therapy.

**PLASMID VECTORS FOR PLANTS** An important vector for carrying new DNA into many types of plants is a plasmid found in *Agrobacterium tumefaciens*. This bacterium lives in the soil, infects plants, and causes a disease called crown gall, which is characterized by the presence of growths, or tumors, in the plant. *A. tumefaciens* contains a plasmid called Ti (for tumor-inducing) (**Figure 18.3B**). The Ti plasmid carries genes that allow the bacterium to infect plant cells and then insert a region of its DNA called the T DNA into the chromosomes of infected cells. The T DNA contains genes that cause the growth of tumors and the production of specific sugars that the bacterium uses as sources of energy. Scientists have exploited this remarkable natural "genetic engineer" to insert foreign DNA into the genomes of plants.

When used as a vector for plant transformation, the tumor-inducing and sugar-producing genes on the T DNA are removed and replaced with foreign DNA. The altered Ti plasmids are first used to transform *Agrobacterium* cells from which the original Ti plasmids have been removed. Then the *Agrobacterium* cells are used to infect plant cells. Whole plants can be regenerated from transgenic cells or, in the case of the model plant *Arabidopsis* (see Section 17.3), the *Agrobacterium* can be used to directly infect germ line cells of whole plants.

## Reporter genes identify host cells containing recombinant DNA

Even when a population of host cells interacts with an appropriate vector, only a small proportion of the cells actually take up the vector. Furthermore, the process of making recombinant DNA is far from perfect. After a ligation reaction, not all the vector copies contain the foreign DNA. How can we identify or select the host cells that contain that sequence?

Selectable markers such as antibiotic resistance genes can be used to select cells containing those genes. Only cells carrying the antibiotic resistance gene can grow in the presence of that antibiotic. If a vector carrying genes for resistance to two different antibiotics is used, one antibiotic can be used to selectively grow cells carrying the vector. If the other antibiotic resistance gene is inactivated by the insertion of foreign DNA, then cells carrying copies of the vector with the inserted DNA can be identified by their sensitivity to that antibiotic (**Figure 18.4**). Since the uptake of recombinant DNA is a rare event (only about 1 cell in 10,000 takes up a plasmid in such experiments), it is vital to be able to select the small number of cells harboring the recombinant DNA.

## TOOLS FOR INVESTIGATING LIFE

### 18.4 Marking Recombinant DNA by Inactivating a Gene

Selectable marker (reporter) genes are used by scientists to select for bacteria that have taken up a plasmid. A second reporter gene allows for the identification of bacteria harboring the recombinant plasmid. The host bacteria in this experiment could display any of the three phenotypes indicated in the table.

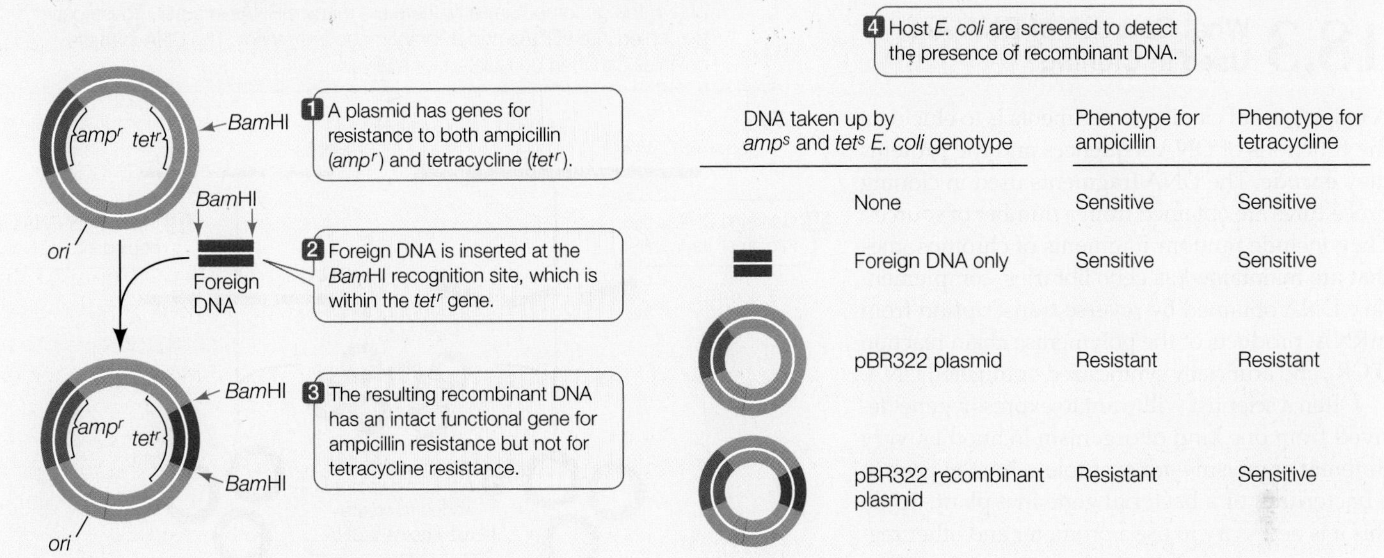

ampʳ
tetʳ
BamHI
ori
1 A plasmid has genes for resistance to both ampicillin (ampʳ) and tetracycline (tetʳ).
BamHI
Foreign DNA
2 Foreign DNA is inserted at the BamHI recognition site, which is within the tetʳ gene.
BamHI
ampʳ
tetʳ
BamHI
ori
3 The resulting recombinant DNA has an intact functional gene for ampicillin resistance but not for tetracycline resistance.
4 Host E. coli are screened to detect the presence of recombinant DNA.

| DNA taken up by *amp*ˢ and *tet*ˢ *E. coli* genotype | Phenotype for ampicillin | Phenotype for tetracycline |
|---|---|---|
| None | Sensitive | Sensitive |
| Foreign DNA only | Sensitive | Sensitive |
| pBR322 plasmid | Resistant | Resistant |
| pBR322 recombinant plasmid | Resistant | Sensitive |

Selectable markers are one type of **reporter gene**, which is any gene whose expression is easily observed. Other reporter genes code for proteins that can be detected visually. For example:

- The β-galactosidase (*lacZ*) gene in the *E. coli lac* operon (see Figure 16.10) codes for an enzyme that can convert the substrate X-Gal into a bright blue product. Many plasmids contain the *lacZ* gene with a multiple cloning site within its sequence. Bacterial colonies containing the plasmid (which also includes an antibiotic resistance gene) are selected on a solid medium containing the antibiotic. X-Gal is also included in the medium, so that bacterial colonies containing the recombinant DNA inserted into the *lacZ* gene produce white, rather than blue, colonies.
- Green fluorescent protein, which normally occurs in the jellyfish *Aequopora victoriana*, emits visible light when exposed to ultraviolet light. The gene for this protein has been isolated and incorporated into vectors. It is now widely used as a reporter gene (**Figure 18.5**).

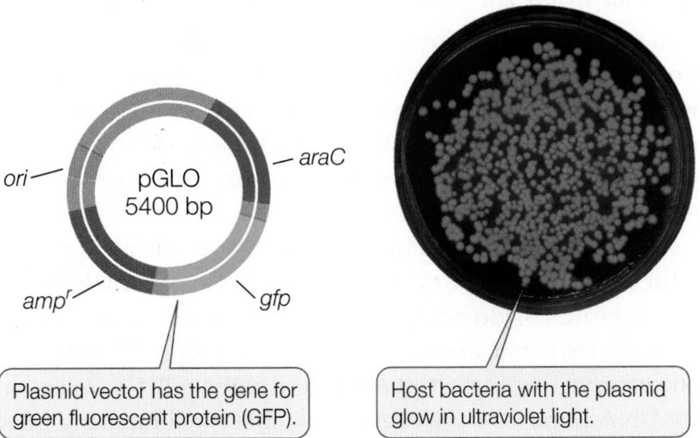

ori
pGLO
5400 bp
araC
ampʳ
gfp
Plasmid vector has the gene for green fluorescent protein (GFP).
Host bacteria with the plasmid glow in ultraviolet light.

**18.5 Green Fluorescent Protein as a Reporter** The presence of a plasmid with the gene for green fluorescent protein is readily apparent in transgenic cells because they glow under ultraviolet light. This allows the identification of cells carrying a plasmid without the use of selection on antibiotics. That is, no cells are killed during the selection process.

Such reporters are not just used to select and identify cells carrying recombinant DNA. They can be attached to promoters in order to study how the promoters function under different conditions or in different tissues of a transgenic multicellular organism. They can also be attached to other proteins, to study how and where those proteins become localized within eukaryotic cells.

## 18.2 RECAP

**Recombinant DNA can be cloned by using a vector to insert it into a suitable host cell. The vector often has a selectable marker or other reporter gene that gives the host cell a phenotype by which transgenic cells can be identified.**

- List the characteristics of a plasmid that make it suitable for introducing new DNA into a host cell. See pp. 389–390
- How are cells harboring a vector that carries recombinant DNA selected? See p. 390 and Figure 18.4

We have described how DNA can be cut or amplified, inserted into a vector, and introduced into host cells. We have also seen how host cells carrying recombinant DNA can be identified. Now let's consider where the genes or DNA fragments used in these procedures come from.

TOOLS FOR INVESTIGATING LIFE

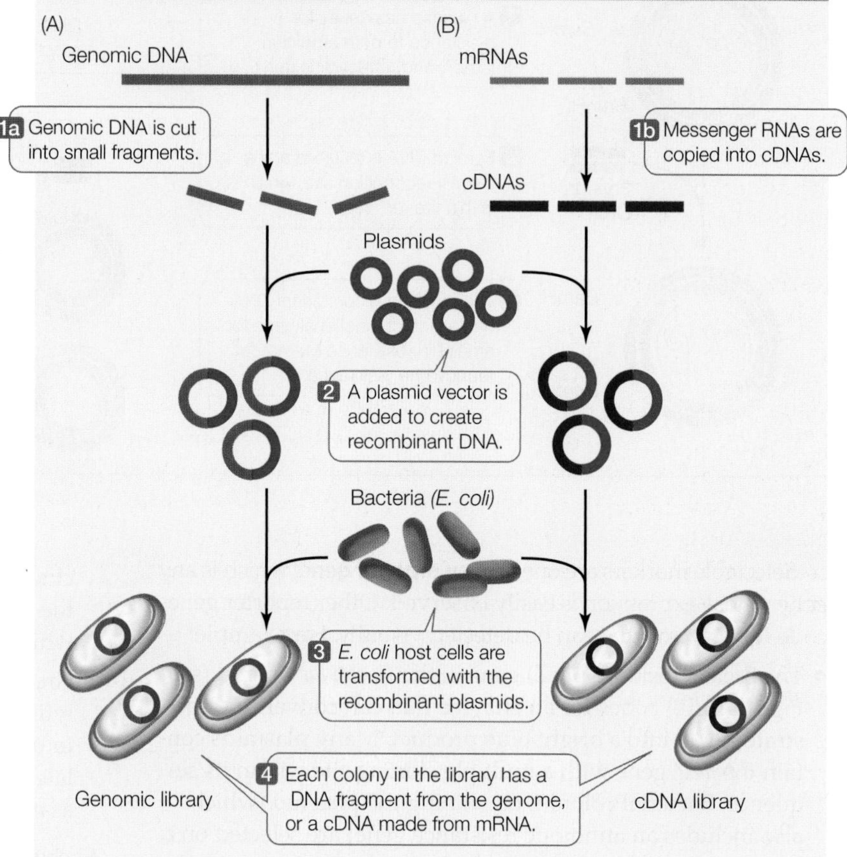
(A)
Genomic DNA
(B)
mRNAs
1a Genomic DNA is cut into small fragments.
1b Messenger RNAs are copied into cDNAs.
cDNAs
Plasmids
2 A plasmid vector is added to create recombinant DNA.
Bacteria (E. coli)
3 E. coli host cells are transformed with the recombinant plasmids.
Genomic library
4 Each colony in the library has a DNA fragment from the genome, or a cDNA made from mRNA.
cDNA library

**18.6 Constructing Libraries** Intact genomic DNA is too large to be introduced into host cells. A genomic library can be made by breaking the DNA into small fragments, incorporating the fragments into a vector, and then transforming host cells with the recombinant vectors. Each colony of cells contains many copies of a small part of the genome. Similarly, there are many mRNAs in a cell. These can be copied into cDNAs and a library made from them. The DNA in these colonies can then be isolated for analysis.

## 18.3 What Sources of DNA Are Used in Cloning?

A major goal of cloning experiments is to elucidate the functions of DNA sequences and the proteins they encode. The DNA fragments used in cloning procedures are obtained from a number of sources. They include random fragments of chromosomes that are maintained as gene libraries, complementary DNA obtained by reverse transcription from mRNA, products of the polymerase chain reaction (PCR), and artificially synthesized or mutated DNA.

Often a scientist will want to express a gene derived from one kind of organism in another, very different organism—for example, a human gene in a bacterium, or a bacterial gene in a plant. To do this it is necessary to use a promoter and other regulatory sequences from the host organism: a bacterial promoter will not function in a plant cell, for example. The coding region of the gene of interest is inserted between a promoter and a transcription termination sequence derived from the host organism, or from one that uses similar mechanisms for gene regulation.

### Libraries provide collections of DNA fragments

In Chapter 17 we introduced the concept of a **genomic library**: a collection of DNA fragments that together comprise the genome of an organism. Now we provide details on how a genomic or other gene library is generated and used.

Restriction enzymes or other means, such as mechanical shearing, can be used to break chromosomes into smaller pieces. These smaller DNA fragments still constitute a genome (**Figure 18.6A**), but the information is now in many smaller "volumes." Each fragment is inserted into a vector, which is then taken up by a host cell. Proliferation of a single transformed cell produces a colony of recombinant cells, each of which harbors many copies of the same fragment of DNA.

When plasmids are used as vectors, about 700,000 separate fragments are required to make a library of the human genome. By using bacteriophage λ, which can carry four times as much DNA as a plasmid, the number of "volumes" in the library can be reduced to about 160,000. Although this seems like a large number, a single petri plate can hold thousands of phage colonies, or plaques, and is easily screened for the presence of a particular DNA sequence by hybridization to an appropriate DNA probe.

### cDNA libraries are constructed from mRNA transcripts

A much smaller DNA library—one that includes only the genes transcribed in a particular tissue—can be made from **complementary DNA**, or **cDNA** (**Figure 18.6B**). This involves isolating mRNA from cells, then making cDNA copies of that mRNA by complementary base pairing. An enzyme, reverse transcriptase, catalyzes this reaction.

A collection of cDNAs from a particular tissue at a particular time in the life cycle of an organism is called a **cDNA library**. Messenger RNAs do not last long in the cytoplasm and are often present in small amounts, so a cDNA library is a "snapshot" that preserves the transcription pattern of the cell. Complementary DNA libraries have been invaluable for comparing gene expression in different tissues at different stages of development. For example, their use has shown that up to one-third of all the genes of an animal are expressed only during development. Complementary DNA is also a good starting point for cloning eukaryotic genes (because the clones contain only the coding sequences of the genes) and genes that are expressed in only a few cell types.

### Synthetic DNA can be made by PCR or by organic chemistry

In Chapter 13 we describe the polymerase chain reaction (PCR), a method of amplifying DNA in a test tube. PCR can begin with as little as $10^{-12}$ g of DNA (a picogram). Any fragment of DNA can be amplified by PCR as long as appropriate primers are available. You will recall that DNA replication (by PCR or any other system) requires not just a template on which DNA polymerase adds complementary nucleotides, but also a short oligonucleotide primer where replication begins (see Figure 13.22). If the appropriate primers (two are needed—one for each strand of DNA) are added to denatured DNA, more than two billion copies of the DNA region between the primers can be produced in just a few hours. This amplified DNA can then be inserted into plasmids to create recombinant DNA and cloned in host cells.

The *artificial synthesis* of DNA by organic chemistry is now fully automated, and a special service laboratory can make short- to medium-length sequences overnight for any number of investigators. Synthetic oligonucleotides (single-stranded DNA fragments of up to 40 bp) are used as primers in PCR reactions. These primers can be designed to create short new sequences at the ends of the PCR products. This might be done to create a mutation in a recombinant gene, or to add restriction enzyme sites at the ends of the PCR product to aid in cloning.

Longer synthetic sequences can be pieced together to construct an artificial gene. If we know the amino acid sequence of the desired protein product, we can use the genetic code to figure out the corresponding DNA sequence. As mentioned above, other sequences must be added, such as the promoter and transcription termination sequences. Appropriate selection of the codon for a given amino acid is another important consideration: many amino acids are encoded by more than one codon (see Figure 14.6), and host organisms vary in their use of synonymous codons.

### DNA mutations can be created in the laboratory

Mutations that occur in nature have been important in demonstrating cause-and-effect relationships in biology. However, mutations in nature are rare events. Recombinant DNA technology allows us to ask "what if" questions by creating mutations artificially. Because synthetic DNA can be made with any desired sequence, it can be manipulated to create specific mutations, the consequences of which can be observed when the mutant DNA is expressed in host cells. These mutagenesis techniques have revealed many cause-and-effect relationships.

For example, consider the experiment illustrated in Figure 14.20. Researchers hypothesized that a nuclear localization signal (NLS) sequence of amino acids is necessary for targeting a protein to the nucleus after it is made at the ribosome. The researchers used recombinant DNA technology to synthesize genes encoding proteins with and without the sequence, which were then used to transform cells. Without the NLS, newly synthesized proteins did not enter the nucleus. Knowing this, the researchers then asked, "Are certain amino acids more functionally important to the NLS than others?" In follow-up experiments, they made a series of mutated genes to test whether certain amino acids were needed at certain locations in the NLS. They found that changing the amino acids at the very beginning or very end of the NLS, but not the middle, abolished its function. This led to a fuller description of the binding of the NLS to its nuclear receptor. Without the ability to generate specific mutations, these experiments would not have been possible.

## 18.3 RECAP

**DNA for cloning can be obtained from genomic libraries, cDNA made from mRNA, or artificially synthesized DNA fragments. Gene function can be investigated by intentionally introducing mutations into natural or synthetic genes and organisms.**

- How are genomic DNA and cDNA libraries made and used? See p. 392 and Figure 18.6
- Explain how recombinant DNA and mutagenesis are used to test "what if" questions in biology. See p. 393

We've explored the various sources of DNA that can be used to make recombinant DNA molecules and the ways the resulting molecules can be used to study the functions of genes and proteins. We now turn to some additional tools that are available for studying DNA.

## 18.4 What Other Tools Are Used to Study DNA Function?

Sections 13.5 and 17.1 describe PCR and DNA sequencing, two important techniques arising from our understanding of DNA replication. In this section we will examine three additional techniques for studying DNA, including homologous recombination to inactivate genes, antisense and RNAi to block gene expression, and DNA microarrays to analyze large numbers of nucleotide sequences.

### Genes can be inactivated by homologous recombination

One way to study a gene or protein in order to understand its function is to inactivate the gene so that it is not transcribed and translated into a functional protein. Such a manipulation is called a **knockout** experiment. In plants, transposons or T DNA insertions can be used to create thousands of knockout mutants, and then the mutants are screened to identify those with altered phenotypes. For example, the mutants can be screened for those that are susceptible to a particular disease. This is an important way to identify genes that are involved in processes such as resistance to disease or other environmental stresses, such as drought and temperature extremes.

A technique called **homologous recombination** is a much more targeted way to produce knockout mutants. In this case, the gene of interest has already been identified, and recombinant DNA technology is used to specifically inactivate that gene. Mice are frequently used in such knockout experiments (**Figure 18.7**). The normal allele of the mouse gene to be tested is inserted into a plasmid. Restriction enzymes are then used to insert a fragment containing a reporter gene into the middle of the normal gene. This addition of extra DNA plays havoc with the targeted gene's transcription and translation; a functional mRNA is seldom made from a gene whose sequence has been thus interrupted.

Once the recombinant plasmid has been made, it is used to transfect mouse embryonic stem cells. (A **stem cell** is an unspecialized cell that divides and differentiates into specialized cells.) Much of the targeted gene is still present in the plasmid (although in two separated regions), and these sequences tend to line up with their homologous sequences in the normal allele on the mouse chromosome. Sometimes recombination occurs, and the plasmid's inactive allele is "swapped" with the functional allele in the host cell. The inactive allele is inserted permanently into the host cell's genome and the normal allele is lost (because the plasmid cannot replicate in mouse cells). The active reporter gene in the insert is used to select those stem cells carrying the inactivated gene.

A transfected stem cell is now transplanted into an early mouse embryo. If the mouse that develops from this embryo has the mutant gene in its germ line cells, its progeny will have the knockout gene in every cell in their bodies. Such mice are inbred to create *knockout mice* carrying the inactivated gene in homozygous form. The mutant mouse can then be observed for phenotypic changes, to find clues about the function of the targeted gene in the normal (wild-type) animal. The knockout technique has been important in assessing the roles of many genes, and has been especially valuable in studying human genetic diseases. Many of these diseases, such as phenylketonuria, have knockout mouse *models*—mouse strains that suffer from an analogous disease—produced by homologous recombination. These models can be used to study a disease and to test potential treatments. Mario Capecchi, Martin Evans, and Oliver Smithies shared the Nobel Prize for developing the knockout mouse technique.

TOOLS FOR INVESTIGATING LIFE

**18.7 Making a Knockout Mouse** Animals carrying mutations are rare. Homologous recombination is used to replace a normal mouse gene with an inactivated copy of that gene, thus "knocking out" the gene. Discovering what happens to a mouse with an inactive gene tells us much about the normal role of that gene.

Target gene
Vector (plasmid)
Reporter gene
1 The targeted gene is inactivated by insertion of the marker gene.
2 The vector is inserted into a mouse stem cell...
Mouse embryonic stem cell
3 ...where the targeted genes on the vector and mouse genome line up via homologous sequence recognition.
Mouse chromosome
4 Recombination occurs. The inactivated gene is now in the mouse genome, and the vector is lost during cell division.
Inactivated mouse gene
5 The stem cell is transplanted into an early mouse embryo, where it replaces most of the embryo's cells during development.
Blastocyst
Development of embryo and birth
6 The resulting mouse is examined for consequences of carrying an inactivated gene.

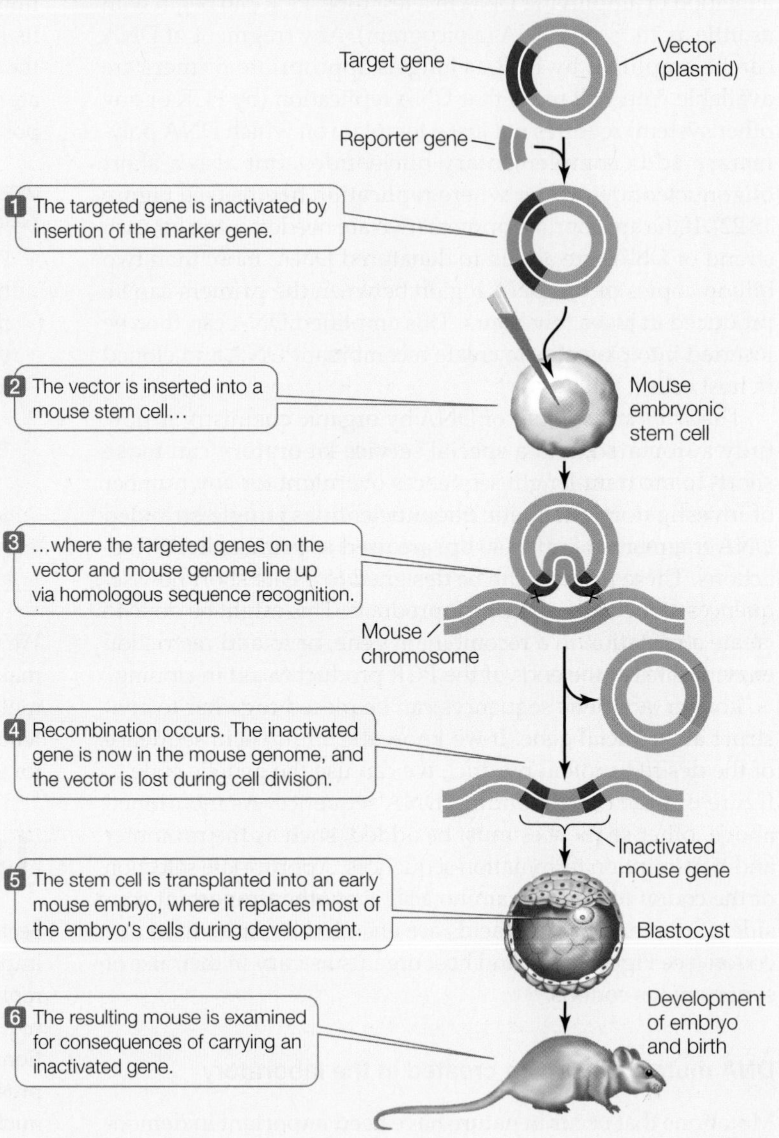

## Complementary RNA can prevent the expression of specific genes

Another way to study the expression of a specific gene is to block the translation of its mRNA. This is an example of scientists imitating nature. As described in Section 16.5, gene expression is sometimes controlled by the production of double-stranded RNA molecules, which are cut up and unwound to produce short, single-stranded RNA molecules (microRNAs) that are complementary to specific mRNA sequences (see Figure 16.23). Such a complementary molecule is called **antisense RNA** because it binds by base pairing to the "sense" bases on the mRNA. The resulting partially double-stranded RNA hybrid inhibits translation of the mRNA, and the hybrid tends to be broken down rapidly in the cytoplasm. Although the gene continues to be transcribed, translation does not take place. After determining the sequence of a gene and its mRNA in the lab-

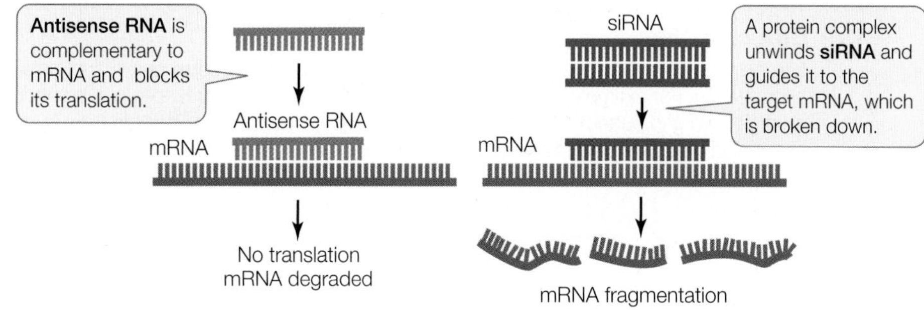


**18.8 Using Antisense RNA and siRNA to Block Translation of mRNA** Once a gene's sequence is known, the synthesis of its protein can be prevented by making either an antisense RNA (left) or a small interfering RNA (siRNA, right) that is complementary to its mRNA.

oratory, scientists can make a specific, single-stranded antisense RNA and add it to a cell to prevent translation of that gene's mRNA (**Figure 18.8, left**).

Several antisense drugs are being developed to reduce the expression of genes involved with cancer. For example, the gene *bcl2* codes for a protein that blocks apoptosis, and in some forms of cancer *bcl2* is activated inappropriately through mutation. These cells fail to undergo apoptosis, continue to divide, and form a tumor. Treatment with oblimersen, an antisense RNA that binds to *bcl2* mRNA, prevents production of the protein, and leads to apoptosis of tumor cells and shrinkage of the tumor.

A related technique to antisense RNA takes advantage of **RNA interference** (**RNAi**), a rare, natural mechanism for inhibiting mRNA translation. In a process similar to that involved in processing microRNAs, a short (about 20 nucleotides) double-stranded RNA is unwound to single strands by a protein complex that guides this RNA to a complementary region on mRNA. The protein complex catalyzes the breakdown of the targeted mRNA. RNAi was not discovered until the late 1990s, but since then scientists have synthesized double-stranded siRNAs to inhibit the expression of known genes (**Figure 18.8, right**). Because these double-stranded siRNAs are more stable than antisense RNAs, the use of siRNAs is the preferred approach for blocking translation. Macular degeneration is an eye disease that results in near-blindness when blood vessels proliferate in the eye. The signaling molecule that stimulates vessel proliferation is a growth factor. An RNAi-based therapy is being developed to target this growth factor's mRNA and shows promise in stopping and even reversing the progress of the disease.

Although medical applications for RNAi are still at the experimental stage, antisense RNA and RNAi have been widely used to test cause-and-effect relationships in biological research. Another powerful research tool with great potential for medicine is the gene chip, or DNA microarray.

### DNA microarrays can reveal RNA expression patterns

The emerging science of genomics has to face two major quantitative realities. First, there are very large numbers of genes in eukaryotic genomes. Second, the pattern of gene expression in different tissues at different times is quite distinctive. For example, a cell from a skin cancer at its early stage may have a unique mRNA "fingerprint" that differs from that of both normal skin cells and the cells of a more advanced skin cancer.

To find such patterns, scientists could isolate mRNA from a cell and test it by hybridization with each gene in the genome, one gene at a time. But that would involve many steps and take a very long time. It is far simpler to do these hybridizations all in one step. This is possible with **DNA microarray** technology, which provides large arrays of sequences for hybridization experiments.

The development of DNA arrays ("gene chips") was inspired by methods used for decades by the semiconductor industry. A silicon microchip consists of an array of microscopic electric circuits etched onto a tiny silicon base, called a chip. In the same way, a series of DNA sequences can be attached to a glass slide in a precise order (**Figure 18.9**). The slide is divided into a grid of microscopic spots, or "wells." Each spot contains thousands of copies of a particular oligonucleotide of 20 or more bases. A computer controls the addition of these oligonucleotide sequences in a predetermined pattern. Each oligonucleotide can hybridize with only one DNA or RNA sequence, and thus is a unique identifier of a gene. Many thousands of different oligonucleotides can be placed on a single microarray.

As we mention in Section 17.4, DNA microarrays can be used to identify specific single nucleotide polymorphisms or other mutations in genomic DNA samples. Or they can be used to analyze RNA from different tissues or cells to identify which genes are expressed in those cell types. If mRNA is to be analyzed, it is usually incubated with reverse transcriptase to make cDNA (see Figure 18.6B). Fluorescent dyes are used to tag the cDNAs from different samples with different colors (usually red and green; see Figure 18.9). The cDNAs are used to probe the DNA on the microarray. Complementary sequences that form hybrids with the DNA on the microarray can be located using a sensitive scanner that detects the fluorescent light.

A clinical use of DNA chips was developed by Laura van 't Veer and her colleagues at the Netherlands Cancer Institute (**Figure 18.10**). Most women with breast cancer are treated with surgery to remove the tumor, and then treated with radiation soon afterward to kill cancer cells that the surgery may have missed. But a few cancer cells may still survive in some patients, and these eventually form tumors in the breast or elsewhere in the body. The challenge for physicians is to develop criteria to identify patients with surviving cancer cells so that they can be treated aggressively with tumor-killing chemotherapy. The scientists in van 't Veer's group followed the medical histories of breast cancer patients to identify those patients whose cancer recurred. They then used a DNA microarray to examine the expression of about 1,000 genes in these patients' original tu-

TOOLS FOR INVESTIGATING LIFE

**18.9 DNA on a Chip** Large arrays of DNA sequences can be used to identify specific sequences in a sample of DNA or RNA by hybridization. For example, thousands of known, synthetic DNA sequences can be attached to a glass slide in an organized grid pattern. This can be hybridized with cDNA samples derived from two different tissues to find out what genes are being expressed in the tissues.

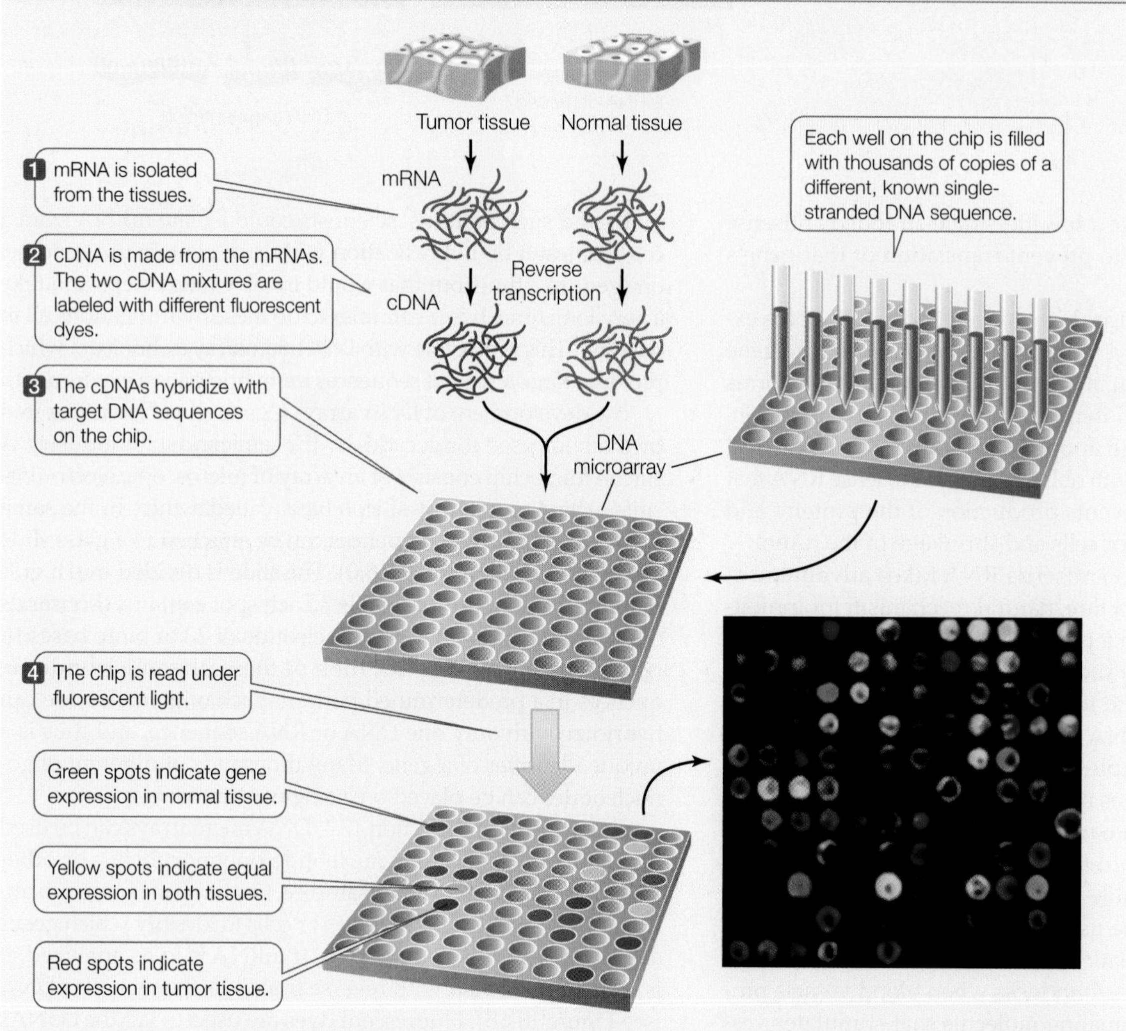


yourBioPortal.com
GO TO Animated Tutorial 18.1 • DNA Chip Technology

mors (which had been stored after their surgical removal) relative to normal tissue. They found 70 genes whose expression differed dramatically between tumors from patients whose cancers recurred and tumors from patients whose cancers did not recur. From this information, the Dutch group was able to identify what is called a *gene expression signature*. This expression pattern is useful in clinical decision-making: patients with a good prognosis can avoid unnecessary chemotherapy, while those with a poor prognosis can receive aggressive treatment.

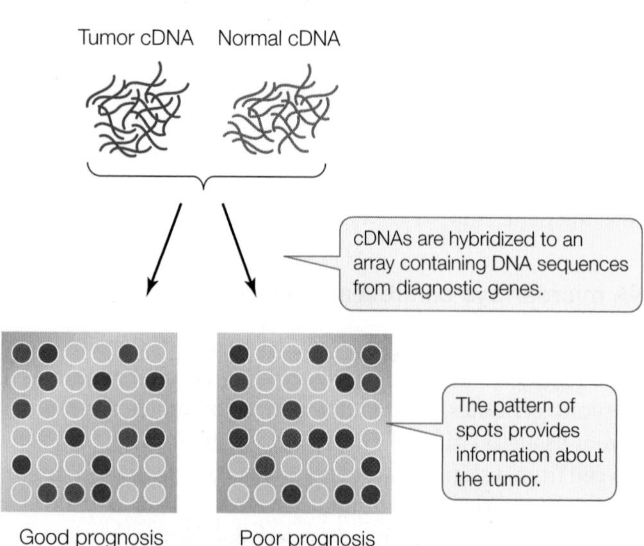

Tumor cDNA
Normal cDNA
cDNAs are hybridized to an array containing DNA sequences from diagnostic genes.
The pattern of spots provides information about the tumor.
Good prognosis
Poor prognosis

**18.10 Using DNA Arrays for Medical Diagnosis** The pattern of expression of 70 genes in tumor tissues indicates whether breast cancer is likely to recur. Actual arrays have more dots than shown here.

**18.4 RECAP**

**Researchers can study the function of a gene by knocking out that gene in a living organism. Antisense RNAs and siRNAs silence genes by selectively blocking mRNA translation. DNA microarrays allow the simultaneous analysis of many different mRNA transcripts.**

- How is a gene "knocked out" in a living organism? See p. 393 and Figure 18.7
- How do antisense RNA and siRNA molecules affect gene expression? See pp. 394–395 and Figure 18.9

Now that you've seen how DNA can be fragmented, recombined, manipulated, and put back into living organisms, let's see some examples of how these techniques are used to make useful products.

# 18.5 What Is Biotechnology?

**Biotechnology** is the use of cells or whole living organisms to produce materials useful to people, such as foods, medicines, and chemicals. People have been doing this for a very long time. For example, the use of yeasts to brew beer and wine dates back at least 8,000 years, and the use of bacterial cultures to make cheese and yogurt is a technique many centuries old. For a long time people were not aware of the molecular basis of each of these biochemical transformations.

About 100 years ago, thanks largely to Louis Pasteur's work, it became clear that specific bacteria, yeasts, and other microbes could be used as biological converters to make certain products. Alexander Fleming's discovery that the mold *Penicillium* makes the antibiotic penicillin led to the large-scale commercial culture of microbes to produce antibiotics as well as other useful chemicals. Today, microbes are grown in vast quantities to make much of the industrial-grade alcohol, glycerol, butyric acid, and citric acid that are used by themselves or as starting materials in the manufacture of other products.

Nevertheless the commercial harvesting of proteins, including hormones and enzymes, was limited by the (often) minuscule amounts that could be extracted from organisms that produce them naturally. Yields were low, and purification was difficult and costly. Gene cloning has changed all this. The ability to insert almost any gene into bacteria or yeasts, along with methods to induce the gene to make its product in large amounts and export it from the cells, has turned these microbes into versatile factories for important products. Today there is interest in producing nutritional supplements and pharmaceuticals in whole transgenic animals and harvesting them in large quantities, for example from the milk of cows or the eggs of chickens. Key to this boom in biotechnology has been the development of specialized vectors that not only carry genes into cells, but also make those cells express them at high levels.

### Expression vectors can turn cells into protein factories

If a eukaryotic gene is inserted into a typical plasmid and used to transform *E. coli*, little if any of the gene product will be made. Other key prokaryotic DNA sequences must be included with the gene. A bacterial promoter, a signal for transcription termination, and a special sequence that is necessary for ribosome binding on the mRNA must all be included in the transformation vector if the gene is to be expressed in the bacterial cell.

To solve this kind of problem, scientists make **expression vectors** that have all the characteristics of typical vectors, as well as the extra sequences needed for the foreign gene (also called a *transgene*) to be expressed in the host cell. For bacterial hosts, these additional sequences include the elements named above (**Figure 18.11**); for eukaryotes, they include the poly A–addition sequence, transcription factor binding sites, and enhancers. An expression vector can be designed to deliver transgenes to

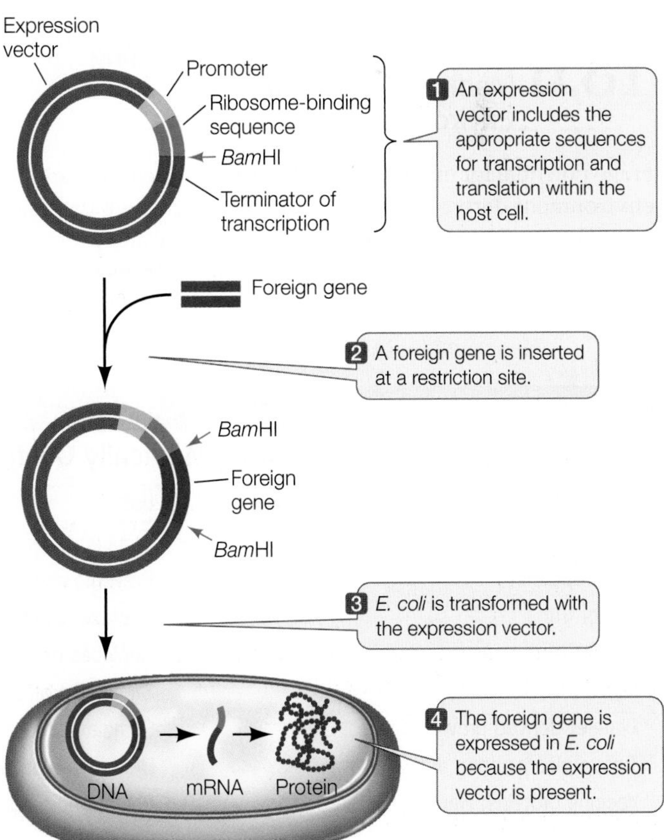


**18.11 Expression of a Transgene in a Host Cell Produces Large Amounts of its Protein Product** To be expressed in *E. coli*, a gene derived from a eukaryote requires bacterial sequences for transcription initiation (promoter), transcription termination, and ribosome binding. Expression vectors contain these additional sequences, enabling the eukaryotic protein to be synthesized in the prokaryotic cell.

**yourBioPortal.com**
GO TO Web Activity 18.1 • Expression Vectors

any class of prokaryotic or eukaryotic host and may include additional features:

- An *inducible promoter*, which responds to a specific signal, can be included. For example, a promoter that responds to hormonal stimulation can be used so that the transgene will be expressed at high levels when the hormone is added.
- A *tissue-specific promoter*, which is expressed only in a certain tissue at a certain time, can be used if localized expression is desired. For example, many seed proteins are expressed only in the plant embryo. Coupling a transgene to a seed-specific promoter will allow it to be expressed only in seeds.
- *Signal sequences* can be added so that the gene product is directed to an appropriate destination. For example, when a protein is made by yeast or bacterial cells in a liquid medium, it is economical to include a signal directing the protein to be secreted into the extracellular medium for easier recovery.

### 18.5 RECAP

**Expression vectors maximize the expression of transgenes inserted into host cells.**

- How do expression vectors work? See pp. 397–398 and Figure 18.11

This chapter has introduced many of the methods that are used in biotechnology. Let's turn now to the ways biotechnology is being applied to meet some specific human needs.

## 18.6 How Is Biotechnology Changing Medicine, Agriculture, and the Environment?

Huge potential for improvements in health, agriculture, and the environment derive from recent developments in biotechnology. We now have the ability to make virtually any protein by recombinant DNA technology and to insert transgenes into many kinds of host cells. With these revolutionary developments in biological capability, concerns have been raised about ethics and safety. We now turn to the promises and problems of biotechnology that uses DNA manipulation.

### Medically useful proteins can be made by biotechnology

Many medically useful products are being made by biotechnology (**Table 18.1**), and hundreds more are in various stages of development. The manufacture of *tissue plasminogen activator* (*TPA*) provides a good illustration of a medical application of biotechnology.

When a wound begins bleeding, a blood clot soon forms to stop the flow. Later, as the wound heals, the clot dissolves. How does the blood perform these conflicting functions at the right times? Mammalian blood contains an enzyme called *plasminogen*. When activated, it becomes *plasmin* and catalyzes the dissolution of the clotting proteins. The conversion of plasminogen to plasmin is catalyzed by the enzyme TPA, which is produced by cells lining the blood vessels:

$$\underset{\text{(inactive)}}{\text{plasminogen}} \xrightarrow{\text{TPA}} \underset{\text{(active)}}{\text{plasmin}}$$

Heart attacks and strokes can be caused by blood clots that form in major blood vessels leading to the heart or the brain, respectively. During the 1970s, a bacterial enzyme called *streptokinase* was found to stimulate the dissolution of clots in some patients. Treatment with this enzyme saved lives, but being a foreign protein, it triggered the body's immune system to react against it. More important, the drug sometimes prevented clotting throughout the entire circulatory system, sometimes leading to a dangerous situation in which blood could not clot where needed.

When TPA was discovered, it had many advantages: it bound specifically to clots, and it did not provoke an immune reaction. But the amounts of TPA that could be harvested from human tissues were tiny, certainly not enough to inject at the site of a clot in the emergency room.

Recombinant DNA technology solved this problem. TPA mRNA was isolated and used to make cDNA, which was then

**TABLE 18.1**
**Some Medically Useful Products of Biotechnology**

| PRODUCT | USE |
|---|---|
| Colony-stimulating factor | Stimulates production of white blood cells in patients with cancer and AIDS |
| Erythropoietin | Prevents anemia in patients undergoing kidney dialysis and cancer therapy |
| Factor VIII | Replaces clotting factor missing in patients with hemophilia A |
| Growth hormone | Replaces missing hormone in people of short stature |
| Insulin | Stimulates glucose uptake from blood in people with insulin-dependent (Type I) diabetes |
| Platelet-derived growth factor | Stimulates wound healing |
| Tissue plasminogen activator | Dissolves blood clots after heart attacks and strokes |
| Vaccine proteins: Hepatitis B, herpes, influenza, Lyme disease, meningitis, pertussis, etc. | Prevent and treat infectious diseases |

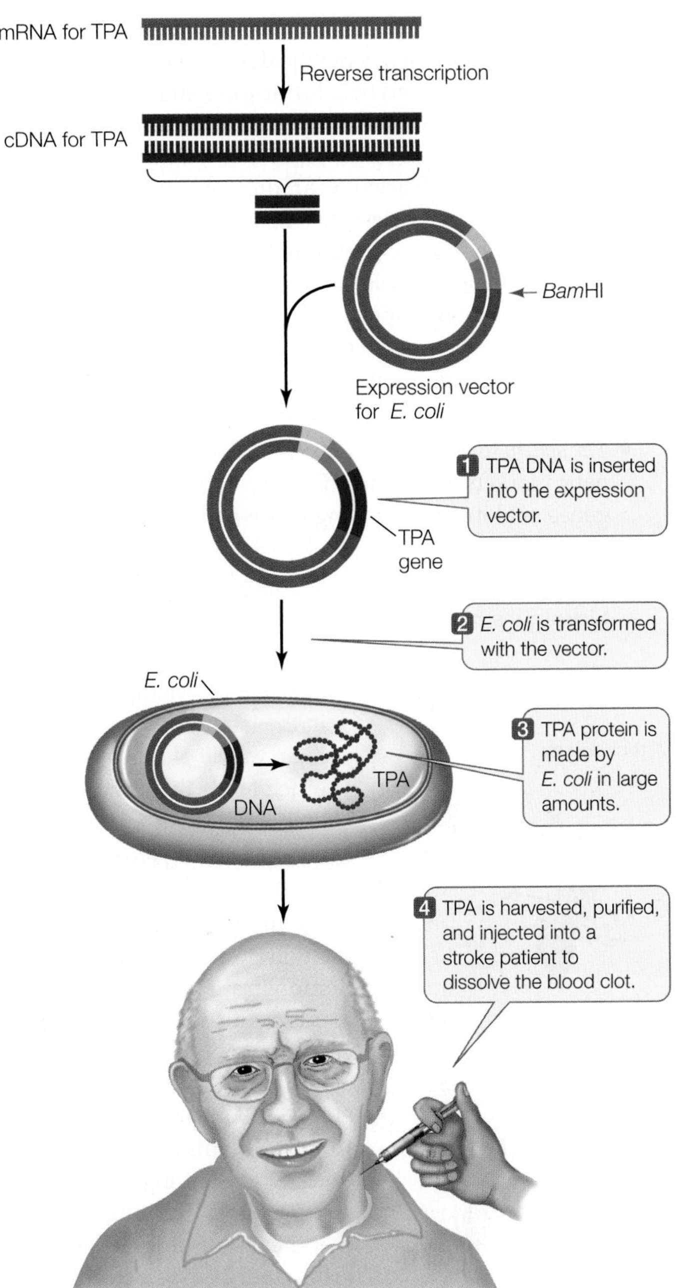


**18.12 Tissue Plasminogen Activator: From Protein to Gene to Drug** TPA is a naturally occurring human protein that dissolves blood clots. It is used to treat patients suffering from blood clotting in heart attacks or strokes, and is manufactured using recombinant DNA technology.

inserted into an expression vector and used to transform *E. coli* (**Figure 18.12**). The transgenic bacteria made the protein in quantity, and it soon became available commercially. This drug has had considerable success in dissolving blood clots in people experiencing strokes and heart attacks.

Another way of making medically useful products in large amounts is **pharming**: the production of pharmaceuticals in farm animals or plants. For example, a gene encoding a useful protein might be placed next to the promoter of the gene that encodes lactoglobulin, an abundant milk protein. Transgenic animals carrying this recombinant DNA will secrete large amounts of the foreign protein into their milk. These natural "bioreactors" can produce abundant supplies of the protein, which can be separated easily from the other components of the milk (**Figure 18.13**).

Human growth hormone is a protein made in the pituitary gland in the brain and has many effects, especially in growing children (see Chapter 41). Children with growth hormone deficiency have short stature as well as other abnormalities. In the past they were treated with protein isolated from the pituitary glands of dead people, but the supply was too limited to meet demand. Recombinant DNA technology was used to coax bacteria to make the protein, but the cost of treatment was high ($30,000 a year). In 2004, a team led by Daniel Salamone at the

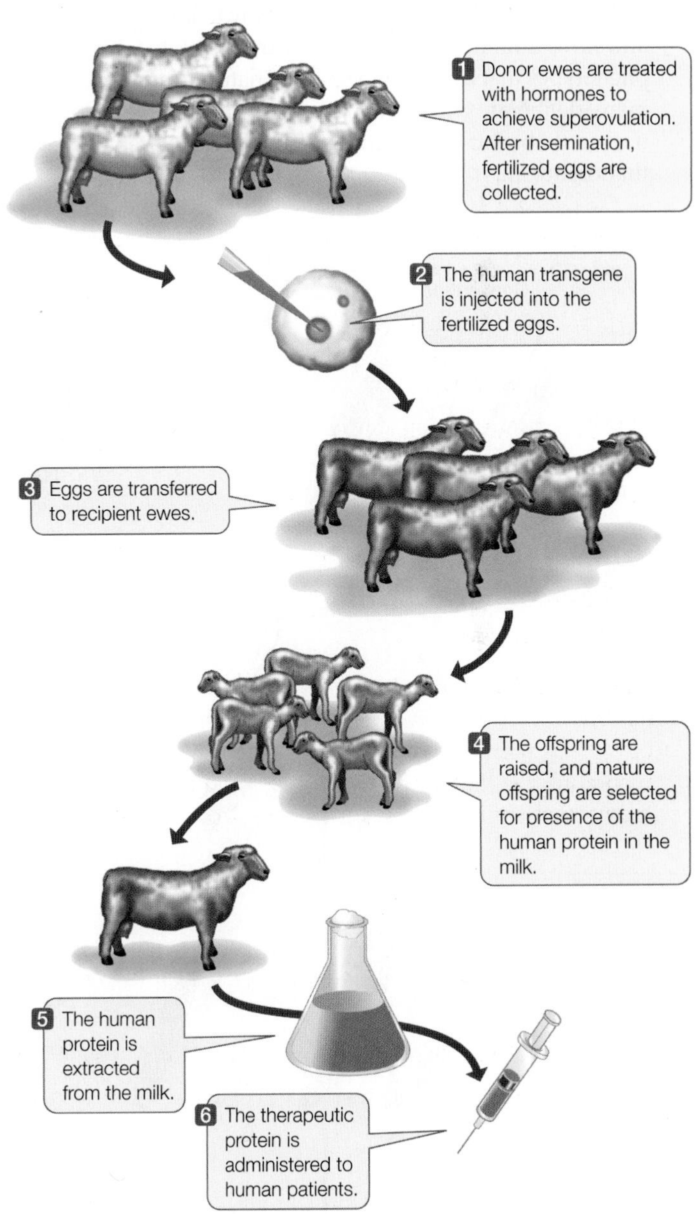


**18.13 Pharming** An expression vector carrying a desired gene can be put into an animal egg, which is implanted into a surrogate mother. The transgenic offspring produce the new protein in their milk. The milk is easily harvested and the protein isolated, purified, and made clinically available to patients.

University of Buenos Aires made a transgenic cow that secretes human growth hormone in her milk. The yield is prodigious: only 15 such cows are needed to meet the needs worldwide of children suffering from this type of dwarfism.

### DNA manipulation is changing agriculture

The cultivation of plants and the husbanding of animals provide the world's oldest examples of biotechnology, dating back more than 10,000 years. Over the centuries, people have adapted crops and farm animals to their needs. Through selective breeding of these organisms, desirable characteristics such as large seeds, high fat content in milk, or resistance to disease have been imparted and improved.

Until recently, the most common way to improve crop plants and farm animals was to identify individuals with desirable phenotypes that existed as a result of natural variation. Through many deliberate crosses, the genes responsible for the desirable trait could be introduced into a widely used variety or breed of that organism.

Despite some spectacular successes, such as the breeding of high-yielding varieties of wheat, rice, and hybrid corn, such deliberate crossing can be a hit-or-miss affair. Many desirable traits are controlled by multiple genes, and it is hard to predict the results of a cross or to maintain a prized combination as a pure-breeding variety year after year. In sexual reproduction, combinations of desirable genes are quickly separated by meiosis. Furthermore, traditional breeding takes a long time: many plants and animals take years to reach maturity and then can reproduce only once or twice a year—a far cry from the rapid reproduction of bacteria.

Modern recombinant DNA technology has several advantages over traditional methods of breeding (**Figure 18.14**):

- *The ability to identify specific genes.* The development of genetic markers allows breeders to select for specific desirable genes, making the breeding process more precise and rapid.
- *The ability to introduce any gene from any organism into a plant or animal species.* This ability, combined with mutagenesis techniques, vastly expands the range of possible new traits.
- *The ability to generate new organisms quickly.* Manipulating cells in the laboratory and regenerating a whole plant by cloning is much faster than traditional breeding.

Consequently, recombinant DNA technology has found many applications in agriculture (**Table 18.2**). We will describe a few examples to demonstrate the approaches that plant scientists have used to improve crop plants.

**PLANTS THAT MAKE THEIR OWN INSECTICIDES** Plants are subject to infections by viruses, bacteria, and fungi, but probably the most important crop pests are herbivorous insects. From the locusts of biblical (and modern) times to the cotton boll weevil, insects have continually eaten the crops people grow.

The development of insecticides has improved the situation somewhat, but insecticides have their own problems. Many, including the organophosphates, are relatively nonspecific and kill beneficial insects in the broader ecosystem as well as crop pests. Some even have

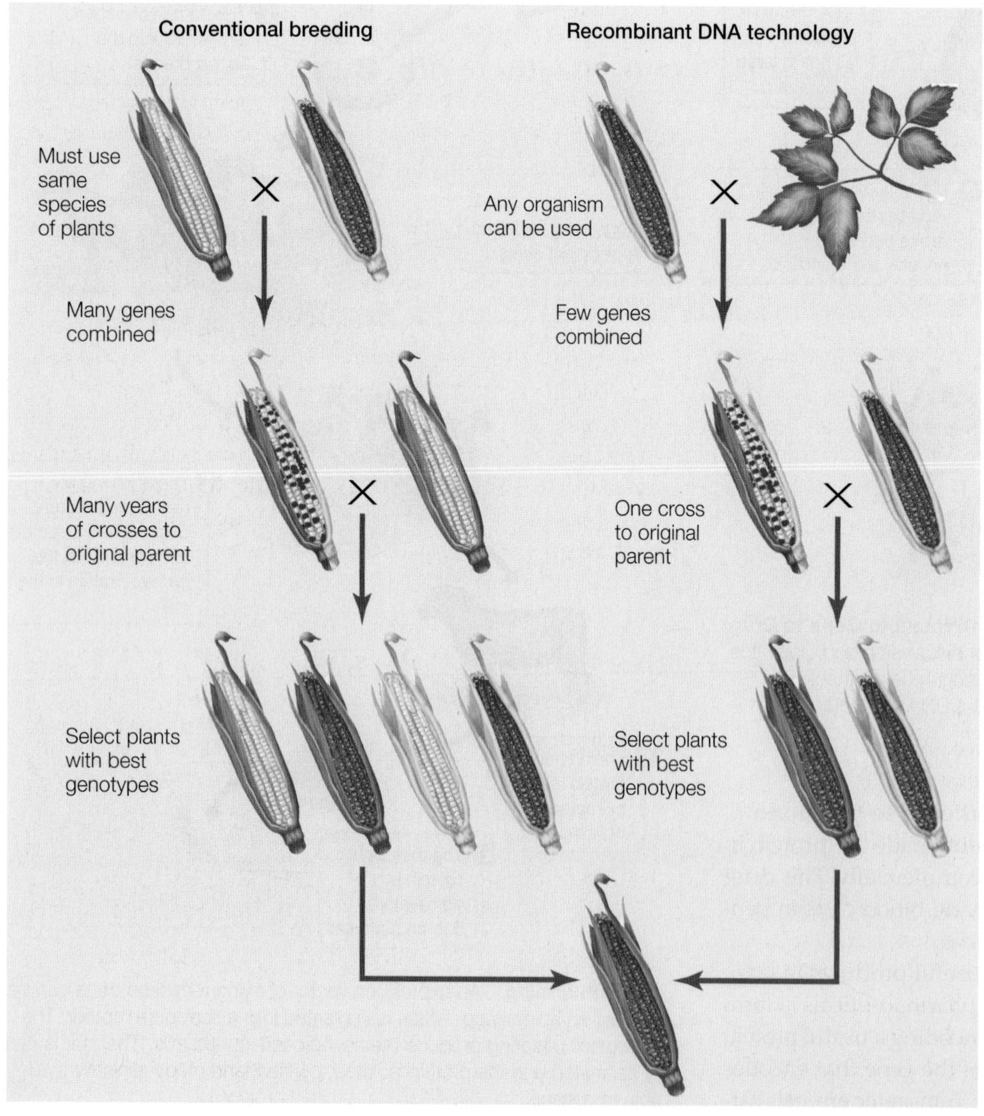


**18.14 Genetically Modified Plants versus Conventional Plant Breeding** Plant biotechnology offers many potential advantages over conventional breeding.

TABLE 18.2
**Agricultural Applications of Biotechnology under Development**

| PROBLEM | TECHNOLOGY/GENES |
|---|---|
| Improving the environmental adaptations of plants | Genes for drought tolerance, salt tolerance |
| Improving nutritional traits | High-lysine seeds; β-carotene in rice |
| Improving crops after harvest | Delay of fruit ripening; sweeter vegetables |
| Using plants as bioreactors | Plastics, oils, and drugs produced in plants |

toxic effects on other groups of organisms, including people. What's more, many insecticides persist in the environment for a long time.

Some bacteria protect themselves by producing proteins that can kill insects. For example, the bacterium *Bacillus thuringiensis* produces a protein that is toxic to the insect larvae that prey on it. The toxicity of this protein is 80,000 times greater than that of a typical commercial insecticide. When a hapless larva eats the bacteria, the toxin becomes activated and binds specifically to the insect's gut, producing holes and killing the insect. Dried preparations of *B. thuringiensis* have been sold for decades as safe insecticides that break down rapidly in the environment. But the biodegradation of these preparations is their limitation, because it means that the dried bacteria must be applied repeatedly during the growing season.

A more permanent approach is to have the crop plants themselves make the toxin, and this is exactly what plant scientists have done. The toxin gene from *B. thuringiensis* has been isolated, cloned, and extensively modified by the addition of a plant promoter and other regulatory sequences. Transgenic corn, cotton, soybeans, tomatoes, and other crops are now being grown successfully with this added gene. Pesticide usage by farmers growing these transgenic crops is greatly reduced.

**CROPS THAT ARE RESISTANT TO HERBICIDES** Herbivorous insects are not the only threat to agriculture. Weeds may grow in fields and compete with crop plants for water and soil nutrients. Glyphosate is a widely used and effective herbicide, or weed killer, that works only on plants. It inhibits an enzyme system in the chloroplast that is involved in the synthesis of amino acids. Glyphosate is a broad-spectrum herbicide that kills most weeds, but unfortunately it also kills crop plants. One solution to this problem is to use it to rid a field of weeds before the crop plants start to grow. But, as any gardener knows, when the crop begins to grow, the weeds reappear. If the crop were not affected by the herbicide, the herbicide could be applied to the field at any time.

Scientists have used expression vectors to make plants that synthesize a different form of the target enzyme for glyphosate that is unaffected by the herbicide. The gene for this enzyme has been inserted into corn, cotton, and soybean plants, making them resistant to glyphosate. This technology has expanded rapidly and a large proportion of cotton and soybean plants now carry this gene.

**GRAINS WITH IMPROVED NUTRITIONAL CHARACTERISTICS** To remain healthy, humans must consume adequate amounts of β-carotene, which the body converts into vitamin A (see Figure 3.21). About 400 million people worldwide suffer from vitamin A deficiency, which makes them susceptible to infections and blindness. One reason is that rice grains, which do not contain β-carotene, make up a large part of their diets. Other parts of the rice plant, and indeed many plants and other organisms, contain enzymes for the biochemical pathway that leads to β-carotene production.

Plant biologists Ingo Potrykus and Peter Beyer isolated one of the genes for the β-carotene pathway from the bacterium *Erwinia uredovora* and another from daffodil plants. They added a promoter and other signals for expression in the developing rice grain, and then transformed rice plants with the two genes. The resulting rice plants produce grains that look yellow because of their high β-carotene content. A newer variety with a corn gene replacing the one from daffodils makes even more β-carotene and is golden in color (**Figure 18.15**). A daily intake of about 150 grams of this cooked rice can supply all the β-carotene a person needs. This new transgenic strain has been crossed with strains adapted for various local environments, in the hope of improving the diets of millions of people.

**CROPS THAT ADAPT TO THE ENVIRONMENT** Agriculture depends on ecological management—tailoring the environment to the needs of crop plants and animals. A farm field is an unnatural, human-designed system that must be carefully managed to maintain optimal conditions for crop growth. For example, excessive irrigation can cause increases in soil salinity. The Fertile Crescent, the region between the Tigris and Euphrates rivers in the Middle East where agriculture probably originated 10,000 years ago, is no longer fertile. It is now a desert, largely because the soil has a high salt concentration. Few plants can grow on salty soils, partly because of osmotic ef-

**18.15 Transgenic Rice Rich in β-Carotene** Right and middle: The grains from these transgenic rice strains are colored because they make the pigment β-carotene, which is converted to vitamin A in the human body. Left: Normal rice grains do not contain β-carotene.

fects that result in wilting, and partly because excess salt ions are toxic to plant cells.

Some plants can tolerate salty soils because they have a protein that transports $Na^+$ ions out of the cytoplasm and into the vacuole, where the ions can accumulate without harming plant growth (see Section 5.3 for a description of the plant vacuole). In many salt-intolerant plants, including *Arabidopsis thaliana*, the gene for this protein exists but is inactive. Recombinant DNA technology has allowed scientists to create active versions of this gene, and to use it to transform crop plants such as rapeseed, wheat, and tomatoes. When this gene was added to tomato plants, they grew in water that was four times as salty as the typical lethal level (**Figure 18.16**). This finding raises the prospect of growing useful crops on what were previously unproductive soils.

This example illustrates what could become a fundamental shift in the relationship between crop plants and the environment. *Instead of manipulating the environment to suit the plant, biotechnology may allow us to adapt the plant to the environment.* As a result, some of the negative effects of agriculture, such as water pollution, could be lessened.

### Biotechnology can be used for environmental cleanup

The thousands of species of bacteria have many unique enzymes and biochemical pathways. Bacteria are nature's recyclers, thriving on many types of nutrients—including what humans refer to as wastes. **Bioremediation** is the use by humans of other organisms to remove contaminants from the environment. Two well-known uses of bacteria for bioremediation are composting and wastewater treatment.

*Composting* involves the use of bacteria to break down large molecules, including carbon-rich polymers and proteins in waste products such as wood chips, paper, straw, and kitchen scraps. For example, some species of bacteria make cellulase, an enzyme that hydrolyzes cellulose. Bacteria are used in *wastewater treatment* to break down human wastes, paper products, and household chemicals.

Transgenic organisms can also be used to clean up environmental contaminants. As we saw at the opening of this chapter, bacteria are being used to help clean up oil spills. As another example, plants that have been modified to take up heavy metals are being explored as a way to remediate contaminated soils, such as mine tailings (see Section 39.4).

### There is public concern about biotechnology

Concerns have been raised about the safety and wisdom of genetically modifying crops and other organisms. These concerns are centered on three claims:

- Genetic manipulation is an unnatural interference with nature.
- Genetically altered foods are unsafe to eat.
- Genetically altered crop plants are dangerous to the environment.

Advocates of biotechnology tend to agree with the first claim. However, they point out that all crops are unnatural in the sense that they come from artificially bred plants growing in a manipulated environment (a farmer's field). Recombinant DNA technology just adds another level of sophistication to these technologies.

To counter the concern about whether genetically engineered crops are safe for human consumption, biotechnology advocates point out that only single genes are added and that these genes are specific for plant function. For example, the *B. thuringiensis* toxin produced by transgenic plants has no effect on people. However, as plant biotechnology moves from adding genes that improve plant growth to adding genes that affect human nutrition, such concerns will become more pressing.

Various negative environmental impacts have been envisaged. There is concern about the possible "escape" of transgenes from crops to other species. If the gene for herbicide resistance, for example, were inadvertently transferred from a crop plant to a closely related weed, that weed could thrive in her-

(A)

(B)

**18.16 Salt-Tolerant Tomato Plants** Transgenic plants containing a gene for salt tolerance thrive in salty water (A), while plants without the transgene die (B). This technology may allow crops to be grown on salty soils.

bicide-treated areas. Another negative impact would be the development of new super-weeds from transgenic crops. For example, a drought tolerant crop plant might spread into, and upset the ecology of, a desert. Or beneficial insects could eat plant materials containing *B. thuringiensis* toxin and die. Transgenic plants undergo extensive field-testing before they are approved for use, but the complexity of the biological world makes it impossible to predict all potential environmental effects of transgenic organisms. In fact, some spreading of transgenes has been detected. Because of the potential benefits of agricultural biotechnology (see Table 18.2), scientists believe that it is wise to proceed with caution.

### 18.6 RECAP

**Biotechnology has been used to produce medicines and to develop transgenic plants with improved agricultural and nutritional characteristics.**

- What are the advantages of using biotechnology for plant breeding compared with traditional methods? **See Figure 18.14**
- What are some of the concerns that people might have about agricultural biotechnology? **See pp. 402–403**

## CHAPTER SUMMARY

### 18.1 What Is Recombinant DNA?

- **Recombinant DNA** is formed by the combination of two DNA sequences from different sources. **Review Figure 18.1**
- Many **restriction enzymes** make staggered cuts in the two strands of DNA, creating fragments that have **sticky ends** with unpaired bases.
- DNA fragments with sticky ends can be used to create recombinant DNA. DNA molecules from different sources can be cut with the same restriction enzyme and spliced together using **DNA ligase**. **Review Figure 18.2**

### 18.2 How Are New Genes Inserted into Cells?

- One goal of recombinant DNA technology is to **clone** a particular gene, either for analysis or to produce its protein product in quantity.
- Bacteria, yeasts, and cultured plant and animal cells are commonly used as hosts for recombinant DNA. The insertion of foreign DNA into host cells is called **transformation** or **transfection** (for animal cells). Transformed or transfected cells are called **transgenic** cells.
- Various methods are used to get recombinant DNA into cells. These include chemical or electrical treatment of the cells, the use of viral vectors, and injection. *Agrobacterium tumefaciens* is often used to insert DNA into plant cells.
- To identify host cells that have taken up a foreign gene, the inserted sequence can be tagged with one or more **reporter genes**, which are genetic markers with easily identifiable phenotypes. **Selectable markers** allow for the selective growth of transgenic cells.
- Replication of the foreign gene in the host cell requires that it become part of a segment of DNA that contains a **replicon** (origin and terminus of replication).
- **Vectors** are DNA sequences that can carry new DNA into host cells. Plasmids and viruses are commonly used as vectors. **Review Figure 18.3**

### 18.3 What Sources of DNA Are Used in Cloning?

- DNA fragments from a genome can be inserted into host cells to create a **genomic library**. **Review Figure 18.6A**
- The mRNAs produced in a certain tissue at a certain time can be extracted and used to create **complementary DNA (cDNA)** by reverse transcription. **Review Figure 18.6B**
- PCR products can be used for cloning.
- Synthetic DNA containing any desired sequence can be made and mutated in the laboratory.

### 18.4 What Other Tools Are Used to Study DNA Function?

- Homologous recombination can be used to **knock out** a gene in a living organism. **Review Figure 18.7**
- Gene silencing techniques can be used to inactivate the mRNA transcript of a gene, which may provide clues to the gene's function. Artificially created **antisense RNA** or **siRNA** can be added to a cell to prevent translation of a specific mRNA. **Review Figure 18.8**
- **DNA microarray** technology permits the screening of thousands of cDNA sequences at the same time. **Review Figure 18.9, ANIMATED TUTORIAL 18.1**

### 18.5 What Is Biotechnology?

- **Biotechnology** is the use of living cells to produce materials useful to people. Recombinant DNA technology has resulted in a boom in biotechnology.
- **Expression vectors** allow a transgene to be expressed in a host cell. **Review Figure 18.11, WEB ACTIVITY 18.1**

### 18.6 How Is Biotechnology Changing Medicine, Agriculture, and the Environment?

- Recombinant DNA techniques have been used to make medically useful proteins. **Review Figure 18.12**
- **Pharming** is the use of transgenic plants or animals to produce pharmaceuticals. **Review Figure 18.13**
- Because recombinant DNA technology has several advantages over traditional agricultural biotechnology, it is being extensively applied to agriculture. **Review Figure 18.14**
- Transgenic crop plants can be adapted to their environments, rather than vice versa.
- **Bioremediation** is the use of organisms, which are often genetically modified, to improve the environment by breaking down pollutants.
- There is public concern about the application of recombinant DNA technology to food production.

## SELF-QUIZ

1. Restriction enzymes
   *a.* play no role in bacteria.
   *b.* cleave DNA at highly specific recognition sequences.
   *c.* are inserted into bacteria by bacteriophage.
   *d.* are made only by eukaryotic cells.
   *e.* add methyl groups to specific DNA sequences.
2. Which of the following is used as a reporter gene in recombinant DNA work with bacteria as host cells?
   *a.* rRNA
   *b.* Green fluorescent protein
   *c.* Antibiotic sensitivity
   *d.* Ability to make ornithine
   *e.* Vitamin synthesis
3. From the list below, select the sequence of steps for inserting a piece of foreign DNA into a plasmid vector, introducing the plasmid into bacteria, and verifying that the plasmid and the foreign gene are present:
   (1) Transform host cells.
   (2) Select for the lack of plasmid reporter gene 1 function.
   (3) Select for the plasmid reporter gene 2 function.
   (4) Digest vector and foreign DNA with a restriction enzyme, which inactivates plasmid reporter gene 1.
   (5) Ligate the digested plasmid together with the foreign DNA.
   *a.* 4, 5, 1, 3, 2
   *b.* 4, 5, 1, 2, 3
   *c.* 1, 3, 4, 2, 5
   *d.* 3, 2, 1, 4, 5
   *e.* 1, 3, 2, 5, 4
4. Possession of which feature is *not* desirable in a vector for gene cloning?
   *a.* An origin of DNA replication
   *b.* Genetic markers for the presence of the vector
   *c.* Many recognition sequences for the restriction enzyme to be used
   *d.* One recognition sequence each for one or more different restriction enzymes
   *e.* Genes other than the target for transfection
5. RNA interference (RNAi) inhibits
   *a.* DNA replication.
   *b.* neither transcription nor translation of specific genes.
   *c.* recognition of the promoter by RNA polymerase.
   *d.* transcription of all genes.
   *e.* translation of specific mRNAs.
6. Complementary DNA (cDNA)
   *a.* is produced from ribonucleoside triphosphates.
   *b.* is produced by reverse transcription.
   *c.* is the "other strand" of single-stranded DNA in a virus.
   *d.* requires no template for its synthesis.
   *e.* cannot be placed into a vector because it has the opposite base sequence of the vector DNA.
7. In a genomic library of frog DNA in *E. coli* bacteria,
   *a.* all bacterial cells have the same sequences of frog DNA.
   *b.* all bacterial cells have different sequences of frog DNA.
   *c.* each bacterial cell has a random fragment of frog DNA.
   *d.* each bacterial cell has many fragments of frog DNA.
   *e.* the frog DNA is transcribed into mRNA in the bacterial cells.
8. An expression vector requires all of the following except
   *a.* genes for ribosomal RNA.
   *b.* a reporter gene.
   *c.* a promoter of transcription.
   *d.* an origin of DNA replication.
   *e.* restriction enzyme recognition sequences.
9. "Pharming" is a term that describes
   *a.* the use of animals in transgenic research.
   *b.* plants making genetically altered foods.
   *c.* synthesis of recombinant drugs by bacteria.
   *d.* large-scale production of cloned animals.
   *e.* synthesis of a drug by a transgenic plant or animal.
10. Which of the following could *not* be used to test whether expression of a particular gene is necessary for a particular biological function?
   *a.* RNAi
   *b.* Knockout technology
   *c.* Antisense
   *d.* Mutant tRNA
   *e.* Transposon mutagenesis

## FOR DISCUSSION

1. Compare PCR (see Section 13.5) and cloning as methods to amplify a gene. What are the requirements, benefits, and drawbacks of each method?
2. As specifically as you can, outline the steps you would take to *(a)* insert and express the gene for a new, nutritious seed protein in wheat, and *(b)* insert and express a gene for a human enzyme in sheep's milk.
3. Compare traditional genetic methods with molecular methods for producing genetically altered plants. For each case, describe *(a)* sources of new genes; *(b)* numbers of genes transferred; and *(c)* how long the process takes.

## ADDITIONAL INVESTIGATION

Green fluorescent protein (GFP) from a jellyfish can be incorporated into a vector as a reporter gene to signal the presence of the vector in a host cell (see Figure 18.5). How would you alter the technique in Figure 18.4 to substitute GFP for one (or both) of the antibiotic resistance markers?

## WORKING WITH DATA (GO TO yourBioPortal.com)

**Recombinant DNA** In 1973, Stanley Cohen and Herbert Boyer pioneered the field of recombinant DNA technology when they demonstrated that biologically functional recombinant bacterial plasmids can be constructed in the laboratory (Figure 18.1). In this exercise, you will examine their original research article and calculations from their data that show that recombinant DNA was made.

# 19 Differential Gene Expression in Development

## On track with stem cells

In horse racing, bettors speak of the "future book" odds on a horse's chances in an upcoming race. On the morning after winning a race in 2005, the future book odds for Greg's Gold did not look good—he was limping because of a shredded tendon in his right front leg. A tendon is like a rubber band connecting muscles and bones, and tendons in the legs store energy when an animal runs. Typically, a damaged tendon is allowed to heal naturally, but scar tissue makes it less flexible, and a horse cannot run as fast as it did before injuring a tendon. So it looked as if Greg's Gold might have to retire from racing.

Greg's Gold's trainer, David Hofmans, decided to try a new therapy. A veterinarian removed a small amount of adipose (fatty) tissue from the horse's hindquarters and sent it to a cell biology laboratory. There, the tissue was treated with enzymes to digest the extracellular molecules that held the cells together. Several cell populations were obtained, among them mesenchymal stem cells. Stem cells are actively dividing, unspecialized cells that have the potential to produce different cell types depending on the signals they receive from the body. Mesenchymal stem cells are able to differentiate into various kinds of connective tissue, including bone, cartilage, blood vessels, tendons, and muscle.

Two days after the tissue was taken, Greg's Gold's veterinarian received the stem cells back from the lab and injected them into the site of the damaged tendon. After several months, the tendon healed with little scar tissue, and Greg's Gold's trainer returned him to the racetrack. Greg's Gold raced for almost two more years, winning over $1 million in purse money before being retired.

The mesenchymal stem cell treatment has been used successfully on several thousand horses, and on dogs with arthritis. Most stem cell therapies for humans are still at the experimental stage, particularly in the United States, where controversy over the use of embryonic stem cells has slowed the progress of research and the adoption of therapeutic techniques. But in Japan, women undergoing reconstructive surgery after the removal of breast cancer have had more favorable outcomes when treated with their own mesenchymal stem cells. Bone marrow transplantation is one form of stem cell therapy that has been used successfully for more than thirty years in the United States, to treat patients with cancers such as leukemia and lymphoma.

**Greg's Gold** Fat stem cells helped repair damage to his tendons and he was able to race—and win—again.

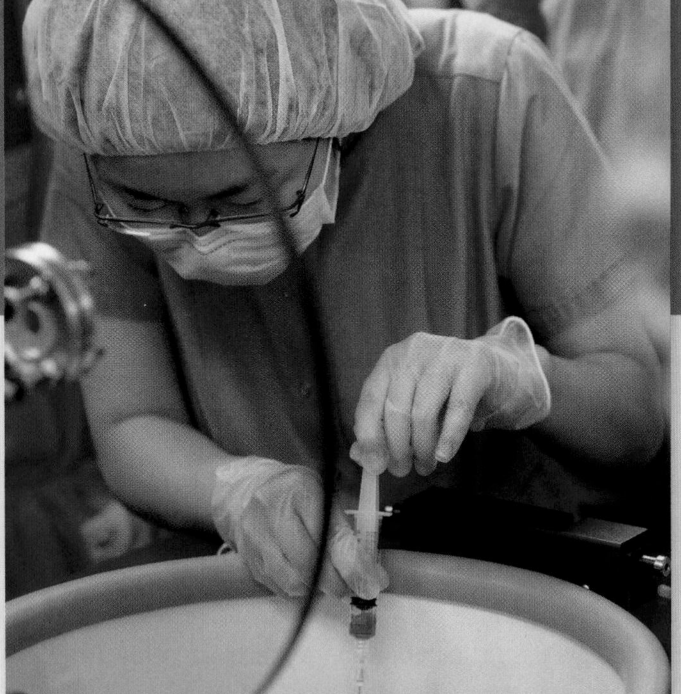

**Fat as a Source of Stem Cells** This centrifuge separates dense fatty tissues from the lighter stem cells. Stem cells from fat have been found to be capable of differentiating into several specialized cell types.

The processes by which an unspecialized stem cell proliferates and forms specialized cells and tissues with distinctive appearances and functions are similar to the developmental processes that occur in the embryo. Much of our knowledge of developmental biology has come from studies on model organisms such as the fruit fly *Drosophila melanogaster*, the nematode worm *Caenorhabditis elegans*, zebrafish, the mouse, and the small flowering plant *Arabidopsis thaliana*. Eukaryotes share many similar genes, and the cellular and molecular principles underlying their development also turn out to be similar. Thus discoveries from one organism can aid us in understanding other organisms, including ourselves.

**IN THIS CHAPTER** we begin by describing how almost every cell in a multicellular organism contains all of the genes present in the zygote that gave rise to that organism. Then we explain how cellular changes during development result from the differential expression of those genes. Finally, we show how the various mechanisms of transcriptional control and chemical signaling that are discussed in previous chapters work together to produce a complex organism.

CHAPTER OUTLINE



## 19.1 What Are the Processes of Development?

**Development** is the process by which a multicellular organism, beginning with a single cell, goes through a series of changes, taking on the successive forms that characterize its life cycle (**Figure 19.1**). After the egg is fertilized, it is called a zygote, and in the earliest stages of development a plant or animal is called an **embryo**. Sometimes the embryo is contained within a protective structure such as a seed coat, an eggshell, or a uterus. An embryo does not photosynthesize or feed itself. Instead, it obtains its food from its mother either directly (via the placenta) or indirectly (by way of nutrients stored in a seed or egg). A series of embryonic stages precedes the birth of the new, independent organism. Many organisms continue to develop throughout their life cycle, with development ceasing only with death.

### Development involves distinct but overlapping processes

The developmental changes an organism undergoes as it progresses from an embryo to mature adulthood involve four processes:

- **Determination** sets the developmental *fate* of a cell—what type of cell it will become—even before any characteristics of that cell type are observable. For example, the mesenchymal stem cells described in the opening story look unspecialized, but their fate to become connective tissue cells has already been determined.
- **Differentiation** is the process by which different types of cells arise, leading to cells with specific structures and functions. For example, mesenchymal stem cells differentiate to become muscle, fat, tendon, or other connective tissue cells.
- **Morphogenesis** (Greek for "origin of form") is the organization and spatial distribution of differentiated cells into the multicellular body and its organs.
- **Growth** is the increase in size of the body and its organs by cell division and cell expansion.

Determination and differentiation occur largely because of differential gene expression. The cells that arise from repeated mitoses in the early embryo may look the same superficially, but they soon begin to differ in terms of which of the thousands of genes in the genome are expressed.

Morphogenesis involves differential gene expression and the interplay of signals between cells. Morphogenesis can occur in several ways:

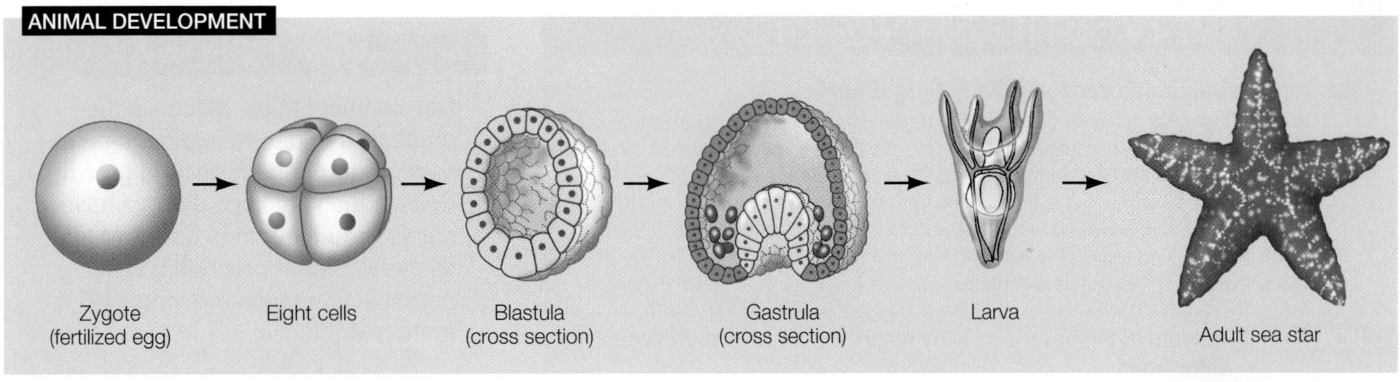


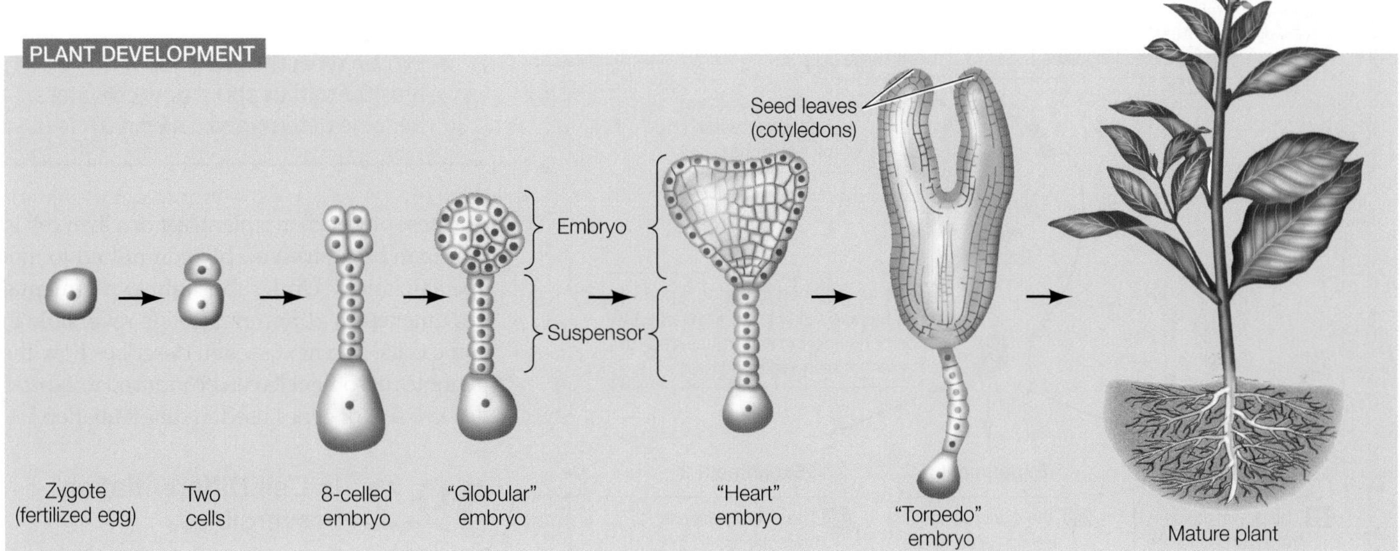


**19.1 From Fertilized Egg to Adult** The stages of development from zygote to maturity are shown for an animal and for a plant. The blastula is a hollow sphere of cells; the gastrula has three cell layers.

**yourBioPortal.com**

GO TO Web Activity 19.1 • Stages of Development

- *Cell division* is important in both plants and animals.
- *Cell expansion* is especially important in plant development, where a cell's position and shape are constrained by the cell wall.
- *Cell movements* are very important in animal morphogenesis (see Section 44.2)
- *Apoptosis* (programmed cell death) is essential in organ development.

Growth can occur by an increase in the number of cells or by the enlargement of existing cells. Growth continues throughout the individual's life in some organisms, but reaches a more or less stable end point in others.

### Cell fates become progressively more restricted during development

During development, each undifferentiated cell will become part of a particular type of tissue—this is referred to as the **cell fate** of that undifferentiated cell. A cell's fate is a function of both differential gene expression and morphogenesis. The role of morphogenesis in determining cell fate was revealed in experiments in which undifferentiated cells were removed from specific locations in early embryos and grafted into new positions on other embryos. The cells were marked with stains so that their development into adult structures could be traced. Such experiments on amphibian embryos indicated that the fates of early embryonic cells are not irrevocably determined, but depend on the cells' environment and stage of development (**Figure 19.2**). In this example, the cells that would have become skin tissue if left in place became brain or notochord tissues, depending on the locations of the grafts.

But as development proceeds from zygote to mature organism, the developmental potential of cells becomes more restricted. For example, if tissue is removed from the brain area of a later-stage frog embryo, it will become brain tissue, even if transplanted to a part of an early-stage embryo that is destined to become another structure.

As we will discuss in this chapter, cell fate determination is influenced by changes in gene expression as well as the extracellular environment. Determination is not something that is visible under the microscope—cells do not change their appearance when they become determined. Determination is followed by differentiation—the actual changes in biochemistry, structure, and function that result in cells of different types. *Determination is a commitment; the final realization of that commitment is differentiation.*

## INVESTIGATING LIFE

**19.2 Developmental Potential in Early Frog Embryos**

In an early embryo, the cells look alike. But marking experiments suggested that the fates of these cells were determined early in development. Was the fate of a cell irrevocable or did it still retain the ability to become a different cell type? To answer this question, biologists transplanted cells from one location in one embryo to a different location in a second embryo. The cells took on the fate of cells at the new location. Therefore, cells in the early embryo retained the ability to form other cell types if placed in the right environment.

**HYPOTHESIS** The fate of the cells in an early amphibian embryo is irrevocably determined.

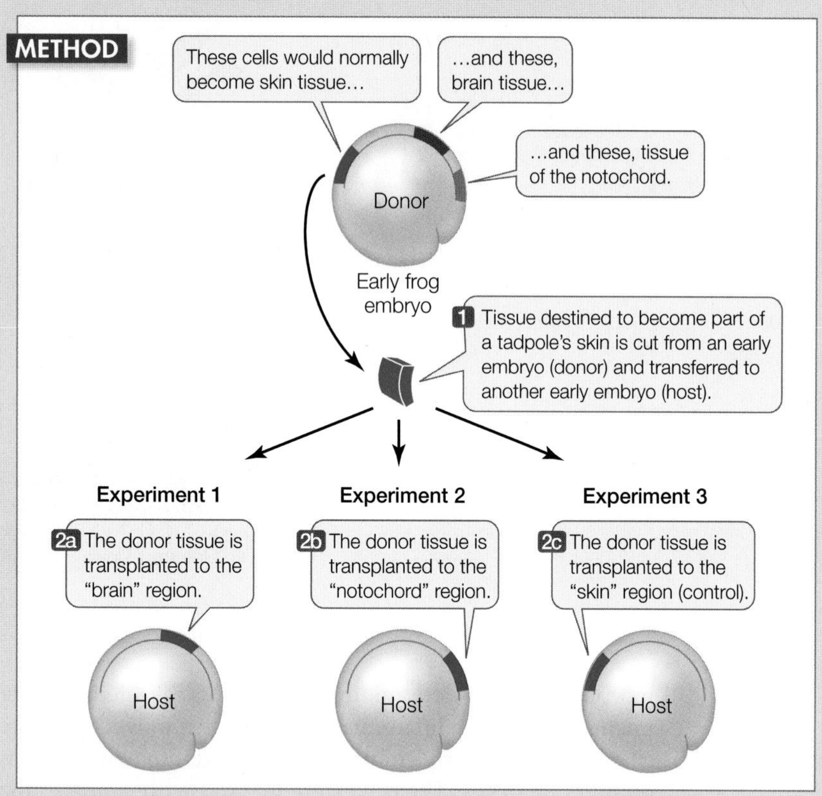
METHOD
These cells would normally become skin tissue...
...and these, brain tissue...
...and these, tissue of the notochord.
Donor
Early frog embryo
1 Tissue destined to become part of a tadpole's skin is cut from an early embryo (donor) and transferred to another early embryo (host).
Experiment 1
2a The donor tissue is transplanted to the "brain" region.
Experiment 2
2b The donor tissue is transplanted to the "notochord" region.
Experiment 3
2c The donor tissue is transplanted to the "skin" region (control).
Host
Host
Host

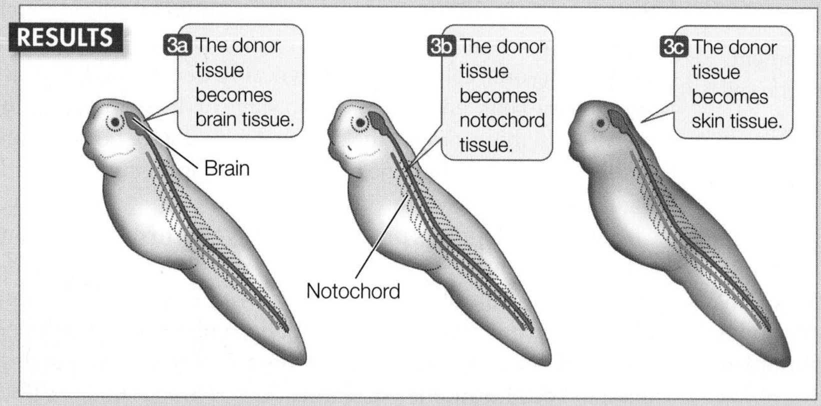
RESULTS
3a The donor tissue becomes brain tissue.
Brain
3b The donor tissue becomes notochord tissue.
Notochord
3c The donor tissue becomes skin tissue.

**CONCLUSION** The hypothesis is rejected. Cell fates in the early embryo are not determined, but can change depending on the environment.

**FURTHER INVESTIGATION:** What would happen if tissue from an adult were transplanted into an early embryo?

Go to **yourBioPortal.com** for original citations, discussions, and relevant links for all INVESTIGATING LIFE figures.

## 19.1 RECAP

**Development takes place via the processes of determination, differentiation, morphogenesis, and growth. Cells in the very early embryo have not yet had their fates determined; as development proceeds, their potential fates become more and more restricted.**

- What are the four processes of development? See p. 405
- Explain what the experiment in Figure 19.2 told us about how cell fates become determined. See p. 407

Is a mesophyll cell in a plant leaf or a liver cell in a human being irrevocably committed to that specialization? Under the right experimental circumstances, differentiation is reversible in some cells. The next section describes how the genomes of some cells can be induced to express *different* sets of genes used in differentiation.

## 19.2 Is Cell Differentiation Irreversible?

A zygote has the ability to give rise to every type of cell in the adult body; in other words, it is **totipotent** (*toti*, "all"; *potent*, "capable"). Its genome contains instructions for all of the structures and functions that will arise throughout the life cycle of the organism. Later in development, the cellular descendants of the zygote lose their totipotency and become determined. These determined cells then differentiate into specialized cells. The human liver cell and the leaf mesophyll cell generally retain their differentiated forms and functions throughout their lives. But this does not necessarily mean that they have irrevocably lost their totipotency. Most of the differentiated cells of an animal or plant have nuclei containing the entire genome of the organism and therefore have the genetic capacity for totipotency. We explore here several examples of how this capacity has been demonstrated experimentally.

### Plant cells can be totipotent

A carrot root cell normally faces a dark future. It cannot photosynthesize and generally does not give rise to new carrot plants. However, in 1958 Frederick Steward at Cornell University showed that if he isolated cells from a carrot root and maintained them in a suitable nutrient medium, he could induce them to dedifferentiate—to lose their differentiated characteristics. The cells could divide and give rise to masses of undifferentiated cells called *calli* (singular *callus*), which could be maintained in culture indefinitely. But, if they were provided with the right chemical cues, the cells could develop into embryos and eventually into complete new plants (**Figure 19.3**). Since the new plants were genetically identical to the cells from which they came, they were clones of the original carrot plant.

The ability to clone an entire carrot plant from a differentiated root cell indicated that the cell contained the entire carrot genome, and that under the right conditions, the cell and its descendants could express the appropriate genes in the right sequence to form a new plant. Many types of cells from other plant species show similar behavior in the laboratory. This ability to generate a whole plant from a single cell has been invaluable in agriculture and forestry. For example, trees from planted forests are used in making paper, lumber, and other products. To replace the trees reliably, forestry companies regenerate new trees from the leaves of selected trees with desirable traits. The characteristics of these clones are more uniform and predictable than those of trees grown from seeds.

### Nuclear transfer allows the cloning of animals

Animal somatic cells cannot be manipulated as easily as plant cells can. However, experiments such as the one shown in Figure 19.2 have demonstrated the totipotency of early embryonic cells from animals. In humans, this totipotency permits both genetic screening (see Section 15.5) and certain assisted reproductive technologies (see Section 43.4). A human embryo can be isolated in the laboratory and one or a few cells removed and examined to determine whether a certain genetic condition is present. Due to their totipotency, the remaining cells can develop into a complete embryo, which can be implanted into the mother's uterus, where it develops into a normal fetus and infant.

Until recently, it was not possible to induce a cell from a fully developed animal to dedifferentiate and then redifferentiate into another cell type. However, nuclear transfer experiments have shown that the genetic information from an animal cell can be used to create cloned animals. Robert Briggs and Thomas King performed the first such experiments in the 1950s using frog embryos. First they removed the nucleus from an unfertilized egg, forming an *enucleated* egg. Then, with a very fine glass needle, they punctured a cell from an early embryo and drew up part of its contents, including the nucleus, which they injected into the enucleated egg. They stimulated the eggs to divide, and many went on to form embryos, and eventually frogs, that were clones from the original implanted nucleus. These experiments led to two important conclusions:

- No information is lost from the nuclei of cells as they pass through the early stages of embryonic development. This fundamental principle of developmental biology is known as **genomic equivalence**.
- The cytoplasmic environment around a cell nucleus can modify its fate.

INVESTIGATING LIFE

**19.3 Cloning a Plant**

When cells were removed from a plant and put into a medium with nutrients and hormones, they lost many of their specialized features—in other words, they dedifferentiated. Did these cells retain the ability to differentiate again? Frederick Steward found that a cultured carrot cell did indeed retain the ability to develop into an embryo and a new plant.

**HYPOTHESIS** Differentiated plant cells can be totipotent and can be induced to generate all types of the plant's cells.

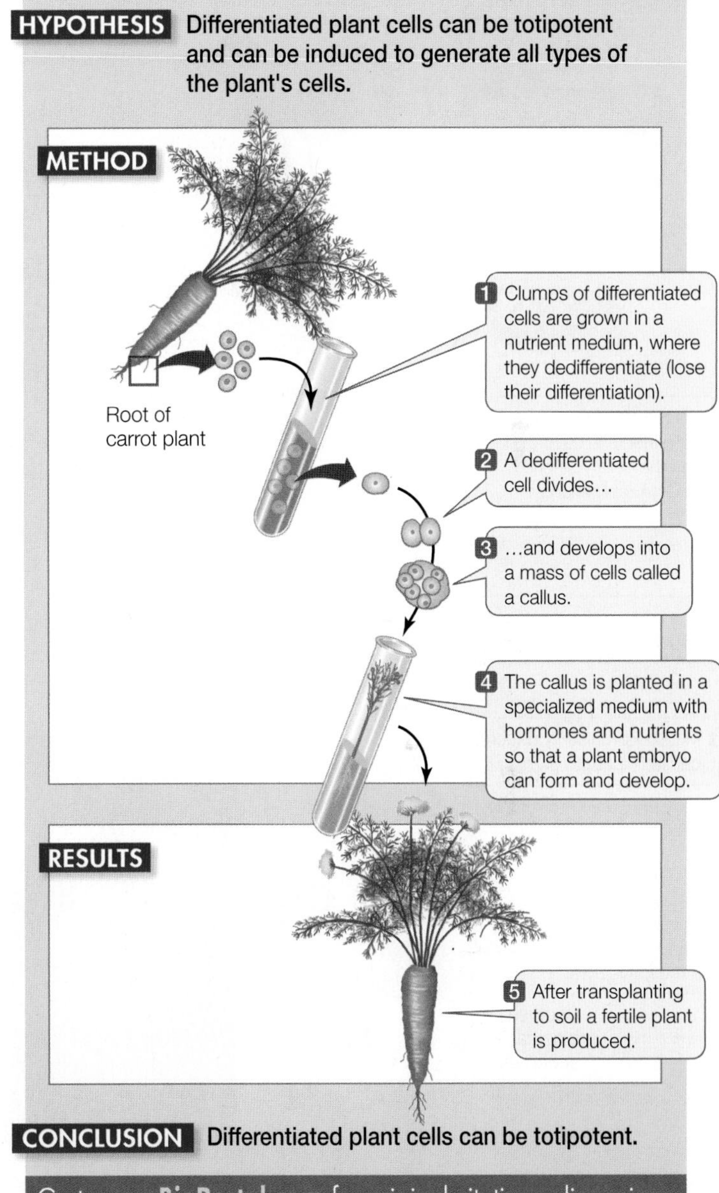

METHOD
Root of carrot plant
1 Clumps of differentiated cells are grown in a nutrient medium, where they dedifferentiate (lose their differentiation).
2 A dedifferentiated cell divides...
3 ...and develops into a mass of cells called a callus.
4 The callus is planted in a specialized medium with hormones and nutrients so that a plant embryo can form and develop.
RESULTS
5 After transplanting to soil a fertile plant is produced.


**CONCLUSION** Differentiated plant cells can be totipotent.

Go to **yourBioPortal.com** for original citations, discussions, and relevant links for all INVESTIGATING LIFE figures.

In 1996, Ian Wilmut and his colleagues in Scotland cloned the first mammal by the cell fusion method. To produce donor cells suitable for nuclear transfer, they took differentiated cells from a ewe's udder and starved them of nutrients for a week, halting the cells in the G1 phase of the cell cycle. One of these cells was fused with an enucleated egg from a different breed

TOOLS FOR INVESTIGATING LIFE

**19.4 Cloning a Mammal**

The experimental procedure described here produced the first cloned mammal, a Dorset sheep named Dolly (shown on the left in the photo). As an adult, Dolly mated and subsequently gave birth to a normal offspring (the lamb on the right), thus proving the genetic viability of cloned mammals.

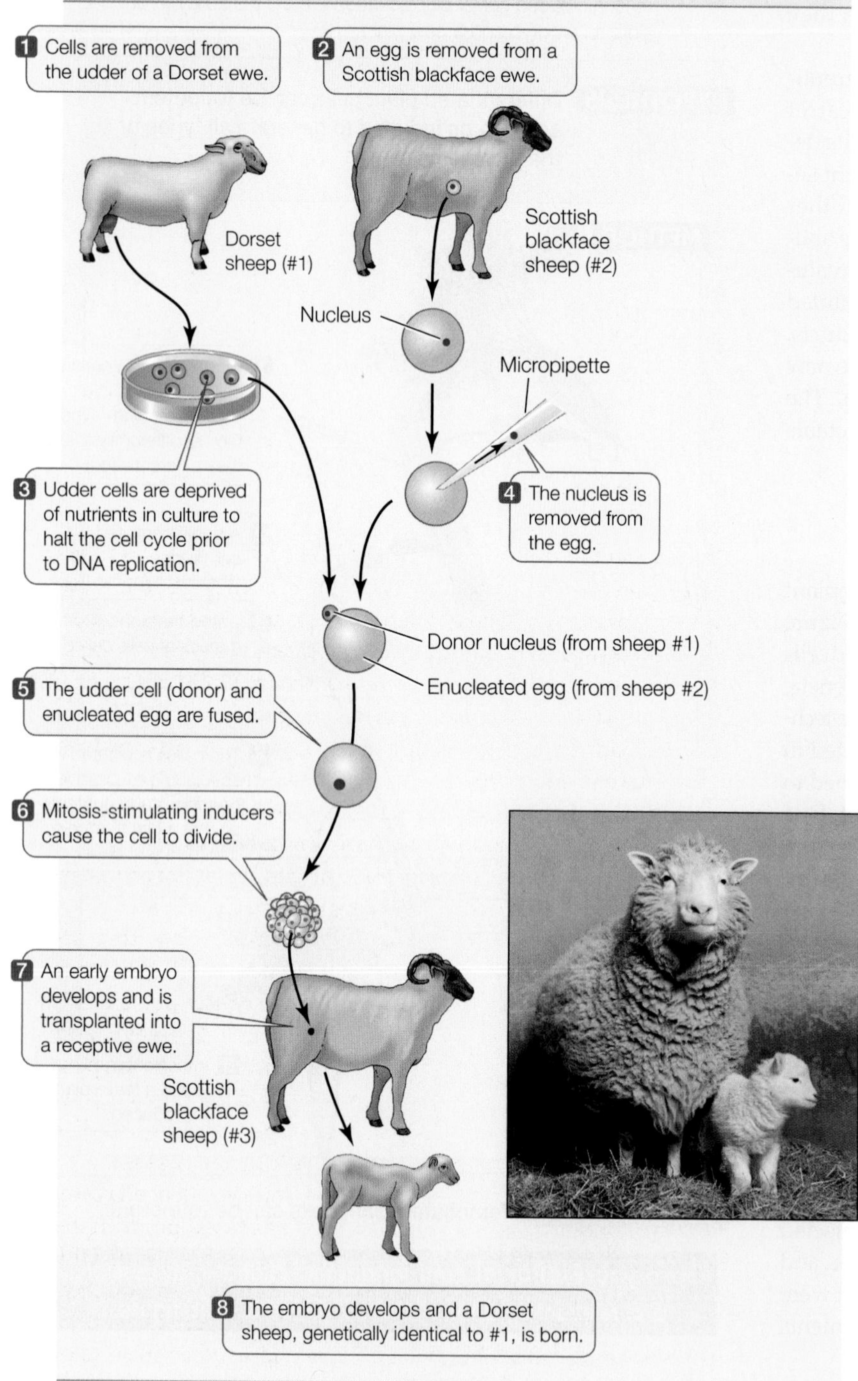

1 Cells are removed from the udder of a Dorset ewe.
2 An egg is removed from a Scottish blackface ewe.
Dorset sheep (#1)
Scottish blackface sheep (#2)
Nucleus
Micropipette
3 Udder cells are deprived of nutrients in culture to halt the cell cycle prior to DNA replication.
4 The nucleus is removed from the egg.
Donor nucleus (from sheep #1)
Enucleated egg (from sheep #2)
5 The udder cell (donor) and enucleated egg are fused.
6 Mitosis-stimulating inducers cause the cell to divide.
7 An early embryo develops and is transplanted into a receptive ewe.
Scottish blackface sheep (#3)
8 The embryo develops and a Dorset sheep, genetically identical to #1, is born.

of ewe. Signals from the egg's cytoplasm stimulated the donor nucleus to enter S phase, and the rest of the cell cycle proceeded normally. After several cell divisions, the resulting early embryo was transplanted into the womb of a surrogate mother (**Figure 19.4**).

Out of 277 successful attempts to fuse adult cells with enucleated eggs, one lamb survived to be born; she was named Dolly, and she became world-famous overnight. DNA analyses confirmed that Dolly's nuclear genes were identical to those of the ewe from whose udder the donor nucleus had been obtained. Dolly grew to adulthood, mated, and produced offspring in the normal manner, thus proving her status as a fully functioning adult animal.

Many other animal species, including cats, dogs, horses, pigs, rabbits, and mice have since been cloned by nuclear transfer. The cloning of animals has practical uses and has given us important information about developmental biology. There are several reasons to clone animals:

- *Expansion of the numbers of valuable animals*: One goal of Wilmut's experiments was to develop a method of cloning transgenic animals carrying genes with therapeutic properties. For example, a cow that was genetically engineered to make human growth hormone in milk has been cloned to produce two more cows that do the same thing. Only 15 such cows would supply the world's need for this medication, which is used to treat short stature due to growth hormone deficiency.
- *Preservation of endangered species*: The banteng, a relative of the cow, was the first endangered animal to be cloned, using a cow enucleated egg and a cow surrogate mother. Cloning may be the only way to save endangered species with low rates of natural reproduction, such as the giant panda.
- *Preservation of pets*: Many people get great personal benefit from pets, and the death of a pet can be devastating. Companies have been set up to clone cats and dogs from cells provided by their owners. Of course, the behavioral characteristics of the beloved pet, which are certainly derived in part from the environment, may not be the same in the cloned pet as in its genetic parent.

### Multipotent stem cells differentiate in response to environmental signals

In plants, the growing regions at the tips of the roots and stems contain *meristems*, which are clusters of undifferentiated, rapidly dividing **stem cells**. These cells can give rise to the specialized cell types that make up the various parts of roots and stems. In general, plants have far fewer (15–20) broad cell types than animals (as many as 200).

In mammals, stem cells are found in adult tissues that need frequent cell replacement, such as the skin, the inner lining of the intestine, and the bone marrow, where blood and other types of cells are formed. Canadian cell biologists Ernest McCulloch and James Till discovered mammalian stem cells in the early 1960s when they injected bone marrow cells into adult mice. They noticed that the recipient mice developed small clumps of tissue in the spleen. When they looked more carefully at the clumps, they found that each was composed of undifferentiated stem cells. Before this, stem cells were believed to be present only in animal embryos.

As they divide, stem cells produce daughter cells that differentiate to replace dead cells and maintain the tissues. These adult stem cells in animals are not totipotent, because their ability to differentiate is limited to a relatively few cell types. In other words, they are **multipotent**. For example, there are two types of multipotent stem cells in bone marrow. One type (called hematopoietic stem cells) produces the various kinds of red and white blood cells, while the other type (mesenchymal stem cells) produces the cells that make bone and surrounding *connective* tissues, such as muscle.

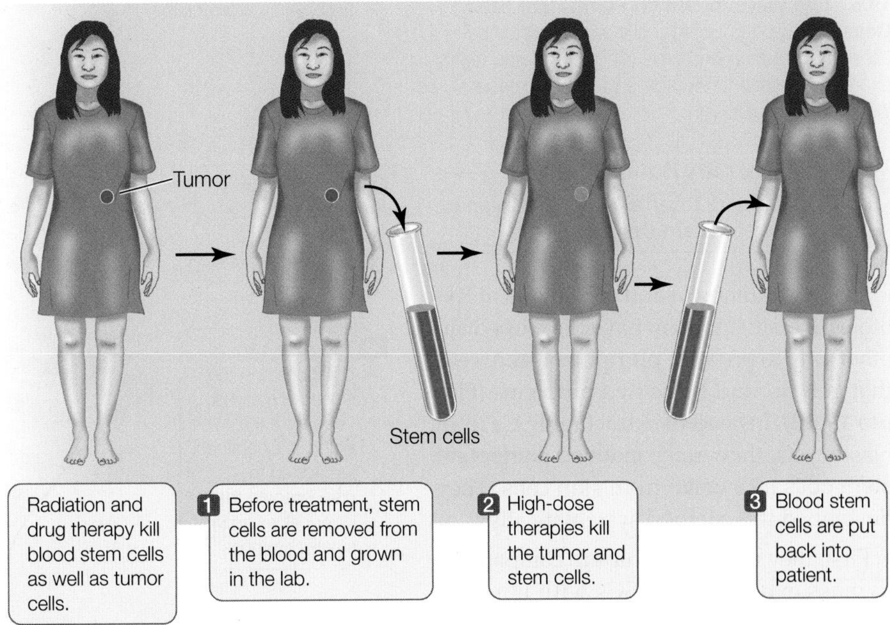

Tumor
Stem cells
Radiation and drug therapy kill blood stem cells as well as tumor cells.
1 Before treatment, stem cells are removed from the blood and grown in the lab.
2 High-dose therapies kill the tumor and stem cells.
3 Blood stem cells are put back into patient.

**19.5 Stem Cell Transplantation** Multipotent blood stem cells can be used in hematopoietic stem cell transplantation, to replace stem cells destroyed by cancer therapy.

The differentiation of multipotent stem cells is "on demand." The blood cells that differentiate in the bone marrow do so in response to specific signals known as growth factors. This is the basis of an important cancer therapy called *hematopoietic stem cell transplantation* (HSCP) (**Figure 19.5**). Because some treatments that kill cancer cells also kill other dividing cells, bone marrow stem cells in patients will die if exposed to these treatments. To circumvent this problem, stem cells are removed from the patient's blood and given growth factors to increase their numbers in the laboratory. The cells are stored during treatment, and then added back to populate the depleted bone marrow when treatment is over. The stored stem cells retain their ability to differentiate in the bone marrow environment. By allowing the use of high doses of treatment to kill tumors, bone marrow transplantation saves thousands of lives each year.

Adjacent cells can also influence stem cell differentiation. We saw this in the opening story of this chapter, in which stem cells from fat differentiated to form cells of the tendon. Bone marrow stem cells that can form muscle will do so if implanted into the heart. Such stem cell transplantation for heart repair has been demonstrated in animals and even in people who had heart attacks, in experiments that used the stem cells to repair a damaged heart. Multipotent stem cells have been found in many organs and tissues, and their use in treating diseases is under intensive investigation.

### Pluripotent stem cells can be obtained in two ways

As stated earlier, totipotent stem cells that can form an entire new animal are found only in very early embryos. In both mice and humans, the earliest embryonic stage before differentiation occurs is called a *blastocyst* (see Figure 44.4). Although they cannot form an entire embryo, a group of cells in the blastocyst still retains the ability to form all of the cells in the body: these cells are **pluripotent** ("pluri," many; "potent," capable). In mice, these **embryonic stem cells** (**ESCs**) can be removed from the blastocyst and grown in laboratory culture almost indefinitely if provided with the right conditions. When cultured mouse ESCs are injected back into a mouse blastocyst, the stem cells mix with the resident cells and differentiate to form all the cell types in the mouse. This indicates that the ESCs do not lose any of their developmental potential while growing in the laboratory.

ESCs growing in the laboratory can also be induced to differentiate in a particular way if the right signal is provided (**Figure 19.6A**). For example, treatment of mouse ESCs with a derivative of vitamin A causes them to form neurons, while other growth factors induce them to form blood cells. Such experiments demonstrate both the cells' developmental potential and the roles of environmental signals. This finding raises the possibility of using ESC cultures as sources of differentiated cells to repair specific tissues, such as a damaged pancreas in diabetes, or a brain that malfunctions in Parkinson's disease.

ESCs can be harvested from human embryos conceived by *in vitro* ("under glass"—in the laboratory) fertilization, with the consent of the donors. Since more than one embryo is usually conceived in this procedure, embryos not used for reproduction might be available for embryonic stem cell isolation. These cells could then be grown in the laboratory and used as sources of tissues for transplantation into patients with tissue damage. There are two problems with this approach:

- Some people object to the destruction of human embryos for this purpose.
- The stem cells, and tissues derived from them, would provoke an immune response in a recipient (see Chapter 42).

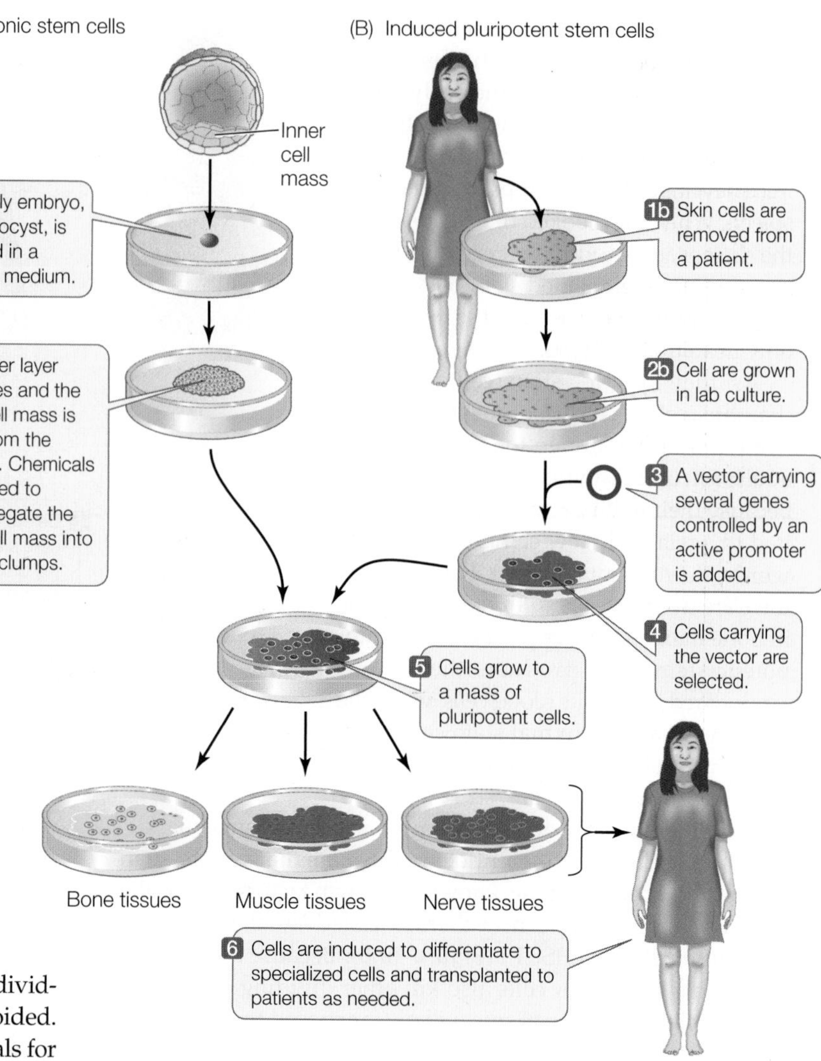

(A) Embryonic stem cells
(B) Induced pluripotent stem cells
Inner cell mass
1a The early embryo, or blastocyst, is cultured in a nutrient medium.
2a The outer layer collapses and the inner cell mass is freed from the embryo. Chemicals are added to disaggregate the inner cell mass into smaller clumps.
1b Skin cells are removed from a patient.
2b Cell are grown in lab culture.
3 A vector carrying several genes controlled by an active promoter is added.
4 Cells carrying the vector are selected.
5 Cells grow to a mass of pluripotent cells.
Bone tissues
Muscle tissues
Nerve tissues
6 Cells are induced to differentiate to specialized cells and transplanted to patients as needed.

**19.6 Two Ways to Obtain Pluripotent Stem Cells** Pluripotent stem cells can be obtained either from human embryos (A) or by adding highly expressed genes to skin cells to transform them into stem cells (B).

**yourBioPortal.com**
GO TO **Animated Tutorial 19.1 • Embryonic Stem Cells**

Shinya Yamanaka and coworkers at Kyoto University in Japan have developed another way to produce pluripotent stem cells that gets around these two problems (**Figure 19.6B**). Instead of extracting ESCs from blastocysts, they make **induced pluripotent stem cells** (**iPS cells**) from skin cells. They developed this method systematically:

1. First, they used gene chips to compare the genes expressed in ESCs with nonstem cells (see Figure 18.9). They found several genes that were uniquely expressed at high levels in ESCs. These genes were believed to be essential to the undifferentiated state and function of stem cells.
2. Next, they isolated the genes and inserted them into a vector for genetic transformation of skin cells (see Section 18.5). They found that the skin cells now expressed the newly added genes at high levels.
3. Finally, they showed that the iPS cells were pluripotent and could be induced to differentiate into many tissues.

Because the iPS cells can be made from skin cells of the individual who is to be treated, an immune response may be avoided. Such cells have already been used for cell therapy in animals for diseases similar to human Parkinson's disease (a brain disorder), diabetes, and sickle cell anemia. Human uses are sure to follow.

## 19.2 RECAP

**Even differentiated cells retain their ability to differentiate into other cell types, given appropriate chemical signals. This has made cloning and stem cell technologies possible.**

- Describe the differences between totipotent, pluripotent, and multipotent cells. **See pp. 408–411**
- How are stem cells found in adult body tissues different from embryonic stem cells? **See pp. 410–411**
- What are the two ways to produce pluripotent stem cells? **See pp. 411–412 and Figure19.6**

Cloning experiments and observations of stem cells have shown that most differentiated cells in an organism share the same genes. But not all genes are expressed in every cell. What turns gene expression on and off as cells differentiate? In the next section we explore several of the mechanisms controlling the changes in gene expression that lead to cell differentiation.

# 19.3 What Is the Role of Gene Expression in Cell Differentiation?

Although every cell contains all the genes needed to produce every protein encoded by its genome, each cell expresses only selected genes. For example, certain cells in hair follicles produce keratin, the protein that makes up hair, while other cell types in the body do not. What determines whether a cell will produce keratin? Chapter 16 describes a number of ways in which cells regulate gene expression and protein production—by controlling transcription, translation, and posttranslational protein modifications. But the mechanisms that control gene expression resulting in cell differentiation generally work at the level of transcription.

### Differential gene transcription is a hallmark of cell differentiation

The gene for β-globin, one of the protein components of hemoglobin, is expressed in red blood cells as they form in the bone marrow of mammals. That this same gene is also present—but unexpressed—in neurons in the brain (which do not make hemoglobin) can be demonstrated by nucleic acid hybridization. Recall that in nucleic acid hybridization, a probe made of single-stranded DNA or RNA of known sequence is added to denatured DNA to reveal complementary coding regions on the DNA template strand (see Figure 15.16). A probe for the β-globin gene can be applied to DNA from brain cells and immature red blood cells (recall that mature mammalian red blood cells lose their nuclei during development). In both cases, the probe finds its complement, showing that the β-globin gene is present in both types of cells. On the other hand, if the β-globin probe is applied to mRNA, rather than DNA, from the two cell types, it finds β-globin mRNA only in the red blood cells, not in the brain cells. This result shows that the gene is expressed in only one of the two cell types.

What leads to this differential gene expression? One well-studied example of cell differentiation is the conversion of undifferentiated muscle precursor cells into cells that are destined to form muscle (**Figure 19.7**). In the vertebrate embryo these cells come from a layer called the *mesoderm* (see Section 44.2). A key event in the commitment of these cells to become muscle is that they stop dividing. Indeed, in many parts of the embryo, *cell division and cell differentiation are mutually exclusive.* Cell signaling activates the gene for a transcription factor called **MyoD** (*myo*blast-determining gene). Recall that transcription factors are DNA binding proteins that regulate the expression of specific genes. In this case, MyoD activates the gene for p21, an inhibitor of cyclin-dependent kinases that normally stimulate the cell cycle at G1 (see Figure 11.6). Expression of the *p21* gene causes the cell cycle to stop, and other transcription factors then enter the picture so that differentiation can begin. Interestingly, *myoD* is also activated in stem cells that are present in adult muscle, indicating a role of this transcription factor in repair of muscle as it gets damaged and worn out.

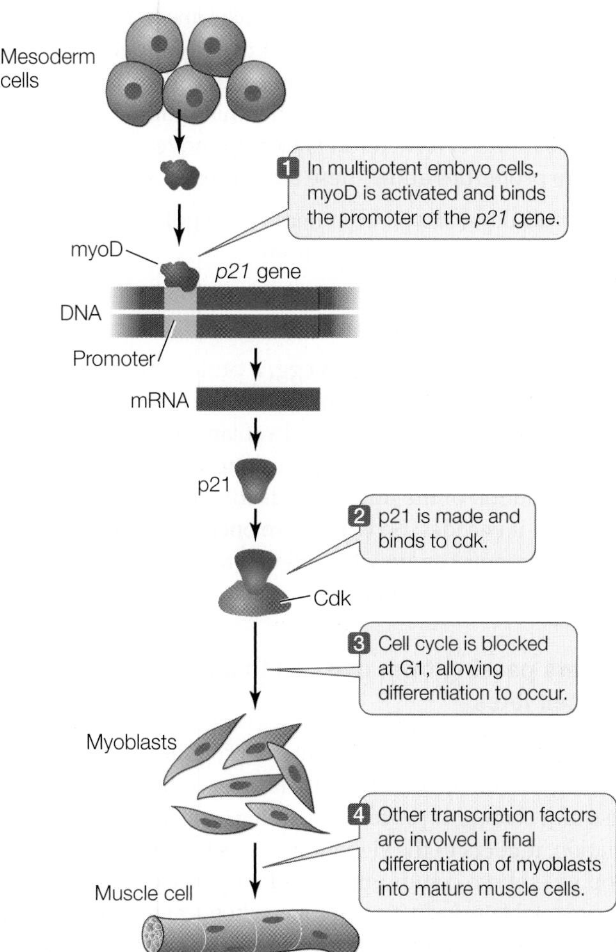

Mesoderm cells
1 In multipotent embryo cells, myoD is activated and binds the promoter of the p21 gene.
myoD
p21 gene
DNA
Promoter
mRNA
p21
2 p21 is made and binds to cdk.
Cdk
3 Cell cycle is blocked at G1, allowing differentiation to occur.
Myoblasts
4 Other transcription factors are involved in final differentiation of myoblasts into mature muscle cells.
Muscle cell

**19.7 Transcription and Differentiation in the Formation of Muscle Cells** Activation of a transcription factor, MyoD, is important in muscle cell differentiation.

Genes such as *myoD* that direct the most fundamental decisions in development (often by regulating other genes on other chromosomes) usually encode transcription factors. In some cases, a single transcription factor can cause a cell to differentiate in a certain way. In others, complex interactions between genes and proteins determine a sequence of transcriptional events that leads to differential gene expression.

## 19.3 RECAP

**Differentiation involves selective gene expression, controlled at the level of transcription by transcription factors.**

- What techniques could you use to identify genes expressed during cell differentiation? See pp. 412–413
- What is the role of transcription factors in controlling differentiation? See p. 413 and Figure 19.7

Cell differentiation involves extensive transcriptional regulation of genes. But what causes a cell to express one set of genes, and not some other set? In other words, how is a cell's fate determined?

# 19.4 How Is Cell Fate Determined?

The fertilized egg undergoes many cell divisions to produce the many differentiated cells in the body (such as liver, muscle, and nerve cells). How can one cell produce so many different cell types? There are two ways that this occurs:

- **Cytoplasmic segregation** (unequal cytokinesis). A factor within an egg, zygote, or precursor cell may be unequally distributed in the cytoplasm. After cell division, the factor ends up in some daughter cells or regions of cells, but not others.
- **Induction** (cell-to-cell communication). A factor is actively produced and secreted by certain cells to induce other cells to become determined.

### Cytoplasmic segregation can determine polarity and cell fate

Some differences in gene expression patterns are the result of *cytoplasmic* differences between cells. One such cytoplasmic difference is the emergence of distinct "top" and "bottom" ends of an organism or structure; such a difference is called **polarity**.

## INVESTIGATING LIFE

### 19.8 Asymmetry in the Early Sea Urchin Embryo

As an embryo develops, cells become determined and their ultimate fate gets more and more narrowly defined. The cells of an eight-celled sea urchin embryo look identical and so might be expected to have the same developmental potential. But do they? Hans Driesch separated different parts of this tiny embryo from one another, to examine their developmental potentials. His experiments showed that even at the eight-cell stage, cell fate determination is underway.

**HYPOTHESIS** **Different regions in the fertilized egg and the embryo have different developmental fates.**

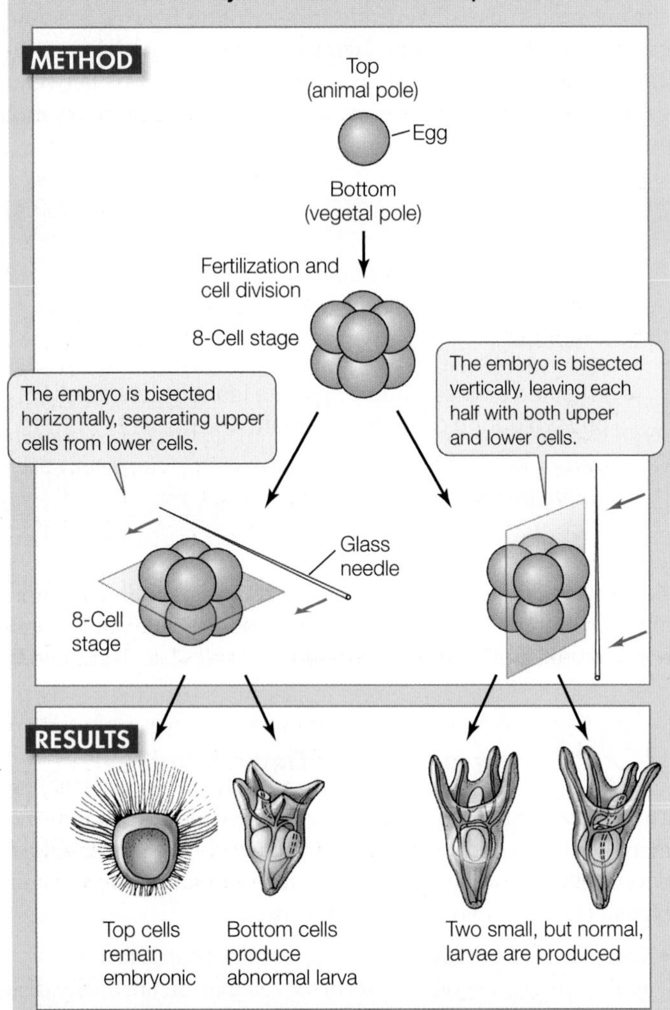
METHOD
Top
(animal pole)
Egg
Bottom
(vegetal pole)
Fertilization and
cell division
8-Cell stage
The embryo is bisected horizontally, separating upper cells from lower cells.
The embryo is bisected vertically, leaving each half with both upper and lower cells.
Glass
needle
8-Cell
stage
RESULTS
Top cells remain embryonic
Bottom cells produce abnormal larva
Two small, but normal, larvae are produced

**CONCLUSION** **The upper and lower halves of a sea urchin embryo differ in their developmental potential.**

Go to **yourBioPortal.com** for original citations, discussions, and relevant links for all INVESTIGATING LIFE figures.

**yourBioPortal.com**

GO TO **Animated Tutorial 19.2 • Early Asymmetry in the Embryo**

Many examples of polarity are observed as development proceeds. Our heads are distinct from our rear ends, and the distal (far) ends of our arms and legs (wrists, ankles, fingers, toes) differ from the proximal (near) ends (shoulders and hips). Polarity may develop early; even within the fertilized egg, the yolk and other factors are often distributed asymmetrically. During early development in animals, polarity is specified by an *animal pole* at the top of the zygote and a *vegetal pole* at the bottom.

In the early twentieth century Hans Driesch at the Marine Biological Station in Naples, Italy, demonstrated the effects of cytoplasmic segregation on development (**Figure 19.8**). Very early development in sea urchins occurs by rapid, equal mitotic divisions of the fertilized egg; there is no increase in size at this stage, just a partitioning of the cells. If an eight-cell embryo is carefully separated vertically into two four-celled halves, both halves develop into normal (albeit small) larvae. But if an eight-cell embryo is cut horizontally, the top half does not develop at all, while the bottom half develops into a small, abnormal larva.

Clearly, then, there must be at least one factor essential for development that is segregated in the vegetal half of the sea urchin egg, such that the bottom cells of the 8-cell embryo get this essential factor and the top cells do not. Experiments have established that certain materials, called **cytoplasmic determinants**, are distributed unequally in the egg cytoplasm. Cytoplasmic determinants play roles in directing the embryonic development of many organisms (**Figure 19.9**). What are these determinants and what accounts for their unequal distribution?

The cytoskeleton contributes to the asymmetric distribution of cytoplasmic determinants in the egg. Recall from Section 5.3 that an important function of the microtubules and microfilaments in the cytoskeleton is to help move materials in the cell. Two properties allow these structures to accomplish this:

- Microtubules and microfilaments have polarity—they grow by adding subunits to the plus end.
- Cytoskeletal elements can bind specific proteins, which can be used in the transport of mRNA.

For example, in the sea urchin egg, a protein binds to both the growing (+) end of a microfilament and to an mRNA encoding a cytoplasmic determinant. As the microfilament grows toward one end of the cell, it carries the mRNA along with it. The asymmetrical distribution of the mRNA leads to a similar distribution of the protein it encodes. So what were once unspecified cytoplasmic determinants can now be defined in terms of cellular structures, mRNAs, and proteins.

### Inducers passing from one cell to another can determine cell fates

The term "induction" has different meanings in different contexts. In biology it can be used broadly to refer to the initiation of, or cause of, a change or process. But in the context of cellular differentiation, it refers to the signaling events by which cells in a developing embryo communicate and influence one another's developmental fate. Induction involves chemical signals and signal transduction mechanisms. We will describe two examples of this form of induction: one in the developing vertebrate eye, and the other in a developing reproductive structure of the nematode *Caenorhabditis elegans*.

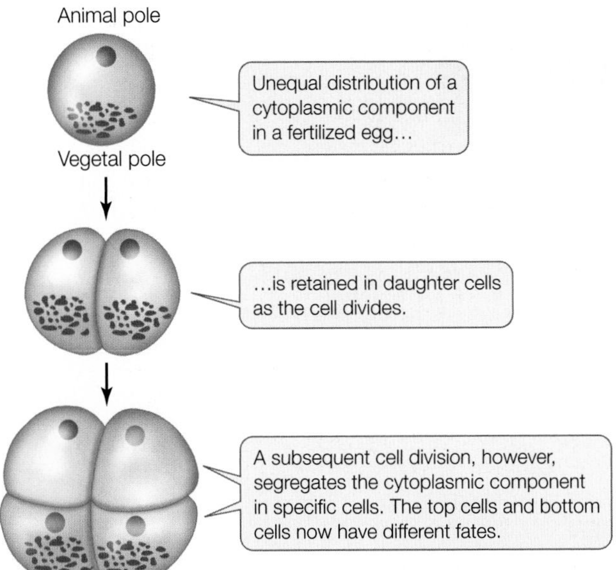

Animal pole
Vegetal pole
Unequal distribution of a cytoplasmic component in a fertilized egg...
...is retained in daughter cells as the cell divides.
A subsequent cell division, however, segregates the cytoplasmic component in specific cells. The top cells and bottom cells now have different fates.

**19.9 The Principle of Cytoplasmic Segregation** The unequal distribution of some component in the cytoplasm of a cell may determine the fates of its descendants.

**LENS DIFFERENTIATION IN THE VERTEBRATE EYE** The development of the lens in the vertebrate eye is a classic example of induction. In a frog embryo, the developing forebrain bulges out at both sides to form the *optic vesicles*, which expand until they come into contact with the cells at the surface of the head (**Figure 19.10**). The surface tissue in the region of contact thickens, forming a *lens placode*—tissue that will ultimately form the lens. The lens placode bends inward, folds over on itself, and ultimately detaches from the surface tissue to produce a structure that will develop into the lens. If the growing optic vesicle is cut away before it contacts the surface cells, no lens forms. Placing an impermeable barrier between the optic vesicle and the surface cells also prevents the lens from forming. These observations suggest that the surface tissue begins to develop into a lens when it receives a signal from the optic vesicle. Such a signal is termed an **inducer**.

Inducers trigger sequences of gene expression in the responding cells. How cells switch on different sets of genes that govern development and direct the formation of body plans is of great interest to developmental and evolutionary biologists. They use model organisms to investigate the major principles governing these processes.

**VULVAL DIFFERENTIATION IN THE NEMATODE** The tiny nematode *Caenorhabditis elegans* is a favorite model organism for studying development. Its genome was one of the first eukaryotic genomes to be sequenced (see Section 17.3). It develops from fertilized egg to larva in only about 8 hours, and the worm reaches the adult stage in just 3.5 days. The process is easily observed using a low-magnification dissecting microscope because the body covering is transparent (**Figure 19.11A**).

The adult nematode is *hermaphroditic*, containing both male and female reproductive organs. It lays eggs through a pore called the *vulva* on the ventral (lower) surface. During development, a single cell, called the *anchor cell*, induces the vulva to form from six cells on the worm's ventral surface. In this case, there are two molecular signals, the *primary inducer* and the *secondary inducer* (or *lateral signal*). Each of the six ventral cells has three possible fates: it may become a primary vulval precursor cell, a secondary vulval precursor cell, or simply become part of the worm's skin—an epidermal cell. You can follow the sequence of events in **Figure 19.11B**. The concentration gradient of the primary inducer, LIN-3, is key: the anchor cell produces LIN-3, which diffuses out of the cell and forms a concentration gradient with respect to adjacent cells. Cells that receive the most LIN-3 become vulval precursor cells; cells slightly farther from the anchor cell receive less LIN-3 and become epidermal cells. Induction involves the activation or inactivation of specific genes through signal transduction cascades in the responding cells (**Figure 19.12**).

Nematode development illustrates the important observation that *much of development is controlled by molecular switches that allow a cell to proceed down one of two alternative tracks*. One challenge for developmental biologists is to find these switches and determine how they work. The primary inducer, LIN-3, released by the *C. elegans* anchor cell is a growth factor homologous to a ver-

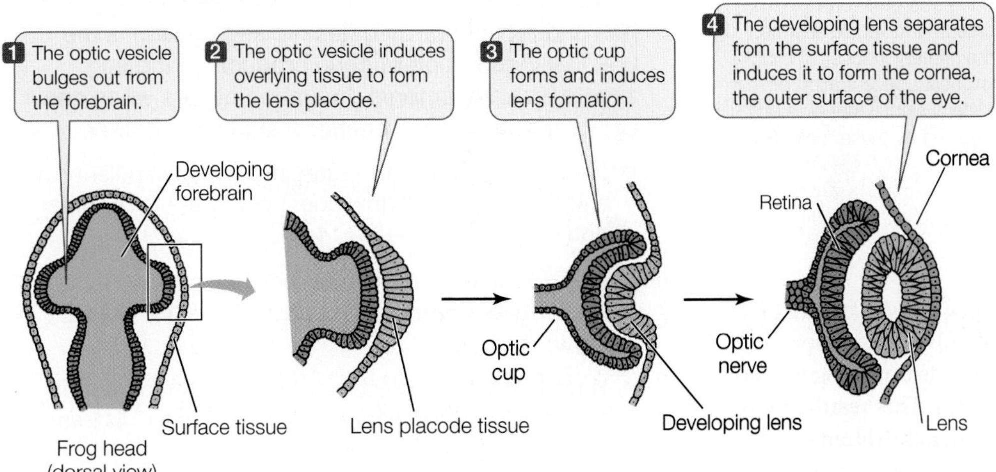

1 The optic vesicle bulges out from the forebrain.
2 The optic vesicle induces overlying tissue to form the lens placode.
3 The optic cup forms and induces lens formation.
4 The developing lens separates from the surface tissue and induces it to form the cornea, the outer surface of the eye.
Developing forebrain
Frog head (dorsal view)
Surface tissue
Lens placode tissue
Optic cup
Developing lens
Retina
Cornea
Optic nerve
Lens

**19.10 Embryonic Inducers in Vertebrate Eye Development** The eye of a frog develops as different cells induce changes in neighboring cells.

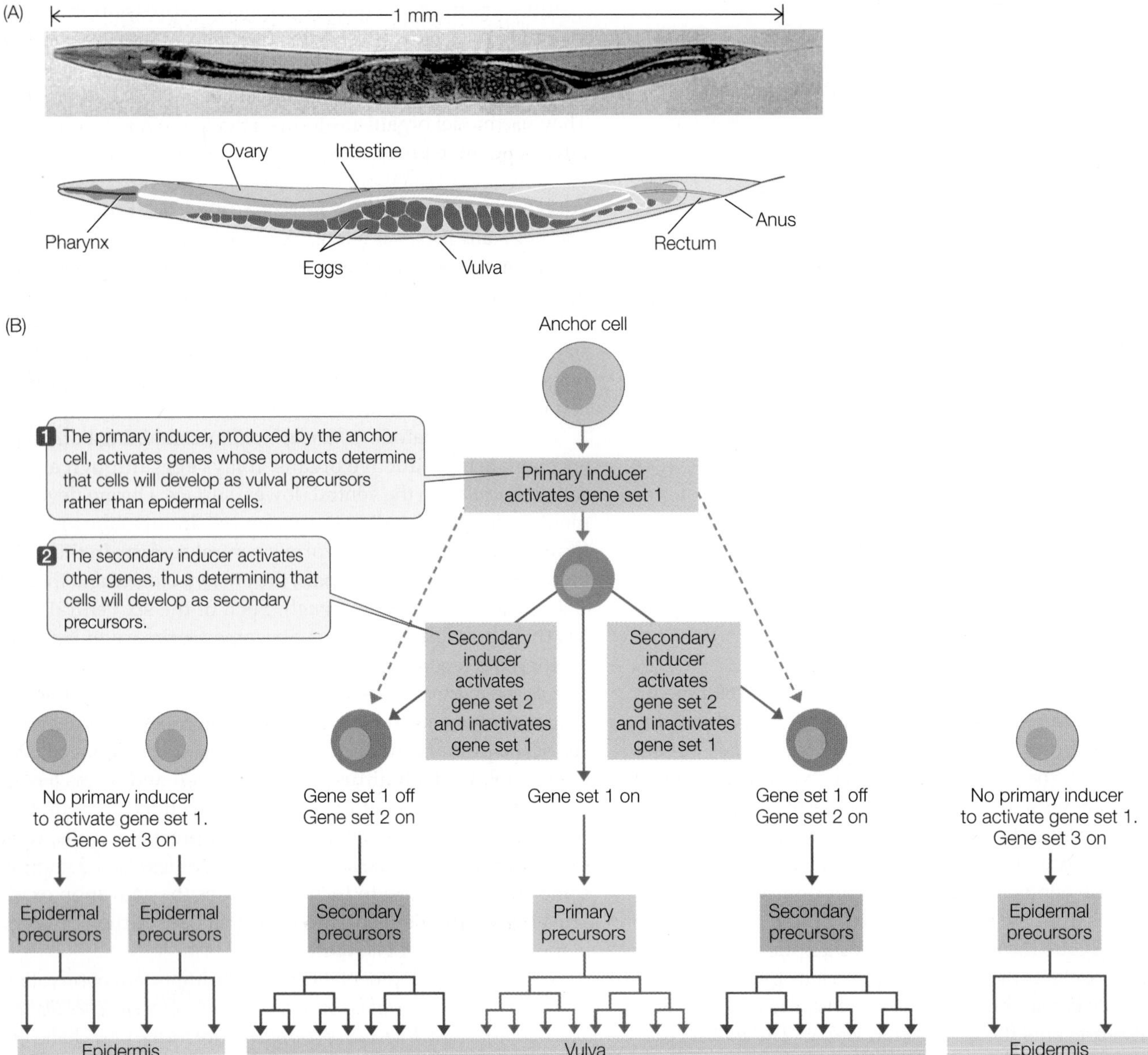


**19.11 Induction during Vulval Development in *Caenorhabditis elegans*** (A) In the nematode *C. elegans*, it has been possible to follow all of the cell divisions from the fertilized egg to the 959 cells found in the fully developed adult. (B) During vulval development, a molecule secreted by the anchor cell (the LIN-3 protein) acts as the primary inducer. The primary precursor cell (the one that received the highest concentration of LIN-3) then secretes a secondary inducer (the lateral signal) that acts on its neighbors. The gene expression patterns triggered by these molecular switches determine cell fates.

tebrate growth factor called EGF (*e*pidermal *g*rowth *f*actor). LIN-3 binds to a receptor on the surfaces of vulval precursor cells, setting in motion a signal transduction cascade involving the Ras protein and MAP kinases (see Figure 7.12). This results in increased transcription of the genes involved in the differentiation of vulval cells.

## 19.4 RECAP

**Cellular differentiation involves cytoplasmic segregation and induction. Cytoplasmic segregation is the unequal distribution of gene products in the egg, zygote, or early embryo. Induction occurs when one cell or tissue sends a chemical signal to another.**

- How does cytoplasmic segregation result in polarity in a fertilized egg, and how does polarity affect cell differentiation? **See pp. 413–414 and Figure 19.9**
- Describe an example of how induction influences tissue formation in the vertebrate eye. **See p. 416 and Figure 19.10**
- How do inducer molecules interact with transcription factors to produce differentiated cells? **See p. 416 and Figure 19.12**

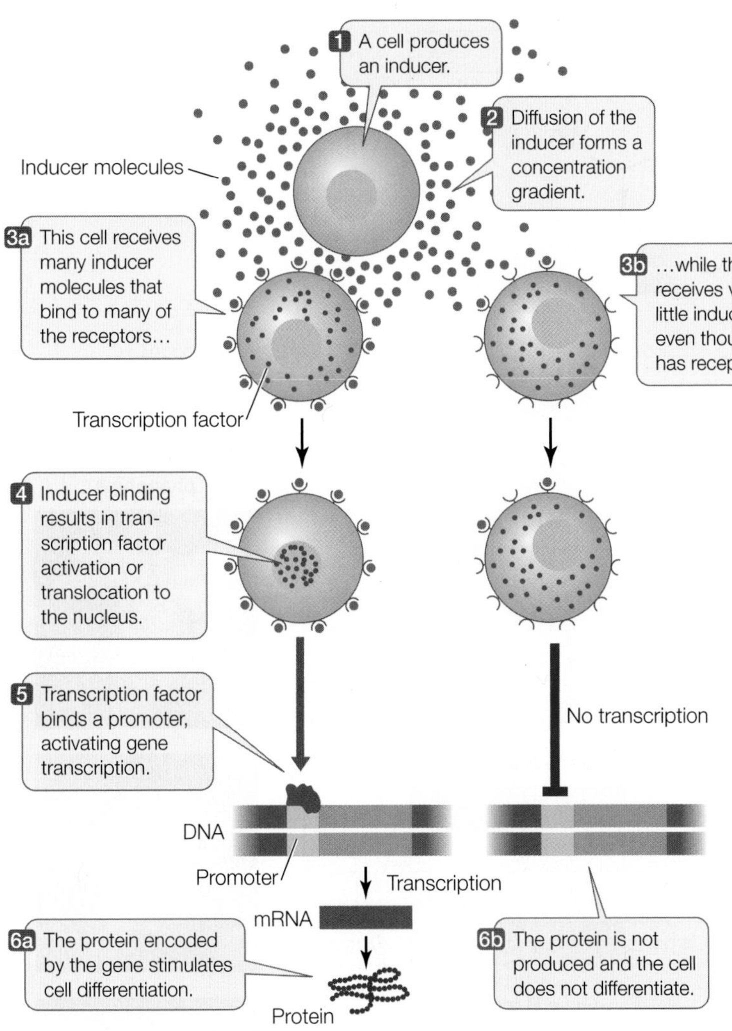


**19.12 Induction** The concentration of an inducer directly affects the degree to which a transcription factor is activated. The inducer acts by binding to a receptor on the target cell. This binding is followed by signal transduction involving transcription factor activation or translocation from the cytoplasm to the nucleus. In the nucleus it acts to stimulate the expression of genes involved in cell differentiation.

We have seen that cytoplasmic segregation and induction lead to cell differentiation, and have seen two examples of how these processes lead to organ formation in developing multicellular organisms. We now take a closer look at how gene expression affects differentiation and development.

## 19.5 How Does Gene Expression Determine Pattern Formation?

**Pattern formation** is the process that results in the spatial organization of a tissue or organism. It is inextricably linked to morphogenesis, the creation of body form. You might expect morphogenesis to involve a lot of cell division, followed by differentiation—and it does. But what you might not expect is the amount of programmed cell death—apoptosis—that occurs during morphogenesis.

### Multiple genes interact to determine developmental programmed cell death

We noted in Section 11.6 that apoptosis is a programmed series of events that leads to cell death. Apoptosis is an integral part of the normal development and life of an organism. For example, in an early human embryo, the hands and feet look like tiny paddles: the tissues that will become fingers and toes are linked by connective tissue. Between days 41 and 56 of development, the cells between the digits die, freeing the individual fingers and toes:

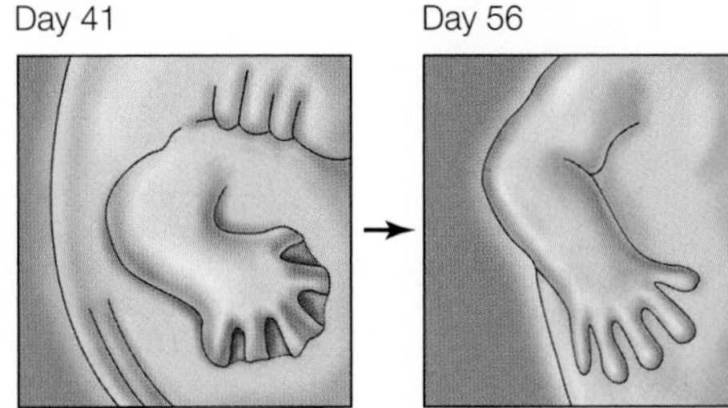


Many cells and structures form and then disappear during development, in processes involving apoptosis.

Model organisms have been very useful in studying the genes involved in apoptosis. For example, the nematode worm *C. elegans* produces precisely 1,090 somatic cells as it develops from a fertilized egg into an adult, but 131 of those cells die (leaving 959 cells in the adult worm). The sequential expression of two genes called *ced-4* and *ced-3* (for *ce*ll *d*eath) appears to control this programmed cell death (**Figure 19.13**).

In the nematode nervous system, 302 neurons come from 405 precursors; thus 103 neural precursor cells undergo apoptosis. If the protein encoded by either *ced-3* or *ced-4* is nonfunctional, *all* 405 cells form neurons, resulting in abnormal brain develop-

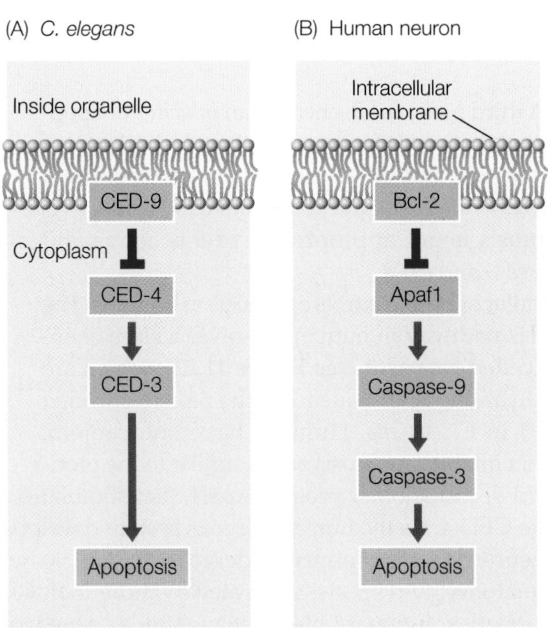


**19.13 Pathways for Apoptosis** In the worm *C. elegans* (A) and humans (B) similar pathways for apoptosis are controlled by genes with similar sequences and functions.

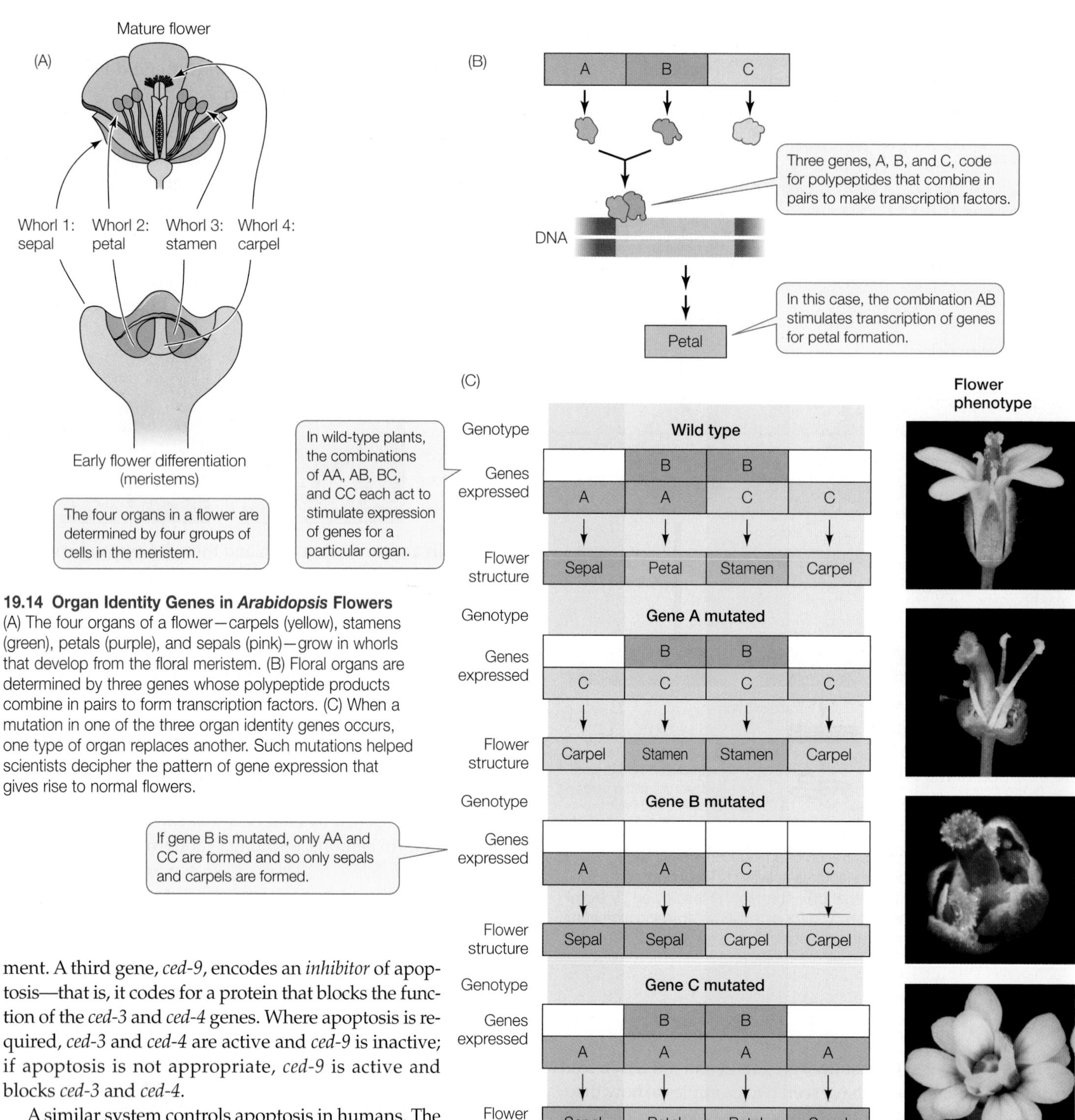


**19.14 Organ Identity Genes in *Arabidopsis* Flowers** (A) The four organs of a flower—carpels (yellow), stamens (green), petals (purple), and sepals (pink)—grow in whorls that develop from the floral meristem. (B) Floral organs are determined by three genes whose polypeptide products combine in pairs to form transcription factors. (C) When a mutation in one of the three organ identity genes occurs, one type of organ replaces another. Such mutations helped scientists decipher the pattern of gene expression that gives rise to normal flowers.

ment. A third gene, *ced-9*, encodes an *inhibitor* of apoptosis—that is, it codes for a protein that blocks the function of the *ced-3* and *ced-4* genes. Where apoptosis is required, *ced-3* and *ced-4* are active and *ced-9* is inactive; if apoptosis is not appropriate, *ced-9* is active and blocks *ced-3* and *ced-4*.

A similar system controls apoptosis in humans. The apoptosis pathway in humans involves a class of enzymes called caspases (see Figure 11.22), which are similar in amino acid sequence to the protein encoded by *ced-3* in *C. elegans*. Humans have one protein, Bcl-2, that inhibits apoptosis and is similar to the product of *ced-9*, and another protein, Apaf1, that stimulates apoptosis like CED-4. As the human nervous system develops, half of the neurons that are formed undergo apoptosis. So humans and nematodes, two species separated by more than 600 million years of evolutionary history, have similar genes controlling programmed cell death (see Figure 19.13). The commonality of this pathway indicates its importance: mutations are harmful and evolution selects against them.

## Plants have organ identity genes

Like animals, plants have organs—for example, leaves and roots. Many plants form flowers, and many flowers are composed of four types of organs: sepals, petals, stamens (male reproductive organs), and carpels (female reproductive organs). These floral organs occur in concentric *whorls*, with groups of each organ type encircling a central axis. The sepals are on the outside and the carpels are on the inside (**Figure 19.14A**).

In the model plant *Arabidopsis thaliana* (thale cress), flowers develop in a radial pattern around the shoot apex as it develops and elongates. The whorls develop from a *meristem* of about 700 undifferentiated cells arranged in a dome, which is at the growing point on the stem. How is the identity of a particular whorl determined? A group of genes called **organ identity genes** encode proteins that act in combination to produce specific whorl features (**Figure 19.14B and C**):

- Genes in class A are expressed in whorls 1 and 2 (which form sepals and petals, respectively).
- Genes in class B are expressed in whorls 2 and 3 (which form petals and stamens).
- Genes in class C are expressed in whorls 3 and 4 (which form stamens and carpels).

Two lines of experimental evidence support this model of organ identity gene function:

- *Loss-of-function mutations*: for example, a mutation in a class A gene results in no sepals or petals. In any organism, the replacement of one organ for another is called *homeosis*, and this type of mutation is a **homeotic mutation** (see Figure 19.14C).
- *Gain-of-function mutations*: for example, a promoter for a class C gene can be artificially coupled to a class A gene. In this case, the class A gene is expressed in all four whorls, resulting in only sepals and petals.

Genes in classes A, B, and C code for transcription factors that are active as dimers, that is, proteins with two polypeptide subunits. Gene regulation in these cases is *combinatorial*—that is, the composition of the dimer determines which genes will be activated. For example, a dimer made up of two class A monomers activates transcription of the genes that make sepals; a dimer made up of A and B monomers results in petals, and so forth. A common feature of the A, B, and C proteins, as well as many other plant transcription factors, is a DNA-binding domain called the **MADS box**. These proteins also have domains that can bind to other proteins in a *transcription initiation complex*. As we discuss in Chapter 16, transcription initiation in eukaryotes is controlled by a complex of proteins that interact with DNA and other proteins at the promoter. The MADS box proteins participate in this complex to control the expression of specific genes.

Some familiar ornamental plants have mutations in floral organ identity genes. For example, many rose varieties have mutations in a C gene, resulting in multiple rows of petals instead of the single set of five petals found in wild roses. An understanding of the molecular basis of floral organ identity may have practical uses. Many of the foods that make up the human diet come from fruits and seeds, which form from parts of the carpel—the female reproductive organ of the flower. Genetically modifying plants to produce more carpels could increase the amount of fruit or grain a crop produces. A genetic system similar to the one described here for *Arabidopsis* controls floral organ formation in rice, humanity's most widely consumed plant. Appropriate mutations in these genes might lead to more grain produced per plant.

Transcription of the floral organ identity genes is controlled by other gene products, including the LEAFY protein. Plants with loss-of-function mutations in the *LEAFY* gene make flowering stems instead of flowers, with increased numbers of modified leaves called bracts. The wild-type LEAFY protein acts as a transcription factor, stimulating expression of the class A, B, and C genes so that they produce flowers. This finding, too, has practical applications. It usually takes 6–20 years for a citrus tree to produce flowers and fruits. Scientists have made transgenic orange trees expressing the *LEAFY* gene coupled to a strongly expressed promoter. These trees flower and fruit years earlier than normal trees.

### Morphogen gradients provide positional information

During development, the key cellular question, "What am I (or what will I be)?" is often answered in part by "Where am I?" Think of the cells in the developing nematode, which develop into different parts of the vulva depending on their positions relative to the anchor cell. This spatial "sense" is called **positional information**. Positional information often comes in the form of an inducer called a **morphogen**, which diffuses from one group of cells to surrounding cells, setting up a concentration gradient. There are two requirements for a signal to be considered a morphogen:

- *It must directly affect target cells, rather than triggering a secondary signal that affects target cells.*
- *Different concentrations of the signal must cause different effects.*

Developmental biologist Lewis Wolpert uses the "French flag model" to explain morphogens (**Figure 19.15**). This model can be applied to the differentiation of the vulva in *C. elegans* (see Figure 19.11) and to the development of vertebrate limbs.

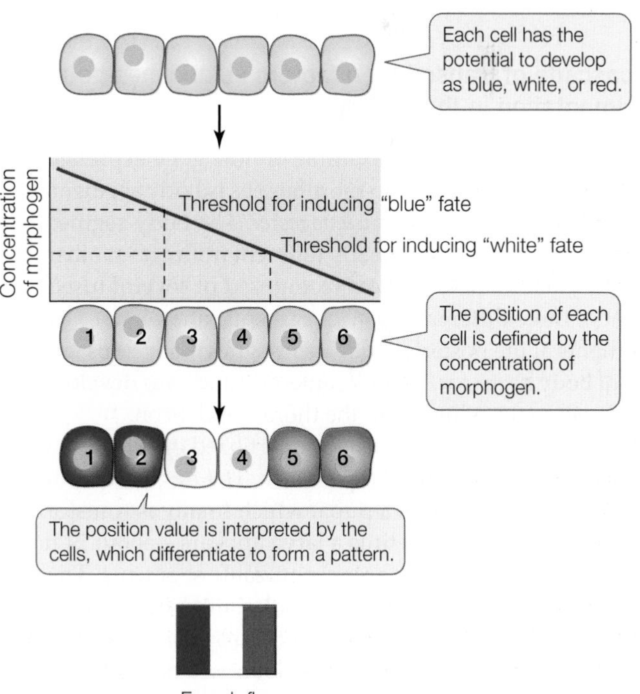


**19.15 The French Flag Model** In the "French flag" model, a concentration gradient of a diffusible morphogen signals each cell to specify its position.

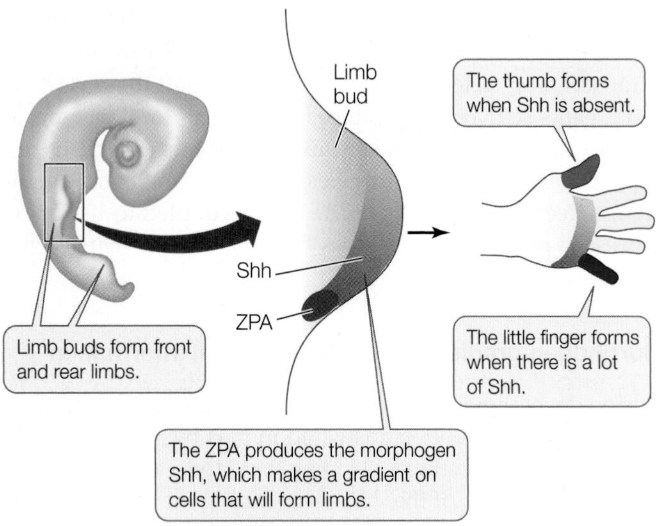
Limb bud
The thumb forms when Shh is absent.
Shh
ZPA
Limb buds form front and rear limbs.
The little finger forms when there is a lot of Shh.
The ZPA produces the morphogen Shh, which makes a gradient on cells that will form limbs.

**19.16 Specification of the Vertebrate Limb and the French Flag Model** The zone of polarizing activity (ZPA) in the limb bud of the embryo secretes the morphogen Sonic *hedge*hog (Shh). Cells in the bud form different digits depending on the concentration of Shh.

The vertebrate limb develops from a paddle-shaped *limb bud* (**Figure 19.16**). The cells that develop into different digits must receive positional information; if they do not, the limb will be totally disorganized (imagine a hand with only thumbs or only little fingers). A group of cells at the posterior base of the limb bud, just where it joins the body wall, is called the *zone of polarizing activity* (ZPA). The cells of the ZPA secrete a morphogen called Sonic *hedge*hog (Shh), which forms a gradient that determines the posterior–anterior (little finger to thumb) axis of the developing limb. The cells getting the highest dose of Shh form the little finger; those getting the lowest dose develop into the thumb. Recall the French flag model when considering the gradient of Shh.

### A cascade of transcription factors establishes body segmentation in the fruit fly

Perhaps the best-studied example of how gene expression affects cell fate in response to morphogens is body segmentation in the fruit fly *Drosophila melanogaster*. The body segments of this model organism are clearly different from one another. The adult fly has an anterior *head* (composed of several fused segments), three different *thoracic* segments, and eight *abdominal* segments at the posterior end. Each segment develops into different body parts: for example, antennae and eyes develop from head segments, wings from the thorax, and so on.

The life cycle of *Drosophila* from fertilized egg to adult takes about 2 weeks at room temperature. The egg hatches into a larva, which then forms a pupa, which finally is transformed into the adult fly. By the time a larva appears—about 24 hours after fertilization—there are recognizable segments. The thoracic and abdominal segments all look similar, but *the fates of the cells to become different adult segments is already determined*. The determination events in the first 24 hours will be our focus here.

Several types of genes are expressed sequentially in the embryo to define these segments:

- First, cells in the mother that are adjacent to the maturing egg release products that set up anterior–posterior and dorsal–ventral axes in the egg.
- Next, a series of gene products in the embryo successively define the position of each cell in a segment relative to these axes. For example, a cell might first be defined as being in the head rather than in the abdomen in the anterior–posterior axis; then it might be defined as being on the ventral (top) side of the head.
- Finally, a set of genes called *Hox genes* control the ultimate identity of each body part; for example, determining that the cells at a particular position in the head will make mouthparts.

The genes involved in each of these steps code for transcription factors, which in turn control the synthesis of other transcription factors acting on the next set of genes. This cascade of events may remind you of a signal transduction cascade (see Section 7.3), only in this case it is a *cascade of events that occurs over time and location*, rather than abruptly and in a single cell. The genes finally expressed are the ones familiar to you: they code for protein kinases, receptors, and other proteins that carry out the functions of the cell.

The description of these events in fruit fly development is one of the great achievements in modern biology. It gave biologists a deep understanding of how the events that specify cell identity unfold. We will only skim the surface of the process here, but keep in mind the basic principle of a transcriptional cascade. As we will see in Chapter 20, the fruit fly has been a true model organism in this case, because these findings have informed research on other organisms, including mammals.

Experimental genetics was used to elucidate the events leading to cell fate determination in *Drosophila*:

- First, developmental mutations were identified. For example, a mutant strain might produce larvae with two heads or no segments.
- Then, the mutant was compared with wild-type flies, and the gene responsible for the developmental mistake, and its protein product (if appropriate), were isolated.
- Finally, experiments with the gene (making transgenic flies) and protein (injecting the protein into an egg or into an embryo) were done to confirm the proposed developmental pathway.

Together, these approaches revealed a sequential pattern of gene expression that results in the determination of each segment within 24 hours after fertilization. Several classes of genes are involved.

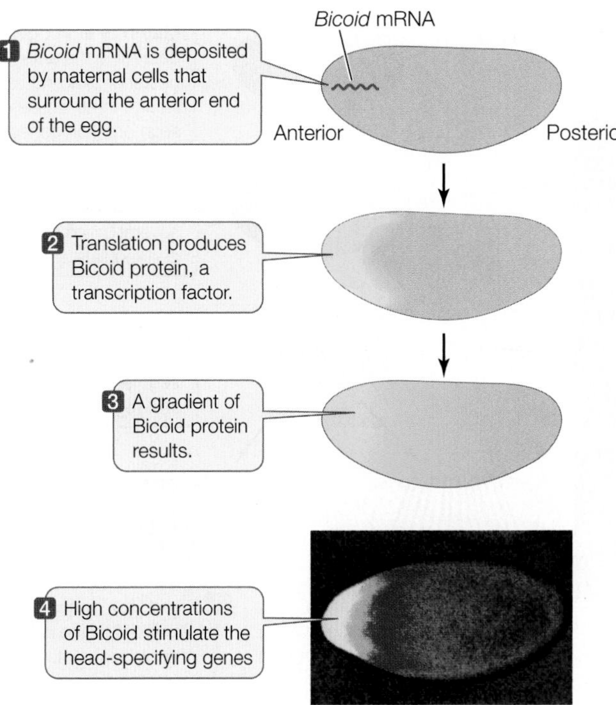
Bicoid mRNA
1 Bicoid mRNA is deposited by maternal cells that surround the anterior end of the egg.
Anterior
Posterior
2 Translation produces Bicoid protein, a transcription factor.
3 A gradient of Bicoid protein results.
4 High concentrations of Bicoid stimulate the head-specifying genes

**19.17 Bicoid Protein Provides Positional Information** The anterior–posterior axis of *Drosophila* arises from the gradient of a morphogen encoded by *Bicoid*, a maternal effect gene. Bicoid protein is also a transcription factor, which activates a gene to specify that the anterior region will become the head of the fly. Other maternal effect genes in the posterior region of the embryo inhibit Bicoid, thus limiting its activity in that region.

**MATERNAL EFFECT GENES** Like the eggs and early embryos of sea urchins, *Drosophila* eggs and larvae are characterized by unevenly distributed cytoplasmic determinants (see Figure 19.9). These molecular determinants, which include both mRNAs and proteins, are the products of specific **maternal effect genes**. These genes are transcribed in the cells of the mother's ovary that surround what will be the anterior portion of the egg. The transcription products are passed to the egg by cytoplasmic bridges. Two maternal effect genes, called *Bicoid* and *Nanos*, help determine the anterior–posterior axis of the egg. (The dorsal–ventral [back–belly] axis is determined by other maternal effect genes that will not be described here.)

The mRNAs for *Bicoid* and *Nanos* diffuse from the mother's cells into what will be the anterior end of the egg. The *Bicoid* mRNA is translated to produce Bicoid protein, which diffuses away from the anterior end, establishing a gradient in the egg cytoplasm (**Figure 19.17**). Where it is present in sufficient concentration, Bicoid acts as a transcription factor to stimulate the transcription of the *Hunchback* gene in the early embryo. A gradient of the Hunchback protein establishes the head, or anterior, region.

Meanwhile, the egg's cytoskeleton transports the *Nanos* mRNA from the anterior end of the egg, where it was deposited, to the posterior end, where it is translated. The Nanos protein forms a gradient with the highest concentration at the posterior end. At that end, the Nanos protein inhibits the translation of *Hunchback* mRNA. Thus, the action of both Bicoid and Nanos establish the Hunchback gradient, which determines the anterior and posterior ends of the embryo.

How did biologists elucidate these pathways? Let's look more closely at the experimental approaches used in this case.

- Females that are homozygous for a particular *bicoid* mutation produce larvae with no head and no thorax; thus the Bicoid protein must be needed for the anterior structures to develop.
- If the eggs of these *bicoid* mutants are injected at the anterior end with cytoplasm from the anterior region of a wild-type egg, the injected eggs develop into normal larvae. This experiment also shows that the Bicoid protein is involved in the development of anterior structures.
- If cytoplasm from the anterior region of a wild-type egg is injected into the posterior region of another egg, anterior structures develop there. The degree of induction depends on how much cytoplasm is injected.
- Eggs from homozygous *nanos* mutant females develop into larvae with missing abdominal segments.
- If cytoplasm from the posterior region of a wild-type egg is injected into the posterior region of a *nanos* mutant egg, it will develop normally.

These and other experiments led scientists to understand the cascade of events that determine cell fates.

The events involving *Bicoid*, *Nanos*, and *Hunchback* begin before fertilization and continue after it, during the multinucleate stage, which lasts a few hours. At this stage, the embryo looks like a bunch of indistinguishable nuclei under the light microscope. But the cell fates have already begun to be determined. After the anterior and posterior ends have been established, the next step in pattern formation is the determination of segment number and locations.

**SEGMENTATION GENES** The number, boundaries, and polarity of the *Drosophila* larval segments are determined by proteins encoded by the **segmentation genes**. These genes are expressed when there are about 6,000 nuclei in the embryo (about three hours after fertilization). Three classes of segmentation genes act one after the other to regulate finer and finer details of the segmentation pattern:

- **Gap genes** organize broad areas along the anterior–posterior axis. Mutations in gap genes result in gaps in the body plan—the omission of several consecutive larval segments.
- **Pair rule genes** divide the embryo into units of two segments each. Mutations in pair rule genes result in embryos missing every other segment.

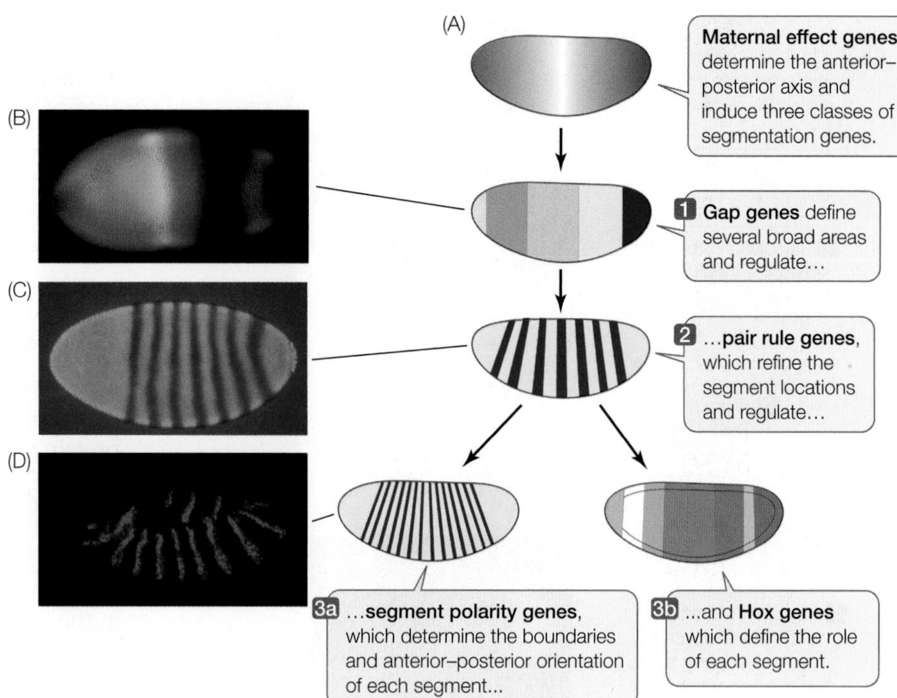

(A)
(B)
(C)
(D)
Maternal effect genes determine the anterior–posterior axis and induce three classes of segmentation genes.
1 Gap genes define several broad areas and regulate...
2 ...pair rule genes, which refine the segment locations and regulate...
3a ...segment polarity genes, which determine the boundaries and anterior–posterior orientation of each segment...
3b ...and Hox genes which define the role of each segment.


**19.18 A Gene Cascade Controls Pattern Formation in the *Drosophila* Embryo** (A) Maternal effect genes (see Figure 19.17) induce gap, pair rule, and segment polarity genes—collectively referred to as segmentation genes. (B) Two gap genes, *Hunchback* (orange) and *Krüppel* (green) overlap; both genes are transcribed in the yellow area. (C) The pair rule gene *Fushi tarazu* is transcribed in the dark blue areas. (D) The segment polarity gene *Engrailed* (bright green) is seen here at a slightly more advanced stage than is depicted in (A). By the end of this cascade, a group of nuclei at the anterior of the embryo, for example, is determined to become the first head segment in the adult fly.

**yourBioPortal.com**
GO TO Animated Tutorial 19.3 • Pattern Formation in the *Drosophila* Embryo

- **Segment polarity genes** determine the boundaries and anterior–posterior organization of the individual segments. Mutations in segment polarity genes can result in segments in which posterior structures are replaced by reversed (mirror-image) anterior structures.

The expression of these genes is sequential (**Figure 19.18**). The maternal effect protein Bicoid, which begins the cascade, acts as a morphogen and transcription factor to stimulate the expression of genes such as *Hunchback* that set up the anterior–posterior axis. As a result, a nucleus in the early embryo "knows" where it is. The Hunchback protein stimulates gap gene transcription, the products of the gap genes activate pair rule genes, and the pair rule gene products activate segment polarity genes. By the end of this cascade, nuclei throughout the embryo "know" which segment they will be part of in the adult fly.

The next set of genes in the cascade determines the form and function of each segment.

**HOX GENES** **Hox genes** encode a family of transcription factors that are expressed in different combinations along the length of the embryo, and help determine cell fate within each segment. Hox gene expression tells the cells of a segment in the head to make eyes, those of a segment in the thorax to make wings, and so on. The *Drosophila* Hox genes occur in two clusters on chromosome 3, in the same order as the segments whose function they determine (**Figure 19.19**). By the time the fruit fly larva hatches, its segments are completely determined. Hox genes are homeotic genes that are shared by all animals, and they are functionally analogous to the organ identity genes of plants. However, they differ from plant homeotic genes in DNA sequence and encoded

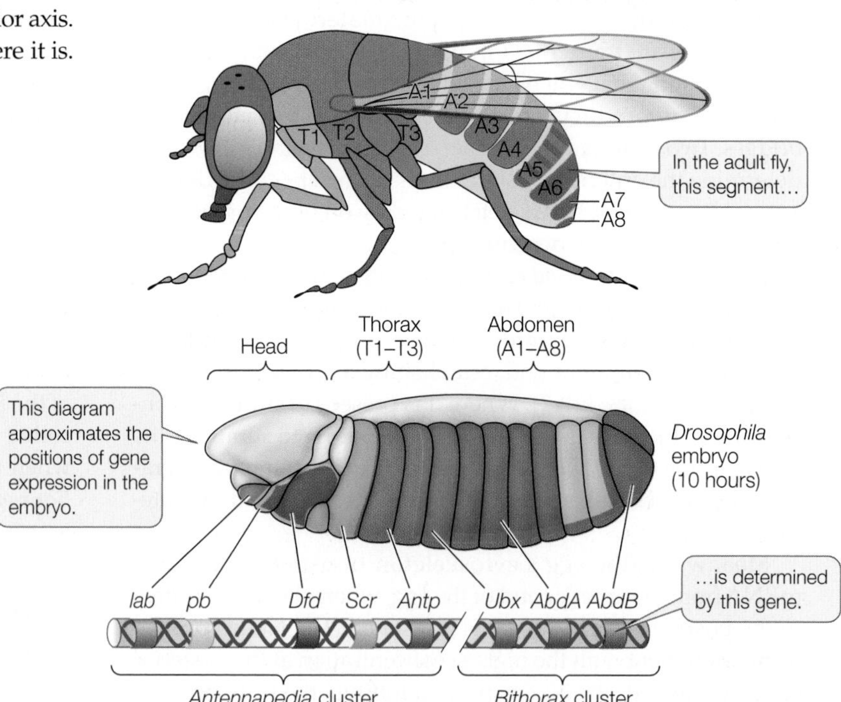


**19.19 Hox Genes in *Drosophila* Determine Segment Identity** Two clusters of Hox genes on chromosome 3 (center) determine segment function in the adult fly (top). These genes are expressed in the embryo (bottom) long before the structures of the segments actually appear.

(A)

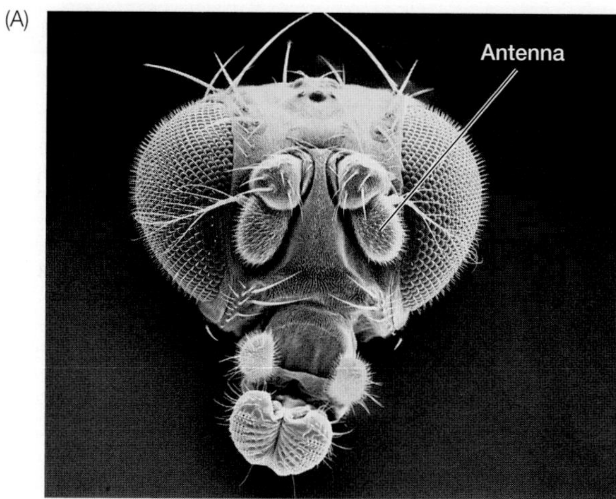


(B)

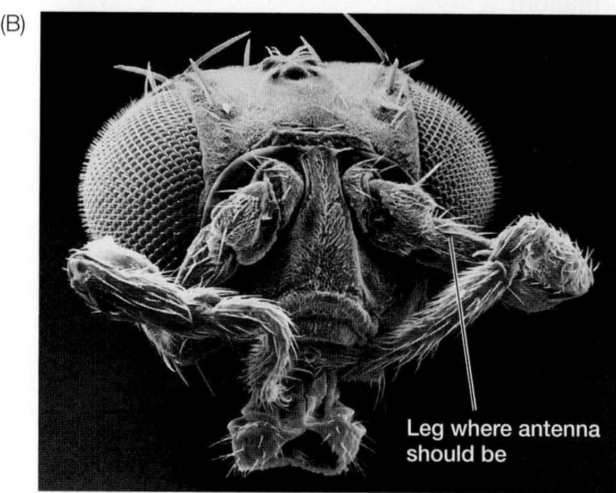


**19.20 A Homeotic Mutation in *Drosophila*** Mutations of the Hox genes cause body parts to form on inappropriate segments. (A) A wild-type fruit fly. (B) An *antennapedia* mutant fruit fly. Mutations such as this reveal the normal role of the *Antennapedia* gene in determining segment function.

protein structure. This is not surprising, given that the last common ancestor of plants and animals was unicellular, and therefore multicellularity evolved independently for plants and animals.

In *Drosophila*, the maternal effect genes, segmentation genes, and Hox genes interact to "build" a larva step by step, beginning with the unfertilized egg. How do we know that the Hox genes determine segment identity? An important clue came from bizarre homeotic mutations observed in *Drosophila*. The *antennapedia* mutation causes legs to grow on the head in place of antennae (**Figure 19.20**), and the *bithorax* mutation causes an extra pair of wings to grow in a thoracic segment where wings do not normally occur (see Figure 20.3). Edward Lewis at Caltech found that *antennapedia* and *bithorax* mutations resulted from changes in Hox genes.

The first cluster of Hox genes—the *Antennapedia* cluster—specifies anterior segments, starting with genes for the different head segments and ending with thoracic segments. The second cluster (*Bithorax*) contains three genes. It begins with a gene specifying the last thoracic segment, followed by a gene for the anterior abdominal segments, and ends with a gene for the posterior abdominal segments (see Figure 19.19). Lewis hypothesized that all of the Hox genes might have come from the duplication of a single gene in an ancestral, unsegmented organism. Since Lewis put forward this hypothesis, molecular research methods became available to test it.

### Hox genes encode transcription factors

Molecular biologists confirmed Lewis's hypothesis using nucleic acid hybridization. Several scientists found that probes for a sequence found in the *Bithorax* cluster could bind to other sequences in both the *Bithorax* and *Antennapedia* clusters. In other words, this DNA sequence is common to all the Hox genes in both clusters. It is also found in several of the segmentation genes, as well as other genes that encode transcription factors.

This 180-base-pair DNA sequence is called the **homeobox**. It encodes a 60-amino acid sequence called the *homeodomain*. The homeodomain recognizes and binds to a specific DNA sequence in the promoters of its target genes. However, this recognition is usually not sufficient to allow the transcription factor to bind fully to a promoter and turn the target gene on or off. Other transcription factors are also involved.

The Hox genes are found in animals with an anterior–posterior axis, where they play a role in development similar to that played by MADS box genes in plants. But the homeobox is found in many different transcription factors, including some from plants. The evolutionary significance of these common pathways for development will be discussed in the next chapter.

## 19.5 RECAP

**A cascade of transcription factors governs pattern formation and the subsequent development of animal and plant organs. Often these transcription factors create or respond to morphogen gradients. In plants, cell fate is often determined by MADS box genes, and in animal embryos, cell fate is determined in part by Hox genes.**

- How is apoptosis crucial in shaping the developing embryo? **See p. 417**
- How do organ identity genes act in *Arabidopsis*? **See pp. 418–419 and Figure 19.14**
- List the key attributes of a morphogen. How does the Bicoid protein fit this definition? **See pp. 420–421 and Figure 19.17**
- How is segment identity established in the *Drosophila* embryo? **Review pp. 421–422 and Figure 19.19**

## CHAPTER SUMMARY

### 19.1 What Are the Processes of Development?

- A multicellular organism begins its development as an embryo. A series of embryonic stages precedes the birth of an independent organism. **Review Figure 19.1, WEB ACTIVITY 19.1**
- The processes of development are **determination**, **differentiation**, **morphogenesis**, and **growth**.
- **Differential gene expression** is responsible for the differences between cell types. **Cell fate** is determined by environmental factors, such as the cell's position in the embryo, as well as by intracellular influences. **Review Figure 19.2**
- Determination is followed by differentiation, the actual changes in biochemistry, structure, and function that result in cells of different types. Determination is a commitment; differentiation is the realization of that commitment.

### 19.2 Is Cell Differentiation Irreversible?

- The zygote is **totipotent**; it is capable of forming every type of cell in the adult body.
- The ability to create **clones** from differentiated cells demonstrates the principle of **genomic equivalence**. **Review Figures 19.3 and 19.4**
- **Stem cells** produce daughter cells that differentiate when provided with appropriate intercellular signals. Some **multipotent** stem cells in the adult body can differentiate into a limited number of cell types to replace dead cells and maintain tissues. **Review Figure 19.5**
- Embryonic stem cells are **pluripotent** and can be cultured in the laboratory. Under suitable environmental conditions, these cells can differentiate into any tissue type. **Induced pluripotent stem cells** have similar characteristics. This has led to technologies to replace cells or tissues damaged by injury or disease. **Review Figure 19.6, ANIMATED TUTORIAL 19.1**

### 19.3 What Is the Role of Gene Expression in Cell Differentiation?

- Differential gene expression results in cell differentiation. Transcription factors are especially important in regulating gene expression during differentiation.
- Complex interactions of many genes and their products are responsible for differentiation during development. **Review Figure 19.7**

### 19.4 How Is Cell Fate Determined?

- **Cytoplasmic segregation**—the unequal distribution of **cytoplasmic determinants** in the egg, zygote, or early embryo—can establish **polarity** and lead to cell fate determination. **Review Figures 19.8 and 19.9, ANIMATED TUTORIAL 19.2**
- **Induction** is a process by which embryonic animal tissues direct the development of neighboring cells and tissues by secreting chemical signals, called **inducers**. **Review Figure 19.10**
- The induction of the vulva in the nematode *Caenorhabditis elegans* offers an example of how inducers act as molecular switches to direct a cell down one of two differentiation paths. **Review Figures 19.11 and 19.12**

### 19.5 How Does Gene Expression Determine Pattern Formation?

- **Pattern formation** is the process that results in the spatial organization of a tissue or organism.
- During development, selective elimination of cells by apoptosis results from the expression of specific genes. **Review Figure 19.13**
- Sepals, petals, stamens, and carpels form in plants as a result of combinatorial interactions between transcription factors encoded by **organ identity genes**. **Review Figure 19.14**
- The transcription factors encoded by floral organ identity genes contain an amino acid sequence called the **MADS box** that can bind to DNA.
- Both plants and animals use **positional information** as a basis for pattern formation. Positional information usually comes in the form of a signal called a **morphogen**. Different concentrations of the morphogen cause different effects. **See Figures 19.15 and 19.16**
- In the fruit fly *D. melanogaster*, a cascade of transcriptional activation sets up the axes of the embryo, the development of the segments, and finally the determination of cell fate in each segment. The cascade involves the sequential expression of maternal effect genes, gap genes, pair rule genes, segment polarity genes, and Hox genes. **Review Figures 19.18 and 19.19, ANIMATED TUTORIAL 19.3**
- Hox genes help to determine cell fate in the embryos of all animals. The **homeobox** is a DNA sequence found in Hox genes and other genes that code for transcription factors. The sequence of amino acids encoded by the homeobox is called the homeodomain.

## SELF-QUIZ

1. Which statement about determination is true?
   *a.* Differentiation precedes determination.
   *b.* All cells are determined after two cell divisions in most organisms.
   *c.* A determined cell will keep its determination no matter where it is placed in an embryo.
   *d.* A cell changes its appearance when it becomes determined.
   *e.* A differentiated cell has the same pattern of transcription as a determined cell.
2. Cloning experiments on sheep, frogs, and mice have shown that
   *a.* nuclei of adult cells are totipotent.
   *b.* nuclei of embryonic cells can be totipotent.
   *c.* nuclei of differentiated cells have different genes than zygote nuclei have.
   *d.* differentiation is fully reversible in all cells of a frog.
   *e.* differentiation involves permanent changes in the genome.

3. The term "induction" describes a process in which a cell or cells
   *a.* influence the development of another group of cells.
   *b.* trigger cell movements in an embryo.
   *c.* stimulate the transcription of their own genes.
   *d.* organize the egg cytoplasm before fertilization.
   *e.* inhibit the movement of the embryo.
4. Stem cells from adult animals
   *a.* are always totipotent.
   *b.* divide when provided with external signals.
   *c.* are not present in bone marrow.
   *d.* are present in an embryo but not an adult.
   *e.* can be turned into differentiated cells with only a few genes.
5. Which statement about cytoplasmic determinants in *Drosophila* is *not* true?
   *a.* They specify the dorsal–ventral and anterior–posterior axes of the embryo.
   *b.* Their positions in the embryo are determined by cytoskeletal action.
   *c.* Some are products of specific genes in the mother fruit fly.
   *d.* They do not produce gradients.
   *e.* They have been studied by the transfer of cytoplasm from egg to egg.
6. In fruit flies, the following genes are used to determine segment polarity: (k) gap genes; (l) Hox genes; (m) maternal effect genes; (n) pair rule genes. In what order are these genes expressed during development?
   *a.* k, l, m, n
   *b.* l, k, n, m
   *c.* m, k, n, l
   *d.* n, k, m, l
   *e.* n, m, k, l
7. Which statement about induction is *not* true?
   *a.* One group of cells induces adjacent cells to develop in a certain way.
   *b.* It triggers a sequence of gene expression in target cells.
   *c.* Single cells cannot form an inducer.
   *d.* A tissue may be induced as well as make an inducer.
   *e.* The chemical identification of specific inducers has not been achieved.
8. In the process of pattern formation in the *Drosophila* embryo,
   *a.* the first steps are specified by Hox genes.
   *b.* mutations in pair rule genes result in embryos missing every other segment.
   *c.* mutations in gap genes result in the insertion of extra segments.
   *d.* segment polarity genes determine the dorsal–ventral axes of segments.
   *e.* all segments develop the same organs.
9. Homeotic mutations
   *a.* are often severe and result in structures at inappropriate places.
   *b.* cause subtle changes in the forms of larvae or adults.
   *c.* occur only in prokaryotes.
   *d.* do not affect the animal's DNA.
   *e.* are confined to the zone of polarizing activity.
10. Which statement about the homeobox is *not* true?
   *a.* It is transcribed and translated.
   *b.* It is found only in animals.
   *c.* Proteins containing the homeodomain bind to DNA.
   *d.* It is a sequence of DNA shared by more than one gene.
   *e.* It occurs in Hox genes.

## FOR DISCUSSION

1. Molecular biologists can attach genes to active promoters and insert them into cells (see Section 18.5). What would happen if the following were inserted and overexpressed? Explain your answers.
   *a.* *ced-9* in embryonic neuron precursors of *C. elegans*
   *b.* *MyoD* in undifferentiated myoblasts
   *c.* the gene for Sonic hedgehog in a chick limb bud
   *d.* *Nanos* at the anterior end of the *Drosophila* embryo
2. A powerful method to test for the function of a gene in development is to generate a "knockout" organism, in which the gene in question is inactivated (see Section 18.4). What do you think would happen in each of the following cases?
   *a.* a knocked-out *ced-9* in *C. elegans*
   *b.* a knocked-out *Nanos* in *Drosophila*
3. If you wanted a rose flower with only petals, what kind of homeotic mutation would you seek in the rose genome?
4. During development, an animal cell's potential for differentiation becomes ever more limited. In the normal course of events, most cells in the adult animal have the potential to be only one or a few cell types. On the basis of what you have learned in this chapter, discuss possible mechanisms for the progressive limitation of the cell's potential.
5. How were biologists able to obtain such a complete accounting of all the cells in *C. elegans*? What major conclusions came from these studies?

## ADDITIONAL INVESTIGATION

Cloning involves considerable reprogramming of gene expression in a differentiated cell so that it acts like an egg cell. How would you investigate this reprogramming?

## WORKING WITH DATA (GO TO yourBioPortal.com)

**Cloning a Mammal** In this hands-on exercise, you will examine the experimental protocol used by Wilmut and colleagues to clone Dolly the sheep (Figure 19.4). You will see the data on the efficiency of this process, as well as the genetic evidence that Dolly was indeed a clone.

# 20 Development and Evolutionary Change

## The eyes have it

Eyes are not essential for survival; many animals and all plants get by just fine without them. However, almost all animals *do* have eyes or some type of light-sensing organs, and having eyes can confer a selective advantage.

About a dozen different kinds of eyes are found among the different animals, including the camera-like eyes of humans and the compound eyes of insects, with their thousands of individual units. In trying to understand the origin of this variety, scientists—starting with Charles Darwin—proposed that eyes evolved independently many times in different animal groups, and that each improvement in the ability of eyes to gather light and form images conferred a selective advantage on their possessor.

A remarkable discovery in the 1990s may have overturned this long-held dogma about the evolution of eyes. Years earlier a mutant fruit fly without eyes was found, and the gene involved—appropriately called *eyeless*—was mapped onto one of its chromosomes. This mutant fly remained a laboratory curiosity until 1994, when the Swiss developmental biologists Rebecca Quiring and Walter Gehring began looking for transcription factors that might be involved in fly development. The gene for one of the proteins they identified mapped to the *eyeless* locus. Thus, the product of the *eyeless* gene is a transcription factor that controls the formation of the eye. Quiring and Gehring demonstrated this by making recombinant DNA constructs that allowed the *eyeless* gene to be expressed in various embryonic tissues of transgenic flies. These experiments resulted in adult flies with extra eyes on various body parts—including on the legs, under the wings, and on the antennae—depending on where the *eyeless* gene was expressed in the embryos.

But the big surprise came when the scientists performed a database search and found that the *eyeless* gene sequence was quite similar to that of *Pax6*, a gene in mice that, when mutated, leads to the development of very small eyes. Could the very different eyes of flies and mice just be variations on a common developmental theme? To test for functional similarity between the insect and mammalian genes, Gehring and colleagues repeated their gene expression experiments using the mouse *Pax6* gene instead of the fly *eyeless* gene. Once again, eyes developed at various sites on the transgenic flies. So a gene whose expression normally leads to the development of a mammalian "camera" eye now led to the development of an insect's "compound"

**Eye of the Fly** Unlike the single-lensed eyes of vertebrates, the compound eyes of flies and other insects are composed of thousands of individual lenses, or ommatidia.

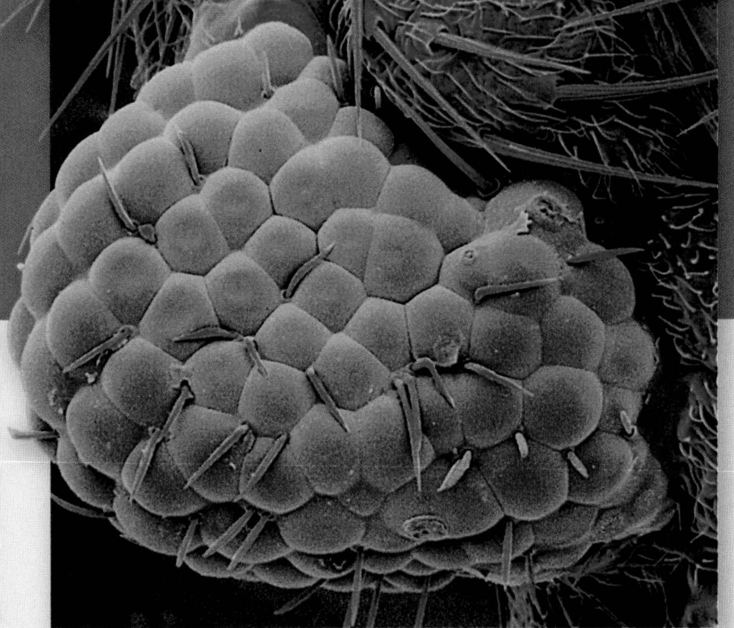

**A Mouse Gene Can Produce a Fly's Eye** When the mouse *Pax6* eye-specifying gene was implanted in the part of the fruit fly embryo that normally produces a limb, ommatidia emerged in place of a leg.

eye—a very different eye type. Thus a single transcription factor appears to function as a molecular switch that turns on eye development. Although eyes evolved many times during animal evolution, all of them depend on the same gene. The special features of the many different eyes in diverse animals all evolved from a common developmental process.

The discovery that the same genes govern development in a wide variety of animals led to the rapid growth of the discipline of evolutionary developmental biology, often known as "evo-devo." Evolutionary developmental biologists compare the genes that regulate development in many different multicellular organisms to understand how a single gene can do so many different things.

**IN THIS CHAPTER** we show that the genes controlling pattern formation, which we introduced in Chapter 19, are shared by a diverse array of organisms. We next describe how changes can occur in some parts of an organism without causing undesirable changes in other parts. We see how a common set of genes can produce a great variety of body forms. We then turn to the ways some organisms can modulate their development by responding to signals from their environment. Finally, we examine how developmental processes constrain evolution.

## CHAPTER OUTLINE



## 20.1 What Is Evo-Devo?

The modern study of evolution and development is called **evolutionary developmental biology**, or **evo-devo**. Its ideas have come from studies of the molecular mechanisms that underlie the development of morphology, and how the genes controlling these mechanisms have evolved. The principles of evo-devo are:

- Many groups of animals and plants, even distantly related ones, share similar molecular mechanisms for morphogenesis and pattern formation. As we saw in the opening essay, some genes that are experimentally swapped from one organism to another can retain similar functions in the new organisms. These mechanisms can be thought of as "toolkits," in the same sense that a few tools in a carpenter's toolkit can be used to build many different structures.
- The molecular pathways that determine different developmental processes, such as anterior–posterior polarity and organ formation in animals, operate independently from one another. This is called **modularity**.
- Changes in the location and timing of expression of particular genes are important in the evolution of new body forms and structures.
- Development produces morphology, and much of morphological evolution occurs by modifications of existing development genes and pathways, rather than the introduction of radically new developmental mechanisms.
- Mechanisms of development have often evolved to be responsive to environmental conditions.

Biologists have long known that the morphological differences between species are due to differences in their genomes. But we have also discussed how the genomes of different species—including distantly related ones—share numerous similar regulatory and coding sequences (see Section 17.3). When developmental biologists began to describe the events of differentiation, morphogenesis, and pattern formation at the molecular level, they found common regulatory genes and pathways in organisms that don't appear similar at all, such as fruit flies and mice.

### Developmental genes in distantly related organisms are similar

In the opening story of this chapter, we describe how a single developmental switch turns on the production of eyes in two widely divergent species—fruit flies and mice—that are only

Mouse *Pax6* gene:

DNA GTATCCAACGGTTGTGTGAGTAAAATTCTGGGCAGGTATTACGAGACTGGCTCCATCAGA

Amino acids V S N G C V S K I L G R Y Y E T G S I R

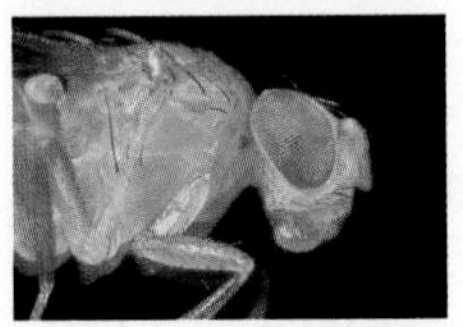

Fly *eyeless* gene:

77% GTATCAAATGGATGTGTGAGCAAAATTCTCGGGAGGTATTATGAAACAGGAAGCATACGA

100% V S N G C V S K I L G R Y Y E T G S I R

Shark eye control gene:

85% GTGTCCAACGGTTCTGTCAGTAAAATCCTGGGCAGATACTATGAAACAGGATCCATCAGA

100% V S N G C V S K I L G R Y Y E T G S I R

Squid eye control gene:

78% GTCTCCAACGGCTGCGTTAGCAAGATTCTCGGACGGTACTATGAGACGGGCTCCATAAGA

100% V S N G C V S K I L G R Y Y E T G S I R

**20.1 DNA Sequence Similarity in Eye Development Genes** Genes controlling eye development contain regions that are highly conserved, even among species with very different eyes. These sequences, from a conserved region of the *Pax6* gene and its homologs in other species, are similar at the DNA level (top sequence in each pair) and identical at the amino acid level (bottom sequence). The percentages beside the sequences represent the percent match with the corresponding DNA and protein sequences in the mouse.

distantly related by evolution. The genes that control this switch, *eyeless* in fruit flies and *Pax6* in mice, contain sequences that are highly conserved in these species and in other animals (**Figure 20.1**). As described in Section 22.1, biologists infer from these similarities that the genes are *homologous*, meaning that they evolved from a gene present in a common ancestor of mice and fruit flies. In recent years, thousands of genes have been found that are homologous across distantly related species.

An even more dramatic example of homology in genes that control development, because it involves a whole set of genes, is the Hox gene cluster. These genes provide positional information and control pattern formation in early *Drosophila* embryos (see Figure 19.19). When scientists looked for similar sequences in the mouse and human genomes, the results were amazing. The Hox genes had homologs in mammals, and what is more, the genes were arranged in similar clusters in the genomes of mammals and fruit flies, and were expressed in similar patterns in their embryos (**Figure 20.2**). Over the millions of years that have elapsed since the common ancestor of these animals, the genes in question have mostly been maintained, suggesting that their functions were favored over many different conditions.

These and other examples have lead evo-devo biologists to the idea that certain developmental mechanisms, controlled by specific DNA sequences, have been conserved over long periods during the evolution of multicellular organisms. These sequences comprise the **genetic toolkit**, which has been modified and reshuffled over the course of evolution to produce the remarkable diversity of plants, animals, and other organisms in the world today.

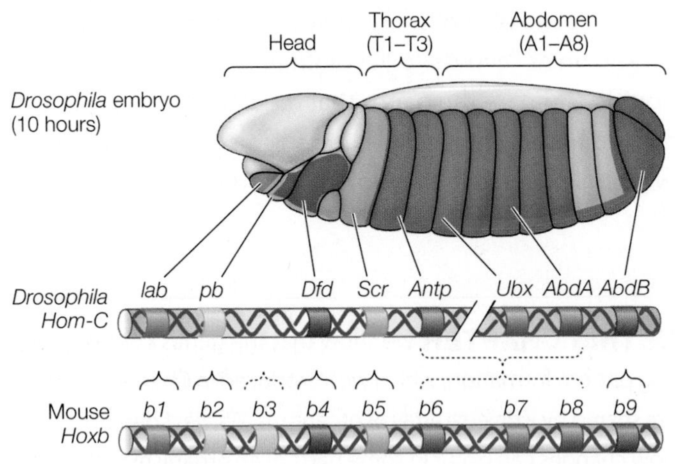


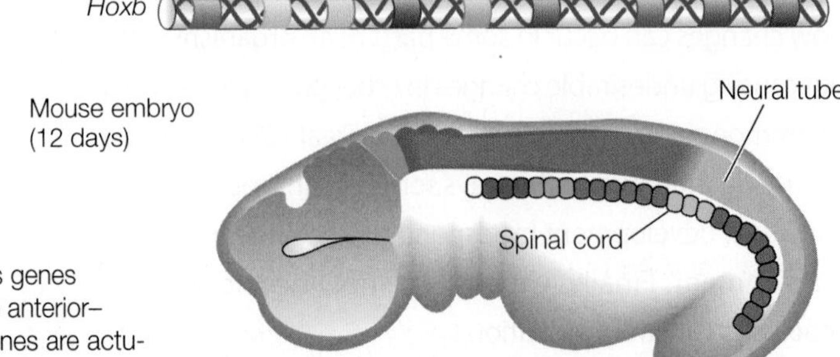


**20.2 Regulatory Genes Show Similar Expression Patterns** Homologous genes encoding similar transcription factors are expressed in similar patterns along the anterior–posterior axes of both insects and vertebrates. The mouse (and human) Hox genes are actually present in multiple copies; this prevents a single mutation from having drastic effects.

## 20.1 RECAP

**Changes in development underlie the changes in morphology that result in evolution of body form. Evo-devo is the study of the evolutionary aspects of development. Evo-devo uses the knowledge gained from, and techniques of, molecular genetics. A genetic toolkit consisting of highly conserved regulatory genes encoding transcription factors governs pattern formation in multicellular organisms.**

- Describe the main ideas of evo-devo. See p. 427
- How does the story of the eye-determining genes *Pax6* and *eyeless* demonstrate the existence of common developmental genes in different organisms? See pp. 427–428 and Figure 20.1
- What is the significance of the similarity between the fruit fly and mouse genes involved in the development of polarity and pattern formation? See p. 428 and Figure 20.2

Many developmental mutations in fruit flies that result in striking abnormalities (e.g., a head segment that forms a leg; see Figure 19.20) affect only a single structure, segment or region. The rest of the embryo is often unaffected. How is this possible?

## 20.2 How Can Mutations With Large Effects Change Only One Part of the Body?

In Chapter 19 we describe how development involves interactions between gene products, which determine a sequence of transcriptional events leading to differential gene expression. On the other hand, the study of homeotic mutations revealed that embryos, like adults, are made up of **developmental modules**—functional entities encompassing genes and various signaling pathways that determine physical structures such as body segments and legs.

The form of each module in an organism may be changed independently of the other modules because some developmental genes exert their effects on only a single module. For example, the form of a developing animal's heart can change independently of changes in its limbs, because some of the genes that govern heart formation do not affect limb formation, and vice versa. If this were not true, a mutation in any developmental gene might result in an adult with multiple, widely different deformities. Such an adult would have difficulty functioning well in any environment.

**yourBioPortal.com**
GO TO **Animated Tutorial 20.1 • Modularity**

### Genetic switches govern how the genetic toolkit is used

Different structures can evolve within a single organism using a common set of genetic instructions because there are mechanisms called **genetic switches** that control how the genetic toolkit is used. These mechanisms involve promoters and the transcription factors that bind them. The signal cascades that converge on and operate these switches determine when and where genes will be turned on and off. Multiple switches control each gene by influencing its expression at different times and in different places. In this way, elements of the genetic toolkit can be involved in multiple developmental processes while still allowing individual modules to develop independently.

Genetic switches integrate positional information in the developing embryo and play key roles in determining the developmental pathways of different modules. For example, each Hox gene codes for a transcription factor that is expressed in a particular segment or appendage of the developing fruit fly. The pattern and functioning of each segment depend on the unique Hox gene or combination of Hox genes that are expressed in the segment.

Consider the formation of fruit fly wings. *Drosophila* has three thoracic segments, the first of which bears no wings. The second segment bears the large forewings, and the third segment bears small hindwings, called *halteres*, that function as balancing organs. Hox proteins are not expressed in forewing cells, but all hind wing cells express the Hox gene *Ultrabithorax* (*Ubx*) because a set of genetic switches activates the *Ubx* gene in the third thoracic segment. Ubx turns off genes that promote the formation of the veins and other structures of the forewing, and it turns on genes that promote the formation of hind wing features (**Figure 20.3**). In butterflies, on the other hand, Ubx influences target genes so that wings develop in the third-segment cells, so full hind wings develop. Therefore, a simple genetic change results in a major morphological difference in the wings of flies and butterflies.

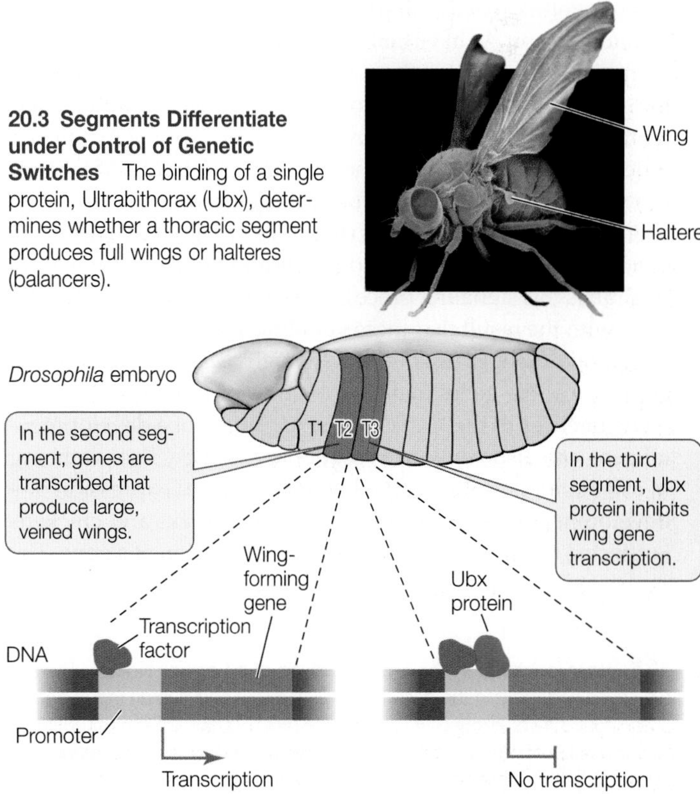


**20.3 Segments Differentiate under Control of Genetic Switches** The binding of a single protein, Ultrabithorax (Ubx), determines whether a thoracic segment produces full wings or halteres (balancers).

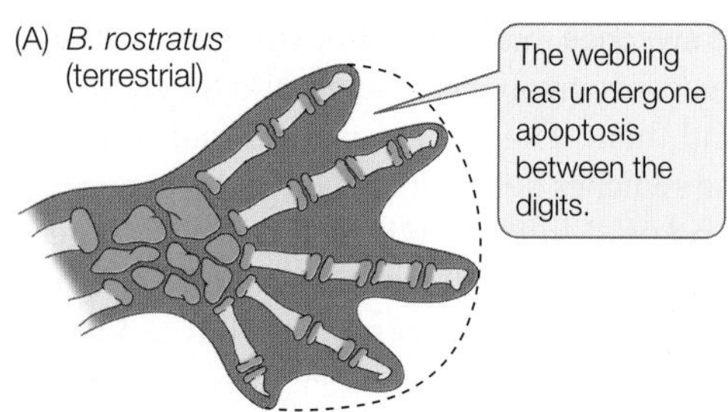
(A) B. rostratus
(terrestrial)
The webbing has undergone apoptosis between the digits.

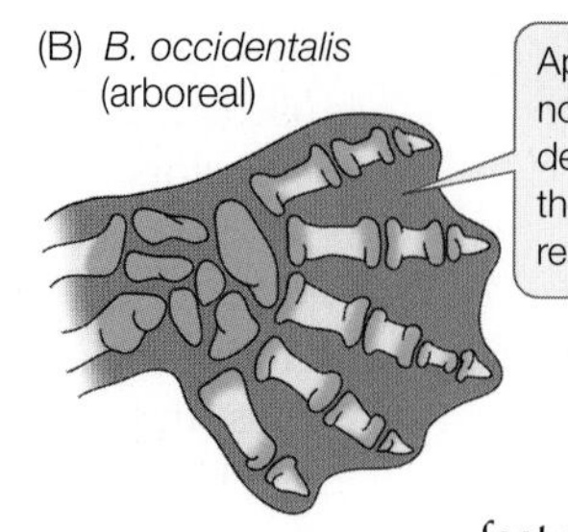
(B) B. occidentalis
(arboreal)
Apoptosis did not occur during development, so the webbing remains.

**20.4 Heterochrony Resulted in the Evolution of a Tree-Climbing Salamander** (A) The foot of an adult *B. rostratus*, a terrestrial salamander. (B) The foot of *B. occidentalis*, a closely related salamander, does not lose its webbing. This species uses the suction of its webbed feet in an arboreal lifestyle.

### Modularity allows for differences in the timing and spatial pattern of gene expression

Modularity allows the relative *timing* of different developmental processes to shift independently of one another, in a process called **heterochrony**. That is, the genes regulating the development of one module (say, the eyes of vertebrates) may be expressed at different developmental stages in different species.

Salamanders of the genus *Bolitoglossa* illustrate how heterochrony can result in major morphological changes. Salamander embryos have webbing between their toes, but in most species of salamanders a particular gene triggers apoptosis in the webbing as the salamanders develop. The resulting independent digits allow the adult salamander to walk more easily than if it had webbed feet. This is the case with *Bolitoglossa rostratus*, a species that lives on the forest floor (**Figure 20.4A**). But in arboreal species such as *Bolitoglossa occidentalis* (**Figure 20.4B**), this gene is not expressed and apoptosis does not occur. The feet of *B. occidentalis* are webbed throughout life, acting like suction cups so the animal can adhere to vertical surfaces such as tree trunks. Thus a simple change in gene expression led to a major morphological change and allowed a new lifestyle.

The evolution of the giraffe's neck provides another example of heterochrony. As in virtually all mammals (with the exception of manatees and sloths) there are seven vertebrae in the neck of the giraffe. So the giraffe did not get a longer neck by adding vertebrae. Instead the cervical (neck) vertebrae of the giraffe are much longer than those of other mammals (**Figure 20.5**). Bones grow due to the proliferation of cartilage-producing cells called chondrocytes. Bone growth is stopped by a signal that results in death of the chondrocytes and calcification of the bone matrix. In giraffes this signaling process is delayed in the cervical vertebrae, with the result that these vertebrae grow longer. Thus, the evolution of longer necks acted through *changes in the timing of expression* of the genes that control bone formation.

Differences in the *spatial expression pattern* of a developmental gene can also result in evolutionary change. Foot webbing in salamanders is determined by the temporal expression of a developmental gene, but foot webbing in ducks and chickens is affected by alterations in the spatial expression of a gene. The feet of all bird embryos have webs of skin that connect their toes. This webbing is retained in adult ducks (and other aquatic birds) but not in adult chickens (and other non-aquatic birds). The loss of webbing is due to a signaling protein called bone morphogenetic protein 4 (BMP4) that instructs the cells in the webbing to undergo apoptosis. The death of these cells destroys the webbing between the toes.

Embryonic duck and chicken hindlimbs both express the *BMP4* gene in the webbing between the toes, but they differ in expression of a gene called *Gremlin*, which encodes a BMP *inhibitor* protein (**Figure 20.6**). In ducks, but not chickens, the *Gremlin* gene is expressed in the webbing cells. The Gremlin protein inhibits the BMP4 protein from signaling for apoptosis, and the result is a webbed foot. If chicken hindlimbs are experimentally exposed to Gremlin during development, apoptosis does not occur, and ducklike webbed feet form on the chicken (**Figure 20.7**).

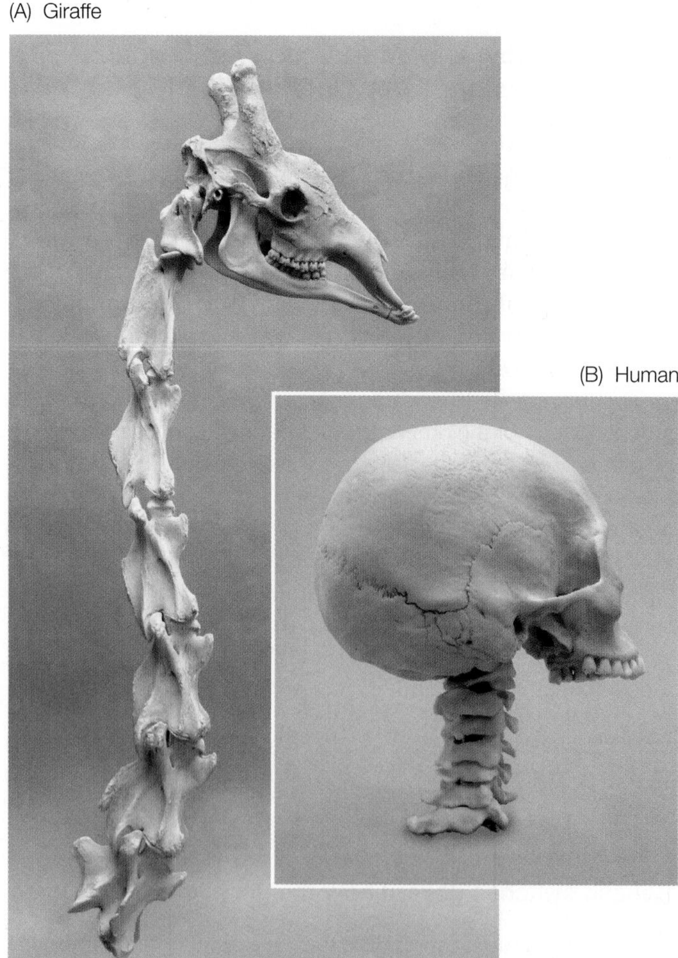
(A) Giraffe
(B) Human

**20.5 Heterochrony in the Development of a Longer Neck** There are seven vertebrae in the neck of the giraffe (left) and human (right; not to scale). But the vertebrae of the giraffe are much longer (25 cm compared to 1.5 cm) because during development, growth continues for a longer period of time. This timing difference is called heterochrony.

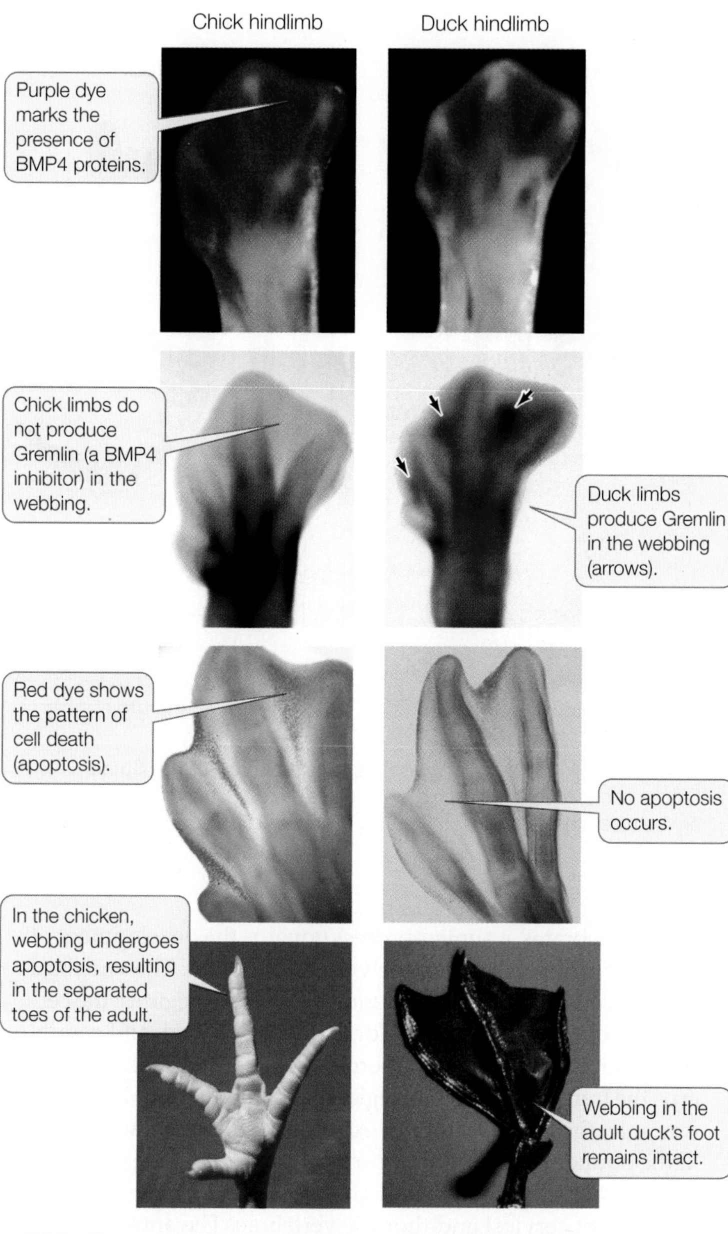

Chick hindlimb
Duck hindlimb
Purple dye marks the presence of BMP4 proteins.
Chick limbs do not produce Gremlin (a BMP4 inhibitor) in the webbing.
Duck limbs produce Gremlin in the webbing (arrows).
Red dye shows the pattern of cell death (apoptosis).
No apoptosis occurs.
In the chicken, webbing undergoes apoptosis, resulting in the separated toes of the adult.
Webbing in the adult duck's foot remains intact.


**20.6 Changes in Gremlin Expression Correlate with Changes in Hindlimb Structure** The left column of photos shows the development of a chicken's foot; the right column shows foot development in a duck. Gremlin protein in the webbing of the duck foot inhibits BMP4 signaling, thus preventing the embryonic webbing from undergoing apoptosis.

## INVESTIGATING LIFE

### 20.7 Changing the Form of an Appendage

Ducks have webbed feet and chickens do not—a major difference in the adaptations of these species. Webbing is initially present in the chick embryo, but undergoes apoptosis that is stimulated by the protein BMP4. In ducks another protein, Gremlin, binds to BMP4 and inhibits it, preventing apoptosis and resulting in webbed feet. J. J. Hurle and colleagues at the Universidad de Cantabria in Spain asked what would happen if Gremlin were put onto a developing chick foot. They hypothesized that apoptosis would be inhibited, and it was: the chick developed webbed feet. Thus, a single developmental switch controls foot shape—an important adaptation to the environment.

**HYPOTHESIS** **Adding Gremlin protein (a BMP4 inhibitor) to a developing chicken foot will transform it into a ducklike foot.**

**METHOD** Chip a small window in chick egg shell and carefully add Gremlin-secreting beads to the webbing of embryonic chicken hindlimbs. Add beads that do not contain Gremlin to other hindlimbs (controls). Close the eggs and observe limb development.

**RESULTS** In the hindlimbs in which Gremlin was secreted, the webbing does not undergo apoptosis, and the hindlimb resembles that of a duck. The control hindlimbs develop the normal chicken form.

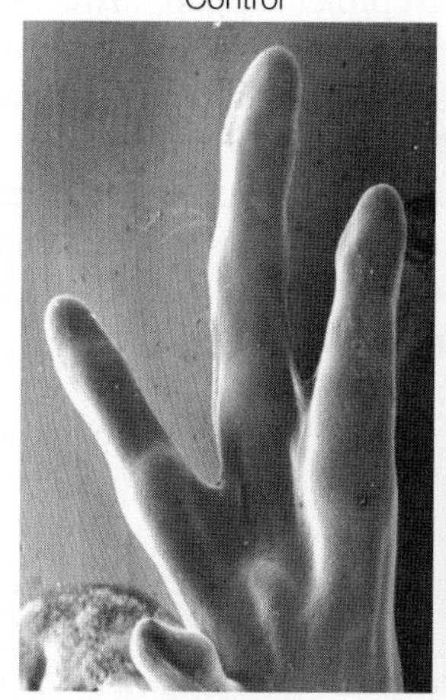


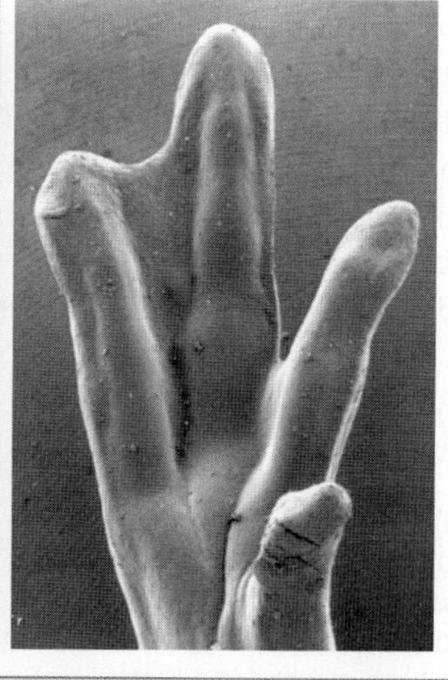


**CONCLUSION** **Differences in *Gremlin* gene expression cause differences in morphology, allowing duck hindlimbs to retain their webbing.**

Go to **yourBioPortal.com** for original citations, discussions, and relevant links for all INVESTIGATING LIFE figures.

## 20.2 RECAP

**Embryos and adult organisms are made up of self-contained units called modules. The form of each module may change independently because some developmental genes exert their effects on only a single module.**

- How do genetic switches control the way a gene is used? **See p. 429 and Figure 20.3**
- Explain how heterochrony can result in evolutionary change. **See p. 430 and Figures 20.4 and 20.5**

Genetic manipulations and studies of pattern formation within embryos have shown that the same signals can control development of different structures in an individual organism. For example, the protein BMP4 promotes apoptosis between developing digits in feet, and then is involved later in the formation of bone. These studies suggest that the processes that generate multiple structures *within* an organism might also explain how different structures develop in different species.

## 20.3 How Can Differences among Species Evolve?

Can the processes that allow different structures to develop in different regions of an embryo also explain major morphological differences among species? Apparently they can. The genetic switches that determine where and when genes will be expressed appear to underlie both the transformation of an individual from egg to adult and the many major differences in body form that exist among species. Arthropods provide good examples of how morphological differences among species can evolve through mutations in the genes that regulate the differentiation of segments.

The arthropods (which include crustaceans, centipedes, spiders, and insects) are segmented, with head, thoracic, and abdominal segments. In centipedes, both thoracic and abdominal segments form legs; but in insects, only the thoracic segments do. Arthropods express a gene called *Distal-less* (*Dll*) that causes legs to form from segments. What shuts down *Dll* expression in insect abdominal segments? The product of the Hox gene *Ubx* is produced in arthropod abdominal segments. But it has very different effects in different organisms. In centipedes, the Ubx protein apparently activates expression of the *Dll* gene to promote the formation of legs. But during the evolution of insects, a change in the *Ubx* gene sequence resulted in a modified Ubx protein that *represses Dll* expression in abdominal segments, so leg formation is inhibited. A phylogenetic tree of arthropods shows that this change in *Ubx* occurred in the ancestor of insects, at the same time that abdominal legs were lost (**Figure 20.8**).

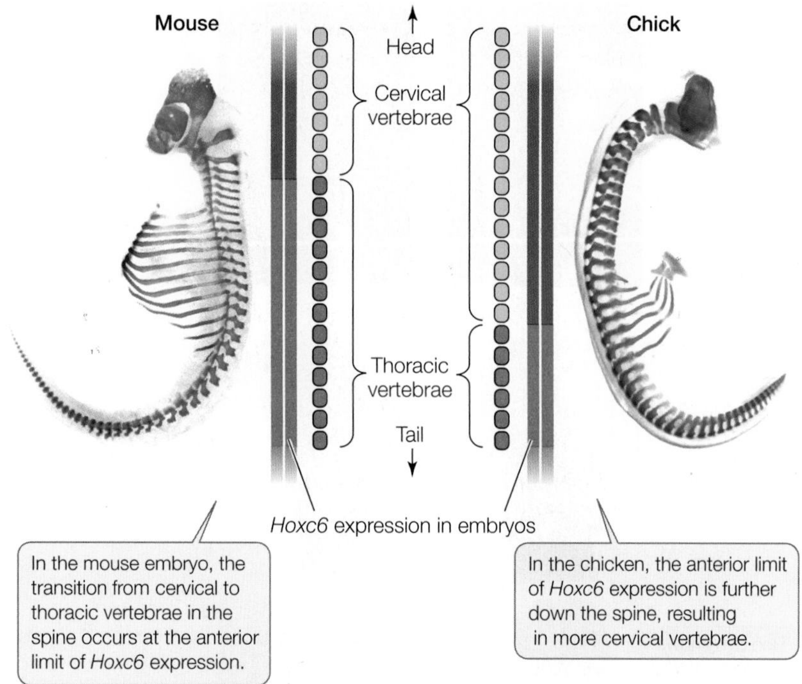

Mouse
Chick
Head
Cervical vertebrae
Thoracic vertebrae
Tail
*Hoxc6* expression in embryos
In the mouse embryo, the transition from cervical to thoracic vertebrae in the spine occurs at the anterior limit of *Hoxc6* expression.
In the chicken, the anterior limit of *Hoxc6* expression is further down the spine, resulting in more cervical vertebrae.

**20.9 Changes in Gene Expression and Evolution of the Spine** Differences in the pattern of *Hoxc6* expression result in a different boundary between the cervical and thoracic vertebrae in mice and chicks.

In vertebrates, a similar process governs the development of differences in segments of the vertebral column. The vertebral column consists of a set of anterior-to-posterior regions: the cervical (neck), thoracic (chest), lumbar (back), sacral, and caudal (tail) regions. The spatial pattern of Hox gene expression governs the transition from one region to another (**Figure 20.9**). For example, the anterior limit of expression of *Hoxc6* always falls at the boundary between the cervical and thoracic vertebrae of mice and chickens, even though these animals have different numbers of cervical and thoracic vertebrae. The anterior-most

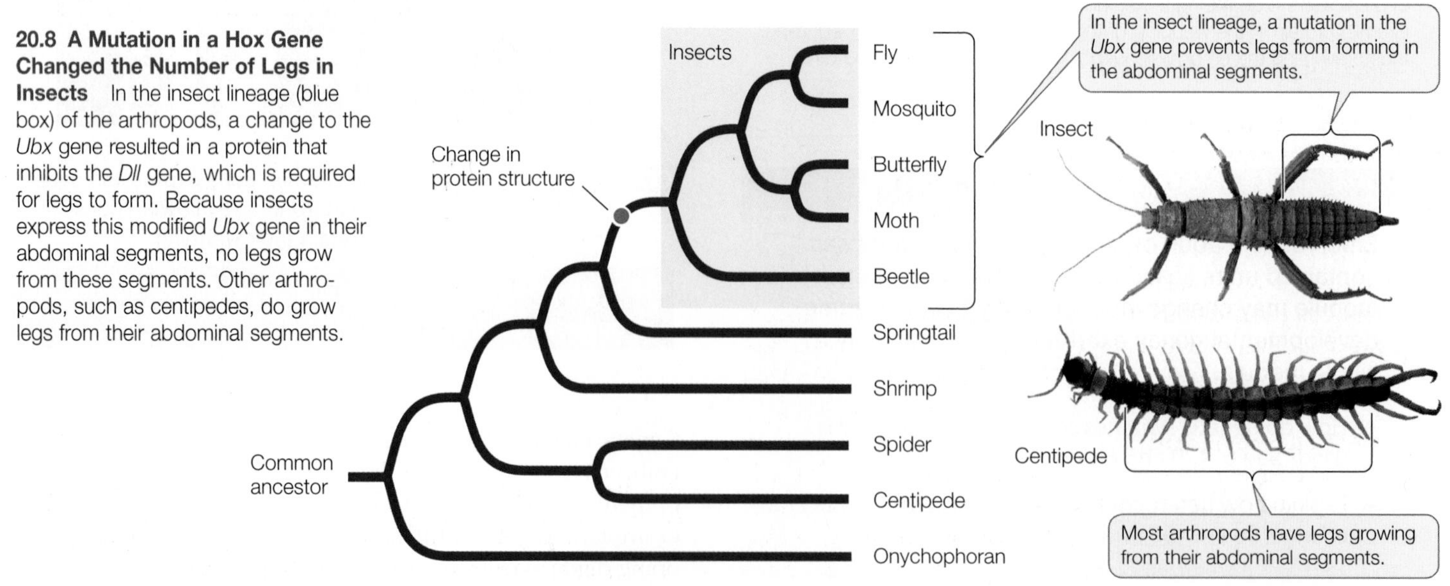


**20.8 A Mutation in a Hox Gene Changed the Number of Legs in Insects** In the insect lineage (blue box) of the arthropods, a change to the *Ubx* gene resulted in a protein that inhibits the *Dll* gene, which is required for legs to form. Because insects express this modified *Ubx* gene in their abdominal segments, no legs grow from these segments. Other arthropods, such as centipedes, do grow legs from their abdominal segments.

segment that expresses *Hoxc6* is the segment where the forelimbs will develop. Thus, genetic changes that expanded or contracted the expression domains of different Hox genes resulted in changes in the characteristic numbers of different vertebrae during evolution.

## 20.3 RECAP

**Changes in the genetic switches that determine where and when genes will be expressed underlie the evolution of differences in form among species.**

- Why do insects, unlike other arthropods, lack abdominal limbs? See p. 432 and Figure 20.8
- How can the evolution of the spinal column be explained by changes in a developmental gene? See p. 432 and Figure 20.9

So far in this chapter, we have focused on how modular genetic signaling cascades control the development of an organism and how changes in genetic switches can produce differences between species. You may have the impression that all of these processes unfold from the genetic information contained in the fertilized egg, but that is not the case. Information from the environment can influence the genetic signaling cascades and thereby alter the form of the organism.

# 20.4 How Does the Environment Modulate Development?

The environment an individual lives in may differ from the one its parents lived in. Some environmental signals can produce developmental changes in an organism. If such changes result in higher reproductive fitness, they will be favored by natural selection. The ability of an organism to modify its development in response to environmental conditions is called **developmental plasticity** or *phenotypic plasticity*. It means that *a single genotype has the capacity to produce two or more different phenotypes.*

### Temperature can determine sex

In Chapter 12 we discuss the genetic mechanisms that determine sex. In mammals there are two sex chromosomes; XX individuals are female and XY individuals are male. But in some reptiles, sex is determined not by genetic differences between individuals, but rather by the temperature at which the eggs are incubated—a remarkable case of developmental plasticity.

Research in the laboratory of David Crews at the University of Texas has shown that if eggs of the red-eared slider turtle are incubated at temperatures below 28.6°C, they will all become males, whereas if the eggs are incubated above 29.4°C, they will all become females. In the less than 1°C range between these two temperatures, a mix of males and females will hatch from the eggs (**Figure 20.10**). In other species with temperature-dependent sex determination, the incubation temperatures that produce males and females may differ from those that produce males and females in the red-eared slider. These different temperature dependencies indicate that the effects of incubation temperature can vary among species. But how does temperature control this developmental plasticity?

In vertebrates, the development of male and female organs in the embryo is controlled by the actions of sex steroid hormones. This is the case whether the organism's sex determination is controlled genetically or by temperature. Sex steroid biosynthesis in both males and females begins with cholesterol, which goes through many chemical reactions to produce the male sex steroids (androgens) and the female sex steroids (estrogens). In this biosynthetic sequence, the step that produces the first androgen—testosterone—precedes the step that produces estrogens; therefore, both males and females produce testosterone.

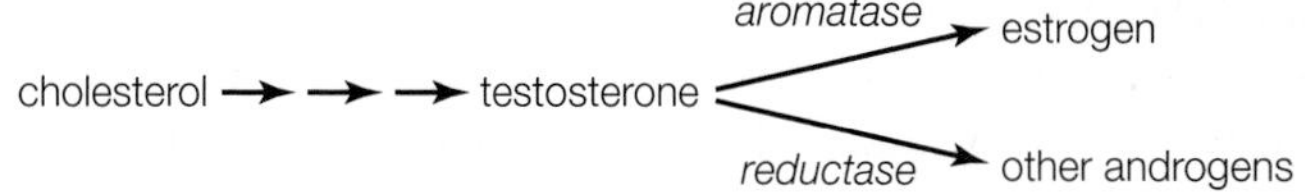


In animals with temperature-controlled sex determination, incubation temperature influences sex development by controlling the expression of the enzyme aromatase, which converts testosterone to estrogen. If aromatase is abundantly expressed, estrogens are dominant and female organs develop. If aromatase is not expressed, testosterone is dominant, and male organs develop. Applying estrogen to eggs results in the development of females, even at the male-inducing temperature.

What is the evolutionary advantage of this sex determination mechanism? It might be that incubation temperature influences the growth rate of the embryo and the time of hatching.

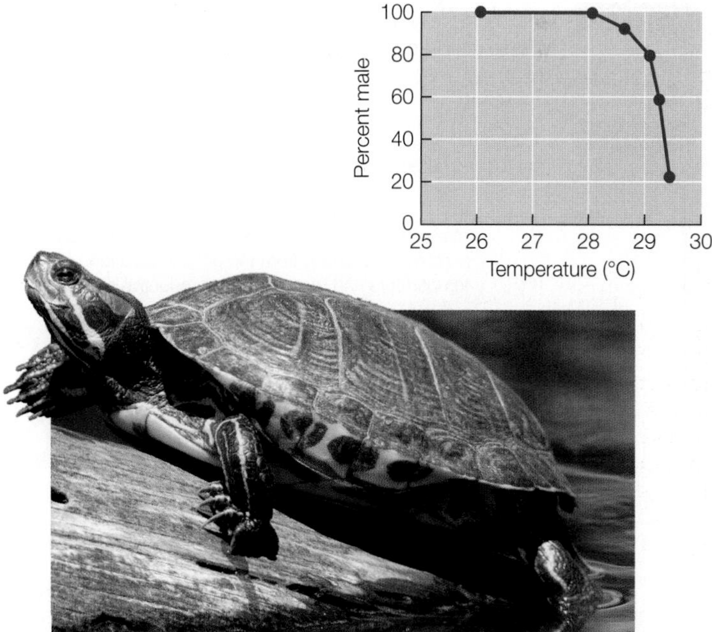


**20.10 Hot Females, Cool Males** Whether the embryo of a red-eared slider turtle develops into a male or a female depends on the temperature at which the egg is incubated. Higher temperatures produce only females and lower temperatures produce only males. This is apparently due to temperature sensitivity in the synthesis of sex hormones.

In species in which males compete for territories and for females, a larger body size would be a benefit for males, but not necessarily for females. Depending on availability of food in the environment, earlier hatching may have a positive or a negative effect on growth rate. For these kinds of reasons, incubation temperature may have a differential effect on the reproductive successes of males and females in a population.

One interesting experiment clearly demonstrated the fitness value of temperature-determined sex (**Figure 20.11**). At the University of Sydney, Daniel Warner and Richard Shine used hormones to manipulate the eggs of a lizard called the jacky dragon (*Amphibolurus muricatus*) to produce males at temperatures that would normally result in females. In this species, females are produced at all incubation temperatures, but males are only produced at incubation temperatures between 27°C and 30°C. The hormonal manipulations allowed the investigators to obtain both males and females from three different incubation temperatures—low, medium, and high—and to compare their subsequent growth characteristics and reproductive success.

The young lizards were released into outdoor enclosures and allowed to behave naturally for the next three years. The males incubated at the medium temperature had higher reproductive success over the three-year study. The reproductive successes of females from the low and medium incubation temperatures did not differ, but were higher for the high-temperature female group. These data support the hypothesis that the incubation temperature differentially affects reproductive success in males and females, and provides an explanation for why there could be selection for temperature-dependent sex determination in some species.

### Organisms use information that predicts future conditions

In many cases of phenotypic plasticity, the adaptation of the different body forms to different but predictable environments is quite obvious. An excellent example is the moth *Nemoria arizonaria*, which produces two generations each year. Caterpillars

## INVESTIGATING LIFE

**20.11 Temperature-Dependent Sex Determination is Associated with Sex-Specific Fitness Differences**

In some reptiles, sex is determined by the incubation temperature of the developing embryo. This led to the hypothesis that male-inducing temperatures during development result in males with higher reproductive fitness. Warner and Shine tested this hypothesis by using a drug to block estrogen synthesis, so that males developed instead of females at high and low temperatures. These males had much lower reproductive fitness than males that developed at the normal male-inducing temperature (which is intermediate). In contrast, females showed highest fitness when they developed from eggs incubated at higher temperatures.

**HYPOTHESIS** **Incubation temperature has a differential effect on reproductive success in lizards.**

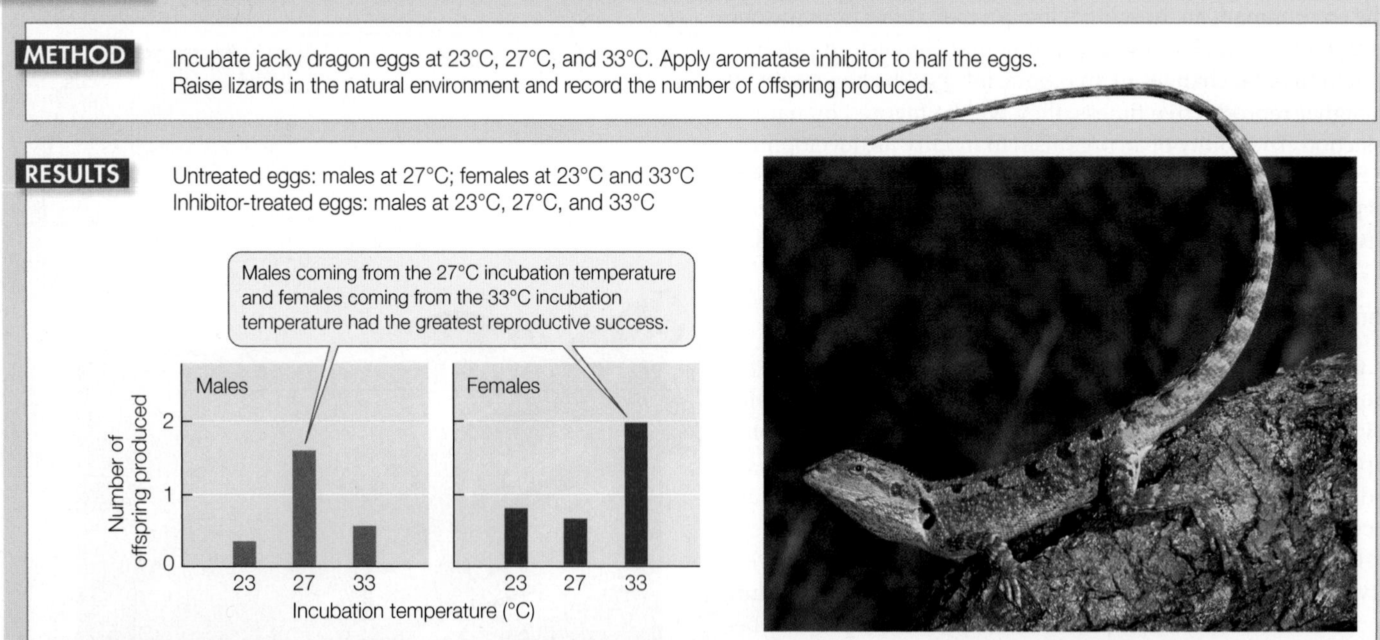

METHOD
Incubate jacky dragon eggs at 23°C, 27°C, and 33°C. Apply aromatase inhibitor to half the eggs.
Raise lizards in the natural environment and record the number of offspring produced.
RESULTS
Untreated eggs: males at 27°C; females at 23°C and 33°C
Inhibitor-treated eggs: males at 23°C, 27°C, and 33°C
Males coming from the 27°C incubation temperature and females coming from the 33°C incubation temperature had the greatest reproductive success.
Males
Females
Number of offspring produced
0
1
2
23
27
33
23
27
33
Incubation temperature (°C)


**CONCLUSION** **Incubation temperature during development has a differential effect on fitness of male and female jacky dragons, and thus could be a selective pressure leading to the pattern of temperature determination of sex in this species.**

Go to **yourBioPortal.com** for original citations, discussions, and relevant links for all INVESTIGATING LIFE figures.

that hatch from eggs in spring feed on oak tree flowers (catkins). These caterpillars complete their development and transform into adult moths in summer. The summer moths lay their eggs on oak leaves, and the caterpillars that hatch eat the leaves. When these caterpillars transform into adult moths, they lay eggs that overwinter and hatch in the spring when the catkins are once again in bloom. Both types of caterpillars are camouflaged in the environments in which they feed. The body form of the spring caterpillars resembles the catkins (**Figure 20.12A**), and the body form of the summer caterpillars resembles small oak branches (**Figure 20.12C**). At the time of hatching, the young caterpillars all look similar, but their diets trigger developmental changes that result in the differences in appearance. The ability to avoid predation by phenotypic plasticity increases evolutionary fitness.

### A variety of environmental signals influence development

In addition to temperature and diet, there are other environmental signals that initiate developmental changes. A ubiquitous and dependable source of environmental information is light, which provides predictive information about seasonal changes. Outside of the equatorial region, lengthening days herald spring and summer while shortening days indicate oncoming fall and winter. Many insects use day length to enter or exit a period of developmental or reproductive arrest called **diapause**, which enables them to better survive harsh conditions. Deer, moose, and elk use day length to time the development and the dropping of antlers, and many organisms use day length to optimize the timing of reproduction. As we discuss in Chapter 38, many plants initiate reproduction in response to the length of the night (an absence of light) and others respond to certain wavelengths of light with developmental changes.

You may wonder why we are mentioning processes like antler growth and seasonal reproduction in a chapter on development. Development encompasses more than the events that occur before an organism reaches maturity. Development includes changes in body form and function that can occur throughout the life of the organism.

Plants provide a particularly clear example of this. Redwood trees that are thousands of years old still have undifferentiated tissues called meristems that produce new differentiated tissues for the tree—stems, leaves, reproductive structures, and so on—throughout its life. These developmental processes are not a simple read-out of a genetic program; they are adjusted to optimize plant form in the environment in which the plant grows. Light, which plants need for photosynthesis, is an important environmental signal in plant development. Dim light stimulates the elongation of stem cells, so that plants growing in the shade become tall and spindly (**Figure 20.13**). This developmental plasticity is adaptive because a spindly plant is more likely to reach a patch of brighter light than a plant that remains compact. In bright light a plant does not need to grow tall, and can put its energy into growing leaves.

yourBioPortal.com
GO TO Web Activity 20.1 • Plant Development

Natural selection can act on any genes or signaling pathways with important developmental functions that can influence reproductive success. Antler growth cycles involve the turning on and off of genes controlling bone growth. Seasonal breeding involves turning on and off the same genes that were involved in sex development and maturation. The evolution of development extends to all stages of life.

**20.12 Spring and Summer Forms of a Caterpillar** (A) Spring caterpillars of the moth *N. arizonaria* resemble the oak catkins on which they feed. They develop into adults (B), which lay eggs on oak leaves. The summer caterpillars (C) of the same species resemble oak twigs.

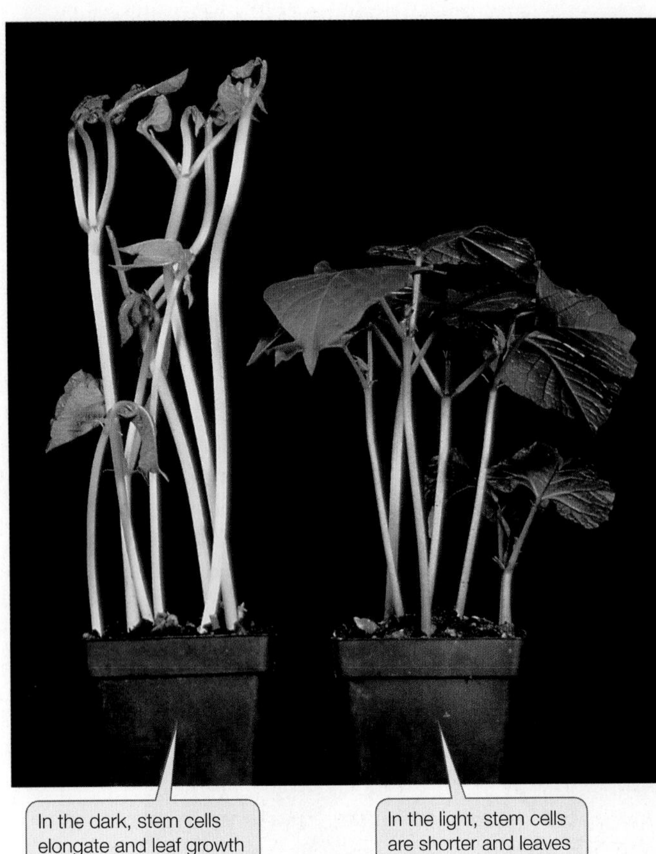

In the dark, stem cells elongate and leaf growth is reduced.
In the light, stem cells are shorter and leaves are bigger.

**20.13 Light Seekers** The bean plants on the left were grown under low light levels. The plant's cells have elongated in response to the low light, and the plants have become spindly. The control plants on the right were grown under normal light conditions.

## 20.4 RECAP

**Developmental plasticity enables developing organisms to adjust their forms to fit the environments in which they live. Organisms respond to environmental signals that are accurate predictors of future conditions. Development continues throughout life, and can result in adaptive changes in the forms and functions of adult organisms.**

- Describe several examples of how an organism's phenotype can be a response to environmental signals. **See pp. 433–435 and Figures 20.10 and 20.12**
- How would you determine whether or not an environmental effect on development is adaptive? **See pp. 434–435 and Figure 20.11**

Appropriate responses to new environmental conditions are likely to evolve over time, but what are the limits of such evolution? Do developmental genes dictate what structures and forms are possible?

# 20.5 How Do Developmental Genes Constrain Evolution?

Four decades ago, the French geneticist François Jacob made the analogy that evolution works like a tinker, assembling new structures by *combining and modifying the available materials*, and not like an engineer, who is free to develop dramatically different designs (say, a jet engine to replace a propeller-driven engine). We have seen that the evolution of morphology has not been governed by the appearance of radically new genes, but by modifications of existing genes and their regulatory pathways. Thus, developmental genes and their expression constrain evolution in two major ways:

- Nearly all evolutionary innovations are modifications of previously existing structures.
- The genes that control development are highly conserved; that is, the regulatory genes themselves change slowly over the course of evolution.

### Evolution proceeds by changing what's already there

The features of organisms almost always evolve from preexisting features in their ancestors. New "wing genes" did not suddenly appear in insects, birds, and bats; instead, wings arose as modifications of existing structures. Wings evolved independently in insects and vertebrates—once in insects, and in three independent instances among the vertebrates (**Figure 20.14**). *In vertebrates, the wings are modified limbs.*

Like limbs, wings have a common structure: a humerus that articulates with the body; two longer bones, the radius and ulna, that project away from the humerus; and then metacarpals and phalanges (digits). During development these bones have different lengths and weights in different organisms.

Developmental controls also influence how organisms lose structures. The ancestors of present-day snakes lost their forelimbs as a result of changes in the segmental expression of Hox genes. The snake lineage subsequently lost its hindlimbs by the loss of expression of the *Sonic hedgehog* gene in the limb bud tissue. But some snake species such as boas and pythons still have rudimentary pelvic bones and upper leg bones.

### Conserved developmental genes can lead to parallel evolution

The nucleotide sequences of many of the genes that govern development have been highly conserved throughout the evolution of multicellular organisms—in other words, these genes exist in similar form across a broad spectrum of species (see Figure 20.2).

The existence of highly conserved developmental genes makes it likely that similar traits will evolve repeatedly, especially among closely related species—a process called **parallel phenotypic evolution**. A good example of this process is provided by a small fish, the three-spined stickleback (*Gasterosteus aculeatus*).

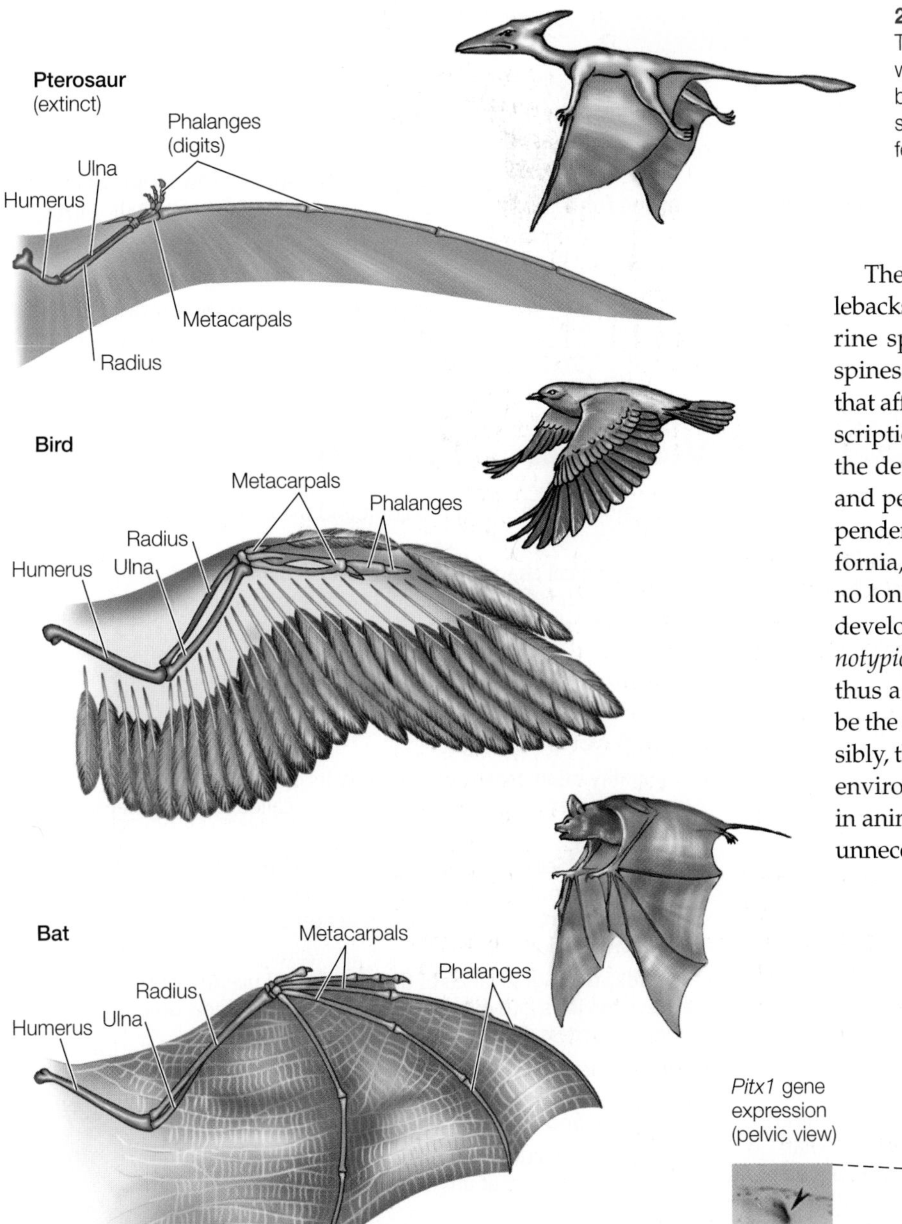


**20.14 Wings Evolved Three Times in Vertebrates** The wings of pterosaurs (the earliest flying vertebrates, which lived from 265 to 220 million years ago), birds, and bats are all modified forelimbs constructed from the same skeletal components. However, the components have different forms in the different groups of vertebrates.

Sticklebacks are widely distributed across the Atlantic and Pacific Oceans and are also found in many freshwater lakes. Marine populations of this species spend most of their lives at sea, but return to fresh water to breed. Members of freshwater populations live in lakes and never journey to salt water. Genetic evidence shows that freshwater populations have arisen independently from marine populations many times, most recently at the end of the last ice age. Marine sticklebacks have several structures that protect them from predators: well-developed pelvic bones with pelvic spines, and bony plates. In the freshwater populations descended from them, this body armor is greatly reduced, and dorsal and pelvic spines are much shorter or even lacking (**Figure 20.15**).

The difference between marine and freshwater sticklebacks is not induced by environmental conditions. Marine species that are reared in fresh water still grow spines. Not surprisingly, the difference is due to a gene that affects development. The *Pitx1* gene codes for a transcription factor that is normally expressed in regions of the developing embryo that form the head, trunk, tail, and pelvis of the marine stickleback. However, in independent populations from Japan, British Columbia, California, and Iceland, the gene has evolved such that it is no longer expressed in the pelvis, and the spines do not develop. *This same gene has evolved to produce similar phenotypic changes in several independent populations*, and is thus a good example of parallel evolution. What could be the common selective mechanism in these cases? Possibly, the decreased predation pressure in the freshwater environment allows for increased reproductive success in animals that invest less energy in the development of unnecessary protective structures.

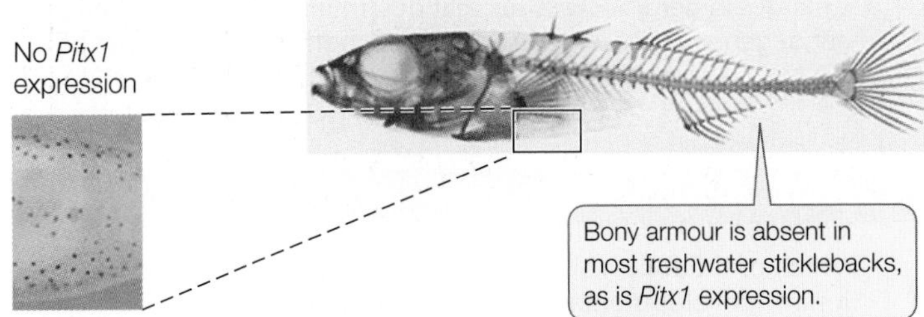


**20.15 Parallel Phenotypic Evolution in Sticklebacks** A developmental gene, *Pitx1*, encodes a transcription factor that stimulates the production of plates and spines. This gene is active in marine sticklebacks, but mutated and inactive in various freshwater populations of the fish. The fact that this mutation is found in geographically distant and isolated freshwater populations is evidence for parallel evolution.

## 20.5 RECAP

**Developmental controls constrain evolution because nearly all evolutionary innovations are modifications of previously existing structures. The conservation of many genes makes it likely that similar traits will evolve repeatedly.**

- How have diverse body forms evolved by means of modifications in the functioning of existing genes? **See p. 436 and Figures 20.14**
- Explain how the differences between marine and freshwater sticklebacks exemplify parallel evolution via changes in gene regulation. **See p. 437 and Figure 20.15**

During the course of evolution, many novel traits have arisen, but failed to persist beyond a single generation. Part Six of this book examines the processes of evolution—the powerful forces that influence the survival and reproductive success of various life forms. We will examine how different adaptations become prevalent in different environments, resulting in the extraordinary diversity of life on Earth today.

## CHAPTER SUMMARY

### 20.1 What Is Evo-Devo?

- **Evolutionary developmental biology (evo-devo)** is the modern study of the evolutionary aspects of development, and it focuses on molecular mechanisms.
- Changes in development underlie evolutionary changes in morphology that produce major differences in body forms.
- Similarities in the basic mechanisms of development between widely divergent organisms reflect common ancestry. Review Figure 20.1
- Evolutionary diversity is produced using a modest number of regulatory genes.
- The transcription factors and chemical signals that govern pattern formation in the bodies of multicellular organisms, and the genes that encode them, can be thought of as a **genetic toolkit**.
- Regulatory genes have been highly conserved during evolution. Review Figure 20.2

### 20.2 How Can Mutations with Large Effects Change Only One Part of the Body?

**SEE ANIMATED TUTORIAL 20.1**

- The bodies of developing and mature organisms are organized into self-contained units called **developmental modules** that can be modified independently. Modularity allows the timing of different developmental processes to shift independently, in a process called **heterochrony**. Review Figure 20.4 and 20.5
- Alterations in the spatial expression patterns of regulatory genes can also result in evolutionary changes. Review Figure 20.6

### 20.3 How Can Differences among Species Evolve?

- Changes in **genetic switches** that determine where and when a set of genes will be expressed underlie both the transformation of an individual from egg to adult and the evolution of differences among species.
- Morphological changes in species can evolve through mutations in the genes that regulate the differentiation of body segments. Review Figure 20.8

### 20.4 How Does the Environment Modulate Development?

**SEE WEB ACTIVITY 20.1**

- The ability of an organism to modify its development in response to environmental conditions is called **developmental plasticity**.
- In many species of reptiles, sex development is determined by incubation temperature, which acts through genes that control the production, modification, and action of sex hormones. Review Figure 20.10
- The adaptive significance of developmental plasticity is not always obvious, but experiments can test for effects on reproductive success. Review Figure 20.11
- Some environmental cues, such as those that anticipate seasons, are highly regular and can reliably drive seasonal adaptations in body form and function. Review Figure 20.12
- Environmental cues that trigger developmental change are diverse and can act at any stage of the life of an organism.

### 20.5 How Do Developmental Genes Constrain Evolution?

- Virtually all evolutionary innovations are modifications of preexisting structures. Review Figure 20.14
- Because many genes that govern development have been highly conserved, similar traits are likely to evolve repeatedly, especially among closely related species. This process is called **parallel phenotypic evolution**. Review Figure 20.15

## SELF-QUIZ

1. Which of the following is *not* one of the principles of evolutionary developmental biology (evo-devo)?
   *a.* Animal groups share similar molecular mechanisms for morphogenesis.
   *b.* Changes in the timing of gene expression are important in the evolution of new structures.
   *c.* Evolution of development is not responsive to the environment.
   *d.* Changes in the locations of gene expression in the embryo can lead to new structures.
   *e.* Evolution occurs by modification of existing developmental genes and pathways.
2. The developmental control pathway that results in polarity and pattern formation in the head–abdomen axis in *Drosophila*
   *a.* has a similar gene sequence and chromosome order in the mouse.

*b.* arose only in insects during evolution.
*c.* determines only the organs that arise in head segments.
*d.* involves only gene products made by the embryo.
*e.* arose through new genes that had not existed before in any form.

3. Which of the following is *not* true of genetic switches?
   *a.* They control how a genetic toolkit is used.
   *b.* They integrate positional information in an embryo.
   *c.* A single switch controls each gene.
   *d.* They allow different structures to develop within an individual organism.
   *e.* They determine when and where a gene is turned on or off.
4. Ducks have webbed feet and chickens do not because
   *a.* ducks need webbed feet to swim, whereas terrestrial chickens do not.
   *b.* both duck and chicken embryos express *BMP4* in the webbing between the toes, but the *Gremlin* gene is expressed in the webbing cells only in ducks.
   *c.* both duck and chicken embryos express *BMP4* in the webbing between the toes, but the *Gremlin* gene is expressed in the webbing cells only in chickens.
   *d.* only duck embryos express *BMP4* in the webbing between the toes.
   *e.* only chicken embryos express *BMP4* in the webbing between the toes.
5. Modularity is important for development because it
   *a.* guarantees that all units of a developing embryo will change in a coordinated way.
   *b.* coordinates the establishment of the anterior–posterior axis of the developing embryo.
   *c.* allows changes in developmental genes to change one part of the body without affecting other parts.
   *d.* guarantees that the timing of gene expression is the same in all parts of a developing embryo.
   *e.* allows organisms to be built up one module at a time.
6. Organisms often respond to environmental signals that accurately predict future conditions by
   *a.* stopping development until the signal changes.
   *b.* altering their development to adapt to the future environment.
   *c.* altering their development such that the resulting adult can produce offspring adapted to the future environment.
   *d.* producing new mutants.
   *e.* developing normally because the predicted conditions may not last long.
7. The process whereby changes in the timing of developmental events can change the form of an organism is called
   *a.* heterochrony.
   *b.* developmental plasticity.
   *c.* adaptation.
   *d.* modularity.
   *e.* mutation.
8. Which of the following is true about temperature determination of sex in some reptiles?
   *a.* It ensures that males and females are produced at different seasons of the year.
   *b.* It ensures that males are faster than females.
   *c.* It acts through the inactivation of the male sex chromosome.
   *d.* There is no evidence that it has evolved because of effects on reproductive success.
   *e.* Temperature effects are due to modifications of concentrations and actions of sex steroids.
9. Which of the following examples of evolutionary change do *not* involve Hox genes?
   *a.* Difference in numbers of legs between bees and centipedes.
   *b.* Difference in number of cervical vertebrae between a goose and a giraffe.
   *c.* Loss of forelegs in snakes.
   *d.* Loss of webbing in the feet of chickens.
   *e.* Location of legs and antennae in *Drosophila*.
10. Parallel phenotypic evolution is likely to occur because
    *a.* closely related organisms typically face similar problems.
    *b.* the conservation of regulatory genes during evolution means that similar traits are likely to evolve repeatedly.
    *c.* many different phenotypes can be produced by a given genotype.
    *d.* phenotypic plasticity, which generates parallel phenotypic evolution, is widespread.
    *e.* evolutionary biologists have looked especially hard to find evidence of it.

## FOR DISCUSSION

1. What environmental influences on development would probably be missed if investigations were confined to unicellular organisms such as bacteria and single-celled eukaryotes?
2. If evolutionary innovations can result from rather simple changes in the timing of expression of a few genes, why have such innovations arisen relatively infrequently during evolution?
3. François Jacob stated that evolution was more like tinkering than engineering. Does the observation that developmental genes have changed little over evolutionary time support his assertion? Why?
4. Despite their major differences, plants and animals share many of the genes that regulate development. What are the implications of this observation for the ways in which humans can respond to the adverse effects of the many substances we release into the environment that cause developmental abnormalities in plants and animals? What kinds of substances are most likely to have such effects? Why?

## ADDITIONAL INVESTIGATION

Figure 20.7 describes an experiment in which the protein Gremlin, which inhibits expression of the *BMP4* gene, was introduced into the foot of a developing chicken. What results would you expect from introducing Gremlin into other parts of a developing chicken? Why? Into what other body parts would it be most informative to introduce Gremlin? If you were particularly interested in parallel phenotypic evolution, what other organisms might you use in these experiments?

# 21 Evidence and Mechanisms of Evolution

## Evolutionary theory leads to better flu vaccines

On November 11, 1918, an armistice agreement signed in France signaled the end of the First World War. But the death toll from four years of war was soon surpassed by the casualties of a massive influenza epidemic that began in the spring of 1918 among soldiers in a U.S. Army barracks. Over the next year and a half, this particular strain of flu virus spread across the globe in a true pandemic that killed more than 50 million people worldwide—more than twice the number of WWI-related combat deaths.

The 1918–1919 pandemic was also noteworthy because the death rate among young adults—who are usually less likely to die from influenza than are the elderly or the very young—was 20 times higher than in influenza epidemics before or since. Why was that particular flu virus so deadly, especially to typically hardy individuals?

Influenza viruses evolve constantly and rapidly. The 1918 strain triggered an especially intense reaction in the human immune system. This overreaction (called a "cytokine storm") meant that people with strong immune systems were likely to be more severely affected. Usually, however, our immune system helps fight flu viruses, and this immune response is the basis of *vaccination* (see Chapter 42). The first flu vaccines were offered in 1945, and since then immunization programs have helped keep the number and severity of outbreaks in check. But flu viruses evolve so rapidly that last year's vaccine is not effective against this year's virus.

Although many of the influenza strains circulating in a given season are closely related, they are not identical. *Genetic variation* ensures that there are always different strains, and these different strains compete with one another. The strains that are best able to escape detection by the immune systems of their hosts are most likely to spread, and thus have an advantage over other strains. But the immune system responds to counteract the virus, and last year's virus loses its advantage. If flu viruses did not evolve, we would become resistant and annual vaccination would be unnecessary.

Our immune system recognizes an influenza infection by detecting a protein called hemagglutinin on the viral surface. New mutations arise rapidly, and genetic sequence changes in the viral genome sometimes result in variation in the structure of hemagglutinin. Variants with altered hemagglutinin structure

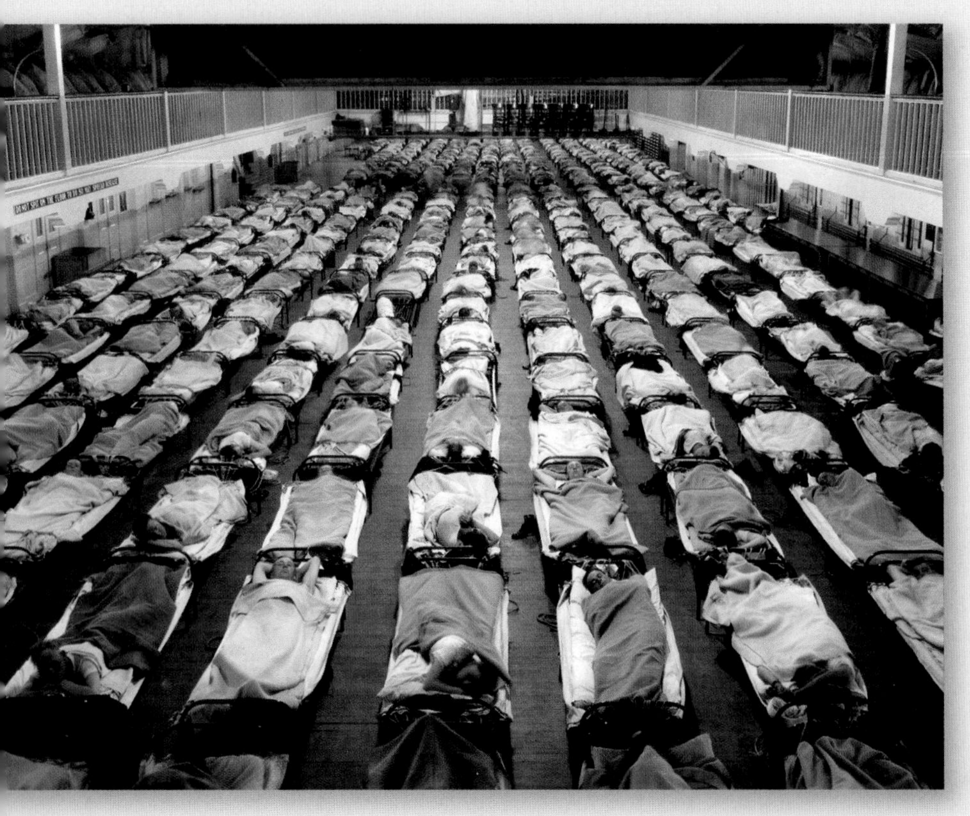

**Deadly Epidemic** So many people were incapacitated during the flu epidemic of 1918–1919 that temporary hospitals had to be set up. Here the beds of flu-stricken patients cover the floor of the Dartmouth College gymnasium during the peak of the epidemic in the United States.

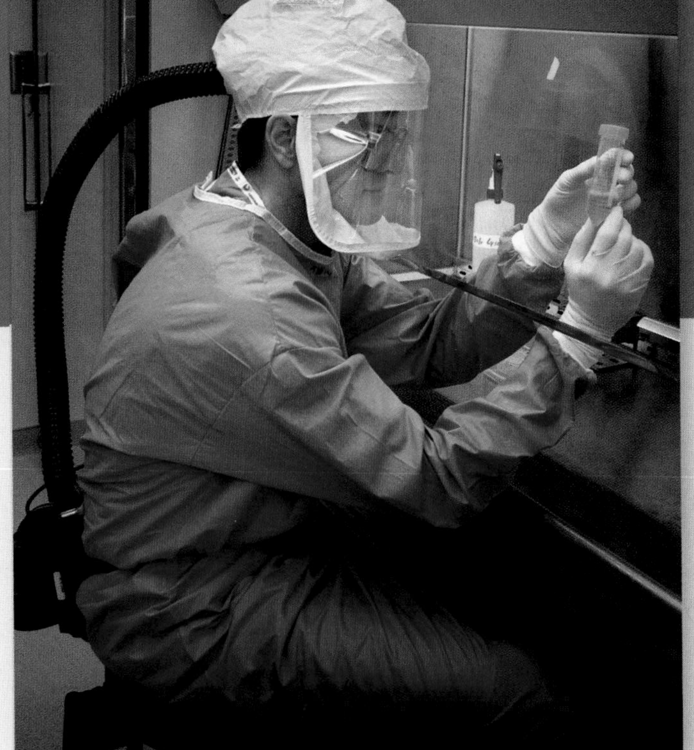

**Reconstructing a Deadly Virus** Terrence Tumpey, of the Centers for Disease Control and Prevention, reconstructed the 1918 influenza virus in his lab to identify the characteristics that made it so deadly.

are more likely to escape detection by our immune system, and thus more likely to survive and replicate. The term *positive selection* describes the evolution of favored changes like those in influenza hemagglutinin protein.

In developing flu vaccines, biologists examine the hemagglutinin gene, particularly at certain sites known to be under positive selection for change. Viruses with the greatest number of changes at these sites are the ones most likely to cause next year's flu epidemic, and therefore are the best targets for new vaccines. Understanding evolutionary theory thus helps us determine the causes and find solutions for a potentially deadly disease.

**IN THIS CHAPTER** we will examine the factual basis of evolution and consider some of the mechanisms that result in evolutionary change. We will see how Charles Darwin developed his ideas on one such mechanism (natural selection). We will discuss the genetic basis of evolution and show how genetic variation in populations is measured. Throughout the chapter, we will discuss ways that evolution can be applied to practical problems and how it helps us understand the diversity of life.

## CHAPTER OUTLINE



## 21.1 What Facts Form the Basis of Our Understanding of Evolution?

The living world is constantly changing. Biologists observe many of these changes directly, both in laboratory experiments and in natural populations. Many other changes are recorded in the fossil record of life. We can measure the rate at which new mutations arise, observe the spread of new genetic variants through a population, and see the effects of genetic change on the form and function of organisms. In other words, evolution is a fact that we can observe directly. Biologists also have accumulated a large body of evidence about *how* evolutionary changes occur, and about *what* evolutionary changes have occurred in the past. The understanding and application of the mechanisms of evolutionary change to biological problems is known as evolutionary theory.

Evolutionary theory has many useful applications, such as the development of influenza vaccines described in the opening of this chapter. We use evolution to study, understand, and treat diseases; to develop better agricultural crops and practices; and to develop industrial processes that produce new molecules with useful properties. Knowledge of evolutionary principles has helped biologists understand how life diversified and how species interact. It also allows us to make predictions about the biological world.

In everyday speech, people tend to use the word "theory" to mean an untested hypothesis, or even a guess. But the term "evolutionary theory" does not refer to any single hypothesis, and it certainly is not guesswork. As used in science, "theory" refers to the entire body of work on the understanding and application of a field of knowledge. When we refer to "gravitational theory," we are not implying that gravity is an untested idea. No one doubts that gravity exists—we can see its effects all around us. Instead, we are referring to our understanding of the mechanisms that result in gravitational pull, and the use of that understanding to make predictions about the interactions of physical objects. Drop this book, and it will fall at a predicable rate, according to gravitational theory.

In a similar manner, when we refer to evolutionary theory, we are referring to our understanding of the mechanisms that result in biological changes in populations over time, and the use of that understanding to interpret changes and interactions of biological organisms. That biological populations evolve through time is not disputed by biologists. We can, and do, observe evolutionary change on a regular basis. We can directly observe the evolution of influenza viruses, but it is evolutionary theory that allows us to apply that information to the problem of developing more effective vaccines.

Several mechanisms of evolutionary change are recognized, and the scientific community is continually expanding its understanding of how and when these mechanisms apply to particular biological problems. Studying the mechanisms of evolution and their innumerable applications constitutes the active and exciting field of evolutionary theory.

### Charles Darwin articulated the principle of natural selection

Today a rich array of geological, morphological, and molecular data support and enhance the factual basis of evolution. In the 1820s, however, it was not yet evident to the young Charles Darwin (or almost anyone else) that life had evolved. Darwin was passionately interested in both geology and natural history—the scientific study of how different organisms function and carry out their lives in nature. Despite these interests, he planned (at his father's behest) to become a doctor. But surgery conducted without anesthesia nauseated Darwin, and he gave up medicine to study at Cambridge University for a career as a clergyman in the Church of England. Always more interested in science than in theology, he gravitated toward scientists on the faculty, especially the botanist John Henslow. In 1831, Henslow recommended Darwin for a position on the H.M.S. *Beagle*, which was preparing for a survey voyage around the world (**Figure 21.1**).

Whenever possible during the 5-year voyage, Darwin (who was often seasick) went ashore to study rocks and to observe and collect plants and animals. He noticed striking differences between the species he saw in South America and those from Europe. He observed that the species of the temperate regions of South America (Argentina and Chile) were more similar to those of tropical South America (Brazil) than they were to temperate European species. When he explored the islands of the

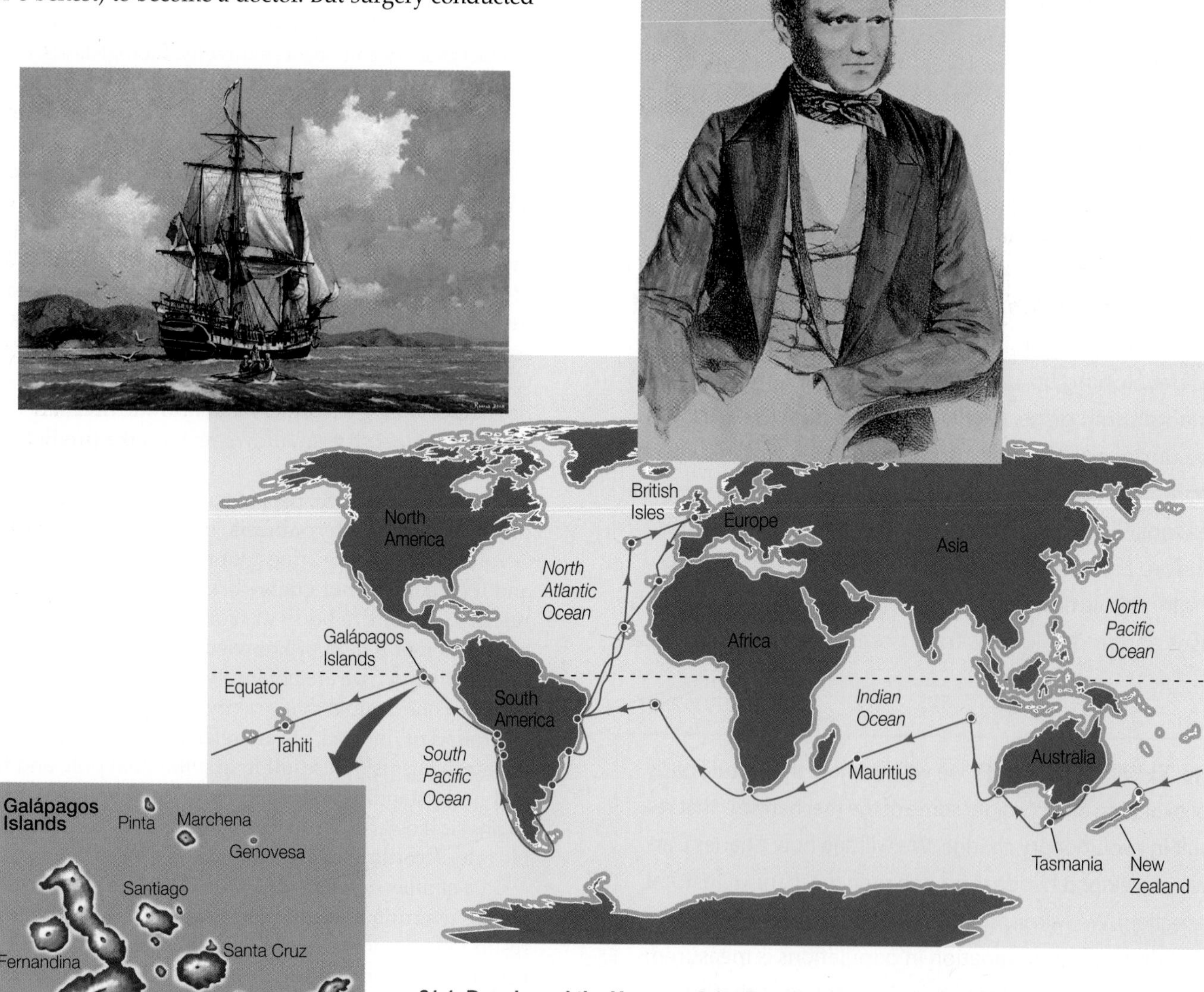


**21.1 Darwin and the Voyage of the Beagle** The mission of H.M.S. *Beagle* was to chart the oceans and collect oceanographic and biological information from around the world. The world map indicates the ship's path; the inset map shows the Galápagos Islands, whose organisms were an important source of Darwin's ideas on natural selection. The portrait is of Charles Darwin at age 24, shortly after the *Beagle* returned to England.

Galápagos Archipelago west of Ecuador, he noted that most of its animal species were found nowhere else, although they were similar to animals found on the mainland of South America. Darwin also observed that the animals of the Galápagos differed from island to island. He postulated that some animals had come to the archipelago from mainland South America and had subsequently undergone different changes on each of the islands. He then wondered what might account for these changes.

When he returned to England in 1836, Darwin continued to ponder his observations. Within a decade he had developed the major elements of an explanatory theory for evolutionary change based on three major propositions:

- Species are not immutable; they change over time.
- Divergent species share a common ancestor.
- The mechanism that produces changes in species is **natural selection**, or the differential survival and reproduction of individuals in a population based on variation in their traits.

The revolutionary assertions in Darwin's first two propositions were that evolution is a historical fact that can be demonstrated to have taken place, and that species are related to one another through common descent. In 1844, he wrote a long essay on natural selection, which he described as the mechanism of evolution (his third proposition), but despite urging from colleagues, he was reluctant to publish it, preferring to assemble more evidence first.

Darwin's hand was forced in 1858 when he received a letter and manuscript from another traveling naturalist, Alfred Russel Wallace, who was studying the biota of the Malay Archipelago. Wallace asked Darwin to evaluate his manuscript, which included an explanation of natural selection almost identical to Darwin's. Darwin was at first dismayed, believing Wallace had preempted his idea. But parts of Darwin's 1844 essay, together with Wallace's manuscript, were presented to the Linnaean Society of London on July 1, 1858, thereby crediting both men for the idea of natural selection. Darwin then worked quickly to finish his own book, *The Origin of Species*, which was published the next year.

Although Darwin and Wallace independently articulated the concept of natural selection, Darwin developed his ideas first. Furthermore, *The Origin of Species* provided exhaustive evidence from many fields to support both natural selection and evolution itself. Thus both concepts are more closely associated with Darwin than with Wallace.

The facts that Darwin used to conceive and develop his explanation of evolution by natural selection were familiar to most contemporary biologists. His insight was to perceive the significance of relationships among these facts. Both Darwin and Wallace were influenced by the ideas of the economist Thomas Malthus, who in 1838 published *An Essay on the Principle of Population*. Malthus argued that because the rate of human population growth is greater than the rate of increase in food production, unchecked growth inevitably leads to famine. Darwin saw parallels throughout nature. He recognized that populations of all species have the potential to rapidly increase in number. To illustrate this point, he used the following example:

> Suppose...there are eight pairs of birds, and that only four pairs of them annually...rear only four young, and that these go on rearing their young at the same rate, then at the end of seven years...there will be 2048 birds instead of the original sixteen.

Such increases are rarely seen in nature, though. Darwin therefore reasoned that death rates in nature must also be high. If they weren't, even the most slowly reproducing species would quickly reach enormous population sizes.

Darwin also observed that although offspring tend to resemble their parents, the offspring of most organisms are not identical to one another or to their parents. He suggested that slight variations among individuals affect the chance that a given individual will survive and reproduce. Darwin called this differential survival and reproduction of individuals natural selection.

Darwin may have used the words "natural selection" because he was familiar with the **artificial selection** of strains with certain desirable traits by animal and plant breeders. Many of Darwin's observations on the nature of variation came from domesticated plants and animals. Darwin was a pigeon breeder, and he knew firsthand the astonishing diversity in color, size, form, and behavior that breeders could achieve (**Figure 21.2**). He recognized

**21.2 Artificial Selection** Charles Darwin raised pigeons as a hobby, and he noted similar forces at work in artificial and natural selection. The "fancy" pigeons shown here represent three of the more than 300 varieties derived from the wild rock dove *Columba livia* (at left) by artificial selection for character traits such as color and feather distribution.

close parallels between selection by breeders and selection in nature. As he argued in *The Origin of Species*,

> How can it be doubted, from the struggle each individual has to obtain subsistence, that any minute variation in structure, habits or instincts, adapting that individual better to the new conditions, would tell upon its vigour and health? In the struggle it would have a better chance of surviving; and those of its offspring which inherited the variation, be it ever so slight, would have a better chance.

That statement, written more than 150 years ago, still stands as a good expression of the process of evolution by natural selection.

It is important to remember that, as Darwin clearly understood, *individuals do not evolve; populations do.* A **population** is a group of individuals of a single species that live and interbreed in a particular geographic area at the same time. A major consequence of the evolution of populations is that their members become adapted to the environments in which they live. But what do biologists mean when they say that an organism is adapted to its environment?

yourBioPortal.com
GO TO Animated Tutorial 21.1 • Natural Selection

### Adaptation has two meanings

In evolutionary biology, **adaptation** refers both to the *processes* by which characteristics that appear to be useful to their bearers evolve—that is, the evolutionary mechanisms that produce them—and to the *characteristics* themselves. With respect to the latter, an adaptation is a phenotypic characteristic that has made it more likely for an organism to survive and reproduce.

Biologists regard an organism as being adapted to a particular environment when they can demonstrate that a slightly different organism reproduces and survives less well in that environment. To understand adaptation, biologists compare the performance of individuals that differ in their traits. For example, biologists can assess the adaptive role of changes to the hemagglutinin protein of influenza viruses, as described in the opening of this chapter. By comparing the survival and proliferation rates of influenza viruses that have different hemagglutinin gene sequences, biologists can study adaptation of the viruses through time.

When Darwin proposed his ideas on evolution by natural selection, he could point to many examples of evolutionary mechanisms operating in nature, but none were supported by experiments. Since then, biologists have conducted thousands of observational and experimental studies that have confirmed the important role of natural selection as a mechanism of evolution. Biologists have also documented changes over time in the genetic composition and morphology of many populations, and our understanding of the mechanisms of inheritance has improved enormously since Darwin's time.

### Population genetics provides an underpinning for Darwin's theory

Darwin had no knowledge of the mechanisms of genetic transmission, and his speculations on the topic proved to be incorrect. Biologists did not have a good understanding of the genetic details of how natural selection works until the field of transmission genetics was established in the early 1900s. At that time, the rediscovery of Gregor Mendel's publications (see Section 12.1) paved the way for the development in the 1930s and 1940s of the field of *population genetics*. As the principles of evolution were integrated with the principles of modern genetics during this period, a new understanding of evolutionary biology—known as the *Modern Synthesis*—emerged. This was when biologists began to study mechanistic aspects of evolution as well as the broad evolutionary patterns that were so evident in nature.

For a population to evolve, its members must possess heritable genetic variation, which is the raw material on which mechanisms of evolution act. In everyday life, we do not directly observe the genetic compositions of organisms. What we see are *phenotypes*, the physical expressions of organisms' genes (including interactions among genes). The features of a phenotype are its *characters*—eye color, for example. The specific form of a character, such as brown eyes, is a *trait*. A **heritable trait** is a characteristic that is at least partly determined by the organism's genes. The genetic constitution that governs a character is called its *genotype*. *A population evolves when individuals with different genotypes survive or reproduce at different rates.*

The field of population genetics has three main goals:

- To explain the patterns and organization of genetic variation
- To explain the origin and maintenance of genetic variation
- To understand the mechanisms that cause changes in allele frequencies in populations

The perspective of population genetics complements the insights into evolutionary processes provided by developmental biology, as described in Chapter 20.

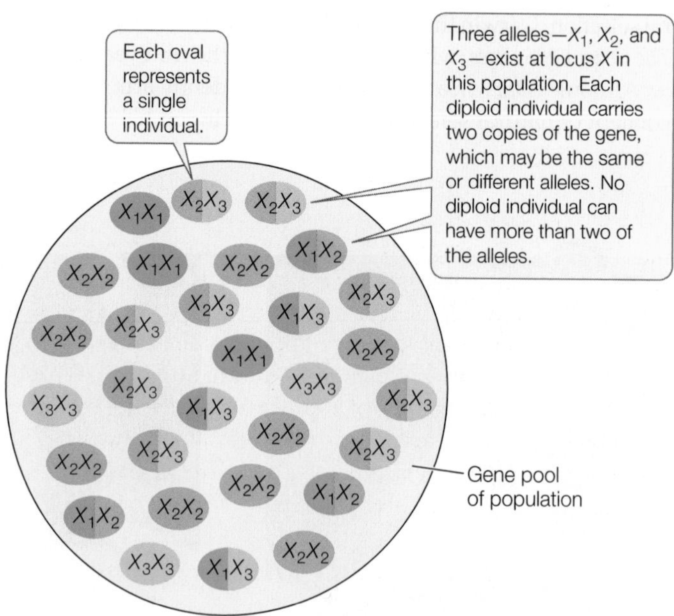


**21.3 A Gene Pool** A gene pool is the sum of all the alleles found in a population, or for a particular locus. This figure shows the gene pool for one locus, $X$. The allele frequencies are 0.20 for $X_1$, 0.50 for $X_2$, and 0.30 for $X_3$.

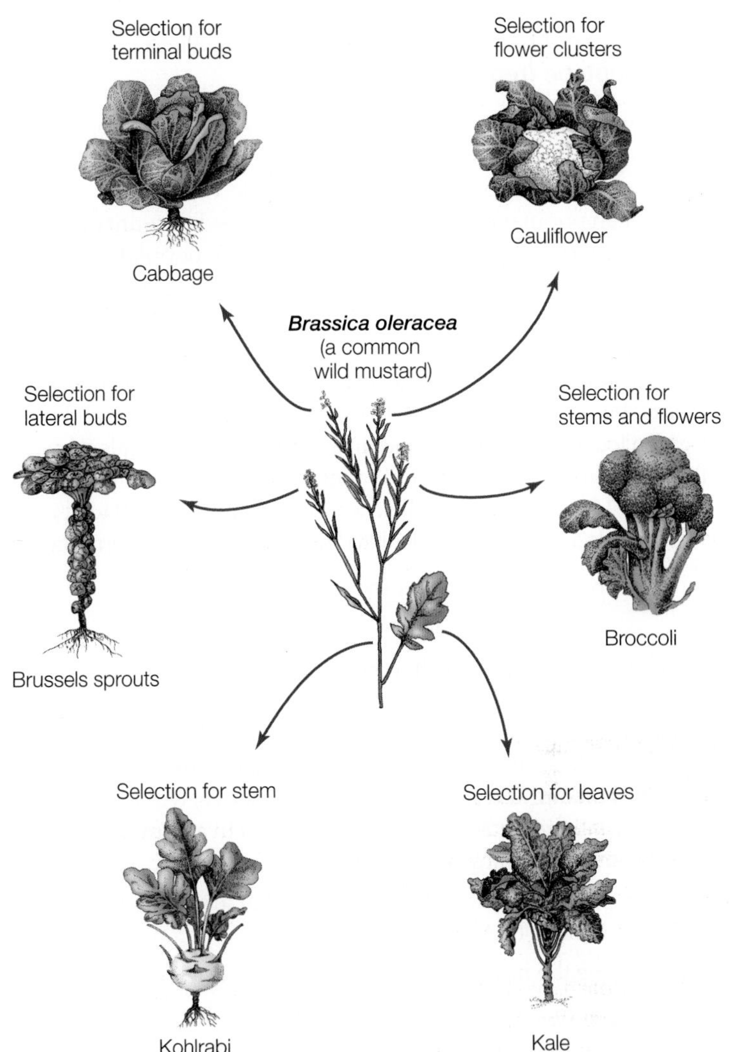


**21.4 Many Vegetables from One Species** All of the crop plants shown here derive from a single wild mustard species. European agriculturalists produced these crop species by selecting and breeding plants with unusually large buds, stems, leaves, or flowers. The results substantiate the vast amount of variation present in a gene pool.

Different forms of a gene, known as **alleles**, may exist at a particular locus. At any particular locus, a single individual has only some of the alleles found in the population to which it belongs (**Figure 21.3**). The sum of all copies of all alleles at all loci found in a population constitutes its **gene pool**. (We can also refer to the "gene pool" for a particular locus or loci.) The gene pool is the source of the genetic variation that produces the phenotypic traits on which natural selection acts. To understand evolution and the role of natural selection, we need to know how much genetic variation populations have, what the sources of that genetic variation are, and how genetic variation changes in populations over space and time.

## Most populations are genetically variable

Nearly all populations have genetic variation for many characters. Artificial selection on different characters in a single European species of wild mustard produced many important crop plants (**Figure 21.4**). Agriculturalists could achieve these results because the original mustard population had genetic variation for the characters of interest.

Laboratory experiments also demonstrate the existence of considerable genetic variation in populations. In one such experiment, investigators attempted to breed populations of the fruit fly *Drosophila melanogaster* with high or low numbers of bristles on their abdomens from an initial population with intermediate numbers of bristles. After 35 generations, all flies in both the high- and low-bristle lineages had bristle numbers that fell well outside the range found in the original population (**Figure 21.5**). Thus there must have been considerable genetic variation in the original fruit fly population on which selection could act.

Studying the genetic basis of natural selection is difficult because genotypes alone do not determine all phenotypes. With dominance, for example, a particular phenotype can be produced by more than one genotype (e.g., *AA* and *Aa* individuals may be phenotypically identical). Also, as we describe in Section 20.5, a given genotype can produce different phenotypes, depending on the environment encountered during development. For example, the cells of all the leaves on a tree or shrub are usually genetically identical, yet leaves of the same plant often differ in shape and size depending, for example, on the amount of ambient light they receive.

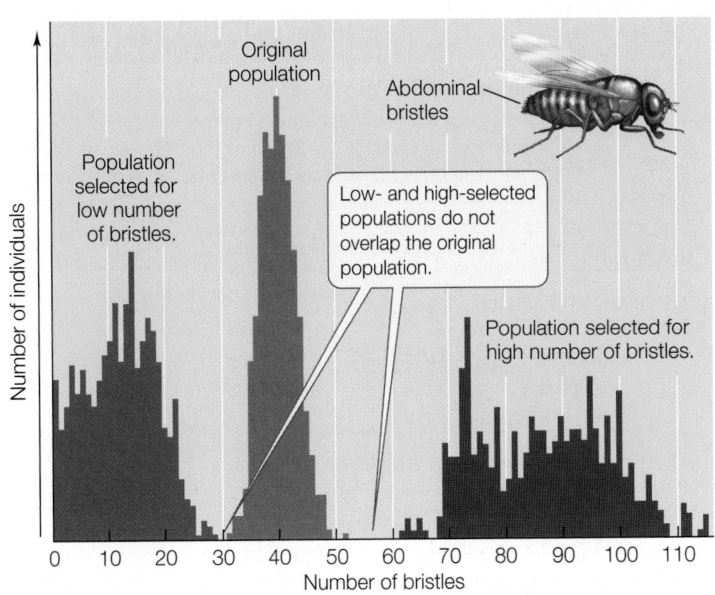


**21.5 Artificial Selection Reveals Genetic Variation** In experiments subjecting *Drosophila melanogaster* to artificial selection for bristle number, this trait evolved rapidly. The graphs show the number of flies with different numbers of bristles after 35 generations of artificial selection, which clearly diverged from the bristle numbers present in the original population (the blue bars in the center).

### Evolutionary change can be measured by allele and genotype frequencies

Allele frequencies are usually estimated in locally interbreeding groups, called Mendelian populations, within a geographic population of a species. To measure allele frequencies in a Mendelian population precisely, we would need to count every allele at every locus in every individual in the population. By doing so, we could determine the frequencies of all alleles in the population. The word *frequency* in this case refers to an allele's proportion in the gene pool at a particular locus.

Fortunately, we do not need to make complete measurements, because we can reliably estimate allele frequencies for a given locus by counting alleles in a sample of individuals from the population. The sum of all allele frequencies at a locus is equal to 1, so measures of allele frequency range from 0 to 1.

An allele's frequency is calculated using the following formula:

$$p = \frac{\text{number of copies of the allele in the population}}{\text{sum of alleles in the population}}$$

If only two alleles (we'll call them $A$ and $a$) for a given locus are found among the members of a diploid population, they may combine to form three different genotypes: $AA$, $Aa$, and $aa$. Such a population is said to be *polymorphic* at that locus, since there is more than one allele. Using the formula above, we can calculate the relative frequencies of alleles $A$ and $a$ in a population of $N$ individuals as follows:

- Let $N_{AA}$ be the number of individuals that are homozygous for the $A$ allele ($AA$).
- Let $N_{Aa}$ be the number that are heterozygous ($Aa$).
- Let $N_{aa}$ be the number that are homozygous for the $a$ allele ($aa$).

Note that $N_{AA} + N_{Aa} + N_{aa} = N$, the total number of individuals in the population, and that the total number of copies of both alleles present in the population is $2N$, because each individual is diploid. Each $AA$ individual has two copies of the $A$ allele, and each $Aa$ individual has one copy of the $A$ allele. Therefore, the total number of $A$ alleles in the population is $2N_{AA} + N_{Aa}$. Similarly, the total number of $a$ alleles in the population is $2N_{aa} + N_{Aa}$.

If $p$ represents the frequency of $A$, and $q$ represents the frequency of $a$, then

$$p = \frac{2N_{AA} + N_{Aa}}{2N}$$

and

$$q = \frac{2N_{aa} + N_{Aa}}{2N}$$

**Figure 21.6** shows how these formulas can be used to calculate allele frequencies in two hypothetical populations, each containing 200 diploid individuals. Population 1 has mostly homozygotes (90 $AA$, 40 $Aa$, and 70 $aa$), whereas population 2 has mostly heterozygotes (45 $AA$, 130 $Aa$, and 25 $aa$).

The calculations in Figure 21.6 demonstrate two important points. First, notice that for each population, $p + q = 1$, which means that $q = 1 - p$. So when there are only two alleles at a given locus in a population, we can calculate the frequency of one allele and then obtain the second allele's frequency by subtraction. If there is only one allele at a given locus in a population, its frequency is 1: the population is then *monomorphic* at that locus, and the allele is said to be *fixed*.

The second thing to notice is that population 1 (consisting mostly of homozygotes) and population 2 (consisting mostly of heterozygotes) have the same allele frequencies for $A$ and $a$. Thus they have the same gene pool for this locus. Because the alleles in the gene pool are distributed differently among individuals, however, the *genotype frequencies* of the two populations differ. Genotype frequencies are calculated as the number of individuals that have a given genotype divided by the total number of individuals in the population. Using the numbers in Figure 21.6, the genotype frequencies in population 1 would be 0.45 $AA$, 0.20 $Aa$, and 0.35 $aa$.

The frequencies of different alleles at each locus and the frequencies of different genotypes in a Mendelian population de-

## TOOLS FOR INVESTIGATING LIFE

**21.6 Calculating Allele Frequencies**

Allele frequencies for any gene pool can be calculated using the equations in panel 1. When these equations are applied to the populations in panel 2, we find that the *frequencies* of alleles A and a in the two populations are the same, but the alleles are distributed differently between heterozygous and homozygous genotypes.

**1** Determine the allele frequencies in the population.

**In any population:**

$$\text{Frequency of allele } A = p = \frac{2N_{AA} + N_{Aa}}{2N} \qquad \text{Frequency of allele } a = q = \frac{2N_{aa} + N_{Aa}}{2N}$$

where $N$ is the total number of individuals in the population.

**2** Compute allele frequencies for different populations.

**For population 1 (mostly homozygotes):**

$N_{AA} = 90$, $N_{Aa} = 40$, and $N_{aa} = 70$

so

$$p = \frac{180 + 40}{400} = 0.55$$

$$q = \frac{140 + 40}{400} = 0.45$$

**For population 2 (mostly heterozygotes):**

$N_{AA} = 45$, $N_{Aa} = 130$, and $N_{aa} = 25$

so

$$p = \frac{90 + 130}{400} = 0.55$$

$$q = \frac{50 + 130}{400} = 0.45$$

scribe that population's **genetic structure**. Allele frequencies measure the amount of genetic variation in a population; genotype frequencies show how a population's genetic variation is distributed among its members. Other measures, such as the proportion of polymorphic loci, are also used to measure variation in populations. With these measurements, it becomes possible to consider how the genetic structure of a population changes or remains the same over generations—that is, to measure evolutionary change.

### The genetic structure of a population changes over time, unless certain restrictive conditions exist

In 1908, the British mathematician Godfrey Hardy and the German physician Wilhelm Weinberg independently deduced the conditions that must prevail if the genetic structure of a population is to remain the same over time. If the conditions they identified do not exist, then evolution will occur. The resulting principle, known as **Hardy–Weinberg equilibrium**, is a cornerstone of population genetics. Hardy–Weinberg equilibrium describes a model in which allele frequencies do not change across generations and genotype frequencies can be predicted from allele frequencies (**Figure 21.7**). The principles of Hardy–Weinberg equilibrium apply only to sexually reproducing organisms. Several conditions must be met for a population to be at Hardy–Weinberg equilibrium:

- *Mating is random.* Individuals do not preferentially choose mates with certain genotypes.
- *Population size is infinite.* The larger a population, the smaller will be the effect of **genetic drift**—random (chance) fluctuations in allele frequencies from one generation to another.
- *There is no* **gene flow** (movement of individuals into or out of the population, or reproductive contact with other populations).
- *There is no mutation.* There is no change to alleles in the population, and no new alleles are added to change the gene pool.
- *Selection does not affect the survival of particular genotypes.* There is no differential survival of individuals with different genotypes.

If these "ideal" conditions hold, two major consequences follow. First, the frequencies of alleles at a locus remain constant from generation to generation. Second, following one generation of random mating, the genotype frequencies occur in the following proportions:

| Genotype | *AA* | *Aa* | *aa* |
|---|---|---|---|
| Frequency | $p^2$ | $2pq$ | $q^2$ |

Consider a population that is *not* in Hardy–Weinberg equilibrium, such as generation I of Figure 21.7. This could occur, for example, if the initial population is founded by migrants from several other populations, thus violating the Hardy–Weinberg assumption of no gene flow. In this example, "generation I" has more homozygous individuals and fewer heterozygous individuals than would be expected under Hardy–Weinberg equilibrium (a condition known as *heterozygote deficiency*).

Generation I (Founder population)

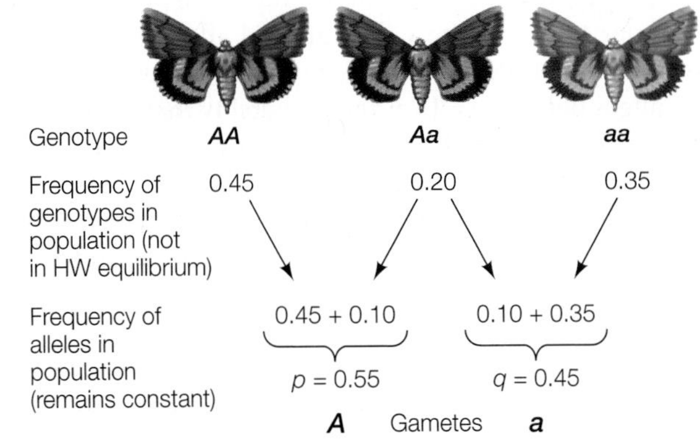


Generation II (Hardy–Weinberg equilibrium restored)

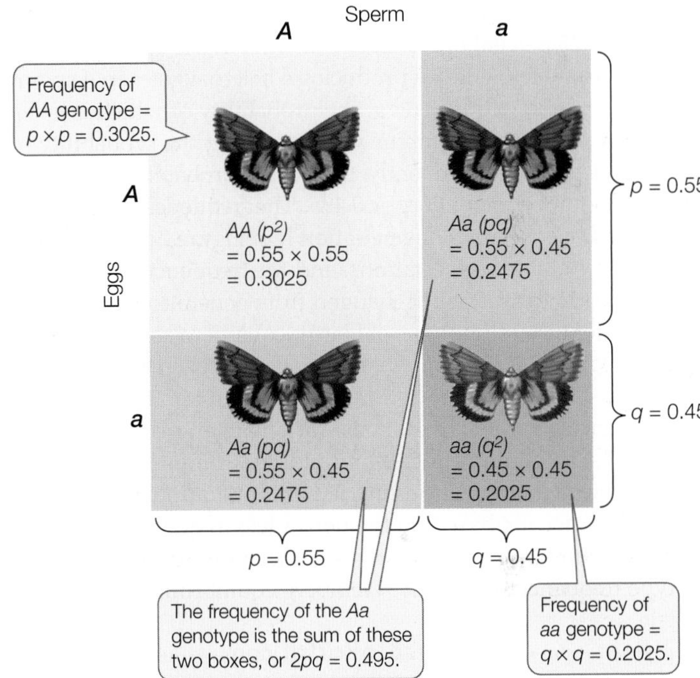


**21.7 One Generation of Random Mating Restores Hardy–Weinberg Equilibrium** Generation I of this example population is founded by migrants from several source populations, and so is not initially in Hardy–Weinberg equilibrium. After one generation of random mating, the allele frequencies are unchanged and the genotype frequencies return to Hardy–Weinberg expectations. The length of the sides of each rectangle are proportional to the allele frequencies in the population; the areas of the rectangles are proportional to the genotype frequencies.

Even with a starting population that is not in Hardy–Weinberg equilibrium, we would predict that after a single generation of random mating, and without violating the other Hardy–Weinberg assumptions, the *allele frequencies* will remain unchanged but the *genotype frequencies* will return to Hardy–Weinberg expectations. Let's explore why this is true.

In generation I of Figure 21.7, the frequency of the *A* allele ($p$) is 0.55. Because we assume that individuals select mates at random, without regard to their genotype, gametes carrying *A* or $a$ combine at random—that is, as predicted by the allele frequencies of $p$ and $q$. Thus in this example, the probability that a particular sperm or egg will bear an $A$ allele is 0.55. In other words, 55 out of 100 randomly sampled sperm or eggs will bear an $A$ allele. Because $q = 1 - p$, the probability that a sperm or egg will bear an $a$ allele is $1 - 0.55 = 0.45$. (You may wish to review the discussion of probability in Section 12.1.)

To obtain the probability of two $A$-bearing gametes coming together at fertilization, we multiply the two independent probabilities of their occurring separately:

$$p \times p = p^2 = (0.55)^2 = 0.3025$$

Therefore, 0.3025, or 30.25 percent, of the offspring in generation II will have homozygous genotype $AA$. Similarly, the probability of bringing together two $a$-bearing gametes is

$$q \times q = q^2 = (0.45)^2 = 0.2025$$

Thus 20.25 percent of generation II will have the $aa$ genotype.

There are two ways of producing a heterozygote: an $A$ sperm may combine with an $a$ egg, the probability of which is $p \times q$; or an $a$ sperm may combine with an $A$ egg, the probability of which is $q \times p$. Consequently, the overall probability of obtaining a heterozygote is $2pq$, or 0.495. The frequencies of the $AA$, $Aa$, and $aa$ genotypes in generation II of Figure 21.7 are now at Hardy–Weinberg expectations, and the frequencies of the two alleles ($p$ and $q$) have not changed from generation I.

Under the assumptions of Hardy–Weinberg equilibrium, allele frequencies $p$ and $q$ remain constant for each generation. If Hardy–Weinberg assumptions are violated and the genotype frequencies in the parental generation are altered (say, by the loss of a large number of $AA$ individuals from the population), then the allele frequencies in the next generation would be altered. However, based on the new allele frequencies, another generation of random mating is sufficient to restore the genotype frequencies to Hardy–Weinberg equilibrium.

**yourBioPortal.com**

GO TO Animated Tutorial 21.2 • Hardy–Weinberg Equilibrium

### Deviations from Hardy–Weinberg equilibrium show that evolution is occurring

You probably have realized that populations in nature never meet the stringent conditions necessary to be at Hardy–Weinberg equilibrium. Why, then, is this model considered so important for the study of evolution? There are two reasons. First, the equation is often useful for predicting the approximate genotype frequencies of a population from its allele frequencies. Second—and crucially—the model describes the conditions required for there to be *no* evolution in a population.

Few if any of the Hardy–Weinberg model's conditions are ever met completely in real populations, and allele frequencies in all populations do in fact change through time—that is, populations *do* evolve. The specific patterns of deviation from Hardy–Weinberg equilibrium can help us identify the various mechanisms of evolutionary change.

### 21.1 RECAP

**Evolutionary change is directly observable in biological populations. Genetic variation is needed for evolutionary change to occur, and natural selection is one of the major mechanisms that acts on that variation and results in evolution.**

- Can you articulate the principle of natural selection? The two meanings of adaptation? See pp. 442–444
- How does calculating allele frequencies allow us to measure evolutionary change? See pp. 446–447 and Figure 21.6
- Why is the concept of Hardy–Weinberg equilibrium important even though the assumptions on which it is based are never completely met in nature? See pp. 447–448

We have briefly outlined Charles Darwin's vision of natural selection and adaptation and explained the mathematical basis of Hardy–Weinberg equilibrium and its importance for studying evolution. We'll now examine some of the forces that cause populations to deviate from equilibrium—the mechanisms of evolutionary change.

## 21.2 What Are the Mechanisms of Evolutionary Change?

Evolutionary mechanisms are forces that change the genetic structure of a population. Hardy–Weinberg equilibrium is a null hypothesis that assumes those forces are absent. The known evolutionary mechanisms include mutation, gene flow, genetic drift, nonrandom mating, and selection—each of which contradicts one of the five basic assumptions of Hardy–Weinberg equilibrium. We have already discussed Darwin's principal explanation for evolution, namely natural selection. Although natural selection is in many cases an important component of evolution, even Darwin recognized that it was not the only mechanism of evolution, and many additional evolutionary forces have been discovered since Darwin's time. Here we discuss some of the other mechanisms that result in evolution.

### Mutations generate genetic variation

The origin of genetic variation is mutation. A **mutation**, as Section 14.6 describes, is any change in the nucleotide sequences of an organism's DNA. The process of DNA replication is not perfect, and changes appear almost every time a genome is replicated. Mutations occur randomly with respect to an organism's adaptive needs; it is selection acting on this random variation that results in adaptation. Most mutations are either harmful to their bearers or neutral. A few are beneficial, however, and previously harmful or neutral alleles may become advantageous if conditions change. In addition, mutations can restore to a population genetic variation that other evolutionary processes have

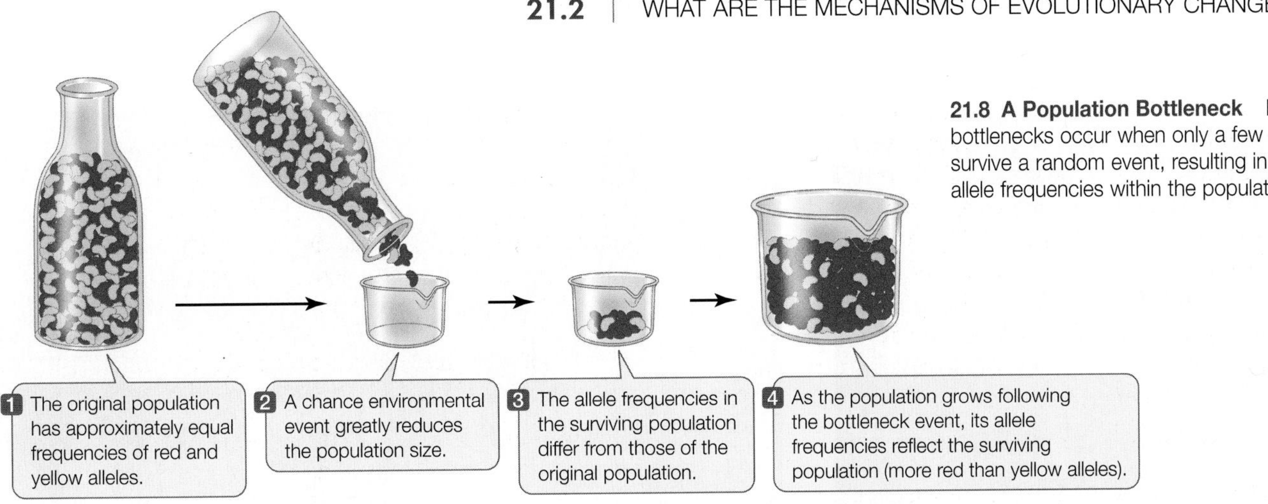

1 The original population has approximately equal frequencies of red and yellow alleles.
2 A chance environmental event greatly reduces the population size.
3 The allele frequencies in the surviving population differ from those of the original population.
4 As the population grows following the bottleneck event, its allele frequencies reflect the surviving population (more red than yellow alleles).


**21.8 A Population Bottleneck** Population bottlenecks occur when only a few individuals survive a random event, resulting in a shift in allele frequencies within the population.

removed. Thus mutations both create and help maintain genetic variation in populations.

Mutation rates can be high, as we saw with the influenza viruses described in the opening of this chapter, but in many organisms the mutation rate is very low (on the order of $10^{-8}$ to $10^{-9}$ changes per base pair of DNA per generation). Even low overall mutation rates, however, are sufficient to create considerable genetic variation, because each of a large number of genes may change, and populations often contain large numbers of individuals. For example, if the probability of a point mutation (an addition, deletion, or substitution of a single base) were $10^{-9}$ per base pair per generation, then each human gamete, the DNA of which contains $3 \times 10^9$ base pairs, would average three new point mutations ($3 \times 10^9 \times 10^{-9} = 3$)—and each zygote would carry an average of six new mutations. The current human population of about 7 billion people would be expected to carry about 42 billion new mutations that were not present one generation earlier. So even though the mutation rate in humans is quite low, human populations still contain enormous genetic variation on which selection can act.

One of the conditions for Hardy–Weinberg equilibrium is that there be no mutation. Although this condition is never strictly met, the rate at which mutations arise at a single locus is usually so low that mutations by themselves result in only small deviations from Hardy–Weinberg equilibrium. If large deviations are found, it is usually appropriate to dismiss mutation as the cause and to look for evidence of other evolutionary mechanisms acting on the population.

### Gene flow may change allele frequencies

Few populations are completely isolated from other populations of the same species. Migration of individuals and movements of gametes between populations—a phenomenon called gene flow—can change allele frequencies in a population. If the arriving individuals survive and reproduce in their new location, they may add new alleles to the population's gene pool, or they may change the frequencies of alleles already present if they come from a population with different allele frequencies. For a population to be at Hardy–Weinberg equilibrium, there must be no gene flow from populations with different allele frequencies.

### Genetic drift may cause large changes in small populations

In small populations, genetic drift—random changes in allele frequencies from one generation to the next—may produce large changes in allele frequencies over time. Harmful alleles may increase in frequency, and rare advantageous alleles may be lost. Even in large populations, genetic drift can influence the frequencies of alleles that do not affect the survival and reproductive rates of their bearers.

As an example, suppose we cross *Aa* × *Aa* fruit flies to produce an $F_1$ population in which $p = q = 0.5$ and in which the genotype frequencies are 0.25 *AA*, 0.50 *Aa*, and 0.25 *aa*. If we randomly select 4 individuals (= 8 copies of the gene) from the $F_1$ population to produce the $F_2$ generation, the allele frequencies in this small sample population may differ markedly from $p = q = 0.5$. If, for example, we happen by chance to draw 2 *AA* homozygotes and 2 heterozygotes (*Aa*), the allele frequencies in the sample will be $p = 0.75$ (6 out of 8) and $q = 0.25$ (2 out of 8). If we replicate this experiment 1,000 times, one of the two alleles will be missing entirely from about 8 of the 1,000 sample populations.

The same principles operate when a population is reduced dramatically in size. Populations that are normally large may occasionally pass through a period in which only a small number of individuals survive, a situation known as a **population bottleneck**. During population bottlenecks, genetic variation can be reduced by genetic drift. This is illustrated in **Figure 21.8**, in which red and yellow beans represent two different alleles of a gene. Most of the "surviving" beans in the small sample taken from the original population are, just by chance, red, so the new population has a much higher frequency of red beans than the previous generation had. In a real population, the allele frequencies would be described as having "drifted."

A population forced through a bottleneck is likely to lose much of its genetic variation. For example, when Europeans first arrived in North America, millions of greater prairie-chickens (*Tympanuchus cupido*) inhabited the prairies. As a result of hunting and habitat destruction by the new settlers, the Illinois population of this species plummeted from about 100 million birds in 1900 to fewer than 50 in the 1990s (**Figure 21.9A**). A comparison of DNA from birds collected in Illinois during the middle of the twentieth century with DNA from the surviving pop-

(A) *Tympanuchus cupido*

(B) *Washingtonia filifera*

**21.9 Species with Low Genetic Variation** (A) Greater prairie-chickens in Illinois lost most of their genetic variation when the population crashed from millions to fewer than 50 individuals. (B) The California fan palm, whose range has been reduced to a small area of southern California and neighboring Mexico, has little genetic variation.

ulation in the 1990s showed that Illinois prairie-chickens have lost most of their genetic diversity. The remaining population is experiencing low reproductive success. Similarly, the California fan palm (*Washingtonia filifera*) was once widespread in California and Mexico; today it is restricted to a few oases in extreme southern California and adjacent Mexico (**Figure 21.9B**). The species has little genetic variation: an average individual is heterozygous at fewer than 1 percent of its loci.

Genetic drift can have similar effects when a few pioneering individuals colonize a new region. Because of its small size, the colonizing population is unlikely to have all the alleles found among members of its source population. The resulting change in genetic variation, called a **founder effect**, is equivalent to that in a large population reduced by a bottleneck. For example, the current population of the pitcher plant *Sarracenia purpurea* on a small island in central Ohio arose from a single individual that was planted there in 1912. Today the population has only one detectable polymorphic locus in its entire genome.

Scientists were given an opportunity to study the genetic composition of founding populations when *Drosophila subobscura*, a well-studied species of fruit fly native to Europe, was discovered near Puerto Montt, Chile (in 1978), and at Port Townsend, Washington (in 1982). The *D. subobscura* founders probably reached Chile from Europe on a ship, and a few flies carried north from Chile on another ship founded the North American population. In both South and North America, populations of the flies grew rapidly and expanded their ranges. Today in North America, *D. subobscura* ranges from British Columbia to central California. In Chile it has spread across 15 degrees of latitude (**Figure 21.10**).

European populations of *D. subobscura* have 80 different *chromosomal inversions*, but the North and South American populations have only a subset of 20 of these inversions—and they are the same 20 on both continents. North and South American populations also have lower allele diversity at certain enzyme-producing genes compared with European populations. Only those alleles that have a frequency higher than 0.10 in European populations are present in the Americas. Thus, as expected for a small founding population, only a small part of the total genetic variation found in Europe reached the Americas. Geneticists estimate that somewhere between 4 and 100 flies founded the North and South American populations.

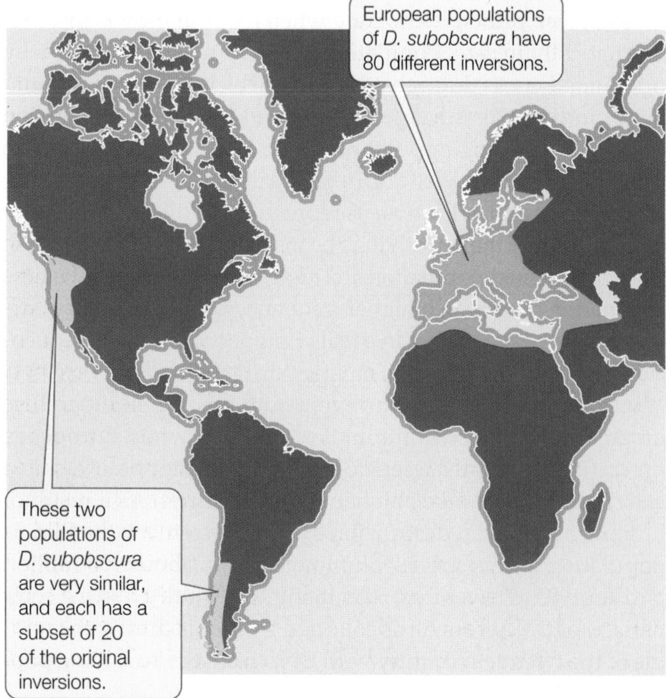

European populations of D. subobscura have 80 different inversions.
These two populations of D. subobscura are very similar, and each has a subset of 20 of the original inversions.


**21.10 A Founder Effect** Populations of the fruit fly *Drosophila subobscura* in North and South America contain less genetic variation than the European populations from which they came, as measured by the number of chromosome inversions in each population. Within two decades of arriving in the Americas, *D. subobscura* populations had increased dramatically and spread widely in spite of their reduced genetic variation.

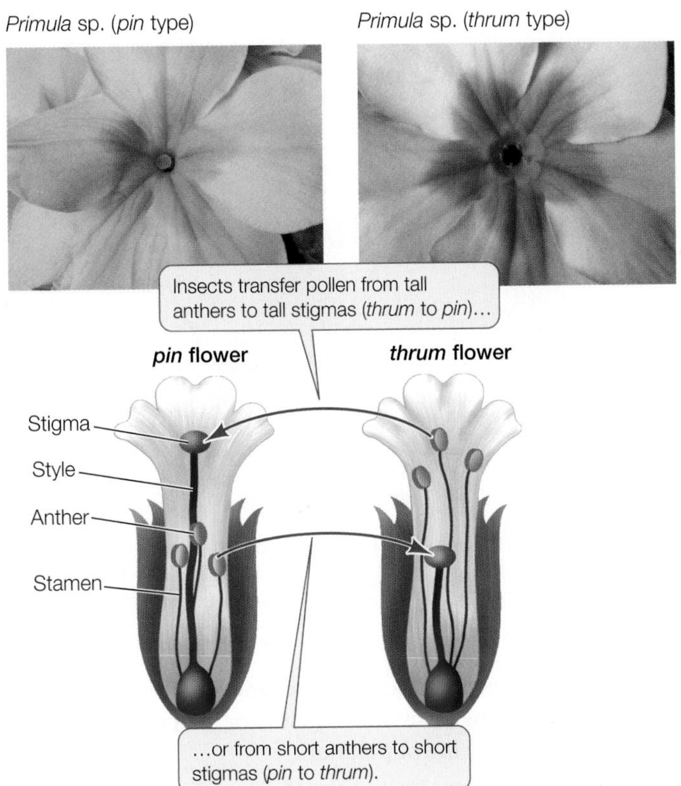
Primula sp. (pin type)
Primula sp. (thrum type)
Insects transfer pollen from tall anthers to tall stigmas (thrum to pin)...
pin flower
thrum flower
Stigma
Style
Anther
Stamen
...or from short anthers to short stigmas (pin to thrum).

**21.11 Flower Structure Fosters Nonrandom Mating** Differing floral structure within the same plant species, as illustrated by this primrose, ensures that pollination usually occurs between individuals of different genotypes.

### Nonrandom mating can change genotype frequencies

Mating patterns may alter genotype frequencies if individuals in a population do not choose mates at random. For example, if they mate preferentially with individuals of the same genotype, then homozygous genotypes will be overrepresented and heterozygous genotypes underrepresented relative to Hardy–Weinberg expectations. Alternatively, individuals may mate primarily or exclusively with individuals of different genotypes.

Nonrandom mating is seen in some plant species, such as primroses (genus *Primula*), in which individual plants bear flowers of only one of two different types. One type, known as *pin*, has a long style (the stalk that supports the stigma, where pollen is received) and short stamens (the stalks ending in anthers, where pollen is produced). The other type, known as *thrum*, has a short style and long stamens (**Figure 21.11**). In many species with this reciprocal arrangement, pollen from one flower type can fertilize only flowers of the other type. Pollen grains from *pin* and *thrum* flowers are deposited on different parts of the bodies of insects that visit the flowers. When the insects visit other flowers, pollen grains from *pin* flowers are most likely to come into contact with stigmas of *thrum* flowers, and vice versa.

Self-fertilization (*selfing*), another form of nonrandom mating, is common in many groups of organisms, especially plants. Selfing reduces the frequencies of heterozygous individuals from Hardy–Weinberg equilibrium and increases the frequencies of homozygotes, but it does not change allele frequencies.

*Sexual selection* is a particularly important form of nonrandom mating that *does* change allele frequencies and also often results in the evolution of significant differences between males and females of a species. We will discuss this important evolutionary mechanism in detail in the next section.

## 21.2 RECAP

**Evolutionary mechanisms are processes that change the genetic structure of a population. Known evolutionary mechanisms include mutation, gene flow, genetic drift, nonrandom mating, and selection.**

- Why do mutations, by themselves, result in only small deviations from Hardy–Weinberg equilibrium? **See pp. 448–449**
- Explain how genetic drift can cause large changes in small populations. **See pp. 449–450 and Figure 21.8**
- Why is it that some types of nonrandom mating alter genotype frequencies but not allele frequencies? **See p. 451**

The evolutionary mechanisms discussed so far influence the frequencies of alleles and genotypes in populations. Although all of these processes influence the course of biological evolution, only natural selection results in adaptation. For adaptation to occur, individuals that differ in heritable traits must survive and reproduce with different degrees of success.

## 21.3 How Does Natural Selection Result in Evolution?

Although evolution is defined as changes in the gene frequencies of a population from one generation to the next, natural selection acts on the *phenotype*—the physical features expressed by an organism with a given genotype—rather than directly on the genotype. The reproductive contribution of a phenotype to subsequent generations relative to the contributions of other phenotypes is called its **fitness**.

Changes in reproductive rate do not necessarily change the genetic structure of a population. For example, if all individuals in a population experience the same increase in reproductive rate (during an environmentally favorable year, for instance), the genetic structure of the population will not change. Changes in numbers of offspring are responsible for increases and decreases in the *size* of a population, but only changes in the *relative* success of different phenotypes in a population lead to changes in allele frequencies from one generation to the next. The fitness of individuals of a particular phenotype is a function of the probability of those individuals surviving multiplied by the average number of offspring they produce over their lifetimes. In other words, the *fitness of a phenotype is determined by the relative rates of survival and reproduction of individuals with that phenotype.*

### Natural selection can change or stabilize populations

To simplify our discussion until now, we have considered only characters influenced by alleles at a single locus. As we describe in Section 12.3, however, most characters are influenced by alleles at more than one locus. Such characters are likely to show quantitative rather than qualitative variation. For example, the distribution of body sizes of individuals in a population, a character that is influenced by genes at many loci as well as by the environment, is likely to resemble the bell-shaped curves shown in the right-hand column of Figure 21.12.

Natural selection can act on characters with quantitative variation in any one of several different ways, producing quite different results:

- **Stabilizing selection** preserves the average characteristics of a population by favoring average individuals.

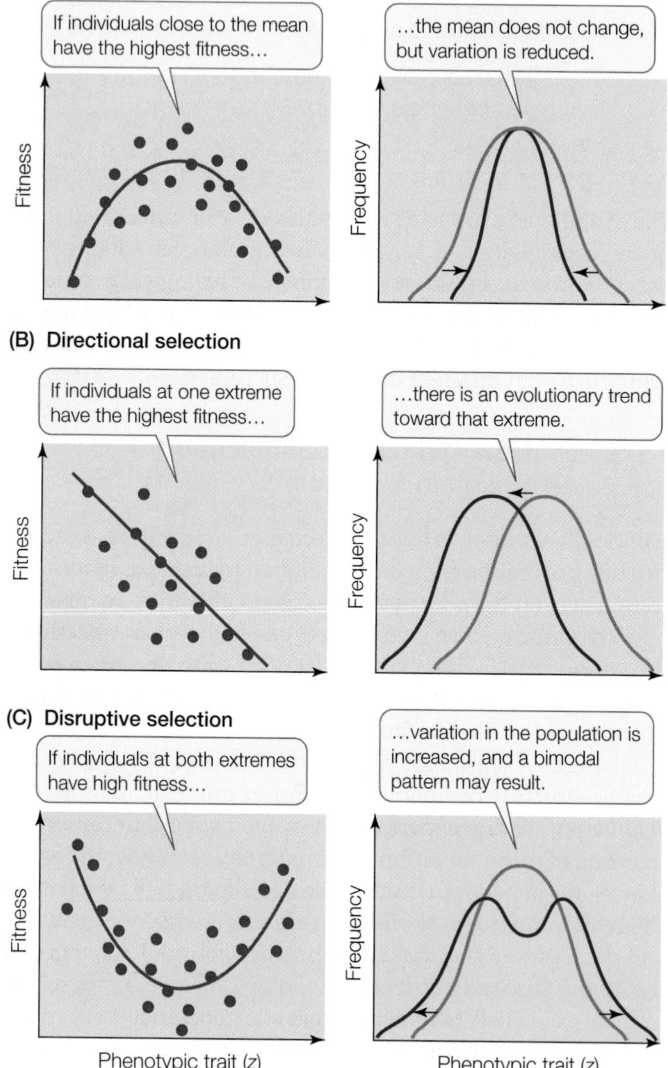


**21.12 Natural Selection Can Operate in Several Ways** The graphs in the left-hand column show the fitness of individuals with different phenotypes of the same trait. The graphs on the right show the distribution of the phenotypes in the population before (light green) and after (dark green) the influence of selection.

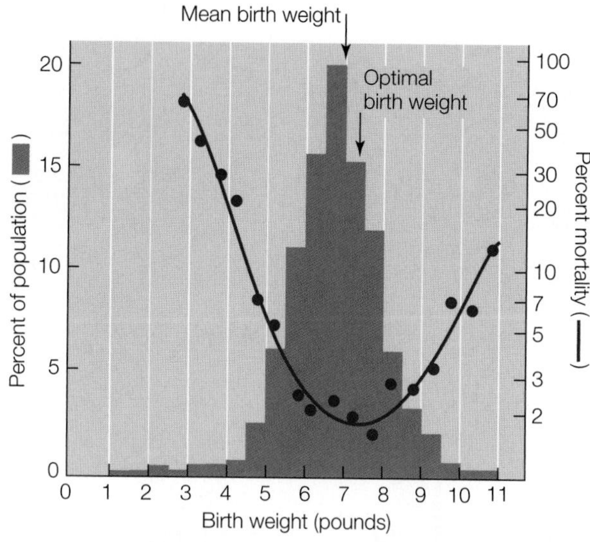


**21.13 Human Birth Weight Is Influenced by Stabilizing Selection** Babies that weigh more or less than average are more likely to die soon after birth than babies with weights close to the population mean.

- **Directional selection** changes the characteristics of a population by favoring individuals that vary in one direction from the mean of the population.
- **Disruptive selection** changes the characteristics of a population by favoring individuals that vary in both directions from the mean of the population.

**STABILIZING SELECTION** If the smallest and largest individuals in a population contribute fewer offspring to the next generation than do individuals closer to the average size, then stabilizing selection is operating on size (**Figure 21.12A**). Stabilizing selection reduces variation in populations, but it does not change the mean. Natural selection frequently acts in this way, countering increases in variation brought about by sexual recombination, mutation, or migration. Rates of evolution in many species are slow because natural selection is often stabilizing. Stabilizing selection operates, for example, on human birth weight. Babies born lighter or heavier than the population mean die at higher rates than babies whose weights are close to the mean (**Figure 21.13**). In discussions of specific genes, stabilizing selection is often called *purifying selection*, because there is selection against any deleterious mutations to the usual gene sequence.

**DIRECTIONAL SELECTION** Directional selection is operating when individuals at one extreme of a character distribution contribute more offspring to the next generation than other individuals do, shifting the average value of that character in the population toward that extreme. In the case of a single gene locus, directional selection may result in favoring a particular genetic variant (known as *positive selection* for that variant). By favoring one phenotype over another, directional selection results in an increase of the frequencies of alleles that produce the favored phenotype (as with the hemagglutinin gene of influenza in the opening of this chapter).

If directional selection operates over many generations, an *evolutionary trend* is seen in the population (**Figure 21.12B**). Directional evolutionary trends often continue for many genera-

**21.14 Texas Longhorns Are the Result of Directional Selection** Longer horns were advantageous for defending young calves from attacks by predators, so feral herds of Spanish cattle developed much longer horns between the early 1500s and the 1860s. The trend has been maintained in modern times by ranchers practicing artificial selection.

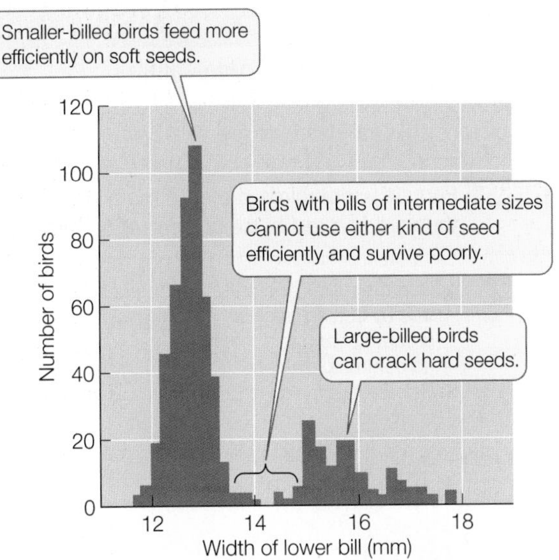
Smaller-billed birds feed more efficiently on soft seeds.
Birds with bills of intermediate sizes cannot use either kind of seed efficiently and survive poorly.
Large-billed birds can crack hard seeds.
Number of birds
0
20
40
60
80
100
120
12
14
16
18
Width of lower bill (mm)

**21.15 Disruptive Selection Results in a Bimodal Distribution** The bimodal distribution of bill sizes in the black-bellied seedcracker of West Africa is a result of disruptive selection, which favors individuals with larger and smaller bill sizes over individuals with intermediate-sized bills.

tions, but they can be reversed if the environment changes and different phenotypes are favored, or halted when an optimal phenotype is reached or when trade-offs oppose further change. The character then falls under stabilizing selection.

Many cases of directional selection have been observed directly, and long-term examples abound in the fossil record. The long horns of Texas Longhorn cattle (**Figure 21.14**) are an example of a trait that has evolved through directional selection. Texas Longhorns are descendants of cattle that Christopher Columbus brought to the New World. Columbus picked up a few cattle in the Canary Islands and brought them to the island of Hispaniola in 1493. The cattle quickly multiplied, and their descendants were taken to the mainland of Mexico. As the Spanish explored what would later become Texas and the southwestern United States, they brought some of these cattle with them, some of which escaped and formed feral herds. Populations of these feral cattle increased greatly over the next few hundred years, but there was heavy predation from bears, mountain lions, and wolves, especially on the young calves. Cows with longer horns were more successful in protecting their calves against attacks, and over the next few hundred years the average horn length of cattle in the feral herds increased considerably. In addition, the cattle evolved resistance to endemic diseases of the Southwest, as well as higher fecundity and longevity. Texas Longhorn cows often live and produce calves well into their twenties, or about twice as long as many breeds of cattle that have been artificially selected by humans for traits such as high fat content or high milk production (which are examples of artificial directional selection).

**DISRUPTIVE SELECTION** When disruptive selection operates, individuals at opposite extremes of a character distribution contribute more offspring to the next generation than do individuals close to the mean, which increases variation in the population (**Figure 21.12C**).

The strikingly bimodal (two-peaked) distribution of bill sizes in the black-bellied seedcracker (*Pyrenestes ostrinus*), a West African finch (**Figure 21.15**), illustrates how disruptive selection can influence populations in nature. The seeds of two types of sedges (marsh plants) are the most abundant food source for these finches during part of the year. Birds with large bills can readily crack the hard seeds of the sedge *Scleria verrucosa*. Birds with small bills can crack *S. verrucosa* seeds only with difficulty; however, they feed more efficiently on the soft seeds of *S. goossensii* than do birds with larger bills.

Young finches whose bills deviate markedly from the two predominant bill sizes do not survive as well as finches whose bills are close to one of the two sizes represented by the distribution peaks. Because there are few abundant food sources in the environment, and because the seeds of the two sedges do not overlap in hardness, birds with intermediate-sized bills are less efficient in using either one of the principal food sources. Disruptive selection therefore maintains a bimodal bill size distribution.

### Sexual selection influences reproductive success

**Sexual selection** acts on characteristics that determine reproductive success. In *The Origin of Species*, Darwin devoted only a few pages to sexual selection, but in 1871 he wrote an entire book about it: *The Descent of Man, and Selection in Relation to Sex*. Sexual selection was Darwin's explanation for the evolution of conspicuous characters that would appear to inhibit survival, such as bright colors, long tails, and elaborate courtship displays in males of many species. He hypothesized that these features either improved the ability of their bearers to compete for access to mates (*intrasexual selection*) or made their bearers more attractive to members of the opposite sex (*intersexual selection*). The concept of sexual selection was either ignored or questioned for many decades, but recent investigations have demonstrated its importance.

INVESTIGATING LIFE

**21.16 Sexual Selection in Male Widowbirds**

The extensive tail of the territorial male African long-tailed widowbird (*Euplectes progne*) actually inhibits its ability to fly. Darwin attributed the evolution of this trait to sexual selection. Behavioral ecologist Malte Andersson tested this hypothesis.

**HYPOTHESIS** **Female widowbirds prefer to mate with the male that displays the longest tail; longer-tailed males thus are favored by sexual selection because they will father more offspring.**

**METHOD**

1. Capture males and artificially lengthen or shorten tails by cutting or gluing on feathers. In a control group, cut and replace tails to their normal length (to control for the effects of tail-cutting).
2. Release the males to establish their territories and mate.
3. Count the nests with eggs or young on each male's territory.

**RESULTS** Male widowbirds with artificially shortened tails established and defended display sites sucessfully but fathered fewer offspring than did control or unmanipulated males. Males with artificially lengthened tales fathered the most offspring.

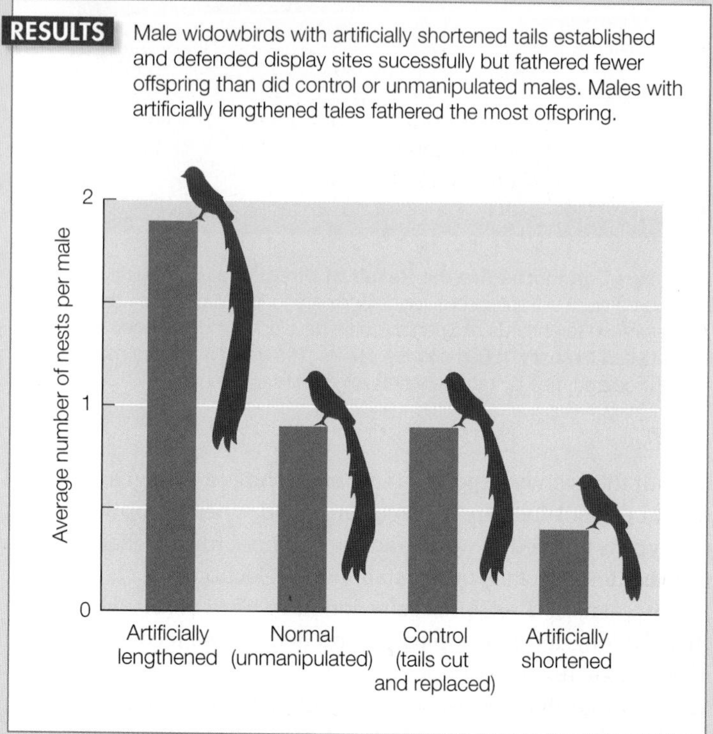


**CONCLUSION** **Sexual selection in *Euplectes progne* has favored the evolution of long tails in the male.**

Go to **yourBioPortal.com** for original citations, discussions, and relevant links for all INVESTIGATING LIFE figures.

Darwin devoted an entire book to sexual selection because he recognized that, whereas natural selection typically favors traits that enhance the survival of their bearers or their descendants, sexual selection is primarily about success in reproduction. Of course, an animal must survive to reproduce, but if it survives and fails to reproduce, it makes no contribution to the next generation. Thus sexual selection may favor traits that enhance an individual's chances of reproduction but reduce its chances of survival. For example, females may be more likely to see or hear males with a given trait (and thus be more likely to mate with those males), even though the favored trait may also increase the chances that the male will be seen or heard by a predator. In other cases, the sexual signal may indicate a successful genotype in the male. In many species of frogs, for example, females prefer males with low-frequency calls. Males' calls vary with body size, and a low-frequency call is indicative of a large-bodied frog. Frogs exhibit indeterminate growth—that is, they continue to grow indefinitely—so a large frog is a long-lived frog, which indicates high survivorship. In this case, the sexual signal represents what is known as an *honest signal* of the male's ability to survive in the local environment.

One example of a trait that Darwin attributed to sexual selection is the remarkable tail of the male African long-tailed widowbird (*Euplectes progne*), which is longer than the bird's head and body combined. Male widowbirds normally select, and defend from other males, a territory where they perform courtship displays to attract females. To investigate whether sexual selection drove the evolution of widowbird tails, Malte Andersson, a behavioral ecologist at Gothenburg University, Sweden, clipped the tails of some captured male widowbirds and lengthened the tails of others by gluing on additional feathers. He then cut and reglued the tail feathers of still other males, which served as controls. Both short- and long-tailed males successfully defended their display territories, indicating that a long tail does not confer an advantage in male–male competition. However, males with artificially elongated tails attracted about four times more females than did males with shortened tails (**Figure 21.16**).

Why do female widowbirds prefer males with long tails? One possibility is that ability to grow and maintain a costly feature such as a long tail may indicate that the male bearing it is vigorous and healthy, even though the tail impairs the bird's ability to fly. If so, then females that are attracted to long tails are in-

## INVESTIGATING LIFE

### 21.17 Do Bright Bills Signal Good Health?

Female zebra finches (*Taeniopygia guttata*) preferentially choose mates with the brightest bill color. Does this preference increase their chances of mating with the healthiest males? This experiment made use of carotenoids (antioxidant pigment molecules believed to boost immune response) to test the hypothesis.

**HYPOTHESIS** **The brightness of a male zebra finch's red bill is correlated with the strength of the bird's immune response and a corresponding likelihood of good health.**

*Taeniopygia guttata*

**METHOD**

1. Provide carotenoids in the drinking water of experimental, but not control, males.
2. Challenge all males immunologically and measure responses.

**RESULTS** Experimental males responded more strongly to the immunological challenge. They also developed brighter bills than control males.

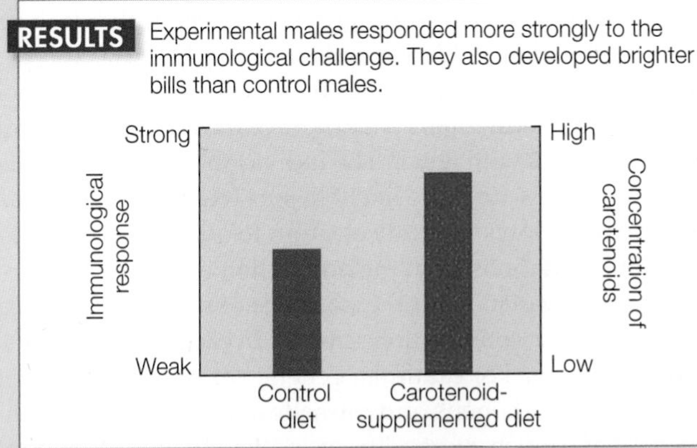


**CONCLUSION** **Bill color is an indication of immunological strength and general health.**

**FURTHER INVESTIGATION:** How would you test this same hypothesis in the field? What would constitute experimental and control birds?

Go to **yourBioPortal.com** for original citations, discussions, and relevant links for all INVESTIGATING LIFE figures.

directly attracted to vigorous, healthy males, which likely carry beneficial genes that would lead to higher survivorship of offspring. Although the manipulated males in Andersson's investigation did not have to pay the price of growing and supporting (except briefly) artificially long tails, the hypothesis that having well-developed ornamental traits signals vigor and health has been tested experimentally with captive zebra finches.

The bright red bills of male zebra finches (*Taeniopygia guttata*) are the result of red and yellow carotenoid pigments. Zebra finches (and most other animals) cannot synthesize carotenoids and must obtain them from their food. In addition to influencing bill color, carotenoids are antioxidants and components of the immune system. Males in good health may need to allocate fewer carotenoids to immune function than males in poorer health. If so, then females can use the brightness of a male's bill to assess his health.

Tim Birkhead and his colleagues at Sheffield University manipulated blood levels of carotenoids in genetically similar male zebra finches by giving experimental males drinking water with added carotenoids; they gave control males only distilled water. All the males had access to the same food. After one month, the experimental males had higher levels of carotenoids in their blood, had much brighter bills than the control males, and were preferred by female zebra finches.

Next, the investigators challenged both groups of males immunologically by injecting phytohemagglutinin (PHA) into their wings. PHA induces a response by T lymphocytes, a type of white blood cell that functions in the immune system to recognize and deactivate foreign substances (see Chapter 42). The injection results in an accumulation of white blood cells and thus a thickening of the skin at the injection site. Experimental males with enhanced carotenoid levels developed thicker skins because they responded more strongly to PHA than control males did, indicating that higher carotenoid levels are associated with stronger immune systems (**Figure 21.17**).

This experiment showed that when a female chooses a male with a bright red bill, she probably gets a mate with a healthy immune system. Such males are less likely to become infected with parasites and diseases, and are better able to assist with parental care.

## 21.3 RECAP

**Variation in genotype can lead to variation in fitness. Fitness refers to the relative reproductive contribution of a phenotype to subsequent generations. Natural and sexual selection can both change and stabilize phenotypes within populations.**

- Explain why natural selection that acts on a phenotype results in changes in genotype frequencies. See p. 451
- Describe the differences between stabilizing, directional, and disruptive selection, giving examples of each. See pp. 452–453 and Figure 21.12
- Why did Darwin devote an entire book to sexual selection? See pp. 453–455

Genetic drift, stabilizing selection, and directional selection all tend to reduce genetic variation within populations. Nevertheless, as we have seen, most populations harbor considerable genetic variation. What processes produce and maintain genetic variation within populations?

## 21.4 How Is Genetic Variation Maintained within Populations?

Genetic variation is the raw material on which mechanisms of evolution act. In this section we will discuss several processes—neutral mutations, sexual recombination, frequency-dependent selection, and heterozygote advantage—that operate to maintain genetic variation in populations, despite the action of other forces (such as genetic drift and many types of selection) that reduce variation. We will also show how genetic variation may be maintained over geographic space.

### Neutral mutations accumulate in populations

As we discuss in Section 14.6, some mutations do not affect the function of the proteins encoded by the mutated genes. An allele that does not affect the fitness of an organism—that is, an allele that is no better or worse than alternative alleles at the same locus—is called a **neutral allele**. Neutral alleles are unaffected by natural selection. Even in large populations, neutral alleles may be lost or may increase in frequency, purely by genetic drift. Neutral alleles are added to a population over time through mutation, providing the population with considerable genetic variation.

Much of the phenotypic variation we are able to observe is not neutral. However, modern techniques enable us to measure neutral variation at the molecular level and provide the means to distinguish it from adaptive variation. Section 24.2 discusses how variation in neutral molecular traits can be used to study divergence among genes, populations, and species.

### Sexual recombination amplifies the number of possible genotypes

In asexually reproducing organisms, each new individual is genetically identical to its parent unless there has been a mutation. When organisms reproduce sexually, however, offspring differ from their parents because of crossing over and independent assortment of chromosomes during meiosis as well as the combination of genetic material from two different gametes, as described in Chapter 11. Sexual recombination generates an endless variety of genotypic combinations that increase the evolutionary potential of populations—a long-term advantage of sex. Although many species may reproduce asexually most of the time, few are strictly asexual over long periods of evolutionary time. Almost all have some means of achieving genetic recombination.

The evolution of the mechanisms of meiosis and sexual recombination were crucial events in the history of life. Exactly how these attributes arose is puzzling, however, because sex has at least three striking disadvantages in the short term:

- Recombination breaks up adaptive combinations of genes.
- Sex reduces the rate at which females pass genes on to their offspring.
- Dividing offspring into separate genders greatly reduces the overall reproductive rate.

To see why this last disadvantage exists, consider an asexual female that produces the same number of offspring as a sexual female. Let's assume that both females produce two offspring, but that the sexual female produces 50 percent males. In the next generation, both asexual $F_1$ females will produce two more offspring, but there is only one sexual $F_1$ female to produce offspring. Thus, the effective reproductive rate of the asexual lineage is twice that of the sexual lineage. The evolutionary problem is to identify the advantages of sex that can overcome such short-term disadvantages.

A number of hypotheses have been proposed for the existence of sex, none of which are mutually exclusive. One is that sexual recombination facilitates repair of damaged DNA, because breaks and other errors in DNA on one chromosome can be repaired by copying the intact sequence from the homologous chromosome.

Another advantage of sexual reproduction is that it permits the elimination of deleterious mutations. As Section 13.4 describes, DNA replication is not perfect. Errors are introduced in every generation, and many or most of these errors result in lower fitness. Asexual organisms have no mechanism to eliminate deleterious mutations. Hermann J. Muller noted that the accumulation of deleterious mutations in a non-recombining genome is like a genetic ratchet. The mutations accumulate—"ratchet up"—at each replication: that is, a mutation occurs and is passed on when the genome replicates, then two new mutations occur in the next replication, so three mutations are passed on, and so on. Deleterious mutations cannot be eliminated except by the death of the lineage or a rare back mutation. This accumulation of deleterious mutations in lineages that lack genetic recombination is known as *Muller's ratchet*.

In sexual species, on the other hand, genetic recombination produces some individuals with more of these deleterious mutations and some with fewer. The individuals with fewer deleterious mutations are more likely to survive. Therefore, sexual reproduction allows natural selection to eliminate particular deleterious mutations from the population over time.

Another explanation for the existence of sex is that the great variety of genetic combinations created in each generation may be advantageous. For example, genetic variation can be a defense against pathogens and parasites. Most pathogens and parasites have much shorter life cycles than their hosts and can rapidly evolve counteradaptations to host defenses. Sexual recombination might give the host's defenses a chance to keep up.

Sexual recombination does not directly influence the frequencies of alleles; rather, *it generates new combinations of alleles on which natural selection can act*. It expands variation in a character influenced by alleles at many loci by creating new genotypes. That is why artificial selection for bristle number in *Drosophila* (see Figure 21.5) resulted in flies that had either more or fewer bristles than the flies in the initial population had.

### Frequency-dependent selection maintains genetic variation within populations

Natural selection often preserves variation as a polymorphism (two or more variants of a trait present in the same population). A polymorphism may be maintained when the fitness of a given phenotype depends on its frequency in a population, a phenomenon known as **frequency-dependent selection**.

A small fish that lives in Lake Tanganyika in East Africa provides an example of frequency-dependent selection. Because of an asymmetrical jaw joint, the mouth of this scale-eating fish, *Perissodus microlepis*, opens either to the right or to the left; the direction is genetically determined (**Figure 21.18**). The scale-eater approaches its prey (another fish) from behind and dashes in to bite off several scales from its flank. "Right-mouthed" individuals always attack from the victim's left, and "left-mouthed" individuals always attack from the victim's right. The distorted mouth enlarges the area of teeth in contact with the prey's flank, but only if the scale-eater attacks from the appropriate side.

Prey fish are alert to approaching scale-eaters, so attacks are more likely to be successful if the prey must watch both flanks. Vigilance by prey thus favors equal numbers of right- and left-mouthed scale-eaters, because if attacks from one side were more common than the other, prey fish would pay more attention to potential attacks from that side. Over an 11-year study of this fish in Lake Tanganyika, the polymorphism was found to be stable: the two forms of *P. microlepis* remained at about equal frequencies.

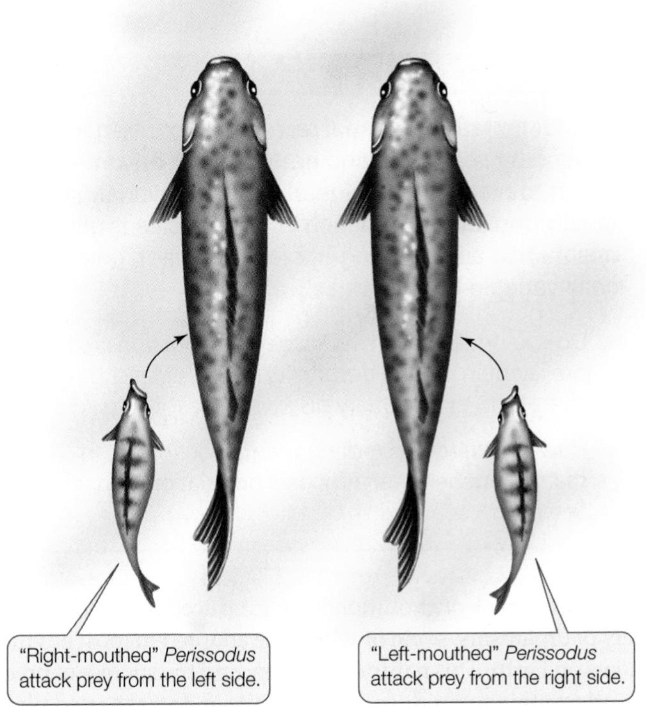

"Right-mouthed" *Perissodus* attack prey from the left side.
"Left-mouthed" *Perissodus* attack prey from the right side.

**21.18 A Stable Polymorphism** Frequency-dependent selection maintains equal proportions of left- and right-mouthed individuals of the scale-eating fish *Perissodus microlepis*.

### Heterozygote advantage maintains polymorphic loci

In many cases, different alleles for a particular gene have advantages under different environmental conditions. Most organisms, however, experience a wide diversity of environments over time. A night is dramatically different from the preceding day. A cold, cloudy day differs from a clear, hot one. Day length and temperature change seasonally. For many genes, a single allele is unlikely to perform well under all these conditions. In such situations, a heterozygous individual (with two different alleles) is likely to outperform individuals that are homozygous for either one of the alleles.

*Colias* butterflies of the Rocky Mountains live in environments where dawn temperatures often are too cold, and afternoon temperatures too hot, for the butterflies to fly. Populations of this butterfly are polymorphic for the enzyme phosphoglucose isomerase (PGI), which influences how well the butterfly flies at different temperatures. Butterflies with certain PGI genotypes can fly better during the cold hours of early morning; others perform better during midday heat. The optimal body temperature for flight is 35°C to 39°C, but some butterflies can fly with body temperatures as low as 29°C or as high as 40°C. During spells of unusually hot weather, heat-tolerant genotypes are favored; during spells of unusually cool weather, cold-tolerant genotypes are favored.

Heterozygous *Colias* butterflies can fly over a greater temperature range than homozygous individuals, which should give them an advantage in foraging and finding mates. A test of this prediction did find a mating advantage in heterozygous males, and further, that this advantage maintains the polymorphism in the population (**Figure 21.19**). Of course, the heterozygotes can never become fixed in the population, because the offspring of two heterozygotes will include both classes of homozygotes in addition to heterozygotes.

### Much genetic variation in species is maintained in geographically distinct populations

Much of the genetic variation in species is preserved as differences among members living in different places (populations). Populations often vary genetically because they are subjected to different selective pressures in different environments. Environments may vary significantly over short distances. For example, in the Northern Hemisphere, temperature and soil moisture differ dramatically between north- and south-facing mountain slopes. In the Rocky Mountains of Colorado, the proportion of ponderosa pines (*Pinus ponderosa*) that are heterozygous for a particular peroxidase enzyme is particularly high on south-facing slopes, where temperatures fluctuate dramatically, often on a daily basis. This heterozygous genotype performs well over a broad range of temperatures. On north-facing slopes and at higher elevations, where temperatures are cooler and fluctuate less strikingly, a peroxidase homozygote, which has a lower optimal temperature, is much more frequent.

Plant species may also vary geographically in the chemicals they synthesize to defend themselves against herbivores. Some individuals of the white clover (*Trifolium repens*) pro-

## INVESTIGATING LIFE

### 21.19 A Heterozygote Mating Advantage

Among butterflies of the genus *Colias*, males that are heterozygous for two alleles of the PGI enzyme can fly farther under a broader range of temperatures than males that are homozygous for either allele. Does this ability give heterozygous males a mating advantage?

**HYPOTHESIS** **Heterozygous male *Colias* will have proportionally greater mating success than homozygous males.**

**METHOD**

1. For each of two *Colias* species, capture butterflies in the field. In the laboratory, determine their genotypes and allow them to mate.
2. Determine the genotypes of the offspring, thus revealing paternity and mating success of the males.

**RESULTS** For both species, the proportion of heterozygous males that mated successfully was higher than the proportion of all males seeking females ("flying").

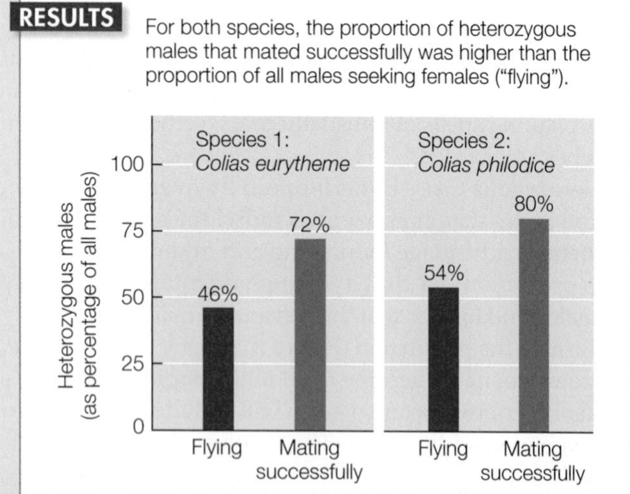


**CONCLUSION** **Heterozygous *Colias* males have a mating advantage over homozygous males.**

Go to **yourBioPortal.com** for original citations, discussions, and relevant links for all INVESTIGATING LIFE figures.

duce the poisonous chemical cyanide. Poisonous individuals are less appealing to herbivores—particularly mice and slugs—than are nonpoisonous individuals. However, clover plants that produce cyanide are more likely to be killed by frost, because freezing damages cell membranes and releases cyanide into the plant's own tissues.

In European populations of *Trifolium repens*, the frequency of cyanide-producing individuals increases gradually from north to south and from east to west (**Figure 21.20**). This gradual change in phenotype across a geographic gradient is known as **clinal variation**. In this cline, poisonous plants make up a large proportion of clover populations only in areas where winters are mild. Cyanide-producing individuals are rare where winters are cold, even though herbivores graze clovers heavily in those areas.

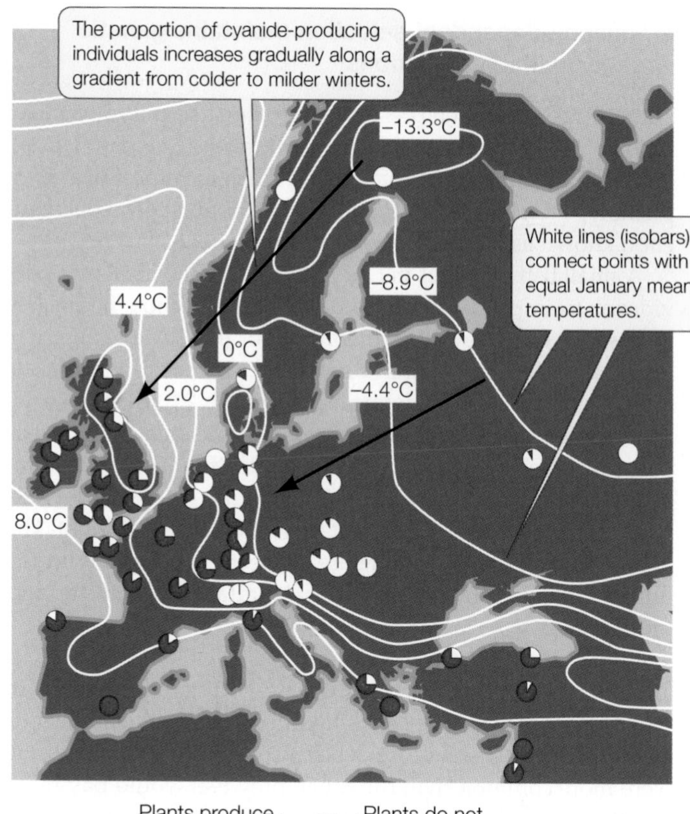


**21.20 Geographic Variation in a Defensive Chemical** The frequency of cyanide-producing individuals in European populations of white clover (*Trifolium repens*) depends on winter temperatures.

## 21.4 RECAP

**Neutral mutations, sexual recombination, frequency-dependent selection, and heterozygote advantage all act to maintain considerable genetic variation in most populations. Variation within species is also maintained among geographically distinct, genetically variable populations.**

- Do you understand why sexual reproduction is so prevalent in nature, despite its having at least three short-term evolutionary disadvantages? See p. 456
- How does frequency-dependent selection act to maintain genetic variation in a population? See p. 457

The mechanisms of evolution have produced a remarkable variety of organisms, some of which are adapted to most environments on Earth. This natural variation, and the success of breeders attempting to produce desired traits in domesticated plants and animals, suggests that evolution can produce a wide variety of adaptive traits. But are there limits to the adaptations evolution can produce?

## 21.5 What Are the Constraints on Evolution?

We would be mistaken to assume that evolutionary mechanisms can produce any trait we might imagine. Evolution is constrained in many ways. Lack of appropriate genetic variation, for example, prevents the development of many potentially favorable traits. If the allele for a given trait does not exist in a population, that trait cannot evolve even if it would be highly favored by natural selection. Most possible combinations of genes and genotypes have never existed in any population, and so have never been tested under natural selection.

Constraints are imposed on organisms by the dictates of physics and chemistry. The size of cells, for example, is constrained by the stringencies of surface area-to-volume ratios (see Section 2.1). The ways in which proteins can fold are limited by the bonding capacities of their constituent molecules (see Section 3.2). And the energy transfers that fuel life must operate within the laws of thermodynamics (see Section 8.1). Keep in mind that evolution works within the boundaries of these universal constraints, as well as the constraints described here.

### Developmental processes constrain evolution

As Section 20.5 explained, developmental constraints on evolution are paramount because *all evolutionary innovations are modifications of previously existing structures*. Human engineers seeking to power an airplane can start "from scratch" to design a completely new type of engine (powered by jet propulsion), to replace the previous type (powered by propellers). Evolutionary changes, however, cannot happen in this way. Current phenotypes of organisms are constrained by historical conditions and past selective pressures.

A striking example of such developmental constraints is provided by the evolution of fish that spend most of their time on the sea bottom. One lineage, the bottom-dwelling skates and rays, share a common ancestor with sharks, whose bodies were already somewhat ventrally flattened and whose skeletal frame is made of flexible cartilage. Skates and rays evolved a body type that further flattened their bellies, allowing them to swim along the ocean floor (**Figure 21.21A**).

By contrast, plaice, sole, and flounder are bottom-dwelling descendants of deep-bellied, laterally flattened ancestors with bony skeletons. The only way these fishes can lie flat is to flop over on their sides. Their ability to swim is thus curtailed, but their bodies can lie still and are well camouflaged. During development, one eye of these flatfishes moves so that both eyes are positioned on the same side of the body (**Figure 21.21B**). Such shifts in eye position have evolved several times, and shifts have happened in both directions (that is, both left- and right-eyed flatfishes have evolved independently). Small shifts in the position of one eye probably helped ancestral flatfishes see better, resulting in the flat body forms found today. This path to producing a flattened body may not be optimal, but the fishes' developmental capabilities constrain the pathways that evolution can take.

(A) *Taeniura lymma*

(B) *Bothus lunatus*

**21.21 Two Solutions to a Single Problem** (A) This stingray, whose ancestors were dorsoventrally flattened, lies on its belly. Stingrays' bodies are symmetrical around the dorsal backbone. (B) This flounder, whose ancestors were laterally flattened, lies on its side. (The backbone of this individual is at the right.) Flounders' eyes migrate during development so that both are on the same side of the body.

### Trade-offs constrain evolution

Adaptations frequently impose both fitness costs and benefits. For an adaptation to evolve, the fitness benefits it confers must exceed the fitness costs it imposes—in other words, the **trade-off** must be worthwhile. For example, there are metabolic costs associated with developing and maintaining certain conspicuous features (such as antlers or horns) that males use to compete with other males for access to females. The fact that these features are common in many species suggests that the benefits derived from possessing them must outweigh the costs.

As a result of trade-offs, many traits that are adaptive in one context may be maladaptive in another. Consider the rough-skinned newt, *Taricha granulosa*, and one of its predators, the common garter snake, *Thamnophis sirtalis* (**Figure 21.22A**). The newt sequesters in its skin a potent neurotoxin called tetrodotoxin (TTX). TTX paralyzes nerves and muscles by blocking

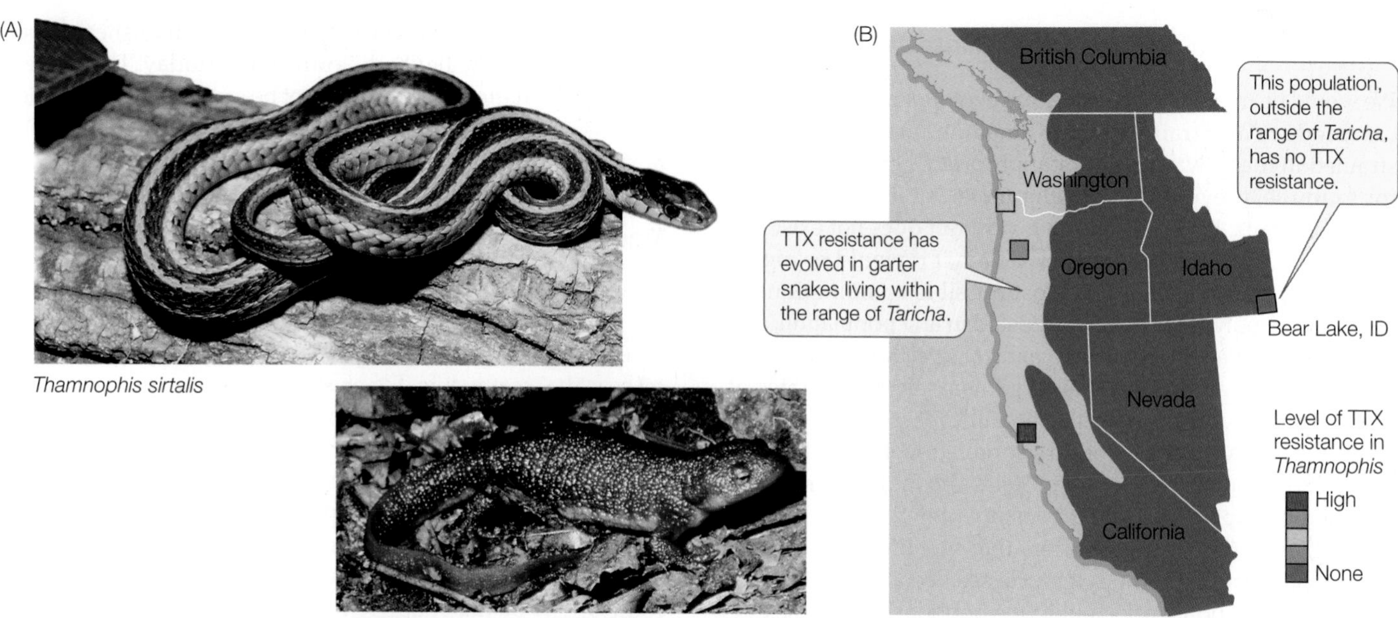


**21.22 Resistance to a Toxin Comes at a Cost** (A) Garter snakes (above) prey on newts (below). Rough-skinned newts counter with the ability to sequester a neurotoxin, TTX, in their skin. In turn, TTX-resistant sodium channels have evolved in some snake populations, allowing the snakes to eat toxic prey but resulting in slower movement by the snakes. (B) High resistance to TTX in garter snakes is only found in regions where snake and newt populations overlap (tan area).

**yourBioPortal.com**
GO TO Animated Tutorial 21.3 • Assessing the Costs of Adaptation

sodium channels (see Section 6.3). Most vertebrates—including many garter snakes—will die if they eat a rough-skinned newt. But some snakes can eat rough-skinned newts and survive. In some populations of garter snakes, TTX-resistant sodium channels have evolved in the nerves and muscles (see Chapter 24 for another example of the evolution of sodium channels). However, the snakes pay a price for this attribute. For several hours after eating a newt, TTX-resistant snakes can move only slowly, and they never move as fast as nonresistant snakes. Thus resistant snakes are more vulnerable to their own predators than are TTX-sensitive snakes that simply don't encounter poisonous newts. Therefore, there is selection against TTX-resistant sodium channels in populations of garter snakes that occur outside the range of rough-skinned newts, but selection for TTX-resistance in many areas where newts are present (**Figure 21.22B**).

### Short-term and long-term evolutionary outcomes sometimes differ

The short-term changes in allele frequencies within populations that we have emphasized in this chapter are an important focus of study for evolutionary biologists. These changes can be observed directly, they can be manipulated experimentally, and they demonstrate the actual processes by which evolution occurs. By themselves, however, they do not enable us to predict long-term evolutionary changes.

Long-term patterns of evolutionary change can be strongly influenced by events that occur so infrequently (a meteorite impact, for example) or so slowly (continental drift) that they are unlikely to be observed during short-term studies. The ways in which evolutionary processes act may change over time with changing environmental conditions. Even among the descendants of a single ancestral species, different lineages may evolve in different directions. Therefore, additional types of evidence, demonstrating the effects of rare and unusual events on trends in the fossil record, must be gathered if we wish to understand the course of evolution over billions of years.

## 21.5 RECAP

**Developmental processes constrain evolution because all evolutionary innovations are modifications of previously existing structures. An adaptation can evolve only if the fitness benefits it confers exceed the fitness costs it imposes.**

- Describe an example of an evolutionary trade-off in which the advantages of an adaptation outweigh its costs in the long run. See pp. 459–460
- Do you see why the presence of a great deal of genetic variation within a population could increase the chances that some members of the population would survive an unprecedented environmental change? Do you also understand why there is no guarantee that this would be the case?

# CHAPTER SUMMARY

## 21.1 What Facts Form the Base of Our Understanding of Evolution?

- Charles Darwin attributed changes in species over time to the possession of advantageous traits by some individuals. He understood that it is not individuals that evolve but **populations**. A population evolves when individuals with favorable **heritable traits** survive and reproduce at higher rates than other members of the population.
- **Adaptation** refers both to characteristics of organisms and the way those characteristics are acquired via **natural selection**. **ANIMATED TUTORIAL 21.1**
- The sum of all copies of all alleles at all loci found in a population constitutes its **gene pool** and represents the genetic variation that results in different phenotypic traits on which natural selection can act. Review Figure 21.3
- **Artificial selection** and laboratory experiments demonstrate the existence of considerable genetic variation in most populations. Review Figure 21.5
- Allele frequencies measure the amount of genetic variation in a population; genotype frequencies show how a population's genetic variation is distributed among its members. Together, allele and genotype frequencies describe a population's **genetic structure**. Review Figure 21.6
- **Hardy–Weinberg equilibrium** predicts the allele frequencies in populations in the absence of evolution. Deviation from these frequencies indicates the work of evolutionary mechanisms. Review Figure 21.7, **ANIMATED TUTORIAL 21.2**

## 21.2 What Are the Mechanisms of Evolutionary Change?

- **Mutation** provides new genetic variants; favored variants increase in populations through natural selection.
- Migration or mating of individuals between populations results in **gene flow**.
- In small populations, **genetic drift**—the random loss of individuals and the alleles they possess—may produce large changes in allele frequencies from one generation to the next and greatly reduce genetic variation. Review Figure 21.8
- **Population bottlenecks** occur when only a few individuals survive a random event, resulting in a drastic shift in allele frequencies within the population and the loss of variation. Similarly, a population established by a small number of individuals colonizing a new region may lose variation via a **founder effect**.
- Nonrandom mating may result in genotype frequencies that deviate from Hardy–Weinberg equilibrium.

## 21.3 How Does Natural Selection Result in Evolution?

- **Fitness** is the reproductive contribution of a phenotype to subsequent generations relative to the contributions of other phenotypes.
- Changes in numbers of offspring are responsible for changes in the absolute size of a population, but only changes in the relative success of different phenotypes within a population lead to changes in allele frequencies.
- Natural selection can act on variable traits in several different ways, resulting in **stabilizing**, **directional**, or **disruptive selection**. Review Figure 21.12
- **Sexual selection** primarily affects success in reproduction, rather than success in survival. Review Figures 21.16 and 21.17

## 21.4 How Is Genetic Variation Maintained within Populations?

- Neutral mutations, sexual recombination, frequency-dependent selection, and heterozygote advantage can all maintain genetic variation within populations.
- **Neutral alleles** do not affect the fitness of an organism, are not affected by natural selection, and may accumulate or be lost by genetic drift.
- Despite short-term disadvantages, sexual reproduction generates countless genotypic combinations that increase the evolutionary potential and survivorship of populations.
- A polymorphism may be maintained by **frequency-dependent selection** when the fitness of a genotype depends on its frequency in a population.
- Genetic variation within species may be maintained by the existence of genetically distinct populations over geographic space. A gradual change in phenotype across a geographic gradient is known as **clinal variation**. Review Figure 21.20

## 21.5 What Are the Constraints on Evolution?

- Developmental processes constrain evolution because all evolutionary innovations are modifications of previously existing structures.
- Most adaptations impose costs. An adaptation can evolve only if the benefits it confers exceed the costs it imposes, a situation that leads to **trade-offs**. Review Figure 21.22, **ANIMATED TUTORIAL 21.3**

# SELF-QUIZ

1. Long-horned cattle have greater difficulty moving through heavily forested areas compared with cattle that have short or no horns, but long-horned cattle are better able to defend their young against predators. This contrast is an example of
   *a.* an adaptation.
   *b.* genetic drift.
   *c.* natural selection.
   *d.* a trade-off.
   *e.* none of the above

2. Which of the following is true?
   *a.* Darwin and Wallace were both influenced by Malthus.
   *b.* Wallace proposed a theory of evolution by natural selection that was similar to Darwin's.
   *c.* Malthus claimed that because human population growth would outstrip any increases in food production, famine was a likely result.
   *d.* Darwin realized that all populations had the capacity to rapidly increase in numbers.
   *e.* All of the above

3. The phenotype of an organism is
   *a.* the type specimen of its species in a museum.
   *b.* its genetic constitution, which governs its traits.
   *c.* the chronological expression of its genes.
   *d.* the physical expression of its genotype.
   *e.* its adult form.
4. The appropriate unit for defining and measuring genetic variation is the
   *a.* cell.
   *b.* individual.
   *c.* population.
   *d.* community.
   *e.* ecosystem.
5. Which statement about allele frequencies is *not* true?
   *a.* The sum of all allele frequencies at a locus is always 1.
   *b.* If there are two alleles at a locus and we know the frequency of one of them, we can obtain the frequency of the other by subtraction.
   *c.* If an allele is missing from a population, its frequency in that population is 0.
   *d.* If two populations have the same allele frequencies at a locus, they must have the same proportion of homozygotes at that locus.
   *e.* If there is only one allele at a locus, its frequency is 1.
6. Which of the following is *not* required for a population at Hardy–Weinberg equilibrium?
   *a.* There is no migration between populations.
   *b.* Natural selection is not acting on the alleles in the population.
   *c.* Mating is random.
   *d.* Multiple alleles must be present at every locus.
   *e.* All of the above.
7. The fitness of a genotype is a function of the
   *a.* average rates of survival and reproduction of individuals with that genotype.
   *b.* individuals that have the highest rates of both survival and reproduction.
   *c.* individuals that have the highest rates of survival.
   *d.* individuals that have the highest rates of reproduction.
   *e.* average reproductive rate of individuals with that genotype.
8. Laboratory selection experiments with fruit flies have demonstrated that
   *a.* bristle number is not genetically controlled.
   *b.* bristle number is not genetically controlled, but changes in bristle number are caused by the environment in which the fly is raised.
   *c.* bristle number is genetically controlled, but there is little variation on which natural selection can act.
   *d.* bristle number is genetically controlled, but selection cannot result in flies having more bristles than any individual in the original population had.
   *e.* bristle number is genetically controlled, and selection can result in flies having more, or fewer, bristles than any individual in the original population had.
9. Disruptive selection maintains a bimodal distribution of bill size in the West African seedcracker because
   *a.* bills of intermediate shapes are difficult to form.
   *b.* the birds' two major food sources differ markedly in size and hardness.
   *c.* males use their large bills in displays.
   *d.* migrants introduce different bill sizes into the population each year.
   *e.* older birds need larger bills than younger birds.
10. Which of the following is *not* a reason why trade-offs constrain evolution?
   *a.* Most adaptations impose both fitness costs and benefits.
   *b.* Structures such a horns and antlers are metabolically costly to produce, but result in more reproduction by the males that possess them.
   *c.* Changes in allele frequencies may be influenced by chance events.
   *d.* Ability to consume toxic prey may reduce mobility.
   *e.* Adaptations can evolve only if the fitness benefits they confer exceed the costs they impose.

## FOR DISCUSSION

1. In what ways does artificial selection by humans differ from natural selection? Was Darwin wise to base so much of his argument for natural selection on the results of artificial selection?
2. In nature, mating among individuals in a population is never truly random, immigration and emigration are common, and natural selection is continuous. Why, then, is Hardy–Weinberg equilibrium, which is based on assumptions known generally to be false, so useful in our study of evolution? Can you think of other models in science that are based on false assumptions? How are such models used?
3. As far as we know, natural selection cannot adapt organisms to future events. Yet many organisms appear to respond to natural events before they happen. For example, many mammals go into hibernation while it is still quite warm. Similarly, many birds leave the temperate zone for their southern wintering grounds long before winter has arrived. How can such "anticipatory" behaviors evolve?
4. Populations of most of the thousands of species that have been introduced to areas where they were previously not found, including those that have become pests, began with a few individuals. Founding populations therefore begin with much less genetic variation than their parental populations have. If genetic variation is generally advantageous, why have so many of these species been successful in their new environments?
5. Why is it important that the ways in which males advertise their health and vigor to females reliably indicate their status?
6. As more humans live longer, many people face degenerative conditions such as Alzheimer's disease that (in most cases) are linked to advancing age. Assuming that some individuals may be genetically predisposed to successfully combat these conditions, is it likely that natural selection alone would act to favor such a predisposition in human populations? Why or why not?

## ADDITIONAL INVESTIGATION

During the past 50 years, more than 200 species of insects that attack crop plants have become highly resistant to DDT and other pesticides. Using your recently acquired knowledge of evolutionary processes, explain the rapid and widespread evolution of resistance. What proposals concerning pesticide use would you make in order to slow down the rate of evolution of resistance? Explain why you think your proposals could work and how you might test them.

## WORKING WITH DATA (GO TO yourBioPortal.com)

**Testing for Significant Differences** In this hands-on exercise based on Figure 21.16, you will use a simple method for randomizing Malte Andersson's data to test for significant differences among the various experimental groups. You will also explore how sample size affects the power to make significant conclusions in experiments.

**Female Mating Preference in Zebra Finches** In this exercise based on Figure 21.17, you will evaluate the data Jonathan Blount and his colleagues used to demonstrate female preference for males with bright bills among zebra finches. You will also consider the limitations to the experiment and explore alternative study designs.

**Determining the Paternity of Butterfly Larvae** In working with a sample of the data collected by Ward et al. for the experiment described in Figure 21.19, you will consider how many larvae from a clutch of butterfly eggs must be examined in order to determine (with a high level of confidence) whether the clutch was fathered by a heterozygous or a homozygous male. You will also consider alternative hypotheses to the authors' conclusions and suggest how these alternative ideas could be tested.

# 22 Reconstructing and Using Phylogenies

## Phylogenetic trees in the courtroom

Transmitting HIV, while irresponsible, is not usually prosecuted as a crime. But in one true-crime case, a woman we'll call "April" went to the police immediately upon learning she was HIV-positive. April believed she was the victim of an attempted murder by "Victor," a physician and her former boyfriend, who had repeatedly threatened violence when she tried to break up with him. April contended that Victor, under the pretense of administering vitamin therapy, had injected her with blood from one of his HIV-infected patients.

Police investigators discovered that Victor had drawn blood from one of his HIV-positive patients just before giving April the injection. The blood draw had no clinical purpose, and Victor had tried to hide the records of it. The police were convinced he might indeed have committed the alleged crime.

The district attorney, however, had to show that April's HIV infection had come from Victor's patient, and from no other source. To reconstruct the history of the infection, the district attorney turned to *phylogenetic analysis*—the study of the evolutionary relationships among a group of organisms.

The district attorney's task was complicated by the nature of HIV. HIV is a retrovirus, in which poor repair of replication errors leads to a high rate of evolution. Once a person is infected with HIV, the virus not only replicates quickly but evolves quickly, so that the infected individual is soon host to a genetically diverse population of viruses. Thus when one person transmits HIV to another, typically very few viral particles (often only one) initiate the infectious event. But the person who is the source of the infection may be host to a large, genetically diverse population of viruses—not just the variant he or she transmits to the recipient.

Enter molecular phylogeny. Samples of HIV from an infected individual can be sequenced to trace their evolutionary lineages back to the originally transmitted virus. The virus that is passed to the recipient will be very closely related to some of the viruses in the source individual and more distantly related to others. A reconstruction of the evolutionary history of the viruses in both individuals is needed to reveal not only whether the two individuals' viruses are closely related, but also who infected whom.

**Human Immunodeficiency Virus** A computer image of the human immunodeficiency virus (HIV), the cause of acquired immunodeficiency syndrome, or AIDS. To combat AIDS, it is also essential to understand the phylogeny of HIV.

**A Source of the Virus** AIDS is a zoonotic disease, meaning that the virus was transferred to humans from another animal. Phylogenetic analyses of immunodeficiency viruses show that humans acquired HIV-1 from chimpanzees (see Figure 22.9). Other forms of the virus have been passed to humans by different simians.

To prove attempted murder, the district attorney needed to demonstrate that April's HIV was more closely related to that of Victor's patient than to other HIV variants in her community. Samples of HIV were isolated from the blood of the patient, from April, and from other HIV-positive individuals in the community. Phylogenetic analysis revealed that April's HIV was indeed closely related to a subset of the patient's HIV, and more distantly related to the other HIV sources in the community. Given this fact, along with other evidence in the case, Victor was convicted of attempted murder.

**IN THIS CHAPTER** we will examine the field of systematics, the scientific study of the diversity of life. We will see how phylogenetic methods are used to reconstruct evolutionary history and to study diversity across genes, populations, species, and larger groups of organisms. We will see how systematists reconstruct the past and use phylogenies to make predictions in biology. We will end the chapter with a look at taxonomy, the theory and practice of classifying organisms.

## CHAPTER OUTLINE



# 22.1 What Is Phylogeny?

**Phylogeny** is the evolutionary history of relationships among organisms or their genes. A **phylogenetic tree** is a diagram that portrays a reconstruction of that history. Phylogenetic trees are commonly used to depict the evolutionary history of species, populations, and genes. Each split (or *node*) in a phylogenetic tree represents a point at which lineages diverged in the past. In the case of species, these splits represent past speciation events, when one lineage divided into two. Thus a phylogenetic tree can be used to trace the evolutionary relationships from the ancient common ancestor of a group of species, through the various speciation events (when lineages split), up to the present populations of the organisms (**Figure 22.1**). Over the past several decades, phylogenetic trees have become important tools for studying and describing evolutionary patterns, and for applying evolutionary theory throughout biology. You will need to understand phylogenetic trees to comprehend many articles and books about biology, including this one.

A phylogenetic tree may portray the evolutionary history of all life forms; of a major evolutionary group (such as the insects); of a small group of closely related species; or in some cases, even the history of individuals, populations, or genes within a species. The common ancestor of all the organisms in the tree forms the *root* of the tree. The phylogenetic trees in this book depict time flowing from left (earliest) to right (most recent) (**Figure 22.2A**). It is also common practice to draw trees with the earliest times at the bottom.

The timing of splitting events in lineages is shown by the position of nodes on a time axis, sometimes called a *divergence axis*. These splits represent events where one lineage diverged into two, such as a speciation event (for a tree of species), a gene duplication event (for a tree of genes), or a transmission event (for a tree of viral lineages transmitted through a host population). The divergence axis may have an explicit scale or simply show the relative timing of splitting events. In this book, the order of nodes along the horizontal (time) axis have meaning, but the vertical distance between the branches does not. Vertical distances are adjusted for legibility and clarity of presentation; they do not correlate with the degree of similarity or difference between groups. Note too that lineages can be rotated around nodes in the tree, so the vertical order of lineages is also largely arbitrary (**Figure 22.2B**). The important information in the tree is the branching order along the horizontal axis, as this indicates when the various lineages last shared a common ancestor.

Any group of species that we designate or name is called a **taxon** (plural *taxa*). Some examples of familiar taxa include humans, primates, mammals, and vertebrates (note that in this

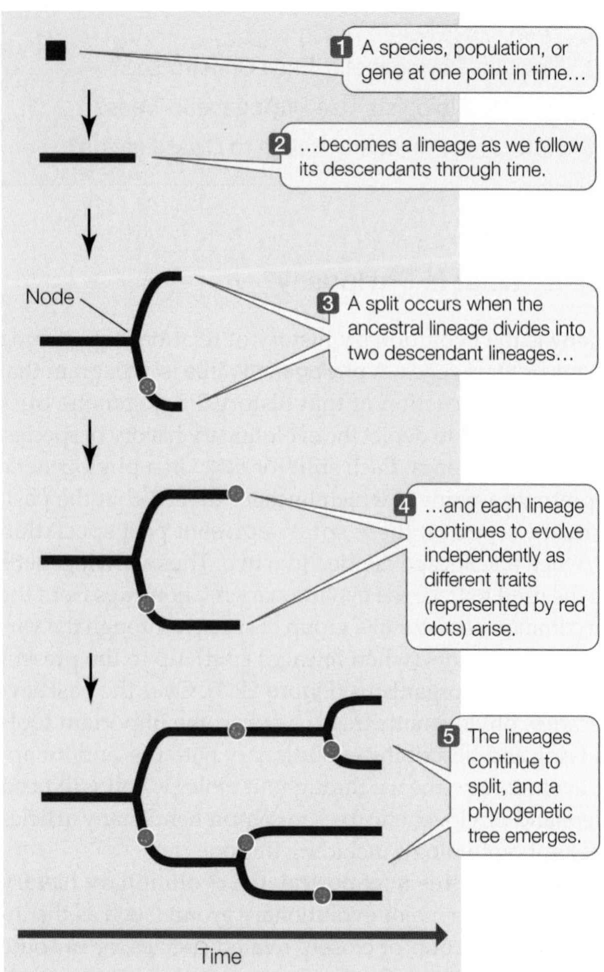

1 A species, population, or gene at one point in time...
2 ...becomes a lineage as we follow its descendants through time.
Node
3 A split occurs when the ancestral lineage divides into two descendant lineages...
4 ...and each lineage continues to evolve independently as different traits (represented by red dots) arise.
5 The lineages continue to split, and a phylogenetic tree emerges.
Time

**22.1 A Phylogenetic Tree** Evolutionary relationships among lineages, as well as the evolution of new traits, can be represented in a treelike diagram.

series, each taxon in the list is also a member of the next, more inclusive taxon). Any taxon that consists of all the evolutionary descendants of a common ancestor is called a **clade**. Clades can be identified by picking any point on a phylogenetic tree and then tracing all the descendant lineages to the tips of the terminal branches (**Figure 22.3**). Two species that are each other's closest relatives are called **sister species**; similarly, any two clades that are each other's closest relatives are called **sister clades**.

Before the 1980s, phylogenetic trees tended to be seen only in the literature on evolutionary biology, especially in the area of **systematics**: the study and classification of biodiversity. But almost every journal in the life sciences published during the last few years contains phylogenetic trees. Trees are widely used in molecular biology, biomedicine, physiology, behavior, ecology, and virtually all other fields of biology. Why have phylogenetic studies become so important?

### All of life is connected through evolutionary history

In biology, we study life at all levels of organization—from genes, cells, organisms, populations, and species to the major divisions of life. In most cases, however, no individual gene or organism (or other unit of study) is exactly like any other gene or organism that we investigate.

Consider the individuals in your biology class. We recognize each person as an individual human, but we know that no two are exactly alike. If we knew everyone's family tree in detail, the genetic similarity of any pair of students would be more predictable. We would find that more closely related students have many more traits in common (from the color of their hair to their susceptibility or resistance to diseases). Likewise, biologists use phylogenies to make comparisons and predictions about shared traits across genes, populations, and species.

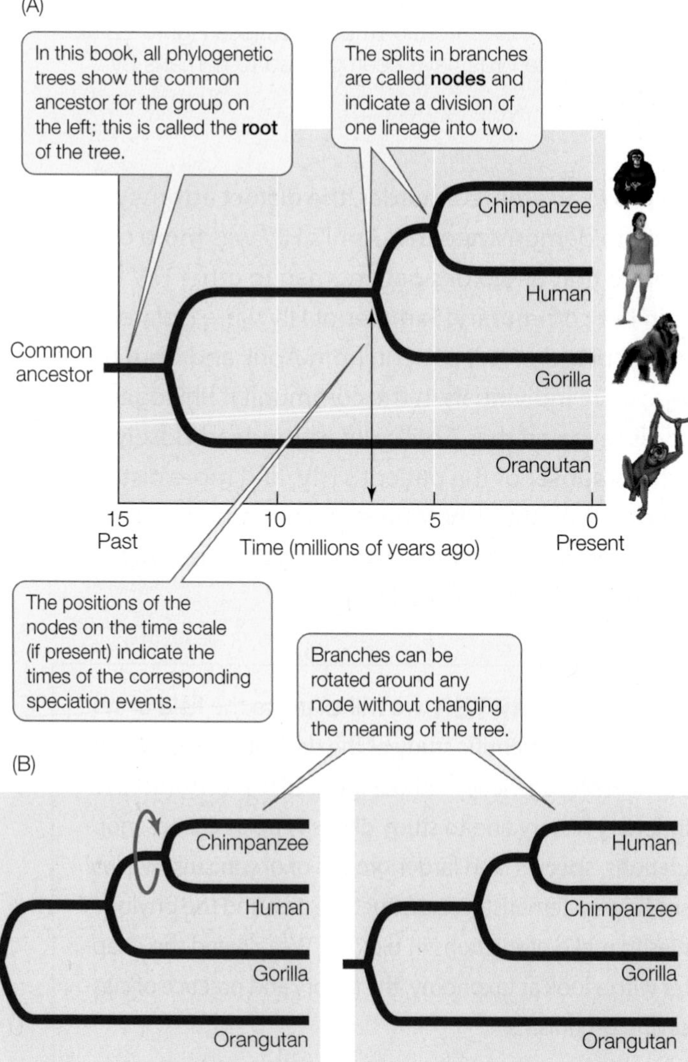


**22.2 How to Read a Phylogenetic Tree** (A) A phylogenetic tree displays the evolutionary relationships among organisms. Such trees can be produced with time scales, as shown here, or with no indication of time. If no time scale is shown, then the branch lengths show relative rather than absolute times of divergence. (B) Lineages can be rotated around a given node, so the vertical order of taxa is largely arbitrary.

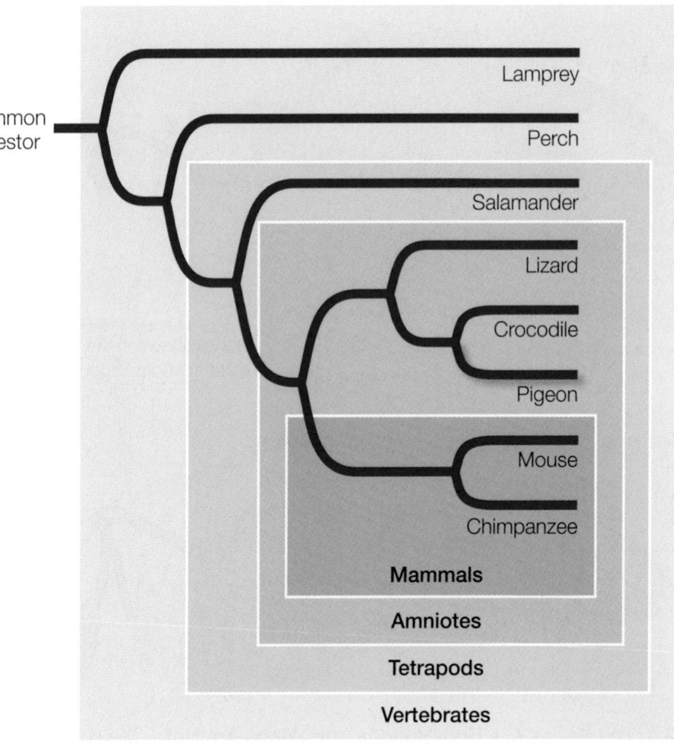


**22.3 Clades Represent All the Descendants of a Common Ancestor** All clades are subsets of larger clades, with all of life as the most inclusive taxon. In this example, the groups called mammals, amniotes, tetrapods, and vertebrates represent successively larger clades. Only a few species within each clade are represented on the tree.

One of the great unifying concepts in biology is that all life is connected through its evolutionary history. The complete evolutionary history of life is known as the **tree of life**. Biologists estimate that there are tens of millions of species on Earth. Only about 1.8 million have been formally described and named. New species are being discovered and named all the time, and phylogenetic analyses reviewed and revised, but our knowledge of the tree of life is far from complete, even for known species. Yet knowledge of evolutionary relationships is essential for making comparisons in biology, so biologists build phylogenies for groups of interest as the need arises. The evolutionary relationships among species, as shown in the tree of life, form the basis for biological classification. This evolutionary framework allows biologists to make many predictions about the behavior, ecology, physiology, genetics, and morphology of species that have not yet been studied in detail.

### Comparisons among species require an evolutionary perspective

When biologists make comparisons among species, they observe traits that differ within the group of interest and try to ascertain when these traits evolved. In many cases, investigators are interested in how the evolution of a trait depends on environmental conditions or selective pressures. For instance, scientists have used phylogenetic analyses to discover changes in the genome of HIV that confer resistance to particular drug treatments. The association of a particular genetic change in HIV with a particular treatment provides a hypothesis about the evolution of resistance that can be tested experimentally.

Any features shared by two or more species that have been inherited from a common ancestor are said to be **homologous**. Homologous features may be any heritable traits, including DNA sequences, protein structures, anatomical structures, and even some behavior patterns. Traits that are shared across a group of interest are likely to have been inherited from a common ancestor. For example, all living vertebrates have a vertebral column, and all known fossil vertebrates had a vertebral column. Therefore, the vertebral column is judged to be homologous in all vertebrates.

In tracing the evolution of a trait, biologists distinguish between *ancestral* and *derived* traits. A trait that was already present in the ancestor of a group is known as an **ancestral trait** for that group. A trait found in a descendent that differs from its ancestral form is called a **derived trait**. Derived traits that are shared among a group of organisms, and are viewed as evidence of the common ancestry of the group, are called **synapomorphies** (*syn*, "shared"; *apo*, "derived"; *morph*, "form," referring to the "form" of a trait). Thus the vertebral column is considered a synapomorphy—a shared, derived trait—of the vertebrates.

Not all similar traits are evidence of relatedness, however. Similar traits in unrelated groups of organisms can develop for either of the following reasons:

- Independently evolved traits subjected to similar selection pressures may become superficially similar, a phenomenon called **convergent evolution**. For example, although the wing bones of bats and birds are homologous, having been inherited from a common ancestor, the wings of bats and the wings of birds are not homologous because they evolved independently from the forelimbs of different nonflying ancestors (**Figure 22.4**).
- A character may revert from a derived state back to an ancestral state in an event called an **evolutionary reversal**. For example, most frogs lack teeth in the lower jaw, but the ancestor of frogs did have such teeth. Teeth have been regained in the lower jaw of one South American species, and thus represent an evolutionary reversal in that species.

Similar traits generated by convergent evolution and evolutionary reversals are called *homoplastic traits* or **homoplasies**.

A particular trait may be ancestral or derived, depending on our point of reference in a phylogeny. For example, all birds have feathers, which are highly modified scales. We infer from this that feathers were present in the common ancestor of modern birds. Therefore, we consider the presence of feathers to be an *ancestral* trait for any particular group of modern birds, such as the songbirds. However, feathers are not present in any other living animals. If we were reconstructing a phylogeny of all

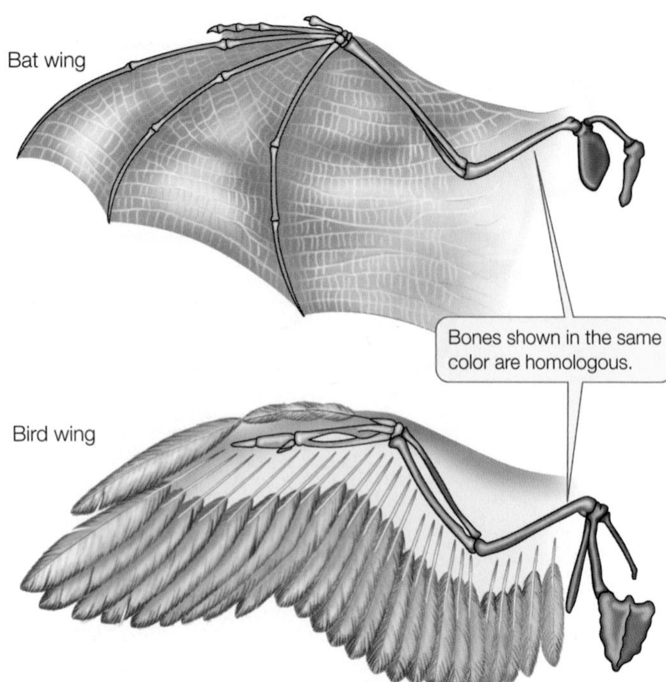


**22.4 The Bones Are Homologous, the Wings Are Not** The supporting bone structures of both bat wings and bird wings are derived from a common four-limbed ancestor and are thus homologous. However, the wings themselves—an adaptation for flight—evolved independently in the two groups.

living vertebrates, the presence of feathers would be a *derived* trait that is found only among birds (and thus a synapomorphy of the birds).

## 22.1 RECAP

**A phylogenetic tree is a description of evolutionary relationships—how a group of genes, populations, or species have evolved from a common ancestor. All living organisms share a common ancestor and are related through the phylogenetic tree of life.**

- Do you understand the different elements of a phylogenetic tree? See pp. 465–466 and Figure 22.2
- Explain the difference between an ancestral and a derived trait. See p. 467
- Do you see how similar traits might arise independently in species that are only distantly related? See p. 467 and Figure 22.4

Phylogenetic analyses have become increasingly important to many types of biological research in recent years, and they are the basis for the comparative nature of biology. For the most part, however, evolutionary history cannot be observed directly. How, then, do biologists reconstruct the past? One way is by using phylogenetic analyses to construct a tree.

## 22.2 How Are Phylogenetic Trees Constructed?

To illustrate how a phylogenetic tree is constructed, let's consider the eight vertebrate animals listed in **Table 22.1**: lamprey, perch, salamander, lizard, crocodile, pigeon, mouse, and chimpanzee. We will assume initially that a given derived trait evolved only once during the evolution of these animals (that is, there has been no convergent evolution), and that no derived traits were lost from any of the descendant groups (there has been no evolutionary reversal). For simplicity, we have selected traits that are either present (+) or absent (–).

In a phylogenetic study, the group of organisms of primary interest is called the **ingroup**. As a point of reference, an ingroup is compared with an **outgroup**. a closely related species or group known to be phylogenetically outside the group of interest. If the outgroup is known to have diverged before the ingroup, the outgroup can be used to determine which traits of the ingroup are derived (evolved within the ingroup) and which are ancestral (evolved before the origin of the ingroup). As we will see in Chapter 33, a group of jawless fishes called the lampreys is thought to have separated from the lineage leading to the other vertebrates before the jaw arose. Therefore, we have included the lamprey as the outgroup for our analysis. Because derived traits are traits acquired by other members of the vertebrate lineage *after* they diverged from the outgroup, any trait that is present in both the lamprey and the other vertebrates is judged to be ancestral.

We begin by noting that the chimpanzee and mouse share two derived traits—mammary glands and fur—that are absent in both the outgroup and in the other species of the ingroup. Therefore, we infer that mammary glands and fur are derived traits that evolved in a common ancestor of chimpanzees and mice after that lineage separated from the lineages leading to the other vertebrates. In other words, we provisionally assume that mammary glands and fur evolved only once among the animals in our ingroup. These characters are synapomorphies that unite chimpanzees and mice (as well as all other mammals, although we have not included other mammalian species in this example). By the same reasoning, we can infer that the other shared derived traits are synapomorphies for the various groups in which they are expressed. For instance, keratinous scales are a synapomorphy of the lizard, crocodile, and pigeon.

Table 22.1 also tells us that, among the animals in our ingroup, the pigeon has a unique trait: the presence of feathers. Feathers are a synapomorphy of birds, but since we only have one bird in this example, the presence of feathers provides no clues concerning relationships among the eight species of vertebrates we have sampled. However, gizzards are found in birds and crocodiles, so this trait is evidence of a close relationship between birds and crocodilians.

By combining information about the various synapomorphies, we can construct a phylogenetic tree. We infer, for example, that mice and chimpanzees, the only two animals that share fur and mammary glands in our example, share a more recent common ancestor with each other than they do with

## TABLE 22.1

### Eight Vertebrates Ordered According to Unique Shared Derived Traits

| | DERIVED TRAIT[a] | | | | | | | |
|---|---|---|---|---|---|---|---|---|
| TAXON | JAWS | LUNGS | CLAWS OR NAILS | GIZZARD | FEATHERS | FUR | MAMMARY GLANDS | KERATINOUS SCALES |
| Lamprey (outgroup) | – | – | – | – | – | – | – | – |
| Perch | + | – | – | – | – | – | – | – |
| Salamander | + | + | – | – | – | – | – | – |
| Lizard | + | + | + | – | – | – | – | + |
| Crocodile | + | + | + | + | – | – | – | + |
| Pigeon | + | + | + | + | + | – | – | + |
| Mouse | + | + | + | – | – | + | + | – |
| Chimpanzee | + | + | + | – | – | + | + | – |

[a] A plus sign indicates the trait is present, a minus sign that it is absent.

pigeons and crocodiles. Otherwise, we would need to assume that the ancestors of pigeons and crocodiles also had fur and mammary glands but subsequently lost them—unnecessary additional assumptions.

**Figure 22.5** shows a phylogenetic tree for the vertebrates in Table 22.1, based on the shared derived traits we examined and the assumption that each derived trait evolved only once. This particular tree was easy to construct because the animals and characters we chose met the assumptions that derived traits appeared only once and were never lost after they appeared. Had we included a snake in the group, our second assumption would have been violated, because we know that the lizard ancestors of snakes had limbs that were subsequently lost. We would need to examine additional characters to determine that the lineage leading to snakes separated from the one leading to lizards long after the lineage leading to lizards separated from

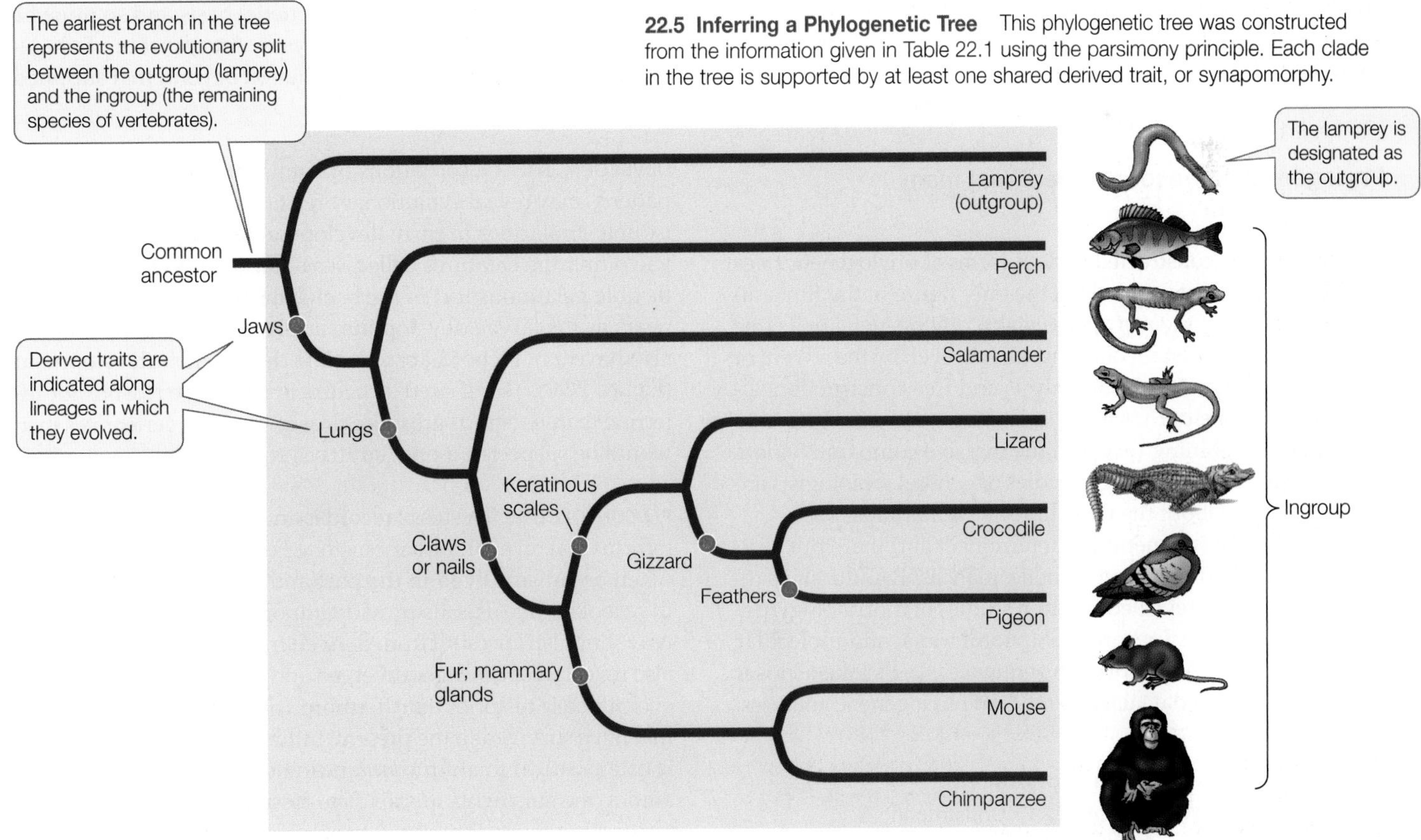
The earliest branch in the tree represents the evolutionary split between the outgroup (lamprey) and the ingroup (the remaining species of vertebrates).
Common ancestor
Jaws
Derived traits are indicated along lineages in which they evolved.
Lungs
Claws or nails
Keratinous scales
Gizzard
Feathers
Fur; mammary glands
Lamprey (outgroup)
Perch
Salamander
Lizard
Crocodile
Pigeon
Mouse
Chimpanzee
The lamprey is designated as the outgroup.
Ingroup

**22.5 Inferring a Phylogenetic Tree** This phylogenetic tree was constructed from the information given in Table 22.1 using the parsimony principle. Each clade in the tree is supported by at least one shared derived trait, or synapomorphy.

the others. In fact, the analysis of several characters shows that snakes evolved from burrowing lizards that became adapted to a subterranean existence.

### Parsimony provides the simplest explanation for phylogenetic data

The phylogenetic tree shown in Figure 22.5 is based on only a very small sample of traits. Typically, biologists construct phylogenetic trees using hundreds or thousands of traits. With larger data sets, we would expect to observe some traits that have changed more than once, and thus we would expect to see some convergence and evolutionary reversal. How do we determine which traits are synapomorphies and which are homoplasies? One way is to invoke the principle of *parsimony*.

In its most general form, the **parsimony principle** states that the preferred explanation of observed data is the simplest explanation. Applying the principle of parsimony to the reconstruction of phylogenies entails minimizing the number of evolutionary changes that need to be assumed over all characters in all groups in the tree. In other words, the best hypothesis under the parsimony principle is one that requires the fewest homoplasies. This application of parsimony is a specific case of a general principle of logic called *Occam's razor*: the best explanation is the one that fits the data best while making the fewest assumptions.

We apply the parsimony principle in constructing phylogenetic trees not because all evolutionary changes always occurred parsimoniously, but because it is logical to adopt the simplest explanation that can account for the observed data. More complicated explanations are accepted only when the evidence requires them. Phylogenetic trees represent our best estimates about evolutionary relationships. They are continually modified as additional evidence becomes available.

### Phylogenies are reconstructed from many sources of data

Naturalists have constructed various forms of phylogenetic trees for more than 150 years. In fact, the only figure in the first edition of Darwin's *Origin of Species* was a phylogenetic tree. Tree construction has been revolutionized, however, by the advent of computer software for trait analysis and tree construction, allowing us to consider far more data than could ever before be processed. Combining this with the massive comparative data sets being generated through studies of genomes, biologists are learning details about the tree of life at a remarkable pace.

Any trait that is genetically determined, and therefore heritable, can be used in a phylogenetic analysis. Evolutionary relationships can be revealed through studies of morphology, development, the fossil record, behavioral traits, and molecular traits such as DNA and protein sequences. Let's take a closer look at the types of data used in modern phylogenetic analyses.

yourBioPortal.com
GO TO Web Activity 22.1 • Constructing a Phylogenetic Tree

**MORPHOLOGY** An important source of phylogenetic information is *morphology*: the presence, size, shape, and other attributes of body parts. Since living organisms have been observed, depicted, and studied for millenia, we have a wealth of recorded morphological data as well as extensive museum and herbarium collections of organisms whose traits can be measured. New technological tools, such as the electron microscope and computed tomography (CT) scans, enable systematists to examine and analyze the structures of organisms at much finer scales than was formerly possible.

Most species are described and known primarily by their morphology, and morphology provides the most comprehensive data set available for many taxa. The features of morphology that are important for phylogenetic analysis are often specific to a particular group of organisms. For example, the presence, development, shape, and size of various features of the skeletal system are important for the study of vertebrate phylogeny, whereas floral structures are important for studying the relationships among flowering plants (*angiosperms*).

Although often useful, morphological approaches to phylogenetic analysis have some limitations. Some taxa exhibit little morphological diversity, despite great species diversity. For example, the phylogeny of the leopard frogs of North and Central America would be difficult to infer from morphological differences alone, because the many species look very similar, despite important differences in their behavior and physiology. At the other extreme, few morphological traits can be compared across distantly related species (consider earthworms and mammals, for instance). Some morphological variation has an environmental (rather than a genetic) basis and so must be excluded from phylogenetic analyses. An accurate phylogenetic analysis often requires information beyond that supplied by morphology.

**DEVELOPMENT** Observations of similarities in developmental patterns may reveal evolutionary relationships. Some organisms exhibit similarities in early developmental stages only. The larvae of marine creatures called sea squirts, for example, have a flexible gelatinous rod in the back—the *notochord*—that disappears as the larvae develop into adults. All vertebrate animals also have a notochord at some time during their development (**Figure 22.6**). This shared structure is one of the reasons for inferring that sea squirts are more closely related to vertebrates than would be suspected if only adult sea squirts were examined.

**PALEONTOLOGY** The fossil record is another important source of information on evolutionary history. Fossils show us where and when organisms lived in the past and give us an idea of what they looked like. Fossils provide important evidence that helps us distinguish ancestral from derived traits. The fossil record can also reveal when lineages diverged and began their independent evolutionary histories. Furthermore, in groups with few species that have survived to the present, information on extinct species is often critical to an understanding of the large divergences among the surviving species. The fossil record does have limitations, however. Few or no fossils have been found for some groups, and the fossil record for many groups is fragmentary.

Sea squirt larva
Neural tube
Notochord
Sea squirt and frog larvae (tadpoles) share several morphological similarities, including the presence of a notochord for body support.
Frog larva
Neural tube
Notochord
Adult
Adult
Despite the similarity of their larvae, the morphology of adult frogs and sea squirts provides little evidence of the common ancestry of these two groups.

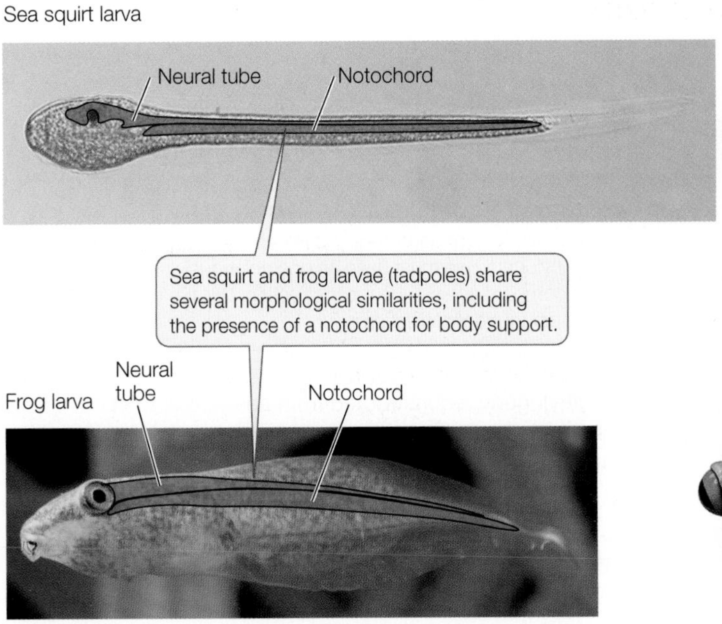

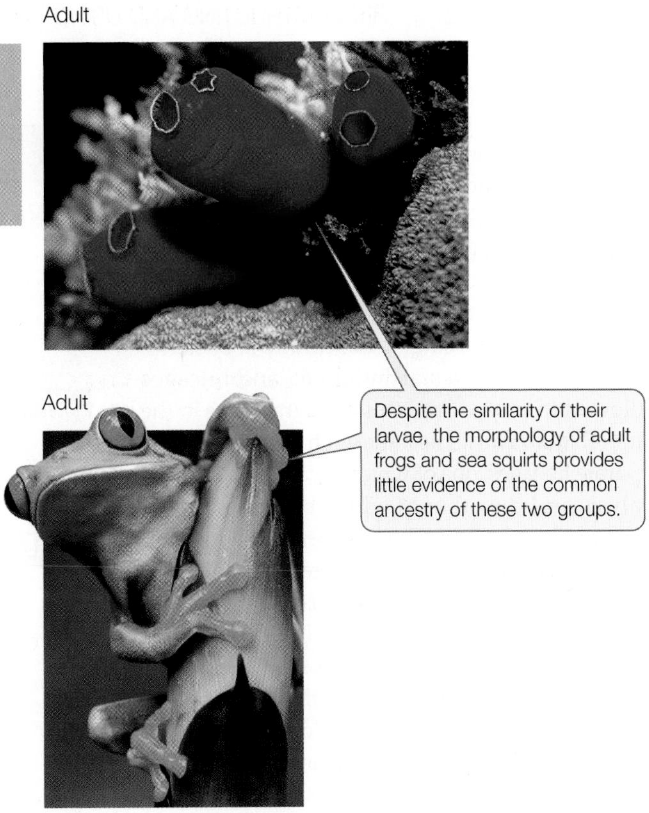

**22.6 The Evolutionary Relationship Between Sea Squirts and Vertebrates** All chordates—a taxonomic group that includes sea squirts and frogs—have notochords at some stage of their development. The larvae share similarities that are not apparent in the adults. Such similarities in development can provide useful evidence of evolutionary relationships. The notochord is lost in adult sea squirts. In adult frogs, as in all vertebrates, the vertebral column replaces the notochord as the support structure.

**BEHAVIOR** Some behavioral traits are culturally transmitted and some are inherited. If a particular behavior is culturally transmitted, it may not accurately reflect evolutionary relationships (but may nonetheless reflect cultural connections). Bird songs, for instance, are often learned and may be inappropriate traits for phylogenetic analysis. Frog calls, however, are genetically determined and appear to be acceptable sources of information for reconstructing phylogenies.

**MOLECULAR DATA** All heritable variation is encoded in DNA, and so the complete genome of an organism contains an enormous set of traits (the individual nucleotide bases of DNA) that can be used in phylogenetic analyses. In recent years, DNA sequences have become among the most widely used sources of data for constructing phylogenetic trees. Comparisons of nucleotide sequences are not limited to the DNA in the cell nucleus. Eukaryotes have genes in their mitochondria as well as in their nuclei; plant cells also have genes in their chloroplasts. The chloroplast genome (cpDNA), which is used extensively in phylogenetic studies of plants, has changed slowly over evolutionary time, so it is often used to study relatively ancient phylogenetic relationships. Most animal mitochondrial DNA (mtDNA) has changed more rapidly, so mitochondrial genes have been used extensively to study evolutionary relationships among closely related animal species (the mitochondrial genes of plants evolve more slowly). Many nuclear gene sequences are also commonly analyzed, and now that several entire genomes have been sequenced, they too are used to construct phylogenetic trees. Information on gene products (such as the amino acid sequences of proteins) is also widely used for phylogenetic analyses, as we discuss in Chapter 24.

### Mathematical models expand the power of phylogenetic reconstruction

As biologists began to use DNA sequences to infer phylogenies in the 1970s and 1980s, they developed explicit mathematical models describing how DNA sequences change over time. These models account for multiple changes at a given position in a DNA sequence. They also take into account different rates of change at different positions in a gene, at different positions in a codon, and among different nucleotides (see Section 24.1). For example, *transitions* (changes between two purines or between two pyrimidines) are usually more likely than are *transversions* (changes between a purine and pyrimidine).

Mathematical models can be used to compute how a tree might evolve given the observed data. A **maximum likelihood** method will identify the tree that most likely produced the observed data, given the assumed model of evolutionary change. Maximum likelihood methods can be used for any kind of characters, but they are most often used with molecular data, for which explicit mathematical models of evolutionary change are easier to develop. The principal advantages to maximum likelihood analyses are that they incorporate more information about evolutionary change than do parsimony methods, and they are easier to treat in a statistical framework. The principal disadvantages are that they are computationally intensive and require explicit models of evolutionary change (which may not be available for some kinds of character change).

### The accuracy of phylogenetic methods can be tested

If phylogenetic trees represent reconstructions of past events, and if many of these events occurred before any humans were around to witness them, how can we test the accuracy of phylogenetic methods? Biologists have conducted experiments both in living organisms and with computer simulations that have demonstrated the effectiveness and accuracy of phylogenetic methods.

In one experiment designed to test the accuracy of phylogenetic analysis, a single viral culture of bacteriophage T7 was used as a starting point, and lineages were allowed to evolve from this ancestral virus in the laboratory (**Figure 22.7**). The initial culture was split into two separate lineages, one of which became the ingroup for analysis and the other of which became the outgroup for rooting the tree. The lineages in the ingroup were split in two after every 400 generations, and samples of the virus were saved for analysis at each branching point. The lineages were allowed to evolve until there were eight lineages in the ingroup. Mutagens were added to the viral cultures to increase the mutation rate so that the amount of change and the degree of homoplasy would be typical of the organisms analyzed in average phylogenetic analyses. The investigators then sequenced samples from the end points of the eight lineages, as well as from the ancestors at the branching points. They then gave the sequences from the end points of the lineages to other investigators to analyze, without revealing the known history of the lineages or the sequences of the ancestral viruses.

After the phylogenetic analysis was completed, the investigators asked two questions: Did phylogenetic methods reconstruct the known history correctly, and were the sequences of the ancestral viruses reconstructed accurately? The answer in both cases was yes: the branching order of the lineages was reconstructed exactly as it had occurred, more than 98 percent of the nucleotide positions of the ancestral viruses were reconstructed correctly, and 100 percent of the amino acid changes in the viral proteins were reconstructed correctly.

**yourBioPortal.com**

**GO TO Animated Tutorial 22.1 • Using Phylogenetic Analysis to Reconstruct Evolutionary History**

The experiment shown in Figure 22.7 demonstrated that phylogenetic analysis was accurate under the conditions tested, but it did not examine all possible conditions. Other experimental studies have taken other factors into account, such as the sensitivity of phylogenetic analysis to convergent environments and highly variable rates of evolutionary change. In addition, computer simulations based on evolutionary models have been used extensively to study the effectiveness of phylogenetic analysis. These studies have also confirmed the accuracy of phylogenetic methods and have been used to refine those methods and extend them to new applications.

## INVESTIGATING LIFE

### 22.7 The Accuracy of Phylogenetic Analysis

To test whether analysis of gene sequences can accurately reconstruct evolutionary phylogeny, we must have an unambiguously known phylogeny to compare against the reconstruction. Will the observed phylogeny match the reconstruction?

**HYPOTHESIS** **A phylogeny reconstructed from analysis of the DNA sequences of living organisms can accurately match the known evolutionary history of the organisms.**

**METHOD** In the laboratory, researchers produced an unambiguous phylogeny of nine viral lineages, enhancing the mutation rate to increase variation among the lineages.

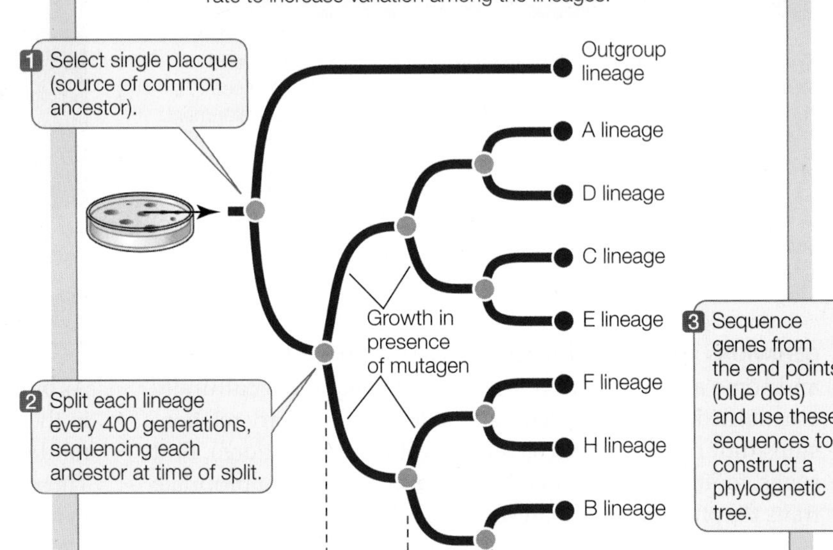
1 Select single placque (source of common ancestor).
Outgroup lineage
A lineage
D lineage
C lineage
E lineage
F lineage
H lineage
B lineage
G lineage
Growth in presence of mutagen
2 Split each lineage every 400 generations, sequencing each ancestor at time of split.
3 Sequence genes from the end points (blue dots) and use these sequences to construct a phylogenetic tree.
400
400
400
Generations

Viral sequences from the end points of each lineage (blue dots) were subjected to phylogenetic analysis by investigators who were unaware of the history of the lineages or the gene sequences of the ancestral viruses. These investigators reconstructed the phylogeny based solely on their analyses of the descendants' genomes.

**RESULTS** The true phylogeny and ancestral DNA sequences were accurately reconstructed solely from the DNA sequences of the viruses at the tips of the tree.

**CONCLUSION** **Phylogenetic analysis of DNA sequences can accurately reconstruct evolutionary history.**

**FURTHER INVESTIGATION:** The lineages in this experiment evolved under similar conditions. How might changing environmental conditions for some of the lineages affect the result?

Go to **yourBioPortal.com** for original citations, discussions, and relevant links for all INVESTIGATING LIFE figures.

## 22.2 RECAP

**Phylogenetic trees can be constructed by using the parsimony principle to find the simplest explanation for the evolution of traits. Maximum likelihood methods incorporate more explicit models of evolutionary change to reconstruct evolutionary history.**

- Do you understand how a phylogenetic tree is constructed? See pp. 468–471 and Figure 22.5
- Is there a way to test whether phylogenetic trees provide accurate reconstructions of evolutionary history? See p. 474 and Figure 22.7

Biologists in many fields now routinely reconstruct phylogenetic relationships. Let's examine some of the many uses of these phylogenetic trees.

# 22.3 How Do Biologists Use Phylogenetic Trees?

Information about the evolutionary relationships among organisms is useful to scientists investigating a wide variety of biological questions. In this section we will illustrate how phylogenetic trees can be used to ask questions about the past, and to compare aspects of the biology of organisms in the present.

### Phylogenies help us reconstruct the past

Most flowering plants reproduce by mating with another individual—a process called *outcrossing*. Many outcrossing species have mechanisms to prevent self-fertilization, and so are referred to as *self-incompatible*. Individuals of some species, however, regularly fertilize themselves with their own pollen; they are termed *selfing* species, which of course requires that they be *self-compatible*. How can we tell how often self-compatibility has evolved in a group of plants? We can do so by conducting a phylogenetic analysis of outcrossing and selfing species and testing the species for self-compatibility.

The evolution of fertilization mechanisms was examined in *Linanthus*, a genus in the phlox family that exhibits a diversity of breeding systems and pollination mechanisms. The outcrossing species of *Linanthus* have long petals, are pollinated by long-tongued flies, and are self-incompatible. The self-pollinating species of *Linanthus*, in contrast, all have short petals and do not require insect pollinators to reproduce successfully. Investigators reconstructed a phylogeny for 12 species in the genus using nuclear ribosomal DNA sequences (**Figure 22.8**). They determined whether each species was self-compatible by artificially pollinating flowers with the plant's own pollen or with pollen from other individuals and observing whether viable seeds formed.

Several lines of evidence suggest that self-incompatibility is the ancestral state in *Linanthus*. Multiple origins of self-incompatibility have not been found in any flowering plant family to date. Self-incompatibility depends on physiological mechanisms in both the pollen and the stigma (the female organ on which pollen lands) and is under the control of least three different alleles. Therefore, a change from self-incompatibility to

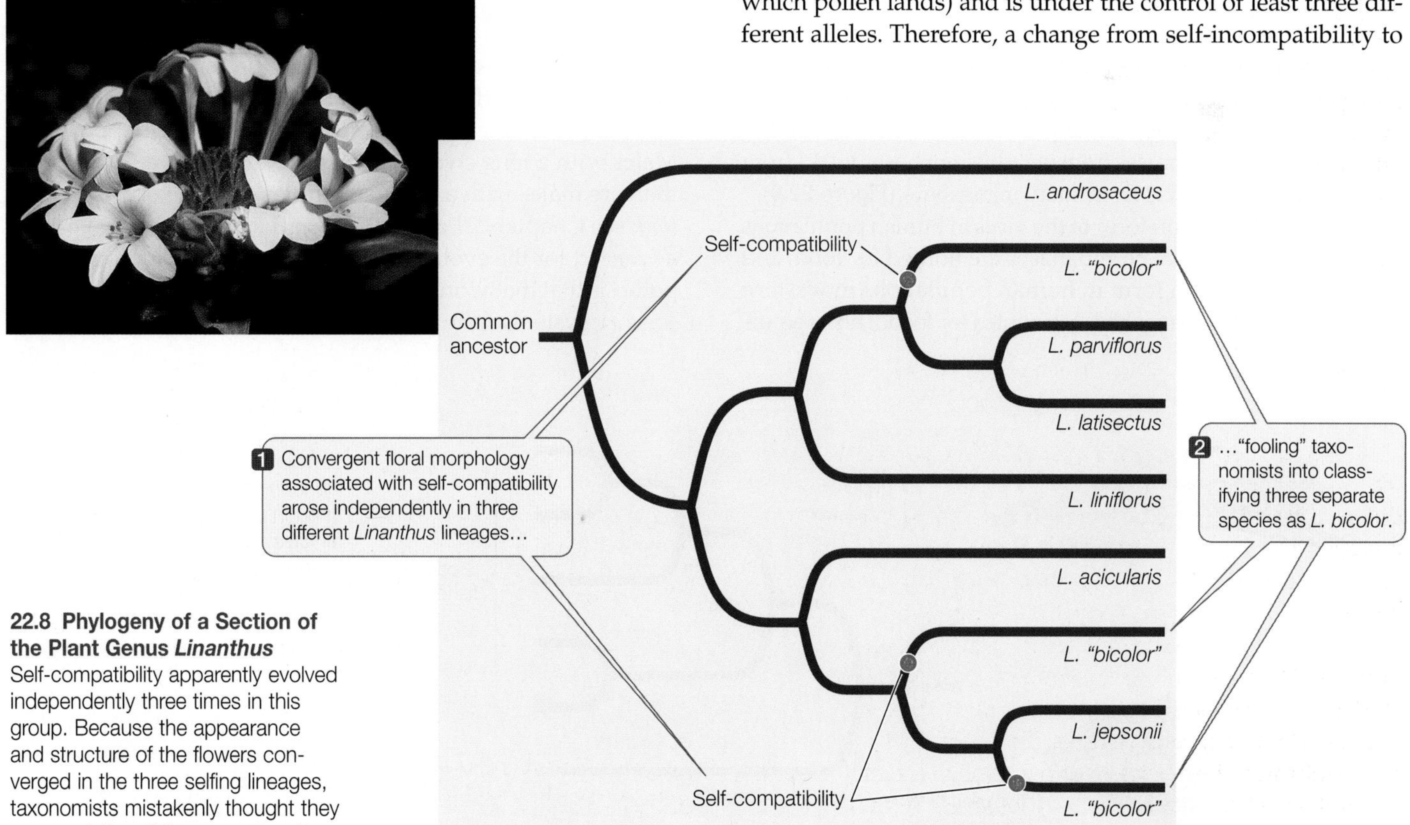


**22.8 Phylogeny of a Section of the Plant Genus *Linanthus*** Self-compatibility apparently evolved independently three times in this group. Because the appearance and structure of the flowers converged in the three selfing lineages, taxonomists mistakenly thought they were varieties of the same species.

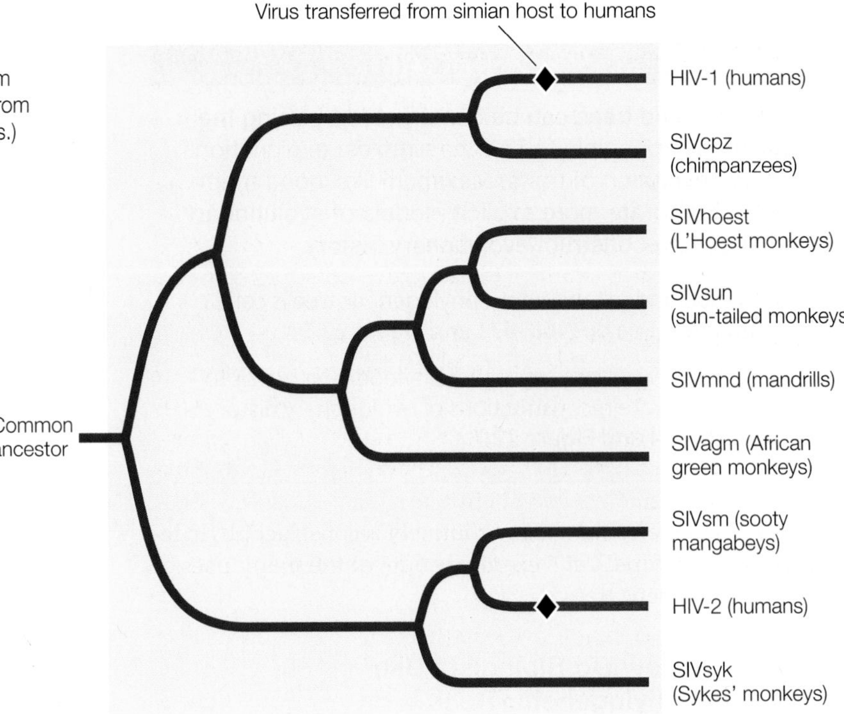


**22.9 Phylogenetic Tree of Immunodeficiency Viruses** Immunodeficiency viruses have been transmitted to humans from two different simian hosts: HIV-1 from chimpanzees and HIV-2 from sooty mangabeys. (SIV stands for simian immunodeficiency virus.)

self-compatibility would be easier than the reverse change. In addition, in all self-incompatible species of *Linanthus*, the site of pollen rejection is the stigma, even though sites of pollen rejection vary greatly among other plant families.

Assuming that self-incompatibility is the ancestral state, the reconstructed phylogeny suggests that self-compatibility evolved three times within this group of *Linanthus* (see Figure 22.8). The change to self-compatibility has been accompanied by the evolution of reduced petal size. Interestingly, the striking similarity of the flowers in the self-compatible groups once led to their being classified as members of a single species. The phylogenetic analysis using ribosomal DNA showed them to be members of three distinct lineages, however.

Reconstructing the past is important for understanding many biological processes. In the case of *zoonotic* diseases (diseases caused by infectious organisms transmitted to humans from another animal host), it is important to understand when, where, and how the disease first entered a human population. Human immunodeficiency virus (HIV) is the cause of such a zoonotic disease, acquired immunodeficiency syndrome, or AIDS. As we described in the opening to this chapter, phylogenetic analyses have become important for studying the transmission of viruses such as HIV. Phylogenies are also important for understanding the present global diversity of HIV and for determining the virus's origins in human populations. A broader phylogenetic analysis of immunodeficiency viruses shows that humans acquired these viruses from two different hosts: HIV-1 from chimpanzees, and HIV-2 from sooty mangabeys (**Figure 22.9**).

HIV-1 is the common form of the virus in human populations in central Africa, where chimpanzees are hunted for food, and HIV-2 is the common form in human populations in western Africa, where sooty mangabeys are hunted for food. Thus it seems likely that these viruses entered human populations through hunters who cut themselves while skinning chimpanzees and sooty mangabeys. The relatively recent global pandemic of AIDS occurred when these infections in local African populations rapidly spread through human populations around the world.

### Phylogenies allow us to compare and contrast living organisms

Male swordtails—a group of fishes in the genus *Xiphophorus*—have a long, colorful tail extension (**Figure 22.10A**), and their reproductive success is closely associated with this appendage. Males with a long sword are more likely to mate successfully than are males with a short sword (an example of *sexual selection*; see Chapters 21 and 23). Several explanations have been advanced for the evolution of this structure, including the hypothesis that the sword simply exploits a preexisting bias in the sensory system of the females. This *sensory exploitation hypoth-*

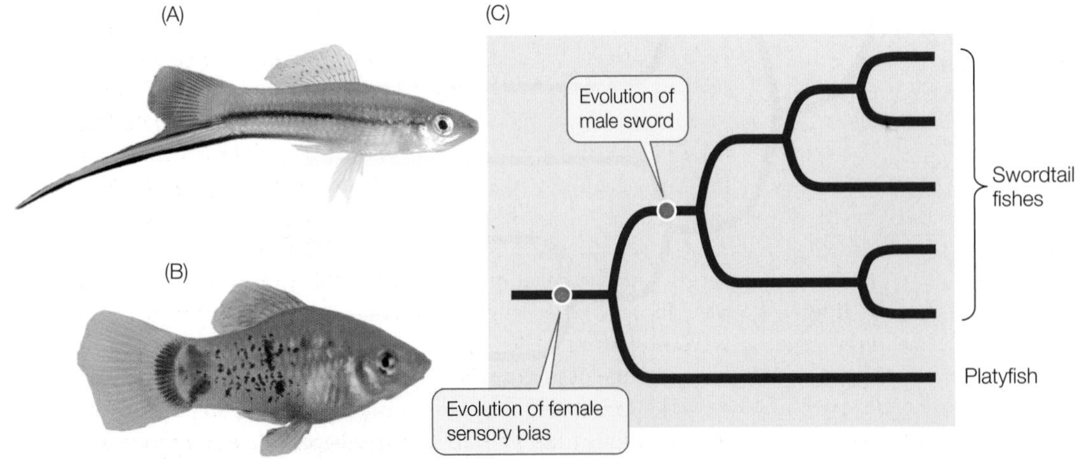


**22.10 The Origin of a Sexually Selected Trait** (A) The large tail of male swordtail fishes (genus *Xiphophorus*) apparently evolved through sexual selection, with females mating preferentially with males with a longer "sword." (B) A male platyfish, member of a related species. (C) Phylogenetic analysis reveals that the platyfishes split from the swordtails before the evolution of the sword. The independent finding that female platyfishes prefer males with an artificial sword further supports the idea that this appendage evolved as a result of a preexisting preference in the females.

*esis* suggests that female swordtails had a preference for males with long tails even before the tails evolved (perhaps because females assess the size of males by their total body length—including the tail—and prefer larger males).

To test the sensory exploitation hypothesis, a phylogeny was used to identify the swordtail relatives that had split most recently from their lineage before the evolution of sword extensions. These closest relatives turned out to be the platyfishes, another group of *Xiphophorus* (**Figure 22.10B**). Even though male platyfishes do not normally have swords, when researchers attached artificial swordlike structures to the tails of some male platyfishes, female platyfishes preferred the males with an artificial sword, thus providing support for the hypothesis that female *Xiphophorus* had a preexisting sensory bias favoring tail extensions even before the trait evolved (**Figure 22.10C**). Thus, a long tail became a sexually selected trait because of the preexisting preference of the females.

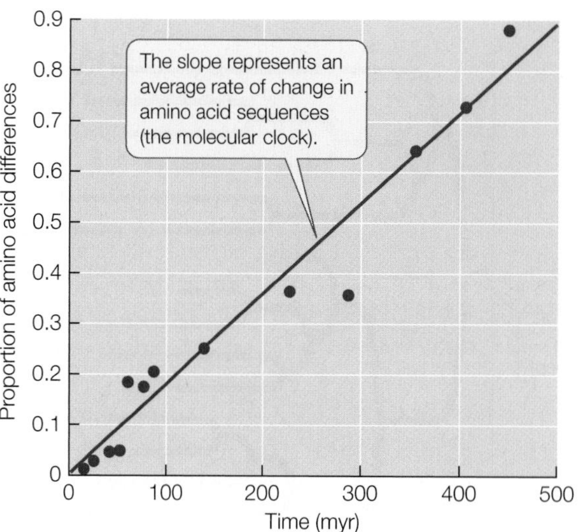


**22.11 A Molecular Clock of the Protein Hemoglobin** Amino acid replacements in hemoglobin have occurred at a relatively constant rate over nearly 500 million years of evolution. The graph shows the relationship between time of divergence and proportion of amino acid change for 13 pairs of vertebrate hemoglobin proteins. The average rate of change represents the molecular clock for hemoglobin in vertebrates.

### Ancestral states can be reconstructed

In addition to using phylogenetic methods to infer evolutionary relationships among lineages, biologists can use them to reconstruct the morphology, behavior, or nucleotide and amino acid sequences of ancestral species (as was demonstrated for the ancestral sequences of bacteriophage T7 in the experiment shown in Figure 22.7). For instance, a phylogenetic analysis was used to reconstruct an opsin protein in the ancestral archosaur (the most recent common ancestor of birds, dinosaurs, and crocodiles). Opsins are pigment proteins involved in vision; different opsins (with different amino acid sequences) are excited by different wavelengths of light. Knowledge of the opsin sequence in the ancestral archosaur would provide clues about the animal's visual capabilities and therefore about some of its probable behaviors. Investigators used phylogenetic analysis of opsin from living vertebrates to estimate the amino acid sequence of the pigment that existed in the ancestral archosaur. A protein with this same sequence was then constructed in the laboratory. The investigators tested the reconstructed opsin and found a significant shift toward the red end of the spectrum in the light sensitivity of this protein compared with that of most modern opsins. Modern species that exhibit similar sensitivity are adapted for nocturnal vision, so the investigators inferred that the ancestral archosaur might have been active at night. Thus, reminiscent of the movie *Jurassic Park*, phylogenetic analyses are being used to reconstruct extinct species, one protein at a time.

### Molecular clocks help date evolutionary events

For many applications, biologists want to know not only the order in which evolutionary lineages split but also the timing of those splits. In 1965, Emile Zuckerkandl and Linus Pauling hypothesized that rates of molecular change were constant enough that they could be used to predict evolutionary divergence times—an idea that has become known as the *molecular clock hypothesis*.

Of course, different genes evolve at different rates, and there are also differences in evolutionary rates among species related to differing generation times, environments, efficiencies of DNA repair systems, and other biological factors. Nonetheless, among closely related species, a given gene usually evolves at a reasonably constant rate. Therefore, the protein encoded by the gene also accumulates amino acid substitutions at a relatively constant rate (**Figure 22.11**). A **molecular clock** uses the average rate at which a given gene or protein accumulates changes to gauge the time of divergence for a particular split in the phylogeny. Molecular clocks must be calibrated using independent data, such as the fossil record, known times of divergence, or biogeographic dates (such as the dates for separations of continents). Using such calibrations, times of divergence have been estimated for many groups of species that have diverged over millions of years.

Molecular clocks are not only used to date ancient events; they are also used to study the timing of comparatively recent events. Most samples of HIV-1 have been collected from humans only since the early 1980s, although a few isolates from medical biopsies are available from as early as the 1950s. But biologists can use the observed changes in HIV-1 over the past several decades to project back to the common ancestor of all HIV-1 isolates, and estimate when HIV-1 first entered human populations from chimpanzees. The clock can be calibrated using the samples from the 1980s and 1990s, and then tested using the samples from the 1950s. As shown in Figure 22.12C, a sample from a 1959 biopsy is dated by molecular clock analysis at 1957 ± 10 years. The molecular clock was also used to project back to the common ancestor of this group of HIV-1 samples. Extrapolation suggests a date of origin for this group of viruses of about 1930. Although AIDS was unknown to Western medicine until the 1980s, this analysis shows that HIV-1 was present (probably at very low frequency) in human populations in Africa for at least a half-century before its emergence as a

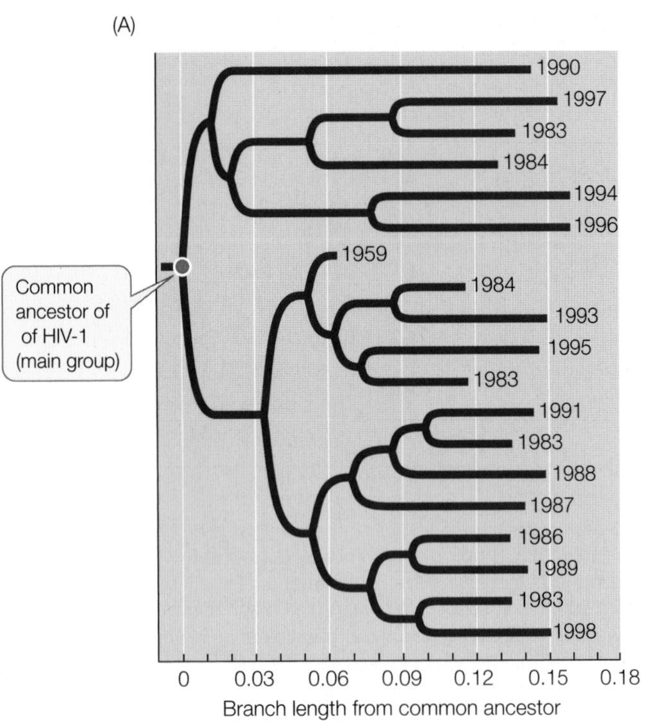


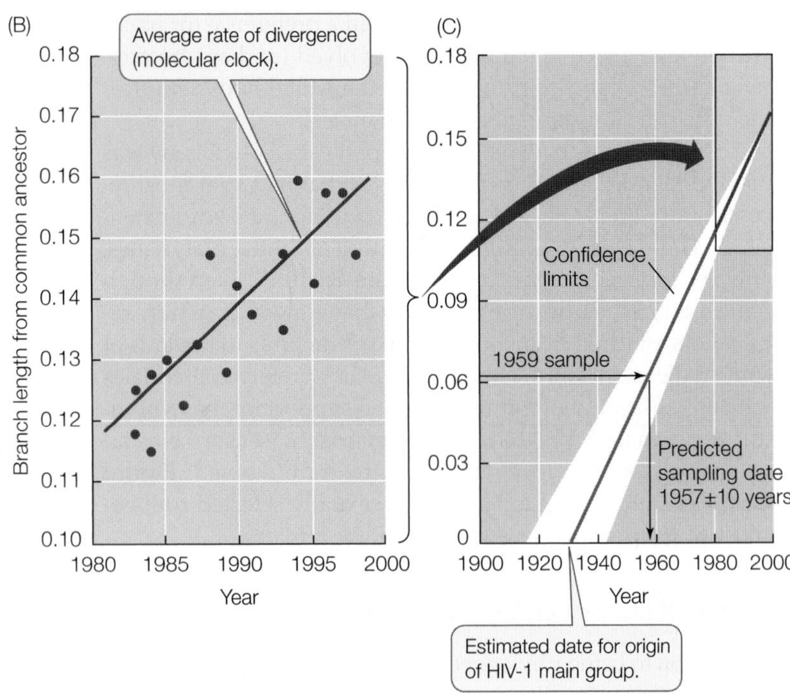


**22.12 Dating the Origin of HIV-1 in Human Populations**
(A) A phylogenetic analysis of the main group of HIV-1 viruses. The dates indicate the years in which samples were taken. (For clarity, only a small fraction of the samples that were examined in the original study are shown.) (B) A plot of year of isolation versus genetic divergence from the common ancestor provides an average rate of divergence, or a molecular clock. (C) The molecular clock is used to date a sample taken in 1959 (as a test of the clock) and the unknown date of origin of the HIV-1 main group (about 1930).

global pandemic (**Figure 22.12**). Biologists have used similar analyses to conclude that immunodeficiency viruses have been transmitted repeatedly into human populations from multiple primates for more than a century (see also Figure 22.9).

## 22.3 RECAP

**Phylogenetic trees are used to reconstruct the past history of lineages, to determine when and where traits arose, and to make relevant biological comparisons among genes, populations, and species. They can also be used to reconstruct ancestral traits and to estimate the timing of evolutionary events.**

- Explain how phylogenetic trees can help determine the number of times a particular trait evolved. See pp. 473–474 and Figure 22.8
- How does the reconstruction of ancestral traits help biologists explain the evolution of visual pigment proteins? See p. 475
- How do molecular clocks add a time dimension to phylogenetic trees? See p. 475 and Figure 22.12

All of life is connected through evolutionary history, and the relationships among organisms provide a natural basis for making biological comparisons. For these reasons, biologists use phylogenetic relationships as the basis for organizing life into a coherent classification system, described in the next section.

## 22.4 How Does Phylogeny Relate to Classification?

The biological classification system in widespread use today is derived from a system developed by the Swedish biologist Carolus Linnaeus in the mid-1700s. Linnaeus developed a naming system called **binomial nomenclature** that has allowed scientists throughout the world to refer unambiguously to the same organisms by the same names (**Figure 22.13**).

Linnaeus gave each species two names, one identifying the species itself and the other the genus to which it belongs. A **genus** (plural, *genera*) is a group of closely related species. Optionally, the name of the taxonomist who first proposed the species name may be added at the end. Thus *Homo sapiens* Linnaeus is the name of the modern human species. *Homo* is the genus to which the species belongs, and *sapiens* identifies the particular species in the genus *Homo;* Linnaeus proposed the species name *Homo sapiens*. You can think of the generic name *Homo* as equivalent to your surname and the specific name *sapiens* as equivalent to your first name. The name of the genus is always capitalized, and the name identifying the species is always lowercased. Both names are italicized, whereas common names of organisms are not. Rather than repeating the name of a genus when it is used several times in the same discussion, biologists often spell it out only once and abbreviate it to the initial letter thereafter (*D. melanogaster* rather than *Drosophila melanogaster*, for example).

(A) *Campanula rotundifolia*

(B) *Endymion non-scriptus*

(C) *Mertensia virginica*

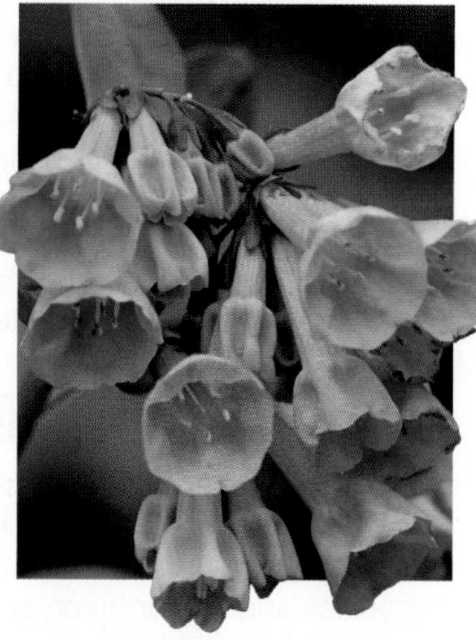

**22.13 Many Different Plants Are Called Bluebells** All three of these distantly related plant species are called "bluebells." Binomial nomenclature allows us to communicate exactly what is being described. (A) *Campanula rotundifolia*, found on the North American Great Plains, belongs to a larger group of bellflowers. (B) *Endymion non-scriptus*, English bluebell, is related to hyacinths. (C) *Mertensia virginica*, Virginia bluebell, belongs in a very different group of plants known as borages.

As we noted earlier, any group of organisms that is treated as a unit in a biological classification system, such as the genus *Drosophila*, or all insects, is called a *taxon*. In the Linnaean system, species and genera are further grouped into a hierarchical system of higher taxonomic categories. The taxon above the genus in the Linnaean system is the **family**. The names of animal families end in the suffix "-idae." Thus Formicidae is the family that contains all ant species, and the family Hominidae contains humans and our recent fossil relatives, as well as our closest living relatives, the chimpanzees and gorillas. Family names are based on the name of a member genus; Formicidae is based on the genus *Formica*, and Hominidae is based on *Homo*. The same rules are used in classifying plants, except that the suffix "-aceae" is used for plant family names instead of "-idae." Thus Rosaceae is the family that includes the genus of roses (*Rosa*) and its close relatives. In the Linnaean system, families are grouped into **orders**, orders into **classes**, and classes into **phyla** (singular *phylum*), and phyla into **kingdoms**. However, Linnaean classification is often subjective; whether a particular taxon is considered, say, an order or a class is often a subjective decision. Today, Linnaean terms are used largely for convenience. Although families are always grouped within orders, orders within classes, and so forth, there is nothing that makes a "family" in one group equivalent (in number of genera or in evolutionary age, for instance) to a "family" in another group.

Linnaeus recognized the overarching hierarchy of life, but he developed his system before evolutionary thought had become widespread. Biologists today recognize the tree of life as the basis for biological classification and often name clades without placing them into any Linnaean rank. But regardless of whether they rank organisms into Linnaean categories or use unranked clade names, modern biologists use evolutionary relationships as the basis for distinguishing biological taxa.

### Evolutionary history is the basis for modern biological classification

Biological classification systems are used to express relationships among organisms. The kind of relationship we wish to express influences which features we use to classify organisms. If, for instance, we were interested in a system that would help us decide what plants and animals were desirable as food, we might devise a classification based on tastiness, ease of capture, and the number of edible parts each organism possessed. Early Hindu classifications of organisms were designed according to these criteria. Such systems served the needs of the people who developed them, but are not adequate for formal scientific classification.

Taxonomists today use biological classifications to express the evolutionary relationships of organisms. Taxa are expected to be **monophyletic**, meaning that the taxon contains an ancestor and all descendants of that ancestor, and no other organisms (**Figure 22.14**). In other words, the taxon is an historical group of related species, or a complete branch on the tree of life (a *clade*). Although biologists seek to describe and name only monophyletic taxa, the detailed phylogenetic information needed to do so is not always available. A group that does not include its common ancestor is called a **polyphyletic** group. A group that does not include all the descendants of a common ancestor is called a **paraphyletic** group.

A true monophyletic group (i.e., a clade) can be removed from a phylogenetic tree by a single "cut" in the tree, as shown in Figure 22.14. Note that there are many monophyletic groups on any phylogenetic tree, and that these groups are successively smaller subsets of larger monophyletic groups. This hierarchy of biological taxa, with all of life as the most inclusive taxon and many smaller taxa within larger taxa, down to the individual species, is the modern basis for biological classification.

Virtually all taxonomists now agree that polyphyletic and paraphyletic groups are inappropriate as taxonomic units, because they do not correctly reflect evolutionary history. The classifications used today still contain such groups because some organisms have not been evaluated phylogenetically. As mistakes in prior classifications are detected, taxonomic names

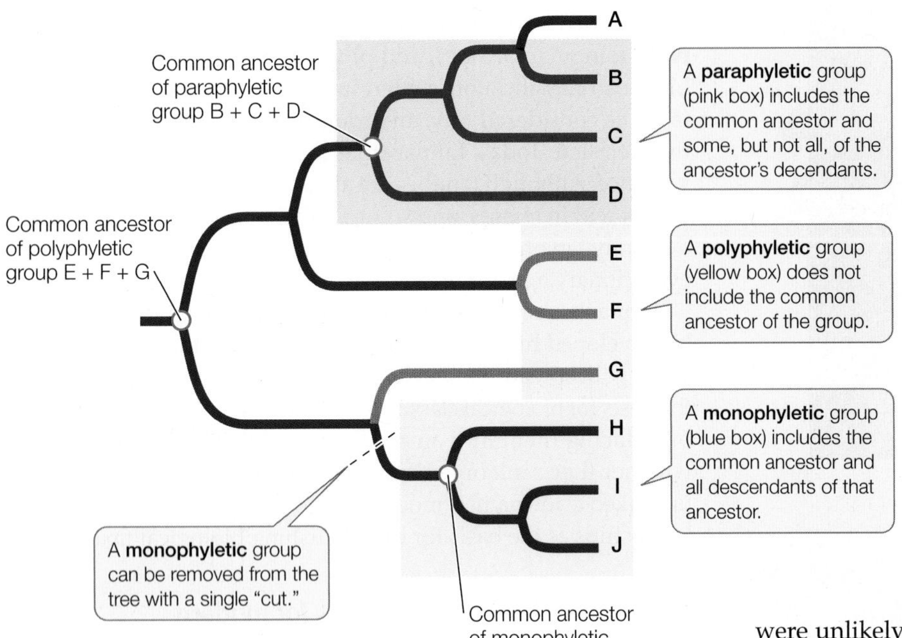

A
B
C
D
E
F
G
H
I
J
Common ancestor of paraphyletic group B + C + D
Common ancestor of polyphyletic group E + F + G
A **paraphyletic** group (pink box) includes the common ancestor and some, but not all, of the ancestor's decendants.
A **polyphyletic** group (yellow box) does not include the common ancestor of the group.
A **monophyletic** group (blue box) includes the common ancestor and all descendants of that ancestor.
A **monophyletic** group can be removed from the tree with a single "cut."
Common ancestor of monophyletic group H + I + J

**22.14 Monophyletic, Polyphyletic, and Paraphyletic Groups** Monophyletic groups are the basis of biological taxa in modern classifications. Polyphyletic and paraphyletic groups do not accurately reflect evolutionary history.

**yourBioPortal.com**
GO TO Web Activity 22.2 • Types of Taxa

are revised and polyphyletic and paraphyletic groups are eliminated from the classifications.

### Several codes of biological nomenclature govern the use of scientific names

Several sets of explicit rules govern the use of scientific names. Biologists around the world follow these rules voluntarily to facilitate communication and dialogue. Although there may be dozens of common names for an organism in many different languages, the rules of biological nomenclature are designed so that there is only one correct scientific name for any single recognized taxon and (ideally) a given scientific name applies only to a single taxon (that is, each scientific name is unique). Sometimes the same species is named more than once (when more than one taxonomist has taken up the task); the rules specify that the valid name is the first name that was proposed. If the same name is inadvertently given to two different species, then a replacement name must be given to the species that was named second.

Because of the historical separation of the fields of zoology, botany (including, originally, the study of fungi), and microbiology, different sets of taxonomic rules were developed for each of these groups. Yet another set of rules for classifying viruses emerged later. This has resulted in many duplicated names in groups that are governed by different sets of rules: *Drosophila*, for instance, is both a genus of fruit flies and a genus of fungi, and there are species in both groups that have identical names. Until recently these duplicated names caused little confusion, since traditionally biologists who studied fruit flies were unlikely to read the literature on fungi (and vice versa). Today, however, given the use of large, universal biological databases (such as GenBank, which includes DNA sequences from across all life), it is increasingly important that each taxon have a unique name. Taxonomists are now working to develop common sets of rules that can be applied across all living organisms.

## 22.4 RECAP

**Biologists organize and classify life by identifying and naming monophyletic groups. Several sets of rules govern the use of scientific names so that each species and higher taxon can be identified and named unambiguously.**

- Explain the difference between monophyletic, paraphyletic, and polyphyletic groups. See p. 477 and Figure 22.14
- Do you understand why biologists prefer monophyletic groups in formal classifications? See p. 477

Now that we have seen how evolution occurs and how phylogenies can be used to study evolutionary relationships, we are ready to consider the process of speciation. Speciation is what leads to the splitting events on the tree of life, and is the process that results in the millions of species that constitute biodiversity.

## CHAPTER SUMMARY

### 22.1 What Is Phylogeny?

- **Phylogeny** is the history of descent of organisms from their common ancestor. Groups of evolutionarily related species are represented as related branches in a **phylogenetic tree**. Review Figure 22.2
- A group of species that consists of all the evolutionary descendants of a common ancestor is called a **clade**. Named clades and species are called **taxa**.
- **Homologies** are similar traits that have been inherited from a common ancestor. Review Figure 22.4

- A trait that is shared by two or more taxa and is derived through evolution from a common ancestral form is called a **synapomorphy**.
- Similar traits may occur among species that do not result from common ancestry. **Convergent evolution** and **evolutionary reversals** can give rise to such traits, which are called **homoplasies**.

## 22.2 How Are Phylogenetic Trees Constructed?

**SEE WEB ACTIVITY 22.1**

- Phylogenetic trees can be inferred from synapomorphies using the principle of **parsimony**. Review Figure 22.5
- Sources of phylogenetic information include morphology, patterns of development, the fossil record, behavioral traits, and molecular traits such as DNA and protein sequences.
- Phylogenetic trees can also be inferred with **maximum likelihood** methods, which calculate the probability that a particular tree will have generated the observed data.

## 22.3 How Do Biologists Use Phylogenetic Trees?

- Phylogenetic trees are used to reconstruct the past and understand the origin of traits. Review Figure 22.8
- Phylogenetic trees are used to make appropriate evolutionary comparisons among living organisms.
- Biologists can use phylogenetic trees to reconstruct ancestral states. **SEE ANIMATED TUTORIAL 22.1**
- Phylogenetic trees may include estimates of times of divergence of lineages determined by **molecular clock** analysis. Review Figure 22.12

## 22.4 How Does Phylogeny Relate to Classification?

- Taxonomists organize biological diversity on the basis of evolutionary history.
- Taxa in modern classifications are expected to be **monophyletic** groups. **Paraphyletic** and **polyphyletic** groups are not considered appropriate taxonomic units. Review Figure 22.14, **WEB ACTIVITY 22.2**
- Several sets of rules govern the use of scientific names, with the goal of providing unique and universal names for biological taxa.

# SELF-QUIZ

1. A *clade* is
   *a.* a type of phylogenetic tree.
   *b.* a group of evolutionarily related species that share a common ancestor.
   *c.* a tool for constructing phylogenetic trees.
   *d.* an extinct species.
   *e.* an ancestral species.
2. Phylogenetic trees may be constructed for
   *a.* genes.
   *b.* species.
   *c.* major evolutionary groups.
   *d.* viruses.
   *e.* All of the above.
3. A shared derived trait, used as the basis for inferring a monophyletic group, is called
   *a.* a synapomorphy.
   *b.* a homoplasy.
   *c.* a parallel trait.
   *d.* a convergent trait.
   *e.* a phylogeny.
4. The parsimony principle can be used to infer phylogenetic trees because
   *a.* evolution is nearly always parsimonious.
   *b.* it is logical to adopt the simplest hypothesis capable of explaining the known facts.
   *c.* once a trait changes, it never reverses condition.
   *d.* all species have an equal probability of evolving.
   *e.* closely related species are always very similar to one another.
5. Convergent evolution and evolutionary reversal are two sources of
   *a.* homology.
   *b.* parsimony.
   *c.* synapomorphy.
   *d.* monophyly.
   *e.* homoplasy.
6. Which of the following are commonly used to infer phylogenetic relationships among plants but not among animals?
   *a.* Nuclear genes
   *b.* Chloroplast genes
   *c.* Mitochondrial genes
   *d.* Ribosomal RNA genes
   *e.* Protein-coding genes
7. Which of the following is *not* true of maximum likelihood or parsimony methods for inferring phylogeny?
   *a.* The maximum likelihood method requires an explicit model of evolutionary character change.
   *b.* The parsimony method is computationally easier than the maximum likelihood method.
   *c.* The maximum likelihood method is easier to treat in a statistical framework.
   *d.* The maximum likelihood method is most often used with molecular data.
   *e.* Parsimony is usually used to infer time on a phylogenetic tree.
8. Taxonomists strive to include taxa in biological classifications that are
   *a.* monophyletic.
   *b.* paraphyletic.
   *c.* polyphyletic.
   *d.* homoplastic.
   *e.* monomorphic.
9. Which of the following groups have separate sets of rules for nomenclature?
   *a.* Animals
   *b.* Plants and fungi
   *c.* Bacteria
   *d.* Viruses
   *e.* All of the above

10. If two scientific names are proposed for the same species, how do taxonomists decide which name should be used?
    *a.* The name that provides the most accurate description of the organism is used.
    *b.* The name that was proposed most recently is used.
    *c.* The name that was used in the most recent taxonomic revision is used.
    *d.* The first name to be proposed is used, unless that name was previously used for another species.
    *e.* Taxonomists use whichever name they prefer.

## FOR DISCUSSION

1. Why are taxonomists concerned with identifying species that share a particular common ancestor?
2. How are fossils used to identify ancestral and derived traits of organisms? How can fossils be integrated into phylogenetic analyses?
3. The parsimony principle is often used to construct phylogenetic trees. What are the limitations of parsimony, and why do some biologists prefer model-based approaches, such as maximum likelihood methods?
4. A student of the evolution of frogs has proposed a strikingly new classification of frogs based on an analysis of a few mitochondrial genes from about 10 percent of frog species. Should frog taxonomists immediately accept the new classification? Why or why not?
5. What are some of the assumptions that go into a molecular clock analysis? How could these assumptions be violated? How could molecular clock analyses be modified to consider these additional sources of variation?
6. Classification systems summarize much information about organisms and enable us to remember the traits of many organisms. From your general knowledge, how many traits can you associate with the following names: conifer, fern, bird, mammal?

## ADDITIONAL INVESTIGATION

West Nile virus kills birds of many species and can cause fatal encephalitis (inflammation of the brain) in humans and horses. The virus was first isolated in Africa (where it is thought to be endemic) in the 1930s, and by the 1990s it had been found throughout much of Eurasia. West Nile virus was not found in North America until 1999, but since that time it has spread rapidly across most of the United States. The genome of West Nile virus evolves quickly. How could you use phylogenetic analysis to investigate the geographic origin of the West Nile virus that was introduced into North America in 1999?

## WORKING WITH DATA (GO TO yourBioPortal.com)

**Constructing a Phylogenetic Tree** In this exercise based on Figure 22.7, you will use a subset of the DNA sequences from the experimental lineages to reconstruct the evolutionary relationships among the viruses. You will also use these data to reconstruct the DNA sequences of the viral ancestors.

# 23 Species and Their Formation

## Catching speciation in the act

When biologists first explored the Cuatro Ciénegas basin of northern Mexico, they found many organisms that are not found anywhere else in the world. So far researchers have described about 150 *species* of plants and animals that are restricted to this small region. Even though Cuatro Ciénegas is in a desert, about 30 of these unique species are aquatic, living in the isolated springs and marshes of the basin. An unusual aquatic box turtle, beautiful cichlid fishes, and tiny crustaceans are among the many aquatic species that are confined to Cuatro Ciénegas. Why are so many different species found here and nowhere else?

Biologists and geologists found that, over the past several million years, this desert oasis has repeatedly been isolated by a succession of geological events that cut it off from the river systems and mountain ranges of northern Mexico. Many different *speciation events* associated with these geological events make Cuatro Ciénegas a natural laboratory for studying speciation by geographic isolation.

Each time gene flow between organisms in the basin and the surrounding areas ceased, populations living inside and outside the basin began to diverge from one another. Over thousands of generations of such isolation, new species developed. These new species no longer share the same gene pool, are adapted to different environments, and look different from one another. And—extremely important—the organisms have diverged to the point that they are no longer capable of reproducing with one another—one of the hallmarks of distinct species.

Although speciation is often studied in natural settings such as Cuatro Ciénegas, some aspects of speciation can be studied in controlled laboratory experiments, using organisms with short generation times. For example, William Rice and George Salt conducted an experiment in which fruit flies were allowed to choose food sources in different habitats, where mating also took place. The habitats were vials in different parts of an experimental cage. The habitats differed in three parameters: (1) light; (2) the direction in which the fruit flies could move (up or down); and (3) concentrations of two aromatic chemicals, ethanol and acetaldehyde. In just 35 generations, two groups of flies were genetically and reproductively isolated from one another because they had evolved distinct preferences for different habitats. In controlled experiments like these, biologists are beginning to study and understand the genetic details of speciation.

**A Natural Laboratory** A swimmer surveys several of the fish species that are isolated in the desert oasis of the Cuatro Ciénegas basin in northern Mexico.

**Experimental Subjects** Fruit flies of the genus *Drosophila* are easily reared in the laboratory. Their short generation time (7–10 days from newly laid egg to reproductive adult) makes them ideal subjects for controlled experiments on speciation.

The *origin of species*—the splitting and diverging of a single lineage into two or more distinct and evolutionarily independent lineages—is one of the most important phenomena in biological science. Charles Darwin recognized its preeminence when he chose the title of *The Origin of Species*. But without the underlying knowledge supplied by the modern science of genetics, Darwin was primarily viewing the consequences of speciation, not its underlying causes. Today biologists are actively searching for and finding answers to the many questions about the process of speciation, something biologists have been known to call "the mystery of mysteries."

**IN THIS CHAPTER** we will describe what species are and discuss how Earth's millions of species came into being. We will examine the mechanisms by which a lineage splits into new species and how such separations are maintained. Finally, we will look at different factors that can make speciation a rapid or a very slow process.

## CHAPTER OUTLINE



# 23.1 What Are Species?

Although "species" is a useful and commonly used term in biology, the concept of "species" sometimes varies among different biologists. Biologists are interested in several different aspects of the divergence of biological lineages. Different biologists think about species differently because they ask different questions about species: How can we recognize and identify species? How do new species arise? How do different species remain separate? Why do rates of speciation differ among groups? In answering these questions, biologists focus on different attributes of species, leading to several different ways of thinking about what species are and how they form. Most of the various *species concepts* proposed by biologists are not mutually exclusive—they are just different ways of approaching the question "What are species?"

### We can recognize many species by their appearance

Biological diversity does not always vary in a smooth, incremental way; groups of organisms often differ in distinct, obvious ways. People have long recognized groups of similar organisms that mate with one another, and there are usually distinct morphological breaks between these groups. Groups of organisms that mate with one another are commonly called *species* (note that this is both the plural and singular form of the word). Someone who is knowledgeable about a group of organisms, such as birds or flowering plants, usually can distinguish the different species found in a particular area simply by looking at them. Standard field guides to birds, mammals, insects, and wildflowers are possible only because many species change little in appearance over large geographic distances. A casual birdwatcher can easily recognize male red-winged blackbirds (*Agelaius phoeniceus*) from the east and west coasts of North America as members of the same species (**Figure 23.1A**).

More than 250 years ago, the Swedish biologist Carolus Linnaeus developed the binomial system of biological nomenclature by which species are named today (see Section 22.4). Linnaeus described thousands of species, and because he knew nothing about genetics or the mating behavior of the organisms he was naming, he classified them only on the basis of their appearance. Linnaeus differentiated species using a **morphological species concept**, a construct that assumes a species consists of individuals that "look alike," and that individuals that don't look alike belong to different species. Although Linnaeus did not know it, members of many of the groups that he classified as species by their appearance look alike because they share many of the alleles that code for their body structures.

**23.1 Members of the Same Species Look Alike—or Not** (A) Both of these male red-winged blackbirds are members of the same species—*Agelaius phoeniceus*—even though one is from the eastern United States and the other is from western Canada. Despite the geographic distance, the two individuals are morphologically very similar. (B) Red-winged blackbirds are sexually dimorphic, which means the female of the species looks quite different from the male.

(A)

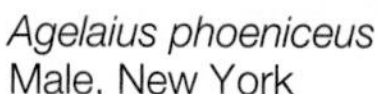

*Agelaius phoeniceus*
Male, New York

*Agelaius phoeniceus*
Male, British Columbia

(B)

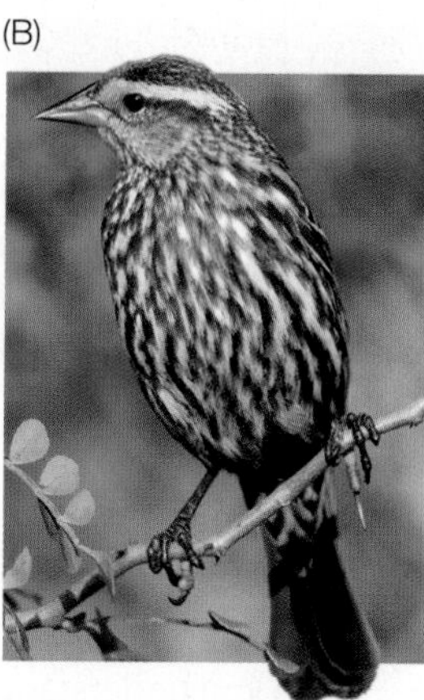

*Agelaius phoeniceus*
Female

The morphological species concept has limitations, however. In some cases, for instance, not all members of a species look alike. For example, males, females, and young individuals do not always resemble one another closely (**Figure 23.1B**). The morphological species concept is of little use in the case of *cryptic species*, instances in which two or more morphologically indistinguishable species do not interbreed. Biologists therefore cannot always rely on appearance alone in determining whether individual organisms are members of the same or different species. Several additional types of information (especially behavioral and genetic data) are used today to help biologists differentiate species.

### Species are reproductively isolated lineages on the tree of life

Evolutionary biologists often think of species as branches on the tree of life, which is known as a **lineage species concept**. Under this concept, each species has a history that starts at a speciation event (one lineage on the tree splits into two) and ends either at extinction or another speciation event, at which time the species produces two daughter species. The process of lineage splitting may be gradual, and thus take thousands of generations to complete. On the other hand, one ancestral lineage may be split into two as a result of a sudden geological event, such as a volcanic eruption that isolates two parts of ancestral species range. Either way, the lineage species concept treats speciation as the process by which one species splits into two or more daughter species, which thereafter evolve as distinct lineages. The gradual nature of some splitting events means that at a single point in time, the final outcome of the process may not be clear. In these cases it is often impractical to try to decide whether the partially isolated populations of the *incipient species* will continue to diverge and become fully isolated from one another, or perhaps merge again in the future.

The most important component that determines long-term isolation of sexually reproducing lineages from one another is the development of **reproductive isolation**, a state in which two populations can no longer exchange genes. If individuals of a population mate and reproduce with one another, but not with individuals of other populations, they constitute a distinct group within which genes recombine. In other words, they comprise an independent evolutionary lineage—a separate branch on the tree of life.

It was recognition of the importance of reproductive isolation that brought Ernst Mayr to propose the **biological species concept**: "Species are groups of actually or potentially interbreeding natural populations which are reproductively isolated from other such groups." The terms "actually" and "potentially" are important elements of the definition. "Actually" says that the individuals live in the same area and interbreed with one another. "Potentially" says that although the individuals do not live in the same area, and therefore cannot interbreed, other information suggests that they would do so if they did get together. This widely used concept of species does not apply to organisms that reproduce asexually, and it is limited to a single point in evolutionary time.

These various concepts of species are not entirely incompatible; they simply emphasize different aspects of species. The morphological species concept emphasizes the practicality of humans recognizing species, although it underestimates or overestimates the actual number of species in some cases. Mayr's biological species concept emphasizes that reproductive isolation is what keeps sexual lineages on the tree of life separated from one another. The lineage species concept embraces the idea that sexual species are maintained by reproductive isolation, while recognizing the existence of asexual species. The lineage species concept also allows biologists to consider species through evolutionary time, which we will discuss in the next section.

Virtually all species exhibit some degree of genetic recombination among individuals, even if recombination is relatively rare (see Section 21.4). Significant reproductive isolation between species is thus necessary for lineages to remain distinct through evolutionary time. Reproductive isolation is also responsible for the morphological distinctiveness of most species, because genetic mutations that result in morphological changes cannot spread between reproductively isolated species. Therefore, no matter which species concept we emphasize, the evolution of reproductive isolation is important for understanding the origin of species.

## 23.1 RECAP

**Species are distinct lineages on the tree of life. Speciation is usually a gradual process as one lineage divides into two. Over time, lineages of sexual species remain distinct from one another because they have become reproductively isolated.**

- Explain how the various species concepts emphasize different attributes of species. See pp. 482–483
- Why is the biological species concept not applicable to asexually reproducing organisms?
- Do you understand why reproductive isolation is an important component of all of the species concepts discussed here? See p. 483

Although Charles Darwin titled his groundbreaking book *The Origin of Species*, it did not extensively discuss the processes of speciation. He devoted most of his attention to demonstrating that individual species are altered by natural selection over time. We will next discuss the many things that biologists have learned about speciation since Darwin's time.

# 23.2 How Do New Species Arise?

Not all evolutionary changes result in new species. A single lineage may change over time without giving rise to a new species. Speciation, in contrast, usually requires the evolution of reproductive isolation within a species whose members formerly exchanged genes. But if a genetic change prevents reproduction between individuals of a species, how can such a change spread through a species in the first place?

### Gene incompatibilities can produce reproductive isolation in two daughter species

If a new allele that causes reproductive incompatibility arises in a population, it cannot spread through the population, because no other individuals are reproductively compatible with the individual that carries the new allele. So how can one reproductively cohesive lineage ever split into two reproductively isolated species? Several early geneticists, including Theodosius Dobzhansky and Hermann Joseph Muller, developed a genetic model to explain this apparent conundrum (**Figure 23.2**).

The Dobzhansky-Muller model is quite simple. First, assume that a single ancestral population is divided into two (by the formation of a new mountain range, for instance). In one of the descendant populations, a new allele (*A*) arises and becomes fixed (see Figure 23.2). In the other population, another new allele (*B*) becomes fixed *at a different gene locus*. Neither new allele at either gene results in any loss of reproductive compatibility. However, the two new forms of these two genes have never occurred together in the same organism. The products of many genes must work together, and it is quite likely that the new protein forms encoded by these two genes are not compatible with one another. If the two populations come back together, they may still be able to interbreed, or **hybridize**. However, the hybrid offspring will have a new combination of genes that may be functionally inferior, or even lethal. Of course, this will not happen with all new combinations of genes, but over time isolated populations will accumulate many allele differences at many gene loci. Thus, we expect that genetic incompatibility will develop through time in the two isolated populations.

Many empirical examples support the Dobzhansky-Muller model, which works not only for pairs of individual genes but also for some kinds of chromosomal rearrangements. For example, bats of the genus *Rhogeessa* exhibit considerable variation in centric fusions of their chromosomes. The chromosomes of the various species contain the same basic chromosomal arms, but in some species two acrocentric (one-armed) chromosomes have fused to form larger, metacentric (two-armed) chromosomes. A polymorphism in this character causes few, if any, problems in meiosis, because the respective chromosomes can still align and assort normally. Therefore, a given centric fusion can become fixed in a population. However, if a *different* centric fusion becomes fixed in a second population, then hybrids between the two populations can no longer produce normal gametes in meiosis (**Figure 23.3**). Most of the closely related species of *Rhogeessa* display different combinations of these centric fusions, and are thereby reproductively isolated from one another.

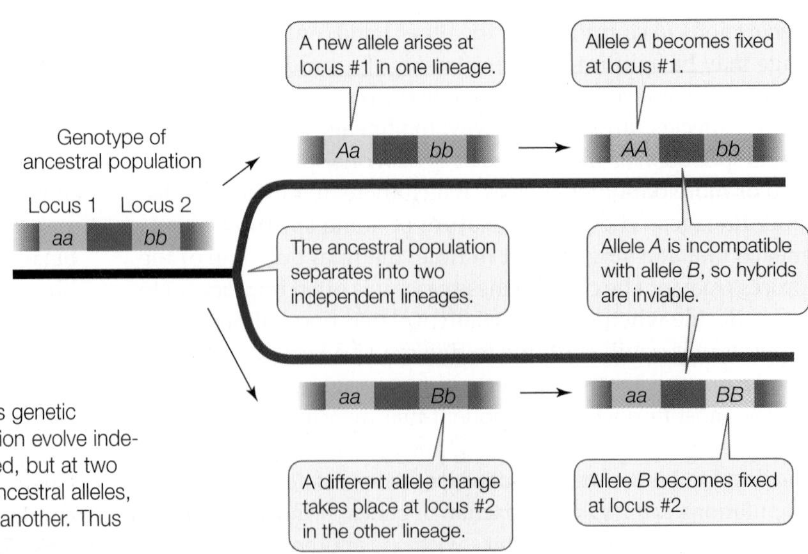


**23.2 The Dobzhansky-Muller Model** In this simple two-locus genetic model, two populations isolated from the same ancestral population evolve independently. In each descendant lineage, a new allele becomes fixed, but at two different loci. Neither of the new alleles is incompatible with the ancestral alleles, but the two new alleles at the two loci are incompatible with one another. Thus the two descendant lineages are reproductively incompatible.

**23.3 Speciation by Centric Fusion** In this chromosomal version of the Dobzhansky–Muller model of speciation, two independent centric fusions of one-armed chromosomes occur in two sister lineages. When polymorphic, neither centric fusion event by itself results in major difficulties at meiosis. However, the independent centric fusions are incompatible at meiosis, because the three different chromosomes involved in these fusions cannot pair normally, leading to sterility of the $F_1$ hybrid. Most of the species in the bat genus *Rhogeessa* differ from one another by such centric fusions.

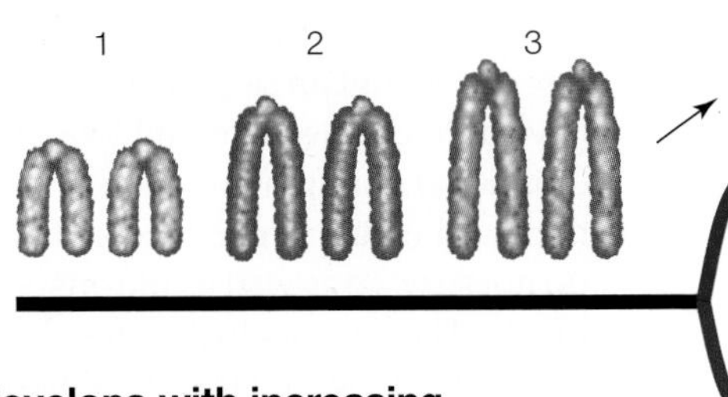
1
2
3

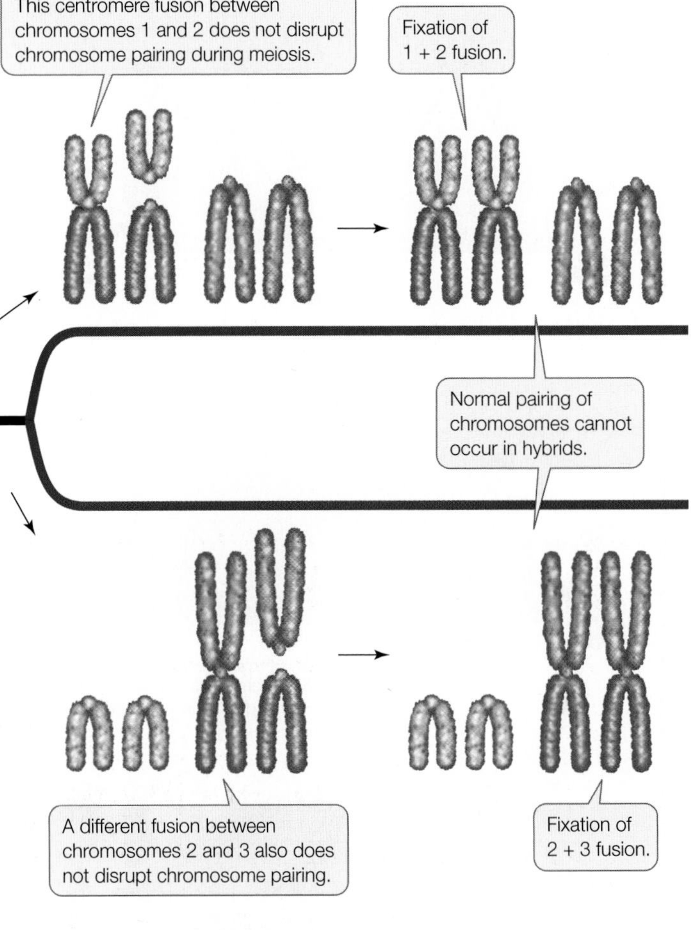
This centromere fusion between chromosomes 1 and 2 does not disrupt chromosome pairing during meiosis.
Fixation of 1 + 2 fusion.
Normal pairing of chromosomes cannot occur in hybrids.
A different fusion between chromosomes 2 and 3 also does not disrupt chromosome pairing.
Fixation of 2 + 3 fusion.

## Reproductive isolation develops with increasing genetic divergence

As pairs of species diverge genetically, they become increasingly reproductively isolated from one another (**Figure 23.4**). Both the rate at which reproductive isolation develops and the mechanisms that produce it vary from group to group. Nonetheless, reproductive incompatibility has been shown to develop gradually in many groups of plants, animals, and fungi, reflecting the slow pace at which incompatible genes accumulate in each lineage. In some cases, complete reproductive isolation may take millions of years to develop. In other cases (as with the chromosomal fusions of *Rhogeessa* described above), reproductive isolation can develop over just a few generations.

*Partial* reproductive isolation has evolved in strains of *Phlox drummondii* artificially isolated by humans. In 1835, Thomas Drummond, after whom this species of garden plant is named, collected seeds in Texas and distributed them to nurseries in Europe. Over the next 80 years, the European nurseries established more than 200 true-breeding strains of this phlox, which differed in flower size and color and plant growth form. The plant breeders did not select directly for reproductive incompatibility between strains, but in subsequent experiments in which fertilization rates were measured and compared, biologists found that reproductive compatibility between strains had been reduced by 14 to 50 percent, depending on the strain.

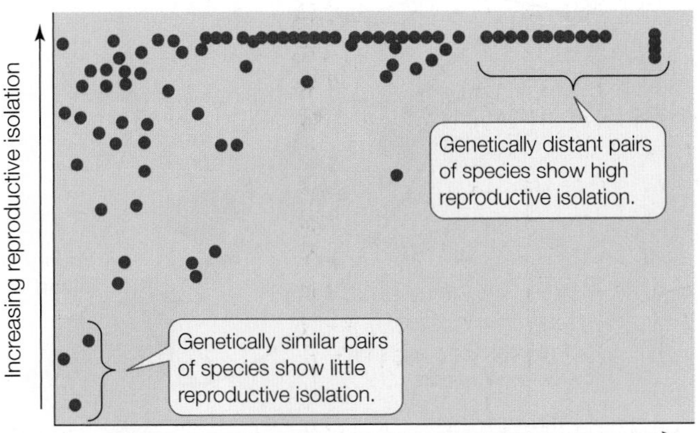
Increasing reproductive isolation
Genetically distant pairs of species show high reproductive isolation.
Genetically similar pairs of species show little reproductive isolation.
Increasing genetic divergence

**23.4 Reproductive Isolation Increases with Time of Separation** This positive relationship between genetic distance and reproductive isolation, shown here for pairs of *Drosophila* species, has been observed in many groups of plants, animals, and fungi (each dot represents a comparison of one species pair).

Many laboratory experiments have demonstrated the gradual evolution of reproductive incompatibility during isolation. We described one such experiment in the opening of this chapter. These laboratory experiments are also consistent with our observations of naturally isolated populations, as described for the unique species of the Cuatro Ciénegas basin in the chapter opening.

## Geographic barriers give rise to allopatric speciation

Speciation that results when a population is divided by a physical barrier is known as **allopatric speciation** (Greek *allos*, "other"; *patra*, "homeland"), also called *geographic speciation* (**Figure 23.5**). Allopatric speciation is thought to be the dominant mode of speciation in most groups of organisms. The physical barrier that divides the range of a species may be a water body, a mountain range, or other inhospitable habitat for terrestrial organisms, or dry land for aquatic organisms. Barriers can form when continents drift, sea levels rise, glaciers advance and retreat, or climates change. These processes continue to generate

Time
A single species is distributed over a broad range.
A barrier separates two populations. Populations adapt to differing environments on opposite sides of the barrier.
The barrier is removed. The populations recolonize the intervening area and mingle, but do not interbreed.
Range of overlap

**23.5 Allopatric Speciation** Allopatric speciation may result when a population is divided into two separate populations by a physical barrier, such as rising sea levels.

physical barriers today. The populations separated by such barriers are often, but not always, initially large. They evolve differences for a variety of reasons, including genetic drift (see Section 21.1), but especially because the environments in which they live are, or become, different.

Allopatric speciation may also result when some members of a population cross an existing barrier and found a new, isolated population. Many of the more than 800 species of *Drosophila* found in the Hawaiian archipelago are restricted to a single island. We know that these species are the descendants of new populations founded by individuals dispersing among the islands because the closest relative of a species on one island is often a species on a neighboring island rather than a species on the same island. Biologists who have studied the chromosomes of these fruit flies estimate that speciation in this group of *Drosophila* has resulted from at least 45 such founder events (**Figure 23.6**).

The 14 species of finches found in the Galápagos, an archipelago 1,000 kilometers off the coast of Ecuador, were generated by allopatric speciation. Darwin's finches (as they are usually called, because Darwin was the first scientist to study them) arose in the Galápagos from a single South American species that colonized the islands. Today the 14 species differ strikingly from their closest mainland relative and from one another (**Figure 23.7**). The islands of the Galápagos archipelago are sufficiently far apart that finches move among them only infrequently. In addition, environmental conditions differ from island to island. Some are relatively flat and arid; others have forested mountain slopes. Finch populations on different islands have differentiated over millions of years to the point that, when occasional immigrants arrive from other islands, they either do not breed with the residents or, if they do, the resulting offspring do not survive as well as the offspring of established island residents. The genetic distinctness and cohesiveness of the different finch species are thus maintained.

**yourBioPortal.com**
GO TO Animated Tutorial 23.1 • Founder Events and Allopatric Speciation

### Sympatric speciation occurs without physical barriers

Although physical isolation is usually required for speciation, under some circumstances speciation can occur without it. A partition of a gene pool without physical isolation is called **sympatric speciation** (Greek *sym*, "together with"). Given that speciation is usually a gradual process, how can reproductive isolation develop when individuals have frequent opportunities

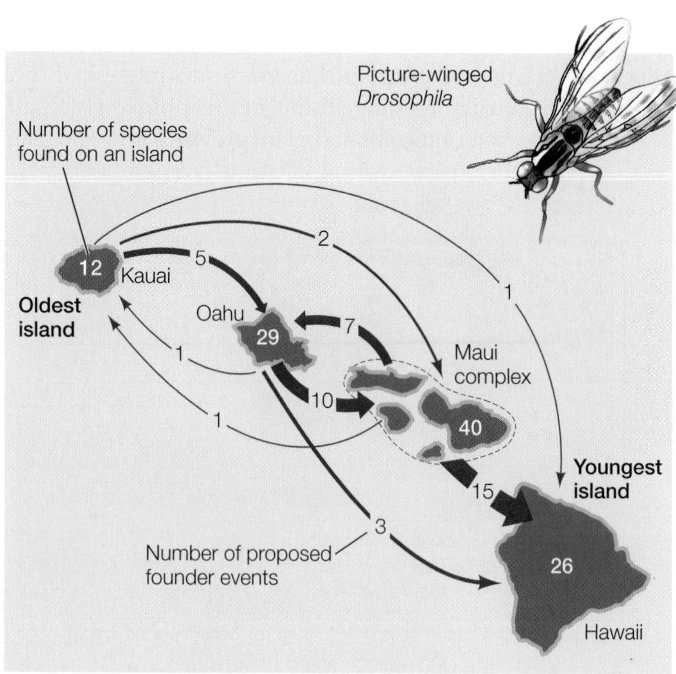


**23.6 Founder Events Lead to Allopatric Speciation** The large number of species of picture-winged *Drosophila* in the Hawaiian Islands is the result of founder events: the founding of new populations by individuals dispersing among the islands. The islands, which were formed in sequence as Earth's crust moved over a volcanic "hot spot," vary in age.

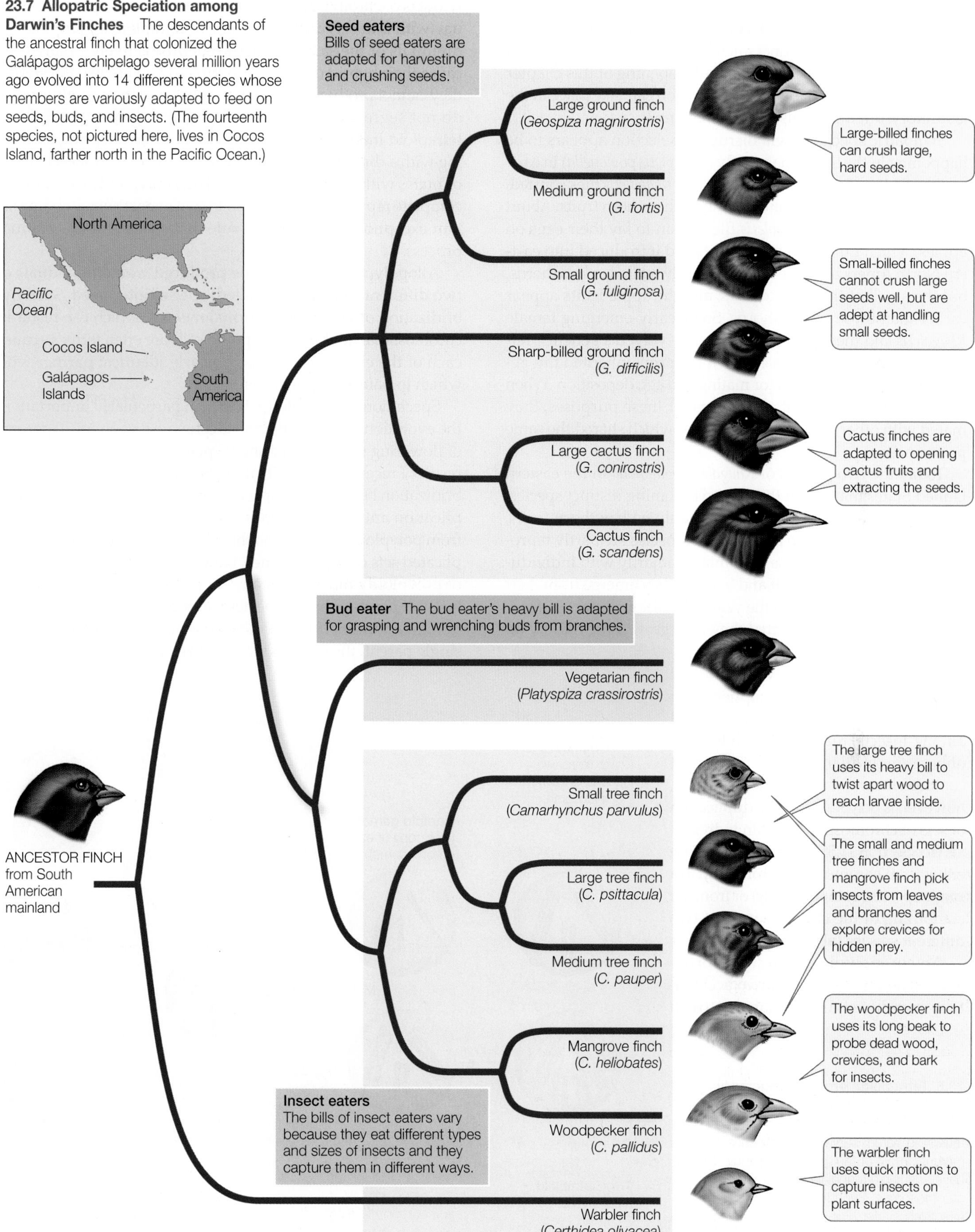


**23.7 Allopatric Speciation among Darwin's Finches** The descendants of the ancestral finch that colonized the Galápagos archipelago several million years ago evolved into 14 different species whose members are variously adapted to feed on seeds, buds, and insects. (The fourteenth species, not pictured here, lives in Cocos Island, farther north in the Pacific Ocean.)

to mate with one another? Sympatric speciation may occur with some form of disruptive selection in which certain genotypes have a preference for distinct microhabitats where mating takes place. The experiment described in the opening of this chapter shows that this kind of disruptive selection can take place in the laboratory, but does it also occur in nature?

Sympatric speciation via disruptive selection appears to be happening in the apple maggot fly (*Rhagoletis pomonella*) in eastern North America. Until the mid-1800s, *Rhagoletis* flies courted, mated, and deposited their eggs only on hawthorn fruits. About 150 years ago, some *Rhagoletis* flies began to lay their eggs on apples, which European immigrants had introduced into eastern North America. Apple trees are closely related to hawthorns, but the smell of the fruits differs, and the apple fruits appear earlier than those of hawthorns. Some early-emerging female *Rhagoletis* laid their eggs on apples and evolved a genetic preference for the smell of apples. Their offspring inherited this genetic preference for apples for mating and egg deposition. When the offspring sought out apple trees for these purposes, they mated with other flies reared on apples, which shared the same preferences.

Today the two groups of *Rhagoletis pomonella* in the eastern United States may be on the way to becoming distinct species. One group mates and lays eggs primarily on hawthorn fruits, the other on apples. The two incipient species are partly reproductively isolated because they mate primarily with individuals raised on the same fruit and because they emerge from their pupae at different times of the year. In addition, the apple-feeding flies have evolved so that they now grow more rapidly on apples than they originally did.

Sympatric speciation via ecological isolation, as appears to be happening in *Rhagoletis pomonella*, may be widespread among insects, many of which feed on only a single plant species. The most common means of sympatric speciation, however, is **polyploidy**, or the duplication of sets of chromosomes within individuals (see Section 11.5). Polyploidy can arise either from chromosome duplication in a single species (**autopolyploidy**) or from the combining of the chromosomes of two different species (**allopolyploidy**).

An autopolyploid individual originates when (for example) two accidentally unreduced diploid gametes (with two sets of chromosomes) combine to form a tetraploid individual (with four sets of chromosomes). Tetraploid and diploid individuals of the same species are reproductively isolated because their hybrid offspring are triploid and are usually sterile; they cannot produce viable gametes because their chromosomes do not segregate evenly during meiosis (**Figure 23.8**). So a tetraploid individual cannot produce viable offspring by mating with a diploid individual—but it *can* do so if it self-fertilizes or mates with another tetraploid. Thus polyploidy can result in complete reproductive isolation in two generations—an important exception to the general rule that speciation is a gradual process.

Allopolyploids may also be produced when individuals of two different (but closely related) species interbreed. Such hybridization often disrupts normal meiosis, which can result in chromosomal doubling. Allopolyploids are often fertile because each of the chromosomes has a nearly identical partner with which to pair during meiosis.

Speciation by polyploidy has been particularly important in the evolution of plants. Botanists estimate that about 70 percent of flowering plant species and 95 percent of fern species are the result of recent polyploidization. Some of these arose from hybridization between two species, followed by chromosomal duplication and self-fertilization. Many other species diverged from polyploid ancestors, so the new species also shared the duplicated sets of chromosomes. New species may arise by means of polyploidy more easily among plants than among animals because plants of many species can reproduce by self-fertilization. In addition, if polyploidy arises in several offspring of a single parent, the siblings can fertilize one another.

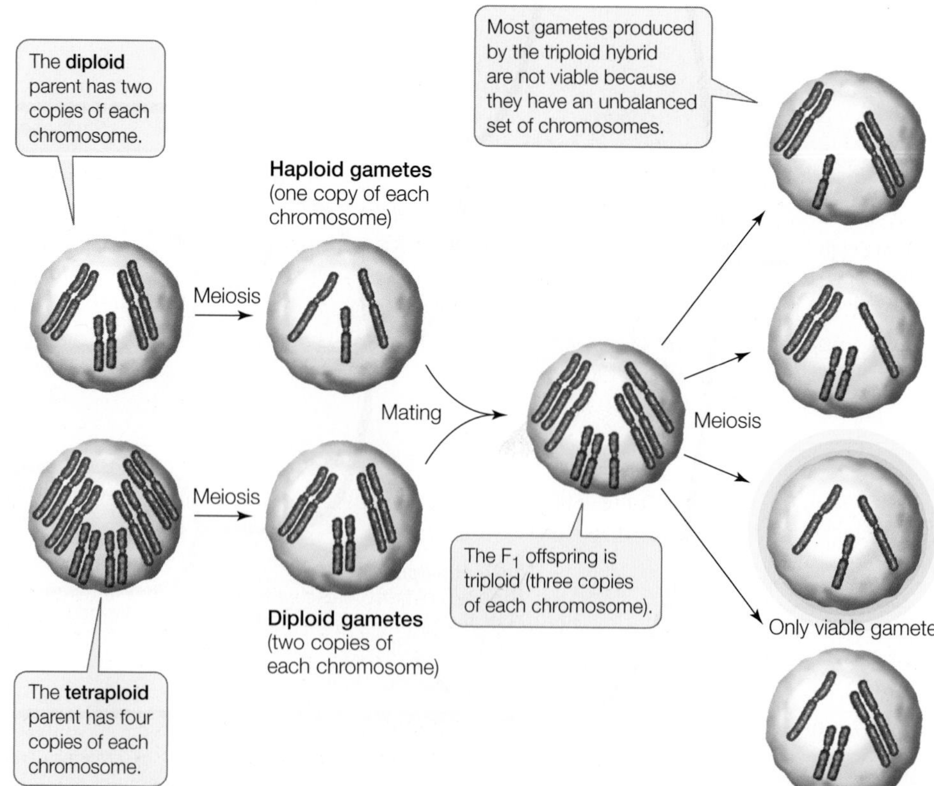
The **diploid** parent has two copies of each chromosome.
Meiosis
Haploid gametes (one copy of each chromosome)
Most gametes produced by the triploid hybrid are not viable because they have an unbalanced set of chromosomes.
Mating
Meiosis
Meiosis
The $F_1$ offspring is triploid (three copies of each chromosome).
Diploid gametes (two copies of each chromosome)
Only viable gamete
The **tetraploid** parent has four copies of each chromosome.

**23.8 Tetraploids Are Reproductively Isolated from Their Diploid Ancestors** Even if the triploid offspring of a diploid and a tetraploid parent survives and reaches sexual maturity, most of the gametes it produces have aneuploid (unbalanced) numbers of chromosomes. Such triploid individuals are effectively sterile. (For simplicity, the diagram shows only three chromosomes; most species have many more than that.)

## 23.2 RECAP

**Allopatric speciation results from the separation of populations by geographic barriers; it is the dominant mode of speciation among most groups of organisms. Sympatric speciation may result from ecological isolation, but among plants and some animals, polyploidy is the most common cause of sympatric speciation.**

- How can speciation via polyploidy happen in two generations? See p. 488
- Explain why an effective barrier to gene flow for one species may not effectively isolate another species.

**yourBioPortal.com**

GO TO Animated Tutorial 23.2 • Speciation Mechanisms

Polyploidy, as we have just seen, can result in a new species that is completely reproductively isolated from its parent species in two generations, but most populations separated by a physical barrier become reproductively isolated only very slowly. Let's see how reproductive isolation may become established once two populations have separated from each other.

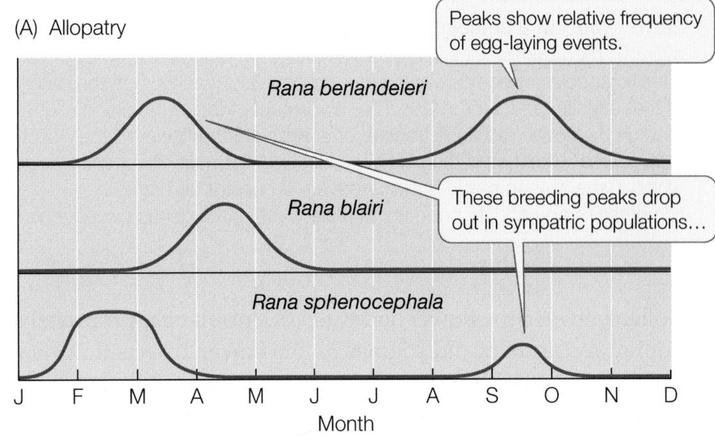

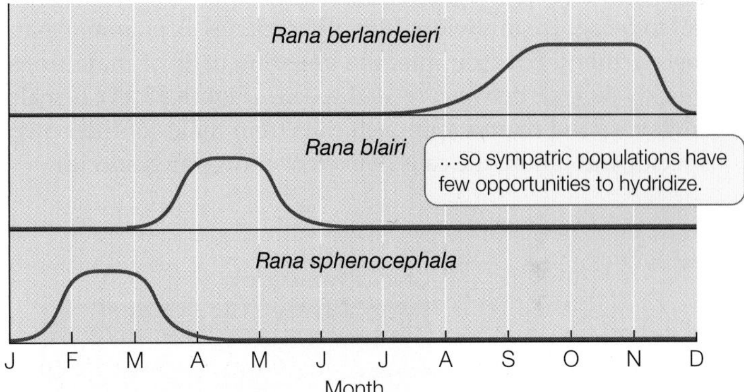


**23.9 Temporal Isolation in the Breeding Seasons of Three Species of Frogs** (A) The peak breeding seasons of three species of *Rana* overlap when the species are physically separated (allopatry). (B) When two or more species of *Rana* occupy the same territory (sympatry), overlap between peak breeding seasons of each species is greatly reduced or eliminated. In areas where only one species is found, the breeding seasons are broader. Selection against hybridization in areas of overlap helps reinforce the prezygotic isolating mechanism.

# 23.3 What Happens When Newly Formed Species Come Together?

As discussed in the previous section, once a barrier to gene flow is established, reproductive isolation can develop through genetic divergence. Over many generations, genetic differences accumulate that reduce the probability that members of the two populations can mate and produce viable offspring. In this way, reproductive isolation evolves as a by-product of the genetic changes in the two populations. What types of mechanisms prevent or reduce gene flow between populations, leading to reproductive isolation? Reproductive isolating mechanisms fall into two major categories: **prezygotic reproductive barriers** act *before* fertilization to prevent individuals of different species or populations from mating, whereas **postzygotic reproductive barriers** act *after* fertilization to prevent the development of viable offspring, or to reduce the offsprings' fertility.

## Prezygotic barriers prevent fertilization

Prezygotic mechanisms come into play before fertilization and can involve several kinds of reproductive isolation.

**HABITAT ISOLATION** When individuals of different species evolve genetic preferences for different habitats in which they live or mate, they may never come into contact during their respective mating periods. The *Rhagoletis* flies in the eastern United States (discussed in Section 23.2) experienced such habitat isolation, as did the *Drosophila* in the experiment described in the opening of this chapter.

**TEMPORAL ISOLATION** Many organisms have distinct mating seasons. If two closely related species breed at different times of the year (or different times of day), the two may never have an opportunity to hybridize. For example, in sympatric populations of three closely related leopard frogs, each species breeds at a different time of year (**Figure 23.9**). Although there is some overlap in the breeding seasons, the opportunities for hybridization are minimized.

**MECHANICAL ISOLATION** Differences in the sizes and shapes of reproductive organs may prevent the union of gametes from different species. With animals, this may involve a match in the shape of reproductive organs between males and females, so that reproduction between species with mismatching structures is not physically possible. In plants, the mechanical isolation between species may involve a pollinator. For example, orchids of the genus *Cryptostylis* produce flowers that look and smell like the females of particular species of wasps (**Figure 23.10**). When a male wasp visits and attempts to mate with the flower (thinking it is a female wasp), the mating action results in transfer of

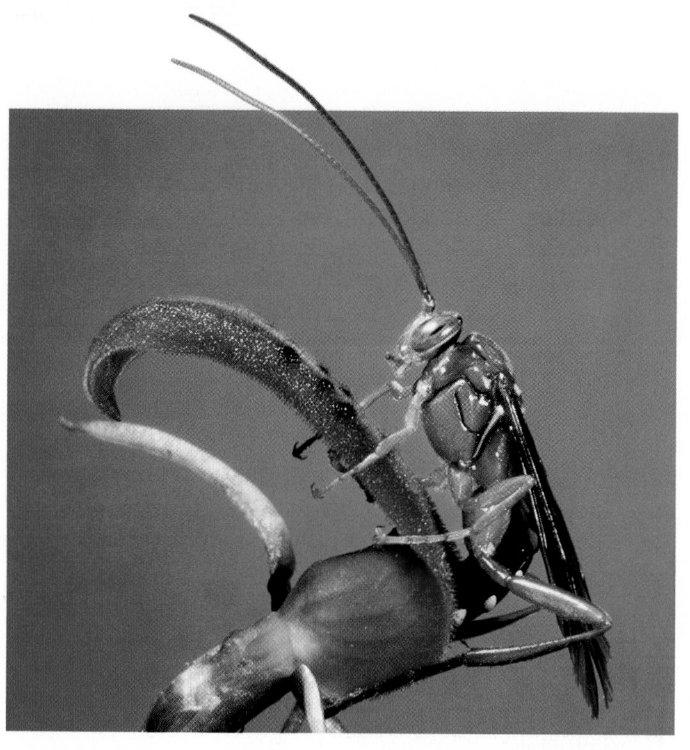

**23.10 Mechanical Isolation through Mimicry** Many orchid species maintain reproductive isolation because their flowers look and smell like a specific species of bee or wasp, inducing copulatory actions on the part of that specific pollinator insect. The placement of the anthers and stigmas on the flower results in transfer of pollen from the flower to the insect that "mates" with it, and then from the insect to the next flower with which it attempts to mate. Shown here are an Australian orchid (*Cryptostylis* sp.) and its pollinator, a male wasp of the genus *Lissopimpla*.

pollen between the flower and wasp as a result of appropriately configured anthers and stigmas on the flower. Insects that visit the flower but do not attempt to mate with it do not trigger the transfer of pollen between the insect and flower.

**BEHAVIORAL ISOLATION** Individuals of a species may reject, or fail to recognize, individuals of other species as potential mating partners. For example, the breeding calls of male frogs quickly diverge between related species (**Figure 23.11**). Female frogs respond to and approach calls from males of their own species, but ignore the calls of even closely related species.

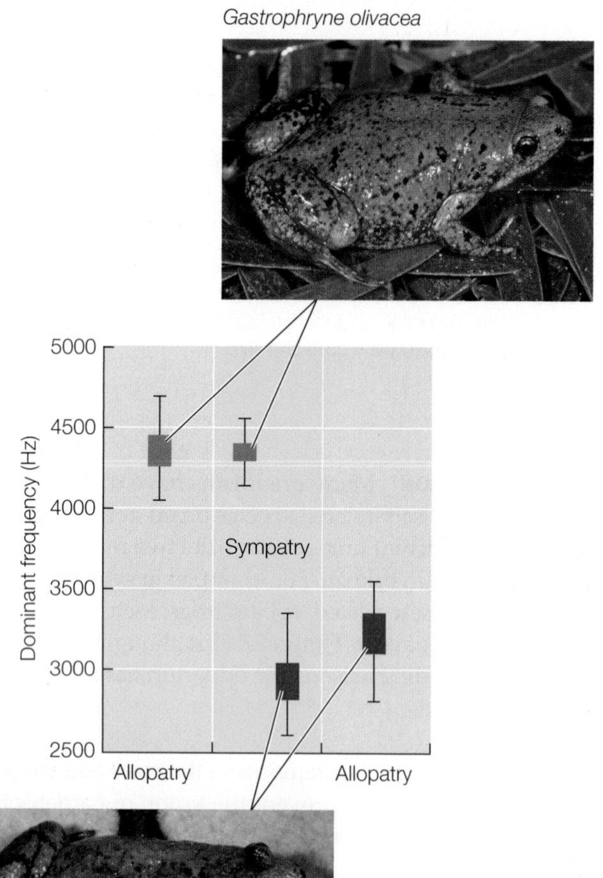


**23.11 Behavioral Isolation in the Mating Calls of Male Frogs** The males of most species of frogs produce species-specific calls. The calls of the two closely related frog species in this figure differ in their dominant frequency (a high-frequency sound wave results in a high-pitched sound; a low frequency results in a low-pitched sound). Female frogs are attracted to the calls of males of their own species. Note that the calls of the two species are more distinct in areas of sympatry than in areas of allopatry (an example of reinforcement).

Sometimes the mate choice of one species is mediated by the behavior of individuals of other species. For example, whether two plant species hybridize may depend on the food preferences of their pollinators. The floral traits of plants, including their color and shape, can enhance reproductive isolation either by influencing which pollinators are attracted to the flowers or by altering where pollen is deposited on the bodies of pollinators. A plant whose flowers are pendant (**Figure 23.12A**) will be pollinated by an animal with different physical characteristics than a plant in which the flowers grow upright (**Figure 23.12B**). Because each pollinator prefers (and is adapted to) a different type of flower, the pollinators rarely transfer pollen from one plant species to the other.

Such isolation by pollinator behavior is seen in the case of two sympatric species of columbines (*Aquilegia*) in the mountains of California that have diverged in flower color, structure, and orientation. *Aquilegia formosa* has pendant flowers with short spurs (spikelike, nectar-containing structures) and is pollinated by hummingbirds (**Figure 23.12C**). *A. pubescens* has upright, lighter-colored flowers with long spurs and is pollinated by hawkmoths (**Figure 23.12D**). The difference in pollinators means that these two species are effectively reproductively isolated even though they populate the same geographic range.

**GAMETIC ISOLATION** Sperm of one species may not attach to the eggs of another species because the eggs do not release the appropriate attractive chemicals, or the sperm may be unable to penetrate the egg because the two gametes are chemically in-

(A)

(B)

(C) *Aquilegia formosa*

(D) *Aquilegia pubescens*

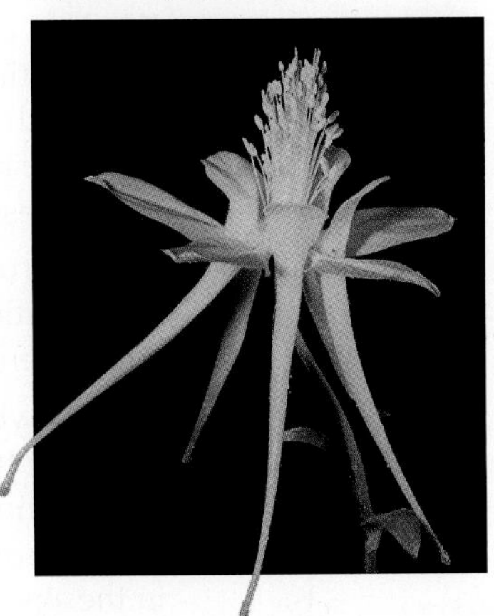

**23.12 Floral Morphology is Associated with Pollinator Morphology** (A) This hummingbird's morphology and behavior are adapted to approach plants whose flowers are pendant (hanging downward). (B) The nectar-extracting proboscis of this hawkmoth is adapted to flowers that grow upright. (C) *Aquilegia formosa* flowers are normally pendant and are pollinated by hummingbirds. (D) Flowers of *A. pubescens* are normally upright, which facilitates pollination by hawkmoths. In addition, the long floral spurs appear to restrict access by some other potential pollinators.

compatible. Thus, even though individuals of the two species may attempt to mate, the gametes never fuse into a zygote. For example, gametic isolation has arisen between species of sea urchins. A protein known as bindin occurs in sea urchin sperm and functions in attaching ("binding") the sperm to eggs. All sea urchins produce this egg-recognition protein, but the gene sequence diverges rapidly between species. The sperm protein evolves so that it will only bind to eggs of the same species, thus preventing interspecific hybridization.

### Postzygotic barriers can isolate species after fertilization

If individuals of two different populations lack complete prezygotic reproductive barriers, postzygotic reproductive barriers may prevent the species from merging. Genetic differences that accumulate while the populations are isolated from each other may reduce the survival and reproduction of hybrid offspring in any of several ways:

- *Low hybrid zygote viability*. Hybrid zygotes may fail to mature normally, either dying during development or developing such severe abnormalities that they cannot mate as adults.
- *Low hybrid adult viability*. Hybrid offspring may simply have lower survivorship than offspring resulting from within-population matings.
- *Hybrid infertility*. Hybrids may mature normally but be infertile. For example, the offspring of matings between horses and donkeys—mules—are healthy but sterile; they produce no descendants.

Although natural selection does not directly favor the evolution of postzygotic reproductive barriers, if hybrid offspring survive poorly, natural selection may favor the evolution of prezygotic barriers. This happens because individuals that mate with individuals of the other population will leave fewer surviving descendants than individuals that mate only within their own population. In this case, individuals that can avoid mating with members of the other population have a selective advantage, and any trait that favors such avoidance will be favored by natural selection. Such strengthening of prezygotic barriers is known as **reinforcement**.

Donald Levin of the University of Texas noticed that individuals of *Phlox drummondii* in most of the range of the species in Texas have pink flowers. Where *P. drummondii* is sympatric with the pink-flowered *P. cuspidata*, however, *P. drummondii* has red flowers. No other *Phlox* species has red flowers. Levin performed an experiment whose results showed that reinforcement might explain the evolution of red flowers where the two species are sympatric (**Figure 23.13**).

Reinforcement can also be detected by comparing sympatric and allopatric populations of potentially hybridizing species. If reinforcement is occurring, then sympatric pairs of closely related species should evolve more effective prezygotic reproductive barriers than do allopatric populations of the same species. The examples of temporal isolation shown in Figure 23.9 and of behavioral isolation shown in Figure 23.11 illustrate reinforcement of prezygotic barriers. The breeding seasons of the sympatric populations of frogs (Figure 23.9) overlap much less than do those of the corresponding allopatric populations. Similarly, the frequencies of the frog mating calls illustrated in Figure 23.11 are more divergent in sympatric populations than in allopatric populations. In both cases, there appears to have been selection against hybrids in areas of sympatry, so individuals that do not produce hybrids are more likely to leave more genes to future generations.

INVESTIGATING LIFE

**23.13 Flower Color Reinforces a Reproductive Barrier in *Phlox*** Most *Phlox drummondii* flowers are pink, but in regions where they are sympatric with *P. cuspidata*—which is always pink—most *P. drummondii* individuals are red. Most pollinators preferentially visit flowers of one color or the other. In this experiment, Donald Levin explored whether flower color reinforces a prezygotic reproductive barrier, lessening the chances of interspecific hybridization.

**HYPOTHESIS** Red-flowered *P. drummondii* are less likely to hybridize with *P. cuspidata* than are pink-flowered *P. drummondii.*

**METHOD**

1. Introduce equal numbers of red- and pink-flowered *P. drummondii* individuals into an area with many pink-flowered *P. cuspidata*.

2. After the flowering season ends, measure hybridization by assessing the genetic composition of the seeds produced by *P. drummondii* plants of both colors.

**RESULTS** Of the seeds produced by pink-flowered *P. drummondii*, 38% were hybrids with *P. cuspidata*. Only 13% of the seeds produced by red-flowered individuals were genetic hybrids.

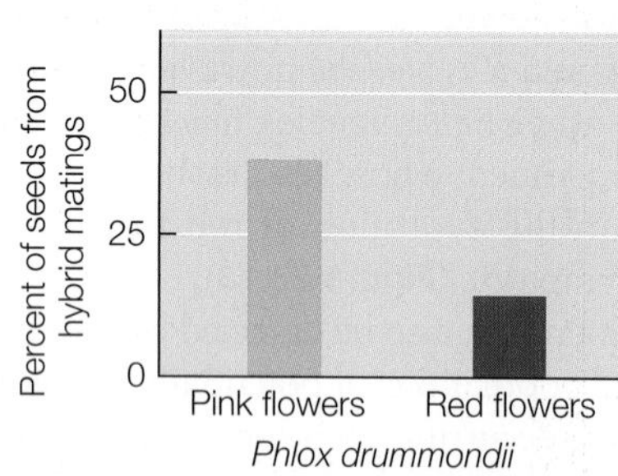


**CONCLUSION** *P. drummondii* and *P. cuspidata* are less likely to hybridize if the flowers of the two species differ in color.

**FURTHER INVESTIGATION:** This experiment did not address the probable reproduction advantages for individual *Phlox* plants of donating and receiving primarily intraspecific pollen. Can you design an experiment to measure such an advantage?

Go to **yourBioPortal.com** for original citations, discussions, and relevant links for all INVESTIGATING LIFE figures.

During the process of reinforcement, closely related species may form hybrids in areas where their ranges overlap, and they may continue to do so for many years. Let's examine what happens when reproductive barriers do not completely prevent individuals from different populations from mating and producing offspring.

### Hybrid zones may form if reproductive isolation is incomplete

If contact is reestablished between formerly isolated populations before complete reproductive isolation has developed, members of the two populations may interbreed. Three outcomes of such interbreeding are possible:

- If hybrid offspring are as fit as those resulting from matings within each population, hybrids will mate with individuals of both parental species. The gene pools will gradually become completely mixed, resulting in one species.
- If hybrid offspring are less fit, complete reproductive isolation may evolve as reinforcement strengthens prezygotic reproductive barriers.
- Even if hybrid offspring are at some disadvantage, a narrow **hybrid zone**—a region in which genetically distinct populations come together and produce offspring of mixed ancestry—may develop in the absence of reinforcement, or before reinforcement is complete.

When a hybrid zone first forms, most hybrids are offspring of crosses between purebred individuals of the two populations. However, subsequent generations include a variety of individuals with different proportions of their genes derived from the original two populations. Thus hybrid zones contain recombinant individuals resulting from many generations of hybridization. Detailed genetic studies can tell us much about why hybrid zones may be narrow and stable for long periods of time.

The hybrid zone between two species of European toads of the genus *Bombina* has been studied intensively. The fire-bellied toad (*B. bombina*) lives in eastern Europe; the closely related yellow-bellied toad (*B. variegata*) lives in western and southern Europe. The ranges of the two species overlap in a long but very narrow zone stretching 4,800 kilometers from eastern Germany to the Black Sea (**Figure 23.14**). Hybrids between the two species suffer from a range of defects, many of which are lethal. Those hybrids that survive often have skeletal abnormalities, such as misshapen mouths, ribs that are fused to vertebrae, and a reduced number of vertebrae. By following the fates of thousands of toads from the hybrid zone, investigators found that a hybrid toad is on average only half as fit as a purebred individual. The

**23.14 Hybrid Zones May Be Long and Narrow** The narrow zone in Europe where fire-bellied toads meet and hybridize with yellow-bellied toads stretches across Europe. This hybrid zone has been stable for hundreds of years and has never expanded, and no reinforcement has evolved.

*B. bombina* (fire-bellied toad)

Hybrid zone

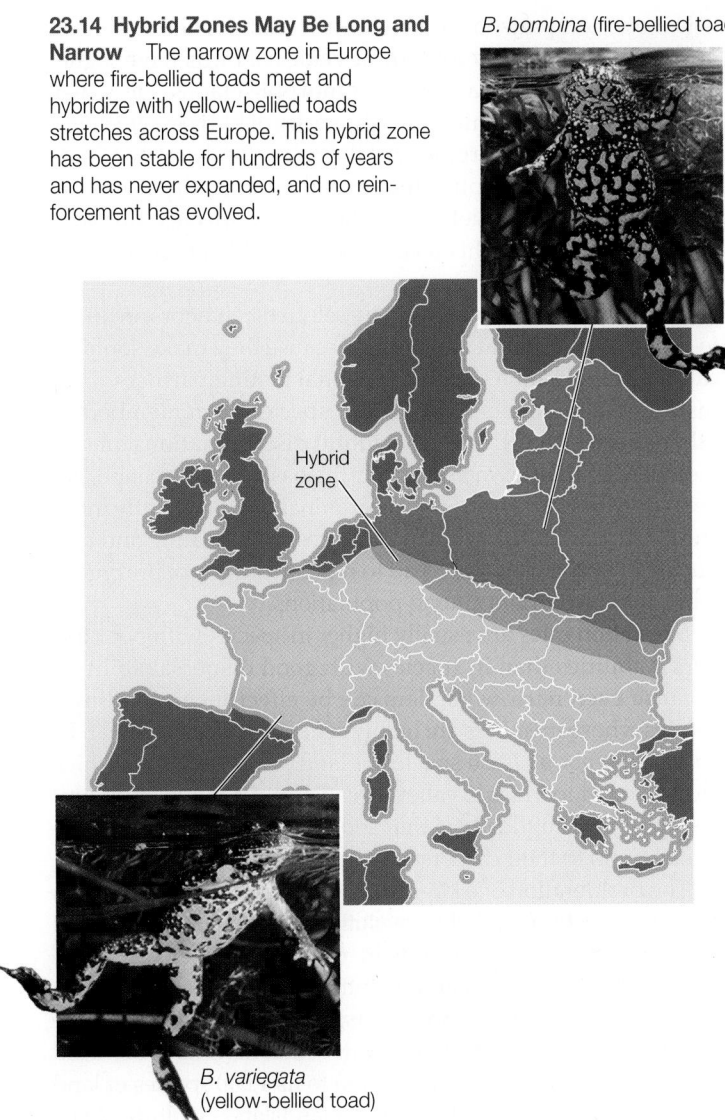

*B. variegata* (yellow-bellied toad)

hybrid zone thus remains narrow because there is strong selection against hybrids, and because adult toads do not move over long distances. The zone has persisted for hundreds of years, however, because many individuals of both species continue to move short distances into it, constantly replenishing the hybrid population.

## 23.3 RECAP

**Reproductive isolation may result from prezygotic or postzygotic reproductive barriers. Lower fitness of hybrids in contact zones can lead to the reinforcement of prezygotic reproductive barriers.**

- Describe various kinds of prezygotic and postzygotic reproductive barriers. **See pp. 489–491**
- Why is reinforcement of prezygotic barriers likely if hybrid offspring survive more poorly than offspring produced by within-population matings? **See p. 491**

Some groups of organisms have many species, others only a few. Hundreds of species of *Drosophila* evolved in the small area of the Hawaiian Islands over about 20 million years. In contrast, there are only a few species of horseshoe crabs in the world, and only one species of ginkgo tree, even though these latter groups have persisted for hundreds of millions of years. Why do different groups of organisms have such different rates of speciation?

## 23.4 Why Do Rates of Speciation Vary?

Rates of speciation (the proportion of existing species that split to form new species over a given period of time) vary greatly because many factors influence the likelihood that a lineage will split to form two or more species. What are some of the factors that influence the probability of a given lineage splitting into two?

Populations of species that have specialized diets may be more likely to diverge than are populations that have generalized diets. To investigate the effects of diet on rates of speciation, Charles Mitter and colleagues compared species richness in some closely related groups of true bugs (hemipterans). The common ancestor of these groups was a predator that fed on other insects, but a dietary shift to herbivory (eating plants) evolved at least twice in the groups under study. The herbivorous groups have many more species than those that are predatory (**Figure 23.15**). Herbivorous bugs typically specialize on one or a few closely related species of plants, whereas predatory bugs tend to feed on many different species of insects. High diversity of host plants can thus lead to a correspondingly high diversity in the herbivorous specialists.

Speciation rates in plants are faster in animal-pollinated than in wind-pollinated plants. Animal-pollinated groups have, on average, 2.4 times as many species as related groups pollinated by wind. Among animal-pollinated plants, speciation rates are correlated with pollinator specialization. In columbines (*Aquilegia*), the rate of evolution of new species has been about three times faster in lineages that have long nectar spurs than in lineages that lack spurs. Why do nectar spurs increase the spe-

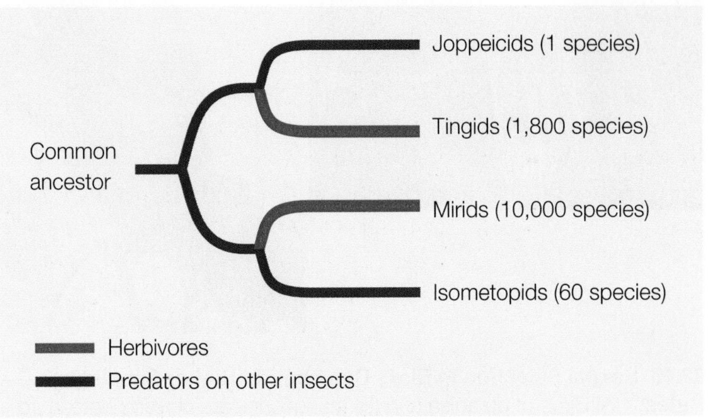


**23.15 Dietary Shifts Can Promote Speciation** Herbivorous groups of hemipteran insects have speciated several times faster than closely related predatory groups.

ciation rate? Apparently it is because having longer spurs restricts the number of pollinator species that visit the flowers, thus increasing opportunities for reproductive isolation (see Figure 23.12).

The mechanisms of sexual selection (see Section 21.2) appear to result in increased rates of speciation. Some of the most striking examples of sexual selection are found in birds with promiscuous mating systems. Bird-watchers travel thousands of miles to Papua New Guinea to witness the mating displays of male birds of paradise, some of which have long, brightly colored tail feathers and look distinctly different than the females (*sexual dimorphism*). In many of these 33 species, males assemble at display grounds, called *leks*, and females come there to choose a male with whom to copulate. After mating, the females leave the display grounds, build their nests, lay their eggs, and feed their offspring with no help from the males. The males remain to court more females (**Figure 23.16A**).

The closest relatives of the birds of paradise are the manucodes. Male and female manucodes differ only slightly in size and plumage (so they are *sexually monomorphic*). They form monogamous pair bonds, and both sexes contribute to raising the young. There are only 5 species of manucodes (**Figure 23.16B**), compared with 33 species of birds of paradise. This is just one comparison, and by itself would not be convincing evidence that sexually dimorphic clades of birds have higher rates of speciation than do monomorphic clades. However, when biologists compare all the examples of birds in which one clade is sexually dimorphic, and the most closely related clade is sexually monomorphic, the sexually dimorphic clades are significantly more likely to contain more species. But why would sexual dimorphism be associated with a higher rate of speciation?

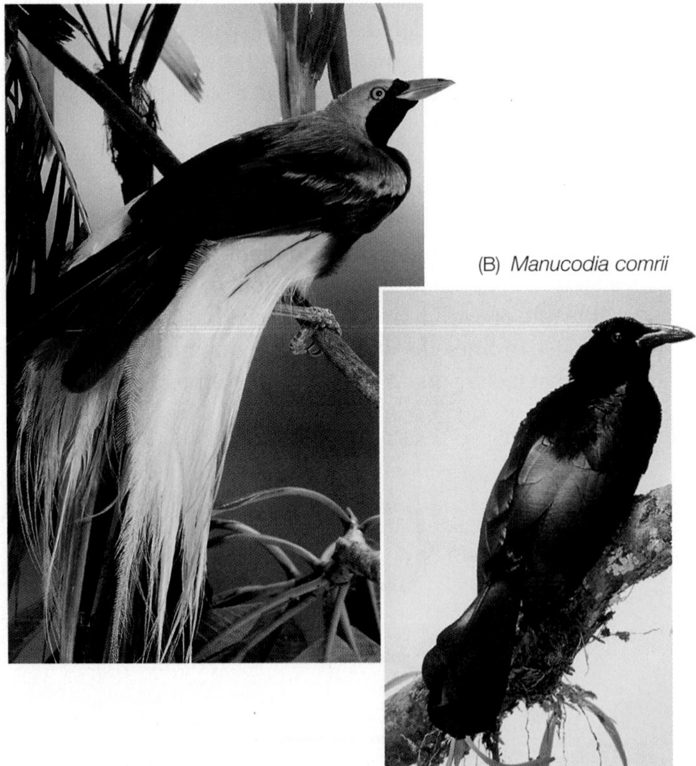


**23.16 Sexual Selection in Birds Can Lead to Higher Speciation Rates** (A) Birds of paradise and (B) manucodes are closely related bird groups of the South Pacific. However, speciation rates are much higher among the sexually dimorphic, polygynous birds of paradise (33 species) than among manucodes (5 species).

Animals with complex sexually selected behaviors are likely to form new species at a high rate because they make sophisticated discriminations among potential mating partners. They distinguish members of their own species from members of other species, and they make subtle discriminations among members of their own species on the basis of size, shape, appearance, and behavior. Such discriminations can greatly influence which individuals are most successful in mating and producing offspring, and may lead to rapid evolution of prezygotic reproductive barriers among populations.

Speciation rates are usually higher in species with poor dispersal abilities than in species with good dispersal abilities, because even narrow barriers can be effective in dividing a species whose members are highly sedentary. The Hawaiian Islands have about 1,000 species of land snails, many of which are restricted to a single valley. Because snails move only short distances, the high ridges that separate the valleys are effective barriers to their dispersal.

The proliferation of a large number of daughter species from a single ancestor is called an **evolutionary radiation**. If the rapid proliferation of species results in an array of species that live in a variety of environments and differ in the characteristics they use to exploit those environments, the radiation is said to be **adaptive**. Several remarkable adaptive radiations have occurred in the Hawaiian Islands. In addition to its 1,000 species of land snails, the native biota of the Hawaiian Islands includes 1,000 species of flowering plants, 10,000 species of insects, and more than 100 bird species. However, there were no amphibians, no terrestrial reptiles, and only one native terrestrial mammal—a bat—on the islands until humans introduced additional species. The 10,000 known native species of insects on Hawaii are believed to have evolved from only about 400 immigrant species; only 7 immigrant species are believed to account for all the native Hawaiian land birds. Similarly, as we saw earlier in this chapter, an adaptive radiation in the Galápagos archipelago resulted in the 14 species of Darwin's finches, which differ strikingly in the size and shape of their bills and, accordingly, in the food resources they use (see Figure 23.7).

The 28 species of Hawaiian sunflowers called silverswords are an impressive example of an adaptive radiation in plants. DNA sequences show that these species share a relatively recent common ancestor with a species of tarweed from the Pacific coast of North America (**Figure 23.17**). Whereas all mainland tarweeds are small, upright herbs (nonwoody plants), the silverswords include prostrate and upright herbs, shrubs, trees, and vines. Silversword species occupy nearly all the habitats of the Hawaiian Islands, from sea level to above timberline

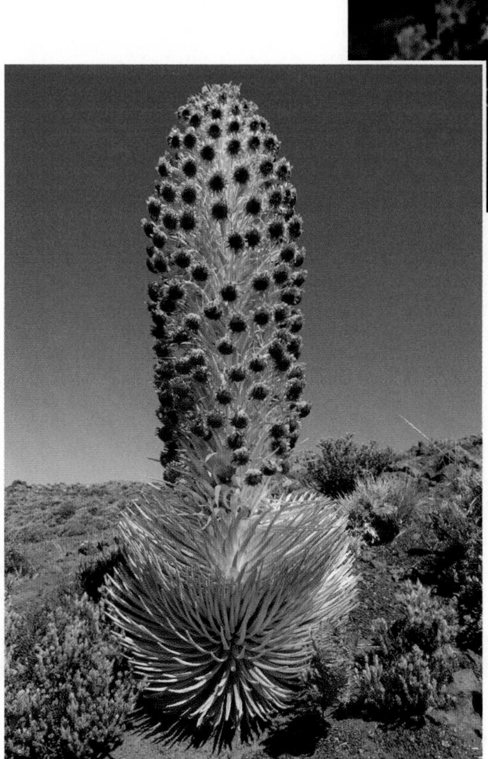

**23.17 Rapid Evolution among Hawaiian Silverswords** The Hawaiian silverswords, three closely related genera of the sunflower family, are believed to have descended from a single common ancestor (a plant similar to the tarweed; upper right) that colonized Hawaii from the Pacific coast of North America. The four plants shown here are more closely related than they appear to be based on their morphology.

in the mountains. Despite their extraordinary morphological diversification, the silverswords are genetically very similar.

The Hawaiian silverswords are more diverse in size and shape than the mainland tarweeds because the tarweed ancestors first arrived on islands that harbored very few plant species. In particular, there were few trees and shrubs, because such large-seeded plants rarely disperse to oceanic islands. Trees and shrubs have evolved from nonwoody ancestors on many oceanic islands. On the mainland, however, tarweeds live in ecological communities that contain many tree and shrub species in lineages with long evolutionary histories. In those environments, opportunities to exploit the "tree" way of life had already been preempted.

## 23.4 RECAP

**Dispersal ability, dietary specialization, and mechanisms of sexual selection affect rates of speciation. Speciation rates in plants can depend on mechanisms of pollination. Open ecological niches present opportunities for evolutionary radiations.**

- Explain how pollinator specialization in plants and sexual selection in animals can increase rates of speciation. See pp. 493–494
- Why do adaptive radiations often occur when a founder species invades an isolated geographic area? See p. 494

The processes described in this chapter, operating over billions of years, have produced a world in which life is organized into millions of species, each adapted to live in a particular environment and to use environmental resources in a particular way. In the next chapter we consider how species evolve at the level of their genes and genomes.

## CHAPTER SUMMARY

### 23.1 What Are Species?

- **Speciation** is the process by which one species splits into two or more daughter species, which thereafter evolve as distinct lineages.
- The **biological species concept** distinguishes species on the basis of **reproductive isolation**.
- The **morphological species concept** distinguishes species on the basis of physical similarities; it often underestimates or overestimates the actual number of reproductively isolated species.
- The **lineage species concept** recognizes evolutionarily independent lineages as species, allowing biologists to consider species over evolutionary time.

### 23.2 How Do New Species Arise?

- Speciation usually results from the interruption of gene flow within a population.
- The Dobzhansky-Muller model describes how reproductive isolation can develop between two descendant species. Review Figure 23.2
- **Allopatric speciation**, which results when populations are separated by a physical barrier, is the dominant mode of speciation. This type of speciation may follow from founder events, in which some members of a population cross a barrier and found a new, isolated population. Review Figure 23.5, **ANIMATED TUTORIAL 23.1**
- **Sympatric speciation** results when the genomes of two groups diverge in the absence of physical isolation. It can result from disruptive selection for two or more distinct microhabitats.
- Sympatric speciation can occur within two generations via **polyploidy**, an increase in the number of chromosomes sets. Polyploidy may arise from chromosome duplications within a species (**autopolyploidy**) or from hybridization that results in combining the chromosomes of two species (**allopolyploidy**). Review Figure 23.8

**SEE ANIMATED TUTORIAL 23.2**

### 23.3 What Happens When Newly Formed Species Come Together?

- **Prezygotic barriers** to reproduction operate before fertilization; **postzygotic barriers** to reproduction operate after fertilization. Prezygotic barriers may be favored by natural selection if postzygotic barriers are incomplete. Review Figures 23.9, 23.11
- **Hybrid zones** may form when previously separated populations come into contact and reproductive isolation is incomplete. Review Figure 23.14

### 23.4 Why Do Rates of Speciation Vary?

- Dispersal ability, dietary specialization, type of pollination, and sexual selection all influence speciation rates. Review Figure 23.15

**SEE WEB ACTIVITY 23.1 for a concept review of this chapter.**

## SELF-QUIZ

1. The biological species concept defines a species as a group of
   *a.* actually interbreeding natural populations that are reproductively isolated from other such groups.
   *b.* potentially interbreeding natural populations that are reproductively isolated from other such groups.
   *c.* actually or potentially interbreeding natural populations that are reproductively isolated from other such groups.
   *d.* actually or potentially interbreeding natural populations that are reproductively connected to other such groups.
   *e.* actually interbreeding natural populations that are reproductively connected to other such groups.
2. Which of the following is *not* a condition expected to favor allopatric speciation?
   *a.* Continents drift apart and separate previously connected lineages.
   *b.* A mountain range separates formerly connected populations.
   *c.* Different environments on two sides of a barrier cause populations to diverge.
   *d.* The range of a species is separated by loss of intermediate habitat.
   *e.* Tetraploid individuals arise in one part of the range of a species.
3. Finches speciated in the Galápagos Islands because
   *a.* the Galápagos Islands are not far from the mainland.
   *b.* the Galápagos Islands are thought to promote sympatric speciation in birds.
   *c.* hybridization across different island populations of finches led to high levels of polyploidy.
   *d.* the islands of the Galápagos archipelago are sufficiently isolated from one another that there is little migration among them.
   *e.* the islands of the Galápagos archipelago are close enough to one another that there is considerable migration among them.
4. Which of the following is *not* a potential prezygotic reproductive barrier?
   *a.* Temporal segregation of breeding seasons
   *b.* Differences in chemicals that attract mates
   *c.* Hybrid infertility
   *d.* Spatial segregation of mating sites
   *e.* Sperm that cannot penetrate an egg
5. A common means of sympatric speciation is
   *a.* polyploidy.
   *b.* hybrid infertility.
   *c.* temporal segregation of breeding seasons.
   *d.* spatial segregation of mating sites.
   *e.* imposition of a geographic barrier.
6. Narrow hybrid zones may persist for long times because
   *a.* hybrids are always at a disadvantage.
   *b.* hybrids have an advantage only in narrow zones.
   *c.* hybrid individuals never move far from their birthplaces.
   *d.* individuals that move into the zone have not previously encountered individuals of the other species, so reinforcement of reproductive barriers has not occurred.
   *e.* Narrow hybrid zones are artifacts because biologists generally restrict their studies to contact zones between species.

7. Which statement about speciation is *not* true?
   *a.* It always takes thousands of years.
   *b.* Reproductive isolation may develop slowly between diverging lineages.
   *c.* Among animals, it usually requires a physical barrier.
   *d.* Among plants, it often happens as a result of polyploidy.
   *e.* It has produced the millions of species living today.
8. Which of the following is often associated with higher rates of speciation?
   *a.* Sexually dimorphic compared with sexually monomorphic birds
   *b.* Insects with specialized diets compared with insects with generalized diets
   *c.* Species with low dispersal ability compared with species with high dispersal ability
   *d.* Plants with animal pollination compared with plants with wind pollination
   *e.* All of the above
9. Evolutionary radiations
   *a.* happen often on continents but rarely on island archipelagoes.
   *b.* characterize birds and plants but not other groups of organisms.
   *c.* have happened on continents as well as on islands.
   *d.* require major reorganizations of the genome.
   *e.* never happen in species-poor environments.
10. Speciation is an important component of evolution because it
   *a.* generates the variation on which natural selection acts.
   *b.* generates the variation on which genetic drift and mutations act.
   *c.* enabled Charles Darwin to perceive the mechanisms of evolution.
   *d.* generates the high extinction rates that drive evolutionary change.
   *e.* has resulted in a world with millions of species, each adapted for a particular way of life.

## FOR DISCUSSION

1. The North American snow goose has two distinct color forms, blue and white. Matings between the two color forms are common. However, blue individuals pair with blue individuals and white individuals pair with white individuals much more frequently than would be expected by chance. Suppose that blue and white snow geese are equally frequent in a population, and that 75 percent of all mated pairs consist of two individuals of the same color. What would you conclude about speciation processes in these geese? What if 100 percent of pairs were the same color?
2. Suppose pairs of snow geese of mixed colors were found only in a narrow zone within the broad Arctic breeding range of the geese, with blue geese found on one side and white geese found on the other side of this narrow zone. Would your answer to Question 1 remain the same?
3. Although many butterfly species are divided into local populations among which there is little gene flow, these species often show relatively little morphological variation among populations. Describe the studies you would conduct to determine what maintains this morphological similarity.
4. Evolutionary radiations are common and easily studied on oceanic islands. In what types of *mainland* situations would you expect to find major evolutionary radiations? Why?
5. Fruit flies of the genus *Drosophila* are distributed worldwide, but 30 to 40 percent of all the species in the genus are found on the Hawaiian Islands (which comprise far less than 1% of Earth's total land area). What might account for this distribution pattern?
6. What factors can cause extinction rates to exceed speciation rates in a clade? Name some clades in which human activities are increasing extinction rates without increasing speciation rates.
7. If it is true that natural selection does not directly favor lower viability of hybrids, why is it that hybrid individuals so often have lowered viability?

## ADDITIONAL INVESTIGATION

In the two *Aquilegia* species shown in Figure 23.12, the orientation of the flowers and the length of flower spurs are associated with the respective pollinator species (hummingbirds and hawkmoths). Columbine flowers vary in other ways as well; for example, they differ in color, and probably in odor. What experiments could you design to determine the traits that various pollinators use to distinguish among the flowers of different columbine species?

## WORKING WITH DATA (GO TO yourBioPortal.com)

**Examining Evidence for Reinforcement of Prezygotic Barriers**
In this exercise based on Figure 23.13, you will examine some of the data collected by Don Levin to study reinforcement of prezygotic reproductive barriers in *Phlox*. You will also critique the study design of the experiment, and consider alternative explanations for the results.

# 24 Evolution of Genes and Genomes

## Shocking evolution

Some fishes, including the famous electric eel of Central and South America, can produce high-voltage discharges of electricity (up to 650 volts) that they use to stun their prey. A variety of other fish species are known to produce somewhat weaker electric discharges. Most of these latter species live in murky water where visual cues are limited; they use electric signals to locate (but not to stun) their prey. Electric signals also allow them to communicate with other individuals of their own species.

Electric organs have evolved independently in several fish lineages. How did these organs evolve? Let's consider first the physical basis of the electrical signal. *Voltage-gated sodium channels* are large proteins that underlie the generation and propagation of rapid electrical signals in nerve, muscle, and heart tissues (see Chapter 45). Electric signals are transmitted along nerves to muscles as the sodium channels embedded in cell membranes are stimulated to open. These channels control the concentration of positively charged sodium ions ($Na^+$) on the inside relative to the outside of cells, resulting in an electric charge that is transmitted across the surface of the muscle, leading to muscular contraction.

Most vertebrates have a number of different copies of the genes encoding the several proteins that make up the sodium channel. These copies arose through a series of *gene duplications* in the distant past of vertebrate genome evolution. Such duplications allowed for the specialization of protein function, making it possible for different sodium channels to exist in different types of tissue. In the case of electric fishes, one of the sodium channel genes ordinarily expressed in muscle diverged and a new functional protein evolved. Changes in a relatively small number of nucleotide positions in the gene resulted in modified sodium channels, allowing the development of a new organ with a unique function—the generation of externally transmitted electric energy.

The "living battery" electric organ differs from skeletal muscle in important ways. The organ is composed of many *electrocytes*, each of which is a derived muscle cell capable of producing a small electric charge. Electrocytes are stacked in series, much like the plates in a car battery. Rather than producing muscle contraction and movement, however, the organ generates an electric discharge. This signal is species-specific, which allows intraspecific communication and also serves as an isolating mechanism between species (see Chapter 23).

The repeated evolution of electric organs from muscle tissue is facilitated by relatively simple molecular changes

**An Electric Fish** The elephant-nose fish (*Gnathonemus petersi*), a river-dwelling species from West Africa, is one of many fishes in which weakly discharging electric organs have evolved via modifications in sodium channel proteins.

**A High-Voltage Electric Fish** This torpedo ray can put out as much as 220 volts of electricity. So far this particular species remains unidentified; it has been found only in a single bay among the islands of Komodo National Park, Indonesia.

in certain genes, changes that result in major functional changes to sodium channels. Gene duplication facilitates the process, since "extra" genes allow for such specialization in protein function. Finally, interspecific differences in sodium channel function result from additional changes in nucleotide sequences of the respective genes. These small differences allow different species of fishes to use different communication signals, which improves intraspecific communication while reducing interspecific interference.

The evolution of sodium channels is just one example of how an understanding of the evolution of genes and genomes helps biologists understand the diversity of life on Earth. Molecular investigations also allow biologists to observe the process of evolution directly in the laboratory, and to use evolutionary principles to produce new molecules with useful functions.

**IN THIS CHAPTER** we will see how molecular biologists infer both the patterns and the causes of molecular evolution from studies of nucleic acids and proteins. We will explore how the functions of molecules change, how genomes change in size, and where new genes come from. Finally, we will explore some practical applications of molecular evolution for producing new molecules with novel functions.

## CHAPTER OUTLINE



## 24.1 How Are Genomes Used to Study Evolution?

An organism's **genome** is the full set of genes it contains, as well as any noncoding regions of the DNA (or in the case of some viruses, RNA). Most of the genes of eukaryotic organisms are found on chromosomes in the nucleus, but genes are also present in chloroplasts and mitochondria. In organisms that reproduce sexually, both males and females transmit nuclear genes, but mitochondrial and chloroplast genes usually are transmitted only via the cytoplasm of one of the two gametes (usually from the female parent).

Genomes must be replicated to be transmitted from parents to offspring. DNA replication does not occur without error, however. Mistakes in DNA replication—mutations—provide much of the raw material for evolutionary change. Mutations are essential for the long-term survival of life, because they are the initial source of the genetic variation that permits organisms to evolve in response to changes in their environment.

A particular copy of a gene will not be passed on to successive generations unless an individual with that copy survives and reproduces. Therefore, the capacity to cooperate with different combinations of other genes is likely to increase the probability that a particular allele will become fixed in a population. Moreover, the degree and timing of a gene's expression are affected by its location in the genome. For these reasons, the genes of an individual organism can be viewed as interacting members of a group, among which there are divisions of labor but also strong interdependencies.

A genome, then, is not simply a random collection of genes in random order along chromosomes. Rather, it is a complex set of integrated genes, regulatory sequences, and structural elements, as well as vast stretches of noncoding DNA that may have little direct function. The positions of genes, as well as their sequences, are subject to evolutionary change, as are the extent and location of noncoding DNA. All of these changes can affect the phenotype of an organism. Biologists have now sequenced the complete genomes of a large number of organisms, including humans. This information is helping us to understand how and why organisms differ, how they function, and how they have evolved.

### Evolution of genomes results in biological diversity

The field of **molecular evolution** investigates the mechanisms and consequences of the evolution of macromolecules. Molecular evolutionists study relationships between the structures of genes and proteins and the functions of organisms. They also examine molecular variation to reconstruct evolutionary history and to study the mechanisms and consequences of evolution. The molecules of special interest to molecular evolutionists are nucleic acids (DNA and RNA) and proteins. Students of this field ask questions such as: What does molecular variation tell us about a gene's function? Why do the genomes of different organisms vary in size? What evolutionary forces shape patterns of variation among genomes? And a crucial question from an evolutionary perspective, How do genomes acquire new functions? Investigations into the evolution of particular nucleic acids and proteins are instrumental in reconstructing the evolutionary histories of genes and in determining which organisms carry them. Ultimately, molecular evolutionists hope to explain the molecular basis of biological diversity.

The evolution of nucleic acids and proteins depends on genetic variation introduced by mutations. One of several ways in which genes evolve is by means of *nucleotide substitutions*. In genes that encode proteins, nucleotide substitutions sometimes result in amino acid replacements that can change the charge, the structure (secondary or tertiary), and other chemical and physical properties of the encoded protein. Phenotypic changes in the protein often affect the way that protein functions in the organism.

Evolutionary changes in genes and proteins can be identified by comparing the nucleotide or amino acid sequences of different organisms. The longer two sequences have been evolving separately, the more differences they accumulate (bearing in mind that different genes in the same species evolve at different rates). Determining when changes in nucleotide or amino acid sequences occurred is a first step toward inferring their causes. Knowledge of the pattern and rate of evolutionary change in a given macromolecule is useful in reconstructing the evolutionary history of groups of organisms.

To compare genes or proteins across different organisms, biologists need a way to identify homologous parts of molecules. (Recall from Section 22.1 that *homologous* features are shared by two or more species and have been inherited from a common ancestor.) Homologous parts of a protein can be traced to homologous amino acid sequences. And, since nucleotide sequences encode amino acid sequences, the concept of homology extends down to the level of individual nucleotide positions. Therefore, one of the first steps in studying the evolution of genes or proteins is to align homologous positions in the nucleotide or amino acid sequence of interest.

### Genes and proteins are compared through sequence alignment

Once the DNA or amino acid sequences of molecules from different organisms have been determined, they can be compared. Homologous positions can be identified only if we first pinpoint the locations of deletions

## TOOLS FOR INVESTIGATING LIFE

### 24.1 Amino Acid Sequence Alignment

Amino acid sequence alignment is a way of arranging protein sequences to identify regions of homology between the sequences. Gaps are inserted between the amino acid residues to align similar residues in columns. Differences and similarities between each pair of aligned sequences are then summarized in a similarity matrix. Homologous DNA sequences can be aligned in a similar manner.

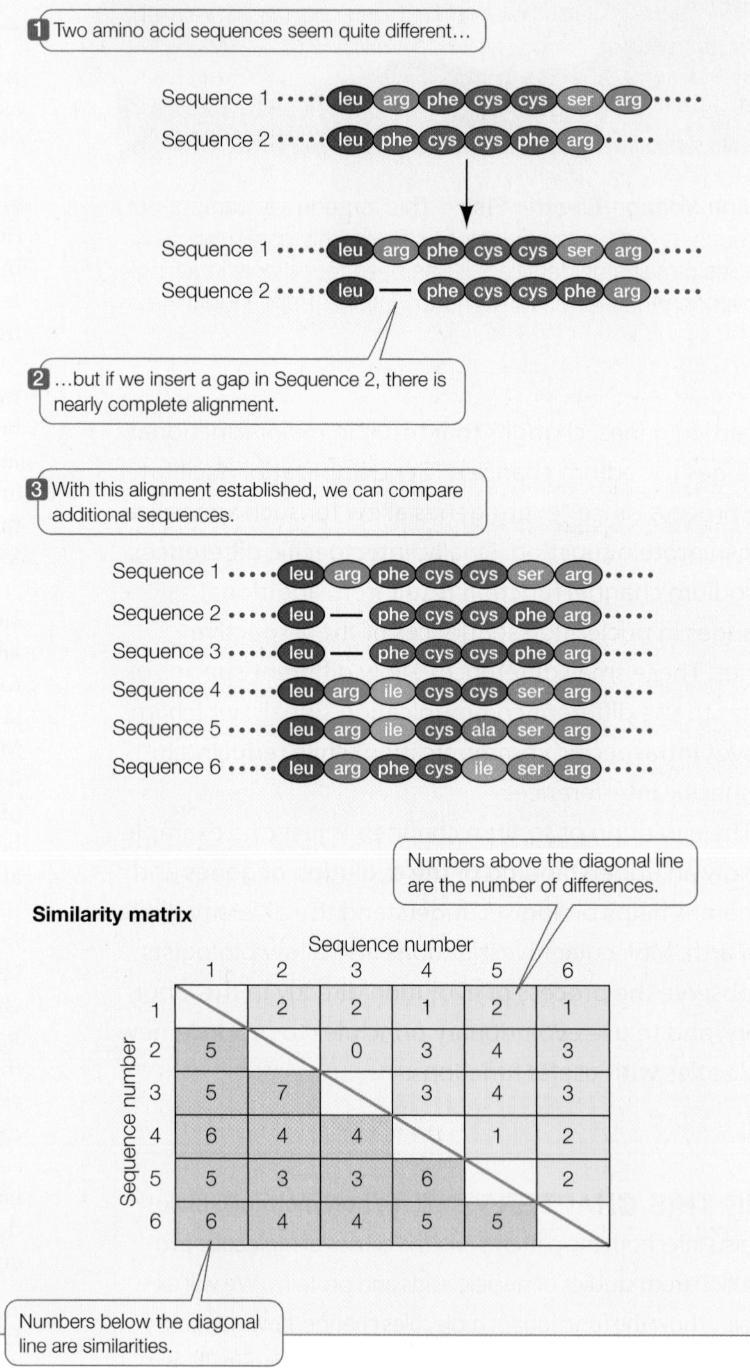


**yourBioPortal.com**
GO TO Web Activity 24.1 • Amino Acid Sequence Alignment

and insertions that have occurred in the molecules of interest in the time since the organisms diverged from a common ancestor. A simple hypothetical example illustrates this **sequence alignment** technique. In **Figure 24.1** we compare two amino acid sequences (1 and 2) from homologous proteins in different organisms. The two sequences at first appear to differ in both the number and identity of their amino acids, but if we insert a gap after the first amino acid in Sequence 2 (after leucine), similarities in the two sequences become more obvious. This gap represents the occurrence of one of two evolutionary events: an insertion of an amino acid in the longer protein, or a deletion of an amino acid in the shorter protein.

Having adjusted for this insertion or deletion event, we can see that the two sequences differ by only one amino acid at position 6 (serine or phenylalanine). Adding a single gap—that is, identifying a deletion or an insertion—*aligns* these sequences. Longer sequences and those that have diverged more extensively require more elaborate adjustments. Explicit models (incorporated into computer algorithms) have been developed to account for the relative probabilities of deletions, insertions, and particular amino acid replacements.

Having aligned the sequences, we can compare them by counting the number of nucleotides or amino acids that differ between them. Summing the number of similar and different amino acids in each pair of sequences allows us to construct a **similarity matrix**, which gives us a measure of the minimum number of changes that have occurred during the divergence between each pair of organisms (see Figure 24.1).

yourBioPortal.com
GO TO Web Activity 24.2 • Similarity Matrix Construction

### Models of sequence evolution are used to calculate evolutionary divergence

The sequence comparison procedure illustrated in Figure 24.1 gives a simple count of the number of differences and similarities between the proteins of two species. In the context of two aligned DNA sequences, we can count the number of differences at homologous nucleotide positions, and this count indicates the minimum number of nucleotide changes that must have occurred since the two sequences diverged from a common ancestral sequence.

Although it is useful in determining a *minimum* number of changes between two sequences, the count provided by sequence alignment almost certainly underestimates the actual number of changes that have occurred since the sequences diverged. Any given change counted in a similarity matrix of DNA sequences may result from multiple substitution events that occurred at a given nucleotide position over time. As illustrated in **Figure 24.2**, any of the following events may have occurred at a given nucleotide position that would not be revealed by a simple count of similarities and differences between two DNA sequences:

- *Multiple substitutions.* More than one change occurs at a given position between the ancestral sequence and at least one of the observed sequences.
- *Coincident substitutions.* At a given position, different substitutions occur between the ancestral sequence and each observed sequence.
- *Parallel substitutions.* The same substitution occurs independently between the ancestral sequence and each observed sequence.
- *Back substitutions* (also called *reversions*). In a variation on multiple substitutions, after a change at a given position, a subsequent substitution changes the position back to the ancestral state.

To correct for undercounting of substitutions, molecular evolutionists have developed mathematical models that describe how DNA (and protein) sequences evolve. These models take into account the relative rates of change from one nucleotide to another; for example, *transitions* (changes between the two purines, A ↔ G, or between the pyrimidines, C ↔ T) are more frequent than *transversions* (a purine is replaced by a pyrimidine, or vice versa). Models also include parameters such as the different rates of substitution across different parts of a gene and the proportions of each nucleotide present in a given sequence. Once such parameters have been estimated, the model

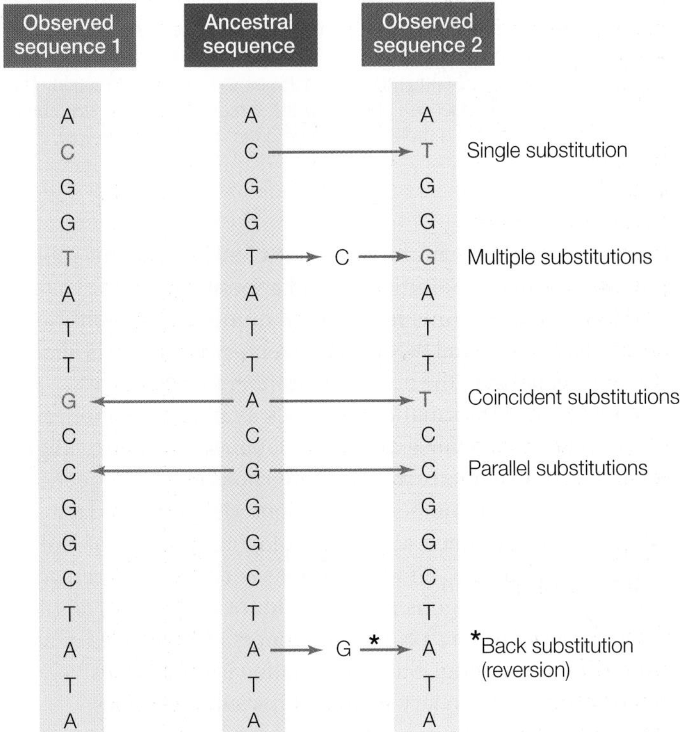


**24.2 Multiple Substitutions Are Not Reflected in Pairwise Sequence Comparisons** Two observed sequences are descended from a common ancestral sequence (center) through a series of substitutions. Although the two observed sequences differ in only three nucleotide differences (colored letters), these three differences resulted from a total of nine substitutions (arrows).

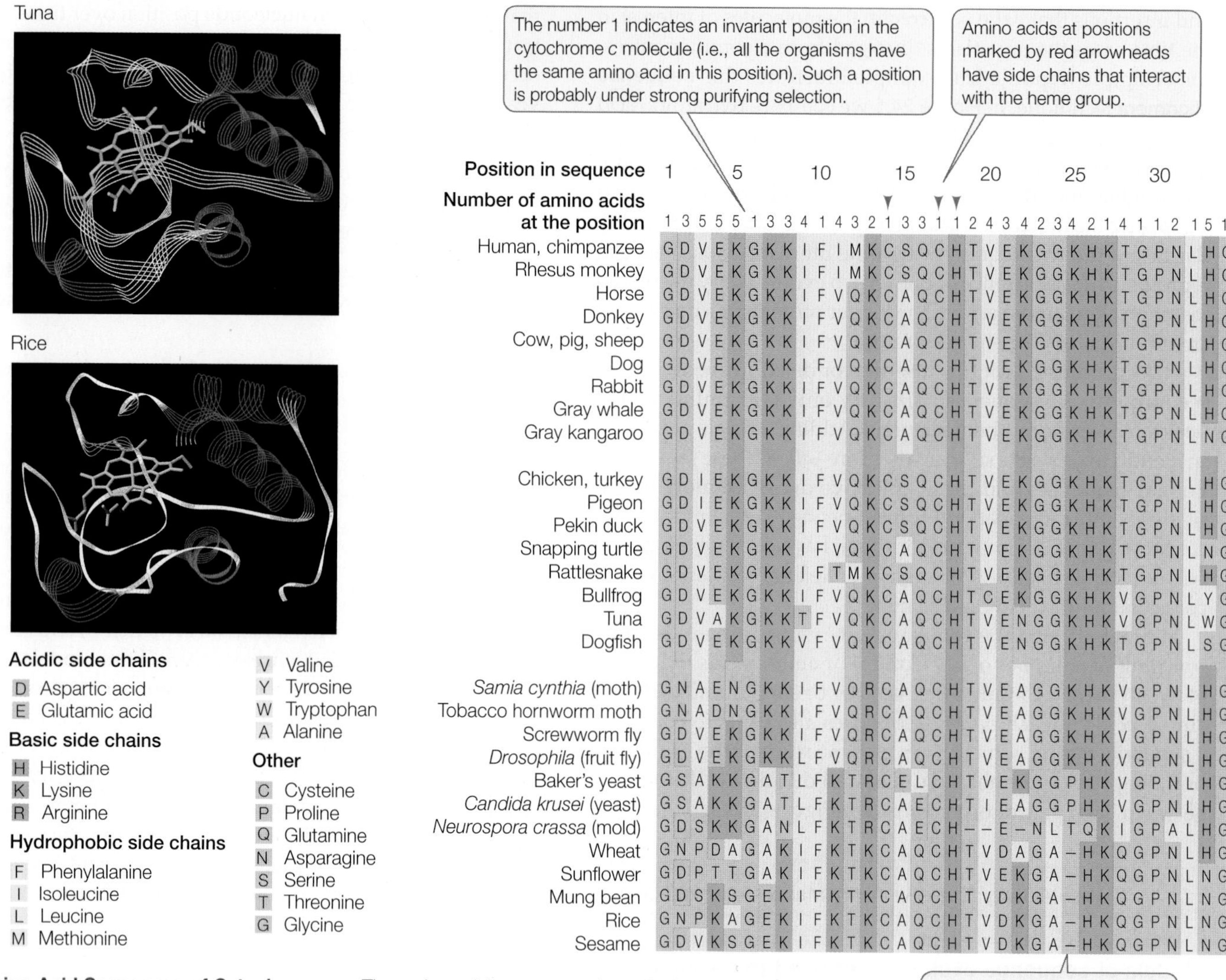


**24.3 Amino Acid Sequences of Cytochrome *c*** The amino acid sequences shown in the table were obtained from analyses of the enzyme cytochrome *c* from 33 species of plants, fungi, and animals. Note the lack of variation across the sequences at positions 70–80, suggesting that this region is under strong stabilizing selection and that changing its amino acid sequence would impair the protein's function. The computer graphics at the upper left are created from these sequences and show the three-dimensional structures of tuna and rice cytochrome *c*. Alpha helixes are in red, and the molecule's heme group is shown in yellow.

is used to correct for multiple substitutions, coincident substitutions, parallel substitutions, and back substitutions. The revised estimate accounts for the *total* number of substitutions likely to have occurred between two sequences, which is almost always greater than the observed number of differences.

As sequence information becomes available for more and more genes in an ever-expanding database, sequence alignments can be extended across multiple homologous sequences, and the minimum number of insertions, deletions, and substitutions can be summed across homologous genes of an entire group of organisms. Similar databases have also been constructed for homologous proteins. **Figure 24.3** shows aligned data for cytochrome *c* protein sequences in 33 species of animals, plants, and fungi. Such information is used extensively in determining evolutionary relationships among species.

### Experimental studies examine molecular evolution directly

Although molecular evolutionists are often interested in naturally evolved genes and proteins, molecular and phenotypic evolution can also be observed directly in the laboratory. Increasingly, evolutionary biologists are studying evolution experimentally. Because substitution rates are related to generation time rather than to absolute time, most of these experiments use unicellular organisms or viruses with short generations. Viruses, bacteria, and unicellular eukaryotes (such as the yeasts) can be cultured in large populations in the laboratory, and many of these organisms can evolve rapidly. In the case of some RNA viruses, the natural substitution rate may be as high as 1 substitution per 1,000 nucleotides per generation. Therefore, in a virus of a few thousand nucleotides, one

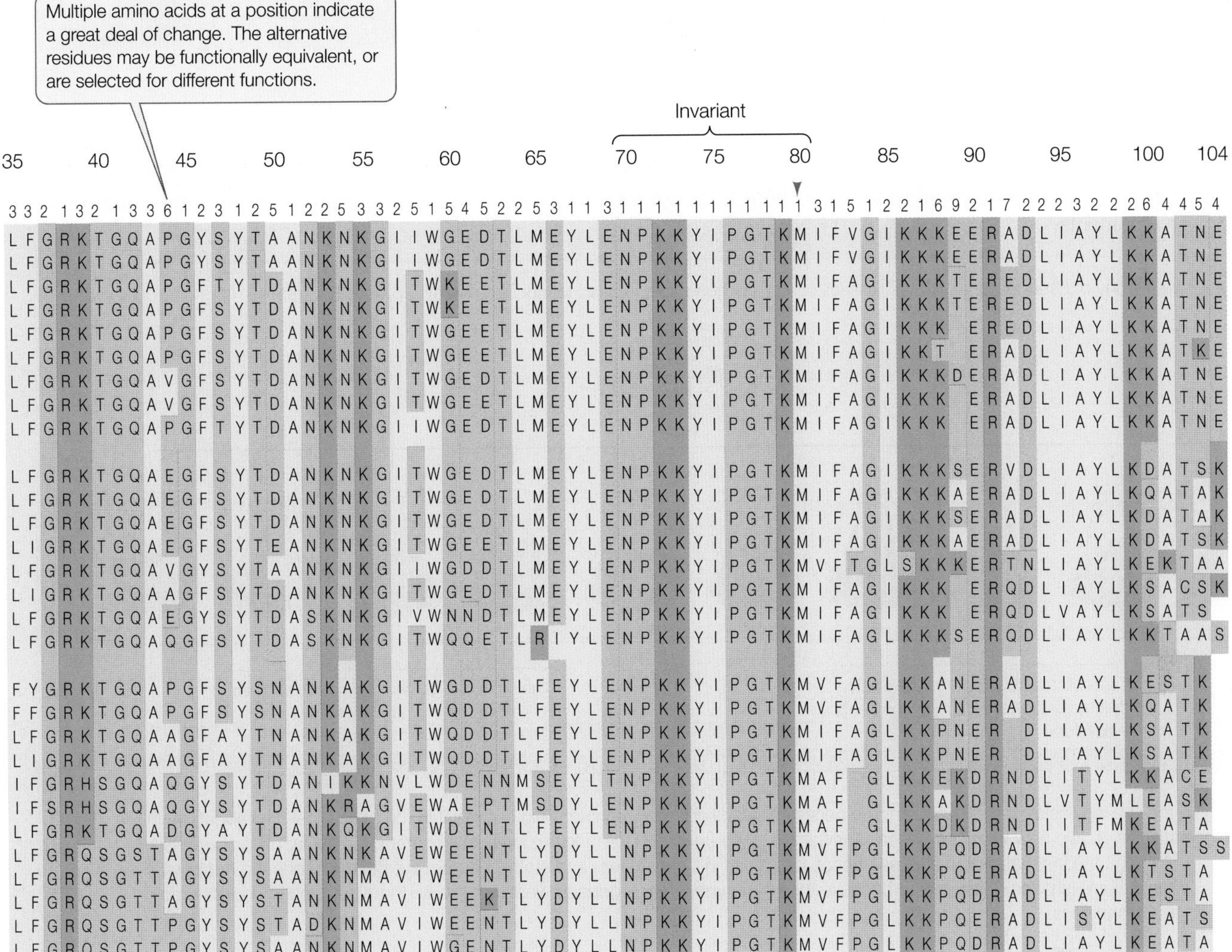


or more substitutions are expected (on average) every generation, and these changes can easily be determined by sequencing the entire genome (because of its small size). Generation time may be only tens of minutes (rather than years or decades, as in humans), so biologists can directly observe substantial molecular evolution in a controlled population over the course of days, weeks, or months.

An example of an experimental evolutionary study is shown in **Figure 24.4**. Paul Rainey and Michael Travisano wanted to examine a potential cause of adaptive radiations, which are a major source of biological diversity (see Section 23.4). For instance, near the beginning of the Cenozoic era, mammals rapidly diversified into species as diverse as elephants, moles, whales, and bats. While Rainey and Travisano clearly couldn't experimentally manipulate mammals over many millions of years, they could test the idea that heterogeneous environments with unoccupied niches lead to adaptive radiation by experimentally manipulating a bacterial lineage.

Rainey and Travisano inoculated several flasks containing culture medium with the same strain of the bacterium *Pseudomonas fluorescens*. They then shook some of the cultures to maintain a constantly uniform environment. Others they left alone (static cultures), allowing them to develop a spatially distinct structure. In the static cultures, the environment on the surface film of the medium differed from that on the walls of the flasks and from parts of the culture not touching any surfaces.

When the cultures were started, the ancestral phenotype of the bacterium produced a smooth colony, which the investigators called a "smooth morph." In just a few days, however, the static cultures consistently and independently developed two other morphs: a "wrinkly spreader" and a "fuzzy spreader." The researchers determined that the two new morphs had a genetic basis and were adaptively superior in some of the environments found within the static cultures. For example, the "wrinkly spreader" cells adhered firmly to one another as well as to surfaces, and thus were able to form a mat across the surface of the medium, where they could compete successfully for oxygen.

DNA sequencing of the genomes of these morphs showed that the same phenotypes had evolved repeatedly, and that many different substitutions could produce the same phenotypes. The homogeneous shaken cultures, in contrast, showed no evolution in phenotype. The same mutations occurred in the

## INVESTIGATING LIFE

### 24.4 Evolution in a Heterogeneous Environment

Rainey and Travisano cultured the rapidly evolving bacterium *Pseudomonas fluorescens* in homogeneous and heterogeneous environments to examine the relationship between phenotypic diversity and environmental variability.

**HYPOTHESIS** **Heterogeneous environments are more conducive to the evolution of phenotypic diversity than are homogeneous environments.**

**METHOD** One colony of *Pseudomonas fluorescens* (all of a single genotype) is used to inoculate many replicate cultures.

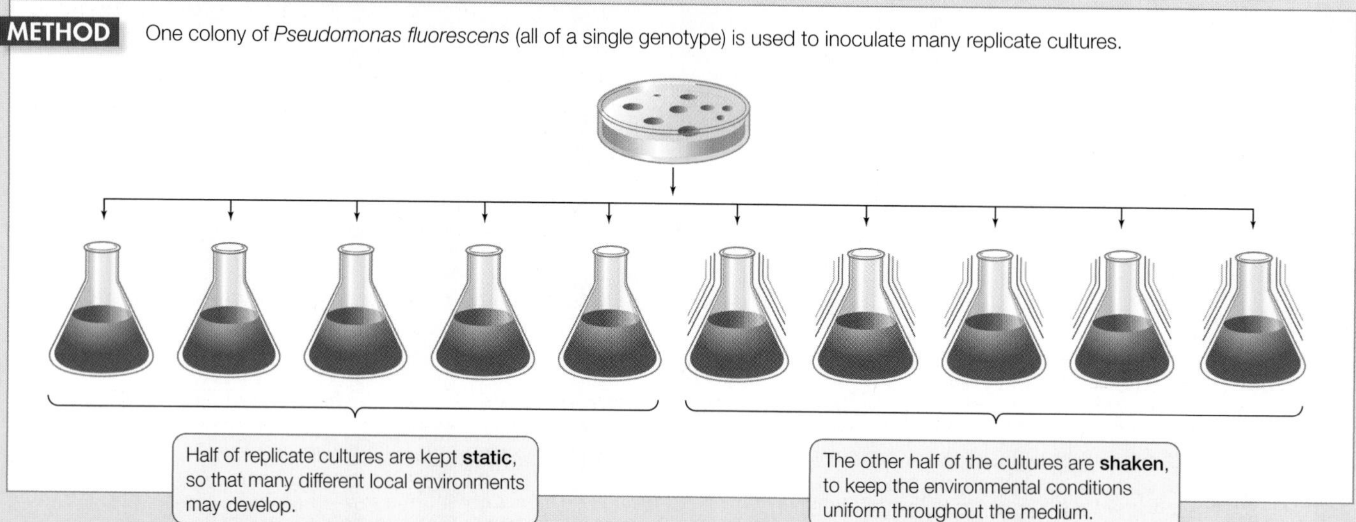

Half of replicate cultures are kept **static**, so that many different local environments may develop.
The other half of the cultures are **shaken**, to keep the environmental conditions uniform throughout the medium.

**RESULTS** In the shaken flasks, the ancestral morph persisted; the uniform environment did not result in morphological diversification. In the static flasks, two new morphotypes regularly arose, each adapted to a different local environment. Molecular analysis revealed that the mutations that produce these phenotypes arose in both shaken and static cultures, but the mutations did not persist in the uniform (shaken) environment because the phenotypes they produced were selectively disadvantageous under homogeneous conditions.

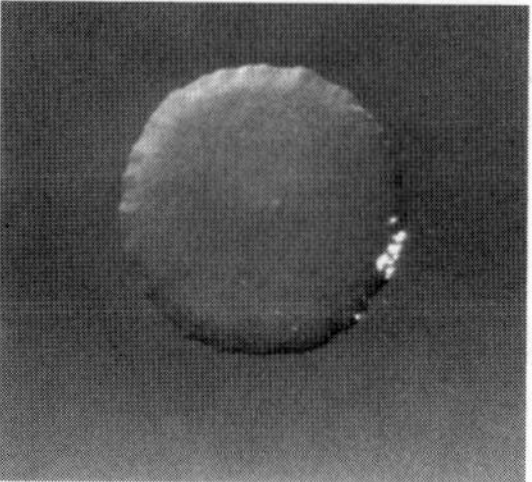

Smooth morph (ancestral)

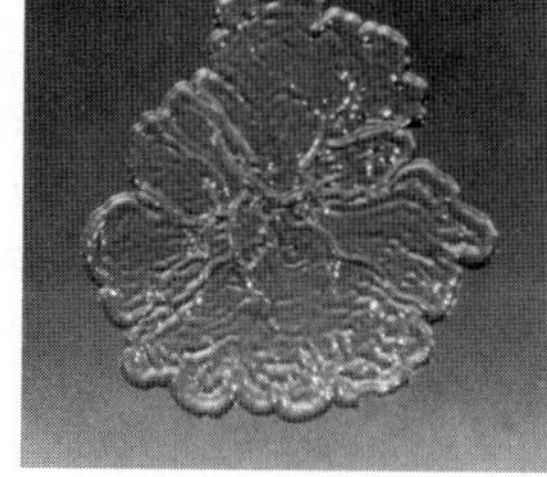
"Wrinkly spreader"

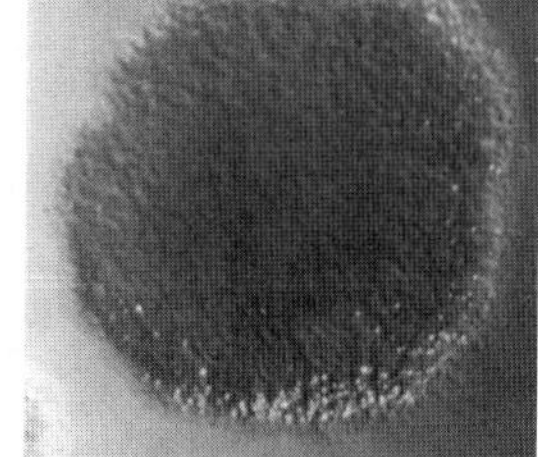
"Fuzzy spreader"

**CONCLUSION** **Phenotypic change and diversification are enhanced in a heterogeneous environment.**

**FURTHER INVESTIGATION:** Do you think the two evolved phenotypes could compete in the homogeneous environment if they were introduced after having become successfully established in the heterogeneous environment? How would you test your hypothesis?

Go to **yourBioPortal.com** for original citations, discussions, and relevant links for INVESTIGATING LIFE figures.

shaken cultures but did not persist, because the novel phenotypes they produced were selectively disadvantageous (i.e., less fit) under the "shaken" environmental conditions.

Experimental molecular evolutionary studies are used for a wide variety of purposes and have greatly expanded the ability of evolutionary biologists to test evolutionary concepts and principles. Biologists now routinely study evolution in the laboratory and, as we will see later in this chapter, use in vitro evolutionary techniques to produce novel molecules that perform new functions with industrial and pharmaceutical uses.

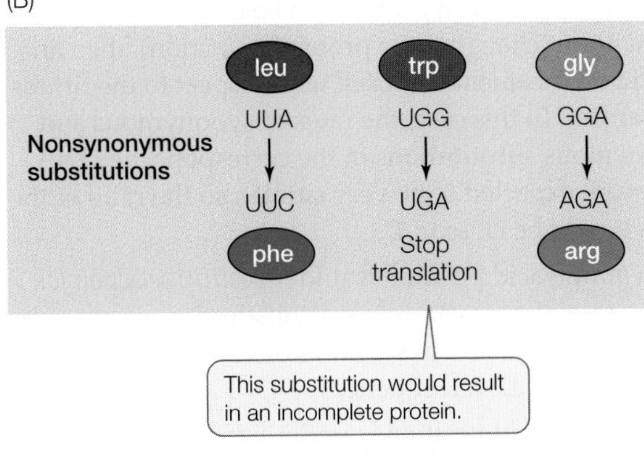


**24.5 When One Nucleotide Does or Doesn't Make a Difference** (A) Synonymous substitutions do not change the amino acid specified and do not affect protein function; such substitutions are less likely to be subject to natural selection. (B) Nonsynonymous substitutions do change the amino acid sequence and are likely to have an effect (often deleterious) on protein function; such substitutions are targets for natural selection.

## 24.1 RECAP

**The genomes of all organisms evolve over time, as can be detected by direct observation in the laboratory, as well as by aligning genes and proteins between species. Experimental studies of molecular evolution allow biologists to study many processes of evolution directly under controlled conditions.**

- How do biologists align nucleotide and amino acid sequences they wish to compare, and how do they calculate the minimum number of changes that have occurred between pairs of aligned sequences? **See pp. 500–501 and Figure 24.1**
- Explain why a simple count of nucleotide differences between two sequences underestimates the actual number of nucleotide substitutions since the sequences diverged. **See Figure 24.2**

We have seen that molecular evolutionists can directly observe the evolution of genomes over time, and can compare the genomes of different organisms and reconstruct the changes that have occurred during their evolution. Let's turn now to the question of how genomes change and examine some of the consequences of those changes.

# 24.2 What Do Genomes Reveal About Evolutionary Processes?

A *mutation*, as we saw in Chapter 15, is any change in the genetic material. A nucleotide substitution is one type of mutation. Many nucleotide substitutions have no effect on phenotype, even if the change occurs in a gene that encodes a protein, because most amino acids are specified by more than one codon (see Figure 14.6). A substitution that does not change the encoded amino acid is known as a **synonymous substitution** or **silent substitution** (**Figure 24.5A**). Synonymous substitutions do not affect the functioning of a protein (although they may have other effects, such as changes in mRNA stability or translation rates; see Section 14.5), and are therefore less likely to be influenced by natural selection.

A nucleotide substitution that *does* change the amino acid sequence encoded by a gene is known as a **nonsynonymous substitution**, also known as a *missense substitution* (**Figure 24.5B**). In general, nonsynonymous substitutions are more likely to be deleterious to the organism. But not every amino acid replacement alters a protein's shape and charge (and hence its functional properties). Therefore, some nonsynonymous substitutions may also be selectively neutral, or nearly so. Conversely, an amino acid replacement that confers an advantage to the organism would result in positive selection for the corresponding nonsynonymous substitution.

The rate of nonsynonymous nucleotide substitutions in several mammalian protein-coding genes is about $3 \times 10^{-9}$ substitutions per position per year. Synonymous substitutions in these genes have occurred about five times more frequently than nonsynonymous substitutions. In other words, *substitution rates are highest at nucleotide positions that do not change the amino acid being expressed* (**Figure 24.6**). The rate of substitution is even higher in **pseudogenes**, which are duplicate copies of genes that are no longer functional.

As we saw Chapter 21, most natural populations harbor far more genetic variation than we would expect to find if genetic

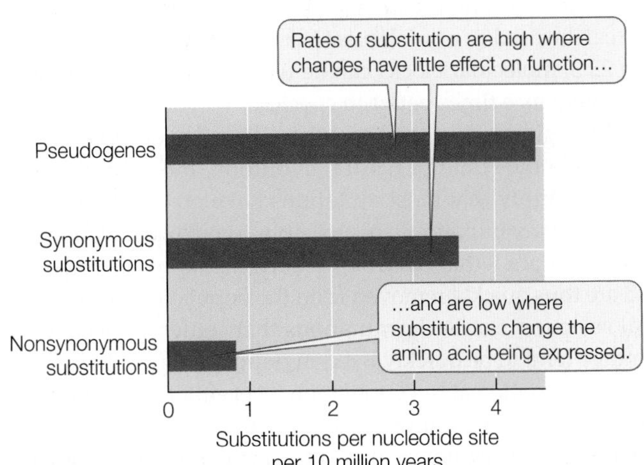


**24.6 Rates of Substitution Differ** Rates of nonsynonymous substitution typically are much lower than rates of synonymous substitution and the substitution rate in pseudogenes. This pattern reflects differing levels of functional constraints.

variation were influenced by natural selection alone. This discovery, combined with the knowledge that many mutations do not change molecular function, stimulated the development of the *neutral theory* of molecular evolution.

### Much of evolution is neutral

In 1968, Motoo Kimura proposed the *neutral theory of molecular evolution*. Kimura suggested that, at the molecular level, the majority of variants we observe in most populations are selectively neutral; that is, they confer neither an advantage nor a disadvantage on their bearers. Therefore, these neutral variants accumulate through genetic drift rather than through positive selection.

The rate of fixation of neutral mutations by genetic drift is independent of population size. To see why this is so, consider a population of size $N$ and a neutral mutation rate $\mu$ (mu) per gamete per generation at a locus. The number of new mutations would be, on average, $\mu \times 2N$, because $2N$ gene copies are available to mutate in a population of diploid organisms. The probability that a given mutation will be fixed by drift alone is its frequency, which equals $1/(2N)$ for a newly arisen mutation. We can multiply these two terms to get the rate of fixation of neutral mutations in a given population of $N$ individuals:

$$2N\mu\frac{1}{2N} = \mu$$

Therefore, the rate of fixation of neutral mutations depends only on the neutral mutation rate $\mu$ and is independent of population size. A given mutation is more likely to appear in a large population than in a small one, but any mutation that does appear is more likely to become fixed in a small population. These two influences of population size cancel each other out.

Therefore, the rate of fixation of neutral mutations is equal to the mutation rate. As long as the underlying mutation rate is constant, macromolecules evolving in different populations should diverge from one another in neutral changes at a constant rate. Empirically, the rate of evolution of particular genes and proteins is often relatively constant over time, and therefore can be used as a "molecular clock." As we described in Section 22.3, molecular clocks can be used to calculate evolutionary divergence times between species.

Although much of the genetic variation we observe in populations is the result of neutral evolution, the neutral theory does not imply that most mutations have no effect on the organism. Many mutations are never observed in populations because they are lethal or strongly detrimental to the organism and are thus quickly removed from the population through natural selection. Similarly, mutations that confer a selective advantage tend to be quickly fixed in populations, so they do not result in variation at the population level either. Nonetheless, if we compare homologous proteins from different populations or species, some amino acid positions will remain constant under purifying selection, others will vary through neutral genetic drift, and still others will differ between species as a result of positive selection for change. How can these evolutionary processes be distinguished?

### Positive and purifying selection can be detected in the genome

As we have just seen, substitutions in a protein-coding gene can be either synonymous or nonsynonymous, depending on whether they change the resulting amino acid sequence of the protein. The relative rates of synonymous and nonsynonymous substitutions are expected to differ in regions of genes that are evolving neutrally, under positive selection for change, or staying unchanged under purifying selection.

- If a given amino acid in a protein can be one of many alternatives (without changing the protein's function), then an amino acid replacement is *neutral* with respect to the fitness of an organism. In this case, the rates of synonymous and nonsynonymous substitutions in the corresponding DNA sequences are expected to be very similar, so the ratio of the two rates would be close to 1.
- If a given amino acid position is under *positive selection for change*, the rate of nonsynonymous substitutions is expected to exceed the rate of synonymous substitutions in the corresponding DNA sequences.
- If a given amino acid position is under *purifying selection*, then the rate of synonymous substitutions in the corresponding DNA sequences is expected to be much higher than the rate of nonsynonymous substitutions.

By comparing the gene sequences that encode proteins from many species, scientists can determine the history and timing of synonymous and nonsynonymous substitutions. This information can be mapped on a phylogenetic tree, as we saw in Chapter 22.

Regions of genes that are evolving under neutral, purifying, or positive selection can be identified by comparing the nature and rates of substitutions across the phylogenetic tree. The evolution of lysozyme illustrates how and why particular positions of a gene sequence might be under different modes of selection.

The enzyme lysozyme (see Figure 3.8) is found in almost all animals. It is produced in the tears, saliva, and milk of mammals and in the albumen (whites) of bird eggs. Lysozyme digests the cell walls of bacteria, rupturing and killing them. As a result, it plays an important role as a first line of defense against invading bacteria. Most animals defend themselves against bacteria by digesting them, which is probably why most animals have lysozyme. Some animals also use lysozyme in the digestion of food.

Among mammals, a mode of digestion called *foregut fermentation* has evolved twice. In mammals with this mode of digestion, the foregut—the posterior esophagus and/or the stomach—has been converted into a chamber in which bacteria break down ingested plant matter by fermentation. Foregut fermenters can obtain nutrients from the otherwise indigestible cellulose that makes up a large proportion of plant tissue. Foregut fermentation evolved independently in ruminants (a group of hoofed mammals that includes cattle) and in certain leaf-eating monkeys, such as langurs. We know these evolutionary events were independent, because both langurs and ruminants have close relatives that are not foregut fermenters.

In both foregut-fermenting lineages, the enzyme lysozyme has been modified to play a new, nondefensive role. This lysozyme ruptures some of the bacteria that live in the foregut, releasing nutrients metabolized by the bacteria, which the mammal then absorbs. How many changes in the lysozyme molecule were needed to allow it to perform this function amid the digestive enzymes and acidic conditions of the mammalian foregut? To answer this question, molecular evolutionists compared the lysozyme-coding sequences in foregut fermenters with those in several of their nonfermenting relatives. They determined which amino acids differed and which were shared among the species (**Figure 24.7A**), as well as the rates of synonymous and nonsynonymous substitutions in lysozyme genes across the evolutionary history of the sampled species.

For many of the amino acid positions of lysozyme, the rate of synonymous substitutions (in the corresponding gene) is much higher than the rate of nonsynonymous substitutions. This observation indicates that many of the amino acids that make up lysozyme are evolving under purifying selection. In other words, there is selection against change in the protein at these positions, and the observed amino acids must therefore be critical for lysozyme function. At other positions, several different amino acids function equally well, and the corresponding regions of the genes have similar rates of synonymous and nonsynonymous substitutions. The most striking finding is that amino acid replacements in lysozyme happened at a much higher rate in the lineage leading to langurs than in any other primates. The high rate of nonsynonymous substitutions in the langur lysozyme gene shows that lysozyme went through a period of rapid change in adapting to the stomachs of langurs. Moreover, the lysozymes of langurs and cattle share five amino acid replacements, all of which lie on the surface of the lysozyme molecule, well away from the enzyme's active site. Several of these shared replacements involve changes from arginine to lysine, which makes the proteins more resistant to attack by the stomach enzyme pepsin. By understanding the functional significance of amino acid replacements, molecular evolutionists can explain the observed changes in amino acid sequences in terms of changes in the functioning of the protein.

A large body of fossil, morphological, and molecular evidence shows that langurs and cattle do not share a recent common ancestor. However, langur and ruminant lysozymes share several amino acids that neither mammal shares with the lysozymes of its own closer relatives. The lysozymes of these two mammals have undergone *evolutionary convergence* at some amino acid positions despite their very different ancestry. The amino acids they share give these lysozymes the ability to lyse the bacteria that ferment plant material in the foregut.

The hoatzin, an unusual leaf-eating South American bird and the only known avian foregut fermenter, offers another remarkable story of the convergent evolution of lysozyme (**Figure 24.7B**). Many birds have an enlarged esophageal chamber called a *crop*. The crop of the hoatzin contains lysozyme and bacteria and acts as a fermenting chamber. Many of the amino acid replacements that occurred in the adaptation of hoatzin crop lysozyme are identical to those that evolved in ruminants and langurs. Thus, even though the hoatzin and foregut-fermenting mammals have not shared a common ancestor in hundreds of millions of years, they have all evolved similar adaptations in their lysozymes that enable them to recover nutrients from their fermenting bacteria.

## Genome size and organization also evolve

We know that genome size varies tremendously among organisms. Across broad taxonomic categories, there is some correlation between genome size and organismal complexity. The

(A)

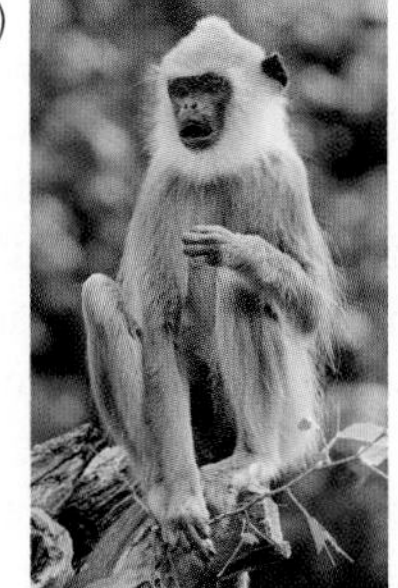

*Semnopithecus* sp.

The lysozymes of langurs and cattle are convergent for 5 amino acid residues, indicative of the independent evolution of foregut fermentation in these two species.

| | Langur | Baboon | Human | Rat | Cattle | Horse |
|---|---|---|---|---|---|---|
| Langur | | 14 | 18 | 38 | 32 | 65 |
| Baboon | 0 | | 14 | 33 | 39 | 65 |
| Human | 0 | 1 | | 37 | 41 | 64 |
| Rat | 0 | 0 | 0 | | 55 | 64 |
| Cattle | 5 | 0 | 0 | 0 | | 71 |
| Horse | 0 | 0 | 0 | 0 | 1 | |

(B) *Opisthocomus hoazin*

**24.7 Convergent Molecular Evolution of Lysozyme** (A) The number of amino acid differences in the lysozymes of several pairs of mammals are shown above the diagonal line; the number of convergent similarities between these same pairs are shown below the diagonal. The two foregut-fermenting species share convergent amino acid replacements related to this digestive adaptation. (B) The hoatzin—the only known foregut-fermenting bird species—has been evolving independently from mammals for hundreds of millions of years but has independently evolved similar modifications to lysozyme.

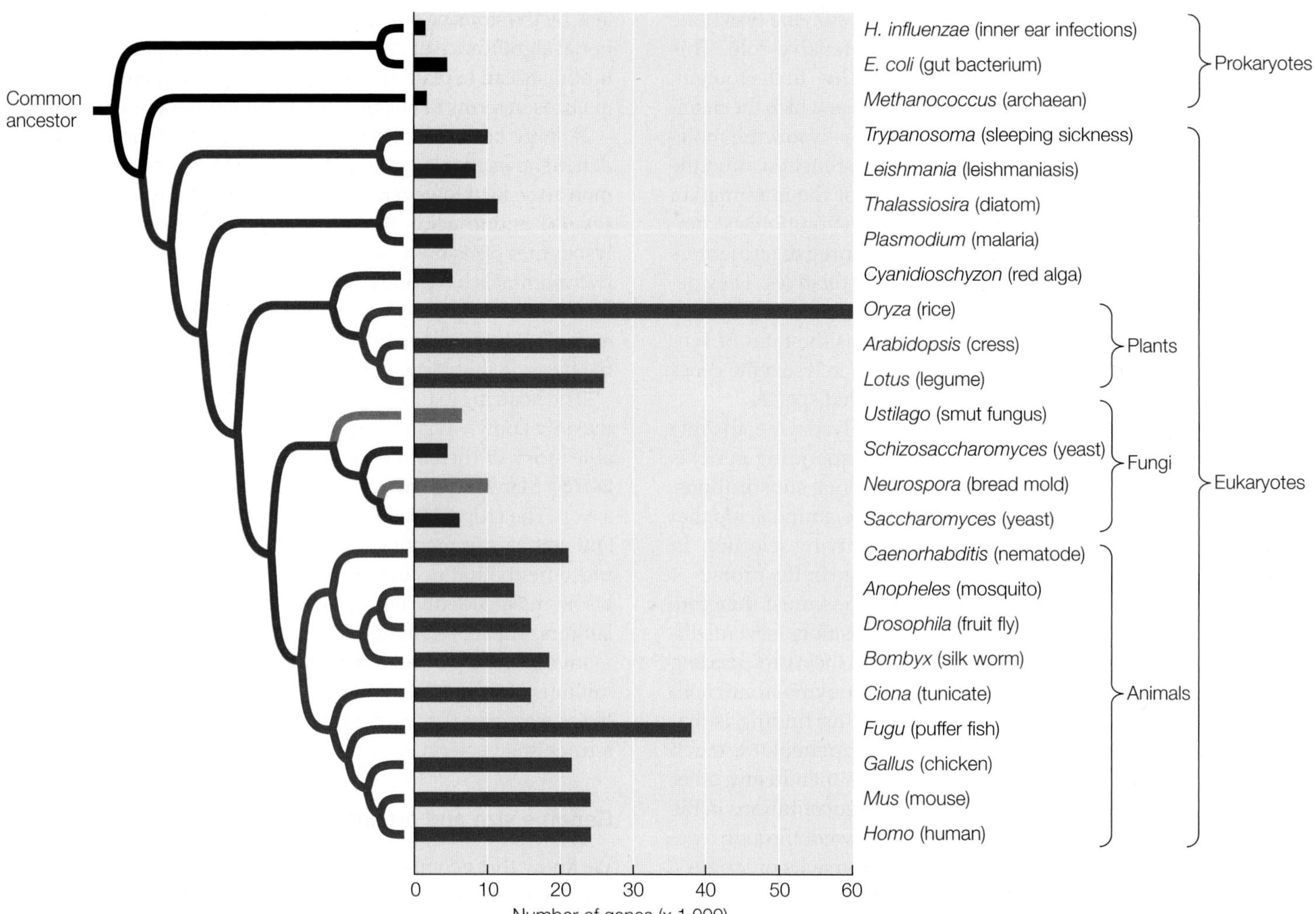


**24.8 Genome Size Varies Widely** This figure shows the number of genes from a sample of organisms with fully sequenced genomes, arranged by their evolutionary relationships. Bacteria and archaea typically have fewer genes than most eukaryotes. Among eukaryotes, multicellular organisms with tissue organization (plants and animals; blue branches) have more genes than single-celled organisms (red branches) or multicellular organisms that lack pronounced tissue organization (green branches).

genome of the tiny bacterium *Mycoplasma genitalium* has only 470 genes. *Rickettsia prowazekii*, the bacterium that causes typhus, has 634 genes. *Homo sapiens*, by contrast, has about 23,000 protein-coding genes. **Figure 24.8** shows the number of genes from a sample of organisms with fully sequenced genomes, arranged by their evolutionary relationships. As this figure reveals, however, a larger genome does not always indicate greater complexity (compare rice to the other plants, for example). It is not surprising that more complex genetic instructions are needed for building and maintaining a large, multicellular organism than a small, single-celled bacterium. What is surprising is that some organisms, such as lungfishes, some salamanders, and lilies, have about 40 times as much DNA as humans do. Structurally, a lungfish or a lily is not 40 times more complex than a human. So why does genome size vary so much?

Differences in genome size are not so great if we take into account only the portion of DNA that actually encodes RNAs or proteins. The organisms with the largest total amounts of nuclear DNA (some ferns and flowering plants) have 80,000 times as much DNA as do the bacteria with the smallest genomes, but no species has more than about 100 times as many protein-coding genes as a bacterium. Therefore, much of the variation in genome size lies not in the number of functional genes but in the amount of noncoding DNA (**Figure 24.9**).

Why do the cells of most eukaryotic organisms have so much noncoding DNA? Does this noncoding DNA have a function, or is it "junk"? Although some of this DNA does not appear to have a direct function, it can alter the expression of the surrounding genes. The degree or timing of gene expression can be changed dramatically depending on the gene's position relative to noncoding sequences. Other regions of noncoding DNA consist of pseudogenes that are simply carried in the genome because the cost of doing so is very small. These pseudogenes may become the raw material for the evolution of new genes with novel functions. Some noncoding sequences function in maintaining chromosomal structure. Still others consist of parasitic transposable elements that spread through populations because they reproduce faster than the host genome.

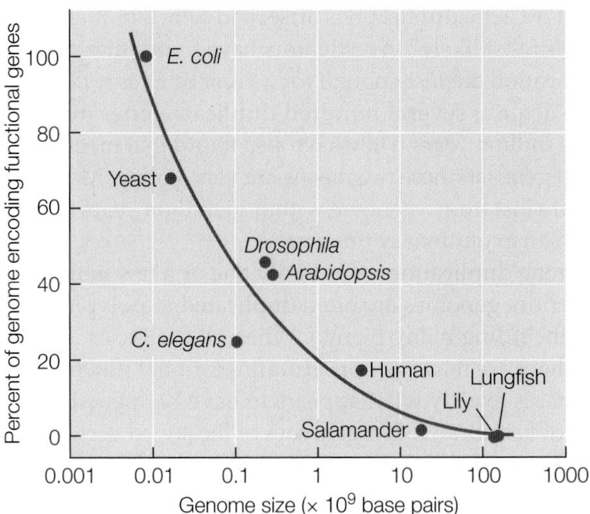


**24.9 A Large Proportion of DNA Is Noncoding** Most of the DNA of bacteria and yeasts encodes RNAs or proteins, but a large percentage of the DNA of multicellular species is noncoding.

Investigators can use retrotransposons to estimate the rates at which species lose DNA. Retrotransposons are transposable elements (see Figure 17.5) that copy themselves through an RNA intermediate. The most common type of retrotransposon carries duplicated sequences at each end, called long terminal repeats, or LTRs. Occasionally, LTRs recombine in the host genome, so that the DNA between them is excised. When this happens, one recombined LTR is left behind. The number of such "orphaned" LTRs in a genome is a measure of how many retrotransposons have been lost. By comparing the number of LTRs in the genomes of Hawaiian crickets (*Laupala*) and fruit flies (*Drosophila*), investigators found that *Laupala* loses DNA more than 40 times more slowly than does *Drosophila*. Therefore, it is not surprising that the genome of *Laupala* is much larger than that of *Drosophila*.

Why do species differ so greatly in the rate at which they gain or lose apparently functionless DNA? One hypothesis is that genome size is related to the rate at which the organism develops, which may be under selection pressure. Large genomes can slow down the rate of development and thus alter the relative timing of expression of particular genes. As discussed in Section 20.2, changes in the timing of gene expression (*heterochrony*) can produce major changes in phenotype. Thus, although some noncoding DNA sequences may have no direct function, they may still affect the development of the organism.

Another hypothesis is that the proportion of noncoding DNA is related primarily to population size. Noncoding sequences that are only slightly deleterious to the organism are likely to be purged by selection most efficiently in species with large population sizes. In species with small populations, the effects of genetic drift can overwhelm selection against noncoding sequences that have small deleterious consequences. Therefore, selection against the accumulation of noncoding sequences is most effective in species with large populations, so such species (such as bacteria or yeasts) have relatively little noncoding DNA compared with species with small populations (see Figure 24.9).

## 24.2 RECAP

**By examining the relative rates of synonymous and nonsynonymous substitutions in genes across evolutionary history, biologists can distinguish the evolutionary mechanisms acting on individual genes. Neutral theory provides an explanation for the relatively constant rate of molecular change seen in many species.**

- Describe how the ratio of synonymous to nonsynonymous substitutions can be used to determine whether a particular gene region is evolving neutrally, under positive selection, or under stabilizing selection. **See pp. 505–506 and Figure 24.6**
- Contrast two hypotheses for the wide diversity of genome sizes among different organisms. **See pp. 508–509**

We have examined some of the ways that biologists can use genomes to study the molecular mechanisms of evolution. But how do organisms gain new functions through time?

# 24.3 How Do Genomes Gain and Maintain Functions?

As we noted in the previous section, most multicellular organisms have many more genes than do most unicellular species. But multicellular organisms evolved from unicellular ancestors. Therefore, some mechanisms must exist that result in the increase of gene numbers within genomes over evolutionary time. There are two primary ways to accomplish this increase: genes can be transferred from other species, or genes can be duplicated within species.

### Lateral gene transfer can result in the gain of new functions

In Chapter 22, we noted that the tree of life is usually visualized as a branching diagram, with each lineage dividing into two (or more) lineages through time, from one common ancestor to the millions of species that are alive today. Chapter 23 described how, in the process of speciation, ancestral lineages divide into descendant lineages, and it is those speciation events that the tree of life captures. However, there are also processes of **lateral gene transfer**, which allow individual genes, organelles, or fragments of genomes to move horizontally from one lineage to another. Some species may pick up fragments of DNA directly from the environment. Other genes may be picked up in a viral genome and then transferred to a new host when the virus becomes integrated in the new host's genome. Hybridization between species also results in the transfer of large numbers of genes.

Lateral gene transfer can be highly advantageous to the species that incorporates novel genes from a distant relative. Genes that confer antibiotic resistance, for example, are commonly transferred among different species of bacteria. Lateral gene transfer is another way, in addition to mutation and recombination, that species can increase their genetic variability. Genetic variability then provides the raw material on which selection acts, resulting in evolution.

A phylogenetic tree constructed from a single laterally transferred genome fragment is likely to reflect only that transfer event, rather than the overall organismal phylogeny (see Section 26.3). Most biologists prefer to build trees from large samples of genes or their products, so that the underlying species tree (as well as any lateral gene transfer events) can be reconstructed. The depiction of lateral gene transfer events on the underlying species tree are known as *reticulations* on the phylogenetic tree.

The degree to which lateral gene transfer events occur in various parts of the tree of life is a matter of considerable current investigation and debate. Lateral gene transfer appears to be relatively uncommon among most eukaryote lineages, although the two major endosymbioses that gave rise to mitochondria and chloroplasts can be viewed as lateral transfers of entire bacteria genomes to the eukaryote lineage. Some groups of eukaryotes, most notably some plants, are subject to relatively high levels of hybridization among closely related species. Hybridization leads to the exchange of many genes among recently separated lineages of plants. The greatest degree of lateral transfer, however, appears to occur among species of bacteria. Many bacteria genes have been transferred repeatedly among lineages of bacteria, to the point that relationships among the bacteria species are often hard to decipher. Nonetheless, the broad relationships of the major groups of bacteria can still be determined (as we will discuss in Part Seven of this book). Lateral transfer of genes makes it difficult to identify the boundaries of bacteria species, which is one reason why fewer bacteria species have been named than are known to exist.

### Most new functions arise following gene duplication

**Gene duplication** is yet another way in which genomes can acquire new functions. When a gene is duplicated, one copy of that gene is potentially freed from having to perform its original function. The identical copies of a duplicated gene can have any one of four different fates:

- Both copies of the gene may retain their original function (which can result in a change in the amount of gene product that is produced by the organism).
- Both copies of the gene may retain the ability to produce the original gene product, but the expression of the genes may diverge in different tissues or at different times of development.
- One copy of the gene may be incapacitated by the accumulation of deleterious substitutions and become a functionless pseudogene.
- One copy of the gene may retain its original function while the second copy accumulates enough substitutions that it can perform a different function.

How often do gene duplications arise, and which of these four outcomes is most likely? Investigators have found that rates of gene duplication are fast enough for a yeast or *Drosophila* population to acquire several hundred duplicate genes over the course of a million years. They have also found that most of the duplicated genes in these organisms are very young. Many extra genes are lost from a genome within 10 million years (which is rapid on an evolutionary time scale).

Many gene duplications affect only one or a few genes at a time, but entire genomes are often duplicated in polyploid organisms (including many plants). When all the genes are duplicated, there are massive opportunities for new functions to evolve. That is exactly what appears to have happened in the evolution of vertebrates. The genomes of the jawed vertebrates appear to have four diploid sets of many major genes, which leads biologists to believe that two genome-wide duplication events occurred in the ancestor of these species. These duplications have allowed considerable specialization of individual vertebrate genes, many of which are now highly tissue-specific in their expression. A good example is the duplication of sodium channel genes, which allowed the evolution of the electric organs of electric fishes described in the opening of this chapter.

Although many extra genes disappear rapidly, some duplication events lead to the evolution of genes with new functions. Several successive rounds of duplication and mutation may result in a **gene family**, a group of homologous genes with related functions, often arrayed in tandem along a chromosome. An example of this process is provided by the *globin gene family* (see Figure 17.10). The globins were among the first proteins to be sequenced and compared. Comparisons of their amino acid sequences strongly suggest that the different globins arose via gene duplications. These comparisons can also tell us how long the globins have been evolving separately, because differences among these proteins have accumulated with time.

Hemoglobin, a tetramer (four-subunit molecule) consisting of two α-globin and two β-globin polypeptide chains, carries oxygen in blood. Myoglobin, a monomer, is the primary $O_2$ storage protein in muscle. Myoglobin's affinity for $O_2$ is much higher than that of hemoglobin, but hemoglobin has evolved to be more diversified in its role. Hemoglobin binds $O_2$ in the lungs or gills, where the $O_2$ concentration is relatively high, transports it to deep body tissues, where the $O_2$ concentration is low, and releases it in those tissues. With its more complex tetrameric structure, hemoglobin is able to carry four molecules of $O_2$, as well as hydrogen ions and carbon dioxide, in the blood.

To estimate the time of the globin gene duplication that gave rise to the α- and β-globin gene clusters, we can create a phylogenetic tree of the gene sequences that encode the various globins (**Figure 24.10**). The rate of molecular evolution of globin genes has been estimated from other studies, using the divergence times of groups of vertebrates that are well documented in the fossil record. These studies indicate an average rate of divergence for globin genes of about 1 nucleotide substitution every 2 million years. Applying this rate to the gene tree, the two globin gene clusters are estimated to have split about 450 million years ago.

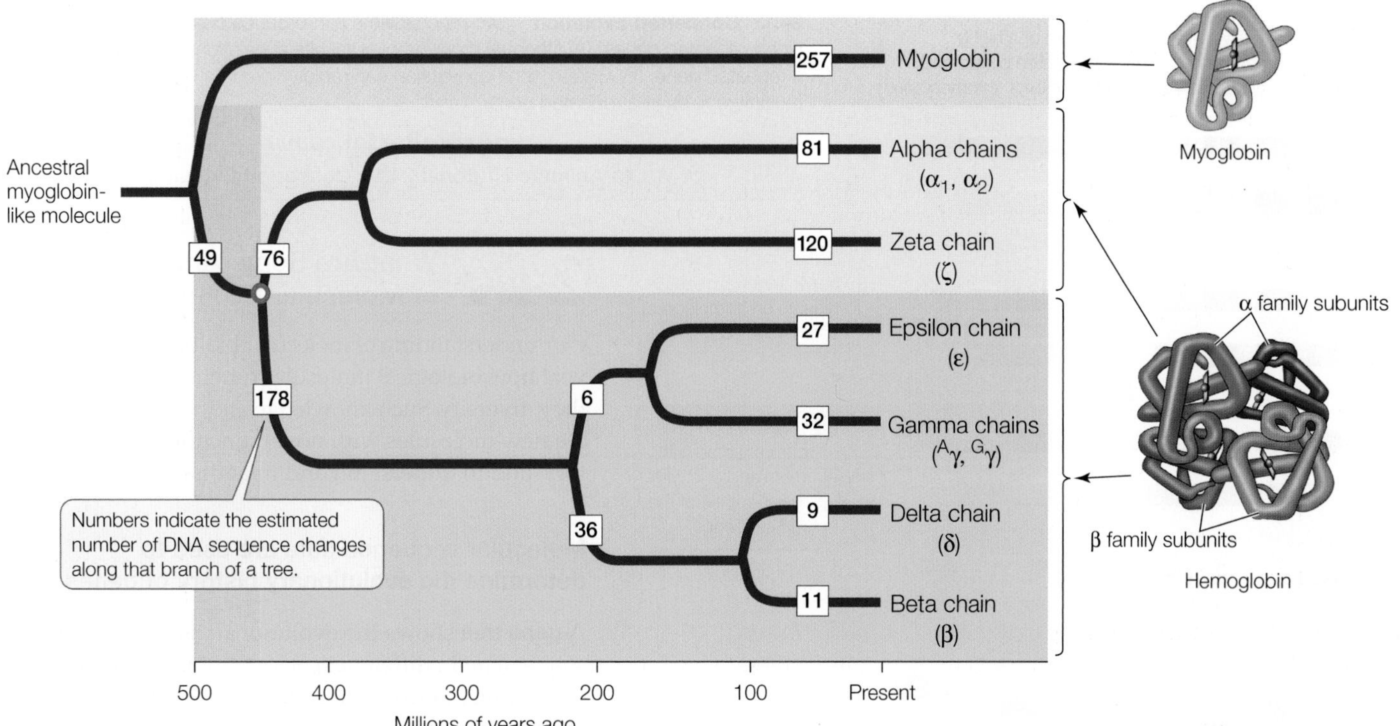


**24.10 A Globin Family Gene Tree** This gene tree suggests that the α-globin and β-globin gene clusters diverged about 450 million years ago (open circle), soon after the origin of the vertebrates.

**yourBioPortal.com**

GO TO **Web Activity 24.3 • Gene Tree Construction**

### Some gene families evolve through concerted evolution

Although the members of the globin gene family have diversified in form and function, the members of many other gene families do not evolve independently of one another. For instance, almost all organisms have many copies (up to thousands) of the ribosomal RNA genes. *Ribosomal RNA* (rRNA) is the principal structural element of ribosomes and, as such, has a primary role in protein synthesis. Every living species needs to synthesize proteins, often in large amounts (especially during early development). Having many copies of the rRNA genes ensures that organisms can rapidly produce many ribosomes and thereby maintain a high rate of protein synthesis.

Like all portions of the genome, ribosomal RNA genes evolve, and differences accumulate in the rRNA genes of different species. But within any one species, the multiple copies of rRNA genes are very similar, both structurally and functionally. This similarity makes sense, because ideally every ribosome in a species should synthesize proteins in the same way. In other words, within a given species, the multiple copies of these rRNA genes are evolving in concert with one another, a phenomenon called **concerted evolution**.

How does concerted evolution occur? There must be one or more mechanisms to cause a substitution in one copy to spread to other copies in a species so that all of the copies remain similar. In fact, two different mechanisms appear to be responsible for concerted evolution. The first of these is *unequal crossing over*. When DNA is replicated during meiosis in a diploid species, the homologous chromosome pairs align and recombine by crossing over (see Section 11.4).

In the case of highly repeated genes, however, it is easy for genes to become displaced in alignment, since so many copies of the same genes are present in the repeats (**Figure 24.11A**). The end result is that one chromosome will gain extra copies of the repeat and the other chromosome will have fewer copies of the repeat. If a new substitution arises in one copy of the repeat, it can spread to new copies (or be eliminated) through unequal crossing over. Thus, over time, a novel substitution will either become fixed or lost entirely from the repeat. In either case, all the copies of the repeat will remain very similar to one another.

The second mechanism that produces concerted evolution is *biased gene conversion*. This mechanism can be much faster than unequal crossing over, and has been shown to be the primary mechanism for concerted evolution of rRNA genes. DNA strands break often, and are repaired by the DNA repair systems of cells (see Section 13.4). At many times during the cell cycle, the genes for ribosomal RNA are clustered close together. If damage occurs to one of the genes, a copy of the rRNA gene on another chromosome may be used to repair the damaged copy, and the sequence that is used as a template can thereby replace the original sequence (**Figure 24.11B**). In many cases, this repair system appears to be biased in favor of using particular sequences as templates for repair, and thus the favored sequence rapidly spreads across all copies of the gene. In this way, changes may appear in a single copy and then rapidly spread to all the other copies.

Regardless of the mechanism responsible, the net result of concerted evolution is that the copies of a highly repeated gene do not evolve independently of one another. Mutations still oc-

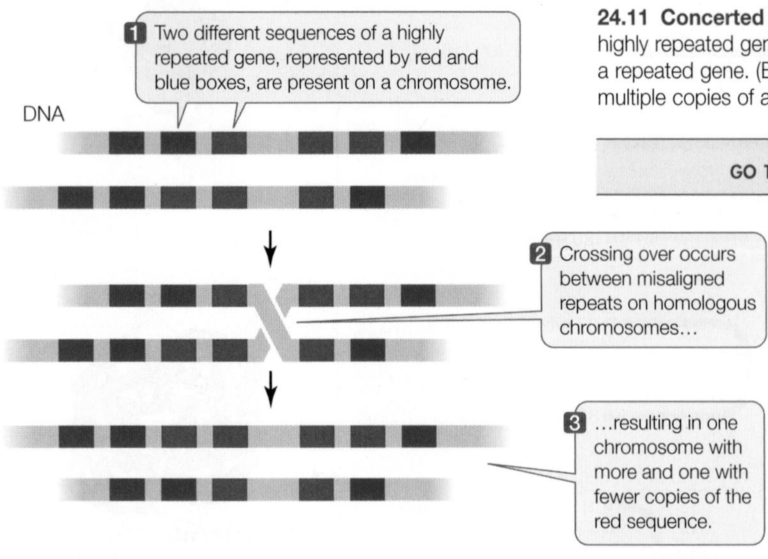


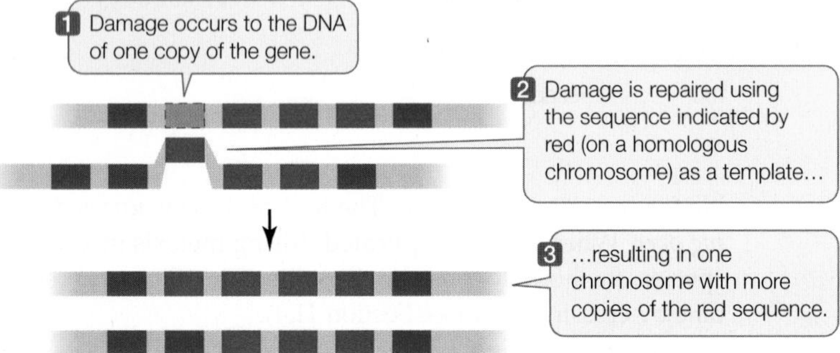

(B) Biased gene conversion
1 Damage occurs to the DNA of one copy of the gene.
2 Damage is repaired using the sequence indicated by red (on a homologous chromosome) as a template...
3 ...resulting in one chromosome with more copies of the red sequence.


**24.11 Concerted Evolution** Two mechanisms can produce concerted evolution of highly repeated genes. (A) Unequal crossing over results in deletions and duplications of a repeated gene. (B) Biased gene conversion can rapidly spread a new variant across multiple copies of a repeated gene.

yourBioPortal.com
GO TO Animated Tutorial 24.1 • Concerted Evolution

cur, but once they arise in one copy, they either spread rapidly across all the copies or are lost from the genome completely. This process allows the products of each copy to remain similar through time in both sequence and function.

## 24.3 RECAP

**Gene duplication can lead to the evolution of new functions. Lateral gene transfer can result in the spread of genetic functions between distantly related species. Some highly repeated genes evolve in concert, which maintains uniform functionality.**

- Explain the potential advantages of lateral gene transfer. See pp. 509–510
- What are four possible outcomes of gene duplication? See p. 510
- Describe the pattern of concerted evolution among highly repeated genes and the mechanisms that lead to concerted evolution. See p. 511 and Figure 24.11

We have seen how the principles and methods of molecular evolution have opened new vistas in evolutionary science. Now let's consider some of the practical applications of this field.

# 24.4 What Are Some Applications of Molecular Evolution?

Our understanding of molecular biology has helped reveal how biological molecules function as well as how they diversify. Such knowledge allows scientists to create new molecules with novel functions in the laboratory, and to understand and treat disease.

### Molecular sequence data are used to determine the evolutionary history of genes

A **gene tree** shows the evolutionary relationships of a single gene in different species or of the members of a gene family (as in Figure 24.10). The methods for constructing a gene tree are the same as those we described in Section 22.2 for building phylogenetic trees of species. The process involves identifying differences between genes and using those differences to reconstruct the evolutionary history of the genes. Gene trees are often used to infer phylogenetic trees of species, but the two types of trees are not necessarily equivalent. Processes such as gene duplication can give rise to differences between the phylogenetic trees of genes and species. From a gene tree, biologists can reconstruct the history and timing of gene duplication events and learn how gene diversification has resulted in the evolution of new protein functions.

All of the genes of a particular gene family have similar sequences because they have a common ancestry. As we discussed in Section 22.1, features that are similar as a result of common ancestry are referred to as *homologs* of one another. When discussing gene trees, however, we usually need to distinguish between two forms of homology. Genes found in different species, and whose divergence we can trace to the speciation events that gave rise to various species, are called **orthologs**. Genes in the same or different species that are related through gene duplication events are called **paralogs**. When we examine a gene tree, the questions we wish to address determine whether we should compare orthologous or paralogous genes. If we wish to reconstruct the evolutionary history of the species that contain the genes, then our comparison should be restricted to orthologs (because they will reflect the history of speciation events). If we are interested in the changes in function that have resulted from gene duplication events, however, then the appropriate comparison is among paralogs (because they will reflect the history of gene duplication events). If our focus is on the diversification of a gene family through both processes, then we will want to include both paralogs and orthologs in our analysis.

**Figure 24.12** depicts a gene tree for the members of a gene family called *engrailed* (its members encode transcription factors

Ancestral *engrailed* gene
Sea urchin *En*
*Amphioxus En*
Lamprey *En*
These species have a single *engrailed* gene.
Zebrafish *Eng2b*
Zebrafish *Eng2a*
Chicken *En2*
Mouse *En2*
Human *En2*
**En2**
**En1**
Zebrafish *Eng1b*
Zebrafish *Eng1a*
Chicken *En1*
Mouse *En1*
Human *En1*
In vertebrates, a gene duplication event resulted in two paralogous *engrailed* genes, *En1* and *En2*.
Additional gene duplications occurred in the zebrafish lineage.
Within orthologous groups of genes, the relationships among the species are the same (compare the relationships of *En1* genes to those of *En2* genes).

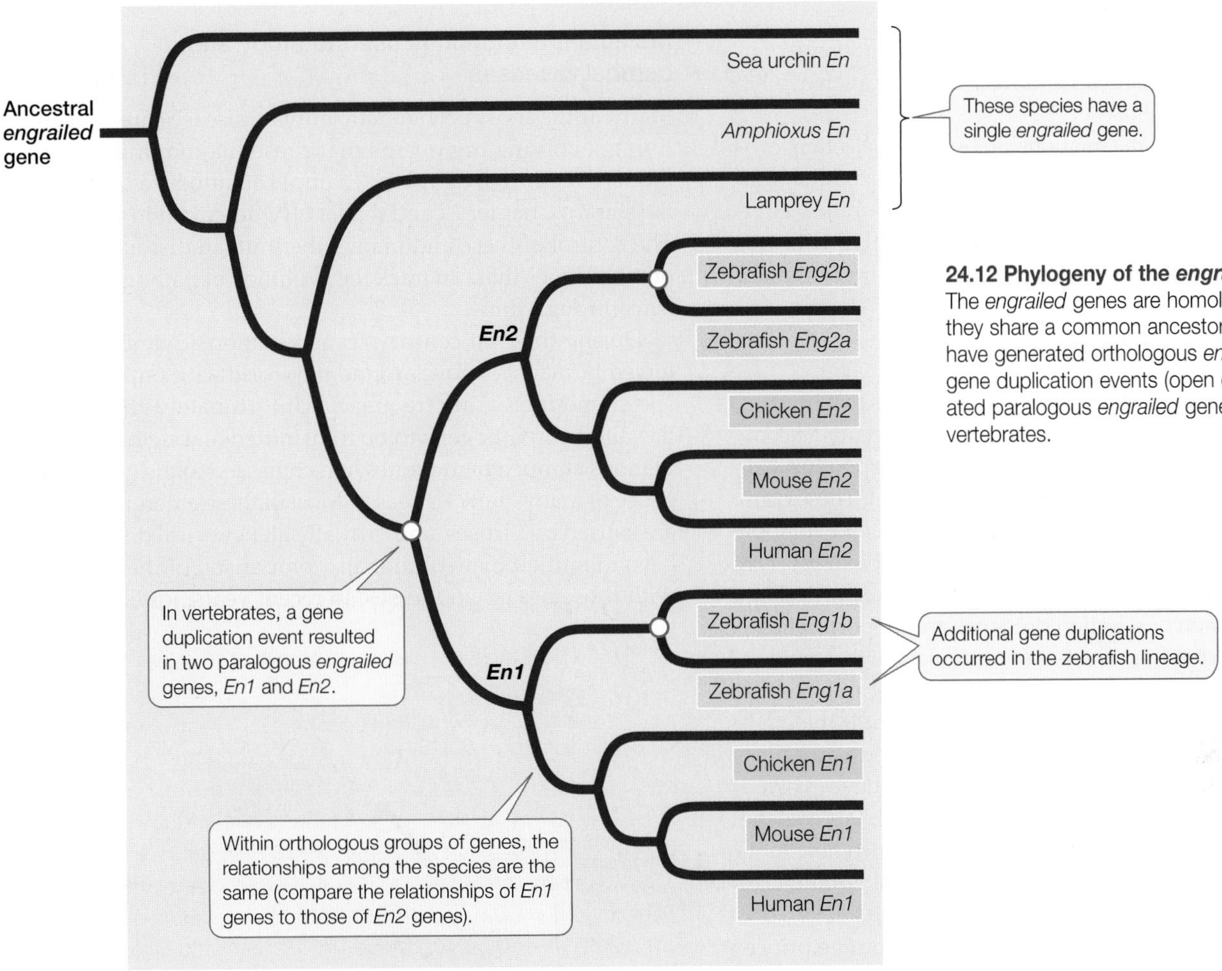

**24.12 Phylogeny of the *engrailed* Genes** The *engrailed* genes are homologous because they share a common ancestor. Speciation events have generated orthologous *engrailed* genes, and gene duplication events (open circles) have generated paralogous *engrailed* genes among bony vertebrates.

that regulate development). At least three gene duplications have occurred in this family, resulting in up to four different *engrailed* genes (*En*) in some vertebrate species (such as the zebrafish). All of the *engrailed* genes are homologs because they have a common ancestor. Gene duplication events have generated paralogous *engrailed* genes in some lineages of vertebrates. We could compare the orthologous sequences of the *En1* group of genes to reconstruct the history of the bony vertebrates (i.e., all the species in Figure 24.12 except the lamprey), or we could use the orthologous sequences of the *En2* group of genes and expect the same answer (because there is only one history of the underlying speciation events). All bony vertebrates have both groups of *engrailed* genes because the two groups arose from a gene duplication event in the common ancestor of bony vertebrates. If we wanted to focus on the diversification that occurred as a result of this duplication, then the appropriate comparison would be between the paralogous genes of the *En1* versus *En2* groups.

### Gene evolution is used to study protein function

Earlier in this chapter we discussed the ways in which biologists can detect regions of genes that are under positive selection for change. What are the practical uses of this information? Consider the evolution of the family of gated sodium channel genes, which we introduced in the opening of this chapter. Sodium channels have many functions, including the control of nerve impulses in the nervous system. Sodium channels can become blocked by various toxins, such as the tetrodotoxin that is present in puffer fishes and many other animals. A human who eats the tissues of a puffer fish that contain tetrodotoxin can become paralyzed and die, because the tetrodotoxin blocks sodium channels and prevents nerves and muscles from functioning.

Puffer fish themselves have sodium channels; so why doesn't the tetrodotoxin cause paralysis in a puffer fish? The sodium channels of puffer fish (and other animals that sequester tetrodotoxin) have evolved to become resistant to the toxin. Nucleotide substitutions in the puffer fish genome have resulted in changes to the proteins that make up sodium channels, and those changes prevent tetrodotoxin from binding to the sodium channel pore and blocking it. Several different substitutions that result in tetrodotoxin resistance have evolved in the various duplicated sodium channel genes of the many species of puffer fish.

Many other changes that have nothing to do with the evolution of tetrodotoxin resistance have occurred in these genes as well. Biologists who study the function of sodium channels can learn a great deal about how the channels work (and about neurological diseases that are caused by mutations in the sodium channel genes) by understanding which changes have been selected for tetrodotoxin resistance. They can do this by comparing the rates of synonymous and nonsynonymous substitutions across the genes in various lineages that have evolved tetrodotoxin resistance. In a similar manner, molecular evolutionary principles are used to understand function and diversification of function in many other proteins.

As biologists studied the relationship between selection, evolution, and function in macromolecules, they realized that molecular evolution could be used in a controlled laboratory environment to produce new molecules with novel and useful functions. Thus were born the applications of in vitro evolution.

### In vitro evolution produces new molecules

Living organisms produce thousands of compounds that humans have found useful. The search for such naturally occurring compounds, which can be used for pharmaceutical, agricultural, or industrial purposes, has been termed *bioprospecting*. These compounds are the result of millions of years of molecular evolution across millions of species of living organisms. Yet biologists can imagine molecules that could have evolved but have not, lacking the right combination of selection pressures and opportunities.

For instance, we might like to have a molecule that binds a particular environmental contaminant so that it can be easily isolated and extracted from the environment. But if the environmental contaminant is synthetic (not produced naturally), then it is unlikely that any living organism would have evolved a molecule with the function we desire. This problem was the inspiration for the field of **in vitro evolution**, in which new molecules are produced in the laboratory to perform novel and useful functions.

The principles of in vitro evolution are based on the principles of molecular evolution that we have learned from the natural world. Consider the evolution of a new RNA molecule that was produced in the laboratory using the principles of mutation and selection. This molecule's intended function was to join two other RNA molecules (acting as a ribozyme with a function similar to that of the naturally occurring DNA ligase described in Section 13.3, but for RNA molecules). The process started with a large pool of random RNA sequences ($10^{15}$ different sequences, each about 300 nucleotides long), which were then selected for any ligase activity (**Figure 24.13**). None were very effective ribozymes for ligase activity, but some were slightly better than others. The best of the ribozymes were selected and reverse-transcribed into cDNA (using the enzyme reverse transcriptase). The cDNA molecules were then amplified using the polymerase chain reaction (PCR; see Figure 13.22).

PCR amplification is not perfect, and it introduced many new mutations into the pool of sequences. These sequences were then transcribed back into RNA molecules using RNA polymerase, and the process was repeated. The ligase activity of the RNAs evolved quickly; after 10 rounds of in vitro evolution, it had increased by about 7 million times (see Figure 24.13). Similar techniques have since been used to create a wide variety of molecules with novel enzymatic and binding functions.

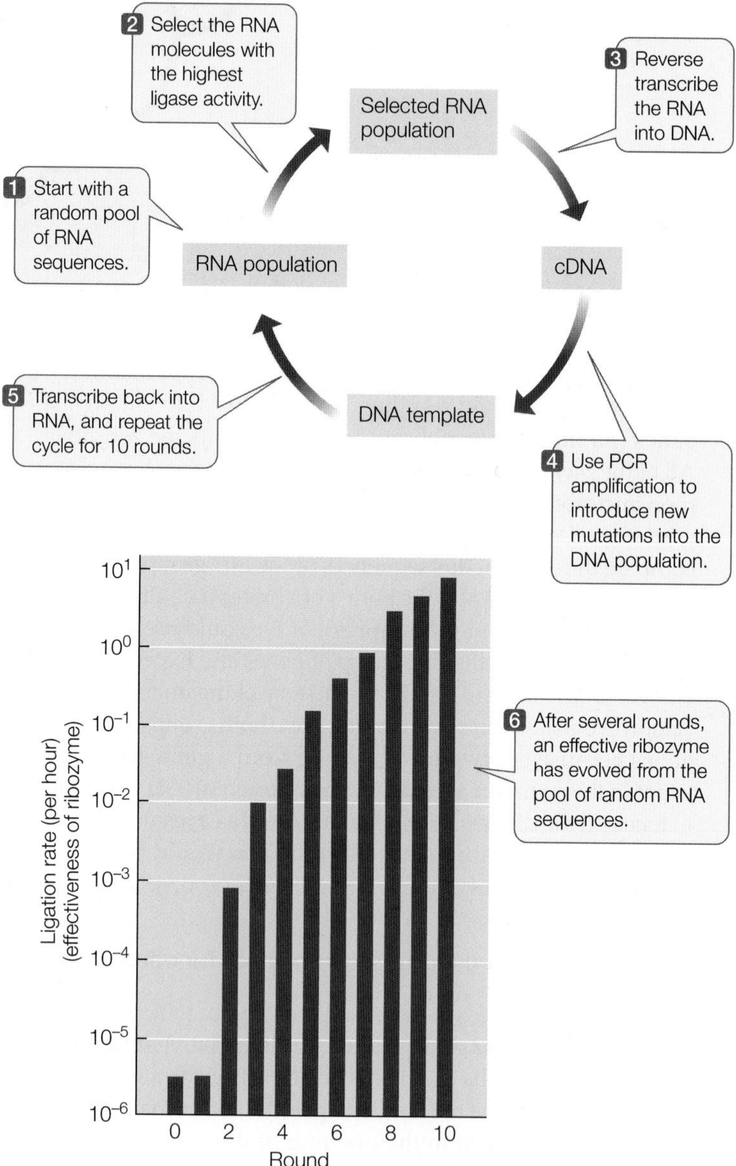


**24.13 In Vitro Evolution** Starting with a large pool of random RNA sequences, Bartel and Szostak produced a new ribozyme through rounds of mutation and selection for the ability to ligate RNA sequences.

### Molecular evolution is used to study and combat diseases

Many of the most problematic human diseases are caused by living, evolving organisms that present a moving target for modern medicine. Recall the example of influenza described at the start of Chapter 21 and that of HIV described in Chapter 22. The control of these and many other human diseases depends on techniques that can track the evolution of pathogenic organisms through time.

During the past century, transportation advances have allowed humans to move around the world with unprecedented speed and increasing frequency. Unfortunately, this mobility has allowed pathogens to be transmitted among human populations at much higher rates, which has led to the global emergence of many "new" diseases. Most of these emerging diseases are caused by viruses, and virtually all new viral diseases have been identified by evolutionary comparison of their genomes with those of known viruses. In recent years, for example, rodent-borne hantaviruses have been identified as the source of widespread respiratory illnesses, and the virus (and its host) that causes Sudden Acute Respiratory Syndrome (SARS) has been identified using evolutionary comparisons of genes. Studies of the origins, the timing of emergence, and the global diversity of many human pathogens depend on the principles of molecular evolution, as do the efforts to develop and use effective vaccines against these pathogens. For example, the techniques to develop polio vaccines, as well as the methods used to track their effectiveness in human populations, rely on molecular evolutionary approaches.

In the future, molecular evolution will become even more critical to the identification of human (and other) diseases. Once biologists have collected data on the genomes of enough organisms, it will be possible to identify an infection by sequencing a portion of the infecting organism's genome and comparing this sequence with other sequences on an evolutionary tree. At present, it is difficult to identify many common viral infections (those that cause "colds," for instance). As genomic databases and evolutionary trees increase, however, automated methods of sequencing and rapid phylogenetic comparison of the sequences will allow us to identify and treat a much wider array of human illnesses.

## 24.4 RECAP

**Molecular evolutionary studies have provided biologists with new tools to understand the functions of macromolecules and how those functions can change through time. Molecular evolution is used to develop synthetic molecules for industrial and pharmaceutical uses and to identify and combat human diseases.**

- Why might a biologist limit a particular investigation to orthologous (as opposed to paralogous) genes? See pp. 512–513
- Explain how gene evolution can be used to study protein function. See pp. 513–514
- Describe the process of in vitro evolution. See p. 514 and Figure 24.13

Now that we have discussed how organisms and biological molecules evolve, we are ready to consider the evolutionary history of the Earth. Chapter 25 describes the long-term evolutionary changes that have given rise to all of life's diversity.

## CHAPTER SUMMARY

### 24.1 How Are Genomes Used to Study Evolution?

SEE WEB ACTIVITY 24.1

- The field of **molecular evolution** concerns relationships between the structures of genes and proteins and the functions of organisms.
- A **genome** is an organism's full set of genes and noncoding DNA. In eukaryotes, the genome includes genetic material in the nucleus of the cell as well as in mitochondria and chloroplasts (where present).
- Nucleotide substitutions may or may not result in amino acid replacements in the encoded proteins.
- The estimated number of substitutions between sequences can be calculated from a **similarity matrix** using models of sequence evolution that account for changes that cannot be observed directly. Review Figure 24.1, WEB ACTIVITY 24.2
- The concept of homology (similarity that results from common ancestry) extends down to the level of particular positions in nucleotide or amino acid sequences. **Sequence alignments** from different organisms allow us to compare the sequences and identify homologous positions. Review Figure 24.3

### 24.2 What Do Genomes Reveal about Evolutionary Processes?

- **Nonsynonymous substitutions** of nucleotides result in amino acid replacements in proteins, but **synonymous substitutions** do not. Review Figure 24.5
- Rates of synonymous substitution are typically higher than rates of nonsynonymous substitution in protein-coding genes (a result of stabilizing selection). Review Figure 24.6
- Much of the molecular change in nucleotide sequences is a result of **neutral evolution**. The rate of fixation of neutral mutations is independent of population size and is equal to the mutation rate.
- Positive selection for change in a protein-coding gene may be detected by a higher rate of nonsynonymous versus synonymous substitutions.
- Genome size evolves by the addition or deletion of genes and noncoding DNA. The total size of genomes varies much more widely across multicellular species than does the number of functional genes. Review Figures 24.8 and 24.9
- Even though many noncoding regions of the genome may not have direct functions, these regions can affect the phenotype of an organism by influencing gene expression.
- Functionless **pseudogenes** can serve as the raw material for the evolution of new genes.

### 24.3 How Do Genomes Gain and Maintain Functions?

- Lateral gene transfer can result in the rapid acquisition of new functions from distantly related species.
- **Gene duplications** can result in increased production of the gene's product, in pseudogenes, or in new gene functions. Several rounds of gene duplication can give rise to multiple genes with related functions, known as a **gene family**.
- Some highly repeated genes evolve by **concerted evolution**: multiple copies within an organism maintain high similarity, while the genes continue to diverge between species. SEE ANIMATED TUTORIAL 24.1

## 24.4 What Are Some Applications of Molecular Evolution?

SEE WEB ACTIVITY 24.3

- **Gene trees** describe the evolutionary history of particular genes or gene families.
- **Orthologs** are genes that are related through speciation events, whereas **paralogs** are genes that are related through gene duplication events. Review Figure 24.12
- Protein function can be studied by examining gene evolution. Detection of positive selection can be used to identify molecular changes that have resulted in functional changes.
- **In vitro evolution** is used to produce synthetic molecules with particular desired functions. Review Figure 24.13
- Many diseases are identified, studied, and combated through molecular evolutionary investigations.

## SELF-QUIZ

1. A higher rate of synonymous relative to nonsynonymous substitutions in a protein-coding gene is expected under
   *a.* purifying selection.
   *b.* positive selection.
   *c.* neutral evolution.
   *d.* concerted evolution.
   *e.* none of the above
2. Before nucleotide and amino acid sequences can be compared in an evolutionary framework, they must be aligned to account for
   *a.* deletions and insertions.
   *b.* selection and neutrality.
   *c.* parallelisms and convergences.
   *d.* gene families.
   *e.* all of the above
3. Models of nucleotide sequence evolution, developed by biologists to estimate sequence divergence, include parameters that account for
   *a.* substitution rates between different nucleotides.
   *b.* differences in substitution rates across different positions in a gene.
   *c.* differences in nucleotide frequencies.
   *d.* all of the above
   *e.* none of the above
4. The rate of fixation of neutral mutations is
   *a.* independent of population size.
   *b.* higher in small populations than in large populations.
   *c.* higher in large populations than in small populations.
   *d.* slower than the rate of fixation of deleterious mutations.
   *e.* none of the above
5. Genome size differs widely among different multicellular species of eukaryotes. What is the greatest contributing cause for these differences?
   *a.* The number of protein-coding genes
   *b.* The amount of noncoding DNA
   *c.* The number of duplicated genes
   *d.* The degree of concerted evolution
   *e.* The amount of positive selection for change in protein-coding genes
6. Which of the following is *not* true of concerted evolution?
   *a.* Concerted evolution refers to the nonindependent evolution of some repeated genes within a species.
   *b.* Unequal crossing over may produce concerted evolution.
   *c.* Biased gene conversion may produce concerted evolution.
   *d.* Ribosomal RNA genes are an example of a gene family that has undergone concerted evolution.
   *e.* Concerted evolution results in divergence of members of a gene family within an organism.
7. When a gene is duplicated, which of the following may occur?
   *a.* Production of the gene's product may increase.
   *b.* The two copies may become expressed in different tissues.
   *c.* One copy of the gene may accumulate deleterious substitutions and become functionless.
   *d.* The two copies may diverge and acquire different functions.
   *e.* All of the above
8. Paralogous genes are genes that trace back to a common
   *a.* speciation event.
   *b.* substitution event.
   *c.* insertion event.
   *d.* deletion event.
   *e.* duplication event.
9. Which of the following is true of in vitro evolution?
   *a.* In vitro evolution refers to bioprospecting for naturally occurring macromolecules.
   *b.* In vitro evolution can produce new molecular sequences not known from nature.
   *c.* In vitro evolution can only produce new proteins.
   *d.* In vitro evolution only selects for changes that were present in the starting pool of molecules, and does not introduce any new mutations.
   *e.* All of the above
10. Which of the following is true of the use of molecular evolutionary studies of human disease?
   *a.* Molecular evolutionary studies are useful for identifying many diseases.
   *b.* Molecular evolutionary studies are often used to determine the origin of emerging diseases.
   *c.* Molecular evolutionary studies are important for developing vaccines against diseases.
   *d.* Molecular evolutionary studies are used to track the effectiveness of polio vaccines in human populations.
   *e.* All of the above

## FOR DISCUSSION

1. Rates of evolutionary change differ among different molecules, and different species differ widely in generation times and population sizes. How does this variation limit how and in what ways we can use the concept of a molecular clock to help us answer questions about the evolution of both molecules and organisms?
2. One hypothesis proposed to explain the existence of large amounts of noncoding DNA is that the cost of maintaining that DNA is so small that natural selection is too weak to reduce it. How could you test this hypothesis against the hypothesis that genome size is functionally related to developmental rate?
3. If fossil evidence and molecular evidence disagree on the date of a major lineage split, which of the two kinds of evidence would you favor? Why?
4. Scientists can produce and release into the wild genetically modified mosquitoes that are unable to harbor and transmit malarial parasites. What ethical issues need to be discussed before such releases are permitted?

## ADDITIONAL INVESTIGATION

Over evolutionary history, many groups of organisms that inhabit caves have lost the organs of sight. For instance, although surface-dwelling crayfishes have functional eyes, several crayfish species that are restricted to underground habitats lack eyes. Opsins are a group of light-sensitive proteins known to have an important function in vision (see Chapter 46), and opsin genes are expressed in eye tissues. Opsin genes are present in the genomes of eyeless, cave-dwelling crayfish. Two alternative hypotheses are (1) the opsin genes are no longer experiencing purifying selection (because there is no longer selection for function in vision); or (2) the opsin genes are experiencing selection for a function other than vision. How would you investigate these alternatives using the sequences of the opsin genes in various species of crayfishes?

# 25 The History of Life on Earth

## When hawk-sized dragonflies ruled the air

Almost anyone who has spent time around fresh water ponds is familiar with dragonflies. Their hovering flight, bright colors, and transparent wings stimulate our visual senses on bright summer afternoons as they fly about their business of devouring mosquitoes, mating, and laying their eggs. The largest dragonflies alive today have wingspans that can be covered by a human hand. Three hundred million years ago, however, dragonflies such as *Meganeuropsis permiana* had wingspans of more than 70 centimeters—well over 2 feet, matching or exceeding the wingspans of many modern birds of prey—and were the largest flying predators on Earth.

No flying insects alive today are anywhere near this size. But during the Carboniferous and Permian geological periods, between 350 and 250 million years ago, many groups of flying insects contained gigantic members. *Meganeuropsis* probably ate huge mayflies and other giant flying insects that shared their home in the Permian swamps. These enormous insects were themselves eaten by giant amphibians.

None of the giant flying insects or amphibians of that time would be able to survive on Earth today. The oxygen concentrations in Earth's atmosphere were about 50 percent higher then than they are now, and those high oxygen levels are thought to have been necessary to support giant insects and their huge amphibian predators.

Paleontologists have uncovered fossils of *Meganeuropsis permiana* in the rocks of Kansas. How do we know the age of these fossils, and how can we know how much oxygen that long-vanished atmosphere contained? The *stratigraphic* layering of the rocks allows us to tell their ages relative to each other, but it does not by itself indicate a given layer's absolute age.

One of the remarkable achievements of twentieth-century scientists was to develop sophisticated techniques that use the decay rates of various radioisotopes, changes in Earth's magnetic field, and the ratios of certain molecules to infer conditions and events in the remote past and to date them accurately. It is those methods that allow us to age the fossils of *Meganeuropsis* and to calculate the concentration of oxygen in Earth's atmosphere at the time.

The development of the science of biology is intimately linked to changing concepts of

**Giant Dragonflies** *Meganeuropsis permiana*, shown here in a reconstruction from fossils, dwarfed modern dragonflies (shown in the inset at the same scale) in size. Otherwise, however, the Permian giant was quite similar to modern dragonflies in general appearance.

**Younger Rocks Lie on Top of Older Rocks** In the Grand Canyon, the Colorado River cut through and exposed many strata of ancient rocks. The oldest rocks visible here formed about 540 million years ago. The youngest, at the top, are about 500 million years old. Knowing the ages of rock strata allows scientists to date the fossils found in each stratum.

time, especially of the age of Earth. About 150 years ago, geologists first provided solid evidence that Earth is ancient; before 1850, most people believed it was no more than a few thousand years old. For many more years, physicists continued to underestimate Earth's age, until an understanding of radioactive decay was developed. Today we know that Earth is about 4.5 billion years old and that life has existed on it for about 3.8 billion of those years. That means human civilizations have occupied Earth for less than 0.0003 percent of the history of life. Discovering what happened before humans were around is an ongoing and exciting area of science.

**IN THIS CHAPTER** we will examine how biologists assign dates to events in the distant evolutionary past, and how such dating allows us to review the major changes in physical conditions on Earth during the past 4 billion years. We will then look at how these changes in physical conditions have influenced the major patterns in the evolution of life, and describe how scientists organize our knowledge of biological diversity based on the relationships among species.

## CHAPTER OUTLINE



## 25.1 How Do Scientists Date Ancient Events?

Many evolutionary changes happen rapidly enough to be studied directly and manipulated experimentally. Plant and animal breeding by agriculturalists and insects' evolution of resistance to pesticides are examples of rapid, short-term evolution. Other changes, such as the appearance of new species and evolutionary lineages, usually take place over much longer time frames.

To understand the long-term patterns of evolutionary change, we must think in time frames spanning many millions of years, and consider events and conditions very different from those we observe today. Earth of the distant past was so unlike the present that it seams like a foreign planet inhabited by strange organisms. The continents were not where they are today, and climates were sometimes dramatically different from those of today.

**Fossils**—the preserved remains of ancient organisms—can tell us a great deal about the body form, or *morphology*, of organisms that lived long ago, as well as how and where they lived. Fossils provide a direct record of evolution. But to understand patterns of evolutionary change, we must also understand how Earth has changed over time.

Earth's history is largely recorded in its rocks. We cannot tell the ages of rocks just by looking at them, but we can determine the ages of rocks relative to one another. The first person to formally recognize that this could be done was the seventeenth-century Danish physician Nicolaus Steno. Steno realized that in undisturbed **sedimentary rocks** (rocks formed by the accumulation of grains on the bottom of bodies of water), the oldest layers of rock, or **strata** (singular *stratum*), lie at the bottom; thus successively higher strata are progressively younger.

Geologists, particularly the eighteenth-century English scientist William Smith, subsequently combined Steno's insight with their observations of fossils contained in sedimentary rocks. They concluded that:

- Fossils of similar organisms are found in widely separated places on Earth.
- Certain fossils are always found in younger rocks, and certain other fossils in older rocks.
- Organisms found in higher, more recent strata are more similar to modern organisms than are those found in lower, more ancient strata.

These patterns revealed much about the relative ages of sedimentary rocks as well as patterns in the evolution of life. But the geologists still could not tell how old the rocks were. A method of dating rocks did not become available until after radioactivity was discovered at the beginning of the twentieth century.

### Radioisotopes provide a way to date rocks

Radioactive isotopes of atoms (see Section 2.1) decay in a predictable pattern over long time periods. During each successive time interval, known as a **half-life**, half of the remaining radioactive material of the radioisotope decays to become a different, stable isotope (**Figure 25.1A**).

To use a radioisotope to date a past event, we must know or estimate the concentration of the isotope at the time of that event. In the case of carbon, the production of new carbon-14 ($^{14}C$) in the upper atmosphere (by the reaction of neutrons with nitrogen-14) just balances the natural radioactive decay of $^{14}C$ to $^{14}N$. Therefore, the ratio of $^{14}C$ to its stable isotope, carbon-12 ($^{12}C$), is relatively constant in living organisms and their environment. As soon as an organism dies, however, it ceases to exchange carbon compounds with its environment. Its decaying $^{14}C$ is no longer replenished, and the ratio of $^{14}C$ to $^{12}C$ in its remains decreases through time. Paleontologists can use the ratio of $^{14}C$ to $^{12}C$ in fossil material to date fossils that are less than 50,000 years old (and thus the sedimentary rocks that contain those fossils). If fossils are older than that, so little $^{14}C$ remains that the limits of detection using this particular isotope are reached.

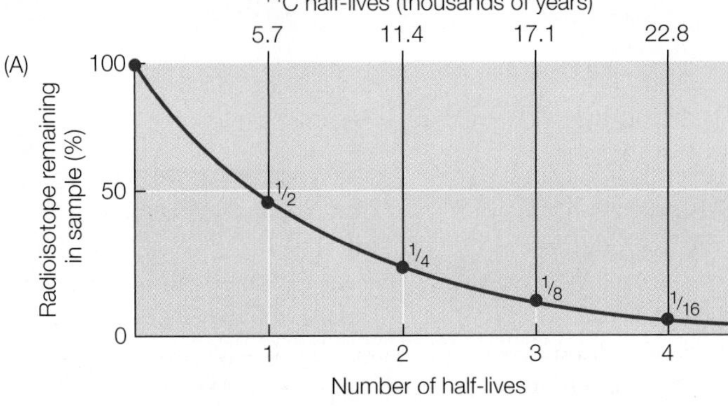


(B)

| Radioisotope | Half-life (years) | Decay product | Useful dating range (years) |
|---|---|---|---|
| Carbon-14 ($^{14}C$) | 5,700 | Nitrogen-14 ($^{14}N$) | 100 – 50,000 |
| Potassium-40 ($^{40}K$) | 1.3 billion | Argon-40 ($^{40}Ar$) | 10 million – 4.5 billion |
| Uranium-238 ($^{238}U$) | 4.5 billion | Lead 206 ($^{206}Pb$) | 10 million – 4.5 billion |

**25.1 Radioactive Isotopes Allow Us to Date Ancient Rocks** The decay of radioactive "parent" atoms into stable "daughter" isotopes happens at a steady rate known as a half-life. (A) The graph demonstrates the principle of half-life using carbon-14 ($^{14}C$) as an example. (B) Radioisotopes have different characteristic half-lives that allow us to measure how much time has elapsed since the rocks containing them were laid down.

## TABLE 25.1

### Earth's Geological History

| RELATIVE TIME SPAN | ERA | PERIOD | ONSET | MAJOR PHYSICAL CHANGES ON EARTH |
|---|---|---|---|---|
| | Cenozoic | Quaternary | 2.6 mya | Cold/dry climate; repeated glaciations |
| | | Tertiary | 65 mya | Continents near current positions; climate cools |
| | Mesozoic | Cretaceous | 145 mya | Northern continents attached; Gondwana begins to drift apart; meteorite strikes Yucatán Peninsula |
| | | Jurassic | 200 mya | Two large continents form: Laurasia (north) and Gondwana (south); climate warm |
| | | Triassic | 251 mya | Pangaea begins to slowly drift apart; hot/humid climate |
| | Paleozoic | Permian | 297 mya | Extensive lowland swamps; $O_2$ levels 50% higher than present; by end of period continents aggregate to form Pangaea, and $O_2$ levels begin to drop rapidly |
| | | Carboniferous | 359 mya | Climate cools; marked latitudinal climate gradients |
| | | Devonian | 416 mya | Continents collide at end of period; meteorite probably strikes Earth |
| | | Silurian | 444 mya | Sea levels rise; two large land masses emerge; hot/humid climate |
| | | Ordovician | 488 mya | Massive glaciation, sea level drops 50 meters |
| | | Cambrian | 542 mya | $O_2$ levels approach current levels |
| Precambrian | Precambrian | | 900 mya | $O_2$ level at >5% of current level |
| | | | 1.5 bya | $O_2$ level at >1% of current level |
| | | | 3.8 bya | $O_2$ first appears in atmosphere |
| | | | 4.5 bya | |

*Note*: mya, million years ago; bya, billion years ago.

### Radioisotope dating methods have been expanded and refined

Sedimentary rocks are formed from materials that existed for varying lengths of time before being transported, sometimes over long distances, to the site of their deposition. Therefore, the inorganic isotopes in a sedimentary rock do not contain reliable information about the date of its formation. Dating rocks more ancient than 50,000 years requires estimating isotope concentrations in *igneous* rocks—rocks formed when molten material cools. To date older sedimentary rocks, geologists search for places where sedimentary rocks show igneous intrusions of volcanic ash or lava flows.

A preliminary estimate of the age of an igneous rock determines which isotope is used to date it (**Figure 25.1B**). The decay of potassium-40 (which has a half-life of 1.3 billion years) to argon-40 has been used to date many of the ancient events in the evolution of life. Fossils in the adjacent sedimentary rock that are similar to those in other rocks of known ages provide additional clues.

Radioisotope dating of rocks, combined with fossil analysis, is the most powerful method of determining geological age. But in places where sedimentary rocks do not contain suitable igneous intrusions and few fossils are present, paleontologists turn to other methods.

One method, known as **paleomagnetic dating**, relates the ages of rocks to patterns in Earth's magnetism, which change over time. Earth's magnetic poles move and occasionally reverse themselves. Because both sedimentary and igneous rocks preserve a record of Earth's magnetic field at the time they were formed, paleomagnetism helps determine the ages of those rocks. Other dating methods use information about continental drift, sea level changes, and molecular clocks (the last of which is described in Section 22.3).

Using these methods, geologists divided the history of life into *eras*, which in turn are subdivided into *periods* (**Table 25.1**). The boundaries between these time frames are based on striking differences scientists have observed in the assemblages of fossil organisms contained in successive layers of rocks. Geologists defined and named these divisions before they were able to establish the ages of fossils, adding and refining the time scales as new methods for geological dating were developed.

## 25.1 RECAP

**Fossils in sedimentary rocks enabled geologists to determine the relative ages of organisms, but absolute dating was not possible until the discovery of radioactivity. Geologists divide the history of life into eras and periods, based on assemblages of fossil organisms found in successive layers of rocks.**

- What observations about fossils suggested to geologists that they could be used to determine the relative ages of rocks? See p. 519
- How is the rate of decay of radioisotopes used to estimate the absolute ages of rocks? See p. 520 and Figure 25.1

The scale at the left of Table 25.1 gives a relative sense of geological time, especially the vast expanse of the Precambrian era, during which early life evolved amid stupendous physical changes of Earth and its atmosphere. During the Precambrian to Cambrian transition, an "explosion" of new life forms took place as representatives of many of the major multicellular groups of life evolved. Earth continued to undergo massive physical changes that influenced the evolution of life, and these events and important milestones are listed in the table. In the next two sections we'll discuss the most important of these changes.

## 25.2 How Have Earth's Continents and Climates Changed over Time?

The globes and maps that adorn our walls, shelves, and books give an impression of a static Earth. It would be easy for us to assume that the continents have always been where they are, but we would be wrong. The idea that Earth's land masses have changed position over the millennia, and that they continue to do so, was first put forth in 1912 by the German meteorologist

| MAJOR EVENTS IN THE HISTORY OF LIFE |
|---|
| Humans evolve; many large mammals become extinct |
| Diversification of birds, mammals, flowering plants, and insects |
| Dinosaurs continue to diversify; mass extinction at end of period (≈76% of species disappear) |
| Diverse dinosaurs; radiation of ray-finned fishes; first fossils of flowering plants |
| Early dinosaurs; first mammals; marine invertebrates diversify; mass extinction at end of period (≈65% of species disappear) |
| Reptiles diversify; giant amphibians and flying insects present; mass extinction at end of period (≈96% of species disappear) |
| Extensive "fern" forests; first reptiles; insects diversify |
| Fishes diversify; first insects and amphibians; mass extinction at end of period (≈75% of species disappear) |
| Jawless fishes diversify; first ray-finned fishes; plants and animals colonize land |
| Mass extinction at end of period (≈75% of species disappear) |
| Rapid diversification of multicellular animals; diverse photosynthetic protists |
| Ediacaran fauna; earliest fossils of multicellular animals |
| Eukaryotes evolve |
| Origin of life; prokaryotes flourish |

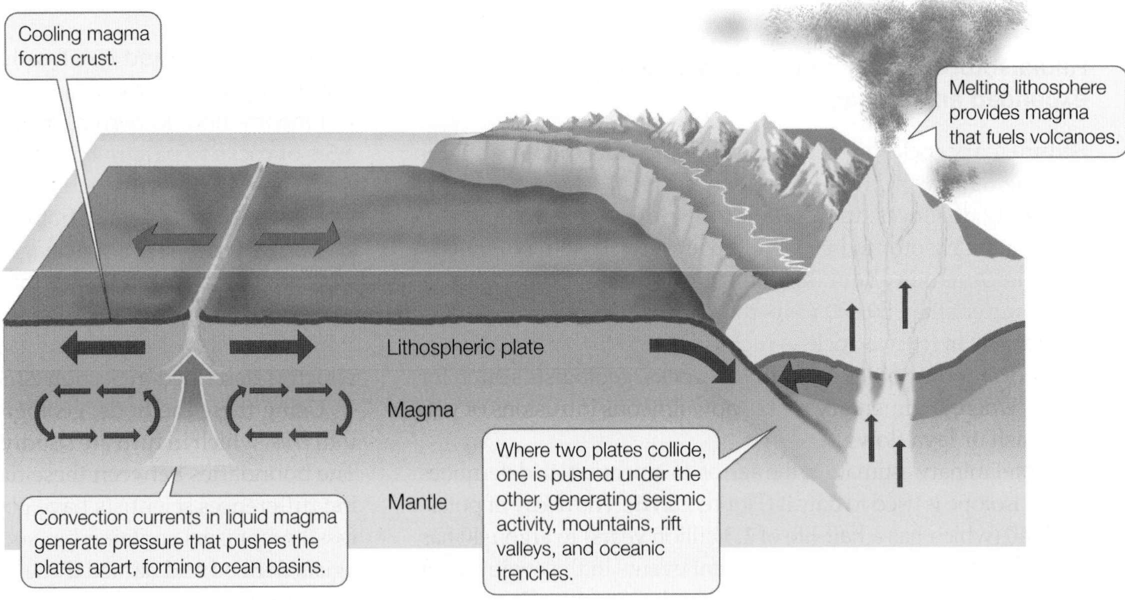

Cooling magma forms crust.
Melting lithosphere provides magma that fuels volcanoes.
Lithospheric plate
Magma
Mantle
Convection currents in liquid magma generate pressure that pushes the plates apart, forming ocean basins.
Where two plates collide, one is pushed under the other, generating seismic activity, mountains, rift valleys, and oceanic trenches.

**25.2 Plate Tectonics and Continental Drift** The heat of Earth's core generates convection currents (arrows) in the magma that push the lithospheric plates, along with the land masses lying on them, together or apart. When lithospheric plates collide, one often slides under the other. The resulting seismic activity can create mountains and deep rift valleys (the latter known as trenches when they occur under ocean basins).

and geophysicist Alfred Wegener. His book *The Origin of Continents and Oceans* was initially met with skepticism and resistance. By the 1960s, however, physical evidence and increased understanding of the geophysics of **plate tectonics**—the study of movement of major land masses—had convinced virtually all geologists of the reality of Wegener's vision.

Earth's crust consists of several solid plates approximately 40 kilometers thick, which collectively make up the lithosphere. The lithospheric plates float on a fluid layer of molten rock, or magma (**Figure 25.2**). Heat produced by radioactive decay deep in Earth's core sets up convection currents in the fluid magma, which then rises and exerts tremendous pressure on the solid plates. When the pressure of the rising magma pushes plates apart, ocean basins may form between them. When plates are pushed together, they either move sideways past each other or one plate slides under the other, pushing up mountain ranges and carving deep rift valleys. When they occur under the water of ocean basins, rift valleys are known as trenches. The movement of the lithospheric plates and the continents they contain is known as **continental drift**.

We now know that at times the drifting of the plates has brought continents together and at other times has pushed them apart (these movements are depicted in Figure 25.12). The positions and sizes of the continents influence oceanic circulation patterns, global climates, and sea levels. Major drops in sea level have usually been accompanied by massive extinctions—particularly of marine organisms, which could not survive the exposure of vast areas of the continental shelves and the disappearance of the shallow seas that covered them (**Figure 25.3**).

yourBioPortal.com
GO TO Animated Tutorial 25.1 • Evolution of the Continents

**25.3 Sea Levels Have Changed Repeatedly** Most mass extinctions of marine organisms (indicated by asterisks) have coincided with periods of low sea levels.

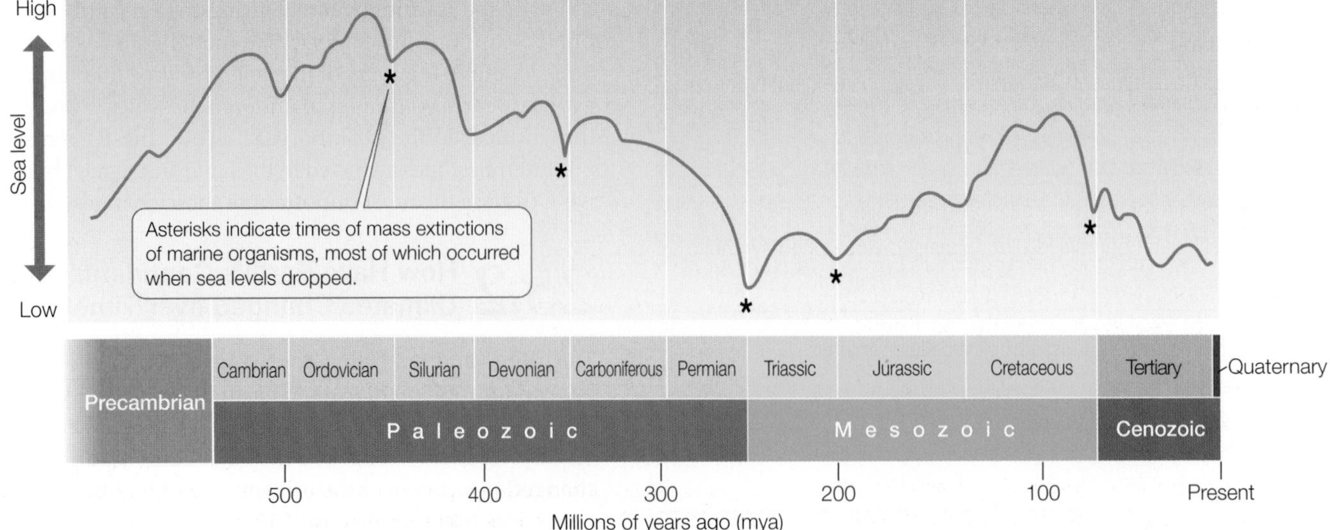

(A)

(B)

**25.4 Stromatolites** (A) A vertical section through a fossil stromatolite. (B) These rocklike structures are living stromatolites that thrive in the very salty waters of Shark Bay in western Australia. Layers of cyanobacteria are found in the uppermost parts of the structures.

### Oxygen concentrations in Earth's atmosphere have changed over time

As the continents have moved over Earth's surface, the world has experienced other physical changes, including large increases and decreases in atmospheric oxygen. The atmosphere of early Earth probably contained little or no free oxygen gas ($O_2$). The increase in atmospheric $O_2$ came in two big steps more than a billion years apart. The first step occurred at least 2.4 billion years ago (bya), when certain bacteria evolved the ability to use water as the source of hydrogen ions for photosynthesis. By chemically splitting $H_2O$, these bacteria generated atmospheric $O_2$ as a waste product. They also made electrons available for reducing $CO_2$ to form organic compounds (see Section 10.3).

One group of $O_2$-generating bacteria, the *cyanobacteria*, formed rocklike structures called *stromatolites*, which are abundantly preserved in the fossil record. Cyanobacteria are still forming stromatolites today in a few very salty places on Earth (**Figure 25.4**). Cyanobacteria liberated enough $O_2$ to open the way for the evolution of oxidation reactions as the energy source for the synthesis of ATP (see Section 9.1).

The evolution of life thus irrevocably changed the physical nature of Earth. Those physical changes, in turn, influenced the evolution of life. When it first appeared in the atmosphere, $O_2$ was poisonous to the anaerobic prokaryotes that inhabited Earth at the time. Over millennia, however, prokaryotes that evolved the ability to metabolize $O_2$ not only survived but gained several advantages. Aerobic metabolism proceeds more rapidly and harvests energy more efficiently than anaerobic metabolism (see Section 9.4), and organisms with aerobic metabolism replaced anaerobes in most of Earth's environments.

An atmosphere rich in $O_2$ also made possible larger cells and more complex organisms. Small unicellular aquatic organisms can obtain enough $O_2$ by simple diffusion even when $O_2$ concentrations are very low. Larger unicellular organisms have lower surface area-to-volume ratios (see Figure 5.2); to obtain enough $O_2$ by simple diffusion, they must live in an environment with a relatively high oxygen concentration. Bacteria can thrive on 1 percent of the current atmospheric $O_2$ levels; eukaryotic cells require levels that are at least 2–3 percent of current concentrations. (For concentrations of dissolved $O_2$ in the oceans to reach these levels, much higher atmospheric concentrations were needed.)

Probably because it took many millions of years for Earth to develop an oxygenated atmosphere, only unicellular prokaryotes lived on Earth for more than 2 billion years. About 1.5 bya, atmospheric $O_2$ concentrations became high enough for large eukaryotic cells to flourish (**Figure 25.5**). Further increases in atmospheric $O_2$ levels 750 to 570 million years ago (mya) enabled several groups of multicellular organisms to evolve.

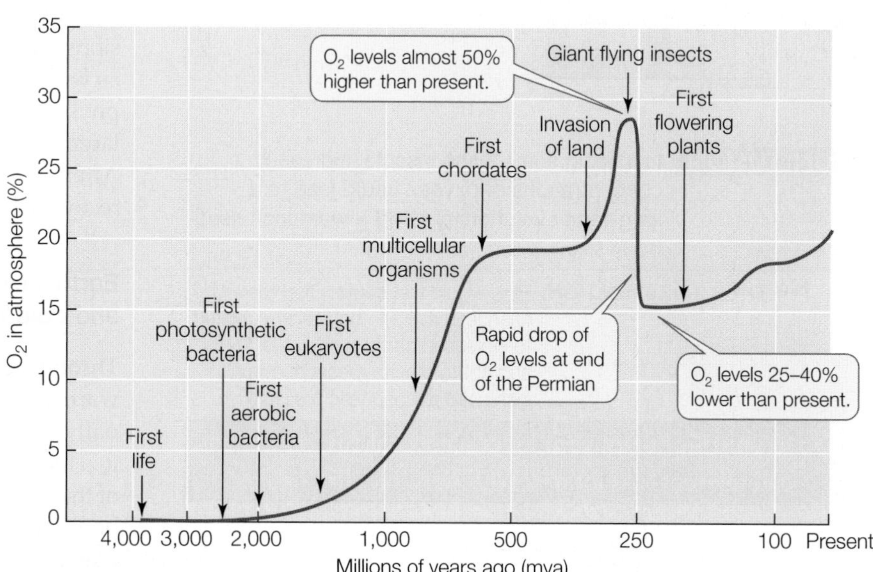


**25.5 Larger Cells, Larger Organisms Need More Oxygen** Changes in oxygen concentrations have strongly influenced, and been influenced by, the evolution of life. (Note that the horizontal axis of the graph is on a logarithmic scale.)

## INVESTIGATING LIFE

### 25.6 Rising Oxygen Levels and Body Size in Insects

In this experiment, flies were raised under hyperbaric conditions (increased atmospheric pressure), thus increasing the partial pressure of $O_2$ in a manner that simulated the greater levels of atmospheric $O_2$ characteristic of the Carboniferous and Permian. Robert Dudley asked if flies raised in hyperbaric conditions would grow larger than their normal counterparts.

**HYPOTHESIS** **Under increased atmospheric pressure, the increased partial pressure of $O_2$ will allow directional selection for increased body size in flying insects.**

**METHOD**

1. Divide a population of fruit flies (*Drosophila melanogaster*) into two lines.
2. Raise one line (the control) at current atmospheric oxygen conditions. Raise the experimental line in hyperbaric conditions (increased partial pressure of $O_2$, simulating increased atmospheric oxygen concentrations). Continue for 5 generations.
3. Raise the $F_6$ offspring of both lines under identical environmental conditions.
4. Weigh all the $F_6$ individuals and test for statistical differences in the average body mass of the flies in each population.

**RESULTS** The average body mass of $F_6$ individuals of both sexes in the experimental line was significantly ($p < 0.0001$) greater than that of insects in the control line.

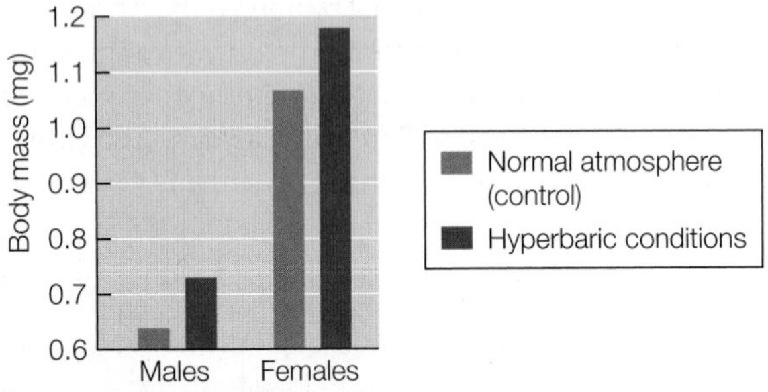

Body mass (mg): 1.2, 1.1, 1.0, 0.9, 0.8, 0.7, 0.6
Males, Females
Normal atmosphere (control)
Hyperbaric conditions

**CONCLUSION** **In at least some flying insects, increased concentrations of oxygen could lead to a long-term evolutionary trend toward increased body size.**

**FURTHER INVESTIGATION:** How would you confirm that the change in average body size is related to increased partial pressure of $O_2$, and not to other aspects of overall increased atmospheric pressure?

Go to **yourBioPortal.com** for original citations, discussions, and relevant links for all INVESTIGATING LIFE figures.

$O_2$ concentrations increased again during the Carboniferous and Permian periods because of the evolution of large vascular plants in the expansive lowland swamps that existed then (see Table 25.1). These swamps resulted in extensive burial of plant debris from vascular plants, which led to the formation of Earth's vast coal deposits. As the buried organic material was not subject to oxidation, and the living plants were producing large quantities of $O_2$, atmospheric $O_2$ increased to concentrations that have not been reached again in Earth's history (see Figure 25.5). As mentioned in the opening of this chapter, high concentrations of atmospheric $O_2$ allowed the evolution of giant flying insects and amphibians that could not survive in today's atmosphere. The drying of the lowland swamps at the end of the Permian reduced global organic burial, and also the production of atmospheric $O_2$, so $O_2$ concentrations dropped rapidly. Over the past 200 million years, with the diversification of flowering plants, $O_2$ concentrations have again increased, but not to the levels that characterized the Carboniferous and Permian periods.

Biologists have conducted experiments that demonstrate the changing selective pressures that can accompany changes in $O_2$ levels. In experimental conditions, an increase in $O_2$ concentration can be simulated by increasing atmospheric pressure in a hyperbaric chamber. Increasing atmospheric pressure increases the *partial pressure of oxygen* (see Chapter 49) in a manner that simulates an increase in $O_2$ concentration at normal atmospheric pressure. When lines of fruit flies (*Drosophila*) are raised in artificial hyperbaric atmospheres (which have higher partial pressure of $O_2$), they quickly evolve larger body sizes over just a few generations (**Figure 25.6**). The current levels of atmospheric $O_2$ appear to constrain body size evolution of these flying insects; increases in $O_2$ appear to relax these constraints. This demonstrates that the *stabilizing selection* on body size at present $O_2$ concentrations can quickly switch to *directional selection* (see Section 21.3) for a change in body size in response to a change in $O_2$ levels. Directional selection over a period of millions of years would be sufficient to account for giant insects such as *Meganeuropsis*, described at the beginning of this chapter.

Many physical conditions on Earth have oscillated in response to the planet's internal processes, such as volcanic activity and continental drift. Extraterrestrial events, such as collisions with meteorites, have also left their mark. In some cases, as we saw earlier and will see again in this chapter, changing physical parameters caused **mass extinctions**, during which a large proportion of the species living at the time disappeared. After each mass extinction, the diversity of life rebounded, but recovery took millions of years.

### Earth's climate has shifted between hot/humid and cold/dry conditions

Through much of its history, Earth's climate was considerably warmer than it is today, and temperatures decreased more gradually toward the poles. At other times, Earth was colder than it is today. Large areas were covered with glaciers near the end of the Precambrian and Ordovician, and during parts of the Carboniferous and Permian periods. These cold periods were separated by long periods of milder climates (**Figure 25.7**). Because we are living in one of the colder periods in Earth's history, it is difficult for us to imagine the mild climates that were found at high latitudes during much of the history of life. During the

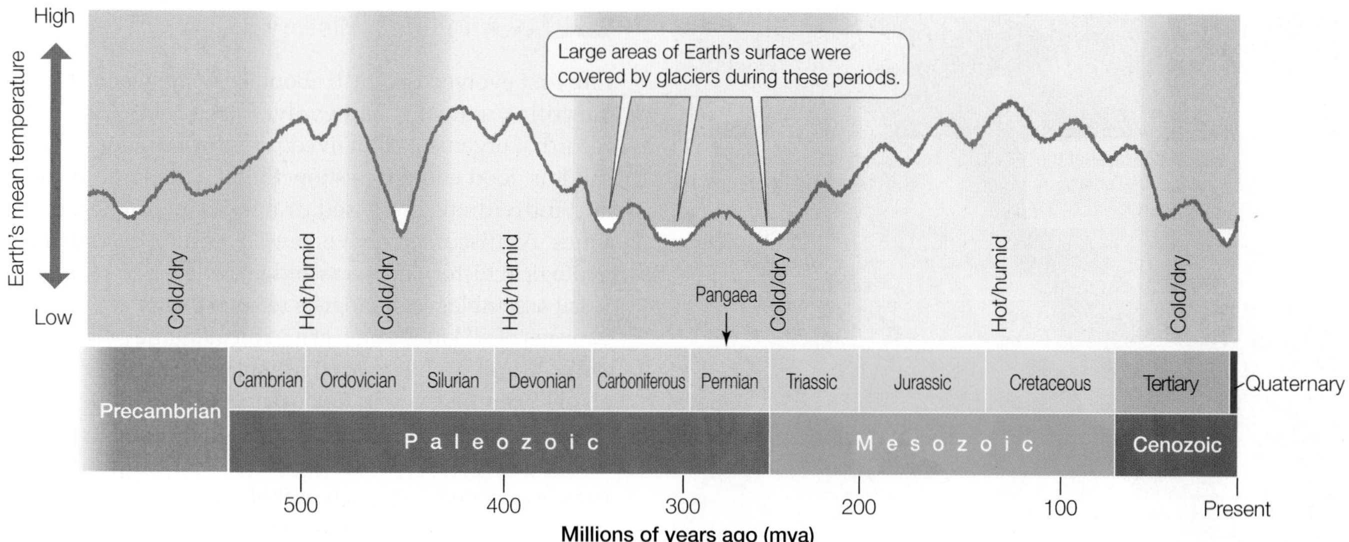


**25.7 Hot/Humid and Cold/Dry Conditions Have Alternated over Earth's History** Throughout Earth's history, periods of cold climates and glaciations (white depressions) have been separated by long periods of milder climates.

Quaternary period there has been a series of glacial advances, interspersed with warmer interglacial intervals during which the glaciers retreated.

"Weather" refers to daily events, such as individual storms. "Climate" refers to long-term average expectations of the various seasons at a given location. Weather often changes rapidly; climates typically change slowly. Major climatic shifts have taken place over periods as short as 5,000 to 10,000 years, primarily as a result of changes in Earth's orbit around the sun. A few climatic shifts have been even more rapid. For example, during one Quaternary interglacial period, the ice-locked Antarctic Ocean became nearly ice-free in less than 100 years. Such rapid changes are usually caused by sudden shifts in ocean currents. Some climate changes have been so rapid that the extinctions caused by them appear to be nearly "instantaneous" in the fossil record.

We are currently living in a time of rapid climate change thought to be caused by a buildup of atmospheric $CO_2$, primarily from the burning of fossil fuels. We are reversing the process of organic burial that occurred (especially) in the Carboniferous and Permian, but we are doing so over a few hundred years rather than the many millions of years over which these deposits accumulated. The current rate of increase of atmospheric $CO_2$ is unprecedented in Earth's history. A doubling of the atmospheric $CO_2$ concentration—which may happen during the current century—is expected to increase the average temperature of Earth, change rainfall patterns, melt glaciers and ice caps, and raise sea level. The possible consequences of such climate changes are discussed in Chapters 58 and 59.

### Volcanoes have occasionally changed the history of life

Most volcanic eruptions produce only local or short-lived effects, but a few large volcanic eruptions have had major consequences for life. When Krakatoa erupted in Indonesia in 1883, it ejected more than 25 cubic kilometers of ash and rock, as well as large quantities of sulphur dioxide gas ($SO_2$). The $SO_2$ was ejected into the stratosphere and then moved by high-level winds around the planet. This led to high concentrations of sulphurous acid ($H_2SO_3$) in high-level clouds, which meant less sunlight got through to Earth's surface. Global temperatures dropped by 1.2°C in the year following the eruption, and global weather patterns showed strong effects for another 5 years. This was all the result of a single volcanic eruption. The collision of continents during the Permian period (about 275 mya) formed a single, gigantic land mass (Pangaea) and caused many massive volcanic eruptions. These eruptions resulted in considerable blockage of sunlight, contributing to the glaciations of that time (see Figure 25.7). Massive volcanic eruptions occurred again as the continents drifted apart during the late Triassic and at the end of the Cretaceous.

### Extraterrestrial events have triggered changes on Earth

At least 30 meteorites between the sizes of baseballs and soccer balls hit Earth each year. Collisions with large meteorites or comets are rare, but such collisions have probably been responsible for several mass extinctions. Several types of evidence tell us about these collisions. Their craters, and the dramatically disfigured rocks that resulted from their impact, are found in many places. Geologists have also discovered compounds in these rocks that contain helium and argon with isotope ratios characteristic of meteorites, which are very different from the ratios found elsewhere on Earth.

A meteorite caused or contributed to a mass extinction at the end of the Cretaceous period (about 65 mya). The first clue that a meteorite was responsible came from the abnormally high concentrations of the element iridium in a thin layer separating rocks deposited during the Cretaceous from those deposited during the Tertiary (**Figure 25.8**). Iridium is abundant in some meteorites, but it is exceedingly rare on Earth's surface. Scientists discovered a circular crater 180 kilometers in diameter buried beneath the northern coast of the Yucatán Peninsula of

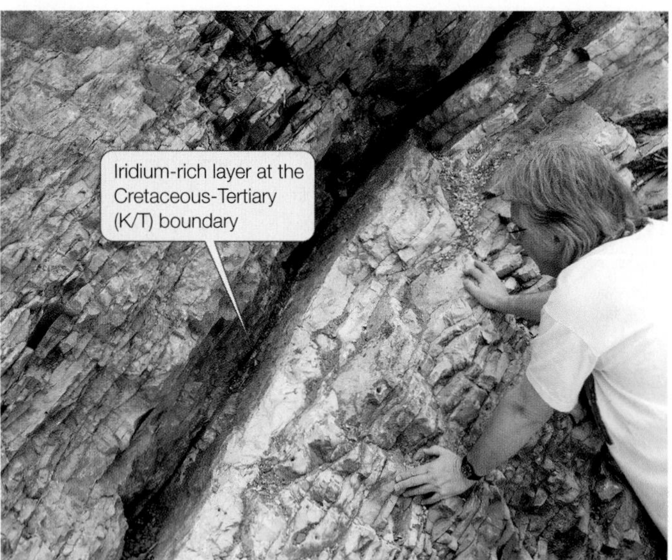


**25.8 Evidence of a Meteorite Impact** The white layers of rock are Cretaceous in age; the layers at the upper left were deposited in the Tertiary. Between the two is a thin, dark layer of clay that contains large amounts of iridium, a metal common in some meteorites but rare on Earth. Its high concentration in sediments deposited about 65 million years ago suggests the impact of a large meteorite.

Mexico. When it collided with Earth, the meteorite released energy equivalent to that of 100 million megatons of high explosives, creating great tsunamis. A massive plume of debris swelled to a diameter of up to 200 kilometers, spread around Earth, and descended. The descending debris heated the atmosphere to several hundred degrees, ignited massive fires, and blocked the sun, preventing plants from photosynthesizing. The settling debris formed the iridium-rich layer. About a billion tons of soot, which has a composition that matches smoke from forest fires, was also deposited. Many fossil species (particularly dinosaurs) that are found in Cretaceous rocks are not found in the Tertiary rocks of the next layer.

## 25.2 RECAP

**Conditions on Earth have changed dramatically over time. Changes in atmospheric concentrations of $O_2$ and in Earth's climate have had major effects on biological evolution. Continental drift, volcanic eruptions, and large meteorite strikes have contributed to climatic changes during Earth's history.**

- Describe how increases in atmospheric concentrations of $O_2$ affected the evolution of multicellular organisms. **See pp. 523–524 and Figure 25.5**
- How have volcanic eruptions and meteorite strikes influenced the course of life's evolution? **See p. 525**

The many dramatic physical events of Earth's history have influenced the nature and timing of evolutionary changes among Earth's living organisms. We now will look more closely at some of the major events that characterize the history of life on Earth.

## 25.3 What Are the Major Events in Life's History?

Life first evolved on Earth about 3.8 bya. By about 1.5 bya, eukaryotic organisms had evolved (see Table 25.1). The fossil record of organisms that lived prior to 550 mya is fragmentary, but it is good enough to show that the total number of species and individuals increased dramatically in late Precambrian times. As discussed above, pre-Darwinian geologists divided geological history into eras and periods based on their distinct fossil assemblages. Biologists refer to the assemblage of all organisms of all kinds living at a particular time or place as a **biota**. All of the plants living at a particular time or place are its **flora**; all of the animals are its **fauna**. Table 25.1 describes some of the physical and biological changes, such as mass extinctions and dramatic increases in the diversity of major groups of organisms, associated with each unit of time.

About 300,000 species of fossil organisms have been described, and the number steadily grows. The number of named species, however, is only a tiny fraction of the species that have ever lived. We do not know how many species lived in the past, but we have ways of making reasonable estimates. Of the present-day biota, nearly 1.8 million species have been named. The actual number of living species is probably well over 10 million, and possibly much higher, because many species have not yet been discovered and described by biologists. So the number of described fossil species is only about 3 percent of the estimated minimum number of living species. Life has existed on Earth for about 3.8 billion years. Many species last only a few million years before undergoing speciation or going extinct; therefore, Earth's biota must have turned over many times during geological history. So the total number of species that have lived over evolutionary time must vastly exceed the number living today. Why have only about 300,000 of these tens of millions of species been described from fossils to date?

### Several processes contribute to the paucity of fossils

Only a tiny fraction of organisms ever become fossils, and only a tiny fraction of fossils are ever discovered by paleontologists. Most organisms live and die in oxygen-rich environments in which they quickly decompose. They are not likely to become fossils unless they are transported by wind or water to sites that lack oxygen, where decomposition proceeds slowly or not at all. Furthermore, geological processes often transform rocks, destroying the fossils they contain, and many fossil-bearing rocks are deeply buried and inaccessible. Paleontologists have studied only a tiny fraction of the sites that contain fossils, but they find and describe many new ones every year.

The fossil record is most complete for marine animals that had hard skeletons (which resist decomposition). Among the nine major animal groups with hard-shelled members, approximately 200,000 species have been described from fossils—roughly twice the number of living marine species in these same groups. Paleontologists lean heavily on these groups in their interpretations of the evolution of life. Insects and spiders are also relatively well represented in the fossil record, because they are numerically abundant and have hard exoskeletons

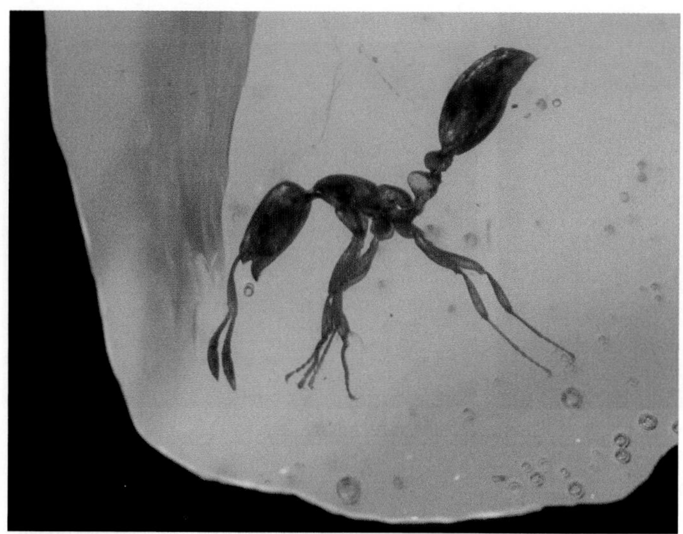

*Solenopsis* sp.

**25.9 Insect Fossils** Chunks of amber—fossilized tree resin—often contain insects that were preserved when they were trapped in the sticky resin. This fire ant fossil is some 30 million years old.

(**Figure 25.9**). The fossil record, though incomplete, is good enough to document clearly the factual history of the evolution of life.

By combining information about geological changes during Earth's history with evidence from the fossil record, scientists have composed portraits of what Earth and its inhabitants may have looked like at different times. We know in general where the continents were and how life changed over time, but many of the details are poorly known, especially for events in the more remote past.

### Precambrian life was small and aquatic

For most of its history, life was confined to the oceans, and all organisms were small. Over the long ages of the **Precambrian era**—more than 3 billion years—the shallow seas slowly began to teem with life. For most of the Precambrian, life consisted of microscopic prokaryotes; eukaryotes evolved about two-thirds of the way through the era (**Figure 25.10**). Unicellular eukaryotes and small multicellular animals fed on floating photosynthetic microorganisms. Small floating organisms, known collectively as *plankton*, were strained from the water and eaten by slightly larger *filter-feeding* animals. Other animals ingested sediments on the seafloor and digested the remains of organisms within them. By the late Precambrian (630–542 mya), many kinds of multicellular soft-bodied animals had evolved. Some of them were very different from any animals living today, and may be members of groups that have no living descendants (**Figure 25.11**).

### Life expanded rapidly during the Cambrian period

The Cambrian period (542–488 mya) marks the beginning of the Paleozoic era. The oxygen concentration in the Cambrian atmosphere was approaching its current level, and the land masses had come together to form several large continents. A geologically rapid diversification of life took place that is sometimes referred to as the **Cambrian explosion** (although in fact it began before the Cambrian, and the "explosion" took millions of years). Several of the major groups of animals that have species living today first evolved during the Cambrian. An overview of the continental and biotic shifts that characterized the Cambrian and subsequent periods is shown in **Figure 25.12** on the following pages.

For the most part, fossils tell us only about the hard parts of organisms, but in three known Cambrian fossil beds—the Burgess Shale in British Columbia, Sirius Passet in northern Greenland, and the Chengjiang site in southern China—the soft parts of many animals were preserved. Crustacean arthropods (crabs, shrimps, and their relatives) are the most diverse group

**25.10 A Sense of Life's Time** The top timeline shows the 4.5 billion year history of life on Earth. Most of this time is accounted for by the Precambrian, a 3.4 billion year era that saw the origin of life and the evolution of cells, photosynthesis, and multicellularity. The final 600 million years are expanded in the second timeline and detailed in Figure 25.12.

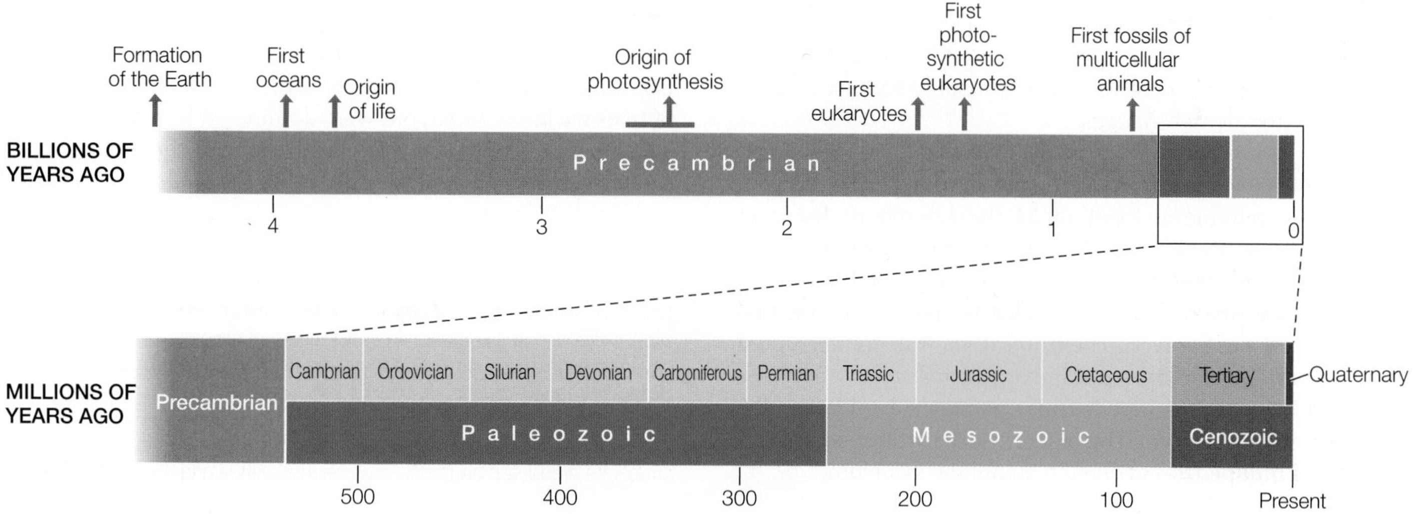

*Spriggina floundersi*

*Mawsonites spriggi*

*Dickinsonia costata*

**25.11 Precambrian Life** These fossils of soft-bodied invertebrates, excavated at Ediacara in southern Australia, were formed about 600 million years ago. Very different from later life forms, they illustrate the diversity of life at the end of the Precambrian era.

in the Chinese fauna; some of them were large carnivores. Multicellular diversity was largely or completely aquatic during the Cambrian. If there was life on land at this time, it was probably restricted to microbial organisms.

## Many groups of organisms that arose during the Cambrian later diversified

Geologists divide the remainder of the Paleozoic era into the Ordovician, Silurian, Devonian, Carboniferous, and Permian periods. Each period is characterized by the diversification of specific groups of organisms. Mass extinctions marked the ends of the Ordovician, Devonian, and Permian.

**THE ORDOVICIAN (488–444 MYA)** During the Ordovician period, the continents, which were located primarily in the Southern Hemisphere, still lacked multicellular plants. Evolutionary radiation of marine organisms was spectacular during the early Ordovician, especially among animals, such as brachiopods and mollusks, that lived on the seafloor and filtered small prey from the water. At the end of the Ordovician, as massive glaciers formed over the southern continents, sea levels dropped about 50 meters and ocean temperatures dropped. About 75 percent of the animal species became extinct, probably because of these major environmental changes.

**THE SILURIAN (444–416 MYA)** During the Silurian period, the continents began to merge together. Marine life rebounded from the mass extinction at the end of the Ordovician. Animals able to swim in open water and feed above the ocean bottom appeared for the first time. Jawless fishes diversified, and the first ray-finned fishes evolved. The tropical sea was uninterrupted by land barriers, and most marine organisms were widely distributed. On land, the first vascular plants evolved late in the Silurian (about 420 mya). The first terrestrial arthropods—scorpions and millipedes—evolved at about the same time.

**THE DEVONIAN (416–359 MYA)** Rates of evolutionary change accelerated in many groups of organisms during the Devonian period. The major land masses continued to move slowly toward each other. In the oceans there were great evolutionary radiations of corals and of shelled, squidlike cephalopod mollusks. Fishes diversified as jawed forms replaced jawless ones and as heavy armor gave way to the less rigid outer coverings of modern fishes.

Terrestrial communities changed dramatically during the Devonian. Club mosses, horsetails, and tree ferns became common; some attained the size of large trees. Their roots accelerated the weathering of rocks, resulting in the development of the first forest soils. The ancestors of gymnosperms—the first plants to produce seeds—appeared in the Devonian. The first known fossils of centipedes, spiders, mites, and insects date to this period, and fishlike amphibians began to occupy the land.

A massive extinction of about 75 percent of all marine species marked the end of the Devonian. Paleontologists are uncertain about its cause, but two large meteorites that collided with Earth at that time (one in present-day Nevada and the other in western Australia) may have been responsible, or at least a contributing factor. The continued coalescence of the continents, with the corresponding reduction in continental shelves, may have also contributed to this mass extinction event.

**THE CARBONIFEROUS (359–297 MYA)** Large glaciers formed over high-latitude portions of the southern land masses during the Carboniferous period, but extensive swamp forests grew on the tropical continents. These forests were not made up of the kinds of trees we know today, but were dominated by giant tree ferns and horsetails with small leaves. Fossilized remains of those forests formed the coal we now mine for energy. In the seas,

**25.12 A Brief History of Life on Earth** The geologically rapid "explosion" of life during the Cambrian saw the rise of several animal groups that have representatives surviving today. The following three pages depict life's history from the Cambrian forward. Movements of the major continents during the past half-billion years are shown in the maps of Earth, and associated biotas for each time period are depicted. The artists' reconstructions are based on fossils such as those shown in the photographs.

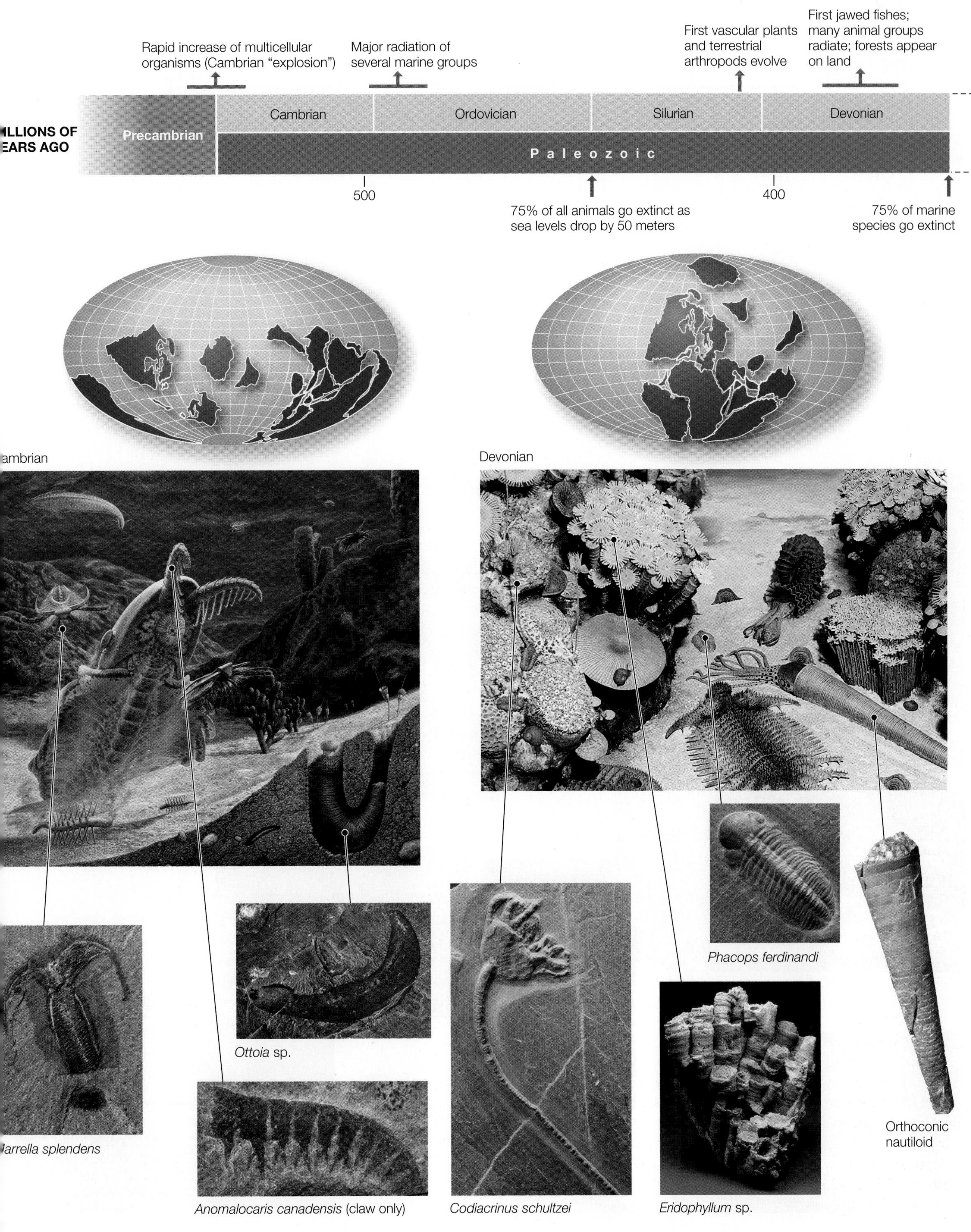

Rapid increase of multicellular organisms (Cambrian "explosion")
Major radiation of several marine groups
First vascular plants and terrestrial arthropods evolve
First jawed fishes; many animal groups radiate; forests appear on land
ILLIONS OF
EARS AGO
Precambrian
Cambrian
Ordovician
Silurian
Devonian
Paleozoic
500
400
75% of all animals go extinct as sea levels drop by 50 meters
75% of marine species go extinct
ambrian
Devonian

Marrella splendens

Anomalocaris canadensis (claw only)

Ottoia sp.

Codiacrinus schultzei

Eridophyllum sp.

Phacops ferdinandi

Orthoconic nautiloid

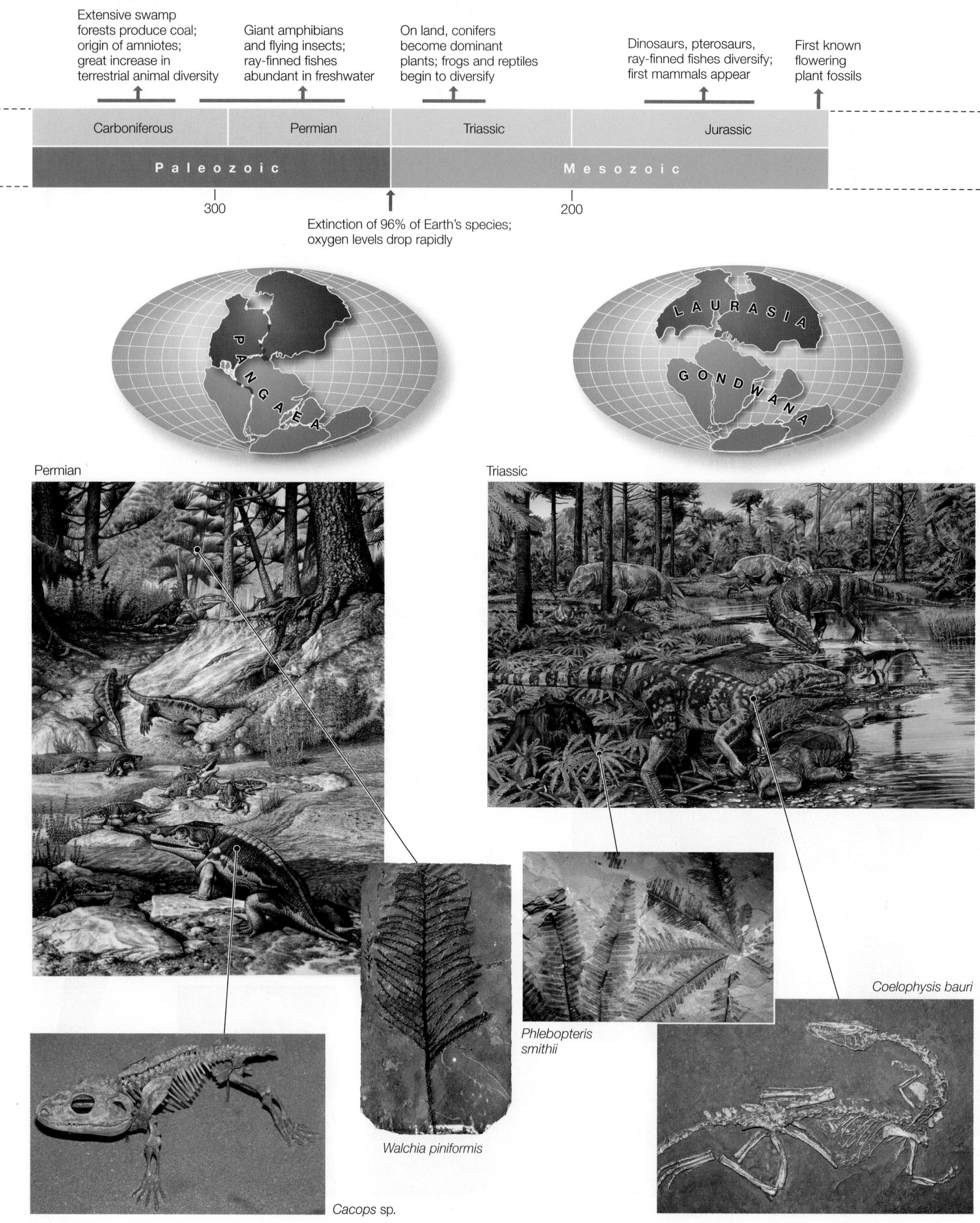

Extensive swamp forests produce coal; origin of amniotes; great increase in terrestrial animal diversity
Giant amphibians and flying insects; ray-finned fishes abundant in freshwater
On land, conifers become dominant plants; frogs and reptiles begin to diversify
Dinosaurs, pterosaurs, ray-finned fishes diversify; first mammals appear
First known flowering plant fossils
Carboniferous
Permian
Triassic
Jurassic
Paleozoic
Mesozoic
300
200
Extinction of 96% of Earth's species; oxygen levels drop rapidly
PANGAEA
LAURASIA
GONDWANA
Permian
Triassic
Cacops sp.
Walchia piniformis
Phlebopteris smithii
Coelophysis bauri

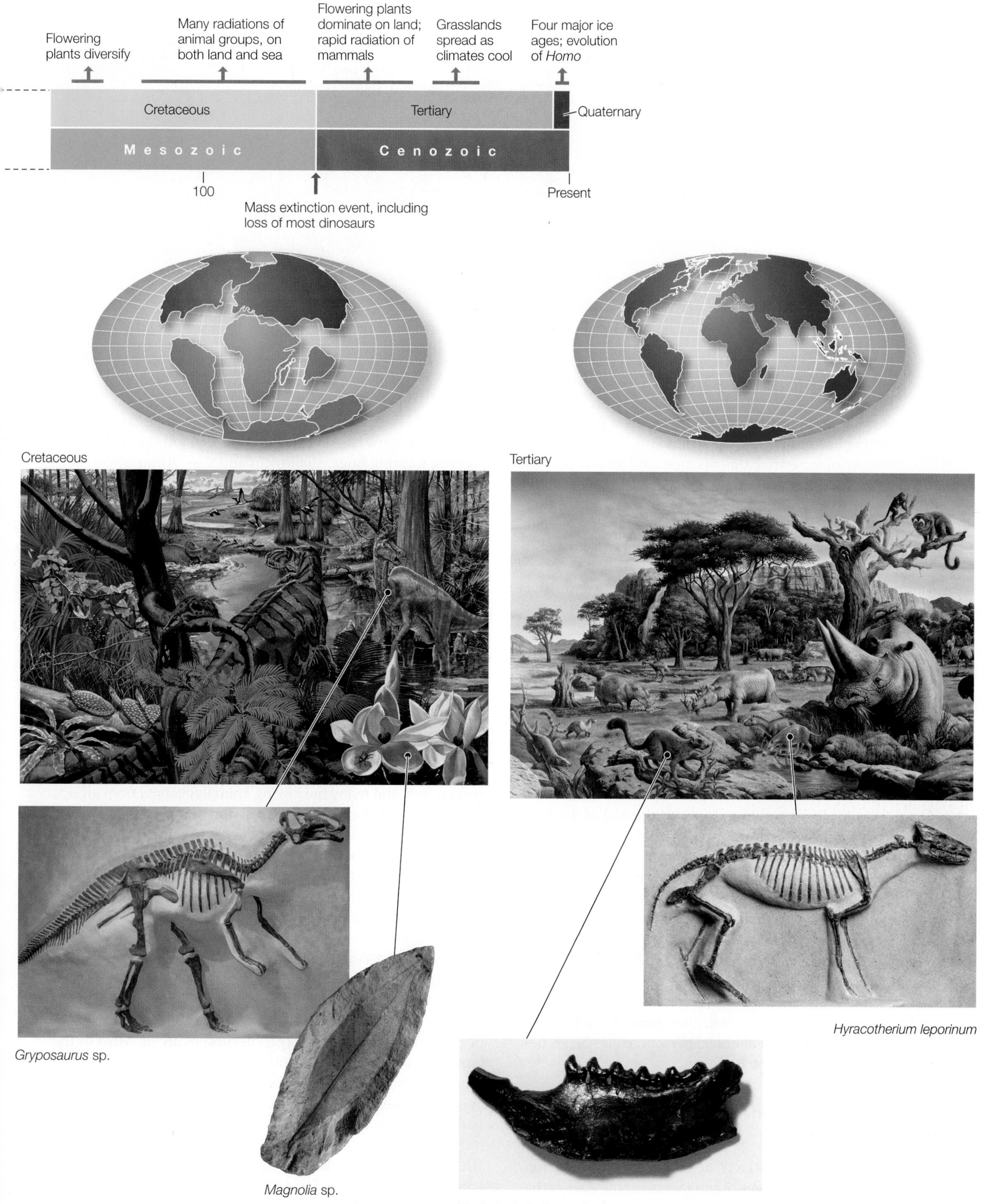
Flowering plants diversify
Many radiations of animal groups, on both land and sea
Flowering plants dominate on land; rapid radiation of mammals
Grasslands spread as climates cool
Four major ice ages; evolution of *Homo*
Cretaceous
Tertiary
Quaternary
Mesozoic
Cenozoic
100
Present
Mass extinction event, including loss of most dinosaurs
Cretaceous
Tertiary

*Gryposaurus* sp.

*Magnolia* sp.

*Hyracotherium leporinum*

*Plesiadapis fodinatus* (jaw)

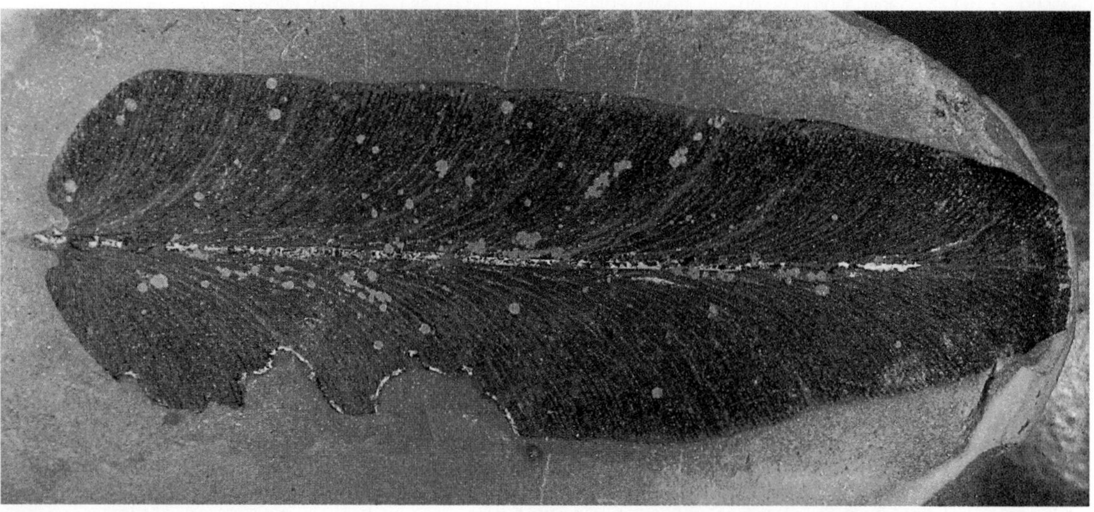

**25.13 Evidence of Insect Diversification** The margins of this fossil fern leaf from the Carboniferous have been chewed by insects.

crinoids (sea lilies and feather stars) reached their greatest diversity, forming "meadows" on the seafloor.

The diversity of terrestrial animals increased greatly during the Carboniferous. Snails, scorpions, centipedes, and insects were abundant and diverse. Insects evolved wings, becoming the first animals to fly. Flight gave herbivorous insects easy access to tall plants; plant fossils from this period show evidence of chewing by insects (**Figure 25.13**). The terrestrial vertebrate lineage split, and amphibians became larger and better adapted to terrestrial existence, while the sister lineage led to the **amniotes**, vertebrates with well-protected eggs that can be laid in dry places.

**THE PERMIAN (297–251 MYA)** During the Permian period, the continents coalesced completely into the supercontinent **Pangaea**. Permian rocks contain representatives of many major groups of insects we know today. By the end of the period, the reptiles split from a second amniote lineage (which would lead to the mammals). Ray-finned fishes became common in the fresh waters of Pangaea.

Conditions for life deteriorated toward the end of the Permian. Massive volcanic eruptions resulted in outpourings of lava that covered large areas of Earth. The ash and gases produced by the volcanoes blocked sunlight and cooled the climate, resulting in the largest glaciers in Earth's history. Atmospheric oxygen concentrations gradually dropped from about 30 to 15 percent. At such low concentrations, most animals would have been unable to survive at elevations above 500 meters; thus about half of the land area would have been uninhabitable at the end of the Permian. The combination of these changes resulted in the most drastic mass extinction event in Earth's history. Scientists estimate that about 96 percent of all species became extinct at the end of the Permian.

## Geographic differentiation increased during the Mesozoic era

The few organisms that survived the Permian mass extinction found themselves in a relatively empty world at the start of the Mesozoic era (251 mya). As Pangaea slowly began to break apart, the oceans rose and once again flooded the continental shelves, forming huge, shallow inland seas. Atmospheric oxygen concentrations gradually rose. Life once again proliferated and diversified, but different groups of organisms came to the fore. The three groups of phytoplankton (floating photosynthetic organisms) that dominate today's oceans—dinoflagellates, coccolithophores, and diatoms—became ecologically important at this time; their remains are the primary origin of the world's oil deposits. Seed-bearing plants replaced the trees that had ruled the Permian forests.

The Mesozoic era is divided into three periods: the Triassic, Jurassic, and Cretaceous. The Triassic and Cretaceous were terminated by mass extinctions, probably caused by meteorite impacts.

**THE TRIASSIC (251–200 MYA)** Pangaea began to break apart during the Triassic period. Many invertebrate groups became more species-rich, and many burrowing animals evolved from groups living on the surfaces of seafloor sediments. On land, conifers and seed ferns were the dominant trees. The first frogs and turtles appeared. A great radiation of reptiles began, which eventually gave rise to crocodilians, dinosaurs, and birds. The end of the Triassic was marked by a mass extinction that eliminated about 65 percent of the species on Earth.

**THE JURASSIC (200–145 MYA)** During the Jurassic period, Pangaea became fully divided into two large continents: **Laurasia** drifted northward and **Gondwana** drifted south. Ray-finned fishes rapidly diversified in the oceans. The first lizards appeared, and flying reptiles (pterosaurs) evolved. Most of the large terrestrial predators and herbivores of the period were dinosaurs. Several groups of mammals made their first appearance, and the earliest known fossils of flowering plants are from late in this period.

**THE CRETACEOUS (145–65 MYA)** By the early Cretaceous period, Laurasia and Gondwana had begun to break apart into the continents we know today. A continuous sea encircled the tropics. Sea levels were high, and Earth was warm and humid. Life proliferated both on land and in the oceans. Marine invertebrates increased in diversity and in number of species. On land, the reptile radiation continued as dinosaurs diversified further and the first snakes appeared. Early in the Cretaceous, flowering plants began the radiation that led to their current dominance of the land. By the end of the period, many groups of mammals had evolved. Most early mammals were small, but one species recently discovered in China, *Repenomamus giganticus*, was large enough to capture and eat young dinosaurs.

As described in Section 25.2, another meteorite-caused mass extinction took place at the end of the Cretaceous (the impact site was near the present day Yucatán Peninsula of Mexico). In the seas, many planktonic organisms and bottom-dwelling invertebrates became extinct. On land, almost all animals larger than about 25 kilograms in body weight became extinct. Many species of insects died out, perhaps because the growth of their food plants was greatly reduced following the impact. Some species in northern North America and Eurasia survived in areas that were not subjected to the devastating fires that engulfed most low-latitude regions.

### Modern biota evolved during the Cenozoic era

By the early Cenozoic era (65 mya), the positions of the continents resembled those of today, but Australia was still attached to Antarctica, and the Atlantic Ocean was much narrower. The Cenozoic was characterized by an extensive radiation of mammals, but other groups were also undergoing important changes.

Flowering plants diversified extensively and came to dominate world forests, except in the coolest regions, where the forests were composed primarily of gymnosperms. Mutations of two genes in one group of plants (the legumes) allowed them to use atmospheric nitrogen directly by forming symbioses with a few species of nitrogen-fixing bacteria (see Section 36.4). The evolution of this symbiosis between certain early Cenozoic plants and these specialized bacteria was the first "green revolution" and dramatically increased the amount of nitrogen available for terrestrial plant growth; the symbiosis remains fundamental to the ecological base of life as we know it today.

The Cenozoic era is divided into the Tertiary and the Quaternary periods. Because both the fossil record and our subsequent knowledge of evolutionary history become more extensive as we approach our own time, paleontologists have subdivided these two periods into *epochs* (**Table 25.2**).

**THE TERTIARY (65–2.6 MYA)** During the Tertiary period, Australia began its northward drift. By 20 mya it had nearly reached its current position. The early Tertiary was a hot and humid time, and the ranges of many plants shifted latitudinally. The tropics were probably too hot for rainforests and were clothed in low-lying vegetation instead. In the middle of the Tertiary, however, Earth's climate became considerably cooler and drier. Many lineages of flowering plants evolved herbaceous (nonwoody) forms, and grasslands spread over much of Earth.

TABLE 25.2

**Subdivisions of the Cenozoic Era**

| PERIOD | EPOCH | ONSET (MYA) |
|---|---|---|
| Quaternary | Holocene[a] | 0.01 (~10,000 years ago) |
| | Pleistocene | 2.6 |
| Tertiary | Pliocene | 5.3 |
| | Miocene | 23 |
| | Oligocene | 34 |
| | Eocene | 55.8 |
| | Paleocene | 65 |

[a]The Holocene is also known as the Recent.

By the start of the Cenozoic era, invertebrate faunas had already come to resemble those of today. It is among the terrestrial vertebrates that evolutionary changes during the Tertiary were most rapid. Frogs, snakes, lizards, birds, and mammals all underwent extensive radiations during this period. Three waves of mammals dispersed from Asia to North America across one of the several land bridges that have intermittently connected the two continents during the past 55 million years. Rodents, marsupials, primates, and hoofed mammals appeared in North America for the first time.

**THE QUATERNARY (2.6 MYA TO PRESENT)** We are living in the Quaternary period. It is subdivided into two epochs, the Pleistocene and the Holocene (the Holocene also being known as the Recent).

The Pleistocene was a time of drastic cooling and climate fluctuations. During 4 major and about 20 minor "ice ages," massive glaciers spread across the continents, and the ranges of animal and plant populations shifted toward the equator. The last of these glaciers retreated from temperate latitudes less than 15,000 years ago. Organisms are still adjusting to these changes. Many high-latitude ecological communities have occupied their current locations for no more than a few thousand years.

It was during the Pleistocene that divergence within one group of mammals, the primates, resulted in the evolution of the hominoid lineage. Subsequent hominoid radiation eventually led to the species *Homo sapiens*—modern humans (see Section 33.5). Many large bird and mammal species became extinct in Australia and in the Americas when *H. sapiens* arrived on those continents about 45,000 and 15,000 years ago, respectively. Many paleontologists believe these extinctions were probably the result of hunting and other influences of *Homo sapiens*.

### The tree of life is used to reconstruct evolutionary events

The fossil record reveals broad patterns in life's evolution. To reconstruct major events in the history of life, biologists also rely on the phylogenetic information in the tree of life (see Chapter 22 and the Tree of Life Appendix). We can use phylogeny (in combination with the paleontological record) to reconstruct the timing of such major events as the acquisition of mitochondria in the ancestral eukaryotic cell, the several independent origins of multicellularity, and the movement of life onto dry land. We can also follow major changes in the genomes of organisms, and even reconstruct many gene sequences of species that are long extinct (see Chapter 24).

Changes to the physical environment on Earth have clearly influenced the great diversity in living organisms we see on the planet today. To study the evolution of that diversity, biologists examine the evolutionary relationships among species. Deciphering these relationships is an important step in understanding how life has diversified on Earth. Part Seven of this book explores the major groups of life and the different solutions that have evolved for major functions such as reproduction, energy acquisition, dispersal, and escape from predation.

**yourBioPortal.com**
GO TO The Interactive Tree of Life

## 25.3 RECAP

**Life evolved in the Precambrian oceans. It diversified as atmospheric oxygen approached its current level and the continents came together to form several large land masses. Numerous climate changes and rearrangements of the continents, as well as meteorite impacts, contributed to five major mass extinctions.**

- Why have so few of the multitudes of organisms that have existed over millennia become fossilized? See pp. 526–527
- What do we mean when we refer to the "Cambrian explosion"? See p. 527
- In what ways has continental drift affected the evolution of life on Earth? See Figure 25.12

## CHAPTER SUMMARY

### 25.1 How Do Scientists Date Ancient Events?

- The relative ages of organisms can be determined by the dating of fossils and the **strata** of **sedimentary rocks** in which they are found.
- Paleontologists use a variety of radioisotopes with different **half-lives** to date events at different times in the remote past. Review Figure 25.1
- Geologists divide the history of life into eras and periods, based on major differences in the fossil assemblages found in successive layers of rocks. Review Table 25.1

### 25.2 How Have Earth's Continents and Climates Changed over Time?

- Earth's crust consists of solid lithospheric plates that float on fluid magma. **Continental drift** caused by convection currents in the magma moves these plates and the continents that lie on top of them. Review Figure 25.2, **ANIMATED TUTORIAL 25.1**
- Conditions on Earth have changed dramatically over time. Increases in atmospheric oxygen and changes in Earth's climate have greatly influenced the evolution of life on Earth. Review Figures 25.5 and 25.7
- Oxygen-generating cyanobacteria liberated enough $O_2$ to open the door to oxidation reactions in metabolic pathways. The aerobic prokaryotes were able to harvest more energy than anaerobic organisms and began to proliferate. Increases in atmospheric $O_2$ levels supported the evolution of large eukaryotic cells.
- Major physical events on Earth, such as the collision of continents that formed the supercontinent **Pangaea**, have affected Earth's surface, climate, and atmosphere. In addition, extraterrestrial events such as meteorite strikes created sudden and dramatic environmental shifts. All of these changes have affected the history of life.

### 25.3 What Are The Major Events in Life's History?

- Paleontologists use fossils and evidence of geological changes to determine what Earth and its **biota** may have looked like at different times.
- During most of its history, life was confined to the oceans. Multicellular life diversified extensively during the **Cambrian explosion**. Review Figure 25.11
- The periods of the Paleozoic era were each characterized by the diversification of specific groups of organisms. **Amniotes**—vertebrates whose eggs can be laid in dry places—first appeared during the Carboniferous period.
- During the Mesozoic era, distinct terrestrial biotas evolved on each continent.
- Five episodes of **mass extinction** punctuated the history of life in the Paleozoic and Mesozoic eras.
- Earth's **flora** has been dominated by flowering plants since the Cenozoic era.
- Phylogenetic trees help reconstruct the timing of evolutionary events and clarify relationships among modern species.

**SEE WEB ACTIVITY 25.1 for a concept review of this chapter.**

## SELF-QUIZ

1. Which of the following is *not* true of the giant flying dragonflies of the Carboniferous and Permian?
   *a.* Some species grew to have wing spans as wide or wider than many modern birds of prey.
   *b.* They were the largest flying predators of the time.
   *c.* Such large flying insects could exist because of the higher concentrations of atmospheric oxygen compared to the present.
   *d.* Their predators were giant reptiles.
   *e.* Fossils of one large species, *Meganeurosis permiana*, have been found in the Permian rocks of Kansas.
2. In undisturbed strata of sedimentary rock, the oldest rocks
   *a.* lie at the top.
   *b.* lie at the bottom.
   *c.* are in the middle.
   *d.* are distributed among the strata of younger rocks.
   *e.* none of the above
3. $^{14}C$ can be used to determine the ages of fossil organisms because
   *a.* all organisms contain many carbon compounds.
   *b.* $^{14}C$ has a regular rate of decay to $^{14}N$.
   *c.* the ratio of $^{14}C$ to $^{12}C$ in living organisms is always the same as that in the atmosphere.
   *d.* the production of new $^{14}C$ in the atmosphere just balances the natural radioactive decay of $^{14}C$.
   *e.* all of the above
4. The concentration of oxygen in the Earth's atmosphere
   *a.* has increased steadily through time.
   *b.* has decreased steadily through time.
   *c.* has been both higher and lower in the past than at present.
   *d.* was lower during most of the Permian than at present.
   *e.* was at its highest levels in the Cambrian.
5. The total of all species of organisms in a given region is known as the region's
   *a.* biota.
   *b.* flora.
   *c.* fauna.
   *d.* flora and fauna.
   *e.* biogeography.
6. The coal beds we now mine for energy are largely the remains of
   *a.* plants that grew in swamps during the Carboniferous period.
   *b.* algae that grew in marshes during the Devonian period.
   *c.* giant insects and amphibians of the Permian period.
   *d.* plants that grew in the oceans during the Carboniferous period.
   *e.* none of the above
7. The mass extinction at the end of the Ordovician period was probably caused by
   *a.* the collision of Earth with a large meteorite.
   *b.* massive volcanic eruptions.
   *c.* massive glaciation on the southern continents and associated climatic changes.
   *d.* the uniting of all continents to form Pangaea.
   *e.* changes in Earth's orbit.
8. The cause of the mass extinction at the end of the Mesozoic era probably was
   *a.* continental drift.
   *b.* the collision of Earth with a large meteorite.
   *c.* changes in Earth's orbit.
   *d.* massive glaciation.
   *e.* changes in the salt concentration of the oceans.
9. Which of the following times was marked by the largest mass extinction of life in the history of Earth?
   *a.* The end of the Cretaceous
   *b.* The end of the Devonian
   *c.* The end of the Permian
   *d.* The end of the Triassic
   *e.* The end of the Silurian
10. Paleontologists have subdivided the Cenozoic era into epochs because
    *a. Homo sapiens* evolved at the start of the Cenozoic.
    *b.* the continents had achieved their present positions.
    *c.* the number of species stopped increasing at this time.
    *d.* our knowledge of the evolutionary events of the Cenozoic is more extensive than for other eras.
    *e.* starting with the Cenozoic, the fossil record is no longer a necessary source of information about evolutionary relationships.

## FOR DISCUSSION

1. Some groups of organisms have evolved to contain large numbers of species; other groups have produced only a few species. Is it meaningful to consider the former groups more successful than the latter? What does the word "success" mean in evolution?
2. Scientists date ancient events using a variety of methods, but nobody was present to witness or record those events. Accepting those dates requires us to understand the accuracy and appropriateness of indirect measurement techniques. What other basic scientific concepts are also based on the results of indirect measurement techniques?
3. Why is it useful to be able to date past events absolutely as well as relatively?
4. If we are living during one of the cooler periods in Earth's history, why should we be concerned about human activities that are thought to contribute to global climate warming?
5. What conditions may have favored the evolution of multicellular groups of organisms near the end of the Precambrian?
6. In what ways do endosymbiotic events (such as the origin of mitochondria and chloroplasts) complicate the classification of the major groups of life?

## ADDITIONAL INVESTIGATION

The experiment in Figure 25.6 showed that body size of insects may evolve quickly following changes in atmospheric oxygen concentrations. What other experiments could you devise to test the effects of changing atmospheric oxygen?

PART SEVEN THE EVOLUTION OF DIVERSITY

# 26 Bacteria and Archaea: The Prokaryotic Domains

## Life on a strange planet

It must have been quite a shock when Thomas "Grif" Taylor's Antarctic exploration team first spotted Blood Falls in 1911. The blood-red falls were certainly a surprise in the snowy, icy terrain. What could possibly cause a red waterfall in Antarctica?

A few million years ago, the Taylor Glacier (which now bears the explorer's name) moved above a pool of salty water, trapping the pool under 400 meters of ice. The harsh environment in the resulting enclosed subglacial sea could hardly seem more hostile to life. It is extremely cold; there is no light and virtually no oxygen; and salt concentrations are several times higher than seawater. In short, it is not a place one might expect to find a diverse living ecosystem.

Some water is able to seep out of this subglacial sea. This water is stained a dark, rusty red, and it spills from the head of Taylor Glacier to form Blood Falls. Taylor speculated that red algae might account for the red coloration, but in the 1960s geologist Robert Black discovered that the water's color arises from iron oxides that come from the underlying bedrock. With the methods then available, biologists could not detect any living organisms in the cold, saline, iron-rich water.

A half-century later, biologists were better equipped to study microscopic life in strange places. By then it was also possible to amplify and sequence genes from single microbes, and to place these gene sequences into the framework of the tree of life to identify and classify the microbes. Microbiologist Jill Mikucki and her colleagues used these techniques on water samples from Blood Falls, and reported in 2009 that the falls contain an unusual ecosystem of at least 17 different species of bacteria. The bacteria survive by metabolizing minute amounts of organic matter trapped in the subglacial sea, using sulfate and iron ions as catalysts and electron acceptors.

The presence of living organisms in Blood Falls confirms that it is hard to find a place on or even near the surface of the Earth that does not contain populations of prokaryotes. There are prokaryotes in volcanic vents, in the clouds, in environments as acidic as battery acid or as alkaline as household ammonia. There are species that can survive below the freezing point and above the boiling point

**A Splash of Color in a Frozen World of White** Antarctica's Blood Falls is the outflow of a subglacial sea that contains an unusual ecosystem of bacteria that rely on sulfate and iron ions for metabolism.

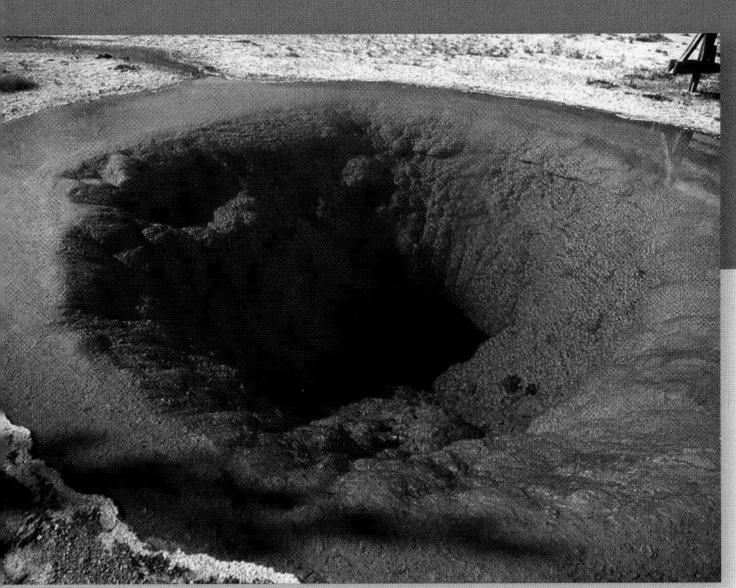

**Prokaryotes Can Take the Heat** Entire ecosystems of prokaryotes create the beauty of Morning Glory Pool, a hot spring in Yellowstone National Park. Cyanobacteria impart the "morning glory blue" color. Archaea live in the intensely hot regions of the pool's interior.

of water. There are more prokaryotes living on and inside our bodies than we have human cells. The prokaryotes are masters of metabolic ingenuity, having developed more ways to obtain energy from the environment than the eukaryotes have. They have been around much longer than other organisms, too.

Prokaryotes are by far the most numerous organisms on Earth. Late in the twentieth century, it became apparent to microbiologists that all prokaryotes are not most closely related to one another. Two prokaryotic lineages diverged early in life's evolution: Bacteria and Archaea. An early merging between members of these two groups is thought to have given rise to the eukaryotic lineage, which includes humans.

**IN THIS CHAPTER** we will discuss the distribution of prokaryotes and examine their remarkable metabolic diversity. We will describe the difficulties involved in determining evolutionary relationships among the prokaryotes and will survey the surprising diversity of organisms in each domain. We will discuss how prokaryotes can have enormous influence on their environments. Finally, we will discuss the evolutionary origin and diversity of viruses and their relationship to the rest of life.

CHAPTER OUTLINE



## 26.1 How Did the Living World Begin to Diversify?

You may think that you have little in common with unicellular prokaryotes. But multicellular eukaryotes like yourself actually share many attributes with Bacteria and Archaea. For example, all three of you:

- conduct glycolysis
- use DNA as the genetic material that encodes proteins
- produce those proteins by transcription and translation using a similar genetic code
- replicate DNA semiconservatively
- have plasma membranes and ribosomes in abundance

These features support the conclusion that all living organisms are related to one another. If life had multiple origins, there would be little reason to expect all organisms to use overwhelmingly similar genetic codes or to share structures as unique as ribosomes. Furthermore, similarities in DNA sequences of universal genes (such as those that encode the structural components of ribosomes) confirm the monophyly of life.

Despite the commonalities found across all three domains, major differences have evolved as well. Let's first distinguish between Eukarya and the two prokaryotic domains. Note that "domain" is a subjective term used for the largest groups of life. There is no objective definition of a domain, any more than there is of a kingdom or a family.

### The three domains differ in significant ways

Prokaryotic cells differ from eukaryotic cells in three important ways:

- *Prokaryotic cells lack a cytoskeleton and a nucleus, so they do not divide by mitosis.* Instead, after replicating their DNA (see Figure 11.2), prokaryotic cells divide by their own method, *binary fission.*
- *The organization of the genetic material differs.* The DNA of the prokaryotic cell is not organized within a membrane-enclosed nucleus. DNA molecules in prokaryotes (both bacteria and archaea) are often circular. Many (but not all) prokaryotes have only one main chromosome and are effectively haploid, although many have additional smaller DNA molecules, called *plasmids*, as well (see Section 12.6).

TABLE 26.1

**The Three Domains of Life on Earth**

| CHARACTERISTIC | BACTERIA | DOMAIN ARCHAEA | EUKARYA |
|---|---|---|---|
| Membrane-enclosed nucleus | Absent | Absent | Present |
| Membrane-enclosed organelles | Absent | Absent | Present |
| Peptidoglycan in cell wall | Present | Absent | Absent |
| Membrane lipids | Ester-linked | Ester-linked | Ester-linked |
| | Unbranched | Branched | Unbranched |
| Ribosomes[a] | 70S | 70S | 80S |
| Initiator tRNA | Formylmethionine | Methionine | Methionine |
| Operons | Yes | Yes | No |
| Plasmids | Yes | Yes | Rare |
| RNA polymerases | One | One[b] | Three |
| Ribosomes sensitive to chloramphenicol and streptomycin | Yes | No | No |
| Ribosomes sensitive to diphtheria toxin | No | Yes | Yes |

[a]70S ribosomes are smaller than 80S ribosomes.
[b]Archaeal RNA polymerase is similar to eukaryotic polymerases.

- *Prokaryotes have none of the membrane-enclosed cytoplasmic organelles—mitochondria, Golgi apparatus, and others—that are found in most eukaryotes.* However, the cytoplasm of a prokaryotic cell may contain a variety of infoldings of the plasma membrane and photosynthetic membrane systems not found in eukaryotes.

A glance at **Table 26.1** will show you that there are also major differences (most of which cannot be seen even under an electron microscope) between the two prokaryotic domains. In some ways archaea are more like eukaryotes; in other ways they are more like bacteria. (Note that we use lowercase when referring to the members of these domains and uppercase when referring to the domains themselves.) The structures of prokaryotic and eukaryotic cells are compared in Chapter 5. The basic unit of an *archaeon* (the term for a single archaeal organism) or *bacterium* (a single bacterial organism) is the prokaryotic cell. Each single-celled organism contains a full complement of genetic and protein-synthesizing systems, including DNA, RNA, and all the enzymes needed to transcribe and translate the genetic information into proteins. The prokaryotic cell also contains at least one system for generating the ATP it needs.

Genetic studies clearly indicate that all three domains had a single common ancestor. For a major portion of their genome, eukaryotes share a more recent common ancestor with Archaea than they do with Bacteria (**Figure 26.1**). However, the mitochondria of eukaryotes (as well as the chloroplasts of photosynthetic eukaryotes, such as plants) originated through the endosymbiosis of a bacterium, as described in Section 5.5. Some biologists prefer to view the origin of eukaryotes as a fusion of two equal partners (one ancestor that was related to modern archaea, and another that was more closely related to modern bacteria). Others view the divergence of the early eukaryotes from the archaea as a separate and earlier event than the later endosymbiosis of the bacterium (the origin of mitochondria). In either case, some genes of eukaryotes are more closely related to those of archaea, while others are more closely related to those of bacteria. The tree of life therefore contains some merging of lineages, as well as the predominant diverging of lineages.

**yourBioPortal.com**

GO TO Animated Tutorial 26.1 • The Evolution of the Three Domains

The common ancestor of all three domains had DNA as its genetic material, and its machinery for transcription and translation produced RNAs and proteins, respectively. This ancestor probably had a circular chromosome.

Three shapes are particularly common among the bacteria: spheres, rods, and curved or helical forms (**Figure 26.2**). A spherical bacterium is called a *coccus* (plural *cocci*). Cocci may live singly or may associate in two- or three-dimensional arrays

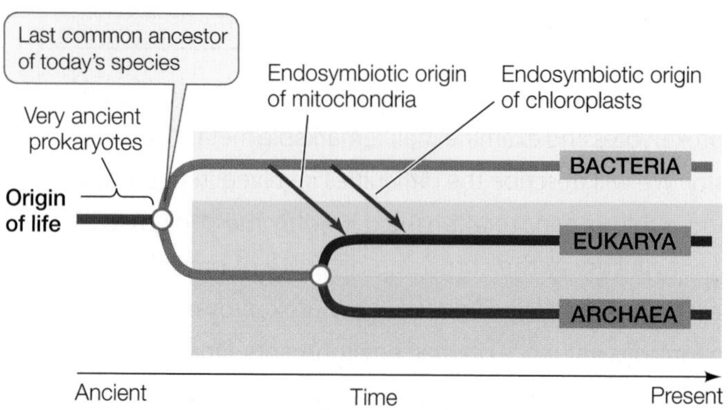


**26.1 The Three Domains of the Living World** All three domains share a common prokaryotic ancestor.

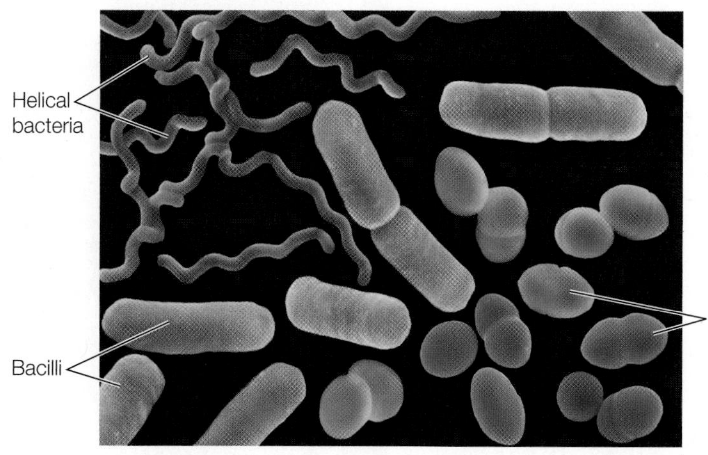


**26.2 Bacterial Cell Shapes** This composite, colorized micrograph shows the three common types of bacterial morphology. Spherical cells are called cocci; the acid-producing cocci shown here in green are a species of *Enterococcus* from the mammalian gut. The rod-shaped bacilli (orange) are represented by *Escherichia coli*, also a resident of the gut. *Leptospira interrogans* is a helical (spiral) bacterium and a human pathogen.

as chains, plates, blocks, or clusters of cells. A rod-shaped bacterium is called a *bacillus* (plural *bacilli*). The spiral form (like a corkscrew), or *helix* (plural *helices*), is the third main bacterial shape. Bacilli and helices may be single, form chains, or gather in regular clusters. Among the other bacterial shapes are long filaments and branched filaments.

Less is known about the shapes of archaea because many of these organisms have never been seen. Many archaea are known only from samples of DNA from the environment, as we describe in Section 26.4. However, the morphology of some species is known, including cocci, bacilli, and even triangular and square-shaped species; the latter grow on surfaces, arranged like sheets of postage stamps.

Archaea, Bacteria, and Eukarya are all products of billions of years of mutation, natural selection, and genetic drift, and they are all well adapted to present-day environments. None is "primitive." Their last common ancestor probably lived 2 to 3 billion years ago. The earliest prokaryotic fossils date back at least 3.5 billion years, and they indicate that there was considerable diversity among the prokaryotes even during the earliest days of life.

## 26.1 RECAP

**Bacteria and archaea are highly divergent from each other and are only distantly related on the tree of life. Eukaryotes received ancient evolutionary contributions from both of these prokaryotic lineages.**

- What are the principal differences between the prokaryotes and the eukaryotes? See pp. 537–538 and Table 26.1
- Why don't we group Bacteria and Archaea together in a single domain? See p. 538 and Table 26.1

The prokaryotes were alone on Earth for a very long time, adapting to new environments and to changes in existing environments. They have survived to this day, in massive numbers and incredible diversity, and they are found everywhere.

## 26.2 What Are Some Keys to the Success of Prokaryotes?

If success is measured by numbers of individuals, the prokaryotes are the most successful organisms on Earth. Individual bacteria and archaea in the oceans number more than $3 \times 10^{28}$. This stunning number is perhaps 100 million times as great as the number of stars in the visible universe. In fact, the bacteria living in your intestinal tract outnumber all the humans who have ever lived.

Prokaryotes are unicellular organisms, although many form multicellular colonies that contain many individual cells. These multicellular associations are not cases of true multicellular organisms, however, because each individual cell is fully viable and independent. These associations arise as cells adhere to one another after reproducing by binary fission. Associations in the form of chains are called **filaments**. Some filaments become enclosed in delicate tubular sheaths.

### Prokaryotes generally form complex communities

Prokaryotic cells and their associations do not usually live in isolation. Rather, they live in communities of many different species of organisms, often including microscopic eukaryotes. (Microscopic organisms are often collectively referred to as *microbes*.) Some microbial communities form layers in sediments, and others form clumps a meter or more in diameter. While some microbial communities are harmful to humans, others provide important services. They help us digest our food, break down municipal waste, and recycle organic matter in the environment.

Many microbial communities tend to form dense **biofilms**. Upon contacting a solid surface, the cells secrete a gel-like sticky polysaccharide matrix that then traps other cells (**Figure 26.3**). Once this biofilm forms, it is difficult to kill the cells. Pathogenic (disease-causing) bacteria are difficult for the immune system—and modern medicine—to combat once they form a biofilm. For example, the film may be impermeable to antibiotics. Worse, some drugs stimulate the bacteria in a biofilm to lay down more matrix, making the film even more impermeable.

Biofilms form on contact lenses, on artificial joint replacements, and on just about any available surface. They foul metal pipes and cause corrosion, a major problem in steam-driven electricity generation plants. The stain on our teeth that we call dental plaque is also a biofilm. Fossil stromatolites—large, rocky structures made up of alternating layers of fossilized microbial biofilm and calcium carbonate—are the oldest remnants of early life on

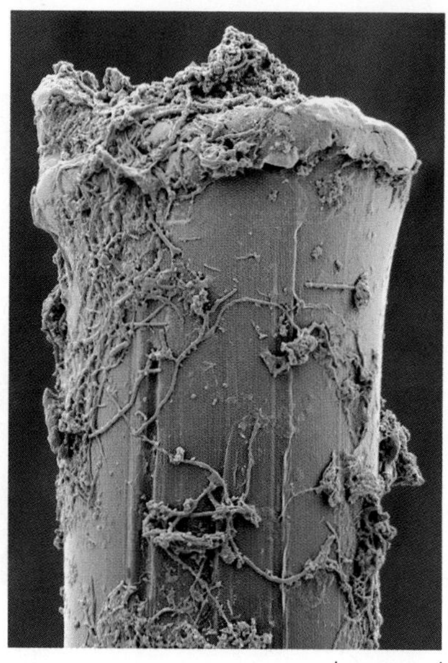


**26.3 Forming a Biofilm** (A) Free-swimming bacteria and archaea readily attach themselves to surfaces and form films stabilized and protected by a surrounding matrix. Once the population size is large enough, the developing biofilm can send chemical signals that attract other microorganisms. (B) Scanning electron micrography reveals a biofilm of plaque on a used toothbrush bristle. The matrix of dental plaque consists of proteins from both bacterial secretions and saliva.

Earth (see Figure 25.4). Stromatolites still form today in some parts of the world.

Biofilms are the subject of much current research. For example, some biologists are studying the chemical signals that bacteria in biofilms use to communicate with one another. By blocking the signals that lead to the production of the matrix polysaccharides, researchers may be able to prevent biofilms from forming.

A team of bioengineers and chemical engineers recently devised a sophisticated technique that enables them to monitor biofilm development in extremely small populations of bacteria, cell by cell. They developed a tiny chip housing six separate growth chambers, or "microchemostats" (**Figure 26.4**). The techniques of *microfluidics* use microscopic tubes and computer-controlled valves to direct fluid flow through complex "plumbing circuits" in the growth chambers.

## TOOLS FOR INVESTIGATING LIFE

**26.4 The Microchemostat**

Using techniques from microfluidic engineering, biologists can monitor the dynamics of extremely small bacterial populations. The photograph shows six microchemostats on a chip. Each of the six is equipped with input ports for growth and flushing media, and a number of output ports (diagram). Tiny valves, controlled by a computer, direct flow. Samples are removed through the output ports and are analyzed to record changes in the bacterial population.

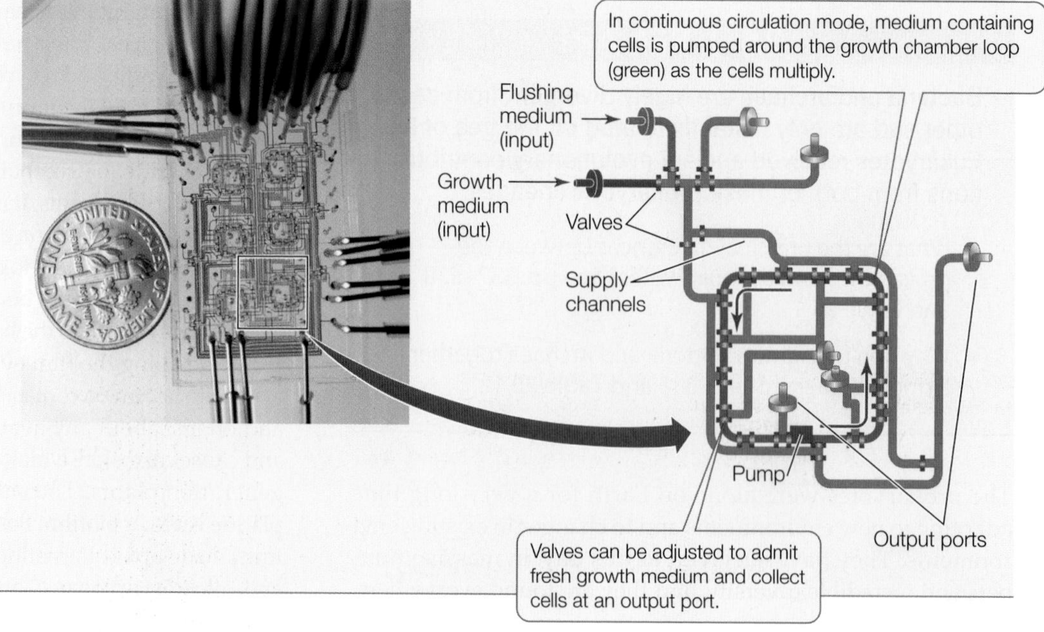

In continuous circulation mode, medium containing cells is pumped around the growth chamber loop (green) as the cells multiply.
Flushing medium (input)
Growth medium (input)
Valves
Supply channels
Pump
Output ports
Valves can be adjusted to admit fresh growth medium and collect cells at an output port.

### Prokaryotes have distinctive cell walls

Many prokaryotes have a thick and relatively stiff cell wall. It is quite different from the cell walls of land plants and algae, which contain cellulose and other polysaccharides, and from those of fungi, which contain chitin. The cell walls of almost all bacteria contain **peptidoglycan** (a cross-linked polymer of amino sugars), which produces a meshlike structure around the cell. Archaeal cell walls are of differing types, but most contain significant amounts of protein. One group of archaea has *pseudopeptidoglycan* in its cell wall; as you can probably guess from the prefix *pseudo*, pseudopeptidoglycan is similar to, but distinctly different from, the peptidoglycan of bacteria. The monomers making up pseudopeptidoglycan differ from and are differently linked than those of peptidoglycan. Peptidoglycan is a substance unique to bacteria; its absence from the walls of archaea is a key difference between the two prokaryotic domains.

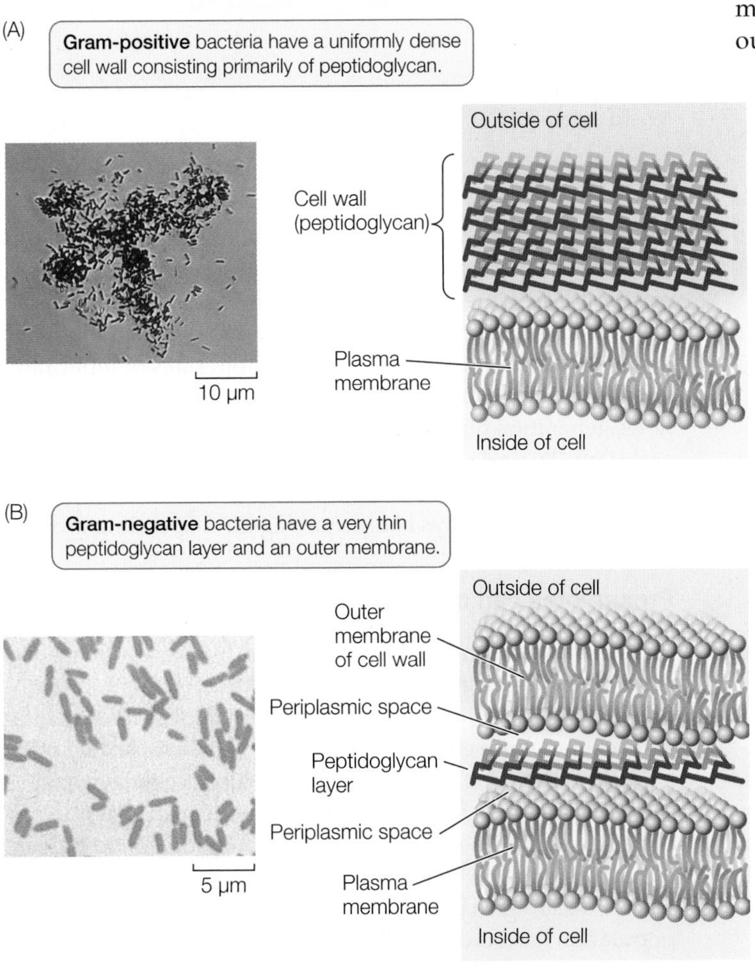

To appreciate the complexity of some bacterial cell walls, consider the reactions of bacteria to a simple staining process. A test called the **Gram stain** separates most types of bacteria into two distinct groups, Gram-positive and Gram-negative. A smear of cells on a microscope slide is soaked in a violet dye and treated with iodine; it is then washed with alcohol and counterstained with a red dye (safranine). **Gram-positive bacteria** retain the violet dye and appear blue to purple (**Figure 26.5A**). The alcohol washes the violet stain out of Gram-negative cells; these cells then pick up the safranine counterstain, so **Gram-negative bacteria** appear pink to red (**Figure 26.5B**).

For most bacteria, the Gram-staining results are determined by the chemical structure of the cell wall. A Gram-negative cell wall usually has a thin peptidoglycan layer, and outside the peptidoglycan layer the cell is surrounded by a second, outer membrane quite distinct in chemical makeup from the plasma membrane (see Figure 26.5B). Between the inner (plasma) and outer membranes of Gram-negative bacteria is a *periplasmic space*. This space contains proteins that are important in digesting some materials, transporting others, and detecting chemical gradients in the environment.

A Gram-positive cell wall usually has about five times as much peptidoglycan as a Gram-negative wall. This thick peptidoglycan layer is a meshwork that may serve some of the same purposes as the periplasmic space of the Gram-negative cell wall.

The consequences of the different features of prokaryotic cell walls are numerous and relate to the disease-causing characteristics of some bacteria. Indeed, the cell wall is a favorite target in medical combat against pathogenic bacteria because it has no counterpart in eukaryotic cells. Antibiotics such as penicillin and ampicillin, as well as other agents that specifically interfere with the synthesis of peptidoglycan-containing cell walls, tend to have little, if any, effect on the cells of humans and other eukaryotes.



**26.5 The Gram Stain and the Bacterial Cell Wall** When treated with Gram stain, the cell walls of different bacteria react in one of two ways. (A) Gram-positive bacteria have a thick peptidoglycan cell wall that retains the violet dye and appears deep blue or purple. (B) Gram-negative bacteria have a thin peptidoglycan layer that does not retain the violet dye but picks up the counterstain and appears pink to red.

**yourBioPortal.com**
GO TO **Web Activity 26.1 • Gram Stain and Bacteria**

### Prokaryotes have distinctive modes of locomotion

Although many prokaryotes cannot move, others are *motile*. These organisms move by one of several means. Some helical bacteria, called *spirochetes*, use a corkscrew-like motion made possible by modified flagella, called *axial filaments*, running along the axis of the cell beneath the outer membrane (**Figure 26.6A**). Many cyanobacteria and a few other groups of bacteria use various poorly understood gliding mechanisms, including rolling. Various aquatic prokaryotes, including some cyanobacteria, can move slowly up and down in the water by adjusting the amount of gas in *gas vesicles* (**Figure 26.6B**). By far the most common type of locomotion in prokaryotes, however, is that driven by flagella.

Prokaryotic **flagella** are slender filaments that extend singly or in tufts from one or both ends of the cell or are distributed all around it (**Figure 26.7**). A prokaryotic flagellum consists of a single fibril made of the protein *fla-*

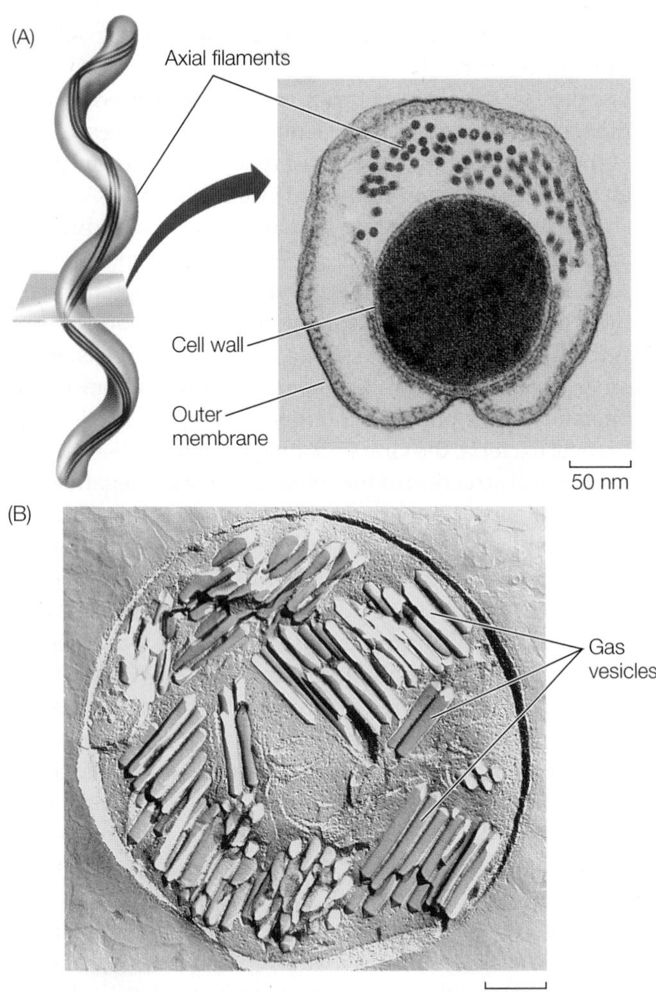


**26.6 Structures Associated with Prokaryote Motility** (A) A spirochete from the gut of a termite, seen in cross section, shows the axial filaments used to produce a corkscrew-like motion. (B) Gas vesicles in a cyanobacterium, visualized by the freeze-fracture technique.

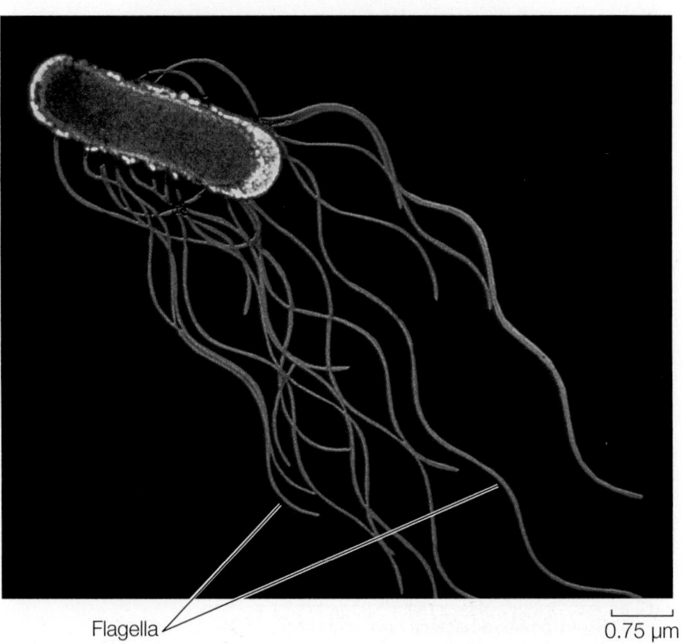


**26.7 Some Prokaryotes Use Flagella for Locomotion** Multiple flagella propel this *Salmonella* bacillus.

*gellin*, projecting from the cell surface, plus a hook and basal body responsible for motion (see Figure 5.5). In contrast, the flagellum of eukaryotes is enclosed by the plasma membrane and usually contains a circle of nine pairs of microtubules surrounding two central microtubules, all containing the protein *tubulin*, along with many other associated proteins. The prokaryotic flagellum rotates about its base, much like a propeller, rather than beating in a whiplike manner, as a eukaryotic flagellum or cilium does.

### Prokaryotes reproduce asexually, but genetic recombination can occur

Prokaryotes reproduce by binary fission, an asexual process (see Figure 11.2). Recall, however, that there are also processes—transformation, conjugation, and transduction—that allow the exchange of genetic information between some prokaryotes without reproduction occurring. So prokaryotes can exchange and recombine their DNA with other individuals (this is sex in the genetic sense of the word), but this genetic exchange is not directly linked to reproduction as it is in most eukaryotes.

If conditions are favorable, some prokaryotes can multiply very rapidly. The shortest known prokaryote generation times are about 10 minutes, although these rapid rates of replication usually are not maintained for long. Under less optimal conditions, generation times often extend to many hours or even several days. Bacteria living deep in Earth's crust may suspend their growth for more than a century without dividing, then multiply for a few days before once again suspending growth.

### Prokaryotes can communicate

Prokaryotes can send and receive signals from one another and from other organisms. One communication channel they employ is chemical. Another is physical, with light as the medium.

Bacteria release chemical substances that are sensed by other bacteria of the same species. They can announce their availability for conjugation, for example, by means of such signals. They can also monitor the density of their population. As the density of bacteria in a particular region increases, the concentration of a chemical signal builds up. When the bacteria sense that their population has become sufficiently dense, they can commence activities that smaller densities could not manage, such as forming a biofilm (see Figure 26.3). This density-sensing technique is called **quorum sensing**.

Like fireflies and many other organisms, some bacteria can emit light by a process called **bioluminescence**. A complex, enzyme-catalyzed reaction requiring ATP causes the emission of light but not heat. Often such bacteria luminesce only when a quorum has been sensed. The bioluminescent spots present in some deep-sea fishes are produced by colonies of biolumines-

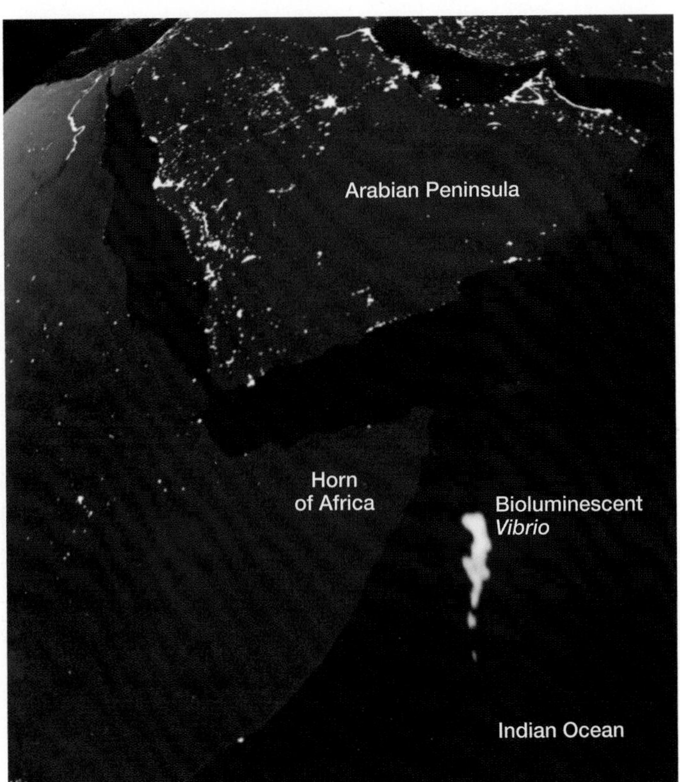


**26.8 Bioluminescent Bacteria Seen from Space** In this satellite photo, legions of bioluminescent *Vibrio harveyi* form a glowing patch thousands of square kilometers in area in the Indian Ocean, off the Horn of Africa. Compare their blue glow with the white light of cities in eastern Africa and the Middle East.

cent bacteria. On land, some soil-dwelling bioluminescent bacteria produce eerily glowing patches of ground at night.

How is bioluminescence useful to a prokaryote? One fairly well understood case is that of some bacteria of the genus *Vibrio*. These bacteria can live freely, but they truly thrive inside the guts of fish. Inside the fish, they may attach to food particles and then can be expelled as waste along with particulate matter. Reproducing on the particles, a bacteria population increases until a glowing particle attracts another fish, which ingests the bacteria along with the particle—giving the bacteria a new home and food source for a while. In this case, *Vibrio* are both communicating with another species and enhancing their own nutritional status. In the Indian Ocean off the eastern coast of Africa, *Vibrio* sometimes concentrate over such a large area (several thousand square kilometers) that their bioluminescence is visible from space (**Figure 26.8**).

### Prokaryotes have amazingly diverse metabolic pathways

Bacteria and archaea outdo the eukaryotes in terms of metabolic diversity. Although much more diverse in size and shape, eukaryotes draw on fewer metabolic mechanisms for their energy needs. In fact, much of the eukaryotes' energy metabolism is carried out in organelles—mitochondria and chloroplasts—that are endosymbiotic descendants of bacteria, as described in Section 5.5.

The long evolutionary history of bacteria and archaea, during which they have had time to explore a wide variety of habitats, has led to the extraordinary diversity of their metabolic "lifestyles"—their use or nonuse of oxygen, their energy sources, their sources of carbon atoms, and the materials they release as waste products.

**ANAEROBIC VERSUS AEROBIC METABOLISM** Some prokaryotes can live only by anaerobic metabolism because molecular oxygen is poisonous to them. These oxygen-sensitive organisms are called **obligate anaerobes**. Other prokaryotes can shift their metabolism between anaerobic and aerobic modes (see Chapter 9) and thus are called **facultative anaerobes**. Many facultative anaerobes alternate between anaerobic metabolism (such as fermentation) and cellular respiration as conditions dictate. **Aerotolerant anaerobes** cannot conduct cellular respiration but are not damaged by oxygen when it is present. By definition, an anaerobe does not use oxygen as an electron acceptor for its respiration.

At the other extreme from the obligate anaerobes, some prokaryotes are **obligate aerobes**, unable to survive for extended periods in the *absence* of oxygen. They require oxygen for cellular respiration.

**NUTRITIONAL CATEGORIES** All living organisms face the same nutritional challenges: they must synthesize energy-rich compounds such as ATP to power their life-sustaining metabolic reactions, and they must obtain carbon atoms to build their own organic molecules. Biologists recognize four broad nutritional categories of organisms: photoautotrophs, photoheterotrophs, chemolithotrophs, and chemoheterotrophs. Prokaryotes are represented in all four groups (**Table 26.2**).

**Photoautotrophs** perform photosynthesis. They use light as their energy source and carbon dioxide ($CO_2$) as their carbon source. Like green plants and other photosynthetic eukaryotes, the cyanobacteria, a group of photoautotrophic bacteria, use chlorophyll *a* as their key photosynthetic pigment and produce

TABLE 26.2

**How Organisms Obtain Their Energy and Carbon**

| NUTRITIONAL CATEGORY | ENERGY SOURCE | CARBON SOURCE |
|---|---|---|
| Photoautotrophs (found in all three domains) | Light | Carbon dioxide |
| Photoheterotrophs (some bacteria) | Light | Organic compounds |
| Chemolithotrophs (some bacteria, many archaea) | Inorganic substances | Carbon dioxide |
| Chemoheterotrophs (found in all three domains) | Organic compounds | Organic compounds |

oxygen gas ($O_2$) as a by-product of noncyclic electron transport (see Section 10.1).

There are other photosynthetic groups among the bacteria, but these use *bacteriochlorophyll* as their key photosynthetic pigment, and they do not release $O_2$. Indeed, some of these photosynthesizers produce particles of pure sulfur, because hydrogen sulfide ($H_2S$) rather than $H_2O$ is their electron donor for photophosphorylation (see Section 10.2). Bacteriochlorophyll molecules absorb light of longer wavelengths than the chlorophyll molecules used by all other photosynthesizing organisms. As a result, bacteria using this pigment can grow in water under fairly dense layers of algae, using light of wavelengths that are not absorbed by the algae (**Figure 26.9**).

**Photoheterotrophs** use light as their energy source but must obtain their carbon atoms from organic compounds made by other organisms. Their "food" consists of organic compounds such as carbohydrates, fatty acids, and alcohols. For example, compounds released from plant roots (as in rice paddies) or from decomposing photosynthetic bacteria in hot springs are taken up by photoheterotrophs and metabolized to form building blocks for other compounds; sunlight provides the necessary ATP through photophosphorylation. The purple nonsulfur bacteria, among others, are photoheterotrophs.

**Chemolithotrophs** (also called chemoautotrophs) obtain their energy by oxidizing inorganic substances, and they use some of that energy to fix $CO_2$. Some chemolithotrophs use reactions identical to those of the typical photosynthetic cycle, but others use alternative pathways to fix $CO_2$. Some bacteria oxidize ammonia or nitrite ions to form nitrate ions. Others oxidize hydrogen gas, hydrogen sulfide, sulfur, and other materials. Many archaea are chemolithotrophs.

Deep-sea hydrothermal vent ecosystems are dependent on chemolithotrophic prokaryotes that are incorporated into large communities of crabs, mollusks, and giant worms, all living at a depth of 2,500 meters—below any hint of sunlight. These bacteria obtain energy by oxidizing hydrogen sulfide and other substances released in the near-boiling water flowing from volcanic vents in the ocean floor.

Finally, **chemoheterotrophs** obtain both energy and carbon atoms from one or more complex organic compounds that have been synthesized by other organisms. Most known bacteria and archaea are chemoheterotrophs—as are all animals and fungi and many protists.

**NITROGEN AND SULFUR METABOLISM** Key metabolic reactions in many prokaryotes involve nitrogen or sulfur. For example, some bacteria carry out respiratory electron transport without using oxygen as an electron acceptor. These organisms use oxidized inorganic ions such as nitrate, nitrite, or sulfate as electron acceptors. Examples include the **denitrifiers**, bacteria that release nitrogen to the atmosphere as nitrogen gas ($N_2$). These normally aerobic bacteria, mostly species of the genera *Bacillus* and *Pseudomonas*, use nitrate ($NO_3^-$) as an electron acceptor in place of oxygen if they are kept under anaerobic conditions:

$$2\ NO_3^- + 10\ e^- + 12\ H^+ \rightarrow N_2 + 6\ H_2O$$

**Nitrogen fixers** convert atmospheric nitrogen gas into a chemical form (ammonia) usable by the nitrogen fixers themselves as well as by other organisms, especially land plants:

$$N_2 + 6\ H \rightarrow 2\ NH_3$$

All organisms require nitrogen in order to build proteins, nucleic acids, and other important compounds. Nitrogen fixation is thus vital to life as we know it. This all-important biochemical process is carried out by a wide variety of archaea and bacteria (including cyanobacteria) but by no other organisms, so we depend on these prokaryotes for our very existence. We describe the details of nitrogen fixation in Chapter 36.

Ammonia is oxidized to nitrate in soil and in seawater by chemolithotrophic bacteria called **nitrifiers**. Bacteria of two genera, *Nitrosomonas* and *Nitrosococcus*, convert ammonia to nitrite ions ($NO_2^-$), and *Nitrobacter* oxidizes nitrite to nitrate ($NO_3^-$).

What do the nitrifiers get out of these reactions? Their metabolism is powered by the energy released by the oxidation of ammonia or nitrite. For example, by passing the electrons from nitrite through an electron transport chain (see Section 9.3), *Nitrobacter* can make ATP, and using some of this ATP, can also make NADH. With this ATP and NADH, the bacterium can convert $CO_2$ and $H_2O$ to glucose.

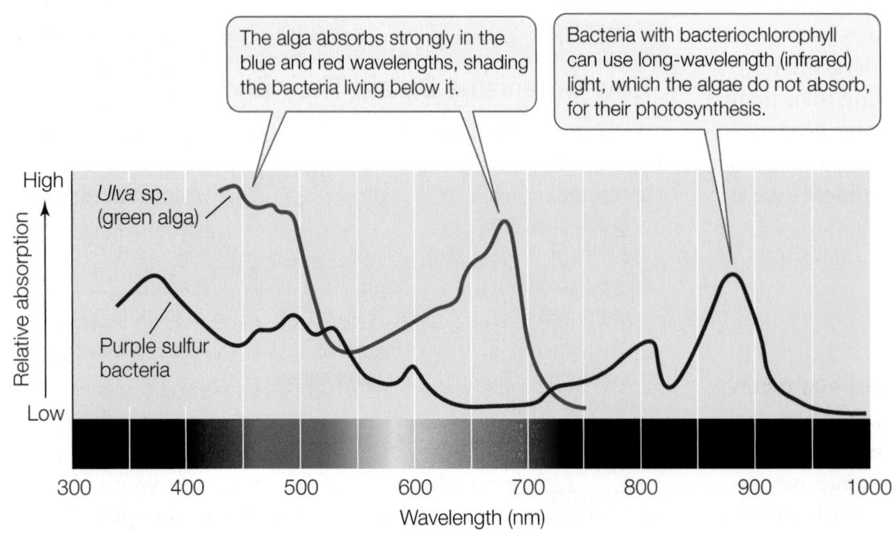
The alga absorbs strongly in the blue and red wavelengths, shading the bacteria living below it.
Bacteria with bacteriochlorophyll can use long-wavelength (infrared) light, which the algae do not absorb, for their photosynthesis.
High
Low
Relative absorption
Ulva sp. (green alga)
Purple sulfur bacteria
300
400
500
600
700
800
900
1000
Wavelength (nm)

**26.9 Bacteriochlorophyll Absorbs Long-Wavelength Light** The chlorophyll in *Ulva*, a green alga, absorbs no light of wavelengths longer than 750 nm. Purple sulfur bacteria, which contain bacteriochlorophyll, can conduct photosynthesis using longer wavelengths.

## 26.2 RECAP

**Prokaryotes have established themselves everywhere on Earth. They may form communities called biofilms that coat materials with a gel-like matrix. Prokaryotes have distinctive cell walls and modes of locomotion, communication, reproduction, and nutrition.**

- How do biofilms form and why are they of special interest to researchers? See pp. 539–540 and Figure 26.3
- Describe bacterial cell wall architecture. See p. 541 and Figure 26.5
- How are the four nutritional categories of prokaryotes distinguished? See pp. 543–544 and Table 26.2
- Explain why nitrogen metabolism in the prokaryotes is vital to other organisms. See p. 544

We noted earlier that only recently have scientists appreciated the huge distinctions between Bacteria and Archaea. How do researchers approach the classification of organisms they can't even see?

# 26.3 How Can We Resolve Prokaryote Phylogeny?

As detailed in Chapter 22, classification schemes serve three primary purposes: to identify organisms, to reveal evolutionary relationships, and to provide universal names. Classifying bacteria and archaea is of particular importance to humans because scientists and medical technologists must be able to identify bacteria quickly and accurately; when the bacteria are pathogenic, lives may depend on it. In addition, many emerging biotechnologies (see Chapter 18) depend on a thorough knowledge of prokaryote biochemistry, and understanding an organism's phylogeny allows biologists to make predictions about the distribution of biochemical processes across the wide diversity of prokaryotes.

### The small size of prokaryotes has hindered our study of their phylogeny

Until about 300 years ago, nobody had even *seen* an individual prokaryote; these organisms remained invisible to humans until the invention of the first simple microscope. Prokaryotes are so small that even the best light microscopes don't reveal much about them. It took the advanced microscopic equipment and techniques of the twentieth century (see Figure 5.3) to open up the microbial world.

Until recently, taxonomists based prokaryote classification on observable phenotypic characters such as shape, color, motility, nutritional requirements, antibiotic sensitivity, and reaction to the Gram stain. When biologists learned how to grow bacteria in pure culture on nutrient media, they learned a great deal about the genetics, nutrition, and metabolism of those species that could be cultured. However, these species represent probably less than 1 percent of living prokaryote species. Furthermore, this work provided little insight into how prokaryotic organisms evolved—a question of great interest to microbiologists and evolutionary biologists. Only recently have systematists developed the appropriate tools to produce classification schemes that make sense in evolutionary terms.

### The nucleotide sequences of prokaryotes reveal their evolutionary relationships

Analyses of nucleotide sequences of ribosomal RNA (rRNA) genes provided the first comprehensive evidence of evolutionary relationships among prokaryotes. For several reasons, rRNA is particularly useful for evolutionary studies of living organisms:

- rRNA is evolutionarily ancient, as it was found in the common ancestor of life.
- No free-living organism lacks rRNA, so rRNA genes can be compared throughout the tree of life.
- rRNA plays a critical role in translation in all organisms, so *lateral transfer* of rRNA genes among distantly related species is unlikely.
- rRNA has evolved slowly enough that gene sequences can be aligned and analyzed among even distantly related species.

Comparisons of rRNA genes from a great many organisms have revealed the probable phylogenetic relationships from throughout the tree of life. Databases such as GenBank contain rRNA gene sequences from hundreds of thousands of species—more than any other gene sequences.

Although these data are helpful, it is clear that even distantly related prokaryotes sometimes exchange genetic material. In some groups of prokaryotes, analyses of multiple gene sequences have suggested several different phylogenetic patterns. How could such differences among different gene sequences arise?

### Lateral gene transfer can lead to discordant gene trees

As noted earlier, prokaryotes reproduce by binary fission. If we could follow these divisions back through evolutionary time, we would be tracing the path of the complete tree of life for bacteria and archaea. This underlying tree of relationships, represented in highly abbreviated form in Appendix A, is called the *organismal* (or *species*) *tree*. Because whole genomes are replicated during asexual binary fission divisions, we expect phylogenetic trees constructed from most gene sequences to reflect these same relationships (see Chapter 22).

From early in evolution to the present day, however, some genes have been moving "sideways" from one prokaryotic species to another, a phenomenon known as **lateral gene transfer**. Mechanisms of lateral gene transfer include transfer by plasmids and viruses and uptake of DNA from the environment by transformation. Lateral gene transfers are well documented, especially among closely related species; some have been documented even across the three primary domains of life.

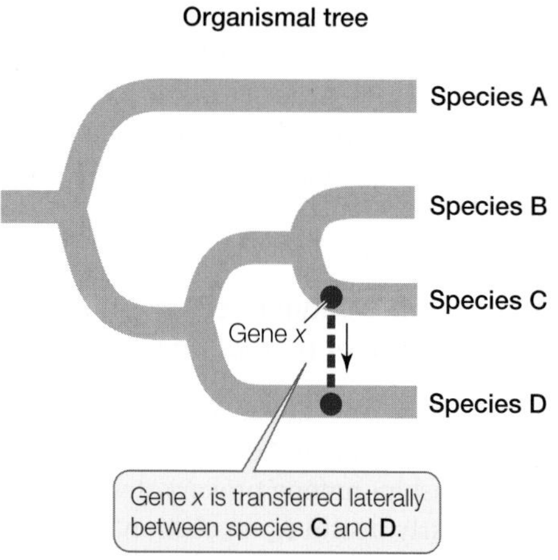
Organismal tree
Species A
Species B
Species C
Gene x
Species D
Gene x is transferred laterally between species C and D.

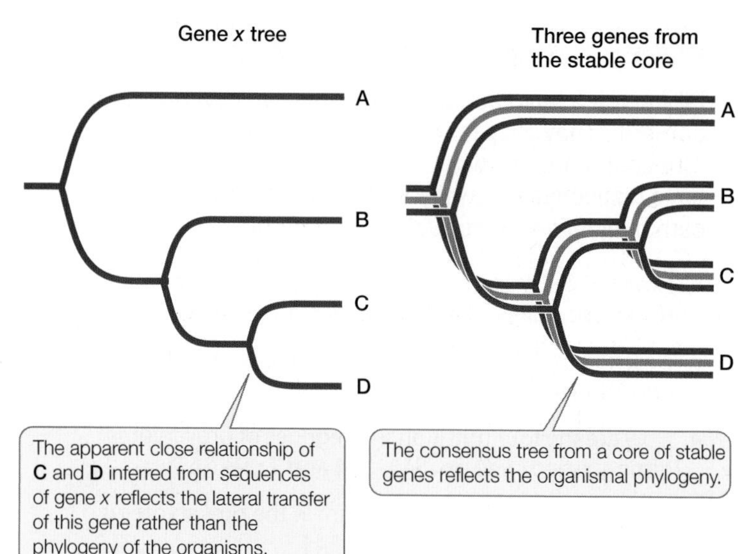
Gene x tree
A
B
C
D
The apparent close relationship of C and D inferred from sequences of gene x reflects the lateral transfer of this gene rather than the phylogeny of the organisms.
Three genes from the stable core
A
B
C
D
The consensus tree from a core of stable genes reflects the organismal phylogeny.

**26.10 Lateral Gene Transfer Complicates Phylogenetic Relationships** (A) The phylogeny of four hypothetical species, with a lateral gene transfer of gene *x*. (B) A tree based only on gene *x* shows the phylogeny of the laterally transferred gene, rather than the organismal phylogeny. (C) In many cases, a "stable core" of prokaryote genes can be used to reconstruct the organismal phylogeny of prokaryotes.

Consider, for example, the genome of *Thermotoga maritima*, a bacterium that can survive extremely high temperatures. In comparing the 1,869 gene sequences of *T. maritima* against sequences for the same proteins in other species, investigators found that some of this bacterium's genes have their closest relationships not with those of other bacterial species, but with the genes of archaeal species that live in similar environments.

When genes involved in lateral transfer events are sequenced and analyzed phylogenetically, the resulting individual gene trees will not match the organismal phylogeny in every respect (**Figure 26.10**). Individual gene trees will vary because the history of lateral gene transfer events is different for each gene. Biologists reconstruct the underlying organismal phylogeny by comparing multiple genes (to produce a *consensus tree*), or by concentrating on genes that are unlikely to be involved in lateral gene transfer events. For example, genes that are involved in fundamental cell processes (such as the rRNA genes discussed above) are unlikely to be replaced by the same genes from other species, since functional, locally adapted copies of these genes are already present.

What kinds of genes are most likely to be involved in lateral gene transfer? Genes that result in a new, adaptive function that will convey higher fitness to a recipient species are most likely to be transferred repeatedly among species. For example, genes that produce antibiotic resistance are often transferred on plasmids among many bacterial species, especially under the strong selective conditions of antibiotic medication by humans. This selection for antibiotic resistance is why informed physicians are now more careful in prescribing antibiotics. Improper or frequent use of antibiotics can lead to selection for resistant strains of bacteria, which are then much harder to treat effectively.

It is debatable whether lateral gene transfer has seriously complicated our attempts to resolve the tree of prokaryotic life. Recent work suggests that it has not—while it complicates studies in some individual species, it need not present problems at higher levels. It is now possible to make nucleotide sequence comparisons involving entire genomes, and these studies are revealing a *stable core* of crucial genes that are uncomplicated by lateral gene transfer. Gene trees based on this stable core more accurately reveal relationships of the organismal phylogeny (see Figure 26.10). The problem remains, however, that only a very small proportion of the prokaryotic world has been described and studied.

### The great majority of prokaryote species have never been studied

Most prokaryotes have defied all attempts to grow them in pure culture, causing biologists to wonder how many species, and possibly even important clades, we might be missing. A window onto this problem was opened with the introduction of a new way to look at nucleic acid sequences. Unable to work with the whole genome of a single species, biologists instead examine sequences in individual genes collected from a random sample of the environment.

Norman Pace of the University of Colorado isolated individual rRNA gene sequences from extracts of environmental samples such as soil and seawater. Comparing such sequences with previously known ones revealed an extraordinary number of new sequences, implying that they came from previously unrecognized species. Biologists have described only about 10,000 species of bacteria and only a few hundred species of archaea (see Figure 1.10). The results of Pace's and similar studies suggest that there may be millions, perhaps hundreds of millions, of prokaryote species on Earth. Other biologists put the estimate much lower, and argue that the high dispersal ability of many bacterial species greatly reduces local endemism (geographically restricted species). Only the magnitude of these estimates differ, however; all sides agree that we have just begun to uncover Earth's bacterial and archaeal diversity.

## 26.3 RECAP

**The study of prokaryote phylogeny and diversity has been inhibited by the organisms' small size, our inability to grow some of them in pure culture, and lateral gene transfer. However, nucleotide sequences of essential genes are providing a much clearer picture of bacterial and archaeal evolutionary relationships.**

- How did biologists classify bacteria before it became possible to determine nucleotide sequences? See p. 545
- Explain why nucleotide sequences of rRNA genes are useful for evolutionary studies. See p. 545
- How does lateral gene transfer complicate evolutionary studies? See p. 545–546 and Figure 26.10

With the advent of sequencing techniques, biologists have made rapid progress in understanding the phylogeny of prokaryotes. In the next section, we identify the characteristics and life history of the major groups.

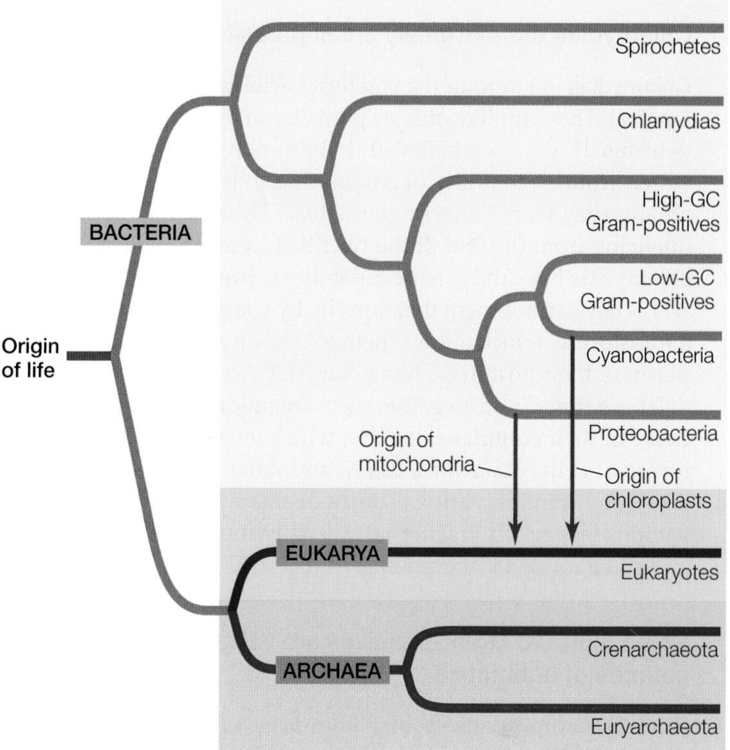


**26.11 Two Domains: A Brief Overview** This abridged summary classification of the domains Bacteria and Archaea shows their relationships to each other and to Eukarya. The relationships among the many clades of bacteria, not all of which are listed here, are incompletely resolved at this time.

## 26.4 What Are the Major Known Groups of Prokaryotes?

Here we use a widely accepted classification scheme that has considerable support from nucleotide sequence data. More than a dozen major clades have been proposed under this scheme, just a few of which we discuss here. We pay the closest attention to six groups that have received the most study: the spirochetes, chlamydias, high-GC Gram-positives, cyanobacteria, low-GC Gram-positives, and proteobacteria (**Figure 26.11**). First, however, a few words about the origins of the prokaryotes are in order.

Several of the earliest branching lineages of bacteria and archaea are **thermophiles** (Greek, "heat-lovers"). This observation is in line with the hypothesis that the first living organisms were thermophiles, given that most environments on early Earth were much hotter than those of today. While additional evidence continues to support this hypothesis, some researchers believe that the various thermophilic groups evolved more recently than did the lineages leading to the spirochetes and chlamydias.

### Spirochetes move by means of axial filaments

**Spirochetes** are Gram-negative, motile, chemoheterotrophic bacteria characterized by unique structures called axial filaments, which are modified flagella running through the periplasmic space (see Figure 26.6A). The cell body is a long cylinder coiled into a helix (**Figure 26.12**). The axial filaments begin at either end of the cell and overlap in the middle. Protein motors connect the axial filaments to the cell wall, enabling rotation of these structures as they do in other prokaryotic flagella. Many spirochetes live in humans as parasites; a few are pathogens, including those that cause syphilis and Lyme disease. Others live free in mud or water.

**26.12 A Spirochete** This corkscrew-shaped bacterium causes syphilis in humans.

### Chlamydias are extremely small parasites

**Chlamydias** are among the smallest bacteria (0.2–1.5 μm in diameter). They can live only as parasites in the cells of other organisms. It was once believed that this obligate parasitism resulted from an inability of chlamydias to produce ATP—that chlamydias were "energy parasites." However, genome sequencing from the end of the twentieth century indicates that chlamydias have the genetic capability to produce at least some ATP. They can augment this capacity by using an enzyme called a translocase, which allows them to take up ATP from the cytoplasm of their host in exchange for ADP from their own cells.

These tiny, Gram-negative cocci are unique prokaryotes because of their complex life cycle, which involves two different forms of cells, *elementary bodies* and *reticulate bodies* (**Figure 26.13**). In humans, various strains of chlamydias cause eye infections (especially trachoma), sexually transmitted diseases, and some forms of pneumonia.

### Some high-GC Gram-positives are valuable sources of antibiotics

**High-GC Gram-positives**, also known as *actinobacteria*, derive their name from the relatively high ratio of G-C to A-T nucleotide base pairs in their DNA. These bacteria develop an elaborately branched system of filaments (**Figure 26.14**) and can resemble the filamentous growth habit of fungi, albeit at a reduced scale. Some high-GC Gram-positives reproduce by forming chains of spores at the tips of the filaments. In species that do not form spores, the branched, filamentous growth ceases, and the structure breaks up into typical cocci or bacilli, which then reproduce by binary fission.

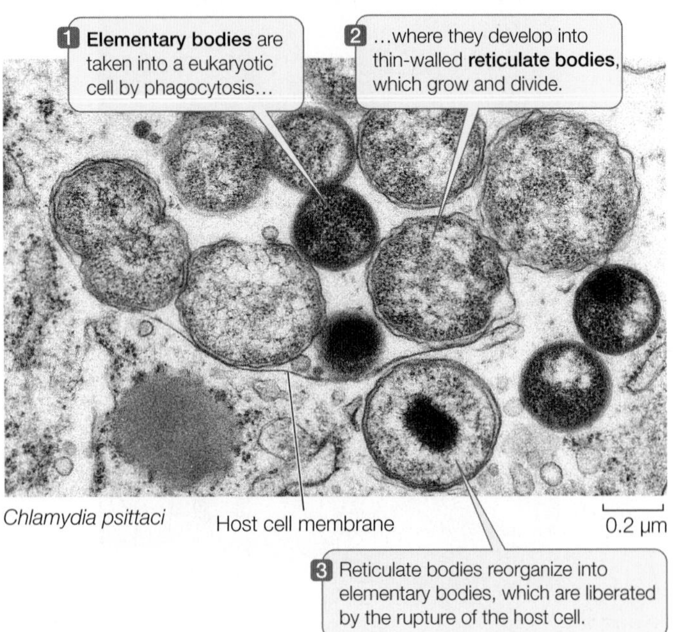

1 **Elementary bodies** are taken into a eukaryotic cell by phagocytosis...
2 ...where they develop into thin-walled **reticulate bodies**, which grow and divide.
Chlamydia psittaci
Host cell membrane
0.2 μm
3 Reticulate bodies reorganize into elementary bodies, which are liberated by the rupture of the host cell.

**26.13 Chlamydias Change Form during their Life Cycle** Elementary bodies and reticulate bodies are the two major phases of the chlamydia life cycle.

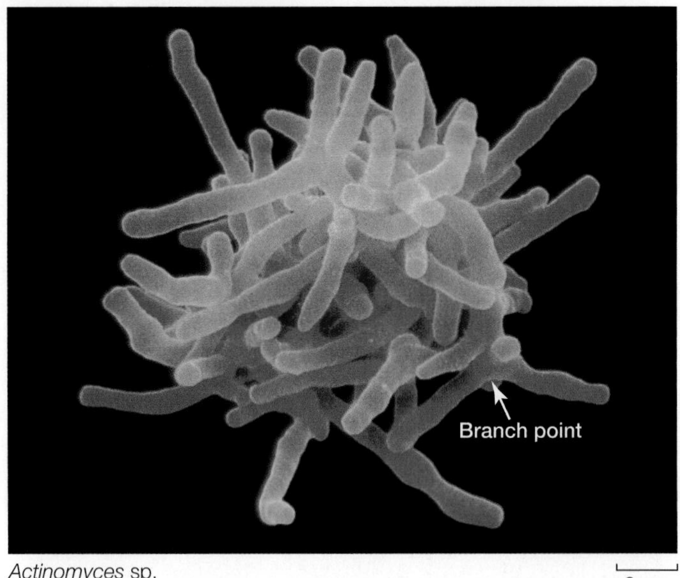


**26.14 Filaments of a High-GC Gram-Positive** The branching filaments seen in this scanning electron micrograph are typical of this medically important bacterial group.

The high-GC Gram-positives include several medically important bacteria. *Mycobacterium tuberculosis* causes tuberculosis, which kills 3 million people each year. Genetic data suggest that this bacterium arose 3 million years ago in East Africa, making it the oldest known human bacterial affliction. *Streptomyces* produce streptomycin as well as hundreds of other antibiotics. We derive most of our antibiotics from members of the high-GC Gram-positives.

### Cyanobacteria are important photoautotrophs

**Cyanobacteria**, sometimes called *blue-green bacteria* because of their pigmentation, are photoautotrophs that require only water, nitrogen gas, oxygen, a few mineral elements, light, and carbon dioxide to survive. They use chlorophyll *a* for photosynthesis and release oxygen gas; many species also fix nitrogen. Their photosynthesis was the basis of the "oxygen revolution" that transformed Earth's atmosphere (see Section 25.3).

Cyanobacteria carry out the same type of photosynthesis that is characteristic of eukaryotic photosynthesizers. They contain elaborate and highly organized internal membrane systems called *photosynthetic lamellae*. The chloroplasts of photosynthetic eukaryotes are derived from an endosymbiotic cyanobacterium.

Cyanobacteria may live free as single cells or associate in colonies. Depending on the species and on growth conditions, colonies may range from flat sheets one cell thick to filaments to spherical balls of cells.

Some filamentous colonies of cyanobacteria differentiate into three cell types: vegetative cells, spores, and heterocysts (**Figure 26.15**). **Vegetative cells** photosynthesize, **spores** are resting stages that can survive harsh environmental conditions and eventually develop into new filaments, and **heterocysts** are cells specialized for nitrogen fixation. All of the known cyanobacteria with heterocysts fix nitrogen. Heterocysts also have a role in reproduction: when filaments break apart to reproduce, the heterocyst may serve as a breaking point.

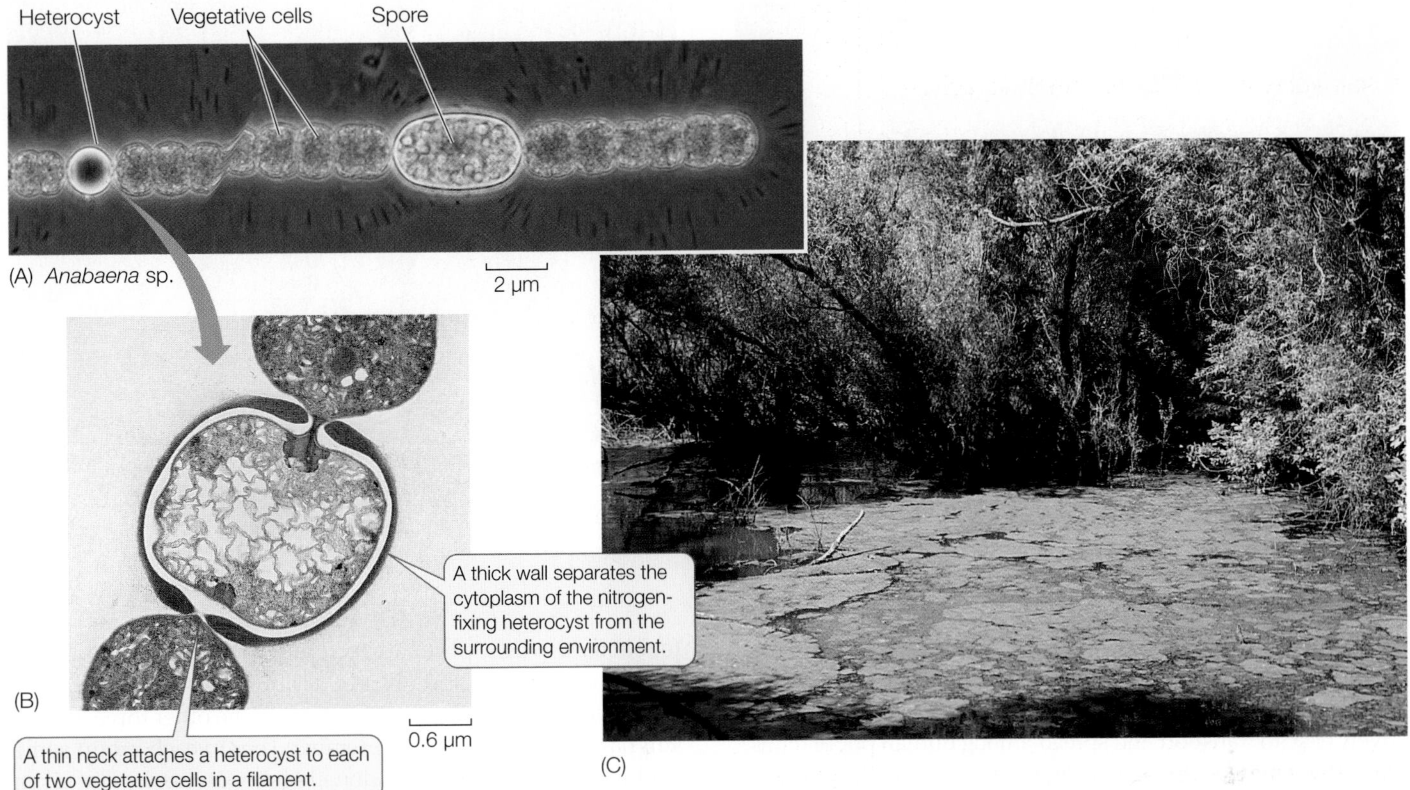
Heterocyst
Vegetative cells
Spore
(A) Anabaena sp.
2 μm
(B)
A thick wall separates the cytoplasm of the nitrogen-fixing heterocyst from the surrounding environment.
A thin neck attaches a heterocyst to each of two vegetative cells in a filament.
0.6 μm
(C)

**26.15 Cyanobacteria** (A) *Anabaena* is a genus of cyanobacteria that form filamentous colonies containing three cell types. (B) Heterocysts are specialized for nitrogen fixation and serve as a breaking point when filaments reproduce. (C) Cyanobacteria appear in enormous numbers in some environments. This California pond has experienced eutrophication: phosphorus and other nutrients generated by human activity have accumulated, feeding an immense green mat (commonly referred to as "pond scum") that is made up of several species of free-living cyanobacteria.

### The low-GC Gram-positives include the smallest cellular organisms

As their name suggests, the **low-GC Gram-positives** have a lower ratio of G-C to A-T nucleotide base pairs than do the high-GC Gram-positives. Some of the low-GC Gram-positives are in fact Gram-negative, and some have no cell wall at all. Despite these differences among the various species, phylogenetic analyses of DNA sequences support the monophyly of this clade.

Some low-GC Gram-positives can produce heat-resistant resting structures called **endospores** (**Figure 26.16**). When a key nutrient such as nitrogen or carbon becomes scarce, the bacterium replicates its DNA and encapsulates one copy, along with some of its cytoplasm, in a tough cell wall heavily thickened with peptidoglycan and surrounded by a spore coat. The parent cell then breaks down, releasing the endospore. Endospore production is not a reproductive process; the endospore merely replaces the parent cell. The endospore, however, can survive harsh environmental conditions that would kill the parent cell, such as high or low temperatures or drought, because it is *dormant*—its normal activity is suspended. Later, if it encounters favorable conditions, the endospore becomes metabolically active and divides, forming new cells that are like the parent cells.

Some endospores can be reactivated after more than 1,000 years of dormancy. There are even credible claims of reactivation of *Bacillus* endospores after millions of years. Members of this endospore-forming group of low-GC Gram-positives include the many species of *Clostridium* and *Bacillus*.

Dormant endospores of *Bacillus anthracis* are the source of an *exotoxin* (see page 555) that causes anthrax. The spores germinate when they sense specific molecules (macrophages; see Chapter 42) in the cytoplasm of mammalian blood cells. However, endospores of other, nonpathogenic *Bacillus* species do not germi-

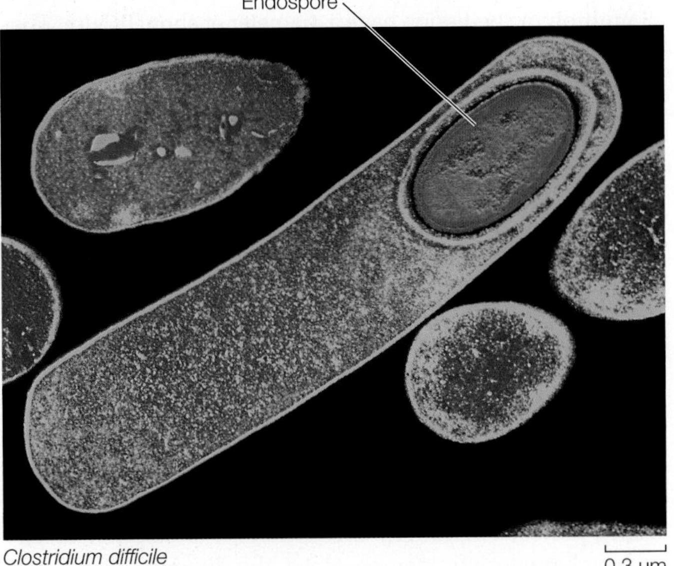
Endospore
Clostridium difficile
0.3 μm

**26.16 A Structure for Waiting Out Bad Times** This low-GC Gram-positive bacterium, which can cause severe colitis in humans, produces endospores as resistant resting structures.

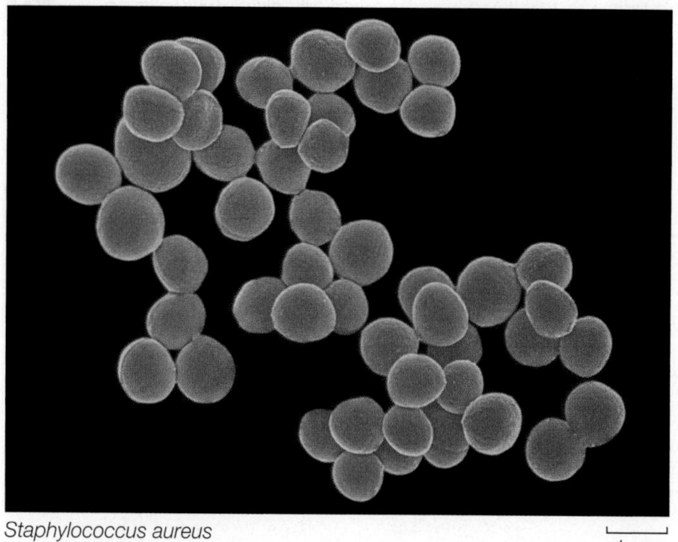


**26.17 Low-GC Gram-Positives** "Grape clusters" are the usual arrangement of staphylococci.

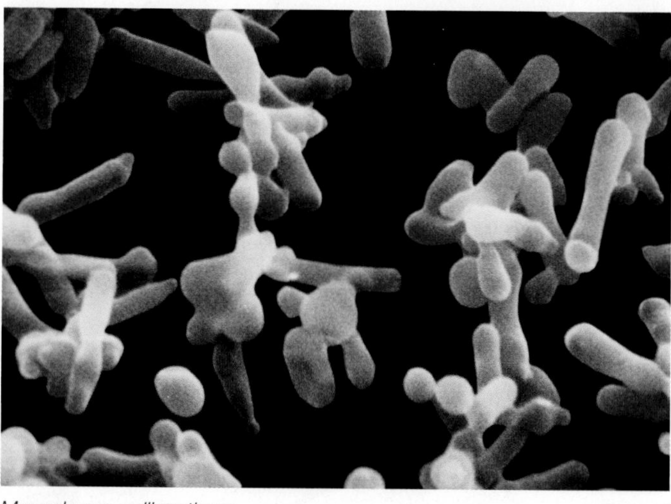


**26.18 The Tiniest Cells** Containing only about one-fifth as much DNA as *E. coli*, mycoplasmas are the smallest known bacteria.

nate in this environment. *B. anthracis* has been used as a bioterrorism agent because large quantities of its endospores are relatively easy to transport and spread among human populations, where they may be inhaled or ingested.

The genus *Staphylococcus*—the **staphylococci**—includes low-GC Gram-positives that are abundant on the human body surface; they are responsible for boils and many other skin problems (**Figure 26.17**). *Staphylococcus aureus* is the best-known human pathogen in this genus; it is found in 20 to 40 percent of normal adults (and in 50 to 70 percent of hospitalized adults). In addition to skin diseases, it can cause respiratory, intestinal, and wound infections.

Another interesting group of low-GC Gram-positives, the **mycoplasmas**, lack cell walls, although some have a stiffening material outside the plasma membrane. The mycoplasmas include the smallest cellular creatures known (**Figure 26.18**)—even smaller than chlamydias. The smallest mycoplasmas capable of multiplying by fission have a diameter of about 0.2 μm. They are small in another crucial sense as well: they have less than half as much DNA as most other prokaryotes. It has been speculated that the amount of DNA in a mycoplasma, which codes for fewer than 500 proteins, may be the minimum amount required to encode the essential properties of a living cell.

### The proteobacteria are a large and diverse group

By far the largest group of bacteria, in terms of number of described species, is the **proteobacteria**. The proteobacteria include many species of Gram-negative, bacteriochlorophyll-containing, sulfur-using photoautotrophs—as well as dramatically diverse bacteria that bear no phenotypic resemblance to the photoautotrophic species. Genetic and morphological evidence indicates that the mitochondria of eukaryotes were derived from a proteobacterium by endosymbiosis (see Section 27.2).

No characteristic demonstrates the diversity of the proteobacteria more clearly than their metabolic pathways (**Figure 26.19**). There are five groups of proteobacteria: alpha, beta, gamma, delta, and epsilon. The common ancestor of all the proteobacteria was a photoautotroph. Early in evolution, two groups of proteobacteria lost their ability to photosynthesize and have been chemotrophs ever since. The other three groups still have photoautotrophic members, but in each group some evolutionary lines have abandoned photoautotrophy and taken up other modes of nutrition. There are chemolithotrophs and chemoheterotrophs in all three groups. Why? One possibility is that each of the trends shown in Figure 26.19 was an evolution-

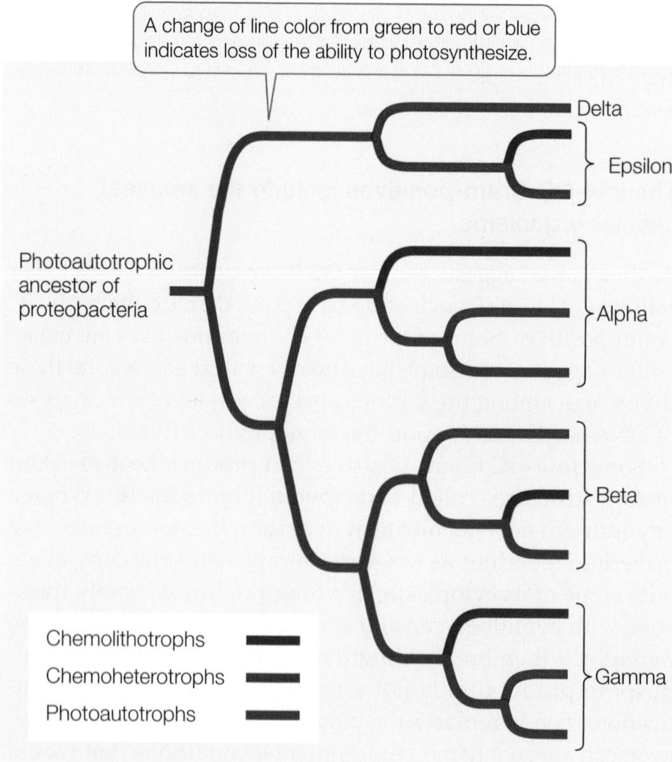


**26.19 Modes of Nutrition in the Proteobacteria** The common ancestor of all proteobacteria was probably a photoautotroph. As they encountered new environments, the delta and epsilon proteobacteria lost the ability to photosynthesize. In the other three groups, some evolutionary lineages became chemolithotrophs or chemoheterotrophs.

ary response to selective pressures encountered as these bacteria colonized new habitats that presented new challenges and opportunities. Lateral gene transfer may have played a role in these responses.

Among the proteobacteria are some nitrogen-fixing genera, such as *Rhizobium* (see Figure 36.9), and other bacteria that contribute to the global nitrogen and sulfur cycles. *Escherichia coli*, one of the most studied organisms on Earth, is a proteobacterium. So, too, are many of the most famous human pathogens, such as *Yersinia pestis* (which causes bubonic plague), *Vibrio cholerae* (cholera), and *Salmonella typhimurium* (gastrointestinal disease).

Although fungi cause most plant diseases, and viruses cause others, about 200 known plant diseases are of bacterial origin. *Crown gall*, with its characteristic tumors (**Figure 26.20**), is one of the most striking. The causal agent of crown gall is *Agrobacterium tumefaciens*, a proteobacterium that harbors a plasmid used in recombinant DNA studies as a vehicle for inserting genes into new plant hosts (see Section 18.2).

We have discussed six clades of bacteria in some detail. Other bacterial clades are well known, and there are probably dozens more waiting to be discovered. This estimate is conservative because so few bacteria have been cultured and studied in the laboratory.

### Archaea differ in several important ways from bacteria

The separation of Archaea from Bacteria and Eukarya was originally based on phylogenetic relationships determined from sequences of rRNA genes. This conclusion was supported when biologists sequenced the first archaeal genome. It consisted of 1,738 genes, more than half of which were unlike any genes ever found in the other two domains. Archaea are well known for living in extreme habitats such as those with high salinity (salt content), low oxygen concentrations, high temperatures, or high or low pH (**Figure 26.21**). However, many archaea are not extremeophiles but live in moderate habitats; they are common in

**26.20 Crown Gall** A crown gall, the type of tumor shown here growing on the stem of a bushy shrub, is caused by the proteobacterium *Agrobacterium tumefaciens*.

## INVESTIGATING LIFE

**26.21 What Is the Highest Temperature Compatible with Life?**

Can any organism thrive at temperatures above 120°C? This is the temperature used for sterilization, known to destroy all previously described organisms. Kazem Kashefi and Derek Lovley isolated an unidentified prokaryote from water samples taken near a hydrothermal vent and found it survived and even grew at 121°C. The organism was dubbed "Strain 121," and its gene sequencing results indicate that it is an archaeal species.

**HYPOTHESIS** **Some prokaryotes survive and even multiply at temperatures above the 120°C threshold of sterilization.**

**METHOD**

1. Seal samples of unidentified, iron-reducing, thermal vent prokaryotes in tubes with a medium containing $Fe^{3+}$ as an electron acceptor. Control tubes contain $Fe^{3+}$ but no organisms.
2. Hold both tubes in a sterilizer at 121°C for 10 hours. If the iron-reducing organisms are metabolically active, they will reduce the $Fe^{3+}$ to $Fe^{2+}$ (as magnetite, which can be detected with a magnet).
3. Isolate any surviving organisms and test for growth at various temperatures.

**RESULTS**

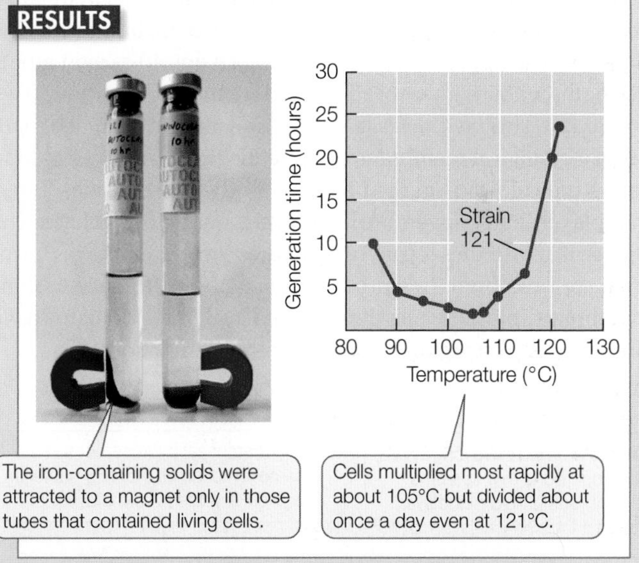
Generation time (hours)
30
25
20
15
10
5
80
90
100
110
120
130
Temperature (°C)
Strain 121
The iron-containing solids were attracted to a magnet only in those tubes that contained living cells.
Cells multiplied most rapidly at about 105°C but divided about once a day even at 121°C.

**CONCLUSION** **Some prokaryotic organisms can survive and grow at temperatures above the previously defined sterilization limit.**

**FURTHER INVESTIGATION:** Note that Strain 121 did not grow during a 2-hour exposure to 130°C, but it did not die, either. How would you demonstrate that it was still alive?

Go to **yourBioPortal.com** for original citations, discussions, and relevant links for all INVESTIGATING LIFE figures.

soil, for example. Perhaps the largest number of archaea live in the ocean depths.

One current classification scheme divides Archaea into two principal groups, **Euryarchaeota** and **Crenarchaeota**. Less is known about two more recently discovered groups, **Korarchaeota** and **Nanoarchaeota**. In fact, we know relatively little about the phylogeny of archaea, in part because the study of archaea is still in its early stages.

Two characteristics shared by all archaea are the absence of peptidoglycan in their cell walls and the presence of lipids of distinctive composition in their cell membranes (see Table 26.1). The unusual lipids in the membranes of archaea are found in all archaea and in no bacteria or eukaryotes. Most bacterial and eukaryotic membrane lipids contain unbranched long-chain fatty acids connected to glycerol molecules by *ester linkages*:

```
 O     H
 ‖     |
—C—O—C—
       |
       H
```

In contrast, some archaeal membrane lipids contain long-chain hydrocarbons connected to glycerol molecules by *ether linkages*:

```
 H     H
 |     |
—C—O—C—
 |     |
 H     H
```

These ether linkages are a synapomorphy of archaea. In addition, the long-chain hydrocarbons of archaea are branched. One class of these lipids, with hydrocarbon chains 40 carbon atoms in length, contains glycerol at *both* ends of the hydrocarbons (**Figure 26.22**). This *lipid monolayer* structure, unique to archaea, still fits in a biological membrane because the lipids are twice as long as the typical lipids in the bilayers of other membranes. Lipid monolayers and bilayers are both found among the archaea. The effects, if any, of these structural features on membrane performance are unknown. In spite of this striking difference in their membrane lipids, the membranes seen in all three domains have similar overall structures, dimensions, and functions.

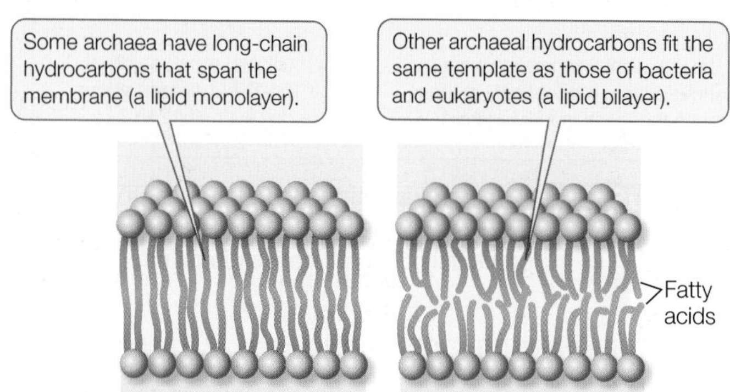

Some archaea have long-chain hydrocarbons that span the membrane (a lipid monolayer).
Other archaeal hydrocarbons fit the same template as those of bacteria and eukaryotes (a lipid bilayer).
Fatty acids


**26.22 Membrane Architecture in Archaea** The long-chain hydrocarbons of many archaeal membranes have glycerol molecules at both ends, so that the membranes consist of a lipid monolayer. In contrast, the membranes of other archaea, bacteria, and eukaryotes consist of a lipid bilayer.

### Most Crenarchaeota live in hot and/or acidic places

Most known Crenarchaeota are either thermophilic (heat loving), acidophilic (acid loving), or both. Members of the genus *Sulfolobus* live in hot sulfur springs at temperatures of 70°C to 75°C. They become metabolically inactive at 55°C (131°F). Hot sulfur springs are also extremely acidic. *Sulfolobus* grows best in the range from pH 2 to pH 3, but some members of this genus readily tolerate pH values as low as 0.9. Most acidophilic thermophiles maintain an internal pH of 5.5 to 7 (close to neutral) in spite of their acidic environment. These and other thermophiles thrive where very few other organisms can even survive. The archaea living in the volcanic vent shown in the opening of this chapter are examples of such thermophiles.

### Euryarchaeota are found in surprising places

Some species of Euryarchaeota share the property of producing methane ($CH_4$) by reducing carbon dioxide. All of these methanogens are obligate anaerobes, and methane production is the key step in their energy metabolism. Comparison of rRNA gene sequences has revealed a close evolutionary relationship among these methanogenic species, which were previously assigned to several different bacterial groups.

Methanogens release approximately 2 billion tons of methane gas into Earth's atmosphere each year, accounting for 80 to 90 percent of the methane in the atmosphere, including the methane produced in some mammalian digestive systems. Approximately a third of this methane comes from methanogens living in the guts of grazing herbivores such as cattle, sheep, and deer, and another large fraction comes from the methanogens that live in the guts of termites and cockroaches. Methane is increasing in Earth's atmosphere by about 1 percent per year and contributes to the greenhouse effect. Part of the increase is due to increases in cattle and rice farming and the methanogens associated with both.

One methanogen, *Methanopyrus*, lives on the ocean bottom near hot hydrothermal vents. *Methanopyrus* can survive and grow at 122°C. It grows best at 98°C and not at all at temperatures below 84°C.

Another group of Euryarchaeota, the **extreme halophiles** (salt lovers), lives exclusively in very salty environments, such as the water of Blood Falls described in the opening of this chapter. Because they contain pink carotenoid pigments, halophiles are sometimes easy to see (**Figure 26.23**). Halophiles grow in the Dead Sea and in brines of all types: the reddish pink spots that can occur on pickled fish are colonies of halophilic archaea. Few other organisms can live in the saltiest of the homes that the extreme halophiles occupy; most would "dry" to death, losing too much water to the hypertonic environment. Extreme halophiles have been found in lakes with pH values as high as 11.5—the most alkaline environment inhabited by living organisms, and almost as alkaline as household ammonia.

Some of the extreme halophiles have a unique system for trapping light energy and using it to form ATP—without using any form of chlorophyll—when oxygen is in short supply. They use the pigment *retinal* (also found in the vertebrate eye) com-

**26.23 Extreme Halophiles** Commercial seawater evaporating ponds (these are in San Francisco Bay) are home to salt-loving archaea, easily visible here because of their carotenoid pigments.

bined with a protein to form a light-absorbing molecule called *bacteriorhodopsin*, and they form ATP by a chemiosmotic mechanism of the kind described in Figure 9.9.

Another member of the Euryarchaeota, *Thermoplasma*, has no cell wall. It is thermophilic and acidophilic, its metabolism is aerobic, and it lives in coal deposits. It has the smallest genome among the archaea, and among the smallest (along with the mycoplasmas) of any free-living organism—1,100,000 base pairs.

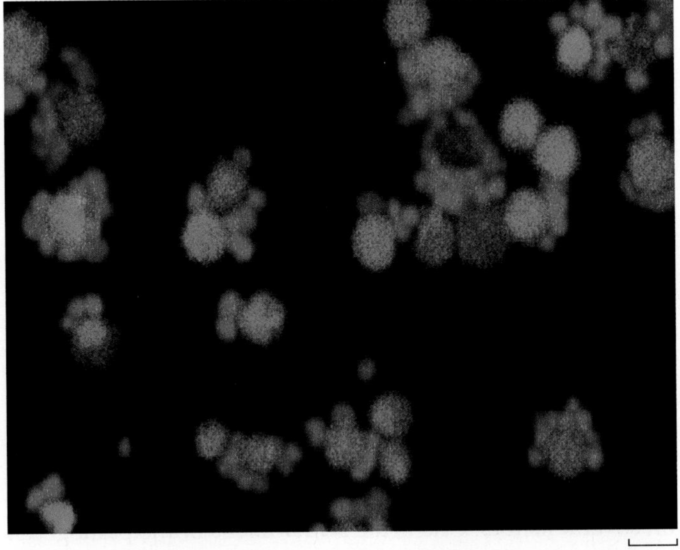


**26.24 A Nanoarchaeote Growing in Mixed Culture with a Crenarchaeote** *Nanoarchaeum equitans* (red), discovered living near deep-ocean hydrothermal vents, is the only representative of the nanoarchaeote group so far discovered. This tiny organism lives attached to cells of the crenarchaeote *Ignicoccus* (green). For this confocal laser micrograph, the two species were visually differentiated by fluorescent dye "tags" that are specific to their separate gene sequences.

### Korarchaeota and Nanoarchaeota are less well known

The Korarchaeota are known only by evidence derived from DNA isolated directly from hot springs. No korarchaeote has been successfully grown in pure culture.

Another distinctive archaeal lineage has been discovered at a deep-sea hydrothermal vent off the coast of Iceland. It is the first representative of a group christened Nanoarchaeota because of their minute size. This organism lives attached to cells of *Ignicoccus*, a crenarchaeote. Because of their association, the two species can be grown together in culture (**Figure 26.24**).

### 26.4 RECAP

**Bacteria and Archaea are highly diverse groups that survive in almost every imaginable habitat on Earth. Many can survive and even thrive in habitats where no eukaryotes can live, including extremely hot, acidic, or alkaline conditions.**

- Can you explain how metabolic diversity could have become so great in the proteobacteria? **See pp. 550–551 and Figure 26.19**
- What makes the membranes of archaea unique? **See p. 552 and Figure 26.22**

Because prokaryotes have so many different metabolic and nutritional capabilities, and because they can live in so many environments, it is reasonable to expect that they affect their environments in many ways. As we are about to see, prokaryotes directly affect humans—in ways both beneficial and harmful.

## 26.5 How Do Prokaryotes Affect Their Environments?

Prokaryotes live in and exploit all kinds of environments and are part of all ecosystems. In this section we examine the roles of prokaryotes that live in soils, in water, and even in other organisms, where they may exist in a neutral, beneficial, or parasitic relationship with their host's tissues. The roles of some prokaryotes living in extreme environments have yet to be determined.

Remember that in spite of our frequent mention of prokaryotes as human pathogens, only a small minority of the known prokaryotic species are pathogenic. Many more prokaryotes play positive roles in our lives and in the biosphere. We make direct use of many bacteria and a few archaea in such diverse applications as cheese production, sewage treatment, and the industrial production of an amazing variety of antibiotics, vitamins, organic solvents, and other chemicals.

### Prokaryotes are important players in element cycling

Many prokaryotes are decomposers—organisms that metabolize organic compounds in dead organisms and other organic material and return the products to the environment as inorganic substances. Prokaryotes, along with fungi, return tremendous quantities of organic carbon to the atmosphere as carbon dioxide, thus carrying out a key step in the carbon cycle. Prokaryotic decomposers also return inorganic nitrogen and sulfur to the environment.

Animals depend on plants and other photosynthetic organisms for their food, directly or indirectly. But plants depend on other organisms—prokaryotes—for their own nutrition. The extent and diversity of life on Earth would not be possible without nitrogen fixation by prokaryotes. Nitrifiers are crucial to the biosphere because they convert the products of nitrogen fixation into nitrate ions, the form of nitrogen most easily used by many plants (see Figure 36.11). Plants, in turn, are the source of nitrogen compounds for animals and fungi. Denitrifiers also play a key role in keeping the nitrogen cycle going. Without denitrifiers, which convert nitrate ions back into nitrogen gas, all forms of nitrogen would leach from the soil and end up in lakes and oceans, making life on land impossible. Other prokaryotes—both bacteria and archaea—contribute to a similar cycle of sulfur.

In the ancient past, the cyanobacteria had an equally dramatic effect on life: their photosynthesis generated oxygen, converting Earth's atmosphere from an anaerobic to an aerobic environment (see Section 25.3). A major result was the wholesale loss of obligate anaerobic species that could not tolerate the oxygen generated by the cyanobacteria. Only those anaerobes that adapted to aerobic conditions or colonized environments that remained anaerobic survived. However, this transformation to aerobic environments made possible the evolution of cellular respiration and the subsequent explosion of eukaryotic life.

### Prokaryotes live on and in other organisms

Prokaryotes work together with eukaryotes in many ways. As we have seen, mitochondria and chloroplasts are descended from what were once free-living bacteria. Much later in evolutionary history, some plants became associated with bacteria to form cooperative nitrogen-fixing nodules on their roots (see Figure 36.9).

Many animals harbor a variety of bacteria and archaea in their digestive tracts. Cattle depend on prokaryotes to perform important steps in digestion. Like most animals, cattle cannot produce cellulase, the enzyme needed to start the digestion of the cellulose that makes up the bulk of their plant food. However, bacteria living in a special section of the gut, called the rumen, produce enough cellulase to process the daily diet for the cattle.

Humans use some of the metabolic products—especially vitamins $B_{12}$ and K—of bacteria living in our large intestine. These and other bacteria and archaea line our intestines with a dense biofilm that is in intimate contact with the mucosal lining of the gut. This biofilm facilitates nutrient transfer from the intestine into the body and induces immunity to the gut contents. The biofilm in the gut is a major part of an "organ" consisting of prokaryotes that is essential to our health. Its makeup varies from time to time and from region to region of the intestinal tract, and it has a complex ecology that scientists have just begun to explore in detail—including the possibility that the species composition of an individual's prokaryote gut fauna may contribute to obesity (or the resistance to it).

We are heavily populated inside and out by bacteria. A 2009 study of bacteria that live on human skin identified more than 1,000 species living on the outside of our bodies, and many of these are thought to be critical to maintaining the skin's health. Although only a small percentage of bacterial species are agents of disease, popular notions of bacteria as "germs" and fear of the consequences of infection arouse our curiosity about those few.

### A small minority of bacteria are pathogens

The late nineteenth century was a productive era in the history of medicine—a time when bacteriologists, chemists, and physicians proved that many diseases are caused by microbial agents. During this time the German physician Robert Koch laid down a set of four rules for establishing that a particular microorganism causes a particular disease:

- The microorganism is always found in individuals with the disease.
- The microorganism can be taken from the host and grown in pure culture.
- A sample of the culture produces the same disease when injected into a new, healthy host.
- The newly infected host yields a new, pure culture of microorganisms identical to those obtained in the second step.

These rules, called **Koch's postulates**, were very important in a time when it was not widely understood that microorganisms cause disease. Although medical science today has more powerful diagnostic tools, the postulates remain useful on occasion. For example, physicians were taken aback in the 1990s when stomach ulcers—long accepted and treated as the result of excess stomach acid—were shown by Koch's postulates to be caused by the bacterium *Helicobacter pylori* (see Figure 51.14).

Only a tiny percentage of all prokaryotes are pathogens, and of those that are known, all are in the domain Bacteria. For an organism to be a successful pathogen, it must:

- arrive at the body surface of a potential host;
- enter the host's body;
- evade the host's defenses;
- multiply inside the host; and finally
- infect a new host.

Failure to successfully complete any of these steps ends the reproductive career of a pathogenic organism. However, in spite of the many defenses available to potential hosts (see Chapter 42), some bacteria are very successful pathogens.

For the host, the consequences of a bacterial infection depend on several factors. One is the **invasiveness** of the pathogen—its ability to multiply in the host's body. Another is its **toxigenicity**—its ability to produce toxins, chemical substances that are harmful to the host's tissues. *Corynebacterium diphtheriae*, the agent that causes diphtheria, has low invasiveness and multiplies only in the throat, but its toxigenicity is so great that the entire body is affected. In contrast, *Bacillus anthracis*, which causes anthrax (a disease primarily of cattle and sheep, but which is also sometimes fatal in humans), has low toxigenicity but is so invasive that the entire bloodstream ultimately teems with the bacteria.

There are two general types of bacterial toxins: exotoxins and endotoxins. **Endotoxins** are released when certain Gram-negative bacteria grow or lyse (burst). Endotoxins are lipopolysaccharides (complexes consisting of a polysaccharide and a lipid component) that form part of the outer bacterial membrane. Endotoxins are rarely fatal; they normally cause fever, vomiting, and diarrhea. Among the endotoxin producers are some strains of the gamma-proteobacteria *Salmonella* and *Escherichia*.

**Exotoxins** are usually soluble proteins released by living, multiplying bacteria, and they may travel throughout the host's body. They are highly toxic—often fatal—to the host, but they do not produce fevers. Human diseases induced by bacterial exotoxins include tetanus (*Clostridium tetani*), cholera (*Vibrio cholerae*), and bubonic plague (*Yersinia pestis*). Anthrax results from three exotoxins produced by *Bacillus anthracis*. Botulism is caused by exotoxins produced by *Clostridium botulinum* that are among the most poisonous ever discovered. The lethal dose of the botulinum A exotoxin for humans is about one-millionth of a gram (1 μg). Nonetheless, much smaller doses of this exotoxin, marketed under various trade names, are used to treat muscle spasms and also for cosmetic purposes (temporary wrinkle reduction in skin).

Pathogenic bacteria are often surprisingly difficult to combat, even with today's arsenal of antibiotics. One source of this difficulty is the ability of prokaryotes to form biofilms.

## 26.5 RECAP

**Prokaryotes play key roles in the cycling of Earth's elements. Many prokaryotes are beneficial and even necessary to other forms of life; others are pathogens.**

- Describe the roles of bacteria in the nitrogen cycle. See p. 554
- What are some of the challenges facing a pathogen? See p. 554

Before moving on to discuss the diversity of eukaryotic life, it is appropriate to consider how viruses are related to the rest of life. Although they are not cellular, viruses are numerically among the most abundant organisms on Earth. Their effects on other organisms are enormous. Where did viruses come from, and how do they fit into the tree of life? Biologists are still working to answer these questions.

# 26.6 Where Do Viruses Fit into the Tree of Life?

Some biologists do not think of viruses as living organisms, primarily because they are not cellular and must depend on cellular organisms for basic life functions such as replication and metabolism. But viruses are derived from the cells of other living organisms. They use the same essential forms of genetic storage and transmission as do cellular organisms. Viruses infect all cellular forms of life, including bacteria, archaea, and eukaryotes. They replicate, mutate, evolve, and interact with other organisms, often causing serious diseases when they infect their hosts. They are also numerically among the most abundant organisms on the planet. And, finally, viruses clearly evolve independently of other organisms, so it is almost impossible not to treat them as a part of life.

Several factors make virus phylogeny difficult to resolve. The tiny size of many viral genomes restricts the phylogenetic analyses that can be conducted to relate viruses to cellular organisms. The rapid mutational rate, which results in rapid evolution of viral genomes, tends to cloud evolutionary relationships across long periods of time. There are no known viral fossils (viruses are too small and delicate to fossilize), so the paleontological record offers no clues as to viral origins. Finally, viruses are highly diverse (see Figure 26.25), and several lines of evidence support the hypothesis that viruses have evolved repeatedly within each of the major groups of life.

## Many RNA viruses probably represent escaped genomic components

Although viruses are now obligate parasites of cellular species, they may once have been cellular components involved in basic cellular functions—that is, they may be "escaped" components of cellular life that now evolve independently of their hosts.

**NEGATIVE-SENSE SINGLE-STRANDED RNA VIRUSES** A case in point is a class of viruses whose genome is composed of single-stranded RNA that is the complement (negative-sense) of the mRNA needed for protein translation. Many of these negative-sense single-stranded RNA viruses have only a few genes, including an RNA-dependent RNA polymerase that allows them to make mRNA from their negative-sense RNA genome. Modern cellular organisms cannot generate mRNA in this manner (at least in the absence of viral infections), but scientists speculate that single-stranded RNA genomes may have been common in the distant past, before DNA became the primary molecule for genetic information storage.

A self-replicating RNA polymerase gene that begins to replicate independently of a cellular genome could conceivably acquire a few additional protein-coding genes through recombination with its host's DNA. If one or more of these genes were to foster the development of a protein coat, the virus might then survive outside the host and infect new hosts. It is believed that this scenario has been repeated many times independently across the tree of life, given that many of the negative-sense single-stranded RNA viruses that infect organisms from bacteria to humans are not closely related to one another. In other words, negative-sense single-stranded RNA viruses do not represent a distinct taxonomic group, but exemplify a particular process of cellular escape that probably happened many different times.

Examples of familiar negative-sense single-stranded RNA viruses include the viruses that cause measles, mumps, rabies, and influenza (**Figure 26.25A**).

(A)

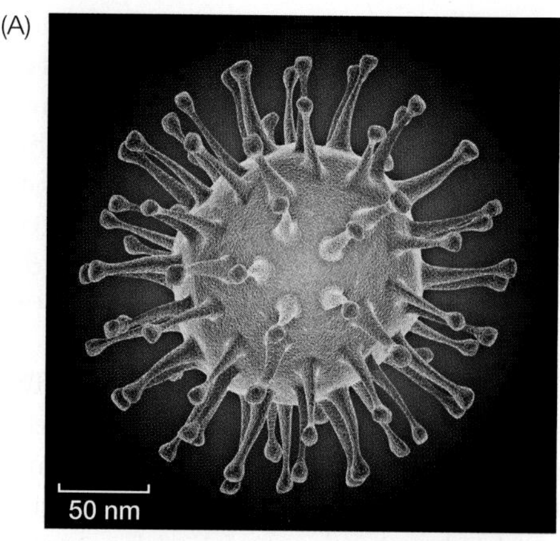


A negative-sense single-stranded RNA virus: Influenza virus H5N1, the "bird flu" virus. Surface view.

(B)

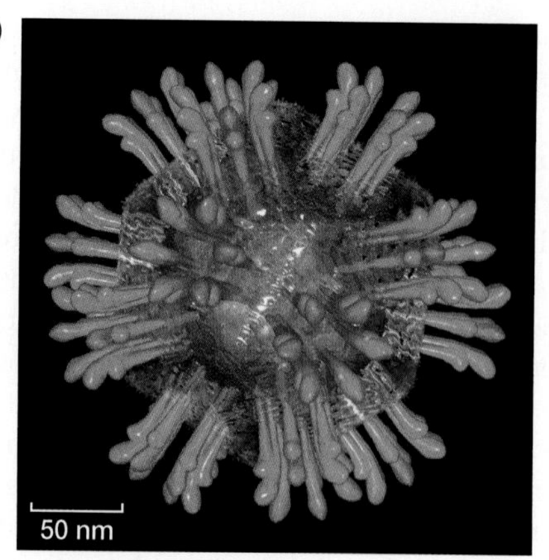


A positive-sense single-stranded RNA virus: Hepatitis C virus, the cause of a human liver disease. Surface view.

(C)

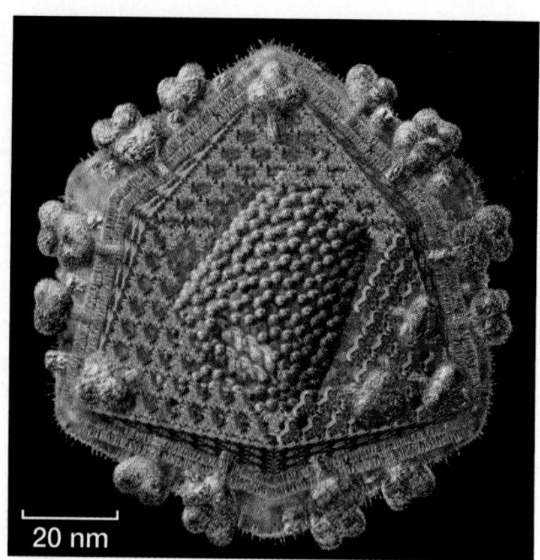


An RNA retrovirus: One of the human immunodeficiency viruses (HIV) that causes AIDS. Cutaway view.

(D)

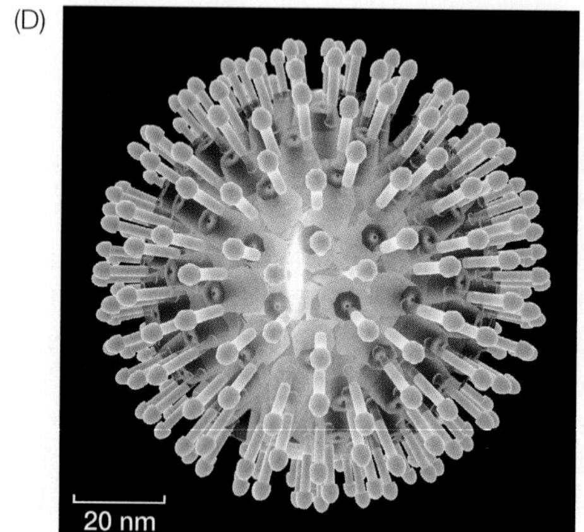


A double-stranded DNA virus: One of the many herpes viruses (Herpesviridae). Different herpes viruses are responsible for many human infections, including chicken pox, shingles, cold sores. and genital herpes (HSV1/2). Surface view.

(E)

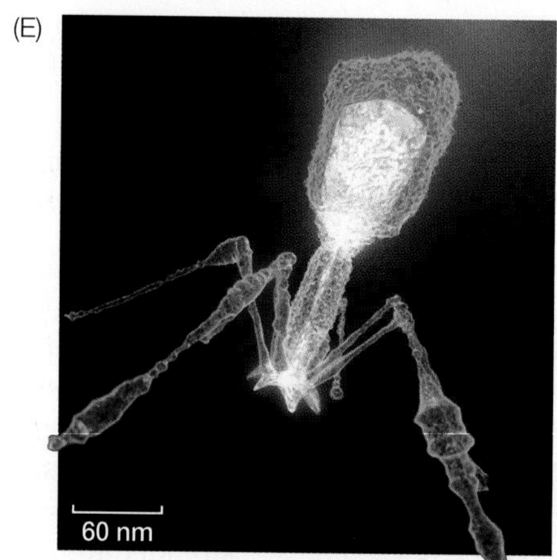


A double-stranded DNA virus: Bacteriophage T4. Viruses that infect bacteria are referred to as bacteriophage (or simply phage). T4 attaches leglike fibers to the outside of its host cell and injects its DNA into the cytoplasm through its "tail" (pink structure in this rendition).

(F)

A double-stranded DNA mimivirus: This *Acanthamoeba polyphaga* mimivirus (APMV) has the largest diameter of all known viruses and a genome larger than some prokaryote genomes. Cutaway view.

**26.25 Viruses Are Diverse** Relatively small genomes and rapid evolutionary rates make it difficult to reconstruct phylogenetic relationships among some classes of viruses. Instead, viruses are classified largely by general characteristics of their genomes. The images here are computer artists' reconstructions based on cryoelectron micrographs.

**POSITIVE-SENSE SINGLE-STRANDED RNA VIRUSES** The genome of another type of single-stranded RNA viruses is composed of positive-sense RNA. Positive-sense genomes are already set for translation; unlike in negative-sense RNA, no replication of the genome into the complement strand is needed before protein translation can take place. Positive-sense single-stranded RNA viruses are the most abundant and diverse class of viruses (**Figure 26.25B**). Most of the viruses that cause diseases in crop plants are in this group. When they infect plants, these viruses kill patches of cells in the leaves or stems of the plants, leaving live cells amid a patchwork of discolored dead plant tissue (giving them the name of *mosaic* or *mottle viruses*; **Figure 26.26**). Other viruses in this group infect bacteria, fungi, and animals. Human diseases caused by positive-sense single-stranded RNA viruses include polio, hepatitis C, and the common cold. As is true of the other functionally defined groups of viruses, these viruses appear to have evolved multiple times across the tree of life from different groups of cellular ancestors.

**RNA RETROVIRUSES** The RNA retroviruses are best known as the group that includes the human immunodeficiency viruses (HIV; **Figure 26.25C**). Like the previous two categories of viruses, RNA retroviruses have genomes composed of single-stranded RNA and likely evolved as escaped cellular components.

Yellow areas are dead leaf cells, killed by the mosaic virus.

**26.26 Mosaic Viruses Are a Problem for Agriculture** Mosaic or "mottle" viruses are the most abundant and diverse class of viruses. This leaf is from an apple tree infected with mosaic virus.

Retroviruses are so named because they regenerate themselves by reverse transcription. When the retroviruses enter the nucleus of their vertebrate host, viral reverse transcriptase produces cDNA from the viral RNA genome and then replicates the single-stranded cDNA to produce double-stranded DNA. Another virally encoded enzyme called integrase catalyzes the integration of the new piece of double-stranded viral DNA into the host's genome. The viral genome is then replicated along with the host cell's DNA; the integrated retroviral DNA is known as a *provirus*. Many components of cellular species (such as retrotransposons; see Section 17.3) resemble components of retroviruses.

Retroviruses are only known to infect vertebrates, although genomic elements that resemble portions of these viruses are a component of the genomes of a wide variety of organisms including bacteria, plants, and many animals. Several retroviruses are associated with the development of various forms of cancer, as cells infected with these viruses are likely to undergo uncontrolled replication.

**DOUBLE-STRANDED RNA VIRUSES** Double-stranded RNA viruses may have evolved repeatedly from single-stranded RNA ancestors—or perhaps vice versa. These viruses, which are not closely related to one another, infect organisms from throughout the tree of life. For example, many important plant diseases are caused by double-stranded RNA viruses, whereas other viruses of this type cause many cases of infant diarrhea in humans.

## Some DNA viruses may have evolved from reduced cellular organisms

Another class of viruses is composed of those that have a double-stranded DNA genome (**Figure 26.25D–F**). This group is also almost certainly polyphyletic (with many independent origins). Many of the common phage that infect bacteria (*bacteriophage*) are double-stranded DNA viruses, as are the viruses that cause smallpox and herpes in humans.

Some biologists think that at least some of the DNA viruses may represent highly reduced parasitic organisms that have lost their cellular structure as well as their ability to survive as free-living species. For example, some of the largest DNA viruses are the mimiviruses (see Figure 26.25F), which have a genome in excess of a million base pairs of DNA. This is double the genome size of some parasitic bacteria such as *Mycoplasma genitalium*. Phylogenetic analyses of these DNA viruses suggest that they have evolved repeatedly from cellular organisms. Furthermore, recombination among different viruses may have allowed the exchange of various genetic modules, further complicating the history and origins of viruses.

### 26.6 RECAP

**Viruses are highly diverse and appear to have evolved independently from many different groups of cellular organisms from throughout the tree of life. Some viruses appear to have evolved from escaped components of cellular organisms, whereas other viruses may have evolved from parasitic cellular ancestors.**

- What are some the reasons that it is difficult to place viruses precisely within the phylogeny of cellular organisms? See p. 555
- Explain the two main hypotheses of viral origins. See p. 555–557

It appears that the enormous diversity of viruses is, at least in part, a result of their multiple origins from many different cellular organisms. Some may have arisen as escaped genetic components, while others may represent reduced and specialized parasites. Some appear to be derived from bacterial species, others to be derived from eukaryotic organisms. It may be best to view viruses as spinoffs from the various branches on the tree of life—sometimes evolving independently of cellular genomes, sometimes recombining with them. One way to think of viruses is as the "bark" on the tree of life: certainly an important component all across the tree, but not quite like the main branches.

## CHAPTER SUMMARY

### 26.1 How Did the Living World Begin to Diversify?

- Two of life's three domains, Bacteria and Archaea, are prokaryotic. They are distinguished from Eukarya in several ways, including their lack of membrane-enclosed organelles. Review Table 26.1
- Eukaryotes are related to both Archaea and Bacteria, and appear to have formed through a merging of members of these two lineages. The last common ancestor of all three domains probably lived 2 to 3 billion years ago. Review Figure 26.1, ANIMATED TUTORIAL 26.1
- Prokaryotes may be single or may form chains or clusters. The three most common bacterial morphologies are **cocci** (spheres), **bacilli** (rods), and **helices** (spirals). Review Figure 26.2
- The cells of some bacteria aggregate, forming **filaments** and other structures.

### 26.2 What Are Some Keys to the Success of Prokaryotes?

- Prokaryotes are the most numerous organisms on Earth. They occur in an enormous variety of habitats, including inside other organisms and deep in Earth's crust.
- Prokaryotes form complex communities, some of which become dense films called **biofilms**. Review Figure 26.3
- Most prokaryotes have cell walls containing molecules important in transport, digestion, and sensing the environment. Almost all bacterial cell walls contain **peptidoglycan**. The cell walls of archaea lack peptidoglycan and instead include pseudopeptidoglycan or proteins. Review Figure 26.5, WEB ACTIVITY 26.1
- Bacteria can be placed into two groups by the **Gram stain**. **Gram-negative bacteria** have a periplasmic space between the inner (plasma) and outer membranes. In **Gram-positive bacteria**, the cell wall usually has about five times as much peptidoglycan as a Gram-negative wall.
- Prokaryotes move by a variety of means, including axial filaments, gas vesicles, and **flagella**. Review Figures 26.6 and 26.7
- Prokaryotes reproduce asexually but may undergo genetic recombination. Reproduction and genetic recombination are not directly linked in prokaryotes.
- Prokaryote metabolism is very diverse. Some prokaryotes are anaerobic, others are aerobic, and yet others can shift between these modes. Prokaryotes are classified as **photoautotrophs**, **photoheterotrophs**, **chemolithotrophs**, or **chemoheterotrophs**. Review Table 26.2
- The metabolic pathways of some prokaryotes involve sulfur or nitrogen. **Nitrogen fixers** convert nitrogen gas into a form organisms can metabolize.

### 26.3 How Can We Resolve Prokaryote Phylogeny?

- Early attempts to classify prokaryotes were hampered by the organisms' small size and difficulties growing them in pure culture. Phylogenetic classification of prokaryotes is now based on nucleotide sequences of rRNA and other slowly evolving genes.
- Although **lateral gene transfer** has occurred throughout prokaryotic evolutionary history, elucidation of prokaryote phylogeny is still possible. Review Figure 26.10
- Only a tiny percentage of all prokaryote species have been described.

### 26.4 What Are the Major Known Groups of Prokaryotes?

- Several major taxonomic groups of prokaryotes have been recognized. The members of a prokaryote clade often differ profoundly from one another. Review Figures 26.11 and 26.19
- Of the major taxonomic groups of Bacteria, the **proteobacteria** embrace the largest number of species. Other important groups include the **cyanobacteria**, **spirochetes**, **chlamydias**, and **low-GC Gram-positives**. Some **high-GC Gram positives** produce important antibiotics.
- The low-GC Gram-positives include the **mycoplasmas**, which are among the smallest cellular organisms ever discovered.
- Ether linkages in the branched long-chain hydrocarbons of cell membranes are a synapomorphy of archaea.
- The best-studied groups of Archaea are **Euryarchaeota** and **Crenarchaeota**. Some species of Euryarchaeota are methanogens; extreme halophiles are also found among the Euryarchaeota. Most known Crenarchaeota are both thermophilic and acidophilic.

### 26.5 How Do Prokaryotes Affect their Environments?

- Prokaryotes play key roles in the cycling of elements such as nitrogen, oxygen, sulfur, and carbon. One such role is as decomposers of dead organisms.
- Nitrogen-fixing bacteria fix the nitrogen needed by all other organisms. **Nitrifiers** convert that nitrogen into forms that can be used by plants, and **denitrifiers** ensure that nitrogen is returned to the atmosphere.
- Oxygen production by early photosynthetic cyanobacteria reconfigured Earth's atmosphere, which made aerobic forms of life possible.
- Prokaryotes inhabiting the guts of many animals help them digest their food.
- **Koch's postulates** establish the criteria by which an organism may be classified as a pathogen. Relatively few bacteria—and no archaea—are known to be pathogens.

### 26.6 Where Do Viruses Fit Into the Tree of Life?

- Viruses have evolved many times from many different groups of cellular organisms. They do not represent a single taxonomic group.
- Some viruses are probably derived from escaped genetic elements of cellular species; others are thought to have evolved as highly reduced parasites.
- Viruses are categorized by the nature of their genomes. Review Figure 26.25

## SELF-QUIZ

1. Most prokaryotes
   *a.* are agents of disease.
   *b.* lack ribosomes.
   *c.* evolved from the most ancient eukaryotes.
   *d.* lack a cell wall.
   *e.* are chemoheterotrophs.
2. The division of the living world into three domains
   *a.* is based on the number of cells in organisms of each group.
   *b.* is based mostly on the major morphological differences between archaea and bacteria.
   *c.* emphasizes the greater importance of eukaryotes.
   *d.* was proposed by the early microscopists.
   *e.* is based on phylogenetic relationships determined from nucleotide sequences of rRNA and other genes.
3. Which statement about archaeal genomes is *true*?
   *a.* They are typically organized in a circular chromosome, like bacterial genomes.
   *b.* They include no rRNA genes.
   *c.* They are always much smaller than bacterial genomes.
   *d.* They are housed in the nucleus of the archaeal cell.
   *e.* No archaeal genome has yet been sequenced.
4. Which statement about nitrogen metabolism is *not* true?
   *a.* Certain prokaryotes reduce atmospheric $N_2$ to ammonia.
   *b.* Some nitrifiers are soil bacteria.
   *c.* Denitrifiers are obligate anaerobes.
   *d.* Nitrifiers obtain energy by oxidizing ammonia and nitrite.
   *e.* Without nitrifiers, terrestrial organisms would lack a nitrogen supply.
5. All photosynthetic bacteria
   *a.* use chlorophyll *a* as their photosynthetic pigment.
   *b.* use bacteriochlorophyll as their photosynthetic pigment.
   *c.* release oxygen gas.
   *d.* produce particles of sulfur.
   *e.* are photoautotrophs.
6. Gram-negative bacteria
   *a.* appear blue to purple following Gram staining.
   *b.* appear pink to red following Gram staining.
   *c.* are all either bacilli or cocci.
   *d.* contain no peptidoglycan in their cell walls.
   *e.* are all photosynthetic.
7. Endospores
   *a.* are produced by viruses.
   *b.* are reproductive structures.
   *c.* are very delicate and easily killed.
   *d.* are resting structures.
   *e.* lack cell walls.
8. Chlamydias
   *a.* are among the smallest archaea.
   *b.* live on the surface of human skin.
   *c.* are never pathogenic to humans.
   *d.* live only as parasites in the cells of other organisms.
   *e.* have a very simple life cycle.
9. Archaea
   *a.* have cytoskeletons.
   *b.* have distinctive lipids in their plasma membranes.
   *c.* survive only at moderate temperatures and near neutrality.
   *d.* all produce methane.
   *e.* have substantial amounts of peptidoglycan in their cell walls.
10. Genetic evidence suggests that viruses
   *a.* are most closely related to Bacteria.
   *b.* are most closely related to Archaea.
   *c.* are most closely related to Eukarya.
   *d.* have evolved multiple times from many different cellular species.
   *e.* evolved from the fusion of a bacterial and an archaeal species.

## FOR DISCUSSION

1. Why do systematic biologists find rRNA sequence data more useful than data on metabolism or cell structure for classifying prokaryotes?
2. How can lateral gene transfer mislead studies about prokaryote phylogeny? How could this issue be addressed?
3. Differentiate among the members of the following sets of related terms:
   *a.* prokaryotic/eukaryotic
   *b.* obligate anaerobe/facultative anaerobe/obligate aerobe
   *c.* photoautotroph/photoheterotroph/chemolithotroph/chemoheterotroph
   *d.* Gram-positive/Gram-negative
4. Why are the endospores of low-GC Gram-positives not considered to be reproductive structures?
5. Originally, the cyanobacteria were called "blue-green algae" and were not grouped with the bacteria. Suggest several reasons for this (abandoned) tendency to separate the cyanobacteria from the bacteria. Why are the cyanobacteria now grouped with the other bacteria?
6. The high-GC Gram-positives are of great commercial interest. Why?
7. Thermophiles are of great interest to molecular biologists and biochemists. Why? What practical concerns might motivate that interest?
8. How can biologists discuss the Korarchaeota when they have never seen one?
9. Do you consider viruses to be living organisms? Why or why not?

## ADDITIONAL INVESTIGATION

Kashefi and Lovley (see Figure 26.21) were able to grow an unnamed archaeal species at temperatures above 120°C only because they used $Fe^{3+}$ as an electron acceptor—no other electron acceptor they tried allowed growth. How might you explore the same or other high-temperature environments for other hyperthermophilic organisms not detected by Kashefi and Lovley using $Fe^{3+}$?

# 27 The Origin and Diversification of Eukaryotes

## How a microbial eukaryote may have changed the course of science

Charles Darwin suffered throughout his adult life from a debilitating illness that was never conclusively diagnosed. Whatever the cause of this illness, it seems to have contributed to the determination with which Darwin pursued his studies. In his autobiography, Darwin wrote that his "chief enjoyment and sole employment throughout life has been scientific work; and the excitement from such work makes me for the time forget, or drives quite away, my daily discomfort." To Darwin, science was a pleasure that provided a distraction from his pain.

One hypothesis is that Darwin suffered from Chagas' disease. Caused by the trypanosome *Trypanosoma cruzi*, one of many microbial eukaryotes, this disease currently affects 16 to 18 million people, primarily in Central and South America. The trypanosome is transmitted to people by trypanosome-infected assassin bugs, which bite to suck blood from their victims and then often defecate near the wound. When people crush the insect or scratch its bite, they can become infected with the trypanosome, which is present in the assassin bug's feces. That Darwin may have been infected in this way is suggested by a journal entry he made during his voyage around the world on the HMS *Beagle*. On March 25, 1835, while visiting Argentina, Darwin wrote:

> *At night I experienced an attack (for it deserves no less a name) of the* Benchuca … *the great black bug of the Pampas. It is most disgusting to feel soft wingless insects, about an inch long, crawling over one's body. Before sucking they are quite thin but afterwards they become round and bloated with blood, and in this state they are easily crushed.*

Darwin's entry describes the experience of being bitten by one of the assassin bugs that carry *Trypanosoma cruzi*, so clearly Darwin had the opportunity to become infected. In fact, he even experimented with the insects to learn more about their habits. "[I]f a finger was presented, the bold insect would immediately protrude its sucker, make a charge, and if allowed, draw blood," he wrote in his journal.

There is neither a vaccine nor any effective drug treatment for Chagas'

**An Assassin with a Deadly Cargo** This member of the insect genus *Rhodnius* is an "assassin bug," one of several species that transmits *Trypanosoma cruzi*, the causative agent of Chagas' disease in humans.

**Thugs in a Huddle** Sometimes trypanosomes, like these *Leishmania major*, form clusters held together by a tangle of mucilage secreted around their flagella; nobody is sure yet why they behave this way.

disease, which kills more than 40,000 people a year. Other trypanosome species cause other debilitating diseases, including leishmaniasis (60,000 deaths/year) and African sleeping sickness (50,000 deaths/year). Since the infective organism in each of these diseases is a eukaryote like us, most of the drugs that are toxic to the trypanosomes are often toxic to humans as well. Although the treatment of bacterial diseases often capitalizes on the many differences between prokaryotes and eukaryotes in order to target only the bacteria, the diseases caused by microbial eukaryotes usually are, like Chagas' disease, much more difficult to treat.

**IN THIS CHAPTER** we describe the origin and early diversification of the eukaryotes and the complexity achieved by some single cells. We then explore some of the diversity of microbial eukaryote body forms and adaptations and present the developing current view of the evolutionary relationships among the major eukaryote groups.

## CHAPTER OUTLINE



# 27.1 How Did the Eukaryotic Cell Arise?

Many members of the domain Eukarya are familiar to us. We have no problem recognizing trees, mushrooms, and insects as plants, fungi, and animals, respectively. However, a dazzling assortment of other eukaryotes—mostly microscopic organisms—do not fit into any of these three groups. Eukaryotes that are not plants, animals, or fungi have traditionally been "dropped" into the category **protists**. Phylogenetic analyses, however, are clear and consistent in showing that many of the groups that fall under the rubric of "protists" in fact are not closely related to one another, but are *paraphyletic* (see Figure 22.14). Thus the word "protist" does not describe a formal taxonomic group, but is a convenience term—a shorthand way of saying "all the eukaryotes that are not plants, animals, or fungi."

### The diversity of protists is reflected in both morphology and phylogeny

In terms of their evolutionary relationships, as well as in many aspects of their biology, protists are more diverse than any of the three better known eukaryote groups. Some protists are motile, while others do not move; some are photosynthetic, others heterotrophic; most are unicellular, but some are multicellular (**Table 27.1**). Most are microscopic, but a few are huge: giant kelps, for example, can grow to be longer than a football field. We refer to the unicellular species of protists as *microbial eukaryotes*, but you should keep in mind that there are large, multicellular protists as well.

The phylogeny of the major eukaryote lineages remains a subject of research and debate. Some groups of protists are closely related to animals and fungi, whereas others are closely related to the land plants, and still others are only distantly related to any of these familiar eukaryotes (**Figure 27.1**).

### Cellular features support the monophyly of eukaryotes

Eukaryotic cells differ in many ways from prokaryotic cells, and these unique characters of eukaryotes lead us to conclude that eukaryotes are monophyletic. In other words, there was a single eukaryotic ancestor which diversified into the many different lineages of protists, as well as plants, animals, and fungi. Given the nature of evolutionary processes, the many synapomorphies of eukaryotes undoubtedly did not arise simultaneously. We can make some reasonable inferences about the most important events that led to the evolution of a new cell type, bearing in mind that the global environment underwent an

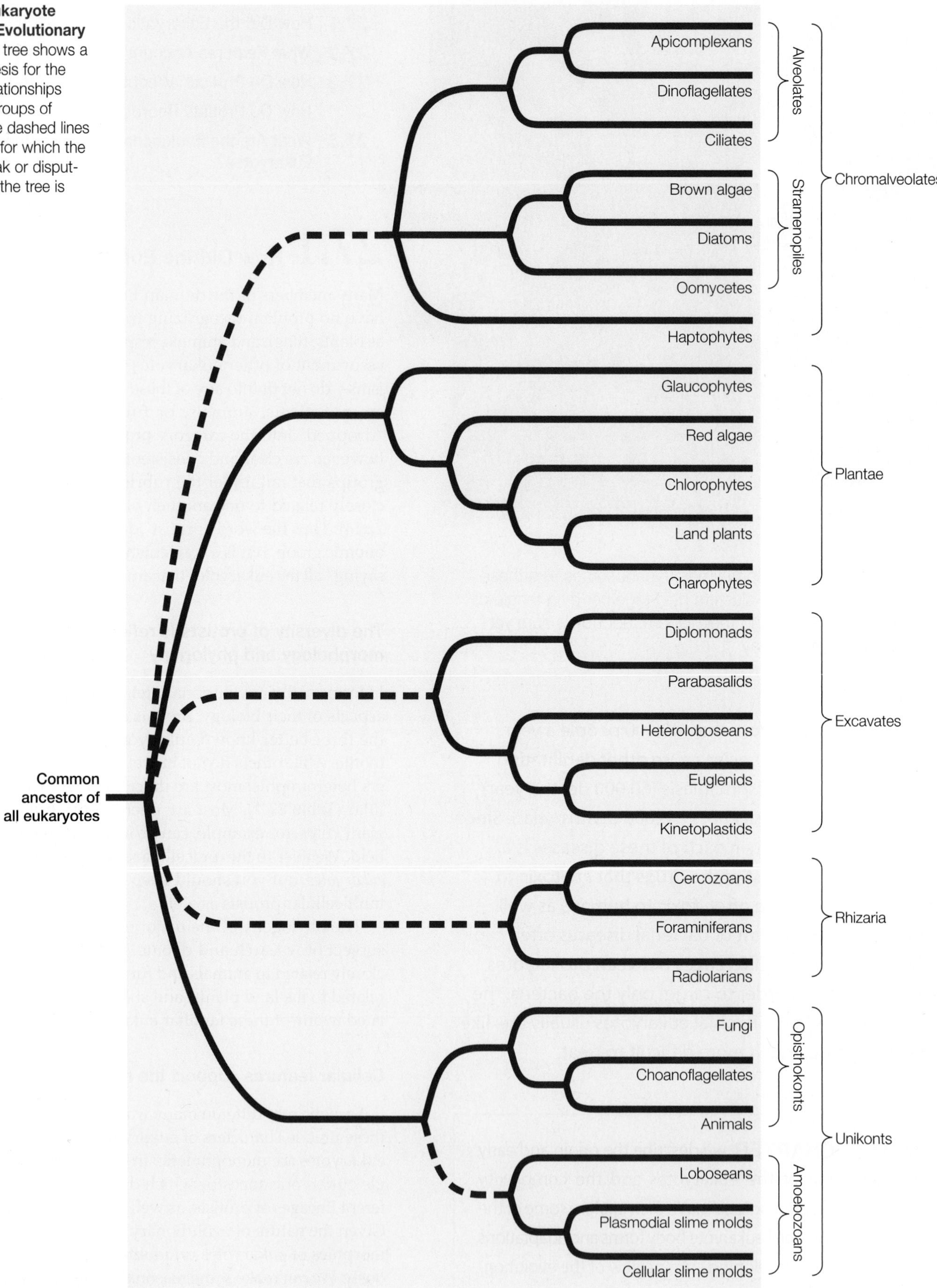


**27.1 Major Eukaryote Groups in an Evolutionary Context** This tree shows a current hypothesis for the evolutionary relationships among major groups of eukaryotes. The dashed lines indicate clades for which the evidence is weak or disputed. The root of the tree is uncertain.

TABLE 27.1

**Major Eukaryote Clades**

| CLADE | ATTRIBUTES | EXAMPLE (GENUS) |
|---|---|---|
| **Chromalveolates** | | |
| Haptophytes | Unicellular, often with calcium carbonate scales | *Emiliania* |
| Alveolates | Sac-like structures beneath plasma membrane | |
| Apicomplexans | Apical complex for penetration of host | *Plasmodium* |
| Dinoflagellates | Pigments give golden-brown color | *Gonyaulax* |
| Ciliates | Cilia; two types of nuclei | *Paramecium* |
| Stramenopiles | Hairy and smooth flagella | |
| Brown algae | Multicellular; marine; photosynthetic | *Macrocystis* |
| Diatoms | Unicellular; photosynthetic; two-part cell walls | *Thalassiosira* |
| Oomycetes | Mostly coenocytic; heterotrophic | *Saprolegnia* |
| **Plantae** | | |
| Glaucophytes | Peptidoglycan in chloroplasts | *Cyanophora* |
| Red algae | No flagella; chlorophyll *a* and *c*; phycoerythrin | *Chondrus* |
| Chlorophytes | Chlorophyll *a* and *b* | *Ulva* |
| *Land plants (Chs. 28–29) | Chlorophyll *a* and *b*; protected embryo | *Ginkgo* |
| Charophytes | Chlorophyll *a* and *b*; mitotic spindle oriented as in land plants | *Chara* |
| **Excavates** | | |
| Diplomonads | No mitochondria; two nuclei; flagella | *Giardia* |
| Parabasalids | No mitochondria; flagella and undulating membrane | *Trichomonas* |
| Heteroloboseans | Can transform between amoeboid and flagellate stages | *Naegleria* |
| Euglenids | Flagella; spiral strips of protein support cell surface | *Euglena* |
| Kinetoplastids | Kinetoplast within mitochondrion | *Trypanosoma* |
| **Rhizaria** | | |
| Cercozoans | Threadlike pseudopods | *Cercomonas* |
| Foraminiferans | Long, branched pseudopods; calcium carbonate shells | *Globigerina* |
| Radiolarians | Glassy endoskeleton; thin, stiff pseudopods | *Astrolithium* |
| **Unikonts** | | |
| Opisthokonts | Single, posterior flagellum | |
| *Fungi (Ch. 30) | Heterotrophs that feed by absorption | *Penicillium* |
| Choanoflagellates | Resemble sponge cells; heterotrophic; with flagella | *Choanoeca* |
| *Animals (Chs. 31–33) | Heterotrophs that feed by ingestion | *Drosophila* |
| Amoebozoans | Amoebas with lobe-shaped pseudopods | |
| Loboseans | Feed individually | *Amoeba* |
| Plasmodial slime molds | Form coenocytic feeding bodies | *Physarum* |
| Cellular slime molds | Cells retain their identity in pseudoplasmodium | *Dictyostelium* |

*Clades marked with an asterisk are made up of multicellular organisms and are discussed in the chapters indicated. All other groups listed are treated here as *protists*.

enormous change—from low to high availability of free oxygen—during the course of these events (see Section 25.3). Keep in mind, however, that these inferences, although reasonable and grounded, are still conjectural; the hypothesis we pursue here is one of a few that biologists are currently considering. We describe here a prominent theory on the origin of the eukaryotic cell as a framework for thinking about the challenging question of eukaryotic origins.

### The modern eukaryotic cell arose in several steps

Several events preceded the origin of the modern eukaryotic cell:

- The origin of a flexible cell surface
- The origin of a cytoskeleton
- The origin of a nuclear envelope, which enclosed a genome organized into chromosomes
- The appearance of *digestive vacuoles*
- The acquisition of certain organelles via *endosymbiosis*

**RAMIFICATIONS OF A FLEXIBLE CELL SURFACE** Many fossil prokaryotes look like rods, and we presume that these ancient organisms, like most present-day prokaryotic cells, had firm cell walls. The first step toward the eukaryotic condition was the loss of the cell wall by an ancestral prokaryotic cell. This wall-less condition is present in some present-day bacteria, although many others have developed new types of cell walls. Let's consider the possibilities open to a flexible cell without a wall.

First, think of cell size. As a cell grows larger, its surface area-to-volume ratio decreases (see Figure 5.2). Unless the surface area can be increased, the cell volume will reach an upper limit. If the cell's surface is flexible, it can fold inward and elaborate itself, creating more surface area for gas and nutrient exchange (**Figure 27.2**).

With a surface flexible enough to allow infolding, the cell can exchange materials with its environment rapidly enough to sustain a larger volume and more rapid metabolism (Figure 27.2, steps 1–2). Furthermore, a flexible surface can pinch off bits of the environment, bringing them into the cell by endocytosis.

**CHANGES IN CELL STRUCTURE AND FUNCTION** Other early steps in the evolution of the eukaryotic cell are likely to have included three advances: the formation of ribosome-studded internal membranes, some of which surrounded the DNA; the appearance of a cytoskeleton; and the evolution of digestive vacuoles (Figure 27.2, steps 3–7).

Prokaryotic cell
Cell wall
DNA
1 The protective cell wall was lost.
2 Infolding of the plasma membrane added surface area without increasing the cell's volume.
3 Cytoskeleton (microfilament and microtubules) formed.
4 Internal membranes studded with ribosomes formed.
5 As DNA attached to the membrane of an infolded vesicle, a precursor of a nucleus formed.
6 Microtubules from the cytoskeleton formed eukaryotic flagellum, enabling propulsion.
7 Early digestive vacuoles evolved into lysosomes using enzymes from the early endoplasmic reticulum.
8 Mitochondria formed through endosymbiosis with a proteobacterium.
9 Endosymbiosis with cyanobacteria led to the development of chloroplasts, which supplied the cell with the means to manufacture materials using solar energy (see Figure 27.4).
Flagellum
Eukaryotic cell
Chloroplast
Mitochondrion
Nucleus

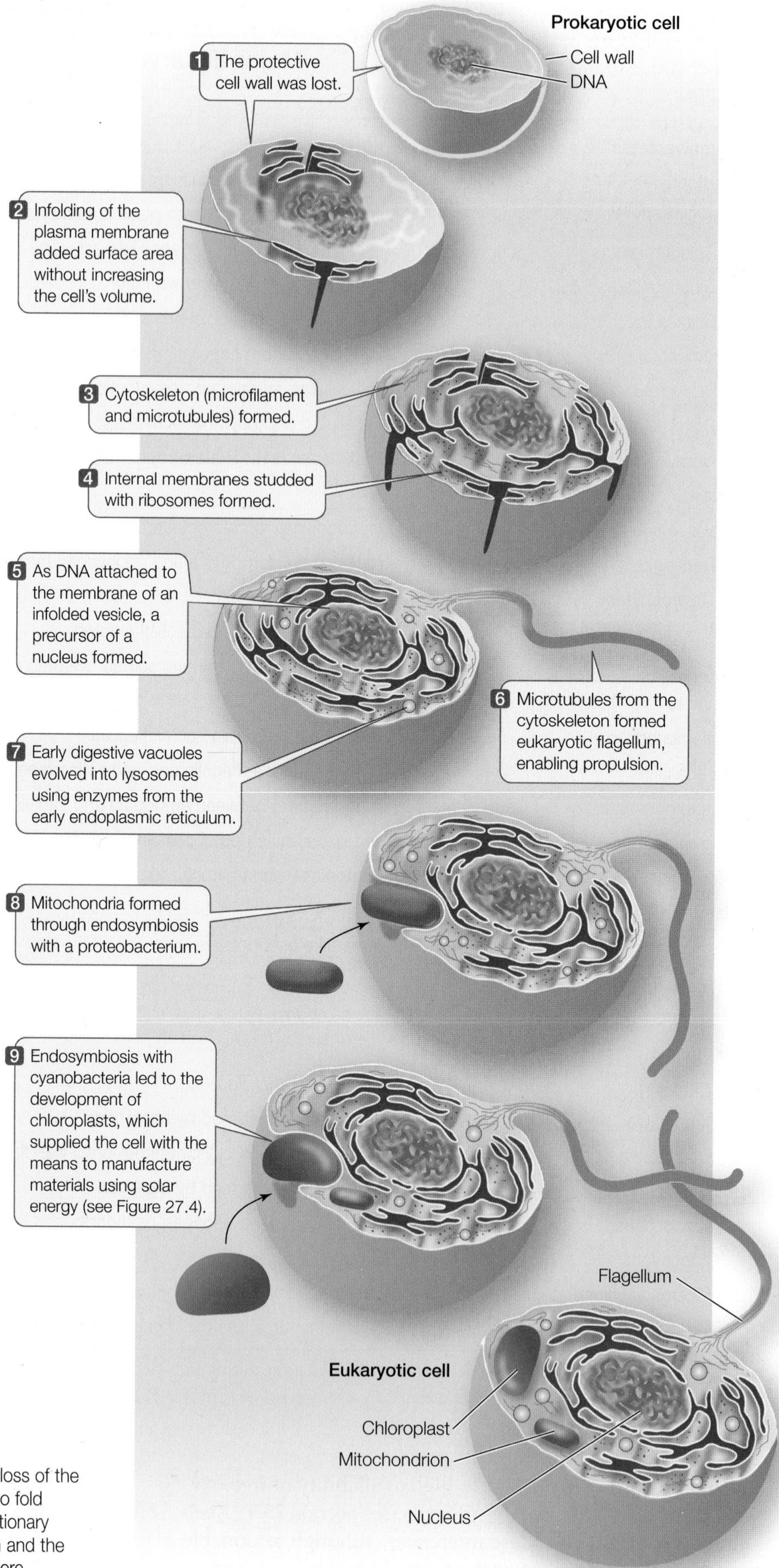

**27.2 From Prokaryotic Cell to Eukaryotic Cell** The loss of the rigid prokaryotic cell wall allowed the plasma membrane to fold inward and create more surface area. One possible evolutionary sequence, which includes the formation of a cytoskeleton and the enclosure of genetic material into the nucleus, is shown here.

A cytoskeleton composed of microfilaments and microtubules would support the cell and allow it to manage changes in shape, to distribute daughter chromosomes, and to move materials from one part of the now much larger cell to other parts. The presence of microtubules in the cytoskeleton could have evolved in some cells to give rise to the characteristic eukaryotic flagellum. The origin of the cytoskeleton is becoming clearer, as homologs of the genes that encode many cytoskeletal proteins have been found in modern prokaryotes.

The DNA of a prokaryotic cell is attached to a site on its plasma membrane. If that region of the plasma membrane were to fold into the cell, the first step would be taken toward the evolution of a *nucleus*, a primary feature of the eukaryotic cell.

From an intermediate kind of cell, the next step was probably *phagocytosis*—the ability to engulf and digest other cells. The cytoskeleton and nuclear envelope appeared early in the eukaryote lineage. Early eukaryotes may also have had an associated endoplasmic reticulum and Golgi apparatus, and perhaps one or more flagella of the eukaryotic type.

**ENDOSYMBIOSIS AND ORGANELLES** At the same time the processes outlined above were taking place, cyanobacteria were generating oxygen gas as a product of photosynthesis. The increasing $O_2$ levels in the atmosphere had disastrous consequences because most organisms of the time (archaea and bacteria) were unable to tolerate the newly oxidizing environment. But some prokaryotes managed to cope with these changes, and—fortunately for us—so did some of the new phagocytic eukaryotes.

At about this time, endosymbiosis might have come into play (Figure 27.2, steps 8–9). Recall that the theory of endosymbiosis proposes that certain organelles are the descendants of prokaryotes engulfed, but not digested, by ancient eukaryotic cells (see Section 5.5). One crucial endosymbiotic event in the history of the Eukarya was the incorporation of a proteobacterium that evolved into the mitochondrion. Initially, the new organelle's primary function was probably to detoxify $O_2$ by reducing it to water. Later, this reduction became coupled with the formation of ATP in cellular respiration (see Chapter 9). Upon completion of this step, the essential modern eukaryotic cell was complete.

Some important eukaryotes are the result of yet another endosymbiotic step, the incorporation of a prokaryote related to today's cyanobacteria, which became the chloroplast.

## Chloroplasts are a study in endosymbiosis

Eukaryotes in several different groups possess chloroplasts, and groups with chloroplasts appear in several distantly related clades. Some of these groups differ in the photosynthetic pigments their chloroplasts contain. And we'll see that not all chloroplasts have a pair of surrounding membranes—in some microbial eukaryotes, they are surrounded by *three or more* membranes. We now view these observations in terms of a remarkable series of endosymbioses, supported by extensive evidence from electron microscopy and nucleic acid sequence comparisons.

All chloroplasts trace their ancestry back to the engulfment of one cyanobacterium by a larger eukaryotic cell (**Figure 27.3A**).

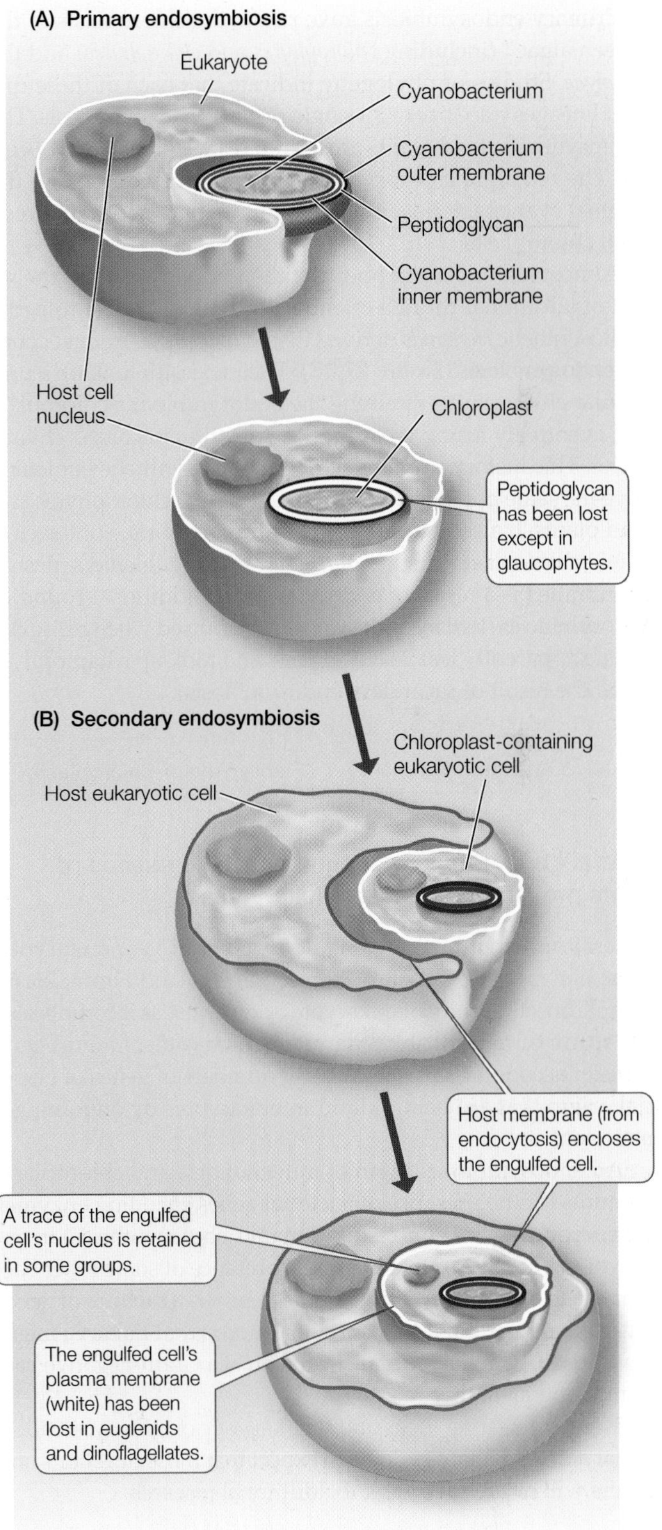
(A) Primary endosymbiosis
Eukaryote
Cyanobacterium
Cyanobacterium outer membrane
Peptidoglycan
Cyanobacterium inner membrane
Host cell nucleus
Chloroplast
Peptidoglycan has been lost except in glaucophytes.
(B) Secondary endosymbiosis
Chloroplast-containing eukaryotic cell
Host eukaryotic cell
Host membrane (from endocytosis) encloses the engulfed cell.
A trace of the engulfed cell's nucleus is retained in some groups.
The engulfed cell's plasma membrane (white) has been lost in euglenids and dinoflagellates.

**27.3 Endosymbiotic Events in the "Family Tree" of Chloroplasts**
(A) A single instance of primary endosymbiosis ultimately gave rise to all of today's chloroplasts. A eukaryotic cell engulfed a cyanobacterium but did not digest it. (B) Secondary endosymbiosis—the uptake and retention of a chloroplast-containing cell by another eukaryotic cell—took place several times, independent of each other.

This event, the step that gave rise to the photosynthetic eukaryotes, is known as **primary endosymbiosis**. The cyanobacterium, a Gram-negative bacterium, had both an inner and outer membrane. Thus the original chloroplasts had two surrounding membranes—the inner and outer membranes of the cyanobacterium. Remnants of the peptidoglycan-containing wall of the bacterium are present in the form of a bit of peptidoglycan between the chloroplast membranes of *glaucophytes*, the first microbial eukaryote group to branch off following primary endosymbiosis of the cyanobacterium.

Primary endosymbiosis gave rise to the chloroplasts of the "green algae" (including *chlorophytes* and *charophytes*) and the *red algae*. Studies of phylogeny indicate that each of these distinct lineages trace back to a single primary endosymbiosis. The photosynthetic land plants arose later from a green algal ancestor. The red algal chloroplast retains certain pigments of the original cyanobacterial endosymbiont that are absent in green algal chloroplasts.

Almost all remaining photosynthetic eukaryotes are the result of additional rounds of endosymbiosis. For example, the photosynthetic *euglenids* derived their chloroplasts from **secondary endosymbiosis** (**Figure 27.3B**). Their ancestor took up a unicellular chlorophyte, retaining the endosymbiont's chloroplast and eventually losing the rest of the constituents of the chlorophyte. This history explains why the photosynthetic euglenids have the same photosynthetic pigments as the chlorophytes and land plants. It also accounts for the third membrane of the euglenoid chloroplast, which is derived from the euglenid's plasma membrane (as a result of endocytosis). An additional round of endosymbiosis (**tertiary endosymbiosis**) occurred when a dinoflagellate apparently lost its chloroplast and took up a haptophyte (itself the result of secondary endosymbiosis).

**yourBioPortal.com**

GO TO Animated Tutorial 27.1 • Family Tree of Chloroplasts

### Lateral gene transfer accounts for the presence of some prokaryotic genes in eukaryotes

Several uncertainties remain about the origins of eukaryotic cells. Lateral gene transfer (see Section 26.4 and Figure 26.10) complicates the study of eukaryote origins, just as it complicates the study of relationships among prokaryotes. Lateral gene transfer accounts for the increasing numbers of genes of bacterial origin that are being found in eukaryotes by ongoing genetic analyses.

An endosymbiotic origin of mitochondria and chloroplasts accounts for the presence of bacterial genes encoding enzymes for energy metabolism (respiration and photosynthesis) in eukaryotes, but it does not explain the presence of some other bacterial genes. The eukaryotic genome clearly is a mixture of genes with different origins. A recent suggestion is that Eukarya might have arisen from the mutualistic fusion of a Gram-negative bacterium and an archaean.

Many interesting ideas about eukaryotic origins await additional data and analysis. We can expect that these questions and others will eventually yield to additional research.

### 27.1 RECAP

**The modern eukaryotic cell likely arose from an ancestral prokaryote in several steps, including endosymbiosis. Primary endosymbiosis involves the engulfment of a bacterium by another free-living organism; the endosymbiosis of a bacterium probably gave rise to mitochondria, whereas the chloroplasts of photosynthesizing eukaryotes are thought to have originated in the endosymbiosis of a cyanobacterium.**

- Explain why protists are described as paraphyletic rather than monophyletic. **See p. 561, Table 27.1, and Figure 27.1**
- Why was the development of a flexible cell surface a key event in eukaryotic cell history? **See p. 564**
- Identify some of the probable events involved in the evolution of the eukaryotic cell from a prokaryotic cell. **See pp. 564–565 and Figure 27.2**
- What is the difference between primary and secondary endosymbiosis? **See p. 566 and Figure 27.3**

Having considered some of the known and suspected steps that led from the prokaryotic to the eukaryotic condition, let's now see what use the early eukaryotes made of their new features.

## 27.2 What Features Account for Protist Diversity?

The eukaryotic cell possesses some very useful features (detailed in Chapter 5). The cytoskeleton allows for various means of locomotion and also manages the controlled movement of cellular constituents (notably the mitotic and meiotic chromosomes). The specialized organelles of eukaryotes support a variety of activities. Given these tools, eukaryotes have been able to explore many environments and have exploited a variety of nutrient sources.

### Protists occupy many different niches

Most protists are aquatic. Some live in marine environments, others in fresh water, and still others in the body fluids of host organisms. Many aquatic protists are plankton: free-floating aquatic organisms. The *slime molds* inhabit damp soil, animal feces, and the moist, decaying bark of rotting trees. Other microbial protists also live in soil water, and some of them contribute to the global nitrogen cycle by preying on soil bacteria and recycling their nitrogen compounds into nitrates.

Many metabolic lifestyles are found among the protists. Some protists are photosynthetic autotrophs, some are heterotrophs, and some switch with ease between the autotrophic and heterotrophic modes of nutrition. Some of the heterotrophs ingest their food; others, including many parasites, absorb nutrients from their environment.

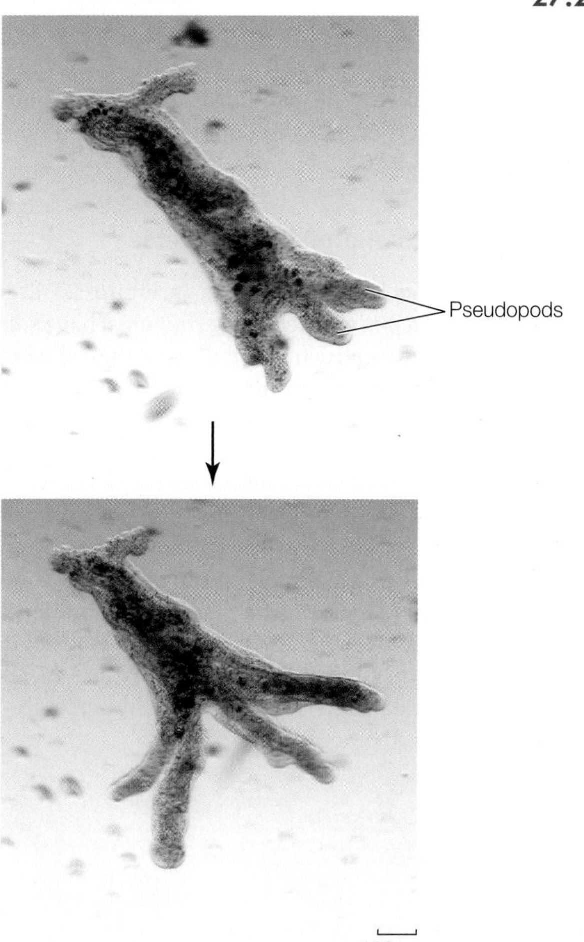
*Peloxima carolinensis*
Pseudopods
120 μm

**27.4 An Amoeba in Motion** Its flowing pseudopods are constantly changing shape as this "chaos amoeba" moves and feeds.

Two general terms are sometimes used to designate two broad categories of protists: the *protozoans* and *algae*. These are convenience terms used to describe categories of distantly related species that have some similar attributes. The term *protozoans* refers to various groups of protists, formerly classified as animals, that often survive by ingesting other species. Likewise, the term *algae* (singular *alga*) refers to several groups of photosynthetic protists. Neither of these terms designate formal taxonomic groups, however.

## Protists have diverse means of locomotion

Although a few protist groups consist entirely of nonmotile organisms, most groups include organisms that move by amoeboid motion, by ciliary action, or by means of flagella. Each of these types of motion is based on activities of the cytoskeleton.

In *amoeboid motion*, the cell forms **pseudopods** ("false feet") which are extensions of its constantly changing cell shape. Cells such as the one shown in **Figure 27.4** simply extend a pseudopod and then flow into it. Regions of the cytoplasm alternate between a more liquid state and a stiffer state, and a network of cytoskeletal microfilaments squeezes the more liquid cytoplasm forward.

The proteins of the eukaryotic cytoskeleton form microtubules which allowed the evolution of different means of locomotion. *Cilia* are tiny, hairlike organelles that beat in a coordinated fashion to move the cell forward or backward. Some ciliated organisms can change direction rapidly in response to their environment. A eukaryotic flagellum moves like a whip; some flagella *push* the cell forward, others *pull* the cell forward. Cilia and eukaryotic flagella are identical in cross section, with a "9 + 2" arrangement of microtubules (see Figure 5.20); they differ only in length.

## Protists employ vacuoles in several ways

Most unicellular organisms are microscopic. As noted above, an important reason that cells are small is that they need enough membrane surface area in relation to their volume to support the exchange of materials required for their existence. Some relatively large unicellular eukaryotes minimize this problem by having membrane-enclosed vacuoles of various types that increase their effective surface area.

Organisms living in fresh water are hypertonic to their environment (see Section 6.3). Many freshwater protists such as *Paramecium* address this problem by means of specialized vacuoles that excrete the excess water they constantly take in by osmosis. Members of several groups have such **contractile vacuoles**. The excess water collects in the contractile vacuoles, which then expel the water from the cell (**Figure 27.5**).

A **digestive vacuole** is a second important type of vacuole found in *Paramecium* and many other protists. These organisms engulf solid food by endocytosis, forming a vacuole within which the food is digested. Smaller vesicles containing digested food pinch away from the digestive vacuole and enter the cy-

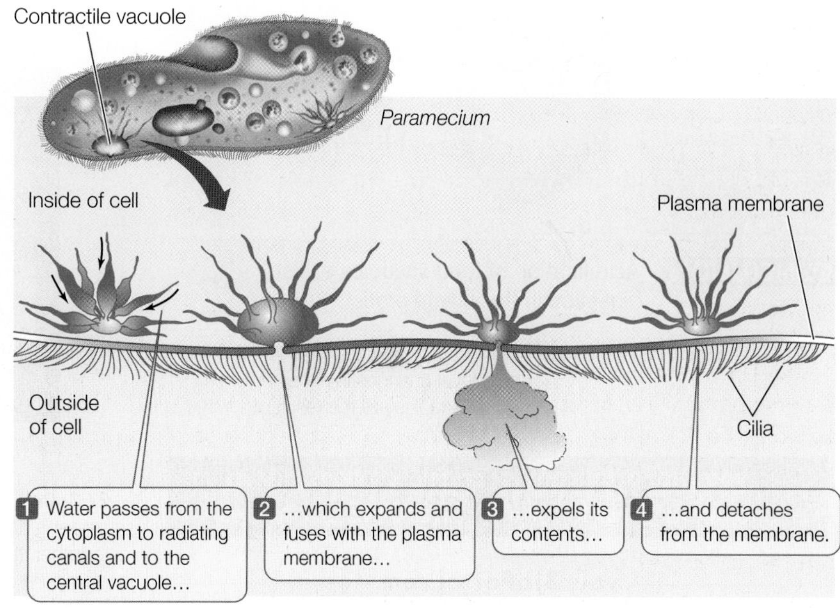
Contractile vacuole
*Paramecium*
Inside of cell
Plasma membrane
Outside of cell
Cilia
1 Water passes from the cytoplasm to radiating canals and to the central vacuole...
2 ...which expands and fuses with the plasma membrane...
3 ...expels its contents...
4 ...and detaches from the membrane.

**27.5 Contractile Vacuoles Bail Out Excess Water** Contractile vacuoles remove the water that constantly enters freshwater protists by osmosis.

INVESTIGATING LIFE

**27.6 The Role of Vacuoles in Ciliate Digestion**
An experiment with the ciliate protist *Paramecium* demonstrates the function of food vacuoles. Given that an acidic environment is known to aid digestion in many organisms, does this microbial eukaryote use acid to obtain nutrients?

**HYPOTHESIS** **The food vacuoles of *Paramecium* produce an acidic environment that allows the organism to digest food particles.**

**METHOD**
1. Feed *Paramecium* yeast cells stained with Congo red, a dye that is red at neutral or basic pH but turns green at acidic pH.
2. Under a light microscope, observe the formation and degradation of food vacuoles within the *Paramecium*. Note time and sequence of color (i.e., acid level) changes.

**RESULTS**

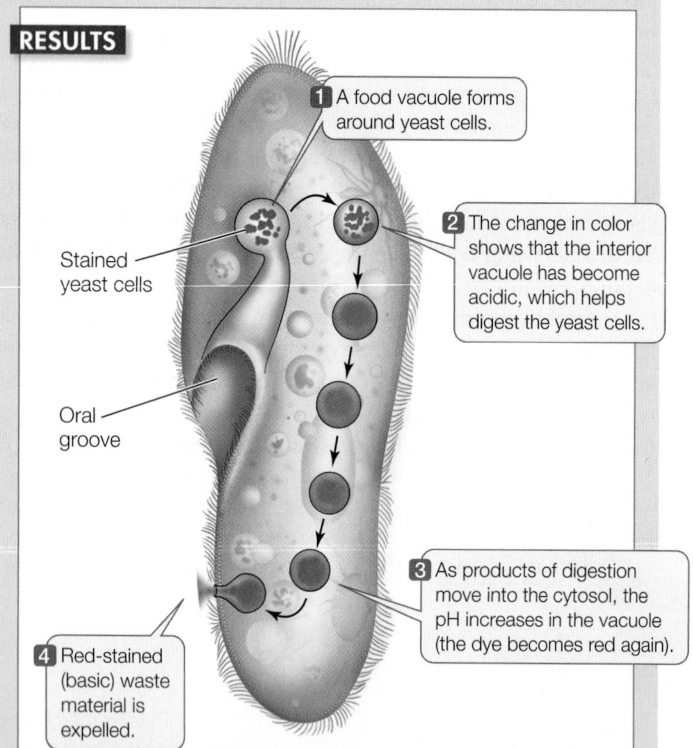
1 A food vacuole forms around yeast cells.
2 The change in color shows that the interior vacuole has become acidic, which helps digest the yeast cells.
3 As products of digestion move into the cytosol, the pH increases in the vacuole (the dye becomes red again).
4 Red-stained (basic) waste material is expelled.
Stained yeast cells
Oral groove

**CONCLUSION** **Acidification of food vacuoles assists digestion in this ciliate protist.**

**FURTHER INVESTIGATION:** How might you determine whether the acid level changes in *Paramecium's* food vacuoles are the result of enzymes?

Go to **yourBioPortal.com** for original citations, discussions, and relevant links for all INVESTIGATING LIFE figures.

**yourBioPortal.com**
GO TO **Animated Tutorial 27.2 • Food Vacuoles Handle Digestion and Excretion**

toplasm. These tiny vesicles provide a large surface area across which the products of digestion may be absorbed by the rest of the cell (**Figure 27.6**).

### The cell surfaces of protists are diverse

A few protists, such as the amoeba in Figure 27.4, are surrounded by only a plasma membrane, but most have stiffer surfaces that maintain the structural integrity of the cell. Many have cell walls, which are often complex in structure, outside the plasma membrane. Other protists that lack cell walls have a variety of ways of strengthening their surfaces.

*Paramecium* has proteins in its cell surface—known as a *pellicle* in this genus (see Figure 27.18)—that make it flexible but resilient. Other groups have external "shells," which the organism either produces itself or makes from bits of sand and thickenings immediately beneath the plasma membrane, as some amoebas do (**Figure 27.7A**). The complex cell walls of diatoms are glassy, based on silica (silicon dioxide; **Figure 27.7B**). Biologists recently measured, at a microscopic scale, the forces

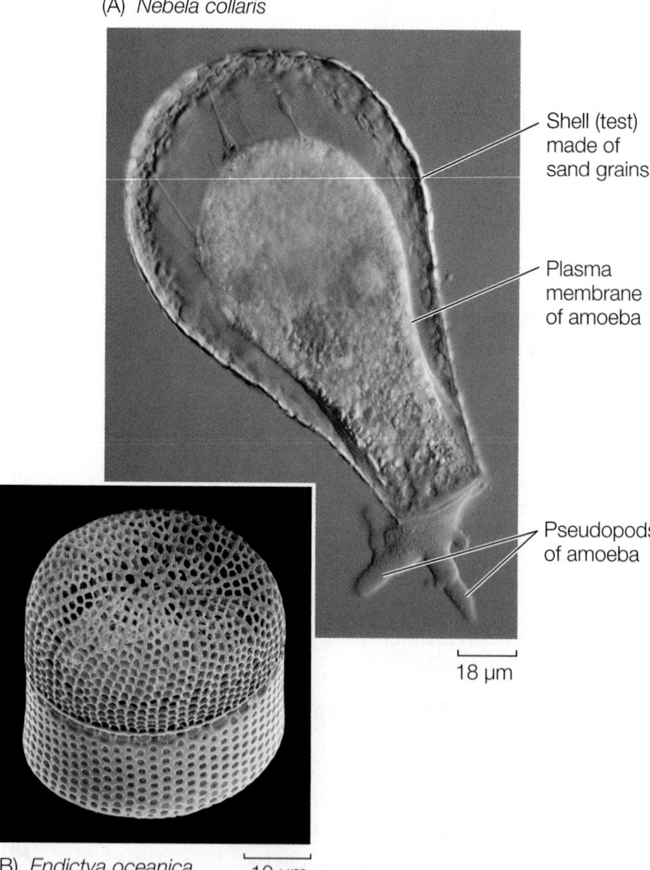
(A) *Nebela collaris*
Shell (test) made of sand grains
Plasma membrane of amoeba
Pseudopods of amoeba
18 μm
(B) *Endictya oceanica*
10 μm

**27.7 Cell Surfaces among Microbial Eukaryotes** (A) This testate amoeba has built a lightbulb-shaped shell, or test, by gluing sand grains together. Its pseudopods extend through the single aperture in the test (compare with Figure 27.4). (B) Scanning electron micrography reveals the intricate patterning of the silica-dense cell walls of this diatom. These spectacular unicellular, photosynthetic eukaryotes dominate the aquatic phytoplankton community (see also Figure 27.19).

needed to break single, living diatoms, and discovered that the glassy cell walls are exceptionally strong. Evolution of these walls by natural selection may have given diatoms an enhanced defense against predators, and thus an edge over competitors.

## 27.2 RECAP

**Protists are diverse in their habitat, nutrition, locomotion, and body form. Some protists move by use of pseudopods, cilia, or flagella. Protists may have cell walls or external coverings that provide structural support and protection.**

- Can you explain the roles of the cytoskeleton in the locomotion of protists? See p. 567
- Do you understand the operation of contractile and food vacuoles in *Paramecium*? See p. 567 and Figures 27.5 and 27.6

The diversity of body form, habitat, nutrition, and locomotion found among the protists reflects the diversity of avenues pursued during the early evolution of eukaryotes. Protists have an enormous effect on biotic and physical environments.

# 27.3 How Do Protists Affect the World Around Them?

Protists are extremely diverse, and their effects on other organisms and on the physical environment are almost as diverse. Some protists are food for marine animals, while others poison the sea; some are packaged as nutritional supplements, and some are pathogens; the remains of some form the sands of many modern beaches, and others are a major source of today's ever more expensive crude oil. Many protists are constituents of the plankton. Photosynthetic members of the plankton are called **phytoplankton**.

A single eukaryote clade, the *diatoms*, is responsible for about a fifth of all photosynthetic carbon fixation on Earth—about the same amount of photosynthesis performed by all of Earth's rainforests. These spectacular unicellular organisms (see Figures 27.7B and Figure 27.19) are the predominant members of the phytoplankton, but other protist clades also include important phytoplanktonic species that contribute heavily to global photosynthesis. Like green plants on land, the phytoplankton serve as a gateway for energy from the sun into the living world; in other words, they are *primary producers*. In turn, they are eaten by heterotrophs, including animals and many other protists. Those consumers are, in turn, eaten by other consumers. Most aquatic heterotrophs (with the exception of some species existing in the deep ocean) depend on the photosynthesis performed by phytoplankton.

### Some protists are endosymbionts

As we have described, endosymbiosis is the condition in which two organisms live together, one inside the other. Endosymbiosis is common among the microbial protists, many of which live within the cells of animals. Members of the *dinoflagellates* are common symbionts in both animals and in other protists; most but not all dinoflagellate endosymbiont species are photosynthetic. Some dinoflagellates live endosymbiotically in the cells of corals, contributing products of their photosynthesis to the partnership. The importance to the coral is demonstrated when the dinoflagellates die as a result of changing environmental conditions, a phenomenon known as coral bleaching (**Figure 27.8A**); the coral is ultimately damaged or destroyed when its nutrient supply is reduced.

Many *radiolarians* also harbor photosynthetic endosymbionts (**Figure 27.8B**). As a result, the radiolarians, which are not photosynthetic themselves, appear greenish or golden, depending on the type of endosymbiont they contain. This arrangement is often mutually beneficial: the radiolarian can make use of the organic

(A)

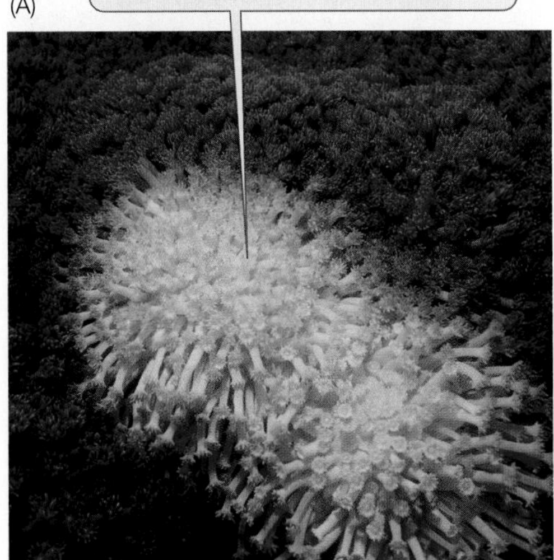

Bleaching occurs when the photosynthetic dinoflagellates living endosymbiotically within this coral die.
*Goniopora* sp.


(B)

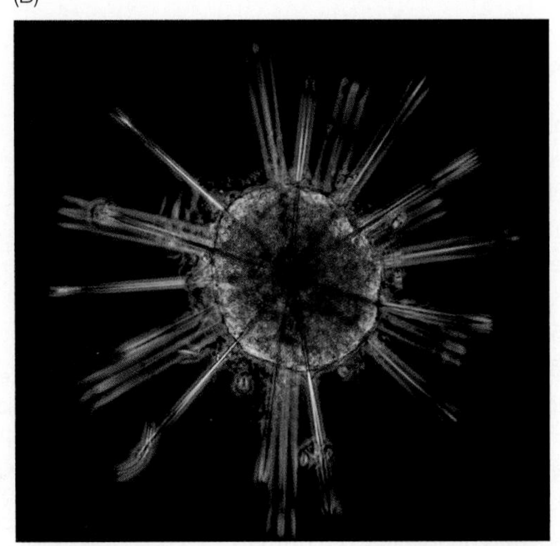


**27.8 Dinoflagellate Endosymbionts are Photosynthesizers** (A) Some corals lose their chief nutritional source when their photosynthetic endosymbionts die, often as a result of changing environmental conditions such as warming water. (B) Dinoflagellates live endosymbiotically inside a radiolarian (another protist), providing organic nutrients for the radiolarian and imparting the golden brown pigmentation seen at the center of its glassy skeleton.

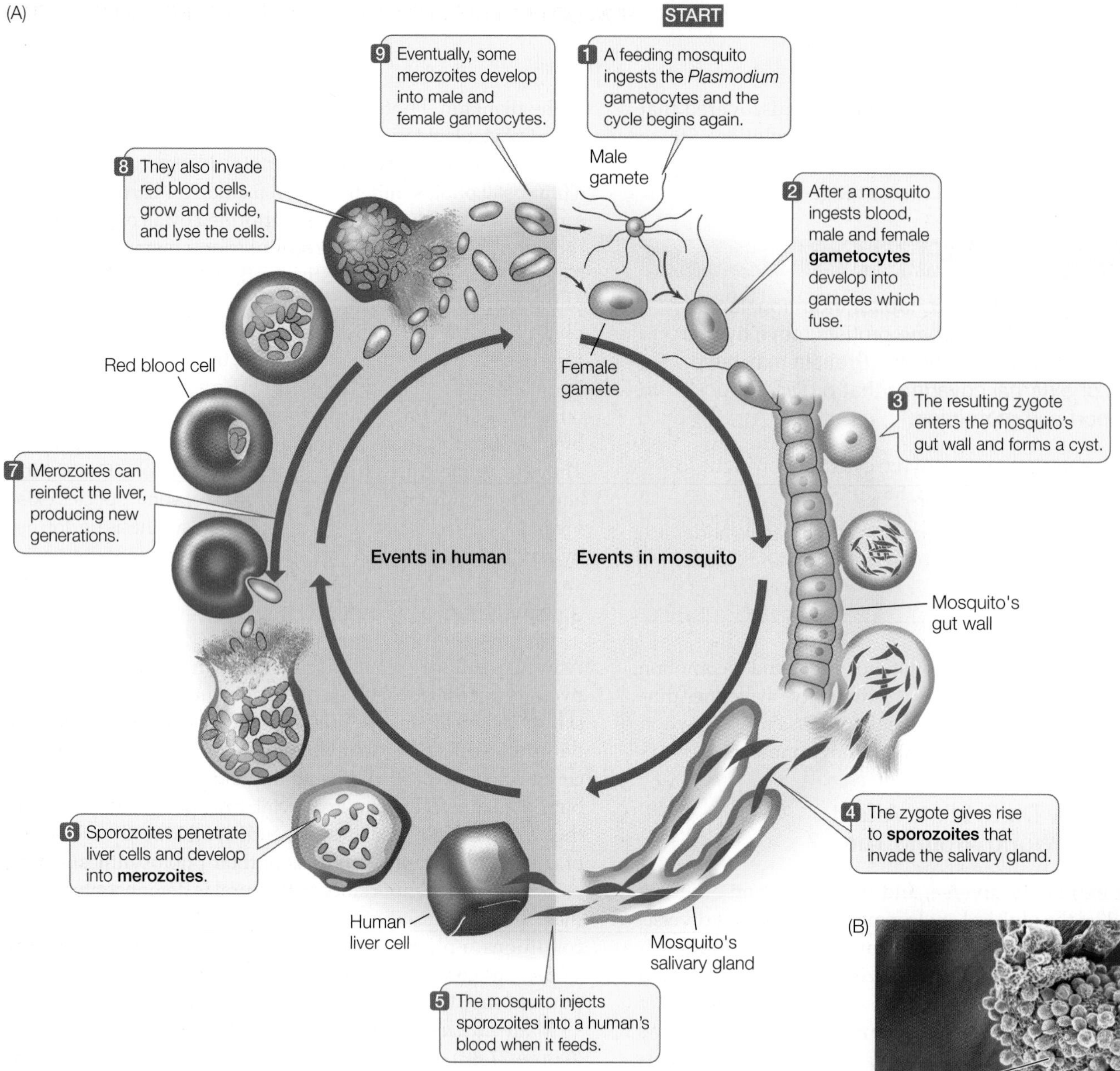

(A)

START

1 A feeding mosquito ingests the *Plasmodium* gametocytes and the cycle begins again.

2 After a mosquito ingests blood, male and female **gametocytes** develop into gametes which fuse.

3 The resulting zygote enters the mosquito's gut wall and forms a cyst.

4 The zygote gives rise to **sporozoites** that invade the salivary gland.

5 The mosquito injects sporozoites into a human's blood when it feeds.

6 Sporozoites penetrate liver cells and develop into **merozoites**.

7 Merozoites can reinfect the liver, producing new generations.

8 They also invade red blood cells, grow and divide, and lyse the cells.

9 Eventually, some merozoites develop into male and female gametocytes.

Male gamete

Female gamete

Red blood cell

Events in human

Events in mosquito

Mosquito's gut wall

Human liver cell

Mosquito's salivary gland

nutrients produced by its photosynthetic guest, and the guest may in turn make use of metabolites made by the host or receive physical protection. In some cases, the guest is exploited for its photosynthetic products while receiving little or no benefit itself.

## Some microbial protists are deadly

The best-known pathogenic microbial protists are members of the genus *Plasmodium*, a highly specialized group of *apicomplexans* that spend part of their life cycle as parasites in human red blood cells, where they are the cause of malaria (**Figure 27.9**). In terms of the number of people affected, malaria is one of the world's three most serious infectious diseases; it kills about 880,000 people each year, out of 250 million infected individuals. On average, someone dies from malaria every 36 seconds—usually in sub-Saharan Africa, although malaria occurs in more than 100 countries.

Female mosquitoes of the genus *Anopheles* transmit *Plasmodium* to humans. The parasite enters the human circulatory system when an infected *Anopheles* mosquito penetrates the human skin in search of blood. The parasites find their way to cells in

(B)

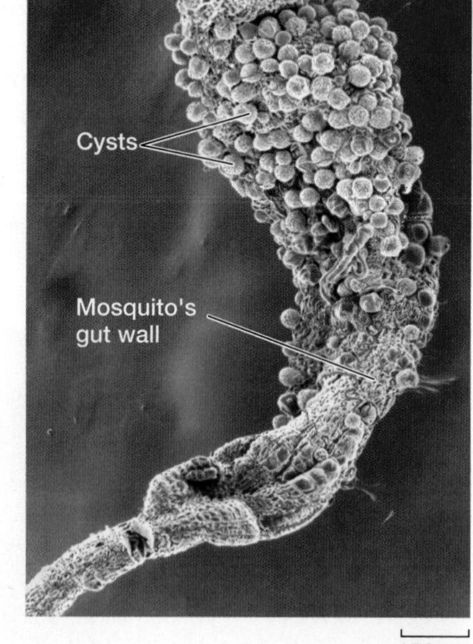


**27.9 Life Cycle of the Malarial Parasite** (A) Like many parasitic species, the apicomplexan *Plasmodium falciparum* has a complex life cycle, part of which is spent in mosquitoes of the genus *Anopheles* and part in humans. The sexual phase (gamete fusion) of this life cycle takes place in the insect, and the zygote is the only diploid stage. (B) Encysted *Plasmodium* zygotes (artificially colored blue) cover the stomach wall of a mosquito. Invasive sporozoites will hatch from the cysts and be transmitted to a human, in whom the parasite causes malaria.

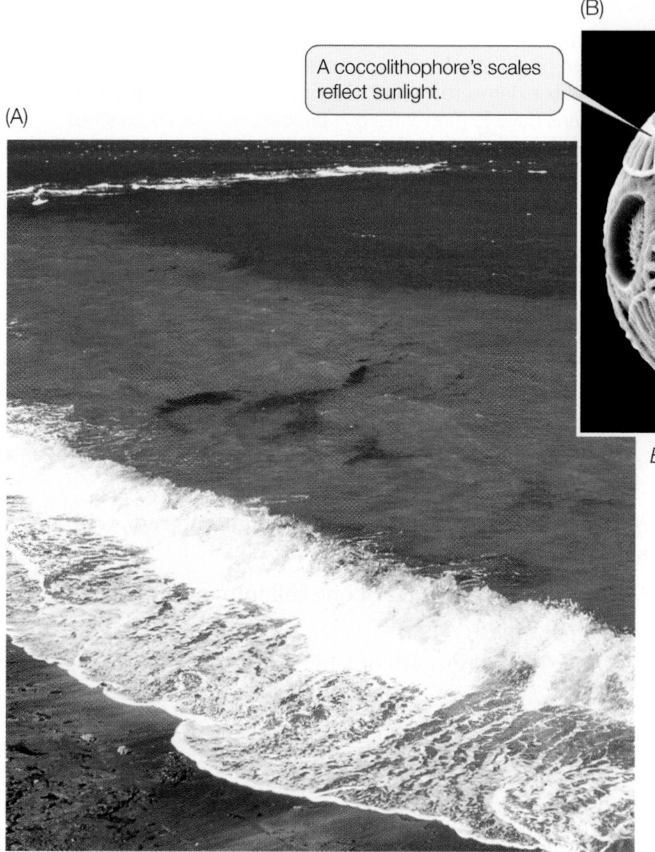


**27.10 Chromalveolates Can Bloom in the Oceans** (A) By reproducing in astronomical numbers, the dinoflagellate *Gonyaulax tamarensis* can cause toxic red tides, such as this one along the coast of Baja California. (B) Massive blooms of this coccolithophore, a tiny haptophyte, can reduce the amount of sunlight able to penetrate to the waters below.

the liver and the lymphatic system, change their form, multiply, and reenter the bloodstream, attacking red blood cells.

The parasites multiply inside the red blood cells, which then burst, releasing new swarms of parasites. If another *Anopheles* bites the victim, the mosquito takes in *Plasmodium* cells along with blood. Some of the ingested cells are gametes that formed in human cells. The gametes unite in the mosquito, forming zygotes that lodge in the mosquito's gut, divide several times, and move into its salivary glands, from which they can be passed on to another human host. Thus *Plasmodium* is an extracellular parasite in the mosquito vector and an intracellular parasite in the human host.

*Plasmodium* has proved to be a singularly difficult pathogen to attack. The complex *Plasmodium* life cycle is best broken by the removal of stagnant water, in which mosquitoes breed. Using insecticides to reduce the *Anopheles* population can be effective, but the benefits must be weighed against the ecological, economic, and health risks posed by the insecticides themselves.

The genomes of one malarial parasite, *Plasmodium falciparum*, and one of the mosquitoes that transmits malaria, *Anopheles gambiae*, have been sequenced. These advances should lead to a better understanding of the biology of malaria and to the development of drugs, vaccines, or other means of dealing with this pathogen or its insect vectors. The opening story of Chapter 30 describes a novel weapon against malaria: mosquito netting containing fungi that attack mosquitoes.

Some *kinetoplastids* are human pathogens, such as the trypanosomes discussed at the opening of this chapter. Recall that trypanosomes cause sleeping sickness, leishmaniasis, and Chagas' disease. The genomes of all three of the trypanosomes responsible for these diseases have now been sequenced.

Some *chromalveolates*, including diatoms, dinoflagellates, and haptophytes, reproduce in enormous numbers in warm and somewhat stagnant waters. The result can be a "red tide," so called because of the reddish color of the sea that results from the dinoflagellates' pigments (**Figure 27.10A**). During a dinoflagellate red tide, the concentration of cells may reach 60 million per liter of ocean water. Some red tide species produce a potent nerve toxin that can kill tons of fish. The genus *Gonyaulax* produces a toxin that can accumulate in shellfish in amounts that, although not fatal to the shellfish, may kill a person who eats the shellfish.

The haptophyte *Emiliania huxleyi* is one of the smallest unicellular protists, but it can form tremendous blooms in ocean waters. This *coccolithophore* ("sphere of stone") has an armored coating that makes the surface water more reflective (**Figure 27.10B**). This reflectivity cools the deeper layers of water below the bloom by reducing the amount of sunlight that penetrates. At the same time, it is possible that *E. huxleyi* contributes to global warming, because its metabolism increases the amount of dissolved $CO_2$ in ocean waters.

### We continue to rely on the products of ancient marine protists

Diatoms are often lovely to look at, but their importance to us goes far beyond aesthetics. They store oil as an energy reserve and to help them float at the correct depth in the ocean. Over millions of years, diatoms have died and sunk to the ocean floor, ultimately undergoing chemical changes and becoming a major source of petroleum and natural gas, two of our most important energy supplies and political concerns.

Other marine protists have also contributed to today's world. Some *foraminiferans*, for example, secrete shells of calcium carbonate. After they reproduce (by mitosis and cytokinesis), the daughter cells abandon the parent shell and make new shells of their own. The discarded shells of ancient foraminiferans make up extensive limestone deposits in various parts of the world, forming a layer hundreds to thousands of meters deep over mil-

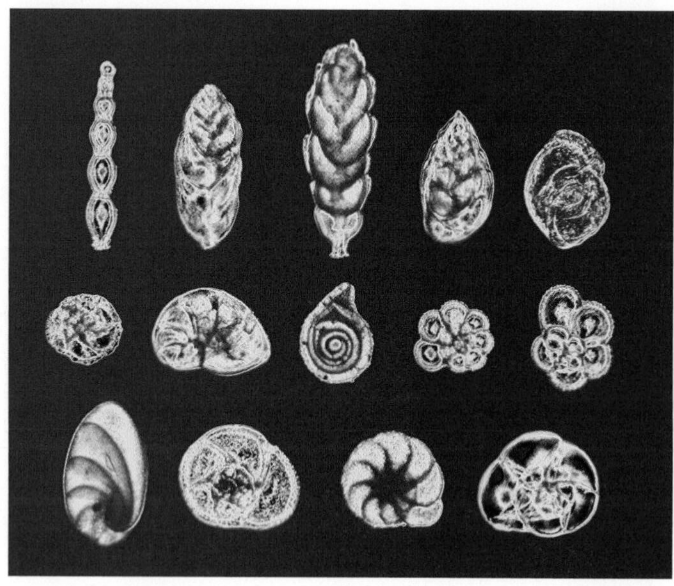


**27.11 Foraminiferan Shells Are Building Blocks** Some foraminiferan shells are made of calcium carbonate that has been mineralized on an organic matrix external to the cell. Over millions of years, the remains of foraminiferans have formed limestone deposits and sandy beaches. Several species are shown in this micrograph.

lions of square kilometers of ocean bottom. Foraminiferan shells also make up much of the sand of some beaches. A single gram of such sand may contain as many as 50,000 foraminiferan shells and shell fragments.

The shells of individual foraminiferans are easily preserved as fossils in marine sediments. The shells of foraminiferan species have distinctive shapes (**Figure 27.11**), and each geological period has a distinctive assemblage of foraminiferan species. For this reason, and because they are so abundant, the remains of foraminiferans are especially valuable in classifying and dating sedimentary rocks, as well as in oil prospecting. Analyses of foraminiferan shells are also used in determining the global temperatures prevalent at the time of their existence.

## 27.3 RECAP

**Protists have many effects, both positive and negative, on other organisms and on global ecosystems. Some species are primary producers, many are endosymbionts, and some are pathogens. They are among the most important producers of petroleum products, and they are important for producing and dating sedimentary rock formations.**

- Can you describe the role of female mosquitoes of the genus *Anopheles* in the transmission of malaria? See pp. 570–571 and Figure 27.9
- Do you understand the role of dinoflagellates in the very different phenomena of coral bleaching and red tides? See pp. 569 and 571

This section has presented a brief overview of the many ways protists interact with one another and with other species. Next we examine their diverse forms of reproduction.

## 27.4 How Do Protists Reproduce?

Although most protists engage in both asexual and sexual reproduction, sexual reproduction has yet to be observed in some groups. As we will see, some protists separate the acts of sex and reproduction, so that the two are not directly linked.

Asexual reproductive processes found among the protists include:

- *Binary fission:* Equal splitting of one cell into two, with mitosis followed by cytokinesis
- *Multiple fission:* Splitting of one cell into multiple (i.e., more than two) cells
- *Budding:* The outgrowth of a new cell from the surface of an old one
- *Spores:* The formation of specialized cells that are capable of developing into new organisms

Sexual reproduction also occurs among the protists, and it takes various forms. In some protists, as in animals, the gametes are the only haploid cells. In others, the zygote is the only diploid cell. In still others, both diploid and haploid cells undergo mitosis, giving rise to *alternation of generations* (the alternation of multicellular diploid and haploid life stages).

### Some protists have reproduction without sex, and sex without reproduction

Members of the genus *Paramecium* are *ciliates,* a protist group characterized by the possession of two types of nuclei in a single cell—commonly one *macronucleus* and from one to several *micronuclei.* The micronuclei, which are typical eukaryotic nuclei, are essential for genetic recombination. The macronucleus is derived from micronuclei. Each macronucleus contains many copies of the genetic information, packaged in units containing very few genes each. The macronuclear DNA is transcribed and translated to regulate the life of the cell. In asexual reproduction, all of the nuclei are copied before the cell divides.

Paramecia also have an elaborate sexual behavior called **conjugation**, in which two paramecia line up tightly against each other and fuse in the oral groove region of the body. Nuclear material is extensively reorganized and exchanged over the next several hours (**Figure 27.12**). Each cell ends up with two haploid micronuclei, one of its own and one from the other cell, which fuse to form a new diploid micronucleus. A new macronucleus develops from the micronucleus through a series of dramatic chromosomal rearrangements. The exchange of nuclei is fully reciprocal—each of the two paramecia gives and receives an equal amount of DNA. The two organisms then separate and go their own ways, each equipped with new combinations of alleles.

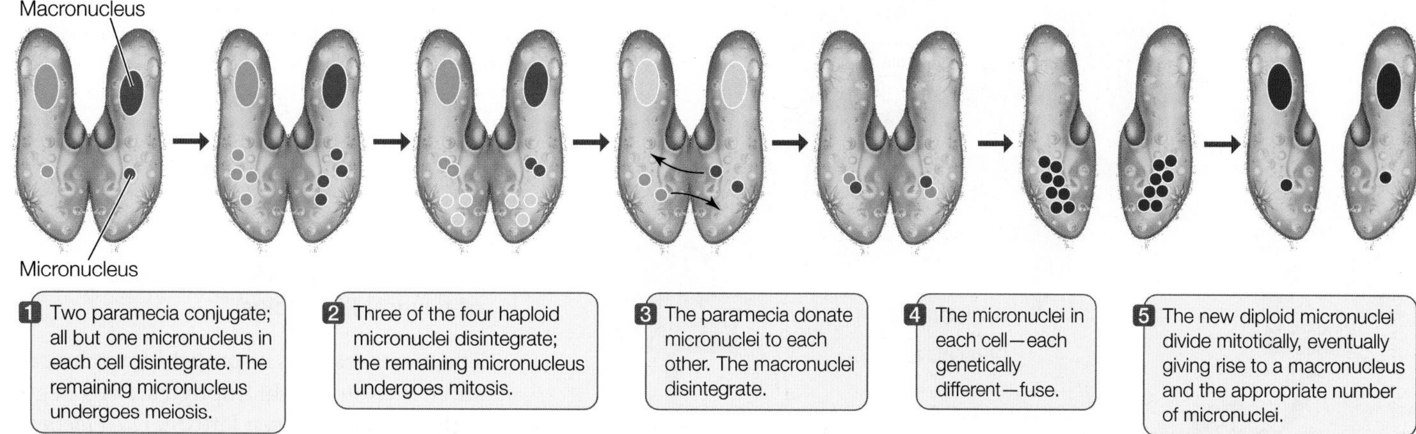

Macronucleus
Micronucleus
1 Two paramecia conjugate; all but one micronucleus in each cell disintegrate. The remaining micronucleus undergoes meiosis.
2 Three of the four haploid micronuclei disintegrate; the remaining micronucleus undergoes mitosis.
3 The paramecia donate micronuclei to each other. The macronuclei disintegrate.
4 The micronuclei in each cell—each genetically different—fuse.
5 The new diploid micronuclei divide mitotically, eventually giving rise to a macronucleus and the appropriate number of micronuclei.

**27.12 Paramecia Achieve Genetic Recombination by Conjugating** The exchange of micronuclei by two conjugating *Paramecium* individuals results in genetic recombination. After conjugation, the cells separate and continue their lives as two individuals.

Conjugation in *Paramecium* is a *sexual* process of genetic recombination, but it is not a *reproductive* process. Two cells begin the process and two cells are there at the end, so no new cells are created. As a rule, each asexual clone of paramecia must periodically conjugate. Experiments have shown that if some species are not permitted to conjugate, the clones can live through no more than approximately 350 cell divisions before they die out.

## Some protist life cycles feature alternation of generations

**Alternation of generations** is a type of life cycle found in many multicellular protists, land plants, and some fungi. The term describes a life cycle in which a multicellular, diploid, spore-producing organism gives rise to a multicellular, haploid, gamete-producing organism. When two haploid gametes fuse (*fertilization*, or *syngamy*), a diploid organism is formed (**Figure 27.13**). The haploid organism, the diploid organism, or both may also reproduce asexually.

The two alternating generations (spore-producing and gamete-producing) differ genetically (one has diploid cells, the other haploid cells), but they may or may not differ morphologically. In **heteromorphic** alternation of generations, the two generations differ morphologically; in **isomorphic** alternation of generations, they do not, despite their genetic difference.

Examples of both heteromorphic and isomorphic alternation of generations are found in both brown algae and green algae. As we discuss the life cycles of land plants and multicellular photosynthetic protists, we will use the terms **sporophyte** ("spore plant") and **gametophyte** ("gamete plant") to refer to the multicellular diploid and haploid generations, respectively.

Gametes are not produced by meiosis because the gametophyte generation is already haploid. Instead, specialized cells of the diploid sporophyte, called **sporocytes**, divide meiotically to produce four haploid spores. The spores may eventually germinate and divide mitotically to produce multicellular haploid gametophytes, which produce gametes by mitosis and cytokinesis.

Gametes, unlike spores, can produce new organisms only by fusing with other gametes. The fusion of two gametes produces a diploid zygote, which then undergoes mitotic divisions to produce a diploid organism: the sporophyte generation. The sporocytes of the sporophyte generation then undergo meiosis and produce haploid spores, starting the cycle anew.

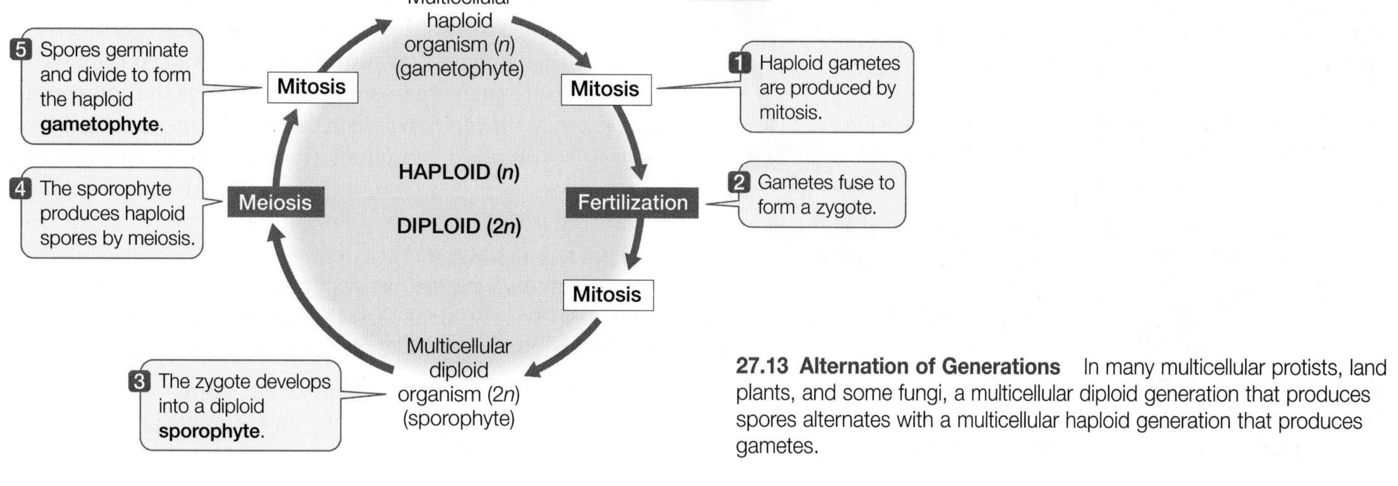


**27.13 Alternation of Generations** In many multicellular protists, land plants, and some fungi, a multicellular diploid generation that produces spores alternates with a multicellular haploid generation that produces gametes.

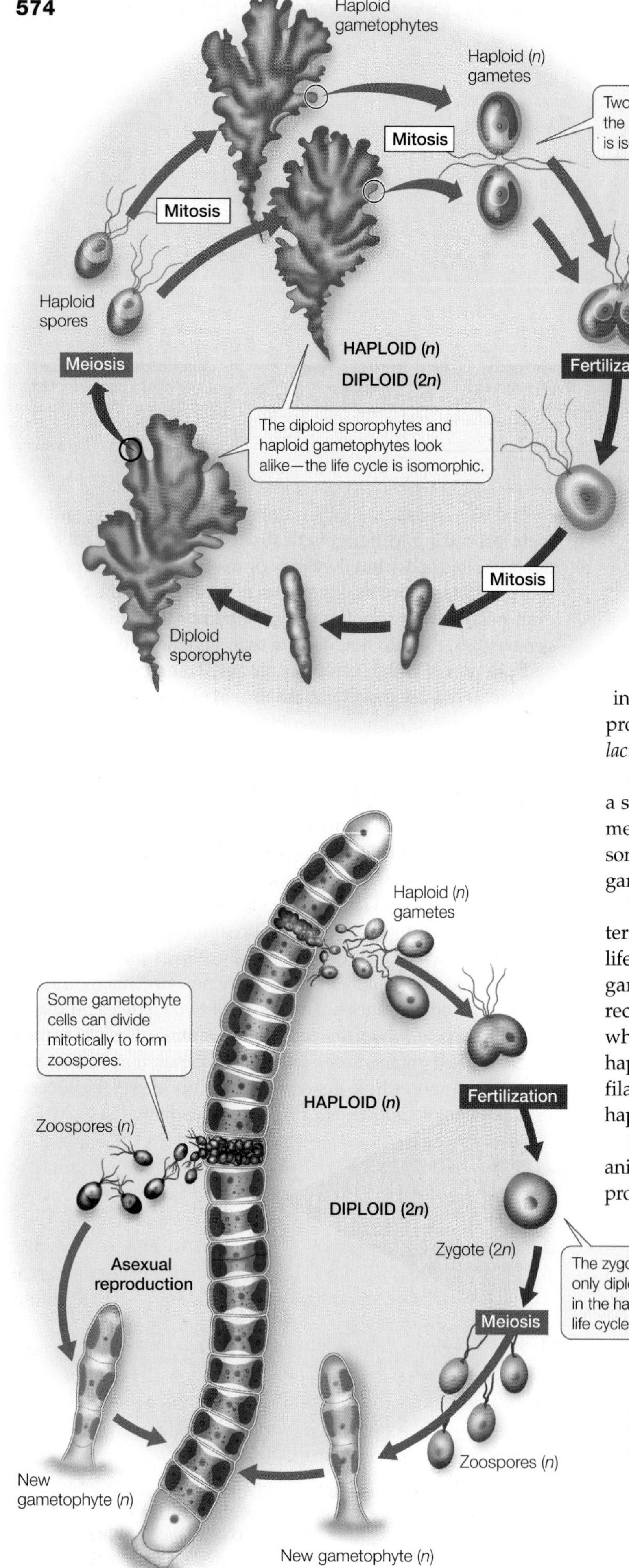


**27.14 An Isomorphic Life Cycle** The sexual life cycle of sea lettuce (*Ulva lactuca*) is an example of isomorphic alternation of generations.

**yourBioPortal.com**
GO TO Web Activity 27.1 • An Isomorphic Life Cycle

## Chlorophytes provide examples of several life cycles

The major features of protist life cycles can all be seen in the chlorophytes. Let's begin with the common sea lettuce, *Ulva lactuca*. Like many chlorophytes, sea lettuce exhibits alternation of generations. The diploid sporophyte of this common multicellular seashore organism is a broad sheet only two cells thick. Some of its cells (sporocytes) differentiate and undergo meiosis and cytokinesis, producing motile haploid spores. These *zoospores* swim away, and some find a suitable place to settle. The zoospores begin to divide mitotically, producing a thin filament that develops into a broad sheet only two cells thick. The gametophyte thus produced looks just like the sporophyte—in other words, *Ulva lactuca* has an isomorphic life cycle (**Figure 27.14**).

In most species of *Ulva*, the gametes are structurally of just a single type, making those species **isogamous**—having gametes of identical appearance. Other chlorophytes, including some other species of *Ulva*, are **anisogamous**—they have female gametes that are distinctly larger than the male gametes.

The life cycles of many other chlorophytes do not feature alternation of generations. Some chlorophytes have a **haplontic** life cycle, in which a multicellular haploid individual produces gametes that fuse to form a zygote. The zygote functions directly as a sporocyte, undergoing meiosis to produce spores, which in turn produce a new haploid individual. In the entire haplontic life cycle, only one cell—the zygote—is diploid. The filamentous organisms of the genus *Ulothrix* are examples of haplontic chlorophytes (**Figure 27.15**).

Some chlorophytes have a **diplontic** life cycle like that of many animals. In a diplontic life cycle, meiosis of diploid sporocytes produces haploid gametes directly; the gametes fuse, and the resulting diploid zygote divides mitotically to form a new multicellular diploid sporophyte. In such organisms, all cells except the gametes are diploid. Between these two extremes are chlorophytes in which the gametophyte and sporo-

**27.15 A Haplontic Life Cycle** In the life cycle of *Ulothrix*, a filamentous, multicellular haploid gametophyte generation alternates with a diploid sporophyte generation consisting of a single cell (the zygote). *Ulothrix* gametophytes can also reproduce asexually.

**yourBioPortal.com**
GO TO Web Activity 27.2 • A Haplontic Life Cycle

phyte generations are both multicellular, but one generation (usually the sporophyte) is much larger and more prominent than the other.

### The life cycles of some protists require more than one host species

The trypanosome diseases discussed at the opening of this chapter share a striking feature with malaria: in each case, the eukaryote pathogen completes part of its life cycle in the human host and part in an insect (see Figure 27.9). Many other protist life cycles require the participation of two different host species.

What might be the advantage of a life cycle with two hosts? This remains an intriguing question. It may be relevant that in the human pathogens described above, the sexual phase of the organism's life cycle—the fusion of gametes into a zygote—takes place in the insect host. Could this imply that the human host is nothing but a copying machine for the products of sexual reproduction in the insect host?

### 27.4 RECAP

**Protists reproduce both asexually and sexually, although sex may occur independently of reproduction in some species. Some multicellular protists exhibit alternation of generations, alternating between multicellular haploid and diploid life stages. Parasitic protists may have complex life cycles in which they infect more than one host species.**

- Why is conjugation between paramecia considered a sexual process but not a reproductive process? See p. 573 and Figure 27.12
- Can you explain the difference between the diplontic human life cycle and a life cycle with alternation of generations? See pp. 573–574

The success of protists' diverse adaptations for nutrition, locomotion, and reproduction is evident from the abundance and diversity of eukaryotes living today. In the next section we survey that diversity.

## 27.5 What Are the Evolutionary Relationships among Eukaryotes?

Biologists used to classify the various groups of protists largely on the basis of life history and reproductive features. However, scientists using electron microscopy and gene sequencing have revealed many new patterns of evolutionary relatedness. Analyses of slowly evolving gene sequences are making it possible to explore evolutionary relationships among eukaryotes in ever greater detail and with greater confidence. Today we recognize great diversity among the many distantly related protist clades, whose members have explored a great variety of lifestyles.

Most eukaryotes can be classified in one of five major hypothesized clades: chromalveolates, Plantae, excavates, rhizaria, or unikonts (see Figure 27.1 and Table 27.1). As we will see, some of these groups consist of organisms with enormously diverse body forms and nutritional lifestyles. The phylogenetic relationships among the major groups of eukaryotes comprise an active area of study, and new data from genomes are rapidly changing our understanding of the evolution of these species. These relationships also help us understand how the major multicellular eukaryotic groups (brown algae, plants, fungi, and animals) originated from the microbial eukaryotes.

#### CHROMALVEOLATES

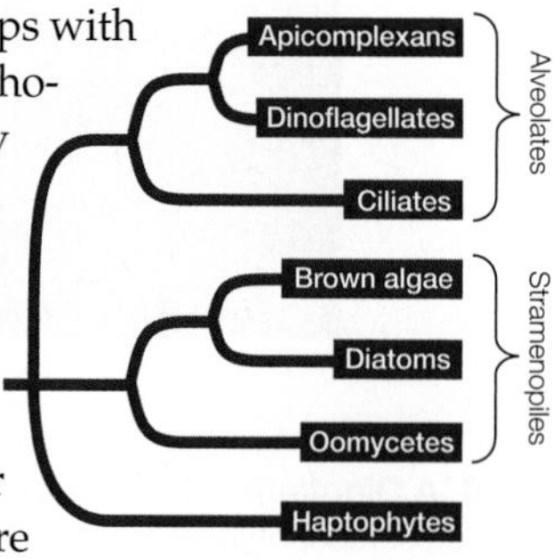

We begin our tour of protist groups with the **chromalveolates**, a group of photosynthetic organisms, usually with cellulose in their cell walls, that includes the haptophytes, alveolates and stramenopiles. The monophyly of the chromalveolates is not yet well established. The **haptophytes** are unicellular organisms with flagella; many are "armored" with elaborate scales (see Figure 27.10B). The alveolates and stramenopiles are large and diverse clades that we explore here in greater detail.

### Alveolates have sacs under their plasma membrane

The synapomorphy that characterizes the **alveolate** clade is the possession of sacs called *alveoli* just below their plasma membranes. The alveoli may play a role in supporting the cell surface. These organisms are all unicellular, but they are diverse in body form. The alveolate groups we consider in detail here are the dinoflagellates, apicomplexans, and ciliates.

**DINOFLAGELLATES** The **dinoflagellates** are of great ecological, evolutionary, and morphological interest. Most dinoflagellates are marine, and they are important primary photosynthetic producers of organic matter in the oceans. A distinctive mixture of photosynthetic and accessory pigments gives dinoflagellate chloroplasts a golden brown color. (Section 27.1 describes the endosymbiotic events that gave rise to dinoflagellates with different numbers of membranes surrounding their chloroplasts.) Some are photosynthetic endosymbionts living in the cells of other organisms, including invertebrate animals (such as corals) and other marine protists (see Figure 27.8). Some are nonphotosynthetic and live as parasites within other marine organisms.

Dinoflagellates have a distinctive appearance. They are unicellular and generally have two flagella, one in an equatorial groove around the cell, the other starting near the same point as the first and passing down a longitudinal groove before extending into the surrounding medium (**Figure 27.16**). Some dinoflagellates can take on different forms, including amoeboid ones, depending on environmental conditions. It has been claimed that the dinoflagellate *Pfiesteria piscicida* can occur in at least two dozen distinct forms, although this claim is highly con-

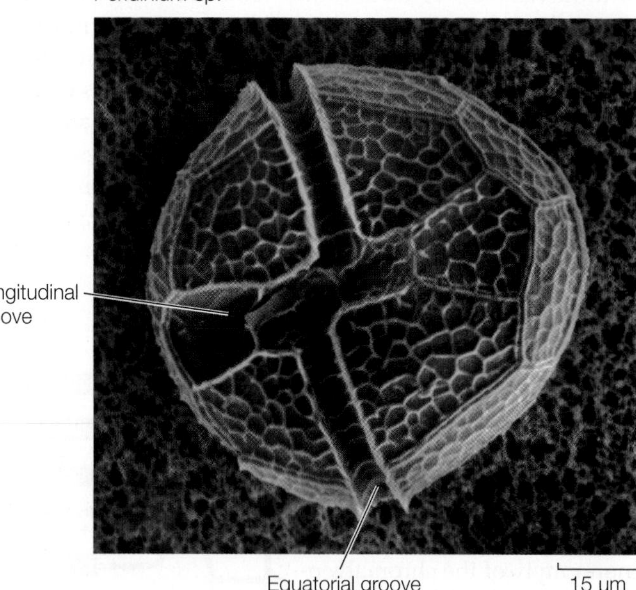


**27.16 A Dinoflagellate** The dinoflagellates are an important group of alveolates. Most of them are photosynthetic and are a crucial component of the world's phytoplankton. They are often endosymbiotic (see Figure 27.8) and can be the agents of deadly ocean "blooms" (see Figure 27.10A).

troversial. In any case, this remarkable dinoflagellate is harmful to fish and can, when present in great numbers, both stun and feed on them.

**APICOMPLEXANS** The exclusively parasitic **apicomplexans** derive their name from the *apical complex*, a mass of organelles contained in the *apical* end (the tip) of a cell. These organelles help the apicomplexan invade its host's tissues. For example, the apical complex enables the merozoites and sporozoites of *Plasmodium*, the causative agent of malaria, to enter their target cells in the human body.

Like many obligate parasites, apicomplexans have elaborate life cycles featuring asexual and sexual reproduction by a series of very dissimilar life stages. Often these stages are associated with two different types of host organisms, as is the case with *Plasmodium*. The apicomplexan *Toxoplasma* alternates between cats and rats to complete its life cycle. A rat infected with *Toxoplasma* loses its fear of cats, making it more likely to be eaten by, and thus transfer the parasite to, a cat.

Apicomplexans lack contractile vacuoles. They contain a much-reduced chloroplast that no longer has a photosynthetic function (derived, like all chromalveolate chloroplasts, from secondary endosymbiosis of a red alga). This chloroplast might be a target for a future antimalarial drug.

**CILIATES** The **ciliates** are so named because they characteristically have numerous hairlike cilia shorter than, but otherwise identical to, eukaryotic flagella. This group is noteworthy for its diversity and ecological importance (**Figure 27.17**). Almost all ciliates are heterotrophic (although a few contain photosynthetic endosymbionts), and they are much more complex in body form than are most other unicellular eukaryotes. The definitive characteristic of ciliates is the possession of two types of nuclei (as seen in the paramecia in Figure 27.12).

*Paramecium*, a frequently studied ciliate genus, exemplifies the complex structure and behavior of ciliates (**Figure 27.18**). The slipper-shaped cell is covered by an elaborate pellicle, a structure composed principally of an outer membrane and an inner layer of closely packed, membrane-enclosed sacs (the alveoli) that surround the bases of the cilia. Defensive organelles

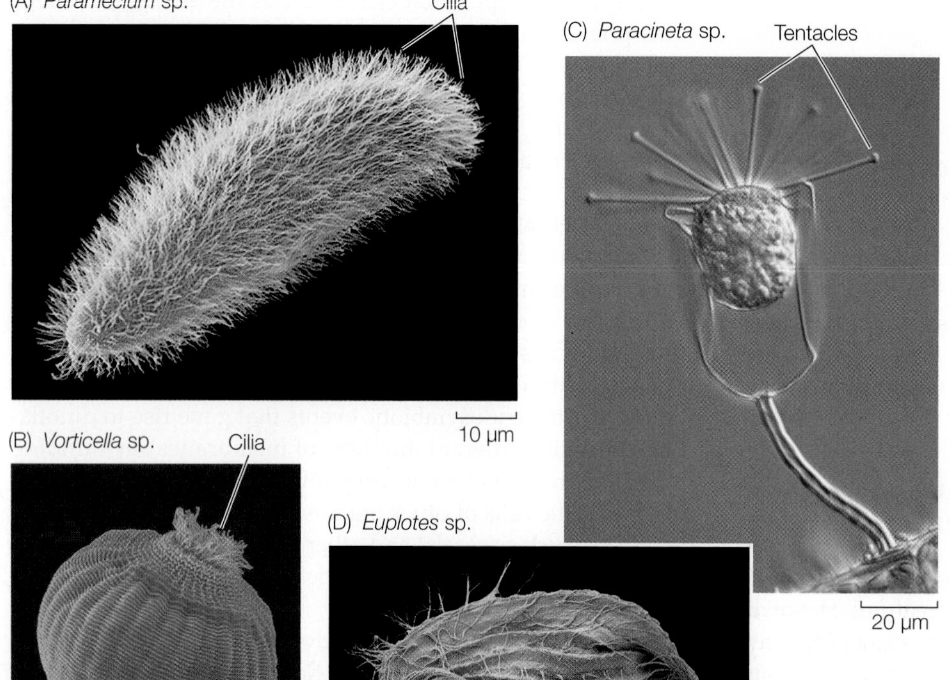


**27.17 Diversity among the Ciliates** (A) A free-swimming organism, this paramecium belongs to a ciliate group whose members have many cilia of uniform length. (B) Members of this group have cilia on their mouthparts. (C) In this group, cilia are replaced by tentacles as development proceeds. (D) Some of the cilia in *Euplotes* are grouped into flat sheets that sweep food particles into the oral groove.

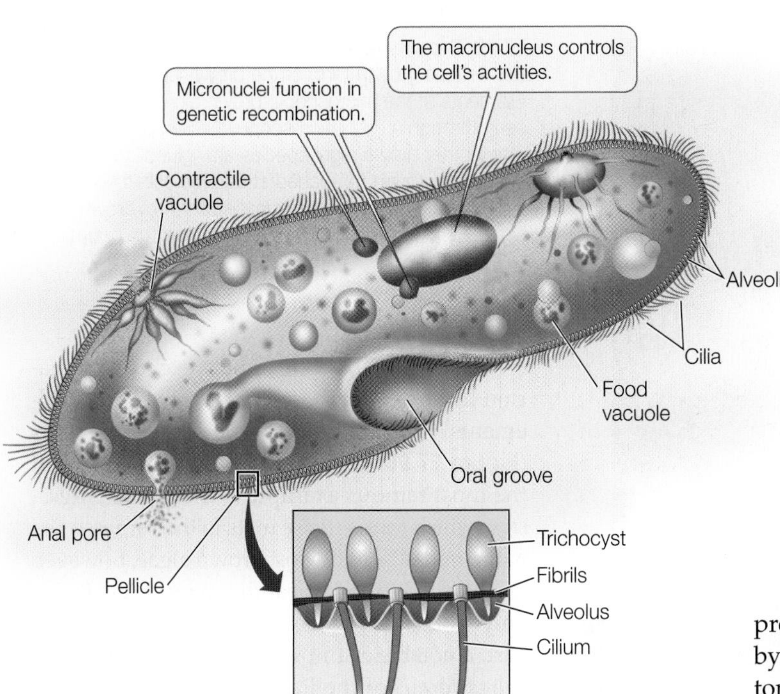


**27.18 Anatomy of *Paramecium*** This diagram shows the complex structure of a typical paramecium, detailing the pellicle (outer covering) with its trichocysts and alveoli.

**yourBioPortal.com**

GO TO Web Activity 27.3 • Anatomy of *Paramecium*

called *trichocysts* are also present in the pellicle. In response to a threat, a microscopic explosion expels the trichocysts in a few milliseconds, and they emerge as sharp darts, driven forward at the tip of a long, expanding filament.

The cilia provide a form of locomotion that is generally more precise than locomotion by flagella or pseudopods. A paramecium can coordinate the beating of its cilia to propel itself either forward or backward in a spiraling manner. It can also back off swiftly when it encounters a barrier or a negative stimulus. The coordination of ciliary beating is probably the result of a differential distribution of ion channels in the plasma membrane near the two ends of the cell.

### Stramenopiles have two unequal flagella, one with hairs

A morphological synapomorphy shared by most **stramenopiles** is the possession of rows of tubular hairs on the longer of their two flagella. Some stramenopiles lack flagella, but they are descended from ancestors that possessed flagella. The stramenopiles include the diatoms and the brown algae, which are photosynthetic, and the oomycetes and slime nets, which are not. Most golden algae are photosynthetic, but nearly all of them become heterotrophic when light intensity is limited or when there is a plentiful food supply; some even feed on diatoms or bacteria. The slime nets (not to be confused with slime molds) are unicellular organisms that produce networks of filaments along which the cells move.

**DIATOMS** All of the **diatoms** are unicellular, although some species associate in filaments. Many have sufficient carotenoids in their chloroplasts to give them a yellow or brownish color. All make carbohydrates and oils as photosynthetic storage products. Diatoms lack flagella except in male gametes.

Architectural magnificence on a microscopic scale is the hallmark of the diatoms. As mentioned earlier, almost all diatoms deposit silica (hydrated silicon dioxide) in their cell walls. The cell wall is constructed in two pieces, with the top overlapping the bottom like the top and bottom of a petri plate. The silica-impregnated walls have intricate patterns unique to each species (**Figure 27.19**). Despite their remarkable morphological diversity, all diatoms are symmetrical—either bilaterally (with "right" and "left" halves) or radially (with the type of symmetry possessed by a circle).

Diatoms reproduce both sexually and asexually. Asexual reproduction is by binary fission and is somewhat constrained by the stiff, silica-containing cell wall. Both the top and the bottom of the "petri plate" become tops of new "plates" without changing appreciably in size; as a result, the new cell made from the former bottom is smaller than the parent cell. If this process continued indefinitely, one cell line would simply vanish, but sexual reproduction largely solves this potential problem. Gametes are formed, shed their cell walls, and fuse. The resulting zygote then increases substantially in size before a new cell wall is laid down.

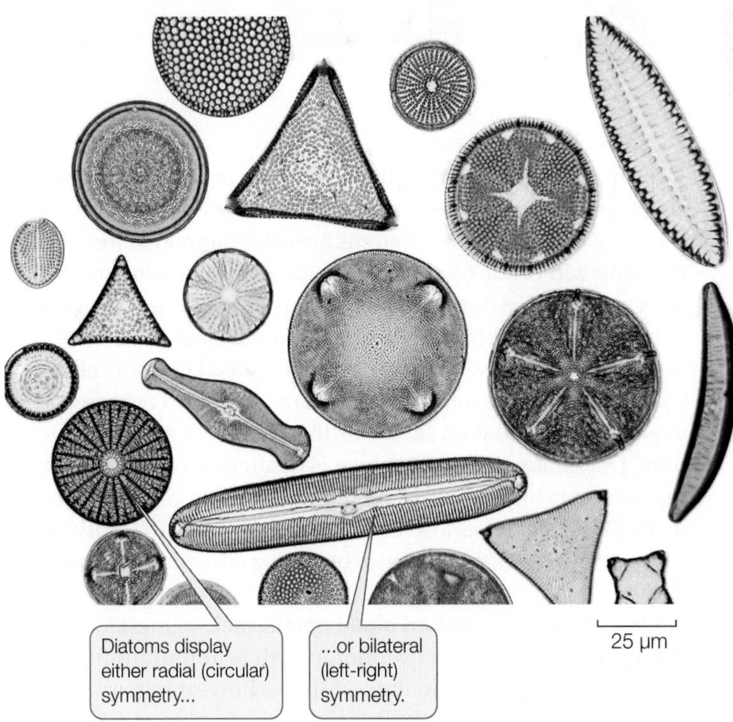


**27.19 Diatom Diversity** This brightfield micrograph of diatoms shows a variety of species-specific forms. Diatoms are photosynthesizers and dominant components of the world's phytoplankton.

(A) *Cystoseira usneoides*

(B) *Ectocarpus* sp.

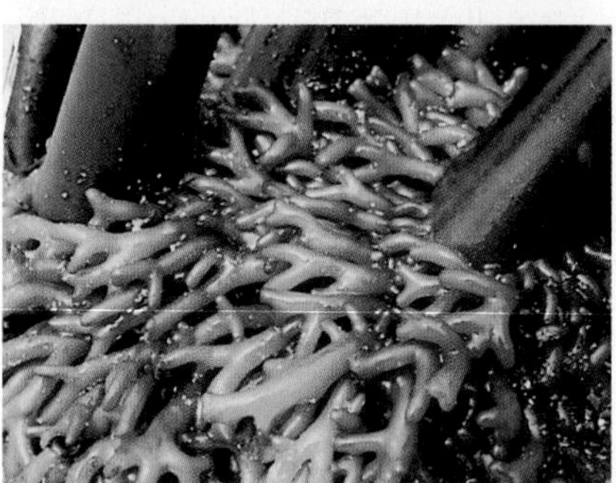

(C) *Postelsia palmaeformis*

(D) *Postelsia palmaeformis*

**27.20 Brown Algae** (A) This seaweed illustrates the filamentous growth form of the brown algae. (B) Filaments of the microscopic brown alga *Ectocarpus* seen through a light microscope. (C) Sea palms and many other brown algal species are "glued" to the substratum by tough, branched structures called holdfasts. (D) Sea palms exemplify the leaflike form of brown algae. Holdfasts allow them to withstand the pounding of the surf.

Diatoms are found in all the oceans and are frequently present in great numbers, making them major photosynthetic producers in coastal waters. Diatoms are also common in fresh water and even occur on the wet surfaces of terrestrial mosses. They are also the dominant organisms in the dense "blooms" of plankton that occasionally appear in the open ocean.

Because the silica-containing walls of dead diatom cells resist decomposition, certain sedimentary rocks are composed almost entirely of diatom skeletons that sank to the seafloor over time. Diatomaceous earth, which is obtained from such rocks, has many industrial uses, such as insulation, filtration, and metal polishing. It has also been used as an "Earth-friendly" insecticide that clogs the tracheae (breathing structures) of insects.

**BROWN ALGAE** The **brown algae** obtain their namesake color from the carotenoid *fucoxanthin*, which is abundant in their chloroplasts. The combination of this yellow-orange pigment with the green of chlorophylls *a* and *c* yields a brownish tinge. All brown algae are multicellular, and some are extremely large. Giant kelps, such as those of the genus *Macrocystis*, may be up to 60 meters long.

The brown algae are almost exclusively marine. They are composed either of branched filaments (**Figure 27.20A,B**) or of leaflike growths (**Figure 27.20D**). Some float in the open ocean; the most famous example is the genus *Sargassum*, which forms dense mats in the Sargasso Sea in the mid-Atlantic. Most brown algae, however, are attached to rocks near the shore. A few thrive only where they are regularly exposed to heavy surf; a notable example is the sea palm *Postelsia palmaeformis* of the Pacific coast. All of the attached forms develop a specialized structure, called a *holdfast*, that literally glues them to the rocks (**Figure 27.20C**). The "glue" of the holdfast is *alginic acid*, a gummy polymer of sugar acids found in the walls of many brown algal cells. In addition to its function in holdfasts, alginic acid cements algal cells and filaments together, and is harvested and used by humans as an emulsifier in ice cream, cosmetics, and other products.

Some brown algae differentiate extensively into specialized organs. Some, like the sea palm, have stemlike stalks and leaflike blades. Some develop gas-filled cavities or bladders that serve as floats. In addition to organ differentiation, the larger brown algae also exhibit considerable tissue differentiation. Most of the giant kelps have photosynthetic filaments only in the outermost regions of their stalks and blades. Within the stalks and blades lie filaments of tubular cells that closely resemble the nutrient-conducting tissue of land plants. Called *trumpet cells* because they have flared ends, these tubes rapidly conduct the products of photosynthesis through the body of the organism.

**OOMYCETES** A nonphotosynthetic stramenopile group called the **oomycetes** consists in large part of the water molds and their terrestrial relatives, such as the downy mildews. Water molds are filamentous and stationary, and they are *absorptive heterotrophs*—that is, they secrete enzymes that digest large food molecules into smaller molecules that the water mold can absorb. If you have seen a whitish, cottony mold growing on dead fish or dead insects in water, it was probably a water mold of the common genus *Saprolegnia* (**Figure 27.21**).

Don't be misled by the "mycete" in the name of this group. That term means "fungus," and it is there because these organisms were once classified as fungi. However, we now know that the oomycetes are more distantly related to the fungi than

*Saprolegnia* sp.

**27.21 An Oomycete** The filaments of a water mold radiate from the carcass of an insect.

are many other eukaryote groups, including ourselves (see Figure 27.1), and the similarity of oomycetes to fungi is only superficial. For example, the cell walls of oomycetes are typically made of cellulose, whereas those of fungi are made of chitin.

Some oomycetes are **coenocytes**, which means they have many nuclei enclosed in a single plasma membrane. Their filaments have no cross-walls to separate the many nuclei into discrete cells. Their cytoplasm is continuous throughout the body of the organism, and there is no single structural unit with a single nucleus, except in certain reproductive stages. A distinguishing feature of the oomycetes is their flagellated reproductive cells. Oomycetes are diploid throughout most of their life cycle.

The water molds, such as *Saprolegnia*, are all aquatic and **saprobic**—meaning they feed on dead organic matter. Some other oomycetes are terrestrial. Although most of the terrestrial oomycetes are harmless or helpful decomposers of dead matter, a few are serious plant parasites that attack crops such as avocados, grapes, and potatoes.

Although their presumed chromalveolate ancestors had chloroplasts and were photosynthetic, the oomycetes lack chloroplasts.

## Plantae



**Plantae** contains mostly photosynthetic species and consists of several major clades, including glaucophytes, red algae, chlorophytes, charophytes, and the land plants, all of which probably trace their chloroplasts back to a single incidence of endosymbiosis (see Section 27.1). It is for this reason that the small clade known as the **glaucophytes**, unicellular organisms that live in fresh water, is of particular interest.

The glaucophytes were likely the first group to diverge after the primary endosymbiosis event. Their chloroplast is unique in containing a small amount of peptidoglycan between its inner and outer membranes—the same arrangement as that found in cyanobacteria (see Figure 27.3A). The presence of peptidoglycan, the characteristic cell wall component of bacteria, suggests that this feature has been retained in glaucophytes but lost in the other Plantae groups.

### Red algae have a distinctive accessory photosynthetic pigment

Almost all **red algae** are multicellular (**Figure 27.22**). Their characteristic color is a result of the accessory photosynthetic pigment *phycoerythrin*, which is found in relatively large amounts in the chloroplasts of many species. In addition to phycoerythrin, red algae contain phycocyanin, carotenoids, and chlorophyll *a*.

The red algae include species that grow in the shallowest tide pools as well as the photosynthesizers found deepest in the ocean (as deep as 260 meters if nutrient conditions are right and the water is clear enough to permit light to penetrate). Very few red algae inhabit fresh water. Most grow attached to a substratum by a holdfast.

In a sense, the red algae are misnamed. They have the capacity to change the relative amounts of their

(A) *Ceramium* sp.

(B) *Calliarthron tuberculatum*

**27.22 Red Algae** (A) Differential contrast light microscopy reveals the rich red color of the pigment phycoerythrin. (B) Coralline red alga, named for its coral-like appearance.

various photosynthetic pigments depending on the light conditions where they are growing. Thus the leaflike *Chondrus crispus*, a common North Atlantic red alga, may appear bright green when it is growing at or near the surface of the water and deep red when growing at greater depths. The ratio of pigments present depends to a remarkable degree on the intensity of the light that reaches the alga. In deep water, where the light is dimmest, the alga accumulates large amounts of phycoerythrin. The algae in deep water have as much chlorophyll as the green ones near the surface, but the accumulated phycoerythrin makes them look red.

In addition to being the only photosynthetic eukaryotes with phycoerythrin among their pigments, the red algae have two other distinctive characteristics:

- They store the products of photosynthesis in the form of *floridean starch*, which is composed of very small, branched chains of approximately 15 glucose monomers.
- They produce no motile, flagellated cells at any stage of their life cycle. The male gametes lack cell walls and are slightly amoeboid; the female gametes are completely immobile.

Some red algal species enhance the formation of coral reefs (see Figure 27.22B). Like coral animals, they possess the biochemical machinery for secreting calcium carbonate, which they deposit both in and around their cell walls. After the deaths of corals and algae, the calcium carbonate persists, sometimes forming substantial rocky masses.

Some red algae produce large amounts of mucilaginous polysaccharide substances, which contain the sugar galactose with a sulfate group attached. This material readily forms solid gels and is the source of agar, a substance widely used in the laboratory for making a solid aqueous medium on which tissue cultures and many microorganisms can be grown.

The distinctive chloroplasts of the photosynthetic chromalveolates are derived by secondary endosymbiosis of a red alga, as discussed in Section 27.1.

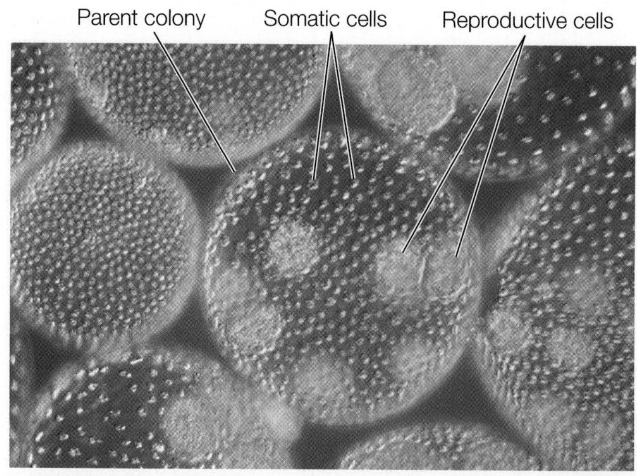

**27.23 Chlorophytes** (A) *Volvox* colonies are precisely spaced arrangements of cells. Specialized reproductive cells produce daughter colonies, which will eventually release new individuals. (B) Sea lettuce grows in ocean tidewaters.

## Chlorophytes, charophytes, and land plants contain chlorophylls a and b

One major clade of "green algae" is the **chlorophytes**. A sister group to the chlorophytes contains another green algal clade—the **charophytes**, or **Charales**—along with the land plants (see Section 28.1). The green algae share several characters that distinguish them from other protists: like the land plants, they contain chlorophylls *a* and *b*, and their reserve of photosynthetic products is stored as starch in chloroplasts. Through secondary endosymbiosis, a chlorophyte became the chloroplast of the euglenids.

There are more than 17,000 species of chlorophytes. Most are aquatic—some are marine, though more are freshwater forms—but others are terrestrial, living in moist environments. The chlorophytes range in size from microscopic unicellular forms to multicellular forms many centimeters long.

The chlorophytes display an incredible variety of shapes and body forms. *Chlamydomonas* is an example of the simplest type: unicellular and flagellated. Surprisingly large and well-formed colonies of cells are found in such freshwater groups as the genus *Volvox* (**Figure 27.23A**). The cells in these colonies are not differentiated into specialized tissues and organs, as in land plants and animals, but the colonies show vividly how the preliminary step of this great evolutionary innovation might have been taken. In *Volvox*, the origins of cell specialization can be seen in certain cells in the colony that are specialized for reproduction.

While *Volvox* is colonial and spherical, *Oedogonium* is multicellular and filamentous, and each of its cells has only one nucleus. *Cladophora* is multicellular, but each cell is multinucleate. *Bryopsis* is tubular and coenocytic, forming cross-walls only when reproductive structures form. *Acetabularia* is a single giant, uninucleate cell a few centimeters in length that becomes multinucleate only at the end of its reproductive stage. *Ulva lactuca* is a thin, membranous sheet a few centimeters across; its distinctive appearance justifies its common name of sea lettuce (**Figure 27.23B**).

As mentioned above, the chlorophytes are the largest clade of green algae, but there are other green algal clades as well. Those clades are branches of a clade that also includes the land plants, which are described in the next chapter.

## EXCAVATES

Diplomonads
Parabasalids
Heteroloboseans
Euglenids
Kinetoplastids

Excavates include a number of diverse clades, several of which lack mitochondria. This absence of mitochondria once led to the view that these groups might represent early diverging groups of eukaryotes that diversified before the evolution of mitochondria. However, the absence of mitochondria seems to be a derived condition, judging in part from the presence of nuclear genes normally associated with mitochondria. Ancestors of these organisms probably possessed mitochondria that were lost or reduced in the course of evolution. The existence of such organisms today shows that eukaryotic life is feasible without mitochondria, and for that reason, these groups are the focus of much attention.

### Diplomonads and parabasalids are excavates that lack mitochondria

The **diplomonads** and **parabasalids**, all of which are unicellular, are distinctive in their lack of mitochondria. *Giardia lamblia*, a diplomonad, is a familiar parasite that contaminates water supplies and causes the intestinal disease giardiasis (**Figure 27.24A**). This tiny organism contains two nuclei bounded by nuclear envelopes, and it has a cytoskeleton and multiple flagella.

*Trichomonas vaginalis* is a parabasalid responsible for a sexually transmitted disease in humans (**Figure 27.24B**). Infection of the male urethra, where it may occur without symptoms, is less common than infection of the vagina. In addition to flagella and a cytoskeleton, the parabasalids have undulating membranes that also contribute to the cell's locomotion.

### Heteroloboseans alternate between amoeboid forms and forms with flagella

The amoeboid body form appears in several protist groups—including the **loboseans** and **heteroloboseans**—that are only distantly related to one another. These groups belong, respectively, to the unikonts and excavates. Amoebas of the free-living heterolobosean genus *Naegleria*, some of which can enter humans and cause a fatal disease of the nervous system, usually have a two-stage life cycle, in which one stage has amoeboid cells and the other flagellated cells.

### Euglenids and kinetoplastids have distinctive mitochondria and flagella

The euglenids and kinetoplastids, both of which are excavates, together constitute a clade of unicellular organisms with flagella. Their mitochondria contain distinctive, disc-shaped cristae, and their flagella contain a crystalline rod not found in other organisms. They reproduce primarily asexually by binary fission.

**EUGLENIDS** The **euglenids** possess flagella arising from a pocket at the anterior end of the cell. Spiraling strips of proteins under their plasma membranes control the cell's shape. Some members of the group are photosynthetic. Euglenids used to be claimed by the zoologists as animals and by the botanists as plants.

**Figure 27.25** depicts a cell of the genus *Euglena*. Like most other euglenids, this common freshwater organism has a complex cell structure. It propels itself through the water with

(A) *Giardia* sp.

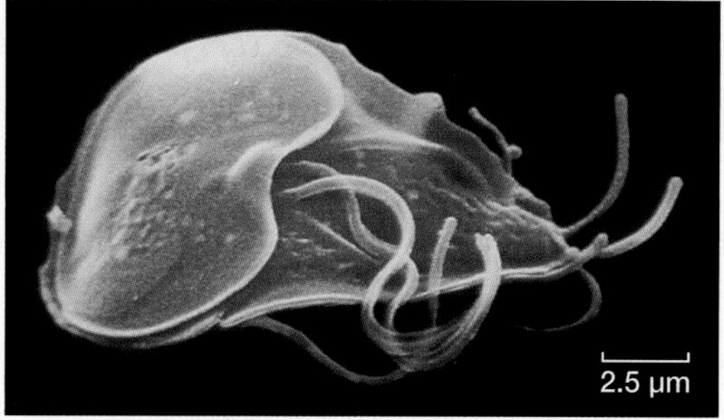


(B) *Trichomonas vaginalis*

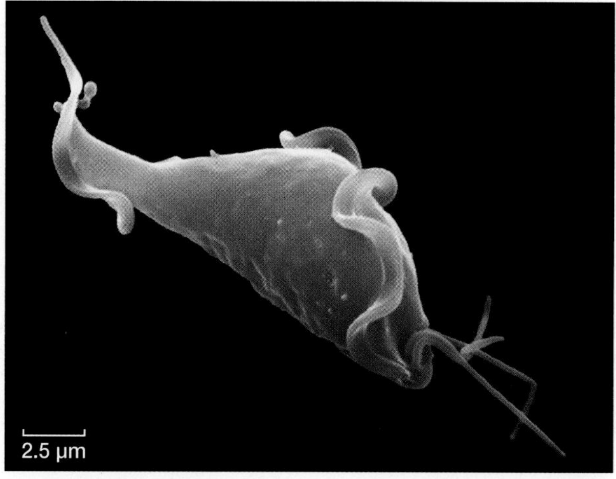


**27.24 Some Excavate Groups Lack Mitochondria** (A) *Giardia*, a diplomonad, has flagella and two nuclei. (B) *Trichomonas*, a parabasalid, has flagella and undulating membranes. Neither of these organisms possesses mitochondria.

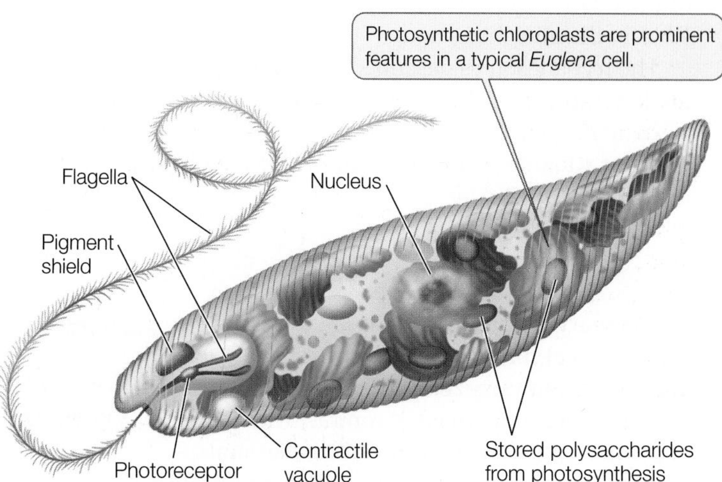


**27.25 A Photosynthetic Euglenid** Several *Euglena* species possess flagella. In this species, the second flagellum is rudimentary.

TABLE 27.2
**A Comparison of Three Kinetoplastid Trypanosomes**

| | *Trypanosoma brucei* | *Trypanosoma cruzi* | *Leishmania major* |
|---|---|---|---|
| Human disease | Sleeping sickness | Chagas' disease | Leishmaniasis |
| Insect vector | Tsetse fly | Assassin bug | Sand fly |
| Vaccine or effective cure | None | None | None |
| Strategy for survival | Changes surface recognition molecules frequently | Causes changes in surface recognition molecules on host cell | Reduces effectiveness of macrophage hosts |
| Site in human body | Bloodstream; attacks nerve tissue in final stages | Enters cells, especially muscle cells | Enters cells, primarily macrophages |
| Approximate number of deaths per year | 50,000 | 43,000 | 60,000 |

the longer of its two flagella, which may also serve as an anchor to hold the organism in place. The second flagellum is often rudimentary.

Euglenids have diverse nutritional requirements. Many species are always heterotrophic. Other species are fully autotrophic in sunlight, using chloroplasts to synthesize organic compounds through photosynthesis. The chloroplasts of euglenids are surrounded by three membranes as a result of secondary endosymbiosis (see Figure 27.3B). When kept in the dark, these euglenids lose their photosynthetic pigment and begin to feed exclusively on dissolved organic material in the water around them. Such a "bleached" *Euglena* resynthesizes its photosynthetic pigment when it is returned to the light and becomes autotrophic again. But *Euglena* cells treated with certain antibiotics or mutagens lose their photosynthetic pigment completely; neither they nor their descendants are ever autotrophs again. However, those descendants function well as heterotrophs.

**KINETOPLASTIDS** The **kinetoplastids** are unicellular parasites with two flagella and a single, large mitochondrion. That mitochondrion contains a *kinetoplast*—a unique structure housing multiple circular DNA molecules and associated proteins. Some of these DNA molecules encode "guides" that edit messenger RNA in the mitochondrion.

The trypanosomes described at the opening of this chapter are kinetoplastids. They are able to change their cell surface recognition molecules frequently, which allows them to evade our best attempts to kill them and eradicate the diseases they cause (**Table 27.2**).

### RHIZARIA

The primary groups of **Rhizaria** are unicellular aquatic eukaryotes. Foraminiferans, radiolarians, and cercozoans typically have long, thin pseudopodia that contrast with the broader, lobelike pseudopodia of the familiar amoebas. These groups have contributed to ocean sediments, some of which have become terrestrial features in the course of geological history.

Cercozoans
Foraminiferans
Radiolarians

The **cercozoans** are a diverse group, with many forms and habitats. Some are amoeboid, while others have flagella. Some are aquatic; others live in soil. One group of cercozoans possesses chloroplasts derived from a green alga by secondary endosymbiosis—and that chloroplast contains a trace of the green alga's nucleus.

#### Foraminiferans have created vast limestone deposits

Some **foraminiferans** secrete external shells of calcium carbonate (see Figure 27.11), which over time have accumulated to produce much of the world's limestone. Some foraminiferans live as plankton; others live on the seafloor. Living foraminiferans have been found at the deepest point in the world's oceans—10,896 meters down in the Challenger Deep, in the western Pacific. At that depth, however, they cannot secrete a normal shell because the surrounding water is too poor in calcium carbonate.

Long, threadlike, branched pseudopods reach out through numerous microscopic apertures in the shell and interconnect to create a sticky, reticulated net, which planktonic foraminifera use to catch smaller plankton. The pseudopods provide locomotion in some species.

#### Radiolarians have thin, stiff pseudopods

The **radiolarians** are recognizable by their thin, stiff pseudopods, which are reinforced by microtubules. These pseudopods:

- greatly increase the surface area of the cell for exchange of materials with the environment
- help the cell float in its marine environment

Found exclusively in marine environments, radiolarians are immediately recognizable by their distinctive radial symmetry (see Figure 27.8B). Almost all radiolarian species secrete glassy *endoskeletons* (internal skeletons). A central capsule lies within the cytoplasm. The skeletons of the different species are as varied as snowflakes, and many have elaborate geometric designs (**Figure 27.26**). A few radiolarians are among the largest of the unicellular eukaryotes, measuring several millimeters across.

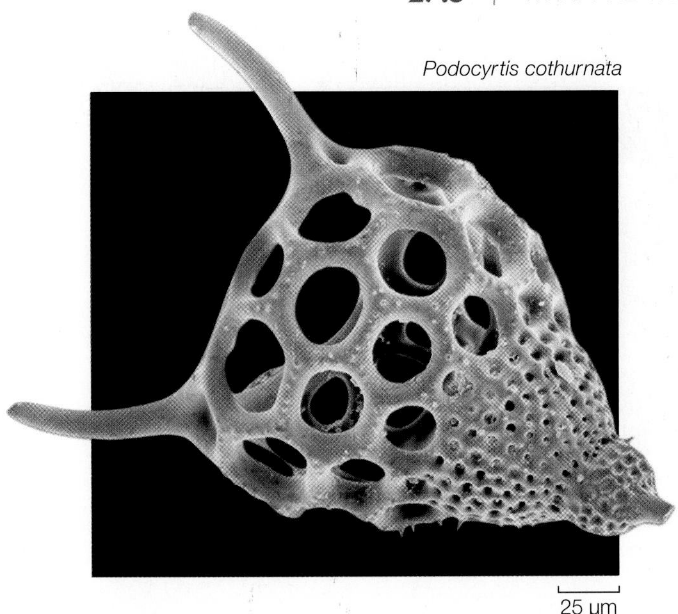


**27.26 A Radiolarian's Glass House** Radiolarians secrete intricate glassy skeletons such as the one shown here. A living radiolarian with its endosymbionts is shown in Figure 27.8B.

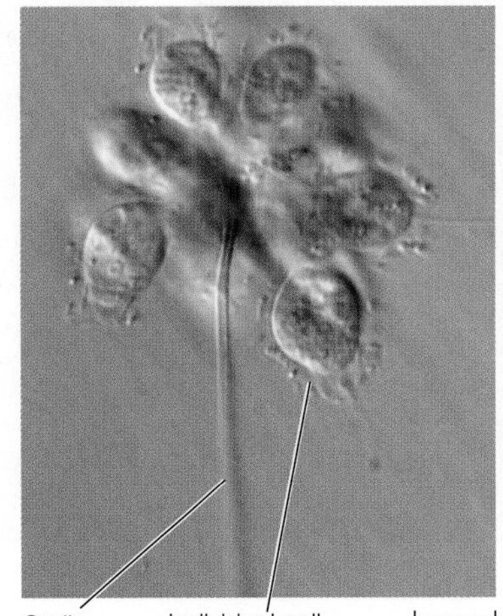


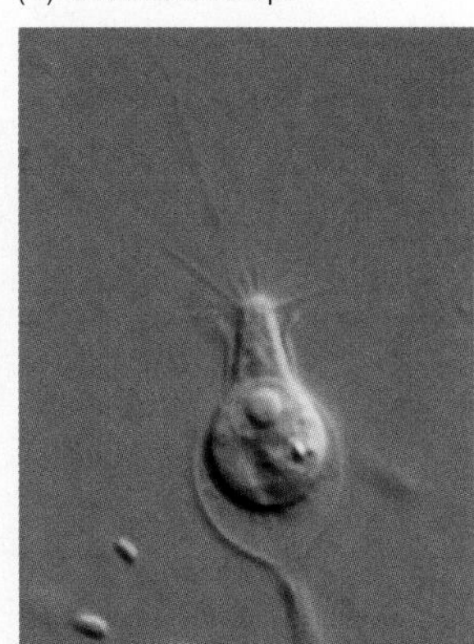


**27.27 A Link to the Animals** Choanoflagellates are the sister group to the animals. (A) The formation of colonies by unicellular organisms, as in this choanoflagellate species, is one route to the evolution of multicellularity. (B) A solitary choanoflagellate illustrates the similarity of this microbial protist group to a cell type present in the multicellular sponges (see Figure 31.7).

## Unikonts

We now consider the **unikonts**, a large clade that may be close to the root of the eukaryote tree. The name *unikont* derives from the Greek for "single cone," and in these eukaryotes, the flagella, if present, are single. Unikonts consist of two major groups, the *opisthokonts* (which include animals and fungi) and the *amoebozoans* (see Figure 27.1).

A morphological synapomorphy of the opisthokonts is that their flagellum, if present, is posterior, as in animal sperm. The flagella of all other eukaryotes are anterior. In addition to the animals and fungi, opisthokonts also include the choanoflagellates. Fungi and animals are discussed in Chapters 30–33. The choanoflagellates, or collar flagellates, are sister to the animals, and the animal–choanoflagellate clade is sister to the fungi.

Some choanoflagellates are colonial (**Figure 27.27A**). They bear a striking resemblance to the most characteristic type of cell found in the sponges (compare **Figure 27.27B** with Figure 31.7).

### Amoebozoans use lobe-shaped pseudopods for locomotion

The lobe-shaped pseudopods used by amoebozoans are a hallmark of the amoeboid body plan. The amoebozoan pseudopod differs in form and function from the slender pseudopods of rhizaria. We consider three amoebozoan groups here: the loboseans and two clades of slime molds.

**LOBOSEANS** A lobosean, such as the *Peloxima carolinensis* shown in Figure 27.4, consists of a single cell. Unlike the cells of the slime molds, loboseans live independently of one another and do not aggregate. A lobosean feeds on small organisms and particles of organic matter by phagocytosis, engulfing them with its pseudopods. Many loboseans are adapted for life on the bottoms of lakes, ponds, and other bodies of water. Their creeping locomotion and their manner of engulfing food particles fit them for life close to a relatively rich supply of sedentary organisms or organic particles. Most loboseans exist as predators, parasites, or scavengers.

Some loboseans are shelled, living in casings of sand grains glued together (see Figure 27.7A). Others have shells secreted by the organism itself.

**SLIME MOLDS** The two major groups of *slime molds* share only general characteristics. All are motile, all ingest particulate food by endocytosis, and all form spores on erect structures called *fruiting bodies*. They undergo striking changes in organization during their life cycles, and one stage consists of isolated cells that take up food particles by endocytosis. Some slime molds may cover areas of 1 meter or more in diameter while in their less aggregated stage. Such a large slime mold may weigh more than 50 grams. Slime molds of both types favor cool, moist habitats, primarily in forests. They range from colorless to brilliant yellow and orange.

**PLASMODIAL SLIME MOLDS** If the nucleus of an amoeba began rapid mitotic division, accompanied by a tremendous increase in cytoplasm and organelles but no cytokinesis, the resulting organism might resemble the multinucleate mass of a plasmodial slime mold. During its vegetative (feeding) phase, a plasmodial slime mold is a wall-less mass of cytoplasm with numerous diploid nuclei. This mass streams very slowly over its substratum in a remarkable network of strands called a *plas-*

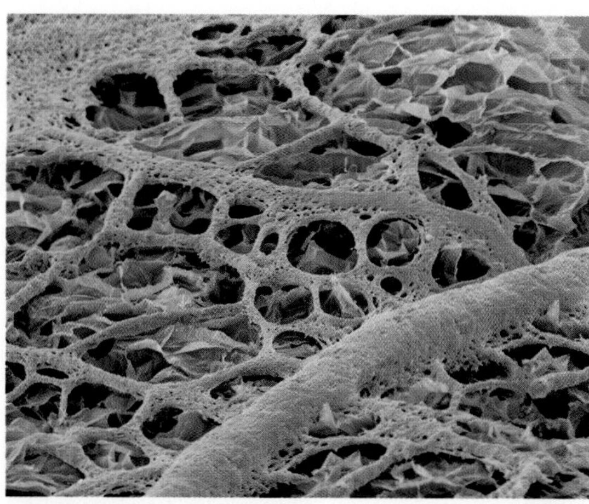


(A) *Physarum polycephalum*

(B) *Physarum polycephalum*

**27.28 Plasmodial Slime Molds** (A) The slime mold *Physarum* is most often seen in its plasmodial form, covering rocks, decaying logs, and other objects as it engulfs bacteria and other food items from the surface. (B) Fruiting structures of *Physarum*.

*modium*. The plasmodium of such a slime mold is another example of a coenocyte, with many nuclei enclosed in a single plasma membrane. The outer cytoplasm of the plasmodium (closest to the environment) is normally less fluid than the interior cytoplasm and thus provides some structural rigidity (**Figure 27.28A**).

Plasmodial slime molds provide a dramatic example of movement by cytoplasmic streaming. The outer cytoplasmic region of the plasmodium becomes more fluid in places, and cytoplasm rushes into those areas, stretching the plasmodium. This streaming somehow reverses its direction every few minutes as cytoplasm rushes into a new area and drains away from an older one, moving the plasmodium over its substratum. Sometimes an entire wave of plasmodium moves across the substratum, leaving strands behind. Microfilaments and a contractile protein called *myxomyosin* interact to produce the streaming movement. As it moves, the plasmodium engulfs food particles by endocytosis—predominantly bacteria, yeasts, spores of fungi, and other small organisms, as well as decaying animal and plant remains.

A plasmodial slime mold can grow almost indefinitely in its plasmodial stage, as long as the food supply is adequate and other conditions, such as moisture and pH, are favorable. However, one of two things can happen if conditions become unfavorable. First, the plasmodium can form an irregular mass of hardened cell-like components called a *sclerotium*. This resting structure rapidly becomes a plasmodium again when favorable conditions are restored.

Alternatively, the plasmodium can transform itself into spore-bearing fruiting structures (**Figure 27.28B**). These stalked or branched structures rise from heaped masses of plasmodium. They derive their rigidity from walls that form and thicken between their nuclei. The diploid nuclei of the plasmodium divide by meiosis as the fruiting structure develops. One or more knobs, called *sporangia*, develop on the end of the stalk. Within a sporangium, haploid nuclei become surrounded by walls and form spores. Eventually, as the fruiting body dries, it sheds its spores.

The spores germinate into wall-less, haploid cells called *swarm cells*, which can either divide mitotically to produce more haploid swarm cells or function as gametes. Swarm cells can live as separate individual cells that move by means of flagella or pseudopods, or they can become walled and resistant resting cysts when conditions are unfavorable; when conditions improve again, the cysts release swarm cells. Two swarm cells can also fuse to form a diploid zygote, which divides by mitosis (but without a wall forming between the nuclei) and thus forms a new, coenocytic plasmodium.

**CELLULAR SLIME MOLDS** Whereas the plasmodium is the basic vegetative (feeding, nonreproductive) unit of the plasmodial slime molds, an amoeboid cell is the vegetative unit of the cellular slime molds. Large numbers of cells called *myxamoebas*, which have single haploid nuclei, engulf bacteria and other food particles by endocytosis and reproduce by mitosis and fission. This simple life cycle stage, consisting of swarms of independent, isolated cells, can persist indefinitely as long as food and moisture are available.

When conditions become unfavorable, the cellular slime molds aggregate and form fruiting structures, as do their plasmodial counterparts. The individual myxamoebas aggregate into a mass called a *slug* or *pseudoplasmodium* (**Figure 27.29**). Unlike the true plasmodium of the plasmodial slime molds, this structure is not simply a giant sheet of cytoplasm with many nuclei; the individual myxamoebas retain their plasma membranes and, therefore, their identity.

A slug may migrate over its substratum for several hours before becoming motionless and reorganizing to construct a delicate, stalked fruiting structure. Cells at the top of the fruiting structure develop into thick-walled spores, which are eventually released. Later, under favorable conditions, the spores germinate, releasing myxamoebas.

The cycle from myxamoebas through slug and spores to new myxamoebas is asexual. Cellular slime molds also have a sex-

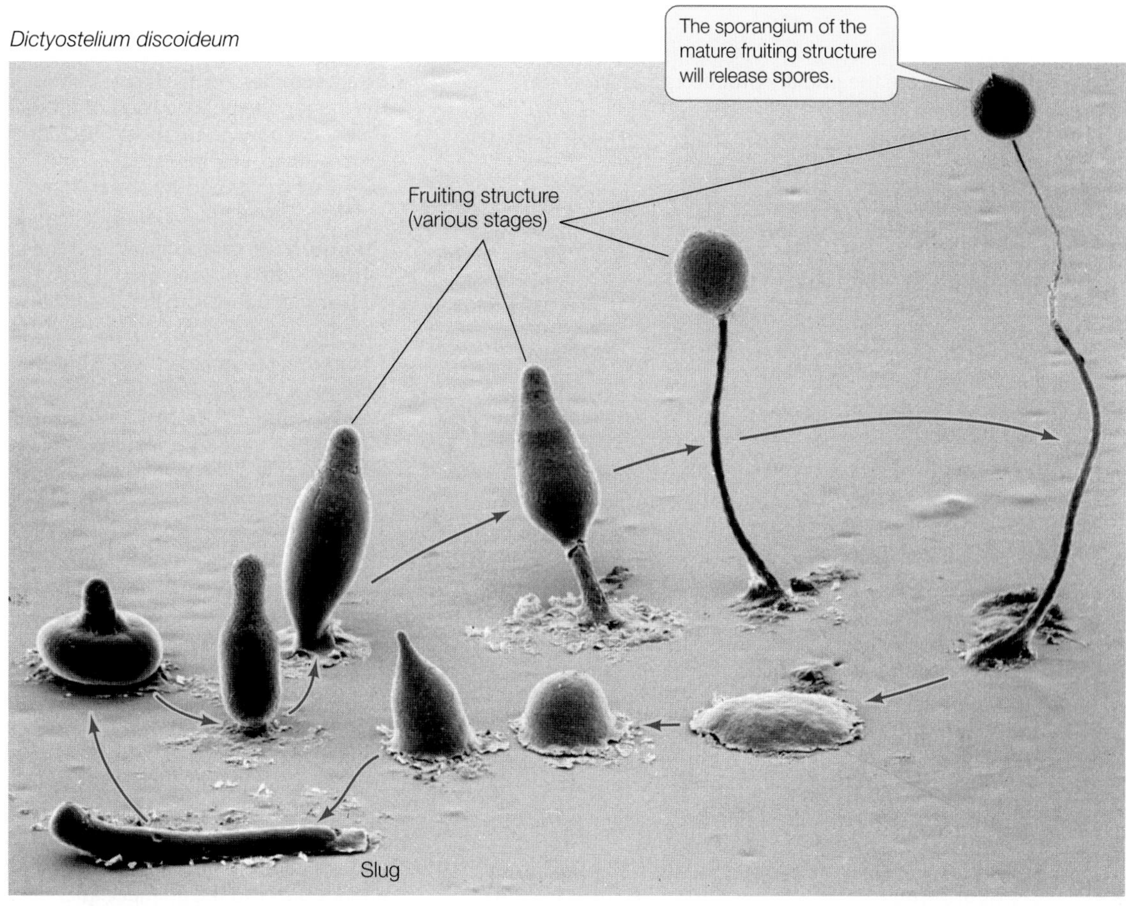


**27.29 A Cellular Slime Mold** The life cycle of the slime mold *Dictyostelium* is shown here in a composite micrograph.

ual cycle, in which two myxamoebas fuse. The product of this fusion develops into a spherical structure that ultimately germinates, releasing new haploid myxamoebas.

In subsequent chapters we explore the three major groups of multicellular eukaryotes. Chapters 28 and 29 describe the origin and diversification of the plants, Chapter 30 presents the fungi, and Chapters 31–33 describe the animals. All three of these groups arose from protist ancestors.

# CHAPTER SUMMARY

## 27.1 How Did the Eukaryotic Cell Arise?

- The modern eukaryotic cell evolved from an ancestral prokaryote. Probable early events in this evolution include the loss of the cell wall and infolding of the plasma membrane. Such infolding probably led to the segregation of the genetic material in a membrane-enclosed nucleus. The development of a cytoskeleton gave the evolving cell increasing control over its shape and distribution of daughter chromosomes. Review Figure 27.2
- Some organelles were acquired by endosymbiosis. Mitochondria evolved from a proteobacterium.
- **Primary endosymbiosis** of a eukaryote and a cyanobacterium gave rise to the chloroplasts, beginning with those of glaucophytes, red algae, green algae, and land plants. **Secondary** and **tertiary endosymbiosis** of these chloroplast-containing eukaryotes within other eukaryotes gave rise to the chloroplasts of euglenids, dinoflagellates, and other groups. Review Figure 27.3, **ANIMATED TUTORIAL 27.1**

## 27.2 What Features Account for Protist Diversity?

- **Protists** are a diverse, paraphyletic assemblage of mostly unicellular eukaryotes. Review Figure 27.1
- Some protists are photosynthetic autotrophs, some are heterotrophs, and some are both.
- The cytoskeleton allows for various means of locomotion. Most protists are motile, moving by amoeboid motion with **pseudopods** or by means of cilia or flagella.
- Some protist cells contain **contractile vacuoles** that pump out excess water, or **digestive vacuoles** where food is digested. Review Figures 27.5 and 27.6, **ANIMATED TUTORIAL 27.2**
- Many protists have protective cell surfaces such as cell walls, external "shells," or shells constructed from sand.

## 27.3 How Do Protists Affect the World Around Them?

- The diatoms, part of the plankton, are responsible for about a fifth of all the photosynthetic carbon fixation on Earth.

- **Phytoplankton** are the primary producers in the marine environment.
- Endosymbiosis is common among protists and often helpful to both partners. Review Figure 27.8
- Pathogenic protists include species of *Plasmodium* and trypanosomes. Review Figure 27.9

## 27.4 How Do Protists Reproduce?

- Most protists reproduce both asexually and sexually.
- **Conjugation** in paramecia is a sexual process but not a reproductive one. Review Figure 27.12
- **Alternation of generations**, which includes a multicellular diploid phase and a multicellular haploid phase, is a feature of many protist life cycles. Review Figure 27.13
- Alternation of generations may be **heteromorphic** or **isomorphic**. The alternating generations are the (diploid) **sporophyte** and (haploid) **gametophyte**. Specialized cells of the sporophyte, called **sporocytes**, divide meiotically to produce haploid spores.
- Depending on whether their gametes look identical or dissimilar, species are termed **isogamous** or **anisogamous**.
- In a **diplontic** life cycle, the gametes are the only haploid cells. In a **haplontic** life cycle, the zygote is the only diploid cell. Review Figures 27.14 and 27.15, **WEB ACTIVITIES 27.1 AND 27.2**
- Some protist life cycles involve more than one host species.

## 27.5 What Are the Evolutionary Relationships Among Eukaryotes?

- Most eukaryotes can be classified in one of five major clades: chromalveolates, Plantae, excavates, rhizaria, or unikonts. Review Table 27.1 and Figure 27.1
- The **chromalveolates** include the haptophytes, alveolates, and stramenopiles.
- **Alveolates** are unicellular organisms with sacs (alveoli) beneath their plasma membranes. Alveolate clades include the marine **dinoflagellates**, the parasitic **apicomplexans**, and the diverse, highly motile **ciliates**. **REVIEW WEB ACTIVITY 27.3**
- **Stramenopiles** typically have two flagella of unequal length, the longer one bearing rows of tubular hairs. Included among the stramenopiles are the unicellular **diatoms**, the multicellular **brown algae**, and the nonphotosynthetic **oomycetes**, including the water molds and downy mildews.
- Clades in the **Plantae** include the **glaucophytes**, **red algae**, **chlorophytes**, **charophytes**, and the land plants. All are photosynthetic and contain chloroplasts. Glaucophyte chloroplasts contain peptidoglycan between their inner and outer membranes.
- The **excavates** include the diplomonads, parabasalids, heteroloboseans, euglenids, and kinetoplastids. The **diplomonads** and **parabasalids** lack mitochondria, having apparently lost them during their evolution. **Heteroloboseans** are amoebas with a two-stage life cycle. **Euglenids** are often photosynthetic and have anterior flagella and strips of protein that support their cell surface. **Kinetoplastids** have a single, large mitochondrion in which mitochondrial mRNA is edited.
- **Rhizaria** are unicellular and aquatic; most are amoeboid. This group includes the **foraminiferans**, whose shells have contributed to great limestone deposits; the **radiolarians** with thin, stiff pseudopods and glassy endoskeletons; and the **cercozoans**, which take many forms and live in diverse habitats.
- The **unikonts** encompass organisms with single flagella on their flagellated cells (if any). They can be divided into two subgroups, the opisthokonts and the amoebozoans.
- In the **opisthokonts**, the flagellum (when present) is posterior. The opisthokont subgroups are the fungi, choanoflagellates, and animals. **Choanoflagellates** resemble the cells of sponges and are sister to the animal clade.
- The **amoebozoans** move by means of lobe-shaped pseudopodia. They comprise the **loboseans**, plasmodial slime molds, and cellular slime molds. A lobosean consists of a single cell; these cells do not aggregate. Plasmodial slime molds are amoebozoans whose feeding phase is coenocytic. In the feeding phase, movement is by cytoplasmic streaming. In cellular slime molds, the individual cells maintain their identity at all times but aggregate to form fruiting bodies.

## SELF-QUIZ

1. Microbial eukaryotes with flagella
   *a.* appear in several clades.
   *b.* are all algae.
   *c.* all have pseudopods.
   *d.* are all colonial.
   *e.* are never pathogenic.
2. Which statement about eukaryotic phytoplankton is *not* true?
   *a.* Some are important primary producers.
   *b.* Some contributed to the formation of petroleum.
   *c.* Some form toxic "red tides."
   *d.* Some are food for marine animals.
   *e.* They constitute a clade.
3. Apicomplexans
   *a.* possess flagella.
   *b.* possess a glassy shell.
   *c.* are all parasitic.
   *d.* are algae.
   *e.* include the trypanosomes that cause sleeping sickness.
4. The ciliates
   *a.* move by means of short flagella.
   *b.* use amoeboid movement.
   *c.* include *Plasmodium*, the agent of malaria.
   *d.* possess both a macronucleus and micronuclei.
   *e.* are autotrophic.
5. The chloroplasts of photosynthetic protists
   *a.* are structurally identical.
   *b.* gave rise to mitochondria.
   *c.* are all descended from a once free-living cyanobacterium.
   *d.* all have exactly two surrounding membranes.
   *e.* are all descended from a once free-living red alga.

6. Which statement about the brown algae is *not* true?
   *a.* They are all multicellular.
   *b.* They use the same photosynthetic pigments as do land plants.
   *c.* They are almost exclusively marine.
   *d.* A few are many meters in length.
   *e.* They are stramenopiles.
7. Which statement about the chlorophytes is *not* true?
   *a.* They use the same photosynthetic pigments as do land plants.
   *b.* Some are unicellular.
   *c.* Some are multicellular.
   *d.* All are microscopic in size.
   *e.* They display a great diversity of life cycles.
8. The red algae
   *a.* are mostly unicellular.
   *b.* are mostly marine.
   *c.* owe their red color to a special form of chlorophyll.
   *d.* have flagella on their gametes.
   *e.* are all heterotrophic.
9. The plasmodial slime molds
   *a.* form a plasmodium that is a coenocyte.
   *b.* lack fruiting bodies.
   *c.* consist of large numbers of myxamoebas.
   *d.* consist at times of a mass called a pseudoplasmodium.
   *e.* possess flagella.
10. The cellular slime molds
    *a.* possess apical complexes.
    *b.* lack fruiting bodies.
    *c.* form a plasmodium that is a coenocyte.
    *d.* have haploid myxamoebas.
    *e.* possess flagella.

## FOR DISCUSSION

1. For each type of organism below, give a single characteristic that may be used to differentiate it from the other, related organism(s) named in parentheses.
   *a.* Foraminiferans (radiolarians)
   *b.* *Euglena* (*Volvox*)
   *c.* *Trypanosoma* (*Giardia*)
   *d.* Plasmodial slime molds (cellular slime molds)
2. In what sense are sex and reproduction independent of each other in the ciliates? What does that suggest about the role of sex in biology?
3. Why are dinoflagellates and apicomplexans placed in one group of microbial eukaryotes and brown algae and oomycetes in another?
4. Unlike many protists, apicomplexans lack contractile vacuoles. Why don't apicomplexans need a contractile vacuole?
5. Giant seaweeds (mostly brown algae) have "floats" that aid in keeping their fronds suspended at or near the surface of the water. Why is it important that the fronds be suspended in this way?
6. Why are algal pigments so much more diverse than those of land plants?
7. Consider the chloroplasts of chlorophytes, euglenids, and red algae. For each of these groups, indicate how many membranes surround their chloroplasts, and offer a reasonable explanation in each case. Why do some dinoflagellates have more membranes around their chloroplasts than other dinoflagellates?

## ADDITIONAL INVESTIGATION

Mitochondrial, chloroplast, and nuclear genomes of eukaryotes each contain ribosomal RNA genes. If the endosymbiotic theory for the origin of mitochondria and chloroplasts is correct, would you expect the ribosomal genes of these organelles to be more closely related to the nuclear ribosomal RNA genes, or to ribosomal RNA genes in proteobacteria and cyanobacteria? How would you test the endosymbiotic theory for the origin of these organelles using a phylogenetic analysis of ribosomal RNA genes?

# 28 Plants without Seeds: From Water to Land

## What surprises lurk in a rock?

John William Dawson, later Sir William Dawson, was an outspoken critic of Charles Darwin and evolutionary theory. Ironically, Dawson made one of the first contributions to our understanding of the early evolution of plants. While surveying the geology of Nova Scotia, this Canadian scientist found a puzzling fossil fragment on the Gaspé Peninsula. It was 1859—the very year in which Darwin published *The Origin of Species*.

What Dawson found was the remains of a remarkable plant, which he named *Psilophyton*, meaning "naked plant." The fossilized plant appeared to have no roots and no leaves, and its stem grew both below and above the ground. The aboveground part of the stem, which branched and ended in spore cases, was about 50 centimeters tall. Could this Devonian fossil plant represent one of the stages in the momentous transition of plants from an aquatic existence to the forests that cover the land today? Dawson presented *Psilophyton* and other finds in a major lecture he delivered in 1870, but his audience was unreceptive, with his fellow scientists chuckling that *Psilophyton* grew nowhere but in Dawson's imagination. His published drawing of the fossil plant was regarded as an amusing curiosity.

Decades later, Dawson's interpretation of *Psilophyton* was vindicated by the work of two British botanists who studied plant fossils. The year 1915 found Robert Kidston and William Lang in the hills near Rhynie, Scotland, where they discovered evidence that a marsh had existed there 400 million years ago. In the intervening hundreds of millions of years, that Devonian marsh had become a flinty, fossil-laden rock called chert. Although the rock now lay in a hilly site some 50 kilometers from the sea, its original location must have been near the shore.

The most startling feature of the Rhynie chert was the great abundance of fossils of a small plant with spore cases but neither roots nor leaves—a plant, in fact, that looked strikingly like Dawson's "imaginary" *Psilophyton*! Kidston and Lang described their find in detail and gave it the genus name *Rhynia* in honor of the site of its discovery.

**Making It on Land** An artist's reconstruction of the earliest vascular plant ancestors, based on the Rhynie chert fossil bed in Scotland. The plants in this scene are rhyniophytes. Heavily mineralized water from hot springs and geysers may have been partly responsible for the remarkable preservation of plant fossils in the Rhynie bed.

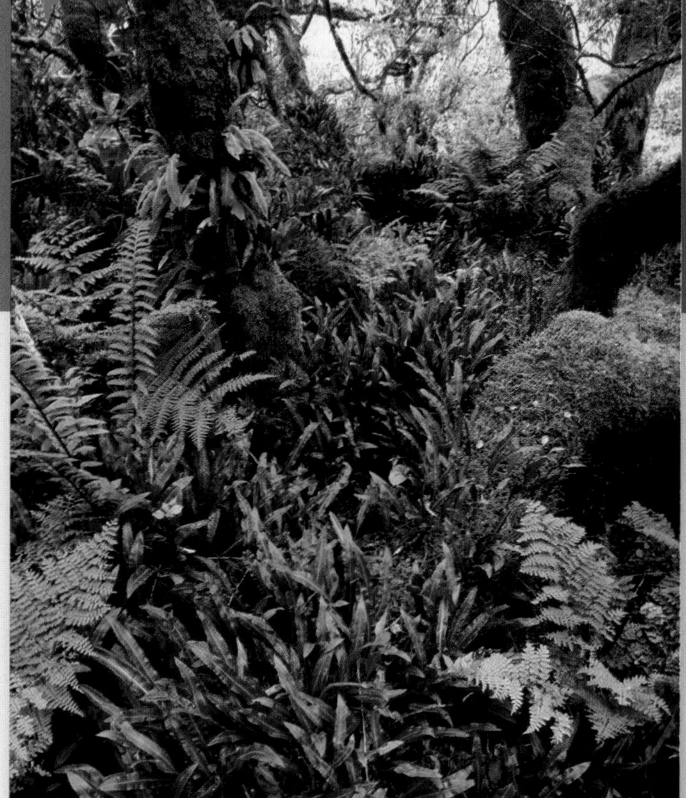

**Green Mansions** Seedless plants—mosses and ferns—dominate the understory of this pristine rainforest on the island of Maui, Hawaii. The only seed plants to be seen are the trees, which are almost obscured by mosses and ferns.

*Rhynia* is just one of several ancient groups of plants that lack seeds and other features of more modern plants. The glory days of the seedless plants are long past, but many of them—most notably the mosses and ferns—are still abundant. We rely on these surviving plants, as well as on fossils, to help us understand the evolution of land plants from aquatic algae growing at the edge of a sea or marsh. Much of the fossilized history, and the great bulk of the mass of seedless plants, was preserved as coal, which today is an economically important substance that we burn as fuel and to produce electricity. What a library of botanical history the fossils in the world's coal beds provide!

**IN THIS CHAPTER** we will see how plants first invaded land, how land plants evolved, and how plant clades diversified, resulting in plants ever better equipped to face the challenges of terrestrial environments. The descriptions here will concentrate on those land plants that lack seeds. Chapter 29 considers the seed plants that dominate the terrestrial scene today.

## CHAPTER OUTLINE



# 28.1 How Did the Land Plants Arise?

The question "Where did plants come from?" embraces two questions: "What were the ancestors of plants?" and "Where did those ancestors live?" We consider both questions in this section.

The **land plants** are monophyletic: all land plants descend from a single common ancestor and form a branch of the evolutionary tree of life. Although we refer to this clade of mostly terrestrial species as land plants, some of them actually live in shallow water. One of the key shared derived traits, or *synapomorphies*, of the land plants is development from an embryo protected by tissues of the parent plant. For this reason, land plants are sometimes called **embryophytes** (*phyton*, "plant"). Land plants retain the derived features they share with the "green algae" described in Chapter 27: the use of chlorophylls *a* and *b* in photosynthesis, and the use of starch as a photosynthetic storage product. Both land plants and green algae have cellulose in their cell walls.

There are several ways to define "plant" and still refer to a clade (**Figure 28.1**; see also Figure 27.1). Throughout this book and in everyday language, the unmodified common name "plants" usually refers to the land plants. Some biologists, however, use the term "plants" to include land plants plus some closely related groups of green algae; this monophyletic group is also known as the **streptophytes**. The addition of the remainder of the green algae to the streptophytes results in a more inclusive clade, commonly called the **green plants**, which encompasses all the groups that possess chlorophyll *b*. Green plants, streptophytes, and land plants each have been called "the plant kingdom" by different authorities; others take an even broader view and include red algae and glaucophytes as "plants." To avoid confusion in this chapter, we will use modifying terms ("land plants" or "green plants," for example) to refer to the various clades shown in Figure 28.1.

### There are ten major groups of land plants

The land plants that exist today fall naturally into ten major clades (**Table 28.1**). Members of seven of those clades possess well-developed vascular systems that transport materials throughout the plant body. We call these seven groups, collectively, the **vascular plants**, or **tracheophytes**, because they all possess fluid-conducting cells called **tracheids**. Taken together, the seven groups of vascular plants constitute a clade.

The remaining three clades (liverworts, hornworts, and mosses) lack tracheids. These three groups are sometimes collectively called *bryophytes*, but in this text we reserve that term for the largest monophyletic group—the mosses—and refer to these three clades collectively as **nonvascular land plants**.

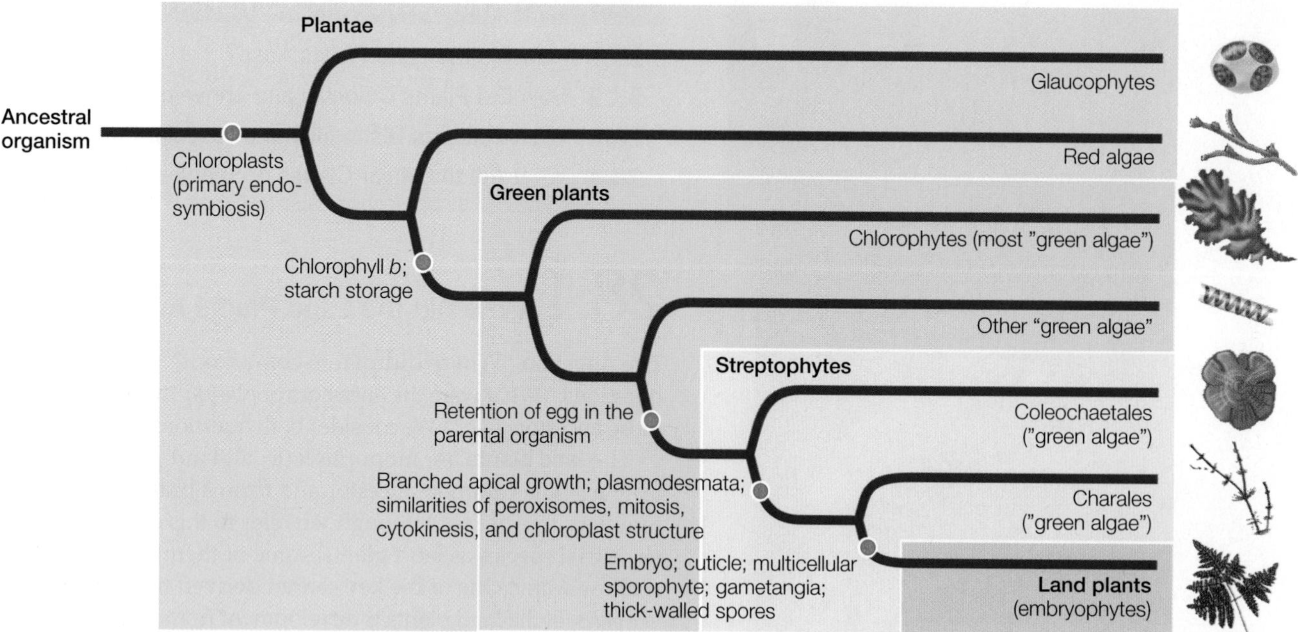


**28.1 What Is a Plant?** In its broadest definition, the term "plant" includes the green plants, red algae, and glaucophytes—all the groups descended from a common ancestor with primary chloroplasts. Some biologists restrict the term to the green plants (those with chlorophyll *b*) or, even more narrowly, to the land plants (embryophytes).

Note, however, that *these three groups do not form a clade.* Some nonvascular land plants have conducting cells, but none have tracheids.

Where did the land plants come from? To determine which living organisms are most closely related to land plants, biologists have considered several of the synapomorphies of land plants and looked for their origins in various other groups.

### The land plants arose from a green algal clade

Several microscopic structural features, backed by clear-cut evidence from molecular studies, indicate that the closest relatives of the land plants are two groups of aquatic green algae, the **Coleochaetales** and the **Charales** (see Figure 28.1). Both of these algal groups retain their eggs in the parental organism, as do land plants. Of these two candidates, Charales is thought to be the sister-group of land plants, based on the following synapomorphies:

- Plasmodesmata join the cytoplasm of adjacent cells (see Figure 34.6)
- Growth is branching and apical (from the tip)

**TABLE 28.1**

**Classification of Land Plants**

| GROUP | COMMON NAME | CHARACTERISTICS |
|---|---|---|
| **NONVASCULAR LAND PLANTS** | | |
| Hepatophyta | Liverworts | No filamentous stage; gametophyte flat |
| Anthocerophyta | Hornworts | Embedded archegonia; sporophyte grows basally (from the ground) |
| Bryophyta | Mosses | Filamentous stage; sporophyte grows apically (from the tip) |
| **VASCULAR PLANTS** | | |
| Lycopodiophyta | Lycophytes: Club mosses and allies | Microphylls in spirals; sporangia in leaf axils |
| Monilophyta | Horsetails, whisk ferns, ferns | Differentiation between main stem and side branches (overtopping growth) |
| SEED PLANTS | | |
| Gymnosperms | | |
| Cycadophyta | Cycads | Compound leaves; swimming sperm; seeds on modified leaves |
| Ginkgophyta | Ginkgo | Deciduous; fan-shaped leaves; swimming sperm |
| Gnetophyta | Gnetophytes | Vessels in vascular tissue; opposite, simple leaves |
| Coniferophyta | Conifers | Seeds in cones; needlelike or scalelike leaves |
| Angiosperms | Flowering plants | Endosperm; carpels; gametophytes much reduced; seeds within fruit |

(A) *Chara vulgaris* (stonewort)

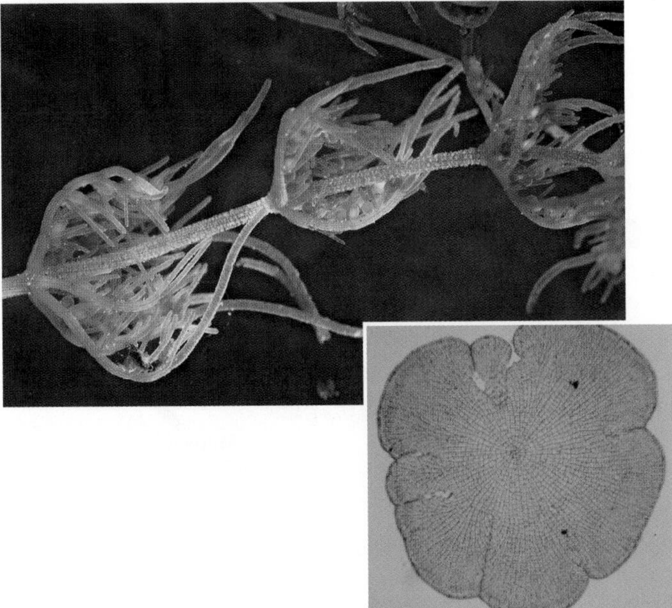

(B) *Coleochaete* sp.

**28.2 The Closest Relatives of Land Plants** (A) The land plants probably evolved from a common ancestor shared with the Charales, a green algal group. (B) This species of Coleochaete is a representative of the Coleochaetales, the sister group of Charales plus land plants.

- Similar peroxisome contents, mechanics of mitosis and cytokinesis, and chloroplast structure

Both of these algal groups, however, have some features that are similar to groups of land plants. The Charales, as represented by stoneworts of the genus *Chara* (**Figure 28.2A**), exhibit the branching growth form found among most land plants. But the flattened growth form of many members of Coleochaetales, as represented by the genus *Coleochaete* (**Figure 28.2B**), is more like the growth form of basal land plants such as liverworts.

## 28.1 RECAP

**Land plants are photosynthetic organisms that develop from embryos protected by parental plant tissue.**

- Explain the different possible uses of the term "plant." See p. 589 and Figure 28.1
- What is the key difference between the vascular plants and the other three clades of land plants? See pp. 589–590 and Table 28.1
- What evidence supports the phylogenetic relationship between land plants and Charales? See p. 590

The green algal ancestors of the land plants lived at the margins of ponds or marshes, ringing them with a green mat. It was from such a marginal habitat, which was sometimes wet and sometimes dry, that early plants made the transition onto land.

# 28.2 How Did Plants Colonize and Thrive on Land?

Land plants, or their immediate ancestors in those ancient green mats, first appeared in the terrestrial environment between 400 and 500 million years ago. How did they survive in an environment that differed so dramatically from the aquatic environment of their ancestors? While the water essential for life is everywhere in the aquatic environment, water is difficult to obtain and retain in the terrestrial environment.

### Adaptations to life on land distinguish land plants from green algae

No longer bathed in fluid, organisms on land faced potentially lethal desiccation (drying). Large terrestrial organisms had to develop ways to transport water to body parts distant from the source. And whereas water provides aquatic organisms with support against gravity, a plant living on land must either have some other support system or sprawl unsupported on the ground. A land plant must also use different mechanisms for dispersing its gametes and progeny than its aquatic relatives, which can simply release them into the water. Survival on land required numerous adaptations. The first colonists—the *nonvascular land plants*—met at least some of these challenges.

Most of the characteristics that distinguish land plants from green algae are evolutionary adaptations to life on land:

- The *cuticle*, a waxy covering that retards water loss
- *Stomata*, small closable openings in leaves and stems that are used to regulate gas exchange (stomata are not present in liverworts)
- *Gametangia*, multicellular organs that enclose plant gametes and prevent them from drying out
- *Embryos*, young plants contained within a protective structure
- Certain *pigments* that afford protection against the mutagenic ultraviolet radiation that bathes the terrestrial environment
- Thick *spore walls* containing a polymer (called sporopollenin) that protects the spores from desiccation and resists decay
- A *mutually beneficial association with a fungus* that promotes nutrient uptake from the soil

The **cuticle** may be the most important—and earliest—of these features. Composed of several unique waxy lipids (see Section 3.4) that coat the leaves and stems of land plants, the cuticle has several functions, the most obvious and important of which is to keep water from evaporating from the plant body.

As ancient plants colonized land, they modified the terrestrial environment by contributing to the formation of soil. Acid secreted by plants helps break down rock, and the organic compounds produced by the breakdown of dead plants contribute to soil structure. Such effects are repeated today as plants grow in new areas.

### Nonvascular land plants usually live where water is readily available

Living species of liverworts, mosses, and hornworts are thought to be similar in many ways to the earliest land plants. Most of these plants grow in dense mats, usually in moist habitats. Even the largest of these species are only about half a meter tall, and most are only a few centimeters tall or long. Why have they not evolved to be taller? The probable answer is that they lack an efficient vascular system for conducting water and minerals from the soil to distant parts of the plant body.

The nonvascular land plants lack the leaves, stems, and roots that characterize the vascular plants, although they have structures analogous to each. Their growth pattern allows water to move through the mats of plants by capillary action. They have leaflike structures that readily catch and hold any water that splashes onto them. They are small enough that minerals can be distributed throughout their bodies by diffusion. As in all land plants, layers of maternal tissue protect their embryos from desiccation. Nonvascular land plants also have a cuticle, although it is often very thin (or even absent in some species) and thus is not highly effective in retarding water loss.

Most nonvascular land plants live on the soil or on vascular plants, but some grow on bare rock, dead and fallen tree trunks, and even on buildings. The ability to grow on such marginal surfaces results from a mutualistic association with the glomeromycetes, a clade of fungi. The earliest association of land plants and fungi dates back at least 460 million years. This mutualism probably facilitated the absorption of water and minerals, especially phosphorus, from the first soils.

Nonvascular land plants are widely distributed over six continents and even exist (albeit very locally) on the coast of the seventh, Antarctica. They are well adapted to their environments. Most are terrestrial. Although a few species live in fresh water, these aquatic forms are descended from terrestrial ones. None live in the oceans.

Land plants and green algae differ not just in structure but also in their life cycles. Differences in the life cycles of land plants and those of their ancestors are partially a function of the relative dependence of these respective plant groups on a water supply for reproduction.

### Life cycles of land plants feature alternation of generations

A universal feature of the life cycles of land plants is alternation of generations (**Figure 28.3**). Recall from Section 27.4 the two hallmarks of alternation of generations:

- The life cycle includes both a multicellular diploid stage and a multicellular haploid stage.
- Gametes are produced by mitosis, not by meiosis. Meiosis produces spores that develop into multicellular haploid organisms.

If we begin looking at the land plant life cycle at the single-cell stage—the diploid zygote—then the first phase of the cycle is the formation, by mitosis and cytokinesis, of a multicellular embryo, which eventually grows into a mature diploid plant. This multicellular diploid plant is the **sporophyte** ("spore plant").

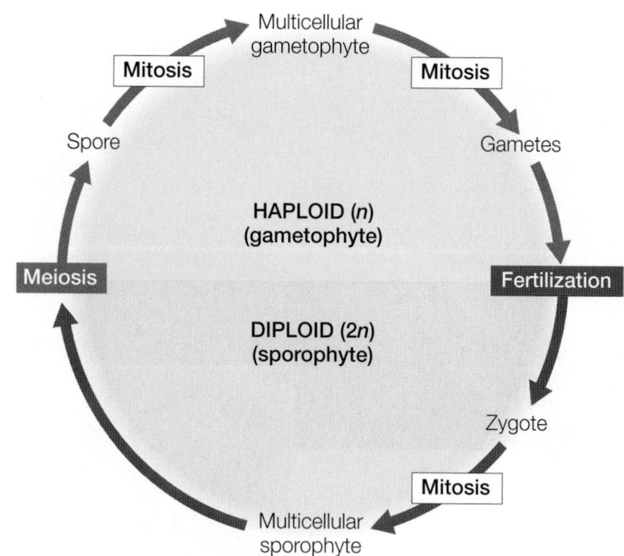


**28.3 Alternation of Generations in Plants** A multicellular diploid sporophyte generation produces spores by meiosis and alternates with a multicellular haploid gametophyte generation that produces gametes by mitosis.

Cells contained within **sporangia** (singular *sporangium*) of the sporophyte undergo meiosis to produce haploid, unicellular spores. By mitosis and cytokinesis, a spore forms a haploid plant. This multicellular haploid plant, called the **gametophyte** ("gamete plant"), produces haploid gametes by mitosis. The fusion of two gametes (*syngamy*, or *fertilization*) forms a single diploid cell—the zygote—and the cycle is repeated (**Figure 28.4**).

The *sporophyte generation* extends from the zygote through the adult multicellular diploid plant and sporangium formation; the *gametophyte generation* extends from the spore through the adult multicellular haploid plant to the gametes. The transitions between the generations are accomplished by fertilization and meiosis. In all land plants, the sporophyte and gametophyte differ genetically: the sporophyte has diploid cells, and the gametophyte has haploid cells.

There is a trend toward reduction of the gametophyte generation in plant evolution. In the nonvascular land plants, the gametophyte is larger, longer-lived, and more self-sufficient than the sporophyte. In those groups that appeared later in plant evolution, however, the sporophyte generation is the larger, longer-lived, and more self-sufficient one. In the seed plants, this evolutionary trend has led to a condition in which water is not required for the sperm to reach the egg.

### The sporophytes of nonvascular land plants are dependent on gametophytes

In nonvascular land plants, the conspicuous green structure visible to the naked eye is the gametophyte (see Figure 28.4). This is in contrast to vascular plants, such as ferns and seed plants, in which the familiar forms are sporophytes. The gametophyte of liverworts, hornworts, and mosses is photosynthetic and

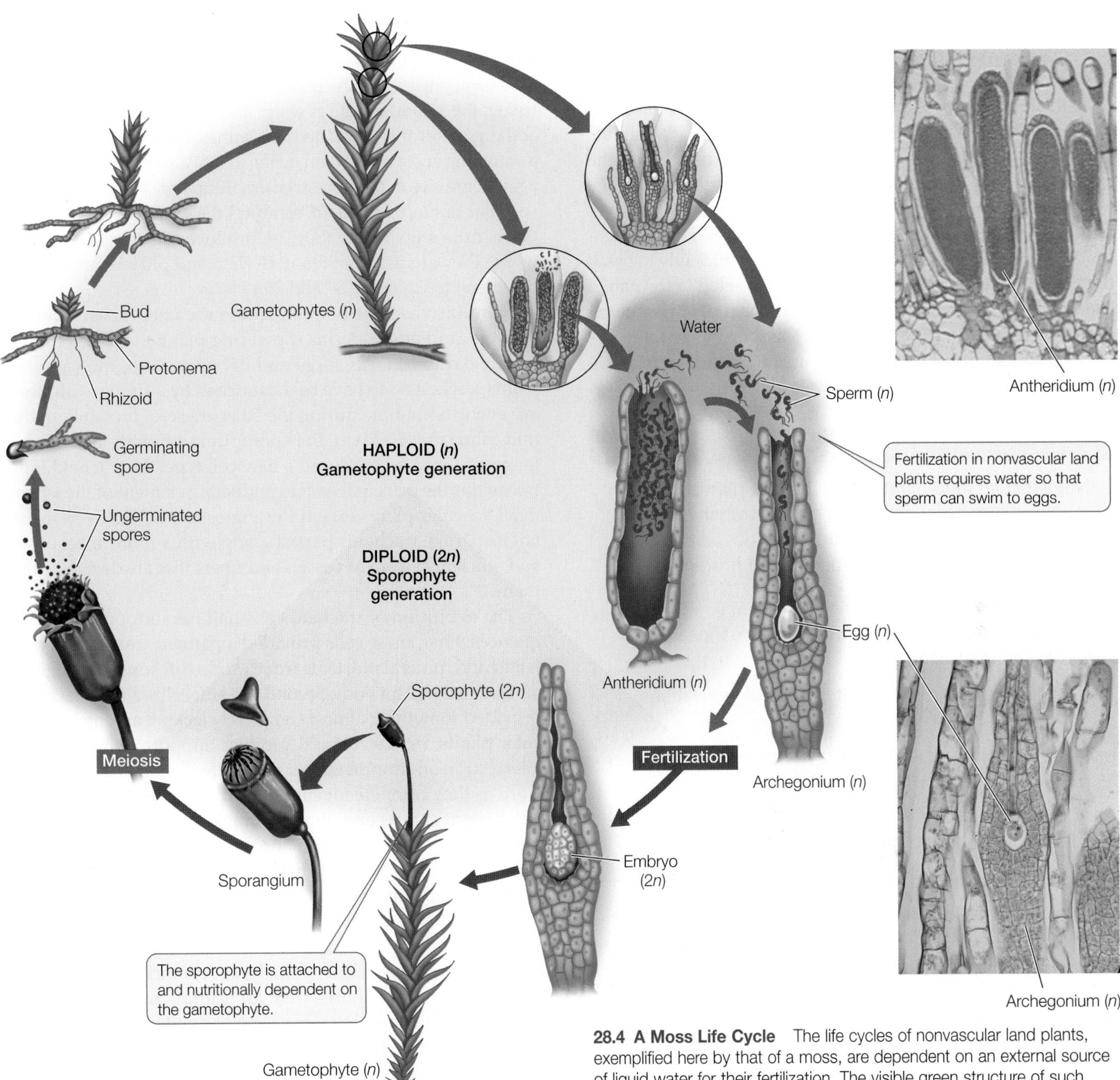


**28.4 A Moss Life Cycle** The life cycles of nonvascular land plants, exemplified here by that of a moss, are dependent on an external source of liquid water for their fertilization. The visible green structure of such plants is the gametophyte.

**yourBioPortal.com**
GO TO Animated Tutorial 28.1 • Life Cycle of a Moss

therefore nutritionally independent; the sporophyte may or may not be photosynthetic, but it is always nutritionally dependent on the gametophyte and remains permanently attached to it.

In nonvascular land plants, a sporophyte produces unicellular haploid spores as products of meiosis within a sporangium. A spore germinates, giving rise to a multicellular haploid gametophyte whose cells contain chloroplasts and are thus photosynthetic. Eventually gametes form within specialized sex organs, the **gametangia**. The **archegonium** is a multicellular, flask-shaped female sex organ with a long neck and a swollen base, which produces a single egg. The **antheridium** is a male sex organ in which sperm, each bearing two flagella, are produced in large numbers (see the insets in Figure 28.4). Both archegonia and antheridia are produced on the same individual, so each individual can have both male and female reproductive structures. Fertilization is often between adjacent individuals, however, which helps maintain genetic diversity in the population.

Once released from the antheridium, the sperm must swim or be splashed by raindrops to a nearby archegonium on the same or a neighboring plant—a constraint that reflects the aquatic origins of the nonvascular land plants' ancestors. The sperm are aided in this task by chemical attractants released by the egg or the archegonium. Before sperm can enter the archegonium, certain cells in the neck of the archegonium must break down, leaving a water-filled canal through which the sperm swim to complete their journey. Note that *all of these events require liquid water.*

On arrival at the egg, the nucleus of a sperm fuses with the egg nucleus to form a diploid zygote. Mitotic divisions of the zygote produce a multicellular, diploid sporophyte embryo. The base of the archegonium grows and surrounds the embryo, which protects the embryo during its early development. Eventually, the developing sporophyte elongates sufficiently to break out of the archegonium, but it remains connected to the gametophyte by a "foot" that is embedded in the parent tissue and absorbs water and nutrients from it. The sporophyte then produces a sporangium, within which meiotic divisions produce spores and thus the next gametophyte generation.

## 28.2 RECAP

**Nonvascular land plants were the first plants to inhabit a terrestrial environment. The transition to land required numerous evolutionary developments, including the cuticle, gametangia, and protected embryos.**

- Describe several adaptations of plants to the terrestrial environment. See p. 591
- Explain what is meant by alternation of generations. See p. 592 and Figure 28.3
- Why do nonvascular land plants require liquid water for fertilization? See pp. 592–594
- In the nonvascular land plants, how is the sporophyte dependent on the gametophyte? See pp. 593–594

Further adaptations to the terrestrial environment appeared as plants continued to evolve. One of the most important of these later adaptations was the appearance of vascular tissues.

## 28.3 What Features Distinguish the Vascular Plants?

The first land plants were nonvascular, lacking both water-conducting and food-conducting tissues. These plants did not leave a very complete fossil record, however, because most nonvascular plant species readily decompose, leaving little or nothing to fossilize.

That the nonvascular land plants evolved tens of millions of years before the earliest vascular plants is supported by extensive phylogenetic and molecular-clock analyses of DNA sequences. The first plants possessing vascular tissue (characterized by fluid-conducting tracheid cells) arose much later (**Figure 28.5**).

### Vascular tissues transport water and dissolved materials

Vascular plants differ from the other land plants in crucial ways, one of which is the possession of a well-developed **vascular system** consisting of tissues specialized for the transport of materials from one part of the plant to another. One type of vascular tissue, the **xylem**, conducts water and minerals from the soil to aerial parts of the plant. Because some of its cell walls contain a stiffening substance called *lignin*, xylem also provides support against gravity in the terrestrial environment. The other type of vascular tissue, the **phloem**, conducts the products of photosynthesis from sites where they are produced or released to sites where they are used or stored. (Xylem and phloem are further discussed in Chapters 34 and 35.)

Familiar vascular plants include the club mosses, ferns, conifers, and angiosperms (flowering plants). Although they are an extraordinarily large and diverse group, the vascular plants can be said to have been launched by a single evolutionary event. Sometime during the Paleozoic era, probably in the mid-Silurian (430 Mya), the sporophyte generation of a now long-extinct plant produced a new cell type, the tracheid. The tracheid is the principal water-conducting element of the xylem in all vascular plants except the angiosperms, and even in the angiosperms, tracheids persist along with a more specialized and efficient system of vessels and fibers that are derived from them.

The evolution of tracheid cells had two important consequences. First, these cells provided a pathway for transport of water and mineral nutrients from a source of supply to regions of need in the plant body. Second, the stiff cell walls of tracheids provided something almost completely lacking among nonvascular plants: rigid structural support. Support is important in a terrestrial environment because it allows plants to grow upward as they compete for sunlight to power photosynthesis. A taller plant can receive direct sunlight and photosynthesize more readily than a shorter plant, whose leaves may be shaded by the taller one. Increased height also improves the dispersal of spores. Thus tracheids set the stage for the complete and permanent invasion of land by plants.

The vascular plants featured another evolutionary novelty: a branching, independent sporophyte. A branching sporophyte can produce more spores than an unbranched body, and it can develop in complex ways. The sporophyte of a vascular plant is nutritionally independent of the gametophyte at maturity. Among the vascular plants, the sporophyte is the large and obvious plant that one normally pays attention to in nature. In contrast, the sporophyte of nonvascular land plants is attached to, dependent on, and usually much smaller than the gametophyte.

The present-day evolutionary descendants of the early vascular plants include the lycophytes, monilophytes, and seed plants (see Figure 28.5; the vascular plant groups are bracketed on the right). Two types of life cycles are seen in the vascular plants: one that involves seeds and another that does not. As we discuss in greater detail in the next chapter, a **seed** consists of a plant embryo, together with a food source, surrounded by a protective coat. The life cycles of the club mosses, ferns, whisk ferns, and horsetails do not involve seeds. We will describe these seedless vascular plant groups in detail after taking a closer look at vascular plant evolution. The major groups of seed plants are described in Chapter 29.

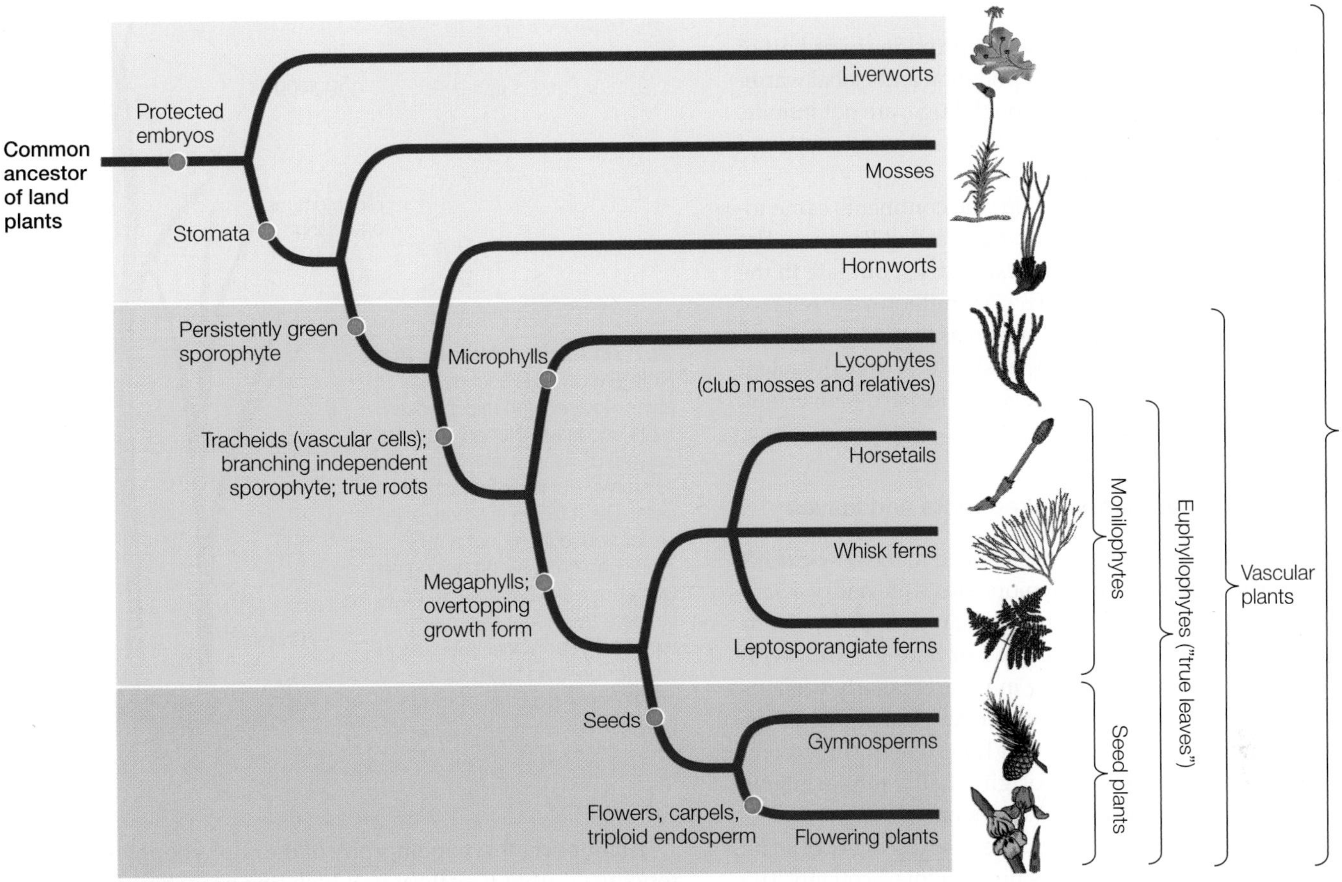

Common ancestor of land plants
Protected embryos
Stomata
Persistently green sporophyte
Tracheids (vascular cells); branching independent sporophyte; true roots
Microphylls
Megaphylls; overtopping growth form
Seeds
Flowers, carpels, triploid endosperm
Liverworts
Mosses
Hornworts
Lycophytes (club mosses and relatives)
Horsetails
Whisk ferns
Leptosporangiate ferns
Gymnosperms
Flowering plants
Monilophytes
Seed plants
Euphyllophytes ("true leaves")
Vascular plants
Land plants (embryophytes)

**28.5 The Evolution of Plants** Three key characteristics that emerged during plant evolution—protected embryos, vascular tissues, and seeds—are adaptations to life in a terrestrial environment.

## Vascular plants have been evolving for almost half a billion years

The evolution of an effective cuticle and protective layers for the gametangia (archegonia and antheridia) helped make the first vascular plants successful, as did the initial absence of herbivores (plant-eating animals) on land. By the late Silurian period (about 425 Mya), vascular plants were being preserved as fossils that we can study today. During the Silurian, the largest vascular plants were only a few centimeters tall, yet fossils uncovered in Wales in 2004 give clear evidence for the earliest known wildfire, which burned vigorously even in the Silurian atmosphere, which had 14 percent less oxygen than today's atmosphere (see Figure 25.6). The small plants must have been abundant to sustain fire in such an atmosphere. Their proliferation made the terrestrial environment more hospitable to animals. Amphibians and insects arrived on land soon after land plants became established.

Trees of various kinds appeared in the Devonian period and dominated the landscape of the Carboniferous period (359–297 Mya). Forests of lycophytes (club mosses) up to 40 meters tall, along with horsetails and tree ferns, flourished in the tropical swamps of what would become North America and Europe (**Figure 28.6**). Plant parts from those forests sank in the swamps and were gradually covered by sediment. Over millions of years, as the buried plant material was subjected to intense pressure and elevated temperatures, it was

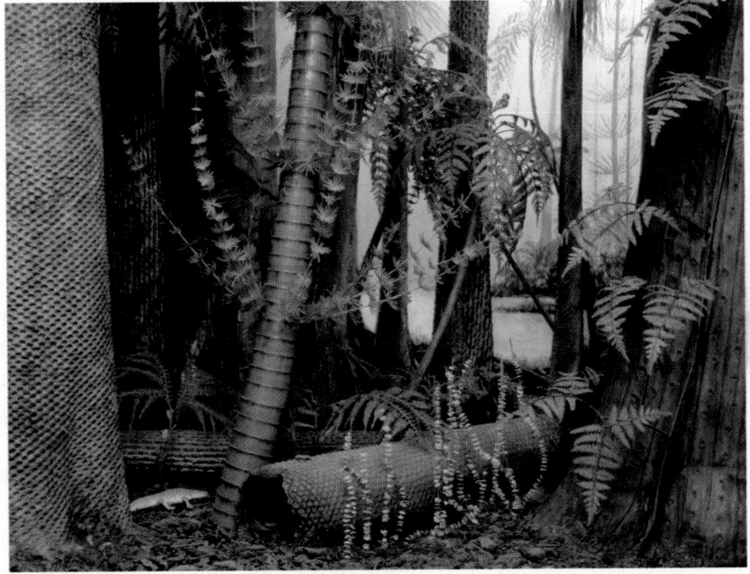

**28.6 Reconstruction of an Ancient Forest** This Carboniferous forest once thrived in what is now Michigan. The "trees" to the left and in the background are lycophytes of the genus *Lepidodendron*; abundant ferns are visible to the right. The plant in the foreground is a relative of the modern horsetails.

transformed into coal. Today that coal provides over half of our electricity (and contributes to air pollution and global warming). The world's coal deposits, although huge, are not infinite, and they cannot be renewed because the conditions that created coal no longer exist.

In the subsequent Permian period, the continents came together to form a single gigantic land mass called Pangaea. The continental interior became warmer and drier, but late in the period glaciation was extensive. The 200-million-year reign of the lycophyte–fern forests came to an end as they were replaced by forests of seed plants (gymnosperms), which were prevalent until a different group of seed plants (angiosperms) overtook the landscape about 65 million years ago.

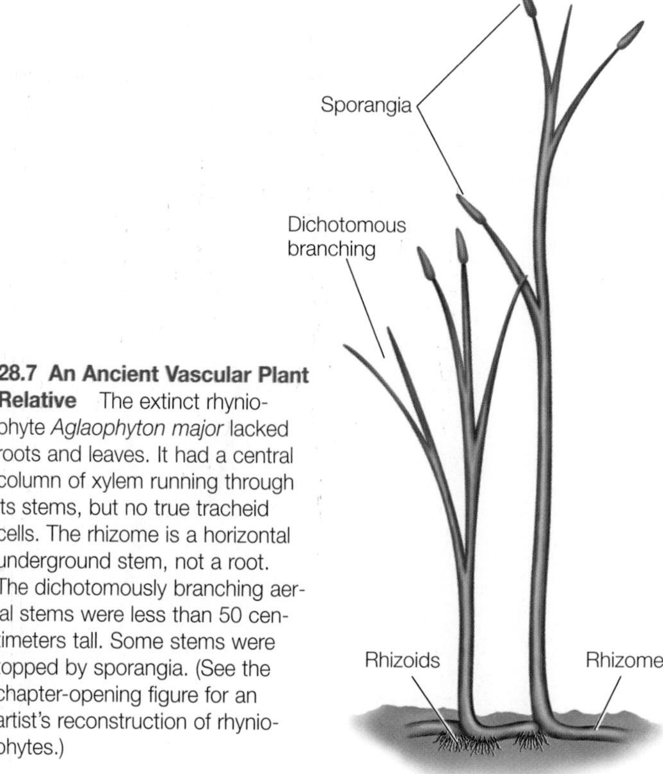


**28.7 An Ancient Vascular Plant Relative** The extinct rhyniophyte *Aglaophyton major* lacked roots and leaves. It had a central column of xylem running through its stems, but no true tracheid cells. The rhizome is a horizontal underground stem, not a root. The dichotomously branching aerial stems were less than 50 centimeters tall. Some stems were topped by sporangia. (See the chapter-opening figure for an artist's reconstruction of rhyniophytes.)

### The earliest vascular plants lacked roots and leaves

The earliest known vascular plants belonged to the now-extinct group called **rhyniophytes**. The rhyniophytes were one of a very few types of vascular plants in the Silurian period. The landscape at that time probably consisted of bare ground, with stands of rhyniophytes in low-lying moist areas. Early versions of the structural features of all the other vascular plant groups appeared in the rhyniophytes of that time. These shared features strengthen the case for the origin of all vascular plants from a common nonvascular plant ancestor.

The beginning of this chapter described the discovery of some important fossils in Devonian rocks near Rhynie, Scotland. The preservation of these plants was remarkable, considering that the rocks were more than 395 million years old. These fossil plants had a simple vascular system of phloem and xylem, but not all had the tracheids characteristic of today's vascular plants.

These plants also lacked roots. Like most modern ferns and lycophytes, they were apparently anchored in the soil by horizontal portions of stem, called **rhizomes**, which bore water-absorbing unicellular filaments called **rhizoids**. These rhizomes also bore aerial branches, and sporangia—homologous to the sporangia of mosses—were found at the tips of those branches. Their branching pattern was *dichotomous;* that is, the apex (tip) of the shoot divided to produce two equivalent new branches, each pair diverging at approximately the same angle from the original stem (**Figure 28.7**). Scattered fragments of such plants had been found earlier, but never in such profusion or as well preserved as those discovered by Kidston and Lang.

Although they were apparently ancestral to the other vascular plant groups, the rhyniophytes themselves are long gone. None of their fossils appear anywhere after the Devonian period.

### The vascular plants branched out

The **lycophytes** (club mosses and their relatives) first appeared in the Silurian period. The **monilophytes** (ferns and fern allies) appeared during the Devonian period. These two groups, both still with us today, arose from rhyniophyte-like ancestors. Important new features of the vascular plants are found in these groups, such as true roots, true leaves, and a differentiation between two types of spores. The monilophytes and seed plants constitute a clade called the **euphyllophytes** (*eu,* "true"; *phyllos,* "leaf").

An important synapomorphy of the euphyllophytes is **overtopping**, a growth pattern in which one branch differentiates from and grows beyond the others. Overtopping growth would have given these plants an advantage in the competition for light for photosynthesis, enabling them to shade their dichotomously growing competitors. And, as we'll see, the overtopping growth of the euphyllophytes enabled a new type of leaf to evolve.

### Roots may have evolved from branches

The rhyniophytes had only rhizoids arising from a rhizome with which to gather water and minerals. How, then, did subsequent groups of vascular plants come to have the complex roots we see today?

It is probable that roots had their evolutionary origins as a branch, either of a rhizome or of the aboveground portion of a stem. That branch presumably penetrated the soil and branched further. The underground portion could anchor the plant firmly, and even in this primitive condition, it could absorb water and minerals. The discovery of several fossil plants from the Devonian period, all having horizontal stems (rhizomes) with both underground and aerial branches, supports this hypothesis.

Underground and aboveground branches, growing in sharply different environments, were subjected to very different selection pressures during the succeeding millions of years. Thus the two parts of the plant body—the aboveground shoot system and the underground root system—diverged in structure and evolved distinct internal and external anatomies. In spite of these differences, scientists believe that the root and shoot systems of vascular plants are homologous—that they were once part of the same organ.

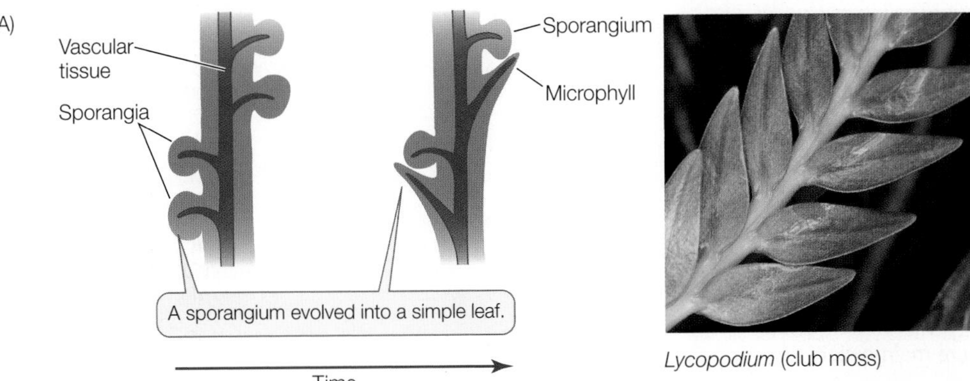
(A)
Vascular tissue
Sporangia
Sporangium
Microphyll
A sporangium evolved into a simple leaf.
Time
*Lycopodium* (club moss)

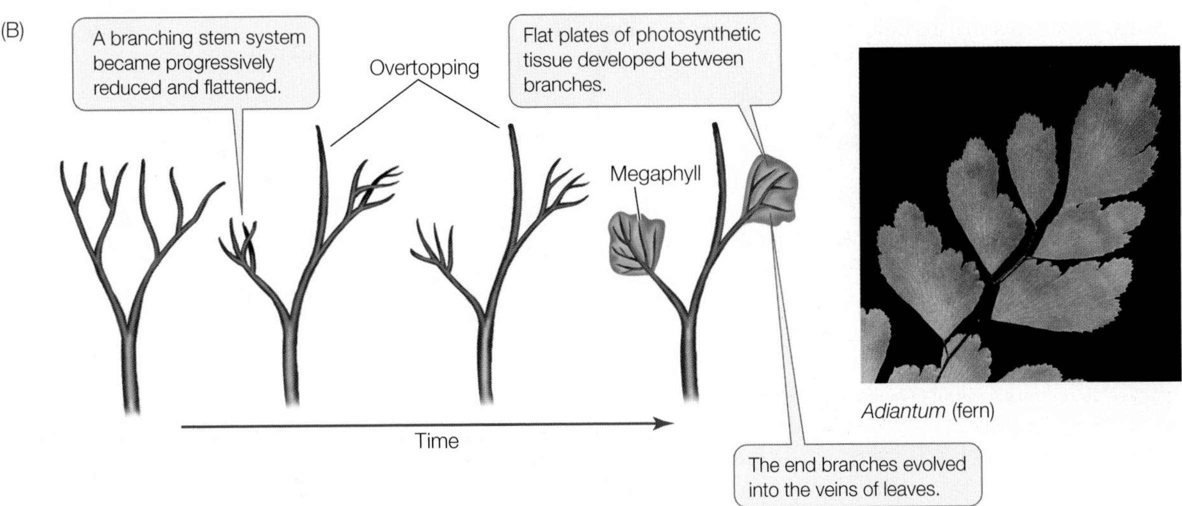
(B)
A branching stem system became progressively reduced and flattened.
Overtopping
Flat plates of photosynthetic tissue developed between branches.
Megaphyll
Time
The end branches evolved into the veins of leaves.
*Adiantum* (fern)

**28.8 The Evolution of Leaves** (A) Microphylls are thought to have evolved from sterile sporangia. (B) The megaphylls of monilophytes and seed plants may have arisen as photosynthetic tissue developed between branch pairs that were "left behind" as dominant branches overtopped them.

### Monilophytes and seed plants have true leaves

Thus far we have used the term "leaf" rather loosely. In the strictest sense, a *leaf* is a flattened photosynthetic structure emerging laterally from a stem or branch and possessing true vascular tissue. Using this precise definition as we take a closer look at true leaves in the vascular plants, we see that there are two different types of leaves, very likely of different evolutionary origins.

The first leaf type, the **microphyll**, is usually small and only rarely has more than a single vascular strand, at least in existing plants. Lycophytes, of which only a few genera survive, have such simple leaves. Some biologists believe that microphylls had their evolutionary origins as sterile sporangia (**Figure 28.8A**). The principal characteristic of this type of leaf is a vascular strand that departs from the vascular system of the stem in such a way that the structure of the stem's vascular system is scarcely disturbed. This was true even in the lycophyte trees of the Carboniferous period, many of which had leaves many centimeters long.

The other leaf type is found in monilophytes and seed plants. This larger, more complex leaf is called a **megaphyll**. The megaphyll is thought to have arisen from the flattening of a dichotomously branching stem system with overtopping growth. This change was followed by the development of photosynthetic tissue between the members of overtopped groups of branches (**Figure 28.8B**), which had the advantage of increasing the photosynthetic surface area of those branches.

The first megaphylls, which were very small, appeared in the Devonian period. We might expect that evolution should have led swiftly to the appearance of more and larger megaphylls because of their greater photosynthetic capacity. However, it took some 50 million years, until the Carboniferous period, for large megaphylls to become common. Why should this have been so, especially given that other advances in plant structure were taking place during that time?

According to one theory, the high concentration of $CO_2$ in the atmosphere during the Devonian period reduced selection for the tiny pores, called **stomata**, that allow a leaf to take up $CO_2$ for use in photosynthesis. With more $CO_2$ available, fewer stomata were needed. Today, when stomata are open, they allow water vapor to escape the leaf and $CO_2$ to enter. In the Devonian, larger leaves would have absorbed heat from sunlight, but they would have been unable to lose heat fast enough by evaporation of water through their limited number of stomata. The resulting overheating would have been lethal. Recent research has supported this hypothesis, indicating that larger megaphylls evolved only as $CO_2$ concentrations dropped over millions of years (**Figure 28.9**).

### Heterospory appeared among the vascular plants

In the lineages of present-day vascular plants that are most similar to their ancestors, the gametophyte and the sporophyte are

## INVESTIGATING LIFE

### 28.9 $CO_2$ Levels and the Evolution of Megaphylls

High concentrations of atmospheric $CO_2$ may have acted as a constraint on the evolution of leaf size. C. P. Osborne and colleagues compared the leaf sizes of fossil plants against estimates of $CO_2$ concentrations in the atmosphere at the time the plants were alive.

**HYPOTHESIS** **High atmospheric $CO_2$ concentrations during the Devonian constrained the evolution of leaf size, with selection pressure maintaining relatively small leaves.**

**METHOD**

1. Analyze 300 plant fossils from the Devonian and Carboniferous periods and calculate the sizes of their leaves.
2. Compare the pattern of change in leaf size with that of the estimated change in atmospheric $CO_2$ concentrations over the same time frame.

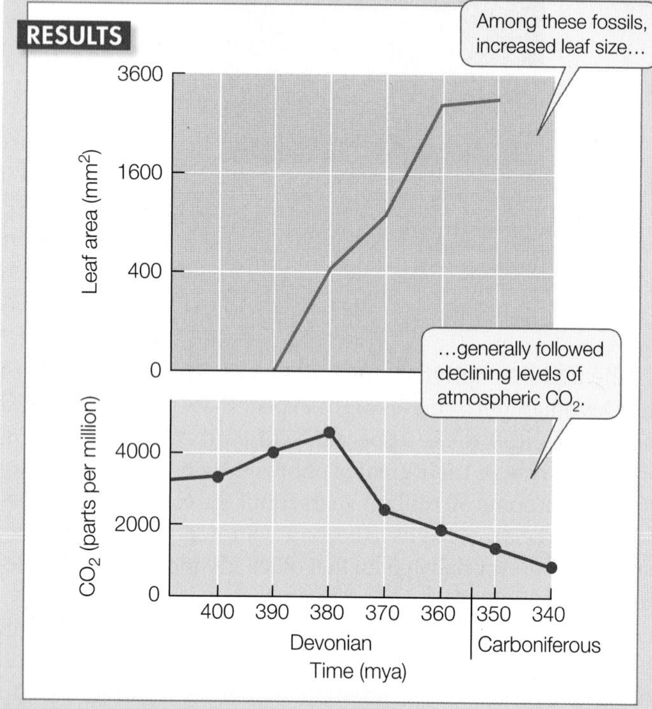


**CONCLUSION** **Leaf size increased as levels of $CO_2$ in the atmosphere decreased.**

**FURTHER INVESTIGATION:** What sort of experiment would you do to determine the effects of stomata on overheating of fern leaves today?

Go to **yourBioPortal.com** for original citations, discussions, and relevant links for all INVESTIGATING LIFE figures.

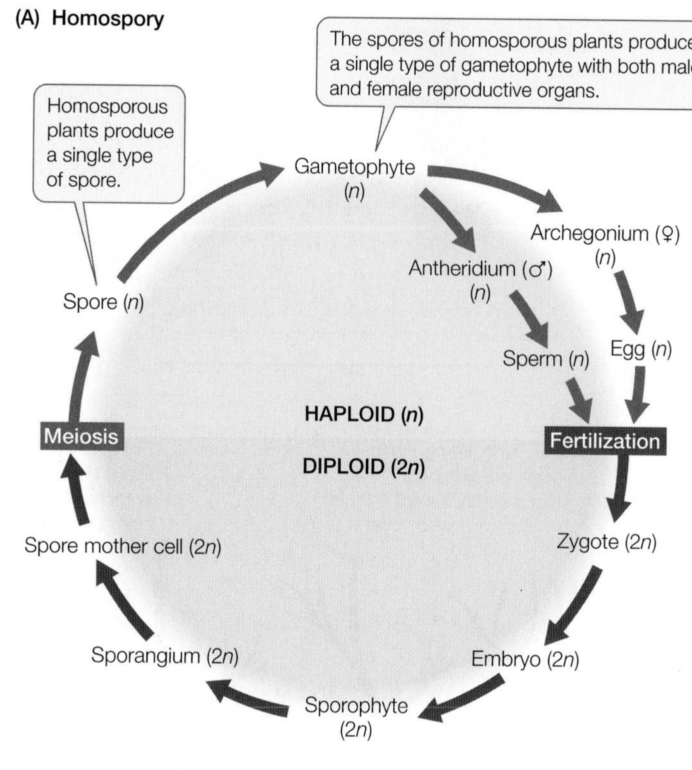


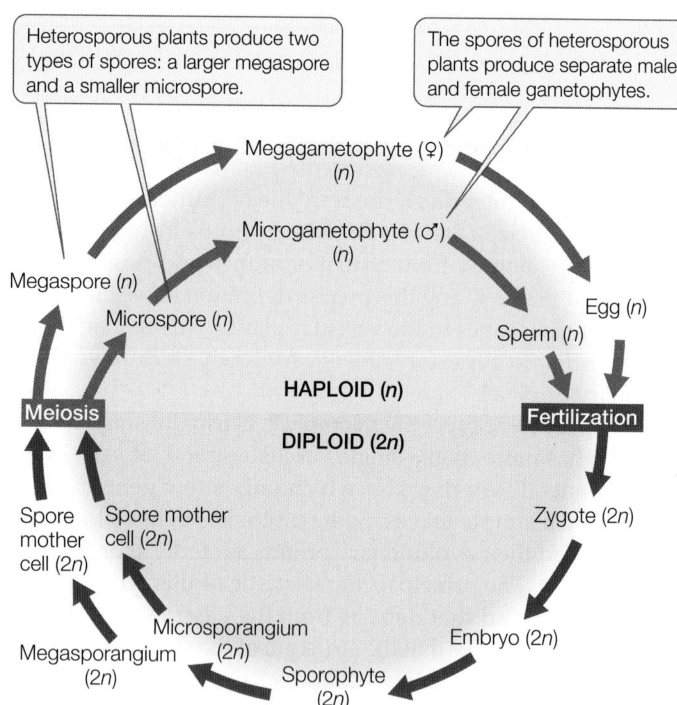


**28.10 Homospory and Heterospory** (A) Homosporous plants bear a single type of spore. Each gametophyte has two types of sex organs, antheridia (male) and archegonia (female). (B) Heterosporous plants bear two types of spores that develop into distinctly male and female gametophytes.

**yourBioPortal.com**
GO TO Web Activities 28.1 and 28.2 • Homospory and Heterospory

independent, and both are usually photosynthetic. Spores produced by the sporophyte are of a single type and develop into a single type of gametophyte that bears both female and male reproductive organs. The female organ is a multicellular archegonium containing a single egg. The male organ is an antheridium, producing many sperm. Such plants, which bear a single type of spore, are said to be **homosporous** (**Figure 28.10A**).

A system with two distinct types of spores evolved somewhat later. Plants of this type are said to be **heterosporous** (**Figure 28.10B**). In heterospory, one type of spore—the **megaspore**—develops into a specifically female gametophyte (a *megagametophyte*) that produces only eggs. The other type, the **microspore**, is smaller and develops into a male gametophyte (a *microgametophyte*) that produces only sperm. The sporophyte produces megaspores in small numbers in *megasporangia*, and microspores in large numbers in *microsporangia*. Heterospory affects not only the spores and the gametophyte but also the sporophyte plant itself, which must develop two types of sporangia.

The most ancient vascular plants were all homosporous, but heterospory evidently evolved several times in the early descendants of the rhyniophytes. The fact that heterospory evolved repeatedly suggests that it affords selective advantages. Subsequent evolution in the land plants featured ever greater specialization of the heterosporous condition.

### 28.3 RECAP

**A new type of cell, the tracheid, marked the origin of the vascular plants. Later evolutionary events included the appearance of roots and leaves.**

- How do the vascular tissues xylem and phloem serve the vascular plants? See p. 594
- Describe the difference between the two leaf types (microphylls and megaphylls). See p. 597 and Figure 28.8
- Explain the concept of heterospory. See p. 599 and Figure 28.10

The liverworts, hornworts, mosses, lycophytes, and monilophytes have come a long way from the aquatic environment to meet the challenges of life on dry land. Let's look at the diversity within these groups.

## 28.4 What Are the Major Clades of Seedless Plants?

Three principal clades of living land plants lack tracheids; these nonvascular land plants are the liverworts, hornworts, and mosses. The structure and growth pattern of the sporophyte differ among the three groups. There are four principal clades of seedless vascular plants—lycophytes (club mosses and their relatives), horsetails, whisk ferns, and leptosporangiate ferns. Most plants known as "ferns" are leptosporangiate ferns, although there are also small groups called "ferns" that are more closely related to horsetails and whisk ferns.

### Liverworts may be the most ancient surviving plant clade



There are about 9,000 species of **liverworts** (Hepatophyta, "liver-plants"). Most liverworts have leafy gametophytes (**Figure 28.11A**). Some have *thalloid* gametophytes—green, leaflike layers that lie flat on the ground (**Figure 28.11B**). The simplest liverwort gametophytes, however, are flat plates of cells, a centimeter or so long, that produce antheridia or archegonia on their upper surfaces and rhizoids on their lower surfaces.

Liverwort sporophytes are shorter than those of mosses and hornworts, rarely exceeding a few millimeters. The liverwort sporophyte has a stalk that connects sporangium and foot. In most species, the stalk elongates by expansion of cells throughout its length. This elongation raises the sporangium above ground level, allowing the spores to be dispersed more widely. The sporangia of liverworts are simple: a globular sporangium wall surrounds a mass of spores. In some species of liverworts, spores are not released by the sporophyte until the surrounding sporangium wall rots. In other liverworts, however, the

**28.11 Liverwort Structures** Liverworts display various characteristic structures. (A) The gametophyte of a leafy liverwort. (B) Gametophytes of a thalloid liverwort. (C) This thalloid liverwort bears archegonia in the structures that look like bunches of bananas. It also bears cups containing gemmae.

(A) *Bazzania trilobata*

(B) *Marchantia* sp.

These cups contain gemmae—small, lens-shaped outgrowths of the plant body, each capable of developing into a new plant.
The banana-like structures bear archegonia.

(C) *Marchantia* sp.

spores are thrown from the sporangium by structures that shorten and compress a "spring" as they dry out. When the stress becomes sufficient, the compressed spring snaps back to its resting position, throwing spores in all directions.

Among the most familiar thalloid liverworts are species of the genus *Marchantia*. *Marchantia* is easily recognized by the characteristic structures on which its male and female gametophytes bear their antheridia and archegonia (**Figure 28.11C**). Like most liverworts, *Marchantia* also reproduces asexually by simple fragmentation of the gametophyte. *Marchantia* and some other liverworts and mosses also reproduce asexually by means of *gemmae* (singular *gemma*), which are lens-shaped clumps of cells. In a few liverworts, the gemmae are held in structures called *gemmae cups*, which promote dispersal of the gemmae by raindrops (see Figure 28.11C).

### Water- and sugar-transport mechanisms first emerged in the mosses

The most familiar of the nonvascular land plants are the **mosses (Bryophyta)**. These hardy little plants, of which there are about 15,000 species, are found in almost every terrestrial environment. They are often found on damp, cool ground, where they form thick mats (**Figure 28.12**). The mosses are the sister lineage to the vascular plants plus the hornworts (see Figure 28.5).

(A) (B) *Polytrichum* sp.

**28.12 Mosses Grow in Dense Mats** (A) Dense layers of moss carpet a field of solidified volcanic lava in Iceland. (B) A close-up view of moss growing on a forest floor in Michigan.

The mosses, along with the hornworts and vascular plants, share an advance over the liverwort clade in their adaptation to life on land: they have stomata, which are important for both water and gas exchange. Stomata are a shared derived trait of mosses and all other land plants except liverworts.

In mosses, the gametophyte begins its development following spore germination as a branched, filamentous structure called a *protonema* (see Figure 28.4). Although the protonema looks a bit like a filamentous green alga, this structure is unique to the mosses. Some of the filaments contain chloroplasts and are photosynthetic; others, called rhizoids, are nonphotosynthetic and anchor the protonema to the substratum. After a period of linear growth, cells close to the tips of the photosynthetic filaments divide rapidly in three dimensions to form *buds*. The buds eventually develop a distinct tip, or apex, and produce the familiar leafy moss shoot with leaflike structures arranged spirally. These leafy shoots produce antheridia or archegonia (see Figure 28.4).

Some moss gametophytes are so large they could not transport enough water solely by diffusion. Gametophytes and sporophytes of many mosses contain a type of cell called a *hydroid*, which dies and leaves a tiny channel through which water can travel. The hydroid is functionally similar to the tracheid, the characteristic water-conducting cell of the vascular plants, but it lacks lignin and the cell-wall structure found in tracheids. The possession of hydroids and of a limited system for transport of sugar by some mosses (via cells called *leptoids*) shows that the term "nonvascular plant" is somewhat misleading when applied to mosses. Despite their simple system of internal transport, however, the mosses are not vascular plants because they lack true xylem and phloem.

Mosses of the genus *Sphagnum* (**Figure 28.13A**) often grow in cool, swampy places, where the plants begin to decompose in the water after they die. Rapidly growing upper layers of moss compress the deeper-lying, decomposing layers. Partially decomposed plant matter is called *peat*. In some parts of the world, people derive the majority of their fuel from peat bogs (**Figure 28.13B**). *Sphagnum*-dominated peatlands cover an area approximately half the size of the United States—more than 1 percent of Earth's surface. Millions of years ago, continued compression of peat composed primarily of other seedless plants gave rise to coal.

### Hornworts have distinctive chloroplasts and sporophytes without stalks

The approximately 100 species of **hornworts** comprise the group Anthocerophyta ("horn plants"), so named because their sporophytes look like little horns (**Figure 28.14**). Hornworts appear at first glance to be liverworts with very simple gametophytes. Their gametophytes are flat plates of cells a few cells thick.

Hornworts have two characteristics that distinguish them from liverworts and mosses. First, the cells of hornworts each

(A)

(B)

**28.13 Sphagnum Moss** (A) *Sphagnum* bogs are extremely dense growths of the moss shown here. These bogs can cover large areas in temperate climates. (B) A farmer mines a bog for peat, a fossil fuel formed from decomposing *Sphagnum* mosses.

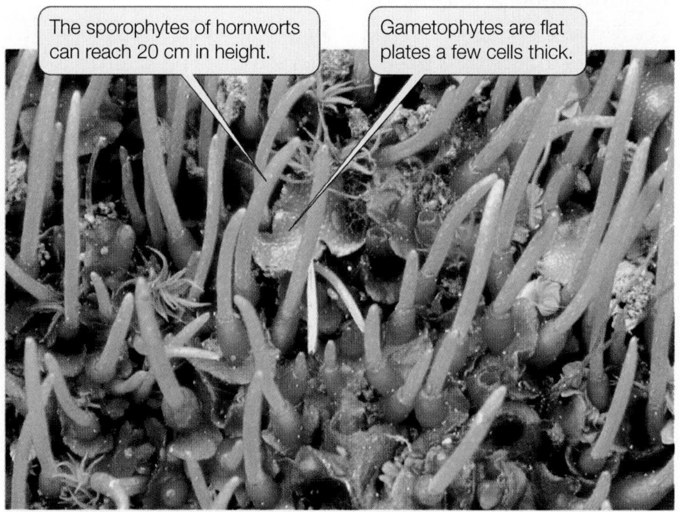

The sporophytes of hornworts can reach 20 cm in height.
Gametophytes are flat plates a few cells thick.
Anthoceros sp.


**28.14 A Hornwort** The sporophytes of many hornworts resemble little horns.

contain a single large, platelike chloroplast, whereas the cells of the other two groups contain numerous small, lens-shaped chloroplasts. Second, of the sporophytes in all three groups, those of the hornworts come closest to being capable of growth without a set limit. Liverwort and moss sporophytes have a stalk that stops growing as the sporangium matures, so elongation of the sporophyte is strictly limited. The hornwort sporophyte, however, has no stalk. Instead, a basal region of the sporangium remains capable of indefinite cell division, continuously producing new spore-bearing tissue above. The sporophytes of some hornworts growing in mild and continuously moist conditions can become as tall as 20 centimeters. Eventually, however, the sporophyte's growth is limited by the lack of a transport system.

Hornworts have evolved a symbiotic relationship that promotes their growth by providing them with greater access to nitrogen, which is often a limiting resource. Hornworts have internal cavities filled with mucilage; these cavities are often populated by cyanobacteria that convert atmospheric nitrogen gas into a form usable by their host plant.

We present the hornworts here as sister to the vascular plants, but this is only one possible interpretation of the current data. The hornworts and vascular plants are united by DNA sequence analyses, and they also share a persistently green sporophyte. The exact evolutionary position of the hornworts is still unclear, and in some morphological analyses they are placed as the sister group to the mosses plus the vascular plants (the two groups that express apical cell division).

### Some vascular plants have vascular tissue but not seeds

The earliest vascular plants did not have seeds, and several major clades of seedless vascular plants have survived to the present. These plants have a large, independent sporophyte and a small gametophyte that is independent of the sporophyte. The gametophytes of the surviving seedless vascular plants are rarely more than 2 centimeters long and are short-lived, whereas their sporophytes are often highly visible and long-lived; the sporophyte of a tree fern, for example, may be up to 20 meters tall and live for many years.

The most prominent resting stage in the life cycle of the seedless vascular plants is the single-celled spore. A spore may "rest" for some time before developing further. This feature makes the life cycle of seedless vascular plants similar to those of the fungi, green algae, and nonvascular land plants, but not, as we will see in the next chapter, to that of the seed plants. Like nonvascular land plants, seedless vascular plants must have an aqueous environment for at least one stage of their life cycle because fertilization is accomplished by flagellated, swimming sperm.

The leptosporangiate ferns are the most abundant and diverse group of seedless vascular plants today, but the club mosses and horsetails were once dominant elements of Earth's vegetation. A fourth group, the whisk ferns, contains only two genera. Let's look at the characteristics of these four groups and at some of the evolutionary advances that appeared in them.

## The lycophytes are sister to the other vascular plants

The club mosses and their relatives, the spike mosses and quillworts, together called **lycophytes**, are the sister-group of the remaining vascular plants. There are relatively few surviving species of lycophytes—just over 1,200.

The lycophytes have roots that branch dichotomously. The arrangement of vascular tissue in their stems is simpler than in the other vascular plants. They bear only microphylls, and these simple leaves are arranged spirally on the stem. Growth in club mosses comes entirely from apical cell division, and branching in the stems is also dichotomous, by a division of the apical cluster of dividing cells.

The sporangia of many club mosses are aggregated in cone-like structures called *strobili* (singular *strobilus;* **Figure 28.15**). The strobilus of a club moss is a cluster of spore-bearing leaves inserted on an axis (linear supporting structure). Other club mosses lack strobili and bear their sporangia on (or adjacent to) the upper surfaces of leaves called *sporophylls*. This placement contrasts with the terminal sporangia of the rhyniophytes. There are both homosporous and heterosporous species of club mosses.

Although they are only a minor element of present-day vegetation, the lycophytes are one of two groups that appear to have been the dominant vegetation during the Carboniferous period. One type of coal (cannel coal) is formed almost entirely from fossilized spores of the tree lycophyte *Lepidodendron*—which gives us an idea of the abundance of this genus in the forests of that time (see Figure 28.6). Other major elements of Carboniferous vegetation included horsetails and ferns.

(A)

(B)

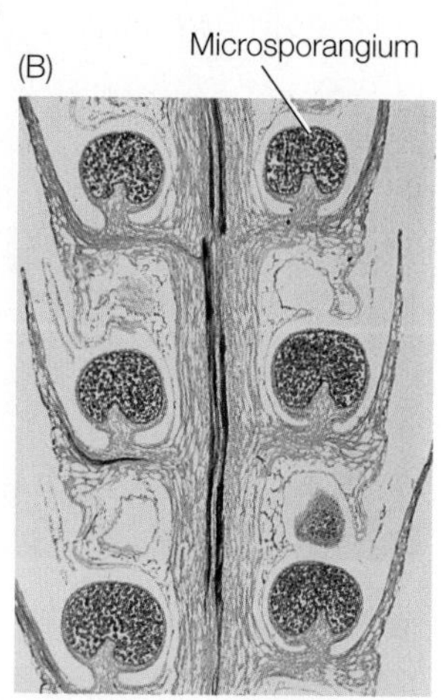


*Lycopodium obscurum*

**28.15 Club Mosses** (A) A strobilus is visible at the tip of this club moss. Club mosses have microphylls arranged spirally on their stems. (B) A thin section through a strobilus of a club moss, showing microsporangia.

## Horsetails, whisk ferns, and ferns constitute a clade

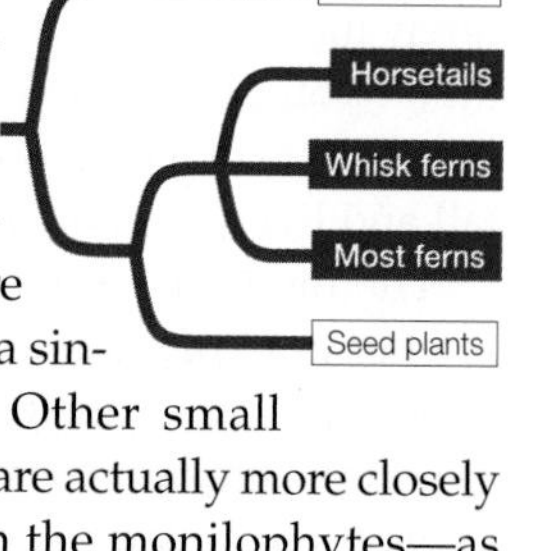


Once thought to be only distantly related, the horsetails, whisk ferns, and ferns form a clade, the monilophytes, or "ferns and fern allies." Within that clade, the whisk ferns and the horsetails are both monophyletic; the ferns are not. However, most ferns do belong to a single clade, the *leptosporangiate ferns*. Other small groups that are commonly called ferns are actually more closely related to horsetails or whisk ferns. In the monilophytes—as in all seed plants—there is differentiation between the main stem and side branches. This pattern contrasts with the dichotomous branching characteristic of the lycophytes and rhyniophytes, in which each split gives rise to two branches of similar size (see Figure 28.7).

**HORSETAILS** Today there are only about 15 species of **horsetails**, all in the genus *Equisetum*. Horsetails have been sometimes called "scouring rushes" because rough silica deposits found in their cell walls made them useful for cleaning. They have true roots that branch irregularly. Horsetails have a large sporophyte and a small gametophyte, both independent.

The small leaves of horsetails are reduced megaphylls and form in distinct whorls (circles) around the stem (**Figure 28.16**). Growth in horsetails originates to a large extent from discs of dividing cells just above each whorl of leaves, so each segment of the stem grows from its base. Such basal growth is uncommon in plants, although it is found in the grasses, a major group of flowering plants, as well as in the hornworts.

(A)

Dark "dots" on the apical structures are sporangia.

*Equisetum pratense*

(B)

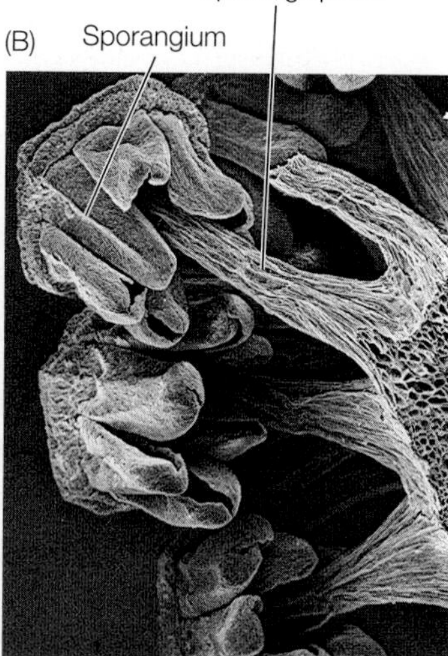

Sporangium
Sporangiophore

*Equisetum arvense*

**28.16 Horsetails** (A) Horsetails have a distinctive growth pattern in which the stem grows in segments above each whorl of leaves. These are fertile shoots, with sporangia-bearing structures at the apex. (B) Scanning electron micrograph of horsetail sporangia.

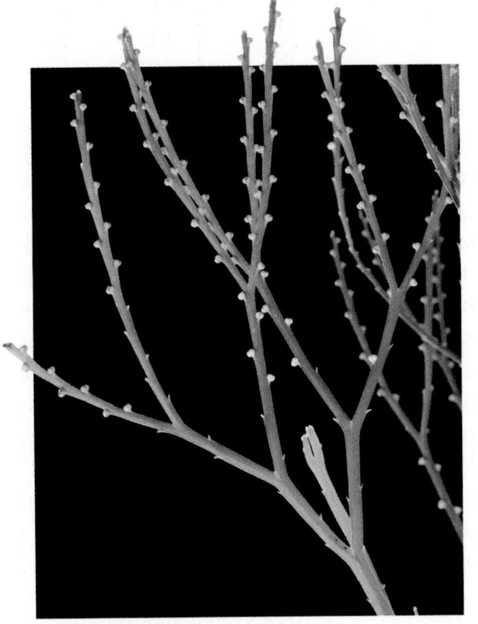


**28.17 A Whisk Fern** DNA sequence data reveals that *Psilotum*—once considered to be a surviving rhyniophyte—is in fact a much more recent genus that arose from a fernlike ancestor. Whisk ferns are now grouped with the monilophytes.

**WHISK FERNS** There once was some disagreement about whether the rhyniophytes are entirely extinct. The confusion arose because of the existence today of about 15 species in two genera of rootless, spore-bearing plants, *Psilotum* and *Tmesipteris*, collectively called the **whisk ferns**. *Psilotum* (**Figure 28.17**) has minute scales instead of true leaves, but plants of the genus *Tmesipteris* have flattened photosynthetic organs—reduced megaphylls—with well-developed vascular tissue. Are these two genera the living relics of the rhyniophytes, or do they have more recent origins?

The whisk ferns once were thought to be evolutionarily ancient descendants of anatomically simple ancestors. That hypothesis was weakened by an enormous hole in the fossil record between the rhyniophytes, which apparently became extinct more than 300 million years ago, and *Psilotum* and *Tmesipteris*, which are modern plants. DNA sequence data finally settled the question in favor of a more recent origin of the whisk ferns from fernlike ancestors. The whisk ferns are a clade of highly specialized plants that evolved fairly recently from anatomically more complex ancestors by loss or reduction of megaphylls and true roots. Whisk fern gametophytes live below the surface of the ground and lack chlorophyll. They depend on fungal partners for their nutrition.

**LEPTOSPORANGIATE FERNS** The first **leptosporangiate ferns** appeared during the Devonian period; today this group comprises more than 12,000 species. The sporangia of leptosporangiate ferns are borne on a stalk and have walls only one cell thick. The name "leptosporangiate" refers to these thin-walled sporangia (*lepton*, "thin"). Two other small groups that are also called ferns are actually more closely related to horsetails and whisk ferns, even though they are superficially similar to the leptosporangiate ferns.

The sporophytes of ferns, like those of the seed plants, have true roots, stems, and leaves. Ferns are characterized by large leaves with branching vascular strands (**Figure 28.18A**). During its development, the fern leaf unfurls from a tightly coiled "fiddlehead" (**Figure 28.18B**). Some fern leaves become climbing organs and may grow to be as long as 30 meters. A few species have small leaves as a result of evolutionary reduction, but even these small leaves have more than one vascular strand, and are thus megaphylls (**Figure 28.18C**).

**28.18 Fern Leaves Take Many Forms** (A) Tree ferns and crown ferns dominate this forest on Stewart Island, New Zealand. (B) The "fiddlehead" (developing leaf) of a common forest fern will unfurl and expand, giving rise to a complex adult leaf such as that of a crown fern. (C) The leaves of two species of water ferns.

**THE FERN LIFE CYCLE** In all ferns, *spore mother cells* inside the sporangia undergo meiosis to form haploid spores. Once shed, the spores may be blown great distances by the wind and eventually germinate to form independent gametophytes far from the parent sporophyte. A case in point is Old World climbing fern (*Lygodium microphyllum*), which is currently spreading disastrously through the Florida Everglades, choking off the growth of other plants. This rapid spread is testimony to the effectiveness of windborne spores. Another example is the remarkable diversity of ferns that have spread through the isolated Hawaiian Islands.

Fern gametophytes have the potential to produce both antheridia and archegonia, although not necessarily at the same time or on the same gametophyte. Sperm swim through water to archegonia—often to those on other gametophytes—where they unite with an egg. The resulting zygote develops into a new sporophyte embryo. The young sporophyte sprouts a root and can thus grow independently of the gametophyte. In the alternating generations of a fern, the gametophyte is small, delicate, and short-lived, but the sporophyte can be very large and can sometimes survive for hundreds of years (**Figure 28.19**).

Because they require liquid water for the transport of the male gametes to the female gametes (as do all other plant groups discussed in this chapter), most ferns inhabit shaded, moist woodlands and swamps. Tree ferns can reach heights of 20 meters. Tree ferns are not as rigid as woody plants, and they have poorly developed root systems. Thus they do not grow in sites exposed directly to strong winds, but rather in ravines or beneath trees in forests. The sporangia of ferns typically are found on the undersurfaces of the leaves, sometimes covering the entire undersurface and sometimes only at the edges. In most species, the sporangia are found in clusters called *sori* (singular *sorus*).

Most ferns are homosporous. However, two groups of aquatic ferns, the Marsileaceae and the Salviniaceae (see Figure 28.18C),

yourBioPortal.com
GO TO Web Activity 28.3 • The Fern Life Cycle

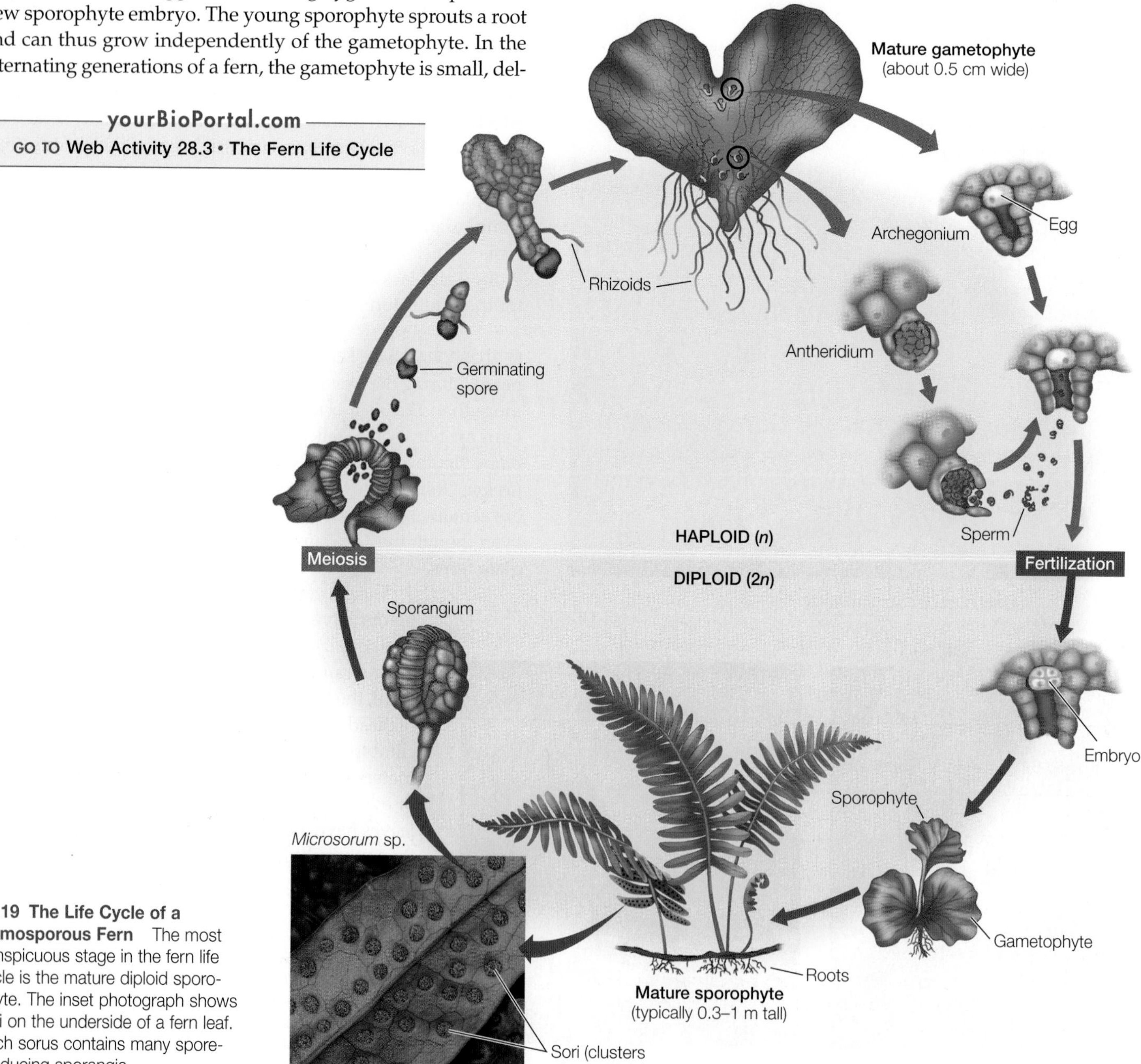


**28.19 The Life Cycle of a Homosporous Fern** The most conspicuous stage in the fern life cycle is the mature diploid sporophyte. The inset photograph shows sori on the underside of a fern leaf. Each sorus contains many spore-producing sporangia.

are derived from a common ancestor in which heterospory evolved. The megaspores and microspores of these plants (which germinate to produce female and male gametophytes, respectively) are produced in different sporangia (megasporangia and microsporangia), and the microspores are always much smaller and greater in number than the megaspores.

A few genera of ferns produce a tuberous, fleshy gametophyte instead of the characteristic flattened, photosynthetic structure produced by most ferns. These tuberous gametophytes depend on a mutualistic fungus for nutrition. In some genera, even the sporophyte embryo must become associated with the fungus before its development can proceed. Chapter 30 discusses many other important plant–fungus mutualisms.

The seedless vascular plants, and especially the ferns, were long considered an evolutionary cul-de-sac—that is, a group with great diversity in the fossil record but less diversity in the present. However, recent DNA-based research has suggested that the diversification of today's ferns took place much more recently than previously thought. The expansion of seed plants and their dominance of forests actually predates the diversification of extant ferns, which presumably took advantage of the new environments created by those forests.

## 28.4 RECAP

**Three clades of land plants lack true vascular systems (liverworts, mosses, and hornworts). The seedless vascular plants include the club mosses, horsetails, whisk ferns, and leptosporangiate ferns.**

- Describe the different branching patterns of lycophytes and monilophytes. See pp. 602–603
- Why were whisk ferns once thought to be close relatives of the rhyniophytes? See p. 603
- Why do most ferns live in moist, shady areas? See p. 604

All of the groups described in this chapter—the nonvascular land plants and the seedless vascular plants—require water at a key stage in their life cycles, because their sperm can reach their eggs only by means of liquid water. In addition, the vascular plant groups we have discussed thus far all disperse by spores. The next chapter describes the *seed plants* that dominate Earth's vegetation today, and in which the seed affords new sporophytes protection unavailable to the progeny of seedless plants.


## CHAPTER SUMMARY

### 28.1 How Did the Land Plants Arise?

- **Land plants**, sometimes referred to as **embryophytes**, are photosynthetic eukaryotes that develop from embryos protected by parental tissue. Review Figure 28.1
- **Streptophytes** include the land plants and certain green algae. **Green plants** include the streptophytes and the remaining green algae.
- Land plants arose from an aquatic green algal ancestor related to today's **Charales**.
- The **vascular plants** have well developed water-conducting tissues with cells including **tracheids**; the three major groups of **nonvascular land plants** do not. Review Table 28.1

### 28.2 How Did Plants Colonize and Thrive on Land?

- The acquisition of a **cuticle**, **stomata**, **gametangia**, a protected embryo, protective pigments, thick spore walls with a protective polymer, and a mutualistic association with a fungus are all adaptations to terrestrial life.
- All land plant life cycles feature alternation of generations, in which a multicellular diploid **sporophyte** alternates with a multicellular haploid **gametophyte**. Review Figure 28.3
- Spores form in **sporangia**; gametes form in gametangia. In seedless land plants, the female and male gametangia are, respectively, an **archegonium** and an **antheridium**.
- In liverworts, hornworts, and mosses, the sporophyte is smaller than the gametophyte and depends on it for water and nutrition. Review Figure 28.4, **ANIMATED TUTORIAL 28.1**

### 28.3 What Features Distinguish the Vascular Plants?

- A **vascular system** consisting of **xylem** and **phloem** conducts water, minerals, and products of photosynthesis through the bodies of vascular plants.
- Among living vascular plant groups, the **lycophytes** (club mosses and relatives) have only small, simple leaves (**microphylls**). Larger, more complex leaves (**megaphylls**) are found in **monilophytes** (horsetails, ferns, and allies) and seed plants. These latter two groups comprise the **euphyllophytes**. Review Figure 28.5
- In vascular plants, the sporophyte is larger than and independent of the gametophyte.
- The **rhyniophytes**, the earliest vascular plants, are known to us only in fossil form. They lacked roots and leaves but possessed **rhizomes** and **rhizoids**. Review Figure 28.7
- Roots may have evolved either from rhizomes or from branches. Microphylls probably evolved from sterile sporangia, and megaphylls may have resulted from the flattening and reduction of an **overtopping** branching stem system. Review Figure 28.8
- Many seedless vascular plants are **homosporous**, but **heterospory**—the production of distinct **megaspores** and **microspores**—evolved several times. Megaspores develop into megagametophytes; microspores develop into microgametophytes. Review Figure 28.10, **WEB ACTIVITIES 28.1 and 28.2**

### 28.4 What Are the Major Clades of Seedless Plants?

- The nonvascular land plant clades are the **liverworts**, the **mosses**, and the **hornworts**. The seedless vascular plant groups are the **lycophytes** (club mosses and their relatives) and the monilophytes (**horsetails**, **whisk ferns**, and **leptosporangiate ferns**). Review Figure 28.5
- Mosses, hornworts, and vascular plants all have surface pores (stomata) in their leaves.
- The gametophyte of ferns is small, delicate, and short-lived, whereas the sporophyte of ferns is typically much larger and longer lived. Review Figure 28.19, **WEB ACTIVITY 28.3**

## SELF-QUIZ

1. Land plants differ from photosynthetic protists in that only the plants
   *a.* are photosynthetic.
   *b.* are multicellular.
   *c.* possess chloroplasts.
   *d.* have multicellular embryos protected by the parent.
   *e.* are eukaryotic.
2. Which statement about alternation of generations in land plants is *not* true?
   *a.* The gametophyte and sporophyte differ in appearance.
   *b.* Meiosis occurs in sporangia.
   *c.* Gametes are always produced by meiosis.
   *d.* The zygote is the first cell of the sporophyte generation.
   *e.* The gametophyte and sporophyte differ in chromosome number.
3. Which statement is *not* evidence for the origin of land plants from the green algae?
   *a.* Some green algae have multicellular sporophytes and multicellular gametophytes.
   *b.* Both plants and green algae have cellulose in their cell walls.
   *c.* The two groups have the same photosynthetic pigments.
   *d.* Both plants and green algae produce starch as their principal storage carbohydrate.
   *e.* All green algae produce large, stationary eggs.
4. Liverworts, mosses, and hornworts
   *a.* lack a sporophyte generation.
   *b.* grow in dense masses, allowing capillary movement of water.
   *c.* possess xylem and phloem.
   *d.* possess true leaves.
   *e.* possess true roots.
5. Which statement is *not* true of the mosses?
   *a.* The sporophyte is dependent on the gametophyte.
   *b.* Sperm are produced in archegonia.
   *c.* There are more species of mosses than of liverworts and hornworts combined.
   *d.* The sporophyte grows by apical cell division.
   *e.* Mosses are probably sister to the vascular plants plus hornworts.
6. Megaphylls
   *a.* probably evolved only once.
   *b.* are found in all the vascular plant groups.
   *c.* probably arose from sterile sporangia.
   *d.* are the characteristic leaves of club mosses.
   *e.* are the characteristic leaves of horsetails and ferns.
7. The rhyniophytes
   *a.* lacked tracheids.
   *b.* possessed true roots.
   *c.* possessed sporangia at the tips of stems.
   *d.* possessed leaves.
   *e.* lacked branching stems.
8. Club mosses and horsetails
   *a.* have larger gametophytes than sporophytes.
   *b.* possess small leaves.
   *c.* are represented today primarily by trees.
   *d.* have never been a dominant part of the vegetation.
   *e.* produce fruits.
9. Which statement about ferns is *not* true?
   *a.* The sporophyte is larger than the gametophyte.
   *b.* Most are heterosporous.
   *c.* The young sporophyte can grow independently of the gametophyte.
   *d.* The leaf is a megaphyll.
   *e.* The gametophytes produce archegonia and antheridia.
10. The leptosporangiate ferns
   *a.* are not a monophyletic group.
   *b.* have sporangia with walls more than one cell thick.
   *c.* constitute a minority of all ferns.
   *d.* are monilophytes.
   *e.* produce seeds.

## FOR DISCUSSION

1. Mosses and ferns share a common trait that makes water droplets a necessity for sexual reproduction. What is that trait?
2. Are the mosses well adapted to terrestrial life? Explain your answer.
3. Ferns display a dominant sporophyte generation (with large leaves). Describe the major advance in anatomy that enables most ferns to grow much larger than mosses.
4. What features distinguish club mosses from horsetails? What features distinguish these groups from rhyniophytes? From ferns?
5. Why did some botanists once believe that the whisk ferns should be classified together with the rhyniophytes?
6. Contrast microphylls with megaphylls in terms of structure, evolutionary origin, and occurrence among plants.

## ADDITIONAL INVESTIGATION

The findings of Osborne and co-workers on the evolution of megaphylls (see Figure 28.9) support the concept that large megaphylls became common only after the atmospheric $CO_2$ level had dropped, so that more stomata were produced, allowing water to evaporate and cool larger leaves. How might you extend that work to confirm the involvement of temperature as a factor limiting leaf size?

# 29 The Evolution of Seed Plants

## A seed from biblical times germinates after 2,000 years

The fruit of the Judean date palm was once much prized. The prophet Muhammad admired its nutritional and medicinal properties; in the Koran, it is associated with heaven and described as a symbol of goodness. The Judean date was the source of the "honey" in the biblical "land of milk and honey." Today that ancient strain of date is extinct. Or is it?

Around two thousand years ago, at the beginning of the Common Era, a seed developed in a fruit on a Judean date palm. The fruit that contained that seed found its way to a storeroom in the fortress Masada in Judea. In 73 C.E. almost a thousand Jewish Zealots involved in a religious revolt against Rome fled to this refuge with their families. Roman legions followed, and the ensuing siege lasted more than two years. In the end, rather than be killed or enslaved by the Roman soldiers, the Zealots are said to have killed themselves and their families in a dramatic mass suicide.

Twenty centuries later, archeologists working in Masada discovered the date seed that was long ago stored in the Masada fortress and confirmed its age. The previous record for seed survival and germination was 1,300 years, held by lotus seeds that recently germinated under the care of scientists in China. But botanist Elaine Solowey succeeded in making the 2,000-year-old date seed germinate! The resulting seedling (nicknamed Methuselah) has continued to thrive and grow. Perhaps the ancient Egyptians knew what they were doing when they placed date seeds in the tombs of the Pharaohs as symbols of immortality.

Seeds are important structures for the survival of plants. They protect the plant embryo within from environmental extremes through what may be a long and stressful resting period—in the case of the Judean date, many centuries in a harsh desert. Seeds of the coconut palm remain dormant for years as they float across vast expanses of ocean, finally washing up on a distant shore, where they germinate and grow. Such hardiness is one of the properties that have contributed to making seed plants the predominant plants on Earth. All of today's forests are dominated by seed plants.

So will the seedling growing under Dr. Solowey's care serve as the parent of a new population of Judean dates, thus resurrecting that genotype from extinction? Unfortunately, it cannot do so alone, because date trees are of two different sexes. However, it may be possible to germinate

**A Refuge** King Herod of Judea fortified Masada and stocked it with water and food—including Judean dates.

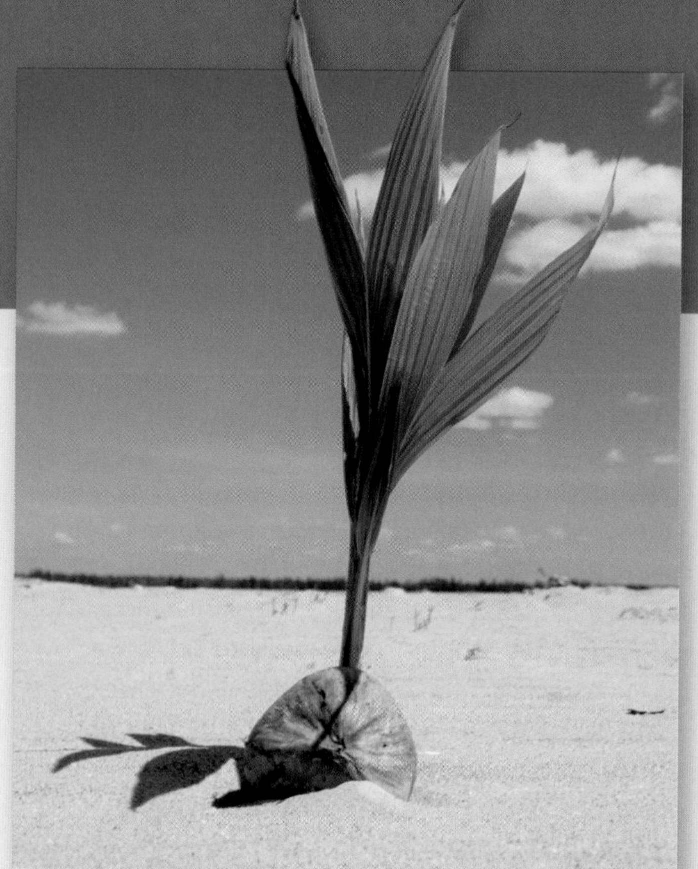

**The Hardy Seed** This coconut seed arrived on a beach, where it germinated successfully. The evolution of seeds was a major factor in the eventual dominance of the seed plants.

other Judean dates, and/or to cross the tree with Moroccan, Egyptian, or Iraqi varieties of date palms. Although these latter varieties differ from Judean dates at almost half of the gene loci that have been examined, crossing these varieties may introduce many beneficial alleles from Judean dates back into modern date plantations.

Today, humans are collecting and storing seeds from many species in seed banks (see Chapter 34). Many plant species or cultivars that would otherwise be lost to extinction are thus being preserved for restoration projects or other uses by future generations.

**IN THIS CHAPTER** we will describe the defining characteristics of the seed plants as a group. Living seed plants include the gymnosperms, which produce seeds not protected by ovary or fruit tissue, and the angiosperms, which are characterized by flowers and fruits. We will cover some of the unsolved problems in seed plant evolution, concluding with a survey of the diversity of living seed plants.

## CHAPTER OUTLINE



## 29.1 How Did Seed Plants Become Today's Dominant Vegetation?

By the late Devonian period, more than 360 million years ago, Earth was home to a great variety of land plants, many of which are discussed in Chapter 28. The land plants shared the hot, humid terrestrial environment with insects, spiders, centipedes, and fishlike amphibians (early tetrapods). The plants and animals affected one another, acting as agents of natural selection.

In the Devonian an innovation had appeared: some plants developed extensively thickened woody stems, which resulted from the proliferation of xylem. This type of growth in the diameter of stems and roots is called **secondary growth**. Among the first plants with this adaptation were seedless vascular plants called *progymnosperms*, all species of which are now extinct.

The earliest fossil evidence of seed plants is found in late Devonian rocks. Like the progymnosperms, these *seed ferns* were woody. They possessed fernlike foliage but had seeds attached to their leaves. By the end of the Permian, other groups of seed plants became dominant (**Figure 29.1**).

The living seed plants fall into two major groups, the **gymnosperms** (such as pines and cycads) and the **angiosperms** (flowering plants). There are several competing phylogenetic hypotheses regarding the major groups of gymnosperms relative to the angiosperms, but **Figure 29.2** shows the most widely supported relationships. All living gymnosperms and many angiosperms show secondary growth. The life cycles of all seed plants also share distinctive features, as we are about to see.

### Features of the seed plant life cycle protect gametes and embryos

Section 28.2 describes a trend in plant evolution: the sporophyte became less dependent on the gametophyte, which became smaller in relation to the sporophyte. This trend continued with the appearance of the seed plants, whose gametophyte generation is reduced even further than it is in the ferns (**Figure 29.3**). The haploid gametophyte develops partly or entirely while attached to and nutritionally dependent on the diploid sporophyte.

Among the seed plants, only the earliest groups of gymnosperms (such as cycads and ginkgos) had swimming sperm. Later groups of gymnosperms and the angiosperms evolved other means of bringing eggs and sperm together. The culmination of this striking evolutionary trend in seed plants was independence from the liquid water that earlier plants needed to assist sperm in reaching the egg. The advent of the seed gave seed plants the opportunity to colonize drier areas and spread over the terrestrial environment.

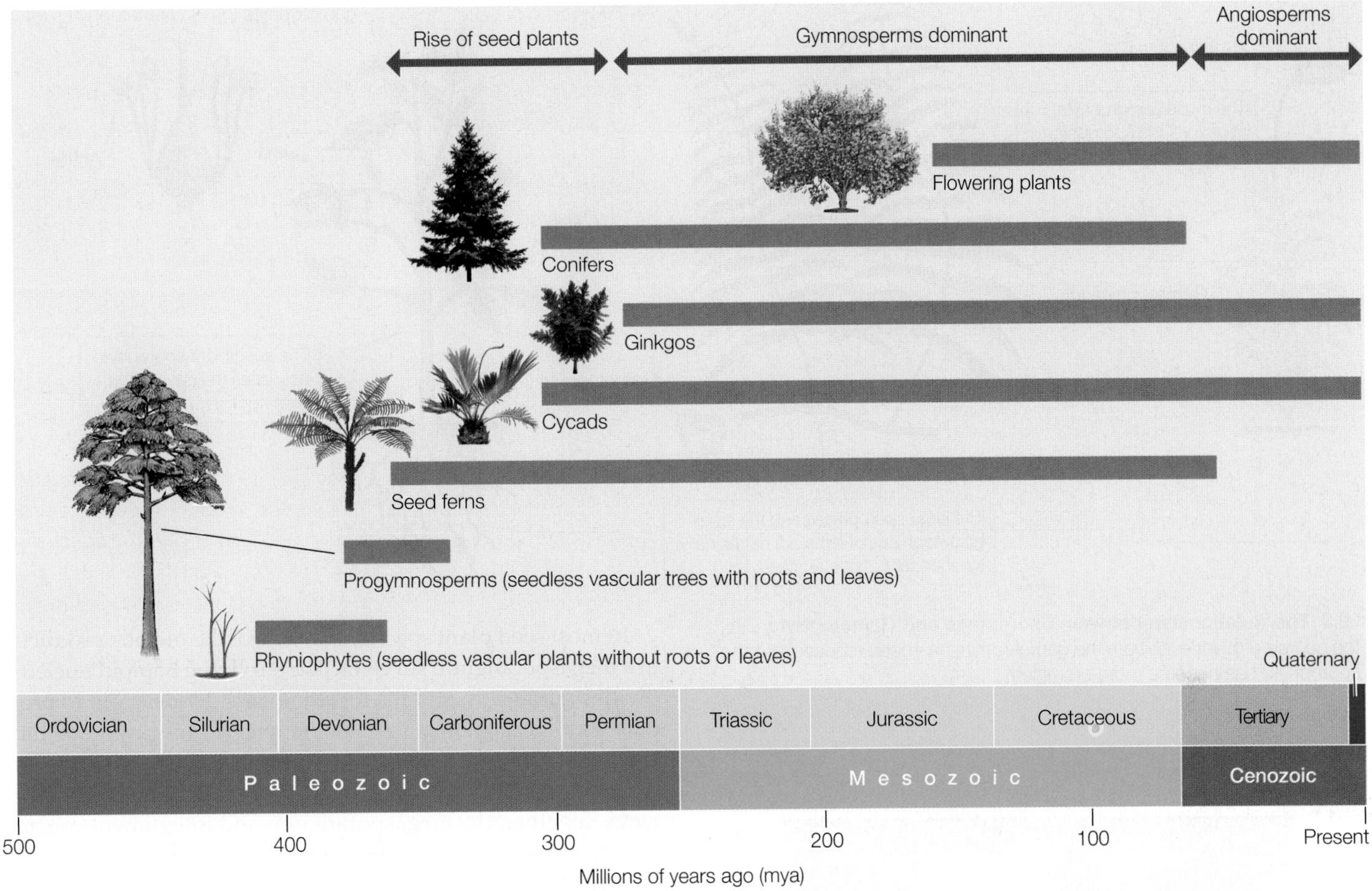

**29.1 The Fossil Record of Seed Plants and Some of Their Extinct Seedless Relatives** Woody growth evolved in the seedless progymnosperms. The now-extinct seed ferns had woody growth, fernlike foliage, and seeds attached to their leaves. New lineages of seed plants arose during the Carboniferous, but the earliest known fossils of flowering plants are from the late Jurassic.

Seed plants are heterosporous (see Figure 28.9B); that is, they produce two types of spores, one that becomes the male gametophyte and one that becomes the female gametophyte. They form separate microsporangia and megasporangia on structures that are grouped on short axes, such as the stamens and pistils of an angioperm flower.

Within the microsporangium, the meiotic products are microspores, which divide mitotically within the spore wall one or a few times to form a multicellular male gametophyte called a **pollen grain**. Pollen grains are released from the microsporangium to be distributed by wind or by an animal pollinator (**Figure 29.4**). The wall of the pollen grain contains *sporopollenin*, the most chemically resistant biological compound known, which protects the pollen grain against dehydration and chemical damage—another advantage in terms of survival in the terrestrial environment. Recall that sporopollenin in spore walls contributed to the successful colonization of the terrestrial environment by the earliest land plants.

In contrast to the pollen grains produced by the microporangia, the megaspores of seed plants are not shed. Instead, they develop into female gametophytes within the megasporangia. These megagametophytes are dependent on the sporophyte for food and water.

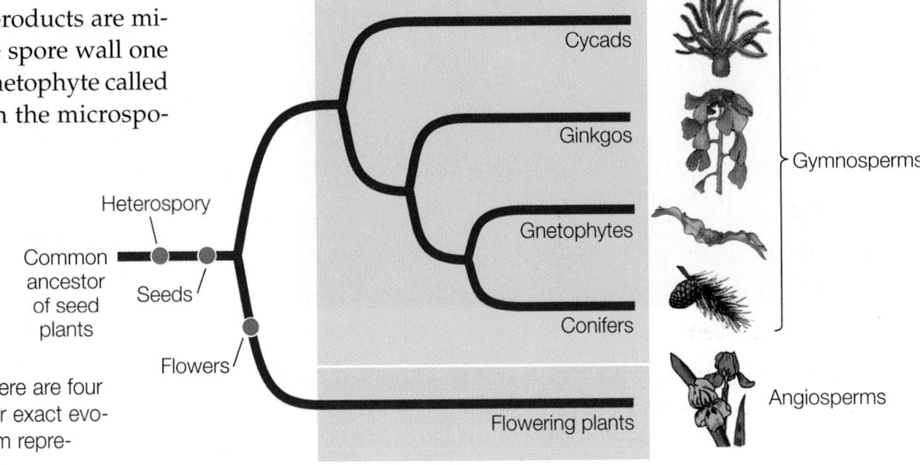


**29.2 The Major Groups of Living Seed Plants** There are four groups of gymnosperms and one of angiosperms. Their exact evolutionary relationship is still uncertain, but this cladogram represents one current interpretation.

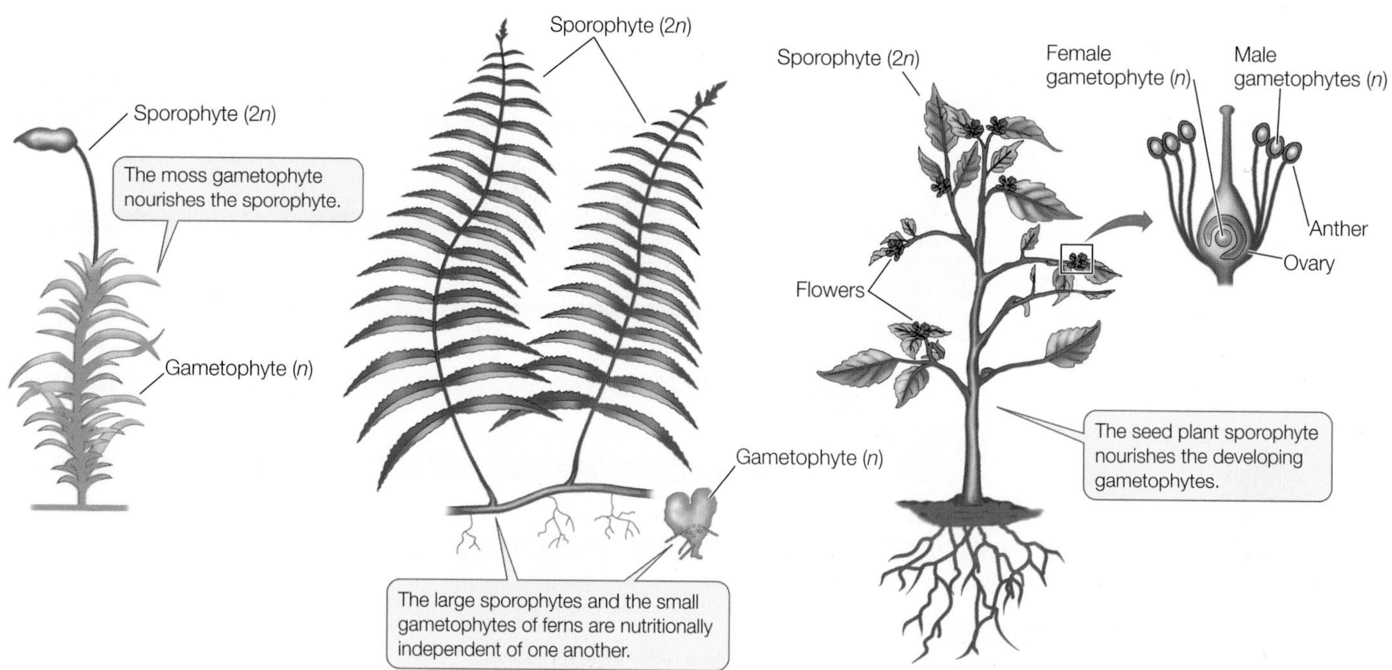


**29.3 The Relationship between Sporophyte and Gametophyte** In the course of plant evolution, the gametophyte has been reduced and the sporophyte has become more prominent.

**29.4 Pollen Grains** Pollen grains are the male gametophytes of seed plants. The pollen of this silver birch is dispersed by the wind, and grains may land near the female gametophytes of the same or other silver birch trees.

In most seed plant species, only one of the meiotic products in a megasporangium survives. The surviving haploid nucleus divides mitotically, and the resulting cells divide again to produce a multicellular female gametophyte. The megasporangium is surrounded by sterile sporophytic structures, which form an **integument** that protects the megasporangium and its contents. Together, the megasporangium and integument constitute the **ovule**, which will develop into a seed after fertilization (**Figure 29.5**).

The arrival of a pollen grain at an appropriate landing point, close to a female gametophyte on a sporophyte of the same species, is called **pollination**. A pollen grain that reaches this point develops further. It produces a slender **pollen tube** that elongates and digests its way toward the megagametophyte (see Figure 29.5). When the tip of the pollen tube reaches the megagametophyte, sperm are released from the tube and fertilization occurs.

The resulting diploid zygote divides repeatedly, forming an embryonic sporophyte. After a period of embryonic development, growth is temporarily suspended (the embryo enters a *dormant* stage). The end product at this stage is a multicellular **seed**.

## The seed is a complex, well-protected package

A seed contains tissues from three generations. A *seed coat* develops from the integument—tissues of the diploid sporophyte parent that surround the megasporangium. Within the megasporangium is the haploid female gametophytic tissue from the next generation, which contains a supply of nutrients for the developing embryo. (This tissue is fairly extensive in most gymnosperm seeds. In angiosperm seeds it is greatly reduced, and nutrition for the embryo is supplied by a tissue called *endosperm*, which is described below.) In the center of the seed is the third generation, the embryo of the new diploid sporophyte.

The seed of a gymnosperm or an angiosperm is a well-protected resting stage. The seeds of some species may remain dormant but stay *viable* (capable of growth and development) for

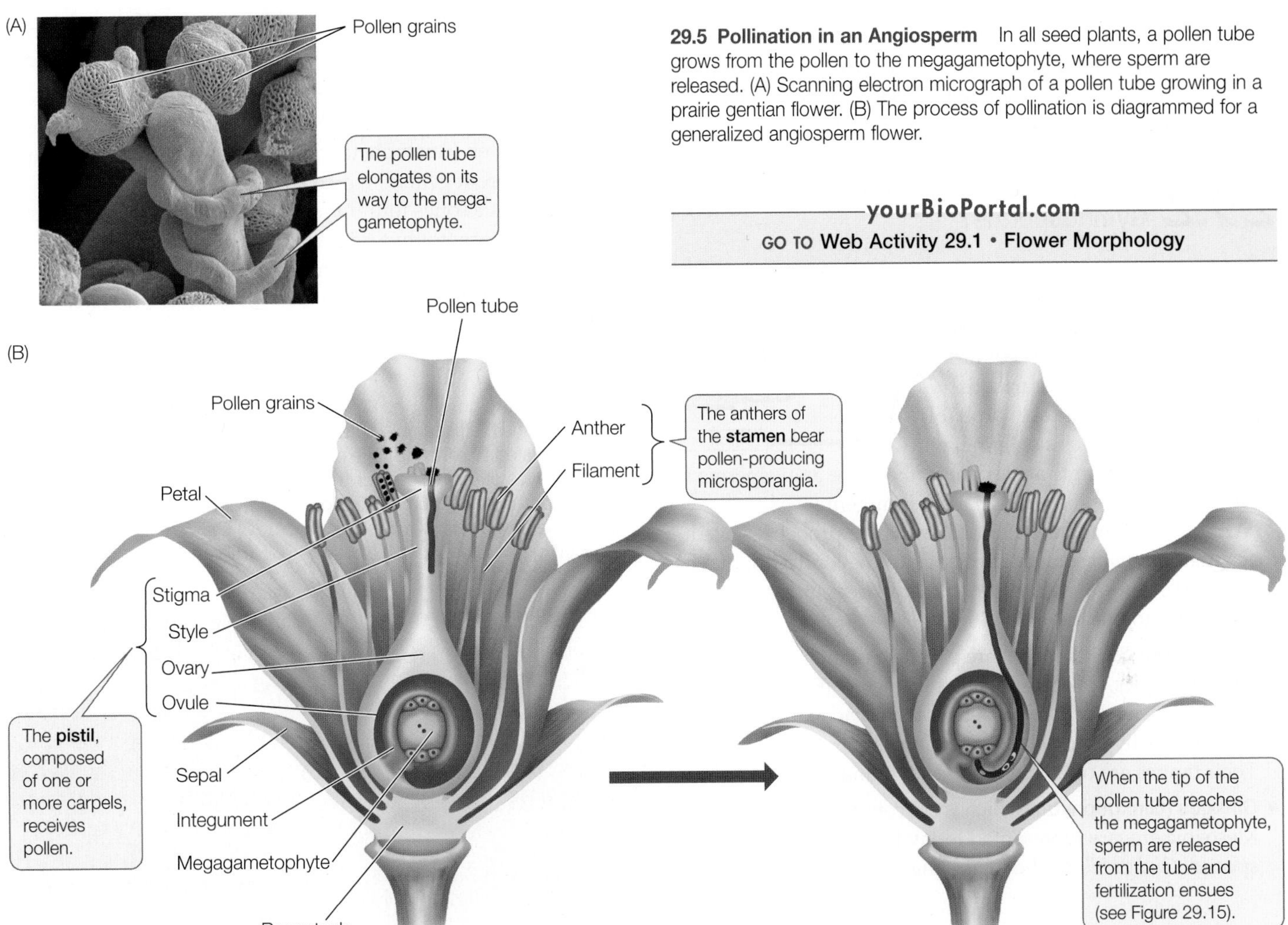
(A)
Pollen grains
The pollen tube elongates on its way to the mega-gametophyte.
(B)
Pollen tube
Pollen grains
Petal
Anther
Filament
The anthers of the **stamen** bear pollen-producing microsporangia.
Stigma
Style
Ovary
Ovule
The **pistil**, composed of one or more carpels, receives pollen.
Sepal
Integument
Megagametophyte
Receptacle
When the tip of the pollen tube reaches the megagametophyte, sperm are released from the tube and fertilization ensues (see Figure 29.15).

**29.5 Pollination in an Angiosperm** In all seed plants, a pollen tube grows from the pollen to the megagametophyte, where sperm are released. (A) Scanning electron micrograph of a pollen tube growing in a prairie gentian flower. (B) The process of pollination is diagrammed for a generalized angiosperm flower.

**yourBioPortal.com**
GO TO **Web Activity 29.1 • Flower Morphology**

many years, germinating only when conditions are favorable for the growth of the sporophyte, as happened with the 2,000-year-old Judean date seed mentioned at the beginning of this chapter. In contrast, the embryos of seedless plants develop directly into sporophytes, which either survive or die, depending on environmental conditions. Spores of some seedless plants may remain dormant and viable for long periods of time, but seeds provide a more secure and lasting dormant stage.

During the dormant stage, the seed coat protects the embryo from excessive drying and may also protect it against potential predators that would otherwise eat the embryo and its nutrient reserves. Many seeds have structural adaptations that promote their dispersal by wind or, more often, by animals. When the young sporophyte resumes growth, it draws on the food reserves in the seed. The possession of seeds is a major reason for the enormous evolutionary success of the seed plants, which are the dominant life forms of most modern terrestrial floras.

### A change in anatomy enabled seed plants to grow to great heights

The most ancient seed plants produced **wood**—extensively proliferated xylem—which gave them the support to grow taller than other plants around them, thus capturing more light for photosynthesis. The younger portion of wood is well adapted for water transport, whereas older wood becomes clogged with resins or other materials. Although no longer functional in transport, the older wood continues to provide support for the plant.

Not all seed plants are woody. In the course of seed plant evolution, many seed plants lost the woody growth habit; however, other advantageous attributes helped them become established in an astonishing variety of places.

## 29.1 RECAP

**Pollen, seeds, and wood are major evolutionary innovations of the seed plants. Protection of the gametes and embryos is a hallmark of seed plants.**

- Distinguish between the roles of the megagametophyte and the pollen grain. See p. 609
- Explain the importance of pollen in freeing seed plants from dependence on liquid water. See pp. 609–610 and Figure 29.5
- What are some of the advantages afforded by seeds? By wood? See pp. 610–611

The seed ferns have long been extinct, but the surviving seed plants have been remarkable successes. Next we examine the gymnosperms, the next group of plants to dominate terrestrial environments.

## 29.2 What Are the Major Groups of Gymnosperms?

Cycads
Ginkgos
Gnetophytes
Conifers
Angiosperms

The gymnosperms are seed plants that do not form flowers. Gymnosperms (which means "naked-seeded") are so named because their ovules and seeds are not protected by ovary or fruit tissue. Although there are probably fewer than 1,200 species of living gymnosperms, these plants are second only to the angiosperms in their dominance of the terrestrial environment.

Although the modern gymnosperms are probably a clade, their monophyly has not been established beyond a doubt. The four major groups of living gymnosperms bear little superficial resemblance to one another.

- The **cycads** (**Cycadophyta**) are palmlike plants of the tropics and subtropics, growing as tall as 20 meters (**Figure 29.6A**). Of the present-day gymnosperms, the cycads are probably the earliest-diverging clade. There are about 300 species. Their tissues are often highly toxic to humans if ingested.
- **Ginkgos** (**Ginkgophyta**), which were common during the Mesozoic era, are represented today by a single genus and species: *Ginkgo biloba*, the maidenhair tree (**Figure 29.6B**). There are both male (microsporangiate) and female (megasporangiate) maidenhair trees. The difference is determined by X and Y sex chromosomes, as in humans; few other plants have distinct sex chromosomes.
- **Gnetophytes** (**Gnetophyta**) number about 90 species in three very different genera, which share certain characteristics analogous to ones found in the angiosperms, as we will see. One of the gnetophytes is *Welwitschia* (**Figure 29.6C**), a long-lived desert plant with just two straplike leaves that sprawl on the sand and can grow as long as 3 meters.

**29.6 Diversity among the Gymnosperms** (A) Many cycads have growth forms that resemble both ferns and palms, but cycads are not closely related to either. (B) The characteristic fleshy seed coat and broad leaves of the maidenhair tree. (C) A gnetophyte growing in the Namib Desert of Africa. Straplike leaves grow throughout the life of the plant, breaking and splitting as they grow. (D) Conifers, like this giant sequoia growing in Sequoia National Park, California, dominate many modern forests.

(A) *Encephalartos villosus*

(B) *Ginkgo biloba*

(C) *Welwitschia mirabilis*

(D) *Sequoiadendron giganteum*

- **Conifers** (**Coniferophyta**) are by far the most abundant of the gymnosperms. There are about 700 species of these cone-bearing plants, including the pines and redwoods (**Figure 29.6D**).

With the exception of the gnetophytes, the living gymnosperm groups have only tracheids as water-conducting and support cells within the xylem. In most angiosperms, cells called vessel elements and fibers (specialized for water conduction and support, respectively) are found alongside tracheids. While the gymnosperm water-transport and support system may thus seem somewhat less efficient than that of the angiosperms, it serves some of the largest trees known. The coastal redwoods of California are the tallest gymnosperms; the largest are well over 100 meters tall.

During the Permian, as environments became warmer and dryer, the conifers and cycads flourished. Gymnosperm forests changed over time as the gymnosperm groups evolved. Gymnosperms dominated the Mesozoic era, during which the continents drifted apart and large dinosaurs walked the Earth. Gymnosperms were the principal trees in all forests until about 65 million years ago, and even down to the present day, conifers are the dominant trees in many forests, especially of higher latitudes and altitudes. The oldest living single organism on Earth today is a gymnosperm in California—a bristlecone pine that germinated about 4,800 years ago, at about the time the ancient Egyptians were just starting to develop writing.

### Conifers have cones but no motile gametes

The great Douglas fir and cedar forests found in the northwestern United States and the massive boreal forests of pine, fir, and spruce of the northern regions of Eurasia and North America, as well as on the upper slopes of mountain ranges everywhere, rank among the great vegetation formations of the world. All these trees belong to one group of gymnosperms, Coniferophyta—the conifers, or cone-bearers.

Male and female **cones** contain the reproductive structures of conifers. The female (seed-bearing) cone is known as a **megastrobilus** (plural: *megastrobili*); this is the familiar woody cone of pine trees. The seeds in a megastrobilus are protected by a tight cluster of woody scales, which are modifications of branches extending from a central axis (**Figure 29.7A**). The typically much smaller male (pollen-bearing) cone is known as a **microstrobilus**. The microstrobilus is typically herbaceous rather than woody, as its scales are composed of modified leaves, beneath which are the pollen-bearing microsporangia (**Figure 29.7B**).

We will use the life cycle of a pine to illustrate reproduction in gymnosperms (**Figure 29.8**). The production of male gametophytes in the form of pollen grains frees the plant completely from its dependence on liquid water for fertilization. Wind, rather than water, assists conifer pollen grains in their first stage of travel from the strobilus to the female gametophyte inside a cone (see Figure 29.4). The pollen tube provides the sperm with the means for the last stage of travel by elongating through maternal sporophytic tissue, as diagrammed for an angiosperm in Figure 29.5. When it reaches the female gametophyte, it releases two sperm, one of which degenerates after the other unites with an egg. Union of sperm and egg results in a zygote; mitotic di-

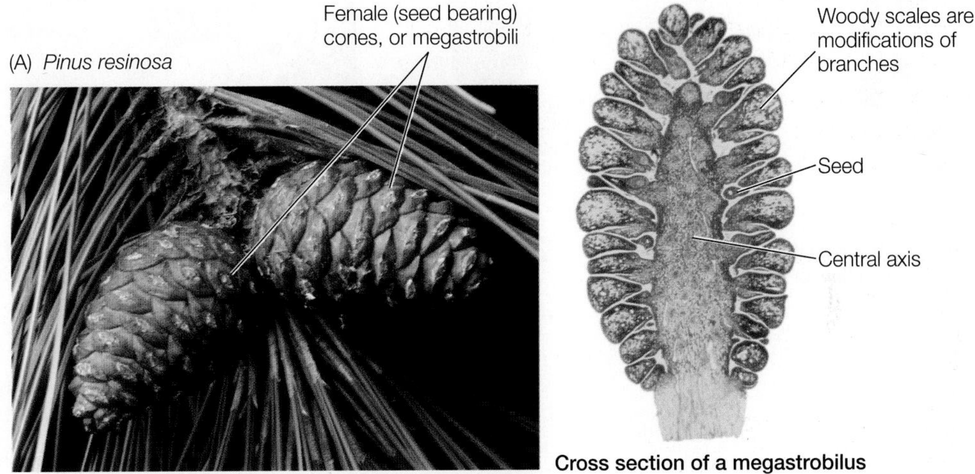

**29.7 Female and Male Cones**
(A) The scales of female cones (megastrobili) are modified branches. (B) The spore-bearing structures in male cones (microstrobili) are modified leaves.

visions and further development of the zygote result in an embryo.

yourBioPortal.com
GO TO Web Activity 29.2 and Animated Tutorial 29.1 • Life Cycle of a Conifer

The megasporangium, in which the female gametophyte will form, is enclosed in a layer of sporophytic tissue—the integument—that will eventually develop into the seed coat that protects the embryo. The integument, the megasporangium inside it, and the tissue attaching it to the maternal sporophyte constitute the ovule. The pollen grain enters through a small opening in the integument at the tip of the ovule, the **micropyle**.

Most conifer ovules (which will develop into seeds after fertilization) are borne exposed on the upper surfaces of the mod-

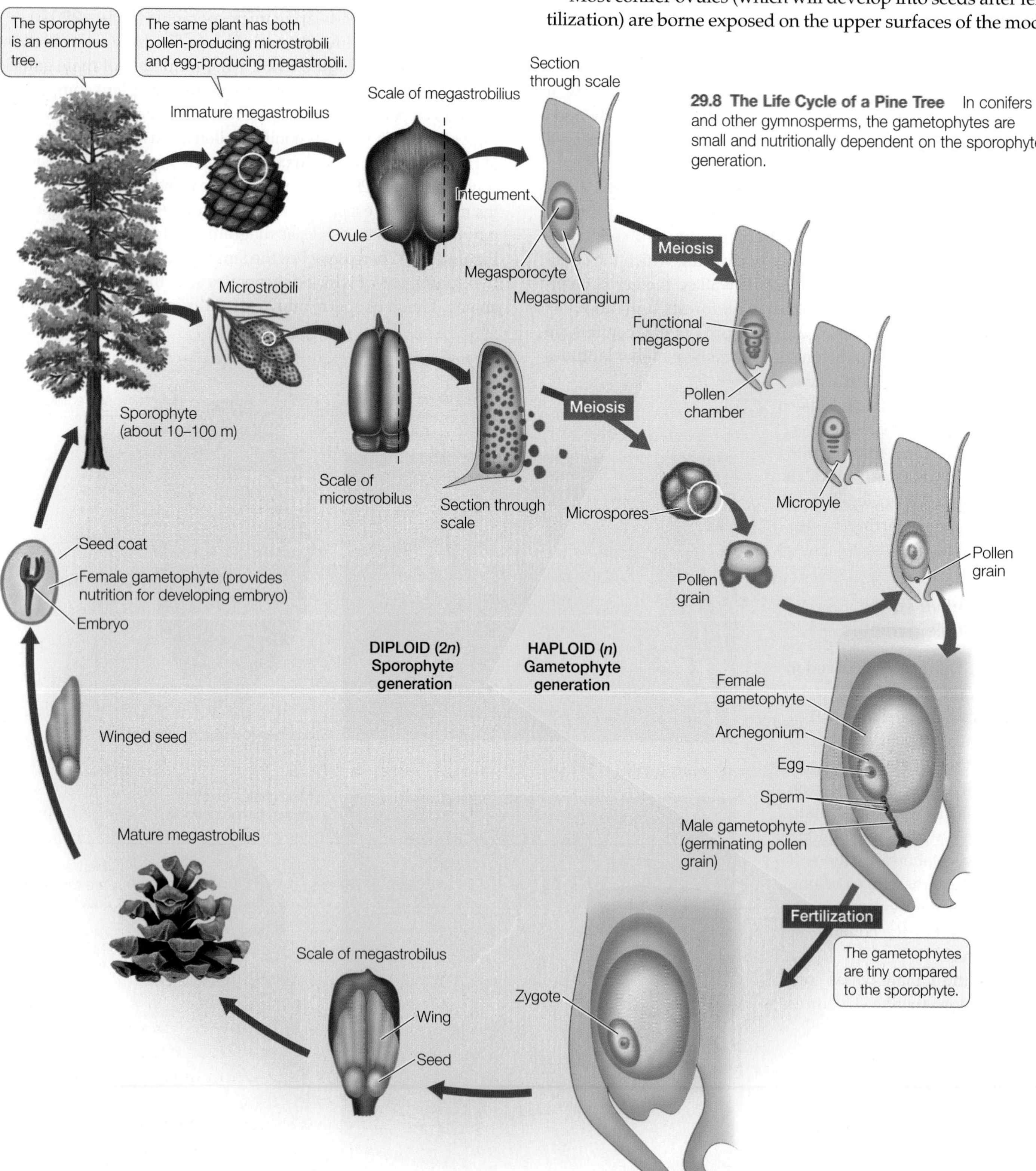


**29.8 The Life Cycle of a Pine Tree** In conifers and other gymnosperms, the gametophytes are small and nutritionally dependent on the sporophyte generation.

**29.9 From Devastation, New Life Emerges** A stand of lodgepole pines in Yellowstone National Park. The mature trees were destroyed by a forest fire in 1988. The fire released large numbers of lodgepole pine seeds from the cones, and now many young lodgepole pine trees are emerging in the burn area.

ified branches that form the scales of the cone (megastrobilus). The only protection of the ovules comes from the scales, which are tightly pressed against one another within the cone. Some pines, such as the lodgepole pine, have such tightly closed cones that only fire suffices to split them open and release the seeds. A fire devastated lodgepole pine forests in Yellowstone National Park in 1988 but also released large numbers of seeds from the cones. As a result, large numbers of lodgepole pine seedlings are now emerging in the burn area (**Figure 29.9**).

About half of all conifer species have soft, fleshy modifications of cones that envelop their seeds; examples are the fruit-like cones or "berries" of juniper and yew. Animals may eat these tissues and then disperse the seeds in their feces, often carrying them considerable distances from the parent plant.

## 29.2 RECAP

**Living gymnosperms include cycads, ginkgos, gnetophytes, and conifers, all of which are woody and have seeds that are not protected by ovaries.**

- Explain the different functions of a megastrobilus and a microstrobilus. See p. 613 and Figure 29.8
- What is the role of the integument in a gymnosperm seed? See p. 614 and Figure 29.8
- Do you understand how fire can be necessary for the survival of some species? See p. 615

The "berries" on some gymnosperms (such as juniper and yew) are not true fruits but rather are fleshy cones. As we will see, true fruits are the ripened ovaries of plants. Ovaries are absent in gymnosperms but are a characteristic of the plant group that is dominant today: the angiosperms. Let's look at other ways in which angiosperms differ from gymnosperms.

## 29.3 What Features Contributed to the Success of the Angiosperms?

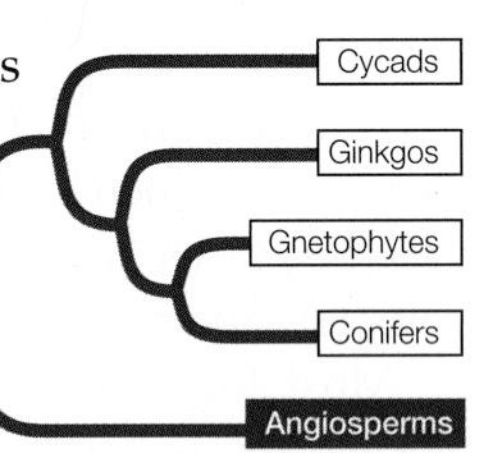


The oldest fossil evidence of angiosperms dates back to the late Jurassic period, about 150 million years ago (see Figure 29.1). The angiosperms radiated explosively in the Tertiary (beginning 65 million years ago) and became the dominant plant life on Earth. Today there are more than a quarter million species of angiosperms.

The female gametophyte of the angiosperms is even more reduced than that of the gymnosperms, usually consisting of just seven cells. Thus the angiosperms represent the current extreme of the trend we have traced throughout the evolution of the vascular plants: the sporophyte generation becomes larger and more independent of the gametophyte, while the gametophyte generation becomes smaller and more dependent on the sporophyte. What else sets the angiosperms apart from other plants?

The major synapomorphies (shared derived traits) that characterize the angiosperms include:

- Double fertilization
- Production of a nutritive tissue called the endosperm
- Ovules and seeds enclosed in a carpel
- Flowers
- Fruits
- Phloem with companion cells
- Reduced gametophytes

In the angiosperms, pollination consists of the arrival of a microgametophyte—a pollen grain—on a receptive surface in a flower (the *stigma*). As in the gymnosperms, pollination is the first in a series of events that result in the formation of a seed. The next is the growth of a pollen tube extending to the megagametophyte (see Figure 29.5). The third event is a fertilization process that, in detail, is unique to the angiosperms.

Double fertilization is often considered the single most reliable distinguishing characteristic of the angiosperms. *Two* male gametes, contained in a single microgametophyte, participate in fertilization events in the megagametophyte of an angiosperm. The nucleus of one sperm combines with that of the egg to produce a diploid zygote, the first cell of the sporophyte generation. In most angiosperms, the other sperm nucleus combines with two other haploid nuclei of the female gametophyte to form a *triploid* ($3n$) nucleus (further discussed below; see Figure 29.15). That nucleus, in turn, divides to form a triploid tissue, the **endosperm**, that nourishes the embryonic sporophyte during its early development. This process, in which two fertilization events take place, is known as **double fertilization**.

Double fertilization occurs in nearly all present-day angiosperms. We are not sure when and how it evolved because there is no known fossil evidence on this point. It may have first resulted in two diploid embryos, as it does in the three existing genera of the gnetophytes.

The name *angiosperm* ("enclosed seed") is drawn from another distinctive characteristic of these plants: the ovules and seeds are enclosed in a modified leaf called a **carpel**. Besides protecting the ovules and seeds, the carpel often interacts with incoming pollen to prevent self-pollination, thus favoring cross-pollination and increasing genetic diversity. Of course, the most obvious diagnostic feature of angiosperms is that they have **flowers**. Production of **fruits** is another of their unique characteristics. As we will see, both flowers and fruits afford major advantages to angiosperms.

Most angiosperms are distinguished by the possession of specialized water-transporting cells called **vessel elements** in their xylem (see Chapter 34). These cells are broad in diameter and connect without obstruction, allowing easy water movement. A second distinctive cell type in angiosperm xylem is the **fiber**, which plays an important role in supporting the plant body. Angiosperm phloem possesses another unique cell type, called a **companion cell**. Like the gymnosperms, woody angiosperms show secondary growth, producing secondary xylem and secondary phloem and growing in diameter.

In the remainder of this section we examine the structure and function of flowers, evolutionary trends in flower structure, the functions of pollen and fruits, and the angiosperm life cycle.

### The sexual structures of angiosperms are flowers

If you examine any familiar flower, you will notice that the outer parts look somewhat like leaves. In fact, all the parts of a flower *are* modified leaves.

We showed a generalized flower (for which there is no exact counterpart in nature) in Figure 29.5. The structures bearing microsporangia are called **stamens**. Each stamen is composed of a **filament** bearing an **anther** that contains pollen-producing microsporangia. The structures bearing megasporangia are the carpels. A structure composed of one carpel or two or more fused carpels is called a **pistil**. The swollen base of the pistil, containing one or more ovules (each containing a megasporangium surrounded by its protective integument), is called the **ovary**. The apical stalk of the pistil is the **style**, and the terminal surface that receives pollen grains is the **stigma**.

In addition, a flower often has several specialized sterile (non-spore-bearing) leaves. The inner ones are called **petals** (collectively, the **corolla**) and the outer ones **sepals** (collectively, the **calyx**). The corolla and calyx (collectively, the *perianth*) can be quite showy and often play roles in attracting animal pollinators to the flower. The calyx more commonly protects the immature flower in bud. From base to apex, the sepals, petals, stamens, and carpels (which are referred to as the *floral organs*; see Figure 19.14) are usually positioned in circular arrangements or whorls and attached to a central stalk called the **receptacle**.

The generalized flower in Figure 29.5 has functional megasporangia and microsporangia; such flowers are referred to as **perfect** (or hermaphroditic). Many angiosperms produce two types of flowers, one with only megasporangia and the other with only microsporangia. Consequently, either the stamens or the carpels are nonfunctional or absent in a given flower, and the flower is referred to as **imperfect**.

Species such as corn or birch, in which both megasporangiate (female) and microsporangiate (male) flowers occur on the same plant, are said to be **monoecious** (meaning "one-housed"—but, it must be added, one house with separate rooms). Complete separation is the rule in some other angiosperm species, such as willows and date palms; in these species, a given plant produces either flowers with stamens or flowers with carpels, but never both. Such species are said to be **dioecious** ("two-housed").

Flowers come in an astonishing variety of forms—just think of some of the flowers you recognize. The generalized flower in Figure 29.5 has distinct petals and sepals arranged in distinct whorls. In nature, however, petals and sepals sometimes are indistinguishable. Such appendages are called **tepals**. In other flowers, petals, sepals, or tepals are completely absent.

Flowers may be single, or they may be grouped together to form an **inflorescence**. Different families of flowering plants have characteristic types of inflorescences, such as the compound umbels of the carrot family (**Figure 29.10A**), the heads of the aster family (**Figure 29.10B**), and the spikes of many grasses (**Figure 29.10C**).

### Flower structure has evolved over time

The flowers of the most basal clades of angiosperms have a large and variable number of tepals (or sepals and petals), carpels, and stamens (**Figure 29.11A**). Evolutionary change within the angiosperms has included some striking modifications of this early condition: reductions in the number of each type of floral organ to a fixed number, differentiation of petals from sepals, and changes in symmetry from radial (as in a lily or magnolia) to bilateral (as in a sweet pea or orchid), often accompanied by an extensive fusion of parts (**Figure 29.11B**).

According to one theory, the first carpels to evolve were leaves with marginal sporangia, folded but incompletely closed. Early in angiosperm evolution, the carpels fused and became progressively more buried in receptacle tissue, forming the ovary (**Figure 29.12A**). In some flowers, the other floral organs are attached at the top of the ovary, rather than at the bottom as in Figure 29.5. The stamens of the most ancient flowers may have appeared leaflike (**Figure 29.12B**), little resembling those of the generalized flower in Figure 29.5.

Why do so many flowers have pistils with long styles and anthers with long filaments? Natural selection has favored length in both of these structures, probably because length increases the likelihood of successful pollination. Long filaments may bring the anthers into contact with insect bodies, or they may place the anthers in a better position to catch the wind. Similar arguments apply to long styles.

A perfect flower represents a compromise of sorts. On the one hand, in attracting a pollinating bird or insect, the plant is attending to both its female and male functions with a single

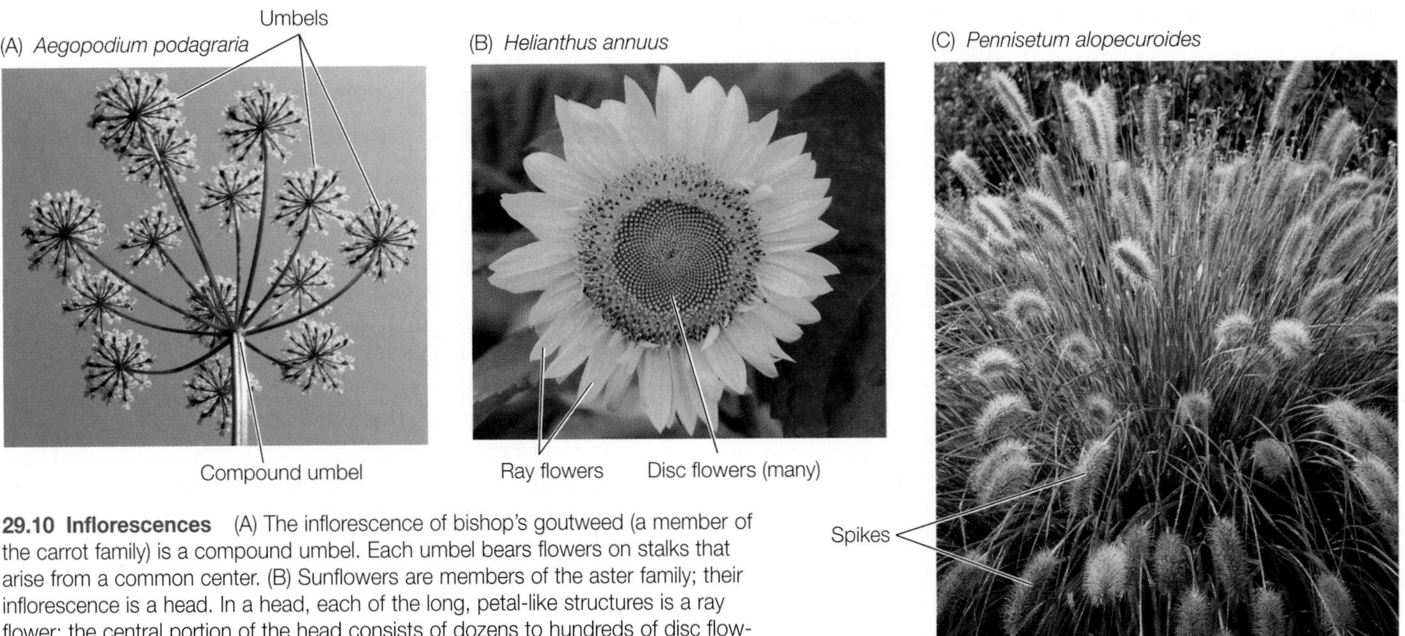


**29.10 Inflorescences** (A) The inflorescence of bishop's goutweed (a member of the carrot family) is a compound umbel. Each umbel bears flowers on stalks that arise from a common center. (B) Sunflowers are members of the aster family; their inflorescence is a head. In a head, each of the long, petal-like structures is a ray flower; the central portion of the head consists of dozens to hundreds of disc flowers. (C) Grasses such as this fountain grass have inflorescences called spikes.

**29.11 Flower Form and Evolution** (A) A magnolia flower shows the major features of early flowers: it is radially symmetrical, and the individual tepals, carpels, and stamens are separate, numerous, and attached at their bases. (B) Orchids, such as this ladyslipper, have a bilaterally symmetrical structure that evolved much later.

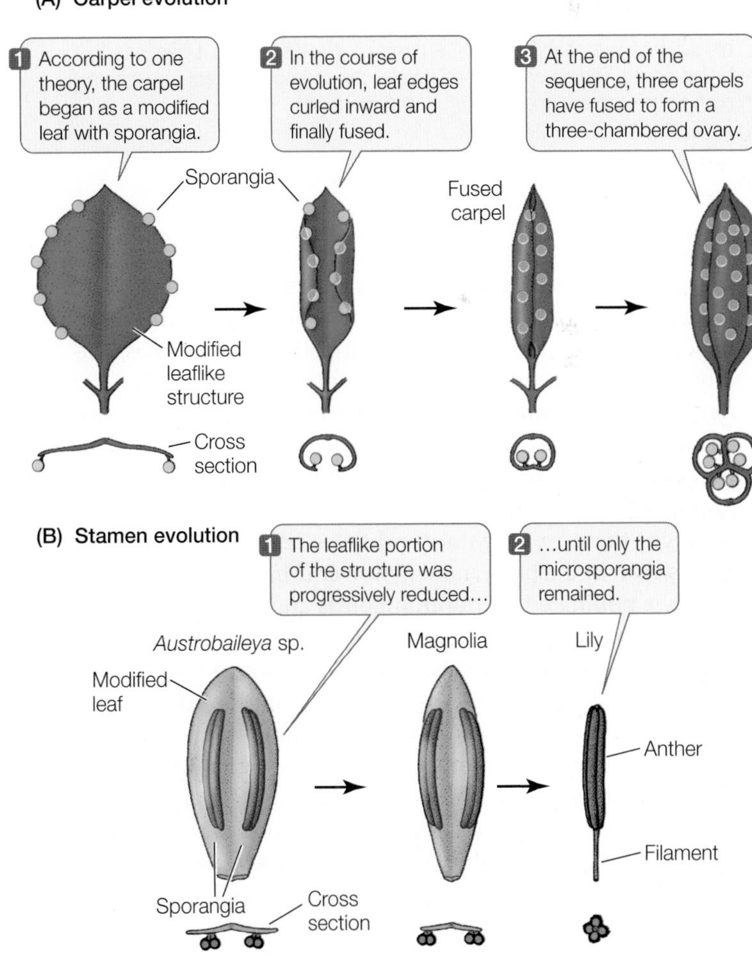


**29.12 Carpels and Stamens Evolved from Leaflike Structures** (A) Possible stages in the evolution of a carpel from a more leaflike structure. (B) The stamens of three modern plants show the various stages in the evolution of that organ. It is *not* implied that these species evolved from each other; their structures simply illustrate the possible stages.

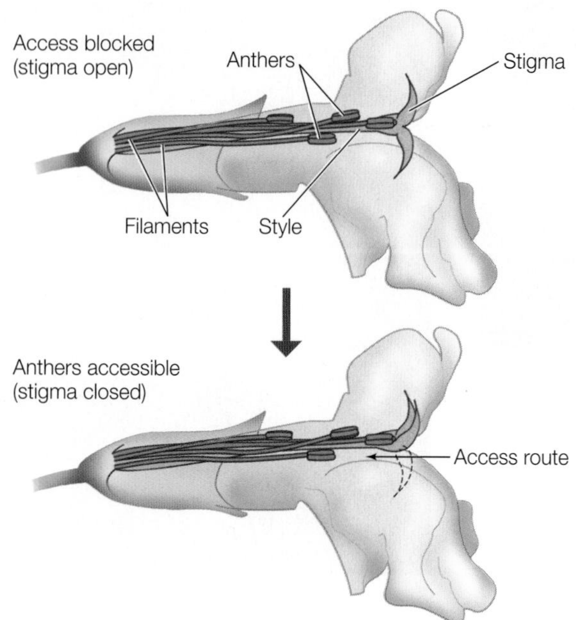


**29.13 An Unusual Solution to Selfing** Both long stamens and long styles facilitate cross pollination, but if these male and female structures are too close to each other, the likelihood of (disadvantageous) self-pollination increases. In *Mimulus aurantiacus*, initially the stigma is open, blocking access to the anthers. A hummingbird's touch as it deposits pollen on the stigma causes one lobe of the stigma to retract, creating a path to the anthers and allowing pollen dispersal.

## INVESTIGATING LIFE

**29.14 Stigma Behavior in *Mimulus* Flowers**

Elizabeth Fetscher's experiments showed that the unusual stigma retraction mechanism in monkeyflowers (see Figure 29.13) enhances the dispersal of pollen from the anthers.

**HYPOTHESIS** **Stigma responses in *M. aurantiacus* increase the likelihood that an individual's pollen will be exported once pollen from another individual has been deposited on the stigma.**

**METHOD**

1. Set up experimental arrays such that only one flower in each array can donate pollen. In control arrays, the pollen donor styles function normally. In another set of arrays, donor stigmas are artificially sealed closed; in a third set, the pollen donor stigmas are permanently propped open.
2. Allow hummingbirds to visit the arrays, then count the pollen grains from each donor on the next flower visited.

**RESULTS**

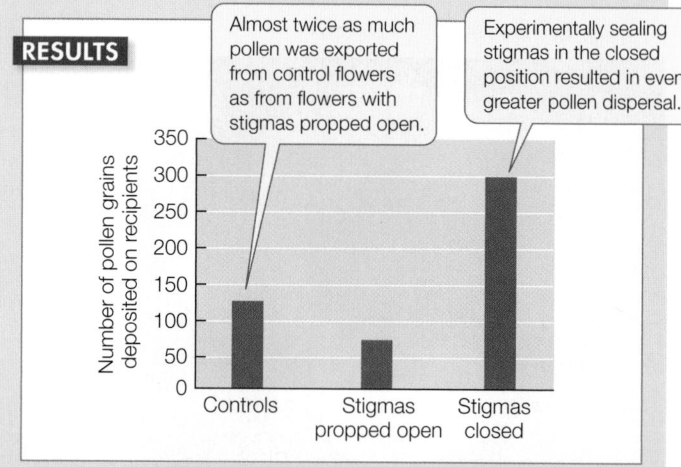


**CONCLUSION** **Stigma responses enhance the male function of the flower (dispersal of pollen) once the female function (receipt of pollen) has been performed.**

**FURTHER INVESTIGATION:** How might you test how this mechanism affects self-pollination of the flower?

Go to **yourBioPortal.com** for original citations, discussions, and relevant links for all INVESTIGATING LIFE figures.

flower type, whereas plants with imperfect flowers must create that attraction twice—once for each type of flower. On the other hand, the perfect flower can favor self-pollination, which is usually disadvantageous. Another potential problem is that the female and male functions might interfere with each other—for example, the stigma might be so placed as to make it difficult for pollinators to reach the anthers, thus reducing the export of pollen to other flowers.

Might there be a way around these problems? One solution is seen in the bush monkeyflower (*Mimulus aurantiacus*), which is pollinated by hummingbirds and has a stigma that initially serves as a screen, hiding the anthers (**Figure 29.13**). Once a hummingbird touches the stigma, one of the stigma's two lobes folds, so that subsequent hummingbirds pick up pollen from the previously screened anthers. The first bird transfers pollen from another plant to the stigma, eventually leading to fertilization. Later visitors pick up pollen from the now-accessible anthers, fulfilling the flower's male function. **Figure 29.14** describes the experiment that revealed the function of this mechanism.

### Angiosperms have coevolved with animals

Whereas many gymnosperms are wind-pollinated, most angiosperms are animal-pollinated. The many different pollination symbioses between plants and animals are vital to both parties. This fascinating coevolutionary field is covered in more detail in Chapter 56, but we mention some aspects here.

Many flowers entice animals to visit them by providing food rewards. Some flowers produce a sugary fluid called nectar, and the pollen grains themselves sometimes serve as food for animals. In the process of visiting flowers to obtain nectar or pollen, animals often carry pollen from one flower to another or from one plant to another. Thus, in their quest for food, the animals contribute to the genetic diversity of the plant population. In-

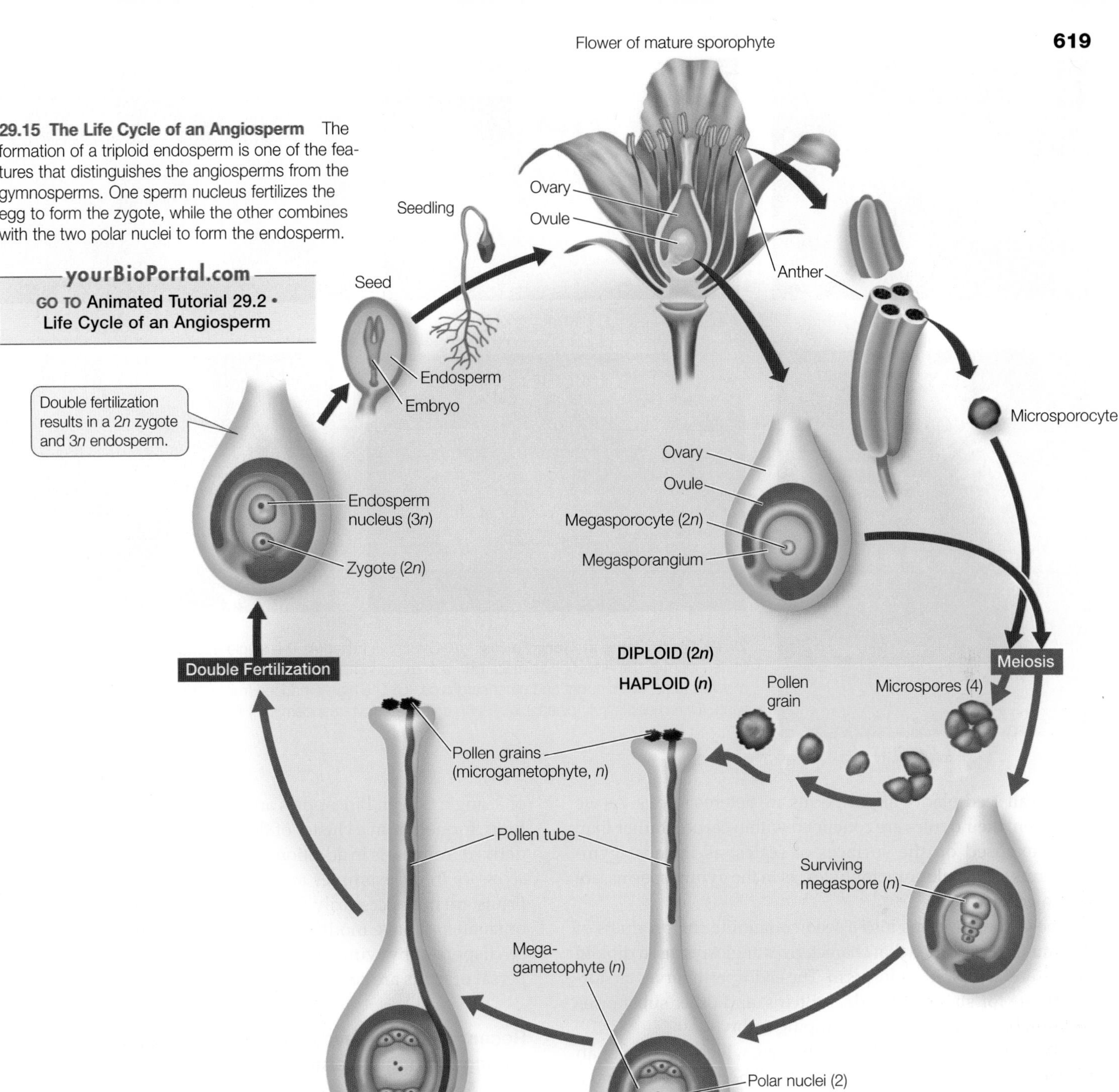


**29.15 The Life Cycle of an Angiosperm** The formation of a triploid endosperm is one of the features that distinguishes the angiosperms from the gymnosperms. One sperm nucleus fertilizes the egg to form the zygote, while the other combines with the two polar nuclei to form the endosperm.

yourBioPortal.com
GO TO Animated Tutorial 29.2 • Life Cycle of an Angiosperm

sects, especially bees, are among the most important pollinators; birds and some species of bats are also major pollinators.

For more than 150 million years, angiosperms and their animal pollinators have coevolved in the terrestrial environment. The animals have affected the evolution of the plants, and the plants have affected the evolution of the animals. Flower structure has become incredibly diverse under these selection pressures. Some of the products of coevolution are highly specific; for example, some yucca species are pollinated by only one species of moth. Pollination by just one or a few animal species provides a plant species with a reliable mechanism for transferring pollen from one of its members to another.

Most plant–pollinator interactions are much less specific; that is, many different animal species pollinate the same plant species, and the same animal species pollinate many different plant species. However, even these less specific interactions have developed some specialization. Bird-pollinated flowers are often red and odorless. Many insect-pollinated flowers have characteristic odors, and bee-pollinated flowers may have conspicuous markings, or *nectar guides*, that may be visible only in the ultraviolet region of the spectrum, where bees have better vision than in the red region (see Table 56.1 and Figure 56.10).

### The angiosperm life cycle features double fertilization

The angiosperm life cycle is considered in detail in Chapter 38, but let's look at it briefly here, comparing **Figure 29.15** with the conifer life cycle shown in Figure 29.8.

**29.16 Fruits Come in Many Forms** (A–C) Simple fruits. (A) The single-seeded drupe of the fleshy-fruited plum is dispersed by animals. (B) Each macadamia seed is covered by a hard, woody fruit that allows it to survive drought and other hardships. (C) The highly reduced fruits of dandelions are dispersed by wind. (D) A multiple fruit (pineapple). (E) An aggregate fruit (raspberry). (F) An accessory fruit (strawberry).

Like all seed plants, angiosperms are heterosporous. As we have seen, their ovules are contained within carpels rather than being exposed on the surfaces of scales, as in most gymnosperms. The male gametophytes, as in the gymnosperms, are pollen grains.

The ovule develops into a seed containing the products of the double fertilization that characterizes angiosperms: a diploid zygote and a triploid endosperm. The endosperm serves as storage tissue for starch or lipids, proteins, and other substances that will be needed by the developing embryo.

The zygote develops into an embryo, which consists of an embryonic axis (the "backbone" that will become a stem and a root) and one or two **cotyledons**, or seed leaves. The cotyledons have three different possible fates in different plants. In many, they serve as absorptive organs that take up and digest the endosperm. In others, they enlarge and become photosynthetic when the seed germinates. Often they play both roles.

### Angiosperms produce fruits

The ovary of a flower (together with the seeds it contains) develops into a fruit after fertilization. The fruit protects the seeds and can also promote seed dispersal by becoming attached to or being eaten by an animal. Other fruits are adapted for dispersal by wind or water. A fruit may consist only of the mature ovary and its seeds, or it may include other parts of the flower or structures associated with it (**Figure 29.16**). A *simple fruit*, such as a plum, is one that develops from a single carpel or several united carpels. A raspberry is an example of an *aggregate fruit*—one that develops from several separate carpels of a single flower. Pineapples and figs are examples of *multiple fruits*, formed from a cluster of flowers (an inflorescence). Fruits derived from parts in addition to the carpel and seeds are called *accessory fruits*; examples are apples, pears, and strawberries. Fruits are not necessarily fleshy; they can be hard and woody or small and have modified structures that allow the seeds to be dispersed by wind. The development, ripening, and dispersal of fruits are considered in Chapters 37 and 38.

### Recent analyses have revealed the oldest split among the angiosperms

**Figure 29.17** shows the relationships among the major angiosperm clades. The two largest clades— the **monocots** and the **eudicots**—include the great majority of angiosperm species. The monocots are so called because they have a single embryonic cotyledon; the eudicots have two. (Chapter 34 describes other differences between these two groups.)

Some familiar angiosperms belong to other clades, including the water lilies, star anise and its relatives, and the magnoliid complex (**Figure 29.18**). The magnoliids are less numerous than the monocots and eudicots, but they include many familiar and useful plants such as magnolias, avocados, cinnamon, and black pepper.

The root of the evolutionary tree of flowering plants was once a matter of great controversy. A fundamental issue was identifying the group that is sister to the remaining angiosperms, and the magnoliid clade was a leading candidate for the postion. At the close of the twentieth century, however, an impressive convergence of molecular and morpholog-

**29.17 Evolutionary Relationships among the Angiosperms** Recent analyses of many angiosperm genes have clarified the relationships among the major groups.

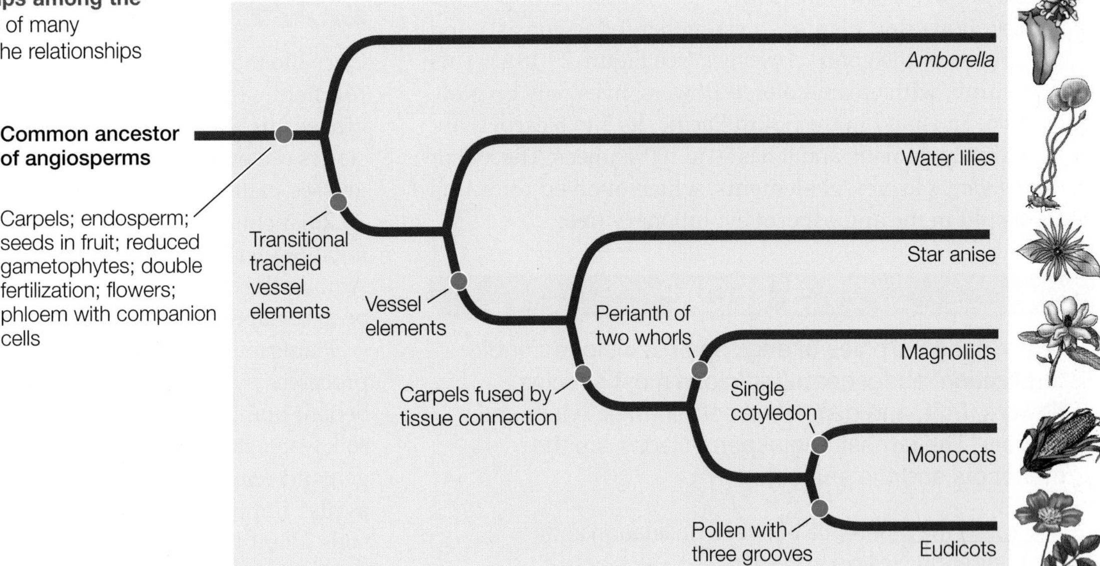

(A) *Amborella trichopoda*

(B) *Nymphaea* sp.

(C) *Illicium anisatum*

(D) *Piper nigrum*

(E) *Aristolochia macrophylla*

(F) *Persea americana*

**29.18 Monocots and Eudicots Are Not the Only Surviving Angiosperms** (A) *Amborella*, a shrub, is the sister to the remaining extant angiosperms. Of interest in this photo of a female flower is the pair of false anthers (male flower parts), possibly serving to lure insects that are searching for pollen. (B) The water lily clade is the next clade to diverge after *Amborella*. (C) Star anise and its relatives belong to another basal clade. (D–F) The largest clade other than the monocots and eudicots is the magnoliid complex, represented here by (D) black pepper, (E) Dutchman's pipe, and (F) avocado. The magnolia in Figure 29.11A is another magnoliid.

ical evidence led to the conclusion that the sister group to the remaining flowering plants is a clade that today consists of a single species of the genus *Amborella* (see Figure 29.18A). This woody shrub, with cream-colored flowers, lives only on New Caledonia, an island in the South Pacific. Its 5 to 8 carpels are in a spiral arrangement, and it has 30 to 100 stamens. The xylem of *Amborella* lacks vessel elements, which evolved after this deepest split in the angiosperm evolutionary tree.

## 29.3 RECAP

**The synapomorphies of angiosperms include double fertilization, endosperm (triploid in most species), flowers, fruit, and distinctive cells in their xylem and phloem. The largest angiosperm clades are the monocots and the eudicots.**

- Explain the difference between pollination and fertilization. See pp. 615–616
- Give some examples of how animals have affected the evolution of the angiosperms. See pp. 618–619
- What are the respective roles of the two sperm in double fertilization? See pp. 619–620 and Figure 29.15
- What are the different functions of flowers, fruits, and seeds?

The remarkable diversity of the seed plants has been shaped in part by the different environments in which these and other plants have evolved. In turn, land plants—and seed plants in particular—affect their environments.

# 29.4 How Do Plants Support Our World?

Once life moved onto land, it was plants that shaped the environment, and today's environment is dominated by angiosperms. Representatives of the two largest angiosperm clades are everywhere. The monocots (**Figure 29.19**) include grasses, cattails, lilies, orchids, and palms. The eudicots (**Figure 29.20**) include the vast majority of familiar seed plants, including most herbs (i.e., nonwoody plants), vines, trees, and shrubs. Among the eudicots are such diverse plants as oaks, willows, beans, snapdragons, and sunflowers.

Plants make profound contributions to ecosystem services—processes by which the environment maintains resources that benefit humans. These benefits include the effects of plants on soil, water, the atmosphere, and the climate. Plants produce oxygen and remove carbon dioxide from the atmosphere, as well as play important roles in forming and renewing the fertility of soils. Plant roots help hold soil in place, providing a defense against erosion by wind and water. They also moderate local climate in various ways, such as by increasing humidity, providing shade, and blocking wind.

### Seed plants are our primary food source

Plants are **primary producers**; that is, their photosynthesis traps energy and carbon, making those resources available not only for their own needs but also for the herbivores and omnivores that consume them, for the carnivores and omnivores that eat the herbivores, and for the prokaryotes and fungi that complete the food chain. The earliest steps in human civilization involved cultivating angiosperms to provide a reliable food supply.

Today, twelve species of seed plants stand between the human race and starvation: rice, coconut, wheat, corn (also called maize), potato, sweet potato, cassava (also called tapioca or man-

(A) *Phoenix dactylifera*

(B) *Triticum aestivum*

(C) *Lilium* sp.

**29.19 Monocots** (A) Palms are among the few monocot trees. Date palms are a major food source in some areas of the world. (B) Grasses such as this cultivated wheat and the fountain grass in Figure 29.10C are monocots. (C) Monocots include popular garden flowers such as these lilies. Orchids (Figure 29.11B) are also highly sought-after monocot flowers.

(A) *Echinocereus reichenbachii*

(B) *Mimosa nuttallii*

(C) *Crataegus viridis*

**29.20 Eudicots** (A) The cactus family is a large group of eudicots, with about 1,500 species in the Americas. This black lace cactus bears large pink flowers for a brief period of the year. (B) This sensitive briar is a legume ("bean and pea family"), an economically important plant group with a large number of species. (C) The green hawthorn is a small eudicot tree and a member of the family that includes roses.

ioc), sugarcane, sugar beet, soybean, common bean (*Phaseolus vulgaris*), and banana. Hundreds of other seed plants are cultivated for food, but none rank with these twelve in importance.

Indeed, more than half of the world's human population derives the bulk of its food energy from the seeds of a single plant: rice, *Oryza sativa*. Rice is particularly important in the Far East, where it has been cultivated for more than 8,000 years. People also use rice straw in many ways, such as thatching for roofs, food and bedding for livestock, and clothing. Rice hulls, too, have many uses, ranging from fertilizer to fuel.

Another vitally important angiosperm is the coconut (*Cocos nucifera*), whose seed is illustrated at the opening of this chapter. In some cultures of the coastal tropics, this monocot tree is known as the "tree of life." Every aboveground part of the plant—its seeds, fruit, stem (trunk), leaves, buds, and even its sap—is of use and value to humans. Millions of people get most of their protein from the "meat" of coconut seeds, and the seed's "milk" is vital in areas where water is scarce or unfit to drink.

### Seed plants have been sources of medicine since ancient times

One of the oldest human professions is that of medicine man or shaman—a person who cures others with medicines derived from seed plants. It is claimed that a legendary Chinese emperor around 2700 B.C.E. knew some 365 medicinal plants. Although we also use medicines derived from fungi, lichens, and actinobacteria, seed plants are the source of many of our medications, just a few of which are shown in **Table 29.1**. Even in synthetic pharmaceuticals, the chemical structures of active ingredients are often based on the biochemistry of substances isolated from plants.

How are plant-based medicines discovered? These days many are found by systematic testing of tremendous numbers of plants from all over the world, a process that began in the 1960s. One example is taxol, an important anticancer drug. Among the myriad plant samples that had been tested by 1962, extracts of the bark of Pacific yew (*Taxus brevifolia*) showed antitumor activity in tests against rodent tumors. The active ingredient, taxol, was isolated in 1971 and tested against human cancers in 1977. After another 16 years, the U.S. Food and Drug Administration approved it for human use, and taxol is now widely used in treating breast and ovarian cancers as well as several other types of cancers.

TABLE 29.1

**Some Medicinal Plants and Their Products**

| PRODUCT | PLANT SOURCE | MEDICAL APPLICATION |
|---|---|---|
| Atropine | Belladonna | Dilating pupils for eye examination |
| Bromelain | Pineapple stem | Controlling tissue inflammation |
| Digitalin | Foxglove | Strengthening heart muscle contraction |
| Ephedrine | *Ephedra* | Easing nasal congestion |
| Menthol | Japanese mint | Relief of coughing |
| Morphine | Opium poppy | Relief of pain |
| Quinine | *Cinchona* bark | Treatment of malaria |
| Taxol | Pacific yew | Treatment of ovarian and breast cancers |
| Tubocurarine | Curare plant | Muscle relaxant in surgery |
| Vincristine | Periwinkle | Treatment of leukemia and lymphoma |

Widespread screening of plant samples eventually was deemphasized in favor of a purely chemical approach. Using automation and miniaturization, pharmaceutical laboratories generate vast numbers of compounds that are screened just as plant materials were screened in the search for taxol and other plant-based medicines. Now, however, the plant screening is getting renewed interest. Both approaches are based on trial and error.

The other leading source of medicinal plants is work by *ethnobotanists*, who study how people use and view plants in their local environments. This work proceeds all over the globe today. An older example is the discovery of quinine as a treatment for malaria. In the sixteenth century, Spanish priests in Peru became aware that the native population used the bark of local *Cinchona* trees to treat fevers. The priests successfully used the bark to treat malaria. Word of the medicine spread to Europe, where it is said to have been in use in Rome by 1631. The active ingredient of *Cinchona* bark—quinine—was identified in 1820, and quinine remained the standard malarial remedy well into the twentieth century.

## CHAPTER SUMMARY

### 29.1 How Did Seed Plants Become Today's Dominant Vegetation?

- Fossils of woody seed ferns are the earliest evidence of vascular seed plants. The surviving groups of seed plants are the **gymnosperms** and **angiosperms**.
- All seed plants are heterosporous, and their gametophytes are much smaller than (and dependent on) their sporophytes. Review Figure 29.3
- An **ovule** consists of the seed plant megagametophyte and the **integument** that protects it. The ovule develops into a **seed**.
- **Pollen grains**, the microgametophytes, do not require liquid water to perform their functions. Following **pollination**, a **pollen tube** emerges from the pollen grain and elongates to deliver gametes to the megagametophyte. Review Figure 29.5, WEB ACTIVITY 29.1
- Seeds are well protected, and they are often capable of long periods of dormancy, germinating when conditions are favorable.

### 29.2 What Are the Major Groups of Gymnosperms?

- The gymnosperms produce seeds that are not protected by ovaries.
- The major gymnosperm groups are the **cycads**, **ginkgos**, **gnetophytes**, and **conifers**. Review Figure 29.2.
- The megaspores of conifers are produced in woody **cones** called **megastrobili**, and microspores are produced in herbaceous cones called **microstrobili**. Pollen reaches the megagametophyte by way of the **micropyle**, an opening in the integument of the ovule. Review Figure 29.8, WEB ACTIVITY 29.2, ANIMATED TUTORIAL 29.1

### 29.3 What Features Contributed to the Success of the Angiosperms?

- Only angiosperms have **flowers** and **fruits**.
- The ovules and seeds of angiosperms are enclosed in and protected by **carpels**. Review Figure 29.12
- Angiosperms have **double fertilization**, resulting in the production of a zygote and **endosperm** (which is triploid in most species). Review Figure 29.15, ANIMATED TUTORIAL 29.2
- The xylem and phloem of angiosperms are more complex and efficient than those of the gymnosperms. **Vessel elements** in the xylem of angiosperms function in water transport. **Fibers** in angiosperm xylem play an important role in structural support.
- The floral organs, from the apex to the base of the flower, are the **pistil**, **stamens**, **petals**, and **sepals**. Stamens bear microsporangia in **anthers**. The pistil (consisting of one or more carpels) includes an **ovary** containing ovules. The **stigma** is the receptive surface of the pistil. The floral organs are borne on the **receptacle**. Review Figure 29.5
- A flower with both megasporangia and microsporangia is **perfect**; all other flowers are **imperfect**. Flowers may be grouped to form an **inflorescence**. Flowers may be pollinated by wind or animals.
- A **monoecious** species has megasporangiate and microsporangiate flowers on the same plant. A **dioecious** species is one in which megasporangiate and microsporangiate flowers occur on different plants.
- The most species-rich angiosperm clades are the **monocots** and the **eudicots**. The magnoliids form the sister group to the monocots and eudicots. Review Figure 29.17
- The oldest evolutionary split among the angiosperms is between the single species in the genus *Amborella* and all the remaining flowering plants.

### 29.4 How Do Plants Support Our World?

- Plants provide **ecosystem services** that affect soil, water, air quality, and climate.
- Plants are **primary producers** and as such are the foundation of the entire terrestrial food web.
- Plants provide many important medicinal products. Review Table 29.1

## SELF-QUIZ

1. Which of the following statements about seed plants is *true*?
   *a.* Seeds are produced only by flowering plants (angiosperms).
   *b.* The sporophyte generation is more reduced in seed plants than in the ferns.
   *c.* The gametophytes of seed plants are independent of the sporophytes.
   *d.* All seed plant species are heterosporous.
   *e.* The zygote of seed plants divides repeatedly to form the gametophyte.
2. The gymnosperms
   *a.* dominate all land masses today.
   *b.* have never dominated land masses.
   *c.* have secondary growth.
   *d.* all have vessel elements.
   *e.* lack sporangia.
3. Conifers
   *a.* produce ovules in microstrobili and pollen in megastrobili.
   *b.* depend on liquid water for fertilization.
   *c.* have triploid endosperm.
   *d.* have pollen tubes that release two sperm.
   *e.* have vessel elements.
4. Most angiosperms
   *a.* have seeds enclosed in a carpel.
   *b.* produce triploid endosperm by the union of two eggs and one sperm.
   *c.* lack secondary growth.
   *d.* bear two kinds of cones.
   *e.* lack flowers.
5. Which statement about flowers is *not* true?
   *a.* Pollen is produced in the anthers.
   *b.* Pollen is received on the stigma.
   *c.* An inflorescence is a cluster of flowers.
   *d.* A species having female and male flowers on the same plant is dioecious.
   *e.* A flower with both megasporangia and microsporangia is said to be perfect.
6. Which statement about fruits is *not* true?
   *a.* They develop from ovaries.
   *b.* They may include other parts of the flower.
   *c.* A multiple fruit develops from several carpels of a single flower.
   *d.* They are produced only by angiosperms.
   *e.* A cherry is a simple fruit.
7. Which statement is *not* true of angiosperm pollen?
   *a.* It is the male gamete.
   *b.* It is haploid.
   *c.* It produces a long tube.
   *d.* It interacts with the carpel.
   *e.* It is produced in microsporangia.
8. Which statement is *not* true of carpels?
   *a.* They are thought to have evolved from leaves.
   *b.* They bear megasporangia.
   *c.* They may fuse to form a pistil.
   *d.* They are floral organs.
   *e.* They are absent in perfect flowers.
9. *Amborella*
   *a.* was the first flowering plant.
   *b.* belongs to the first gymnosperm clade.
   *c.* is the sister group of all other living angiosperms.
   *d.* is a eudicot.
   *e.* has vessel elements in its xylem.
10. The eudicots
    *a.* include many herbs, vines, shrubs, and trees.
    *b.* and the monocots are the only extant angiosperm clades.
    *c.* are not a clade.
    *d.* include the magnolias.
    *e.* include orchids and palm trees.

## FOR DISCUSSION

1. In most seed plant species, only one of the products of meiosis in the megasporangium survives. How might this be advantageous?
2. Suggest an explanation for the great success of the angiosperms in occupying terrestrial habitats.
3. In many locales, large gymnosperms predominate over large angiosperms. Under what conditions might gymnosperms have the advantage, and why?
4. Not all flowers possess all of the following floral organs: sepals, petals, stamens, and carpels. Which floral organ or organs do you think might be found in the flowers that have the smallest number of floral organ types? Discuss the possibilities, both for a single flower and for a species.
5. The origin of the angiosperms has long been "an abominable mystery," as Charles Darwin once put it. The earliest known angiosperm fossils are from the late Jurassic, but fossils of their sister-group, the gymnosperms, are known from as early as the late Carboniferous (about 150 million years earlier). Given that these two sister-groups are thought to have arisen at the same time from a single split in the seed plant tree, what might explain the lack of earlier angiosperms fossils?

## ADDITIONAL INVESTIGATION

The flower of a particular species of orchid has a long, spurlike tube into which a pollinating insect can insert its proboscis to suck nectar. The spurs are of different lengths in different habitats, apparently correlated with the lengths of the probosces of the local pollinators. How might you test the hypothesis that this correlation increases reproductive success, in terms of pollen transfer to the flower?

# 30 Fungi: Recyclers, Pathogens, Parasites, and Plant Partners

## A fungus battles witchweed

More than 300 million Africans in 25 countries are suffering because their crops have been invaded by witchweed (*Striga*), a parasitic flowering plant. Witchweed has attacked more than two-thirds of the sorghum, corn, and millet crops in sub-Saharan Africa. Reduced crop yields cost an estimated U.S. $7 billion each year.

A team of Canadian scientists set out to find a biological solution to the *Striga* problem. Their strategy was to look for an organism that would destroy witchweed in the fields. They succeeded in isolating a strain of a fungus—the mold *Fusarium oxysporum*—that has several outstanding properties. First, it grows on *Striga* and kills a high percentage of these parasitic plants. Second, the fungus does not attack the crop plants that *Striga* parasitizes. And finally, *F. oxysporum* is not toxic to humans. In subsequent fieldwork, the scientists established specific techniques for applying *F. oxysporum* to witchweed. Farmers who apply the fungus to their crops are rewarded by greatly increased crop yields as *Striga* is held in check.

It may be possible to repeat the *Striga* story—the use of a fungus to wipe out a particular type of flowering plant—in a very different context. A different strain of *F. oxysporum* preferentially attacks coca plants, the source of cocaine. A controversial proposal to use *F. oxysporum* to wipe out the coca plantations in Andean South America and other parts of the world has been proposed; however, the specificity of the fungus to infect only coca is not clear. Establishing the degree of host specificity is crucial, because some naturally occurring strains of *F. oxysporum* attack important crops in various parts of the world. Introducing this plant pathogen on a widespread basis could have unintended consequences for many non-target plant species.

Fungi can also be used to battle animal pests. Research teams reported in 2005 that two fungi, *Beauveria bassiana* and *Metarhizium anisopliae*, killed malaria-carrying mosquitoes when applied to mosquito netting. Certain fungi are already used against other insect pests, notably termites and aphids. In

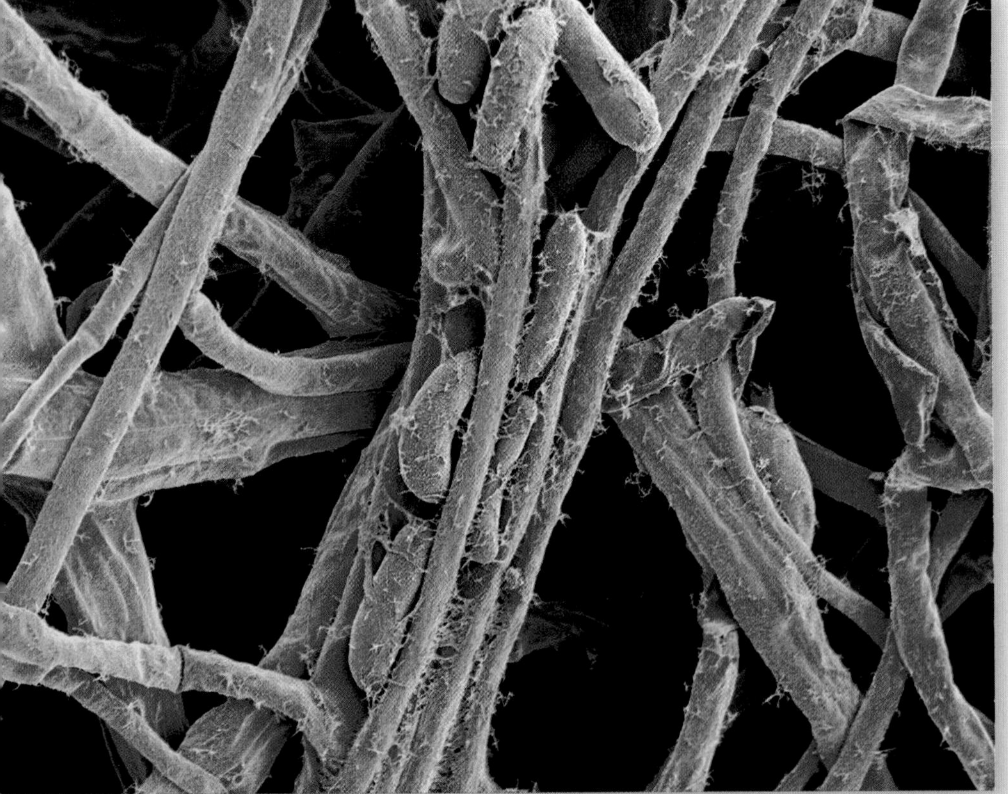

**Pathogenic Fungus, Parasitic Plant** The fungus *Fusarium oxysporum* is a potent pathogen of witchweed (*Striga*), a parasitic plant that attacks crops. The fungal spores are shown in blue; fungal filaments are in tan. Both colors were added to enhance this scanning electron micrograph.

**An Alien Meal** The tropical fungus whose fruiting body is growing on the stalk projecting from this ant's carcass developed internally in the host, from a spore ingested by the ant.

research conducted thus far, no insect pests have yet been known to develop resistance to fungi (as they can to DDT and other pesticides).

Of course, fungi don't need human help to find organisms to grow on. Fungal spores ingested by suitable animal hosts such as ants can germinate in the host's gut and develop into internal parasites. Absorbing nutrients from their unwitting hosts, some ingested fungi may eventually kill the host, producing new spores to infect new hosts as they do so.

Fungi interact with other organisms in many different ways, some of which are beneficial and some harmful to the other organisms. As we begin our study, recall that the fungi are more closely related to animals than to plants. That means molds and mushrooms are more closely related to you than they are to the plants discussed in Chapter 29.

**IN THIS CHAPTER** we will see that fungi differ from other eukaryotes in some very interesting ways. We will explore the diversity of body forms, reproductive structures, and life cycles that have evolved among six major groups of fungi. We will also examine the mutually beneficial associations of certain fungi with other organisms.

CHAPTER OUTLINE



## 30.1 What Is a Fungus?

Modern fungi are believed to have evolved from a unicellular protist ancestor that had a flagellum. The probable common ancestor of the animals was also a flagellated microbial eukaryote that may have been similar to the existing choanoflagellates (see Figure 27.27). Current evidence suggests that today's choanoflagellates, fungi, and animals share a common ancestor not shared by other eukaryotes, and thus the three lineages are often grouped together as the *opisthokonts*. Synapomorphies that distinguish the fungi among the opisthokonts include absorptive heterotrophy and the presence of chitin in their cell walls (**Figure 30.1**). Molecular sequences of many genes also support these relationships among the opisthokonts. Thus, fungi represent one of four large, independent evolutionary origins of multicellular organisms (plants, brown algae, and animals are the other three).

The fungi live by **absorptive heterotrophy**: they secrete digestive enzymes outside their bodies to break down large food molecules in the environment, then absorb the breakdown products through the plasma membranes of their cells. Absorptive heterotrophy is successful in virtually every conceivable environment. Many fungi are **saprobes**, which absorb nutrients from dead organic matter. Others are **parasites**, which absorb nutrients from living hosts (such as the ant shown at the opening of this chapter). Still others are **mutualists** living in intimate associations with other organisms that benefit both partners.

We will discuss six major groups of fungi (**Table 30.1**): microsporidia, chytrids, zygospore fungi (Zygomycota), arbuscular mycorrhizal fungi (Glomeromycota), sac fungi (Ascomycota), and club fungi (Basidiomycota). The chytrids and zygospore fungi are not thought to represent monophyletic groups, but instead consist of several distantly related lineages that retain some ancestral features. Nonetheless, these groupings represent convenient categories for a general introduction to the fungi. The clades that are thought to be monophyletic within these two paraphyletic groupings are listed in Table 30.1. Major fungal groups were originally defined by their methods and structures for sexual reproduction and also, to a lesser extent, by other morphological differences. More recently, evidence from DNA analyses has established the placement of microsporidia among the fungi, the paraphyly of chytrids and zygospore fungi, the independence of arbuscular mycorrhizal fungi from the other fungal groups, and the monophyly of sac fungi and club fungi (**Figure 30.2**). The chytrids are almost all aquatic, but the other groups are mostly terrestrial.

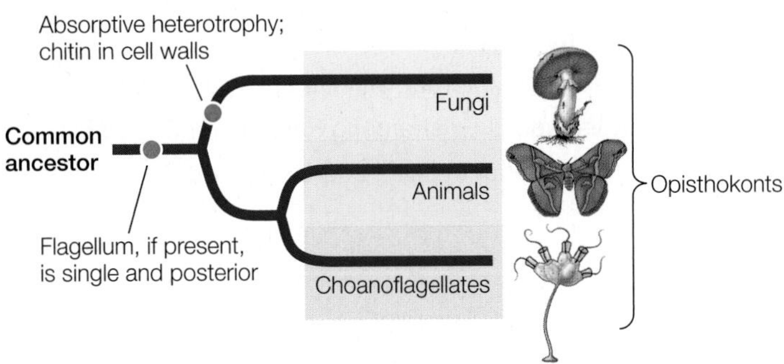


**30.1 Fungi in Evolutionary Context** Absorptive heterotrophy and the presence of chitin in their cell walls distinguish the fungi from other opisthokonts.

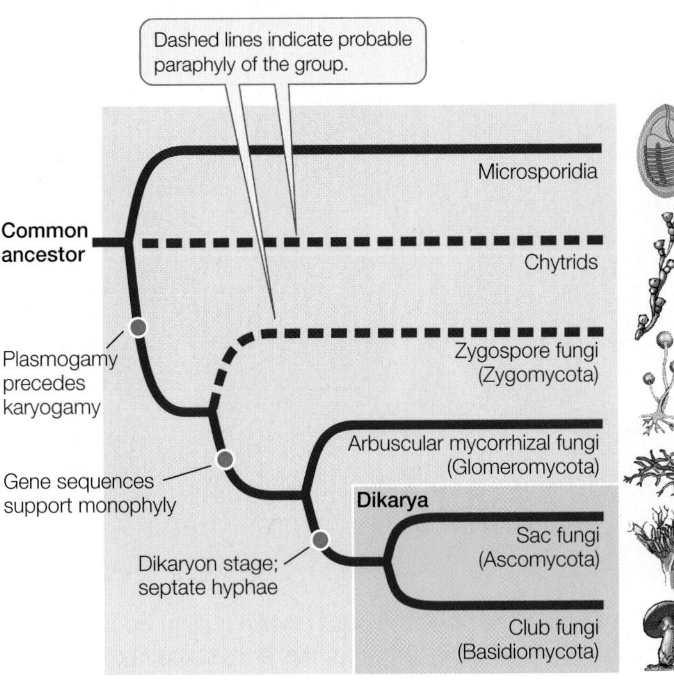


**30.2 Phylogeny of the Fungi** Microsporidia are reduced, parasitic fungi whose relationships among the fungi are uncertain. They may be the sister group of most other fungi or more closely related to particular groups of chytrids or zygospore fungi. The dashed lines indicate that chytrids and zygospore fungi are thought to be paraphyletic; the relationships of the lineages within these two informal groups (see Table 30.1) are not yet well resolved. The sac fungi and club fungi together form the clade Dikarya.

yourBioPortal.com
GO TO Web Activity 30.1 • Fungal Phylogeny

## Unicellular fungi are known as yeasts

Most fungi are multicellular, but single-celled species are found in most of the fungal groups. Unicellular forms of zygospore fungi, sac fungi, and club fungi are called **yeasts** (**Figure 30.3**). Some fungi that have yeast stages also include a filamentous stage. Yeasts live in liquid or moist environments and absorb nutrients directly across their cell surfaces. The term "yeast" does not refer to a single taxonomic group of organisms but rather to a lifestyle that has evolved multiple times. The ease with which many yeast species are cultured, combined with their rapid growth rates, has made them ideal model organisms for study in the laboratory. They present many of the same advantages to laboratory investigators as do many bacteria, but since they are eukaryotes, they have genomic structure and cells that are much more like those of humans and other eukaryotes.

TABLE 30.1

**Classification of the Fungi**

| GROUP | COMMON NAME | FEATURES |
|---|---|---|
| Microsporidia | Microsporidia | Intracellular parasites of animals; greatly reduced, among smallest eukayotes known; polar tube used to infect hosts |
| Chytrids (paraphyletic)[a]<br>Chytridiomycota<br>Neocallimastigomycota<br>Blastocladiomycota | Chytrids | Mostly aquatic and microscopic; zoospores have flagella |
| Zygomycota (paraphyletic)[a]<br>Entomophthoromycotina<br>Kickxellomycotina<br>Mucoromycotina<br>Zoopagomycotina | Zygospore fungi | Reproductive structure is a unicellular zygospore with many diploid nuclei in a zygosporangium; no regularly occurring septa; usually no fleshy fruiting body |
| Glomeromycota | Arbuscular mycorrhizal fungi | Form arbuscular mycorrhizae on plant roots; only asexual reproduction is known |
| Ascomycota | Sac fungi | Sexual reproductive saclike structure known as an ascus, which contains haploid ascospores; perforated septa; dikaryon |
| Basidiomycota | Club fungi | Sexual reproductive structure is a basidium, a swollen cell at the tip of a specialized hypha that supports haploid basidiospores; perforated septa; dikaryon |

[a]The formally named groups within the chytrids and Zygomycota are each thought to be monophyletic, but their relationships to one another (and to Microsporidia) are not yet well resolved.

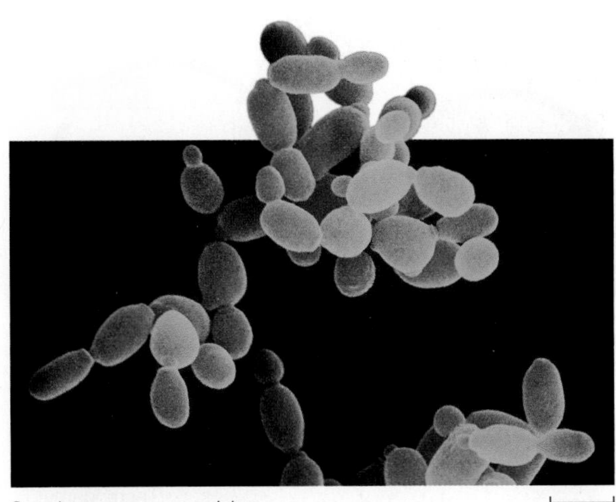
Saccharomyces cerevisiae
10 µm

**30.3 Yeasts Are Unicellular Fungi** Unicellular species of the zygospore fungi, sac fungi, and club fungi are known as yeasts. Many yeasts reproduce by budding—mitosis followed by asymmetrical cell division—as those shown here are doing.

### The body of a multicellular fungus is composed of hyphae

The body of a multicellular fungus is called a **mycelium** (plural *mycelia*). A mycelium is composed of a mass of individual tubular filaments called **hyphae** (singular *hypha*; **Figure 30.4A,B**). The cell walls of the hyphae are greatly strengthened by microscopic fibrils of *chitin*, a nitrogen-containing polysaccharide. In some species of fungi, the hyphae are subdivided into cell-like compartments by *incomplete* cross-walls called **septa** (singular *septum*); these hyphae are referred to as **septate**. Septa do not completely close off compartments in the hyphae. Gaps in the septa known as *pores* allow organelles—sometimes even nuclei—to move in a controlled way between compartments (**Figure 30.4C**). In other species of fungi, the hyphae lack septa but may contain hundreds of nuclei; these hyphae are referred to as **coenocytic**. The coenocytic condition results from repeated nuclear divisions without cytokinesis.

Certain modified hyphae, called *rhizoids*, anchor some fungi to their substratum (the dead organism or other matter on which they feed). These rhizoids are not homologous to the rhizoids of plants, and they are not specialized to absorb nutrients and water. Parasitic fungi, however, may possess modified hyphae that take up nutrients from their host.

The total hyphal growth of a fungal mycelium (not the growth of an individual hypha) may exceed 1 kilometer a day! The hyphae may be widely dispersed to forage for nutrients over a large area, or they may clump together in a cottony mass to exploit a rich nutrient source. In some members of certain fungal groups, when sexual spores are produced, portions of the mycelium become reorganized into a reproductive *fruiting body*, such as a mushroom. The mycelial mass is often far larger than the mushroom alone. The mycelium of one individual fungus in Michigan covers 15 hectares underground and weighs

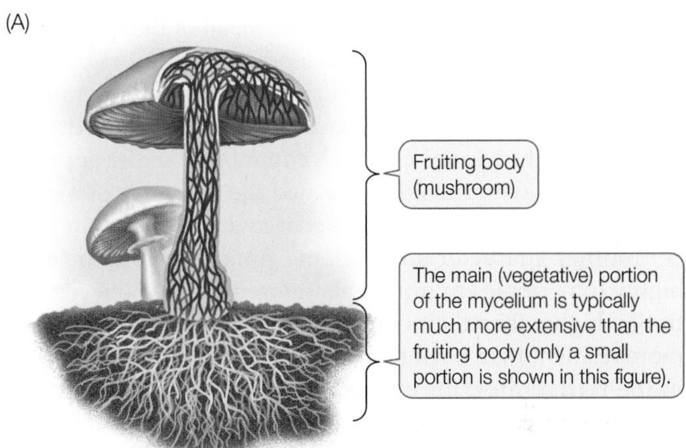
(A)
Fruiting body (mushroom)
The main (vegetative) portion of the mycelium is typically much more extensive than the fruiting body (only a small portion is shown in this figure).

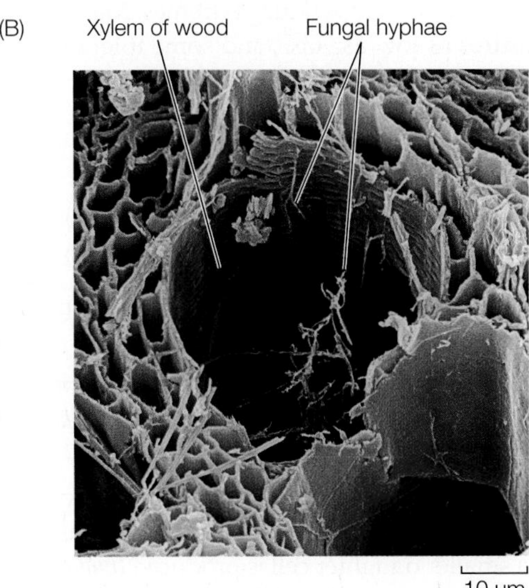
(B)
Xylem of wood
Fungal hyphae
10 µm

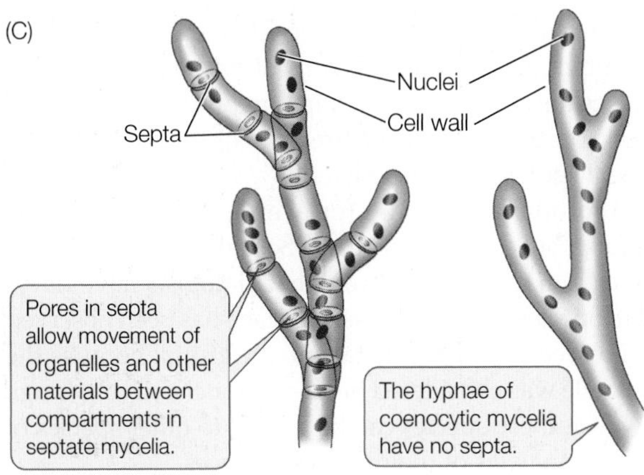
(C)
Nuclei
Cell wall
Septa
Pores in septa allow movement of organelles and other materials between compartments in septate mycelia.
The hyphae of coenocytic mycelia have no septa.

**30.4 Mycelia Are Made Up of Hyphae** (A) The fruiting body of a club fungus is transient, but the filamentous, nutrient-absorbing mycelium can be long-lived and cover large areas. (B) The minute individual hyphae of fungal mycelia can penetrate small spaces. In this artificially colored micrograph, hyphae (yellow structures) of a dry-rot fungus are penetrating the xylem cavities and other woody tissues of a log. (C) The hyphae of septate fungal species are divided into organelle-containing compartments by porous septa, while coenocytic hyphae have no septa.

more than a blue whale. Aboveground, this individual is evident only as isolated clumps of mushrooms.

### Fungi are in intimate contact with their environment

The filamentous hyphae of a fungus give it a unique relationship with its physical environment. The fungal mycelium has an enormous surface area-to-volume ratio compared with that of most large multicellular organisms. This large ratio is a marvelous adaptation for absorptive heterotrophy. Throughout the mycelium (except in fruiting structures), all of the hyphae are very close to their environmental food source.

The downside of the great surface area-to-volume ratio of the mycelium is its tendency to lose water rapidly in a dry environment. Thus fungi are most common in moist environments. You have probably observed the tendency of molds, toadstools, and other fungi to appear in damp places.

Another characteristic of some fungi is a tolerance for highly hypertonic environments (those with a solute concentration higher than their own; see Section 6.3). Many fungi are more resilient than bacteria in hypertonic surroundings. Jelly in the refrigerator, for example, will not become a growth medium for bacteria because it is too hypertonic to those organisms, but it may eventually harbor mold colonies. This presence of fungi in the refrigerator illustrates yet another trait of many fungi: tolerance of temperature extremes. Many fungi tolerate temperatures as low as –6°C, and some tolerate temperatures above 50°C.

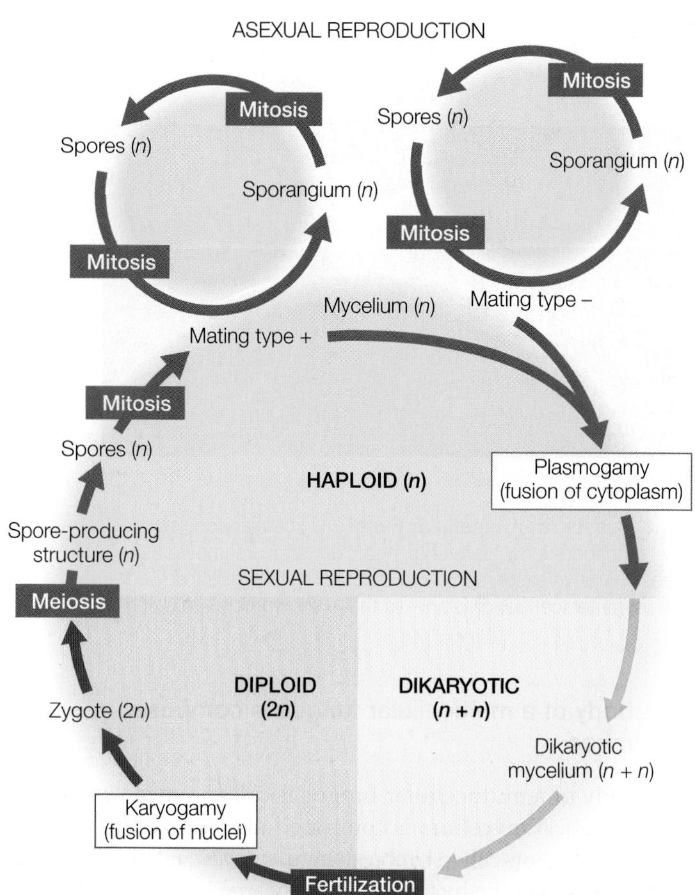


**30.5 Asexual and Sexual Reproduction in a Fungal Life Cycle** Environmental conditions may determine which mode of reproduction takes place at a given time. In the sexual phase, many fungi are characterized not by male and female individuals but by genetically distinct mating types (top of figure).

### Fungi reproduce both sexually and asexually

Both asexual and sexual reproduction occur among the fungi (**Figure 30.5**). Asexual reproduction takes several forms:

- The production of (usually) haploid spores within structures called *sporangia*
- The production of haploid spores (not enclosed in sporangia) at the tips of hyphae; such spores are called *conidia* (Greek *konis*, "dust")
- Cell division by unicellular fungi—either a relatively equal division of one cell into two (*fission*) or an asymmetrical division in which a smaller daughter cell is produced (*budding*)
- Simple breakage of the mycelium

Asexual reproduction in fungi can be spectacular in terms of spore quantity. A 2.5-centimeter colony of *Penicillium*, the mold that produces the antibiotic penicillin, can produce as many as 400 million conidia. The air we breathe contains as many as 10,000 fungal spores per cubic meter.

Sexual reproduction is rare (or even unknown) in some groups of fungi but common in others. Sexual reproduction may not occur, or it may occur so rarely that biologists have never observed it. Species in which no sexual stage has been observed were once placed in a separate taxonomic group, because the sexual life cycle was considered necessary for classifying fungi. Now, however, these species can be related to other species of fungi through analysis of their DNA sequences.

When it does occur, sexual reproduction in many fungi features an interesting twist. There is often no morphological distinction between female and male structures, or between female and male individuals. Rather, there is a genetically determined distinction between two *or more* **mating types**. Individuals of the same mating type cannot mate with one another, but they can mate with individuals of another mating type within the same species, thus preventing self-fertilization. Individuals of different mating types differ genetically but are often visually and behaviorally indistinguishable. Many protists also have mating type systems.

Fungi reproduce sexually when hyphae (or, in one fungal group, motile cells) of different mating types meet and fuse. In many fungi, the zygote nuclei formed by sexual reproduction are the only diploid nuclei in the life cycle. These nuclei undergo meiosis, producing haploid nuclei that become incorporated into spores. Haploid fungal spores, whether produced sexually in this manner or asexually, germinate, and their nuclei divide mitotically to produce haploid hyphae.

We will discuss reproduction in fungi in more detail in Section 30.3.

## 30.1 RECAP

**The rapid growth and large surface area-to-volume ratio of fungal hyphae allow fungi to practice absorptive heterotrophy efficiently in moist environments. Both asexual and sexual reproduction occur in fungi; sexual reproduction occurs between different mating types.**

- Describe the relationship between fungal structure and absorptive heterotrophy. **See p. 627**
- Explain the concept of mating types. How is this different from male and female sexes? **See p. 630**

Fungi are important components of healthy ecosystems. Fungi interact with other organisms in many ways, some of which are harmful and some beneficial to other organisms.

# 30.2 How Do Fungi Interact with Other Organisms?

Without the fungi, our planet would be very different. Picture Earth with only a few stunted plants and watery environments choked with the remains of dead organisms. Fungi do much of Earth's garbage disposal. Fungi that absorb nutrients from dead organisms not only help clean up the landscape and form soil but also play a key role in recycling mineral elements. Furthermore, the colonization of the terrestrial environment was made possible in large part by associations fungi formed with other organisms.

### Saprobic fungi are critical to the planetary carbon cycle

Saprobic fungi (those that consume nonliving organic matter), along with bacteria, are the major decomposers on Earth, contributing to decay and thus to recycling the elements used by living things. In forests, for example, the mycelia of fungi absorb nutrients from fallen trees, thus decomposing their wood. Fungi are the principal decomposers of cellulose and lignin, the main components of plant cell walls (most bacteria cannot break down these materials). Other fungi produce enzymes that decompose keratin and thus break down animal structures such as hair and nails.

Were it not for the fungal decomposers, Earth's carbon cycle would fail: great quantities of carbon atoms would remain trapped forever on forest floors and elsewhere (see Chapter 58). Instead, those carbon atoms are returned to the atmosphere in the form of respiratory $CO_2$, available for photosynthesis by plants.

In fact, there was a time when populations of saprobic fungi declined significantly. During the Carboniferous period, plants in the vast tropical swamps died and began to form peat (see Section 28.4). Peat formation led to acidification of the swamps; that acidity, in turn, drastically reduced the fungal population. The result? With the decomposers largely absent, large quantities of peat remained on the swamp floor and over time were converted into coal.

In contrast to their decline during the Carboniferous, fungi did very well at the end of the Permian, a quarter of a billion years ago, when the aggregation of continents produced volcanic eruptions that triggered a planetwide extinction event of many other organisms (see Chapter 25). The fossil record shows that even though 96 percent of all species became extinct, fungi flourished, demonstrating both their hardiness and their role in recycling the elements in the dead bodies of plants and animals.

Many saprobic fungi can be grown on artificial media, and it is relatively easy to perform experiments to determine their exact nutritional requirements. Sugars are the favored source of carbon for saprobic fungi. Most fungi obtain nitrogen from proteins or the products of protein breakdown. Many fungi can use nitrate ($NO_3^-$) or ammonium ($NH_4^+$) ions as their sole source of nitrogen. No known fungus can get its nitrogen directly from inorganic nitrogen gas as can some bacteria and plant–bacteria associations (that is, fungi cannot "fix" nitrogen; see Section 36.4). Nutritional studies also reveal that most fungi are unable to synthesize certain vitamins and must absorb them from their environment. However, fungi can synthesize some vitamins that animals cannot. Like all organisms, fungi also require some mineral elements.

What happens when a fungus faces a dwindling food supply? A common strategy is to reproduce rapidly and abundantly. When conditions are good, fungi produce great quantities of spores, but the rate of spore production is commonly even higher when nutrient supplies go down. The spores may then remain dormant until conditions improve, or may be dispersed to areas where nutrient supplies are higher.

Not only are fungal spores abundant in number, but they are extremely tiny and easily spread by wind or water (**Figure 30.6**). This virtually assures that the individual that produced them

*Lycoperdon pyriforme*

**30.6 Spores Galore** Puffballs (a type of club fungus) disperse trillions of spores in great bursts. Few of the spores travel very far, however; some 99 percent of them fall within 100 meters of the parent puffball.

will have many progeny, which may be scattered over great distances. No wonder we find fungi just about everywhere.

### Fungi may engage in parasitic and predatory interactions

Whereas saprobic fungi obtain their energy, carbon, and nitrogen directly from dead organic matter, other species of fungi obtain their nutrition from parasitic—and even predatory—interactions.

**PARASITIC FUNGI** Biologists distinguish between two classes of parasitic fungi, based on the degree of dependence on their host species. *Facultative* parasites can attack living organisms but can also grow by themselves, including on artificial media. *Obligate* parasites can grow only on their specific living host, usually a plant species. Because their growth depends on a living host, obligate parasites have specialized nutritional requirements.

The filamentous structure of fungal hyphae is especially well suited to a life of absorbing nutrients from plants. The slender hyphae of a parasitic fungus can invade a plant through stomata, through wounds, or in some cases, by direct penetration of epidermal cell walls (**Figure 30.7A**). Once inside the plant, the hyphae branch out to expand the mycelium. Some hyphae produce **haustoria**, branching projections that push through cell walls into living plant cells, absorbing the nutrients within those cells. The haustoria do not break through the plant cell plasma membranes inside the cell walls; they simply invaginate into the membranes, with the plasma membrane fitting them like a glove (**Figure 30.7B**). Fruiting structures may form, either within the plant body or on its surface in a symbiotic relationship that is usually not lethal to the plant. Some parasitic fungi, however, are *pathogenic*, sickening or even killing the hosts from which they derive nutrition.

**PATHOGENIC FUNGI** Although most human diseases are caused by bacteria or viruses, fungal pathogens are a major cause of death among people with compromised immune systems. Most people with AIDS die of fungal diseases, such as the pneumonia caused by *Pneumocystis jirovecii* or incurable diarrhea caused by other fungi. *Candida albicans* and certain other yeasts also cause severe diseases, such as esophagitis (which impairs swallowing), in individuals with AIDS and in individuals taking immunosuppressive drugs. Fungal diseases are a growing international health problem, requiring vigorous research. Our limited understanding of the basic biology of these fungi still hampers our ability to treat the diseases they cause. Various fungi cause other, less threatening human diseases, such as ringworm and athlete's foot.

The worldwide decline of amphibian species has been linked to the spread of a chytrid fungus, *Batrachochytrium dendrobatidis*. Genetic analyses indicate that the fungus populations attacking amphibian populations around the world are genetically almost identical, which suggests a recent introduction of the fungus across the globe. This chytrid appears to be endemic to southern Africa, and its spread around the world may have initiated in the 1930s with exports of the African clawed frog (*Xenopus laevis*), which was once widely used in human pregnancy tests.

Fungi are by far the most important plant pathogens, causing crop losses amounting to billions of dollars. Bacteria and viruses are less important as plant pathogens. Major fungal diseases of crop plants include black stem rust of wheat and other diseases of wheat, corn, and oats. The agent of black stem rust is *Puccinia graminis*, which has a complicated life cycle that involves two plant hosts (wheat and barberry). In an epidemic in 1935, *P. graminis* was responsible for the loss of about one-fourth of the entire wheat crop in Canada and the United States. However, as we saw at the beginning of this chapter, pathogenic fungi that kill certain weed species can be a boon to agriculture.

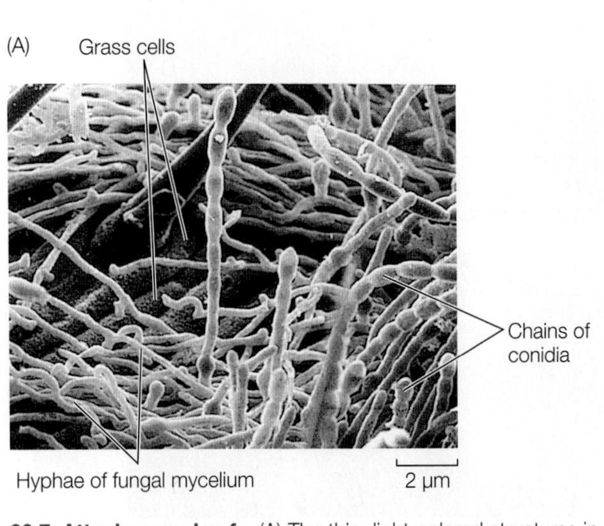
(A)
Grass cells
Chains of conidia
Hyphae of fungal mycelium
2 μm

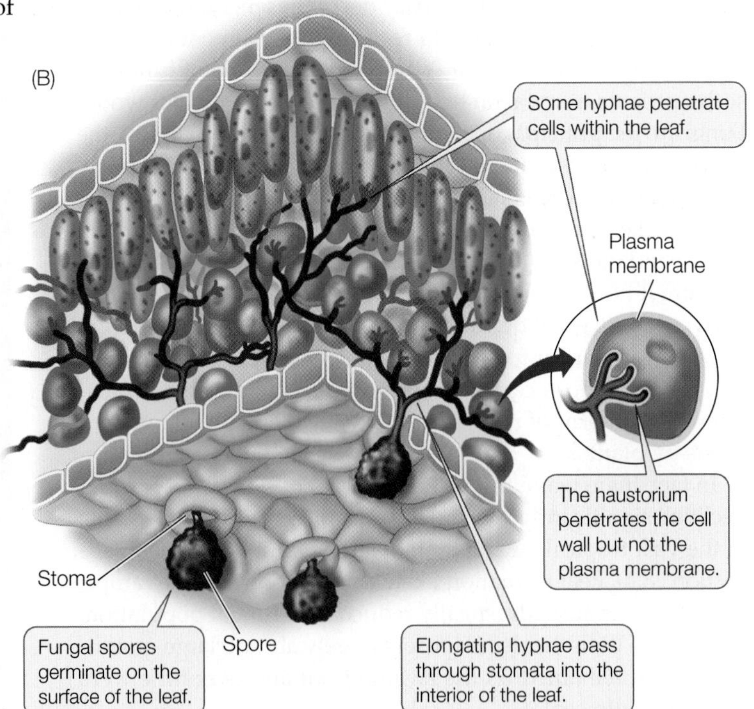
(B)
Some hyphae penetrate cells within the leaf.
Plasma membrane
The haustorium penetrates the cell wall but not the plasma membrane.
Stoma
Fungal spores germinate on the surface of the leaf.
Spore
Elongating hyphae pass through stomata into the interior of the leaf.

**30.7 Attacks on a Leaf** (A) The thin, light-colored structures in the micrograph are hyphae of the parasitic fungus *Blumeria graminis* growing on the dark surface of the leaf of a grass. (B) Haustoria are fungal hyphae that push into the living cells of plants, from which they absorb nutrients.

Nematode
Fungal hyphae
20 µm

**30.8 Some Fungi Are Predators** A nematode is trapped by hyphal rings of the soil-dwelling fungus *Arthrobotrys dactyloides*.

**PREDATORY FUNGI** Some fungi have adaptations that enable them to function as active predators, trapping nearby microscopic protists or animals. The most common predatory strategy seen in fungi is to secrete sticky substances from the hyphae so that passing organisms stick tightly to them. The hyphae then quickly invade the prey, growing and branching within it, spreading through its body, absorbing nutrients, and eventually killing it.

A more dramatic adaptation for predation is the constricting ring formed by some species of *Arthrobotrys*, *Dactylaria*, and *Dactylella* (**Figure 30.8**). All of these fungi grow in soil. When nematodes (tiny roundworms) are present in the soil, these fungi form three-celled rings with a diameter that just fits a nematode. A nematode crawling through one of these rings stimulates the fungus, causing the cells of the ring to swell and trap the worm. Fungal hyphae quickly invade and digest the unlucky victim.

### Some fungi engage in relationships beneficial to both partners

Certain kinds of relationships between fungi and other organisms have nutritional consequences for both partners. Two of these relationships are highly specific and are **symbiotic** (the partners live in close, permanent contact with one another) as well as **mutualistic** (the relationship benefits both partners; see Chapter 56).

**Lichens** are associations of a fungus with a cyanobacterium, a unicellular photosynthetic alga, or both. **Mycorrhizae** (singular *mycorrhiza*) are associations between fungi and the roots of plants. In these associations, the fungus obtains organic compounds from its photosynthetic partner and provides it with minerals and water in return, so that the partner's nutrition is also promoted. In fact, many plants grow very poorly without their fungal partners.

**LICHENS** A lichen is not a single organism but rather a meshwork of two radically different organisms: a fungus and a photosynthetic microorganism. Together the organisms constituting a lichen can survive some of the harshest environments on Earth. The biota of Antarctica, for example, features more than a hundred times as many species of lichens as of plants.

In spite of their hardiness, lichens are highly sensitive to air pollution because they are unable to excrete any toxic substances they absorb. This sensitivity means that lichens are good biological indicators of air pollution levels. It also explains why they are not commonly found in heavily industrialized regions or large cities.

The fungal components of most lichens are sac fungi (Ascomycota), which also include various cup fungi, yeasts, and molds such as the *Fusarium* mentioned at the beginning of the chapter. The photosynthetic component of a lichen is most often a unicellular green alga, but it can be a cyanobacterium, or can include both. Relatively little experimental work has focused on lichens, perhaps because they grow so slowly—typically less than 1 centimeter in a year.

There are nearly 30,000 described "species" of lichens, each of which is assigned the name of its fungal component. These fungal components may constitute as many as 20 percent of all fungal species. Some of these fungi are able to grow independently without a photosynthetic partner, but most have never been observed in nature other than in a lichen association.

Lichens are found in all sorts of exposed habitats: on tree bark, on open soil, and on bare rock. Reindeer moss (not a moss at all, but the lichen *Cladonia subtenuis*) covers vast areas in Arctic, sub-Arctic, and boreal regions, where it is an important part of the diets of reindeer and other large mammals. Lichens come in various forms and colors. *Crustose* (crustlike) lichens look like colored powder dusted over their substratum (**Figure 30.9A**); *foliose* (leafy) and *fruticose* (shrubby) lichens may have complex forms (**Figure 30.9B,C**).

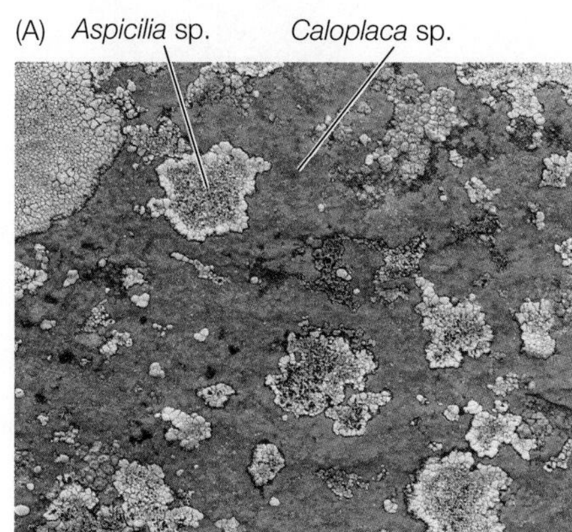


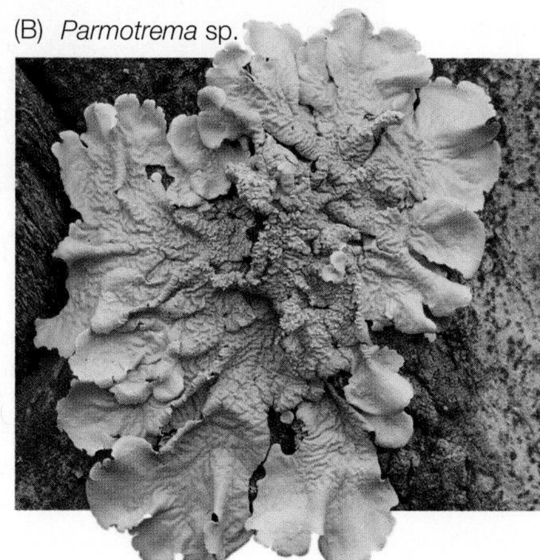

**30.9 Lichen Body Forms** Lichens fall into three principal classes based on their body form. (A) Two crustose lichens are growing on the surface of exposed rock. (B) Foliose lichens have a leafy appearance. (C) The brown and orange growth is of "shrubby" fruticose lichens.

The most widely held interpretation of the lichen relationship is that it is a mutually beneficial symbiosis (a *mutualism*). The hyphae of the fungal mycelium are tightly pressed against the algal or cyanobacterial cells and sometimes even invade them without breaching the plasma membrane (as we described earlier for haustoria in parasitic fungi of plants; see Figure 30.7). The bacterial or algal cells not only survive these indignities but continue their growth and photosynthesis. In fact, the algal cells in a lichen "leak" photosynthetic products at a greater rate than do similar cells growing on their own, and photosynthetic cells from lichens grow more rapidly on their own than when associated with a fungus. On this basis, we could consider lichen fungi to be parasitic on their photosynthetic partners. In many places where lichens grow, however, the photosynthetic cells would not grow at all on their own.

Lichens can reproduce simply by fragmentation of the vegetative body (the *thallus*), or by means of specialized structures called *soredia* (singular *soredium*). Soredia consist of one or a few photosynthetic cells bound by fungal hyphae. The soredia become detached from the lichen, are dispersed by air currents, and upon arriving at a favorable location, develop into a new lichen thallus. Alternatively, the fungal partner may go through its sexual cycle, producing haploid spores. When these spores are discharged, however, they disperse alone, unaccompanied by the photosynthetic partner.

Visible in a cross section of a typical foliose lichen are a tight upper region of fungal hyphae, a layer of photosynthetic cyanobacteria or algae, a looser hyphal layer, and finally hyphal rhizoids that attach the entire structure to its substratum (**Figure 30.10**). The meshwork of fungal hyphae takes up some nutrients needed by the photosynthetic cells and provides a suitably moist environment for them by holding water tenaciously. The fungi derive fixed carbon from the photosynthetic products of the algal or cyanobacterial cells.

A **soredium** consists of one or a few photosynthetic cells surrounded by fungal hyphae.
Soredia detach from the parent lichen and travel in air currents, founding new lichens when they settle in a suitable environment.
Upper layer of hyphae
Photosynthetic cell layer
Loose layer of hyphae
Lower level of hyphal rhizoids

**30.10 Lichen Anatomy** Cross section showing the layers of a foliose lichen and the release of soredia.

Lichens are often the first colonists on new areas of bare rock. They get most of the nutrients they need from the air and rainwater, augmented by minerals absorbed from dust. A lichen begins to grow shortly after a rain, as it begins to dry. As it grows, the lichen acidifies its environment slightly, and this acidity contributes to the slow breakdown of rocks, an early step in soil formation. After further drying, the lichen's photosynthesis ceases. The water content of the lichen may drop to less than 10 percent of its dry weight, at which point it becomes highly insensitive to extremes of temperature.

**MYCORRHIZAE** Many vascular plants depend on a symbiotic association with fungi. Unassisted, the root hairs of such plants often do not take up enough water or minerals to sustain growth. However, their roots usually do become infected with fungi, forming an association called a mycorrhiza. Mycorrhizae are of two types, based on whether or not the fungal hyphae penetrate the plant cell walls.

In *ectomycorrhizae*, the fungus wraps around the root, and its mass is often as great as that of the root itself (**Figure 30.11A**). The fungal hyphae wrap around individual cells in the root but do not penetrate the cell walls. An extensive web of hyphae penetrates the soil in the area around the root, so that up to 25 percent of the soil volume near the root may be fungal hyphae. The hyphae attached to the root increase the surface area for the absorption of water and minerals, and the mass of the mycorrhiza in the soil, like a sponge, holds water efficiently in the neighborhood of the root. Infected roots characteristically are short, swollen, and club-shaped, and they lack root hairs.

The fungal hyphae of *arbuscular mycorrhizae* enter the root and penetrate the cell wall of the root cells, forming arbuscular (treelike) structures inside the cell wall but outside the plasma membrane. These structures, like the haustoria of par-

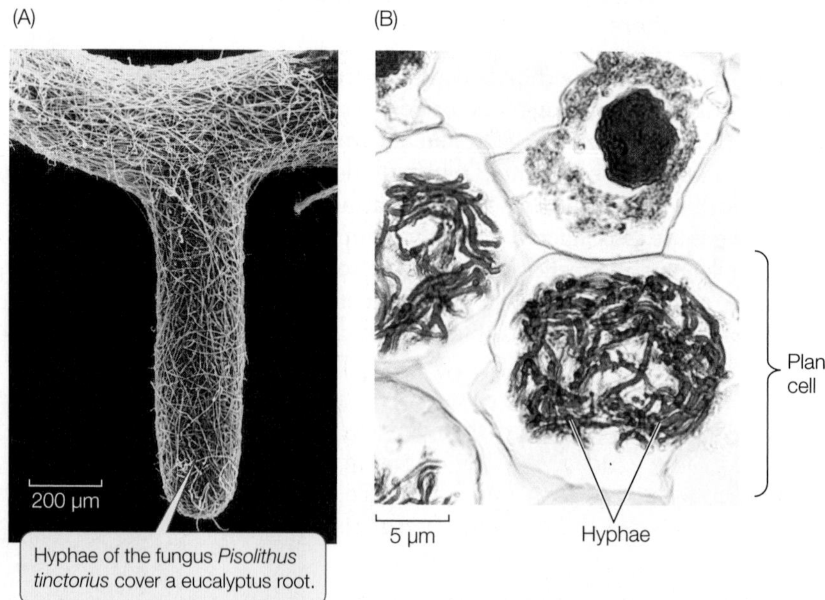


**30.11 Mycorrhizal Associations** (A) Ectomycorrhizal fungi wrap themselves around a plant root, increasing the area available for absorption of water and minerals. (B) Hyphae of arbuscular mycorrhizal fungi infect the root internally and penetrate the root cell walls, branching within the cells and forming treelike (arbuscular) structures that provide the plant with nutrients. Hyphae fill much of the cell outside the nucleus and invaginate the plasma membrane without puncturing it.

asitic fungi and the contact regions of fungi and algal cells in lichens, become the primary site of exchange between plant and fungus (**Figure 30.11B**). As in the ectomycorrhizae, the fungus forms a vast web of hyphae leading from the root surface into the surrounding soil.

The mycorrhizal association is important to both partners. The fungus obtains needed organic compounds, such as sugars and amino acids, from the plant. In return, the fungus, because of its very high surface area-to-volume ratio and its ability to penetrate the fine structure of the soil, greatly increases the plant's ability to absorb water and minerals (especially phosphorus). The fungus may also provide the plant with certain growth hormones and may protect it against attack by disease-causing microorganisms. Plants that have active arbuscular mycorrhizae typically are a deeper green and may resist drought and temperature extremes better than plants of the same species that have little mycorrhizal development. Attempts to introduce some plant species to new areas have failed until a bit of soil from the native area (presumably containing the fungus necessary to establish mycorrhizae) was provided. Trees without ectomycorrhizae will not grow well in the absence of abundant nutrients and water, so the health of our forests depends on the presence of ectomycorrhizal fungi.

The partnership between plant and fungus results in a plant that is better adapted for life on land. It has been suggested that the evolution of mycorrhizae was the single most important step in the colonization of the terrestrial environment by living things. Fossils of mycorrhizal structures 460 million years old have been found. Some liverworts, which are representatives of one of the oldest lineages of terrestrial plants (see Section 28.4), form mycorrhizal associations with fungi.

Certain plants that live in nitrogen-poor habitats, such as cranberry bushes and orchids, invariably have mycorrhizae. Orchid seeds will not germinate in nature unless they are already infected by the fungus that will form their mycorrhizae. Plants that lack chlorophyll always have mycorrhizae, which they often share with the roots of green, photosynthetic plants. In effect, these plants without chlorophyll are feeding on nearby green plants, using the fungus as a bridge.

Biologists had long suspected that roots secrete a chemical signal that enables fungi to find and invade them to form arbuscular mycorrhizae. This was proved to be correct in 2005 when researchers succeeded in isolating the signaling compound. Might the compound also be used by parasitic plants to attack their host plants? Indeed it is. *Striga*, discussed at the beginning of this chapter, turns out to be one of the parasitic plants that use exactly this signal. Thus, in attracting its helper fungus, a plant may also attract a dangerous parasite. The *Striga* story is continued at the end of Chapter 36.

### Endophytic fungi protect some plants from pathogens, herbivores, and stress

In a tropical rainforest, 10,000 or more fungal spores land on a single leaf each day. Some are plant pathogens, some do not attack the plant at all, and some invade the plant in a beneficial way. Fungi that live within aboveground parts of plants without causing obvious deleterious symptoms are called **endophytic fungi**. Recent research has shown that endophytic fungi are abundant in plants in all terrestrial environments.

Grasses with endophytic fungi are more resistant to pathogens and to insect and mammalian herbivores than are grasses lacking endophytes. The fungi produce alkaloids (nitrogen-containing compounds) that are toxic to animals. The alkaloids do not harm the host plant; in fact, some plants produce alkaloids (such as nicotine) themselves. The fungal alkaloids also increase the ability of host plants to resist stress of various types, including drought (water shortage) and salty soils. Such resistance is useful in agriculture.

The role, if any, of endophytic fungi in most broad-leafed plants is unclear, however. They may convey protection against pathogens or simply occupy space within leaves, without conferring any benefit but also without doing harm. The benefit, in fact, might be all for the fungus.

### 30.2 RECAP

**Fungi interact with other organisms in many ways, both harmful and beneficial. Lichens are mutualistic associations of a fungus with an alga and/or a cyanobacterium. Mycorrhizae are associations of fungi and the roots of plants; they are essential for the survival of most plant species.**

- What is the role of fungi in Earth's carbon cycle? See p. 631
- Describe the nature and benefits of the lichen association. See pp. 633–635
- Why do plants grow better in the presence of mycorrhizal fungi? See p. 635

One of the most important criteria for assigning fungi to taxonomic groups, before molecular techniques clarified phylogenetic relationships among fungi, was the nature of their life cycles.

## 30.3 What Variations Exist among Fungal Life Cycles?

Different fungal groups have different life cycles. Some chytrids feature alternation of generations, a type of life cycle found in all plants and some protists. The life cycles of sac fungi and club fungi feature a unique stage called a *dikaryon*, in which a single hypha has two genetically distinct nuclei. In this section we examine the diverse life cycles of major groups of fungi.

### Alternation of generations is seen among some aquatic chytrids

The alternation between multicellular haploid ($n$) and multicellular diploid ($2n$) generations that evolved in plants and certain protist groups (see Section 28.2) is seen in some chytrids as well (**Figure 30.12A**). Alternation of generations is not usual for the life cycles of other fungal groups. The basal chytrids, which are

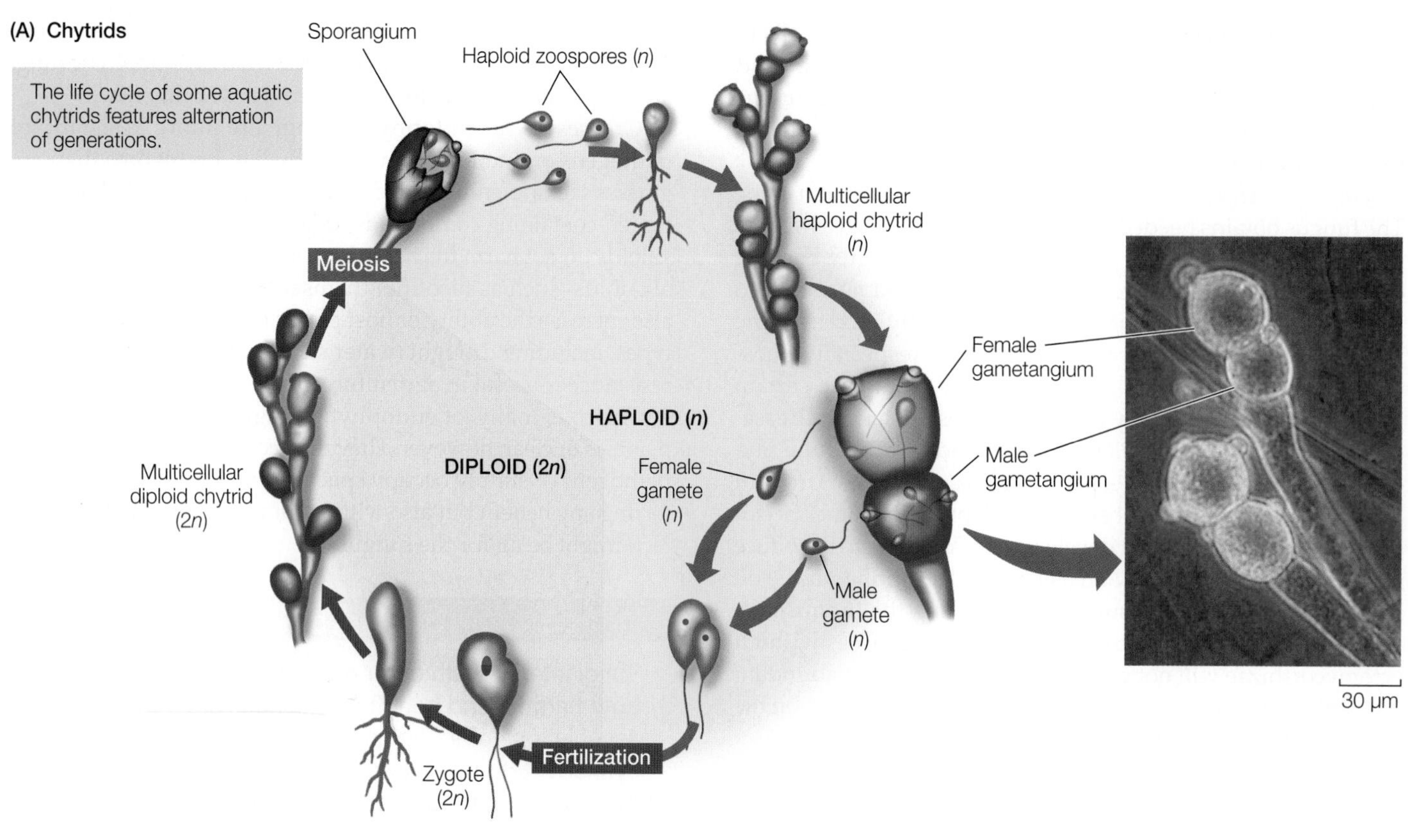

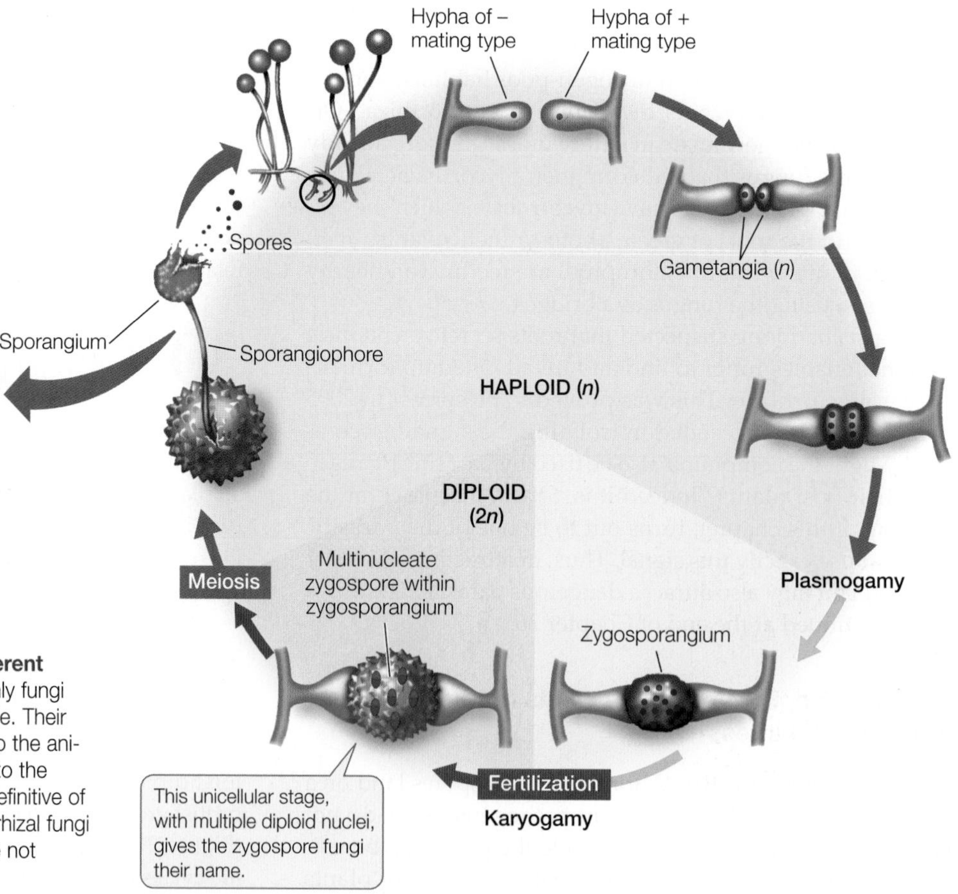

**30.12 Sexual Life Cycles Vary among Different Groups of Fungi** (A) The chytrids are the only fungi that posses flagella at any stage of the life cycle. Their flagellated gametes and zoospores link them to the animals. (B) A multinucleate zygospore is unique to the zygospore fungi. (C,D) The dikaryon stage is definitive of the sac fungi (C) and the club fungi (D). Mycorrhizal fungi are only known to reproduce asexually and are not depicted here.

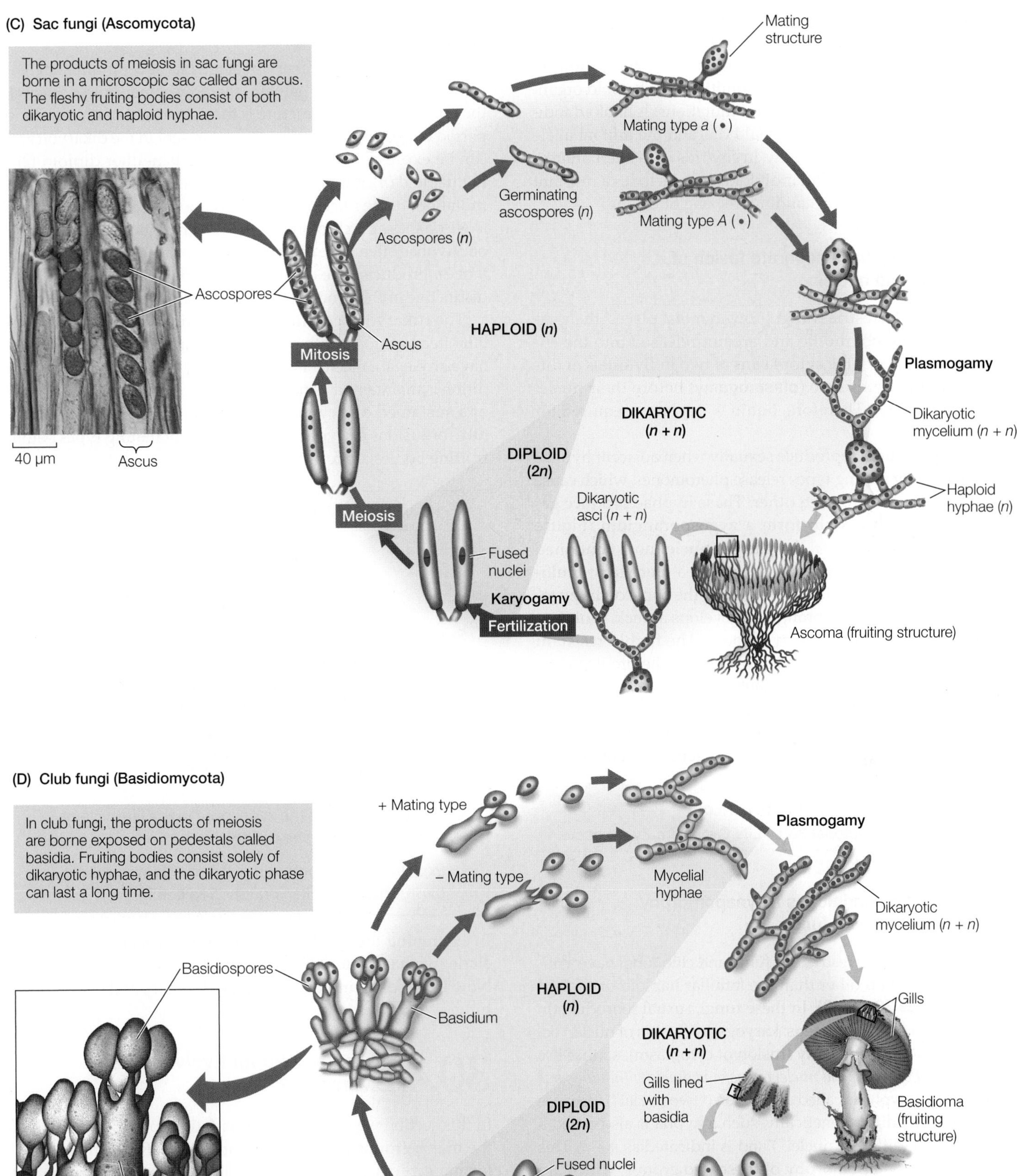

(C) Sac fungi (Ascomycota)
The products of meiosis in sac fungi are borne in a microscopic sac called an ascus. The fleshy fruiting bodies consist of both dikaryotic and haploid hyphae.
Mating structure
Mating type a (•)
Germinating ascospores (n)
Mating type A (•)
Ascospores (n)
Ascospores
Ascus
Mitosis
HAPLOID (n)
Plasmogamy
Dikaryotic mycelium (n + n)
DIKARYOTIC (n + n)
DIPLOID (2n)
Haploid hyphae (n)
40 µm
Ascus
Dikaryotic asci (n + n)
Meiosis
Fused nuclei
Karyogamy
Fertilization
Ascoma (fruiting structure)
(D) Club fungi (Basidiomycota)
In club fungi, the products of meiosis are borne exposed on pedestals called basidia. Fruiting bodies consist solely of dikaryotic hyphae, and the dikaryotic phase can last a long time.
+ Mating type
Plasmogamy
– Mating type
Mycelial hyphae
Dikaryotic mycelium (n + n)
Basidiospores
HAPLOID (n)
Basidium
Gills
DIKARYOTIC (n + n)
Gills lined with basidia
DIPLOID (2n)
Basidioma (fruiting structure)
Fused nuclei
Meiosis
Developing basidium (n + n)
Nuclei
50 µm
Basidium
Karyogamy
Fertilization

aquatic, possess flagellated gametes and flagellated spores. Flagella have been lost in the terrestrial fungi.

What are the consequences of alternation of generations in the chytrids? It is possible that the multicellular haploid organisms serve as a "filter" for harmful mutations. A haploid individual with such a mutation would die, and the mutant allele would not be passed to progeny. In chytrids with alternation of generations, the multicellular diploid stage includes a resistant structure capable of withstanding freezing or drying.

### Terrestrial fungi have separate fusion of cytoplasms and nuclei

Although the terrestrial fungi grow in moist places, their gamete nuclei are not motile and are not released into the environment. Instead, the cytoplasms of two individuals of different mating types fuse (**plasmogamy**) before their nuclei fuse (**karyogamy**). Therefore, liquid water is not required for fertilization.

Zygospore fungi reproduce sexually when adjacent hyphae of two different mating types release pheromones, which cause them to grow toward each other. These hyphae produce gametangia, which fuse to form a zygosporangium (**Figure 30.12B**). Sometime later, the gamete nuclei now contained within the zygosporangium fuse to form a unicellular multinucleate **zygospore**, which is the basis of the name of the zygospore fungi. The zygosporangium develops a thick, multilayered wall that protects the zygospore. The highly resistant zygospore may remain dormant for months before its nuclei undergo meiosis and a *sporangiophore* sprouts, bearing a *sporangium*. The sporangium contains the products of meiosis: haploid nuclei that are incorporated into spores. These spores disperse and germinate to form a new generation of haploid hyphae.

yourBioPortal.com
GO TO Animated Tutorial 30.1 • Life Cycle of a Zygomycete

### The dikaryotic condition is a synapomorphy of sac fungi and club fungi

Certain hyphae of terrestrial sac fungi and club fungi have a nuclear configuration other than the familiar haploid or diploid states (**Figures 30.12C,D**). In these fungi, sexual reproduction begins in two distinctive steps: karyogamy (fusion of nuclei) occurs long after plasmogamy (fusion of cytoplasm), so that *two genetically different haploid nuclei coexist and divide within the same hypha*. Such a hypha is called a **dikaryon** ("two nuclei"). Because the two nuclei differ genetically, such a hypha is also called a *heterokaryon* ("different nuclei") and is indicated as $n + n$. This dikaryon is a synapomorphy of these two groups, which are placed together in the clade called Dikarya.

Eventually, specialized fruiting structures form, within which the pairs of genetically dissimilar nuclei—one from each parent—fuse, giving rise to zygotes long after the original "mating." The diploid zygote nucleus undergoes meiosis, producing four haploid nuclei. The mitotic descendants of those nuclei become spores, which germinate to give rise to the next generation—usually of hyphae.

A life cycle with a dikaryon stage has several unusual features. First, there are no gamete *cells*, only gamete *nuclei*. Second, the only true diploid structure is the zygote, although for a long period the genes of both parents are present in the dikaryon and can be expressed. In effect, the hypha is neither diploid ($2n$) nor haploid ($n$); rather, it is *dikaryotic* ($n + n$). A harmful recessive mutation in one nucleus may be compensated for by a normal allele on the same chromosome in the other nucleus, and dikaryotic hyphae often have characteristics that are different from their $n$ or $2n$ products. The dikaryotic condition is perhaps the most distinctive of the genetic peculiarities of the fungi.

The dikaryotic condition is short-lived in most sac fungi but often lasts for months or even years in club fungi. Club fungi have an elegant mechanism that ensures that the dikaryotic condition is maintained as new cells are formed. One consequence of a sustained dikaryotic condition is an increased opportunity for fusions of hyphae of two different mating types before fruiting bodies are formed.

yourBioPortal.com
GO TO Web Activity 30.2 • Life Cycle of a Dikaryotic Fungus

### 30.3 RECAP

**Some chytrid fungi exhibit alternation of generations between haploid and diploid multicellular states. The life cycle of zygospore fungi includes a resistant-spore stage with many diploid nuclei. The sac fungi and club fungi share a derived dikaryotic condition, in which two genetically different haploid nuclei coexist in the same cell.**

- What is the role of the zygospore in the life cycle of zygospore fungi? See p. 638
- Explain the phenomenon of dikaryosis in terms of plasmogamy and karyogamy. See p. 638

In examining the most important properties of the fungi as a clade, we have often drawn examples from specific groups. Now let's look in more detail at the diversity and evolution of fungi.

## 30.4 How Have Fungi Evolved and Diversified?

In this section we examine representative species from each of the major groups of fungi (see Table 30.1). Recall that the informal groups called chytrids and zygospore fungi are not clades, but the various lineages within these informal groups shown in Table 30.1 are each thought to be monophyletic. In addition, Microsporidia, Glomeromycota (arbuscular mycorrhizal fungi), Ascomycota (sac fungi), and Basidiomycota (club fungi) are all clades, and the latter two groups form a monophyletic group called Dikarya.

### Microsporidia are highly reduced, parasitic fungi

**Microsporidia** are unicellular fungi with walls that contain chitin. They are among the smallest eukaryotes known, with infective spores that are only 1–40 μm in diameter. About 1,500 species have been described, but many more species are thought to exist. Their relationships among the eukaryotes have puzzled biologists for many decades.

Microsporidia lack true mitochondria, although they have reduced structures known as *mitosomes* that are derived from mitochondria. Unlike mitochondria, however, mitosomes contain no DNA; the mitochondrial genome has been completely transferred to the nucleus. Because microsporidia lack mitochondria, biologists initially suspected that they represent an early lineage of eukaryotes that branched before the endosymbiotic event from which mitochondria evolved. The presence of the mitosome, however, indicates that this hypothesis is incorrect. DNA sequence analysis has confirmed that microsporidia are in fact highly reduced, parasitic fungi, although the exact placement of microsporidia among the fungal lineages is still being investigated.

Microsporidia are obligate intracellular parasites of animals, especially of insects, crustaceans, and fishes. Some species are known to infect mammals, including humans. Most infections by microsporidia cause chronic diseases in the hosts, with effects that include weight loss, reduced fertility, and shortened life span. The host cell is penetrated by a *polar tube* of the microsporidian spore, and the contents of the spore are injected into the host (**Figure 30.13**). The sporoplasm then replicates within the host cell and produces new infective spores. The life cycle of some species is complex and involves multiple hosts, whereas other species infect a single host. In some insects, the parasitic microsporida are transmitted vertically (i.e., from parent to offspring).

### Chytrids are the only fungi with flagella

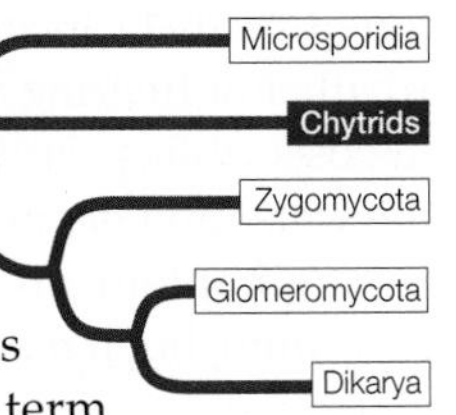
Microsporidia
Chytrids
Zygomycota
Glomeromycota
Dikarya

**Chytrids** include several distinct lineages of aquatic microorganisms once classified with the protists. However, morphological evidence (cell walls that consist primarily of chitin) and molecular evidence support their classification as basal fungi. In this book we use the term "chytrid" to refer to all three of the formally named clades shown as chytrids in Table 30.1, but some mycologists use this term to refer to one particular clade, the Chytridiomycota. There are fewer than 1,000 described species among the three taxonomic groups of chytrids.

Like the animals (and many other eukaryotes), most chytrids possess flagellated gametes. The retention of this character reflects the aquatic environment in which fungi first evolved. Chytrids are the only fungi that have flagella at any life cycle stage.

Chytrids may be parasitic (on organisms such as algae, mosquito larvae, and nematodes) or saprobic. Some have complex mutualistic relationships in the compound stomachs of foregut-fermenting animals such as cattle and deer. Many chytrids live in freshwater habitats or in moist soil, but some are marine. Some chytrids are unicellular, others have rhizoids (**Figure 30.14**), and still others have coenocytic hyphae. Chytrids reproduce both sexually and asexually, but they do not have a dikaryon stage.

*Allomyces*, a well-studied genus of chytrids, is a member of the group of chytrids (**Blastocladiomycota**; see Table 30.1) that displays alternation of generations. Both female and male gametes have flagella. The motile female gamete produces a *pheromone*, a chemical signal that attracts the swimming male gamete. The two gametes fuse, and then their nuclei fuse to form a diploid zygote that germinates to form a diploid mycelium. This is the diploid generation. Mitosis and cytokine-

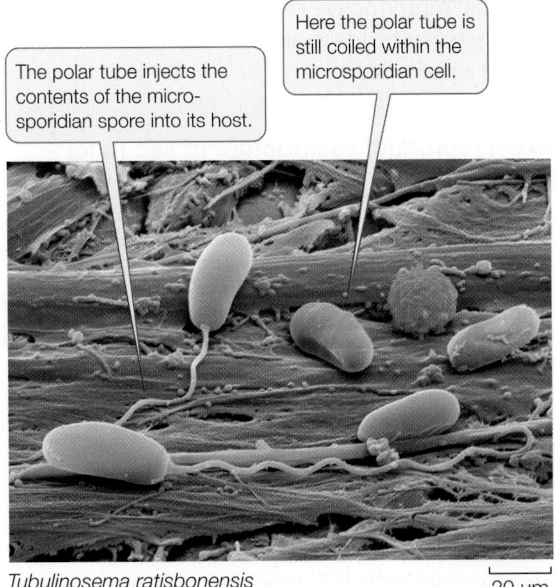
Here the polar tube is still coiled within the microsporidian cell.
The polar tube injects the contents of the microsporidian spore into its host.
*Tubulinosema ratisbonensis*
20 μm

**30.13 Spores of Microsporidia Inject Host Cells** The polar tubes of microsporidia spores transfer their contents into the host's cells. This species infects many animals, including humans.

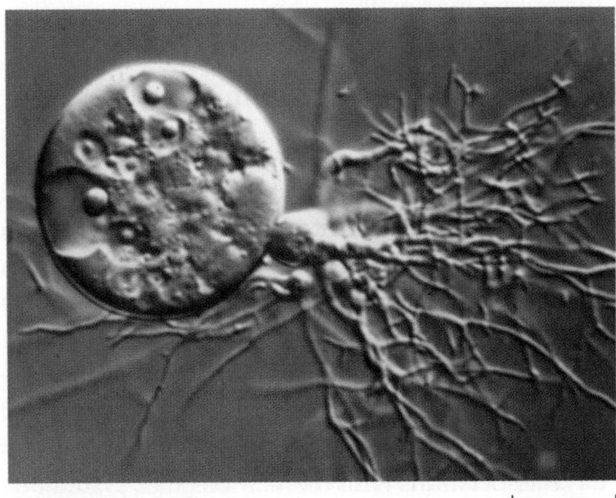
*Chytriomyces hyalinus*
25 μm

**30.14 A Chytrid** Branched rhizoids emerge from the sporangium of a mature chytrid.

sis in the zygote give rise to a small, multicellular, diploid organism, which produces numerous diploid flagellated zoospores. These diploid zoospores disperse and germinate to form more diploid organisms. Eventually, the diploid organisms produce a second kind of zoosporangium, a thick-walled resting sporangium, that can survive unfavorable conditions such as dry weather or freezing. Nuclei in the resting sporangium undergo meiosis, giving rise to haploid zoospores that are released into the water and begin the haploid stage of the life cycle.

The haploid mycelium is similar to that of the diploid mycelium, but it matures to produce female and male *gametangia* (gamete cases; see Figure 30.12A). Note that meiosis occurred in the earlier diploid generation and not immediately before gamete formation, as it does in animals. This life cycle with meiosis in spore formation is similar to the alternation of generations seen in plants.

### Zygospore fungi are terrestrial saprobes, parasites, and mutualists

Zygomycota
Glomeromycota
Ascomycota
Basidiomycota

**Zygospore fungi** (Zygomycota) include four major lineages of terrestrial fungi that live on soil as saprobes, as parasites of insects and spiders, or as mutualists of other fungi and invertebrate animals. They produce no cells with flagella, and only one diploid cell—the zygote—appears in the entire life cycle. Their hyphae are coenocytic. The mycelium spreads over its substratum, growing forward by means of vegetative hyphae. Most species do not form a fleshy fruiting structure; rather, the hyphae spread in an apparently random fashion, with occasional stalked **sporangiophores** reaching up into the air (**Figure 30.15**). These reproductive structures may bear one or many sporangia.

More than a thousand species of zygospore fungi have been described. One species you may have seen is *Rhizopus stolonifer*, the black bread mold. *Rhizopus* produces many stalked sporangiophores, each bearing a single sporangium containing hundreds of minute spores (see Figure 30.12B). As in other filamentous fungi, the spore-forming structure is separated from the rest of the hypha by a wall.

### Arbuscular mycorrhizal fungi form symbioses with plants

**Arbuscular mycorrhizal fungi** (Glomeromycota) are terrestrial fungi that associate with plant roots in a close symbiotic relationship (see Figure 30.11B). As we noted earlier in this chapter, these associations are important for most species of plants, which benefit from absorption of water and mineral nutrients through the large surface area of the fungal mycelium. Many of the fungi found in soils are arbuscular mycorrhizal fungi. Fewer than 200 species have been described, but 80 to 90 percent of all plants have associations with them. Molecular systematic studies have suggested that arbuscular mycorrhizal fungi are the sister group to Dikarya (sac fungi and club fungi), although this position is still subject to some debate.

**30.15 Zygospore Fungi Produce Sporangiophores** These transparent structures are sporangiophores (spore-bearing hyphae) growing on decomposing animal dung. Sporangiophores grow toward the light and end in tiny sporangia, which the filamentous sporangiophores can eject as far as 2 meters. Animals ingest those sporangia that land on grass and then disseminate the spores in their feces.

The hyphae of arbuscular mycorrhizal fungi are coenocytic. These fungi use glucose from their plant partners as their primary energy source, converting the glucose to other, fungus-specific sugars that cannot return to the plant. Arbuscular mycorrhizal fungi reproduce asexually; there is not yet any direct evidence that they reproduce sexually.

The next two fungal clades that we'll discuss, the sac fungi and the club fungi, are related groups with many similarities, including a dikaryon stage and septate hyphae. The two clades differ in sexual reproductive structure; in the sac fungi the sexual spores are borne inside a sac, whereas in the club fungi they are borne on a pedestal.

### The sexual reproductive structure of sac fungi is the ascus

Zygomycota
Glomeromycota
Ascomycota
Basidiomycota

The **sac fungi** (Ascomycota) are a large and diverse group of fungi found in marine, freshwater, and terrestrial habitats. There are approximately 64,000 known species, nearly half of which are the fungal partners in lichens. The hyphae of sac fungi are segmented by more or less regularly spaced septa. A pore in each septum permits extensive movement of cytoplasm and organelles (including nuclei) from one segment to the next.

Sac fungi are distinguished by the production of sacs called **asci** (singular *ascus*), which after meiosis and cytoplasmic cleavage contain sexually produced haploid *ascospores* (see Figure 30.12C). The ascus is the characteristic sexual reproductive structure of the sac fungi. The sac fungi contain many diverse groups and in the past were divided on the basis of whether or not the asci are contained within a specialized fruiting structure known as an **ascoma** (plural *ascomata*), and on the morphology of this fruiting structure. DNA sequence analyses have resulted in a revision of these traditional groupings, however.

Some species of sac fungi are unicellular yeasts. Perhaps the best known of the 800 or so species of yeasts in this group is baker's, or brewer's, yeast (*Saccharomyces cerevisiae*; see Figure 30.3). These yeasts are among the most important domesticated fungi. *S. cerevisiae* metabolizes glucose obtained from its environment to ethanol and carbon dioxide by fermentation. It forms carbon dioxide bubbles in bread dough and gives baked bread its light texture. Although they are baked away in bread making (which produces the pleasant aroma of baking bread), the ethanol and carbon dioxide are both retained when yeast ferments grain into beer. Other yeasts live on fruits such as figs and grapes and play an important role in the making of wine. Many other yeasts are associated with insects; in the guts of some insects, yeasts provide enzymes for digestion of refractory materials, especially cellulose.

Sac fungus yeasts reproduce asexually by budding. Sexual reproduction takes place when two adjacent haploid cells of opposite mating types fuse. In some species, the resulting zygote buds to form a diploid cell population. In others, the zygote nucleus undergoes meiosis immediately; when this happens, the entire cell becomes an ascus. Depending on whether the products of meiosis then undergo mitosis, a yeast ascus contains either eight or four ascospores. The ascospores germinate to become haploid cells. The sac fungus yeasts have lost the dikaryon stage.

Most sac fungi are filamentous species, such as the cup fungi (**Figure 30.16**), in which the ascomata are cup-shaped and can be as large as several centimeters across (although most are much smaller). The inner surfaces of the cups, which are covered with a mixture of specialized hyphae and asci, produce huge numbers of spores. The edible ascomata of some species, including morels and truffles, are regarded by humans as gourmet delicacies (and can sell at prices higher than gold). The underground ascomata of truffles have a strong odor that attracts mammals such as pigs, which then eat and disperse the fungus.

The sac fungi also include many of the filamentous fungi known as molds. Many of these species are parasites of flowering plants. Chestnut blight and Dutch elm disease are both caused by filamentous molds. Between its introduction to the United States in the 1890s and 1940, the chestnut blight fungus destroyed the American chestnut as a commercial species. Before the blight, this species accounted for more than half the trees in the forests of the eastern United States. Another familiar story is that of the American elm. Sometime before 1930, the Dutch elm disease fungus (first discovered in the Netherlands but native to Asia) was introduced into the United States on infected elm logs from Europe. Spreading rapidly—sometimes by way of connected root systems—the fungus destroyed great numbers of American elm trees.

Other plant pathogens among the sac fungi include the powdery mildews that infect cereal grains, lilacs, and roses, among many other plants. Mildews can be a serious problem to farmers and gardeners, and a great deal of research has focused on ways to control these agricultural pests.

Brown molds of the genus *Aspergillus* are important in some human diets. *A. tamarii* acts on soybeans in the production of soy sauce, and *A. oryzae* is used in brewing the Japanese alcoholic beverage sake. Some species of *Aspergillus* that grow on grains and on nuts such as peanuts and pecans produce extremely carcinogenic (cancer-inducing) compounds called *aflatoxins*. In the United States and most other industrialized countries, moldy grain infected with *Aspergillus* is thrown out. In Africa, where food is scarcer, the grain gets eaten, moldy or not, and causes severe health problems, including high levels of certain cancers.

*Penicillium* is a genus of green molds, of which some species produce the antibiotic penicillin, presumably for defense against competing bacteria. Two species, *P. camembertii* and *P. roquefortii*, are the organisms responsible for the characteristic strong flavors of Camembert and Roquefort cheeses, respectively.

The filmentous sac fungi reproduce asexually by means of conidia that form at the tips of specialized hyphae (**Figure 30.17**). Small chains of conidia are produced by the millions and can survive for weeks in nature. The conidia are what give molds their characteristic colors. *Fusarium oxysporum*, the plant pathogen mentioned at the beginning of this chapter, is a filamentous sac fungus with no known sexual stage. It produces conidia in abundance.

The sexual reproductive cycle of filamentous sac fungi includes the formation of a dikaryon, although this stage is relatively brief compared with that in club fungi. Many filamentous sac fungi form mating structures of two different mating types (types *A* and *a*; see Figure 30.12C). A nucleus from one of the mating structures (specialized hyphae) on type *a* enters the

(A) *Sarcoscypha coccinea*

(B) *Morchella* sp.

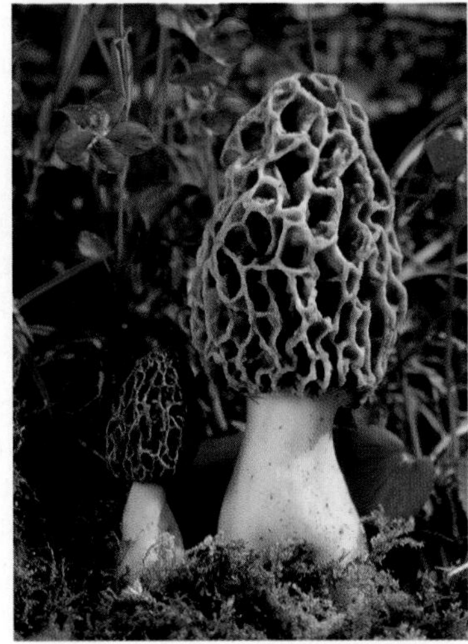

**30.16 Cup Fungi** (A) These brilliant red cups are the ascomata of a cup fungus. (B) Morels, which have a spongelike ascoma and a subtle flavor, are considered a culinary delicacy by humans.

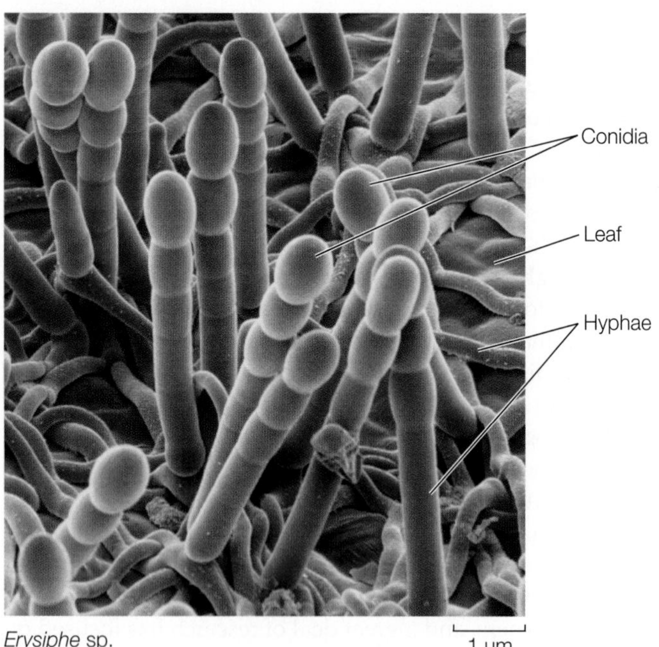


**30.17 Conidia** Chains of conidia are developing at the tips of specialized hyphae arising from this powdery mildew growing on a leaf.

hypha of type *A* (or vice versa). *Ascogenous* (ascus-forming) hyphae develop from the now dikaryotic hypha. The mating type *a* and *A* nuclei divide simultaneously with the host nuclei, and eventually asci form at the tips of the ascogenous hyphae. Only with the formation of asci do the nuclei finally fuse within the cell that now is defined as the ascus. Both nuclear fusion and the subsequent meiosis to produce the haploid ascospores take place within individual asci. The meiotic products are incorporated into ascospores that are ultimately released (sometimes shot off forcefully) by the ascus to begin the new haploid generation.

### The sexual reproductive structure of club fungi is a basidium

Zygomycota
Glomeromycota
Ascomycota
Basidiomycota

**Club fungi** (Basidiomycota) produce some of the most spectacular fruiting structures found anywhere among the fungi. These fruiting structures, called **basidiomata** (singular, basidioma), include puffballs (see Figure 30.6), some of which may be more than half a meter in diameter, mushrooms of all kinds, and the bracket fungi often encountered on trees and fallen logs in a damp forest. About 30,000 species of club fungi have been described. These include about 4,000 species of mushrooms, including the familiar *Agaricus bisporus* you may enjoy on your pizza, as well as poisonous species such as some members of the genus *Amanita* (**Figure 30.18A**). Bracket fungi (**Figure 30.18B**) do great damage both to cut lumber and timber stands; they also play an important role in returning carbon to the carbon cycle. Some of the most damaging plant pathogens are club fungi, including the rust fungi and the smut fungi that parasitize cereal grains. In contrast, other club fungi contribute to the survival of plants as fungal partners in ectomycorrhizae.

The hyphae of club fungi characteristically have septa with small, distinctive pores. The **basidium** (plural *basidia*), a swollen cell at the tip of a specialized hypha, is the characteristic sexual reproductive structure of the club fungi. In mushroom-forming club fungi, the basidia form on specialized structures known as gills or pores. The basidium is the site of nuclear fusion and meiosis and thus plays the same role in the club fungi as the ascus does in the sac fungi and the zygosporangium does in the zygospore fungi.

As seen in Figure 30.12D, after nuclei fuse in the basidium, the resulting diploid nucleus undergoes meiosis, and the four resulting haploid nuclei are incorporated into haploid *basidiospores*, which form on tiny stalks on the outside of the basidium. A single basidioma of the common bracket fungus *Ganoderma applanatum* can produce as many as 4.5 *trillion* basidiospores in one growing season. Basidiospores typically are forcibly discharged from their basidia and then germinate, giving rise to hyphae with haploid nuclei. As these hyphae grow, haploid hyphae of different mating types meet and fuse, forming dikaryotic hyphae, each cell of which contains two nuclei, one from each parent hypha. The dikaryotic mycelium grows and eventually, when triggered by rain or another environmental cue, produces a basidioma. The dikaryon stage may persist for years—some club fungi live for decades or even centuries. This pattern contrasts with the life cycle of the sac fungi, in which the dikaryon is found only in the stages leading up to formation of the asci.

## 30.4 RECAP

**The ancestor of all fungi was probably aquatic like the chytrids, but fungi have diversified to become important components of terrestrial ecosystems. Sac fungi and club fungi (which together form the clade Dikarya) contain the largest number of species.**

- Explain how the microsporidia infect the cells of their animal hosts. See p. 639 and Figure 30.13
- What feature of the chytrids suggests an aquatic ancestor to the fungi? See p. 639
- What distinguishes the fruiting bodies of sac fungi from those of club fungi? See pp. 640–642 and Figures 30.12C,D

Whether living on their own or in symbiotic associations, fungi have spread successfully over much of Earth since their origin from a protist ancestor. That ancestor also gave rise to the choanoflagellates and the animals, as we describe in Chapter 31.

(A) *Amanita muscaria*

(B) *Laetiporus sulphureus*

**30.18 Club Fungus Fruiting Structures** The basidiomata of the club fungi are probably the most familiar of all fungal structures. (A) These highly poisonous mushrooms were produced by a fungus in the genus *Amanita*, which forms ectomycorrhizal relationships with trees. (B) This edible bracket fungus is an agent of decay on dead wood.

# CHAPTER SUMMARY

## 30.1 What Is a Fungus?

- Fungi are opisthokonts with **absorptive heterotrophy** and with chitin in their cell walls. Fungi have various nutritional modes: some are **saprobes**, others are **parasites**, and some are **mutualists**. **Yeasts** are unicellular fungi. **SEE WEB ACTIVITY 30.1**
- The body of a multicellular fungus is a **mycelium**—a meshwork of **hyphae** that may be **septate** (having **septa**) or **coenocytic**. **Review Figure 30.4**
- Fungi are tolerant of hypertonic environments, and many are tolerant of low or high temperatures.
- Many species of fungi reproduce both sexually and asexually. Sexual reproduction occurs between individuals of different **mating types**.

## 30.2 How Do Fungi Interact with Other Organisms?

- Saprobic fungi, as decomposers, make crucial contributions to the recycling of elements, especially carbon. Certain fungi have relationships with other organisms that are both **symbiotic** and **mutualistic**.
- Many fungi are parasitic plant pathogens, harvesting nutrients from plant cells by means of **haustoria**. **Review Figure 30.7**
- Some fungi associate with cyanobacteria and/or green algae to form **lichens**, which live on many exposed surfaces of rocks, trees, and soil. **Review Figure 30.10**
- **Mycorrhizae** are mutualistic associations of fungi with plant roots. They improve a plant's ability to take up nutrients and water. **Review Figure 30.11**
- **Endophytic fungi** live within plants and can provide protection to their hosts from herbivores.

## 30.3 What Variations Exist among Fungal Life Cycles?

- Some chytrids have a life cycle that includes alternation of generations. **Review Figure 30.12A**
- In the sexual reproduction of terrestrial fungi, hyphae fuse, allowing "gamete" nuclei to be transferred. **Plasmogamy** (fusion of cytoplasm) precedes **karyogamy** (fusion of nuclei).
- Zygospore fungi have a resistant-spore stage with many diploid nuclei, known as a **zygospore**. **Review Figure 30.12B, ANIMATED TUTORIAL 30.1**
- In sac fungi and club fungi, a **dikaryon** is formed. The dikaryotic ($n + n$) condition is unique to the fungi. **Review Figures 30.12C,D, WEB ACTIVITY 30.2**

## 30.4 How Have Fungi Evolved and Diversified?

- The relationships of microsporidia, chytrids, and zygospore fungi are not well resolved, but these groups diversified early in fungal evolution. The mycorrhizal fungi, sac fungi, and club fungi form a monophyletic group, and the latter two groups form the clade Dikarya. **Review Figure 30.2 and Table 30.1**
- **Microsporidia** are reduced, intracellular parasitic species of fungi that infect several animal groups, especially insects, crustaceans, and fishes.
- The three distinct lineages of **chytrids** have flagellated gametes. **Review Figure 30.12A**
- The four distinct lineages of **zygospore fungi** have coenocytic hyphae and a zygospore in their life cycle. Their fruiting structures are simple stalked **sporangiophores**. **Review Figure 30.12B**
- **Arbuscular mycorrhizal fungi** form symbiotic associations with plant roots. These mycorrhizae increase water and mineral uptake by the plants and provide a carbon source to the fungi. They are only known to reproduce asexually. Their hyphae are coenocytic.
- **Sac fungi** have septate hyphae; their sexual reproductive structures are **asci**. Many sac fungi are partners in lichen and endophytic associations. Filimentous sac fungi produce fleshy fruiting bodies called **ascomata**. The dikaryon stage in the sac fungus life cycle is relatively brief. **Review Figure 30.12C**
- **Club fungi** have septate hyphae. Many of the species are plant pathogens, although mushroom-forming species are more familiar to most people. Their fruiting bodies are called **basidiomata**, and their sexual reproductive structures are **basidia**. The dikaryon stage may last for years. **Review Figure 30.12D**

## SELF-QUIZ

1. Which statement about fungi is *not* true?
   *a.* A multicellular fungus has a body called a mycelium.
   *b.* Hyphae are composed of individual mycelia.
   *c.* Many fungi tolerate highly hypertonic environments.
   *d.* Many fungi tolerate low temperatures.
   *e.* Some fungi are anchored to their substrate by rhizoids.
2. The absorptive heterotrophy of fungi is aided by
   *a.* dikaryon formation.
   *b.* spore formation.
   *c.* the fact that they are all parasites.
   *d.* their large surface area-to-volume ratio.
   *e.* their possession of chloroplasts.
3. Which statement about fungal nutrition is *not* true?
   *a.* Some fungi are active predators.
   *b.* Some fungi form mutualistic associations with other organisms.
   *c.* All fungi require mineral nutrients.
   *d.* Fungi can make some of the compounds that are vitamins for animals.
   *e.* Facultative parasites can grow only on their specific hosts.
4. Which statement about dikaryosis is *not* true?
   *a.* The cytoplasm of two cells fuses before their nuclei fuse.
   *b.* The two haploid nuclei are genetically different.
   *c.* The two nuclei are of the same mating type.
   *d.* The dikaryon stage ends when the two nuclei fuse.
   *e.* Not all fungi have a dikaryon stage.
5. Reproductive structures consisting of one or more photosynthetic cells surrounded by fungal hyphae are called
   *a.* ascospores.
   *b.* basidiospores.
   *c.* conidia.
   *d.* soredia.
   *e.* gametes.
6. Members of the zygospore fungi
   *a.* have hyphae without regularly occurring septa.
   *b.* produce motile gametes.
   *c.* form fleshy fruiting bodies.
   *d.* are haploid throughout their life cycle.
   *e.* have sexual reproductive structures similar to those of the sac fungi.
7. Which statement about sac fungi is *not* true?
   *a.* Some species are yeasts.
   *b.* They form reproductive structures called asci.
   *c.* Their hyphae are segmented by septa.
   *d.* Many of their species have a dikaryotic state.
   *e.* All have fruiting structures called ascomata.
8. Club fungi
   *a.* often produce fleshy fruiting structures.
   *b.* have hyphae without septa.
   *c.* have no sexual stage.
   *d.* produce basidia within basidiospores.
   *e.* form diploid basidiospores.
9. Microsporidia
   *a.* lack true mitochondria.
   *b.* are parasites of animals.
   *c.* contain mitosomes.
   *d.* are among the smallest eukaryotes known.
   *e.* all of the above
10. Which statement about lichens is *not* true?
    *a.* They can reproduce by fragmentation of the vegetative body.
    *b.* They are often the first colonists in a new area.
    *c.* They render their environment more basic (alkaline).
    *d.* They contribute to soil formation.
    *e.* They may contain less than 10 percent water by weight.

## FOR DISCUSSION

1. You are shown an object that looks superficially like a pale green mushroom. Describe at least three criteria (including anatomical and chemical traits) that would enable you to tell whether the object is a piece of a plant or a piece of a fungus.
2. Differentiate among the members of the following pairs of related terms:
   *a.* hypha/mycelium
   *b.* ascus/basidium
   *c.* ectomycorrhiza/arbuscular mycorrhiza
3. For each type of organism listed below, give a single characteristic that may be used to differentiate it from the other, related organism(s) in parentheses.
   *a.* zygospore fungi (sac fungi)
   *b.* sac fungi (club fungi)
   *c.* baker's yeast (*Penicillium*)
4. Many fungi are dikaryotic during part of their life cycle. Why are dikaryons described as $n + n$ instead of $2n$?
5. If all the fungi on Earth were suddenly to die, how would the surviving organisms be affected?
6. How might the first mycorrhizae have arisen?
7. What attributes might account for the ability of lichens to withstand the intensely cold environment of Antarctica? Be specific in your answer.
8. What factors must be taken into account in using fungi to combat agricultural pests?

## ADDITIONAL INVESTIGATION

We noted that lichens are highly sensitive to air pollution. How could you use lichen diversity and abundance to measure air quality? How would you expect lichen diversity to vary with respect to distance from major metropolitan areas? Would you expect prevailing wind direction to be a factor in this pattern?

# 31 Animal Origins and the Evolution of Body Plans

## Getting back to our roots

In 1883, the zoologist Franz Schulze noticed something unusual in his Austrian laboratory: transparent, flattened organisms were crawling on the sides of his saltwater aquarium. Collected accidentally along with the sponges that were Schulze's primary interest, these organisms were unlike any previously described animals—especially since they continually changed shape as they moved.

Schulze's examination of the new organisms revealed that they were animals. But structurally they were among the simplest animals that he—or anyone else—had ever observed, with only four types of cells. He named the new species *Trichoplax adhaerens*, which means "sticky hairy plate," and argued that the new species had no close relationships with other major animal groups. For decades, however, most biologists dismissed Schulze's findings, insisting that the transparent organisms must be larval forms of other, well-known, animals.

In the 1960s, new and more detailed studies confirmed Schultze's findings and the distinctive nature of *Trichoplax*. Even then, this odd animal continued to be known almost exclusively from aquariums. Only in the past decade have biologists been able to locate and study natural field populations of *Trichoplax adhaerens*. A few additional closely related species have been discovered, and they are collectively known as placozoans (Greek, "flat animals").

The more biologists have studied *Trichoplax*, the odder this animal appears. It has the smallest genome of any animal studied to date. The mature stages lack body symmetry and have no mouth, gut, or nervous system. Is *Trichoplax* a relict representative of a group of animals that appeared early in animal evolution? Indeed, some recent phylogenetic analyses support the possibility that *Trichoplax* is a representative of the most divergent group of animals.

Although biologists agree on which groups of organisms are animals, the root of the animal tree is a subject of considerable investigation and debate. Traditionally, the first split is thought to have been between the sponges and all other animals, and most evidence still favors that view. Some gene sequence analyses, however, suggest that the major groups of sponges are not even each other's closest relatives, and that the glass sponges alone split with all the remaining animals (including other sponges)

**Are Placozoans at the Base of Animal Phylogeny?** Is the simplicity of placozoans an ancestral feature, or did these organisms descend from ancestors with more complex body plans? Answering that question requires an understanding of animal phylogeny.

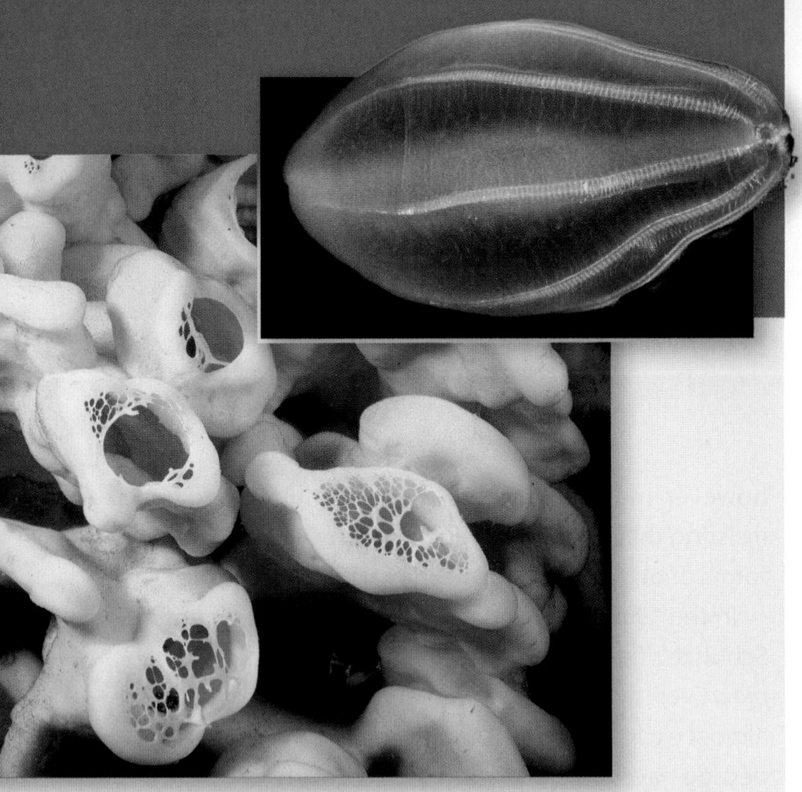

**Alternative Candidates** Some studies place either sponges (left) or ctenophores (upper right) rather than placozoans as the sister group of all other multicellular animals. Most evidence favors sponges as the most divergent animal group.

at the base of the animal tree. Other genomic investigations have suggested that the ctenophores—comb jellies—may be the sister group to all other animals. Ctenophores are beautiful, transparent organisms whose sticky tentacles capture planktonic prey.

Newly acquired insights into how animal genomes are structured and have evolved play an ever more important role in understanding the relationships among the major animal groups. This chapter explores the earliest branches on the animal tree, and how a few fundamental "body plans" have been modified to yield the remarkable variety of animal forms described in this and the following two chapters.

**IN THIS CHAPTER** we will review the evidence that has led biologists to conclude that the animals are monophyletic and will then present the best-supported current hypotheses of animal phylogeny. We will describe how the diverse animal forms are derived from a small array of body plans. We will discuss various animal strategies for obtaining food and describe the amazingly varied life cycles of animals—how they are born, grow, disperse, and reproduce. Finally, we will describe members of several basal clades of animals.

## CHAPTER OUTLINE



# 31.1 What Characteristics Distinguish the Animals?

How do we recognize an organism as an animal? That may seem obvious for many familiar animals, but less so for groups like sponges, which were once thought to be plants. Some of the general aspects that we associate with animals include:

- *Multicellularity.* In contrast to the Bacteria, Archaea, and most protists (see Chapters 26 and 27), all animals are multicellular. Animal life cycles feature complex patterns of development from a single-celled zygote into a multicellular adult.
- *Heterotrophic metabolism.* In contrast to most plants (see Chapters 28 and 29), all animals are heterotrophs. Animals are able to synthesize very few organic molecules from inorganic chemicals, so they must take in nutrients from their environment (either through their own actions, or in some cases with the aid of symbiotic species).
- *Internal digestion.* Although the fungi are also heterotrophs (see Chapter 30), animals digest their food differently. Whereas fungi rely on external digestion, most animals use internal processes to break down materials from their environment into the organic molecules they need most. Most animals ingest food into an internal gut that is continuous with the outside environment and in which digestion takes place.
- *Movement.* In contrast to the majority of plants and fungi, most animals can move. Animals must move to find food or bring food to them. Muscle tissue is unique to animals, and many animal body plans are specialized for movement.

Although these general features help us recognize animals, none is diagnostic for all animals. Some animals do not move, at least during certain life stages, and some plants and fungi do have limited movement. Some animals lack a gut. Many multicellular organisms are not animals. So what is the evidence that groups all animals together in a single clade?

### Animal monophyly is supported by gene sequences and morphology

The most convincing evidence that all the organisms considered to be animals share a common ancestor comes from phylogenetic analyses of their gene sequences. Relatively few complete animal genomes are available, but many more genomes are being sequenced each year. Analyses of these genomes, as well as many individual gene sequences, have shown that all animals are indeed *monophyletic;* a currently well-supported phylogenetic tree of the animals is shown in **Figure 31.1**.

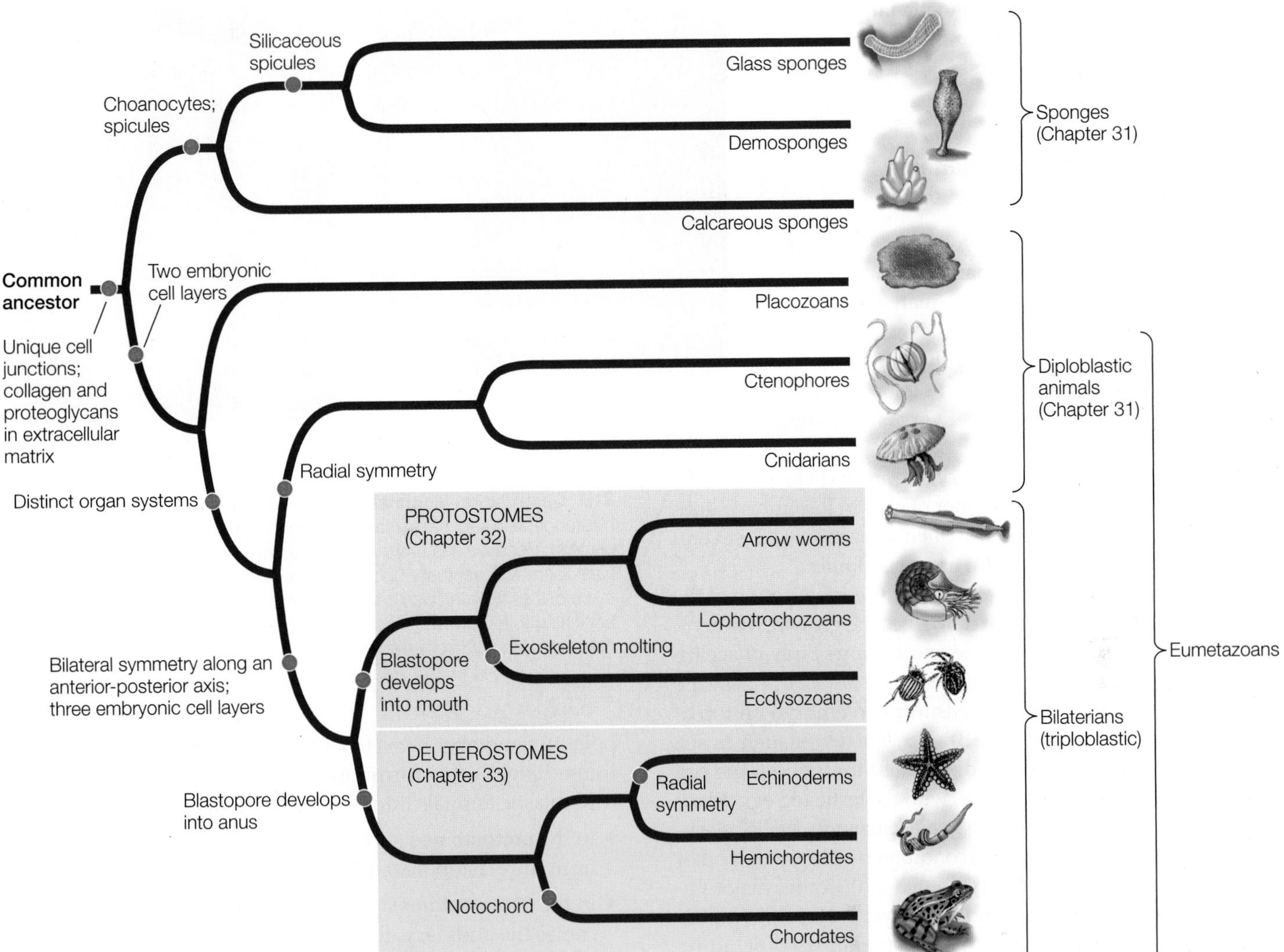


**31.1 The Phylogeny of Animals** This tree presents the best supported current hypotheses of evolutionary relationships among major groups of animals. The traits highlighted by red circles will be explained as you read this chapter; you should review this figure closely after you complete your reading.

yourBioPortal.com
GO TO Web Activity 31.1 • Sponge and Diploblast Classification

Although animals were considered to belong to a single clade long before gene sequencing became possible, surprisingly few morphological features are shared across all species of animals. These morphological *synapomorphies* include:

- Unique types of junctions between their cells (tight junctions, desmosomes, and gap junctions; see Figure 6.7).
- A common set of extracellular matrix molecules, including collagen and proteoglycans (see Figure 5.25).

Although some animals in a few groups lack one or another of these characteristics, it is believed that these traits were possessed by the ancestor of all animals and subsequently lost in those groups. Similarities in the organization and function of Hox and other developmental genes (see Chapter 20) provide additional evidence of developmental mechanisms shared by a common animal ancestor. The Hox genes specify body pattern and axis formation, leading to developmental similarities across animals.

The common ancestor of animals was probably a colonial flagellated protist similar to existing colonial choanoflagellates (see Figure 27.27). The most reasonable current scenario postulates a choanoflagellate lineage in which certain cells in the colony began to be specialized—some for movement, others for nutrition, others for reproduction, and so on. Once this *functional specialization* had begun, cells could have continued to differentiate. Coordination among groups of cells could have improved by means of specific regulatory and signaling molecules that guided differentiation and migration of cells in developing embryos. Such coordinated groups of cells eventually evolved into the larger and more complex organisms that we call animals.

More than a million living animal species have been named and described, and millions of additional animal species await discovery. Clues to the evolutionary relationships among animal groups can be found in fossils, in patterns of embryonic development, in the morphology and physiology of living animals, in the structure of animal proteins, and in gene sequences.

Increasingly, studies of higher-level relationships have come to depend on genomic sequence comparisons, as genomes are ultimately the source of all inherited trait information.

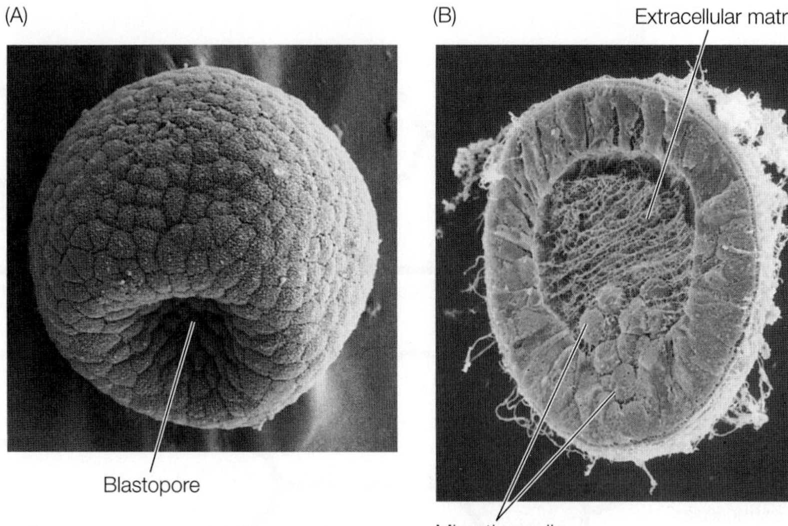


**31.2 Gastrulation Illuminates Evolutionary Relationships** (A) The blastopore is clear in this scanning electron micrograph of a sea urchin gastrula. Because sea urchins (echinoderms) are deuterostomes, this blastopore will eventually become the anal end of the animal's gut. (B) In this cross section through a later-stage sea urchin gastrula, the cells are beginning to look different from one another. The molecules of the extracellular matrix guide cell movement.

### A few basic developmental patterns differentiate major animal groups

Differences in patterns of embryonic development have until recently provided many of the most important clues to animal phylogeny. Analyses of gene sequences, however, are now showing that some developmental patterns are more evolutionarily variable than previously thought. We describe here the basic developmental patterns that vary among the major animal clades.

The first few cell divisions of a zygote are known as **cleavage**. In general, the number of cells in the embryo doubles with each cleavage. As described in Section 44.1, several different **cleavage patterns** exist among animals.

Cleavage patterns are influenced by the configuration of the *yolk*, the nutritive material that nourishes the growing embryo. In reptiles, for example, the presence of a large body of acellular yolk within the fertilized egg creates an *incomplete* cleavage pattern in which the dividing cells form an embryo on top of the yolk mass (see Figure 44.3B). In *echinoderms* such as sea urchins, limited yolk is evenly distributed throughout the egg cytoplasm, so cleavage is *complete*, with the fertilized egg dividing in an even pattern known as **radial cleavage**. Radial cleavage is thought to be the ancestral condition for the animals other than sponges, as it is widely distributed in the other major lineages. **Spiral cleavage**—a complicated permutation of radial cleavage—is found among many *lophotrochozoans*, including earthworms and clams. Lophotrochozoans with spiral cleavage are thus sometimes known as *spiralians*. The early branches of the *ecdysozoans* (molting animals, such as insects and nematodes) have radial cleavage, although most ecdysozoans have an idiosyncratic cleavage pattern that is neither radial nor spiral in organization (see Figure 44.3C).

Distinct layers of cells form during the early development of most animals. These cell layers differentiate into specific organs and organ systems as development continues. The embryos of **diploblastic** animals have two cell layers: an outer *ectoderm* and an inner *endoderm*. Embryos of **triploblastic** animals have, in addition to ectoderm and endoderm, a third distinct cell layer, *mesoderm*, between the ectoderm and the endoderm. The existence of three cell layers in embryos is a synapomorphy of triploblastic animals, whereas the diploblastic animals (placozoans, ctenophores, and cnidarians) exhibit the ancestral condition. Some biologists consider sponges to be diploblastic, but since they do not have clearly differentiated tissue types or embryonic cell layers, the term is not usually applied to them.

During early development in many animals, in a process known as *gastrulation*, a hollow ball one cell thick indents to form a cup-shaped structure. The opening of the cavity formed by this indentation is called the *blastopore* (**Figure 31.2**). The process of gastrulation is covered in detail in Section 44.2; the point to remember here is that the *overall pattern* of gastrulation immediately after formation of the blastopore divides the triploblastic animals into two major groups:

- In the **protostomes** (Greek, "mouth first"), the mouth arises from the blastopore, and the anus forms later.
- In the **deuterostomes** ("mouth second"), the blastopore becomes the anus, and the mouth forms later.

Although the developmental patterns of animals are more varied than suggested by this simple dichotomy, sequencing data indicate that the protostomes and deuterostomes represent distinct animal clades. Together, these two groups are known as the **bilaterians** (named for their usual bilateral symmetry), and they account for the vast majority of animal species.

## 31.1 RECAP

**The animals are thought to be monophyletic because they share several derived traits, especially among their gene sequences. Major developmental differences also provide evidence of evolutionary relationships, although phylogenetic analyses of gene sequences have shown that the evolutionary history of these features is more complex than was once thought.**

- What general features of animals distinguish this group from other living organisms? See p. 646
- Describe the difference between diploblastic and triploblastic embryos, and between protostomes and deuterostomes. See p. 648

We devote Chapter 32 to the protostomes and Chapter 33 to the deuterostomes. Later in this chapter, we describe several groups of animals that diverged before the origin of the bilaterians. We begin our exploration of animal diversity by discussing general features of animal body plans.

## 31.2 What Are the Features of Animal Body Plans?

The general structure of an animal, the arrangement of its organ systems, and the integrated functioning of its parts are referred to as its **body plan**. As Chapter 20 describes, the regulatory and signaling genes that govern the development of body symmetry, body cavities, segmentation, and appendages are widely shared among the different animal groups. Thus we might expect animals to share body plans. Although animal body plans vary tremendously, they can be seen as variations on four key features:

- The *symmetry* of the body
- The structure of the *body cavity*
- The *segmentation* of the body
- *External appendages* that are used for sensing, chewing, locomotion, mating, and other functions

Each of these features affects how an animal moves and interacts with its environment.

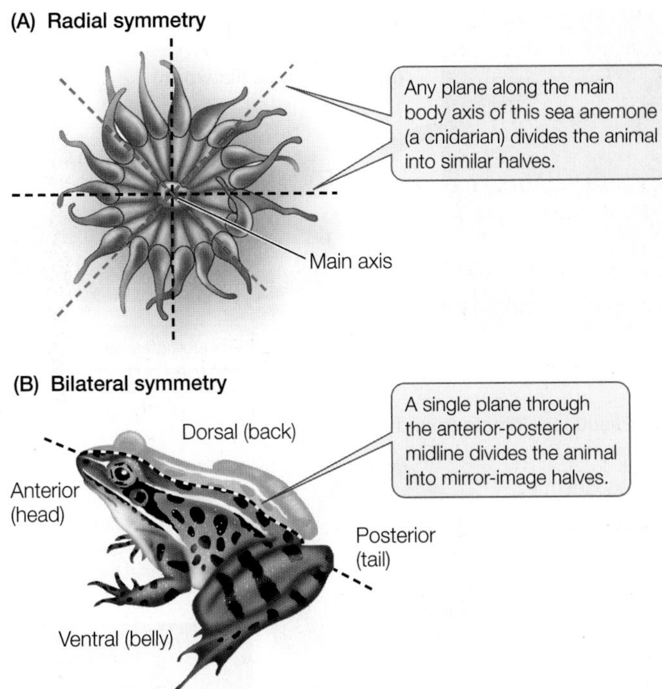
(A) Radial symmetry
Any plane along the main body axis of this sea anemone (a cnidarian) divides the animal into similar halves.
Main axis
(B) Bilateral symmetry
Dorsal (back)
A single plane through the anterior-posterior midline divides the animal into mirror-image halves.
Anterior (head)
Posterior (tail)
Ventral (belly)

**31.3 Body Symmetry** Most animals are either radially or bilaterally symmetrical.

### Most animals are symmetrical

The overall shape of an animal can be described by its **symmetry**. An animal is said to be *symmetrical* if it can be divided along at least one plane into similar halves. Animals that have no plane of symmetry are said to be *asymmetrical*. Placozoans and many sponges are asymmetrical, but most other animals have some kind of symmetry, which is governed by the expression of regulatory genes during development.

The simplest form of symmetry is **spherical symmetry**, in which body parts radiate out from a central point. An infinite number of planes passing through the central point can divide a spherically symmetrical organism into similar halves. Spherical symmetry is widespread among unicellular protists, but most animals possess other forms of symmetry.

In organisms with **radial symmetry**, body parts are arranged around one main axis at the body's center (**Figure 31.3A**). *Ctenophores* are radially symmetrical, as are many *cnidarians* and *echinoderms*. A perfectly radially symmetrical animal can be divided into similar halves by any plane that contains the main axis. However, most radially symmetrical animals—including the adults of echinoderms such as sea stars and sand dollars—are slightly modified so that fewer planes can divide them into identical halves. Some radially symmetrical animals are sessile (sedentary) or drift with water currents. Others move slowly but can move equally well in any direction.

**Bilateral symmetry** is characteristic of animals that have a distinct front end, which typically precedes the rest of the body as the animal moves. A bilaterally symmetrical animal can be divided into mirror-image (left and right) halves by a single plane that passes through the midline of its body (**Figure 31.3B**). This plane runs from the front, or **anterior**, end of the body, to its rear, or **posterior**, end.

A plane at right angles to the midline divides the body into two dissimilar sides. The back of a bilaterally symmetrical animal is its **dorsal** surface; the underside is its **ventral** surface.

Bilateral symmetry is strongly correlated with **cephalization**, which is the concentration of sensory organs and nervous tissues in a head at the anterior end of the animal. Cephalization has been evolutionarily favored because the anterior end of a bilaterally symmetrical animal typically encounters new environments first.

### The structure of the body cavity influences movement

Animals can be divided into three types—*acoelomate*, *pseudocoelomate*, and *coelomate*—based on the presence and structure of an internal, fluid-filled **body cavity**. The structure of an animal's body cavity strongly influences the ways in which it can move.

**Acoelomate** animals such as flatworms lack an enclosed, fluid-filled body cavity. Instead, the space between the gut (derived from endoderm) and the muscular body wall (derived from mesoderm) is filled with masses of cells called *mesenchyme* (**Figure 31.4A**). These animals typically move by beating cilia.

Body cavities come in two types. Some lie between mesoderm and endoderm, and others are enclosed completely within mesoderm.

- **Pseudocoelomate** animals have a body cavity called a *pseudocoel*, a fluid-filled space in which many of the internal

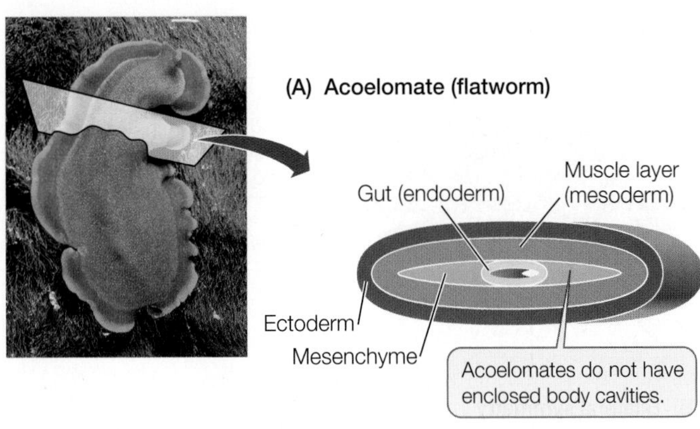


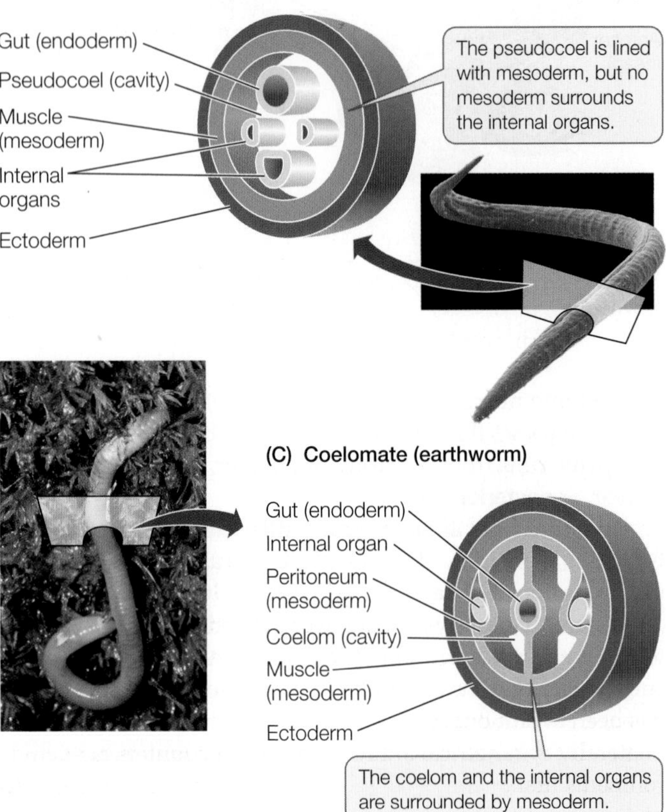


**31.4 Animal Body Cavities** (A) Acoelomates do not have enclosed body cavities. (B) Pseudocoelomates have a body cavity bounded by endoderm and mesoderm. (C) Coelomates have a peritoneum surrounding the internal organs in a region bounded by mesoderm.

**yourBioPortal.com**
GO TO Web Activity 31.2 • Animal Body Cavities

organs are suspended. A pseudocoel is enclosed by muscles (mesoderm) only on its outside; there is no inner layer of mesoderm surrounding the internal organs (**Figure 31.4B**).

- **Coelomate** animals have a *coelom*, a body cavity that develops within the mesoderm. It is lined with a layer of muscular tissue called the *peritoneum*, which also surrounds the internal organs. The coelom is thus enclosed on both the inside and the outside by mesoderm (**Figure 31.4C**). A coelomate animal has better control over the movement of the fluids in its body cavity than a pseudocoelomate animal does.

The body cavities of many animals function as **hydrostatic skeletons**. Fluids are relatively incompressible, so when the muscles surrounding them contract, fluids shift to another part of the cavity. If the body tissues around the cavity are flexible, fluids squeezed out of one region can cause some other region to expand. The moving fluids can thus move specific body parts. (You can see how a hydrostatic skeleton works by watching a snail emerge from its shell.) An animal with both *circular muscles* (encircling the body cavity) and *longitudinal muscles* (running along the length of the body) has even greater control over its movement.

In terrestrial environments, the hydrostatic function of fluid-filled body cavities applies mostly to relatively small, soft-bodied organisms. Most larger animals (as well as many smaller ones) have hard skeletons that provide protection and facilitate movement. Muscles are attached to those firm structures, which may be inside the animal or on its outer surface (in the form of a shell or cuticle).

### Segmentation improves control of movement

Many animal bodies are divided into segments. **Segmentation** facilitates specialization of different body regions. It also allows an animal to alter the shape of its body in complex ways and to control its movements precisely. If an animal's body is segmented, muscles in each individual segment can change the shape of that segment independently of the others. In only a few segmented animals is the body cavity separated into discrete compartments, but even partly separated compartments allow better control of movement. As we see in Chapters 32 and 33, segmentation occurs in several groups of protostomes and deuterostomes.

In some animals, segments are not apparent externally (as with the segmented vertebrae of vertebrates). In other animals, such as annelids, similar body segments are repeated many times (**Figure 31.5A**). And in yet other animals, including most arthropods, segments are visible but differ strikingly (**Figure 31.5B**). As described in Chapter 32, the dramatic evolutionary radiation of the arthropods (including the insects, spiders, centipedes, and crustaceans) was based on changes in a segmented body plan that features muscles attached to the inner surface of an external skeleton, including a variety of external appendages that move these animals.

### Appendages have many uses

Getting around under their own power is important to many animals. It allows them to obtain food, to avoid predators, and to find mates. Even some sedentary species, such as sea anemones, have larval stages that use cilia to swim, thus increasing the animal's chances of finding a suitable habitat.

Appendages that project externally from the body greatly enhance an animal's ability to move around. Many echinoderms, including sea urchins and sea stars, have myriad *tube feet* that allow them to move slowly across the substratum. Highly con-

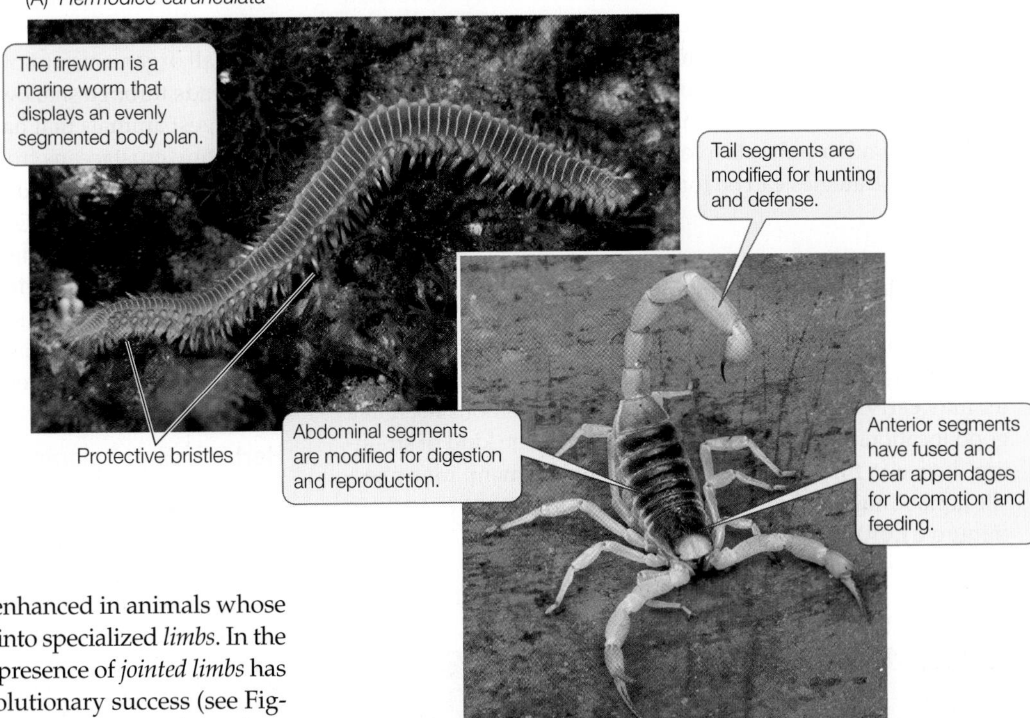

(A) Hermodice carunculata
The fireworm is a marine worm that displays an evenly segmented body plan.
Tail segments are modified for hunting and defense.
Protective bristles
Abdominal segments are modified for digestion and reproduction.
Anterior segments have fused and bear appendages for locomotion and feeding.
(B) Hadrurus arizonensis

**31.5 Segmentation** The body cavities of many animals are segmented. (A) All of the segments of this marine fireworm, an annelid, are similar. Its appendages are tipped with bristles (setae) that are used for locomotion and (in this species) for protection—the setae contain a noxious toxin. (B) Segmentation allows the evolution of differentiation among the segments. The segments of this scorpion, an arthropod, differ in their form, function, and the appendages they bear.

trolled, rapid movement is greatly enhanced in animals whose appendages have become modified into specialized *limbs*. In the arthropods and the vertebrates, the presence of *jointed limbs* has been a prominent factor in their evolutionary success (see Figure 31.5B). In four independent instances—among the arthropod insects and among the vertebrate pterosaurs, birds, and bats—body plans emerged in which limbs were modified into wings, allowing these animals to take to the air.

Appendages also include many structures that are not used for locomotion. Many animals have antennae, which are specialized appendages used for sensing the environment. Other appendages (such as claws and mouth parts of many arthropods) are adaptations for capturing prey or chewing food. In some species, appendages are used for reproductive purposes, such as sperm transfer or egg incubation.

## 31.2 RECAP

**The body plans of animals are variations on patterns of symmetry, body cavities, segmentation, and appendages. All four can affect movement and locomotion, which are important aspects of the animal way of life.**

- Describe the main types of symmetry found in animals. How can an animal's symmetry influence the way it moves? **See p. 649 and Figure 31.3**
- Explain several ways in which body cavities and segmentation improve control over movement. **See pp. 649–650**

Many of the modifications to animal body plans affect ways of finding, capturing, and processing food. Evolutionary changes in symmetry, body cavities, appendages, and segmentation have played key roles in enabling animals to obtain food from their environment, as well as helping them avoid becoming food for other animals.

# 31.3 How Do Animals Get Their Food?

As noted in Section 31.1, animals are heterotrophs, or "other feeders." Although many animals rely on photosynthetic endosymbionts for nutrition (see Figures 5.14C and 27.8), most animals must actively obtain an outside source of nutrition, otherwise known as food. The need to locate food has favored the evolution of sensory structures that can provide animals with detailed information about their environment, as well as nervous systems that can receive, process, and coordinate that information.

To acquire food, most animals must expend energy, either to move through the environment to where food is located or to move the environment and the food it contains to them. Animals that can move from one place to another are **motile**; animals that stay in one place are **sessile**.

The principal feeding strategies that animals use fall into five broad categories:

- *Filter feeders* capture small organisms delivered to them by their environment.
- *Herbivores* eat plants or parts of plants.
- *Predators* capture and eat other animals that typically are relatively large.
- *Parasites* live in or on other, generally much larger, organisms from which they obtain energy and nutrients.
- *Detritivores* actively feed on dead organic material.

Each of these strategies can be found in many different animal groups, and none of them is limited to a single group. In addition, individuals of some species may employ more than one feeding strategy, and some animals employ different feeding strategies at different points in their life cycle. The constant and ongoing need to obtain food, the variety of nutrient sources available in any given environment, and the necessity of competing with other animals to obtain food means that a variety of feeding strategies can be found among all the major animal groups.

### Filter feeders capture small prey

Air and water often contain small organisms and organic molecules that are potential food for animals. Moving air and water may carry those items to an animal that positions itself in a good location. These **filter feeders** then use some kind of straining device to filter the food from the environment. Many sessile aquatic animals rely on water currents to bring prey to them (**Figure 31.6A**).

Motile filter feeders bring the nutrient-containing medium to them. Flamingos, for example, uses their serrated beak to filter small organisms out of the muddy mixture they pick up as they wade through shallow water (**Figure 31.6B**). Blue whales—the largest animals that have ever lived—are filter feeders that strain tiny crustaceans from the water column as they swim.

Some sessile filter feeders expend energy to move water past their food-capturing devices. Sponges, for example, bring water into their body by beating the flagella of their specialized feeding cells, called **choanocytes** (**Figure 31.7**). These flagellated feeding cells of sponges are similar in structure to protists known as choanoflagellates, which provides evidence for the close relationship of choanoflagellates to animals (see Section 27.5).

### Herbivores eat plants

Animals that eat plants are called **herbivores**. An individual plant has many different structures—leaves, wood, sap, flowers, fruits, nectar, and seeds—that animals can consume. Not surprisingly, then, many different kinds of herbivores may feed on a single kind of plant, consuming different parts of the plant or eating the same part in different ways. An individual animal that is captured by a predator is likely to die, but herbivores often feed on plants without killing them.

Animals do not need to expend energy subduing and killing plants. However, they do need to digest them, and animals must expend energy to detoxify plants' defensive chemicals. Digestion can pose challenges to terrestrial herbivores because the dominant land plants tend to have several different kinds of tissues, many of which are tough or fibrous. Herbivorous animals typically have long, complex guts to accomplish the tasks involved in digesting plants (see Section 51.2).

### Predators capture and subdue large prey

**Predators** possess features that enable them to capture and subdue relatively large animals (referred to as their **prey**). Many vertebrate predators have sensitive sensory organs that enable them to locate prey, as well as sharp teeth or claws that allow them to capture and subdue large prey (**Figure 31.8**). Predators may stalk and pursue their prey, or wait (often camouflaged) for their prey to come to them.

Another weapon of predators (as well as of prey) is toxins. We are all aware of the dangers of encountering the toxins of a venomous snake. Toxins often have both a defensive role as well as a role in the capture and sometimes the digestion of prey. Cnidarians (jellyfishes and their relatives) are one of many animal groups that use toxins to capture and subdue prey. The cnidarians' tentacles are covered with specialized cells that con-

**31.6 Filter-Feeding Strategies** (A) Sessile marine filter feeders such as this "Christmas tree worm," a polychaete, allow the ocean currents to bring their food—plankton—to them. (B) The greater flamingo of South America is a motile filter feeder, using its appendages (legs) to stir up mud as it wades through ocean lagoons and salty lakes. The bird then uses its beak (close-up) to strain small organisms out of the muddy mixture.

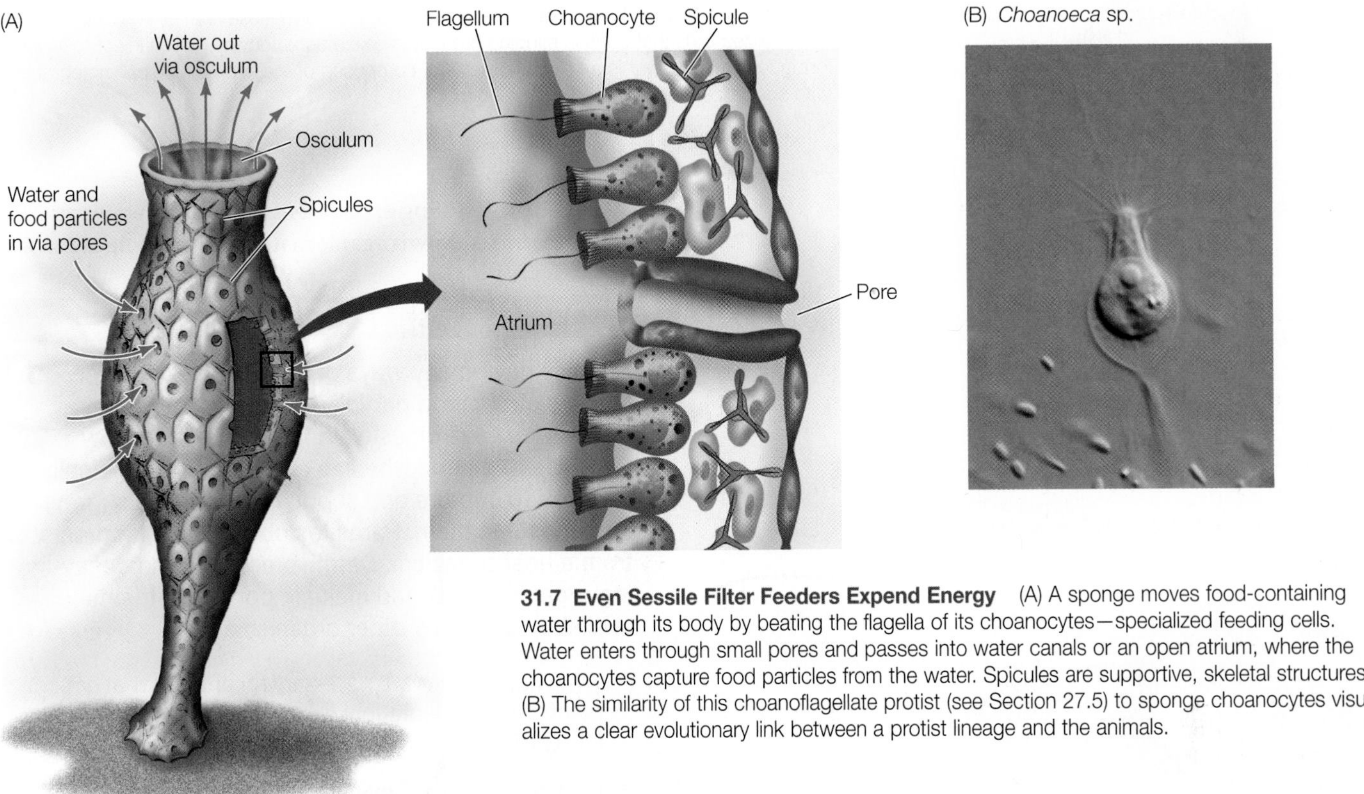


**31.7 Even Sessile Filter Feeders Expend Energy** (A) A sponge moves food-containing water through its body by beating the flagella of its choanocytes—specialized feeding cells. Water enters through small pores and passes into water canals or an open atrium, where the choanocytes capture food particles from the water. Spicules are supportive, skeletal structures. (B) The similarity of this choanoflagellate protist (see Section 27.5) to sponge choanocytes visualizes a clear evolutionary link between a protist lineage and the animals.

tain stinging organelles called **nematocysts**, which inject toxins into the prey (**Figure 31.9**).

**Omnivores** are animals, such as raccoons and humans, that eat both plants and other animals. The diet of some omnivores differs at different life stages; many songbirds, for example, eat fruit or seeds as adults but feed insects to their young.

### Parasites live in or on other organisms

**Parasites** are animals that live in or on another organism—called a *host*—and obtain their nutrients from that host. Some parasites consume parts of the host itself (such as ticks that suck body fluids); others highjack nutrients the host would otherwise consume (such as tapeworms that may live in our intestines). Most animal parasites are much smaller than their hosts, and many parasites can consume parts of their host without killing it. To reside within a host, a parasite must first overcome the host's defenses. Parasites often have complex life cycles that rely on multiple hosts, as we detail in Section 31.4.

**31.8 Tooth and Claw** (A) The teeth of the predatory gray wolf are adapted for killing prey and shearing meat. (B) The appendages (legs and wings) of the bald eagle, along with its strong beak, are adaptations to the life of a predatory hunter.

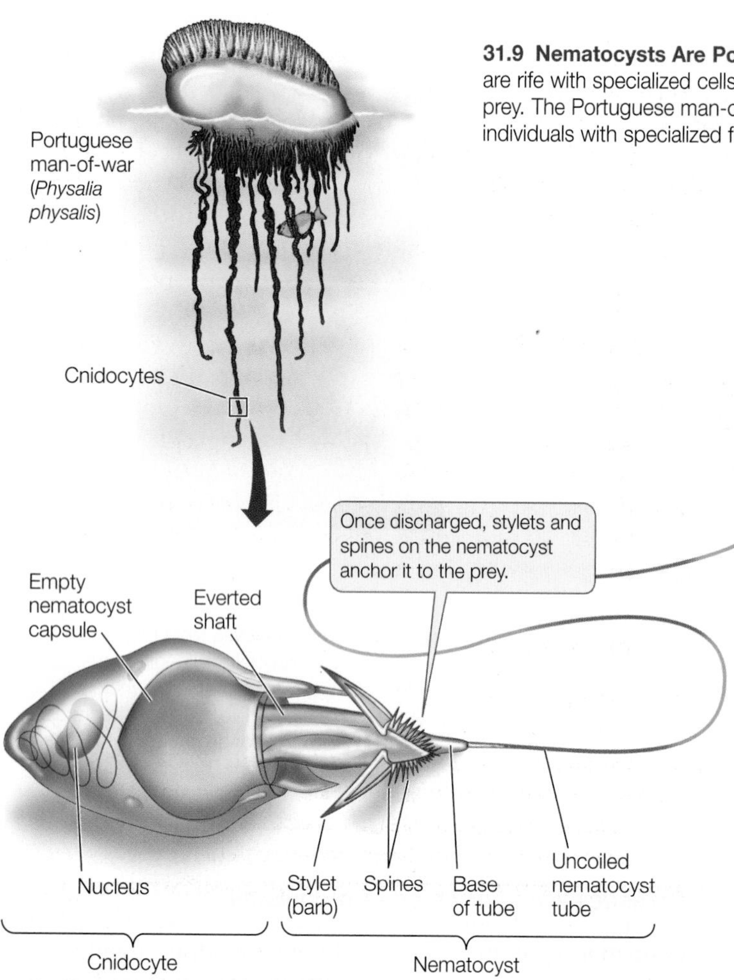


**31.9 Nematocysts Are Potent Weapons** The tentacles of the Portuguese man-of-war, a cnidarian, are rife with specialized cells that contain stinging organelles called nematocysts, which inject toxin into prey. The Portuguese man-of-war is a colonial organism, composed of many physiologically integrated individuals with specialized functions.

Parasites that live inside their hosts are called *endoparasites*, and they are often morphologically very simple. They can often function without a digestive system because they absorb food directly from the host's gut or bodily tissues. Many flatworms are endoparasites of humans and other mammals, as described in Chapter 32.

Parasites that live outside their hosts are called *ectoparasites*, and they are generally more complex morphologically than endoparasites. Ectoparasites have digestive tracts and mouthparts that enable them to pierce the host's tissues or suck on their body fluids. Fleas and ticks are widely known ectoparasitic arthropods that many humans have unfortunately experienced.

### Detritivores live off the remains of other organisms

**Detritivores** feed on decomposing organic matter, or *detritus*. In so doing, they perform an important ecosystem function by returning nutrients to the environment in a state that can be used by other organisms. Detrivores are common in any soil with high organic content, as well as on the ocean floors. Well-known detrivores include earthworms and other annelids, millipedes, and many insects and crustaceans.

Charles Darwin became fascinated with the action of earthworms, and wrote a book called *The Formation of Vegetable Mould Through the Action of Worms*. He was particularly impressed by the importance of earthworms in soil formation. Darwin conducted many interesting experiments to establish how quickly earthworms break down organic matter and build up rich soils.

## 31.3 RECAP

**Animals have many ways of acquiring food. Filter feeders strain food particles from the water or air. Herbivores have digestive adaptations that allow them to eat plants, whereas predators are physically adapted to capture and subdue other animals (prey) and consume them. Parasites obtain their nutrition from a host organism. Detritivores consume decaying organic matter and make important nutrients available for use by other organisms.**

- What adaptations are necessary for animals that eat plants? What adaptations are needed for a predatory lifestyle? See pp. 652–653
- Looking at the brief overview of some of the diverse feeding modes of animals presented here, and using whatever you already know about different animals, how useful do you think feeding behavior would be as a criterion for grouping animals into phylogenetic categories?

As an animal grows from a single cell into a larger, more complex adult, its body structure, its diet, and the environment in which it lives may all change. In the next section we describe some animal life cycles and discuss why they are so varied.

## 31.4 How Do Life Cycles Differ among Animals?

The **life cycle** of an animal encompasses its embryonic development, birth, growth to maturity, reproduction, and death. During its life an individual animal ingests food, grows, interacts with other individuals of the same and other species, and reproduces.

In some groups of animals, newborns bear many similarities to adults (a pattern called *direct development*). Newborns of most species, however, differ dramatically from adults. Consider, for example a **larva** (plural *larvae*), the immature life stage that some organisms take early in their life cycle before assuming an adult form. Some of the most striking life cycle changes are found among insects such as beetles, flies, moths, butterflies, and bees, which undergo radical changes (called *metamorphosis*) between their larval and adult stages (**Figure 31.10**). In these animals, one stage may be specialized for feeding and the other for reproduction. Adults of most moth species, for example, do not eat. In some animals species, individuals eat during all life cycle stages, but what they eat changes with the stage. For exam-

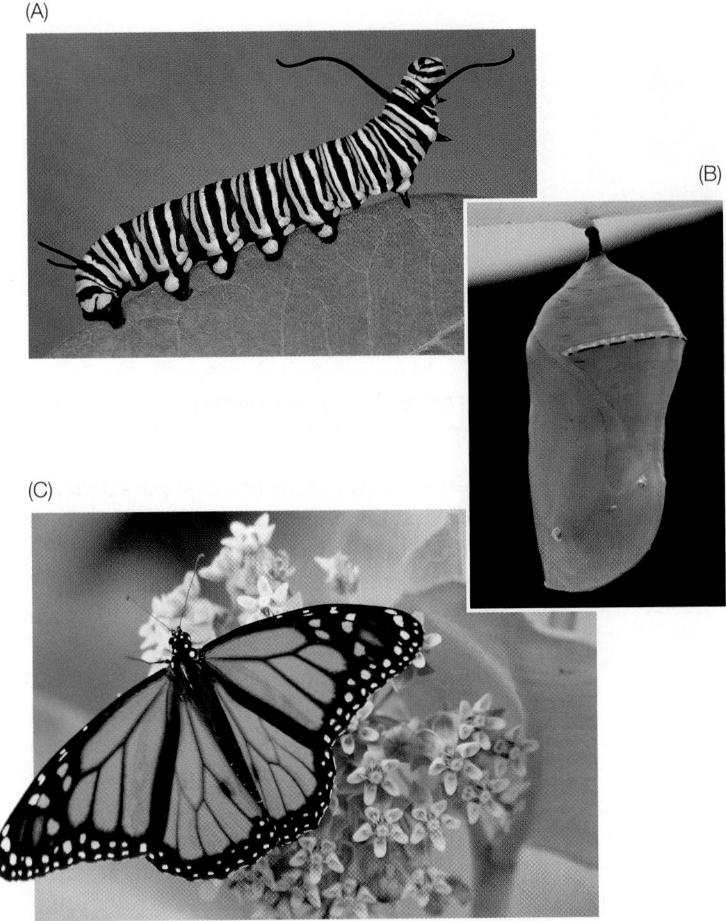


**31.10 A Life Cycle with Metamorphosis** (A) The larval stage (caterpillar) of the monarch butterfly (*Danaus plexippus*) is specialized for feeding. (B) The pupa is the stage during which the transformation to the adult form occurs. (C) The adult butterfly is specialized for dispersal and reproduction.

ple, butterfly larvae, known as *caterpillars*, eat leaves and flowers, whereas most adult butterflies eat only nectar. Having different life cycle stages that are specialized for different activities may increase the efficiency with which an animal performs particular tasks.

## Most animal life cycles have at least one dispersal stage

At some time during its life, an animal moves, or is moved, so few animals die exactly where they were born. Movement of organisms from a parent organism or from an existing population is called **dispersal**.

Animals that are sessile as adults typically disperse as eggs or larvae. Most sessile marine animals discharge their small eggs and sperm into the water, where fertilization takes place. A larva soon hatches and floats freely in the plankton, where it filters small prey from the water.

Many animals that live on the seafloor, including polychaete worms and mollusks, have a radially symmetrical larval form known as a **trochophore** (**Figure 31.11A**); others, such as crustaceans, have a bilaterally symmetrical larval form called a **nauplius** (**Figure 31.11B**). Both types of larvae feed for some time in the plankton before settling on a substratum and transforming into adults.

Although the main dispersal phase of many animals occurs early in the life cycle, some species that are motile as adults disperse when they are mature. A caterpillar, for example, may spend its entire larval stage feeding on a single plant, but after it metamorphoses into a flying adult—a butterfly—it may fly to and lay eggs on other plants located far from the one where it spent its caterpillar days. In some species, individuals disperse during several different life cycle stages.

## No life cycle can maximize all benefits

The common saying "a jack-of-all-trades is master of none" suggests why there are constraints on the evolution of life cycles. The characteristics an animal has in any one life cycle stage may improve its performance in one activity but reduce its performance in another—a situation known as a *trade-off*. An animal that is good at filtering small food particles from the water, for example, probably cannot capture large prey. Similarly, energy devoted to building protective structures such as shells cannot be used for growth.

Some major trade-offs can be seen in animal reproduction. Some animals produce large numbers of small eggs, each with

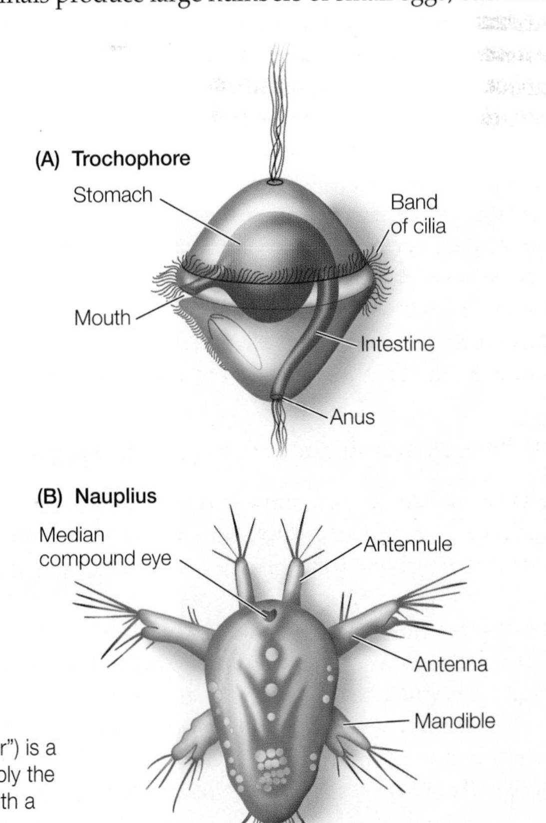


**31.11 Planktonic Larval Forms of Marine Animals** (A) The trochophore ("wheel-bearer") is a distinctive larval form found in several marine animal clades with spiral cleavage, most notably the polychaete worms and the mollusks. (B) This nauplius larva will mature into a crustacean with a segmented body and jointed appendages.

(A) *Rana sylvatica*

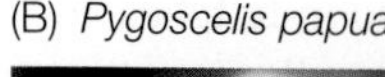

(B) *Pygoscelis papua*

**31.12 Many Small or Few Large** Allocation of energy to eggs requires trade-offs. (A) This wood frog has divided her reproductive energy among a large number of small eggs. (B) This gentoo penguin invested all of her reproductive energy in one large egg.

a small energy store (**Figure 31.12A**). Other animals produce a small number of large eggs, each with a large energy store (**Figure 31.12B**). With a fixed amount of available energy, a female animal can produce many small eggs or a few large eggs, but she cannot produce many large eggs. Thus there is a trade-off between the number of offspring produced and the energy resources each offspring receives from its mother.

The larger the energy store in an egg, the longer an offspring can develop before it must either find its own food or be fed by its parents. Birds of all species lay relatively small numbers of relatively large eggs, but incubation periods vary. In some species, eggs hatch when the young are still helpless (**Figure 31.13A**). Such *altricial* young must be fed and cared for until they can feed themselves; parents can provide for only a small number of altricial offspring. In contrast, some bird species incubate their eggs longer, and the hatchlings are developed to a point that they are able to forage for themselves almost immediately (**Figure 31.13B**). The young of such species are called *precocial*.

### Parasite life cycles evolve to facilitate dispersal and overcome host defenses

Animals that live as internal parasites are bathed in the nutritious tissues of their host or in the digested food that fills their host's digestive tract. Thus they may not need to exert much energy to obtain food, but to survive they must overcome the host's defenses. Furthermore, either they or their offspring must disperse to new hosts while their host is still living, because they die when their host dies.

The fertilized eggs of some parasites are voided with the host's feces and later ingested directly by other host individuals. Most parasite species, however, have complex life cycles involving one or more intermediate hosts and several larval stages (**Figure 31.14**). Some intermediate hosts transport individual parasites directly between other hosts. Others house and support the parasite until another host ingests it. Complex life cycles may thus facilitate the transfer of individual parasites among hosts.

### Colonial organisms are composed of genetically identical, physiologically integrated individuals

Most people tend to view the distinction between individuals and populations as clear-cut. However, in several groups of animals, asexual reproduction without fission can lead to colonies

(A) *Parus caeruleus*

(B) *Branta canadensis*

**31.13 Helpless or Independent** (A) The altricial young of the blue tit are essentially helpless when they hatch. Their parents feed and care for them for several weeks. (B) Canada goose hatchlings are precocial, ready to swim and feed independently almost immediately after hatching.

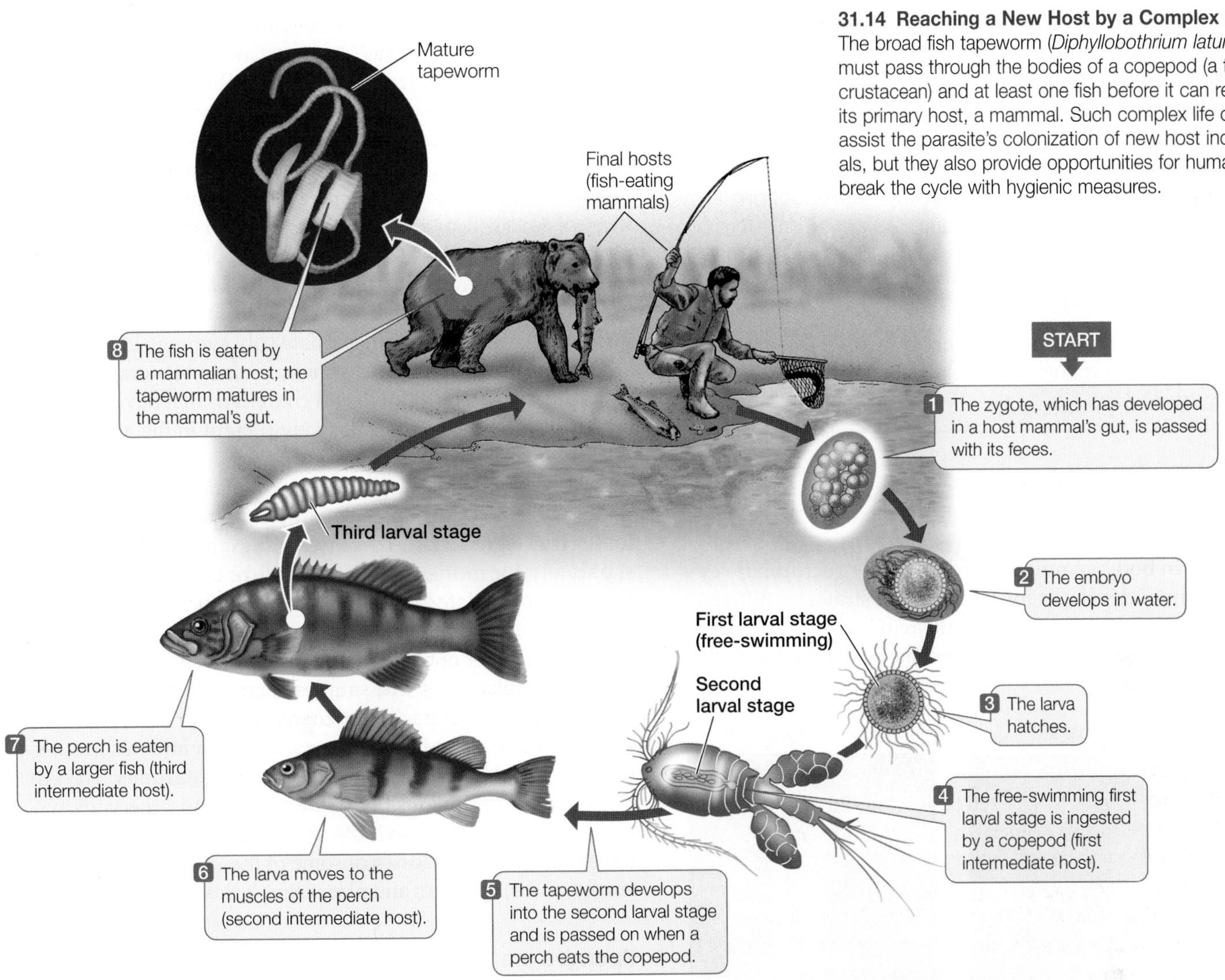

**31.14 Reaching a New Host by a Complex Route** The broad fish tapeworm (*Diphyllobothrium latum*) must pass through the bodies of a copepod (a type of crustacean) and at least one fish before it can reinfect its primary host, a mammal. Such complex life cycles assist the parasite's colonization of new host individuals, but they also provide opportunities for humans to break the cycle with hygienic measures.

Mature tapeworm

Final hosts (fish-eating mammals)

START

1 The zygote, which has developed in a host mammal's gut, is passed with its feces.

2 The embryo develops in water.

First larval stage (free-swimming)

3 The larva hatches.

Second larval stage

4 The free-swimming first larval stage is ingested by a copepod (first intermediate host).

5 The tapeworm develops into the second larval stage and is passed on when a perch eats the copepod.

6 The larva moves to the muscles of the perch (second intermediate host).

7 The perch is eaten by a larger fish (third intermediate host).

Third larval stage

8 The fish is eaten by a mammalian host; the tapeworm matures in the mammal's gut.

of organisms composed of many physiologically integrated individuals, which at first appearance may look much like a single integrated organism. The individuals in a colony are clonal copies of one another, so they are genetically homogeneous. Coloniality has arisen several times among animal groups, with widely varying levels of integration and specialization among the individuals. In some species, colonies are composed of loosely connected but integrated individuals that all function alike (**Figure 31.15**). In other colonial species, the individuals may become specialized for different functions, just as different cell types in multicellular organisms have different functions. The Portuguese man-of-war (see Figure 31.9) is an example of such a colonial animal, as it is composed of many individuals of four specialized body forms, all integrated and functioning together. The individuals in the colony are themselves multicellular, however, unlike the cells of a single multicellular organism.

*Diaperoecia californica*

The individual animals...
...secrete a gelatinous matrix that brings the colony together.

**31.15 Colonial Animals** This colonial bryozoan consists of many asexually reproducing, genetically homogeneous, physiologically interacting individuals. The colony looks much like a single individual with many parts, but in fact it is many individuals acting together.

## 31.4 RECAP

**Many animals have a larval stage that clearly differs from the adult in morphology. In some animals, the larval form is a dispersal stage; in other species, the adults are more likely to disperse than are larvae. In several groups of organisms, asexual reproduction without fission leads to coloniality.**

- How do trade-offs constrain the evolution of life cycles? **See pp. 655–656**
- Do you understand the differences between a single multicellular organism and coloniality? **See p. 657 and Figure 31.15**

## 31.5 What Are the Major Groups of Animals?

Variations in body symmetry, body cavity structure, life cycles, patterns of development, and survival strategies differentiate millions of animal species. In the remainder of this chapter and in Chapters 32 and 33, we become acquainted with the major animal groups and learn how the general characteristics described in this chapter apply to each of them.

**Table 31.1** summarizes the living members of the major animal groups. **Bilateria** is a large monophyletic group embracing all animals other than sponges, placozoans, ctenophores, and cnidarians. Some major traits that support the monophyly of bilaterians (in addition to genomic analyses) are the presence of three distinct cell layers in embryos (triploblasty) and the presence of at least seven Hox genes (see Chapters 19 and 20). Although bilateral symmetry is often viewed as a synapomorphy of bilaterians (and the trait gives the group its name), some groups of cnidarians are also bilaterally symmetrical. Recent studies have shown that the genetic basis of bilateral symmetry is the same in bilaterians and cnidarians that have bilateral symmetry, so this feature was likely present in the ancestor of these two groups.

Bilaterian animals comprise the two major categories mentioned earlier in this chapter, the protostomes and the deuterostomes (see Figure 31.1). These two groups have been evolving separately for more than 500 million years—since the early Cambrian or the late Precambrian. We describe the protostomes in Chapter 32 and the deuterostomes in Chapter 33.

The remainder of this chapter describes those animal groups that are not bilaterians. The simplest animals, the sponges, have no distinct tissue types. Placozoans have four cell types and weakly differentiated tissue layers. All other animals groups, including the bilatarians, are known as **eumetazoans**. The eumetazoans have obvious body symmetry, a gut, a nervous system, and tissues organized into distinct organs (although there have been secondary losses of some of these structures in some eumetazoans). Sponges and placozoans lack all of these features.

### Sponges are loosely organized animals

**Sponges** are the simplest animals. Although they have some specialized cells, they have no distinct embryonic cell layers and no true organs. Early naturalists thought sponges were plants because they were sessile and lacked body symmetry.

Sponges have hard skeletal elements called **spicules**, which may be small and simple or large and complex. There are three major groups of sponges, which separated soon after the split between sponges and the rest of the animals. Members of two groups (*glass sponges* and *demosponges*) have skeletons composed of silicaceous spicules made of hydrated silicon dioxide (**Figure 31.16A,B**). These spicules are remarkable in having greater flexibility and toughness than synthetic glass rods of similar length. Members of the third group, the *calcareous sponges*, take their name

(A) *Xestospongia testudinaria*

(B) *Euplectella aspergillum*

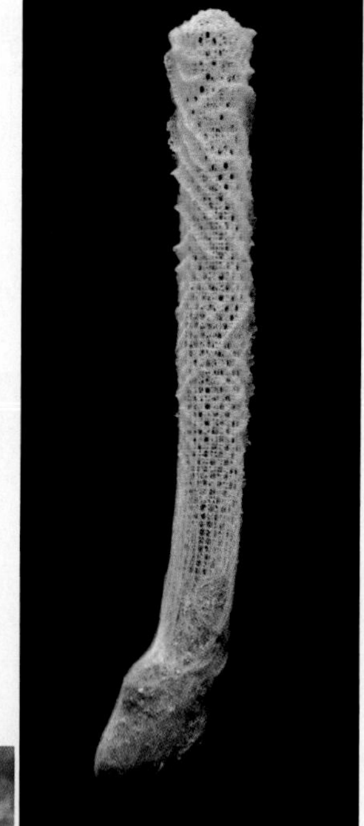

(C) *Leucilla nuttingi*

**31.16 Sponge Diversity** (A) The great majority of sponge species are demosponges, such as these Pacific barrel sponges. The system of pores and water canals "typical" of the sponge body plan is apparent. (B) The supporting structures of both demosponges and glass sponges are silicaceous spicules, seen here in the skeleton of a glass sponge. (C) The skeletons of calcareous sponges are made of calcium carbonate.

TABLE 31.1

**Summary of Living Members of the Major Animal Groups**

| | APPROXIMATE NUMBER OF LIVING SPECIES DESCRIBED | MAJOR GROUPS |
|---|---|---|
| Sponges | 9,000 | Demosponges, glass sponges, calcareous sponges |
| Placozoans | 2 | |
| Ctenophores | 150 | |
| Cnidarians | 11,000 | Anthozoans: Corals, sea anemones |
| | | Hydrozoans: Hydras and hydroids |
| | | Scyphozoans: Jellyfishes |
| **PROTOSTOMES** | | |
| Arrow worms | 100 | |
| **Lophotrochozoans** | | |
| Bryozoans | 4,500 | |
| Flatworms | 25,000 | Free-living flatworms; flukes and tapeworms (all parasitic); monogeneans (ectoparasites of fishes) |
| Rotifers | 1,800 | |
| Ribbon worms | 1,000 | |
| Phoronids | 20 | |
| Brachiopods | 335 | |
| Annelids | 16,500 | Polychaetes (all marine) |
| | | Clitellates: Earthworms, freshwater worms, leeches |
| Mollusks | 100,000 | Monoplacophorans |
| | | Chitons |
| | | Bivalves: Clams, oysters, mussels |
| | | Gastropods: Snails, slugs, limpets |
| | | Cephalopods: Squids, octopuses, nautiloids |

| | APPROXIMATE NUMBER OF LIVING SPECIES DESCRIBED | MAJOR GROUPS |
|---|---|---|
| **Ecdysozoans** | | |
| Kinorhynchs | 150 | |
| Loriciferans | 100 | |
| Priapulids | 16 | |
| Horsehair worms | 320 | |
| Nematodes | 25,000 | |
| Onychophorans | 150 | |
| Tardigrades | 800 | |
| Arthropods: | | |
| Crustaceans | 52,000 | Crabs, shrimps, lobsters, barnacles, copepods |
| Hexapods | 1,000,000 | Insects and relatives |
| Myriapods | 14,000 | Millipedes, centipedes |
| Chelicerates | 98,000 | Horseshoe crabs, arachnids (scorpions, harvestmen, spiders, mites, ticks) |
| **DEUTEROSTOMES** | | |
| Echinoderms | 7,000 | Crinoids (sea lilies and feather stars); brittle stars; sea stars; sea daisies; sea urchins; sea cucumbers |
| Hemichordates | 100 | Acorn worms and pterobranchs |
| Urochordates | 3,000 | Ascidians (sea squirts) |
| Cephalochordates | 30 | Lancelets |
| Vertebrates | 62,000 | Hagfish; lampreys |
| | | Cartilaginous fishes |
| | | Ray-finned fishes |
| | | Coelacanths; lungfishes |
| | | Amphibians |
| | | Reptiles (including birds) |
| | | Mammals |

from their calcium carbonate skeletons (**Figure 31.16C**). There is some question about the monophyly of sponges. Analyses of some gene sequences suggest that calcareous sponges are actually more closely related to the eumetazoans than to the other groups of sponges. However, genomic analyses that combine information from many genes do support the monophyly of sponges.

The body plan of sponges of all three groups—even large ones, which may reach a meter or more in length—is an aggregation of cells built around a water canal system. Water, along with any food particles it contains, enters the sponge by way of small pores and passes into the water canals or a central atrium, where choanocytes capture food particles (see Figure 31.7).

A skeleton of simple or branching spicules, and often a complex network of elastic fibers, supports the body of most sponges. Sponges also have an extracellular matrix, composed of collagen, adhesive glycoproteins, and other molecules, that holds the cells together. Most species are filter feeders; a few species are carnivores that trap prey on hook-shaped spicules that protrude from the body surface.

Most of the 9,000 species of sponges are marine animals; only about 50 species live in fresh water. Sponges come in a wide variety of sizes and shapes that are adapted to different movement patterns of water. Sponges living in intertidal or shallow subtidal environments with strong wave action are firmly attached to the substratum. Most sponges that live in slowly flowing water are flattened and are oriented at right angles to the direction of current flow. They intercept water and the prey it contains as it flows past them.

Sponges reproduce both sexually and asexually. In most species, a single individual produces both eggs and sperm, but individuals do not self-fertilize. Water currents carry sperm from one individual to another. Asexual reproduction is by budding and fragmentation.

### Placozoans are abundant but rarely observed



As discussed in the opening of this chapter, **placozoans** are structurally very simple animals with only a few distinct cell types (see the photograph on p. 645). Individuals in the mature, asymmetrical life stage are usually observed adhering to surfaces (such as the glass of aquariums, where they were first discovered, or to rocks and other hard substrates in nature). Their structural simplicity—they have no mouth, gut, or nervous system—initially led biologists to suspect they might be the sister group of all other animals. Most phylogenetic analyses have not supported this hypothesis, however, and some aspects of the placozoans' structural simplicity may be secondarily derived. They are generally considered to have a diploblastic body plan, with upper and lower epithelial (surface) layers that sandwich a layer of contractile fiber cells.

Recent studies have found that placozoans have a pelagic (open-ocean) stage that is capable of swimming (**Figure 31.17**), but the life history of placozoans is incompletely known. Most studies have focused on the larger adherent stages that are usually found in aquariums, where they appear after being inadvertently collected with other marine organisms. The transparent nature and small size of placozoans make them very difficult to observe in nature. Nonetheless, it is known that placozoans can reproduce both asexually as well as sexually, although the details of their sexual reproduction are mostly unknown. As we noted in the opening story, placozoans have mostly been studied in aquariums, although we now know that pelagic-stage placozoans are abundant in warm seas around the world.

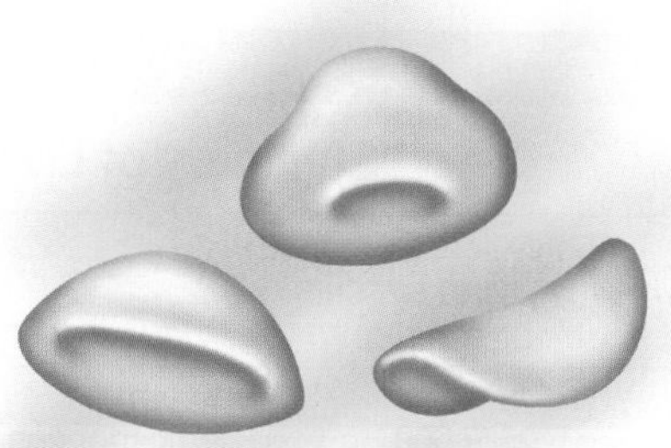

**31.17 A Swimming Stage in the Life of a Placozoan** Placozoans are tiny and transparent, and thus difficult to observe in nature. Recent studies have found a small, weakly swimming pelagic stage of placozoan to be abundant in many warm tropical and subtropical seas.

### Ctenophores are radially symmetrical and diploblastic



**Ctenophores**, also known as *comb jellies*, lack most of the Hox genes found in all other eumetazoans. Ctenophores have a radially symmetrical, diploblastic body plan. The two cell layers are separated by an inert, gelatinous extracellular matrix called **mesoglea**. Ctenophores have a *complete gut*: food enters through a mouth, and wastes are eliminated through two anal pores.

Ctenophores move by beating cilia rather than muscular contractions. Most of the 150 known species have eight comblike rows of cilia-bearing plates, called **ctenes** (**Figure 31.18**). The feeding tentacles of ctenophores are covered with cells that discharge adhesive material when they contact prey. After capturing its prey, a ctenophore retracts its tentacles to bring the food to its mouth. In some species, the entire surface of the body is coated with sticky mucus that captures prey. Most ctenophores

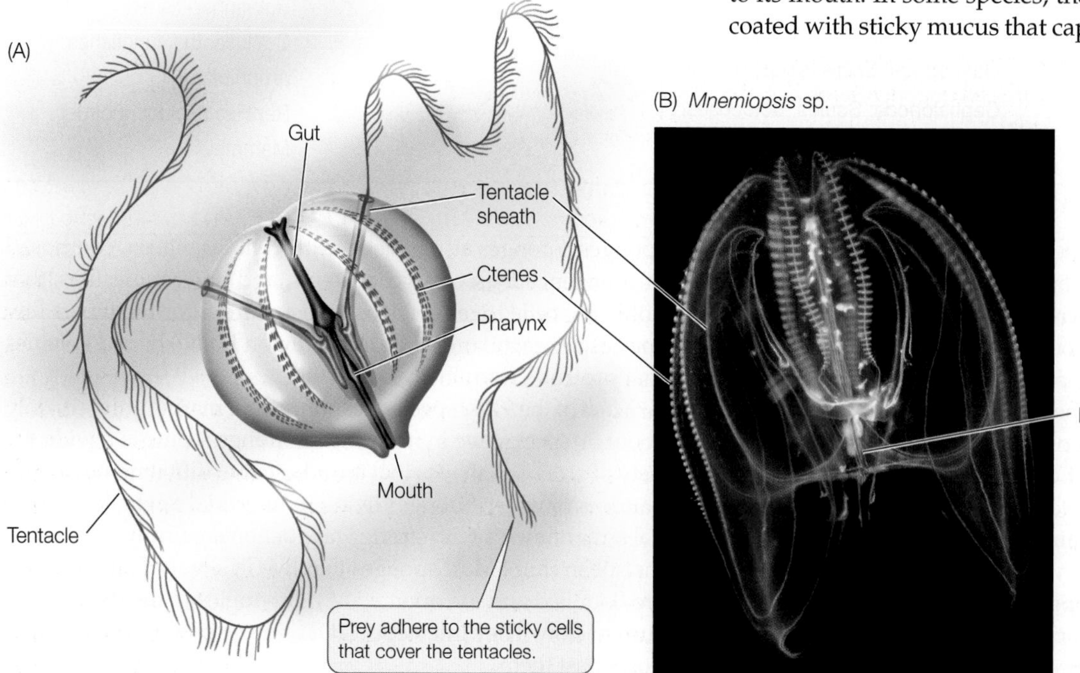


**31.18 Comb Jellies Feed with Tentacles** (A) The body plan of a typical ctenophore. The long, sticky tentacles sweep through the water, efficiently harvesting small prey. (B) This comb jelly, photographed in Sydney Harbor, Australia, has short tentacles.

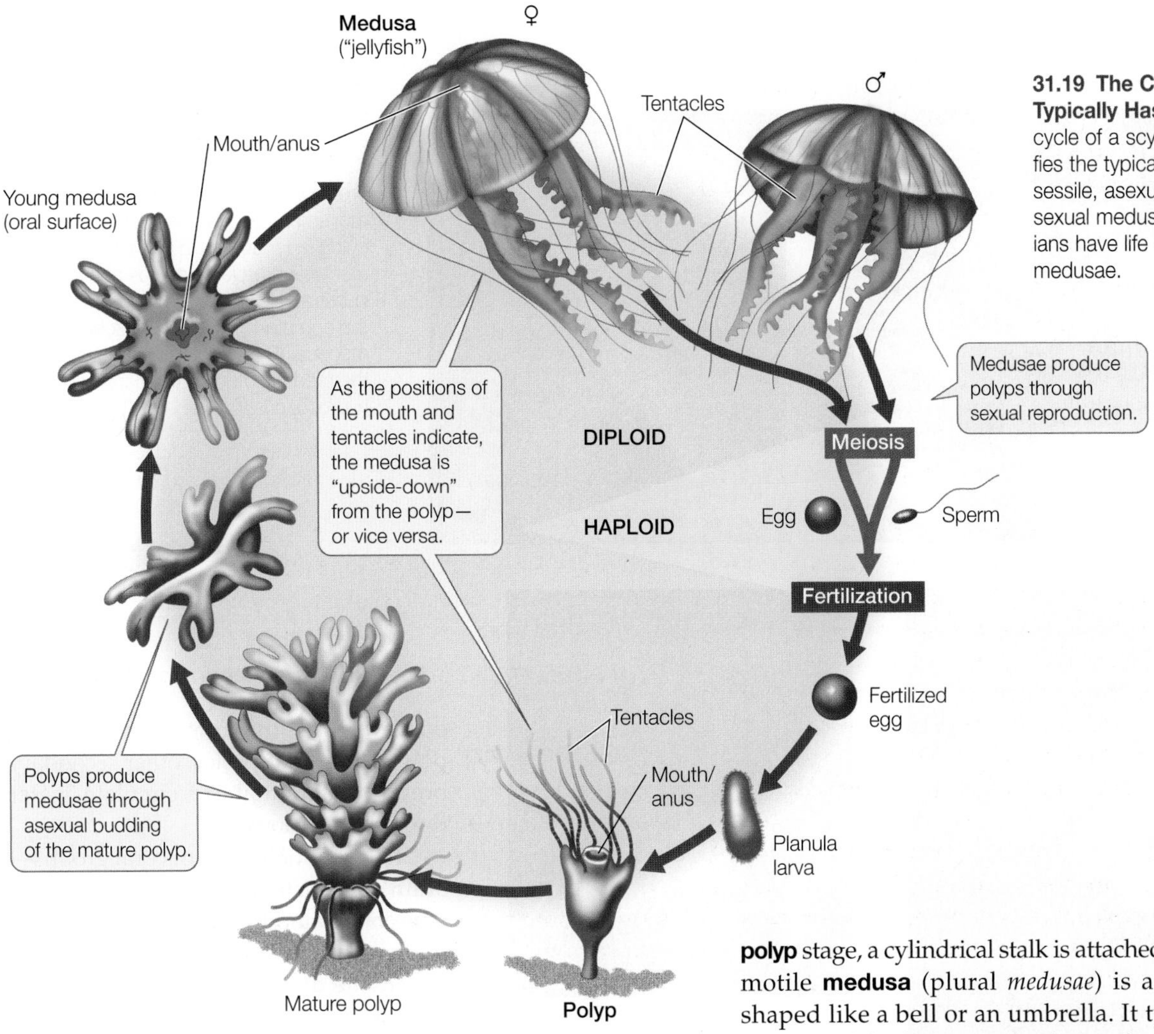

**31.19 The Cnidarian Life Cycle Typically Has Two Stages** The life cycle of a scyphozoan (jellyfish) exemplifies the typical cnidarian body forms: the sessile, asexual polyp; and the motile, sexual medusa. Some species of cnidarians have life cycles that lack polyps or medusae.

eat small planktonic organisms, although some eat other ctenophores. They are common in open seas and can become abundant in protected bodies of water, where large populations of ctenophores can damage local ecosystems.

Ctenophore life cycles are uncomplicated. Gametes are released into the body cavity and then discharged through the mouth or the anal pores. Fertilization takes place in open seawater. In nearly all species, the fertilized egg develops directly into a miniature ctenophore that gradually grows into an adult.

### Cnidarians are specialized carnivores

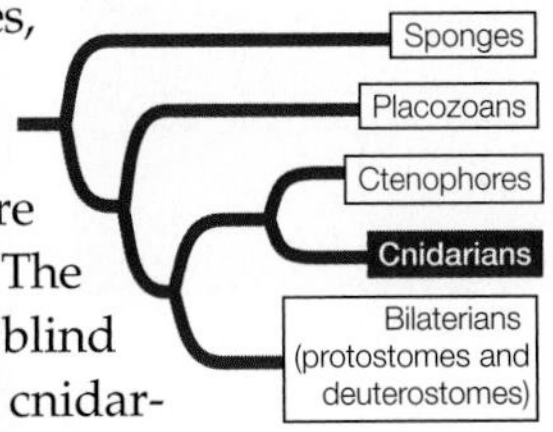

The **cnidarians** (jellyfishes, sea anemones, corals, and hydrozoans) may be the sister group of the ctenophores, although some biologists think they are more closely related to the bilaterians. The mouth of a cnidarian is connected to a blind sac called the **gastrovascular cavity** (a cnidarian thus does not have a complete gut). The gastrovascular cavity functions in digestion, circulation, and gas exchange, and it also acts as a hydrostatic skeleton. The single opening serves as both mouth and anus.

The life cycle of many cnidarians has two distinct stages, one sessile and the other motile (**Figure 31.19**), although one or the other of these stages is absent in some groups. In the sessile **polyp** stage, a cylindrical stalk is attached to the substratum. The motile **medusa** (plural *medusae*) is a free-swimming stage shaped like a bell or an umbrella. It typically floats with its mouth and feeding tentacles facing downward. Mature polyps produce medusae by asexual budding. Medusae then reproduce sexually, producing eggs or sperm by meiosis and releasing the gametes into the water. A fertilized egg develops into a free-swimming, ciliated larva called a **planula**, which eventually settles to the bottom and develops into a polyp.

**yourBioPortal.com**
GO TO Animated Tutorial 31.1 • Life Cycle of a Cnidarian

Cnidarians have epithelial cells with muscle fibers whose contractions enable the animals to move, as well as simple *nerve nets* that integrate their body activities. They also have specialized structural molecules (collagen, actin, and myosin). They are specialized carnivores, using the toxin in their nematocysts to capture relatively large and complex prey (see Figure 31.9). Some cnidarians, including many corals and anemones, gain additional nutrition from photosynthetic endosymbionts that live in their tissues. Cnidarians, like ctenophores, are largely made up of inert mesoglea. They have low metabolic rates and can survive in environments where they encounter prey only infrequently.

Of the roughly 11,000 living cnidarian species, all but a few live in the oceans (**Figure 31.20**). The smallest cnidarians can hardly be seen without a microscope. The largest known jellyfish is 2.5 meters in diameter, and some colonial siphonophores

(A) *Anthopleura elegantissima*

(B) *Pteroeides* sp.

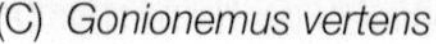

(C) *Gonionemus vertens*

(D) *Polyorchis penicillatus*

**31.20 Diversity among Cnidarians** (A) The nematocyst-studded tentacles of this sea anemone from British Columbia are poised to capture large prey carried to the animal by water movement. (B) The sea pen is a colonial cnidarian that lives in soft bottom sediments and projects polyps above the substratum. (C) This jellyfish illustrates the complexity of a scyphozoan medusa. (D) The internal structure of the medusa of a North Atlantic colonial hydrozoan is visible here.

(which include the Portuguese man-of-war; see Figure 31.9) can reach lengths in excess of 30 meters. Here we describe three clades of cnidarians that have many species: anthozoans, scyphozoans, and hydrozoans.

**ANTHOZOANS** Members of the **anthozoan** clade include sea anemones, sea pens, and corals. Sea anemones (see Figure 31.20A), all of which are solitary, are widespread in both warm and cold ocean waters. Sea pens (see Figure 31.20B), by contrast, are colonial. Each colony consists of two or more different kinds of polyps. The primary polyp has a lower portion anchored in the bottom sediment and a branched upper portion that projects above the substratum. Along the upper portion, the primary polyp produces smaller secondary polyps by budding. Some of these secondary polyps differentiate into feeding polyps; in some species, other secondary polyps differentiate to circulate water through the colony.

The common names of coral groups—brain corals, staghorn corals, and organ pipe corals, among others—often describe their appearance (**Figure 31.21A**). Corals are sessile and colonial. The polyps of

(A) *Diploria labyrinthiformis*

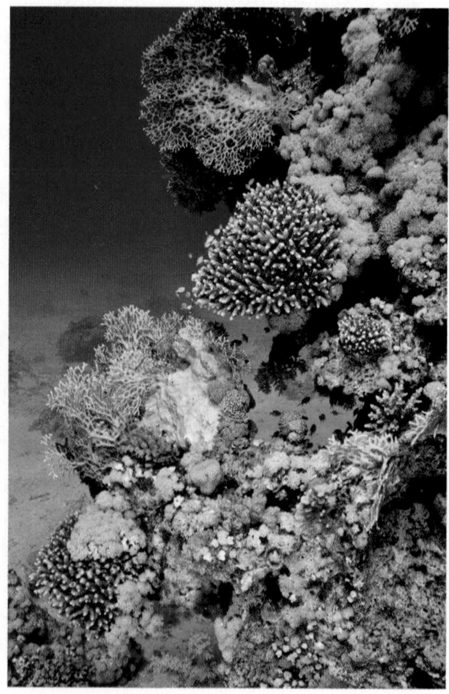

**31.21 Corals** (A) The descriptive common name of this Caribbean coral is "brain coral." (B) Many different coral species form this reef in the Red Sea between Egypt and the Arabian Peninsula.

♀ Medusa
♂
Oral surface
Gonad
Medusae develop asexually within an enlarged polyp.
Meiosis
DIPLOID
HAPLOID
Egg
Sperm
Fertilization
Eggs produced by medusae are fertilized in the open water by sperm produced by other medusae.
Fertilized egg
Planula larva
The polyps of the hydrozoan *Obelia* are interconnected and share a gastrovascular cavity.
Larvae settle on the substratum and grow into polyps.

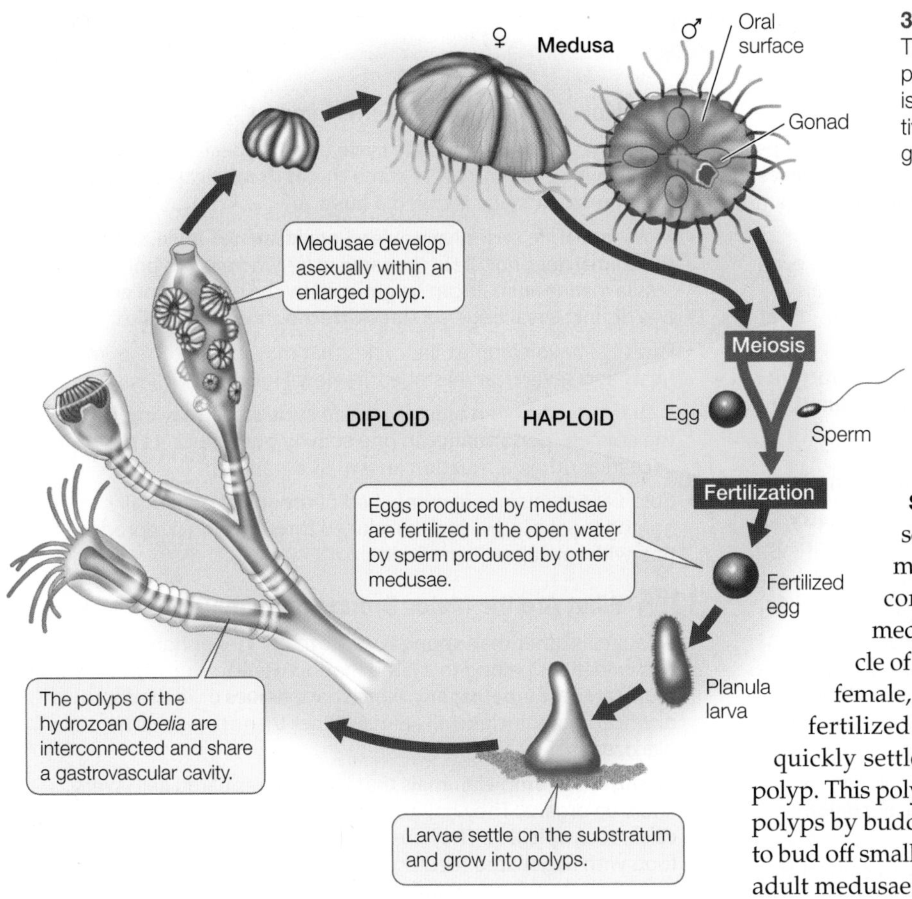

**31.22 Hydrozoans Often Have Colonial Polyps** The polyps in a hydrozoan colony may differentiate to perform specialized tasks. In the species whose life cycle is diagrammed here, the medusa is the sexual reproductive stage, producing eggs and sperm in organs called gonads.

most species form a skeleton by secreting a matrix of organic molecules on which they deposit calcium carbonate, which forms the eventual skeleton of the coral colony. As the colony grows, old polyps die but their calcium carbonate skeletons remain. Living corals form a layer on top of a growing bank of skeletal remains, eventually forming chains of islands and reefs (**Figure 31.21B**). The Great Barrier Reef along the northeastern coast of Australia is a system of coral formations more than 2,000 kilometers long—about the distance from New York City to St. Louis. A single coral reef in the Red Sea has been calculated to contain more material than all the buildings in the major cities of North America combined.

Corals flourish in clear, nutrient-poor tropical waters. They grow well in such environments because unicellular photosynthetic dinoflagellates live endosymbiotically within their cells. These dinoflagellates provide the corals with products of photosynthesis; the corals, in turn, provide the dinoflagellates with nutrients and a place to live. This endosymbiotic relationship explains why reef-forming corals are restricted to clear surface waters, where light levels are high enough to support photosynthesis.

Coral reefs throughout the world are threatened both by global warming, which is raising the temperatures of shallow tropical ocean waters, and by polluted runoff from development on adjacent shorelines. Warming can lead to the loss of coral endosymbionts (*coral bleaching*; see Figure 27.8A), and an overabundance of nitrogen in runoff gives an advantage to algae, which overgrow and eventually smother the corals.

**SCYPHOZOANS** The several hundred species of scyphozoans are all marine. The mesoglea of their medusae is thick and firm, giving rise to their common name—jellyfishes or sea jellies. The medusa rather than the polyp dominates the life cycle of scyphozoans. An individual medusa is male or female, releasing eggs or sperm into the open sea. The fertilized egg develops into a small planula larva that quickly settles on a substratum and develops into a small polyp. This polyp feeds and grows and may produce additional polyps by budding. After a period of growth, the polyp begins to bud off small medusae, which feed, grow, and transform into adult medusae (see Figures 31.19 and 31.20C).

**HYDROZOANS** Hydrozoans have diverse life cycles. The polyp typically dominates the life cycle, but some species have only medusae and others have only polyps. Most hydrozoans are colonial (see Figure 31.20D). A single larval planula eventually gives rise to a colony of many polyps, all interconnected and sharing a continuous gastrovascular cavity (**Figure 31.22**). Within such a colony, some polyps have tentacles with many nematocysts; they capture prey for the colony. Other individuals lack tentacles and are unable to feed but are specialized for the asexual production of medusae. Still others are fingerlike and defend the colony with their nematocysts.

## 31.5 RECAP

**Bilaterian animals are in one of two major clades, protostomes or deuterostomes. The non-bilaterian animals—the sponges, placozoans, ctenophores, and cnidarians—have diverse life cycles, feeding strategies, and growth forms.**

- Why are sponges are considered to be animals, even though they lack the complex body structures found among most other animal groups? **See pp. 658–659 and Figure 31.7**
- Describe some major features of the following groups: sponges, placozoans, ctenophores, and cnidarians. **See pp. 658–663 and Figures 31.16–31.19**

# CHAPTER SUMMARY

## 31.1 What Characteristics Distinguish the Animals?

- Animals share a set of derived traits not found in other groups of organisms. These traits include similarities in the sequences of many of their genes, the structure of their cell junctions, and the components of their extracellular matrix.
- Patterns of embryonic development provide clues to the evolutionary relationships among animals. **Diploblastic** animals develop two embryonic cell layers; **triploblastic** animals develop three cell layers.
- Differences in their patterns of early development characterize two major clades of triploblastic animals, the **protostomes** and the **deuterostomes**.

## 31.2 What Are the Features of Animal Body Plans?

- Animal **body plans** can be described in terms of **symmetry**, **body cavity** structure, **segmentation**, and type of appendages.
- A few animals have no symmetry, but most animals have either **radial symmetry** or **bilateral symmetry**. Review Figure 31.3
- Most animals with radial symmetry move slowly or not at all, whereas most animals with bilateral symmetry are able to move more rapidly. Many bilaterally symmetrical animals exhibit **cephalization**, with sensory and nervous tissues in an anterior head.
- On the basis of their body cavity structure, animals can be described as **acoelomates**, **pseudocoelomates**, or **coelomates**. Review Figure 31.4
- Segmentation, which takes many forms, improves control of movement, especially if the animal also has appendages.

## 31.3 How Do Animals Get Their Food?

- **Motile** animals can move to find food; **sessile** animals stay in one place and capture food by filter feeding or through interaction with endosymbionts.
- **Filter feeders** strain small organisms and organic molecules from their environment.
- **Predators** have morphological features such as sharp teeth, beaks, and claws that enable them to capture and subdue animal **prey**.
- **Herbivores** consume plants, usually without killing them.
- **Parasites** live in or on other organisms and obtain nutrition from these host individuals.
- **Detritivores** consume decaying organic matter and return nutrients to the ecosystem.

## 31.4 How Do Life Cycles Differ among Animals?

- The stages of an animal's **life cycle** may be specialized for different activities. An immature stage that is dramatically different from the adult stage is called a **larva**.
- Most animal life cycles have at least one **dispersal** stage, so that the animal does not die in the same place where it was born. Many sessile marine animals can be grouped by the presence of one of two distinct larval dispersal stages: **trochophore** or **nauplius**.
- Parasites have complex life cycles that may involve one or more hosts and several larval stages. Review Figure 31.14
- A characteristic of an animal or a life cycle stage may improve the animal's performance in one activity but reduce its performance in another, a situation known as a trade-off.
- Colonial organisms are composed of groups of genetically homogeneous individuals produced through clonal reproduction without subsequent fission.

## 31.5 What Are the Major Groups of Animals?

- All animals other than sponges, placozoans, ctenophores, and cnidarians belong to a large monophyletic group called the **Bilateria**. **Eumetazoans**, which have tissues organized into distinct organs, include all animals other than sponges and placozoans.
- **Sponges** are simple animals that lack differentiated cell layers and true organs. They have skeletons made up of silicaceous or calcareous **spicules**. They create water currents and capture food with flagellated feeding cells called choanocytes. Choanocytes are an evolutionary link between the animals and the choanoflagellate protists. Review Figure 31.7
- **Placozoans** have only a few cell types and lack true organs, although their simplicity may be secondarily derived. They are abundant in warm seas of the world, but their transparent form and small size make them difficult to observe.
- **Ctenophores** and many **cnidarians** are radially symmetrical, although some cnidarians are bilaterally symmetrical. The mature adherent stage of placozoans is asymmetrical.
- The two cell layers of ctenophores are separated by an inert extracellular matrix called **mesoglea**.
- Ctenophores move by beating fused plates of cilia called **ctenes**. Review Figure 31.18
- The life cycle of most cnidarians has two distinct stages: a sessile **polyp** stage and a motile **medusa**. A fertilized egg develops into a free-swimming larval **planula**, which settles to the bottom and develops into a polyp. Review Figures 31.19 and 31.22, **ANIMATED TUTORIAL 31.1**

**SEE WEB ACTIVITIES 31.1 and 31.2 for a concept review of this chapter.**

# SELF-QUIZ

1. The body plan of an animal is
   *a.* its general structure.
   *b.* the integrated functioning of its parts.
   *c.* its general structure and the integrated functioning of its parts.
   *d.* its general structure and its evolutionary history.
   *e.* the integrated functioning of its parts and its evolutionary history.
2. A bilaterally symmetrical animal can be divided into mirror images by
   *a.* any plane through the midline of its body.
   *b.* any plane from its anterior to its posterior end.
   *c.* any plane from its dorsal to its ventral surface.
   *d.* any plane through the midline of its body from its anterior to its posterior end.
   *e.* a single plane through the midline of its body from its dorsal to its ventral surface.

3. Among protostomes, cleavage of the fertilized egg is
   *a.* delayed while the egg continues to mature.
   *b.* always radial.
   *c.* spiral, radial, or idiosyncratic.
   *d.* triploblastic.
   *e.* diploblastic.
4. Many parasites evolved complex life cycles because
   *a.* they are too simple to disperse readily.
   *b.* they are poor at recognizing new hosts.
   *c.* they were driven to it by host defenses.
   *d.* complex life cycles increase the probability of a parasite's transfer to a new host.
   *e.* their ancestors had complex life cycles and they simply retained them.
5. Bilateral symmetry
   *a.* is found only among bilaterians.
   *b.* is characteristic of all sponges.
   *c.* is characteristic of all ctenophores.
   *d.* is characteristic of all cnidarians.
   *e.* none of the above
6. In the common ancestor of protostomes and deuterostomes, the pattern of early cleavage was
   *a.* spiral.
   *b.* radial.
   *c.* biradial.
   *d.* deterministic.
   *e.* haphazard.
7. A fluid-filled body cavity can function as a hydrostatic skeleton because
   *a.* fluids are moderately compressible.
   *b.* fluids are highly compressible.
   *c.* fluids are incompressible at physiological pressures.
   *d.* fluids have the same density as body tissues.
   *e.* fluids can be moved by ciliary action.
8. Which of the following is *not* a feature that enables some animals to capture large prey?
   *a.* Sharp teeth
   *b.* Claws
   *c.* Toxins
   *d.* A filtering device
   *e.* Tentacles with stinging cells
9. The sponge body plan is characterized by
   *a.* a mouth and digestive cavity but no muscles or nerves.
   *b.* muscles and nerves but no mouth or digestive cavity.
   *c.* a mouth, digestive cavity, and spicules.
   *d.* muscles and spicules but no digestive cavity or nerves.
   *e.* the lack of a mouth, digestive cavity, muscles, or nerves.
10. The endosymbiotic dinoflagellates present in many corals
    *a.* provide the corals with the products of photosynthesis.
    *b.* allow corals to grow rapidly in clear, nutrient-poor tropical waters
    *c.* can be lost when environmental conditions change.
    *d.* all of the above
    *e.* none of the above

## FOR DISCUSSION

1. Differentiate among the members of each of the following sets of related terms:
   *a.* radial symmetry/bilateral symmetry
   *b.* protostome/deuterostome
   *c.* diploblastic/triploblastic
   *d.* coelomate/pseudocoelomate/acoelomate
2. In this chapter we listed some of the traits shared by all animals that convince most biologists that all animals are descendants of a single common ancestral lineage. In your opinion, which of these traits provides the most compelling evidence that animals are monophyletic? If morphological and molecular data do not agree, should one type of evidence be given greater weight? If so, which one?
3. Describe some features that allow animals to capture prey that are larger and more complex than they themselves are.
4. Why is bilateral symmetry strongly associated with cephalization, the concentration of sensory organs in an anterior head?
5. How does a slow metabolic rate enable an animal to live in an unproductive environment?

## ADDITIONAL INVESTIGATION

The discoveries that pelagic (swimming in open ocean) stages of placozoans are abundant in warm seas, and that the mature stages settle onto smooth surfaces, suggest how these organisms might be collected and surveyed. What sampling procedures might you use to discover whether placozoans occur at a particular location along a coast?

## WORKING WITH DATA (GO TO yourBioPortal.com)

**Reconstructing Animal Phylogeny** In this exercise, you will analyze a subset of protein sequences that provide evidence about the evolutionary relationships of some major animal groups. Using the parsimony method described in Chapter 22, you will build a phylogenetic tree of some of the major animal lineages depicted in Figure 31.1.

# 32 Protostome Animals

## Tiny parasites exert mind control

Most people have never heard of strepsipterans, and even those who *have* heard of them have probably never seen one. They don't know what they are missing! These tiny insects—there are some 600 species of them—parasitize hundreds of other insect species, including bees, wasps, ants, grasshoppers, and cockroaches. Males and females of most strepsipteran species parasitize the same host species, although in one clade males parasitize ants whereas females parasitize grasshoppers.

Strepsipteran males and females are often so different that even determining that they are members of the same species requires DNA analysis. They also have some of the strangest life cycles of any animals. Once grown to maturity within their hosts (whose internal organs they consume), the males of most species emerge looking like a "typical" insect. The females also consume the host from within, but they usually remain inside the host. A mature female extrudes her head and part of her body from the body of her host. The extruded body part contains an opening that receives sperm from a male. Much later, this opening becomes an exit for the strepsipteran larvae. The host insects are left dead or severely damaged and produce no offspring of their own.

Strepsipterans dramatically change the behavior of their hosts in ways that help them complete their life cycles at the expense of their hosts' reproduction. For example, when wasps—a typical host—are parasitized by strepsipterans, the parasites generate signals that induce the wasps to leave their nest and form a mating aggregation. This aggregation, however, serves the strepsipterans, not the wasps. Once the wasps aggregate, the male strepsipterans emerge from their hosts to search for and mate with female strepsiterans, whose heads are now poking out of the bodies of other wasps.

Adult male strepsipterans live only a few hours, during which they must find a female and mate. Because the protruding part of a female's body is barely visible, the males have unusually large eyes,

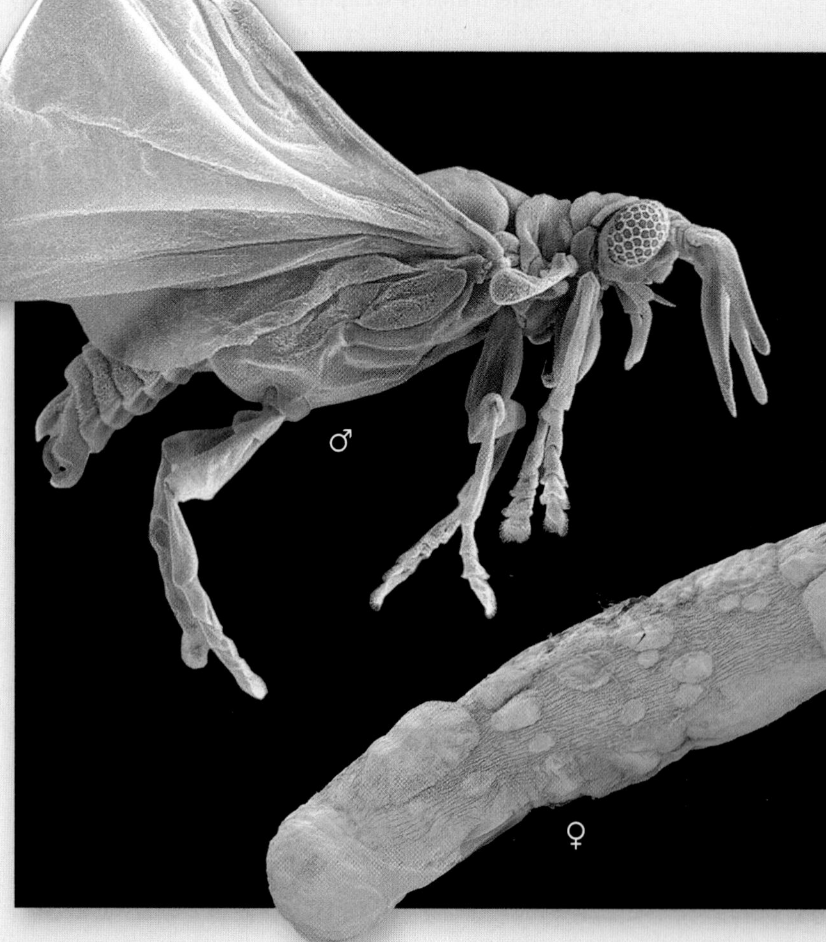


**Same Parasite, Different Lifestyles**
Strepsipterans are parasitic insects that grow to maturity within host insects. Most male strepsipterans look insectlike upon reaching maturity, when they leave their host to find and mate with a very different-appearing female strepsipteran. Females remain inside their host.

**A Host Insect** Strepsipterans parasitize many different insect species; wasps of the genus *Polistes* (paper wasps) are common hosts. Three female strepsipterans (arrows) can be seen on this *P. dorsalis* individual.

and about 75 percent of their brain cells are allocated to vision. This highly developed sensory system evolved as an adaptation to help the male find a female.

Strepsipterans and their hosts are all insects, and insects account for more than half of all described species on Earth. Other protostome groups, such as mollusks, nematodes, crabs, spiders, and ticks, are also species-rich. Many protostome species are parasites. Parasites often live within their hosts and absorb nutrients through their body walls. Some parasites, including the strepsipterans, can more technically be described as *parasitoids*, which consume the host's tissues as they develop from eggs laid on or in the host's body, ultimately killing the host.

**IN THIS CHAPTER** we will describe the characteristics of protostome animals and describe the members of two major protostome clades, the lophotrochozoans and the ecdysozoans. We will give particular attention to the arthropods, a species-rich group of ecdysozoans with rigid exoskeletons and jointed appendages.

## CHAPTER OUTLINE



## 32.1 What Is a Protostome?

You may recall from Chapter 31 that the embryos of diploblastic animals have two cell layers: an outer ectoderm and an inner endoderm (see Section 31.1). Sometime after the origin of the diploblastic animals (the placozoans, ctenophores, and cnidarians), a third embryological germ layer evolved—the mesoderm, which lies between the ectoderm and the endoderm. Mesoderm is found in the two major triploblastic animal clades, the protostomes and the deuterostomes. If we were to judge solely on the basis of numbers, both of species and of individuals, the protostomes would emerge as by far the more successful of the two groups.

As noted in Chapter 31, the name *protostome* means "mouth first." In protostomes that have an embryonic blastopore, this opening becomes the mouth as the animal develops. In contrast, in deuterostomes ("mouth second"), the blastopore becomes the anal opening of the digestive tract (see Figure 31.2). The protostomes can be further divided into two major clades—the *lophotrochozoans* and the *ecdysozoans*—largely on the basis of DNA sequence analysis (**Figure 32.1**).

The protostomes are extremely varied, but they are all bilaterally symmetrical animals whose bodies exhibit two major derived traits:

- An anterior *brain* that surrounds the entrance to the digestive tract
- A ventral *nervous system* consisting of paired or fused longitudinal nerve cords

Other aspects of protostome body organization differ widely from group to group (**Table 32.1**). Although the common ancestor of the protostomes had a *coelom*, subsequent modifications of the coelom distinguish many protostome lineages. In at least one protostome lineage (the flatworms), the coelom has been lost (that is, the flatworms reverted to an *acoelomate* state). Some lineages are characterized by a *pseudocoelom*, which you may recall is a body cavity lined with mesoderm in which the internal organs are suspended (see Figure 31.4). In two of the most prominent protostome clades, the coelom has been highly modified:

- The *arthropods* lost the ancestral condition of the coelom over the course of evolution. Their internal body cavity has become a *hemocoel*, or "blood chamber," in which fluid from an open circulatory system bathes the internal organs before returning to blood vessels.

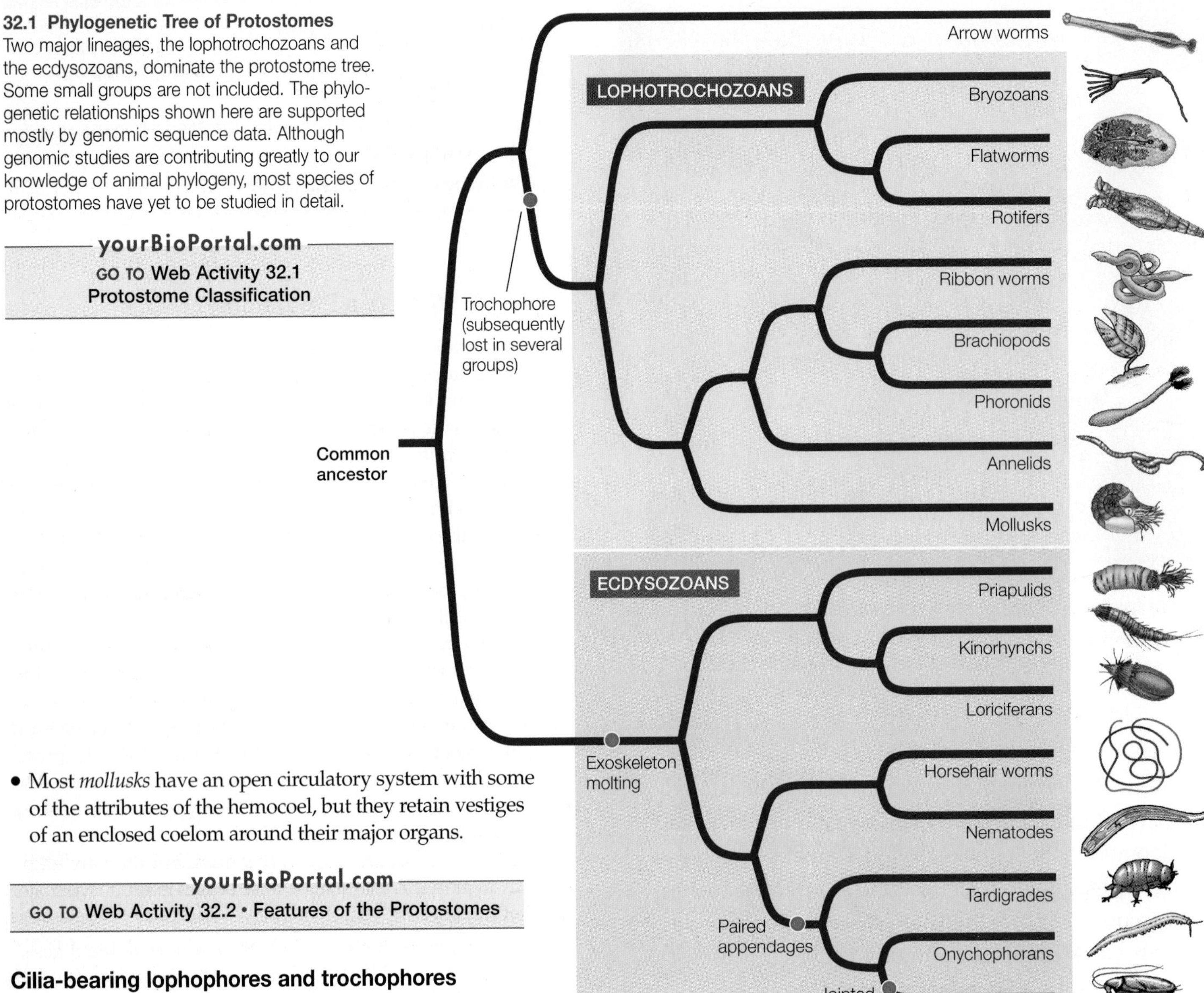


**32.1 Phylogenetic Tree of Protostomes** Two major lineages, the lophotrochozoans and the ecdysozoans, dominate the protostome tree. Some small groups are not included. The phylogenetic relationships shown here are supported mostly by genomic sequence data. Although genomic studies are contributing greatly to our knowledge of animal phylogeny, most species of protostomes have yet to be studied in detail.

**yourBioPortal.com**
GO TO **Web Activity 32.1 Protostome Classification**

- Most *mollusks* have an open circulatory system with some of the attributes of the hemocoel, but they retain vestiges of an enclosed coelom around their major organs.

**yourBioPortal.com**
GO TO **Web Activity 32.2 • Features of the Protostomes**

### Cilia-bearing lophophores and trochophores evolved among the lophotrochozoans

**Lophotrochozoans** derive their name from two different features that involve cilia: a feeding structure known as a *lophophore* and a free-living larva known as a *trochophore*. Neither a lophophore nor a trochophore is universal for all lophotrochozoans, however.

Several distantly related groups of lophotrochozoans (including bryozoans, brachiopods, and phoronids) have a **lophophore**, a circular or U-shaped ring of ciliated, hollow tentacles around the mouth (**Figure 32.2**). This complex structure is an organ for both food collection and gas exchange. Biologists once grouped taxa that have lophophores together as *lophophorates*, but it is now clear that they are not one another's closest relatives. The lophophore appears to have evolved independently several times, or else it is an ancestral feature of lophotrochozoans and has been lost in many groups. Nearly all animals with a lophophore are sessile as adults. They use the tentacles and cilia of the lophophore to capture small floating organisms from the water. Other sessile lophotrochozoans use less well developed tentacles for the same purpose.

Some lophotrochozoans, especially in their larval form, use cilia for locomotion. The larval form known as a **trochophore** moves by beating a band of cilia (see Figure 31.11A). This movement of cilia also brings plankton closer to the larva, where the plankton can be captured and ingested (similar in function to the cilia of the lophophore). Trochophore larvae are found among many of the major groups of lophotrochozoans, including the mollusks, annelids, ribbon worms, and bryozoans. This larval form was probably present in the common ancestor of lophotrochozoans, although it has been subsequently lost in several lineages.

As discussed Chapter 31, some lophotrochozoans (including flatworms, ribbon worms, annelids, and mollusks) exhibit a form of cleavage in early development known as *spiral cleavage*. Some biologists group these taxa together as *spiralians*, although phylogenetic analyses of gene sequences do not support the

**TABLE 32.1**

**Anatomical Characteristics of Some Major Protostome Groups[a]**

| GROUP | BODY CAVITY | DIGESTIVE TRACT | CIRCULATORY SYSTEM |
|---|---|---|---|
| Arrow worms | Coelom | Complete | None |
| **LOPHOTROCHOZOANS** | | | |
| Flatworms | None | Dead-end sac | None |
| Rotifers | Pseudocoelom | Complete | None |
| Bryozoans | Coelom | Complete | None |
| Brachiopods | Coelom | Complete in most | Open |
| Phoronids | Coelom | Complete | Closed |
| Ribbon worms | Coelom | Complete | Closed |
| Annelids | Coelom | Complete | Closed or open |
| Mollusks | Reduced coelom | Complete | Open except in cephalopods |
| **ECDYSOZOANS** | | | |
| Horsehair worms | Pseudocoelom | Greatly reduced | None |
| Nematodes | Pseudoceolom | Complete | None |
| Arthropods | Hemocoel | Complete | Open |

[a] Note that all protostomes have bilateral symmetry.

Bryozoans can oscillate, rotate, and retract their lophophore tentacles.
Plumatella repens
100 µm


**32.2 Bryozoans Use Their Lophophore to Feed** The extended lophophore dominates the anatomy of the colonial bryozoans. This species inhabits fresh water, although most bryozoans are marine. Bryozoan colonies can grow to contain more than a million individuals, all stemming from the asexual reproduction of the colony's founder.

species with spiral cleavage as monophyletic. Nonetheless, spiral cleavage may have been present in the lophotrochozoan ancestor, with several subsequent losses of this cleavage pattern.

Many lineages of lophotrochozoans are *wormlike*, which means they are bilaterally symmetrical, legless, soft-bodied, and at least several times longer than they are wide. A wormlike body form enables animals to burrow efficiently through marine sediment or soil. However, as we will describe later in this chapter, the *mollusks*—the most familiar of the lophotrochozoans to many people—have a very different body organization.

### Ecdysozoans must shed their cuticles

**Ecdysozoans** have an external covering, or **cuticle**, which is secreted by the underlying *epidermis* (the outermost cell layer). The cuticle provides these animals with both protection and support. Once formed, however, the cuticle cannot grow. How, then, can ecdysozoans increase in size? They do so by shedding, or **molting**, the cuticle and replacing it with a new, larger one. This molting process gives the clade its name (Greek *ecdysis*, "to get out of").

A recently discovered fossil of a Cambrian soft-bodied arthropod, preserved in the process of molting, shows that molting evolved more than 500 million years ago (**Figure 32.3A**). An increasingly rich array of molecular and genetic evidence, including a set of Hox genes shared by all ecdysozoans, suggests they have a single common ancestor. Thus molting of a cuticle is a trait that may have evolved only once during animal evolution.

Before an ecdysozoan molts, a new cuticle is already forming underneath the old one. Once the old cuticle is shed, the new one expands and hardens. Until it has hardened, though, the animal is vulnerable to its enemies, both because its outer surface is easy to penetrate and because an animal with a soft cuticle moves slowly or not at all (**Figure 32.3B**).

In many ecdysozoans that have wormlike bodies, the cuticle is relatively thin and flexible; it offers the animal some protection but provides only modest body support. A thin cuticle allows the exchange of gases, minerals, and water across the body surface, but it restricts the animal to moist habitats. Many species of ecdysozoans with thin cuticles live in marine sediments from which they obtain prey, either by ingesting sediments and extracting organic material from them, or by capturing larger prey using a toothed *pharynx* (a muscular organ at the anterior end of the digestive tract). Some freshwater species absorb nutrients directly through their thin cuticles, as do parasitic species that live within their hosts. Many wormlike ecdysozoans are predators, eating protists and small animals.

The cuticles of other ecdysozoans, mainly arthopods, function as external skeletons, or **exoskeletons**. These exoskeletons are thickened by layers of protein and a strong, waterproof polysaccharide called **chitin**. An animal with a rigid, chitin-reinforced exoskeleton can neither move in a wormlike manner nor use cilia for locomotion. A hard exoskeleton also impedes the passage of oxygen and nutrients into the animal, presenting new

(A)

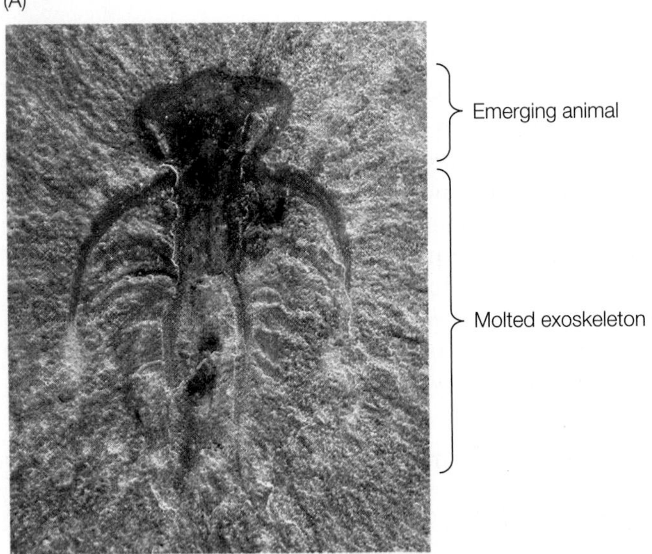


(B) *Phrynus parvulus*

**32.3 Molting: Past and Present** (A) This 500-million-year-old fossil from the Cambrian captured an individual of a long-extinct arthropod species in the process of molting and shows that the molting process is an evolutionarily ancient trait. (B) This whip spider has just emerged from its discarded exoskelton and will be highly vulnerable until its new cuticle has hardened.

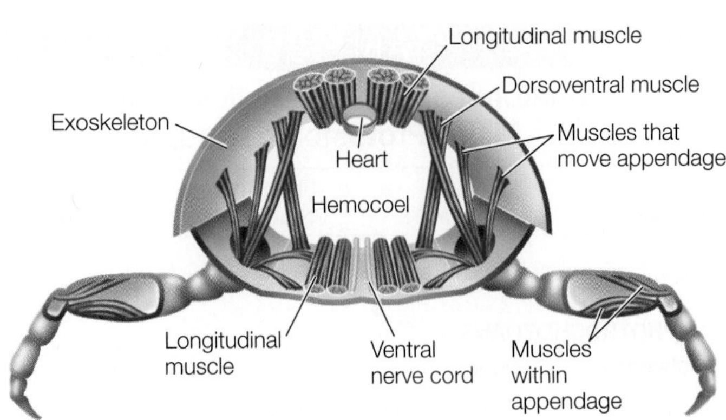


**32.4 Arthropod Skeletons Are Rigid and Jointed** This cross section through a thoracic segment of a generalized arthropod illustrates the arthropod body plan, which is characterized by a rigid exoskeleton with jointed appendages.

challenges in other areas besides growth. New mechanisms of locomotion and gas exchange evolved in those ecdysozoans with hard exoskeletons.

To move rapidly, an animal with a rigid exoskeleton must have body extensions that can be manipulated by muscles. Such *appendages* evolved in the late Precambrian, leading to the **arthropod** ("jointed foot") clade. Arthropod appendages exist in an amazing variety of forms. They serve many functions, including walking and swimming, gas exchange, food capture and manipulation, copulation, and sensory perception. Arthropods grasp food with their mouths and associated appendages and digest it internally. Their muscles are attached to the inside of the exoskeleton. Each segment has muscles that operate that segment and the appendages attached to it (**Figure 32.4**).

The arthropod exoskeleton has had a profound influence on the evolution of these animals. Encasement within a rigid body covering provides support for walking on dry land, and the waterproofing provided by chitin keeps the animal from dehydrating in dry air. Aquatic arthropods were, in short, excellent candidates to invade terrestrial environments. As we will see, they did so several times.

### Arrow worms retain some ancestral developmental features

Nearly all triploblastic animal groups can be readily classified as either protostomes or deuterostomes, but the evolutionary relationships of one small group, the **arrow worms**, were debated for many years. Although the early development of arrow worms seems similar to that of deuterostomes, it is now known that arrow worms merely retain developmental features that are ancestral to triploblastic animals in general. Recent studies of gene sequences clearly identify arrow worms as protostomes. There is still some question as to whether they are the closest relatives of the lophotrochozoans (as shown in Figure 32.1), or possibly the sister group of all other protostomes.

The arrow worm body is divided into three compartments: head, trunk, and tail (**Figure 32.5**). The body is transparent or translucent. Most arrow worms swim in the open sea. A few species live on the seafloor. Their abundance as fossils indicates that they were common more than 500 million years ago. The 100 or so living species of arrow worms are small enough—ranging from 3 millimeters to less than 12 centimeters in length—that their gas-exchange and waste-excretion requirements are met by diffusion through the body surface. They lack a circulatory system; wastes and nutrients are moved around the body in the coelomic fluid, which is propelled by cilia that line the coelom. There is no distinct larval stage. Miniature adults hatch directly from eggs that are fertilized internally following elaborate courtship between two hermaphroditic in-

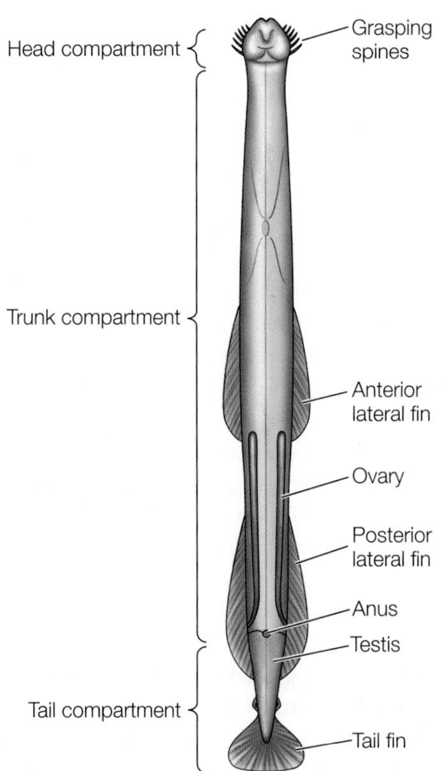


**32.5 An Arrow Worm** Arrow worms have a three-part body organization. Their fins and grasping spines are adaptations for a predatory lifestyle. Individuals are hermaphroditic, producing both eggs and sperm (ovary and testis).

dividuals (each arrow worm produces both male and female gametes).

Arrow worms are stabilized in the water by means of one or two pairs of lateral fins and a tail fin. They are major predators of planktonic organisms in the open ocean, ranging in size from small protists to young fish as large as the arrow worms themselves. An arrow worm typically lies motionless in the water until water movement signals the approach of prey. The arrow worm then darts forward and uses the stiff spines adjacent to its mouth to grasp its prey.

## 32.1 RECAP

**The shared derived traits of protostomes include a blastopore that develops into a mouth, as well as the structure of the anterior brain and ventral nervous system. Several lophotrochozoan groups are characterized by a filter-feeding structure known as a lophophore and/or by cilia-bearing larvae known as trochophores. Ecdysozoans, which have a body covering known as a cuticle, must molt periodically in order to grow.**

- How does an animal's body covering influence the way it breathes, feeds, and moves? See pp. 669–670
- What features made arthropods well adapted for colonizing terrestrial environments? See p. 670

# 32.2 What Features Distinguish the Major Groups of Lophotrochozoans?

Lophotrochozoans come in a variety of sizes and shapes, ranging from relatively simple animals with a blind gut (that is, a gut with only one opening) and no internal transport system to animals with a complete gut (having separate entrance and exit openings) and a complex internal transport system. They include some highly species-rich groups, such as flatworms, annelids, and mollusks. A number of these groups exhibit wormlike bodies, but the lophotrochozoans encompass a wide diversity of morphologies, including a few groups with external shells. Some lophotrochozoan groups have only recently been discovered by biologists.

### Bryozoans live in colonies

The 4,500 species of **bryozoans** ("moss animals") are colonial animals that live in a "house" made of material secreted by the external body wall. Almost all bryozoans are marine, although a few species occur in fresh or brackish water. A bryozoan colony consists of many small (1–2 mm) individuals connected by strands of tissue along which nutrients can be moved (**Figure 32.6**). The colony is created by the asexual reproduction of its founding member, and a single colony may contain as many as 2 million individuals. Rocks in coastal regions in many parts of the world are covered with luxuriant growths of bryozoans. Some bryozoans create miniature reefs in shallow waters. In some species, the individual colony members are differentially specialized for feeding, reproduction, defense, or support.

Bryozoans
Flatworms
Rotifers
Ribbon worms
Brachiopods
Phoronids
Annelids
Mollusks

*Sertella septentrionalis*

**32.6 A Bryozoan Colony** The rigid orange tissue of this bryozoan colony connects and supplies nutrients to thousands of individual animals (see Figure 32.2).

Individual bryozoans in a colony are able to oscillate their lophophore to increase contact with prey. They can also retract it into their "house" (see Figure 32.2). Bryozoans can reproduce sexually by releasing sperm into the water, which carries the sperm to other individuals. Eggs are fertilized internally; developing embryos are brooded before they exit as larvae to seek suitable sites for attachment to the substratum.

### Flatworms and rotifers are structurally diverse relatives

Flatworms and rotifers compose a structurally diverse group, and only recently have their relationships to one another been hypothesized. Before gene sequences were available for phylogenetic analysis, biologists considered the structure of the body cavity to be a critical feature in animal classification. If recent genomic studies prove correct, however, this monophyletic animal group includes both acoelomate subgroups (e.g., the flatworms) and pseudocoelomate subgroups (e.g., the rotifers), and yet the closest relatives of these two groups—the bryozoans—are coelomate. Thus, systematists today conclude that body cavity forms have undergone considerable convergence in the course of animal evolution (see Table 32.1).

**Flatworms** lack specialized organs for transporting oxygen to their internal tissues. Lacking a gas transport system, each cell must be near a body surface, a requirement met by the dorsoventrally flattened body form. The digestive tract of a flatworm consists of a mouth opening into a blind sac. The sac is often highly branched, forming intricate patterns that increase the surface area available for the absorption of nutrients. Some small free-living flatworms are cephalized, with a head bearing chemoreceptor organs, two simple eyes, and a tiny brain composed of anterior thickenings of the longitudinal nerve cords. Free-living flatworms glide over surfaces, powered by broad bands of cilia (**Figure 32.7A**).

Although many flatworms are free-living, most of the species are parasites. Of the parasitic species, most are internal parasites. There are also flatworms that feed externally on animal tissues (living or dead), and some graze on plants. A likely evolutionary transition was from feeding on dead organisms to feeding on the body surfaces of dying hosts to invading and consuming parts of healthy hosts.

Most of the 25,000 species of living flatworms are *tapeworms* and *flukes*; members of these two groups are internal parasites, particularly of vertebrates (**Figure 32.7B**). Because they absorb digested food from the digestive tracts of their hosts, many parasitic flatworms lack digestive tracts of their own. Some cause serious human diseases, such as schistosomiasis, which is common in parts of Asia, Africa, and South America. The species that causes this devastating disease has a complex life cycle involving both freshwater snails and mammals as hosts. Members of another flatworm group, the *monogeneans*, are external parasites of fishes and other aquatic vertebrates. The *turbellarians* include most of the free-living species.

Most **rotifers** are tiny (50–500 μm long)—smaller than some ciliate protists—but they have specialized internal organs (**Figure 32.8**). A complete gut passes from an anterior mouth to a posterior anus; the body cavity is a pseudocoel that functions as a hydrostatic skeleton. Rotifers typically propel themselves through the water by means of rapidly beating cilia rather than by muscular contraction.

The most distinctive organ of rotifers is a conspicuous ciliated organ called the *corona*, which surmounts the head of many species. Coordinated beating of the cilia sweeps particles of

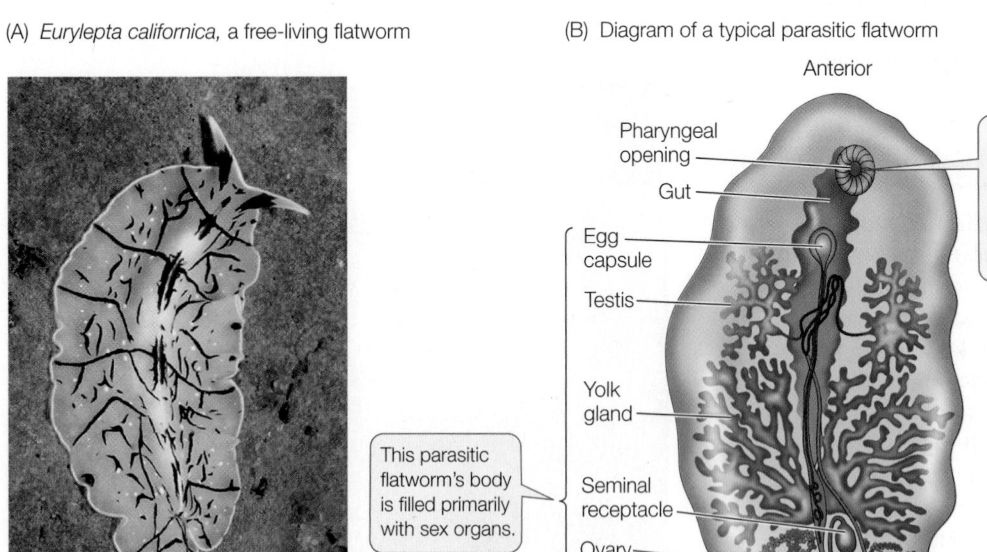


**32.7 Flatworms May Live Freely or Parasitically** (A) Some flatworm species, such as this Pacific marine flatworm, are free-living. (B) The fluke diagrammed here lives parasitically in the gut of sea urchins and is representative of parasitic flatworms. Because their hosts provide all the nutrition they need, these internal parasites do not require elaborate feeding or digestive organs and can devote most of their bodies to reproduction.

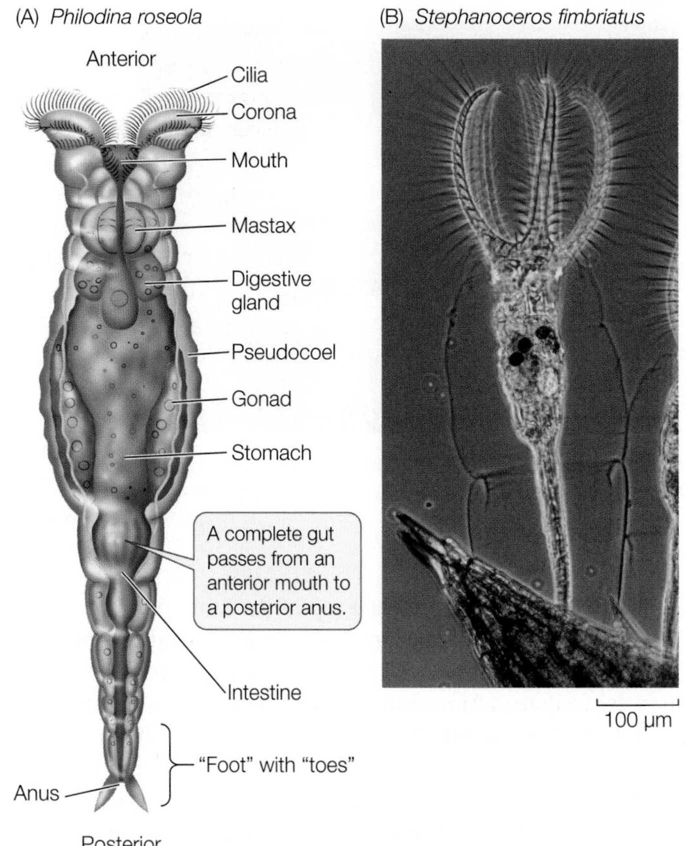


**32.8 Rotifers** (A) The individual diagrammed here reflects the general structure of many rotifers. (B) A micrograph reveals the internal complexity of a rotifer that has ingested photosynthetic protists. This species, one of the larger rotifers, is found in the Great Lakes of North America.

organic matter from the water into the animal's mouth and down to a complicated structure called the *mastax*, in which food is ground into small pieces. By contracting muscles around the pseudocoel, a few rotifer species that prey on protists and small animals can protrude the mastax through their mouth and seize small objects with it.

Most of the 1,800 known species of rotifers live in fresh water. Some species rest on the surfaces of mosses or lichens in a desiccated, inactive state until it rains. When rain falls, they absorb water and become mobile, feeding in the films of water that temporarily cover the plants. Most rotifers live no longer than a few weeks.

Both males and females are found in some species of rotifers, but only females are known among the *bdelloid rotifers* (the *b* in *bdelloid* is silent). Biologists have concluded that the bdelloid rotifers may have existed for tens of millions of years without regular sexual reproduction. In general, lack of genetic recombination leads to the buildup of deleterious mutations, so long-term asexual reproduction typically leads to extinction (see Section 21.4). Recent studies, however, have indicated that bdelloid rotifers may avoid this problem by picking up fragments of genes from their environment during the desiccation–rehydration cycle, which allows genetic recombination among individuals in the absence of direct sexual exchange.

### Ribbon worms have a long, protrusible feeding organ

Bryozoans
Flatworms
Rotifers
Ribbon worms
Brachiopods
Phoronids
Annelids
Mollusks

**Ribbon worms** (nemerteans) have simple nervous and excretory systems similar to those of flatworms. Unlike flatworms, however, they have a complete digestive tract with a mouth at one end and an anus at the other. Small ribbon worms move slowly by beating their cilia. Larger ones employ waves of muscle contraction to move over the surface of sediments or to burrow into them.

Within the body of nearly all of the 1,000 species of ribbon worms is a fluid-filled cavity called the *rhynchocoel*, within which lies a hollow, muscular *proboscis*. The proboscis, which is the worm's feeding organ, may extend much of the length of the body. Contraction of the muscles surrounding the rhynchocoel causes the proboscis to evert explosively through an anterior pore (**Figure 32.9A**). The proboscis may be armed with sharp stylets that pierce prey and discharge paralysis-causing toxins into the wound.

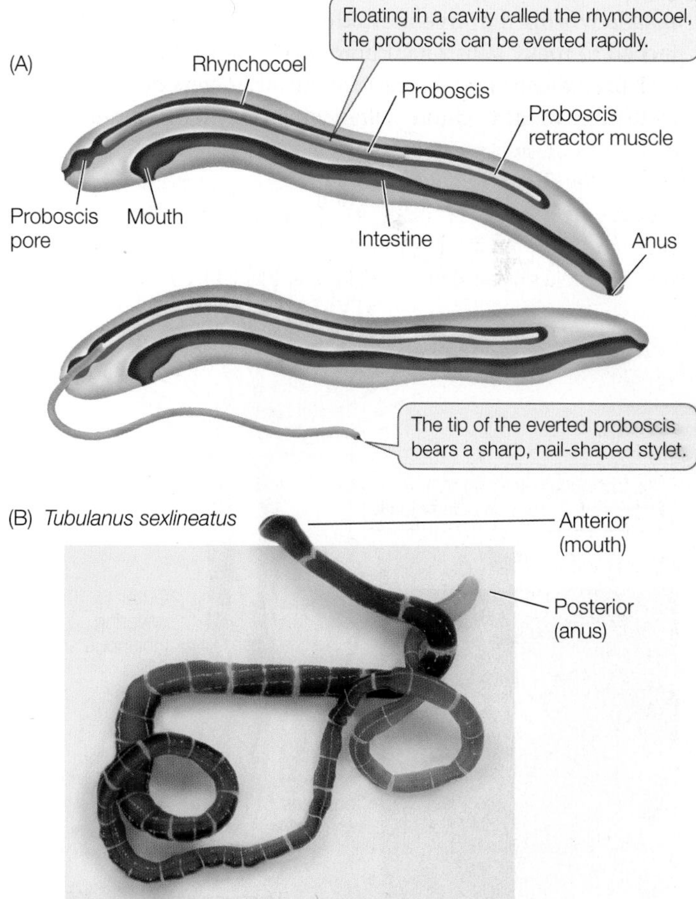


**32.9 Ribbon Worms** (A) The proboscis is the ribbon worm's feeding organ. (B) This large marine nemertean is found in harbors and bays along the Pacific Coast of North America. Its proboscis is not everted in this photograph.

Ribbon worms are largely marine, although there are species that live in fresh water or on land. Most species are less than 20 centimeters long, but individuals of some species reach 20 *meters* or more. Some genera feature species that are conspicuous and brightly colored (**Figure 32.9B**). Recent molecular analyses suggest that ribbon worms may be most closely related to the phoronids and brachiopods.

### Phoronids and brachiopods use lophophores to extract food from the water

Recall that the bryozoans use a lophophore to feed. Phoronids and brachiopods also feed using a lophophore, but this structure may have evolved more than once among these groups. Although neither the phoronids nor the brachiopods are represented by many living species, the brachiopods (which have shells and thus leave an excellent fossil record) are known to have been much more abundant during the Paleozoic and Mesozoic eras.

Bryozoans
Flatworms
Rotifers
Ribbon worms
Brachiopods
Phoronids
Annelids
Mollusks

The 20 known species of **phoronids** are small (5–25 cm long), sessile worms that live in muddy or sandy sediments or attached to rocky substrata. Phoronids are found in marine waters, from the intertidal zone to about 400 meters deep. They secrete tubes made of chitin, within which they live (**Figure 32.10**).

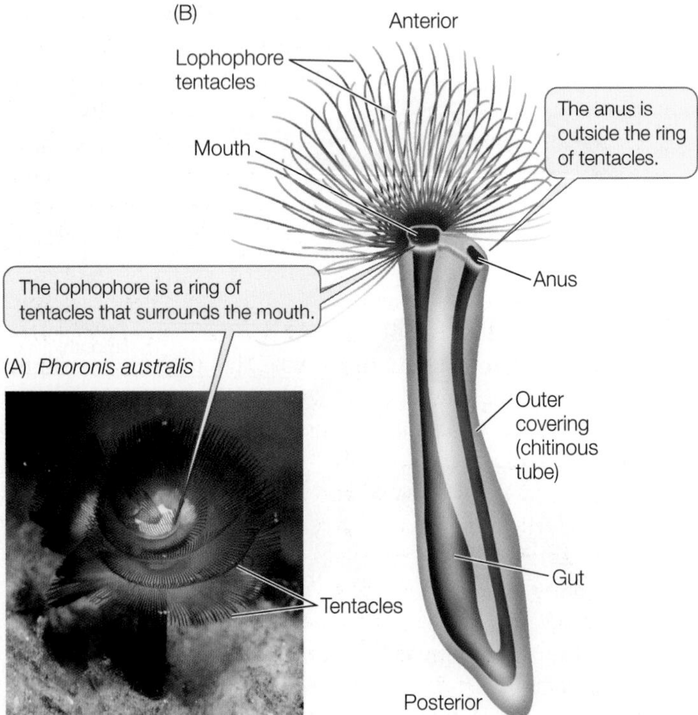


**32.10 Phoronids** (A) The tentacles of this phoronid's lophophore form a spiral. (B) The phoronid gut is U-shaped, as seen in this generalized diagram.

**32.11 A Brachiopod's Lophophore** The lophophore of this North Pacific brachiopod can be seen between the valves of its shell.

Their cilia drive water into the top of the lophophore, and the water exits through the narrow spaces between the tentacles. Suspended food particles are caught and transported to the mouth by ciliary action. Some species release eggs into the water, where they are fertilized, but other species produce large eggs that are fertilized internally and retained in the parent's body, where they are brooded until they hatch.

**Brachiopods** are solitary marine animals. They have a rigid shell that is divided into two parts connected by a ligament (**Figure 32.11**). The two halves can be pulled shut to protect the soft body. Brachiopods superficially resemble bivalve mollusks, but shells have evolved independently in the two groups. The two halves of the brachiopod shell are dorsal and ventral, rather than lateral as in bivalves. The lophophore is located within the shell. The beating of cilia on the lophophore draws water into the slightly opened shell. Food is trapped in the lophophore and directed to a ridge, along which it is transferred to the mouth. Most brachiopods are 4–6 centimeters long.

Brachiopods live attached to a solid substratum or embedded in soft sediments. Most species are attached by means of a short, flexible stalk that holds the animal above the substratum. Gases are exchanged across body surfaces, especially the tentacles of the lophophore. Most brachiopods release their gametes into the water, where they are fertilized. The larvae remain among the plankton for only a few days before they settle and develop into adults.

Brachiopods reached their peak abundance and diversity in Paleozoic and Mesozoic times. More than 26,000 fossil species have been described. Only about 335 species survive, but they remain common in some marine environments.

### Annelids have segmented bodies

The bodies of **annelids** are clearly segmented. As discussed in Section 31.2, segmentation allows an animal to move different parts of its body independently of one another, giving it much better control of its movement. The earliest segmented worms, preserved as fossils from the middle Cambrian, were burrowing marine annelids.

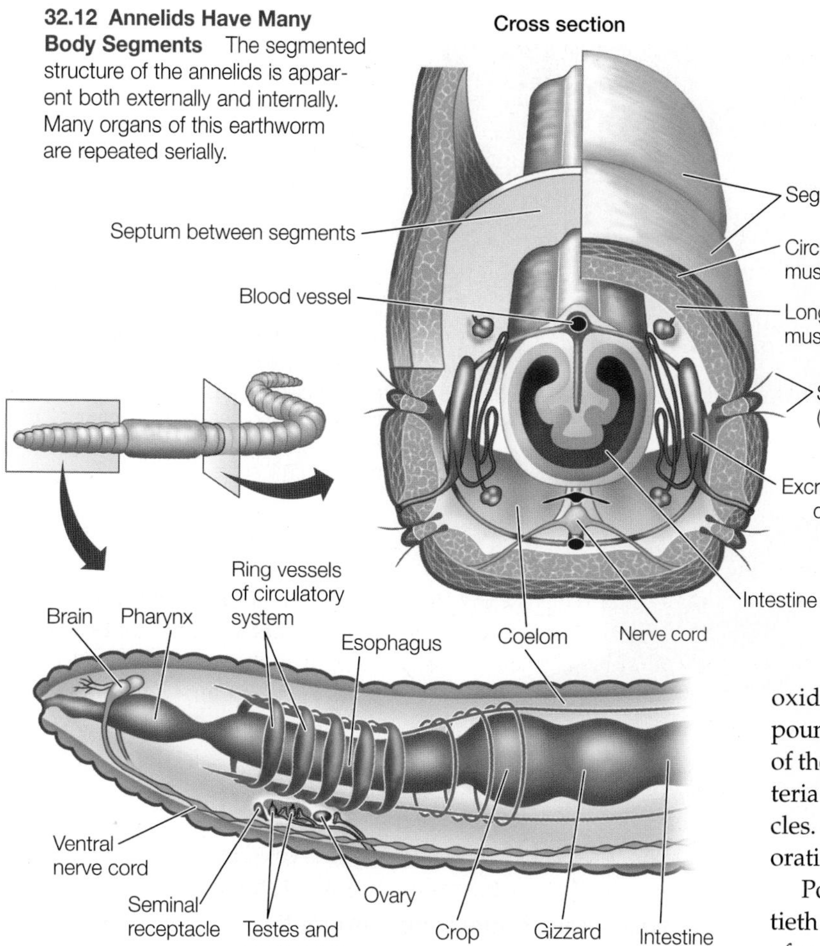


**32.12 Annelids Have Many Body Segments** The segmented structure of the annelids is apparent both externally and internally. Many organs of this earthworm are repeated serially.

In most large annelids, the coelom in each segment is isolated from those in other segments (**Figure 32.12**). A separate nerve center called a *ganglion* controls each segment; nerve cords that connect the ganglia coordinate their functioning. Most annelids lack a rigid external protective covering; instead, they have a thin, permeable body wall that serves as a general surface for gas exchange. Most annelids are thus restricted to moist environments because they lose body water rapidly in dry air. The approximately 16,500 described species live in marine, freshwater, and moist terrestrial environments.



**POLYCHAETES** More than half of all annelid species are commonly known as *polychaetes* ("many hairs"), although this is a descriptive term rather than the name of a single clade. Most polychaetes are marine, and many live in burrows in soft sediments. Most of them have one or more pairs of eyes and one or more pairs of tentacles, with which they capture prey or filter food from the surrounding water, at the anterior end of the body (**Figure 32.13A**; see also Figure 31.6A). In some species, the body wall of most segments extends laterally as a series of thin outgrowths called *parapodia*. The parapodia function in gas exchange, and some species use them to move. Stiff bristles called *setae* protrude from each parapodium, forming temporary contact with the substratum and preventing the animal from slipping backward when its muscles contract. Recent molecular studies indicate that polychaetes are paraphyletic with respect to the remaining annelids.

Members of one polychaete clade, the *pogonophorans*, secrete tubes made of chitin and other substances, in which they live (**Figure 32.13B**). Pogonophorans have lost their digestive tract (they have no mouth or gut). So how do they obtain nutrition? Part of the answer is that pogonophorans can take up dissolved organic matter directly from the sediments in which they live or from the surrounding water. Much of their nutrition, however, is provided by endosymbiotic bacteria that live in a specialized organ known as the *trophosome*. These bacteria oxidize hydrogen sulfide and other sulfur-containing compounds, fixing carbon from methane in the process. The uptake of the hydrogen sulfide, methane, and oxygen used by the bacteria is facilitated by hemoglobin in the pogonophorans' tentacles. It is this hemoglobin that gives the tentacles their red coloration (see Figure 32.13B).

Pogonophorans were not discovered until early in the twentieth century, when the first species were discovered at depths of up to a few hundred meters. In recent decades, deep-sea explorers have found them living many thousands of meters below the ocean surface. In these deep oceanic sediments, they may reach densities of many thousands per square meter. About 160 species have been described. The largest and most remarkable pogonophorans are 2 meters or more in length and live near deep-sea *hydrothermal vents*—volcanic openings in the seafloor through which hot, sulfide-rich water pours. The methane and hydrogen sulfide from these vents provide the raw materials for carbon fixation by the pogonophorans' endosymbiotic bacteria.

**CLITELLATES** The approximately 3,000 described species of this well-supported clade within the annelids are found in freshwater, marine, or terrestrial environments. The clitellates appear to be phylogenetically nested among various groups of polychaetes, although the exact relationships are not yet clear. There are two major groups of clitellates, the oligochaetes and the leeches.

*Oligochaetes* ("few hairs") have no parapodia, eyes, or anterior tentacles, and they have only four pairs of setae bundles per segment. Earthworms—the most familiar oligochaetes—burrow in and ingest soil, from which they extract food particles. All oligochaetes are *hermaphroditic*; that is, each individual is both male and female. Sperm are exchanged simultaneously between two copulating individuals (**Figure 32.13C**). Eggs and sperm are deposited in a cocoon outside the adult's body. Fertilization occurs within the cocoon after it is shed, and when development is complete, miniature worms emerge and immediately begin independent life.

(A) *Eudistylia* sp.

(C) *Lumbricus terrestris*

(B) *Riftia* sp.

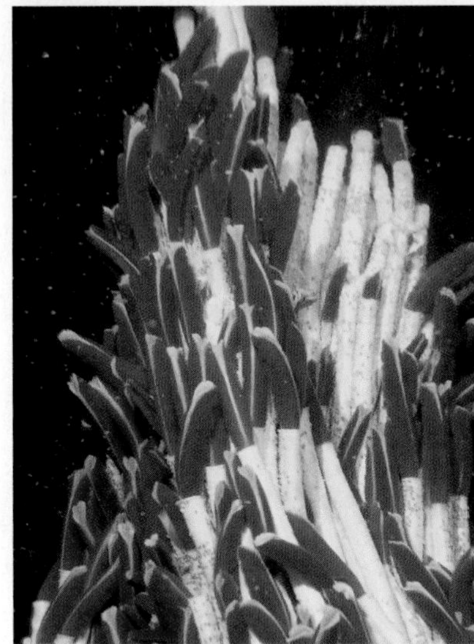

(D) *Hirudo medicinalis*

**32.13 Diversity among the Annelids** (A) "Fan worms" or "feather duster worms" are sessile marine polychaetes that grow in masses, filtering food from the water with their tentacles. This individual has been removed from its chitinous tube. (B) Pogonophorans live around hydrothermal vents deep in the ocean. Their tentacles can be seen protruding from their chitinous tubes. (C) Earthworms are hermaphroditic; when they copulate, each individual donates and receives sperm. (D) The medicinal leech has been a tool of physicians and healers for centuries. Even today, leeches have uses in modern clinical practice.

*Leeches*, like oligochaetes, lack parapodia and tentacles. The coelom of leeches is not divided into compartments; the coelomic space is largely filled with undifferentiated tissue. Groups of segments at each end of the body are modified to form suckers, which serve as temporary anchors that aid the leech in movement. With its posterior sucker attached to a substratum, the leech extends its body by contracting its circular muscles. The anterior sucker is then attached, the posterior one detached, and the leech shortens itself by contracting its longitudinal muscles. Leeches live in freshwater or terrestrial habitats.

A leech makes an incision in its host, from which blood flows. It can ingest so much blood in a single feeding that its body may enlarge several-fold. The leech secretes an anticoagulant into the wound that keeps the host's blood flowing. For centuries, medical practitioners employed leeches to draw blood to treat diseases they believed were caused by an excess of blood or by "bad blood." Although most leeching practices (such as inserting a leech in a person's throat to alleviate swollen tonsils) have been abandoned, *Hirudo medicinalis* (the medicinal leech; **Figure 32.13D**) is used today to reduce fluid pressure and prevent blood clotting in damaged tissues, to eliminate pools of coagulated blood, and to prevent scarring. The anticoagulants of certain other leech species that also contain anesthetics and blood vessel dilators are being studied for possible medical uses.

### Mollusks have undergone a dramatic evolutionary radiation

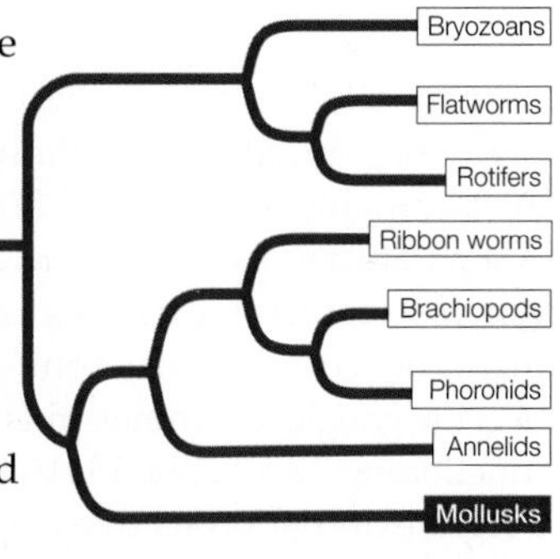


**Mollusks** are the most diverse group of lophotrochozoans, both in numbers of species and in environments they occupy. Although the major groups of mollusks differ dramatically in morphology, they all share the same three major body components: a foot, a visceral mass, and a mantle (**Figure 32.14**).

- The molluscan *foot* is a large, muscular structure that originally was both an organ of locomotion and a support for the internal organs. In squids and octopuses, the foot has been modified to form arms and tentacles borne on a head with complex sensory organs. In other groups, such as clams, the foot is a burrowing organ. In some groups the foot is greatly reduced.
- The heart and the digestive, excretory, and reproductive organs are concentrated in a centralized, internal *visceral mass*.
- The *mantle* is a fold of tissue that covers the organs of the visceral mass. The mantle secretes the hard, calcareous shell that is typical of many mollusks.

In most mollusks, the mantle extends beyond the visceral mass to form a *mantle cavity*. Within this cavity lie *gills* that are used for gas exchange. When cilia on the gills beat, they create a cur-

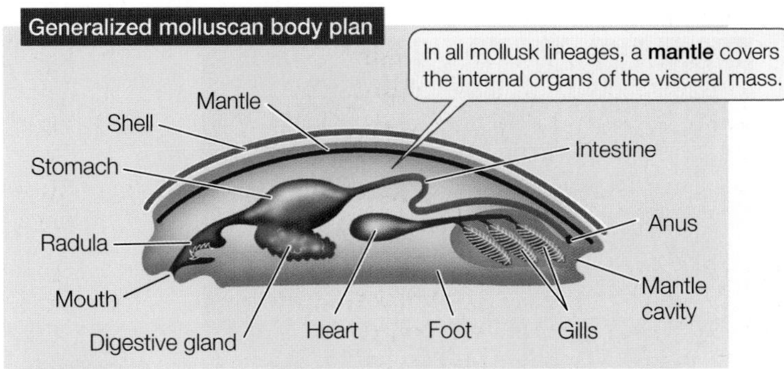


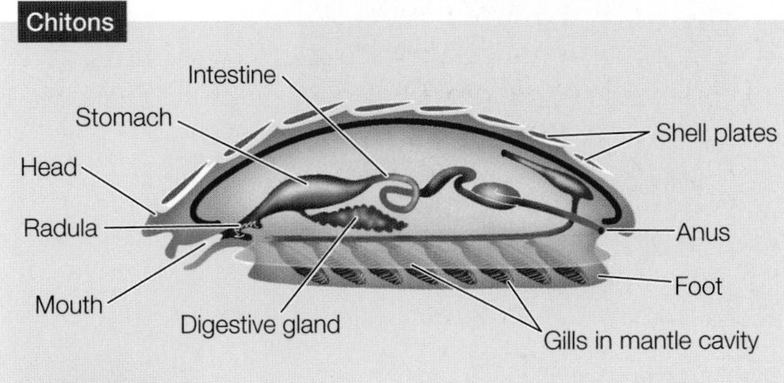


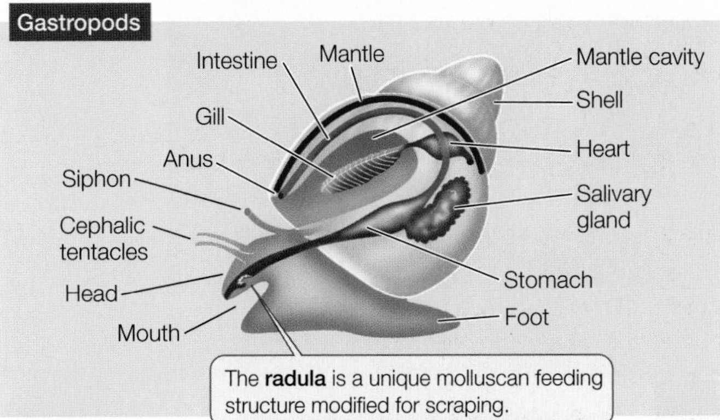


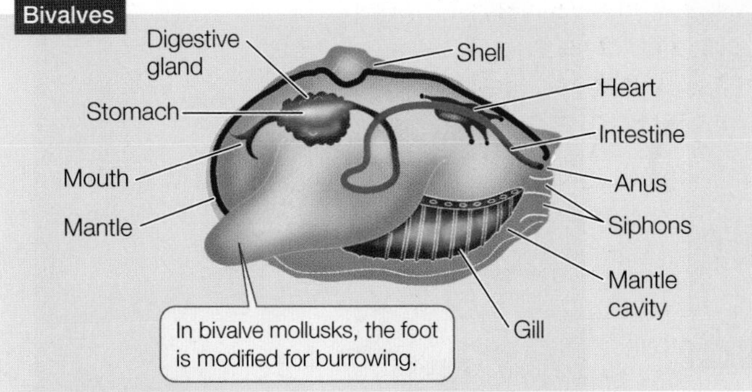


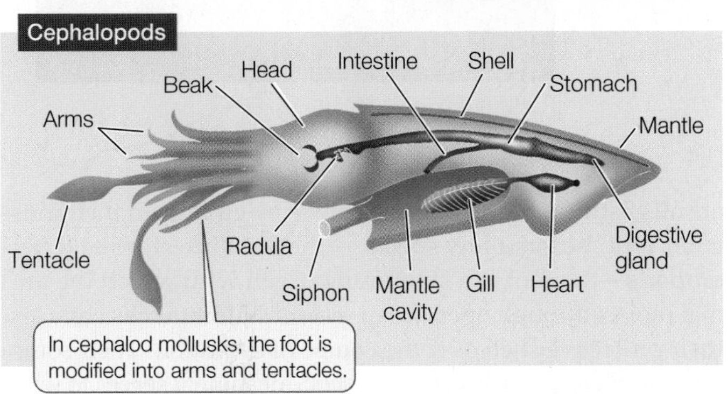


**32.14 Organization of Molluscan Bodies** The major molluscan groups display different variations on a general body plan that includes three major components: a foot, a visceral mass of internal organs, and a mantle. The mantle may secrete a calcareous shell, as in the gastropods and bivalves.

rent of water. The tissue of the gills, which is highly *vascularized* (containing many blood vessels), takes up oxygen from the water and releases carbon dioxide. Many mollusk species use their gills as filter-feeding devices, whereas others feed using a rasping structure known as the *radula* to scrape algae from rocks. In some mollusks, such as the marine cone snails, the radula has been modified into a drill or poison dart.

Molluscan blood vessels do not form a closed system. Blood and other fluids empty into a large, fluid-filled *hemocoel*, through which fluids move around the animal and deliver oxygen to the internal organs. Eventually, the fluids reenter the blood vessels and are moved by a heart.

**Monoplacophorans** were the most abundant mollusks during the Cambrian period, 500 million years ago, but only a few species survive today. Unlike in all other living mollusks, in monoplacophorans the gas exchange organs, muscles, and excretory pores are repeated over the length of the body.

Figure 32.15 illustrates the four major clades of living mollusks: chitons, bivalves, gastropods, and cephalopods. The species shown here are a tiny sample of the approximately 100,000 living species of mollusks.

**CHITONS** Eight overlapping calcareous plates, surrounded by a structure known as the girdle, protect the internal organs and muscular foot of **chitons** (**Figure 32.15A**). The chiton body is bilaterally symmetrical, and the internal organs, particularly the digestive and nervous systems, are relatively simple. Most chitons are marine omnivores that scrape algae, bryozoans, and other organisms from rocks with their sharp radula. An adult chiton spends most of its life clinging tightly to rock surfaces with its large, muscular, mucus-covered foot. It moves slowly by means of rippling waves of muscular contraction in the foot. Fertilization in most chitons takes place in the water, but in a few species fertilization is internal and embryos are brooded within the body. There are approximately 1,000 living species of chitons.

**BIVALVES** Clams, oysters, scallops, and mussels are all familiar **bivalves**. The 30,000 living species are found in both marine and freshwater environments. Bivalves have a very small head and a hinged, two-part shell that extends over the sides of the body as well as the top (**Figure 32.15B**). Many clams use their foot to burrow into mud and sand. Bivalves feed by taking in water through an opening called an *incurrent siphon* and filtering food from the water with their large gills, which are also the main sites of gas exchange. Water and gametes exit through the *excurrent siphon*. Fertilization takes place in open water in most species.

**GASTROPODS** **Gastropods** are the most species-rich and widely distributed mollusks, with nearly 70,000 living species. Snails, whelks, limpets, slugs, nudibranchs (sea slugs), and abalones

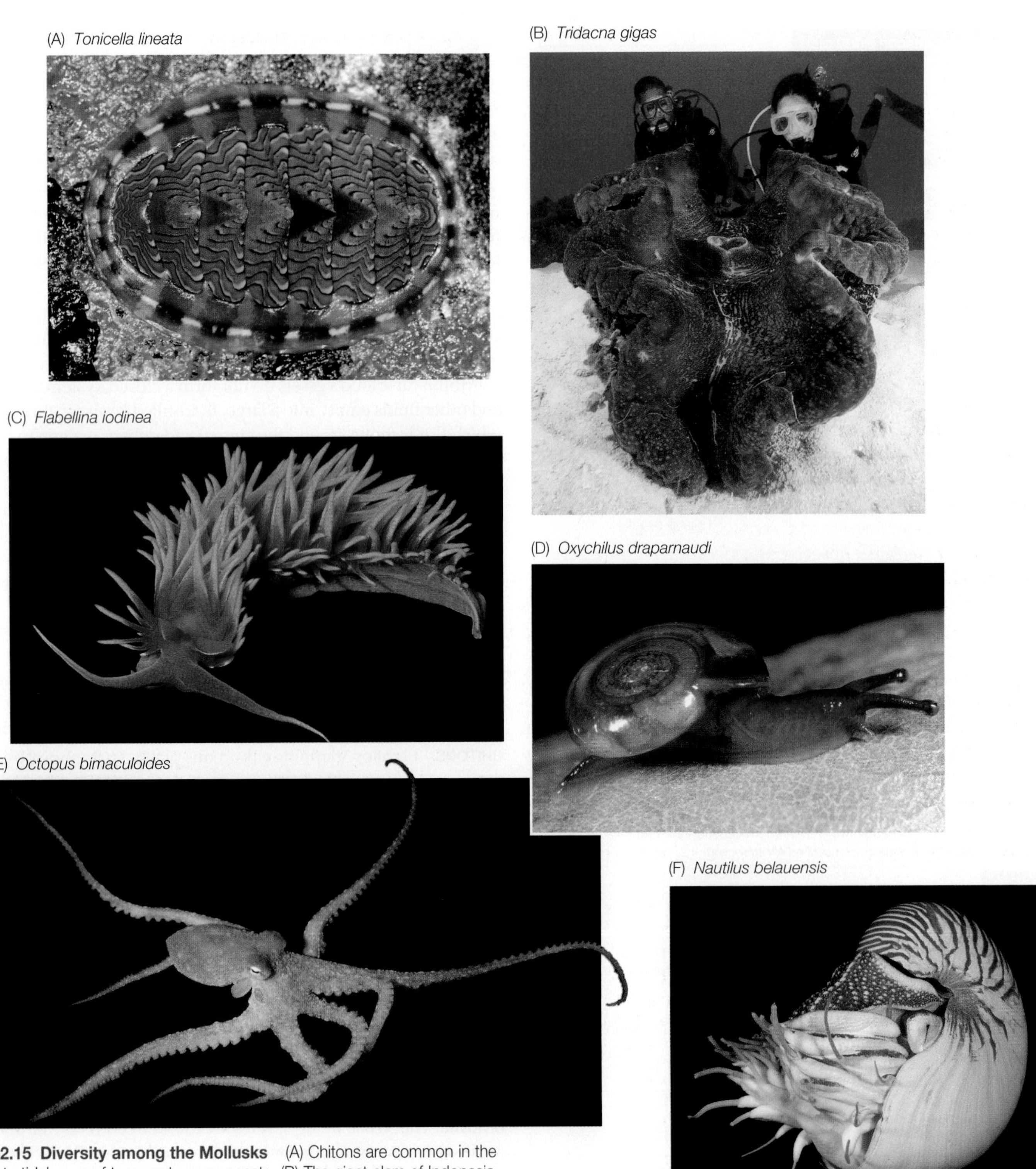


**32.15 Diversity among the Mollusks** (A) Chitons are common in the intertidal zone of temperate-zone coasts. (B) The giant clam of Indonesia is among the largest of the bivalve mollusks. (C) This shell-less nudibranch ("naked gills"), or sea slug, is brightly colored, a signal of its toxicity. (D) Pulmonate land snails, such as this dark-bodied glass snail, have a vascularized lung that allows them to survive in terrestrial environments. (E) Cephalopods such as the octopuses are active predators. (F) The boundaries of chambers are clearly visible on the outer surfaces of this shelled cephalopod, one of the chambered nautiluses.

are all gastropods. Most species move by gliding on their muscular foot, but in a few species—the sea butterflies and heteropods—the foot is a swimming organ with which the animal moves through open ocean waters. Nudibranchs have lost their protective shell over the course of evolution. Their sometimes brilliant coloration is *aposematic*, meaning it serves to warn potential predators of toxicity (**Figure 32.15C**). Other nudibranch species exhibit camouflaged coloration.

Shelled gastropods have one-piece shells. The only mollusks that live in terrestrial environments—land snails and slugs—are gastropods (**Figure 32.15D**). In these terrestrial species, the mantle tissue is modified into a highly vascularized lung.

**CEPHALOPODS** The **cephalopods**—squids, octopuses, and nautiluses—first appeared near the beginning of the Cambrian period. By the Ordovician period a variety of types were present. Today there are about 800 living species.

In the cephalopods, the excurrent siphon is modified to allow the animal to control the water content of the mantle cavity. The modification of the mantle into a device for forcibly ejecting water from the cavity through the siphon enables these animals to move rapidly by "jet propulsion" through the water. With their greatly enhanced mobility, cephalopods became the major predators in the open waters of the Devonian oceans. They remain important marine predators today.

Cephalopods capture and subdue prey with their tentacles; octopuses also use their tentacles to move over the substratum (**Figure 32.15E**). As is typical of active, rapidly moving predators, cephalopods have a head with complex sensory organs, most notably eyes that are comparable to those of vertebrates in their ability to resolve images. The head is closely associated with a large, branched foot that bears the tentacles and a siphon. The large, muscular mantle provides a solid external supporting structure. The gills hang in the mantle cavity. Many cephalopods have elaborate courtship behavior, which may involve striking color changes.

Many early cephalopods had a chambered shell divided by partitions penetrated by tubes through which gases and liquids could be moved to control the animal's buoyancy. Nautiluses (genus *Nautilus*) are the only surviving cephalopods that have such external chambered shells (**Figure 32.15F**).

## 32.2 RECAP

**Lophotrochozoans include animals with diverse body types. Wormlike forms include some flatworms, ribbon worms, phoronids, and annelids. There has been convergent evolution of lophophores (in bryozoans versus brachiopods and phoronids) and of external two-part shell coverings (in brachiopods versus bivalve mollusks).**

- Can you explain how flatworms can survive without a gas transport system? **See p. 672 and Figure 32.7**
- Do you know why most annelids are restricted to moist environments? **See p. 675**
- Briefly describe how the basic body organization of mollusks has been modified to yield a wide diversity of animals. **See pp. 676–679 and Figure 32.14**

The second of the two major protostome clades, the ecdysozoans, contains the vast majority of Earth's animal species. What evolutionary innovations led to this massive diversity?

# 32.3 What Features Distinguish the Major Groups of Ecdysozoans?

Many ecdysozoans are wormlike in form, although the *arthropods, onychophorans*, and *tardigrades* have limbs. In this section we will look at the two clades of wormlike ecdysozoans: the priapulids, kinorhynchs, and loriciferans in one group, and the horsehair worms and nematodes in the other. Section 32.4 is devoted to the most diverse ecdysozoans—the arthropods and their relatives—and the many forms their appendages take.

## Several marine groups have relatively few species

Members of several species-poor groups of wormlike marine ecdysozoans—the priapulids, kinorhynchs, and loriciferans—have relatively thin cuticles that are molted periodically as the animals grow to full size. In 2004, embryos of a fossil species related to these ecdysozoans were discovered in sediments laid down in China about 500 million years ago. This remarkable discovery shows that the ancestors of these animals developed directly from an egg to the adult form, as their modern descendants do.



The 16 species of **priapulids** are cylindrical, unsegmented, wormlike animals with a three-part body plan consisting of a proboscis, trunk, and caudal appendage ("tail"). It should be clear from their appearance why they were named after the Greek fertility god Priapus (**Figure 32.16A**). Priapulids range in length from 0.5 millimeters to 20 centimeters. They burrow in fine marine sediments and prey on soft-bodied invertebrates such as polychaetes, which they capture with a toothed, muscular pharynx that they evert through their mouth and then withdraw into their body together with the grasped prey. Fertilization is external, and most species have a larval form that also lives in the mud.

About 150 species of **kinorhynchs** have been described. They live in marine sands and muds and are virtually microscopic; no kinorhynchs are longer than 1 millimeter. Their bodies are divided into 13 segments, each with a separate cuticular plate (**Figure 32.16B**). These plates are periodically molted during growth. Kinorhynchs feed by ingesting sediments through their retractable proboscis (the name means "movable snout"). They then digest the organic material found in the sediment, which may include living algae as well as dead matter. Kinorhynchs have no distinct larval stage; fertilized eggs develop directly into juveniles, which emerge from their egg cases with 11 of the 13 body segments already formed.

**Loriciferans** are also minute animals less than 1 millimeter long. They were not discovered until 1983. About 100 living species are known to exist, although many of these are still being described. The body is divided into a head, neck, thorax, and abdomen and is covered by six plates, from which the loriciferans get their name (Latin *lorica*, "corset"). The plates around the base of the neck bear anterior-directed spines of unknown

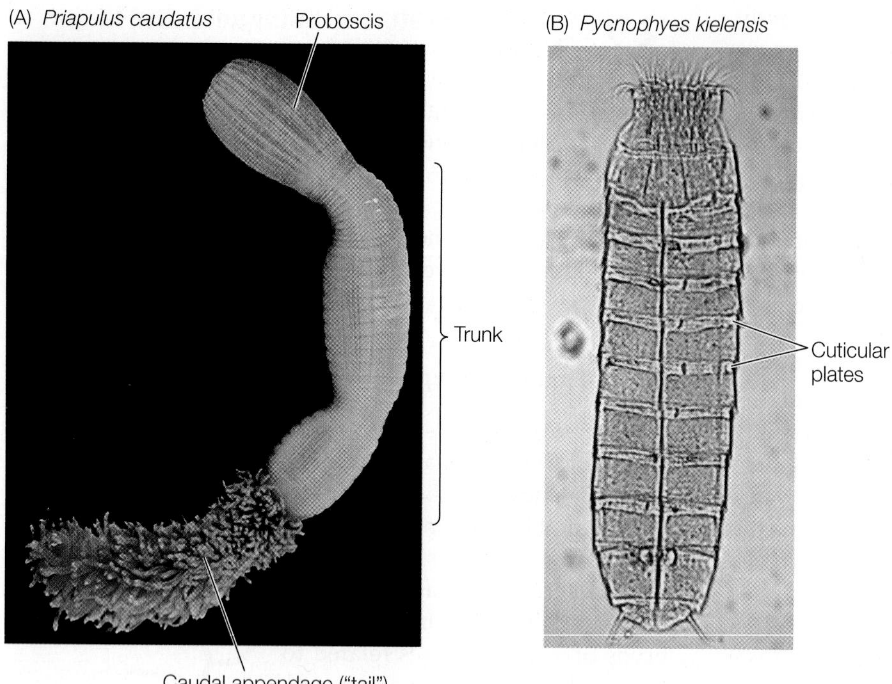


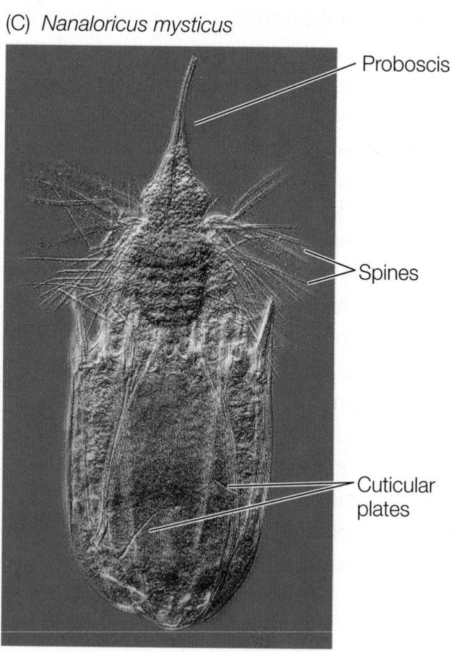


**32.16 Wormlike Marine Ecdysozoans** Members of these groups are marine bottom-dwellers. (A) Most priapulid species live in burrows on the ocean floor, extending the proboscis for feeding. (B) Kinorhynchs are virtually microscopic. The cuticular plates that cover their bodies are molted periodically. (C) Six cuticular plates form a "corset" around the minute loriciferan body.

function (**Figure 32.16C**). Loriciferans live in coarse marine sediments. Little is known about what they eat, but some species apparently eat bacteria.

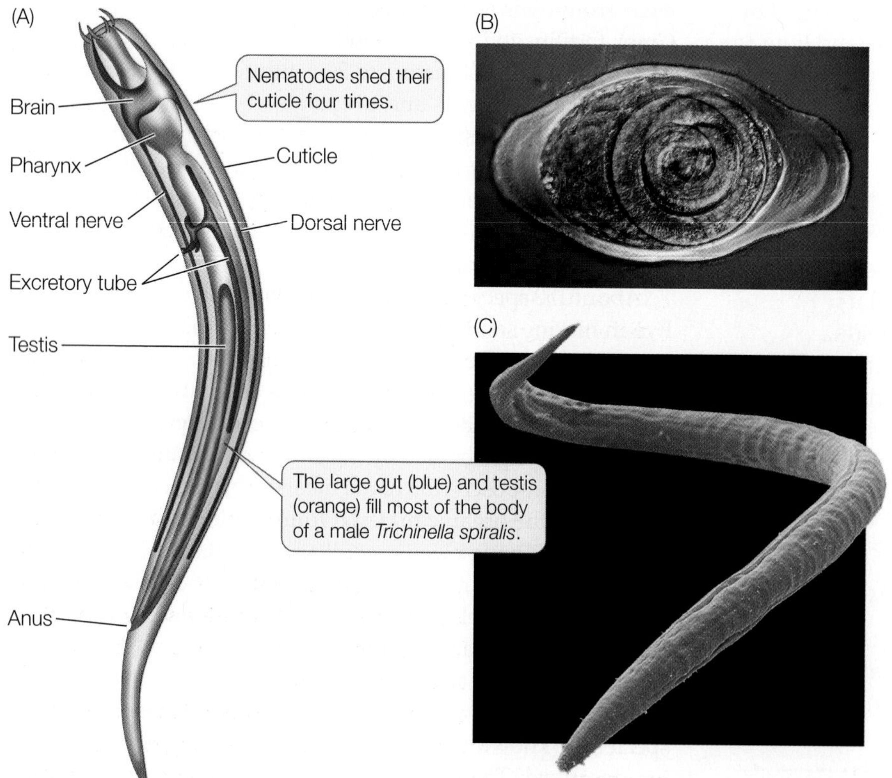


**32.17 Nematodes** (A) *Trichinella spiralis*, a parasitic nematode that causes trichinosis. (B) This polarized light micrograph shows a cyst of *T. spiralis* in the muscle tissue of a host. (C) This free-living nematode lives in freshwater environments.

### Nematodes and their relatives are abundant and diverse

**Nematodes** (roundworms) have a thick, multilayered cuticle that gives their unsegmented body its shape (**Figure 32.17**). As a nematode grows, it sheds its cuticle four times. Nematodes exchange oxygen and nutrients with their environment through both the cuticle and the gut, which is only one cell layer thick. Materials are moved through the gut by rhythmic contraction of a highly muscular organ, the *pharynx*, at the worm's anterior end. Nematodes move by contracting their longitudinal muscles.

Nematodes are probably the most abundant and universally distributed of all animal groups. Many are microscopic; the largest known nematode, which reaches a length of 9 meters, is a parasite in the placentas of sperm whales. About 25,000 species have been described, but the actual number of living species may be more than a million. Countless nematodes live as scavengers in the upper layers of the soil, on the bottoms of lakes and streams, and in marine sediments. The topsoil of rich farmland may contain from 3 to 9 billion nematodes per acre. A single rotting apple may contain as many as 90,000 individuals.

One soil-inhabiting nematode, *Caenorhabitis elegans*, serves as a model organism in the laboratories of geneticists and developmental biologists. It is ideal for such research because it is easy to cultivate, matures in 3 days, and has a fixed number of body cells. Its genome has been completely sequenced.

Many nematodes are predators, feeding on protists and small animals (including other roundworms). Most significant to humans, however, are the many species that parasitize plants and animals. The nematodes that parasitize humans (causing serious diseases such as trichinosis, filariasis, and elephantiasis), domestic animals, and economically important plants have been studied intensively in an effort to find ways of controlling them.

The structure of parasitic nematodes is similar to that of free-living species, but the life cycles of many parasitic species have special stages that facilitate the transfer of individuals among hosts. *Trichinella spiralis*, the species that causes the human disease trichinosis, has a relatively simple life cycle. A person may become infected by eating the flesh of an animal (usually a pig) that has *Trichinella* larvae encysted in its muscles (see Figure 32.17B). The larvae are activated in the person's digestive tract, emerge from their cysts, and attach to the intestinal wall, where they feed. Later, they bore through the intestinal wall and are carried in the bloodstream to muscles, where they form new cysts. If present in great numbers, these cysts can cause severe pain or death.

About 320 species of the unsegmented **horsehair worms** have been described. As their name implies, these animals are extremely thin in diameter; horsehair worms range from a few millimeters up to a meter in length. Most adult worms live in fresh water, among leaf litter and algal mats near the edges of streams and ponds. A few species live in damp soil.

Horsehair larvae are internal parasites of freshwater crayfish and of terrestrial and aquatic insects (**Figure 32.18**). An adult horsehair worm has no mouth, and its gut is greatly reduced and probably nonfunctional. Some species may feed only as larvae, absorbing nutrients from their hosts across the body wall. But other species continue to grow after they have left their hosts, shedding their cuticles, suggesting that adult worms may also absorb nutrients from their environment.

An adult horsehair worm exits the wood cricket it parasitized during its larval development.



**32.18 Horsehair Worm Larvae Are Parasitic** Like the strepsipteran insects described at the start of this chapter, the larvae of this horsehair worm (*Paragordius tricuspidatus*) can manipulate its host's behavior. The hatching worm causes the cricket to jump into water, where the worm will continue its life cycle as a free-living adult. The insect, having delivered its parasitic burden, drowns.

### 32.3 RECAP

**Priapulids, kinorhynchs, and loriciferans are relatively small, poorly known groups of wormlike marine ecdysozoans. Horsehair worms and nematodes are also wormlike. Nematodes are probably the most abundant and universal animal group. The biology of some nematodes has been studied extensively, particularly that of *Caenorhabitis elegans*, which serves as a model laboratory organism.**

- Describe at least three different ways in which nematodes have a significant impact on humans. See pp. 680–681

We now turn to the animals that not only dominate the ecdysozoan clade but are also the most diverse animals on Earth.

## 32.4 Why Are Arthropods So Diverse?



Arthropods and their relatives are ecdysozoans with paired appendages. Arthropods are the most diverse group of animals in numbers of species (about a million have been described, and many more remain to be discovered). Furthermore, the number of individual arthropods alive at any one time is estimated to be about $10^{18}$, or a billion billion. Among the animals, only the nematodes are thought to exist in greater numbers.

Several key features have contributed to the success of the arthropods. Their bodies are segmented, and their muscles are attached to the inside of their rigid exoskeletons. Each segment has muscles that operate that segment and the jointed appendages attached to it (see Figure 32.4). Jointed appendages permit complex movements, and different appendages are specialized for different functions. Encasement of the body within a rigid exoskeleton provides the animal with support for walking in the water or on dry land and provides some protection against predators. The waterproofing provided by chitin keeps the animal from dehydrating in dry air.

Representatives of the four major arthropod groups living today are all species-rich: the *crustaceans* (including shrimps, crabs, and barnacles), *hexapods* (insects and their relatives), *myriapods* (millipedes and centipedes), and *chelicerates* (including the arachnids—spiders, scorpions, mites, and their relatives). Phylogenetic relationships among arthropod groups are currently being reexamined in light of a wealth of new information, much of it based on gene sequences. These studies show close relationships between the myriapods and chelicerates in one clade and between the crustaceans and hexapods in another, with strong support for the monophyly of arthropods as a whole.

The jointed appendages of arthropods gave the clade its name, from the Greek words *arthron*, "joint," and *podos*, "foot" or "limb." Arthropods evolved from ancestors with simple, unjointed appendages. The exact forms of those ancestors are unknown, but some arthropod relatives with segmented bodies and unjointed appendages survive today. Before we describe the modern arthropods, we will discuss those arthropod relatives, as well as an early clade that went extinct but left an important fossil record.

### Arthropod relatives have fleshy, unjointed appendages

Until fairly recently, biologists debated whether the **onychophorans** (velvet worms) were more closely related to annelids or arthropods, but molecular evidence clearly links them to the latter. Indeed, with their soft, fleshy, unjointed, claw-bearing legs, onychophorans may be similar in appearance to the ancestors of arthropods (**Figure 32.19A**). The 150 species of onychophorans live in leaf litter in humid tropical environments. They have soft, segmented bodies that are covered by a thin, flexible cuticle that contains chitin. They use their fluid-filled body cavities as hydrostatic skeletons. Fertilization is internal, and the large, yolky eggs are brooded within the body of the female.

**Tardigrades** (water bears) also have fleshy, unjointed legs and use their fluid-filled body cavities as hydrostatic skeletons (**Figure 32.19B**). Tardigrades are extremely small (0.1–0.5 millimeters long) and lack both a circulatory system and gas-exchange organs. The 800 known extant species live in marine sands and on temporary water films on plants. When these films dry out, the animals also lose water and shrink to small, barrel-shaped objects that can survive for at least a decade in a dormant state. Tardigardes have been found in densities as high as 2 million per square meter of moss.

### Jointed appendages first appeared in the trilobites

The **trilobites** flourished in Cambrian and Ordovician seas, but they disappeared in the great Permian extinction at the close of the Paleozoic era (251 mya). Because they had heavy exoskeletons that readily fossilized, they left behind an abundant record of their existence (**Figure 32.20**). About 10,000 species have been described.

The body segmentation and appendages of trilobites followed a relatively simple, repetitive plan, but some of their

(A) *Peripatus* sp.

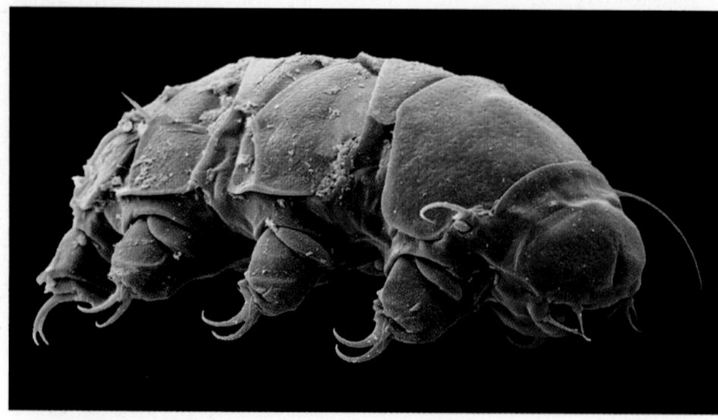

(B) *Echiniscus* sp. 50 μm

**32.19 Arthropod Relatives with Unjointed Appendages**
A) Onychophorans have unjointed legs and use the body cavity as a hydrostatic skeleton. (B) Tardigrades can be abundant on the wet surfaces of mosses and plants, and in temporary pools of water.

*Acanthopyge* sp.

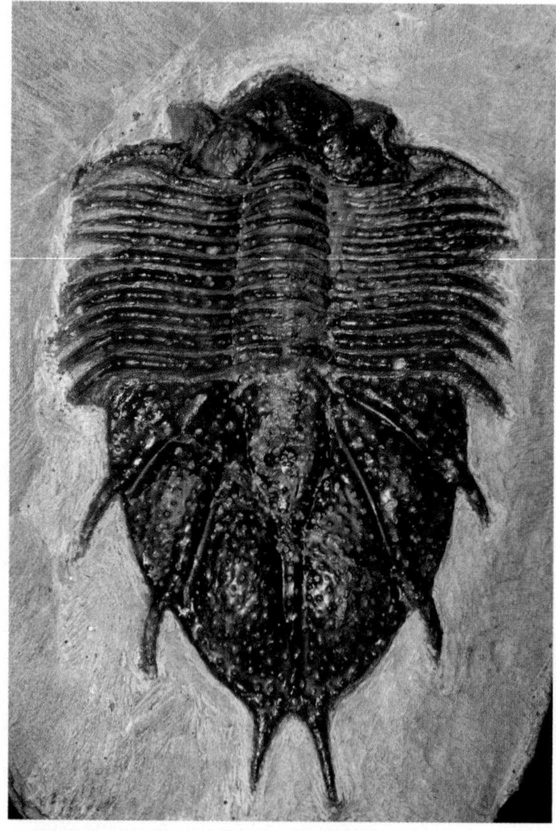

**32.20 A Trilobite** The relatively simple, repetitive segments of the now-extinct trilobites are illustrated by a fossil trilobite from the shallow seas of the Devonian period, some 400 million years ago.

jointed appendages were modified for different functions. This specialization of appendages is a theme in the continuing evolution of the arthropods.

### Myriapods have many legs

Members of two arthropod groups, the myriapods and the chelicerates, have a body with just two regions: a head and a trunk. This contrasts with the hexapods and crustaceans, most of which have a three-part body, with a head, thorax, and abdomen.

The **myriapods** comprise the centipedes, millipedes, and their close relatives. Centipedes and millipedes have a well-formed head and a long, flexible, segmented trunk that bears many pairs of legs (**Figure 32.21**). Centipedes, which have one pair of legs per segment, prey on insects and other small animals. In millipedes, two adjacent segments are fused so that each fused segment has two pairs of legs. Millipedes scavenge and eat plants. More than 3,000 species of centipedes and 11,000 species of millipedes have been described; many more species probably remain unknown. Although most myriapods are less than a few centimeters long, some tropical species are ten times that size.

### Most chelicerates have four pairs of legs

In the two-part body of **chelicerates**, the head bears two pairs of appendages modified to form mouthparts. In addition, many chelicerates have four pairs of walking legs. The 98,000 described species are usually placed in three major clades: pycnogonids, horseshoe crabs, and arachnids.

The *pycnogonids*, or sea spiders, are a poorly known group of about 1,000 marine species (**Figure 32.22A**). Most are small, with leg spans less than 1 centimeter, but some deep-sea species have leg spans up to 60 centimeters. A few pycnogonids eat algae, but most are carnivorous, eating a variety of small invertebrates.

There are four living species of *horseshoe crabs*, but many close relatives are known from fossils. Horseshoe crabs, which have changed very little morphologically during their long fossil history, have a large horseshoe-shaped covering over most of the body. They are common in shallow waters along the eastern

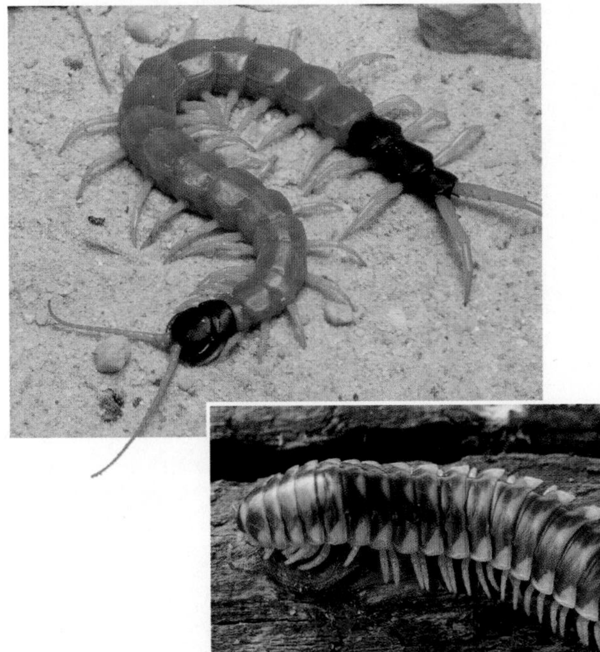

(A) *Scolopendra heros*

(B) *Sigmoria trimaculata*

**32.21 Myriapods** (A) Centipedes have modified appendages that function as poisonous fangs for capturing active prey and one pair of legs per segment. (B) Millipedes, which are scavengers and plant eaters, have smaller jaws and legs than centipedes do. They have two pairs of legs per segment.

(A) *Endeis* sp.

(B) *Limulus polyphemus*

**32.22 Two Small Chelicerate Groups** (A) Although they are not spiders, it is easy to see why sea spiders were given their common name. (B) This spawning aggregation of horseshoe crabs was photographed on a sandy beach of Delaware Bay. Horseshoe crabs are an ancient group that has changed very little in morphology over time; such species are sometimes referred to as "living fossils."

coast of North America and the southern and eastern coasts of Asia, where they scavenge and prey on bottom-dwelling animals. Periodically they crawl into the intertidal zone in large numbers to mate and lay eggs (**Figure 32.22B**).

*Arachnids* are abundant in terrestrial environments. Most arachnids have a simple life cycle in which miniature adults hatch from internally fertilized eggs and begin independent lives almost immediately. Some arachnids retain their eggs during development and give birth to live young.

The most species-rich and abundant arachnids are the spiders, scorpions, harvestmen, mites, and ticks (**Figure 32.23**). More than 50,000 described species of mites and ticks live in soil, leaf litter, mosses, and lichens, under bark, and as parasites of plants and animals. Mites are vectors for wheat and rye mosaic viruses; they cause mange in domestic animals and skin irritation in humans.

Spiders, of which 38,000 species have been described, are important terrestrial predators. Some have excellent vision that enables them to chase and seize their prey. Others spin elaborate webs made of protein threads in which they snare prey. The threads are produced by modified abdominal appendages connected to internal glands that secrete the proteins, which solidify on contact with air. The webs of different groups of spiders are strikingly varied, and this variation enables the spiders to position their snares in many different environments for many different types of prey.

### Crustaceans are diverse and abundant

**Crustaceans** are the dominant marine arthropods today, and they are also common in freshwater and some terrestrial environments. The most familiar crustaceans are the shrimps, lobsters, crayfishes, and crabs (all *decapods*; **Figure 32.24A**) and the sow bugs (*isopods*; **Figure 32.24B**). Additional species-rich groups of crustaceans include *amphipods*, *ostracods*, and *branchiopods* (**Figure 32.24C**), all of which are found in freshwater and marine environments. (Don't confuse the branchiopods with the similarly named brachiopods, described in Section 32.2.) *Krill* are small but abundant oceanic crustaceans that are important food items for a variety of large vertebrates, including baleen whales. Also

**32.23 Arachnid Diversity** (A) "Tarantulas" encompass many free-living species as well as several hundred species of hairy, ground-dwelling spiders, some of which can grow to the size of a dinner plate. Their venomous bite, although painful, is not usually dangerous to humans. (B) Scorpions are nocturnal predators. (C) Harvestmen, also called daddy longlegs, are scavengers. (D) Mites include many free-living species as well as blood-sucking external parasites.

(A) *Poecilotheria metallica*

(B) *Pseudouroctonus minimus*

(C) *Leiobunum rotundum*

(D) *Brevipalpus phoenicis*

(A) *Randallia ornata*

(C) *Triops longicaudatus*

(B) *Armadillium vulgare*

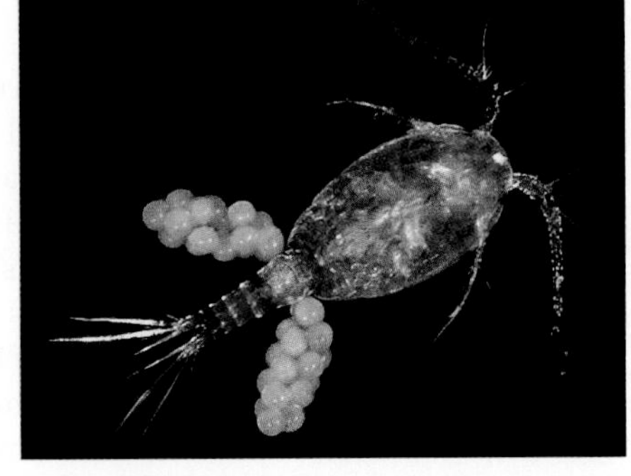

(D) Cyclopoid copepod

(E) *Lepas pectinata*

**32.24 Crustacean Diversity** (A) This decapod crustacean, a purple sand crab, is also referred to as a "globe crab" based on its globelike body shape. (B) Isopods are sometimes referred to as "rock lice," a name derived from their preferred habitat. (C) This tadpole shrimp, a branchiopod, is common in seasonal pools of the southwestern United States. Molecular studies suggest that branchiopods may be more closely related to hexapods than to other crustaceans. (D) This minute copepod from a freshwater pond is brooding eggs. (E) Gooseneck barnacles attach to a substratum by their muscular stalks and feed by protruding and retracting their feeding appendages.

included in the crustacean clade are the small *copepods* (**Figure 32.24D**), which are also abundant in the open oceans.

*Barnacles* are unusual crustaceans that are sessile as adults (**Figure 32.24E**). Adult barnacles look more like mollusks than like other crustaceans, but as the zoologist Louis Agassiz remarked more than a century ago, a barnacle is "nothing more than a little shrimp-like animal, standing on its head in a limestone house and kicking food into its mouth."

Recent molecular studies suggest that crustaceans may not be monophyletic. In analyses of DNA sequences, the branchiopods (which include brine shrimp, fairy shrimp, water fleas, and tadpole shrimp; see Figure 32.24D) appear to be more closely related to the hexapods than to other crustaceans.

Most of the 52,000 described species of crustaceans have a body that is divided into three regions: head, thorax, and abdomen (**Figure 32.25**). The segments of the head are fused together, and the head bears five pairs of appendages. Each of the multiple thoracic and abdominal segments usually bears one pair of appendages. The appendages on different parts of the body are specialized for different functions, such as gas exchange, chewing, capturing food, sensing, walking, and swimming. In some cases, the appendages are branched, with different branches

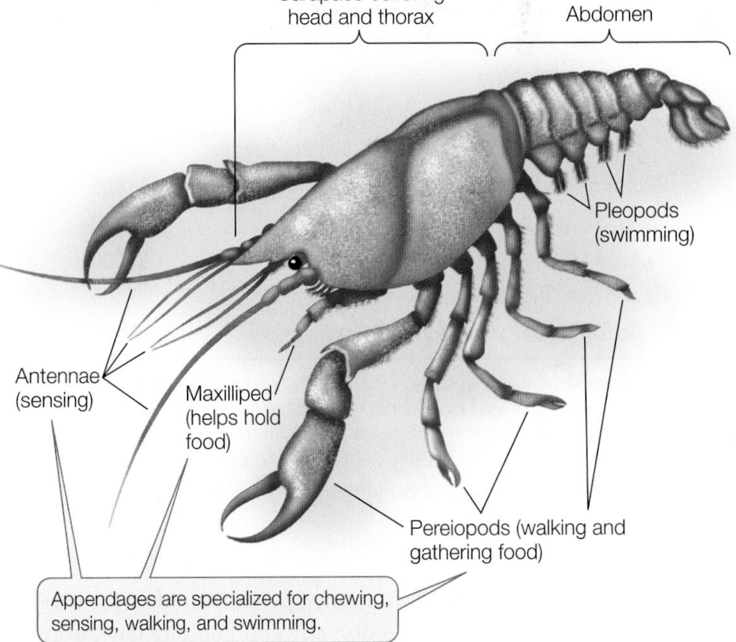


**32.25 Crustacean Body Plan** The bodies of crustaceans are divided into three regions: the head, thorax, and abdomen. Each region bears specialized appendages, and a shell-like carapace covers the head and thorax.

TABLE 32.2

**The Major Insect Groups**[a]

| GROUP | APPROXIMATE NUMBER OF DESCRIBED LIVING SPECIES |
|---|---|
| Jumping bristletails (Archaeognatha) | 300 |
| Silverfish (Thysanura) | 370 |
| **PTERYGOTE (WINGED) INSECTS (PTERYGOTA)** | |
| Mayflies (Ephemeroptera) | 2,000 |
| Dragonflies and damselflies (Odonata) | 5,000 |
| **Neopterans (Neoptera)**[b] | |
| Ice-crawlers (Grylloblattodea) | 25 |
| Gladiators (Mantophasmatodea) | 15 |
| Stoneflies (Plecoptera) | 1,700 |
| Webspinners (Embioptera) | 300 |
| Angel insects (Zoraptera) | 30 |
| Earwigs (Dermaptera) | 1,800 |
| Grasshoppers and crickets (Orthoptera) | 20,000 |
| Stick insects (Phasmida) | 3,000 |
| Cockroaches (Blattodea) | 3,500 |
| Termites (Isoptera) | 2,750 |
| Mantids (Mantodea) | 2,300 |
| Booklice and barklice (Psocoptera) | 3,000 |
| Thrips (Thysanoptera) | 5,000 |
| Lice (Phthiraptera) | 3,100 |
| True bugs, cicadas, aphids, leafhoppers (Hemiptera) | 80,000 |
| **Holometabolous neopterans (Holometabola)**[c] | |
| Ants, bees, wasps (Hymenoptera) | 125,000 |
| Beetles (Coleoptera) | 375,000 |
| Strepsipterans (Strepsiptera) | 600 |
| Lacewings, ant lions, dobsonflies (Neuropterida) | 4,700 |
| Scorpionflies (Mecoptera) | 600 |
| Fleas (Siphonaptera) | 2,400 |
| True flies (Diptera) | 120,000 |
| Caddisflies (Trichoptera) | 5,000 |
| Butterflies and moths (Lepidoptera) | 250,000 |

[a] The hexapod relatives of insects include the springtails (Collembola; 3,000 spp.), two-pronged bristletails (Diplura; 600 spp.), and proturans (Protura; 10 spp.). All are wingless and have internal mouthparts.

[b] Neopteran insects can tuck their wings close to their bodies

[c] Holometabolous insects are neopterans that undergo complete metamorphosis.

serving different functions. In many species, a fold of the exoskeleton, the *carapace*, extends dorsally and laterally back from the head to cover and protect some of the other segments.

The fertilized eggs of most crustacean species are attached to the outside of the female's body, where they remain during their early development. At hatching, the young of some species are released as larvae; those of other species are released as juveniles that are similar in form to the adults. Still other species release eggs into the water or attach them to an object in the environment. The typical crustacean larva, called a *nauplius*, has three pairs of appendages and one central eye (see Figure 31.11B).

### Insects are the dominant terrestrial arthropods

During the Devonian period, more than 400 million years ago, some arthropods colonized terrestrial environments. Of the several groups that successfully colonized the land, none is more prominent today than the six-legged *hexapods*: the **insects** and their relatives (**Table 32.2**). Insects are abundant and diverse in terrestrial and freshwater environments; only a few live in salt water.

Insects, like crustaceans, have a body with three regions—head, thorax, and abdomen. In addition, insects have a unique mechanism for gas exchange in air: a system of air sacs and tubular channels called *tracheae* (singular *trachea*) that extend from external openings called *spiracles* inward to tissues throughout the body. Insects have a single pair of antennae on the head and three pairs of legs attached to the thorax (**Figure 32.26**). Unlike the other arthropods, insects have no appendages growing from their abdominal segments. Insects can be distinguished from other hexapods by their external mouthparts and paired antennae that contain a motion-sensitive receptor called *Johnston's organ*.

The wingless relatives of the insects—the springtails (**Figure 32.27**), two-pronged bristletails, and proturans—are probably the most similar of living forms to insect ancestors. These relatives of insects have a simple life cycle; they hatch from eggs as miniature adults. They differ from insects in having inter-

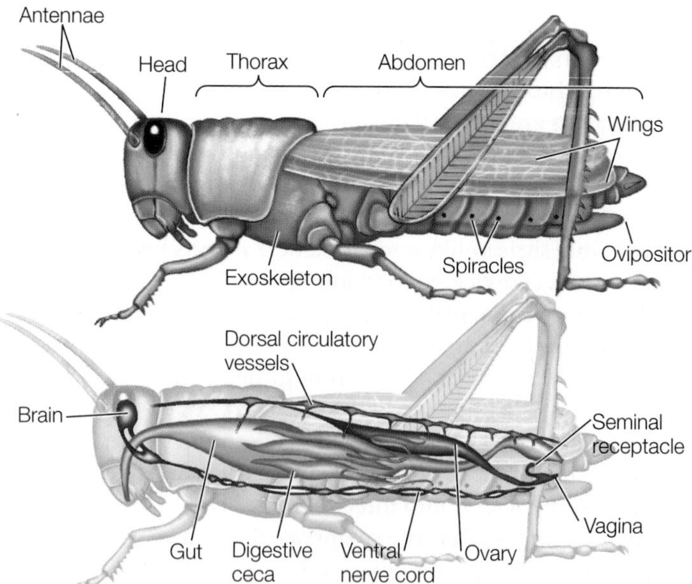


**32.26 Body Plan of an Insect** Insects, like crustaceans, have a three-part body plan with a head, thorax, and abdomen. The middle region, the thorax, bears three pairs of legs and, in most groups, two pairs of wings. Unlike other arthropods, the abdomen of an insect bears no appendages (see Figure 20.8).

*Sminthurides aquaticus*

**32.27 Wingless Hexapods** The wingless hexapods, such as this springtail, have a simple life cycle. They hatch looking like miniature adults, then grow by successive molts of the cuticle.

nal mouthparts. Springtails can be extremely abundant (up to 200,000 per square meter) in soil, leaf litter, and on vegetation, and are the most abundant hexapods in the world in terms of number of individuals (rather than number of species).

Insects use nearly all species of plants and many species of animals as food. Herbivorous insects can consume massive amounts of plant matter. Gypsy moth caterpillars, for example, have been known to denude entire forests of their hardwood tree leaves, and the cyclical depredations wrought by locust hordes can be of biblical proportions. Many insects are predatory, feeding on small animals or on other insects. *Detritivorous* insects such as dung beetles are important in the recycling of chemicals through ecosystems. Some insects are internal parasites of plants or animals; others suck their host's blood or consume surface body tissues.

As recently as the 1980s, most biologists thought that about half of existing insect species had been described, but today they think that the approximately 1 million species of described insects are a small fraction of the total number of living species. Why did they change their minds?

A simple but important field study published in 1988 suggested that the number of existing insect species had been significantly underestimated. Knowing that the insects of tropical forests (the most species-rich habitat on Earth) were poorly known, entomologist Terry Erwin decided to make a comprehensive sample of beetles in the canopies of a single species of tropical forest tree, *Luehea seemannii*. Erwin fogged the canopies of 19 large *L. seemannii* trees with a pesticide and collected the insects that fell from the trees in collection nets (**Figure 32.28**). His sample contained about 1,000 species of beetles—many of them undescribed—from this one species of tree. He then used a set of assumptions to estimate the total number of insect species in tropical evergreen forests. Erwin's assumptions included estimates of the number of species of host trees in these forests, the proportion of beetles that specialized on a specific species of host tree, the proportion of beetles relative to other insect groups, and the proportion of beetles that live in trees versus leaf litter, among several other factors. From this and similar studies, Erwin estimated that there may be as many as 50 million species of insects on Earth! This estimate was surprising to many biologists, since it is about 50 times higher than the known species diversity of insects.

The assumptions Erwin used to make his calculations are open to debate, and Erwin did not intend for his calculations to provide more than a rough estimate of undiscovered biodiversity. If different assumptions are used, estimates of the number of species of insects change dramatically. For example, one of Erwin's assumptions involved an estimate of the proportion of insect species thought to be *specialists*—species that specialize on a particular host plant (in this case, *Luehea seemannii*). If the proportion of specialist species was overestimated, then the total number of species was likely overestimated; if the proportion of specialists was underestimated, the total number of species may have been underestimated. Erwin's extrapolations also assumed that his samples were representative of other forest trees, and that his sampling areas were representative of other tropical forests, among many other assumptions. But even if his estimates were off by several fold, Erwin's pioneering study highlighted the fact that we live on a poorly known planet, most of whose species have yet to be named and described.

Two groups of insects—the jumping bristletails and silverfish—are wingless and have simple life cycles, like the springtails and other close insect relatives. The remaining insects are the *pterygote insects*. Pterygotes have two pairs of wings, except in

**32.28 Erwin's Research Project** In an important study, entomologist Terry Erwin fogged tree canopies in a Panamanian rainforest with insecticide. Using the number of species represented among the fallen insects as a base, he extrapolated to estimate the total number of insect species on Earth.

**32.29 Diverse Winged Insects** (A) Unlike most flying insects, a dragonfly cannot fold its wings over its back. (B) The plains lubber, a grasshopper, is an orthopteran. (C) Milkweed bugs are true bugs (hemipterans). The plants they feed on (*Asclepias* spp.) are toxic to many species; its coloring serves as a signal to predators that the bug also toxic. (D–H) Holometabolous insects undergo complete metamorphosis. (D) A larval caddisfly (right) emerges from its dark pupal case. (E) Coleopterans (beetles) such as this New Guinean weevil account for more than half of all holometabolous species and are the largest insect group. (F) The comet moth of Madagascar is a lepidopteran. (G) Flies such as this green bottle fly are dipterans. (H) Many hymenopteran genera, such as these yellowjacket wasps, are social.

some groups where one or both pairs of wings have been secondarily lost. Secondarily wingless groups include the parasitic lice and fleas, some beetles, and the worker individuals in many ants.

Hatchling pterygote insects do not look like adults, and they undergo substantial changes at each molt. The immature stages of insects between molts are called **instars**. As Section 31.4 describes, a substantial change that occurs between one developmental stage and another is called **metamorphosis**. If the changes between its instars are gradual, an insect is said to have **incomplete metamorphosis**. If the change between at least some instars is dramatic, an insect is said to have **complete metamorphosis**.

The most familiar examples of species with complete metamorphosis are butterflies (see Figure 31.10). The wormlike butterfly larva, called a *caterpillar*, transforms itself during a specialized phase, the **pupa**, in which larval tissues are broken down and the adult form develops. In many insects with complete metamorphosis, the different life stages are specialized for different environments and use different food sources. In many species, the larvae are adapted for feeding and growing, whereas the adults are specialized for reproduction and dispersal.

Pterygote insects were the first animals in evolutionary history to achieve the ability to fly. Flight opened up many new lifestyles and feeding opportunities that only the insects could exploit, and it is almost certainly one of the reasons for the remarkable numbers of insect species and individuals, and for their unparalleled evolutionary success.

The adults of most flying insects have two pairs of stiff, membranous wings attached to the thorax. True flies, however, have one pair of wings and a pair of stabilizers called *haltares*. In winged beetles, one pair of wings—the forewings—forms heavy, hardened wing covers. Flying insects are important pollinators of flowering plants.

Two groups of pterygotes, the mayflies and dragonflies (**Figure 32.29A**), cannot fold their wings against their bodies. This is the ancestral condition for pterygote insects, and these two groups are not closely related to one another. Members of these groups have predatory or herbivorous aquatic larvae that transform into flying adults after they crawl out of the water. Dragonflies (and their relatives the damselflies) are active predators as adults. In contrast, adult mayflies lack functional digestive tracts. Mayflies live only about a day, just long enough to mate and lay eggs.

All other pterygote insects—the *neopterans*—can tuck their wings out of the way upon landing and crawl into crevices and other tight places. Some neopteran groups undergo incomplete metamorphosis, so hatchlings of these insects are sufficiently similar in form to adults to be recognizable. Examples include the grasshoppers (**Figure 32.29B**), roaches, mantids, stick insects, termites, stoneflies, earwigs, thrips, true bugs (**Figure 32.29C**), aphids, cicadas, and leafhoppers. They acquire adult organ systems, such as wings and compound eyes, gradually through several juvenile instars.

More than 80 percent of all insects belong to a subgroup of the neopterans called the *holometabolous* insects (see Table 32.2), which undergo complete metamorphosis (**Figure 32.29D**). The many species of beetles account for almost half of this group (**Figure 32.29E**). Also included are lacewings and their relatives; caddisflies; butterflies and moths (**Figure 32.29F**); sawflies; true flies (**Figure 32.29G**); and bees, wasps, and ants, some species of which display unique and highly specialized social behaviors (**Figure 32.29H**).

Molecular data suggest that insects began to diversify about 450 million years ago, about the time of the appearance of the first land plants. These early hexapods evolved in a terrestrial environment that lacked any other similar organisms, which in part accounts for their remarkable success. But the success of the insects is also due to their wings. Homologous genes control the development of insect wings and crustacean appendages, suggesting that the insect wing evolved from a dorsal branch of a crustacean-like limb (**Figure 32.30**). The dorsal limb branch of crustaceans is used for gas exchange. Thus the insect wing probably evolved from a gill-like structure that had a gas exchange function.

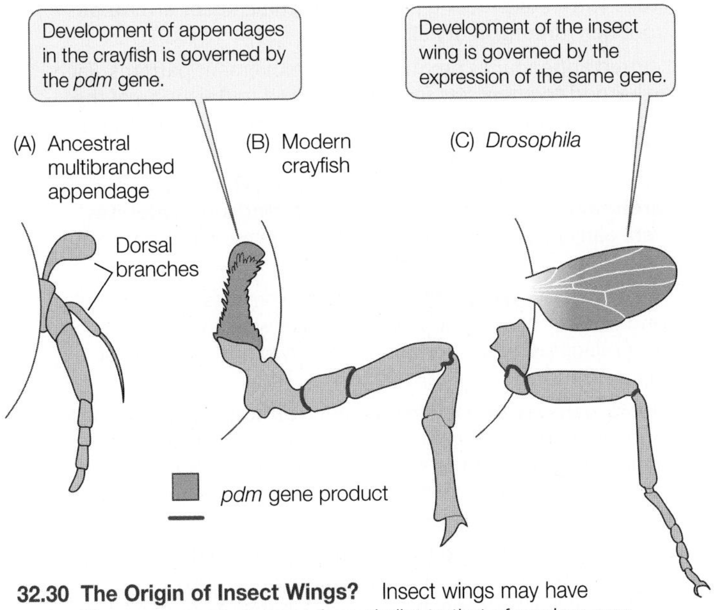

Development of appendages in the crayfish is governed by the *pdm* gene.

Development of the insect wing is governed by the expression of the same gene.

(A) Ancestral multibranched appendage

Dorsal branches

(B) Modern crayfish

(C) *Drosophila*

*pdm* gene product

**32.30 The Origin of Insect Wings?** Insect wings may have evolved from an ancestral appendage similar to that of modern crustaceans. (A) A diagram of the ancestral, multibranched arthropod limb. (B, C) The *pdm* gene, a Hox gene, is expressed throughout the dorsal limb branch and walking leg of the thoracic limb of a crayfish (B) and in the wings and legs of *Drosophila* (C).

## 32.4 RECAP

**All arthropods have segmented bodies. Muscles in each segment operate that segment and the appendages attached to it. These jointed, specialized appendages permit complex patterns of movement, including, in insects, the ability to fly. With flight, insects took advantage of new feeding and lifestyle opportunities, which contributed to the unparalleled evolutionary success of this group.**

- What features have contributed to making arthropods among the most abundant animals on Earth, both in number of species and number of individuals? **See p. 681**
- Describe the difference between incomplete and complete metamorphosis. **See p. 688**
- Can you think of some possible reasons why there are so many species of insects? **See pp. 688–689**

The protostomes encompass the majority of Earth's animal species, so it is not surprising that the different protostome groups display a huge variety of different characteristics.

### An Overview of Protostome Evolution

The myriad species of protostomes encompass a staggering number of different body forms and life styles. The following aspects of protostome evolution have contributed to this enormous diversity:

- The evolution of *segmentation* permitted some groups of protostomes to move different parts of the body independently of one another. Species in some groups gradually evolved the ability to move rapidly over and through the substratum, through water, and through air.

- *Complex life cycles* with dramatic changes in form between one stage and another allow individuals of different stages to specialize on different resources.
- *Parasitism* has evolved repeatedly, and many protostome groups parasitize plants and animals.
- The evolution of *diverse feeding structures* allowed protostomes to specialize on many different food sources. Specialization on food sources undoubtedly contributed to species isolation and further diversification.
- Predation was a major selection pressure favoring the development of *hard external body coverings* (exoskeletons and shells). Such coverings evolved independently in many lophotrochozoan and ecdysozoan groups. In addition to providing protection, these coverings became key elements in the development of new systems of locomotion.
- *Better locomotion* permitted prey to escape from predators, but also allowed predators to pursue their prey more effectively. Thus the evolution of animals has been, and continues to be, a complex "arms race" among predators and prey.

Many major evolutionary trends among the protostomes are shared by the deuterostomes, which includes the chordates, the group to which humans belong. We turn to the deuterostomes in the next chapter.

## CHAPTER SUMMARY

### 32.1 What Is a Protostome?

- Protostomes ("mouth first") are bilaterally symmetrical animals with an anterior brain that surrounds the entrance to the digestive tract and a ventral nervous system. The blastopore of protostomes (when present in development) develops into a mouth. Protostomes comprise two major clades, the lophotrochozoans and the ecdysozoans. **Review Figure 32.1, WEB ACTIVITIES 32.1 and 32.2**
- **Lophotrochozoans** include a wide diversity of animals. Within this group evolved **lophophores** (a complex organ for both food collection and gas exchange), free-living **trochophore** larvae, and spiral cleavage. Some of these features were subsequently lost in some lineages.
- **Ecdysozoans** have a body covering known as the **cuticle**, which they must **molt** in order to grow. Some ecdysozoans have a relatively thin cuticle. Others, especially the **arthropods**, have a rigid cuticle reinforced with **chitin**. This rigid cuticle functions as an **exoskeleton**. New mechanisms of locomotion and gas exchange evolved among the arthropods. **Review Figure 32.4**

### 32.2 What Features Distinguish the Major Groups of Lophotrochozoans?

- Lophotrochozoans range from animals having only one entrance to the digestive tract and no oxygen transport system to animals with complete digestive tracts and complex internal transport systems.
- Lophophores, wormlike body forms, and external shells are each found in many unrelated groups of lophotrochozoans.
- The most species-rich groups of lophotrochozoans are the **flatworms**, **annelids**, and **mollusks**.
- Annelids are a diverse group of segmented worms that live in moist terrestrial and aquatic environments. **Review Figure 32.12**
- Mollusks underwent a dramatic evolutionary radiation based on a body plan consisting of three major components: a foot, a mantle, and a visceral mass. The four major living molluscan clades—**chitons**, **bivalves**, **gastropods**, and **cephalopods**—demonstrate the diversity that evolved from this three-part body plan. **Review Figure 32.14**

### 32.3 What Features Distinguish the Major Groups of Ecdysozoans?

- Many ecdysozoan groups are wormlike in form. Members of several species-poor groups of wormlike marine ecdysozoans—**priapulids**, **kinorhynchs**, and **loriciferans**—have thin cuticles.
- **Horsehair worms** are extremely thin; many are internal parasites as larvae.
- **Nematodes**, or roundworms, have a thick, multilayered cuticle. Nematodes are among the most abundant and universally distributed of all animal groups. **Review Figure 32.17**
- One major ecdysozoan clade, the arthropods, has evolved jointed, paired appendages that have a wide diversity of functions. Collectively, arthropods are the dominant animals on Earth in number of described species, and among the most abundant in number of individuals.

### 32.4 Why Are Arthropods So Diverse?

- Encasement within a rigid exoskeleton provides arthropods with support for walking as well as some protection from predators. The waterproofing provided by chitin keeps arthropods from dehydrating in dry air.
- Jointed appendages permit complex movement patterns. Each arthropod segment has muscles attached to the inside of the exoskeleton that operate that segment and the appendages attached to it.
- Two groups of arthropod relatives, the **onychophorans** and the **tardigrades**, have simple, unjointed appendages. **Trilobites** were early marine arthropods that disappeared in the Permian extinction.
- The bodies of **myriapods** have only two regions, a head, and a long trunk with many segments, each of which carries appendages. **Chelicerates** also have a two-part body; most chelicerates have four pairs of walking legs.
- **Crustaceans** are the dominant marine arthropods, and are also found in many freshwater and some terrestrial environments. Their segmented bodies are divided into three regions (head, thorax, and abdomen) with different, specialized appendages in each region. **Review Figure 32.25**
- Hexapods—**insects** and their relatives—are the dominant terrestrial arthropods. They have the same three body regions as crustaceans, but no appendages form in their abdominal segments. Wings and the ability to fly first evolved among the insects, allowing them to exploit new lifestyles. **Review Figure 32.26**

## SELF-QUIZ

1. Members of which groups have lophophores?
   *a.* Phoronids, brachiopods, and nematodes
   *b.* Phoronids, brachiopods, and bryozoans
   *c.* Brachiopods, bryozoans, and flatworms
   *d.* Phoronids, rotifers, and bryozoans
   *e.* Rotifers, bryozoans, and brachiopods
2. Which of the following is *not* part of the molluscan body plan?
   *a.* Mantle
   *b.* Foot
   *c.* Radula
   *d.* Visceral mass
   *e.* Jointed skeleton
3. Nautiluses control their buoyancy by
   *a.* adjusting salt concentrations in their blood.
   *b.* forcibly expelling water from the mantle.
   *c.* pumping water and gases in and out of internal chambers.
   *d.* using the complex sensory organs in their heads.
   *e.* swimming rapidly.
4. The outer covering of ecdysozoans
   *a.* is always hard and rigid.
   *b.* is always thin and flexible.
   *c.* is hard and rigid in larvae but thin in adults.
   *d.* ranges from very thin to hard and rigid depending on the species.
   *e.* grows throughout life to accommodate a growing body.
5. Nematodes are abundant and diverse because
   *a.* they are both parasitic and free-living and eat a wide variety of foods.
   *b.* they are able to molt their exoskeleton.
   *c.* their thick cuticle enables them to move in complex ways.
   *d.* their body cavity is a pseudocoelom.
   *e.* their segmented body enables them to live in many different places.
6. The arthropod exoskeleton is composed of a
   *a.* mixture of several kinds of polysaccharides.
   *b.* mixture of several kinds of proteins.
   *c.* single complex polysaccharide called chitin.
   *d.* single complex protein called arthropodin.
   *e.* mixture of layers of proteins and a polysaccharide called chitin.
7. Which groups are arthropod relatives with unjointed legs?
   *a.* Trilobites and onychophorans
   *b.* Onychophorans and tardigrades
   *c.* Trilobites and tardigrades
   *d.* Onychophorans and chelicerates
   *e.* Tardigrades and chelicerates
8. The body plan of insects is composed of which of the three following regions?
   *a.* Head, abdomen, and trachea
   *b.* Head, abdomen, and cephalothorax
   *c.* Cephalothorax, abdomen, and trachea
   *d.* Head, thorax, and abdomen
   *e.* Abdomen, trachea, and mantle
9. Insects whose hatchlings are sufficiently similar in form to adults to be recognizable are said to have
   *a.* instars.
   *b.* neopterous development.
   *c.* accelerated development.
   *d.* incomplete metamorphosis.
   *e.* complete metamorphosis.
10. Factors that may have contributed to the remarkable evolutionary success of insects include
    *a.* the lack of any other similar organisms in the terrestrial environments colonized by insects.
    *b.* the ability to fly.
    *c.* complete metamorphosis.
    *d.* a new mechanism for delivering oxygen to their internal tissues.
    *e.* all of the above

## FOR DISCUSSION

1. Segmentation has either arisen several times during animal evolution, or else it arose early in animal evolution and was subsequently lost multiple times. What advantages does segmentation provide? Given these advantages, why might some animals have lost their segmentation?
2. Major structural novelties have arisen only infrequently during the course of evolution. Which of the features of protostomes do you think are major evolutionary novelties? Can you think of morphological features that may have led to major evolutionary radiations?
3. There are more described and named species of insects than of all other species on Earth combined. However, only a very few insect species live in marine environments, and those species are restricted to the intertidal zone or the ocean surface. What factors may have contributed to this lack of success by the insects in the oceans?

## ADDITIONAL INVESTIGATION

If you were given funding to carry out studies to improve our understanding of Earth's biodiversity, how would you spend it? How would your investigation differ if you targeted particular major groups of animals, such as insects or nematodes? Which animal groups do you think should receive highest priority for such studies, and why?

## WORKING WITH DATA (GO TO yourBioPortal.com)

**Estimating the Number of Species of Insects** In this exercise based on the study illustrated in Figure 32.28, you will use the data collected by Terry Erwin to estimate the number of insects in a hectare (10,000 $m^2$) of Panamanian rainforest, as well as to estimate the number of insect species found on Earth. You will also examine Erwin's assumptions and consider how they might be modified and tested.

# 33 Deuterostome Animals

## Good parents, or cannibals?

Why would a frog swallow its own offspring? That may not sound like a good parenting skill, but female gastric brooding frogs of Australia shut down their digestive system to brood their tadpoles in their stomach. This provides the tadpoles with a safe haven from predators, making it much more likely that they will survive to metamorphosis. That was the case, at least, until gastric brooding frogs went extinct in the early 1980s, after humans introduced pathogenic fungi into their native range.

Not all frogs are as involved in raising their young as were gastric brooding frogs, however. Bullfrogs, for example, lay thousands of eggs each year and provide no parental care for their offspring. The eggs are fertilized by a male bullfrog and are then left to develop on their own. The eggs hatch into tadpoles and transform into tiny frogs—if they aren't eaten first; most of these offspring don't survive. In the water, the tadpoles are eaten by many species of fishes, turtles, birds, snakes, aquatic insects, and other predators. As the young tadpoles transform into frogs, they are likely to be eaten by predators hunting at the margins of the pond. Out of the tens of thousands of tadpoles an adult bullfrog produces in its lifetime, an average of only two offspring are expected to survive to reproduce themselves.

In many species of frogs, complex behaviors associated with parental care change these long odds. Rather than producing huge numbers of offspring, each with a minimal chance of surviving, some frogs invest more energy in each offspring and care for the young as they grow. This increases the chances that any one offspring will survive and reproduce—but it also means far fewer offspring can be produced, because of the greater parental investment per offspring.

Strategies for parental care include many behaviors in addition to gastric brooding. The females of many species guard a terrestrial egg clutch until it hatches. Other females carry their developing embryos or tadpoles around with them on their backs, or even in special brood pouches (as in the marsupial frogs of South America). Parental care often involves males as well. Males of many frog species guard egg masses or carry young, sometimes in unusual ways. In Darwin's frog

**Eating One's Offspring?** Young frogs emerge from the digestive tract of a female *Rheobatrachus silus*, one of the now-extinct gastric brooding frogs of Australia. In this unique form of gestation, eggs hatch and tadpoles develop within the protected environment of the mother's stomach.

**Riding Piggyback** The young of this female *Flectonotus pygmaeus* are developing within a pouch on her back. The marsupial frogs of South America take their name from another vertebrate group—marsupial mammals such as kangaroos—whose females also protect their developing young in pouches.

of southern South America, for example, the tadpoles develop within the male's vocal sacs.

Parental care can extend beyond protection to feeding the young. Some female dart-poison frogs of the tropical Americas carry each individual tadpole to one of the tiny pools of water that collect in bromeliad plants growing on the trees. The female then returns to each bromeliad "pond" and lays unfertilized eggs as food for the single tadpole developing there.

Frogs and other vertebrates constitute one of the major groups of deuterostome animals. Although there are far fewer species of deuterostomes than of protostomes, deuterostomes are of particular interest to biologists because of their importance in many ecosystems, their often complex behaviors, and their widespread use as models in developmental biology and genetics.

**IN THIS CHAPTER** we will introduce the deuterostomes and describe the principal animal groups: the echinoderms, hemichordates, and chordates. We will then discuss the evolution of the vertebrates within the chordates, including the features that allowed vertebrates to colonize most habitats on Earth. We will look especially closely at the primate lineage, which includes our own species.

## CHAPTER OUTLINE



# 33.1 What Is a Deuterostome?

It may surprise you to learn that both you and a sea urchin are deuterostomes. Adult sea stars, sea urchins, and sea cucumbers—the most familiar echinoderms—look so different from adult vertebrates (fishes, frogs, lizards, birds, and mammals) that it may be difficult to believe all these animals are closely related. The evidence that all deuterostomes share a common ancestor that is not shared with the protostomes includes common early developmental patterns and phylogenetic analysis of gene sequences, factors that are not apparent in the forms of the adult animals.

Historically, the deuterostomes were characterized by three early developmental patterns:

- Radial cleavage
- Formation of the mouth at the opposite end of the embryo from the blastopore (the pattern that gives the deuterostomes their name); in deuterostomes, the blastopore develops into the anus (see Figure 31.2)
- Development of a coelom from mesodermal pockets that bud off from the cavity of the gastrula rather than by splitting of the mesoderm, as occurs among protostomes

Radial cleavage is not exclusive to deuterostomes, and as noted in Chapter 31, it is now thought to be the ancestral condition for bilaterians. In fact, some of the groups now known to be protostomes were once thought to be deuterostomes because they have developmental patterns similar to those of echinoderms and chordates. The development of the blastopore into an anus may also be the ancestral condition for bilaterians, rather than a derived feature of deuterostomes. Today, the strongest support for the shared evolutionary relationships of echinoderms, hemichordates, and chordates (the groups that now compose the deuterostomes) comes from phylogenetic analyses of DNA sequences of many different genes.

Although there are far fewer species of deuterostomes than of protostomes (see Table 31.1), we have a special interest in deuterostomes, in part because we are members of that clade. The deuterostomes are also of interest because they include many large animals that strongly influence the characteristics of ecosystems. Many deuterostome species have been intensively studied in all fields of biology. Complex behaviors, such as the parenting behaviors described in the opening of this chapter, are especially well developed among some deuterostomes.

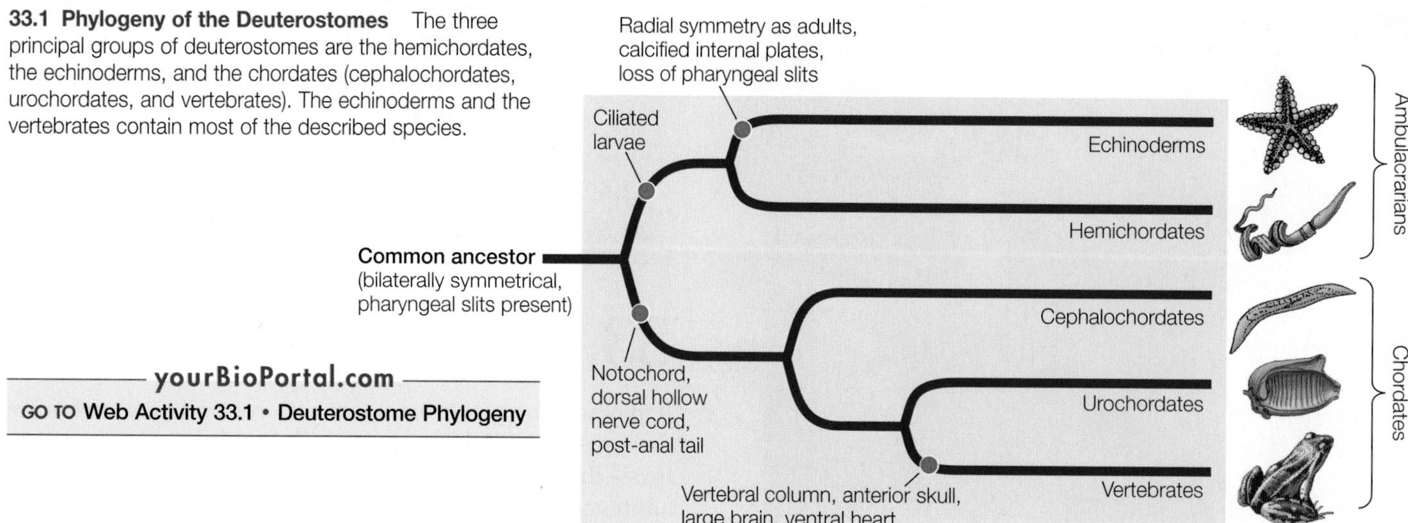


**33.1 Phylogeny of the Deuterostomes** The three principal groups of deuterostomes are the hemichordates, the echinoderms, and the chordates (cephalochordates, urochordates, and vertebrates). The echinoderms and the vertebrates contain most of the described species.

**yourBioPortal.com**
GO TO **Web Activity 33.1 • Deuterostome Phylogeny**

Living deuterostomes comprise three distinct clades (**Figure 33.1**):

- *Echinoderms*: sea stars, sea urchins, and their relatives
- *Hemichordates*: acorn worms and pterobranchs
- *Chordates*: sea squirts, lancelets, and vertebrates

All deuterostomes are triploblastic, coelomate animals (see Figure 31.4C). Skeletal support features, where present, are internal rather than external. Some species have segmented bodies, but the segments are less obvious than those of annelids and arthropods.

Scientists are learning much about the ancestors of modern deuterostomes from fossils recently discovered in 520-million-year-old rocks in China. Some of these early deuterostomes had a skeleton similar to that of a modern echinoderm, but they had pharyngeal slits and bilateral symmetry. Another group of early deuterostomes, the *yunnanozoans*, were discovered in China's Yunnan Province. These well-preserved animals had a large mouth, six pairs of external gills, and a segmented posterior body section bearing a light cuticle (**Figure 33.2**).

The features of these fossil animals support the findings from phylogenetic analyses of living species in showing that the earliest deuterostomes were bilaterally symmetrical, segmented animals with a pharynx that had slits through which water flowed. Echinoderms evolved their adult forms with unique symmetry (in which the body parts are arranged along five radial axes) much later, whereas other deuterostomes retained the ancestral bilateral symmetry.

*Yunnanozoon lividum*

**33.2 Ancestral Deuterostomes Had External Gills** The extinct yunnanozoans may be ancestral to all deuterostomes. This fossil, which dates from the Cambrian, shows the six pairs of external gills and segmented posterior body that characterized these animals.

## 33.1 RECAP

**The deuterostomes include the echinoderms, hemichordates, and chordates. The common ancestry of these groups is supported by developmental similarities and by phylogenetic analyses of DNA sequences.**

- What are three developmental patterns the earliest deuterostomes had in common? See p. 693
- Why is radial cleavage no longer considered to be evidence for the monophyly of deuterostomes? See p. 693

## 33.2 What Are the Major Groups of Echinoderms and Hemichordates?

About 13,000 species of echinoderms in 23 major groups have been described from their fossil remains. They are probably only a small fraction of those that actually lived. Only 6 of the 23 major groups known from fossils are represented by species that survive today; many clades became extinct during the periodic mass extinctions that have occurred throughout Earth's history. Nearly all of the 7,000 extant species of echinoderms live only in marine environments. There are far fewer species of living hemichordates, with only about 100 known species.

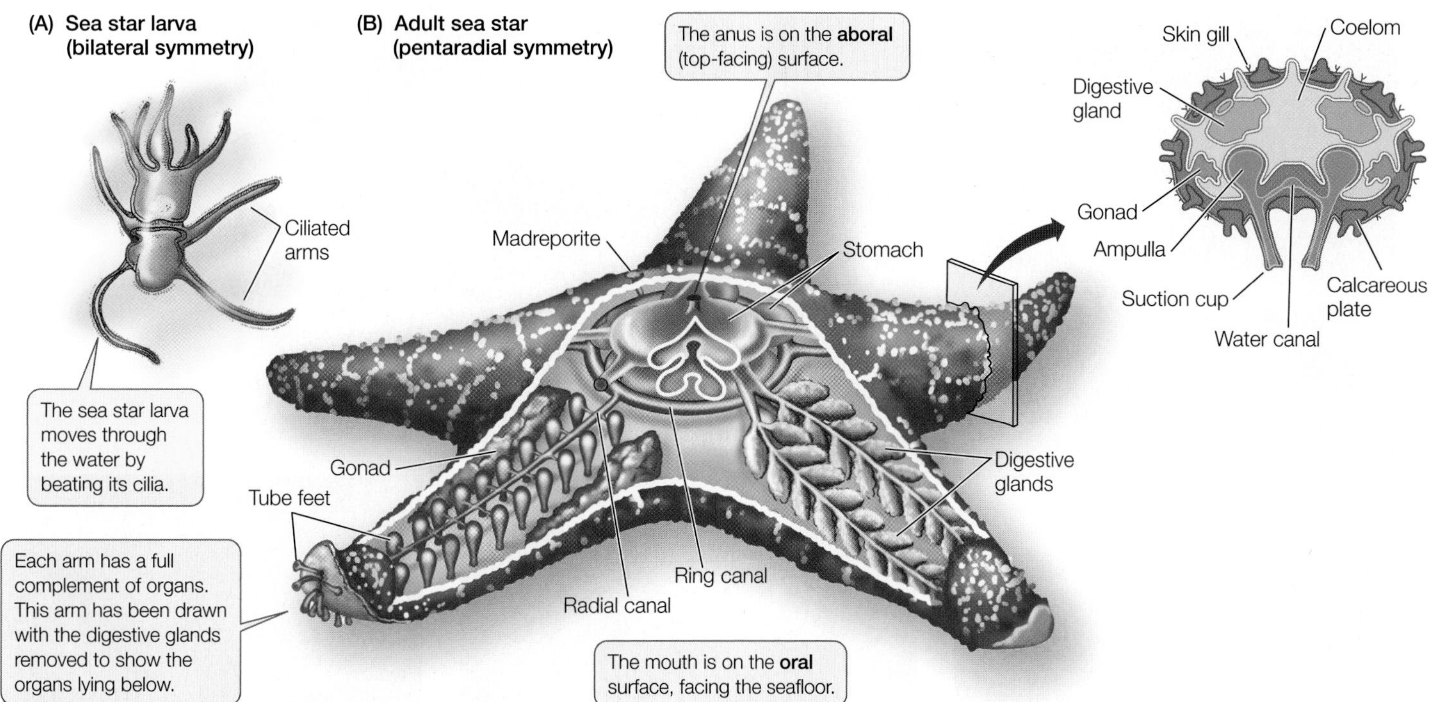
(A) Sea star larva (bilateral symmetry)
(B) Adult sea star (pentaradial symmetry)
The anus is on the **aboral** (top-facing) surface.
Ciliated arms
The sea star larva moves through the water by beating its cilia.
Each arm has a full complement of organs. This arm has been drawn with the digestive glands removed to show the organs lying below.
Madreporite
Stomach
Gonad
Tube feet
Digestive glands
Ring canal
Radial canal
The mouth is on the **oral** surface, facing the seafloor.
Skin gill
Coelom
Digestive gland
Gonad
Ampulla
Suction cup
Calcareous plate
Water canal

**33.3 Echinoderms Are Bilaterally Symmetrical as Larvae but Radially Symmetrical as Adults** (A) The ciliated larva of a sea star has bilateral symmetry. Hemichordates have a similar larval form. (B) The radially symmetrical adult sea star displays the canals and tube feet of the echinoderm water vascular system, as well as the calcified internal skeleton. The body's orientation is oral–aboral rather than anterior–posterior.

The echinoderms and hemichordates (together known as *ambulacrarians*) have a bilaterally symmetrical, ciliated larva (**Figure 33.3A**). Adult hemichordates also are bilaterally symmetrical. Echinoderms, however, undergo a radical change in form as they develop into adults (**Figure 33.3B**), changing from a bilaterally symmetrical larva to an adult with **pentaradial symmetry** (symmetry in five or multiples of five). As is typical of animals with radial symmetry, echinoderms have no head, and they move slowly and equally well in many directions. Rather than having an anterior–posterior (head–tail) and dorsal–ventral (back–belly) body organization, echinoderms have an *oral* side (containing the mouth) and an opposite *aboral* side (containing the anus).

## Echinoderms have unique structural features

Echinoderms
Hemichordates
Chordates

In addition to having pentaradial symmetry, adult echinoderms have two unique structural features. One is a system of calcified internal plates covered by thin layers of skin and some muscles. The calcified plates of most echinoderms are thick, and they fuse inside the entire body, forming an *internal skeleton*. The other unique feature is a **water vascular system**, a network of water-filled canals leading to extensions called **tube feet**. This system functions in gas exchange, locomotion, and feeding (see Figure 33.3B). Seawater enters the system through a perforated structure called a *madreporite*. A calcified canal leads from the madreporite to the *ring canal*, which surrounds the *esophagus* (the tube leading from the mouth to the stomach). *Radial canals* branch off from the ring canal, extending through the arms (in species that have arms) and connecting with the tube feet. Echinoderms use their tube feet in many different ways to move and to capture prey. These structural innovations have been modified in many ways, resulting in a striking array of very different animals.

Members of one major extant echinoderm clade, the *crinoids* (sea lilies and feather stars), were more abundant and species-rich 300 to 500 million years ago than they are today. There are some 80 described living sea lily species, most of which are sessile organisms attached to the substratum by a stalk. Feather stars (**Figure 33.4A**) grasp the substratum with flexible appendages that allow for limited movement. About 600 living species of feather stars have been described.

Unlike the mostly sessile crinoids, most surviving echinoderms are motile. The two main groups of motile echinoderms are the *echinozoans* (sea urchins and sea cucumbers) and *asterozoans* (sea stars and brittle stars). Sea urchins are hemispherical in shape and lack arms (**Figure 33.4B**). They are covered with spines that are attached to the underlying skeleton with ball-and-socket joints. These joints enable the spines to be moved so they can converge toward a point that has been touched. The spines vary in size and shape and can be used for locomotion; a few produce toxic substances. They provide effective protection for the urchin, as many a scuba diver has found out the hard way. Sand dollars are flattened, disc-shaped relatives of sea urchins.

Sea cucumbers also lack arms, and their bodies are oriented in an atypical manner for an echinoderm (**Figure 33.4C**). The mouth is anterior and the anus is posterior (front and rear), not oral and aboral (top and bottom) as in other echinoderms.

(A) *Oxycomanthus bennetti*

(B) *Sphaerechinus granularis*

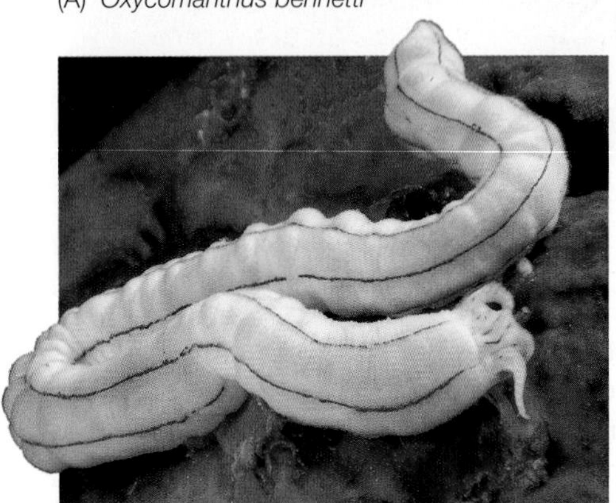

(C) *Synaptula* sp.

(D) *Asterias rubens*

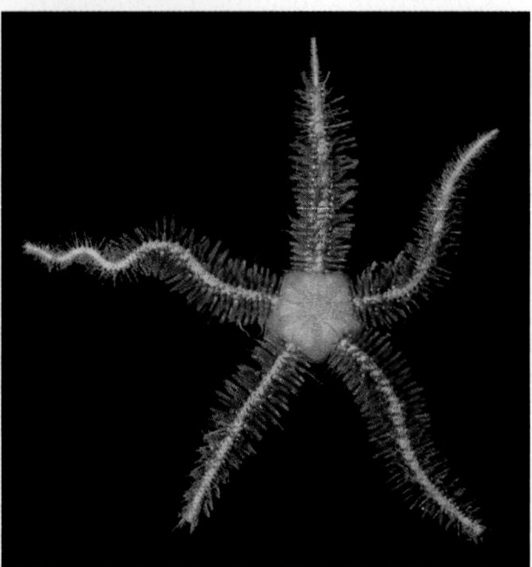

(E) *Ophiothrix spiculata*

**33.4 Echinoderm Diversity** (A) The flexible arms of this golden feather star are clearly visible. (B) Sea urchins are important grazers on algae in the intertidal zones of the world's oceans. (C) Sea cucumbers are unique among echinderms in having an anterior–posterior rather than an oral–aboral orientation of the mouth and anus. (D) Sea stars are important predators on bivalve mollusks such as mussels and clams. Suction tips on its tube feet allow a sea star to grasp both shells of the bivalve and pull them open. (E) The arms of the brittle star are composed of hard but jointed plates.

Sea cucumbers use most of their tube feet primarily for attaching to the substratum rather than for moving.

Sea stars, popularly called starfish, are the most familiar echinoderms (**Figure 33.4D**). Their gonads and digestive organs are located in the arms, as seen in Figure 33.3B. Their tube feet serve as organs of locomotion, gas exchange, and attachment. Each tube foot of a sea star consists of an internal *ampulla* connected by a muscular tube to an external suction cup that can stick to the substratum. The tube foot is moved by expansion and contraction of the circular and longitudinal muscles of the tube. Brittle stars are similar in structure to sea stars, but their flexible arms are composed of jointed, hard plates (**Figure 33.4E**).

Echinoderms use their tube feet in a great variety of ways to capture prey. Sea lilies, for example, feed by orienting their arms in passing water currents. Food particles then strike and stick to the tube feet, which are covered with mucus-secreting glands. The tube feet transfer these particles to grooves in the arms, where ciliary action carries the food to the mouth. Sea cucumbers capture food with their anterior tube feet, which are modified into large, feathery, sticky tentacles that can be protruded from the mouth. Periodically, a sea cucumber withdraws the tentacles, wipes off the material that has adhered to them, and digests it.

Many sea stars use their tube feet to capture large prey such as polychaetes, gastropod and bivalve mollusks, small crustaceans such as crabs, and fishes. With hundreds of tube feet acting simultaneously, a sea star can grasp a bivalve in its arms, anchor the arms with its tube feet, and by steady contraction of the muscles in its arms, gradually exhaust the muscles the bivalve uses to keep its shell closed (see Figure 33.4D). To feed on a bivalve, a sea star can push its stomach out through its mouth and then through the narrow space between the two halves of the bivalve's shell. The sea star's stomach then secretes enzymes that digest the prey.

Most sea urchins eat algae, which they catch with their tube feet from the plankton or scrape from rocks with a complex rasping structure. Most of the 2,000 species of brittle stars ingest particles from the upper layers of sediments and assimilate the organic material from them, although some species filter suspended food particles from the water and others capture small animals.

### Hemichordates are wormlike marine deuterostomes

Hemichordates—acorn worms and pterobranchs—have a body organized in three major parts, consisting of a *proboscis*, a *collar* (which bears the mouth), and a *trunk* (which contains the other body parts). The 70 known species of *acorn worms* range up to 2 meters in length (**Figure 33.5A**). They live in burrows in muddy and sandy marine sediments. The digestive tract of an acorn worm consists of a mouth behind which are a muscular *pharynx* and an *intestine*. The pharynx opens to the outside through a number of *pharyngeal slits* through which water can exit. Highly vascularized tissue surrounding the pharyngeal slits serves as a gas-exchange apparatus. Acorn worms breathe by pumping water into the mouth and out through the pharyngeal slits. They capture prey with the large proboscis, which is coated with sticky mucus to which small organisms in the sediment stick.

(A)

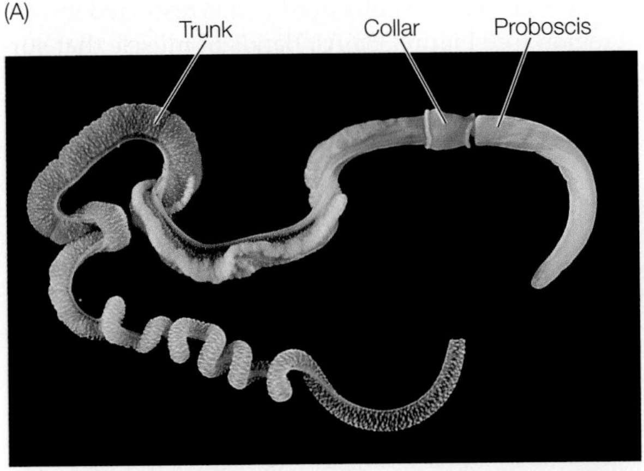


(B)

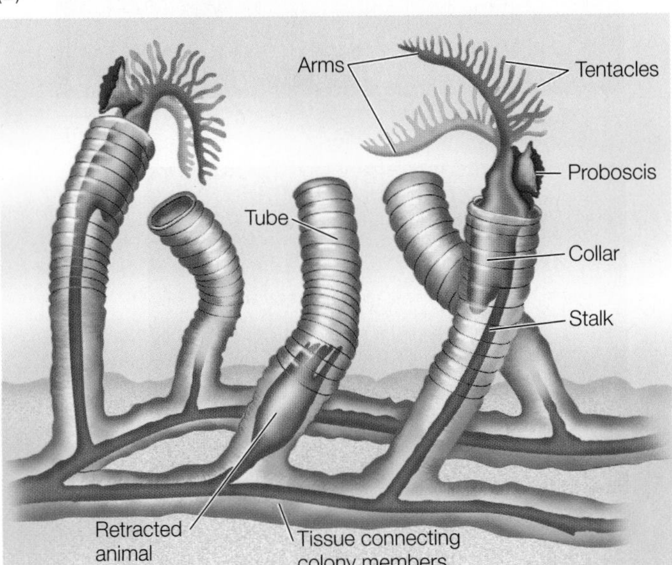


**33.5 Hemichordates** (A) The proboscis of an acorn worm is modified for burrowing. (B) The structure of a colonial pterobranch.

The mucus and its attached prey are conveyed by cilia to the mouth. In the esophagus, the food-laden mucus is compacted into a ropelike mass that is moved through the digestive tract by ciliary action.

The 30 living species of *pterobranchs* are sedentary marine animals up to 12 millimeters long that live in a tube secreted by the proboscis. Some species are solitary; others form colonies of individuals joined together (**Figure 33.5B**). Behind the proboscis is a collar with anywhere from one to nine pairs of arms. The arms bear long tentacles that capture prey and function in gas exchange.

#### 33.2 RECAP

**Echinoderms have an internal skeleton of calcified plates and a unique water vascular system. Hemichordates have a bilaterally symmetrical body divided into three parts: proboscis, collar, and trunk.**

- What are some of the ways that echinoderms use their tube feet to obtain food? See p. 696
- How do hemichordates obtain food? See p. 697

We have described the deuterostome groups that are most distantly related to us. Now we turn our attention to the chordates, the clade to which humans belong.

## 33.3 What New Features Evolved in the Chordates?

It is not obvious from examining adult animals that echinoderms and chordates share a common ancestor. The evolutionary relationships among some chordate groups are not immediately apparent, either. The features that reveal the evolutionary relationships both among the chordates and between chordates and echinoderms are seen primarily in the larvae—in other words, it is during the early developmental stages that their evolutionary relationships are evident.

There are three principal chordate clades: the **cephalochordates**, the **urochordates**, and the **vertebrates** (see Figure 33.1). There are about 3,000 living species of urochordates and 62,000 living species of vertebrates, but only about 30 living species of cephalochordates.

Adult chordates vary greatly in form, but all chordates display the following derived structures at some stage in their development (**Figure 33.6**):

- A dorsal hollow nerve cord
- A tail that extends beyond the anus
- A dorsal supporting rod called the notochord

The **notochord** is the most distinctive derived chordate trait. It is composed of a core of large cells with turgid fluid-filled vacuoles, which make it rigid but flexible. In the urochordates the notochord is lost during metamorphosis to the adult stage. In most vertebrate species, it is replaced during development by skeletal structures that provide support for the body.

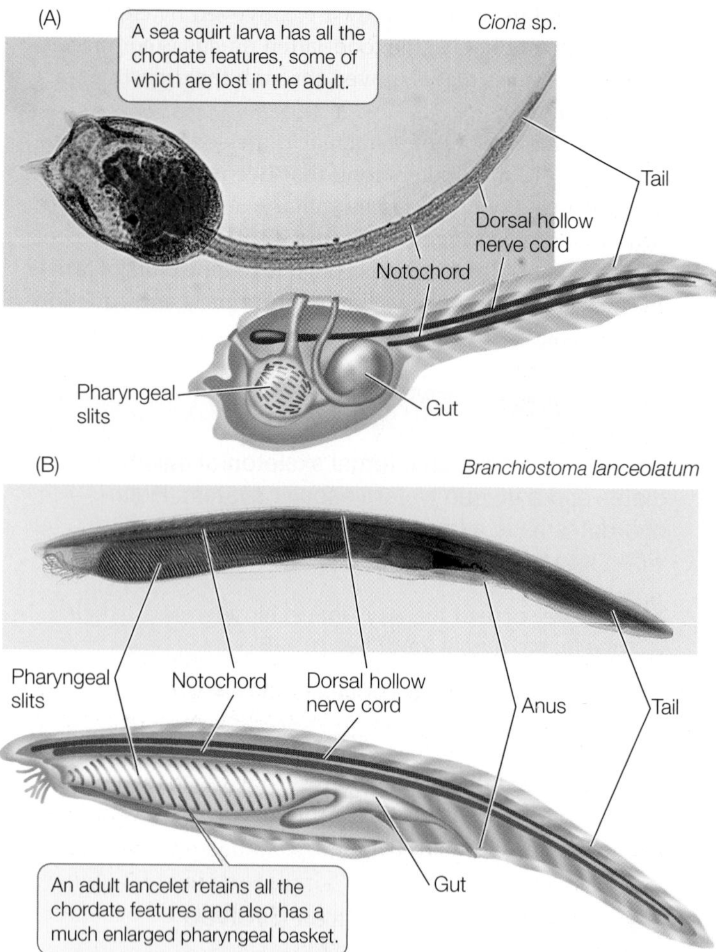


**33.6 The Key Features of Chordates Are Most Apparent in Early Developmental Stages** The pharyngeal slits of both the urochordate sea squirt and the cephalochordate lancelet develop into a pharyngeal basket. (A) The sea squirt larva (but not the adult) has all three chordate features: dorsal hollow nerve cord, postanal tail, and a notochord. (B) All three chordate synapomorphies are retained in the adult lancelet.

The ancestral pharyngeal slits (not a derived feature of this group) are present at some developmental stage of chordates but are often lost in adults. The *pharynx*, which develops around the pharyngeal slits, functioned in chordate ancestors as the site for oxygen uptake and the elimination of carbon dioxide and water (as in acorn worms). The pharynx is much enlarged in some chordate species (as in the *pharyngeal basket* of the lancelet in Figure 33.6B) but has been lost in others.

## Adults of most cephalochordates and urochordates are sessile

The 30 species of cephalochordates, or *lancelets*, are small animals that rarely exceed 5 centimeters in length. The notochord, which provides body support, extends the entire length of the body throughout their lives (see Figure 33.6B). Lancelets are found in shallow marine and brackish waters worldwide. Most of the time they lie covered in sand with their head protruding above the sediment, but they can swim. They filter prey from the water with their pharyngeal basket. During the reproductive season, the gonads of the males and females greatly enlarge. At spawning, the walls of the gonads rupture, releasing eggs and sperm into the water column, where fertilization takes place.

All members of the three major urochordate groups—the ascidians, thaliaceans, and larvaceans—are marine animals. More than 90 percent of the known species of urochordates are *ascidians* (sea squirts). Individual ascidians range in length from less than 1 millimeter to 60 centimeters. Some ascidians form colonies by asexual budding from a single founder. Colonies may measure several meters across. The baglike body of an adult ascidian is enclosed in a tough tunic, leading to its alternate name of "tunicate" (**Figure 33.7A**). The tunic is composed of proteins and a complex polysaccharide secreted by epidermal cells. The ascidian pharynx is enlarged into a pharyngeal basket that filters prey from the water passing through it.

In addition to its pharyngeal slits, an ascidian larva has a dorsal hollow nerve cord and a notochord that is restricted mostly to the tail region (see Figure 33.6A). Bands of muscle that surround the notochord provide support for the body. After a short

(A) *Clavelina dellavallei* 1 cm

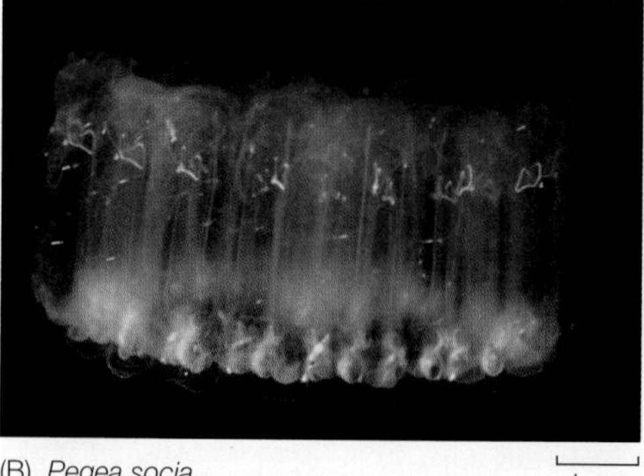

(B) *Pegea socia* 1 cm

**33.7 Adult Urochordates** (A) The transparent tunic and the pharyngeal basket are clearly visible in this ascidian (sea squirt, also known as a tunicate). (B) A chainlike colony of thaliaceans (salps) floats in tropical waters.

time swimming in the plankton, the larvae of most species settle on the seafloor and transform into sessile adults. The swimming, tadpolelike larvae suggest a close evolutionary relationship between ascidians and vertebrates (see Figure 22.6).

*Thaliaceans* (salps) can live singly or in chainlike colonies up to several meters long (**Figure 33.7B**). They float in tropical and subtropical oceans at all depths down to 1,500 meters. *Larvaceans* are solitary planktonic animals that retain their notochord and dorsal hollow nerve cord throughout their lives. Most larvaceans are less than 5 millimeters long, but some species that live near the bottom of deep ocean waters build delicate casings of mucus that may be more than a meter wide. They snare sinking organic particles (their primary food source) with elaborate filters built into their mucus "houses." When the old "house" gets clogged with excess debris, the animals build a new one.

### A dorsal supporting structure replaces the notochord in vertebrates

In one chordate group, a new dorsal supporting structure evolved. The **vertebrates** take their name from the jointed, dorsal **vertebral column** that replaces the notochord during early development as their primary supporting structure (**Figure 33.8**).



All of the non-vertebrate deuterostomes (the hemichordates, echinoderms, cephalochordates, and urochordates) are exclusively or primarily marine. The lineage that led to the vertebrates is also thought to have evolved in the oceans, although probably in an estuarine environment (where fresh water meets salt water). Vertebrates have since radiated into marine, freshwater, terrestrial, and aerial environments worldwide.

The *hagfishes* are thought by many biologists to be the sister group to the remaining vertebrates (as shown in Figure 33.8). Hagfishes (**Figure 33.9A**) have a weak circulatory system, with three small accessory hearts (rather than a single, large heart), a partial *cranium* or skull (containing no *cerebrum* or *cerebellum*, two main regions of the brain of other vertebrates), and no jaws or stomach. They also lack separate, jointed vertebrae and have a skeleton composed of cartilage. Thus, some biologists do not consider hagfishes vertebrates, and use instead the term *craniates* to refer collectively to the hagfishes and the vertebrates. Some analyses of gene sequences suggest, however, that hagfishes may be more closely related to the vertebrate lampreys (**Figure 33.9B**); in this phylogenetic arrangement, the two groups are collectively called the *cyclostomes* ("circle mouths"). If in fact the hagfishes and lampreys do form a monophyletic group, then hagfishes must have secondarily lost many of the major vertebrate morphological features during their evolution.

The 58 known species of hagfishes are unusual marine animals that produce copious quantities of slime as a defense. They are virtually blind and rely largely on the four pairs of sensory tentacles around their mouth to detect food. Although they have no jaws, hagfishes have a tonguelike structure equipped with toothlike rasps that they can use to tear apart dead organisms and to capture their principal prey, polychaete worms. Hagfishes have direct development (no larvae), and individuals may actually change sex from year to year (from male to female and vice versa).

The nearly 50 species of lampreys either live in fresh water, or they live in coastal salt water and move into fresh water to breed. Although the lampreys and hagfishes may look super-

**33.8 Phylogeny of the Living Vertebrates** This phylogenetic tree shows the evolution of some of the key innovations among the major groups of vertebrates.

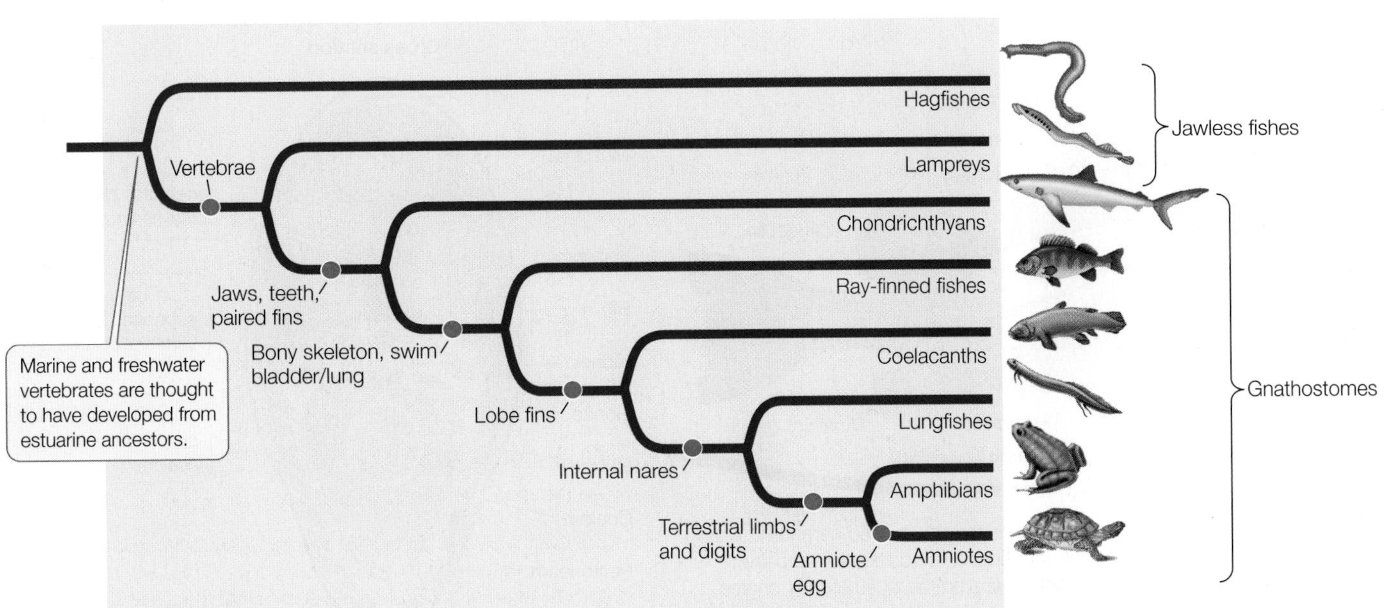

*Eptatretus stoutii*

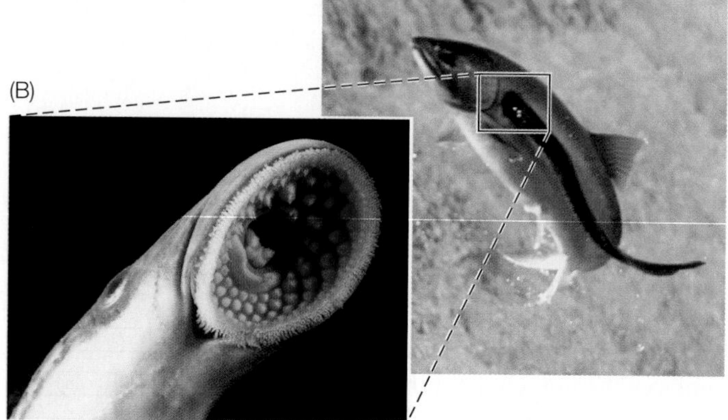

*Petromyzon marinus*

**33.9 Modern Jawless Fishes** (A) Hagfishes burrow in the ocean mud, from which they extract small prey. They also scavenge on dead or dying fish. Hagfishes have degenerate eyes, which has led to their being miscalled "blind eels." (B) Sea lampreys are ectoparasites that attach to the bodies of living fish and use their large, jawless mouths to suck blood and flesh. They can survive in both fresh and salt water, as this individual attached to a salmon returning to its spawning ground will do.

ficially similar (with elongate eel-like bodies and no paired fins), they differ greatly in their biology.

Lampreys have a complete braincase and distinct and separate (although rudimentary) vertebrae, all cartilaginous rather than bony. Lampreys undergo a complete metamorphosis from filter-feeding larvae known as *ammocoetes*, which are morphologically quite similar in general structure to adult lancelets. The adults of many species of lampreys are parasitic, although several lineages of lampreys evolved to become nonfeeding as adults. These nonfeeding adult lampreys survive only a few weeks after metamorphosis—just long enough to breed. In the species that are parasitic as adults, the round mouth is a rasping and sucking organ that is used to attach to their prey and rasp at the flesh (see Figure 33.9B). Some species of lampreys are critically endangered because of recent habitat changes and losses.

### The vertebrate body plan can support large, active animals

Four key features characterize the vertebrates:

- An anterior *skull* with a large brain
- A rigid internal *skeleton* supported by the vertebral column
- Internal organs *suspended in a coelom*
- A well-developed *circulatory system*, driven by contractions of a ventral *heart*

This organization of the vertebrate body is exemplified by the bony fish diagrammed in **Figure 33.10**. Many kinds of jawless fishes were found in the seas, estuaries, and fresh waters of the Devonian period, but hagfishes and lampreys are the only jawless fishes that survived beyond the Devonian. During that period, the *gnathostomes* (Greek, "jaw mouths") evolved jaws via modifications of the skeletal arches that supported the gills (**Figure 33.11A**). Jaws greatly improved feeding efficiency, as an animal with jaws can grasp, subdue, and swallow large prey.

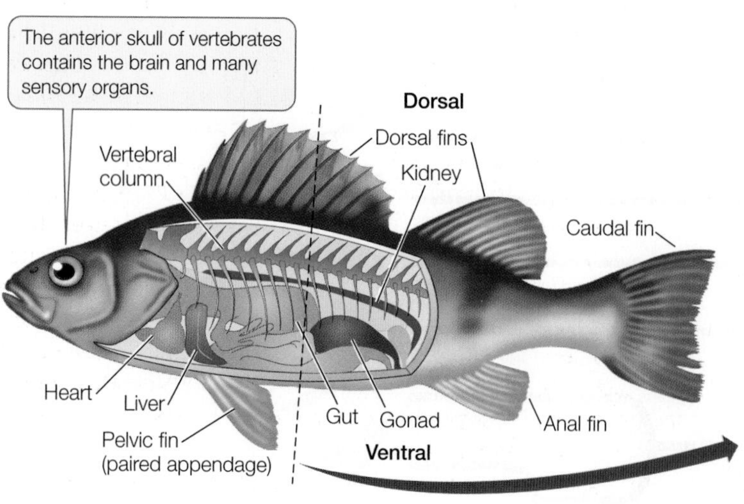


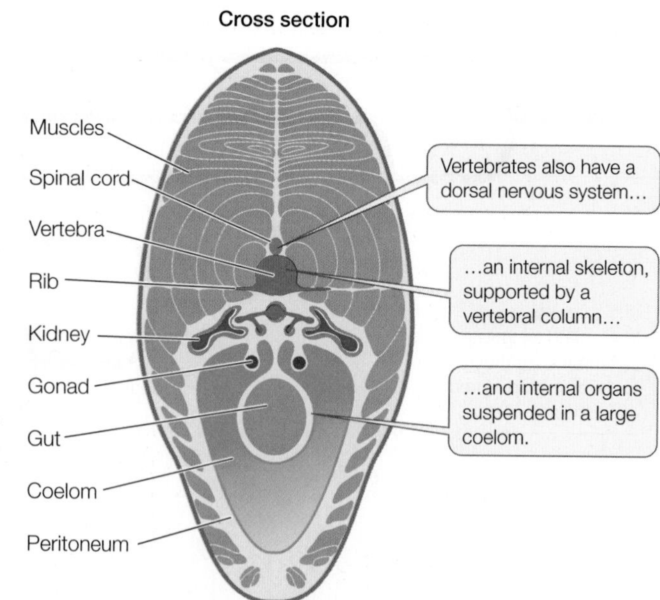


**33.10 The Vertebrate Body Plan** A ray-finned fish is used here to illustrate the structural elements common to all vertebrates. In addition to the paired pelvic fins, these fishes have paired pectoral fins on the sides of their bodies (not seen in this cutaway view).

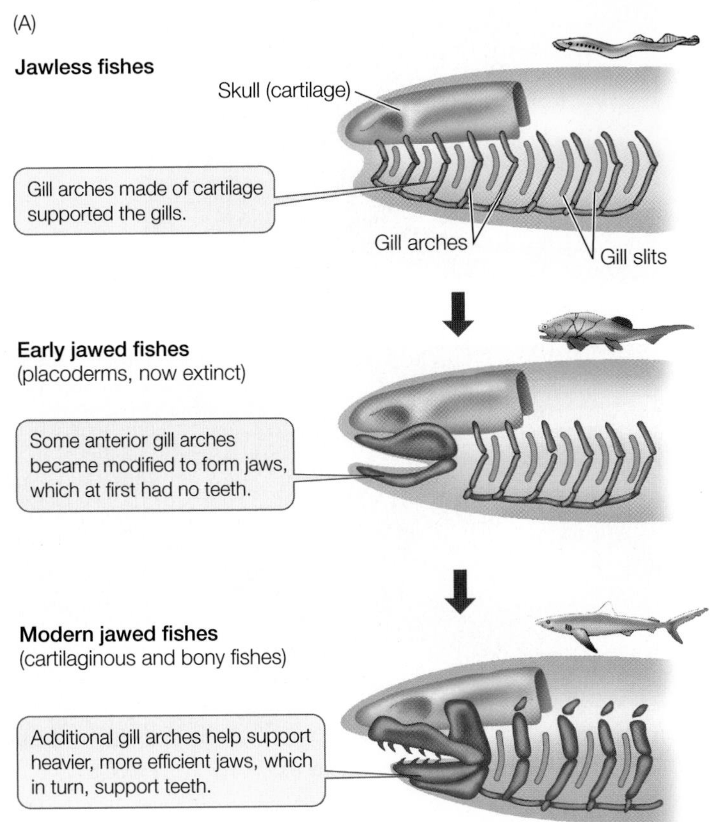
(A)
Jawless fishes
Skull (cartilage)
Gill arches made of cartilage supported the gills.
Gill arches
Gill slits
Early jawed fishes
(placoderms, now extinct)
Some anterior gill arches became modified to form jaws, which at first had no teeth.
Modern jawed fishes
(cartilaginous and bony fishes)
Additional gill arches help support heavier, more efficient jaws, which in turn, support teeth.

(B)

**33.11 Jaws and Teeth Increased Feeding Efficiency** (A) The diagrams illustrate one probable scenario for the evolution of jaws from the anterior gill arches of jawless fishes. (B) Jaws of the extinct giant shark (*Carcharodon megalodon*) display the teeth that indicate an extreme predatory lifestyle.

The earliest jaws were simple, but the evolution of *teeth* made predators more effective (**Figure 33.11B**). In predators, teeth function crucially both in grasping and in breaking up prey. In both predators and herbivores, teeth enable an animal to chew both soft and hard body parts of their food. Chewing also aids chemical digestion and improves an animal's ability to extract nutrients from its food, as we describe in Chapter 51. Vertebrates are remarkable in the diversity of their jaws and teeth.

## Fins and swim bladders improved stability and control over locomotion

Paired fins stabilize the position of jawed fishes in water (and in some cases, help propel them). Most aquatic gnathostomes have a pair of pectoral fins just behind the gill slits, and a pair of pelvic fins anterior to the anal region (see Figures 33.10 and 33.12). Median dorsal and anal fins stabilize the fish, or may be used for propulsion in some species. In many fishes, the caudal fin helps propel the animal and enables it to turn rapidly.

Hagfishes
Lampreys
Cartilaginous fishes
Ray-finned fishes
Sacopterygians

Several groups of gnathostomes became abundant during the Devonian. Among them were the **chondrichthyans**—sharks, skates, and rays (940 living species) and chimaeras (40 living species). Like hagfishes and lampreys, these fishes have a skeleton composed entirely of firm but pliable cartilage. Their skin is flexible and leathery, sometimes bearing scales that give it the consistency of sandpaper. Sharks move forward by means of lateral undulations of their body and caudal (tail) fin (**Figure 33.12A**). Skates and rays propel themselves by means of vertical undulating movements of their greatly enlarged pectoral fins (**Figure 33.12B**).

Most sharks are predators, but some feed by straining plankton from the water. Most skates and rays live on the ocean floor, where they feed on mollusks and other animals buried in the sediments. Nearly all cartilaginous fishes live in the oceans, but a few are estuarine or migrate into lakes and rivers. One group of stingrays is found in river systems of South America. The less familiar chimaeras (**Figure 33.12C**) live in deep-sea or cold waters.

In the ancestor of the bony vertebrates, gas-filled sacs supplemented the gas-exchange function of the gills by giving the animals access to atmospheric oxygen. These features enabled those fishes to live where oxygen was periodically in short supply, as it often is in freshwater environments. These lunglike sacs evolved into *swim bladders*, which are organs of buoyancy, as well as into the lungs of tetrapods. By adjusting the amount of gas in its swim bladder, a fish can control the depth at which it remains suspended in the water while expending very little energy to maintain its position.

**Ray-finned fishes**, and most remaining groups of vertebrates, have internal skeletons of calcified, rigid *bone* rather than flexible cartilage. The outer body surface of most species of ray-finned fishes is covered with flat, thin, lightweight scales that provide some protection or enhance movement through the wa-

(A) *Carcharhinus plumbeus*

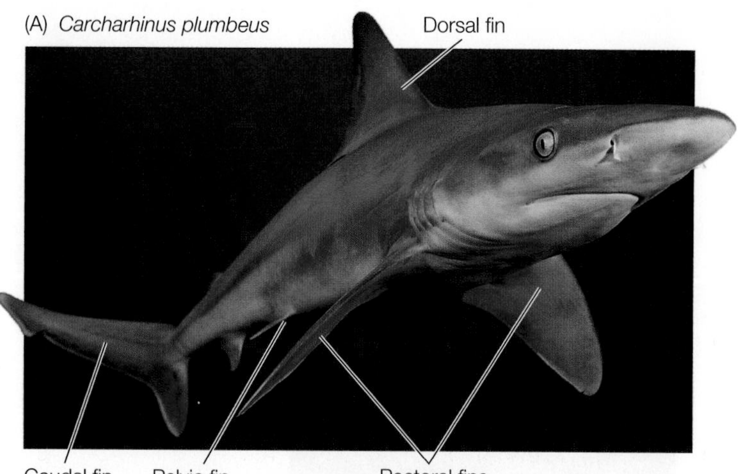


(B) *Myliobatis australis*

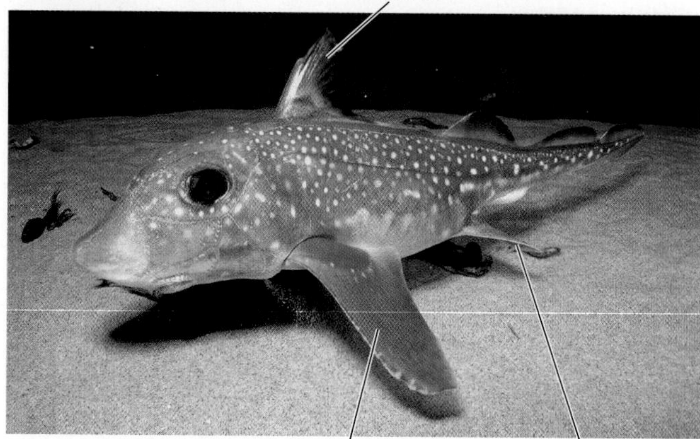


(C) *Hydrolagus colliei*

**33.12 Chondrichthyans** (A) Most sharks, such as this sandbar shark, are active marine predators. (B) Skates and rays, represented here by an eagle ray, feed on the ocean bottom. Their modified pectoral fins are used for propulsion; their other fins are greatly reduced. (C) A chimaera, or ratfish. These deep-sea fishes often possess modified dorsal fins that contain toxins.

**33.13 Diverse Ray-Finned Fishes** (A) Eels such as this moray have the large teeth and powerful jaws typical of predatory fishes. (B) The wrasses contain more than 500 described species. Many species, such as the flame fairy wrasse seen here, inhabit coral reefs. (C) Another large ray-fin clade, the serranids, includes the sea basses and groupers. Panther groupers such as this one are endangered by the loss of Pacific coral reef habitat. (D) A unique structure that resembles a fishing lure has evolved among the anglerfishes. Deep-sea anglerfishes such as this one live below the level of light penetration; their lures are bioluminescent.

(A) *Gymnothorax favagineus*

(B) *Cirrhilabrus jordani*

(C) *Cromileptes altivelis*

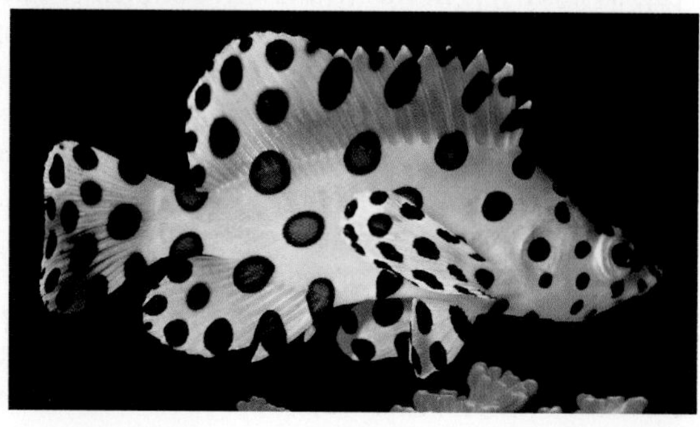

(D) *Gigantactis vanhoeffeni* luring prey

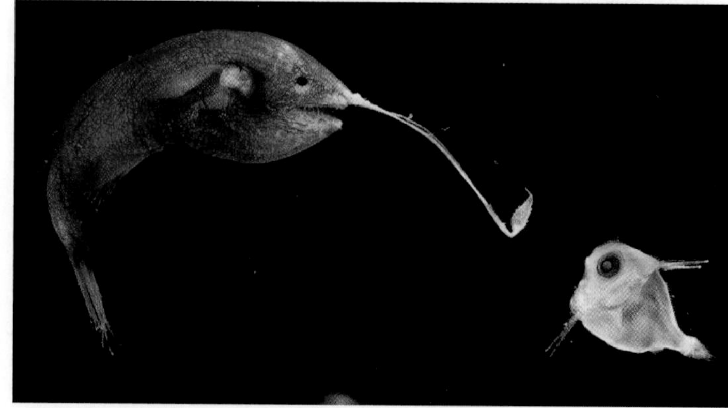

ter. The gills of ray-finned fishes open into a single chamber covered by a hard flap, called an *operculum*. Movement of the operculum improves the flow of water over the gills, where gas exchange takes place.

Ray-finned fishes radiated extensively in the Tertiary. Today there are about 30,000 known living species, encompassing a remarkable variety of sizes, shapes, and lifestyles (**Figure 33.13**). The smallest are less than 1 centimeter long as adults; the largest weigh as much as 900 kilograms. Ray-finned fishes exploit nearly all types of aquatic food sources. In the oceans they filter plankton from the water, rasp algae from rocks, eat corals and other soft-bodied colonial animals, dig animals from soft sediments, and prey on virtually all kinds of other fishes. In fresh water they eat plankton, devour insects, eat fruits that fall into the water in flooded forests, and prey on other aquatic vertebrates and, occasionally, terrestrial vertebrates. Many ray-finned fishes are solitary, but in open water others form large aggregations called *schools*. Many species perform complicated behaviors to maintain schools, build nests, court and choose mates, and care for their young.

Although ray-finned fishes can readily control their position in open water using their fins and swim bladder, their eggs tend to sink. Some species produce small eggs that are buoyant enough to complete their development in the open water, but many marine fishes move to food-rich shallow waters to lay their eggs. That is why coastal waters and estuaries are so important in the life cycles of many marine fishes. Some, such as salmon, are *anadromous*; these species leave salt water when they breed, ascending rivers to spawn in freshwater streams and lakes (see Figure 33.9B).

## 33.3 RECAP

**Chordates are characterized by a dorsal hollow nerve chord, a post-anal tail, and a dorsal supporting rod called a notochord at some point during the life cycle. Specialized structures for support (such as vertebrae), locomotion (such as fins), and feeding (such as jaws and teeth) evolved among the vertebrates, which allowed them to colonize and adapt to most of Earth's environments.**

- What synapomorphies respectively characterize the chordates and the vertebrates? **See pp. 697–700 and Figures 33.6 and 33.10**
- How do the hagfishes differ from the lampreys in morphology and life history? Do you see why some biologists do not consider the hagfishes to be vertebrates? **See p. 699**

In some fishes, the lunglike sacs that gave rise to swim bladders became specialized for another purpose: breathing air. That adaptation set the stage for the vertebrates to move onto the land.

# 33.4 How Did Vertebrates Colonize the Land?

The evolution of lunglike sacs in fishes set the stage for the invasion of the land. Some early ray-finned fishes probably used those sacs to supplement their gills when oxygen levels in the water were low, as lungfishes and many groups of ray-finned fishes do today. But with their unjointed fins, those fishes could only flop around on land. Changes in the structure of the fins first allowed some fishes to support themselves better in shallow water and, later, to move better on land.

### Jointed fins enhanced support for fishes



Two pairs of muscular, jointed fins evolved in the ancestor of the **sarcopterygians**, which include coelacanths, lungfishes, and tetrapods. Each of the jointed appendages of sarcopterygians is joined to the body by a single enlarged bone. The coelacanths flourished from the Devonian until about 65 million years ago, when they were thought to have become extinct. However, in 1938 a commercial fisherman caught a living coelacanth off South Africa. Since that time, hundreds of individuals of this extraordinary fish, *Latimeria chalumnae*, have been collected. A second species, *L. menadoensis*, was discovered in 1998 off the Indonesian island of Sulawesi. *Latimeria*, a predator of other fishes, reaches a length of about 1.8 meters and weighs up to 82 kilograms (**Figure 33.14A**). Its skeleton is composed mostly of cartilage, not bone. A cartilaginous skeleton is a derived feature in this clade because it had bony ancestors.

**Lungfishes**, which also have jointed fins that are connected to the body by a single enlarged bone, were important predators in shallow-water habitats in the Devonian, but most lineages died out. The six surviving species live in stagnant swamps and muddy waters in South America, Africa, and Australia (**Figure 33.14B**). Lungfishes have lungs derived from the lunglike sacs of their ancestors as well as gills. When ponds dry up, individuals of most species can burrow deep into the mud and survive for many months in an inactive state while breathing air.

It is believed that some early aquatic sarcopterygians began to use terrestrial food sources, became more fully adapted to life on land, and eventually evolved to become ancestral **tetrapods** (four-legged vertebrates).

How was the transition from an animal that swam in water to one that walked on land accomplished? Early in 2006, scientists reported the discovery of a Devonian fossil they believe represents an intermediate between the finned appendages of fishes and the limbs of terrestrial tetrapods (**Figure 33.14C**). It appears that limbs able to prop up a large fish with the front-to-rear movement necessary for walking evolved while the animals still lived in water. These limbs appear to have functioned in holding the animals upright in shallow water, perhaps even allowing them to hold their head above the water's surface. These same structures were then co-opted for movement on land, at first probably for foraging on brief trips out of water.

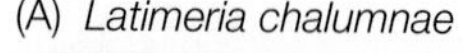

(A) *Latimeria chalumnae*

(B) *Protopterus annectens*

*Tiktaalik*'s pectoral fins show some of the skeletal structures of tetrapod limbs.


(C) *Tiktaalik roseae*

**33.14 The Closest Relatives of Tetrapods** (A) The African coelacanth, discovered in deep waters of the Indian Ocean, represents one of two surviving species of a group that was once thought to be extinct. (B) All surviving lungfish species, such as this African lungfish, live in the Southern Hemisphere. (C) This fossil from the Devonian is believed to represent a transitional species intermediate between the finned fishes and the limbed tetrapods.

## Amphibians adapted to life on land

Lungfishes
Amphibians
Amniotes

During the Devonian, the first tetrapods arose from an aquatic ancestor. In this lineage, legs capable of movement on land evolved from the short, stubby fins of their aquatic ancestors. The basic elements of those legs have remained throughout the evolution of terrestrial vertebrates, although they have changed considerably in their form.

Most modern **amphibians** are confined to moist environments because they lose water rapidly through the skin when exposed to dry air. In addition, their eggs are enclosed within delicate membranous envelopes that cannot prevent water loss in dry conditions. In some amphibian species, adults live mostly on land but return to fresh water to lay their eggs, which are usually fertilized outside the body (**Figure 33.15**). The fertilized eggs give rise to larvae that live in water until they undergo metamorphosis to become terrestrial adults. However, many amphibians (especially those in tropical and subtropical areas) have evolved a wide diversity of additional reproductive modes and types of parental care, as described in the opening of this chapter. Internal fertilization evolved many times among the amphibians.

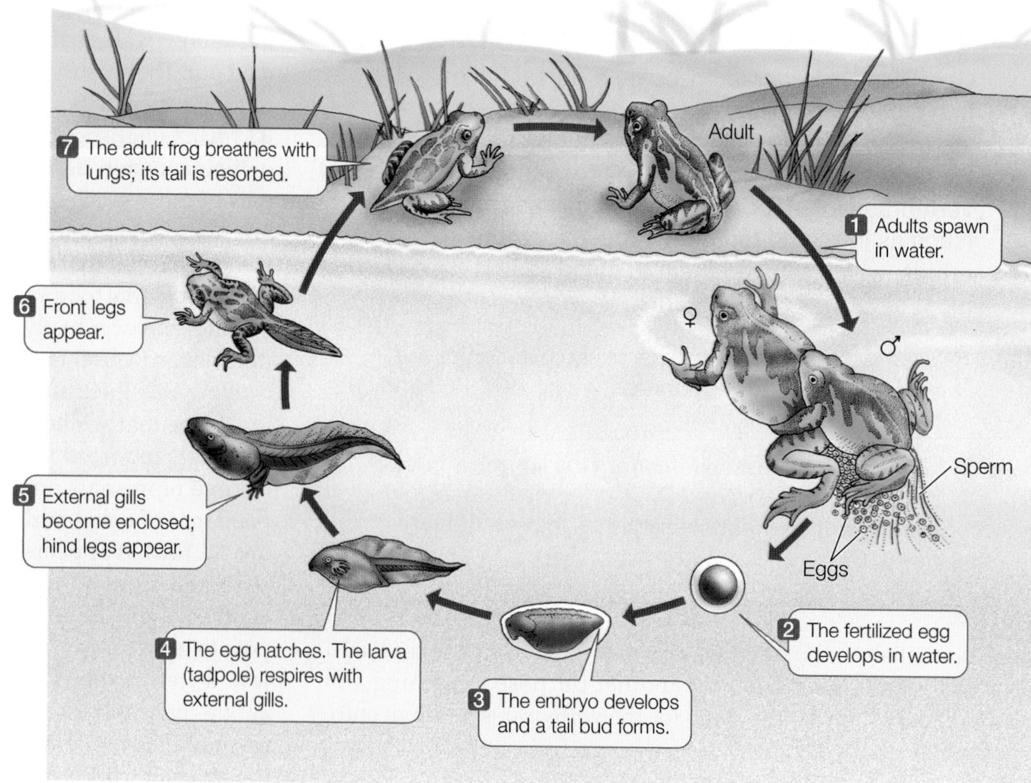


1 Adults spawn in water.

4 The egg hatches. The larva (tadpole) respires with external gills.
5 External gills become enclosed; hind legs appear.
6 Front legs appear.
7 The adult frog breathes with lungs; its tail is resorbed.

**33.15 In and Out of the Water** Most early stages in the life cycle of many amphibians take place in water. The aquatic tadpole transforms into a terrestrial adult through metamorphosis. Some species of amphibians, however, have direct development (with no aquatic larval stage), and others are aquatic throughout life.

**yourBioPortal.com**
GO TO **Animated Tutorial 33.1 • Life Cycle of a Frog**

Many species develop directly into adultlike forms from fertilized eggs laid on land or carried by the parents. Other species of amphibians are entirely aquatic, never leaving the water at any stage of their lives, and many of these species retain a larval-like morphology.

The more than 6,500 known species of amphibians living on Earth today belong to three major groups: the wormlike, limbless, tropical, burrowing or aquatic *caecilians* (**Figure 33.16A**), the tail-less frogs and toads (collectively called *anurans;* **Figure 33.16B**), and the tailed *salamanders* (**Figure 33.16C and D**).

Anurans are most diverse in wet tropical and warm temperate regions, although a few are found at very high latitudes. There are far more anurans than any other amphibians, with about 5,800 described species and more being discovered every year. Some anurans have tough skins and other adaptations that enable them to live for long periods in very dry deserts, whereas others live in moist terrestrial and arboreal environments. Some species are completely aquatic as adults. All anurans have a very short vertebral column, with a strongly modified pelvic region that is modified for leaping, hopping, or propelling the body through water by kicking the hind legs.

The approximately 600 described species of salamanders are most diverse in temperate regions of the Northern Hemisphere, but many species are also found in cool, moist environments in the mountains of Central America, and a few species penetrate into the South American tropics. Many salamanders live in rotting logs or moist soil. One major group has lost lungs, and these species exchange gases entirely through the skin and mouth lining—body parts that all amphibians use in addition to their lungs. Through *paedomorphosis* (retention of the juvenile state; see Chapter 20), a completely aquatic lifestyle has evolved several times among the salamanders (see Figure 33.16D). Most species of salamanders have internal fertilization, which is usually achieved through the transfer of a small jellylike, sperm-embedded capsule called a *spermatophore*.

Many amphibians have complex social behaviors. Most male anurans utter loud, species-specific calls to attract females of their own species (and sometimes to defend breeding territories), and they compete for access to females that arrive at the breeding sites. Many amphibians lay large numbers of eggs, which they abandon once they are deposited and fertilized. As described in the opening of this chapter, some amphibians lay only a few eggs, which are fertilized and then guarded in a nest, or carried on the backs, in the vocal pouches, or even in the stomachs of one of the parents. A few species of frogs, salamanders, and caecilians are *viviparous*, meaning they give birth to well-developed young that have received nutrition from the female during gestation.

**33.16 Diversity among the Amphibians** (A) Burrowing caecilians superficially look more like worms than amphibians. (B) Male golden toads in the cloud forest of Monteverde, Costa Rica. This species has recently become extinct, one of many amphibian species to do so in the past few decades. (C) An adult barred tiger salamander. (D) This Austin blind salamander's life cycle remains aquatic; it has no adult terrestrial stage. The eyes of this cave dweller have become greatly reduced.

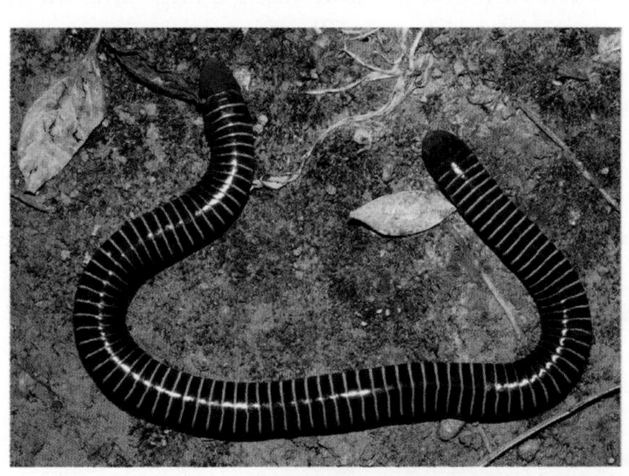
(A) *Siphonops annulatus*

(B) *Bufo periglenes*

(C) *Ambystoma mavortium*

(D) *Eurycea waterlooensis*

Amphibians are the focus of much attention today because populations of many species are declining rapidly, especially in mountainous regions of western North America, Central and South America, and northeastern Australia. Worldwide, about one-third of amphibian species are now threatened with extinction or have disappeared completely in the last few decades (as happened with the gastric brooding frogs described at the beginning of this chapter). Scientists are investigating several hypotheses to account for these population declines, including the adverse effects of habitat alteration by humans, increased solar radiation caused by destruction of Earth's ozone layer, pollution from urban and industrial areas and airborne agricultural pesticides and herbicides (see Chapter 1), and the spread of a pathogenic chytrid fungus that attacks amphibians. Scientists have documented the spread of the chytrid fungus through Central America, where many species of amphibians have become extinct (including Costa Rica's golden toad; see Figure 33.16B).

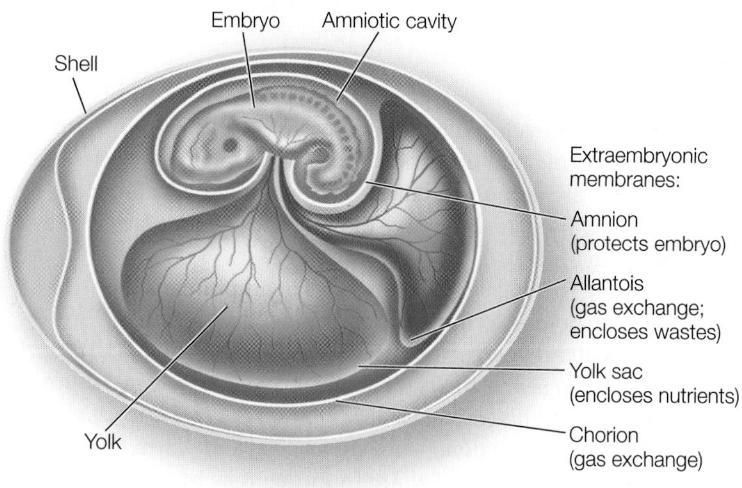


**33.17 An Egg for Dry Places** The evolution of the amniote egg, with its water-retaining shell, four extraembryonic membranes, and embryo-nourishing yolk, was a major step in the colonization of the terrestrial environment.

yourBioPortal.com
GO TO Web Activity 33.2 • The Amniote Egg

### Amniotes colonized dry environments

Several key innovations contributed to the ability of members of one clade of tetrapods to exploit a wide range of terrestrial habitats. The animals that evolved these water-conserving traits are called **amniotes**.

The **amniote egg** (which gives the group its name) is relatively impermeable to water and allows the embryo to develop in a contained aqueous environment (**Figure 33.17**). The leathery or brittle, calcium-impregnated shell of the amniote egg retards evaporation of the fluids inside but permits passage of oxygen and carbon dioxide. The egg also stores large quantities of food in the form of *yolk*, allowing the embryo to attain a relatively advanced state of development before it hatches. Within the shell are *extraembryonic membranes* that protect the embryo from desiccation and assist its gas exchange and excretion of waste nitrogen.

In several different groups of amniotes, the amniote egg became modified, allowing the embryo to grow inside (and receive nutrition from) the mother. For instance, the mammalian egg lost its shell while the functions of the extraembryonic membranes were retained and expanded; we examine the roles of these membranes in detail in Section 44.4.

Other innovations evolved in the organs of terrestrial adults. A tough, impermeable skin, covered with scales or modifications of scales such as hair and feathers, greatly reduced water loss. Adaptations of the vertebrate excretory organs, the kidneys, allowed amniotes to excrete concentrated urine, ridding the body of waste nitrogen without losing a large amount of water in the process (see Chapter 52).

During the Carboniferous, amniotes split into two major groups, the mammals and reptiles (**Figure 33.18**). More than

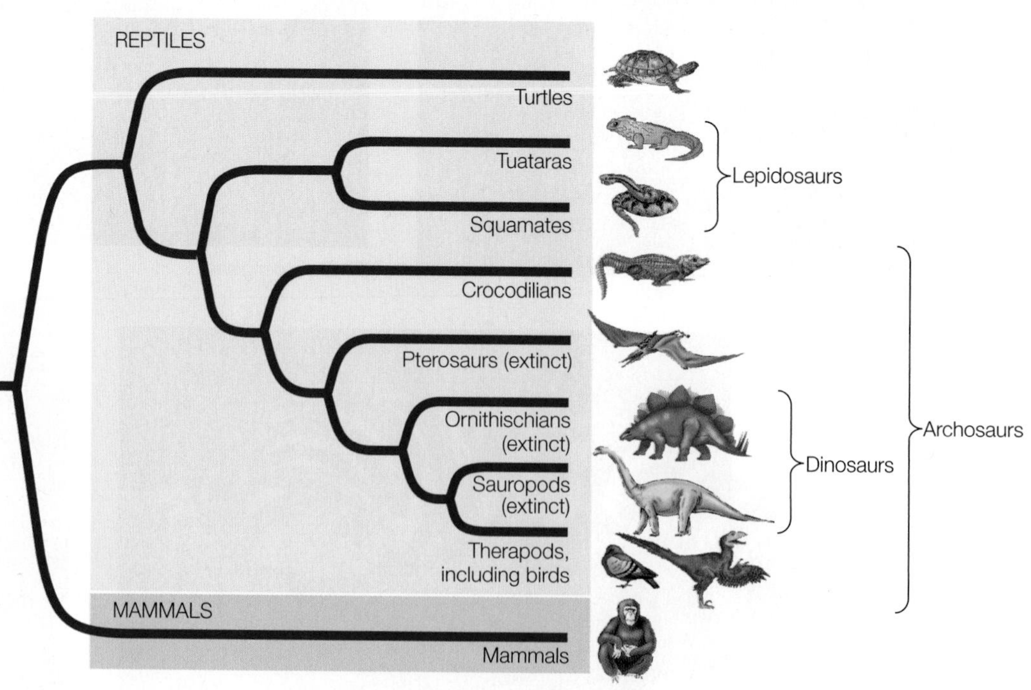


**33.18 Phylogeny of Amniotes** This tree of amniote relationships shows the primary split between mammals and reptiles. The reptile portion of the tree shows a lineage that led to the turtles, another to the lepidosaurs (snakes, lizards, and tuataras), and a third branch that includes all the archosaurs (crocodiles, several extinct groups, and the birds). There is some uncertainty in the placement of the turtle lineage; some data support a relationship between turtles and archosaurs.

18,700 species of **reptiles** exist today, over half of which are *birds*. Birds are the only living members of the otherwise extinct *dinosaurs*, the dominant terrestrial predators of the Mesozoic.

### Reptiles adapted to life in many habitats

Amphibians
Reptiles
Mammals

The lineage leading to modern reptiles began to diverge from other amniotes about 250 million years ago. One reptilian group that has changed very little over the intervening millennia is the *turtles*. The dorsal and ventral bony plates of turtles form a shell into which the head and limbs can be withdrawn in many species (**Figure 33.19A**). The dorsal shell is an expansion of the ribs, and it is a mystery how the pectoral girdles evolved to be inside the ribs of turtles, making them unlike any other vertebrates. Most turtles live in aquatic environments, but several groups, such as tortoises and box turtles, are terrestrial. Sea turtles spend their entire lives at sea except when they come ashore to lay eggs. Human exploitation of sea turtles and their eggs has resulted in worldwide declines of these species, all of which are now endangered. A few species of turtles are strict herbivores or carnivores, but most species are omnivores that eat a variety of aquatic and terrestrial plants and animals.

The *lepidosaurs* constitute the second-most species-rich clade of living reptiles. This group is composed of the *squamates* (lizards, snakes, and amphisbaenians—the last a group of mostly legless, wormlike, burrowing reptiles with greatly reduced eyes) and the *tuataras*, which superficially resemble lizards but differ from them in tooth attachment and several internal anatomical features. Many species related to the tuataras lived during the Mesozoic era, but today only two species, restricted to a few islands off New Zealand, survive (**Figure 33.19B**).

The skin of a lepidosaur is covered with horny scales that greatly reduce loss of water from the body surface. These scales, however, make the skin unavailable as an organ of gas exchange. Gases are exchanged almost entirely via the lungs, which are proportionally much larger in surface area than those of amphibians. A lepidosaur forces air into and out of its lungs by bellowslike movements of its ribs. The three-chambered lepidosaur heart partially separates oxygenated blood from the lungs from deoxygenated blood returning from the body. With this type of heart, lepidosaurs can generate high blood pressure and can sustain a relatively high metabolism.

Most lizards are insectivores, but some are herbivores; a few prey on other vertebrates. The largest lizard, which grows as

**33.19 Reptilian Diversity** (A) Green sea turtles are widely distributed in tropical oceans. (B) This tuatara represents one of only two surviving species in a lineage that diverged long ago. (C) The leopard gecko, a desert dweller native to Afghanistan, Pakistan, and northwestern India. (D) The ringneck snake of North America is nonvenomous. It coils its tail to reveal a bright orange underbelly, which distracts potential predators from the vital head region.

(A) *Chelonia mydas*

(B) *Sphenodon punctatus*

(C) *Eublepharis macularius*

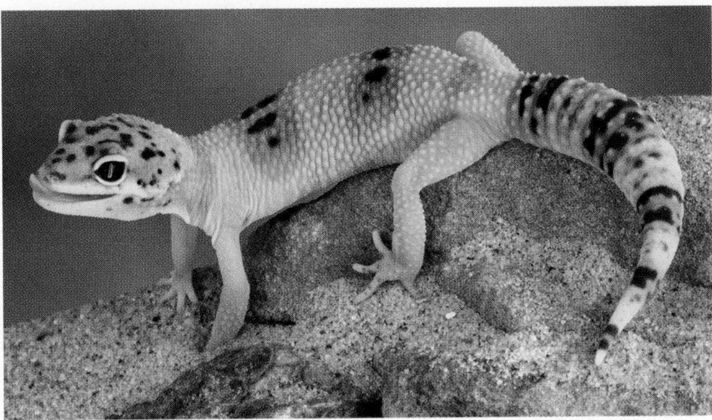

(D) *Diadophis punctatus*

long as 3 meters and can weigh more than 150 kilograms, is the predaceous Komodo dragon of the East Indies. Most lizards walk on four limbs (**Figure 33.19C**), although limblessness has evolved repeatedly in the group, especially in burrowing and grassland species. One major group of limbless squamates is the snakes (**Figure 33.19D**). All snakes are carnivores, and many can swallow objects much larger than themselves. Several snake groups evolved venom glands and the ability to inject venom rapidly into their prey.

### Crocodilians and birds share their ancestry with the dinosaurs

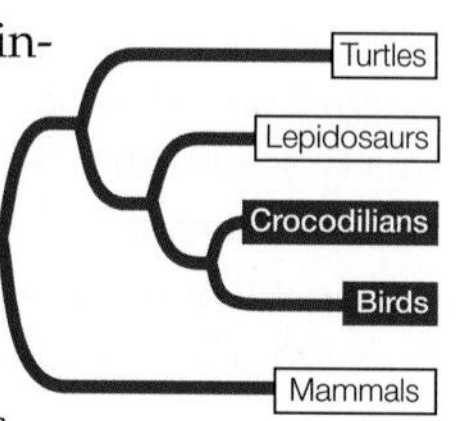


Another reptilian group, the *archosaurs*, includes the crocodilians, dinosaurs, and birds. *Dinosaurs* rose to prominence about 215 million years ago and dominated terrestrial environments for about 150 million years; only one group of dinosaurs, the *birds*, survived the mass extinction event at the Cretaceous–Tertiary boundary. During the Mesozoic, most terrestrial animals more than a meter long were dinosaurs. Many were agile and could run rapidly; they had special muscles that enabled the lungs to be filled and emptied while the limbs moved. We can infer the existence of such muscles in dinosaurs from the structure of the vertebral column in fossils. Some of the largest dinosaurs weighed as much as 70,000 kilograms.

Modern *crocodilians*—crocodiles, caimans, gharials, and alligators—are confined to tropical and warm temperate environments (**Figure 33.20A**). Crocodilians spend much of their time in water, but they build nests on land or on floating piles of vegetation. The eggs are warmed by heat generated by decaying organic matter that the female places in the nest. Typically, the female guards the eggs until hatching, and she often facilitates hatching. In some species, the female continues to guard and communicate with her offspring after they hatch. All crocodilians are carnivorous. They eat vertebrates of all kinds, including large mammals.

Biologists have long accepted the phylogenetic position of birds among the reptiles, although birds clearly have many unique, derived morphological features. In addition to the strong morphological evidence for the placement of birds among the reptiles, fossil and molecular data emerging over the last few decades have provided definitive supporting evidence. Birds are thought to have emerged among the *theropods*, a group of predatory dinosaurs that share such traits as bipedal stance, hollow bones, a *furcula* ("wishbone"), elongated metatarsals with three-fingered feet, elongated forelimbs with three-fingered hands, and a pelvis that points backward.

The living bird species fall into two major groups that diverged during the late Cretaceous, about 80 to 90 million years ago, from a flying ancestor. The few modern descendants of one lineage include a group of secondarily flightless and weakly flying birds, some of which are very large. This group, called the *palaeognaths*, includes the South and Central American tinamous and several large flightless birds of the southern continents—

**33.20 Archosaurs** (A) Crocodiles, alligators, and their relatives live in tropical and warm temperate climates. This crocodile lives in saltwater and estuarine environments along Australia's coast. (B) Birds are the other living archosaur group, represented here by the winged but flightless ostrich.

the rheas, emu, kiwis, cassowaries, and the world's largest bird, the ostrich (**Figure 33.20B**). The second lineage, the *neognaths*, has left a much larger number of descendants, most of which have retained the ability to fly.

### The evolution of feathers allowed birds to fly

Fossil dinosaurs discovered recently in early Cretaceous deposits in Liaoning Province, in northeastern China, show that the scales of some small predatory dinosaurs were highly modified to form *feathers*. The feathers of one of these dinosaurs, *Microraptor gui*, were structurally similar to those of modern birds (**Figure 33.21A**).

During the Mesozoic era, about 175 million years ago, a lineage of theropods gave rise to the birds. The oldest known avian fossil, *Archaeopteryx*, which lived about 150 million years ago, had teeth, but it was covered with feathers that are virtually identical to those of modern birds (**Figure 33.21B**). It also had well-developed wings, a long tail, and a furcula to which some of the flight muscles were probably attached. *Archaeopteryx* had clawed fingers on its forelimbs, but it also had typical perching bird claws on its hindlimbs. It probably lived in trees and shrubs and used the fingers to assist it in clambering over branches.

The evolution of feathers was a major force for diversification. Feathers are lightweight but are strong and structurally complex (**Figure 33.22**). The large quills of the flight feathers on

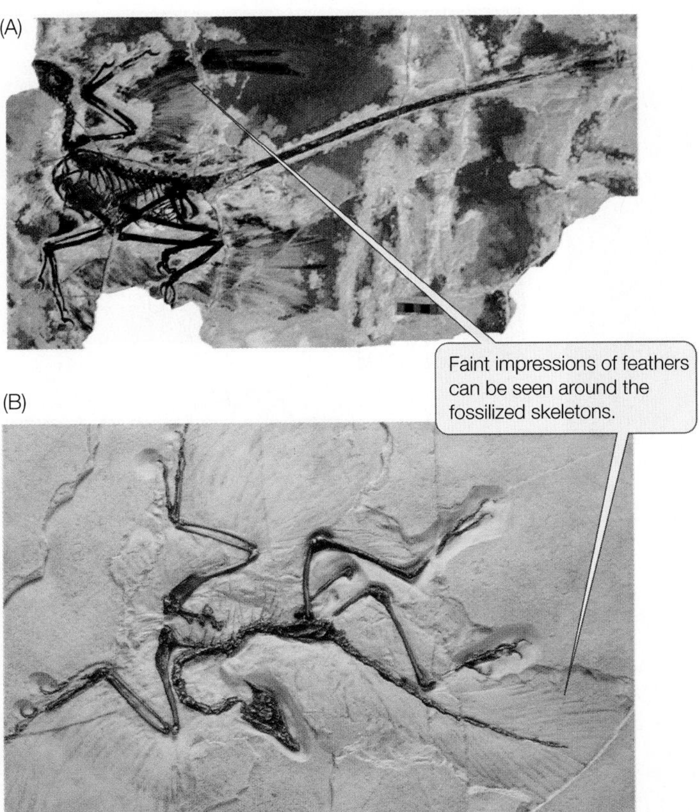
(A)
(B)
Faint impressions of feathers can be seen around the fossilized skeletons.

**33.21 Mesozoic Bird Relatives** Fossil remains demonstrate the evolution of birds from other dinosaurs. (A) *Microraptor gui* was a feathered dinosaur from the early Cretaceous (about 140 mya). (B) Dating from roughly the same timeframe, *Archaeopteryx* is the oldest known birdlike fossil.

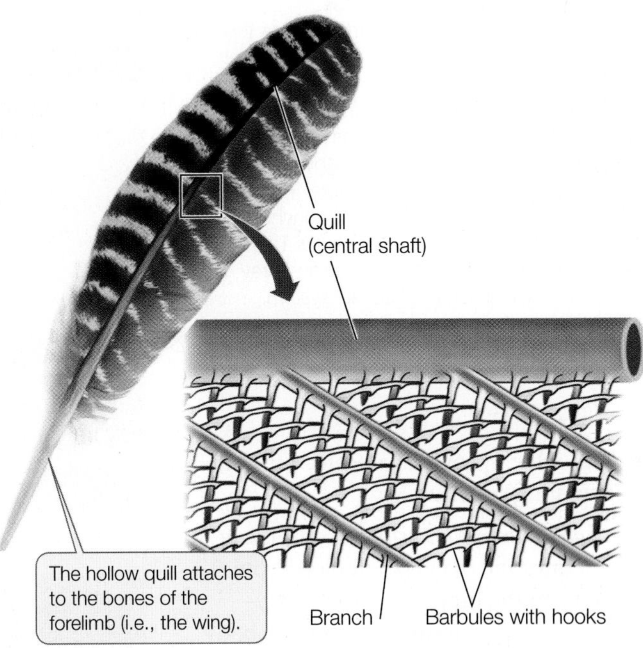
Quill (central shaft)
The hollow quill attaches to the bones of the forelimb (i.e., the wing).
Branch
Barbules with hooks

**33.22 Feathers Represent a Major Evolutionary Innovation** The flight feathers of birds are attached to the forelimb (wing) bones by a hollow central shaft. Fine branches with interlocking hooks and barbs radiate from the shaft, creating a strong, lightweight surface that enables flight.

the wings arise from the skin of the forelimbs to create the flying surfaces. Other strong feathers sprout like a fan from the shortened tail and serve as stabilizers during flight. The feathers that cover the body, along with an underlying layer of down feathers, provide birds with insulation that helps them to survive in virtually all of Earth's climates.

The bones of theropod dinosaurs, including birds, are hollow with internal struts that increase their strength. Hollow bones would have made early theropods lighter and more mobile; later they facilitated the evolution of flight. The sternum (breastbone) of flying birds forms a large, vertical keel to which the flight muscles are attached.

Flight is metabolically expensive. A flying bird consumes energy at a rate about 15 to 20 times faster than a running lizard of the same weight. Because birds have such high metabolic rates, they generate large amounts of heat. They control the rate of heat loss using their feathers, which may be held close to the body or elevated to alter the amount of insulation they provide. The lungs of birds allow air to flow through unidirectionally rather than by pumping air in and out (see Section 49.2). This flow-through structure of the lungs increases the efficiency of gas exchange and thereby supports an increased metabolic rate.

There are about 10,000 species of living birds, which range in size from the 150-kilogram ostrich to a tiny hummingbird weighing only 2 grams. The teeth so prominent among other dinosaurs were secondarily lost in the ancestral birds, but birds nonetheless eat almost all types of animal and plant material. Insects and fruits are the most important dietary items for terrestrial species. Birds also eat seeds, nectar and pollen, leaves and buds, carrion, and other vertebrates. By eating the fruits and seeds of plants, birds serve as major agents of seed dispersal. **Figure 33.23** shows representatives of a few of the major groups of birds.

### Mammals radiated after the extinction of dinosaurs

Small and medium-sized **mammals** coexisted with the dinosaurs for millions of years. After the non-avian dinosaurs disappeared during the mass extinction at the close of the Mesozoic era, mammals increased dramatically in numbers, diversity, and size. Today mammals range in size from tiny shrews and bats weighing only about 2 grams to the blue whale, the largest animal on Earth, which measures up to 33 meters long and can weigh as much as 160,000 kilograms. Mammals have far fewer, but more highly differentiated, teeth than do fishes, amphibians, or reptiles. Differences among mammals in the number, type, and arrangement of teeth reflect their varied diets (see Figure 51.6).

Four key features distinguish the mammals:

- *Sweat glands*, which secrete sweat that evaporates and thereby cools an animal
- *Mammary glands*, which in females secrete a nutritive fluid (milk) on which newborn individuals feed

(A) *Bombycilla cedrorum*

(B) *Aix galericulata*

(C) *Frigata minor*

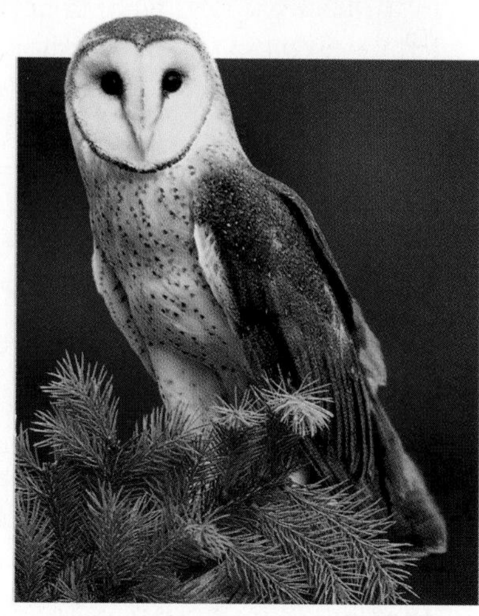

(D) *Tyto alba*

**33.23 Diversity among the Birds** (A) Perching, or passeriform, birds such as this cedar waxwing comprise the most species-rich of all bird groups. (B) This bright-plumaged male Mandarin duck is a member of the group that includes ducks, geese, and swans. (C) Great frigatebirds are among the ocean-going birds that are often found miles from shore. This male is in display mode, with an inflated pouch that advertises his presence to females. (D) This barn owl is a nighttime predator that can find prey using its sensitive auditory "sonar" system.

- *Hair*, which provides a protective and insulating covering
- A *four-chambered heart* that completely separates the oxygenated blood coming from the lungs from the deoxygenated blood returning from the body (this last characteristic is convergent with the archosaurs, including modern birds and crocodiles)

Mammalian eggs are fertilized within the female's body, and the embryos undergo a period of development in the female's body in an organ called the *uterus* prior to being born. Most mammals have a covering of hair (fur), which is luxuriant in some species but has been greatly reduced in others, including the cetaceans (whales and dolphins) and humans. Thick layers of insulating fat (blubber) replace hair as a heat-retention mechanism in the cetaceans; humans learned to use clothing for this purpose when they dispersed from warm tropical areas.



The approximately 5,000 species of living mammals are divided into two primary groups: the *prototherians* and the *therians*. Only five species of prototherians are known, and they are found only in Australia and New Guinea. These mammals, the duck-billed platypus and four species of echidnas, differ from other mammals in lacking a placenta, laying eggs, and having sprawling legs (**Figure 33.24**). Prototherians supply milk for their young, but they have no nipples on their mammary glands; the milk simply oozes out and is lapped off the fur by the offspring.

(A) *Tachyglossus aculeatus*

(B) *Ornithorhynchus anatinus*

**33.24 Prototherians** (A) The short-beaked echidna is one of the four surviving species of echidnas. (B) The duck-billed platypus is another surviving prototherian.

## Most mammals are therians

Members of the *therian* clade are further divided into the *marsupials* and the *eutherians*. Females of most marsupial species have a ventral pouch in which they carry and feed their offspring (see Figure 33.25A). Gestation (pregnancy) in marsupials is brief; the young are born tiny but with well-developed forelimbs, with which they climb to the pouch. They attach to

(A) *Macropus giganteus* (B) *Sarcophilus harrisii* (C) *Marmosa murina*

**33.25 Marsupials** (A) Australia's eastern gray kangaroo is among the largest living marsupials. This female carries her young offspring in the characteristic marsupial pouch. (B) The carnivorous Tasmanian devil is found only in Tasmania, an island off southern Australia. (C) This arboreal opossum is a South American marsupial.

a nipple but cannot suck. The mother ejects milk into the tiny offspring until it grows large enough to suckle. Once her offspring have left the uterus, a female marsupial may become sexually receptive again. She can then carry fertilized eggs that are capable of initiating development and can replace the offspring in her pouch should something happen to them.

At one time marsupials were found on all continents, but the approximately 350 living species are now restricted to Australasia (**Figure 33.25A and B**) and the Americas (especially South America; **Figure 33.25C**). One species, the Virginia opossum, is widely distributed in North America. Marsupials radiated to become herbivores, insectivores, and carnivores, but no marsupials live in the oceans. None can fly, although some *arboreal* (tree-dwelling) marsupials are gliders. The largest living marsupials are the kangaroos of Australia, which can weigh up to 90 kilograms. Much larger marsupials existed in Australia until humans exterminated them soon after reaching that continent about 40,000 years ago.

Eutherians include the majority of mammals. Eutherians are sometimes called *placental mammals*, but this name is inappropriate because some marsupials also have placentas. Eutherians are more developed at birth than are marsupials; no external pouch houses them after they are born.

The approximately 5,000 species of living eutherians are divided into 20 major groups (**Table 33.1**). The largest group is the rodents, with about 2,300 species. Rodents are traditionally defined by the unique morphology of their teeth, which are adapted for gnawing through substances such as wood. The next largest group comprises the approximately 1,100 bat species—the flying mammals. The bats are followed by the moles and shrews, with about 430 species. The relationships of the major groups of eutherians to one another have been difficult to determine, because most of the major groups diverged in a short period of time during an explosive adaptive radiation.

Eutherians are extremely varied in their form and ecology (**Figure 33.26**). The extinction of the non-avian dinosaurs at the end of the Cretaceous may have made it possible for them to diversify and radiate into a large range of ecological *niches*. Many eutherian species grew to become quite large in size, and some assumed the role of dominant terrestrial predators previously occupied by the large dinosaurs. Among these predators, social hunting behavior evolved in several species, including members of the canid (dog), felid (cat), and primate lineages.

*Grazing* and *browsing* by members of several eutherian groups helped transform the terrestrial landscape. Herds of grazing herbivores feed on open grasslands, whereas browsers feed on shrubs and trees. The effects of herbivores on plant life favored the evolution of the spines, tough leaves, and difficult-to-eat growth forms found in many plants. In turn, adaptations to the teeth and digestive systems of many herbivore lineages allowed these species to consume many plants despite such defenses—a striking example of coevolution. A large animal can survive on food of lower quality than a small animal can, and large size evolved in several groups of grazing and browsing mammals (see Figure 33.26C). The evolution of large herbivores, in turn, favored the evolution of large carnivores able to attack and overpower them.

Several lineages of terrestrial eutherians subsequently returned to the aquatic environments their ancestors had left behind (see Figure 33.26D). The completely aquatic cetaceans—whales and dolphins—evolved from artiodactyl ancestors (whales are closely related to the hippopotamuses). The seals, sea lions, and walruses also returned to the marine environment, and their limbs became modified into flippers. Weasel-like otters retain their limbs but have also returned to aquatic environments, colonizing both fresh and salt water. The manatees and dugongs colonized estuaries and shallow seas.

(A) *Erithizon dorsatum*

(B) *Artibeus lituratus*

(C) *Rangifer tarandus*

(D) *Stenella longirostris*

**33.26 Diversity among the Eutherians**
(A) The North American porcupine, a large rodent covered with sharp, protective quills. Almost half of all eutherians are rodents. (B) Flight evolved in the ancestor of bats. This fruit-eating bat ranges from Central to South America. Virtually all bat species are nocturnal. (C) Large hoofed mammals are important herbivores in terrestrial environments. Although this bull is grazing by himself, caribou are usually found in huge herds. (D) Spinner dolphins are cetaceans, a cetartiodactyl group that returned to the marine environment.

TABLE 33.1

**Major Groups of Living Eutherian Mammals**

| GROUP | APPROXIMATE NUMBER OF LIVING SPECIES | EXAMPLES |
|---|---|---|
| Gnawing mammals (Rodentia) | 2,300 | Rats, mice, squirrels, woodchucks, ground squirrels, beaver, capybara |
| Flying mammals (Chiroptera) | 1,100 | Bats |
| Soricomorph insectivores (Soricomorpha) | 430 | Shrews, moles |
| Even-toed hoofed mammals and cetaceans (Cetartiodactyla) | 320 | Deer, sheep, goats, cattle, antelopes, giraffes, camels, swine, hippopotamus, whales, dolphins |
| Carnivores (Carnivora) | 290 | Wolves, dogs, bears, cats, weasels, pinnipeds (seals, sea lions, walruses) |
| Primates (Primates) | 235 | Lemurs, monkeys, apes, humans |
| Lagomorphs (Lagomorpha) | 80 | Rabbits, hares, pikas |
| African insectivores (Afrosoricida) | 50 | Tenrecs, golden moles |
| Spiny insectivores (Erinaceomorpha) | 24 | Hedgehogs |
| Armored mammals (Cingulata) | 21 | Armadillos |
| Tree shrews (Scandentia) | 20 | Tree shrews |
| Odd-toed hoofed mammals (Perissodactyla) | 20 | Horses, zebras, tapirs, rhinoceroses |
| Long-nosed insectivores (Macroscelidea) | 16 | Elephant shrews |
| Pilosans (Pilosa) | 10 | Anteaters, sloths |
| Pholidotans (Pholidota) | 8 | Pangolins |
| Sirenians (Sirenia) | 5 | Manatees, dugongs |
| Hyracoids (Hyracoidea) | 4 | Hyraxes, dassies |
| Elephants (Proboscidea) | 3 | African and Indian elephants |
| Dermopterans (Dermoptera) | 2 | Flying lemurs |
| Aardvark (Tubulidentata) | 1 | Aardvark |

## 33.4 RECAP

**The vertebrate colonization of dry land was facilitated by the evolution of an impermeable body covering, efficient kidneys, and the amniote egg—a structure that resists desiccation and provides an aqueous internal environment in which the embryo grows. Major amniote groups include turtles, lepidosaurs, archosaurs (crocodilians and birds), and mammals.**

- In the not-too-distant past, the idea that birds were reptiles met with skepticism. Explain how fossils, morphology, and molecular evidence now support the position of birds among the reptiles. See pp. 708–709
- In reviewing the discussion of the various vertebrate groups, identify several reasons why tooth structure is such an important area of study.

The evolutionary history of one eutherian group—the primates—is of special interest to us because it includes the human lineage. The primates have been the subject of extensive research in most aspects of their biology, including behavior, ecology, physiology, and molecular biology. Let's take a closer look at the characteristics and evolutionary history of the primates.

## 33.5 What Traits Characterize the Primates?

The eutherian **primates** underwent extensive evolutionary radiation from an ancestral small, arboreal, insectivorous mammal. A nearly complete fossil of an early primate, *Carpolestes*, was found in Wyoming and dated at 56 million years ago; it had grasping feet with an opposable big toe that had a nail rather than a claw. Grasping limbs with opposable digits are one of the major adaptations to arboreal life that distinguish primates from other mammals.

Early in their evolutionary history, the primates split into two main clades, the prosimians and the anthropoids (**Figure 33.27**). *Prosimians*—lemurs, lorises, and their close relatives—once lived on all continents, but today they are restricted to Africa, Madagascar, and tropical Asia. All mainland prosimian species are arboreal and nocturnal (**Figure 33.28**). On the island of Madagascar, however, the site of a remarkable radiation of lemurs, there are also diurnal and terrestrial species.

A second primate lineage, the *anthropoids*—tarsiers, New World monkeys, Old World monkeys, and apes—evolved about 65 million years ago in Africa or Asia. New World monkeys diverged from Old World monkeys and apes at a slightly later date, but early enough that they might have reached South America from Africa when those two continents were still close to each other. All New World monkeys are arboreal (**Figure 33.29A**). Many of them have a long, prehensile tail with which they can grasp branches. Many Old World monkeys are arboreal as well, but a number of species are terrestrial (**Figure 33.29B**). No Old World primate has a prehensile tail.

About 35 million years ago, a lineage that led to the modern apes separated from the Old World monkeys. Between 22 and 5.5 million years ago, dozens of species of apes lived in Europe, Asia, and Africa. The Asian apes—gibbons and orangutans (**Figure 33.30A and B**)—descended from two of these ape lineages. Another extinct genus, *Dryopithecus*, is the sister group of the modern African apes—gorillas and chimpanzees (**Figure 33.30C and D**)—and of humans.

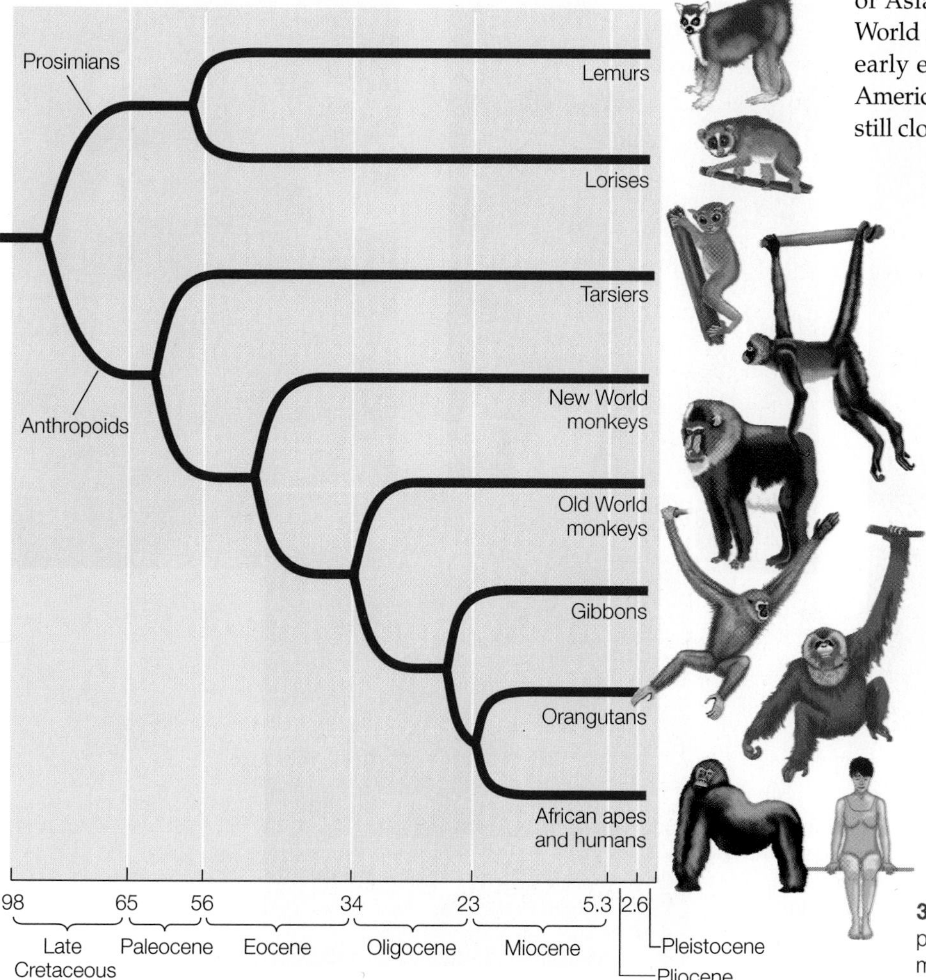


**33.27 Phylogeny of the Primates** The phylogeny of primates is among the best studied of any major group of mammals. This tree is based on evidence from many genes, morphology, and fossils.

*Eulemur coronatus*

33.28 **A Prosimian** The crowned lemur is one of many lemur species found in Madagascar, where it is part of a unique assemblage of endemic plants and animals.

(A) *Ateles geoffroyi*

(B) *Mandrillus sphinx*

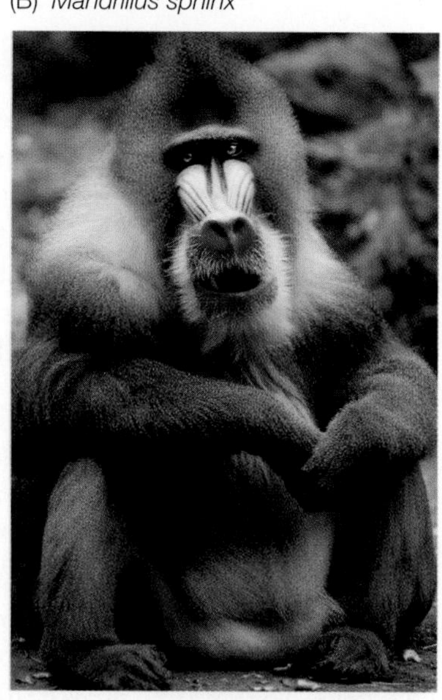

**33.29 Monkeys** (A) The spider monkeys of Central America are typical of the New World monkeys, all of which are arboreal. (B) Although many Old World monkey species are arboreal, the mandrills are among the many terrestrial groups.

(A) *Hylobates lar*

(B) *Pongo pygmaeus*

(C) *Gorilla gorilla*

(D) *Pan troglodytes*

**33.30 Apes** (A) The several genera of gibbons are all smaller in size than the other apes. Gibbons are found throughout southeastern Asia. (B) Orangutans are also native ot Asia, living in the forests of Sumatra and Borneo. (C) Gorillas—the largest apes—are restricted to humid African forests. This male is a lowland gorilla. (D) Chimpanzees, our closest relatives, are found in forested regions of Africa.

### Human ancestors evolved bipedal locomotion

About 6 million years ago in Africa, a lineage split occurred that would lead to the chimpanzees on the one hand and to the *hominid* clade that includes modern humans and their extinct close relatives on the other.

The earliest protohominids, known as *ardipithecines*, had distinct morphological adaptations for *bipedal locomotion* (walking on two legs). Bipedal locomotion frees the forelimbs to manipulate objects and to carry them while walking. It also elevates the eyes, enabling the animal to see over tall vegetation to spot predators and prey. Bipedal locomotion is also energetically more economical than quadrupedal locomotion. All three advantages were probably important for the ardipithecines and their descendants, the australopithecines.

The first *australopithecine* skull was found in South Africa in 1924. Since then australopithecine fossils have been found at many sites in Africa. The most complete fossil skeleton of an australopithecine yet found was discovered in Ethiopia in 1974. The skeleton was approximately 3.5 million years old and was that of a young female who has since become known to the world as "Lucy." Lucy was assigned to the species *Australopithecus afarensis*, and her discovery captured worldwide interest. Fossil remains of more than 100 *A. afarensis* individuals have since been discovered, and there have been recent discoveries of fossils of other australopithecines who lived in Africa 4 to 5 million years ago.

Experts disagree over how many species are represented by australopithecine fossils, but it is clear that multiple species of hominids lived together over much of eastern Africa several million years ago (**Figure 33.31**). A lineage of larger species (weighing about 40 kilograms) is represented by *Paranthropus robustus* and *P. boisei*, both of which died out between 1 and 1.5 million years ago. Members of a smaller lineage of australopithecines gave rise to the genus *Homo*.

Early members of the genus *Homo* lived contemporaneously with *Paranthropus* in Africa for about a million years. Some 2-million-year-old fossils of an extinct species called *H. habilis* were discovered in the Olduvai Gorge, Tanzania. Other fossils of *H. habilis* have been found in Kenya and Ethiopia. Associated with the fossils are tools that these early hominids used to obtain food.

Another extinct hominid species, *Homo erectus*, evolved in Africa about 1.6 million years ago. Soon thereafter it had spread as far as eastern Asia, becoming the first hominid to leave Africa. Members of *H. erectus* were nearly as large as modern people, but their brains were smaller and they had comparatively thick skulls. The cranium, which had thick, bony walls, may have

**33.31 A Phylogenetic Tree of Hominids** At times in the past, more than one species of hominid lived on Earth at the same time. Originating in Africa, hominids spread to Europe and Asia multiple times. All these closely related species are now extinct, while modern *Homo sapiens* have colonized nearly every corner of the planet.

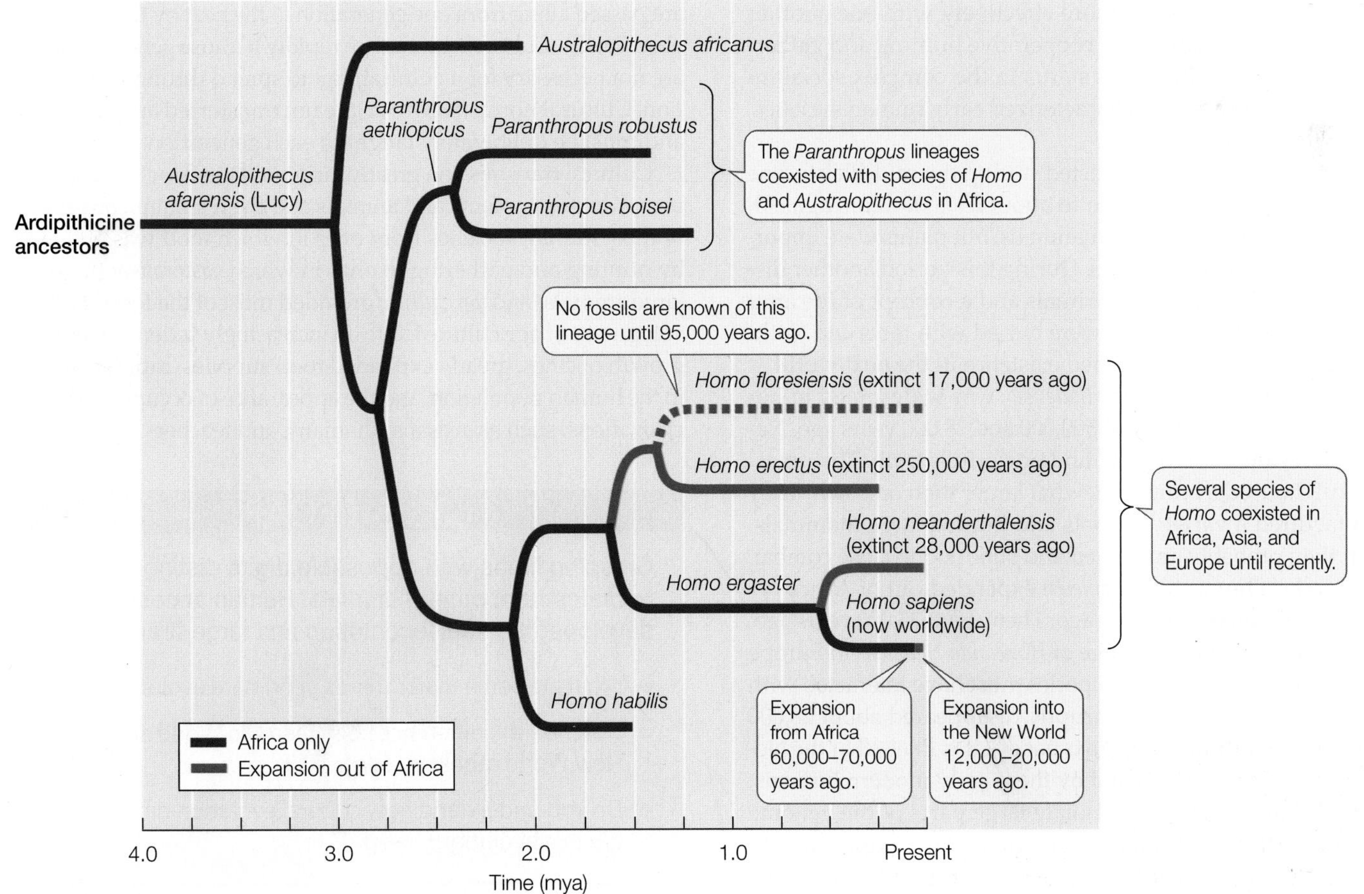

been an adaptation to protect the brain, ears, and eyes from impacts caused by a fall or a blow from a blunt object. What would have been the source of such blows? Fighting with other *H. erectus* individuals is a possible answer.

*Homo erectus* used fire for cooking and for hunting large animals, and made characteristic stone tools that have been found in many parts of the Old World. Populations of *H. erectus* survived until at least 250,000 years ago, although more recent fossils may also be attributable to this species. In 2004 some 18,000-year-old fossil remains of a small *Homo* were found on the island of Flores in Indonesia. Since then, numerous additional fossils of this diminutive hominid have been found on Flores, dating from 95,000 to 17,000 years ago. Many anthropologists think that this small species, named *H. floresiensis*, was most closely related to *H. erectus*.

### Human brains became larger as jaws became smaller

In the hominid lineage leading to *Homo sapiens* and *H. neanderthalensis*, the brain increased rapidly in size. At the same time, the powerful jaw muscles of our ancestors dramatically decreased in size. These two changes were simultaneous, suggesting they might have been functionally correlated. A mutation in a regulatory gene that is expressed only in the head may have removed a barrier that had previously prevented this remodeling of the human cranium.

The striking enlargement of the brain relative to body size in the hominid lineage was probably favored by an increasingly complex social life. Any features that allowed group members to communicate more effectively with one another would have been valuable in cooperative hunting and gathering and for improving one's status in the complex social interactions that must have characterized early human societies, just as they do in ours today.

Several *Homo* species coexisted during the mid-Pleistocene epoch, from about 1.5 million to about 250,000 years ago. All were skilled hunters of large mammals, but plants were important components of their diets. During this period another distinctly human trait emerged: rituals and a concept of life after death. Deceased individuals were buried with tools and clothing, supplies for their presumed existence in the next world.

One species, *Homo neanderthalensis*, was widespread in Europe and Asia between about 500,000 and 28,000 years ago. Neanderthals were short, stocky, and powerfully built. Their massive skull housed a brain somewhat larger than our own. They manufactured a variety of tools and hunted large mammals, which they probably ambushed and subdued in close combat. Early modern humans (*H. sapiens*) expanded out of Africa between 70,000 and 60,000 years ago. Then about 35,000 years ago, *H. sapiens* moved into the range of *H. neanderthalensis* in Europe and western Asia, so the two species must have interacted with one another. Neanderthals abruptly disappeared about 28,000 years ago. Many anthropologists believe it is likely that the Neanderthals were exterminated by these early modern humans. Scientists have been able to isolate large parts of the genome of *H. neanderthalensis* from recent fossils and to compare it with our own. These studies suggest little or no interbreeding between the two species while they occupied the same range in Europe and western Asia.

Early modern humans made and used a variety of sophisticated tools. They created the remarkable paintings of large mammals, many of them showing scenes of hunting, found in European caves. The animals depicted were characteristic of the cold steppes and grasslands that occupied much of Europe during periods of glacial expansion. Early modern humans also spread across Asia, reaching North America perhaps as early as 20,000 years ago, although the date of their arrival in the New World is still uncertain. Within a few thousand years, they had spread southward through North America to the southern tip of South America.

### Humans developed complex language and culture

As our ancestors evolved larger brains, their behavioral capabilities increased, especially the capacity for language. Most animal communication consists of a limited number of signals, which refer mostly to immediate circumstances and are associated with charged emotional states induced by those circumstances. Human language is far richer in its symbolic character than other animal vocalizations. Our words can refer to past and future times and to distant places. We are capable of learning thousands of words, many of them referring to abstract concepts. We can rearrange words to form sentences with complex meanings.

The expanded mental abilities of humans enabled the development of a complex **culture**, in which knowledge and traditions are passed along from one generation to the next by teaching and observation. Cultures can change rapidly because genetic changes are not necessary for a cultural trait to spread through a population. Cultural norms, however, are not transferred automatically and must be deliberately taught to each generation.

Cultural transmission greatly facilitated the development and use of domestic plants and animals and the resultant conversion of most human societies from ones in which food was obtained by hunting and gathering to ones in which *pastoralism* (herding large animals) and *agriculture* provided most of the food. The development of agriculture led to an increasingly sedentary life, the growth of cities, greatly expanded food supplies, rapid increases in the human population, and the appearance of occupational specializations, such as artisans, shamans, and teachers.

### 33.5 RECAP

**Grasping limbs with opposable digits distinguish primates from other mammals. Human ancestors developed bipedal locomotion and large brains.**

- What are some major trends in primate evolution?
- Describe the differences between Old World and New World monkeys. See p. 713
- Do you understand how cultural evolution differs from genetic evolution? See p. 716

## CHAPTER SUMMARY

### 33.1 What Is a Deuterostome?

- Deuterostomes vary greatly in adult form, but based on the distinctive patterns of early development they share and on phylogenetic analyses of gene sequences, they are judged to be monophyletic. There are far fewer species of deuterostomes than of protostomes, but many deuterostomes are large and ecologically important. Review Figure 33.1, **WEB ACTIVITY 33.1**

### 33.2 What Are the Major Groups of Echinoderms and Hemichordates?

- Echinoderms and hemichordates both have bilaterally symmetrical, ciliated larvae.
- Adult echinoderms have **pentaradial symmetry**. Echinoderms have an internal skeleton of calcified plates and a unique **water vascular system** connected to extensions called **tube feet**. Review Figure 33.3
- Hemichordate adults are bilaterally symmetrical and have a three-part body that is divided into a proboscis, collar, and trunk. They include the acorn worms and the pterobranchs. Review Figure 33.5

### 33.3 What New Features Evolved In the Chordates?

- Chordates fall into three principal subgroups: **cephalochordates**, **urochordates**, and **vertebrates**.
- At some stage in their development, all chordates have a dorsal hollow nerve cord, a post-anal tail, and a **notochord**. Review Figure 33.6
- Urochordates include the ascidians (sea squirts), the larvaceans, and the thaliaceans (salps). Cephalochordates are the lancelets, which live buried in the sand of shallow marine and brackish waters.
- The vertebrate body is characterized by a rigid internal skeleton, which is supported by a **vertebral column** that replaces the notochord, internal organs suspended in a coelom, a ventral heart, and an anterior skull with a large brain. Review Figure 33.10
- The evolution of jaws from gill arches enabled individuals to grasp large prey and, together with teeth, cut them into small pieces. Review Figure 33.11
- **Chondrichthyans** have skeletons of cartilage; almost all species are marine. The skeletons of **ray-finned fishes** are made of bone; these fishes have colonized all aquatic environments.

### 33.4 How Did Vertebrates Colonize the Land?

- Lungs and jointed appendages enabled vertebrates to colonize the land. The earliest split in the **tetrapod** tree is between the **amphibians** and the **amniotes** (reptiles and mammals).
- Most modern amphibians are confined to moist environments because they and their eggs lose water rapidly. Review Figure 33.16, **ANIMATED TUTORIAL 33.1**
- An impermeable skin, efficient kidneys, and an egg that could resist desiccation evolved in the amniotes. Review Figure 33.17, **WEB ACTIVITY 33.2**
- The major living **reptile** groups are the turtles, the lepidosaurs (tuataras, lizards, snakes, and amphisbaenians), and the archosaurs (crocodilians and birds). Review Figure 33.18
- **Mammals** are unique among animals in supplying their young with a nutritive fluid (milk) secreted by mammary glands. There are two primary mammalian clades: the prototherians (of which there are only five species) and the species-rich therians. The therian clade is further subdivided into the marsupials and the eutherians. Review Table 33.1

### 33.5 What Traits Characterize the Primates?

- Grasping limbs with opposable digits distinguish **primates** from other mammals. The prosimian clade includes the lemurs and lorises; the anthropoid clade includes monkeys, apes, and humans. Review Figure 33.27
- Hominid ancestors were terrestrial primates that developed efficient bipedal locomotion. In the lineage leading to *Homo*, brains became larger as jaws became smaller; the two events may have been developmentally linked. Review Figure 33.31

**SEE WEB ACTIVITY 33.3 for a concept review of this chapter.**

## SELF-QUIZ

1. Which of the following are *not* deuterostomes?
   *a.* Acorn worms
   *b.* Sea stars
   *c.* Urochordates
   *d.* Brachiopods
   *e.* Lancelets
2. The structure used by adult urochordates to capture food is a
   *a.* pharyngeal basket.
   *b.* proboscis.
   *c.* lophophore.
   *d.* mucus net.
   *e.* radula.
3. The pharyngeal gill slits of chordate ancestors functioned as sites for
   *a.* uptake of oxygen only.
   *b.* release of carbon dioxide only.
   *c.* both uptake of oxygen and release of carbon dioxide.
   *d.* removal of small prey from the water.
   *e.* forcible expulsion of water to move the animal.
4. A bony skeleton evolved in the most recent common ancestor of
   *a.* gnathostomes.
   *b.* vertebrates and hagfishes.
   *c.* craniates and urochordates.
   *d.* ray-finned fishes and sarcopterygians.
   *e.* amniotes and amphibians.
5. In most ray-finned fishes, lunglike sacs evolved into
   *a.* pharyngeal gill slits.
   *b.* true lungs.
   *c.* coelomic cavities.
   *d.* swim bladders.
   *e.* none of the above

6. Many amphibians return to water to lay their eggs because
   *a.* water is isotonic to egg fluids.
   *b.* adults must be in water while they guard their eggs.
   *c.* there are fewer predators in water than on land.
   *d.* amphibians get their nutrition from water.
   *e.* amphibian eggs quickly lose water and desiccate if their surroundings are dry.
7. The horny scales that cover the skin of reptiles prevent them from
   *a.* using their skin as an organ of gas exchange.
   *b.* sustaining high levels of metabolic activity.
   *c.* laying their eggs in water.
   *d.* flying.
   *e.* crawling into small spaces.
8. Which statement about bird feathers is *not* true?
   *a.* They are highly modified reptilian scales.
   *b.* They provide insulation for the body.
   *c.* They exist in two layers.
   *d.* They help birds fly.
   *e.* They are important sites of gas exchange.
9. Prototherians differ from other mammals in that they
   *a.* do not produce milk.
   *b.* lack body hair.
   *c.* lay eggs.
   *d.* live in Australia.
   *e.* have a pouch in which the young are raised.
10. Bipedalism is believed to have evolved in the hominid lineage because bipedal locomotion is
    *a.* more efficient than quadrupedal locomotion.
    *b.* more efficient than quadrupedal locomotion and frees the forelimbs to manipulate objects.
    *c.* less efficient than quadrupedal locomotion but frees the forelimbs to manipulate objects.
    *d.* less efficient than quadrupedal locomotion, but bipedal animals can run faster.
    *e.* less efficient than quadrupedal locomotion, but natural selection does not act to improve efficiency.

## FOR DISCUSSION

1. In what animal groups has the ability to fly evolved? How do the structures used for flying differ among these animals?
2. Extracting suspended food from the water is a common mode of feeding among animals. Which groups contain species that extract prey from the air? Why is this mode of obtaining food so much less common than extracting prey from the water?
3. What risks and benefits are posed by large size?
4. Amphibians have survived and prospered for many millions of years, but today many species are disappearing, and populations of others are declining seriously. What features of their life histories might make amphibians especially vulnerable to the kinds of environmental changes now happening on Earth?
5. The body plan of most vertebrates is based on four appendages. What are the varied forms that these appendages take, and how are they used?
6. Compare the ways in which different animal lineages colonized the land. How were those ways influenced by the body plans of animals in the different groups?

## ADDITIONAL INVESTIGATION

A mutation in the gene that encodes the myosin heavy chain (see Chapter 48) decreased the size of the jaw muscles in human ancestors. This mutation may have enabled a restructuring of the cranium and larger brain size. How could we determine when this mutation arose in the phylogenetic history of the primates?

# 34 The Plant Body

## The doomsday vault

Carved into a sandstone mountain, surrounded by permafrost on the Arctic island of Spitsbergen, Norway, the Svalbard vault is almost the size of a soccer field. It is stable enough to withstand a major earthquake, even though earthquakes are very unlikely at that location. It is high enough in elevation to remain above sea level even if all the polar ice caps were melted by global warming. A cooling system run on energy from local coal holds its interior at –18°C, and even if this system fails, insulation and the cold weather outside ensure that it will be weeks before the interior temperature rises. It is surrounded by a technologically advanced security system.

The 120 nations that participated in establishing this facility agree that the vault at Svalbard needs to be very secure. Does it contain gold bars? No, it contains seeds, the carriers of plant embryos. As photosynthesizers, plants are the keys to the biosphere. They have also been the mainstay of human survival. Of the approximately 300,000 species of plants, humans depend directly on only a few dozen. You can name some of them—wheat, rice, and corn for food; cotton for fiber; forest trees for paper. Millennia ago, humans selected certain plants growing in the wild for their own uses and began to cultivate them. After many generations of artificial selection, the plants looked very distinct from their wild relatives. In addition, because these plants were grown in different environments and for different purposes, different genetic strains of plants came into being.

By the twentieth century, there were thousands of genetically distinct varieties of crop plants—an amazing 100,000 varieties of rice alone. A crop plant variety, also called a *cultivar*, is a member of a species that has been artificially selected for one or more of its useful traits. Although only a small number of these varieties are in use at a given time, plant biologists realized that the genetic diversity of the remaining species should not be allowed to disappear. Because plant seeds are generally quite hardy

**The Svalbard Global Seed Vault** The vault is located 390 feet inside a sandstone mountain on the Norwegian island of Spitsbergen, about 700 miles from the North Pole. The location was chosen for secure, long-term storage of seeds because earthquakes are unlikely there and the ground is permeated with permafrost.

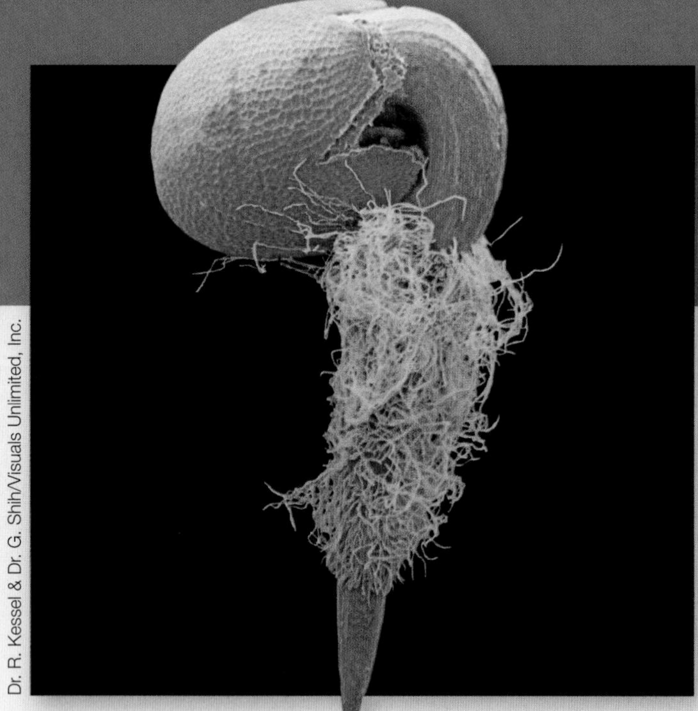

Dr. R. Kessel & Dr. G. Shih/Visuals Unlimited, Inc.

**Plant in Storage** Every seed contains an embryo with the means to create all parts of the plant body.

and relatively easy to store for long periods, seed banks were established. One of the largest is the National Seed Bank in Colorado, USA, where over half a million plant varieties are stored as seeds. When a plant breeder wants to develop a new genetic strain of corn—for example, one that is naturally resistant to a fungal disease—samples of seed can be withdrawn from the seed bank and used for cross-breeding.

The newly established seed vault in Norway is not so much a bank as a safe deposit box. Seed banks from all over the world are depositing samples of seeds for safekeeping in the vault in the event of a disaster that might destroy the seed banks. Such destruction does happen; for example, two valuable seed banks in Iraq and Afghanistan were destroyed during recent wars. As the existence of the Svalbard vault clearly demonstrates, seeds, and the plants that they form, are vital to humanity.

**IN THIS CHAPTER** we will examine plant structure at the levels of cells, tissues, organs, and tissue systems. We will see how that structure arises from clusters of undifferentiated cells, called meristems, that permit continuous growth throughout a plant's lifetime. The chapter concludes with a look at how humans have altered plant form through crop domestication.

## CHAPTER OUTLINE



## 34.1 What Is the Basic Body Plan of Plants?

Plants live by harvesting energy from sunlight and by collecting water and mineral nutrients from the soil. These resources, however, are incredibly sparse in the environment, so plants face the challenge of collecting them from huge areas, both above and below ground. Another challenge plants face is their inability to move. Plants cannot relocate themselves from, say, a dry, shady location to one that is wet and sunny.

The plant body plan allows plants to respond to these challenges. Stems, leaves, and roots enable plants anchored to one spot to capture scarce resources effectively, both above and below the ground. More important, to compensate for their inability to move, plants can grow throughout their lifetimes. Thus, while plants cannot move to a new water source or a new sunny clearing, they can respond to environmental cues by redirecting their growth to exploit opportunities that arise in their immediate environment.

In Chapters 28 and 29 we saw how modern plants arose from aquatic ancestors, giving rise to simple land plants and then vascular plants. Despite their obvious differences in size and form, all vascular plants have essentially the same simple structural organization. This chapter describes the basic architecture of the largest group of vascular plants, the angiosperms (flowering plants), and shows how so much diversity can literally grow out of such a simple basic form.

As we saw in Figure 29.1, angiosperms first appeared about 140 million years ago, radiated explosively over a period of about 60 million years, and became the dominant plant life on this planet. There are over 250,000 angiosperm species today. Flowers, the angiosperms' devices for sexual reproduction and their main distinguishing feature, consist of modified leaves and stems and will be considered in detail in Section 38.1. In this chapter we'll focus on the three kinds of vegetative (nonsexual) organs angiosperms possess: roots, stems, and leaves. Each of these vegetative organs can be understood in terms of its structure. By *structure* we mean both its overall form, called its *morphology*, and its internal component cells and tissues and their arrangement, known as its *anatomy*.

Plant organs are organized into two systems (**Figure 34.1**):

- The **root system** anchors the plant in place, absorbs water and dissolved minerals, and stores the products of photosynthesis from the shoot system. The extreme branching of plant roots and their high surface area-to-volume ratios allow them to absorb water and mineral nutrients from the soil efficiently.

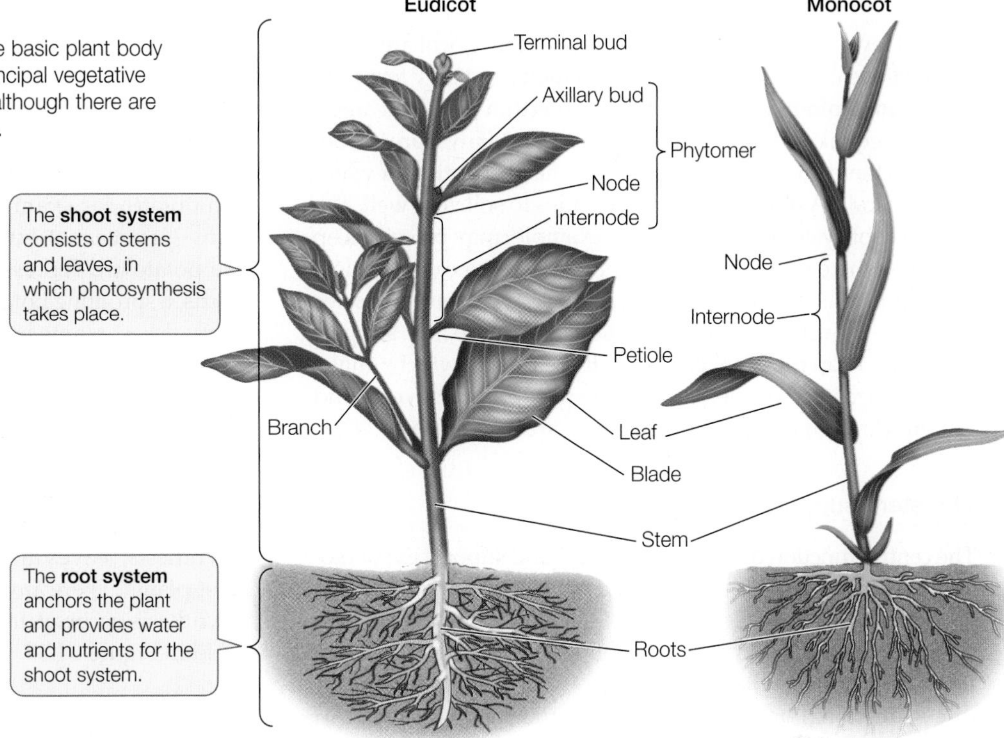
Eudicot
Monocot
Terminal bud
Axillary bud
Phytomer
Node
Internode
Node
Internode
Petiole
Branch
Leaf
Blade
Stem
Roots
The shoot system consists of stems and leaves, in which photosynthesis takes place.
The root system anchors the plant and provides water and nutrients for the shoot system.

**34.1 Vegetative Organs and Systems** The basic plant body plan, with root and shoot systems, and the principal vegetative organs are similar in eudicots and monocots, although there are also some differences between the two clades.

- The **shoot system** of a plant consists of the stems, leaves, and flowers. Broadly speaking, the **leaves** are the chief organs of photosynthesis. The **stems** hold and display the leaves to the sun and provide connections for the transport of materials between roots and leaves.

As we saw in Section 29.3, most angiosperms belong to one of two major clades. *Monocots* are generally narrow-leaved flowering plants such as grasses, lilies, orchids, and palms. *Eudicots* are broad-leaved flowering plants such as soybeans, roses, sunflowers, and maples. These two clades, which account for 97 percent of flowering plant species, differ in several important basic characteristics (see Figure 34.1). Let's take a closer look at how the root and shoot systems are elaborated in eudicots and monocots.

## The root system anchors the plant and takes up water and dissolved minerals

Water and minerals enter most plants through the root system, which is located in the soil. Because light does not penetrate the soil, roots typically lack the capacity for photosynthesis. Although hidden from view, the root system is often larger than the visible shoot system. For example, the root system of a 4-month-old winter rye plant (*Secale cereale*) was found to be 130 times larger than the shoot system, with almost 13 million branches that had a cumulative length of over 500 km!

The root system of angiosperms originates in an embryonic root called the *radicle*. From this common starting point, the root systems of monocots and eudicots develop differently. Following seed germination, the radicle of most eudicots develops as a primary root (called the **taproot**), which extends downward by tip growth and outward by initiating **lateral roots**. The taproot and the lateral roots form a **taproot system**, which can take a variety of forms. For example, the taproot itself often functions as a nutrient storage organ, as in carrots (*Daucus carota*), sugar beets (*Beta vulgaris*), and sweet potato (*Ipomoea batatas*) (**Figure 34.2A**).

In contrast, the primary root of monocots (and some eudicots) is short-lived. Because they originate from the stem at

(A) Taproots

(B) Fibrous root system

(C) Prop roots

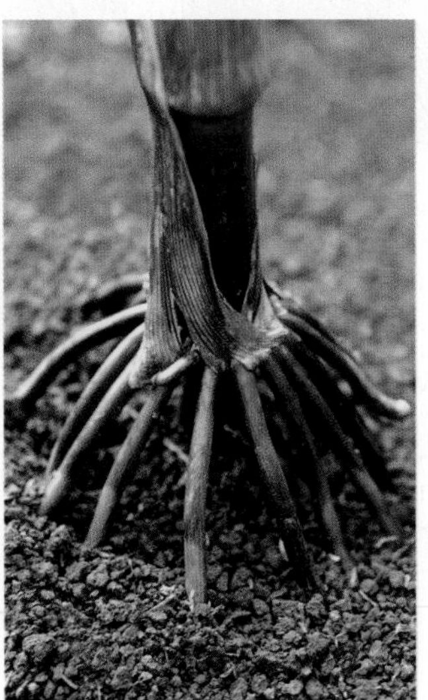

**34.2 Root Systems of Eudicots and Monocots** (A) The taproot systems of eudicots, such as carrots, sugar beets, and sweet potato, contrast with (B) the fibrous root system of a leek and (C) the adventitious prop roots of corn.

ground level or below, the roots of a typical monocot are called **adventitious** ("arriving from outside") **roots**, and they form a **fibrous root system** composed of numerous thin roots that are all roughly equal in diameter (**Figure 34.2B**). Many fibrous root systems have a large surface area for the absorption of water and minerals. A fibrous root system clings to soil very well. The fibrous root systems of grasses, for example, may protect steep hillsides where runoff from rain would otherwise cause erosion.

In some monocots—corn, banyan trees, and some palms, for example—adventitious roots function as props to help support the shoot. **Prop roots** are critical to these plants, which, unlike most eudicot tree species, are unable to support aboveground growth through the thickening of their stems (**Figure 34.2C**).

### The stem supports leaves and flowers

The central function of stems is to elevate and support the photosynthetic organs (leaves) as well as the reproductive organs (flowers).

Unlike roots, stems bear buds of various types. A *bud* is an undeveloped shoot that may or may not develop further to produce additional branches or leaves. Shoots are composed of repeating modules called **phytomers** (see Figure 34.1). A phytomer includes one or more leaves, which are attached to the stem at a **node**; an **internode** (the interval of stem between two nodes); and one or more **axillary buds**, which form in the angle (*axil*) where each leaf meets the stem. The axillary buds are distinguished from the bud at the end of a stem or branch, which is called a **terminal bud**. If it becomes active, an axillary bud can develop into a new *branch*, or extension of the shoot system. The arrangement of leaves along the stem (called the *phyllotaxy*) is often characteristic of the plant species.

Various modifications of stems are seen in nature. The *tuber* of a potato, for example—the part of the plant eaten by humans—is not a root, but rather an underground stem. The "eyes" of a potato are depressions containing axillary buds—in other words, a sprouting potato is just a branching stem (**Figure 34.3A**). Many desert plants have enlarged, water-retaining stems (**Figure 34.3B**). The *runners* of strawberry plants are horizontal stems from which roots grow at frequent intervals (**Figure 34.3C**). If the links between the rooted portions are broken, independent plants can develop on each side of the break—a form of vegetative (asexual) reproduction (see Section 38.3).

Although most young stems are green and capable of photosynthesis, leaves are the principal sites of photosynthesis in most plants. There are, however, exceptions: for example, photosynthesis occurs primarily in the stem of the barrel cactus featured in Figure 34.3B.

### Leaves are the primary sites of photosynthesis

In gymnosperms and in most flowering plants, the leaves are responsible for most of the plant's photosynthesis. Leaves are marvelously adapted for gathering light. Typically, the **blade** of a leaf is a thin, flat structure attached to the stem by a stalk called a **petiole** (see Figure 34.1). In many plants, the leaf blade is held by its petiole at an angle almost perpendicular to the rays of the sun. This orientation, with the leaf surface facing the sun, maximizes the amount of light available for photosynthesis. Some leaves track the sun over the course of the day, moving so that they constantly face it.

In some plant species, leaves are highly modified for special functions. For example, some modified leaves serve as storage depots for energy-rich molecules, as in the bulbs of onions. In other species, such as succulents, the leaves store water. The protective spines of cacti are modified leaves (see Figure 34.3B). Other plants, such as peas, have modified portions of leaves called *tendrils* that support the plant by wrapping around other structures or plants.

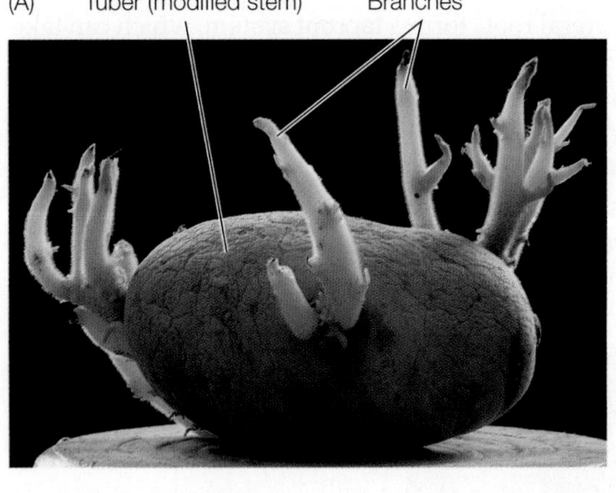

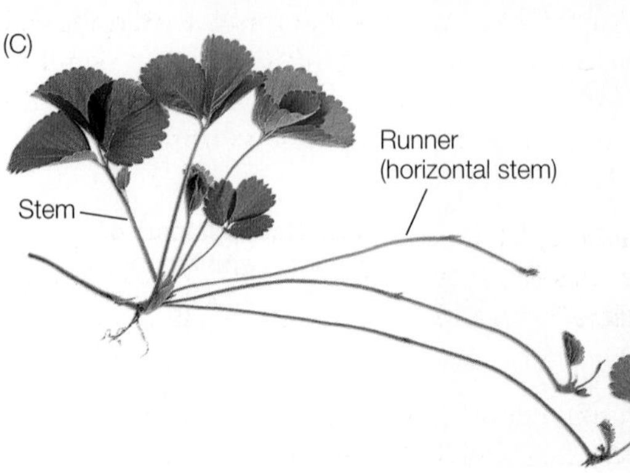


**34.3 Modified Stems** (A) A potato is a modified stem called a tuber; the sprouts that grow from its eyes are shoots, not roots. (B) The stem of this barrel cactus is enlarged to store water. Its highly modified leaves serve as thorny spines. Most of this plant's photosynthesis occurs in the stem. (C) The runners of beach strawberry are horizontal stems that produce roots and shoots at intervals. Rooted portions of the plant can live independently if the runner is cut.

## 34.1 RECAP

**The basic vegetative plant body plan consists of a root system and a shoot system. Stems and leaves, which are part of the shoot system, may be highly modified.**

- How do plants explore their environment for resources even though they cannot move? **See p. 720**
- How would you distinguish between a piece of a root and a piece of a shoot? **See pp. 721–722**
- What are the differences between the root systems of eudicots and monocots? **See p. 721 and Figure 34.2**

Our closer examination of the plant body begins with its fundamental building blocks: its cells. Although plant cells share many features with animal cells, including nucleus, mitochondria, plasma membrane, and Golgi apparatuses (see Section 5.3), it is their distinguishing features that we will consider in this chapter.

# 34.2 How Does the Cell Wall Support Plant Growth and Form?

Plant cells have the essential organelles that are shared by all eukaryotes (see Figure 5.7), but certain additional structures and organelles distinguish them from many other eukaryotic cells:

- *Chloroplasts* or other plastids
- A central *vacuole*
- Rigid, cellulose-containing *cell walls*

As mentioned earlier, plant form is dictated in part by the need to collect energy for photosynthesis, which takes place in the chloroplasts (see Section 10.2). Less obvious is the importance of vacuoles and cell walls in determining plant form.

### Cell walls and vacuoles help determine plant form

Mature plant cells usually contain a single **central vacuole**, which may account for a staggering 90 percent of its volume (see Figure 5.16). The vacuole is a watery sac containing a high concentration of solutes, including enzymes, amino acids, and sugars produced by photosynthesis. Many of these solutes are pumped into the vacuole by transporter proteins located in the **tonoplast**, the vacuolar membrane. This active accumulation of solutes provides the osmotic force for water uptake into the vacuole (as we will see in Section 35.1). As the vacuole expands, it exerts turgor pressure on the cell wall (see Figure 6.10). Turgor pressure not only keeps plants upright, but also is essential for plant growth.

### The structure of cell walls allows plants to grow

Cell walls are a feature of bacteria, fungi, algae, and plants. They serve to regulate cell volume, determine cell shape, and protect the cell contents. Plant cell walls have unique features that derive from their chemical composition. Furthermore, most of the carbon in terrestrial ecosystems is sequestered in the molecules that make up plant cell walls. As such, it is worth taking a closer look at their formation and structure.

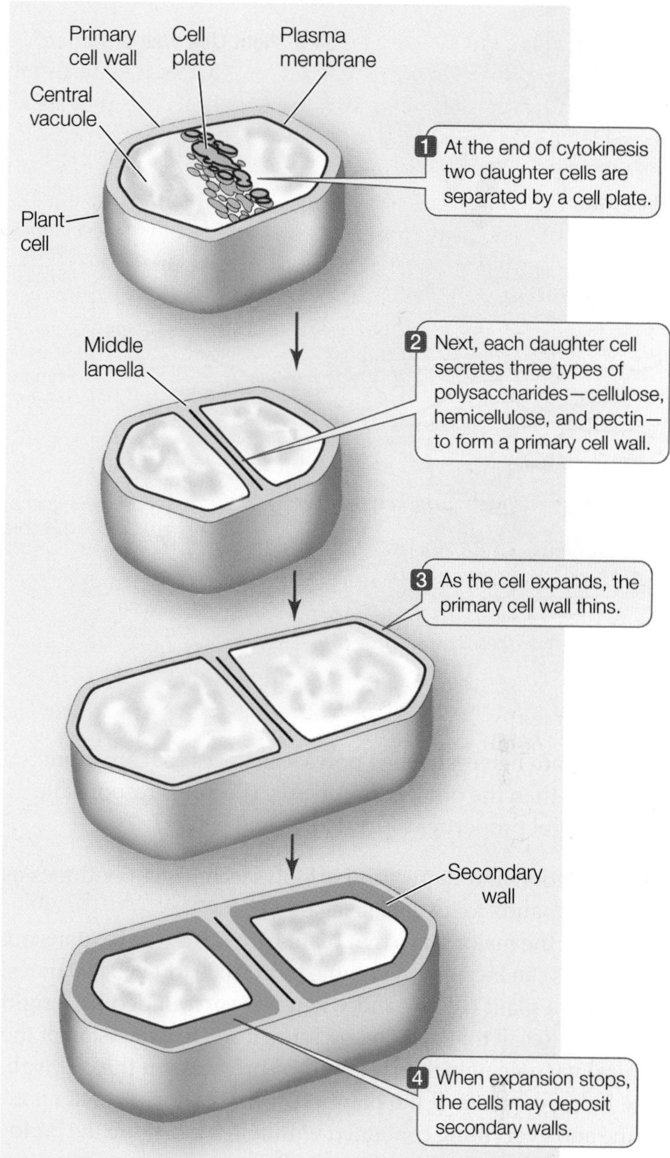

Primary cell wall
Cell plate
Plasma membrane
Central vacuole
Plant cell
1 At the end of cytokinesis two daughter cells are separated by a cell plate.
Middle lamella
2 Next, each daughter cell secretes three types of polysaccharides—cellulose, hemicellulose, and pectin—to form a primary cell wall.
3 As the cell expands, the primary cell wall thins.
Secondary wall
4 When expansion stops, the cells may deposit secondary walls.


**34.4 Plant Cell Wall Formation** Plant cell walls form as the final step in cell division.

The cytokinesis of a plant cell is completed when the two daughter cells are separated by a cell plate (**Figure 34.4**; see also Figure 11.13). A gluelike substance that forms within the cell plate constitutes the **middle lamella**, which persists as a thin layer between the walls of the two daughter cells. Each daughter cell then secretes three types of polysaccharides to form a **primary cell wall** (**Figure 34.5**):

- *Cellulose* is made up of linear polymers of thousands of glucose molecules (see Figure 3.16) that are organized into bundles of **microfibrils**, which form a lattice within the cell wall.
- *Hemicelluloses* are highly branched polysaccharide chains that extensively cross-link the cellulose microfibrils.

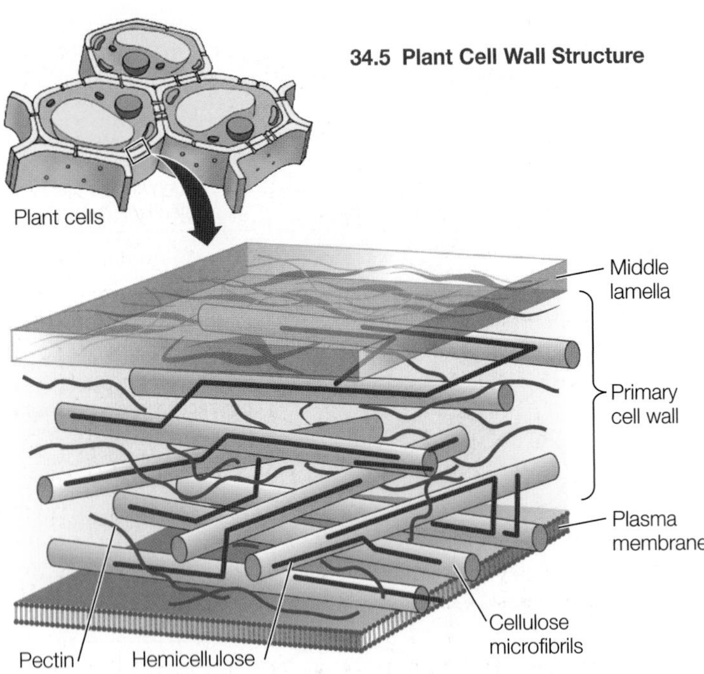


**34.5 Plant Cell Wall Structure**

- *Pectins* are heterogeneous polysaccharides that are more soluble than the other components. (Pectin is responsible for the gel properties of fruit jams and jellies.)

This secretion and deposition of polysaccharides continues as the cell expands to its final size.

One of the major ways that plants grow is by cell expansion. Some cells can increase in volume by 100,000 to 1,000,000 times! How can a plant cell expand when it is surrounded by a rigid cell wall? Recall that osmotic pressure leads to expansion of the central vacuole, which exerts turgor pressure on the cell wall. The living contents of the plant cell—that is, the plasma membrane and everything contained within it—constitute the **protoplast**. The cell wall responds to the increasing size of the protoplast by loosening the linkages between cellulose microfibrils. A class of proteins that reside in the cell wall, called *expansins*, are thought to assist in cell wall loosening by disrupting the noncovalent bonds that link the hemicelluloses and pectins to the cellulose microfibrils. To prevent the cell wall from becoming too thin (so that it does not blow out like an overinflated balloon), new cell wall components are synthesized and integrated.

When cell expansion stops, some types of plant cells deposit one or more additional cellulosic layers to form a thick **secondary cell wall** internal to the primary cell wall (see Figure 34.4). Secondary cell walls provide the mechanical support that allows some plants to produce large stems. Like the primary wall, the secondary wall contains layers of ordered cellulose microfibrils. However, rather than being embedded in pectins, the microfibrils are embedded in a remarkable substance called **lignin**. When secondary walls become lignified, the primary wall and even the middle lamella are also lignified. Lignin is a complex, carbon-containing polymer that forms a hydrophobic matrix that is strong, waterproof, and resistant to digestion by animals. After cellulose, lignin is the most abundant biological polymer on Earth, accounting for 20–35 percent of the dry weight of wood.

Scientists have just begun to dissect the complexity and dynamics of plant cell walls. Their basic components—celluloses, hemicelluloses, pectins, and lignins—are classes of molecules that can be built from several components and modified in a variety of ways. Thus the composition of plant cell walls varies among different types of plant cells. In addition, the composition of the wall of a single plant cell may not be uniform. For example, it is possible that directional growth reflects the deposition of cell wall components that are more easily loosened at one end of the cell. One measure of how much remains unknown is the finding that the genome of the tiny plant *Arabidopsis thaliana* contains more than a thousand genes related to cell wall biosynthesis, the functions of only a small fraction of which are currently known.

Building a plant body requires cooperation between groups of cells. Although they may appear to be isolated by their cell walls, plant cells interact in two ways to build and maintain a complex organism. First, in most areas, the cell wall is permeable to water and mineral ions and allows small molecules to

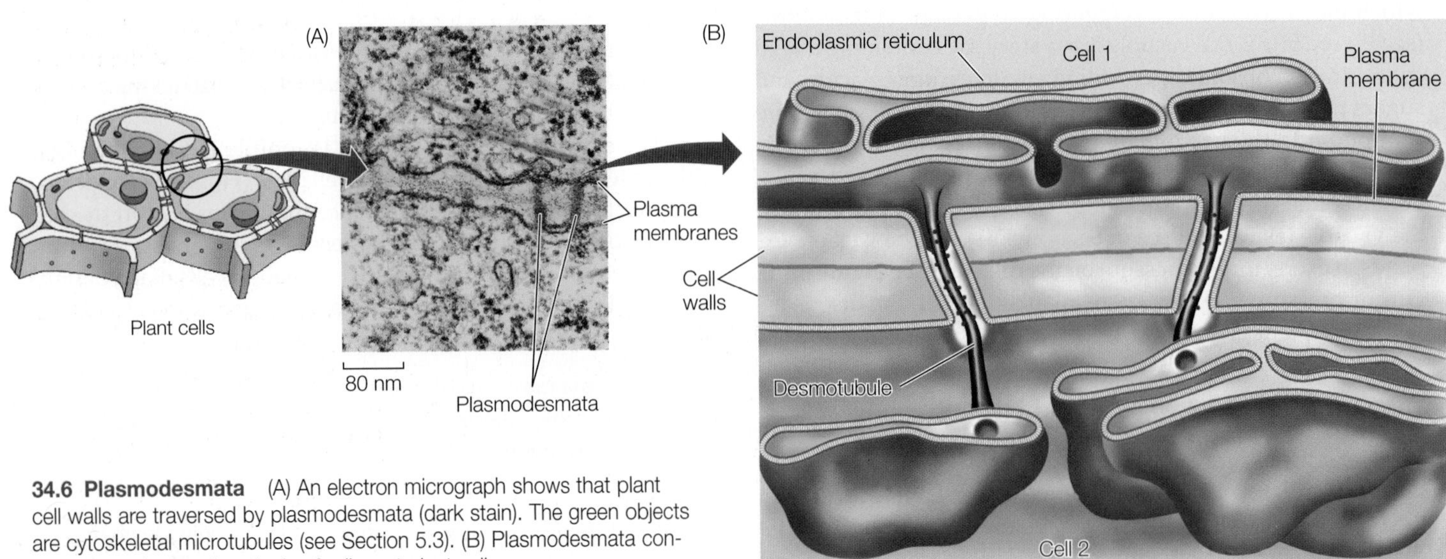


**34.6 Plasmodesmata** (A) An electron micrograph shows that plant cell walls are traversed by plasmodesmata (dark stain). The green objects are cytoskeletal microtubules (see Section 5.3). (B) Plasmodesmata connect the endoplasmic reticula of adjacent plant cells.

reach the plasma membrane. Second, the endoplasmic reticula (ER) of adjacent cells are connected by cytoplasm-filled canals called **plasmodesmata** that pass through the primary wall, allowing direct communication between plant cells (**Figure 34.6**). A single plant cell may be connected to its neighbors by up to a thousand plasmodesmata, which permit the movement of proteins and even RNAs from cell to cell. Some of these plasmodesmata are formed during cytokinesis when the cell plate is deposited.

Evolution has given some plant viruses a clever way to use this intercellular highway to their advantage. Tobacco mosaic virus (TMV), for example, encodes a protein called movement protein, or MP, that helps the virus spread throughout the plant. Without MP, the RNA genome of TMV cannot move from cell to cell. However, in some unknown way, the MP–RNA complex is able to move easily from cell to cell via plasmodesmata.

## 34.2 RECAP

**Plants synthesize a primary cell wall during cell division and cell expansion. In some types of plant cells, a secondary cell wall, reinforced with lignin, forms within the primary cell wall when cell expansion stops, providing additional structural support.**

- What are the components of plant cell walls in which most of the carbon in terrestrial ecosystems is sequestered? See pp. 723–724
- How do plant cell walls accommodate an expanding protoplast? See p. 724 and Figure 34.5
- Describe two features that allow plant cells to interact with one another. See pp. 724–725 and Figure 34.6

That there are dramatic differences between plant and animal body plans should not be surprising, since the multicellular forms of plants and animals evolved independently from entirely distinct protist ancestors (see Figure 1.10). In the next two sections we will look more closely at the unique characteristics of the plant body by following its development from a zygote into an adult.

# 34.3 How Do Plant Tissues and Organs Originate?

How does a single plant cell (a zygote) divide and grow into an organism like a redwood tree, which may grow continuously for over a thousand years to a height of over 100 meters? While still in the seed, a plant establishes the basic body plan for its mature form.

Two patterns that contribute to the plant body plan are established in the embryo:

- The *basal–apical* axis: the arrangement of cells and tissues along the main axis from root to shoot
- The *radial axis*: the concentric arrangement of the tissue systems

In addition, two clusters of undifferentiated cells form at the tips of the embryonic root and shoot. These clusters, called **meristems** (from the Greek *merizein*, "to divide"), will orchestrate all postembryonic development and allow the plant to form organs throughout its lifetime.

Both axes and meristems are best understood in developmental terms. We focus here on embryogenesis (embryo formation) in the model eudicot *Arabidopsis thaliana*, in which the process has been most intensively studied.

The first step in the formation of a plant embryo is a mitotic division of the zygote that gives rise to two daughter cells (**Figure 34.7, step 1**). These two cells face different fates (see Section 19.4). An asymmetrical (uneven) distribution of cytoplasm within the zygote causes one daughter cell to produce the embryo proper and the other daughter cell to produce a supporting structure, the **suspensor** (**Figure 34.7, step 2**). This asymmetrical division of the zygote establishes polarity as well as the basal–apical axis of the new plant. A long, thin suspensor and a more spherical or globular embryo are distinguishable after just four mitotic divisions. The suspensor soon ceases to elongate.

In eudicots, the initially globular embryo develops into the characteristic *heart stage* as the **cotyledons** ("seed leaves") start to grow (**Figure 34.7, step 3**). Further elongation of the cotyledons and of the main axis of the embryo gives rise to the *torpedo stage*, during which some of the internal tissues begin to differentiate (**Figure 34.7, step 4**). Between the cotyledons is the **shoot apical meristem**; at the other end of the axis is the **root apical meristem**. Each of these regions contains undifferentiated cells that will continue to divide to give rise to the organs developing over the life of the plant.

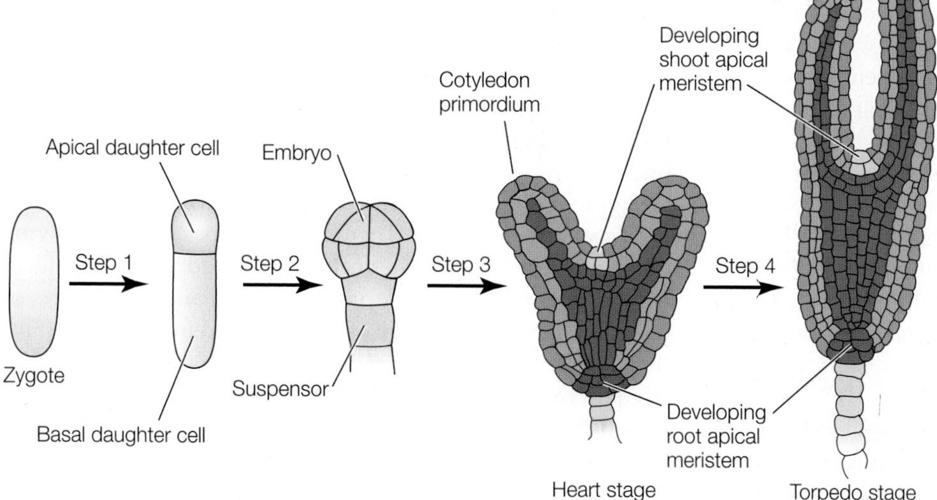


**34.7 Plant Embryogenesis** The basic body plant of the model eudicot (*Arabidopsis thaliana*) is established in several steps. By the heart stage, the three tissue systems are established: the dermal (gold), ground (light green), and vascular (blue) tissue systems.

By the end of embryogenesis, the radial axis of the plant has also been established. The embryonic plant contains three tissue systems, arranged concentrically, that will give rise to the tissues of the adult plant body.

### The plant body is constructed from three tissue systems

A *tissue* is an organized group of cells that have features in common and that work together as a structural and functional unit. In plants, tissues, in turn, are grouped into **tissue systems**. Despite their structural diversity, all vascular plants are constructed from three tissue systems: *dermal*, *vascular*, and *ground*. These three tissue systems are established during embryogenesis and ultimately extend throughout the plant body in a concentric arrangement (**Figure 34.8**). Each tissue system has distinct functions and is composed of different mixtures of cell types.

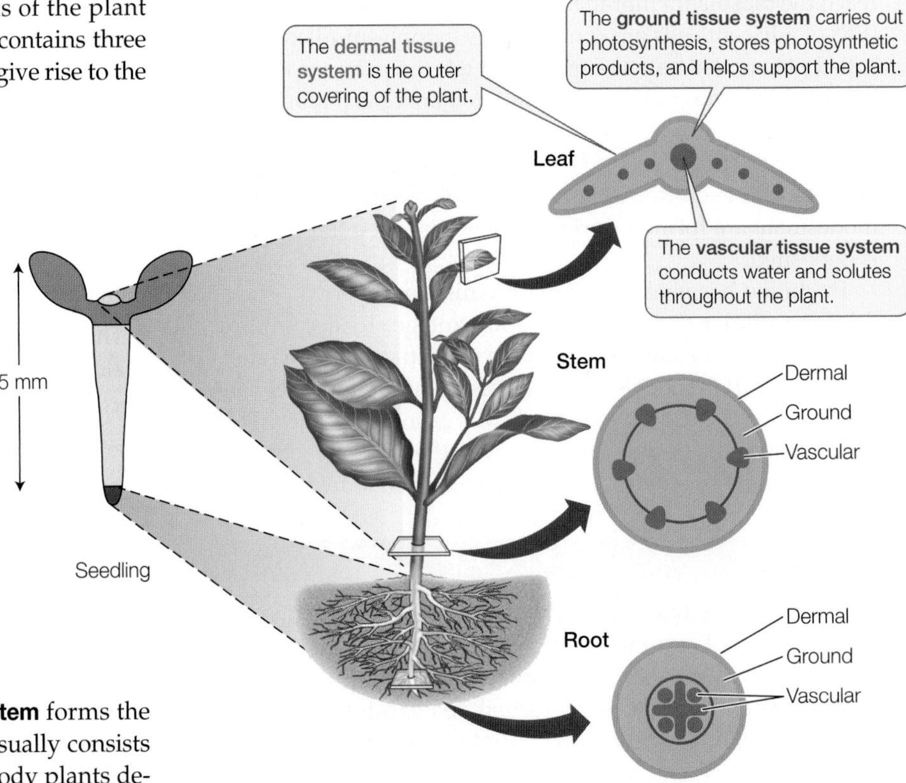

The dermal tissue system is the outer covering of the plant.
The ground tissue system carries out photosynthesis, stores photosynthetic products, and helps support the plant.
Leaf
The vascular tissue system conducts water and solutes throughout the plant.
5 mm
Seedling
Stem
Dermal
Ground
Vascular
Root
Dermal
Ground
Vascular

**34.8 Three Tissue Systems Extend Throughout the Plant Body** The arrangement shown here is typical of eudicots, but the three tissue systems are continuous in the bodies of all vascular plants.

**DERMAL TISSUE SYSTEM** The **dermal tissue system** forms the *epidermis*, or outer covering, of a plant, which usually consists of a single cell layer. The stems and roots of woody plants develop a dermal tissue called *periderm*.

During plant development, the epidermis must grow to cover the expanding plant body. The cells of the epidermis are small and round and usually have a small central vacuole or none at all. Once cell division ceases in the epidermis of an organ, the epidermal cells expand. Some epidermal cells differentiate to form one of three specialized structures:

- *Stomatal guard cells*, which form stomata (pores) for gas exchange in leaves
- *Trichomes*, or leaf hairs, which provide protection against insects and damaging solar radiation
- *Root hairs*, which greatly increase root surface area, thus providing more surface for the uptake of water and mineral nutrients.

Aboveground epidermal cells secrete a protective extracellular **cuticle** made of *cutin* (a polymer composed of long chains of fatty acids), a complex mixture of waxes, and cell wall polysaccharides. The cuticle limits water loss, reflects potentially damaging solar radiation, and serves as a barrier against pathogens.

**GROUND TISSUE SYSTEM** Virtually all the tissue lying between dermal tissue and vascular tissue in both shoots and roots is part of the **ground tissue system**, which therefore makes up most of the plant body. Ground tissue functions primarily in storage, support, and photosynthesis. To fulfill these diverse functions, ground tissues incorporate three cell types that are classified according to their cell wall structure: *parenchyma*, *collenchyma*, and *schlerenchyma*.

The most common cell type in plants is the **parenchyma** cell (**Figure 34.9A**). Parenchyma cells have large vacuoles and thin walls consisting only of a primary wall and the shared middle lamella. They play important roles in photosynthesis (in leaves) and in the storage of, for example, protein (in fruits) and starch (in roots). Many retain the capacity to divide and hence may give rise to new cells, as when a wound results in cell proliferation.

**Collenchyma** cells resemble parenchyma cells that have been modified to provide flexible support. Their primary walls are characteristically thick at the corners of the cells (**Figure 34.9B**). Collenchyma cells are generally elongated. In these cells, the primary wall thickens in part due to the deposition of pectins, but no secondary wall forms. Collenchyma provides support to leaf petioles, nonwoody stems, and growing organs. Tissue made of collenchyma cells is flexible, permitting stems and petioles to sway in the wind without snapping. The familiar "strings" in celery consist primarily of collenchyma cells.

**Sclerenchyma** cells have thickened secondary walls that perform their major function: support. Many sclerenchyma cells undergo programmed cell death (apoptosis; see Section 11.6) after lignifying their cell walls, and thus perform their supporting function when dead. There are two types of sclerenchyma cells: elongated **fibers** and variously shaped **sclereids**. Fibers provide relatively rigid support to wood and other parts of the plant, in which they are often organized into bundles (**Figure 34.9C**). The bark of trees owes much of its mechanical strength to long fibers. Sclereids may pack together densely, as in a nut's shell or in some seed coats (**Figure 34.9D**). Isolated clumps of sclereids, called *stone cells*, in pears and some other fruits give them their characteristic gritty texture.

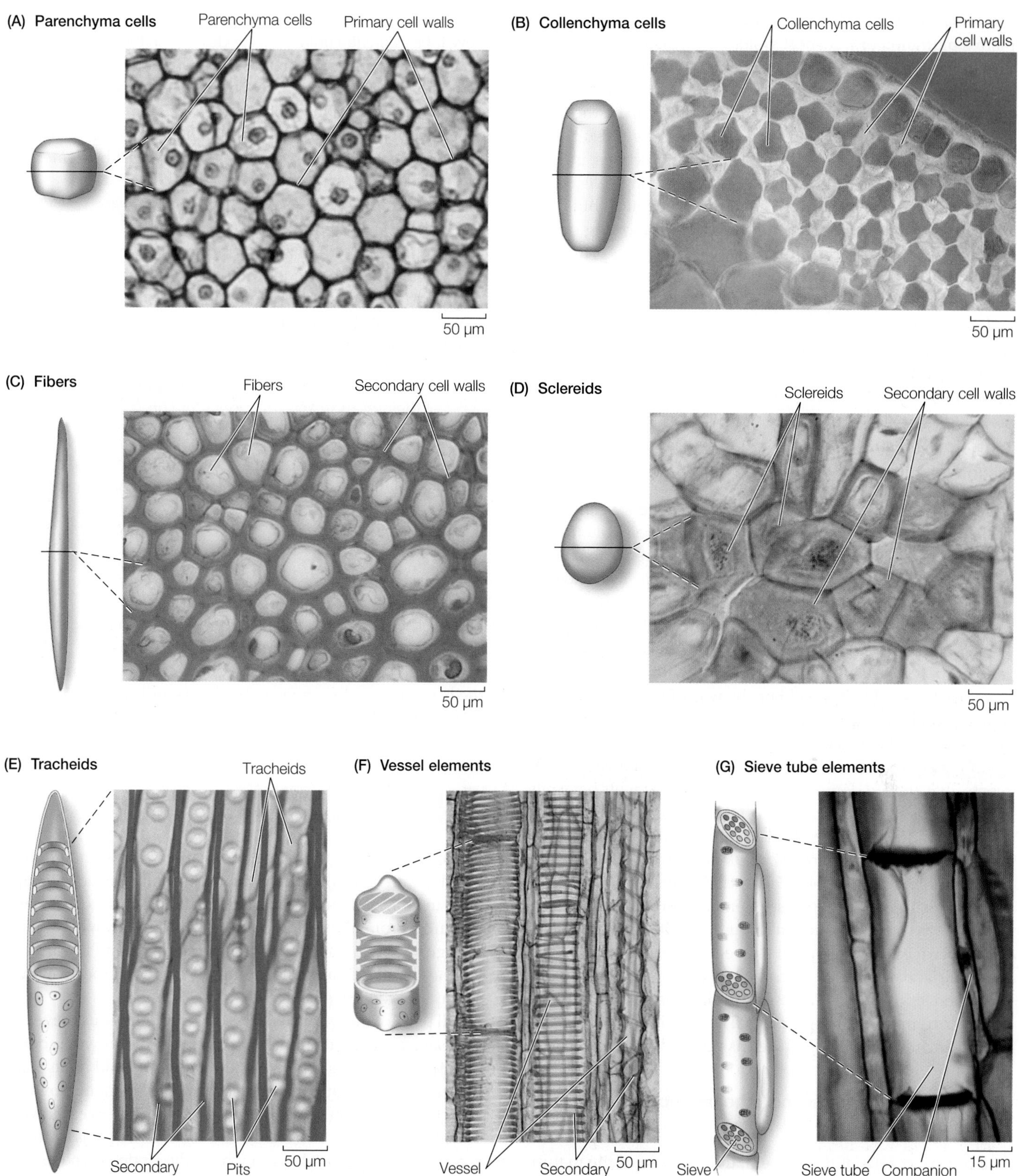


**34.9 Plant Cell Types** (A) Parenchyma cells in the petiole of *Coleus*. Note the thin, uniform cell walls. (B) Collenchyma cells make up the five outer cell layers of this spinach leaf vein. Their walls are thick at the corners of the cells and thin elsewhere. (C) Sclerenchyma: fibers in a sunflower stem (*Helianthus*). The thick secondary walls are stained red. (D) Sclerenchyma: sclereids. The extremely thick secondary walls of sclereids are laid down in layers. They provide support and a hard texture to structures such as nuts and seeds. (E, F) Tracheary elements: (E) Tracheids in pinewood. The thick secondary walls are stained dark red. (F) Vessel elements in the stem of a squash. The secondary walls are stained red; note the different patterns of thickening, including rings and spirals. (G) Sieve tube elements and companion cells in the stem of a cucumber.

**VASCULAR TISSUE SYSTEM** The **vascular tissue system** is the plant's plumbing or transport system—the distinguishing feature of vascular plants. Its two constituent tissues, the xylem and phloem, distribute materials throughout the plant. The **xylem** distributes water and mineral ions taken up by the roots to all the cells of the stems and leaves. **Phloem**, as a result of its cellular complexity, can perform a variety of functions, including transport, support, and storage. All the living cells of the plant body require a source of energy and chemical building blocks. The phloem meets these needs by transporting carbohydrates from sites of production (called *sources*, primarily leaves) to sites of utilization or storage (called *sinks*, such as growing tissue, storage tubers, and developing flowers).

Let's take a closer look at the structure of the diverse cell types that make up these vascular tissues. In Chapter 35 we will see how they transport water and materials throughout the plant body.

### Cells of the xylem transport water and dissolved minerals

Xylem contains conducting cells called **tracheary elements**, which undergo apoptosis before assuming their function of transporting water and dissolved minerals. There are two types of tracheary elements: *tracheids* and *vessel elements*. **Tracheids** are spindle-shaped cells, found in gymnosperms and other vascular plants, that are evolutionarily more ancient than vessel elements (**Figure 34.9E**). When the protoplast disintegrates upon cell death, water and minerals can move with little resistance from one tracheid to its neighbors by way of *pits*, interruptions in the secondary wall that leave the primary wall unobstructed.

Flowering plants evolved a water-conducting system made up of *vessels*, formed from individual cells, called **vessel elements**, that are laid down end-to-end. Vessel elements have pits in their cell walls, as do tracheids, but are generally larger in diameter than tracheids. Vessel elements secrete lignin into their secondary walls, then partially break down their end walls before undergoing apoptosis. The result is a continuous hollow tube consisting of many vessel elements, providing an open pipeline for water conduction (**Figure 34.9F**). In the course of angiosperm evolution, vessel elements have become shorter, and their end walls have become less and less obliquely oriented and less obstructed, presumably increasing the efficiency of water transport through them. The xylem of many angiosperms includes tracheids as well as vessel elements.

### Cells of the phloem transport the products of photosynthesis

The transport cells of the phloem, unlike those of the mature xylem, are living cells. In flowering plants, the characteristic cells of the phloem are **sieve tube elements** (**Figure 34.9G**). Like vessel elements, these cells meet end-to-end. They form long *sieve tubes*, which transport carbohydrates and many other materials from their sources to tissues that consume or store them. In plants with mature leaves, for example, products of photosynthesis move from leaves to root tissues.

Unlike vessel elements, which break down their end walls, sieve tube elements contain plasmodesmata in their end walls that enlarge to form pores. The result is end walls that look like sieves, called *sieve plates*. Although the sieve tube elements remain alive, some components of the protoplast break down. They are closely connected to *companion cells* that retain all their organelles and function as a "life support system" for the sieve tube elements.

#### 34.3 RECAP

**Plant embryos have an embryonic root and shoot containing three concentric tissue systems: dermal, ground, and vascular. These tissue systems carry out different functions through their unique combinations of specialized cell types.**

- What distinguishes the three tissue systems in terms of their location and functions? See pp. 726–728 and Figure 34.8
- What structural differences make tissues made of collenchyma cells more flexible than those consisting primarily of sclerenchyma cells? See p. 726
- Outline the differences between tracheids and vessel elements. See p. 728 and Figure 34.9

By the end of embryogenesis, the plant embryo is encased in a seed and is ready to germinate. We will return to seeds and their germination in the chapters that follow. For now, let's consider the beginning of life as faced by plants and see how the cells and tissues we have just described allow the embryo to build an adult plant body.

## 34.4 How Do Meristems Build a Continuously Growing Plant?

As noted at the beginning of this chapter, plants and animals develop and function differently. While animals use their mobility to forage for food, plants are sessile and must collect scarce resources from above and below ground by growing. Plants thus grow in two directions—toward sunlight, and toward water and dissolved minerals in the soil—through the growth of shoots and roots, respectively.

All parts of the animal body grow as an individual develops from embryo to adult, but in most animals this growth is **determinate**—that is, the growth of the individual and all its parts ceases when the adult state is reached. Determinate growth is also characteristic of some plant organs, such as leaves, flowers, and fruits. The growth of shoots and roots, however, is a lifelong process. Such open-ended growth is called **indeterminate**.

### Plants increase in size through primary and secondary growth

Plants increase their surface area above and below ground by growing. All plants experience **primary growth**, which is characterized by the lengthening of roots and shoots and by the proliferation of new roots and shoots through branching. In addition, many gymnosperms and eudicots, especially trees, experience **secondary growth**, by which they increase in girth.

Primary and secondary growth lead to distinctive traits in the plant body. Primary growth develops what is called the **primary plant body**, while secondary growth develops the **secondary plant body**. All seed plants have a primary plant body, which consists of all the *nonwoody* parts of the plant. Many herbaceous plants—monocots in particular—consist entirely of a primary plant body. *Woody* plants, such as trees and shrubs, have, in addition to the primary plant body, a secondary plant body consisting of wood and bark. As the tissues of the secondary plant body are laid down, the stems and roots thicken. The secondary plant body continues to grow and thicken throughout the life of the plant. The primary plant body also continues to grow, lengthening and branching the shoot and root systems and forming new leaves.

### A hierarchy of meristems generates the plant body

Meristems, as we have seen, are localized regions of undifferentiated cells that are the source of all new organs in the adult plant. Even before seed germination, the plant embryo has two meristems: a shoot apical meristem at the end of the embryonic shoot, and a root apical meristem near the end of the embryonic root (see Figure 34.7).

Meristematic cells are small and closely packed, with very small central vacuoles and a very thin primary cell wall. Meristematic cells are undifferentiated and forever young, retaining the ability to produce new cells indefinitely. The cells that perpetuate the meristems, called **initials**, are comparable to animal stem cells (discussed in Section 19.2). When an initial divides, one daughter cell develops into another meristem cell the size of its parent, while the other daughter cell differentiates into a more specialized cell.

While the plant embryo experiences primary growth through the activities of the root and shoot apical meristems, growth of the adult plant reflects the activity of additional meristem types. Our discussion of postembryonic plant growth begins with a closer look at how the adult plant grows throughout its lifetime and the critical role of meristems in that growth.

Two types of meristems contribute to the growth and development of the adult plant (**Figure 34.10**):

- **Apical meristems** orchestrate primary growth, giving rise to the primary plant body. This growth is characterized by cell division followed by cell enlargement (vertical elongation).
- **Lateral meristems** orchestrate secondary growth. Two lateral meristems, *vascular cambium* and *cork cambium*, contribute to the secondary plant body.

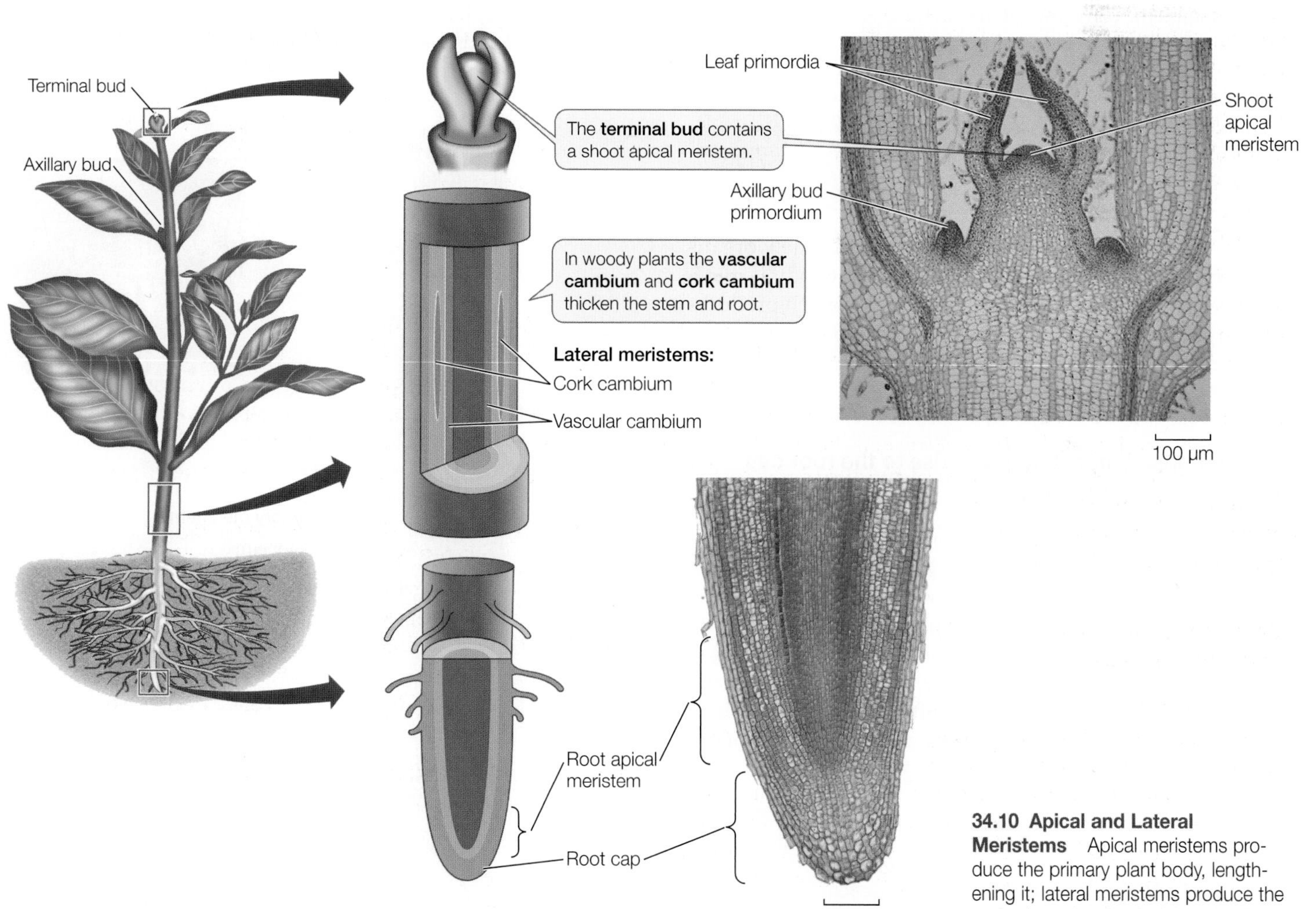


**34.10 Apical and Lateral Meristems** Apical meristems produce the primary plant body, lengthening it; lateral meristems produce the secondary plant body, thickening it.

### Indeterminate primary growth originates in apical meristems

Because apical meristems can perpetuate themselves indefinitely, a shoot or root can continue to lengthen and grow indefinitely; in other words, growth of the shoot or root is indeterminate. Primary growth leads to elongation of shoots and roots and formation of organs (see Figure 34.10). All plant organs arise ultimately from cell divisions in apical meristems, followed by cell expansion and differentiation. Several types of apical meristems play roles in organ formation:

- *Shoot apical meristems* supply the cells that extend stems and branches, allowing more leaves to form and photosynthesize. Apical meristems that form leaves are called *vegetative meristems*. Flowers are formed by apical meristems that become *inflorescence meristems* (see Section 38.2 for more on floral development).
- *Root apical meristems* supply the cells that extend roots, enabling the plant to penetrate and explore the soil for water and minerals.

Apical meristems in both the shoot and the root give rise to a set of cylindrical **primary meristems**, which produce the tissues of the primary plant body. From the outside to the inside of the root or shoot, which are both cylindrical organs, the primary meristems are the **protoderm**, the **ground meristem**, and the **procambium**. These meristems, in turn, give rise to the three tissue systems:

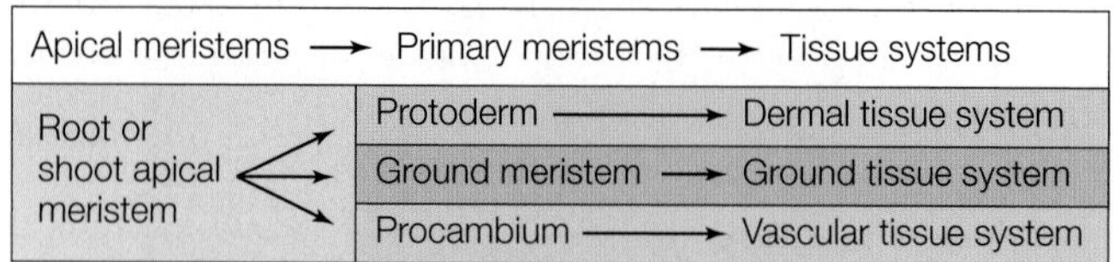

| Apical meristems → | Primary meristems → | Tissue systems |
|---|---|---|
| Root or shoot apical meristem | Protoderm → | Dermal tissue system |
| | Ground meristem → | Ground tissue system |
| | Procambium → | Vascular tissue system |

Because meristems can continue to produce new organs throughout the lifetime of the plant, the plant body is much more variable in form than the animal body, which produces each organ only once. To see how meristems function, let's look more closely at how the root apical meristem produces the root system.

### The root apical meristem gives rise to the root cap and the root primary meristems

The root apical meristem produces all the cells that contribute to growth in the length of a root (**Figure 34.11A**). Some of the daughter cells from the apical (tip) end of the root apical meristem contribute to a **root cap**, which protects the delicate growing region of the root as it pushes through the soil. The root cap secretes a mucopolysaccharide (slime) that acts as a lubricant. Even so, the cells of the root cap are often damaged or scraped away and must therefore be replaced constantly. The root cap is also the structure that detects the pull of gravity and thus controls the downward growth of roots.

In the middle of the root apical meristem is a *quiescent center*, in which cell divisions are rare. The quiescent center can become more active when needed—following injury, for example. The daughter cells produced above the quiescent center (that is, away from the root cap) become the three cylindrical primary meristems: the protoderm, the ground meristem, and the procambium.

The apical and primary meristems constitute the **zone of cell division**, the source of all the cells of the root's primary tissues. Just above this zone is the **zone of cell elongation**, where the newly formed cells are elongating and thus pushing the root farther into the soil. Above that zone is the **zone of maturation**, where the cells are differentiating, taking on specialized forms and functions. These three zones grade imperceptibly into one another; there is no abrupt line of demarcation.

### The products of the root's primary meristems become root tissues

The products of the three primary meristems (the protoderm, ground meristem, and the procambium) are the tissue systems of the mature root (**Figure 34.12**).

The protoderm gives rise to the **epidermis**, an outer layer of cells that is adapted for protection of the root and absorption of mineral ions and water. Many of the epidermal cells become

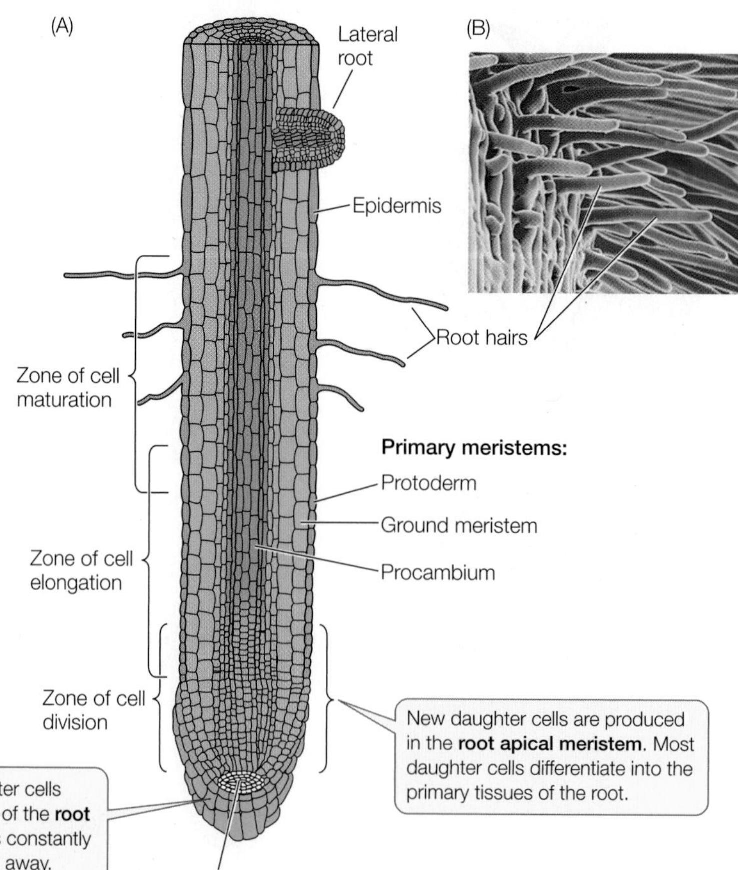
(A)
Lateral root
(B)
Epidermis
Root hairs
Zone of cell maturation
Primary meristems:
Protoderm
Ground meristem
Procambium
Zone of cell elongation
Zone of cell division
New daughter cells are produced in the root apical meristem. Most daughter cells differentiate into the primary tissues of the root.
Some daughter cells become part of the root cap, which is constantly being eroded away.
Quiescent center

**34.11 Tissues and Regions of the Root Tip**
(A) Extensive cell division creates the complex structure of the root. (B) Root hairs, seen with a scanning electron microscope.

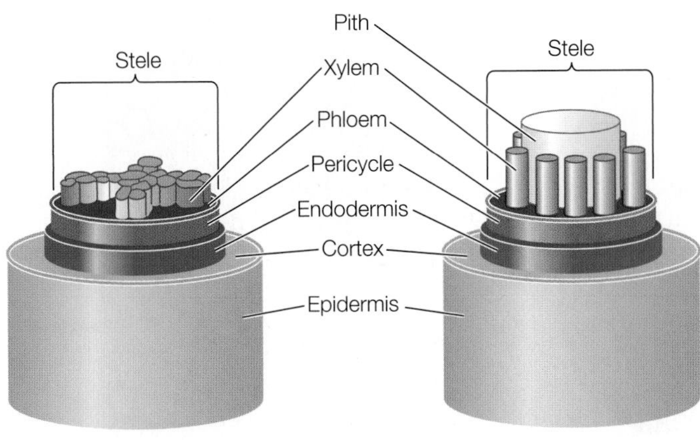


**34.12 Products of the Root's Primary Meristems** The protoderm gives rise to the outermost layer (epidermis). The ground meristem produces the cortex, the innermost layer of which is the endodermis. The primary vascular tissues of the root are found in the stele, which is the product of the procambium. The arrangement of tissues in the stele differs in the roots of eudicots and monocots.

**yourBioPortal.com**
GO TO **Web Activity 34.1 • Eudicot Root**
AND **Web Activity 34.2 • Monocot Root**

long, delicate **root hairs**, which vastly increase the surface area of the root (**Figure 34.11B**). Root hairs grow out among the soil particles, probing nooks and crannies and taking up water and minerals.

Internal to the epidermis, the ground meristem gives rise to a region of ground tissue that is many cells thick, called the **cortex**. The cells of the cortex are relatively unspecialized and often serve as storage depots. The innermost layer of the cortex is the **endodermis**. Unlike those of other cortical cells, the cell walls of the endodermal cells contain *suberin*, a waterproof substance. Strategic placement of suberin in only certain parts of the cell wall enables the cylindrical ring of endodermal cells to control the movement of water and dissolved mineral ions into the vascular tissue system.

Moving inward past the endodermis, we enter the vascular cylinder, or **stele**, produced by the procambium. The stele consists of three tissues: pericycle, xylem, and phloem. The **pericycle** consists of one or more layers of relatively undifferentiated cells. It has three important functions:

- It is the tissue within which lateral roots arise (**Figure 34.13A**).
- It can contribute to secondary growth by giving rise to lateral meristems that thicken the root.
- Its cells contain membrane transport proteins that export nutrient ions into the cells of the xylem.

At the very center of the root of a eudicot lies the xylem. Seen in cross section, it typically has the shape of a star with a variable number of points (**Figure 34.13B**). Between the points are bundles of phloem. In monocots, a region of parenchyma cells, called the **pith**, typically lies in the center of the root, surrounded by xylem and phloem (**Figure 34.13C**). Pith, which often stores carbohydrate reserves, is also found in the stems of both eudicots and monocots.

### The products of the stem's primary meristems become stem tissues

Recall that shoots are composed of repeating modules called phytomers, each consisting of a node with its attached leaf or leaves, the internode between nodes, and axillary buds in the angle between each leaf and the stem (see Figure 34.1). Shoots grow by adding new phytomers. Those new phytomers originate from cells in shoot apical meristems, which are formed at the tips of stems and in axillary buds.

The shoot apical meristem, like the root apical meristem, forms three primary meristems: protoderm, ground meristem, and procambium. These primary meristems, in turn, give rise to the three shoot tissue systems. The shoot apical meristem

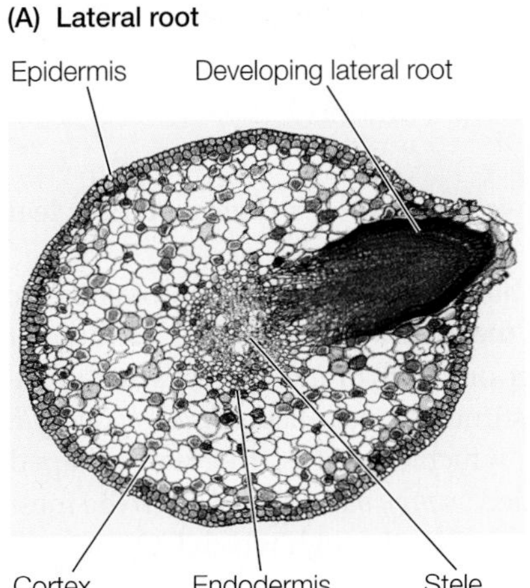


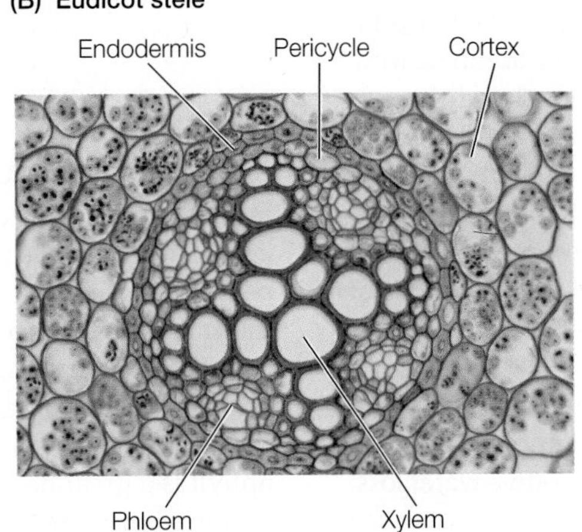


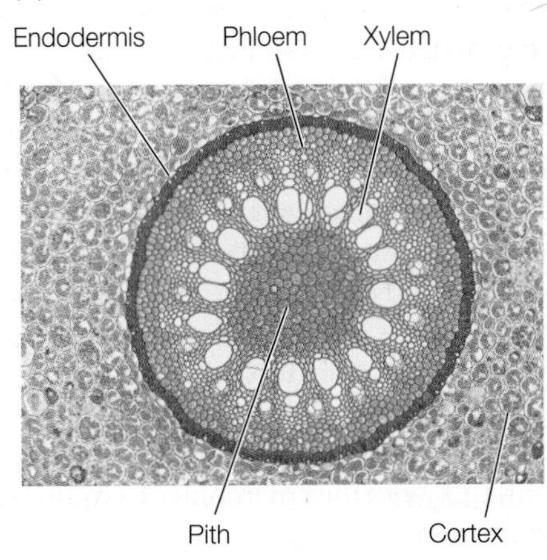


**34.13 Root Anatomy** (A) Cross section through the tip of a lateral root in a willow tree. Cells in the pericycle divide and the products differentiate, forming the tissues of a lateral root. (B, C) Cross sections of the stele of (B) a representative eudicot (the buttercup, *Ranunculus*) and (C) a representative monocot (corn, *Zea mays*), showing the arrangement of the primary root tissues.

repetitively lays down the beginnings of leaves and axillary buds. Leaves arise from bulges called *leaf primordia*, which form as cells divide on the sides of the shoot apical meristem (see Figure 34.10). *Bud primordia* form at the bases of the leaf primordial and where they may become new apical meristems and initiate new shoots. The growing stem has no protective structure analogous to the root cap, but the leaf primordia can act as a protective covering for the shoot apical meristem.

The plumbing of stems differs from that of roots. In a root, the vascular tissue lies deep in the interior, with the xylem at or near the center (see Figure 34.13B and C). The vascular tissue of a young stem, however, is divided into discrete **vascular bundles** (**Figure 34.14**). Each vascular bundle contains both xylem and phloem. In eudicots, the vascular bundles generally form a cylinder, but in monocots, they are seemingly scattered throughout the stem.

In addition to the vascular tissues, the stem contains other important storage and supportive tissues. In eudicots, the pith lies to the inside of the ring of vascular bundles and also extends between them, forming regions called *pith rays*. To the outside lies the cortex, which may contain supportive collenchyma cells with thickened walls. The pith and cortex constitute the ground tissue system of the stem. The outermost cell layer of the young stem is the epidermis.

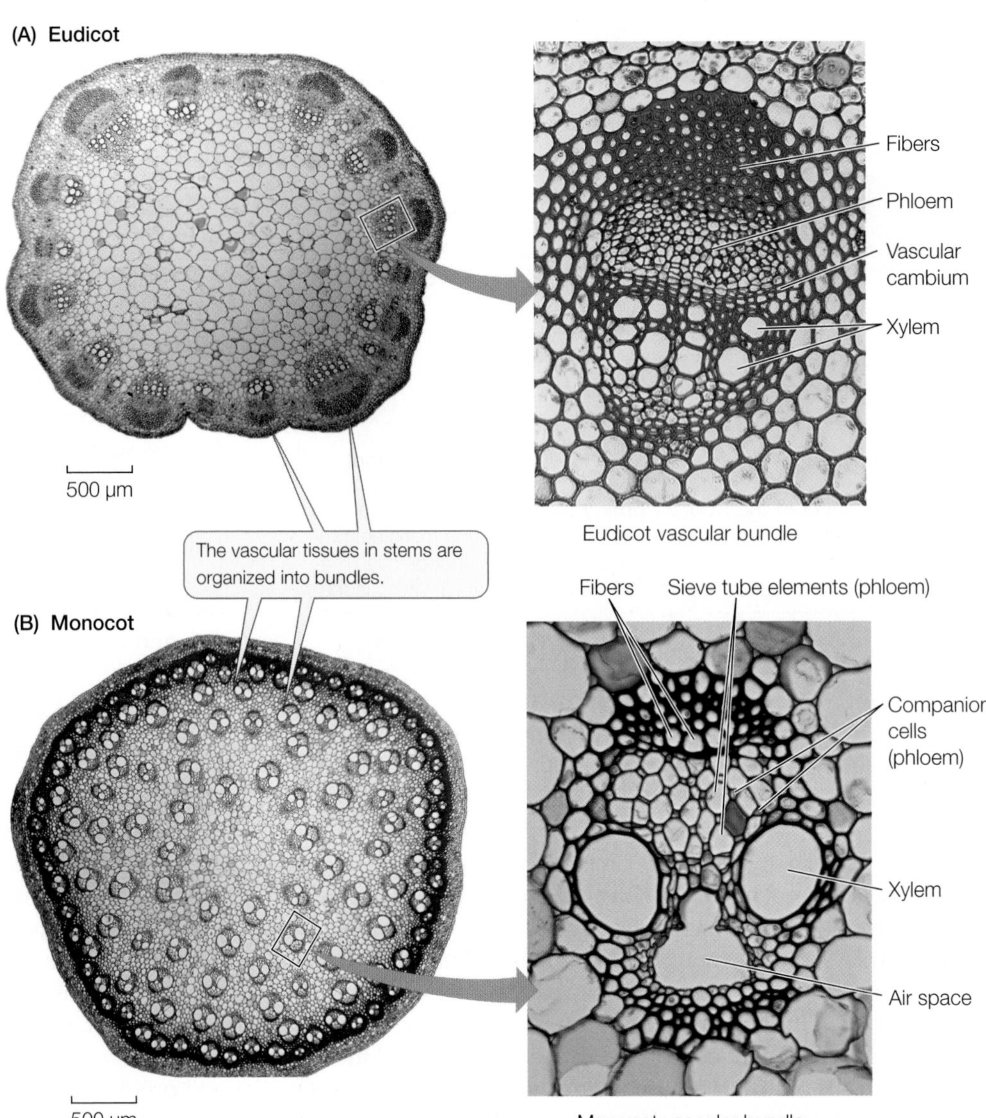


**34.14 Vascular Bundles in Stems** (A) In herbaceous eudicot stems, the vascular bundles are arranged in a cylinder, with pith in the center and the cortex outside the cylinder. (B) A scattered arrangement of vascular bundles is typical of monocot stems.

**yourBioPortal.com**
GO TO Web Activity 34.3 • Eudicot Stem
AND Web Activity 34.4 • Monocot Stem

### Leaves are determinate organs produced by shoot apical meristems

For most of its life a plant produces leaves from apical meristems. Apical meristems that produce leaves are called **vegetative meristems**. As shown in Figure 34.10, leaves originate from the edges of the apical meristem as initial cells that differentiate into leaf primordia. A highly simplified way to think of the development of the leaf from the leaf primordia is to imagine leaves as flattened stems. However, there are two important differences. First, unlike the indeterminate growth of the stem, the growth of a leaf is determinate. Second, while the tissues of the stem are arranged in a radial pattern, the leaf, as a flat organ, has a distinct top side and bottom side.

Leaf anatomy is beautifully adapted to carry out photosynthesis and to support that process by exchanging the gases $O_2$ and $CO_2$ with the environment, limiting evaporative water loss, and exporting the products of photosynthesis to the rest of the plant. **Figure 34.15A** shows a section of a typical eudicot leaf in three dimensions.

Most eudicot leaves have two zones of photosynthetic parenchyma tissue called **mesophyll** (which means "middle of the leaf"). The upper layer or layers of mesophyll, which consist of elongated cells, constitute a zone called *palisade mesophyll*. The lower layer or layers, which consist of irregularly shaped cells, constitute a zone called *spongy mesophyll*. Within the mesophyll is a great deal of air space through which $CO_2$ can diffuse to photosynthesizing cells.

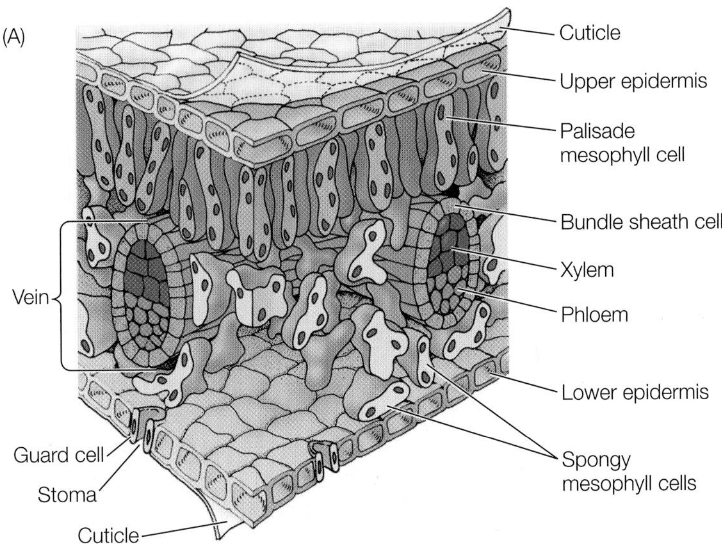


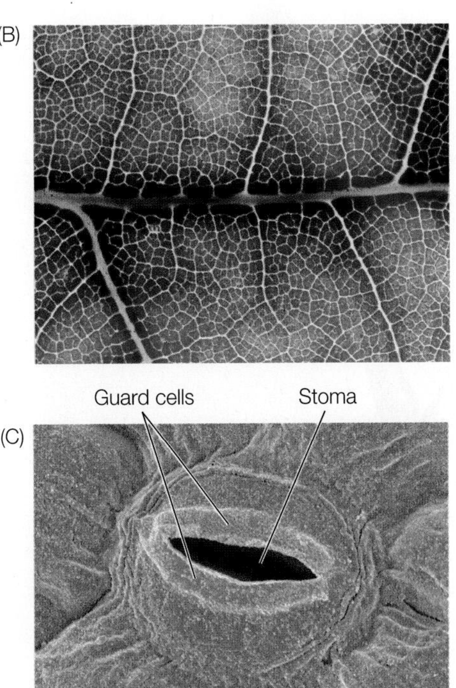


**34.15 The Eudicot Leaf** (A) This three-dimensional diagram shows a section of a eudicot leaf. (B) The network of fine veins in this maple leaf carries water to the mesophyll cells and carries photosynthetic products away from them. (C) Carbon dioxide enters the leaf through stomata like this one on the epidermis of a eudicot leaf.

**yourBioPortal.com**
GO TO **Web Activity 34.5 • Eudicot Leaf**

Vascular tissue branches extensively throughout the leaf, forming a network of *veins* (**Figure 34.15B**). Veins extend to within a few cell diameters of all the cells of the leaf, ensuring that the mesophyll cells are well supplied with water and minerals. The products of photosynthesis are loaded into the veins for export to the rest of the plant.

Covering virtually the entire leaf on both its upper and lower surfaces is a layer of nonphotosynthetic cells, the epidermis. The epidermal cells have an overlying waxy cuticle that is impermeable to water. Although this impermeability prevents excessive water loss, it also poses a problem: while the epidermis keeps water in the leaf, it also keeps out $CO_2$—the other raw material of photosynthesis.

The problem of balancing water retention and carbon dioxide availability is solved by an elegant regulatory system that will be discussed in more detail in Section 35.3. Stomatal *guard cells* are modified epidermal cells that can change their shape, thereby opening or closing pores called *stomata* (singular *stoma*), which serve as passageways between the environment and the leaf's interior (**Figure 34.15C**). When the stomata are open, carbon dioxide can enter and oxygen can leave, but water can also be lost.

## Many eudicot stems and roots undergo secondary growth

As we have seen, the roots and stems of some eudicots develop a secondary plant body, the tissues of which we commonly refer to as *wood* and *bark*. These tissues are derived by secondary growth from the two lateral meristems, the vascular cambium and the cork cambium.

The **vascular cambium** is a cylindrical tissue consisting predominantly of elongated cells that divide frequently. It supplies the cells of the secondary xylem and secondary phloem, which eventually become wood and bark. The **cork cambium** produces mainly waxy-walled protective cells. It supplies some of the cells that become bark.

Each year, deciduous trees lose their leaves, leaving bare branches and twigs in winter. These twigs illustrate both primary and secondary growth (**Figure 34.16**). The apical meristems of the twigs are enclosed in buds protected by *bud scales*. When the buds begin to grow in spring, the scales fall away, leaving scars, which show us where the bud was and allow us to identify each year's growth. The dormant twig shown in Figure 34.16 is the product of primary and secondary growth. Only the buds consist entirely of primary tissues.

The vascular cambium is initially a single layer of cells lying between the primary xylem and the primary phloem within the vascular bundles (see Figure 34.16). The root or stem increases in diameter when the cells of the vascular cambium divide, producing secondary xylem cells toward the inside of the root or stem and producing secondary phloem cells toward the outside (**Figure 34.17**). In the stem, cells in the pith rays between the vascular bundles also divide, forming a continuous cylinder of vascular cambium running the length of the stem. This cylinder, in turn, gives rise to complete cylinders of secondary xylem (**wood**) and secondary phloem, which contributes to the bark. It also produces vascular rays for lateral transport, a structure not found in primary xylem and phloem. The principal cell products of the vascular cambium are vessel elements, tracheids, and supportive fibers in the secondary xylem, and sieve tube elements, companion cells, fibers, and parenchyma cells in the secondary phloem.

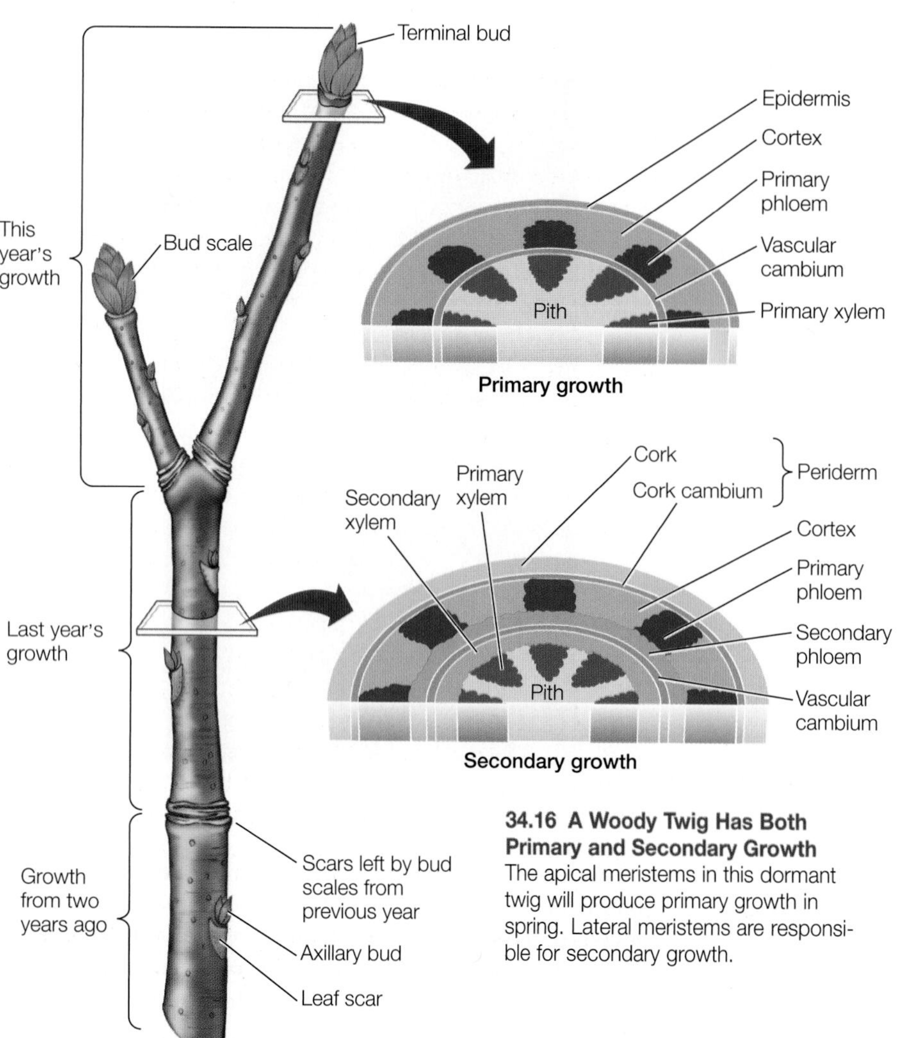


**34.16 A Woody Twig Has Both Primary and Secondary Growth** The apical meristems in this dormant twig will produce primary growth in spring. Lateral meristems are responsible for secondary growth.

As secondary growth of stems or roots continues, the expanding vascular tissue stretches and breaks the epidermis and the outer layers of the cortex, which ultimately flake away. Tissue derived from the secondary phloem then becomes the outermost part of the stem. Before the dermal tissues are broken away, cells lying near the surface of the secondary phloem begin to divide, forming a cork cambium. This meristematic tissue produces layers of *cork*, a protective tissue composed of cells with thick walls waterproofed with suberin. The cork soon becomes the outermost tissue of the stem or root (see Figure 34.16). Without the activity of the cork cambium, the sloughing off of the outer primary tissues would expose the plant to potential damage, such as excessive water loss or invasion by microorganisms. Sometimes the cork cambium produces cells to the inside as well as to the outside; these cells constitute a tissue known as the *phelloderm*.

The cork cambium, cork, and phelloderm constitute a secondary dermal tissue called **periderm**. As the vascular cambium continues to produce secondary vascular tissue, these corky layers are lost, but the continuous formation of new cork cambia in the underlying secondary phloem gives rise to new corky layers. The periderm and the secondary phloem—that is, all the tissues external to the vascular cambium—constitute the **bark**.

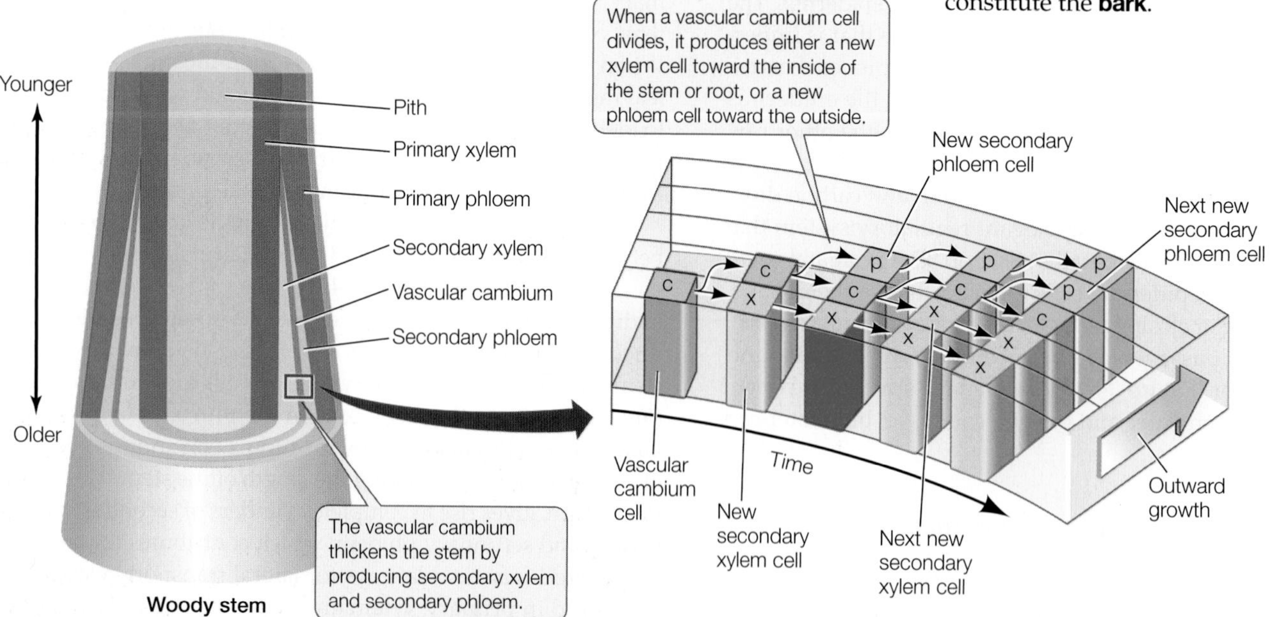


**34.17 Vascular Cambium Thickens Stems and Roots** Stems and roots grow thicker because a thin layer of cells, the vascular cambium, remains meristematic. These highly diagrammatic images emphasize the pattern of deposition of secondary xylem and phloem by the vascular cambium.

**yourBioPortal.com**
GO TO **Animated Tutorial 34.1 • Secondary Growth: The Vascular Cambium**

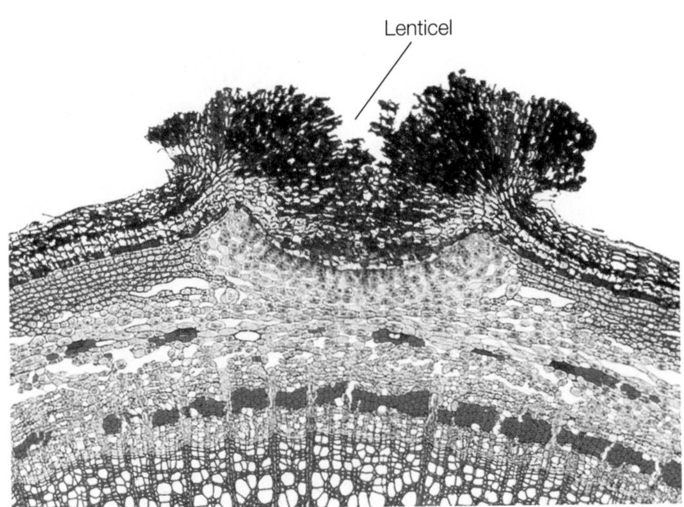


**34.18 Lenticels Allow Gas Exchange through the Periderm** The region of periderm that appears broken open is a lenticel in a year-old elderberry (*Sambucus*) twig; note the spongy tissue that constitutes the lenticel.

When periderm forms on stems or roots, the underlying tissues still need to release carbon dioxide and take up oxygen for cellular respiration. *Lenticels* are spongy regions in the periderm that allow such gas exchange (**Figure 34.18**).

Cross sections of most trunks (mature stems) of trees in temperate-zone forests show annual rings of wood (**Figure 34.19**), which result from seasonal environmental conditions. In spring, when water is relatively plentiful, the tracheids or vessel elements produced by the vascular cambium tend to be large in diameter and thin-walled. Such wood is well adapted for transporting water and minerals. As water becomes less available during the summer, narrower cells with thicker walls are produced, making this summer wood darker and perhaps more dense than the wood formed in spring. Thus each growing season is usually recorded in a tree trunk by a clearly visible annual ring. Trees in the moist tropics do not undergo seasonal growth, so they do not lay down such obvious regular rings. Variations in temperature or water supply can lead to the formation of more than one "annual" ring in a single year, but commonly each year brings a new annual ring and a new batch of leaves.

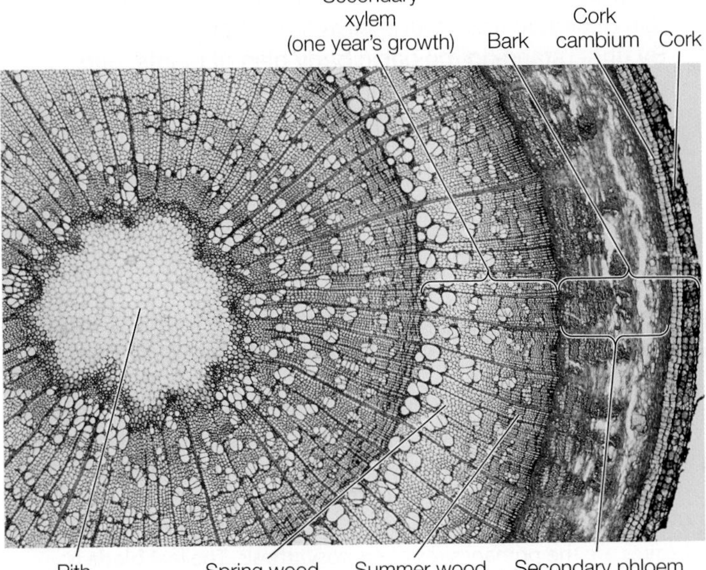


**34.19 Annual Rings** Rings of secondary xylem are the most noticeable feature of this cross section from a tree trunk.

Only eudicots and other non-monocot angiosperms, along with many gymnosperms, have a vascular cambium and a cork cambium and thus undergo secondary growth. The few monocots that form thickened stems—palms, for example—do so without secondary growth. Palms have a very wide apical meristem that produces a wide stem, and dead leaf bases add to the diameter of the stem. All monocots grow in essentially this way, as do other angiosperms that lack secondary growth.

## 34.4 RECAP

**Meristems are localized regions of cell division that are the source of all new organs in the adult plant. Apical meristems are responsible for primary growth, which is associated with the lengthening and branching of shoots and roots. Lateral meristems increase plant girth and form wood and bark in many eudicots.**

- Explain how an apical meristem can be maintained for years while continuing to form leaves. See p. 729 and Figure 34.10
- What cells are derived from the root apical meristem and what is the general process of root growth? See pp. 730-731 and Figure 34.11
- How does the vascular cambium give rise to thicker stems and roots? See p. 733 and Figure 34.16

The building of the plant body by meristems allows a plant to respond to its environment by redirecting its growth. Thus individual plants of the same species can vary greatly in form. What underlies this variation, and how have humans used it to our advantage?

## 34.5 How Has Domestication Altered Plant Form?

We have seen in this chapter that a very simple plant body plan—with roots, stems, leaves, meristems, and relatively few tissue and cell types—underlies the remarkable diversity of the flowering plants that cover our planet. However, while a difference in plant form between members of different species is expected, members of the same species can be remarkably diverse in form as well. From a genetic perspective, this observation suggests that minor differences in gene content or gene regulation can underlie dramatic differences in plant form. (Nevertheless, different plant species do differ greatly sometimes in gene content and genome organization.)

Let's return to the Doomsday Vault. We saw at the opening of this chapter that the vault will be used as a backup for seed banks around the world. Many of these seed banks concentrate on seed from a particular crop species, such as the Maize Stock Center at the University of Illinois (corn) or the Genetic Resources Center of the International Rice Research Institute in the Philippines (rice). In addition to containing large collections of cultivated varieties, these seed banks also contain seeds from populations of their wild relatives.

Why maintain collections of seed from both cultivated crops and their wild relatives? Despite sometimes vast morphological differences, crops and their wild relatives are still members of the same species. As such, when they are crossbred, they can produce viable progeny. These progeny will carry new combinations of their parents' traits.

It is hard to believe that modern corn was domesticated from the wild grass teosinte, which still grows in the hills of Mexico (**Figure 34.20**). One of the most conspicuous differences is that teosinte, like other wild grasses, is highly branched, with many shoots, while domesticated corn has a single shoot. This morphological difference is due in large part to the activity of a single gene called *teosinte branched 1* (*tb1*). The protein product of *tb1* regulates the growth of axillary buds (see Figure 34.1). The allele of *tb1* in domesticated corn represses branching, while the allele in teosinte permits branching.

Even harder to believe is that a single species, *Brassica oleracea* (wild mustard), is the ancestor of so many familiar and morphologically diverse crops such as kale, broccoli, Brussels sprouts, and cabbage (see Figure 21.4). An understanding of how the basic body plan of plants arises makes it possible to appreciate how each of these crops was domesticated. Starting with morphologically diverse populations of the wild ancestor, humans selected and planted the seed from variants with the trait they found desirable. Many generations of such artificial selection produced the crops that fill the produce section of the supermarket or the stands of the farmers' market.

Just as they were for ancient farmers, the genomes of plants are priceless resources today. The genetic variation in crop plants and their wild relatives can be used to improve our crop plants or adapt them to changing conditions. The improvement of crop plants is a work in progress that is being carried out in plant breeding programs worldwide. In fact, these programs are more important than ever. Increased human activity is dramatically changing our planet and leading to the extinction of more and more plant species. When seen in this light, the Doomsday Vault is an insurance policy for our crop plants against the loss of our most valuable resource, the genetic diversity underlying plant form and growth.

**34.20 Modern Corn Was Domesticated from the Wild Grass Teosinte** Beginning more than 8,000 years ago in Mexico, farmers favored plants with minimal branching. Reducing the number of branches results in fewer ears per plant, but allows each ear to grow larger and produce more seeds.

## 34.5 RECAP

**Crop domestication involves artificial selection of certain desirable traits found in wild plant populations. By understanding the basic body plan of plants, one can more easily understand the morphological relationship between a crop plant and its wild relatives.**

- Why is seed from wild relatives of crop plants valuable? See p. 736 and Figure 34.20

# CHAPTER SUMMARY

## 34.1 What Is the Basic Body Plan of Plants?

- The vegetative organs of flowering plants are roots, which form a **root system**, and stems and leaves, which form a **shoot system**. Review Figure 34.1
- The two major clades of flowering plants, eudicots and monocots, differ from each other in a number of structural respects. Most eudicots have a **taproot system**, and most monocots have a **fibrous root system**. Review Figure 34.2
- **Stems** bear undeveloped shoots called buds. **Axillary buds** can develop into new branches. A **terminal bud** is found at the end of a shoot.
- **Leaves** are the primary sites of photosynthesis. The leaf **blade** is attached to the stem by a **petiole**.

## 34.2 How Does the Cell Wall Support Plant Growth and Form?

- Plant cells differ from other eukaryotic cells in having chloroplasts or other plastids, large **central vacuoles**, and cellulose-containing cell walls.
- Once cytokinesis of a plant cell is complete, each daughter plant cell produces a **primary cell wall**. The walls of the two cells are separated by a **middle lamella**. Review Figure 34.4
- The primary cell wall is made up of bundles of cellulose **microfibrils** cross-linked by hemicellulose and pectin. Review Figure 34.5
- The primary cell wall is rigid but dynamic. By loosening the linkages between microfibrils, the cell wall can expand in volume by up to a million times.
- Some cells produce a thick **secondary cell wall**. **Lignin** in the secondary cell wall offers exceptional structural support.
- **Plasmodesmata** connect adjacent plant cells and allow direct communication between them. Review Figure 34.6

## 34.3 How Do Plant Tissues and Organs Originate?

- During embryogenesis, the basal–apical axis and the radial axis of the plant body are established. Review Figure 34.7
- The **shoot apical meristem** and the **root apical meristem** are also established during embryogenesis. These clusters of undifferentiated cells will orchestrate all postembryonic development.
- Three **tissue systems**, arranged concentrically, extend throughout the plant body: the vascular tissue, dermal tissue, and ground tissue systems. Review Figure 34.8
- The **dermal tissue system** protects the plant body surface. Dermal cells form the epidermis and, in woody plants, the periderm.
- The **ground tissue system** contains cells of three types. Some **parenchyma** cells carry out photosynthesis; others store starch. **Collenchyma** cells provide flexible support. **Sclerenchyma** cells include **fibers** and **sclereids** that provide strength and mechanical support. Review Figure 34.9
- The **vascular tissue system** includes **xylem**, which conducts water and minerals absorbed by the roots, and **phloem**, which conducts the products of photosynthesis throughout the plant body.
- **Tracheary elements** include **tracheids** and **vessel elements**, which are the conducting cells of the xylem. **Sieve tube elements** are the conducting cells of the phloem.

## 34.4 How Do Meristems Build a Continuously Growing Plant?

- All seed plants possess a **primary plant body** consisting of nonwoody tissues. Woody plants also possess a **secondary plant body** consisting of wood and bark. **Apical meristems** generate the primary plant body, and **lateral meristems** generate the secondary plant body. Review Figure 34.10
- Apical meristems are responsible for **primary growth** (lengthening of roots and shoots). Apical meristems at the tips of stems and roots give rise to three **primary meristems** (**protoderm**, **ground meristem**, and **procambium**), which in turn produce the three tissue systems of the primary plant body
- The root apical meristem gives rise to the **root cap** and to three primary meristems. Root tips have overlapping **zones of cell division**, **elongation**, and **maturation**. Review Figure 34.11
- The vascular tissue of roots is contained within the **stele**. It is arranged differently in eudicot and monocot roots. Review Figures 34.12 and 34.13, **WEB ACTIVITIES 34.1 and 34.2**
- In nonwoody stems, the vascular tissue is divided into **vascular bundles**, each containing both xylem and phloem. Review Figure 34.14, **WEB ACTIVITIES 34.3 and 34.4**
- Eudicot leaves have two zones of photosynthetic **mesophyll** that are supplied by **veins** with water and minerals. Veins also carry the products of photosynthesis to other parts of the plant body. A waxy **cuticle** limits water loss from the leaf. **Guard cells** control openings (stomata) in the leaf that allow $CO_2$ to enter, but also allow some water to escape. Review Figure 34.15, **WEB ACTIVITY 34.5**
- Two lateral meristems, the **vascular cambium** and **cork cambium**, are responsible for secondary growth. The vascular cambium produces secondary xylem (wood) and secondary phloem. The cork cambium produces a protective tissue called **cork**. Review Figures 34.16 and 34.17, **ANIMATED TUTORIAL 34.1**

## 34.5 How Has Domestication Altered Plant Form?

- The plant body plan is simple, yet it can be changed dramatically by minor differences in genes, as evidenced by the natural diversity of wild plants.
- Crop domestication involves artificial selection of certain desirable traits found in wild populations. Review Figure 34.20

# SELF-QUIZ

1. Which of the following is a difference between monocots and eudicots?
   *a.* Only eudicots have phytomers.
   *b.* Only monocots have shoot and root apical meristems.
   *c.* Monocot stems do not undergo secondary growth.
   *d.* The vascular bundles of monocot stems are commonly arranged as a cylinder.
   *e.* Eudicot embryos commonly have one cotyledon.
2. Roots
   *a.* always form a fibrous root system that holds the soil.
   *b.* possess a root cap at their tip.
   *c.* form branches from axillary buds.
   *d.* are commonly photosynthetic.
   *e.* do not show secondary growth.
3. The primary plant cell wall
   *a.* lies immediately inside the plasma membrane.
   *b.* is an impermeable barrier between cells.
   *c.* is always waterproofed with either lignin or suberin.
   *d.* always consists of a primary wall and a secondary wall, separated by a middle lamella.
   *e.* contains cellulose and other polysaccharides.
4. Which statement about parenchyma cells is *not* true?
   *a.* They are alive when they perform their functions.
   *b.* They typically lack a secondary wall.
   *c.* They often function as storage depots.
   *d.* They are the most common cell type in the plant body.
   *e.* They are found only in stems and roots.

5. Tracheids and vessel elements
   *a.* must die to become functional.
   *b.* are important constituents of all seed plants.
   *c.* have no secondary cell wall.
   *d.* are always accompanied by companion cells.
   *e.* are found only in the secondary plant body.
6. Which statement about meristems is *not* true?
   *a.* They are formed during embryogenesis.
   *b.* They have secondary cell walls.
   *c.* Their cells have small central vacuoles.
   *d.* They are clusters of undifferentiated cells.
   *e.* They retain the ability to produce new cells indefinitely.
7. The pericycle
   *a.* is the innermost layer of the cortex.
   *b.* is the tissue within which lateral roots arise.
   *c.* consists of highly differentiated cells.
   *d.* forms a star-shaped structure at the very center of the root.
   *e.* is waterproofed by suberin.
8. Which of these statements is true of secondary growth, but *not* primary growth?
   *a.* It occurs in eudicots and monocots.
   *b.* It involves the proliferation of roots and shoots through branching.
   *c.* It derives from the vascular cambium and the cork cambium.
   *d.* It occurs in palms.
   *e.* It derives from the shoot apical meristem
9. Periderm
   *a.* contains lenticels that allow for gas exchange.
   *b.* is produced during primary growth.
   *c.* is permanent; it lasts as long as the plant does.
   *d.* is the innermost part of the plant.
   *e.* contains vascular bundles.
10. Which statement about leaf anatomy is *not* true?
    *a.* Opening of stomata is controlled by guard cells.
    *b.* The cuticle is secreted by the epidermis.
    *c.* The veins contain xylem and phloem.
    *d.* The cells of the mesophyll are packed together, minimizing air space.
    *e.* The spines of cacti are actually modified leaves.

## FOR DISCUSSION

1. When a young oak was 5 m tall, a thoughtless person carved his initials in its trunk at a height of 1.5 m above the ground. Today that tree is 10 m tall. How high above the ground are those initials? Explain your answer in terms of plant growth.
2. Distinguish between the primary cell wall and the secondary cell wall. When do secondary walls form? What cell types lack secondary walls?
3. Distinguish between sclerenchyma cells and collenchyma cells in terms of structure and function.
4. Distinguish between primary and secondary growth. Do all angiosperms undergo secondary growth? Explain.
5. What anatomical features make it possible for a plant to retain water? Describe the plant tissues involved and how and when they form.
6. The Doomsday Vault contains the seeds of both domesticated and wild plants. Why is it important to preserve collections of seeds of both domesticated and wild plants? What kinds of situations would necessitate the withdrawal of seeds from the Doomsday vault?
7. Take a walk through a farmer's market or the produce section of a supermarket. Use your knowledge of plant growth and form to figure out what desirable trait was selected to produce some of your favorite vegetables.

## ADDITIONAL INVESTIGATION

Of the approximately 20,000 genes in the sequenced genome of *Arabidopsis thaliana*, over 1,000 are involved in cell wall biosynthesis. Based on the composition, growth, and functions of cell walls, what types of proteins would you predict some of these genes encode?

# 35 Transport in Plants

## Engineering water-conserving crops

Everyone knows that plants need water to grow. However, it may come as a surprise that the cultivation of crop plants consumes far more water than all other human activities combined. Worldwide demand for water is increasing at the same time that supplies are declining. This situation makes it imperative that we understand how plants use water so that we can breed plants that use it more efficiently.

The question of just how much water plants use while they grow was addressed in 1690 by John Woodward, a professor at Cambridge University. He reported that a plant that gained just 1 g in weight used 76,000 g of water over 77 days. He proposed that most of the water taken up by plants was "drawn off and conveyed through the pores of the leaves and exhaled into the atmosphere."

We know now, of course, that much of the mass plants acquire as they grow is due to net fixation of atmospheric $CO_2$ into carbohydrates through photosynthesis. But Woodward nevertheless articulated a crucial insight: plants need to take up a lot of water to grow. Plant biologists have a name for the ratio of net photosynthetic carbon fixation to water uptake: *water-use efficiency*.

Droughts and a dwindling water supply are challenging farmers all over the world. One of the least water-efficient of all crop plants is, unfortunately, one of our most important: rice. Rice plants use up to 3 times more water per unit of growth than other crops such as wheat and maize (corn). The precariousness of heavily water-dependent rice farming was dramatically demonstrated in eastern India between 1997 and 2003, when drought reduced rice production by over 5 million tons—some farmers lost up to 50 percent of their crop.

Clearly, a strain of rice that needs less water would not only make the world supply of rice less vulnerable to drought but also help conserve water for other uses. A team of molecular biologists, plant physiologists, and crop scientists led by Andrew Pereira at Virginia Polytechnic began their quest for such a strain of rice by studying an entirely different plant—the model organism *Arabidopsis thaliana* (thale cress). They searched for genetic variants of *Arabidopsis* that had improved

**Thirsty Rice** Cultivation of rice, the most important food crop in Asia, requires large quantities of water.

**A Need for Improved Water-Use Efficiency in Plants** Rice that could use water more efficiently would be less vulnerable to drought and might help maintain or even increase crop yields.

water-use efficiency. One variant they chose to study was particularly hard to pull out of the ground because of its extensive root system (indicating more capacity for water uptake) and had thick leaves with abundant photosynthetic tissue (indicating prolific photosynthesis). Molecular and physiological characterization of this *Arabidopsis* strain showed that its improved water usage was linked to a mutation in a single gene that codes for a transcription factor. When this gene (called *HARDY*) was isolated and put into rice plants using recombinant DNA technology, the rice plants also were more efficient, and indeed tolerated dry soil much better than their normal counterparts. While the *HARDY* gene may or may not lead to crops with higher water-use efficiency, many laboratories around the world are using *Arabidopsis* to isolate genes that can be used to improve water usage and other important characteristics of crop plants.

**IN THIS CHAPTER** we will consider the uptake of water and minerals from the soil and the transport of these materials up the plant in the xylem. We will also look at the control of evaporative water loss from leaves, and the translocation (movement from one location to another) of dissolved substances in the phloem.

## CHAPTER OUTLINE



## 35.1 How Do Plants Take Up Water and Solutes?

Terrestrial plants must obtain both water and mineral nutrients from the soil, usually through their roots. The roots, in turn, obtain carbohydrates and other important materials from the leaves (**Figure 35.1**). Water is required for carbohydrate production by photosynthesis in leaves (see Section 10.1), for transporting solutes between plant organs, for cooling the plant, and for developing the internal pressure that supports the plant body.

As our opening story conveys, plants lose large quantities of water to evaporation. To balance this loss, an equally large amount of water must be absorbed through the roots, continue up the stem, and be transported into the leaves. The minerals that a plant needs are transported along with the water. Several steps in water and mineral transport will be considered in this chapter. In this section we will focus on the first part of the journey—the uptake of water and minerals into the roots and their transport into the xylem.

### Water potential differences govern the direction of water movement

The process of water uptake by plants requires water to move through at least one, and usually many cell membranes. Accordingly, we will begin our discussion of water transport by examining the rules that govern the movement of water across membranes. As described in Section 6.3, the movement of water through a membrane in accordance with the laws of diffusion is called *osmosis*.

The overall tendency of a solution to take up water from pure water, across a membrane, is called its **water potential** and is represented as $\psi$, the Greek letter *psi* (pronounced "sigh"). The water potential of a solution is measured as the sum of its (negative) solute potential ($\psi_s$) and its (usually positive) pressure potential ($\psi_p$):

$$\psi = \psi_s + \psi_p$$

Whenever water moves by osmosis, the following important rule applies: *water always moves across a selectively permeable membrane toward the region of lower (more negative) water potential.*

We can measure solute potential, pressure potential, and water potential in *megapascals* (MPa), a unit of pressure. Atmospheric pressure, "one atmosphere," is about 0.1 MPa, or 14.7 pounds per square inch; a typical pressure in an automobile tire is about 0.2 MPa.

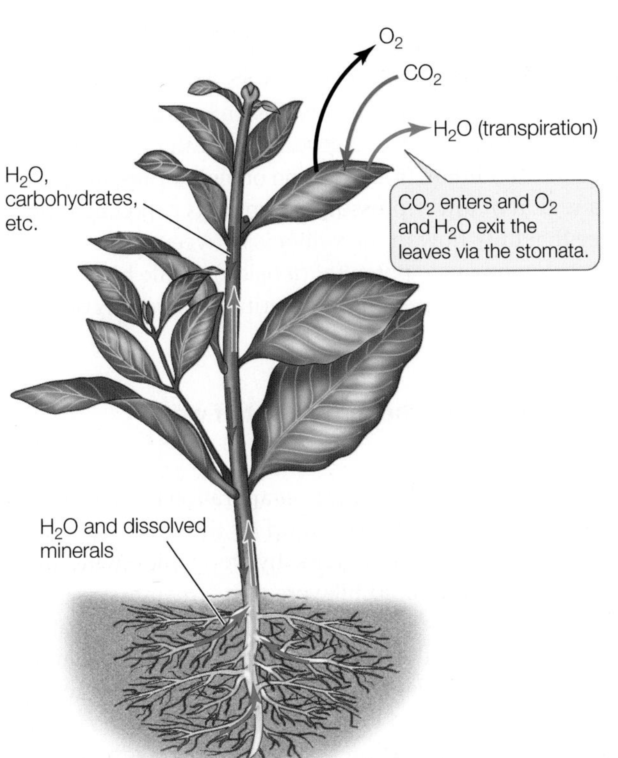


**35.1 The Pathways of Water and Solutes in a Plant** Water travels from the soil to the atmosphere, with only a small fraction used within the plant.

We explore the meaning of the water potential equation in **Figure 35.2**, which assigns values to $\psi_s$ and $\psi_p$ and illustrates how changes in the values of these parameters alter the water potential ($\psi$) and determine the direction of water movement between two compartments (for example, the inside and outside of a plant cell) separated by a semipermeable membrane.

The **solute potential** ($\psi_s$, also called the *osmotic potential*) of a solution is a measure of the effect of dissolved solutes on the osmotic behavior of the solution. The addition of solutes removes free water from the solution because the solute molecules bind water molecules to their surfaces. This is reflected in a more negative value for $\psi_s$ (–0.4 MPa in our example; see Figure 35.2A), which lowers the water potential ($\psi$ = –0.4 MPa) and leads to the movement of water through the membrane to the region of lower $\psi$. Equilibrium is reached when there is no difference in $\psi$ on either side of the membrane (see Figure 35.2B). Now let's see how these same forces determine the direction of water flow through plant cells.

Mature plant cells usually contain a large central vacuole filled with solutes, which are often pumped into the vacuole by transporter proteins. The active accumulation of solutes provides the osmotic force for water uptake into the vacuole. Plant cells are surrounded by a relatively rigid cell wall that resists the expansion of the underlying protoplast. The pressure exerted by the cell wall is equivalent to the positive pressure exerted by the piston on the water column in the idealized example shown in Figure 35.2C. When the pressure potential equals the solute potential, there is no net movement of water through the membrane.

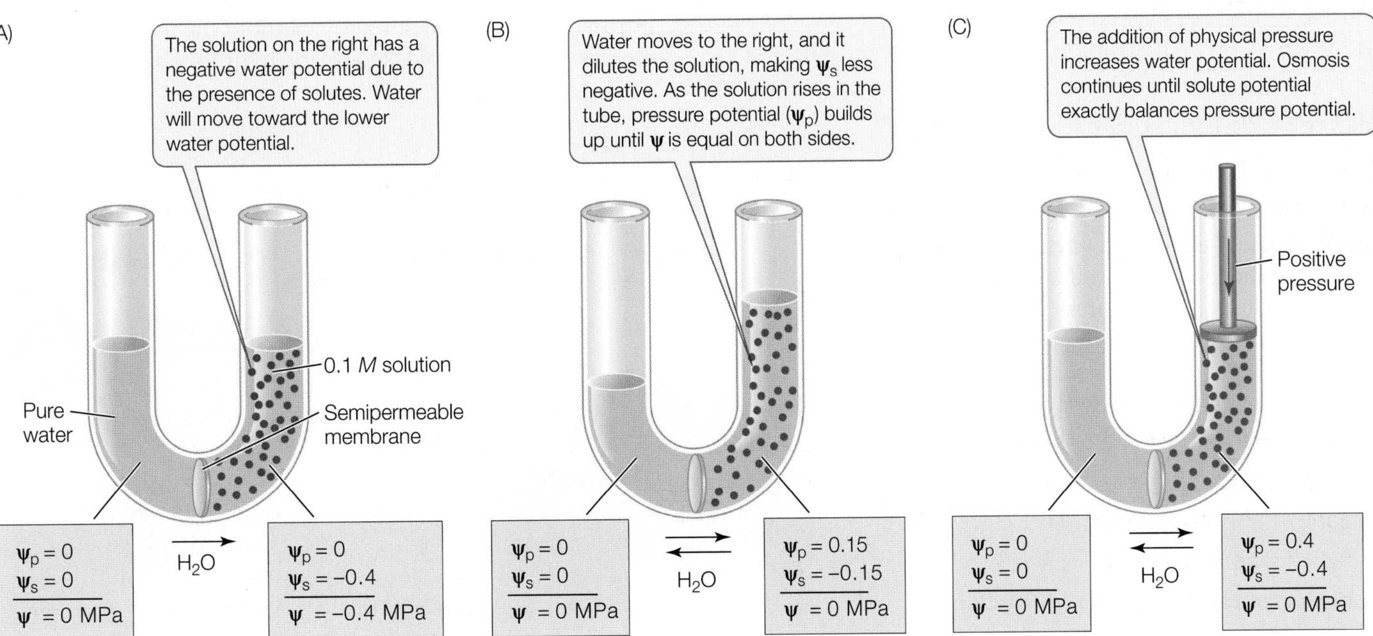


**35.2 Water Potential, Solute Potential, and Pressure Potential** As can be seen in these idealized examples, water flows towards regions of lower water potential ($\psi$), which is the sum of the solute potential ($\psi_s$) and the pressure potential ($\psi_p$). For pure water under no applied pressure, all three of these parameters are equal to zero.

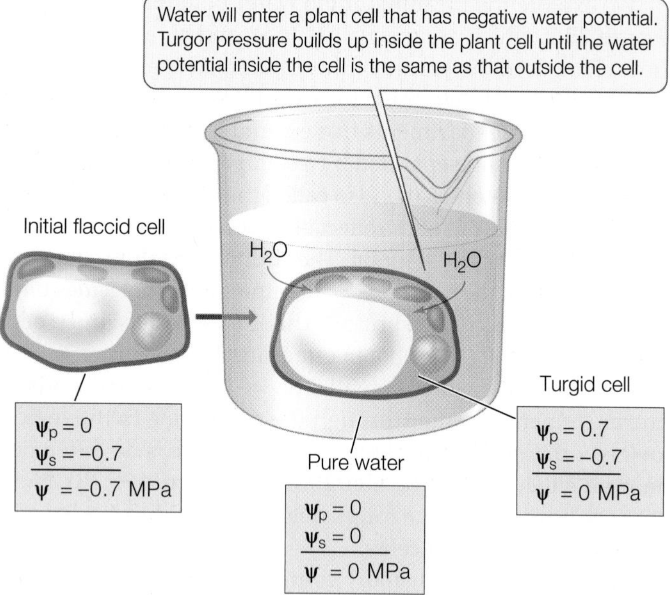


**35.3 Turgor** Turgor pressure builds up inside the cell as the cell wall resists further expansion of the cell.

When the wall of a plant cell is exerting no pressure on the underlying protoplast, the cell is said to be *flaccid* (**Figure 35.3**, left). In this situation,

$$\Psi_s = -0.7 \text{ MPa and } \Psi_p = 0 \text{ MPa}$$

$$\text{So, } \Psi = \Psi_s + \Psi_p = -0.7 \text{ MPa}$$

When a flaccid cell is placed in pure water, water initially moves into the cell due to its negative solute potential (Figure 35.3, right). However, the cell can't expand because it is contained by the cell wall; thus, as water enters, the cell's internal pressure increases and resists the further entry of water.

$$\Psi_s = -0.7 \text{ MPa and } \Psi_p = 0.7 \text{ MPa}$$

$$\text{So, } \Psi = 0$$

This opposing pressure is called **turgor pressure** in plants and is equivalent to the pressure potential ($\psi_p$) exerted by the piston in Figure 35.2C. Water will enter plant cells by osmosis until the pressure potential exactly balances the solute potential. At this point, the cell is **turgid**; that is, it has a significant positive pressure potential. The physical structure of many plants is maintained by the (positive) pressure potential of their cells; if the pressure potential drops (for example, if the plant does not have enough water), the plant *wilts* (**Figure 35.4**).

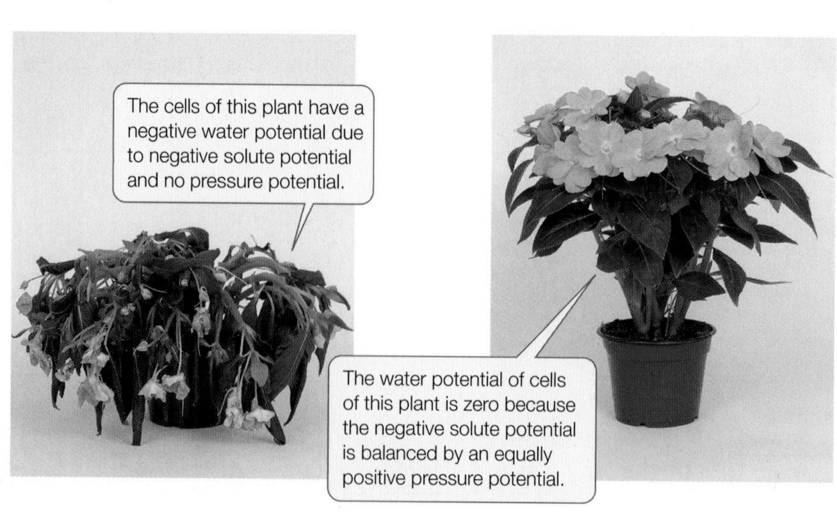

The cells of this plant have a negative water potential due to negative solute potential and no pressure potential.
The water potential of cells of this plant is zero because the negative solute potential is balanced by an equally positive pressure potential.

**35.4 A Wilted Plant** A plant wilts when the pressure potential of its cells is zero.

Within living plant tissues, the movement of water from cell to cell follows a gradient of water potential. Over long distances, in unobstructed tubes such as xylem vessels and phloem sieve tubes, the flow of water and dissolved solutes is driven by a *gradient of pressure potential*, not a gradient of water potential. The movement of a solution from a region of higher pressure potential to a region of lower pressure potential is called **bulk flow**. We'll see that bulk flow in the xylem is between regions of differing *negative* pressure potential (tension), while bulk flow in the phloem is between regions of differing *positive* (turgor) pressure potential.

### Aquaporins facilitate the movement of water across membranes

The large quantities of water lost to evaporation from the leaves must be balanced by water taken up by the roots. Yet only a trickle of water can pass through the hydrophobic environment created by the phospholipid bilayers of cell membranes. How do plants turn this trickle into a gusher? The answer is that water diffuses through cell membranes mainly through channels called **aquaporins** (see Figure 6.13), which are located in both the plasma membrane and the tonoplast (vacuolar membrane) of plant cells. Aquaporins allow water to move rapidly from environment to cell and from cell to cell. The abundance of aquaporins in a plant cell varies with environmental conditions, depending on a cell's need to obtain and retain water. The permeability of some aquaporins also can be regulated. Alterations in aquaporin abundance and permeability change the *rate* of osmosis across the membrane. Note, however, that water movement through aquaporins is always passive, so the *direction* of water movement is unchanged.

### Uptake of mineral ions requires membrane transport proteins

Although water molecules can cross membranes through aquaporins, mineral ions generally cannot. The ions, which carry electric charges, are blocked by the hydrophobic interior of the membrane, and they are too large to pass through aquaporins. Instead, mineral ions generally cross membranes through transport proteins, including ion channels and carrier proteins (see Sections 6.3 and 6.4).

We have just seen that water moves through a water-permeable membrane in response to a water potential gradient. Other molecules and ions also follow their own concentration gradients, as permitted by the characteristics of the membrane. When the concentration of charged ions in the soil is greater than that in the plant, transport proteins can move them into the plant by facilitated diffusion, which is a passive process. The concentrations of most ions in the soil solution, however, are lower than those required inside the plant. In these cases, the plant must actively take up ions *against* their concentration gradients—a process that requires energy.

Electric charge differences also play a role in the uptake of mineral ions. For example, a negatively charged ion that moves into a negatively charged compartment is moving against an *electrical gradient*, and this requires energy. Concentration and electrical gradients combine to form an *electrochemical gradient*. Uptake against an electrochemical gradient involves *active transport*, which requires specific transport proteins and is fueled by ATP generated by cellular respiration.

Unlike animals, plants do not have a sodium–potassium pump (see Section 6.4) for active transport. Rather, plants have a **proton pump**, which uses energy obtained from ATP to move protons out of the cell against a proton concentration gradient (**Figure 35.5, step 1**). Because protons ($H^+$) are positively charged, their accumulation outside the cell has two results:

- An electrical gradient is created such that the region outside the cell becomes more positively charged relative to the region inside.
- A proton concentration gradient develops, with more protons outside the cell than inside.

Each of these results has consequences for the movement of other ions. Because the inside of the cell is now more negative than the outside, cations (positively charged ions) such as potassium ($K^+$) move into the cell by facilitated diffusion through their specific membrane channels (**Figure 35.5, step 2**). In addition, the proton concentration gradient can be harnessed to drive secondary active transport, in which anions (negatively charged ions) such as chloride ($Cl^-$) are moved into the cell against an electrochemical gradient by a symport protein that couples their movement with that of $H^+$ (**Figure 35.5, step 3**).

In sum, there is vigorous traffic of water molecules and mineral ions across plant cell membranes. This traffic involves specific membrane channels and transport proteins, and both active and passive processes. Now we will step back and see how these membrane transport processes participate in the journey of water and nutrients from the soil to the xylem.

### Water and ions pass to the xylem by way of the apoplast and symplast

The journey from the soil through the roots to the xylem occurs primarily by one of two pathways, the fast lane (called the *apoplast*) and the slow(er) lane (called the *symplast*) (**Figure 35.6**):

- The **apoplast** (Greek *apo*, "away from"; *plast*, "living material") consists of the cell walls, which lie outside the plasma membranes, and the intercellular spaces (spaces between cells) that are common in many plant tissues. The apoplast is a continuous meshwork through which water and dissolved substances can flow without ever having to cross a membrane. Movement of materials through the apoplast is thus unregulated and rapid—until it reaches the *Casparian strips* of the endodermis.
- The **symplast** (Greek *sym*, "together with") passes through the continuous cytoplasm of the living cells connected by plasmodesmata. The selectively permeable plasma membranes of the root cells control access to the symplast, so movement of water and dissolved substances into the symplast is tightly regulated.

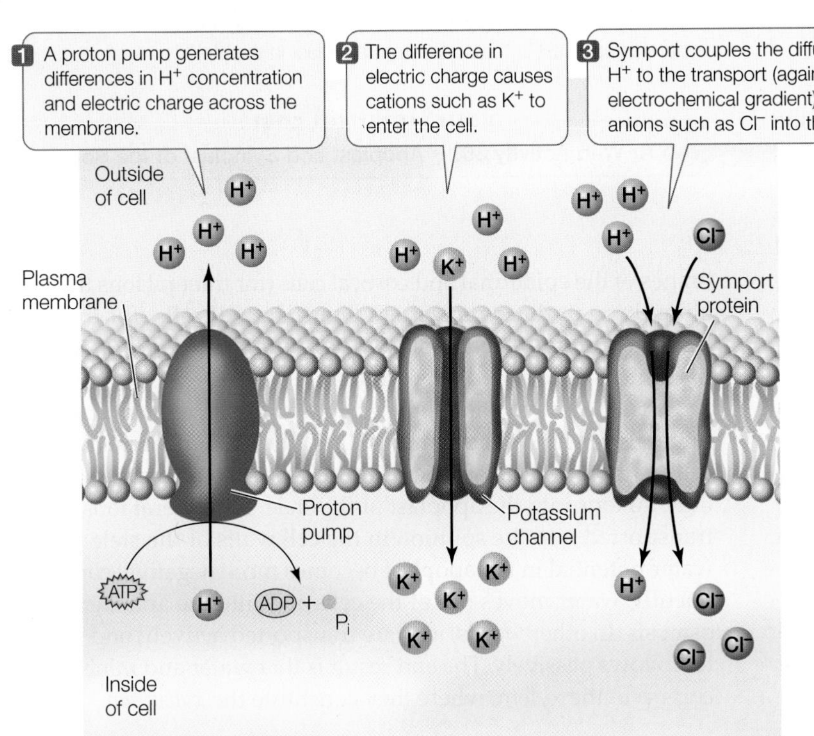

1 A proton pump generates differences in H+ concentration and electric charge across the membrane.
2 The difference in electric charge causes cations such as K+ to enter the cell.
3 Symport couples the diffusion of H+ to the transport (against an electrochemical gradient) of anions such as Cl− into the cell.
Outside of cell
Plasma membrane
Symport protein
Proton pump
Potassium channel
ATP
ADP + Pi
Inside of cell

**35.5 The Proton Pump in Transport of $K^+$ and $Cl^-$**
The active transport of hydrogen ions ($H^+$) out of the cell by the proton pump (1) drives the movement of both cations (2) and anions (3) into the cell.

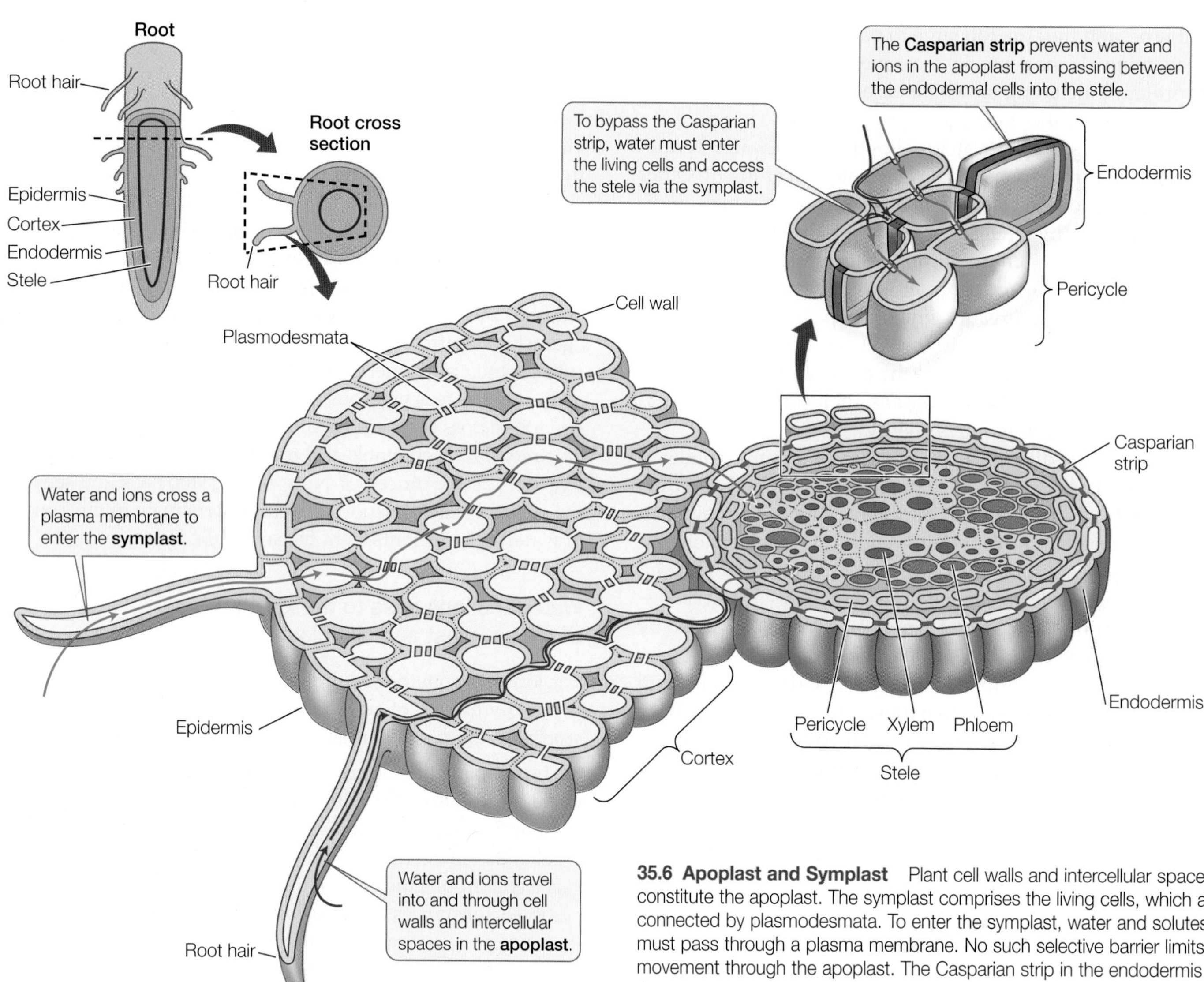


**35.6 Apoplast and Symplast** Plant cell walls and intercellular spaces constitute the apoplast. The symplast comprises the living cells, which are connected by plasmodesmata. To enter the symplast, water and solutes must pass through a plasma membrane. No such selective barrier limits movement through the apoplast. The Casparian strip in the endodermis of the cortex is impregnated with the water-repelling substance suberin and separates apoplast in the cortex from apoplast in the stele.

Water and minerals that pass from the soil solution through the apoplast can travel as far as the endodermis, the innermost layer of the root cortex. The endodermis is distinguished from the rest of the ground tissue by the presence of the **Casparian strip**. This waxy, suberin-impregnated region of the endodermal cell wall forms a water-repelling (hydrophobic) belt around each endodermal cell where it is in contact with other endodermal cells. The Casparian strip acts as a seal that prevents water and ions from moving between the cells (see Figure 35.6).

The Casparian strip of the endodermis completely separates the apoplast of the cortex from the apoplast of the stele. Accordingly, the only way water and ions can enter the stele is by way of the symplast—that is, by entering and passing through the cytoplasm of the endodermal cells. Water and ions already in the symplast can enter the endodermal cells through plasmadesmata, but those in the apoplast must cross the plasma membranes of the endodermal cells (this is possible because the Casparian strip does not obstruct the inner or outer faces of endodermal cells). Thus transport proteins in the plasma membranes of the epidermal and cortical cells (for mineral ions traveling through the symplast) and endodermal cells (for those traveling through the apoplast) determine which mineral ions pass into the stele, and at what rates.

yourBioPortal.com
GO TO Web Activity 35.1 • Apoplast and Symplast of the Root

Once they have passed the endodermal barrier, water and minerals remain in the symplast until they reach parenchyma cells in the pericycle or xylem. These cells then actively export mineral ions into the apoplast of the stele. As mineral ions are transported into the solution in the cell walls of the stele, the water potential in the apoplast becomes more negative; consequently, water moves out of the cells and into the apoplast by osmosis. In other words, ions are transported actively, and water follows passively. The end result is that water and minerals end up in the xylem, where they constitute the *xylem sap*.

## 35.1 RECAP

**Differences in water potential govern the osmotic flow of water from the soil into the plant stele; this is a passive process. Uptake of minerals from the soil occurs along an electrochemical gradient and is therefore an active process requiring energy and membrane transport proteins. Water and minerals can move into the root through either the apoplast or the symplast, but must enter and leave the symplast to reach the xylem.**

- What distinguishes water potential, solute potential, and pressure potential? **See pp. 740–741 and Figure 35.2**
- Explain why the cell wall is so important in determining the direction of water movement. **See pp. 741–742**
- What are aquaporins? Why are they needed? **See p. 742**
- Describe the differences between the apoplast and the symplast. **See p. 743 and Figure 35.6**

So far we've described the movement of water and minerals into plant roots and their entry into the root xylem. How does the xylem sap move once it is in the xylem?

## 35.2 How Are Water and Minerals Transported in the Xylem?

Water has arrived in the xylem—it is all uphill from there! Before we consider the ascent to the leaves, let's revisit the cells that make up the xylem, the xylem vessels (see Figure 34.9E and F). In this section you will learn that the properties of xylem vessels make it possible for water and solutes to be transported efficiently over long distances. Recall that xylem vessels are dead and lack all cell contents. When fused end to end, the xylem vessels form a long tubular "straw" of lignified cell walls that provide both structural support and the rigidity needed to maintain a gradient of pressure.

Consider the magnitude of what xylem accomplishes in transporting a *large amount of water over a great distance* within the plant. A single maple tree 15 meters tall has been estimated to have some 177,000 leaves, with a total leaf surface area of 675 square meters—half again the area of a basketball court. During a summer day, that tree loses 220 liters of water *per hour* to the atmosphere by evaporation from the leaves. So to prevent wilting, the xylem needs to transport 220 liters of water from the roots to the leaves every hour. (By comparison, a 50-gallon drum holds 189 liters.) The tallest leaves can be quite far from the root. The tallest gymnosperm, the coast redwood *Sequoia sempervirens*, and the tallest angiosperm, the Australian *Eucalyptus regnans*, are more than 110 meters tall. Any successful explanation of water transport in the xylem must account for the transport of water to these great heights.

Scientists have proposed various models to explain the ascent of xylem sap. We begin by reviewing some illuminating experiments that ruled out early models, and then turn to evidence in support of the current model.

### Xylem sap is not pumped by living cells

Some of the earliest attempts to explain the rise of sap in the xylem were based on the hypothesis that a pumping action by living cells in the stem might push the sap upward. However, in 1893 the German botanist Eduard Strasburger conducted and published experiments that definitively ruled out such models.

Strasburger worked with trees about 20 meters tall. He sawed through the trunk of each tree at its base and plunged the cut end into a solution of a poison, such as copper sulfate. The solution rose through the trunk, as was evident from the progressive death of the bark higher and higher up. When the solution reached the leaves, the leaves died too, at which point the movement of the solution stopped (as shown by the liquid level in the bucket, which stopped dropping).

This simple experiment established three important points:

- Living, "pumping" cells were not responsible for the upward movement of the solution, because the solution itself killed all living cells with which it came in contact.
- The leaves played a crucial role in transport. As long as they were alive, the solution continued to move upward; when the leaves died, movement ceased.
- The roots did not cause the movement, because the trunk had been completely separated from the roots.

### Root pressure alone does not account for xylem transport

A second hypothesis about xylem transport involved **root pressure**—pressure exerted by the root tissues that would force liquid up the xylem. The basis for root pressure is a higher solute concentration, and accordingly a more negative water potential, in the xylem sap than in the soil solution. This water potential draws water into the stele; once there, the water has nowhere to go but up, so it rises in the xylem vessels.

Root pressure certainly exists—for example, it is responsible for the phenomenon of *guttation*, in which liquid water is forced out through special openings at the margins of leaves. Guttation occurs only when atmospheric humidity is high and soil water is plentiful, conditions that occur most commonly at night. Root pressure is also the source of the sap that oozes from the cut stumps of some plants when their tops are cut off.

Root pressure, however, cannot account for the ascent of sap in trees. Root pressure seldom exceeds 0.1–0.2 MPa (1–2 atmospheres). If root pressure were driving sap up the xylem, we would observe a positive pressure potential in the xylem at all times. In fact, as we are about to see, the xylem sap in most trees is under *tension*—has a *negative* pressure potential—when it is ascending. Furthermore, as Strasburger had already shown, materials can be transported upward in the xylem even when the roots have been removed. If the roots are not pushing the xylem sap upward, what causes it to rise?

### The transpiration–cohesion–tension mechanism accounts for xylem transport

The current model of xylem transport relies on an alternative to pushing: pulling. The evaporative loss of water from the leaves indirectly generates a pulling force—**tension**—on the water in the apoplast of the leaves, which pulls the xylem sap upward. Hydrogen bonding between water molecules makes the sap in the xylem cohesive enough to withstand the tension and rise by bulk flow. Let's see how this process works (**Figure 35.7**).

The concentration of water vapor in the atmosphere is lower than that in the leaf. Because of this difference, water vapor diffuses from the intercellular spaces of the leaf, through the *stomata* (which we will consider in more detail later) to the outside air, in a process called **transpiration**. Within the leaf blade, water evaporates from the moist walls of the mesophyll cells and enters the intercellular spaces. As water evaporates from the aqueous film coating each cell, the film shrinks back into tiny spaces in the cell walls, increasing the curvature of the water surface and thus increasing its surface tension. This increased tension (negative pressure potential) in the surface film draws more water into the cell walls, replacing that which was lost. The resulting tension in the mesophyll draws water from the xylem of the nearest vein into the apoplast surrounding the mesophyll cells. The removal of water from the veins, in turn, establishes tension on the entire column of water contained within the xylem, so that the column is drawn upward all the way from the roots.

Water can be pulled upward through tiny tubes because of the remarkable **cohesion** of water—the tendency of water molecules to stick to one another by hydrogen bonding (see Section 2.4 and Figure 35.7, step 6). The narrower the tube, the greater the tension the water column can withstand without breaking. The integrity of the column is also maintained by the *adhesion* of water to the xylem walls.

In summary, the key elements of water transport in the xylem are:

- *Transpiration* of water molecules from the leaves by evaporation
- *Tension* in the xylem sap resulting from transpiration from the leaves
- *Cohesion* of water molecules in the xylem sap, from the leaves to the roots

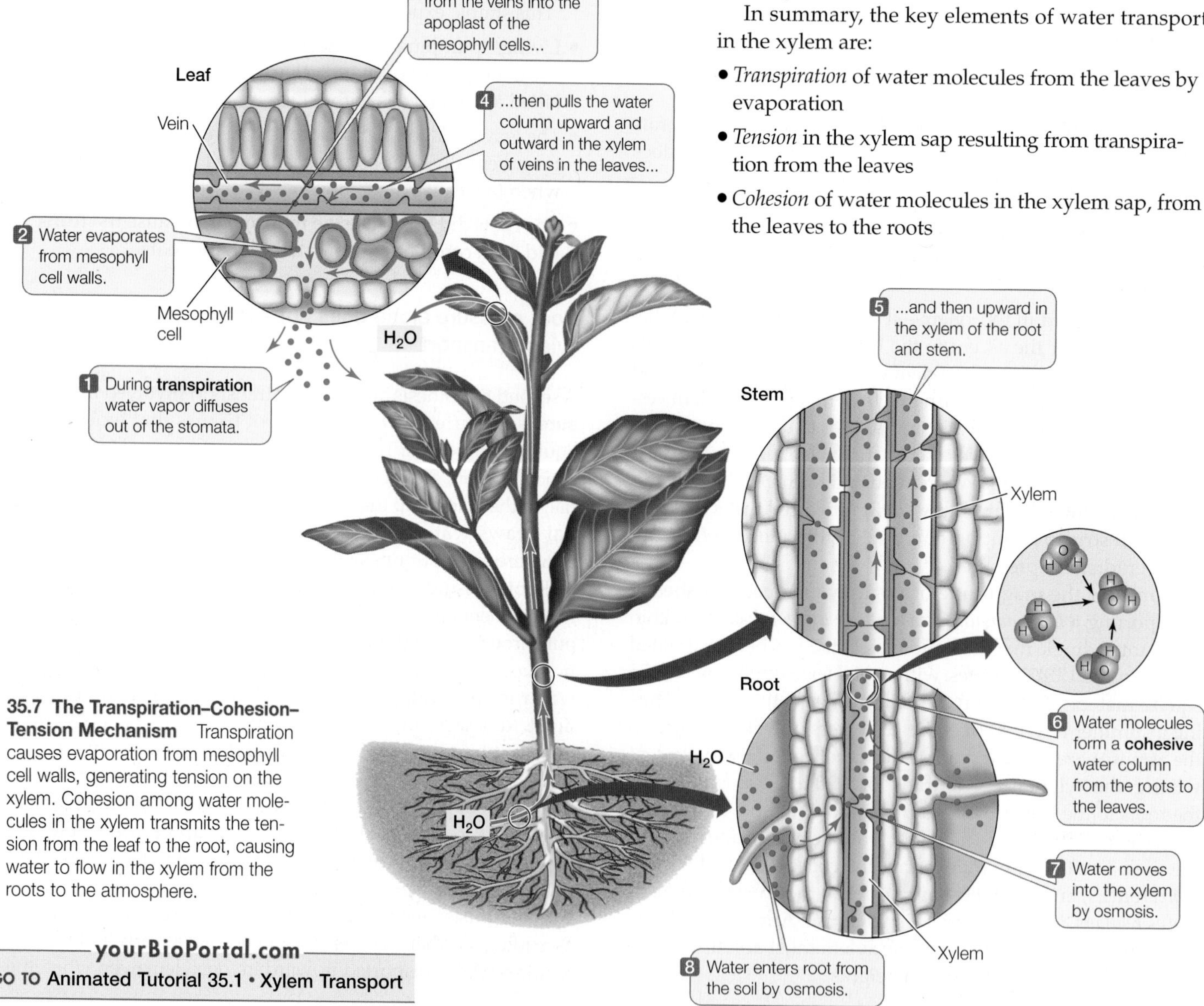
3 Tension pulls water from the veins into the apoplast of the mesophyll cells...
Leaf
Vein
4 ...then pulls the water column upward and outward in the xylem of veins in the leaves...
2 Water evaporates from mesophyll cell walls.
Mesophyll cell
H2O
1 During transpiration water vapor diffuses out of the stomata.
5 ...and then upward in the xylem of the root and stem.
Stem
Xylem
Root
H2O
H2O
6 Water molecules form a cohesive water column from the roots to the leaves.
7 Water moves into the xylem by osmosis.
Xylem
8 Water enters root from the soil by osmosis.

**35.7 The Transpiration–Cohesion–Tension Mechanism** Transpiration causes evaporation from mesophyll cell walls, generating tension on the xylem. Cohesion among water molecules in the xylem transmits the tension from the leaf to the root, causing water to flow in the xylem from the roots to the atmosphere.

yourBioPortal.com
GO TO Animated Tutorial 35.1 • Xylem Transport

The water transport process we have described, called the **transpiration–cohesion–tension mechanism**, *requires no work (that is, no expenditure of energy) on the part of the plant*. At each step between soil and atmosphere, water moves passively; first toward a region with lower water potential, and then to a region of lower pressure potential. Dry air has the most negative water potential (–95 MPa at 50% relative humidity), and the soil solution has the least negative water potential (between –0.01 and –3 MPa). Xylem sap has a water potential more negative than that of cells in the cortex of the root, but less negative than that of mesophyll cells in the leaf.

In the tallest trees, such as a 110-meter *Sequoia*, the difference in pressure potential between the top and the bottom of the column may be as great as 3 MPa. Compare this to root pressure and the pressure in a typical automobile tire, which seldom exceed 0.2 MPa. The cohesion of water in the xylem is great enough to withstand the huge tensions that develop in the tallest trees.

Mineral ions contained in the xylem sap rise passively with water as it ascends from root to leaf. In this way the nutritional needs of the shoot are met. Some of the mineral elements brought to the leaves are subsequently redistributed to other parts of the plant by way of the phloem, but the initial delivery from the roots is through the xylem.

In addition to promoting the transport of minerals, transpiration has an added benefit of cooling a plant's leaves. The evaporation of water from mesophyll cells consumes heat, thereby decreasing the leaf temperature. A farmer can hold a leaf between thumb and forefinger to estimate its temperature; if the leaf doesn't feel cool, that means that transpiration is not occurring and it must be time to water.

The cooling effect of evaporation (also evident in the cooling of our skin when we sweat) may also be important in enabling plants to live in hot environments. However, while transpiration may lead to the cooling of the leaf, this effect is a consequence of the need to transpire, not a reason for it.

## TOOLS FOR INVESTIGATING LIFE

**35.8 Measuring the Pressure of Xylem Sap with a Pressure Chamber** Xylem sap pulls away from a cut stem because the pressure in the intact xylem is lower than that of the atmosphere. The negative pressure potential originally present in the plant can be measured in a pressure chamber in which the pressure can be raised. The cut surface remains outside the chamber. As gas pressure increases, the xylem sap is pushed back to the cut surface. When the sap first becomes visible again at the cut surface, the pressure in the chamber is recorded. This pressure is equal in magnitude but opposite in sign to the tension (negative pressure potential) originally present in the xylem.

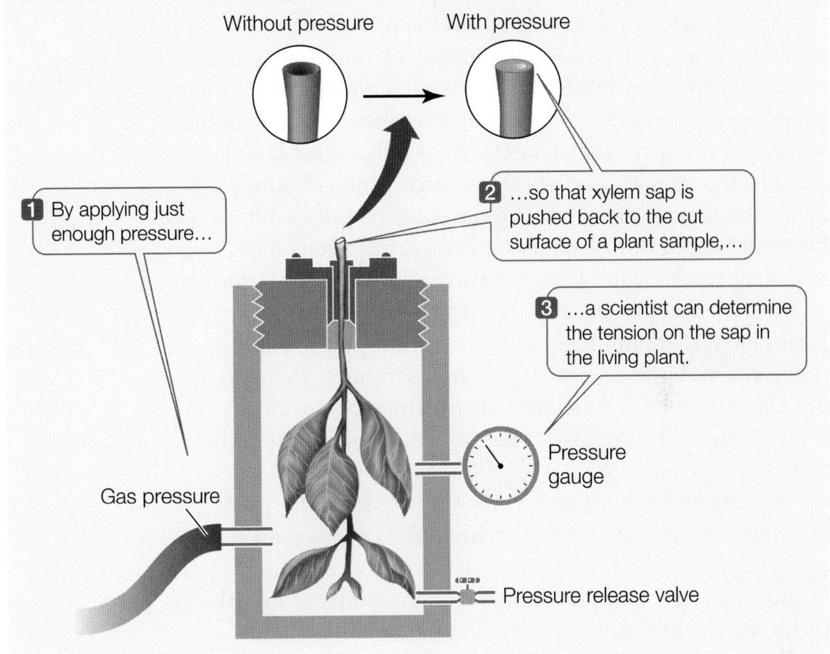

Without pressure
With pressure
1 By applying just enough pressure...
2 ...so that xylem sap is pushed back to the cut surface of a plant sample,...
3 ...a scientist can determine the tension on the sap in the living plant.
Pressure gauge
Gas pressure
Pressure release valve


### A pressure chamber measures tension in the xylem sap

The transpiration–cohesion–tension model holds true only if the column of sap in the xylem is under tension (has a negative pressure potential). The most elegant demonstrations of this tension, and of its adequacy to account for the ascent of xylem sap in tall trees, were performed by the biologist Per Scholander, who measured tension in stems with an instrument called a **pressure chamber** (**Figure 35.8**).

Scholander used the pressure chamber to study dozens of plant species from diverse habitats, growing under a variety of conditions. The rate at which xylem sap ascends is not the same at all times. No flow of xylem sap takes place at night, when there is little or no transpiration. By day, when the sap is ascending, the rate of ascent depends on several factors. These include temperature, light intensity, and wind velocity, all of which affect the transpiration rate, and hence the rate of sap flow. In addition, Scholander found that the xylem sap in developing vines was not under tension until leaves formed. Once leaves developed, the tension increased and transport in the xylem began.

## 35.2 RECAP

**The transpiration–cohesion–tension mechanism explains the ascent of xylem sap. Transpiration draws water out of leaves, resulting in tension that pulls water from the xylem. Because of cohesion between water molecules, water is pulled passively through the xylem vessels in continuous columns, always toward a region with lower pressure potential.**

- What are the roles of transpiration, cohesion, and tension in xylem transport? **See p. 746 and Figure 35.7**
- What properties of the water molecule contribute to cohesion and tension? **See p. 746**
- Describe how mineral ions get from the roots to the leaves. **See p. 747**

Although transpiration provides the driving force for the transport of water and minerals in the xylem, it also results in the loss of tremendous quantities of water from the plant. How plants control this loss is the subject of the next section.

## 35.3 How Do Stomata Control the Loss of Water and the Uptake of $CO_2$?

The epidermis of leaves and stems minimizes transpirational water loss by secreting a waxy cuticle, which is impermeable to water. However, the cuticle is also impermeable to carbon dioxide. This poses a problem: how can the plant balance its need to retain water with its need to obtain $CO_2$ for photosynthesis?

An elegant compromise has evolved in plants in the form of pores called **stomata** (singular *stoma*) in the epidermis of their leaves. A pair of specialized epidermal cells, called **guard cells**, controls the opening and closing of each stoma (**Figure 35.9A**). When the stomata are open, $CO_2$ can enter the leaf by diffusion—but water vapor diffuses out of the leaf at the same time. Closed stomata prevent water loss, but also exclude $CO_2$ from the leaf.

Most plants open their stomata only when the light intensity is sufficient to maintain a moderate rate of photosynthesis. At night, when darkness precludes photosynthesis, their stomata are closed; no $CO_2$ is needed at this time, and water is conserved. Even during the day, the stomata close if water is being lost at too great a rate.

Stomata are ancient structures that are found in plant fossils that are over 400 million years old. For this reason, they are thought to predate the evolution of leaves. Stomata are found in all vascular plants and in many nonvascular plants, including mosses (but not liverworts; see Chapter 28).

The stoma and guard cells seen in Figure 35.9A are typical of eudicots. Monocots typically have specialized epidermal cells associated with their guard cells. However, the principle of operation, which we will now describe in more detail, is the same for both monocot and eudicot stomata.

### The guard cells control the size of the stomatal opening

Light causes the stomata of most plants to open, admitting $CO_2$ for photosynthesis. Another cue for stomatal opening is the level of $CO_2$ in the intercellular spaces inside the leaf. A low level favors opening of the stomata, thus allowing the uptake of more $CO_2$.

Stomata can respond to changes in light and $CO_2$ in a matter of minutes. How can such an important biological process happen so rapidly? The answer is that the opening and closing of stomata is controlled by turgor pressure changes in the guard cells. Changes in turgor pressure are in turn driven by changes in $K^+$ concentration in the guard cells. Blue light, absorbed by a pigment in the guard cell plasma membrane, activates a proton pump, which actively transports $H^+$ out of the guard cells and into the apoplast of the surrounding epidermis. The resulting electrochemical gradient drives $K^+$ into the guard cells, where it accumulates (**Figure 35.9B**). The increased internal concentration of $K^+$ makes the water potential of the guard cells more negative. Negatively charged chloride ions and organic ions also move into and out of the guard cells along with the potassium ions, maintaining electrical balance and contributing to the change in the solute potential of the guard cells. Water enters by osmosis (guard cell membranes are particularly

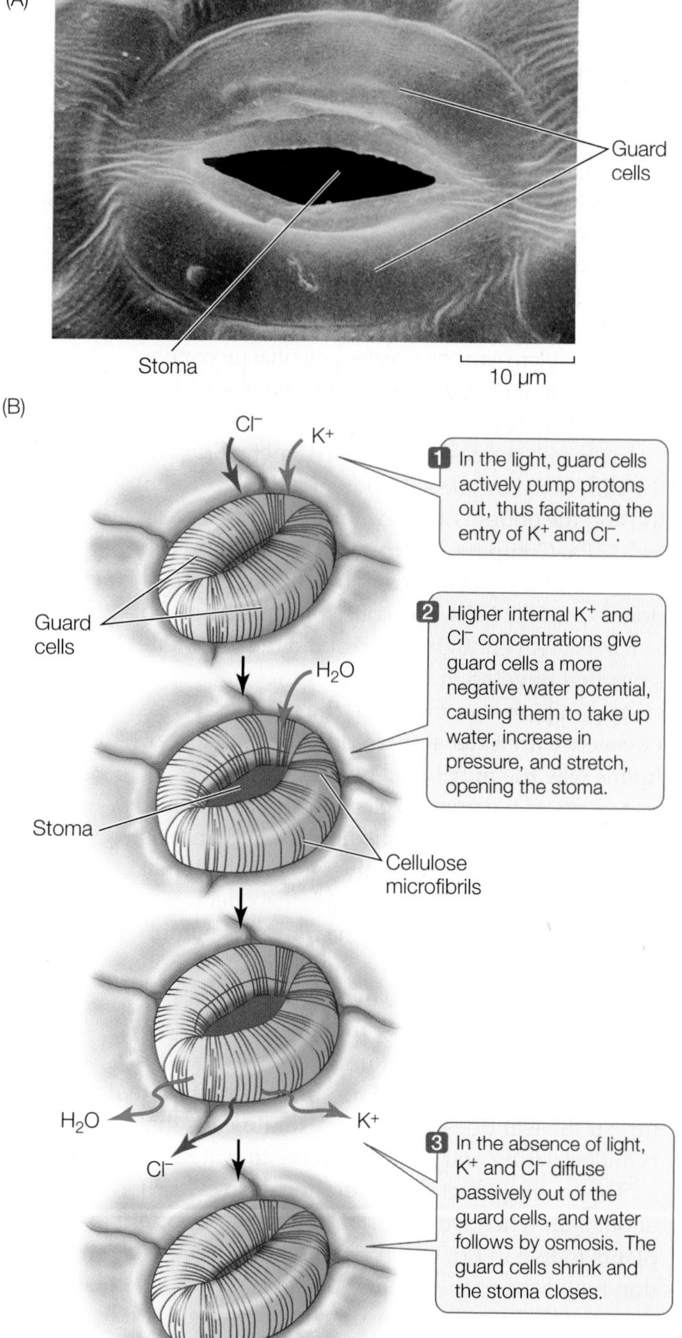

(A)
Guard cells
Stoma
10 μm
(B)
Cl⁻
K⁺
1 In the light, guard cells actively pump protons out, thus facilitating the entry of K⁺ and Cl⁻.
Guard cells
H₂O
2 Higher internal K⁺ and Cl⁻ concentrations give guard cells a more negative water potential, causing them to take up water, increase in pressure, and stretch, opening the stoma.
Stoma
Cellulose microfibrils
H₂O
K⁺
Cl⁻
3 In the absence of light, K⁺ and Cl⁻ diffuse passively out of the guard cells, and water follows by osmosis. The guard cells shrink and the stoma closes.

**35.9 Stomata** (A) Scanning electron micrograph of an open stoma formed by two sausage-shaped guard cells. (B) Potassium ion concentrations affect the water potential of the guard cells, controlling the opening and closing of stomata. Negatively charged ions (e.g., $Cl^-$) that accompany $K^+$ maintain electrical balance and contribute to the changes in water potential that open and close the stomata.

rich in aquaporin protein channels), increasing the pressure potential of the guard cells. The arrangement of the cellulose microfibrils in their cell walls (see Figure 34.5) is such that the guard cells change shape in response to the increase in pressure potential, so that a gap—the stoma—appears between them.

The stoma closes in the absence of blue light. The proton pump becomes less active, potassium ions diffuse passively out of the guard cells, water follows by osmosis, the pressure potential decreases, and the guard cells sag together and seal off the stoma.

INVESTIGATING LIFE

**35.10 Measuring Potassium Ion Concentration in Guard Cells**

G. D. Humble and Klaus Raschke used the electron probe microanalyzer to examine individual stomata of the broad bean. In electron probe microanalysis, electron bombardment of the sample causes it to emit X rays. The wavelength and intensity of the lines in the X-ray spectrum can be analyzed to identify the elements present in the specimen and estimate their concentrations.

**HYPOTHESIS** **Guard cells of open stomata contain more potassium ions than do those of closed stomata.**

**METHOD**

1. Peel strips of epidermis from leaves of broad beans in the dark (closed stomata) and in the light (open stomata).
2. Examine the strips to locate stomata.
3. Scan across guard cells with the electron probe microanalyzer set to measure $K^+$ concentration.

**RESULTS**

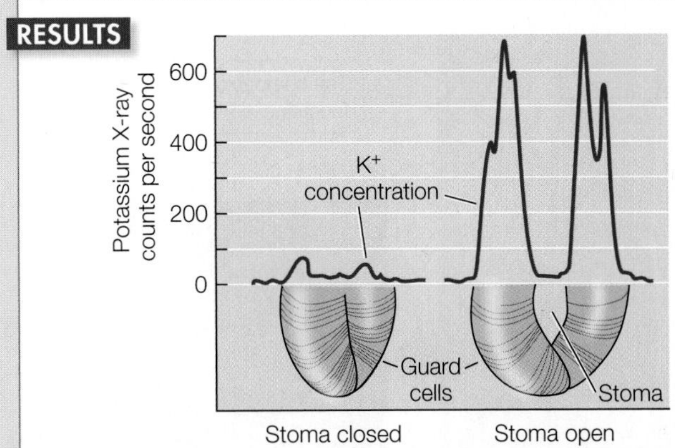


**CONCLUSION** **$K^+$ concentration within the guard cells surrounding an open stoma was much greater than that in the guard cells surrounding a closed stoma.**

**FURTHER INVESTIGATION:** What other ion or ions would you study in order to further explore the mechanism of stomatal opening?

Go to **yourBioPortal.com** for original citations, discussions, and relevant links for all INVESTIGATING LIFE figures.

Showing how much potassium moves into the guard cells to open a stoma was a difficult feat. A typical guard cell is very small, with a total volume of less than 0.03 nanoliters when the stoma is closed and almost 0.05 nanoliters when it is open. The scientists who solved the problem used an *electron probe microanalyzer*, an instrument normally used by metallurgists to study the fine structure of alloys (**Figure 35.10**).

Individual stomata are tiny and plants lose large amounts of water (a single corn plant can lose 2 quarts per day); even so, you may find it surprising that there can be up to 250,000 stomata per square inch of leaf surface! To survive, plants limit water loss by controlling stomata in two very different ways:

- By regulating stomatal opening and closing
- By controlling the total number of stomata

The opening and closing of stomata is regulated by several environmental and endogenous factors. For example, water stress is a common problem for plants, especially on hot, windy days. In response to these conditions, plants will close their stomata, even when the sun is shining. The water potential of the mesophyll cells is the cue for this protective response. If the mesophyll is too dehydrated—that is, if its water potential is too negative—its cells release a plant hormone called *abscisic acid*. Abscisic acid causes the guard cells to close the stomata and prevent further drying of the leaf. This response reduces the rate of photosynthesis, but it protects the plant. Stomata also close in most plants when $CO_2$ levels in the mesophyll spaces are high. Can you think of a reason why this makes good biological sense?

A plant can also reduce its total number of stomata when water is in short supply. Trees can do this by losing some of their leaves. Other plants reduce the number of stomata on the new leaves they produce. For example, if the model plant *Arabidopsis* is exposed to high $CO_2$ levels, the new leaves that form on the plant have fewer stomata than they would have had under normal conditions. Why do you think this might be advantageous?

## 35.3 RECAP

**Leaf pores called stomata admit the $CO_2$ needed for photosynthesis and also permit the exit of water by transpiration. Stomata can be opened or closed by guard cells to regulate water loss.**

- What is the role of $K^+$ ions in the functioning of guard cells? **See p. 748 and Figure 35.9**
- Describe how the external environment (including $CO_2$ level and light intensity) can affect stomatal function and number during the life of a plant. **See p. 749**

Stomata are normally open during daylight hours, allowing $CO_2$ to be fixed and converted to the products of photosynthesis. In the next section we'll see how these products are delivered to other parts of the plant, supporting plant growth.

## 35.4 How Are Substances Translocated in the Phloem?

Photosynthesis occurs primarily in the leaf (see Figure 10.1). The carbohydrate products of photosynthesis (mainly sucrose) diffuse to the nearest small vein (composed of xylem and phloem), where they are actively transported into sieve tube elements. The movement of carbohydrates and other solutes through the plant in the phloem is called **translocation**. Phloem content has several names, including *phloem sap*, *photosynthate*, and *assimilates*.

Substances in the phloem are translocated from *sources* to *sinks*.

- A **source** is an organ (such as a mature leaf or a storage root) that *produces* (by photosynthesis or by digestion of stored reserves) more sugars than it requires.
- A **sink** is an organ (such as a root, flower, developing fruit or tuber, or immature leaf) that *consumes* sugars for its own growth and storage needs.

Sugars (primarily sucrose), amino acids, some minerals, and a variety of other solutes are translocated between sources and sinks in the phloem. However, an organ that is a sink can turn into a source. For example, storage roots (such as sweet potatoes) are sinks when they accumulate carbohydrates but are sources when the stored reserves are needed to nourish other organs in the plant.

How do we know that such organic solutes are translocated in the phloem, rather than in the xylem? Over 300 years ago, the Italian scientist Marcello Malpighi performed a classic experiment in which he removed a ring of bark (containing the phloem) from the trunk of a tree—that is, he *girdled* the tree (**Figure 35.11**). Over time, the bark in the region above the girdle swelled. We now know that the swelling resulted from the accumulation of organic solutes that came from higher up the tree and could no longer continue downward because of the disruption of the phloem. Later, the bark below the girdle died because it no longer received sugars from the leaves. Eventually the roots, and then the entire tree, died.

Any explanation of the translocation of organic solutes must account for a few important observations:

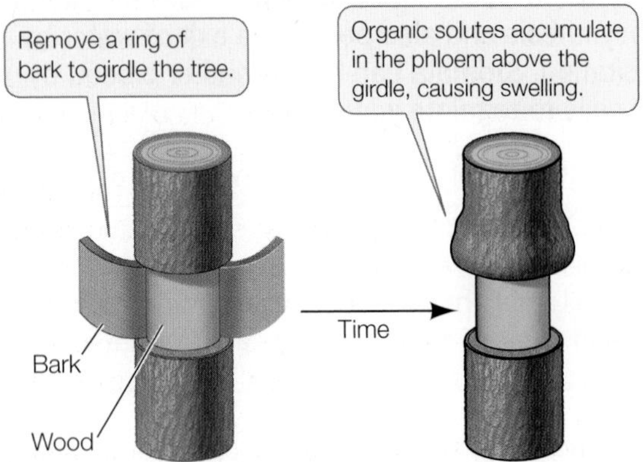


**35.11 Girdling Blocks Translocation in the Phloem** By girdling—removing a ring of bark containing the phloem—Malpighi blocked the translocation of organic solutes in a tree. Bark below the girdle died because it no longer received nutrients; eventually the entire tree died.

- Translocation stops if the phloem tissue is killed by heating or other methods; thus the mechanism must be different from that of transport in the xylem. Recall that xylem is composed of dead cells.
- Translocation often proceeds in both directions—with some phloem transporting up the stem and parallel phloem transporting down the stem. The direction depends on the location of sources and sinks.
- Translocation is inhibited by compounds that inhibit respiration and thus limit the ATP supply in the source. Thus transport in the phloem, unlike the xylem, depends on the input of energy.

Let's first revisit the structure of the phloem to find clues to how it functions. Recall from Chapter 34 that the characteristic cells of the phloem are **sieve tube elements** (see Figure 34.9G). Like vessel elements, these cells meet end-to-end. However, unlike vessel elements, which break down their end walls as they mature, sieve tube elements contain plasmodesmata in the end walls. During sieve tube development, the diameter of these plasmodesmata increases 10- to 100-fold to form pores, enhancing the connection between neighboring cells. The result is end walls that look like sieves, called **sieve plates** (**Figure 35.12**).

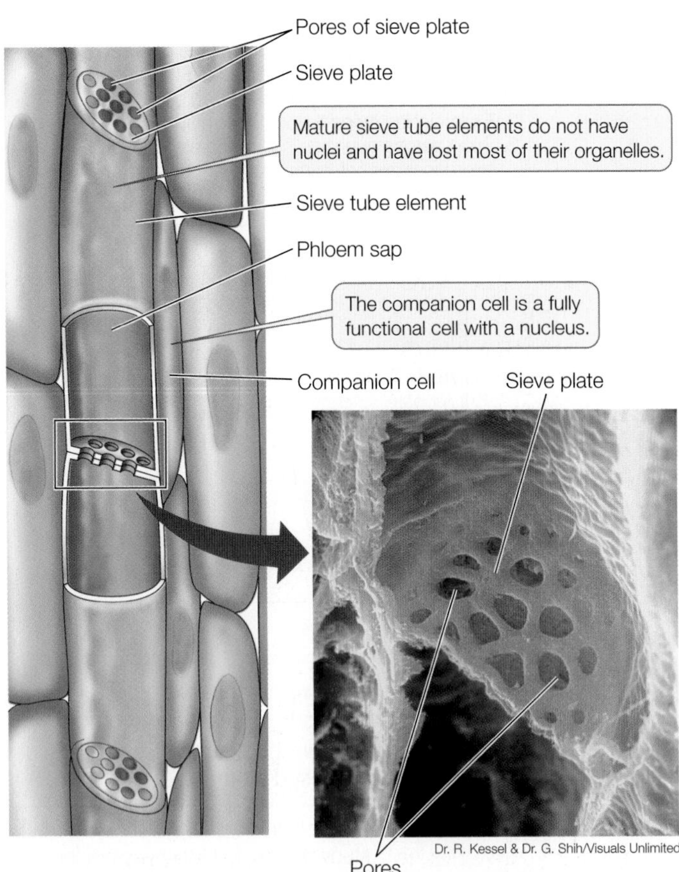


**35.12 Sieve Tubes** Individual sieve tube elements join together to form long tubes that transport carbohydrates and other nutrient molecules throughout the plant body in the phloem. Sieve plates form at the ends of each sieve tube element, and phloem sap passes through the pores in the sieve plate.

What happens next is truly remarkable and makes sieve tube elements among the most unusual cell types in nature. As the holes in the sieve plates expand, most of the cell contents are lost, including the nucleus, Golgi apparatus, and most of the ribosomes and cytoskeleton. Despite this, sieve tube elements live for an entire growing season in trees and for decades in some cases. How can they live for so long with no nucleus? The answer is that each sieve tube element has one or more **companion cells** (see Figure 35.12), produced as daughter cells along with the sieve tube element when a parent cell divides. Numerous plasmodesmata link a companion cell with its sieve tube element. Companion cells retain all their organelles and, through the activities of their nuclei, provide all the functions needed to maintain the sieve tube elements—they may be thought of as the "life support systems" of the sieve tube elements.

**35.13 Aphids Collect Sap** Aphids feed on sap drawn from a sieve tube, which they penetrate with a modified feeding organ, the stylet. Pressure inside the sieve tube forces sap through the aphid's digestive tract.

A mature sieve tube element is filled with phloem sap consisting of water, dissolved sugars, and other solutes. This solution moves through the sieve tube within the symplast, moving from cell to cell through enlarged plasmodesmata. However, because of the unique structural features of the living sieve tubes, phloem sap is able to move rapidly by bulk flow (like xylem sap).

Now that we have a better picture of the structure of the phloem, let's consider the experiments that led to our current understanding of phloem function.

To investigate translocation, plant physiologists needed to obtain samples of pure sieve tube sap from individual sieve tube elements. This difficult task was simplified when it was discovered that a common garden pest, the aphid, feeds on plants by drilling into a sieve tube with its specialized feeding organ, the *stylet*. The pressure within the sieve tube is higher than outside the plant, so the nutritious sieve tube sap is forced through the stylet and into the aphid's digestive tract. So great is the pressure that sugary liquid is forced through the insect's body and out the anus (**Figure 35.13**). This works because the phloem sap is under strongly positive pressure, unlike the negative pressure potential in the xylem. (We will discuss the forces underlying phloem movement a bit later in this section.)

Plant physiologists use aphids to collect phloem sap. When liquid appears on the aphid's abdomen, indicating that the insect has connected with a sieve tube, the physiologist quickly freezes the aphid and cuts its body away from the stylet, which remains in the sieve tube element and can exude phloem sap for hours. Chemical analysis of phloem sap collected from such a stylet reveals the contents of a single sieve tube element over time. Physiologists can also infer the rates at which different substances are translocated by measuring how long it takes for radioactive tracers administered to a leaf to appear at stylets at different distances from the leaf.

Recall that different substances can move in opposite directions in the phloem of a stem. Experiments with aphid stylets have shown that all the contents of any given sieve tube element move in the same direction. Thus, to account for bidirectional translocation in the phloem, different sieve tubes must be conducting sap in opposite directions. These and other experiments led to the general adoption of the *pressure flow model* as an explanation for translocation in the phloem.

### The pressure flow model appears to account for translocation in the phloem

As briefly noted above, phloem sap flows under positive pressure through the sieve tubes, moving from one sieve tube element to the next by bulk flow through the sieve plates, without crossing a membrane. We need to understand how this pressure is generated in order to understand translocation in the phloem.

Two steps in translocation require metabolic energy:

- Transport of sucrose and other solutes from sources into the sieve tubes; called *loading*
- Removal of the solutes from the sieve tubes into sinks; called *unloading*

According to the **pressure flow model** of translocation in the phloem, sucrose is actively transported into sieve tube elements at a source, giving those cells a greater sucrose concentration than the surrounding cells (think lower solute potential—more negative $\psi_s$). Water therefore enters the sieve tube elements from xylem vessels by osmosis. The entry of this water causes a greater pressure potential (turgor pressure) at the source end of the sieve tube, so that the entire fluid content of the sieve tube is pushed toward the sink end of the tube—in other words, the sap moves by bulk flow in response to a pressure gradient (**Figure 35.14**). In the sink, the sucrose is unloaded both passively and by active transport and water moves back to xylem vessels. In this way the gradient of solute potential and pressure potential needed for movement of phloem sap is maintained.

The pressure flow model of translocation in the phloem is contrasted with the transpiration–cohesion–tension model of xylem transport in **Table 35.1**.

TABLE 35.1

**Mechanisms of Sap Flow in Plant Vascular Tissues**

| | XYLEM | PHLOEM |
|---|---|---|
| Driving force for bulk flow | Transpiration from leaves | Active transport of sucrose at source |
| Site of bulk flow | Nonliving vessel elements and tracheids | Living sieve tube elements |
| Pressure potential in sap | Negative (pull from top; tension) | Positive (push from source; pressure) |

### The pressure flow model has been experimentally tested

Even though the pressure flow model was first proposed more than half a century ago, some of its features are still being debated. Other mechanisms have been proposed to account for translocation in sieve tubes, but some have been disproved, and others do not have as much support as the pressure flow model.

Two requirements must be met in order for the pressure flow model to be valid:

- The sieve plates must be unobstructed, so that bulk flow from one sieve tube element to the next is possible.
- There must be an effective method for loading sucrose and other solutes into the phloem in source tissues and unloading them in sink tissues.

Let's see whether these requirements are met.

**ARE THE SIEVE PLATES OPEN?** Early electron microscopic studies of phloem samples cut from plants seemed to contradict the pressure flow model. The pores in the sieve plates always appeared to be plugged with masses of a fibrous protein, suggesting that sieve tube sap could not flow freely.

However, the fibrous protein seen in early electron micrographs turned out to be a normal plant response to wounding. When phloem is cut, the contents of the sieve elements are forced out through the sieve plate pores. To prevent massive losses of phloem sap, a fibrous protein, called P-protein, is synthesized in companion cells, enters the sieve elements through plasmodesmata, moves to the sieve plate pores, and seals them off. This short-term solution gives the plant time to repair the wound site by synthesizing a polysaccharide called *callose*, which is thought to strengthen the cell wall.

More recent studies using rapid freezing, which prevents the wounding response, show open pores (see Figure 35.13) in the sieve plates.

**HOW ARE SIEVE TUBE ELEMENTS LOADED AND UNLOADED?** If the pressure flow model is correct, there must be mechanisms for loading sugars and other solutes into the phloem in source regions and for unloading them in sink regions.

Two general routes can be taken by sugars and other solutes as they move from the mesophyll cells to the phloem: apoplastic and symplastic. The exact details vary widely among plant species. In many plants, sugars and other solutes follow the *apoplastic pathway;* they leave the mesophyll cells and enter the apoplast before they reach the sieve elements. Specific sugars and amino acids are then actively transported into cells of the phloem, and in this way reenter the symplast. Because the solutes cross at least one selectively permeable membrane in the apoplastic pathway, selective transport can be used to regulate which specific substances enter the phloem. In other plants, solutes follow a *symplastic pathway;* the solutes remain within the symplast all the way from the mesophyll cells to the seive tube cells. Because no membranes are crossed in the symplastic pathway, a mechanism that does not involve membrane transport is used to load sucrose into the phloem.

In the apoplastic pathway, sucrose is actively loaded into the companion cells and sieve tubes by sucrose–proton symport, a secondary active transport mechanism. A proton pump actively pumps protons out of the phloem cells, increasing the concentration of protons in the apoplast. The protons then diffuse back

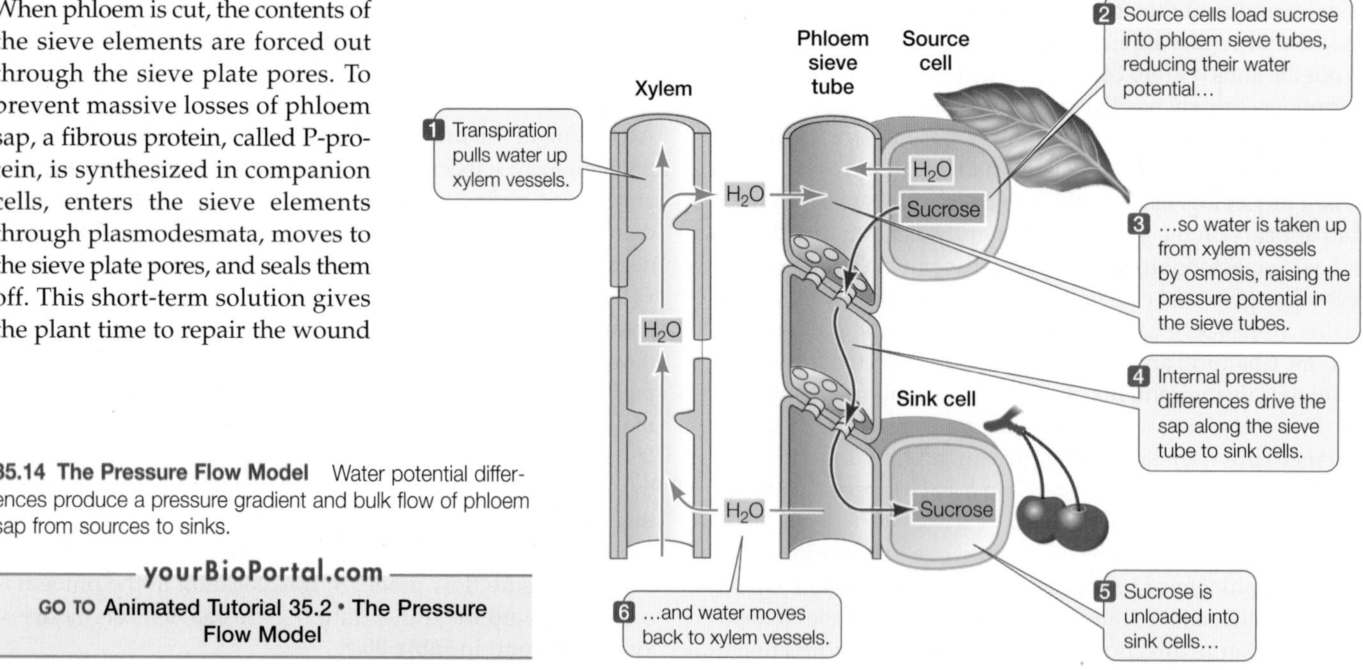
Xylem
Phloem sieve tube
Source cell
1 Transpiration pulls water up xylem vessels.
2 Source cells load sucrose into phloem sieve tubes, reducing their water potential...
3 ...so water is taken up from xylem vessels by osmosis, raising the pressure potential in the sieve tubes.
4 Internal pressure differences drive the sap along the sieve tube to sink cells.
5 Sucrose is unloaded into sink cells...
6 ...and water moves back to xylem vessels.
$H_2O$
Sucrose
Sink cell

**35.14 The Pressure Flow Model** Water potential differences produce a pressure gradient and bulk flow of phloem sap from sources to sinks.

yourBioPortal.com

GO TO Animated Tutorial 35.2 • The Pressure Flow Model

into the phloem cells through sucrose–proton symport proteins, bringing sucrose with them.

In sink regions, the solutes are actively transported *out* of the sieve tube elements and into the surrounding tissues. This unloading serves two purposes: it helps maintain the gradient of solute potential and hence of pressure potential in the sieve tubes; and it helps build up high concentrations of sugars and starch in storage organs, such as developing fruits and seeds. Thus the second requirement of the pressure flow model is met, and the model is supported.

## 35.4 RECAP

**Carbohydrates produced by photosynthesis are translocated from source to sink through the phloem by a pressure flow mechanism.**

- Explain the difference between a source and a sink. See p. 750
- How does loading of sucrose at the source result in bulk flow toward the sink? See p. 751 and Figure 35.14

## CHAPTER SUMMARY

### 35.1 How Do Plant Cells Take Up Water and Solutes?

- Water moves through biological membranes by osmosis, always moving toward regions with a more negative water potential. The **water potential** ($\psi$) of a cell or solution is the sum of the **solute potential** and the **pressure potential**. Review Figure 35.2
- **Turgid** plant cells have significant positive pressure potential because the rigid cell wall pushes back on the protoplast. This positive pressure (**turgor pressure**) maintains the physical structure of many plant cells; if the pressure potential drops, the plant *wilts*. Review Figures 35.3 and 35.4
- The movement of a solution due to a difference in pressure potential between two parts of a plant is called **bulk flow**.
- **Aquaporins** are channel proteins that facilitate movement of water molecules through biological membranes.
- Mineral uptake requires transport proteins. Some minerals enter the plant passively by facilitated diffusion; others enter by active transport. A **proton pump** provides energy for the active transport of many mineral ions across membranes in plants. Review Figure 35.5
- Water and minerals pass from the soil into the root by way of the **apoplast** and **symplast**, but must pass through the symplast to cross the endodermis and enter the xylem. The **Casparian strip** in the endodermis blocks movement of water and minerals through the apoplast. Review Figure 35.6, WEB ACTIVITY 35.1

### 35.2 How Are Water and Minerals Transported in the Xylem?

- **Root pressure** is responsible for guttation and for the oozing of sap from cut stumps, but it cannot account for the ascent of xylem sap in trees.
- Water transport in the xylem results from the combined effects of **transpiration**, **cohesion**, and **tension**—the **transpiration–cohesion–tension mechanism**. Evaporation from the leaf produces tension in the mesophyll cells, which pulls a column of water—held together by cohesion—up through the xylem from the root. Review Figure 35.7, ANIMATED TUTORIAL 35.1
- Transport in the xylem is by bulk flow. It does not require the expenditure of energy. Dissolved minerals are carried passively in the water.

### 35.3 How Do Stomata Control the Loss of Water and the Uptake of $CO_2$?

- **Stomata** allow a compromise between water retention and carbon dioxide uptake.
- A pair of **guard cells** controls the size of the stomatal opening. A light-activated proton pump, moves protons out of the guard cells to the walls of surrounding epidermal cells, setting up an electrochemical gradient that drives the transport of potassium ions into the guard cells. Water follows osmotically, swelling the guard cells and opening the stomata. Review Figures 35.9 and 35.10
- When threatened by dehydration, mesophyll cells release abscisic acid, which causes guard cells to close the stomata, even in the light.

### 35.4 How Are Substances Translocated in the Phloem?

- Products of photosynthesis, as well as some minerals, are translocated through sieve tubes in the phloem by way of living **sieve tube elements**. Review Figure 35.12
- **Translocation** in the phloem can proceed in both directions in the stem, although it only goes one way in any given sieve tube. Translocation requires a supply of ATP.
- Translocation in the phloem is explained by the **pressure flow model**: the difference in solute concentration between **sources** and **sinks** creates a difference in (positive) pressure potential along the sieve tubes, resulting in bulk flow. Review Figure 35.14 and Table 35.1, ANIMATED TUTORIAL 35.2

## SELF-QUIZ

1. Osmosis
   *a.* requires ATP.
   *b.* results in the bursting of plant cells placed in pure water.
   *c.* can cause a cell to become turgid.
   *d.* is independent of solute concentrations.
   *e.* continues until the pressure potential equals the water potential.

2. Water potential
   *a.* is the difference between the solute potential and the pressure potential.
   *b.* is analogous to the air pressure in an automobile tire.
   *c.* is the movement of water through a membrane.
   *d.* determines the direction of water movement between cells.
   *e.* is defined as 1.0 MPa for pure water under no applied pressure.

3. Which statement about aquaporins is *not* true?
   *a.* They are membrane transport proteins.
   *b.* Water movement through aquaporins is always active.
   *c.* The permeability of some aquaporins is subject to regulation.
   *d.* They vary in abundance depending on environmental conditions.
   *e.* They enable water to pass through the phospholipid bilayer without encountering a hydrophobic environment.
4. Which statement about proton pumping across the plasma membrane of plants is *not* true?
   *a.* It requires ATP.
   *b.* The region inside the membrane becomes positively charged with respect to the region outside.
   *c.* It enhances the movement of $K^+$ ions into the cell.
   *d.* It pushes protons out of the cell against a proton concentration gradient.
   *e.* It can drive the secondary active transport of negatively charged ions.
5. Which statement is *not* true?
   *a.* The symplast consists of the interconnected cytoplasm of living cells.
   *b.* Water can enter the stele without entering the symplast.
   *c.* The Casparian strips prevent water from moving between endodermal cells.
   *d.* The endodermis is a cell layer in the cortex.
   *e.* Water can move freely in the apoplast without entering cells.
6. In the xylem,
   *a.* the products of photosynthesis travel down the stem.
   *b.* living, pumping cells push the sap upward.
   *c.* the driving force is in the roots.
   *d.* the sap is often under tension.
   *e.* the sap must pass through sieve plates.
7. Which of the following is *not* part of the transpiration–cohesion–tension mechanism?
   *a.* Water evaporates from the walls of mesophyll cells.
   *b.* Removal of water from the xylem exerts a pull on the water column.
   *c.* Water is remarkably cohesive.
   *d.* The wider the tube, the greater the tension its water column can withstand.
   *e.* At each step, water moves to a region with a more strongly negative water potential.
8. Stomata
   *a.* control the opening of guard cells.
   *b.* release less water to the environment than do other parts of the epidermis.
   *c.* open when $CO_2$ levels inside the leaf are high.
   *d.* do not respond to light.
   *e.* close when water is being lost at too great a rate.
9. Which statement about phloem transport is *not* true?
   *a.* It takes place in sieve tubes.
   *b.* It depends on mechanisms for loading solutes into the phloem at sources.
   *c.* It stops if the phloem is killed by heat.
   *d.* A high pressure potential is maintained in the sieve tubes.
   *e.* At sinks, solutes are actively transported into sieve tube elements.
10. The fibrous protein in sieve tube elements
    *a.* may plug leaks when a plant is damaged.
    *b.* clogs the sieve plates at all times.
    *c.* never clogs the sieve plates.
    *d.* serves no known function.
    *e.* provides the driving force for transport in the phloem.

## FOR DISCUSSION

1. Epidermal cells protect against excess water loss. How do they perform this function? What differences might you expect to find in the structure of the epidermis in stems, roots, and leaves?
2. Phloem transports material from sources to sinks. Give examples of each. How might the distribution of sources and sinks change in the course of a year?
3. What is the minimum number of plasma membranes a water molecule would have to cross in order to get from the soil solution to the atmosphere by way of the stele? To get from the soil solution to a mesophyll cell in a leaf?
4. Transpiration exerts a powerful pulling force on the water column in the xylem. When would you expect transpiration to proceed most rapidly? Why? Describe the source of the pulling force.

## ADDITIONAL INVESTIGATION

In the story that opened this chapter we saw that a mutation in the *HARDY* gene resulted in *Arabidopsis* plants with a more extensive root system and thicker leaves than wild-type plants. When the *HARDY* gene was isolated, it was found to encode a transcription factor that stimulates expression of genes for increased water use efficiency. What type of mutation in the *HARDY* gene would cause *Arabidopsis* plants to use water more efficiently? How would you investigate the effect of the *HARDY* mutation on stomata? What results would you expect?

# 36 Plant Nutrition

## When the land blew away

Two of the greatest disasters of recent times occurred because of the mismanagement of soil resources. One was in North America in the 1930s and the other is ongoing in Haiti.

Beginning in the early 1930s, a prolonged drought, combined with a culturally modified landscape, turned the central plains of North America into the Dust Bowl. The native vegetation in the Plains States in the nineteenth century was long grass in the east and short grass in the west. Cattlemen moved in to where their herds could eat a seemingly endless supply of food. But in fact the supply wasn't endless. As one area was overgrazed, the cattle were moved on to new areas, leaving damaged soil in their wake. Settlers followed, "busted the sod" with plows, planted crops, and disrupted vegetation cycles that had persisted for centuries.

In the early twentieth century, farmers in the Plains States began growing wheat, plowing both good and marginal soils. But wheat requires more water than did the native grasses, and rainfall was irregular throughout the 1920s. In 1932 rainfall was almost nonexistent and did not return at adequate levels until 1939.

Without water, crops failed—if they even started to grow. The U.S. plains are windy regions, and without plant roots there was nothing to hold the soil in place when the winds blew. The farms literally blew away. Farmers spent the last of their money on seeds, but dry year followed dry year. Destitute farmers migrated westward, along with others whose livelihoods had depended on the farmers. But they only encountered more difficulties, as these events took place during the Great Depression.

Unfortunately Dust Bowl conditions are not just a thing of the past. Today many countries around the world are struggling with land use issues. The consequences of poor land management, principally deforestation, are dramatically visible on the Caribbean island of Hispaniola. The island is shared by the countries of Haiti, which has mismanaged its soil resources, and the Dominican Republic, which hasn't. From an airplane window it is easy to see the 120-mile-long bor-

**Dreams Disappeared in a Cloud of Dust** This photograph of a family displaced by the Dust Bowl was taken by Dorothea Lange in the winter of 1936. The family had traveled to northern California looking for work as migrant farm labor. In Lange's words, "I saw and approached the hungry and desperate mother…I did not ask her name or her history. …She said that they had been living on frozen vegetables from the surrounding fields, and birds that the children killed. She had just sold the tires from her car to buy food."

**A Border Marks Life and Death** An aerial view of the border between Haiti (left) and the Dominican Republic (right) provides a dramatic illustration of the extent of deforestation in Haiti.

der between these two countries. The Dominican side is verdant and forested, while the Haitian side is devoid of plant life and most of its soil is gone.

Haiti's population, in addition to living in the poorest nation in the Western Hemisphere, has paid a huge price for the loss of its trees and soil. Because there are few plant roots to stabilize the soil in mountainous areas, rain washes the soil into the sea. Not only is this soil loss detrimental to agriculture, but the runoff hurts offshore reefs and the fisheries they support. To make matters worse, tropical storms all too often result in devastating landslides like the one that killed 2,600 Haitians in 2004.

As human activity increases, ecological disasters become all too common. Today, parts of sub-Saharan Africa's farmland are losing their topsoil as a result of poor land management, swelling populations, and a challenging climate. Crop failures, starvation, and large-scale human displacements are inevitable consequences.

**IN THIS CHAPTER** we consider the nutritional conditions that foster healthy and sustained plant growth. We identify nutrients that are essential to plants and how plants acquire them. Because most plant nutrients come from the soil, we discuss the formation of soils and the effects of plants on soils. We devote a section to the role played by fungi and bacteria in the uptake of phosphorus and nitrogen by plants, and we conclude with a look at carnivorous and parasitic plants.

## CHAPTER OUTLINE



# 36.1 How Do Plants Acquire Nutrients?

Every living thing—and plants are no exception—must obtain raw materials from its environment. These **nutrients** include the major ingredients of macromolecules: carbon, hydrogen, oxygen, and nitrogen. Plants are *autotrophs*, and obtain carbon from atmospheric carbon dioxide through the carbon-fixing reactions of photosynthesis (see Chapter 10). Hydrogen and oxygen come mainly from water, so these elements are plentiful with an adequate water supply. Nitrogen, as you will see later in this chapter, enters plants primarily through the activities of bacteria.

Living organisms need other **mineral nutrients** as well, which most plants obtain from the soil. For example, proteins contain sulfur (S), nucleic acids contain phosphorus (P), chlorophyll contains magnesium (Mg), cytochromes contain iron (Fe), and cellular signaling can involve calcium (Ca). Within the soil, these and other minerals dissolve in water as ions, forming a solution—called the **soil solution**—that contacts the roots of plants.

### How does a stationary organism find nutrients?

Many organisms can move from place to place to find the nutrients they need. An organism that cannot move, termed a *sessile* organism, must obtain nutrients from sources that are somehow brought to it. With the exception of the carbon and oxygen in $CO_2$, a plant's supply of nutrients is strictly local, and a plant may use up the water and mineral nutrients in its local environment as it grows. How does a plant cope with the problem of scarce nutrient supplies?

As discussed in Chapter 34, plants differ fundamentally from animals in that they grow throughout their lifetimes. In fact, growth is a plant's version of movement. For example, roots obtain most of the mineral nutrients plants need. By growing through the soil, roots mine it for new sources of mineral nutrients and water. The growth of stems and leaves helps a plant secure light and carbon dioxide, which in turn allows the roots to continue their growth through the soil.

As it grows, a plant—or even a single root—must deal with a variable environment. Animal droppings create high local concentrations of nitrogen. A particle of calcium carbonate may make a tiny area of the soil alkaline, while dead organic matter may make a nearby area acidic. Such microenvironments encourage or discourage the proliferation of a root system and help direct its growth.

## 36.1 RECAP

**Plants are autotrophs that obtain carbon by photosynthesis, and mineral nutrients and water from the soil.**

- Why do plants need phosphorus? Why do they need nitrogen? See p. 756
- How does the ability to grow throughout their lifetime allow plants to seek out nutrients? See p. 756

We know that plants need nutrients to support their growth. Let's look in more detail at the specific mineral nutrients they need.

# 36.2 What Mineral Nutrients Do Plants Require?

Plants require many mineral nutrients (**Table 36.1**). Except for nitrogen, all mineral nutrients derive from rock and are usually taken up from the soil solution. A nutrient is called an **essential element** if its absence causes severe disruption of normal plant growth and reproduction. An essential element cannot be replaced by another element.

Essential elements fall roughly into two categories—*macronutrients* and *micronutrients* (see Table 36.1)—based on the amounts required by plants.

- A plant needs **macronutrients** in concentrations of at least 1 gram per kilogram of the plant's dry matter.
- A plant needs **micronutrients** in concentrations of less than 100 milligrams per kilogram of the plant's dry matter.

How do we know if a plant is getting enough of a particular nutrient?

### Deficiency symptoms reveal inadequate nutrition

Before a plant that is deficient in an essential element dies, it usually displays characteristic **deficiency symptoms**. Table 36.1 lists the symptoms of some common mineral deficiencies, one of which is also shown in **Figure 36.1**. Such symptoms help horticulturists diagnose mineral nutrient deficiencies in plants. With proper diagnosis, the missing nutrient(s) can be provided in the form of a **fertilizer** (an added source of mineral nutrients).

We know that the elements listed in Table 36.1 are essential to the life of all plants. How did biologists discover which elements are essential?

## TABLE 36.1
### Mineral Elements Required by Plants

| ELEMENT | ABSORBED FORM | MAJOR FUNCTIONS | DEFICIENCY SYMPTOMS |
|---|---|---|---|
| **MACRONUTRIENTS** | | | |
| Nitrogen (N) | $NO_3^-$ and $NH_4^+$ | In proteins, nucleic acids | Oldest leaves turn yellow and die prematurely; plant is stunted |
| Phosphorus (P) | $H_2PO_4^-$ and $HPO_4^{2-}$ | In nucleic acids, ATP, phospholipids | Plant is dark green with purple veins and is stunted |
| Potassium (K) | $K^+$ | Enzyme activation; water balance; ion balance; stomatal opening | Older leaves have dead edges |
| Sulfur (S) | $SO_2^{4-}$ | In proteins and coenzymes | Young leaves are yellow to white with yellow veins |
| Calcium (Ca) | $Ca^{2+}$ | Affects the cytoskeleton, membranes, and many enzymes; second messenger | Growing points die back; young leaves are yellow and crinkly |
| Magnesium (Mg) | $Mg^{2+}$ | In chlorophyll; required by many enzymes; stabilizes ribosomes | Older leaves have yellow stripes between veins |
| **MICRONUTRIENTS** | | | |
| Iron (Fe) | $Fe^{2+}$ and $Fe^{3+}$ | In active site of many redox enzymes and electron carriers; chlorophyll synthesis | Young leaves are white or yellow |
| Chlorine (Cl) | $Cl^-$ | Photosynthesis; ion balance | Leaf tips wilt; leaves turn yellow and die |
| Manganese (Mn) | $Mn^{2+}$ | Activation of many enzymes | Younger leaves are pale with green veins |
| Boron (B) | $B(OH)_3$ | Required for proper cell wall formation and expansion | Poor growth of leaves and roots |
| Zinc (Zn) | $Zn^{2+}$ | Enzyme activation; auxin synthesis | Young leaves are abnormally small; older leaves have many dead spots |
| Copper (Cu) | $Cu^{2+}$ | In active site of many redox enzymes and electron carriers | New leaves are dark green, may have dead spots |
| Nickel (Ni) | $Ni^{2+}$ | Activation of the enzyme urease | Leaf tips die; deficiency is rare |
| Molybdenum (Mo) | $MoO_4^{2-}$ | Nitrate reduction | Leaves turn yellow between veins; older leaves die |

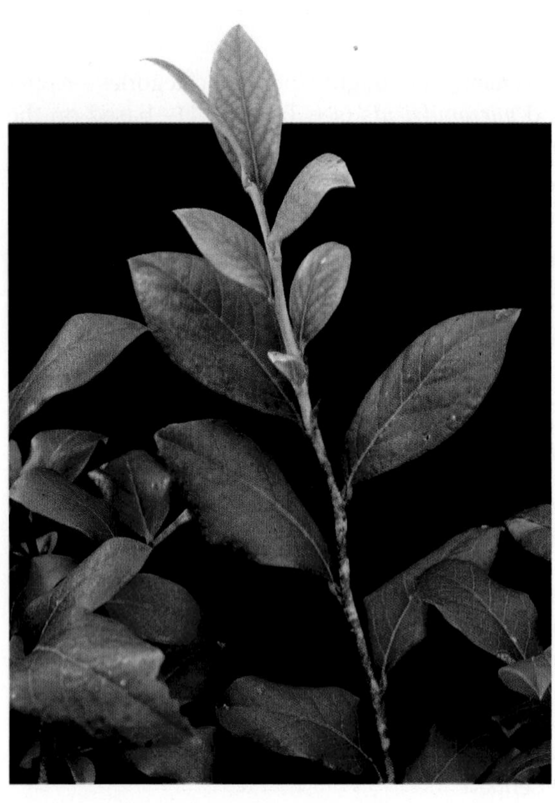

**36.1 Iron Deficiency Symptoms** In crop plants, mineral deficiencies can often be detected in leaves, as in these blueberry leaves. As is typical of iron deficiency, the younger leaves are yellow, whereas the older leaves look normal.

### Hydroponic experiments identified essential elements

An element is considered essential to plants if a plant fails to complete its life cycle or grows abnormally when that element is absent or insufficient. The essential elements for plants were identified by growing plants **hydroponically**—that is, with their roots suspended in nutrient solutions instead of soil (**Figure 36.2**). Growing plants in this manner allows for greater control of nutrient availability than is possible in a complex medium like soil. In the first successful experiments of this type, performed a century and a half ago, plants grew seemingly normally in solutions containing only calcium nitrate [$Ca(NO_3)_2$], magnesium sulfate ($MgSO_4$), and potassium phosphate ($KH_2PO_4$). A solution missing any of these compounds could not support normal growth. Tests with other compounds that included various combinations of these elements soon established six macronutrients—calcium, nitrogen, magnesium, sulfur, potassium, and phosphorus—as essential elements.

Identifying essential *micronutrients* by this experimental approach proved to be more difficult. The chemicals used in nineteenth-century experiments on plant nutrition were sometimes so impure that they provided micronutrients that investigators thought they had excluded. Furthermore, some micronutrients are required in such tiny amounts that a seed may contain enough to supply the embryo and the resultant second-generation plant throughout its lifetime. There might even be enough left over to pass on to third-generation plants. Because of such difficulties, nutrition experiments must be performed in tightly controlled laboratories with special air filters that exclude microscopic salt particles in the air, and must use only the purest available chemicals.

Iron was the first micronutrient to be clearly established as essential, in the 1840s. The last micronutrient to be listed as essential was nickel, in 1983 (the experiment is described in **Figure 36.3**).

yourBioPortal.com
GO TO **Animated Tutorial 36.1 • Nitrogen and Iron Deficiencies**

TOOLS FOR INVESTIGATING LIFE

**36.2 Growing Plants Hydroponically**

Hydroponics is used to grow plants without soil. It is a classic procedure for identifiying nutrients essential to plants. Nitrogen is used here as an example.

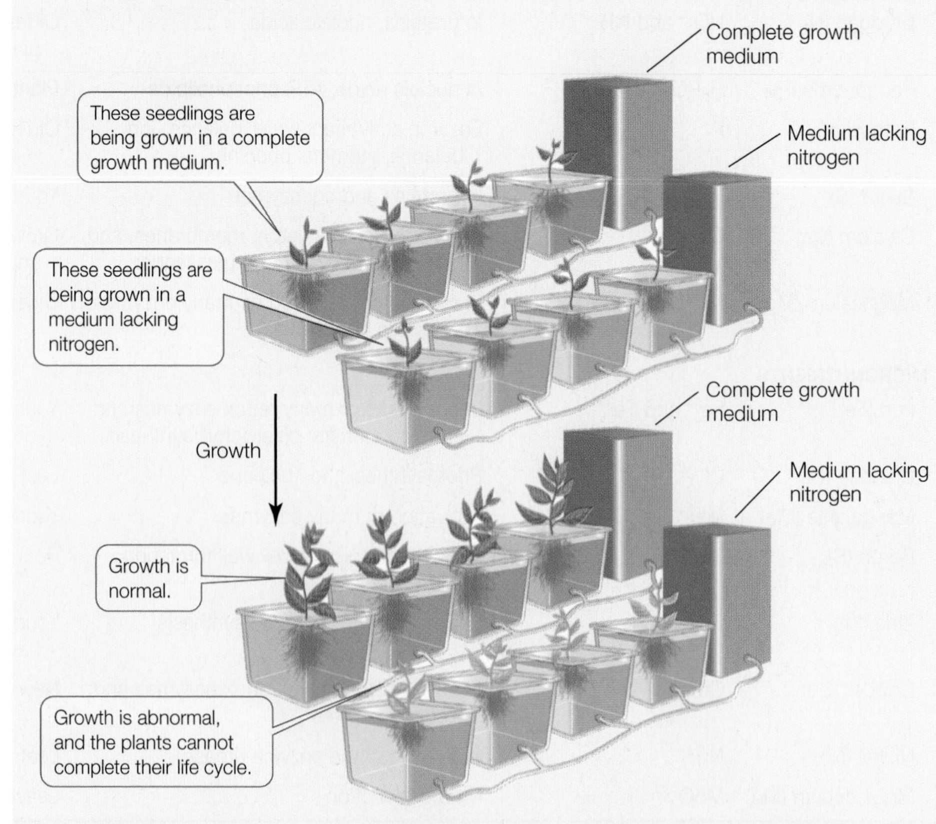
Complete growth medium
These seedlings are being grown in a complete growth medium.
Medium lacking nitrogen
These seedlings are being grown in a medium lacking nitrogen.
Complete growth medium
Growth
Medium lacking nitrogen
Growth is normal.
Growth is abnormal, and the plants cannot complete their life cycle.

## INVESTIGATING LIFE

### 36.3 Is Nickel an Essential Element for Plant Growth?

Using highly purified salts in growth media, Patrick Brown and his colleagues tested whether barley can complete its life cycle in the absence of nickel. Other investigators showed that no other element could substitute for nickel.

**HYPOTHESIS** **Nickel is an essential element for a plant to complete its life cycle.**

**METHOD**

1. Grow barley plants for 3 generations in nutrient solutions containing 0, 0.6, and 1.0 $\mu M$ $NiSO_4$.
2. Harvest seeds from 5–6 third-generation plants in each of the groups.
3. Determine the nickel concentration in seeds from each plant.
4. Germinate other seeds from the same plants on nickel-free medium and plot the success of germination against nickel concentration.

**RESULTS** There was a positive correlation between seed germination and seed nickel concentration. There was significantly less germination at the lowest nickel concentrations.

● 0 $\mu M$ $NiSO_4$
● 0.6 $\mu M$ $NiSO_4$
● 1.0 $\mu M$ $NiSO_4$

Percent germination: 0, 20, 40, 60, 80, 100

Nickel concentration in seeds (ng/g): 0, 50, 100, 150, 200, 250

**CONCLUSION** **Barley seeds require nickel in order to germinate and thereby complete the life cycle.**

Go to **yourBioPortal.com** for original citations, discussions, and relevant links for all INVESTIGATING LIFE figures.

## 36.2 RECAP

**Mineral nutrients required by plants are classified as macronutrients and micronutrients, depending on the amount needed. Micronutrients are often needed in such minute amounts that only sophisticated chemical experiments can determine their essentiality.**

- What are some specific mineral deficiency symptoms seen in plants? **See p. 757 and Table 36.1**
- Outline an experimental method for determining whether an element is essential to a plant. **See Figures 36.2 and 36.3**

As we have seen, all plants require nutrients for growth. Plants get nutrients in two ways. For plants growing in natural settings such as forests and fields, nutrients are derived from minerals in the soil. Crop plants, which often need large quantities of nutrients to support rapid growth, may be given nutrient supplements in the form of fertilizer. However, in either case, nutrition is not the only role the soil plays in the life of plants.

# 36.3 How Does Soil Structure Affect Plants?

Most terrestrial plants grow in soil. Soils provide:

- mechanical support
- mineral nutrients and water from the soil solution
- $O_2$ for root respiration

Soils also harbor many bacteria and other organisms; some of these are beneficial to plant life, but others are harmful. Some soils contain toxic levels of metal ions such as cadmium, chromium, and lead (see Chapter 39).

Soils are modified by natural phenomena, such as rain, temperature extremes, and the activities of plants and animals, and by the practices of humans—particularly agriculture. In this section, we examine the composition, structure, and formation of soils, as well as their role in plant nutrition.

### Soils are complex in structure

**Soils** have living and nonliving components (**Figure 36.4**). The living components include plant roots as well as populations of bacteria, fungi, protists, and animals such as earthworms and insects. The nonliving portion of the soil includes rock fragments

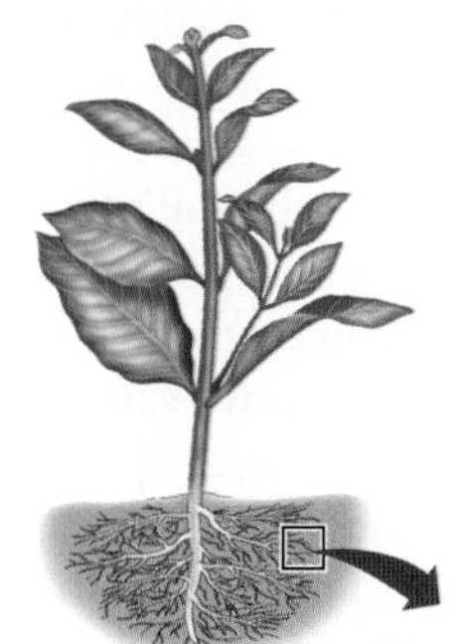

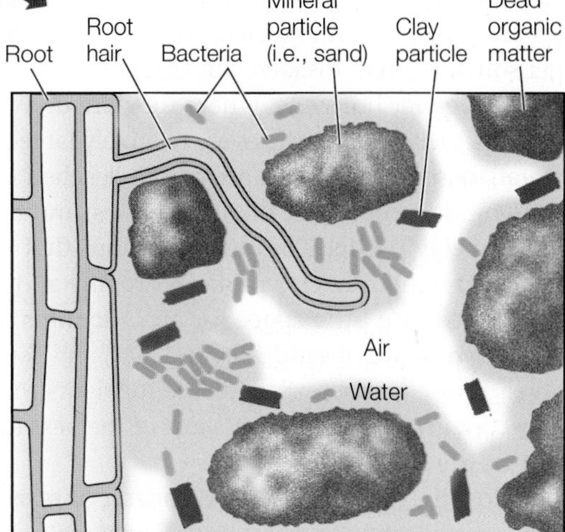


**36.4 The Complexity of Soil** Soils favorable for plant growth contain both clay and larger mineral particles, as well as water, air, and organic matter. Other organisms are also present.

A horizon
Topsoil

B horizon
Subsoil

C horizon
Weathering parent rock (bedrock)

**36.5 A Soil Profile** The A, B, and C horizons can sometimes be seen in road cuts such as this one in Australia. The dark upper layer (the A horizon) is home to most of the living organisms in the soil.

ranging in size from large stones to sand to silt and finally to tiny particles of *clay* that are 2 μm or less in diameter. Soil also contains water and dissolved mineral nutrients, air spaces, and dead organic matter. The air spaces in soil contain $O_2$.

Although soils vary greatly, almost all of them have a *soil profile* consisting of several recognizable horizontal layers, called **horizons**, lying on top of one another. Soil scientists recognize three major *horizons*—termed A, B, and C—in the profile of a typical soil (**Figure 36.5**).

**Topsoil** is the **A horizon**, which contains most of the soil's living and dead organic matter. Successful agriculture depends on the presence of a suitable A horizon; the A horizon is what blew away from the U.S. plains during the Dust Bowl (as we saw at the beginning of this chapter).

Topsoils vary greatly in their proportions of sand, silt, and clay, and this influences their ability to support plant growth. For example, mineral nutrients tend to be **leached** from the upper soil horizons—dissolved in rain or irrigation water and carried to deeper horizons, where they are unavailable to plant roots. Because sand particles are relatively large and cannot hold water, dissolved minerals are readily leached from sandy soil. Clay binds more water than sand does, and the charged surfaces of clay particles bind mineral ions that plant roots ultimately take up. But clay particles are tiny and pack tightly together, leaving little space for air. A **loam** is a soil that is an optimal mixture of sand, silt, and clay, and thus has sufficient levels of air, water, and available nutrients for plants. Loams also contain organic matter. Most of the best topsoils for agriculture are loams.

Below the A horizon is the **B horizon**, or **subsoil**, which is the zone of infiltration and accumulation of materials leached from above. Farther down, the **C horizon** is the **parent rock**, also called bedrock, that is breaking down to form soil. Some deep-growing roots extend into the B horizon to obtain water and nutrients, but roots rarely enter the C horizon.

### Soils form through the weathering of rock

Rocks are broken down into soil particles (**weathered**) in two ways: First, there is *mechanical weathering*, which is the physical breakdown of materials by wetting, drying, and freezing. Second there is *chemical weathering*, the alteration of the chemistry of the materials in the rocks. Several types of chemical weathering occur, all of which influence the availability of mineral nutrients:

- *Oxidation* by atmospheric oxygen
- *Hydrolysis* (reaction with water)
- Reaction with *acids* (particularly carbonic acid)

The parent rock and the weathering it undergoes determine the basic structure and chemical composition of a soil. However, a key soil characteristic for plants is the *availability* of nutrients, which must be dissolved in the soil solution for uptake by the plant. Chemical weathering results in clay particles covered with negatively charged chemical groups, which bind positively charged mineral nutrients (**Figure 36.6**). How do roots obtain these mineral nutrients?

### Soils are the source of plant nutrition

Negatively charged clay particles form ionic bonds (see Section 2.2) with the positively charged ions (cations) of many minerals that are important for plant nutrition, such as potassium ($K^+$), magnesium ($Mg^{2+}$), and calcium ($Ca^{2+}$). To become available to plants or other organisms, these cations must be detached from the clay particles.

Recall that the root surface is covered with root hair cells (see Figure 34.11). Protein transporters in the plasma membrane of these cells actively pump protons ($H^+$) out of the cell. In addition, cellular respiration in the roots releases $CO_2$, which dissolves in the soil water and reacts with it to form carbonic acid. This acid ionizes to form bicarbonate and free protons:

$$CO_2 + H_2O \rightleftharpoons H_2CO_3 \rightleftharpoons H^+ + HCO_3^-$$

Proton-pumping by the root and ionization of carbonic acid both act to increase the proton concentration in the soil surrounding the root. The protons bind more strongly to clay particles than do mineral cations; in essence, they trade places with the cations in a process called **ion exchange** (see Figure 36.6). Ion exchange releases important cations into the soil solution, where they are available to be taken up by the roots. The capacity of a soil to support plant growth, called *soil fertility*, is determined in part by its ability to provide nutrients in this manner.

There is no comparable mechanism for binding and releasing negatively charged ions. As a result, important anions such as nitrate ($NO_3^-$) and sulfate ($SO_4^{2-}$)—direct sources of nitrogen and sulfur, respectively—may leach rapidly from the A horizon.

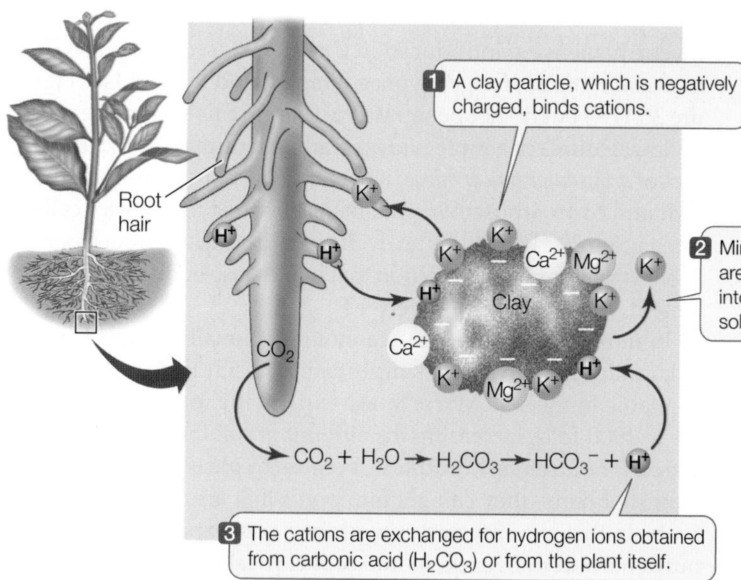

1 A clay particle, which is negatively charged, binds cations.
Root hair
Clay
2 Mineral cations are released into the soil solution.
$CO_2 + H_2O \rightarrow H_2CO_3 \rightarrow HCO_3^- + H^+$
3 The cations are exchanged for hydrogen ions obtained from carbonic acid ($H_2CO_3$) or from the plant itself.


**36.6 Ion Exchange** Plants obtain mineral nutrients from the soil primarily in the form of positive ions; potassium ($K^+$) is the example shown here.

### Fertilizers and lime are used in agriculture

Agricultural soils are often deficient in one or more essential elements. Irrigation and rainwater leach mineral nutrients from the soil, and the harvesting of crops removes the nutrients that the crops took up from the soil during their growth. Unless the soil is replenished, crop yields will decrease. Mineral nutrients may be replaced by adding **fertilizers**: inorganic fertilizers of various types; or organic fertilizers, such as compost or rotted manure.

**INORGANIC AND ORGANIC FERTILIZERS** The three elements most commonly added to agricultural soils are nitrogen (N), phosphorus (P), and potassium (K). Commercial inorganic fertilizers are characterized by their "N-P-K" percentages. A 5-10-10 fertilizer, for example, contains 5 percent nitrogen, 10 percent phosphate ($P_2O_5$), and 10 percent potash ($K_2O$) by weight (of the nutrient-containing compound, not as weights of the elements N, P, and K). Sulfur, in the form of ammonium sulfate, is also occasionally added to soils.

Organic fertilizers such as manure or crop residues can also be used to supply mineral nutrients. Organic fertilizers have both advantages and disadvantages over inorganic fertilizers. Among the advantages:

- Organic fertilizers release nutrients slowly, which results in less leaching than occurs with a one-time application of an inorganic fertilizer.
- They contain residues of plant or animal materials that improve the structure of the soil, providing spaces for air movement, root growth, and drainage.

However, the nutrients in organic fertilizers are not in a form that is immediately available for absorption, as are the nutrients in inorganic fertilizers. Furthermore, unlike organic fertilizers, inorganic fertilizers can be formulated to meet the specific requirements of a particular soil and a particular crop.

**pH EFFECTS ON NUTRIENTS** The availability of nutrient ions, whether they are naturally present in the soil or added as fertilizer, depends on soil pH. The proton concentration can affect the binding of nutrient cations to clay particles, as we saw earlier, and can also affect the solubility of other nutrients, such as iron, in the soil solution. The optimal soil pH for most crops is about 6.5, but so-called acid-loving crops such as blueberries prefer a pH closer to 4.

Rainfall and decomposition of organic substances lower the pH of soil, sometimes making it so acidic that plant growth is inhibited. Such acidification can be reversed by **liming**—the application of compounds commonly known as *lime*, such as calcium carbonate, calcium hydroxide, or magnesium carbonate. The addition of these compounds removes $H^+$ from the soil, and also increases the availability of calcium to plants.

Sometimes, on the other hand, a soil is not acidic enough for a crop. In this case, sulfur can be added in the form of elemental sulfur, which soil bacteria convert to sulfuric acid. Iron and some other elements are more available to plants at a slightly acidic pH. Because soil pH is so important for soil fertility, measuring pH is often the first step in deciding which amendments to add to soils for home gardens and agriculture.

**SPRAY APPLICATION OF NUTRIENTS** Spraying leaves with a nutrient solution is another effective way to deliver some essential elements to growing plants. Plants take up copper, iron, and manganese more readily from foliar (leaf) sprays than from the soil. Many foliar applications contain chemicals that partially dissolve the protective covering of leaf cells (the cuticle) to increase nutrient uptake.

### Plants affect soil fertility and pH

The relationship between plants and soils is not a one-way affair—soils affect plants, but plants also affect soils. The soil that forms in a particular place depends not only on the underlying parent rock, mechanical weathering, and other such factors, but also on the particular plants that grow there. For example, dead plant matter provides most of the carbon-rich materials that break down to form **humus**—a dark-colored, organic soil component, each particle of which is too small to be recognizable with the naked eye. Soil bacteria and fungi produce humus by breaking down plant litter (such as fallen leaves and dead roots), animal feces, dead organisms, and other organic material. Humus is rich in mineral nutrients, especially nitrogen (from animal excrement). Humus also favors plant growth by trapping supplies of water and oxygen for absorption by roots.

Plants also affect the pH of the soil in which they grow. Roots maintain a balance of electric charges. If roots absorb more cations than anions, they excrete $H^+$, thus lowering the soil pH. If they absorb more anions than cations, they excrete $OH^-$ or $HCO_3^-$, raising the soil pH. Roots can also actively change the pH in their immediate vicinity by exuding organic acids, such as citric acid and malic acid, that acidify the soil, making it easier to take up certain ions such as ferric iron ($Fe^{3+}$). Looking at the big picture, we see that successful plant growth can help create conditions that favor further plant growth.

### 36.3 RECAP

**Land plants live anchored in the soil and obtain water and mineral nutrients from it. Plants and soil interact in many ways. Plants can affect many aspects of the soil in which they grow, including mineral content and availability, pH, and the amount of humus. Many of these effects are beneficial to future plant growth.**

- Explain how mechanical and chemical weathering form soil from rock. **See p. 760**
- How is soil fertility enhanced by the process of ion exchange? **See p. 760 and Figure 36.6**

Thus far we have focused on the uptake of nutrients in the soil by plant roots. An understanding of how plants acquire nutrients from the soil would be incomplete, however, without taking into account the involvement of *soil microbes*, including fungi and bacteria. In the next section we will focus on the intimate interactions of plants with these organisms, which are essential to the success of most terrestrial plants.

## 36.4 How Do Fungi and Bacteria Increase Nutrient Uptake by Plant Roots?

One gram of soil contains 6,000–50,000 bacterial *species* and up to 200 *meters* of fungal hyphae (the long branching cells of fungi), although both are largely invisible to the naked eye. In Chapter 39 we describe the strategies plants use to prevent infection by harmful soil microbes. It may surprise you that plants actively encourage a few species of fungi and bacteria to infect their roots and even invade root cells. In this section we describe the resulting "intracellular trading posts," where products are exchanged to the mutual benefit of plants and a few very special soil microbes.

### Mycorrhizae expand the root system of plants

The association of fungi with roots is so prevalent that it has its own name: **mycorrhizae** (singular, *mycorrhiza*) (from the Greek *mycos*, "fungus," and *rhiza*, "root"). Recall from Chapter 30 that a multicellular fungus is called a **mycelium** (plural, *mycelia*) and that it is composed of rapidly growing individual tubular filaments called **hyphae** (singular, *hypha*). Two types of mycorrhizae were introduced in Chapter 30 (see Section 30.2). In *ectomycorrhizae*, fungal hyphae wrap around the root (see Figure 30.11A) but do not penetrate the cells. In this section we will review features of a more widespread and intimate association: that of *arbuscular mycorrhizae*, where the fungal hyphae enter the root and form arbuscular (treelike) structures inside root cells (see Figure 30.11B). This is an evolutionarily ancient association. What is it about this interaction that makes it so enduring? What benefit does each partner derive?

In most cases, roots alone cannot nutritionally support vascular plant growth—they simply cannot reach all the nutrients available in the soil. Mycorrhizae expand the root surface area 10- to 1000-fold, increasing the amount of soil that can be scavenged for nutrients. In addition, because hyphae are much finer than root hairs, they can get into pores that are inaccessible to roots. In this way, mycorrhizae probe a vast expanse of soil for nutrients and deliver them into root cortical cells.

The primary nutrient that the plant obtains from a mycorrhizal interaction is phosphorus. In exchange, the fungus obtains an energy source, largely in the form of simple sugars. In fact, up to 20 percent of the photosynthate (the product[s] of photosynthesis) of terrestrial plants is directed to and consumed by arbuscular mycorrhiza fungi. Such associations are excellent examples of *mutualism*, an interaction between two species in which both species benefit (further discussed in Chapter 56). They are also examples of *symbiosis*, in which two different species live in close contact for a significant portion of their life cycles.

The events in the formation of arbuscular mycorrhiza are shown in **Figure 36.7**. Plant roots produce compounds called **strigolactones** that stimulate rapid growth of fungal hyphae toward the root. (We will return to strigolactones at the end of this chapter.) In response, fungi produce signals that stimulate expression of plant symbiosis-related genes. The products of some of these genes give rise to the *prepenetration apparatus* (PPA), which guides the growth of the fungal hyphae into the root cortex. The sites of nutrient exchange between fungus and plant are the *arbuscules*, which form within root cortical cells. Despite the intimacy of this association, the plant and fungal cytoplasms never mix—they are separated by two membranes, the fungal plasma membrane and the *periarbuscular membrane* (PAM), which is continuous with the plant plasma membrane. We will return to this structure and the features it shares with bacteria-induced root nodules in the next section.

### Soil bacteria are essential in getting nitrogen from air to plant cells

The essential mineral nutrient most commonly in short supply, in both natural and agricultural situations, is nitrogen. This is surprising because elemental nitrogen ($N_2$) makes up almost four-fifths of Earth's atmosphere. However, plants cannot use $N_2$ directly as a nutrient. The triple bond linking the two nitrogen atoms is extremely stable, and a great deal of energy is required to break it; thus $N_2$ is a highly unreactive substance. How, then, do plants obtain usable nitrogen for the synthesis of proteins and nucleic acids?

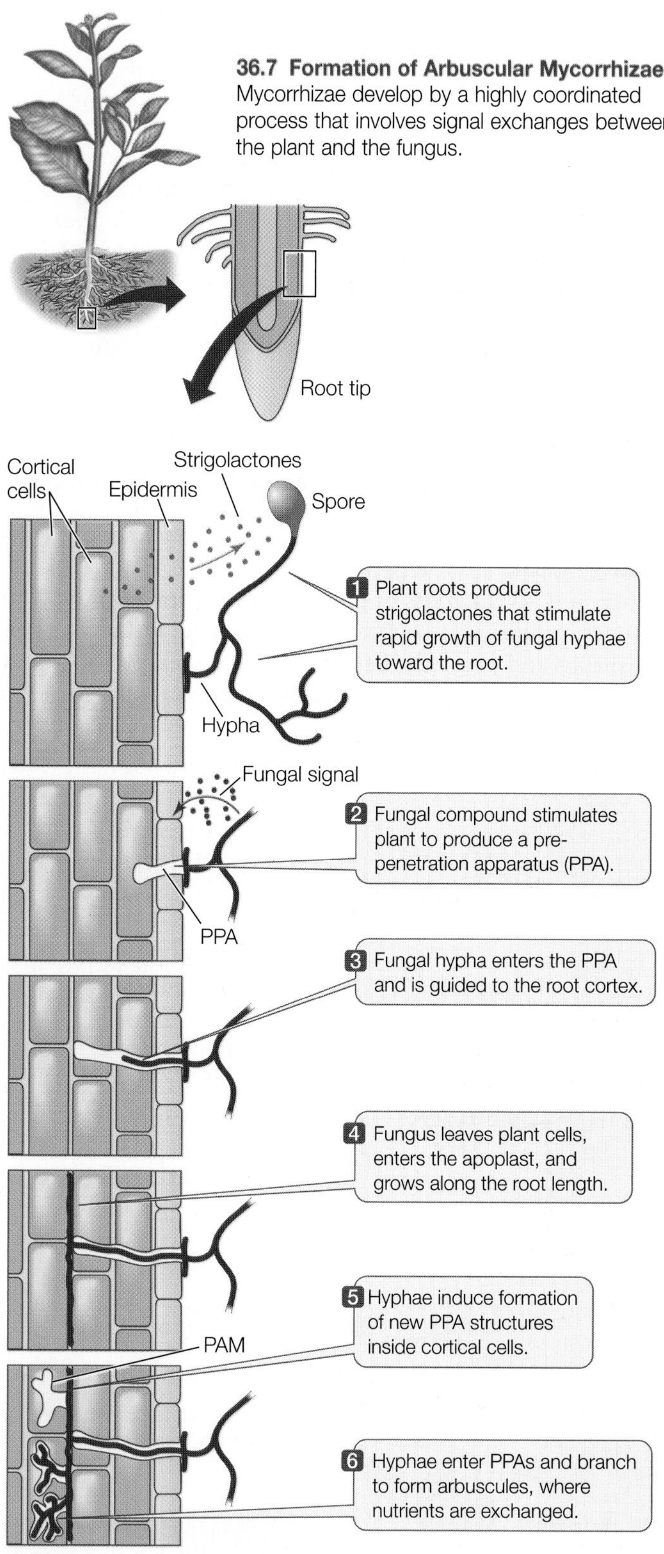

Root tip
Cortical cells
Epidermis
Strigolactones
Spore
1 Plant roots produce strigolactones that stimulate rapid growth of fungal hyphae toward the root.
Hypha
Fungal signal
2 Fungal compound stimulates plant to produce a pre-penetration apparatus (PPA).
PPA
3 Fungal hypha enters the PPA and is guided to the root cortex.
4 Fungus leaves plant cells, enters the apoplast, and grows along the root length.
5 Hyphae induce formation of new PPA structures inside cortical cells.
PAM
6 Hyphae enter PPAs and branch to form arbuscules, where nutrients are exchanged.


**36.7 Formation of Arbuscular Mycorrhizae** Mycorrhizae develop by a highly coordinated process that involves signal exchanges between the plant and the fungus.

A few species of bacteria have an enzyme that enables them to convert $N_2$ into a more reactive and biologically useful form by a process called **nitrogen fixation**. These prokaryotic organisms—*nitrogen fixers*—convert $N_2$ to ammonia ($NH_3$). Although there are relatively few species of nitrogen fixers, and their biomass is small compared to that of the organisms that depend on them, these talented prokaryotes are essential to the biosphere as we know it.

### Nitrogen fixers make all other life possible

By far the greatest share of total world nitrogen fixation is performed biologically by **nitrogen-fixing bacteria**, which fix approximately 170 million metric tons of nitrogen per year. About 80 million metric tons is fixed industrially by humans. Smaller amounts of nitrogen, about 20 million metric tons per year, are fixed in the atmosphere by nonbiological means such as lightning, volcanic eruptions, and forest fires. Rain brings these atmospherically formed products to the ground.

Several groups of bacteria fix nitrogen. In the oceans, various photosynthetic bacteria, including cyanobacteria, fix nitrogen. In fresh water, cyanobacteria are the principal nitrogen fixers. On land, free-living soil bacteria make some contribution to nitrogen fixation, but they fix only what they need for their own use and release the fixed nitrogen only when they die.

Important groups of nitrogen-fixing bacteria live in close association with plant roots. The plant obtains fixed nitrogen from the bacterium, and the bacterium obtains energy sources from the plant. As with arbuscular mycorrhizae, the relationship nitrogen-fixing bacteria have with plants is both mutualistic and symbiotic.

We will look at nitrogen-fixing symbioses in more detail later. But first: how does biological nitrogen fixation work?

### Nitrogenase catalyzes nitrogen fixation

Nitrogen fixation is the *reduction* of nitrogen gas (see Section 9.1). It proceeds by the stepwise addition of three pairs of hydrogen atoms to $N_2$ (**Figure 36.8**). In addition to $N_2$, these reactions require three things:

- A strong reducing agent to transfer hydrogen atoms (protons and electrons) to $N_2$ and to the intermediate products of the reaction
- A great deal of energy, which is supplied by ATP
- The enzyme **nitrogenase**, which catalyzes the reaction

Depending on the species of nitrogen fixer, either respiration or photosynthesis provides the necessary reducing agent and ATP.

Nitrogenase is strongly inhibited by oxygen, and many nitrogen fixers are anaerobes that live in environments with little or no $O_2$. But rhizobia are aerobic and fix nitrogen in aerobic plant roots. How can nitrogenase function under these circumstances?

Plants typically house nitrogen-fixing bacteria in special root structures called **nodules**. Within a nodule, $O_2$ is maintained at a low level that is sufficient to support respiration, but not so high as to inactivate nitrogenase. This is possible because the cytoplasm of nodule cells contains a plant-produced protein called **leghemoglobin**, which is an $O_2$ carrier. Leghemoglobin is a close relative of hemoglobin, the red, oxygen-carrying pigment of animals, and is thus an evolutionarily ancient molecule. Some plant nodules contain enough of it to be bright pink inside. Leghemoglobin, with its iron-containing heme groups, transports enough oxygen to the nitrogen-fixing bacteria to support their respiration, while keeping free oxygen concentrations low enough to protect nitrogenase.

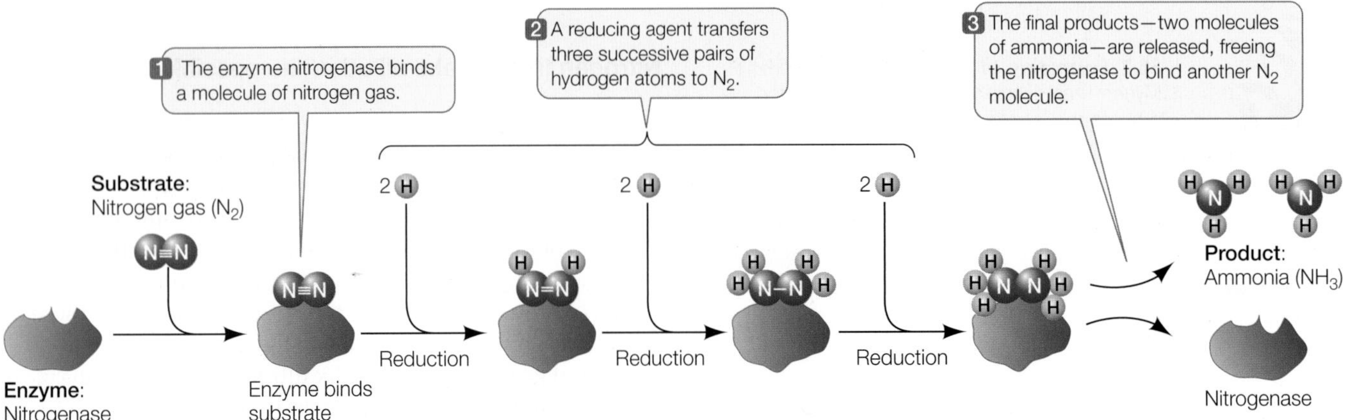


**36.8 Nitrogenase Fixes Nitrogen** Throughout the chemical reactions of nitrogen fixation, the reactants are bound to the enzyme nitrogenase. A reducing agent transfers hydrogen atoms to nitrogen, and eventually the final product—ammonia—is released. This reaction requires a large input of energy: about 16 ATPs are consumed per reaction.

### Some plants and bacteria work together to fix nitrogen

Bacteria of several different genera, collectively known as **rhizobia** (singular, *rhizobium*), fix nitrogen in close, mutualistic association with the roots of plants in the legume family. The legumes include peas, soybeans, clover, alfalfa, and many tropical shrubs and trees. The bacteria infect the plant's roots, and in response the roots develop nodules that house the bacteria.

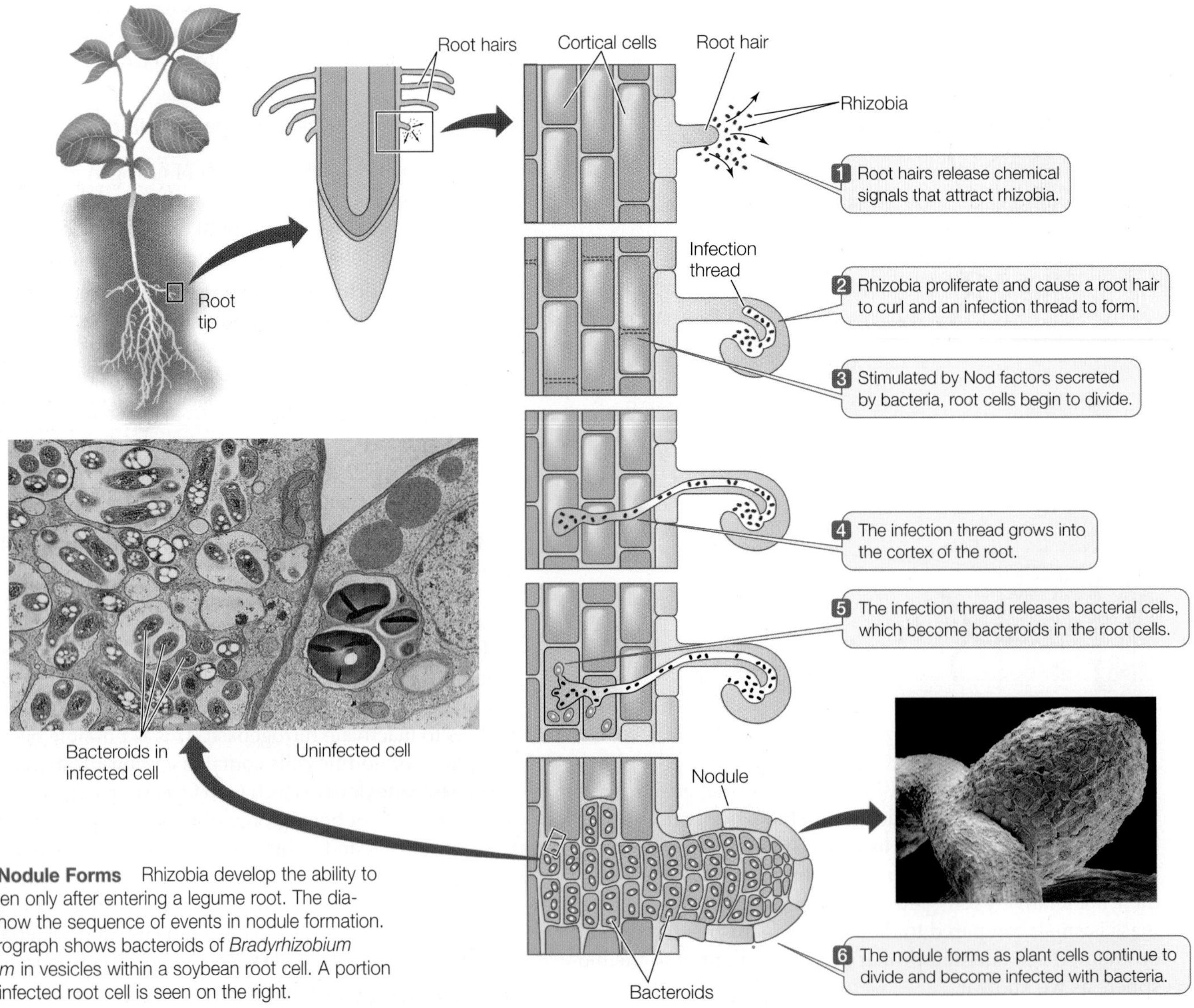

Root hairs
Cortical cells
Root hair
Rhizobia
Root tip
1 Root hairs release chemical signals that attract rhizobia.
Infection thread
2 Rhizobia proliferate and cause a root hair to curl and an infection thread to form.
3 Stimulated by Nod factors secreted by bacteria, root cells begin to divide.
4 The infection thread grows into the cortex of the root.
5 The infection thread releases bacterial cells, which become bacteroids in the root cells.
Bacteroids in infected cell
Uninfected cell
Nodule
Bacteroids
6 The nodule forms as plant cells continue to divide and become infected with bacteria.

**36.9 A Nodule Forms** Rhizobia develop the ability to fix nitrogen only after entering a legume root. The diagrams show the sequence of events in nodule formation. The micrograph shows bacteroids of *Bradyrhizobium japonicum* in vesicles within a soybean root cell. A portion of an uninfected root cell is seen on the right.

Because of their ability to form nitrogen-fixing associations, legumes are often used in crop rotations; for example, farmers might plant clover or alfalfa occasionally to increase the available nitrogen content of the soil.

The legume–rhizobium association is not the only nitrogen-fixing symbiosis. Some cyanobacteria fix nitrogen in association with fungi (in lichens) or with ferns, cycads, or nonvascular plants. Rice farmers can increase crop yields by growing the water fern *Azolla*, with its symbiotic nitrogen-fixing cyanobacterium, in the flooded fields where rice is grown. Another group of bacteria, the filamentous actinobacteria, fix nitrogen in association with woody species such as alder and mountain lilacs.

### Legumes and rhizobia communicate using chemical signals

Neither free-living rhizobia nor uninfected legumes can fix nitrogen. Only when the two are closely associated in root nodules does the reaction take place. The establishment of this symbiosis between a rhizobium and a legume requires a complex series of steps, with active contributions by both the bacterium and the plant root (**Figure 36.9**). First the root releases flavonoids and other chemical signals that attract soil-living rhizobia to the vicinity of the root. Flavonoids trigger the transcription of bacterial *nod* genes, the products of which synthesize Nod (nodulation) factors. These factors, secreted by the bacteria, cause cells in the root cortex to divide, leading to the formation of a primary nodule meristem. This meristem gives rise to the plant tissue that constitutes the root nodule.

The bacteria enter the root via an infection thread and eventually reach cells in the interior of the root nodule. There the bacteria are released into the cytoplasm of the nodule cells, enclosed in plant-derived membrane vesicles. Inside the vesicles, the bacteria differentiate into **bacteroids**—the form of bacteria that can fix nitrogen.

The legume–rhizobium interaction is very specific. For example, only one species of rhizobium will form a nitrogen-fixing symbiosis with alfalfa; another rhizobium will only infect clover. The specificity of the interaction is determined in part by the specificity of the chemical signals exchanged by the plant and bacterium. The soil may not have the correct bacterium for a given legume crop, so farmers and gardeners often coat legume seeds with the appropriate rhizobium before planting.

There is increasing evidence that nodule formation depends on some of the same genes and mechanisms that allow mycorrhizae to develop. For example, both processes involve invagination of the plasma membrane to allow entry of the fungal hypha or rhizobia. The similarities of the structures formed during the development of mycorrhizae and nodules are especially striking considering that the symbioses involve members of two different *kingdoms* (fungi and bacteria) (**Figure 36.10**).

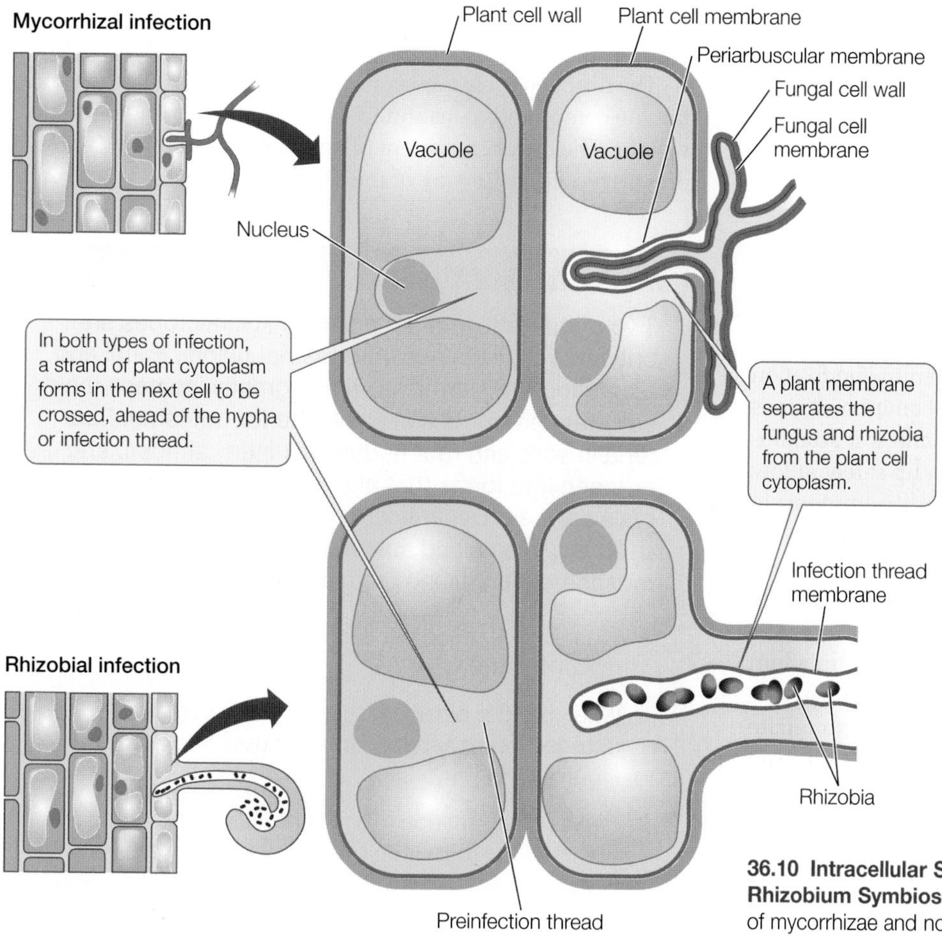

**36.10 Intracellular Structures in Plant–Fungus and Plant–Rhizobium Symbioses** Several steps in the development of mycorrhizae and nodules involve similar structures.

### Biological nitrogen fixation does not always meet agricultural needs

Bacterial nitrogen fixation is not always sufficient to support the needs of agriculture. Traditional farmers used to plant dead fish along with corn; the decaying fish released nitrogen that the developing corn could use. Today farmers use inorganic nitrogen fertilizers produced through industrial nitrogen fixation to meet the food needs of a rapidly expanding population.

Most industrial nitrogen fixation is done by the *Haber process*, a chemical reduction that requires a great deal of energy. (Recall that biological nitrogen fixation consumes a lot of ATP—about 16 ATP per N fixed; see Figure 36.8.) At present in the United States, the manufacture of nitrogen-containing fertilizer takes more energy—primarily natural gas and hydroelectric—than does any other aspect of crop production. The rising cost and dwindling supply of energy

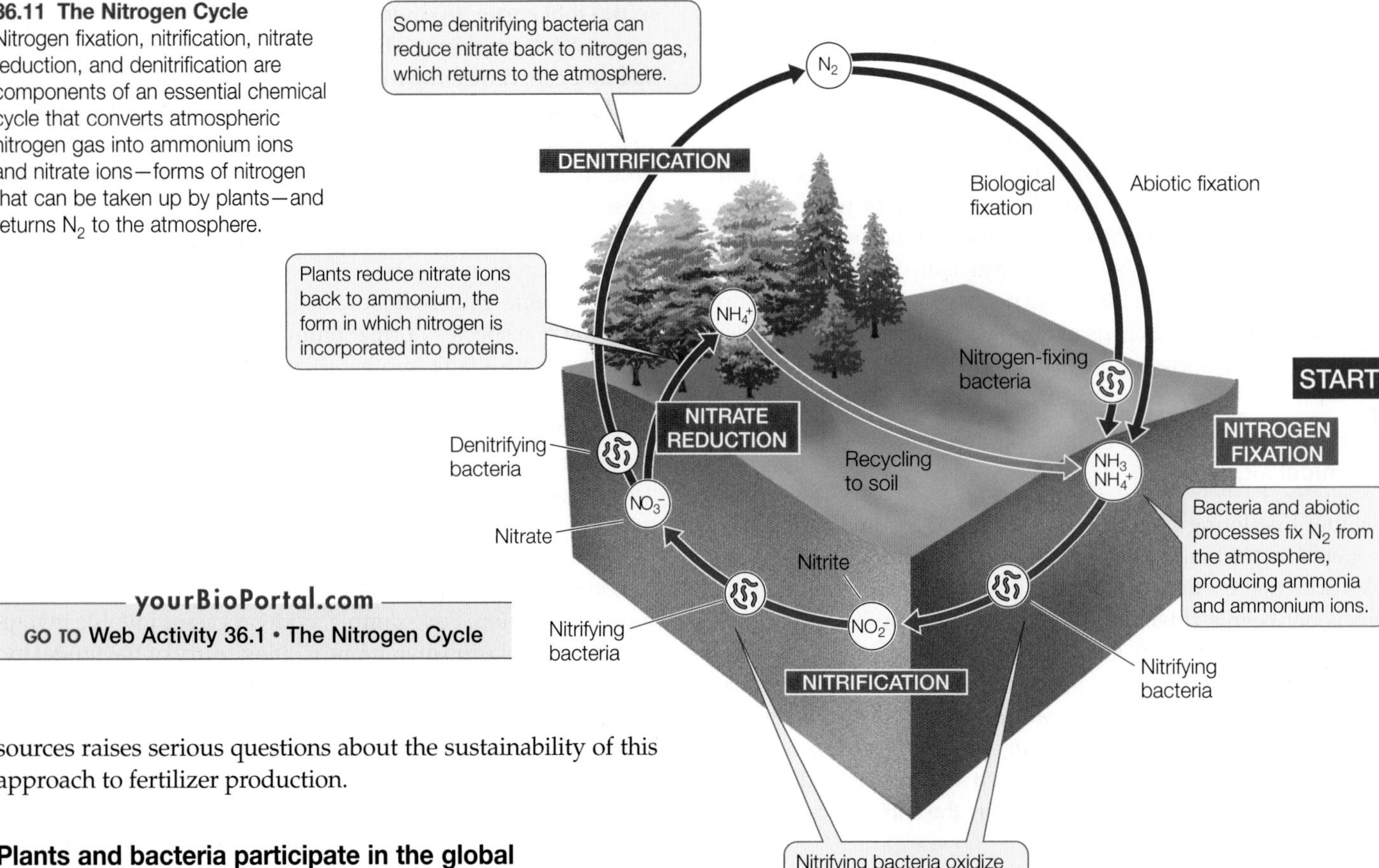

Some denitrifying bacteria can reduce nitrate back to nitrogen gas, which returns to the atmosphere.
DENITRIFICATION
$N_2$
Biological fixation
Abiotic fixation
Plants reduce nitrate ions back to ammonium, the form in which nitrogen is incorporated into proteins.
$NH_4^+$
Nitrogen-fixing bacteria
START
NITRATE REDUCTION
Denitrifying bacteria
Recycling to soil
$NH_3$ $NH_4^+$
NITROGEN FIXATION
$NO_3^-$
Nitrate
Bacteria and abiotic processes fix $N_2$ from the atmosphere, producing ammonia and ammonium ions.
Nitrite
$NO_2^-$
Nitrifying bacteria
NITRIFICATION
Nitrifying bacteria
Nitrifying bacteria oxidize ammonia to nitrate ions.

**36.11 The Nitrogen Cycle**
Nitrogen fixation, nitrification, nitrate reduction, and denitrification are components of an essential chemical cycle that converts atmospheric nitrogen gas into ammonium ions and nitrate ions—forms of nitrogen that can be taken up by plants—and returns $N_2$ to the atmosphere.

**yourBioPortal.com**
GO TO **Web Activity 36.1 • The Nitrogen Cycle**

sources raises serious questions about the sustainability of this approach to fertilizer production.

### Plants and bacteria participate in the global nitrogen cycle

Nitrogen moves through the biosphere in a **global nitrogen cycle** (**Figure 36.11**), which includes four key steps:

1. *Fixation* of atmospheric $N_2$ to $NH_3$ and $NH_4^+$ by bacteria and by abiotic processes
2. *Nitrification* of these molecules to nitrate by bacteria
3. *Nitrate reduction* by plants
4. *Denitrification* of nitrate by bacteria back to $N_2$, which is then released to the atmosphere to begin another cycle

The nitrogen released into the soil as a result of nitrogen fixation is primarily in the form of ammonia ($NH_3$) and ammonium ions ($NH_4^+$). Although ammonia can be toxic to plants if it accumulates in tissues, ammonium ions can be taken up safely at low concentrations. Soil bacteria called **nitrifiers** oxidize ammonia to nitrate ions ($NO_3^-$)—another form that plants can take up—by the process of **nitrification**. Soil pH affects which form of nitrogen is taken up by plants: nitrate ions are taken up preferentially under more acidic conditions, ammonium ions under more basic ones. To use nitrate, a plant must first reduce it to ammonium in a process called **nitrate reduction**. This occurs in two enzyme-catalyzed steps. The first step, from nitrate ($NO_3^-$) to nitrite ($NO_2^-$), takes place in the cytoplasm; the second, from nitrite ($NO_2^-$) to ammonia ($NH_4^+$), in the plastids. The plant uses the ammonia to manufacture amino acids, from which the plant's proteins and all its other nitrogen-containing compounds are formed. Animals cannot reduce nitrogen, and they depend on plants to supply them with reduced nitrogenous compounds.

The nitrogen cycle is essential for life on Earth: nitrogen-containing compounds constitute 5–30 percent of a plant's total dry weight. The nitrogen content of animals is even higher, and all of it arrives there by way of the plant kingdom.

## 36.4 RECAP

**Two mutualistic interactions with soil microbes are critical to the success of terrestrial plants. Fungi and plants form mycorrhizae, which greatly increase the soil area that roots can scavenge for nutrients. Bacteria in soils and root nodules fix inert, atmospheric nitrogen into forms that plants and ultimately animals can use. Denitrification returns nitrogen from dead organisms and animal waste back to the atmosphere, continuing the global nitrogen cycle.**

- What is exchanged between plants and fungi in mycorrhizae? **See p. 762**
- What, besides nitrogenase, is required to reduce nitrogen gas to a form plants can use? **See p. 763 and Figure 36.8**
- How is the formation of a root nodule on a legume similar to the formation of an arbuscular mychorriza? **See p. 765 and Figure 36.10**

Let's turn now to some special mechanisms for obtaining nutrients that have evolved in plant species with unusual lifestyles.

# 36.5 How Do Carnivorous and Parasitic Plants Obtain a Balanced Diet?

Most plants obtain their mineral nutrients from the soil solution (with the help of fungi), but some use other sources. Carnivorous and parasitic plants are examples of such plants.

## Carnivorous plants supplement their mineral nutrition

Some plants augment their nitrogen supply by capturing and digesting flies and other insects. There are about 500 of these **carnivorous plant** species, the best known of which are Venus flytraps (genus *Dionaea*; **Figure 36.12A**), sundews (genus *Drosera*; **Figure 36.12B**), and pitcher plants (genus *Sarracenia*).

Carnivorous plants are typically found in boggy habitats that are acidic and nutrient deficient. To obtain extra nitrogen, these plants capture animals, digest their proteins, and absorb the amino acids. Pitcher plants have pitcher-shaped leaves that collect small amounts of rainwater. Insects and even small rodents are lured into the pitchers by bright colors or attractive scents and are prevented from leaving by stiff, downward-pointing hairs. The animals eventually die and are digested by a combination of plant enzymes and bacteria in the water. Sundews have leaves covered with hairs that secrete a clear, sticky, sugary liquid. Insects become stuck to these hairs, and more hairs curve over to further entrap them. Enzymes secreted by the plant digest the insects. Venus flytraps have specialized leaves with two halves that fold together. When an insect touches trigger hairs on a leaf, its two halves quickly come together, their spiny margins interlocking and imprisoning the insect before it can escape. The leaf then secretes enzymes that digest its prey.

The closing of the Venus flytrap's leaf is one of the fastest movements in the plant world, requiring only 0.1 sec. To find out how this happens, Dr. Lakshminarayanan Mahadevan and colleagues painted fluorescent dots on the surface of the flytrap's leaf surface and used high-speed cameras to record the trap snapping shut when its trigger hairs were touched. They then used computer analysis of the recorded dot movements to generate a mathematical model to help explain the movement. The researchers found that the first step is the elongation of cells on the outer surface of the leaf. The expansion of only one side of the leaf causes it to snap from a convex into a concave shape, much like a contact lens flipping inside out.

Carnivorous plants do not need to feed on insects, but doing so helps them grow faster in their natural habitats. They use the additional nitrogen from the insects to make more proteins, chlorophyll, and other nitrogen-containing compounds.

(A) *Dionaea muscipula*

(B) *Drosera rotundifolia*

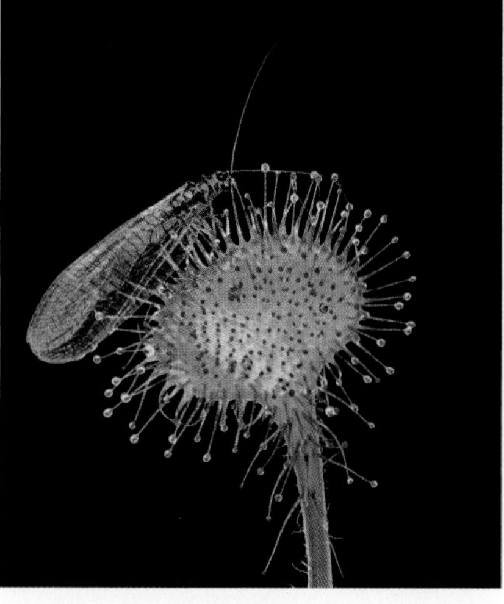

**36.12 Carnivorous Plants** Some plants have adapted to nitrogen-poor environments by becoming carnivorous. (A) The Venus flytrap obtains nitrogen from the bodies of insects trapped inside the plant when its hinges snap shut. (B) Sundews trap insects on sticky hairs. Secreted enzymes will digest the carcass externally.

## Parasitic plants take advantage of other plants

Approximately 1 percent of flowering plant species derive some or all of their water, nutrients, and sometimes even photosynthate from other plants. In these **parasitic plants**, absorptive organs called **haustoria** have evolved that invade the host and tap into the vascular tissues in the root or stem.

Parasitic plants are divided into two broad classes based on their nutritional interactions with their hosts. **Hemiparasites** can still photosynthesize, but derive water and mineral nutrients from the living bodies of other plants. Perhaps the most familiar hemiparasites are the several genera of mistletoes. Mistletoes are green and carry on some photosynthesis, but they parasitize other plants for water and mineral nutrients and may derive photosynthetic products from them as well. Dwarf mistletoe (*Arceuthobium americanum*) is a serious parasite in forests of the western United States, destroying more than 3 billion board feet of lumber per year.

**Holoparasites** are completely parasitic and do not perform photosynthesis. They are taxonomically and morphologically diverse. Some, such as members of the dodder family, are plant-like in appearance, with small leaf remnants and flowers (**Figure 36.13**). Some holoparasites do not have leaves or stems because they spend most of their life cycle underground and only break the surface to flower.

Several parasitic plant species lack many of the genes normally present in the chloroplast genome (which in turn is only a remnant of the genome in the original endosymbiont from which the chloroplast evolved; see Sections 5.5 and 27.1). These genes, which are needed for photosynthesis, have been lost because there is no evolutionary pressure to retain them. Thus, while the parasitic lifestyle can be viewed as a free ride, for some

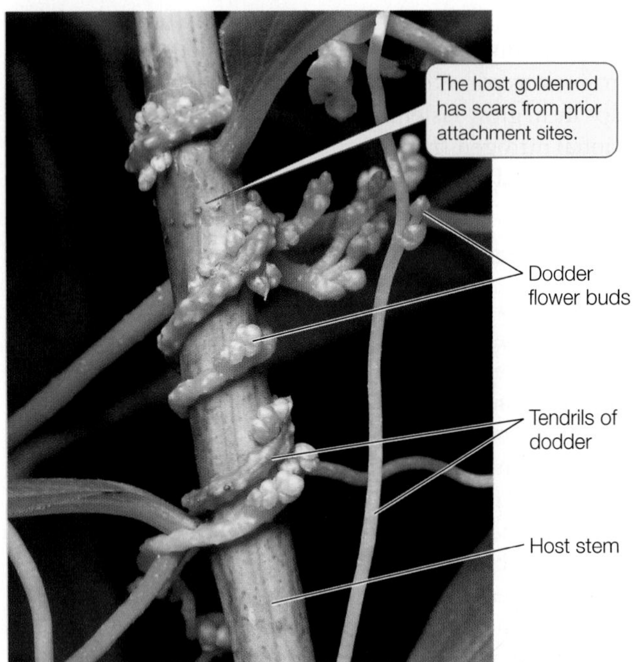


**36.13 A Parasitic Plant** Tendrils of dodder (genus *Cuscuta*) wrap around a goldenrod (genus *Solidago*). The parasitic dodder obtains water, sugars, and other nutrients through tiny, rootlike protuberances that penetrate the surface of the host plant.

plants it is also a one-way ticket, with no possibility of return to self-sufficiency.

### The plant–parasite relationship is similar to plant–fungi and plant–bacteria associations

Plant–bacteria and plant–fungus relations both involve reciprocal signaling between the two species (see Figures 36.7 and 36.9). Parasitic plants also need to detect nearby plants so they can obtain their nutrients, but obviously this is to the disadvantage of the potential host plant. In one interesting case, a mechanism has evolved in a parasitic plant to recognize the chemical signals produced by plants to attract beneficial fungi.

The holoparasite *Striga* (witchweed) is a serious pest of cereal crops in Africa (see the opening story of Chapter 30). Earlier in this chapter, you learned that arbuscular fungi are attracted to plant roots by compounds called strigolactones. One of these same molecules was discovered over 60 years ago as an inducer of germination of some parasitic plants, including *Striga*. Scientists strongly suspect that this is no coincidence. The mycorrhizal interaction is ancient (over 400 million years old) and predates the evolution of parasitic plants. For this reason scientists hypothesize that a mechanism evolved in the ancestors of modern *Striga* to recognize a compound that was already produced by plants to attract soil microbes.

In *Striga* we thus find an example of "opportunistic evolution"—that is, the repurposing of preexisting processes rather than the invention of new processes from scratch. This is not the first time we have encountered this phenomenon in this chapter. Recall that the formation of nodules by rhizobia uses some of the same mechanisms utilized by arbuscular fungi to establish residence inside plant cells (see Figure 35.10), implying an evolutionary connection between the two symbioses.

## 36.5 RECAP

**Carnivorous plants supplement their nutrition by extracting materials from animals. Rapid reflexes have evolved in some of these plants for trapping their prey. Parasitic plants, on the other hand, get some or all of their sustenance from other plants. Extreme holoparasites cannot function as autotrophs, having lost chloroplast genes coding for photosynthetic machinery. At least one parasitic plant responds to the same signaling molecule that the host plant uses to attract beneficial fungi.**

- Why do carnivorous plants eat animals? See p. 767
- How do the needs of holoparasitic plants differ from those of carnivorous plants? See p. 767
- What characteristics are shared among plant–parasite, plant–fungus, and plant–bacteria associations?

## CHAPTER SUMMARY

### 36.1 How Do Plants Acquire Nutrients?

- Plants are photosynthetic autotrophs that can produce all their organic molecules from carbon dioxide, water, and minerals, including a nitrogen source.
- **Mineral nutrients** are obtained from the **soil solution**.
- Root growth allows plants, which are sessile, to search for mineral resources.
- Microenvironments within the soil, such as acidic or alkaline areas, affect the direction of root growth.

### 36.2 What Mineral Nutrients Do Plants Require?

- Plants require 14 **essential elements**. Of these, six are **macronutrients** and eight are **micronutrients**. **Deficiency symptoms** suggest what essential element a plant lacks. Review Table 36.1 and Figure 36.1
- The essential elements were discovered by growing plants on **hydroponic solutions** that lacked individual elements. Review Figures 36.2 and 36.3, **ANIMATED TUTORIAL 36.1**

### 36.3 How Does Soil Structure Affect Plants?

- **Soils** contain water, air, and inorganic and organic substances. Soils have living (biotic) and nonliving (abiotic) components. Review Figure 36.4

- A soil typically consists of two or three horizontal zones called **horizons**. **Topsoil** forms the uppermost or **A horizon**. Topsoil tends to lose mineral nutrients through **leaching**. **Loams** are excellent agricultural topsoils, with a good balance of sand, silt, clay, and organic matter. Review Figure 36.5
- Soils form by mechanical and chemical **weathering** of rock. Chemical weathering imparts mineral nutrients to **clay** particles. Plant litter and other organic matter decomposes to form **humus**. Plants obtain some mineral nutrients through **ion exchange** between the soil solution and the surface of clay particles. Review Figure 36.6
- Farmers use **fertilizers** to make up for deficiencies in soil mineral nutrient content. **Liming** can reverse acidification.
- Plants can influence the characteristics, including the pH, of the soil in which they grow.

## 36.4 How Do Fungi and Bacteria Increase Nutrient Uptake by Plant Roots?

- **Mycorrhizae** are symbiotic root–fungus associations that greatly increase a plant's absorption of water and minerals, especially phosphorus. They occur in 80 percent of plant species.
- The fungal **mycelia** invade root cortex cells and form arbuscules, which are the sites of nutrient exchange between the fungus and plant. Review Figure 36.7
- In the earliest stages of mycorrhiza formation, the **hyphae** of arbuscular fungi grow toward **strigolactones**, compounds that are produced by the plant roots.
- Some **nitrogen-fixing bacteria** live free in the soil; others live symbiotically as **bacteroids** within plant roots. In **nitrogen fixation**, nitrogen gas ($N_2$) is reduced to ammonia ($NH_3$) or ammonium ions ($NH_4^+$) in a reaction catalyzed by **nitrogenase**. Review Figure 36.8
- Nitrogenase requires anaerobic conditions, but the bacteroids in root **nodules** require oxygen, which is maintained at the proper level by **leghemoglobin**.
- The formation of a root nodule requires interaction between the root system of a legume and a **rhizobium**. Review Figure 36.9
- Several steps in the formation of root nodules and arbuscules are similar and probably involve some of the same plant genes. Review Figure 36.10
- In agriculture, biological nitrogen fixation must usually be supplemented with commercial nitrogen fertilizers made by the Haber process.
- Plants and bacteria interact in the **global nitrogen cycle**, which involves a series of reductions and oxidations of nitrogen-containing molecules. Review Figure 36.11, **WEB ACTIVITY 36.1**
- **Nitrification** by bacteria converts ammonia to nitrate ions in the soil. **Nitrate reduction** is carried out by plant enzymes, enabling plants to form their own nitrogen compounds. **Denitrification** returns nitrogen from animal wastes and dead organisms to the atmosphere.

## 36.5 How Do Carnivorous and Parasitic Plants Obtain a Balanced Diet?

- **Carnivorous plants** are autotrophs that supplement a low nitrogen supply by feeding on insects or other small animals.
- **Parasitic plants** draw on other plants to meet their needs, which may include minerals, water, or the products of photosynthesis.
- **Hemiparasites**, such as mistletoes, can still photosynthesize. Extreme **holoparasites** cannot function as auxotrophs because they have lost chloroplast genes that code for components of the photosynthetic apparatus (which they no longer need).
- A strigolactone—a compound in the same category of compounds plants use to attract mycorrhizal fungi—also induces the germination of some parasitic plants, including *Striga*.
- Scientists hypothesize that a mechanism evolved in the ancestors of modern *Striga* to recognize a compound that was already produced by plants to attract arbuscular fungi.

## SELF-QUIZ

1. Macronutrients
   *a.* are so called because they are more essential than micronutrients.
   *b.* include manganese, boron, and zinc, among others.
   *c.* function as catalysts.
   *d.* are required in concentrations of at least 1 gram per kilogram of plant dry matter.
   *e.* are obtained by the process of photosynthesis.
2. Which of the following is *not* an essential mineral element for plants?
   *a.* Potassium
   *b.* Magnesium
   *c.* Calcium
   *d.* Lead
   *e.* Phosphorus
3. Fertilizers
   *a.* are often characterized by their N-P-O percentages.
   *b.* are not required if crops are removed frequently enough.
   *c.* restore needed mineral nutrients to the soil.
   *d.* are needed to provide carbon, hydrogen, and oxygen to plants.
   *e.* are needed to destroy soil pests.
4. In a typical soil,
   *a.* the topsoil tends to lose mineral nutrients by leaching.
   *b.* there are four or more horizons.
   *c.* the C horizon consists primarily of loam.
   *d.* the dead and decaying organic matter gathers in the B horizon.
   *e.* more clay means more air space and thus more oxygen for roots.
5. Which of the following is *not* true for arbuscules?
   *a.* They are an ancient association between plants and fungi.
   *b.* They expand the effective root area of plants and allow more efficient water uptake.
   *c.* They are a significant source of fixed nitrogen for plants.
   *d.* They are a significant source of phosphorous for plants.
   *e.* Most land plants have them.
6. Nitrogen fixation is
   *a.* performed only by plants.
   *b.* the oxidation of nitrogen gas.
   *c.* catalyzed by the enzyme nitrogenase.
   *d.* a single-step chemical reaction.
   *e.* possible because $N_2$ is a highly reactive substance.

7. Nitrification is
   *a.* performed only by plants.
   *b.* the reduction of ammonium ions to nitrate ions.
   *c.* the reduction of nitrate ions to nitrogen gas.
   *d.* catalyzed by the enzyme nitrogenase.
   *e.* performed by certain bacteria in the soil.
8. Which of the following is an early step in the formation of *both* arbuscules and root nodules?
   *a.* Invasion of a plant root by a fungus
   *b.* Invasion of a plant root by a bacterium
   *c.* Strigolactones produced by the root are recognized by the microbe
   *d.* Root cells are invaded but there is no direct contact between plant and microbe cell contents
   *e.* Root cells are invaded and there is direct contact between plant and microbe cell contents
9. Which of the following is a parasite?
   *a.* Venus flytrap
   *b.* Pitcher plant
   *c.* Sundew
   *d.* Dodder
   *e.* Tobacco
10. All carnivorous plants
   *a.* are parasites.
   *b.* depend on animals as a source of carbon.
   *c.* are incapable of photosynthesis.
   *d.* depend on animals as their sole source of phosphorus.
   *e.* obtain supplemental nitrogen from animals.

## FOR DISCUSSION

1. Methods for determining whether a particular element is essential have been known for more than a century. Since these methods are so well established, why was the essentiality of some elements discovered only recently?
2. If a Venus flytrap were deprived of soil sulfates and hence made unable to synthesize the amino acids cysteine and methionine, would it die from lack of protein? Explain.
3. Soils are dynamic systems. What changes might result when land is subjected to heavy irrigation for agriculture after being relatively dry for many years? What changes in the soil might result when a virgin deciduous forest is cut down and replaced by crops that are harvested each year? Even though the countries share the same island, why are hurricanes frequently accompanied by the loss of life in Haiti but not in the Dominican Republic?
4. We mentioned that important positively charged ions are held in the soil by clay particles, but other, equally important, negatively charged ions are leached deeper into the soil's B horizon. Why doesn't leaching cause an electrical imbalance in the soil? (Hint: think of the ionization of water.)
5. The biosphere of Earth as we know it depends on the existence of a few species of nitrogen-fixing prokaryotes. What do you think might happen if one of these species were to become extinct? If all of them were to disappear?
6. Holoparasitic plants have lost many of the morphological and genetic traits necessary for an autotrophic lifestyle. From an evolutionary point of view, how do you think this happened? (Hint: think about selection pressures.)

## ADDITIONAL INVESTIGATION

Some mutant *Arabidopsis* plants that are very bushy (their shoots are more highly branched than wild-type plants) cannot make strigolactones because of a mutation in a gene necessary for strigolactone biosynthesis. If an investigator applies strigolactones to the plants, they grow normally. What does this experiment suggest about the role of strigolactones in plant growth? How does this add to the story of strigolactones as signals for arbuscules and parasitic plants?

## WORKING WITH DATA (GO TO yourBioPortal.com)

**Is Nickel an Essential Element for Plant Growth?** In this hands-on exercise, you will critically examine the experimental approach used by Patrick Brown and his colleagues to show that barley plants require nickel for their life cycle (Figure 36.3). Analyzing data from the original paper, you will calculate the critical value, the tissue concentration below which growth is significantly reduced.

# 37 Regulation of Plant Growth

## Saving millions of lives by regulation of plant development

There is no Nobel Prize for agriculture as there is for medicine, the other branch of applied biology. This is not because plant biology is unimportant; it is because agriculture was not mentioned in the will of the prize's benefactor, Alfred Nobel. But a Nobel prize was awarded in this field: plant geneticist Norman Borlaug received the Peace Prize for research on wheat that has been estimated to have saved a billion lives.

In their constant search for ways to help farmers produce more food for a growing population, biologists have developed crop plants whose physiology allows them to produce more grain per plant (higher yield). However, when a plant produces a lot of seeds, the sheer weight of the load may cause the stem to bend over, or even break. This makes harvesting the seeds impossible: think of how hard it would be to get enough food for your family if you had to pick up seeds on the ground, some of which had already sprouted.

In 1945, the U.S. Army temporarily occupied Japan, which it had defeated in World War II. During the war, Japan, an island nation, was blockaded and could not import food. How had they been able to grow enough grain to feed their people? The answer lay in the fields: the Japanese had bred genetic strains of rice and wheat with short, strong stems that could bear a high yield of grain without bending or breaking. This innovation made an impression on an agricultural advisor who happened to be among the first wave of U.S. occupiers, and seeds of the Japanese strains were sent back to the U.S.

A decade later, Borlaug, who was working in Mexico at the time, began genetic crosses of what were known as semi-dwarf wheat plants from Japan with varieties that had genes conferring rapid growth, adaptability to varying climates, and resistance to fungal diseases. The results were genetic strains of wheat that gave record yields, first in Mexico and then in India and Pakistan in the 1960s. At about the same time and using a similar strategy, scientists in the Philippines developed semi-dwarf rice with equally spectacular results. People who had lived on the edge of starvation now produced enough food. Countries that had been relying on food aid from

**Norman Borlaug** Seen here in a field of semi-dwarf wheat, plant geneticist Norman Borlaug carried out a program of genetic crosses that led to high-yielding varieties and saved millions from starvation.

**Semi-Dwarf Rice** The short variety of rice can give higher yields of grain than its taller counterpart (right). The difference is that the latter can respond to the hormone gibberellin.

other countries were now growing so much grain that they could export the surplus. The development of these semi-dwarf grains began what was called the "Green Revolution."

It is only recently that plant biologists have discovered why semi-dwarf wheat and rice have short stems. In normal plants a hormone called gibberellin stimulates stem elongation. But in the semi-dwarf plants, a mutation affects the signal transduction mechanism for gibberellin so that the stem cells do not respond to it and growth is reduced. The lives of countless people have been saved by a disruption of hormone signaling.

**IN THIS CHAPTER** we will give a brief overview of the life of a flowering plant and its developmental stages. We will explore the nature of the environmental cues, photoreceptors, and hormones (including gibberellin) that regulate plant growth and development. We will also consider the multiple roles and interactions of these different elements.

CHAPTER OUTLINE



## 37.1 How Does Plant Development Proceed?

As Chapter 34 describes, plants are sessile organisms that must seek out resources above and below the ground. Features that maximize the ability of plants to obtain the resources that they need to grow and reproduce include:

- *Meristems*. Plants have permanent collections of stem cells (undifferentiated, constantly dividing cells) that allow them to continue growing throughout their lifetimes (see Section 34.4).
- *Post-embryonic organ formation*. Unlike animals, plants can initiate development of new organs such as leaves and flowers throughout their lifetimes.
- *Differential growth*. Plants can allocate their resources so that they grow more of the organs that will benefit them most; for example, more leaves to harvest more sunlight or more roots to obtain more water and nutrients.

To use growth for maximal advantage, plants must continuously monitor their environment and redirect their growth as appropriate. Under normal circumstances a plant's environment is never completely stable. For example, the amount of light changes from day to night and from season to season. In addition, other plants are often vying for what light there is, and plants modulate their growth to compete with their neighbors for this precious resource. As you will see in this chapter, several mechanisms have evolved in plants to sense their environment and trigger appropriate growth responses.

The *development* of a plant—the series of progressive changes that take place throughout its life—is regulated in many ways. Key factors involved in regulating plant growth and development are:

- *Environmental cues*, such as day length
- *Receptors* that allow a plant to sense environmental cues, such as photoreceptors that absorb light, and chemoreceptors that signal the presence of pathogens (see Chapter 39)
- *Hormones*—chemical signals that mediate the effects of the environmental cues, including those sensed by receptors
- The plant's *genomes*, which encode regulatory proteins and enzymes that catalyze the biochemical reactions of development

We will explore these regulatory mechanisms in more detail later in this chapter. But first let's look at the initial steps of plant development—from seed to seedling—and the types of internal and external cues that guide them.

### In early development, the seed germinates and forms a growing seedling

If all developmental activity is suspended in a seed, even when conditions appear to be suitable for its growth, the seed is said to be **dormant**. Cells in dormant seeds do not divide, expand, or differentiate. For the embryo to begin developing, seed dormancy must be broken by one of the mechanisms discussed later in this section.

As the seed begins to **germinate**—to develop into a seedling—it takes up water. The growing embryo then obtains chemical building blocks—carbohydrate, amino acid, and lipid monomers—for its development by digesting the polysaccharides, fats, and proteins stored in the seed. As we will see later, the embryos of some plant species secrete hormones that direct the mobilization of these reserves. Germination is completed when the **radicle** (embryonic root) emerges from the seed coat. The plant is then called a **seedling**.

If the seed germinates underground, the new seedling must elongate rapidly (in the right direction!) and cope with a period of life in darkness or dim light. A series of photoreceptors direct this stage of development and prepare the seedling for growth in the light.

Early shoot development varies among the flowering plants. **Figure 37.1** shows the shoot development patterns of monocots and eudicots.

### Environment cues can initiate seed germination

The seeds of some plant species are capable of germinating as soon as they have matured. All they need for germination is water. But the seeds of many species are dormant at maturity. Seed storage and dormancy may last for weeks, months, years, or even centuries, as we saw in the opening story of Chapter 29: in 2005 a botanist was able to germinate a date palm seed recovered from a 2000-year-old storage bin at Masada in Israel. The mechanisms that maintain seed dormancy are numerous and diverse, but three principal strategies dominate:

- *Exclusion of water or oxygen* from the embryo by an impermeable seed coat
- *Mechanical restraint* of the embryo by a tough seed coat
- *Chemical inhibition* of germination

Seed dormancy must be broken before germination can begin. The dormancy of seeds with impermeable seed coats can be broken if the coat is abraded as the seed tumbles across the ground or through a creek bed or passes through the digestive tract of an animal. Cycles of freezing and thawing can also aid in making the seed coat permeable, as can soil microorganisms. Fire can end seed dormancy by melting waterproof wax in seed coats, allowing water to reach the embryo. Fire can also release mechanical restraint by cracking the seed coat. *Leaching*—the dissolving and diffusing away of water-soluble chemical inhibitors by prolonged exposure to water—is another way in which dormancy can be broken.

yourBioPortal.com
GO TO Web Activity 37.1 • Monocot Shoot Development
AND Web Activity 37.2 • Eudicot Shoot Development

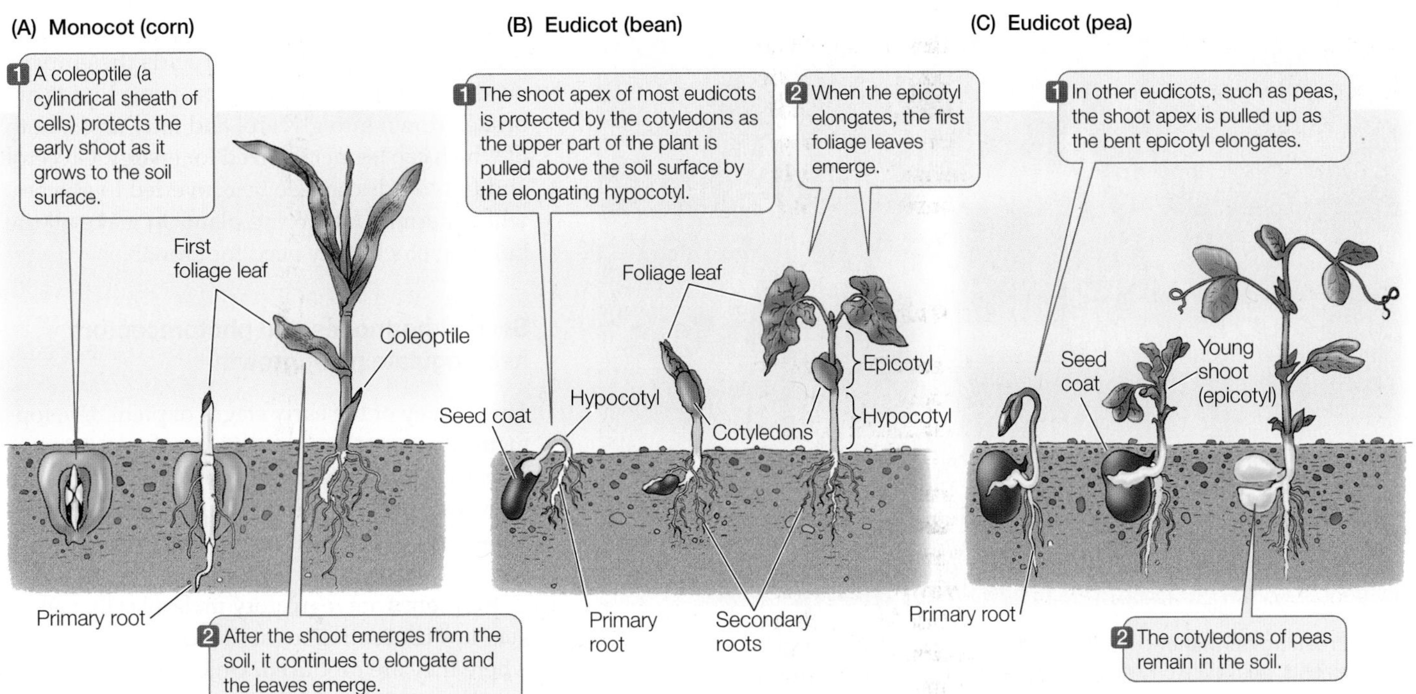
(A) Monocot (corn)
1 A coleoptile (a cylindrical sheath of cells) protects the early shoot as it grows to the soil surface.
First foliage leaf
Coleoptile
Primary root
2 After the shoot emerges from the soil, it continues to elongate and the leaves emerge.
(B) Eudicot (bean)
1 The shoot apex of most eudicots is protected by the cotyledons as the upper part of the plant is pulled above the soil surface by the elongating hypocotyl.
2 When the epicotyl elongates, the first foliage leaves emerge.
Foliage leaf
Epicotyl
Hypocotyl
Seed coat
Hypocotyl
Cotyledons
Primary root
Secondary roots
(C) Eudicot (pea)
1 In other eudicots, such as peas, the shoot apex is pulled up as the bent epicotyl elongates.
Seed coat
Young shoot (epicotyl)
Primary root
2 The cotyledons of peas remain in the soil.

**37.1 Patterns of Early Shoot Development** (A) In grasses and some other monocots, growing shoots are protected by a coleoptile until they reach the soil surface. (B) In most eudicots, the growing point of the shoot is protected within the cotyledons. (C) In some eudicots, the cotyledons remain in the soil, and the apex is protected by the first true leaves.

### Seed dormancy affords adaptive advantages

What are the potential advantages of seed dormancy? For many plant species, dormancy ensures survival during unfavorable conditions and results in germination when conditions are more favorable for growth. To avoid germination in the dry days of late summer, for example, some seeds require exposure to a long cold period (winter) before they will germinate. Other seeds will not germinate until a certain amount of time has passed, regardless of how they are treated. Among other things, this strategy prevents germination while the seeds are still attached to the parent plant. Plants whose seeds need fire to break dormancy avoid competition with other plants by germinating only where an area has been cleared by fire. Dormancy also helps seeds to survive long-distance dispersal, allowing plants to colonize new territory.

The dormancy of some seeds is broken by exposure to light. These seeds, which germinate only at or near the surface of the soil, are generally tiny and have few food reserves. Seedlings from such seeds would not have enough food to reach the light if they germinated deep in the soil. Conversely, the germination of some other seeds is inhibited by light; these seeds germinate only when deeply buried. Light-inhibited seeds are often large and well-stocked with nutrients.

Dormancy may also increase the likelihood of a seed germinating in a favorable ecological setting. Some cypress trees, for example, grow in standing water, and their seeds germinate only if germination inhibitors are leached away by water (**Figure 37.2**).

### Seed germination begins with the uptake of water

Seeds begin to germinate when dormancy is broken and environmental conditions are satisfactory. The first step in germination is the uptake of water, called **imbibition** (from *imbibe*, "to drink in"). A dormant seed contains very little water; only 5 to 15 percent of a seed's weight is water compared to 80 to 95 percent for most other plant parts. Seeds also contain polar macromolecules, such as cellulose and starch, that attract and bind polar water molecules. Consequently a seed has a very negative water potential (see Section 35.1), and will take up water if the seed coat is permeable. The force exerted by imbibing seeds, which expand several-fold in volume, demonstrates the magnitude of seed water potential. Imbibing cocklebur seeds can exert a pressure of up to 1,000 atmospheres (approximately 100 kilopascals or 15,000 pounds per square inch)!

As a seed takes up water, it undergoes metabolic changes: enzymes are activated upon hydration, RNA and then proteins are synthesized, the rate of cellular respiration increases, and other metabolic pathways are activated. In many seeds, there is no initiation of the cell division cycle during the early stages of germination. Instead, growth results solely from the expansion of small preformed cells. DNA is synthesized only after the radicle begins to grow and ruptures the seed coat.

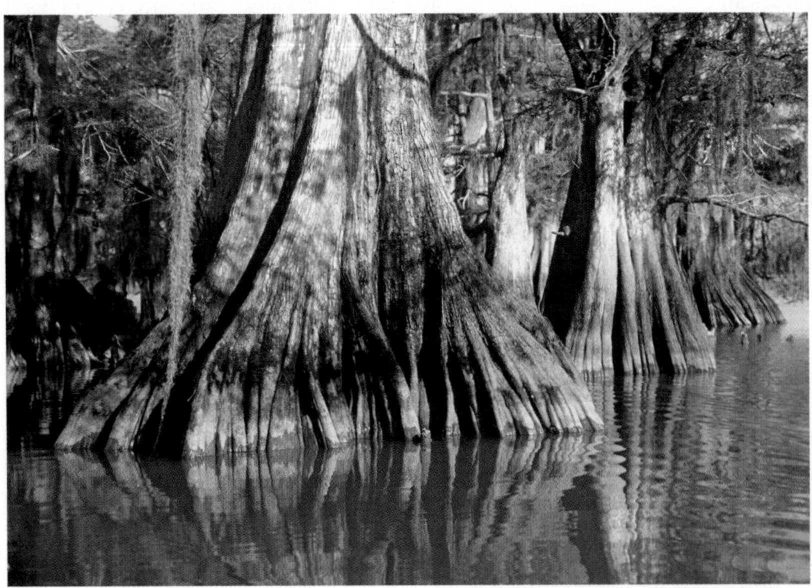

**37.2 Breaking dormancy** The seeds of bald cypress, a tree adapted to moist or wet environments, germinate only after being leached by water, which increases the chances that they will germinate in a location suitable for their growth.

### The embryo must mobilize its reserves

To fuel its metabolic activities, the embryo must use the reserves of energy and raw materials stored in the seed. Until the young plant is able to photosynthesize, it depends on these reserves, which are stored in the **cotyledons** (see Figure 37.1) or in the **endosperm** (the specialized nutritive tissue) of the seed. The principal reserve of energy and carbon in many seeds, such as wheat, is starch. Other seeds, such as sunflower, store fats or oils. Usually the seed holds amino acid reserves in the form of proteins, rather than as free amino acids.

Before they can be used to support growth of the embryo, starch, lipids, and proteins must be broken down by enzymes into monomers. Starch breakdown yields glucose for energy metabolism and for the synthesis of cellulose and other cell wall constituents. Digestion of stored proteins provides the amino acids the embryo needs to synthesize its own proteins. Lipids are broken down into glycerol and fatty acids, both of which can be metabolized for energy. Glycerol and fatty acids can also be converted to glucose, which permits fat-storing plants to make all the building blocks they need for growth.

### Several hormones and photoreceptors help regulate plant growth

This survey of the early stages of plant development illustrates the many internal and external cues that influence plant growth. A plant's responses to these cues are initiated and maintained by two types of regulators: *hormones* and *photoreceptors*.

**Hormones** are regulatory molecules that act at very low concentrations at sites often distant from where they are produced. Unlike animals, which usually produce each hormone in specific cells within the body, plants produce hormones in many types of cells. Each plant hormone plays multiple regulatory roles, affecting several different aspects of plant development (**Table 37.1**). Interactions among these hormones can be complex. Several hormones regulate plant growth from seedling to adult. Other hormones are involved in the plant's defenses against herbivores and microorganisms (discussed in Chapter 39).

**Photoreceptors**, like hormones, are involved in many developmental processes in plants. However, unlike plant hormones, which are small molecules, plant photoreceptors consist of *pigments* (molecules that absorb light) associated with proteins. Light acts directly on photoreceptors, which in turn regulate developmental processes that need to be responsive to light, such as the many changes that occur as a young seedling germinates and emerges from the soil.

### Signal transduction pathways are involved in all stages of plant development

Plants, like other organisms, make extensive use of *signal transduction pathways*, sequences of biochemical reactions by which a cell generates a response to a stimulus (see Chapter 7). Cell signaling in plant development generally involves a receptor (for a hormone or for light) and a signal transduction pathway, and concludes with a cellular response that is relevant to development. Protein kinase cascades often amplify responses to signals in plants, as they do in other organisms (see Figure 7.12). We will look at several plant signal transduction pathways in more detail in the remaining sections of this chapter.

No matter what cues regulate development, the plant's genome ultimately determines the limits of plant development. The genome encodes the master plan, but its interpretation depends on conditions in the environment. For several decades biologists focused on identifying the hormones and photoreceptors that control plant development, but recent advances in molecular genetics have now made it possible to explore the underlying processes that regulate development, such as signal transduction pathways.

TABLE 37.1

**Plant Growth Hormones**

| HORMONE | STRUCTURE | TYPICAL ACTIVITIES |
|---|---|---|
| Abscisic acid* | [structure: $H_3C$, $CH_3$, $CH_3$, OH, O, $CH_3$, COOH] | Maintains seed dormancy; closes stomata |
| Auxins (e.g., indole-3-acetic acid) | [structure: OH, O, N, H] | Promote stem elongation, adventitious root initiation, and fruit growth; inhibit axillary bud outgrowth, leaf abscission, and root elongation |
| Brassinosteroids | [structure: OH, OH, HO, HO, H, O, O] | Promote stem and pollen tube elongation; promote vascular tissue differentiation |
| Cytokinins | [structure: H, $CH_2OH$, C=C, HN—$CH_2$, $CH_3$, N, N, N, N, H] | Inhibit leaf senescence; promote cell division and axillary bud outgrowth; affect root growth |
| Ethylene | [structure: H, H, C=C, H, H] | Promotes fruit ripening and leaf abscission; inhibits stem elongation and gravitropism |
| Gibberellins | [structure: O, CO, H, H, =$CH_2$, $CO_2H$, $CH_3$] | Promote seed germination, stem growth, and fruit development; break winter dormancy; mobilize nutrient reserves in grass seeds |

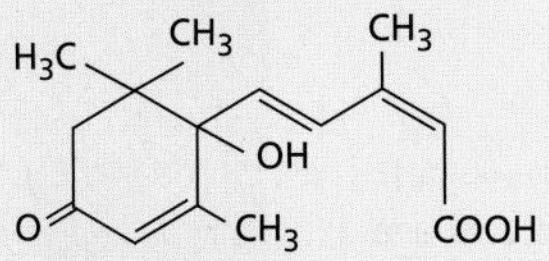

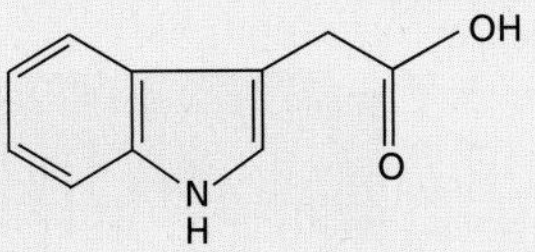

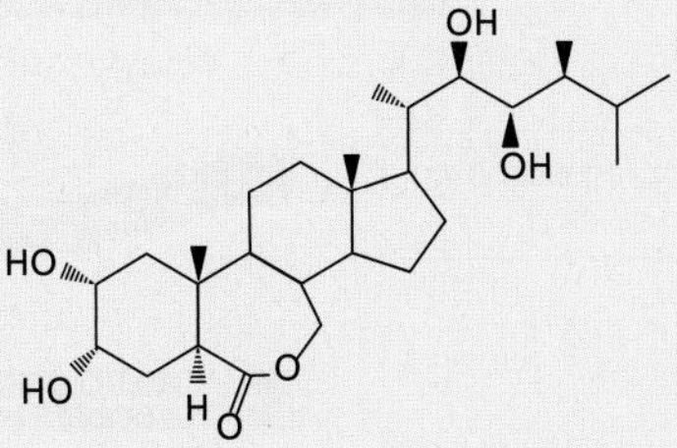

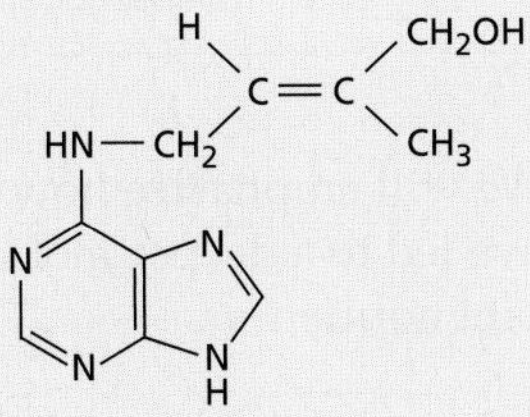

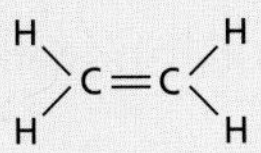

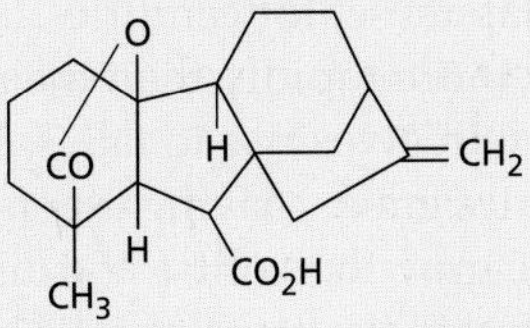

*See Chapter 38.

### Studies of *Arabidopsis thaliana* have increased our understanding of plant signal transduction

Many recent advances in understanding plant growth and development have come from work with *Arabidopsis thaliana*, a weed in the mustard family. This plant is used as a model organism by researchers because its body and seeds are tiny, its nuclear genome is unusually small for a flowering plant (125 million base pairs), and it flowers and forms many seeds (up to 10,000 per plant) within weeks of germination. Furthermore, its genomes (nuclear, plastid, and mitochondrial) are fully sequenced, so researchers have an accounting of all genes in the plant.

In Chapter 19, we describe how genetics can be used to identify the steps along a developmental pathway. You will recall the theme of these experiments: *if a mutation for a certain biochemical process disrupts a developmental event, then the biochemical process must be essential for that developmental event.* Similarly, genetics can be used to dissect pathways for receptor activation and signal transduction in plants: if proper signaling does not occur in a mutant strain, the mutant gene must be involved in the signal transduction process. Mapping the mutant gene and characterizing its molecular phenotype is a starting point for understanding the signaling pathway.

One technique for identifying genes involved in a plant signal transduction pathway is illustrated in **Figure 37.3**. Called a **genetic screen**, the process involves creating a collection of mutants and identifying those individuals that are likely to have a defect in the pathway being studied. Genes can be randomly mutated in two ways:

- Insertion of a transposon (see Section 17.2)
- Point mutation by a chemical mutagen, usually ethyl methane sulfonate

In both cases, a large number of mutated plants are then examined for a specific phenotype, usually a characteristic that is easy to see or measure (e.g., height). The growth conditions and plant characteristics used for the screen are carefully chosen to maximize the chances that the selected plants will have a defect in the pathway of interest. Once mutant plants have been selected, their genotypes and phenotypes are compared to those of wild-type plants. *Arabidopsis* mutants with altered developmental patterns have provided a wealth of new information about the hormones present in plants and the mechanisms of hormone and photoreceptor action.

TOOLS FOR INVESTIGATING LIFE

**37.3 A Genetic Screen**

Genetics of the model plant *Arabidopsis thaliana* can be used to identify the steps of a signal transduction pathway. If a mutant strain does not respond to a hormone (in this case, ethylene), the corresponding wild-type gene must be essential for the pathway (in this case, ethylene response). This method has been instrumental to scientists in understanding plant growth regulation.

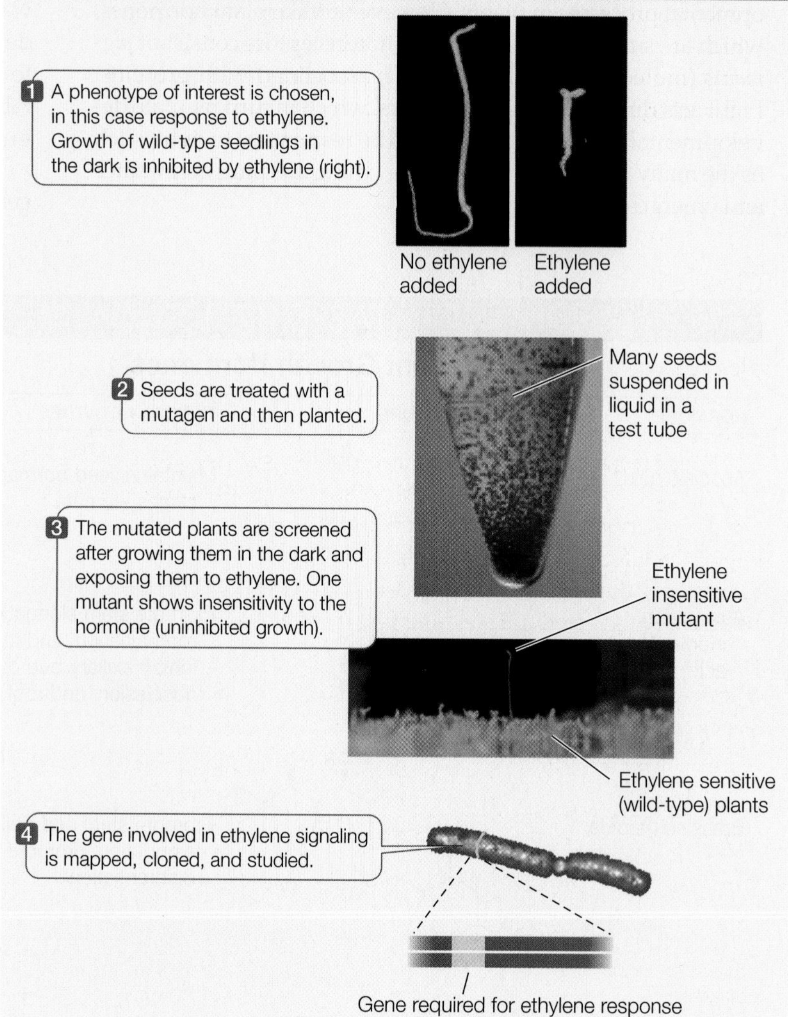

1 A phenotype of interest is chosen, in this case response to ethylene. Growth of wild-type seedlings in the dark is inhibited by ethylene (right).
No ethylene added
Ethylene added
2 Seeds are treated with a mutagen and then planted.
Many seeds suspended in liquid in a test tube
3 The mutated plants are screened after growing them in the dark and exposing them to ethylene. One mutant shows insensitivity to the hormone (uninhibited growth).
Ethylene insensitive mutant
Ethylene sensitive (wild-type) plants
4 The gene involved in ethylene signaling is mapped, cloned, and studied.
Gene required for ethylene response

## 37.1 RECAP

**Plant development is under the control of external cues in the environment as well as internal hormones. In both cases, signal transduction pathways regulate plant development. Genetic screens have been useful in describing signal transduction pathways in the model plant *Arabidopsis thaliana.* Seed dormancy often precedes seed germination.**

- Describe how monocots and eudicots differ in early development. See p. 773 and Figure 37.1
- Under what circumstances is seed dormancy advantageous? See p. 774
- What fuels the metabolic activities of a young plant embryo before it is able to commence photosynthesis? See p. 774
- What is a genetic screen and how can it be used to analyze the regulation of plant development? See p. 776 and Figure 37.3

You have now seen the early stages of plant development and growth, and how the environment influences these processes. Plant hormones are central to the internal regulation of development, a subject to which we now turn. We will describe how hormones were discovered and what physiological effects they have on plants. We will emphasize how genetic screens and other methods have led to a deeper molecular understanding of the action of plant hormones.

## 37.2 What Do Gibberellins Do?

In Asia, rice farmers have known about "foolish seedling disease" (*bakanae* in Japanese) for centuries. Seedlings affected by this disease grow more rapidly than their healthy neighbors, but this rapid growth gives rise to tall, spindly plants that die before producing rice grains, having expended most of their energy on vegetative growth. At first, the disease was attributed to an inherited defect in the plants themselves. But by the twentieth century it was clear that it is caused by infection with the ascomycete fungus *Gibberella fujikuroi*. How does the infection cause the disease?

The mystery of foolish seedling disease was solved in 1925 by Japanese biologist Eiichi Kurosawa. He hypothesized that the fungus must release a molecule that overstimulates plant growth. To isolate it, he grew the fungus in a liquid medium and then removed the fungus from the medium by filtration. He heated the filtered medium to kill any remaining fungus, but found that the resulting heat-treated filtrate was still capable of inducing rapid growth in rice seedlings. Medium that had never contained the fungus did not have this effect. This experiment established that *G. fujikuroi* produces a growth-promoting chemical substance, which Kurosawa called a **gibberellin**. Soon, gibberellin was isolated and its chemical nature described (see Table 37.1).

## Gibberellins are plant hormones

Once externally applied gibberellin was shown to affect rice plant growth, a question arose: *does a plant make the same or similar molecules to regulate its own growth?* Biologists used a genetic approach to answer this question, by studying mutant strains of corn and tomato that were dwarfs—they had abnormally short stems. The stems hardly grew, even though other parts of the plants appeared normal. Because this was the exact opposite of the effect of *too much* gibberellin (in the fungus-infected rice plants), the biologists hypothesized that the dwarf mutants had *too little* gibberellin. This hypothesis was tested in two ways:

- Gibberellin was applied to the dwarf plants. As a result, they grew to normal height (**Figure 37.4**).
- The levels of gibberellin were measured in wild-type and dwarf plants, and the wild-type plants had much more gibberellin.

These experiments clearly showed that gibberellin is made by plants and acts to stimulate stem elongation. Numerous chemically related gibberellins exist, all belonging to a family of common plant metabolites called *diterpenoids*.

## Gibberellins have many effects on plant growth and development

The functions of gibberellins can be inferred from the effects of experimentally decreasing gibberellins or blocking their action at various points in plant development. Such experiments reveal that gibberellins have multiple roles in regulating plant growth.

**FRUIT GROWTH** Gibberellins and other hormones regulate the growth of fruits. Grapevines that produce seedless grapes develop smaller fruit than varieties that produce seed-bearing grapes. Biologists wanting to explain this phenomenon removed seeds from immature seeded grapes and found that this prevented normal fruit growth, suggesting that the seeds are sources of a growth regulator. Biochemical studies showed that developing seeds produce gibberellins, which diffuse out into the immature fruit tissue. Spraying young seedless grapes with a gibberellin solution causes them to grow as large as seeded ones, and this is now a standard commercial practice (**Figure 37.5**).

**MOBILIZATION OF SEED RESERVES** Early in seed germination hydrolytic enzymes are produced to break down stored reserves of starch, proteins, and lipids. Just after imbibition in germinating seeds of barley and other cereals, the embryo secretes gibberellins. The hormones diffuse through the endosperm to a surrounding tissue called the **aleurone layer**, which lies underneath the seed coat. The gibberellins trigger a cascade of events in the aleurone layer, causing it to syn-

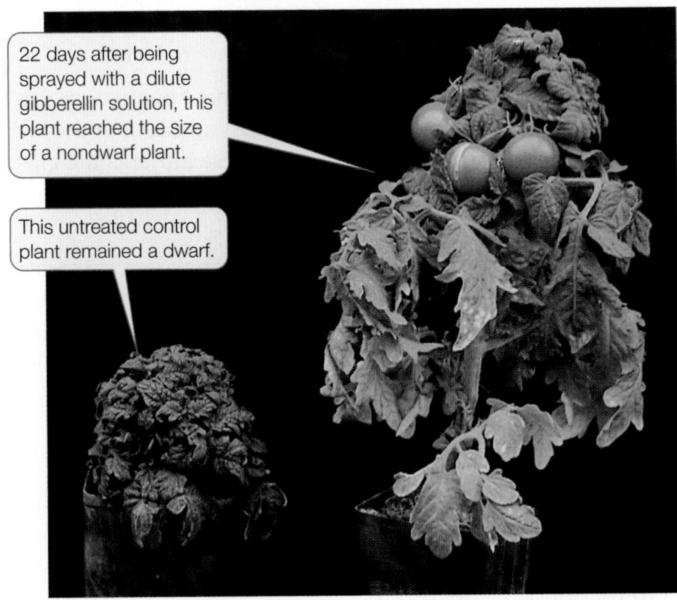

22 days after being sprayed with a dilute gibberellin solution, this plant reached the size of a nondwarf plant.
This untreated control plant remained a dwarf.

**37.4 The Effect of Gibberellins on Dwarf Plants** Both of the dwarf tomato plants in this photograph were the same size when the one on the right was treated with gibberellins.

**37.5 Gibberellin and Fruit Growth** Spraying developing seedless grapes with gibberellins (right) increases their size compared to untreated fruit (left).

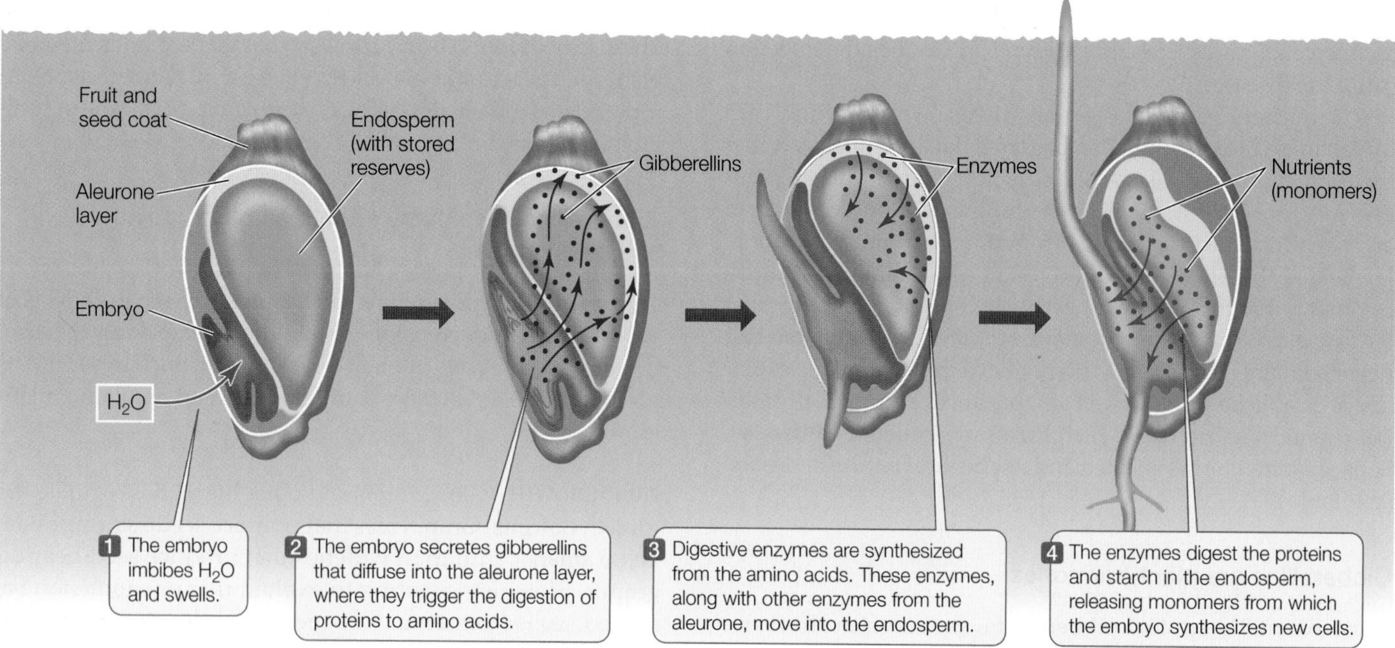

Fruit and seed coat
Endosperm (with stored reserves)
Aleurone layer
Embryo
$H_2O$
Gibberellins
Enzymes
Nutrients (monomers)
1 The embryo imbibes $H_2O$ and swells.
2 The embryo secretes gibberellins that diffuse into the aleurone layer, where they trigger the digestion of proteins to amino acids.
3 Digestive enzymes are synthesized from the amino acids. These enzymes, along with other enzymes from the aleurone, move into the endosperm.
4 The enzymes digest the proteins and starch in the endosperm, releasing monomers from which the embryo synthesizes new cells.


**37.6 Embryos Mobilize Their Reserves** During seed germination in cereal grasses, gibberellins trigger a cascade of events that result in the conversion of starch and protein reserves into monomers that can be used by the developing embryo.

**yourBioPortal.com**
GO TO **Web Activity 37.3 • Events of Seed Germination**

thesize and secrete enzymes that digest proteins and starch stored in the endosperm (**Figure 37.6**). These observations have practical importance: in the beer brewing industry, gibberellins are used to enhance the "malting" (germination) of barley and the breakdown of its endosperm, producing sugar that is fermented to alcohol.

**STEM ELONGATION** The effects of gibberellins on wild-type plants are not as dramatic as those seen on dwarf plants. However, gibberellins are indeed active in wild-type plants, because inhibitors of gibberellin synthesis cause a reduction in stem elongation. Such inhibitors can be put to practical uses. For example, plants such as chrysanthemums that are grown in greenhouses tend to get too tall; but leggy plants, unfortunately, do not appeal to consumers. Flower growers thus spray such plants with gibberellin synthesis inhibitors to control their height. Some wheat crops are similarly sprayed to keep them short, so they do not fall over when they produce grain—this is essentially a chemically induced version of the semi-dwarf genetic varieties described in the opening story of this chapter. In some plants, such as cabbage, the normal growth habit is to be a squat, leafy head near the ground. When environmental signals are right, however, the plant "bolts," quickly producing a tall stem with flowers. This response is mediated by gibberellins.

## Gibberellins act by initiating the breakdown of transcriptional repressors

The molecular mechanisms underlying gibberellin action have been worked out with the help of genetic screens. Biologists started by identifying mutant plants whose growth and development are *insensitive* to gibberellins; that is, they are *not* affected by added gibberellins. Several such mutants have been found—both natural mutant strains and induced mutants selected from genetic screens—and they fall into two general categories:

- *Excessively tall plants*. These plants resemble wild-type plants given an excess of gibberellin, and get no taller when given extra gibberellin. They are also tall even when treated with inhibitors of gibberellin synthesis. Their gibberellin response is always "on," even in the absence of the hormone. It is presumed that the normal allele for the mutant gene codes for an *inhibitor* of the gibberellin signal transduction pathway. In wild-type plants, the pathway is "off" but in the mutant plants, the pathway is "on" and the plant grows tall.
- *Dwarf plants*. These plants resemble dwarf tomato or maize plants that are deficient in gibberellin synthesis, but they do not respond to added gibberellin. In these mutants the gibberellin response is always "off," regardless of the presence of the hormone.

The two types of mutations described above *affect the same protein*, which turns out to be a *repressor* of a transcription factor that stimulates the expression of growth-promoting genes. The repressor protein has two important domains, explaining how mutations in the same protein can have seemingly opposite effects:

- *One region of the repressor protein binds to the transcription complex to inhibit transcription*. This is the region mutated in the excessively tall plants: the growth-promoting genes are always "on" because the repressor does not bind to the transcription complex.
- *Another part of the repressor protein causes it to be removed from the transcription complex*. This is the region mutated in the

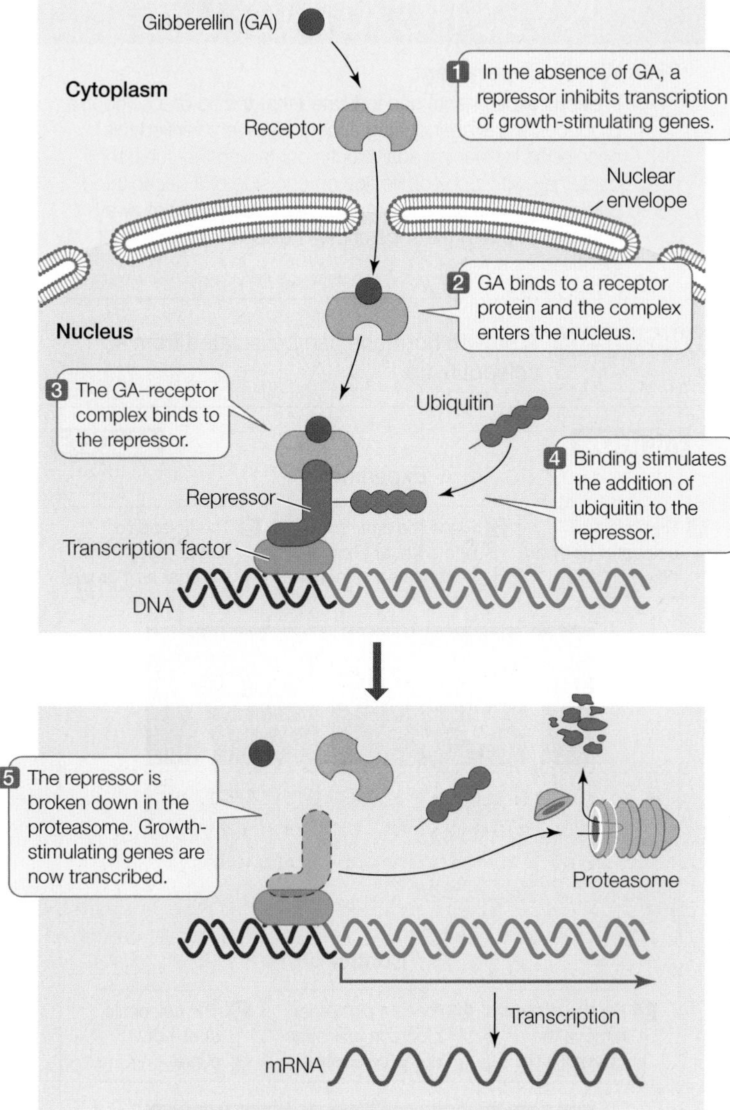

Gibberellin (GA)
Cytoplasm
Receptor
1 In the absence of GA, a repressor inhibits transcription of growth-stimulating genes.
Nuclear envelope
Nucleus
2 GA binds to a receptor protein and the complex enters the nucleus.
3 The GA–receptor complex binds to the repressor.
Ubiquitin
4 Binding stimulates the addition of ubiquitin to the repressor.
Repressor
Transcription factor
DNA
5 The repressor is broken down in the proteasome. Growth-stimulating genes are now transcribed.
Proteasome
Transcription
mRNA

**37.7 How Gibberellin Works** Gibberellin acts to stimulate gene transcription by inactivating a repressor protein, a common mechanism for plant hormone action.

dwarf plants: the growth-promoting genes are always "off" because the repressor is always bound to the complex.

How does gibberellin act in this system? *Gibberellin acts by removing the repressor from the transcription complex* (**Figure 37.7**). It does this by binding to a receptor protein, which in turn binds to the repressor. Binding of the gibberellin–receptor complex stimulates poly-ubiquitination of the repressor, targeting it for breakdown in the proteasome (see Figure 16.24). The gibberellin receptor contains a region called an **F-box** that facilitates protein–protein interactions necessary for protein breakdown. While animal genomes have few F-box containing proteins, plant genomes have hundreds, an indication that this type of gene regulation is common in plants. As you will see, this regulatory mechanism underlies the effects of another important plant hormone: auxin.

## 37.2 RECAP

**Gibberellins are plant hormones that affect stem growth, fruit size, seed germination, and many other aspects of plant development; the effects vary from species to species.**

- How were gibberellins shown to be plant hormones? See p. 777
- How do gibberellins contribute to the germination of barley seeds? See Figure 37.6
- Explain how gibberellins work at the molecular level. See pp. 778–779 and Figure 37.7

Most other hormones, like the gibberellins, have multiple effects within the plant, and they often interact with one another to regulate developmental processes. In controlling stem elongation, for example, gibberellins interact with another hormone, auxin, to which we now turn.

# 37.3 What Does Auxin Do?

**Auxin** was discovered as a result of investigations into **phototropism**, the growth of plants toward light (as in shoots) or away from it (as in roots). Ever the curious biologist, Charles Darwin wanted to know how plants bend toward a source of light. In 1880, he and his son Francis published the results of their experiments on what part of a plant senses light.

yourBioPortal.com
GO TO Animated Tutorial 37.1 • Tropisms

The Darwins worked with canary grass (*Phalaris canariensis*) seedlings grown in the dark. A young grass seedling has a **coleoptile**—a cylindrical sheath a few cells thick that protects the delicate shoot as it pushes through the soil (see Figure 37.1A). When the seedling breaks through the soil surface, the coleoptile soon stops growing, and the shoot emerges unharmed. The coleoptiles of grasses are phototropic—they grow toward the light.

To find the light-receptive region of the coleoptile, the Darwins "blindfolded" the coleoptiles of dark-grown canary grass seedlings in various places, and then illuminated them from one side (**Figure 37.8**). The coleoptile grew toward the light whenever its tip was exposed. If the top millimeter or more of the coleoptile was covered, however, the coleoptile showed no phototropic response. Thus, the Darwins were able to conclude that the tip contains the photoreceptor that responds to light. The actual bending toward the light, however, takes place in a growing region a few millimeters below the tip. Therefore, the Darwins reasoned, *some type of signal must travel from the tip of the coleoptile to the growing region.* Later, others demonstrated that this signal is a chemical substance by showing that it can move through certain permeable materials, such as gelatin, but not through impermeable materials, such as a metal sheet.

# INVESTIGATING LIFE

## 37.8 The Darwins' Phototropism Experiment

Charles Darwin and his son Francis wanted to know how plants bend toward the light. They grew canary grass seedlings (coleoptiles) in the dark. To discover what part of the coleoptile responds to light, they covered up ("blindfolded") different regions of each coleoptile and then exposed the seedlings to light from one side. The Darwins discovered that the tip of the seedling senses the light and that growth occurs below the tip. Their observations led them to hypothesize the existence of a growth-promoting signal produced by the coleoptile tip.

**HYPOTHESIS** Only part of the coleoptile senses the light that triggers phototropism.

**METHOD**

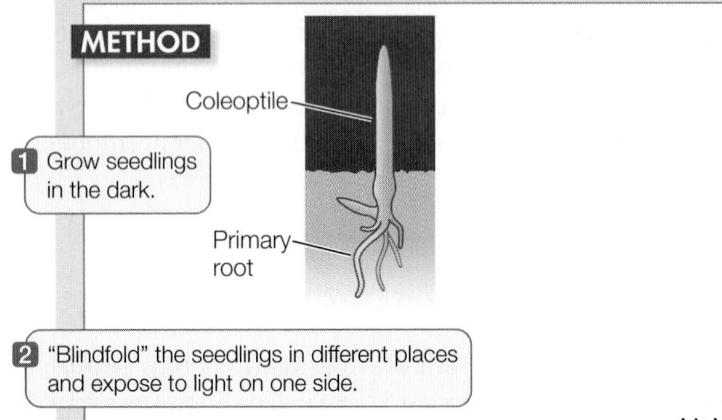

Coleoptile
1 Grow seedlings in the dark.
Primary root

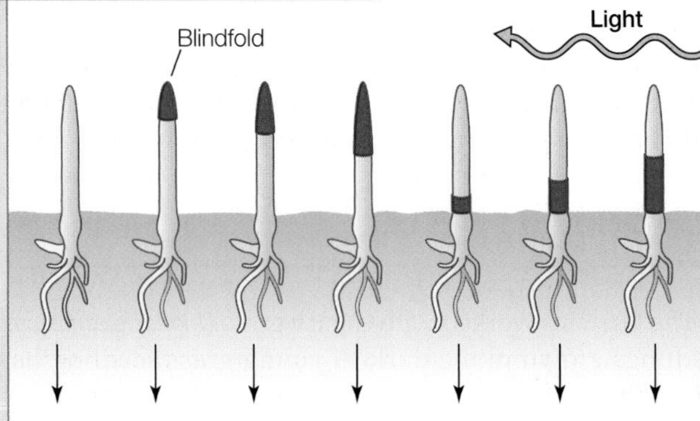

2 "Blindfold" the seedlings in different places and expose to light on one side.
Blindfold
Light

**RESULTS**

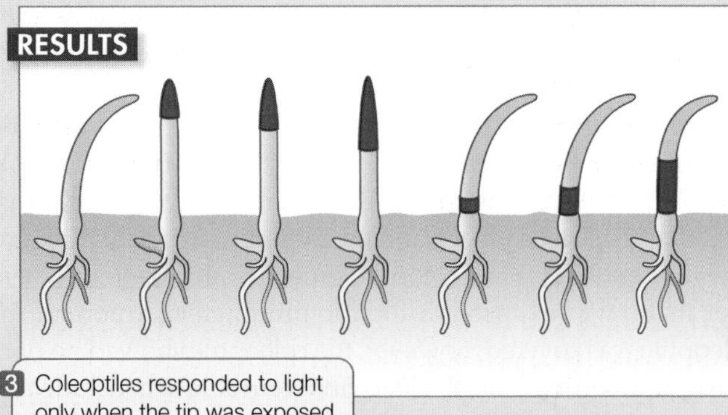

3 Coleoptiles responded to light only when the tip was exposed.

**CONCLUSION** The part of the coleoptile that senses light is in the tip, and it sends a signal from the tip to the growing region.

Go to **yourBioPortal.com** for original citations, discussions, and relevant links for all INVESTIGATING LIFE figures.

# INVESTIGATING LIFE

## 37.9 Went's Experiment

Previous experiments had indicated that the tip of a coleoptile produces a growth-inducing substance. Went verified this conclusion by placing agar blocks containing the substance contained within coleoptile tips on one side of a decapitated coleoptile. In the absence of light, the coleoptile bent away from the side with the substance. The substance was later identified as auxin.

**HYPOTHESIS** A growth hormone can be isolated from a coleoptile tip.

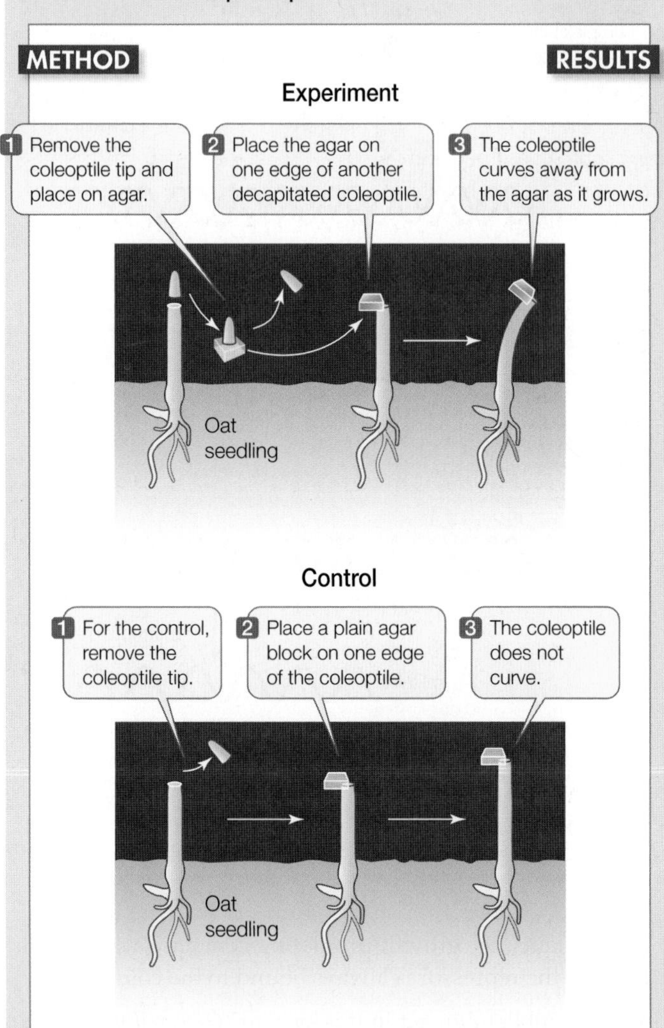

METHOD
RESULTS
Experiment
1 Remove the coleoptile tip and place on agar.
2 Place the agar on one edge of another decapitated coleoptile.
3 The coleoptile curves away from the agar as it grows.
Oat seedling
Control
1 For the control, remove the coleoptile tip.
2 Place a plain agar block on one edge of the coleoptile.
3 The coleoptile does not curve.
Oat seedling

**CONCLUSION** A growth hormone diffused from the tip into the agar, and from the agar into another plant. It had an effect on the growth of the plant similar to that of a coleoptile tip.

Go to **yourBioPortal.com** for original citations, discussions, and relevant links for all INVESTIGATING LIFE figures.

**yourBioPortal.com**
GO TO Animated Tutorial 37.2 • Went's Experiment

In the 1920s, the Dutch botanist Frits Went followed up on the Darwins' experiment. He removed coleoptile tips and placed their cut surfaces on a block of agar. Then he placed pieces of that agar on decapitated coleoptiles—positioned to cover only one side (**Figure 37.9**). As they grew, the coleoptiles curved away from the side with the agar, showing that the agar contained a substance that stimulated elongation of cells on that side of the coleoptile. This substance had diffused into the agar block from the isolated coleoptile tips. Eventually, the hormone indole-3-acetic acid (see Table 37.1) was isolated from similar agar blocks and was nicknamed *auxin* (from the Latin "to increase").

### Auxin transport is polar and requires carrier proteins

The experiments we have just described showed that in coleoptiles auxin movement is strictly *polar*—that is, it is unidirectional along a line from apex to base. Auxin transport is polar in other organs as well. For example, in a leaf petiole, which connects the leaf blade to the stem, auxin moves from the leaf blade end toward the stem. In roots, auxin moves unidirectionally toward the root tip. How does this directional transport occur?

Polar transport depends on four biochemical conditions that should be familiar from earlier chapters (**Figure 37.10**):

- *Diffusion across a plasma membrane*. Polar molecules diffuse across the plasma membrane less readily than nonpolar molecules.
- *Membrane protein asymmetry*. Active transport carriers for auxin are located only at the basal (bottom) end of the plasma membrane.

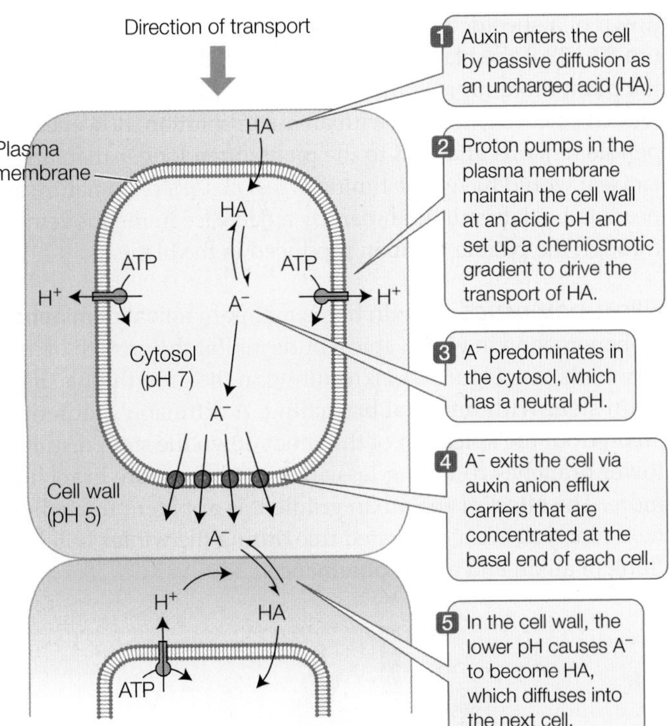


**37.10 Polar Transport of Auxin** Proton pumps set up a chemiosmotic gradient directing $A^-$ toward the basally placed auxin active transport carriers, leading to a net movement of auxin in a basal direction.

- *Proton pumping/chemiosmosis*. An ATP-driven proton pump removes $H^+$ from the cell, thereby increasing the intracellular pH and decreasing the pH in the cell wall. Proton pumping also sets up an electrochemical gradient, with potential energy to drive the transport of auxin by the carriers noted above.
- *The ionization of a weak acid*. Indole-3-acetic acid (which, recall, is the chemical name for auxin; see Table 37.1) is a weak acid:

$$A^- + H^+ \rightleftharpoons HA$$

When the pH is low, the increased $H^+$ concentration drives this reaction to the right and HA (non-ionized auxin) is the predominant form. When the pH is higher, there is more $A^-$ (ionized auxin).

### Auxin transport mediates responses to light and gravity

While polar auxin transport distributes the hormone along the longitudinal axis of the plant, *lateral* (side-to-side) redistribution of auxin is responsible for plant movements. This redistribution is carried out by auxin carrier proteins that move from the base of the cell to one side; because of this, auxin exits the cell only on that side of the cell, rather than at the base, and moves sideways within the tissue.

When light strikes a grass coleoptile on one side, auxin at the tip moves laterally toward the shaded side. The asymmetry thus established is maintained as polar transport moves auxin down the coleoptile, so that in the growing region below, the auxin concentration is highest on the shaded side. Cell elongation is thus speeded up on that side, causing the coleoptile to bend toward the light (**phototropism**; **Figure 37.11A**). If you have noticed a houseplant bending toward a window, you have observed phototropism.

Light is not the only signal that can cause redistribution of auxin. Auxin moves to the lower side of a shoot that has been tipped sideways, causing more rapid growth in the lower side and, hence, an upward bending of the shoot. Such growth in a direction determined by gravity is called **gravitropism** (**Figure 37.11B**). The upward gravitropic response of shoots is defined as *negative gravitropism*; that of roots, which bend downward, is *positive gravitropism*.

How does a plant cell sense light and gravity and respond with an asymmetric distribution of auxin? Different mechanisms have been proposed:

- *The phototropic response*. As you will see later in the chapter, plants have a membrane receptor called phototropin that absorbs blue light. This receptor was discovered in a genetic screen in *Arabidopsis* for mutant plants that failed to bend toward light. When the receptor is activated, a signal transduction pathway results in redistribution of auxin transport carriers so that the hormone is transported to the cells on the shaded side. This results in bending toward light (see Figure 37.11).
- *The gravitropic response*. Some types of plant cells contain starch that is stored in large plastids called amyloplasts. These plastids tend to settle on the downward side of a cell

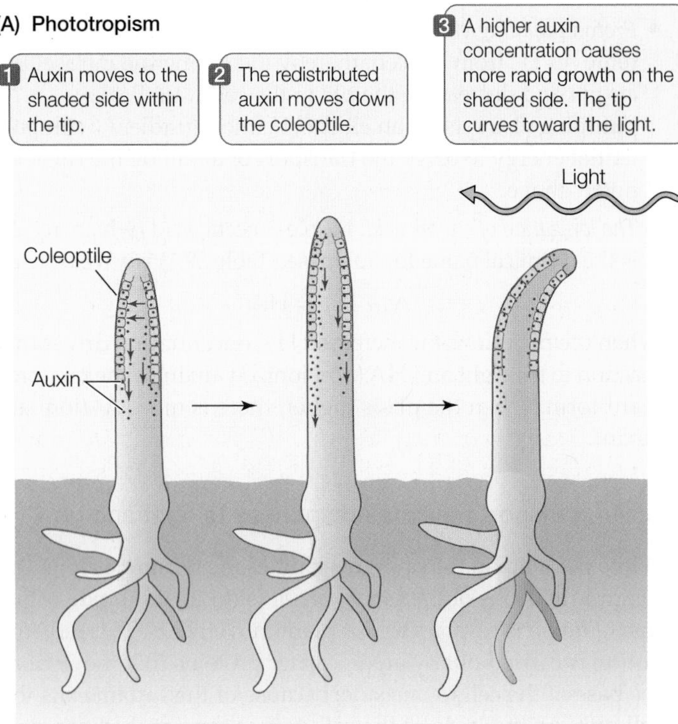


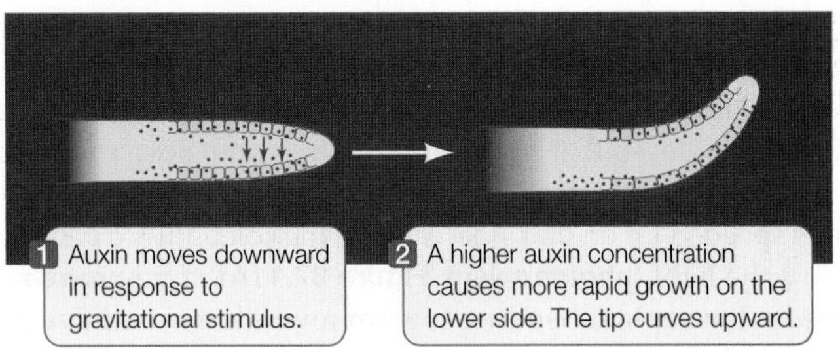

(B) Negative gravitropism of shoot
1 Auxin moves downward in response to gravitational stimulus.
2 A higher auxin concentration causes more rapid growth on the lower side. The tip curves upward.


**37.11 Plants Respond to Light and Gravity** (A) Phototropism and (B) gravitropism occur in shoot apices in response to a redistribution of auxin.

in response to gravity. How gravity-induced plastid movement is sensed is not well understood, but it may be through disturbance of endoplasmic reticulum membranes on the downward side of the cell. This in turn triggers auxin transport to the bottom side of the root or shoot, which causes bending in the appropriate direction.

### Auxin affects plant growth in several ways

Like the gibberellins, auxin has many roles in plant development. It affects the vegetative and reproductive growth of plants in a number of ways.

**ROOT INITIATION** Cuttings from the shoots of some plants can produce roots and develop into entire new plants. For this to occur, certain undifferentiated cells in the interior of the shoot, originally destined to function only in food storage, must set off on a new mission: they must change their cell fate and become organized into the apical meristem of a new root. These changes are similar to those that take place in the pericycle of a root when a lateral root forms (see Figure 34.13A). Shoot cuttings of many species can be made to develop roots by dipping the cut surfaces into an auxin solution. These observations suggest that in an intact plant the plant's own auxin plays a role in the initiation of lateral roots. Commercial preparations that enhance the rooting of plant cuttings typically contain synthetic auxins.

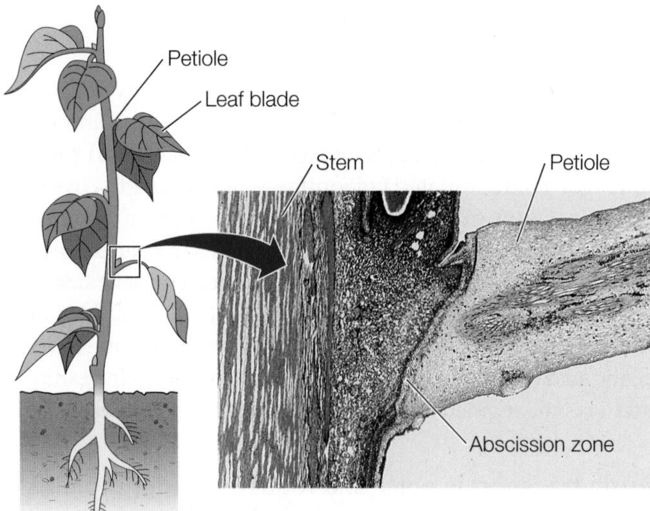


**37.12 Changes Occur when a Leaf Is About to Fall**
The breakdown of cells in the abscission zone of the petiole causes the leaf to fall.

**LEAF ABSCISSION** In contrast to its stimulatory effect on root initiation, auxin inhibits the detachment of old leaves from stems. This detachment process, called **abscission**, is the cause of autumn leaf fall. Most leaves consist of a blade and a petiole that attaches the blade to the stem. Abscission results from the breakdown of a specific part of the petiole, the *abscission zone* (**Figure 37.12**). If the blade of a leaf is cut off, the petiole falls from the plant more rapidly than if the leaf had remained intact. If the cut surface is treated with an auxin solution, however, the petiole remains attached to the plant, often longer than an intact leaf would have. The timing of leaf abscission in nature appears to be determined in part by a decrease in the movement through the petiole of auxin produced in the blade.

**APICAL DOMINANCE** Auxin helps maintain **apical dominance**, a phenomenon in which apical buds inhibit the growth of axillary buds (see Figure 34.1), resulting in the growth of a single main stem with minimal branching. A diffusion gradient of auxin from the apical tip of the shoot down the stem results in lower branches receiving less auxin and therefore branching more. The effect of the auxin gradient is apparent in conifers: next time you see a decorated tree during the winter holidays, think of auxin and apical dominance.

Apical dominance can be demonstrated by an experiment with a young seedling. If the plant remains intact, the stem elongates and the axillary buds remain inactive. Removal of the apical bud—the major site of auxin production—results in growth of the axillary buds. If the cut surface of the stem is treated with auxin, however, the axillary buds do not grow. The apical buds of branches also exert apical dominance: the axillary buds on the branch are inactive unless the apex of the branch is removed. That is why gardeners prune shrubs to encourage branching.

In the experiments on leaves and stems just discussed, removal of a particular part of the plant elicits a response—abscission or loss of apical dominance—and that response is prevented by treatment with auxin. These results are consistent with other data showing that the excised part of the leaf or stem is an auxin source and that auxin in the intact plant delays the abscission of leaves and helps maintain apical dominance.

**FRUIT DEVELOPMENT** Fruit development normally depends on prior fertilization of the ovule (egg), but in many species treatment of an unfertilized ovary with auxin or gibberellins causes **parthenocarpy**—fruit formation without fertilization. Parthenocarpic fruits form spontaneously in some cultivated varieties of plants, including seedless grapes, bananas, and some cucumbers.

**CELL EXPANSION** The expansion of plant cells is what causes plant growth. Because the plant cell wall normally prevents expansion of the protoplast (see Section 35.1), the cell wall plays a key role in controlling the rate and direction of plant cell growth. Auxin acts on cell walls to regulate this process.

The expansion of a plant cell is driven primarily by the uptake of water, which enters the cytoplasm of the cell and accumulates in its central vacuole (see Section 35.1). Growth of the vacuole accounts for most of the increase in volume of a growing cell, and the vacuole often makes up more than 90 percent of the volume of a mature cell. As the vacuole expands, it presses the cytoplasm against the cell wall, and the wall resists this force (the basis of turgor pressure). The cell wall is an extensively cross-linked network of polysaccharides and proteins, dominated by cellulose fibrils (see Figure 34.5). If the cell is to expand, some adjustments must be made in the wall structure to allow the wall to "give" under turgor pressure. Think of a balloon (the cell surrounded by a membrane) inside a box (the cell wall). How does the cell wall "box" loosen to allow expansion?

The **acid growth hypothesis** offers a possible explanation for auxin-induced cell expansion (**Figure 37.13**). The hypothesis holds that protons ($H^+$) are pumped from the cytoplasm into the cell wall, lowering the pH of the wall and activating enzymes called *expansins* that catalyze changes in the cell wall structure such that the polysaccharides adhere to each other less strongly. This loosens the cell wall, making it easier to stretch as the cell expands. Auxin is believed to have two roles in this process: to increase the synthesis of the proton pumps, and to guide their insertion into the plasma membrane. Auxin may also increase the activity of proton pump proteins already in the plasma membrane. Several lines of evidence support the acid growth hypothesis. For example, adding acid to the cell wall to lower the pH stimulates cell expansion even in the absence of auxin. Conversely, when a buffer is used to prevent the wall from becoming more acidic, auxin-induced cell expansion is blocked. The model works more or less well depending on species; in some plants auxin stimulates secretion of new cell wall components quickly enough to account for even rapid changes in growth rate.

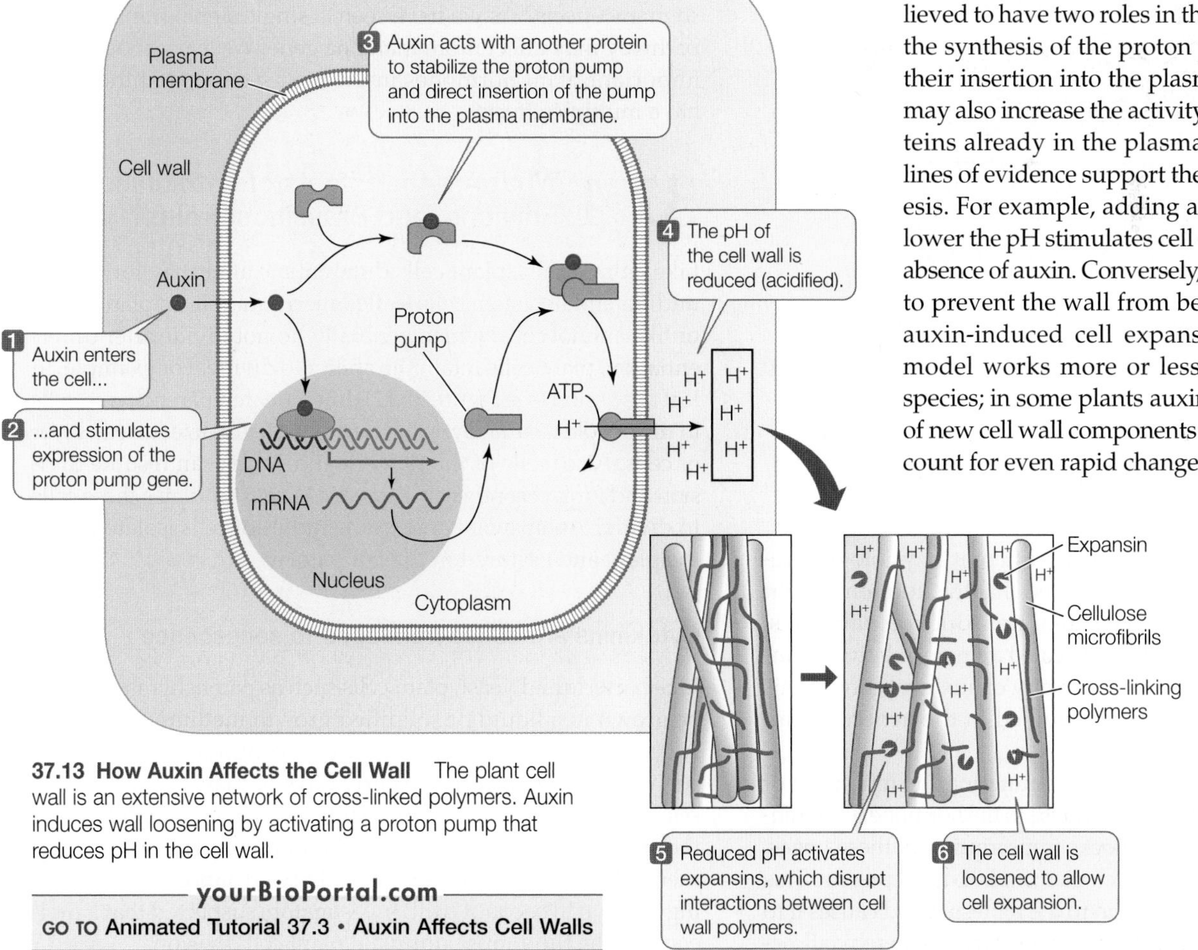
Plasma membrane
Cell wall
Auxin
1 Auxin enters the cell...
2 ...and stimulates expression of the proton pump gene.
3 Auxin acts with another protein to stabilize the proton pump and direct insertion of the pump into the plasma membrane.
Proton pump
DNA
mRNA
Nucleus
Cytoplasm
ATP
H+
4 The pH of the cell wall is reduced (acidified).
Expansin
Cellulose microfibrils
Cross-linking polymers
5 Reduced pH activates expansins, which disrupt interactions between cell wall polymers.
6 The cell wall is loosened to allow cell expansion.

**37.13 How Auxin Affects the Cell Wall** The plant cell wall is an extensive network of cross-linked polymers. Auxin induces wall loosening by activating a proton pump that reduces pH in the cell wall.

yourBioPortal.com
GO TO Animated Tutorial 37.3 • Auxin Affects Cell Walls

(A) Repression: Auxin absent

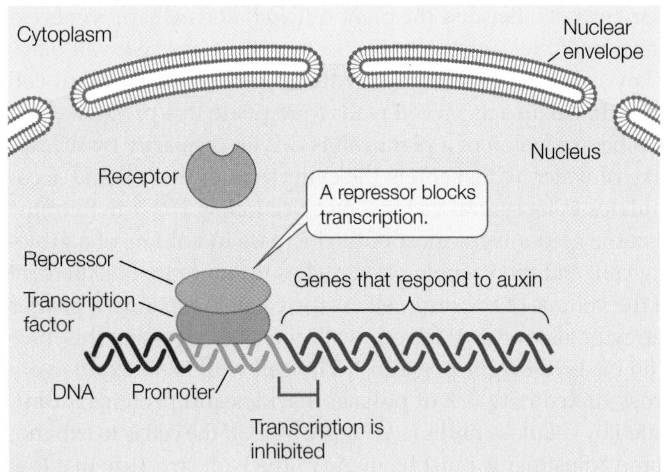


(B) Activation: Auxin present

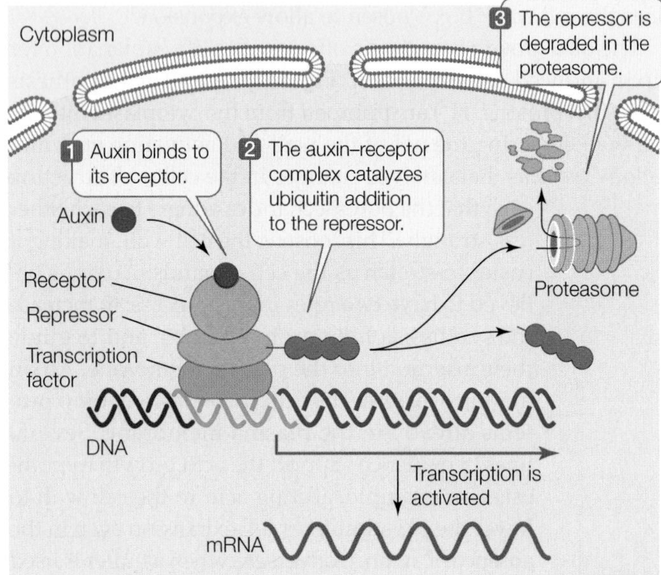


**37.14 Signal Transduction Pathway for Auxin** As with gibberellins (see Figure 37.7), the auxin response involves the release from inhibition of transcription.

### At the molecular level, auxin and gibberellins act similarly

Given that auxin induces so many different physiological responses in plants, biologists expected that the hormone might act through several different signal transduction pathways. However, genetic screens have revealed a single, relatively simple mechanism that accounts for many of auxin's actions. This mechanism is similar to that involved in the action of gibberellins (see Figure 37.7).

Like gibberellin, auxin acts by countering the inhibition of genes involved in the cellular response to the hormone. In the absence of auxin, a repressor blocks transcription of these genes. Auxin binds to a receptor, which then binds to the repressor. This stimulates addition of ubiquitin to the repressor and causes it to be degraded by the proteasome (**Figure 37.14**), thereby allowing transcription. The molecular response takes longer, and is longer-lasting than the rapid acid growth response in the cell wall.

### 37.3 RECAP

**Auxin regulates stem elongation (cell expansion) and mediates phototropism and gravitropism; it also plays roles in apical dominance, leaf abscission, and root initiation. The acid growth hypothesis explains auxin-induced cell wall loosening. Similar molecular mechanisms explain the effects of auxin and gibberellin on gene expression.**

- What is the evidence for polar transport of auxin and how does it occur? **See pp. 779–780 and Figures 37.9 and 37.10**
- Explain why, even though auxin moves *away* from the lighted side of a coleoptile tip, the coleoptile bends *toward* the light. **See p. 781 and Figure 37.11**
- How does auxin cause cell wall loosening? **See p. 783 and Figure 37.13**
- What are the similarities between the signal transduction pathways for auxin and gibberellin? **See p. 784 and Figures 37.7 and 37.14**

How can a single hormone, such as auxin or a gibberellin, have so many effects? As we have seen, a single signal transduction pathway may affect more than one gene. We learn about other important plant hormones in the next section, and they, too, have multiple effects.

## 37.4 What Are the Effects of Cytokinins, Ethylene, and Brassinosteroids?

Like animal cells, plant cells differentiate after they form from undifferentiated stem cells (called meristem cells in plants). But unlike animal cells, which generally do not divide after differentiation, plant cells retain the ability to divide. For example, in leaf abscission (see Figure 37.12) differentiated parenchyma cells in the petiole resume division, forming a specialized, weak layer of cells. Also, cells of the phloem and cortex can resume division and form secondary meristems. What stimulates these cells to divide? An answer came from studies of cells isolated from the plant and cultured in the laboratory.

### Cytokinins are active from seed to senescence

Like bacteria and yeast, plant cells such as parenchyma cells can be grown in a liquid or solidified growth medium containing sugars and salts. The cells will divide continuously until they run out of nutrients. In the early days of plant cell culturing, scientists experimented with many supplements to determine the optimal chemical environment for growth. The best supplement was coconut milk, the fluid that surrounds the developing embryo in coconut fruit. Investigators suspected that a molecule in the fluid must stimulate plant cell division.

A clue to the identity of the molecule came when Folke Skoog at the University of Wisconsin tested various pure substances that might substitute for coconut milk. DNA was among the substances tested, and it did not work; however, heating DNA at high pressure in an autoclave produced a mixture that strongly promoted plant cell division. A derivative of adenine called *kinetin* was identified as the active ingredient. Because it stimulated cell division (cytokinesis), it was called a **cytokinin**.

Kinetin does not exist in cells, but it gave scientists a hint as to what type of molecule might be the active ingredient in coconut milk. In 1963, an adenine derivative called **zeatin** was extracted from corn endosperm, the "coconut milk of corn" (see Table 37.1). Since then, over 150 different cytokinins have been isolated, and most are derivatives of adenine.

Cytokinins have a number of different effects, in many cases interacting with auxin:

- Adding an appropriate combination of auxin and cytokinins to a growth medium induces rapid proliferation of cultured plant cells.
- Cytokinins can cause certain light-requiring seeds to germinate even when kept in constant darkness.
- In cell cultures, a high cytokinin-to-auxin ratio promotes the formation of shoots; a low ratio promotes the formation of roots.
- Cytokinins usually inhibit the elongation of stems, but they cause lateral swelling of stems and roots (the fleshy roots of radishes are an extreme example).
- Cytokinins stimulate axillary buds to grow into branches; the auxin-to-cytokinin ratio controls the extent of branching (bushiness) of a plant.
- Cytokinins delay the senescence of leaves. If leaf blades are detached from a plant and floated on water or a nutrient solution, they quickly turn yellow and show other signs of senescence. If instead they are floated on a solution containing a cytokinin, they remain green and senesce much more slowly. Roots contain abundant cytokinins, and cytokinin transport to the leaves delays senescence.

Cytokinin signaling appears to act through a pathway that includes proteins with amino acid sequences similar to proteins in *two-component systems* in bacteria (see Figure 7.3). Indeed, this system was one of the first of its kind discovered in eukaryotes. The two components in such a system are:

- A *receptor* that can act as a protein kinase, phosphorylating itself as well as a target protein
- A *target protein*, generally a transcription factor, that can act as an *effector*

Genetic screens in *Arabidopsis* for abnormalities in the response to cytokinin have identified the receptor (AHK; *Arabidopsis* histidine kinase) and target effector (ARR; *Arabidopsis* response regulator), the latter acting as a transcription factor when phosphorylated. The cytokinin signal transduction pathway also includes a third protein (AHP; *Arabidopsis* histidine phosphotransfer protein), which transfers phosphates from the receptor to the effector (**Figure 37.15**). The plant genome has over 20 genes that are expressed in response to this signaling pathway.

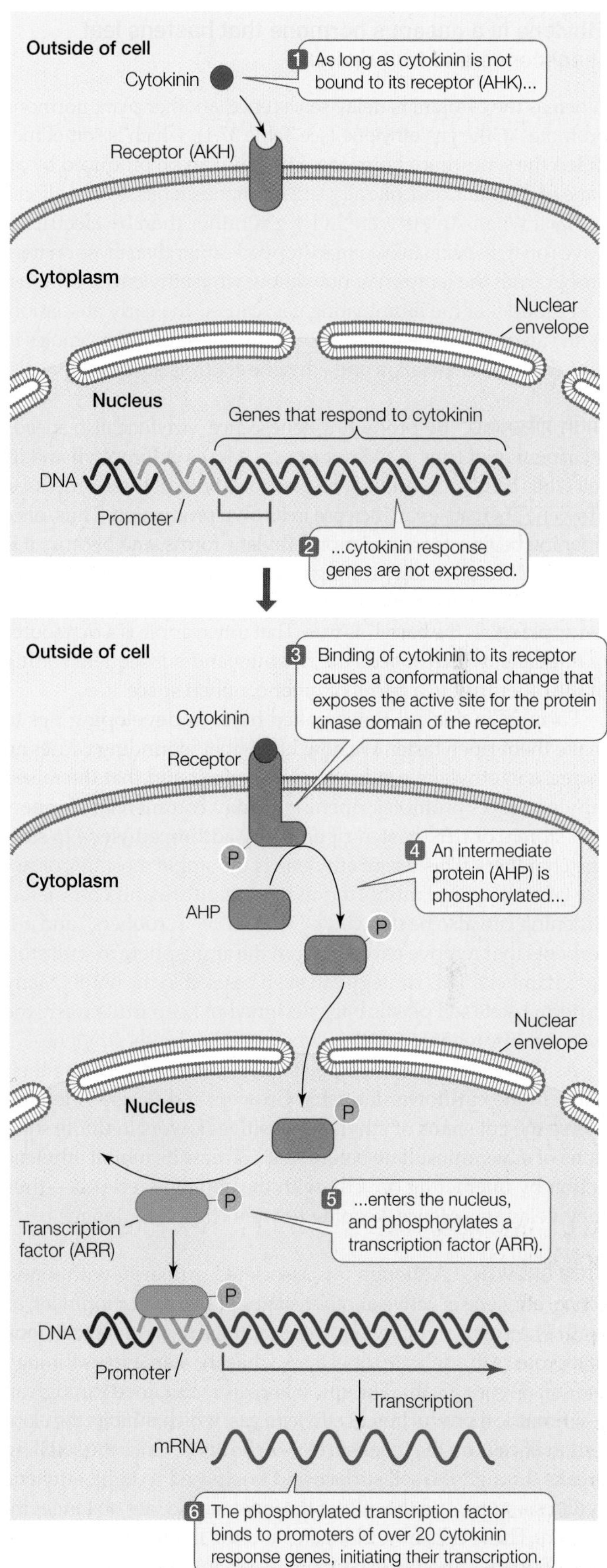


**37.15 The Cytokinin Response Pathway** Plant cells respond to cytokinins using a signal transduction pathway related to bacterial two-component systems.

## Ethylene is a gaseous hormone that hastens leaf senescence and fruit ripening

Whereas the cytokinins delay senescence, another plant hormone promotes it: the gas **ethylene** (see Table 37.1), which is sometimes called the senescence hormone. Ethylene can be produced by all parts of the plant, and, like all plant hormones, it has several effects.

Back when streets were lit by gas rather than by electricity, leaves on trees near street lamps dropped earlier than those on trees farther from the lamps. We now know why: ethylene, a combustion product of the illuminating gas, caused the early abscission. While auxin delays leaf abscission, ethylene strongly promotes it; thus the balance of auxin and ethylene controls abscission.

**FRUIT RIPENING** By promoting senescence, ethylene also speeds the ripening of fruit. As a fruit ripens, it loses chlorophyll and its cell walls break down; ethylene promotes both of these processes. Ethylene also causes an increase in its own production. Thus, once ripening begins, more and more ethylene forms, and because it is a gas, it diffuses readily throughout the fruit and even to neighboring fruits on the same or other plants. The old saying "one rotten apple spoils the barrel" is true. That rotten apple is a rich source of ethylene, which speeds the ripening and subsequent rotting of the other fruit in a barrel or other confined space.

Farmers in ancient times poked holes in developing figs to make them ripen faster. We now know that wounding causes an increase in ethylene production by the fruit and that the raised ethylene level promotes ripening. Today commercial shippers and storers of fruit hasten ripening by adding ethylene to storage chambers. This use of ethylene is the single most important use of a natural plant hormone in agriculture and commerce. Ripening can also be delayed by the use of "scrubbers" and adsorbents that remove ethylene from the atmosphere in fruit storage chambers. This strategy can even be used in the home. Many supermarkets sell plastic bags designed to keep fruits fresh; the bags are impregnated with a substance that binds ethylene.

As flowers senesce, their petals may abscise, decreasing their value in the cut-flower industry. Growers and florists often immerse the cut stems of ethylene-sensitive flowers in dilute solutions of silver thiosulfate before sale. Silver salts inhibit ethylene action by interacting directly with the ethylene receptor—thus they delay senescence, keeping flowers "fresh" for longer.

**STEM GROWTH** Although it is associated primarily with senescence, ethylene is active at other stages of plant development, as well. The stems of many eudicot seedlings form an **apical hook** that protects the delicate shoot apex while the stem grows through the soil (**Figure 37.16**). The apical hook is maintained through an asymmetrical production of ethylene gas, which inhibits the elongation of cells on the inner surface of the hook. Once the seedling breaks through the soil surface and is exposed to light, ethylene synthesis stops, and the cells of the inner surface are no longer inhibited. These cells now elongate, and the hook unfolds, raising the shoot apex and the expanding leaves into the sun.

Ethylene also inhibits stem elongation in general, promotes lateral swelling of stems (as do the cytokinins), and decreases the sensitivity of stems to gravitropic stimulation. Together, these three phenomena constitute the *triple response*, a well-characterized stunted growth habit observed when plants are treated with ethylene.

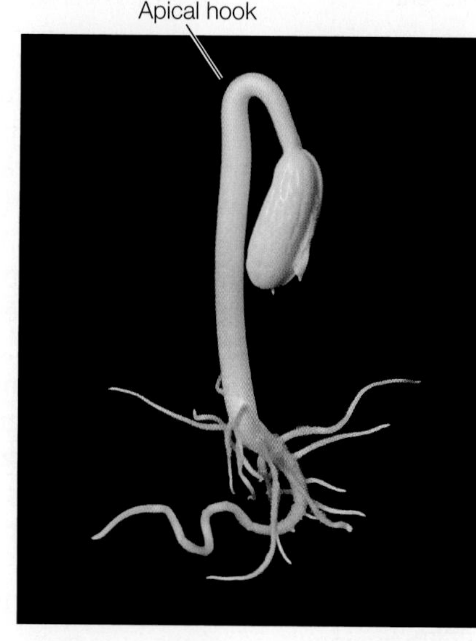


**37.16 The Apical Hook of a Eudicot** Asymmetrical production of ethylene is responsible for the apical hook of this bean seedling. The ethylene concentration was highest on the right side, so more rapid growth on the left caused and maintained the hook.

**THE ETHYLENE SIGNAL TRANSDUCTION PATHWAY** The mechanism of ethylene action has been worked out by analyzing *Arabidopsis* mutants that have ethylene-related defects. Some of these mutants do not respond to applied ethylene, and others act as if they have been exposed to ethylene even though they have not. Researchers studied the mutant genes and compared their protein products to other known proteins; thus they worked out some of the details of the signal transduction pathway through which ethylene acts (**Figure 37.17**).

The pathway includes two membrane proteins in the endoplasmic reticulum. The first is an ethylene receptor (labeled A in the figure) and the second is a channel protein (C). In the absence of ethylene, a protein kinase (B) keeps C inactive by phosphorylation. When receptor A binds ethylene it inactivates B. Without B to inactivate it, C activates a transcription factor (D), which then moves into the nucleus, where it turns on the genes that produce ethylene's effects in the cell. In other words, ethylene turns off the "off" signal.

## Brassinosteroids are plant steroid hormones

In animals, steroid hormones such as cortisol and estrogen are formed from cholesterol (see Figure 3.22). Animal steroids are widespread and have been well studied for many decades. In contrast, plant steroid hormones are a relatively recent discovery. In the 1970s, biologists isolated a steroid (see Table 37.1) from the pollen of rape, a member of the Brassicaceae (mustard family). When applied to various plant tissues, this **brassinosteroid** stimulated cell elongation, pollen tube elongation, and vascular tissue differentiation, but it inhibited root elongation. Since then, dozens of chemically related, growth-affecting brassinosteroids have been found in plants.

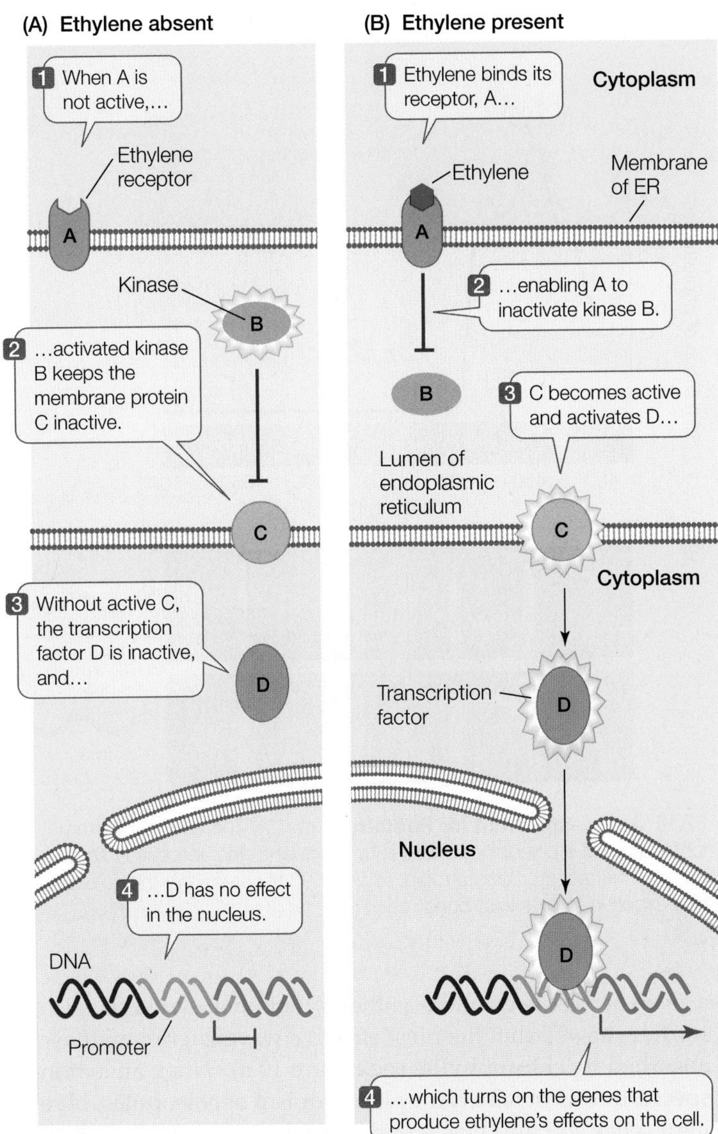


**37.17 The Signal Transduction Pathway for Ethylene** This diagram shows the roles of four proteins (A, B, C, and D) in the signal transduction pathway through which ethylene exerts its many effects.

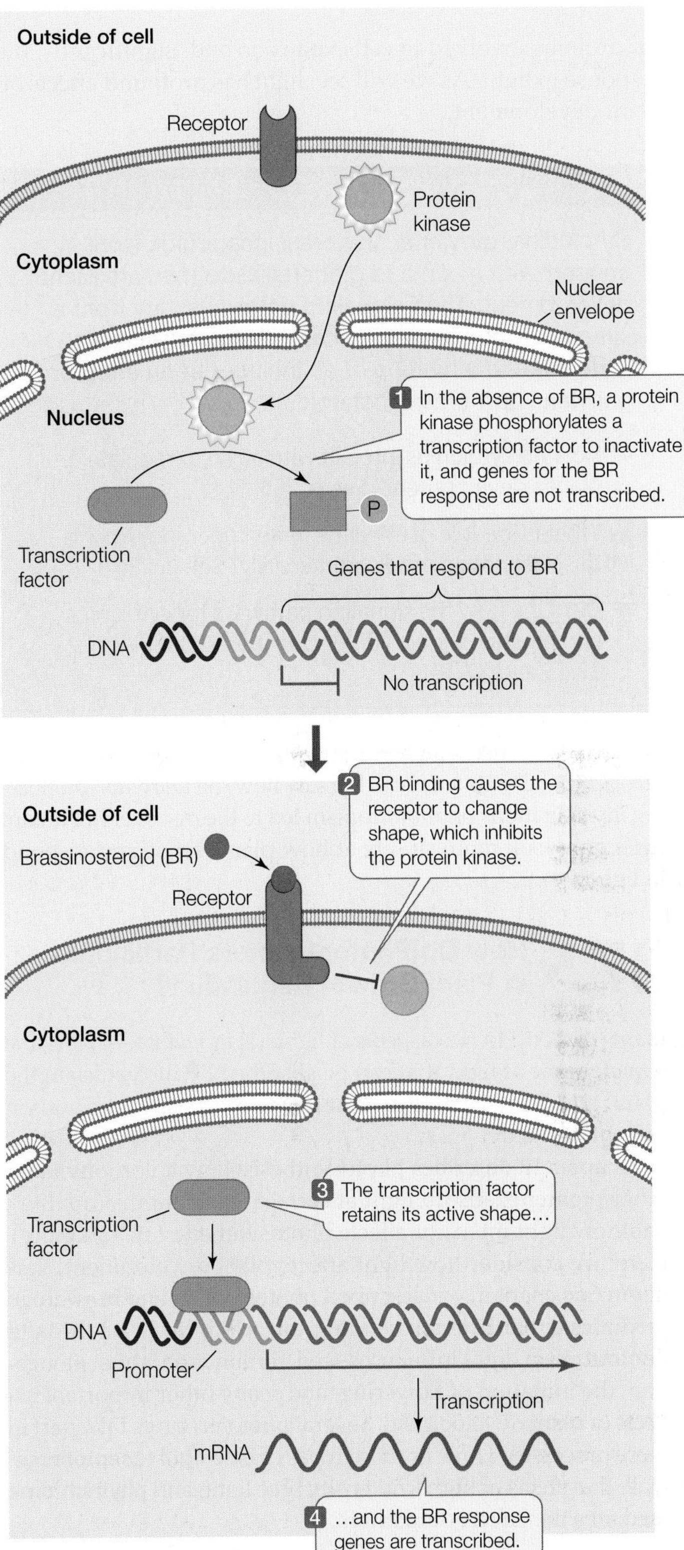


**37.18 The Brassinosteroid Signal Transduction Pathway Begins at the Plasma Membrane** Unlike the receptors for animal steroid hormones, the brassinosteroid (BR) receptor is a membrane protein. The signal transduction pathway concludes by activating certain genes.

Mutant plants that either do not make brassinosteroids or have defects in brassinosteroid reception and signal transduction are usually dwarf, infertile, and slow to develop. These effects can be reversed by adding small amounts of brassinosteroids, indicating that brassinosteroids are true hormones. These hormones have diverse effects, which vary among plants. Brassinosteroids can:

- enhance cell elongation and cell division in shoots
- promote xylem differentiation
- promote growth of pollen tubes during reproduction
- promote seed germination
- promote apical dominance and leaf senescence

The signaling pathway for these plant steroids differs sharply from those for steroid hormones in animals. In animals, steroids diffuse through the plasma membrane and bind to receptors in the cytoplasm. In contrast, the receptor for brassinosteroids is an integral protein in the plasma membrane (**Figure 37.18**). Binding of a brassinosteroid by the receptor inactivates a protein kinase that would otherwise inactivate a transcription factor. The genes activated by this pathway code for proteins involved in cell expansion and, significantly, the response to light. As we will see, light has profound effects on plant development.

## RECAP 37.4

**Cytokinins, ethylene, and brassinosteroids work in concert with auxin and gibberellins to mediate plant development. Their signaling pathways vary from a simple two-component receptor–effector system (cytokinin) to inhibition of an inhibitor of an effector (ethylene and brassinosteroids).**

- How do cytokinins interact with auxin to regulate a plant's development? **See p. 785**
- What is the role of ethylene in fruit ripening? How is this knowledge used commercially? **See p. 786**
- Describe how the signaling pathways for cytokinins and brassinosteroids differ. **See p. 787 and Figures 37.15 and 37.18**

A plant's response to light—the energy source for photosynthesis—is crucial to its survival. We saw how the Darwins' pioneering investigations of phototropism led to the discovery of auxin. Let's now look more closely at how plants sense and respond to light.

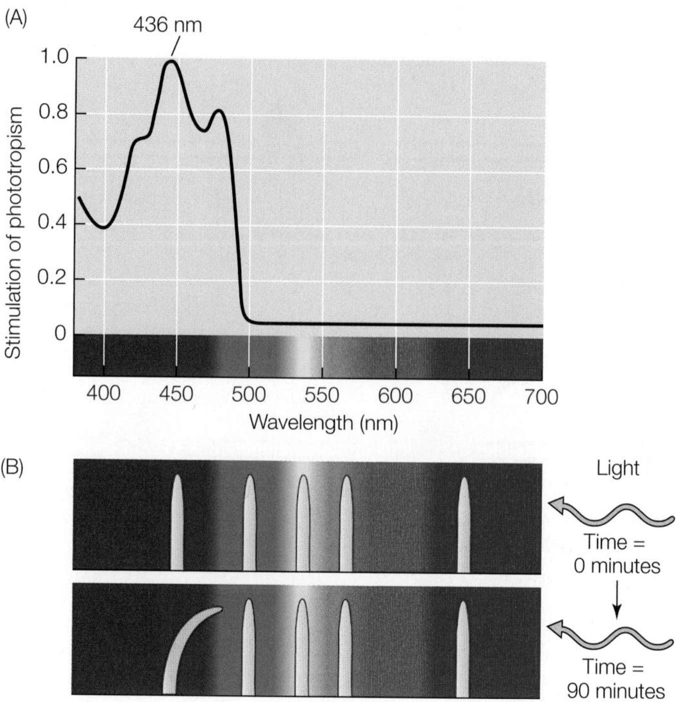


**37.19 Action Spectrum for Phototropism** (A) The action spectrum for bending of a coleoptile toward light is similar to the absorption spectrum for the receptor, phototropin. (B) After 90 minutes, only the coleoptiles exposed to blue light bend.

# 37.5 How Do Photoreceptors Participate in Plant Growth Regulation?

Plants respond to two aspects of light: (1) its *quality*—that is, the wavelengths of light that can be absorbed by molecules in the plant; and (2) its *quantity*—that is, the intensity and duration of light exposure.

Chapter 10 describes photosynthesis: how chlorophyll and other pigments absorb light at certain wavelengths (quality), and how light intensity affects photosynthetic rate (quantity). Here, we consider how light affects plant development. Earlier in this chapter, we described phototropism and how auxin mediates a plant stem's bending toward light. In addition to phototropism, light influences seed germination, shoot elongation, the initiation of flowering, and many other important aspects of plant development. Several photoreceptors take part in these processes. Three or more types of **blue-light receptors** mediate the effects of higher-intensity blue light, and **phytochrome** mediates the effects of red light.

## Phototropins, cryptochromes, and zeaxanthin are blue-light receptors

Charles and Francis Darwin showed that the apical tip of a growing coleoptile receives light as a signal and then redistributes auxin to stimulate cell elongation below the tip on the shaded side. You may recall from Chapter 10 that an *action spectrum* involves exposing plants to different wavelengths of light to determine what wavelengths are most effective in driving a given process (e.g., photosynthesis). For photosynthesis, such studies showed that the most effective wavelengths are those absorbed by chlorophylls (see Figure 10.6). When an action spectrum was obtained for phototropism of coleoptiles, blue light (peak 436 nm) was found to be the most effective at inducing the coleoptile to curve (**Figure 37.19**). What is the blue-light-absorbing receptor/pigment? Biologists have used a genetic approach to answer this question, once again employing the model plant *Arabidopsis.*

Researchers recovered blue-light-insensitive *Arabidopsis* mutants from a genetic screen and identified the gene for a blue-light receptor protein located in the plasma membrane called **phototropin**. Phototropin protein has a flavin mononucleotide associated with it that absorbs blue light, leading to a change in the shape of the protein. This change exposes an active site for a protein kinase, which in turn initiates a signal transduction cascade that ultimately results in stimulation of cell elongation by auxin.

Phototropin is also involved in chloroplast movements in relation to light, and participates with another type of blue-light receptor, the plastid pigment **zeaxanthin**, in the light-induced opening of stomata (see Figure 35.9).

Yet another class of blue-light receptors is the **cryptochromes**, which absorb blue and ultraviolet light. These yellow pigments are located primarily in the plant cell nucleus and affect seedling development and flowering. The exact mechanism of cryptochrome action is not yet known. Strong blue light inhibits cell elongation through the action of cryptochromes, although the most rapid responses are mediated by phototropins.

### Phytochromes mediate the effects of red and far-red light

A number of physiological and developmental events in plants are controlled by light, a process called **photomorphogenesis**. For example:

- A bean seedling germinating below ground has an elongated stem, a pale yellow, folded leaf, and a hook that protects the leaf (see Figures 37.2 and 37.16)—it is **etiolated**. As the seedling reaches the surface of the soil, it undergoes several light-induced changes: the apical hook straightens, the rudimentary leaf unfolds, and chlorophyll is made so that photosynthesis can begin. Even very dim light will induce these changes.
- Lettuce seeds spread on the soil will germinate only in response to light. Even just a flash of dim light will suffice.
- Adult cocklebur plants flower when they are exposed to long nights. If there is a brief light flash in the middle of the night, they do not flower.

Action spectra of the above processes show that they are induced by red light (650–680 nm). This indicates that plants must have a photoreceptor pigment that absorbs red light and initiates photomorphogenesis.

What is especially remarkable about these red light responses is that *they are reversible by far-red light* (710–740 nm). For example, if lettuce seeds are exposed to brief, alternating periods of red and far-red light in close succession, they respond only to the final exposure. If it is red, they germinate; if it is far-red, they remain dormant (**Figure 37.20**). This reversibility of the effects of red and far-red light regulates many other aspects of plant development, including flowering and seedling growth.

The basis for the effects of red and far-red light resides in a bluish photoreceptor pigment protein in the cytosol of plants called **phytochrome**. Phytochrome exists in two interconvertible "isoforms" or states. The molecule undergoes a conformational change upon absorbing light at particular wavelengths. The default or "ground" state, which absorbs principally red light, is called $\mathbf{P_r}$. When $P_r$ absorbs a photon of red light it is converted into $\mathbf{P_{fr}}$. The $P_{fr}$ form preferentially absorbs far-red light; when it does so, it is converted back to $P_r$.

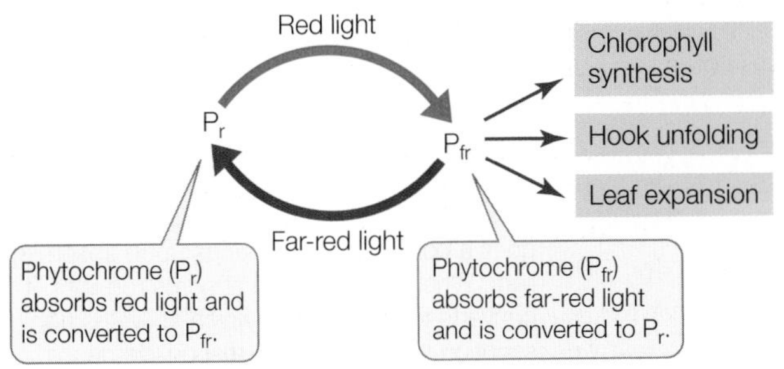


INVESTIGATING LIFE

**37.20 Sensitivity of Seeds to Red and Far-Red Light** Lettuce seeds will germinate if exposed to a brief period of light. An action spectrum indicated that red light was most effective in promoting germination, but far-red light would reverse the stimulation if presented right after the red light flash. Harry Borthwick and his colleagues asked what would be the effect of repeated alternating flashes of red and far-red light. In each case, the final exposure determined the germination response. This observation led to the conclusion that a single, photoreversible molecule was involved. That molecule turned out to be phytochrome.

**HYPOTHESIS** **The effects of red and far-red light on lettuce seed germination are mutually reversible.**

**METHOD** Expose lettuce seeds to alternate periods of red light R for 1 minute and far-red light FR for 4 minutes.

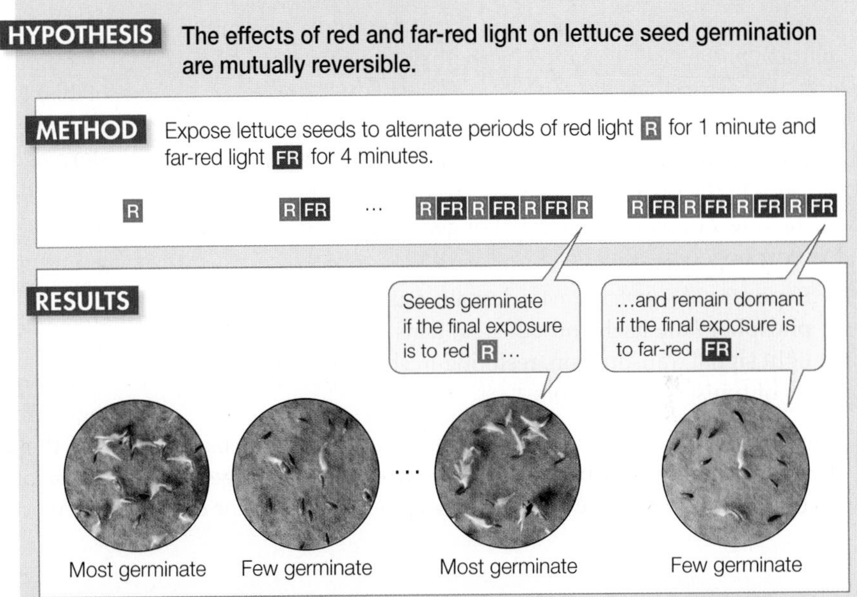


**CONCLUSION** **Red light and far-red light reverse each other's effects.**

Go to **yourBioPortal.com** for original citations, discussions, and relevant links for all INVESTIGATING LIFE figures.

$P_{fr}$, not $P_r$, is the active form of phytochrome—the form that triggers important biological processes in various plants. As we have seen, these processes include seed germination, shoot development after etiolation, and flowering.

For a plant in nature, the ratio of red to far-red light determines whether a phytochrome-mediated response will occur. For example, during daylight, the ratio is about 1.2:1; because there is more red than far-red light, the $P_{fr}$ form predominates. But for a plant growing in the shade of other plants, the ratio is as low as 0.13:1, and phytochrome is mostly in the $P_r$ form. The low ratio of red to far-red light in the shade results from absorption of red light by chlorophyll in the leaves overhead, so less of the red light gets through to the plants below. Shade-intolerant species respond by stimulating cell elongation in the stem and thus growing taller to escape the shade. Shade cast by other plants also prevents germination of seeds that require red light to germinate (see Figure 37.20). The reflective properties of the soil can also affect the red to far-red ratio—and thus plant behavior. For example, cotton seedlings grow more slowly on soils (such as clay) that reflect more red than far-red light.

### Phytochrome stimulates gene transcription

How does phytochrome, or more specifically, $P_{fr}$, work? Phytochrome is a cytoplasmic protein composed of two subunits (**Figure 37.21**). Each subunit has a protein chain and a nonprotein pigment from the plastid called a *chromophore.* In *Arabidopsis*, there is a gene family that encodes five slightly different phytochromes, each functioning in different photomorphogenic responses.

Gene transcription is stimulated when $P_r$ is converted to the $P_{fr}$ isoform. When $P_r$ absorbs red light, the chromophore changes shape, which leads to a change in the conformation of the protein itself from the $P_r$ form to the $P_{fr}$ form. Conversion to the $P_{fr}$ form exposes two important regions of the phytochrome protein (see Figure 37.21), both of which affect transcriptional activity:

- Exposure of a *nuclear localization sequence* (see Figure 14.20) results in movement of $P_{fr}$ from the cytosol to the nucleus. Once in the nucleus, $P_{fr}$ binds to transcription factors and thereby stimulates expression of genes involved in photomorphogenesis.
- Exposure of a *protein kinase* domain causes $P_{fr}$ protein to phosphorylate itself and other proteins involved in red-light signal transduction, resulting in changes in the activity of transcription factors.

The effect of activating these transcription factors is quite large: In *Arabidopsis*, phytochrome affects an amazing 2,500 genes (10 percent of the entire genome!) by either increasing or decreasing their expression. Some of these genes are related to other hormones. For example, when $P_{fr}$ is formed in seed germination, genes for gibberellin synthesis are activated and genes for gibberellin breakdown are repressed. As a result, gibberellins accumulate and seed reserves are mobilized.

### Circadian rhythms are entrained by light reception

The timing and duration of biological activities in living organisms are governed in all eukaryotes and some prokaryotes by what is commonly called a "biological clock"—an oscillator within cells that alternates back and forth between two states at roughly 12-hour intervals. The major outward manifestations of this clock are known as **circadian rhythms** (Latin *circa*, "about," and *dies*, "day"). Think of your own life: in all probability you sleep at night, and you are awake during the day. In plants, circadian rhythms influence, for example, the opening (during the day) and closing (at night) of stomata in *Arabidopsis*, and the raising toward the sun (during the day) and lowering (at night) of leaves in bean plants. From these two examples, it is obvious that circadian rhythms are ecologically useful adaptations, in that they relate the plant's physiology to its environment.

Two qualities characterize circadian rhythms, as well as other regular biological cycles: the **period** is the length of one cycle, and the **amplitude** is the magnitude of the change over the course of a cycle. The circadian rhythms of plants have several noteworthy characteristics:

- The period of a circadian rhythm is remarkably insensitive to temperature, although lowering the temperature may drastically reduce the amplitude.
- Circadian rhythms are highly persistent; they may continue for days, even in the absence of environmental cues, such as light–dark periods.
- Circadian rhythms can be *entrained*, within limits, by light–dark cycles that do not exactly correspond to 24 hours. That is, the period of a rhythm can be made to coincide (within limits) with that of the light–dark cycle to which the organism is exposed.

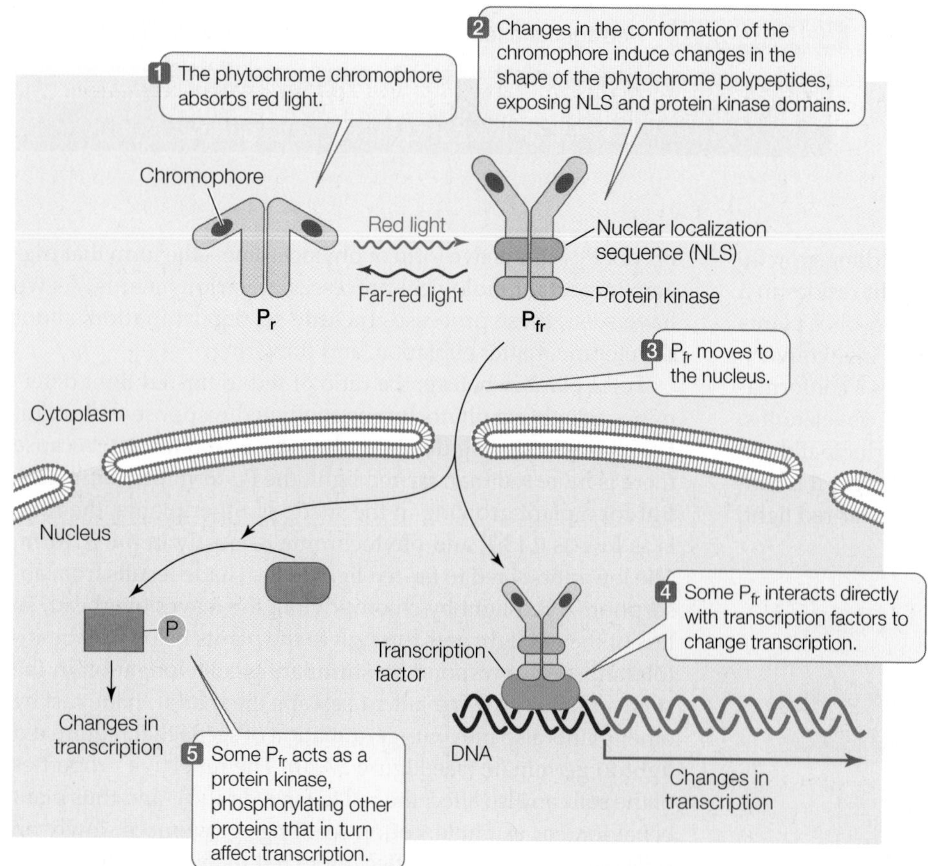

1 The phytochrome chromophore absorbs red light.
2 Changes in the conformation of the chromophore induce changes in the shape of the phytochrome polypeptides, exposing NLS and protein kinase domains.
Chromophore
Red light
Far-red light
Nuclear localization sequence (NLS)
Protein kinase
$P_r$
$P_{fr}$
3 $P_{fr}$ moves to the nucleus.
Cytoplasm
Nucleus
P
Changes in transcription
Transcription factor
4 Some $P_{fr}$ interacts directly with transcription factors to change transcription.
DNA
Changes in transcription
5 Some $P_{fr}$ acts as a protein kinase, phosphorylating other proteins that in turn affect transcription.


**37.21 Phytochrome Stimulates Gene Transcription** Phytochrome is composed of two polypeptide chains, each with a chromophore pigment. This pair of polypeptides undergoes a conformational change upon absorbing light. When phytochrome absorbs red light, it converts to the $P_{fr}$ form, which activates transcription of phytochrome-responsive genes.

Consider what happens when a person abruptly moves across many time zones: what was the night becomes the day, and gradually the person's sleep–wakefulness circadian rhythm entrains to the new environmental cues. Similar entrainment occurs in plants adapting to day length as the seasons progress during the year. The action spectrum for plant entrainment indicates that phytochrome (and to a lesser extent, blue-light receptors) is very likely involved. At sundown phytochrome is mostly in the active $P_{fr}$ form. But as the night progresses, $P_{fr}$ gradually gets converted back to the inactive $P_r$ form. By dawn phytochrome is mostly in the $P_r$ state, but as daylight begins, it rapidly converts to $P_{fr}$. The switch to the $P_{fr}$ state resets the plant's biological clock. However long the night, the clock is still reset at dawn every day. Thus while the total period measured by the clock is consistent, the clock adjusts to changes in day length over the course of the year.

## 37.5 RECAP

**Light controls a number of physiological and developmental events in plants, a process called photomorphogenesis. Pigment photoreceptors such as phototropin, cryptochromes, and phytochrome mediate the effects of light on plant growth and development. Phytochrome exists in two interconvertible states; conversion from one state to the other is controlled by the ratio of red to far-red light. Circadian rhythms are influenced by light reception.**

- Give the evidence for blue-light receptors in plants. See p. 788 and Figure 37.19
- Why does red light affect seed germination differently from far-red light? See p. 789 and Figure 37.20
- What are circadian rhythms? How are they related to photoreception? See p. 790

Photoreceptors also play a regulatory role in flowering. In addition to light, another environmental cue—temperature—regulates flowering. We will examine these topics and others in the next chapter, which focuses on reproduction in flowering plants.

## CHAPTER SUMMARY

### 37.1 How Does Plant Development Proceed?

- As sessile organisms, plants maximize their ability to grow by using meristems, forming new organs and growing throughout life.
- The environment, photoreceptors, hormones, and the plant's genome all regulate plant development.
- Seed **dormancy**, which has adaptive advantages, is maintained by a variety of mechanisms. In nature, dormancy is broken by, for example, abrasion, fire, leaching, and low temperatures. When dormancy ends and the seed **imbibes** water, it **germinates** and develops into a **seedling**. Review Figure 37.1, **WEB ACTIVITIES 37.1 and 37.2**
- Plants have several hormones, each of which regulates multiple aspects of development. Interactions among these hormones are often complex. Review Table 37.1
- **Hormones** and **photoreceptors** act through signal transduction pathways to regulate seedling development. Before the germinating embryo can begin photosynthesis, it relies on energy reserves in the **cotyledons** or the **endosperm**.
- **Genetic screens** using the model organism *Arabidopsis thaliana* have contributed greatly to our understanding of signaling in plants. Review Figure 37.3

### 37.2 What Do Gibberellins Do?

- The embryos of cereal seeds secrete **gibberellins**, which cause the **aleurone layer** to synthesize and secrete digestive enzymes that break down macromolecules stored in the endosperm. Review Figure 37.6, **WEB ACTIVITY 37.3**
- Dozens of gibberellins exist. These hormones regulate the growth of stems and some fruits.
- Gibberellins act through the breakdown of transcriptional repressors. Review Figure 37.7

### 37.3 What Does Auxin Do?

**SEE ANIMATED TUTORIAL 37.1**

- In **coleoptiles**, **auxin** is made in cells at the tip and moves down to the growing region. Review Figures 37.8 and 37.9, **ANIMATED TUTORIAL 37.2**
- Auxin transport is polar. Auxin active transport carriers—membrane proteins confined to the basal ends of cells—cause auxin to move from the tip to the base of the shoot. Review Figure 37.10
- Lateral movement of auxin, mediated by auxin transport carriers, is responsible for **phototropism** and **gravitropism**. Review Figure 37.11
- Auxin plays roles in root formation, leaf **abscission**, **apical dominance**, and **parthenocarpic fruit development**. Certain synthetic auxins are used as selective herbicides.
- The **acid growth hypothesis** explains how auxin promotes cell expansion by disrupting interactions between cell wall microfibrils. Review Figure 37.13, **ANIMATED TUTORIAL 37.3**
- The molecular mechanism underlying auxin action is similar to that of gibberellin; as long as auxin is not bound to its receptor, transcription is repressed. When the auxin–receptor complex binds to a transcriptional repressor, the repressor is degraded and transcription is initiated. Review Figure 37.14

### 37.4 What Are the Effects of Cytokinins, Ethylene, and Brassinosteroids?

- **Cytokinins** are adenine derivatives. They promote plant cell division, promote seed germination in some species, inhibit stem elongation, promote lateral swelling of stems and roots, stimulate the growth of axillary buds, promote the expansion of leaf tissue, and delay leaf senescence.

- **Cytokinins** act on plant cells by a signal transduction pathway that is similar to bacterial two-component systems. **Review Figure 37.15**
- A balance between auxin and **ethylene** controls leaf abscission. Ethylene promotes senescence and fruit ripening. It causes the formation of a protective **apical hook** in eudicot seedlings. In stems, it inhibits elongation, promotes lateral swelling, and causes a loss of gravitropic sensitivity.
- Ethylene acts on cells by a protein kinase pathway located in the endoplasmic reticulum. **Review Figure 37.17**
- Dozens of different **brassinosteroids** affect cell elongation, pollen tube elongation, vascular tissue differentiation, and root elongation. Some effects of light are mediated by changes in the action and levels of brassinosteroids. These steroids act at a plasma membrane receptor. **Review Figure 37.18**

### 37.5 How Do Photoreceptors Participate in Plant Growth Regulation?

- **Phototropins** are blue-light photoreceptors for phototropism and chloroplast movements. **Zeaxanthin** acts in conjunction with the phototropins to mediate the light-induced opening of stomata. **Cryptochromes** are blue-light photoreceptors that control seedling development, stem elongation, and floral initiation.
- **Phytochromes** exist in the cytosol in two interconvertible forms, $\mathbf{P_r}$ and $\mathbf{P_{fr}}$. The relative amounts of these two forms are a function of the ratio of red to far-red light. Phytochromes affect seedling growth, flowering, and etiolation. **Review Figure 37.20**
- The phytochrome signal transduction pathway affects transcription in two different ways; the $P_{fr}$ form interacts directly with some transcription factors, and influences transcription indirectly through interactions with protein kinases. **Review Figure 37.21**
- **Circadian rhythms** are activities that occur on a near-24-hour cycle. Light can entrain these activities through photoreceptors such as phytochrome.

## SELF-QUIZ

1. Which of the following is *not* an advantage of seed dormancy?
   *a.* It makes the seed more likely to be digested by birds that disperse it.
   *b.* It counters the effects of year-to-year variations in the environment.
   *c.* It increases the likelihood that a seed will germinate in the right place.
   *d.* It favors dispersal of the seed.
   *e.* It may result in germination at a favorable time of year.
2. Which of the following does *not* occur in seed germination?
   *a.* Imbibition of water
   *b.* Metabolic changes
   *c.* Growth of the radicle
   *d.* Mobilization of nutrient reserves
   *e.* Extensive mitotic divisions
3. To mobilize its nutrient reserves, a germinating barley seed
   *a.* becomes dormant.
   *b.* undergoes senescence.
   *c.* secretes gibberellins into its endosperm.
   *d.* converts glycerol and fatty acids into lipids.
   *e.* takes up proteins from the endosperm.
4. The gibberellins
   *a.* are responsible for phototropism and gravitropism.
   *b.* are gases at room temperature.
   *c.* are produced only by fungi.
   *d.* cause flowering in plants.
   *e.* inhibit the synthesis of digestive enzymes by barley seeds.
5. In coleoptile tissue, auxin
   *a.* is transported from base to tip.
   *b.* is transported from tip to base.
   *c.* can be transported toward either the tip or the base, depending on the orientation of the coleoptile with respect to gravity.
   *d.* is transported by simple diffusion, with no preferred direction.
   *e.* is not transported, because auxin is used where it is made.
6. Which process is *not* directly affected by auxin?
   *a.* Apical dominance
   *b.* Leaf abscission
   *c.* Synthesis of digestive enzymes by barley seeds
   *d.* Root initiation
   *e.* Cell elongation
7. Signal transduction for both auxin and gibberellins involves
   *a.* binding of the hormone to a nuclear receptor.
   *b.* degradation of a repressor of gene transcription.
   *c.* production of a small molecule second messenger.
   *d.* light absorption followed by chemical changes.
   *e.* breakdown of the hormone.
8. Which statement about cytokinins is *not* true?
   *a.* They promote cell division in tissue cultures.
   *b.* They delay the senescence of leaves.
   *c.* They usually promote the elongation of stems.
   *d.* They act by a receptor with protein kinase activity.
   *e.* They were discovered as a breakdown product of DNA.
9. Ethylene
   *a.* causes the triple response in seedlings growing underground.
   *b.* is liquid at room temperature.
   *c.* delays the ripening of fruits.
   *d.* generally promotes stem elongation.
   *e.* inhibits the swelling of stems, in opposition to cytokinin's effects.
10. Phytochrome
    *a.* is the only photoreceptor pigment in plants.
    *b.* exists in two forms interconvertible by light.
    *c.* is a pigment that is colored red or far-red.
    *d.* is a green-light receptor.
    *e.* is the photoreceptor for phototropism in coleoptiles.

## FOR DISCUSSION

1. Describe the circumstances under which it would be advantageous for a species to have the dormancy of its seeds broken by fire.
2. Cocklebur fruits contain two seeds each that are kept dormant by two different mechanisms. Why might having two mechanisms of dormancy be advantageous to cockleburs?
3. Supermarkets sell plastic bags that are impregnated with activated charcoal, which binds gases. The bags are designed to keep fruit fresh. How do they work?
4. Corn stunt virus causes a great reduction in the growth rate of infected corn plants. Diseased plants take on a dwarfed form. Since their appearance is reminiscent of the genetically dwarfed corn, you suspect that the virus may inhibit the synthesis of gibberellins by corn plants. Describe two experiments you might conduct to test this hypothesis, only one of which should require chemical measurement.

## ADDITIONAL INVESTIGATION

The semi-dwarf wheat and rice plants that led to the Green Revolution described in the chapter opening have mutations in the signal transduction pathway for gibberellins. You wish to use genetic engineering to make corn plants that are semi-dwarf.

*a.* How would you do a genetic screen to identify the genes in corn involved in gibberellin signaling?

*b.* Assuming that the signal transduction pathway is similar to that in *Arabidopsis*, what gene would you select for inactivation?

*c.* Besides short stature, what other effects would you expect for the signal transduction mutant strain? How would you use other hormones to overcome them?

## WORKING WITH DATA (GO TO yourBioPortal.com)

**The Darwins' Phototropism Experiment** In this exercise based on Figure 37.8, you will read excerpts from a book by Charles Darwin, *The Power of Movement in Plants*, in which he describes the experiments he and his son Francis undertook that ultimately led to the isolation by others of the plant hormone auxin. You will see how they planned their experiments and controls, and analyze the results.

# 38 Reproduction in Flowering Plants

## The language of flowers

In the recent film *Kate and Leopold*, a Victorian English nobleman named Leopold is transplanted to modern-day New York, where he meets and falls in love with Kate. At one point, Leopold sees a bundle of flowers at a florist' shop and is amazed that this bouquet would be given to a woman. It's all wrong, he explains: the lavender implies distrust; the orange lily stands for extreme hatred. Better to send amaryllis, which symbolizes great beauty.

During the Victorian era in England (1837–1891) floral symbolism reached its peak of popularity. Social convention discouraged open displays of emotion, so flowers were often used to convey messages people dared not speak aloud. This botanical language was so elaborate that dictionaries were written to describe the specific "meanings" of flowers and their colors.

A student at Cambridge University might "tell" a woman that she was beautiful with a calla lily. He might indicate he would be patient by presenting her with a daisy. If the woman found her suitor attractive, she could tell him so with a camellia; a geranium, on the other hand, would say, "Let's just be friends." Colors had meaning, too. A red rose symbolized love, while yellow was associated with jealousy and white with innocence.

By the early twentieth century the rules of social communication were sufficiently relaxed that intricate floral communication was no longer necessary. Nevertheless, certain flowers continue to have symbolic meaning. Poppies are worn in the British Commonwealth to memorialize soldiers who died in battle. Lilies are often used at funerals to symbolize life and, for Christians, resurrection. The Hindu god Vishnu is often shown with a lotus flower, symbolizing that he is the pure source of all creation.

Floral symbolism flourishes even in the United States. Consider the poinsettia, *Euphorbia pulcherrima*, a bright red shrub native to Central America that was used by the Aztecs as a source of red dye. The plant was brought to the U.S. by the first U.S. ambassador to Mexico, John Roberts Poinsett, an amateur botanist. Some years later a much shorter strain of the plant was developed by a Californian plant breeder named Paul Ecke. By 1950, his son, Paul Ecke, Jr., began promoting this now portable plant as a holiday decoration, blanketing television specials with offers of free plants during the period between Thanksgiving and Christmas. The campaign was successful: over 100 million poinsettia plants are now sold in the U.S. during the winter holidays every year, making it the best-selling potted plant.

**Floral Message** A girl holds a single flower, perhaps wondering what message it conveys.

**Flowers Have Diverse Forms and Meanings**
The language of flowers still had some popularity in the early twentieth century, as demonstrated by these Edwardian postcards.

You may be surprised to learn that the brightly colored poinsettia "flowers" are not flowers at all. The red (or sometimes pale yellow) parts of the plant that we most notice and appreciate are actually leaves. The poinsettia has a single tiny yellow female flower, without petals, surrounded by male flowers.

The main task of flowers is not to convey messages to humans. Flowers are reproductive equipment: they produce gametophytes, female and male, which in turn produce the gametes that give rise to the next sporophyte generation. Wildflowers (those not "improved" by plant breeders) may have pleasing shapes and colors, but these are in aid not of poetry but of pollination, which is crucial to angiosperm reproduction.

> **IN THIS CHAPTER** we contrast sexual and asexual reproduction in plants, focusing on the details of sexual reproduction. We consider angiosperm gametophytes, pollination, double fertilization, embryo development, and the roles of fruits in seed dispersal. We examine the transition from the vegetative state to the flowering state, a key event in angiosperm development. We conclude by considering the role of asexual reproduction in nature and in agriculture.

## CHAPTER OUTLINE



# 38.1 How Do Angiosperms Reproduce Sexually?

Flowers—the hallmark of angiosperms—contain sex organs; thus it is no surprise that almost all angiosperms reproduce sexually. But many reproduce asexually as well; some even reproduce asexually most of the time. What are the advantages and disadvantages of these two kinds of reproduction?

The relative benefits of sexual versus asexual production are a matter of whether genetic recombination will be advantageous. As we have seen, sexual reproduction produces new combinations of genes and diverse phenotypes (see Section 11.4). Asexual reproduction, in contrast, produces a clone of genetically identical individuals.

Many plants can reproduce either sexually or asexually. For example, strawberry plants can reproduce perfectly well by flowers and seeds (sexual reproduction), but they also reproduce asexually by a stem called a *runner* that spreads over the surface of the soil, sprouting new plants at intervals. For the strawberry plant it might be advantageous to reproduce sexually when possible; this generates genetic diversity, and the seeds that are produced facilitate dispersal to far-flung sites. However, too much diversity can be a drawback for farmers, and they generally propagate this crop asexually to deliver predictably plump and tasty strawberries to the market.

We will return to asexual reproduction later in this chapter. Our concern for now is sexual reproduction.

### The flower is an angiosperm's structure for sexual reproduction

Sexual reproduction involves mitosis and meiosis, and the alternation of haploid and diploid generations (see Chapter 11):

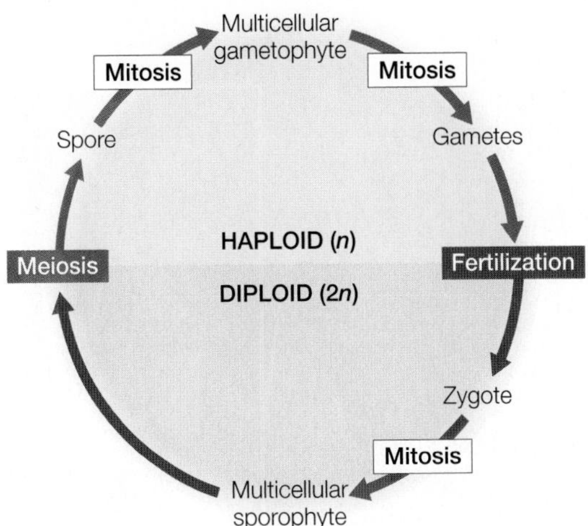


In angiosperms, the plant that we see in nature is a sporophyte and male and/or female gametophytes are contained in the flowers. A complete flower consists of four concentric groups of organs arising from modified leaves: the *carpels*, *stamens*, *petals*, and *sepals*.

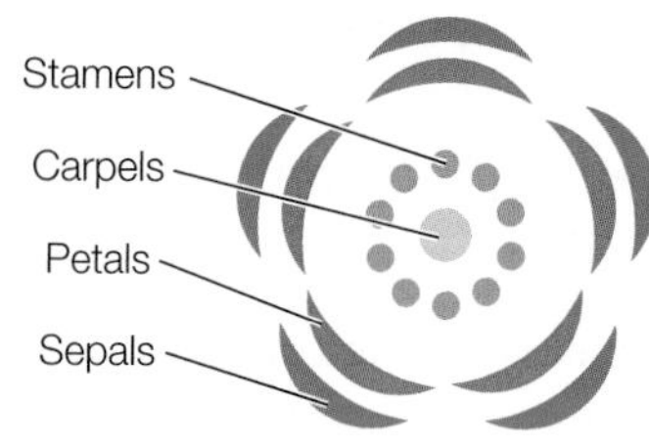


The parts of the flower are usually borne on a stem tip, and derive from a meristem. The differentiation of the meristem into the various organs of the flower is controlled by specific transcription factors (see Figure 19.14). As we discussed in the introductory essay for this chapter, flower parts are very diverse in form.

The carpels and stamens are, respectively, the female and male sex organs. Flowers usually have both stamens and carpels; such flowers are termed *perfect* (**Figure 38.1A**). *Imperfect* flowers, on the other hand, are those with only male or only female sex organs. Male flowers have stamens but not carpels, and female flowers have carpels but not stamens. Some plants, such as corn, bear both male and female flowers on an individual plant; such species are called **monoecious** ("one house") (**Figure 38.1B**). In **dioecious** species, on the other hand, individual plants bear either male-only or female-only flowers; an example is bladder campion (**Figure 38.1C**).

## Flowering plants have microscopic gametophytes

**Figure 38.2** offers a detailed look at the gametophytes central to angiosperm reproduction. The haploid gametophytes—the gamete-producing structures—develop from haploid spores in the flower:

- Female gametophytes (megagametophytes), which are called **embryo sacs**, develop in megasporangia.
- Male gametophytes (microgametophytes), which are called **pollen grains**, develop in microsporangia.

**FEMALE GAMETOPHYTE** Locate the ovule in the flower shown in Figure 38.2. Within the ovule, a megasporocyte—a cell within the megasporangium—divides meiotically to produce four haploid megaspores. In most flowering plants, all but one of these megaspores then undergo apoptosis. The surviving megaspore usually goes through three mitotic divisions without cytokinesis, producing eight haploid nuclei, all initially contained within a single cell—three nuclei at one end, three at the other, and two in the middle. Subsequent cell wall formation leads to an elliptical, seven-celled megagametophyte with a total of eight nuclei:

- At one end of the elliptical megagametophyte are three tiny cells: the **egg** and two cells called *synergids*. The egg is the female gamete, and the synergids participate in fertilization

(A) Perfect: lily

(B) Imperfect monoecious: corn

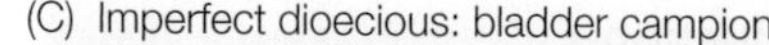

**38.1 Perfect and Imperfect Flowers** (A) A lily is an example of a perfect flower, meaning one that has both male and female sex organs. (B) Imperfect flowers are either male or female. Corn is a monoecious species: both types of imperfect flowers are borne on the same plant. (C) Bladder campion is a dioecious species; some bladder campion plants bear male imperfect flowers while others bear female imperfect flowers.

(C) Imperfect dioecious: bladder campion

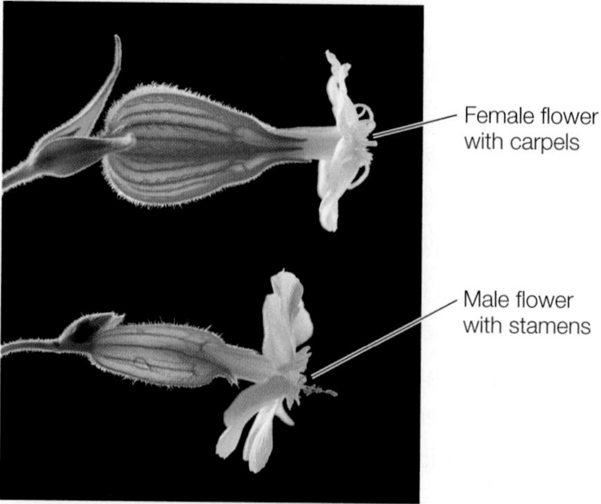

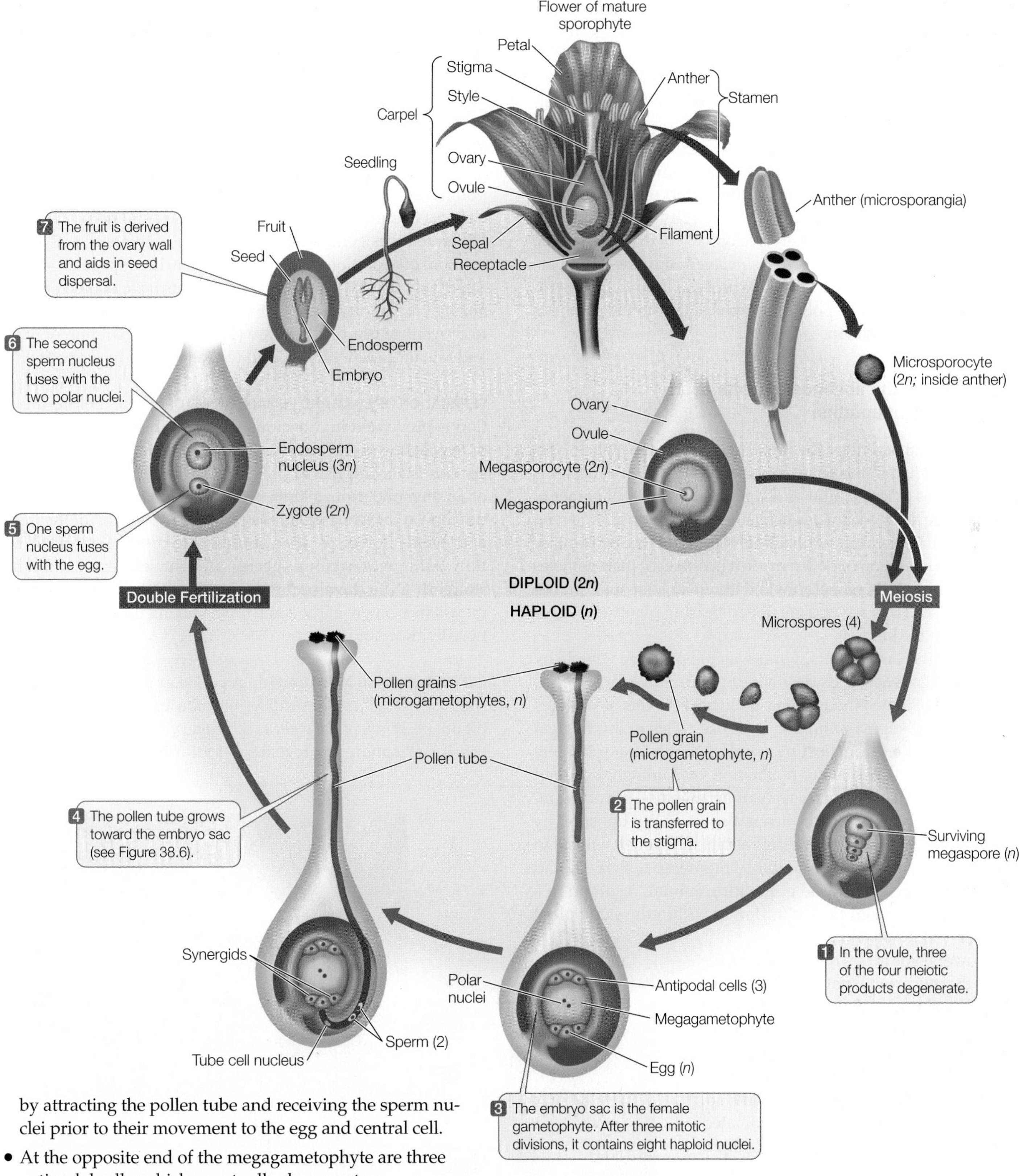


**38.2 Sexual Reproduction in Angiosperms** The embryo sac is the female gametophyte; the pollen grain is the male gametophyte. The male and female nuclei meet and fuse within the embryo sac. Angiosperms have double fertilization, in which a zygote and an endosperm nucleus form from separate fusion events—the zygote from one sperm and the egg, and the endosperm from the other sperm and two polar nuclei.

by attracting the pollen tube and receiving the sperm nuclei prior to their movement to the egg and central cell.

- At the opposite end of the megagametophyte are three antipodal cells, which eventually degenerate.
- In the large central cell are two **polar nuclei**, which together combine with a sperm nucleus.

**yourBioPortal.com**

GO TO **Animated Tutorial 38.1 • Double Fertilization**

The embryo sac (megagametophyte) is the entire seven-cell, eight-nucleus structure.

**MALE GAMETOPHYTE** The pollen grain (microgametophyte) consists of fewer cells and nuclei than the embryo sac. The development of a pollen grain begins when a microsporocyte within the anther divides meiotically. Each resulting haploid microspore develops a spore wall, within which it normally undergoes one mitotic division before the anthers open and release these two-celled pollen grains. The two cells are the tube cell and the generative cell. Further development of the pollen grain, which we will describe shortly, is delayed until the pollen arrives at a stigma (the receptive part of the carpel). In angiosperms, the transfer of pollen from the anther to the stigma is referred to as **pollination**.

### Pollination in the absence of water is an evolutionary adaptation

As Chapter 28 describes, the union of gametes in aquatic plants is accomplished in the water. Fertilization of mosses and ferns also requires at least a film of water for movement of gametes. While there are mechanisms to ensure fertilization if and when the two gametes meet, fertilization is clearly a low-probability event. The evolution of pollen made it possible for male gametes to reach the female gametophyte without an aqueous conduit. With this selective advantage, pollen-bearing plants were able to colonize the land.

In the first land plants, wind was the primary vehicle by which pollen reached its destination, and many plant species are wind-pollinated today. Wind-pollinated flowers have sticky or featherlike stigmas, and they produce pollen grains in great numbers. Pollen transport by wind is, however, a relatively chancy means of achieving pollination, explaining why about 75 percent of all angiosperms rely upon animals—including insects, birds, and bats—for pollen transport. Pollen transport by animals greatly increases the probability that pollen will get to the female gametophyte. Suitably pigmented, shaped, and scented flowers attract the pollinating animal, resulting in a pollen transfer from flower to flower within the same plant species (**Figure 38.3**).

Flower color is one of several adaptations that attract pollinators. Bees, for example, are attracted to blue and yellow flowers (bees cannot sense red but are attracted to patterns exhibited by pigments visible in ultraviolet light; see Figure 56.10). Many birds, on the other hand, are attracted to red flowers (bird-pollinated plants also are often shaped to fit their pollinator's beak.) In both cases, the animals may derive nutrition from the flowers in the form of carbohydrate-rich nectar and/or pollen—a mutually beneficial situation.

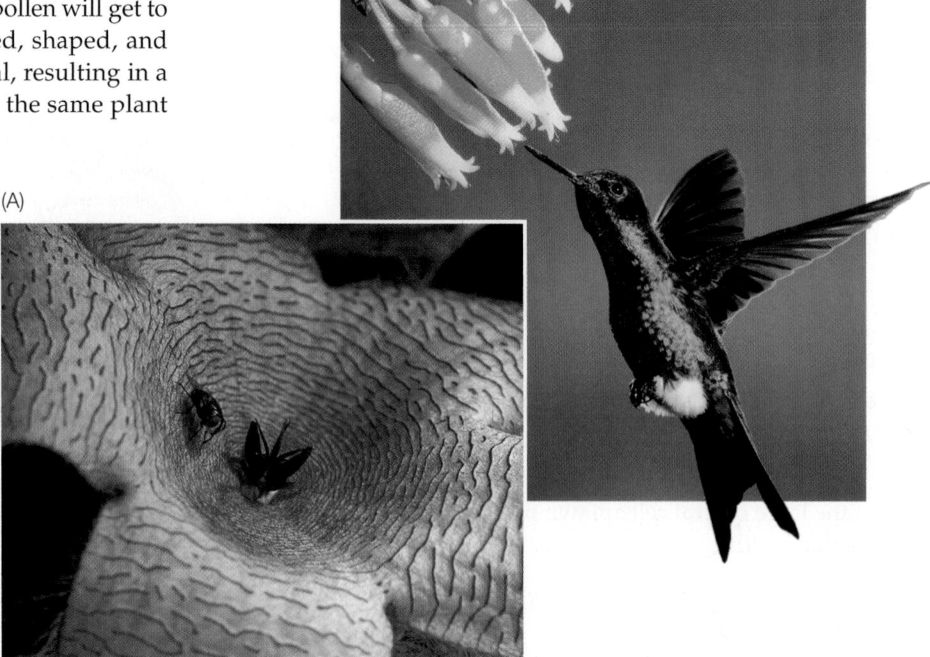


**38.3 Flowers and Pollinators** (A) Flies are attracted to some flowers (in this case, the tropical plant *Stapelia gigantea*) by chemicals emitted from the flower. (B) Other flowers, such as these *Cavendishia* sp. flowers, have red pigments and a shape that attracts certain birds.

### Flowering plants prevent inbreeding

You may recall from discussions of Mendel's work (see Section 12.1) that some plants can reproduce sexually by both cross-pollination and self-pollination. Self-pollination increases the chances of successful pollination, but leads to homozygosity, which reduces genetic diversity. Because diversity is the raw material of evolution by natural selection, homozygosity can be selectively disadvantageous. Most plants have evolved mechanisms that prevent self-fertilization. The two primary means to prevent self-fertilization are (1) physical separation of male and female gametophytes, and (2) genetic self-incompatibility.

**SEPARATION OF MALE AND FEMALE GAMETOPHYTES** Self-fertilization is prevented in dioecious species, which bear only male or female flowers on a particular plant. Pollination in dioecious species is accomplished only when one plant pollinates another. In monoecious plants, which bear both male and female flowers on the same plant, the physical separation of the male and female flowers is often sufficient to prevent self-fertilization. Some monoecious species prevent self-fertilization by staggering the development of male and female flowers so they do not bloom at the same time, making these species functionally dioecious.

**GENETIC SELF-INCOMPATIBILITY** A pollen grain that lands on the stigma of the same plant will fertilize the female gamete (review Figure 38.2) *only if the plant is self-compatible,* meaning capable of self-pollination. To prevent self-fertilization, many plants are

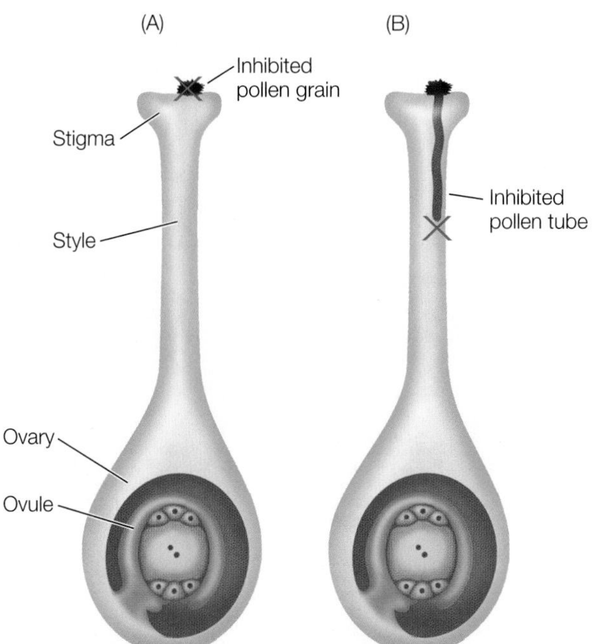


**38.4 Self-Incompatibility** In a self-incompatible plant, pollen is rejected if it expresses an *S* allele that matches one of the *S* alleles of the stigma and style. Self pollen may (A) fail to germinate or (B) its pollen tube may die before reaching an ovule. In either case, the egg cannot be fertilized by a sperm from the same plant.

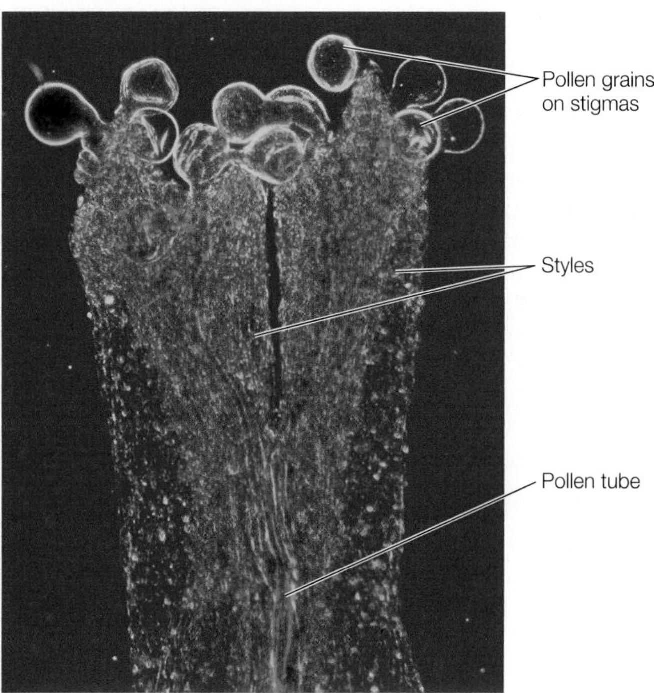


**38.5 Pollen Tubes Begin to Grow** Staining pollen with a fluorescent dye allows them to be seen through a fluorescence microscope. These pollen grains have landed on the stigmas of a crocus.

**self-incompatible**, which depends upon the ability of a plant to determine whether pollen is genetically similar or genetically different from "self." Rejection of "same-as-self" pollen prevents self-fertilization. How does it occur?

Self-incompatibility in plants is controlled by a cluster of tightly linked genes called the *S* locus (for self-incompatibility). The *S* locus encodes proteins in the pollen and style that interact during the recognition process. A self-incompatible species typically has many alleles of the *S* locus, and when the pollen carries an allele that matches one of the alleles of the recipient pistil, the pollen is rejected. Depending on the type of self-incompatibility system, the rejected pollen either fails to germinate or is prevented from growing through the style (**Figure 38.4**); either way, self-fertilization is prevented.

### A pollen tube delivers sperm cells to the embryo sac

When a functional pollen grain lands on the stigma of a compatible pistil, it germinates. A key event is water uptake by pollen from the stigma: pollen loses most of its water as it matures. Germination involves the development of a **pollen tube** (**Figure 38.5**). The pollen tube either traverses the spongy tissue of the style or, if the style is hollow, grows on the inner surface of the style until it reaches an ovule. The pollen tube typically grows at the rate of 1.5–3 mm/hr, taking just an hour or two to reach its destination, the female gametophyte.

The growth of the pollen tube is guided in part by a chemical signal in the form of a small protein produced by the synergids within the ovule. If one synergid is destroyed, the ovule still attracts pollen tubes, but destruction of both synergids renders the ovule unable to attract pollen tubes, and fertilization does not occur. The attractant appears to be species-specific: in some cases, isolated female gametophytes attract only pollen tubes of the same species.

### Angiosperms perform double fertilization

In most angiosperm species, the mature pollen grain consists of two cells, the tube cell and the generative cell. The larger tube cell encloses the much smaller generative cell. Guided by the tube cell nucleus, the pollen tube eventually grows through the style tissue and reaches the embryo sac. The generative cell, meanwhile, has undergone one mitotic division and cytokinesis to produce two haploid **sperm cells** (**Figure 38.6, steps 1 and 2**).

Two fertilization events now occur. One of the two synergids degenerates when the pollen tube arrives and the two sperm cells are released into its remains. (**Figure 38.6, step 3**). Each sperm cell then fuses with a different cell of the embryo sac (**Figure 38.6, steps 4 and 5**). One sperm cell fuses with the egg cell, producing the diploid zygote. The nucleus of the other fuses with the two polar nuclei in the central cell, forming a **triploid** (**3n**) **nucleus**. While the zygote nucleus begins mitotic division to form the new sporophyte embryo, the triploid nucleus undergoes rapid mitosis to form a specialized nutritive tissue, the **endosperm**. The endosperm will later be digested by the developing embryo as a source of nutrients, energy, and carbon-based anabolic building blocks (since it often begins its development

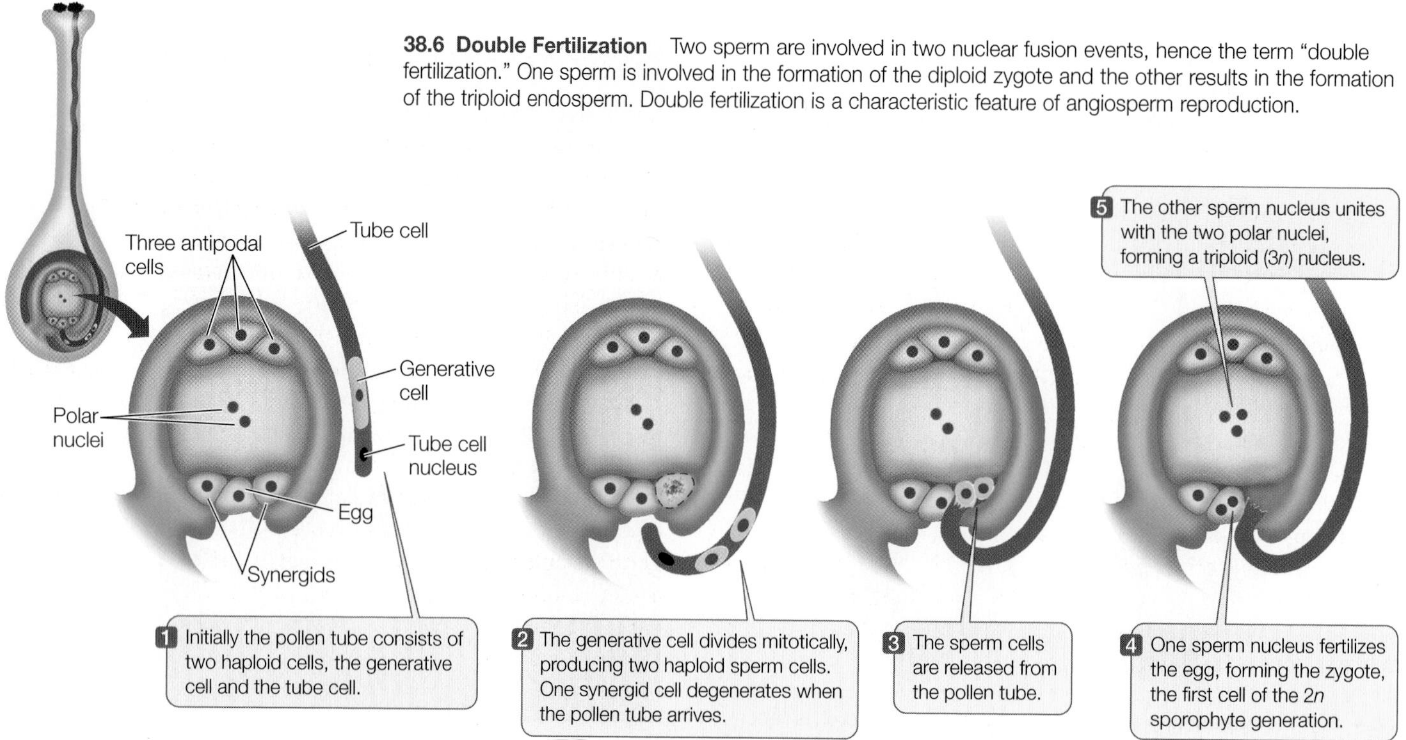

Three antipodal cells
Tube cell
Generative cell
Polar nuclei
Tube cell nucleus
Egg
Synergids
1 Initially the pollen tube consists of two haploid cells, the generative cell and the tube cell.
2 The generative cell divides mitotically, producing two haploid sperm cells. One synergid cell degenerates when the pollen tube arrives.
3 The sperm cells are released from the pollen tube.
4 One sperm nucleus fertilizes the egg, forming the zygote, the first cell of the 2n sporophyte generation.
5 The other sperm nucleus unites with the two polar nuclei, forming a triploid (3n) nucleus.

**38.6 Double Fertilization** Two sperm are involved in two nuclear fusion events, hence the term "double fertilization." One sperm is involved in the formation of the diploid zygote and the other results in the formation of the triploid endosperm. Double fertilization is a characteristic feature of angiosperm reproduction.

underground and thus cannot perform photosynthesis right away). The remaining cells of the male and female gametophytes, the antipodal cells, and the remaining synergid eventually degenerate, as does the pollen tube nucleus.

**Double fertilization** is so named because it involves two nuclear fusion events:

- One sperm nucleus fuses with the egg cell nucleus.
- The other sperm nucleus fuses with the two polar nuclei.

The fusion of a sperm cell nucleus with the two polar nuclei to form endosperm is one of the defining characteristics of angiosperms.

### Embryos develop within seeds

Fertilization initiates the highly coordinated growth and development of the embryo, endosperm, integuments, and carpel. The *integuments*—tissue layers immediately surrounding the megasporangium—develop into the seed coat, and the carpel ultimately becomes the wall of the fruit that encloses the seed.

The first step in the formation of the embryo is a mitotic division of the zygote that gives rise to two daughter cells. These two cells face different fates. An asymmetrical (uneven) distribution of cytoplasm within the zygote causes one daughter cell to produce the embryo proper and the other daughter cell to produce a supporting structure, the **suspensor** (**Figure 38.7**). The suspensor pushes the embryo against or into the endosperm, thereby facilitating the transfer of nutrients from the endosperm into the embryo.

yourBioPortal.com
GO TO **Web Activity 38.1 • Early Development of a Eudicot**

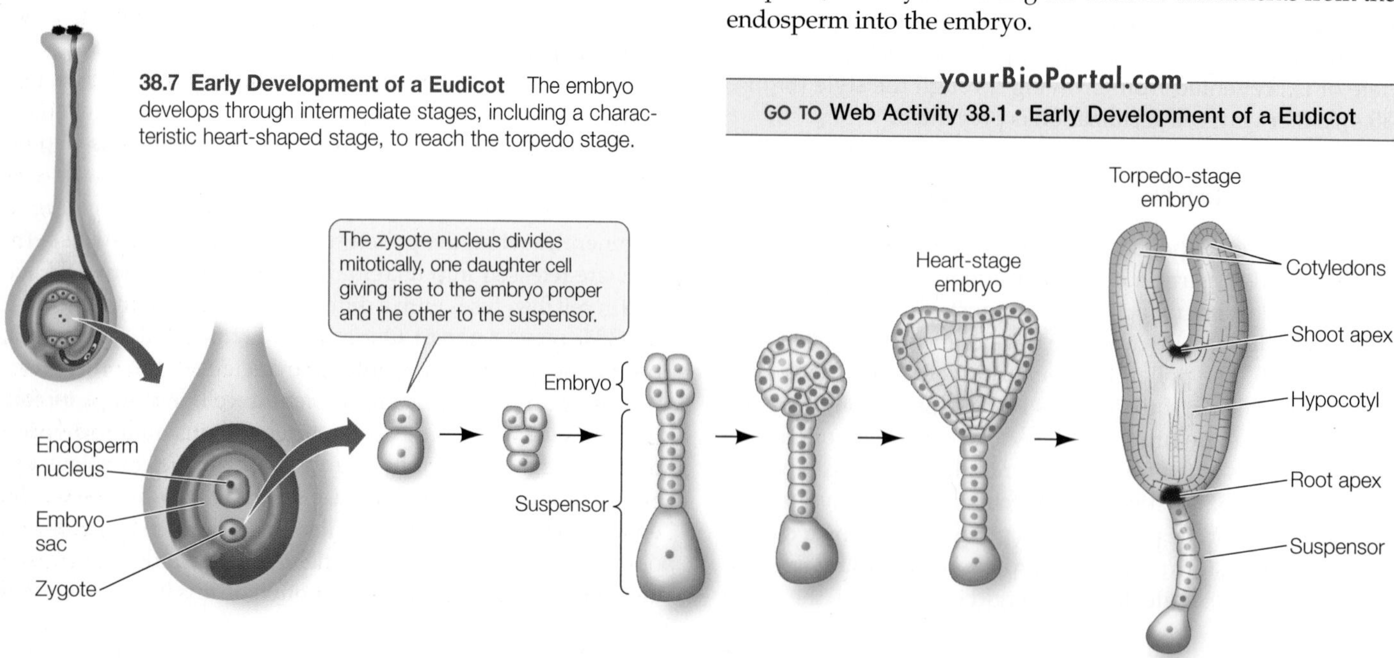

The zygote nucleus divides mitotically, one daughter cell giving rise to the embryo proper and the other to the suspensor.
Endosperm nucleus
Embryo sac
Zygote
Embryo
Suspensor
Heart-stage embryo
Torpedo-stage embryo
Cotyledons
Shoot apex
Hypocotyl
Root apex
Suspensor

**38.7 Early Development of a Eudicot** The embryo develops through intermediate stages, including a characteristic heart-shaped stage, to reach the torpedo stage.

The asymmetrical division of the zygote establishes polarity as well as the longitudinal axis of the new plant. A long, thin suspensor and a more spherical or globular embryo are distinguishable after just four mitotic divisions. The suspensor soon ceases to elongate, and the primary meristems and first organs begin to form within the embryo.

In eudicots, the initially globular embryo develops into the characteristic heart stage as the cotyledons ("seed leaves") start to grow. Further elongation of the cotyledons and of the main axis of the embryo gives rise to the torpedo stage, during which some of the internal tissues begin to differentiate (see Figure 34.7). Between the cotyledons is the shoot apex; at the other end of the axis is the root apex. Each of the apical regions contains a cluster of meristematic cells that continue to divide to give rise to new organs throughout the life of the plant.

During seed development, large amounts of nutrients are moved in from other parts of the parent plant, and the endosperm accumulates starch, lipids, and proteins. In many species, the cotyledons absorb the nutrient reserves from the surrounding endosperm and grow very large in relation to the rest of the embryo (**Figure 38.8A**). In others, the cotyledons remain thin (**Figure 38.8B**) and draw on the reserves in the endosperm as needed when the seed germinates.

In the late stages of embryonic development, the seed loses water—sometimes as much as 95 percent of its original water content. This helps the seed remain viable during the time between the seed's dispersal from the parent plant and its eventual germination.

What keeps seeds viable when they have lost water? It appears that as water leaves, sugars and certain protective proteins become more concentrated inside the seeds, creating a very viscous fluid similar to glass. The membranes and proteins of the cells inside the seed retain their integrity in this viscous state. Once the embryo has become desiccated, it is incapable of further development; it remains dormant until internal and external conditions are right for germination (as we saw in Section 37.1).

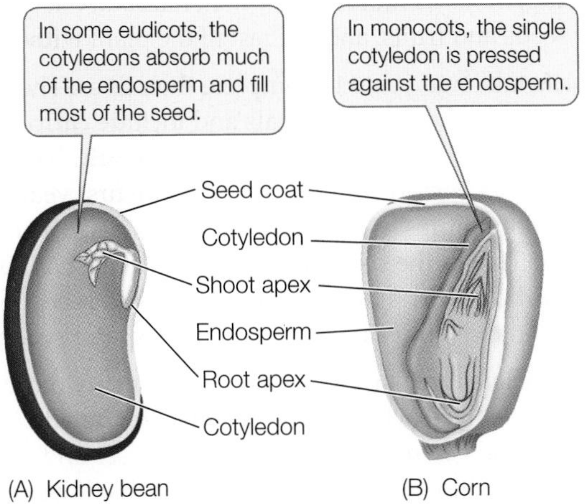


**38.8 Variety in Angiosperm Seeds** In some seeds, such as kidney beans (A), the nutrient reserves of the endosperm are absorbed by the cotyledons. In others, such as corn (B), the reserves in the endosperm will be drawn upon after germination.

## Seed development is under hormonal control

Chapter 37 describes the role of the hormone gibberellin in the mobilization of stored macromolecules in the seed endosperm during germination. The development of seeds is under the control of a different hormone, **abscisic acid** (**ABA**). Most plant tissues make this hormone, and like other plant hormones it has multiple effects (see Table 37.1). (Unfortunately, its name is misleading, because it does not directly control leaf abscission.) During early seed development the ABA level is low, and it rises as the seed matures. This increase stimulates the endosperm to synthesize seed storage proteins. It also stimulates the synthesis of proteins that prevent cell death as the seeds dry.

ABA also keeps the developing seed from germinating on the plant before it dries. Premature germination, termed **vivipary**, is undesirable in seed crops (such as wheat) because the grain is damaged if it has started to sprout. Viviparous seedlings are also unlikely to survive if they remain attached to the parent plant and are unable to establish themselves in the soil. Mutants of corn that are insensitive to ABA have viviparous seeds, indicating the importance of ABA in preventing precocious germination.

The general effect of ABA in preventing germination extends to seed dormancy. Seeds stay dormant if their ABA level is high and germinate when the level goes down, as usually occurs as dormancy is broken.

## Fruits assist in seed dispersal

In angiosperms the ovary wall—together with its seeds—develops into a fruit after fertilization has occurred. Fruits have two main functions:

- They protect the seed from damage by animals and infection by microbial diseases
- They aid in seed dispersal

A **fruit** may consist of only the mature ovary and seeds, or it may include other parts of the flower. Some species produce fleshy, edible fruits such as peaches and tomatoes, while the fruits of other species are dry or inedible.

Fruits are clearly important for carrying seeds, with their embryos, away from the parent plant. Why has this characteristic been selected for during evolution? As products of sexual reproduction, seeds are genetically diverse, and dispersal spreads this diversity around. But if a plant has successfully grown to reproduce, its environment would presumably be favorable for the next generation, too. Some offspring do indeed stay near the parent, as is the case in many tree species, where the seeds essentially fall to the ground. However, this strategy has several disadvantages. If the species is a perennial, offspring that germinate near their parent will be competing with their parent for resources, which may be too limited to support a dense population. Furthermore, even though the local conditions were good enough for the parent to produce at least some seed, there is no guarantee that conditions will still be good the next year, or that they won't be even better elsewhere. Thus, in many cases, seed dispersal is vital to a species' survival.

(A) *Asclepias syriaca*

(B) *Arctium* sp.

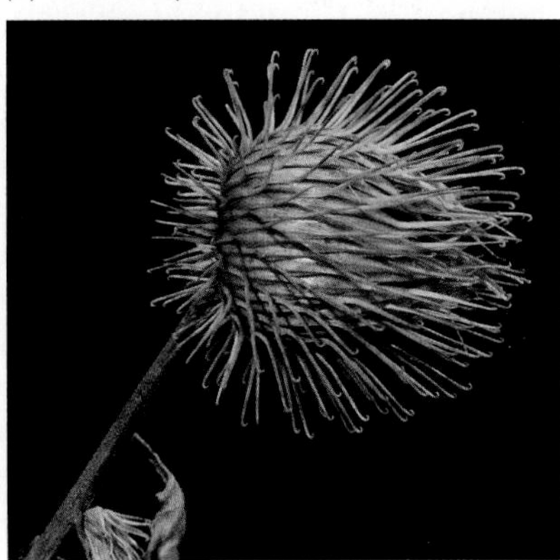

**38.9 Dispersing Fruit** (A) A milkweed seed pod. Silky filaments catch the wind currents and carry the brown seeds with them. (B) Animals who rub up against the "hook-and-loop" surface of burdock fruit walk away with it attached to their fur, thus making the animals unwitting agents of dispersal. This feature of the fruit is said to have inspired the invention of Velcro.

Some fruits help disperse seeds over substantial distances, increasing the probability that at least a few of the many seeds produced by a plant will find suitable conditions for germination and growth to sexual maturity. Various plants, including milkweed and dandelion, produce a fruit with a "parachute" that may be blown some distance from the parent plant by the wind (**Figure 38.9A**). Still other fruits move by hitching rides with animals—either on them, as with burrs stuck to an animal's fur (or to your hiking socks) (**Figure 38.8B**), or inside them, as with berries eaten by birds. Water disperses some fruits; coconuts have been known to travel thousands of miles between islands. Seeds swallowed whole along with fruits such as berries travel through the animal's digestive tract and are deposited some distance from the parent plant. In some species, seeds must pass through an animal in order to break dormancy.

## 38.1 RECAP

**Flowers contain the organs for sexual reproduction in angiosperms. Plants that use pollen for reproduction have several selective advantages, among them the ability to accomplish fertilization without water, which allowed plants to colonize land. After fertilization, the flower develops into seed(s) and fruit. The selective advantages of seeds and fruits include long-term viability and multiple modes of dispersal.**

- What are the relationships between an ovule and an ovary, and between a fruit and a seed? **See p. 796 and Figure 38.2**
- How do plants prevent self-pollination? **See pp. 798–799 and Figure 38.4**
- Describe the roles of the two sperm nuclei in double fertilization. **See p. 799 and Figure 38.6**
- How is plant development controlled by the hormone abscisic acid? **See p. 801**

We have now traced the sexual life cycle of angiosperms from the flower, to the fruit, to the dispersal of seeds. Seed germination and the vegetative development of the seedling are presented in Chapter 37. The next section covers the rest of the angiosperm life cycle—the transition from the vegetative to the flowering state—and how this transition is regulated.

## 38.2 What Determines the Transition from the Vegetative to the Flowering State?

The act of flowering is one of the major events in a plant's life. It represents a reallocation of energy and materials away from making more plant parts (vegetative growth) to making flowers and gametes (reproductive growth). Once a plant is old enough, it can respond to internal or external signals to initiate reproduction. This can happen right at maturity as part of a predetermined developmental program (as in a dandelion plant in the summer) or in response to environmental cues such as light or temperature (as with most ornamental flowers).

Plants fall into three categories depending upon when they mature and initiate flowering, and what happens after they flower:

- **Annuals** complete their lives in one year. This class includes many crops important to the human diet, such as corn, wheat, rice, and soybean. When the environment is suitable, they grow rapidly, with little or no secondary growth. After flowering, they use most of their materials and energy to develop seeds and fruits, and the rest of the plant withers away.
- **Biennials** take two years to complete their lives. They are much less common than annuals and include carrots, cabbage, onions, and Queen Anne's lace. Typically, biennials produce just vegetative growth during the first year and store carbohydrates in underground roots (carrot) and stems (onion). In the second year, they use most of the stored carbohydrates to produce flowers and seeds rather than vegetative growth, and the plant dies after seeds form.
- **Perennials** live three or more—sometimes many more—years. Maple trees, whose leaves symbolize Canada, can live up to 400 years. Perennials include many trees and shrubs, as well as wildflowers. Typically these plants flower every year, but stay alive and keep growing for another season; the reproductive cycle repeats each year. However, some perennials (e.g., century plant) grow vegetatively for many years, flower once, and die.

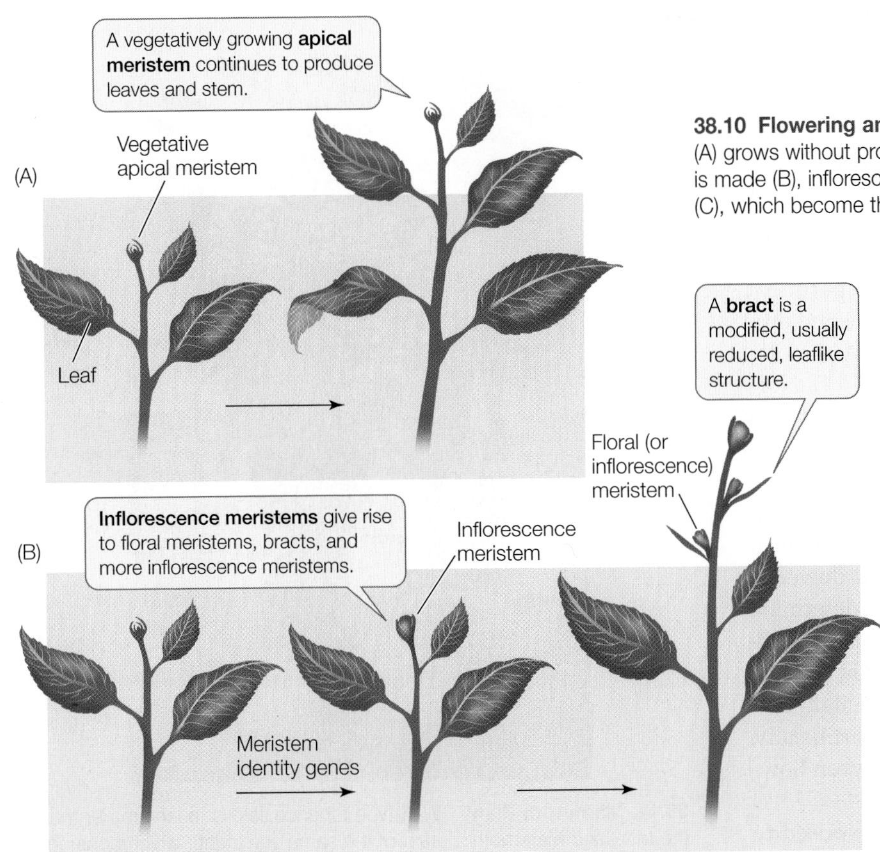
(A)
A vegetatively growing **apical meristem** continues to produce leaves and stem.
Vegetative apical meristem
Leaf
(B)
**Inflorescence meristems** give rise to floral meristems, bracts, and more inflorescence meristems.
Inflorescence meristem
Meristem identity genes
A **bract** is a modified, usually reduced, leaflike structure.
Floral (or inflorescence) meristem
(C)
A **floral meristem** gives rise to a flower.
Floral meristem
Floral identity genes
Carpel
Stamen
Petal
Sepal

**38.10 Flowering and the Apical Meristem** A vegetative apical meristem (A) grows without producing flowers. Once the transition to the flowering state is made (B), inflorescence meristems give rise to bracts and to floral meristems (C), which become the flowers.

No matter what type of life cycle they have, angiosperms all make the transition to flowering. This transition entails significant developmental changes, to which we now turn.

### Apical meristems can become inflorescence meristems

The first visible sign of a transition to the flowering state may be a change in one or more apical meristems in the shoot system. As described in Chapter 34, meristems have a pool of undetermined cells. During vegetative growth, an apical meristem continually produces leaves, axillary buds, and stem tissues (**Figure 38.10A**) in a kind of unrestricted growth called *indeterminate growth* (see Section 34.4).

Flowers may appear singly or in an orderly cluster that constitutes an **inflorescence**. If a vegetative apical meristem becomes an **inflorescence meristem**, it ceases production of leaves and axillary buds and produces other structures: smaller leafy structures called bracts, as well as new meristems in the angles between the bracts and the stem (**Figure 38.10B**). These new meristems may also be inflorescence meristems, or they may be **floral meristems**, each of which gives rise to a flower.

Each floral meristem typically produces four consecutive whorls or spirals of organs—the sepals, petals, stamens, and carpels discussed earlier in the chapter—separated by very short internodes, keeping the flower compact (**Figure 38.10C**). In contrast to vegetative apical meristems and some inflorescence meristems, floral meristems are responsible for *determinate growth*—growth of limited duration, like that of leaves.

### A cascade of gene expression leads to flowering

How do apical meristems become floral meristems or inflorescence meristems, and how do inflorescence meristems give rise to floral meristems? How does a floral meristem give rise, in short order, to four different floral organs (sepals, petals, stamens, and carpels)? How does each flower come to have the correct number of each of the floral organs? Numerous genes are expressed and interact to produce these results. We'll refer here to some of the genes whose actions have been most thoroughly studied in *Arabidopsis* and snapdragons (*Antirrhinum*) (see Figure 38.10):

- Expression of a group of **meristem identity genes** initiates a cascade of further gene expression that leads to flower formation. The expression of the genes *LEAFY* and *APETALA1* is both necessary and sufficient for flowering. How do we know this? There are two types of evidence, genetic and molecular. For example, a mutated allele of the gene *APETALA1* leads to continued vegetative growth, even if all other conditions are suitable for flowering. On the other hand, if the wild-type *APETALA1* gene is coupled to an active promoter and introduced into an apical meristem, the plant will flower regardless of the environment. This is powerful evidence that *APETALA1* plays a role in switching meristem cells from a vegetative to a reproductive fate.
- Meristem identity gene products trigger the expression of **floral organ identity genes**, which work in concert to specify the successive whorls of the flower (see Figure 19.14). Floral identity genes are homeotic genes whose products are transcription factors that determine whether cells in the floral meristem will be sepals, petals, stamens, or carpels. An example is the gene *AGAMOUS*, which causes florally determined cells to form stamens and carpels in the "ABC" system described in Section 19.5.

How is this cascade of events initiated? Depending on the species, plants respond to either internal or external cues. Among external clues, the best studied are photoperiod (day length) and temperature. We begin with photoperiod, as it has a fascinating history and clear experimental support.

### Photoperiodic cues can initiate flowering

In 1920, W. W. Garner and H. A. Allard of the U.S. Department of Agriculture studied the behavior of a newly discovered mutant tobacco plant. The mutant, named Maryland Mammoth, had large leaves and exceptional height (**Figure 38.11**). Normally tobacco is an annual that flowers in the summer and then stops growing. In contrast, Maryland Mammoth plants remained vegetative and continued to grow.

Garner and Allard now tried to figure out why the mutant plants did not flower in the summer. It wasn't that they *could not flower*: the scientists found that the plants would flower in December in the greenhouse under natural light. To determine what induces flowering in December, they tested several likely environmental variables, such as temperature. The key variable proved to be day length. By moving plants between light and dark rooms at different times to vary the day length artificially, the scientists were able to establish a direct link between flowering and day length.

Maryland Mammoth plants did not flower if exposed to more than 14 hours of light per day, but flowering commenced once the daylight period became shorter than 14 hours, as in December. Thus the **critical day length** for Maryland Mammoth tobacco is 14 hours (**Figure 38.12**). Control of an organism's responses by the length of day or night is called **photoperiodism**.

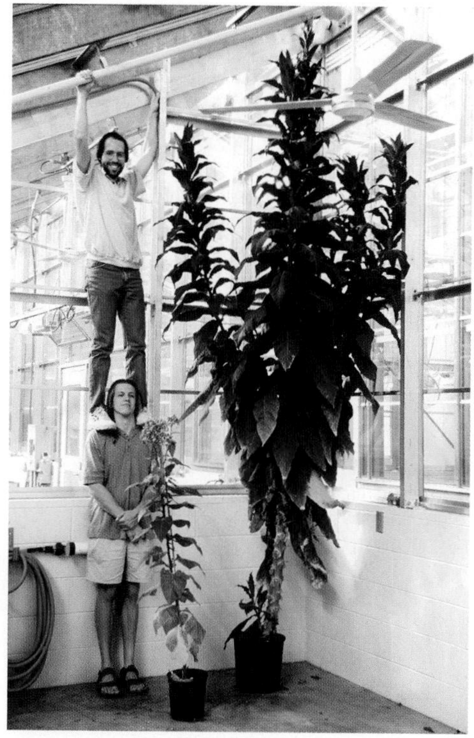

**38.11 Mammoth Plant** Wild-type tobacco (left) is much smaller than the Maryland Mammoth mutant of the same age (right), which does not respond to an environmental cue to stop growing and flower.

### Plants vary in their responses to photoperiodic cues

Plants that flower in response to photoperiodic stimuli fall into two main classes:

- **Short-day plants (SDPs)** flower only when the day is shorter than a critical maximum. They include poinsettias and chrysanthemums, as well as Maryland Mammoth tobacco. Thus, for example, we see chrysanthemums in nurseries in the fall, and poinsettias in winter, as noted in the opening of this chapter.
- **Long-day plants (LDPs)** flower only when the day is longer than a critical minimum. Spinach and clover are examples of LDPs. For example, spinach tends to flower and become bitter in the summer, and is therefore normally planted in early spring.

While there are variations on these two patterns, photoperiodic control of flowering serves an important role: it synchronizes the flowering of plants of the same species in a local population, and this promotes cross-pollination and successful reproduction.

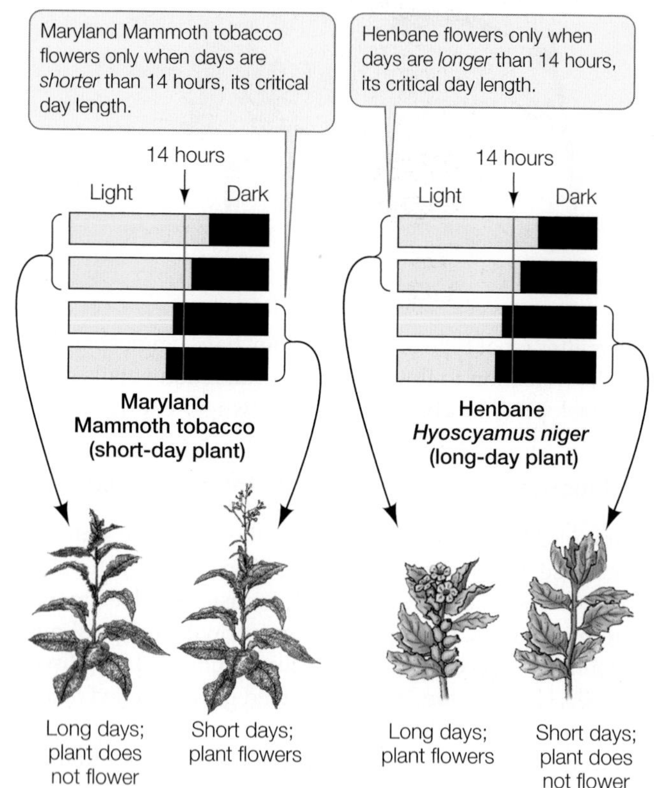

Maryland Mammoth tobacco flowers only when days are *shorter* than 14 hours, its critical day length.
Henbane flowers only when days are *longer* than 14 hours, its critical day length.
14 hours
Light
Dark
14 hours
Light
Dark
Maryland Mammoth tobacco (short-day plant)
Henbane *Hyoscyamus niger* (long-day plant)
Long days; plant does not flower
Short days; plant flowers
Long days; plant flowers
Short days; plant does not flower

**38.12 Day Length and Flowering** By artificially varying the day length in a 24-hour period, Garner and Allard showed that the flowering of Maryland Mammoth tobacco is initiated when the days become shorter than a critical length. Maryland Mammoth tobacco is thus called a short-day plant. Henbane, a long-day plant, shows an inverse pattern of flowering.

### The length of the night is the key photoperiodic cue determining flowering

The terms "short-day plant" and "long-day plant" became entrenched before scientists determined that *photoperiodically sen-*

*sitive plants actually measure the length of the night (darkness), rather than the length of the day.* This was demonstrated by Karl Hamner of the University of California at Los Angeles and James Bonner of the California Institute of Technology (**Figure 38.13**).

INVESTIGATING LIFE

**38.13 Night Length and Flowering**
Short-day plants (SDP) flower only when the day is shorter than a critical maximum. But what environmental cue initiates SDP flowering: day length or night length? To find out, Karl Hamner and James Bonner carried out greenhouse experiments using cocklebur, a SDP.

**HYPOTHESIS** **Short-day plants measure day length.**

**METHOD** Divide plants into two groups. Expose groups to different light conditions: one group to a constant daylight period and varied periods of darkness, the other to varied periods of daylight and fixed periods of darkness.

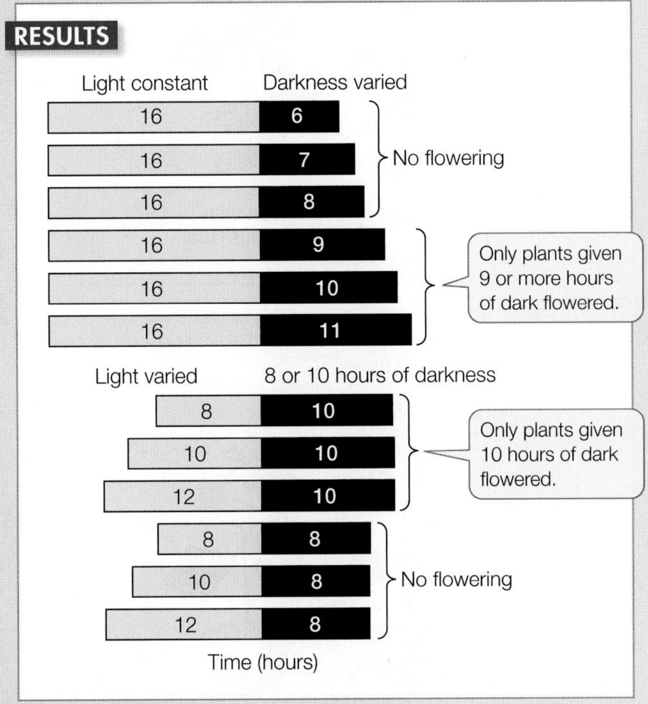


**CONCLUSION** **The data do not support the hypothesis. Short-day plants measure the length of the night and thus could more accurately be called long-night plants.**

**FURTHER INVESTIGATION:** How would you perform these experiments using long-day plants and what would be the results?

Go to **yourBioPortal.com** for original citations, discussions, and relevant links for all INVESTIGATING LIFE figures.

Working with cocklebur, an SDP, Hamner and Bonner ran a series of experiments using two sets of conditions:

- One group of plants was exposed to a constant light period—either shorter or longer than the critical day length—and the dark period was varied.
- A second group of plants was exposed to a constant dark period—and the light period was varied.

Plants flowered under all treatments in which the dark period exceeded 9 hours, regardless of the length of the light period. Hamner and Bonner thus concluded that the length of the *night* is critical to flowering. For cocklebur, the **critical night length** is about 9 hours. It is thus more accurate to call cocklebur a "long-night plant" than a short-day plant.

In cocklebur, a single long night is sufficient to trigger full flowering some days later, even if the intervening nights are short. Most plants are less sensitive than cocklebur and require from two to several nights of appropriate length to induce flowering. For some plants a single shorter night in a series of long ones inhibits flowering, even if the short night comes only one day before flowering would have commenced.

Through other experiments Hamner and Bonner gained some insight into how plants measure night length. They grew SDPs and LDPs under a variety of light/dark conditions. In some experiments, the dark period was interrupted by a brief exposure to light; in others, the light period was interrupted briefly by darkness. Interruptions of the light period by darkness had no effect on the flowering of either short-day or long-day plants. Even a brief interruption of the dark period by light, however, completely nullified the effect of a long night. An SDP flowered *only if the long nights were uninterrupted*. The investigators hypothesized that something must accumulate during that long night that could be broken down by a flash of light in the middle of the night.

To find out what that "something" might be, Hamner and Bonner tested flashes of interrupting light at various wavelengths. You may recall from Section 37.5 that several photoreceptors play roles in regulating plant growth, and that these are sensitive to different wavelengths. In the interrupted-night experiments, the most effective wavelengths of light were in the red range (**Figure 38.14**), and the effect of a red-light interruption of the night could be fully reversed by a subsequent exposure to far-red light, indicating that a phytochrome is the photoreceptor. Where does this occur and what happens downstream from the reception? Once again, elegant experiments provided the answer.

### The flowering stimulus originates in a leaf

Early experiments indicated that reception of the photoperiodic stimulus occurs within the leaf. For example, in spinach, an LDP, flowering would occur if the leaves were exposed to long-day periods of light while the bud meristem was masked to simulate short days. Flowering could *not* occur when its leaves were masked to simulate short days while the bud was exposed to long-day periods of light.

## INVESTIGATING LIFE

**38.14 Interrupting the Night**

Knowing that plants measure night duration, the question became whether the dark hours to which a plant is exposed must be continuous. Using SDPs and LDPs as test subjects, Hamner and Bonner interrupted the night with light of different wavelengths.

**HYPOTHESIS** Red light participates in the photoperiodic timing mechanism.

**METHOD** Grow plants under short-day conditions, but interrupt the night with light of different wavelengths.

| Short-day plants | Light/dark combinations | Long-day plants |
|---|---|---|
| Flowering | (uninterrupted night) | No flowering |
| | (night interrupted with light) | |

**RESULTS**

| Short-day plants | Light | Long-day plants |
|---|---|---|
| No flowering | R | Flowering |
| Flowering | FR | No flowering |
| Flowering | R FR | No flowering |
| No flowering | R FR R | Flowering |
| Flowering | R FR R FR | No flowering |

**CONCLUSION** When plants are exposed to red (R) and far-red (FR) light in alternation, the final treatment determines the effect. Phytochrome is the photoreceptor.

**FURTHER INVESTIGATION:** How would you show that interrupting the day with a brief period of darkness had no effect on flowering?

Go to **yourBioPortal.com** for original citations, discussions, and relevant links for all INVESTIGATING LIFE figures.

**yourBioPortal.com**

GO TO **Animated Tutorial 38.2 • The Effect of Interrupted Days and Nights**

## INVESTIGATING LIFE

**38.15 The Flowering Signal Moves from Leaf to Bud**

The receptor for photoperiod, phytochrome, is in the leaf but flowering occurs in the bud meristem. To investigate whether there is a diffusible substance that travels from leaf to bud, James Knott exposed only the leaf to the photoperiodic stimulus.

**HYPOTHESIS** The leaves measure the photoperiod.

**METHOD** Grow cocklebur plants under long days and short nights. Mask a leaf on some plants and see if flowering occurs.

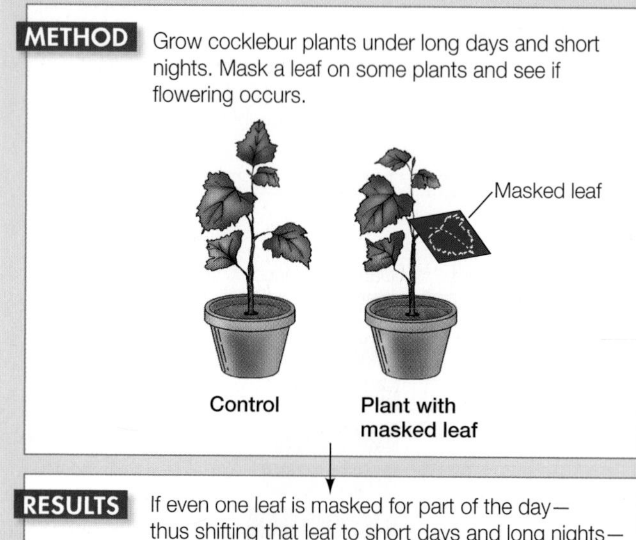


**RESULTS** If even one leaf is masked for part of the day—thus shifting that leaf to short days and long nights—the plant will flower.

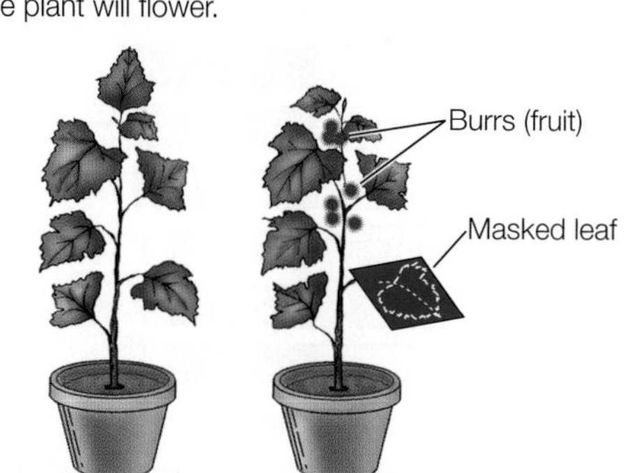


**CONCLUSION** The leaves measure the photoperiod. Therefore, some signal must move from the induced leaf to the flowering parts of the plant.

**FURTHER INVESTIGATION:** How would you show experimentally that the flowering signal is the same in different species of plants?

Go to **yourBioPortal.com** for original citations, discussions, and relevant links for all INVESTIGATING LIFE figures.

These "masking" experiments were extended to SDP plants as well (**Figure 38.15**). Because the receptor of the stimulus (in the leaf) is physically separated from the tissue on which the stimulus acts (the bud meristem), the inference can be drawn that a systemic signal travels from the leaf through the plant's tissues to the bud meristem. Other evidence that a diffusible chemical travels from the leaf to the bud meristem signal includes the following:

- If a photoperiodically induced leaf is immediately removed from a plant after the inductive dark period, the plant does not flower. If, however, the induced leaf remains attached to the plant for several hours, the plant will flower. This result suggests that something is synthesized in the leaf in response to the inductive dark period, and then moves out of the leaf to induce flowering.
- If two or more cocklebur plants are grafted together and if one plant is exposed to inductive long nights and its graft

partners are exposed to noninductive short nights, all the plants flower.

- In several species, if an induced leaf from one species is grafted onto another, noninduced plant of a different species, the recipient plant flowers.

Although the transmissible signal was long ago given a name, **florigen** ("flower inducing"), the nature of the signal has only recently been explained.

### Florigen is a small protein

The characterization of florigen was made possible by genetic and molecular studies of the model organism *Arabidopsis,* an LDP. Three genes are involved (**Figure 38.16**):

- ***FT*** **(*FLOWERING LOCUS T*)** *codes for florigen.* A small protein (20 kDa molecular weight), FT can travel through plasmodesmata. FT is synthesized in phloem companion cells of the leaf and diffuses into the adjacent sieve elements, where it enters the phloem flow to the apical meristem. If *FT* is coupled to an active promoter and expressed at high levels in the shoot meristem, flowering is induced even in the absence of an appropriate photoperiodic stimulus.
- ***CO*** **(*CONSTANS*)** *codes for a transcription factor that activates the synthesis of FT.* Like *FT, CO* is expressed in leaf companion cells. If *CO* is experimentally overexpressed in the leaf, flowering is induced. Overexpression of *CO* in the apical meristem does not, however, induce flowering, indicating that its essential activity is in the leaf. CO protein is expressed all the time but is unstable; an appropriate photoperiodic stimulus stabilizes CO so that there is enough to turn on FT synthesis.
- ***FD*** **(*FLOWERING LOCUS D*)** *codes for a protein that binds to FT protein when it arrives in the apical meristem.* The FD protein is a transcription factor that when complexed with FT protein, activates promoters for meristem identity genes, such as *APETALA1* (see Figure 38.10). The expression of FD primes meristem cells to change from a vegetative fate to a reproductive fate once florigen arrives.

Before florigen was isolated, grafting experiments indicated that many different plant species could be induced to flower by the same chemical signal. A photoperiod-induced leaf from one species can induce flowering when grafted onto an uninduced plant of another species. Results of molecular experiments confirm that the *FT* gene is involved in photoperiod signaling in many species:

- Transgenic plants (e.g., tobacco and tomato) that express the *Arabidopsis FT* gene at high levels flower regardless of day length.
- Transgenic *Arabidopsis* plants that express high levels of *FT* homologs from other plants (e.g., rice and tomato) flower regardless of day length.

While the molecular basis of the action of florigen has been elucidated, commercial applications of this knowledge have been harder to realize. It was hoped that florigen might be a very small molecule, like auxin or gibberellin that could be sprayed on economically important plants to induce flowering

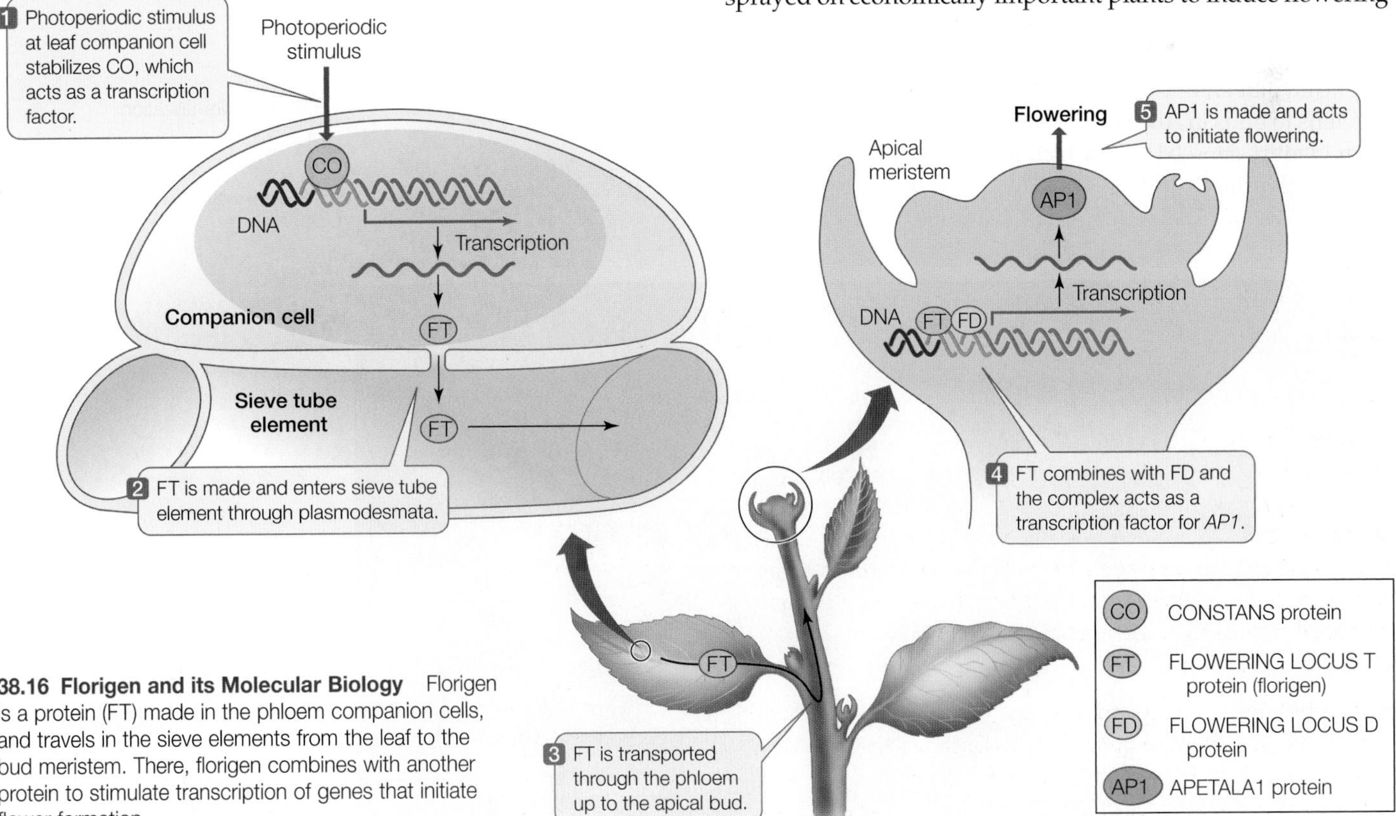

1 Photoperiodic stimulus at leaf companion cell stabilizes CO, which acts as a transcription factor.
Photoperiodic stimulus
CO
DNA
Transcription
Companion cell
FT
Sieve tube element
FT
2 FT is made and enters sieve tube element through plasmodesmata.
3 FT is transported through the phloem up to the apical bud.
FT
Flowering
5 AP1 is made and acts to initiate flowering.
Apical meristem
AP1
Transcription
DNA
FT
FD
4 FT combines with FD and the complex acts as a transcription factor for AP1.
CO CONSTANS protein
FT FLOWERING LOCUS T protein (florigen)
FD FLOWERING LOCUS D protein
AP1 APETALA1 protein

**38.16 Florigen and its Molecular Biology** Florigen is a protein (FT) made in the phloem companion cells, and travels in the sieve elements from the leaf to the bud meristem. There, florigen combines with another protein to stimulate transcription of genes that initiate flower formation.

at will. The fact that florigen is a protein that cannot readily enter cells from the outside environment makes the development of commercial florigen treatments unlikely.

We have considered the photoperiodic regulation of flowering, from photoreceptors in the leaf to florigen that travels from the induced leaf to the sites of flower formation. In some plant species, however, flowering is induced by other stimuli.

### Flowering can be induced by temperature or gibberellin

**TEMPERATURE** In some plant species, notably certain cereal grains, the environmental signal for flowering is cold temperature, a phenomenon called **vernalization** (Latin *vernus*, "spring"). In both wheat and rye, we distinguish two categories of flowering behavior. Spring wheat, for example, is a typical annual plant: it is sown in the spring and flowers in the same year. Winter wheat is sown in the fall, grows to a seedling, overwinters (often covered by snow), and flowers the following summer. If winter wheat is not exposed to cold in its first year, it will not flower normally the next year.

How vernalization leads to flowering has been elucidated from model organisms such as *Arabidopsis*. In strains of *Arabidopsis* that require vernalization to flower (**Figure 38.17**), a gene called *FLC* (FLOWERING LOCUS C) encodes a transcription factor that blocks the FT–FD florigen pathway (see Figure 38.16) by inhibiting expression of FT and FD. Cold temperature inhibits the synthesis of FLC protein, allowing FT and FD proteins to be expressed and flowering to proceed. Similar proteins control some steps in vernalization in cereals.

**GIBBERELLIN** *Arabidopsis* plants do not flower if they are genetically deficient in the hormone gibberellin, or if they are treated with an inhibitor of gibberellin synthesis. These observations implicate gibberellins in flowering. Direct application of gibberellins to buds in *Arabidopsis* results in activation of the meristem identity gene *LEAFY*, which in turn promotes the transition to flowering.

### Some plants do not require an environmental cue to flower

A number of plant species and strains do not require a photoperiod, vernalization, or gibberellin to flower, but instead flower on cue from an "internal clock." For example, flowering in some strains of tobacco will be initiated in the terminal bud when the stem has grown four phytomers in length (recall that stems are composed of repeating units called phytomers; see Figure 34.1). If such a bud and a single adjacent phytomer is removed and planted, the cutting will flower because the bud has already received the cue for flowering. But the rest of the shoot below the bud that has been removed will not flower because it is only three phytomers long. After it grows an additional phytomer, it flowers. These results suggest that there is something about the *position* of the bud (atop four phytomers of stem) that determines its transition to flowering.

The bud might "know" its position by the concentration of some substance that forms a positional gradient along the length of the plant. Such a gradient could be formed if the root makes a diffusible inhibitor of flowering whose concentration diminishes with plant height. When the plant reaches a certain height, the concentration of the inhibitor would become sufficiently low at the tip of the shoot to allow flowering. What this inhibitor might be is unclear, but there is evidence that it acts by decreasing the amount of FLC, allowing the FT–FD pathway to proceed (just as cold acts on FLC in vernalization). A positional gradient that acts on FLC would be consistent with other mechanisms affecting flowering, which all converge on *LEAFY* and *APETALA1*:

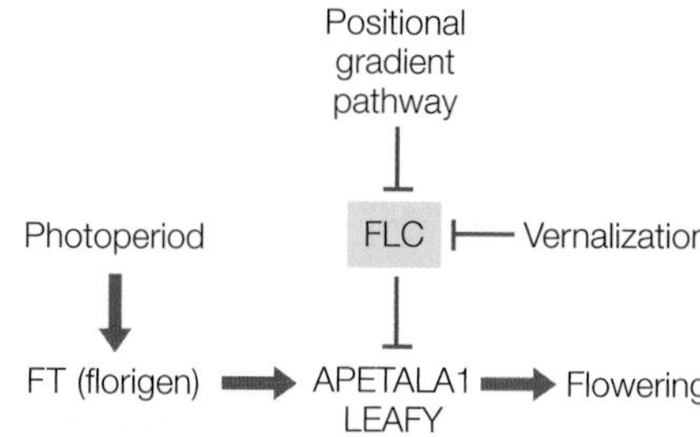

**38.17 Vernalization** A genetic strain of *Arabidopsis* (winter-annual *Arabidopsis*) requires vernalization for flowering. Without it, the plant is large and vegetative (left), but with the cold period it is smaller and flowers (right).

## 38.2 RECAP

**Flowering of some angiosperms is controlled by night length, a phenomenon called photoperiodism. Gibberellins can induce flowering in some species, as can exposure to low temperatures (vernalization). Some species flower when their stems have grown by a certain amount, independent of environmental cues. All pathways to flowering converge on the meristem identity genes.**

- What are the differences between apical meristems, inflorescence meristems, and floral meristems? What genes control the transitions between them? **See p. 803 and Figure 38.10**
- Explain why "short-day plant" is a misleading term. **See p. 805 and Figure 38.13**
- What is the evidence for florigen? What is its molecular mechanism of action? **See p. 807 and Figures 38.15 and 38.16**

We have seen how environmental factors interact with genes to control flowering in angiosperms. The function of flowers is sexual reproduction, which maintains beneficial genetic variation in a population. Many angiosperms, however, also benefit from being able to reproduce asexually.

# 38.3 How Do Angiosperms Reproduce Asexually?

Although sexual reproduction takes up most of the space in this chapter, asexual reproduction accounts for many of the individual plants present on Earth. This fact suggests that in some circumstances asexual reproduction must be advantageous.

We have noted that genetic recombination is one of the advantages of sexual reproduction. Self-fertilization is a form of sexual reproduction, but offers fewer opportunities for genetic recombination than does cross-fertilization. A diploid, self-fertilizing plant that is heterozygous for a certain locus can produce both kinds of homozygotes for that locus plus the heterozygote among its progeny, but it cannot produce any progeny carrying alleles that it does not itself possess. Nevertheless many self-fertilizing plant species produce viable and vigorous offspring.

Asexual reproduction eliminates genetic recombination altogether. A plant that reproduces asexually produces progeny genetically identical to the parent (clones). What, then, is the advantage of asexual reproduction? If a plant is well adapted to its environment, asexual reproduction allows it to pass on to all its progeny a superior combination of alleles, which might otherwise be separated by sexual recombination.

### Many forms of asexual reproduction exist

Stems, leaves, and roots are considered *vegetative organs* and are distinguished from flowers, the reproductive parts of the plant. Asexual reproduction is often accomplished through the modification of a vegetative organ, which is why the term **vegetative reproduction** is sometimes used to describe asexual reproduction in plants. Another type of asexual reproduction, *apomixis*, involves flowers but no fertilization.

**VEGETATIVE REPRODUCTION** Often the stem is the organ that is modified for vegetative reproduction. As noted earlier, strawberries produce horizontal stems, called *stolons* or runners, which grow along the soil surface, form roots at intervals, and establish potentially independent plants. Asexual reproduction by *tip layers* is accomplished when the tips of upright branches sag to the ground and develop roots, as in blackberry and forsythia.

Some plants, such as potatoes, form enlarged fleshy tips of underground stems, called *tubers*, that can produce new plants (from the "eyes"). *Rhizomes* are horizontal underground stems that can give rise to new shoots. Bamboo is a striking example of a plant that reproduces vegetatively by means of rhizomes. A single bamboo plant can give rise to a stand—even a forest—of plants constituting a single, physically connected entity.

Whereas stolons and rhizomes are horizontal stems, bulbs and corms are short, vertical, underground stems. Lilies and onions form bulbs (**Figure 38.18A**), short stems with many fleshy, highly modified

(A) *Allium* sp.

Storage leaves grow in layers from the stem of this onion.

The short stem is visible at the bottom of the bulb.

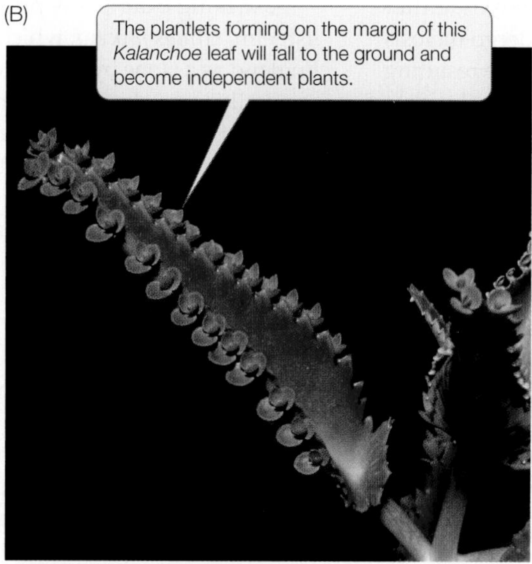

(B)
The plantlets forming on the margin of this *Kalanchoe* leaf will fall to the ground and become independent plants.


**38.18 Vegetative Organs Modified for Reproduction** (A) Bulbs are short stems with large leaves that store nutrients and can give rise to new plants. (B) In *Kalanchoe*, new plantlets can form on leaves.

leaves that store nutrients. These storage leaves make up most of the bulb. Bulbs are thus large underground buds. They can give rise to new plants by dividing or by producing new bulbs from axillary buds. Crocuses, gladioli, and many other plants produce corms, underground stems that function very much as bulbs do. Corms are disclike and consist primarily of stem tissue; they lack the fleshy modified leaves that are characteristic of bulbs.

Stems are not the only vegetative organs modified for asexual reproduction. Leaves may also be the source of new plantlets, as in some succulent plants of the genus *Kalanchoe* (**Figure 38.18B**). Many kinds of angiosperms, ranging from grasses to trees such as aspens and poplars, form interconnected, genetically homogeneous populations by means of suckers—shoots produced by roots. What appears to be a whole stand of aspen trees, for example, may be a clone derived from a single tree by suckers. This is why the leaves of a whole stand of aspens typically turn yellow at the same time.

Plants that reproduce vegetatively often grow in physically unstable environments such as eroding hillsides. Plants with stolons or rhizomes, such as beach grasses, rushes, and sand verbena, are common pioneers on coastal sand dunes. Rapid vegetative reproduction enables these plants, once introduced, not only to multiply but also to survive burial by the shifting sand; in addition, the dunes are stabilized by the extensive network of rhizomes or stolons that develops. Vegetative reproduction is also common in some deserts, where the environment is often not suitable for seed germination and the establishment of seedlings.

**APOMIXIS** Some plants produce flowers but use them to reproduce asexually rather than sexually. Dandelions, blackberries, some citrus, and some other plants reproduce by the asexual production of seeds, called **apomixis**. As described earlier, in alternation of generations meiosis typically reduces the number of chromosomes in gametes by half, and fertilization restores the sporophytic (diploid) number of chromosomes in the zygote. In a female gametophyte undergoing apomixis, either meiosis begins and the chromosomes do not undergo meiosis II, or meiosis does not occur at all. In either case, the resulting gamete is diploid. Cells within the ovule simply develop into the embryo and the ovary wall develops into a fruit. The result of apomixis is a fruit with seeds that are genetically identical to the parent plant.

Apomixis would be considered an oddity of the plant reproductive world were it not for its potential use in propagating crop plants. You may recall from Chapter 12 that many crop plants (such as corn) are grown as hybrids because the progeny of a cross between two inbred lines are often superior to either of their parents. The explanation for this phenomenon, called *hybrid vigor*, is not completely understood. One hypothesis attributes the superiority of hybrids to the suppression of undesirable recessive alleles from one parent by dominant alleles from the other. Another hypothesis states that certain advantageous combinations of alleles can be obtained by crossing two inbred strains.

Unfortunately, once a farmer has obtained a hybrid with desirable characteristics, (s)he cannot use those plants for further crosses with themselves (selfing) to get more seeds for the next generation. You can imagine the genetic chaos when a hybrid, which is heterozygous at many of its loci (e.g., *AaBbCcDdEe*, etc.), is crossed with itself: there will be many new combinations of alleles (e.g., *AabbCCDdee*, etc.), resulting in highly variable progeny. The only way to reliably reproduce the hybrid is to maintain populations of the original parents to cross again each year.

However, if a hybrid carried a gene for apomixis, it could reproduce asexually, and its offspring would be genetically identical to itself. So the search is on for a gene for apomixis that could be introduced into desirable crops and allow them to be propagated indefinitely. (A recently published detective novel, *Day of the Dandelion*, explores this idea.)

Researchers recently found a strain of *Arabidopsis* that exhibits apomixis as a result of a mutation in a single gene called *dyad*. In normal plants, *dyad* is essential for chromosome organization, specifically synapsis, during meiosis I (see Figure 11.17). In the apomictic strain, meiosis I resembles mitosis, and the chromosomes replicate again before what would be meiosis II. The result is diploid cells that are genetically identical to the parent instead of the genetically recombined haploid gametes that normally result from meiosis. Scientists are trying to isolate and transfer such apomictic genes into corn and other cereal crops with the hope that plant breeders can use apomixis to propagate plants with desirable hybrid traits (such as high yields, and disease- and insect-resistance) without compromising their hybrid vigor.

### Vegetative reproduction has a disadvantage

Vegetative reproduction is highly efficient in an environment that is stable over the long term. A change in the environment, however, can leave an asexually reproducing species at a disadvantage.

A striking example is provided by the demise of the English elm, *Ulmus procera*, which was apparently introduced into England as a clone by the ancient Romans. This tree reproduces asexually by suckers and is incapable of sexual reproduction. In 1967, Dutch elm disease first struck the English elms. After two millennia of clonal growth, the population lacked genetic diversity, and no individuals carried genes that would protect them against the disease. Today the English elm is all but gone from England.

### Vegetative reproduction is important in agriculture

Farmers and gardeners take advantage of some natural forms of vegetative reproduction. They have also developed new types of asexual reproduction by manipulating plants. One of the oldest methods of vegetative reproduction used in agriculture consists of simply making cuttings of stems, inserting them in soil, and waiting for them to form roots and thus become autonomous plants. The cuttings are usually encouraged to root

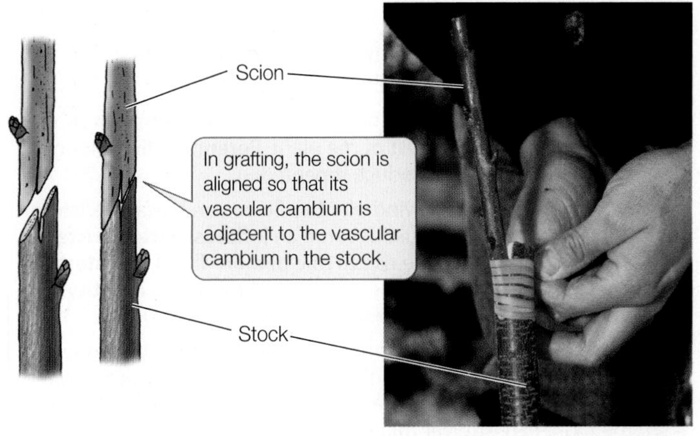

Scion
In grafting, the scion is aligned so that its vascular cambium is adjacent to the vascular cambium in the stock.
Stock

**38.19 Grafting** Grafting—attaching a piece of a plant to the root or root-bearing stem of another plant—is a common horticultural technique. The "host" root or stem is the stock; the upper grafted piece is the scion. In the photo, a Bing cherry scion is being grafted onto a hardier stock.

by treatment with a plant hormone, auxin, as described in Section 37.3.

Horticulturists reproduce many woody plants by **grafting**—attaching a bud or a piece of stem from one plant to the root or root-bearing stem of another plant. The part of the resulting plant that comes from the root-bearing "host" is called the **stock**; the part grafted on is the **scion** (**Figure 38.19**).

For a graft to succeed, the vascular cambium of the scion must associate with that of the stock. By cell division, both cambiums form masses of wound tissue. If the two masses meet and connect, the resulting continuous cambium can produce xylem and phloem, allowing transport of water and minerals to the scion and of photosynthate to the stock. Grafts are most often successful when the stock and scion belong to the same or closely related species. Much fruit grown for market in the United States is produced on grafted trees. Another example is wine grapes. The roots of most grape strains are susceptible to soil pests, and so grape varieties are grafted onto root stocks that have pest resistance.

Scientists in universities and commercial laboratories have been developing new ways to produce useful plants through tissue culture. Because many plant cells are totipotent, cultures of undifferentiated tissue can give rise to entire plants, as can small pieces of tissue cut directly from a parent plant. Tissue cultures sometimes are used commercially to produce new plants. This is common in the forestry industry, where uniformity of trees is desirable.

Culturing tiny bits of apical meristem can produce plants free of viruses. Because apical meristems lack developed vascular tissues, viruses tend not to enter them. Treatment with hormones causes a single apical meristem to give rise to 20 or more shoots; thus, a single plant can give rise to millions of genetically identical plants within a year by repeated meristem culturing. Using this approach, strawberry and potato producers are able to start each year's crop from virus-free plants.

## 38.3 RECAP

**Angiosperms may reproduce asexually by means of modified stems, roots, or leaves, or by apomixis. Asexual reproduction is advantageous when a plant has a superior genotype well adapted to its environment, but decreases the genetic diversity of plant populations.**

- How does apomixis differ from sexual reproduction? See p. 810
- Explain how vegetative reproduction of plants is advantageous to humans. See pp. 810–811

We have seen how angiosperms reproduce sexually and asexually. A disadvantage of asexual reproduction is that its genetic inflexibility may leave a population unable to cope with new challenges. In the next chapter we focus on the mechanisms that have evolved in plants to cope with biological and physical challenges in their environment.

# CHAPTER SUMMARY

## 38.1 How Do Angiosperms Reproduce Sexually?

- Sexual reproduction promotes genetic diversity in a population. The flower is an angiosperm's structure for sexual reproduction.
- Flowering plants have microscopic gametophytes. The megagametophyte is the **embryo sac**, which typically contains eight nuclei in a total of seven cells. The microgametophyte is the **pollen grain**, which usually contains two cells. Review Figure 38.2, **ANIMATED TUTORIAL 38.1**
- Following **pollination**, the pollen grain delivers **sperm cells** to the embryo sac by means of a **pollen tube**.
- Most angiosperms exhibit **double fertilization**: one sperm nucleus fertilizes the **egg**, forming a zygote, and the other sperm nucleus unites with the two **polar nuclei** to form a triploid **endosperm**. Review Figure 38.6
- Plants have both physical and genetic methods of preventing inbreeding. Physical separation of the gametophytes and genetic **self-incompatibility** prevent self-pollination.
- The zygote develops into an embryo (with an attached **suspensor**), which remains quiescent in the seed until conditions are right for germination. Review Figure 38.7, **WEB ACTIVITY 38.1**
- Ovules develop into seeds, and the ovary wall and the enclosed seeds develop into a **fruit**.
- The hormone **abscisic acid** promotes seed development and dormancy.

## 38.2 What Determines the Transition from the Vegetative to the Flowering State?

- In **annuals** and **biennials**, flowering and seed formation usually leads to death of the rest of the plant. **Perennials** live a long time and typically reproduce repeatedly.
- For a vegetatively growing plant to flower, an apical meristem in the shoot system must become an **inflorescence meristem**, which in turn must give rise to one or more **floral meristems**. These events are under the influence of **meristem identity genes** and **floral organ identity genes**. Review Figure 38.10
- Some plants flower in response to **photoperiod**. **Short-day plants** flower when the nights are longer than a **critical night length** specific to each species; **long-day plants** flower when the nights are shorter than a critical night length. Review Figure 38.13
- The mechanism of photoperiodic control involves phytochromes and a biological clock. Review Figure 38.14, ANIMATED TUTORIAL 38.2
- A flowering signal, called **florigen**, is formed in a photoperiodically induced leaf and is translocated to the sites where flowers will form. Review Figures 38.15 and 38.16
- In some angiosperm species, exposure to low temperatures—**vernalization**—is required for flowering; in others internal signals (one of which is gibberellin in some plants) induce flowering. All of these stimuli converge on the meristem identity genes.

## 38.3 How Do Angiosperms Reproduce Asexually?

- Asexual reproduction allows rapid multiplication of organisms that are well suited to their environment.
- **Vegetative reproduction** involves the modification of a vegetative organ—usually the stem—for reproduction.
- Some plant species produce seeds asexually by **apomixis**.
- Horticulturists often **graft** different plants together to take advantage of favorable properties of both **stock** and **scion**. Review Figure 38.19

## SELF-QUIZ

1. Sexual reproduction in angiosperms
   *a.* is by way of apomixis.
   *b.* requires the presence of petals.
   *c.* can be accomplished by grafting.
   *d.* gives rise to genetically diverse offspring.
   *e.* cannot result from self-pollination.
2. The typical angiosperm female gametophyte
   *a.* is called a microspore.
   *b.* has eight nuclei.
   *c.* has eight cells.
   *d.* is called a pollen grain.
   *e.* is carried to the male gametophyte by wind or animals.
3. Pollination in angiosperms
   *a.* always requires wind.
   *b.* never occurs within a single flower.
   *c.* always requires help by animal pollinators.
   *d.* is also called fertilization.
   *e.* makes most angiosperms independent of external water for reproduction.
4. Which statement about double fertilization is *not* true?
   *a.* It is found in most angiosperms.
   *b.* It takes place in the microsporangium.
   *c.* One of its products is a triploid nucleus.
   *d.* One sperm nucleus fuses with the egg nucleus.
   *e.* One sperm nucleus fuses with two polar nuclei.
5. The suspensor
   *a.* gives rise to the embryo.
   *b.* is heart-shaped in eudicots.
   *c.* separates the two cotyledons of eudicots.
   *d.* ceases to elongate early in embryonic development.
   *e.* is larger than the embryo.
6. Which statement about photoperiodism is *not* true?
   *a.* It is related to the biological clock.
   *b.* A phytochrome plays a role in the timing process.
   *c.* It is based on measurement of the length of the night.
   *d.* Some plants do not flower in response to photoperiod.
   *e.* It is limited to plants.
7. Florigen is
   *a.* produced in the leaves and transported to the apical bud.
   *b.* produced in the roots and transported to the shoots.
   *c.* produced in the coleoptile tip and transported to the base.
   *d.* the same as gibberellin.
   *e.* activated by prolonged (more than a month) high temperature.
8. Which statement about vernalization is *not* true?
   *a.* It decreases the abundance of an inhibitor of flowering.
   *b.* Vernalization involves exposure to cold temperatures.
   *c.* It only occurs in crop plants such as cereals.
   *d.* In the vernalized state, the synthesis of FLC protein is inhibited.
   *e.* If winter wheat is not exposed to cold, it will not flower.
9. Which of the following does *not* participate in asexual reproduction?
   *a.* Stolon
   *b.* Rhizome
   *c.* Zygote
   *d.* Tuber
   *e.* Corm
10. Apomixis involves
    *a.* sexual reproduction.
    *b.* complete meiosis.
    *c.* fertilization.
    *d.* a diploid embryo.
    *e.* no production of a seed.

## FOR DISCUSSION

1. Which method of reproduction might a farmer prefer for a crop plant that reproduces both sexually and asexually? Why?
2. Thompson Seedless grapes are produced by vines that are triploid. Think about the consequences of this chromosomal condition for meiosis in the flowers. Why are these grapes seedless? Describe the role played by the flower in fruit formation when no seeds are being formed. How do you suppose Thompson Seedless grapes are propagated?
3. Poinsettias are popular ornamental plants that typically bloom just before Christmas. Their flowering is photoperiodically controlled. Are they long-day or short-day plants? Explain.
4. You plan to induce the flowering of a crop of long-day plants in the field by using artificial light. Is it necessary to keep the lights on continuously from sundown until the point at which the critical day length is reached? Why or why not?

## ADDITIONAL INVESTIGATION

The isolation of *dyad*, the *Arabidopsis* gene that controls apomixis, offers possibilities for crop plant breeding. How would you investigate the possibility of using the mutant allele of this gene to produce hybrid corn plants that can be propagated and retain their hybrid nature?

# 39 Plant Responses to Environmental Challenges

## Sharing plants' defensive strategies

The tropical rainforest of the eastern slopes of the Andes teems with plant life. This region of the Amazon Basin is host to about 40 species of *Cinchona*, a genus of trees that grow to a height of about 20 meters. In this moist environment, *Cinchona* trees grow rapidly and thrive, despite the many natural enemies that threaten their survival and growth.

Unlike animals, which can sometimes escape their enemies, plants must confront their enemies in place. Over evolutionary time, plants with the ability to fight off attackers have survived and passed on that ability to their offspring. In many instances, that ability comes in the form of defensive chemicals. In the case of *Cinchona*, one of those chemicals is quinine, a bitter molecule that is toxic to insects. It may be familiar to you as an ingredient in tonic water.

The Quechua, a group of people native to the Andes forests, have a long history of putting local plants to medicinal use. In the tropics, malaria has long been, and still is, a common and lethal disease. Even today, about 400 million cases arise worldwide, and over 1.5 million people die from malaria each year. Centuries ago, the Quechua found that a tea made from the bark of *Cinchona* trees was highly effective in treating malaria. Legend has it that the tree got its name from the Peruvian countess of Cinchon, who was cured of malaria in 1638 when her physician got some bark from the Quechua. Use of the bark extract quickly spread around the world.

In 1820, the active ingredient of *Cinchona* bark, quinine, was isolated and became the mainstay of malaria treatment. Malaria is caused by an apicomplexan parasite that infects red blood cells. Quinine kills the parasite by interfering with its ability to break down hemoglobin. Unfortunately, mutations render some parasites resistant to quinine and its chemical derivatives. Over time, treatment of billions of people with quinine drugs has selected for parasites with genes for quinine resistance. An urgent need for alternative treatments has led scientists to another plant and its defensive chemical.

*Artemisia annua*, or sweet wormwood, grows in forests all over the world. It synthesizes a molecule called artemisinin that helps defend the plant against insects. For over 3000 years, people in Asia have made a curative tea from sweet wormwood. In 1972 Japanese chemists isolated artemisinin and found that it works by reacting with iron in red blood cells to form free radicals, which damage lipids and DNA in the infecting par-

**The Source of Quinine** *Cinchona* trees from the Amazon rainforest synthesize a defensive chemical, quinine, that has been used to treat people with malaria.

**A Substitute for Quinine** Sweet wormwood (*Artemisia annua*) grows in forests throughout the world. It synthesizes a defensive chemical, called artemisinin, that is now being used to treat people with malaria whose infection is resistant to quinine drugs.

asite. Since this mechanism of action differs from that of quinine, it was thought that artemisinin might be effective in treating quinine-resistant malaria—and it is. Indeed, during the Vietnam War, drinking *Artemisia* tea helped Vietnamese soldiers cope with the quinine-resistant malaria that struck American soldiers. Since 2000, artemisinin has become a mainstay of malaria treatment. Millions of people take it every day.

Until recently, hundreds of such plant chemicals were contemplated only in the context of plant biochemistry. Today we view them as adaptations arising from a plant's interactions with its environment.

**IN THIS CHAPTER** we describe the biological (biotic) and nonbiological (abiotic) environmental challenges faced by plants and how plants deal with them. We begin by examining interactions between plants and plant pathogens, such as fungi, and then consider plant interactions with herbivores. The chapter concludes by considering plant adaptations to abiotic factors such as water, temperature, salinity, and heavy metals.

CHAPTER OUTLINE



## 39.1 How Do Plants Deal with Pathogens?

Botanists know of dozens of diseases, with many different genetic strains, that can kill a wheat plant, each of them caused by a different pathogen. Plant pathogens—which include bacteria, fungi, protists, and viruses—are part of nature, and for that reason alone they merit our study in biology. Because we humans depend on plants for our food, however, the stakes in our effort to understand plant pathology are especially high. That is why, just as medical schools have departments of pathology, many universities in agricultural regions have departments of plant pathology.

Successful infection by a pathogen can have significant effects on a plant, reducing photosynthesis and causing massive cell and tissue death. Like the responses of the human immune system (see Chapter 42), the responses by which plants fight off disease are varied and fascinating. Plants and pathogens have evolved together in a continuing "arms race": pathogens have evolved mechanisms with which to attack plants, and plants have evolved mechanisms for defending themselves against those attacks. Each set of mechanisms uses information from the other. For example, the pathogen's enzymes may break down the plant's cell walls, and the breakdown products may signal to the plant that it is under attack. In turn, the plant's defenses alert the pathogen that it, too, is under attack.

An arms race of global importance is under way to combat wheat rust, a fungus that can devastate wheat crops (**Figure 39.1**). In 1999, a new genetic strain of the wheat rust fungus *Puccinia graminis* was identified in a wheat field in Uganda. Although many strains of wheat have natural resistance to other strains of rust, the new fungal strain, called Ug99, overcomes resistance in almost all of them. It has spread to the Middle East, and there is a good probability that it has already reached Asia. Over 90 percent of the wheat strains in its path are susceptible. Scientists are racing to discover wheat genes that confer resistance to Ug99 and to implement genetic crosses to get this resistance into the wheat strains under widespread cultivation. Failure in this arms race could have disastrous consequences for the global food supply.

What determines the outcome of a battle between a plant and a pathogen? The key to success for the plant is to respond to information from the pathogen quickly and massively. Plants use both mechanical and chemical defenses in this effort. These defenses can either be **constitutive**, always present in the plant, or **induced**, produced in reaction to damage or stress.

### Mechanical defenses include physical barriers

A plant's first line of defense is its outer surfaces, which can prevent the entry of pathogens. As Section 34.3 describes, the or-

**39.1 Wheat Rust** A field of wheat infected with the wheat rust fungus (right) has a much reduced growth rate and has produced much less grain than an uninfected field of wheat that is resistant to the fungus (left). Inset: Wheat rust causes cell death in leaves.

gans of a growing plant that are exposed to the outside environment are covered with cutin, suberin, and waxes. These substances not only prevent water loss by evaporation, but can also prevent fungal spores and bacteria from entering the underlying tissues. Some fungi get around this defense, however, by secreting enzymes that hydrolyze components of these substances, breaking them down to gain entry.

Much more important to the plant are the induced resistance mechanisms initiated when a pathogen lands on a plant. As we discuss these mechanisms, refer to the overview in **Figure 39.2**.

### Plants can seal off infected parts to limit damage

While animals generally repair tissues that have been damaged by pathogens, plants do not. Instead, plants seal off and sacrifice damaged tissues so that the rest of the plant does not become infected. Plants have the option of discarding damaged tissues because most plants, unlike most animals, can replace damaged parts by growing new ones.

Before we look at the details of the defensive process, note that a key response by plant cells to invasion by pathogens is the rapid deposition of additional polysaccharides, as well as a cell wall protein called *extensin*, on the inside of the cell wall. These macromolecules not only reinforce the mechanical barrier formed by the cell wall, but also block the plasmodesmata,

yourBioPortal.com
GO TO **Animated Tutorial 39.1 • Signaling between Plants and Pathogens**

Pathogen
1 Some elicitors from the pathogen are recognized directly.
2 When certain pathogenic enzymes attack the plant cell wall, the breakdown products are recognized as elicitors by a membrane receptor.
4 Defensive molecules such as phytoalexins and PR proteins attack the pathogen directly.
Polysaccharide
Extensin
Receptors in plasma membrane
5 Some PR proteins serve as "alarm signals" to cells that have not yet been attacked.
Phytoalexins
3 Signaling molecules trigger cellular responses, including the production of defensive molecules.
PR proteins
Nucleus
Polysaccharides
Plasmodesma
6 Polysaccharides and extensin strengthen the cell wall and block plasmodesmata.
Cell wall
Plant cell

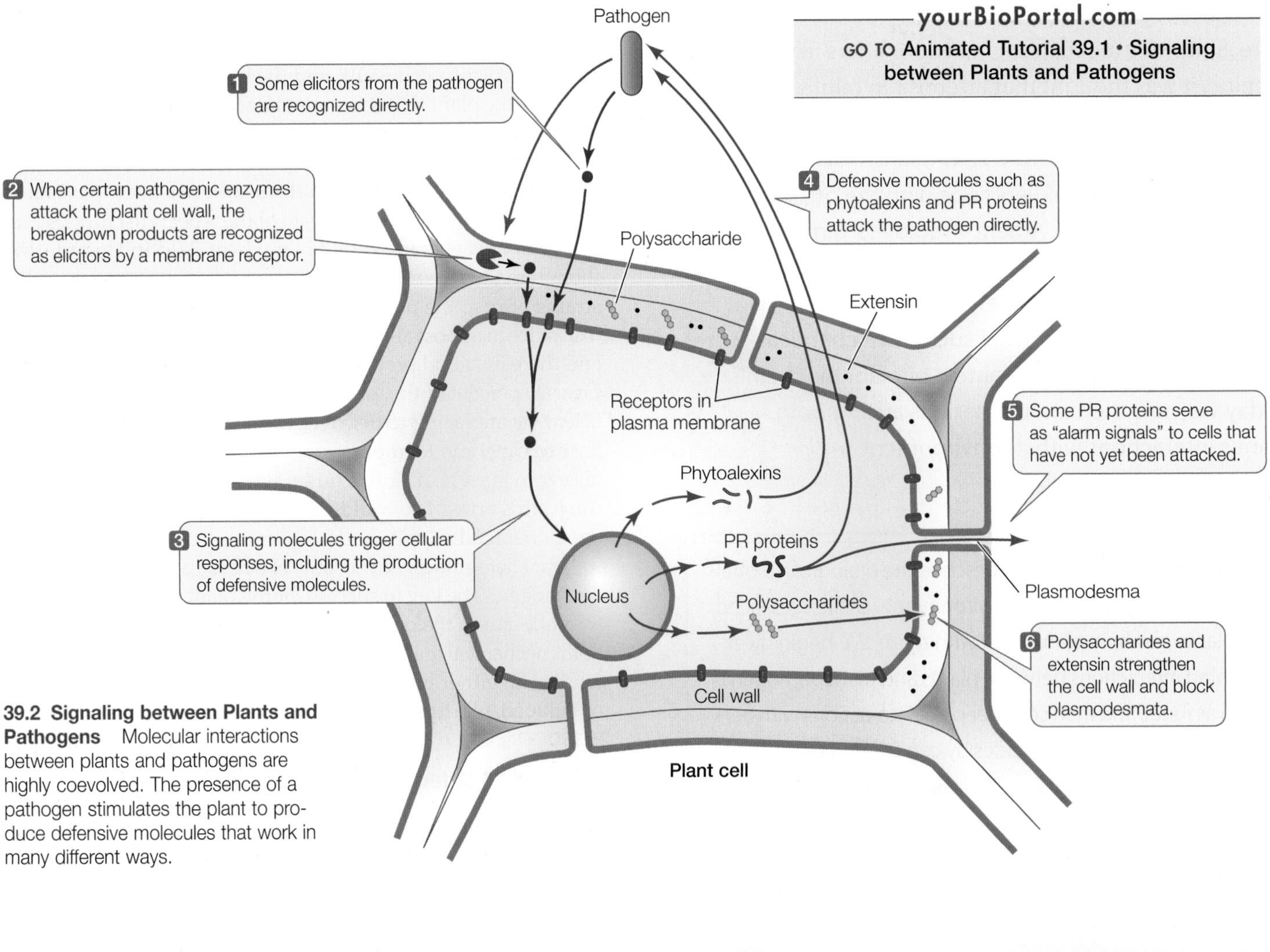

**39.2 Signaling between Plants and Pathogens** Molecular interactions between plants and pathogens are highly coevolved. The presence of a pathogen stimulates the plant to produce defensive molecules that work in many different ways.

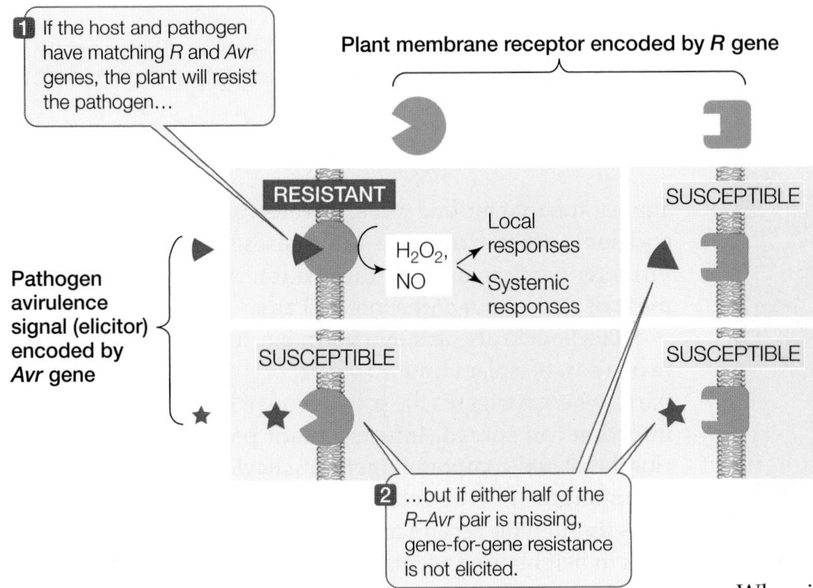

1 If the host and pathogen have matching *R* and *Avr* genes, the plant will resist the pathogen...
Plant membrane receptor encoded by *R* gene
RESISTANT
Local responses
$H_2O_2$, NO
Systemic responses
SUSCEPTIBLE
SUSCEPTIBLE
SUSCEPTIBLE
Pathogen avirulence signal (elicitor) encoded by *Avr* gene
2 ...but if either half of the *R–Avr* pair is missing, gene-for-gene resistance is not elicited.

**39.3 Gene-for-Gene Resistance** If a gene in a pathogen that codes for an elicitor "matches" a gene in a plant that codes for a receptor, the receptor binds the elicitor, and a defensive response results.

limiting the ability of viral pathogens to move from cell to cell. The polysaccharides also serve as a base on which lignin may be laid down. Lignin enhances the mechanical barrier, and the toxicity of lignin precursor chemicals makes the cell inhospitable to some pathogens. These lignin building blocks are only one example of the toxic substances that plants use as chemical defenses.

## Plant responses to pathogens may be genetically determined

Plant pathogens cause the host plant to activate various chemical defense responses. Several distinctive molecules called *elicitors* have been identified that trigger these plant defenses. These molecules vary in character, from peptides made by bacteria to cell wall fragments from fungi. Elicitors can also be derived from fragments of plant cell wall components broken down by pathogens (see Figure 39.2, labels 1 and 2). Pathogen genes that code for elicitors are called **avirulence genes (*Avr*)**; there are hundreds of such genes. When an elicitor enters a plant cell, it may encounter a receptor protein in the cytoplasm. If the receptor binds to the elicitor, a signal transduction pathway is set in motion that leads to the plant's defensive response. If the plant has no receptor to bind to the elicitor, the plant does not defend itself.

Over 50 years ago, plant pathologist Harold Flor at North Dakota State University studied the susceptibility of various genetic strains of cereal grain plants to various strains of rust fungus. He proposed that susceptibility is determined by a genetic relationship between the pathogen and the plant (**Figure 39.3**). Today we know that plants have **resistance genes (*R*)**, and we know that these resistance genes code for receptors. We also know that pathogens have *Avr* genes that code for elicitors. If the *R* and *Avr* genes match, there is resistance: the molecules they encode will bind to each other and set off a response. Flor's idea, known as **gene-for-gene resistance**, has been borne out by molecular studies, and it has had a great influence on crop plant breeding for resistance to pathogens.

The signal transduction pathway started by receptor–elicitor binding is mediated by the production of nitric oxide and the toxic peroxide $H_2O_2$. Together, these substances initiate local defenses and, later, systemic defenses (defenses in parts of the plant distant from the attack site).

## Receptor–elicitor binding evokes the hypersensitive response

When infected by certain fungi and bacteria, plants try to contain the infection locally by what is called the **hypersensitive response**. This three-pronged response involves the production of defensive compounds as well as physical isolation of the infection site (see Figure 39.2, labels 3 and 4).

**PHYTOALEXINS** **Phytoalexins** are antibiotics produced by infected plants that are toxic to many fungi and bacteria. Most are small molecules, and each is made by only a few plant species. They are produced by infected cells, and by their immediate neighbors, within hours of the onset of infection. Because their antimicrobial activity is nonspecific, phytoalexins can destroy many species of fungi and bacteria in addition to the one that originally triggered their production. Phytoalexins also kill the plant cells that produced them, thus sealing off the infection.

Phytoalexins are an example of an induced plant defense: they are not normally present in plants, but are synthesized rapidly when a bacterial or fungal infection occurs. Physical injuries and viral infections can also induce the production of phytoalexins.

Neither the mechanism of phytoalexin induction by pathogens, nor the specific effects of phytoalexins on invading organisms, is clear. Camalexin, a phytoalexin made by the model organism *Arabidopsis thaliana*, is being used to investigate these phenomena. It is synthesized by the plant from the amino acid tryptophan:

COOH
N S
N H → N H
Tryptophan Camalexin

**PATHOGENESIS-RELATED PROTEINS** Plants produce several types of **pathogenesis-related proteins**, or **PR proteins**. Some are enzymes that break down the cell walls of pathogens. Chitinase, for example, is a PR protein that breaks down chitin, which is found in many fungal cell walls. In some cases, the breakdown

**39.4 Sealing Off the Pathogen and the Damage** These necrotic lesions on the leaves of a broad bean plant are a response to "chocolate spot" fungus, *Botrytis fabae*.

products of the pathogen's cell walls serve as elicitors that trigger further defensive responses. Other PR proteins may serve as alarm signals to plant cells that have not yet been attacked (Figure 39.2, label 5). In general, PR proteins appear not to be rapid-response weapons; rather, they act more slowly, perhaps after other mechanisms have blunted the pathogen's attack.

**PHYSICAL ISOLATION** A third component of the hypersensitive response seals off the damage and the pathogen from the rest of the plant. Cells around the site of infection undergo apoptosis, preventing the spread of the pathogen by depriving it of nutrients. Some of these cells produce phytoalexins and other chemicals before they die. The dead tissue, called a *necrotic lesion*, contains and isolates what is left of the infection (**Figure 39.4**). The rest of the plant remains free of the infecting pathogen.

### Systemic acquired resistance is a form of long-term "immunity"

The hypersensitive response is not the only defensive response initiated by receptor–elicitor binding. **Systemic acquired resistance** is a general increase in the resistance of the entire plant to a wide range of pathogens. It is not limited to the pathogen that originally triggered it, or to the site of the original infection, and its effect may last as long as an entire growing season.

This defensive response is initiated by salicylic acid, a defensive chemical produced during the local hypersensitive response. Since ancient times, people in Asia, Europe, and the Americas have used willow (*Salix*) leaves and bark to relieve pain and fever. The active ingredient in willow is salicylic acid, the substance from which aspirin is derived. It now appears that all plants contain at least some salicylic acid.

Systemic acquired resistance is accompanied by the synthesis of PR proteins. Treatment of plants with salicylic acid or aspirin leads to the production of PR proteins and to resistance to pathogens. It provides substantial protection, for example, against tobacco mosaic virus (a well-studied plant pathogen) and some other viruses. In some cases salicylic acid inhibits virus replication, and in others it interferes with the movement of viruses out of the infected area.

Salicylic acid also acts as a hormone. In some cases, infection in one part of a plant leads to the export of salicylic acid to other parts, where it triggers the production of PR proteins before the infection can spread. Infected plant parts also produce the closely related compound methyl salicylate (also known as oil of wintergreen). This volatile substance travels to other plant parts through the air and may trigger the production of PR proteins in neighboring plants that have not yet been infected.

### Plants develop specific immunity to RNA viruses

Before we leave the topic of plant defenses against pathogens, let's consider a recently discovered defense mechanism directed against a specific pathogen type: RNA viruses (viruses that have RNA instead of DNA as their hereditary material). The plant uses its own enzymes to convert some of the single-stranded RNA of the invading virus into *double-stranded RNA* (dsRNA) and to chop that dsRNA into small pieces called *small interfering RNAs* (siRNAs). Some of the viral RNA is transcribed, forming mRNAs that advance the infection. However, the siRNAs interact with another cellular component to degrade those mRNAs, blocking viral replication. This phenomenon is an example of *RNA interference* (RNAi) (see Section 18.4).

The immunity conferred by RNAi spreads quickly throughout the entire plant through plasmodesmata. However, the establishment of immunity depends on the extent of the original infection and the speed of the plant's response. Plant viruses are continuing their side of the arms race: most have evolved mechanisms to confound RNA interference.

## 39.1 RECAP

**In the hypersensitive response to infection by pathogens, plants produce two types of chemical defenses and seal off infected areas. Systemic acquired resistance, providing a longer-lasting, more general immunity, may follow.**

- Name two types of defensive compounds produced by plant cells when they are infected by bacteria or fungi. **See p. 817 and Figure 39.2**
- How do *R* and *Avr* genes determine which pathogens a plant can resist? **See p. 817 and Figure 39.3**
- How do infected plant cells signal infection to other parts of the plant, or other plants? **See p. 818**

Not all biological threats to plants come from pathogens. Another threat comes from the many animals, from inchworms to elephants, that eat plants.

## 39.2 How Do Plants Deal with Herbivores?

Herbivores—animals that eat plants—depend on plants for energy and nutrients. Their foraging activities cause physical damage to plants, and they often spread disease among plants as well. While the majority of herbivores are insects (**Figure 39.5**), every major class of vertebrates includes at least a few herbivores (see also Section 56.2, which discusses herbivory in the ecological context of species interactions). Plants cannot evade their consumers by running away, but they have many other ways of protecting themselves against herbivory.

While in most cases, the physical damage caused by herbivores is severe, sometimes limited herbivory may be harmless or even beneficial. Before turning to the ways that plants resist herbivory, let's examine those few cases in which herbivory enhances plant growth.

### Herbivory increases the growth of some plants

How detrimental is herbivory to plants? How well have plants adapted to their place in the food web? Like plants and pathogens, plants and herbivores have evolved together, each acting as an agent of natural selection on the other. This coevolution has led to arms races in some cases, but it has favored increased photosynthetic production in some plant species subjected to herbivory.

Removal of some leaves from a plant usually increases the rate of photosynthesis in the remaining leaves for several reasons:

- Nitrogen obtained from the soil by the roots no longer needs to be divided among as many leaves.
- The export of sugars and other photosynthetic products from the leaves may be enhanced because the demand for those products in the roots is undiminished, while the sources for those products—leaves—have been decreased.
- The removal of older or dead leaves makes more light available to the younger, more active leaves or leaf parts.

Grasses are especially tolerant of herbivory because, unlike most other plants, which grow from shoot apical meristems, grasses grow from the base of the shoot and leaf, so their growth is not cut short by grazing.

In western North America, mule deer and elk graze many plants, including a wildflower called scarlet gilia (*Ipomopsis aggregata*). Although grazing removes about 95 percent of the aboveground plant, the scarlet gilia quickly regrows not one, but up to four replacement stems (**Figure 39.6**). Grazed individuals produce three times as many fruits by the end of the growing season as do ungrazed plants.

### Mechanical defenses against herbivores are widespread

All parts of the plant body offer some resistance against herbivores. In addition, plants have a number of constitutive anatom-

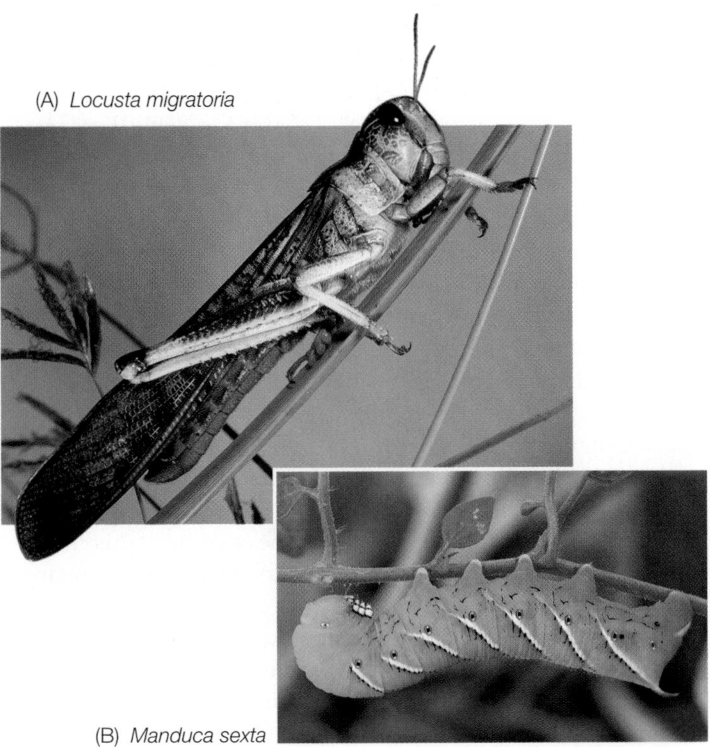


**39.5 Insect Herbivores** The great majority of herbivores are insects. (A) Some herbivores, such as this locust, are generalists that will attack nearly any plant. (B) Others are specialists, like this tobacco hornworm, which feeds only on tobacco plants.

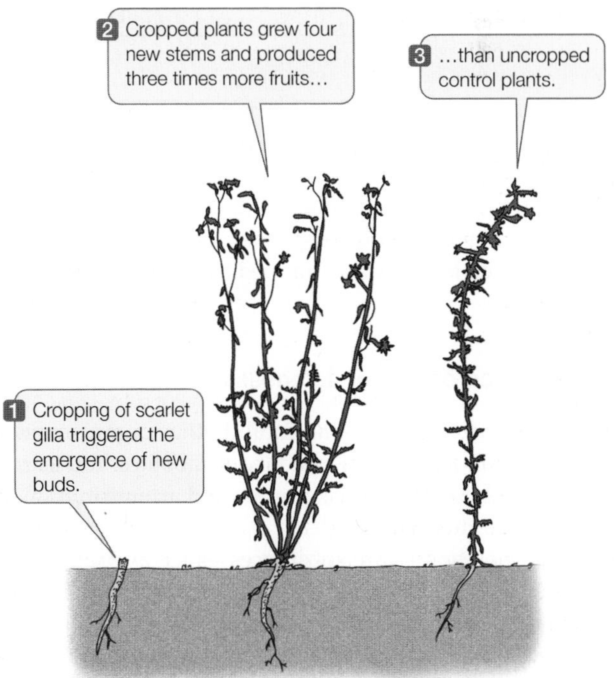

1 Cropping of scarlet gilia triggered the emergence of new buds.
2 Cropped plants grew four new stems and produced three times more fruits...
3 ...than uncropped control plants.

**39.6 Overcompensation for Being Eaten** Experiments confirm that herbivory increases the growth of some plants.

TABLE 39.1

**Secondary Metabolites Used in Defense**

| CLASS | TYPE | ROLE | EXAMPLE |
|---|---|---|---|
| Nitrogen-containing | Alkaloids | Neurotoxin | Nicotine in tobacco |
| 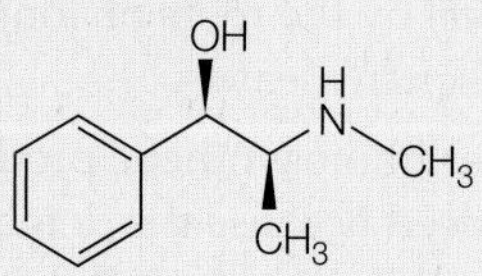 | Glycosides | Inhibit electron transport | Dhurrin in sorghum |
| Ephedrine (an alkaloid) | Nonprotein amino acids | Disrupt protein structure | Canavanine in jack bean |
| Nitrogen–sulfur-containing | Glucosinolates | Inhibit respiration | Methylglucosinolate in cabbage |
| 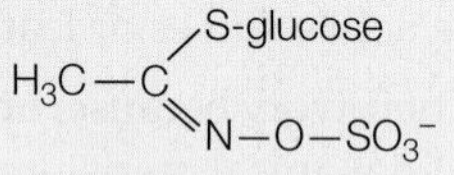 Methylglucosinolide | | | |
| Phenolics | Coumarins | Block cell division | Umbelliferone in carrots |
| 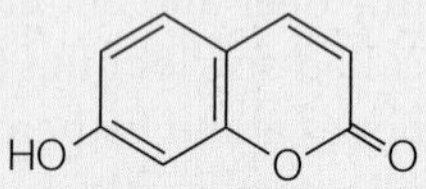 | Flavonoids | Phytoalexins | Capsidol in peppers |
| Umbelliferone | Tannins | Inhibit enzymes | Gallotannin in oak trees |
| Terpenes | Monoterpenes | Neurotoxins | Pyrethrin in chrysanthemums |
| 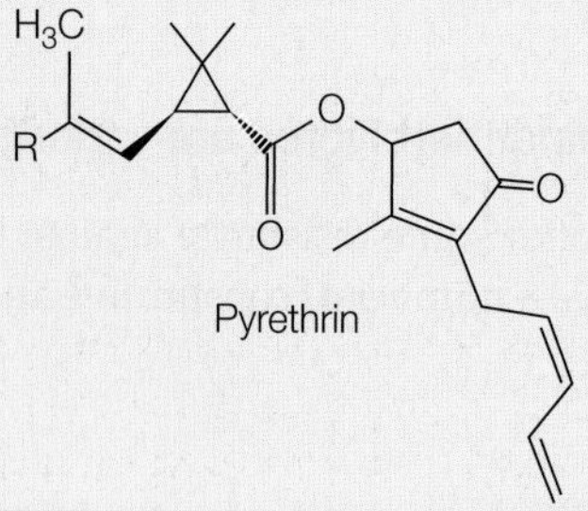 | Diterpenes | Disrupt reproduction and muscle function | Gossypol in cotton |
| | Triterpenes | Inhibit ion transport | Digitalis in foxglove |
| | Sterols | Block animal hormones | Spinasterol in spinach |
| Pyrethrin | Polyterpenes | Deter feeding | Latex in *Euphorbia* |

ical features, such as trichomes, thorns, spines, or hairs, that are specialized for defense. An example of an induced mechanical defense is the production of latex. When they are injured by an herbivore attack, some plants, such as *Euphorbia* species, produce a thick, white aqueous suspension of cellular debris, oils, and resins called *latex*. Insects trapped by this sticky substance starve to death.

### Plants produce chemical defenses against herbivores

Many plants attract, resist, and inhibit other organisms with special chemicals known as secondary metabolites. You learned about two of these chemicals, quinine and artemisinin, in the opening of this chapter.

Primary metabolites are substances—such as proteins, nucleic acids, carbohydrates, lipids, and their building blocks—that are produced and used by all living organisms, including plants. They and their metabolic products are used in basic cellular processes such as photosynthesis, respiration, and nutrient uptake. **Secondary metabolites** are substances that are not used for basic cellular processes. Each is found in only certain plants or plant groups.

The more than 10,000 known secondary metabolites range in molecular mass from about 70 to more than 390,000 daltons, but most have a low molecular mass (**Table 39.1**). Some are produced by only a single plant species, while others are characteristic of entire genera or even families. The effects of defensive secondary metabolites on animals are diverse. Some act on the nervous systems of herbivorous insects, mollusks, or mammals. Others mimic the natural hormones of insects, causing some larvae to fail to develop into adults. Still others damage the digestive tracts of herbivores. Some secondary metabolites are toxic to fungal pests. As we saw at the opening of this chapter, humans make use of many secondary metabolites as pharmaceuticals and pesticides.

The secondary metabolite nicotine was one of the first insecticides to be used by farmers and gardeners. This molecule kills insects by acting as an inhibitor of nervous system function. Yet commercial varieties of tobacco and related plants that produce nicotine are still attacked, with moderate damage, by pests such as the tobacco hornworm (see Figure 39.5B). Given that observation, does nicotine really deter herbivores? Biologists answered this question conclusively with a study that used tobacco plants in which an enzyme involved in nicotine biosynthesis had

## INVESTIGATING LIFE

### 39.7 Nicotine Is a Defense against Herbivores

The secondary metabolite nicotine, made by tobacco plants, is an insecticide, yet most commercial varieties of tobacco are susceptible to insect attack. Ian Baldwin demonstrated that a tobacco strain with a reduced nicotine concentration was more susceptible to insect damage.

**HYPOTHESIS** Nicotine helps protect tobacco plants against insects.

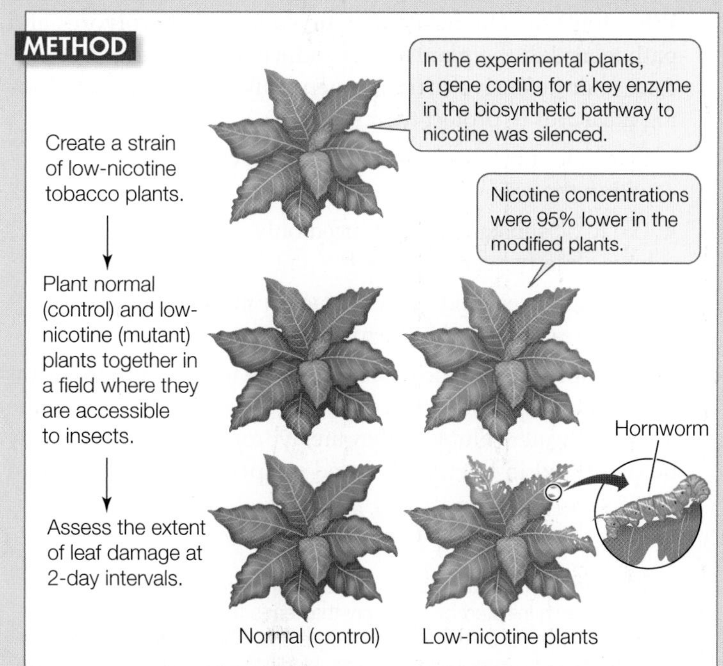

METHOD
Create a strain of low-nicotine tobacco plants.
In the experimental plants, a gene coding for a key enzyme in the biosynthetic pathway to nicotine was silenced.
Nicotine concentrations were 95% lower in the modified plants.
Plant normal (control) and low-nicotine (mutant) plants together in a field where they are accessible to insects.
Hornworm
Assess the extent of leaf damage at 2-day intervals.
Normal (control)
Low-nicotine plants


**RESULTS** The low-nicotine plants suffered more than twice as much leaf damage as did the wild-type controls.

Leaf area damaged (% of total): 0, 5, 10, 15, 20
Low-nicotine plant
Normal (control)
Days after planting: 4, 6, 8, 10, 12, 14, 16

**CONCLUSION** Nicotine provides tobacco plants with at least some protection against insects.

**FURTHER INVESTIGATION:** Treatment of tobacco plants with jasmonate (a hormone) elicits the production of nicotine and other compounds. How would you modify this experiment to determine whether nicotine is the only insecticidal compound produced by tobacco?

Go to **yourBioPortal.com** for original citations, discussions, and relevant links for all INVESTIGATING LIFE figures.

been silenced, lowering the nicotine concentration in the plants by more than 95 percent. These low-nicotine plants suffered much more damage from insect herbivory than normal plants did (**Figure 39.7**).

### Some secondary metabolites play multiple roles

Canavanine is a secondary metabolite that has two important roles in the plants that produce it. The first is as a nitrogen-storing compound in seeds. The second role is defensive, and is based on its chemical structure. Canavanine is an amino acid that is not found in proteins, but is very similar to the amino acid arginine, which is found in almost all proteins:

| Arginine | Canavanine |
|---|---|
| $NH_2$ | $NH_2$ |
| C=NH | C=NH |
| N—H | N—H |
| H—C—H | O |
| H—C—H | H—C—H |
| H—C—H | H—C—H |
| $H_2N$—C—H | $H_2N$—C—H |
| C—OH | C—OH |
| O | O |

A seemingly slight chemical difference... ...produces an inactive protein.

When an insect larva consumes canavanine-containing plant tissue, the canavanine is incorporated into the insect's proteins in some of the places where the mRNA codes for arginine, because the enzyme that charges the tRNA specific for arginine fails to discriminate accurately between the two amino acids (see Section 14.5). The structure of canavanine, however, is different enough from that of arginine that some of the resulting proteins end up with a modified tertiary structure, and hence reduced biological activity. These defects in protein structure and function lead to developmental abnormalities that kill the insect.

In plants that produce them, canavanine and other secondary metabolites are constitutive defenses—that is, they are present regardless of whether the plant is under attack. Other chemical defenses come into play only when an herbivore strikes.

### Plants respond to herbivory with induced defenses

The first step in a plant's response to herbivory is to somehow sense the event. Two mechanisms for plant perception of herbivore damage have been described: *membrane signaling* and *chemical signaling*.

**MEMBRANE SIGNALING** The plasma membrane is the part of the plant cell that is in contact with the environment. Within the first minute after an herbivore strikes, changes in the electrical potential of the plasma membrane occur in the damaged area. As you will see later when you study the animal nervous system (see Section 45.2), such changes can be rapidly transmitted as a signal along the plasma membrane. In the case of plants responding to herbivory, the continuity of the symplast (see Figure 35.6) ensures that the signal travels over much of the plant within 10 minutes.

**CHEMICAL SIGNALING** When an insect chews on a plant, substances in the insect's saliva combine with fatty acids derived from the consumed plant tissue. The resulting compounds act as elicitors (see Section 39.1) to trigger both local and systemic responses to the herbivore. In corn, the herbivore-produced elicitor has been named volicitin for its ability to induce production of volatile signals that can travel to other plant parts—and to neighboring corn plants—and simulate their defense responses.

**SIGNAL TRANSDUCTION PATHWAY** The perception of damage from herbivory initiates a signal transduction pathway in the plant that involves the plant hormone *jasmonic acid* (*jasmonate*) (**Figure 39.8**):

COOH

Jasmonic acid

When the plant senses an herbivore-produced elicitor, it makes jasmonate, which triggers many plant defenses, including the synthesis of a protease inhibitor. The inhibitor, once in an insect's gut, interferes with the digestion of proteins and thus stunts the insect's growth. Jasmonates also "call for help" by triggering the formation of volatile compounds that attract insects that prey on the herbivores attacking it.

### Why don't plants poison themselves?

Why don't the chemicals that are so toxic to herbivores and pathogens kill the plants that produce them? Plants that produce toxic defensive chemicals generally use one of the following measures to protect themselves:

- The toxic substance is isolated in a special compartment within the cell.
- The toxic substance is produced only after the plant's cells have already been damaged.
- The plant uses modified enzymes or modified receptors that do not react with the toxic substance.

Isolation of the toxic substance is the most common means of avoiding exposure. Plants store their toxins in vacuoles if the toxins are water-soluble. If they are hydrophobic, the toxins may be dissolved in latex and stored in specialized tubes called

**39.8 A Signal Transduction Pathway for Induced Defenses** The chain of events initiated by herbivory that leads to the production of a defensive chemical can consist of many steps. These steps may include the synthesis of one or two hormones, binding of receptors, gene activation, and, finally, synthesis of defensive compounds.

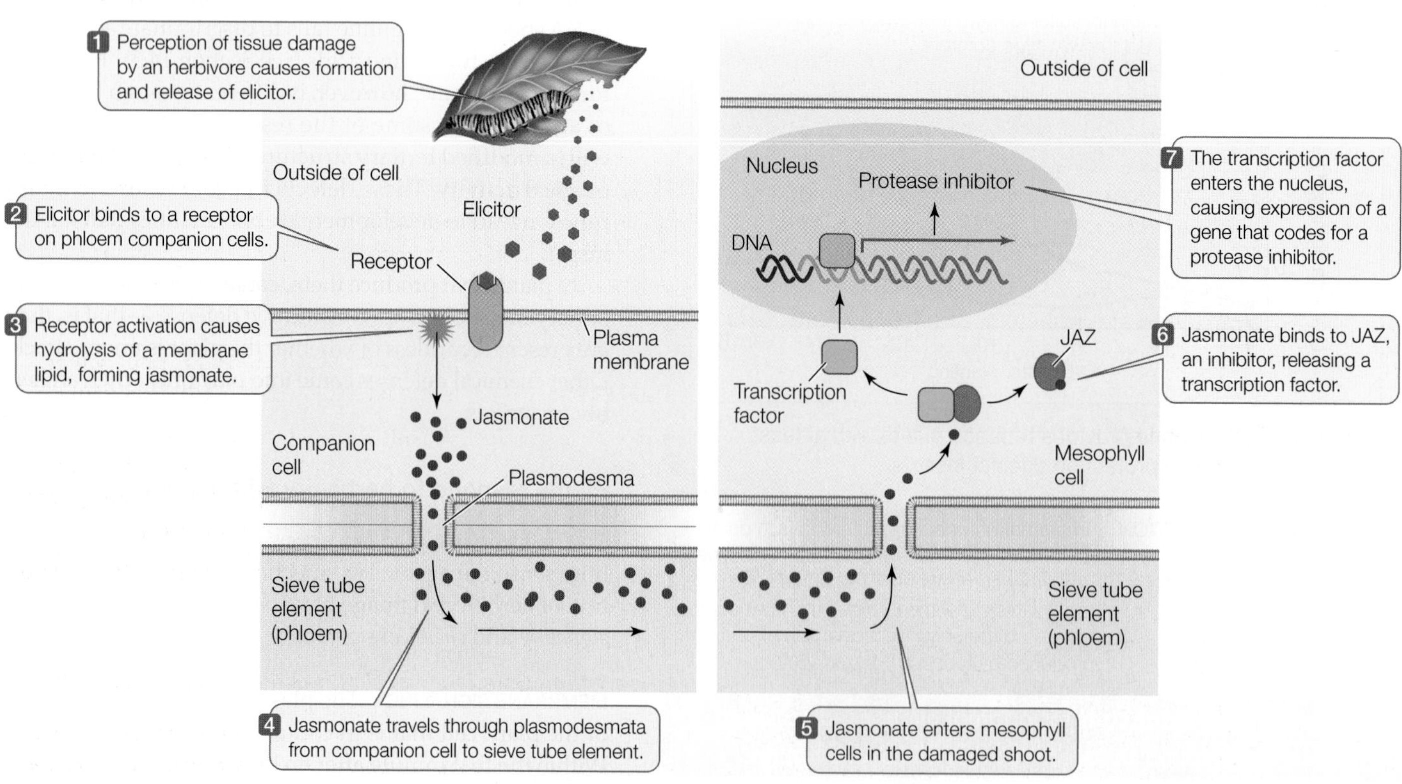
1 Perception of tissue damage by an herbivore causes formation and release of elicitor.
Outside of cell
2 Elicitor binds to a receptor on phloem companion cells.
Elicitor
Receptor
3 Receptor activation causes hydrolysis of a membrane lipid, forming jasmonate.
Plasma membrane
Jasmonate
Companion cell
Plasmodesma
Sieve tube element (phloem)
4 Jasmonate travels through plasmodesmata from companion cell to sieve tube element.
Outside of cell
Nucleus
Protease inhibitor
DNA
7 The transcription factor enters the nucleus, causing expression of a gene that codes for a protease inhibitor.
JAZ
6 Jasmonate binds to JAZ, an inhibitor, releasing a transcription factor.
Transcription factor
Mesophyll cell
Sieve tube element (phloem)
5 Jasmonate enters mesophyll cells in the damaged shoot.

**39.9 Disarming a Plant's Defenses** This beetle is inactivating a milkweed's defense system by cutting its laticifer supply lines.

**laticifers**, or dissolved in waxes on the epidermal surface. Such compartmentalized storage keeps the toxins away from the mitochondria, chloroplasts, and other parts of the plant's metabolic machinery.

Some plants store the precursors of toxic substances in one type of tissue, such as the epidermis, and store the enzymes that convert those precursors into the active toxin in another type, such as the mesophyll. When an herbivore chews part of the plant, cells are ruptured, the enzymes come into contact with the precursors, and the toxin is produced. The only part of the plant that is damaged by the toxin is that which was already damaged by the herbivore. Plants such as sorghum and some legumes, which respond to herbivory by producing cyanide—an inhibitor of cellular respiration—are among those that use this type of protective measure.

The third protective measure is used by the canavanine-producing plants described earlier. In these plants, unlike most other plants, the enzyme that charges the arginine tRNA discriminates correctly between arginine and canavanine, so canavanine is not incorporated into their proteins. Some herbivores, however, can evade canavanine poisoning in a similar manner, demonstrating that no plant defense is perfect. Like plants and pathogens, plants and herbivores evolve together in a continuing arms race, which the plants don't always win.

### The plant doesn't always win the arms race

Milkweeds such as *Asclepias syriaca* store their defensive chemicals in laticifers. When damaged, a milkweed releases copious amounts of toxic latex from its laticifers, which run alongside the veins in its leaves. Field studies have shown that most insects that feed on neighboring plants of other species do not attack laticiferous plants, but there are exceptions. One population of beetles that feeds on *A. syriaca* exhibits a remarkable prefeeding behavior: these beetles cut a few veins in the leaves before settling down to dine (**Figure 39.9**). Cutting the veins, with their adjacent laticifers, causes massive latex leakage and interrupts the latex supply to a downstream portion of the leaf. The beetles then move to the relatively latex-free portion and eat their fill.

Does this behavior of the beetles negate the adaptive value of latex protection? Not entirely. Great numbers of potential insect pests are still effectively deterred by the latex. And evolution proceeds. Over time, milkweed plants producing higher concentrations of toxins may be selected by virtue of their ability to kill even the beetles that cut their laticifers.

### 39.2 RECAP

**Many plants use secondary metabolites as defenses against herbivory. Other defenses are induced by herbivory through a signal transduction pathway involving the hormone jasmonate.**

- Describe one example of a secondary metabolite and how it affects herbivores. **See pp. 820–821**
- What role does jasmonate play in plant defense? **See p. 822 and Figure 39.8**
- What are three ways in which a plant avoids being poisoned by its own defensive chemicals? **See pp. 822–823**

A plant's survival depends not only on successful defenses against pathogens and herbivores, but also on coping with a sometimes hostile physical environment. In the next section we consider how plants deal with climate-imposed stresses.

## 39.3 How Do Plants Deal with Climatic Extremes?

Plants are threatened by many aspects of the physical environment, such as drought, waterlogged soils, and extreme temperatures. How do plants survive these environmental challenges? Plants cope with environmental stresses through adaptation or acclimation.

- *Adaptation* is genetically encoded resistance to stress. A plant may have structures or biochemical properties that aid in its survival in the face of environmental challenges.
- *Acclimation* is increased tolerance for environmental extremes because of prior exposure to them. An individual plant previously exposed to extreme cold, for example, may be more likely to survive the subsequent winter.

Let's begin by describing some adaptations of plants to extremes of water availability and temperature.

### Desert plants have special adaptations to dry conditions

Many plants, especially those living in deserts, must cope with extremely limited water supplies. Some desert plants have no spe-

**39.10 Desert Annuals Avoid Drought** The seeds of many desert annuals lie dormant for long periods awaiting conditions appropriate for germination. When they do receive enough moisture to germinate, they grow and reproduce rapidly before the short wet season ends. During the long dry spells, only dormant seeds remain alive.

**39.12 Succulence** The *Aloe* plant stores water in its fleshy leaves.

cial structural adaptations for water conservation. Instead, these desert annuals, called *drought avoiders*, simply evade periods of drought. Drought avoiders carry out their entire life cycle—from seed to seed—during a brief period in which rainfall has made the surrounding desert soil sufficiently moist (**Figure 39.10**). Deciduous plants, particularly in Africa and South America, shed their leaves in response to drought as a way to conserve water.

Stomata

A section through a leaf's surface shows stomata sunken in crypts protected by hairs.

Protective hairs

Lower surface of leaf

**39.11 Stomatal Crypts** Stomata in the leaves of some xerophytes are located in sunken cavities called stomatal crypts. The hairs covering these crypts trap moist air.

**LEAF STRUCTURES** Most desert plants grow in their dry environment year-round. Plants adapted to dry environments are called **xerophytes** (from the Greek *xeros*, "dry"). Three structural adaptations are found in the leaves of many xerophytes:

- Specialized leaf anatomy that reduces water loss
- A thick cuticle and a profusion of trichomes over the leaf epidermis, which retard water loss
- Diffraction and reflection of sunlight by trichomes, which decrease the intensity of the light impinging on the leaves, thus decreasing the risk of damage to the photosynthetic apparatus by excess light

In some species the stomata are strategically located in sunken cavities below the leaf surface (known as **stomatal crypts**), where they are sheltered from the drying effects of air currents (**Figure 39.11**). Hairs surrounding the stomata slow air currents as well.

Cacti and similar plants have spines rather than typical leaves, and photosynthesis is confined to the fleshy stems. The spines may help plants cope with desert condidtions by reflecting incident radiation, or by dissipating heat.

**WATER-STORING STRUCTURES** **Succulence**—the possession of fleshy, water-storing leaves or stems—is an adaptation to dry environments (**Figure 39.12**). Other adaptations of succulents include a reduced number of stomata and CAM photosynthesis, which separates the light-requiring and $CO_2$-assimilating reactions of photosynthesis to conserve water (see Section 10.4).

**39.13 Mining Water with Deep Taproots** In Death Valley, California, the root of this mesquite tree must reach far beneath the dunes for its water supply.

**ROOT SYSTEMS THAT MAXIMIZE WATER UPTAKE** Roots may also be adapted to dry environments. Cacti have shallow but extensive fibrous root systems that effectively intercept water at the soil surface following even light rains. Mesquite trees (*Prosopis*; **Figure 39.13**) obtain water through taproots that grow to great depths, reaching water supplies far underground, as well as from condensation on their leaves. The Atacama Desert in northern Chile often goes several years without measurable rainfall, but the landscape there has many surprisingly large mesquite trees.

One of the most successful desert plants of the southwestern United States, creosote bush (*Larrea tridentata*), displays a range of xerophytic features. It has a deep taproot, a shallow and extensive root system that can absorb water quickly after rare rain events, and small wax-covered leaves. The plant owes its name to its ability to produce noxious resins that smell like the wood preservative creosote. These natural resins not only help to seal in water, but also render the leaves virtually indigestible to browsing mammals—another adaptation to the stresses of desert life.

**CHANGES IN WATER POTENTIAL** Xerophytes and other plants that must cope with inadequate water supplies may accumulate high concentrations of the amino acid proline or of secondary metabolites in their vacuoles. These accumulations lower the water potential in the plant's cells below that in the soil, which results in the uptake of water by the cells via osmosis (see Section 35.1). Plants living in salty environments share this and several other adaptations with xerophytes, as we will see shortly.

## In water-saturated soils, oxygen is scarce

For some plants, the environmental challenge is the opposite of that faced by xerophytes: too much water. Some plants live in environments so wet that the diffusion of oxygen to their roots is severely limited. Since most plant roots require oxygen to support respiration and ATP production, most plants cannot tolerate saturated soil conditions for long.

Some species, however, are adapted to life in a water-saturated habitat. Their roots grow slowly and hence do not penetrate deeply. Because the oxygen concentration in saturated soil is too low to support aerobic respiration, their roots carry on alcoholic fermentation (an anaerobic process; see Section 9.4), which provides ATP for the activities of the root system. This adaptation explains why their growth is slow: fermentation is much less efficient in producing ATP than aerobic respiration.

The root systems of some plants adapted to swampy environments, such as cypresses and some mangroves, have **pneumatophores**, which are extensions that grow out of the water and up into the air (**Figure 39.14**). Pneumatophores have lenticels that allow oxygen to diffuse through them, aerating the submerged parts of the root system.

Pneumatophores are root extensions that grow out of the water, under which the rest of the roots are submerged.

**39.14 Coming Up for Air** The roots of these mangroves obtain oxygen through pneumatophores.

Large air spaces are found in the leaf parenchyma and in the petioles of many submerged or partly submerged aquatic plants. Tissue containing such air spaces is called **aerenchyma** (**Figure 39.15**). Aerenchyma stores oxygen produced by photosynthesis and permits its ready diffusion to parts of the plant where it is needed for cellular respiration. Aerenchyma also imparts buoyancy. Furthermore, because it contains far fewer cells than most other plant tissues, metabolism in aerenchyma proceeds at a lower rate, so the need for oxygen is much reduced.

Many plants, rather than facing continual water deficits or excesses, live in fluctuating environments with unpredictable rainfall. We now turn to the mechanisms plants use to respond to those challenges.

## Plants can acclimate to drought stress

When the weather is abnormally dry, the water content of the soil is reduced, and less water is available to plants. Water deficits in plant cells have two major effects: a reduction in membrane integrity as the polar–nonpolar forces that orient the lipid bilayer proteins are reduced, and changes in the three-dimensional structures of proteins. Plant growth is reduced when the structure of plant cells is compromised in these ways. Indeed, inadequate water supply is the single most important factor that limits production of our most important food crops.

Plants can, however, acclimate to drought stress to maintain their structure and function. How do they do it? When plants sense a water deficit in their roots, a signal transduction pathway is set in motion that initiates several measures to conserve water and maintain cellular integrity. This signal transduction pathway begins with the production of a hormone, abscisic acid, in the roots. This hormone travels from the roots to the shoot, where it causes stomatal closure and initiates gene transcription that leads to other physiological events that conserve water and cellular integrity (**Figure 39.16**).

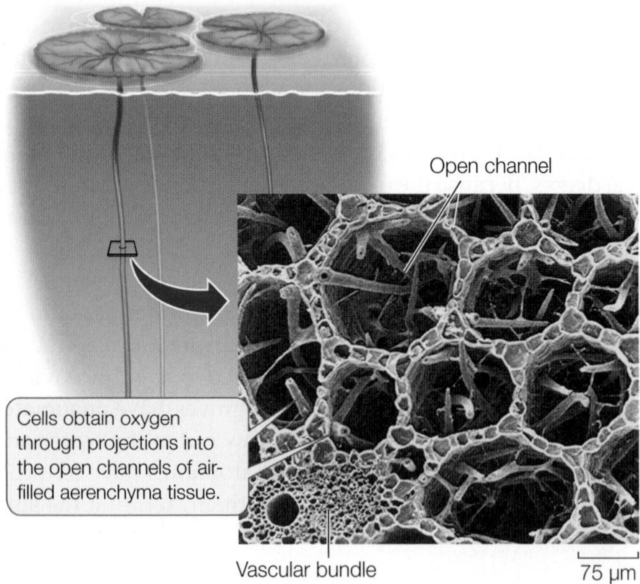

Open channel
Cells obtain oxygen through projections into the open channels of air-filled aerenchyma tissue.
Vascular bundle
75 µm

**39.15 Aerenchyma Lets Oxygen Reach Submerged Tissues** This scanning electron micrograph of a cross section of a petiole of the yellow water lily shows the structure of the air-filled channels that make up aerenchyma tissue.

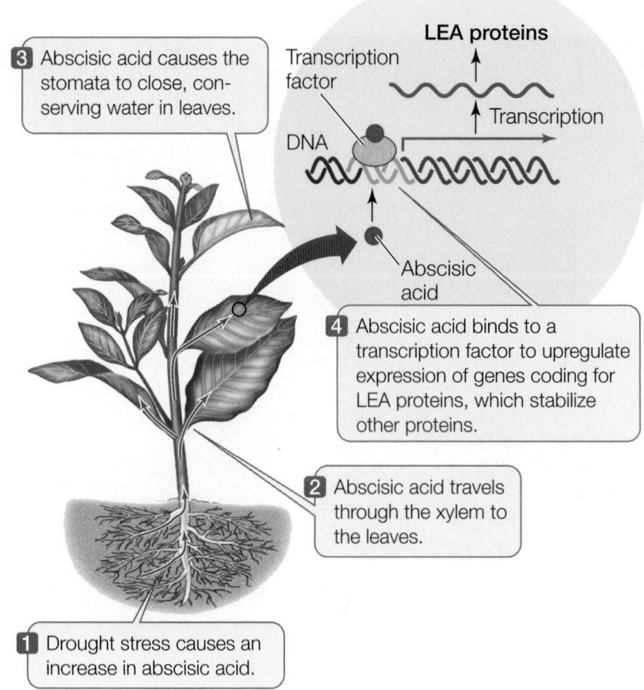

3 Abscisic acid causes the stomata to close, conserving water in leaves.
LEA proteins
Transcription factor
Transcription
DNA
Abscisic acid
4 Abscisic acid binds to a transcription factor to upregulate expression of genes coding for LEA proteins, which stabilize other proteins.
2 Abscisic acid travels through the xylem to the leaves.
1 Drought stress causes an increase in abscisic acid.

**39.16 A Signal Transduction Pathway in Response to Drought Stress** Acclimation to drought stress begins in the root with the production of the hormone abscisic acid.

Plant genes whose expression is altered by drought stress have been identified, largely through research using DNA microarrays (see Figure 18.9), proteomics (see Figure 17.17), and other molecular approaches. One group of proteins whose production is upregulated during drought stress is the *l*ate *e*mbryogenesis *a*bundant (LEA; pronounced "lee-yuh") proteins. These hydrophobic proteins accumulate in maturing seeds as they dry out. The LEA proteins bind to membranes and other cellular proteins to stabilize them, preventing their aggregation during desiccation. The importance of LEA proteins in coping with drought stress was demonstrated by Ray Wu and his colleagues at Cornell University, who showed that transgenic rice plants expressing a high level of a LEA protein in their leaves and roots grew better than normal plants under drought conditions (**Figure 39.17**). Genes that code for LEA proteins are widely distributed among plants, bacteria, and invertebrates; in all these organisms, high-level expression results in drought tolerance. As described at the opening of Chapter 35, which discusses the discovery of *HARDY*, another gene that promotes water use efficiency in plants, genetic research holds the promise of solutions to many agricultural problems, including those that may be presented by global climate change.

## Plants have ways of coping with temperature extremes

Temperatures that are too high or too low can stress plants and even kill them. Plants differ in their sensitivity to heat and cold, but all plants have their limits. Any temperature extreme can damage cellular membranes.

## INVESTIGATING LIFE

### 39.17 A Molecular Response to Drought Stress

Understanding the responses of plants to drought conditions is vital for agriculture. Ray Wu and colleagues transformed rice cells with a gene that codes for a LEA protein that is expressed in seeds as they mature and dry out. The investigators then measured the response of the transgenic rice plants to 28 days of drought.

**HYPOTHESIS** LEA proteins protect plants from the effects of drought stress.

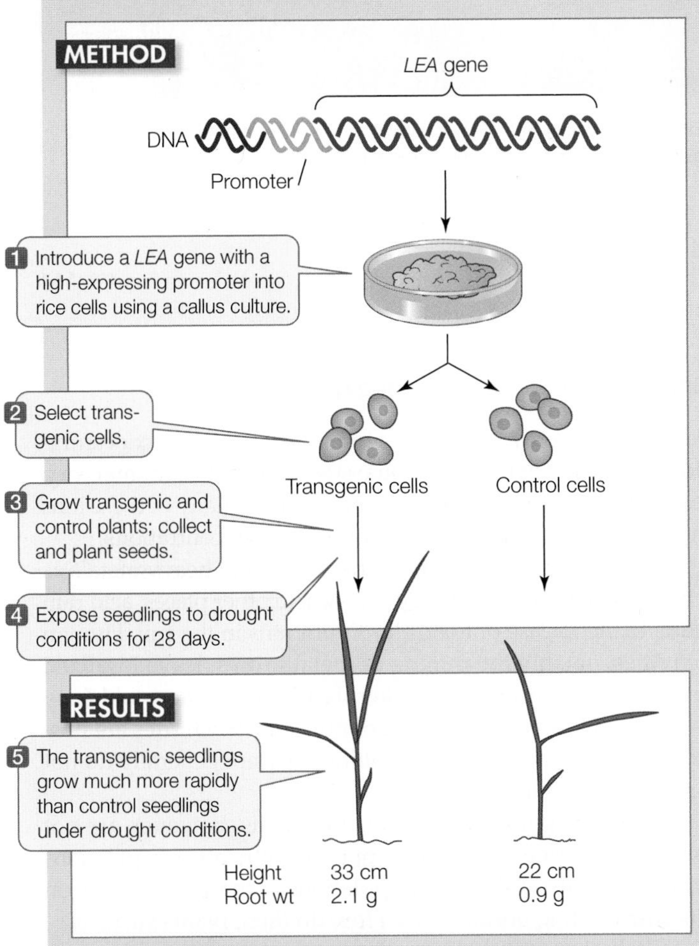

**METHOD**

*LEA* gene
DNA
Promoter
1 Introduce a *LEA* gene with a high-expressing promoter into rice cells using a callus culture.
2 Select transgenic cells.
Transgenic cells
Control cells
3 Grow transgenic and control plants; collect and plant seeds.
4 Expose seedlings to drought conditions for 28 days.

**RESULTS**

5 The transgenic seedlings grow much more rapidly than control seedlings under drought conditions.

| | Transgenic | Control |
|---|---|---|
| Height | 33 cm | 22 cm |
| Root wt | 2.1 g | 0.9 g |

**CONCLUSION** Plants with higher concentrations of a LEA protein grow better under drought conditions.

Go to **yourBioPortal.com** for original citations, discussions, and relevant links for all INVESTIGATING LIFE figures.

- High temperatures destabilize membranes and denature many proteins, especially some of the enzymes of photosynthesis.
- Low temperatures cause membranes to lose their fluidity and alter their permeabilities to solutes.
- Freezing temperatures may cause ice crystals to form, damaging membranes.

Therefore, it is not surprising that many plants living in hot environments have adaptations similar to those of xerophytes. These adaptations include hairs and spines that dissipate heat, leaf forms that intercept less direct sunlight, and CAM photosynthesis, which allows plants to perform some metabolic processes in the cool of night (see Section 10.4).

The plant response to heat stress is similar to the response to drought stress in that new proteins are made, often under the direction of an abscisic acid–mediated signal transduction pathway. Within minutes of experimental exposure to raised temperatures (typically a 5°C–10°C increase), plants synthesize several kinds of **heat shock proteins**. Among these proteins are chaperonins (see Figure 3.12), which help other proteins maintain their structures and avoid denaturation. Threshold temperatures for the production of heat shock proteins vary, but 39°C is sufficient to induce them in most plants.

Low temperatures above the freezing point can cause *chilling injury* in many plants, including crops such as rice, corn, and cotton, as well as tropical plants such as bananas. Many plant species can acclimate to cooler temperatures through a process called **cold-hardening**, which requires repeated exposure to cool, but not injurious, temperatures over many days. A key change during the hardening process is an increase in the proportion of unsaturated fatty acids in membranes, which allows them to retain their fluidity and function normally at cooler temperatures. Plants have a greater ability to modify the degree of saturation of their membrane lipids than animals do. In addition, low temperatures induce the formation of proteins similar to heat shock proteins, which protect against chilling injury.

If ice crystals form within plant cells, they can kill the cells by puncturing organelles and plasma membranes. Furthermore, the growth of ice crystals outside the cells can draw water from the cells and dehydrate them. Freeze-tolerant plants have a variety of adaptations to cope with these problems, including the production of *antifreeze proteins* that slow the growth of ice crystals.

## 39.3 RECAP

**Plants that live in continually dry or water-saturated environments have structural adaptations to cope with those conditions. Mechanisms that protect plants from drought stress are initiated by a signal transduction pathway involving abscisic acid. Heat shock proteins help plants acclimate to high and low temperatures.**

- Describe two structural adaptations for growth in water saturated soils. **See pp. 825–826**
- What is the role of abscisic acid in acclimation to drought stress? **See p. 826 and Figure 39.16**
- What environmental conditions induce the formation of heat shock proteins, and what functions do those proteins serve? **See p. 827**

Just as climatic extremes can limit plant growth, the presence of certain substances, such as salt and heavy metals, can make an environment inhospitable to plant growth.

## 39.4 How Do Plants Deal with Salt and Heavy Metals?

A number of toxic solutes are found in soils, but worldwide, no toxic substance restricts angiosperm growth more than salt (sodium chloride). Saline—salty—habitats support, at best, limited types of vegetation. Saline habitats are found in diverse locales, from hot, dry, deserts to moist, cool coastal marshes. Along the seashore, saline environments are created by ocean spray. The ocean itself is a saline environment, as are estuaries, where fresh and salt water meet and mingle. Salinization of agricultural land is an increasing global problem. Even where crops are irrigated with fresh water, sodium ions from the water accumulate in the soil to ever greater concentrations as the water evaporates.

Saline environments pose an osmotic challenge for plants. Because of its high salt concentration, a saline environment has an unusually negative water potential. To obtain water from such an environment, a plant must have an even more negative water potential; otherwise, water will diffuse out of its cells, and the plant will wilt and die. Plants in saline environments are also challenged by the potential toxicity of sodium, which inhibits enzymes and protein synthesis.

### Most halophytes accumulate salt

**Halophytes**—plants adapted to saline habitats—are found in a wide variety of flowering plant groups. Most halophytes share one adaptation: they take up sodium and, usually, chloride ions and transport those ions to their leaves. The accumulated ions are stored in the central vacuoles of leaf cells, away from more sensitive parts of the cells. Nonhalophytes accumulate relatively little sodium, even when placed in a saline environment; of the sodium that is absorbed by their roots, very little is transported to the shoot. The increased salt concentration in the tissues of halophytes lowers their water potential and allows them to take up water from their saline environment.

Some halophytes have other adaptations to life in saline environments. Some, for example, have **salt glands** in their leaves. These glands excrete salt, which collects on the leaf surface until it is removed by rain or wind (**Figure 39.18**). This adaptation, which reduces the danger of poisoning by accumulated salt, is found in some desert plants, such as *Frankenia palmeri*, and in some mangroves growing in seawater.

Salt glands can play multiple roles, as in the desert shrub *Atriplex halimus*. This shrub has glands that secrete salt into small bladders on the leaves. By lowering the water potential of the leaves, this salt not only helps them obtain water from the roots, but also reduces their transpirational loss of water to the atmosphere.

The adaptations we have just discussed are specific to halophytes. Several other adaptations are shared by halophytes and xerophytes, including thick cuticles, succulence, and CAM photosynthesis.

**39.18 Excreting Salt** This saltwater mangrove plant has special salt glands that excrete salt, which appears here as crystals on the leaves.

### Some plants can tolerate heavy metals

Salt is not the only toxic solute found in soils. High concentrations of some heavy metal ions, such as chromium, mercury, lead, and cadmium, are toxic to most plants; many of these ions are more toxic than sodium at equivalent concentrations.

Some geographic sites are naturally rich in heavy metals as a result of normal geological processes. In other places, acid rain leads to the release of toxic aluminum ions in the soil. Human activities, notably the mining of metallic ores, leave localized areas—known as tailings—with high concentrations of heavy metals and low concentrations of nutrients. Such sites are hostile to most plants, and seeds falling on them generally do not produce adult plants.

Most mine tailings rich in heavy metals, however, are not completely barren. They may support healthy plant populations that differ genetically from populations of the same species on the surrounding normal soils. How do these plants survive?

Initially, botanists believed that some plants were able to tolerate heavy metals by excluding them: that by not taking up the metal ions, the plants avoided being poisoned. Further investigations have shown, however, that tolerant plants growing on mine tailings do take up heavy metals, accumulating concentrations that would kill most plants. Over 200 plant species have been identified as **hyperaccumulators** that store large quantities of metals such as arsenic (As), cadmium (Cd), nickel (Ni), aluminum (Al), and zinc (Zn).

Perhaps the best-studied hyperaccumulator is alpine pennycress (*Thlaspi caerulescens*). Before the advent of chemical analysis, miners used to use the presence of this plant as an indicator of mineral-rich deposits. A *Thlaspi* plant may accumulate as much as 30,000 ppm Zn (most plants contain 100 ppm) and 1,500 ppm Cd (most plants contain 1 ppm). Studies of *Thlaspi*

**39.19 Phytoremediation** Plants that accumulate heavy metals can be used to clean up contaminated soils. Here, poplars are being used to remove contaminants from an air base.

and other hyperaccumulators have revealed the presence of several common mechanisms:

- Increased ion transport into the roots
- Increased rates of translocation of ions to the leaves
- Accumulation of ions in vacuoles in the shoot
- Resistance to the ions' toxicity

Knowledge of these hyperaccumulation mechanisms and the genes underlying them has led to the emergence of **phytoremediation**, a form of bioremediation (see Section 18.6) that uses plants to clean up environmental pollution. Some phytoremediation projects use natural hyperaccumulators, while others use genes from hyperaccumulators to create transgenic plants that grow more rapidly and are better adapted to a particular polluted environment. In either case, the plants are grown in the contaminated soil, where they act as natural "vacuum cleaners" by taking up the contaminants (**Figure 39.19**). The plants are then harvested and disposed of to remove the contaminants. Perhaps the most dramatic use of phytoremediation occurred after an accident at the nuclear power plant at Chernobyl, Ukraine (then part of the Soviet Union), in 1986, when sunflower plants were used to remove uranium from the nearby soil. Phytoremediation is now widely used in cleaning up land after strip mining.

After finding plants that accumulate valuable metals such as Ni, cobalt (Co), and silver (Ag), some scientists have proposed using those plants for *phytomining*. As in phytoremediation, the plants would be used to take up metals from the soil, but the metals would be extracted from the plants after they are harvested.

## 39.4 RECAP

**Halophytes have a number of adaptations to saline habitats, most of which involve mechanisms that lower their water potential. Some plants can tolerate heavy-metal-rich soils that are toxic to most other plants.**

- What are some of the roles of salt glands in halophyte leaves? See p. 828
- How are plants used for phytoremediation? See p. 829

# CHAPTER SUMMARY

### 39.1 How Do Plants Deal with Pathogens?

- Plants and pathogens have evolved together in a continuing "arms race": pathogens have evolved mechanisms for attacking plants, and plants have evolved mechanisms for defending themselves against those attacks. Review Figure 39.2, ANIMATED TUTORIAL 39.1
- Plants strengthen their cell walls and block plasmodesmata when attacked, limiting the ability of viral pathogens to move from cell to cell.
- **Gene-for-gene resistance** depends on a match between a plant's **resistance (*R*) genes** and a pathogen's **avirulence (*Avr*) genes**. Review Figure 39.3
- In the **hypersensitive response** to infection by bacteria or fungi, cells produce two kinds of defensive molecules: **phytoalexins** and **PR proteins**. Some cells around the infected area die, sealing off the pathogens and the damage they have caused.
- The hypersensitive response is often followed by **systemic acquired resistance**, in which salicylic acid activates further synthesis of defensive compounds.
- Plants use RNA interference to develop specific immunity to invading RNA viruses.

### 39.2 How Do Plants Deal with Herbivores?

- Herbivory increases the productivity of some plants. Review Figure 39.6
- Some plants produce **secondary metabolites** as defenses against herbivores. Review Table 39.1, Figure 39.7
- Hormones, including **jasmonates**, participate in signal transduction pathways leading to the production of defensive compounds. Review Figure 39.8
- Plants protect themselves against their own toxic defensive chemicals by isolating them in specialized compartments, by producing them only after the plant has already been damaged, or by using modified enzymes or receptors that are not affected by the toxic substance.

## 39.3 How Do Plants Deal with Climatic Extremes?

- Plants cope with environmental stresses by **adaptation** (genetically encoded resistance) or **acclimation** (increased tolerance).
- **Xerophytes** are plants that are adapted to dry environments.
- Some xerophytic adaptations are structural, including thickened cuticles, specialized trichomes, **stomatal crypts, succulence**, and long taproots.
- Some plants accumulate solutes, making their water potential lower so they can tolerate drought.
- Adaptations to water-saturated habitats include **pneumatophores**, extensions of roots allow oxygen uptake from the air, and **aerenchyma**, in which oxygen can be stored and diffuse throughout the plant.
- A signal transduction pathway involving abscisic acid initiates a plant's response to drought stress. **Review Figures 39.16 and 39.17**
- Membranes and proteins can be damaged by extremely high or low temperatures. Plants respond to extreme temperatures by producing **heat shock proteins**.
- Some plants undergo **cold-hardening**, an acclimation process that includes changes in membrane lipids and production of heat shock proteins.
- Some plants resist freezing by producing antifreeze proteins.

## 39.4 How Do Plants Deal with Salt and Heavy Metals?

- Most **halophytes** accumulate salt. Some have **salt glands** that excrete salt to the leaf surface.
- Some plants living in soils that are rich in heavy metals are **hyperaccumulators** that take up large amounts of those metals into their tissues.
- **Phytoremediation** is the use of hyperaccumulating plants or their genes to clean up environmental pollution.

**SEE WEB ACTIVITY 39.1 for a concept review of this chapter.**

## SELF-QUIZ

1. Which of the following is *not* a common defense against bacteria and fungi?
   *a.* Lignin formation
   *b.* Phytoalexins
   *c.* A waxy covering
   *d.* The hypersensitive response
   *e.* Mycorrhizae
2. Plants sometimes protect themselves from their own toxic secondary metabolites by
   *a.* producing special enzymes that destroy the toxin.
   *b.* storing precursors of the toxic substances in one compartment and the enzymes that convert those precursors to toxic products in another compartment.
   *c.* storing the toxic substances in mitochondria or chloroplasts.
   *d.* distributing the toxic substances to all cells of the plant.
   *e.* performing crassulacean acid metabolism.
3. Herbivory
   *a.* is an attack by plants on animals.
   *b.* always reduces plant growth.
   *c.* usually increases the rate of photosynthesis in the remaining leaves.
   *d.* reduces the rate of transport of photosynthetic products from the remaining leaves.
   *e.* is always lethal to the grazed plant.
4. Which statement about secondary metabolites is *not* true?
   *a.* They may be used in defense against fungi.
   *b.* Some are poisonous to herbivores.
   *c.* Some are amino acids that are normally part of proteins.
   *d.* Water soluble molecules are stored in vacuoles.
   *e.* Some mimic the hormones of animals.
5. Which statement about latex is *not* true?
   *a.* It is sometimes contained in laticifers.
   *b.* It is typically white.
   *c.* It is often toxic to insects.
   *d.* It is a rubbery solid.
   *e.* Milkweeds produce it.
6. Which of the following is *not* an adaptation to dry environments?
   *a.* Increased solute concentration in the vacuoles
   *b.* Hairy leaves
   *c.* A heavier cuticle over the leaf epidermis
   *d.* Sunken stomata
   *e.* A root system that grows each rainy season and dies back when it is dry
7. Some plants adapted to swampy environments meet the oxygen needs of their roots by means of a specialized tissue called
   *a.* parenchyma.
   *b.* aerenchyma.
   *c.* collenchyma.
   *d.* sclerenchyma.
   *e.* chlorenchyma.
8. Halophytes
   *a.* may accumulate abscisic acid in their vacuoles.
   *b.* may have water potentials that are lower than those of other plants.
   *c.* only accumulate sodium.
   *d.* have low root-to-shoot ratios.
   *e.* rarely accumulate sodium.
9. Which of the following is *not* involved in the response to drought stress?
   *a.* Abscisic acid
   *b.* Closing of aquaporins
   *c.* LEA gene expression
   *d.* Closing of stomata
   *e.* Jasmonate
10. Plants that tolerate heavy metals commonly
    *a.* differ genetically from other members of their species.
    *b.* do not take up the heavy metals.
    *c.* are tolerant to all heavy metals.
    *d.* are slow to colonize an area rich in heavy metals.
    *e.* weigh more than plants that are sensitive to heavy metals.

## FOR DISCUSSION

1. How might plant adaptations affect the evolution of herbivores? How might the adaptations of herbivores affect plant evolution?
2. The stomata of the common oleander (*Nerium oleander*) are located in sunken crypts in its leaves. Whether or not you know what an oleander is, you should be able to describe an important feature of its natural habitat. What is that feature?
3. In the coming decades, climate change may have significant effects on the growth and productivity of plants, in particular the crop plants on which we depend for our food. Discuss the physiological effects, and possible genetic responses in terms of plant breeding, of the following:
   *a.* In Pakistan, reduced rainfall causes a reduction in wheat yields.
   *b.* In the Mekong Delta of Vietnam, rising sea level inundates rice fields, causing a drastic reduction in yields.
   *c.* Increased temperature and humidity in western Canada causes an increase in wheat rust.

## ADDITIONAL INVESTIGATION

The tobacco hornworm (*Manduca sexta*) is adapted to feeding on nicotine-producing plants. Using the genetically modified tobacco plants described in Figure 39.7, how might you test the hypothesis that dietary nicotine protects the tobacco hornworm against its parasite *Cotesia congregata*?

## WORKING WITH DATA (GO TO yourBioPortal.com)

**Nicotine Is a Defense against Herbivores** To test the hypothesis that nicotine is a plant defense chemical, Baldwin and colleagues generated transgenic plants that expressed a low amount of nicotine (Figure 39.7). In this exercise, you will analyze data from the original research paper, which included the effects of the hormone jasmonic acid.

PART NINE ANIMALS: FORM AND FUNCTION

# 40 Physiology, Homeostasis, and Temperature Regulation

## Cool it!

"A new world record!" These words convey the thrill of world-class athletic competition. But as records are broken by mere centimeters or by fractions of a second, are we reaching absolute limits to human performance? We can assess many physiological limits to extreme performance—maximum breathing rate, for example, or the maximum rate at which the heart can supply blood to the muscles. A less obvious physiological limit is temperature.

The 2008 New York City Marathon took place on a cold, clear, windy day in November. For the third time, the first-place woman in this 41-km race was world record holder Paula Radcliffe. Radcliffe had also been expected to win the women's marathon back in the 2004 Olympics. But that race took place on a hot, humid day in Athens. Overcome by heat stress, Radcliffe collapsed 6 km from the finish line. The critical difference in the two races was probably temperature.

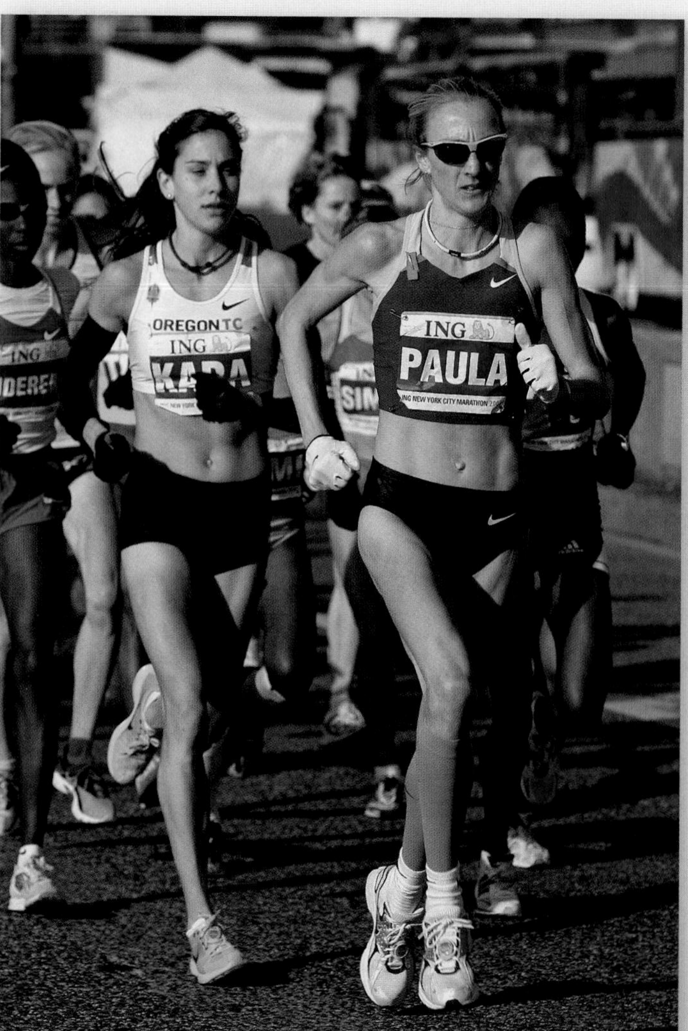


**Limits to Performance** Paula Radcliffe, photographed here during her winning performance at the 2008 New York City Marathon, collapsed from heat stress during the 2004 Olympic marathon. When the body's internal temperature is subjected to extreme heat, its homeostatic mechanisms may fail.

Thermal stress can have more serious consequences than losing a race. Every year some athletes die of heat stroke, which can occur when internal body temperature exceeds 41°C. This elevated internal temperature results in the failure of major organs and, in more than 20 percent of cases, death. Soldiers in the desert are at extreme risk of heat stroke, as are firefighters. Agricultural, industrial, and construction workers are also subject to the adverse affects of heat. Biologists at Stanford University developed a technology to cool individuals in such situations, and in the process discovered a way to enhance athletic performance.

Working muscles produce heat, which is carried by the blood to skin surfaces, where it is lost to the environment. Not all skin surfaces are equally good at dissipating heat, however. Because fur impedes heat loss, mammals evolved efficient bare-skin heat-loss portals such as the nose, tongue, footpads, and parts of the face. These areas have specialized blood vessels that can act like radiators to disperse heat or close down to conserve heat. Humans are not furred, but we retain these ancestral blood vessels in our hands, feet, and face (which is why we blush). The Stanford team designed a device to amplify heat extraction from these areas.

The heat extractor is a chamber that encases the hand and is sealed at the wrist. The hand is in contact with a

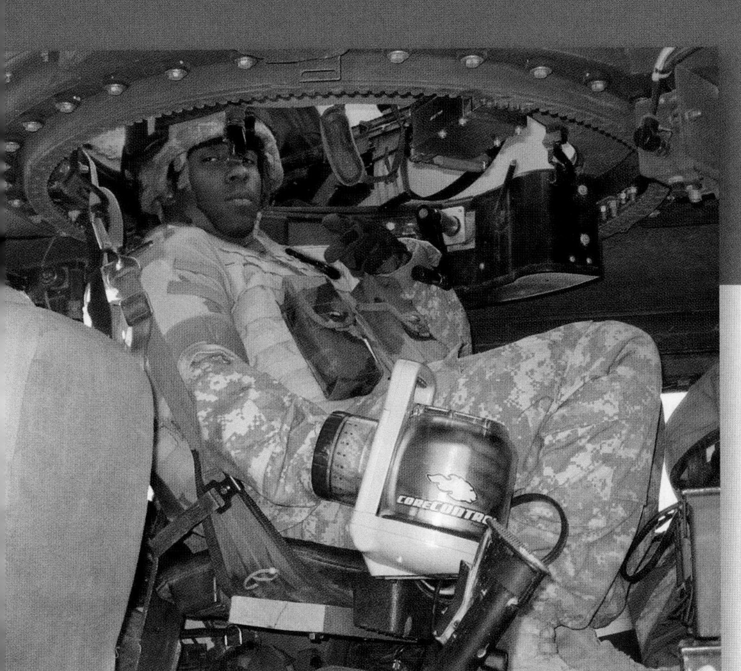

**A Cooling Glove** The heat extractor increases heat loss and allows the body to perform at a higher level in severe conditions.

cooled surface. A mild vacuum in the chamber pulls more blood into the hand, enabling the cool surface to extract more heat. With this device, an active individual's body temperature rises more slowly, and cools more rapidly during rest. An unexpected benefit is that cooling reduces fatigue and greatly increases exercise capacity. In one study, college freshmen improved their push-up performance at a rate of 5 push-ups a day without cooling, but 9 push-ups a day with cooling. Some men and women in the study achieved more than 800 push-ups in a workout session.

Human beings survive in environments that are extremely hot or extremely cold because we have the physiological and behavioral means of regulating our internal body temperature—an example of *homeostasis*, the maintenance of a "steady state" in our internal environment.

**IN THIS CHAPTER** we will explore the internal environment that serves the needs of all of the body's cells. We will survey the cell and tissue types that make up physiological systems and discuss how these systems maintain the internal environment within certain physiological limits, a condition called homeostasis. Homeostasis will be described using one important example, the regulation of body temperature.

CHAPTER OUTLINE



## 40.1 How Do Multicellular Animals Supply the Needs of Their Cells?

All animal cells need nutrients and oxygen from the environment and must eliminate carbon dioxide and other waste products of metabolism to the environment. The cells of very small or very thin aquatic animals meet these needs by direct exchanges with the external environment. In such animals, no cell is far from direct contact with the water it lives in; the water contains nutrients, absorbs waste, and provides a relatively unchanging physical environment. Most cells of larger animals do not have direct contact with the external environment, and their needs must be served by an environment that is wholly internal to the animal.

### An internal environment makes complex multicellular animals possible

The cells of multicellular animals exist within an **internal environment** of extracellular fluid (ECF). A human, for example, is about 60 percent water. Two-thirds of that water is contained within cells, and one-third makes up the ECF that is our internal environment. About 20 percent of that extracellular fluid, or 3 liters, is the blood plasma that circulates in our blood vessels. The rest—about 11 liters—is the **interstitial fluid** that bathes every cell of the body (**Figure 40.1**). Individual cells get their nutrients from this interstitial fluid and dump their waste products into it. As long as conditions in this internal environment are held within certain limits, the cells are protected from changes or harsh conditions in the external environment. A stable internal environment makes it possible for an animal to occupy habitats that would kill its cells if they were exposed to it directly. How is the internal environment kept constant?

As multicellular organisms evolved, cells became specialized for maintaining specific aspects of the internal environment. In turn, the internal environment enabled these specializations, since each cell did not have to provide for all of its own needs. Some cells evolved to be the interface between the internal and the external environments and to provide the necessary transport functions to get nutrients in and move wastes out. Other cells became specialized to provide internal functions such as circulation of the extracellular fluids, energy storage, movement, and information processing. The evolution of physiological systems to maintain the internal environment made it possible for multicellular animals to become larger, thicker, and more complex, and to occupy many different habitats.

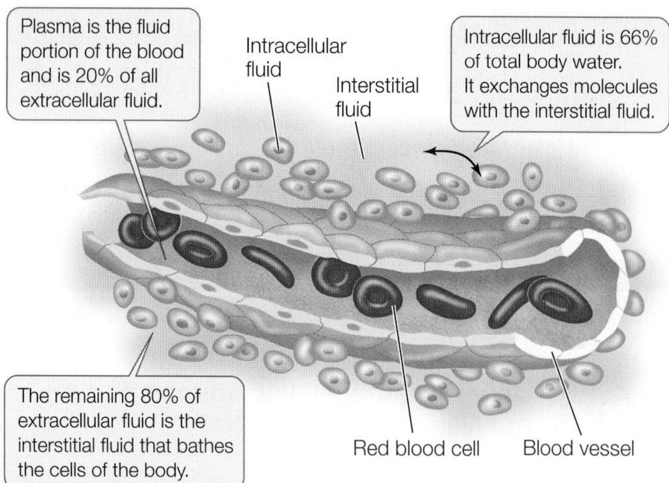

Plasma is the fluid portion of the blood and is 20% of all extracellular fluid.
Intracellular fluid
Interstitial fluid
Intracellular fluid is 66% of total body water. It exchanges molecules with the interstitial fluid.
The remaining 80% of extracellular fluid is the interstitial fluid that bathes the cells of the body.
Red blood cell
Blood vessel


**40.1 The Internal Environment** The "internal environment" is the extracellular fluid, or ECF. ECF, which accounts for about one-third of total body water, is made of the blood plasma and the interstitial fluid. The physiological composition of the ECF must remain stable within narrow limits, and maintaining that stability is the job of the body's organ systems.

The composition of the internal environment is constantly being challenged by the external environment and by the metabolic activity of the cells of the body. The maintenance of stable conditions (within a narrow range) in the internal environment is called **homeostasis**. If a physiological system fails to function properly, homeostasis is compromised and cells are damaged and can die. To avoid the loss of homeostasis, physiological systems must be controlled and regulated in response to changes in both the external and internal environments.

### Physiological systems maintain homeostasis

The activities of all physiological systems are controlled—speeded up or slowed down—by actions of the nervous and endocrine systems. But to *regulate* the internal environment, information is required. As an analogy, think of driving a car (**Figure 40.2**). To regulate the speed of your car, you have to know both how fast you are going and how fast you want to go. The desired speed is a **set point**, or reference point; the reading on your speedometer is **feedback information**. Any difference between the set point and the feedback information is an **error signal**. Error signals suggest corrective actions, such as stepping on the accelerator or brake.

Some components of physiological systems are called **effectors** because they *effect* changes in the internal environment. Effectors are **controlled systems** because their activities are controlled by commands from regulatory systems. **Regulatory systems**, in contrast, obtain, process, and integrate information, then issue commands to controlled systems.

Important components of any regulatory system are the **sensors** that provide the feedback information to be compared with the internal set point. How is information from the sensors used?

**Negative feedback** is the most common use of sensory information in regulatory systems. Negative feedback information is used to counteract the influence that created an error signal. Whatever force is pushing the system away from its set point must be "negated." In our car analogy, the recognition that you are going too fast is negative feedback that causes you to release the accelerator and press the brake.

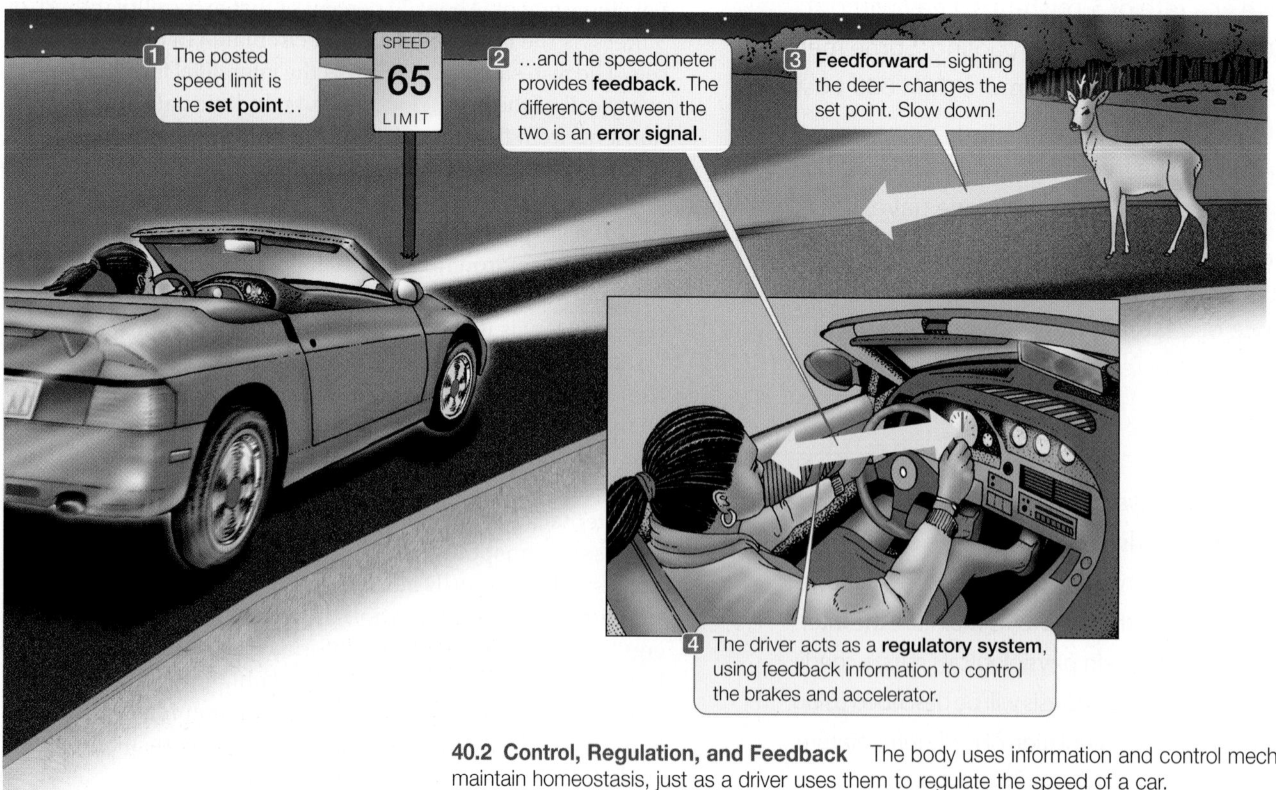


3 **Feedforward**—sighting the deer—changes the set point. Slow down!

**40.2 Control, Regulation, and Feedback** The body uses information and control mechanisms to maintain homeostasis, just as a driver uses them to regulate the speed of a car.

Although not as common as negative feedback, **positive feedback** is also seen in physiological systems. Rather than returning a system to a set point, positive feedback amplifies a response (i.e., it *increases* the deviation from the set point). Examples of regulatory systems that use positive feedback are the responses that empty body cavities, such as urination and defecation. Another example is sexual behavior, in which a little stimulation causes more behavior, which causes more stimulation, and so on. Positive feedback responses tend to reach a limit and terminate rapidly. The birth process is a good example. Contractions of the uterus stretch the birth canal, which stimulates more and stronger contractions until the baby is delivered, at which time contractions cease.

**Feedforward information** is another feature of regulatory systems. The function of feedforward information is to change the set point. Seeing a deer in the road when you are driving is an example of feedforward information; this information takes precedence over the posted speed limit, and you slow down. Before the start of a race, hearing the command "on your mark" is feedforward information that raises your heart rate before you begin to run. Feedforward information predicts a change in the internal environment before that change occurs.

These principles of control and regulation help organize our thinking about physiological systems. Once we understand how a system works, we can then ask how it is regulated. The example we will explore in this chapter is the regulation of body temperature. But first we need to become acquainted with the important structural features that all physiological systems have in common.

### Cells, tissues, organs, and systems are specialized to serve homeostatic needs

Each physiological system is composed of discrete organs, such as the liver, heart, lungs, and kidneys. These organs are made up of assemblages of cells known as **tissues**. Although there are many specialized cell types, there are only four kinds of tissues: *epithelial*, *muscle*, *connective*, and *nervous*.

The word "tissue" is often used in a general way to refer to a piece of an organ, such as "lung tissue" or "kidney tissue." As we will see, an organ always consists of more than one of the four kinds of tissues.

**EPITHELIAL TISSUES** **Epithelial tissues** are sheets of densely packed, tightly connected *epithelial cells* (**Figure 40.3**). Epithelial cells create boundaries between the inside and the outside of the body and between body compartments; they line the blood vessels and make up various ducts and tubules.

Filtration and transport are important functions of epithelial cells; they both act as barriers and provide transport across those barriers. They control what molecules and ions can move between the blood and the interstitial fluid. They can selectively transport ions and molecules from one side of an epithelial membrane to the other. Examples are the absorption of nutrient molecules from your gut and the secretion of acid into your stomach. Some epithelial cells, like those in the lungs or at the skin's surface, are extremely thin (*squamous*) to facilitate movement of substances across them.

The skin is epithelial tissue that receives much wear and tear. Accordingly, epithelial cells in the deepest layer of the skin have a high rate of cell division, producing new cells that move progressively to the skin surface, die, and are shed. A cross section of the skin reveals the layering of cells, from the newly formed ones on the innermost germinal layer to the dead ones on the surface. Because of this appearance, the skin is called a stratified epithelium (see Figure 40.3A). In contrast, gut epithelium consists of a single layer of tall, closely packed cells called a simple columnar epithelium. The epithelial cells of your gut are replaced about every 5 days; those in your skin are renewed every 1 to 2 months.

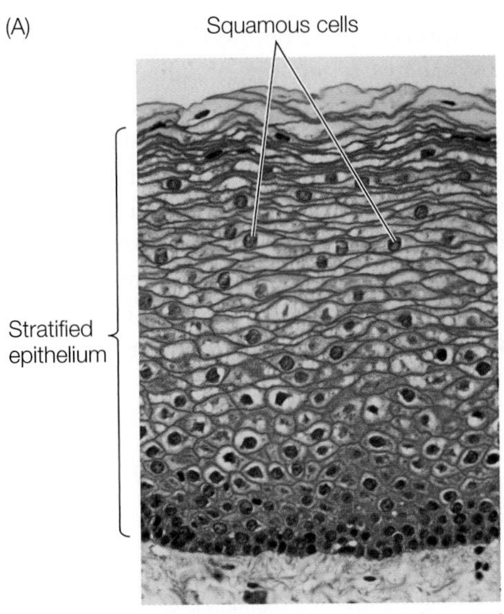


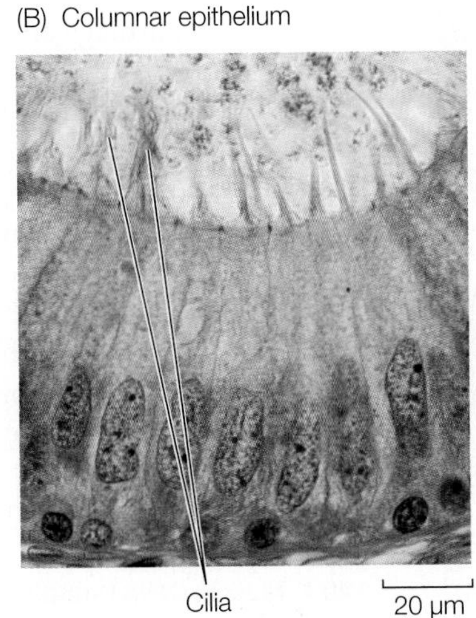


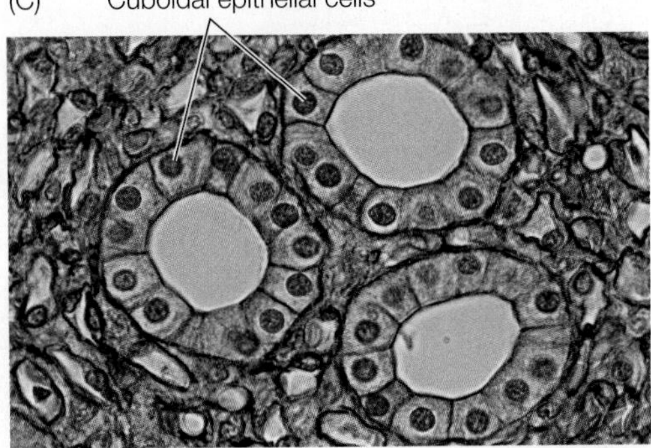


**40.3 Epithelial Tissue** (A) Epithelial cells make up the outer layers of skin. They are stratified, from extremely thin (squamous) older cells at the surface to rapidly dividing new layers that will rise to the surface as older cells are shed. (B) Ciliated columnar epithelium from the male reproductive duct (the vas deferens). (C) A single layer of cuboidal epithelial cells forms a tubule in the kidney. These cells have many molecular transport functions.

Epithelial cells have many other roles. Some secrete hormones, milk, mucus, digestive enzymes, or sweat. Others have cilia that move substances over surfaces or through tubes (see Figure 40.3B). Epithelial cells can also provide information to the nervous system. Smell and taste receptors, for example, are epithelial cells that detect specific chemicals.

**MUSCLE TISSUES** **Muscle tissues** consist of elongated cells that contract to generate forces and cause movement. Muscle tissues are the most abundant tissues in the body, and they use most of the energy produced in the body. All muscle cells contain long protein polymers called actin and myosin which interact to cause muscle cells to contract and exert force. There are three types of muscle tissues (**Figure 40.4**).

(A)

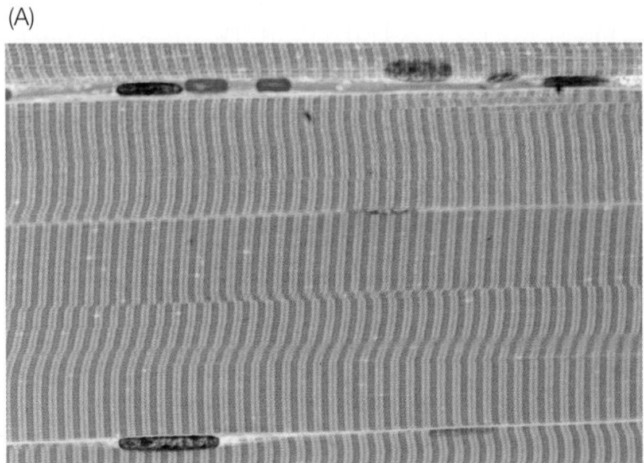

15 µm

(B)

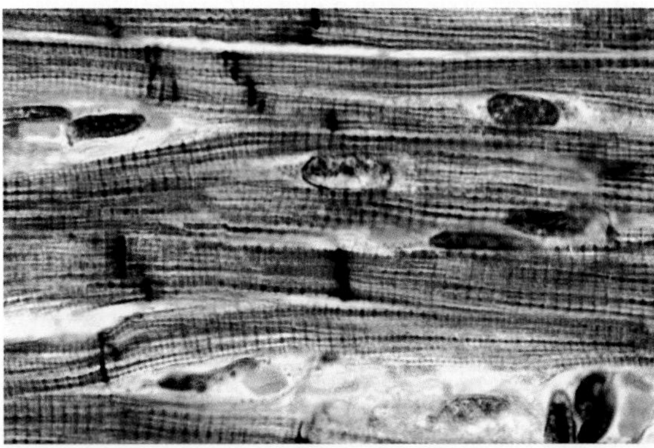

15 µm

(C)

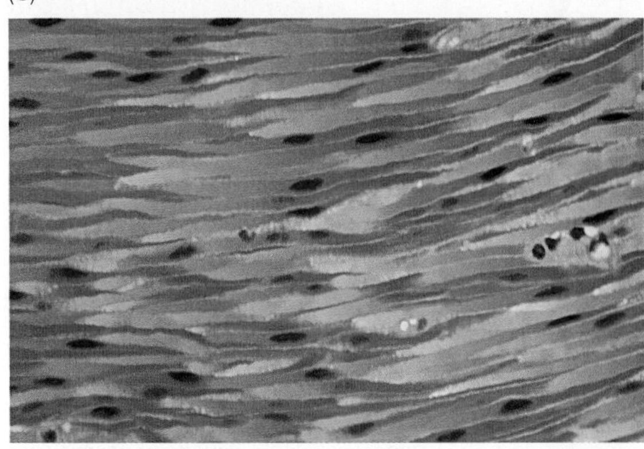

30 µm

**40.4 Muscle Cells Contain Protein Filaments** The filaments of two specific proteins—actin and myosin—interact to cause contraction and generate force in muscle tissue. (A) The regular arrangement of actin and myosin filaments results in the striated (striped) appearance of skeletal muscle. (B) The individual cells of cardiac muscle are branched and form a strong structural meshwork. (C) The actin and myosin filaments of smooth muscle are not regularly arranged and thus it does not have a striated appearance.

- **Skeletal muscles** (so named because they mostly attach to bones) are responsible for locomotion and other body movements such as facial expressions, shivering, and breathing.
- **Cardiac muscle** makes up the heart and is responsible for the beating of the heart and the pumping of blood. Individual cardiac muscle cells are branched, and the interweaving of these branches gives heart muscle structural strength.
- **Smooth muscle** is responsible for involuntary generation of forces in many hollow internal organs such as the gut, bladder, and blood vessels.

Skeletal muscles are under both voluntary and involuntary control, as will be described in detail in Section 48.1. Cardiac and smooth muscles are under involuntary control; they are controlled by physiological regulatory systems.

**CONNECTIVE TISSUES** In contrast to densely packed epithelial and muscle tissues, **connective tissues** are generally dispersed populations of cells embedded in an *extracellular matrix* that they secrete (**Figure 40.5**). The composition and properties of the matrix differ among types of connective tissues.

Protein fibers are an important component of the extracellular matrix of connective tissue cells. The dominant protein in the extracellular matrix is *collagen* (see Figure 5.25), which makes up about 25 percent of total body protein. Collagen fibers are strong and resistant to stretch, giving strength to the skin and to the connections between bones and between bones and muscles. The fibers provide a netlike framework for organs, giving them shape and structural strength.

*Elastin* is another type of protein fiber in the extracellular matrix of connective tissues. It is so named because it can be stretched to several times its resting length and then recoil. Fibers composed of elastin are most abundant in tissues that are regularly stretched, such as the walls of the lungs and the large arteries.

Cartilage and bone are connective tissues that provide rigid structural support. In *cartilage*, a network of collagen fibers is embedded in a flexible matrix consisting of a protein–carbohydrate complex, along with a specific type of cell called a *chondrocyte*. Cartilage, which lines the joints of vertebrates, is resistant to compressive forces. Since it is flexible, it provides structural support for flexible structures such as external ears and noses. The extracellular matrix in *bone* also contains many collagen fibers, but it is hardened by the deposition of the mineral calcium phosphate. We discuss cartilage and bone in greater detail in Section 48.3.

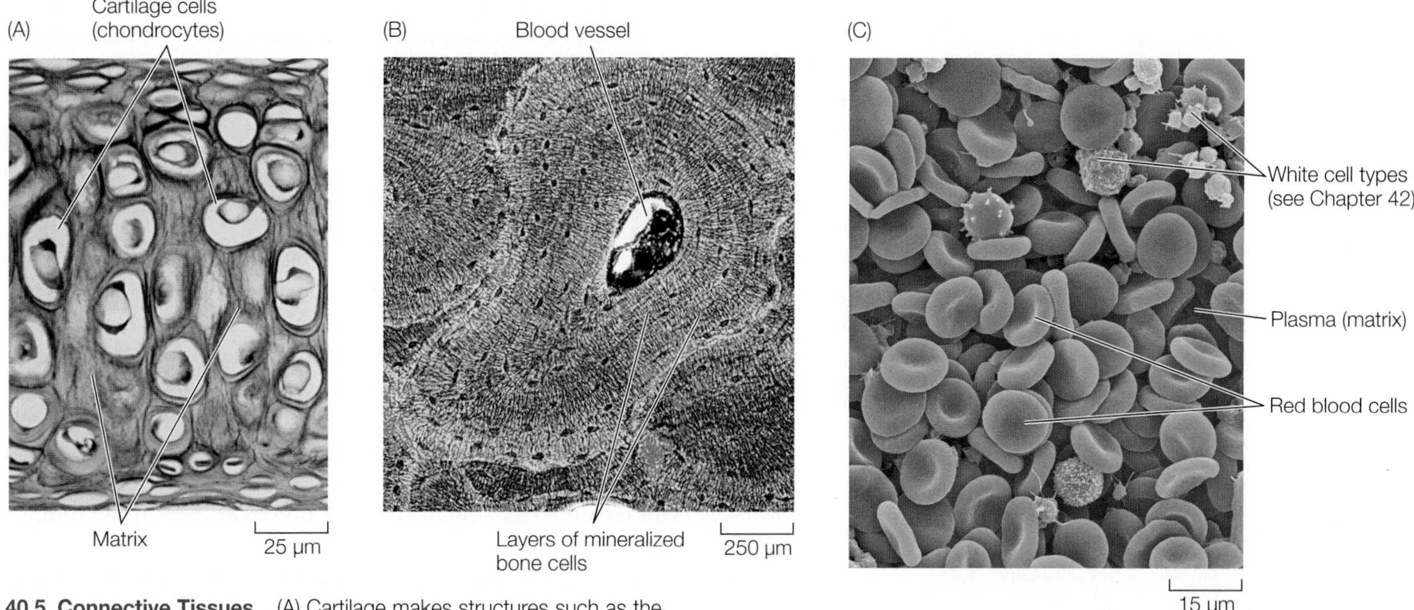


**40.5 Connective Tissues** (A) Cartilage makes structures such as the ear stiff but flexible. Cartilage cells, or chondrocytes, secrete an extracellular matrix rich in collagen and elastin fibers. In this micrograph, the elastin fibers are stained dark blue. (B) Bone is the mineral-rich connective tissue of the vertebrate skeleton. (C) Blood is unique among the connective tissues, consisting of blood cells floating in an extracellular matrix of plasma.

*Adipose* cells form loose connective tissue that stores lipids. Adipose tissue, or "fat," is a major source of stored energy. It also cushions organs, and layers of adipose tissue under the skin can provide a barrier to heat loss.

*Blood* is a connective tissue consisting of cells dispersed in an extensive liquid extracellular matrix, the blood *plasma*. We present many of the proteins and cellular elements of blood in Section 42.1, and we will discuss blood again in Section 50.4.

**NERVOUS TISSUES** The two basic cell types in **nervous tissues** are *neurons* and *glial cells* (**Figure 40.6**). Neurons come in many shapes and sizes, and all neurons encode information as electrical signals. These signals can travel over long extensions called *axons* to communicate with other neurons, muscle cells, or secretory cells through the release of chemicals called *neurotransmitters.* Neurons control the activities of most organ systems. Glial cells do not generate or conduct electrical signals, but they provide a variety of supporting functions for neurons. There are more glial cells than neurons in the nervous system.

Chapters 45, 46, and 47 detail the many fascinating properties of nervous tissues.

### Organs consist of multiple tissues

**Organs** include more than one kind of tissue, and most organs include all four (**Figure 40.7**). The wall of the gut is a good example. Its inner surface is lined with a sheet of columnar epithelial cells. Different epithelial cells secrete mucus, enzymes, or stomach acid. Beneath the epithelial lining is connective tissue. Within this connective tissue are blood vessels, neurons, and glands (clusters of secretory epithelial

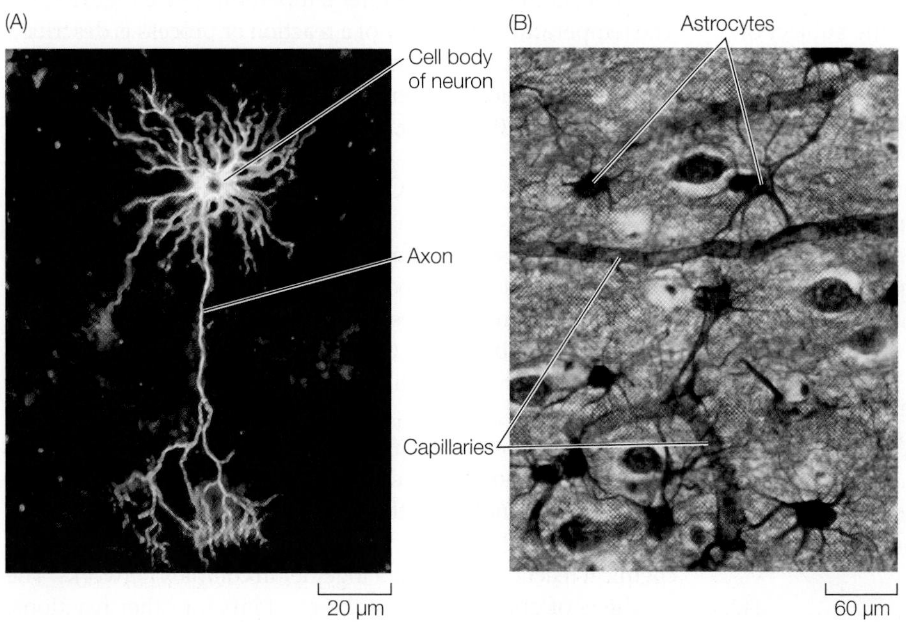


**40.6 Nervous Tissue Includes Neurons and Glial Cells** (A) This human neuron consists of a cell body, a number of processes that receive input from other neurons, and one long axon that sends information to other cells. (B) A section through human brain tissue shows astrocytes, a type of glial cell. Glial cells provide support and protection for neurons, including creating a barrier that protects the brain from many chemicals circulating in the blood.

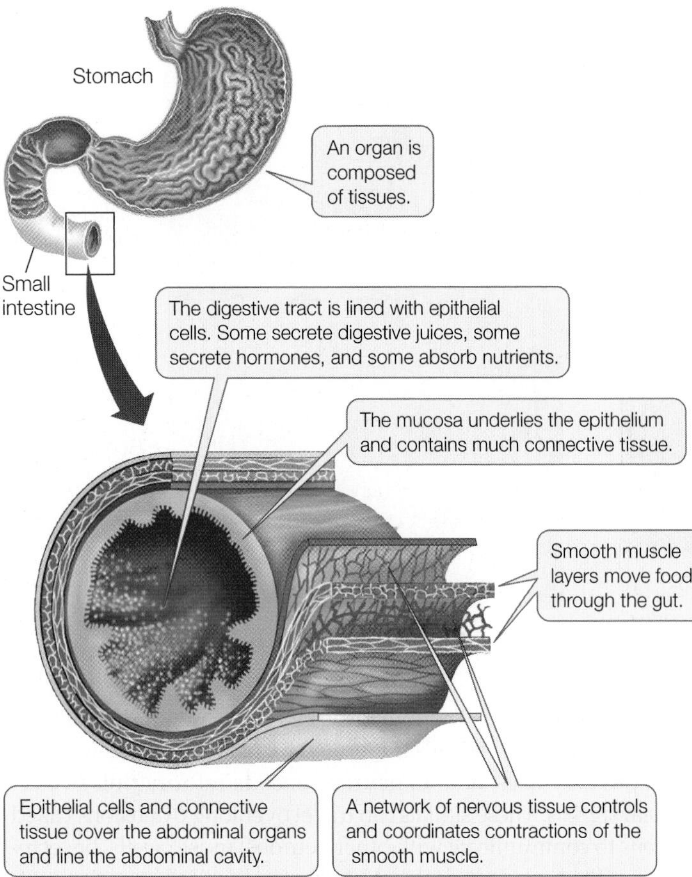

Stomach
An organ is composed of tissues.
Small intestine
The digestive tract is lined with epithelial cells. Some secrete digestive juices, some secrete hormones, and some absorb nutrients.
The mucosa underlies the epithelium and contains much connective tissue.
Smooth muscle layers move food through the gut.
Epithelial cells and connective tissue cover the abdominal organs and line the abdominal cavity.
A network of nervous tissue controls and coordinates contractions of the smooth muscle.

**40.7 Tissues Form Organs** Most organs contain more than one of the four tissue types. The organs of the human digestive system, such as the stomach and small intestine, are made up of all four.

cells). Concentric layers of smooth muscle tissue enable the gut to contract to mix food with digestive juices. A network of neurons between the muscle layers controls these movements.

An individual organ is usually part of an **organ system**—a group of organs that work together to carry out certain functions. The stomach and small intestine, for example, are part of the digestive system. The digestive system is the subject of Chapter 51.

## 40.1 RECAP

**The internal environment provides for the needs of all the cells that make up a complex animal. Organs and organ systems control the composition of the internal environment. The activities of organs and organ systems are regulated to maintain homeostasis of the internal environment.**

- Explain the difference between negative and positive feedback control mechanisms. **See pp. 834–835 and Figure 40.2**
- Describe a key function of each of the four kinds of tissue found in animals. **See pp. 835–837 and Figures 40.3–40.7**

Subsequent chapters will describe each of the organ systems mentioned above in much greater detail. The remainder of this chapter focuses on the mechanisms of homeostasis, using one important variable of the internal environment—its temperature—as our example.

## 40.2 How Does Temperature Affect Living Systems?

Temperatures vary enormously over the face of Earth, from the boiling hot springs of Yellowstone National Park to the interior of Antarctica, where the temperature can fall below –80°C. Cells, however, can function over only a narrow range of temperatures. If cells cool below 0°C, ice crystals form and damage cell structures. Some animals have adaptations, such as antifreeze molecules in their blood, that help them resist freezing; others can survive freezing. Generally, however, cells must remain above 0°C to stay alive.

The upper temperature limit for survival in most cells is about 45°C (although some specialized algae can grow in hot springs at 70°C, and some archaea live at near 100°C). In general, proteins begin to denature and lose their function as temperatures rise above 40°C. Therefore, most cellular functions are limited to the range between 0°C and 40°C, which approximates the thermal limits for life. Each particular species, however, usually has much narrower limits. To stay within those limits in spite of environmental conditions, animals have evolved thermoregulatory adaptations that give them certain thermal tolerances that determine their distribution ranges. When environments change rapidly, as may be happening with global climate warming, animals may find themselves in situations that exceed their thermal tolerances.

### $Q_{10}$ is a measure of temperature sensitivity

Even between 0°C and 40°C, changes in tissue temperature create problems for animals. Most physiological processes, like the biochemical reactions that constitute them, are temperature-sensitive, going faster at higher temperatures (see Figure 8.21). The temperature sensitivity of a reaction or process is described in terms of $\mathbf{Q_{10}}$, a factor calculated by dividing the rate of a process or reaction at a certain temperature, $R_T$, by that rate at a temperature 10°C lower, $R_{T-10}$:

$$Q_{10} = \frac{R_T}{R_{T-10}}$$

$Q_{10}$ can be measured for a simple enzymatic reaction or for a complex physiological process, such as rate of oxygen consumption. If a reaction or process is not temperature-sensitive, it has a $Q_{10}$ of 1. Most biological $Q_{10}$ values are between 2 and 3. A $Q_{10}$ of 2 means that the reaction rate doubles as temperature increases by 10°C, and a $Q_{10}$ of 3 indicates a tripling of the rate over a 10° temperature range (**Figure 40.8**).

Changes in body temperature can disrupt an animal's physiology because not all of the biochemical reactions that constitute the metabolism of an animal have the same $Q_{10}$. These biochemical reactions are linked together in complex networks. The products of one reaction are the reactants for other reactions.

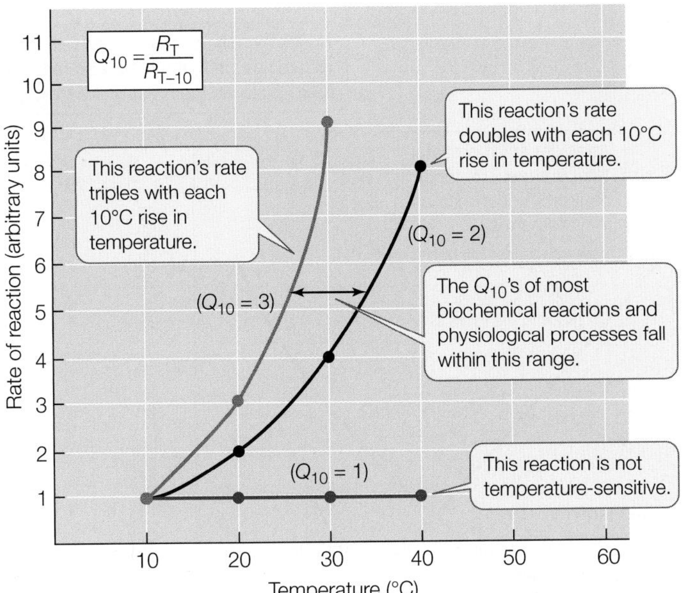

$Q_{10} = \frac{R_T}{R_{T-10}}$
This reaction's rate triples with each 10°C rise in temperature.
This reaction's rate doubles with each 10°C rise in temperature.
($Q_{10}$ = 3)
($Q_{10}$ = 2)
The $Q_{10}$'s of most biochemical reactions and physiological processes fall within this range.
($Q_{10}$ = 1)
This reaction is not temperature-sensitive.
Rate of reaction (arbitrary units)
Temperature (°C)

**40.8 $Q_{10}$ and Reaction Rate** The larger the $Q_{10}$ of a reaction or process, the faster its rate rises in response to an increase in temperature.

Because different reactions have different $Q_{10}$'s, changes in tissue temperature will shift the rates of some reactions more than others, disrupting the overall network. To maintain homeostasis, organisms must be able to compensate for or prevent changes in body temperature.

### Animals acclimatize to seasonal temperatures

The body temperature of some animals (especially aquatic animals) is coupled to environmental temperature. The body temperature of a fish in a pond, for example, will be the same as the water temperature, which might range from 4°C in winter to 24°C in summer. If we bring that fish into the laboratory in the summer and measure its metabolism at different temperatures, we will demonstrate a $Q_{10}$ relationship and can predict what the fish's metabolic rate will be in its pond in the winter. However, if we bring that fish back into the laboratory in the winter and measure its metabolic rate at winter pond temperature, we will find the rate to be much higher than we predicted. The fish's biochemistry and physiology have *acclimatized* to the seasonal change in water temperature so that the fish can remain active at winter temperatures. For example, it may express isozymes with different temperature optima. The ability to acclimatize means that metabolic functions are less sensitive to long-term changes in temperature than to short-term changes.

## 40.2 RECAP

**Cells can survive only within a narrow range of temperatures, but even changes within that range can be disruptive because different physiological processes have different temperature sensitivities.**

- Plot a $Q_{10}$ = 2.5 curve for a physiological process. **See p. 838 and Figure 40.8**
- Explain how a change in body temperature can disrupt physiological processes. **See pp. 838–839**

We have seen how animals are affected by the temperature of their environment. Now let's take a look at the adaptations that allow animals to control and regulate their body temperatures.

## 40.3 How Do Animals Alter Their Heat Exchange with the Environment?

Many of us learned to think of animals as being either "cold-blooded" or "warm-blooded," which implies a comparison with our own body temperature and sets mammals and birds apart from other animals. This simple classification breaks down when we realize that mammals that hibernate become cold, and that many reptiles and insects can be quite warm when they are active. Physiologists sometimes classify animals according to whether they have a constant body temperature (homeotherms) or a variable body temperature (poikilotherms). But a deep-sea fish has a constant body temperature. Should it be classified with mammals?

A thermal classification system that avoids such irrational results is one based on the source of heat that predominantly determines the temperature of the animal. **Ectotherms** are animals whose body temperatures are determined primarily by external sources of heat. **Endotherms** regulate their body temperatures by producing heat metabolically or by using active mechanisms of heat loss.

Mammals and birds are endotherms most of the time; other animals are ectotherms most of the time. Like the homeotherm/poikilotherm classification, the endotherm/ectotherm scheme is not perfect. Therefore we have a third category; a **heterotherm** is an animal that behaves sometimes as an endotherm and other times as an ectotherm. For example, a mammal that hibernates is a perfect endotherm over the summer, but during the winter it has bouts of hibernation during which its internal heat production falls and it behaves much like an ectotherm. At times some ectotherms can produce internal heat and act like endotherms.

### Endotherms produce heat metabolically

Section 8.1 described how transfers of energy in biological systems are always inefficient. With every transfer of energy—from food molecules to ATP, from ATP to biological work—some of the energy is lost as heat. This is true for both ectotherms and endotherms, so why do endotherms produce more heat? The answer is that the cells of endotherms are less efficient at using energy than are the cells of ectotherms.

In a resting endotherm, most of the energy expended goes into pumping ions across membranes. $K^+$ is the dominant positive ion inside cells, and $Na^+$ is the dominant positive ion outside cells. To the extent that cell membranes permit, these ions diffuse down their concentration gradients. To maintain their proper concentrations inside and outside cells, the ions must be transported back "uphill," which requires expending energy.

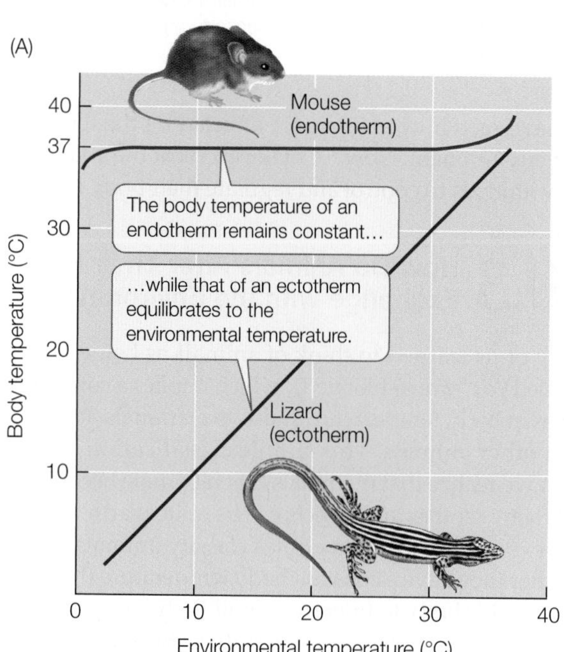


**40.9 Ectotherms and Endotherms React Differently to Environmental Temperatures** (A) At the same environmental temperature, an ectotherm and an endotherm of approximately the same body size (here, a lizard and a mouse) have different body temperatures. (B) The metabolic rates of the lizard and mouse react in opposite manners to cooler temperatures. (The mouse's metabolic rate rises again at higher temperatures because, after a certain point, it takes metabolic energy to dissipate heat by sweating or panting.)

While this is true for both ectotherms and endotherms, the cells of endotherms tend to be more "leaky" to ions than are those of ectotherms. Thus endotherms must expend more energy (and thus release more heat) than do ectotherms to maintain ion concentration gradients. This is akin to running on a treadmill: the faster the treadmill goes (analogous to leaking ions), the faster you have to run (analogous to pumping ions) to remain in the same position.

We can speculate that a mutation allowing seemingly faulty or leaky ion channels may have led to the evolution of endothermy. Such a mutation in a small ectotherm may have promoted sufficient heat production to allow this ectotherm to remain active for a longer time after the sun went down. Thus, for the first endotherms an entirely new nocturnal world of ecological opportunities opened, one in which there was less competition from similar-sized ectotherms. Major differences between endotherms and ectotherms are their resting metabolic rates—the sum total of all energy expenditures in their bodies when at rest—and their responses to changes in environmental temperature.

## Ectotherms and endotherms respond differently to changes in temperature

Let's compare how two similar-sized animals, a lizard (an ectotherm) and a mouse (an endotherm) respond to changes in temperature. We put each animal in a closed chamber and measure its body temperature and metabolic rate as we change the temperature of the chamber from 37°C to 0°C.

The body temperature of the lizard equilibrates with that of the chamber, whereas the body temperature of the mouse remains stable (**Figure 40.9A**). The metabolic rate of the lizard (already much lower than that of the mouse) decreases as the temperature is lowered (**Figure 40.9B**). In contrast, the mouse's metabolic rate increases as the chamber temperature falls below 25°C. The increase in the mouse's metabolism produces enough heat to prevent its body temperature from falling. In other words, the mouse can regulate its body temperature by increasing its metabolic rate; the lizard cannot.

This experiment might lead us to conclude that the ectotherm cannot regulate its body temperature, but observations of the lizard in nature do not support this conclusion. In nature, unlike in the laboratory, the lizard's body temperature is sometimes considerably different than the environmental temperature. The desert habitat where the lizard lives can fluctuate by 40°C in a few hours. During its daily activities, however, the lizard maintains a fairly stable body temperature by using behavior to alter its heat exchange with the environment (**Figure 40.10A**). Its behavioral strategies include spending time in a burrow, basking in the sun, seeking shade, climbing vegetation, and changing its orientation with respect to the sun. While the lizard can regulate its body temperature quite well, it does so by behavioral mechanisms rather than by altering its internal metabolic heat production.

Behavioral thermoregulation is not the exclusive domain of ectotherms (**Figure 40.10B**). Endotherms usually select the most comfortable thermal environment possible. They may change posture, orient to the sun, move between sun and shade, and move between still air and moving air, the same as the ectotherm in our field experiment. Examples of more complex

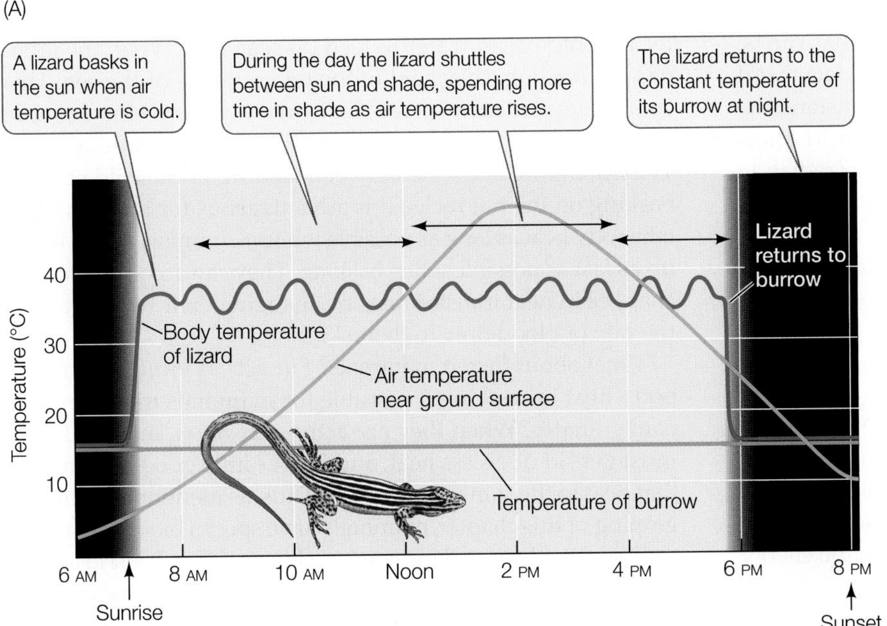


The lizard returns to the constant temperature of its burrow at night.
Lizard returns to burrow

(B) *Loxodonta africana*

(C) *Maccaca fuscata*

**40.10 Using Behavior to Regulate Body Temperature** (A) The body temperature of a lizard (an ectotherm) depends on environmental heat, but the lizard can regulate its temperature by moving from place to place within its environment. (B) When air temperatures on the African savanna soar, an elephant (an endotherm) may thermoregulate by showering itself with water. (C) Japanese macaques are social primates and will huddle together for warmth.

thermoregulatory behaviors include nest construction and social behaviors such as huddling. Humans put on or remove clothing and burn fossil fuels to generate the energy to heat or cool buildings.

## Energy budgets reflect adaptations for regulating body temperature

Both ectotherms and endotherms can influence their body temperatures by altering four avenues of heat exchange between their bodies and the environment (**Figure 40.11**):

- **Radiation**: Heat transfers from warmer objects to cooler ones via the exchange of infrared radiation (what you feel when you stand in front of a fire).
- **Convection**: Heat transfers to a surrounding medium such as air or water as that medium flows over a surface (the wind-chill factor).
- **Conduction**: Heat transfers directly when objects of two different temperatures come into contact (think of putting an icepack on a sprained ankle).
- **Evaporation**: Heat transfers away from a surface when water evaporates on that surface (the effect of sweating).

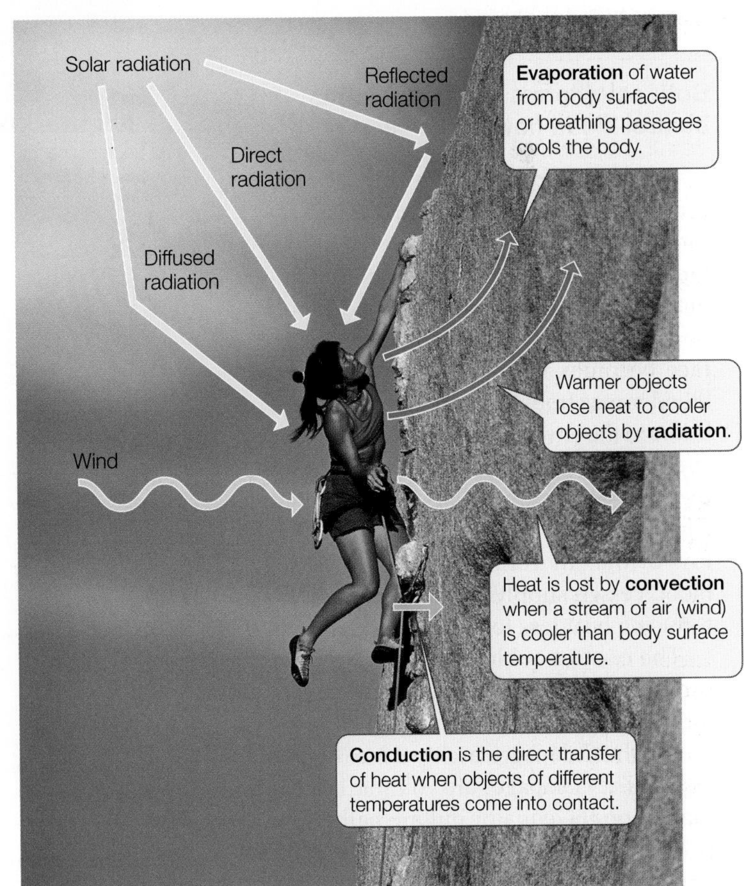

Solar radiation
Reflected radiation
Direct radiation
Diffused radiation
**Evaporation** of water from body surfaces or breathing passages cools the body.
Warmer objects lose heat to cooler objects by **radiation**.
Wind
Heat is lost by **convection** when a stream of air (wind) is cooler than body surface temperature.
**Conduction** is the direct transfer of heat when objects of different temperatures come into contact.


**40.11 Animals Exchange Heat with the Environment** An animal's body temperature is determined by the balance between internal heat production and four avenues of heat exchange with the environment: radiation, convection, conduction, and evaporation.

The total balance of heat production and heat exchange can be expressed as an **energy budget**, based on the simple fact that if the body temperature of an animal is to remain constant, the heat entering the animal must equal the heat leaving it. The heat coming in usually comes from metabolism and solar radiation ($R_{abs}$, for radiation absorbed). Heat leaves the body via the four mechanisms listed above: radiation emitted ($R_{out}$), convection, conduction, and evaporation. The energy budget takes the following form:

$$\overbrace{\text{metabolism} + R_{abs}}^{\text{heat}_{\text{in}}} = \overbrace{R_{out} + \text{convection} + \text{conduction} + \text{evaporation}}^{\text{heat}_{\text{out}}}$$

Anyone who has experienced a very hot environment knows that heat can also *enter* the body through convection (e.g., the hot desert wind) and conduction (e.g., a hot car seat). In that case, the values of those factors become negative in the energy budget equation.

The energy budget is a useful concept because any adaptation that influences the ability of an animal to deal with its thermal environment must affect one or more components of the budget. So the energy budget gives us the ability to quantify and compare the thermal adaptations of animals. One interesting observation is that all of the components on the right side of the energy budget equation—that is, the heat-loss side—depend on the surface temperature of the animal. One way surface temperature can be controlled is by altering the flow of blood to the skin.

### Both ectotherms and endotherms control blood flow to the skin

Heat exchange between the internal environment and the skin occurs largely through blood flow. As described at the beginning of this chapter, when body temperature rises because of exercise, blood flow to the skin increases, and the skin surface becomes warm. The heat that the blood brings from the body core to the skin is lost to the environment through the four avenues listed above, which helps bring the body temperature back to normal. In contrast, when body temperature is too low or the environment is too cold, the blood vessels supplying the skin constrict, reducing heat loss to the environment.

The control of blood flow to the skin can be an important adaptation for an ectotherm such as the marine iguana (a reptile) of the Galápagos archipelago (**Figure 40.12**). The Galápagos are volcanic islands that lie on the equator but are bathed by cold ocean currents. The iguanas bask on hot black lava rocks on the shore, then enter the cold ocean water to feed on seaweed. When the iguanas are feeding, they cool to the temperature of the sea. This cooling lowers their metabolism, making them slower, more vulnerable to predators, and incapable of efficient digestion. They therefore alternate between feeding in the cold sea and basking on the hot rocks. It is advantageous for iguanas to retain body heat as long as possible while swimming and to warm up as fast as possible when basking. They can accomplish these changes in heat transfer rates by changing their heart rate and the rate of blood flow to their skin.

What about furred mammals? Fur acts as *insulation* to keep body heat in, making it possible for mammals to live in very cold climates. When they are active, however, mammals still must get rid of excess heat, and it does little good to transport that heat to the skin under the fur. Thus, as mentioned at the beginning of this chapter, mammals have special blood vessels for transporting heat to their hairless skin surfaces. Heat loss from these areas of skin is tightly controlled by the opening and closing of these special blood vessels. When you are cold, the blood flow to your hands and feet decreases and they can feel very cold, but when you exercise, these same surfaces can get very hot quickly.

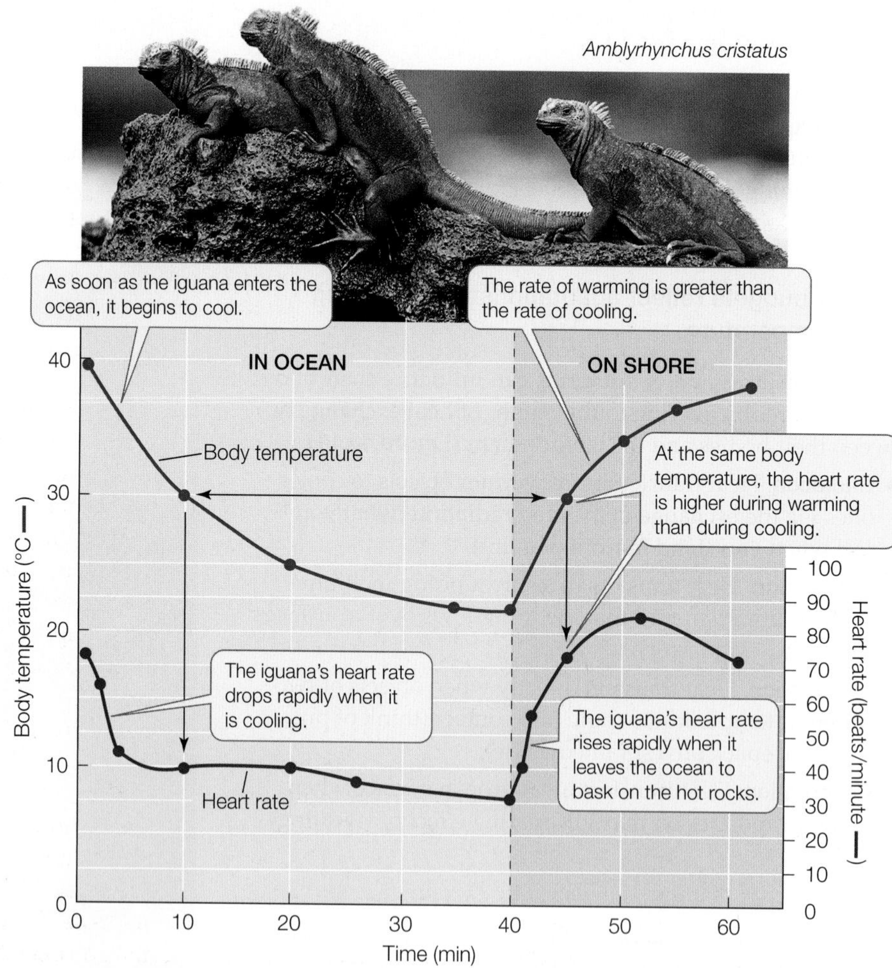
Amblyrhynchus cristatus
As soon as the iguana enters the ocean, it begins to cool.
The rate of warming is greater than the rate of cooling.
IN OCEAN
ON SHORE
Body temperature
At the same body temperature, the heart rate is higher during warming than during cooling.
The iguana's heart rate drops rapidly when it is cooling.
The iguana's heart rate rises rapidly when it leaves the ocean to bask on the hot rocks.
Heart rate
Body temperature (°C —)
Heart rate (beats/minute —)
Time (min)
0
10
20
30
40
50
60
100
90
80
70
60
50
40
30
20
10

**40.12 Some Ectotherms Regulate Blood Flow to the Skin** Galápagos marine iguanas control blood flow to the skin to alter their heating and cooling rates.

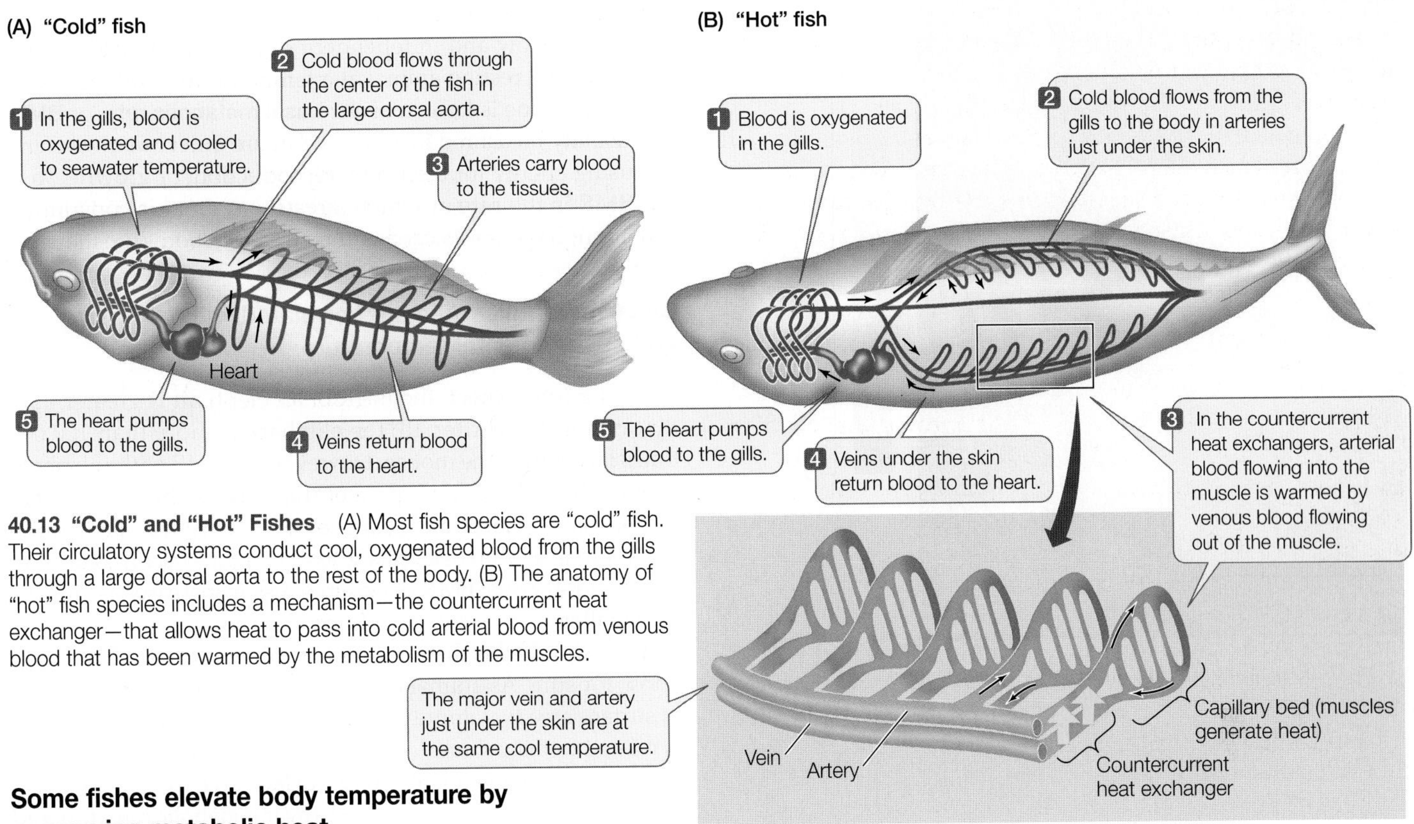
(A) "Cold" fish
1 In the gills, blood is oxygenated and cooled to seawater temperature.
2 Cold blood flows through the center of the fish in the large dorsal aorta.
3 Arteries carry blood to the tissues.
Heart
4 Veins return blood to the heart.
5 The heart pumps blood to the gills.
(B) "Hot" fish
1 Blood is oxygenated in the gills.
2 Cold blood flows from the gills to the body in arteries just under the skin.
3 In the countercurrent heat exchangers, arterial blood flowing into the muscle is warmed by venous blood flowing out of the muscle.
4 Veins under the skin return blood to the heart.
5 The heart pumps blood to the gills.
The major vein and artery just under the skin are at the same cool temperature.
Vein
Artery
Capillary bed (muscles generate heat)
Countercurrent heat exchanger

**40.13 "Cold" and "Hot" Fishes** (A) Most fish species are "cold" fish. Their circulatory systems conduct cool, oxygenated blood from the gills through a large dorsal aorta to the rest of the body. (B) The anatomy of "hot" fish species includes a mechanism—the countercurrent heat exchanger—that allows heat to pass into cold arterial blood from venous blood that has been warmed by the metabolism of the muscles.

### Some fishes elevate body temperature by conserving metabolic heat

Active fishes can produce substantial amounts of metabolic heat, but they have difficulty retaining any of that heat. Blood pumped from the heart goes directly to the gills, where it comes very close to the surrounding water to exchange respiratory gases. So any heat that the blood picks up from metabolically active muscles is lost to the surrounding water as it flows through the gills. It is thus surprising that some large, rapidly swimming fishes, such as bluefin tuna and great white sharks, can maintain temperature differences as great as 10° to 15°C between their bodies and the surrounding water. The heat comes from their powerful swimming muscles, and the ability of these "hot" fishes to conserve that heat is based on the remarkable arrangements of their blood vessels.

In the usual ("cold") fish circulatory system, oxygenated blood from the gills collects in a large dorsal vessel, the aorta, which travels through the center of the fish, distributing blood to all organs and muscles (**Figure 40.13A**). "Hot" fishes have a smaller central dorsal aorta, and most of their oxygenated blood is transported in large vessels just under the skin (**Figure 40.13B**). The cold blood from the gills is thus kept close to the surface of the fish. Smaller vessels transporting this cold blood into the muscle mass run parallel to vessels transporting warm blood from the muscle mass back toward the heart. Since the vessels carrying the cold blood into the muscle are in close contact with the vessels carrying warm blood away, heat flows from the warm to the cold blood by conduction and is therefore retained in the muscle mass.

Because heat is exchanged between blood vessels carrying blood in opposite directions, this adaptation is called a **countercurrent heat exchanger**. It keeps the heat within the muscle mass, enabling these fishes to have an internal body temperature considerably higher than the water temperature. Why is it advantageous for the fish to be warm? Each 10°C rise in muscle temperature increases the fish's sustainable power output almost threefold, giving it a faster foraging capability!

### Some ectotherms regulate heat production

Some ectotherms raise their body temperature by producing heat. For example, the powerful flight muscles of many insects must reach 35° to 40°C before the insects can fly, and they must maintain these high temperatures during flight. Such insects produce the required heat by contracting their flight muscles in a manner analogous to shivering in mammals. The heat-producing ability of insects can be quite remarkable. Probably the most impressive case is a species of scarab beetle that lives mostly underground in mountains north of Los Angeles, California. To mate, these beetles come aboveground, and males fly in search of females. They undertake this mating ritual at night, in winter, and only during snowstorms.

Honey bees regulate temperature as a group. They live in large colonies consisting mostly of female worker bees that maintain the hive and rear the larval offspring of the single queen bee. During winter, worker bees cluster around the brood of larvae. They adjust their individual metabolic heat production and density of clustering so that the brood temperature remains remarkably constant, at about 34°C, even as the outside air temperature drops below freezing (**Figure 40.14**).

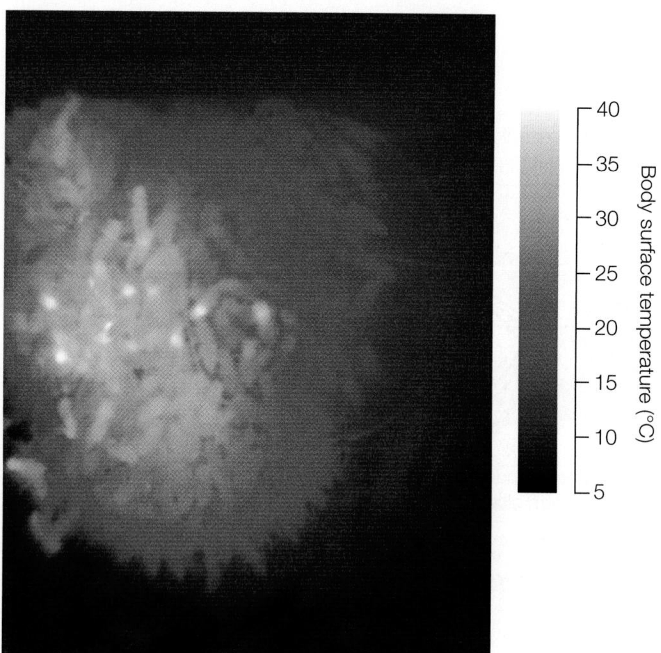


**40.14 Bees Keep Warm in Winter** Honey bee colonies survive winter cold because workers generate metabolic heat. In this infrared photograph of the center of an overwintering hive, individual bees are discernible by the heat their bodies produce as they cluster around their queen.

## 40.3 RECAP

**Animals that metabolically produce their own heat are called endotherms. Those that depend on environmental sources of heat are called ectotherms. Heat exchange between an animal and its environment occurs via radiation, convection, conduction, and evaporation.**

- In terms of the energy budget relationship, why is the control of blood flow to the skin so important for thermoregulation? See p. 842
- Explain how countercurrent heat exchange makes it possible for some fishes to have a body temperature higher than that of the surrounding water. See p. 843 and Figure 40.13

Endotherms must keep their body temperatures within a critical physiological range. Let's look more closely at the evolutionary adaptations that enable endothermic mammals to maintain this optimal temperature range.

# 40.4 How Do Mammals Regulate Their Body Temperatures?

As we saw in Figure 40.9, endotherms can respond to changes in environmental temperature by changing their metabolic rate. Physiologists determine metabolic rate by measuring the rate at which an animal consumes $O_2$ and produces $CO_2$. Within a narrow range of environmental temperatures, called the **thermoneutral zone** (see Figure 40.9B), the metabolic rate of endotherms is low and independent of temperature. The metabolic rate of a resting animal at a temperature within the thermoneutral zone is known as the **basal metabolic rate**, or **BMR**. It is usually measured in animals that are quiet but awake and not using energy for digestion, reproduction, or growth. Thus the BMR is the rate at which a resting animal is consuming just enough energy to carry out its minimal body functions.

### Basal metabolic rates are correlated with body size and environmental temperature

As you might expect, the BMR of an elephant is greater than that of a mouse. After all, the elephant is more than 100,000 times larger than the mouse. However, the BMR of the elephant is only about 7,000 times greater than that of the mouse. That means that a gram of mouse tissue uses energy at a rate 15 times greater than a gram of elephant tissue (**Figure 40.15**). Across all of the endotherms, BMR per gram of tissue increases as animals get smaller.

Why should this disproportionate difference exist? We don't know for sure. As animals get bigger, they have a smaller ratio of surface area to volume (see Figure 5.2). Since heat production is related to the volume, or mass, of the animal, but its capacity to dissipate heat is related to its surface area, it was once reasoned that larger animals evolved lower metabolic rates to avoid overheating. This explanation is insufficient because the relationship between body mass and metabolic rate holds for even very small organisms and for ectotherms, in which overheating is not a problem. Other hypotheses have also been advanced. For example, a larger animal has a greater proportion of support tissue (skin, bone), which is not as metabolically active. The real answer is probably a mixture of different causative factors.

In an endotherm, the metabolic rate versus environmental temperature curve represents the integrated response of all of the animal's thermoregulatory adaptations (**Figure 40.16**). The

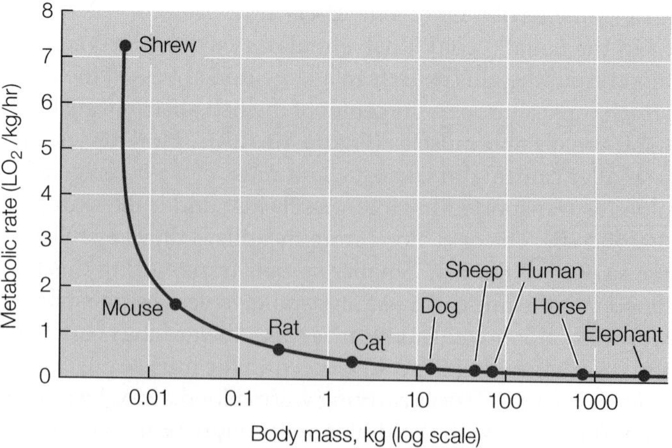


**40.15 The Mouse-to-Elephant Curve** On a weight-specific basis, the metabolic rate of small endotherms is much greater than that of larger endotherms. This graph plots $O_2$ consumption per kilogram of body mass (a measure of metabolic rate) against a logarithmic plot of body mass.

Body temperature

Within the **thermoneutral zone**, body temperature is regulated by passive mechanisms, such as heat loss through the skin.

Metabolic rate

Basal metabolic rate (BMR)

Lower critical temperature

Upper critical temperature

Ambient temperature (°C)

Below the lower critical temperature, the animal expends energy to produce metabolic heat.

Above the upper critical temperature, the animal must expends energy to lose heat by panting or sweating.

**40.16 Environmental Temperature and Mammalian Metabolic Rates** Outside the thermoneutral zone, maintaining a constant body temperature requires expending energy. Outside extreme limits (0°C and 40°C in this instance), the animal cannot maintain its body temperature and dies.

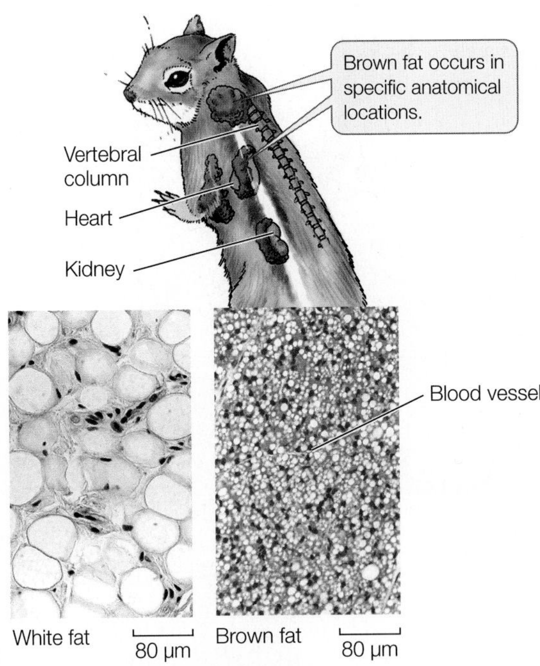


**40.17 Brown Fat** In many mammals, specialized brown fat tissue produces heat. When looking through a microscope at similar magnifications, we see that white fat cells (left) contain large droplets of lipid but have few organelles and limited blood supply, whereas brown fat cells (right) are packed with mitochondria and richly supplied with blood.

thermoneutral zone is bounded by a *lower* and an *upper critical temperature*. Within its thermoneutral zone, an endotherm's thermoregulatory adaptations do not require much energy and could be considered passive; such adaptations include changing posture, fluffing fur, and controlling blood flow to the skin. Outside its thermoneutral zone, however, an endotherm's thermoregulatory responses are active and require considerable metabolic energy, as shown on the left side of Figure 40.16.

**yourBioPortal.com**

GO TO **Web Activity 40.1 • Thermoregulation in an Endotherm**

## Endotherms respond to cold by producing heat and adapt to cold by reducing heat loss

When environmental temperatures fall below the lower critical temperature, endotherms must produce heat to compensate for the heat they lose to the environment. Mammals can accomplish this in two ways: shivering and nonshivering heat production. Birds use only shivering heat production.

*Shivering* uses the contractile machinery of skeletal muscles to consume ATP without causing large movements. Shivering muscles pull against each other so that little movement other than a tremor results. The energy from the conversion of ATP to ADP in this process is released as heat. "Shivering heat production" is perhaps too narrow a term, however; increased muscle tone and increased body movements also contribute to increased heat production in cold environments.

Most *nonshivering heat production* occurs in a specialized adipose tissue called **brown fat** (**Figure 40.17**). This tissue looks brown because of its abundant mitochondria and rich blood supply. In brown fat cells, a protein called *thermogenin* uncouples proton movement from ATP production, allowing protons to leak across the inner mitochondrial membrane rather than having to pass through the ATP synthase and generate ATP (review the discussion of the chemiosmotic mechanism and Figure 9.9). As a result, metabolic fuels are consumed without producing ATP, but heat is still released. Brown fat is abundant in newborns of many mammalian species (including humans), in some adult mammals that are small and acclimatized to cold, and in mammals that hibernate.

In spite of their ability to produce heat, endotherms in cold climates have evolved adaptations to reduce their heat loss and therefore remain within their thermoneutral zones as much as possible. Heat is lost from the body surface, and cold-climate species have anatomical adaptations that give them smaller surface-to-volume ratios than their warm-climate relatives. These adaptations include rounder body shapes and shorter appendages (**Figure 40.18**).

The most significant means of decreasing heat loss is to increase thermal insulation. Animals adapted to cold climates have much thicker layers of fur, feathers, or fat than do their warm-climate relatives. Fur and feathers are good insulators because they trap a layer of still, warm air close to the skin surface. If that air is displaced by water, insulation is drastically reduced. In many species, oil secretions spread through fur or

**40.18 Adaptations to Hot and Cold Climates** (A) The antelope jackrabbit is found in the Sonoran Desert of Arizona. Its large ears serve as heat exchangers, passing heat from the animal's blood to the surrounding air. (B) The thick fur of the Arctic hare provides insulation in the frigid winter. This species' ears and extremities are smaller than those of the jackrabbit.

feathers by grooming are critical for resisting wetting and maintaining a high level of insulation.

Decreasing blood flow to the skin is an important thermoregulatory adaptation in the cold. Constriction of blood vessels in the skin, and especially in the appendages, greatly improves an animal's ability to conserve heat. Countercurrent heat exchange like we saw in the "hot" fishes is also an important adaptation in the appendages of endotherms. Blood flowing out to the paw of a wolf, the hoof of a caribou, or the foot of a bird parallels the flow of the blood returning. Heat is transferred from the outgoing to the returning blood, thus retaining heat in the animal's core.

### Evaporation of water can dissipate heat, but at a cost

As environmental temperature rises within an endotherm's thermoneutral zone, the animal dissipates more of its metabolic heat by increasing blood flow to the skin. When the temperature exceeds the upper critical temperature, however, overheating becomes a problem. For an exercising animal, overheating can occur even at low environmental temperatures. Large mammals, especially those in hot habitats such as elephants, rhinoceroses, and water buffaloes, have little or no insulating fur and seek out water to wallow in when the air temperature is high (see Figure 40.10B). Having water in contact with the skin greatly increases heat loss because the heat-absorbing capacity of water is much greater than that of air.

Evaporation from external or internal body surfaces through sweating or panting can also cool an endotherm. A gram of water absorbs about 580 calories of heat when it evaporates. If this evaporation occurs on the skin, most of that heat is absorbed from the skin and the underlying blood. Sweat and saliva that fall off of the body, however, provide no cooling. Thus when the need for heat loss is greatest, water from the internal environment can be squandered with no cooling benefit. Water is heavy, so animals do not carry an excess supply of it, and many hot environments are also arid. In habitats that are both hot and dry, sweating and panting are cooling adaptations of last resort.

Sweating and panting are *active* processes that require expending metabolic energy. That is why the metabolic rate increases when the upper critical temperature is exceeded (see Figure 40.16). A sweating or panting animal is generating heat in the process of dissipating heat, which can be a losing battle.

### The mammalian thermostat uses feedback information

The thermoregulatory mechanisms and adaptations we have just discussed work through a regulatory system that integrates information from environmental and physiological sources and then issues commands that control body temperature. Such a regulatory system is based on feedback information, and can be thought of as a thermostat like the one in your home.

The major thermoregulatory integrative center of mammals is at the bottom of the brain in a structure called the **hypothalamus**. If you slide your tongue back as far as possible along the roof of your mouth, it will be just a few centimeters below your hypothalamus. The hypothalamus is a key part of many regulatory systems, including thermoregulation in all vertebrates.

In many vertebrates and all mammals, the temperature of the hypothalamus itself is the major negative feedback signal, and damage to the hypothalamus can disrupt thermoregulation. The hypothalamus generates a set point like a setting on a home thermostat. When the temperature of the hypothalamus exceeds or drops below that set point, thermoregulatory responses (the controlled system) are activated to reverse the direction of temperature change.

In mammals, experiments show that directly cooling the hypothalamus increases metabolic heat production and stimulates constriction of the blood vessels that supply the skin, thus causing body temperature to rise. Conversely, mild warming of the hypothalamus stimulates dilation of the blood vessels, while stronger hypothalamic heating stimulates sweating or panting. Consequently, heating the hypothalamus causes the overall body temperature to fall (**Figure 40.19**).

yourBioPortal.com
GO TO Animated Tutorial 40.1 • The Hypothalamus

The mammalian thermoregulatory system has adjustable set points and integrates sources of information in addition to hypothalamic temperature. For example, temperature sensors in the skin register environmental temperature; change in skin tem-

INVESTIGATING LIFE

**40.19 The Hypothalamus Regulates Body Temperature** In this laboratory experiment, a mammal's hypothalamus was subjected directly to temperature manipulation. The body's responses to the experimenters' manipulations were as expected if the hypothalamus is indeed the mammalian "thermostat."

**HYPOTHESIS** Heating or cooling the mammalian hypothalamus results in corresponding and predictable changes in body temperature.

**METHOD** 1. Implant a probe that directly heats or cools the hypothalamus.

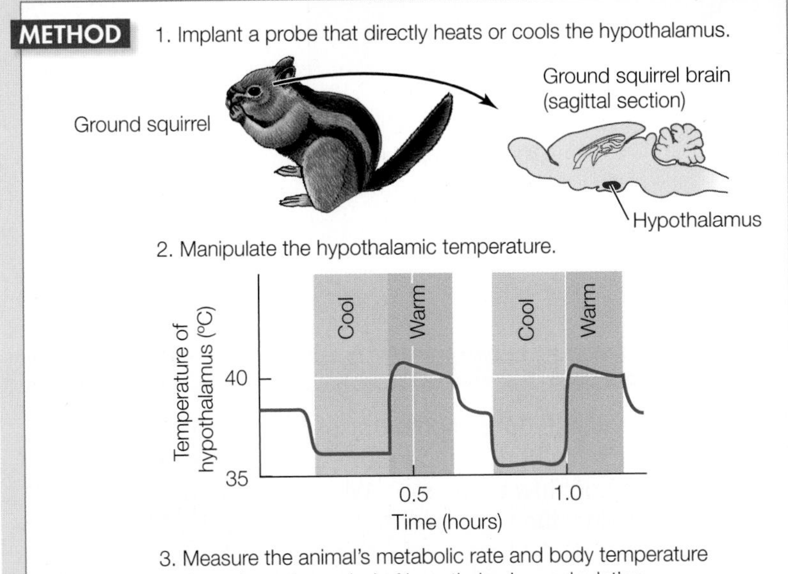


3. Measure the animal's metabolic rate and body temperature throughout the period of hypothalamic manipulation.

**RESULTS**

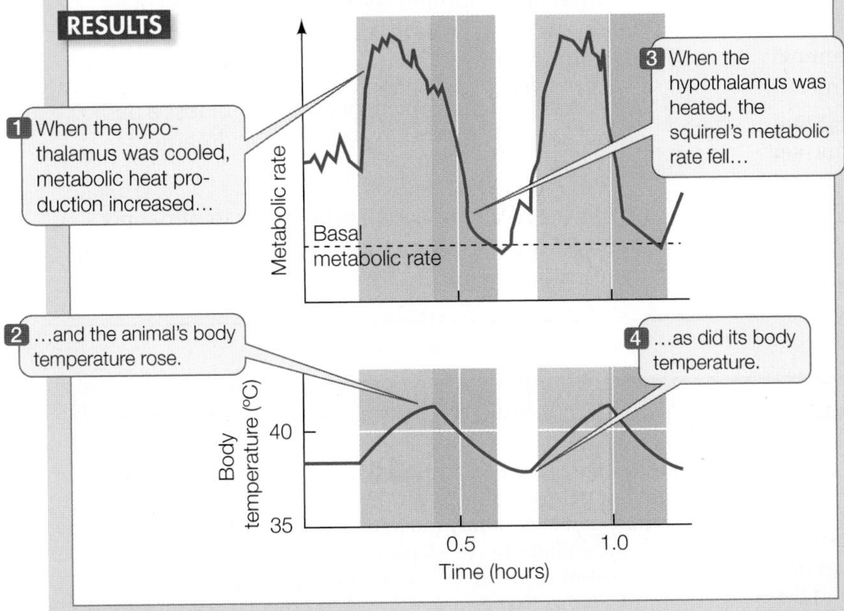


**CONCLUSION** The ground squirrel's hypothalamus acts as a thermostat. When cooled it activates metabolic heat production; when warmed, it suppresses metabolic heat production and favors heat loss.

Go to **yourBioPortal.com** for original citations, discussions, and relevant links for all INVESTIGATING LIFE figures.

perature is feedforward information that shifts the hypothalamic set point for thermoregulatory responses. The set point for metabolic heat production is higher when skin is cold and lower when skin is warm.

Hypothalamic set points are higher during wakefulness than during sleep, and they are higher during the active part of the daily cycle than during the inactive part, even if the animal is awake at both times. Even when an endotherm is kept under constant environmental conditions, its body temperature displays a daily cycle of changes in set point. This kind of cycle is controlled by an internal *circadian rhythm;* we discuss these endogenous bodily rhythms in Chapter 54.

### Fever helps the body fight infections

Fever is an adaptive response that helps the body fight pathogens. A fever is a rise in body temperature in response to molecules called **pyrogens**. *Exogenous pyrogens* come from foreign substances such as bacteria or viruses that invade the body. *Endogenous pyrogens* are produced by cells of the immune system in response to infection.

The presence of a pyrogen in the body causes a rise in the hypothalamic set point for the metabolic heat production response. As a result, you shiver, put on a sweater, or crawl under a blanket, and your body temperature rises until it matches the new set point. At the higher body temperature you no longer feel cold, and you may not feel hot, but someone touching your forehead will say that you are "burning up." Taking aspirin lowers your set point to normal. Now you feel hot, take off clothes, and even sweat until your elevated body temperature returns to normal. Although modest fevers help the body fight infections, extreme fevers can be dangerous and must be controlled, usually with fever-reducing drugs.

### Turning down the thermostat

**Hypothermia** is a below-normal body temperature. It can result from starvation (lack of metabolic fuel), exposure to extreme cold, serious illness, or anesthesia. In each of these cases, the drop in body temperature is unregulated. However, many birds and mammals undergo regulated drops in body temperature to survive periods of cold and food scarcity, an adaptation known as *regulated hypothermia.*

Hummingbirds, for example, are very small endotherms with a high metabolic rate. Just getting through a single day without food could exhaust their metabolic reserves. Hummingbirds and other small endotherms can extend the pe-

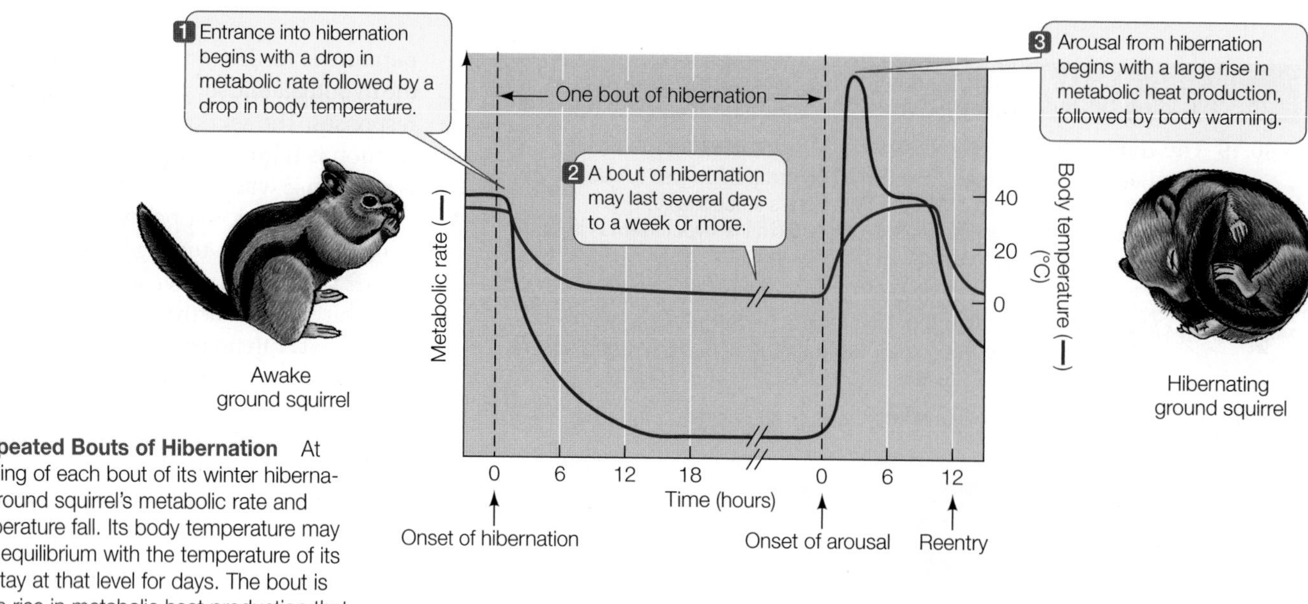


**40.20 Repeated Bouts of Hibernation** At the beginning of each bout of its winter hibernation, the ground squirrel's metabolic rate and body temperature fall. Its body temperature may come into equilibrium with the temperature of its nest and stay at that level for days. The bout is ended by a rise in metabolic heat production that returns body temperature to a normal level.

riod over which they can survive without food by dropping their body temperature during the portion of day or night when they are normally inactive. This adaptive hypothermia is called **daily torpor**. Body temperature can drop 10° to 20°C during daily torpor, lowering metabolic rate and saving energy.

Regulated hypothermia that lasts for days or even weeks, with body temperature falling close to the ambient temperature, is called **hibernation** (**Figure 40.20**). Many species of mammals, including bats, bears, and ground squirrels, hibernate, but only one species of bird (the poorwill) has been shown to hibernate. The metabolic rate needed to sustain a hibernating animal may be only one-fiftieth its basal metabolic rate, and many animals maintain body temperatures close to the freezing point. Arousal from hibernation occurs when the hypothalamic set point returns to the normal level for a mammal.

The ability of hibernators to reduce their thermoregulatory set point so dramatically probably evolved as an extension of the set point decrease that accompanies sleep even in nonhibernating species of mammals and birds.

## 40.4 RECAP

**Within the thermoneutral zone, an endotherm controls its body temperature by altering insulation and blood flow to the skin. When the temperature drops below the thermoneutral zone, the animal increases metabolic heat production. Above this zone, it dissipates heat by panting or sweating.**

- Describe how endotherms produce heat. How does heat production change with body size? See p. 844 and Figure 40.15
- Why is dependence on evaporative water loss a dangerous strategy for dealing with hot environments? See p. 846
- What is the nature of negative feedback information and feedforward information used by the mammalian thermostat? See pp. 847–848 and Figure 40.19

## CHAPTER SUMMARY

### 40.1 How Do Multicellular Animals Supply the Needs of Their Cells?

- Multicellular animals provide for the needs of all their cells by maintaining a stable **internal environment**, which consists of the two types of extracellular fluid: the interstitial fluid, and the plasma of blood. Review Figure 40.1
- The regulation of physiological systems is mostly through **negative feedback**. **Feedforward information** functions to change **set points**. Review Figure 40.2
- The four types of **tissues** are assemblages of cells. **Epithelial tissues** provide barriers and have secretory and transport functions. The three types of **muscle tissue** (skeletal, cardiac, and smooth muscle) are able to contract and are the source both voluntary and involuntary movement. **Connective tissues**, including cartilage, bone, adipose tissue, and blood, are supportive tissues made up of cells embedded in an extracellular matrix. **Nervous tissues** process and communicate information; they contain two cell types, neurons and glial cells.

**Organs** are made up of tissues, and most organs contain all four kinds of tissue. Organs are grouped into **organ systems**. Review Figure 40.7

## 40.2 How Does Temperature Affect Living Systems?

- Life is sustained within a narrow range of environmental temperatures. $\mathbf{Q_{10}}$ is a measure of the sensitivity of a life process to temperature. A $Q_{10}$ of 2 means that the reaction rate doubles as temperature increases by 10°C. Review Figure 40.8
- Animals can acclimatize to seasonal changes in temperature through biochemical and physiological adaptations.

## 40.3 How Do Animals Alter Their Heat Exchange with the Environment?

- **Endotherms** can produce considerable metabolic heat to compensate for heat loss to the environment. **Ectotherms** generally do not. Review Figure 40.9
- **Energy budgets** describe all pathways for heat exchange between an organism and its environment. The four avenues of heat exchange are **radiation**, **convection**, **conduction**, and **evaporation**. Review Figure 40.11
- Skin temperature is an important variable, and it can be influenced by blood flow. Circulatory system adaptations such as **countercurrent heat exchange** can conserve metabolic heat. Review Figures 40.12 and 40.13

## 40.4 How Do Mammals Regulate Their Body Temperatures?

- Within the **thermoneutral zone**, mammals have a **basal metabolic rate** (BMR) that scales with body size. Review Figures 40.15 and 40.16, **WEB ACTIVITY 40.1**
- In mammals, control of thermoregulatory effectors relies on commands from a regulatory center in the **hypothalamus**. This thermostat uses its own temperature as a major negative feedback signal, and skin temperature as a feedforward signal. Review Figure 40.19, **ANIMATED TUTORIAL 40.1**

## SELF-QUIZ

1. Which of the following characterizes the protein elastin?
   *a.* It functions predominantly in muscle tissue to resist excess stretching.
   *b.* It is found predominantly in epithelial tissue.
   *c.* It is found in the extracellular matrix of connective tissue.
   *d.* It is the most abundant protein in the body.
   *e.* It is responsible for the elasticity of the long extensions of neurons.
2. If the $Q_{10}$ of the metabolic rate of an animal is 2, then
   *a.* the animal is better acclimatized to a cold environment than if its $Q_{10}$ is 3.
   *b.* the animal is an ectotherm.
   *c.* the animal consumes half as much oxygen per hour at 20°C as it does at 30°C.
   *d.* the animal's metabolic rate is not at basal levels.
   *e.* the animal produces twice as much heat at 20°C as it does at 30°C.
3. Which statement about brown fat is true?
   *a.* It produces heat without producing ATP.
   *b.* It insulates animals acclimatized to cold.
   *c.* It is a major source of heat production for birds.
   *d.* It is found only in hibernators.
   *e.* It provides fuel for muscle cells.
4. Which of the following is the most important and most general characteristic of endotherms adapted to cold climates compared with those adapted to warm climates?
   *a.* Higher basal metabolic rates
   *b.* Higher $Q_{10}$ values
   *c.* Brown fat
   *d.* Greater insulation
   *e.* Ability to hibernate
5. Which of the following would cause a decrease in the hypothalamic temperature set point for metabolic heat production?
   *a.* Entering a cold environment
   *b.* Taking an aspirin when you have a fever
   *c.* Arousing from hibernation
   *d.* Getting an infection that causes a fever
   *e.* Cooling the hypothalamus
6. Mammalian hibernation
   *a.* occurs when animals run out of metabolic fuel.
   *b.* is a regulated decrease in body temperature.
   *c.* is less common than hibernation in birds.
   *d.* can occur at any time of year.
   *e.* lasts for several months, during which body temperature remains close to the environmental temperature.
7. Which of the following is an important difference between an ectotherm and an endotherm of similar body size?
   *a.* An ectotherm has higher $Q_{10}$ values.
   *b.* Only an ectotherm uses behavioral thermoregulation.
   *c.* Only an endotherm can constrict and dilate the blood vessels to the skin to alter heat flow.
   *d.* Only an endotherm can have a fever.
   *e.* At a body temperature of 37°C, an ectotherm has a lower metabolic rate than an endotherm.
8. How would you describe the role of skin temperature in the human thermoregulatory system?
   *a.* It provides feedforward information.
   *b.* It acts as a set point for metabolic heat production.
   *c.* It provides positive feedback information.
   *d.* It provides an error signal.
   *e.* It provides negative feedback information.
9. What is the biggest difference between a "cold" fish such as a trout and a "hot" fish such as a tuna?
   *a.* The temperature of the blood leaving the heart
   *b.* The temperature of the blood entering the gills
   *c.* The arrangement of blood vessels in the gills
   *d.* The temperature of the brain
   *e.* The volume of blood flowing in arteries just under the skin
10. Which of the following statements about the thermoneutral zone is true?
    *a.* Metabolic heat production is variable.
    *b.* Skin blood flow is variable.
    *c.* The environmental temperature equals body temperature.
    *d.* The lower boundary (lower critical temperature) is lower for small than for large endotherms.
    *e.* It is the range of hypothalamic temperatures that do not alter metabolic heat production.

## FOR DISCUSSION

1. What is the advantage of feedforward information for homeostasis? Can you suggest what some sources of feedforward information could be for regulation of breathing, blood pressure, secretion of digestive juices, and elimination of wastes?
2. In some epithelial tissues there are "tight junctions" between the individual cells that prevent anything from passing between them (see Figure 6.7); in other cases the junctions between epithelial cells are quite loose. What are the possible advantages in different organs of loose versus tight junctions between epithelial cells? Give some examples in which these differences would be important.
3. Newton's law of cooling describes how a physical object comes into thermal equilibrium with its environment. The law is expressed

   $$HL = K(T_o - T_a)$$

   *HL* is the rate of heat loss, K is the thermal conductance constant (how easily an object loses heat), $T_o$ is the temperature of the object, and $T_a$ is the ambient temperature. Compare this expression with the metabolic rate/temperature curve for endotherms. In Newton's law of cooling, K is a constant reflecting the properties of the object. What would K represent for an endotherm? Using a version of Newton's law that replaces $T_o$ with $T_b$ (body temperature), explain why the metabolic rate curve projects to zero at an ambient temperature that equals body temperature.
4. The range of temperatures compatible with life is about 0°C to 40°C. Endotherms have regulated body temperatures much closer to the upper limit of this range than to the lower. What are the advantages of living so close to the upper limit?
5. Discuss what it means when we say that the metabolic rate of mammals scales to the ¾ power of body mass. In contrast, heart size of mammals scales according to the first power of body mass. What does this difference imply for the functions of the hearts of mammals of different sizes?

## ADDITIONAL INVESTIGATION

1. The text described the drop in body temperature of a hibernator as regulated hypothermia—a turning down of the thermostat. Yet we also saw that if we put an ectotherm in a cold environment, its body temperature will fall. What experiment could you do to prove that the mammalian hibernator did not just simply turn off or inactivate its thermoregulatory system in order to behave like an ectotherm in the cold?
2. The observations on the Galápagos marine iguana showed that its body temperature rose faster in air than it fell in water. The inference was that the iguana was influencing its gain or loss of heat by altering the blood flow to its skin. However, the thermal properties of air and water are different, and in the case described in the text, the animal was breathing when in air but not when diving in the water. What experiment could you do to strengthen the argument that the iguana was actively altering the flow of heat across its skin?

## WORKING WITH DATA (GO TO yourBioPortal.com)

**A Hibernator's Thermostat** In this exercise based on the experiments outlined in Figure 40.19, you will plot data gathered from hibernating and non-hibernating ground squirrels to graph the relationship between hypothalamic temperature and metabolic rate. You will then analyze these data and draw conclusions about the role of the hypothalamus in mammalian hibernation.

# 41 Animal Hormones

## Juiced

The use of performance-enhancing drugs—particularly *anabolic steroids*—has become a scandal in athletics. Olympic champions have lost medals, professional athletes have been suspended, coaches have lost their jobs, and exceptional performances have been expunged from the record books. The recent history of baseball in the United States has been termed "the steroid era" because of the huge impact the extensive use of performance-enhancement drugs has had on the game and the controversy this has created. Gains in performance and new records raise the question of whether an aspiring athlete can succeed without using performance-enhancing drugs. The U.S. Congress has passed laws against nonmedical use of steroids, and Major League Baseball has instituted penalties for players who break the law. To date, over 100 Major League players have admitted to or been implicated in the use of steroids, and many have been suspended for up to 80 games.

You have probably heard of one anabolic steroid: testosterone, the male sex hormone. Shortly before puberty, the male reproductive system increases its production of this important chemical signal. Testosterone enters cells, where it binds to receptors and alters gene expression. Cells that have these receptors are those involved in the development of male secondary sexual characteristics, such as a deep voice, facial and body hair, and increased muscle and bone mass. Anabolic steroids are used therapeutically to treat conditions such as delayed puberty, erectile dysfunction, and the loss of muscle mass that occurs with certain diseases.

When a muscle is exercised, an interaction between the exercise and the steroids results in growth of that muscle. Body builders who abuse anabolic steroids typically use them in doses 10 to100 times greater than normal levels or therapeutic doses. The resulting extreme growth of skeletal muscle mass occurs in women as well as men. Both sexes have receptors for testosterone, but women normally have much lower concentrations of testosterone in their blood than men do. When female body builders use these hormones, they develop male muscle patterns. They also develop deep voices and body and facial hair, and because these steroids generate negative feedback

**How Baseball Got Big** Jose Canseco was the American League Rookie of the Year in 1986. In 1988, he became the first major league player to hit 40 home runs and steal 40 bases in a season and was named the American League's Most Valuable Player. In 2005 he wrote a book called *Juiced: Wild Times, Rampant 'Roids, Smash Hits, and How Baseball Got Big* in which he admitted to using anabolic steroids and implicated many other players.

**Anabolic Steroids Build Big Muscles** Anabolic steroids greatly enhance the development of skeletal muscle in response to exercise. Steroids have this effect on women as well as men.

information that controls female reproductive physiology, their breast tissue diminishes, they stop menstruating, and they become infertile. Similar negative feedback in males causes infertility. Behavioral changes—"roid rage"—are also common. The most serious side effects, for both men and women, are greatly increased risks of cancer and of heart, liver, and kidney diseases.

Despite the risks, athletes seeking an advantage have frequently turned to anabolic steroids. Athletic governing organizations administer blood and urine tests to detect their use; in their turn, illicit drug makers constantly seek to design new forms of anabolic steroids that produce the desired physical results but are not detectable.

**IN THIS CHAPTER** we will examine how hormones control and regulate anatomical, developmental, physiological, and behavioral changes in animals. First we examine hormonal control of invertebrate life cycles. Next we discuss the general characteristics of hormones and their receptors. Then we describe the functions, control, and mechanisms of action of mammalian hormones.

## CHAPTER OUTLINE



## 41.1 What Are Hormones and How Do They Work?

In multicellular animals, physiological control and regulation require information and cell-to-cell communication. Most intercellular communication is by means of chemical signals that bind to receptors, as described in Chapter 7. **Hormones** are chemical signals that are released by certain types of cells and that influence the activities of other cells at a distance. In this and subsequent chapters you will learn about hormones and other examples of chemical signals, including growth factors, morphogens, cytokines, and neurotransmitters. These general names come from the context in which the chemical signal operates—endocrine system, growth and development, immune system, or nervous system—but the principles of their function are the same: one cell releases a chemical signal that travels to and binds to a receptor, causing a cellular response.

The information that animals use to develop, grow, and function comes from four major sources: the genome, the endocrine system, the immune system, and the nervous system. In each of these systems, information is encoded in the specificity of chemical signals and their receptors. In earlier chapters we learned a lot about genetic information. In this and following chapters we discuss the endocrine, immune, and nervous systems. Lest you think that all signaling is chemical, however, keep in mind that there are also receptors in the nervous system that encode physical sources of information, such as temperature, pressure, and light. And the nervous system uses electrical signals called action potentials to get information from place to place in the body. Regardless of the system, the processing of information depends on which cells have receptors for the signals and how those cells respond and interact with other cells.

Some analogies might help distinguish how the immune, endocrine, and nervous informational systems operate. The immune system (the topic of Chapter 42) operates like an army of private security guards. The various cellular agents make their rounds of the body, and when they detect a security breach, they sound their alarms—cytokines—which activate the body's defenses. The nervous system (see Chapters 45–47) operates like a telephone system with a central integration and command center that sends signals along specific wires to specific receivers. The endocrine system is more like a radio or TV network that broadcasts signals that can be picked up by anyone who has an appropriate receiver that is turned on and tuned in.

In this chapter we focus on the **endocrine system**, which includes a variety of cells that produce and release hormonal chemical signals into the extracellular fluid.

## Chemical signals can act locally or at a distance

**Endocrine cells** secrete chemical signals; **target cells** have receptors for those signals. Chemicals secreted into the extracellular fluid diffuse locally and may diffuse into the blood. Endocrine signals that enter the blood are called hormones, and they can activate target cells far from their site of release (**Figure 41.1A**). Testosterone is an example of a **hormone**.

Some endocrine signals are released in such tiny quantities, or are so rapidly inactivated by enzymes, or are taken up so efficiently by local cells that they never diffuse into the blood in sufficient amounts to act on distant cells. Because these signals affect only target cells near their release site, they are called **paracrines** (*para*, "near"; **Figure 41.1B**). An example of a paracrine is histamine, one of the mediators of inflammation. The most local action an endocrine signal can have is when it binds to receptors on or in the same cell that secreted it. When a chemical signal influences the cell that secreted it, it has **autocrine** function (**Figure 41.1C**). Hormones and paracrines can have autocrine functions as a means of providing negative feedback to control their rates of secretion.

Some endocrine cells exist as single cells within a tissue. Hormones of the digestive tract, for example, are secreted by isolated endocrine cells in the walls of the stomach and small intestine. Many hormones, however, are secreted by aggregations of endocrine cells forming secretory organs called **endocrine glands**. The name "endocrine" reflects the fact that these glands secrete their products directly into the extracellular fluid, which they pass into the blood. In contrast, **exocrine glands,** such as sweat glands or salivary glands, have ducts that carry their products to the surface of the skin or into a body cavity such as the gut. A single endocrine gland may secrete multiple hormones.

To complete our overview of intercellular chemical communication, we must mention neurotransmitters, which we discuss in detail in Chapters 45–47, and pheromones, which we discuss in Chapter 54. Neurons, the cells of the nervous system, conduct information over long distances as electrical signals, but where a neuron communicates that information to another cell, be it another neuron, a muscle cell, or a secretory cell, it does so by releasing chemical signals called **neurotransmitters**. Most neurotransmitters act very locally and frequently act on the neuron that released them. Some neurotransmitters, however, diffuse into the blood and are therefore referred to as **neurohormones**. **Pheromones** are chemical signals that an animal releases into the environment to communicate information to other individuals of the same species.

## Hormonal communication has a long evolutionary history

Intercellular chemical signaling was critical for the evolution of multicellularity. A protist, the slime mold *Dictyostelium*, uses a chemical signal (cAMP) to coordinate the aggregation of individual cells to form a multicellular fruiting structure (see Figure 27.32). The most primitive of the multicellular animals—the sponges—do not have nervous systems, but they do have intercellular chemical communication. And as discussed in Chapter 37, plant growth is regulated by a variety of hormones.

Studying the evolution of hormonal signaling reveals an interesting generalization: the signal molecules themselves are highly conserved. We find the same chemical compounds over broad groups of organisms, but their functions differ. As organisms have evolved to occupy different environments and have different lifestyles, hormone–receptor systems have evolved to serve different functions—for example, the hormone prolactin (**Figure 41.2**). Another important example is the hormonal control of molting and metamorphosis—critical events in the lives of arthropods, the most diverse animal group on Earth (see Chapter 32). The hormones involved represent an ancient system of hormonal communication that is genetically related to the anabolic steroid system discussed at the opening of this chapter.

### HORMONAL CONTROL OF MOLTING IN ARTHROPODS

Insects, like all arthropods, have a rigid exoskeleton. Therefore, their growth is episodic, punctu-

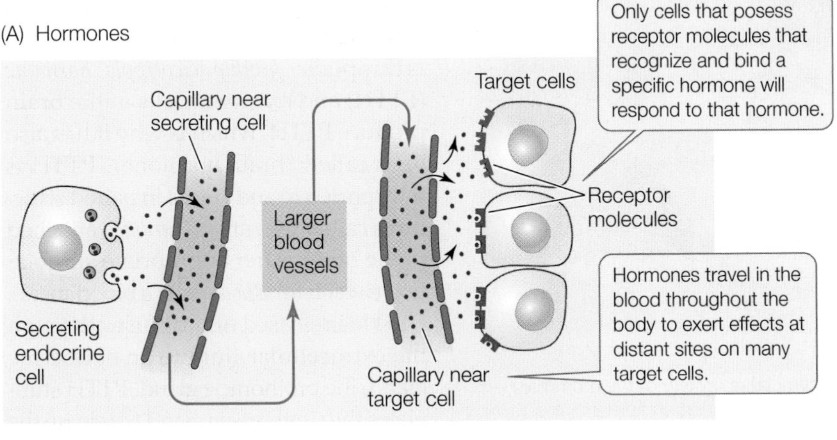
(A) Hormones
Only cells that posess receptor molecules that recognize and bind a specific hormone will respond to that hormone.
Target cells
Capillary near secreting cell
Receptor molecules
Larger blood vessels
Hormones travel in the blood throughout the body to exert effects at distant sites on many target cells.
Secreting endocrine cell
Capillary near target cell

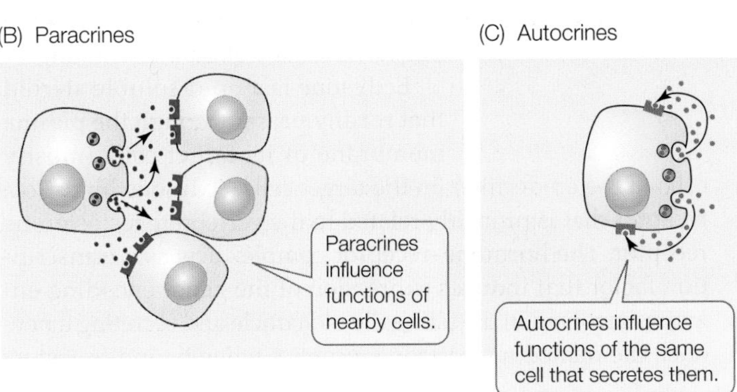
(B) Paracrines
(C) Autocrines
Paracrines influence functions of nearby cells.
Autocrines influence functions of the same cell that secretes them.

**41.1 Chemical Signaling Systems** Hormones (A) are distributed throughout the body by the blood. Paracrines and autocrines do not enter the bloodstream; paracrines (B) simply diffuse to nearby cells, while autocrines (C) influence the cells that release them.

**Fish**
Required for osmoregulation in freshwater species. In saltwater species that return to fresh water to spawn (e.g., salmon), prolactin production in adults may play a role in generating the drive to return to natal streams.

**Amphibians**
Alters the osmoregulatory properties of the skin for animals that enter fresh water. In some species, creates a "water drive" that returns adults to breeding locations. Stimulates oviduct development and production of egg jelly in females. In some species, controls development of sexual characteristics.

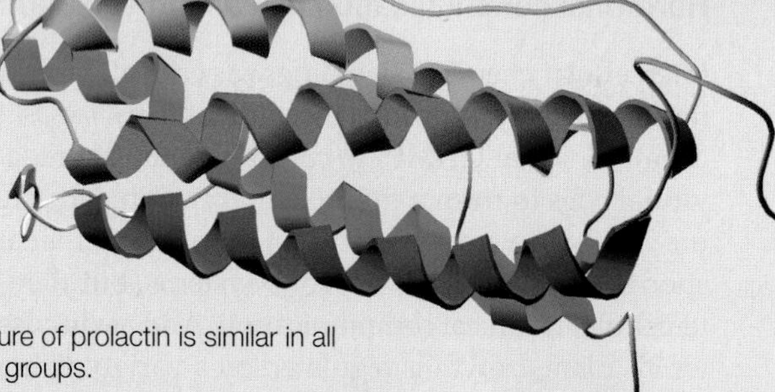

The structure of prolactin is similar in all vertebrate groups.

**Birds**
In some species, stimulates nesting activity, incubation behavior, and parental care in both sexes. Stimulates the epithelial cells of the upper GI tract to proliferate and slough off to form "crop milk" to nourish the young.

**Mammals**
In females, stimulates growth of the mammary glands and milk production. In humans, it is responsible for the sensation of sexual gratification as well as the male refractory period following sexual intercourse.

**41.2 Prolactin's Structure Is Conserved, but Its Functions Have Evolved** The hormone prolactin is found in all vertebrate groups and has a long evolutionary history. Its probable function in early vertebrates was in regulating the body's salt and water balance (osmoregulation). It maintains this function in some species, and has evolved in others to control a number of physiological processes, most of which are associated with reproduction.

ated with *molts* (shedding of the exoskeleton). Each growth stage between two molts is called an *instar*.

The British physiologist Sir Vincent Wigglesworth was a pioneer in the study of the hormonal control of growth and development in insects. Wigglesworth conducted a series of experiments on the bloodsucking bug *Rhodnius prolixus*. Upon hatching, *Rhodnius* looks like a miniature adult but lacks certain adult features. The juvenile bug molts five times before developing into a mature adult; a blood meal triggers each episode of molting and growth.

*Rhodnius* is an amazingly hardy experimental animal—it survives for quite a long time even after its head is cut off. Wigglesworth's studies revealed that, if decapitated within an hour after a blood meal, *Rhodnius* can survive for up to a year, but it never molts. If decapitated a week after its blood meal, however, it does molt. Wigglesworth hypothesized that the time lag meant that the substance that triggers molting diffuses slowly from the head. He tested this hypothesis with the experiment described in **Figure 41.3**.

We now know that two hormones working in sequence regulate molting in arthropods: *prothoracicotropic hormone* (PTTH) and *ecdysone*. Cells in the brain produce PTTH, which is why it has also been called "brain hormone." PTTH is transported to and stored in paired structures called the *corpora cardiaca* attached to the brain. After appropriate stimulation (which for *Rhodnius* is a blood meal), PTTH is released and diffuses through the extracellular fluid to an endocrine gland, the prothoracic gland. PTTH stimulates the prothoracic gland to release the hormone ecdysone. Ecdysone diffuses to target tissues and stimulates molting.

Ecdysone is a lipid-soluble steroid that readily passes through the plasma membrane of its target cells (mostly cells of the epidermis). In the target cells, ecdysone binds to a receptor that is probably related to the vertebrate testosterone receptor. The hormone–receptor complex acts as a transcription factor that induces expression of the genes encoding enzymes involved in digesting the old cuticle and secreting a new one. The related testosterone receptor, when bound to testos-

## INVESTIGATING LIFE

**41.3 A Diffusible Substance Triggers Molting**
The bloodsucking bug *Rhodnius prolixus* develops from hatchling to adult in a series of five molts (instars) that are triggered by ingesting blood. Sir Vincent Wigglesworth's experiments demonstrated that a blood meal stimulates production of some molt-inducing substance in the insect's head.

**HYPOTHESIS** **The substance that controls molting in *R. prolixus* is produced in the head segment and diffuses slowly through the body.**

**OBSERVATION**

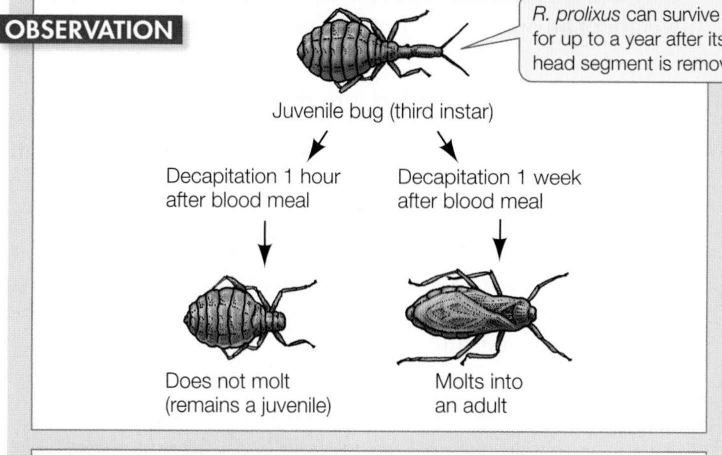
R. prolixus can survive for up to a year after its head segment is removed.
Juvenile bug (third instar)
Decapitation 1 hour after blood meal
Decapitation 1 week after blood meal
Does not molt (remains a juvenile)
Molts into an adult

**METHOD** 1. Decapitate third-instar juveniles at different times after blood meal.

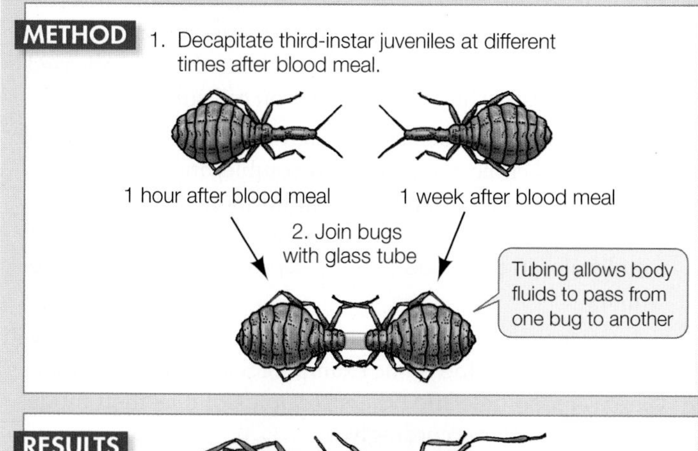
1 hour after blood meal
1 week after blood meal
2. Join bugs with glass tube
Tubing allows body fluids to pass from one bug to another

**RESULTS**

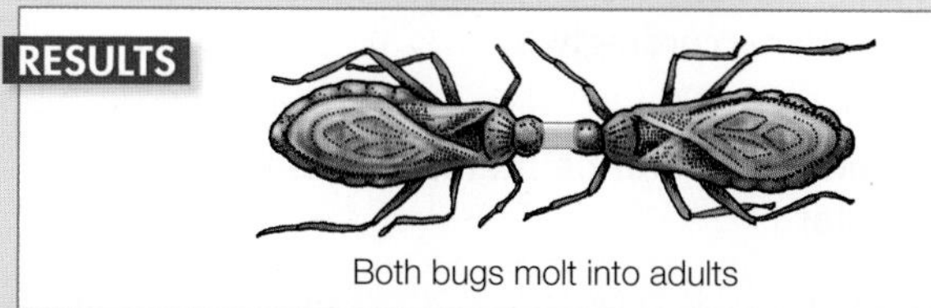
Both bugs molt into adults

**CONCLUSION** **A blood meal stimulates production of some substance within the insect's head that then diffuses slowly through the body, triggering a molt.**

Go to **yourBioPortal.com** for original citations, discussions, and relevant links for all INVESTIGATING LIFE figures.

terone, also plays important roles in development and growth, illustrating the evolutionary conservatism of both the chemical signal and its general domain of function.

The control of molting by PTTH and ecdysone is a general arthropod hormonal control mechanism, and it exemplifies how the endocrine system works with the nervous system to integrate diverse information and induce long-term effects. The nervous system of an arthropod receives various types of information about the animal's environment (such as day length, temperature, social cues, and nutrition) that help determine the optimal timing for the stages of growth and development. When conditions are right, the brain (part of the nervous system) signals the prothoracic gland (part of the endocrine system), which produces the hormone (ecdysone) that orchestrates the physiological processes involved in development and molting. Later in this chapter we will see similar links between the nervous system and endocrine glands in vertebrates.

#### HORMONAL CONTROL OF MATURATION IN ARTHROPODS

Wigglesworth's experiments with *Rhodnius* yielded another curious result: regardless of the instar used, decapitated bugs that molted always molted directly into an adult. Additional experiments by Wigglesworth demonstrated that another hormone determines whether a bug molts into another juvenile instar or into an adult.

Because the head of *Rhodnius* is long, it is possible to remove just the front part of it, which contains the brain, while leaving the rear part intact. When fourth-instar bugs that had had a blood meal a week earlier were partly decapitated in this way, they molted into fifth-star juveniles, not into adults.

This experiment was followed by more experiments using glass tubes to connect individual bugs. When an unfed, completely decapitated fifth-instar bug was connected to a fed, partly decapitated fourth-instar bug (with only the front part of its head removed), both bugs molted into juvenile forms. A substance from the rear part of the head of the fourth-instar bug prevented both bugs from molting into adults.

The substance responsible for preventing maturation is **juvenile hormone**, which is released continuously from the *corpora allata* (structures that are attached to the corpora cardiaca, which release PTTH). As long as juvenile hormone is present, *Rhodnius* molts into another juvenile instar. Normally *Rhodnius* stops producing juvenile hormone during the fifth instar, then molts into an adult.

The control of development by juvenile hormone is more complex in insects, such as butterflies, that undergo complete metamorphosis. These animals undergo dramatic developmental changes in their life cycles. The fertilized egg hatches into a *larva*, which feeds and molts several times, becoming bigger each time. After a fixed number of molts, it enters an inactive stage called *pupation*. The pupa undergoes major body reorganization and finally emerges as an adult.

An excellent example of complete metamorphosis is provided by the silkworm moth *Hyalophora cecropia*

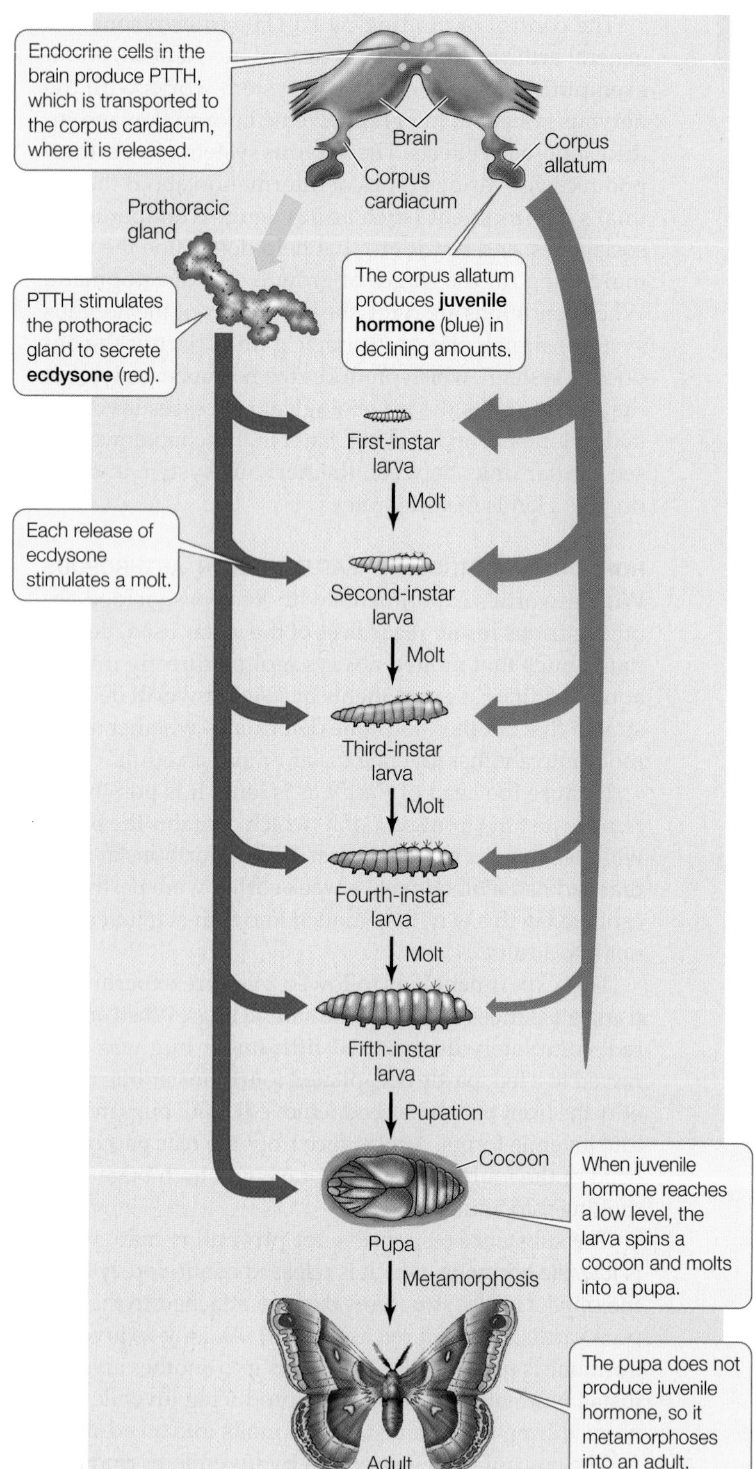
Endocrine cells in the brain produce PTTH, which is transported to the corpus cardiacum, where it is released.
Brain
Corpus allatum
Corpus cardiacum
Prothoracic gland
PTTH stimulates the prothoracic gland to secrete **ecdysone** (red).
The corpus allatum produces **juvenile hormone** (blue) in declining amounts.
First-instar larva
Molt
Each release of ecdysone stimulates a molt.
Second-instar larva
Molt
Third-instar larva
Molt
Fourth-instar larva
Molt
Fifth-instar larva
Pupation
Cocoon
When juvenile hormone reaches a low level, the larva spins a cocoon and molts into a pupa.
Pupa
Metamorphosis
The pupa does not produce juvenile hormone, so it metamorphoses into an adult.
Adult

**41.4 Hormonal Control of Metamorphosis** Three hormones control molting and metamorphosis in the silkworm moth *Hyalophora cecropia*.

yourBioPortal.com
GO TO Animated Tutorial 41.1 • Complete Metamorphosis

(**Figure 41.4**). As long as juvenile hormone is present in high concentrations, larvae molt into larger larvae. When the level of juvenile hormone falls, larvae spin cocoons and molt into pupae. Because no juvenile hormone is produced in pupae, they molt into adults.

The existence and function of insect hormones were experimentally demonstrated many years before the hormones were identified chemically. That is not surprising when you consider the tiny amounts of certain hormones that exist in an organism. In one of the earliest studies of ecdysone, biochemists produced only 250 milligrams of pure ecdysone (about one-fourth the weight of an apple seed) from 4 tons of silkworms!

### Hormones can be divided into three chemical groups

Now that we have seen examples of the roles hormones can play in long-term physiological and developmental processes, we can step back and ask some general questions about them. What kinds of hormones exist? What is their chemical nature, and what is their mode of action? There is enormous diversity in the chemical structure of hormones, but most of them can be classified into three groups:

- The majority of hormones are *peptides* or *proteins*. These hormones (insulin is an example) are water-soluble and thus are easily transported in the blood without carrier molecules. Peptide and protein hormones are packaged in vesicles within the cells that make them; they are released by exocytosis. Their receptors are on cell surfaces.
- *Steroid hormones* (such as testosterone) are synthesized from the steroid cholesterol, are lipid-soluble, and easily pass through cell membranes. Steroid hormones diffuse out of the cells that make them and are usually bound to carrier molecules in the blood. Their receptors are mostly intracellular
- *Amine hormones* are mostly synthesized from the amino acid tyrosine (thyroxine is one example). Some amine hormones are water-soluble and others are lipid-soluble; their modes of release differ accordingly.

### Hormone receptors can be membrane-bound or intracellular

Water-soluble hormones cannot pass readily through plasma membranes, so their receptors are located on the surfaces of target cells. These receptors are large transmembrane glycoprotein complexes with three domains: a *binding domain* that projects outside the plasma membrane, a *transmembrane domain* that anchors the receptor in the membrane, and a *cytoplasmic domain* that extends into the cytoplasm of the cell. When a hormone binds to the binding domain, the cytoplasmic domain initiates the target cell's response through a second messenger-activated cascade, eventually activating protein kinases or protein phosphatases (see Figures 7.7 and 7.8). In most cases these protein kinases and phosphatases activate

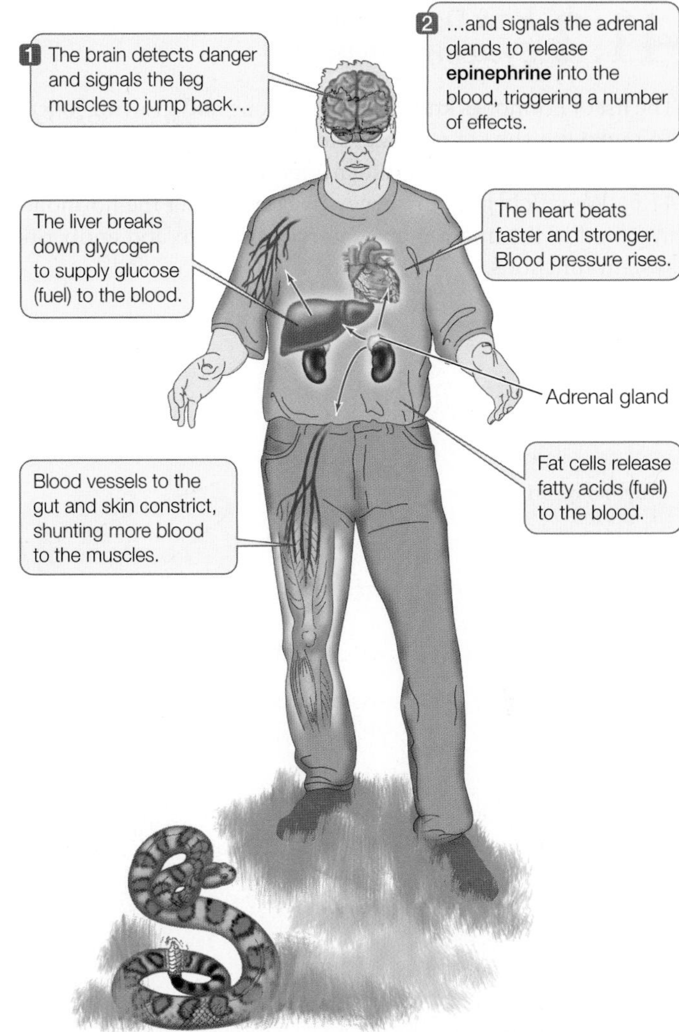

1 The brain detects danger and signals the leg muscles to jump back...
2 ...and signals the adrenal glands to release **epinephrine** into the blood, triggering a number of effects.
The liver breaks down glycogen to supply glucose (fuel) to the blood.
The heart beats faster and stronger. Blood pressure rises.
Adrenal gland
Fat cells release fatty acids (fuel) to the blood.
Blood vessels to the gut and skin constrict, shunting more blood to the muscles.

**41.5 The Fight-or-Flight Response** When a person is suddenly faced with a threatening situation, the brain sends a signal to the adrenal glands, which almost instantaneously release the hormone epinephrine. Epinephrine circulates around the body and induces the various components of the fight-or-flight response in different tissues.

or inactivate enzymes in the cytoplasm, which leads to the cell's response, but the signaling cascade initiated by the receptor can also generate signals that enter the nucleus and alter gene expression (see Figure 7.12).

Lipid-soluble hormones can diffuse through plasma membranes, and therefore their receptors are usually inside cells, in either the cytoplasm or the nucleus (although some membrane-bound receptors for lipid-soluble hormones have recently been described). In most cases, the complex formed by the lipid-soluble hormone and its receptor acts by altering gene expression in the cell's nucleus (see Figure 7.9).

### Hormone action depends on the nature of the target cell and its receptors

Wherever a hormone encounters a cell with a receptor to which it can bind, it binds to that receptor and triggers a response. The nature of the response depends on the responding cell and its receptors. Thus the same hormone can cause different responses in different types of cells.

Consider the amine hormone *epinephrine*. Suppose you are walking in the forest and almost trip over a rattlesnake. You jump back. Your heart starts to thump, and an entire set of protective actions are set in motion. The jump and the initial heart thumping are driven by your nervous system, which reacts very quickly. Simultaneously with these muscular responses, however, your nervous system stimulates endocrine cells in the adrenal glands just above your kidneys to secrete epinephrine. Within seconds, epinephrine is diffusing into your blood and circulating around your body to activate the many components of the **fight-or-flight response** (**Figure 41.5**).

Epinephrine binds to receptors in your heart, causing a faster and stronger heartbeat. Your heart is now pumping more blood. Epinephrine also binds to receptors in certain blood vessels. By causing constriction of blood vessels supplying your skin, kidneys, and digestive tract (digesting lunch can wait!), the hormone diverts more blood to the muscles needed for your escape from danger.

Epinephrine also binds to cells in the liver and to receptors on fat cells. In the liver, epinephrine stimulates the breakdown of glycogen into glucose for a quick energy supply (see Figure 7.20). In fatty tissue, it stimulates the breakdown of fats to yield fatty acids—another source of energy. These are just some of the many actions that are triggered by one hormone; they all contribute to increasing your chances of surviving a dangerous situation. In each case the cellular response depends on the cell's receptors and associated intracellular signaling cascade.

## 41.1 RECAP

**Hormones are chemical signals released by endocrine cells into the extracellular fluid, where they diffuse into the blood and travel to distant target cells. The receptors for water-soluble hormones are on the surfaces of target cells; receptors for most lipid-soluble hormones are inside the target cells.**

- What is the role of juvenile hormone in metamorphosis? **See pp. 855–856 and Figures 41.3 and 41.4**
- Describe the different methods by which water-soluble and lipid-soluble hormones reach their receptors. **See p. 856**
- Do you understand why a single hormone can have diverse effects in the body?

Since the nervous system and the endocrine system are the two major information systems of the body, it is not surprising that their activities are coordinated. Let's look next at how this coordination works.

## 41.2 How Do the Nervous and Endocrine Systems Interact?

The list of hormones known to exist is long and growing longer. To make the subject manageable, we will focus primarily on the endocrine system of humans and other mammals (**Figure 41.6**). We will begin our survey by considering the hormones involved in the integration of nervous system and endocrine system functions.

### The pituitary connects the nervous and endocrine systems

The **pituitary gland** sits in a depression at the bottom of the skull, just over the back of the roof of the mouth (**Figure 41.7A**). It is attached by a stalk to a part of the brain called the **hypothalamus,** which is involved in many physiological regulatory systems. Through its close connection with the hypothalamus, the pituitary serves as the interface between the nervous system and the endocrine system and is involved in the hormonal control of many physiological processes.

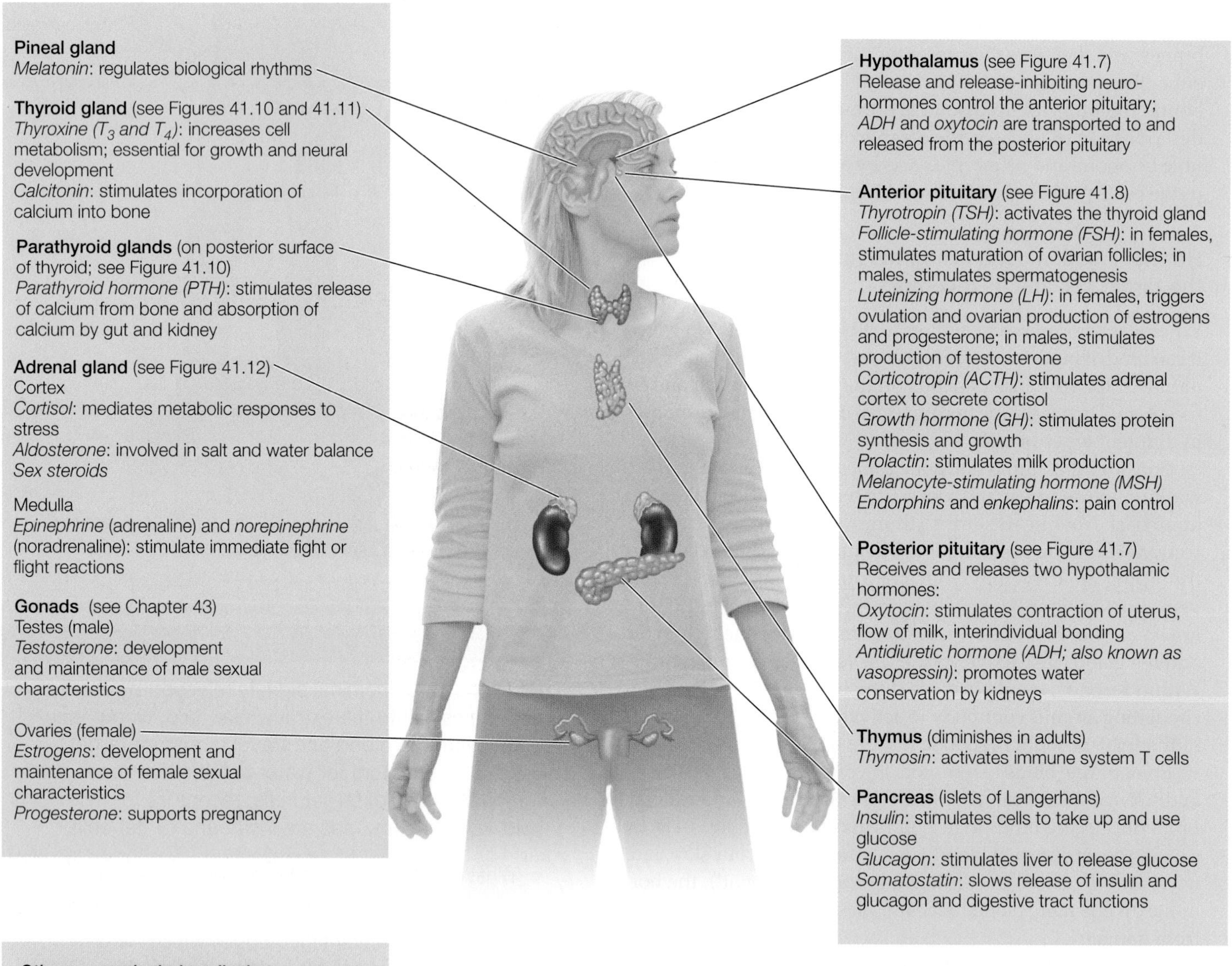

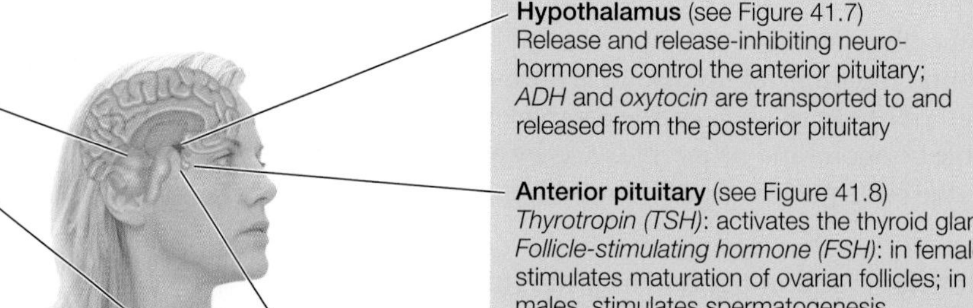
**Pineal gland**
*Melatonin*: regulates biological rhythms
**Thyroid gland** (see Figures 41.10 and 41.11)
*Thyroxine ($T_3$ and $T_4$)*: increases cell metabolism; essential for growth and neural development
*Calcitonin*: stimulates incorporation of calcium into bone
**Parathyroid glands** (on posterior surface of thyroid; see Figure 41.10)
*Parathyroid hormone (PTH)*: stimulates release of calcium from bone and absorption of calcium by gut and kidney
**Adrenal gland** (see Figure 41.12)
Cortex
*Cortisol*: mediates metabolic responses to stress
*Aldosterone*: involved in salt and water balance
*Sex steroids*
Medulla
*Epinephrine* (adrenaline) and *norepinephrine* (noradrenaline): stimulate immediate fight or flight reactions
**Gonads** (see Chapter 43)
Testes (male)
*Testosterone*: development and maintenance of male sexual characteristics
Ovaries (female)
*Estrogens*: development and maintenance of female sexual characteristics
*Progesterone*: supports pregnancy
**Hypothalamus** (see Figure 41.7)
Release and release-inhibiting neurohormones control the anterior pituitary;
*ADH* and *oxytocin* are transported to and released from the posterior pituitary
**Anterior pituitary** (see Figure 41.8)
*Thyrotropin (TSH)*: activates the thyroid gland
*Follicle-stimulating hormone (FSH)*: in females, stimulates maturation of ovarian follicles; in males, stimulates spermatogenesis
*Luteinizing hormone (LH)*: in females, triggers ovulation and ovarian production of estrogens and progesterone; in males, stimulates production of testosterone
*Corticotropin (ACTH)*: stimulates adrenal cortex to secrete cortisol
*Growth hormone (GH)*: stimulates protein synthesis and growth
*Prolactin*: stimulates milk production
*Melanocyte-stimulating hormone (MSH)*
*Endorphins* and *enkephalins*: pain control
**Posterior pituitary** (see Figure 41.7)
Receives and releases two hypothalamic hormones:
*Oxytocin*: stimulates contraction of uterus, flow of milk, interindividual bonding
*Antidiuretic hormone (ADH; also known as vasopressin)*: promotes water conservation by kidneys
**Thymus** (diminishes in adults)
*Thymosin*: activates immune system T cells
**Pancreas** (islets of Langerhans)
*Insulin*: stimulates cells to take up and use glucose
*Glucagon*: stimulates liver to release glucose
*Somatostatin*: slows release of insulin and glucagon and digestive tract functions

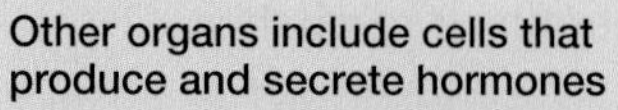

**Other organs include cells that produce and secrete hormones**

| Organ | Hormone |
|---|---|
| Adipose tissue | Leptin |
| Heart | Atrial natriuretic peptide |
| Kidney | Erythropoietin |
| Stomach | Gastrin |
| Intestine | Secretin, cholecystokinin |
| Skin | Vitamin D (cholecalciferol) |
| Liver | Somatomedins, insulin-like growth factors |

**41.6 The Endocrine System of Humans** Cells that produce and secrete hormones may be organized into discrete endocrine glands, or they may be embedded in the tissues of other organs, such as the digestive tract or kidneys. The hypothalamus is part of the brain, but it includes cells that secrete neurohormones into the extracellular fluid.

**yourBioPortal.com**
GO TO Web Activity 41.1 • The Human Endocrine Glands

The pituitary has two parts with different developmental origins. The *anterior pituitary* originates as an outpocketing of the roof of the embryonic mouth cavity, whereas the *posterior pituitary* originates as an outpocketing of the floor of the developing brain. Thus the anterior pituitary originates from gut epithelial tissue and the posterior pituitary from neural tissue. Both parts interact with the nervous system but in different ways. The anterior pituitary contains endocrine cells controlled by neurohormones secreted by the hypothalamus. The posterior pituitary contains axons from hypothalamic neurons.

yourBioPortal.com

GO TO Animated Tutorial 41.2 • The Hypothalamic–Pituitary–Endocrine Axis

**THE POSTERIOR PITUITARY** Long axons extend into the **posterior pituitary** from neurons in the hypothalamus. The ends (*terminals*) of those axons release two hormones produced by the neurons, antidiuretic hormone and oxytocin (**Figure 41.7B**). These neurohormones are packaged in vesicles that are transported down the axons. The vesicles are stored until an action potential stimulates their release.

The main action of **antidiuretic hormone** (**ADH**) in mammals and birds is to increase the amount of water conserved by the kidneys. When ADH secretion is high, the kidneys produce only a small volume of highly concentrated urine. When ADH secretion is low, the kidneys produce a large volume of dilute urine. The posterior pituitary increases its release of ADH when blood pressure falls or the blood becomes too salty. ADH is also known as *vasopressin* because at high concentrations it causes the constriction of peripheral blood vessels as a means of elevating blood pressure.

When a woman is about to give birth, her posterior pituitary releases **oxytocin**, which stimulates the uterine contractions that deliver the baby (see Figure 44.16). Oxytocin also brings about the flow of milk from the mother's breasts. The baby's suckling stimulates neurons in the mother's brain that cause the secretion of oxytocin. Even the sight and sound of a baby can cause a nursing mother to secrete oxytocin and release breast milk. This is a good example of how the nervous system integrates information and contributes to the control of hormonally mediated processes.

Hormones, in turn, can influence the nervous system. Oxytocin, for example, promotes bonding (see the story that opens Chapter 7). If oxytocin release is experimentally blocked, mammalian mothers, from rats to sheep, will reject their newborn offspring, but if a virgin rat is given a dose of oxytocin, she will adopt strange pups as if they were her own. Oxytocin promotes pair bonding and trust in a variety of animals. In humans, its secretion rises with intimate sexual contact. Not surprisingly, oxytocin has been nicknamed the "cuddle hormone."

**THE ANTERIOR PITUITARY** The **anterior pituitary** releases four peptide and protein hormones that act as **tropic hormones**, meaning they control the activities of other endocrine glands

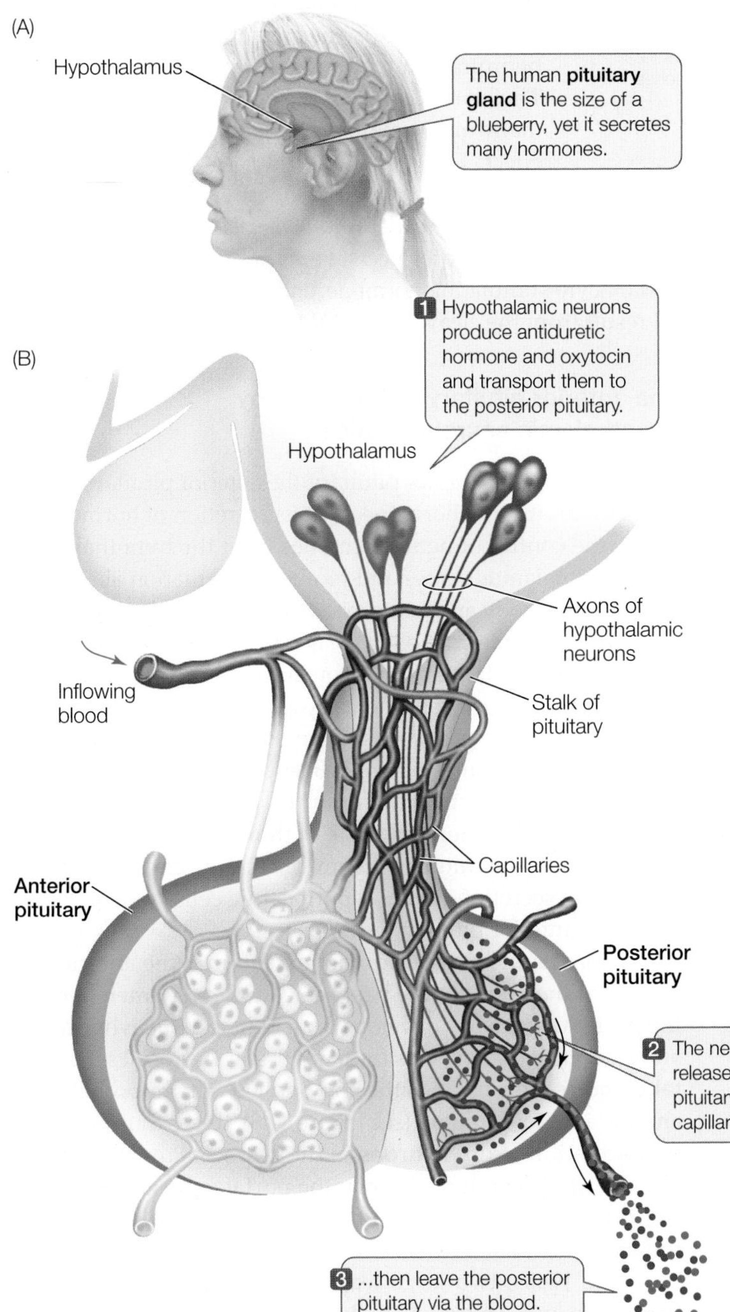
(A)
Hypothalamus
The human **pituitary gland** is the size of a blueberry, yet it secretes many hormones.
(B)
1 Hypothalamic neurons produce antidiuretic hormone and oxytocin and transport them to the posterior pituitary.
Hypothalamus
Axons of hypothalamic neurons
Inflowing blood
Stalk of pituitary
Capillaries
Anterior pituitary
Posterior pituitary
2 The neurohormones are released in the posterior pituitary and diffuse into capillaries...
3 ...then leave the posterior pituitary via the blood.

**41.7 The Posterior Pituitary Releases Neurohormones**
Neurons in the hypothalamus produce two peptide neurohormones, which are stored and released by the posterior pituitary.

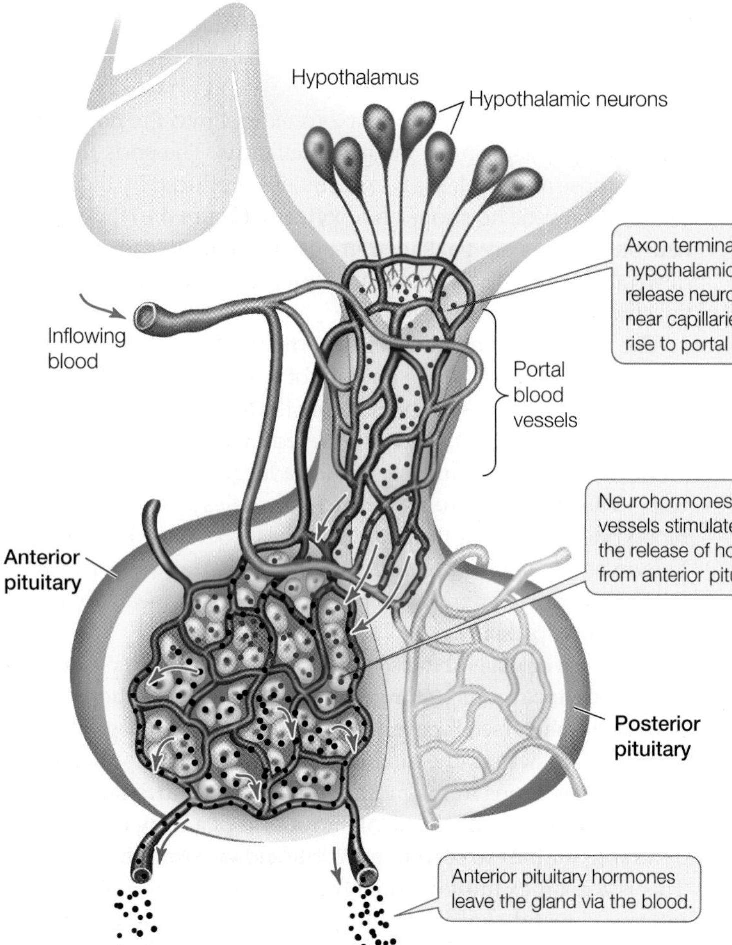

Hypothalamus
Hypothalamic neurons
Axon terminals of hypothalamic neurons release neurohormones near capillaries that give rise to portal vessels.
Inflowing blood
Portal blood vessels
Neurohormones from portal vessels stimulate or inhibit the release of hormones from anterior pituitary cells.
Anterior pituitary
Posterior pituitary
Anterior pituitary hormones leave the gland via the blood.

**41.8 The Anterior Pituitary Is Controlled by the Hypothalamus** Cells of the anterior pituitary produce four tropic hormones that control other endocrine glands, as well as several other peptide and protein hormones. These cells are controlled by neurohormones produced in the hypothalamus and delivered through portal blood vessels that run between the hypothalamus and the anterior pituitary through the pituitary stalk.

(**Figure 41.8**). These four hormones are *thyrotropin, luteinizing hormone, follicle-stimulating hormone,* and *corticotropin*. Each tropic hormone is produced by a different type of pituitary cell. We will say more about the tropic hormones when we describe their target glands (thyroid, testes, ovaries, and adrenal cortex) later in this chapter and in Chapter 43.

Other peptide and protein hormones produced by the anterior pituitary are *growth hormone, prolactin* (see Figure 41.2) *melanocyte-stimulating hormone, enkephalins*, and *endorphins*.

**Growth hormone** (**GH**) acts on a wide variety of tissues to promote growth. One of its important effects is to stimulate cells to take up amino acids. Growth hormone also promotes growth by stimulating the liver to produce chemical signals called *somatomedins* or *insulin-like growth factors* (*IGFs*), which stimulate the growth of bone and cartilage. Thus growth hormone can be considered a tropic hormone because it stimulates endocrine cells in the liver.

Overproduction of growth hormone in children causes *gigantism,* in which affected individuals may grow to nearly 8 feet tall. Underproduction causes *pituitary dwarfism*, in which individuals fail to reach normal adult height. Beginning in the late 1950s, children with serious growth hormone deficiencies were treated with growth hormone extracted from pituitaries of human cadavers. The treatment was successful in stimulating substantial growth, but a year's supply of the hormone for one individual required up to 50 cadaver pituitaries. In the 1980s, scientists using recombinant DNA technology isolated the gene for human growth hormone and introduced it into bacteria that could be grown in large quantities, making it possible to purify enough of the hormone to make it widely available.

**Endorphins** and **enkephalins** are the body's natural painkillers. In the brain, these molecules act as neurotransmitters in pathways that control pain. Their production in the anterior pituitary is normally quite small and probably has no significant effect. They are a by-product of the production of two other anterior pituitary hormones. One gene encodes a large parent molecule called *pro-opiomelanocortin*, or POMC, which is cleaved to produce several peptides. Corticotropin, melanocyte-stimulating hormone, endorphins, and enkephalins all result from the cleavage of POMC.

### The anterior pituitary is controlled by hypothalamic neurohormones

In contrast to the posterior pituitary, the anterior pituitary makes and secretes its own hormones, but its secretion of hormones is under the control of neurohormones from the hypothalamus. The hypothalamus senses and receives information about conditions in the body and in the external environment, and it communicates that information to the anterior pituitary by releasing neurohormones. If the connection between the hypothalamus and the pituitary is experimentally cut, the release of pituitary hormones no longer changes when conditions in the internal or external environment change. In experiments in which pituitary cells were maintained in culture, adding extracts of hypothalamic tissue stimulated some of those cells to release their hormones into the culture medium. Therefore, scientists hypothesized that secretions from hypothalamic cells control the activities of anterior pituitary cells.

Hypothalamic neurons do not extend into the anterior pituitary as they do into the posterior pituitary. Remember that the posterior pituitary develops from neural tissue whereas the anterior pituitary develops from gut tissue. A special set of **portal blood vessels** bridges the gap between the hypothalamus and the anterior pituitary (see Figure 41.8). It was thus proposed that secretions from neurons in the hypothalamus enter the blood and are conducted down the portal vessels to the anterior pituitary, where they stimulate the release of anterior pituitary hormones.

In the 1960s, two large teams of scientists, led by Roger Guillemin and Andrew Schally, initiated the search for these hypothalamic secretions. Because the amounts of such neurohormones in any individual mammal would be tiny, massive numbers of hypothalami from pigs and sheep were collected from slaughterhouses and shipped to laboratories in refrigerated trucks. One extraction effort began with the hypothalami from 270,000 sheep and yielded only 1 milligram of purified **thyrotropin-releasing hormone** (**TRH**). TRH was the first hypothalamic *releasing hormone* (that is, release-stimulating hormone) to be isolated and characterized. It turned out to be a simple tripeptide consisting of glutamine, histidine, and proline. It causes certain anterior pituitary cells to release the tropic hormone thyrotropin, which in turn stimulates the activity of the thyroid gland.

Soon after discovering TRH, Guillemin's and Schally's teams identified **gonadotropin-releasing hormone** (**GnRH**), which stimulates certain anterior pituitary cells to release the tropic hormones that control the activity of the gonads (the ovaries and the testes). For these discoveries, Guillemin and Schally received the 1977 Nobel Prize in Medicine. Many other hypothalamic neurohormones, including both releasing hormones and release-inhibiting hormones, are now known. The major hypothalamic neurohormones that control anterior pituitary function are:

- Thyrotropin-releasing hormone
- Gonadotropin-releasing hormone
- Prolactin-releasing and release-inhibiting hormones
- Growth hormone–releasing hormone
- Growth hormone release-inhibiting hormone (somatostatin)
- Corticotropin-releasing hormone

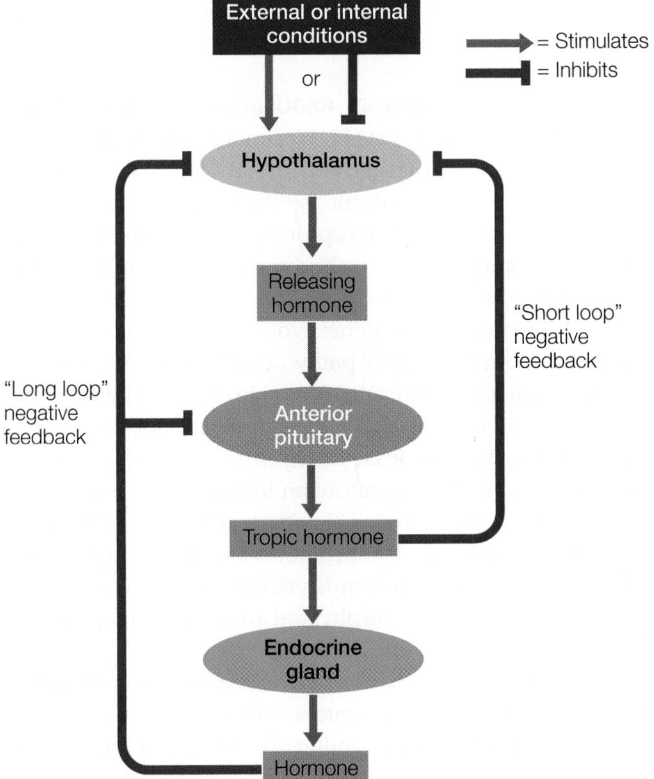


**41.9 Multiple Feedback Loops Control Hormone Secretion** Multiple negative feedback loops regulate the chain of command from hypothalamus to anterior pituitary to endocrine glands.

### Negative feedback loops regulate hormone secretion

As well as being controlled by hypothalamic releasing and release-inhibiting hormones, the endocrine cells of the anterior pituitary are also under direct and indirect negative feedback control by the hormones of the target glands they stimulate (**Figure 41.9**). For example, the hormone cortisol, produced by the adrenal gland in response to corticotropin secreted by the anterior pituitary, reaches the pituitary in the circulating blood and inhibits further release of that tropic hormone. Cortisol also acts as a negative feedback signal to the hypothalamus, inhibiting the release of corticotropin-releasing hormone. In some cases, a tropic hormone also exerts negative feedback control on the hypothalamic cells that produce the corresponding releasing hormone.

### 41.2 RECAP

**The pituitary is the interface between the nervous system and the endocrine system. The posterior pituitary releases two neurohormones. The anterior pituitary, under the control of other neurohormones from the hypothalamus, releases hormones that control other endocrine glands.**

- Describe the anatomical and functional relationships between the brain and the two parts of the pituitary.
- What are the tropic hormones of the anterior pituitary, and how do they influence the endocrine system? **See pp. 859–860 and Figure 41.8**

Now that we know some of the mechanisms by which endocrine systems are controlled, we will take a more detailed look at the functions of the major endocrine glands of the mammalian body.

## 41.3 What Are the Major Mammalian Endocrine Glands and Hormones?

Hormones help regulate functions in all mammalian physiological systems. In this section we will examine a few major examples of hormonal action in physiological processes. We will see many more in the chapters that follow.

### The thyroid gland secretes thyroxine

The **thyroid gland** wraps around the front of the windpipe (*trachea*) and expands into a lobe on either side (see Figure 41.6). There are two cell types in the thyroid gland, each of which produces a specific hormone. Thyroxine is produced, stored, and released by epithelial cells that make up round, colloid-contain-

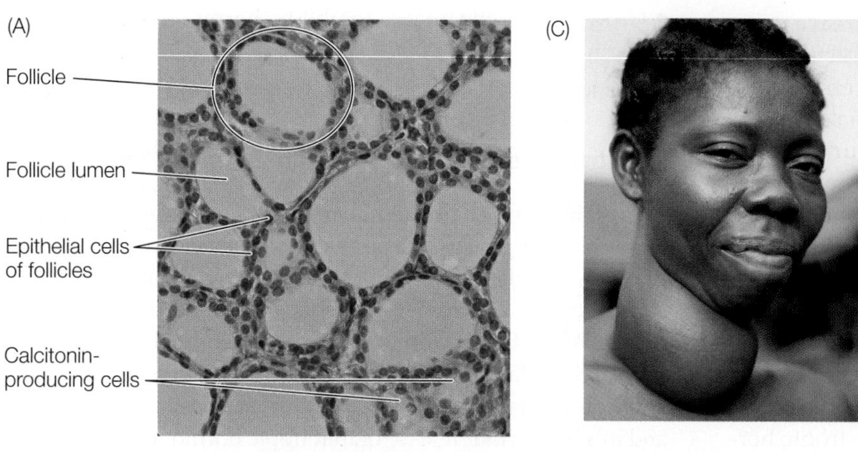


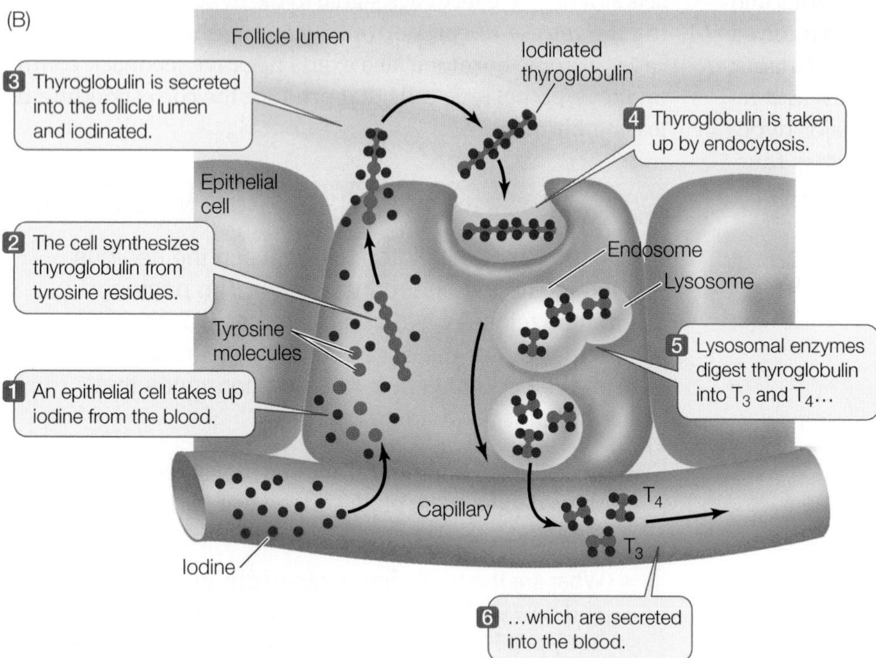


**41.10 The Thyroid Gland Consists of Many Follicles** (A) Cross section through a thyroid gland, showing numerous follicles bounded by epithelial cells. Calcitonin-secreting cells are located in the spaces between the follicles. (B) The epithelial cells of the follicle synthesize thyroglobulin and secrete it into the lumen of the follicle, where it is iodinated and stored until it is processed by the epithelial cells to generate $T_3$ and $T_4$. (C) Iodine deficiency can result in hypothyroid goiter. In this condition, a lack of functional thyroxine results in oversynthesis of thyroglobulin and subsequent enlarged follicles.

ing structures called *follicles* (**Figure 41.10A and B**). Calcitonin is produced in cells in spaces between the follicles.

**Thyroxine** begins as the glycoprotein **thyroglobulin**, which the follicle cells synthesize as long chains consisting largely of tyrosine residues. As these thyroglobulin molecules are exported into the follicle for storage, the tyrosine residues are iodinated with one or two atoms of iodine. When the thyroid gland is stimulated to release thyroxine, the follicle cells take up thyroglobulin from the follicle by endocytosis. These bits of thyroglobulin are then cleaved to form smaller molecules consisting of only two tyrosine residues, and these molecules leave the follicle cells. If these molecules are iodinated at the maximum of four sites on the tyrosine residues, the hormone is tetraiodothyronine, or $T_4$;

[Structure of $T_4$: HO–(ring with two I)–O–(ring with two I)–$CH_2$–CH($NH_2$)–C(=O)–OH]

and if they are iodinated at only three sites, they are triiodothyronine, or $T_3$:

[Structure of $T_3$: HO–(ring with one I)–O–(ring with two I)–$CH_2$–CH($NH_2$)–C(=O)–OH]

The thyroid usually releases about four times as much $T_4$ as $T_3$; however, $T_3$ is the more active hormone in the cells of the body. Circulating $T_4$ can be converted to $T_3$ by an enzyme within target cells. Therefore, each target cell can set its own sensitivity to thyroid hormones by controlling the conversion of $T_4$ to the more active $T_3$. When you read about thyroxine, keep in mind that the actions discussed are primarily those of $T_3$.

#### THYROXINE REGULATES CELL METABOLISM

Thyroxine in mammals plays many roles in regulating cell metabolism. Thyroxine is lipid-soluble, so it enters cells readily and binds to receptors in the nucleus. The receptor is found in most cells of the body, and when combined with thyroxine, it acts as a transcription factor that stimulates the transcription of numerous genes whose products are enzymes involved in energy metabolic pathways, transport proteins, and structural proteins. As a result, thyroxine elevates the metabolic rates of most cells and tissues. Exposure to cold for several days leads to an increased release of thyroxine, an increased conversion of $T_4$ to $T_3$, and therefore an increased basal metabolic rate (see Section 40.4). Thyroxine is especially crucial during development and growth, as it promotes amino acid uptake and protein synthesis. Insufficient thyroxine in a human fetus or growing child greatly retards physical and mental development, resulting in a condition known as *cretinism*.

The tropic hormone **thyrotropin**, or **thyroid-stimulating hormone** (**TSH**), produced by the anterior pituitary, activates the thyroxine-producing follicle cells in the thyroid. Thyrotropin-releasing hormone (TRH), produced in the hypothalamus and transported to the anterior pituitary through the portal blood vessels, activates the TSH-producing pituitary cells. The hypothalamus uses environmental information, such as temperature or day length, to determine whether to increase or decrease its secretion of TRH. This sequence of steps is regulated by a negative feedback loop like the one described earlier for cortisol (see Figure 41.9). Circulating thyroxine inhibits the response of pituitary cells to TRH, so less TSH is released when thyroxine levels are high, and more TSH is released when thyroxine levels are low. Circulating thyroxine also exerts negative feedback on the production and release of TRH by the hypothalamus.

A *goiter* is an enlarged thyroid gland (**Figure 41.10C**) that can be associated with either **hyperthyroidism** (excess production of thyroxine) or **hypothyroidism** (thyroxine deficiency). The negative feedback loop whereby thyroxine controls TSH release helps explain how two very different conditions can result in the same symptom.

The most common cause of *hyperthyroid* goiter is an autoimmune disease involving an antibody to the TSH receptor. This antibody binds to and activates the TSH receptors on the follicle cells, causing uncontrolled production and release of thyroxine. Blood levels of TSH are quite low because of the negative feedback from high levels of thyroxine, but the thyroid remains maximally stimulated and grows bigger. People with hyperthyroidism have high metabolic rates, are jumpy and nervous, usually feel hot, and may develop a buildup of fat behind the eyeballs, which causes their eyes to bulge.

*Hypothyroid* goiter results when there is not enough circulating thyroxine to turn off TSH production. The most common cause of this condition is a deficiency of dietary iodine, without which the follicle cells cannot make thyroxine. Without sufficient thyroxine, TSH levels remain high, and the thyroid continues to produce large amounts of thyroglobulin. Because sufficient iodine is not available, however, the thyroglobulin is poorly iodinated. When it is broken down by the follicle cells, it does not yield functional thyroxine ($T_3$ or $T_4$). TSH levels remain high and stimulate more and more synthesis of thyroglobulin, and the thyroid gets bigger. The symptoms of hypothyroidism are low metabolism, intolerance of cold, and general physical and mental sluggishness.

Goiter affects about 5 percent of the world's population. The addition of iodine to table salt has greatly reduced the incidence of hypothyroid goiter in industrialized nations, but the condition is still common in the other parts of the world.

## Three hormones regulate blood calcium concentrations

The regulation of calcium concentration in the blood is a crucial and difficult task. It is crucial because shifts in blood calcium concentration above or below a narrow range can cause serious problems. When blood calcium falls below this range, the nervous system becomes overly excited, resulting in muscle spasms and even seizures. When blood calcium rises above this range, the nervous system becomes depressed and muscles—including the heart—weaken. Regulation of blood calcium is difficult because only about 0.1 percent of the calcium in the body is located in the extracellular fluid. About 1 percent is in cells, and almost 99 percent is in the bones. Therefore, the body must maintain a tiny pool of calcium in the blood at a precise concentration, and that tiny pool can be influenced greatly by relatively small shifts in the much larger pools of calcium in the cells and bones.

The body has multiple mechanisms for changing blood calcium levels, including:

- Deposition or absorption of bone
- Excretion or retention of calcium by the kidneys
- Absorption of calcium from the digestive tract

These mechanisms are controlled by the hormones calcitonin, parathyroid hormone, and calcitriol (synthesized from vitamin D).

**CALCITONIN REDUCES BLOOD CALCIUM** **Calcitonin** released by the thyroid lowers the concentration of calcium in the blood, mainly by regulating bone turnover (**Figure 41.11**). Bone is continuously remodeled through a dynamic process that involves

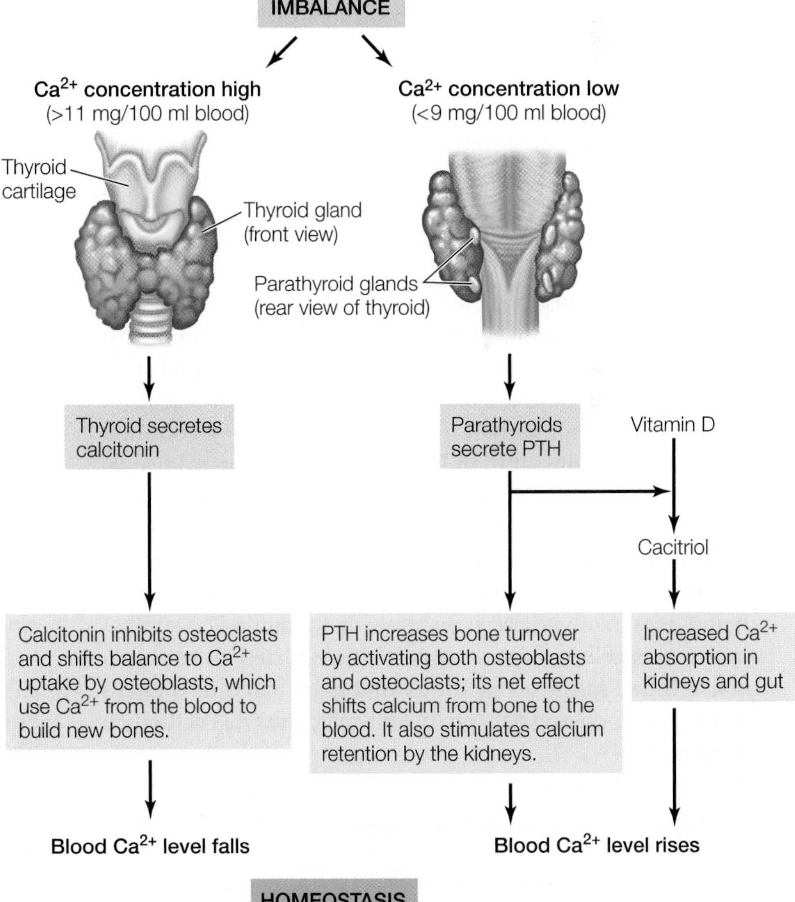
IMBALANCE
Ca2+ concentration high
(>11 mg/100 ml blood)
Ca2+ concentration low
(<9 mg/100 ml blood)
Thyroid cartilage
Thyroid gland
(front view)
Parathyroid glands
(rear view of thyroid)
Thyroid secretes calcitonin
Parathyroids secrete PTH
Vitamin D
Cacitriol
Calcitonin inhibits osteoclasts and shifts balance to Ca2+ uptake by osteoblasts, which use Ca2+ from the blood to build new bones.
PTH increases bone turnover by activating both osteoblasts and osteoclasts; its net effect shifts calcium from bone to the blood. It also stimulates calcium retention by the kidneys.
Increased Ca2+ absorption in kidneys and gut
Blood Ca2+ level falls
Blood Ca2+ level rises
HOMEOSTASIS
Ca2+ concentration between
9 and 11 mg/100 ml blood

**41.11 Hormonal Regulation of Calcium** Calcitonin, parathyroid hormone (PTH), and vitamin D regulate calcium levels in the blood.

yourBioPortal.com
GO TO Animated Tutorial 41.3 • Hormonal Regulation of Calcium

both resorption of old bone and synthesis of new bone, as we discuss in Section 48.3. Cells called *osteoclasts* break down bone and release calcium into the blood, and cells called *osteoblasts* take up calcium from the blood and deposit it in new bone. Calcitonin decreases the activity of osteoclasts and thereby favors removal of calcium from the blood and its deposition in bone by osteoblasts. The turnover of bone in adult humans is not very high, so calcitonin does not play a major role in calcium homeostasis in adults. It is probably more important in young individuals whose bones are actively growing.

**PARATHYROID HORMONE INCREASES BLOOD CALCIUM** The **parathyroid glands** are four tiny structures embedded in the posterior surface of the thyroid gland (see Figure 41.11). Their single hormone product, **parathyroid hormone** (also called **PTH** or *parathormone*), is the most important hormone in the regulation of blood calcium levels. Circulating calcium activates receptors in the plasma membrane of the parathyroid cells. When these receptors are active, they inhibit the synthesis and release of PTH. A fall in blood calcium removes this inhibition and triggers the synthesis and release of PTH.

PTH stimulates bone turnover by actions on both osteoclasts and osteoblasts. The end result of these actions of PTH is a net increase of calcium in the blood.

Another mechanism by which PTH raises blood concentrations of calcium is by stimulating the kidneys to reabsorb it rather than excrete it in the urine. In addition, PTH activates the synthesis of calcitriol from vitamin D, which in turn causes the digestive tract to increase its absorption of calcium from food.

**VITAMIN D IS A HORMONE** A *vitamin* is a substance that the body requires in small quantities but cannot synthesize, and must therefore obtain from the diet. By this definition, **vitamin D** is not a vitamin, because the body can and does synthesize it.

It had long been known that fragile bones were common among people living at high latitudes, where winter days are short and the winter diet often lacked meat, fish, dairy products, and fresh vegetables. Since the condition could be reversed by taking cod-liver oil (which, as it turned out, contains large amounts of vitamin D), scientists assumed that a dietary vitamin was involved. We now know that vitamin D is synthesized in skin cells, where cholesterol is converted into vitamin D (also called *calciferol*) by ultraviolet light. Vitamin D circulates in the blood and acts on distant cells; thus it is actually a hormone.

### PTH lowers blood phosphate levels

Bone minerals are made up primarily of calcium and phosphate. Thus, when PTH stimulates the release of calcium from bone, it also causes the release of phosphate. Excessive blood concentrations of both elements can be dangerous. The normal concentrations of calcium and phosphate in the blood are just below the levels at which they precipitate out of solution as calcium phosphate salts. Thus even a small rise may cause such precipitation, leading to maladies such as kidney stones and calcium deposits in the arteries (hardening of the arteries). To reduce this risk, PTH acts on the kidneys to increase the elimination of phosphate via the urine.

### Insulin and glucagon regulate blood glucose concentrations

Before the 1920s, *diabetes mellitus* was a fatal disease, characterized by weakness, lethargy, and a dramatic loss of body mass. The disease was known to be connected somehow with the **pancreas**—a large gland located just below the stomach (see Figure 41.6)—and with abnormal glucose metabolism, but the links were not clear.

Today we know that diabetes mellitus is caused by a lack of the protein hormone **insulin** (in type I or juvenile-onset diabetes) or by a lack of insulin responsiveness in target tissues (in type II or adult-onset diabetes). Glucose enters cells by diffusion, not active transport, but cell membranes are not very permeable to glucose. Therefore there are proteins called glucose transporters in cell membranes that facilitate the movement of glucose into cells. The glucose transporters that are most common in muscle and adipose tissue are controlled by insulin. When insulin binds to its receptor on the cell membrane, it causes these glucose transporters to move from cytoplasmic vesicles to the cell membrane, making the cell more permeable to glucose. When insulin is not present, these transporters are returned to the cytoplasmic pool through endocytosis.

In the absence of insulin or insulin responsiveness, glucose entry into cells is impaired, and so much glucose accumulates in the blood that it starts to spill over into the urine; urine production also goes up. A high concentration of glucose in the blood increases urine output by two mechanisms. First, it causes water to move from cells into the blood by osmosis, and this increase in blood volume results in increased urine production. Second, the increased glucose in the tubules of the kidneys pulls more water into the urine by osmosis. Thus diabetic individuals can become dehydrated, but more importantly they suffer from a lack of metabolic fuel. Because glucose uptake by muscle and adipose tissue is impaired in the absence of insulin, muscle cells must depend on fat and protein for fuel and adipose tissue cannot replenish its stores of triglycerides. As a result, the body of the untreated diabetic can waste away.

For centuries, the prospects for diabetics were bleak. A change came almost overnight in 1921, when the physician Frederick Banting and a medical student, Charles Best, at the University of Toronto, discovered that they could reduce the symptoms of diabetes by injecting an extract prepared from pancreatic tissue. The active component of this extract was found to be a small protein hormone—insulin—consisting of just 51 amino acids. In the United States today, insulin replacement therapy, using manufactured insulin, allows 1.5 million people with type I diabetes to lead almost normal lives.

**ISLETS OF LANGERHANS** Insulin is produced in clusters of endocrine cells in the pancreas. These clusters are called **islets of Langerhans** after the German medical student who discovered them. They contain three types of cells, each of which produces a specific hormone:

- Beta (β) cells produce and secrete insulin.
- Alpha (α) cells produce and secrete glucagon, a hormone that has effects mostly opposite from those of insulin.
- Delta (δ) cells produce a hormone called somatostatin.

The rest of the pancreas is made up of exocrine tissue, which produces enzymes and other secretions that travel through ducts to the gut, where they participate in digestion.

After a meal, the concentration of glucose in the blood rises as glucose is absorbed from the food in the gut. This increase stimulates the β cells of the islets to release insulin, which causes target cells throughout the body to use the circulating glucose as fuel and to convert it into storage products, such as glycogen and fat. When the gut contains no more food, the glucose concentration in the blood falls and the islets stop releasing insulin. As a result, most cells shift to using glycogen and fat, rather than glucose, as fuel. If the concentration of glucose in the blood falls substantially below normal, the islet α cells release **glucagon**, which stimulates the liver to break down stored glycogen and release glucose into the blood. These actions are discussed in greater detail in Section 51.4.

**SOMATOSTATIN** **Somatostatin** is released from the δ cells of the pancreas in response to rapid increases of glucose and amino acids in the blood. This hormone has paracrine functions within the islets: it inhibits the release of both insulin and glucagon. Its actions outside the pancreas slow the digestive activities of the gut, extending the period during which nutrients are absorbed. Somatostatin is also produced in very small amounts by cells in the hypothalamus. This somatostatin is transported in the portal blood vessels to the anterior pituitary, where it acts as a neurohormone to inhibit the release of growth hormone and thyrotropin.

### The adrenal gland is two glands in one

An **adrenal gland** sits above each kidney, just below the middle of your back. Functionally and anatomically, each adrenal gland consists of a gland within a gland (**Figure 41.12**). The core, or **adrenal medulla**, produces two hormones: **epinephrine** (also known as *adrenaline*) and, to a lesser degree, **norepinephrine** (or *noradrenaline*). The medulla develops from nervous tissue and is under the control of the nervous system. Surrounding the medulla is the **adrenal cortex**, which produces steroid hormones. The cortex is under hormonal control, largely by corticotropin produced by the anterior pituitary.

**THE ADRENAL MEDULLA** The adrenal medulla produces epinephrine and norepinephrine in response to stressful situations, arousing the body to action. As we saw earlier in this chapter, epinephrine increases heart rate and blood pressure and diverts blood flow to active muscles and away from the gut and skin. Norepinephrine has similar functions.

Epinephrine and norepinephrine are both water-soluble, and both bind to the same set of receptors on the surfaces of target cells. These *adrenergic receptors* are of two general types, *α-adrenergic* and *β-adrenergic*; each type stimulates different actions within cells (**Figure 41.13**). The α-adrenergic receptors are the most common on target cells of the sympathetic nervous system, and they respond more strongly to norepinephrine than to epinephrine. The β-adrenergic receptors respond about equally to both epinephrine and norepinephrine. One class of the β-adrenergic receptors are found on cells not innervated by the sympathetic fibers, so they are positioned to respond to circulating epinephrine. Therefore, drugs called *beta blockers* (so named because they selectively block β-adrenergic receptors) can reduce the fight-or-flight response to epinephrine without disrupting the physiological regulatory functions of norepinephrine. Beta blockers are commonly prescribed to reduce the symptoms of anxiety, such as dry mouth and elevated heart rate (palpitations).

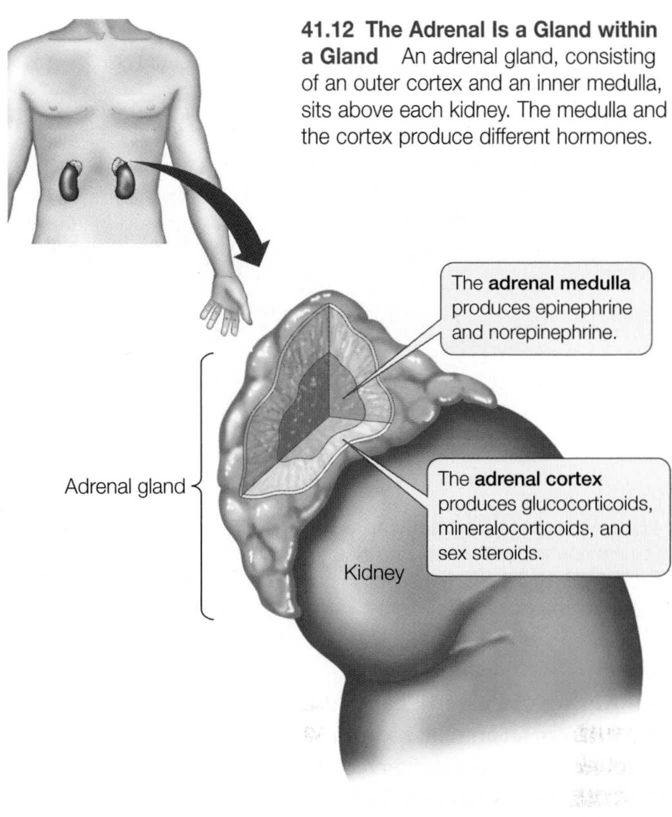
The **adrenal medulla** produces epinephrine and norepinephrine.
The **adrenal cortex** produces glucocorticoids, mineralocorticoids, and sex steroids.
Adrenal gland
Kidney

**41.12 The Adrenal Is a Gland within a Gland** An adrenal gland, consisting of an outer cortex and an inner medulla, sits above each kidney. The medulla and the cortex produce different hormones.

**THE ADRENAL CORTEX** The cells of the adrenal cortex use cholesterol to produce three classes of steroid hormones, collectively called **corticosteroids**:

- The *glucocorticoids* influence blood glucose concentrations as well as other aspects of fat, protein, and carbohydrate metabolism.
- The *mineralocorticoids* influence the salt and water balance of the extracellular fluid.
- The *sex steroids* play roles in sexual development, sexual behavior, and anabolism.

In adults, the adrenal cortex secretes sex steroids in only negligible amounts. The major producers of sex steroids are the gonads, as we will see in the following section.

**Aldosterone**, the main mineralocorticoid (**Figure 41.14A**), stimulates the kidneys to conserve sodium and excrete potas-

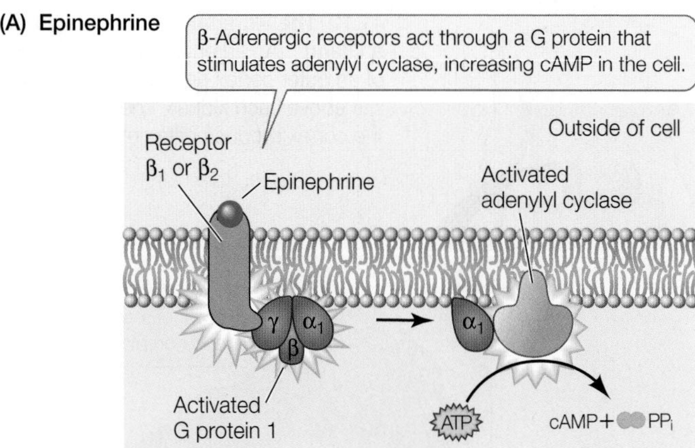


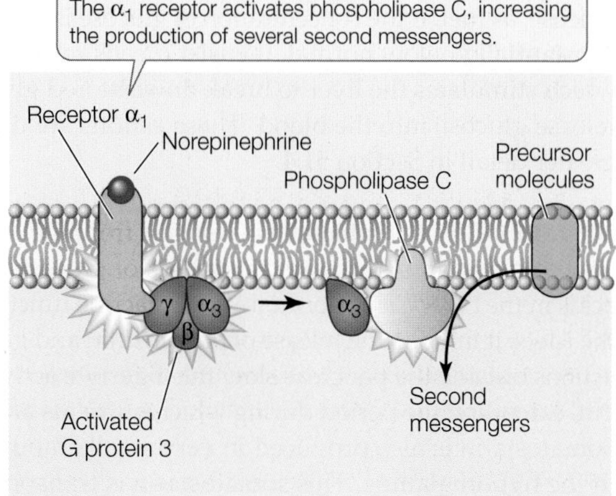


**41.13 Hormones Can Activate a Variety of Signal Transduction Pathways** Epinephrine and norepinephrine bind to G protein-linked adrenergic receptors that act through different signal transduction pathways. Epinephrine acts equally on both α- and β-adrenergic receptors; norepinephrine acts mostly on α-adrenergic receptors.

sium, as we discuss in Chapter 51. If the adrenal glands are removed from an animal, sodium must be added to its diet, or its sodium will be depleted and it will die.

The main glucocorticoid in humans is **cortisol** (**Figure 41.14B**), which is critical for mediating the body's metabolic responses to stress. Within minutes of a stressful stimulus (one provoking fear or anger, for example), your blood cortisol level rises. Cells not critical for your fight-or-flight responses are stimulated by cortisol to decrease their use of blood glucose and shift instead to using fats and proteins for energy. This is no time to feel sick, have allergic reactions, or heal wounds, so cortisol also blocks immune system reactions. That is why cortisol and drugs that mimic its action are useful for reducing inflammation and allergic responses.

Cortisol release is controlled from the anterior pituitary by **corticotropin** (also called *adrenocorticotropic hormone*, or **ACTH**), whose release is controlled in turn by **corticotropin-releasing hormone** from the hypothalamus. Because the cortisol response to stress has this chain of steps, each involving secretion, diffusion, circulation, and cell activation, it is much slower than the epinephrine response to stress. Furthermore, in many cases cortisol acts as a transcription factor to change gene expression, and those changes take time.

Turning off the responses to stress activated by cortisol is as important as turning them on. A study of stress in rats showed that old rats could turn on these stress responses as effectively as young rats, but they had lost the ability to turn them off as rapidly. As a result, they suffered from the well-known consequences of stress seen in humans: digestive system problems, cardiovascular

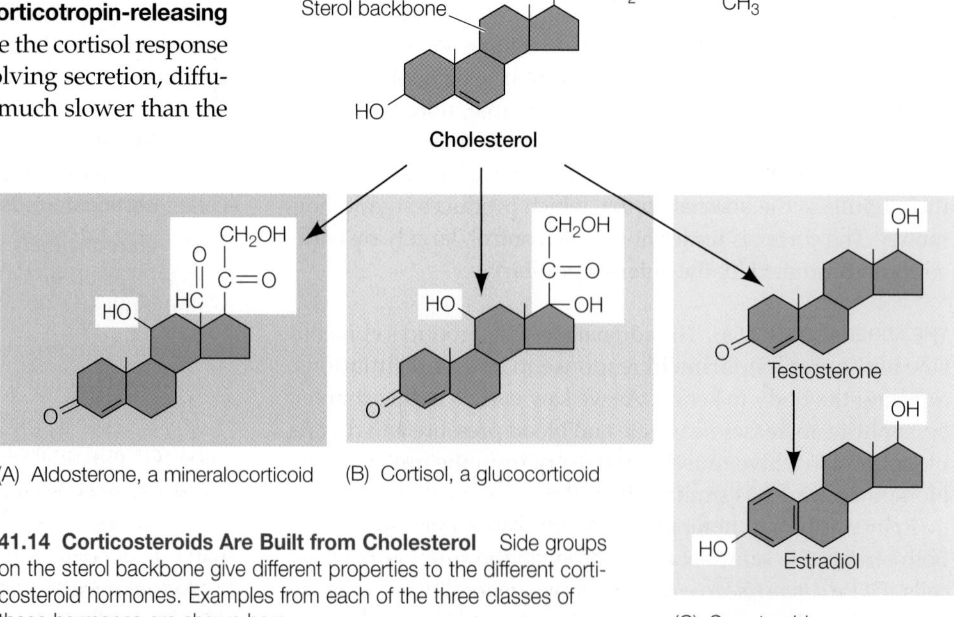


**41.14 Corticosteroids Are Built from Cholesterol** Side groups on the sterol backbone give different properties to the different corticosteroid hormones. Examples from each of the three classes of these hormones are shown here.

problems, strokes, impaired immune system function, and increased susceptibility to cancers and other diseases. Further research showed that these stress responses are turned off by the negative feedback action of cortisol on the hypothalamus, which causes a decrease in the release of corticotropin-releasing hormone (see Figure 41.9). Repeated activation of this negative feedback mechanism, either through repeated stress or prolonged medical use of cortisol, leads to a gradual loss of cortisol-sensitive cells in the hypothalamus, and therefore to a decreased ability to terminate the stress response.

### Sex steroids are produced by the gonads

The **gonads**—the testes of the male and the ovaries of the female—produce hormones as well as sperm and ova. The male steroids are collectively called **androgens**, and the dominant one is *testosterone*. The female steroids are **estrogens** and **progesterone**. The dominant estrogen is *estradiol*, which is synthesized from testosterone. Males and females both synthesize testosterone, but females have an enzyme (aromatase) that converts testosterone to estradiol (**Figure 41.14C**).

The sex steroids have important developmental effects: they determine whether a mammalian embryo develops into a phenotypic female or male. After birth, the sex steroids control the maturation of the reproductive organs and the development and maintenance of secondary sexual characteristics, such as breasts and facial hair. The roles they play in adult sexual behavior and reproduction are described in Chapter 43.

**EARLY DEVELOPMENT** Sex steroids begin to exert their effects in the human embryo in the seventh week of development. Until that time, the embryo has the potential to develop into either sex. In mammals and birds, the instructions for sex determination reside in the genes. In mammals, individuals that receive two X chromosomes normally become females, and individuals that receive an X and a Y chromosome normally become males (see Section 12.4).

These genetic instructions are carried out through the production and action of the sex steroids. In humans, the presence of a Y chromosome normally causes the undifferentiated embryonic gonads to begin producing androgens in the seventh week of development. In response to those androgens, the reproductive system develops into that of a male. If androgens are not produced at that time, female reproductive structures develop (**Figure 41.15**). In other words, androgens are required to trigger male development in humans, and the default condition is female.

**INITIATION OF PUBERTY** Sex steroids have dramatic effects at **puberty**—the time of sexual maturation in humans. Sex steroids are produced at low levels by the juvenile gonads, but their production increases rapidly at the beginning of puberty—around the age of 12 to 13 years. Why does this sudden increase occur?

In the juvenile, as in the adult, the production of sex steroids by the gonads is controlled by the tropic hormones **luteinizing hormone** (**LH**) and **follicle-stimulating hormone** (**FSH**), which together are called the **gonadotropins**. The production of these tropic hormones by the anterior pituitary is under the control of the hypothalamic neurohormone gonadotropin-releasing hormone (GnRH). Before puberty, the gonads can respond to gonadotropins and the anterior pituitary can respond to GnRH, but the hypothalamus produces only very low levels of GnRH. Puberty is initiated by a reduction in the sensitivity of hypothalamic GnRH-producing cells to negative feedback from sex steroids and from gonadotropins. As a result, GnRH release increases, stimulating increased production of gonadotropins and hence increased production of sex steroids.

In females, increasing levels of LH and FSH at puberty stimulate the ovaries to increase their production of the female sex hormones. The increased circulating levels of these hormones initiate the development of the traits of a sexually mature

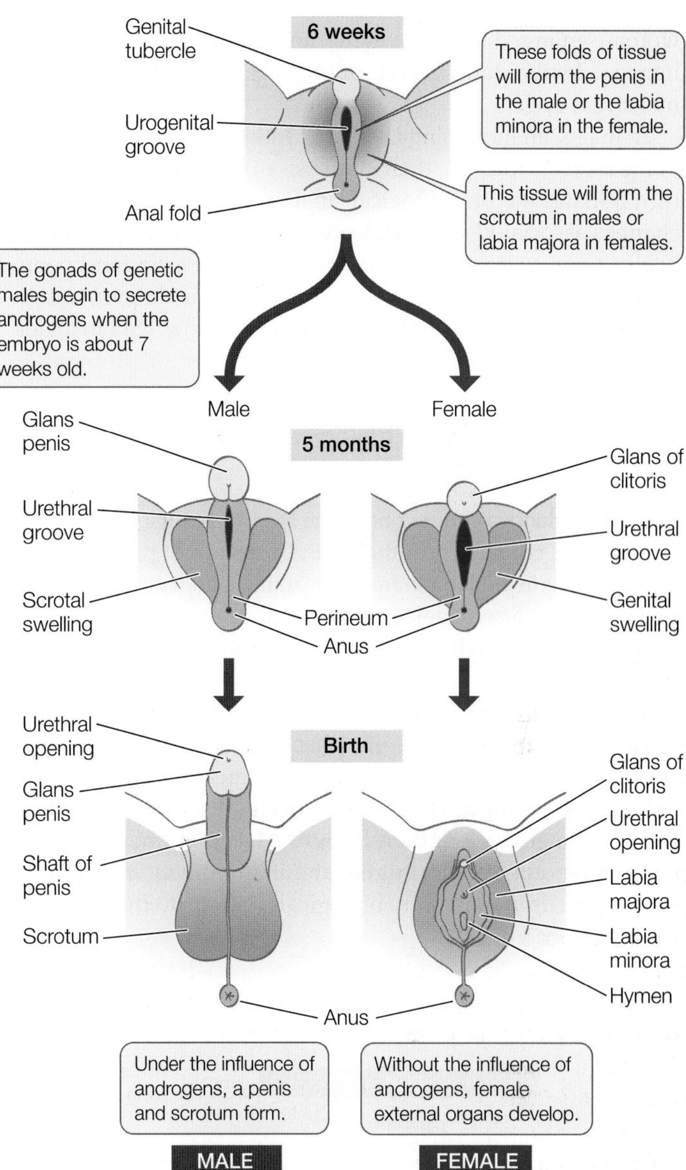


**41.15 Sex Steroids Direct the Development of Human Sex Organs** The sex organs of early human embryos are undifferentiated. Androgens promote the development of male sex organs. In the absence of androgens, female sex organs form.

woman: enlarged breasts, vagina, and uterus; broadened hips; increased subcutaneous fat; pubic hair; and initiation of the menstrual cycle.

In males, an increasing level of LH stimulates groups of cells in the testes to increase their synthesis of testosterone, which in turn initiates the physiological, anatomical, and psychological changes associated with adolescence. The voice deepens, hair begins to grow on the face and body, and the testes and penis grow larger. As we saw at the beginning of this chapter, testosterone also help bones and skeletal muscles grow. FSH in males stimulates production of sperm.

### Melatonin is involved in biological rhythms and photoperiodicity

The **pineal gland** is situated between the two hemispheres of the brain and is connected to the brain by a stalk. It produces the amine hormone **melatonin** from the amino acid tryptophan. The pineal gland releases melatonin in the dark and therefore marks the length of the night. Exposure to light inhibits the release of melatonin.

In vertebrates, melatonin is involved in biological rhythms, including **photoperiodicity**—the phenomenon whereby seasonal changes in day length cause physiological changes. Many species, for example, come into reproductive condition when the days begin to lengthen (**Figure 41.16**). Humans are not strongly photoperiodic, but melatonin in humans may play a role in synchronizing daily biological rhythms to the daily cycle of light and dark.

(A)

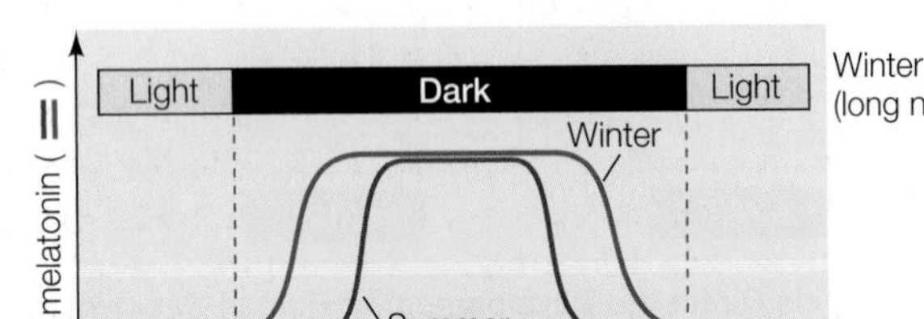


(B)

**41.16 Melatonin Regulates Seasonal Changes** (A) Melatonin release occurs in the dark and is inhibited by exposure to light. The duration of daily melatonin release thus changes as day length (photoperiod) changes, inducing dramatic seasonal physiological changes in some animals. (B) In winter, these Siberian hamsters are white and do not reproduce. In summer, they are mottled brown and breed.

### Many chemicals may act as hormones

We have discussed the major endocrine glands and their hormones in this chapter, but many more hormones exist. As we discuss the organ systems of the body in the chapters that follow, we will frequently describe hormones that their tissues produce as well as hormones that control their functions.

## 41.3 RECAP

**In mammals, the major endocrine glands include the hypothalamus, pituitary gland, thyroid gland, parathyroid glands, pancreas, adrenal glands, gonads, and pineal gland. Each of these glands secretes, and responds to, hormones that play crucial roles in controlling physiology and development.**

- Describe how thyroxine is produced and how its production and release are controlled. **See pp. 861–862 and Figure 41.10**
- How is the concentration of calcium in the blood regulated? **See pp. 863–864 and Figure 41.11**
- What are the hormonal bases for the two forms of diabetes? **See p. 864**
- What changes in the feedback control of sex steroids result in puberty? **See pp. 867–868**

Studying endocrine systems is not easy. Many hormones are released in very small quantities, and some disappear from the extracellular fluid rapidly. A hormone's receptors may be found on diverse cells around the body, and those cells can respond in different ways to the same hormone. How have we overcome these difficulties to learn how hormones work?

## 41.4 How Do We Study Mechanisms of Hormone Action?

We can break the study of hormone actions into different sets of problems. First, we must be able to detect, identify, and measure hormones. Second, we must be able to identify and characterize the receptors for hormones. Third, we must understand the signal transduction pathways activated by hormones in different tissues.

### Hormones can be detected and measured with immunoassays

As we have seen, testosterone has many dramatic and diverse effects, yet its concentration in the blood of adult human males is only about 30 to 100 *billionths of a gram* per milliliter. Measuring hypothalamic neurohormones requires calibrations in the range of *trillionths* of a gram per milliliter.

The ability to detect and measure minute quantities of hormone was an important breakthrough. Rosalyn Yalow developed a method called *radioimmunoassay* because it used radioactive la-

## TOOLS FOR INVESTIGATING LIFE

**41.17 An Immunoassay Allows Measurement of Small Concentrations** An immunoassay uses labeled and unlabeled samples of a purified hormone (or other substance to be measured) and an antibody to that hormone to develop a standard curve against which a sample of unknown concentration can be measured.

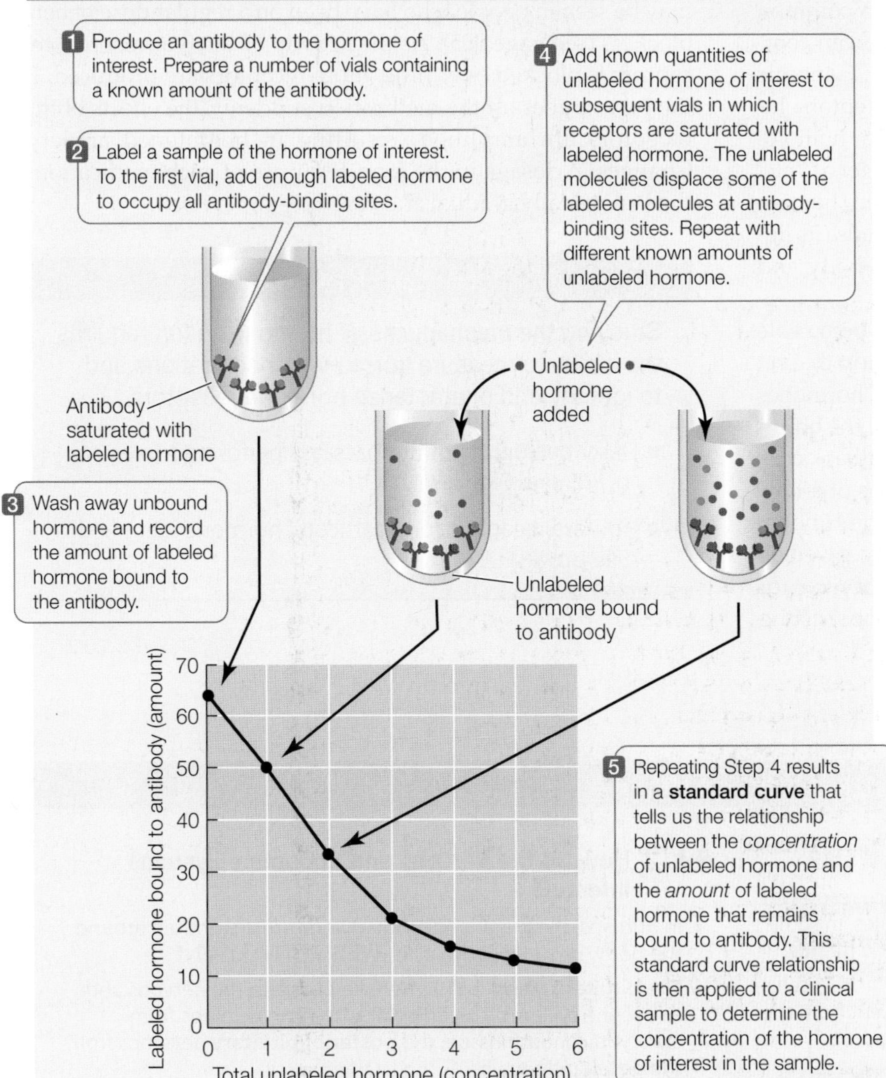
1 Produce an antibody to the hormone of interest. Prepare a number of vials containing a known amount of the antibody.
2 Label a sample of the hormone of interest. To the first vial, add enough labeled hormone to occupy all antibody-binding sites.
4 Add known quantities of unlabeled hormone of interest to subsequent vials in which receptors are saturated with labeled hormone. The unlabeled molecules displace some of the labeled molecules at antibody-binding sites. Repeat with different known amounts of unlabeled hormone.
Antibody saturated with labeled hormone
Unlabeled hormone added
3 Wash away unbound hormone and record the amount of labeled hormone bound to the antibody.
Unlabeled hormone bound to antibody
Labeled hormone bound to antibody (amount)
Total unlabeled hormone (concentration)
0 10 20 30 40 50 60 70
0 1 2 3 4 5 6
5 Repeating Step 4 results in a **standard curve** that tells us the relationship between the *concentration* of unlabeled hormone and the *amount* of labeled hormone that remains bound to antibody. This standard curve relationship is then applied to a clinical sample to determine the concentration of the hormone of interest in the sample.

bels (she used radioactive isotopes of iodine) to track interactions between an antigen (the hormone or other substance of interest to be measured) and an antibody made to that antigen. Today we are more likely to use nonradioactive labels, so the technique is called simply **immunoassay** (**Figure 41.17**). Being able to measure hormones in the blood made it possible to study many important hormonal mechanisms.

Hormones are not simple on–off switches, and an important characteristic of a hormone is the time course over which it acts. This time course can be measured by the hormone's *half-life* in the blood (defined as the length of time it takes for one-half of the hormone molecules to disappear). Soon after endocrine cells are stimulated to secrete their hormone, the hormone reaches its maximum concentration in the blood. By taking a series of blood samples and using immunoassays, researchers can determine how long it takes for the circulating hormone to drop to half of that maximum concentration. The fight-or-flight response to epinephrine, as we have seen, is relatively quick in its onset and termination; the half-life of epinephrine in the blood is only 1 to 3 minutes. The effects of other hormones, such as cortisol and thyroxine, are expressed over much longer periods, and their half-lives are on the order of days or weeks.

Immunoassays have also facilitated the measurement of dose–response relationships. To evaluate a drug or a natural hormone for therapeutic use, it is critical to know the sensitivity of the body to that drug or hormone. Being able to measure the concentrations of drugs or hormones in the blood makes it possible to construct dose–response curves that help physicians adjust dosages appropriately (**Figure 41.18**).

### A hormone can act through many receptors

Different receptors may be involved in mediating the actions of a single hormone. Because there are slight differences among the receptors for a particular hormone, it is possible to create drugs that are very selective in blocking or stimulating specific responses. A number of receptors have been identified, isolated, and purified through biochemical separation techniques. For example, a hormone can be bound to a substrate such as

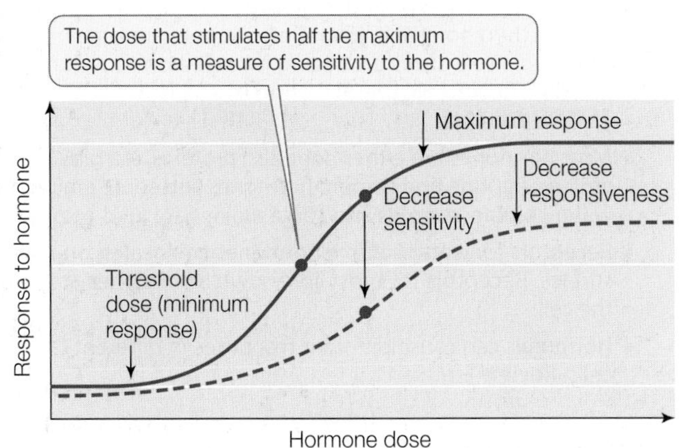
The dose that stimulates half the maximum response is a measure of sensitivity to the hormone.
Maximum response
Decrease responsiveness
Decrease sensitivity
Threshold dose (minimum response)
Response to hormone
Hormone dose

**41.18 Dose-Response Curves Quantify the Body's Response to a Hormone** Between the threshold and maximum doses, a dose-response curve frequently has an S shape. Anything that changes the responsiveness of a system—such as an increase or decrease in the number of receptors in target cells—affects the position of the curve.

resin beads packed into a glass column. When an extract of cells suspected of containing receptors to that hormone is added to the column, the receptors bind to the hormone molecules on the beads. The hormone–receptor complexes can then be washed off the beads and the receptors isolated. This technique is called **affinity chromatography**.

As more receptors are isolated and characterized, researchers discover that they frequently exist in families with common structural features. These common features result from common nucleotide sequences in the receptors' genes. Genomic analyses have led to the discovery of many new receptors. Investigators "scan" the genome for sequences that bear homologies to known receptor gene sequences; when they get a "hit," they have found a candidate gene for a new receptor. They can then identify the molecule to which that receptor binds (its ligand), describe where the receptor occurs within the body, and characterize the receptor's physiological effects.

Knowing the molecular identity of a receptor and being able to measure its concentration makes it possible to study that receptor's regulation. We saw above that the release of hormones can be under negative feedback control. Similarly, the abundance of receptors for a hormone can be under feedback control. In some cases, continuous high concentrations of a hormone can decrease the number of its receptors, a process known as **downregulation**. **Upregulation** of receptors can occur when hormone secretion is suppressed. The regulation of receptor abundance is an important mechanism controlling the sensitivity of the body to hormonal signaling.

An example of downregulation occurs in type II diabetes mellitus, which is characterized by elevated insulin concentrations in the blood and a loss of insulin receptors. Although genetic factors are probably involved, a possible immediate cause of the disease is an overstimulation of pancreatic release of insulin by excessive carbohydrate intake, which leads to downregulation of the insulin receptors. An example of upregulation may be seen in people who have been on a regular dose of beta blockers (see page 865). As the activity of the β-adrenergic receptors is blocked over time, more receptors are produced. If the person goes off the medication suddenly, the effects of the receptors are amplified, resulting in heightened anxiety. Changes in dosage in the long-term use of such medications thus are usually gradual and carefully supervised.

## 41.4 RECAP

**Studying the mechanisms of hormone action requires the ability to measure hormone concentrations and to identify and characterize hormone receptors.**

- Describe how an immunoassay is performed. See p. 869 and Figure 41.17
- How are receptors for a particular hormone identified? See pp. 869–870

## CHAPTER SUMMARY

### 41.1 What Are Hormones and How Do They Work?

- **Endocrine cells** secrete chemical signals that induce responses in other cells that have receptors for those molecules. In some cases endocrine cells are aggregated into **endocrine glands**.
- **Hormones** are endocrine signals that are secreted from a cell, circulate in the blood, and bind to **target cells** distant from the secreting cell. Review Figure 41.1
- The chemical structures of hormones are highly conserved, but through evolution, hormones have acquired different functions in different animal groups. Review Figure 41.2
- Two hormones, PTTH and ecdysone, control molting in arthropods. A third hormone, **juvenile hormone**, prevents maturation. When an insect stops producing juvenile hormone, it molts into an adult. Review Figures 41.3 and 41.4, **ANIMATED TUTORIAL 41.1**
- Most hormones are either peptides, proteins, steroids, or amines. Peptide and protein hormones and some amines are water-soluble; steroids and some amines are lipid-soluble.
- Receptors for water-soluble hormones are located on the cell surface. Receptors for most lipid-soluble hormones are inside the cell.
- Hormones can cause different responses in different target cells. Review Figure 41.5

### 41.2 How Do the Nervous and Endocrine Systems Interact?

- In humans, the major endocrine glands are distributed around the body. Review Figure 41.6, **WEB ACTIVITY 41.1**
- The **pituitary gland** is the interface between the nervous and endocrine systems. The **anterior pituitary** develops from embryonic mouth tissue; the **posterior pituitary** develops from the developing brain.
- The posterior pituitary secretes two **neurohormones**: **antidiuretic hormone** (**ADH**) and **oxytocin**. The anterior pituitary secretes **tropic hormones** (thyrotropin, corticotropin, luteinizing hormone, and follicle-stimulating hormone) as well as **growth hormone**, **prolactin**, melanocyte-stimulating hormone, **endorphins**, and **enkephalins**.
- The anterior pituitary is controlled by neurohormones produced by cells in the **hypothalamus** and transported through **portal blood vessels** to the anterior pituitary. Review Figures 41.7 and 41.8, **ANIMATED TUTORIAL 41.2**
- Hormone release is controlled by negative feedback loops. Review Figure 41.9

## 41.3 What Are the Major Mammalian Endocrine Glands and Hormones?

- The **thyroid gland** is controlled by **thyrotropin** and secretes **thyroxine**, which controls cell metabolism. Review Figure 41.10
- The level of calcium in the blood is regulated by three hormones. **Calcitonin** from the thyroid lowers blood calcium by promoting bone deposition. **Parathyroid hormone** (**PTH**) raises blood calcium by promoting bone turnover and decreasing calcium excretion. **Vitamin D** promotes calcium absorption from the digestive tract. Review Figure 41.11, **ANIMATED TUTORIAL 41.3**
- The **pancreas** secretes three hormones. **Insulin** stimulates glucose uptake by cells and lowers blood glucose, **glucagon** raises blood glucose, and **somatostatin** slows the rate of nutrient processing.
- The **adrenal gland** has two portions, one within the other. The inner portion, the **adrenal medulla**, releases **epinephrine** and **norepinephrine** in response to stress. The outer portion, the **adrenal cortex,** produces three classes of **corticosteroids**: glucocorticoids, mineralocorticoids, and small amounts of sex steroids. Review Figures 41.12 and 41.13
- **Aldosterone** is a mineralocorticoid that stimulates the kidneys to conserve sodium and excrete potassium. **Cortisol** is a glucocorticoid that is released in response to stressful stimuli but acts more slowly than the hormones of the adrenal medulla. Review Figure 41.14
- Sex hormones (**androgens** in males, **estrogens** and **progesterone** in females) are produced by the **gonads** in response to tropic hormones. Sex hormones control sexual development, secondary sexual characteristics, and reproductive functions. Review Figures 41.14 and 41.15
- The **pineal gland** releases **melatonin**, which is involved in controlling biological rhythms. Review Figure 41.16

## 41.4 How Do We Study Mechanisms of Hormone Action?

- **Immunoassay** techniques are used to measure concentrations of hormones and other substances. Review Figure 41.17
- The sensitivity of a cell to hormones can be altered by **down-regulation** or **upregulation** of the receptors in that cell.

**SEE WEB ACTIVITY 41.2 for a concept review of this chapter.**

# SELF-QUIZ

1. Before puberty
   *a.* the pituitary secretes luteinizing hormone and follicle-stimulating hormone, but the gonads are unresponsive.
   *b.* the hypothalamus does not secrete much gonadotropin-releasing hormone.
   *c.* males can stimulate massive muscle development through a vigorous training program.
   *d.* testosterone plays no role in development of the male sex organs.
   *e.* genetic females will develop male genitals unless estrogen is present.
2. Both epinephrine and cortisol are secreted in response to stress. Which of the following statements is also true for *both* of these hormones?
   *a.* They act to increase blood glucose availability.
   *b.* Their receptors are on the surfaces of target cells.
   *c.* They are secreted by the adrenal cortex.
   *d.* Their secretion is stimulated by corticotropin.
   *e.* They are secreted into the blood within seconds of the onset of stress.
3. Growth hormone
   *a.* can cause adults to grow taller.
   *b.* stimulates protein synthesis.
   *c.* is released by the hypothalamus.
   *d.* can be obtained only from cadavers.
   *e.* is a steroid.
4. PTH
   *a.* stimulates osteoblasts to lay down new bone.
   *b.* reduces blood calcium levels.
   *c.* stimulates calcitonin release.
   *d.* is produced by the thyroid gland.
   *e.* is released when blood calcium levels fall.
5. Steroid hormones
   *a.* are produced only by the adrenal cortex.
   *b.* have only cell surface receptors.
   *c.* are water-soluble.
   *d.* act by altering the activity of proteins in the target cell.
   *e.* act by altering gene expression in the target cell.
6. The hormone ecdysone
   *a.* is released from the posterior pituitary.
   *b.* stimulates molting in insects.
   *c.* maintains an insect in larval stages unless PTTH is present.
   *d.* stimulates the secretion of juvenile hormone from the prothoracic glands.
   *e.* keeps the insect exoskeleton flexible to permit growth.
7. The posterior pituitary
   *a.* synthesizes oxytocin.
   *b.* is under the control of hypothalamic releasing neurohormones.
   *c.* secretes tropic hormones.
   *d.* secretes neurohormones.
   *e.* is under feedback control by thyroxine.
8. Which of the following contributes to the development of goiter?
   *a.* Inadequate iodine in the diet
   *b.* Autoimmune antibodies that stimulate the TSH receptor
   *c.* Lack of feedback from circulating $T_3$ and $T_4$
   *d.* Overproduction of thyroglobulin
   *e.* All of the above
9. Which of the following is a likely cause of diabetes?
   *a.* Overproduction of insulin by $\beta$ cells of the pancreas
   *b.* Loss of $\alpha$ cells of the pancreas
   *c.* Loss of insulin receptors
   *d.* Overproduction of glucagon
   *e.* Loss of receptors for somatostatin
10. Which statement is true of all hormones?
   *a.* They are secreted by glands.
   *b.* They have receptors on cell surfaces.
   *c.* They may stimulate different responses in different cells.
   *d.* There is a gene that codes for each hormone.
   *e.* When the same hormone occurs in different species, it has the same action.

## FOR DISCUSSION

1. Explain how both hyperthyroidism and hypothyroidism can cause goiter. Refer to the roles of the hypothalamus and the pituitary in your answer.
2. Neurons, the cells of the nervous system, do not require insulin. Why might this be so, and why is it important?
3. Various side effects of anabolic steroid abuse were mentioned at the opening of this chapter. Some of these effects are due to the direct action of the steroid, but others are due to the negative feedback action of the steroid. Discuss an example of each and explain possible mechanisms.
4. Compare the characteristics you would expect of a hormone signaling system that controls a short-term process, such as digestion, with the characteristics you would expect of a hormone signaling system that controls a long-term process, such as development.

## ADDITIONAL INVESTIGATION

Each spring, male deer grow antlers that they use in male–male competition for mates. Each fall they shed those antlers. Although females of most deer species do not grow antlers, the caribou are an exception. Female caribou grow antlers in the spring but do not shed them in the fall. (Over the winter, they use their antlers to defend patches of food from males.) In addition, newborn caribou begin to grow antlers after birth. Assuming that antler growth is controlled by hormones, how would you investigate the control of antler growth in caribou? What hormonal assays might you want to use? What hypotheses would you test with respect to the relationship between time of year and antler growth?

# 42 Immunology: Animal Defense Systems

## The plague of Athens

*As a rule, however, there was no ostensible cause; but people in good health were all of a sudden attacked by violent heats in the head, and redness and inflammation in the eyes, the inward parts, such as the throat or tongue, becoming bloody and emitting an unnatural and fetid breath.*

The Athenian historian Thucydides wrote these words in 430 BCE. He was describing a rapidly spreading infectious disease, or plague, that was sweeping the city of Athens in ancient Greece, and would ultimately kill about one-third of its inhabitants.

One year earlier, in 431 BCE, the Spartans had surrounded Athens. Sparta was a rival state with a highly disciplined army, while Athens was near a port and had a powerful navy. The Athenian leader Pericles decided to take advantage of his naval supremacy. He instructed his soldiers to abandon the countryside to the Spartans, relying on the navy to keep the enemies out of the city and to keep the Athenians supplied with food. People living in the countryside poured into Athens.

As Athens became more crowded, declining sanitation and close living conditions resulted in the city becoming an incubator for infectious diseases. In his famous *History of the Peloponnesian War*, Thucydides speculated that the plague came to Athens from Africa. His precise clinical description of the disease has provoked great debate among medical historians regarding its nature. In 1994, a burial ground was found near Athens that apparently contained the remains of several dozen people who died in the plague. PCR (polymerase chain reaction) analyses of dental pulp from the skeletons revealed the presence of DNA sequences from the bacterium that causes typhoid fever, and this has been proposed as the cause of the plague. Other scientists disagree, using arguments based on a comparison of Thucydides' account with the clinical description of typhoid fever today.

Whatever the cause, another passage from Thucydides (who was one of the few lucky survivors of the plague) bears noting:

*Yet it was with those who had recovered from the disease that the sick and the dying found most compassion. These knew what it was from experience and now had no fear for themselves; for the same man was never attacked twice, never at least fatally.*

Two questions arise from this passage: First, how could a person survive the disease? And second, how could the survivors be resistant to further infection?

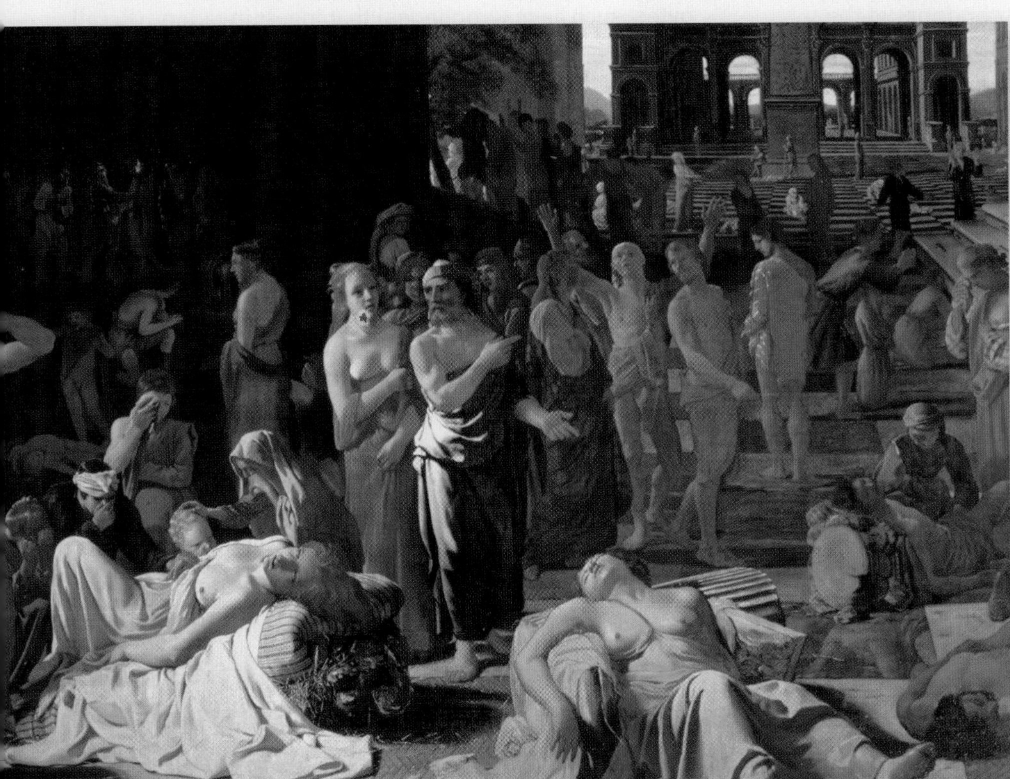

**The Peloponnesian War** The blockade of Sparta and its Peloponnesian allies on Athens made the latter crowded and vulnerable to the spread of an infectious disease.

**Thucydides** The Athenian general wrote a detailed history of the Peloponnesian war and the spread of an infectious disease that killed 1 in 3 Athenians. He survived the disease, presumably because of his immune system.

Almost 2,500 years after Thucydides recorded his observations, we have some answers to these questions. These answers come from studies of the mammalian immune system, which recognizes and fights off pathogens. When Thucydides caught the disease, specialized white blood cells called T cells recognized proteins of the infecting agent on the surfaces of his body's cells. This initiated a series of events that resulted in the killing of infected cells and destruction of the disease agent by antibodies, specific proteins made by a different set of white blood cells called B cells. Some descendants of the T and B cells persisted in his body as "memory cells," which defended him when he was exposed to the disease later. The operation of the mammalian immune system and how its components defend the body are the focus of this chapter.

**IN THIS CHAPTER** we first describe how the innate, nonspecific immune system prevents most pathogens from entering the body and deals with those that do. Then we look at how the specific, adaptive immune system targets invaders such as the typhoid bacterium, which may have killed the ancient Athenians. We see how the reshuffling of genetic material helps animals fight off the staggering diversity of potential invaders, and finally, we see what happens when this complex and crucial system malfunctions.

## CHAPTER OUTLINE



## 42.1 What Are the Major Defense Systems of Animals?

Animals have a number of ways of defending themselves against **pathogens**—harmful organisms and viruses that can cause disease. These defense systems are based on the distinction between *self*—the animal's own molecules—and *nonself*, or foreign, molecules. The defensive response involves three phases:

- *Recognition phase*. The organism must be able to discriminate between self and nonself.
- *Activation phase*. The recognition event leads to a mobilization of cells and molecules to fight the invader.
- *Effector phase*. The mobilized cells and molecules destroy the invader.

There are two general types of defense mechanisms:

- **Nonspecific defenses**, or *innate defenses*, provide the first line of defense against pathogens. They typically act very rapidly, and include barriers such as the skin, molecules that are toxic to invaders, and phagocytic cells (**phagocytes**, such as **macrophages**) that ingest invaders. (Recall from Section 6.5 that phagocytosis is a form of endocytosis, in which a cell engulfs a large particle or cell.) This system recognizes broad classes of organisms or molecules and gives a quick response, within minutes or hours. Most animals have nonspecific defenses.
- **Specific defenses** are *adaptive* mechanisms aimed at specific pathogens. For example, a specific defense system can make an antibody protein that will recognize, bind to, and aid in the destruction of a certain virus if that virus ever enters the bloodstream. These systems recognize specific configurations of atoms in a molecule and are typically slow to develop and long-lasting. Specific defense mechanisms are found in vertebrate animals.

Mammals have both kinds of defense mechanism and are the focus of this chapter. In mammals and other vertebrates, the nonspecific and specific mechanisms operate together as a *coordinated* defense system. The nonspecific defenses are the body's first line of defense because the specific defenses often require days or even weeks to become effective.

### Blood and lymph tissues play important roles in defense

The components of the mammalian defense system are dispersed throughout the body and interact with almost all of its other tissues and organs. The *lymphoid tissues*, which include the thymus, bone marrow, spleen, and lymph nodes, are essential parts of the defense system (**Figure 42.1**). The blood and lymph are complex systems with nondefensive functions that are discussed in Chapter 50. They each have central roles in defense as well.

The blood and lymph both consist of liquids in which cells are suspended:

- **Blood plasma** is a yellowish solution containing ions, small molecular solutes, and soluble proteins. Suspended in the plasma are red blood cells, white blood cells, and platelets (cell fragments essential to blood clotting). While red blood cells are normally confined to the closed circulatory system (the heart, arteries, capillaries, and veins), white blood cells and platelets are also found in the lymph.
- **Lymph** is a fluid derived from the blood and other tissues that accumulates in intercellular spaces throughout the body. From these spaces, the lymph moves slowly into the vessels of the lymphatic system. Tiny lymph capillaries conduct this fluid to larger ducts that eventually join together, forming one large vessel, the thoracic duct, which joins a major vein (the left subclavian vein) near the heart. By this system of vessels, the lymph is eventually returned to the blood and the circulatory system.

At many sites along the lymph vessels are small, roundish structures called **lymph nodes**, which contain a type of white blood cell called a **lymphocyte**. As lymph passes through a lymph node, the lymphocytes encounter foreign cells and molecules that have entered the body, and if they are recognized, an immune response is initiated.

### White blood cells play many defensive roles

One milliliter of human blood typically contains about *5 billion* red blood cells and 7 *million* of the larger white blood cells. All of these cells originate from multipotent stem cells (constantly dividing undifferentiated cells that can form several different cell types) in the bone marrow. Examine **Figure 42.2** and you will see that there are two major families of **white blood cells** (also called *leukocytes*): *phagocytes* and *lymphocytes*. Lymphocytes, which include **B cells** and **T cells**, are smaller than phagocytes and are not phagocytic. Each family contains different types of cells with specialized functions. Natural killer cells and some kinds of phagocytes are also referred to collectively as *granulocytes* because they contain numerous granules (vesicles containing defensive enzymes). Defensive proteins and signals play fundamental roles in the interactions and functioning of these cells.

### Immune system proteins bind pathogens or signal other cells

Similar to the actions of hormones, which we discuss in Chapter 41, the cells that defend mammalian bodies work together, interacting with one another and with the cells of invading pathogens. These cell–cell interactions are accomplished by a variety of key proteins, including receptors, other cell surface proteins, and signaling molecules. Four of the major players are listed here, and will be discussed in more detail later in the chapter.

Thoracic duct
Lymph ducts conduct lymph.
T cells mature in the thymus.
In the lymph nodes, lymph is filtered and white blood cells inspect it for pathogens.
The spleen filters circulating blood.
B cells mature in the bone marrow.

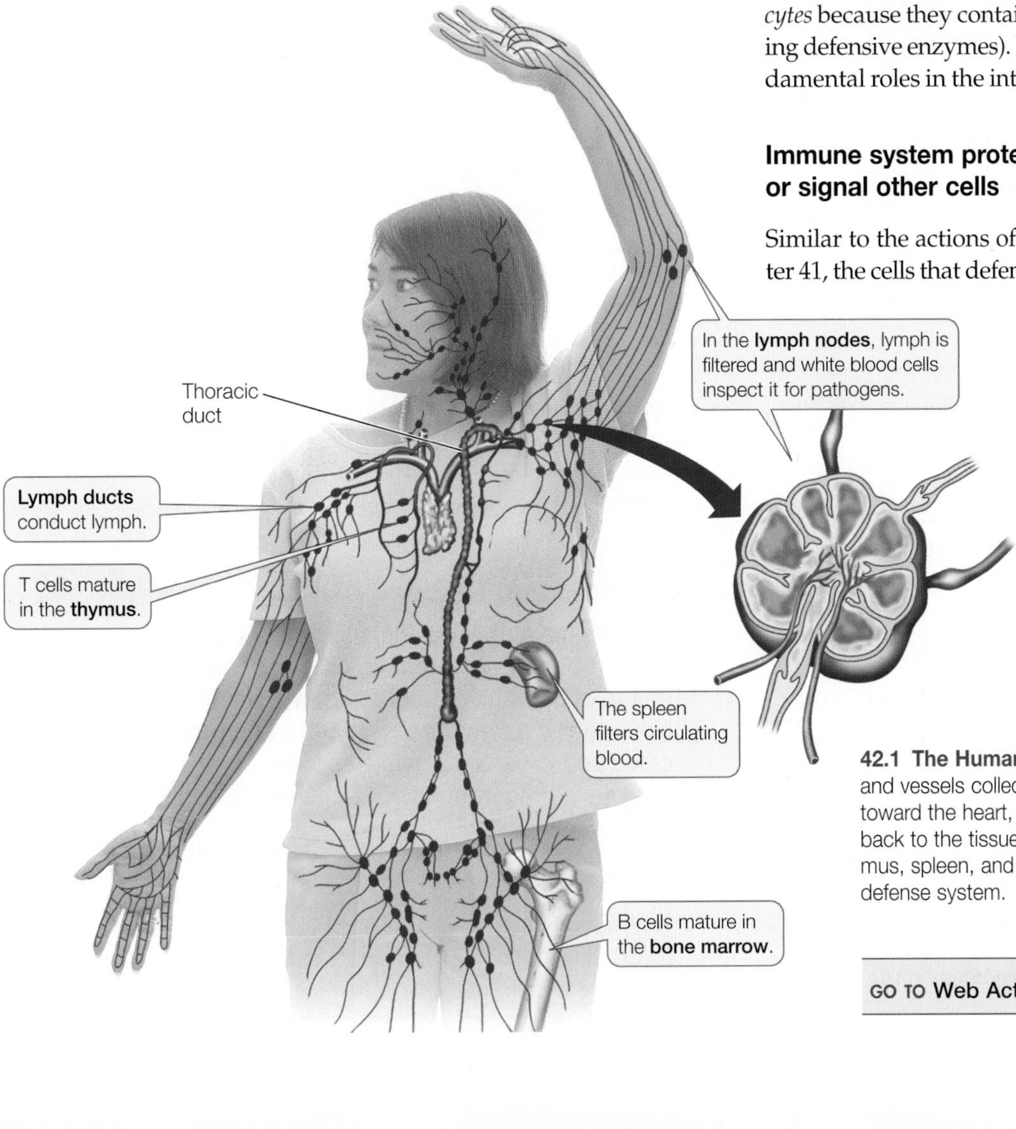

**42.1 The Human Lymphatic System** A network of ducts and vessels collects lymph from body tissues and carries it toward the heart, where it mixes with blood to be pumped back to the tissues. Other lymphoid tissues, including the thymus, spleen, and bone marrow, are also essential to the body's defense system.

yourBioPortal.com
GO TO Web Activity 42.1 • The Human Defense System

**42.2 Blood Cells** Multipotent stem cells in the bone marrow can differentiate into red blood cells, platelets, and the various types of white blood cells.

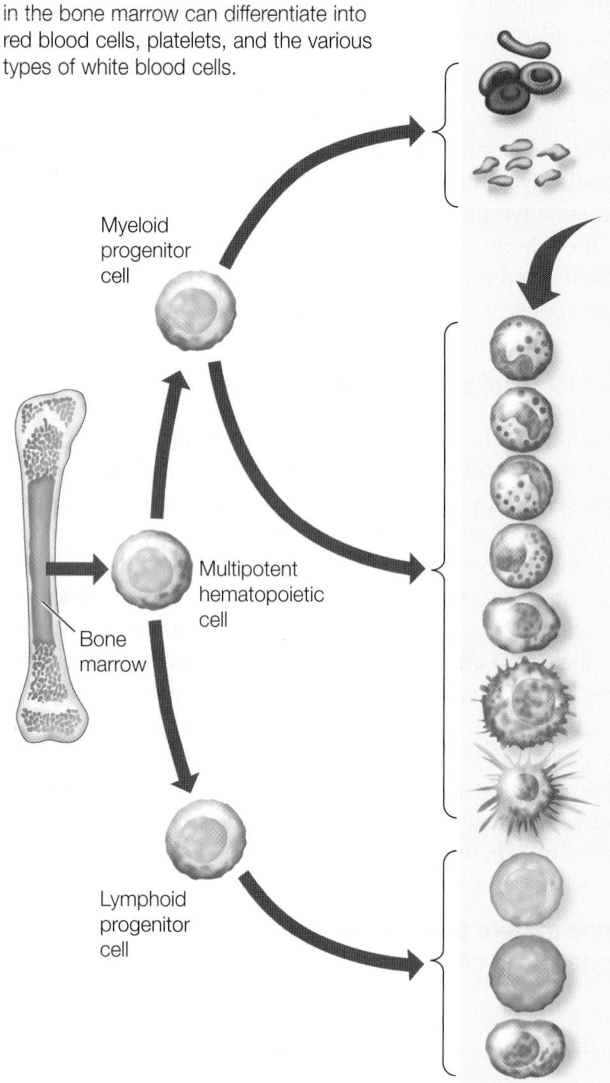

Bone marrow

Multipotent hematopoietic cell

Myeloid progenitor cell

Lymphoid progenitor cell

| TYPE OF CELL | FUNCTION |
|---|---|
| Red blood cells (erythrocytes) | Transport oxygen and carbon dioxide |
| Platelets (cell fragments without nuclei) | Initiate blood clotting |
| White blood cells (leukocytes) | |
| GRANULAR CELLS | |
| Basophils | Release histamine; may promote development of T cells |
| Eosinophils | Kill antibody-coated parasites |
| Neutrophils | Stimulate inflammation |
| Mast cells | Release histamine when damaged |
| AGRANULAR CELLS | |
| Monocytes | Develop into macrophages and dendritic cells |
| Macrophages | Engulf and digest microorganisms; activate T cells |
| Dendritic cells | Present antigens to T cells |
| B lymphocytes | Differentiate to form antibody-producing cells and memory cells |
| T lymphocytes | Kill virus-infected cells; regulate activities of other white blood cells |
| Natural killer cells | Attack and lyse virus-infected or cancerous body cells |

yourBioPortal.com

GO TO Animated Tutorial 42.1 • Cells of the Immune System

- **Antibodies** are proteins that bind specifically to certain substances identified by the immune system as nonself or *altered self* (a major histocompatibility protein that has been altered by a pathogen). This binding can inactivate viruses and toxins, and it can act as tag on nonself cells, making them easier for the immune system cells to attack. Antibodies are produced by B cells.
- **Major histocompatibility complex (MHC)** proteins are found in two classes. MHC I proteins are found on the surfaces of most cells in the mammalian body. MHC II proteins are found on most immune system cells. MHC proteins are important self-identifying labels and play a major role in coordinating interactions between lymphocytes and macrophages.
- **T cell receptors** are integral membrane proteins on the surfaces of T cells. They recognize and bind to nonself substances presented with self MHC molecules on the surfaces of other cells.
- **Cytokines** are soluble signaling proteins released by many cell types. They bind to cell surface receptors and alter the behavior of their target cells. Various cytokines activate or inactivate B cells, macrophages, and T cells.

## 42.1 RECAP

**Animals have nonspecific and specific defenses against pathogens. Both kinds of mechanism are based on the ability to differentiate self from nonself. Nonspecific defenses target a broad range of molecules and organisms, while specific defenses target specific configurations of atoms on molecules.**

- List the differences between specific and nonspecific defenses. See p. 874
- What are the two classes of white blood cells, and how do they function in vertebrate defense systems? See p. 875 and Figure 42.2

The outcome of a disease—the life or death of the host—often depends on the success of both rapid, nonspecific responses and long-lasting, specific responses to invading pathogens. We turn now to the nonspecific defenses that protect vertebrates from disease.

## 42.2 What Are the Characteristics of the Nonspecific Defenses?

Nonspecific defenses are general protection mechanisms that attempt to stop pathogens from invading the body. As noted above, they provide first line of defense, in both time and location. Essentially, they are "ready to go," in contrast to specific immunity, which takes time to develop after a pathogen or toxin has been recognized as nonself. In mammals, nonspecific mechanisms include physical barriers as well as cellular and chemical defenses (**Table 42.1**).

### Barriers and local agents defend the body against invaders

Consider a pathogenic bacterium that causes disease in internal organs and that lands on human skin. The challenges faced by the bacterium just to reach its target are formidable. First, there is the physical barrier of the skin. Bacteria rarely penetrate intact skin; by the same token, broken skin increases the risk of infection. The saltiness and dryness of skin may not be hospitable to the growth of the invader. On the other hand, there are some bacteria and fungi that normally live and sometimes reproduce in great numbers on our body surfaces without causing disease. They are referred to as **normal flora**. These natural occupants of our bodies form a nonspecific defense because they compete with pathogens for space and nutrients.

Consider a pathogen that lands on the surface of the nose. The mucous membranes found at the inner surfaces of the nose (as well as the digestive, respiratory, and urogenital systems) contain the enzyme **lysozyme**, which attacks the cell walls of many bacteria, causing them to *lyse* (burst open). Mucus in the nose traps airborne microorganisms, and most of those that get past this filter end up trapped in mucus deeper in the respiratory tract. Mucus and trapped pathogens are removed by the beating of cilia in the respiratory passageway, which continuously move a sheet of mucus and its trapped debris up toward the nose and mouth where it can be expelled or swallowed. Sneezing is another way to remove microorganisms from the respiratory tract.

TABLE 42.1

**Human Nonspecific Defenses**

| DEFENSIVE MECHANISM | FUNCTION |
|---|---|
| **Surface barriers** | |
| Skin | Prevents entry of pathogens and foreign substances |
| Acid secretions | Inhibit bacterial growth on skin |
| Mucus | Prevents entry of pathogens; produces defensins that kill pathogens |
| Mucous secretions | Trap bacteria and other pathogens in digestive and respiratory tracts |
| Nasal hairs | Filter bacteria in nasal passages |
| Cilia | Move mucus and trapped materials away from respiratory passages |
| Gastric juice | Concentrated HCl and proteases destroy pathogens in stomach |
| Acid in vagina | Limits growth of fungi and bacteria in female reproductive tract |
| Tears, saliva | Lubricate and cleanse; contain lysozyme, which destroys bacteria |
| **Nonspecific cellular, chemical, and coordinated defenses** | |
| Normal flora | Compete with pathogens; may produce substances toxic to pathogens |
| Fever | Body-wide response inhibits microbial multiplication and speeds body repair processes |
| Coughing, sneezing | Expels pathogens from upper respiratory passages |
| Inflammatory response (involves leakage of blood plasma and phagocytes from vessels) | Limits spread of pathogens to neighboring tissues; concentrates defenses; digests pathogens and dead tissue cells; releases chemical mediators that attract phagocytes and lymphocytes to site |
| Phagocytes (macrophages and neutrophils) | Engulf and destroy pathogens that enter body |
| Natural killer cells | Attack and lyse virus-infected or cancerous body cells |
| Antimicrobial proteins | |
| Interferons | Released by virus-infected cells to protect healthy tissue from viral infection; mobilize specific defenses |
| Complement proteins | Lyse microorganisms, enhance phagocytosis, and assist in inflammatory and antibody responses |

Finally, the mucous membranes produce **defensins**, peptides that are 18–45 amino acids long and contain hydrophobic domains. They are toxic to a wide range of pathogens, including bacteria, microbial eukaryotes, and enveloped viruses. Defensins insert themselves into the plasma membranes of these organisms and make the membranes permeable, thus killing the invaders. Defensins are also produced in phagocytes, where they kill pathogens trapped by phagocytosis. Plants also produce defensins in response to pathogen exposure (see Chapter 39).

Consider a bacterium that lands in the mouth. If it survives the lysozyme in saliva and enters the digestive tract (stomach, small intestine, and large intestine) it is met by other defenses. The gastric juice in the stomach is a deadly environment for many bacteria because of the hydrochloric acid and proteases that are secreted into it. Bacteria cannot normally penetrate the lining of the small intestine, and some pathogens are killed by bile salts secreted into this part of the digestive tract. Any bacteria that survive these lines of defense (which include normal flora) and enter the large intestine are removed quickly with the feces.

All of these barriers and local agents are nonspecific defenses because they act on all invading pathogens in the same way. Most are properties of the tissue, called *epithelial tissue*, that covers surfaces on and within organs. More complex nonspecific defenses await any pathogens that manage to elude this first set of defenses.

### Other nonspecific defenses include specialized proteins and cellular processes

Pathogens that penetrate the body's outer and inner surfaces encounter more complex nonspecific defenses. These include the activation of defensive cells and the secretion of various defensive proteins, such as *complement* and *interferon* proteins.

**COMPLEMENT PROTEINS** Vertebrate blood contains more than 20 different proteins that make up the antimicrobial **complement system**. This system can be activated by various mechanisms, including both nonspecific and specific defense responses. The proteins act in a characteristic sequence, or cascade, with each protein activating the next:

- First, they attach to specific components on the surface of a microbe or to an antibody that has already bound to the microbe's surface. In either case, binding helps phagocytes recognize and destroy the microbe.
- Then, they activate the inflammation response and attract phagocytes to the site of infection.
- Finally, they lyse invading cells (such as bacteria).

**INTERFERONS** When a cell is infected by a pathogen, it produces small amounts of signaling proteins called **interferons** that increase the resistance of neighboring cells to infection. Interferons are a class of cytokines and have been found in many vertebrates. Various molecules, including double stranded (viral) RNA, induce the production of interferons. Thus interferons are particularly important as a first line of nonspecific defense against viruses. Interferons bind to receptors on the plasma membranes of uninfected cells, stimulating a signaling pathway that inhibits viral reproduction if the cells are subsequently infected. In addition, interferons stimulate the cells to hydrolyze bacterial or viral proteins to peptides, an initial step in specific immunity (see Section 42.3).

**PHAGOCYTES** Some phagocytes travel freely in the circulatory and lymphatic systems; others can move out of blood vessels and adhere to certain tissues. Pathogenic cells, viruses, or fragments of these invaders can be recognized by phagocytes, which then ingest them by phagocytosis. Defensins, nitric oxide, and reactive oxygen intermediates inside these phagocytes then kill the pathogens.

**NATURAL KILLER CELLS** One class of lymphocytes, known as **natural killer cells**, can distinguish virus-infected cells and some tumor cells from their normal counterparts and initiate the apoptosis of these target cells. In addition to this nonspecific action, natural killer cells interact with the specific defense mechanisms by lysing antibody-labeled target cells.

### Inflammation is a coordinated response to infection or injury

When tissue is damaged because of infection or injury, the body responds with **inflammation**. This response can happen almost anywhere in the body, internally as well as on the surface. Inflammation is an important phenomenon: it isolates the damaged area to stop the spread of the damage; it recruits cells and molecules to the damaged location to kill the invader; and it promotes healing. The first responders to tissue damage are **mast cells**, which adhere to the skin and the linings of organs, and release numerous chemical signals including:

- **tumor necrosis factor**, a cytokine protein that kills target cells and activates immune cells.
- **prostaglandins**, fatty acid derivatives that play roles in various responses including the widening of blood vessels. Prostaglandins interact with nerve endings and are partly responsible for the pain caused by inflammation.
- **histamine**, an amino acid derivative that leads to itchy, watery eyes, and rashes seen with some types of allergic reactions.

The redness and heat of inflammation result from the dilation and leakiness of blood vessels in the infected or injured area (**Figure 42.3**). Phagocytes enter the inflamed area where they engulf the invaders and dead tissue cells. Phagocytes are responsible for most of the healing associated with inflammation. They produce several cytokines, which (among other functions) signal the brain to produce a fever. This rise in body temperature accelerates lymphocyte production and phagocytosis, thereby speeding the immune response. In some cases, pathogens are temperature-sensitive, and their growth is inhibited. The pain of inflammation results from increased pressure due to swelling, the action of leaked enzymes on nerve endings, and the action of prostaglandins, which increase the sensitivity of the nerve endings to pain.

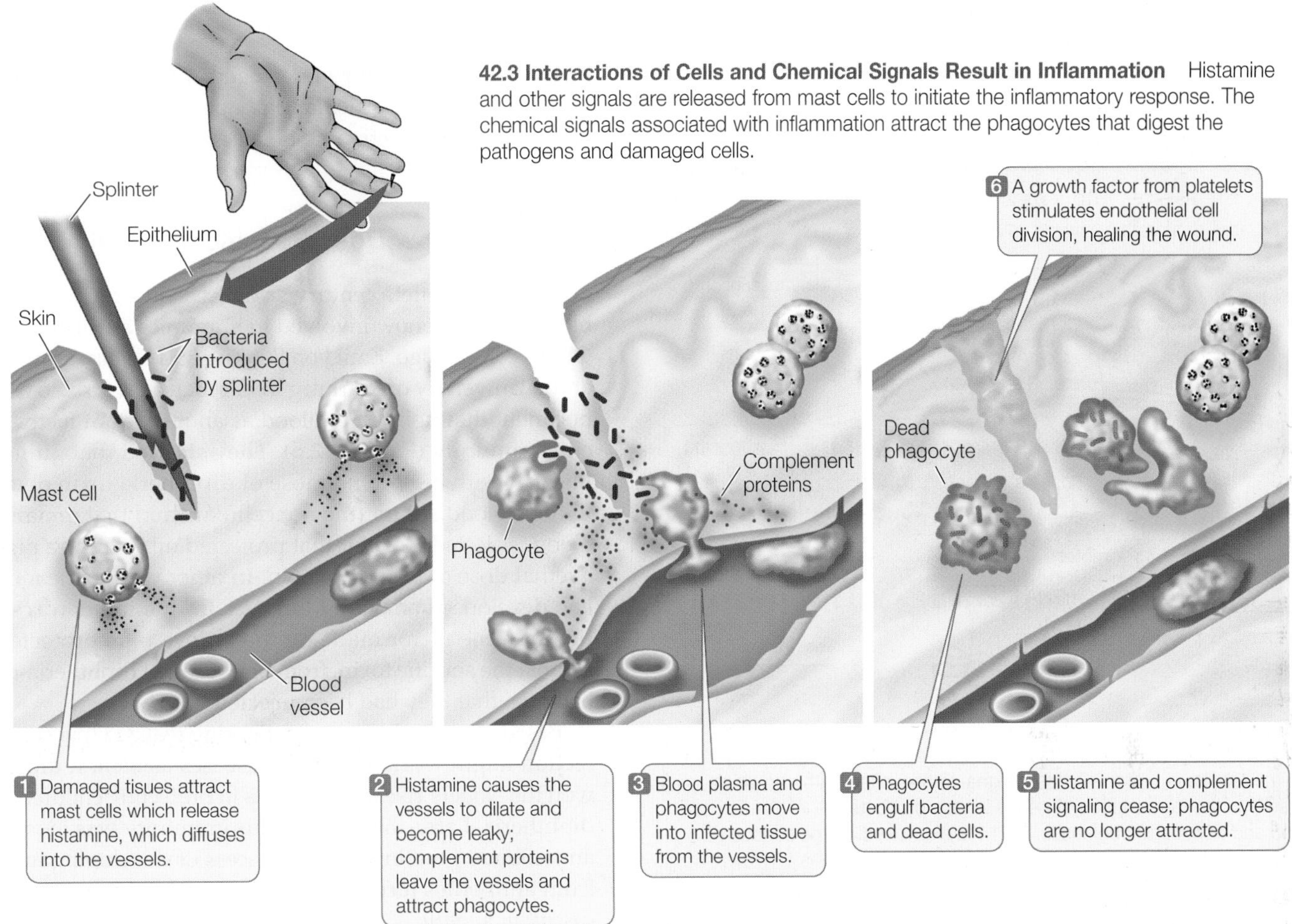

Splinter
Epithelium
Skin
Bacteria introduced by splinter
Mast cell
Blood vessel
Phagocyte
Complement proteins
Dead phagocyte
6 A growth factor from platelets stimulates endothelial cell division, healing the wound.
1 Damaged tisues attract mast cells which release histamine, which diffuses into the vessels.
2 Histamine causes the vessels to dilate and become leaky; complement proteins leave the vessels and attract phagocytes.
3 Blood plasma and phagocytes move into infected tissue from the vessels.
4 Phagocytes engulf bacteria and dead cells.
5 Histamine and complement signaling cease; phagocytes are no longer attracted.


**42.3 Interactions of Cells and Chemical Signals Result in Inflammation** Histamine and other signals are released from mast cells to initiate the inflammatory response. The chemical signals associated with inflammation attract the phagocytes that digest the pathogens and damaged cells.

yourBioPortal.com
GO TO Web Activity 42.2 • Inflammation Response

Following inflammation, pus may accumulate. Pus is a mixture of leaked fluid and dead cells: bacteria, neutrophils (the most abundant white blood cells—see Figure 42.2), and damaged body cells. Pus is a normal result of inflammation, and is gradually consumed and further digested by phagocytes called macrophages.

### Inflammation can cause medical problems

While inflammation is generally a good thing, sometimes the inflammatory response is inappropriately strong, resulting in some allergies, cases of autoimmunity, and sepsis. In these cases, the response causes more damage than was originally there. We will discuss allergy and autoimmune diseases in Section 42.7. In some cases of severe bacterial infection, the inflammation response does not remain local. Instead, it extends throughout the bloodstream in a condition called *sepsis*. As in a local infection or injury, blood vessels dilate, but they do so throughout the body. The lowering of blood pressure that results is a medical emergency and can be lethal.

The symptoms of swelling, pain, and fever caused by excessive inflammation can be bothersome to the point of incapacitation. Diseases such as rheumatoid arthritis and chronic obstructive pulmonary disease, and accidents such as athletic injuries result in tissue damage and an inflammatory response. In order to manage excessive inflammation, drugs have been developed that act on the various cytokines and signal pathways to reduce inflammation and its symptoms. For example, the bark of the willow tree, *Salix*, has been known for millennia to contain a potent antidote to fever and swelling. In the nineteenth century the active ingredient, salicylic acid, was isolated and a better version, acetylsalicylic acid, was made. This drug—also called aspirin—works by inhibiting an enzyme on the pathway to the synthesis of prostaglandins. Other anti-inflammatory drugs act on the prostaglandin pathway, on the actions of tumor necrosis factor, and on the actions of histamine.

### Cell signaling pathways stimulate the body's defenses

An invading pathogen can be regarded as a signal. In response to that signal, the body produces molecules (complement proteins, interferons, and other cytokines) that regulate phagocytosis and other defense processes. Not surprisingly, the link between signal and response is a signal transduction pathway, similar to the ones we considered in Section 7.3. A key group of receptors is called the **toll-like receptors**. The toll protein was first identified in insects, where it is involved in development and in sensing infection. Comparative genomics has revealed at least ten similar receptors in humans.

Toll-like receptors are part of a protein kinase cascade that ultimately results in the transcription of at least 40 genes in-

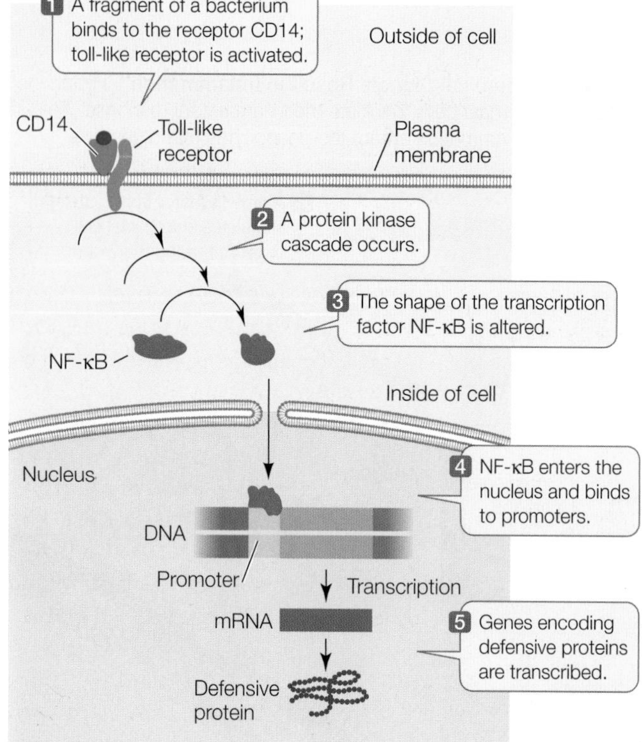
1 A fragment of a bacterium binds to the receptor CD14; toll-like receptor is activated.
Outside of cell
CD14
Toll-like receptor
Plasma membrane
2 A protein kinase cascade occurs.
3 The shape of the transcription factor NF-κB is altered.
NF-κB
Inside of cell
Nucleus
4 NF-κB enters the nucleus and binds to promoters.
DNA
Promoter
Transcription
mRNA
5 Genes encoding defensive proteins are transcribed.
Defensive protein

**42.4 Cell Signaling and Defense** Binding of a pathogenic molecule to the toll-like receptor initiates a signal transduction pathway which results in the transcription of genes whose products are involved in specific and nonspecific defenses. CD14 is expressed on the surface of white blood cells, including macrophages and monocytes.

volved in both nonspecific and specific defenses (**Figure 42.4**). The CD14 protein (for *c*luster of *d*ifferentiation) is a membrane protein that binds to the toll-like receptor. Among the molecules that stimulate this pathway are bacterial and fungal cell wall fragments. Binding of these fragment molecules to the receptor sets in motion a cascade of molecular changes that results in a change in the three-dimensional structure of the transcription factor NF-κB. (NF-κB stands for *n*uclear *f*actor *kappa* light chain enhancer of activated *B* cells. It is a key transcription factor in the activation of both the nonspecific and specific immune systems.) The shape change allows NF-κB to enter the nucleus, bind to the promoters of genes, and activate the transcription of genes encoding defensive proteins.

## 42.2 RECAP

**Nonspecific defenses are the first line of defense against pathogens. These include physical barriers such as the skin; defensive proteins; and coordinated responses such as inflammation.**

- How do complement proteins and interferons defend the body against microbes? See p. 578
- Describe the inflammation response. See pp. 878–879 and Figure 42.3

Often the innate immune system, with its nonspecific defenses, is adequate to prevent or fight off a pathogenic infection. But in many cases, this system works together with specific immunity, which adapts to a particular pathogen. We now turn to the development and functioning of specific immunity.

# 42.3 How Does Specific Immunity Develop?

Before the twentieth century, scientists had long suspected that blood was somehow involved in immunity against pathogens. Over a century ago, Emil von Behring and Shibasaburo Kitasato at the University of Marburg in Germany performed a key experiment that pointed to blood as an important factor in specific immunity (**Figure 42.5**). They showed that guinea pigs injected with a sublethal dose of diphtheria toxin developed in their blood serum (the noncellular fluid that remains after blood is clotted) a factor that protected other guinea pigs from a lethal dose of the same toxin. In other words, the recipients had developed **immunity**. Moreover, the immunity was specific: the immune factor made by the guinea pigs protected only against the specific toxin, from one strain of diphtheria-causing bacteria, that they had been injected with.

Based on the animal model, Behring realized that serum protection might work in human diseases as well. It did, and he won the Nobel Prize for his efforts in protecting children against diphtheria. Later, the agent of this immunity was identified as an antibody protein, and the process of developing immunity from antibodies received from another individual was called *passive immunity*.

In this section we outline the main features of the specific immune system, much of which does indeed occur in blood serum. We will consider the two major types of specific responses: the *humoral immune response*, which produces antibodies; and the *cellular immune response*, which destroys infected cells.

**yourBioPortal.com**
GO TO Animated Tutorial 42.2 • Pregnancy Test

## Adaptive immunity has four key features

Four important features of the adaptive immune system are: specificity; the ability to distinguish self from nonself; the ability to respond to an enormous diversity of nonself molecules; and immunological memory.

**SPECIFICITY** Lymphocytes (B and T cells) are crucial components of specific immunity. T cell receptors and the antibodies produced by B cells recognize and bind to specific nonself or altered self substances (**antigens**), and this interaction initiates a specific immune response. The specific sites on antigens that the immune system recognizes are called **antigenic determinants** or *epitopes*:

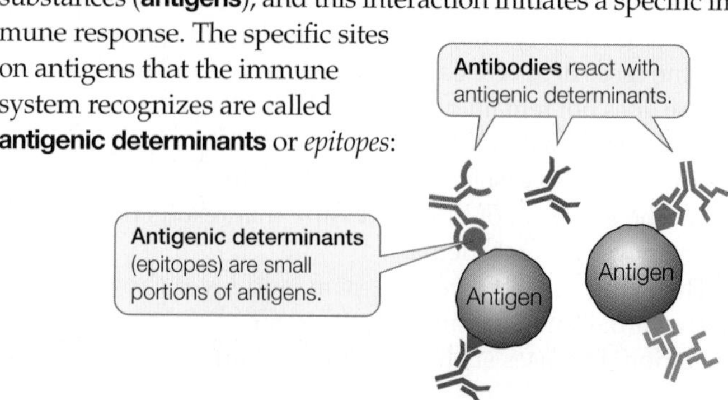
**Antibodies** react with antigenic determinants.
**Antigenic determinants** (epitopes) are small portions of antigens.
Antigen
Antigen

INVESTIGATING LIFE

**42.5 The Discovery of Specific Immunity**

Until the twentieth century, most people did not survive an attack of the bacterium that causes diphtheria, but a few did. Emil von Behring and Shibasaburo Kitasato performed a key experiment using an animal model, and demonstrated that the factor(s) responsible for immunity against diphtheria were in blood serum.

**HYPOTHESIS** Serum from guinea pigs injected with a sublethal dose of diphtheria toxin protects other guinea pigs that are exposed to a lethal dose of the toxin.

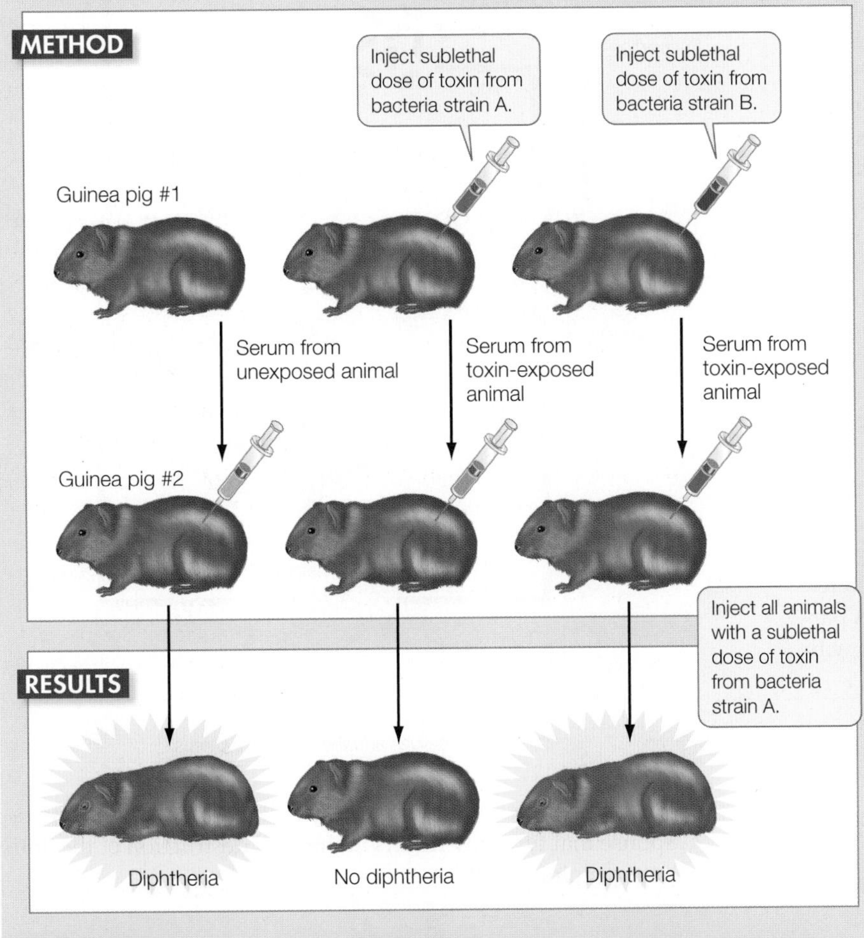
METHOD
Inject sublethal dose of toxin from bacteria strain A.
Inject sublethal dose of toxin from bacteria strain B.
Guinea pig #1
Serum from unexposed animal
Serum from toxin-exposed animal
Serum from toxin-exposed animal
Guinea pig #2
Inject all animals with a sublethal dose of toxin from bacteria strain A.
RESULTS
Diphtheria
No diphtheria
Diphtheria

**CONCLUSION** Serum of toxin-exposed guinea pigs is protective against later exposure to a lethal dose of the toxin from the same genetic strain of bacteria, but not a different strain.

Go to **yourBioPortal.com** for original citations, discussions, and relevant links for all INVESTIGATING LIFE figures.

An antigenic determinant is a specific portion of a large molecule, such as a certain sequence of amino acids that may be present in a protein. Antigens are usually proteins or polysaccharides, and there can be multiple antigens on a single invading bacterium. A single antigenic molecule can have multiple, different antigenic determinants. The host animal responds to the presence of an antigen with highly specific defenses involving T cell receptors and antibodies. These receptors and soluble proteins bind to the antigenic determinants. Each T cell and each antibody is specific for a single antigenic determinant.

For the remainder of the chapter, we will refer to antigenic determinants simply as "antigens."

**DISTINGUISHING SELF FROM NONSELF** The human body contains tens of thousands of different proteins, each with a specific three-dimensional structure capable of generating immune responses. Thus every cell in the body bears a tremendous number of antigens. A crucial requirement of an individual's immune system is that it recognize the body's own antigens and not attack them. This is accomplished by clonal deletion, which you will encounter in a few pages.

**DIVERSITY** Challenges to the immune system are numerous. Pathogens take many forms: viruses, bacteria, protists, fungi, and multicellular parasites. Furthermore, each pathogenic species usually exists as many subtly different genetic strains, and each strain possesses multiple surface features. Estimates vary, but a reasonable guess is that humans can respond specifically to 10 million different antigens. Upon recognizing an antigen, the immune system responds by activating lymphocytes of the appropriate specificity. This capacity is accomplished by a special genetic recombination mechanism described in Section 42.6.

**IMMUNOLOGICAL MEMORY** After responding to a particular type of pathogen once, the immune system "remembers" that pathogen and can usually respond more rapidly and powerfully to the same threat in the future. This **immunological memory** usually saves us from repeats of childhood diseases such as chicken pox. Vaccination against specific diseases works because the immune system "remembers" the antigens that were introduced into the body.

All four of these features of specific immune defense characterize both the humoral immune response and the cellular immune response.

### Two types of specific immune responses interact: an overview

The specific immune system mounts two types of responses against invaders: the humoral immune response and the cellular immune response. B cells that make antibodies are the workhorses of the humoral immune response, and cytotoxic

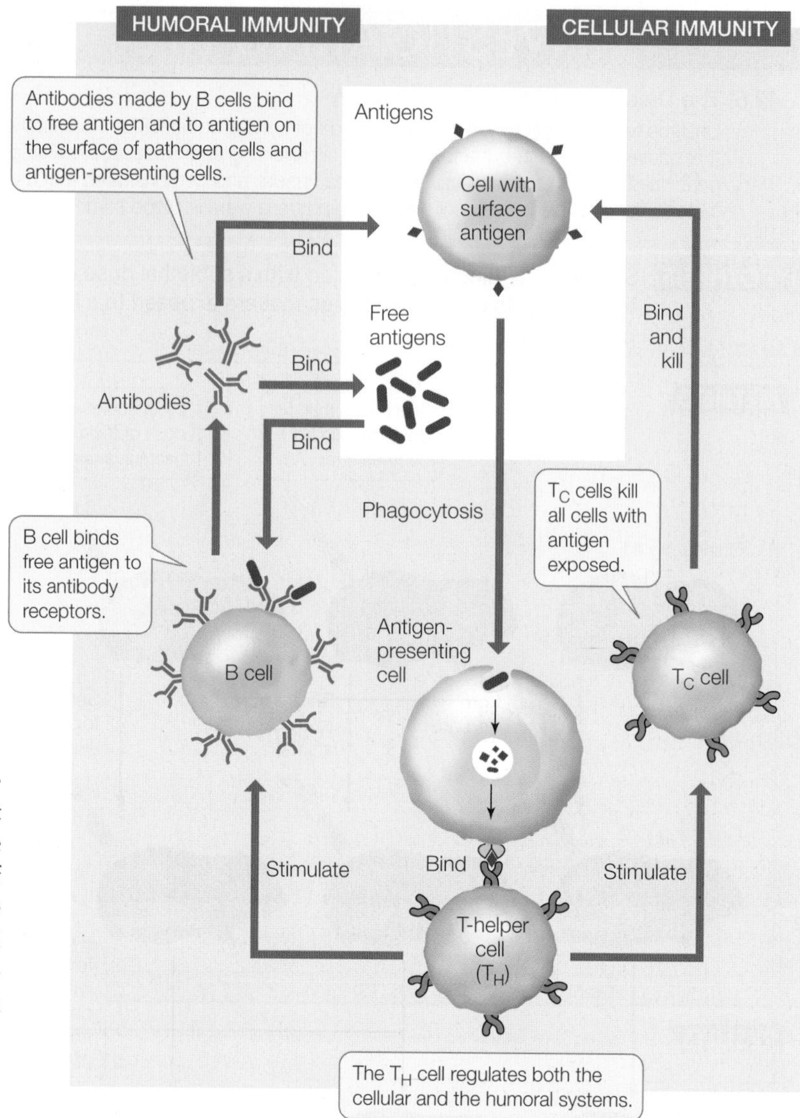


**42.6 The Specific Immune System** Humoral immunity involves the production of antibodies by B cells. Cellular immunity involves the activation of cytotoxic T cells that bind to cells expressing the antigen. For further details, see Figure 42.13.

(killer) T cells are the workhorses of the cellular immune response. These two responses operate simultaneously and cooperatively, sharing many mechanisms. A key event early in these two processes is the exposure of the antigen's three-dimensional structure to the immune system. This occurs when an antigenic molecule, or a fragment of the molecule, is inserted into the plasma membrane of a cell and the unique epitope structure protrudes from the membrane, where it is exposed to nearby T or B cells. Many different types of cell can "present" the antigen to the immune system in this way, and they are collectively referred to as **antigen-presenting cells**. **Figure 42.6** provides an overview of antigen presentation and the roles of T and B cells in the immune response. The key player integrating the two responses is the *T-helper* ($T_H$) *cell*. By binding to the antigen on a presenting cell, the $T_H$ cell stimulates events in both responses.

**HUMORAL IMMUNE RESPONSE** In the **humoral immune response** (from the Latin *humor*, "fluid"), antibodies react with antigens on pathogens in blood, lymph, and tissue fluids. An animal can produce a staggering diversity of antibodies capable of binding to almost any conceivable antigen the animal encounters. Antibodies are secreted by B cells and travel free in the blood and lymph. A particular B cell also possesses receptors on its surface with the same specificity as the antibodies it produces.

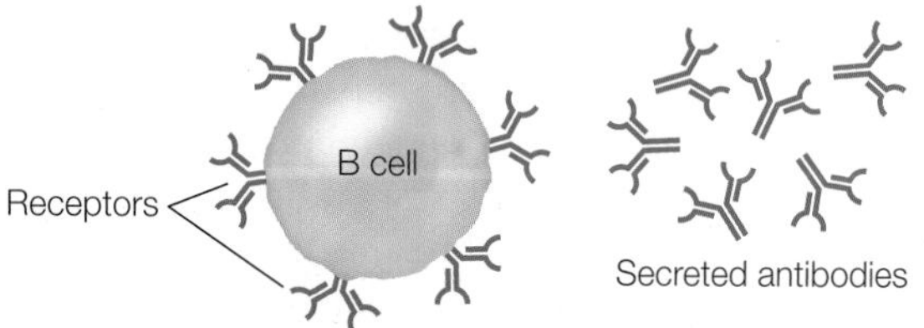


The first time a specific antigen invades the body, it may be presented and then detected by binding to a B cell receptor. This binding activates the B cell, which makes and secretes multiple copies of the antibody.

**CELLULAR IMMUNE RESPONSE** The **cellular immune response** is directed against antigens that have become established within a cell of the host animal. It detects and destroys virus-infected or mutated cells, such as cancer cells expressing unique proteins caused by mutations.

T cells within the lymph nodes, the bloodstream, and the intercellular spaces carry out the cellular immune response. These T cells have integral membrane proteins—T cell receptors—that recognize and bind to antigens. T cell receptors are rather similar to antibodies in structure and function, each including specific molecular configurations that bind to specific antigens.

Once a T cell is bound to an antigen, it initiates an immune response that typically results in the total destruction of the antigen-containing cell.

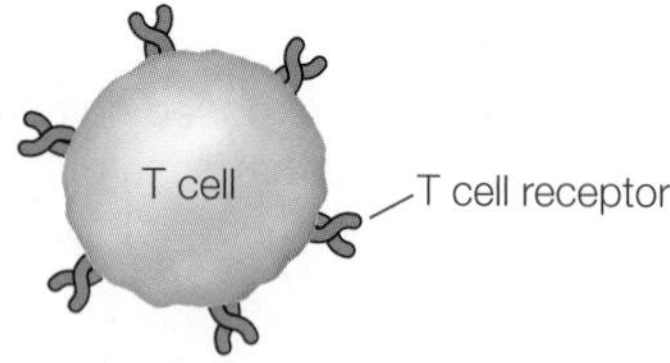


## Genetic changes and clonal selection generate the specific immune response

Before the reactions just described for the humoral and cellular immune responses can take place, the body needs to generate a vast diversity of lymphocytes that have the ability to bind different antigens. How does this tremendous diversity arise? How do lymphocytes specific for certain antigens proliferate? The answers lie in the processes of DNA rearrangement and **clonal selection**. These processes generate the enormous numbers of different receptors with antibody-like specificities in the

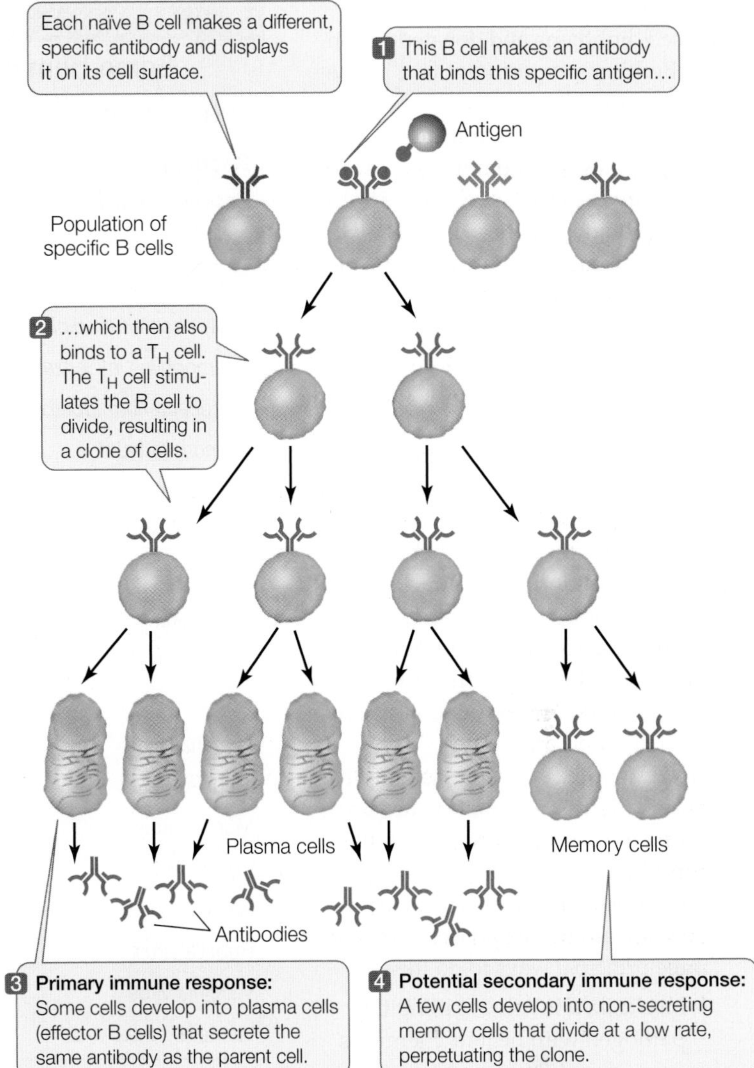

Each naïve B cell makes a different, specific antibody and displays it on its cell surface.
1 This B cell makes an antibody that binds this specific antigen...
Antigen
Population of specific B cells
2 ...which then also binds to a $T_H$ cell. The $T_H$ cell stimulates the B cell to divide, resulting in a clone of cells.
Plasma cells
Memory cells
Antibodies
3 Primary immune response: Some cells develop into plasma cells (effector B cells) that secrete the same antibody as the parent cell.
4 Potential secondary immune response: A few cells develop into non-secreting memory cells that divide at a low rate, perpetuating the clone.


**42.7 Clonal Selection in B Cells** The binding of an antigen to a specific receptor on the surface of a B cell stimulates that cell to divide, producing a clone of genetically identical cells to fight that invader.

B and T lymphocytes, even before the body encounters antigens for these receptors.

- *Diversity is generated primarily by DNA changes*—chromosomal rearrangements and other mutations—that occur just after the B and T cells are formed in the bone marrow. Each B cell is able to produce only one kind of antibody; thus there are millions of different B cells. Similarly, there are millions of different T cells with specific T cell receptors. We will describe the mechanisms for the formation of these antibodies and receptors later in the chapter. For now, keep in mind that the adaptive immune system is "predeveloped"—*all of the machinery available to respond to an immense diversity of antigens is already there, even before the antigens are ever encountered.*
- *Antigen binding "selects" a particular B or T cell for proliferation*. When an antigen fits the surface receptor on a B cell and binds to it, that B cell is activated. It divides to form a clone of cells (a genetically identical group derived from a single cell), all of which produce and secrete antibodies with the same specificity as the receptor (**Figure 42.7**). Binding, activation, and proliferation also apply to T lymphocytes. When a foreign or abnormal cell antigen binds to a T cell receptor, that binding activates the proliferation of T cells with that particular receptor. Binding and activation select a particular lymphocyte, while proliferation generates the clone, hence the term "clonal selection."

## Immunity and immunological memory result from clonal selection

The first time a vertebrate animal is exposed to a particular antigen there is a time lag (usually several days) before the B cell-produced antibody molecules and T cells specific to that antigen slowly increase. But for years afterward—sometimes for life—the immune system "remembers" that particular antigen. How does this happen?

The answer lies in the fact that activated lymphocytes divide and differentiate to produce *two types* of daughter cells: effector cells and memory cells.

- **Effector cells** carry out the attack on the antigen. Effector B cells, called **plasma cells**, secrete antibodies. Effector T cells release cytokines and other molecules that initiate reactions that destroy nonself or altered cells. Effector cells live only a few days.
- **Memory cells** are long-lived cells that retain the ability to start dividing on short notice to produce more effector and more memory cells. Memory B and T cells may survive in the body for decades, rarely dividing.

These two types of lymphocytes can respond to an antigen in two different ways:

- When the body first encounters a particular antigen, a **primary immune response** is activated, in which the "naïve" (previously unexposed to antigen) lymphocytes that recognize that antigen proliferate to produce clones of effector and memory cells.
- After a primary immune response to a particular antigen, subsequent encounters with the same antigen will trigger a much more rapid and powerful **secondary immune response**. The memory cells that bind with that antigen proliferate, launching a huge army of plasma cells and effector T cells.

## Vaccines are an application of immunological memory

You will recall Behring's experiment on using the serum of diphtheria-exposed animals to protect other animals from the disease (see Figure 42.5). As we noted, he later used this method on people to cause passive immunity. Now we can contrast this with *active immunity*, in which the immune system develops a specific response, including memory cells that lead to long-term protection. The animals that survived diphtheria and donated serum in Berhing's experiment had active immunity.

Thanks to active immunity and immunological memory, exposure to many diseases, such as chicken pox, provides a natural immunity to those diseases. Furthermore, it is possible to provide artificial immunity against many life-threatening diseases by **vaccination**: the introduction of antigen into the body in a form that does not cause disease.

Vaccination initiates a primary immune response, generating memory cells without making the person ill. Later, if a pathogen carrying the same antigen attacks, specific memory cells already exist. They recognize the antigen and quickly overwhelm the invaders with a massive production of lymphocytes and antibodies.

Because the antigens used for immunization or vaccination are produced by pathogenic organisms, they must be altered so that they cannot cause disease but are still able to provoke an immune response. There are three principal ways to do this:

- *Inactivation* involves killing the pathogen with heat or chemicals.
- *Attenuation* involves reducing the virulence of a virus by repeatedly infecting cells with it in the laboratory; this results in mutations in the virus that render it nontoxic but still recognized as nonself.
- *Recombinant DNA technology* can be used to produce peptide fragments that bind to and activate lymphocytes but do not have the harmful part of a protein toxin.

For most of the 70 or so bacteria, viruses, fungi, and parasites known to cause serious human diseases, vaccines are already available or will be in the next few years (**Table 42.2**). Vaccination has completely or almost completely wiped out some deadly diseases, including smallpox, diphtheria, and polio, in industrialized countries.

TABLE 42.2

**Some Human Pathogens for which Vaccines are Available**

| INFECTIOUS AGENT | DISEASE | VACCINATED POPULATION |
|---|---|---|
| **Bacteria** | | |
| *Bacillus anthracis* | Anthrax | Those at risk in biological warfare |
| *Bordetella pertussis* | Whooping cough | Children and adults |
| *Clostridium tetani* | Tetanus | Children and adults |
| *Corynebacterium diphtheriae* | Diphtheria | Children |
| *Haemophilus influenzae* | Meningitis | Children |
| *Mycobacterium tuberculosis* | Tuberculosis | All people |
| *Salmonella typhi* | Typhoid fever | People in areas exposed to agent |
| *Streptococcus pneumoniae* | Pneumonia | Elderly people |
| *Vibrio cholerae* | Cholera | People in areas exposed to agent |
| **Viruses** | | |
| Adenovirus | Respiratory disease | Military personnel |
| Hepatitis A | Liver disease | People in areas exposed to agent |
| Hepatitis B | Liver disease, cancer | All people |
| Influenza virus | Flu | All people |
| Measles virus | Measles | Children and adolescents |
| Mumps virus | Mumps | Children and adolescents |
| Poliovirus | Polio | Children |
| Rabies virus | Rabies | People exposed to agent |
| Rubella virus | German measles | Children |
| Vaccinia virus | Smallpox | Laboratory workers, military personnel |
| Varicella-zoster virus | Chicken pox | Children |

### Animals distinguish self from nonself and tolerate their own antigens

Normally, the body is *tolerant* of its own molecules—the same molecules that would generate an immune response in another individual. This occurs primarily during the early differentiation of T and B cells, when they encounter self antigens. Any immature B or T cell that shows the potential to mount an immune response against self antigens undergoes programmed cell death (apoptosis) within a short time. This process is called **clonal deletion**.

42.3 RECAP

**The specific immune system reacts against nonself or altered self molecules called antigens. The system generates amazing diversity. Immunological memory arises from clonal selection.**

- How does an antigen initiate a specific immune response? See pp. 881–882
- Describe clonal selection. How does it contribute to immunological memory? See pp. 882–883 and Figure 42.7
- How do vaccines make use of immunological memory? See pp. 883–884

Now that we understand the general features of the specific immune system, let's focus in more detail on the B lymphocytes and the humoral immune response.

# 42.4 What Is the Humoral Immune Response?

Every day, billions of B cells survive the test of clonal deletion and are released from the bone marrow into the circulation. B cells are the basis for the humoral immune response.

yourBioPortal.com
GO TO Animated Tutorial 42.3 • Humoral Immune Response

## Some B cells develop into plasma cells

A B cell begins by making a receptor protein on its cell surface. As we have seen, if a B cell is activated by antigen binding to this receptor, it gives rise to clones of plasma cells and memory cells. The plasma (effector B) cells secrete antibodies into the bloodstream (see Figure 42.7).

Usually, for a naïve B cell (one that has not yet been exposed to antigen) to develop into an antibody-secreting plasma cell, a T-helper cell ($T_H$) with the same specificity must also bind to the antigen (see Figure 42.6). The division and differentiation of the B cell is stimulated by the receipt of chemical signals from the $T_H$ cell. Thus, as we will see in Section 42.5, the B cell also functions as an antigen-presenting cell.

As plasma cells develop, the number of ribosomes and the amount of endoplasmic reticulum in their cytoplasms increase greatly (**Figure 42.8**). These increases allow the cells to synthesize and secrete large amounts of antibody proteins—up to 2,000 per second! *All the plasma cells arising from a given B cell produce antibodies that are specific for the antigen that originally bound to the parent B cell.* Thus antibody specificity is maintained as B cells proliferate.

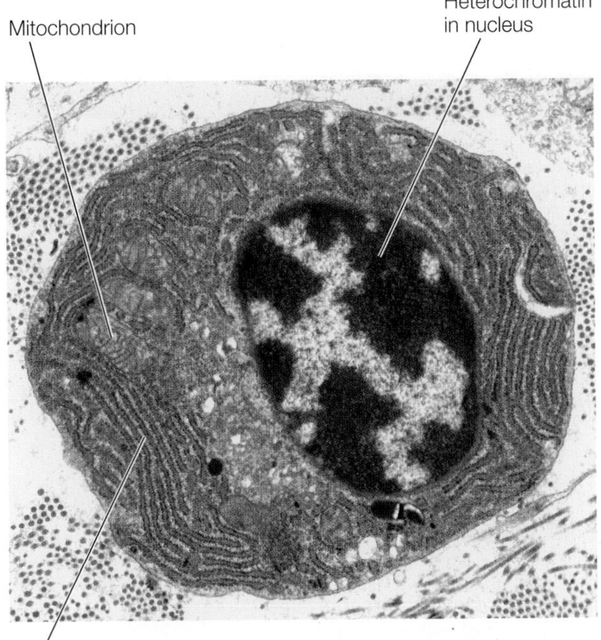


**42.8 A Plasma Cell** A prominent nucleus with a large amount of heterochromatin (dark brown) and a cytoplasm crowded with rough endoplasmic reticulum (purple) are features of a cell that is actively synthesizing and exporting proteins. In this case, a B cell is making a specific antibody. Whole blocks of genes not needed for this specialized function are kept inactive in the heterochromatin.

## Different antibodies share a common structure

Antibodies belong to a class of proteins called **immunoglobulins**. There are several types of immunoglobulins, but all contain a tetramer consisting of four polypeptide chains (**Figure 42.9**). In each immunoglobulin molecule, two of these polypeptides are identical *light chains*, and two are identical *heavy chains*.

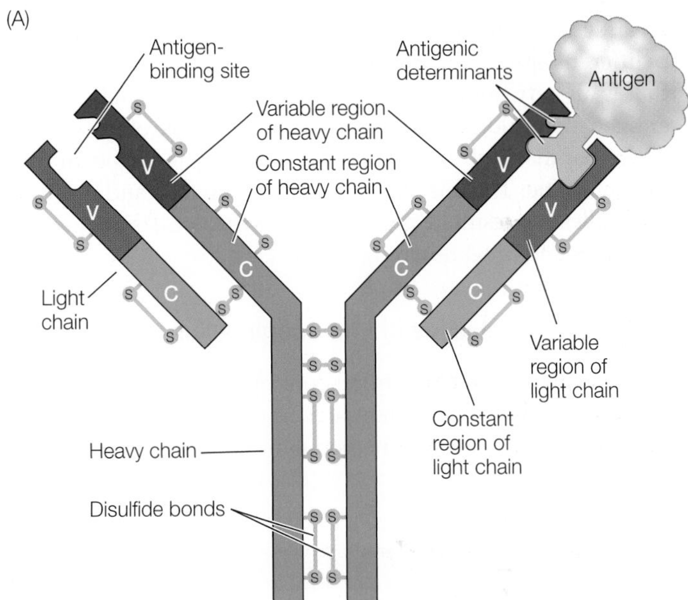


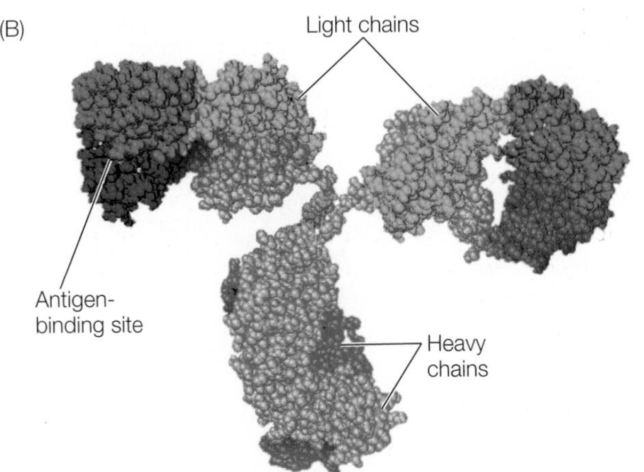


**42.9 The Structure of An Immunoglobulin** Four polypeptide chains (two light, two heavy) make up an immunoglobulin molecule. Here we show both diagrammatic (A) and space-filling (B) representations of immunoglobulin.

yourBioPortal.com
GO TO Web Activity 42.3 • Immunoglobulin Structure

Disulfide bonds hold the chains together. Each polypeptide chain has a constant region and a variable region:

- The amino acid sequences of the **constant regions** are similar among the immunoglobulins. They determine the destination and function—the class—of each immunoglobulin.
- The amino acid sequences of the **variable regions** are different for each specific immunoglobulin. Their three-dimensional antigen-binding sites are determined by their secondary structures, and are responsible for antibody specificity.

The two antigen-binding sites on each immunoglobulin molecule are identical, making the antibody *bivalent* (*bi*, "two"; *valent*, "binding"). This ability to bind two antigen molecules at once, along with the existence of multiple epitopes on each antigen, permits antibodies to form large complexes with the antigens. For example, one antibody might bind two molecules of an antigen. Another antibody might bind the same antigen at a different epitope. It may bind one of the antigen molecules that is already bound to the first antibody, along with a third antigen molecule. A third antibody may bind one of the antigen molecules that is already part of the complex, and a fourth antigen molecule. This binding of multiple antigens and multiple antibodies can result in large complexes that are easy targets for ingestion and breakdown by phagocytes.

### There are five classes of immunoglobulins

While the variable regions are responsible for the specificity of an immunoglobulin, the constant regions of the heavy chain determine the class of the immunoglobulin—for example, whether it will be an integral membrane receptor or a soluble antibody that is secreted into the bloodstream. The five immunoglobulin classes are described in **Table 42.3**. The most abundant class is IgG; these soluble antibody proteins make up about 80 percent of the total immunoglobulin content of the bloodstream. They are made in greatest quantity during a secondary immune response. IgG molecules defend the body in several ways. For example, after some IgG molecules bind to antigens, they become attached by their heavy chains to macrophages. This attachment permits the macrophages to destroy the antigens by phagocytosis.

### Monoclonal antibodies have many uses

The specificity of antibodies suggested to scientists that they might be useful for detecting specific substances in the laboratory. Suppose that a physician wishes to measure the level of the hormone estrogen in a woman's blood. Because hormones are present in such minute amounts, their concentrations cannot usually be measured effectively by conventional chemical methods. A more sensitive method is needed. One way would be to add an antibody specific for estrogen to a sample of the patient's blood and observe how much antigen–antibody complex formed.

However, the immune response to a complex antigen is polyclonal—that is, most antigens carry many different antigenic determinants and will produce a complex mixture of antibodies, each made by a different clone of B cells. Furthermore, as emphasized in our study of biochemistry, many biological molecules share regions of similar structure—all human steroid hormones, for example, have a similar multi-ring structure (see Fig-

**TABLE 42.3**
**Antibody Classes**

| CLASS | GENERAL STRUCTURE | LOCATION | FUNCTION |
|---|---|---|---|
| IgG | Monomer | Free in blood plasma; about 80 percent of circulating antibodies | Most abundant antibody in primary and secondary immune responses; crosses placenta and provides passive immunization to fetus |
| IgM | Pentamer | Surface of B cell; free in blood plasma | Antigen receptor on B cell membrane; first class of antibodies released by B cells during primary response |
| IgD | Monomer | Surface of B cell | Cell surface receptor of mature B cell; important in B cell activation |
| IgA | Dimer | Saliva, tears, milk, and other body secretions | Protects mucosal surfaces; prevents attachment of pathogens to epithelial cells |
| IgE | Monomer | Secreted by plasma cells in skin and tissues lining gastrointestinal and respiratory tracts | Binds to mast cells and basophils to sensitize them to subsequent binding of antigen, which triggers release of histamine that contributes to inflammation and some allergic responses |

ure 3.22). A polyclonal group of antibodies targeted to estrogen might be uninformative because some of the antibodies would bind to any steroid hormone present in the blood sample. More useful would be a clone of B cells that produce large amounts of an antibody that binds to only one specific epitope—a **monoclonal antibody**. How could such a clone be produced?

A clone of cells that produce a single antibody can be made artificially by fusing a single B cell (which has a finite lifetime and makes a lot of antibody) with a tumor cell (which has an infinite lifetime and can be grown in culture). The resulting hybrid cell, called a **hybridoma**, makes a specific monoclonal antibody and proliferates in culture (**Figure 42.10**). In order to produce useful monoclonal antibodies by this approach, the hybridoma clones must be screened to isolate those producing antibodies that are specific for the target molecule. Recently, new methods have been developed for producing monoclonal antibodies using recombinant DNA technology. These approaches do not require the injection of live animals with antigen, the isolation of their B cells, or the production and screening of hybridomas.

Monoclonal antibodies have many applications:

- *Immunoassays* use monoclonal antibodies to detect tiny amounts of molecules in tissues and fluids. For example, this technique is used in pregnancy tests to detect human chorionic gonadotropin, the hormone made by the developing embryo.
- *Immunotherapy* uses monoclonal antibodies targeted against antigens on the surfaces of cancer cells. The coupling of a radioactive ligand or toxin to the antibody makes it into a medical "smart bomb." In a related approach, binding of the antibody itself is enough to trigger a cellular immune response that destroys the cancer. This is the case with trastuzumab (Herceptin®), a monoclonal antibody that binds to a growth factor receptor on some breast cancer cells.

## TOOLS FOR INVESTIGATING LIFE

**42.10 Creating Hybridomas for the Production of Monoclonal Antibodies**

Cancerous myeloma cells and normal B cells can be fused so that the proliferative properties of the myeloma cells are combined with the ability to produce specific antibodies.

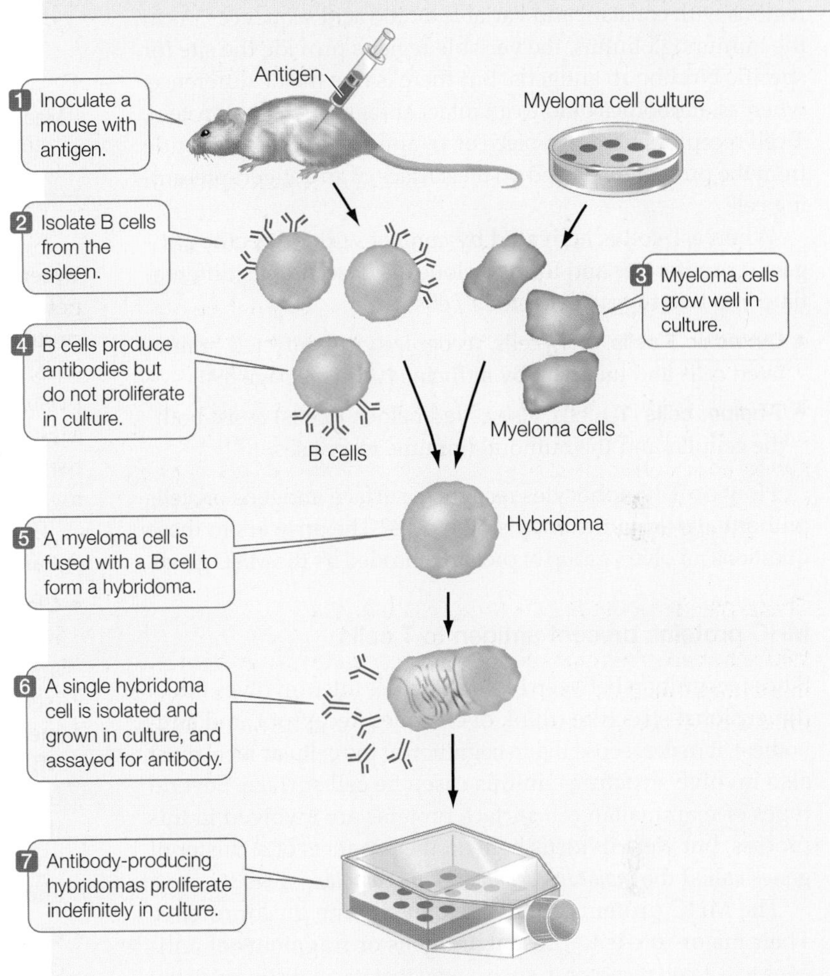


## 42.4 RECAP

**The humoral immune response is based on the synthesis by B cells of specific antibodies directed against antigens. The specificity of an antibody derives from the amino acid sequence of its variable regions. Monoclonal antibodies are specific to one epitope on an antigen and can be produced artificially for use in diagnostics and therapy.**

- How does a B cell respond to an antigen? See p. 885
- How is the structure related to the function of an antibody molecule? See pp. 885–886 and Figure 42.9
- What are monoclonal antibodies? How are they used? See pp. 886–887 and Figure 42.10

By making antibodies, B cells are the major players in the humoral immune response. We now turn to the cellular immune response, where T cells are active at all stages.

# 42.5 What Is the Cellular Immune Response?

Two types of effector T cells (T-helper cells and cytotoxic T cells) are involved in the cellular immune response, along with proteins of the major histocompatibility complex, the MHC proteins, which underlie the immune system's tolerance for the body's own cells.

yourBioPortal.com
GO TO Animated Tutorial 42.4 • Cellular Immune Response

### T cell receptors bind to antigens on cell surfaces

Like B cells, T cells possess specific membrane receptors. The T cell receptor is not an immunoglobulin, however, but a glycoprotein with a molecular weight of about half that of an IgG. It is made up of two polypeptide chains, each encoded by a separate gene (**Figure 42.11**). Thus the two chains have distinct regions with constant and variable amino acid sequences. As in the immunoglobulins, the variable regions provide the site for specific binding to antigens. But there is one major difference: whereas antibodies bind to an intact antigen such as a protein, T cell receptors bind to a piece of an antigen, such as a peptide from the protein displayed on the surface of an antigen-presenting cell.

When a T cell is activated by contact with a specific antigen, it proliferates and forms a clone. Its descendants differentiate into two types of effector T cells:

- **Cytotoxic T cells**, or $T_C$ cells, recognize virus-infected or mutated cells and kill them by inducing lysis (see page 890).
- **T-helper cells** ($T_H$ cells, also called helper T cells) assist both the cellular and the humoral immune responses.

How do T lymphocytes recognize surface antigens on cells without attacking the body's own cells? The answers to these questions involve a group of proteins encoded by the *MHC* genes.

### MHC proteins present antigen to T cells

Since recognition between biological molecules involves three-dimensional structure (think of enzymes, receptors, and antibodies), it makes sense that recognition at the cellular level must also involve structure—in this case, the cell surface. Several types of mammalian cell surface proteins are involved in this process, but we will focus here on the products of a cluster of genes called the *major histocompatibility complex*, or MHC.

The MHC proteins are plasma membrane glycoproteins. Their major role is to present antigens or fragments of antigens to a T cell receptor in such a way that it can distinguish between self and nonself antigens. In humans, there are three genetic loci for MHC class I proteins and three for MHC class II proteins. Each of these six loci has as many as 100 different alleles. With so many possible allele combinations, it is not surprising that different people are very likely to have different MHC genotypes. So the unique three-dimensional structure of the MHC protein is important in self-tolerance: the T cell receptor binds to the MHC and the antigen together.

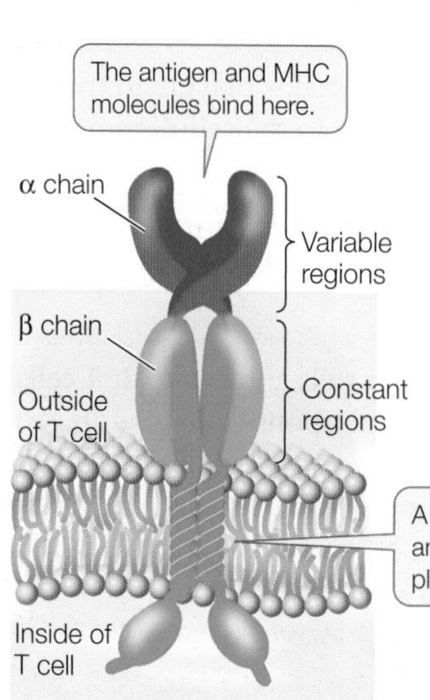


**42.11 A T Cell Receptor** The receptors on T lymphocytes are smaller than those on B lymphocytes, but their two polypeptides contain both variable and constant regions. As with the B cell receptors, the constant region fixes the receptor in the plasma membrane, while the variable regions establish the specificity for binding to antigen.

**TABLE 42.4**

**The Interaction between T Cells and Antigen-Presenting Cells**

| PRESENTING CELL TYPE | ANTIGEN PRESENTED | MHC CLASS | T CELL TYPE | T CELL SURFACE PROTEIN |
|---|---|---|---|---|
| Any cell | Intracellular protein fragment | Class I | Cytotoxic T cell ($T_C$) | CD8 |
| Macrophages and B cells | Fragments from extracellular proteins | Class II | Helper T cell ($T_H$) | CD4 |

There are two classes of MHC proteins. Both function to present antigens to the different T lymphocytes:

- **Class I MHC** proteins are present on the surface of every nucleated cell in the animal body. When cellular proteins are degraded into small peptide fragments by a proteasome (see Figure 16.24), an MHC I protein may bind to a fragment and carry it to the plasma membrane. There, the MHC I protein is oriented in such a way that the bound fragment is exposed to the outside of the cell, where both the MHC and the fragment can interact with $T_C$ cells. The $T_C$ cells have a surface protein called CD8 that recognizes and binds to MHC I.
- **Class II MHC** proteins are found mostly on the surfaces of B cells, macrophages, and and other antigen-presenting cells (see Figure 42.2). When one of these cells ingests a nonself antigen such as a virus, the antigen is broken down in a phagosome. An MHC II molecule may bind to one of the fragments and carry it to the cell surface, where it is presented to a $T_H$ cell (**Figure 42.12**). $T_H$ cells have a surface protein called CD4 that recognizes and binds to MHC II.

The information on MHC proteins, the cellular origins of antigens, and T lymphocytes is summarized in **Table 42.4**. To accomplish their role in antigen presentation, both MHC I and MHC II proteins have an antigen-binding site, which can hold a peptide of about 10–20 amino acids. The T cell receptor recognizes not just the antigenic fragment, but the MHC I or MHC II molecule to which the fragment is bound.

MHC proteins play a vital role in the selection of T cells during their development:

1. *Binding to self MHC molecules.* The ultimate goal is for the T cell receptor to bind not to the antigen alone but to the antigen–MHC complex. Here, there is positive selection for T

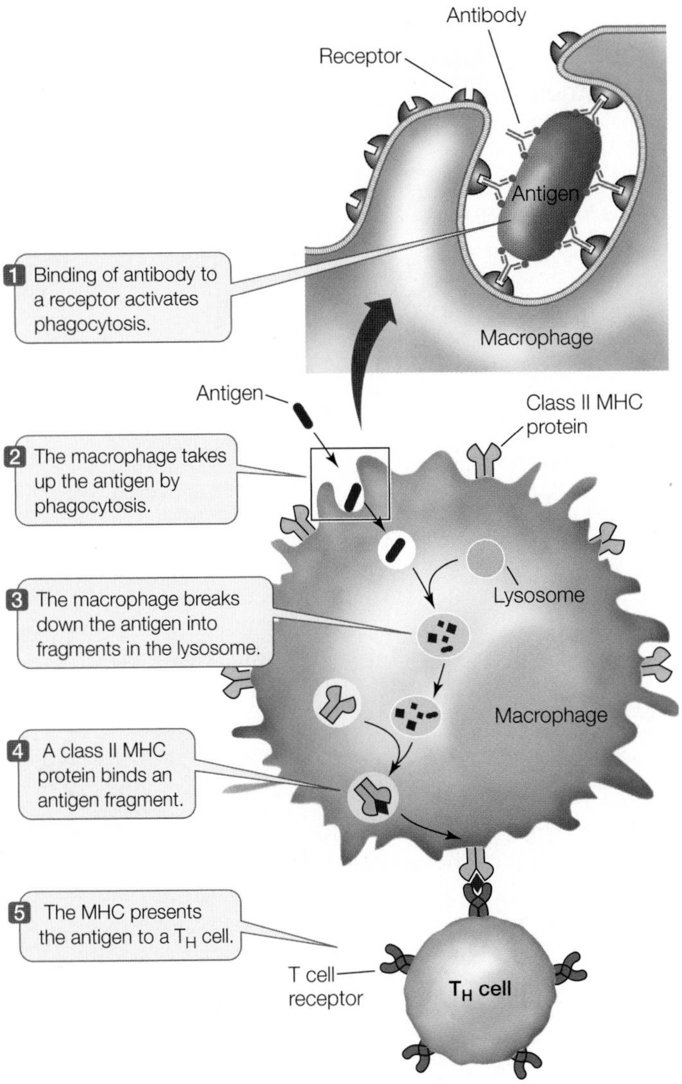

**42.12 Macrophages Are Antigen-Presenting Cells** A fragment of an antigen is displayed by MHC II on the surface of a macrophage. T cell receptors on a specific T-helper cell can then bind to and interact further with the antigen–MHC II complex.

cells that bind. Any T cells that do not recognize self MHC (do not bind to the antigen-presenting cell) are eliminated soon after they develop; the rest of the T cells go on to the next selection step.

2. *Binding to self peptides bound to self MHC.* Here, there is negative selection for T cells that bind, thereby eliminating the further production of T cells that react to self antigens. Both of these selection events occur in the thymus gland (see Figure 41.6).

### T-helper cells and MHC II proteins contribute to the humoral immune response

When a $T_H$ cell survives the selection processes and binds to an antigen-presenting macrophage, it releases cytokines, which activate the $T_H$ cell to proliferate, producing a clone of $T_H$ cells with the same specificity. The steps to this point constitute the activation phase of the humoral immune response, and they occur in the lymphoid tissues. Next comes the effector phase, in which the $T_H$ cells activate naïve B cells with the same specificity to produce antibodies.

B cells are also antigen-presenting cells. B cells take up antigens bound to their surface immunoglobulin receptors by endocytosis, break them down, and display antigenic fragments on class II MHC proteins. When a $T_H$ cell binds to the displayed antigen–MHC II complex, it releases cytokines that cause the B cell to produce a clone of plasma cells (**Figure 42.13A**). Finally, the plasma cells secrete antibodies, completing the effector phase of the humoral immune response.

### Cytotoxic T cells and MHC I proteins contribute to the cellular immune response

Class I MHC proteins play a role in the cellular immune response that is similar to the role played by class II MHC proteins in the humoral immune response. In a virus-infected or mutated cell, foreign or abnormal proteins or peptide fragments combine with MHC I molecules. The resulting complex is displayed on the cell surface and presented to $T_C$ cells. When a $T_C$ cell recognizes and binds to this antigen–MHC I complex, it is activated to proliferate (**Figure 42.13B**).

In the effector phase of the cellular immune response, $T_C$ cells recognize and bind to cells bearing the same antigen–MHC I complex. These bound $T_C$ cells produce a substance called *perforin*, which lyses the bound target cell. In addition, the $T_C$ cells can bind to a specific receptor (called Fas) on the target cell that initiates apoptosis in that cell. These two mechanisms, cell lysis and programmed cell death, work in concert to eliminate the antigen-containing host cell. Because $T_C$ cells recognize self MHC proteins complexed with nonself antigens, they help rid the body of its own virus-infected cells.

In addition to the binding of an antigen–MHC complex to its receptors, a T cell must receive a second signal for activation. This *co-stimulatory* signal occurs after the initial specific binding. It involves the interaction between another receptor on the T cell surface and a protein called B7 on certain antigen-presenting cells. This second binding event leads to T cell activation, including cytokine production and proliferation.

### Regulatory T cells suppress the humoral and cellular immune responses

As we describe in Chapter 40, *homeostasis*—the maintenance of internal constancy—is a hallmark of animal biology. A third class of T cells called **regulatory T cells** (**Tregs**) ensures that the immune response does not spiral out of control. Like $T_H$ and $T_C$ cells, Tregs are made in the thymus gland, express the T cell receptor, and become activated if they bind to antigen–MHC complexes. But Tregs are different in one important way: the antigens that Tregs recognize are *self antigens*. The activation of Tregs causes them to secrete the cytokine *interleukin 10*, which blocks T cell activation and leads to apopto-

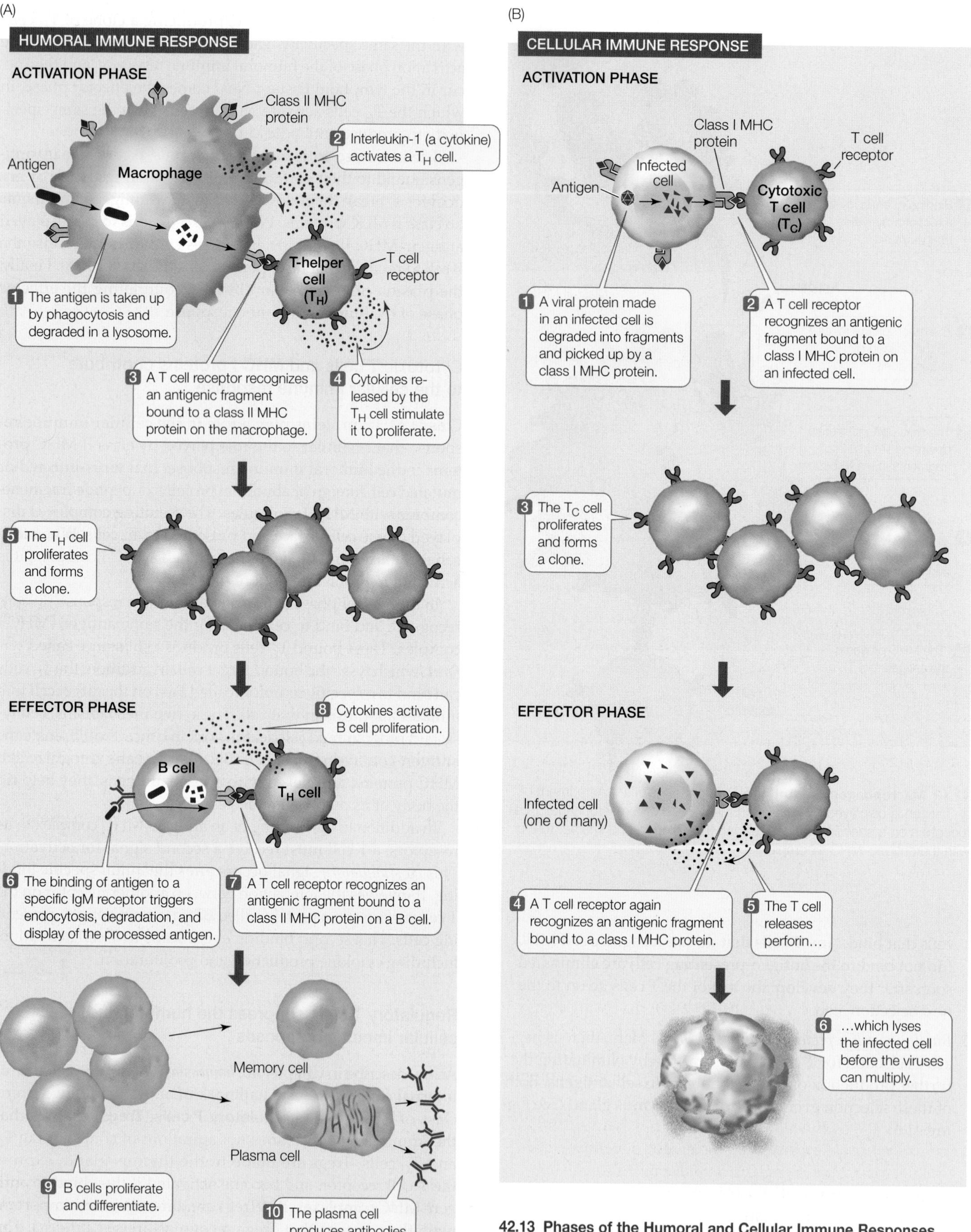
(A)
HUMORAL IMMUNE RESPONSE
ACTIVATION PHASE
Class II MHC protein
Antigen
Macrophage
2 Interleukin-1 (a cytokine) activates a $T_H$ cell.
T-helper cell ($T_H$)
T cell receptor
1 The antigen is taken up by phagocytosis and degraded in a lysosome.
3 A T cell receptor recognizes an antigenic fragment bound to a class II MHC protein on the macrophage.
4 Cytokines released by the $T_H$ cell stimulate it to proliferate.
5 The $T_H$ cell proliferates and forms a clone.
EFFECTOR PHASE
8 Cytokines activate B cell proliferation.
B cell
$T_H$ cell
6 The binding of antigen to a specific IgM receptor triggers endocytosis, degradation, and display of the processed antigen.
7 A T cell receptor recognizes an antigenic fragment bound to a class II MHC protein on a B cell.
Memory cell
Plasma cell
9 B cells proliferate and differentiate.
10 The plasma cell produces antibodies.
(B)
CELLULAR IMMUNE RESPONSE
ACTIVATION PHASE
Class I MHC protein
T cell receptor
Infected cell
Antigen
Cytotoxic T cell ($T_C$)
1 A viral protein made in an infected cell is degraded into fragments and picked up by a class I MHC protein.
2 A T cell receptor recognizes an antigenic fragment bound to a class I MHC protein on an infected cell.
3 The $T_C$ cell proliferates and forms a clone.
EFFECTOR PHASE
Infected cell (one of many)
4 A T cell receptor again recognizes an antigenic fragment bound to a class I MHC protein.
5 The T cell releases perforin...
6 ...which lyses the infected cell before the viruses can multiply.

**42.13 Phases of the Humoral and Cellular Immune Responses** Both the humoral and the cellular immune responses have activation and effector phases, all of which involve T cells.

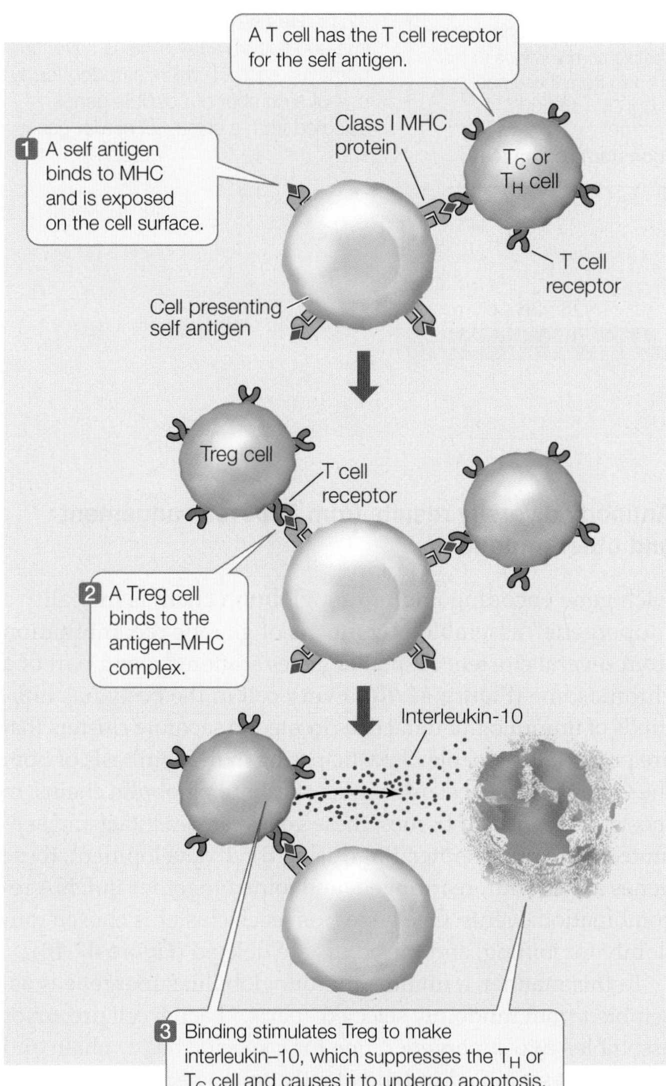
A T cell has the T cell receptor for the self antigen.
Class I MHC protein
1 A self antigen binds to MHC and is exposed on the cell surface.
$T_C$ or $T_H$ cell
T cell receptor
Cell presenting self antigen
Treg cell
T cell receptor
2 A Treg cell binds to the antigen–MHC complex.
Interleukin-10
3 Binding stimulates Treg to make interleukin-10, which suppresses the $T_H$ or $T_C$ cell and causes it to undergo apoptosis.

**42.14 Tregs and Tolerance** A special class of T cells called regulatory T cells (Tregs) inhibit the activation of the immune system in response to self antigens.

sis of the $T_C$ and $T_H$ cells that are bound to the same antigen-presenting cell (**Figure 42.14**).

The important role of Tregs is to maintain homeostasis by mediating tolerance to "self" antigens. Thus they constitute one of the mechanisms for distinguishing self from nonself. How do we know this? There are two lines of evidence for the role of Tregs. As in many other biological studies, the cause-and-effect relationships were worked out using experimental manipulations and genetics:

- If Tregs are experimentally destroyed in the thymus of a mouse, the mouse grows up with an out-of-control immune system, mounting strong immune responses to self antigens (*autoimmunity*—see Section 42.7).
- In humans, a rare X-linked inherited disease occurs when a gene critical to Treg function is mutated. An infant with this disease, called IPEX (*i*mmune dysregulation, *p*olyendocrinopathy and *e*nteropathy, *X*-linked), mounts an immune response that attacks the pancreas, thyroid, and intestine. Most affected individuals die within the first few years of life.

### MHC proteins are important in tissue transplants

In humans, one consequence of the major histocompatibility complex became important with the development of organ transplant surgery. Because the proteins produced by the MHC are specific to each individual, they act as nonself antigens if transplanted into another individual. An organ or a piece of tissue transplanted from one person to another is recognized as nonself by the host body and soon provokes an immune response; the tissue is then killed, or "rejected," by the host's cellular immune system. But if the transplant is performed immediately after birth, or if it comes from a genetically identical person (an identical twin), the material is recognized as self and is not rejected.

The rejection problem can be overcome by treating a patient with a drug, such as cyclosporin, that suppresses the immune system. Cyclosporin blocks the activation of a transcription factor that is essential for T cell development. However, this approach compromises the ability of transplant recipients to defend themselves against pathogens. This problem must be managed by the use of antibiotics and other drugs to combat infections that develop.

### 42.5 RECAP

**The cellular immune response acts against antigens expressed on the surfaces of virus-infected or mutated body cells. Specific receptors on T cells bind to antigen–MHC complexes displayed on the cell surface. During development, T cells are selected that recognize self MHC proteins. Tregs suppress immune responses to self antigens.**

- What are the roles of a T cell receptor in cellular immunity? See p. 888
- What parts do MHC proteins play in the cellular immune response? See pp. 888–889
- What occurs during the cellular immune response to a virus-infected cell? See p. 889 and Figure 42.13

We have alluded to genes that encode various components of the immune system and to the tremendous diversity in these components. We will now consider the genetic mechanisms that make this diversity possible.

## 42.6 How Do Animals Make So Many Different Antibodies?

Each mature B cell makes one—and only one—specific antibody with a specific amino acid sequence targeted to a single antigen. As we have seen, there are millions of possible epitopes to which a human is exposed or can be exposed. With millions of possible amino acid sequences in immunoglobu-

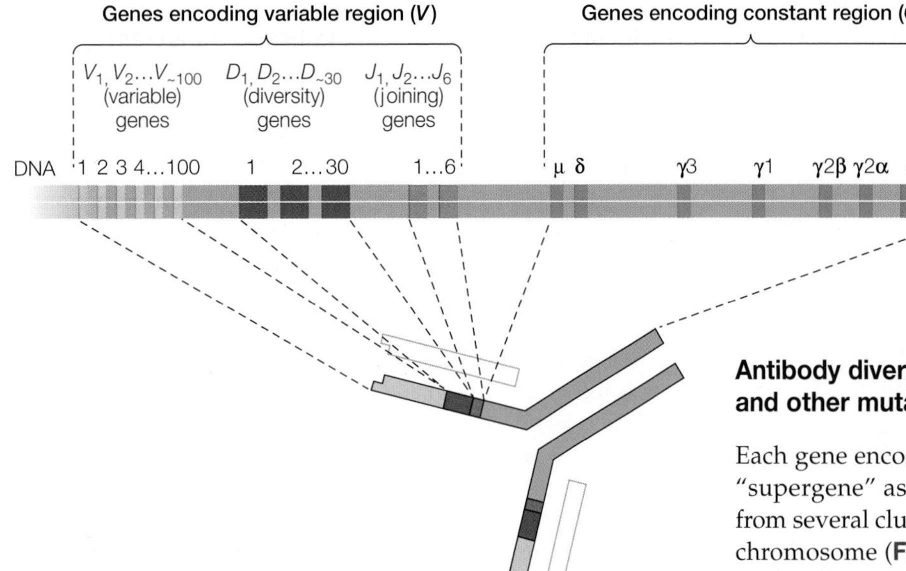
The **variable region** for the heavy chain of a specific antibody is encoded by one *V* gene, one *D* gene, and one *J* gene. Each of these genes is taken from a pool of like genes.
The **constant region** is selected from another pool of genes. The number of possible combinations to make an immunoglobulin heavy chain from these pools of genes is (100 *V*)(30 *D*)(6 *J*)(8 *C*) = 144,000.
Genes encoding variable region (*V*)
Genes encoding constant region (*C*)
$V_1, V_2...V_{\sim 100}$ (variable) genes
$D_1, D_2...D_{\sim 30}$ (diversity) genes
$J_1, J_2...J_6$ (joining) genes
DNA
1 2 3 4...100
1 2...30
1...6
μ δ
γ3
γ1
γ2β γ2α ε α

**42.15 Heavy-Chain Genes** Mouse immunoglobulin heavy chains have four domains, each of which is coded for by one of a number of possible genes selected from a cluster of similar genes.

lins, the molecular genetic explanation would be that there are millions of genes, each one coding for one antibody molecule. A simple calculation using approximate numbers shows that this is impossible:

$$\text{one immunoglobulin heavy chain} = \text{500 amino acids coded for by 1500 bp DNA}$$

$$\text{one immunoglobulin light chain} = \text{200 amino acids coded for by 600 bp DNA}$$

therefore,

$$\text{one antibody (2 each identical light and heavy chains)} = \text{2100 bp DNA}$$

$$\text{10 million different antibodies} = \text{21 billion bp DNA}$$

This is 7 times the size of the entire human genome! There must be another way to generate antibody diversity.

It turns out that instead of a single gene encoding each immunoglobulin, the genome of the differentiating B cell has a limited number of alleles for several *regions* or *domains* of the protein, and that *combinations of these alleles* generate diversity. In this section we will describe the unusual process of shuffling this genetic deck to generate the enormous immunological diversity that characterizes each individual mammal. Then we will see how similar events produce five classes of immunoglobulins with different cellular locations or functions in the body.

**yourBioPortal.com**

GO TO Animated Tutorial 42.5 • A B Cell Builds an Antibody

### Antibody diversity results from DNA rearrangement and other mutations

Each gene encoding an immunoglobulin chain is in reality a "supergene" assembled by means of genetic recombination from several clusters of smaller genes scattered along part of a chromosome (**Figure 42.15**). Every cell in the body has hundreds of immunoglobulin genes located in separate clusters that are potentially capable of participating in the synthesis of both the variable and constant regions of immunoglobulin chains. In most body cells and tissues, these genes remain intact and separated from one another. But during B cell development, these genes are cut out, rearranged, and joined together in DNA recombination events. One gene from each cluster is chosen randomly for joining, and the others are deleted (**Figure 42.16**).

In this manner, a unique immunoglobulin supergene is assembled from randomly selected "parts." Each B cell precursor assembles two supergenes, one for a specific heavy chain and the other, assembled independently, for a specific light chain. This remarkable example of irreversible cell differentiation generates an enormous diversity of immunoglobulins from the same starting genome. It is a major exception to the generalization that all somatic cells derived from the fertilized egg have identical DNA.

In both humans and mice, the gene clusters encoding immunoglobulin heavy chains are on one pair of chromosomes and those for light chains are on two others. Two families of genes encode the variable region of the light chain, and three families encode the variable region of the heavy chain.

Figure 42.15 illustrates the gene families that encode the constant and variable regions of the heavy chain in mice. There are multiple genes that encode each of three parts of the variable region: 100 *V*, 30 *D*, and 6 *J* genes. Each B cell randomly selects one gene from each of these clusters to make the final coding sequence (*VDJ*) of the heavy-chain variable region. So the number of different heavy chains that can be made through this random recombination process is quite large:

$$100\ V \times 30\ D \times 6\ J = 18{,}000 \text{ possible combinations}$$

Now consider that the light chains are similarly constructed, with a similar amount of diversity made possible by random recombination. If we assume that the degree of potential light-

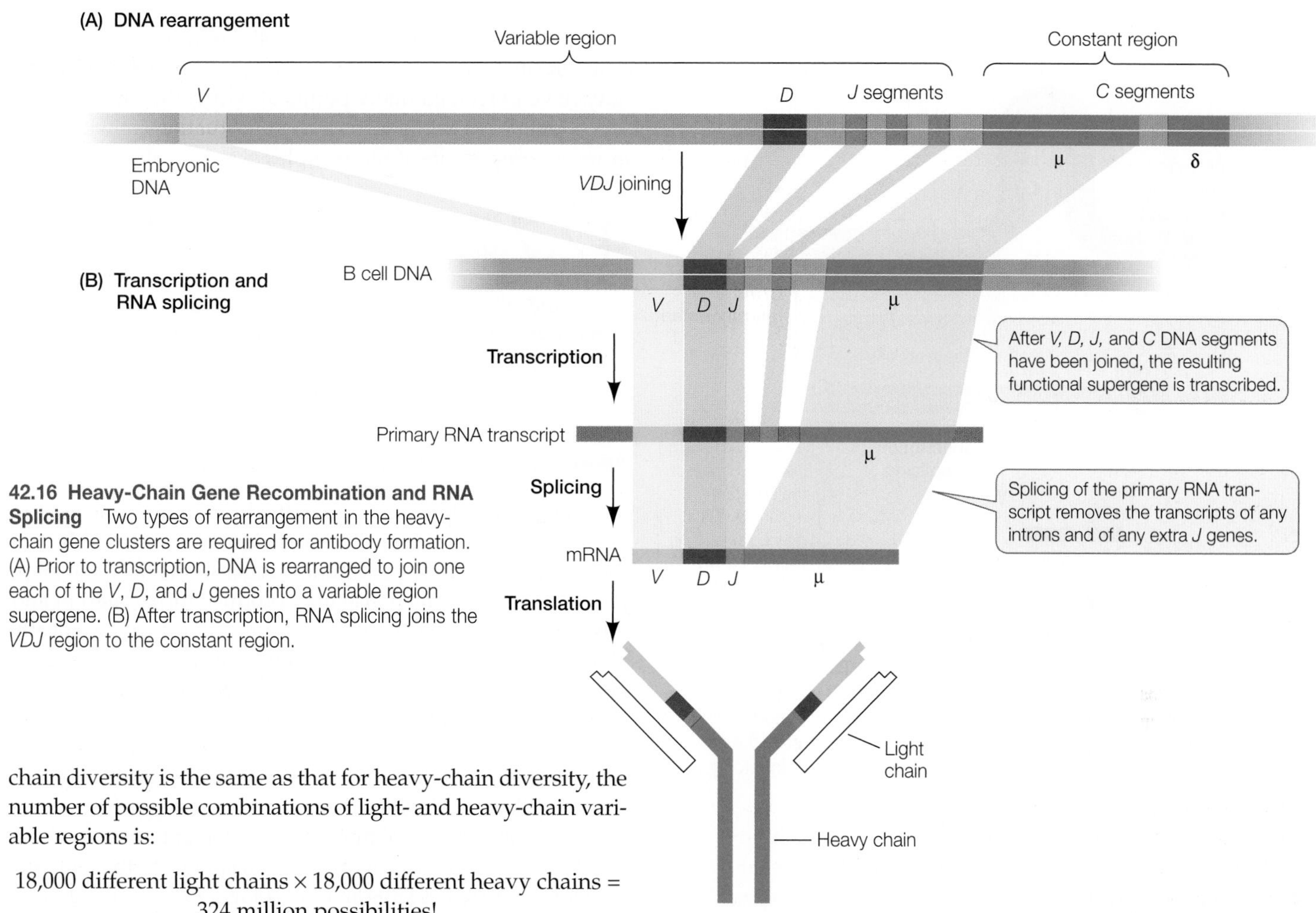


**42.16 Heavy-Chain Gene Recombination and RNA Splicing** Two types of rearrangement in the heavy-chain gene clusters are required for antibody formation. (A) Prior to transcription, DNA is rearranged to join one each of the *V*, *D*, and *J* genes into a variable region supergene. (B) After transcription, RNA splicing joins the *VDJ* region to the constant region.

chain diversity is the same as that for heavy-chain diversity, the number of possible combinations of light- and heavy-chain variable regions is:

18,000 different light chains × 18,000 different heavy chains = 324 million possibilities!

There are other mechanisms that generate even more diversity:

- When the DNA sequences that encode the *V*, *D*, and *J* regions are rearranged so that they are next to one another, the recombination event is not precise, and errors occur at the junctions. This *imprecise recombination* can create frameshift mutations, generating new codons at the junctions, with resulting amino acid changes.
- After the DNA sequences are cut out and before they are joined, the enzyme *terminal transferase* often adds some nucleotides to the free ends of the DNA pieces. These additional bases create insertion mutations.
- There is a relatively high *spontaneous mutation rate* in immunoglobulin genes. Once again, this process creates many new alleles and adds to antibody diversity.

When we include these possibilities with the millions of combinations that can be made by random DNA rearrangements, it is not surprising that the immune system can mount a response to almost any natural or artificial substance.

Once the pretranscriptional processing is completed, each supergene is transcribed and then translated to produce an immunoglobulin light chain or heavy chain. These combine to form an active immunoglobulin protein.

This genetic system is capable of still other kinds of changes. The B cell or plasma cell can switch the immunoglobulin class it produces while retaining its antigen specificity.

### The constant region is involved in immunoglobulin class switching

Table 42.3 describes the different classes of immunoglobulins and their functions. Generally, a B cell makes only one class at a time. But **class switching** can occur, in which a B cell changes the immunoglobulin class it synthesizes. For example, a B cell making IgM can switch to making IgG.

Early in its life, a B cell produces IgM molecules, which are the receptors responsible for its recognition of a specific antigen. At this time, the constant region of the heavy chain is encoded by the first constant region gene, the μ gene (see Figure 42.16). If the B cell later becomes a plasma cell during a humoral immune response, another deletion occurs in the cell's DNA, positioning the variable region genes (consisting of the same *V*, *D*, and *J* genes) next to a constant region gene farther away on the original DNA molecule (**Figure 42.17**). Such a DNA deletion results in the production of a new immunoglobulin with a different constant region of the heavy chain, and therefore a different function (see Table 42.3). However, this immunoglobulin has the same variable regions, and therefore the same antigen specificity, as the IgM produced by the parent B cell. The new immunoglobulin protein falls into one of the four classes (IgA, IgD, IgE, or IgG), depending on which of the constant region genes is placed adjacent to the variable region genes.

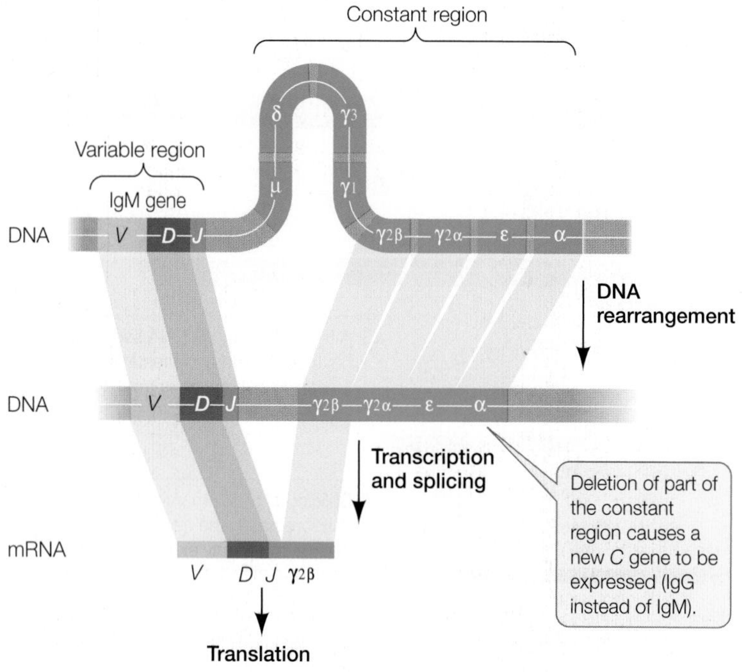

**42.17 Class Switching: Exchanging C Regions** The supergene produced by joining *V*, *D*, *J*, and *C* genes (see Figure 42.16) may later be modified, causing a different *C* region to be transcribed. This modification, known as class switching, is accomplished by deletion of part of the constant region gene cluster. Shown here is class switching from IgM to IgG.

What triggers class switching, and what determines the class to which a given B cell will switch? T-helper ($T_H$) cells direct the course of an immune response and determine the nature of the attack on the antigen. These T cells induce class switching by sending cytokine signals. The cytokines bind to receptors on the target B cells, generating signal transduction cascades that result in recombination and altered expression of the immunoglobulin genes.

## 42.6 RECAP

**The immune system can make millions of immunoglobulins with different specificities by rearranging the genes that encode the variable regions of the heavy and light chains. Additional mechanisms create mutations that provide for further diversity in the immunoglobulins. The class of the immunoglobulin molecule can be changed by recombination of the genes coding for the constant region of the heavy chain.**

- How can millions of antibodies with different specificities be generated from a relatively small number of genes? See pp. 892–893 and Figures 42.15 and 42.16
- What is the role of the constant region of the immunoglobulin in class switching? See pp. 893–894 and Figure 42.17

Given the numerous and complex cellular interactions that activate the immune system and generate antibody diversity, you may have perceived many points at which the immune system could fail. We now turn to several situations in which one or more components of this complex system malfunction.

## 42.7 What Happens When the Immune System Malfunctions?

Sometimes the immune system fails us in one way or another. It may overreact, as in an allergic reaction; it may attack self antigens, as in an autoimmune disease; or it may function weakly or not at all, as in an immune deficiency disease.

### Allergic reactions result from hypersensitivity

An **allergic reaction** arises when the human immune system overreacts to (is hypersensitive to) a dose of antigen. Although the antigen itself may present no danger to the host, the inappropriate immune response may produce inflammation and other symptoms, which can cause serious illness or even death. Allergic reactions are the most familiar examples of this phenomenon. Two types of allergic reactions involve immediate hypersensitivity and delayed hypersensitivity.

**IMMEDIATE HYPERSENSITIVITY** **Immediate hypersensitivity** arises when an allergic individual is exposed to an antigen (in this case referred to as an allergen) from the environment, such as a food, pollen, or the venom of an insect. In response to the allergen, the individual makes large amounts of IgE. When this happens, mast cells in tissues and basophils in the blood bind the constant end of the IgE. If that individual is exposed to the same allergen again, binding of the allergen to the IgE causes the mast cells and basophils to rapidly release a large amount of histamine (**Figure 42.18**). This results in symptoms such as dilation of blood vessels, inflammation, and difficulty breathing. If not treated with antihistamines, a severe allergic reaction can lead to death. It is not known why some people produce excessive amounts of IgE in response to allergens. There is some evidence for genetic factors predisposing people to allergic responses.

Allergy to pollen can be treated using a process called *desensitization*. The process involves injecting small amounts of the allergen (typically just an extract of the offending plant tissue) into the skin—enough to stimulate IgG production but not enough to stimulate IgE production. The next time the person is exposed to the allergen, IgG binds to it, tying it up before IgE can bind it and exert its harmful effects.

Desensitization does not work for food allergens because the IgE response to those substances is so strong that even a small amount of antigen provokes it. The best approach for those with food allergies—there are an estimated 3 million in the U.S.—is to avoid foods containing the allergens. This can be difficult, but food labels listing all the ingredients are helpful. Molecular biologists are beginning to identify the antigens that act as allergens, with the hope of developing vaccines or genetically modified foods that lack the allergenic epitopes.

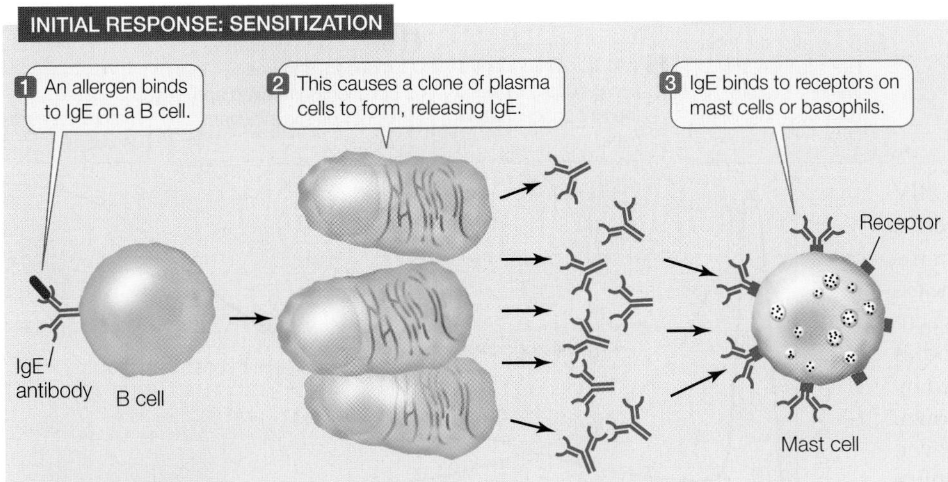

INITIAL RESPONSE: SENSITIZATION
1 An allergen binds to IgE on a B cell.
2 This causes a clone of plasma cells to form, releasing IgE.
3 IgE binds to receptors on mast cells or basophils.
Receptor
IgE antibody
B cell
Mast cell


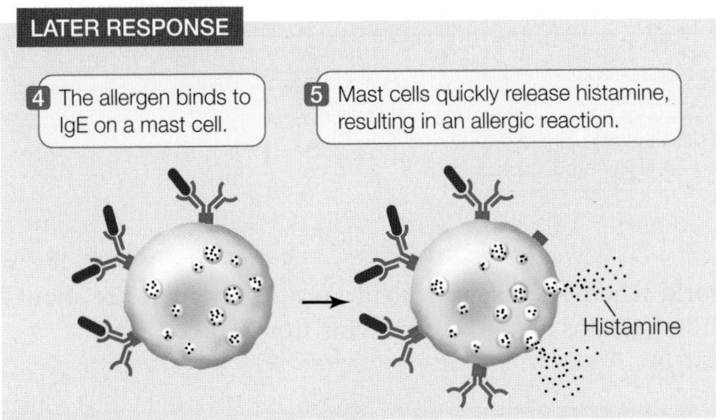

LATER RESPONSE
4 The allergen binds to IgE on a mast cell.
5 Mast cells quickly release histamine, resulting in an allergic reaction.
Histamine


**42.18 An Allergic Reaction** An allergen is an antigen that stimulates B cells to make large amounts of IgE antibodies, which bind to mast cells and basophils. When the body encounters the allergen again, these cells produce large amounts of histamine, which have harmful physiological effects.

**DELAYED HYPERSENSITIVITY** **Delayed hypersensitivity** is an allergic reaction that does not begin until hours after exposure to an antigen. In this case, the antigen is taken up by antigen-presenting cells and a T cell response is initiated. An example is the rash that develops after exposure to poison ivy.

### Autoimmune diseases are caused by reactions against self antigens

Errors in the selection of T cells in the thymus can result in T cells that bind to antigen–MHC complexes that carry self antigens. Although the precise origin of **autoimmunity** is not known, there are several hypotheses:

- *Failure of clonal deletion*. A clone of lymphocytes making antibodies against self antigens that should have been destroyed by Tregs and apoptosis is not destroyed.
- *Molecular mimicry*. T cells that recognize a nonself antigen, such as a virus, also recognize something on the self antigen that has a similar structure.

Autoimmunity does not always result in disease, but a number of autoimmune diseases are common:

- People with *systemic lupus erythematosis* (SLE) have antibodies to many cellular components, including DNA and nuclear proteins. These antinuclear antibodies can cause serious damage when they bind to normal tissue antigens and form large circulating antigen–antibody complexes, which become stuck in tissues and provoke inflammation.
- People with *rheumatoid arthritis* have difficulty in shutting down a T cell response. An inhibitory protein called CTLA4 blocks T cells from reacting to self antigens. People with rheumatoid arthritis may have low CTLA4 activity, which results in inflammation of the joints due to the infiltration of excess white blood cells.
- *Hashimoto's thyroiditis* is the most common autoimmune disease in women over 50. Immune cells attack thyroid tissue, resulting in fatigue, depression, weight gain, and other symptoms.
- *Insulin-dependent diabetes mellitus*, or type I diabetes, occurs most often in children. It is caused by an immune reaction against several proteins in the cells of the pancreas that manufacture the protein hormone insulin. This reaction kills the insulin-producing cells, so people with type I diabetes must take insulin daily in order to survive.

### AIDS is an immune deficiency disorder

There are a number of inherited and acquired *immune deficiency disorders*. In some individuals, T or B cells never form; in others, B cells lose the ability to give rise to plasma cells. In either case, the affected individual is unable to mount an immune response and thus lacks a major line of defense against pathogens. The $T_H$ cell is perhaps the most central component of the immune system because of its essential roles in both the humoral and cellular immune responses (see Figure 42.6). This cell is the target of **human immunodeficiency virus** (**HIV**), the retrovirus that results in **acquired immune deficiency syndrome** (**AIDS**).

HIV can be transmitted from person to person in body fluids containing the virus (such as blood, semen, vaginal fluid, or breast milk). The recipient tissue is either blood (by transfusion) or a mucous membrane lining an organ (the mucus contains a high concentration of lymphocytes). HIV initially infects macrophages, $T_H$ cells, and antigen-presenting dendritic cells in the blood and tissues. At first there is an immune response to the viral infection, and $T_H$ cells are activated. But because HIV

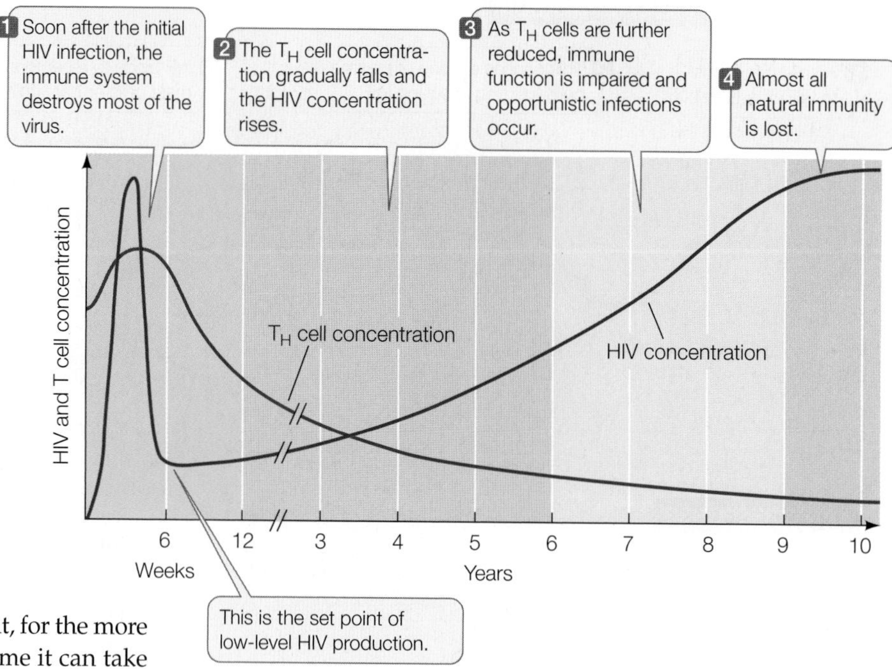
1 Soon after the initial HIV infection, the immune system destroys most of the virus.
2 The $T_H$ cell concentration gradually falls and the HIV concentration rises.
3 As $T_H$ cells are further reduced, immune function is impaired and opportunistic infections occur.
4 Almost all natural immunity is lost.
HIV and T cell concentration
$T_H$ cell concentration
HIV concentration
6
12
3
4
5
6
7
8
9
10
Weeks
Years
This is the set point of low-level HIV production.

**42.19 The Course of an HIV Infection** An HIV infection may be carried, unsuspected, for many years before the onset of symptoms.

infects the $T_H$ cells, they are killed both by HIV itself and by $T_C$ cells that lyse infected $T_H$ cells. Consequently, $T_H$ cell numbers decline after the first month or so of infection. Meanwhile, the extensive production of HIV by infected cells activates the humoral immune system. Antibodies bind to HIV and the complexes are removed by phagocytes. The HIV level in blood goes down. There is still a low level of infection, however, because of the depletion of $T_H$ cells (**Figure 42.19**). This process reaches a low, steady-state level called the "set point." This point varies among individuals and is a strong predictor of the rate of progression of the disease. For most people it takes 8–10 years, even without treatment, for the more severe manifestations of AIDS to develop. In some it can take as little as a year; in others, 20 years.

During this dormant period, people carrying HIV generally feel fine, and their $T_H$ cell levels are adequate for them to mount immune responses. Eventually, however, the virus destroys the $T_H$ cells, and their numbers fall to the point where the infected person is susceptible to infections that the $T_H$ cells would normally eliminate. These infections result in conditions such as Kaposi's sarcoma, a skin tumor caused by a herpesvirus; pneumonia caused by the fungus *Pneumocystis jirovecii*; and lymphoma tumors caused by the Epstein–Barr virus. These conditions result from *opportunistic infections* because the pathogens take advantage of the crippled immune system of the host. They lead to death within a year or two.

The molecular biology of HIV and its life cycle have been intensively studied (see Figure 16.6). Drug treatments are focused on inhibiting the viral proteins, such as *reverse transcriptase* that makes cDNA from the viral RNA, and *viral protease* that cuts the large precursor viral protein into its final active proteins. Combinations of such drugs result in long-term survival. Unfortunately, like many medical treatments, HIV drugs are not available to all who need them—particularly in poor regions of the world where AIDS is prevalent. As a result, there are about 2 million deaths per year worldwide from AIDS.

## 42.7 RECAP

**Failures of the immune system include allergic reactions (caused by hypersensitivity to antigens), autoimmune diseases (caused by reactions against self antigens), and immune deficiency disorders.**

- How does immediate hypersensitivity develop? See p. 894 and Figure 42.18
- What is an autoimmune disease? Give an example. See p. 895
- Describe the course of events in the human immune system during HIV infection. See pp. 895–896 and Figure 42.19

## CHAPTER SUMMARY

### 42.1 What Are the Major Defense Systems of Animals?

- Animal defenses against pathogens are based on the body's ability to distinguish between self and nonself.
- **Nonspecific (innate) defenses** are inherited mechanisms that protect the body from many kinds of pathogens. They typically act rapidly.
- **Specific defenses** are adaptive mechanisms that respond to specific pathogens. They develop more slowly than nonspecific defenses but are long-lasting.
- Many defenses are implemented by cells and proteins carried in the blood plasma and lymph. Review Figure 42.1, **WEB ACTIVITY 42.1**
- White blood cells fall into two broad groups. **Phagocytes** include **macrophages** that engulf pathogens by phagocytosis. **Lymphocytes**, which include **B cells** and **T cells**, participate in specific responses. Review Figure 42.2, **ANIMATED TUTORIAL 42.1**

## 42.2 What Are the Characteristics of the Nonspecific Defenses?

- An animal's nonspecific defenses include physical barriers such as the skin, and competing resident microorganisms known as **normal flora**. Review Table 42.1
- The **complement system** consists of more than 20 different antimicrobial proteins that act to alter membrane permeability and kill targeted cells.
- Circulating defensive cells, such as phagocytes and **natural killer cells**, eliminate invaders.
- The **inflammation response** activates several types of cells and proteins that act against invading pathogens. **Mast cells** release **histamines**, which cause blood vessels to dilate and become "leaky." Review Figure 42.3, WEB ACTIVITY 42.2
- A cell signaling pathway involving the **toll-like receptor** stimulates the body's defenses. Review Figure 42.4

## 42.3 How Does Specific Immunity Develop?

SEE ANIMATED TUTORIAL 42.2

- The specific immune response recognizes specific **antigens**, responds to an enormous diversity of **antigenic determinants**, distinguishes self from nonself, and remembers the antigens it has encountered.
- Each antibody and each T cell is specific for a single antigenic determinant. **T cell receptors** bind to antigens on the surfaces of virus-infected cells and abnormal cells.
- The **humoral immune response** is directed against pathogens in the blood, lymph, and tissue fluids. The **cellular immune response** is directed against an antigen established within a host cell. Both responses are mediated by antigenic fragments being presented on a cell surface along with the proteins of the **major histocompatibility complex**.
- **Clonal selection** accounts for the specificity and diversity of the immune response and for **immunological memory**. Review Figure 42.7
- An activated B or T lymphocyte produces **effector cells** that attack the antigen, and **memory cells** that are long-lived and rarely divide. Effector B cells are called **plasma cells** and secrete specific **antibodies**.
- **Vaccination** is inoculation with modified pathogens or antigens that provoke an immune response but are not pathogenic.

## 42.4 What Is the Humoral Immune Response?

SEE ANIMATED TUTORIAL 42.3

- B cells are the basis of the humoral immune response. Naïve B cells are activated by binding of antigen and by stimulation by $T_H$ cells with the same specificity, and then form plasma cells. These cells synthesize and secrete specific antibodies.
- The basic unit of an **immunoglobulin** is a tetramer of four polypeptides: two identical light chains and two identical heavy chains, each consisting of a **constant region** and a **variable region**. Review Figure 42.9, WEB ACTIVITY 42.3
- The variable regions determine the specificity of an immunoglobulin, and the constant regions of the heavy chain determine its **class**. There are five classes of immunoglobulins with different body locations and functions. Review Table 42.3
- A **monoclonal antibody** can be made by fusing a B cell with a tumor cell to form a **hybridoma**. Review Figure 42.10

## 42.5 What Is the Cellular Immune Response?

SEE ANIMATED TUTORIAL 42.4

- T cells are the effectors of the cellular immune response. T cell receptors are somewhat similar in structure to the immunoglobulins, having variable and constant regions. Review Figure 42.11
- There are three types of T cells. **Cytotoxic T cells** recognize and kill virus-infected cells or mutated cells. **T-helper cells** direct both the cellular and humoral immune responses. **Regulatory T cells** inhibit the other T cells from mounting an immune response to self antigens.
- The genes of the major histocompatibility complex (MHC) encode membrane proteins that bind antigenic fragments and present them to T cells. Review Figures 42.12 and 42.13
- Organ transplants are rejected when the host's immune system recognizes MHC proteins on transplanted tissue as nonself and initiates an immune defense attacking the foreign tissue.

## 42.6 How Do Animals Make So Many Different Antibodies?

SEE ANIMATED TUTORIAL 42.5

- B cell genomes undergo random recombinations of genes coding for regions of the immunoglobulin polypeptide chains so that each cell can produce a specific antibody protein. The immunoglobulin chains derive from "supergenes" that are constructed from different combinations of *V*, *D*, *J*, and *C* genes. This DNA rearrangement and rejoining yields millions of different immunoglobulin chains. Review Figures 42.15 and 42.16
- Once a B cell becomes a plasma cell, it may undergo **class switching**, in which a deletion of one or more constant region genes results in the production of an immunoglobulin with a different constant region and a different function. Review Figure 42.17

## 42.7 What Happens When the Immune System Malfunctions?

- An **allergic reaction** is an inappropriate immune response caused by immediate or delayed **hypersensitivity** to certain antigens. Review Figure 42.18
- Autoimmune diseases result when the immune system produces B and T cells that attack self antigens.
- Immune deficiency disorders result from failure of one or another part of the immune system. **AIDS** is an acquired immune deficiency disorder arising from depletion of the $T_H$ cells as a result of infection with HIV. Review Figure 42.19

## SELF-QUIZ

1. Phagocytes kill harmful bacteria by
   *a.* endocytosis.
   *b.* producing antibodies.
   *c.* complement proteins.
   *d.* T cell stimulation.
   *e.* inflammation.
2. Which statement about immunoglobulins is true?
   *a.* They help antibodies do their job.
   *b.* They recognize and bind antigenic determinants.
   *c.* They encode some of the most important genes in an animal.
   *d.* They are the chief participants in nonspecific defense mechanisms.
   *e.* They are a specialized class of white blood cells.
3. Which statement about an antigenic determinant is *not* true?
   *a.* It is a specific chemical grouping.
   *b.* It may be part of many different molecules.
   *c.* It is the part of an antigen to which an antibody binds.
   *d.* It may be part of a cell.
   *e.* A single protein has only one on its surface.
4. T cell receptors
   *a.* are the primary receptors for the humoral immune system.
   *b.* are carbohydrates.
   *c.* cannot function unless the animal has previously encountered the antigen.
   *d.* are produced by plasma cells.
   *e.* are important in combating viral infections.
5. According to the clonal selection theory,
   *a.* an antibody changes its shape to match the antigen it meets.
   *b.* an individual animal contains only one type of B cell.
   *c.* an individual animal contains many types of B cells, each producing one kind of antibody.
   *d.* each B cell produces many types of antibodies.
   *e.* many clones of antiself lymphocytes appear in the bloodstream.
6. Immunological tolerance
   *a.* depends on exposure to an antigen.
   *b.* develops late in life and is usually life-threatening.
   *c.* disappears at birth.
   *d.* results from the activities of the complement system.
   *e.* results from DNA splicing.
7. The extraordinary diversity of antibodies results in part from
   *a.* the action of monoclonal antibodies.
   *b.* the splicing of protein molecules.
   *c.* the action of cytotoxic T cells.
   *d.* the rearrangement of genes.
   *e.* their remarkable nonspecificity.
8. Which of the following play(s) no role in the B cell response?
   *a.* T-helper cells
   *b.* Growth factors
   *c.* Macrophages
   *d.* Reverse transcriptase
   *e.* Products of class II MHC genes
9. The major histocompatibility complex
   *a.* codes for specific proteins found on the surfaces of cells.
   *b.* plays no role in T cell immunity.
   *c.* plays no role in antibody responses.
   *d.* plays no role in skin graft rejection.
   *e.* is encoded by a single locus with multiple alleles.
10. Which of the following is important in the immune deficiency caused by HIV?
    *a.* Pneumonia infection by bacteria
    *b.* Activated $T_H$ cell infection by HIV
    *c.* Circulating free HIV in blood plasma
    *d.* Antibodies made against HIV-infected cells
    *e.* Increase in T cells

## FOR DISCUSSION

1. Describe the part of an antibody molecule that interacts with an antigen. How is it similar to the active site of an enzyme? How does it differ from the active site of an enzyme?
2. Contrast immunoglobulins and T cell receptors with respect to their structures and functions.
3. Discuss the diversity of antibody specificities in an individual in relation to the diversity of enzymes. Does every cell in an animal contain genetic information for all the organism's enzymes? Does every cell contain genetic information for all the organism's immunoglobulins?
4. The gene family determining MHC on the cell surface in humans is on a single chromosome. A father's MHC type is A1, A3, B5, B7, D9, D11. A mother's genotype is A2, A4, B6, B7, D11, D12. Their child is A1, A4, B6, B7, D11, D12. What are the parents' haplotypes—that is, which alleles are linked on each of the two chromosomes of each parent? Assuming there is no recombination among the genes determining the MHC type, can these same two parents have a child who is A1, A2, B7, B8, D10, D11?

## ADDITIONAL INVESTIGATION

Development of an effective HIV vaccine requires that the person being vaccinated develop both cellular and humoral immunity against HIV. Explain what this means in terms of studies you would do on people given a potential new vaccine.

# 43 Animal Reproduction

## Explosive sex

Producer of valuable honey and pollinator of many crucial plants, the common honey bee, *Apis mellifera*, has been the subject of study and fascination for humans throughout recorded history. The unique sex life of these social insects is among their most intriguing aspects.

More than 99 percent of female honey bees do not reproduce. They exist to help just one female—the queen—reproduce. At any given time, there is only one queen in a hive. She lays all of the eggs, which is a full-time job. She also determines whether or not an egg will be fertilized. Fertilized eggs develop into females; unfertilized eggs develop into males. Wait, you say, it's the *male* that fertilizes the egg! Yes, the male's sperm fertilizes the egg, but in this case the female—the queen bee—stores sperm after mating and controls its delivery to her eggs.

Eventually, every hive must have a new queen. Sometimes the old queen dies. Other times, in a phenomenon known as *swarming*, the queen and a retinue of workers leave their current hive to start a new one. Whether the queen dies or leaves, a new queen must be produced. The remaining worker bees enlarge a few cells in the honeycomb that contain fertilized eggs laid by the old queen. The larvae that hatch from those eggs are fed special food that stimulates their growth and development into prospective queens.

The first queen to pupate and emerge from her royal chamber kills any other aspiring queens. She then leaves the hive for her mating flight. Males from all around get the message that a virgin queen is available and congregate around her. While in flight, she mates with 15 to 20 males, and each coupling is an event. After a male manages to insert his penis into her vagina, he literally explodes, leaving inside the female not only his sperm but also his sex organs (the latter will drop out later). This process is repeated with other males, which also die as a result. The queen returns to the hive with a lifetime supply of sperm and sets about laying eggs.

The queen lives for about 2 years and lays as many as 3,000 eggs each day. If she releases sperm, the eggs will be fertilized and develop into sterile females that will devote their lives to feeding her, maintaining the hive, foraging for nectar and pollen, and raising their sisters. Unfertilized eggs develop into males, which hang around the hive doing nothing useful until they take off to search for a receptive queen—an unlikely event. It is thus in the

**A Unique Reproductive Strategy** Among honey bees and other hymenopteran insects, the only reproductive female is the queen, seen here in the center as she deposits eggs in the comb. The female workers that attend her are sterile.

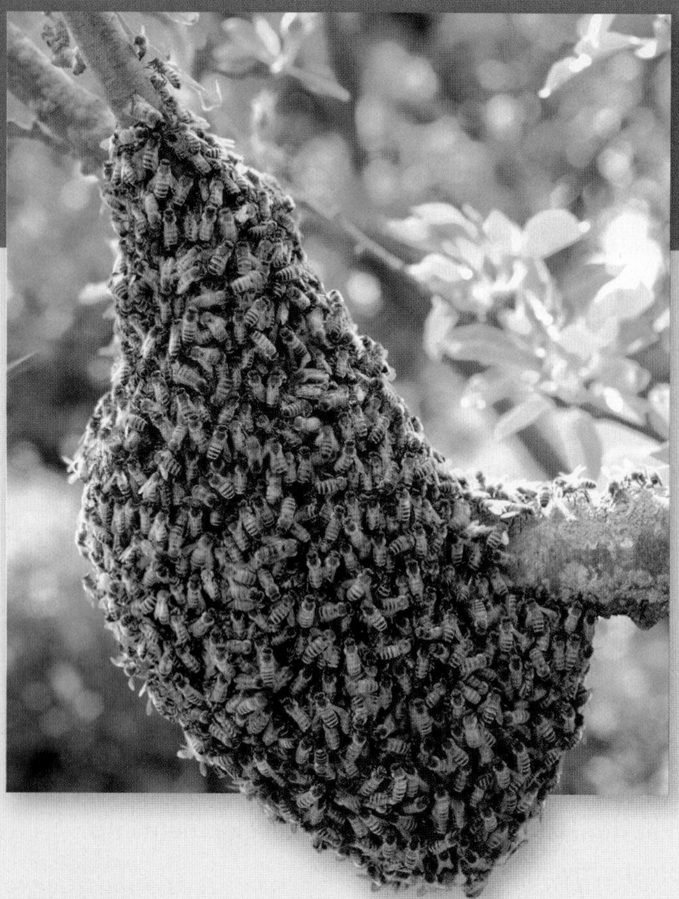

**Swarming Means a New Queen Bee** When a honey bee colony swarms, the queen leaves the hive and takes a retinue of workers with her. A new queen will emerge to take over the old hive, at which time a few males will get to perform their one brief function—fertilizing a virgin queen—before they die.

queen's best interest to limit the number of males she produces.

Natural selection has resulted in some amazing adaptations, none more so than those involved in reproduction. Sexual or asexual, bizarre or otherwise, the anatomy, physiology, biochemistry, and behavior surrounding the urge to propagate one's species are fascinating fields of study.

**IN THIS CHAPTER** we will examine the diverse ways in which animals produce offspring. First we examine asexual reproduction, in which only a single parent is involved. We then turn to sexual reproduction, in which an egg and a sperm unite to create a new diploid individual. Finally, we focus on the anatomy, function, and endocrine control of the human reproductive system, as well as technologies used to limit or enhance human fertility.

CHAPTER OUTLINE



## 43.1 How Do Animals Reproduce without Sex?

Sexual reproduction is a nearly universal trait in animals, although many species can also reproduce asexually and some reproduce only asexually. Offspring produced asexually are genetically identical to one another and to their parents. Asexual reproduction is efficient because no time or energy is wasted on mating and every member of the population can convert resources into offspring. However, asexual reproduction does not generate genetic diversity as sexual reproduction does, and this diversity is the raw material that enables natural selection to shape adaptations in response to environmental change. When such changes occur, the lack of genetic diversity can be disadvantageous to a species.

A variety of animals, mostly invertebrates, reproduce asexually. They tend to be species that are attached to their substrate and cannot search for mates or that live in sparse populations and rarely encounter potential mates. Asexually reproducing species are likely to be found in relatively constant environments where genetic diversity is less important for species success. In fact, asexual reproduction is a good way to preserve a genotype that is successful in a particular environment, as long as that environment does not change. Three common modes of asexual reproduction are *budding, regeneration,* and *parthenogenesis.*

### Budding and regeneration produce new individuals by mitosis

Many simple multicellular animals produce offspring by **budding**. New individuals form as outgrowths or buds from the bodies of older animals. A bud grows by mitotic cell division, and the cells differentiate before the bud breaks away from the parent (**Figure 43.1A**). The bud is genetically identical to the parent, and it may grow as large as the parent before it becomes independent.

**Regeneration** is usually thought of as the replacement of damaged tissues or lost limbs, but in some cases pieces of an organism can regenerate complete individuals. Echinoderms, for example, have remarkable abilities to regenerate. If sea stars (starfishes) are cut into pieces, each piece that includes an arm and a portion of the central disc can grow into a new animal (**Figure 43.1B**). In the early 1900s, oyster fishermen in Narragansett Bay tried to eliminate the sea stars that were preying on their oysters. Whenever they encountered a sea star, they chopped it up with knives and threw it back into the water. As a result, the sea star population increased explosively.

Regeneration can occur when an animal is broken by an outside force such as wave action in the intertidal zone. In some

(A) *Hydra* sp.

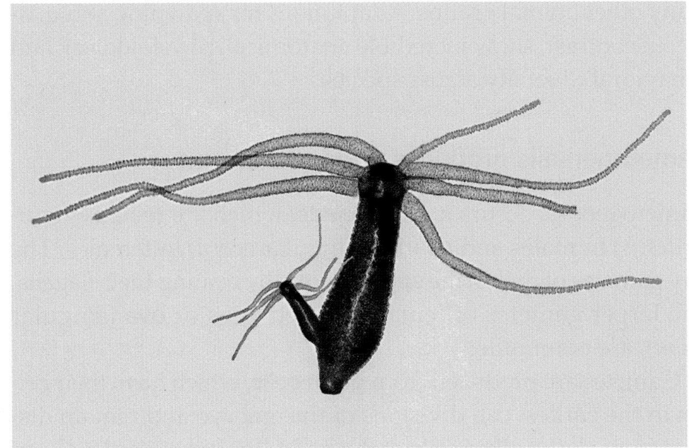

(B) *Fromia* sp.

**43.1 Two Forms of Asexual Reproduction** (A) Budding: A new individual forms as an outgrowth from an adult hydra. (B) Regeneration: A single severed arm and a piece of the central disc of a mature sea star can regenerate into an entire animal.

cases, breakage occurs in the absence of external forces. Some species of segmented marine worms develop segments with rudimentary heads bearing sensory organs. The segments then break apart, and each one forms a new worm.

## Parthenogenesis is the development of unfertilized eggs

Not all eggs must be fertilized to develop. A common mode of asexual reproduction in arthropods is the development of offspring from unfertilized eggs. This phenomenon, called **parthenogenesis**, also occurs in some species of fishes, amphibians, and reptiles. Most species that reproduce parthenogenetically also engage in sexual reproduction or at least sexual behavior at other times.

In some species, parthenogenesis is part of the mechanism that determines sex. As we saw at the beginning of this chapter, in honey bees (as well as in most ants and wasps), males develop from unfertilized eggs and are haploid. Females develop from fertilized eggs and are diploid.

Parthenogenetic reproduction in some species requires sexual behavior even though sperm are not delivered to the female reproductive tract and eggs are not fertilized. One case that has been investigated extensively by David Crews and his students at the University of Texas is parthenogenetic reproduction in a species of whiptail lizard. This species has no males, but females can act as males, engaging in all aspects of courtship display and mating, although no sperm are produced or transferred (**Figure 43.2**). Whether a specific female acts as a female or as a male depends on cyclical hormonal states. When estrogen levels are high, she acts as a female. When her progesterone level peaks, she acts as a male. The stimulation resulting from the sexual activity triggers the release of eggs from the ovaries.

(A)

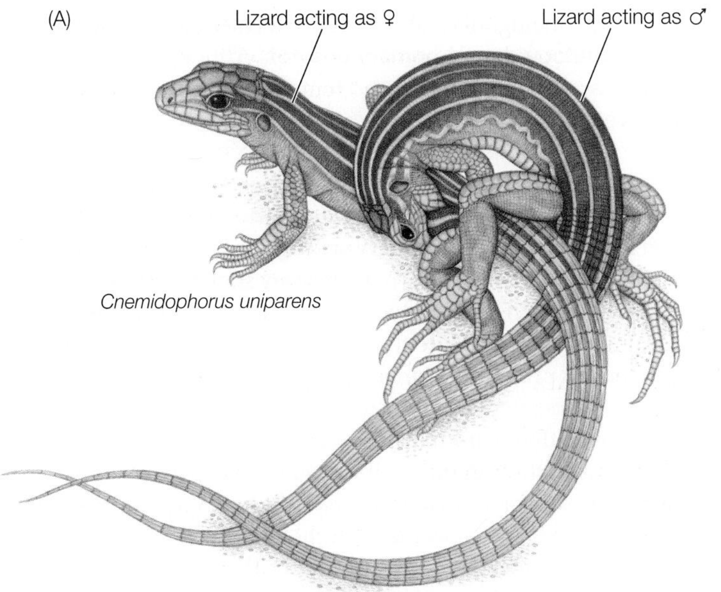


(B)

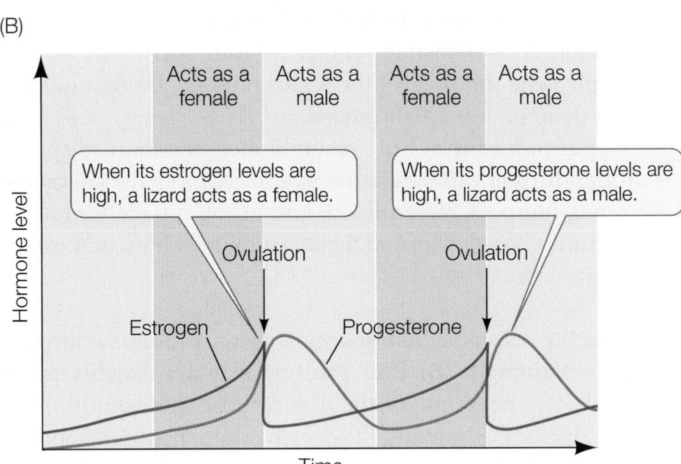


**43.2 Sexual Behavior May Be Required for Asexual Reproduction** (A) Parthenogenetic whiptail lizards are all females, but they take turns acting the male role in reproductive behavior. The stimulation from sexual behavior is necessary for ovulation to occur. (B) The stage of the ovarian cycle determines the role an individual whiptail plays.

## 43.1 RECAP

**Most animals reproduce sexually, but many can also or can only reproduce asexually, through budding, regeneration, or parthenogenesis.**

- Explain why asexual reproduction might be disadvantageous for an animal living in a changing environment. See p. 900
- How is parthenogenesis related to sex determination in honey bees? See p. 901

Asexual reproduction is an efficient way to use resources. However, the fact that sexual reproduction produces genetic diversity must be a tremendous advantage, because most species reproduce sexually.

# 43.2 How Do Animals Reproduce Sexually?

Given the efficiency of asexual reproduction in perpetuating an organism's genome, the prevalence of sexual reproduction is somewhat surprising. Even the evolution of meiosis—an extremely complicated process in comparison to mitosis—has been the subject of much speculation and debate among evolutionary biologists. And of course, mating behaviors involve costs and risks. Costs include time and energy for finding, attracting, and competing for a mate as well as the opportunity costs of detracting from other activities such as feeding and caring for existing offspring. Risks include increased exposure to predation and the potential for physical damage. Despite these disadvantages, most eukaryotic organisms reproduce sexually. Thus it would seem that the production of genetic diversity is an evolutionary advantage that overwhelms "the cost of sex" (see Section 21.4).

Sexual reproduction requires the joining of two haploid sex cells to form a diploid individual. These haploid cells, or *gametes*, are produced through **gametogenesis**, a process that involves meiotic cell divisions. Two events in meiosis contribute to genetic diversity: *crossing over* between homologous chromosomes and the *independent assortment* of chromosomes (see Sections 11.5 and 12.1). Sexual reproduction itself also contributes to genetic diversity. The genetic variation among the gametes of a single individual and the genetic variation between any two parents produce an enormous potential for genetic variation between any two offspring of a sexually reproducing pair of individuals.

Sexual reproduction in animals consists of three fundamental steps:

- *Gametogenesis*: making gametes
- *Spawning* or *mating*: bringing gametes together
- *Fertilization*: fusing gametes

The process of gametogenesis is very similar across sexually reproducing animal species. Processes of fertilization are also rather similar in widely different species. Therefore, while our discussion of gametogenesis will focus generally on mammals, and our discussion of fertilization will feature sea urchins, the facts would not be dramatically different were we to consider many other animal groups. Adaptations for spawning and mating, in contrast, show incredible anatomical, physiological, and behavioral diversity across species.

## Gametogenesis produces eggs and sperm

Gametogenesis occurs in the **gonads**, which are **testes** (singular *testis*) in males and **ovaries** (singular *ovary*) in females. The tiny gametes of males, the **sperm**, move by beating their flagella. The larger gametes of females, called eggs or **ova** (singular *ovum*), are nonmotile.

Gametes are produced from **germ cells**, which have their origin in the earliest cell divisions of the embryo and remain distinct from all the other cells of the body (the *somatic cells*). Germ cells are sequestered in the body of the embryo until its gonads begin to form. The germ cells then migrate to the developing gonads, where they take up residence and proliferate by mitosis, producing **spermatogonia** (singular *spermatogonium*) in males and **oogonia** (singular *oogonium*) in females. Spermatogonia and oogonia are diploid and multiply by mitosis.

In the next step in gametogenesis, meiotic cell division reduces the chromosomes to the haploid number (see Section 11.5). The spermatogonia and oogonia that enter meiosis are **primary spermatocytes** and **primary oocytes**. Although the steps of meiosis are similar in males and females, gametogenesis differs between the sexes.

**SPERMATOGENESIS** The initial proliferation of male germ cells into spermatogonia proceeds by mitosis in the embryo. As illustrated in **Figure 43.3A**, primary spermatocytes then undergo the first meiotic division to form **secondary spermatocytes**. The second meiotic division produces four haploid **spermatids** for each primary spermatocyte that enters meiosis. In mammals, the progeny of primary spermatocytes remain connected by *cytoplasmic bridges* after each division.

One reason that mammalian spermatocytes remain in cytoplasmic contact throughout their development is the asymmetry of sex chromosomes in males. Half the secondary spermatocytes receive an X chromosome, the other half a Y chromosome. The Y chromosome contains fewer genes than the X chromosome, and some of the products of genes found only on the X chromosome are essential for spermatocyte development. By remaining in cytoplasmic contact, all four spermatocytes can share the gene products of the X chromosomes, although only half of them have an X chromosome.

A spermatid bears little resemblance to a mature sperm. Through further differentiation (*spermiogenesis*), the spermatid becomes compact, streamlined, and grows a flagellum to become motile. We will look at the production of human sperm in the next section.

**OOGENESIS** Oogonia, like spermatogonia, proliferate through mitosis (**Figure 43.3B**). The resulting primary oocytes immediately enter prophase of the first meiotic division. In many species, including humans, the oocyte experiences developmental arrest at this point and may remain in that state for days,

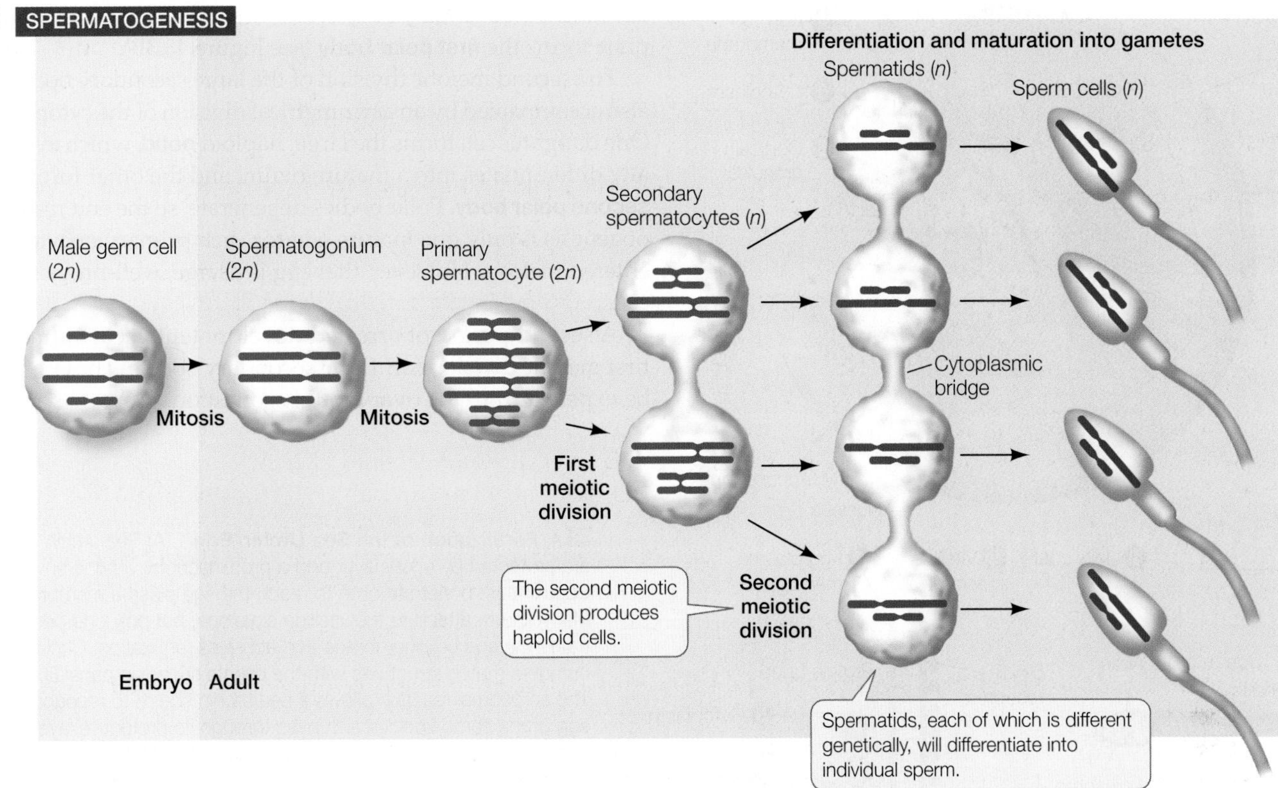


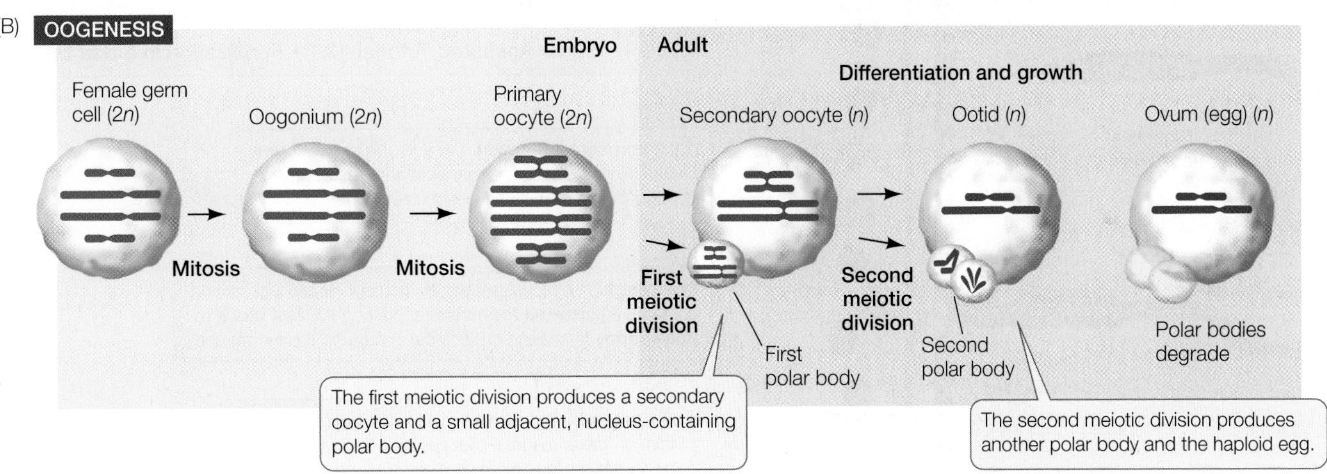


**43.3 Gametogenesis** Male and female germ cells proliferate by mitosis and produce diploid spermatogonia and oogonia that mature into primary spermatocytes and oocytes before entering meiosis. (A) Spermatogonia continue to divide by mitosis in adults, producing a steady supply of spermatocytes that divide meiotically to produce haploid spermatids, which differentiate into sperm. In many species, the progeny of spermatocytes remain in contact through cytoplasmic bridges until the sperm mature. (B) In mammals, oogonia cease division in the embryo, and primary oocytes remain arrested in prophase I of meiosis until they are ovulated and fertilized. Each oocyte will produce one haploid ootid which matures into an ovum.

months, or years. In the human female, this period of arrest is at least 10 years (i.e., until puberty), and some primary oocytes remain in prophase I for up to 50 years (i.e., until menopause). In contrast, spermatogenesis continues, uninterupted, to completion once the primary spermatocyte has differentiated.

During this prolonged prophase I, or shortly before it ends, the primary oocyte grows larger through increased production of ribosomes, RNA, cytoplasmic organelles, and energy stores. At this point, the primary oocyte acquires all the energy, raw materials, and RNA that the ovum will need to survive its first cell divisions after fertilization. In fact, the nutrients in the egg must maintain the embryo until it is either nourished by the maternal circulatory system or can feed on its own.

When a primary oocyte resumes meiosis, its nucleus completes the first meiotic division near the surface of the cell. The daughter cells of this division receive grossly unequal shares of cytoplasm. This asymmetry represents another major difference from spermatogenesis, in which cytoplasm is apportioned equally. The daughter cell that receives almost all the cytoplasm becomes the **secondary oocyte**, and the one that receives almost none forms the **first polar body** (see Figure 43.3B).

The second meiotic division of the large secondary oocyte is also accompanied by an asymmetrical division of the cytoplasm. One daughter cell forms the large, haploid **ootid**, which eventually differentiates into a mature ovum, and the other forms the **second polar body**. Polar bodies degenerate, so the end result of oogenesis is only one mature egg for each primary oocyte that entered meiosis. However, that egg is a large, well-provisioned cell.

A second period of arrested development occurs after the first meiotic division forms the secondary oocyte. The egg may be expelled from the ovary in this condition. In many species,

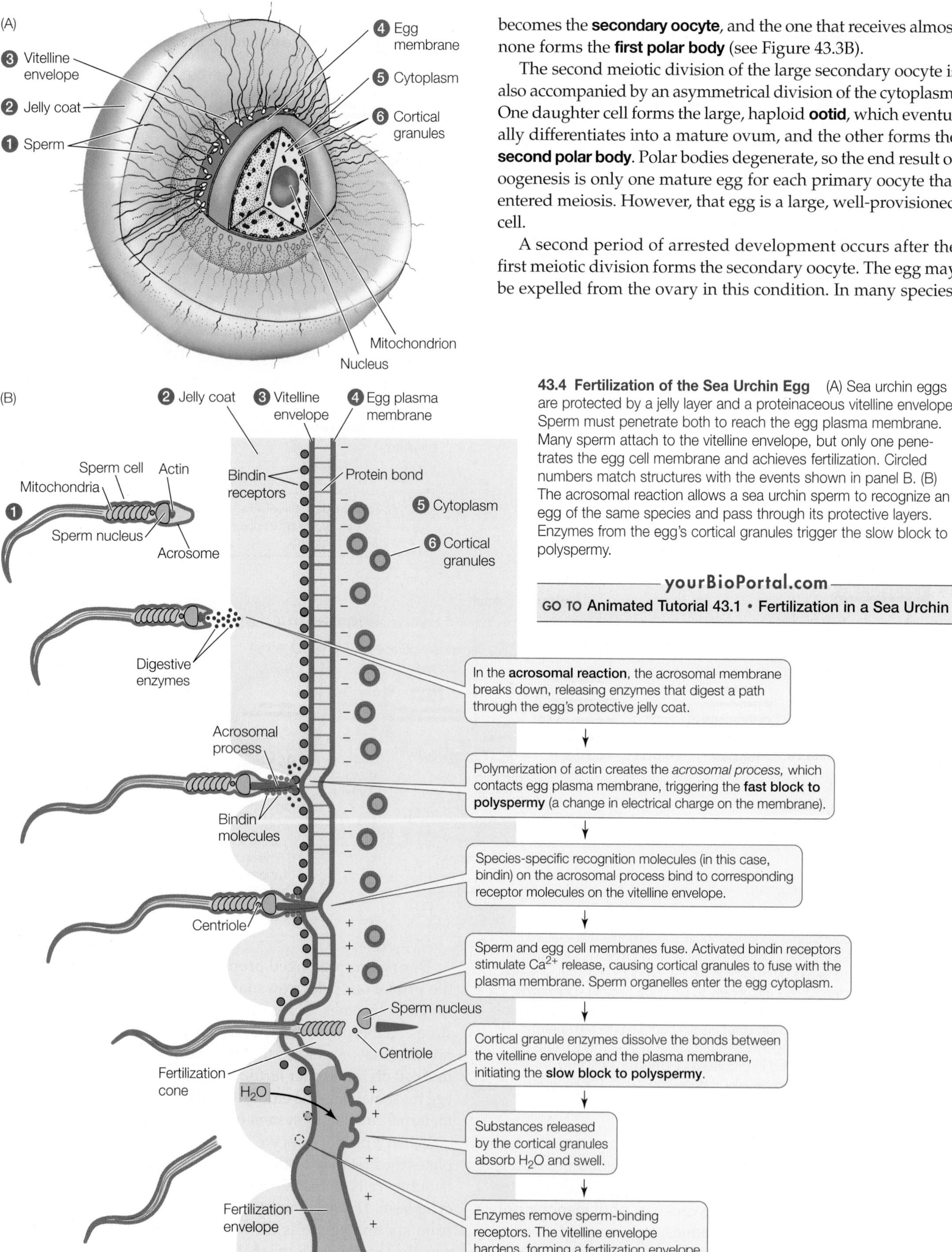


**43.4 Fertilization of the Sea Urchin Egg** (A) Sea urchin eggs are protected by a jelly layer and a proteinaceous vitelline envelope. Sperm must penetrate both to reach the egg plasma membrane. Many sperm attach to the vitelline envelope, but only one penetrates the egg cell membrane and achieves fertilization. Circled numbers match structures with the events shown in panel B. (B) The acrosomal reaction allows a sea urchin sperm to recognize an egg of the same species and pass through its protective layers. Enzymes from the egg's cortical granules trigger the slow block to polyspermy.

**yourBioPortal.com**
GO TO Animated Tutorial 43.1 • Fertilization in a Sea Urchin

including humans, the second meiotic division is not completed until the egg is fertilized by a sperm.

### Fertilization is the union of sperm and egg

The union of the haploid sperm and the haploid egg in **fertilization** creates a single diploid cell, called a **zygote**, which will develop into an embryo. Fertilization does more, however, than just restore the full genetic complement of the animal. The processes associated with fertilization help the egg and sperm get together, prevent the union of the sperm and egg of different species, and guarantee that only one sperm will enter and activate the egg metabolically. Fertilization involves a complex series of events:

- The sperm and the egg recognize each other.
- The sperm is *activated*, enabling it to gain access to the plasma membrane of the egg.
- The plasma membrane of the egg fuses with the plasma membrane of a single sperm.
- The egg blocks entry of additional sperm.
- The egg is metabolically activated and stimulated to start development.
- The egg and sperm nuclei fuse to create the diploid nucleus of the zygote.

**SPECIFICITY IN SPERM–EGG INTERACTIONS** Specific recognition molecules mediate interactions between sperm and eggs. These molecules ensure that the activities of sperm are directed toward eggs and not other cells, and they help prevent eggs from being fertilized by sperm from the wrong species. The latter function is particularly important in aquatic species that release eggs and sperm into the surrounding water, as the eggs of such animals may readily be exposed to sperm of other species. The sea urchin is a good example of such a species, and its mechanisms of fertilization have been well studied.

The eggs of sea urchins release chemical attractants that increase the motility of sperm and cause them to swim toward the egg. These chemical attractants are species-specific. For example, eggs of one species of sea urchin release a specific peptide consisting of 14 amino acids. As this peptide diffuses from the egg, it binds to receptors on the sperm of the same species. The sperm respond by increasing their mitochondrial respiration and motility. Before exposure to the peptide, the sperm swim in tight little circles, but after binding to the peptide, they swim energetically up the concentration gradient of the peptide until they reach the egg that is releasing it.

When sperm reach an egg, they must get through two protective layers before they can fuse with the egg plasma membrane. The eggs of sea urchins are covered with a **jelly coat**, which surrounds a proteinaceous **vitelline envelope** (**Figure 43.4A**). The success of a sperm's assault on these protective layers depends on a membrane-enclosed structure at the front of the sperm head called an **acrosome**.

The acrosome, which contains enzymes and other proteins, forms a cap over the sperm nucleus. When the sperm makes contact with an egg of its own species, substances in the jelly coat trigger an *acrosomal reaction*, which begins with the breakdown of the plasma membrane covering the sperm head and the underlying acrosomal membrane (**Figure 43.4B**). The acrosomal enzymes are released and digest a hole through the jelly coat.

As a result of the polymerization of actin triggered by the acrosomal reaction, a structure called the *acrosomal process* extends out of the head of the sperm. The acrosomal process is coated with species-specific recognition molecules called *bindin*, and there are bindin receptors on the vitelline envelope of the egg. The interaction of these two molecules enables the sperm to contact the egg plasma membrane. That contact results in fusion of the sperm and egg plasma membranes and the formation of a *fertilization cone* that engulfs the sperm head, bringing it into the egg cytoplasm. The sperm mitochondrion, which largely constitutes the mid-piece of the sperm, is also drawn into the egg cytoplasm, but it degrades and disappears so that the mitochondria and mitochondrial genes of the new urchin are derived only from the egg.

In animals that practice internal fertilization, mating behaviors help guarantee species specificity, but egg–sperm recognition mechanisms still exist. The mammalian egg is surrounded by a thick layer called the **cumulus**, which consists of a loose assemblage of maternal cells in a gelatinous matrix (**Figure 43.5**). Beneath the cumulus is a glycoprotein envelope called the **zona pellucida**, which is functionally similar to the vitelline envelope of sea urchin eggs. When mammalian sperm are deposited in the female reproductive tract, they are metabolically activated and made capable of an acrosomal reaction

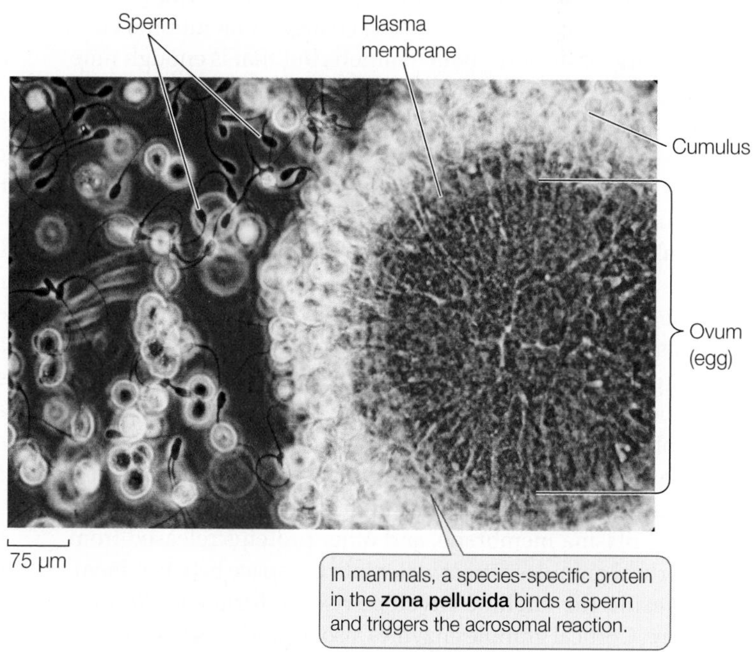
Sperm
Plasma membrane
Cumulus
Ovum (egg)
75 µm
In mammals, a species-specific protein in the **zona pellucida** binds a sperm and triggers the acrosomal reaction.

**43.5 Barriers to Sperm** This human egg, like other mammalian eggs, is surrounded by the cumulus and zona pellucida, both of which a sperm must penetrate to fertilize the egg. Only one sperm will penetrate the zona pellucida and fuse with the plasma membrane.

should they encounter an egg. An activated sperm can penetrate the cumulus and interact with the zona pellucida.

Unlike the jelly coat of sea urchin eggs, the cumulus of mammalian eggs does not trigger the acrosomal reaction. When sperm make contact with the zona pellucida, a species-specific glycoprotein binds to recognition molecules on the head of the sperm. This binding triggers the acrosomal reaction, releasing acrosomal enzymes that digest a path through the zona pellucida. When the sperm head reaches the egg plasma membrane, other proteins facilitate its adhesion to and fusion with the egg plasma membrane.

The importance of the zona pellucida and its sperm-binding molecules as a species-specific recognition mechanism was revealed in experiments on mammalian eggs and sperm in culture dishes. When the zona was stripped from human eggs and the eggs were exposed to hamster sperm, fertilization took place, resulting in a hamster–human hybrid zygote. The hybrid zygote did not survive its first cell division, but the experiment demonstrated that a recognition mechanism in mammalian species resides in the zona pellucida.

**BLOCKS TO POLYSPERMY** The fusion of the sperm and egg plasma membranes and the entry of the sperm into the egg initiate a programmed sequence of events. The first responses to sperm entry are **blocks to polyspermy**—that is, mechanisms that prevent more than one sperm from entering the egg. If more than one sperm enters the egg, the embryo is unlikely to survive.

Blocks to polyspermy have been studied extensively in sea urchin eggs, which can be fertilized in a dish of seawater. Within seconds after the sperm membrane contacts the egg membrane, an influx of sodium ions changes the electric charge difference across the egg's plasma membrane. This *fast block to polyspermy* prevents the fusion of any other sperm with the egg plasma membrane, but it is transient. The change in membrane electrical charge lasts only about a minute, but that is enough time to allow a slower block to sperm entry to develop.

The *slow block to polyspermy* involves converting the vitelline envelope to a physical barrier that sperm cannot penetrate. Before fertilization, the vitelline envelope is bonded to the egg plasma membrane. Just under the plasma membrane are vesicles called *cortical granules* (see Figure 43.4), which contain enzymes and other proteins.

The sea urchin egg, like all animal cells, contains calcium ions sequestered in its endoplasmic reticulum. Sperm entry into the egg stimulates the release of calcium ions from the endoplasmic reticulum and into the egg cytoplasm. This increase in cytosolic calcium causes the egg's cortical granules to fuse with the plasma membrane and release their contents. Cortical granule enzymes break the bonds between the vitelline envelope and the plasma membrane, and other proteins released from the cortical granules attract water into the space between them. As a result, the vitelline envelope rises to form a *fertilization envelope*. Cortical granule enzymes also degrade sperm-binding molecules on the surface of the fertilization envelope and cause it to harden, thus preventing additional sperm from contacting the egg's plasma membrane.

In mammals, sperm entry does not cause a rapid change in membrane potential, but it does trigger a release of calcium from the endoplasmic reticulum. As in the sea urchin, the increased calcium causes the cortical granules to fuse with the egg plasma membrane. A fertilization envelope does not form around the mammalian egg, but the cortical granule enzymes destroy the sperm-binding molecules in the zona pellucida. The rise in cytosolic calcium also signals the egg to complete meiosis. The stage is set for the first cell division.

### Getting eggs and sperm together

As we have just seen, sexual reproduction requires the production of haploid gametes (gametogenesis) and the joining together of those gametes to form a diploid zygote (fertilization). Spawning and mating behaviors get eggs and sperm close enough together that fertilization can occur. Fertilization can occur externally or internally.

**EXTERNAL FERTILIZATION** In an aquatic environment, animals can bring their gametes together by simply releasing them into the water. This practice, called spawning, results in **external fertilization**. Many aquatic animals are not very mobile, but they produce huge numbers of gametes that can travel far from the point of release. A female oyster, for example, will release millions of eggs when she spawns, and the number of sperm produced by a male oyster is astronomical.

Numbers alone, however, do not guarantee that gametes will meet. The reproductive activities of the males and females of a population must be synchronized, since released gametes have a limited life span. Seasonal breeders may use day length, changes in temperature, or changes in weather to time the production and release of their gametes. Mutual stimulation is also important. Release of gametes into the water by one individual can stimulate others to spawn.

Behavior can play an important role in bringing gametes together even when fertilization is external. Many species travel great distances to congregate with potential mates and release their gametes at the same time in a suitable environment. Salmon are an extreme example. They hatch and develop in freshwater streams and then migrate to the ocean where they remain for years. When they are grown, they travel hundreds of miles to spawn back in the stream where they hatched. Males and females expend great amounts of energy to swim up the streams to the spawning grounds, where they pair up, prepare a depression in the streambed gravel, and together release their sperm and eggs. As the gametes drift down into the gravel, fertilization occurs.

**INTERNAL FERTILIZATION** Terrestrial animals cannot simply release their gametes into the environment. Sperm can move only through liquid, and delicate gametes released into air would dry out and die. Terrestrial animals avoid these problems by **internal fertilization**, the release of sperm into the female reproductive tract. Some aquatic animals also practice internal fertilization, but it is ubiquitous in terrestrial animals.

Animals have evolved an astonishing diversity of behavioral and anatomical adaptations for internal fertilization. As we saw above, gametogenesis occurs in the gonads, which are the *primary sex organs*. All additional anatomical components of an animal's reproductive system are called *accessory sex organs*. An obvious accessory sex organ in males of many species is the **penis**, which enables the male to deposit sperm in the female's **vagina**, the entry to her reproductive tract. Accessory sex organs include a variety of glands, tubules, ducts, and other structures.

**Copulation** is the physical joining of male and female accessory sex organs. Transfer of sperm in internal fertilization can also be indirect. Males of many invertebrate species (for example, mites and scorpions) and a few vertebrates (salamanders) deposit *spermatophores*—packets of sperm—in the environment. The packets protect the sperm from desiccation. When a female mite encounters a spermatophore from a potential mate, she straddles it and opens a pair of plates in her abdomen so that the tip of the spermatophore enters her reproductive tract and allows the sperm to enter.

Male squids and spiders play a more active role in spermatophore transfer. The male spider secretes a drop containing sperm onto a bit of web, then uses a special structure on his foreleg to pick up the sperm-containing web and insert it through the female's genital opening. Male squids use one specialized tentacle to pick up a spermatophore and insert it into the female's genital opening.

Most male insects copulate and transfer sperm to the female's vagina through a penis. The **genitalia**—external sex organs—of insects often have species-specific shapes that ensure that the male and female genitalia match in a lock-and-key fashion. This mechanism ensures a tight, secure fit between the mating pair during the prolonged period of sperm transfer. In some insect species in which females mate with more than one male, the males have elaborate structures on their penises that can scoop sperm deposited by other males out of a female's reproductive tract, replacing it with their own.

### An individual animal can function as both male and female

In most species, gametes are produced by individuals that are either male or female. Species that have separate male and female members are called **dioecious** species (from the Greek for "two houses"). In some species, however, a single individual may produce both sperm and eggs. Such species are called **monoecious** ("one house") or **hermaphroditic**, species.

Almost all invertebrate groups contain some hermaphroditic species. An earthworm is an example of a *simultaneous hermaphrodite*, meaning an individual is both male and female at the same time. When two earthworms mate, they exchange sperm, and as a result, the eggs of each are fertilized (see Figure 32.13C). Some vertebrates are *sequential hermaphrodites*, meaning that an individual may function as a male or a female at different times in its life. An example is the anemone fish, or clown fish (**Figure 43.6**), a species that lives in small groups within large sea anemones. All anemone fish are born male. The largest one in a group becomes a functional female. If that fish is removed from the group, the next largest male becomes a female. Also, the second-largest anemone fish in the group is the only male in breeding condition.

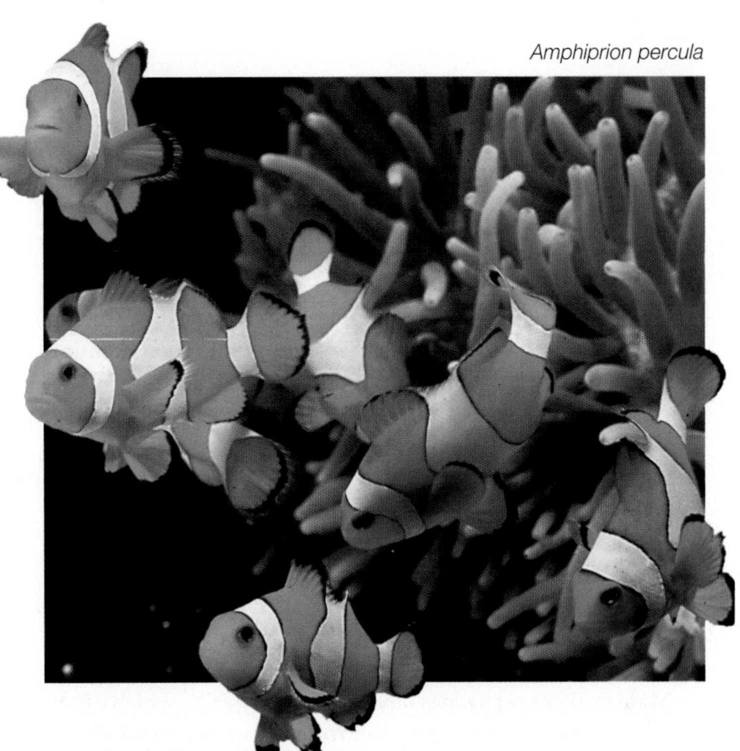


**43.6 When Size Determines Sex** Anemone fish (also known as clown fish) live in groups of about a dozen centered on a single sea anemone. All anemone fish are born male, and the largest one in the group becomes a functional female. Thus any one fish may function as a male and as a female at different times in its life.

What is the evolutionary advantage of hermaphroditism? Some simultaneous hermaphrodites, such as parasitic tapeworms, have a low probability of meeting a potential mate—it may be the only tapeworm in the host. Although tapeworms can fertilize their own eggs, most simultaneous hermaphrodites must mate with another individual; but because every member of the population is both male and female, the probability of encountering a possible mate is double what it would be in strictly monoecious species. In some sequential hermaphrodites, all siblings are either male or female at the same time, thus reducing the incidence of inbreeding.

### The evolution of vertebrate reproductive systems parallels the move to land

The earliest vertebrates evolved in aquatic environments. The closest living relatives of those earliest vertebrates are modern-day fishes. They remain exclusively aquatic animals, and most practice external fertilization. The most primitive of the fishes, the lampreys and hagfishes, simply release their gametes into the environment. In most fishes, however, mating behaviors bring females and males into close proximity at the time of gamete release. In sharks and rays, fins have evolved into claspers that hold the male and female together and enable sperm to be transferred directly into the female reproductive tract.

Amphibians were the first vertebrates to live in terrestrial environments. They dealt with the challenge of a dry environment by returning to water to reproduce, as most amphibians still do today.

Reptiles were the first vertebrate group to solve the problem of reproduction in the terrestrial environment. Their solution, the **amniote egg**, is shared with the birds (see Section 33.4). A chicken egg is a good example of an amniote egg. It contains a supply of food (yolk) and water for the developing embryo. A hard shell protects the embryo and impedes water loss while allowing the diffusion of oxygen into the egg and carbon dioxide out of the egg (**Figure 43.7A**). The eggshell creates an obvious problem for fertilization. Sperm cannot penetrate the shell, so they must reach the egg before the shell forms. Hence internal fertilization and the evolution of accessory sex organs were necessary for the evolution of the amniote egg.

Male snakes and lizards have paired *hemipenes*, which can be filled with blood and thereby extruded from the male's body. Only one hemipenis is inserted into the female's reproductive tract at a time. It is usually rough or spiny at the end to achieve a secure hold while sperm are transferred down a groove on its surface. Retractor muscles pull the hemipenis back into the male's body when mating is completed. Some evolutionarily ancient bird species have erectile penises that channel sperm along a groove into the female's reproductive tract. Birds with more recent evolutionary origins, however, do not have erectile penises; instead, the male and female simply bring their genital openings close together to transfer sperm. Usually this involves the male standing on the female's back (**Figure 43.7B**).

All mammals practice internal fertilization, and with the exception of the prototherian mammals, the developing embryo is retained for some time in the female reproductive tract. Prototherian mammals (the monotremes; see Figure 33.24) lay eggs. The other mammals (the *therians*) vary enormously as to the developmental stage of their offspring at the time of birth.

### Animals with internal fertilization are distinguished by where the embryo develops

Two patterns of care and nurture of the embryo have evolved in animals: *oviparity* (egg laying) and *viviparity* (live bearing).

**Oviparous** animals lay eggs in the environment, and their embryos develop outside the mother's body. Oviparous terrestrial animals such as insects, reptiles, and birds protect their eggs from desiccation with waterproof membranes or shells. Oviparity is possible because eggs are stocked with abundant nutrients to supply the needs of the embryo. Some oviparous animals engage in various forms of parental behavior to protect their eggs, but until the eggs hatch, the embryos depend entirely on the nutrients stored in the egg.

**Viviparous** animals retain the embryo within the mother's body during its early developmental stages. Although examples of viviparity exist in all vertebrate groups except the crocodiles, turtles, and birds (even some sharks retain fertilized eggs in their bodies and give birth to free-living offspring), there is a big difference between viviparity in mammals and in other species.

All mammals except the prototherians are viviparous and have a specialized portion of the female reproductive tract, the **uterus** or *womb*, that holds the embryo and interacts with it to produce a **placenta**, which enables the exchange of nutrients and wastes between the blood of the mother and that of the embryo. Very few non-mammalian species have evolved such a connection between the embryo and the mother.

In most non-mammalian viviparous animals, such as garter snakes and the well-known aquarium fish the guppy, fertilized

(A) *Chelonia mydas*

(B) *Merops apiaster*

**43.7 The Shelled Egg** The shelled egg was a major evolutionary step that allows reptiles and birds to reproduce in the terrestrial environment. (A) A female green sea turtle has deposited her eggs in the sand. (B) The shelled egg requires that sperm meet egg before the shell forms. Terrestrial animals thus must practice internal fertilization, as these European bee-eaters are doing.

eggs are retained in the mother's body until they hatch. These embryos still receive nutrition from stores in the egg, so this reproductive adaptation is called **ovoviviparity**.

## 43.2 RECAP

**Sexual reproduction involves gametogenesis, mating, and fertilization. Fertilization can be external or internal and involves mechanisms for ensuring that only one sperm from the right species enters the egg.**

- Describe the steps by which a sea urchin sperm penetrates the egg. **See Figure 43.4**
- Explain how polyspermy is prevented and why it is crucial to do so. **See p. 906 and Figure 43.4**
- What reproductive adaptations made life on land possible? **See p. 908**

Now that we have covered some of the general aspects of gametogenesis and fertilization and have briefly discussed the great diversity of mating systems, we will next consider the human reproductive systems in detail.

# 43.3 How Do the Human Male and Female Reproductive Systems Work?

In this section we describe the structures and functions of the male and female reproductive systems in mammals, using human beings as our prime example. We also discuss the hormonal regulation of both male and female systems. Our discussion includes the primary sex organs (testes in males and ovaries in females) that produce gametes and serve endocrine functions. It also includes the accessory sex organs: the ducts through which the gametes pass, the various glands that empty into those ducts, and the external genitalia. We also discuss *secondary sexual characteristics,* which are not directly involved in reproduction but are responsible for the major differences in external appearance of men and women and are important in mating.

## Male sex organs produce and deliver semen

**Semen** is the product of the male reproductive system. Besides sperm, semen contains a complex mixture of fluids and molecules that support the sperm and facilitate fertilization. Sperm make up less than 5 percent of the volume of the semen.

The male reproductive organs are diagrammed in **Figure 43.8**. Sperm are produced in the testes, the paired male gonads. The testes of most mammals are located outside the body cavity in a pouch of skin called the **scrotum**.

Why should the testes be located outside the body cavity? The optimal temperature for spermatogenesis in most mammals is slightly lower than the normal body temperature. The scrotum

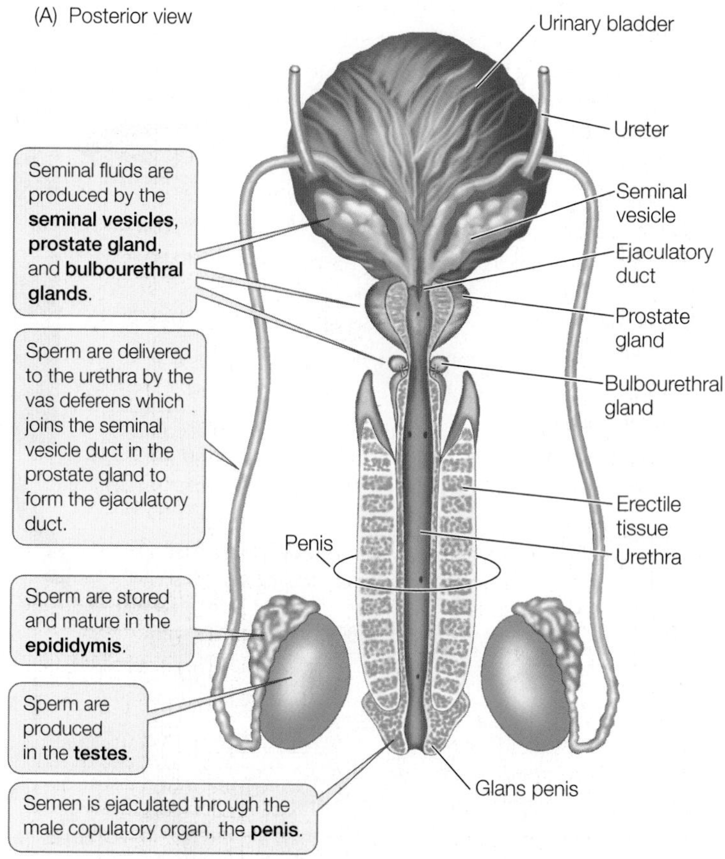


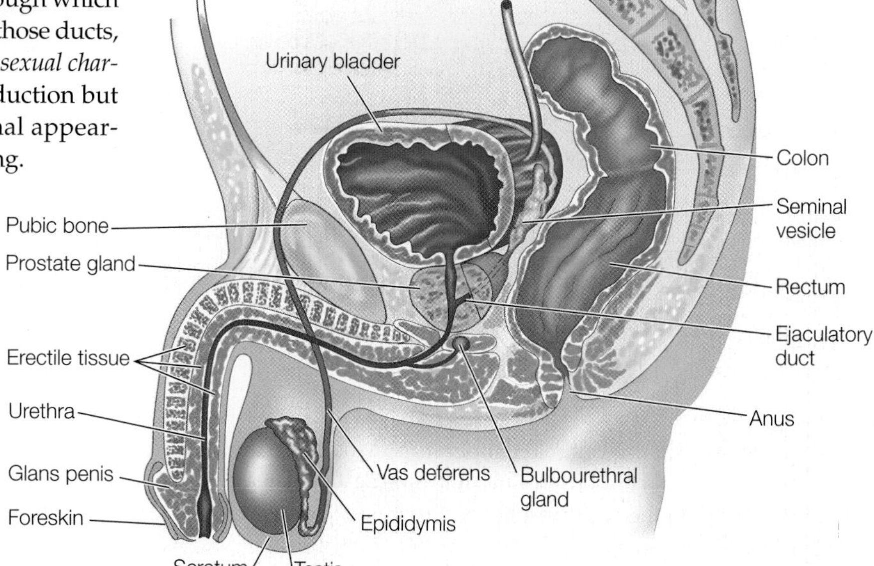


**43.8 Reproductive Tract of the Human Male** The male reproductive organs are shown (A) from the rear and (B) from the side.

**yourBioPortal.com**
GO TO **Web Activity 43.1 • The Human Male Reproductive Tract**

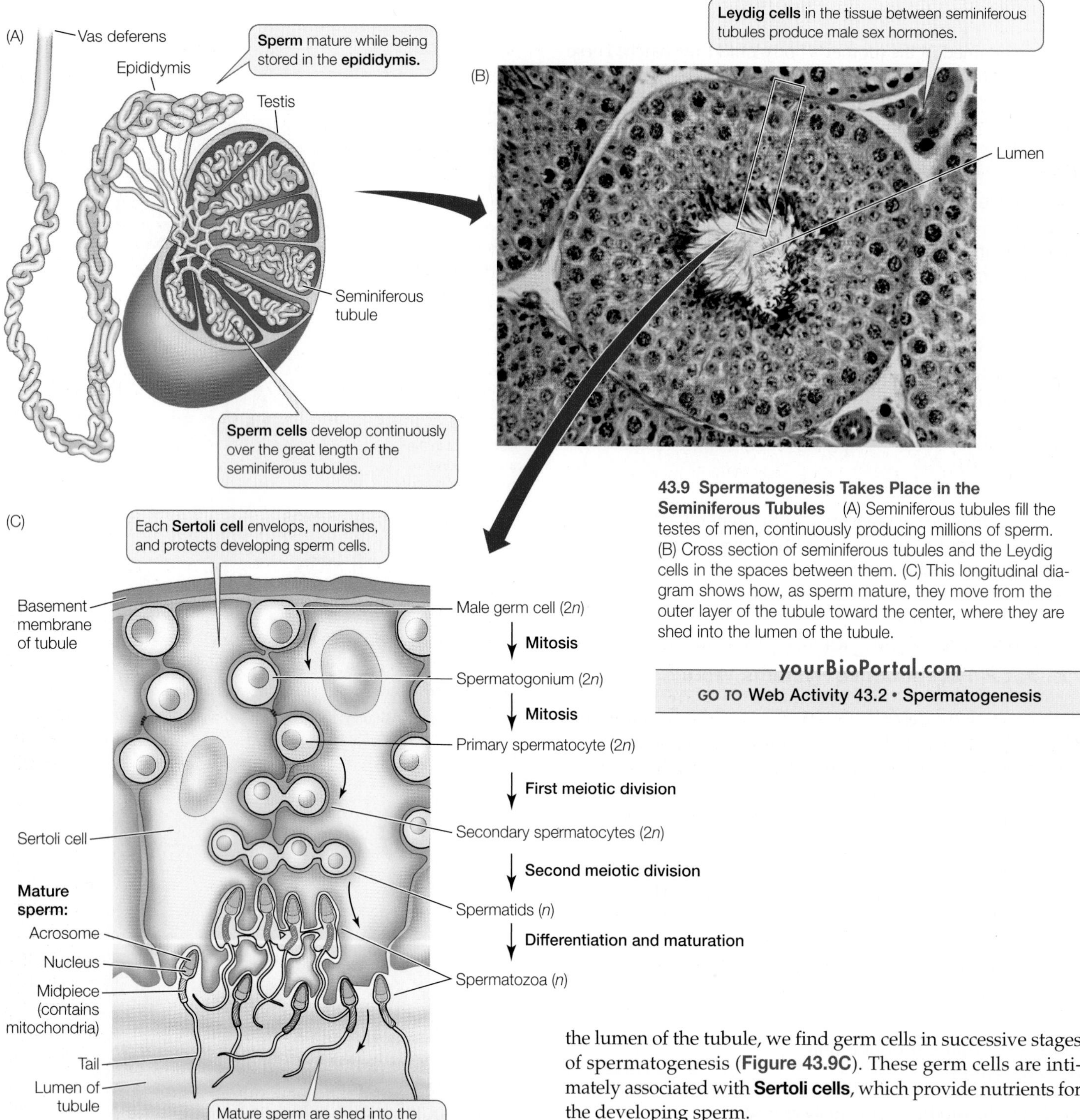


**43.9 Spermatogenesis Takes Place in the Seminiferous Tubules** (A) Seminiferous tubules fill the testes of men, continuously producing millions of sperm. (B) Cross section of seminiferous tubules and the Leydig cells in the spaces between them. (C) This longitudinal diagram shows how, as sperm mature, they move from the outer layer of the tubule toward the center, where they are shed into the lumen of the tubule.

yourBioPortal.com
GO TO Web Activity 43.2 • Spermatogenesis

keeps the testes at this optimal temperature. Muscles in the scrotum contract in a cold environment, bringing the testes closer to the warmth of the body; in a hot environment they relax, cooling the testes by suspending them farther from the body.

Spermatogenesis takes place within the **seminiferous tubules**, which are tightly coiled in each testis (**Figure 43.9A**). Between the seminiferous tubules are clusters of *Leydig cells*, or *interstitial cells*, which produce testosterone (**Figure 43.9B**). Spermatogonia reside in the outer regions of the tubule, just under the basement membrane. Moving inward from these outer layers toward the lumen of the tubule, we find germ cells in successive stages of spermatogenesis (**Figure 43.9C**). These germ cells are intimately associated with **Sertoli cells**, which provide nutrients for the developing sperm.

When the second meiotic division is complete, each primary spermatocyte has given rise to four spermatids (see Figure 43.3A), which develop into spermatozoa as they continue to migrate toward the lumen of the seminiferous tubule. The nucleus becomes compact, and the surrounding cytoplasm is lost. A flagellum, or sperm tail, develops. The mitochondria, which will provide energy for tail motility, become condensed into a midpiece between the head and the tail. An acrosome forms over the nucleus in the head of the sperm. Immature sperm are shed into the lumen of the seminiferous tubule.

From the seminiferous tubules, sperm move into the **epididymis** (see Figure 43.8), where they mature and become motile. The epididymis connects to the **urethra** via the **vas deferens** (plural *vasa deferentia*) and the **ejaculatory duct**. The urethra originates in the bladder, runs through the penis, and opens to the outside of the body at the tip of the penis. It serves as the common final duct for the urinary and reproductive systems. The components of the semen other than sperm come from several accessory glands. About 60 percent of the volume of semen is secreted by the paired **seminal vesicles**, which empty into the vas deferens just before it joins the urethra. Seminal fluid is thick because it contains mucus and fibrinogen, a protein also found in the blood, where it can polymerize to form blood clots. Seminal fluid also contains the monosaccharide fructose, an energy source for the sperm.

The **prostate gland** surrounds the urethra and contributes about 30 percent of the volume of the semen. Prostate fluid is alkaline, so it neutralizes the acidity in the male and female reproductive tracts and makes those environments more hospitable to sperm. The prostate also secretes a clotting enzyme that causes the fibrinogen from the seminal vesicles to convert the semen into a coagulum (gelatinous mass), facilitating its propulsion into and retention in the upper regions of the female reproductive tract. Another enzyme in the prostate fluid, profibrinolysin, is inactive when secreted but is activated shortly after it enters the female reproductive tract. Active fibrinolysin dissolves the clotted semen and liberates the sperm.

The **bulbourethral glands** produce a small volume of an alkaline, mucoid secretion that helps neutralize acidity in the urethra and lubricate it to facilitate the passage of semen at the climax of sexual intercourse. Secretions of the bulbourethral glands precede the climax of the sex act and can carry with them residual sperm from prior sexual activity. Therefore, it is possible for pregnancy to occur even if the penis is withdrawn from the vagina just before climax (a rather ineffective birth control practice known as *coitus interruptus*).

The penis and the scrotum are the male genitalia. The shaft of the penis is covered with normal skin, but the highly sensitive tip, the **glans penis**, is covered with thinner, more sensitive skin that is especially responsive to sexual stimulation. A fold of skin called the *foreskin* covers the glans of the human penis. The procedure known as *circumcision* removes a portion of the foreskin.

Sexual stimulation triggers responses in the nervous system that result in penile **erection**. Nerve endings release a gaseous neurotransmitter, nitric oxide (NO), onto blood vessels leading into the penis. NO stimulates production of the second messenger cGMP (see Figure 7.17), which causes these vessels to dilate. The increased blood flow that results fills and swells shafts of spongy, erectile tissue located along the length of the penis. The enlargement of these blood-filled cavities compresses the vessels that normally carry blood out of the penis. As a result, the erectile tissue becomes more and more engorged with blood. The penis becomes hard and erect, facilitating its insertion into the vagina. Many species of mammals, but not humans, have a bone in the penis, but these species still depend on erectile tissue for copulation.

At the climax of copulation, about 2 to 6 milliliters of semen are propelled through the vasa deferentia and the urethra in two steps, emission and ejaculation. During **emission**, rhythmic contractions of smooth muscles in the vasa deferentia and accessory glands move the semen into the urethra at the base of the penis. **Ejaculation** is caused by contractions of other muscles at the base of the penis surrounding the urethra. These contractions force the coagulum of semen through the urethra and out of the penis. The muscle contractions of ejaculation are accompanied by feelings of intense pleasure known as **orgasm**. They are also accompanied by transient increases in heart rate, blood pressure, breathing, and skeletal muscle contractions throughout the body.

After ejaculation, NO release decreases and enzymes break down cGMP, causing the blood vessels flowing into the penis to constrict. The blood pressure in the erectile tissue decreases, relieving the compression of the blood vessels leaving the penis, and the erection declines.

*Erectile dysfunction* (*ED*), or *impotence*, is the inability to achieve or sustain an erection. ED may have different causes, including cardiovascular disease. Drugs used to treat ED act by inhibiting the breakdown of cGMP, thus enhancing the effect of NO released in the penis, which improves the ability to achieve and maintain an erection.

### Male sexual function is controlled by hormones

Spermatogenesis and maintenance of male secondary sexual characteristics such as facial hair and a deep voice depend on testosterone produced by the Leydig cells of the testes. As described in Section 41.3, increased production of testosterone at puberty results from an increased release of gonadotropin-releasing hormone (GnRH) by the hypothalamus, which stimulates anterior pituitary cells to increase their secretion of luteinizing hormone (LH) and follicle-stimulating hormone (FSH) (**Figure 43.10**). Higher levels of LH stimulate the Leydig cells

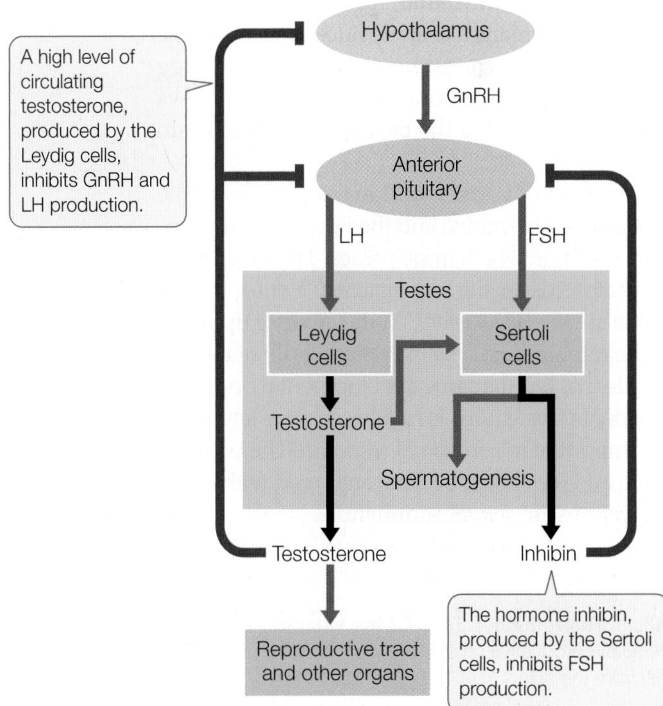


**43.10 Male Reproductive Hormones** The male reproductive system is under hormonal control by the hypothalamus and the anterior pituitary. Red lines indicate inhibition; green lines indicate stimulation.

to increase their production and release of testosterone. Testosterone exerts negative feedback on the anterior pituitary and the hypothalamus. At the time of puberty, the sensitivity of the hypothalamus to negative feedback from testosterone declines, and the level of circulating testosterone increases.

Increased testosterone in pubertal boys causes the development of pubic and facial hair, a deeper voice, enlarged genitals, and an increased growth rate. Testosterone also promotes increased muscle mass and maturation of the testes. Continued production of testosterone after puberty is essential for the maintenance of secondary sexual characteristics and the production of sperm.

Spermatogenesis is controlled by the influence of FSH and testosterone on the Sertoli cells in the seminiferous tubules. The Sertoli cells also produce a hormone called *inhibin*, which exerts negative feedback on the anterior pituitary cells that produce and secrete FSH.

### Female sex organs produce eggs, receive sperm, and nurture the embryo

When a mammalian egg matures, it is released from the ovary directly into the body cavity. But the egg does not go far. The ovaries are close to the *fimbria*, the undulating, fringed openings of the **oviducts** (also known as the *Fallopian tubes*). The fimbria draw the released egg into an oviduct (**Figure 43.11**), where fertilization takes place. Whether or not the egg is fertilized, cilia lining the oviduct propel the egg slowly toward the uterus, a muscular, thick-walled cavity shaped in humans like an upside-down pear. The uterus is where the embryo develops if the egg is fertilized. At the bottom, the uterus narrows into a region called the **cervix**, which leads into the vagina.

In humans, two sets of skin folds surround the opening of the vagina and the opening of the urethra, through which urine passes. The inner, more delicate folds are the *labia minora*; the outer, thicker folds are the *labia majora*. At the anterior tip of the labia minora is the *clitoris*, a small bulb of erectile tissue that has the same developmental origins as the penis. The clitoris is highly sensitive and plays an important role in sexual response. The labia minora and the clitoris become engorged with blood in response to sexual stimulation.

The external opening of an infant's vagina is usually, but not always, partly covered by a thin membrane, the *hymen*. Eventually the hymen can be torn by vigorous physical activity or by first sexual intercourse; it can sometimes make first intercourse difficult or painful for the woman.

To fertilize an egg, sperm deposited in the vagina swim and are propelled by contractions of the female reproductive tract through the cervical opening, the uterus, and most of the oviduct. The egg (actually a secondary oocyte) is fertilized in the upper region of the oviduct. Fertilization stimulates the completion of the second meiotic division, after which the haploid

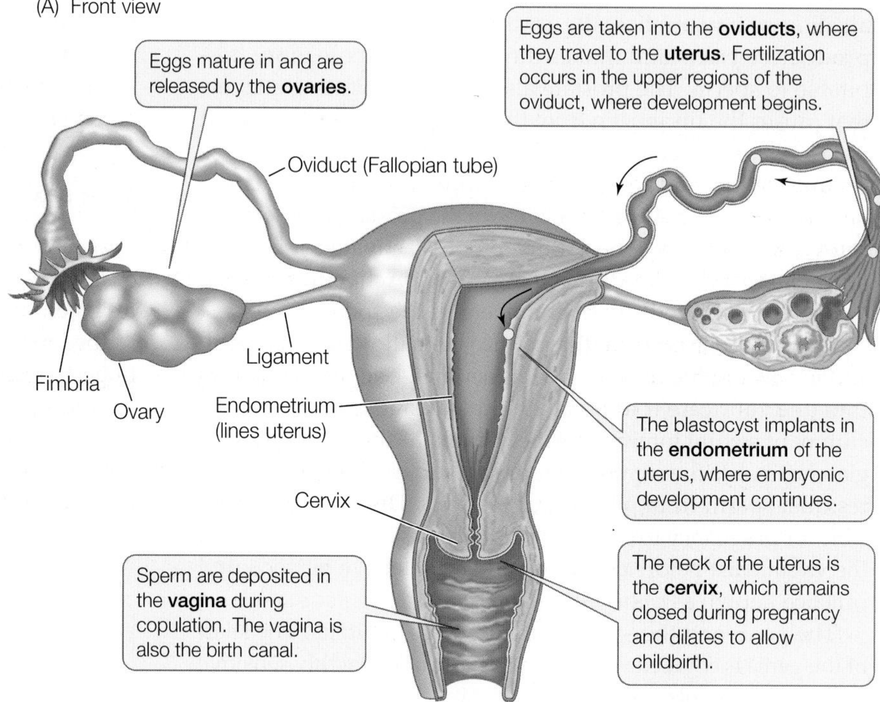
(A) Front view
Eggs mature in and are released by the **ovaries**.
Eggs are taken into the **oviducts**, where they travel to the **uterus**. Fertilization occurs in the upper regions of the oviduct, where development begins.
Oviduct (Fallopian tube)
Ligament
Fimbria
Ovary
Endometrium (lines uterus)
The blastocyst implants in the **endometrium** of the uterus, where embryonic development continues.
Cervix
Sperm are deposited in the **vagina** during copulation. The vagina is also the birth canal.
The neck of the uterus is the **cervix**, which remains closed during pregnancy and dilates to allow childbirth.

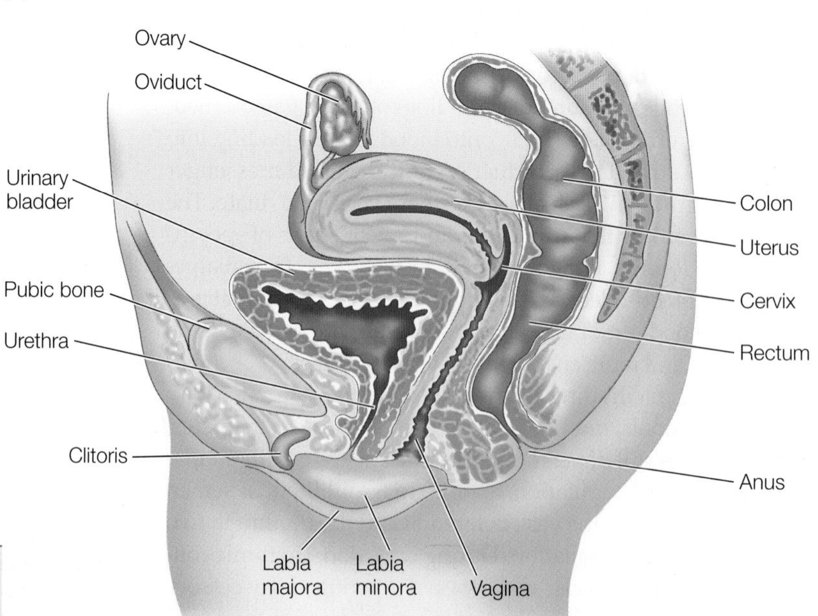
(B) Side view
Ovary
Oviduct
Urinary bladder
Pubic bone
Urethra
Clitoris
Labia majora
Labia minora
Vagina
Colon
Uterus
Cervix
Rectum
Anus

**43.11 Reproductive Tract of the Human Female** The female reproductive organs are shown (A) from the front and (B) from the side.

**yourBioPortal.com**
GO TO Web Activity 43.3 • The Human Female Reproductive Tract

nuclei of the sperm and the egg can fuse to produce a diploid zygote nucleus.

Still in the oviduct, the zygote undergoes its first few cell divisions to become a **blastocyst**. The blastocyst moves down the oviduct to the uterus, where it attaches itself to the epithelial lining of the uterus, or **endometrium**. Once attached, the blastocyst burrows into the endometrium—a process called *implantation*—and interacts with it to form the placenta, as we will see in Chapter 44. The placenta nurtures the embryo and produces hormones that help sustain pregnancy.

As an egg matures in the ovary, the endometrium thickens. If no blastocyst arrives in the uterus, the endometrium regresses or is sloughed off. Thus the female reproductive cycle actually consists of two linked cycles: an *ovarian cycle* that produces eggs and hormones, and a *uterine*, or *menstrual*, *cycle* that prepares the endometrium for the arrival of a blastocyst.

### The ovarian cycle produces a mature egg

An **ovarian cycle** is about 28 days long in the human female, but it varies considerably among individuals. During the first half of each cycle, at least one primary oocyte matures into a secondary oocyte (egg) and is expelled from the ovary (*ovulation*). During the second half of the cycle, cells in the ovary that were associated with the maturing oocyte develop endocrine functions and then regress if the egg is not fertilized. The progression of these events is shown diagrammatically in **Figure 43.12A**.

A newborn baby girl has about a million primary oocytes in each ovary. By the time she reaches puberty, she has only about 200,000; the rest have degenerated. During a woman's fertile years, her ovaries go through about 450 ovarian cycles. During each cycle, several oocytes begin to mature, but usually only one matures completely and is released; the others degenerate. At around the age of 50, a woman reaches **menopause**—the end of fertility—and may have few if any oocytes left in each ovary.

Each primary oocyte in the ovary is surrounded by a layer of ovarian cells. An oocyte and its surrounding cells constitute the functional unit of the ovary, the **follicle** (**Figure 43.12B**). Between puberty and menopause, 6 to 12 follicles begin to mature each month. In each follicle, the oocyte enlarges and the surrounding follicular cells proliferate. After about a week, one follicle is larger than the rest, and it continues to grow, while the others cease to develop. In the enlarged follicle, the follicular cells nurture the growing egg, supplying it with nutrients, growth factors, and hormonal stimulation.

In humans, after 2 weeks of follicular growth, **ovulation** occurs: the follicle ruptures and the egg is released. Following ovulation, the follicle cells that remain in the ovary continue to proliferate and form a mass of endocrine tissue about the size of a marble. This structure is the *corpus luteum* (plural *corpora lutea*). It functions as an endocrine gland, producing estrogen and progesterone for about 2 weeks. It then degenerates, unless a blastocyst implants in the endometrium.

### The uterine cycle prepares an environment for the fertilized egg

The **uterine cycle** parallels the ovarian cycle and consists of a buildup and then a breakdown of the endometrium (**Figure 43.13**). About 5 days into the ovarian cycle, the endometrium starts to grow in preparation for receiving a blastocyst. The uterus

**43.12 The Ovarian Cycle** (A) The ovarian cycle progresses from the development of a follicle to ovulation and finally to growth and degeneration of the corpus luteum. (B) This micrograph shows a mature mammalian follicle; the oocyte is in the center.

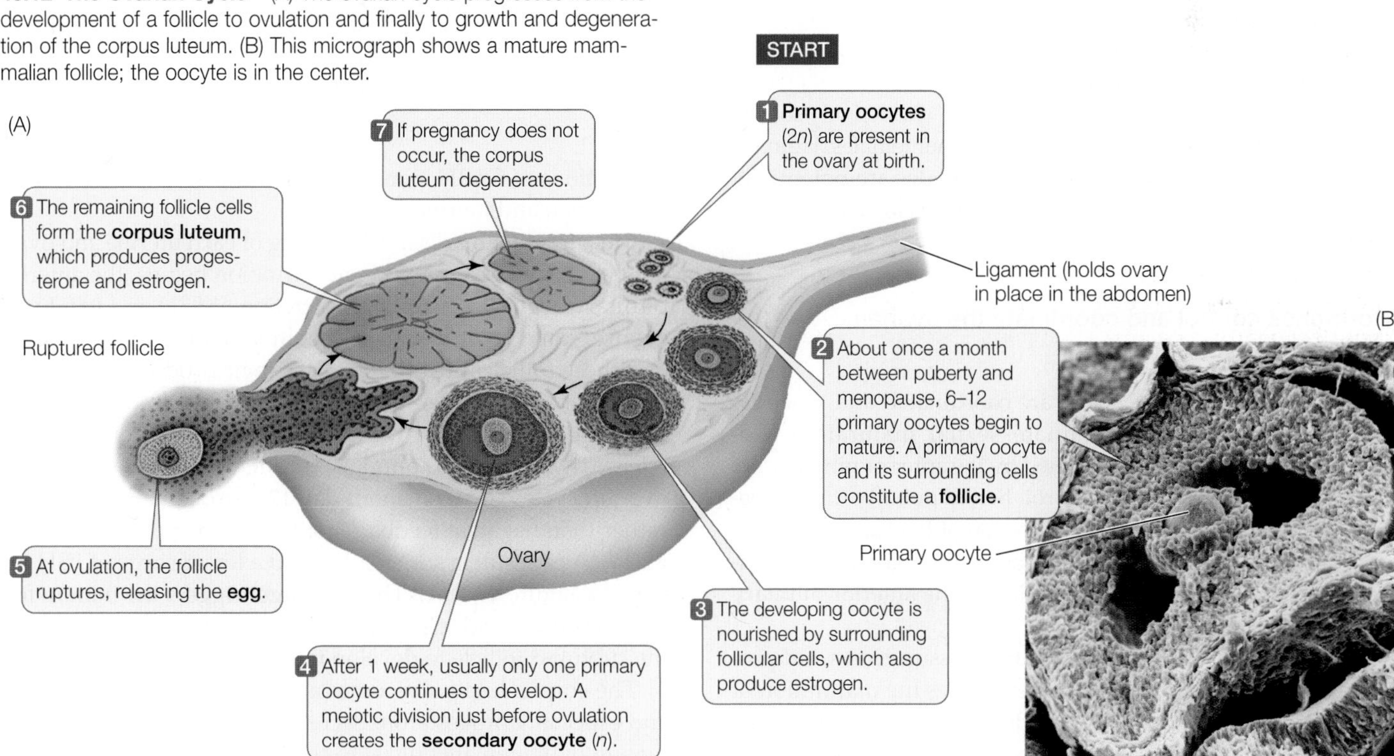

(A)
START
1 Primary oocytes (2n) are present in the ovary at birth.
Ligament (holds ovary in place in the abdomen)
2 About once a month between puberty and menopause, 6–12 primary oocytes begin to mature. A primary oocyte and its surrounding cells constitute a follicle.
3 The developing oocyte is nourished by surrounding follicular cells, which also produce estrogen.
4 After 1 week, usually only one primary oocyte continues to develop. A meiotic division just before ovulation creates the secondary oocyte (n).
5 At ovulation, the follicle ruptures, releasing the egg.
6 The remaining follicle cells form the corpus luteum, which produces progesterone and estrogen.
7 If pregnancy does not occur, the corpus luteum degenerates.
Ruptured follicle
Ovary
(B)
Primary oocyte

**43.13 The Ovarian and Uterine Cycles** During a woman's ovarian and uterine cycles, coordinated changes occur in (A) gonadotropin release by the anterior pituitary, (B) the ovary, (C) the release of female sex steroids, and (D) the uterus. The cycles begin with the onset of menstruation; ovulation is at midcycle (yellow bar).

**yourBioPortal.com**
**GO TO Animated Tutorial 43.2 • The Ovarian and Uterine Cycles**

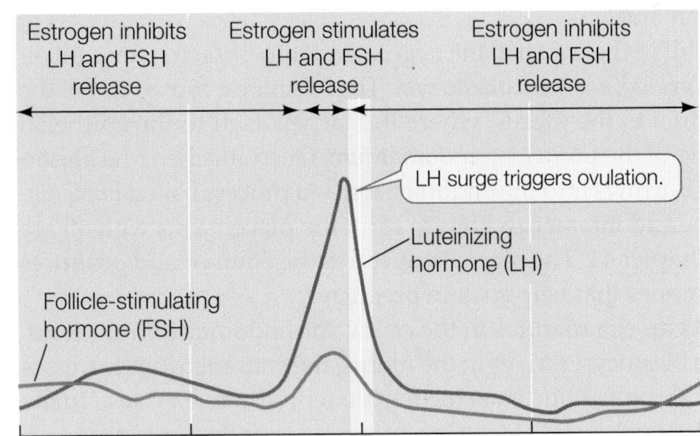
FSH and LH are under control of GnRH from the hypothalamus and the ovarian hormones estrogen and progesterone (part C).
(A) Gonadotropins (from anterior pituitary)
Estrogen inhibits LH and FSH release
Estrogen stimulates LH and FSH release
Estrogen inhibits LH and FSH release
LH surge triggers ovulation.
Luteinizing hormone (LH)
Follicle-stimulating hormone (FSH)

FSH stimulates the development of follicles; the LH surge causes ovulation and then the development of the corpus luteum.

(B) Events in ovary (ovarian cycle)

Oocyte maturation
Developing follicle
Ovulation (day 14)
Corpus luteum
Developing oocyte

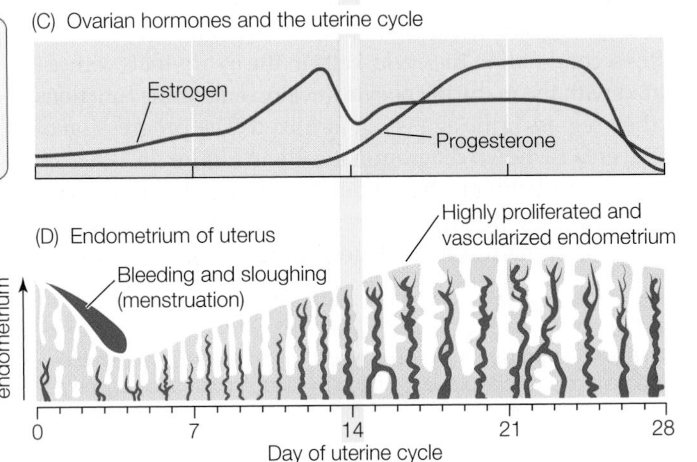
Estrogen and progesterone stimulate the development of the endometrium in preparation for pregnancy.
(C) Ovarian hormones and the uterine cycle
Estrogen
Progesterone
(D) Endometrium of uterus
Highly proliferated and vascularized endometrium
Bleeding and sloughing (menstruation)
Thickness of endometrium
0
7
14
21
28
Day of uterine cycle

attains its maximal state of preparedness about 5 days after ovulation and remains in that state for another 9 days. If a blastocyst has not arrived by that time, the endometrium begins to break down, and the sloughed-off tissue, including blood, flows from the body through the vagina—the process of **menstruation** (from *menses*, the Latin word for "months").

The uterine cycles of most mammals other than humans do not include menstruation; instead, the uterine lining typically is resorbed. In these species, the most obvious correlate of the ovarian cycle is a state of sexual receptivity called **estrus** around the time of ovulation. You may be aware of the bloody discharge that occurs in dogs at the time of estrus. This discharge is not the same as menstruation—in fact it is exactly the opposite. Bleeding in dogs occurs during the *proliferation* of the uterine lining, which occurs just before ovulation. When the female mammal comes into estrus, or "heat," she actively solicits male attention and may be aggressive to other females. Humans are unusual among mammals in that females are potentially sexually receptive throughout their ovarian cycles and at all seasons of the year.

### Hormones control and coordinate the ovarian and uterine cycles

The ovarian and uterine cycles are coordinated and timed by the same hormones that initiate sexual maturation. Gonadotropins (FSH and LH) secreted by the anterior pituitary are the central elements of this control. Before puberty (that is, before about 11 years of age), the secretion of FSH and LH is low and the ovaries are inactive. At puberty, the hypothalamus increases its release of GnRH, stimulating the anterior pituitary to secrete FSH and LH.

In response to FSH and LH, ovarian tissue grows and produces estrogen. The rise in estrogen causes the maturation of the accessory sex organs and the development of female secondary sexual characteristics. Between puberty and menopause, interactions of GnRH, gonadotropins, and sex steroids control the ovarian and uterine cycles.

Menstruation marks the beginning of each uterine and ovarian cycle. A few days before menstruation begins, the anterior pituitary begins to increase its secretion of FSH and LH. In response, some 10 to 20 follicles begin to mature in the ovaries, and these follicles steadily increase their production of estrogen. After about a week, all but one of the follicles wither away.

Estrogen exerts negative feedback control on gonadotropin release by the anterior pituitary during the first 12 days of the ovarian cycle. Then, on about day 12, estrogen exerts positive rather than negative feedback control on the pituitary (**Figure 43.14**). As a result, a surge of LH and a lesser surge of FSH occur (see Figure 43.13A). The LH surge triggers the mature follicle to rupture and release its egg, and it stimulates the cells of the ruptured follicle to develop into a corpus luteum.

The corpus luteum becomes an endocrine gland. Estrogen and especially progesterone secreted by the corpus luteum following ovulation are crucial to continued growth and mainte-

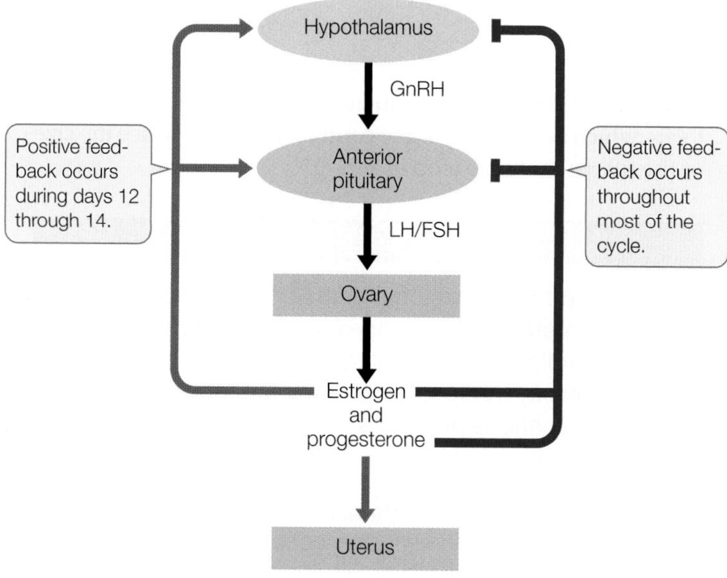


**43.14 Hormones Control the Ovarian and Uterine Cycles** The ovarian and uterine cycles are under a complex series of positive and negative feedback controls involving several hormones.

nance of the endometrium. In addition, these sex steroids exert negative feedback control on the pituitary, inhibiting gonadotropin release and thus preventing new follicles from beginning to mature.

If the egg is not fertilized, the corpus luteum degenerates on about day 26 of the cycle. Without production of progesterone by the corpus luteum, the endometrium sloughs off and menstruation occurs. The decrease in circulating steroids also releases the hypothalamus and pituitary from negative feedback control, so GnRH, FSH, and LH all begin to increase. The increase in these hormones induces the next round of follicle development, and the ovarian cycle begins again.

### In pregnancy, hormones from the extraembryonic membranes take over

If the egg is fertilized and a blastocyst arrives in the uterus and implants in the endometrium, a new hormone comes into play. A layer of cells covering the blastocyst begins to secrete **human chorionic gonadotropin**, or **hCG** (**Figure 43.15A**). This gonadotropin, a molecule similar to LH, stimulates the corpus luteum to continue to produce estrogen and progesterone to support the growth and maintenance of the endometrium and thereby prevent menstruation. Because it is present only in the blood of pregnant women, the presence of hCG is the basis for pregnancy testing. Pregnancy tests use an antibody to detect hCG in urine; they take only minutes, and can be done at home.

(A)

Increase in E/P ratio
Human chorionic gonadotropin (hCG)
Estrogen
Progesterone
Plasma hormone concentration
0 1 2 3 4 5 6 7 8 9 10
Months since start of last menstrual period
Delivery

(B)

The release of oxytocin by the posterior pituitary increases the contractions of the uterus during labor and birth (a positive feedback loop).
Increased ratio of estrogen to progesterone at the end of the third trimester
Growth of fetus
Increased contractility of uterine muscle
Increased stretch of uterus
Increased uterine contractions
Increased pressure of fetal head on cervix
Nervous system communicates uterine stretching to hypothalamus
Increased oxytocin secretion

**43.15 Pregnancy and Childbirth** (A) When a fertilized ovum implants in the uterus, cells surrounding it produce human chorionic gonadotropin, which acts like LH and keeps the corpus luteum functioning as an endocrine gland. The ovarian and uterine cycles are put on hold for the duration of pregnancy. (B) Both mechanical and hormonal signals are involved in stimulating the uterine contractions of labor and delivery. (C) A new person is delivered into the world headfirst.

(C)

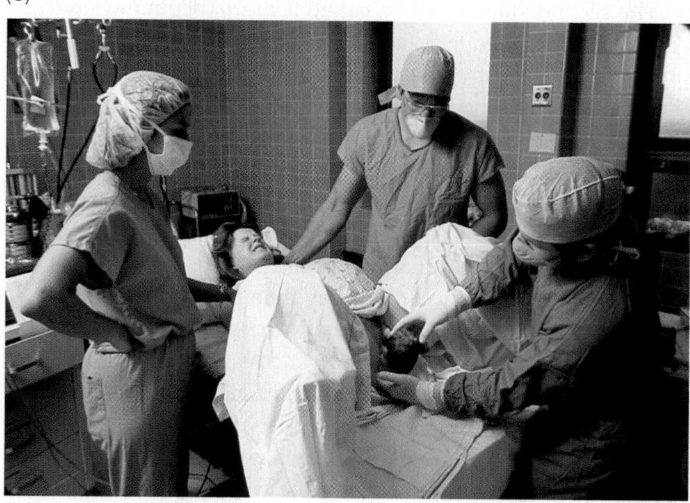

Blastocyst and endometrial tissues form the placenta, which produces estrogen and progesterone (eventually replacing the corpus luteum as the most important source of these sex steroids). Continued high levels of estrogen and progesterone prevent the pituitary from secreting gonadotropins; thus, the ovarian cycle ceases for the duration of pregnancy. (This mechanism underlies the action of birth control pills, which contain synthetic hormones resembling estrogen and progesterone that exert negative feedback control on the hypothalamus and pituitary.)

### Childbirth is triggered by hormonal and mechanical stimuli

Throughout pregnancy, the muscles of the uterine wall periodically undergo slow, weak, rhythmic contractions called *Braxton Hicks contractions*. These contractions become stronger during the third trimester of pregnancy and are sometimes called *false labor contractions*. True labor contractions usually mark the beginning of childbirth. Both hormonal and mechanical stimuli contribute to the onset of labor.

Progesterone inhibits and estrogen stimulates contractions of uterine muscle. Toward the end of the third trimester, the estrogen–progesterone ratio shifts in favor of estrogen. The onset of labor is marked by increased secretion of the hormone oxytocin by the posterior pituitaries of both mother and fetus. Oxytocin is a powerful stimulant of uterine muscle contraction. Manufactured oxytocin is used to induce labor when that is necessary.

Mechanical stimuli come from the stretching of the uterus by the fully grown fetus and the pressure of the fetal head on the cervix. These mechanical stimuli increase the release of oxytocin by the mother's posterior pituitary, which in turn increases the activity of uterine muscle, which causes even more pressure on the cervix. This positive feedback loop converts the weak, slow, rhythmic Braxton Hicks contractions into stronger labor contractions (**Figure 43.15B**).

In the early stage of labor, hormonal changes and pressure created by the contractions cause the cervix to dilate (expand) until it is large enough to allow the baby to pass through. Gradually the contractions become more frequent and more intense. This stage of labor lasts an average of 12 to 15 hours in a first pregnancy; it is usually 8 hours or less in subsequent ones.

The second stage of labor, called *delivery*, begins when the cervix is fully dilated to a diameter of about 10 centimeters (**Figure 43.15C**). The baby's head moves into the vagina; passage of the fetus through the vagina is assisted by the mother's bearing down ("pushing") with her abdominal and other muscles. Once the head and shoulders of the baby clear the cervix, the rest of its body eases out rapidly, but it is still connected to the placenta by the umbilical cord. Once the baby clears the birth canal, it starts breathing and is independent of its mother's circulation. The umbilical cord may then be clamped and cut. The segment still attached to the baby dries up and sloughs off in a few days, leaving behind its distinctive signature, the belly button—more properly called the *umbilicus*. The detachment and expulsion of the placenta and fetal membranes take from a few minutes to an hour, and may be accompanied by uterine contractions. If the baby suckles at the breast immediately following birth, its suckling stimulates additional secretion of oxytocin, which augments uterine contractions that reduce the size of the uterus and help stop bleeding.

## 43.3 RECAP

**The reproductive systems of men and women produce gametes and hormones, and these functions are controlled by hypothalamic and anterior pituitary hormones. In women, the hormonal control of reproductive functions produces linked ovarian and uterine cycles.**

- Describe the path the human sperm and ovum take in moving from their respective gonads to the point at which fertilization occurs. See Figures 43.8 and 43.11
- In males, increased production of GnRH at puberty stimulates the release of what two hormones of the anterior pituitary? What effect do these hormones have? See p. 911
- Explain the events in the ovarian cycle that result in release of a single ovum each month. What events prepare the uterus to receive the egg? See Figures 43.12 and 43.13

Understanding the physiology of human reproduction has led to numerous methods and technologies for controlling it, either to prevent unwanted pregnancies or to overcome infertility.

## 43.4 How Can Fertility Be Controlled?

Sexual issues and sexual behavior are dominant aspects of our society, and reproductive technologies have had huge impacts on our sexual and reproductive lives.

### Human sexual responses have four phases

The responses of both women and men to sexual stimulation consist of four phases: excitement, plateau, orgasm, and resolution. As sexual *excitement* begins in a woman, her heart rate and blood pressure rise, muscular tension increases, breasts swell, and nipples become erect. Her external genitals, including the sensitive clitoris, swell as they become filled with blood, and the walls of the vagina secrete lubricating fluid that facilitates copulation. In the *plateau* phase, her blood pressure and heart rate rise further and her breathing becomes rapid. The sensitivity once focused in the clitoris spreads over the external genitals, and the clitoris itself becomes even more sensitive. *Orgasm* may last as long as a few minutes, and unlike men, some women can experience several orgasms in rapid succession. During the *resolution* phase, blood drains from the genitals, and body physiology returns to close to normal.

In the man, the excitement phase is marked by an increase in blood pressure, heart rate, and muscle tension (just like in the woman) and by penile erection. In the plateau phase, breathing becomes rapid, the diameter of the glans increases, and a few drops of a lubricating fluid from the bulbourethral gland may ooze from the penis. Continued stimulation of the nerve endings in the glans and in the skin along the shaft of the penis eventually trigger orgasm. Massive spasms of the muscles in the genital area and contractions in the accessory reproductive organs result in ejaculation.Within a few minutes after ejaculation, the penis shrinks to its former size and body physiology returns to resting conditions.

The male sexual response also includes a *refractory period* immediately after orgasm. During this period, which may last from minutes to hours, a man cannot achieve a full erection or another orgasm, regardless of the intensity of sexual stimulation. This refractory period is believed to be controlled by the anterior pituitary hormone prolactin (see Figure 41.2), which is released during orgasm. We say "believed" because the study shown in **Figure 43.16** is not conclusive; it compares the data from a single subject who displayed an unusual physiological characteristic with averaged data from a group of typical individuals. In Chapter 1 we learned the importance of statistical tests to establish whether an observed result could be due to chance alone, but we can't perform the necessary tests with a sample size of only one. This problem frequently crops up in medical science when there is an effort to find an explanation for a highly unusual trait or disease. In such cases, data from even a single individual is of interest and can lead to additional studies and experiments.

INVESTIGATING LIFE

**43.16 Prolactin and the Male Refractory Period**

Men experience a refractory period immediately after achieving orgasm, during which most men cannot achieve a full erection. A study done in Germany on a single individual who was able to achieve multiple orgasms indicated that the hormone prolactin may control the onset of refraction.

**HYPOTHESIS** **Release of prolactin during copulation induces a refractory period in males.**

**METHOD**

1. Catheterize subjects and withdraw blood samples (9 normal subjects; 1 multiorgasmic subject).
2. Induce orgasms 30 minutes apart.
3. Withdraw blood samples following orgasms.
4. Analyze all blood samples for levels of prolactin.

**RESULTS**

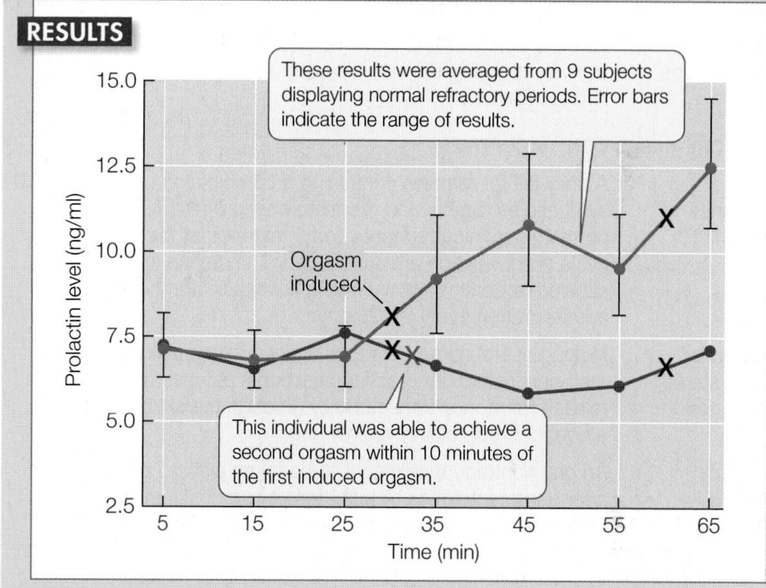

These results were averaged from 9 subjects displaying normal refractory periods. Error bars indicate the range of results.
Orgasm induced
This individual was able to achieve a second orgasm within 10 minutes of the first induced orgasm.
Prolactin level (ng/ml)
15.0
12.5
10.0
7.5
5.0
2.5
5
15
25
35
45
55
65
Time (min)

**CONCLUSION** **In normal men, blood levels of prolactin rise following ejaculation. The multiorgasmic male showed no increase in prolactin levels, which may explain the lack of a refractory period in this individual.**

Go to **yourBioPortal.com** for original citations, discussions, and relevant links for all INVESTIGATING LIFE figures.

### Humans use a variety of methods to control fertility

According to a recent study, almost half of the more than 6 million pregnancies that occur in the United States each year are unintended. For women of college age, a single act of unprotected intercourse in the 2 days prior to ovulation carries a chance of conception as high as 50 percent.

The only failure-proof methods of preventing pregnancy are complete abstinence from sexual activity or the surgical removal of the gonads. Since those options are not acceptable to most people, they turn to other methods to prevent pregnancy. Many of these methods prevent fertilization or implantation (*conception*) and are therefore referred to as **contraception**. Most methods are used by the woman, some are used by the man; they also vary enormously in their effectiveness. **Table 43.1** lists some of the most commonly used contraceptive methods and their relative failure rates.

Once a fertilized egg is successfully implanted in the uterus, any termination of the pregnancy is called an **abortion**. A *spontaneous abortion* is the medical term for what most people call a miscarriage. Spontaneous abortions are common early in pregnancy and are usually the result of either a chromosomal abnormality in the fetus or to a breakdown in the process of implantation. Many spontaneous abortions occur before the woman even realizes she is pregnant.

Abortions that result from medical intervention may be performed either for therapeutic purposes or for fertility control. A therapeutic abortion may be necessary to protect the health of the mother, or it may be performed because prenatal testing reveals that the fetus has a severe defect. Of the approximately 3 million unintended pregnancies in the United States each year, almost half are ended by abortion.

TABLE 43.1

**Methods of Contraception**

| METHOD | MODE OF ACTION | FAILURE RATE[a] | COMMENTS |
|---|---|---|---|
| Unprotected | No form of birth control | 85 | High risk of pregnancy, especially for women 15–30. |
| **NONTECHNOLOGICAL METHODS** | | | |
| Rhythm method | The couple abstains from intercourse between days 10 and 20 of the ovarian cycle (peak fertility). | 15–35 | High failure rate due to miscalculation and/or variation of individual cycles. |
| Coitus interruptus | The man withdraws his penis prior to ejaculation, so no sperm is deposited in the vagina. | 20–40 | Requires self-control, especially by the male partner. Very high failure rate. |
| **BARRIER METHODS[b]** | | | |
| Condom | A sheath of impermeable material (often latex) is fitted over the erect penis. Semen is trapped in the condom, so no sperm is deposited in the vagina. | 15 | If fitted correctly, an intact condom can prevent pregnancy and provide protection against sexually transmitted diseases (STDs), including HIV (AIDS). |
| Spermicidal jellies | Applied inside the vagina, these chemical compounds kill or immobilize sperm. | 25 | Used alone, spermicidal compounds have a fairly high failure rate. |
| Diaphragms, cervical caps | Inserted by the woman prior to intercourse, these items work by blocking the cervix so that sperm cannot pass into the uterus. | 10–15 | Approximately the same failure rate as condom use by males, but do not protect against STDs. Can be used in conjunction with spermicidal jelly for extra protection. |
| **HORMONE-BASED CONTRACEPTIVES** | | | |
| Oral hormones ("the pill") | A daily pill for females containing a combination of synthetic estrogens and progesterone (progestin). These hormones mimic pregnancy to the extent that the ovarian cycle and ovulation are suspended. The uterine cycle is allowed to continue by including a week of non-hormone administration every 21–28 days. | 0–3 | Requires medical consultation and prescription. Taken correctly, oral contraceptives are extremely effective. In the U.S., more than 12 million women use them each year; they are sometimes prescribed to treat menstrual disorders. |
| Non-orally administered hormones | Making use of same hormonal actions as the pill, these methods include long-acting injections, patches that release hormones transdermally (through the skin), and a hormone-containing vaginal ring. | <1 | Same as oral contraceptives. A slightly lower failure rate because the woman does not have to remember to take a daily pill. |
| Progestin-only pill (Plan B®) | An oral contraceptive meant to be taken within 72 hours *after* unprotected sex. A high dose of progestin in two pills prevents ovulation in the same manner birth control pills do. | 5-40[c] | Not an "abortion pill," this drug will not terminate an existing pregnancy. Currently available to women over 17 without a prescription. |
| **IMPLANTATION BLOCKERS** | | | |
| Intrauterine device (IUD) | A medical professional inserts a small plastic or metal device into the uterus. The resulting inflammation reaction (see Chapter 42) releases prostaglandins, which prevent implantation of the fertilized egg. | 0.5–5 | A highly effective contraceptive, it is the most widely used birth control device in China (and hence the world). With medical monitoring, can remain in place for several years. |
| Mifepristone (RU-486) | Also known as the "morning after" or "abortion pill," this drug blocks progesterone receptors necessary to maintain the endometrium during implantation and pregnancy. | 0.5–6 | Prevents implantation when taken up to several days after unprotected intercourse. Can end a pregnancy up to the time of the first missed menstrual period. In the U.S., available from specialized providers. |
| **STERILIZATION** | | | |
| Vasectomy | The vasa deferentia (see Figure 43.8A) are cut and tied off so that sperm can no longer pass into the urethra. Sperm continue to be produced but are reabsorbed by the man's body. Male hormone levels and sexual responses are not affected. | 0–0.15 | A simple surgical procedure performed under local anesthetic in a doctor's office. Although it can theoretically be reversed, vasectomy should be considered permanent. |
| Tubal ligation | The oviducts (see Figure 43.11A) are tied off so that eggs cannot reach the uterus and sperm cannot reach the egg. As with vasectomy, hormone levels and sexual responses are not affected. | 0–0.05 | This surgical procedure is somewhat more complex than vasectomy. It is often performed in conjunction with childbirth when a woman has decided that her family is complete. |

[a] "Failure rate" refers to the number of pregnancies per 100 women per year.

[b] All of these barrier methods are routinely available without medical prescription.

[c] Failure rate varies widely depending on when taken.

In a medical abortion, the cervix is dilated and some of the endometrium, along with the implanted fetus, is removed from the uterus. When performed in the first trimester of a pregnancy, a medical abortion carries less risk of death to the mother than a full-term pregnancy. The risk rises after the first 12 weeks of pregnancy but remains less than that of a full-term pregnancy through the second trimester.

### Reproductive technologies help solve problems of infertility

About 15 percent of the couples in the United States are infertile; they can't have children. There are many reasons for infertility, and they are equally distributed between men and women. A number of technologies have been developed to overcome barriers to both conceiving and bearing a child.

The simplest treatment available is **artificial insemination**, in which the physician positions sperm in the female's reproductive tract. This technique is useful if the male's sperm count is low, if his sperm lack motility, or if problems in the female's reproductive tract prevent the normal movement of sperm up to and through the oviducts. Artificial insemination is used widely in the production of domesticated animals such as cattle.

More recent advances, called **assisted reproductive technologies**, or **ARTs**, involve procedures that remove unfertilized eggs from the ovary, combine them with sperm outside the body, and then place fertilized eggs or egg–sperm mixtures in the appropriate location in the female's reproductive tract for development to take place. The first successful ART was *in vitro fertilization* (IVF). In IVF, the female is treated with hormones that stimulate many follicles in her ovaries to mature. Eggs are collected from these follicles, and sperm are collected from the male. Eggs and sperm are combined in a culture medium outside the body, where fertilization takes place. The resulting embryos can be injected into the mother's uterus in the blastocyst stage or kept frozen for implantation later. The first "test-tube baby" resulting from IVF was born in England in 1978. Since then, more than 3 million babies have been produced by this ART.

A major cause of failure of IVF is failure of sperm to gain access to the egg plasma membrane (see Figure 43.4). To solve this problem, methods have been developed to inject a sperm cell directly into the cytoplasm of an egg. In *intracytoplasmic sperm injection* (*ICSI*), an egg is held in place by suction applied to a polished glass pipette. A slender, sharp pipette is then used to penetrate the egg and inject a sperm (**Figure 43.17**). This ART was used successfully for the first time in 1992 by researchers in Belgium; now thousands of these procedures are performed in U.S. clinics each year, with a success rate of about 25 percent.

IVF, coupled with techniques of genetic analysis, can eliminate the risk that adults who are carriers of genetic diseases will produce affected children. It is now possible to take a cell from a human embryo at the 4- or 8-cell stage (see Figure 44.4) without damaging its developmental potential. The sampled cell can be subjected to molecular analysis to determine whether it carries the harmful gene. This procedure, called preimplantation genetic diagnosis (PGD), makes it possible to determine whether an embryo produced by IVF carries the genetic defect of concern.

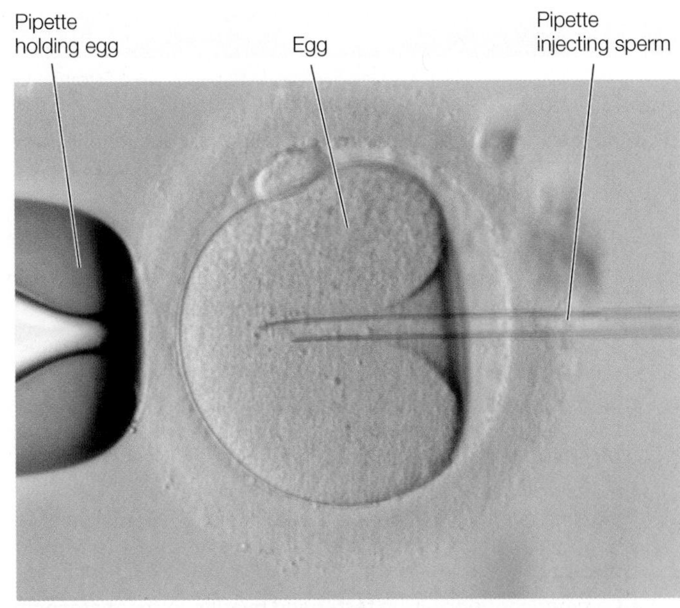


**43.17 Intracytoplasmic Sperm Injection** In this procedure, sperm are injected directly into a mature egg cell. The fertilized egg is then placed in the female reproductive tract, where it can implant and develop into a fetus.

## 43.4 RECAP

**Controlling fertility is an important aspect of modern human life. Decreasing the probability of pregnancy is achieved through methods that prevent sperm and egg from meeting and from preventing implantation. Pregnancies can be facilitated through medical technology.**

- What are the four phases of the human sexual response? See p. 916
- Which method of contraception is the only one to offer protection against sexually transmitted diseases (STDs)? See Table 43.1

The fertilized egg of a sexually reproducing organism is a single cell containing all the genetic information needed to create a new organism. Chapters 19 and 20 introduced some of the molecular aspects of the process of development in multicellular animals. Chapter 44 describes the physical events of animal development.

## CHAPTER SUMMARY

### 43.1 How Do Animals Reproduce Without Sex?

- Asexual reproduction produces offspring that are genetically identical to their parent and to one another; it produces no genetic diversity.
- Means of asexual reproduction include **budding**, **regeneration**, and **parthenogenesis**. Review Figures 43.1 and 43.2

### 43.2 How Do Animals Reproduce Sexually?

- Sexual reproduction consists of three basic steps: gametogenesis; spawning and mating; and fertilization.
- **Gametogenesis** and fertilization are similar in all animals, but spawning and mating includes a great variety of anatomical, physiological, and behavioral adaptations.
- In sexually reproducing species, genetic diversity is created by crossing over and independent assortment of chromosomes during gametogenesis. Fertilization also contributes to genetic diversity.
- Gametogenesis occurs in **testes** and **ovaries**. In spermatogenesis (the production of **sperm**) and oogenesis (the production of eggs), the **germ cells** proliferate mitotically, undergo meiosis, and mature into gametes.
- Each **primary spermatocyte** can produce four haploid sperm through the two divisions of meiosis. Review Figure 43.3A
- **Primary oocytes** immediately enter prophase of the first meiotic division, and in many species, including humans, their development is arrested at this point. Each **oogonium** produces only one egg. Review Figure 43.3B
- Fertilization involves sperm activation, species-specific binding of sperm to egg, the acrosomal reaction, digestion of a path through the protective coverings of the egg, and fusion of sperm and egg plasma membranes. Fusion of these two membranes triggers **blocks to polyspermy**, which prevent additional sperm from entering the egg, and in mammals, signal the egg to complete meiosis and begin development. Review Figure 43.4, **ANIMATED TUTORIAL 43.1**
- **External fertilization** is common in aquatic species. **Internal fertilization** is necessary in terrestrial species and usually involves **copulation**.
- The shelled egg of amniotes is an important evolutionary adaptation that allows reptiles and birds to reproduce in the terrestrial environment.
- **Hermaphroditic** or **monoecious** species have both male and female reproductive systems in the same individual, either sequentially or simultaneously. **Dioecious** species have separate male and female individuals.
- Animals can be classified as **oviparous** or **viviparous**, depending on whether the early stages of development occur outside or inside the mother's body.

### 43.3 How Do the Human Male and Female Reproductive Systems Work?

- Men produce **semen** and deliver it into the woman's reproductive tract. Semen consists of sperm suspended in seminal fluid, which nourishes the sperm and facilitates fertilization.
- Sperm are produced in the **seminiferous tubules** of the testes, mature in the **epididymis**, and are delivered to the **urethra** through the **vasa deferentia**. Other components of semen are produced in the **seminal vesicles**, **prostate gland**, and **bulbourethral gland**. Review Figures 43.8 and 43.9, **WEB ACTIVITY 43.1**
- All components of the semen join in the urethra at the base of the **penis** and are ejaculated through the erect penis by muscle contractions at the culmination of copulation.
- Spermatogenesis depends on testosterone secreted by the Leydig cells of the testes, which are under the control of hormones produced in the anterior pituitary and the hypothalamus. The production of these hormones is controlled by negative feedback from testosterone and another hormone, inhibin, produced by the **Sertoli cells** of the testes. Review Figure 43.10, **WEB ACTIVITY 43.2**
- Eggs mature in the woman's ovaries and are released into the **oviducts**. Sperm deposited in the **vagina** during copulation move up through the **cervix** and **uterus** into the oviducts. Fertilization occurs in the upper regions of the oviducts. Review Figure 43.11, **WEB ACTIVITY 43.3**
- The maturation and release of eggs constitute an **ovarian cycle**. In women, this cycle takes about 28 days. The **uterine cycle** prepares the uterus for receipt of a blastocyst. If no blastocyst is implanted, the lining of the uterus sloughs off in the process of **menstruation**. Review Figure 43.13, **ANIMATED TUTORIAL 43.2**
- Both the ovarian and the uterine cycles are under the control of hypothalamic and pituitary hormones, which in turn are under the feedback control of estrogen and progesterone. Review Figure 43.14
- Childbirth is initiated by hormonal and mechanical stimuli that increase the contraction of uterine muscle. Review Figure 43.15

### 43.4 How Can Fertility Be Controlled?

- Human sexual responses consist of four phases: excitement, plateau, orgasm, and resolution. In addition, men have a refractory period during which renewed excitement is not possible. Review Figure 43.16
- Methods of **contraception** include abstention from copulation and the use of technologies that decrease the probability of fertilization. Review Table 43.1
- **Assisted reproductive technologies** (**ARTs**) have been developed to increase fertility.

## SELF-QUIZ

1. A species in which the individual possesses both male and female reproductive systems is termed
   *a.* dioecious.
   *b.* parthenogenetic.
   *c.* hermaphroditic.
   *d.* monoecious.
   *e.* ovoviviparous.
2. The major advantage of internal fertilization is that
   *a.* it ensures paternity.
   *b.* it permits the fertilization of many gametes.
   *c.* it reduces the incidence of destructive competitive interactions among the members of a group.
   *d.* it increases the number of sperm having access to each egg.
   *e.* it gives the developing organism a greater degree of protection during the early phases of development.

3. Which statement about human oocytes is *true*?
   *a.* By birth, the human female infant has produced a lifetime supply of oocytes.
   *b.* At the onset of puberty, ovarian follicles produce new oocytes in response to hormonal stimulation.
   *c.* At the onset of menopause, a woman stops producing oocytes.
   *d.* Oocytes are produced by a woman throughout adolescence.
   *e.* Oocytes produced by a woman are stored in the oviducts.
4. Spermatogenesis and oogenesis differ in that
   *a.* spermatogenesis produces gametes with greater stores of raw materials than those produced by oogenesis.
   *b.* spermatocytes remain in prophase of the first meiotic division longer than oocytes.
   *c.* oogenesis produces four equally functional haploid cells per meiotic event and spermatogenesis does not.
   *d.* spermatogenesis produces many gametes with meager energy reserves, whereas oogenesis produces relatively few, well-provisioned gametes.
   *e.* spermatogenesis begins before birth in humans, whereas oogenesis does not start until the onset of puberty.
5. Semen contains all of the following except
   *a.* fructose.
   *b.* mucus.
   *c.* clotting enzymes.
   *d.* substances to lower the pH of the uterine environment.
   *e.* an active clot-dissolving enzyme.
6. During oogenesis in mammals, the second meiotic division occurs
   *a.* in the formation of the primary oocyte.
   *b.* in the formation of the secondary oocyte.
   *c.* before ovulation.
   *d.* after fertilization.
   *e.* after implantation.
7. One of the major differences between the sexual responses of men and women is
   *a.* the increase in blood pressure in men.
   *b.* the increase in heart rate in women.
   *c.* the presence of a refractory period in women.
   *d.* the presence of a refractory period in men after orgasm.
   *e.* the increase in muscle tension in men.
8. Which of the following statements about the ovarian and uterine cycles is *false*?
   *a.* Falling estrogen and progesterone levels induce menstruation.
   *b.* A sudden rise in LH induces ovulation.
   *c.* Estrogen levels rise in the first half of the ovarian cycle and progesterone levels rise in the second half.
   *d.* If fertilization occurs, the corpus luteum secretes hCG.
   *e.* Estrogen is produced by follicle cells
9. Contractions of muscles in the uterine wall and milk let-down are stimulated by
   *a.* progesterone.
   *b.* estrogen.
   *c.* prolactin.
   *d.* oxytocin.
   *e.* human chorionic gonadotropin.
10. Which of the following methods of contraception is *most likely* to fail?
    *a.* Rhythm method
    *b.* Birth control pills
    *c.* Diaphragm
    *d.* Vasectomy
    *e.* Condom

## FOR DISCUSSION

1. In the deep ocean, there are species of fish in which the male is much smaller than the female and actually lives attached to her body. In terms of the selective pressures that operate on sexual and asexual reproduction and in terms of the deep-sea environment, what factors do you think resulted in the evolution of this extreme sexual dimorphism?
2. What are two main differences between the immediate products of the first and second meiotic divisions in spermatogenesis and oogenesis? Why do these differences exist?
3. At the beginning of each ovarian cycle in humans, 6 to 12 follicles begin to develop in response to rising levels of FSH, but after a week, only one follicle continues to develop, and the others wither away. Given that follicles produce estrogen, that estrogen stimulates follicle cells to produce FSH receptors, and that estrogen exerts negative feedback on FSH production in the pituitary, can you explain how a single follicle is "selected" to grow?
4. Compare the actions of LH and FSH in the ovaries and testes.
5. Ovarian and uterine events during the month following ovulation differ depending on whether fertilization occurs. Describe the differences and explain their hormonal controls.

## ADDITIONAL INVESTIGATION

No male contraceptive methods exist other than vasectomy or the condom. How would you go about developing a pharmacological method to block sperm production that could lead to a male pill without affecting the maintenance of male secondary sexual characteristics or male sexual behavior?

# 44 Animal Development

## Go with the flow

Place your hand over your heart. Now place it over your liver. Next over your appendix. Surely you put your hand first on the left side of your chest, then on your right side just under your ribs, and finally on the right side of your lower abdomen. But in Chapter 31 you learned that vertebrates (including you) are bilaterally symmetrical (left arm/right arm, left kidney/right kidney, and so on). Clearly, however, our bilateral symmetry is not absolute.

Some of our internal organs—including the heart, liver, appendix, stomach, and the lobes of the lungs—are oriented differently with respect to the left and right sides of the body. How does a developing embryo know which side is left and which is right?

Clues to answering this question came from the fact that in about 1 out of every 7,000 people, the arrangement of the internal organs is reversed, a condition known as *situs inversus*, Latin for "location inverted." This developmental difference arises when the very early embryo goes from being a single layer of cells to multiple layers of cells.

As you will learn in this chapter, to get from a single layer of cells to the next stage with two layers of cells, a pore or slit forms as cells in one area of the embryo migrate inward from the surface. Other cells from the surface migrate toward and through this opening to take up positions underneath. The place where the inward movement of cells starts is called the node. Cells of the node have motile cilia that sweep extracellular fluid through the opening.

Cells bordering the node also have one nonmotile cilium each—a primary cilium. When the primary cilia are bent by the flow of extracellular fluid, they initiate signaling cascades that determine the pattern of internal organ development. Since the fluid driven by the motile cilia tends to flow from right to left, the signaling cascades are not expressed symmetrically, and this initiates the left–right organization of organ development.

Among individuals carrying a mutation that eliminates motility of the nodal cilia, half have the normal orientation of the internal organs and half have *situs inversus*. Most people with this condition lead normal, healthy lives. They may not even know about it

**Go with the Flow** The internal organs of humans are not all symmetrical, and some individuals are born with the mirror-image pattern of what is seen in most people—a condition called *situs inversus*. The left–right asymmetry of the internal organs is initiated by asymmetrical stimulation of primary cilia at a very early stage in development.

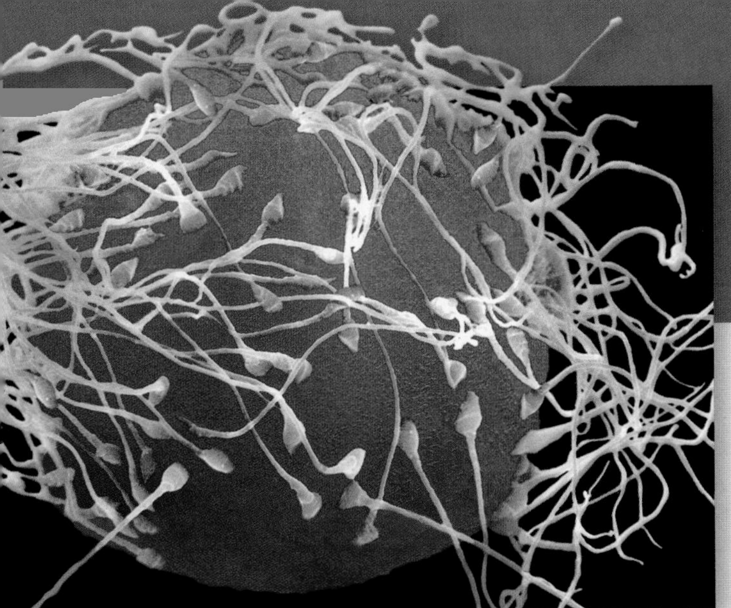

**When Sperm Meet Egg** Development begins with the fertilization of an egg by a sperm. Once one sperm fuses with the egg, all other sperm are blocked. An animal egg typically is much larger than the sperm; the egg cytoplasm is loaded with informational molecules and nutrients that will direct development and nourish the growing embryo.

unless a routine physical exam reveals that the organs are not where they should be.

As frequently happens in biology, we just pushed the question back one more step. Why do the nodal cilia beat in such a way that the extracellular fluid flows from right to left? The root cause of this asymmetry likely originates in some of the early cell divisions of the fertilized egg. The fertilized egg goes through an initial series of cell divisions that subdivide the egg cytoplasm into a mass of undifferentiated cells. Although this mass of cells shows no hints of the eventual body plan, an uneven distribution of molecules from the cytoplasm of the fertilized egg can provide information that directs the fates of cells and sets up the body plan.

**IN THIS CHAPTER** we will see how a single cell becomes a multicellular animal through orderly cell movements that create multiple layers and set up cell–cell interactions. The regional and temporal differences in gene expression that control cell differentiation, described in Chapters 19 and 20, lead to the emergence of the body plan of the animal. We will discuss these early developmental steps in four organisms that have been studied extensively: sea urchins, frogs, chickens, and humans.

## CHAPTER OUTLINE



## 44.1 How Does Fertilization Activate Development?

Fertilization is the joining of sperm and egg. You might therefore think of it as the event that begins development. Keep two things in mind, however. First, in animals that reproduce asexually, development proceeds without fertilization. And second, in animals where fertilization does occur, it is preceded by critical events in the maturing egg that will influence subsequent development. Thus, in studying fertilization we are really asking how it activates or restarts multicellular development in sexually reproducing animals.

Fertilization does far more than just restore a full diploid complement of maternal and paternal genes. The fusion of sperm and egg plasma membranes accomplishes several things:

- It sets up blocks to the entry of additional sperm.
- It stimulates ion fluxes across the egg membrane.
- It changes the egg's pH.
- It increases egg metabolism and stimulates protein synthesis.
- It initiates the rapid series of cell divisions that produce a multicellular embryo.

Section 43.2 described the mechanisms of fertilization. Here we take a closer look at the cellular and molecular interactions of sperm and egg that initiate the first steps of development.

### The sperm and the egg make different contributions to the zygote

In most species, eggs are much larger than sperm. Egg cytoplasm is well stocked with organelles, nutrients, and a variety of molecules, including transcription factors and mRNAs. The sperm is little more than a DNA delivery vehicle. Nearly everything the embryo needs during its early stages of development comes from the mother. In addition to providing its haploid nucleus, the sperm makes another important contribution to the zygote in most species—a centriole.

The centriole becomes the centrosome of the zygote, which organizes the mitotic spindles for subsequent cell divisions (see Figure 11.10). The centriole is also the origin of the primary cilia of cells, which are important in cell signaling, as we saw in the opening story about *situs inversus*.

Cytoplasmic factors in the egg play important roles in setting up the signaling cascades that orchestrate the major events of development: *determination, differentiation, and morphogenesis*.

### Rearrangements of egg cytoplasm set the stage for determination

The unique attributes of amphibian eggs make them ideal models for illustrating how rearrangements of egg cytoplasm set the stage for determination. The molecules in the cytoplasm of the amphibian egg are not homogeneously distributed. The entry of the sperm into the egg stimulates rearrangements of the egg cytoplasm that introduce additional organization to the egg cytoplasm. This rearrangement establishes the polarity of the zygote, and when cell divisions begin, the informational molecules that will guide development are not divided equally among daughter cells.

Rearrangement of egg cytoplasm following fertilization is easily observed in some frog species because of pigments in the cytoplasm. The nutrients in an unfertilized frog egg are dense yolk granules that are concentrated by gravity in the lower half of the egg, called the **vegetal hemisphere**. The haploid nucleus of the egg is located at the opposite end, in the **animal hemisphere**. The outermost (*cortical*) cytoplasm of the animal hemisphere is heavily pigmented, and the underlying cytoplasm has more diffuse pigmentation. The vegetal hemisphere is not pigmented. Because of these differences, it is easy to observe how the cytoplasm is rearranged when a frog egg is fertilized.

The frog egg is radially symmetrical. You can turn it on its vegetal–animal pole axis, and all sides are the same. Sperm-binding sites are localized on the surface of the animal hemisphere, so that is where the sperm enters the egg. When a sperm enters the egg, bilateral symmetry is imposed by creating an anterior–posterior axis. Cortical cytoplasm rotates toward the site of sperm entry. This rotation brings different regions of cytoplasm into contact with each other on opposite sides of the egg, producing a band of diffusely pigmented cytoplasm on the side opposite the site of sperm entry. This band, called the **gray crescent**, marks the location of important developmental events in some species of amphibians (**Figure 44.1**).

The one non-nuclear organelle that the sperm contributes to the egg—the centriole—initiates the cytoplasmic reorganization revealed by the appearance of the gray crescent. The centriole organizes the microtubules in the vegetal hemisphere cytoplasm into a parallel array that guides the movement of the cortical cytoplasm. These microtubules also appear to be directly responsible for movement of specific organelles and proteins, because these organelles and proteins move from the vegetal

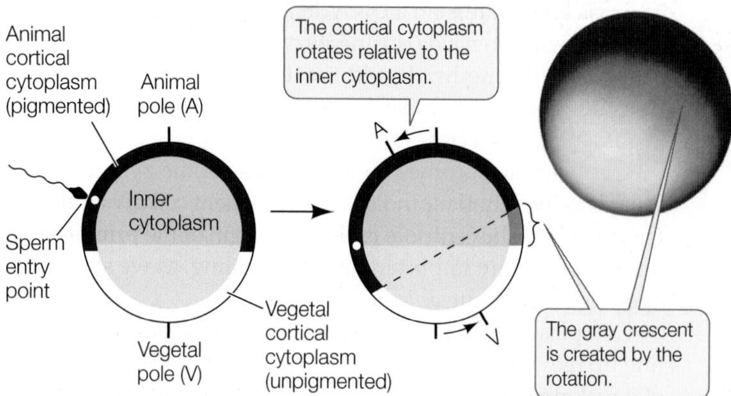


**44.1 The Gray Crescent** Rearrangement of the cytoplasm of frog eggs after fertilization creates the gray crescent.

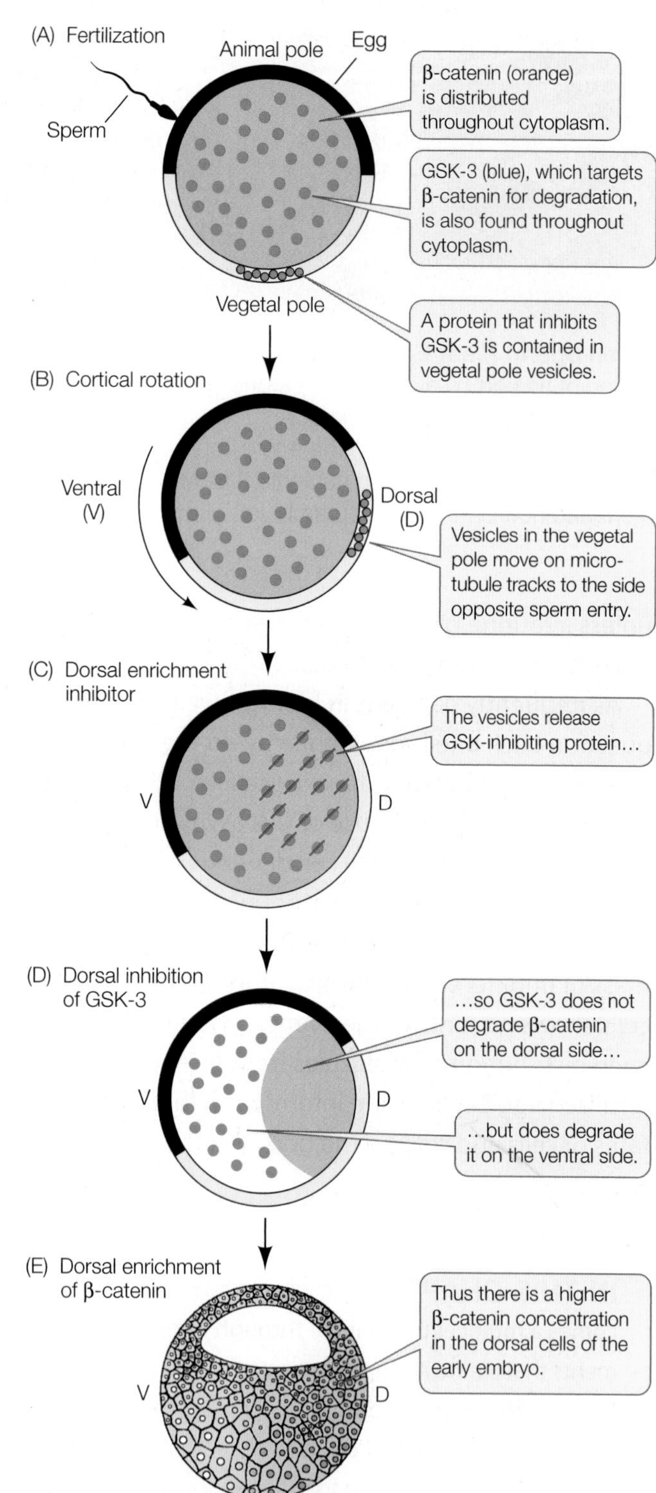

(A) Fertilization
Animal pole
Egg
Sperm
β-catenin (orange) is distributed throughout cytoplasm.
GSK-3 (blue), which targets β-catenin for degradation, is also found throughout cytoplasm.
Vegetal pole
A protein that inhibits GSK-3 is contained in vegetal pole vesicles.
(B) Cortical rotation
Ventral (V)
Dorsal (D)
Vesicles in the vegetal pole move on microtubule tracks to the side opposite sperm entry.
(C) Dorsal enrichment inhibitor
The vesicles release GSK-inhibiting protein...
V
D
(D) Dorsal inhibition of GSK-3
...so GSK-3 does not degrade β-catenin on the dorsal side...
...but does degrade it on the ventral side.
V
D
(E) Dorsal enrichment of β-catenin
Thus there is a higher β-catenin concentration in the dorsal cells of the early embryo.
V
D

**44.2 Cytoplasmic Factors Set Up Signaling Cascades** Cytoplasmic movement changes the distributions of critical developmental signals. In the frog zygote, the interaction of the protein kinase GSK-3, its inhibitor, and the protein β-catenin are crucial in specifying the dorsal–ventral axis of the embryo.

hemisphere to the gray crescent region even faster than the cortical cytoplasm rotates.

The movement of cytoplasm, proteins, and organelles changes the distribution of critical developmental signals. A key transcription factor in early development is β-catenin, which is produced from maternal mRNA (mRNA produced and stored in the egg while it was maturing in the ovary). Beta-catenin is found throughout the egg cytoplasm. Also present throughout the egg cytoplasm is a protein kinase called glycogen synthase kinase-3 (GSK-3), which phosphorylates and thereby targets β-catenin for degradation. An inhibitor of GSK-3 is segregated in the vegetal cortex of the egg. After sperm entry, this inhibitor is moved along microtubules to the gray crescent, where it prevents the degradation of β-catenin. As a result, the concentration of β-catenin is higher on the dorsal than on the ventral side of the developing embryo (**Figure 44.2**).

Beta-catenin plays a major role in the cell–cell signaling cascade that begins the process of cell determination and the formation of the embryo. But before cell–cell signaling can occur, multiple cells must be in place. Let's turn to the early series of cell divisions that transform the zygote into a multicellular embryo.

### Cleavage repackages the cytoplasm

Transformation of the diploid zygote into a mass of cells occurs through a rapid series of cell divisions called **cleavage**. Because the cytoplasm of the zygote is not homogeneous, these first cell divisions result in the differential distribution of nutrients and cytoplasmic determinants in the early embryo.

In most animals, cleavage proceeds with rapid DNA replication and mitosis but with no cell growth and little gene expression. The embryo becomes a solid ball of smaller and smaller cells. Eventually, this ball forms a central fluid-filled cavity called a **blastocoel**, at which point the embryo is called a **blastula**. Its individual cells are called **blastomeres**. The pattern of cleavage in different species influences the form of their blastulas.

- **Complete cleavage** occurs in most eggs that have little **yolk** (stored nutrients). In this pattern, early cleavage furrows divide the egg completely and the blastomeres are of similar size. The frog egg undergoes complete cleavage, but because its vegetal pole contains more yolk, the division of the cytoplasm is unequal and the blastomeres in the animal hemisphere are smaller than those in the vegetal hemisphere (**Figure 44.3A**).

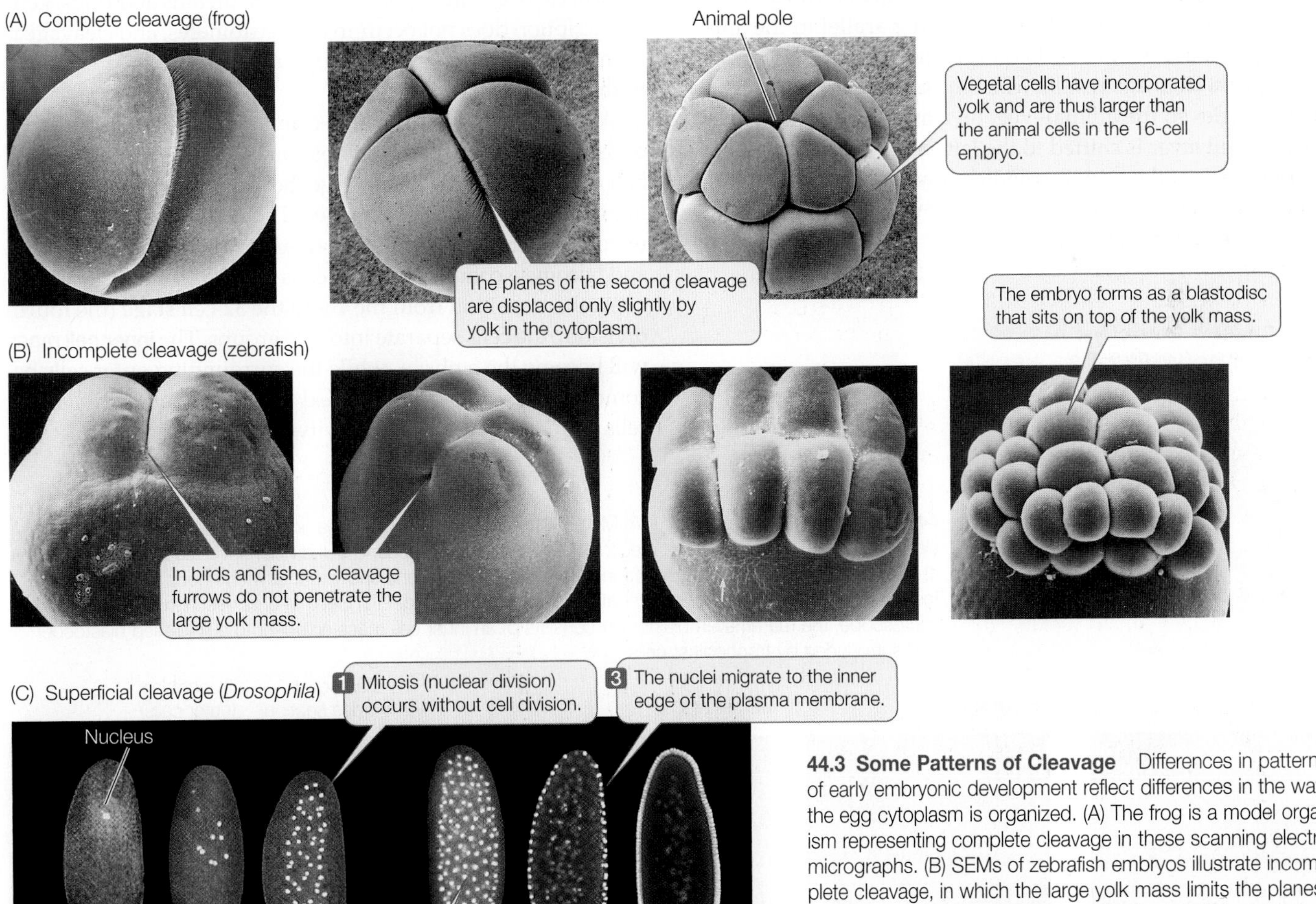
(A) Complete cleavage (frog)
Animal pole
Vegetal cells have incorporated yolk and are thus larger than the animal cells in the 16-cell embryo.
The planes of the second cleavage are displaced only slightly by yolk in the cytoplasm.
(B) Incomplete cleavage (zebrafish)
The embryo forms as a blastodisc that sits on top of the yolk mass.
In birds and fishes, cleavage furrows do not penetrate the large yolk mass.
(C) Superficial cleavage (*Drosophila*)
1 Mitosis (nuclear division) occurs without cell division.
3 The nuclei migrate to the inner edge of the plasma membrane.
Nucleus
2 A syncitium—a single cell with many nuclei—is produced.
4 Cellularization occurs, creating a blastoderm

**44.3 Some Patterns of Cleavage** Differences in patterns of early embryonic development reflect differences in the way the egg cytoplasm is organized. (A) The frog is a model organism representing complete cleavage in these scanning electron micrographs. (B) SEMs of zebrafish embryos illustrate incomplete cleavage, in which the large yolk mass limits the planes of cleavage. (C) Nuclear staining reveals the syncitial nuclei characteristic of the early embryo of a fruit fly. These nuclei migrate to the periphery. Cleavage furrows then move inward to separate the nuclei into individual cells, forming the blastoderm.

- **Incomplete cleavage** occurs in many species in which the egg contains a lot of yolk and the cleavage furrows do not penetrate it all. **Discoidal cleavage** is a type of incomplete cleavage common in fishes, reptiles, and birds, the eggs of which contain a dense yolk mass. The embryo forms as a disc of cells, called a **blastodisc**, that sits on top of the yolk mass (**Figure 44.3B**).
- **Superficial cleavage** is a variation of incomplete cleavage that occurs in insects such as the fruit fly (*Drosophila*). Early in development, cycles of mitosis occur without cell division, producing a *syncytium*—a single cell with many nuclei (**Figure 44.3C**). The nuclei eventually migrate to the periphery of the egg, after which the plasma membrane of the egg grows inward, partitioning the nuclei into individual cells surrounding a core of yolk.

The positions of the mitotic spindles during cleavage are not random but are defined by cytoplasmic factors produced from the maternal genome and stored in the egg (see Section 19.4). The orientation of the mitotic spindles can determine the planes of cleavage and the arrangement of the blastomeres.

In complete cleavage, if the mitotic spindles of successive cell divisions form parallel or perpendicular to the animal–vegetal axis of the zygote, a pattern of **radial cleavage** occurs as seen in the frog: the first two cell divisions are parallel to the animal–vegetal axis and the third is perpendicular to it (see Figure 44.3A). **Spiral cleavage** results when the mitotic spindles are at oblique angles to the animal–vegetal axis. In spiral cleavage, each new cell layer is shifted to the left or right, depending on the orientation of the mitotic spindles. Most mollusks have spiral cleavage, reflected in some species by a coiling shell pattern (as seen in snails).

## Early cell divisions in mammals are unique

Several features of early cell divisions in placental mammals (eutherians) are so different from those seen in other animal groups that some biologists think it is inappropriate to call it cleavage. But whether you call it cleavage or not, it is still the sequence of early cell divisions that produces a body of undifferentiated cells that will become the embryo. This process in mammals is very slow. Cell divisions are 12 to 24 hours apart, compared with tens of minutes to a few hours in non-mammalian species. Also, the cell divisions of mammalian blastomeres are not in synchrony with each other. Because the blastomeres do not undergo mitosis at the same time, the number of cells in the embryo does not increase in the regular (2, 4, 8, 16, 32, etc.) progression typical of other species.

The pattern of mammalian cleavage is *rotational*: the first cell division is parallel to the animal–vegetal axis, yielding two blastomeres. In the second cell division, those two blastomeres divide at right angles to one other: one divides parallel to the animal–vegetal axis, while the other divides perpendicular to it (**Figure 44.4A**).

Another unique feature of the slow, rotational mammalian cleavage is that gene products expressed during cleavage play roles in cleavage. In animals such as sea urchins and frogs, gene transcription does not occur in the blastomeres, and cleavage is therefore directed exclusively by molecules that were present in the egg before fertilization.

As in other animals that have complete cleavage, the early cell divisions in a mammalian zygote produce a loosely associated ball of cells. After the 8-cell stage, however, the behavior of the mammalian blastomeres changes. They change shape to maximize their surface contact with one another, form tight junctions, and become a compact mass of cells (**Figure 44.4B**).

At the transition from the 16- to the 32-cell stage (the fourth division), the cells separate into two groups. The **inner cell mass** will become the embryo, while the surrounding outer cells become an encompassing sac called the **trophoblast**. Trophoblast cells secrete fluid, creating a cavity—the *blastocoel*—with the in-

(A)

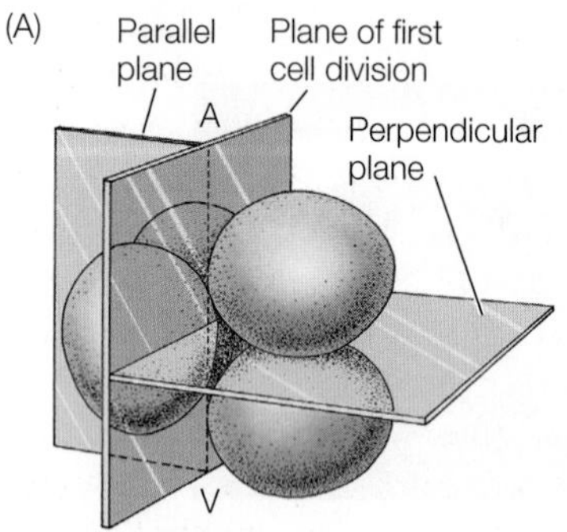


(B)

8-cell stage

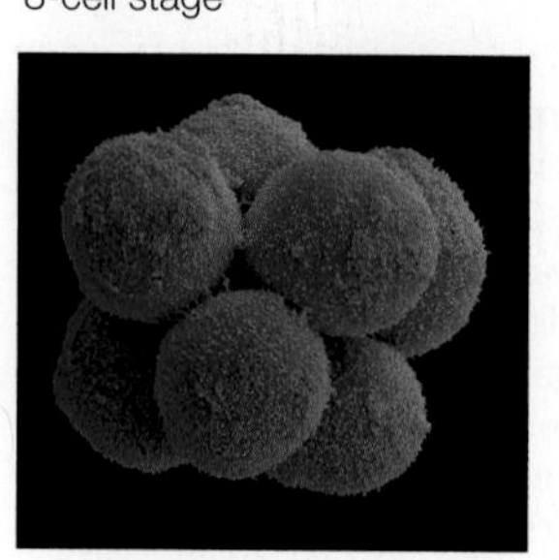

16-cell stage

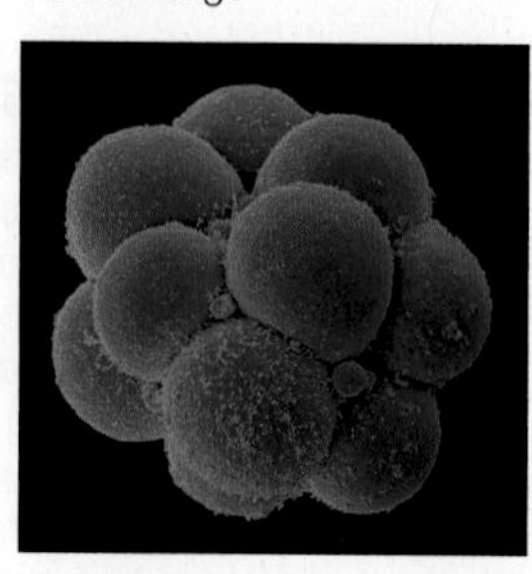

32-cell stage

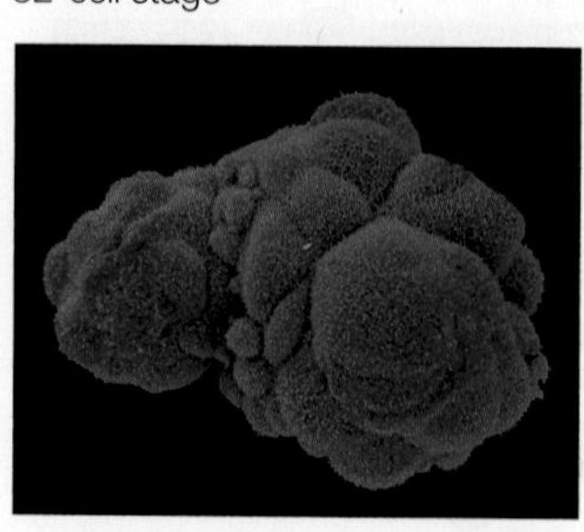

(C)

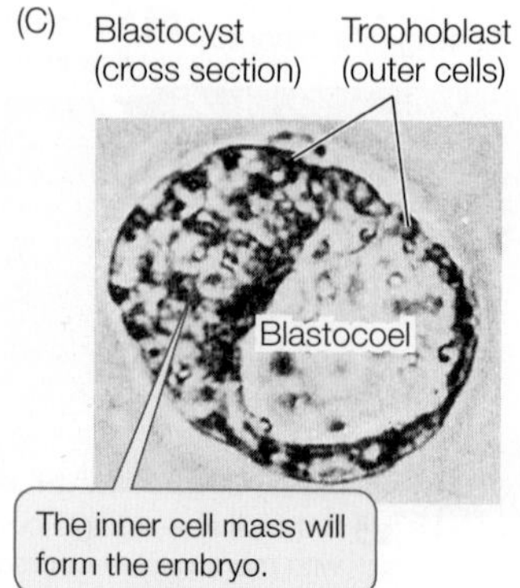


**44.4 Becoming a Blastocyst** (A) Mammals have rotational cleavage, in which the plane of the first cleavage is parallel to the animal–vegetal (A–V) axis, but the second cell division involves two planes (beige) at right angles to each other. (B) Scanning electron micrographs show that asynchronous cell division results in an asymmetrical blastocyst at about the 32-cell stage. (C) Seen in cross section under a light microscope, the mammalian blastocyst consists of an inner cell mass adjacent to a fluid-filled blastocoel and surrounded by trophoblast cells.

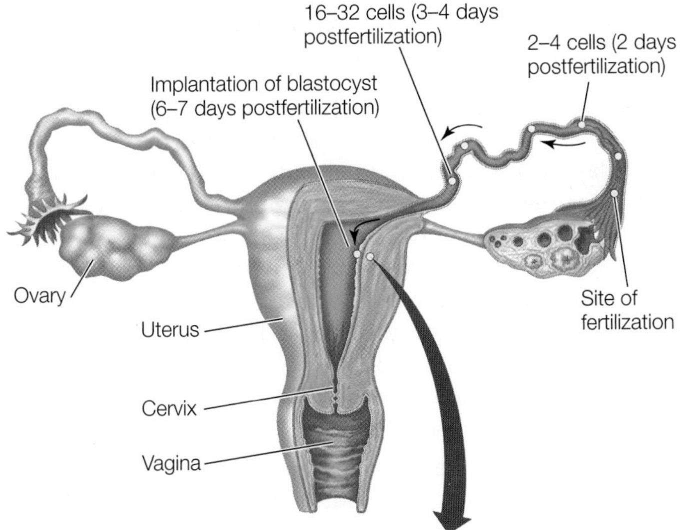


**44.5 A Human Blastocyst at Implantation** Adhesion molecules and proteolytic enzymes secreted by trophoblast cells allow the blastocyst to burrow into the endometrium. Once the blastocyst is implanted in the wall of the uterus, the trophoblast cells send out numerous projections—the chorionic villi—which increase the embryo's area of contact with the mother's bloodstream.

ner cell mass at one end (**Figure 44.4C**). At this stage, the mammalian embryo is called a **blastocyst**, distinguishing it from the blastulas of other animal groups.

Why is mammalian cleavage so different? A key factor is that mammalian eggs contain no yolk and must derive all nutrients from the mother. Mammals are *viviparous*: the embryo develops within the uterus of the mother. To support the developing embryo, a connection develops between the circulatory systems of the embryo and the mother. As we will see later in this chapter, the structures that provide this connection are the placenta and the umbilical cord. Thus, the mammalian blastocyst must produce both the embryo (from the inner cell mass) and its support structures (from the trophoblast).

Fertilization in mammals occurs in the upper reaches of the mother's oviduct, and cleavage occurs as the zygote travels down the oviduct to the uterus. When the blastocyst arrives in the uterus, the trophoblast adheres to the lining of the uterus (the *endometrium*), beginning the process of **implantation**. In humans, implantation begins about 6 days after fertilization and is aided by adhesion molecules and enzymes secreted by the trophoblast (**Figure 44.5**).

As the blastocyst moves down the oviduct to the uterus, it must not embed itself in the oviduct (Fallopian tube) wall, or the result will be an *ectopic*, or *tubal, pregnancy*—a very dangerous condition. Early implantation is prevented by the zona pellucida, which surrounded the egg and remains around the cleaving ball of cells (see Section 43.2). At about the time the blastocyst reaches the uterus, it hatches from the zona pellucida, and implantation can occur.

## Specific blastomeres generate specific tissues and organs

Cleavage results in a repackaging of the egg cytoplasm into a large number of small cells surrounding the fluid-filled blastocoel. Except in mammals, little cell differentiation and little if any gene expression occur during cleavage. Nevertheless, cells in different regions of the blastula possess different complements of the nutrients and cytoplasmic determinants that were present in the egg.

The blastocoel prevents cells from different regions of the blastula from coming into contact and interacting, but that will soon change. During the next stage of development, the cells of the blastula will move around and come into new associations with one another, communicate instructions to one another, and begin to differentiate. In many animals, these movements of the blastomeres are so regular and well orchestrated that it is possible to label a specific blastomere with a dye and identify the tissues and organs that form from its progeny. Such labeling experiments produce **fate maps** of the blastula (**Figure 44.6**).

Blastomeres become **determined**—committed to specific fates—at different times in different species. In some species,

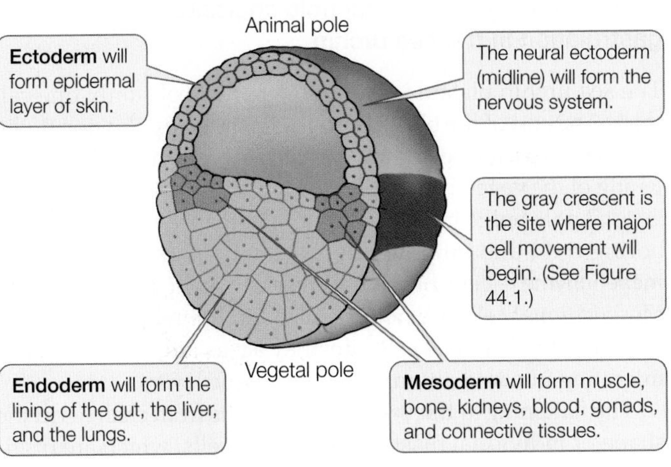


**44.6 Fate Map of a Frog Blastula** Colors indicate the portions of the blastula that will form the three germ layers and subsequently the frog's tissues and organs.

such as roundworms, the fates of blastomeres are restricted as early as the two-cell stage. If one of these blastomeres is experimentally removed, a particular portion of the embryo will not form. This type of development has been called **mosaic development** because each blastomere seems to contribute a specific set of "tiles" to the final "mosaic" that is the adult animal.

In contrast to mosaic development, the loss of some cells during cleavage in **regulative development** does not affect the developing embryo, because the remaining cells compensate for the loss. Regulative development is typical of many vertebrate species, including humans. The pluripotent cells of the mammalian blastocyst (the inner cell mass) are known as *embryonic stem cells* and are the subject of much research, particularly because of their therapeutic potential (see Section 19.2).

If some blastomeres can change their fate to compensate for the loss of other cells during cleavage and blastula formation, can those cells form an entire embryo? To a certain extent, yes. During cleavage or early blastula formation in mammals, for example, if the blastomeres are physically separated into two groups, both groups can produce complete embryos. Since the two embryos come from the same zygote, they will be *monozygotic twins*—genetically identical.

Non-identical twins occur when two separate eggs are fertilized by two separate sperm. Thus, while identical twins are always of the same sex, non-identical twins have a 50 percent chance of being the same sex. In about 1 out of 50,000 human pregnancies, genetic or environmental factors cause the inner cell mass to split partially. The result is twins that are *conjoined* at some point on their bodies, usually sharing some of their organs and limbs.

### 44.1 RECAP

**The egg is stocked with nutrients and informational molecules that power and direct the early stages of development. Fertilization activates the egg and stimulates rearrangement of the cytoplasm, setting up the body axes and positional information that initiate signaling cascades, which control determination and differentiation.**

- Explain how β-catenin becomes concentrated in only certain blastomeres. See p. 925 and Figure 44.2
- In general terms, describe the difference between complete and incomplete cleavage. See pp. 925–926 and Figure 44.3
- What does a fate map tell us? How are fate maps constructed? See pp. 927–928 and Figure 44.6

Of the next stage of development—gastrulation—the developmental biologist Louis Wolpert once said, "It is not birth, marriage, or death, but gastrulation which is the most important time in your life." During gastrulation, cell movements create new cell-to-cell contacts, which in turn sets up signaling cascades. Signaling cascades initiate the differentiation of cells and tissues and set the stage for the emergence of the body plan.

## 44.2 How Does Gastrulation Generate Multiple Tissue Layers?

The blastula is typically a fluid-filled ball of cells. How does this simple ball of cells become an embryo made up of multiple tissue layers with head and tail ends and dorsal and ventral sides? **Gastrulation** is the process whereby the blastula is transformed by massive movements of cells into an embryo with multiple tissue layers and distinct body axes. The resulting spatial relationships between tissues make possible the inductive interactions between cells that trigger differentiation and organ formation (see Figure 19.10).

During gastrulation, three **germ layers** (also called *cell layers* or *tissue layers*) form (see Figure 44.6):

- The **endoderm** is the innermost germ layer, created as some blastomeres move to the inside of the embryo. The endoderm gives rise to the lining of the digestive tract, respiratory tract, pancreas, and liver.
- The **ectoderm** is the outer germ layer, formed from those cells remaining on the outside of the embryo. The ectoderm gives rise to the nervous system, including the eyes and ears; and to the epidermal layer of the skin and structures derived from skin, such as hair, feathers, nails or claws, sweat glands, oil glands, and even teeth and other tissues of the mouth.
- The **mesoderm** is the middle layer and is made up of cells that migrate between the endoderm and the ectoderm. The mesoderm contributes tissues to many organs, including the heart, blood vessels, muscles, and bones.

Some of the most interesting and important challenges in animal development have dealt with two related questions: what directs the cell movements of gastrulation, and what is responsible for the resulting patterns of cell differentiation and organ formation? Scientists have made significant progress in answering both these questions at the molecular level. In the following discussion, we will begin with sea urchin gastrulation because it is the simplest to conceptualize in spatial terms. We will then describe the more complex pattern of gastrulation in frogs, which in turn will help elucidate the still more complex patterns in reptiles, birds, and mammals.

### Invagination of the vegetal pole characterizes gastrulation in the sea urchin

The sea urchin blastula is a hollow ball of cells only one cell thick. The end of the blastula stage is marked by slowing of the rate of mitosis; the beginning of gastrulation is marked by a flattening of the vegetal hemisphere (**Figure 44.7**). Some cells at the vegetal pole bulge into the blastocoel, break away from neighboring cells, and migrate into the cavity. These cells become **mesenchyme** —cells of the middle germ layer, the mesoderm. Mesenchymal cells are not organized in tightly packed sheets or tubes like epithelial cells are; they act as independent units, migrating into and among the other tissue layers.

The flattening at the vegetal pole results from changes in the shape of individual blastomeres. These cells, which are origi-

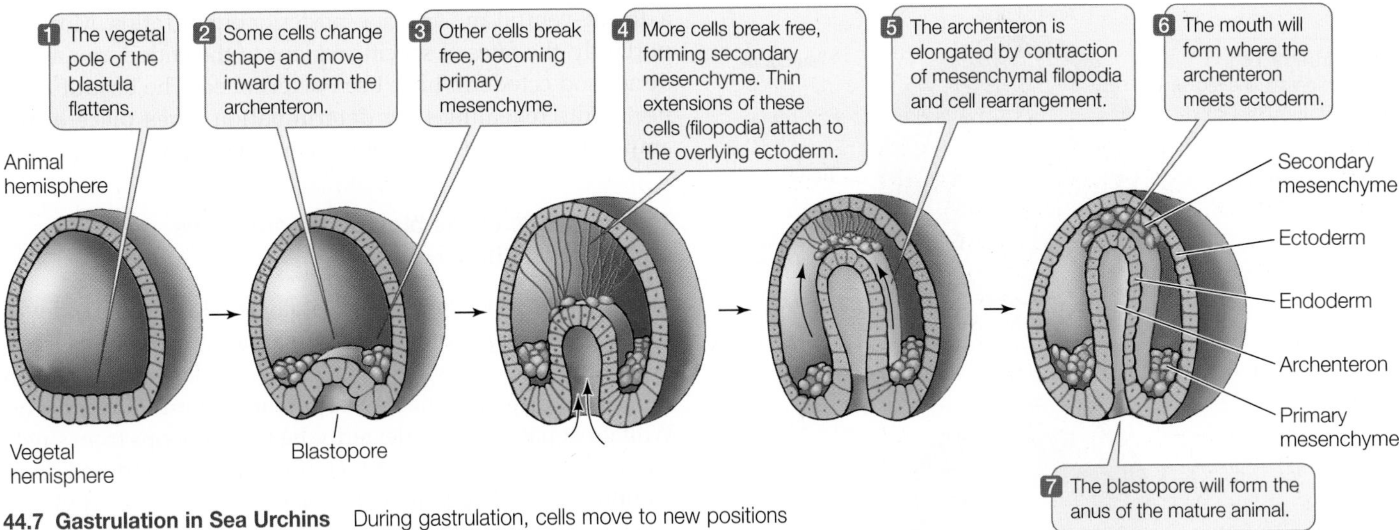


**44.7 Gastrulation in Sea Urchins** During gastrulation, cells move to new positions and form the three germ layers from which differentiated tissues develop.

yourBioPortal.com
GO TO Animated Tutorial 44.1 • Gastrulation

nally rather cuboidal, become wedge-shaped, with smaller outer edges and larger inner edges. As a result, the vegetal pole bulges inward, or *invaginates*, as if someone were poking a finger into a hollow ball (see Figure 44.7). The invaginating cells become endoderm and form the primitive gut, called the **archenteron**. At the tip of the archenteron, more cells enter the blastocoel to form more mesoderm.

Changes in cell shapes cause the initial invagination of the archenteron, but eventually it is pulled by the mesenchyme cells. These cells, attached to the tip of the archenteron, send out extensions called filopodia that adhere to the overlying ectoderm. When the filopodia contract, they pull the archenteron toward the ectoderm at the opposite end of the embryo from where the invagination began. The mouth of the animal forms where the archenteron makes contact with this overlying ectoderm. The opening created by the invagination of the vegetal pole is called the **blastopore**; it will become the anus of the animal.

What mechanisms control the various cell movements of sea urchin gastrulation? The immediate answer is that specific properties of particular blastomeres change. For example, some vegetal cells change shape and bulge into the blastocoel, and these cells become mesenchyme. Once they lose contact with their neighboring cells on the surface of the blastula, they send out filopodia that then move along an extracellular matrix of proteins laid down by the cells lining the blastocoel.

A deeper understanding of gastrulation requires that we discover the molecular mechanisms whereby different blastomeres develop different properties. Cleavage systematically divides up the cytoplasm of the egg. The sea urchin blastula at the 64-cell stage is radially symmetrical, but it has *polarity*, as described in Section 19.4. It consists of tiers of cells. As in the frog blastula, the top is the animal pole and the bottom the vegetal pole.

If different tiers of blastula cells are separated, they show different developmental potentials; only cells from the vegetal pole are capable of initiating the development of a complete larva (see Figure 19.8). It has been proposed that these differences are due to uneven distribution of various transcriptional regulatory proteins in the egg cytoplasm. As cleavage progresses, these proteins end up in different groups of cells. Therefore, specific sets of genes are activated in different cells, determining their different developmental capacities.

Let's turn now to gastrulation in the frog, in which several key signaling molecules have been identified.

### Gastrulation in the frog begins at the gray crescent

Amphibian blastulas have considerable yolk and are more than one cell thick; therefore, gastrulation is more complex in amphibians than in sea urchins. Variation is considerable among different species of amphibians, but in this brief account we will use results from studies done on different species to produce a generalized picture of amphibian development.

Amphibian gastrulation begins when certain cells in the gray crescent region change their shapes and cell adhesion properties. These cells bulge inward toward the blastocoel while they remain attached to the outer surface of the blastula by slender necks. Because of their shape, these cells are called *bottle cells*.

Bottle cells mark the spot where the **dorsal lip** of the blastopore will form (**Figure 44.8**). As the bottle cells move inward, the dorsal lip is created, and a sheet of cells moves over it into the blastocoel. This process is called **involution**. One group of involuting cells is the prospective endoderm; these cells form the primitive gut, or archenteron. Another group will move between the endoderm and the outermost cells to form the mesoderm. These rearrangements are due to changes in cell properties called **convergent extension**. The cells elongate in the direction of movement, but they also intercalate (move in between each other). If they just elongated, the migrating group of cells would become much narrower; by intercalating, they maintain the width of the migrating cell group.

As gastrulation proceeds, cells from the animal hemisphere flatten and move toward the site of involution in a process called **epiboly**. The blastopore lip widens and eventually forms a com-

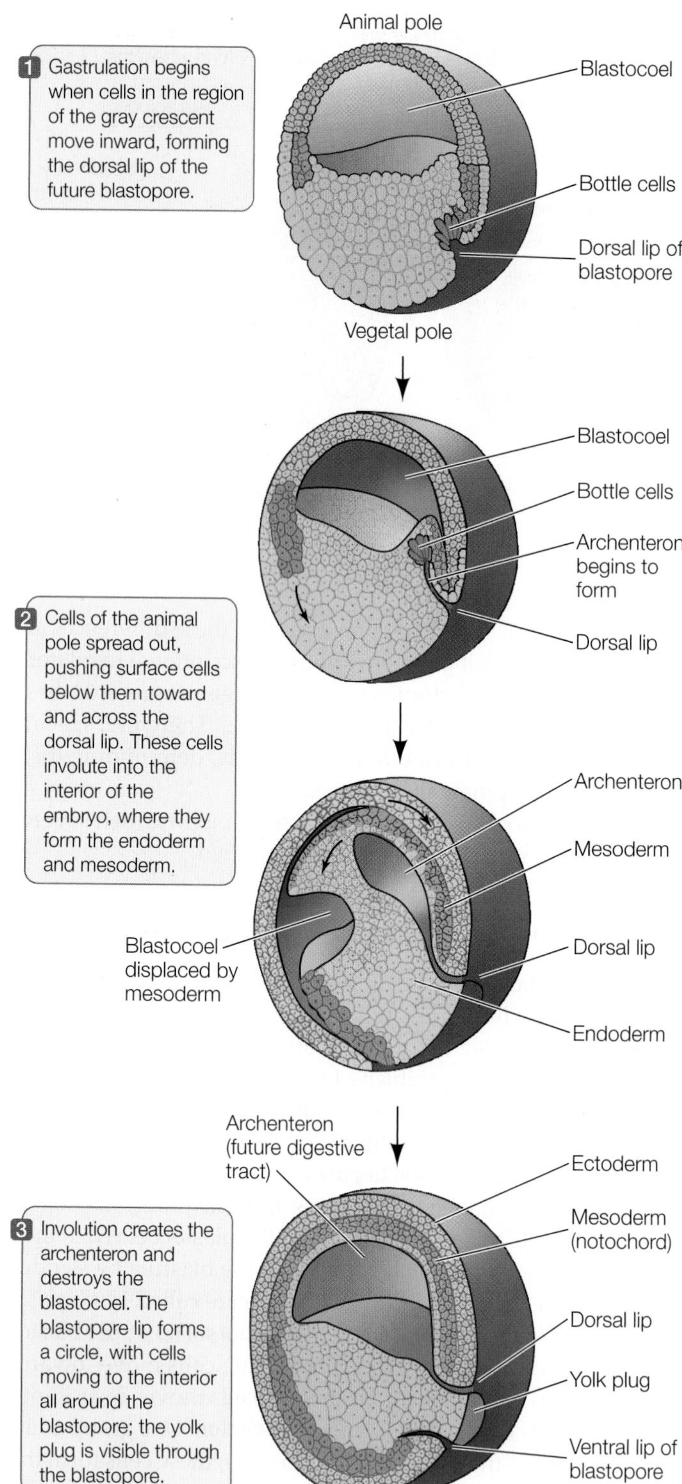

Animal pole
1 Gastrulation begins when cells in the region of the gray crescent move inward, forming the dorsal lip of the future blastopore.
Blastocoel
Bottle cells
Dorsal lip of blastopore
Vegetal pole
Blastocoel
Bottle cells
Archenteron begins to form
Dorsal lip
2 Cells of the animal pole spread out, pushing surface cells below them toward and across the dorsal lip. These cells involute into the interior of the embryo, where they form the endoderm and mesoderm.
Archenteron
Mesoderm
Blastocoel displaced by mesoderm
Dorsal lip
Endoderm
Archenteron (future digestive tract)
Ectoderm
Mesoderm (notochord)
3 Involution creates the archenteron and destroys the blastocoel. The blastopore lip forms a circle, with cells moving to the interior all around the blastopore; the yolk plug is visible through the blastopore.
Dorsal lip
Yolk plug
Ventral lip of blastopore


**44.8 Gastrulation in the Frog Embryo** The colors in this diagram are matched to those in Figure 44.6, the frog fate map.

plete circle surrounding a "plug" of yolk-rich cells. As cells continue to move inward through the blastopore, the archenteron grows, gradually displacing the blastocoel.

As gastrulation comes to an end, the amphibian embryo consists of three germ layers: ectoderm on the outside, endoderm on the inside, and mesoderm in between. The embryo also has a dorsal–ventral and anterior–posterior organization. Most importantly, the fates of specific regions of the endoderm, mesoderm, and ectoderm have been determined. The beautiful experiments revealing how determination takes place in the amphibian embryo are an old but exciting story.

### The dorsal lip of the blastopore organizes embryo formation

In the early 1900s, the German biologist Hans Spemann was studying the development of salamander eggs. He was interested in finding out whether the nuclei of blastomeres remain capable of directing the development of complete embryos. With great patience and dexterity, he formed loops from single hairs taken from a baby (in fact, his daughter) and tied them around fertilized eggs along the plane of the first cell division, effectively dividing the eggs in half, with the nucleus restricted to one side. That side went through cell divisions and developed into a salamander; the other half simply degenerated. Up until the 16-cell stage, if one nucleus escaped to the other side of the constriction, twin salamanders could develop. Thus, each of the nuclei of the blastula (at least up to the 16-cell stage) was capable of directing and supporting development of the whole organism.

But, as often happens in science, Spemann's bisection experiments revealed a new phenomenon. Sometimes the half of the blastula receiving an escaped nucleus did not develop. When his loops bisected the gray crescent, both halves of the zygote developed into a complete embryo. When he tied the loops so the gray crescent was on only one side of the constriction, however, only that half of the zygote developed into a complete embryo (**Figure 44.9**). The half lacking gray crescent material underwent cell division, but even if it contained a nucleus, it became a clump of undifferentiated cells that Spemann called a "belly piece." Spemann hypothesized that cytoplasmic factors unequally distributed in the fertilized egg were necessary for gastrulation and the development of a normal salamander.

To further test the hypothesis that cells receiving different complements of cytoplasmic factors had different developmental fates, Spemann transplanted pieces of early gastrulas to various locations on other gastrulas. Guided by fate maps (see Figure 44.6), he was able to take a piece of ectoderm he knew would develop into skin and transplant it to a region that normally becomes part of the nervous system, and vice versa.

When he performed these transplants in early gastrulas—when the blastopore was just beginning to form—the transplanted pieces always developed into tissues that were appropriate for the location where they were placed. Transplanted cells destined to become epidermis in their original location developed into nervous system tissue, and transplanted cells destined to become nervous system tissue in their original location developed into host epidermis. Thus, Spemann learned that the fates of the transplanted cells had not been determined before the transplantation (see Figure 19.2).

In late gastrulas, however, the same experiment yielded opposite results. Transplanted cells destined to become epidermis in their original location produced patches of skin cells in the host

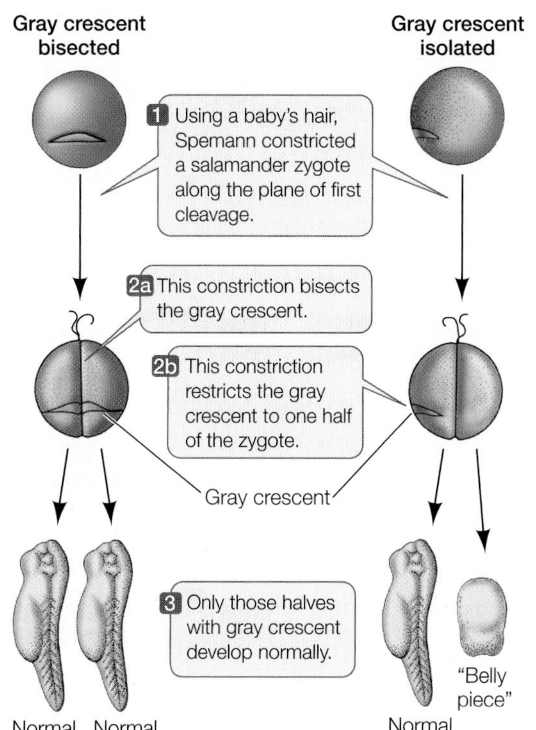
Gray crescent bisected
Gray crescent isolated
1 Using a baby's hair, Spemann constricted a salamander zygote along the plane of first cleavage.
2a This constriction bisects the gray crescent.
2b This constriction restricts the gray crescent to one half of the zygote.
Gray crescent
3 Only those halves with gray crescent develop normally.
"Belly piece"
Normal
Normal
Normal

**44.9 Gastrulation and the Gray Crescent** Spemann's research revealed that gastrulation and subsequent normal development in salamanders depends on cytoplasmic determinants localized in the gray crescent.

# INVESTIGATING LIFE

## 44.10 The Dorsal Lip Induces Embryonic Organization

In a classic experiment, Hans Spemann and Hilde Mangold transplanted the dorsal blastopore lip mesoderm of an early gastrula stage salamander embryo. The results showed that the cells of this embryonic region, which they dubbed "the organizer," could direct the formation of an entire embryo.

**HYPOTHESIS** **Cytoplasmic factors in the early dorsal blastopore lip organize cell differentiation in amphibian embryos.**

**METHOD**

1. Excise a patch of mesoderm tissue from above the dorsal blastopore lip of an early gastrula stage salamander embryo (the donor).
2. Transplant the donor tissue onto a recipient embryo at the same stage. The donor tissue is transplanted onto a region of ectoderm that should become epidermis (skin).

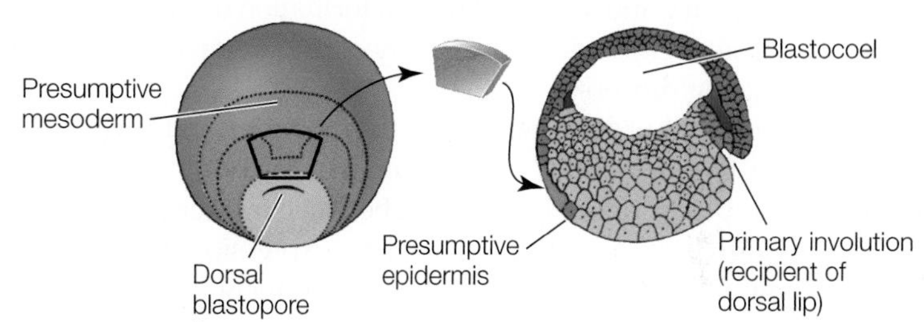
Presumptive mesoderm
Blastocoel
Dorsal blastopore lip
Presumptive epidermis
Primary involution (recipient of dorsal lip)

**RESULTS**

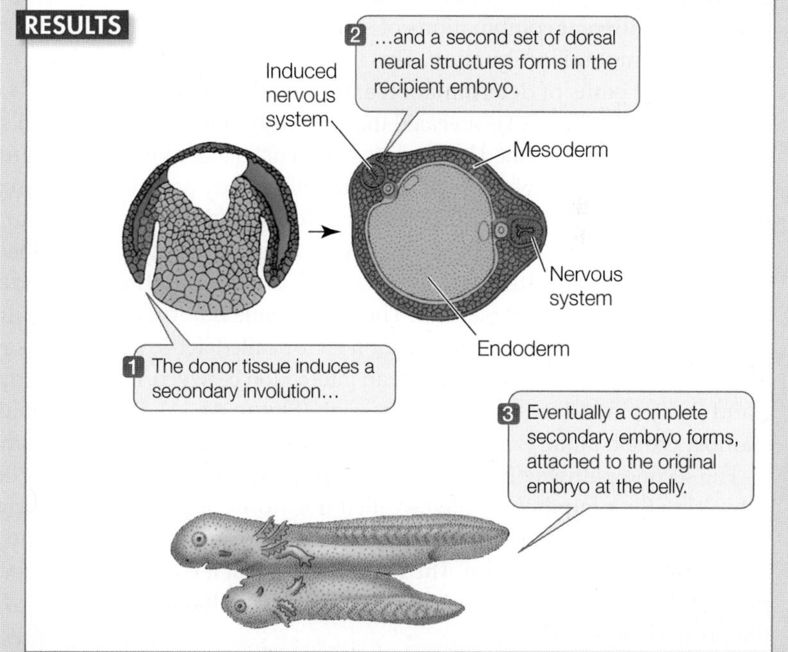
2 ...and a second set of dorsal neural structures forms in the recipient embryo.
Induced nervous system
Mesoderm
Nervous system
Endoderm
1 The donor tissue induces a secondary involution...
3 Eventually a complete secondary embryo forms, attached to the original embryo at the belly.

**CONCLUSION** **The cells of the dorsal blastopore lip can induce other cells to change their developmental fates.**

Go to **yourBioPortal.com** for original citations, discussions, and relevant links for all INVESTIGATING LIFE figures.

nervous system, and the transplanted cells from regions that would develop into nervous system tissue produced neural tissue in the skin of the recipient. At some point during gastrulation, the fates of the embryonic cells had become determined.

Spemann's next experiment, done with his student Hilde Mangold, produced momentous results: they transplanted the dorsal lip of the blastopore (**Figure 44.10**). When this small piece of tissue was transplanted into the presumptive belly area of another gastrula, it stimulated a second site of gastrulation—and a second complete embryo formed belly-to-belly with the original embryo. Because the dorsal lip of the blastopore was apparently capable of inducing the host tissue to form an entire embryo, Spemann and Mangold dubbed the dorsal lip tissue the **primary embryonic organizer**, or simply the **organizer**. For more than 80 years, the organizer has been an active area of research.

**yourBioPortal.com**

**GO TO Animated Tutorial 44.2 • Tissue Transplants Reveal the Process of Determination**

### Transcription factors underlie the organizer's actions

With the advent of modern molecular methods, the primary embryonic organizer has been studied intensively to discover the molecular mechanisms involved in its action. The distribution of the transcription factor β-catenin in the late blastula corresponds to the location of the organizer in the early gastrula, so β-catenin is a candidate for the initiator of organizer activity. To prove that a protein is an inductive signal, it has to be shown that it is both *necessary* and *sufficient* for the proposed effect. In other words, the effect should not occur if the candidate protein is not present (necessity), and the candidate protein should be capable of inducing the effect where it would otherwise not occur (sufficiency).

The criteria of necessity and sufficiency have been satisfied for β-catenin. If β-catenin mRNA transcripts are depleted by injections of antisense RNA into the egg (see Section 18.4), gastrulation does not occur. If β-catenin is experimentally overexpressed in another region of the blastula, it can induce a second axis of embryo formation, as the transplanted dorsal lip did in the Spemann–Mangold experiments. Thus, β-catenin appears to be both necessary and sufficient for the formation of the primary embryonic organizer—but it is only one component of a complex signaling process.

How the presence of β-catenin creates the organizer, and how the organizer then induces the beginnings of the body plan, involves a complex series of interactions between transcription factors and growth factors that control gene expression. What follows is only a portion of this complex and still emerging story. What you should take from this description is not the names of the genes and gene products involved. Rather, we hope you will gain a basic appreciation for how signaling molecules interact to produce different combinations of signals that convey positional and temporal information. This information guides cells into different paths of determination and differentiation.

Studies of early gastrulas revealed that primary embryonic organizer activity is generated by the interaction of β-catenin with signals coming from the vegetal cells. Together, they activate the expression of the transcription factor **Goosecoid**. Expression of the *goosecoid* gene depends on two signaling pathways.

The first of these pathways involves a *goosecoid*-promoting transcription factor called Siamois. The *siamois* gene is normally repressed by a ubiquitous transcription factor called Tcf-3, but in cells in which β-catenin is present, an interaction between Tcf-3 and β-catenin induces *siamois* expression (**Figure 44.11**). But Siamois protein alone is not sufficient for *goosecoid* expression.

The second pathway involves mRNAs from the original egg cytoplasm for a family of proteins called transforming growth factor-β (TGF-β). TGF-β interacts with the Siamois protein to control *goosecoid* transcription. Thus, you can see that it is a complex combination of factors that determines which cells become the primary organizer.

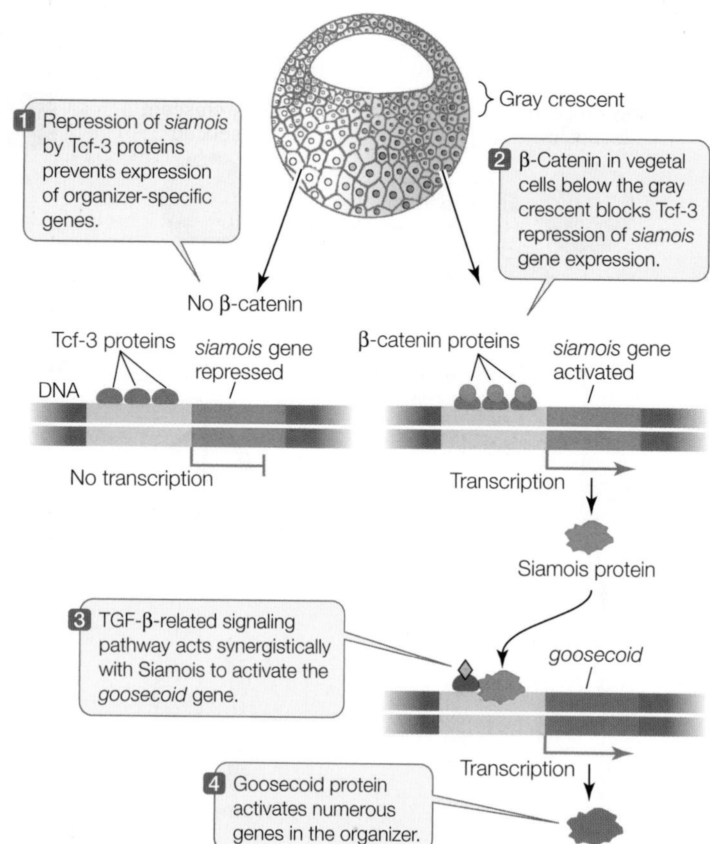


**44.11 Molecular Mechanisms of the Organizer** In amphibians, the organizing potential of the gray crescent depends on the activity of the *goosecoid* gene, which in turn is activated by signaling pathways set up in the vegetal cells below the gray crescent.

### The organizer changes its activity as it migrates from the dorsal lip

Organizer cells begin the process of formation of the dorsal lip of the blastopore. Specifically, these cells are at the center of the dorsal lip and involute, moving forward on the midline (i.e., the middle of the anterior–posterior axis). The first organizer cells to enter the embryo migrate anteriorly to become the head endoderm and head mesoderm. Here, they induce neighboring cells to participate in making structures of the head. Later organizer cells that involute into the embryo will induce structures of the trunk, and the last of the organizer cells to move inward from the dorsal lip will induce structures of the tail. How does the nature of the organizer cells change to enable them to induce head, trunk, or tail structures?

Inductive tissue interactions can suppress as well as activate. As we learned above, the early organizer cells express the transcription factor Goosecoid, which activates genes encoding soluble signals. As these cells move forward in the blastocoel, they come into contact with new populations of cells that produce a number of different growth factors. For head structures to form, certain of these growth factors have to be suppressed. The anteriormost organizer cells, under the influence of Goosecoid, produce and release antagonists to those growth factors.

The induction of trunk structures requires suppression of a different set of growth factors. In organizer cells that involute later than the head organizers, Goosecoid is no longer the dominant transcription factor, and these cells express different growth factor antagonists. The induction of tail structures requires still different activities of the organizer cells that involute last. Thus, the organizer cells express appropriate sets of growth factor antagonists at the right times to achieve different patterns of differentiation on the anterior–posterior axis.

The initiation of the development of the nervous system also involves a suppressive tissue interaction. For a long time it was thought that the involuting organizer cells actively induced the

INVESTIGATING LIFE

**44.12 Differentiation Can Be Due to Inhibition of Transcription Factors**

When organizer cells involute to underlie dorsal ectoderm along the embryo midline, that overlying ectoderm becomes neural tissue rather than skin (epidermis). But do the organizer cells cause dorsal ectoderm to become neural tissue, or do they *prevent* this ectoderm from becoming skin?

**HYPOTHESIS** **The default state of amphibian dorsal ectoderm is neural; it is induced by underlying mesoderm to become epidermis.**

**METHOD**

1. Excise the animal caps of late-stage frog blastulas and disperse the cells in culture medium so there is no cell-to-cell contact. From the culture, extract molecules of BMP4 (secreted by mesoderm cells) and molecules of an inhibitor of BMP4.

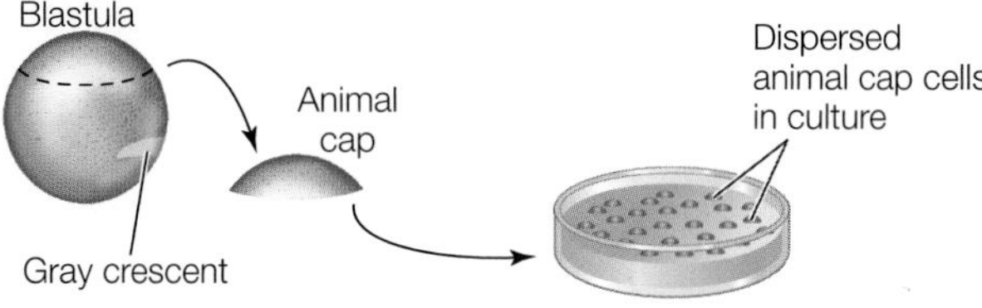


2. Prepare four separate cultures of embryonic ectodermal cells. Incubate with no additions (control); with BMP4 from step 1; with BMP4 inhibitor from step 1; and with both molecules.

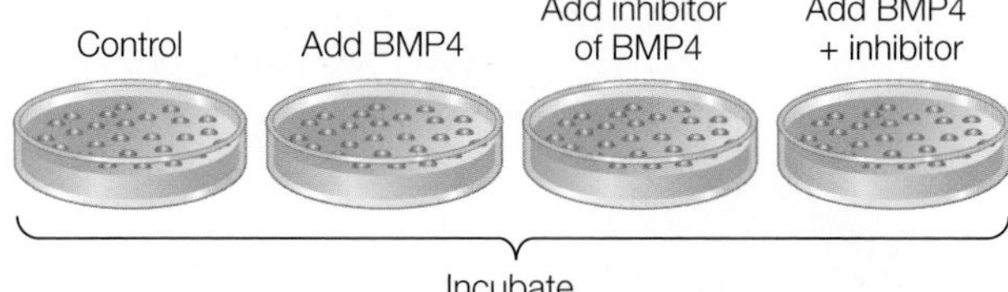


3. After incubation, extract mRNAs from the ectodermal cells and analyze for the presence of mRNAs for marker proteins NCAM (neural cell adhesion molecule, a neural protein) and/or keratin (an epidermal protein).

**RESULTS** The control ectoderm (no inductive factors added) expresses the neural marker. In the presence of mesodermal BMP4, ectoderm expresses the epidermal marker. If BMP4 is inhibited, ectoderm expresses the neural marker.

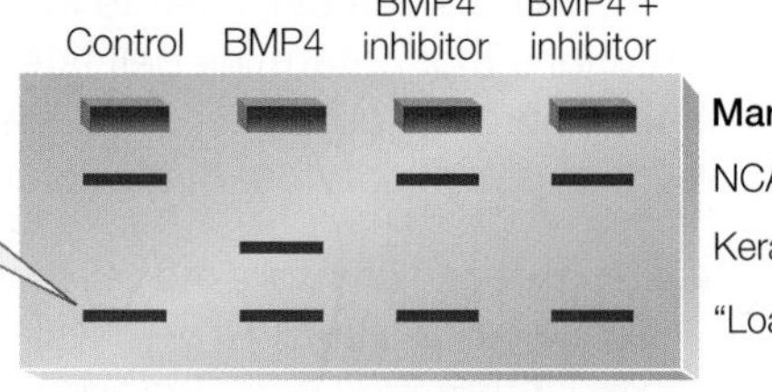


**CONCLUSION** **The default state of amphibian dorsal ectoderm is neural. BMP4 protein from mesoderm can induce ecotoderm cells to differentiate into epidermis. Thus the organizer cells must secrete an inhibitor of BMP4.**

Go to **yourBioPortal.com** for original citations, discussions, and relevant links for all INVESTIGATING LIFE figures.

overlying ectoderm to form neural tissue rather than becoming epidermis. We now know, however, that epidermis is not the default state of the dorsal ectoderm. Rather, the underlying mesoderm secretes factors called BMP proteins that induce the ectoderm to become epidermis. The role of the involuting organizer cells is to block that induction, allowing the overlying ectodermal cells to follow what is really their default pathway—differentiation into neural tissue (**Figure 44.12**).

### Reptilian and avian gastrulation is an adaptation to yolky eggs

The eggs of reptiles and birds contain a mass of yolk, and the blastulas of these groups develop as a disc of cells on top of the yolk (see Figure 44.3B). We will use the chicken egg to show how gastrulation proceeds in a flat disc of cells rather than in a ball of cells.

Cleavage in the chick results in a flat, circular layer of cells called a blastodisc (**Figure 44.13**). Between the blastodisc and the yolk mass is a fluid-filled space. Some cells from the blastodisc break free and move into this space. These cells come together to form a continuous layer called the **hypoblast**, which will later contribute to *extraembryonic membranes* that will support and nourish the developing embryo. The overlying cells make up the **epiblast**, from which the embryo proper will form. Thus, the avian blastula is a flattened structure consisting of an upper epiblast and a lower hypoblast, which are joined at the margins of the blastodisc. The blastocoel is the fluid-filled space between the epiblast and hypoblast.

Gastrulation begins with a thickening in the posterior region of the epiblast, caused by the movement of cells toward the midline and then forward along the midline (see Figure 44.13). The result is a midline ridge called the *primitive streak*. A depression called the *primitive groove* forms along the length of the primitive streak. The primitive groove functions as the blastopore, and cells migrate through it into the blastocoel to become endoderm and mesoderm.

In the chick embryo, no archenteron forms, but the endoderm and mesoderm migrate forward to form the gut and other structures. At the anterior end of the primitive groove is a thickening called **Hensen's node**, which in birds, reptiles, and mammals is the equivalent of the dorsal lip of the amphibian blastopore. Many signaling molecules that have been identified in the frog organizer are also expressed in Hensen's node. Some cells that pass over

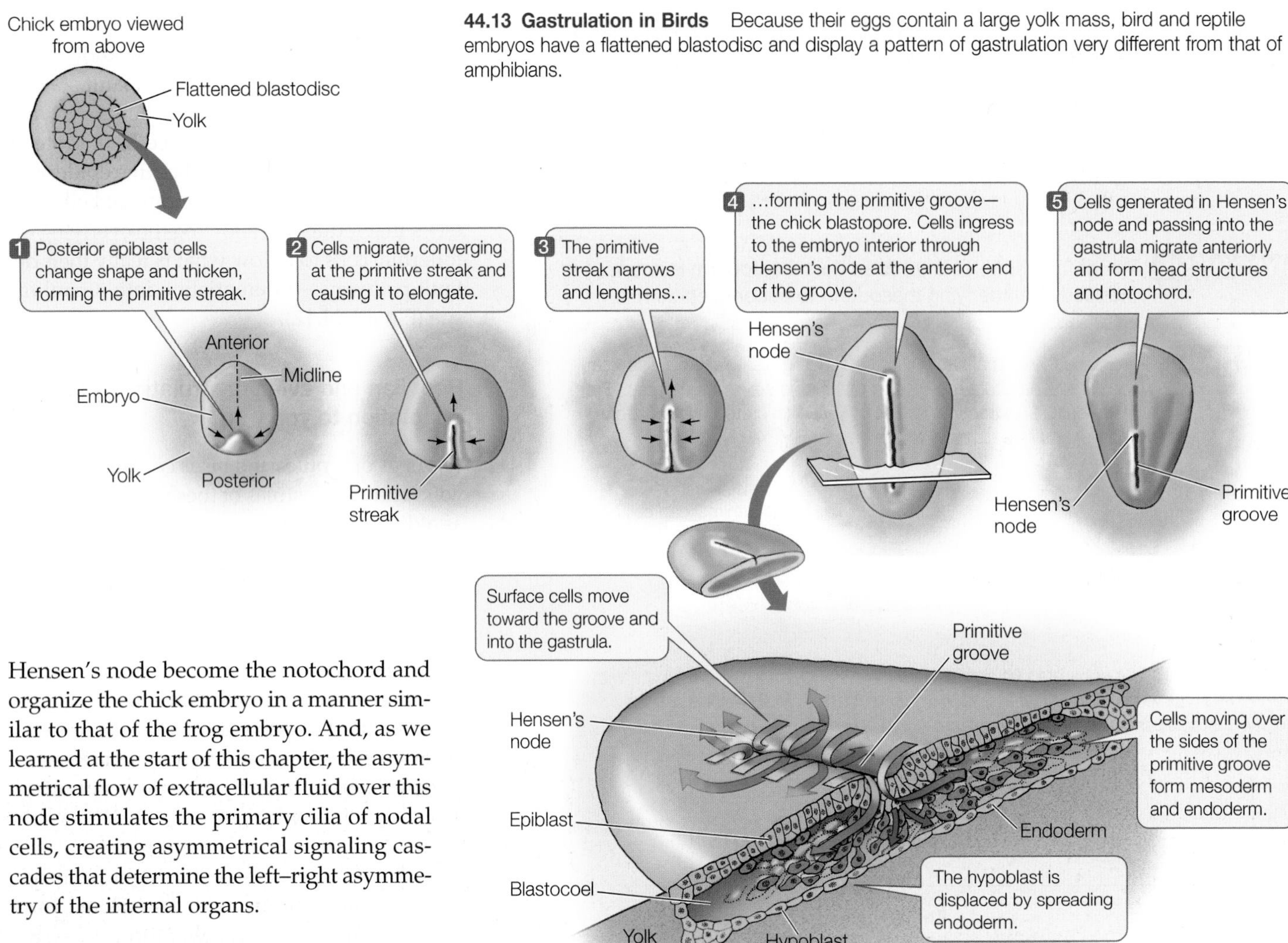

Chick embryo viewed from above
Flattened blastodisc
Yolk
1 Posterior epiblast cells change shape and thicken, forming the primitive streak.
2 Cells migrate, converging at the primitive streak and causing it to elongate.
3 The primitive streak narrows and lengthens…
4 …forming the primitive groove—the chick blastopore. Cells ingress to the embryo interior through Hensen's node at the anterior end of the groove.
5 Cells generated in Hensen's node and passing into the gastrula migrate anteriorly and form head structures and notochord.
Anterior
Midline
Embryo
Yolk
Posterior
Primitive streak
Hensen's node
Hensen's node
Primitive groove
Surface cells move toward the groove and into the gastrula.
Primitive groove
Hensen's node
Epiblast
Blastocoel
Yolk
Hypoblast
Endoderm
Cells moving over the sides of the primitive groove form mesoderm and endoderm.
The hypoblast is displaced by spreading endoderm.
Cross section through chick embryo

**44.13 Gastrulation in Birds** Because their eggs contain a large yolk mass, bird and reptile embryos have a flattened blastodisc and display a pattern of gastrulation very different from that of amphibians.

Hensen's node become the notochord and organize the chick embryo in a manner similar to that of the frog embryo. And, as we learned at the start of this chapter, the asymmetrical flow of extracellular fluid over this node stimulates the primary cilia of nodal cells, creating asymmetrical signaling cascades that determine the left–right asymmetry of the internal organs.

### Placental mammals retain the avian–reptilian gastrulation pattern but lack yolk

Mammalian embryos (with the exception of monotremes) derive their nourishment from the maternal circulation, and therefore mammalian eggs do not have large amounts of yolk to constrain their patterns of cleavage and early development. Nevertheless, mammals and birds evolved from reptilian ancestors, so it is not surprising that they share certain patterns of early development. Earlier we described the development of the mammalian inner cell mass (the equivalent of the avian epiblast) and the outer trophoblast.

As in avian development, in placental mammals the inner cell mass splits into an upper layer called the epiblast and a lower layer called the hypoblast. The embryo forms from the epiblast, while the hypoblast contributes to the extraembryonic membranes that will encase the developing embryo and help form the placenta (see Figure 44.5). The epiblast also contributes to the extraembryonic membranes; specifically, it splits off an upper layer of cells that will form the amnion. The amnion will grow to surround the developing embryo as a membranous sac filled with amniotic fluid. Gastrulation occurs in the mammalian epiblast just as it does in the avian epiblast. A primitive groove forms, and epiblast cells migrate through the groove to become layers of endoderm and mesoderm.

### 44.2 RECAP

**The cell movements of gastrulation convert the blastula into an embryo with three tissue layers. New contacts between cells set up inductive signaling interactions that determine cell fates. Dorsal lip tissue is the source of organizer cells that induce development of preliminary head, trunk, and tail structures.**

- Describe and compare the cell movements that occur during gastrulation in a sea urchin, a frog, and a bird. **See Figures 44.7, 44.8, and 44.13**
- Explain the molecular basis for the inductive capabilities of the organizer. **See pp. 931–932 and Figures 44.11 and 44.12**

We have described how the fertilized egg develops into an embryo with three germ layers and how cellular signals trigger different patterns of differentiation. In the next section we will describe how organs and organ systems develop.

# 44.3 How Do Organs and Organ Systems Develop?

Gastrulation produces an embryo with three germ layers that are positioned to influence one another through inductive tissue interactions. During the next phase of development, called **organogenesis**, many organs and organ systems develop simultaneously and in coordination with one another. An early process of organogenesis in chordates that is directly related to gastrulation is neurulation. **Neurulation** is the initiation of the nervous system. We will examine neurulation in the amphibian embryo, but it occurs in a similar fashion in reptiles, birds, and mammals.

## The stage is set by the dorsal lip of the blastopore

As we learned in the previous section, one group of cells that passes over the dorsal lip of the blastopore moves anteriorly and becomes the endodermal lining of the digestive tract. The other group of cells that involutes over the dorsal lip becomes *chordamesoderm*, so named because it forms a rod of mesoderm—the **notochord**—that extends down the center of the embryo. These cells also have important organizer functions (see Figure 44.8). The notochord gives structural support to the developing embryo and is eventually replaced by the vertebral column. The organizing capacity of the chordamesoderm enables the overlying ectoderm to become neural ectoderm (see Figure 44.12). It does this by expressing signaling molecules (one appropriately called Noggin and another one called Chordin) that initiate differentiation of the different divisions of the nervous system.

Neurulation involves the formation of an internal neural tube from an external sheet of cells. The first signs of neurulation are flattening and thickening of the ectoderm overlying the notochord; this thickened area forms the *neural plate* (**Figure 44.14A**). The edges of the neural plate that run in an anterior–posterior direction continue to thicken to form ridges or folds. Between these neural folds, a groove forms and deepens as the folds roll over it to converge on the midline. The folds fuse, forming a cylinder, the **neural tube**, and a continuous overlying layer of epidermal ectoderm (**Figure 44.14B–D**).

Cells from the most lateral portions of the neural plate do not become part of the neural tube, but disassociate from it and come to lie between the neural tube and the overlying epidermis. These **neural crest cells** migrate outward to lead the development of the connections between the central nervous system (brain and spinal cord) and the rest of the body.

The neural tube develops bulges at the anterior end, which become the major divisions of the brain; the rest of the tube becomes the spinal cord. In humans, failure of the neural folds to fuse in this posterior region results in a birth defect known as *spina bifida*. If the folds fail to fuse at the anterior end, an infant can develop without a forebrain—a condition called *anencephaly*. Although several genetic factors can cause these defects, other factors are environmental, including maternal diet. The incidence of neural tube defects in the United States in the early 1900s was as high as 1 in 300 live births; today it is less than 1 in 1,000. A major factor in this improvement has been the inclusion of folic acid (a B vitamin, also known as folate) in the mother's diet. It is essential for pregnant women to ingest sufficient folic acid.

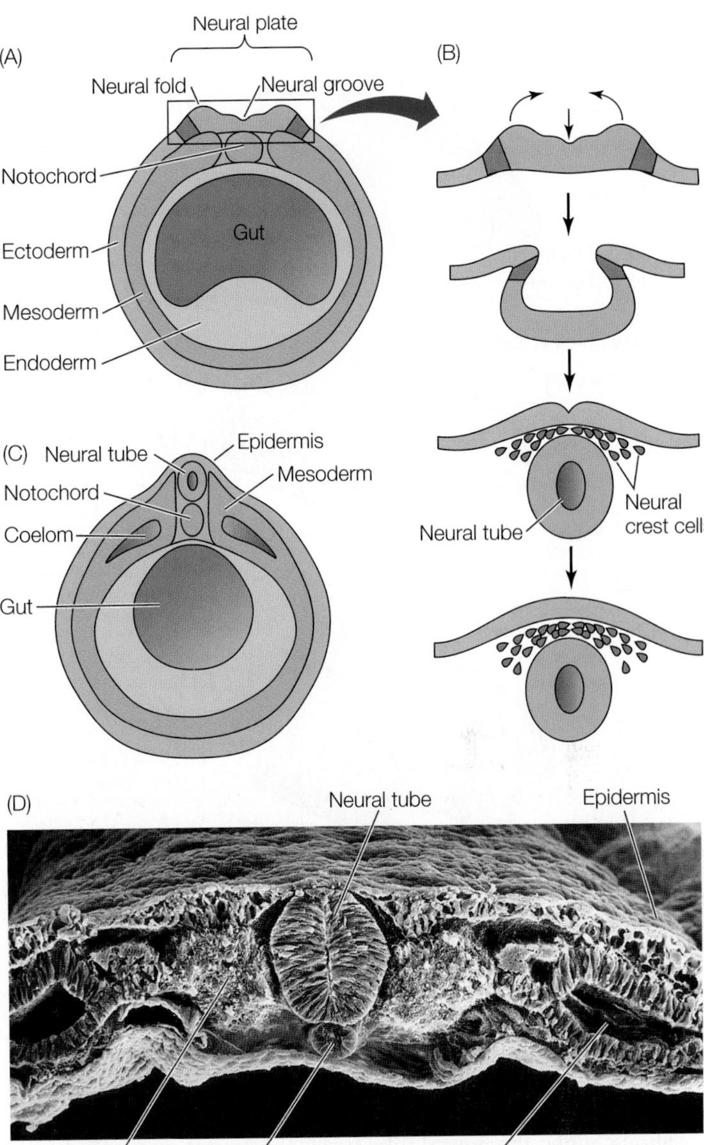


**44.14 Neurulation in a Vertebrate** (A) At the start of neurulation, the ectoderm of the neural plate (green) is flat. (B) The neural plate invaginates and folds, forming a tube. (C,D) The completely formed neural tube seen in (C) diagrammatic form and (D) in a scanning electron micrograph of a chick embryo.

## Body segmentation develops during neurulation

The vertebrate body plan, like that of arthropods, consists of repeating segments that are modified during development. These segments are most evident as the repeating patterns of vertebrae, ribs, nerves, and muscles along the anterior–posterior axis.

As the neural tube forms, mesodermal tissues gather along the sides of the notochord to form separate, segmented blocks

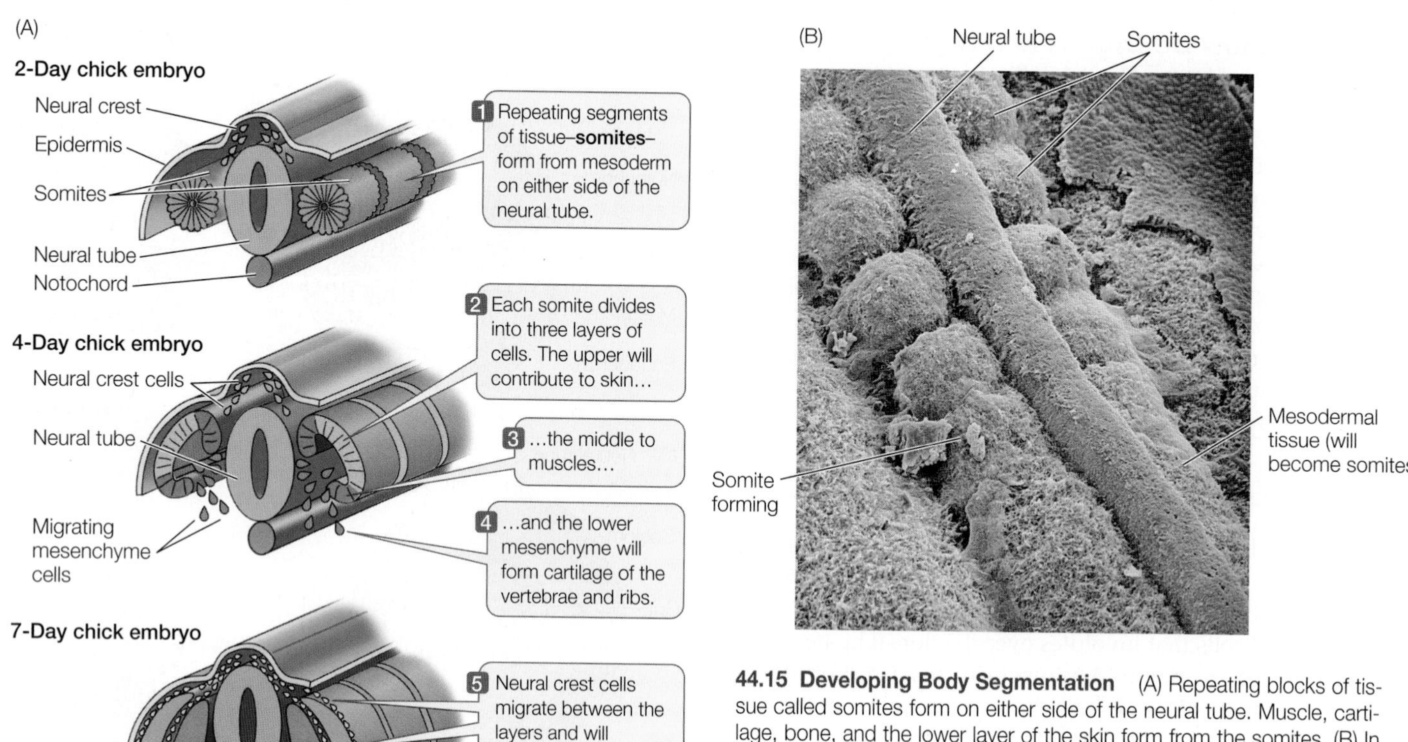
(A)
2-Day chick embryo
Neural crest
Epidermis
Somites
Neural tube
Notochord
1 Repeating segments of tissue–**somites**–form from mesoderm on either side of the neural tube.
4-Day chick embryo
Neural crest cells
Neural tube
Migrating mesenchyme cells
2 Each somite divides into three layers of cells. The upper will contribute to skin...
3 ...the middle to muscles...
4 ...and the lower mesenchyme will form cartilage of the vertebrae and ribs.
7-Day chick embryo
5 Neural crest cells migrate between the layers and will produce nerves and other tissue.
(B)
Neural tube
Somites
Somite forming
Mesodermal tissue (will become somites)

**44.15 Developing Body Segmentation** (A) Repeating blocks of tissue called somites form on either side of the neural tube. Muscle, cartilage, bone, and the lower layer of the skin form from the somites. (B) In this SEM of somite formation in a chick embryo, the overlying ectoderm has been removed and the neural tube and somites are seen from above.

of cells called **somites** (**Figure 44.15**). The somites produce cells that will become the vertebrae, ribs, muscles of the trunk and limbs, and the lower layer of the skin.

Nerves that connect the brain and spinal cord with tissues and organs throughout the body are also arranged segmentally. The somites help guide the organization of these peripheral nerves, but the nerves are not of mesodermal origin. As we saw above, when the neural tube fuses, the neural crest cells break loose and migrate inward between the epidermis and the somites and through the somites. These neural crest cells have diverse fates, including the development of peripheral nerves.

As development progresses, the different segments of the body change. Regions of the spinal cord differ, regions of the vertebral column differ in that some vertebrae grow ribs of various sizes and others do not, forelegs arise in the anterior part of the embryo, and hind legs arise in the posterior region.

## Hox genes control development along the anterior–posterior axis

How is mesoderm in the anterior part of a mouse embryo programmed to produce forelegs rather than hind legs? In Section 19.5, we saw how homeotic genes control body segmentation in *Drosophila*. We also learned that all homeotic genes contain a DNA sequence called the *homeobox*. Some of the genes directing gastrulation in the frog are homeobox genes—for example, *goosecoid* and *siamois*. In vertebrates, the homeotic genes that control differentiation along the anterior–posterior body axis are called **Hox genes**.

In mammals, four Hox gene complexes reside on different chromosomes in clusters of about 10 genes each. Remarkably, the temporal and spatial expression of these genes follows the same pattern as their linear order on their chromosome. That is, the Hox genes closest to the 3′ end of each gene complex are expressed first and in the anterior of the embryo. The Hox genes at the 5′ end of the gene complex are expressed later and in a more posterior part of the embryo. As a result, different segments of the embryo receive different combinations of Hox gene products, which serve as transcription factors (**Figure 44.16**; see also Figure 20.2).

Whereas Hox genes give cells information about their position on the anterior–posterior body axis, other genes provide information about their dorsal–ventral position. Tissues in each segment of the body differentiate according to their dorsal–ventral location. The notochord provides many of these signals. One example of a dorsal–ventral difference is seen in the spinal cord; sensory nerve connections develop in the dorsal region, and motor nerve connections in the ventral region. The protein Sonic hedgehog (named for a video-game character), which is expressed in the mammalian notochord, induces cells in the overlying neural tube (i.e., the ventralmost cells of the neural tube) to become motor neurons.

After the development of body segmentation, the formation of organs and organ systems progresses rapidly. The development of an organ involves extensive inductive interactions of the kind we saw in the example of the vertebrate eye (see Figure 19.10). These inductive interactions are a current focus of study for developmental biologists.

The genes closest to the 3′ end are expressed in the anteriormost positions...
...and those closest to the 5′ end are expressed more posteriorly.
b1 b2 b3 b4 b5 b6 b7 b8 b9
3′
5′
Hoxb
Expression gradients from anterior to posterior of embryo
For example, Hoxb1 is expressed in the hindbrain...
...and Hoxb9 in the spinal cord.
Hindbrain
Spinal cord
Midbrain
Forebrain
Cervical
Thoracic
Lumbar
Mouse embryo

**44.16 Hox Genes Control Body Segmentation** Hox genes are expressed along the anterior–posterior axis of the embryo in the same order as their arrangement between the 3′ and 5′ ends of the gene complex. As a result of gene duplication during evolution, vertebrates have four copies of the Hox gene complex shown.

## 44.3 RECAP

**Gastrulation sets up tissue interactions that initiate organogenesis. Neurulation is initiated by organizer mesoderm that forms the notochord.**

- Describe the formation of the neural tube in vertebrates. See p. 935 and Figure 44.14
- How do somites relate to segmentation of the body axis? See pp. 935–936 and Figure 44.15
- Using information from this chapter and from Chapters 19 and 20, explain what Hox genes are and how they instruct patterns of differentiation along the body axis. See Figures 19.19, 20.1, and 44.16

You may be aware that in mammals the circulatory systems of the fetus and mother are separate and that nourishment reaches the fetus through the placenta and the umbilical cord. In the next section we will examine the developmental events that result in the creation of the placenta.

# 44.4 How is the Growing Embryo Sustained?

There is more to a developing reptile, bird, or mammal than the embryo itself. As mentioned earlier, the embryos of these vertebrates are surrounded by several **extraembryonic membranes**, which originate from the embryo but are not part of it. Extraembryonic membranes function in nutrition, gas exchange, and waste removal. In mammals, they interact with tissues of the mother to form the placenta.

### Extraembryonic membranes form with contributions from all germ layers

The chicken provides a good example of how extraembryonic membranes form from the germ layers created during gastrulation. In the chick, four membranes form—the *yolk sac*, the *allantoic membrane*, the *amnion*, and the *chorion*. The **yolk sac** is the first to form, and it does so by extension of the hypoblast layer along with some adjacent mesoderm. The yolk sac grows to enclose the entire body of yolk in the egg (**Figure 44.17**). It constricts at the top to create a tube that is continuous with the gut of the embryo. However, yolk does not pass through this tube. Yolk is digested by the cells of the yolk sac, and the nu-

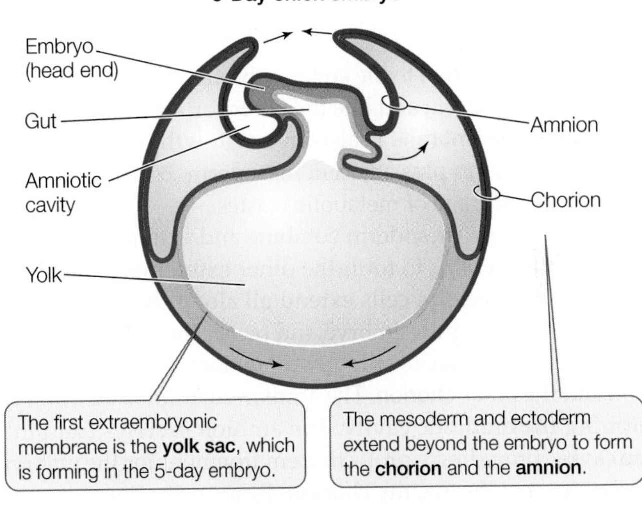
5-Day chick embryo
Embryo (head end)
Gut
Amniotic cavity
Yolk
Amnion
Chorion
The first extraembryonic membrane is the yolk sac, which is forming in the 5-day embryo.
The mesoderm and ectoderm extend beyond the embryo to form the chorion and the amnion.

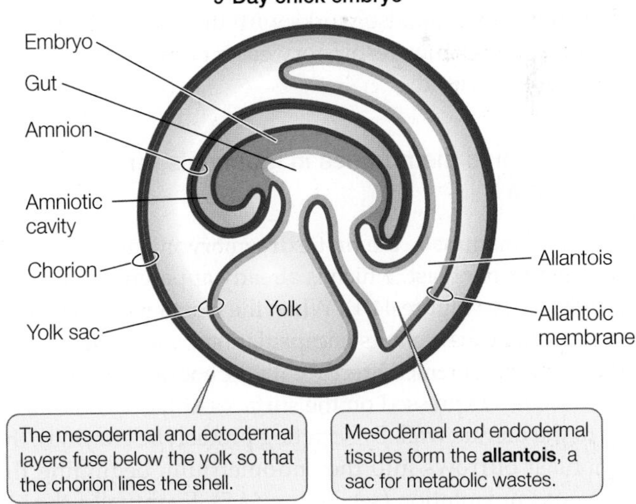
9-Day chick embryo
Embryo
Gut
Amnion
Amniotic cavity
Chorion
Yolk sac
Yolk
Allantois
Allantoic membrane
The mesodermal and ectodermal layers fuse below the yolk so that the chorion lines the shell.
Mesodermal and endodermal tissues form the allantois, a sac for metabolic wastes.

**44.17 The Extraembryonic Membranes** In birds, reptiles, and mammals, the embryo constructs four extraembryonic membranes. The yolk sac encloses the yolk, and the amnion and chorion enclose the embryo. Fluids secreted by the amnion fill the amniotic cavity, providing an aqueous environment for the embryo. The chorion, along with the allantoic membrane, mediates gas exchange between the embryo and its environment. The allantois stores the embryo's waste products.

**yourBioPortal.com**
GO TO **Web Activity 44.1 • Extraembryonic Membranes**

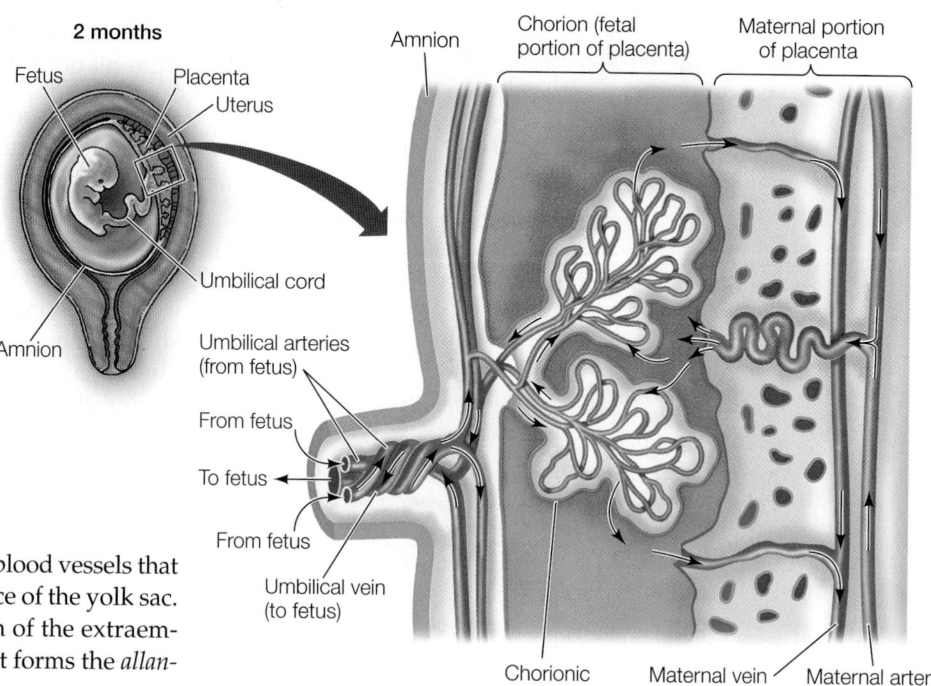


**44.18 The Mammalian Placenta** In humans and most other mammals, nutrients and wastes are exchanged between maternal and fetal blood in the placenta, which forms from the chorion and tissues of the uterine wall. The embryo is attached to the placenta by the umbilical cord. Embryonic blood vessels invade the placental tissue to form fingerlike chorionic villi. Maternal blood flows into the spaces surrounding the villi, and placental blood flows through the villi so nutrients and respiratory gases can be exchanged between the maternal and fetal blood.

trients are transported to the embryo through blood vessels that form from mesoderm and line the outer surface of the yolk sac. The **allantoic membrane** is also an outgrowth of the extraembryonic endoderm plus adjacent mesoderm. It forms the *allantois*, a sac for storage of metabolic wastes.

Ectoderm and mesoderm combine and extend beyond the limits of the embryo to form the other extraembryonic membranes. Two layers of cells extend all along the inside of the eggshell, both over the embryo and below the yolk sac. Where they meet, they fuse, forming two membranes, the inner **amnion** and the outer **chorion**. The amnion surrounds the embryo, forming the amniotic cavity. The amnion secretes fluid into the cavity, providing a protective environment for the embryo. The outer membrane, the chorion, forms a continuous membrane just under the eggshell (see Figure 44.17). It limits water loss from the egg and also works with the enlarged allantoic membrane to exchange respiratory gases between the embryo and the outside world.

### Extraembryonic membranes in mammals form the placenta

In placental mammals, the first extraembryonic membrane to form is the trophoblast, which is already apparent by the fifth cell division (see Figure 44.4). When the blastocyst reaches the uterus and hatches from its encapsulating zona pellucida, the trophoblast cells interact directly with the endometrium. Adhesion molecules expressed on the surfaces of these cells attach them to the uterine wall. By secreting proteolytic enzymes, the trophoblast burrows into the endometrium, beginning the process of implantation (see Figure 44.5). Eventually, the entire trophoblast is within the wall of the uterus. The trophoblast cells then send out numerous projections, or villi, to increase the surface area of contact with maternal blood.

Meanwhile, the hypoblast cells proliferate to form what in the bird would be the yolk sac. But there is no yolk in eggs of placental mammals, so the yolk sac contributes mesodermal tissues that interact with trophoblast tissues to form the chorion. The chorion, along with tissues of the uterine wall, produces the **placenta**, the organ that exchanges nutrients, respiratory gases, and metabolic wastes between the mother and the embryo (**Figure 44.18**).

At the same time the yolk sac is forming from the hypoblast, the epiblast produces the amnion, which grows to enclose the entire embryo in a fluid-filled amniotic cavity. The rupturing of the amnion and chorion and the loss of the **amniotic fluid** (a process called "water breaking") herald the onset of labor in humans.

An allantois also develops in mammals, but its importance depends on how well nitrogenous wastes can be transferred across the placenta. In humans the allantois is minor; in pigs it is important. In humans and other mammals, allantoic tissues contribute to the formation of the umbilical cord, by which the embryo is attached to the chorionic placenta. It is through the blood vessels of the umbilical cord that nutrients and oxygen from the mother reach the developing fetus, and wastes, including carbon dioxide and urea, are removed (see Figure 44.18).

## 44.4 RECAP

**The extraembryonic membranes of reptiles, birds, and mammals sustain the growing embryo. In reptiles and birds, these membranes surround the embryo within the shelled egg. In mammals the extraembryonic membranes form the placenta, an organ that exchanges nutrients, respiratory gases, and metabolic wastes between the mother and the embryo.**

- Describe each of the four extraembryonic membranes and their functions in the developing chick egg. See pp. 937–938 and Figure 44.17
- Explain the role of the trophoblast in the early development of a mammalian embryo. See p. 938

## 44.5 What Are the Stages of Human Development?

In humans, **gestation**, or pregnancy, lasts about 266 days, or 9 months. In smaller mammals gestation is shorter—for example, 21 days in mice—and in larger mammals it is longer—for example, 330 days in horses and 600 days in elephants. The events of human gestation can be divided into three periods of roughly 3 months each, called *trimesters*.

### Organ development begins in the first trimester

Implantation of the human blastocyst begins about 6 days after fertilization. After implantation, gastrulation occurs, tissues differentiate, the placenta forms, and organs begin to develop. The heart begins to beat during week 4, and limbs are formed by week 8 (**Figure 44.19 A,B**). By the end of the first trimester, most organs have started to form. The embryo is about 8 centimeters long and weighs about 40 grams (less than 2 ounces); it would fit neatly in a teaspoon. At about this point in time, the human embryo is medically and legally referred to as a **fetus**. (This distinction is not made for other mammals; developing mice, for example, remain embryos until they are born.)

The first trimester is a time of rapid cell division and tissue differentiation. Signal transduction cascades and the resulting branching sequences of developmental processes are in their early stages. Therefore, the first trimester is the period during which the embryo is most sensitive to damage from radiation, drugs, chemicals, and pathogens that can cause birth defects. An embryo can be damaged before the mother even knows she is pregnant. A classic and tragic case is that of thalidomide, a drug widely prescribed in Europe in the late 1950s to treat nausea. Women who took this drug in the fourth and fifth weeks of pregnancy, when the embryo's limbs are beginning to form, gave birth to children with missing or severely malformed arms and legs.

### Organ systems grow and mature during the second and third trimesters

During the second trimester the fetus grows rapidly to a weight of about 600 grams. The limbs of the fetus elongate, and the fingers, toes, and facial features become well formed (**Figure 44.19C**). Eyebrows and fingernails grow and the fetus's nervous system develops rapidly. Fetal movements are first felt by the mother early in the second trimester, and they become progressively stronger and more coordinated.

The fetus grows rapidly during the third trimester (**Figure 44.19D**). As the trimester approaches its end, internal organs mature. The digestive system begins to function, the liver stores glycogen, the kidneys produce urine, and the brain undergoes cycles of sleep and waking. A human infant is born as soon as the last of its critical organs—the lungs—mature.

Although the first-trimester embryo is the most susceptible to adverse effects of drugs, chemicals, and diseases, the potential for serious effects from exposure to environmental factors

(A) 4 weeks

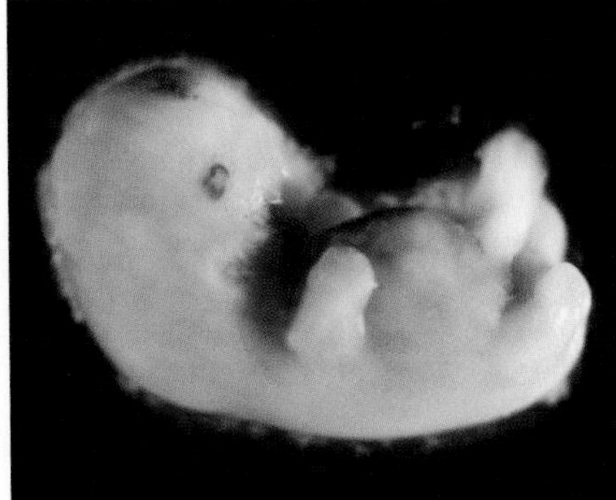

Actual length ~0.4 cm (4 mm)

(B) 8 weeks

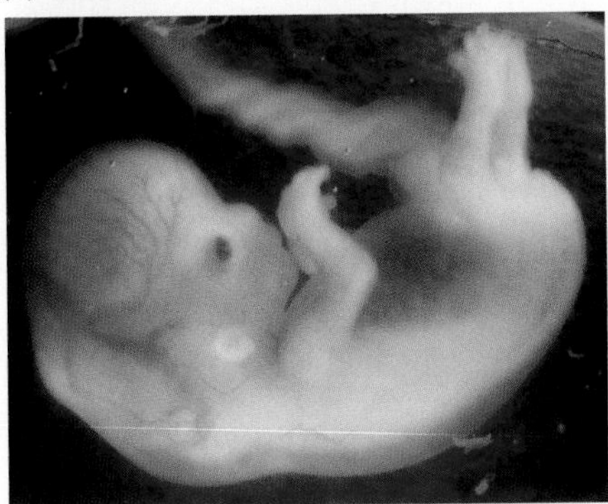

Actual length ~3 cm

(C) 4 months

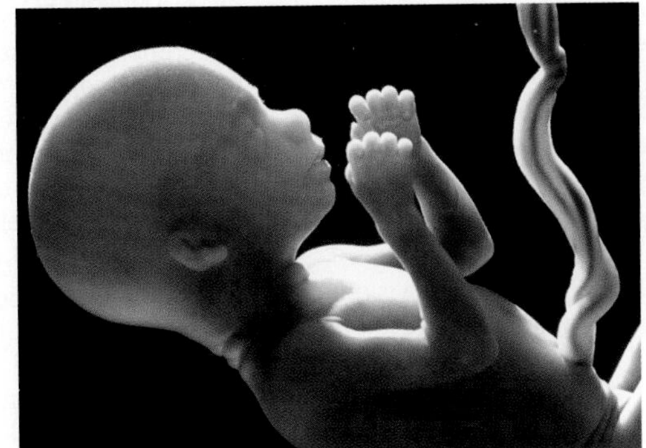

Actual length ~10 cm

(D) 9 months

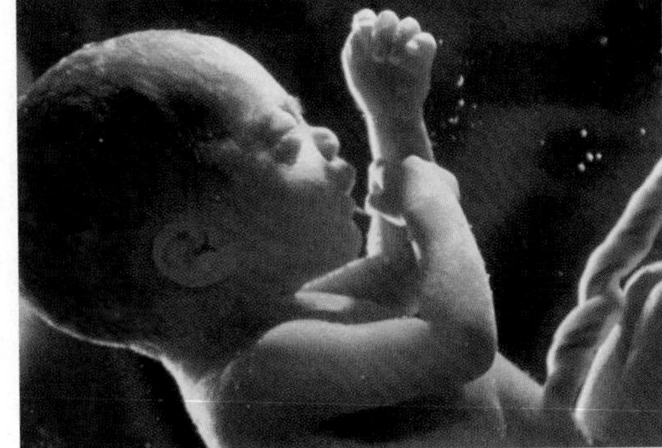

Actual length ~40 cm

**44.19 Stages of Human Development** (A) At 4 weeks of gestation, most of the embryo's organ systems have been formed and the heart is beating. (B) The body structures of this 8-week-old embryo are forming rapidly, and it is visibly a male. The umbilical cord attaches the embryo to the placenta (upper left). (C) At 4 months, the fetus has fully formed limbs with fingers and toes, and moves freely within the amniotic cavity. (D) This fetus is well along in its ninth month. Soon its lungs will be mature enough to trigger the onset of contractions and birth.

exists throughout pregnancy and continues after birth. Severe protein malnutrition, alcohol, and cigarette smoke are examples of factors that can cause low birth weight, mental retardation, and other developmental complications.

### Developmental changes continue throughout life

Development does not end with birth. Obviously, growth continues until adult size is reached, and even when growth stops, organs of the body continue to repair and renew themselves through cycles of cell replacement by the progeny of undifferentiated stem cells. In humans especially, enormous developmental changes occur in the brain in the years between birth and adolescence. Especially in the early years, there is a great deal of plasticity in the organization of the nervous system as the connections between neurons develop.

For example, a child born with misaligned eyes (a condition known as *strabismus*) will use mostly one eye. The connections to the brain from that eye will become strong while connections from the other eye remain weak, and the child will develop with reduced visual acuity and depth perception. If eye alignment is corrected in the first 3 years of life, the connections between the eyes and the brain can improve and the child is likely to develop normal vision. After the age of 3, correcting the connections between the eyes and the brain is less likely to result in improvement, and visual impairments may persist. Thus plasticity in human visual system development declines during early childhood. However, recent data indicate that it is not lost entirely and may be reactivated even in adulthood.

### 44.5 RECAP

**Human gestation lasts 9 months and can be divided into 3 trimesters. At the end of the first trimester, the fetus is very small but most of its organs have begun to form. In the second trimester, limbs elongate and the fetus moves. By the end of the third trimester, most organs have begun to function.**

- Why is a first-trimester embryo particular sensitive to environmental risks? See p. 939

## CHAPTER SUMMARY

### 44.1 How Does Fertilization Activate Development?

- The sperm and the egg contribute differentially to the zygote. The sperm contributes a haploid nucleus and, in most species, a centriole. The egg contributes a haploid nucleus, nutrients, ribosomes, mitochondria, mRNAs, and proteins.
- In amphibians, the cytoplasmic contents of the egg are not distributed homogeneously, and they are rearranged after fertilization to set up the major axes of the future embryo. The nutrient molecules are generally found in the **vegetal hemisphere**, whereas the nucleus is found in the **animal hemisphere**. Review Figures 44.1 and 44.2
- **Cleavage** is a period of rapid cell division. Except in mammals, little if any gene expression occurs during cleavage. Cleavage can be complete or incomplete, and the pattern of cell divisions depends on the orientation of the mitotic spindles. The result of cleavage is a ball or mass of cells called a **blastula**. Review Figure 44.3
- Early cell divisions in mammals are unique in being slow and allowing for gene expression early in the process. These cell divisions produce a **blastocyst** composed of an **inner cell mass** that becomes the embryo and an outer cell mass that becomes the **trophoblast**. At the time of **implantation**, the trophoblast secretes molecules that help the blastocyst implant in the uterine wall. Review Figures 44.4 and 44.5
- A **fate map** can be created by labeling specific **blastomeres** and observing what tissues and organs are formed by their progeny. Review Figure 44.6
- Some species undergo **mosaic development**, in which the fate of each cell is determined during early divisions. Other species, including vertebrates, undergo **regulative development**, in which remaining cells can compensate for cells lost in early cleavages.

### 44.2 How Does Gastrulation Generate Multiple Tissue Layers?

- **Gastrulation** involves massive cell movements that produce three **germ layers** and place cells from various regions of the blastula into new associations with one another. Review Figure 44.7, **ANIMATED TUTORIAL 44.1**
- The initial step of sea urchin and amphibian gastrulation is inward movement of certain blastomeres. The site of inward movement becomes the **blastopore**. Cells that move into the blastula become the **endoderm** and **mesoderm**; cells remaining on the outside become the **ectoderm**. Cytoplasmic factors in the vegetal pole cells are essential to initiate development. Review Figures 44.7 and 44.8
- The **dorsal lip** of the amphibian blastopore is a critical site for cell determination. It has been called the **primary embryonic organizer** because it induces determination in cells that pass over it during gastrulation. Review Figures 44.8, 44.9, and 44.10, **ANIMATED TUTORIAL 44.2**
- The protein β-catenin activates a signaling cascade that induces the primary embryonic organizer and sets up the anterior–posterior body axis. Review Figures 44.2, and 44.11
- Gastrulation in reptiles and birds differs from that in sea urchins and frogs because the large amount of yolk in reptile and bird eggs causes the blastula to form a flattened disc of cells. Review Figure 44.13
- Although their eggs have no yolk, placental mammals have a pattern of gastrulation similar to that of reptiles and birds.

### 44.3 How Do Organs and Organ Systems Develop?

- Gastrulation is followed by **organogenesis**, the process whereby tissues interact to form organs and organ systems.

- In the formation of the vertebrate nervous system, one group of cells that migrates over the blastopore lip is determined to become the **notochord**. The notochord organizes the overlying ectoderm to thicken, form parallel ridges, and fold in on itself to form a **neural tube** below the epidermal ectoderm. The nervous system develops from this neural tube. Review Figure 44.14
- The notochord and **neural crest cells** participate in the segmental organization of mesoderm into structures called **somites** along the body axis. Rudimentary organs and organ systems form during these stages. Review Figure 44.15
- In vertebrates, **Hox genes** determine the pattern of anterior–posterior differentiation along the body axis in mammals. Other genes, such as *sonic hedgehog*, contribute to dorsal–ventral differentiation. Review Figure 44.16

## 44.4 How is the Growing Embryo Sustained?

- The embryos of reptiles, birds, and mammals are protected and nurtured by four **extraembryonic membranes**. In birds and reptiles, the **yolk sac** surrounds the yolk and provides nutrients to the embryo, the **chorion** lines the eggshell and participates in gas exchange, the **amnion** surrounds the embryo and encloses it in an aqueous environment, and the **allantois** stores metabolic wastes. Review Figure 44.17, **WEB ACTIVITY 44.1**
- In mammals, the chorion and the trophoblast cells interact with the maternal uterus to form a **placenta**, which provides the embryo with nutrients and gas exchange. The amnion encloses the embryo in an aqueous environment. Review Figure 44.18

## 44.5 What Are the Stages of Human Development?

- Human pregnancy, or **gestation**, can be divided into three trimesters. The embryo forms in the first trimester; during this time, it is most vulnerable to environmental factors that can lead to birth defects. During the second and third trimesters the **fetus** grows, the limbs elongate, and the organ systems mature.
- Development continues throughout childhood and throughout life.

# SELF-QUIZ

1. Fertilization involves all of the following *except*
   *a.* equal contributions of cell organelles from sperm and egg.
   *b.* joining of sperm and egg haploid nuclei.
   *c.* induction of rearrangements of the egg cytoplasm.
   *d.* sperm binding to specific sites on the egg surface.
   *e.* metabolic activation of the egg.
2. Which of the following does *not* occur during cleavage in frogs?
   *a.* A high rate of mitosis
   *b.* Reduction in the size of cells
   *c.* Expression of genes critical for blastula formation
   *d.* Orientation of cleavage planes at right angles
   *e.* Unequal division of cytoplasmic determinants
3. How does cleavage in mammals differ from cleavage in frogs?
   *a.* Slower rate of cell division
   *b.* Formation of tight junctions
   *c.* Expression of the embryo's genome
   *d.* Early separation of cells that will not contribute to the embryo
   *e.* All of the above
4. Which statement about gastrulation is *true*?
   *a.* In frogs, gastrulation begins in the vegetal hemisphere.
   *b.* In sea urchins, gastrulation produces the notochord.
   *c.* In birds, cells from the surface of the blastodisc move down through the primitive groove to form the hypoblast.
   *d.* In mammals, gastrulation occurs in the hypoblast.
   *e.* In sea urchins, gastrulation produces only two germ layers.
5. Which of the following was a conclusion from the experiments of Spemann and Mangold?
   *a.* Cytoplasmic determinants of development are homogeneously distributed in the amphibian zygote.
   *b.* In the late blastula, certain regions of cells are determined to form skin or nervous tissue.
   *c.* The dorsal lip of the blastopore can be isolated and will form a complete embryo.
   *d.* The dorsal lip of the blastopore can initiate gastrulation.
   *e.* The dorsal lip of the blastopore gives rise to the neural tube.
6. Which of the following is true of human development?
   *a.* Most organs begin to form during the second trimester.
   *b.* Gastrulation takes place in the oviducts.
   *c.* Genetic diseases can be detected by sampling cells from the chorion.
   *d.* Implantation occurs through interactions of the zona pellucida with the uterine lining.
   *e.* Exposure to drugs and chemicals is most likely to cause birth defects when it occurs in the third trimester.
7. Which of the following characterizes neurulation?
   *a.* The notochord forms a neural tube.
   *b.* The neural tube is formed from ectoderm.
   *c.* A neural tube forms around the notochord.
   *d.* The neural tube forms somites.
   *e.* In birds, the neural tube forms from the primitive groove.
8. Which statement about trophoblast cells is *true*?
   *a.* They are capable of producing monozygotic twins.
   *b.* They are derived from the hypoblast of the blastocyst.
   *c.* They are endodermal cells.
   *d.* They secrete proteolytic enzymes.
   *e.* They prevent the zona pellucida from attaching to the oviduct.
9. Which of the following is part of the embryonic contribution to placenta formation?
   *a.* Amnion
   *b.* Chorion
   *c.* Ectoderm
   *d.* Allantois
   *e.* Zona pellucida
10. When is the developing human most susceptible to the occurrence of birth defects from radiation or chemical insults?
    *a.* At the time of birth
    *b.* During the third trimester
    *c.* During the first trimester
    *d.* When it is a zygote
    *e.* During the final stages of organ formation

## FOR DISCUSSION

1. If you found a protein that was localized to a small group of cells in the frog blastula, how would you determine whether that protein played a role in development? Address the issues of sufficiency and necessity.
2. During gastrulation in birds, the *sonic hedgehog* gene is expressed only on the left side of Hensen's node. What might be the cause of this expression pattern, and what is its significance?
3. Much of the early work of describing animal development was done on sea urchins, amphibians, and chickens. Most of the recent work on the molecular mechanisms of animal development has been done on nematodes, fruit flies, zebrafish, and mice. Why do you think there has been a shift in the animal models used by developmental biologists?
4. If all the mitochondria and mitochondrial DNA in the embryo come from the egg, what implications does this have for using mitochondrial DNA for molecular evolutionary studies?
5. There is currently much controversy over therapeutic cloning as a way of obtaining embryonic stem cells to treat diseases. Given that human development is regulative—in other words, twinning can occur if an early blastocyst is divided into two cell masses—can you think of a way to guarantee a source of isogenic (i.e., identically matching a person's own body) stem cells for an individual without resorting to therapeutic cloning? Assume isolated cells can be preserved indefinitely in a frozen state.

## ADDITIONAL INVESTIGATION

It is hypothesized that the differential development of the different body segments is due to the differential expression of Hox genes along the anterior–posterior body axis. For example, in mammals, ribs develop in anterior body trunk segments but not in posterior segments. Using the mouse as a model, how would you test this hypothesis?

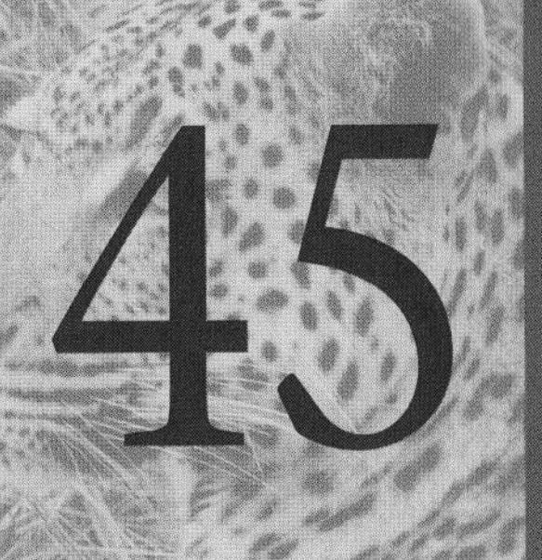

# 45 Neurons and Nervous Systems

## Fear and survival in the brain

Charles Whitman was a normal and responsible child. He became the youngest Eagle Scout in the country. He was a fine son and husband and received commendations as a U.S. Marine. While in the service, however, he began having unexplained fits of anger and other personality disorders. He was discharged from the Marines and entered the University of Texas. Several times he visited campus doctors and complained about having violent thoughts. Then, on August 1, 1966, after killing his wife and mother, Whitman barricaded himself inside the top floor of the clock tower on campus along with several high-powered rifles. From this vantage point, he killed 14 people and wounded 38 others before being shot and killed by Austin police. In his suicide note, he requested that proceeds from his insurance be donated to a mental health foundation. He also requested an autopsy, and that autopsy revealed a tumor pressing on his amygdala.

The *amygdala* (Latin for "almond") is the brain's center for the emotion and memory of fear. When the cells of this structure are activated, your heart beats faster, your breathing becomes rapid and shallow, and your hands get cold and clammy. If you encounter a threatening situation, your amygdala is activated. If you are alone at night and hear a strange noise, your amygdala is activated. If you are faced with attempting something physically dangerous, your amygdyla is activated. Without an amygdala, you would never be scared—and *not* being scared could be hazardous to your health.

A rare case of brain damage left a woman without a functional amygdala. When shown pictures of faces registering different emotions, she could not pick out the ones that were threatening or scary. She could not recall ever having a frightening experience. In tests where she was administered mild electrical shocks, she developed no anticipatory fear; even though she knew that seeing a red card meant she was about to receive a shock, she never reacted to the red card. In real life, she would not have the reflex to pull away from a threat.

People with a damaged amygdala frequently have trouble engaging in normal social relationships. They cannot "read" the nature, mood, or intentions of other people by looking at their faces. The pres-

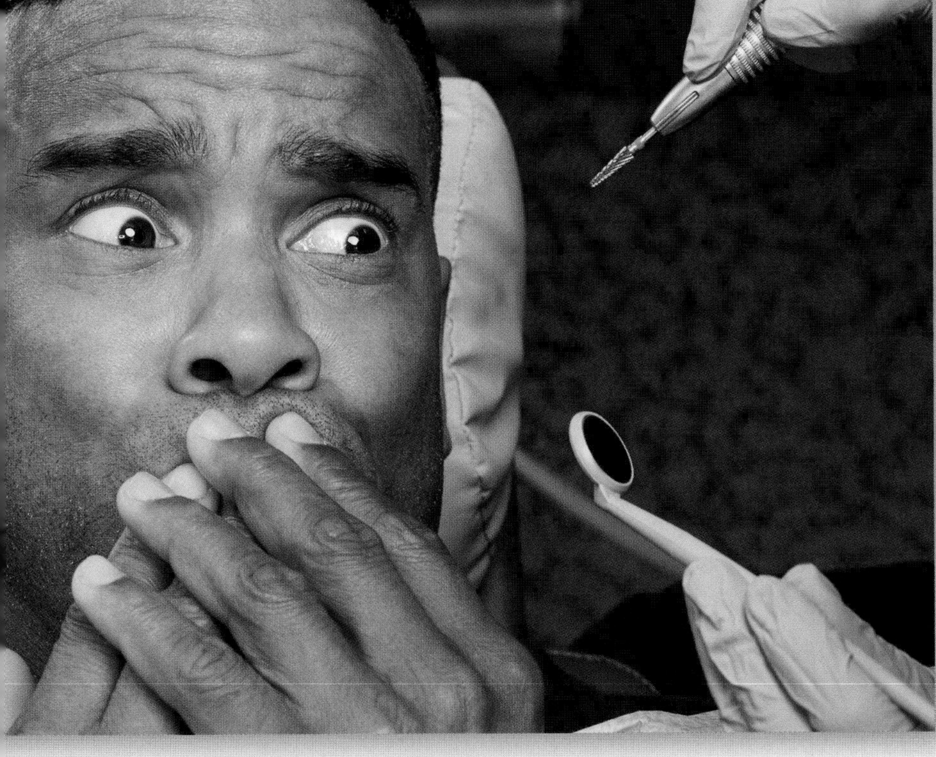

**Fear Factor** The fear response—muscle tension, racing heart, cold sweat—kicks in when you are faced with potential pain, even if that experience ultimately will be beneficial.

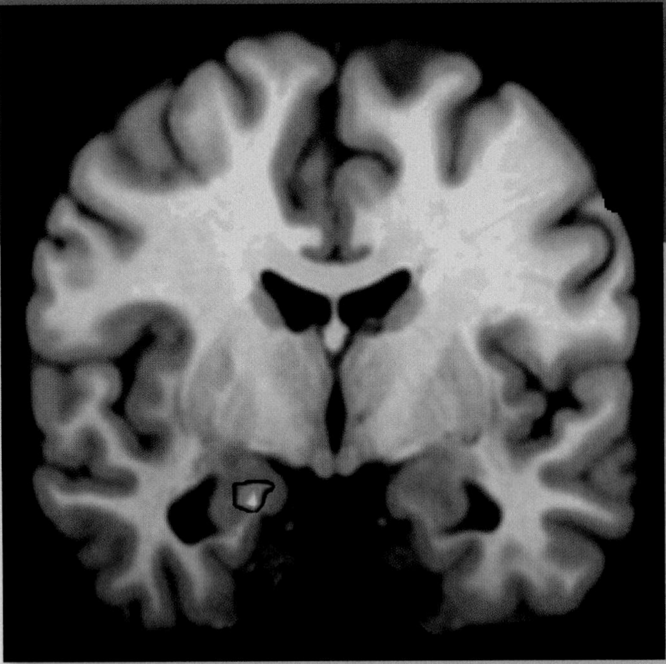

**Source of the Fear Response** Frightening situations—or even memories of such a situation—activate a brain region called the amygdala, as shown in this functional magnetic resonance image (fMRI) of the brain of a person experiencing fear.

sure on Charles Whitman's amygdala may have been a factor in the emotions that drove him to mass murder; this diagnosis remains a matter for medical speculation.

Our nervous systems enable us to experience the world around us and to react to it. But in between sensing and reacting, there is much interpretation based on memory, learning, emotions, and beliefs—all of which are based on the activities of cells in the nervous system. To understand how our eyes see, how our fingers play the piano, or how our emotions affect our behavior, we have to understand how cells in different parts of our nervous systems work and interact.

**IN THIS CHAPTER** we will first describe the cells that make up nervous systems—neurons and glia —and explain how neurons process and transmit information by generating and transmitting electrical signals called action potentials. We will examine the membrane properties and events underlying action potentials and the mechanisms of cell-to-cell communication where neurons come together. Finally, we will consider how changes in the properties of neurons might explain learning and memory.

CHAPTER OUTLINE



## 45.1 What Cells Are Unique to the Nervous System?

Nervous systems are composed of two types of cells: *nerve cells*, or **neurons,** and *glial cells*, or **glia** (see Figure 40.6). Neurons are *excitable*: they can generate and transmit electrical signals, which are known as nerve impulses, or **action potentials**. Many neurons have a long extension called an **axon** that enables them to conduct action potentials over long distances. Glia do not conduct action potentials; rather, they support neurons physically, immunologically, and metabolically. A **nerve** (as distinct from a neuron) is a bundle of axons that come from many different neurons. Many axons are wrapped by glia to electrically isolate them and increase their speed of conduction of action potentials.

Nervous systems can process information because their neurons are organized into **neural networks**. These networks include three functional categories of neurons, which can be thought of as being involved with input, output, and integration:

- **Afferent neurons** carry sensory information into the nervous system. That information comes from specialized **sensory neurons** that transduce (convert) various kinds of sensory stimuli (e.g., light, heat, pressure) into action potentials.
- **Efferent neurons** carry commands to physiological and behavioral *effectors* such as muscles and glands.
- **Interneurons** integrate and store information and communicate between afferent and efferent neurons.

Neural networks can be simple, like the one that causes your leg to kick when a physician taps your knee; or they can be exceedingly complex, like the network that enables you to read and remember this chapter.

### Neural networks range in complexity

Simple animals such as cnidarians (e.g., sea anemones) can process information with simple neural networks that do little more than provide direct lines of communication from sensory cells to effectors; there is little or no integration or processing of signals (**Figure 45.1A**). The cnidarian's *nerve net* is most developed around the tentacles and the oral opening, where it facilitates detection of food or danger and causes tentacles to extend or retract.

Animals that are more complex and actively move about in search for food and mates need to process and integrate larger amounts of information. Even earthworms fit this description, and their increased need for information processing is met by higher numbers of neurons organized into clusters called **ganglia**. Ganglia serving different functions may be distributed around the body, as in earthworms or squid (**Figure 45.1B,C**).

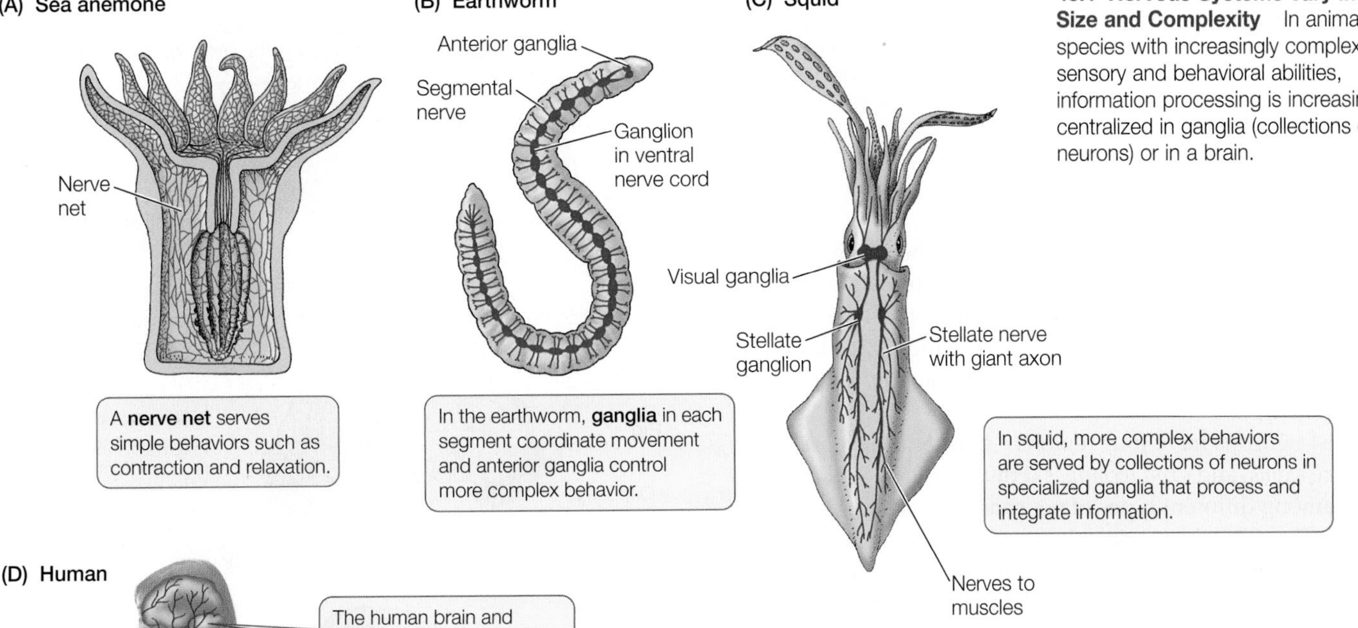
(A) Sea anemone
Nerve net
A **nerve net** serves simple behaviors such as contraction and relaxation.
(B) Earthworm
Anterior ganglia
Segmental nerve
Ganglion in ventral nerve cord
In the earthworm, **ganglia** in each segment coordinate movement and anterior ganglia control more complex behavior.
(C) Squid
Visual ganglia
Stellate ganglion
Stellate nerve with giant axon
Nerves to muscles
In squid, more complex behaviors are served by collections of neurons in specialized ganglia that process and integrate information.
(D) Human
The human brain and spinal cord are the **central nervous system**...
...which communicates to the cells and organs of the body via the **peripheral nervous system**.

**45.1 Nervous Systems Vary in Size and Complexity** In animal species with increasingly complex sensory and behavioral abilities, information processing is increasingly centralized in ganglia (collections of neurons) or in a brain.

In animals that are bilaterally symmetrical, ganglia frequently come in pairs, one on each side of the body. Also, as animals increase in complexity, some ganglia may become enlarged or fused together at the anterior end, forming a larger, centralized integrative center, or **brain**. The small nervous systems of invertebrates can be remarkably complex. Consider the nervous systems of spiders, which have programmed within them the thousands of precise movements necessary to construct a beautiful web without prior experience or opportunities to learn the specific web architecture of their species.

In vertebrates, most cells of the nervous system are found in the brain and the **spinal cord**, the sites of most information processing, storage, and retrieval (**Figure 45.1D**). Therefore, the brain and spinal cord are called the **central nervous system** (**CNS**). Information is transmitted from sensory cells to the CNS and from the CNS to effectors via neurons that extend or reside outside the brain and the spinal cord; these neurons and their supporting cells are called the **peripheral nervous system** (**PNS**). Vertebrates differ greatly in their behavioral complexity and in their physiological specializations, and their nervous systems reflect this diversity. **Figure 45.2** shows the brains of four vertebrate species of similar body mass drawn to the same scale.

The human nervous system contains an estimated $10^{11}$ neurons. Information is passed from one neuron to another where they come into close proximity at structures called **synapses**. The neuron sending the information is the **presynaptic neuron**, and the neuron receiving the information is the **postsynaptic neuron**. A given neuron in the brain can have 1,000 or more synapses. Thus the human brain can contain $10^{14}$ synapses ($10^{11}$ neurons $\times$ $10^3$ synapses per neuron).

Synapses are not constant but instead can be highly plastic. They can increase or decrease in number and size. They can become more or less sensitive. Therein lies the incredible ability of the human brain to process information, to learn, to do complex tasks, to remember, and to have emotions. The astronomical number of neurons and synapses is divided into thousands of distinct but interacting networks that function in parallel. Before we can understand how even one of these circuits works, however, we must understand the properties of individual neurons.

## Neurons are the functional units of nervous systems

Nervous systems of different species vary enormously in structure and function, but neurons behave similarly in animals as different as squid and humans. Their plasma membranes generate action potentials and conduct these signals from one lo-

**45.2 Brains Vary in Size and Complexity** The brains of four vertebrate species—all of which may have a similar body mass—show immense differences. Note that the brainstem, which is involved in physiological regulation and stereotypic behavior, differs less among these species than does the cerebrum, which in higher vertebrates is responsible for complex behavior and learning.

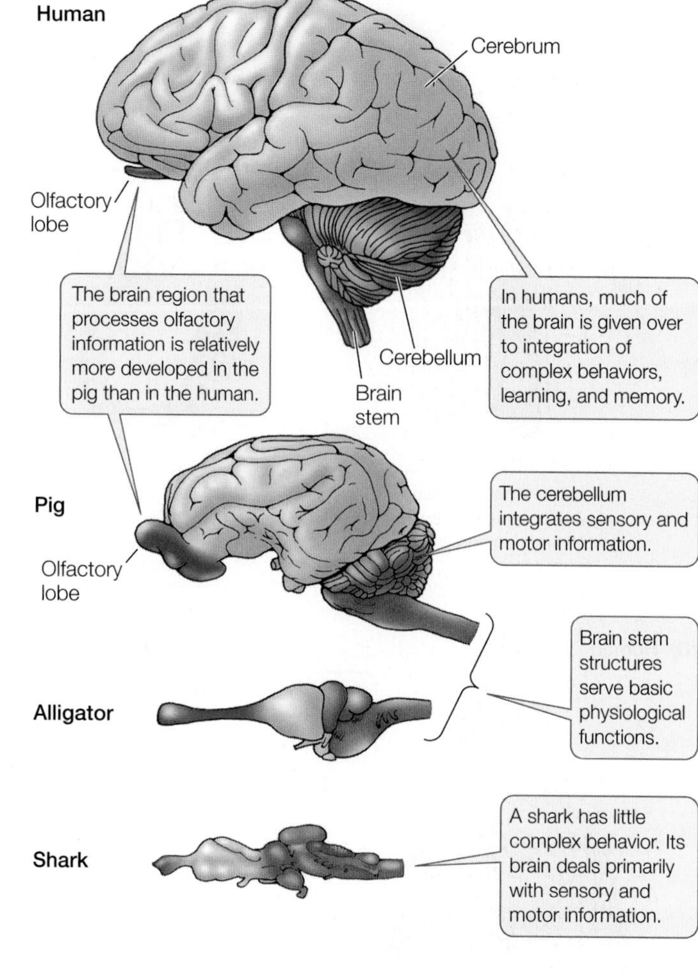
Human
Cerebrum
Olfactory lobe
The brain region that processes olfactory information is relatively more developed in the pig than in the human.
Cerebellum
Brain stem
In humans, much of the brain is given over to integration of complex behaviors, learning, and memory.
Pig
The cerebellum integrates sensory and motor information.
Olfactory lobe
Brain stem structures serve basic physiological functions.
Alligator
Shark
A shark has little complex behavior. Its brain deals primarily with sensory and motor information.

cation on a neuron to the most distant reaches of that cell—a distance that can be more than a meter in a human and many meters in a whale. Moreover, this transmission of action potentials can be rapid—up to 100 meters/sec, which is 360 km/hr—making it possible to sense, process, and act on information very quickly.

Most neurons have four regions—a *cell body*, *dendrites*, one or more *axons*, and *axon terminals* (**Figure 45.3A**)—but the variation among different types of neurons is considerable (**Figure 45.3B**). The *cell body* contains the nucleus and most of the cell's organelles. Many projections may sprout from the cell body. Most of these projections are shrublike **dendrites** (from the Greek *dendron*, "tree"), which bring information from other neu-

**45.3 Neurons** (A) A generalized diagram of a neuron. (B) Neurons from different parts of the mammalian nervous system are specifically adapted to their functions.

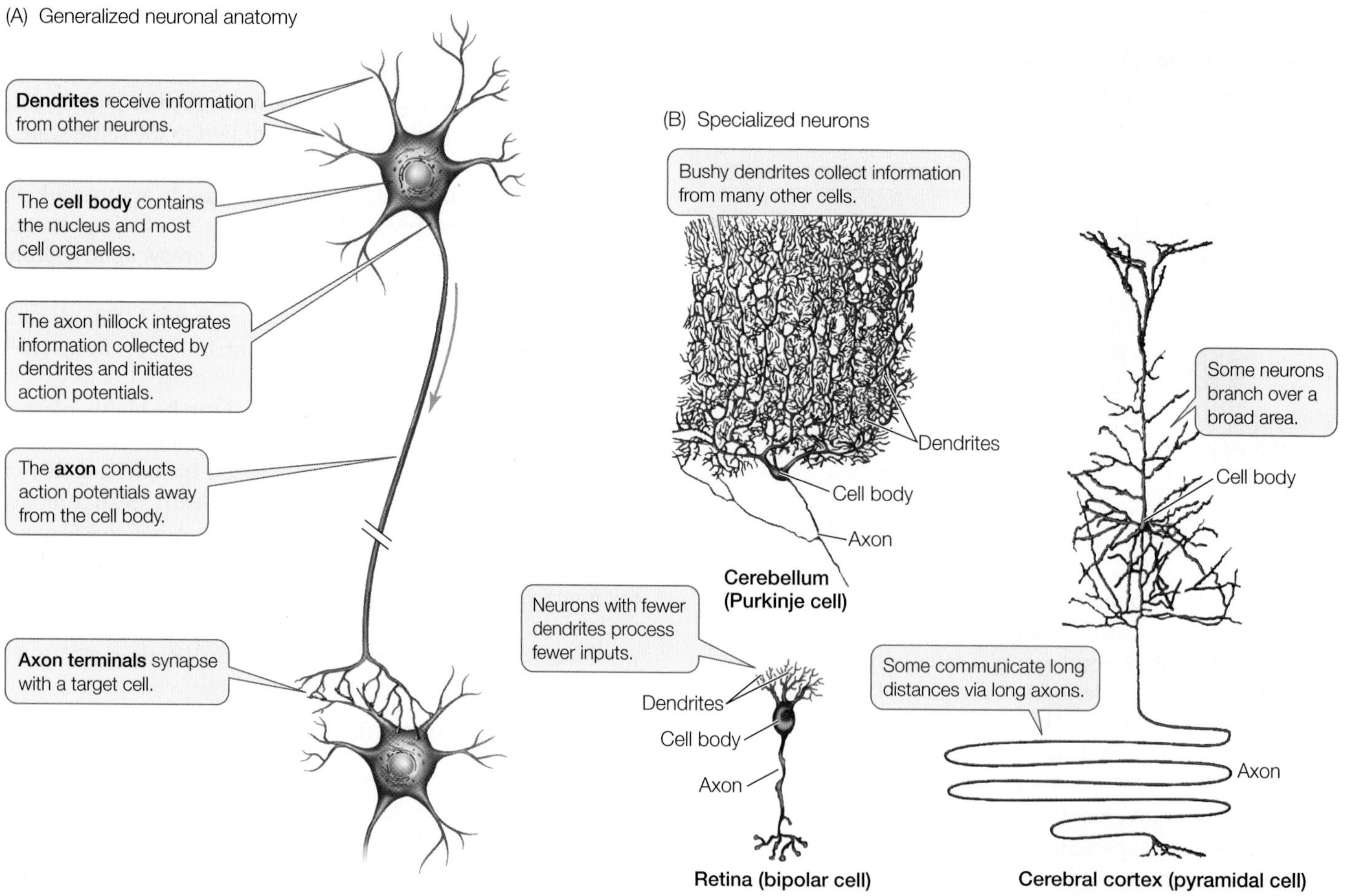

(A) Generalized neuronal anatomy

**Dendrites** receive information from other neurons.

The **cell body** contains the nucleus and most cell organelles.

The axon hillock integrates information collected by dendrites and initiates action potentials.

The **axon** conducts action potentials away from the cell body.

**Axon terminals** synapse with a target cell.

(B) Specialized neurons

Bushy dendrites collect information from many other cells.

Dendrites

Cell body

Axon

**Cerebellum (Purkinje cell)**

Neurons with fewer dendrites process fewer inputs.

Dendrites

Cell body

Axon

**Retina (bipolar cell)**

Some neurons branch over a broad area.

Cell body

Some communicate long distances via long axons.

Axon

**Cerebral cortex (pyramidal cell)**

rons or sensory cells to the cell body. The degree of branching of the dendrites differs among different types of neurons.

In most neurons, one projection—the axon—is much longer than the others. Axons usually carry action potentials away from the cell body. The length of the axon also differs among different types of neurons—some axons are remarkably long, such as those that run from the spinal cord to the toes. Axons are the "telephone lines" of the nervous system. Information received by dendrites can cause the axon to generate an action potential that is conducted down the axon toward its target cell. At the target cell, the axon divides into a spray of fine nerve endings. At the tip of each of these tiny nerve endings is a swelling, called an **axon terminal**, that comes very close to the membrane of the target cell to form a synapse.

As we will discuss in Section 45.3, synapses can be either chemical or electrical. Most synapses in vertebrates are chemical synapses. At chemical synapses a space only about 25 nanometers wide (about 1/2000[th] of a human hair) separates the *presynaptic* and *postsynaptic membranes*. An action potential arriving at an axon terminal causes it to release chemical messenger molecules called **neurotransmitters**. The neurotransmitters diffuse across the space and bind to receptors on the plasma membrane of the postsynaptic or target cell. This binding alters the activity of the postsynaptic neuron. Some neurotransmitter–receptor combinations inhibit activity of the postsynaptic neuron, and other neurotransmitter–receptor combinations excite it. Neurons integrate information by summing excitatory and inhibitory inputs.

### Glia are also important components of nervous systems

*Glia* are another class of nervous system cells. There are many more glia than neurons in the human brain. Like neurons, glia come in several forms and have diverse functions. They do not generate or transmit electrical signals, but they can release neurotransmitters. Some glia physically support and orient the neurons and help them make the right contacts during embryonic development. Others supply neurons with nutrients, maintain the extracellular environment, and insulate axons. Still others consume debris and foreign particles and provide immune functions for the nervous system.

In the CNS, glia called **oligodendrocytes** wrap around the axons of neurons, covering them with concentric layers of insulating plasma membrane. In the PNS, glia called **Schwann cells** perform this function (**Figure 45.4**). **Myelin** is the covering produced by oligodendrocytes and Schwann cells, and it gives many parts of the nervous system a glistening white appearance. Not all axons are myelinated, but those that are can conduct action potentials more rapidly than axons that are not myelinated.

Diseases that affect myelin can be devastating because they impair conduction of action potentials. The most common demyelinating disease is *multiple sclerosis*—meaning literally "multiple scars"—which occurs in about 1 in 700 people in the United States. Individuals with this autoimmune disease produce antibodies to proteins in the myelin in the CNS. The symptoms and damage from the disease depend on where in the CNS the antibody attacks take place. Motor impairment is common. An example of a demyelinating disease that affects the PNS is Guillain-Barre Syndrome. Environmental factors such as pesticide exposure can also damage myelin. There are no known cures for demyelinating diseases.

Glia called **astrocytes** (because they look like stars; see Figure 40.6B) contribute to the **blood–brain barrier**, which protects the brain from toxic chemicals in the blood. Blood vessels throughout the body are very permeable to many chemicals, including toxic ones, which would reach the brain if this barrier did not exist. Astrocytes help form the blood–brain barrier by surrounding the smallest, most permeable blood vessels in the brain. The barrier is not perfect, however. Since it consists of plasma membranes, it is permeable to fat-soluble substances such as anesthetics and alcohol, which explains why these substances have such rapid and marked effects on the nervous system.

The blood–brain barrier usually prevents antibodies in the general circulation from entering the CNS. To provide the CNS with immune defenses, **microglia**, which originate during development from stem cells in the bone marrow, come to reside

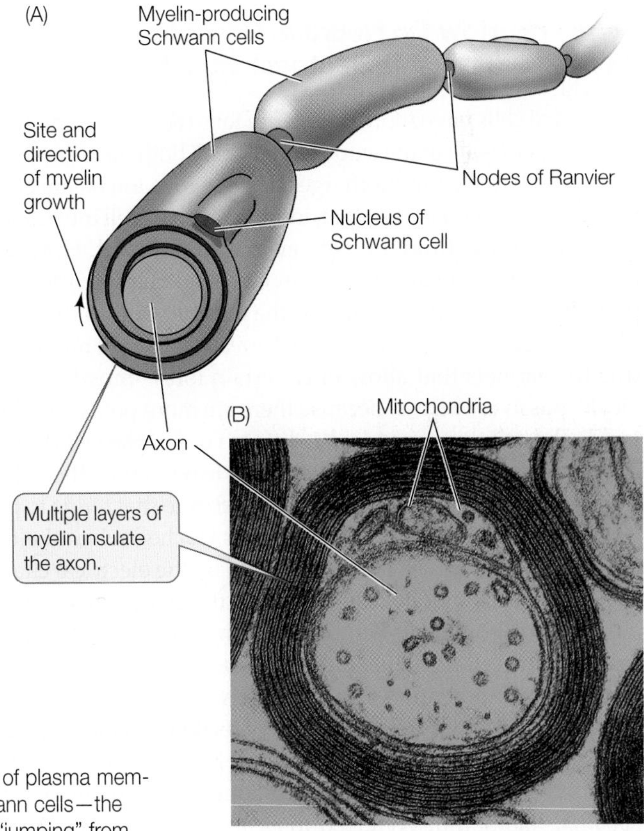


**45.4 Wrapping Up an Axon** (A) Schwann cells produce layers of myelin, a type of plasma membrane that provides electrical insulation to the axon. At the intervals between Schwann cells—the nodes of Ranvier—the axon is exposed. Action potentials travel along the axon by "jumping" from node to node. (B) A myelinated axon, seen in cross section through an electron microscope.

in the CNS and act as macrophages and mediators of inflammatory responses.

## 45.1 RECAP

**Nervous systems consist of neurons and glia. Neurons are arranged in circuits. Sensory neurons convert information from the external or internal environment into electrical signals; interneurons integrate and process information; and efferent neurons carry commands to target tissues. Glia have a wide variety of supporting roles, including contributing to the blood–brain barrier.**

- Comparing nervous systems of anemones, worms, fishes, and humans, what trends seem to reflect increasing complexity? See pp. 945–946 and Figures 45.1 and 45.2
- Describe the different parts of neurons and their functions. See pp. 946–947 and Figure 45.3
- What are some types of glia, and what are their functions? See p. 947

The one feature common to all nervous systems is that they process information in the form of action potentials. In the next section we will focus on how action potentials are generated and transmitted by nervous systems.

# 45.2 How Do Neurons Generate and Transmit Electrical Signals?

All animal cells have more potassium ions ($K^+$) inside and more sodium ions ($Na^+$) in the extracellular fluid. Both inside and outside the cell, the positive charges of these ions are balanced by negatively charged ions so that, individually, the cell interior and the extracellular fluid are both electrically neutral. However, *across the cell membrane* there is an electrical charge difference, with the inside being negative to the outside. Why is this so?

The reason is that there are *leak currents* in the cell membrane due to channels that allow only certain ions—mostly $K^+$—to "leak" passively across. Because there are more potassium ions inside the cell than outside, $K^+$ diffuses out of the cell down a concentration gradient. But when $K^+$ leaks out of the cell, it leaves behind an unbalanced negative charge that tends to pull $K^+$ back into the cell. An equilibrium is reached when the tendency for $K^+$ to diffuse out is countered by the electrical charge pulling $K^+$ back in. The result is an electric charge difference—a **membrane potential**—across the plasma membrane, with the inside of the cell negative to the outside.

Membrane potentials exist in all cells. In neurons, the steady state membrane potential is called the **resting potential**, because neurons can also be active, generating rapid, large changes in membrane potential. These sudden large shifts are called *action potentials* (sometimes referred to as *nerve impulses*). An action potential is generated by sudden openings and rapid closings of ion channels. Before describing the properties of ion channels and action potentials in detail, a review of some simple concepts of electricity may be useful.

### Simple electrical concepts underlie neural function

**Voltage** (electric potential difference) is a force that causes electrically charged particles to move between two points. Voltage is to the flow of electrically charged particles as pressure is to the flow of water. If the negative and positive poles of a battery are connected by a wire, an electric current will flow through the wire because there is a voltage difference between the two poles. This flow of electric current can be used to do work, just as a current of water can be used to do work.

In wires, electric current is carried by electrons, but in solutions and across cell membranes, electric current is carried by ions. The major ions that carry electric charges across the plasma membranes of neurons are sodium ($Na^+$), potassium ($K^+$), calcium ($Ca^{2+}$), and chloride ($Cl^-$). Recall that ions with opposite charges attract one another and that those with like charges repel one another. How do these basic principles of bioelectricity establish the resting potential of the neural plasma membrane? And how is the flow of ions through membrane channels turned on and off to generate action potentials? We address these questions next.

### Membrane potentials can be measured with electrodes

We can record electrical events in a cell using electrodes. **Figure 45.5** shows how this technique is applied across an unstimulated axon to measure the resting potential, which is usually between –60 and –70 mV. The minus sign indicates that the inside of the cell is electrically negative compared with the outside.

The resting potential provides a means for neurons to respond to a stimulus. Because of the voltage difference across the membrane, and the different ion concentrations on either side of the membrane, ions would cross the membrane if they could. For example, $Na^+$ ions are more abundant outside the cell than inside and the inside of the resting cell is negatively charged. Therefore, if the membrane suddenly became permeable to $Na^+$, those positively charged ions would rush into the cell. Any chemical or physical stimulus that changes the permeability of the plasma membrane to ions will produce a change in the cell's membrane potential. The most extreme change in membrane potential is the action potential, a sudden and rapid reversal in the voltage across a portion of the plasma membrane. For one or two milliseconds, positively charged ions flow into the cell, making the inside of the cell *more positive* than the outside.

yourBioPortal.com
GO TO Animated Tutorial 45.1 • The Resting Membrane Potential

### Ion transporters and channels generate membrane potentials

The plasma membranes of neurons, like those of all other cells, are lipid bilayers that are impermeable to ions but contain many

## TOOLS FOR INVESTIGATING LIFE

### 45.5 Measuring the Membrane Potential

An electrode can be made from a glass pipette with a very sharp tip filled with a solution that conducts electric charges. If one electrode is placed inside the plasma membrane of an axon and another is placed just outside the axon, the difference in voltage can be measured.

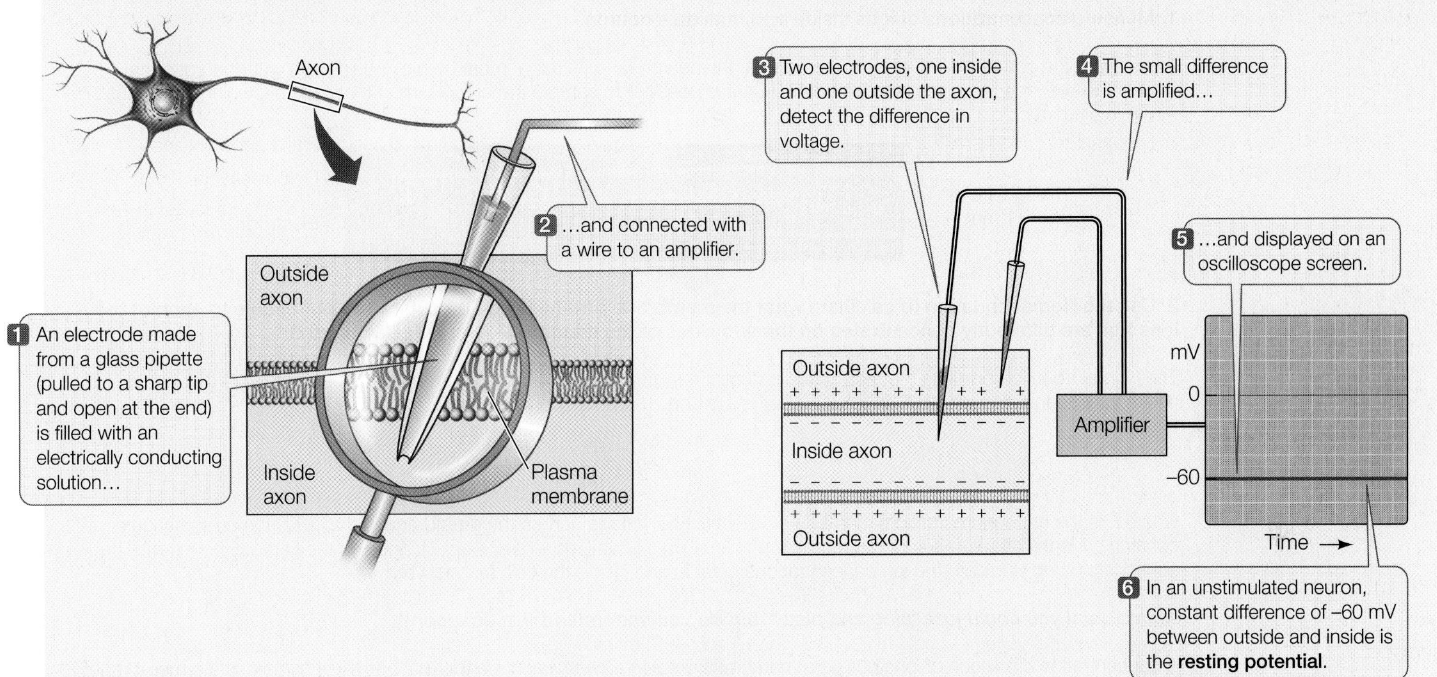

Axon
1 An electrode made from a glass pipette (pulled to a sharp tip and open at the end) is filled with an electrically conducting solution...
Outside axon
Inside axon
Plasma membrane
2 ...and connected with a wire to an amplifier.
3 Two electrodes, one inside and one outside the axon, detect the difference in voltage.
Outside axon
Inside axon
Outside axon
4 The small difference is amplified...
Amplifier
5 ...and displayed on an oscilloscope screen.
mV
0
–60
Time →
6 In an unstimulated neuron, constant difference of –60 mV between outside and inside is the resting potential.

protein molecules that serve as ion transporters and channels. Ion transporters and channels are responsible for the distribution of charges across the membrane that create resting and action potentials.

Ion transporters require energy to move ions against their concentration or electrical gradients and are therefore called ion pumps. A major ion transporter in the plasma membranes of neurons (and all other cells) is the **sodium–potassium pump**, so called because it actively expels $Na^+$ ions from inside the cell, exchanging them for $K^+$ ions from outside the cell (**Figure 45.6A**). The $Na^+$–$K^+$ pump is an *antiporter* (see Section 6.3) and is also known as *sodium–potassium ATPase*, a term emphasizing that it is an enzyme complex requiring ATP to do its work. The $Na^+$–$K^+$ pump keeps the concentration of $K^+$ inside the cell greater than that of the extracellular fluid, and the concentration of $Na^+$ inside the cell less than that of the extracellular fluid. The concentration differences established by this antiporter mean that $K^+$ would diffuse out of the cell and $Na^+$ would dif-

(A) $Na^+$–$K^+$ pump (ATPase)

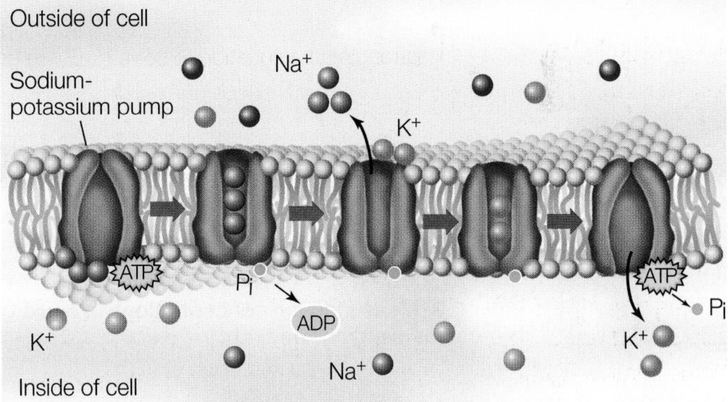


(B) $Na^+$–$K^+$ channels

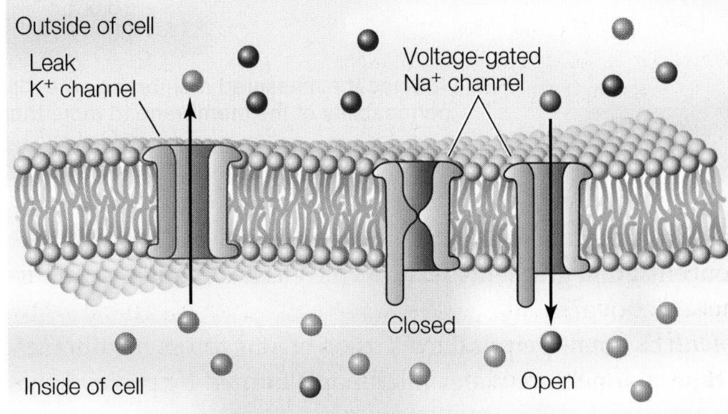


**45.6 Ion Transporters and Channels** (A) The sodium–potassium pump is in an antiporter that actively moves $K^+$ to the inside of a neuron and $Na^+$ to the outside. (B) Ion channels allow specific ions to diffuse down their concentration gradients; $K^+$ tends to leave neurons when potassium channels are open, and $Na^+$ tends to enter neurons when sodium channels are open. Leak channels like the $K^+$ channel shown are always open and create the resting membrane potential. Gated channels like the $Na^+$ channels shown are opened by chemical or electrical stimulation.

## TOOLS FOR INVESTIGATING LIFE

### 45.7 Using the Nernst Equation

The Nernst equation calculates membrane potential when only one type of ion can cross a membrane that separates solutions with different concentrations of that ion.

**1. Measure concentrations of ions inside and outside a neuron.**

To measure the concentration of ions in a neuron, the neuron (and its axon) must be big. Squid have giant neurons that control their escape response (see Figure 45.1C). It is possible to sample the cytoplasm of these axons, which are about 1 mm in diameter.

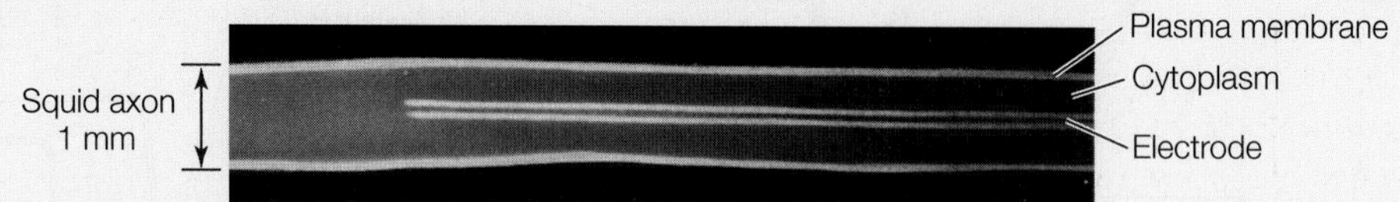


**2. Use the Nernst equation to calculate what the membrane potential would be if it were permeable to each of the ions that are differently concentrated on the two sides of the membrane: $Na^+$, $K^+$, $Ca^{2+}$, and $Cl^-$.**

The Nernst equation predicts the membrane potential resulting from membrane permeability to a single type of ion that differs in concentration on the two sides of the membrane. The equation is written

$$E_{ion} = 2.3\frac{RT}{zF}\log\frac{[ion]_o}{[ion]_i}$$

where $E$ is the equilibrium (resting) membrane potential (the voltage across the membrane in mV), $R$ is the universal gas constant, $T$ is the absolute temperature, $z$ is the charge on the ion (+1, +1, +2, or −1), and $F$ is the Faraday constant. The subscripts o and i indicate the ion concentrations outside and inside the cell, respectively.

***At this point you could just "plug and play," but do you understand this equation?***

A concentration difference of ions across a membrane creates a *chemical* force that pushes the ions across the membrane; however, the resulting unbalanced *electrical charges* will pull the ions back the other way. At *equilibrium*, the work done moving ions in each direction will be the same.

The *chemical* energy pushing the ions will equal $2.3\,RT\log[ion]_o/[ion]_i$
The *electrical* energy pulling the ions will equal $zEF$. So, at equilibrium:

$$zEF = 2.3\,RT\log\frac{[ion]_o}{[ion]_i}$$

Rearranging the equation to solve for $E$, we get the Nernst equation:

$$E_{ion} = 2.3\frac{RT}{zF}\log\frac{[ion]_o}{[ion]_i}$$

We can simplify the equation by picking a temperature—let's use "room temperature," or 20°C—and solving for $2.3\,RT/F$. At 20°C, $2.3\,RT/F$ equals 58. Thus:

$$E_{ion} = 58/z\log\frac{[ion]_o}{[ion]_i}$$

**3. Measuring ion concentrations in squid giant axon cytoplasm and in seawater, then solving the Nernst equation for each ion, we find:**

| | Ion concentration (m*M*) | | Predicted membrane |
|---|---|---|---|
| Ion | in squid axon | in seawater | potential (mV) |
| $K^+$ | 400 | 20 | −75 |
| $Na^+$ | 50 | 460 | +56 |
| $Ca^{2+}$ | 0.5 | 10 | +38 |
| $Cl^-$ | 50 | 560 | −60 |

**4. Since the measured membrane potential is −66 mV, it is clear that the resting potential of the axon is due to permeability of the membrane to more than just one type of ion.**

fuse in if the ions could cross the lipid bilayer. How do these concentration gradients relate to the electric gradients we discussed above?

Ion channels permit the diffusion of ions across membranes. These channels are water-filled pores formed by proteins that cross the lipid bilayer and are generally *selective*—they allow some types of ions to pass through more easily than others (**Figure 45.6B**). Thus, there are potassium channels, sodium channels, chloride channels, and calcium channels, and there are different kinds of each. Ions can diffuse through these channels in either direction. The direction and magnitude of the net movement of ions through a channel depend on the concentration gradient of that ion type across the plasma membrane, as well as on as the voltage difference across that membrane. These two motive forces acting on an ion are termed its **electrochemical gradient**. Although the electrochemical gradient drives the movement of ions through channels across bilayers, movement through channels can be modified by gates that open and close channels.

Potassium channels are the most common open, or leak, channels in the plasma membranes of resting (non-stimulated) neurons. As a consequence, resting neurons are more permeable to $K^+$ than to any other ion. Thus, open potassium channels are largely responsible for the resting membrane potential. Because the potassium channels make the plasma membrane permeable to $K^+$, and because the $Na^+$–$K^+$ pump keeps the concentration of $K^+$ inside the cell much higher than that outside the cell, $K^+$ tends to diffuse down its electrochemical gradient, out of the cell, through the channels. As these positively charged potassium ions diffuse out of the cell, they leave behind unbalanced negative charges, generating an electric potential across the membrane that tends to pull $K^+$ back into the cell.

The membrane potential at which the net diffusion of $K^+$ out of the cell ceases (that is, the point at which $K^+$ diffusion out due to the concentration gradient is balanced by its movement in due to the negative electric potential) is the **potassium equilibrium potential**. The value of the potassium equilibrium potential can be calculated from the concentrations of $K^+$ on the two sides of the membrane using the **Nernst equation** (**Figure 45.7**). This equation, developed in the late 1800s, illustrates that the nature of ion channels in neural membranes was hypothesized long before their specific structures and properties were described.

In the late 1940s, A. L. Hodgkin and A. F. Huxley at the University of Cambridge set out to study the electrical properties of axonal membranes. With the techniques available at that time, the necessary measurements could be made only if you had a very large axon to work with. Such an axon exists in nature, in the huge neuron that controls the escape response of squid. Hodgkin and Huxley used electrodes to measure the voltage across the plasma membrane of this large axon, as seen in Figure 45.7, and to pass electric current into it to change its resting potential. They also changed the concentrations of $Na^+$ and $K^+$ both inside and outside the squid axon and measured the resulting changes in membrane potential. On the basis of their many careful experiments, Hodgkin and Huxley developed virtually all of our basic concepts about the electrical properties of neurons, and received the Nobel prize in 1963.

We now know that, in general, the resting potential is less negative than the Nernst equation predicts because resting neurons are also slightly permeable to other ions, such as $Na^+$ and $Cl^-$. Another equation, called the Goldman equation, takes all of the ions that can cross the membrane into account and therefore can calculate the membrane potential accurately.

### Ion channels and their properties can now be studied directly

Because Hodgkin and Huxley were working long before there were laboratory techniques that could investigate ion channels, they could only hypothesize their properties. These hypotheses could not be tested until the late 1970s, when B. Sakmann and E. Neher developed a technique called **patch clamping**, for which they won the Nobel prize in 1991. Patch clamping, described in **Figure 45.8**, is widely used by neurobiologists, en-

TOOLS FOR INVESTIGATING LIFE

**45.8 Patch Clamping**

The patch clamp is a glass micropipette filled with an electrically conductive solution that has the same composition as extracellular fluids. When this pipette/electrode is positioned against the membrane of a cell and slight suction is applied, a seal forms. If a single ion channel (or a few ion channels) are within the patch of membrane bounded by the seal, the openings and closings of individual channels can be recorded by the electrode. If the pipette is retracted, it can tear the patched membrane away from the cell, and the activities of the ion channels in the patch can continue to be recorded.

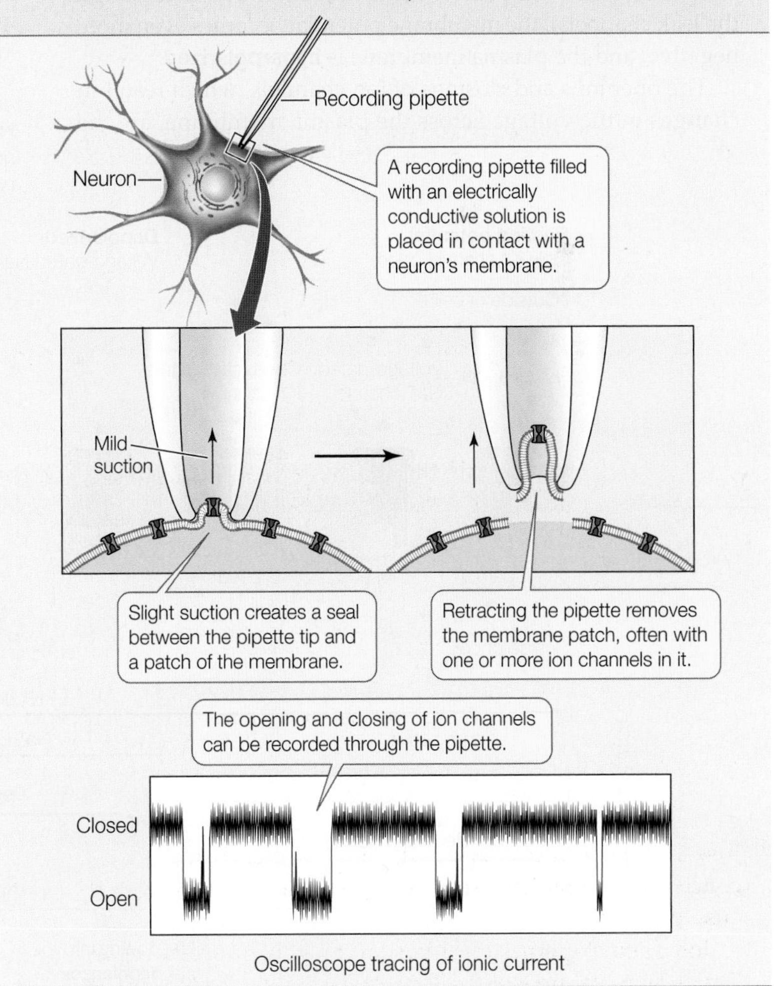

Recording pipette
Neuron
A recording pipette filled with an electrically conductive solution is placed in contact with a neuron's membrane.
Mild suction
Slight suction creates a seal between the pipette tip and a patch of the membrane.
Retracting the pipette removes the membrane patch, often with one or more ion channels in it.
The opening and closing of ion channels can be recorded through the pipette.
Closed
Open
Oscilloscope tracing of ionic current

abling them to record in real time the tiny electrical currents caused by the openings and closings of single ion channels.

## Gated ion channels alter membrane potential

The ion channels called leak channels are always open, but other ion channels in the plasma membranes of neurons behave as if they contain "gates"; they are open under some conditions and closed under other conditions. **Voltage-gated channels** open or close in response to a change in the voltage across the plasma membrane. **Chemically gated channels** open or close depending on the presence or absence of a specific molecule that binds to the channel protein, or to a separate receptor that in turn alters the channel protein. **Mechanically gated channels** open or close in response to mechanical force applied to the plasma membrane. Gated channels play important roles in neural function.

Openings and closings of gated channels alter the resting potential. Imagine what happens, for example, if sodium channels in the plasma membrane open. $Na^+$ diffuses into the neuron down its electrochemical gradient. As a result, the inside of the cell becomes less negative. When the inside of a neuron becomes less negative (or more positive) in comparison to its resting condition, its plasma membrane is **depolarized** (**Figure 45.9**).

An opposite change in the resting potential occurs if gated $K^+$ channels open. When $K^+$ efflux from the neuron increases over the normal leak current (the movement of K+ through the leak channels), the membrane potential becomes even more negative, and the plasma membrane is **hyperpolarized**.

The openings and closings of ion channels, which result in changes in the voltage across the plasma membrane, are the basic mechanisms by which neurons respond to stimuli, be they electrical, chemical, or mechanical. How do such local changes in membrane potential get communicated to other parts of the cell?

A local change in membrane potential causes a flow of ions that spreads the change in membrane potential to adjacent regions of the membrane. For example, when $Na^+$ enters a neuron through open sodium channels at one location, those positively charged ions are attracted to adjacent areas on the inside of the membrane that are more negative, and thus there is a rapid flow of electric current (movement of ions) away from the site of the open $Na^+$ channels. However, this local flow of electric current decays as it spreads and therefore does not spread very far. Electric currents do not spread far in cells because cell membranes are not completely impermeable to ions. An electric current traveling along a membrane is like water flowing through a leaky hose.

## Graded changes in membrane potential can integrate information

Even though the flow of electric current along plasma membranes can only extend over short distances, it can cause graded changes in membrane potentials locally. A **graded membrane**

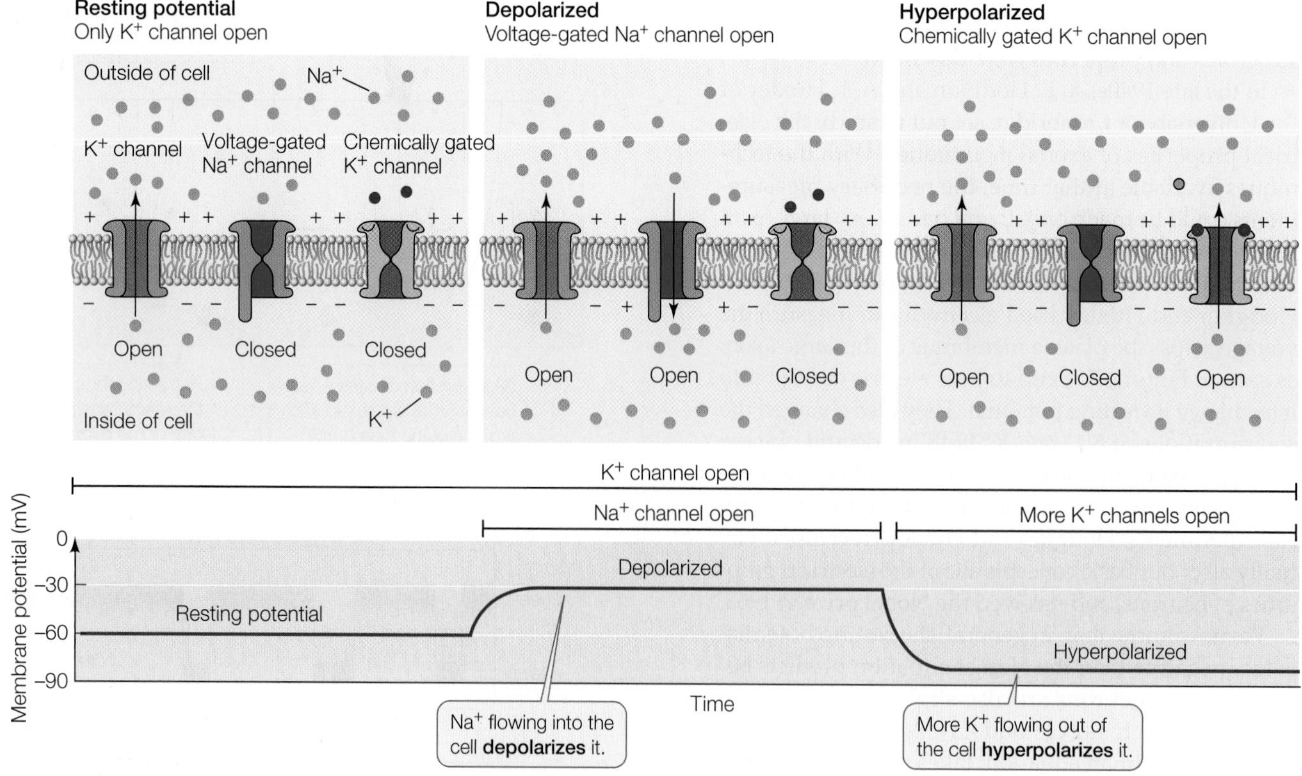


**45.9 Membranes Can Be Depolarized or Hyperpolarized** The resting potential is produced by leak $K^+$ channels. A shift from the resting potential to a less negative membrane potential, as occurs when $Na^+$ enters the cell through a gated sodium channel, is called depolarization. Hyperpolarization occurs when the membrane potential becomes more negative, as when additional $K^+$ leaves the cell through gated $K^+$ channels, which occurs extensively in your brain when you fall asleep.

**potential** is a change from the resting potential. Such changes can be due to chemical or mechanical influences on ion channels. Graded potentials are a means of integrating inputs to a cell because the membrane can respond to those inputs with proportional amounts of depolarization or hyperpolarization.

Graded potentials can transmit signals over very short distances and play an important role at the neuromuscular junction (see Section 45.3). In the next chapter we will learn how they play important roles in sensory systems. However, axons are too long to transmit information as a continuous flow of electric current (as telephone wires do). Therefore axons code information as discrete action potentials that travel along their membranes. Graded potentials, however, play an important role in the generation of action potentials.

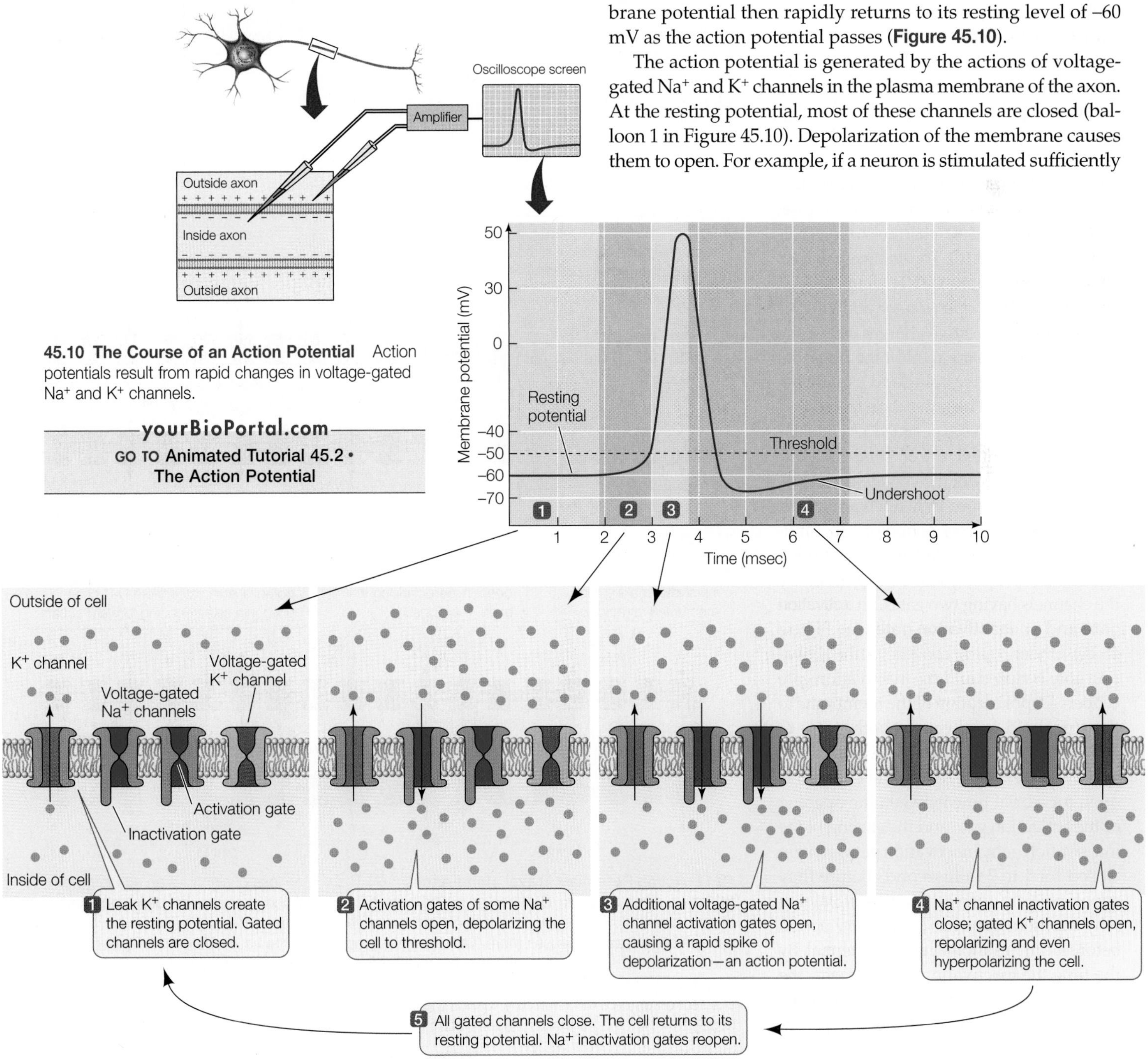

**45.10 The Course of an Action Potential** Action potentials result from rapid changes in voltage-gated $Na^+$ and $K^+$ channels.

yourBioPortal.com
GO TO **Animated Tutorial 45.2 • The Action Potential**

1 Leak $K^+$ channels create the resting potential. Gated channels are closed.

2 Activation gates of some $Na^+$ channels open, depolarizing the cell to threshold.

3 Additional voltage-gated $Na^+$ channel activation gates open, causing a rapid spike of depolarization—an action potential.

4 $Na^+$ channel inactivation gates close; gated $K^+$ channels open, repolarizing and even hyperpolarizing the cell.

5 All gated channels close. The cell returns to its resting potential. $Na^+$ inactivation gates reopen.

## Sudden changes in $Na^+$ and $K^+$ channels generate action potentials

Action potentials are sudden, transient, large changes in membrane potential. In unmyelinated axons they can be conducted at speeds of up to 2 meters per second, but in myelinated axons the conduction velocity can be 100 meters per second. Think of running the 100-meter dash—the world record is slightly under 10 seconds.

If we place the tips of a pair of electrodes on either side of the plasma membrane of a resting axon and measure the voltage difference, the reading might be about –60 mV, as we saw in Figure 45.5. If these electrodes are in place when an action potential travels down the axon, they register a rapid change in membrane potential, from –60 mV to about +50 mV. The membrane potential then rapidly returns to its resting level of –60 mV as the action potential passes (**Figure 45.10**).

The action potential is generated by the actions of voltage-gated $Na^+$ and $K^+$ channels in the plasma membrane of the axon. At the resting potential, most of these channels are closed (balloon 1 in Figure 45.10). Depolarization of the membrane causes them to open. For example, if a neuron is stimulated sufficiently to cause the plasma membrane of its cell body to depolarize, that depolarization can spread by local current flow to the **axon hillock**, the region of the cell body at the base of the axon (see Figure 45.3). Voltage-gated $Na^+$ channels are concentrated in the axon hillock. When the plasma membrane in this area depolarizes, some of these voltage-gated channels open briefly—for less than a millisecond (balloon 2 in Figure 45.10). When these channels open, $Na^+$ rushes into the axon and depolarizes the membrane even more, causing more $Na^+$ channels to open—a *positive feedback* effect. When the membrane is depolarized about 5 to 10 mV above the resting potential, a **threshold** is reached; a large number of sodium channels open (balloon 3 in Figure 45.10), and the membrane potential becomes positive—an action potential. The rising phase of the action potential halts abruptly in 1 to 2 milliseconds, and the membrane potential rapidly becomes negative once again.

What causes the axon to return to resting potential? There are two contributing factors: the voltage-gated $Na^+$ channels close, and voltage-gated $K^+$ channels open (balloon 4 in Figure 45.10). The voltage-gated $K^+$ channels open more slowly than the $Na^+$ channels and stay open longer, allowing $K^+$ to carry excess positive charges out of the axon. As a result, the membrane potential returns to a negative value and usually becomes even more negative than the resting potential until the voltage-gated $K^+$ channels close (balloon 5 in Figure 45.10).

Another feature of the voltage-gated $Na^+$ channels is that once they open and close, they have a **refractory period** of 1 to 2 milliseconds during which they cannot open again. This property can be explained by the channels having two gates, an **activation gate** and an **inactivation gate** (see Figure 45.10). Under resting conditions, the activation gate is closed and the inactivation gate is open. Depolarization of the membrane to the threshold level causes both gates to change state, but the activation gate responds faster. As a result, the channel is open for a brief time between the opening of the activation gate and the closing of the inactivation gate. Inactivation gates remain closed for 1 to 2 milliseconds before they spontaneously open again, thus explaining why the membrane has a refractory period before it can fire another action potential. By the time the inactivation gate reopens, the activation gate is closed, and the membrane is poised to generate another action potential. Another contribution to the refractory period is the duration of the opening of the voltage-gated $K^+$ channels, as we saw above. The dip in the membrane potential following an action potential is called the *after-hyperpolarization* or *undershoot*.

The difference in the concentration of $Na^+$ across the plasma membrane and the negative resting potential constitute the "bat-

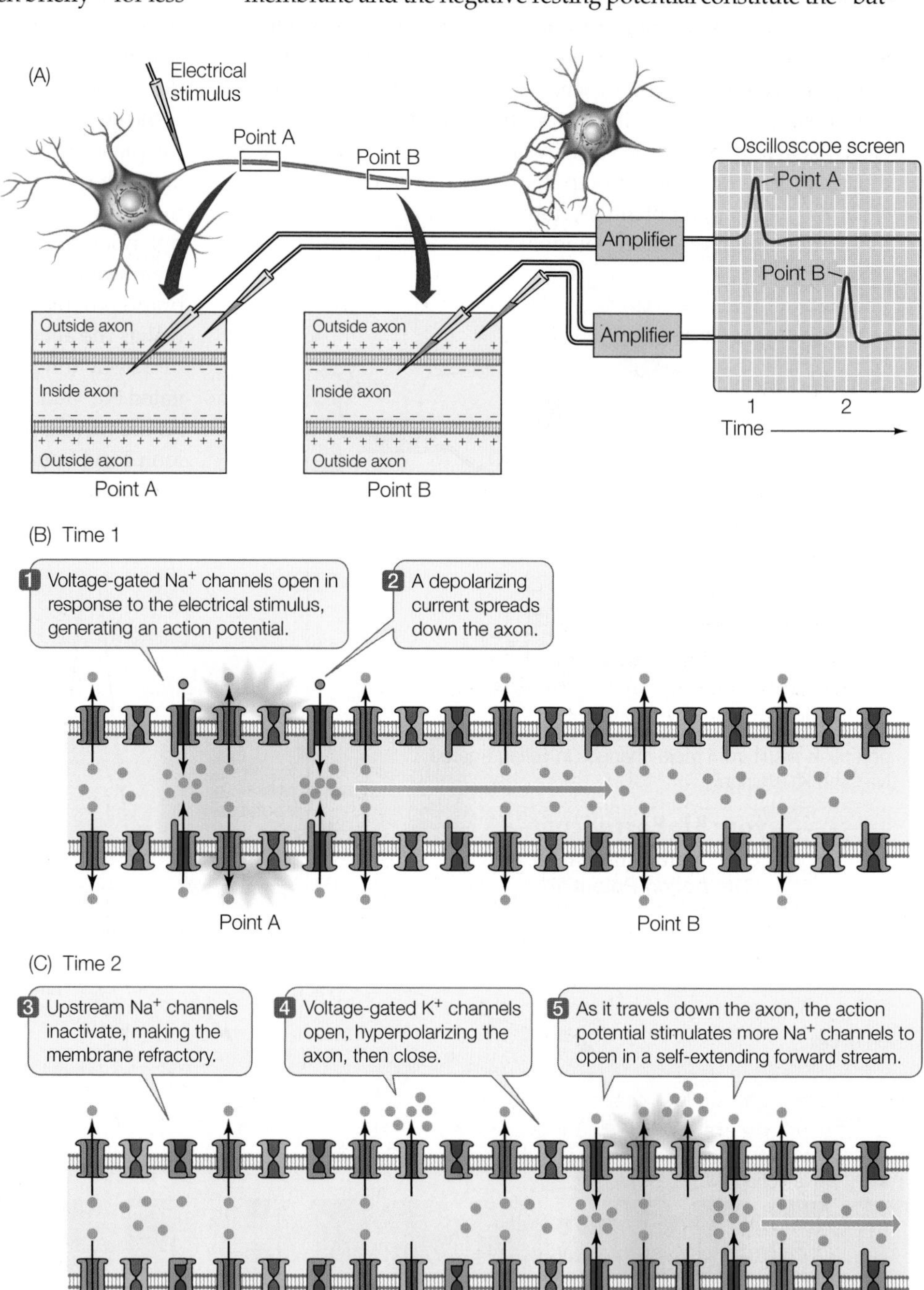


**45.11 Action Potentials Travel along Axons** (A) There is no loss of signal as an action potential travels along an axon. (B) When an action potential is stimulated in one region of membrane, electric current flows to adjacent areas of membrane and depolarizes them. (C) The advancing wave of depolarization causes more $Na^+$ channels to open, and the action potential is generated anew in the next section of membrane. Meanwhile, in the region where the action potential has just fired, the $Na^+$ channels are inactivated and the voltage-gated $K^+$ channels are still open, rendering this section of the axon refractory. Hence the action potential cannot "back up," but moves continuously forward along the axon, regenerating itself as it goes.

tery" that drives action potentials. How rapidly does the battery run down? It might seem that a substantial number of ions would have to cross the membrane for the membrane potential to change from –60 mV to +50 mV and back to –60 mV again. In fact, only a vanishingly small percentage of the $Na^+$ concentrated outside the plasma membrane moves through the channels during the passage of an action potential. Thus the effect of a single action potential on the concentration gradients of $Na^+$ and $K^+$ is very small, and it is possible in most cases for the sodium–potassium pump to keep the "battery" charged, even when the neuron is generating many action potentials every second.

### Action potentials are conducted along axons without loss of signal

Action potentials can travel over long distances with no loss of signal. If we place two pairs of electrodes at two different locations along an axon, we can record an action potential at those two locations as it travels along the axon (**Figure 45.11A**). The magnitude of the action potential does not change between the two recording sites. This constancy is possible because an action potential is an all-or-none, self-regenerating event.

- An action potential is *all-or-none* because of the interaction between the voltage-gated $Na^+$ channels and the membrane potential. If the membrane is depolarized slightly, some voltage-gated $Na^+$ channels open. Some sodium ions cross the plasma membrane and depolarize it even more, opening more voltage-gated $Na^+$ channels, and so on, until the membrane reaches threshold and generates an action potential. This positive feedback mechanism ensures that action potentials always rise to their maximum value.
- An action potential is *self-regenerating* because it spreads by local current flow to adjacent regions of the plasma membrane. The resulting depolarization brings those neighboring areas of membrane to threshold. So when an action potential occurs at one location on an axon, it stimulates the adjacent region of axon to generate an action potential, and so on down the length of the axon.

We can use an electrode to stimulate an axon, causing it to depolarize and to fire an action potential that is then conducted along the axon. **Figure 45.11B** shows the changes in the ion channels in the membrane that are responsible for conducting the action potential along the axon without a reduction in amplitude. Normally, an action potential is propagated in only one direction—away from the body. It cannot reverse itself because the voltage-gated $Na^+$ channels in the region of the membrane it came from are still in their refractory period.

Action potentials do not travel along all axons at the same speed. They travel faster in large-diameter axons than in small-diameter axons because the resistance to current flow decreases as an axon's diameter gets bigger. They travel faster in myelinated than in nonmyelinated axons because they can "jump" from one node to another without traversing the intervening space (**Figure 45.12**). Invertebrates mostly depend on increased axon diameter for fast conduction, but vertebrates mostly depend on myelination of axons to increase conduction velocity.

### Action potentials can jump along axons

In vertebrate nervous systems, increasing the speed of action potentials by increasing the diameter of axons is not feasible because of the huge number of axons involved. Each of our eyes,

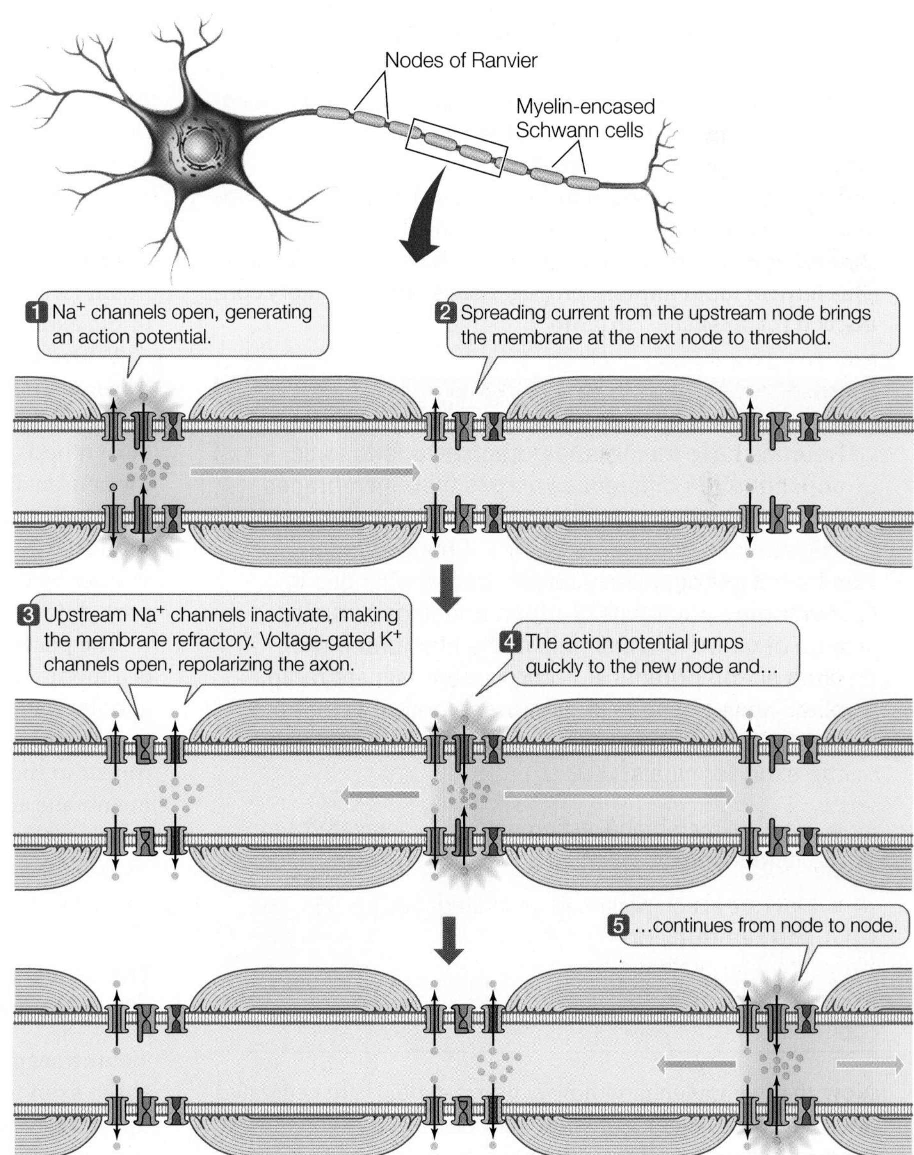


**45.12 Saltatory Action Potentials** Action potentials appear to jump from node to node in myelinated axons.

for example, has about a million axons connecting it to the brain. These axons conduct action potentials at about the same speed as does the squid giant axon—about 20 meters per second—yet the diameter of each is 200 times smaller than the squid axon's diameter. Vertebrates have evolved a different way of increasing conduction velocity of axons, and that adaptation is *myelination*.

When glia wrap themselves around axons, covering them with concentric layers of myelin (see Figure 45.4), they leave regularly spaced gaps, called **nodes of Ranvier**, where the axon is not covered (see Figure 45.12). The leakage of ions across the regions of the plasma membrane that are wrapped in myelin is reduced, so electric current can spread farther along the inside of a myelinated axon than it can along a nonmyelinated axon. Additionally, voltage-gated ion channels are clustered at the nodes of Ranvier. Thus an axon can fire action potentials only at nodes, and those action potentials cannot be propagated through the adjacent patch of membrane covered with myelin. The positive charges that flow into the axon at the node do, however, flow down the inside of the axon in the form of electric current. When the current reaches the next node, the plasma membrane at that node is depolarized to threshold and fires another action potential. Action potentials therefore appear to jump from node to node along the axon.

The speed of conduction is increased in these myelin-wrapped axons because electric current flows much faster through the cytoplasm than ion channels can open and close. This form of rapid impulse propagation is called **saltatory conduction** (Latin *saltare*, "to jump").

### 45.2 RECAP

**Neurons have membrane potentials due to ionic concentration differences across their membranes and because leak channels make the membrane differentially permeable to ions. Changes in ion channel permeabilities cause graded changes in membrane potentials. Sudden openings and closings of gated ion channels in the membrane produce action potentials. Action potentials are rapid, all-or-none changes in membrane potential that are conducted along axons from the neuron body to the axon terminals.**

- How are membrane resting potentials generated and altered? See pp. 949–950 and Figures 45.6 and 45.9
- How are action potentials generated? See pp. 953–955 and Figure 45.10
- How are action potentials transmitted along axons? See pp. 955–956 and Figures 45.11 and 45.12

Now that we understand how action potentials are generated and transmitted along axons, let's address the question of what happens when the action potential gets to the axon terminals. How is it communicated to the next cell, which could be another neuron, a muscle cell, or perhaps a secretory cell?

## 45.3 How Do Neurons Communicate with Other Cells?

Neurons communicate with each other and with other cells at synapses. The most common type of synapse in the nervous system is the **chemical synapse**, in which neurotransmitters released from a presynaptic cell induce changes in a postsynaptic cell. In **electrical synapses** the action potential spreads directly from presynaptic to postsynaptic cell. We will begin this section with a discussion of the synapses between neurons and muscle cells. We will then consider the diversity in synapses, how they integrate information, and how they are involved in learning and memory.

### The neuromuscular junction is a model chemical synapse

**Neuromuscular junctions** are synapses between neurons and the skeletal muscle cells they innervate. They are excellent models for how chemical synaptic transmission works. Like other neurons, a motor neuron has only one axon, but close to its target cell that axon can branch into numerous axon terminals that form many synapses with muscle cells. At each axon terminal an enlarged knob or buttonlike structure contains vesicles filled with neurotransmitter molecules. The neurotransmitter used by all vertebrate neuromuscular synapses is **acetylcholine** (**ACh**). ACh is released by exocytosis when the membrane of a vesicle fuses with the presynaptic membrane of the axon terminal.

Where does the neurotransmitter come from? Some neurotransmitters, such as ACh, are synthesized in the axon terminal and packaged in vesicles. The enzymes required for ACh biosynthesis, however, are produced in the cell body of the motor neuron and are transported along microtubules down the axon to the terminals. In contrast, peptide neurotransmitters are produced in the cell body and packaged into membrane-bound vesicles by the Golgi apparatus. These vesicles are rapidly transported down the axon to the terminals.

The postsynaptic membrane of the neuromuscular junction is a modified part of the muscle cell plasma membrane called a **motor end plate** (**Figure 45.13**). It appears as a depression in the muscle cell membrane, and the terminals of the motor neuron sit in the depression. The space between the presynaptic membrane and the postsynaptic membrane is the **synaptic cleft**, which in chemical synapses is about 20 to 40 nanometers wide. ACh released into the cleft by the presynaptic cell diffuses across to the postsynaptic membrane.

### The arrival of an action potential causes the release of neurotransmitter

Neurotransmitter is released when an action potential arrives at the axon terminal and causes the opening of voltage-gated $Ca^{2+}$ channels in the presynaptic membrane. Because the $Ca^{2+}$ concentration is greater outside the cell than inside, $Ca^{2+}$ enters the axon terminal near the sites of vesicle exocytosis. The increase in $Ca^{2+}$ inside the axon terminal causes the vesicles con-

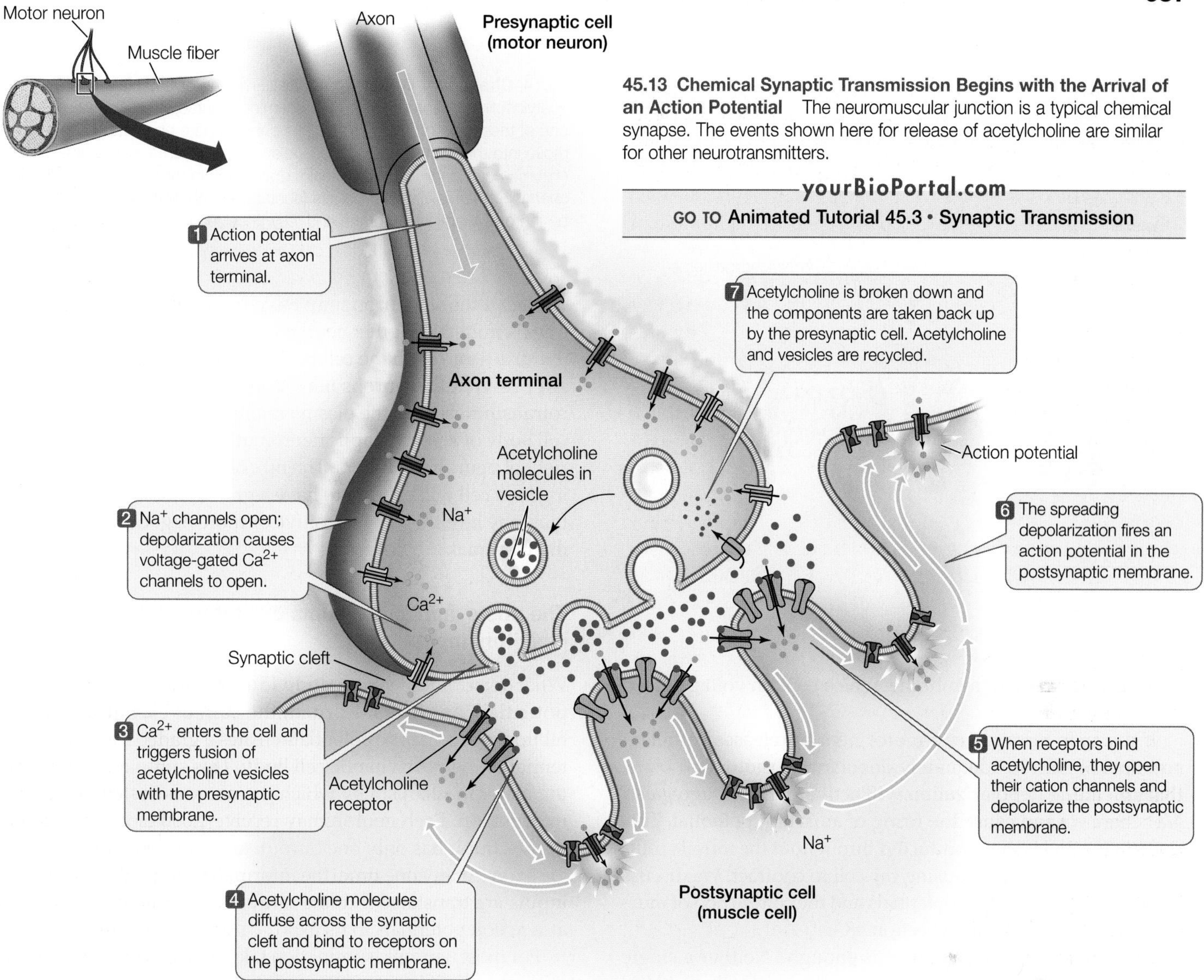
Motor neuron
Muscle fiber
Axon
Presynaptic cell
(motor neuron)
1 Action potential arrives at axon terminal.
Axon terminal
Acetylcholine molecules in vesicle
Na+
2 Na+ channels open; depolarization causes voltage-gated Ca2+ channels to open.
Ca2+
Synaptic cleft
3 Ca2+ enters the cell and triggers fusion of acetylcholine vesicles with the presynaptic membrane.
Acetylcholine receptor
4 Acetylcholine molecules diffuse across the synaptic cleft and bind to receptors on the postsynaptic membrane.
7 Acetylcholine is broken down and the components are taken back up by the presynaptic cell. Acetylcholine and vesicles are recycled.
Action potential
6 The spreading depolarization fires an action potential in the postsynaptic membrane.
5 When receptors bind acetylcholine, they open their cation channels and depolarize the postsynaptic membrane.
Na+
Postsynaptic cell
(muscle cell)

**45.13 Chemical Synaptic Transmission Begins with the Arrival of an Action Potential** The neuromuscular junction is a typical chemical synapse. The events shown here for release of acetylcholine are similar for other neurotransmitters.

yourBioPortal.com

GO TO Animated Tutorial 45.3 • Synaptic Transmission

taining neurotransmitter to fuse with the presynaptic membrane and empty their contents into the synaptic cleft.

In neuromuscular synapses, vesicle fusion and emptying is all-or-none. The vesicle membrane is incorporated into the presynaptic membrane, which actually gets larger as a result—at least until the extra membrane is recycled through endocytosis. The recycled membrane is processed through endosomes to become new vesicles that are then refilled with neurotransmitter.

### Synaptic functions involve many proteins

The description above of the release of neurotransmitter from the presynaptic membrane may seem simple, but it involves hundreds of proteins that are responsible for various aspects of the process: vesicle formation, transport of neurotransmitter into vesicles, anchoring of vesicles to cytoskeletal elements, docking of the vesicles with the presynaptic membrane, fusion of the vesicular and cell membranes, and endocytosis of the vesicle membrane for recycling.

Some of these proteins are the targets of toxins. For example, toxins from bacteria of the genus *Clostridium* are proteases that destroy several of the proteins necessary for the docking of vesicles to the presynaptic membrane. These toxins cause botulism and tetanus, frequently fatal diseases that involve muscle impairment because of loss of transmitter release. Poisons can also become medicines, however. You have surely heard about the use of botulinum toxin (Botox) for decreasing muscle spasms for cosmetic (removal of wrinkles) or therapeutic purposes.

### The postsynaptic membrane responds to neurotransmitter

When ACh is released at a synapse, some of it diffuses across the synaptic cleft and binds to ACh receptors on the postsynaptic membrane (**Figure 45.14**). The postsynaptic membrane of the motor end plate is highly folded. ACh receptors are on the crests of the folds, and voltage-gated cation channels are at the bottoms of the folds and in the surrounding muscle cell membrane (see Figure 45.13). The ACh receptors are channels that allow both $Na^+$ and $K^+$ to flow through, but since the electrochemical gradients favor a net influx of $Na^+$, the response of the motor end plate to ACh is to depolarize. That graded potential reflecting the number of receptors activated spreads to the depths of the folds of the motor end plate membrane and to sur-

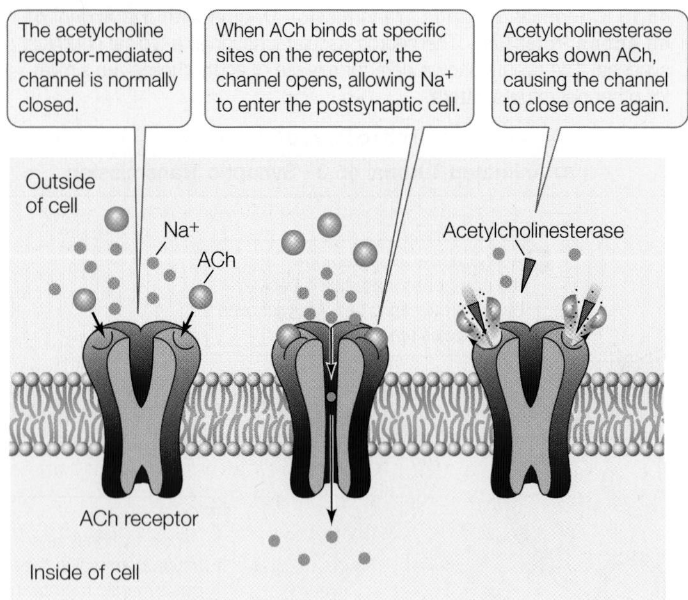

The acetylcholine receptor-mediated channel is normally closed.
When ACh binds at specific sites on the receptor, the channel opens, allowing Na+ to enter the postsynaptic cell.
Acetylcholinesterase breaks down ACh, causing the channel to close once again.
Outside of cell
Na+
ACh
Acetylcholinesterase
ACh receptor
Inside of cell


**45.14 Chemically Gated Channels** The motor end plate contains acetylcholine receptors, which are chemically gated ion channels. When one of these receptors binds ACh, its channel pore opens and $Na^+$ ions move into the postsynaptic cell, depolarizing its plasma membrane. The enzyme acetylcholinesterase breaks down ACh in the synapse, closing the channel; the breakdown products (acetate and choline) are then taken up by the presynaptic membrane and resynthesized into more ACh.

rounding muscle cell membrane, which contain voltage-gated $Na^+$ channels.

If the axon terminal of a motor neuron releases sufficient amounts of ACh to adequately depolarize a motor end plate, that spreading depolarization will activate the voltage-gated $Na^+$ channels and cause the firing of an action potential. This action potential is then conducted throughout the muscle cell's system of membranes, causing the cell to contract. We discuss muscle membrane action potentials and the contraction of muscle cells in greater detail in Section 48.1.

How much neurotransmitter is enough? Neither a single ACh molecule nor the contents of an entire vesicle (about 10,000 ACh molecules) will bring the plasma membrane of a muscle cell to threshold. However, a single action potential in an axon terminal releases the contents of about 100 vesicles, which is more than enough to fire an action potential in the muscle cell and cause it to contract.

### Synapses between neurons can be excitatory or inhibitory

In vertebrates, the synapses between motor neurons and muscle cells are always **excitatory**; that is, motor end plates always respond to ACh with a graded potential that is less negative than the resting potential. However, synapses between neurons can also be inhibitory. For example, recall that there are more chloride ions ($Cl^-$) outside the cell than inside it. If the receptor on the postsynaptic membrane is a $Cl^-$ channel opened by a neurotransmitter, the effect of that neurotransmitter will be to cause $Cl^-$ ions to enter the postsynaptic cell and hyperpolarize it. Hyperpolarization will take the membrane farther from the threshold potential for the voltage-gated $Na^+$ channels, and therefore make it less likely that the cell will fire action potentials. A synapse that causes hyperpolarization of the postsynaptic membrane is **inhibitory**.

Recall that a neuron may have many dendrites. Axon terminals from many other neurons may form synapses with those dendrites and with the cell body. The axon terminals of different presynaptic neurons may store and release different neurotransmitters, and the plasma membrane of the dendrites and cell body of a postsynaptic neuron may have receptors for a variety of neurotransmitters. The mix of synaptic activity impinging on a cell will cause it to have a graded membrane potential that may be either more positive or more negative than its resting potential.

### The postsynaptic cell sums excitatory and inhibitory input

What determines when an individual neuron will fire an action potential? As we just learned, the sum of excitatory and inhibitory postsynaptic potentials creates a graded membrane potential in the postsynaptic cell body. This summation ability is the major mechanism by which the nervous system integrates information. Each neuron may receive 1,000 or more synaptic inputs, but it has only one output: an action potential in a single axon. At any one time, the information from all of the active inputs are translated into the rate at which that neuron generates action potentials in its axon.

For most neurons, summation takes place in the axon hillock at the base of the axon. The plasma membrane of the axon hillock is not insulated by glia and has many voltage-gated $Na^+$ channels. Excitatory and inhibitory postsynaptic potentials from synapses anywhere on the dendrites or the cell body may spread to the axon hillock by local current flow. If the resulting graded potential depolarizes the axon hillock to threshold, it fires an action potential. Because postsynaptic potentials decrease in strength as they spread from the site of the synapse, a synapse at the tip of a dendrite has less influence than a synapse on the cell body, near the axon hillock.

Excitatory and inhibitory postsynaptic potentials are summed over space and over time. **Spatial summation** adds up the simultaneous influences of synapses at different sites on the postsynaptic cell (**Figure 45.15A**). **Temporal summation** adds up postsynaptic potentials generated at the same site in a rapid sequence (**Figure 45.15B**).

### Synapses can be fast or slow

Most neurotransmitter receptors induce changes in postsynaptic cells by opening or closing ion channels. How they do so is the basis for grouping receptors into two general categories:

- **Ionotropic receptors** are ion channels themselves. Neurotransmitter binding to an ionotropic receptor causes a direct

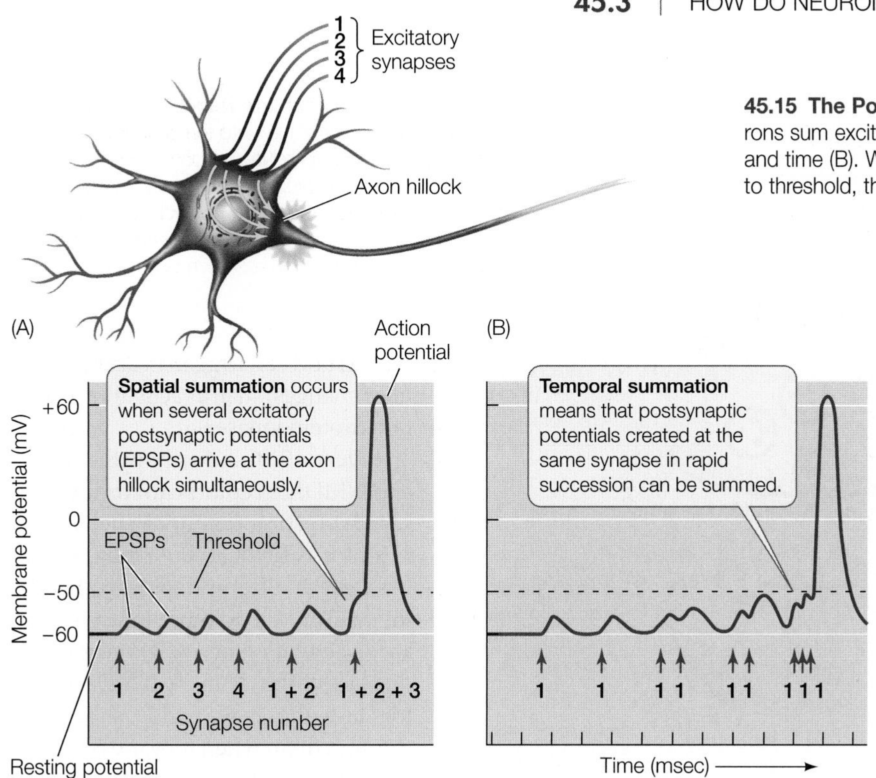


**45.15 The Postsynaptic Neuron Sums Information** Individual neurons sum excitatory and inhibitory postsynaptic potentials over space (A) and time (B). When the sum of the potentials depolarizes the axon hillock to threshold, the neuron generates an action potential.

change in ion movement across the plasma membrane of the postsynaptic cell. These proteins enable fast, short-lived responses.

The ACh receptor of the motor end plate is an example of an ionotropic receptor. It consists of five subunits, each of which extends through the plasma membrane. When assembled, the subunits create a central pore that allows ions to pass through (see Figure 45.14). Of several different kinds of subunits, only one kind has the ability to bind ACh. Each functional receptor has two of the ACh-binding subunits and three other subunits.

- **Metabotropic receptors** are not ion channels, but they induce signaling cascades in the postsynaptic cell that secondarily lead to changes in ion channels (see Figure 7.10A). Postsynaptic cell responses mediated by metabotropic receptors are generally slower and longer-lived than those induced by ionotropic receptors.

Metabotropic receptors are also transmembrane proteins, but instead of acting as ion channels, they initiate an intracellular signaling process that can result in the opening or closing of an ion channel. An example is shown in Figure 7.19.

### Electrical synapses are fast but do not integrate information well

Electrical synapses are different from chemical synapses because they couple neurons electrically. Electrical synapses contain numerous *gap junctions* (see Figure 7.21A). At these synapses, the presynaptic and postsynaptic cell membranes are separated by a space of only 2 to 3 nanometers, and membrane proteins called *connexons* link the two neurons by forming pores that connect the cytoplasm of the two cells. Ions and small molecules can pass directly from cell to cell through these pores. Transmission at electrical synapses is very fast and can proceed in either direction, whereas transmission at chemical synapses is slower and unidirectional.

Electrical synapses are less common in the nervous systems of vertebrates than are chemical synapses for several reasons. First, electrical continuity between neurons does not allow temporal summation of synaptic inputs. Second, an effective electrical synapse requires a large area of contact between the presynaptic and postsynaptic cells. This condition rules out the possibility of thousands of synaptic inputs to a single neuron—which is the norm in complex nervous systems. Third, electrical synapses cannot be inhibitory. Thus, electrical synapses are useful for rapid communication, but they are less useful for processes of integration and learning.

### The action of a neurotransmitter depends on the receptor to which it binds

More than 50 neurotransmitters are now recognized, and more will surely be discovered. ACh, as we have seen, is an important neurotransmitter because it is how the nervous system commands muscles to contract. ACh also plays roles in certain synapses between neurons in the CNS, but it accounts for only a small percentage of the total neurotransmitter content of the CNS.

The workhorse neurotransmitters of the CNS are simple amino acids: glutamate (excitatory) and glycine and γ-aminobutyrate (GABA) (inhibitory). Another important group of neurotransmitters in the CNS is the monoamines, which are derivatives of amino acids. They include dopamine and norepinephrine (derivatives of tyrosine) and serotonin (a derivative of tryptophan). Peptides also function as neurotransmitters; for example, *endorphins* and *enkephalins* are the body's opiates and modulate the sensation of pain. Another peptide, *substance P*, transmits pain sensations. Even a gas, nitric oxide, is used by neurons as an intercellular messenger (see Figure 7.17).

Neurotransmission is complex in part because each neurotransmitter has multiple receptor types. ACh, for example, has two receptor types: *nicotinic receptors*, which are ionotropic, and *muscarinic receptors*, which are metabotropic. Both types of ACh receptors are found in the CNS, where nicotinic receptors tend to be excitatory and muscarinic receptors tend to be inhibitory. ACh actions can differ outside the CNS as well. ACh acting through nicotinic receptors causes the smooth muscle of the gut to increase its motility, but ACh acting through muscarinic receptors causes cardiac muscle to hyperpolarize and therefore to slow down. We could give many more examples of neurotrans-

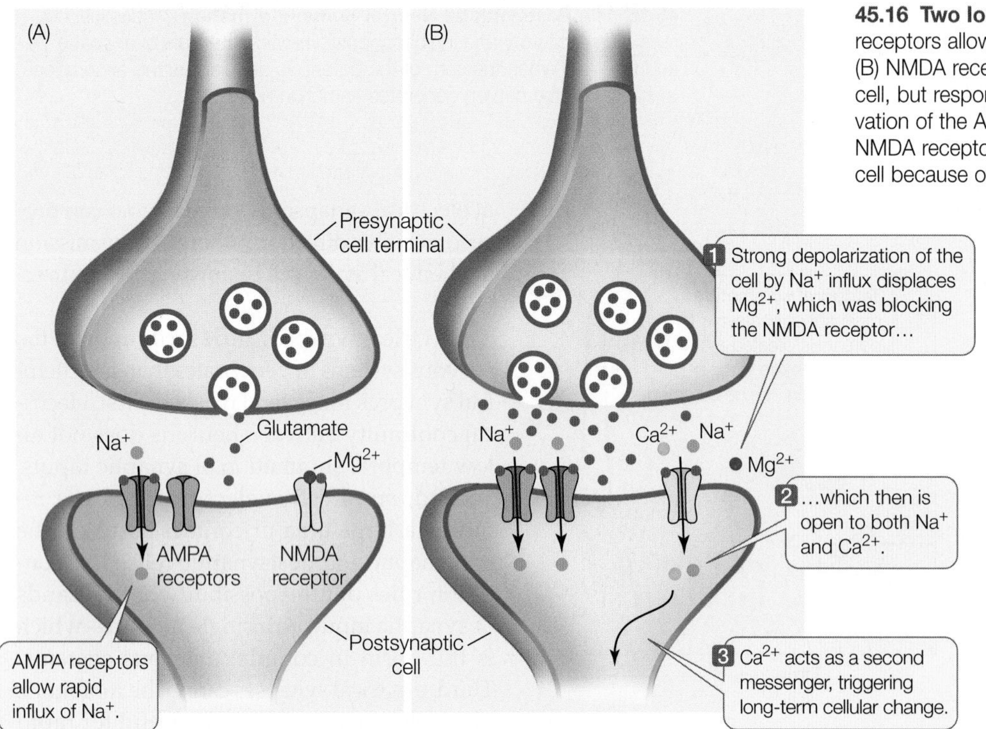
(A)
(B)
Presynaptic cell terminal
Glutamate
Na+
Mg2+
AMPA receptors
NMDA receptor
AMPA receptors allow rapid influx of Na+.
Postsynaptic cell
Na+
Ca2+
Na+
Mg2+
1 Strong depolarization of the cell by Na+ influx displaces Mg2+, which was blocking the NMDA receptor...
2 ...which then is open to both Na+ and Ca2+.
3 Ca2+ acts as a second messenger, triggering long-term cellular change.

**45.16 Two Ionotropic Glutamate Receptors** (A) AMPA receptors allow rapid influx of $Na^+$ into the postsynaptic cell. (B) NMDA receptors allow both $Na^+$ and $Ca^{2+}$ to enter the cell, but respond to synaptic input more slowly. Strong activation of the AMPA receptor leads to unblocking of the NMDA receptor, which results in longer-term effects on the cell because of $Ca^{2+}$ entry.

mitters that have different effects in different tissues, but the important thing to remember is that the action of a neurotransmitter depends on the receptor to which it binds.

**yourBioPortal.com**
GO TO Web Activity 45.1 • Neurotransmitters

## Glutamate receptors may be involved in learning and memory

Glutamate is a neurotransmitter that can bind to a variety of receptors, including both metabotropic and ionotropic receptors. The glutamate receptors are divided into several classes because they can be differentially activated by other chemicals that mimic the action of glutamate. One class of ionotropic glutamate receptors is the *NMDA receptors*, which can be activated by the chemical *N*-methyl-D-aspartate. Another class of ionotropic glutamate receptors is activated by a different chemical, abbreviated as *AMPA* (which stands for α-amino-3-hydroxy-5-methyl-4-isoxazole propionate, demonstrating why biologists are fond of abbreviations).

Glutamate is an excitatory neurotransmitter, so activation of ionotropic glutamate receptors always results in $Na^+$ entry into the neuron and depolarization. But the *timing* of the response to activation by these different types of receptors differs significantly. The AMPA receptors allow a rapid influx of $Na^+$ into the postsynaptic cell, whereas the NMDA receptors allow a slower and longer-lasting influx of $Na^+$. The NMDA receptors require that the cell be somewhat depolarized through the action of other receptors before their pores will open and permit $Na^+$ influx. When they do open, these receptors also allow $Ca^{2+}$ as well as $Na^+$ to enter the cell. Calcium ions act as second messengers in the cell and can trigger a variety of cellular changes, such as activation of certain protein kinases.

**Figure 45.16** shows how the AMPA and NMDA receptors can work in concert. At resting potential, the NMDA receptor is blocked by a magnesium ion ($Mg^{2+}$). Strong depolarization of the neuron due to other inputs—such as the activation of AMPA receptors—displaces $Mg^{2+}$ from the NMDA receptors and allows $Na^+$ and $Ca^{2+}$ to pass through them when they are activated by glutamate. These special properties of the NMDA receptor are probably involved in learning and memory.

Most of the synaptic events we have studied so far happen very quickly. It is therefore a special challenge to understand how the messages carried by action potentials can result in long-term events such as learning and memory. Our understanding of these processes has been greatly enhanced by investigation of a phenomenon called **long-term potentiation** (**LTP**), which has been studied extensively by neurobiologists working with slices of brain kept alive in dishes of culture medium. Using these brain slice preparations, it is possible to stimulate and record from specific brain regions, and even specific neurons.

In the studies leading to the discovery of LTP, experimenters repeatedly stimulated synaptic inputs to a particular neuron and observed the usual action potential response. When the neuron was stimulated many times in rapid succession, however, they found that the properties of the neuron changed. The magnitude of the postsynaptic response was enhanced, or *potentiated*, and this change lasted for days or weeks.

How does potentiation of a synapse occur? The answer, at least for some areas of the brain, now seems clear. With low-frequency stimulation, the glutamate released by presynaptic cells activates only AMPA receptors, and the postsynaptic membrane simply responds with action potentials. With higher frequency stimulation, however, NMDA receptors are also activated, allowing both $Na^+$ and $Ca^{2+}$ ions to enter the postsynaptic neuron. The $Ca^{2+}$ ions induce long-term changes in the postsynaptic membrane that make it more sensitive to synaptic input (**Figure 45.17**).

## To turn off responses, synapses must be cleared of neurotransmitter

Turning off the action of neurotransmitters is as important as turning it on. If released neurotransmitter molecules simply remained in the synaptic cleft, the postsynaptic membrane would become saturated with neurotransmitter, and receptors would be constantly activated. As a result, the postsynaptic cell would remain hyperpolarized or depolarized and would be unresponsive to short-term changes in the presynaptic cell. The more rapidly neurons can respond to input, the more information they can process in a given time. Thus, neurotransmitter must be cleared from the synaptic cleft shortly after it is released by the axon terminal.

Neurotransmitter action may be terminated in several ways. First, enzymes may destroy the neurotransmitter. ACh, for example, is rapidly destroyed by the enzyme acetylcholinesterase, which is present in the synaptic cleft in close association with the ACh receptors on the postsynaptic membrane (see Figure 45.14). Some of the most deadly nerve gases developed for chemical warfare work by inhibiting acetylcholinesterase. As a result, ACh lingers in the synaptic cleft, causing the victim to die of spastic (contracted) muscle paralysis. Some agricultural insecticides, such as malathion, also inhibit acetylcholinesterase and can poison farm workers if used without safety precautions.

Neurotransmitter also may simply diffuse away from the cleft, or be taken up via active transport by nearby cell membranes. The drug commonly prescribed under the brand name Prozac to treat depression slows the reuptake of the neurotransmitter serotonin, thus enhancing its activity at the synapse.

### The diversity of receptors makes drug specificity possible

Many drugs used to treat the nervous system act by modulating specific synaptic interactions. Drugs that mimic or potentiate the effect of a neurotransmitter are called *agonists;* those that block the actions of neurotransmitters are called *antagonists*. For example, morphine is an agonist at the endorphin receptor and therefore blocks pain. Propranolol, known as a β-blocker, is an antagonist of certain adrenergic receptors and therefore decreases panic attacks and anxiety. A major emphasis in neurobiology is to identify neurotransmitter receptor subtypes and design drugs that selectively bind to them to have highly specific effects on nervous system activity.

## INVESTIGATING LIFE

**45.17 Repeated Stimulation Can Cause Long-Term Potentiation**

When a cell receives low-frequency synaptic input, the resulting postsynaptic response remains constant. If that same synaptic pathway is stimulated briefly at a high frequency, however, the subsequent sensitivity of the postsynaptic cell to the original level of synaptic inputis potentiated for a longer period.

**HYPOTHESIS** **Repeated stimulation can change the properties of a synapse.**

**METHOD** Stimulate a cell through two separate input pathways (two sources of stimulus) and record responses.

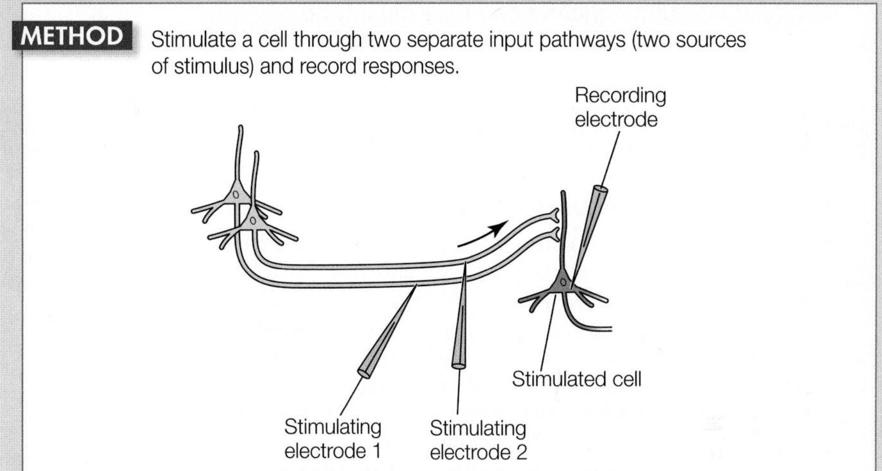


**RESULTS**

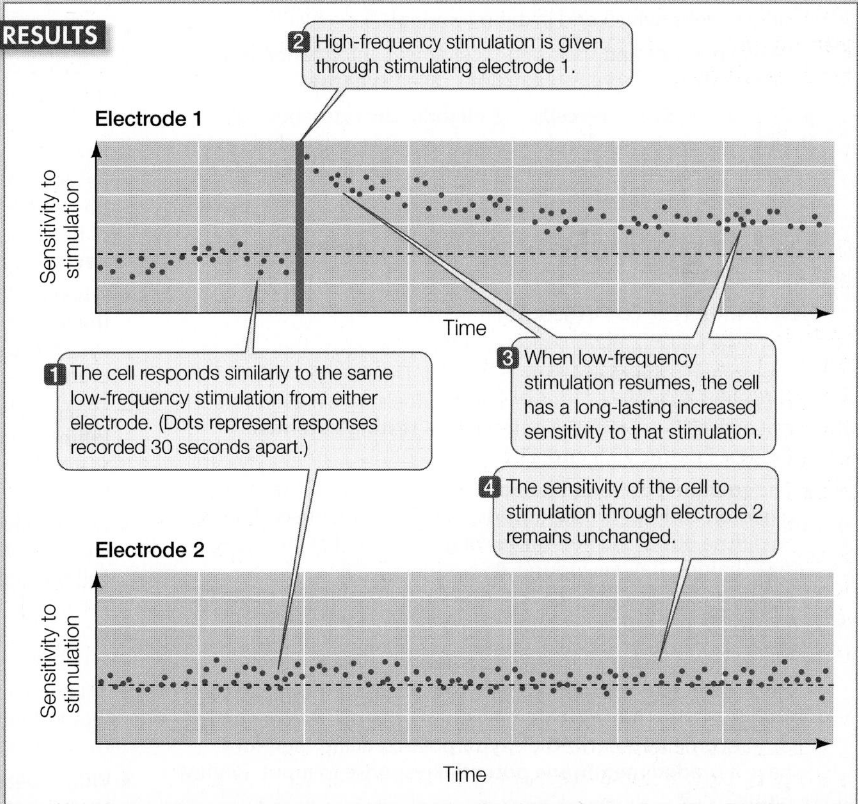

2 High-frequency stimulation is given through stimulating electrode 1.
Electrode 1
Sensitivity to stimulation
Time
1 The cell responds similarly to the same low-frequency stimulation from either electrode. (Dots represent responses recorded 30 seconds apart.)
3 When low-frequency stimulation resumes, the cell has a long-lasting increased sensitivity to that stimulation.
4 The sensitivity of the cell to stimulation through electrode 2 remains unchanged.
Electrode 2
Sensitivity to stimulation
Time


**CONCLUSION** **High-frequency stimulation of synapses can result in a long-lasting change in the sensitivity of these synapses (long-term potentiation, or synaptic memory).**

Go to **yourBioPortal.com** for original citations, discussions, and relevant links for all INVESTIGATING LIFE figures.

## 45.3 RECAP

Chemical synapses involve the release of neurotransmitter molecules stored in vesicles in the presynaptic terminal. Action potentials reaching that terminal cause the fusion of vesicles with the presynaptic membrane, releasing neurotransmitter that can then bind to receptors on the postsynaptic membrane and influence its membrane potential. There is a great diversity of neurotransmitters and their receptors.

- Describe the role of $Ca^{2+}$ channels in synaptic events. See pp. 956–957 and Figure 45.13
- How can some synapses be excitatory and others inhibitory? See p. 958
- How do neurons integrate the input from various synapses? See p. 958 and Figure 45.15

## CHAPTER SUMMARY

### 45.1 What Cells Are Unique to the Nervous System?

- Nervous systems include **neurons** and **glia**. Neurons are organized in circuits with sensory inputs, integration, and outputs to effectors. Glia serve support functions. **Review Figures 45.1 and 45.3**
- In vertebrates, the **brain** and **spinal cord** form the **central nervous system** (**CNS**), which communicates with the rest of the body via the **peripheral nervous system** (**PNS**). The CNS increases in complexity from invertebrates to vertebrates and from fishes to mammals. **Review Figures 45.1 and 45.2**
- Neurons generally receive information via their **dendrites**, of which there can be many, and transmit information via their single **axons**, which end in **axon terminals**. **Review Figure 45.3**
- Where neurons and their target cells meet, information is transmitted across specialized junctions called **synapses**.
- Glia include **Schwann cells** and **oligodendrocytes,** both of which generate **myelin** sheets on axons. Glia also include **astrocytes**, which support neurons metabolically and contribute to the **blood–brain barrier**. **Review Figure 45.4**

### 45.2 How Do Neurons Generate and Transmit Electrical Signals?

**SEE ANIMATED TUTORIAL 45.1**

- Neurons have an electric charge difference across their plasma membranes, the **membrane potential**. The membrane potential is created by ion transporters and channels. When a neuron is not active, its membrane potential is a **resting potential**. **Review Figures 45.5 and 45.6**
- The **sodium–potassium pump** concentrates $K^+$ on the inside of a neuron and $Na^+$ on the outside. Potassium channels allow $K^+$ to diffuse out of the neuron, leaving behind unbalanced negative charges. **Review Figures 45.6 and 45.7**
- **Patch clamping** allows the study of single ion channels. **Review Figure 45.8**
- The resting potential is perturbed when ion channels open or close, changing the permeability of the plasma membrane to charged ions. Through this mechanism, the plasma membrane can become **depolarized** or **hyperpolarized** and therefore have a **graded membrane potential** response to input. **Review Figure 45.9**
- An **action potential** is a rapid reversal in charge across a portion of the plasma membrane resulting from the sequential opening and closing of **voltage-gated channels** of $Na^+$ and $K^+$. These changes in voltage-gated channels occur when the plasma membrane depolarizes to a **threshold** level. **Review Figure 45.10, ANIMATED TUTORIAL 45.2**
- Action potentials are all-or-none, self-regenerating events. They are conducted down axons because local current flow depolarizes adjacent regions of membrane and brings them to threshold. **Review Figure 45.11**
- In myelinated axons, action potentials appear to jump between **nodes of Ranvier**, patches of axonal plasma membrane that are not covered by myelin. **Review Figure 45.12**

### 45.3 How Do Neurons Communicate with Other Cells?

- Neurons communicate with each other and with other cells by transmitting information over **chemical synapses** (with neurotransmitters) or **electrical synapses**.
- The neuromuscular junction is a well-studied chemical synapse between a motor neuron and a skeletal muscle cell. Its neurotransmitter is **acetylcholine** (**ACh**), which causes depolarization of the postsynaptic membrane when it binds to its receptor. **Review Figure 45.13, ANIMATED TUTORIAL 45.3**
- When an action potential reaches an axon terminal, it causes the release of neurotransmitters, which diffuse across the **synaptic cleft** and bind to receptors on the postsynaptic membrane. **Review Figures 45.13 and 45.14**
- Synapses between neurons can be either **excitatory** or **inhibitory**. A postsynaptic neuron integrates information by summing excitatory and inhibitory postsynaptic potentials in both space and time. **Review Figure 45.15**
- **Ionotropic receptors** are ion channels or directly influence ion channels. **Metabotropic receptors** influence the postsynaptic cell through various signal transduction pathways and result in the opening or closing of ion channels. The actions of ionotropic synapses are generally faster than those of metabotropic synapses.
- There are many different neurotransmitters and even more types of receptors. The action of a neurotransmitter depends on the receptor to which it binds. **SEE WEB ACTIVITY 45.1**
- With repeated stimulation, a neuron can become more sensitive to its inputs through **long-term potentiation** (**LTP**). The properties of the NMDA glutamate receptor appear to explain LTP. **Review Figures 45.16 and 45.17**

## SELF-QUIZ

1. The rising phase of an action potential is due to the
   *a.* closing of $K^+$ channels.
   *b.* opening of chemically gated $Na^+$ channels.
   *c.* closing of voltage-gated $Ca^{2+}$ channels.
   *d.* opening of voltage-gated $Na^+$ channels.
   *e.* spread of positive current along the plasma membrane.
2. The resting potential of a neuron is due mostly to
   *a.* local current spread.
   *b.* open $Na^+$ channels.
   *c.* synaptic summation.
   *d.* open $K^+$ channels.
   *e.* open $Cl^-$ channels.
3. Which statement about synaptic transmission is *not* true?
   *a.* The synapses between neurons and skeletal muscle cells use ACh as their neurotransmitter.
   *b.* A single vesicle of neurotransmitter can cause a muscle cell to contract.
   *c.* The release of neurotransmitter at the neuromuscular junction causes the motor end plate to depolarize.
   *d.* In vertebrates, the synapses between motor neurons and muscle fibers are always excitatory.
   *e.* Inhibitory synapses cause the resting potential of the postsynaptic membrane to become more negative.
4. Which statement accurately describes an action potential?
   *a.* Its magnitude increases along the axon.
   *b.* Its magnitude decreases along the axon.
   *c.* All action potentials in a single neuron are of the same magnitude.
   *d.* During an action potential, the membrane potential of a neuron remains constant.
   *e.* An action potential permanently shifts a neuron's membrane potential away from its resting value.
5. A neuron that has just fired an action potential cannot be immediately restimulated to fire a second action potential. The short interval of time during which restimulation is not possible is called
   *a.* hyperpolarization.
   *b.* the resting potential.
   *c.* depolarization.
   *d.* repolarization.
   *e.* the refractory period.
6. Graded membrane potentials
   *a.* can be more negative than resting potential.
   *b.* can be less negative than resting potentials.
   *c.* integrate the many synaptic inputs to a cell.
   *d.* are important means of summing sensory inputs.
   *e.* are all of the above.
7. The binding of an inhibitory neurotransmitter to the postsynaptic receptors results in
   *a.* depolarization of the membrane.
   *b.* generation of an action potential.
   *c.* hyperpolarization of the membrane.
   *d.* increased permeability of the membrane to sodium ions.
   *e.* increased permeability of the membrane to calcium ions.
8. The difference between slow and fast synapses is
   *a.* the width of the synaptic cleft.
   *b.* the size of the synapse.
   *c.* whether or not the neurotransmitter acts directly on ion channels.
   *d.* the density of receptors on the postsynaptic membrane.
   *e.* the amount of neurotransmitter that is released.
9. Whether a synapse is excitatory or inhibitory depends on the
   *a.* type of neurotransmitter.
   *b.* presynaptic axon terminal.
   *c.* size of the synapse.
   *d.* nature of the postsynaptic receptors.
   *e.* concentration of neurotransmitter in the synaptic space.
10. Which of the following is a likely mechanism for long-term potentiation?
    *a.* When glutamate binds to postsynaptic AMPA receptors, it activates intracellular changes.
    *b.* When glutamate binds to NMDA receptors, it allows magnesium ions to enter the cell, which initiate intracellular changes.
    *c.* When sufficient glutamate is released by the presynaptic neuron, it dislodges $Mg^{2+}$ from the AMPA receptors and allows $Ca^{2+}$ to leave the cell.
    *d.* When sufficient glutamate is released, both AMPA and NMDA receptors are activated, and NMDA receptors allow $Ca^{2+}$ as well as $Na^+$ to enter the cell, thus initiating intracellular changes.
    *e.* When both glutamate and ACh are released together, they create a long-lasting depolarization of the postsynaptic cell.

## FOR DISCUSSION

1. The language of the nervous system consists of one "word," the action potential. How can this single message convey a diversity of information, how can that information be quantitative, and how can it be integrated?
2. If you stimulate an axon in the middle, action potentials are conducted in both directions. Yet when an action potential is generated at the axon hillock, it goes only toward the axon terminals and does not backtrack. Explain why action potentials are bidirectional in the first example and unidirectional in the second.
3. The nature of synapses presents various opportunities for plasticity in the nervous system. Discuss at least four synaptic mechanisms that could be altered to change the response of a neuron to a specific input.
4. Benzodiazepines are drugs that potentiate the effects of GABA on its receptor. What effects would you expect these drugs to have?

## ADDITIONAL INVESTIGATION

The patch clamping technique shown in Figure 45.8 produces an "inside-out" patch because the cytoplasmic side of the membrane faces away from the pipette tip and the surface side of the membrane is exposed to the solution in the pipette. Another patch preparation is the "outside-out patch" in which the cytoplasmic side of the membrane is exposed to the solution in the pipette. Describe what kind of experiments you could conduct with the "inside-out" patch and what kind you could conduct with the "outside-out" patch.

# 46 Sensory Systems

## Out of range

A rattlesnake can see to strike and kill a running rodent in complete darkness. How can this be, when "seeing" means using the eyes to detect light, and "complete darkness" means the absence of any light? It is possible because these definitions are based on human capabilities. What we call "light" is actually only a small portion (red, orange, yellow, green, blue, indigo, and violet) of the spectrum of electromagnetic radiation. Other animals see wavelengths we cannot. Insects and birds see patterns on flowers that reflect ultraviolet wavelengths invisible to us. Similarly, rattlesnakes can "see" infrared wavelengths that are invisible to us (although at high enough levels of intensity, humans feel these wavelengths as heat).

It is not the snake's eyes that perceive infrared wavelengths. Rattlesnakes and their relatives have *pit organs* containing high densities of infrared-sensitive neurons. The two pits, located between the nostrils and the eyes on each side of the skull, are positioned in such a way that sensory receptor cells in the pits receive directional information. The fields of "view" of the bilateral pits are overlapping and thus convey a three-dimensional perspective. Information from the pit organs goes to the same region of the brain as information from the eyes, so rattlesnakes actually do "see" the world in a range of electromagnetic radiation that is different from the human visual spectrum.

Our definition of silence is as arbitrary as our definition of darkness. "Sound" is actually pressure waves in the environment, and many animals are sensitive to pressure waves with frequencies (*pitches*) we cannot hear. Elephants communicate in sound waves that are below human hearing range; such long waves travel great distances, an advantage to large animals that roam over extensive areas.

Bats emit incredibly loud, brief sound pulses, the frequencies of which are above our range of hearing, and a flying bat hears echoes of these pulses bouncing off objects in the environment. The pulses are so loud and the echoes so weak that it is rather like a construction worker trying to overhear

**Sensing Infrared Radiation** The "hole" to the right of this diamondback rattlesnake's eye is one of its bilateral pit organs. Pit organs detect infrared radiation from the snake's preferred prey—small rodents—with unerring precision, even in total darkness. The forked tongue also provides positional information, picking up molecular signals that are transmitted to the brain by a specialized organ in the roof of the snake's mouth.

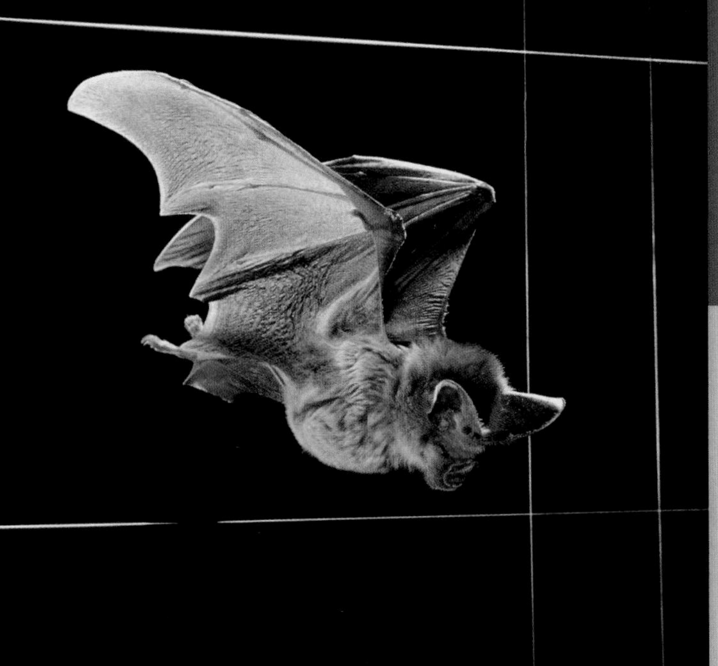

**Echolocating around an Obstacle Course** A bat's ability to echolocate using sound waves is so precise that, in a totally dark room strung with fine wires, bats can capture small insects while avoiding the wires.

a whispered conversation while using a pneumatic drill. Why don't the loud pulses "drown out" the weak echoes for the bat? Small muscles in the bat's ears contract to dampen their hearing sensitivity while the sounds are being emitted, but relax in time for the bat to hear the echo—a truly remarkable ability, since the pulses are emitted at rates of 20 to 80 per second.

Our senses are our windows on the world. "Reality" is what our eyes see, our ears hear, our noses smell, and what we touch and taste. But human beings sense only a limited range of the information available. Animals with different ranges of sensitivity process different sources of information and may perceive "reality" quite differently.

**IN THIS CHAPTER** we will examine how sensory receptor cells convert environmental stimuli into the electrochemical signals of nervous systems. We will examine in detail the diversity of cells responsible for our senses and see how they are incorporated into sensory systems that provide the central nervous system with information about the world around and within us. We will also learn about the unusual sensory abilities of other animals.

## CHAPTER OUTLINE



## 46.1 How Do Sensory Cells Convert Stimuli into Action Potentials?

*Sensory receptor cells*, sometimes simply called *sensors* or *receptors*, transduce (convert) physical and chemical stimuli such as light and sound waves, touch, and odorant and taste molecules into neural signals. These signals are then transmitted to the central nervous system for processing and interpretation. We will examine several sensory systems. In each case, we ask the same general question: How do sensory receptor cells transduce energy from a stimulus into a change in membrane potential?

### Sensory receptor proteins act on ion channels

As discussed in Section 45.2, cells use energy to create gradients of charged ions across their plasma membranes. In this way, cells are like batteries: batteries store potential energy by separating electrical charges between their poles, and cells store potential energy in the ionic gradients across their plasma membranes. If the two poles of a battery are connected by a wire and a switch, current flows through the wire when the switch is closed. Ion channels are like switches. When they open, charged ions can flow down their electrochemical gradient. When the ion channel is selective for one type of ion, the flow of that charged ion creates a change in the electrical potential across the membrane.

**Sensory transduction** begins with a **receptor protein** or some other mechanism in a sensory cell that can detect a specific stimulus modality such as heat, light, chemicals, mechanical force (including sound waves), or electrical fields. The receptor protein then directly or indirectly opens or closes ion channels in that sensory cell, leading either to an action potential or to the release of neurotransmitter.

Section 45.3 notes that synaptic receptor proteins are either *ionotropic* or *metabotropic*; the same distinction can be applied to sensory receptor proteins. Ionotropic sensory receptor proteins are either ion channels themselves or directly affect the opening of an ion channel. Examples are receptors that respond to physical force (*mechanoreceptors*) and those that respond to temperature (*thermoreceptors*). *Electrosensors* most likely have no receptor protein at all, but they are grouped with the ionotropic receptors; the plasma membrane of electrosensory cells is sensitive to the voltage across it and releases neurotransmitter in response to slight changes in membrane potential. Metabotropic sensory receptor proteins influence ion channels indirectly, through G proteins and second messengers as described in Chapter 7. Examples are most *chemoreceptors* and *photoreceptors* (**Figure 46.1**).

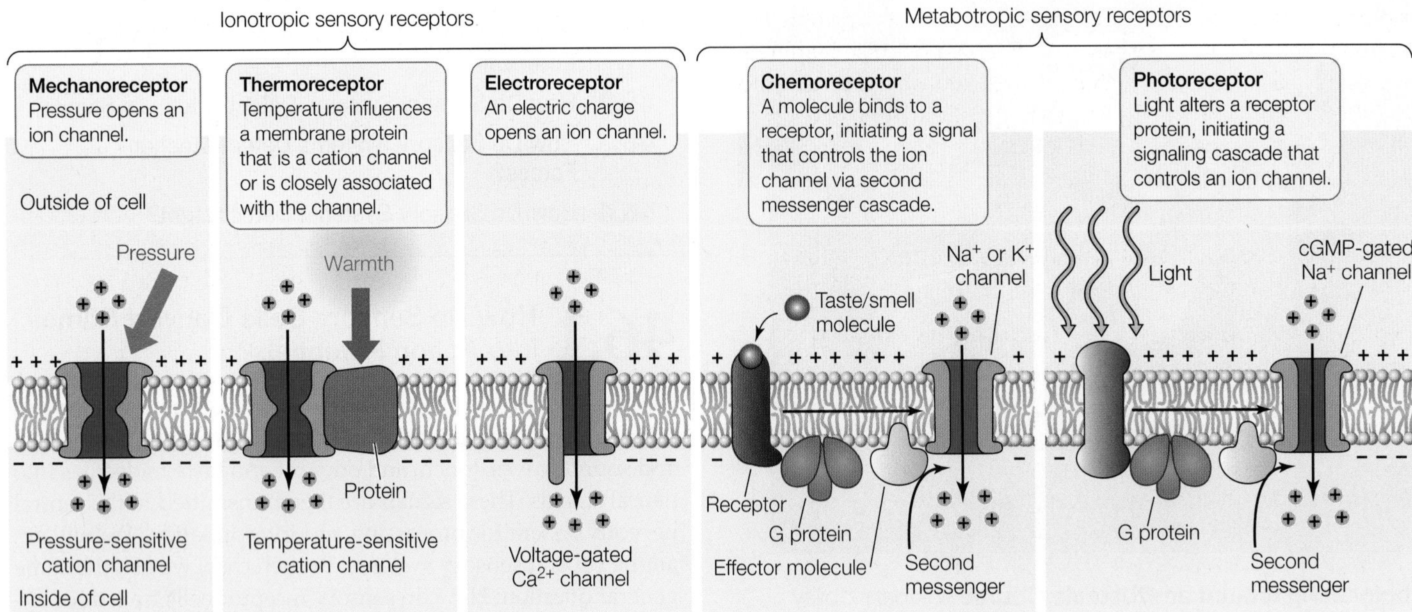


**46.1 Sensory Cell Membrane Receptor Proteins Respond to Stimuli** The receptor proteins in mechanoreceptors are ion channels. The activated receptor proteins of metabotropic chemoreceptors and photoreceptors initiate signal transduction cascades that eventually open or close ion channels.

## Sensory transduction involves changes in membrane potentials

A change in the resting membrane potential of a sensory receptor cell in response to a stimulus is called a **receptor potential**. Receptor potentials are graded membrane potentials that spread over only short distances. To travel long distances in the nervous system, receptor potentials must generate action potentials, which they can do in two ways:

- The receptor potential may generate action potentials within the receptor cell itself.
- The receptor potential may trigger the release of a neurotransmitter that induces a postsynaptic neuron to generate action potentials.

A good example of a sensory receptor cell that can generate action potentials is the stretch receptor of a crayfish (**Figure 46.2**). By placing an electrode in the cell body of a crayfish stretch receptor, we can record the receptor potential that results from stretching the muscle to which the cell's dendrites are attached. These receptor potentials spread to the base of the cell's axon (the axon hillock), which contains voltage-gated $Na^+$ channels. Action potentials generated here travel down the axon to the central nervous system (CNS). The rate at which action potentials are fired by the axon depends on the magnitude of the receptor potential; that, in turn, depends on how much the muscle is stretched.

In a receptor cell that does not fire action potentials, the spreading receptor potential reaches a presynaptic patch of plasma membrane and induces the release of a neurotransmitter. The intensity of the stimulus influences how much neurotransmitter is released. That neurotransmitter binds to receptor proteins on an associated sensory neuron, altering its membrane

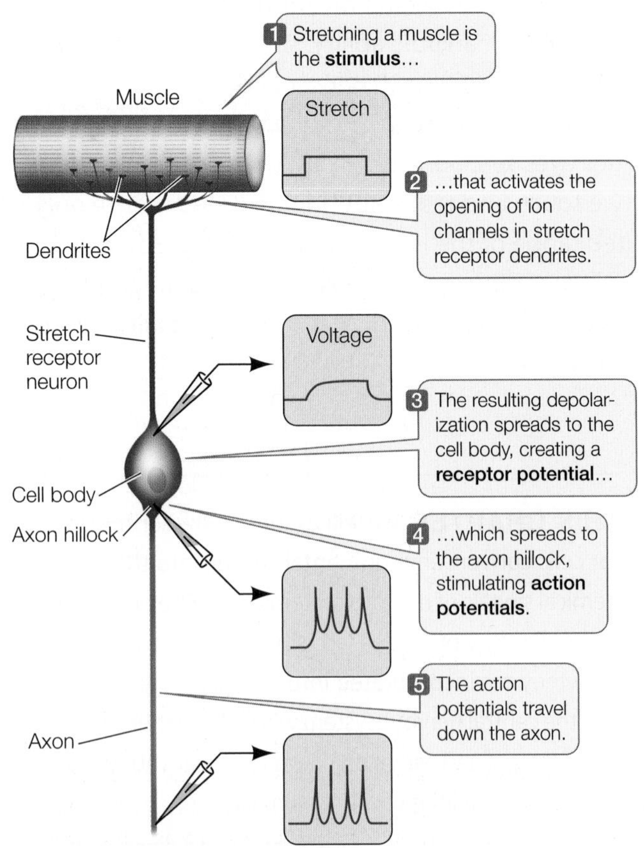

1 Stretching a muscle is the **stimulus**...
Muscle
Stretch
2 ...that activates the opening of ion channels in stretch receptor dendrites.
Dendrites
Stretch receptor neuron
Voltage
3 The resulting depolarization spreads to the cell body, creating a **receptor potential**...
Cell body
Axon hillock
4 ...which spreads to the axon hillock, stimulating **action potentials**.
5 The action potentials travel down the axon.
Axon


**46.2 Stimulating a Sensory Cell Produces a Receptor Potential** Signal transduction in the stretch receptor of a crayfish can be investigated by measuring the membrane potential at different places on the stretch receptor neuron while stretching the muscle innervated by that sensory neuron.

potential and causing it to increase or decrease its rate of firing action potentials. In a few cases, this second cell also responds by changing the rate at which it releases neurotransmitter onto another neuron. Eventually, however, the stimulation of the sensory cell is coded as a change in firing of action potentials in a sensory circuit.

### Sensation depends on which neurons receive action potentials from sensory cells

All sensory systems process information in the form of action potentials. But the sensations we perceive—such as heat, pressure, light, smell, and sound—differ because the messages from different kinds of sensory cells arrive at different places in the CNS. Action potentials arriving in the visual cortex of the brain are interpreted as light, in the auditory cortex as sound, in the olfactory bulb as smell, and so forth.

A small patch of skin on your arm contains some sensory receptor cells that increase their firing rates when the skin is warmed and others that increase their activity when the skin is cooled. Other types of sensory cells in the same patch of skin respond to touch, movement of hairs, irritants such as mosquito bites, and painful stimuli. These receptor cells transmit their messages through axons that enter the CNS at the spinal cord. The synapses made by those axons in the spinal cord and the subsequent pathways of transmission determine whether the stimulation of the patch of skin on your arm is perceived as warmth, cold, touch, tickle, itch, or pain. So even though the action potentials carried by all of these sensory axons look the same, the connectivity of each axon is specific for a given sensory modality.

How is the intensity of the stimulus encoded if, as Section 45.2 describes, each action potential is an all-or-none event? Intensity of sensation is coded as the frequency of the action potentials.

Some sensory cells transmit information about internal conditions in the body, but we may not be consciously aware of that information. The brain continuously receives information about body temperature, blood carbon dioxide and oxygen concentrations, arterial pressure, muscle tension, and the positions of the limbs—all of which are important for homeostasis. All sensory cells produce information that the nervous system can use, but that information does not always result in conscious sensation.

Some sensory receptor cells are assembled with other types of cells into *sensory organs*, such as eyes, ears, and noses, that enhance the ability of the sensory cells to collect, filter, and amplify stimuli. We therefore refer to **sensory systems**, which include the sensory cells, the associated structures, and the neural networks that process the information.

### Many receptors adapt to repeated stimulation

Some sensory cells give gradually diminishing responses to maintained or repeated stimulation. This phenomenon is known as **adaptation**, and it enables an animal to ignore background or unchanging conditions while remaining sensitive to changes or to new information. (Note that this use of the term "adaptation" is different from its application in an evolutionary context.) When you dress, you feel each item of clothing touch your skin, but the sensation of clothes touching your skin is not constantly on your mind throughout the day. You are immediately aware, however, when a seam rips, your shoe comes untied, or someone touches your back.

Animals can discriminate between continuous and changing stimuli partly because some sensory cells adapt; it is also a result of information processing by the CNS. Some sensory cells adapt very little or very slowly; examples are some types of pain receptors and the mechanoreceptors for balance. You do not want to ignore pain that is signaling that something is wrong in your body, and to maintain equilibrium you must continuously know the tensions and forces on all of your joints and muscles.

### 46.1 RECAP

**Sensory receptor cells have receptor proteins that respond to specific stimuli from the external or internal environment by opening or closing ion channels, which results in the generation of action potentials in sensory neurons.**

- Explain the difference between ionotropic and metabotropic sensory receptor proteins. **See p. 965 and Figure 46.1**
- How are we able to perceive action potentials—which are all essentially the same—as different sensations? **See p. 967**

Now that we have a general view of how sensory systems code and process information, we will discuss how sensory systems gather and filter stimuli, transduce specific stimuli into action potentials, and transmit action potentials to the central nervous system. We will look in more depth at specific sensory modalities, beginning with chemosensation, the basis of smell and taste.

## 46.2 How Do Sensory Systems Detect Chemical Stimuli?

A colony of corals responds to a small amount of meat extract in seawater by extending bodies and tentacles and searching for food; a solution of a single amino acid can stimulate this response. Conversely, a small amount of seawater in which corals were crushed will stimulate a defensive retraction of the coral polyps. Humans also react strongly to certain chemical stimuli. When we smell freshly baked bread, we salivate and feel hungry; when we smell rotting meat, we feel nauseated.

All animals receive information about chemical stimuli through **chemoreceptors**, which are receptor proteins that bind to various molecules — their *ligands*—and are responsible for smell and taste. Chemoreceptors are also responsible for monitoring aspects of the internal environment such as the level of carbon dioxide in the blood. Information from chemoreceptors can cause powerful physiological and behavioral responses during activities such as feeding, mating, fighting, and recognizing individuals.

### Arthropods are good models for studying chemoreception

A chemical signal used in communication among members of a species is a **pheromone**. Arthropods use pheromones to attract mates. These signals demonstrate the sensitivity of chemosensory systems, and one of the best-studied examples is that of the silkworm moth *Bombyx mori*.

To attract a mate, the female silkworm moth releases a pheromone called *bombykol* from a gland at the tip of her abdomen. The male silkworm moth has receptors for this molecule on his antennae (**Figure 46.3**). Each feathery antenna carries about 10,000 bombykol-sensitive hairs. A single molecule of bombykol may be sufficient to generate action potentials in the antennal nerve that transmits the signal to the CNS. Because of the male's great sensitivity, the sexual message of a female moth is likely to reach any male within a huge downwind area. When approximately 200 hairs per second are activated, the male orients upwind in search of the female. Because the rate of firing in the male's sensory nerves is proportional to the bombykol concentration in the air, he can follow the airborne concentration gradient and home in on the signaling female.

### Olfaction is the sense of smell

The sense of smell, **olfaction**, depends on chemoreceptors. In vertebrates, the olfactory sensors are neurons embedded in a layer of epithelial tissue at the top of the nasal cavity. The axons from these neurons extend to the olfactory integration area of the brain (the olfactory bulb), whereas their dendrites end in olfactory cilia on the surface of the nasal epithelium. A protective layer of mucus covers the epithelium. Molecules from the environment must diffuse through this mucus to reach the receptor proteins on the olfactory cilia. When you have a cold, the amount of mucus in your nose increases, and the epithelium swells. With this in mind, study **Figure 46.4**, and you will easily understand why respiratory infections can cause you to lose your sense of smell.

Humans have a sensitive olfactory system, but we are unusual among mammals in that we depend more on vision than on olfaction. A typical dog's nasal epithelium is 15 to 20 times larger than a human's and has around 1 billion receptors, compared with around 20 million in humans. For some scents, the threshold sensitivity of the dog is 100 million times lower. A dog's nose reveals a huge amount of information not available to us.

An **odorant** is a molecule that activates an olfactory receptor protein. Odorants bind to receptor proteins on the olfactory cilia of the olfactory neurons. Olfactory receptor proteins are specific for particular odorant molecules. When an odorant molecule binds to its receptor on an olfactory neuron, it activates a G protein. The G protein in turn activates an enzyme that causes an increase of a second messenger (cAMP in vertebrates) in the cytoplasm (see Figure 7.19). The second messenger binds to cation channels in the sensory cell's plasma membrane and opens them, causing an influx of $Na^+$. The sensory neuron depolarizes to threshold and fires action potentials.

The olfactory world has an enormous number of odors, and accordingly, there are a large number of olfactory receptor proteins. In the 1990s, Linda Buck and Richard Axel discovered in mice a family of about 1,000 genes (about 3 percent of the genome) that code for olfactory receptor proteins. Humans have about one-third that number of functional olfactory receptor genes. Each receptor protein that is expressed is found in a limited number of receptor cells in the olfactory epithelium, and each cell expresses just one receptor type. Using a combination of patch clamping (see Figure 45.8) and molecular techniques, the investigators were able to match specific gene prod-

**46.3 Some Scents Travel Great Distances** Mating in silkworm moths of the genus *Bombyx* is coordinated by a pheromone called bombykol.

(A)

The female moth releases a pheromone from a gland at the tip of her abdomen. The pheromone can travel thousands of meters downwind.

(B)

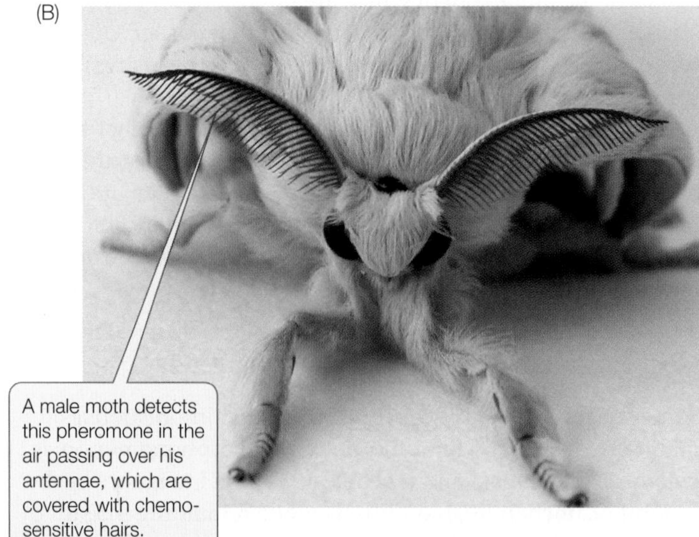

A male moth detects this pheromone in the air passing over his antennae, which are covered with chemo-sensitive hairs.

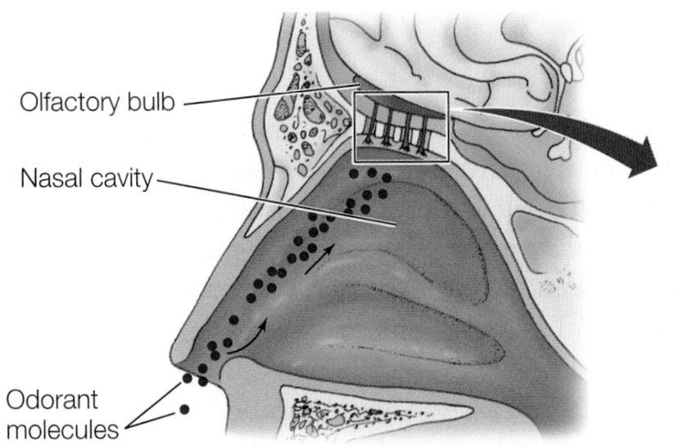
Olfactory bulb
Nasal cavity
Odorant molecules

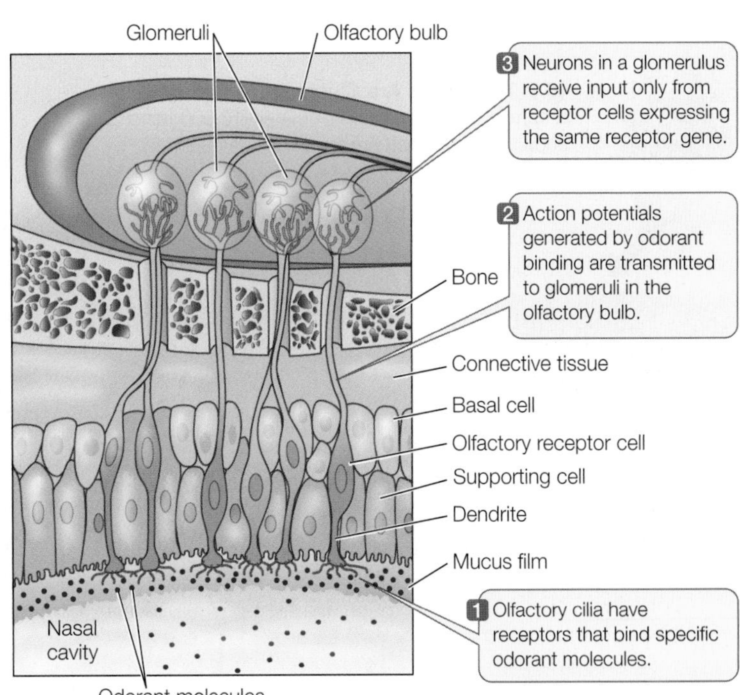
Glomeruli
Olfactory bulb
3 Neurons in a glomerulus receive input only from receptor cells expressing the same receptor gene.
2 Action potentials generated by odorant binding are transmitted to glomeruli in the olfactory bulb.
Bone
Connective tissue
Basal cell
Olfactory receptor cell
Supporting cell
Dendrite
Mucus film
1 Olfactory cilia have receptors that bind specific odorant molecules.
Nasal cavity
Odorant molecules

**46.4 Olfactory Receptors Communicate Directly with the Brain** The receptor cells of the human olfactory system are embedded in epithelial tissues lining the nasal cavity and send their axons to the olfactory bulb of the brain.

ucts with the odorants they detect. For their discoveries of the molecular nature of the olfactory system, Buck and Axel received the Nobel prize in 2004.

Olfactory sensitivity enables discrimination of many more odorants than there are olfactory receptors. An odorant molecule can be quite complex, and different regions of that molecule may bind to different receptor proteins. The next stage of processing olfactory information is in the olfactory bulb, where axons from neurons expressing the same receptor protein cluster together on olfactory bulb neurons, forming structures called *glomeruli* (see Figure 46.4). Therefore, a complex odorant molecule can activate a unique combination of glomeruli in the olfactory bulb, so an olfactory system with hundreds of different receptor proteins can discriminate an astronomically large number of smells. How does the olfactory receptor cell signal the intensity of a smell? The more odorant molecules that bind to receptors, the greater the frequency of action potentials and the greater the intensity of the perceived smell.

In vertebrates, odorants and other chemoreceptor molecules can stimulate strong responses. Many vertebrate species have separate olfactory epithelia that are dedicated to the detection of pheromones.

### The vomeronasal organ contains chemoreceptors

The **vomeronasal organ** (**VNO**) is a small, paired tubular structure embedded in the nasal epithelium of amphibians, reptiles, and many mammals (although probably not humans). In mammals, the VNO is located on the septum dividing the two nostrils (see Figure 53.5).

The vomeronasal organ has a pore that opens into the nasal cavity. When the animal sniffs, the VNO pulsates and draws a sample of nasal fluid over the chemoreceptors embedded in its walls. The information from these chemoreceptors goes to an accessory olfactory bulb in the brain, and information from there goes to brain regions involved in sexual and other instinctive behaviors.

In snakes, the VNO opens into the roof of the mouth cavity. Each time the snake's forked tongue darts in and out, the forks fit into the VNO openings and present the chemoreceptors located there with a sample of molecules from the surrounding air (see the chapter-opening photo). Thus the snake uses its tongue to smell its environment, not to taste it. Why doesn't the snake simply use the flow of air to and from its lungs, as we do, to smell the environment? In reptiles, air flows to and from the lungs slowly (and can even stop entirely for long periods of time), but the tongue can dart in and out many times in a second. It is a quick source of olfactory information.

Studies on mice have led to the hypothesis that the mammalian VNO is a specialized olfactory organ that detects pheromones. Lawrence Katz and his colleagues at Duke University recorded the activity of neurons in the mouse accessory olfactory bulb, which receives input from chemosensors in the VNO. These accessory olfactory neurons were activated when a mouse attached to recording electrodes sniffed another mouse placed in the same cage. However, the neurons fired differentially, depending on the gender and strain of the "intruder" mouse. Other studies on mouse behavior have supported a role for the VNO in gender identification and sexual behaviors that are linked to pheromone perception (see Section 53.2).

### Gustation is the sense of taste

The sense of taste, or **gustation**, in humans and other vertebrates depends on clusters of chemoreceptors called **taste buds**. The taste buds of terrestrial vertebrates are confined to the mouth cavity, but some fishes have taste buds in the skin that enhance their ability to sense their environment. Some fishes living in murky water are very sensitive to small amounts of amino acids in the water around them and can find food without the use of vision. The duck-billed platypus, a prototherian mammal (see Figure 33.24), has similar talents as a result of taste buds on the sensitive skin of its bill.

A human tongue has approximately 10,000 taste buds. The taste buds are embedded in the epithelium, and most are found on the papillae (**Figure 46.5**). (Look at your tongue in a mirror—the papillae make it look fuzzy.) Each papilla has many taste

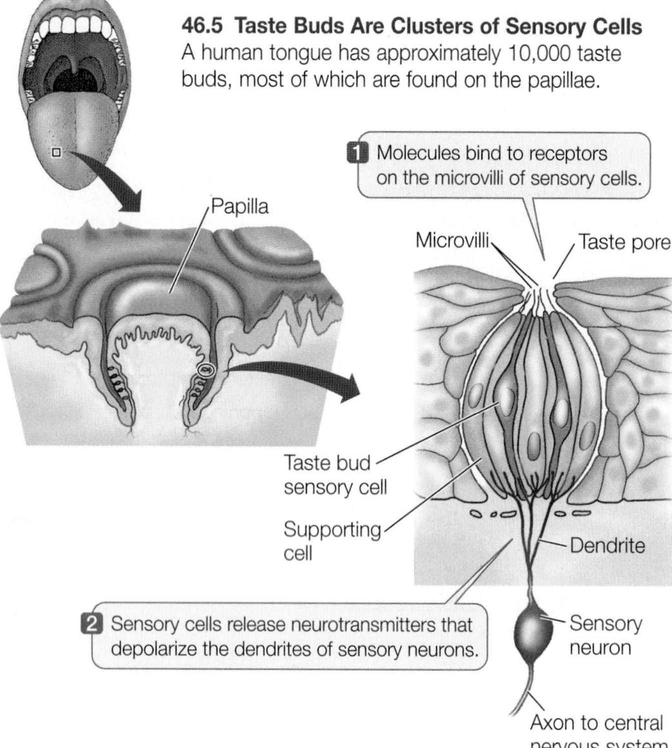
1 Molecules bind to receptors on the microvilli of sensory cells.
Papilla
Microvilli
Taste pore
Taste bud sensory cell
Supporting cell
Dendrite
2 Sensory cells release neurotransmitters that depolarize the dendrites of sensory neurons.
Sensory neuron
Axon to central nervous system

**46.5 Taste Buds Are Clusters of Sensory Cells** A human tongue has approximately 10,000 taste buds, most of which are found on the papillae.

buds, mostly on the sides. The outer surface of a taste bud has a pore that exposes the tips of the sensory receptor cells. Microvilli (tiny hairlike projections) increase the surface area of these cells where their tips converge at the pore. These sensory cells generate action potentials and release neurotransmitter at their bases, where they form synapses with sensory neurons that convey the signals into the central nervous system.

The tongue does a lot of hard work, so its epithelium, along with cells of its taste buds, are shed and replaced at a rapid rate. Individual taste bud cells last about 10 days before they are replaced, but the sensory neurons associated with them live on, always forming new synapses as new taste buds form.

You may have heard that humans can perceive only four tastes: sweet, salty, sour, and bitter. However, taste buds can distinguish among a variety of sweet-tasting molecules and a variety of bitter-tasting molecules. Recently, small families of genes for receptor proteins responding to sweet and bitter tastes have been discovered. In addition, we now recognize a fifth taste, *umami*, which is a savory, meaty taste that originates from receptors for amino acids, including monosodium glutamate (MSG). The full complexity of the chemosensitivity that enables us to enjoy the subtle flavors of food comes from the combined activation of gustatory and olfactory receptors, which is why you lose much of your sense of taste when you have a cold.

Gustation begins with receptor proteins in the membranes of the microvilli of the taste bud sensory cells (see Figure 46.5). The nature of these proteins and the mechanisms by which they depolarize the sensory receptor cell differ for the different basic tastes. Saltiness receptors are ionotropic and simply respond to $Na^+$ diffusing through open $Na^+$ channels depolarizing the sensory cell. Sourness receptors are also probably ionotropic. Depolarization of these receptors is similarly due to a direct effect of $H^+$ ions on $Na^+$ channels. In contrast, sweetness and bitterness involve families of receptor proteins similar to those involved in olfaction, and they are metabotropic. The bitter taste probably evolved as a protective mechanism enabling animals to detect and avoid toxic plant compounds such as quinine, caffeine, and nicotine. Since plants have evolved many such molecules to repel herbivorous predators, a variety of receptors is essential. Similarly, a large number of molecules in food could indicate nutritional value, so a variety of receptors is of value. The diversity of sweet receptors helps explain why it has been possible to invent many different artificial sweeteners.

Regardless of the mechanism of taste transduction by the sensory cells of the taste buds, all these cells release neurotransmitter onto sensory neurons. These neurons then conduct that information to the CNS, where it is interpreted as specific taste sensations.

## 46.2 RECAP

**All animals receive information about chemical stimuli through chemoreceptors, which have diverse structures and bind to a tremendous variety of stimulus molecules. Chemoreceptors are the basis of the sensations of olfaction and gustation and the reception of pheromones. They also monitor some aspects of an animal's internal environment.**

- Why are we able to distinguish so many different smells? Why do some people and some animals experience more or different odors than others? See pp. 968–969 and Figure 46.4
- Describe how different substances in food are transduced into action potentials in taste buds. See pp. 969–970 and Figure 46.5

## 46.3 How Do Sensory Systems Detect Mechanical Forces?

Let's turn now to **mechanoreceptors**, the sensory cells that respond to mechanical forces. Physical distortion of a mechanoreceptor's plasma membrane causes ion channels to open, altering the membrane potential of the cell to create a graded potential, which in turn leads to the release of neurotransmitter or the generation of action potentials. The rate of action potentials tells the CNS the strength of the stimulus to the mechanoreceptor. A considerable diversity of mechanosensory cells and mechanisms has evolved. Involved in many sensory systems, their functions range from interpreting skin sensations to sensing blood pressure to hearing and maintaining balance.

### Many different cells respond to touch and pressure

Human skin (and that of other mammals) is packed with diverse mechanoreceptors that generate varied sensations (**Figure 46.6**). The most important tactile receptors, found in both hairy and nonhairy skin, are *Merkel's discs*, which adapt rather

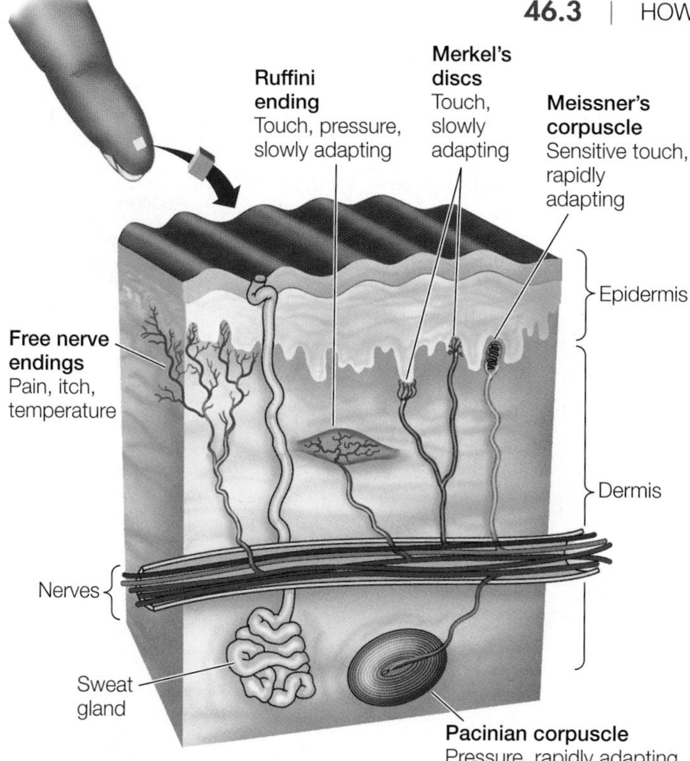


**46.6 The Skin Feels Many Sensations** Even a very small patch of skin contains a variety of sensory cells, making the skin a multi-modal receptor that can sense temperature, pressure, texture, pain, touch, and itch.

slowly and provide continuous information about things touching the skin. *Meissner's corpuscles*, found primarily in nonhairy skin, are very sensitive but adapt rapidly, so they provide information about *changes* in things touching the skin. The rapid adaptation of Meissner's corpuscles is why you roll a small object between your fingers (rather than holding it still) to discern its shape and texture: as you roll it, the object continues to stimulate Meissner's corpuscles.

Two other kinds of mechanoreceptors are found deeper in the skin. *Ruffini endings* adapt slowly and are good at providing information about vibrating stimuli of low frequencies. *Pacinian corpuscles*, which adapt rapidly, provide information about vibrating stimuli of higher frequencies. Even deeper in the skin, dendrites of sensory neurons wrap around hair follicles. When the surface hairs are displaced, those neurons are stimulated.

A *two-point spatial discrimination test* reveals that the density of tactile mechanoreceptor cells varies across the body's surface. By touching someone's skin with two toothpicks simultaneously, you can determine how far apart two stimuli have to be before the person can tell whether the sensations are produced by one toothpick or by two. On the back, the stimuli have to be relatively far apart before they are perceived as two discrete stimuli. But when the same test is applied to the person's lips or fingertips, the person can identify as separate two stimuli that are quite close together, meaning receptor density is much greater in these regions.

### Mechanoreceptors are found in muscles, tendons, and ligaments

An animal receives information from mechanoreceptors about the position of its limbs and the stresses on its muscles and joints. These mechanoreceptors supply information continuously to the CNS, and this information is essential for postural control and the coordination of movements.

The mechanoreceptors found in skeletal muscle are called *muscle spindles*. These are **stretch receptors**, modified muscle cells that are embedded in connective tissue in muscles and innervated by sensory neurons (**Figure 46.7**). Whenever the muscle is stretched, muscle spindles are also stretched, and the neurons

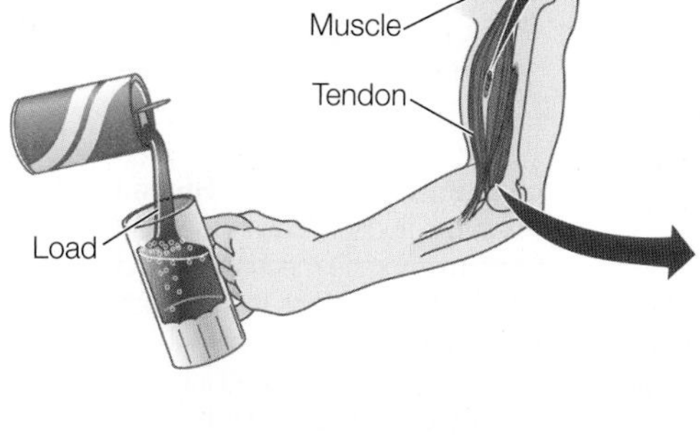

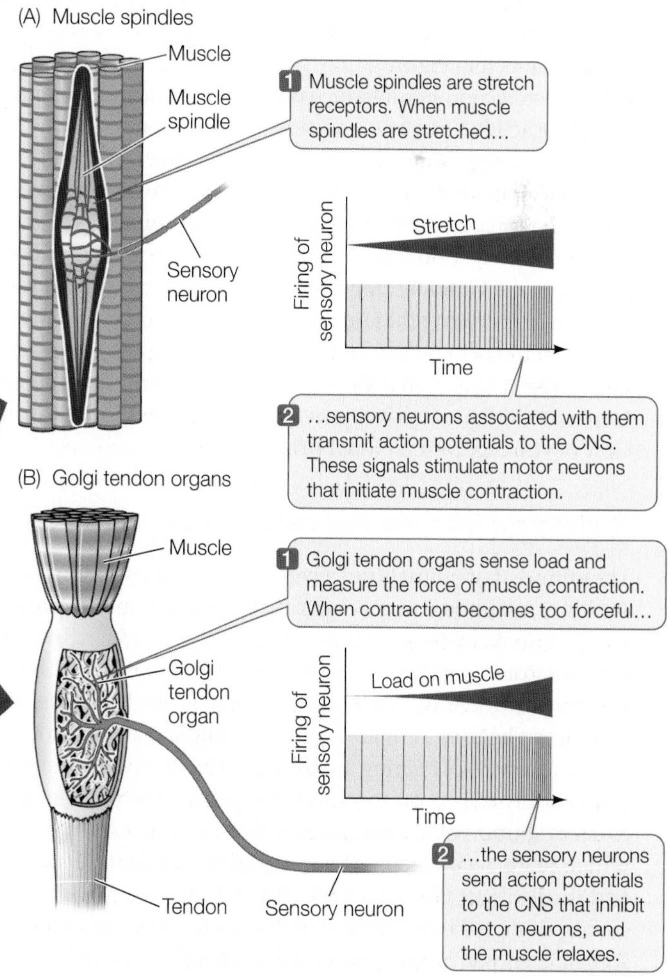

(A) Muscle spindles
Muscle
Muscle spindle
Sensory neuron
1 Muscle spindles are stretch receptors. When muscle spindles are stretched...
Stretch
Firing of sensory neuron
Time
2 ...sensory neurons associated with them transmit action potentials to the CNS. These signals stimulate motor neurons that initiate muscle contraction.
(B) Golgi tendon organs
Muscle
Golgi tendon organ
Tendon
Sensory neuron
1 Golgi tendon organs sense load and measure the force of muscle contraction. When contraction becomes too forceful...
Load on muscle
Firing of sensory neuron
Time
2 ...the sensory neurons send action potentials to the CNS that inhibit motor neurons, and the muscle relaxes.
Muscle
Tendon
Load

**46.7 Stretch Receptors** Stretch receptors provide information about the stresses on muscles and joints in an animal's limbs. (A) Signals from muscle spindles to the CNS initiate muscle contraction. (B) Golgi tendon organs in tendons and ligaments inhibit a contraction that becomes too forceful, triggering a reduction in muscle tension and protecting the muscle from tearing.

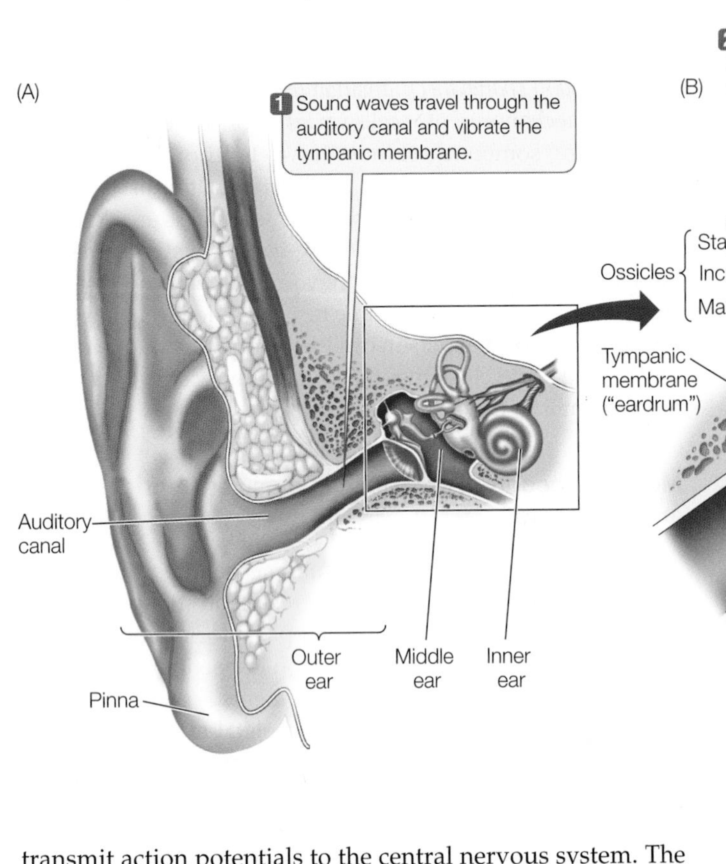

(A)
1 Sound waves travel through the auditory canal and vibrate the tympanic membrane.
Auditory canal
Outer ear
Middle ear
Inner ear
Pinna

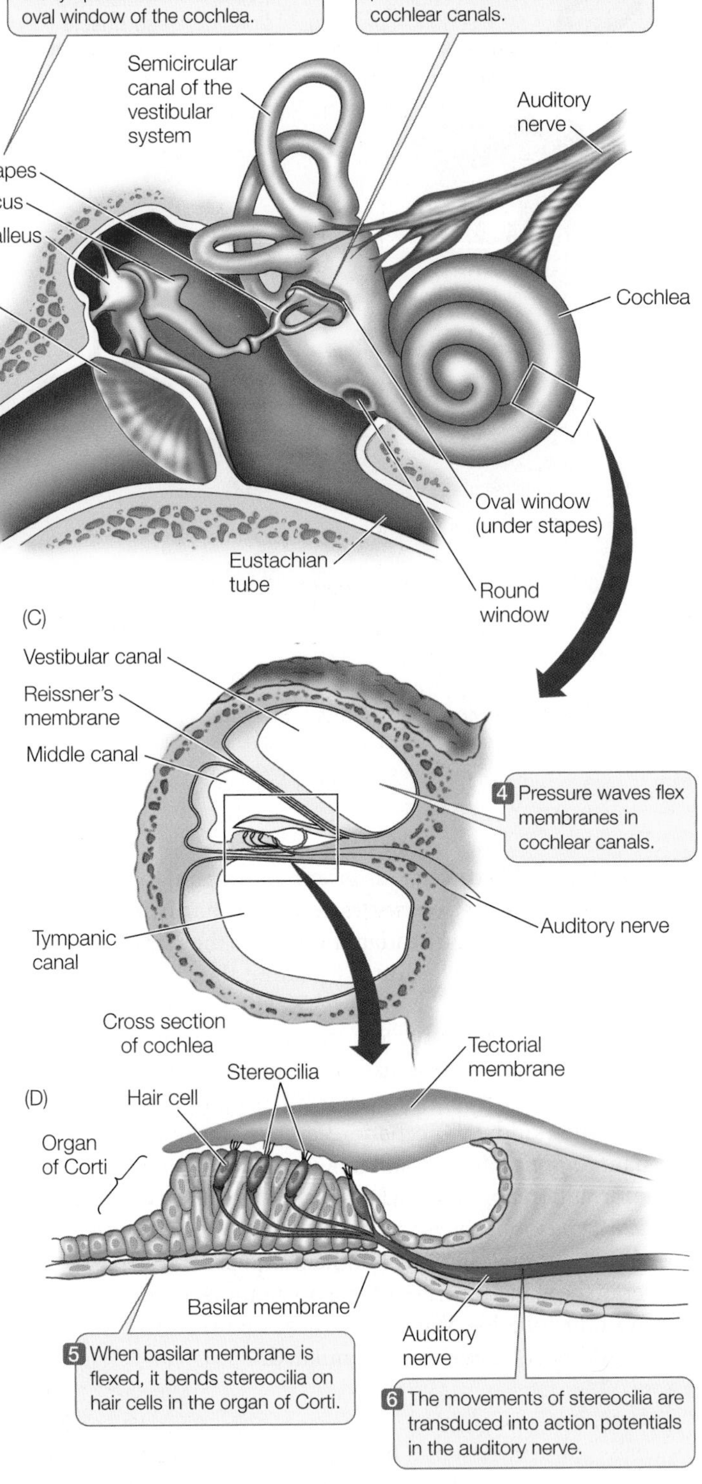

(B)
2 The ossicles transmit vibrations of the tympanic membrane to the oval window of the cochlea.
3 Vibrations at oval window create pressure waves in fluid-filled cochlear canals.
Semicircular canal of the vestibular system
Auditory nerve
Ossicles
Stapes
Incus
Malleus
Tympanic membrane ("eardrum")
Cochlea
Oval window (under stapes)
Eustachian tube
Round window
(C)
Vestibular canal
Reissner's membrane
Middle canal
4 Pressure waves flex membranes in cochlear canals.
Auditory nerve
Tympanic canal
Cross section of cochlea
(D)
Hair cell
Stereocilia
Tectorial membrane
Organ of Corti
Basilar membrane
Auditory nerve
5 When basilar membrane is flexed, it bends stereocilia on hair cells in the organ of Corti.
6 The movements of stereocilia are transduced into action potentials in the auditory nerve.

**46.8 Structures of the Human Ear** (A) The pinnae direct sound waves down the auditory canal to impinge on the tympanic membrane. The tympanic membrane mechanically transmits these pressure waves into movements of the ossicles in the middle ear. (B) The ossicles transmit their movement into pressure waves in the fluid of the cochlea at the oval window. (C) The cochlea is divided into fluid-filled chambers; pressure waves from the ossicles cause the membranes between the chambers to flex. (D) Flexing of the basilar membrane bends stereocilia on hair cells in the organ of Corti.

transmit action potentials to the central nervous system. The CNS uses this information to adjust the strength of the muscle contraction to match the load put on the muscle. Thus a bartender can hold a beer mug in the same position as he fills it from the tap, as seen in Figure 46.7A. In Figure 46.2, we saw how crayfish stretch receptors transduce physical force into action potentials. The actions of muscle spindles are similar.

Another type of mechanoreceptor, the *Golgi tendon organ*, is found in tendons and ligaments and provides information about the force generated by a contracting muscle (see Figure 46.7B). When a contraction becomes too forceful, action potentials from the Golgi tendon organ inhibit the spinal cord motor neurons innervating that muscle, causing it to relax and protecting it from tearing. (You may recall a cell organelle called the Golgi apparatus. What these two cellular structures have in common is their discovery by the Italian anatomist Camillo Golgi, who received a Nobel prize in 1906.)

### Auditory systems use hair cells to sense sound waves

The stimuli that animals perceive as sounds are pressure waves. **Auditory systems** use mechanoreceptors to convert pressure waves into receptor potentials. Auditory systems include special structures that gather sound waves, direct them to the sensory organ, and amplify their effect on the mechanoreceptors.

Human hearing provides a good example of an auditory system. The organs of hearing are the ears. The two prominent structures on the sides of our heads are the *pinnae*. The pinna of an ear collects sound waves and directs them into the *auditory canal*, which leads to the actual hearing apparatus in the *middle ear* and the *inner ear* (**Figure 46.8A**). If you have ever watched a cat or dog change the orientation of its ears to focus on a particular sound, then you have witnessed the role of pinnae in hearing.

yourBioPortal.com
GO TO Web Activity 46.1 • Structures of the Human Ear

The eardrum, or **tympanic membrane**, covers the end of the auditory canal. The tympanic membrane vibrates in response to pressure waves traveling down the auditory canal. The middle ear, an air-filled cavity, lies on the other side of the tympanic membrane.

**MIDDLE EAR** The middle ear is open to the throat at the back of the mouth through the *eustachian tube*. Because the eustachian tube is also filled with air, pressure equilibrates between the middle ear and the outside world. When you have a cold or allergy, the tube can become blocked by mucus or by tissue swelling, so you have difficulty "clearing your ears," or equilibrating the pressure in the middle ear with the outside air pressure, which you have to do when ascending or descending in a plane or when scuba diving.

The middle ear contains three delicate bones called the **ossicles**, individually named the *malleus* (Latin, "hammer"), *incus* ("anvil"), and *stapes* ("stirrup") (**Figure 46.8B**). The ossicles transmit the vibrations of the tympanic membrane to another flexible membrane called the **oval window**. The ossicles act as a lever—like a hammer pulling out a nail—translating a large movement of the tympanic membrane into a smaller movement of the oval window, but a movement of greater force. Also, because the oval window is much smaller than the tympanic membrane, the pressure the stapes transmits to the oval window is more than 20 times greater than the pressure exerted by the sound wave on the tympanic membrane. Behind the oval window lies the fluid-filled inner ear. Movements of the oval window impart pressure changes to that enclosed fluid. These pressure waves are transduced into action potentials. To see how, let's take a closer look at the inner ear.

**INNER EAR** The inner ear is a bony structure consisting of two sets of canals. One is the organ of balance, the **vestibular system**, and the other is the organ of hearing, the **cochlea**. The cochlea (Latin and Greek, "snail" or "spiral shell") is a long, tapered, coiled canal. A cross section of the cochlea reveals that it is composed of three parallel canals separated by two membranes: **Reissner's membrane** and the **basilar membrane** (**Figure 46.8C**). Sitting on the basilar membrane is the **organ of Corti**, which transduces pressure waves into action potentials. The organ of Corti contains *hair cells* with *stereocilia*. The tips of the hair cells are embedded in a gelatinous overhanging shelf called the *tectorial membrane* (**Figure 46.8D**).

Stereocilia are fingerlike extensions of the cell plasma membrane that are stiffened by cross-linked actin filaments. They are not motile, but because their tips are attached to the more rigid tectorial membrane, stereocilia bend when the basilar membrane flexes. The response of the hair cell is a graded membrane potential. The hair cells do not fire action potentials, but the changes in their membrane potential alter the rate at which the hair cells release neurotransmitter onto sensory neurons whose axons make up the auditory nerve and transmit action potentials to the brain.

What causes the basilar membrane to flex, and how does this mechanism distinguish sounds of different frequencies? In **Figure 46.9**, the cochlea is shown uncoiled to make it easier to understand its structure and function. The upper and lower canals separated by the basilar membrane are joined at the distal end of the cochlea (the end farthest from the oval window), making one continuous canal that turns back on itself. Just as the oval window is a flexible membrane at the beginning of the upper canal of the cochlea, the **round window** is a flexible membrane at the end of the lower canal.

Air is highly compressible but fluids are not. Therefore, a pressure wave can travel through air without much displacement of the air, whereas a pressure wave in fluid displaces that

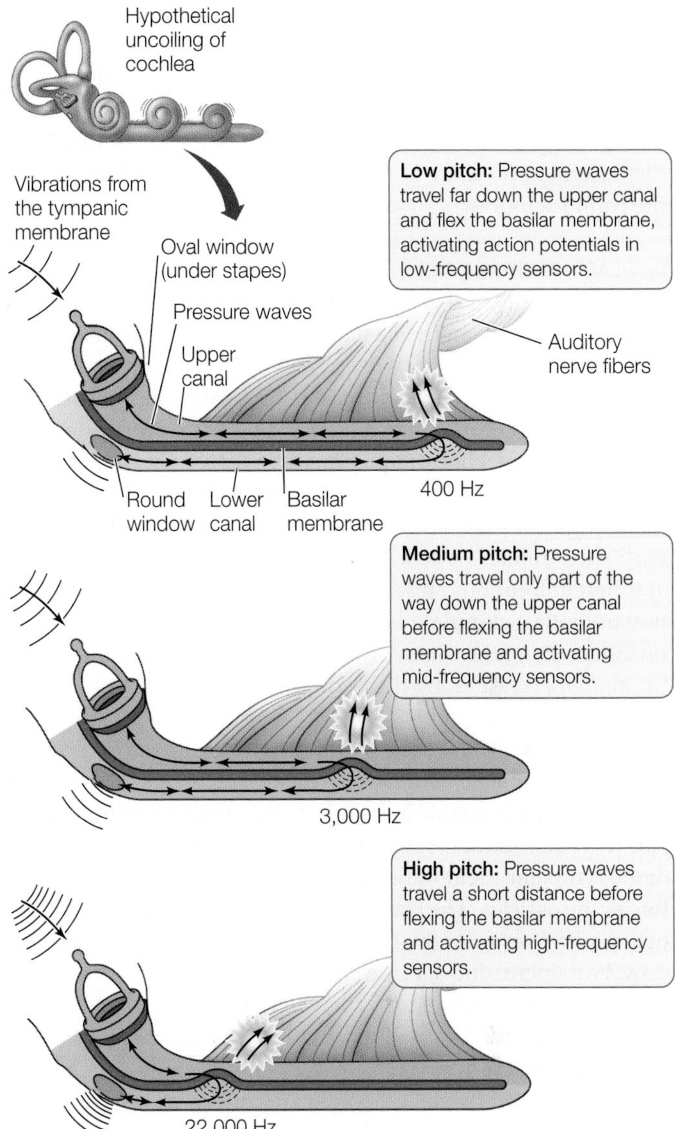
Hypothetical uncoiling of cochlea
Vibrations from the tympanic membrane
Oval window (under stapes)
Pressure waves
Upper canal
Low pitch: Pressure waves travel far down the upper canal and flex the basilar membrane, activating action potentials in low-frequency sensors.
Auditory nerve fibers
Round window
Lower canal
Basilar membrane
400 Hz
Medium pitch: Pressure waves travel only part of the way down the upper canal before flexing the basilar membrane and activating mid-frequency sensors.
3,000 Hz
High pitch: Pressure waves travel a short distance before flexing the basilar membrane and activating high-frequency sensors.
22,000 Hz

**46.9 Sensing Pressure Waves in the Inner Ear** Pressure waves of different frequencies flex the basilar membrane at different locations. Information about sound frequency is specified by which hair cells are activated. For simplicity, this representation illustrates the cochlea as uncoiled, and leaves out the middle ear.

**yourBioPortal.com**
GO TO Animated Tutorial 46.1 • Sound Transduction in the Human Ear

fluid. When the stapes pushes on the oval window, the fluid in the upper canal of the cochlea is displaced. If this movement of the oval window occurs slowly, the cochlear fluid pressure wave travels down the upper canal, around the bend, and back through the lower canal. At the end of the lower canal, the displacement pressure is dissipated by the outward bulging of the round window.

**BASILAR MEMBRANE AND PITCH** If the oval window vibrates in and out rapidly, the waves of fluid pressure create traveling waves, or *flexions*, in the basilar membrane. The basilar membrane is not uniform; it is thicker and stiffer at its base and wider and thinner at its apical end. Pressure waves in the cochlear fluid have different frequencies and set up different patterns of traveling waves. High-frequency waves cause maximal flexion at the basal end of the basilar membrane, whereas low-frequency pressure waves result in maximal flexion at its apical end. Thus different pitches of sound flex the basilar membrane at different locations and activate different sets of hair cells. Action potentials stimulated by the mechanoreceptors at different positions along the organ of Corti travel to different regions of the *auditory cortex* along the auditory nerve.

**HEARING LOSS** There are two general types of hearing loss, or deafness. *Conduction deafness* is caused by the loss of function of the tympanic membrane and/or the ossicles of the middle ear. Repeated infections of the middle ear can cause scarring of the tympanic membrane and stiffening of the connections between the ossicles. The consequence is less efficient conduction of sound waves from the tympanic membrane to the oval window. With increasing age, the ossicles stiffen, resulting in a gradual loss of the ability to hear high-frequency sounds.

*Nerve deafness* is caused by damage to the inner ear or the auditory pathways. A common cause of nerve deafness is damage to the hair cells of the delicate organ of Corti by exposure to loud sounds such as jet engines, pneumatic drills, or highly amplified music. Consistent exposure to sounds above 85 decibels can damage hearing; this damage is cumulative and irreversible. Even using earphones can put you at risk for hearing loss because they generate high-pressure sound waves close to the tympanic membrane. Personal stereo earphones can reach 120 decibels, and people commonly use them at 100 decibels (equivalent to being at a rock concert).

## Hair cells are sensitive to being bent

**Hair cells** are the mechanoreceptors in both the organs of hearing and equilibrium. Stereocilia project from the surface of each hair cell like a set of organ pipes (**Figure 46.10A**). The bending of stereocilia alters ion channels in the hair cell's plasma membrane. Bending in one direction causes the plasma membrane to depolarize, and bending in the other direction causes it to hyperpolarize (**Figure 46.10B**). Graded potentials in the hair cell plasma membrane control its release of neurotransmitter onto the sensory neuron associated with it, and the sensory neuron sends action potentials to the CNS.

How does the bending of the stereocilia open ion channels? The ion channels that are opened are at the ends of the stereocilia. This was discovered by exploring the areas around the stereocilia with microelectrodes, and seeing that local currents were created near the tips of the stereocilia when they were bent. Then, careful electron microscopic work revealed minute filaments that connected the tip of each stereocilium to its taller neighbor. It is hypothesized that these filaments are fine molecular attachments to the ion channels in the stereocilium plasma membrane, and they act like springs that open the channels. If the taller neighboring stereocilium is bent away, the spring tightens and the ion channel is opened. If the taller neighbor bends toward its shorter neighbor, the spring is relaxed and the channel closes (see Figure 46.10B).

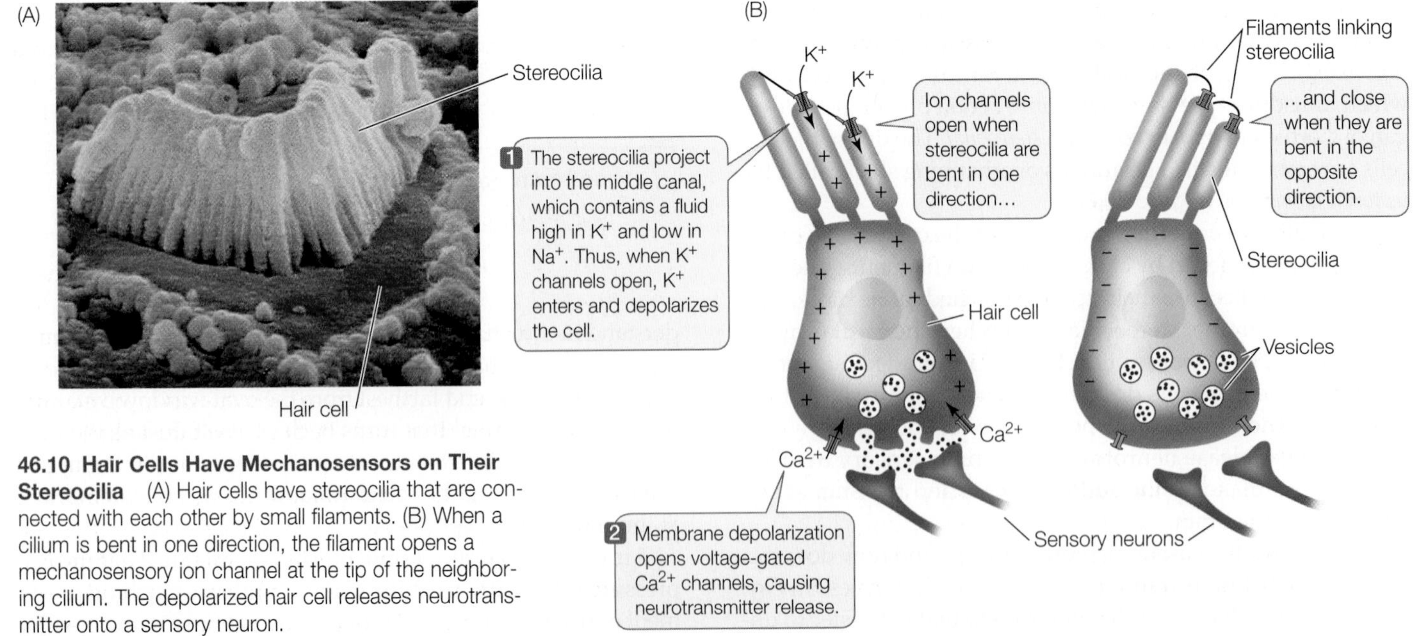
(A)
Stereocilia
Hair cell
1 The stereocilia project into the middle canal, which contains a fluid high in K+ and low in Na+. Thus, when K+ channels open, K+ enters and depolarizes the cell.
(B)
K+
K+
Ion channels open when stereocilia are bent in one direction...
Filaments linking stereocilia
...and close when they are bent in the opposite direction.
Stereocilia
Hair cell
Vesicles
Ca2+
Ca2+
Sensory neurons
2 Membrane depolarization opens voltage-gated Ca2+ channels, causing neurotransmitter release.

**46.10 Hair Cells Have Mechanosensors on Their Stereocilia** (A) Hair cells have stereocilia that are connected with each other by small filaments. (B) When a cilium is bent in one direction, the filament opens a mechanosensory ion channel at the tip of the neighboring cilium. The depolarized hair cell releases neurotransmitter onto a sensory neuron.

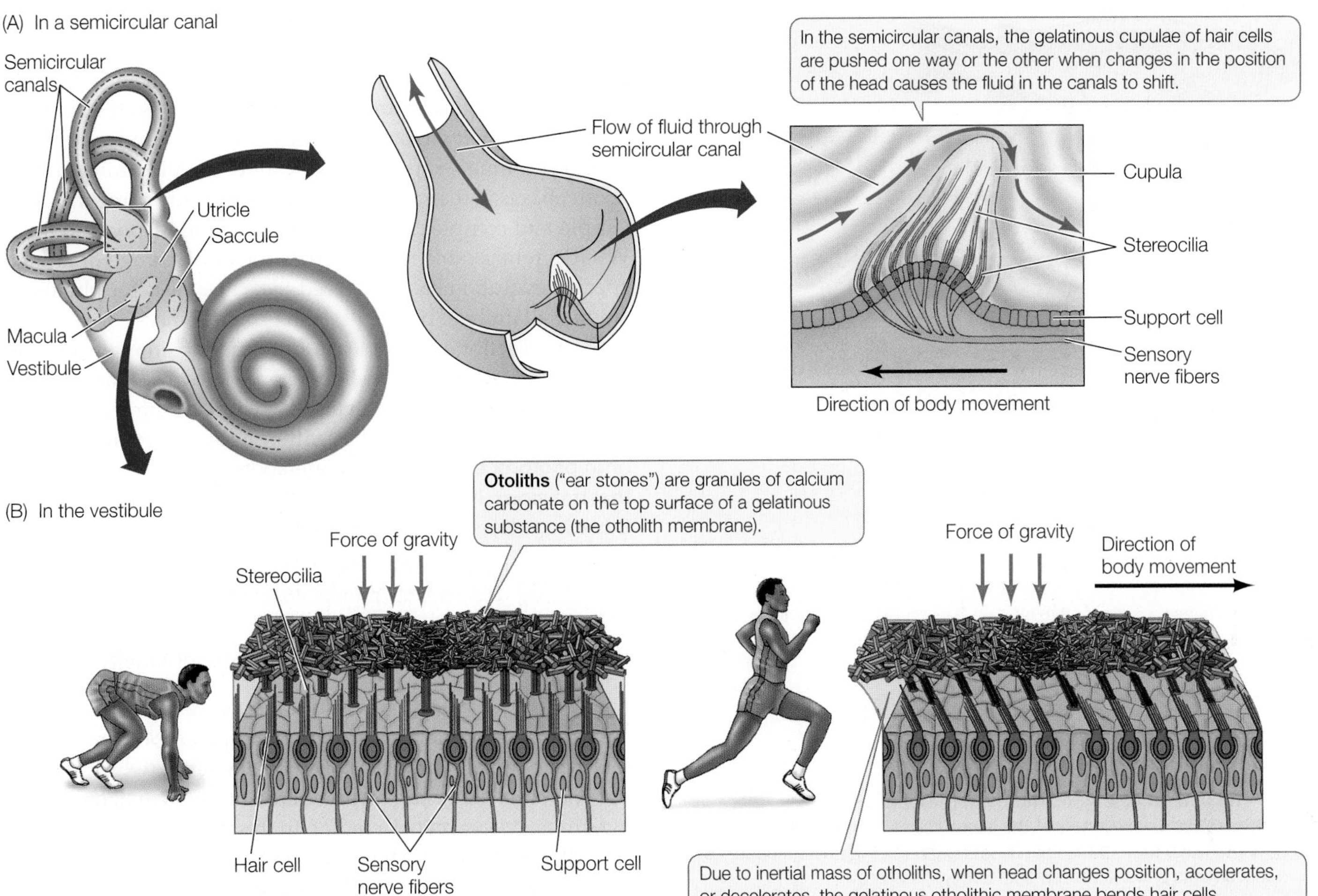

(A) In a semicircular canal
Semicircular canals
Utricle
Saccule
Macula
Vestibule
Flow of fluid through semicircular canal
In the semicircular canals, the gelatinous cupulae of hair cells are pushed one way or the other when changes in the position of the head causes the fluid in the canals to shift.
Cupula
Stereocilia
Support cell
Sensory nerve fibers
Direction of body movement
(B) In the vestibule
Otoliths ("ear stones") are granules of calcium carbonate on the top surface of a gelatinous substance (the otholith membrane).
Force of gravity
Stereocilia
Hair cell
Sensory nerve fibers
Support cell
Force of gravity
Direction of body movement
Due to inertial mass of otholiths, when head changes position, accelerates, or decelerates, the gelatinous otholithic membrane bends hair cells.

**46.11 Organs of Equilibrium** The vestibular system consists of bony chambers and fluid-filled canals. (A) Each semicircular canal has a cupula containing stereocilia. When fluid moves against the cupula, the stereocilia bend. (B) In the saccule and utricle, stereocilia are bent by gravitational forces on the otoliths.

## Hair cells detect forces of gravity and momentum

In the mammalian inner ear, the vestibular system consists of three bony **semicircular canals** and two bony chambers called the *saccule* and the *utricle*. Hair cells in the vestibular system detect the position and movement of the head—information that is essential for maintaining balance (equilibrium). The information from the vestibular system is also crucial for the control of eye movements. When you look at something, you can move your head while staying focused on the object because of your vestibulo-ocular reflex.

The entire vestibular system is filled with a fluid called endolymph. In the semicircular canals, the endolymph shifts when the head changes position. Since the three semicircular canals have different orientations, they respond differentially to the direction of movement. Projecting into the base of each canal is a gelatinous swelling called a *cupula* (plural *cupulae*) that encloses a cluster of hair cell stereocilia. When the shifting endolymph pushes on the cupulae, it bends the stereocilia and causes a graded potential in their hair cell plasma membranes (**Figure 46.11A**). The stereocilia in the saccule and utricle are bent in a different way. These stereocilia are embedded in *otoliths* (Latin for "ear stones"), gelatinous structures containing crystals of calcium carbonate. When the head changes position or when it accelerates or decelerates, gravitational forces are exerted on the otoliths and bend the stereocilia (**Figure 46.11B**).

As in the cochlea, the hair cells of the vestibular system do not fire action potentials, but they release neurotransmitter at synapses with sensory neurons, which in turn fire action potentials.

## Hair cells are evolutionarily conserved

An early evolutionary use of hair cells to measure movement can be seen in the **lateral line** sensory system of fishes. The lateral line is a canal just under the surface of the skin that runs down each side of the fish (**Figure 46.12**). Hair cells line the canal, and their stereocilia, protected by cupulae, project into the stream of water that flows through the lateral line canal when the fish moves through the water. Forward movement of the fish puts pressure on the cupulae, causing the stereocilia to bend in the direction that depolarizes the hair cells. Since water is incompressible, disturbances in the water around the fish are translated into pressure waves that can be picked up by the lateral line stereocilia. Thus, the lateral line system provides

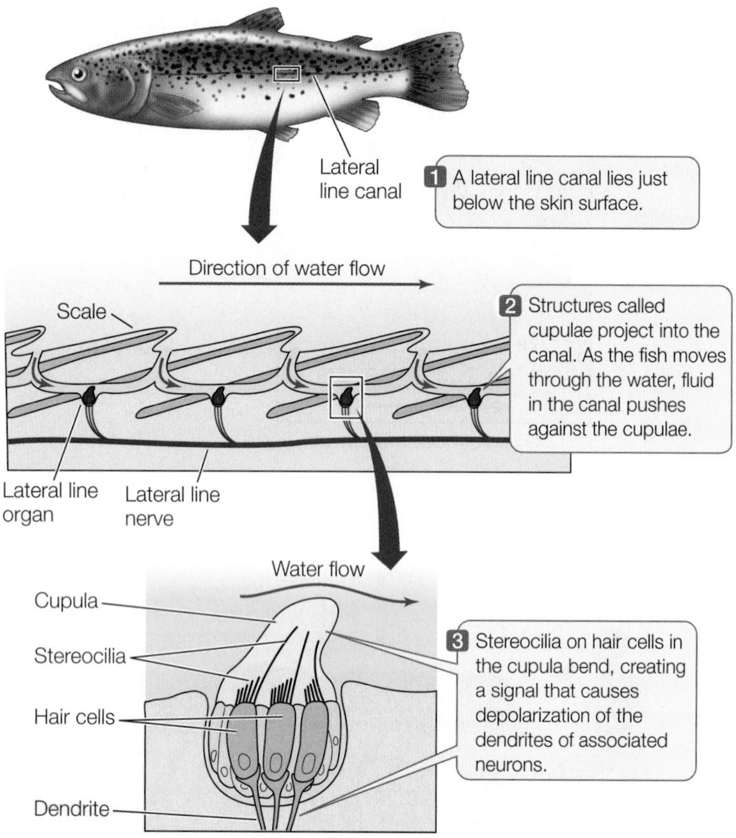

**46.12 The Lateral Line Acoustic System Contains Mechanosensors** Hair cells in the lateral line of a fish detect movement of the water around the animal, giving the fish information about its own movements and the movements of objects nearby.

information about the fish's movements as well as about other moving objects, such as predators or prey.

## 46.3 RECAP

**Sensations that derive from mechanoreceptors include touch, tickle, pressure, joint position, muscle load, hearing, and equilibrium.**

- Describe some of the different mechanoreceptors in the skin and their properties. See pp. 970–971 and Figure 46.6
- How do different frequencies of sound result in action potentials being fired in different acoustic neurons? See pp. 972–973 and Figures 46.8 and 46.9
- How do hair cells transduce force into action potentials? See p. 974 and Figure 46.10

Chemoreception gave us good examples of metabotropic sensory receptors, and mechanoreception has given us good examples of ionotropic sensory receptors. Now let's turn to another example of metabotropic sensory reception—one in which light is the stimulus. We will see how light energy is converted into action potentials.

# 46.4 How Do Sensory Systems Detect Light?

Sensitivity to light—**photosensitivity**—confers on the simplest animals the ability to orient to the sun and sky and gives more complex animals rapid and extremely detailed information about objects in their environment. It is not surprising that both simple and complex animals can sense and respond to light. What is remarkable is that across the entire range of animal species, evolution has conserved the same basis for photosensitivity: a family of pigments called **rhodopsins**.

In this section we will learn how rhodopsin molecules respond when stimulated by light energy and how that response is transduced into neural signals. We will also examine the structures of eyes, the organs that gather light energy and focus it onto **photoreceptor cells**, the metabotropic sensory receptors that transform light energy into action potentials, and the routes those impulses travel to the brain.

yourBioPortal.com
GO TO Animated Tutorial 46.2 • Photosensitivity

## Rhodopsins are responsible for photosensitivity

Photosensitivity depends on the ability of rhodopsins to absorb photons of light and to undergo a change in conformation. A rhodopsin molecule consists of a protein, **opsin** (which alone is not photosensitive), and an associated nonprotein light-absorbing group, **11-*cis*-retinal**, cradled in the center of the opsin and bound covalently to it. The entire rhodopsin molecule sits within the plasma membrane of a photoreceptor cell.

When 11-*cis*-retinal absorbs a photon of light energy, it changes into a different isomer of retinal, called all-*trans*-retinal. This change puts a strain on the bonds between retinal and opsin, changing the conformation of opsin. This change signals the detection of light. In vertebrate eyes, the retinal and the opsin eventually separate from each other—a process called *bleaching*, which causes the molecule to lose its photosensitivity. A series of enzymatic reactions is then required to return the all-*trans*-retinal to the 11-*cis* isomer, which then recombines with opsin so that it once again becomes the photosensitive pigment rhodopsin (**Figure 46.13**).

How does the conformational change of rhodopsin transduce light into a cellular response? After retinal is converted from the 11-*cis* to the all-*trans* form, its interactions with opsin pass through several unstable intermediate stages. One of these stages triggers a cascade of reactions involving a G protein signaling mechanism that results in the alteration of membrane potential that is the photoreceptor cell's response to light.

## Rod cells respond to light

To get a better idea of how rhodopsin alters the membrane potential of a photoreceptor cell and how that photoreceptor cell signals that it has been stimulated by light, let's look at one type of vertebrate photoreceptor cell, the rod cell. The rod cell, named for its shape, is a modified neuron that does not produce action potentials. Rod cells release neurotransmitter from their bases

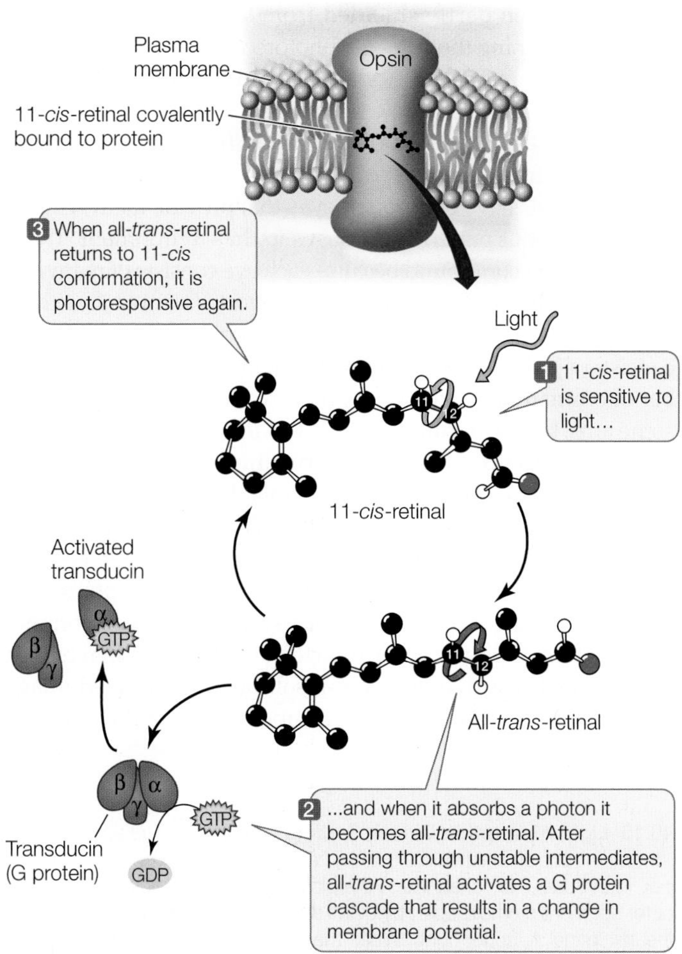

Plasma membrane
Opsin
11-*cis*-retinal covalently bound to protein
3 When all-*trans*-retinal returns to 11-*cis* conformation, it is photoresponsive again.
Light
1 11-*cis*-retinal is sensitive to light...
11-*cis*-retinal
Activated transducin
GTP
All-*trans*-retinal
Transducin (G protein)
GDP
2 ...and when it absorbs a photon it becomes all-*trans*-retinal. After passing through unstable intermediates, all-*trans*-retinal activates a G protein cascade that results in a change in membrane potential.


**46.13 Light Changes the Conformation of Rhodopsin** The light-absorbing molecule 11-*cis*-retinal bonds with the protein opsin to form the pigment rhodopsin, the molecular agent of photosensitivity.

## INVESTIGATING LIFE

### 46.14 A Rod Cell Responds to Light

The plasma membrane of a rod cell hyperpolarizes—becomes more negative—in response to a flash of light. Rod cells do not fire action potentials, but in response to the absorption of light energy, the neuron experiences a change in membrane potential.

**HYPOTHESIS** **When a rod cell absorbs photons (light energy), its membrane potential changes in proportion to the strength of the light stimulus.**

**METHOD**

1. Record membrane potentials from the inner segment of a rod cell.
2. Stimulate the rod cells with light flashes of varying intensity and record the results.

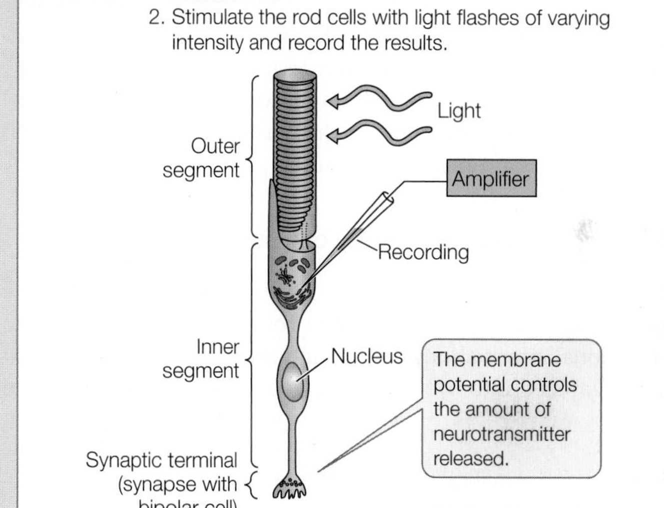


**RESULTS**

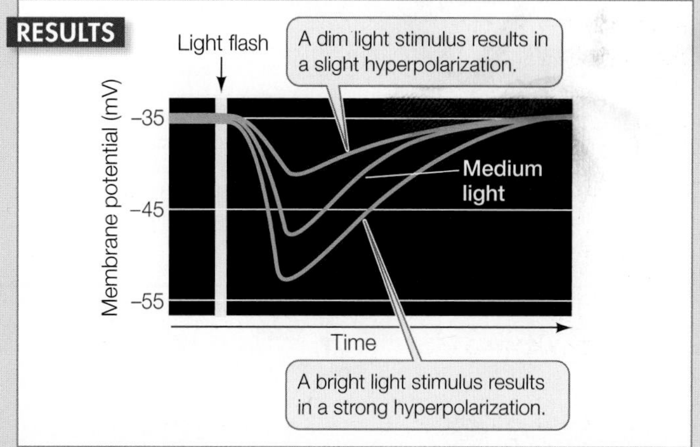


**CONCLUSION** **The membrane potential of rod cells is depolarized in the dark and hyperpolarizes (becomes more negative) in response to light.**

**FURTHER INVESTIGATION:** How would you investigate the effect of background illumination on the rod cell's response to light?

Go to **yourBioPortal.com** for original citations, discussions, and relevant links for all INVESTIGATING LIFE figures.

where they form synapses with the next neurons in the visual pathway (**Figure 46.14**). Each rod cell has an outer segment, an inner segment, and a synaptic terminal. The outer segment is highly specialized and contains a stack of discs of plasma membrane densely packed with rhodopsin. The function of the discs is to capture photons of light passing through the rod cell. The inner segment contains the cell nucleus, mitochondria, and other organelles. The synaptic terminal is where the rod cell communicates with other neurons.

To see how a rod cell responds to light, we can penetrate a single rod cell with an electrode and record its membrane potential in the dark and in the light. From what we have learned about other types of sensory receptors, we might expect stimulation of the rod cell by light would make its membrane potential less negative. But the opposite is true—it becomes more negative.

When a rod cell is kept in the dark, it has a relatively depolarized resting potential compared with other neurons. In fact, the plasma membrane of the rod cell is almost as permeable to $Na^+$ as to $K^+$. In the dark, $Na^+$ continually enters the outer segment of the cell—the dark current. When light is flashed on the dark-adapted rod cell, its membrane potential becomes more negative—it hyperpolarizes (see Figure 46.14). The rate of neurotransmitter release changes as membrane potential changes. As the rod cell hyperpolarizes, its release of neurotransmitter decreases.

How does the absorption of light by rhodopsin hyperpolarize the rod cell? When rhodopsin is excited by light, it initiates a cascade of events. The dark-adapted rod cell has open $Na^+$ channels, allowing a depolarizing dark current (**Figure 46.15A**). Light photoexcites rhodopsin, which activates a G protein called *transducin*. Activated transducin in turn activates a phosphodiesterase (PDE) (**Figure 46.15B**). Activated PDE converts cyclic GMP (cGMP) to GMP, which causes the $Na^+$ channels to close. $Na^+$ is pumped out and the cell hyperpolarizes (**Figure 46.15C**).

This mechanism may seem like a roundabout way of doing business, but its advantage is its enormous amplification ability. Each molecule of photoexcited rhodopsin can activate several hundred transducin molecules, thus activating a large number of PDE molecules. The catalytic capacity of PDE is great: one molecule can hydrolyze several hundred molecules of cGMP per second. The bottom line is that a single photon of light can cause a huge number of $Na^+$ channels to close.

### Invertebrates have a variety of visual systems

Photoreceptors using rhodopsin are incorporated into a variety of visual systems, from simple to complex. Flatworms obtain directional information about light from photoreceptor cells that are organized into **eye cups**. The eye cups are paired bilateral structures, each partly shielded from light by a layer of pigmented cells lining the cup. The photoreceptors on the two sides of the animal are unequally stimulated unless the animal is facing directly toward or away from a light source. The flatworm generally uses directional information from the eye cups to move away from light.

Arthropods have **compound eyes** that provide them with information about patterns or images in the environment. These eyes are called compound because each eye consists of many optical units called **ommatidia** (singular *ommatidium*), each with its own narrow-angle lens (**Figure 46.16**). In contrast, a vertebrate eye consists of just one optical unit with a wide-angle lens. The number of ommatidia in a compound eye varies from only a few in some ants, to 800 in fruit flies, to 30,000 in some dragonflies.

Each ommatidium has a lens structure that directs light onto photoreceptor cells. Flies, for example, have eight elongated photoreceptors in each ommatidium. The inner borders of the photoreceptors are covered with microvilli that contain rhodopsin and trap light. Axons from the photoreceptors send the light information to the nervous system. Since each ommatidium of a compound eye is directed at a slightly different part of the visual world, only a low-resolution or pixillated image can be communicated from the compound eye to the CNS.

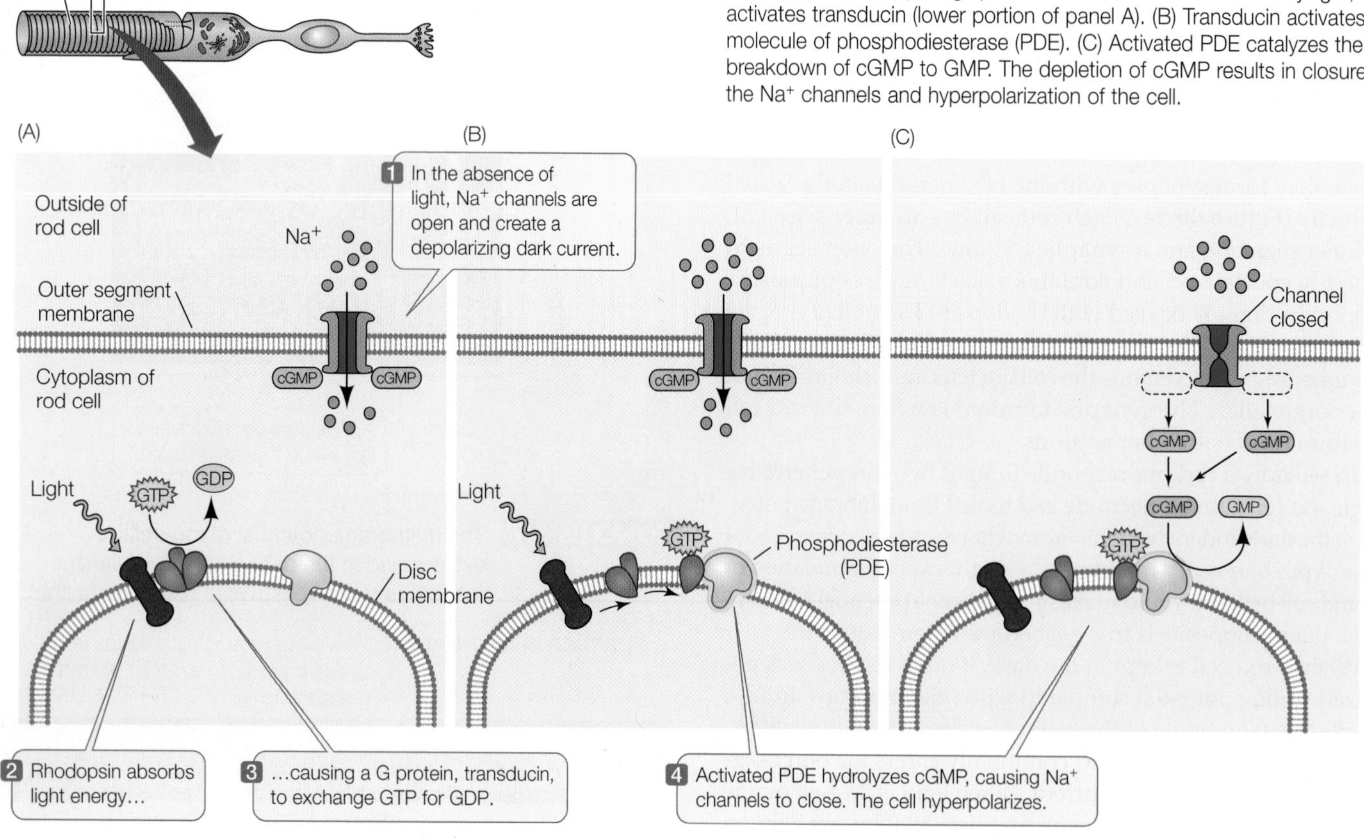
Disc
Rod cell
(A)
(B)
(C)
Outside of rod cell
Na+
1 In the absence of light, Na+ channels are open and create a depolarizing dark current.
Outer segment membrane
Channel closed
Cytoplasm of rod cell
cGMP
cGMP
GMP
Light
GTP
GDP
Phosphodiesterase (PDE)
Disc membrane
2 Rhodopsin absorbs light energy…
3 …causing a G protein, transducin, to exchange GTP for GDP.
4 Activated PDE hydrolyzes cGMP, causing Na+ channels to close. The cell hyperpolarizes.
Dark response
Light response

**46.15 Light Absorption Closes Sodium Channels** The absorption of light by rhodopsin initiates a signaling cascade that hyperpolarizes the rod cell. In the dark, $Na^+$ channels in the plasma membrane of the rod cell's outer segment are held open by cGMP, allowing positive charges to enter the cell (panel A, upper right). When the rod cell is stimulated by light, it activates transducin (lower portion of panel A). (B) Transducin activates a molecule of phosphodiesterase (PDE). (C) Activated PDE catalyzes the breakdown of cGMP to GMP. The depletion of cGMP results in closure of the $Na^+$ channels and hyperpolarization of the cell.

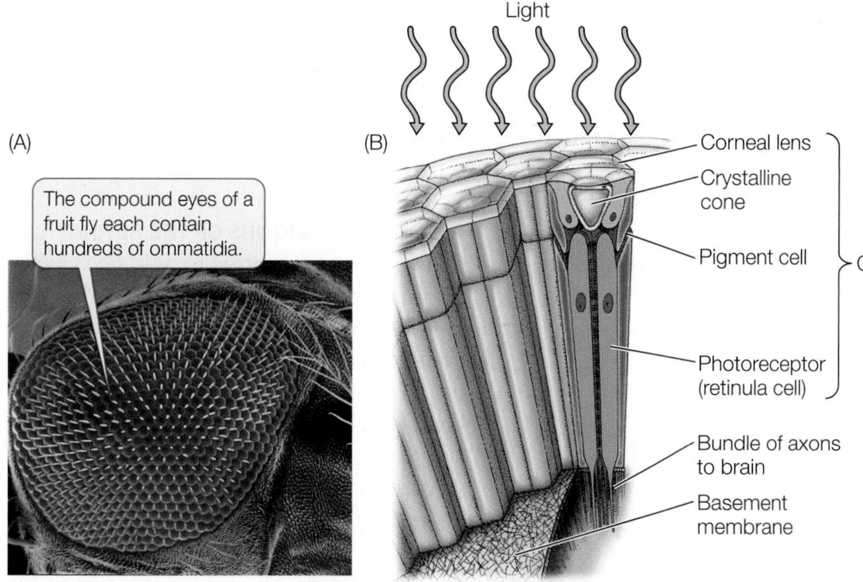


**46.16 Ommatidia: The Functional Units of Insect Eyes** (A) The micrograph shows the compound eye of a fruit fly. (B) The rhodopsin-containing retinula cells are the photoreceptors in ommatidia.

## Image-forming eyes evolved independently in vertebrates and cephalopods

Both vertebrates and cephalopod mollusks have *image-forming eyes*—eyes with exceptional abilities to form detailed images of the visual world. Like cameras, both these eye types focus inverted images on an internal surface that is sensitive to light. Considering that they evolved independently, their degree of similarity is remarkable (**Figure 46.17**).

The vertebrate eye is a spherical, fluid-filled structure bounded by a tough connective tissue layer called the *sclera*. At the front of the eye, the sclera forms the transparent **cornea**, through which light passes to enter the eye. Just inside the cornea is the pigmented **iris**, which gives the eye its color. The iris controls the amount of light that reaches the photoreceptor cells at the back of the eye, just as the diaphragm of a camera controls the amount of light reaching the film. The central opening of the iris is the **pupil**. The iris is under neural control. In bright light, the iris constricts and the pupil is very small. As light levels fall, the iris relaxes and the pupil enlarges.

**yourBioPortal.com**
GO TO **Web Activity 46.2 • Structure of the Human Eye**

Behind the iris is the crystalline protein **lens**, which makes fine adjustments in the focus of images falling on the photosensitive layer—the **retina**—at the back of the eye. The cornea and the fluids within the eye bend light rays passing through them so that they are focused on the retina. The lens makes fine adjustments to the focus and allows the eye to *accommodate*—that is, to focus on objects at various locations in the near visual field. To focus a camera on objects close at hand, you adjust the distance between the lens and the internal surface sensitive to light. Fishes, amphibians, and reptiles accommodate in a similar manner, moving the lenses of their eyes closer to or farther from their

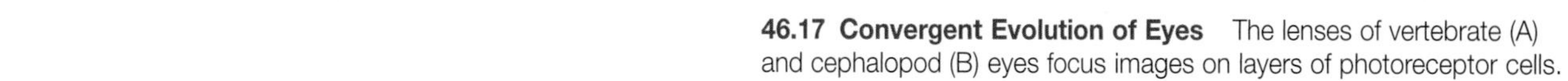

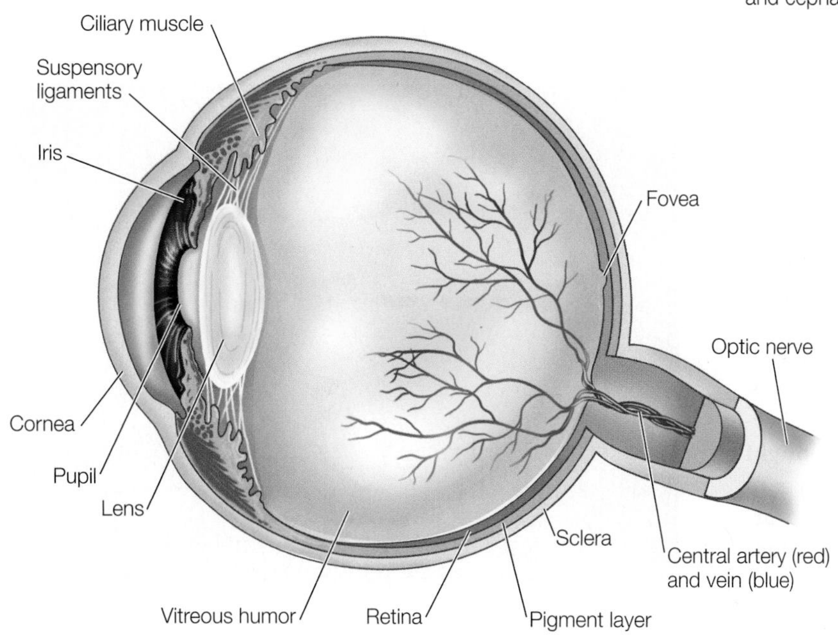


**46.17 Convergent Evolution of Eyes** The lenses of vertebrate (A) and cephalopod (B) eyes focus images on layers of photoreceptor cells.

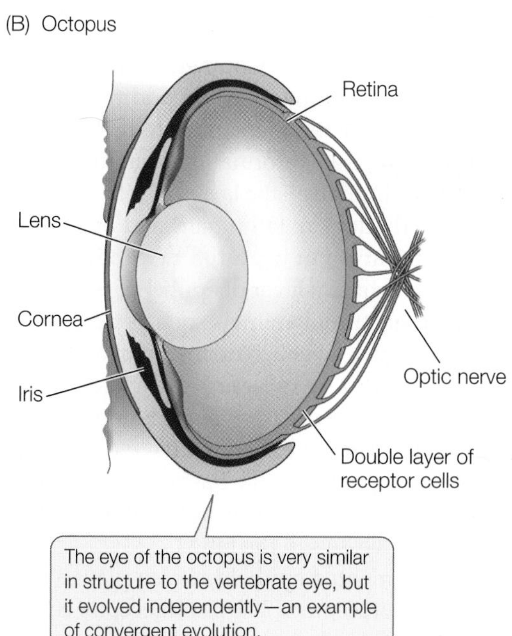

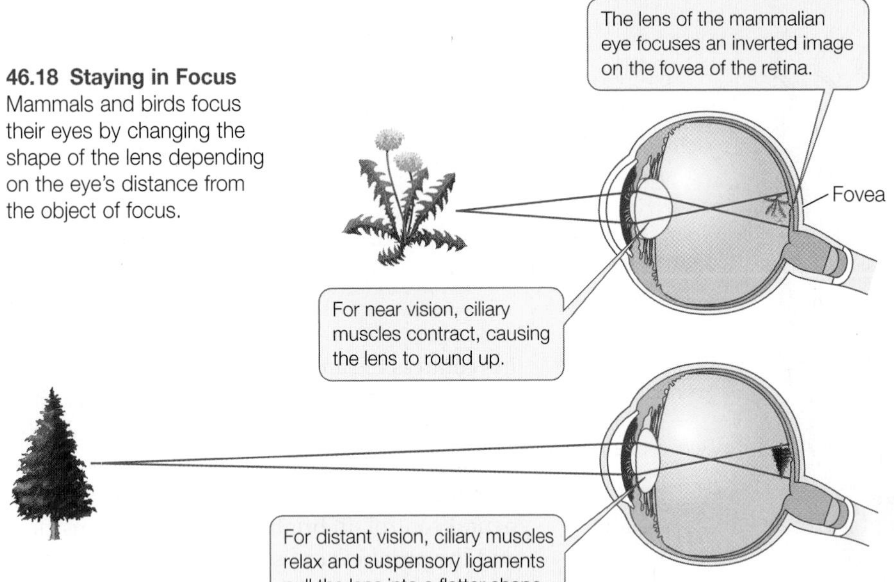

The lens of the mammalian eye focuses an inverted image on the fovea of the retina.
Fovea
For near vision, ciliary muscles contract, causing the lens to round up.
For distant vision, ciliary muscles relax and suspensory ligaments pull the lens into a flatter shape.

**46.18 Staying in Focus** Mammals and birds focus their eyes by changing the shape of the lens depending on the eye's distance from the object of focus.

retinas. Mammals and birds use a different method: they alter the shape of the lens.

The mammalian lens is contained in a connective tissue sheath that tends to keep it in a spherical shape, but it is attached to suspensory ligaments that pull it into a flatter shape. Circular *ciliary muscles* counteract the pull of the suspensory ligaments, permitting the lens to round up. When the ciliary muscles are at rest, the flatter lens has the correct optical properties to focus distant images on the retina. Contracting the ciliary muscles rounds up the lens, changing its light-bending properties to bring close images into focus (**Figure 46.18**).

Lenses become less elastic with age, so we lose the ability to focus on objects close at hand without the help of corrective lenses. Most people over the age of 45 need the assistance of reading glasses or bifocal lenses.

## The vertebrate retina receives and processes visual information

During embryonic development, neural tissue grows out from the brain to form the retina. In addition to a layer of photoreceptor cells, the retina includes four additional layers of cells that process visual information from the photoreceptors. Light must pass through all the layers of retinal cells before being captured by rhodopsin. In humans and other day-active animals, the light that is not captured by rhodopsin is absorbed by a black pigment in a layer of epithelial cells behind the retina. In contrast, nocturnal animals such as deer and raccoons have a white reflective layer behind their retinas that maximize the capture of photons by reflecting them back onto the photoreceptors. Therefore, a deer in the headlights appears to have bright white eyes. Since we do not have the white reflective layer in our retinas, photographic flashes produce the red eye appearance on photos because the light is being reflected by the abundant blood vessels in the retina.

The pigmented epithelium also plays a role in the renewal of the photoreceptors. The photoreceptor cells are always shedding discs from their distal ends as new ones are being generated by the inner segments of those cells. The pigmented epithelial cells phagocytose the shed discs. Each outer segment is totally renewed about every two weeks.

**THE PHOTORECEPTORS OF THE RETINA** Until now we have referred to only one kind of photoreceptor—the rod cell. The other major class of vertebrate photoreceptor, the **cone cell**, is also named for its shape (**Figure 46.19**). Whereas rod cells are responsible for highly sensitive black-and-white vision, cone cells are responsible for the less sensitive color vision. A human retina has about 5 million cones and about 100 million rods, but their density is not the same across the entire retina.

In humans, light coming from the center of the visual field falls on the **fovea**, where the density of cone cells is highest. The human fovea has about 160,000 cones per mm$^2$. The fovea of a hawk has almost twice that number, making the hawk's vision much sharper than ours. The hawk also has *two* foveae in each eye. One receives light from straight ahead, the other from a more lateral field of vision. The forward-looking foveae make binocular vision possible, while the lateral-looking foveae provide high-acuity vision. Birds use both sets of foveae by frequently turning their heads slightly; they cannot move their eyes in the sockets as humans can.

Cones have low sensitivity to light and contribute little to night vision. Night vision depends mostly on rod cells, and therefore vision in dim light is mostly in shades of gray and acuity is low. You may have trouble seeing a small object such as a keyhole at night when you are looking straight at it—that is, when its image is falling on your fovea. If you look a little to the side, so that the image falls on a rod-rich area of your retina, you

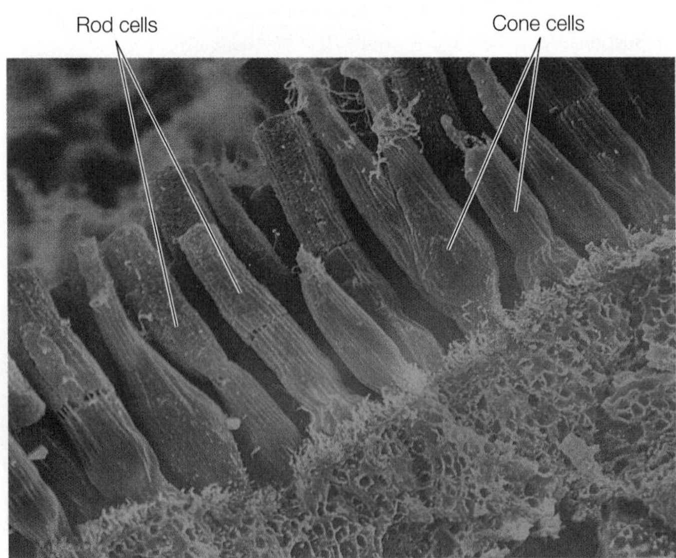


**46.19 Rods and Cones** This scanning electron micrograph of photoreceptors in the retina of a mud puppy (an amphibian) shows cylindrical rods and tapered cones.

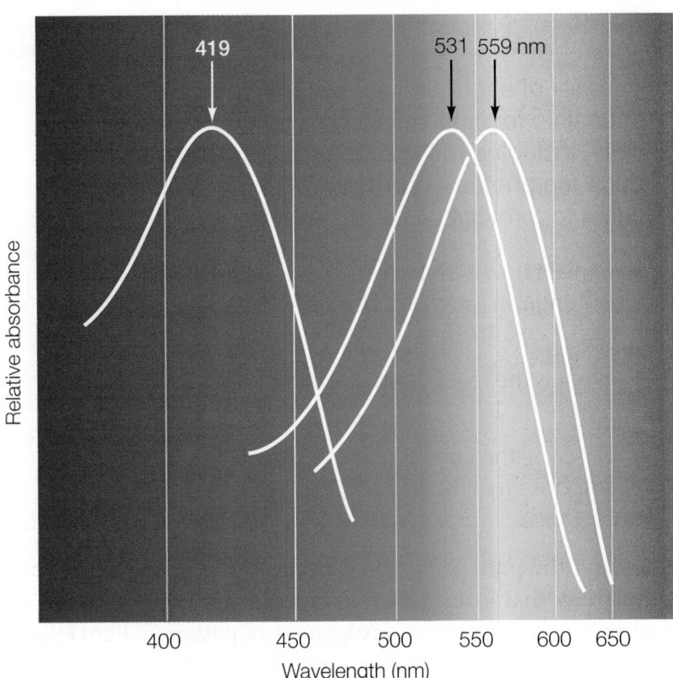
419
531 559 nm
Relative absorbance
400
450
500
550
600 650
Wavelength (nm)

**46.20 Absorption Spectra of Cone Cells** The three kinds of cone cells contain slightly different opsin molecules, which absorb different wavelengths of light.

see the object better (astronomers looking for faint objects in the sky learned this trick a long time ago). The retinas of nocturnal animals, such as flying squirrels, contain a high percentage of rods. By contrast, some animals that are active only during the day (such as chipmunks) have mostly cones in their retinas.

**COLOR VISION** The human retina has three kinds of cone cells, each containing slightly different opsin molecules, which differ in the wavelengths of light they absorb best. Although the same 11-*cis*-retinal group is the light absorber in all three kinds of cones (see Figure 46.13), its molecular interactions with opsin determine the spectral sensitivity of the rhodopsin molecule as a whole (**Figure 46.20**). Because different wavelengths of light are differentially absorbed by the different visual pigments, the brain can interpret the relative inputs from the different classes of cones as a full range of color. Some mammals have only one or two classes of cones, whereas birds have four.

*Color blindness* results from the absence or dysfunction of one or more classes of cone cells. By far the most common form is red–green color blindness, which occurs in about 10 percent of men of European descent. A genetic condition that affects a person's perception of red and green, its inheritance is sex-linked because the defective gene is on the X chromosome (see Figure 12.24).

**INFORMATION FLOW IN THE RETINA** The human retina is organized into five layers of neurons (including the photoreceptor cells) that receive visual information and process it before sending it to the brain (**Figure 46.21**). A first step in understanding how the retina tells the brain what it is seeing is to study how these layers are interconnected and how they influence one another. From our discussion of rod cells, we know that the photoreceptor cells at the back of the retina hyperpolarize in response to light and do not generate action potentials. The cells at the front of the retina (the cells closest to the lens) are **ganglion cells**. They do fire action potentials, and their axons form the optic nerve that travels to the brain. The layers of cells between the photoreceptors and the ganglion cells process information about the visual field.

The photoreceptors and ganglion cells are connected by *bipolar cells*. Changes in the membrane potential of rods and cones in response to light alter the rates at which the rods and cones release neurotransmitter at their synapses with the bipolar cells. In response to this neurotransmitter, the membrane potentials

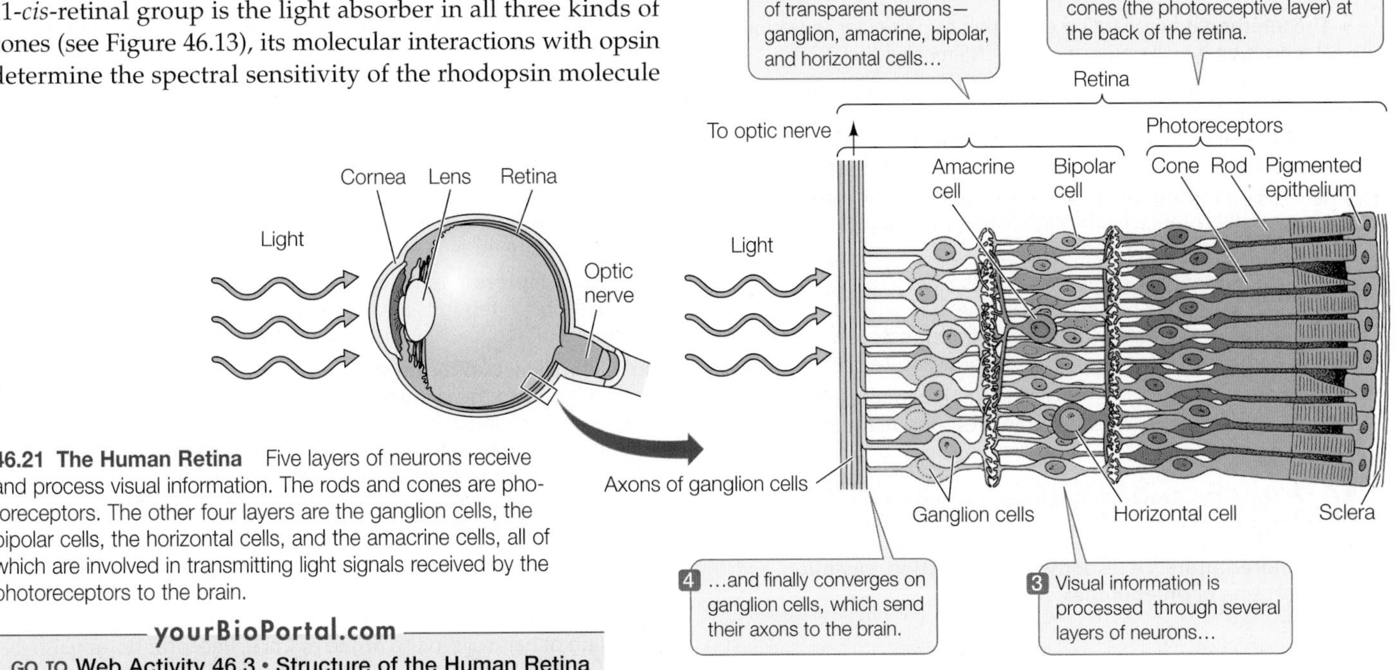
1 Light travels through layers of transparent neurons—ganglion, amacrine, bipolar, and horizontal cells…
2 …and is absorbed by the rods and cones (the photoreceptive layer) at the back of the retina.
Retina
To optic nerve
Photoreceptors
Amacrine cell
Bipolar cell
Cone
Rod
Pigmented epithelium
Cornea
Lens
Retina
Light
Optic nerve
Light
Axons of ganglion cells
Ganglion cells
Horizontal cell
Sclera
4 …and finally converges on ganglion cells, which send their axons to the brain.
3 Visual information is processed through several layers of neurons…

**46.21 The Human Retina** Five layers of neurons receive and process visual information. The rods and cones are photoreceptors. The other four layers are the ganglion cells, the bipolar cells, the horizontal cells, and the amacrine cells, all of which are involved in transmitting light signals received by the photoreceptors to the brain.

**yourBioPortal.com**

GO TO Web Activity 46.3 • Structure of the Human Retina

of the bipolar cells change, altering the rate at which they release neurotransmitter onto ganglion cells. The rate of neurotransmitter release from the bipolar cells determines the rate at which the ganglion cells fire action potentials. Thus, the direct flow of information in the retina is from photoreceptor to bipolar cell to ganglion cell. The ganglion cells send the information to the brain.

The other two cell layers, the horizontal cells and the amacrine cells, consist of interneurons that communicate laterally across the retina. *Horizontal cells* form synapses with neighboring photoreceptors and bipolar cells. Thus, light falling on one photoreceptor can influence the sensitivity of its neighbors to light. This lateral flow of information enables the retina to sharpen the perception of contrast between light and dark patterns. *Amacrine cells* form local interconnections between bipolar cells and ganglion cells. Some amacrine cell types are highly sensitive to changing illumination or to motion. Others assist in adjusting the sensitivity of the eyes according to the overall level of light falling on the retina. When background light levels change, amacrine cell connections to the ganglion cells help the ganglion cells remain sensitive to temporal changes in stimulation. Thus, even with large changes in background illumination, the eyes are sensitive to smaller, more rapid changes in the pattern of light falling on the retina.

## 46.4 RECAP

**A family of photopigments called rhodopsins are responsible for light sensitivity in all animals. Receptor cells, including rod and cone cells in humans, transduce the photosensitivity of rhodopsins to light and use it to form images of the environment.**

- Explain how a photon of light affects the membrane potential in a rod cell. **See pp. 976–978 and Figures 46.14 and 46.15**
- What is the mechanism of color vision? **See p. 981 and Figure 46.20**
- Describe the flow of signals that occurs in the eye in response to light. **See pp. 981–982 and Figure 46.21**

Knowing the path of information in the retina still does not tell us how that information is processed by the brain. What does the eye tell the brain in response to a pattern of light falling on the retina? In Chapter 47 we will describe how the brain reassembles that information into our view of the world.

## CHAPTER SUMMARY

### 46.1 How Do Sensory Cells Convert Stimuli into Action Potentials?

- Sensory receptor cells, also known as sensors or receptors, transduce information about an animal's external and internal environment into action potentials. Some sensors do not fire action potentials, but respond with graded membrane potentials that control release of neurotransmitter onto sensory neurons that do fire action potentials.
- The interpretation of action potentials as particular sensations depends on which neurons in the CNS receive them.
- Sensory receptor cells have membrane **receptor proteins** that cause ion channels to open or close, affecting the resting potential of the cell. Metabotropic receptors act through signal transduction pathways to generate **receptor potentials.** Mechanoreceptors are ionotropic sensory receptors that open ion channels physically through forces such as pressure or stretch. **Review Figure 46.1**
- Receptor potentials initiated by a sensory cell can spread to regions of the cell's plasma membrane that generate action potentials, or they can release neurotransmitter in response to changes in membrane potential. **Review Figure 46.2**
- **Adaptation** enables the nervous system to ignore irrelevant or continuous stimuli while remaining responsive to relevant or new stimuli.

### 46.2 How Do Sensory Systems Detect Chemical Stimuli?

- **Chemoreceptors** are responsible for **olfaction**, **gustation**, and the sensing of **pheromones**.
- Mammalian olfactory sensors project directly to the olfactory bulb of the brain. Sensors for the same **odorant** project to the same area of the olfactory bulb.
- Each olfactory receptor cell expresses one receptor protein that can bind a specific molecule or ion. Binding causes a second messenger to open ion channels, which creates an action potential. **Review Figure 46.4**
- In vertebrates, **taste buds** in the mouth cavity are responsible for gustation. The five basic tastes are sweet, salty, sour, bitter, and umami. **Review Figure 46.5**

### 46.3 How Do Sensory Systems Detect Mechanical Forces?

- The skin contains a variety of ionotropic **mechanoreceptors** that respond to touch and pressure. The density of mechanoreceptors in any skin area determines the sensitivity of that area. **Review Figure 46.6**
- **Stretch receptors** in muscle spindles and in the Golgi tendon organ found in tendons and ligaments inform the CNS of the positions of and loads on parts of the body. **Review Figure 46.7**
- In mammalian **auditory systems**, ear pinnae collect and direct sound waves to the **tympanic membrane**, which vibrates in response to sound waves. The movements of the tympanic membrane are amplified through a chain of **ossicles** that conduct the vibrations to the **oval window**. Movements of the oval window create pressure waves in the fluid-filled **cochlea**. **Review Figure 46.8, WEB ACTIVITY 46.1**
- The **basilar membrane** running down the center of the cochlea is distorted by pressure waves at specific locations that depend on the frequency of the wave. These distortions cause the bending of hair cells in the **organ of Corti**. Receptor potentials

in hair cells cause them to release neurotransmitter, which creates action potentials in the **auditory nerve**. Review Figure 46.9, **ANIMATED TUTORIAL 46.1**

- **Hair cells** are also mechanoreceptors. The bending of their stereocilia alters receptor proteins and therefore their membrane potentials. Hair cells are found in the auditory organs and organs of equilibrium such as the **lateral line** system of fishes and the **semicircular canals** and **vestibular system** of mammals. Review Figures 46.10, 46.11, and 46.12

## 46.4 How Do Sensory Systems Detect Light?

- **Photosensitivity** depends on the absorption of photons of light by **rhodopsin**, a photoreceptor molecule that consists of a protein called **opsin** and a light-absorbing group called retinal. Absorption of light by retinal is the first step in a cascade of intracellular events leading to a change in the membrane potential of the **photoreceptor cell**. Review Figure 46.13, **ANIMATED TUTORIAL 46.2**
- When excited by light, vertebrate photoreceptor cells hyperpolarize and release less neurotransmitter onto the neurons with which they form synapses. They do not fire action potentials. Review Figures 46.14 and 46.15
- Visual systems range from the simple **eye cups** of flatworms, which sense the direction of a light source, to the **compound eyes** of arthropods, which detect shapes and patterns, to the **image-forming eyes** of vertebrates and cephalopods. Review Figures 46.16 and 46.17, **WEB ACTIVITY 46.2**
- Vertebrate and cephalopod eyes focus detailed images of the visual field onto dense arrays of photoreceptors that transduce the visual image into neural signals. Review Figure 46.18
- Vertebrates have two types of photoreceptors, **rod cells** and **cone cells**. In humans, the **fovea** contains almost exclusively cone cells, which are responsible for color vision but are not very sensitive in dim light. Color vision arises from three types of cone cells with different spectral absorption properties. Review Figures 46.19 and 46.20
- The vertebrate **retina** consists of five layers of neurons lining the back of the eye. The light-absorbing photoreceptor cells are at the back of the retina. Review Figure 46.21, **WEB ACTIVITY 46.3**
- The axons of the **ganglion cells**, in the innermost layer of the retina, are bundled together in the optic nerve. Between the photoreceptors and the ganglion cells are neurons that process information from the photoreceptors.

# SELF-QUIZ

1. Which statement about sensory systems is *not* true?
   *a.* Sensory transduction involves the conversion (direct or indirect) of a physical or chemical stimulus into changes in membrane potentials.
   *b.* In general, a stimulus causes a change in the flow of ions across the plasma membrane of a sensory receptor cell.
   *c.* The term "adaptation" refers to the process by which a sensory system becomes insensitive to a continuing source of stimulation.
   *d.* The more intense a stimulus, the greater the magnitude of each action potential fired by a sensory neuron.
   *e.* Sensory adaptation plays a role in the ability of organisms to discriminate between important and unimportant information.
2. The female silkworm moth releases a chemical called bombykol from a gland at the tip of her abdomen. Bombykol is
   *a.* a sex hormone.
   *b.* detected by the male only when present in large quantities.
   *c.* not species-specific.
   *d.* detected by hairs on the antennae of male silkworm moths.
   *e.* a chemical basic to the taste process in arthropods.
3. Which statement about olfaction is *not* true?
   *a.* In general, mammals depend more on vision than on olfaction as their dominant sensory modality.
   *b.* Olfactory stimuli are recognized by the interaction between odorant molecules and receptor proteins on olfactory hairs.
   *c.* The more odorant molecules that bind to receptors, the more action potentials are generated.
   *d.* The greater the number of action potentials generated by an olfactory receptor, the greater the intensity of the perceived smell.
   *e.* The perception of different smells results from the activation of different combinations of olfactory receptors.
4. In general, the touch receptors located close to the surface of both hairy and nonhairy skin
   *a.* are relatively insensitive to light touch.
   *b.* adapt very quickly to stimuli.
   *c.* are uniformly distributed throughout the surface of the body.
   *d.* are called Pacinian corpuscles.
   *e.* adapt slowly and provide almost continuous information.
5. The membrane that is most directly responsible for the ability to discriminate different pitches of sound is the
   *a.* round window.
   *b.* oval window.
   *c.* tympanic membrane.
   *d.* tectorial membrane.
   *e.* basilar membrane.
6. Which statement is *not* true?
   *a.* The transmembrane potential of a rod cell becomes more negative when the rod cell is exposed to light.
   *b.* A photoreceptor releases the most neurotransmitter when in total darkness.
   *c.* Whereas in vision the intensity of a stimulus is encoded by the degree of hyperpolarization of photoreceptors, in hearing the intensity of a stimulus is encoded by changes in firing rates of sensory neurons.
   *d.* Stiffening of the ossicles in the middle ear can lead to deafness.
   *e.* The interaction among hammer (malleus), anvil (incus), and stirrup (stapes) conducts sound waves across the fluid-filled middle ear.
7. In humans, the region of the retina where the central part of the visual field falls is the
   *a.* central ganglion cell.
   *b.* fovea.
   *c.* optic nerve.
   *d.* cornea.
   *e.* pupil.

8. Which of the following statements about information flow in the vertebrate visual system is *true*?
   *a.* Action potentials in bipolar cells cause the release of neurotransmitter onto ganglion cells.
   *b.* Amacrine cells integrate the activity of neighboring rod and cone cells.
   *c.* When photons of light enter they eye, the first cells in the retina they encounter are ganglion cells.
   *d.* The highest density of rod cells in the human retina is centrally located in the fovea, resulting in high acuity dim light vision.
   *e.* Pigmented epithelial cells at the back of the retina provide information about the level of ambient light for contrast adjustments.
9. Which statement about the cone cells in a human eye is *not* true?
   *a.* They are responsible for our sharpest vision.
   *b.* They are responsible for color vision.
   *c.* They are more sensitive to light than rods are.
   *d.* They are fewer in number than rods.
   *e.* They exist in high numbers at the fovea.
10. The color in color vision results from the
   *a.* ability of each cone cell to absorb all wavelengths of light equally.
   *b.* lens of the eye acting like a prism and separating the different wavelengths of light.
   *c.* differential absorption of wavelengths of light by different kinds of rod cells.
   *d.* three different isomers of opsin in cone cells.
   *e.* absorption of different wavelengths of light by amacrine and horizontal cells.

## FOR DISCUSSION

1. Compare and contrast the functioning of olfactory receptors and photoreceptors. How do these sensory cells enable the central nervous system to discriminate between an apple and an orange?
2. Amplification of signal is an important feature of sensory systems. Compare the mechanisms of amplification found in olfactory, visual, and auditory systems.
3. If you were blindfolded and sitting in a wheelchair, how would you know if you were being pushed forward or backward?
4. Animals can use visual, olfactory, tactile, and auditory signals to communicate. From what you know about these sensory systems, discuss the relative advantages and disadvantages of these systems for communication.

## ADDITIONAL INVESTIGATION

A certain region of the brain contains neurons that are sensitive to the osmolarity of the blood. You could imagine that these cells are sensitive to the concentration of a particular solute such as NaCl or to tension in their plasma membranes if they take up or lose water osmotically. Using a slice of this brain region in a culture dish, and patch pipettes, how could you investigate the mechanism of signal transduction in these neurons?

## WORKING WITH DATA (GO TO yourBioPortal.com)

**Rod Cell Response** In this exercise based on the experiments illustrated in Figure 46.14, you will analyze the data generated by electrode recordings from rod cells stimulated by light of varying intensities. You will then use the data to generate a maximum-response curve for the rod cell.

# 47 The Mammalian Nervous System: Structure and Higher Function

## Drive a cab, grow your brain

Go to Google Maps and compare London with New York City at the same scale. In which city is it easier to drive a taxi? In the city with numbered avenues and streets at right angles to each other, or the city with a maze of named streets going in every which direction?

Eleanor Maguire at University College London was so impressed with the navigational abilities of London taxi drivers, she decided to see if there was anything "special" about their brains. Using the brain imaging technique known as MRI (*m*agnetic *r*esonance *i*maging), Maguire and her colleagues examined the brains of taxi drivers with varying years of experience and compared them with each other and with the brains of control subjects who were not taxi drivers. The studies revealed significant changes in the anatomy of the hippocampus among taxi drivers.

The hippocampus is an area of the brain involved in learning and memory. The posterior hippocampus in particular is implicated in the memory of spatial relationships among objects in the environment. Maguire found that the posterior hippocampi of taxi drivers was larger than that of the control subjects and that, among the cab drivers themselves, there was a positive correlation between the size of the posterior hippocampus and years of driving experience.

At the Massachusetts Institute of Technology, Matt Wilson and his colleagues can record the hippocampal activity of lab rats as the animals navigate a maze. By placing electrodes in the rats' hippocampi, Wilson has located specific neurons, sometimes referred to as "place cells," that fire only when the rat is at a particular location in the maze. In a sense he can even see what they are thinking, because when the rats are not moving through the maze, the neurons will occasionally fire in the same pattern as when the animal is running the maze—or they will fire in the reverse sequence, representing where the animal has just been. In a maze that had a choice point (go right or go left), the firing pattern seen when the animal was held at the start position predicted which direction it would turn. Similar firing patterns occur when the rats are sleeping. Are they dreaming about the maze, or are they transferring memory of today's experience into long-term memory?

Neural firing patterns that correlate with the hippocampal sequences are also seen in the visual cortex at the back of the brain

**A Mind-Expanding Maze** The extraordinary ability of taxicab drivers in London to navigate its mazelike warren of streets and byways prompted a study that revealed London cabbies to have larger than normal posterior hippocampi—a brain region implicated in the memory of spatial relationships in the environment.

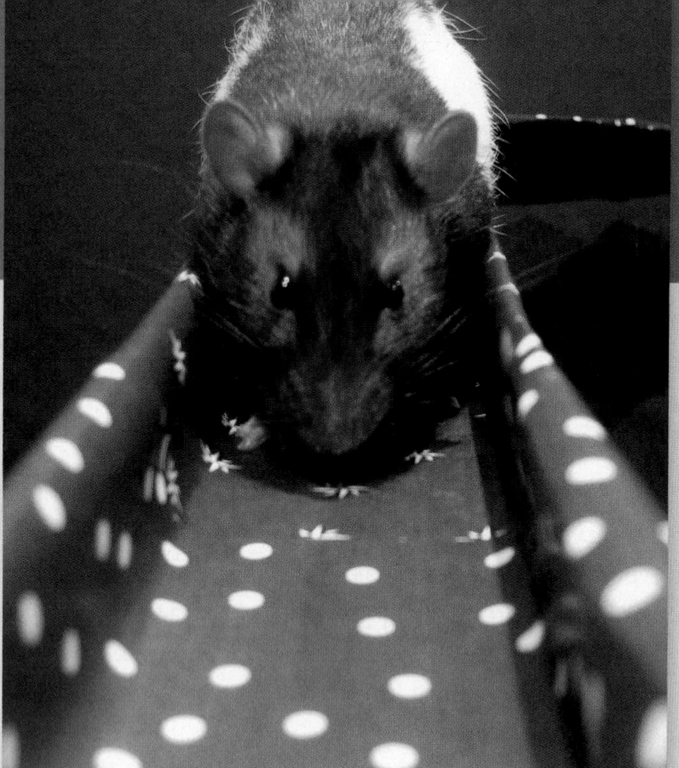

**Seeing What a Rat Thinks** By recording from many neurons in the hippocampus of a rat while it runs a maze, a sequence of "place cell" activities that correspond to different locations in the maze can be described. After the rat has learned the maze, a similar sequence of neural activity can be recorded from its visual cortex.

and in the prefrontal cortex at the front of the brain. In humans, the prefrontal cortex is involved in planning and decision-making.

The extensive ability of humans to remember, to learn, to communicate, and to feel emotions—the "higher functions" of the mammalian brain—is a fascinating subject that brings together the fields of neurobiology, psychology, and medicine. Are these functions simply the product of molecular changes in brain cells? To what extent can we reverse damage to the brain that interferes with higher functioning? These and countless other questions are only beginning to be articulated and studied.

**IN THIS CHAPTER** we will explore the cellular basis of important subsystems of the human nervous system. The human brain has about 100 billion neurons and probably a thousand times as many synapses, which account for its ability to handle vast amounts of information. But the ability of the brain to carry out specific functions—evaluating sensory input, controlling emotions, generating motor output, learning and remembering—arises from functional subsystems.

## CHAPTER OUTLINE



## 47.1 How Is the Mammalian Nervous System Organized?

We can describe the organization of the mammalian nervous system anatomically or functionally. In anatomical terms, all vertebrate nervous systems consist of three parts: a brain, a spinal cord, and a set of peripheral nerves that reach to all parts of the body. As discussed in Section 45.1, the brain and spinal cord are referred to as the *central nervous system*, or *CNS*, and the cranial and spinal nerves that connect the CNS to all the tissues of the body are referred to as the *peripheral nervous system*, or *PNS*. An additional division of the nervous system exists in the gut: the enteric nervous system, which we discuss in Chapter 51.

Recall from Section 45.1 that a *neuron* is an electrically excitable cell that communicates via an axon, and a *nerve* is a bundle of axons that carries information about many things simultaneously. Some axons in a nerve may be carrying information to the CNS, while other axons in the same nerve are carrying information from the CNS to the body's organs. In this chapter, we will further divide the anatomy of the brain, spinal cord, and PNS into smaller functional units.

### A functional organization of the nervous system is based on flow and type of information

**Figure 47.1** illustrates the major avenues of information flow through the nervous system. The *afferent* portion of the PNS carries sensory information to the CNS. We are conscious of much of this information (e.g., light, sound, skin temperature, the position of limbs), but we are usually not conscious of the information involved in physiological regulation (e.g., blood pressure, deep body temperature, blood oxygen levels).

The *efferent* portion of the PNS carries information from the CNS to the muscles and glands of the body. Efferent pathways can be divided into a voluntary division, which executes our conscious movements, and an involuntary, or *autonomic*, division, which controls physiological functions.

In addition to the neural information it receives from the PNS, the CNS receives chemical information from hormones circulating in the blood. In turn, *neurohormones* released by neurons enter the circulation and affect neurons and other cells distant from the site of release (see Figure 41.1).

### The vertebrate CNS develops from the embryonic neural tube

Early in the development of a vertebrate embryo, a tube of neural tissue forms (see Section 44.3). At its anterior end, this *neural tube* forms three swellings that become the **hindbrain**, **midbrain**, and **forebrain**. The rest of the neural tube becomes the

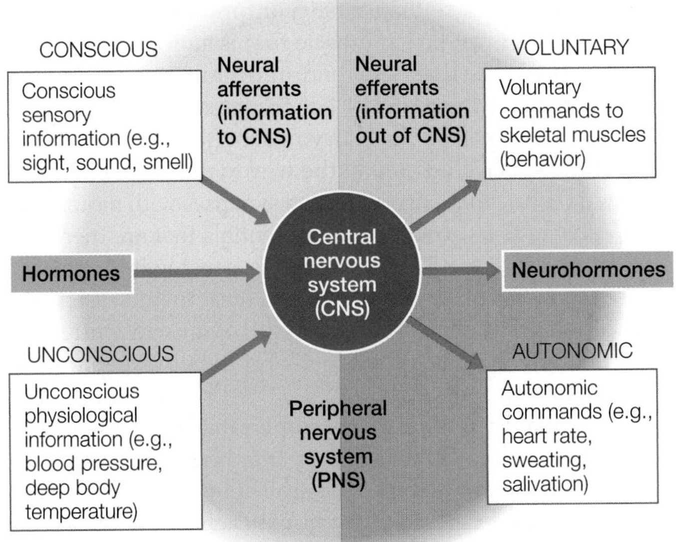


**47.1 Organization of the Nervous System** The peripheral nervous system (beige, green) carries information both to (afferent) and from (efferent) the central nervous system (purple). The CNS also receives hormonal inputs and produces hormonal outputs (lavender).

spinal cord. Peripheral nerves sprout from the midbrain and hindbrain (the cranial nerves) and from the spinal cord (the spinal nerves). From these early stages we see the linear axis of information flow in the nervous system. Although the developing brain will fold and become a complex structure, the information flow in the adult nervous system will follow paths that emerge from the simple linear neural tube.

Each of the three regions of the embryonic brain develops into several structures in the adult brain (**Figure 47.2**). From the embryonic midbrain come structures that process aspects of visual and auditory information. From the hindbrain come the **medulla**, the **pons**, and the **cerebellum**. The medulla is continuous with the spinal cord, the pons is in front of the medulla, and the cerebellum is a dorsal outgrowth of the pons. The medulla and pons contain distinct groups of neurons involved in controlling physiological functions such as breathing and circulation and basic motor patterns such as swallowing and vomiting. All information traveling between the spinal cord and higher brain areas must pass through the pons, the medulla, and the midbrain, which are collectively known as the **brainstem**.

The cerebellum is involved in coordinating muscle activity and maintaining balance. It is like the conductor of an orchestra; the cerebellum receives "copies" of the commands going to the muscles from higher brain areas, and it receives information coming up the spinal cord from the joints and muscles. It compares the motor "score" with the actual behavior of the muscles and refines the motor commands accordingly.

The embryonic forebrain develops a central region called the **diencephalon** and a surrounding structure called the **telencephalon**. The diencephalon is the core of the forebrain and consists of an upper structure called the **thalamus** and a lower structure called the *hypothalamus*. The thalamus is the final relay station for sensory information going to the telencephalon. The

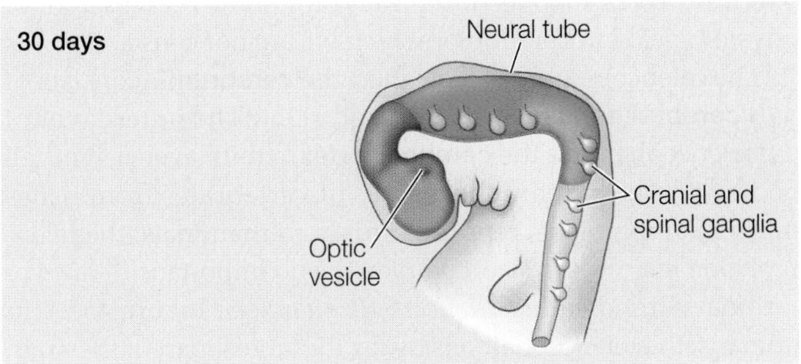


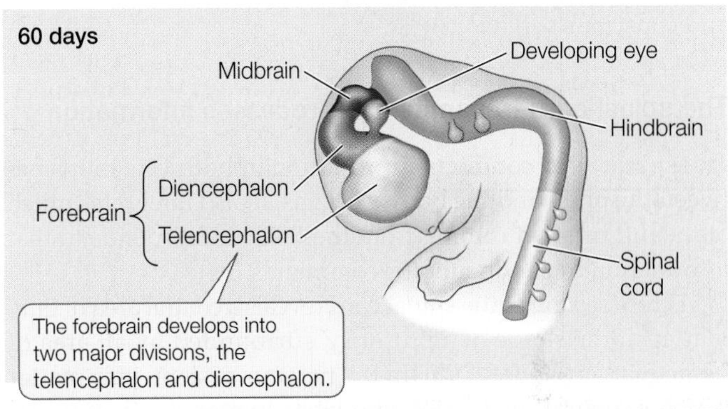


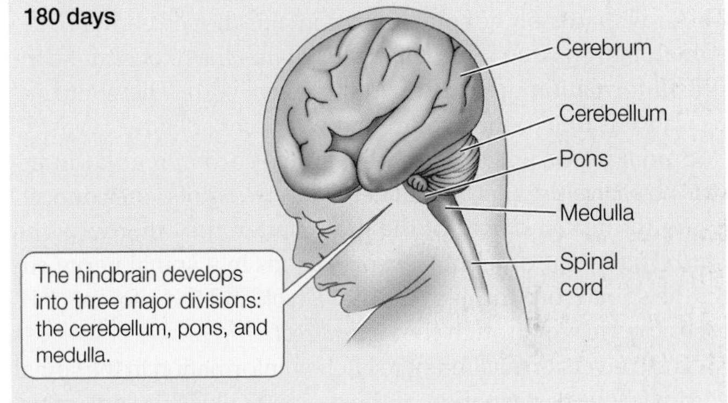


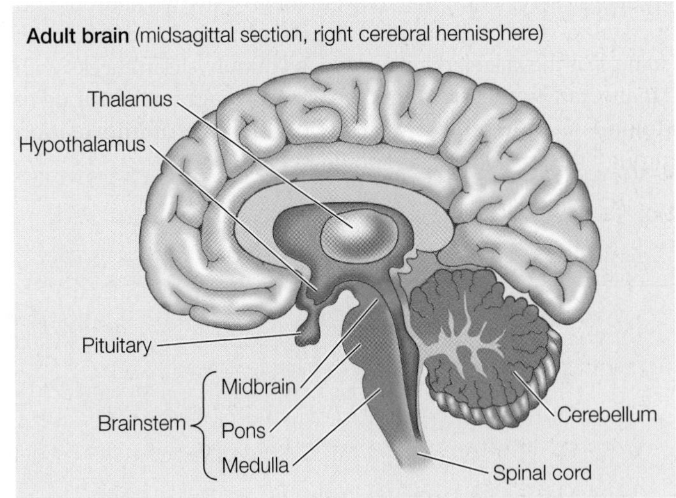


**47.2 Development of the Central Nervous System** In vertebrate embryos, the anterior end of the hollow neural tube differentiates into forebrain, midbrain, and hindbrain. Each of these regions develops into several structures in the adult brain. The remainder of the neural tube becomes the spinal cord.

hypothalamus regulates many physiological functions and biological drives such as hunger and thirst; it receives a lot of physiological information of which we are not conscious.

The telencephalon—also called the **cerebrum**—consists of two **cerebral hemispheres**, left and right. The outer layer of the telencephalon is the **cerebral cortex**, a thin layer rich in cell bodies. If we compare the classes of vertebrates from fishes through amphibians, reptiles, birds, and mammals, the telencephalon increases in size, complexity, and importance—an evolutionary trend called *telencephalization* (see Figure 45.2). In humans, the telencephalon is by far the largest part of the brain and plays major roles in sensory perception, learning, memory, and conscious behavior.

### The spinal cord transmits and processes information

The spinal cord conducts information in both directions between the brain and the body's organs. It also integrates much of the information coming from the PNS and responds to that information by issuing motor commands.

A cross section of the spinal cord reveals a central area of gray matter in the shape of a butterfly, surrounded by an area of white matter (**Figure 47.3**). In the nervous system, **gray matter** is rich in neural cell bodies, and **white matter** contains axons. The gray matter of the spinal cord contains the cell bodies of the spinal neurons; the white matter contains the axons that conduct information up and down the spinal cord. The white appearance is due to the myelin that wraps most of the axons.

Spinal nerves extend from the spinal cord at regular intervals on each side. Each spinal nerve has two roots, one connecting with the *dorsal horn* of the gray matter, the other with the *ventral horn*. The afferent (sensory) axons in a spinal nerve enter the spinal cord through the *dorsal root*, and the efferent (motor) axons leave through the *ventral root*.

The conversion of afferent to efferent information in the spinal cord without participation of the brain is called a **spinal reflex**. The simplest type of spinal reflex involves only two neurons and one synapse and is therefore called a **monosynaptic reflex**. An example is the knee-jerk reflex, which your physician checks with a mallet tap just below your knee. We can diagram the wiring of a monosynaptic reflex by following the flow of information through the spinal cord, as shown in Figure 47.3.

In the knee-jerk reflex, sensory information comes from stretch receptors in the leg muscle that is suddenly stretched when the mallet strikes the tendon running over the knee. Each stretch receptor initiates action potentials that are conducted by the axon of a sensory neuron through the dorsal horn of the spinal cord and all the way to the ventral horn. In the ventral horn, the sensory neuron synapses with motor neurons, causing them to fire action potentials that are then conducted back to the leg extensor muscle, causing it to contract. The function of this simple circuit is to sense an increased load on the limb and to increase the strength of muscle contraction to compensate for the added load and thereby keep muscle length constant.

Most spinal circuits are more complex than this monosynaptic reflex, as we can demonstrate by building on the circuit we just traced. Limb movement is controlled by *antagonistic* sets of muscles—muscles that work against each other. When one member of an antagonistic set of muscles contracts, it bends, or flexes, the limb; it is therefore called a *flexor*. The antagonist to this muscle, the *extensor*, straightens, or extends, the limb. For a limb to move, one muscle of the pair must relax while the other contracts. Thus, sensory input that activates the motor neuron of one muscle also inhibits its antagonist. This coordination is achieved by an *interneuron*, which makes an inhibitory synapse onto the motor neuron of the antagonistic

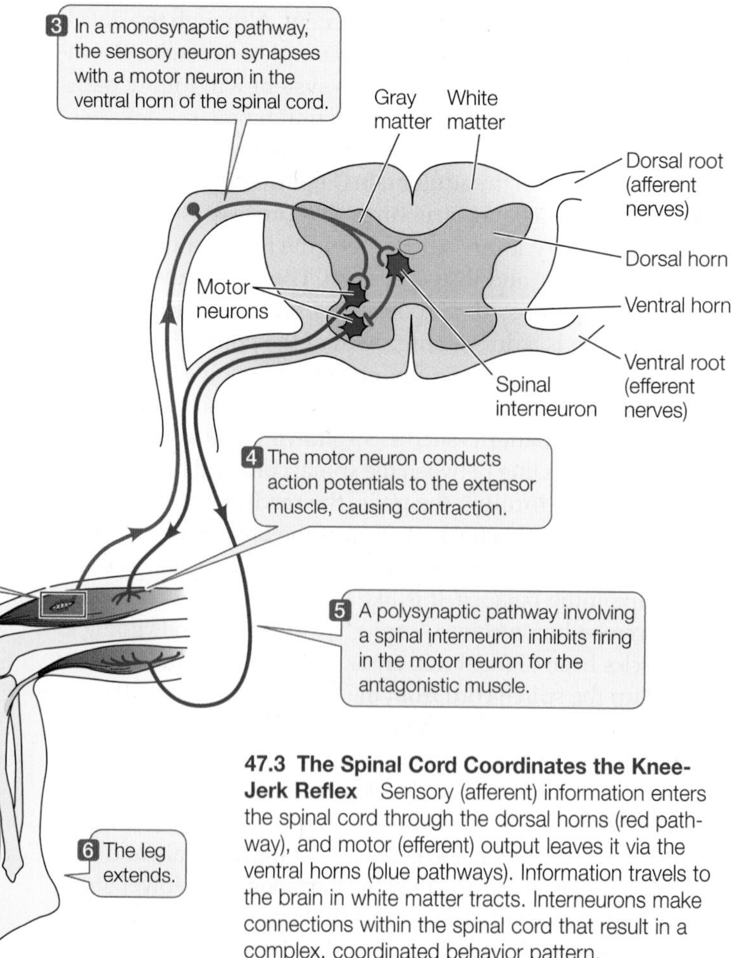
3 In a monosynaptic pathway, the sensory neuron synapses with a motor neuron in the ventral horn of the spinal cord.
Gray matter
White matter
Dorsal root (afferent nerves)
Dorsal horn
Motor neurons
Ventral horn
Ventral root (efferent nerves)
Spinal interneuron
4 The motor neuron conducts action potentials to the extensor muscle, causing contraction.
2 Stretch receptors fire action potentials.
5 A polysynaptic pathway involving a spinal interneuron inhibits firing in the motor neuron for the antagonistic muscle.
1 A hammer tap stretches the tendon in the knee, stretching receptors in the extensor muscle.
6 The leg extends.

**47.3 The Spinal Cord Coordinates the Knee-Jerk Reflex** Sensory (afferent) information enters the spinal cord through the dorsal horns (red pathway), and motor (efferent) output leaves it via the ventral horns (blue pathways). Information travels to the brain in white matter tracts. Interneurons make connections within the spinal cord that result in a complex, coordinated behavior pattern.

yourBioPortal.com
GO TO Animated Tutorial 47.1 • Information Processing in the Spinal Cord

muscle (see Figure 47.3). Thus the reciprocal inhibition of antagonistic muscles involves an interneuron between the sensory cell and the motor neuron of the inhibited muscle, and therefore at least two synapses.

The withdrawal reflex is an example of a polysynaptic spinal reflex that involves many interneurons. When you step on a tack, you immediately pull back your foot: the tack stimulates pain receptors in the foot, and the sensory neurons transmit action potentials into the dorsal horn of the spinal cord on the same side of the body. In the dorsal horn, these neurons synapse with interneurons that send information through their axons to the brain, resulting in the conscious sensation of pain. Before the brain is aware of the pain, however, synapses of the sensory neurons with other interneurons stimulate and inhibit a variety of different motor neurons in the spinal cord. Interneurons on the same side of the spinal cord coordinate the activity of the muscles that withdraw the foot and leg. To pull away, however, the other leg has to extend, and balance must be shifted. The coordination of these actions involves interneurons that make connections across the spinal cord to motor neurons on the opposite side. Thus a rather complex suite of movements is coordinated in the spinal cord without the brain's participation.

### The reticular system alerts the forebrain

Sensory information ascending the spinal cord to final destinations in the forebrain passes through the brainstem. Many sensory axons give off branches in the brainstem that form synapses with a network of brainstem neurons called the **reticular system**. Within this highly complex network of axons and dendrites are many discrete groups of neurons that share a common characteristic such as the neurotransmitter they produce and release. Such an anatomically distinct group of neurons in the CNS is called a **nucleus** (not to be confused with the nucleus of a single cell).

As axons carrying sensory information ascend through the reticular formation, they make connections with nuclei that are involved in controlling many functions of the body. Information from joints and muscles, for example, is directed to nuclei in the pons and cerebellum that are involved in balance and coordination. Sensory information also goes to reticular formation nuclei that control sleep and wakefulness. High reticular formation activity produces waking; in the absence of such stimulation, sleep occurs. Therefore, the reticular core of the brainstem is called the *reticular activating system*.

### The core of the forebrain controls physiological drives, instincts, and emotions

As mentioned above, the diencephalon consists of the thalamus and the hypothalamus. The thalamus communicates sensory information to the cerebral cortex; the hypothalamus receives information about physiological conditions in the body and regulates many homeostatic functions. Section 40.4 describes how the hypothalamus is involved in regulating body temperature, and Section 41.2 discusses the intimate association between the hypothalamus and the pituitary gland and its control of many homeostatic functions.

Surrounding the diencephalon of all vertebrates are phylogenetically older structures of the telencephalon called the **limbic system** (**Figure 47.4**). The limbic system is responsible for basic physiological drives such as hunger and thirst, instincts, long-term memory formation, and emotions such as fear. Within the limbic system are areas that, when stimulated with small electric currents, can cause intense sensations of pleasure, pain, or rage. If a rat is given the opportunity to stimulate its own pleasure centers by pressing a switch, it will ignore food, water, and even sex, pushing the switch until it is exhausted. Pleasure and pain centers in the limbic system are believed to play roles in learning and in physiological drives.

As described in the introduction to Chapter 45, one component of the limbic system—the **amygdala**—is involved in fear and fear memory. If a certain portion of the amygdala is damaged or chemically blocked, an animal cannot learn to be afraid of a stimulus or a situation that would normally induce a strong fear reaction.

Another part of the limbic system, the **hippocampus**, is necessary in humans for the transfer of short-term memory to long-term memory, as we'll discuss in Section 47.3.

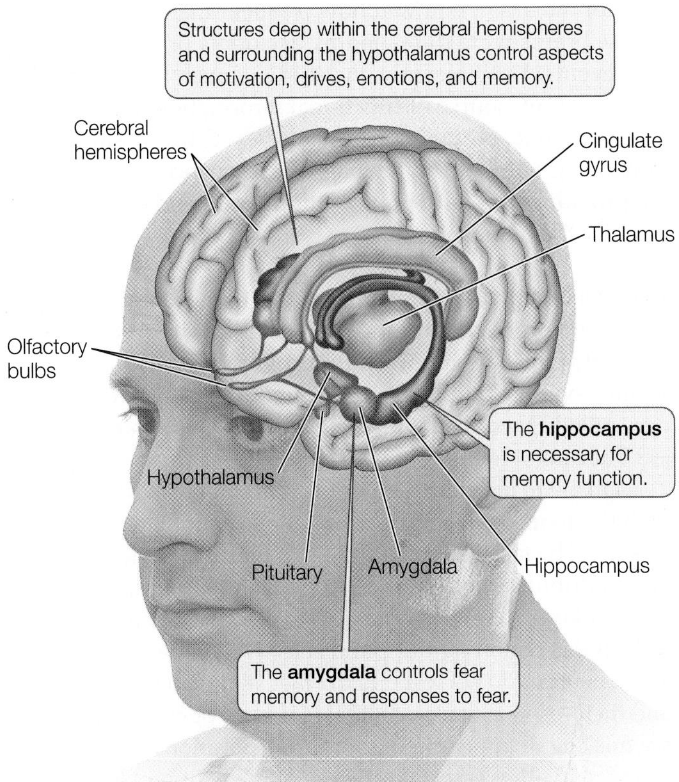

Structures deep within the cerebral hemispheres and surrounding the hypothalamus control aspects of motivation, drives, emotions, and memory.
Cerebral hemispheres
Cingulate gyrus
Thalamus
Olfactory bulbs
The **hippocampus** is necessary for memory function.
Hypothalamus
Pituitary
Amygdala
Hippocampus
The **amygdala** controls fear memory and responses to fear.

**47.4 The Limbic System** The evolutionarily primitive parts of the telencephalon are referred to as the limbic system. The hippocampus is involved in forming long-term memory. The amygdala triggers fear emotions and fear memories (see p. 943).

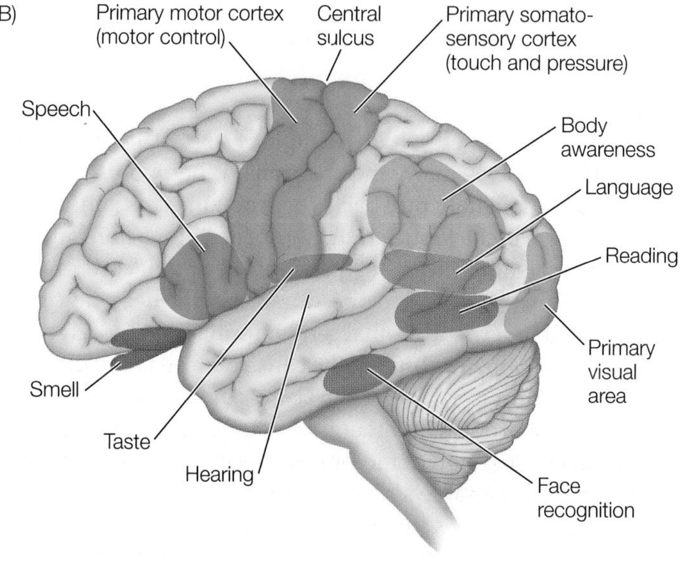


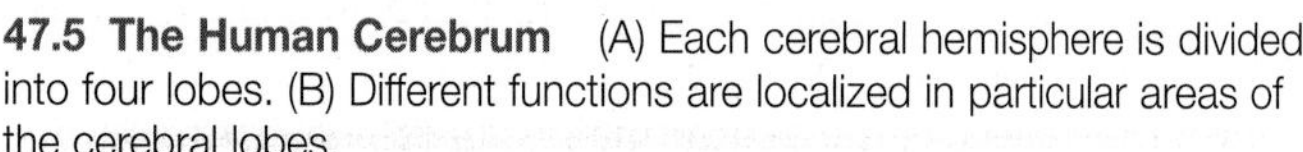
**47.5 The Human Cerebrum** (A) Each cerebral hemisphere is divided into four lobes. (B) Different functions are localized in particular areas of the cerebral lobes.

**yourBioPortal.com**
GO TO **Web Activity 47.1 • The Human Cerebrum**

## Regions of the telencephalon interact to produce consciousness and control behavior

The cerebrum is the dominant structure in the mammalian brain. In humans, it is so large that it covers all other parts of the brain except the cerebellum (**Figure 47.5A**). The cerebral cortex covering the cerebrum is only about 4 millimeters thick, but it covers a surface area larger than 1 $m^2$ because it is folded into ridges (*gyri*; singular *gyrus*) and valleys (*sulci*; singular *sulcus*). These foldings, or *convolutions*, allow the cortex to fit into the skull.

A curious feature of our nervous system is that the left side of the body is served (in both sensory and motor aspects) mostly by the right side of the brain, and the right side of the body is served mostly by the left side of the brain. Thus, sensory input from the right hand goes to the left cerebral hemisphere, and sensory input from the left hand goes to the right cerebral hemisphere. The exception is the head, where the left side is controlled by the left cerebral hemisphere and the right side by the right cerebral hemisphere. The two hemispheres are not symmetrical with respect to all functions. Language abilities, for example, reside predominantly in the left hemisphere.

Different regions of the cerebral cortex have specific functions (**Figure 47.5B**). Some of those functions are easily defined, such as receiving and processing sensory information or generating motor commands, but most of the cortex is involved in higher-order information processing that is less easy to define. These latter areas are given the general name of **association cortex**, so named because they integrate, or *associate*, information from different sensory modalities and from memory.

To understand the cerebral cortex, it helps to have an anatomical road map. Viewed from the left side, the left cerebral hemisphere looks like a boxing glove for the right hand with the fingers pointing forward, the thumb pointing out, and the wrist at the rear. The "thumb" area is the **temporal lobe**, the fingers the **frontal lobe**, the back of the hand the **parietal lobe**, and the wrist the **occipital lobe** (see Figure 47.5A). The right cerebral hemisphere shows a mirror image of this arrangement. Let's look at each lobe separately.

As we explore the functions of the cerebral cortex and other parts of the brain, you will note frequent mention of persons whose brains were damaged by accidents or other unfortunate events. Until recently, the study of such individuals has been the main source of functional information about the human brain, but new imaging technologies such as positron emission tomography (PET) and magnetic resonance imaging (MRI) are providing a wealth of new information and opportunities to study the human brain.

**THE TEMPORAL LOBE** The upper region of the temporal lobe receives and processes auditory information. The association areas of this lobe are involved in recognizing, identifying, and naming objects. Damage to the temporal lobe results in disorders called *agnosias*, in which the individual is aware of an object but cannot identify it.

Damage to a certain area of the temporal lobe results in the inability to recognize faces. Even old acquaintances cannot be identified by facial features, although they may be identified by other attributes such as voice, body features, and posture. Using monkeys, it has been possible to record the activity of neurons in this region that respond selectively to faces in general. These neurons do not respond to other stimuli in the visual field, and their responsiveness decreases if some of the facial features are missing or appear in inappropriate locations (**Figure 47.6**). Damage to other association areas of the temporal lobe causes deficits in understanding spoken language, although speaking, reading, and writing abilities may be intact.

This neuron responds maximally to a complete face viewed from the front.
Firing rate of neuron

**47.6 "Face Neurons" in One Region of the Temporal Lobe** The electrode traces represent the firing rate of a neuron in the temporal lobe of a monkey in response to the pictures shown below them. Highest firing is stimulated by the appearance of a complete face.

**THE FRONTAL LOBE** The frontal and parietal lobes are separated by a deep valley called the *central sulcus*. A strip of the frontal lobe cortex just in front of the central sulcus is called the **primary motor cortex** (see Figure 47.5B). The neurons in this region control muscles in specific parts of the body; the parts of the body have been mapped onto the primary motor cortex largely during neurosurgical procedures. As part of these procedures, electrodes were used to stimulate small areas of cortex. In the area just anterior to the central sulcus, stimulation causes specific muscles to contract. Parts of the body with fine motor control, such as the face and hands, have disproportionate representation (**Figure 47.7A**). Stimulation of neurons in the primary motor cortex causes twitches of muscles, not coordinated movements.

The association functions of the frontal lobe are diverse and are best described as having to do with feeling and planning. They contribute significantly to personality. People with frontal lobe damage have drastic alterations of personality and difficulty planning future events. A dramatic case of frontal lobe damage is the story of Phineas Gage, who in 1848 was an industrious and responsible young railroad construction foreman. Then a blasting accident shot a meter-long, 3-centimeter-wide iron tamping rod through his brain. The rod entered Gage's head below his left eye, passed through his frontal lobe, and exited the top of his head (**Figure 47.8**).

Remarkably, Gage survived, but he was a completely different person. He was quarrelsome, impatient, obstinate, and used profane language, which he had never done before. He lost his railroad job and spent his days as a drifter, earning money by telling his story and exhibiting his scars (and the tamping iron). He died of a seizure in 1860, at the age of 38. If you are in Boston, you can pay him a visit—his skull, death mask, and the tamping iron are on display in the Warren Anatomical Museum of Harvard Medical School.

**THE PARIETAL LOBE** The strip of parietal lobe cortex just behind the central sulcus is the **primary somatosensory cortex** (see Figure 47.5B). This area receives touch and pressure information relayed from the body through the thalamus.

The entire body surface can be mapped onto the primary somatosensory cortex (**Figure 47.7B**). Areas of the body that have a high density of tactile mechanoreceptors and are capable of making fine discriminations in touch (such as the lips and fingers) have disproportionately large representation. If a very small area of the primary somatosensory cortex is stimulated electrically, the subject reports feeling specific sensations, such as touch, in a localized part of the body.

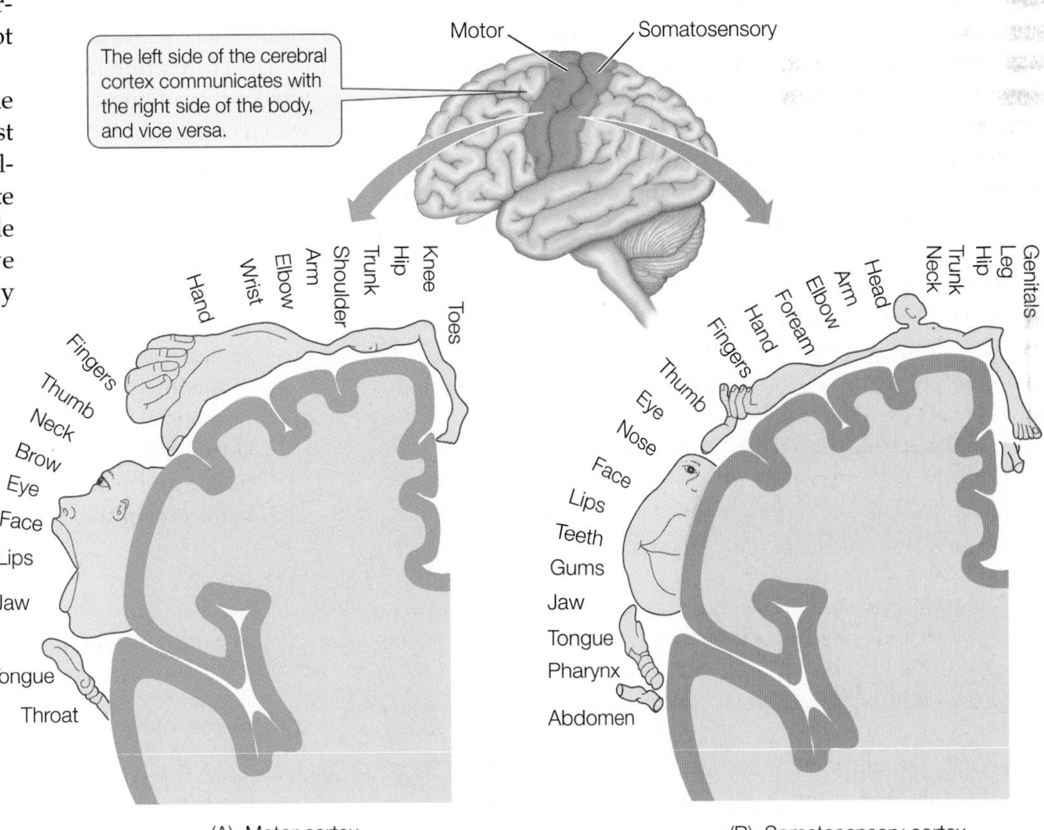


**47.7 The Body Is Represented in Primary Motor and Primary Somatosensory Cortex** Neurons in the primary motor cortex (A) control muscles in specific parts of the body, while neurons in the primary somatosensory cortex (B) receive information from specific parts of the body. The locations of these neurons in the cortex correspond to "maps" on which regions of the body are represented in proportion to the amount of primary cortical area devoted to them.

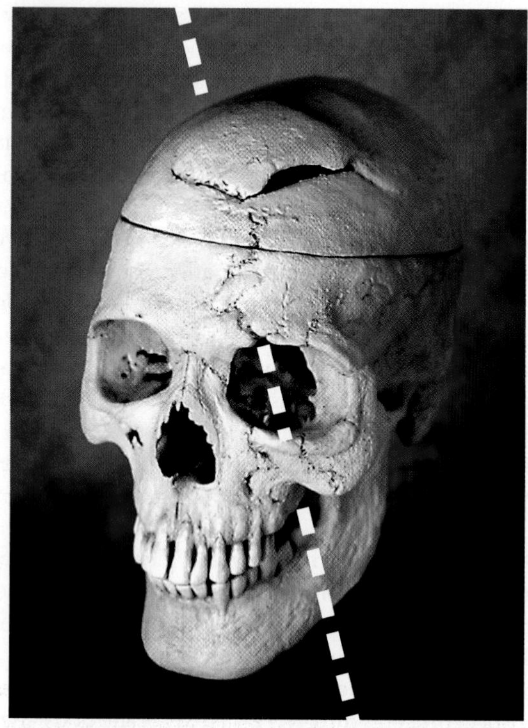

**47.8 A Mind-Altering Experience** Phineas Gage miraculously survived a nineteenth-century railroad construction accident in which an explosion blew an iron rod through his brain. His personality, however, was permanently altered. The path of the iron rod through Gage's brain is superimposed on this reconstruction of his skull.

A major association function of the parietal lobe is attending to complex stimuli. Damage to the right parietal lobe causes a condition called *contralateral neglect syndrome*, in which the individual tends to ignore stimuli from the left side of the body or the left visual field. Such individuals have difficulty performing complex tasks, such as dressing the left side of the body; an afflicted man may not be able to shave the left side of his face. When asked to copy simple drawings, a person who exhibits this syndrome can do well with the right side of the drawing but not the left.

The parietal cortex is not symmetrical with respect to its role in attention, however. Damage to the left parietal cortex does not cause the same degree of neglect of the right side of the body. We will see similar asymmetries in cortical function later in the chapter when we discuss language.

**THE OCCIPITAL LOBE** The occipital lobe receives and processes visual information. The association areas of the occipital cortex are essential for making sense of the visual world and translating visual experience into language. Some deficits resulting from damage to these areas are specific. In one case, a woman with limited damage was unable to see motion. Her vision was intact, but she could see a waterfall only as a still image, and an approaching car only as a series of a stationary objects at different distances.

### The human brain is off the curve

Humans are sometimes called "big-brain primates," and that is an accurate characterization. Across vertebrate species there is a correlation between body size and brain size (**Figure 47.9**). Higher primates such as chimpanzees, baboons, and gorillas all fall above this regression line, but humans stand out because they are so far above the regression line. Gorillas are much larger than humans, but they have smaller brains. Elephants and whales have large brains, but they fall closer to the regression line. Dolphins and humans stand out as having bigger brains than would be predicted by their body sizes.

The correlation of brain size to body size does not tell the whole story of human brain evolution, however. In Figure 45.2, which compares the brains of four vertebrates, we see that the forebrain is larger than other brain regions, and in mammals this is seen as an elaboration of the cerebral cortex. If we look just at mammals, another feature is the degree of *convolution* of the cortex. Since the cortex is a layered, two-dimensional array of neurons, the area of cortex is increased by convolutions, which are greatest in humans. And finally, the percentage of the cortex that is association cortex (i.e., devoted to the integration of information) is by far the greatest in humans. It is these evolutionary changes, primarily in the cortex, that provide the resources for the great intellectual capacity of humans—a topic to which we will return at the end of the chapter.

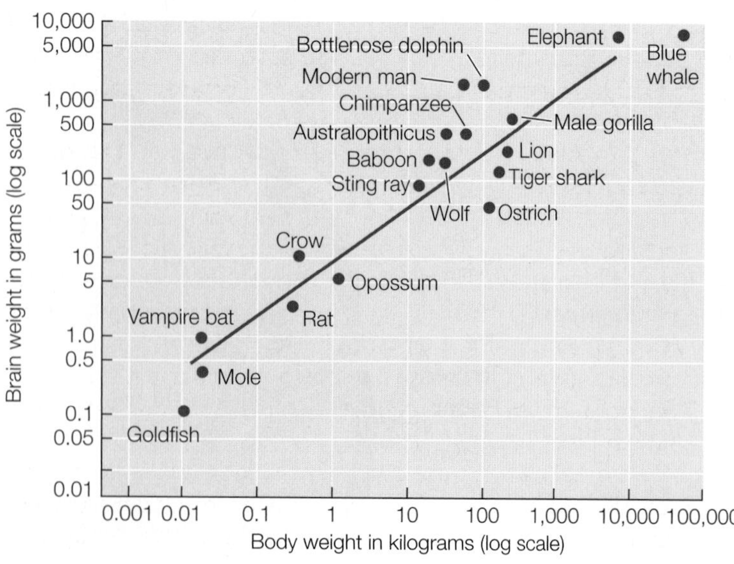


**47.9 Evolution of the Human Brain** Brain size scales to body size across a wide range of vertebrates. The higher primates have larger brains than predicted by the correlation, and humans stand outside this relationship with much bigger brains. The increase in brain size in humans is mostly due to an increase in the cerebral cortex. The human brain is also highly convoluted, and more of it is devoted to associative functions.

## 47.1 RECAP

**The central nervous system communicates with the rest of the body through the peripheral nervous system. We are conscious of some sensory information coming into the CNS, but are not conscious of other afferent information used in physiological regulation. Different regions of the brain have specific functions. Evolution of the human brain has resulted in a greatly increased cerebral cortex devoted to integration of information.**

- Explain how the major functional divisions of the nervous system relate to their origins in the embryonic neural tube. **See pp. 986–988 and Figure 46.2**
- Trace the information flow that results in a spinal reflex. **See pp. 988–989 and Figure 46.3**
- Describe the spatial relations and functions of the major divisions of the telencephalon. **See pp. 990–992 and Figure 47.5**
- What distinguishes the human brain from the brains of other mammals? **See p. 992 and Figure 47.9**

Now that you have some knowledge of the structure and function of different regions of the nervous system, let's explore some examples of how information is processed in the neural circuitry in some specific regions.

## 47.2 How Is Information Processed by Neural Networks?

Specific functions are localized in specific parts of the nervous system and depend on the neural circuits, or networks, in those structures. A major focus of modern neuroscience is understanding how the various functions of the nervous system, ranging from simple reflexes to complex learning and memory, are accomplished by the interactions of neurons in circuits. Let's look at two examples of how neural networks process information: the autonomic nervous system (an output pathway) and the visual system (an input pathway).

### The autonomic nervous system controls involuntary physiological functions

The **autonomic nervous system**, or **ANS**, comprises the output pathways of the CNS that control involuntary functions, such as heart rate, sweating, and some functions of the gut. Its control of diverse organs and tissues is crucial to homeostasis. The ANS has two divisions, **sympathetic** and **parasympathetic**, that work in opposition to each other in their effects on most organs: one division causes an increase in an activity and the other a decrease. The sympathetic and parasympathetic divisions of the ANS are easily distinguished by their anatomy, their neurotransmitters, and their actions (**Figure 47.10**).

The best-known functions of the ANS are those of the sympathetic division that produce the fight-or-flight response: increasing heart rate, blood pressure, and cardiac output and preparing the body for emergencies (see Figure 41.5). In contrast, the parasympathetic division slows the heart and lowers blood pressure; its actions have been characterized as "rest and digest." It is tempting to think of the sympathetic division as speeding things up and the parasympathetic division as slowing things down, but it is not that simple; for example, the sympathetic division slows down the digestive system whereas the parasympathetic division accelerates it.

Whether sympathetic or parasympathetic, every autonomic efferent pathway begins with a *cholinergic neuron* (one that uses acetylcholine as its neurotransmitter) that has its cell body in the brainstem or spinal cord. These cells are called *preganglionic neurons* because the second neuron in the pathway with which they synapse resides in a collection of neurons outside the CNS called a *ganglion*. The second neuron is called a *postganglionic neuron* because its axon extends out from the ganglion. The axon of the postganglionic neuron synapses with cells in the target organs (see Figure 47.10).

The postganglionic neurons of the sympathetic division are called *noradrenergic* because they use norepinephrine (also known as noradrenaline) as their neurotransmitter. In contrast, the postganglionic neurons of the parasympathetic division are mostly cholinergic. In organs that receive both sympathetic and parasympathetic input, the target cells respond in opposite ways to norepinephrine and to acetylcholine. This happens, for example, in a region of the heart called the *pacemaker*, which generates the heartbeat. Stimulating the sympathetic nerve to the heart or dripping norepinephrine onto the pacemaker cells increases their firing rate and causes the heart to beat faster. In contrast, stimulating the parasympathetic nerve to the heart or dripping acetylcholine onto the pacemaker cells decreases their firing rate and causes the heart to beat more slowly.

The sympathetic and parasympathetic divisions of the ANS can also be distinguished by anatomy. The preganglionic neurons of the parasympathetic division come from the brainstem and the *sacral* region; those of the sympathetic division come from the *thoracic* and *lumbar* regions (see Figure 47.10). Most of the ganglia of the sympathetic division are lined up in two chains, one on either side of the spinal cord. The parasympathetic ganglia are close to the target organs.

A specialization of the sympathetic division is its innervation of the adrenal gland, which is critical for the fight-or-flight response. The preganglionic sympathetic neuron sends its axon all the way out to the adrenal gland. The medulla (core) of the gland is composed of hormone-secreting cells that are really developmentally modified postganglionic sympathetic neurons that have lost their axons, and they secrete their "neurotransmitters" (epinephrine and norepinephrine) into the extracellular fluid, from which they enter the circulation and act as hormones (see Section 41.1).

The ANS is an important link between the CNS and many physiological functions. Its control of diverse organs and tissues is crucial to homeostasis. Despite its complexity, work by neu-

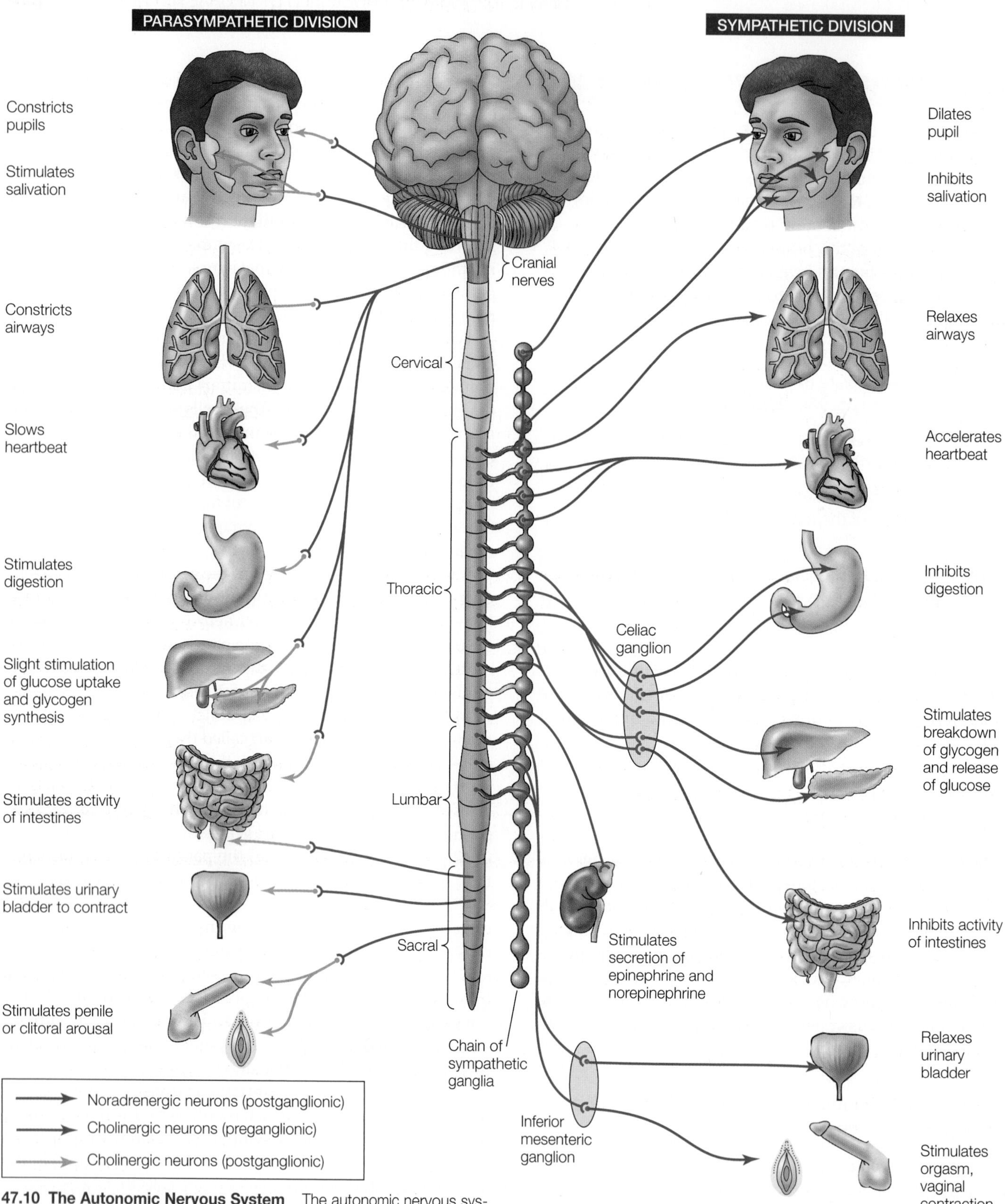


**47.10 The Autonomic Nervous System** The autonomic nervous system is divided into the sympathetic and parasympathetic divisions. The two divisions work in opposition to each other in their effects on most organs; one results in an increase and the other a decrease in activity.

robiologists and physiologists over many decades has made it possible to understand its functions in terms of neuronal properties and circuits.

### Patterns of light falling on the retina are integrated by the visual cortex

The visual system is one of the most studied input pathways to the central nervous system. In Section 46.4 we described how light falling on the retina produces signals that are transmitted through the cellular circuits of the retina, resulting in action po-

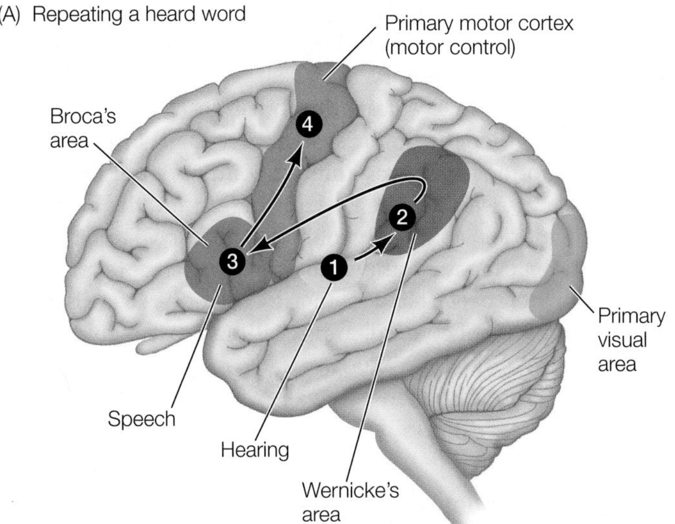


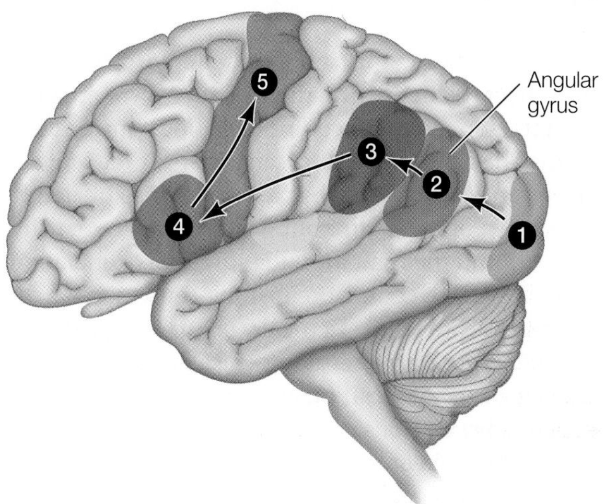


**47.15 Language Areas of the Cortex** Different regions of the left cerebral cortex participate in the processes of (A) repeating a word that is heard and (B) speaking a written word.

**yourBioPortal.com**
GO TO Web Activity 47.2 • Language Areas of the Cortex

experienced. Even very simple animals can learn and remember, but these two abilities are most highly developed in humans. Consider the amount of information associated with learning a language. The capacity of memory and the rate at which memories can be retrieved are remarkable features of the human nervous system.

**LEARNING** Learning that leads to long-term memory and modification of behavior must involve long-lasting synaptic changes. A phenomenon that may explain how long-term synaptic changes might arise is long-term potentiation, or LTP (see Figure 45.18). LTP results from high-frequency electrical stimulation of certain identifiable circuits that makes these circuits more sensitive to subsequent stimulation. In contrast, continuous, repetitive, low-level stimulation of these same circuits reduces their responsiveness, a phenomenon that has been called **long-term depression** (**LTD**). LTP and LTD may be fundamental cellular or molecular mechanisms involved in learning and memory.

Several kinds of learning exist. A form that is widespread among animal species is **associative learning**, in which two unrelated stimuli become linked to the same response. The simplest example of associative learning is the **conditioned reflex**, described by the Russian physiologist Ivan Pavlov. Pavlov observed that a dog salivates at the sight or smell of food—a simple autonomic reflex. He discovered that if he rang a bell just before food was presented to the dog, after a few trials the dog would salivate at the sound of the bell, even if no food followed. The salivation reflex was conditioned to be associated with the sound of a bell, a stimulus that normally is unrelated to feeding and digestion. Experiments in the laboratory of Richard Thompson, now at University of Southern California, located for the first time the circuitry for a conditioned reflex, and it is in the cerebellum.

More complex forms of learning, often referred to as "observational learning" are the foundation of human intelligence. The general pattern of successful observational learning has three elements:

- We pay attention to another person's behavior.
- We retain a memory of what we have observed.
- We try to copy or use that information.

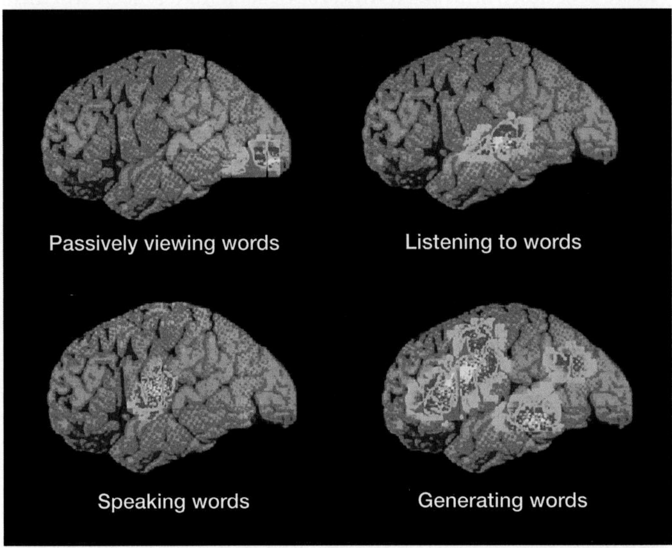


**47.16 Imaging Techniques Reveal Active Parts of the Brain** Positron emission tomography (PET) scanning reveals the brain regions activated by different aspects of language use. Radioactively labeled glucose is given to the subject. Brain areas take up radioactivity in proportion to their metabolic use of glucose. The PET scan visualizes levels of radioactivity in specific brain regions when a particular activity is performed. The red and white areas are the most active.

A key to this scheme of learning is the way in which we create and recall memories.

**MEMORY** Some of the first insights into memory processes came from clinical treatment of patients with severe epilepsy, a disorder characterized by uncontrollable, local increases in neural activity. The resulting *seizures* can endanger the afflicted individual. Serious cases of epilepsy are sometimes treated by destroying the part of the brain from which the surge of activity originates. To find the right area, the surgery is done under local anesthesia, with the patient remaining conscious. As different regions of the brain are electrically stimulated with electrodes, the patient reports the resulting sensations. Stimulation of some regions of the association cortex elicit recall of vivid memories. Such observations provided the first evidence that specific areas in the brain are associated with specific memories and that memory can be attributed to networks of neurons. Destroying a small area of the brain does not completely erase a memory, however, so it is postulated that memory is a function distributed over many brain regions and can be stimulated via many different routes.

You experience several forms of memory everyday. You have **immediate memory** for events that are happening now. Immediate memory is almost perfectly photographic, but it lasts only seconds. **Short-term memory** contains less information but lasts longer—on the order of 10 to 15 minutes. If you are introduced to a group of new people, you may remember most of their names for 5 or 10 minutes, but you will have forgotten them in an hour or so if you have not repeated them, written them down, or used them in a conversation. Repetition, use, or reinforcement by something that gets your attention (such as the title "president") facilitates the transfer of short-term memory to **long-term memory**, which can last for days, months, years, or a lifetime.

Knowledge about neural mechanisms for the transfer of short-term memory to long-term memory has come from observations of persons who have lost parts of the limbic system, notably the hippocampus. A famous case is that of the man identified as H.M., whose hippocampus on both sides of the brain was removed in 1953 in an effort to control severe epilepsy. After the surgery, H.M. was unable to transfer information to long-term memory. If someone was introduced to him, had a conversation with him, and then left the room for several minutes, when that person returned, H.M did not recognize him—it was as if the conversation had never taken place. Up until his death 55 years later, H.M. remembered events that happened before his surgery but could not remember postsurgery events for more than 10 or 15 minutes.

Memory of people, places, events, and things is called **declarative memory** because you can consciously recall and describe them. **Procedural memory** cannot be consciously recalled and described; it is the memory of how to perform a motor task. When you learn to ride a bicycle, ski, or use a computer keyboard, you form procedural memories. Although H.M. was incapable of forming declarative memories, he could form procedural memories. When taught a motor task day after day, he could not recall the lessons of the previous day, yet his performance steadily improved. Thus procedural learning and memory must involve mechanisms different from those used in declarative learning and memory.

Memories can have considerable emotional content. As mentioned earlier, the limbic system plays a major role in controlling emotions. The amygdala, a component of the limbic system, is necessary for the emotion of fear and the formation of fear memories. As we saw in the opening to Chapter 45, patients with a damaged amygdala do not associate fear reactions with their declarative memories. Memories can also have positive emotional content, and recalling those memories activates parts of the brain known to be associated with pleasurable sensations and reward, as revealed by brain imaging technologies.

### We still cannot answer the question "What is consciousness?"

This chapter has only scratched the surface of the organization and functions of the human brain. Even with all of our knowledge of the human brain, and with all of the sophisticated new research tools, we still cannot answer the question "What is consciousness?"

The word "consciousness" is used in everyday language to refer to being awake in contrast to being asleep or in a coma. Here we are referring to the deeper meaning of being mentally aware of yourself, your environment, and events going on around you in such a way that you can plan for future events and make decisions based on experience, evidence, value systems, and predicted consequences. Speculations about consciousness have been the realm of philosophers, but we are getting closer to a neurobiological understanding.

The central requirement for conscious experience is a perception of self that can be integrated with information from the physical and social environment and information from past experience. The basis for a perception of self derives from the huge amount of somatosensory and visceral information that comes from all parts of the body. In the CNS of all vertebrates, this information is used for motor control and for homeostatic regulation. It enables animals to find food, seek mates, seek warmth, avoid cold, avoid danger, and so on. This afferent information goes to appropriate control and regulatory systems in the brainstem and forebrain.

In addition, some of this information goes to somatosensory areas of the cerebral cortex, so the animal is aware of certain information in the sense that it responds to it behaviorally. Visceral afferent information goes beyond its regulatory and control centers in the brainstem and hypothalamus to an area deep within the forebrain called the *insular cortex*, or *insula*. The insula appears to integrate physiological information from all over the body to create a sensation of how the body "feels." Thus, when an animal's actions restore homeostasis, it "feels" better, and this is motivation to do the right thing for well-being.

In humans and the great apes, the insula is greatly expanded and has even acquired new types of spindle-shaped neurons not seen in other animals. The circuitry involving the insula has also evolved to communicate with parts of the brain that are involved in planning and decision making. In imaging studies, the insula is seen to be active in a great diversity of situations that involve strong feelings, be they pleasure, disgust, humor, pain, lust, craving, humiliation, guilt, or empathy. Damage to the insula causes apathy, loss of ability to enjoy music, loss of sexual response, and even loss of the ability to distinguish good food from spoiled food. Humans and the few other species that have expanded insulas and the new spindle cells are the only species that can recognize themselves in a mirror. Could it be that this very discrete part of our brains and its circuitry are the neurobiological bases for self-awareness and conscious experience?

## 47.3 RECAP

**Even complex functions of the nervous system are beginning to be understood in terms of the properties of neurons and neural networks.**

- What events in the brain are associated with wakefulness and the stages of sleep? See p. 999
- Why do some neurobiologists think that the insular cortex might be involved in conscious experience? See p. 1002–1003

## CHAPTER SUMMARY

### 47.1 How Is the Mammalian Nervous System Organized?

- The brain and spinal cord make up the central nervous system (CNS); the cranial and spinal nerves make up the peripheral nervous system (PNS).
- The nervous system can be modeled conceptually in terms of the direction of information flow and whether we are conscious of the information. The afferent component carries information from the PNS to the CNS, and the efferent component directs information from the CNS to the peripheral parts of the body. Review Figure 47.1
- The vertebrate nervous system develops from a hollow dorsal neural tube. The brain forms from three swellings at the anterior end of the neural tube, which become the **hindbrain**, the **midbrain**, and the **forebrain**. The forebrain develops into the **cerebral hemispheres** (the **telencephalon**, or **cerebrum**) and the underlying **thalamus** and hypothalamus (which together compose the **diencephalon**). The midbrain and hindbrain develop into the brainstem and the **cerebellum**. Review Figure 47.2
- The spinal cord communicates information between the brain and the rest of the body. It can issue some commands to the body without input from the brain. Review Figure 47.3, **ANIMATED TUTORIAL 47.1**
- The **reticular system** is a complex network that directs incoming information to appropriate brainstem **nuclei** that control autonomic functions, and transmits the information to the forebrain that results in conscious sensation. The reticular system controls the level of arousal of the nervous system, including sleep and wakefulness.
- The **limbic system** is an evolutionarily primitive part of the telencephalon that is involved in emotions, physiological drives (such as hunger or thirst), instincts, and memory. Review Figure 47.4
- The cerebral hemispheres are the dominant structures of the human brain. Their surfaces are layers of neurons called the **cerebral cortex**. The cerebral hemispheres can be divided into the **temporal**, **frontal**, **parietal**, and **occipital lobes**. Many motor functions are localized in parts of the frontal lobe. Information from many sensory receptors projects to a region of the parietal lobe. Visual information projects to the occipital lobe, and auditory information projects to a region of the temporal lobe. Review Figures 47.5, 47.6, and 47.7, **WEB ACTIVITY 47.1**

### 47.2 How Is Information Processed by Neural Networks?

- The **autonomic nervous system** (**ANS**) consists of efferent pathways that control the physiological function of organs and organ systems. Its **sympathetic** and **parasympathetic** divisions are characterized by their anatomy, neurotransmitters, and effects on target tissues. Review Figure 47.10
- The neural network of vision involves patterns of light falling on **receptive fields** in the retina. Receptive fields have a center and a surround, which have opposing effects on ganglion cell firing. Review Figure 47.11, **ANIMATED TUTORIAL 47.2**
- Information from retinal ganglion cells is communicated via the optic nerve to the thalamus and then to the visual cortex. The visual cortex seems to assemble an image of the visual world by analyzing edges of patterns of light. Review Figure 47.12
- **Binocular vision** is possible because information from both eyes is communicated to binocular cells in the visual cortex. These cells interpret distance by measuring the disparity between where the same stimulus falls on the two retinas. Review Figure 47.13

### 47.3 Can Higher Functions Be Understood in Cellular Terms?

- Humans have a daily cycle of sleep and waking. Sleep can be divided into **rapid-eye-movement** (**REM**) **sleep** and **non-REM sleep.** Deep non-REM sleep is known as slow-wave sleep because of its characteristic EEG patterns. Review Figure 47.14
- Language abilities are localized mostly in the left cerebral hemisphere, a phenomenon known as lateralization. Different areas of the left hemisphere—including **Broca's area**, **Wernicke's area**, and the **angular gyrus**—are responsible for different aspects of language. Review Figure 47.16, **WEB ACTIVITY 47.2**
- Some learning and memory processes have been localized to specific brain areas. Long-lasting changes in synaptic properties referred to as **long-term potentiation** (**LTP**) and **long-term depression** (**LDP**) may be involved in learning and memory.
- Complex memories can be elicited by stimulating small regions of association cortex. Damage to the hippocampus can destroy the ability to form long-term **declarative memory** but not **procedural memory**.
- A sense of the physiological state of the body may be created in the insular cortex from visceral afferent information. Evolution of this integrative function in higher primates and humans could be the basis for conscious experience.

**SEE WEB ACTIVITY 47.3 for a concept review of this chapter.**

## SELF-QUIZ

1. Which of the following describes the route of sensory information from the foot to the brain?
   *a.* Ventral horn, spinal cord, medulla, cerebellum, midbrain, thalamus, parietal cortex
   *b.* Dorsal horn, spinal cord, medulla, pons, midbrain, hypothalamus, frontal cortex
   *c.* Dorsal horn, spinal cord, medulla, pons, midbrain, thalamus, parietal cortex
   *d.* Ventral horn, spinal cord, pons, cerebellum, midbrain, thalamus, parietal cortex
   *e.* Dorsal horn, spinal cord, medulla, pons, midbrain, thalamus, frontal cortex
2. Which statement about the reticular system is *not* true?
   *a.* Increased activity in the reticular system induces sleep.
   *b.* The reticular system is located in the brainstem.
   *c.* Damage to the reticular system in the midbrain can result in coma.
   *d.* Information from the spinal cord is routed to different nuclei in the reticular system and to the forebrain.
   *e.* There are groups of neurons called nuclei in the reticular system.
3. Which statement about afferent and efferent pathways is *not* true?
   *a.* Sympathetic and parasympathetic pathways carry only efferent information.
   *b.* Visceral afferents carry information about physiological functions of which we are not consciously aware.
   *c.* The voluntary division of the efferent portion of the peripheral nervous system executes conscious movements.
   *d.* The cranial nerves and spinal nerves are part of the peripheral nervous system.
   *e.* Afferent and efferent axons never travel in the same nerve.
4. Which statement about the limbic system is *not* true?
   *a.* Damage to one structure in the limbic system makes it impossible to form a fear memory.
   *b.* The limbic system is involved in basic physiological drives,
   instincts, and emotions.
   *c.* The limbic system consists of primitive forebrain structures.
   *d.* In humans, the limbic system is the largest part of the brain.
   *e.* In humans, a part of the limbic system is necessary for the transfer of short-term memory to long-term memory.
5. Which of the following represents the largest portion of the human cerebral cortex?
   *a.* The frontal lobes
   *b.* The primary somatosensory cortex
   *c.* The temporal cortex
   *d.* The association cortex
   *e.* The occipital cortex
6. Which statement about the autonomic nervous system is *true*?
   *a.* The sympathetic division is afferent, and the parasympathetic division is efferent.
   *b.* The transmitter norepinephrine is always excitatory, and acetylcholine is always inhibitory.
   *c.* Each pathway in the autonomic nervous system includes two neurons, and the neurotransmitter of the first neuron is acetylcholine.
   *d.* The cell bodies of many sympathetic preganglionic neurons are in the brainstem.
   *e.* The cell bodies of most parasympathetic postganglionic neurons are in or near the thoracic and lumbar spinal cord.
7. Which statement about cells in the visual cortex is *not* true?
   *a.* Many cortical cells receive inputs directly from single retinal ganglion cells.
   *b.* Many cortical cells respond most strongly to bars of light falling at specific locations on the retina.
   *c.* Some cortical cells respond most strongly to bars of light falling anywhere over large areas of the retina.
   *d.* Some cortical cells receive input from both eyes.
   *e.* Some cortical cells respond most strongly to an object when it is a certain distance from the eyes.
8. Which of the following characterizes non-REM sleep?
   *a.* Dreaming
   *b.* Paralysis of skeletal muscles
   *c.* EEG slow waves
   *d.* Rapid and jerky eye movements
   *e.* It makes up about 20 percent of total sleep time.
9. Which conclusion was supported by experiments on split-brain patients?
   *a.* Language abilities are localized mostly in the left cerebral hemisphere.
   *b.* Language abilities require both Wernicke's area and Broca's area.
   *c.* The ability to speak depends on Broca's area.
   *d.* The ability to read depends on Wernicke's area.
   *e.* The left hand is served by the left cerebral hemisphere.
10. In the withdrawal reflex,
   *a.* action potentials in a pain sensory neuron enter the spinal cord through the ventral root on the same side of the body.
   *b.* the axon from a pain sensory neuron makes inhibitory synapses with motor neurons on the same side of the spinal cord and inhibitory synapses with motor neurons on the other side of the spinal cord.
   *c.* coordinated escape reactions can be initiated before the brain registers the painful sensation.
   *d.* axons from the pain sensory neuron make synapses with sympathetic preganglionic neurons in the spinal cord.
   *e.* axons from motor neurons on the same side of the spinal cord that receives the input from the pain receptors cross the spinal cord to stimulate muscles on the other side of the body.

## FOR DISCUSSION

1. A person receives a stab wound to the left side of his neck. Miraculously, blood vessels are spared. Afterward, however, the man's left pupil remains more constricted than his right pupil, and he drools out of the left side of his mouth. How can you explain these symptoms?
2. The stretch receptors in muscles are modified muscle cells, and they have their own motor neurons. What is the function of those motor neurons? In thinking about this question, remember that the function of the monosynaptic reflex is to adjust muscle tension to a change in load so that the position of the limb does not change.
3. A patient is unable to speak coherently. She can read and write and has no obvious loss of muscle function. Where would you expect to find an abnormality if you did brain scans of this patient?
4. How can you investigate how visual images falling on the retina are communicated to the brain? Start with the retina, and imagine how you would continue your investigation in visual areas of the brain.
5. We described the organization of the visual cortex as columns of cells that alternately receive input from the left and right eye. If a young kitten is allowed to see light out of only one eye for a day, more synapses are maintained in the cortical columns receiving input from that eye, while synapses decrease in the intervening columns. This redistribution of synapses does not occur, however, if the kitten is not allowed to sleep. What hypotheses could you propose on the basis of these results?

## ADDITIONAL INVESTIGATION

The images in Figure 47.16 were made using technology that enables us to image regional activity in the brain. How would you use this imaging methodology to investigate a higher brain function that is of interest to you? First, formulate your hypothesis. Then design your investigation, keeping in mind that your subject has to lie motionless in a large machine during the imaging. You will have to plan how you will deliver the appropriate stimulus to elicit the response or activity you are interested in. Also, describe what controls you will use to draw clear conclusions about the stimulus–response relationships you expect to observe.

# 48 Musculoskeletal Systems

## Champion jumpers

The Olympic record for the women's long jump is 7.4 meters, set in 1988 by Jackie Joyner-Kersee. Another world-record long jump that still stands was set two years earlier by Rosie the Ribeter, who jumped 6.5 meters. Rosie was a frog competing in the Calaveras County Jumping Frog Contest. In some ways, Rosie's jump is more impressive: while Jackie's jump was about 5 times her body length (i.e., her height), Rosie's was about *20 times* her body length.

Both jumps were powered by skeletal muscle. Muscle tissue responds to commands from the nervous system. The cellular mechanisms of muscle contraction are essentially the same in the frog and the human, so why is the frog's jump so much more impressive? The answer involves the concept of *leverage*, which depends on the muscles and skeletal elements working together.

Both frog and human jumping muscles pull on bones that are connected at joints to make levers. A lever makes it possible for the same force to move a large mass a small distance or a small mass a large distance. The ratio of a frog's leg length to its body mass is simply greater than that in a human. Thus the frog's legs are better at moving a small mass a long distance than are the human's legs.

Let's add a flea to our interspecies competition. The flea can jump more than 200 times its body length. This incredible performance is not due to feats of leverage, because no muscle can contract fast enough to explain the take-off velocity of the flea. A different mechanism evolved in the flea—a kind of slingshot action. At the base of the flea's jumping legs is an elastic material that is compressed by muscles while the flea is resting. When a trigger mechanism is released, the elastic material recoils and "fires" the flea into the air.

In a contest of jumping endurance, the uncontested champion would be the kangaroo. As a human runs faster, the number of strides and the energy expended per minute increase rapidly. Neither is true for the kangaroo. When moving at speeds from about 5 to 25 kilometers per hour, the kangaroo takes the same number of strides per minute and its metabolic rate does not increase. Why is this?

In kangaroos, as in frogs and humans, the muscles used to jump are attached to bones by tendons. Like the material at the base of the flea's

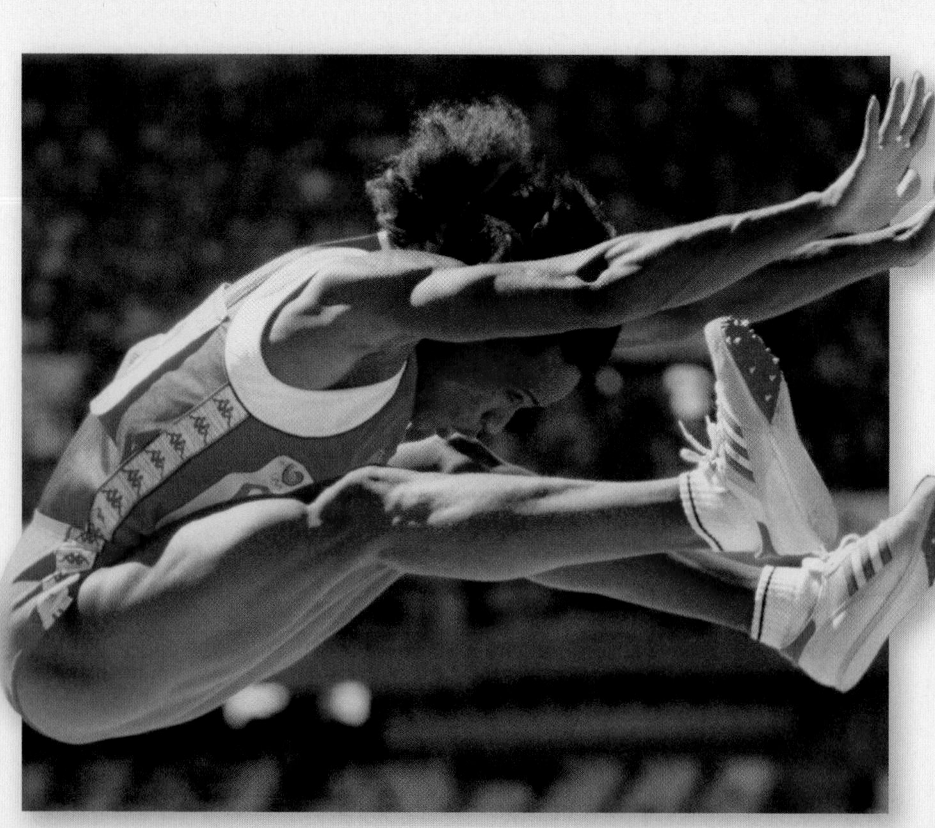

**A Champion Jumper** Jackie Joyner-Kersee set an Olympic record for the women's long jump at the Seoul Olympics in 1988.

**Champion Jumpers** Relative to their size, many animals have more impressive jumping skills than humans. This leopard frog (*Rana pipiens*) can leap distances up to 20 times its body length.

legs, tendons can be elastic. The kangaroo's tendons stretch when it lands, and their recoil helps power the next jump—similar to the action of a pogo stick. To move faster, the kangaroo simply increases the length of its stride, thereby increasing both the stretch on its tendons each time it lands and the magnitude of the recoil at the initiation of each jump.

As discussed in Chapter 31, the ability to move is one of the things that distinguishes animals from the other multicellular organisms. It is our muscles and skeleton—the *musculoskeletal system*—that allow us to move.

**IN THIS CHAPTER** we will describe muscle structure and the molecular mechanisms of contraction. We will discuss the properties that enable different types of muscles to perform different kinds of tasks. To generate specific kinds of movement, muscles must have something to pull on; thus we will describe skeletal systems and their roles in generating movement.

CHAPTER OUTLINE



# 48.1 How Do Muscles Contract?

Most behavior and many physiological actions, such as beating of the heart and moving of food through the digestive tract, depend on muscle contraction. Wherever tissues contract, muscle cells are responsible. As introduced in Section 40.1 and shown in Figure 40.4, there are three types of vertebrate muscle:

- **Skeletal muscle** is responsible for all voluntary movements, such as running or playing a piano. It is also involved in some involuntary actions, such as breathing, shivering and maintaining posture.
- **Cardiac muscle** is responsible for the beating action of the heart.
- **Smooth muscle** creates the movement in many hollow internal organs, such as the gut, bladder, and blood vessels, and is under the control of the autonomic (involuntary) nervous system.

All three muscle types use the same *sliding filament contractile mechanism*, and we begin our study of musculoskeletal movement by describing its underlying molecular mechanisms. We will use vertebrate skeletal muscle as our primary example. Later we will discuss the differences in cardiac and smooth muscle that adapt them to their particular functions.

## Sliding filaments cause skeletal muscle to contract

Skeletal muscle is also called *striated muscle* because of its striped appearance (see Figure 40.4A). Skeletal muscle cells, called **muscle fibers**, are large and have many nuclei. These multinucleate cells form in development through the fusion of many individual embryonic muscle cells called *myoblasts*. A specific muscle such as your biceps (which bends your arm) is composed of hundreds or thousands of muscle fibers bundled together by connective tissue (**Figure 48.1**).

Muscle contraction is due to the interaction between the contractile proteins **actin** and **myosin**. Within muscle cells, actin and myosin molecules are organized into filaments consisting of many molecules. Actin filaments are also called *thin filaments*, and myosin filaments are *thick filaments*. The two kinds of filaments lie parallel to each other. When muscle contraction is triggered, the actin and myosin filaments slide past each other in a telescoping fashion.

What is the relationship between a skeletal muscle fiber and the actin and myosin filaments responsible for its contraction? Each muscle fiber is packed with **myofibrils**—bundles of thin actin and thick myosin filaments arranged in orderly fashion. In most regions of the myofibril, each thick myosin filament is surrounded by six thin actin filaments, and each thin actin filament sits within a triangle of three thick myosin filaments.

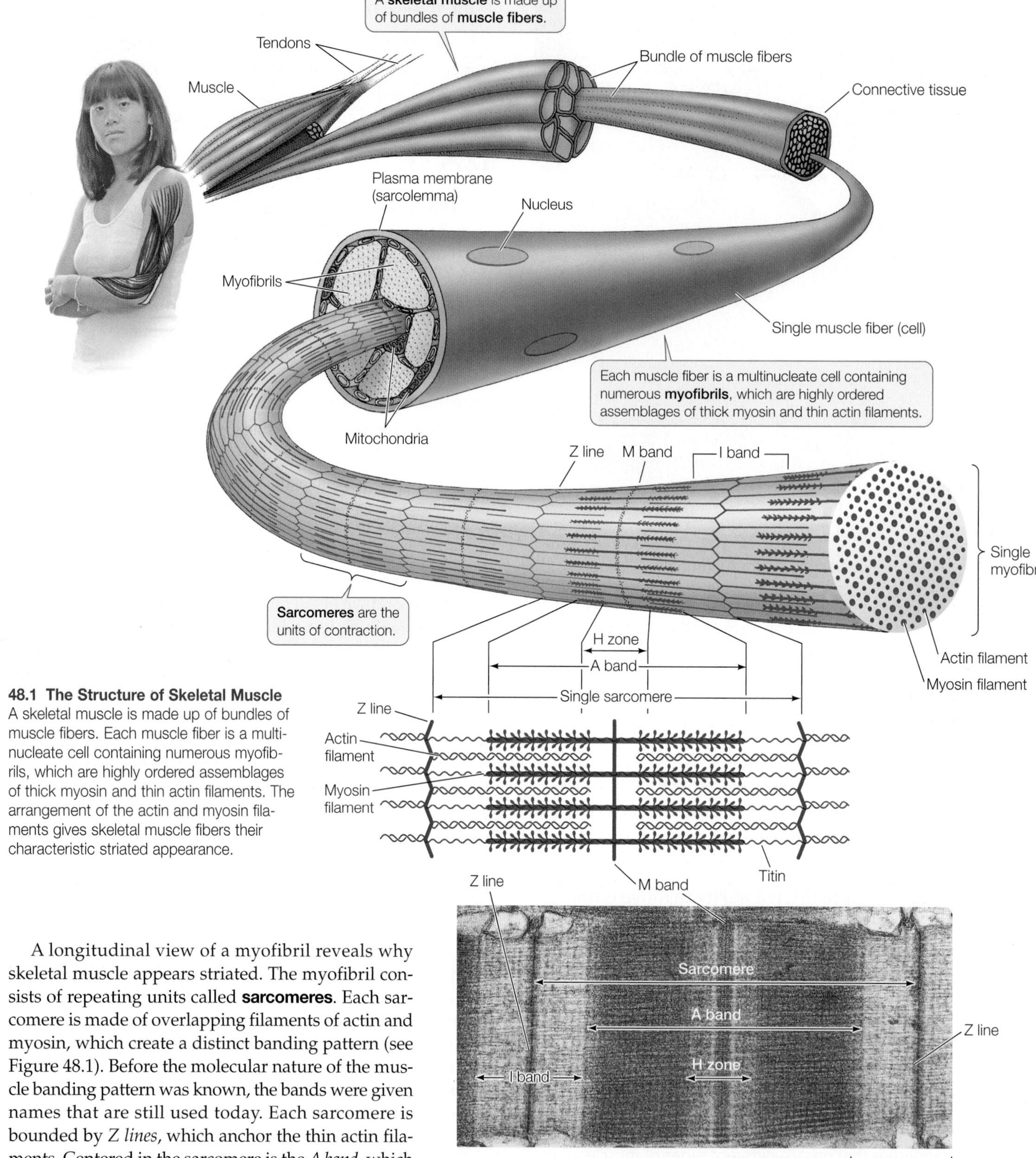


**48.1 The Structure of Skeletal Muscle**
A skeletal muscle is made up of bundles of muscle fibers. Each muscle fiber is a multinucleate cell containing numerous myofibrils, which are highly ordered assemblages of thick myosin and thin actin filaments. The arrangement of the actin and myosin filaments gives skeletal muscle fibers their characteristic striated appearance.

A longitudinal view of a myofibril reveals why skeletal muscle appears striated. The myofibril consists of repeating units called **sarcomeres**. Each sarcomere is made of overlapping filaments of actin and myosin, which create a distinct banding pattern (see Figure 48.1). Before the molecular nature of the muscle banding pattern was known, the bands were given names that are still used today. Each sarcomere is bounded by *Z lines*, which anchor the thin actin filaments. Centered in the sarcomere is the *A band*, which contains all the myosin filaments. The *H zone* and the *I band*, which appear light, are regions where actin and myosin filaments do not overlap in the relaxed muscle. The dark stripe within the H zone is called the *M band*; it contains proteins that help hold the myosin filaments in their regular arrangement.

yourBioPortal.com
GO TO Web Activity 48.1 • The Structure of a Sarcomere

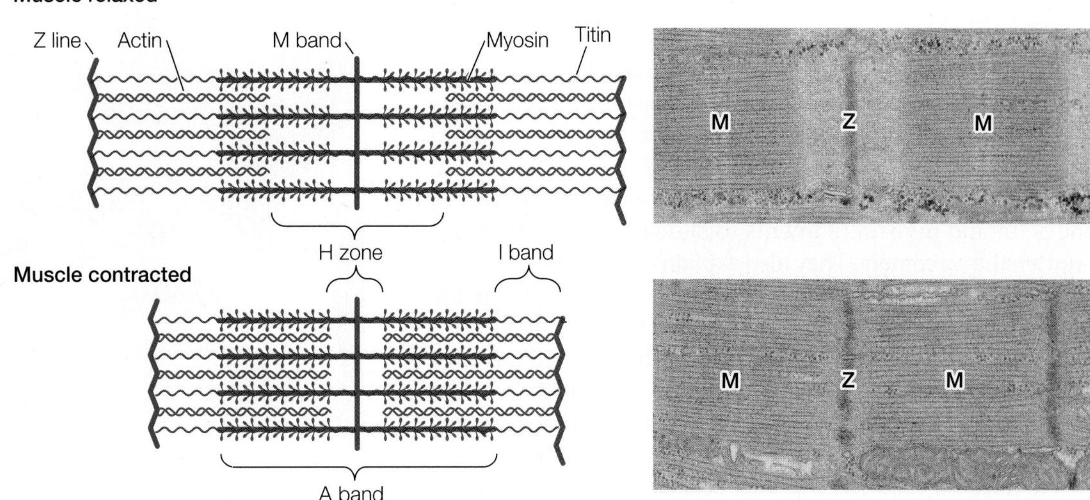


**48.2 Sliding Filaments** The banding pattern of the sarcomere changes as it shortens. Observations of electron micrographs such as those on the right led to the sliding filament hypothesis of muscle contraction.

The bundles of myosin filaments are held in a centered position within the sarcomere by a protein called **titin**. Titin is the largest protein in the body; it runs the full length of the sarcomere from Z line to Z line. Each titin molecule runs right through a myosin bundle. Between the ends of the myosin bundles and the Z lines, titin molecules are very stretchable, like bungee cords. In a relaxed skeletal muscle, resistance to stretch is mostly due to the elasticity of the titin molecules.

As the muscle contracts, the sarcomeres shorten and the band pattern changes. The H zone and the I band become much narrower, and the Z lines move toward the A band as if the actin filaments were sliding into the H zone, the region occupied by the myosin filaments (**Figure 48.2**). In the mid 1950s, this observation independently led two teams of British biologists to propose the **sliding filament model** of muscle contraction.

It is not uncommon in science for critical breakthroughs to be made simultaneously in different laboratories, but in this case the coincidences are remarkable. The leaders of the two teams were named Hugh Huxley and Andrew Huxley—but they were not related. Working in separate Cambridge University labs, the two groups proposed the sliding filament model at the same time, and both papers were published in the same issue of the journal *Nature.*

**yourBioPortal.com**

GO TO **Animated Tutorial 48.1 • Molecular Mechanisms of Muscle Contraction**

## Actin–myosin interactions cause filaments to slide

To understand how the sliding filament model explains muscle contraction, we must first examine the structures of actin and myosin (**Figure 48.3**). A myosin molecule consists of two long polypeptide chains coiled together, each ending in a large globular head. A myosin filament is made up of many myosin molecules arranged in parallel, with their heads projecting sideways at each end of the filament.

An actin filament consists of actin monomers polymerized into a long molecule that looks like two strands of pearls twisted together. Twisting around the actin chains is another protein, *tropomyosin*, and attached to tropomyosin at intervals are molecules of *troponin*. We'll discuss the latter two proteins in the next section.

The myosin heads can bind specific sites on actin, forming cross-bridges between the myosin and the actin filaments. Moreover, when a myosin head binds to an actin filament, the head's conformation changes, and it bends and exerts a tiny force that causes the actin filament to move 5 to 10 nanometers relative to the myosin filament. The myosin heads also have ATPase ac-

**48.3 Actin and Myosin Filaments Overlap to Form Myofibrils** Myosin filaments are bundles of molecules with globular heads and polypeptide tails; the protein titin holds these filaments centered within the sarcomeres. Actin filaments consist of two chains of actin monomers twisted together. They are wrapped by chains of the polypeptide tropomyosin and are studded at intervals with another protein, troponin.

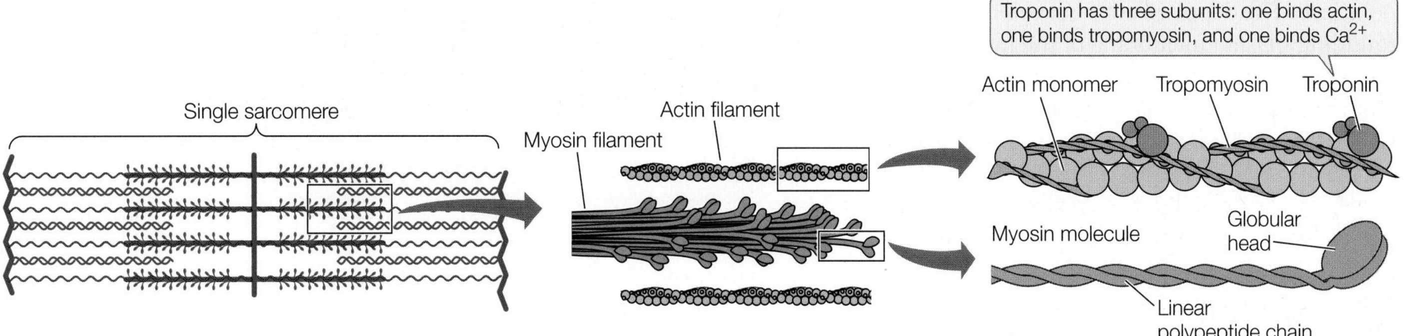

tivity; when they are bound to actin, they can bind and hydrolyze ATP. The energy released when this happens changes the conformation of the myosin head, causing it to release the actin and return to its extended position, from which it can bind to actin again.

Together, these details explain the cycle of events that cause the actin and myosin filaments to slide past each other and shorten the sarcomere. They also explain *rigor mortis*—the stiffening of muscles soon after death. ATP is needed to break the actin–myosin bonds, so when ATP production ceases with death, the actin–myosin bonds cannot be broken and the muscles stiffen. Eventually, however, the proteins begin to lose their integrity, and the muscles soften. The timing of these events helps a medical examiner estimate the time of death.

We have been discussing the cycle of contraction in terms of a single myosin head. Remember that each myosin filament has many myosin heads at both ends and is surrounded by six actin filaments; thus the contraction of the sarcomere involves a great many cycles of interaction between actin and myosin molecules. That is why when a single myosin head breaks its contact with actin, the actin filaments do not slip backward.

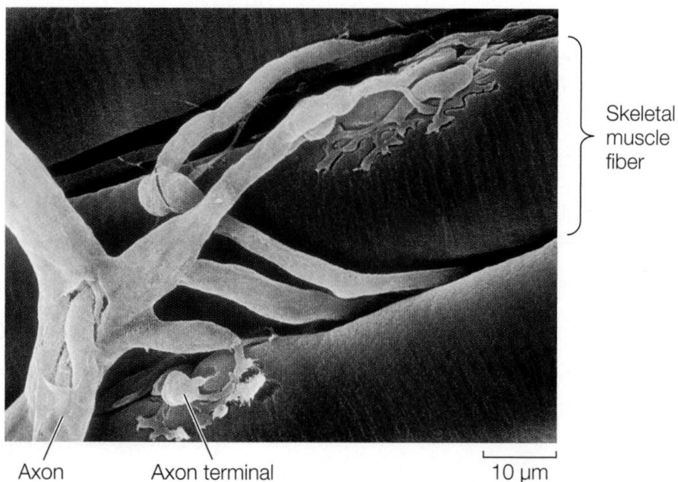


**48.4 The Neuromuscular Junction** Axons branching from a single motor neuron end in terminals that innervate multiple skeletal muscle fibers.

### Actin–myosin interactions are controlled by calcium ions

Like neurons, muscle cells are *excitable*—that is, their plasma membranes can generate and conduct action potentials. In skeletal muscle fibers, action potentials are initiated by motor neurons arriving at a *neuromuscular junction*. The axon terminals of motor neurons are generally highly branched and form synapses with hundreds of muscle fibers (**Figure 48.4**). A motor neuron and all of the fibers with which forms synapses constitute a **motor unit**. The fibers contract simultaneously when its motor neuron fires. A muscle can consist of many motor units. Thus, there are two ways to increase a muscle's strength of contraction: increase the firing rate of an individual motor neuron, or recruit more motor neurons.

As described in Section 45.3, when an action potential arrives at a neuromuscular junction, the neurotransmitter acetylcholine is released from the motor neuron terminals, diffuses across the synaptic cleft, binds to receptors in the postsynaptic membrane, and causes ion channels in the motor end plate to open (see Figures 45.13 and 45.14). Most of the ions that flow through these channels are $Na^+$, and therefore the motor end plate is depolarized. The depolarization spreads to the surrounding plasma membrane of the muscle fiber, which contains voltage-gated sodium channels. When threshold is reached, the plasma membrane fires an action potential that is conducted rapidly to all points on the surface of the muscle fiber.

An action potential in a muscle fiber also travels deep within the cell. The plasma membrane is continuous with a distribution system of tubules that descend into the muscle fiber cytoplasm (also called the **sarcoplasm**). The action potential that spreads over the plasma membrane also spreads through this system of transverse tubules, or **T tubules** (**Figure 48.5**).

The T tubules come very close to the endoplasmic reticulum (ER) of the muscle cell. In muscle cells, the ER is called the **sar-**

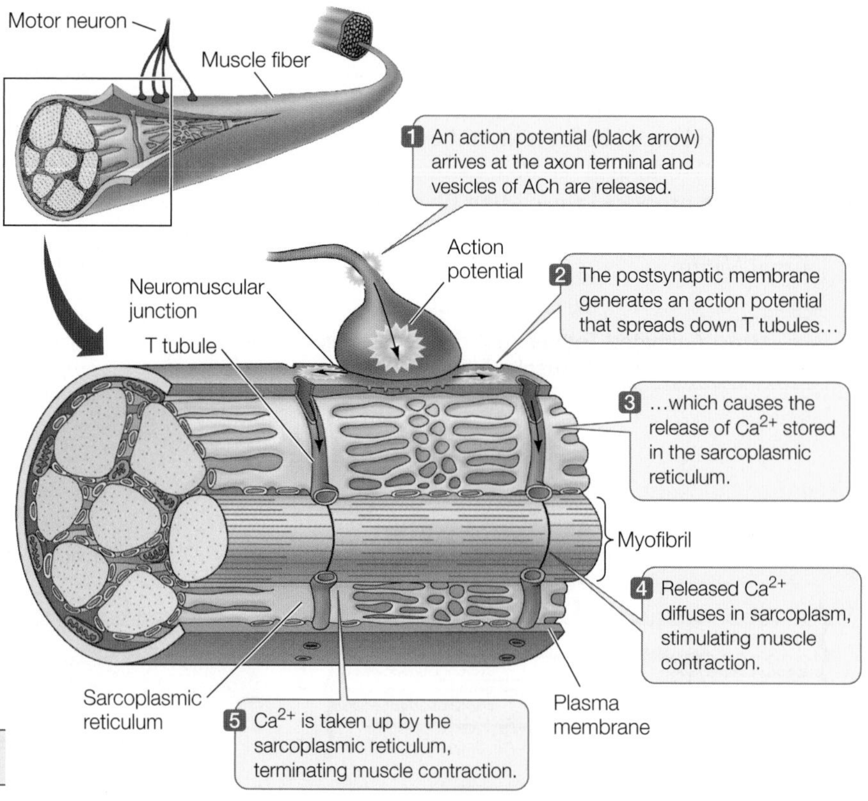

Motor neuron
Muscle fiber
1 An action potential (black arrow) arrives at the axon terminal and vesicles of ACh are released.
Action potential
2 The postsynaptic membrane generates an action potential that spreads down T tubules...
Neuromuscular junction
T tubule
3 ...which causes the release of $Ca^{2+}$ stored in the sarcoplasmic reticulum.
Myofibril
4 Released $Ca^{2+}$ diffuses in sarcoplasm, stimulating muscle contraction.
Sarcoplasmic reticulum
5 $Ca^{2+}$ is taken up by the sarcoplasmic reticulum, terminating muscle contraction.
Plasma membrane


**48.5 T Tubules Spread Action Potentials into the Fiber** An action potential at the neuromuscular junction spreads throughout the muscle fiber via a network of T tubules, triggering the release of $Ca^{2+}$ from the sarcoplasmic reticulum.

yourBioPortal.com
GO TO Web Activity 48.2 • The Neuromuscular Junction

**coplasmic reticulum**, and it is a closed compartment surrounding every myofibril. Calcium pumps in the sarcoplasmic reticulum take up $Ca^{2+}$ ions from the sarcoplasm. Therefore, when the muscle fiber is at rest, there is a higher concentration of $Ca^{2+}$ in the sarcoplasmic reticulum and a lower concentration in the sarcoplasm.

Spanning the space between the membranes of the T tubules and the membranes of the sarcoplasmic reticulum are two proteins. One protein, the *dihydropyridine* (*DHP*) *receptor*, is located in the T tubule membrane; it is voltage-sensitive and changes its conformation when an action potential reaches it. The other protein, the *ryanodine receptor*, is located in the sarcoplasmic reticulum membrane; it is a $Ca^{2+}$ channel. These two proteins are physically connected. When the DHP receptor is activated by an action potential, it changes conformation, causing the ryanodine receptor to allow $Ca^{2+}$ to leave the sarcoplasmic reticulum. $Ca^{2+}$ ions diffuse into the sarcoplasm surrounding the actin and myosin filaments and trigger the interaction of actin and myosin and the sliding of the filaments. How do the $Ca^{2+}$ ions do this?

An actin filament, as we have seen, is a helical arrangement of actin monomers. Twisted around the actin filament are two strands of the protein **tropomyosin** (**Figure 48.6**; see also Figure 48.3). At regular intervals, the filament also includes a globular protein, **troponin**. The troponin molecule has three subunits: one binds actin, one binds tropomyosin, and one binds $Ca^{2+}$.

When the muscle is at rest, the tropomyosin strands are positioned so that they block the sites on the actin filament where myosin heads can bind. When $Ca^{2+}$ is released into the sarcoplasm, it binds to troponin, changing its conformation. Because the troponin is bound to the tropomyosin, this conformational change twists the tropomyosin enough to expose the actin–myosin binding sites. Thus the cycle of making and breaking actin–myosin bonds is initiated, the filaments are pulled past each other, and the muscle fiber contracts. When the calcium pumps remove the $Ca^{2+}$ ions from the sarcoplasm, the conformation of the tropomyosin returns to the state in which it blocks the binding of myosin heads to actin, and the muscle fiber returns to its resting condition. Figure 48.6 summarizes this cycle.

**48.6 Release of $Ca^{2+}$ from the Sarcoplasmic Reticulum Triggers Muscle Contraction** When $Ca^{2+}$ binds to troponin, it exposes myosin-binding sites on the actin. As long as binding sites and ATP are available, the cycle of actin and myosin interactions continues and the filaments slide past each other.

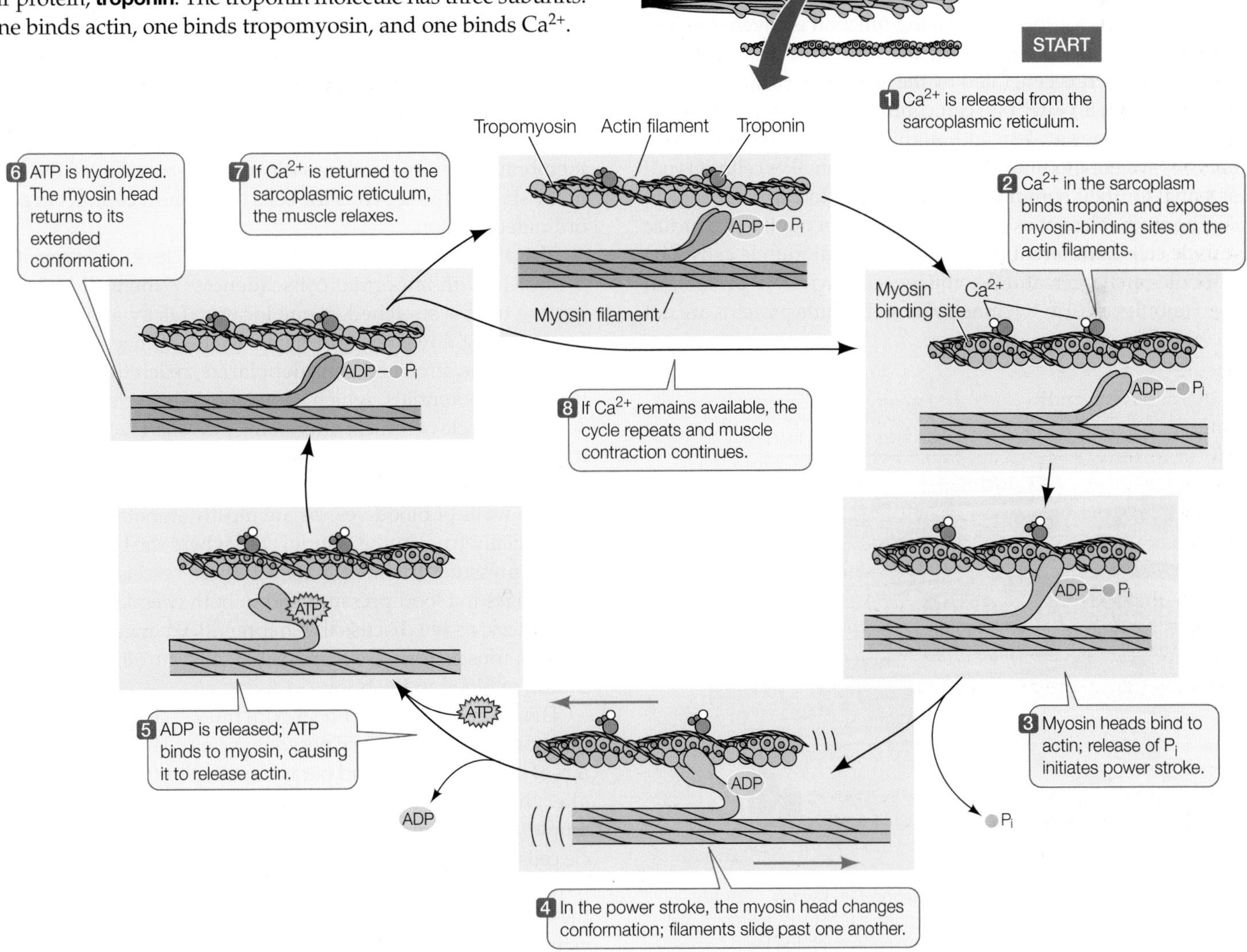

START
1 Ca2+ is released from the sarcoplasmic reticulum.
Tropomyosin
Actin filament
Troponin
ADP–Pi
Myosin filament
2 Ca2+ in the sarcoplasm binds troponin and exposes myosin-binding sites on the actin filaments.
Myosin binding site
Ca2+
3 Myosin heads bind to actin; release of Pi initiates power stroke.
Pi
4 In the power stroke, the myosin head changes conformation; filaments slide past one another.
ADP
ATP
5 ADP is released; ATP binds to myosin, causing it to release actin.
6 ATP is hydrolyzed. The myosin head returns to its extended conformation.
7 If Ca2+ is returned to the sarcoplasmic reticulum, the muscle relaxes.
8 If Ca2+ remains available, the cycle repeats and muscle contraction continues.

### Cardiac muscle is similar to and different from skeletal muscle

Like skeletal muscle, cardiac muscle appears striated because of the regular arrangement of actin and myosin filaments into sarcomeres (**Figure 48.7**). The difference between cardiac and skeletal muscle is that cardiac muscle cells are much smaller and have only one nucleus each (*uninucleate*). Cardiac muscle cells branch, and the branches of adjoining cells interdigitate into a meshwork that is resistant to tearing. As a result, the heart walls can withstand high pressures while pumping blood, without the danger of developing leaks. Adding to the strength of cardiac muscle are *intercalated discs* that provide strong mechanical adhesions between adjacent cells. Gap junctions—protein structures that allow cytoplasmic continuity between cells—in the intercalated discs offer low-resistance pathways for ionic currents to flow between cells (see Figure 7.21A and Section 45.3). Therefore, cardiac muscle cells are electrically coupled. An action potential initiated at one point in the heart spreads rapidly through a large mass of cardiac muscle.

Certain cardiac muscle cells are specialized for generating and conducting electrical signals. These *pacemaker* and *conducting cells* have a low density of actin and myosin filaments, but they initiate and coordinate the rhythmic contractions of the heart. (The molecular basis for this pacemaking function is covered in Section 50.3.) Pacemaker cells make the vertebrate heartbeat *myogenic*, meaning it is generated by the heart muscle itself. A heart removed from a vertebrate can continue to beat with no input from the nervous system; although input from the autonomic nervous system modifies the *rate* of the pacemaker cells, it is not essential for their continued rhythmic function.

The mechanism of excitation–contraction coupling in cardiac muscle cells is different from that in skeletal muscle cells. The T tubules are larger, and the voltage-sensitive DHP proteins in the T tubules are $Ca^{2+}$ channels. These T tubule proteins are not

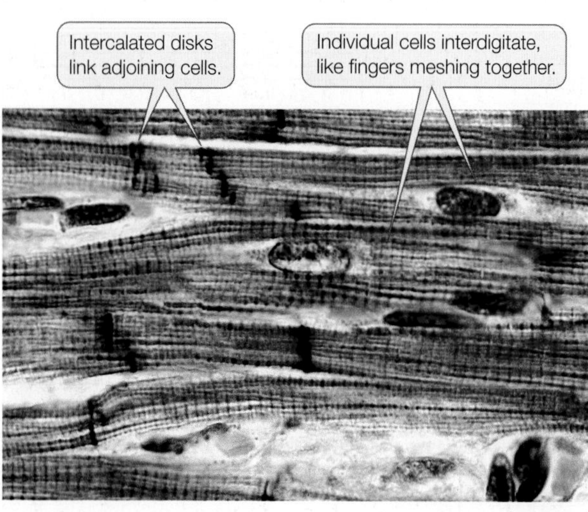
Intercalated disks link adjoining cells.
Individual cells interdigitate, like fingers meshing together.
15 μm

**48.7 Cardiac Muscle Cells Form a Strong Meshwork** Cardiac muscle cells branch and interdigitate, forming a tear-resistant mesh that can withstand the pressure of blood pumping through the heart.

physically connected with the ryanodine receptors in the sarcoplasmic reticulum. Instead, the ryanodine receptors are ion-gated $Ca^{2+}$ channels that are sensitive to $Ca^{2+}$. When an action potential spreads down the T tubules, it causes the voltage-gated channels to open, allowing extracellular $Ca^{2+}$ to flow into the sarcoplasm. This slight rise in sarcoplasmic $Ca^{2+}$ concentration opens the $Ca^{2+}$ channels in the sarcoplasmic reticulum, which in turn causes a huge rise in sarcoplasmic $Ca^{2+}$ concentration, resulting in fiber contraction. This mechanism is called *$Ca^{2+}$-induced $Ca^{2+}$ release.*

### Smooth muscle causes slow contractions of many internal organs

Smooth muscle provides the contractile force for most of our internal organs, which are under the control of the autonomic nervous system. Smooth muscle moves food through the digestive tract, controls the flow of blood through blood vessels, and empties the urinary bladder. Structurally, smooth muscle cells are the simplest muscle cells. They are smaller than skeletal muscle cells, usually long and spindle-shaped, and each has a single nucleus. They are "smooth" because the actin and myosin filaments are not as regularly arranged as they are in skeletal and cardiac muscle, and so do not produce the striated appearance.

Some smooth muscle tissue, such as that from the wall of the digestive tract, has interesting properties. The cells are arranged in sheets, and individual cells in a sheet are in electrical contact with one another through gap junctions, as they are in cardiac muscle. As a result, an action potential generated in the membrane of one smooth muscle cell can spread to all the cells in the sheet of tissue. Thus the cells in the sheet contract in a coordinated fashion.

The plasma membranes of smooth muscle cells are sensitive to stretch, with important consequences. If the wall of the digestive tract is stretched in one location (as by a mouthful of food passing down the esophagus to the stomach), the membranes of the stretched cells depolarize, reach threshold, and fire action potentials, which cause the cells to contract. Thus, smooth muscle contracts after being stretched, and the harder it is stretched, the stronger it contracts. This behavior of smooth muscle is important for moving food through the gut.

The walls of blood vessels are mostly smooth muscle. This is especially true on the arterial side where the blood is under higher pressure. The muscle tone in these vessels is responsive to changes in blood pressure and to both chemical and neural influences, as we discuss in Chapter 50. Changes in vascular smooth muscle tone are responsible for controlling the distribution of blood in the body.

The neural influences on smooth muscle come from the two divisions of the autonomic nervous system. The neurotransmitters of the sympathetic and parasympathetic postganglionic cells alter the membrane potential of smooth muscle cells. For example, in the digestive tract, acetylcholine causes smooth muscle cells to depolarize, making them more likely to fire action potentials and contract. Antagonistically, norepinephrine causes these muscle cells to hyperpolarize and thus be less likely to fire action potentials and contract (**Figure 48.8**). In contrast, norep-

INVESTIGATING LIFE

**48.8 Neurotransmitters Alter the Membrane Potential of Smooth Muscle Cells**

Earlier experiments showed that stretching the smooth muscles of the gut (as in the stretch applied by a full stomach) depolarizes the membranes, causing action potentials that activate the contractile mechanism. This follow-up experiment showed that the parasympathetic neurotransmitter acetylcholine and the sympathetic neurotransmitter norepinephrine act antagonistically to alter the membrane potential of smooth muscle.

**HYPOTHESIS** Stimulation from neurotransmitters of the autonomic nervous system induce contractions in the smooth muscles of the gut.

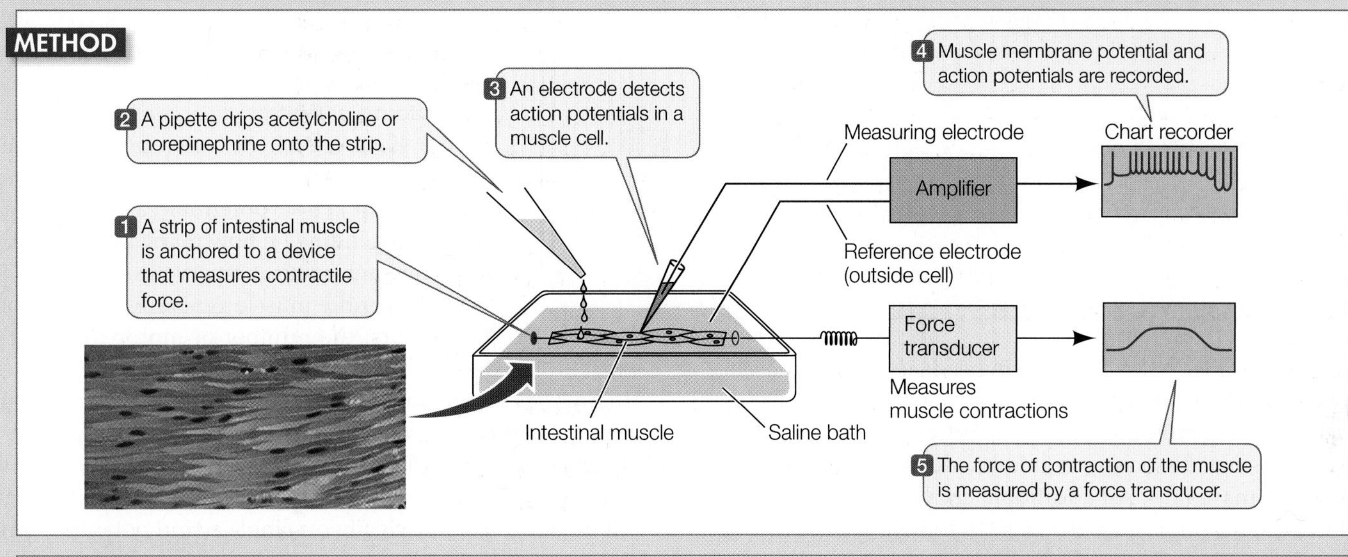

METHOD
1 A strip of intestinal muscle is anchored to a device that measures contractile force.
2 A pipette drips acetylcholine or norepinephrine onto the strip.
3 An electrode detects action potentials in a muscle cell.
4 Muscle membrane potential and action potentials are recorded.
5 The force of contraction of the muscle is measured by a force transducer.
Measuring electrode
Amplifier
Chart recorder
Reference electrode (outside cell)
Force transducer
Measures muscle contractions
Intestinal muscle
Saline bath


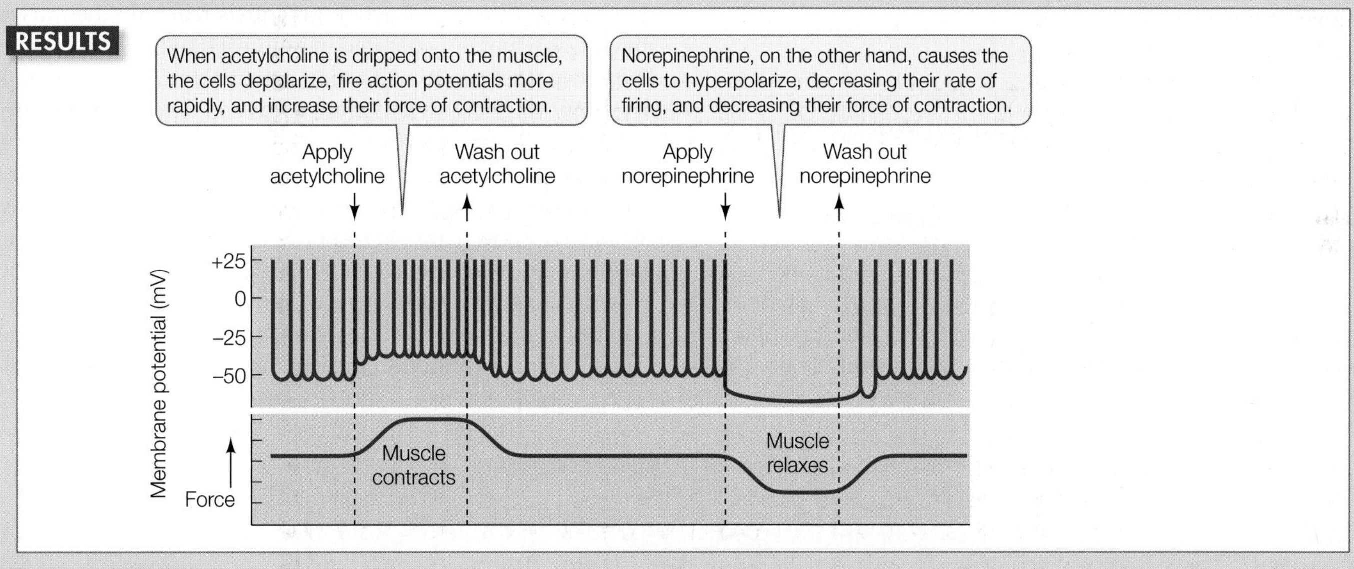


**CONCLUSION** ANS neurotransmitters can alter membrane resting potentials and affect the rate at which smooth muscle cells fire action potentials, thus controlling smooth muscle contraction.

Go to **yourBioPortal.com** for original citations, discussions, and relevant links for all INVESTIGATING LIFE figures.

inephrine acting through G protein-coupled receptors causes the smooth muscle in arteries serving the gut to contract. Remember that the action of the neurotransmitters depends on the receptors in the target tissues. Sympathetic activity is high in a fight-or-flight situation; in an emergency you don't need to digest your lunch, but you do need to send blood to the tissues critical for survival.

Although smooth muscle cell contraction is not controlled by the troponin–tropomyosin mechanism, calcium still plays a critical role. A $Ca^{2+}$ influx into the sarcoplasm of a smooth muscle cell can be stimulated by action potentials, hormones, or stretching. The $Ca^{2+}$ that enters the sarcoplasm combines with a protein called *calmodulin*. The calmodulin–$Ca^{2+}$ complex activates an enzyme called myosin kinase, which can phosphory-

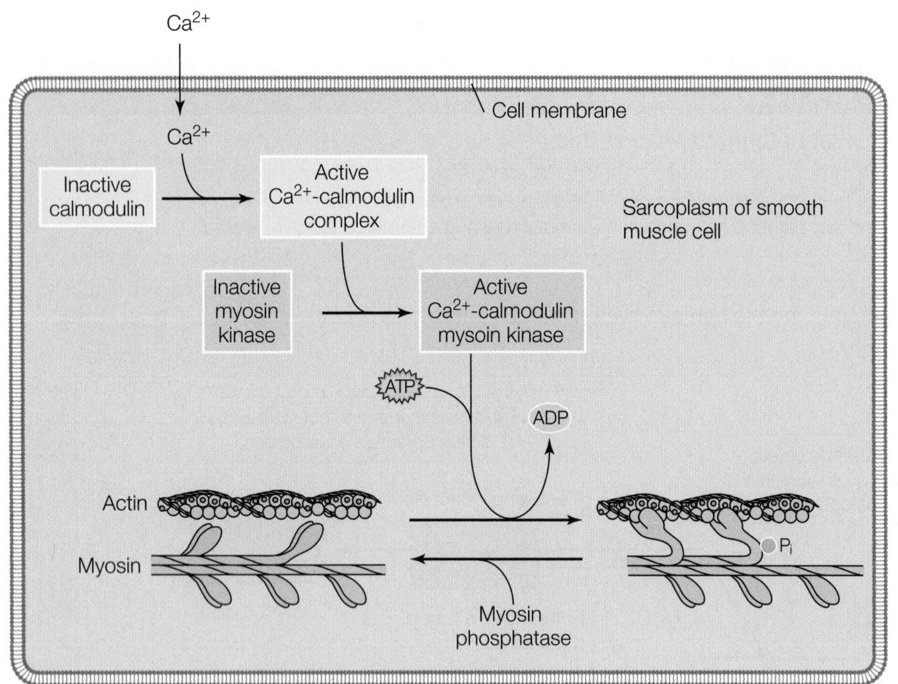


**48.9 The Role of $Ca^{2+}$ in Smooth Muscle Contraction** When a smooth muscle cell is stimulated by neurotransmitter, $Ca^{2+}$ enters the sarcoplasm and binds to calmodulin, which in turn activates an enzyme that phosphorylates the myosin heads, causing them to bind to actin. As long as the myosin remains phosphorylated, actin and myosin go through cycles of binding and release. Thus in smooth muscle, the $Ca^{2+}$-mediated change is on myosin, whereas in skeletal and cardiac muscle it is on the actin-tropomyosin filament.

late myosin heads. When the myosin heads in smooth muscle are phosphorylated, they undergo cycles of binding and releasing actin, causing muscle contraction. As $Ca^{2+}$ is removed from the sarcoplasm, it dissociates from calmodulin, and the activity of myosin kinase falls. An additional enzyme, myosin phosphatase, dephosphorylates the myosin to help reduce actin–myosin interactions (**Figure 48.9**).

**yourBioPortal.com**
GO TO **Animated Tutorial 48.2 • Smooth Muscle Action**

### Single skeletal muscle twitches are summed into graded contractions

In skeletal muscle, the arrival of an action potential at a neuromuscular junction causes an action potential in a muscle fiber. The spread of that action potential through the muscle fiber's T tubule system causes a minimum unit of contraction, called a **twitch**. A twitch can be measured in terms of the *tension*, or force, it generates (**Figure 48.10A**). A single action potential stimulates a single twitch, but the ultimate force generated by an action potential can vary enormously depending on how many muscle fibers are in the motor unit it innervates. The level of tension an entire muscle generates depends on two factors: the number of motor units activated, and the frequency at which the motor units fire

In muscles responsible for fine movements, such as those of the fingers, a motor neuron may innervate only one or a few muscle fibers, but in a muscle that produces large forces, such as the biceps, a motor neuron innervates a large number of muscle fibers.

At the level of a muscle fiber, a single action potential stimulates a single twitch. If action potentials reaching the muscle fiber are adequately separated in time, each twitch is a discrete, all-or-none phenomenon. If action potentials are fired more rapidly, however, new twitches are triggered before the myofibrils have a chance to return to their resting condition. As a result, the twitches sum, and the tension generated by the fiber increases and becomes more sustained. Thus an individual muscle fiber can show a graded response to increased levels of stimulation by its motor neuron.

Twitches sum at high levels of stimulation because the calcium pumps in the sarcoplasmic reticulum are not able to clear the $Ca^{2+}$ ions from the sarcoplasm between action potentials.

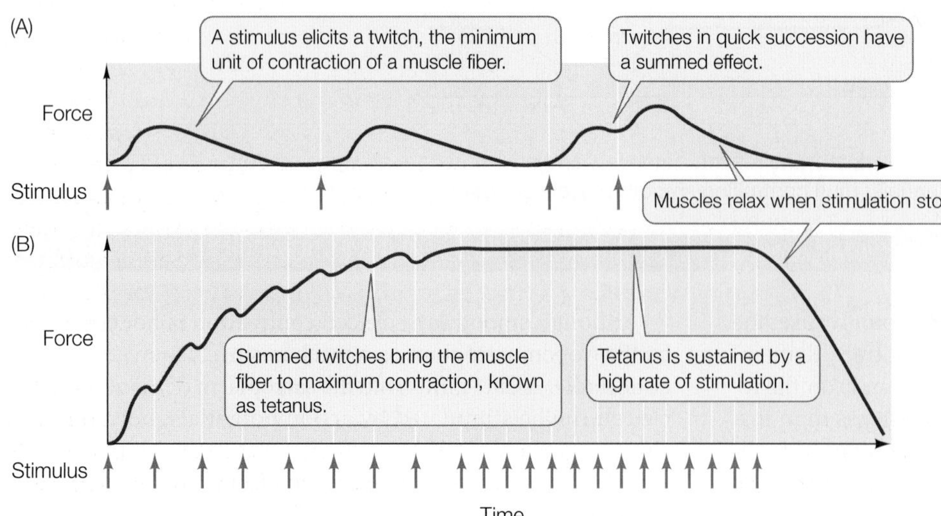


**48.10 Twitches and Tetanus** (A) Action potentials from a motor neuron cause a muscle fiber to twitch. Twitches in quick succession can be summed. (B) Summation of many twitches can bring the muscle fiber to the maximum level of contraction, known as tetanus.

Eventually a stimulation frequency can be reached that results in continuous presence of $Ca^{2+}$ in the sarcoplasm at high enough levels to cause continuous activation of the contractile machinery—a condition known as **tetanus** (**Figure 48.10B**). (Do not confuse this condition with the disease called tetanus, which is caused by a bacterial toxin and is characterized by spastic contractions of skeletal muscles.)

How long a muscle fiber can maintain a tetanic contraction depends on its supply of ATP. Eventually the fiber will become fatigued. It may seem paradoxical that the *lack* of ATP causes fatigue, since the action of ATP is to break actin–myosin bonds. But remember that the energy released from the hydrolysis of ATP "re-cocks" the myosin heads, allowing them to cycle through another power stroke. When a muscle is contracting against a load, the cycle of making and breaking actin–myosin bonds must continue, to prevent the load from stretching the muscle. The situation is like rowing a boat upstream: you cannot maintain your position relative to the stream bank by just holding the oars out against the current; you have to keep rowing. Likewise, actin–myosin bonds have to keep cycling to maintain tension in the muscle.

Many muscles of the body maintain a low level of tension even when the body is at rest. For example, the muscles of the neck, trunk, and limbs that maintain our posture against the pull of gravity are always working, even when we are standing or sitting still. *Muscle tone* comes from the activity of a small but changing number of motor units in a muscle; at any one time, some of the muscle's fibers are contracting and others are relaxed. The nervous system is constantly readjusting muscle tone.

### 48.1 RECAP

**The contractile ability of muscle derives from interactions between actin and myosin filaments. The three types of muscle are skeletal, cardiac, and smooth. Contraction in all three depends on control of $Ca^{2+}$ in the sarcoplasm. Tropomyosin and troponin are controlling elements in skeletal and cardiac muscle. Calmodulin is the controlling element in smooth muscle.**

- Explain how the cellular and subcellular structure of skeletal muscle relate to the sliding filament theory of muscle contraction. **See pp. 1007–1009 and Figures 48.1, 48.2, and 48.3**
- What is the role of $Ca^{2+}$ in the contractile mechanism of skeletal, cardiac, and smooth muscle? **See pp. 1010–1011 and Figures 48.5 and 48.6**
- What role does ATP play in the actin and myosin interactions that produce contraction? **See Figure 48.6**

Now that we understand how muscles generate force, we can look at what determines the characteristics of a muscle, and how individual muscles can change their characteristics with regular use and conditioning.

## 48.2 What Determines Muscle Performance?

The functions that different muscles perform place different demands on them. Some muscles, such as postural muscles, must sustain a load continuously over long periods of time. Other muscles, such as those that control your fingers, generally do not have to sustain long contractions, but they must be able to contract quickly. And what is "quick" for humans doesn't begin to compare with insect flight muscles that can contract as fast as 1,000 times per second. How are muscles adapted to specific functions and demands?

### Muscle fiber types determine endurance and strength

Not all skeletal muscle fibers are alike, and a single muscle often contains more than one type of fiber. The two major types of skeletal muscle fibers express different genes for their myosin molecules, and these myosin variants have different rates of ATPase activity. Those with high ATPase activity can recycle their actin–myosin cross-bridges rapidly and are therefore called fast-twitch fibers. Slow-twitch fibers have lower ATPase activity; they develop tension more slowly but can maintain it longer.

**Slow-twitch fibers** are also called *oxidative* or *red muscle* because they contain the oxygen-binding protein *myoglobin*, have many mitochondria, and are well supplied with blood vessels. These characteristics both increase the fibers' capacity for oxidative metabolism and result in their red appearance. The maximum tension a slow-twitch fiber produces is low and develops slowly but is highly resistant to fatigue. Slow-twitch fibers have substantial reserves of fuel (glycogen and fat), so they can maintain steady, prolonged production of ATP as long as oxygen is available. Muscles with high proportions of slow-twitch fibers are good for long-term *aerobic* work (that is, work that requires oxygen). Long-distance runners, swimmers, cyclists, and other athletes whose activities require endurance have leg and arm muscles consisting mostly of slow-twitch fibers (**Figure 48.11**).

Some **fast-twitch fibers** are also called *glycolytic* or *white muscle* because, compared with slow-twitch fibers, they have few mitochondria, little or no myoglobin, and fewer blood vessels; thus they look pale. Fast-twitch glycolytic fibers can develop maximum tension more rapidly than slow-twitch fibers can, and that maximum tension is greater. However, fast-twitch fibers fatigue rapidly. The myosin of these fibers puts the energy of ATP to work very rapidly, but the fibers cannot replenish ATP quickly enough to sustain contraction for a long time. Fast-twitch fibers are especially good for short-term work that requires maximum strength. Weight lifters and sprinters have leg and arm muscles with high proportions of fast-twitch fibers.

The types of fibers that make up a muscle influence the performance properties of that muscle. A sprinter benefits from muscles that generate maximum force rapidly, but they do not need to sustain a particular load for a long time. By contrast, a marathon runner benefits from muscles with maximum endurance.

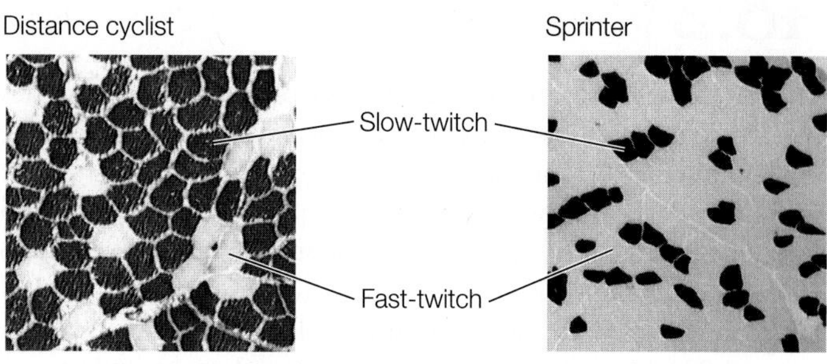


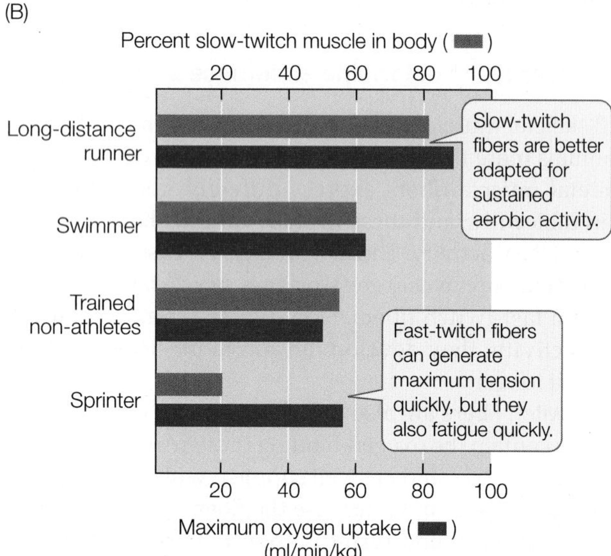


**48.11 Slow- and Fast-Twitch Muscle Fibers** The skeletal muscles in the micrographs were stained with a reagent that shows slow-twitch fibers as dark and fast-twitch muscle as light. Athletes in different sports have different distributions of muscle fiber types.

What can you do to optimize your performance in a particular activity? To a limited extent, you can alter the properties of your muscle fibers through training. There are fast-twitch fibers that are somewhat oxidative and therefore intermediate in their properties between slow-twitch and fast glycolytic fibers. These intermediate fibers can become more oxidative with endurance training and more glycolytic with strength training. However, the most important determinant of your muscle fiber types is your genetic heritage. There is some truth to the statement that champions are born, not made. A person born with a high proportion of fast-twitch fibers is unlikely to become a champion marathon runner, and one born with a high proportion of slow-twitch fibers is unlikely to become a champion sprinter.

### A muscle has an optimal length for generating maximum tension

If you have ever done a pull-up, you know that two parts of this exercise are especially difficult. When you are hanging from the bar with your arms fully extended, it is hard to get the pull-up started; and when your chin has just about reached the bar, pulling yourself up the last small distance is difficult. Why is this? Part of the explanation comes from the lever properties of the muscle–joint interaction that we discuss in the next section, and part comes from the structure of the sarcomere.

When a muscle is stretched and the sarcomeres are lengthened, there is less overlap between the actin and myosin filaments; therefore, fewer cross-bridges can form, and less force can be produced. In fact, if the sarcomeres are stretched too much, actin and myosin do not overlap and no force can be produced. How would a muscle recover from such a difficult situation? The bungee cord–like titin molecules create enough elastic recoil to pull the actin and myosin fibrils back into an overlapping arrangement.

When the muscle is fully contracted, the actin and myosin filaments overlap so much that the myosin bundles are pressed up against the Z lines. Because they have no place to go, additional shortening is difficult. You can see the relationship between the length of a muscle fiber and its ability to develop tension in **Figure 48.12**.

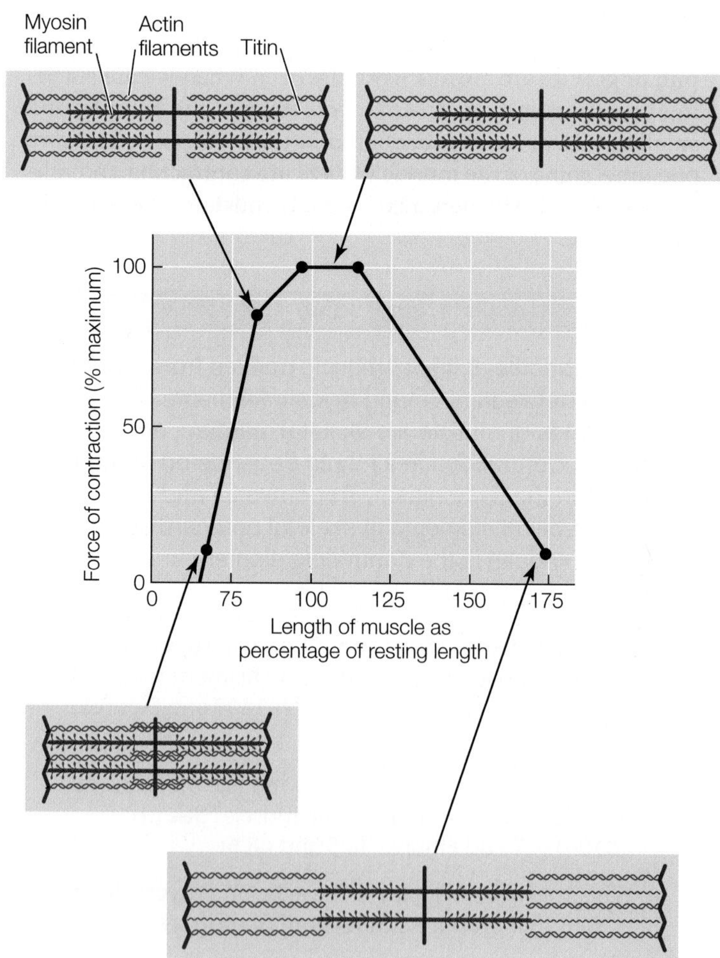


**48.12 Force and Length** The amount of force a sarcomere can generate depends on its resting length. When a muscle is stretched, the sarcomeres lengthen, there is less overlap between the actin and myosin filaments, and less force is produced. Overstretched sarcomeres produce no force because there is no overlap between the actin and myosin.

## Exercise increases muscle strength and endurance

Different types of exercise produce different physical conditioning responses. In general, anaerobic activities, such as weight lifting, increase strength, and aerobic activities, such as jogging, increase endurance. Strength is the maximum force that a muscle can exert, and endurance is work capacity or how long a given workload can be sustained. What are the physiological bases for these differences?

Strength is a function of the cross-sectional area of muscles: the more actin and myosin filaments in a muscle fiber, and the more muscle fibers in a muscle, the more tension it can produce. When athletes undertake strength training, they use weights or exercises such as pull-ups to repeatedly contract specific muscles under heavy loads. Repetitions are usually done until the muscle is completely fatigued. Such stress on a muscle does minor tissue damage—hence the soreness the day after a hard workout—but it also induces the formation of new actin and myosin filaments in existing muscle fibers. The muscle fibers, and hence the muscles, get bigger and stronger. In extreme cases, and after serious muscle damage, new muscle fibers can also be produced from stem cells called *satellite cells* in the muscle. In general, however, the major effect of strength training is to produce bigger, rather than more, muscle fibers.

Aerobic exercise has a completely different effect on muscles: it enhances their oxidative capacity. This effect comes from increases in the number of mitochondria, in enzymes involved in energy use, and in the density of capillaries that deliver oxygen to the muscle. Myoglobin also increases in skeletal muscle cells. **Myoglobin** is an oxygen-binding protein similar to hemoglobin in red blood cells. However, myoglobin has a higher affinity for oxygen than does hemoglobin. Therefore, myoglobin accepts oxygen from the blood, facilitates the diffusion of oxygen throughout the muscle, and provides a store of oxygen for use when oxygen delivery by the blood is insufficient. By increasing the capacity of muscle to use oxygen to produce ATP, aerobic training increases the work load that muscles can sustain through time.

## Muscle ATP supply limits performance

Muscles have three systems for supplying the ATP they need for contraction:

- The *immediate system* uses preformed ATP and creatine phosphate.
- The *glycolytic system* metabolizes carbohydrates to lactate and pyruvate.
- The *oxidative system* metabolizes carbohydrates or fats all the way to $H_2O$ and $CO_2$.

The capacity of these three systems and the rates at which they can produce ATP determine both work capacity and endurance (**Figure 48.13**).

ATP is present in muscles in very small amounts. However, muscle fibers also contain a storage compound called *creatine phosphate* (CP). This molecule stores energy in a phosphate bond, which it can transfer to ADP. The total energy available in all the muscles of your body in the form of ATP and CP—the immediate energy system—is only about 10 kilocalories. When at rest, you metabolize a kilocalorie of energy in less than a minute. Even though the energy available from ATP and CP is limited, it is available immediately, and it enables fast-twitch fibers to generate a lot of force quickly. During burst activity, the immediate system is exhausted in seconds.

The glycolytic system activates within a few seconds to replace the ATP depleted at the onset of muscle activity. The glycolytic enzymes are located in the cytoplasm of the muscle fiber, and therefore the ATP they generate is rapidly available to the myosin filaments. However, as noted in Chapter 9, glycolysis alone is an inefficient way to produce ATP, and it leads to the accumulation of lactic acid, which slows the process. Thus, the glycolytic system and the immediate system together can provide most of the energy for active muscles for less than a minute (see Figure 48.13).

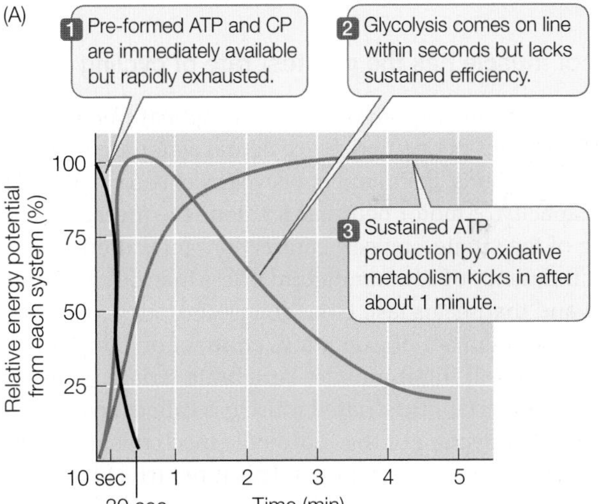


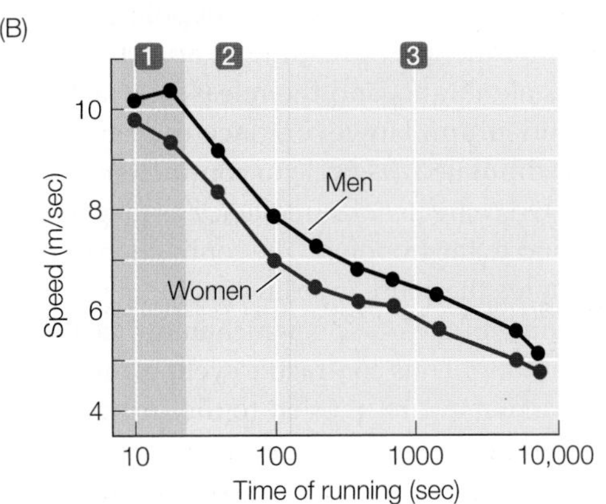


**48.13 Supplying Fuel for High Performance** (A) Muscles have three systems for obtaining the ATP they need for contraction during exertion such as running. (B) Looking at a plot of world-record times for running events of different durations, you can see that the performance of the athletes corresponds to the time courses of the three energy systems.

Oxidative metabolism becomes fully active in about a minute, producing relatively huge amounts of ATP because it can completely metabolize carbohydrates and fats. However, it requires many reactions (see Chapter 9), and it takes place in the mitochondria, so $O_2$ and substrate must diffuse into the mitochondria, and the formed ATP must diffuse from the mitochondria to the myosin filaments in the muscle. These processes are not instantaneous, so the rate at which oxidative metabolism can make ATP available to do work is slower than the rate at which the other two systems can supply ATP.

The fuel supply available to the muscles influences how long someone can sustain a high level of aerobic exercise. From the circulating blood, muscle receives glucose and free fatty acids, which it can metabolize to generate ATP. At high levels of aerobic exercise, however, most of the fuel used by muscles to produce ATP comes from the reserve of glycogen stored in the muscle itself. Depletion of muscle glycogen results in fatigue.

The rate at which muscle glycogen is replenished depends on diet: it is high with a high-carbohydrate diet, low with a high-fat diet, and intermediate with a mixed diet. This fact is the basis for a practice called "carbo-loading." For 3 to 5 days, athletes exercise at a level that depletes muscle glycogen. Then, 2 or 3 days before the event, they taper down their level of training and eat a diet rich in complex carbohydrates. The result can be glycogen supercompensation, in which the restoration of muscle glycogen stores "overshoots" and reaches above-normal levels.

### Insect muscle has the greatest rate of cycling

Insect flight muscle can produce a wingbeat frequency of up to 1,000 cycles per second. Since neural action potentials last 1 to 3 milliseconds, that number of cycles per second would push the capacity of motor neurons, let alone the mechanism of cycling of the striated muscle contraction/relaxation. The wingbeat frequency of a hummingbird is less than 50 cycles per second, and that seems fast. How do insects do it?

The mechanism of excitation/contraction coupling is different in insect flight muscle. Vertebrate striated muscle and much of invertebrate striated muscle is called "synchronous" because the cycling of the contractile mechanism is linked to the firing of the motor neurons. This is not true of insect flight muscle, which is therefore called "asynchronous" muscle. The firing of action potentials in the insect flight motor neurons is not particularly fast, but it does cause depolarization of the muscle cell membrane, the spreading of an action potential throughout the membrane, and the release of $Ca^{2+}$ from the sarcoplasmic reticulum. However, once the asynchronous muscle fiber is stimulated, its cycle of contraction/relaxation proceeds at its own characteristic frequency as long as the $Ca^{2+}$ is available to bind to the troponin. The contraction of the muscle fiber deactivates the actin–myosin binding, which in turn permits a stretching of the muscle, which in turn activates the actin–myosin binding. Thus contractile cycling and the resulting wingbeat frequency are not tied to the firing rate of the flight motor neurons.

## 48.2 RECAP

**Depending on the function a muscle serves, it may need to generate maximum force rapidly, sustain activity for a long period, or contract and relax at a very rapid rate. Properties of muscles can facilitate these types of activity.**

- Describe the differences between slow-twitch and fast-twitch fibers. **See p. 1015 and Figure 48.11**
- How does exercise influence muscle strength and endurance? **See p. 1017**
- How do the different sources of ATP influence performance in different types of exercise? **See pp. 1017–1018 and Figure 48.13**
- Explain how insects can beat their wings 10 times faster than hummingbirds can. **See p. 1018**

Regardless of how much force a muscle can generate, how long it can sustain a work load, or how fast it can contract and relax, a muscle needs something to pull on; otherwise it would just be a lump of pulsating or quivering tissue. Let's look now at how skeletal systems help generate movement.

## 48.3 How Do Skeletal Systems and Muscles Work Together?

Muscles can contract and exert force, or they can relax. To create significant movement, they must have something to pull on and something that stretches the muscle back to a longer position. In some cases, muscles pull on each other, as in the trunk of an elephant or the arms of an octopus. In most cases, however, **skeletal systems** are the rigid supports against which muscles pull to create directed movement. In this section, we examine the three types of skeletal systems: hydrostatic skeletons, exoskeletons, and endoskeletons.

### A hydrostatic skeleton consists of fluid in a muscular cavity

Cnidarians, annelids, and other soft-bodied invertebrates have **hydrostatic skeletons** consisting of a volume of fluid enclosed in a body cavity surrounded by muscle (see Section 31.2). When muscles oriented in one direction contract, the fluid-filled body cavity bulges out in the opposite direction.

An earthworm uses its hydrostatic skeleton to crawl. The earthworm's body cavity is divided into many separate segments, each of which contains a compartment filled with extracellular fluid. The body wall surrounding each segment has two muscle layers: a circular layer and a longitudinal layer. If the circular muscles in a segment contract, the compartment in that segment narrows and elongates. If the longitudinal muscles in a segment contract, the compartment shortens and bulges outward. Alternating contractions of the earthworm's circular and longitudinal muscles create waves of narrowing and widening,

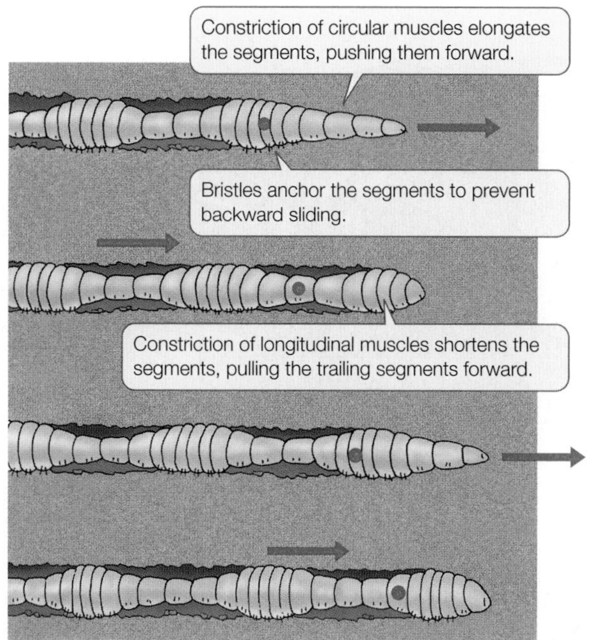

Constriction of circular muscles elongates the segments, pushing them forward.
Bristles anchor the segments to prevent backward sliding.
Constriction of longitudinal muscles shortens the segments, pulling the trailing segments forward.

**48.14 A Hydrostatic Skeleton** Alternating waves of muscle contraction move the earthworm through the soil. The red dot enables you to follow the changes in one segment as the worm moves forward.

lengthening and shortening, that travel down the body. Bulging, shortened segments serve as anchors as long, narrow segments project forward and longitudinal contractions pull other segments forward. Bristles help the widest parts of the body to hold firm against the substratum (**Figure 48.14**).

## Exoskeletons are rigid outer structures

An *exoskeleton* is a hardened, rigid outer surface to which muscles can be attached. Contractions of the muscles cause jointed segments of the exoskeleton to move relative to each other. The simplest example of an exoskeleton is the shell of a mollusk. Some marine mollusks, such as clams, have shells composed of protein strengthened by crystals of calcium carbonate (a rock-hard material). These shells can be massive, affording significant protection against predators. The shells of land mollusks (snails) generally lack the hard mineral component and are much lighter (see Figure 32.15B,D).

The most complex exoskeletons are found among the arthropods. An exoskeleton, or *cuticle*, covers all the outer surfaces of the arthropod's body and all its appendages. It is made up of plates secreted by a layer of cells just below the exoskeleton. The cuticle contains stiffening materials everywhere except at the joints, where flexibility must be retained. Muscles attached to the inner surfaces of the arthropod exoskeleton move its parts around the joints (see Figure 32.4).

A drawback of the rigid arthropod exoskeleton is that it cannot expand. Therefore, if the animal is to become larger, it must *molt*, shedding its exoskeleton and forming a new, larger one. A molting animal is vulnerable because the new exoskeleton takes time to harden. The animal's body is temporarily unprotected, and without a firm exoskeleton against which its muscles can exert maximum tension, it is unable to move rapidly. Soft-shelled crabs, a gourmet delicacy, are crabs caught when they are molting.

## Vertebrate endoskeletons consist of cartilage and bone

The **endoskeleton** of vertebrates is an internal scaffolding. Muscles are attached to it and pull against it. Endoskeletons are composed of rodlike, platelike, and tubelike bones connected to one another at a variety of joints that allow a wide range of movements. An advantage of endoskeletons over the exoskeletons of arthropods is that bones in the body can grow without the animal shedding its skeleton.

The human skeleton consists of 206 bones, some of which are shown in **Figure 48.15**. It can be divided into an *axial skeleton*, which includes the skull, vertebral column, sternum, and ribs; and an *appendicular skeleton*, which includes the pectoral girdle, pelvic girdle, and bones of the arms, legs, hands, and feet.

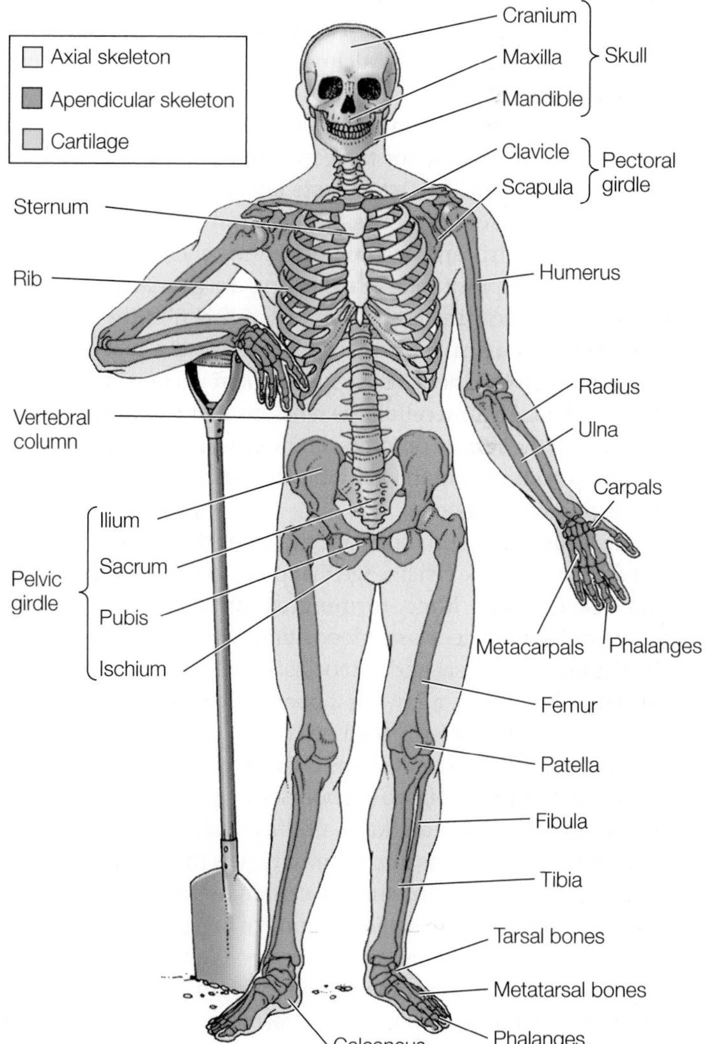


**48.15 The Human Endoskeleton** Cartilage and bone make up the internal skeleton of a human being.

The vertebrate endoskeleton consists of two kinds of connective tissue, *cartilage* and *bone*, which are produced by two kinds of connective tissue cells. *Cartilage cells* produce an extracellular matrix that is a tough, rubbery mixture of polysaccharides and proteins—mainly fibrous collagen. Collagen fibers run in all directions like reinforcing cords through the gel-like matrix and give it the well-known strength and resiliency of "gristle." This matrix, called **cartilage**, is found in parts of the endoskeleton where both stiffness and resiliency are required, such as on the surfaces of joints where bones move against one another. Cartilage is also the supportive tissue in stiff but flexible structures such as the larynx (voice box), nose, and ear pinnae. Sharks and rays are called *cartilaginous fishes* because their skeletons are composed entirely of cartilage. In most other vertebrates, cartilage is the principal component of the embryonic skeleton, but during development most of it is gradually replaced by bone.

**Bone** also contains collagen fibers, but it gets its rigidity and hardness from an extracellular matrix of insoluble calcium phosphate crystals. Bone serves as a reservoir of calcium for the rest of the body and is in dynamic equilibrium with soluble calcium in the extracellular fluids of the body. This equilibrium is under the control of calcitonin and parathyroid hormone (see Figure 41.11). If too much calcium is taken from the skeleton, the bones are seriously weakened.

The living cells of bone—*osteoblasts*, *osteocytes*, and *osteoclasts*—are responsible for the constant dynamic remodeling of bone (**Figure 48.16**). **Osteoblasts** lay down new matrix material on bone surfaces. These cells gradually become surrounded by matrix and eventually become enclosed within the bone, at which point they cease laying down matrix but continue to exist within small lacunae (cavities) in the bone. In this state they are called **osteocytes**. Despite the vast amounts of matrix between them, osteocytes remain in contact with one another through long cellular extensions that run through tiny channels in the bone. Communication between osteocytes is important in controlling the activities of the cells that are laying down or removing bone.

The cells that resorb bone are the **osteoclasts**. They are derived from the same cell lineage that produces white blood cells. Osteoclasts erode bone, forming cavities and tunnels. Osteoblasts follow osteoclasts, depositing new bone. Thus the interplay of osteoblasts and osteoclasts constantly replaces and remodels the bones, allowing a bone to recover from damage and adjust to the forces placed on it.

How the activities of the bone cells are coordinated is not understood, but stress placed on bones somehow provides them with information. A remarkable finding in studies of astronauts who spent long periods in zero gravity was that their bones decalcified. Conversely, in athletes, certain bones thicken during training. Both thickening and thinning of bones are experienced by anyone who has had a leg in a cast for a long time: the bones of the uninjured leg carry the person's weight and thicken while the bones of the inactive leg in the cast thin.

Because of the positive effects of physical stress on bone deposition, weight-bearing exercise is effective in preventing and treating the loss of bone density (and hence strength) known as *osteoporosis*. More than 25 million people in the United States suffer from this debilitating condition. Although osteoporosis is most commonly a problem for postmenopausal women, it can occur in younger people as a result of malnutrition. For example, the condition known as *female athlete triad* includes eating disorders, cessation of menstrual cycling, and osteoporosis. These are interactive conditions in which the eating disorder and excessive training lead to malnutrition that can result in endocrine disruption and osteoporosis. Excessive training and malnutrition can lead to bone loss in males as well.

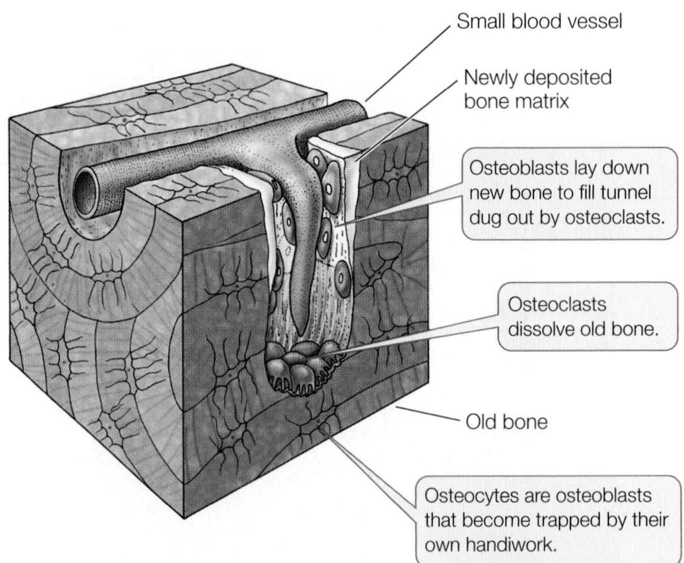
Small blood vessel
Newly deposited bone matrix
Osteoblasts lay down new bone to fill tunnel dug out by osteoclasts.
Osteoclasts dissolve old bone.
Old bone
Osteocytes are osteoblasts that become trapped by their own handiwork.

**48.16 Bone Is Living Tissue** Bones are constantly being remodeled by osteoblasts, which lay down bone, and osteoclasts, which resorb bone.

### Bones develop from connective tissues

Bones are divided into two types on the basis of how they develop. **Membranous bone** forms on a scaffold of connective tissue membrane. **Cartilage bone** forms first as a cartilaginous structure resembling the future mature bone, then gradually hardens, or *ossifies*, to become bone. The outer bones of the skull are membranous bones; the bones of the limbs are cartilage bones.

Cartilage bones can grow throughout the ossification process. The long bones of the legs and arms, for example, ossify first at the centers and later at each end (**Figure 48.17**). Growth can continue until these areas of ossification join. The membranous bones forming the skull cap grow until their edges meet. The soft spot on the top of a baby's head (the fontanelle) is the point at which the skull bones have not yet joined.

The structure of bone may be **compact** (solid and hard) or **cancellous** (having numerous internal cavities that make it appear spongy, although it is rigid). The architecture of a specific bone depends on its position and function, but most bones have both compact and cancellous regions. The shafts of the long bones of the limbs, for example, are cylinders of compact bone surrounding central cavities that contain the bone marrow, where the cellular elements of the blood are made. The ends of the long bones are cancellous (see Figure 48.17). Cancellous bone is lightweight because of its numerous cavities, but it is also strong because its internal meshwork constitutes a support system. It can withstand considerable forces of compression. The

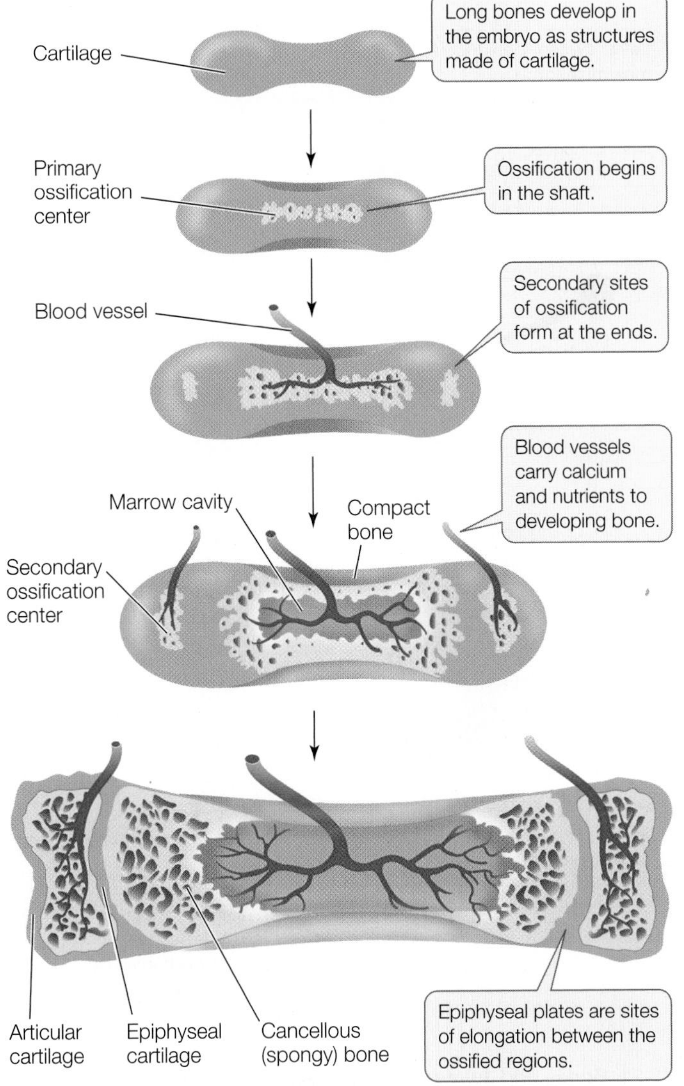


**48.17 The Growth of Long Bones** In the long bones of human limbs, ossification occurs first at the centers and later at each end.

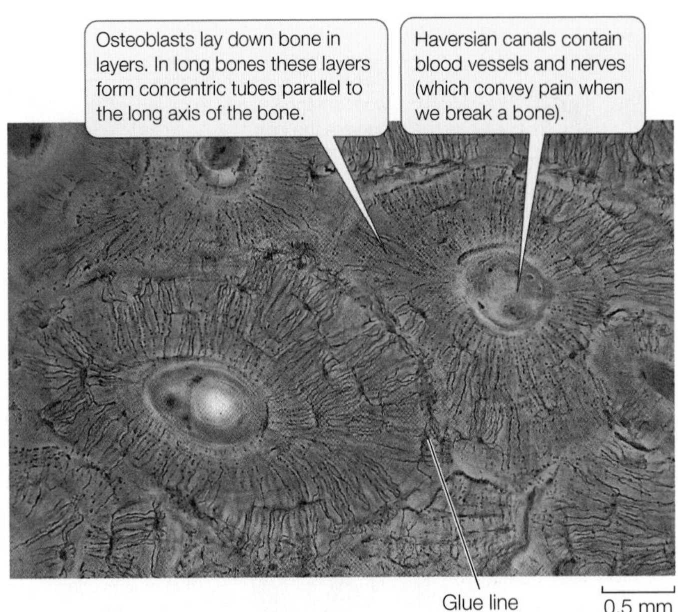

Osteoblasts lay down bone in layers. In long bones these layers form concentric tubes parallel to the long axis of the bone.
Haversian canals contain blood vessels and nerves (which convey pain when we break a bone).
Glue line
0.5 mm


**48.18 Most Compact Bone Is Composed of Haversian Systems** A micrograph of a section of a long bone shows Haversian systems with their central canals. Glue lines separate Haversian systems.

rigid, tubelike shaft of compact bone can withstand compression and bending forces. Architects and nature alike use hollow tubes as lightweight structural elements.

Most of the compact bone in mammals is called *Haversian bone* because it is composed of structural units called **Haversian systems** (**Figure 48.18**). Each Haversian system is a set of thin, concentric bony cylinders, between which are the osteocytes in their lacunae. Through the center of each Haversian system runs a narrow canal containing blood vessels and nerves. Adjacent Haversian systems are separated by boundaries called *glue lines*. Haversian bone is resistant to fracturing because cracks tend to stop at glue lines.

## Bones that have a common joint can work as a lever

Muscles and bones work together around **joints**, where two or more bones come together. Different kinds of joints allow motion in different directions (**Figure 48.19**), but muscles can exert force in only one direction. Therefore, muscles create move-

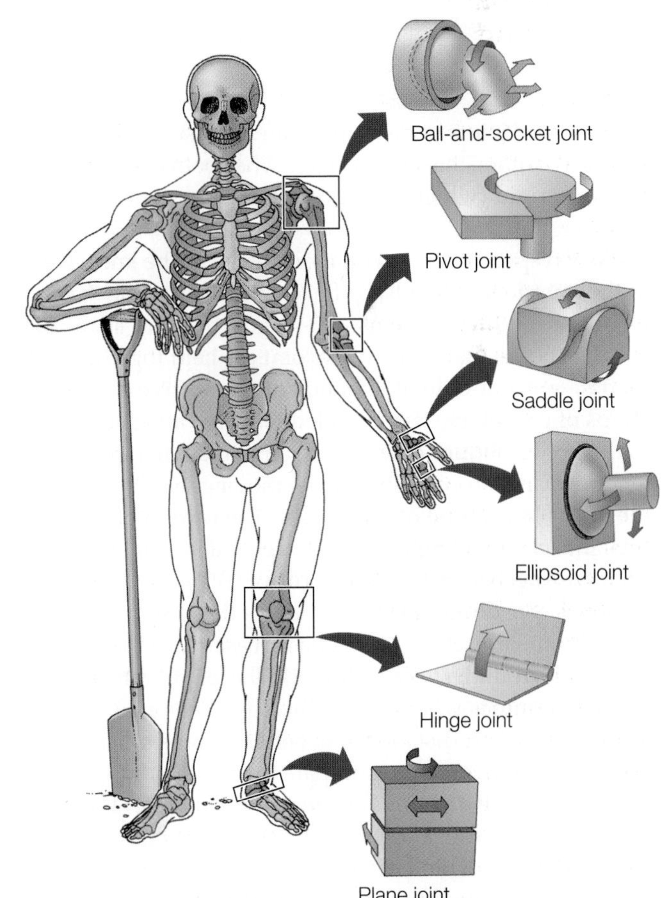


**48.19 Types of Joints** The designs of joints are similar to mechanical counterparts and enable a variety of movements.

**yourBioPortal.com**
GO TO **Web Activity 48.3 • Joints**

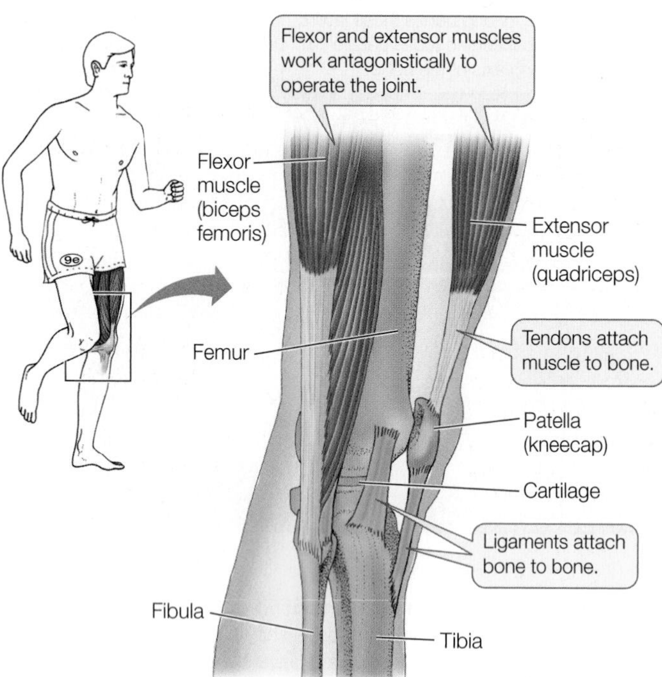


**48.20 Joints, Ligaments, and Tendons** A side view of the knee shows the interactions of muscle, bone, cartilage, ligaments, and tendons at this crucial and vulnerable human joint.

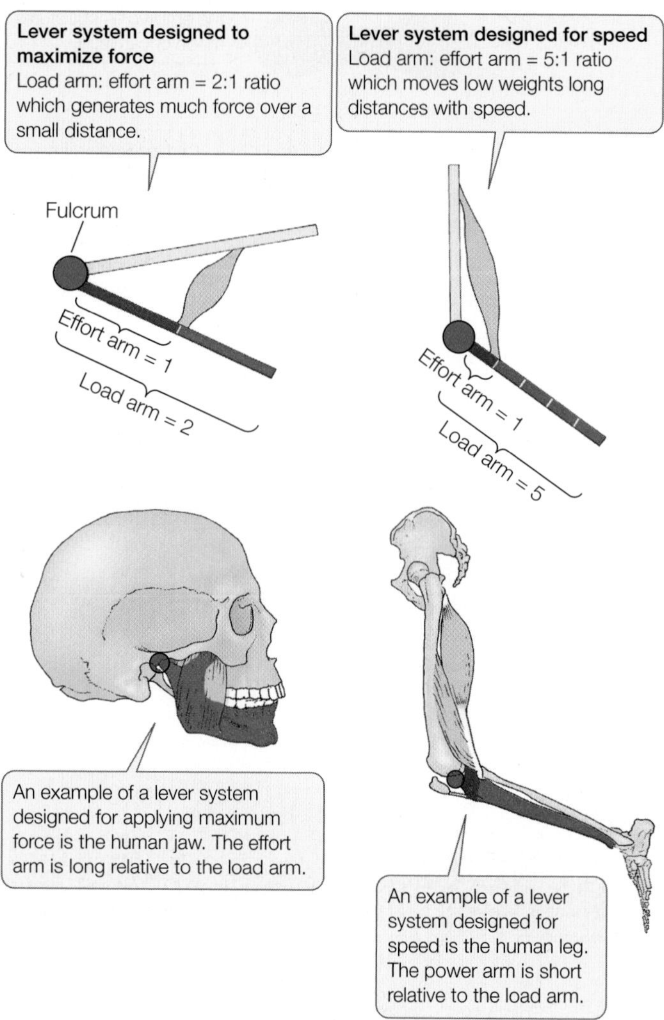


**48.21 Bones and Joints Work Like Systems of Levers** A lever system can be designed for maximizing either force or speed.

ment around joints by working in antagonistic pairs: when one muscle contracts, the other relaxes. When both contract, the joint becomes rigid (which is important for maintaining posture, for example).

With respect to a particular joint, such as the knee, we refer to the muscle that bends, or flexes, the joint as the **flexor**, and the muscle that straightens, or extends, the joint as the **extensor**. The bones that meet at the joint are held together by **ligaments**, which are flexible bands of connective tissue. Other straps of connective tissue, called **tendons**, attach the muscles to the bones (**Figure 48.20**). In many kinds of joints, only the tendon spans the joint, sometimes moving over the surfaces of the bones like a rope over a pulley. The tendon of the quadriceps muscle traveling over the knee joint is what is tapped to elicit the knee-jerk reflex (see Figure 47.3).

Bones constitute a system of levers that are moved around joints by the muscles. A lever has an *effort arm* and a *load arm* that work around a *fulcrum* (pivot). The length ratio of the two arms determines whether a particular lever can exert a lot of force over a short distance or is better at translating force into large or fast movements. Compare the jaw and knee joints, for example (**Figure 48.21**). The effort arm of the jaw is long relative to the load arm, allowing the jaw to apply great force over a small distance. Think of the powerful jaws of carnivores that can easily crack bones. The effort arm of the lower leg, by contrast, is short relative to the load arm, so you can run fast, jump high, and deliver swift kicks.

## 48.3 RECAP

**Muscles can only contract and relax; to achieve organized movement, they must pull against rigid structures—other muscles, hydrostatic skeletons, exoskeletons, or endoskeletons.**

- How do the muscles and fluid-filled body cavity of an earthworm interact to enable the animal to crawl? See pp. 1018–1019 and Figure 48.14
- Describe the differences between membranous and cartilaginous bone and between compact and cancellous bone. See p. 1020 and Figure 48.17
- In terms of levers, explain how specific joints can produce maximum force versus maximum speed. See pp. 1021–1022 and Figure 48.21

# CHAPTER SUMMARY

## 48.1 How Do Muscles Contract?

- **Skeletal muscle** consists of bundles of **muscle fibers**. Each skeletal muscle fiber is a large cell containing multiple nuclei.
- Skeletal muscles contain numerous **myofibrils**, which are bundles of **actin** and **myosin** filaments. The regular, overlapping arrangement of the actin and myosin filaments into **sarcomeres** gives skeletal muscle its striated appearance. Review Figure 48.1, WEB ACTIVITY 48.1
- The changes in the banding patterns of sarcomeres led to the **sliding filament model** of muscle contraction. Review Figure 48.2, ANIMATED TUTORIAL 48.1
- The molecular mechanism of muscle contraction involves the binding of the globular heads of myosin molecules to actin. Upon binding, the myosin head changes its conformation, causing the two filaments to move past each other (the sliding filament mechanism). Release of the myosin heads from actin and their return to their original conformation requires ATP. Review Figures 48.3 and 48.6
- All the fibers activated by a single motor neuron constitute a **motor unit**. Each nerve ending of the motor neuron forms a synapse with the muscle cell membrane. When neurotransmitter is released at the synapse, the muscle cell membrane is depolarized and action potentials are generated. Action potentials spread across the plasma membrane and through the **T tubules**, causing $Ca^{2+}$ to be released from the **sarcoplasmic reticulum**. Review Figure 48.5, WEB ACTIVITY 48.2
- $Ca^{2+}$ binds to **troponin** and changes its conformation, pulling the **tropomyosin** strands away from the myosin-binding sites on the actin filament. The muscle fiber continues to contract until the $Ca^{2+}$ is returned to the sarcoplasmic reticulum. Review Figure 48.6
- **Cardiac muscle** cells are striated, uninucleate, branching, and electrically connected by gap junctions, so that action potentials spread rapidly throughout sheets of cardiac muscle and cause coordinated contractions. Some cardiac muscle cells are pacemaker cells that generate and conduct electrical signals.
- **Smooth muscle** provides contractile force for internal organs. Smooth muscle cells respond to stretch and to neurotransmitters from the autonomic nervous system. Review Figure 48.8, ANIMATED TUTORIAL 48.2
- In skeletal muscle, a single action potential causes a minimum unit of contraction called a **twitch**. Twitches occurring in rapid succession can be summed, thus increasing the strength of contraction. Maximum sustained tension is known as **tetanus**. Review Figure 48.10

## 48.2 What Determines Muscle Performance?

- **Slow-twitch fibers** facilitate extended, aerobic work; **fast-twitch fibers** generate maximum forces for short periods of time. The ratio of slow-twitch to fast-twitch fibers in the muscles of an individual is largely genetically determined. Review Figure 48.11
- The force that a muscle fiber can produce depends on its initial state of extension or contraction. Review Figure 48.12
- Anaerobic exercise stimulates the enlargement of muscle fibers through production of new microfilaments. Aerobic exercise stimulates greater oxidative capacity of muscle fibers.
- Muscle performance depends on a supply of ATP. Available ATP and creatine phosphate can fuel maximum tension instantaneously but are exhausted within seconds. Glycolysis can regenerate ATP rapidly, but this process is inhibited by accumulation of lactic acid. Oxidative metabolism delivers ATP more slowly but can continue to do so for a long time. Review Figure 48.13

## 48.3 How Do Skeletal Systems and Muscles Work Together?

- **Skeletal systems** provide supports against which muscles can pull.
- **Hydrostatic skeletons** are fluid-filled body cavities that can be squeezed by muscles. Review Figure 48.14
- Exoskeletons are hardened outer surfaces to which internal muscles are attached.
- **Endoskeletons** are internal systems of rigid rodlike, platelike, and tubelike supports, consisting of **bone** and **cartilage** to which muscles are attached. Review Figure 48.15
- Bone is continually remodeled by **osteoblasts**, which lay down new bone, and **osteoclasts**, which erode bone. Review Figure 48.16
- Bones develop from connective tissue membranes (**membranous bone**) or from cartilage (**cartilage bone**) through ossification. Cartilage bone can grow until centers of ossification meet. Review Figure 48.17
- Bone can be **compact** (solid and hard), or **cancellous** (containing numerous internal spaces). Most of the compact bone of mammals is composed of **Haversian systems**. Review Figure 48.18
- **Joints** enable muscles to power movements in different directions. Muscles and bones work together around joints as systems of levers. Review Figures 48.19, 48.21, WEB ACTIVITY 48.3
- **Tendons** connect muscles to bones; **ligaments** connect bones to one another. Review Figure 48.20

# SELF-QUIZ

1. Smooth muscle differs from both cardiac and skeletal muscle in that
   *a.* it can act as a pacemaker for rhythmic contractions.
   *b.* contractions of smooth muscle are not due to interactions between neighboring microfilaments.
   *c.* neighboring cells are electrically connected by gap junctions.
   *d.* neighboring cells are tightly coupled by intercalated discs.
   *e.* the membranes of smooth muscle cells are depolarized by stretching.

2. Fast-twitch fibers differ from slow-twitch fibers in that
   *a.* they are more common in the leg muscles of champion sprinters than marathon runners.
   *b.* they have more mitochondria.
   *c.* they fatigue less rapidly.
   *d.* their abundance is more a product of training than of genetics.
   *e.* they are more common in postural muscles than in finger muscles.

3. The role of $Ca^{2+}$ in the control of muscle contraction is to
   *a.* cause depolarization of the T tubule system.
   *b.* change the conformation of troponin, thus exposing myosin-binding sites.
   *c.* change the conformation of myosin heads, thus causing microfilaments to slide past each other.
   *d.* bind to tropomyosin and break actin–myosin cross-bridges.
   *e.* block the ATP-binding site on myosin heads, enabling muscles to relax.
4. Fifteen minutes into a 10-k run, what is the major energy source of the leg muscles?
   *a.* Preformed ATP
   *b.* Glycolysis
   *c.* Oxidative metabolism
   *d.* Pyruvate and lactate
   *e.* High-protein drink consumed right before the race
5. Which statement about skeletal muscle contraction is *not* true?
   *a.* A single action potential at the neuromuscular junction is sufficient to cause a muscle to twitch.
   *b.* Once maximum muscle tension is achieved, no ATP is required to maintain that level of tension.
   *c.* An action potential in the muscle cell activates contraction by releasing $Ca^{2+}$ into the sarcoplasm.
   *d.* Summation of twitches leads to a graded increase in the tension that can be generated by a single muscle fiber.
   *e.* The tension generated by a muscle can be varied by controlling how many of its motor units are active.
6. Which statement about the structure of skeletal muscle is *true*?
   *a.* The light bands of the sarcomere are the regions where actin and myosin filaments overlap.
   *b.* When a muscle contracts, the A bands of the sarcomere lengthen.
   *c.* The myosin filaments are anchored in the Z lines.
   *d.* When a muscle contracts, the H zone of the sarcomere shortens.
   *e.* The sarcoplasm of the muscle cell is contained within the sarcoplasmic reticulum.
7. Insects can beat their wings at exceptionally high frequencies because
   *a.* their wing muscles have mostly fast-twitch fibers.
   *b.* their motor neurons can fire action potentials at a very high frequency.
   *c.* their wings have exoskeletal supports.
   *d.* their wing muscles have extensive sarcoplasmic reticulum that cycles $Ca^{2+}$ very fast.
   *e.* their wing muscles can generate a rapid oscillation of contraction asynchronous with motor neuron firing.
8. The long bones of our arms and legs are strong and can resist both compressional and bending forces because
   *a.* they are solid rods of compact bone.
   *b.* their extracellular matrix contains crystals of calcium carbonate.
   *c.* their extracellular matrix consists mostly of collagen and polysaccharides.
   *d.* they have a high density of osteoclasts.
   *e.* they consist of lightweight cancellous bone with an internal meshwork of supporting elements.
9. If we compare the jaw and knee joints as lever systems,
   *a.* the jaw joint can apply greater compressional forces.
   *b.* their ratios of power arm to load arm are about the same.
   *c.* the knee joint has greater rotational abilities.
   *d.* the knee joint has a greater ratio of power arm to load arm.
   *e.* only the jaw is a hinged joint.
10. Which statement about skeletons is *true*?
   *a.* They can consist of mostly cartilage.
   *b.* Hydrostatic skeletons cannot be used for locomotion.
   *c.* An advantage of exoskeletons is that they can continue to grow throughout the life of the animal.
   *d.* External skeletons must remain flexible, so they never include calcium carbonate crystals, as bones do.
   *e.* Internal skeletons consist of four different types of bone: compact, cancellous, membranous, and Haversian.

## FOR DISCUSSION

1. You can see from the structure of a sarcomere that it can shorten only by a certain percentage of its resting length. Yet muscles can cause a wide variety of ranges of movement—compare the range of movement of a toe and a leg. What are two adaptive design features of muscles and skeletons that can maximize a muscle's ability to cause a greater range of movement of an appendage?
2. If athletes train up until a day before competition and then take a day of rest, performance in which types of events will be most affected by what they eat during the rest day?
3. Wombats are powerful digging animals, and kangaroos are powerful jumping animals. How do you think their leg structures compare in terms of their designs as lever systems?
4. Why are ducks better long-distance fliers than chickens?
5. If an adolescent breaks a leg bone close to the ankle joint, after the break heals, that leg may not grow as long as the other one. Why?

## ADDITIONAL INVESTIGATION

For aerobic exercise endurance, the most important limiting factor is muscle glycogen reserve, but the muscles also use glucose from the blood. How would you design an experiment to test whether or not carbohydrate intake during exercise improves endurance, and if it does, what its effect on the sparing of muscle glycogen is?

# 49 Gas Exchange in Animals

## Why do elephants have long noses?

Elephants use their trunks in a variety of ways. They pluck leaves from trees and pull up plants from the ground; they pick up objects ranging in size from trees to seeds; they suck in water and squirt it into their mouths; they spray water and dust over their bodies; they smell the air around them; they communicate by touch; and mothers can wallop their calves when they misbehave.

Of course, elephants also use their trunks to breathe. This function takes on special interest when elephants go into deep water—for example, to cross rivers. Elephants can swim or walk on the bottom. In either case, they use their trunks as snorkels. As long as the tip of the trunk is above water, an elephant can breathe.

You may be familiar with diving using a snorkel about a foot long. When you want to dive deeper, you hold your breath. Water is heavy, and as you swim deeper, water pressure increases and presses on your body. The air in your lungs is compressible, but the tissues of your body are not. So as you swim deeper, the volume of your lungs decreases.

Think what would happen if you tried to breathe through a long snorkel in deep water. Because the snorkel would be open to the air above, the pressure in your lungs would be at water surface level pressure, but the surrounding water would be pressing on your body. To expand your lungs, you would have to exert enough force to push against the surrounding water. Not so easy! Unlike elephants, we solve this problem with scuba equipment; by breathing air from a pressurized tank with a regulator valve, we keep the air in our lungs at the same pressure as the surrounding water.

The chest of a snorkeling elephant can be 3 meters below the surface—and for every 3 meters, the water pressure increases by 0.3 atmospheres. Even if an elephant's respiratory muscles were strong enough to work against this pressure, there would be a serious problem for the blood vessels lining the cavity in which the lungs are suspended. The pressure difference across the walls of those blood vessels is the difference between the pressure in the vessels (blood pressure plus water pressure) and the air pressure at the surface (where the trunk opens to the air). If the blood vessels couldn't withstand that pressure difference, they would rupture and fill the chest cavity with blood.

**Proboscis as Snorkel** Elephants can cross deep bodies of water either swimming or by walking on the bottom—whichever will allow it to keep its trunk above water.

**Breathing Under Water** Adapted to the aquatic life, fishes such as this whitetip shark have gills that extract respiratory gases from the water. The human cannot "breathe" water, so to enter this environment, he must carry a supply of respiratory gases in a scuba tank.

How does a snorkeling elephant avoid such damage? In all mammals except elephants, the surfaces of the lungs and the chest cavity move easily against each other. As the lungs inflate and deflate, their surface slips and slides against that of the chest cavity. In elephants, however, dense connective tissue attaches the surface of the lungs to the surface of the chest cavity. This connective tissue acts as reinforcement for the fragile blood vessels on the surface of the chest cavity and prevents them from rupturing and filling the cavity with blood, which would destroy the function of the lungs—exchange of oxygen and carbon dioxide.

**IN THIS CHAPTER** we will explore adaptations of the respiratory systems of both water and air breathers for exchanging oxygen and carbon dioxide with the environment. We will first discuss the physical factors that limit these gas exchange systems and identify those factors that natural selection has optimized. We will then examine the gas exchange systems of insects, fishes, birds, and humans and describe the adaptations of the blood for transporting respiratory gases. Finally, we will see how respiratory gas exchange systems are controlled and regulated.

CHAPTER OUTLINE



## 49.1 What Physical Factors Govern Respiratory Gas Exchange?

Gas exchange systems are made up of gas exchange surfaces and the mechanisms that *ventilate* and *perfuse* those surfaces. The **respiratory gases** that organisms must exchange are oxygen ($O_2$) and carbon dioxide ($CO_2$). Cells need to obtain $O_2$ from the environment to produce an adequate supply of ATP by cellular respiration (see Chapter 9). $CO_2$ is an end product of cellular respiration, and it must be removed from the body to prevent toxic effects.

Diffusion is the only means by which respiratory gases are exchanged between an animal's internal body fluids and the outside medium (air or water). There are no active transport mechanisms to move respiratory gases across biological membranes. Because diffusion is a physical process, knowing what physical factors influence rates of diffusion helps us understand the diverse adaptations of gas exchange systems. (You should review the discussion about the physical nature of diffusion in Section 6.3.)

### Diffusion is driven by concentration differences

Diffusion results from the random motion of molecules, and the net movement of a molecule is down its concentration gradient. Concentrations in solutions are easy to think about because they are simply the amount of solute per volume of solution. Concentrations of gases are more complicated because the total number of gas molecules in a specified volume depends on pressure; there are twice as many gas molecules in a liter of gas under 2 atmospheres of pressure as there are in a liter under 1 atmosphere of pressure. Biologists express the concentrations of different gases in a mixture by the *partial pressures* of those gases.

First we have to know what the total pressure is, and we measure that with an instrument called a barometer. There are many types of barometers, but the classical one is a glass tube closed at one end and filled with mercury. This is inverted over a pool of mercury with the open end of the tube under the surface of the mercury. At sea level, the pressure exerted by the atmosphere supports, and therefore equals, a column of mercury in the tube that is about 760 mm high (depending on the weather). Therefore, **barometric pressure** (atmospheric pressure) at sea level is 760 mm of mercury (mm Hg). Because dry air is 20.9 percent $O_2$, the **partial pressure of oxygen** ($P_{O_2}$) at sea level is 20.9 percent of 760 mm Hg, or about 159 mm Hg. If two gas mixtures are separated by a membrane permeable to $O_2$, $O_2$ will diffuse from the mixture where its partial pressure is higher to the mixture where its partial pressure is lower.

Describing the concentration of respiratory gases in a liquid such as water is more complicated because another factor is involved—the solubility of the gas in the liquid. Thus, the actual amount of a gas in a liquid depends on the partial pressure of that gas in the gas phase in contact with the liquid as well as on the solubility of that gas in that liquid. However, the diffusion of the gas between the gaseous phase and the liquid still depends on the partial pressures of the gas in the two phases. Whether in air or water, the diffusion rate of a substance depends on its concentration gradient and on other factors that are expressed in *Fick's law of diffusion*.

### Fick's law applies to all systems of gas exchange

Diffusion is a physical phenomenon that can be described quantitatively with a simple equation called **Fick's law of diffusion**. All environmental variables that limit respiratory gas exchange and all adaptations that maximize respiratory gas exchange are reflected in one or more components of this equation. Fick's law is written as

$$Q = DA\frac{P_1 - P_2}{L}$$

where

- $Q$ is the rate at which a gas such as $O_2$ diffuses between two locations.
- $D$ is the *diffusion coefficient*, which is a characteristic of the diffusing substance, the medium, and the temperature. For example, perfume has a higher $D$ than motor oil vapor, and all substances diffuse faster at higher temperatures and faster in air than in water. Temperature is not expressed explicitly in Fick's law because the diffusion coefficient is usually determined at room temperature (about 20°C).
- $A$ is the cross-sectional area through which the gas is diffusing.
- $P_1$ and $P_2$ are the partial pressures of the gas at the two locations.
- $L$ is the path length, or distance, between the two locations.

Therefore, $(P_1 - P_2)/L$ is a *partial pressure gradient*.

Animals can maximize $D$ for respiratory gases by using air rather than water as their gas exchange medium whenever possible. All other adaptations for maximizing respiratory gas exchange must influence the surface area ($A$) for gas exchange or the partial pressure gradient across that surface area.

### Air is a better respiratory medium than water

The slow diffusion of $O_2$ molecules in water affects both air- and water-breathing animals. Eukaryotic cells carry out cellular respiration in their mitochondria, which are located in the cytoplasm—an aqueous medium. Cells are bathed in extracellular fluid—also an aqueous medium. In addition, all respiratory surfaces must be protected from drying out by a thin film of fluid through which $O_2$ must diffuse. Even in air-breathing animals, the slow rate of $O_2$ diffusion in water limits the efficiency of $O_2$ distribution from gas exchange surfaces to the sites of cellular respiration.

Diffusion of $O_2$ in water is so slow that even animal cells with low rates of metabolism can be no more than a few millimeters away from a good source of environmental $O_2$. Therefore, there are severe size and shape limits on the many species of invertebrates that lack internal systems for distributing $O_2$. Most of these species are very small, but some have grown larger by evolving a flat, thin body with a large external surface area (**Figure 49.1A**). Others have a very thin body built around a central cavity through which water can circulate (**Figure 49.1B**). A critical factor enabling larger, more complex animal bodies has been the evolution of specialized respiratory systems with large surface areas for enhancing gas exchange (**Figure 49.1C**).

$O_2$ can be obtained more easily from air than from water for several reasons:

- The $O_2$ content of air is much higher than the $O_2$ content of an equal volume of water. The maximum $O_2$ content of a bubbling stream in equilibrium with air is less than 10 ml of $O_2$ per liter of water. The $O_2$ content of the air over the stream is about 200 ml of $O_2$ per liter of air.

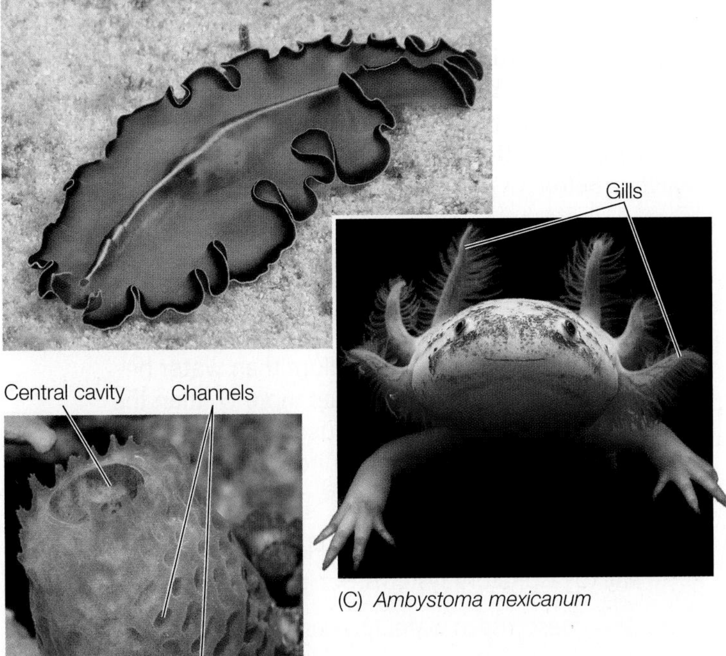


**49.1 Keeping in Touch with the Medium** (A) No cell in the leaflike body of this marine flatworm is more than a millimeter away from seawater. (B) Sponges have body walls perforated by many channels, which allow water to flow between the outside world and a central cavity. No cell in the sponge is more than a millimeter away from seawater. (C) A feathery fringe of gills on this aquatic larval salamander provides a large surface area for gas exchange. Blood circulating through the gills comes into close contact with the respiratory medium.

- $O_2$ diffuses about 8,000 times more rapidly in air than in water. That is why the $O_2$ content of a stagnant pond can be zero only a few millimeters below the surface.
- An animal has to do work to move water or air over its gas exchange surfaces. More energy is required to move water than air because water is 800 times denser than air and about 50 times more viscous. You can appreciate how important these facts were for the movement of life to the terrestrial environment—there were fewer constraints on the evolution of higher metabolic rates.

### High temperatures create respiratory problems for aquatic animals

Animals that use water for their respiratory exchange medium are in a double bind when environmental temperatures rise. Most water-breathing animals are *ectotherms*—their body temperatures are closely tied to the temperature of the water around them. As the water temperature rises, an ectotherm's body temperature and metabolic rate rise (see Figure 40.8). Thus, water breathers need more $O_2$ as the water gets warmer. But warm water holds less dissolved gas than cold water does (think of the gases that escape when you open a warm bottle of soda). In addition, if an animal performs work to move water across its gas exchange surfaces (as fishes do), it must expend more energy to breath as water temperature rises. Therefore, as water temperature goes up, a water-breathing animal must extract more and more $O_2$ from an environment that is increasingly $O_2$ deficient, and a lower percentage of that $O_2$ is available to support activities other than breathing (**Figure 49.2**).

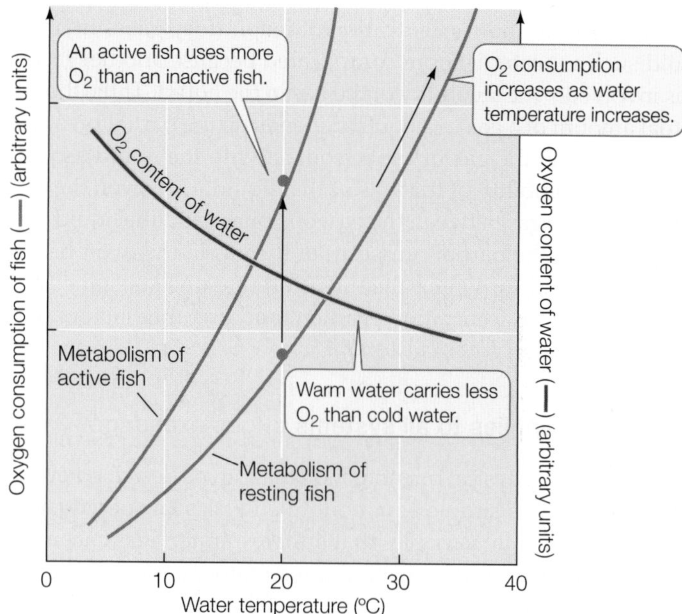


**49.2 The Double Bind of Water Breathers** Fishes need more $O_2$ when the water is warmer, but warm water carries less $O_2$ than cold water.

### $O_2$ availability decreases with altitude

Just as a rise in water temperature reduces the supply of $O_2$ available to water-breathing animals, an increase in altitude reduces the $O_2$ supply for air breathers. At all altitudes, $O_2$ makes up 20.9 percent of the dry air; however, as you go up in altitude, the total amount of gas per unit volume decreases, as reflected in the barometric pressure. For example, at 5,800 m, barometric pressure is only half what it is at sea level, so the $P_{O_2}$ at that altitude is only about 80 mm Hg. At the summit of Mount Everest (8,850 m), $P_{O_2}$ is only about 50 mm Hg—roughly one-third what it is at sea level.

Because the movement of $O_2$ across respiratory gas exchange surfaces and into the body depends on diffusion, its rate of movement depends on the $P_{O_2}$ difference between the air and the body fluids. Therefore, the drastically reduced $P_{O_2}$ in the air at high altitudes constrains $O_2$ uptake. Because of this, mountain climbers attempting the highest peaks usually breathe $O_2$ from pressurized bottles.

### $CO_2$ is lost by diffusion

Respiratory gas exchange is a two-way process: $CO_2$ diffuses out of the body as $O_2$ diffuses in. The direction and rate of diffusion of the respiratory gases across the exchange surfaces depend on the partial pressure gradients of the gases. The partial pressure gradients of $O_2$ and $CO_2$ across these exchange surfaces are quite different. The amount of $CO_2$ in the atmosphere is extremely low (0.03%), so for air-breathing animals there is always a large concentration gradient for diffusion of $CO_2$ from the body to the environment. Whereas the partial pressure gradient for $O_2$ decreases with increasing altitude, the gradient driving $CO_2$ out of the body hardly changes. The partial pressure of $CO_2$ in the atmosphere is close to zero both at sea level and atop Mount Everest.

In general, getting rid of $CO_2$ is not a problem for water-breathing animals because $CO_2$ is much more soluble in water than is $O_2$. Even in stagnant water, where the $P_{CO_2}$ is higher than in moving water, the lack of $O_2$ becomes a problem for an animal long before $CO_2$ exchange difficulties arise.

## 49.1 RECAP

**Respiratory gases are exchanged only by diffusion. Air is a better respiratory medium than water because a given volume of air has more $O_2$ than the same volume of water. $O_2$ diffuses faster in air than in water, and less work is required to move air over respiratory exchange surfaces.**

- Describe how the variables in Fick's law of diffusion relate to respiratory systems. **See p. 1027**
- Why does a rise in water temperature create a double-bind situation for water-breathing animals? **See p. 1028 and Figure 49.2**
- Explain the concept of partial pressures of gases and how they relate to diffusion rates of $O_2$ and $CO_2$ at different altitudes. **See p. 1028**

Now that we have discussed the physical factors that influence diffusion rates of respiratory gases between animals and their environments, let's look at some of the adaptations that have evolved for maximizing respiratory gas exchange.

## 49.2 What Adaptations Maximize Respiratory Gas Exchange?

As you might expect from the components of Fick's law of diffusion, adaptations to maximize respiratory gas exchange can be categorized as those that:

- Increase the surface area for gas exchange
- Maximize the partial pressure difference driving diffusion
- Minimize the diffusion path length
- Minimize the diffusion that takes place in an aqueous medium

### Respiratory organs have large surface areas

A variety of anatomical adaptations maximize the specialized body surface area (*A*) over which respiratory gases can diffuse. Water-breathing animals generally have *gills*, and air-breathing animals have *tracheae* or *lungs*. **External gills** are highly branched and folded extensions of the body surface that provide a large surface area for gas exchange with water (**Figure 49.3A**; see also Figure 49.1C). External gills are found in larval amphibians and in the larvae of many insects. Because they consist of thin, delicate tissues, external gills minimize the path length (*L*) traversed by diffusing molecules of $O_2$ and $CO_2$. External gills are vulnerable to damage, however, and are tempting morsels for predators, so in many animals protective body cavities for gills have evolved. Such **internal gills** are found in most mollusks and arthropods and in all fishes (**Figure 49.3B**).

**Lungs** are internal cavities for respiratory gas exchange with air (**Figure 49.3C**). Their structure is quite different from that of gills. Lungs have a large surface area because they are highly divided; and because they are elastic, they can be inflated with air and deflated.

Insects have a respiratory gas exchange system consisting of a network of air-filled tubes called **tracheae** that branch through all tissues of the insect's body (**Figure 49.3D**). The terminal branches of these tubes are so numerous that they have an enormous surface area compared with the external surface area of the insect's body.

### Transporting gases to and from exchange surfaces optimizes partial pressure gradients

Partial pressure gradients $[(P_1 - P_2)/L]$ drive diffusion across gas exchange surfaces; the larger the gradient, the greater the rate of gas exchange. These gradients can be maximized in several ways:

- *Minimization of path length*: Very thin tissues in gills and lungs reduce the diffusion path length (*L*).
- *Ventilation*: Actively moving the external medium over the gas exchange surfaces (i.e., breathing) regularly exposes those surfaces to fresh respiratory medium containing maximum $O_2$ and minimum $CO_2$ concentrations. This maximizes the concentration gradient.
- *Perfusion*: Actively moving the internal medium (e.g., blood) over the internal side of the exchange surfaces transports $CO_2$ to those surfaces and $O_2$ away from them, thus maximizing the concentration gradients driving diffusion.

This chapter describes four gas exchange systems. First we will look at the unique gas exchange system of insects. Then we will describe two highly efficient gas exchange systems, fish gills and bird lungs. Lastly we examine the structure and function of human lungs.

### Insects have airways throughout their bodies

The tracheal system that enables insects to exchange respiratory gases extends to all tissues in the insect body. Thus, respiratory gases diffuse through air most of the way to and from every cell. The insect respiratory system communicates with the outside environment through gated openings called *spiracles* in the sides of the abdomen (**Figure 49.4A,B**). The spiracles open to allow gas exchange and then close to decrease water loss. They open into tubes called *tracheae* that branch into even finer tubes,

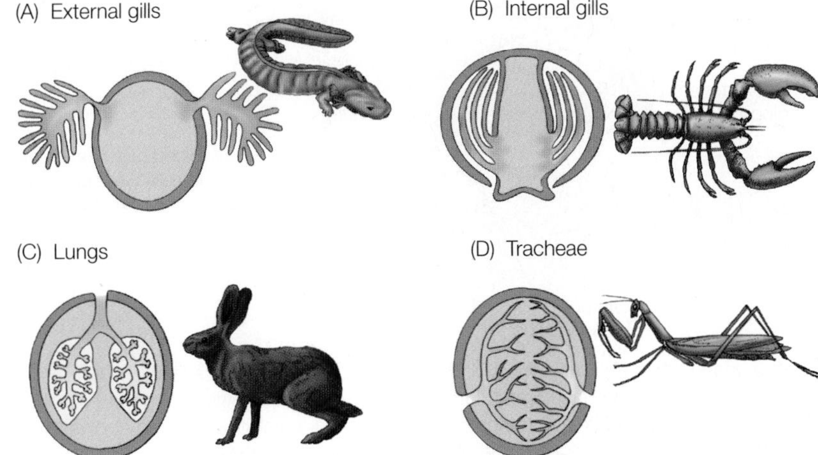


**49.3 Gas Exchange Systems** Large surface areas (blue in these diagrams) for the diffusion of respiratory gases are common features of animals. External (A) and internal (B) gills are adaptations for gas exchange with water. Lungs (C) and tracheae (D) are organs for gas exchange with air.

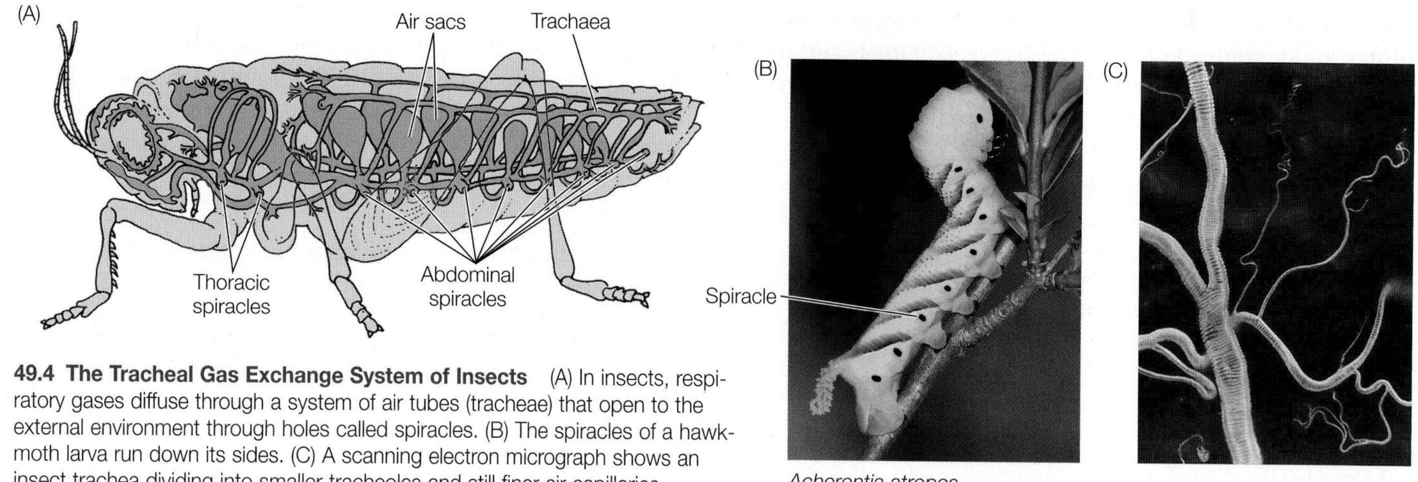


**49.4 The Tracheal Gas Exchange System of Insects** (A) In insects, respiratory gases diffuse through a system of air tubes (tracheae) that open to the external environment through holes called spiracles. (B) The spiracles of a hawkmoth larva run down its sides. (C) A scanning electron micrograph shows an insect trachea dividing into smaller tracheoles and still finer air capillaries.

or *tracheoles*, which end in tiny *air capillaries* that are the actual gas exchange surfaces (**Figure 49.4C**). In the insect's flight muscles and other highly active tissues, every mitochondrion is close to an air capillary.

## Fish gills use countercurrent flow to maximize gas exchange

The internal gills of fishes are supported by *gill arches* that lie between the mouth cavity and the protective *opercular flaps* on the sides of the fish just behind the eyes (**Figure 49.5A**). Water flows *unidirectionally* into the fish's mouth, over the gills, and out from under the opercular flaps. Thus, the gills are continuously bathed with fresh water. This constant, one-way flow

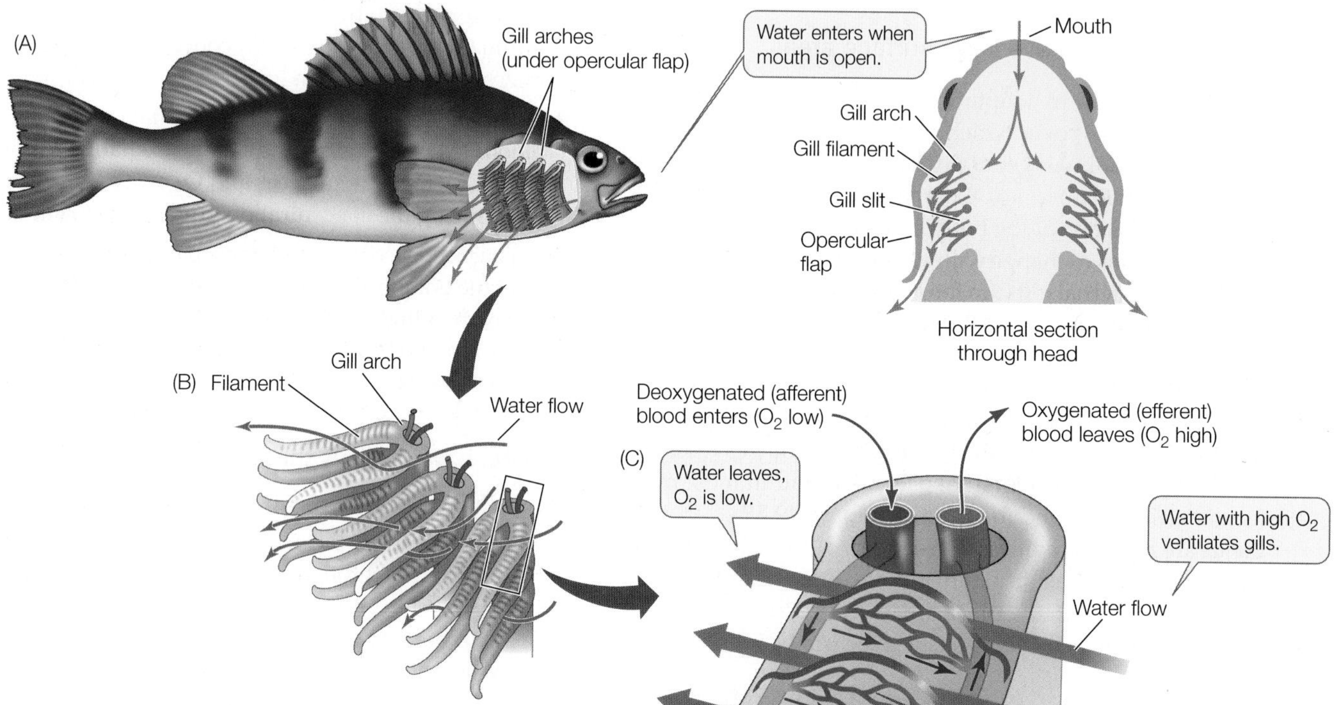

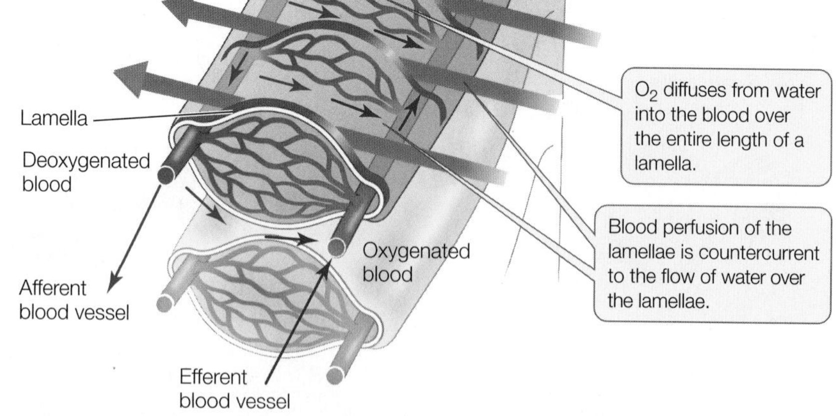

**49.5 Fish Gills** (A) Water flows unidirectionally over the gills of a fish. (B) Gill filaments have a large surface area and thin tissues. (C) Blood flows through the lamellae in the direction opposite (left to right, in this depiction; small blue and red arrows) to the flow of water (right to left; large blue arrows) over the lamellae.

of water over the gills maximizes the $P_{O_2}$ on the external gill surfaces. On the internal side of the gill membranes, the circulation of blood minimizes the $P_{O_2}$ by sweeping $O_2$ away as rapidly as it diffuses across.

Gills have an enormous surface area for gas exchange because they are so highly divided. Each gill consists of hundreds of ribbonlike *gill filaments* (**Figure 49.5B**). The upper and lower flat surfaces of each gill filament are covered with rows of evenly spaced folds, or *lamellae*. The lamellae are the actual gas exchange surfaces. Because they are exceedingly thin, the path length (*L*) for diffusion of gases between blood and water is minimized. The surfaces of the lamellae consist of highly flattened epithelial cells, so the water and the fish's red blood cells are separated by little more than 1 or 2 micrometers.

The flow of blood perfusing the inner surfaces of the lamellae, like the flow of water over the gills, is unidirectional. *Afferent* blood vessels bring deoxygenated blood to the gills, while *efferent* blood vessels take oxygenated blood away from the gills (**Figure 49.5C**). Blood flows through the lamellae in the direction opposite to the flow of water over the lamellae. This **countercurrent flow** optimizes the $P_{O_2}$ gradient between water and blood, making gas exchange more efficient than it would be in a system using concurrent (parallel) flow (**Figure 49.6**).

Some fishes, including anchovies, tuna, and certain sharks, ventilate their gills by swimming almost constantly with their mouths open. Most fishes, however, ventilate their gills by means of a two-pump mechanism. The closing and contracting of the mouth cavity pushes water over the gills, and the expansion of the opercular cavity prior to opening of the opercular flaps pulls water over the gills.

These adaptations for maximizing the surface area (*A*) for diffusion, minimizing the path length (*L*) for diffusion, and maximizing the $P_{O_2}$ gradient allow fishes to extract an adequate supply of $O_2$ from meager environmental sources.

### Birds use unidirectional ventilation to maximize gas exchange

Birds are remarkable for their ability to sustain high levels of activity for a long time—for example, on long-distance flights—even at high altitudes where mammals cannot even survive. The first team to climb Mount Everest was surprised to see birds flying over the mountain when they themselves could barely move without supplemental $O_2$. The highest recorded flight of a bird is from a vulture that collided with an airliner at 11,278 meters; a human could not exist at that altitude without supplemental $O_2$. Yet the lungs of a bird are smaller than the lungs of a similar-sized mammal, and bird lungs expand and contract less during a breathing cycle than do mammalian lungs. Furthermore, bird lungs *are compressed* during inhalation and *expand* during exhalation.

The structure of bird lungs allows air to flow unidirectionally through the lungs, rather than bidirectionally through all the same airways, as it does in mammals. Because mammalian lungs are never completely emptied of air during exhalation, there is always some lung volume that is not ventilated with fresh air. The air remaining in lungs and airways after exhalation is called **dead space**. Bird lungs, by contrast, have very little dead space, and the fresh incoming air is not mixed with stale air. In this way, a high $P_{O_2}$ gradient is maintained.

In addition to lungs, birds have **air sacs** at several locations in their bodies (**Figure 49.7A**). The air sacs are interconnected with each other, with the lungs, and with air spaces in some of the bones. The air sacs receive inhaled air, but they are not gas exchange surfaces. As in other air-breathing vertebrates, air enters and leaves a bird's gas exchange system through the **trachea** (commonly known as the *windpipe*, and not to be confused with the air-conducting tracheae of insects), which divides into smaller airways called **bronchi** (singular *bronchus*).

The bronchi divide into tubelike **parabronchi** that run parallel to one another through the lungs (**Figure 49.7B**). Branching off the parabronchi are numerous tiny *air capillaries*. Air flows through the parabronchi and diffuses into the air capillaries, which are the gas exchange surfaces. They are so numerous that they provide an enormous surface area for gas exchange. The parabronchi coalesce into larger bronchi that take the air out of the lungs and back to the trachea. Thus the anatomy of a bird's airways allows air to flow unidirectionally and continuously through the lungs.

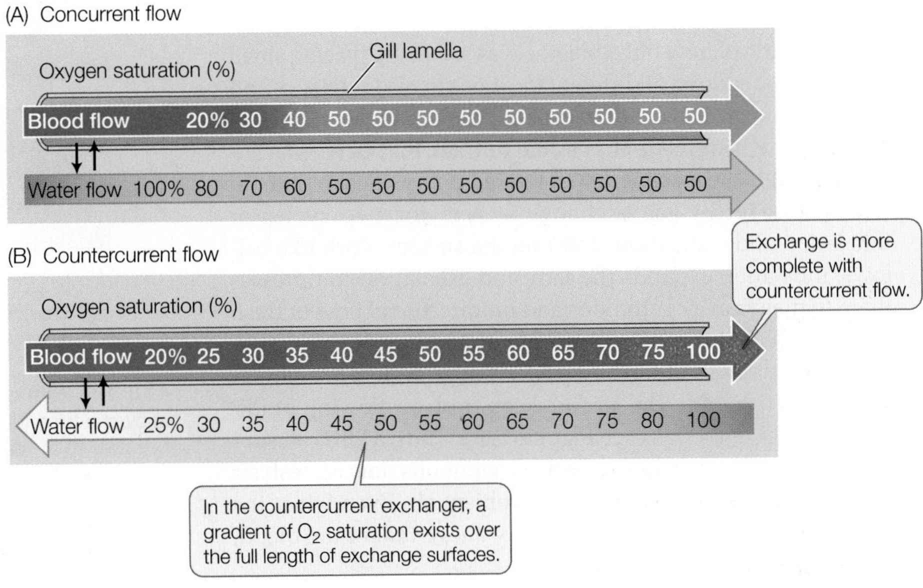


**49.6 Countercurrent Exchange Is More Efficient** In these models of concurrent and countercurrent gas exchange, the numbers represent the $O_2$ saturation percentages of blood and water. (A) In a concurrent exchanger, the saturation percentages of blood and water reach equilibrium halfway across the exchange surface. (B) A countercurrent exchanger allows more complete gas exchange because the water is always more $O_2$-saturated than the blood; thus a gradient of $O_2$ saturation is maintained.

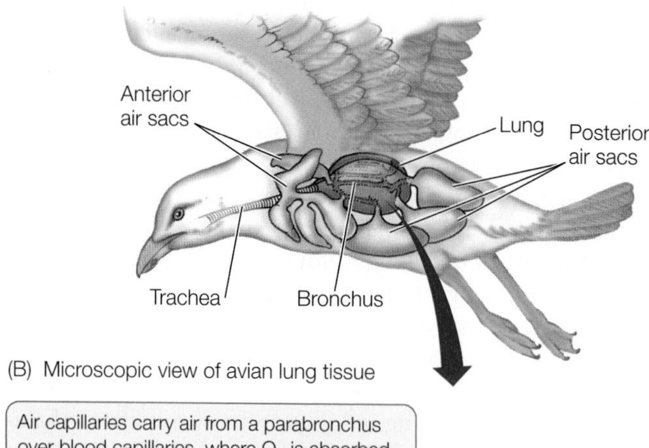


**49.7 The Respiratory System of a Bird** (A) Air sacs and air spaces in the bones are unique to birds. (B) Air flows through bird lungs unidirectionally in parabronchi. Air capillaries, the site of gas exchange, branch off the parabronchi.

The puzzle of how birds breathe was solved by placing small $O_2$ electronic sensors at different locations in birds' air sacs and airways. The birds were then exposed to pure $O_2$ for just a single breath, which made it possible to track that particular inhalation. The experiment demonstrated that a single breath remains in a bird's gas exchange system for two cycles of inhalation and exhalation, and that the air sacs work like bellows; inhalation expands the sacs, and exhalation compresses them to maintain a continuous and unidirectional flow of fresh air through the lungs (**Figure 49.8**).

The advantages of the bird gas exchange system are similar to those of fish gills. The air sacs keep fresh air flowing unidirectionally over the gas exchange surfaces. Thus, a bird can supply its gas exchange surfaces with a continuous flow of fresh air that has a $P_{O_2}$ close to that of the ambient air. Even when the $P_{O_2}$ of the ambient air is only slightly above that of the blood, $O_2$ can diffuse from air to blood.

## Tidal ventilation produces dead space that limits gas exchange efficiency

Lungs evolved in the first "air-gulping" vertebrates as outpocketings of the digestive tract. Although their structure has evolved considerably, lungs remain dead-end sacs in all air-breathing vertebrates except birds. Because of this, ventilation cannot be constant and unidirectional but must be **tidal**: air flows in and exhaled gases flow out by the same route. Since the lungs and airways can never be completely emptied of air, they always contain dead space. We can easily measure the volumes of air exchanged during breathing, but we have to use an indirect method to measure the dead space contained in the lungs and airways.

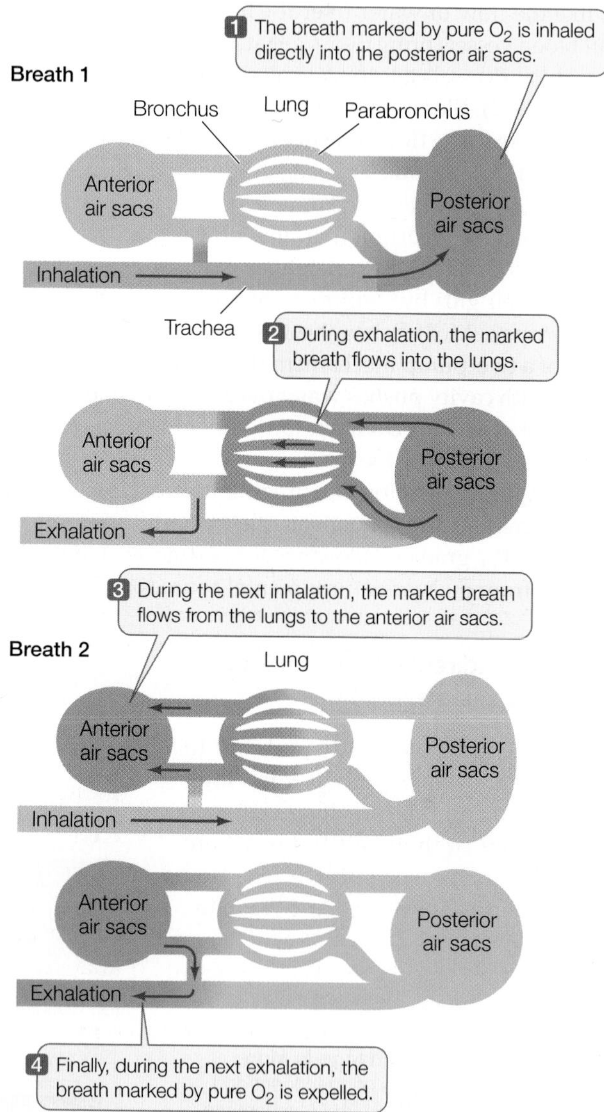


**49.8 The Path of Air Flow through Bird Lungs** The air a bird takes in by breathing (blue) travels through the lungs in one direction, from the posterior to the anterior air sacs. Each breath of air remains in the system for two breathing cycles.

**yourBioPortal.com**
GO TO Animated Tutorial 49.1 • Airflow in Birds

A *spirometer* is a device that measures the volume of air breathed in and out (**Figure 49.9**). Using a human as an example, the amount of air that moves in and out per breath when at rest is called the **tidal volume** (**TV**) (about 500 ml for an average human adult). When we breathe in as much as possible, the additional volume is the **inspiratory reserve volume** (**IRV**). Conversely, if we forcefully exhale as much air as possible, the additional amount of air expelled is the **expiratory reserve volume** (**ERV**). The maximum capacity for air exchange in one breath, or the **vital capacity** (**VC**), is the sum of TV + IRV + ERV. The vital capacity of an athlete is generally greater than that of a non-athlete, and vital capacity decreases with age because of stiffening of the lung tissue.

## TOOLS FOR INVESTIGATING LIFE

### 49.9 Measuring Lung Ventilation

A spirometer is a device that measures the volume of air a person breathes through a mouthpiece. The combined tidal volume, inspiratory reserve volume, and expiratory reserve volume are the lungs' vital capacity.

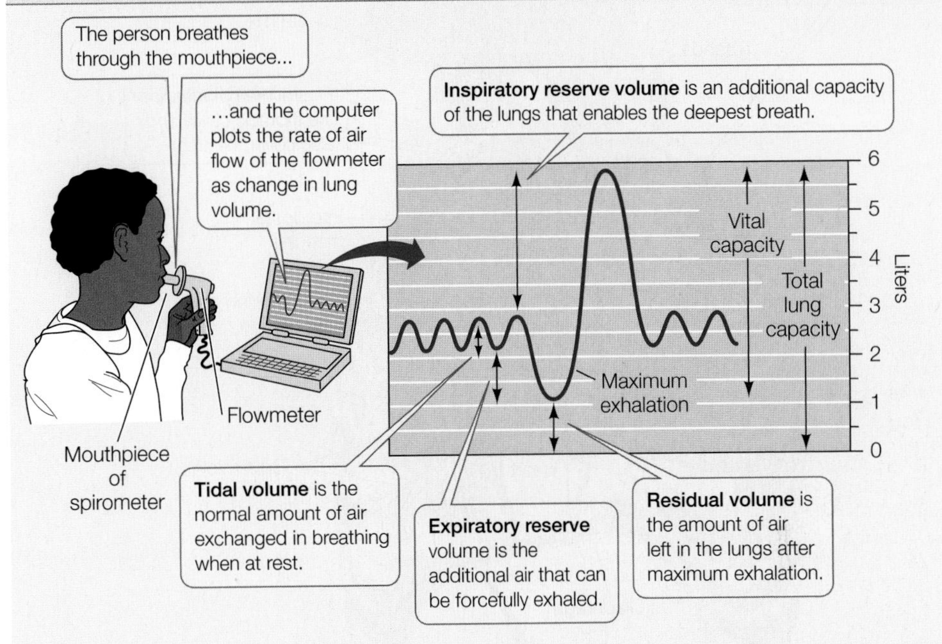

The person breathes through the mouthpiece...
...and the computer plots the rate of air flow of the flowmeter as change in lung volume.
**Inspiratory reserve volume** is an additional capacity of the lungs that enables the deepest breath.
Vital capacity
Total lung capacity
Liters
6
5
4
3
2
1
0
Maximum exhalation
Flowmeter
Mouthpiece of spirometer
**Tidal volume** is the normal amount of air exchanged in breathing when at rest.
**Expiratory reserve** volume is the additional air that can be forcefully exhaled.
**Residual volume** is the amount of air left in the lungs after maximum exhalation.


Vital capacity is not the entire lung capacity because, as mentioned above, there is dead space, also called the **residual volume** (**RV**). We can't measure RV directly with the spirometer, but we can measure it indirectly using the *helium dilution method.* Briefly, a person breathes from a reservoir with a known volume of air containing a known amount of helium (He). The helium is not absorbed from the lungs, so it becomes evenly distributed between the lungs and the reservoir as the subject inhales and exhales. Because the fixed amount of He becomes dispersed in a larger volume of air, its concentration decreases. That decrease in He concentration enables us to calculate the subject's **functional residual volume** (**FRV**), which is the ERV + RV. Since we *can* measure the ERV with the spirometer, we can subtract it and obtain the RV.

**yourBioPortal.com**
GO TO **Working with Data • Calculating the Functional Residual Volume**

Why is the RV important? Referring to Figure 49.9, you will see that for a normal person the ERV is about 1000 ml and the RV is 1000 ml. Thus the FRV is 2000 ml, but the tidal volume is only 500 ml. Thus, the air that reaches the alveoli with each breath consists of only 500 cc of fresh air diluted by 2000 cc of stale air. The maximum $P_{O_2}$ in this mixed air is much below the $P_{O_2}$ of the outside air, and because of the tidal ventilation pattern, the $P_{O_2}$ in the alveoli is steadily dropping during the breathing cycle. The RV is important because it contributes to the FRC and to the dilution of the $O_2$ in the inhaled air. Any disease or condition that increases the RV (such as emphysema or pulmonary fibrosis) compromises a patients respiratory ability. Similarly, considering the mixing of fresh air with the FRV, you can understand why reductions in tidal volume can be a problem—and therefore why patients recovering from surgery are encouraged to breathe deeply, even if it hurts.

Offsetting the inefficiencies of tidal breathing, mammalian lungs have some design features to maximize the rate of gas exchange: an enormous surface area and a very short path length for diffusion.

## 49.2 RECAP

**The major adaptations that increase animals' efficiency of respiratory gas exchange are a large surface area for exchange and a maximized concentration gradient across that surface.**

- Describe three different ways that the concentration gradient for gas exchange is maximized across fish gills. **See pp. 1030–1031 and Figures 49.5 and 49.6**
- What respiratory adaptations enable birds to fly at extremely high altitudes? **See pp. 1031–1032 and Figures 49.7 and 49.8**
- Explain why residual volume limits the efficiency of tidal breathing. **See pp. 1032–1033 and Figure 49.9**

Despite their limitations, mammalian lungs serve the respiratory needs of mammals well. Let's look at the human respiratory system as an example.

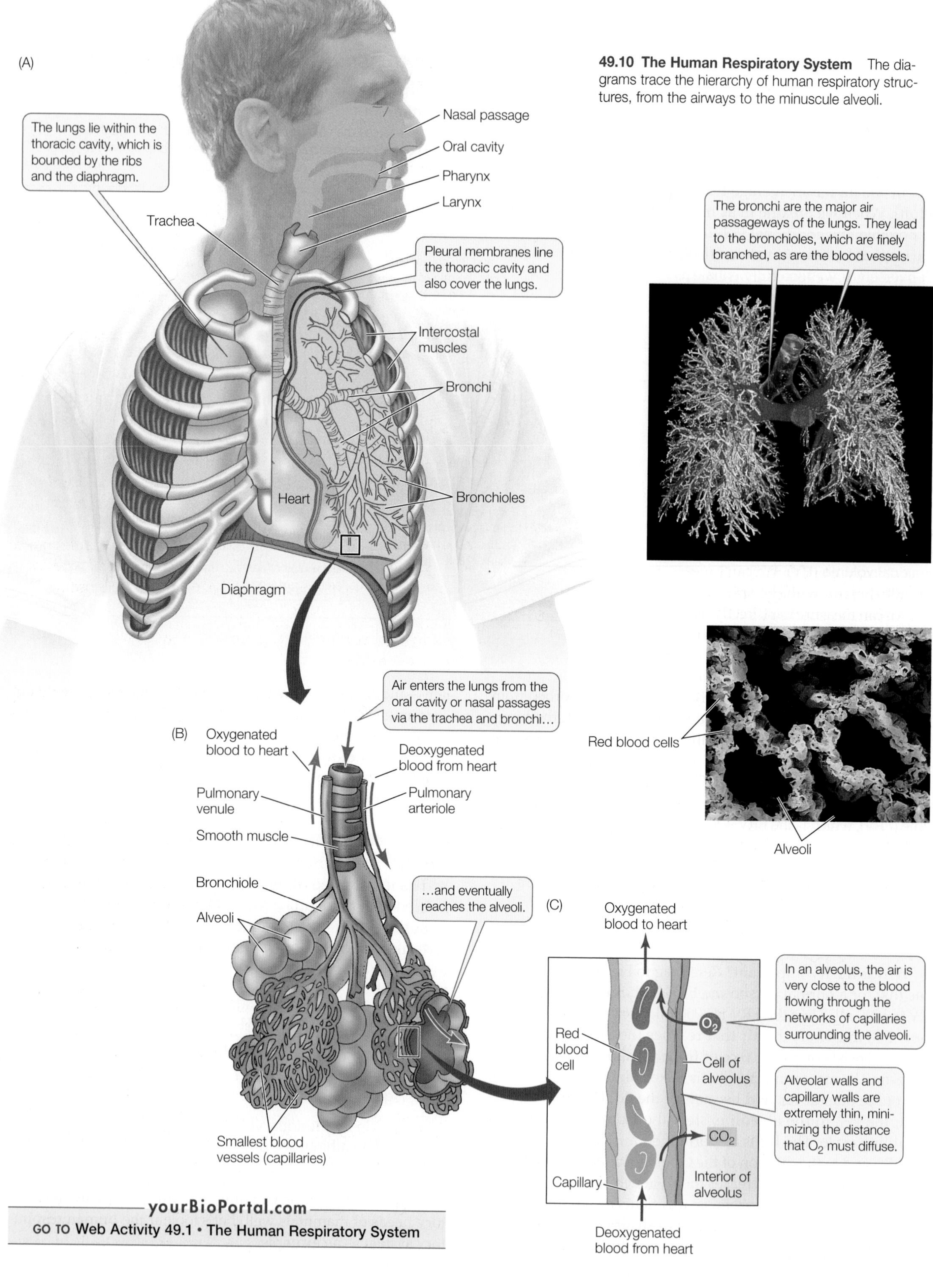
(A)
The lungs lie within the thoracic cavity, which is bounded by the ribs and the diaphragm.
Nasal passage
Oral cavity
Pharynx
Larynx
Trachea
Pleural membranes line the thoracic cavity and also cover the lungs.
Intercostal muscles
Bronchi
Heart
Bronchioles
Diaphragm
The bronchi are the major air passageways of the lungs. They lead to the bronchioles, which are finely branched, as are the blood vessels.
Red blood cells
Alveoli
Air enters the lungs from the oral cavity or nasal passages via the trachea and bronchi…
(B)
Oxygenated blood to heart
Deoxygenated blood from heart
Pulmonary venule
Pulmonary arteriole
Smooth muscle
Bronchiole
Alveoli
…and eventually reaches the alveoli.
Smallest blood vessels (capillaries)
(C)
Oxygenated blood to heart
Red blood cell
$O_2$
Cell of alveolus
$CO_2$
Capillary
Interior of alveolus
Deoxygenated blood from heart
In an alveolus, the air is very close to the blood flowing through the networks of capillaries surrounding the alveoli.
Alveolar walls and capillary walls are extremely thin, minimizing the distance that $O_2$ must diffuse.

**49.10 The Human Respiratory System** The diagrams trace the hierarchy of human respiratory structures, from the airways to the minuscule alveoli.

yourBioPortal.com
GO TO Web Activity 49.1 • The Human Respiratory System

# 49.3 How Do Human Lungs Work?

In humans, air enters the lungs through the oral cavity or nasal passage, which join together in the *pharynx* (**Figure 49.10A**). Below the pharynx, the esophagus conducts food to the stomach, and the trachea leads to the lungs. At the beginning of this airway is the *larynx*, or voice box, which houses the vocal cords. The larynx is the "Adam's apple" that you can see or feel on the front of your neck. The trachea is about 2 cm in diameter. Its thin walls are prevented from collapsing by C-shaped bands of cartilage as air pressure changes during the breathing cycle. If you run your fingers down the front of your neck just below your larynx, you can feel a few of these bands of cartilage.

The trachea branches into two bronchi, one leading to each lung. The bronchi branch repeatedly to generate a treelike structure of progressively smaller airways extending to all regions of the lungs. After four branchings, the cartilage supports disappear, marking the transition to **bronchioles**. After about 16 branchings, the bronchioles are less than a millimeter in diameter, and tiny, thin-walled air sacs called **alveoli** begin to appear. Alveoli are the sites of gas exchange. After alveoli begin to appear, about six more branchings of the airways occur that end in clusters of alveoli (**Figure 49.10B**). Because the airways conduct air only to and from the alveoli and do not themselves participate in gas exchange, their volume is dead space.

Human lungs have about 300 million alveoli. Although each alveolus is very small, their combined surface area for diffusion of respiratory gases is about 70 square meters—about one-fourth the size of a basketball court. Each alveolus is made of very thin cells. Between and surrounding the alveoli are networks of capillaries whose walls are also made up of exceedingly thin cells. Where capillary meets alveolus, very little tissue separates them (**Figure 49.10C**), so the length of the diffusion path between air and blood is less than 2 micrometers.

*Emphysema*, a condition in which inflammation damages and eventually destroys the walls of the alveoli, is the fourth leading cause of death in the United States. As a result of this disease the lungs have fewer but larger alveoli, the RV increases, and the lungs loses elasticity. Although genetic factors can contribute to emphysema, the principal cause of the disease is smoking.

### Respiratory tract secretions aid ventilation

Mammalian lungs produce two secretions that do not directly influence their gas exchange but do affect the process of ventilation: mucus and surfactant.

Many cells lining the airways produce sticky mucus that captures bits of dirt and microorganisms that are inhaled. Other cells lining the airways have cilia whose beating continually sweeps the mucus, with its trapped debris, up toward the pharynx, where it can be swallowed or spit out. This phenomenon, called the *mucus escalator*, can be adversely affected by inhaled pollutants. Smoking one cigarette can immobilize the cilia of the airways for hours. A smoker's cough results from the need to clear the obstructing mucus from the airways when the mucus escalator is out of order. The genetic disease *cystic fibrosis* causes respiratory problems by affecting the respiratory mucus (see Figure 17.3B).

A **surfactant** is a substance that reduces the surface tension of a liquid. **Surface tension** gives the surface of a liquid such as water the properties of an elastic membrane, and it is why certain insects, such as water-striders, can walk on water. As discussed in Section 2.4, surface tension is the result of chemical forces of attraction between water molecules. The attractive forces working on the water molecules at the surface pull from below and from the sides but not from above. This imbalance of forces creates surface tension. The thin film of fluid covering the air-facing surfaces of the alveoli has surface tension that contributes to the lungs' elasticity. To inflate the lungs, enough force has to be generated to overcome both the elasticity of the lung tissue and the surface tension in the alveoli.

Lung surfactant is a fatty, detergent-like substance that is critical for reducing the work necessary to inflate the lungs. Certain cells in the alveoli release surfactant molecules when they are stretched. If a baby is born more than a month prematurely, these cells may not have developed the ability to produce surfactant. A baby with this condition, known as *respiratory distress syndrome*, will have great difficulty breathing and may die from exhaustion and lack of $O_2$. Common treatments for premature babies have been to put them on respirators to assist their breathing and to give them hormones to speed lung development. A new approach is to apply surfactant to the lungs via an aerosol.

### Lungs are ventilated by pressure changes in the thoracic cavity

Human lungs are suspended in the **thoracic cavity**, a closed compartment bounded on the bottom by a sheet of muscle called the **diaphragm** (see Figure 49.10A). Each lung is covered by a continuous sheet of tissue called the **pleural membrane** that also lines the thoracic cavity adjacent to that lung. There is no real space between the pleural membranes of the lung and the thoracic cavity, but there is a thin film of fluid. This fluid lubricates the inner surfaces of the pleural membranes so they can slip and slide against each other during breathing movements. Just as we mentioned above in the explanation of surface tension, there are forces of attraction between the molecules of fluid in the pleural membranes. As a result, it is difficult to pull the pleural membranes apart. Think of two wet panes of glass or two wet microscope slides; you can slide them past each other, but it is difficult to separate them. While the inner surfaces of the pleural membranes are "stuck" to each other by surface tension, the outer surfaces are attached to the wall of the thoracic cavity and to the surface of the lung.

Inhalation and exhalation involve changes in the volume of the thoracic cavity (**Figure 49.11**). Because the pleural membranes are attached to the walls of the thoracic cavity, and because the pleural membranes covering the cavity wall and the lung surface are stuck to each other by surface tension, any attempt to increase the volume of the thoracic cavity increases the tension between the pleural membranes. Even between breaths, there is tension between the pleural membranes because the rib cage is pulling outward and the elasticity of the

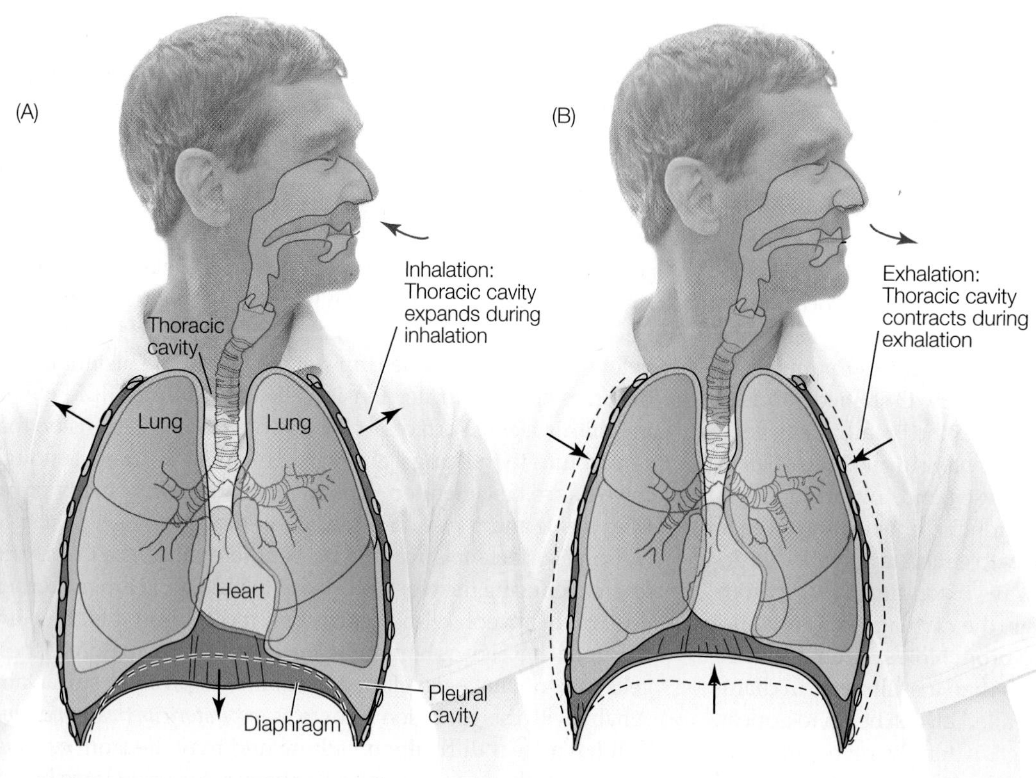


During **inhalation**:
- Diaphragm contracts
- Thoracic cavity expands
- Intrapleural pressure becomes more negative
- Lungs expand
- Air rushes in

During **exhalation**:
- Diaphragm relaxes
- Thoracic cavity contracts
- Intrapleural pressure becomes less negative
- Lungs contract
- Gases in lungs are expelled

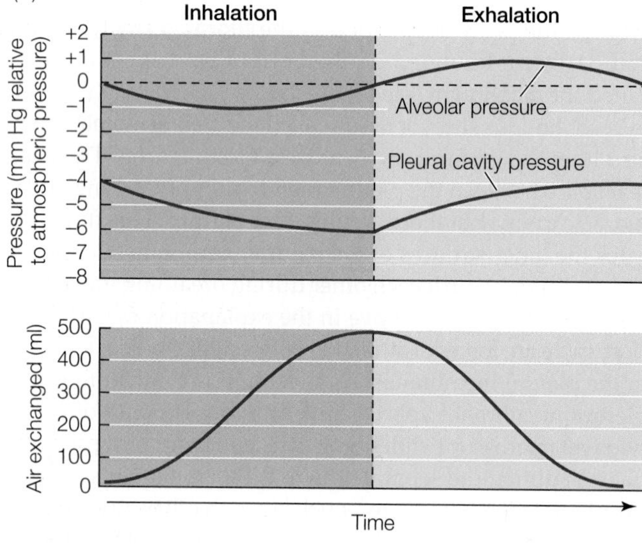


**49.11 Into the Lungs and Out Again** (A) Inhalation is an active process spurred by contraction of the diaphragm. (B) Exhalation generally is a passive process as the diaphragm relaxes. (C) There is always negative pressure in the pleural cavity—the space between the pleural membranes. Variations in that negative pressure cause the lungs to inflate and deflate during the breathing cycle.

**yourBioPortal.com**
GO TO **Animated Tutorial 49.2 • Airflow in Mammals**

lung tissue is pulling inward. This slight negative pressure keeps the alveoli partly inflated even at the end of an exhalation. If the thoracic cavity is punctured—by a knife wound, for example—air can leak into the space between the pleural membranes and cause the lung to deflate. If the wound is not sealed, breathing movements pull air in between the pleural membranes rather than into the lung (a "sucking chest wound"), and there is no ventilation of the alveoli in that lung—a condition called "collapsed lung."

At rest, inhalation is initiated by contraction of the muscular diaphragm (see Figure 49.11A). As the domed diaphragm contracts, it pulls down, expanding the thoracic cavity and pulling on the pleural membranes. Since the pleural membranes cannot separate because of the surface tension of the thin film of fluid between them, they pull on the lungs. The lungs are not a closed cavity; they have an airway to the atmosphere, and they can expand. When the diaphragm contracts, air rushes in through the trachea from the outside and the lungs expand. Exhalation begins when contraction of the diaphragm ceases. As the diaphragm relaxes, the elastic recoil of the lung tissues pulls the diaphragm up and pushes air out through the airways (see Figure 49.11B). When a person is at rest, inhalation is an active process and exhalation is a passive process.

The diaphragm is not the only muscle that can change the volume of the thoracic cavity. Between the ribs are two sets of **intercostal muscles**. The *external intercostal muscles* expand the thoracic cavity by lifting the ribs up and outward. The *internal intercostal muscles* decrease the volume of the thoracic cavity by pulling the ribs down and inward. During strenuous exercise, the external intercostal muscles increase the volume of air inhaled, making use of the inspiratory reserve volume, and the internal intercostal muscles increase the amount of air exhaled, making use of the expiratory reserve volume. The abdominal muscles can also aid in breathing. When they contract, they cause the abdominal contents to push up on the diaphragm and thereby contribute to the expiratory reserve volume.

Remember that ventilation and perfusion work together to maximize the partial pressure gradients across the gas exchange surface. Ventilation delivers $O_2$ to the environmental side of the exchange surface, where it diffuses across and is swept away by the perfusing blood, which carries it to the tissues that need it. The reverse is true for the exchange of $CO_2$. Perfusion delivers $CO_2$ to the exchange surface, where it diffuses out and is swept away by ventilation.

### 49.3 RECAP

**The mammalian respiratory system consists of a highly branching system of airways that lead to blind end sacs called alveoli which are the gas exchange surfaces. Respiratory muscles ventilate the alveoli by creating pressure differences between the lungs and the outside air. $CO_2$ and $O_2$ are exchanged across thin capillary and alveoli walls by diffusion.**

- Describe the path that a breath of air takes from the nose to the gas exchange surfaces. **See p. 1035 and Figure 49.10**
- What roles do mucus and surfactant play in maintaining the function of the mammalian respiratory system? **See p. 1035**
- Explain the anatomical and functional relationships between the thoracic cavity, the pleural membranes, and the lungs. **See pp. 1035–1036 and Figure 49.11**

Now that we have discussed how respiratory gases get to and from the environmental side of the gas exchange membranes through ventilation, we can look at how these gases get to and from the internal side of the gas exchange membranes through perfusion.

## 49.4 How Does Blood Transport Respiratory Gases?

Perfusion of the lungs is one of the functions of the circulatory system. The circulatory system uses a pump (the heart) and a network of vessels to transport blood around the body. Circulatory systems are the subject of Chapter 50, so here we will discuss only one aspect of perfusion: how blood transports respiratory gases.

The liquid part of blood, the *blood plasma*, carries some $O_2$ in solution, but its ability to transport this nonpolar molecule is limited. The blood plasma of a human can contain in solution only about 0.3 ml of $O_2$ per 100 ml of plasma, which is inadequate to support even basal metabolism. However, the blood of most animals, vertebrate and invertebrate, contains molecules that reversibly bind $O_2$ and thus augment its transport capacity. These molecules pick up $O_2$ where $P_{O_2}$ is high and release it where $P_{O_2}$ is lower. There are many $O_2$ transport molecules in the animal kingdom, but in vertebrates this role is played by the protein hemoglobin contained in red blood cells. Hemoglobin increases the capacity of blood to transport $O_2$ by about 60-fold, making high levels of metabolism possible.

### Hemoglobin combines reversibly with $O_2$

Red blood cells contain enormous numbers of hemoglobin molecules. **Hemoglobin** is a protein consisting of four polypeptide subunits (see Figure 3.9), each of which surrounds a *heme group*—an iron-containing ring structure that can reversibly bind a molecule of $O_2$. Thus, each hemoglobin molecule can bind up to four $O_2$ molecules, enabling the blood to carry a large amount of $O_2$ to the body's tissues.

Hemoglobin's ability to pick up or release $O_2$ depends on the $P_{O_2}$ in its environment. When the $P_{O_2}$ of the blood plasma is high, as it usually is in the lung capillaries, each hemoglobin molecule can carry its maximum load of four $O_2$ molecules. As the blood circulates through the rest of the body, it releases some of the $O_2$ it is carrying when it encounters lower $P_{O_2}$ values (**Figure 49.12**).

The relation between $P_{O_2}$ and the amount of $O_2$ bound to hemoglobin is not linear but S-shaped (sigmoidal). The hemoglobin–oxygen binding curve in Figure 49.12 reflects interactions between the four subunits of the hemoglobin molecule. At low $P_{O_2}$ values, only one subunit will bind an $O_2$ molecule. When it does so, the shape of that subunit changes, altering the quaternary structure of the whole hemoglobin molecule. That structural change makes it easier for the other subunits to bind an $O_2$ molecule; that is, their $O_2$ *affinity* is increased. Therefore, a smaller increase in $P_{O_2}$ is necessary to get the hemoglobin molecules to bind a second $O_2$ molecule (that is, to become 50% saturated) than was necessary to get them to bind one $O_2$ molecule

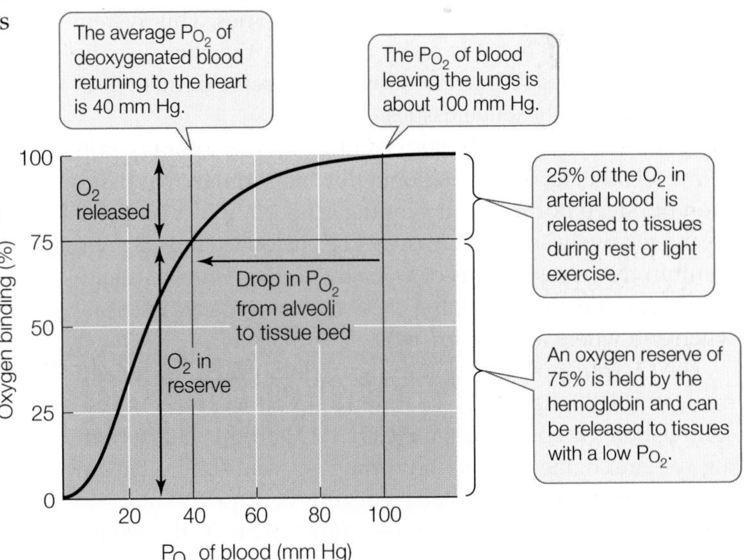


**49.12 Binding of $O_2$ to Hemoglobin Depends on $P_{O_2}$** Hemoglobin in blood leaving the lungs is 100 percent saturated (four $O_2$ molecules are bound to each hemoglobin molecule). Most hemoglobin molecules drop only one of their four $O_2$ molecules as they circulate through the body and are still 75 percent saturated when the blood returns to the lungs. The steep portion of this $O_2$-binding curve comes into play when tissue $P_{O_2}$ falls below the normal 40 mm Hg, at which point hemoglobin "unloads" its $O_2$ reserves.

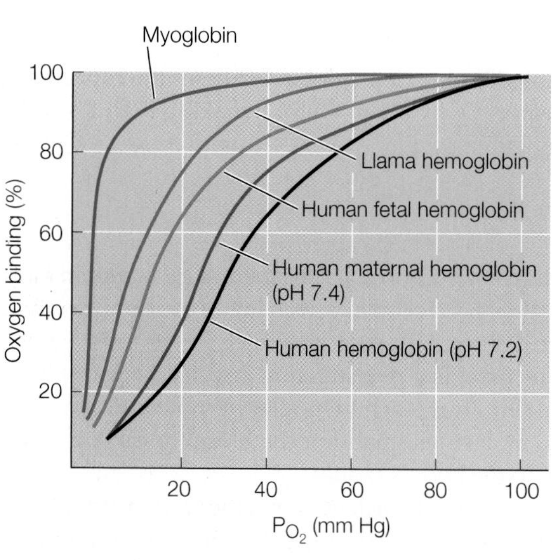


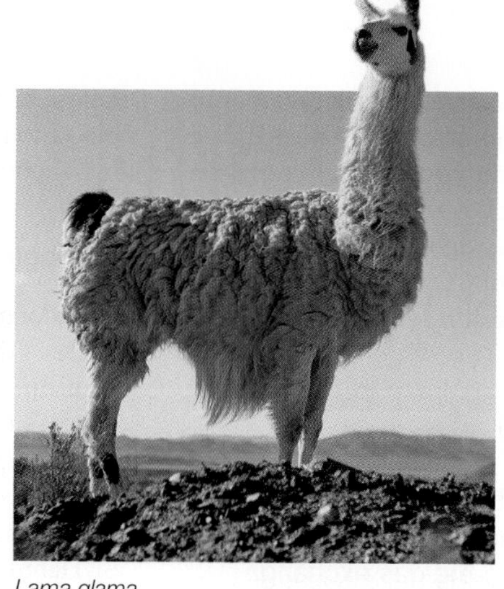

*Lama glama*

**49.13 Oxygen-Binding Adaptations** Myoglobin and the different hemoglobins have different $O_2$-binding properties adapted to different circumstances. The hemoglobin of South American camelids, for example, is adapted for binding $O_2$ at high altitudes, where $P_{O_2}$ is low. Fetal hemoglobin has a higher affinity for $O_2$ than does maternal hemoglobin, facilitating $O_2$ transfer in the placenta. When high metabolism lowers the pH of the blood, hemoglobin releases more of its $O_2$.

yourBioPortal.com
GO TO Web Activity 49.2 • Oxygen-Binding Curves

(to become 25% saturated). This influence of $O_2$ binding by one subunit on the $O_2$ affinity of the other subunits is called **positive cooperativity**.

Once the third $O_2$ molecule is bound, the relationship seems to change, as a larger increase in $P_{O_2}$ is required for the hemoglobin to reach 100 percent saturation. This upper bend of the sigmoid curve is due to a probability phenomenon. The closer we get to having all subunits occupied, the less likely it is that any particular $O_2$ molecule will find a place to bind. Therefore, it takes a relatively greater $P_{O_2}$ to achieve 100 percent saturation.

The $O_2$-binding/dissociation properties of hemoglobin help get $O_2$ to the tissues that need it most. In the lungs, where the $P_{O_2}$ is about 100 mm Hg, hemoglobin is 100 percent saturated. The $P_{O_2}$ in blood returning to the heart from the body is usually about 40 mm Hg. You can see in Figure 49.12 that at this $P_{O_2}$, the hemoglobin is still about 75 percent saturated. This means that as the blood circulates around the body, it releases only about one in four of the $O_2$ molecules it carries. This system seems inefficient, but it is really quite adaptive, because the hemoglobin keeps 75 percent of its $O_2$ in reserve to meet peak demands of highly active tissues.

When a tissue becomes starved of $O_2$ and its local $P_{O_2}$ falls below 40 mm Hg, the hemoglobin flowing through that tissue is on the steep portion of its binding/dissociation curve. That means relatively small decreases in $P_{O_2}$ below 40 mm Hg will result in the release of lots of $O_2$ to the tissue. Thus hemoglobin is very effective in making $O_2$ available to tissues precisely when and where it is needed most.

The $O_2$ transport function of hemoglobin can rapidly and tragically be disrupted by a common by-product of incomplete combustion: carbon monoxide (CO). If CO from a faulty heating system, engine exhaust, or burning charcoal accumulates in a closed space, the results can be deadly. Because CO binds to hemoglobin with a 240-fold higher affinity than $O_2$, it prevents hemoglobin from transporting $O_2$. In the United States, more than 5,000 people die each year from CO poisoning.

### Myoglobin holds an $O_2$ reserve

Muscle cells have their own $O_2$-binding molecule, *myoglobin*. Myoglobin consists of just one polypeptide chain associated with an iron-containing ring structure that can bind one $O_2$ molecule. Myoglobin has a higher affinity for $O_2$ than hemoglobin does, so it picks up and holds $O_2$ at $P_{O_2}$ values at which hemoglobin is releasing its bound $O_2$ (**Figure 49.13**).

Myoglobin facilitates the diffusion of $O_2$ in muscle cells and provides an $O_2$ reserve for times when metabolic demands are high and blood flow is interrupted. Interruption of blood flow in muscles is common because contracting muscles squeeze blood vessels. When tissue $P_{O_2}$ values are low and hemoglobin can no longer supply more $O_2$, myoglobin releases its bound $O_2$. Diving mammals such as seals have high concentrations of myoglobin in their muscles, which is one reason they can stay under water for so long. (We discuss more adaptations for diving in Chapter 50.) Even in nondiving animals, muscles called on for extended periods of work frequently have more myoglobin than muscles that are used for short, intermittent periods, as noted in Section 48.2.

### Hemoglobin's affinity for $O_2$ is variable

Various factors influence the $O_2$-binding/dissociation properties of hemoglobin, thereby influencing $O_2$ delivery to tissues. Three of those factors are the chemical composition of the hemoglobin, the blood pH, and the presence of 2,3-bisphosphoglyceric acid (BPG) in red blood cells.

**HEMOGLOBIN COMPOSITION** There is more than one type of hemoglobin, because the chemical composition of the polypeptide chains that form the hemoglobin molecule varies. The normal hemoglobin of adult humans has two each of two kinds of polypeptide chains—two α-globin chains and two β-globin chains—and the $O_2$-binding characteristics shown in Figure 49.13.

Before birth, the human fetus has a different form of hemoglobin, consisting of two α-globin and two γ-globin chains. The functional difference between fetal and adult hemoglobin is that fetal hemoglobin has a higher affinity for $O_2$. Therefore, the fetal hemoglobin–oxygen binding/dissociation curve is shifted to the left compared with that for adults (see Figure 49.13). You can see from these curves that if both types of hemoglobin are at the same $P_{O_2}$ (as they are in the placenta), fetal hemoglobin will pick up $O_2$ that the maternal hemoglobin releases. This difference in $O_2$ affinities enables the efficient transfer of $O_2$ from the mother's blood to the fetus's blood.

South American camelids—llamas, alpacas, guanacos, and vicuñas—are native to the Andes Mountains. In the natural habitat of these mammals, more than 5,000 m above sea level, the $P_{O_2}$ is below 85 mm Hg, and the $P_{O_2}$ in their lungs is about 50 mm Hg. To increase their chances of survival in their low $O_2$ environment, the hemoglobins of these camelids have $O_2$-binding/dissociation curves to the left of those of most other mammals—in other words, their hemoglobin can become saturated with $O_2$ at lower $P_{O_2}$ values than those of other mammals.

**HEMOGLOBIN AND pH** The $O_2$-binding properties of hemoglobin are also influenced by physiological conditions. The influence of pH on the function of hemoglobin is known as the **Bohr effect**. As blood passes through metabolically active tissue such as exercising muscle, it picks up acidic metabolites such as lactic acid, fatty acids, and $CO_2$. As a result, blood pH falls. The excess $H^+$ binds preferentially to deoxygenated hemoglobin and decreases its affinity for $O_2$ and the $O_2$-binding/dissociation curve of hemoglobin shifts to the right (see Figure 49.13). This shift means the hemoglobin will release more $O_2$ in tissues where pH is low—another way that $O_2$ is supplied where and when it is most needed.

**2,3-BISPHOSPHOGLYCERIC ACID** BPG is a metabolite of glycolysis (see Figure 7.5). Mammalian red blood cells respond to low $P_{O_2}$ by increasing their rate of glycolysis and thus producing more BPG, which is an important regulator of hemoglobin function. BPG, like excess $H^+$, reversibly combines with deoxygenated hemoglobin and lowers its affinity for $O_2$. The result is that at any $P_{O_2}$, hemoglobin releases more of its bound $O_2$ than it otherwise would. In other words, BPG shifts the $O_2$-binding/dissociation curve of mammalian hemoglobin to the right.

When humans go to high altitudes, or when they cease being sedentary and begin to exercise, their red blood cells are exposed to a lower $P_{O_2}$ and their level of BPG goes up, making it easier for hemoglobin to deliver more $O_2$ to tissues. The reason fetal hemoglobin has a left-shifted $O_2$-binding/dissociation curve is that its γ-globin chains have a lower affinity for BPG than do the β-globin chains of adult hemoglobin.

## $CO_2$ is transported as bicarbonate ions in the blood

Delivering $O_2$ to tissues is only half the respiratory function of blood. Blood also must take $CO_2$, a metabolic waste product, away from tissues (**Figure 49.14**). $CO_2$ is highly soluble and readily diffuses through cell membranes, moving from its site of production in the tissues into the blood, where the partial pressure of $CO_2$ ($P_{CO_2}$) is lower. However, very little dissolved $CO_2$ is transported by the blood. Most $CO_2$ produced by the tissues is transported to the lungs in the form of **bicarbonate ions**, $HCO_3^-$. $CO_2$ is converted to $HCO_3^-$, transported to the lungs, and then converted back to $CO_2$ in several steps.

When $CO_2$ dissolves in water, some of it slowly reacts with the water molecules to form carbonic acid ($H_2CO_3$), some of which then dissociates into a proton ($H^+$) and a bicarbonate ion ($HCO_3^-$). This reversible reaction is expressed as follows:

$$CO_2 + H_2O \rightleftharpoons H_2CO_3 \rightleftharpoons H^+ + HCO_3^-$$

In the extracellular fluid, the reaction between $CO_2$ and $H_2O$ proceeds slowly. But it is a different story in the endothelial cells of the capillaries and in the red blood cells, where the enzyme *carbonic anhydrase* speeds up the conversion of $CO_2$ to $H_2CO_3$.

**49.14 Carbon Dioxide Is Transported as Bicarbonate Ions** Carbonic anhydrase in capillary endothelial cells and in red blood cells facilitates conversion of $CO_2$ produced by tissues into bicarbonate ions carried by the plasma. In the lungs, the process is reversed as $CO_2$ is exhaled.

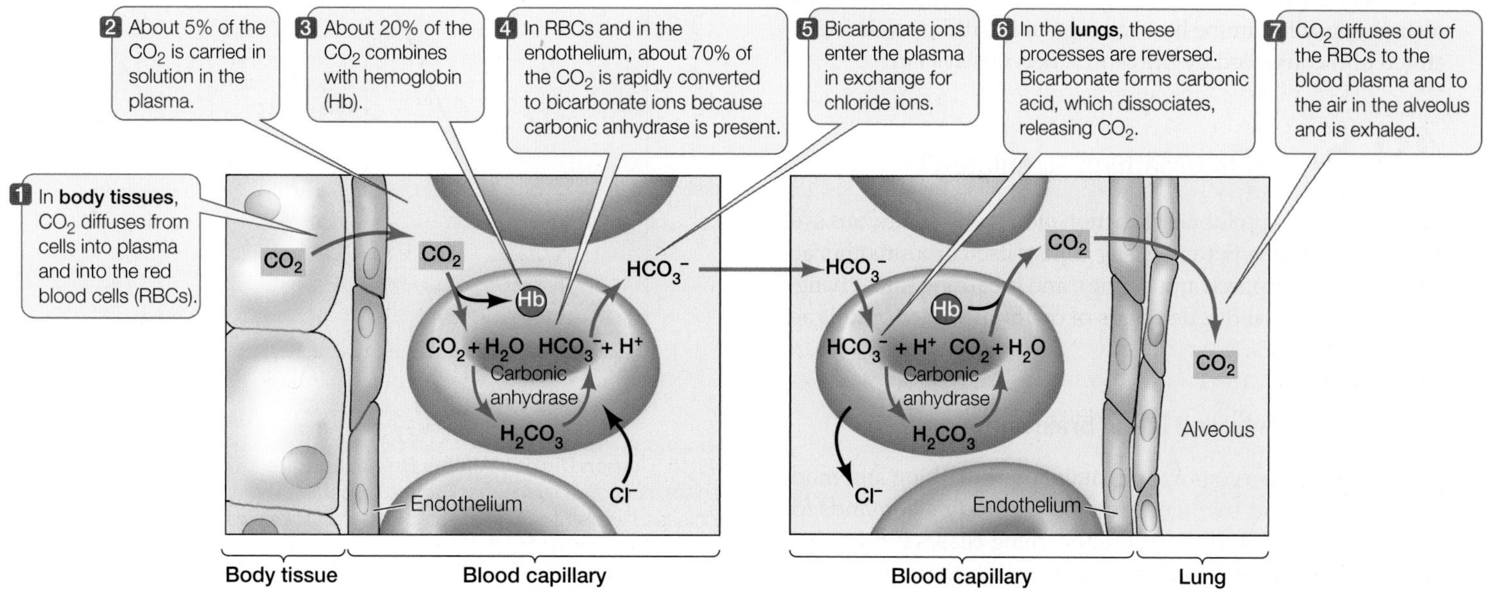
1 In body tissues, CO2 diffuses from cells into plasma and into the red blood cells (RBCs).
2 About 5% of the CO2 is carried in solution in the plasma.
3 About 20% of the CO2 combines with hemoglobin (Hb).
4 In RBCs and in the endothelium, about 70% of the CO2 is rapidly converted to bicarbonate ions because carbonic anhydrase is present.
5 Bicarbonate ions enter the plasma in exchange for chloride ions.
6 In the lungs, these processes are reversed. Bicarbonate forms carbonic acid, which dissociates, releasing CO2.
7 CO2 diffuses out of the RBCs to the blood plasma and to the air in the alveolus and is exhaled.
CO2
CO2
Hb
HCO3−
CO2 + H2O HCO3− + H+
Carbonic anhydrase
H2CO3
Cl−
Endothelium
HCO3−
Hb
CO2
HCO3− + H+ CO2 + H2O
Carbonic anhydrase
H2CO3
Cl−
Endothelium
CO2
Alveolus
Body tissue
Blood capillary
Blood capillary
Lung

The newly formed $H_2CO_3$ dissociates, and the resulting bicarbonate ions enter the plasma in exchange for $Cl^-$ (see Figure 49.14). By converting $CO_2$ to $H_2CO_3$, carbonic anhydrase reduces the $P_{CO_2}$ in these cells and in the plasma, facilitating the diffusion of $CO_2$ from tissue cells to endothelial cells, plasma, and red blood cells. Some $CO_2$ is also carried in chemical combination with hemoglobin.

In the lungs, the reactions involving $CO_2$ and bicarbonate ions are reversed. Remember that an enzyme such as carbonic anhydrase only speeds up a reversible reaction; it does not determine its direction. The direction is determined by concentrations of reactants and products (see Section 6.1). Ventilation keeps the $CO_2$ concentration in the alveoli low, so $CO_2$ diffuses from the blood plasma into the alveoli, lowering the $CO_2$ concentration in the blood, which favors the conversion of $HCO_3^-$ into $CO_2$.

## 49.4 RECAP

**$O_2$ is transported from the lungs to the body's tissues in reversible combination with hemoglobin. Each hemoglobin molecule can reversibly combine with four $O_2$ molecules; the saturation of the binding sites is a function of the $P_{O_2}$ in the hemoglobin's environment.**

- Explain the advantage of having hemoglobin hold on to three $O_2$ molecules at the usual $P_{O_2}$ of mixed venous blood. **See pp. 1037–1038 and Figure 49.12**
- How is the $O_2$-binding/dissociation curve of hemoglobin influenced by pH? By BPG? By development from fetus to newborn infant? **See p. 1039 and Figure 49.13**
- How is $CO_2$ transported in the blood? **See pp. 1039–1040 and Figure 49.14**

We must breathe every minute of our lives, but we don't usually worry about it, or even think about it very often. In the next section we will examine how the regular breathing cycle is generated and controlled by the central nervous system.

# 49.5 How Is Breathing Regulated?

Breathing is an involuntary function of the central nervous system. The breathing pattern easily adjusts itself around other activities (such as speech and eating), and breathing rates change to match the metabolic demands of our bodies. How is this accomplished?

### Breathing is controlled in the brainstem

The autonomic nervous system maintains breathing and modifies its depth and frequency to meet the body's demands for $O_2$ supply and $CO_2$ elimination. Breathing ceases if the spinal cord is severed in the neck region, showing that breathing is generated in the brain. If the brainstem is cut just above the medulla (the segment of the brainstem just above the spinal cord), an irregular breathing pattern remains (**Figure 49.15**).

Groups of respiratory motor neurons in the medulla increase their firing rates just before an inhalation begins. As more and more of these neurons fire—and fire faster and faster—the diaphragm contracts. All of a sudden, the neurons stop firing, the

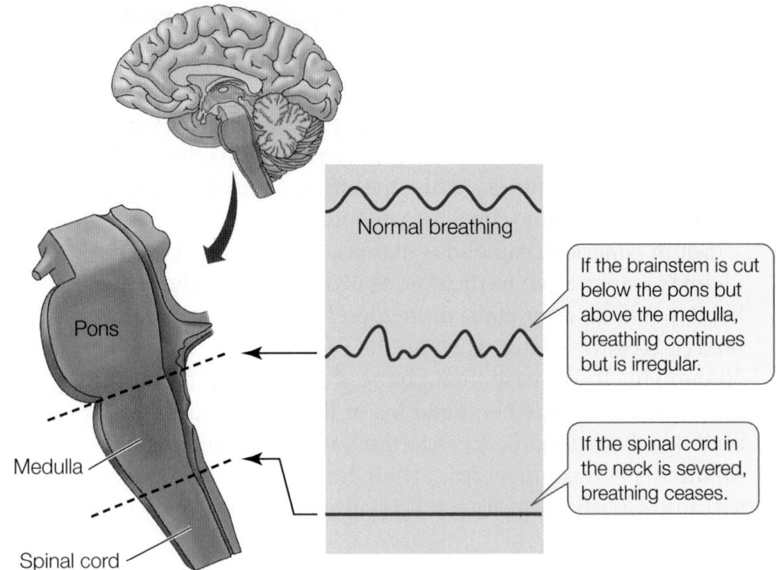


**49.15 Breathing Is Controlled in the Brainstem** Basic breathing rhythm is generated in the medulla and is modified by neurons in or above the pons.

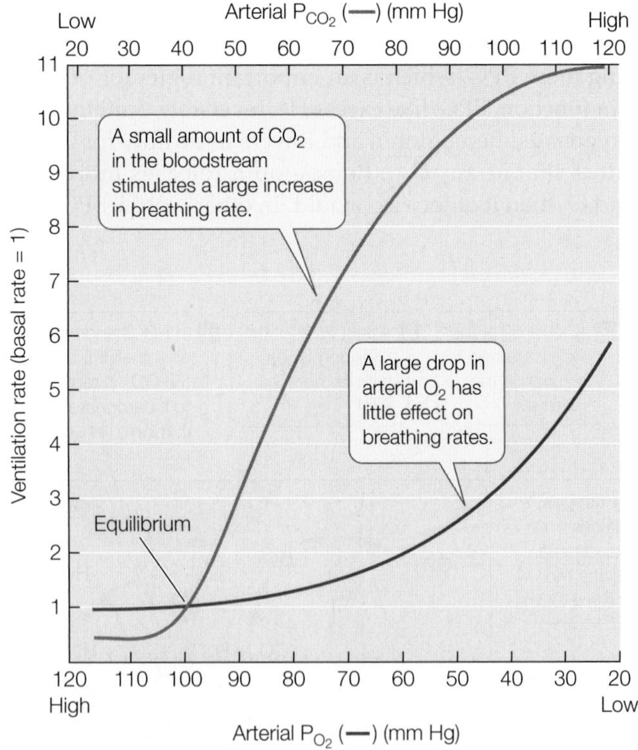


**49.16 Carbon Dioxide Affects Breathing Rate** The breathing mechanism, controlled by feedback to the medulla, is more sensitive to increased levels of $CO_2$ in arterial blood than to decreased amounts of $O_2$. The intersection point occurs at normal (equilibrium) ventilation rate and blood gas levels.

diaphragm relaxes, and exhalation begins. Exhalation is usually a passive process that depends on the elastic recoil of the lung tissues. When breathing demand is high, however, as during strenuous exercise, motor neurons for the intercostal muscles are recruited to a greater extent than during tidal breathing, increasing both the inhalation and the exhalation volumes. Brain areas above the medulla modify breathing to accommodate speech, ingestion of food, coughing, and emotional states.

### Regulating breathing requires feedback information

When breathing or metabolism changes, it alters the $P_{O_2}$ and $P_{CO_2}$ in the blood. We should therefore expect the blood levels of one or both of these gases to provide feedback information to the breathing rhythm generator in the medulla. Experiments in which subjects breathe air with different $P_{O_2}$ and $P_{CO_2}$ concentrations lead us to conclude that humans (and other mammals) are remarkably insensitive to falling levels of $O_2$ in arterial blood but are extremely sensitive to increases in $CO_2$. That is, arterial $P_{O_2}$ can deviate considerably from normal without causing much of an increase in ventilation rate, but even a small rise in arterial $P_{CO_2}$ causes a large increase in ventilation (**Figure 49.16**). This relationship is reversed for water-breathing animals, in which $O_2$ is the primary feedback stimulus for gill ventilation.

We might ask whether it is an increase in the $P_{CO_2}$ of the blood that stimulates increased breathing when we exercise. To answer this question, C. R. Bainton observed dogs running on treadmills at different speeds. As the speed of the treadmill increased, the dogs' respiratory gas exchange rate increased but their blood $P_{CO_2}$ remained constant. Before concluding that blood $P_{CO_2}$ does not control breathing rate, Bainton changed the experiment. Instead of increasing the speed of the treadmill, he gradually increased its slope so that the dogs were running at the same speed but were working harder because they were running uphill (**Figure 49.17**).

In this second experiment, the $P_{CO_2}$ of the blood increased as the slope of the treadmill increased and as the respiratory gas exchange rate increased. Bainton concluded that the $P_{CO_2}$ of the blood is the primary metabolic feedback information for breathing. However, when an animal starts to run or changes its running speed, additional feedback information from receptors in muscles and joints changes its sensitivity to $CO_2$—an example of feedforward information. As noted in Section 40.1, feedforward information can change the sensitivity or the set point of a regulatory system.

Where are partial pressures of gases in the blood sensed? The major site of $P_{CO_2}$ sensitivity is an area on the ventral surface of the medulla, not far from the groups of neurons that generate the breathing rhythm. The primary sensitivity of these chemosensitive cells is not to $CO_2$, however. Rather, they are stimulated by $H^+$ ions. The $H^+$ ion concentration, or pH, in the environment of these cells is a direct reflection of the $P_{CO_2}$ of the

**INVESTIGATING LIFE**

**49.17 Sensitivity of the Respiratory Control System Changes with Exercise** Experiments with dogs running on a treadmill show that sensitivity of the respiratory system to $CO_2$ changes when speed of running changes, but not when workload changes without a change in speed.

**HYPOTHESIS** **Rising levels of blood $CO_2$ during exercise is the feedback signal that stimulates an increase in respiratory rate.**

**METHOD**

1. Dogs are trained to run on a treadmill.
2. The trained dogs are equipped with instruments that measure their respiratory rate and with catheters that withdraw blood samples for measurement of $CO_2$ levels.
3. The dog runs with the treadmill set first at different speeds, and then with the slope of the treadmill elevated (which increases the workload).
4. Respiratory rate is plotted as a function of arterial $CO_2$ concentration.

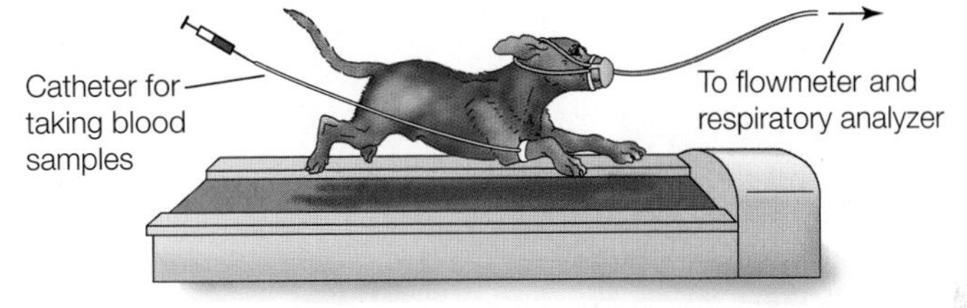


**RESULTS**

**CHANGES IN SPEED**
Respiratory rate changes with running speed but $CO_2$ levels do not. The hypothesis is not supported.

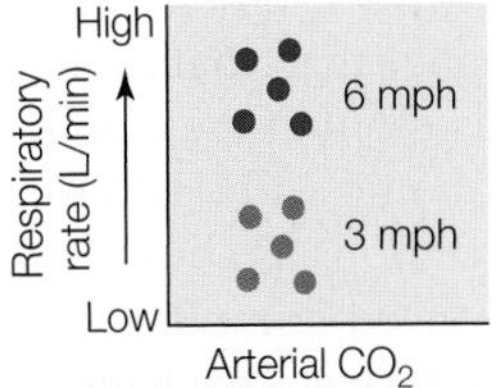


**CHANGES IN SLOPE**
Respiratory rate and $CO_2$ levels both rise as workload increases. The hypothesis is supported.

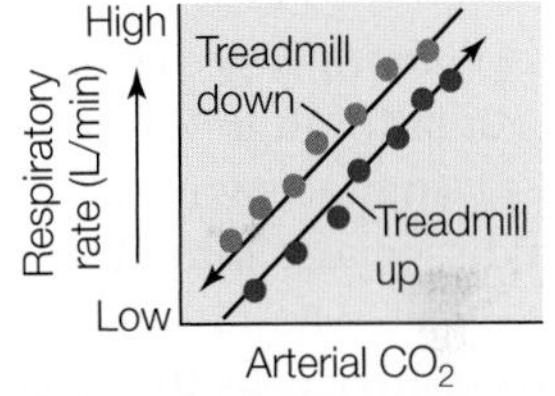


**CONCLUSION** **Arterial $CO_2$ level is the metabolic feedback signal that regulates respiration in response to workload.**

**FURTHER INVESTIGATION:** Exposure to cold is another variable that increases respiration. How would you test whether cold-induced respiration is driven by elevated blood $CO_2$ levels?

Go to **yourBioPortal.com** for original citations, discussions, and relevant links for all INVESTIGATING LIFE figures.

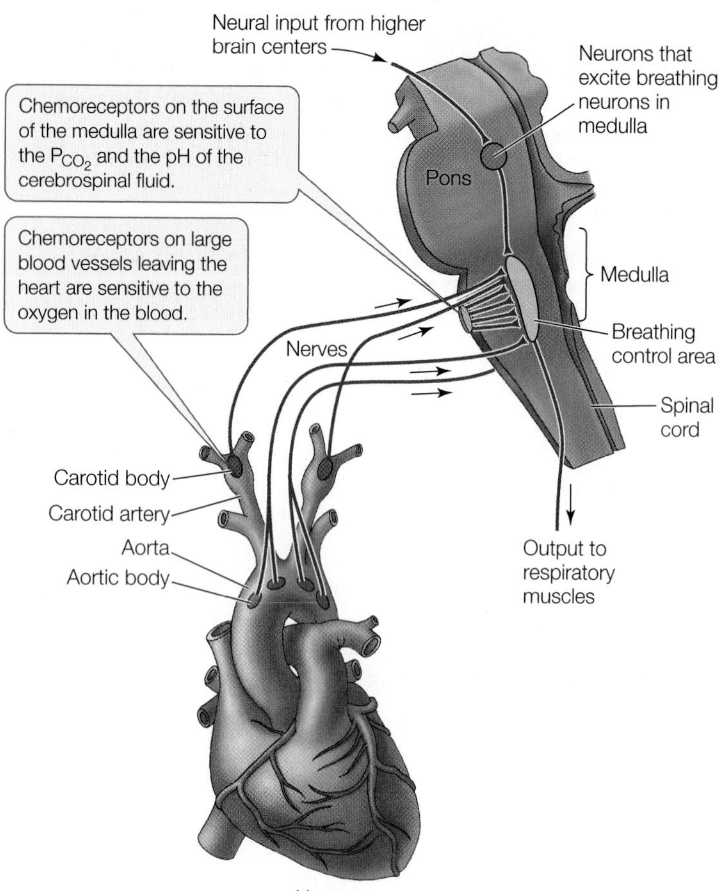


**49.18 Feedback Information Controls Breathing** The body uses feedback information from chemosensors in the heart and brain to match breathing rate to metabolic demand.

blood. When the $P_{CO_2}$ of the blood is higher than that of the extracellular fluid in this area, $CO_2$ diffuses out of the blood. That $CO_2$ interacts with $H_2O$ to form carbonic acid ($H_2CO_3$), which dissociates into $H^+$ ions and $HCO_3^-$ ions as shown in Figure 49.14. The $H^+$ ions that are produced stimulate the chemosensitive cells that increase respiratory gas exchange.

Thus, even though we measure blood $P_{CO_2}$ as the stimulus for breathing, the real stimulus is pH.

Sensitivity to blood $P_{O_2}$ resides in nodes of neural tissue on the large blood vessels leaving the heart: the aorta and the carotid arteries (**Figure 49.18**). These **carotid** and **aortic bodies** are chemosensors. If the blood supply to these structures decreases, or if the blood $P_{O_2}$ falls dramatically, the chemosensors are activated and send nerve impulses to the breathing control center. Although we are not very sensitive to changes in blood $P_{O_2}$, the carotid and aortic bodies can stimulate increases in breathing during exposure to very high altitudes or when blood volume or blood pressure is very low. Also, there is a synergism between $CO_2$ and $O_2$ sensing. When blood $P_{CO_2}$ increases, there is an increased sensitivity to low $O_2$ and vice versa.

## 49.5 RECAP

**The rhythmic contractions of the respiratory muscles that drive breathing are generated by neurons in the brainstem.**

- What is the primary chemical stimulus for controlling the respiratory rate, and where is it sensed? **See p. 1040 and Figures 49.16 and 49.18**
- Explain what feedforward information is. Can you give an example of feedforward information in the respiratory control system. **See Figure 49.17**
- What are the functions of the carotid and aortic bodies? **See p. 1042 and Figure 49.18**

# CHAPTER SUMMARY

## 49.1 What Physical Factors Govern Respiratory Gas Exchange?

- Most cells require a constant supply of $O_2$ and continuous removal of $CO_2$. These **respiratory gases** are exchanged between an animal's body fluids and its environment by diffusion.
- **Fick's law of diffusion** shows how various physical factors influence the diffusion rate of gases. Adaptations to maximize respiratory gas exchange influence one or more variables of Fick's law.
- In water-breathing animals, gas exchange is limited by the low diffusion rate and low amount of $O_2$ in water. If water temperature rises, water-breathing animals face a double bind in that the amount of $O_2$ in water decreases, but their metabolism and the amount of work required to move water over the gas exchange surfaces increase. **Review Figure 49.2**
- In air, the **partial pressure of oxygen ($P_{O_2}$)** decreases with altitude.

## 49.2 What Adaptations Maximize Respiratory Gas Exchange?

- Adaptations to maximize gas exchange include increasing the surface area for gas exchange, maximizing partial pressure gradients across those exchange surfaces, ventilating the outer surface with the respiratory medium, and perfusing the inner surface with blood.
- Insects distribute air throughout their bodies in a system of **tracheae**, **tracheoles**, and air capillaries. **Review Figure 49.4**
- The **gills** of fishes have large gas exchange surface areas that are ventilated continuously and unidirectionally with water. The **countercurrent flow** of blood helps increase the efficiency of gas exchange. **Review Figures 49.5 and 49.6**

- The gas exchange system of birds includes **air sacs** that communicate with the lungs but are not used for gas exchange. Air flows unidirectionally through bird lungs; gases are exchanged in air capillaries that run between **parabronchi**. Review Figure 49.7
- Each breath of air remains in a bird's respiratory system for two breathing cycles. The air sacs work as bellows to supply the air capillaries with a continuous unidirectional flow of fresh air. Review Figure 49.8, **ANIMATED TUTORIAL 49.1**
- In all air-breathing vertebrates except birds, breathing is **tidal**. This is a less efficient form of gas exchange than that of fishes and birds. Although the volume of air exchanged with each breath can vary considerably in tidal breathing, the inhaled air is always mixed with stale air. Review Figure 49.9

## 49.3 How Do Human Lungs Work?

- In mammalian lungs, the gas exchange surface area provided by the millions of **alveoli** is enormous, and the diffusion path length between the air and perfusing blood is short. **Surface tension** in the alveoli would make inflation of the lungs difficult if the alveoli did not produce **surfactant**. Review Figure 49.10, **WEB ACTIVITY 49.1**
- Inhalation occurs when contractions of the **diaphragm** pull on the **pleural membranes** and reduce the pressure in the **thoracic cavity**. Relaxation of the diaphragm increases pressure in the thoracic cavity and results in exhalation. Review Figure 49.11, **ANIMATED TUTORIAL 49.2**
- During periods of heavy metabolic demands such as strenuous exercise, the **intercostal muscles**, located between the ribs, increase the volume of air inhaled and exhaled.

## 49.4 How Does Blood Transport Respiratory Gases?

- $O_2$ is reversibly bound to **hemoglobin** in red blood cells. Each hemoglobin molecule can carry a maximum of four $O_2$ molecules. Because of **positive cooperativity**, hemoglobin's affinity for $O_2$ depends on the $P_{O_2}$ to which the hemoglobin is exposed. Therefore, hemoglobin picks up $O_2$ as it flows through respiratory exchange structures and gives up $O_2$ in metabolically active tissues. Review Figure 49.12
- Myoglobin serves as an $O_2$ reserve in muscle.
- There is more than one type of hemoglobin. Fetal hemoglobin has a higher affinity for $O_2$ than does maternal hemoglobin, allowing fetal blood to pick up $O_2$ from the maternal blood in the placenta. Review Figure 49.13, **WEB ACTIVITY 49.2**
- $CO_2$ is transported in the blood principally as **bicarbonate ions** ($HCO_3^-$). Review Figure 49.14

## 49.5 How Is Breathing Regulated?

- The breathing rhythm is an autonomic function generated by neurons in the medulla and modulated by higher brain centers. The most important feedback stimulus for breathing is the level of $CO_2$ in the blood. Review Figures 49.16 and 14.17
- The breathing rhythm is sensitive to feedback from chemoreceptors on the ventral surface of the medulla and in the **carotid** and **aortic bodies** on the large vessels leaving the heart. Review Figure 49.18

**SEE WEB ACTIVITY 49.3 for a concept review of this chapter.**

## SELF-QUIZ

1. Which of the following statements is *not* true?
   *a.* Respiratory gases are exchanged only by diffusion.
   *b.* $O_2$ has a lower rate of diffusion in water than in air.
   *c.* The $O_2$ content of water falls as the temperature of water rises.
   *d.* The amount of $O_2$ in the atmosphere decreases with increasing altitude.
   *e.* Birds have evolved active transport mechanisms to augment their respiratory gas exchange.
2. Which statement about the gas exchange system of birds is *not* true?
   *a.* Respiratory gases are not exchanged in the air sacs.
   *b.* A bird can achieve more complete exchange of $O_2$ from air to blood than humans can.
   *c.* Air passes through birds' lungs in only one direction.
   *d.* The gas exchange surfaces in bird lungs are the alveoli.
   *e.* A breath of air remains in the system for two breathing cycles.
3. Which statement about gas exchange in fish is *true*?
   *a.* Blood flows over the gas exchange surfaces in a direction opposite to the flow of water.
   *b.* Gases are exchanged across the gill arches.
   *c.* Ventilation of the gills is tidal in fast-swimming fishes.
   *d.* Less work is needed to ventilate gills in warm water than in cold water.
   *e.* The path length for diffusion of respiratory gases is determined by the length of the gill filaments.
4. In the human gas exchange system,
   *a.* the lungs and airways are completely collapsed after a forceful exhalation.
   *b.* the average $P_{O_2}$ concentration of air inside the lungs is always lower than that in the air outside the lungs.
   *c.* the $P_{O_2}$ of the blood leaving the lungs is greater than that of the exhaled air.
   *d.* the amount of air that is moved per breath during normal, at-rest breathing is termed the total lung capacity.
   *e.* $O_2$ and $CO_2$ are actively transported across the alveolar and capillary membranes.
5. Which statement about the human gas exchange system is *not* true?
   *a.* During inhalation, a negative pressure exists in the space between the lung and the thoracic wall.
   *b.* Smoking one cigarette can immobilize the cilia lining the airways for hours.
   *c.* The respiratory control center in the medulla responds more strongly to changes in arterial $O_2$ concentration than to changes in arterial $CO_2$ concentration.
   *d.* Without surfactant, the work of breathing is greatly increased.
   *e.* The diaphragm contracts during inhalation and relaxes during exhalation.

6. The hemoglobin of a human fetus
   *a.* is the same as that of an adult.
   *b.* has a higher affinity for $O_2$ than adult hemoglobin has.
   *c.* has only two protein subunits instead of four.
   *d.* is supplied by the mother's red blood cells.
   *e.* has a higher affinity for BPG than adult hemoglobin has.
7. The amount of $O_2$ carried by hemoglobin depends on the $P_{O_2}$ in the blood. Hemoglobin in active muscles
   *a.* becomes saturated with $O_2$.
   *b.* takes up only a small amount of $O_2$.
   *c.* readily unloads $O_2$.
   *d.* tends to decrease the $P_{O_2}$ in the muscle tissues.
   *e.* is denatured.
8. Most $CO_2$ in the blood is carried
   *a.* in the cytoplasm of red blood cells.
   *b.* as $CO_2$ dissolved in the plasma.
   *c.* in the plasma as bicarbonate ions.
   *d.* bound to plasma proteins.
   *e.* in red blood cells bound to hemoglobin.
9. Myoglobin
   *a.* binds $O_2$ at $P_{O_2}$ values at which hemoglobin is releasing its bound $O_2$.
   *b.* has a lower affinity for $O_2$ than hemoglobin does.
   *c.* consists of four polypeptide chains, just as hemoglobin does.
   *d.* provides an immediate source of $O_2$ for muscle cells at the onset of activity.
   *e.* can bind four $O_2$ molecules at once.
10. When the level of $CO_2$ in the bloodstream increases,
   *a.* the rate of respiration decreases.
   *b.* the pH of the blood rises.
   *c.* the respiratory centers become dormant.
   *d.* the rate of respiration increases.
   *e.* the blood becomes more alkaline.

## FOR DISCUSSION

1. The blood of a certain species of fish that lives in Antarctica has no hemoglobin. What anatomical and behavioral characteristics would you expect to find in this fish, and why is its distribution limited to the waters of Antarctica?
2. Blood banks store whole blood for a much shorter period than they store blood plasma. This is because when blood that has been stored for too long is infused into a patient, it can actually decrease the $O_2$ available to the patient's tissues. Why is this so? Explain in terms of the different physiological functions of BPG.
3. In 1875 three French physiologists went up in a hot air balloon called Zenith to see the effects of reduced air pressure on breathing. As they went higher and higher, their writing became more illegible and nonsensical, but they did not record any sense of danger. They continued to throw out ballast until they all were unconscious. When the balloon returned to the surface, only one of the scientists recovered; the other two were dead. From your knowledge of respiratory gas exchange and regulation of respiration, how would you explain this tragic episode? Why do you think the three men continued tossing out ballast to ascend higher and higher, without realizing they were in mortal danger of asphyxiation?
4. In patients who have emphysema, the fine structures of alveoli break down, resulting in the formation of larger air cavities in the lungs. Also, the lung tissue becomes less elastic. Give at least two explanations for why such patients have a low tolerance for exercise.
5. A condition called "the bends" affects scuba divers who surface too quickly after spending an extended period in deep water, where they have been breathing pressurized air. The cause of the bends is tiny bubbles of nitrogen coming out of solution in the blood plasma. Seals spend much more time under water and at deeper depths than scuba divers, yet they do not suffer the bends. Why?

## ADDITIONAL INVESTIGATION

When you suddenly travel to a location at high altitude, you notice an unusual breathing pattern when you are resting. For a while you stop breathing completely; then suddenly you start breathing rapidly for a short time; then you stop breathing again. This can go on and on in a cyclical pattern called Cheyne-Stokes breathing. It generally goes away after spending a few days at altitude. Can you hypothesize what causes Cheyne-Stokes breathing and design an experiment to test your hypothesis?

## WORKING WITH DATA (GO TO yourBioPortal.com)

**Describing Air Flow in Bird Lungs** The path of avian air flow described in Figure 49.8 was established in experiments by Bretz and Schmidt-Nielsen at Duke University. This exercise presents their methods and data, along with questions to be considered in light of their results.

**Calculating the Functional Residual Volume** A spirometer is used to measure volumes of air exchanged during breathing (see Figure 49.9), but a spirometer cannot measure directly the "dead space," or *residual volume*, of the lungs. The residual volume influences the degree to which incoming fresh air is diluted by the stale air not expelled by the lungs during exhalation. A technique called the *helium dilution method* is used to measure residual volume. In this exercise, you will use data from a helium dilution experiment to calculate the functional residual volume (FRV) of a subject. You will also determine the effect of an increase in FRV on the maximum $P_{O_2}$ in the alveoli.

# 50 Circulatory Systems

## You gotta have heart

On April 29, 1993, the Boston Celtics met the Charlotte Hornets in a National Basketball Association playoff game at the Boston Garden. The Celtics captain and star, 27-year-old Reggie Lewis, had just scored 10 points in 3 minutes when he suddenly slumped forward and fell to the floor. He was examined by the team doctor, who allowed him to return to the game. But Lewis's legs were "wobbly," and he played only briefly.

Lewis had been experiencing dizzy spells for about a month before the playoff incident. Afterward he underwent rigorous testing by cardiologists, who diagnosed him as having a dangerous arrhythmia (irregular heartbeat) caused by cardiomyopathy (diseased cardiac muscle). Accepting this diagnosis would mean the end of his professional athletic career, and Lewis sought a second opinion.

After another battery of tests, a second medical team concluded that Lewis had undergone a transient irregular heartbeat attributable to normal athletic stresses. The condition was thought to be due to Lewis's enlarged heart—a condition common in high-performing athletes. In July 1993, after an hour spent shooting baskets in a pick-up game, Reggie Lewis collapsed and died of heart failure.

Your heart is a muscular pump that, at rest, beats an average of 60 to 70 times per minute. With each beat, it circulates about 70 milliliters of blood through the body. Without taking work or exercise into account, that is 300 liters per hour, 7,200 liters per day, 2.6 million liters per year—no time-outs.

Heart failure is the leading cause of death in the United States, accounting for some one-fourth (about 600,000) of the deaths each year. Heart failure most commonly results from blockage of the vessels that supply the heart muscle with blood, and its risk increases with age. But heart failure is also the leading cause of death among young athletes.

In athletes, heart failure is usually due not to vessel blockage but to defects of the heart or blood vessels. The most common defect is hypertrophic cardiomyopathy (HCM), caused by a mutation that affects the contractile proteins of heart muscle. The heart is good at compensating for these mutant proteins. About 0.5 percent of the population has HCM, and most of them are unaware of it; they can live their entire lives without symptoms. With

**An Athlete's Heart** Basketball star Reggie Lewis died of heart failure at the age of 27. The exact medical situation underlying his heart problems remains clouded in controversy.

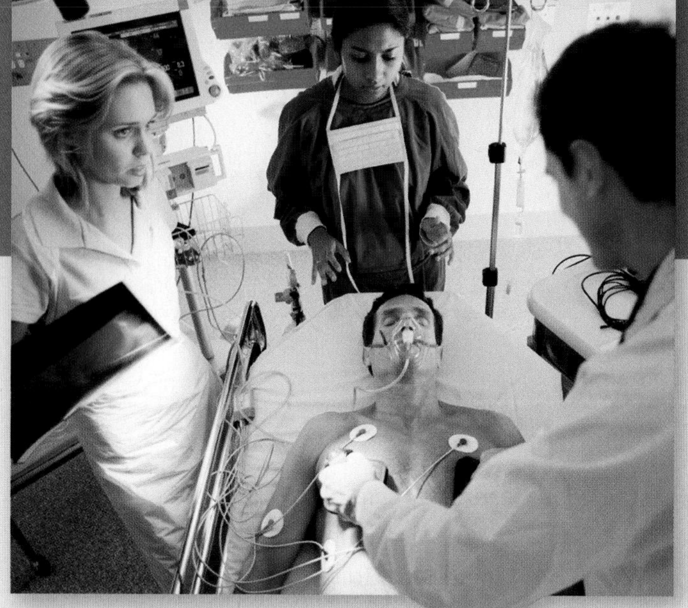

**Clear!** Hospital emergency rooms deal with heart attacks on a daily basis. Advances in heart surgery and emergency resuscitation techniques have saved many lives, but circulatory system failure remains the number one cause of death in Europe, Canada, and the United States.

continued heavy exercise, however, the heart compensates by getting larger. Eventually, this increase in size can disrupt the electrical impulses that coordinate contractions of the heart muscle. When heavy demand is placed on such an enlarged heart, muscle fiber contractions can suddenly become uncoordinated and render the heart incapable of pumping blood. The lack of prior symptoms is why the condition often goes undiagnosed.

Diagnosis is getting better. In December 2008, 33-year-old Cuttino Mobley, who had just joined the New York Knicks, announced his retirement from basketball after learning that his career could kill him. Like Reggie Lewis, Mobley has an enlarged heart. When a physical exam showed arrhythmia, Mobley's physician ordered an MRI that revealed the thickening of the heart walls characteristic of HCM. "Getting the MRI saved my life," Mobley said.

**IN THIS CHAPTER** we will learn about the adaptations of circulatory systems that enable them to match blood supply with demand in a variety of species. Taking the human circulatory system as a model, we will explore the mechanics of the beating heart and the characteristics of the arteries, capillaries, and veins of the vascular system. We will then describe the features of blood before ending with a discussion of the hormonal and neural regulation of the mammalian circulatory system.

CHAPTER OUTLINE



## 50.1 Why Do Animals Need a Circulatory System?

A **circulatory system** consists of a muscular pump (the heart), a fluid (blood), and a series of conduits (blood vessels) through which the fluid can be pumped around the body. Heart, blood, and vessels are also known collectively as a **cardiovascular system** (Greek *kardia*, "heart," and Latin *vasculum*, "vessel"). The function of a circulatory system is to transport things around the body. Preceding chapters discuss how circulatory systems transport heat, hormones, respiratory gases, blood cells, platelets, and cells and molecules of the immune system. Succeeding chapters add nutrients and waste products to that list. In this section we will describe the general types of circulatory systems found in animals.

### Some animals do not have a circulatory system

Single-celled organisms serve all of their needs through direct exchanges with the environment. Such organisms are found mostly in aquatic or very moist terrestrial environments. Similarly, many multicellular aquatic organisms are small or thin enough that all their cells are close to the external environment. Such species may not have a circulatory system because nutrients, respiratory gases, and wastes can diffuse directly between the cells of their bodies and the environment.

The cells of some larger aquatic multicellular animals without a circulatory system are served by highly branched central cavities called *gastrovascular systems* which essentially bring the external environment into the animal. All the cells of a sponge, for example, are in contact with, or very close to, the water that surrounds the animal and circulates through its central cavity (see Figure 49.1B). Very small animals without a circulatory system can maintain high levels of metabolism and activity, but larger animals without a circulatory system such as sponges, jellyfishes, and flatworms tend to be inactive, slow, or even sedentary. Large, active animals need a circulatory system.

### Circulatory systems can be open or closed

The cells of large, mobile animals are supported by the extracellular fluid. All nutrients—oxygen, fuel, essential molecules—come from that fluid, and the waste products of cell metabolism go into it. Circulatory systems have muscular chambers, or *hearts,* that move the extracellular fluid around the body. In *open circulatory systems*, extracellular fluid is the same as the fluid in the circulatory system and is called *hemolymph*. This fluid leaves the vessels of the circulatory system, percolates between cells and through tissues, and then flows back into the heart or vessels of the circulatory system to be pumped out again. In contrast, *closed circulatory systems* completely contain the circulating fluid (*blood*) in a continuous system of vessels. Blood cells and large molecules stay within the system, but water and low-molecular-weight solutes leak out of the smallest vessels, the *capillaries*, which are highly permeable.

In animals with a closed circulatory system, *extracellular fluid* refers to both the fluid in the circulatory system and the fluid outside it. The fluid in the circulatory system is the *blood plasma*; the fluid around the cells is the *interstitial fluid* (see Figure 40.1). A 70-kilogram person has a total extracellular fluid volume of about 14 liters. Less than a quarter of it—about 3 liters—is the blood plasma.

Circulatory systems control the distribution of blood to the body's tissues and organs, with two reciprocal and complementary functions: to maintain the blood composition by picking up nutrients and eliminating wastes, and to supply nutrients to and remove wastes from the body's tissues.

## Open circulatory systems move extracellular fluid

**Open circulatory systems** are found in arthropods, mollusks, and some other invertebrate groups. In these systems, a muscular pump, or heart, helps move the hemolymph through vessels leading to different regions of the body. The fluid leaves the vessels to trickle through the tissues before returning to the heart. In the generalized arthropod shown in **Figure 50.1A**, the fluid returns directly to the heart through openings called *ostia.* Ostia have valves that allow hemolymph to enter the relaxed heart but prevent it from flowing in the reverse direction when the heart contracts. In mollusks (**Figure 50.1B**), open vessels collect hemolymph from different regions of the body and return it to the heart.

## Closed circulatory systems circulate blood through a system of blood vessels

In **closed circulatory systems**, a system of vessels keeps circulating blood separate from the interstitial fluid. Blood is pumped through this *vascular system* by one or more muscular hearts, and some components of the blood never leave the vessels. Closed circulatory systems characterize vertebrates and some invertebrate groups, among them annelids.

A simple example of a closed circulatory system is that of the earthworm (**Figure 50.1C**). One large ventral blood vessel carries blood from the worm's anterior end to its posterior end. Smaller vessels branch off and transport the blood to even smaller vessels serving the tissues in each body segment. In the smallest vessels, respiratory gases, nutrients, and metabolic wastes diffuse between the blood and interstitial fluid. The blood then flows from these vessels into larger vessels that lead into one large dorsal vessel, which carries the blood from the posterior to the anterior end of the body. Five pairs of muscular vessels connect the large dorsal and ventral vessels in the anterior end, thus completing the circuit. The dorsal vessel and the five connecting vessels serve as hearts for the earthworm; their contractions keep the blood circulating. The direction of circulation is determined by one-way valves in the dorsal and connecting vessels.

(A) Arthropod

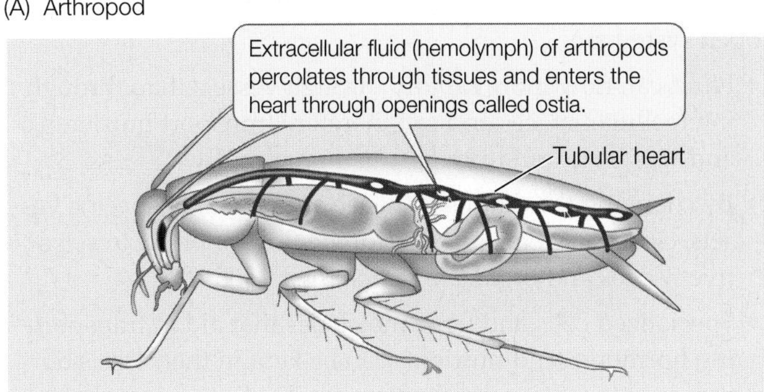


(B) Mollusk

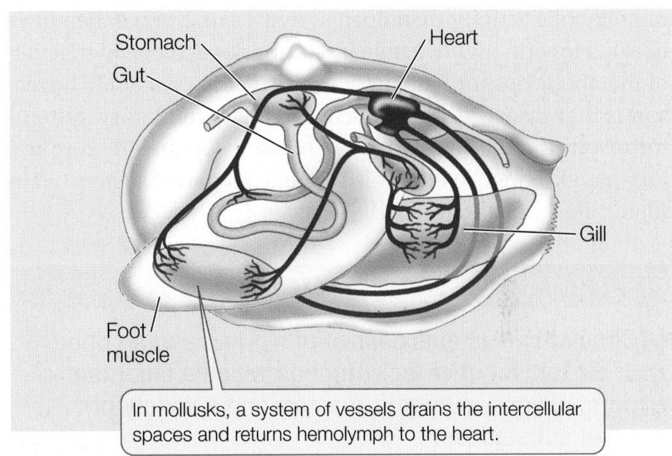


(C) Annelid worm

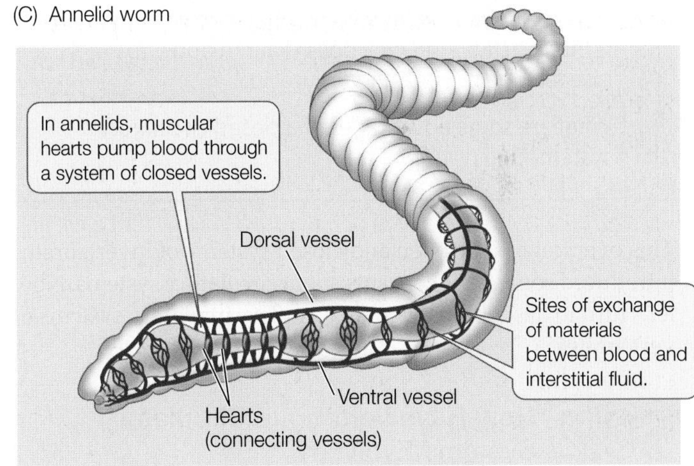


**50.1 Circulatory Systems** Arthropods, illustrated here by an insect (A), and mollusks such as clams (B) have an open circulatory system. Hemolymph is pumped by a tubular heart and directed to different regions of the body through vessels that open into intercellular spaces. (C) Annelids such as earthworms have a closed circulatory system, in which the cellular and macromolecular elements of the blood are confined in a system of vessels and the blood is pumped through those vessels by one or more muscular hearts.

Closed circulatory systems have several advantages over open systems:

- Fluid can flow more rapidly through vessels than through intercellular spaces and can therefore transport nutrients and wastes to and from tissues more rapidly.
- By changing the diameter (hence resistance) of specific vessels, closed systems can control the flow of blood to selective tissues and organs to match their needs.
- Specialized cells and large molecules that aid in transporting hormones and nutrients can be kept in the vessels but can drop their cargo in the tissues where it is needed.

It seems logical to accept the premise that in all but very small animals, closed circulatory systems can support higher levels of metabolic activity than open systems can. But we then have to ask: How do highly active insect species achieve high levels of metabolic output with an open circulatory system? The reason is that insects do not depend on their circulatory systems for respiratory gas exchange. As noted in Chapter 49, respiratory gas exchange in insects is through a separate system of air-filled tubes (see Figure 49.4).

### 50.1 RECAP

**Circulatory systems consist of a pump and an open or closed set of vessels through which a fluid transports oxygen, nutrients, wastes, and a variety of other substances.**

- Why are most animals with an open circulatory system rather inactive, and why does that not apply to insects? **See p. 1047 and Figure 50.1**
- What are some advantages of a closed circulatory system? **See p. 1048**

This overview of the open and closed systems of invertebrates introduced some basic concepts about circulatory systems. Now let's turn to describing the more complex circulatory systems of vertebrates.

## 50.2 How Have Vertebrate Circulatory Systems Evolved?

Vertebrates have a closed circulatory system and a heart with two or more chambers. When a heart chamber contracts, it squeezes the blood, putting it under pressure. Blood then flows out of the heart and into vessels, where pressure is lower. Valves prevent the backflow of blood as the heart cycles between contraction and relaxation.

As we explore the features of the circulatory systems of different classes of vertebrates, a general evolutionary theme will emerge: as circulatory systems become more complex, the blood that flows to the gas exchange organs (gills or lungs; see Chapter 49) becomes more completely separated from the blood that flows to the rest of the body.

In fishes, the phylogenetically oldest vertebrates, blood is pumped from the heart to the gills and then to the tissues of the body and back to the heart—a single circuit. In birds and mammals, blood is pumped from the heart to the lungs and back to the heart in a **pulmonary circuit**, and then from the heart to the rest of the body and back to the heart in a **systemic circuit**. In all other vertebrates we see various adaptations for separating the blood flow into pulmonary and systemic circuits.

Both pulmonary and systemic circuits begin with vessels called **arteries** that carry blood away from the heart. Arteries give rise to smaller vessels called **arterioles**, which feed blood into *capillary beds*. **Capillaries** are the tiny, thin-walled vessels where materials are exchanged between the blood and the tissue fluid. Small vessels called **venules** drain capillary beds. The venules join to form larger vessels called **veins**, which deliver blood back to the heart.

We can trace the evolutionary history of vertebrate circulatory systems by comparing the circulatory systems of fishes, lungfishes, amphibians, ectothermic reptiles, and birds and mammals.

### Fishes have a two-chambered heart

The fish heart has two chambers. An **atrium** (plural *atria*) receives blood from the body and pumps it into a more muscular chamber, the **ventricle**. The ventricle pumps the blood to the gills, where gases are exchanged. Blood leaving the gills collects in a large dorsal artery, the **aorta**, which distributes blood to smaller arteries and arterioles leading to all the organs and tissues of the body. In the tissues, blood flows through beds of tiny capillaries, collects in venules and veins, and eventually returns to the atrium of the heart.

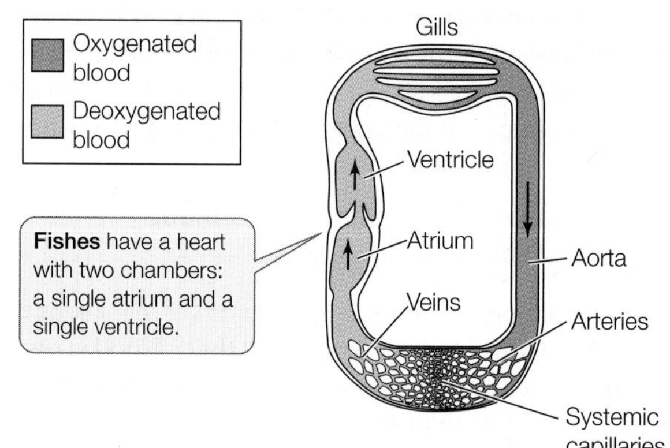


Most of the pressure imparted to the blood by the contraction of the ventricle is lost as a result of the blood entering the many narrow spaces in the gill lamellae. Therefore, blood leaving the gills and entering the aorta is under low pressure, limiting the maximum capacity of the fish circulatory system to supply the tissues with oxygen and nutrients. Yet this limitation on arterial blood pressure does not seem to hamper the performance of many rapidly swimming species, such as tuna and marlin.

The evolutionary transition from breathing water to breathing air had important consequences for the vertebrate circulatory system. An example of how the system changed to serve a primitive lung can be seen in the African lungfish.

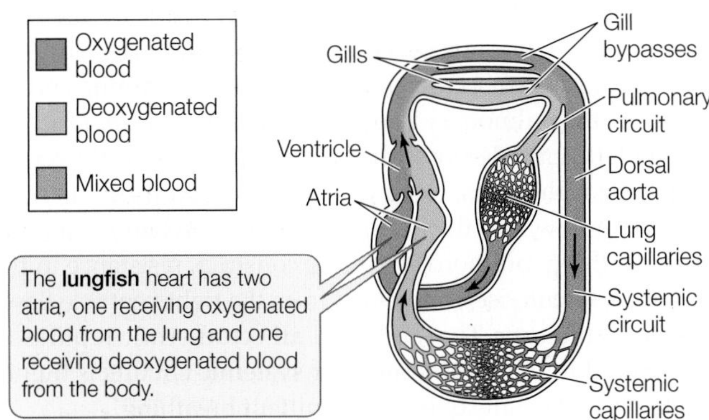


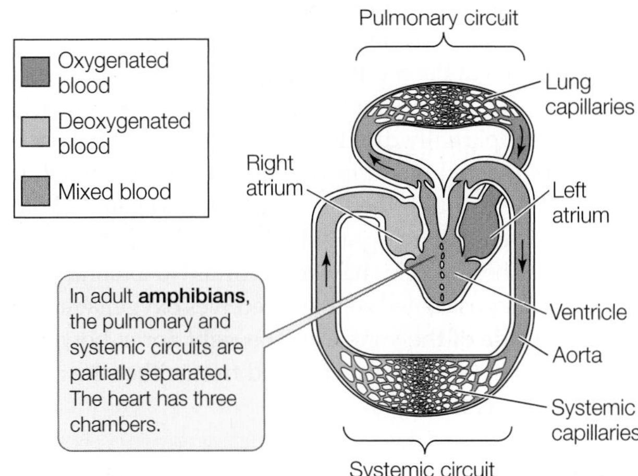


Lungfish are periodically exposed to water with low oxygen content or to situations in which their aquatic environment dries up. The adaptation that deals with these conditions is an outpocketing of the gut that serves as a lung. The lung contains many thin-walled blood vessels, so blood flowing through those vessels can pick up oxygen from air gulped into the lung.

How does the lungfish circulatory system take advantage of this new organ? In fishes, the gills are arranged on supportive gill arches (see Figure 49.5). Blood flows into the gill arch in an afferent arteriole and leaves in an efferent arteriole. In lungfishes, the blood vessels in the posterior pair of gill arteries have been modified to be a low-resistance conduit for blood to the lung, and a new vessel carries oxygenated blood from the lung back to the heart. In addition, two anterior gill arches have lost their gills, and their blood vessels deliver blood from the heart directly to the dorsal aorta. Because a few of the gill arches retain gills, the African lungfish can breathe either air or water.

The lungfish heart partially separates its flow of blood into pulmonary and systemic circuits; it has a partially divided atrium. The left side receives oxygenated blood from the lungs, and the right side receives deoxygenated blood from the other tissues. These two bloodstreams stay mostly separate as they flow through the ventricle and the large vessel leading to the gill arches. As a result, oxygenated blood goes mostly to the anterior gill arteries leading to the dorsal aorta, and deoxygenated blood goes mostly to the other gill arches that have functional gills as well as to the gill arteries that serve the lung.

We can conclude that the lungfish lung evolved as a means of supplementing oxygen uptake from the gills. When the water is fully oxygenated, the lungfish can depend on its gills; but in oxygen-depleted water, it can augment its oxygen intake by gulping air. Associated modifications of the lungfish vascular system set the stage for the evolution of separate pulmonary and systemic circulations in higher vertebrates.

### Amphibians have a three-chambered heart

Pulmonary and systemic circulations are partially separated in adult amphibians. A single ventricle pumps blood to the lungs and the rest of the body. Two atria receive blood returning to the heart: one receives oxygenated blood from the lungs, while the other receives deoxygenated blood from the body. Because both atria deliver blood to the same ventricle, the oxygenated and deoxygenated blood could mix, so that blood going to the tissues would not carry a full load of oxygen. Mixing is limited, however, because anatomical features of the ventricle direct the flow of deoxygenated blood from the right atrium to the pulmonary circuit and the flow of oxygenated blood from the left atrium to the aorta. Partial separation of pulmonary and systemic circulation has the advantage of allowing blood destined for the tissues to sidestep the large pressure drop that occurs in the gas exchange organ. Blood leaving the amphibian heart for the tissues moves directly to the aorta, and hence to the body, at a higher pressure than if it had first flowed through the lungs.

Amphibians have another adaptation for oxygenating their blood: they can pick up a considerable amount of oxygen in blood flowing through small blood vessels in their skin.

### Reptiles have exquisite control of pulmonary and systemic circulation

Reptiles include turtles, snakes, lizards, crocodilians, and birds (see Figure 33.18). Crocodilians and birds have cardiovascular systems with two completely separated ventricles, creating a four-chambered heart. All other reptiles have a three-chambered heart because their ventricles are not completely separated into left and right chambers.

Consider the behavior, ecology, and physiology of *ectothermic* reptiles. Many are active, powerful, fast animals. But their activity comes in bursts that are interspersed with long periods of inactivity, during which these animals' metabolic rates are much lower than the resting metabolic rates of the *endothermic* birds and mammals. So enormous is the range of metabolic demand in ectothermic reptiles that they do not need to breathe continuously. Some species are accomplished divers and spend long periods under water, where they cannot breathe air.

When these animals are not breathing, it would be a waste of energy for them to pump blood through their lungs. Thus, they have evolved the capability to send blood to the lungs and the rest of the body when they are breathing, but when they are not breathing, they can bypass the pulmonary circuit and pump all the blood to the body. How do they do this?

Let's look first at the ectothermic reptiles with a three-chambered heart—the turtles, snakes, and lizards. The ventricle in these species is partially divided into left and right halves by a septum. Oxygenated blood from the lungs enters the left side of the ventricle through the left atrium. Deoxygenated blood from the body enters the right side of the ventricle through the right atrium. These species have two aortas—a left and a right. The left aorta is positioned so that it receives oxygenated blood from the left side of the ventricle. The right aorta, however, is positioned so that it can receive blood from either the right or left side of the ventricle.

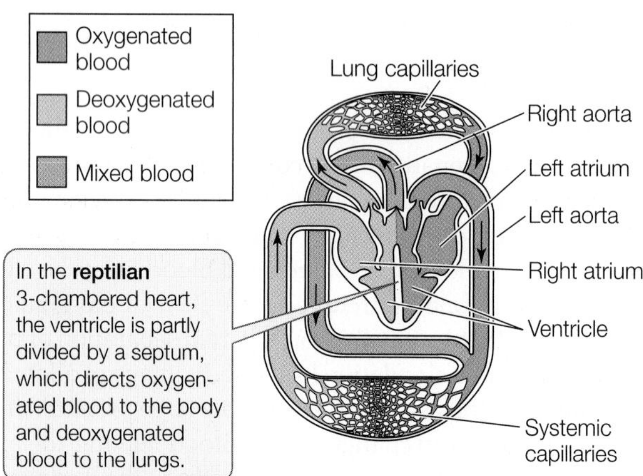
Oxygenated blood
Deoxygenated blood
Mixed blood
Lung capillaries
Right aorta
Left atrium
Left aorta
Right atrium
Ventricle
Systemic capillaries
In the **reptilian** 3-chambered heart, the ventricle is partly divided by a septum, which directs oxygenated blood to the body and deoxygenated blood to the lungs.

When the animal is breathing air, the resistance in the pulmonary circuit is lower than the resistance in the systemic circuit, so blood from the right side of the ventricle tends to flow into the pulmonary artery rather than the right aorta. When the animal is not breathing, pulmonary vessels constrict, resistance in the pulmonary circuit goes up, and blood from the right side of the ventricle tends to flow into the right aorta. As a result, blood from both sides of the ventricle flows through both aortas to the systemic circuit.

Crocodilians, like birds, have two completely separated ventricles. Unlike birds, they have two aortas, one originating in each ventricle. There is a connection between the two aortas just

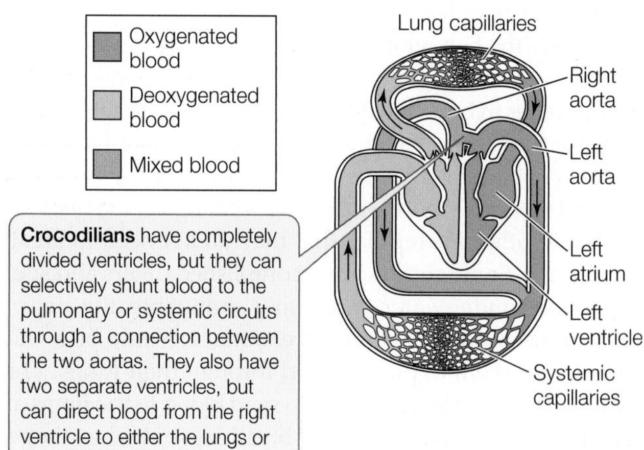
Oxygenated blood
Deoxygenated blood
Mixed blood
Lung capillaries
Right aorta
Left aorta
Left atrium
Left ventricle
Systemic capillaries
**Crocodilians** have completely divided ventricles, but they can selectively shunt blood to the pulmonary or systemic circuits through a connection between the two aortas. They also have two separate ventricles, but can direct blood from the right ventricle to either the lungs or the systemic circuit.

as they leave the heart, and this connection enables them to alter the proportions of blood going to their pulmonary and systemic circuits. When a crocodile or alligator is breathing and resistance in the pulmonary circuit is low, backpressure from the stronger left ventricle closes the valve between the right ventricle and the right aorta, forcing all of the blood from the right ventricle to flow into the pulmonary circuit. When the animal stops breathing, pulmonary vessels constrict, resistance in the pulmonary circuit rises, and blood from the right ventricle flows into the right aorta. This ability of all ectothermic reptiles to direct blood to their pulmonary or systemic circuits is highly adaptive for their lifestyle of intermittent breathing.

### Birds and mammals have fully separated pulmonary and systemic circuits

The four-chambered hearts of birds and mammals have completely separate pulmonary and systemic circuits. Separate circuits have several advantages for these active animals with continuously high metabolic rates:

- Oxygenated and deoxygenated blood cannot mix; therefore, the systemic circuit always receives blood with the highest oxygen content.
- Respiratory gas exchange is maximized because the blood with the lowest oxygen content and highest $CO_2$ content is sent to the lungs.
- Separate systemic and pulmonary circuits can operate at different pressures.

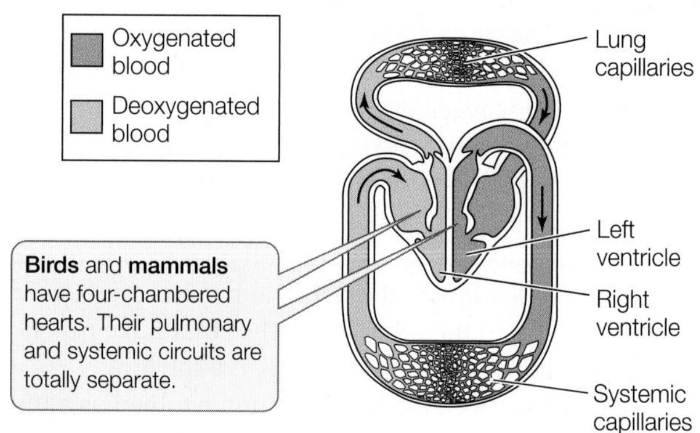
Oxygenated blood
Deoxygenated blood
Lung capillaries
Left ventricle
Right ventricle
Systemic capillaries
**Birds** and **mammals** have four-chambered hearts. Their pulmonary and systemic circuits are totally separate.

The tissues of birds and mammals have high nutrient demands and thus a very high density of blood vessels, requiring the heart to generate a high blood pressure to perfuse all the vessels of the systemic circuit. The pulmonary circuit of these animals receives a blood flow equal to that of the systemic circuit, but the lungs have far fewer blood vessels. Thus, the pulmonary circuit of birds and mammals can function at lower pressures, and the four-chambered heart makes that possible.

**yourBioPortal.com**

GO TO Web Activity 50.1 • Vertebrate Circulatory Systems

## 50.2 RECAP

**The closed circulatory system of vertebrates has evolved from a two-chambered heart in fishes to a four-chambered heart in crocodilians, birds, and mammals.**

- Explain why fishes cannot supply blood to their tissues at high pressure. See p. 1048
- By comparing lungfish and amphibian circulatory systems, explain how a three-chambered heart could have evolved. See pp. 1048–1049
- What are some of advantages of a four-chambered heart? See p. 1050

## 50.3 How Does the Mammalian Heart Function?

Humans have a typical mammalian heart, characterized by four chambers—a right and a left atrium and a right and a left ventricle (**Figure 50.2**). The right ventricle pumps blood through the pulmonary circuit, and the left ventricle pumps blood through the systemic circuit.

One-way valves between the atria and ventricles, the **atrioventricular (AV) valves**, prevent backflow of blood into the atria when the ventricles contract. The **pulmonary valve** and **aortic valve**, one-way valves between the ventricles and the major arteries, prevent backflow of blood into the ventricles when they relax.

In this section, we will describe the flow of blood through the heart and body and examine the unique electrical properties of cardiac muscle that result in the heartbeat.

### Blood flows from right heart to lungs to left heart to body

The heart's right atrium receives deoxygenated blood from the **superior** (upper) **vena cava** and the **inferior** (lower) **vena cava** (see Figure 50.2), large veins that collect blood returning to the

yourBioPortal.com
GO TO Web Activity 50.2 • The Human Heart

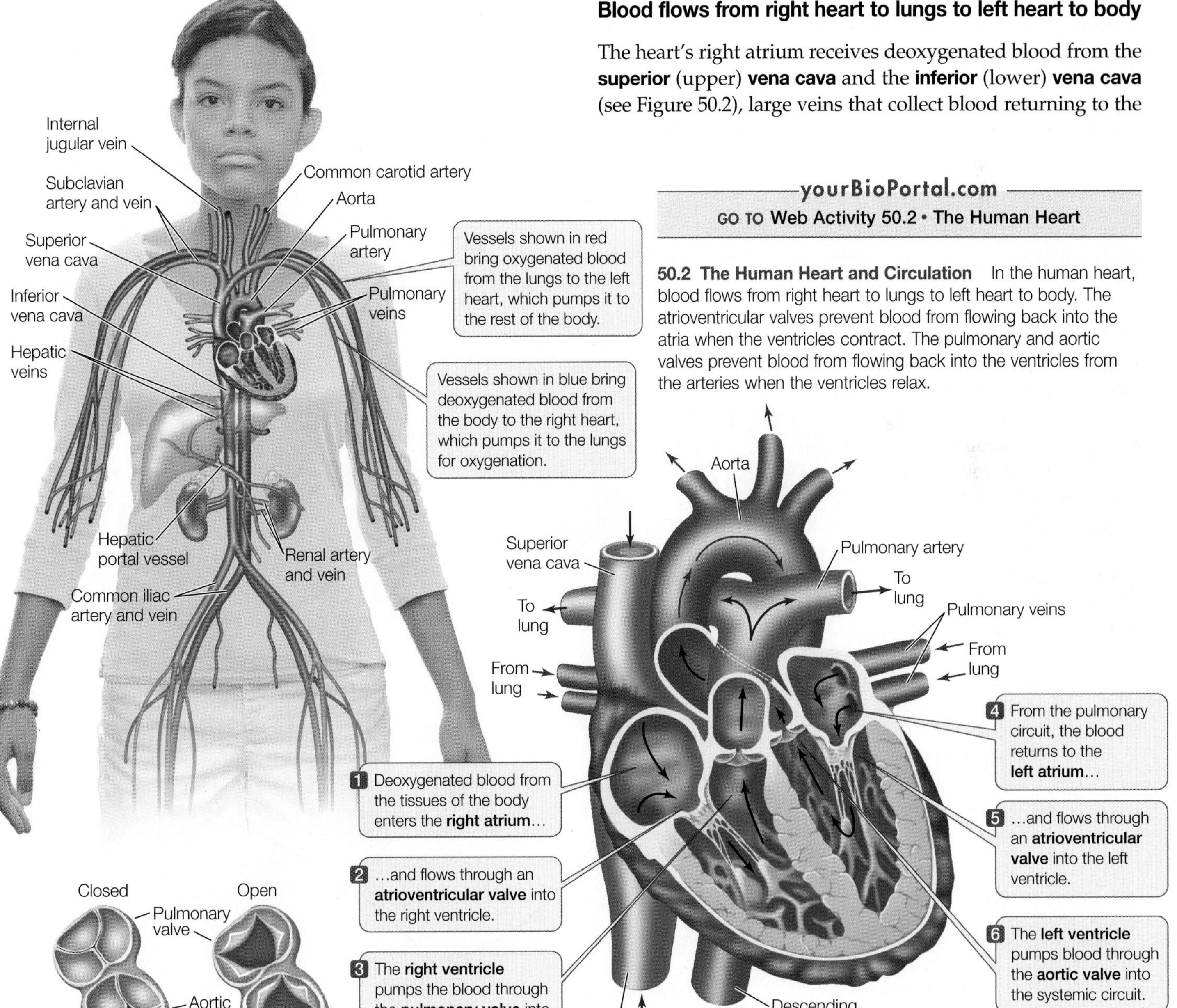
Internal jugular vein
Subclavian artery and vein
Superior vena cava
Inferior vena cava
Hepatic veins
Common carotid artery
Aorta
Pulmonary artery
Pulmonary veins
Vessels shown in red bring oxygenated blood from the lungs to the left heart, which pumps it to the rest of the body.
Vessels shown in blue bring deoxygenated blood from the body to the right heart, which pumps it to the lungs for oxygenation.
Hepatic portal vessel
Renal artery and vein
Common iliac artery and vein
Closed
Open
Pulmonary valve
Aortic valve
Aorta
Superior vena cava
Pulmonary artery
To lung
To lung
Pulmonary veins
From lung
From lung
1 Deoxygenated blood from the tissues of the body enters the right atrium...
2 ...and flows through an atrioventricular valve into the right ventricle.
3 The right ventricle pumps the blood through the pulmonary valve into the pulmonary circuit.
4 From the pulmonary circuit, the blood returns to the left atrium...
5 ...and flows through an atrioventricular valve into the left ventricle.
6 The left ventricle pumps blood through the aortic valve into the systemic circuit.
Inferior vena cava
Descending aorta

**50.2 The Human Heart and Circulation** In the human heart, blood flows from right heart to lungs to left heart to body. The atrioventricular valves prevent blood from flowing back into the atria when the ventricles contract. The pulmonary and aortic valves prevent blood from flowing back into the ventricles from the arteries when the ventricles relax.

heart from the upper and lower body, respectively. The veins of the heart itself also drain into the right atrium. From the right atrium, the blood flows through an AV valve into the right ventricle. Most of the filling of the ventricle results from passive flow while the heart is relaxed between beats. Just at the end of this period of passive ventricular filling, the atrium contracts and adds a little more blood to the ventricular volume. The right ventricle then contracts, causing the AV valve to close and pumping the blood into the **pulmonary artery** leading to the lungs.

After gas exchange occurs in the lungs, **pulmonary veins** return oxygenated blood from the lungs to the left atrium, from which the blood enters the left ventricle through another AV valve. As on the right side of the heart, most left ventricular filling is passive, but ventricular filling is completed when the atria contract.

The walls of the left ventricle are powerful muscles that contract around the blood with a wringing motion starting from the bottom. When pressure in the left ventricle is high enough to push open the aortic valve, blood rushes into the aorta to begin its circulation throughout the body. In Figure 50.2, observe that the walls of the left ventricle are thicker than those of the right ventricle. The left ventricle has to propel blood through many more kilometers of blood vessels than does the right ventricle, and must therefore push against more resistance, even though both ventricles pump the same volume of blood.

Both sides of the heart contract at the same time. Contraction of the two atria, followed by contraction of the two ventricles and then relaxation, is the **cardiac cycle**. The cardiac cycle is divided into two phases: **systole** (pronounced sís-toll-ee), when the ventricles contract, and **diastole** (die-ás-toll-ee), when the ventricles relax (**Figure 50.3**). Just at the end of diastole, the atria contract and top off the volume of blood in the ventricles.

The sounds of the cardiac cycle, the "lub-dup" heard through a stethoscope placed on the chest or the back, are created by the heart valves slamming shut. The closing and opening of these valves are simple mechanical events resulting from pressure differences on the two sides of the valves. As the ventricles begin to contract, the pressure in them rises above the pressure in the atria, so the AV valves close ("lub"). When the ventricles be-

yourBioPortal.com
GO TO Animated Tutorial 50.1 • The Cardiac Cycle

**50.3 The Cardiac Cycle** The rhythmic contraction (systole) and relaxation (diastole) of the ventricles is called the cardiac cycle. The representation below shows pressure and volume changes during the cardiac cycle for the left ventricle only.

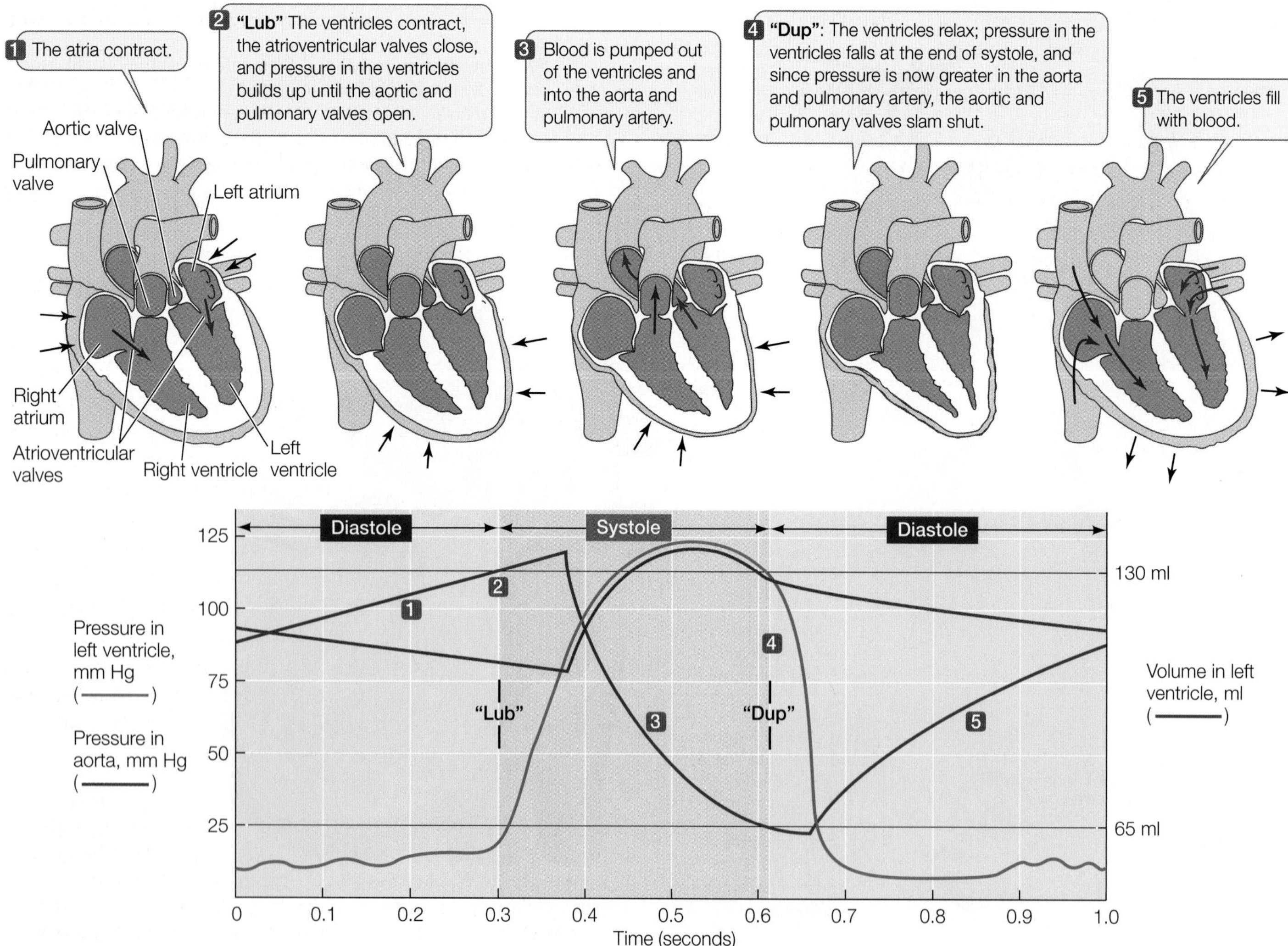
1 The atria contract.
2 "Lub" The ventricles contract, the atrioventricular valves close, and pressure in the ventricles builds up until the aortic and pulmonary valves open.
3 Blood is pumped out of the ventricles and into the aorta and pulmonary artery.
4 "Dup": The ventricles relax; pressure in the ventricles falls at the end of systole, and since pressure is now greater in the aorta and pulmonary artery, the aortic and pulmonary valves slam shut.
5 The ventricles fill with blood.
Aortic valve
Pulmonary valve
Left atrium
Right atrium
Atrioventricular valves
Right ventricle
Left ventricle
Diastole
Systole
Diastole
125
100
75
50
25
130 ml
65 ml
Pressure in left ventricle, mm Hg
Pressure in aorta, mm Hg
Volume in left ventricle, ml
"Lub"
"Dup"
0
0.1
0.2
0.3
0.4
0.5
0.6
0.7
0.8
0.9
1.0
Time (seconds)

gin to relax, the high pressure in the aorta and pulmonary artery closes the aortic and pulmonary valves ("dup").

Defective valves that do not close completely produce turbulent blood flow and the sounds known as *heart murmurs*. For example, if an AV valve does not close completely, blood will flow back into the atrium with a "whoosh" following the "lub."

You can feel the rhythm of the cardiac cycle in the pulsation of arteries such as the one that supplies blood to your hand. You've probably taken your pulse by placing two fingers from one hand lightly over the wrist of your other hand just below the thumb. During systole, the pressure wave created by the contraction of the left ventricle surges through the arteries of your arms; this pressure wave is what you feel as the pulsing of the artery in your wrist.

Blood pressure changes associated with the cardiac cycle can be measured in the large artery in your arm by using an inflatable pressure cuff and a pressure gauge, together called a *sphygmomanometer*, and a stethoscope (**Figure 50.4**). This method measures the minimum pressure necessary to compress an artery so blood does not flow through it at all (the systolic value) and the minimum pressure that causes intermittent flow through the artery (the diastolic value). A conventional blood pressure reading is expressed as the systolic value placed over the diastolic value. Healthy values for a young adult might be 120 millimeters of mercury (mm Hg) during systole and 70 mm Hg during diastole, or 120/70.

### The heartbeat originates in the cardiac muscle

Cardiac muscle (see Section 48.1) has unique adaptations that enable it to function as a pump. First, cardiac muscle cells are in electrical contact with one another through gap junctions, which enable *action potentials* (see Section 45.2) to spread rapidly from cell to cell. Because a spreading action potential stimulates contraction, large groups of cardiac muscle cells contract in unison. This coordinated contraction is essential for pumping blood effectively.

Second, some cardiac muscle cells are **pacemaker cells** that can initiate action potentials without stimulation from the nervous system. When they fire action potentials, they stimulate neighboring cells to contract. The primary pacemaker of the heart is a group of modified cardiac muscle cells, the **sinoatrial node**, located at the junction of the superior vena cava and right atrium (see Figure 50.6). The *resting membrane potentials* of these cells are less negative than those of other cardiac muscle cells and are not stable, but they gradually become even less negative until they reach threshold for initiating an action potential. The action potentials of pacemaker cells are very different from those of neurons and other muscle cells (see Figure 45.10). They are slower to rise; they are broader; and they are slower to return to resting potential. These properties of pacemaker cells are due to the ion channels in their membranes.

Pacemaker potentials involve $Na^+$, $Ca^{2+}$, and $K^+$ channels. As we discuss in Section 45.2, when $Na^+$ or $Ca^{2+}$ channels open, positive charges flow into the cell and the membrane potential becomes less negative. When $K^+$ channels open, positive charges flow out of the cell and the membrane potential becomes more negative. Because the $Na^+$ channels of pacemaker cells are more open than are those of other cardiac muscle cells, the pacemaker resting potential is less negative. The action potential of pacemaker cells is due to voltage-gated $Ca^{2+}$ channels rather than voltage-gated $Na^+$ channels as in neurons, skeletal muscle, and other cardiac muscle cells. These $Ca^{2+}$ ion channels open and close more slowly than voltage-gated $Na^+$ channels, explaining the shape of pacemaker action potentials.

The unstable resting potential of pacemaker cells is due to the behavior of cation channels. As in neurons and skeletal muscle cells, there are voltage-gated $K^+$ channels that open on the rising phase of the action potential. The opening of these channels allows $K^+$ ions to leave the cell and restores the negative charge on the cell membrane. The hyperpolarization (negative membrane potential) causes the opening of a unique class of voltage-gated cation channels that mostly conduct $Na^+$. At the same time, the voltage-gated $K^+$ channels that opened during the action potential are slowly closing. The result is that there are more $Na^+$ ions coming into the cell than there are $K^+$ ions leaving, and the cell membrane potential gradually becomes less negative.

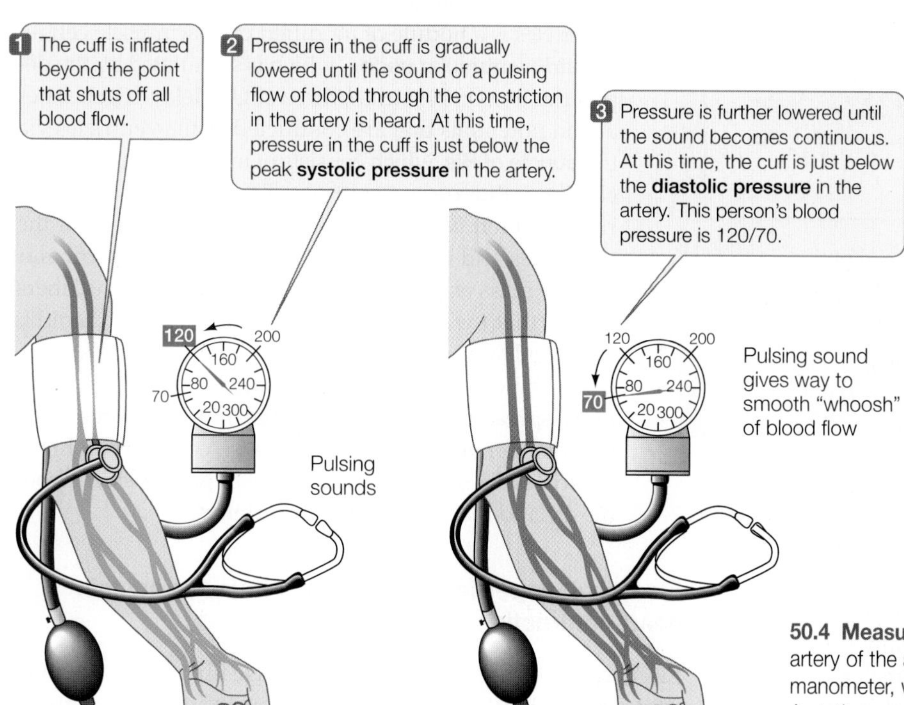

1 The cuff is inflated beyond the point that shuts off all blood flow.
2 Pressure in the cuff is gradually lowered until the sound of a pulsing flow of blood through the constriction in the artery is heard. At this time, pressure in the cuff is just below the peak **systolic pressure** in the artery.
3 Pressure is further lowered until the sound becomes continuous. At this time, the cuff is just below the **diastolic pressure** in the artery. This person's blood pressure is 120/70.
120
200
160
80
240
70
20
300
Pulsing sounds
Pulsing sound gives way to smooth "whoosh" of blood flow

**50.4 Measuring Blood Pressure** Blood pressure in the major artery of the arm can be measured with a device called a sphygmomanometer, which combines an inflatable cuff and a pressure gauge. A stethoscope is also used to detect sounds created by the blood vessels in response to changes in pressure during the cardiac cycle.

# INVESTIGATING LIFE

## 50.5 The Autonomic Nervous System Controls Heart Rate

The membrane potentials of pacemaker cells spontaneously depolarize until action potential threshold is reached. Neurotransmitter signals from the two divisions of the autonomic nervous system speed up and slow down the rate at which the pacemaker membrane potential drifts upward, thereby controlling the rate at which pacemaker cells fire action potentials.

**HYPOTHESIS** **The autonomic nervous system neurotransmitters norepinephrine and acetylcholine influence the membrane potentials of pacemaker cells.**

**METHOD**

1. Culture living sinoatrial node tissue in a dish. Insert an intracellular recording electrode into pacemaker cells.
2. Measure the membrane potential of pacemaker cells during a resting heartbeat (the control) and after applications of the ANS neurotransmitters norepinephrine (sympathetic) and acetylcholine (parasympathetic).

**RESULTS**

Control recording shows that the membrane potential of pacemaker cells gradually depolarizes after an action potential is fired.

When norepinephrine is applied, the rate of depolarization of the membrane potential increases. Time between action potentials decreases and the heart rate increases.

When acetylcholine is applied, the membrane potential is more negative following an action potential and the rate of depolarization is slower. Time between action potentials increases and the heart rate slows down.

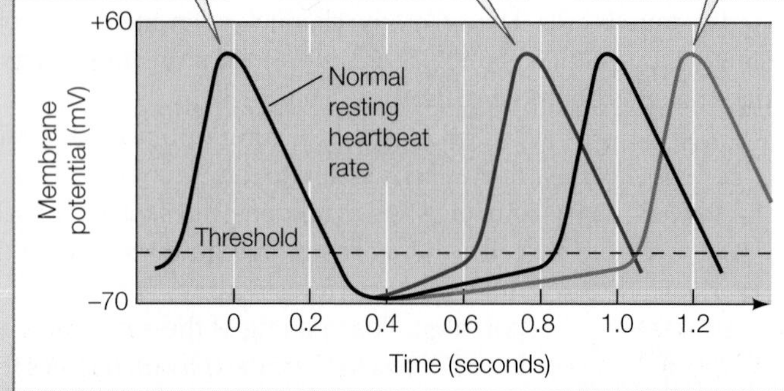


**CONCLUSION** **The ANS neurotransmitters norepinephrine and ACh control heart rate by altering the properties of the pacemaker cell membrane potentials.**

Go to **yourBioPortal.com** for original citations, discussions, and relevant links for all INVESTIGATING LIFE figures.

The gradual rise in membrane potential closes the channels that are allowing $Na^+$ to move into the cell, but as the membrane becomes less negative, some $Ca^{2+}$ channels open, causing the membrane to continue its gradual rise. Eventually this rising membrane potential reaches threshold for the major voltage-gated $Ca^{2+}$ channels, and another action potential is generated. The intricate interaction of these ion channels through their effects on membrane potential causes the rhythmic generation of action potentials that characterizes the pacemaker cells.

The nervous system controls the heartbeat (speeds it up or slows it down) by influencing the rate at which the pacemaker cell resting potentials drift upward (**Figure 50.5**). Norepinephrine released onto pacemaker cells by sympathetic nerves increases the permeability of the $Na^+$ channels and the $Ca^{2+}$ channels. The result is that the resting potential of the pacemaker cells drifts up more rapidly, the interval between action potentials is decreased, and the heart beats faster. Conversely, the parasympathetic neurotransmitter acetylcholine has the opposite effect. ACh increases the permeability of $K^+$ channels so that the membrane potential becomes even more negative following an action potential and rises more slowly. ACh also decreases the permeability of the $Ca^{2+}$ channels so that the rate of rise of the membrane potential slows, the interval between pacemaker action potentials lengthens, and the heart slows down.

### A conduction system coordinates the contraction of heart muscle

A normal heartbeat begins with an action potential in the sinoatrial node (**Figure 50.6**). This action potential spreads rapidly throughout the electrically coupled cells of the atria, causing them to contract in unison. Because there are no gap junctions between the cells of the atria and those of the ventricles, the action potential does not spread directly to the ventricles. Therefore, the ventricles do not contract in unison with the atria.

How does the action potential move from the atria to the ventricles? Situated at the junction of the atria and the ventricles is a nodule of modified cardiac muscle cells—the **atrioventricular node**—which is stimulated by the depolarization of the atria. With a slight delay, it generates action potentials that are conducted to the ventricles via the **bundle of His**, which consists of modified cardiac muscle fibers that do not contract. These fibers divide into right and left *bundle branches* that run to the tips of the ventricles and then spread throughout the ventricular muscle mass as **Purkinje fibers**. These conducting fibers ensure that the cardiac action potential spreads rapidly and evenly throughout the ventricular muscle mass, starting at the very bottom of the ventricles. The short delay in the spread of the action potential imposed by the atrioventricular node ensures that the atria contract before the ventricles do, so that the blood passes progressively from the atria to the ventricles to the arteries.

### Electrical properties of ventricular muscles sustain heart contraction

Electrical properties of ventricular muscle fibers allow them to contract for about 300 milliseconds—much longer than those of skeletal muscle fibers. As in neuronal and skeletal muscle

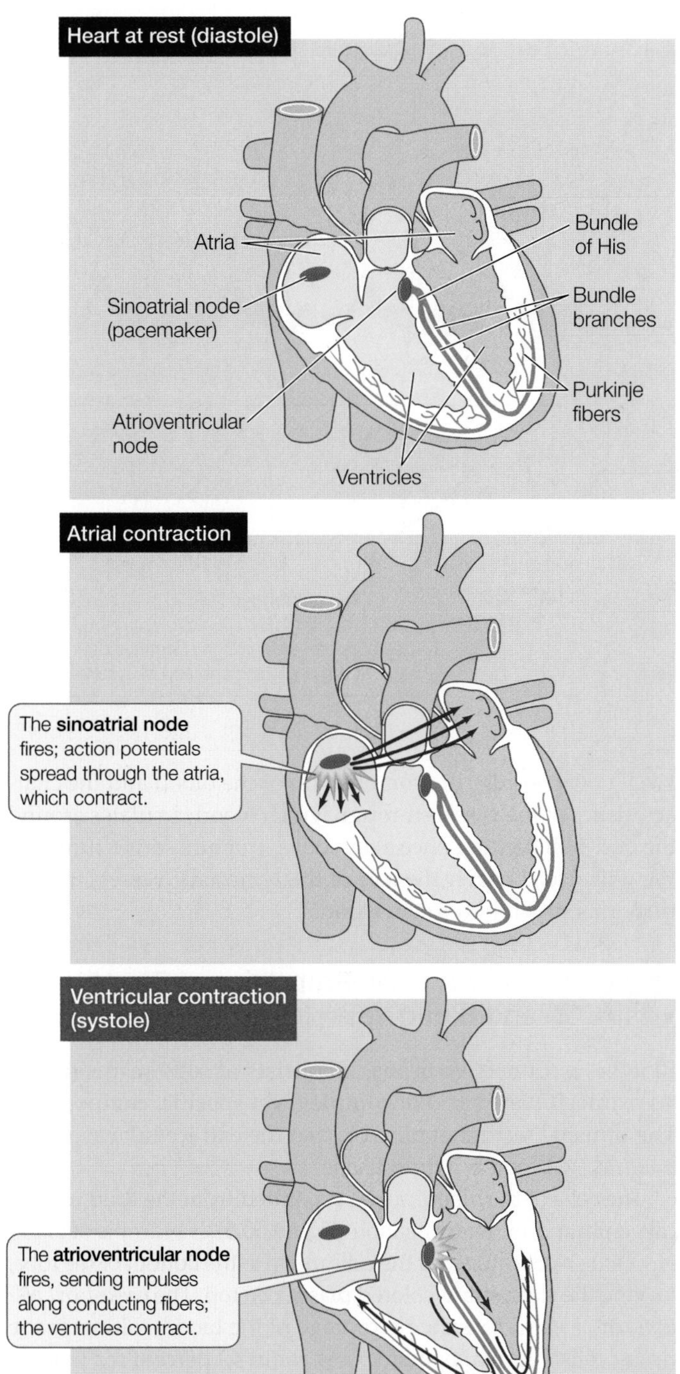
Heart at rest (diastole)
Atria
Sinoatrial node (pacemaker)
Atrioventricular node
Ventricles
Bundle of His
Bundle branches
Purkinje fibers
Atrial contraction
The **sinoatrial node** fires; action potentials spread through the atria, which contract.
Ventricular contraction (systole)
The **atrioventricular node** fires, sending impulses along conducting fibers; the ventricles contract.

**50.6 The Heartbeat** Pacemaker cells in the sinoatrial node initiate the heartbeat by firing action potentials that spread through the electrically coupled atrial muscle. The atrial action potential eventually spreads to the atrioventricular node that, with a delay, conducts it through the Bundle of His and Purkinje fibers to the cells of the ventricles.

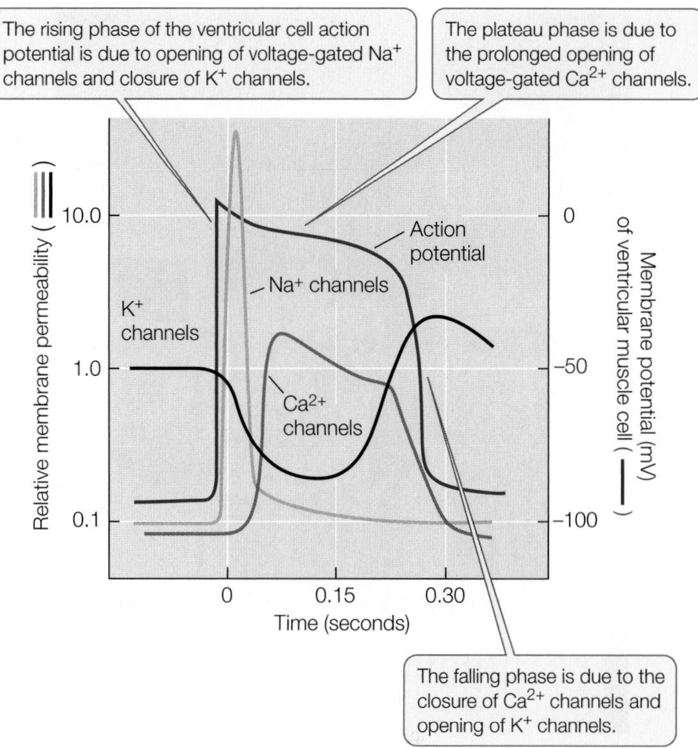
The rising phase of the ventricular cell action potential is due to opening of voltage-gated $Na^+$ channels and closure of $K^+$ channels.
The plateau phase is due to the prolonged opening of voltage-gated $Ca^{2+}$ channels.
Relative membrane permeability
10.0
1.0
0.1
$K^+$ channels
$Na^+$ channels
$Ca^{2+}$ channels
Action potential
0
-50
-100
Membrane potential (mV) of ventricular muscle cell
0
0.15
0.30
Time (seconds)
The falling phase is due to the closure of $Ca^{2+}$ channels and opening of $K^+$ channels.

**50.7 The Action Potential of Ventricular Muscle Fibers** The rising phase of the action potential of ventricular muscle fibers (red) is due to the opening of voltage-gated $Na^+$ channels (gold). However, the membrane remains in a depolarized state for a prolonged time because of the opening of voltage-gated $K^+$ channels (black).

action potentials, the rising phase of the ventricular muscle cell action potentials is due to the opening of voltage-gated $Na^+$ channels. Unlike neurons and skeletal muscle fibers, however, ventricular muscle cells remain depolarized for a long time. This extended plateau of the action potential is due to sustained opening of voltage-gated $Ca^{2+}$ channels (**Figure 50.7**). Like other muscle, cardiac muscle is stimulated to contract when $Ca^{2+}$ is available to bind with troponin (see Figure 48.6). As long as $Ca^{2+}$ remains in the sarcoplasm, the ventricular muscle cells continue to contract.

The drug digitalis has been used since the late 1700s to treat weakened hearts or hearts with irregular patterns of contraction. Digitalis strengthens and slows the heartbeat by slowing the reuptake of $Ca^{2+}$ by the sarcoplasmic reticulum and thereby increasing the concentration of $Ca^{2+}$ in the sarcoplasm. Before being introduced into the practice of medicine, digitalis prepared from the purple foxglove plant, *Digitalis purpurea,* was a folk remedy for heart problems.

### The ECG records the electrical activity of the heart

Electrical events in the cardiac muscle during the cardiac cycle can be recorded by electrodes placed on the surface of the body. Such a recording is an **electrocardiogram**, or **ECG**. **EKG** is also used because German physicians who invented the method used the Greek word for heart and called it the *electrokardiogramm.* The ECG is an important tool for diagnosing heart problems (**Figure 50.8A**).

The action potentials that sweep through the muscles of the atria and ventricles before they contract are such massive, localized electrical events that they cause electric currents to flow throughout the body. Electrodes placed at different locations on

(A)

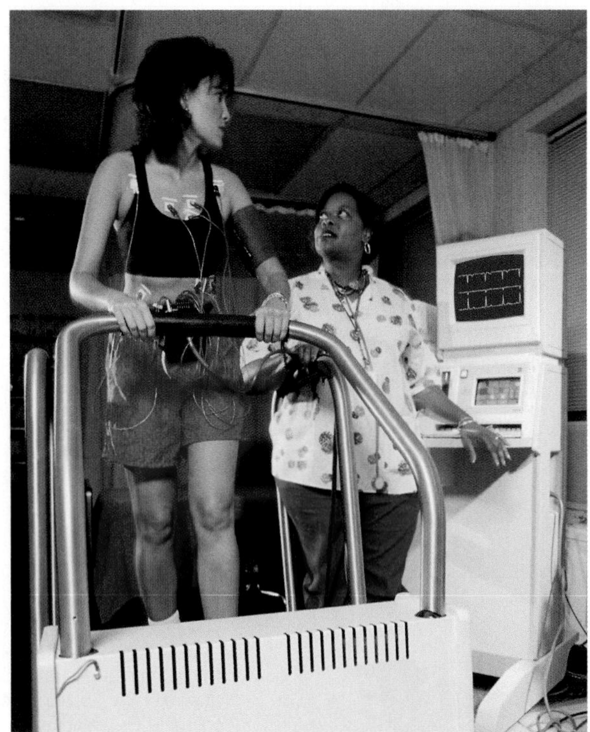

(B)

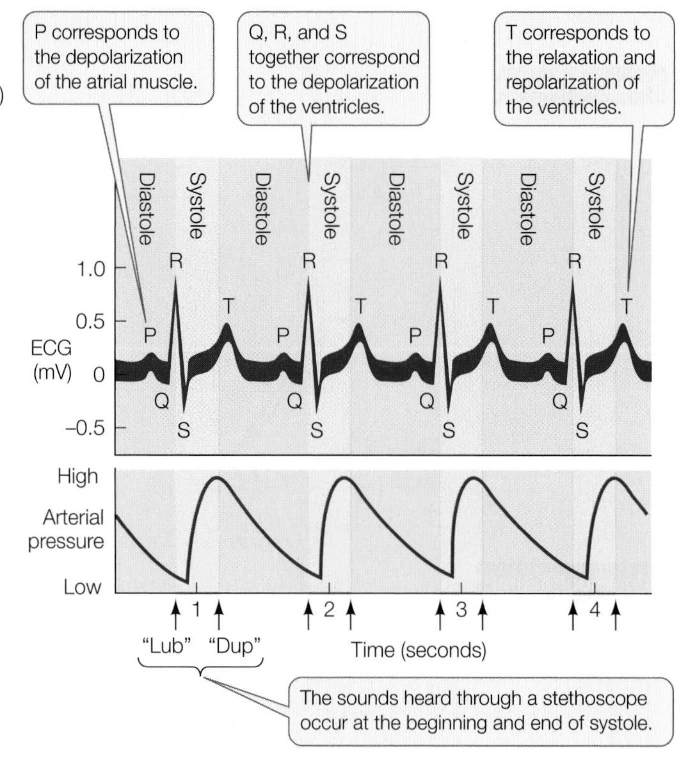
P corresponds to the depolarization of the atrial muscle.
Q, R, and S together correspond to the depolarization of the ventricles.
T corresponds to the relaxation and repolarization of the ventricles.
Diastole
Systole
ECG (mV)
1.0
0.5
0
–0.5
P
Q
R
S
T
High
Arterial pressure
Low
1
2
3
4
"Lub" "Dup"
Time (seconds)
The sounds heard through a stethoscope occur at the beginning and end of systole.

**50.8 The Electrocardiogram** (A) An electrocardiogram (abbreviated as ECG or EKG) is used to monitor heart function. Electrodes attached to the person on the treadmill record an ECG that is amplified and displayed on a monitor. (B) Variations from the normal pattern shown here can be used to diagnose heart problems.

the skin detect those currents at different times and register a voltage difference between them. The appearance of the ECG depends on the placement of the electrodes. Electrodes placed on the right wrist and left ankle produced the normal ECG shown in **Figure 50.8B**. The wave patterns of the ECG are designated P, Q, R, S, and T, each letter representing a particular event in the cardiac muscle, as shown in the figure.

## 50.3 RECAP

**The mammalian heart has two atria and two ventricles. Modified cardiac muscle tissue in the right atrium functions to spontaneously generate pacemaker action potentials. Other modified cardiac muscle tissue between the atria and ventricles and throughout the ventricles conducts those signals and coordinates the heart contraction.**

- Trace the path of blood through both sides of the heart, naming the major blood vessels and heart valves. See pp. 1051–1052 and Figure 50.2
- Differentiate systole and diastole and describe the events of the cardiac cycle. See p. 1052 and Figure 50.3
- How do cells of the sinoatrial node generate the heartbeat? See pp. 1053–1054 and Figures 50.5 and 50.6

We'll now consider the composition of the blood and the characteristics of the vessels through which blood circulates around the body, illustrating once again how structure serves function. We will also consider the role of the lymphatic vessels that return interstitial fluid to the blood.

# 50.4 What Are the Properties of Blood and Blood Vessels?

Blood is a connective tissue. It consists of cells suspended in an extracellular matrix of complex, yet specific, composition. The unusual feature of blood is that the extracellular matrix is a liquid, so blood is a fluid tissue.

The cells of the blood can be separated from the fluid matrix, called **plasma**, by centrifugation (**Figure 50.9**). If a sample of blood is spun in a centrifuge, all the cells move to the bottom of the tube, leaving the clear, straw-colored plasma on top. The *packed-cell volume*, or *hematocrit*, is the percentage of the blood volume made up by cells. Normal hematocrit is about 42 percent for women and 46 percent for men, but these values can vary considerably. They are usually higher, for example, in people who live and work at high altitudes, because the low oxygen concentrations there stimulate the production of more red blood cells.

Here we will consider two elements in blood: the *red blood cells* and the *platelets*, which are pinched-off fragments of cells. *White blood cells*, or *leukocytes*, are the cells of the immune system, which we discussed in Chapter 42.

### Red blood cells transport respiratory gases

Most of the cells in the blood are **erythrocytes**, or red blood cells. Mature red blood cells are biconcave, flexible discs packed with hemoglobin. Their function is to transport respiratory gases. Their shape gives them a large surface area for gas exchange,

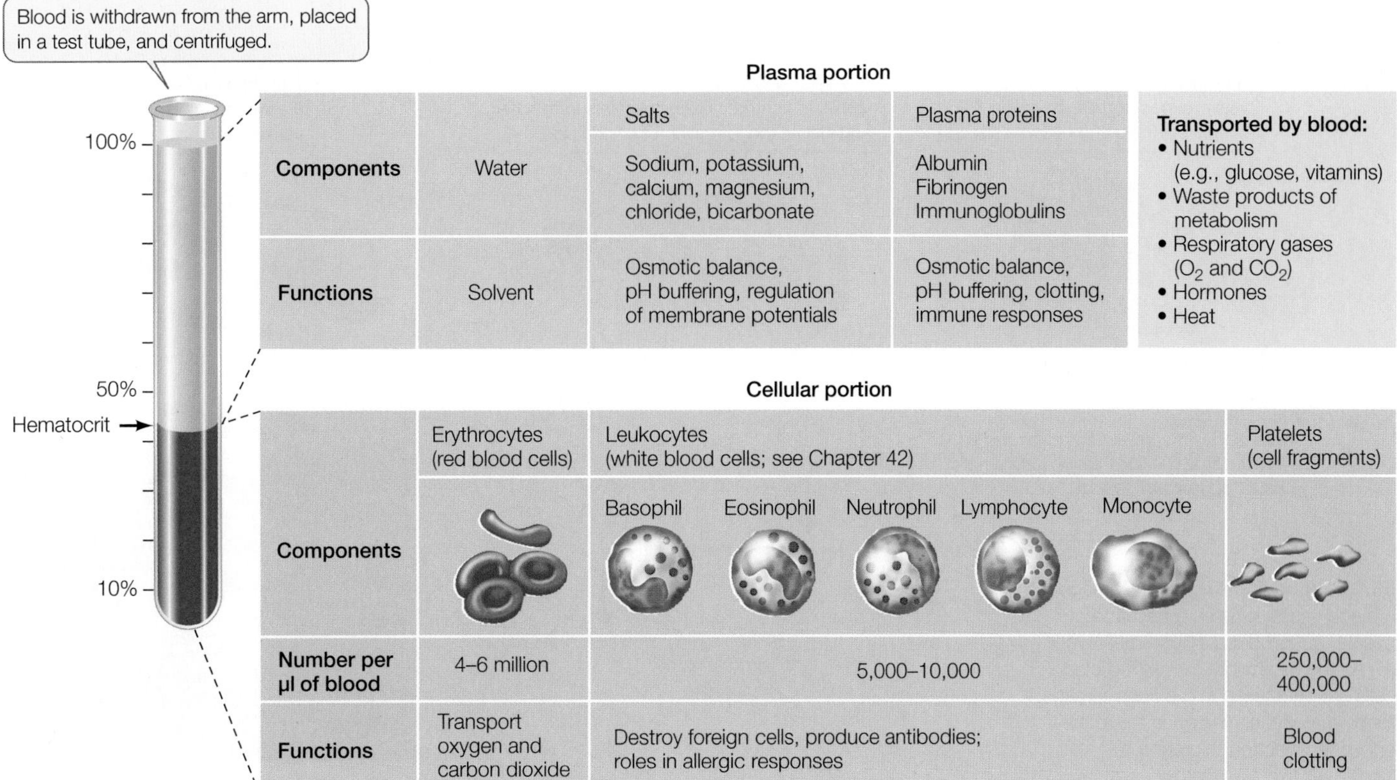

Blood is withdrawn from the arm, placed in a test tube, and centrifuged.
100%
50%
Hematocrit
10%
Plasma portion
Salts
Plasma proteins
Components
Water
Sodium, potassium, calcium, magnesium, chloride, bicarbonate
Albumin
Fibrinogen
Immunoglobulins
Functions
Solvent
Osmotic balance, pH buffering, regulation of membrane potentials
Osmotic balance, pH buffering, clotting, immune responses
Transported by blood:
• Nutrients (e.g., glucose, vitamins)
• Waste products of metabolism
• Respiratory gases ($O_2$ and $CO_2$)
• Hormones
• Heat
Cellular portion
Erythrocytes (red blood cells)
Leukocytes (white blood cells; see Chapter 42)
Platelets (cell fragments)
Basophil
Eosinophil
Neutrophil
Lymphocyte
Monocyte
Components
Number per µl of blood
4–6 million
5,000–10,000
250,000–400,000
Functions
Transport oxygen and carbon dioxide
Destroy foreign cells, produce antibodies; roles in allergic responses
Blood clotting


**50.9 The Composition of Blood** Blood consists of a complex aqueous solution (the plasma) and numerous cell types and cell fragments. The hematocrit (arrow) is a measure of the cellular portion as a percentage of total blood volume.

and their flexibility enables them to squeeze through narrow capillaries. Men have 4.5 to 6.0 million red blood cells per microliter of blood, and women have 3.5 to 5.0 million.

Red blood cells, as well as all the other cellular components of blood, are generated by stem cells in the bone marrow, particularly in the ribs, breastbone, pelvis, vertebrae, and the long bones of the limbs. Red blood cell production is controlled by a hormone, **erythropoietin**, which is released by cells in the kidneys in response to insufficient oxygen—**hypoxia**. Many tissues respond to hypoxia by expressing a transcription factor called *hypoxia-inducible factor 1* (*HIF-1*). When the kidneys become hypoxic and express HIF-1, one of the actions of the transcription factor is to activate the gene encoding erythropoietin. Increased circulating erythropoietin extends the lives of mature red blood cells and stimulates production of new red blood cells in the bone marrow.

Under normal conditions, your bone marrow produces about 2 million red blood cells every second. Developing, immature red blood cells divide many times while still in the bone marrow, and during this time they produce hemoglobin. When the hemoglobin content of a red blood cell approaches about 30 percent, the nucleus, endoplasmic reticulum, Golgi apparatus, and mitochondria of the cell begin to break down. This process is almost complete when the new red blood cell squeezes between the endothelial cells of blood vessels in the bone marrow and enters the circulation. Loss of nuclei from the red blood cells occurs in most mammalian species, but the red cells of some mammals and of all other vertebrates are nucleated.

Each red blood cell circulates for about 120 days. As it gets older, its membrane becomes less flexible and more fragile. Therefore, old red blood cells can rupture as they bend to fit through narrow capillaries. One place where they are really squeezed is in the **spleen**, an organ that sits near the stomach in the upper left side of the abdominal cavity. The spleen has many venous cavities, or sinuses, that serve as a reservoir for red blood cells, but to get into the sinuses, the red blood cells must squeeze between spleen cells. When old red blood cells are ruptured by this squeezing, their remnants are taken up and degraded by macrophages (a class of white blood cells that ingest debris and foreign materials).

### Platelets are essential for blood clotting

Besides producing erythrocytes and leukocytes, the bone marrow stem cells described in Section 42.1 also produce cells called *megakaryocytes*. Megakaryocytes are large cells that remain in the bone marrow and release cell fragments called **platelets** into the circulation. A platelet is just a tiny fragment of a cell without cell organelles, but it is packed with enzymes and chemicals necessary for its function: sealing leaks in blood vessels and initiating **blood clotting** (**Figure 50.10**).

Damage to a blood vessel exposes collagen fibers. An encounter with collagen fibers activates a platelet. The platelet swells, becomes irregularly shaped and sticky, and releases chemicals that activate other platelets and initiate the clotting of blood. The sticky platelets also form a plug at the damaged site.

Blood clotting requires many steps and many clotting factors, most of which are circulating in the blood in an inactive form. The absence of any one of these proteins can impair clotting and cause excessive bleeding. Because the liver produces

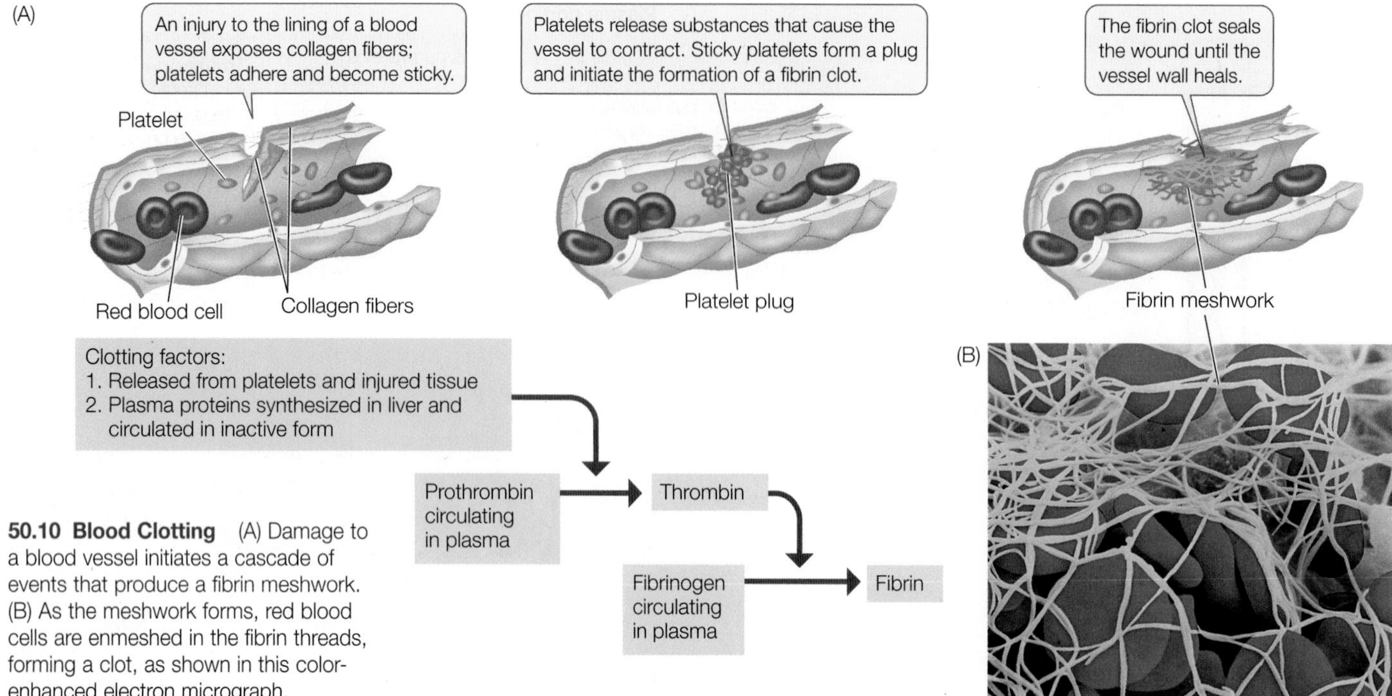
(A)
An injury to the lining of a blood vessel exposes collagen fibers; platelets adhere and become sticky.
Platelets release substances that cause the vessel to contract. Sticky platelets form a plug and initiate the formation of a fibrin clot.
The fibrin clot seals the wound until the vessel wall heals.
Platelet
Red blood cell
Collagen fibers
Platelet plug
Fibrin meshwork
Clotting factors:
1. Released from platelets and injured tissue
2. Plasma proteins synthesized in liver and circulated in inactive form
Prothrombin circulating in plasma
Thrombin
Fibrinogen circulating in plasma
Fibrin
(B)

**50.10 Blood Clotting** (A) Damage to a blood vessel initiates a cascade of events that produce a fibrin meshwork. (B) As the meshwork forms, red blood cells are enmeshed in the fibrin threads, forming a clot, as shown in this color-enhanced electron micrograph.

most of the clotting factors, liver diseases such as hepatitis and cirrhosis can result in excessive bleeding. People with hemophilia experience uncontrolled bleeding because of a genetic inability to produce one of the clotting factors.

Blood clotting factors participate in a cascade of chemical activations of other substances circulating in the blood. The cascade begins with blood vessel and other tissue damage that exposes the blood to proteins such as collagen that are normally separated from the blood by endothelial cells lining the blood vessels. This exposure activates platelets and begins the clotting factor cascade. The end result of this cascade is to convert an inactive circulating enzyme, **prothrombin**, to its active form, **thrombin**. Thrombin cleaves molecules of **fibrinogen**, a plasma protein, forming insoluble threads of **fibrin**. The fibrin threads form the meshwork that binds platelets, seals the vessel, and provides a scaffold for the formation of scar tissue (see Figure 50.10).

### Blood circulates throughout the body in a system of blood vessels

As we mentioned in Section 50.1, blood circulates through the vertebrate body in a system of closed vessels. Blood leaves the heart in arteries and is distributed throughout the body's tissues in arterioles, which feed capillary beds. Exchanges of nutrients, wastes, respiratory gases, and hormones occur in the capillaries. Blood leaving capillary beds collects in venules, which empty into veins that conduct the blood back to the heart.

The walls of the large arteries have many extracellular collagen and elastin fibers, which enable them to withstand the high blood pressures generated by the heart (**Figure 50.11A**). These elastic tissues have another important function: they are stretched during systole, and thereby store some of the energy imparted to the blood by the heart. Elastic recoil during diastole returns this energy to the blood by squeezing it and pushing it forward. As a result, even though pressure in the arteries pulsates with the beating of the heart, the flow of blood is smoother than it would be through a system of rigid pipes.

Smooth muscle cells in the walls of the arteries and arterioles constrict or dilate those vessels. When the diameter of the vessels changes, their resistance to blood flow also changes, and the amount of blood flowing through them changes as a result. Neural and hormonal mechanisms act on smooth muscle cells in the walls of the arteries and arterioles, controlling the flow of blood through these vessels. The arterioles are referred to as *resistance vessels* because their resistance can vary to control the blood flow to specific tissues.

### Materials are exchanged in capillary beds by filtration, osmosis, and diffusion

Beds of capillaries lie between arterioles and venules (**Figure 50.11B**). Few cells are more than a few cell diameters away from a capillary. (Notable exceptions include developing oocytes and the cells of the lens and cornea.) The cells' needs are served by the exchange of materials between blood and interstitial fluid across the capillary walls. This exchange is facilitated by the capillaries' thin, permeable walls, as well as by the slow flow of blood through the capillaries.

It may seem strange that blood flows through the large arteries rapidly at high pressures, but when it reaches the small capillaries, the pressure and rate of flow decrease (**Figure 50.11C**). When you restrict the diameter of a garden hose by placing your thumb over the opening, the pressure in the hose increases, which in turn increases the velocity of the water spraying out of the hose. But keep in mind that the arteries branch into many arterioles, which give rise to a huge number of capillaries. Even

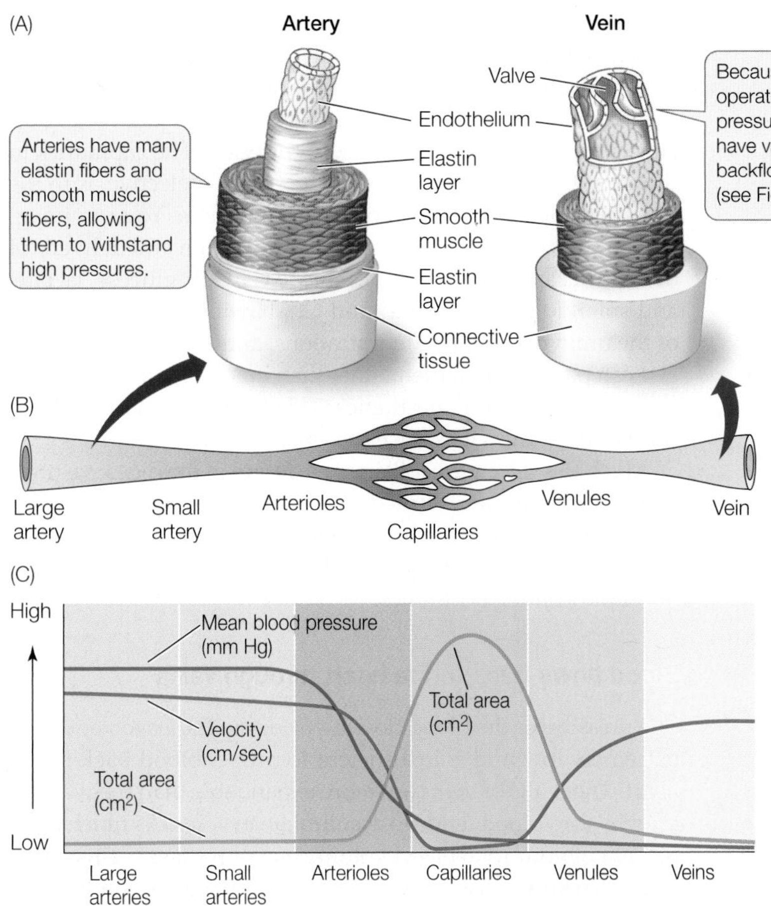


**50.11 Anatomy of Blood Vessels** (A) The different anatomical characteristics of arteries and veins match their functions. (B) Blood from the arterial system feeds into capillary beds, where exchanges with the interstitial fluid occur. The venous system returns the blood to the heart. (C) The area encompassed by each vessel type is graphed along with the pressure and velocity of the blood within them.

yourBioPortal.com
GO TO Web Activity 50.3 • Structure of a Blood Vessel

though each capillary has a diameter so small that red blood cells must pass through in single file, there are so many capillaries that their total cross-sectional area is much greater than that of any other class of vessels. As a result, all the capillaries together have a much greater capacity for blood than do the arterioles. Returning to our garden hose analogy, if we connected the hose to an increasing number of lawn sprinklers, eventually the pressure and flow in each sprinkler would be low.

Capillary walls consist of a single layer of thin endothelial cells (**Figure 50.12**). In most tissues of the body other than the brain, capillaries have tiny holes, or *fenestrations* (*fenestra*, "windows"). Capillaries are permeable to water, some ions, and some small molecules but not to large molecules such as proteins. At the arterial (high pressure) end, blood pressure squeezes water and some small solutes out of the capillaries and into the surrounding intercellular spaces. Why don't water and small-molecular-weight solutes collect in the intercellular spaces? How is the blood volume maintained if fluid is continuously leaking out of the capillaries?

An answer to this question was put forth more than 100 years ago by the physiologist E. H. Starling. Starling suggested that water movement across capillary walls is a result of two opposing forces, which are now known as **Starling's forces**:

- Blood pressure squeezes water and small solutes out of the capillaries.
- Osmotic pressure pulls water back into the capillaries.

Blood pressure is high at the arterial end of a capillary bed and steadily drops as blood moves towards the venous end (**Figure 50.13**). The osmotic pressure is due to the large protein molecules that cannot leave the capillaries, and it is relatively constant along the capillaries. As long as the blood pressure is above the osmotic pressure, fluid leaves the capillaries. At the venule end of most capillaries, blood pressure falls below the osmotic pressure, so fluid returns to the capillaries. The actual numbers for a normal capillary bed in a resting person suggest that there would be a *slight* net loss of fluid to the intercellular spaces. This loss, about 4 liters per day, percolates between cells as the interstitial fluid before it returns to the venous blood via the lymphatic system, which we will discuss later in this chapter.

Several observations supported Starling's model. In people with severe liver disease or protein starvation, a fall in blood protein concentration leads to an accumulation of fluid in the extracellular spaces, which results in tissue swelling, or *edema*. Edema is also characteris-

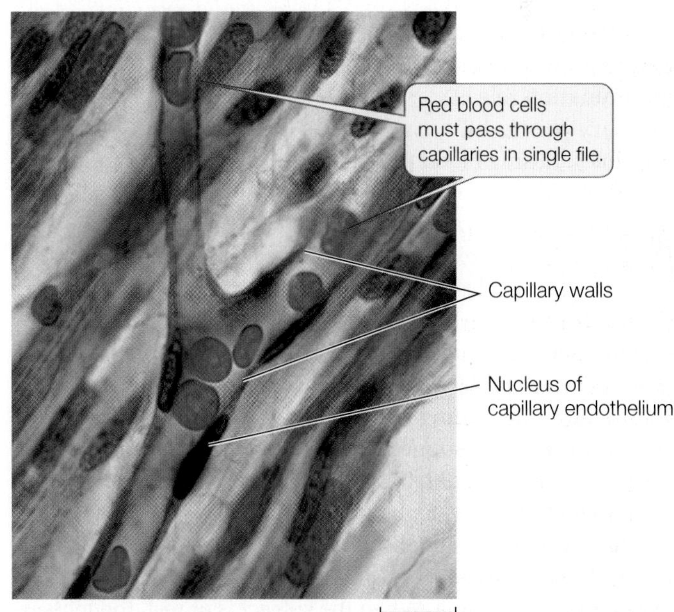


**50.12 A Narrow Lane** Capillaries have a very small diameter, and blood flows through them slowly.

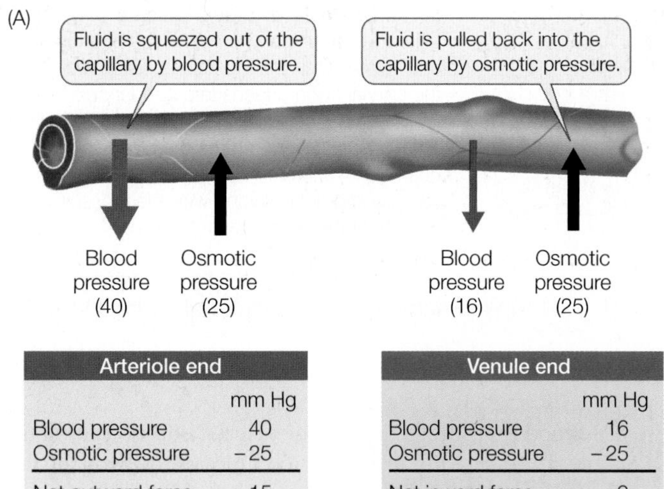


| Arteriole end | mm Hg |
|---|---|
| Blood pressure | 40 |
| Osmotic pressure | –25 |
| Net outward force | 15 |

| Venule end | mm Hg |
|---|---|
| Blood pressure | 16 |
| Osmotic pressure | –25 |
| Net inward force | –9 |

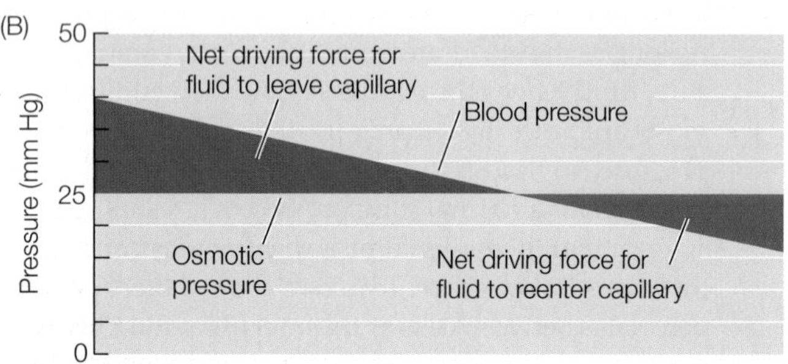


**50.13 Starling's Forces** Starling's model explains how blood volume is maintained in the capillary beds. (A) When blood pressure is greater than osmotic pressure, fluid leaves the capillaries; when blood pressure falls below osmotic pressure, fluid returns to the capillaries. (B) The balance of these two forces changes over the capillary bed as blood pressure falls.

tic of the inflammation response accompanying tissue damage or allergic responses (see Figure 42.3). *Histamine*, a mediator of inflammation released by certain white blood cells, increases capillary permeability and relaxes the smooth muscles of the arterioles, raising blood pressure in the capillaries and leading to fluid leakage into tissues.

A few situations are not explained by Starling's hypothesis. During strenuous exercise, the blood pressure in the arterioles serving the muscles rises substantially but does not result in edema. In birds, the blood pressure in arterioles is much higher than in mammals, and the osmotic pressure is lower. If edema is not a chronic problem in exercising muscles and in birds, what is missing from Starling's model?

Recent research suggests that bicarbonate ions ($HCO_3^-$) in the blood plasma contribute significantly to the osmotic attraction that draws water back into the capillaries. The $CO_2$ produced by cellular metabolism diffuses into the endothelial cells lining the capillaries, where it is converted into $HCO_3^-$ and released into the plasma. When the subject is at rest, the increasing $HCO_3^-$ concentration can cause the osmotic pressure of the blood at the venous end to be 30 mm Hg higher than at the arterial end, and during strenuous exercise this difference can be much higher. Thus it appears that $CO_2$ and $HCO_3^-$ are major factors that pull water back into the capillaries.

Lipid-soluble substances and many small solute molecules can easily pass through capillary walls from an area of higher concentration to one of lower concentration. The capillaries in different tissues, however, are differentially selective as to the sizes of molecules that can pass through them. You can imagine that this is an important issue in the design and delivery of drugs. All capillaries are permeable to $O_2$, $CO_2$, glucose, lactate, and small ions such as $Na^+$ and $Cl^-$. However, the capillaries of the brain do not have fenestrations, and therefore not much else can pass through them other than lipid-soluble substances, such as alcohol and anesthetics. This high selectivity of brain capillaries is known as the *blood–brain barrier.*

Much less selective capillaries are found in the digestive tract, where nutrients are absorbed, and in the kidneys, where wastes are filtered. Even in the brain there are specific regions where the capillaries are more permeable, enabling the brain to detect non-lipid soluble hormones.

### Blood flows back to the heart through veins

The pressure of the blood flowing from capillaries to venules is extremely low and is insufficient to propel blood back to the heart. The walls of veins are more expandable than the walls of arteries, and blood tends to accumulate in veins. As much as 60 percent of your total blood volume may be in your veins when you are resting. Because of their high capacity to stretch and store blood, veins are called *capacitance vessels*.

Blood flow through veins that are above the level of the heart is assisted by gravity. Below the level of the heart, however, venous return is against gravity. The most important force propelling blood from these regions is the squeezing of the veins by the contractions of surrounding skeletal muscles. As muscles contract, the vessels are compressed and blood is squeezed through them. Blood flow may be temporarily obstructed during a prolonged muscle contraction, but when muscles relax, blood is free to move again. One-way valves in the veins of the extremities prevent backflow of blood. Thus, whenever a vein is squeezed, blood is propelled forward toward the heart (**Figure 50.14**).

In a resting person, gravity causes blood accumulation in the veins of the lower body and exerts backpressure on the capillary beds. This backpressure shifts the balance between blood pressure and osmotic pressure, causing increased loss of fluid to the intercellular spaces. That is why your feet swell during a long airline flight.

Because of the one-way valves in the veins of the legs, the contractions of leg muscles act as auxiliary vascular pumps when an animal walks or runs and facilitate the return of blood to the heart from the veins of the lower body. As a greater volume of blood is returned to the heart, the heart contracts more forcefully and its pumping action is enhanced. The heartbeat gets stronger because of a property of cardiac muscle cells described by the **Frank–Starling law**: if the cardiac muscle cells are stretched, as they are when the volume of returning blood increases, they contract more forcefully.

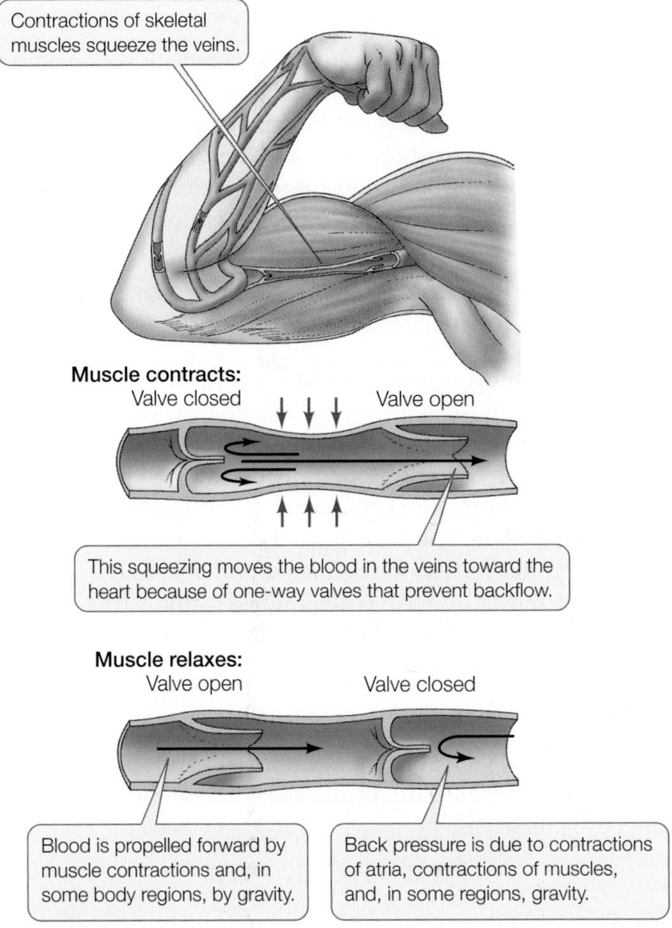

Contractions of skeletal muscles squeeze the veins.
Muscle contracts:
Valve closed
Valve open
This squeezing moves the blood in the veins toward the heart because of one-way valves that prevent backflow.
Muscle relaxes:
Valve open
Valve closed
Blood is propelled forward by muscle contractions and, in some body regions, by gravity.
Back pressure is due to contractions of atria, contractions of muscles, and, in some regions, gravity.

**50.14 One-Way Flow** Veins have valves that prevent blood from flowing backward, and contractions of skeletal muscle help move blood toward the heart.

The actions of breathing also help return venous blood to the heart. The muscles involved in inhalation create negative pressure that pulls air into the lungs (see Figure 49.11), and this negative pressure also pulls blood toward the chest, increasing venous return to the right atrium. In addition, some of the largest veins closest to the heart contain smooth muscle that contracts at the onset of exercise. Contraction of veins can rapidly increase venous return and stimulate the heart in accord with the Frank–Starling law, increasing cardiac output.

### Lymphatic vessels return interstitial fluid to the blood

The interstitial fluid contains water and small molecules, but no red blood cells, and less protein than found in plasma. A separate system of vessels—the **lymphatic system**—returns interstitial fluid to the blood. Each capillary bed contains at least one blind-ended lymph capillary.

Once it enters the lymphatic vessels, the interstitial fluid is called **lymph**. Fine lymphatic capillaries merge into progressively larger vessels and ultimately into two lymphatic vessels—the **thoracic ducts**—that empty into large veins at the base of the neck (see Figure 42.1). The left thoracic duct carries most of the lymph from the lower part of the body and is much larger than the right thoracic duct. Lymph, like blood, is propelled toward the heart by skeletal muscle contractions and breathing movements, and lymphatic vessels, like veins, have one-way valves that keep the lymph flowing toward the thoracic duct.

Mammals and birds have **lymph nodes** along the major lymphatic vessels. Lymph nodes are a major site of lymphocyte production and of the phagocytic action that removes microorganisms and other foreign materials from the circulation (see Section 42.1).

### Vascular disease is a killer

As mentioned at the start of this chapter, cardiovascular disease is responsible for about one-fourth of all deaths each year in the United States, and the same is true for Europe. The immediate cause of most of these deaths is not a defect in heart muscle, as it is in athletes. Instead, the cause is mostly heart attack or stroke, which are usually the end result of a disease called **atherosclerosis** ("hardening of the arteries") that begins many years before symptoms are detected.

Healthy arteries have a smooth internal lining of endothelial cells (**Figure 50.15A**) that can be damaged by chronic high blood pressure, smoking, a high-fat diet, or microorganisms. Deposits called **plaque** begin to form at sites of endothelial damage. First, the damaged endothelial cells attract certain white blood cells to the site. These cells are then joined by smooth muscle cells migrating from the deeper layers of the arterial wall. Lipids, especially cholesterol, are deposited in these cells, so that the developing plaque becomes fatty. Fibrous connective tissue made by the invading smooth muscle cells in the plaque, along with

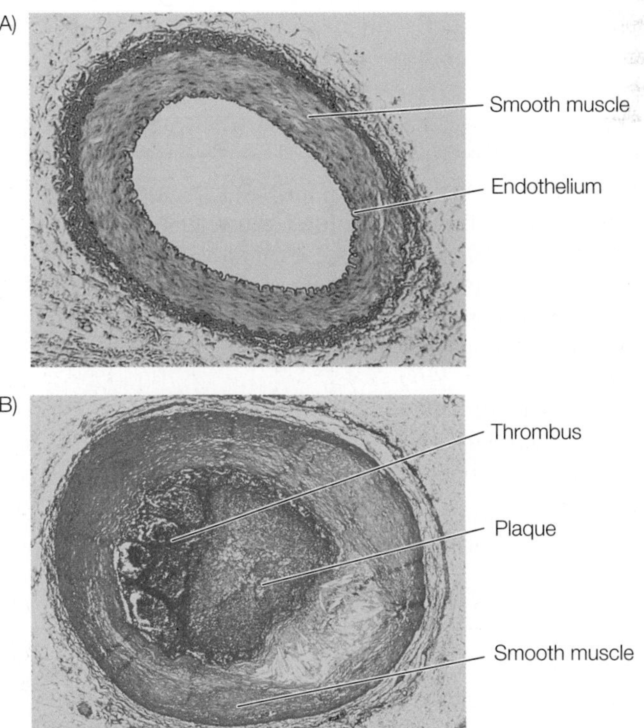


**50.15 Atherosclerotic Plaque** (A) A healthy, clear artery. (B) An atherosclerotic artery, clogged with plaque and a thrombus.

deposits of calcium, make the artery wall less elastic—hence, "hardening of the arteries." The growing plaque deposit narrows the artery and causes turbulence in the blood flow. Blood platelets stick to the plaque (see Figure 50.10) and initiate formation of an intravascular blood clot, a **thrombus**, which can block the artery (**Figure 50.15B**).

The blood supply to the heart muscle flows through the **coronary arteries** which are highly susceptible to atherosclerosis. As these arteries narrow, blood flow to the heart muscle decreases, causing the symptoms of chest pain and shortness of breath during mild exertion. A person with atherosclerosis is at high risk of forming a thrombus in a coronary artery. This condition, called **coronary thrombosis**, can totally block the vessel, causing a *heart attack*, or **myocardial infarction**.

A piece of a thrombus that breaks loose, called an **embolus**, is likely to travel to and become lodged in a vessel of smaller diameter, blocking its flow (an **embolism**). Arteries already narrowed by plaque formation are likely places for an embolism. An embolism in an artery in the brain causes the cells fed by that artery to die. This event is a **stroke**. The specific damage resulting from a stroke, such as memory loss, speech impairment, or paralysis, depends on the location of the blocked artery.

Probably the most important determinants of whether you will get atherosclerosis are your genetic predisposition and your age. Environmental risk factors play a large role, however. If you do have a genetic predisposition to atherosclerosis, it is even more important to minimize environmental risk factors. These include high-fat and high-cholesterol diets, smoking, and a sedentary lifestyle. Certain untreated medical conditions such as *hypertension* (high blood pressure), obesity, and diabetes are also risk factors for atherosclerosis. Changes in diet and behavior and treatment of predisposing medical conditions can prevent and reverse early atherosclerosis and help fend off this silent killer.

## 50.4 RECAP

**Blood is a fluid tissue with cellular components that play roles in transport of respiratory gases, immune system function, and blood clotting. Blood is distributed throughout the body in a system of vessels. Exchanges between the blood and interstitial fluids occur in the smallest of those vessels, the capillaries.**

- How are the structural differences between the various classes of vessels related to their functions? See pp. 1058–1061 and Figure 50.11
- Why are arterioles called resistance vessels and veins called capacitance vessels? See pp. 1058 and 1060
- What factors control the movement of fluids between the vascular and extravascular spaces? See pp. 1059–1060 and Figure 50.13
- What propels blood from the lower part of the body back to the heart? See pp. 1060–1061 and Figure 50.14

Every tissue in the body requires an adequate flow of oxygen-saturated blood that carries essential nutrients and is relatively free of waste products. Blood flow depends on the maintenance of an appropriate blood pressure, and the distribution of blood flow throughout the body depends on control of the resistance in the blood vessels supplying different tissues.

# 50.5 How Is the Circulatory System Controlled and Regulated?

When we investigate how a physiological process is regulated, we start by identifying the critical components of that process, how they can be controlled, and the information used to govern that control. Because blood flow depends on pressure, we can identify the pressure in the aorta as a critical variable of the circulatory system. The pressure in the aorta oscillates between systole and diastole, so we define our variable as the *mean arterial pressure* (MAP). MAP is determined by the cardiac output (CO) and the resistance to flow in the blood vessels, or *total peripheral resistance* (TPR):

$$\text{MAP} = \text{CO} \times \text{TPR}$$

Since CO is a function of the heart rate (HR) and stroke volume (SV), however, the critical relationships we have to understand can be expressed as:

$$\text{MAP} = \text{HR} \times \text{SV} \times \text{TPR}$$

HR, SV, and TPR are controlled by neural and hormonal mechanisms at both the local and systemic levels. At the local level, each tissue controls its own blood flow through **autoregulatory mechanisms** that cause the arterioles supplying that tissue to constrict or dilate.

The collective autoregulatory actions in the capillary beds in all tissues of the body determine TPR and therefore MAP. If many arterioles suddenly dilate, TPR goes down and MAP falls. If many arterioles constrict, TPR goes up and MAP goes up. Changes in MAP provide information about changing needs of the body. As blood flows through capillary beds, its composition changes; its $CO_2$ content goes up and its $O_2$ content goes down. Thus, blood composition also provides information the body uses to regulate the circulatory system.

The nervous and endocrine systems respond to changes in MAP and blood composition by changing breathing rate, heart rate, stroke volume, and peripheral resistance to match the metabolic needs of the body. We will now see how these mechanisms work.

### Autoregulation matches local blood flow to local need

The amount of blood that flows through a capillary bed is controlled by the smooth muscle of the arteries and arterioles feeding that bed. **Figure 50.16** illustrates the flow of blood in a typical capillary bed. Blood flows into the bed from an arteriole. Smooth muscle "cuffs," or **precapillary sphincters**, on the arteriole can shut off the supply of blood to the capillary bed. When the precapillary sphincters are relaxed and the arteriole is open, the arterial blood pressure pushes blood into the capillaries.

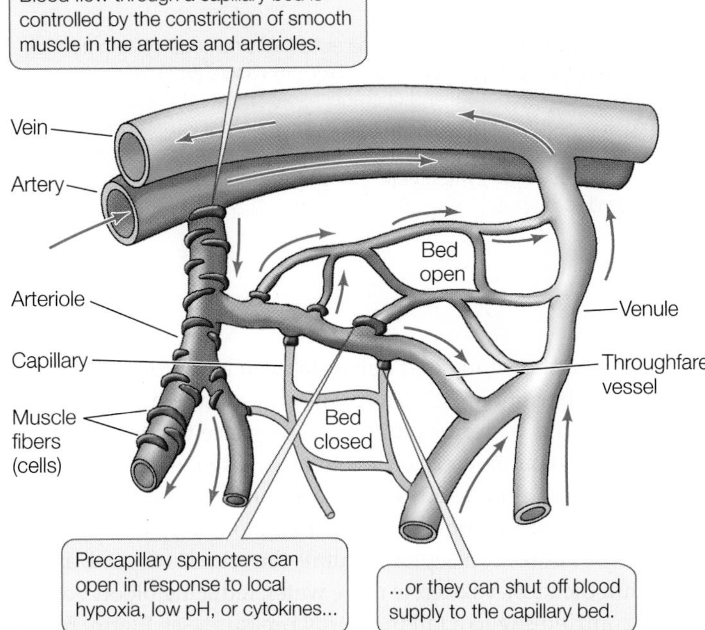
Blood flow through a capillary bed is controlled by the constriction of smooth muscle in the arteries and arterioles.
Vein
Artery
Arteriole
Capillary
Muscle fibers (cells)
Bed open
Bed closed
Venule
Throughfare vessel
Precapillary sphincters can open in response to local hypoxia, low pH, or cytokines...
...or they can shut off blood supply to the capillary bed.

**50.16 Local Control of Blood Flow** Low $O_2$ concentrations or high levels of metabolic by-products cause the smooth muscle of the arteries and arterioles to relax, thus increasing the supply of blood to the capillary bed.

Autoregulation depends on the sensitivity of the smooth muscle to its local chemical environment. Low $O_2$ concentrations and high $CO_2$ concentrations cause the smooth muscle to relax, thus increasing the supply of blood, which brings in more $O_2$ and carries away $CO_2$—a response known as **hyperemia**, which means "excess blood." Increases in other by-products of metabolism, such as lactic acid, hydrogen ions, potassium, and adenosine (all of which increase in exercising muscle), promote hyperemia through the same mechanism. Hence, activities that increase the metabolism of a tissue also induce hyperemia in that tissue.

### Arterial pressure is regulated by hormonal and neural mechanisms

Control and regulation of the circulatory system begins with the local autoregulatory mechanisms that alter the resistance of arteries and arterioles feeding capillary beds. The demands of the capillary beds influence MAP and blood composition. Both of these provide information for the control of endocrine and neural responses that act to return blood pressure and composition to normal. Thus circulatory functions are matched to the regional and overall needs of the body.

Arteries and arterioles are innervated by the autonomic nervous system, particularly the sympathetic division. The sympathetic postganglionic neurotransmitter norepinephrine binds to receptors in smooth muscle in blood vessels in the gut and other tissues not essential for "fight or flight" and causes these vessels to constrict, resulting in reduced blood flow through them and an elevation in MAP. In skeletal muscle, however, specialized sympathetic neurons release acetylcholine, causing the smooth muscle of these arterioles to relax and the vessels to dilate, increasing blood to flow to the muscle. As we discussed earlier in this chapter, increased sympathetic activity increases heart rate, and by increasing the strength of the cardiac muscle contraction, it also increases stroke volume.

Hormones also play a role in regulating arterial pressure. *Epinephrine* has actions similar to those of norepinephrine and is released from the adrenal medulla during massive sympathetic activation stimulated by a fall in arterial pressure or by activation of the fight-or-flight response to a dangerous threat. Another hormone, *angiotensin,* is produced when blood pressure to the kidneys falls (**Figure 50.17**). These hormones influence arterioles located for the most part in peripheral tissues (extremities) or in tissues whose functions need not be maintained continuously (such as the gut). By reducing blood flow in those arterioles, these hormones increase central blood pressure and blood flow to essential organs such as the heart, brain, and kidneys.

The autonomic nervous system activity that controls heart rate and constriction of blood vessels originates in a cardiovascular control center in the medulla. Many inputs converge on this central integrative network and influence the commands it issues via parasympathetic and sympathetic nerves (**Figure 50.18**). Of special importance is incoming information about changes in blood pressure and composition from both **baroreceptors** (stretch receptors) and *chemoreceptors* in the walls of the large arteries leading to the brain—the aorta and the carotid arteries.

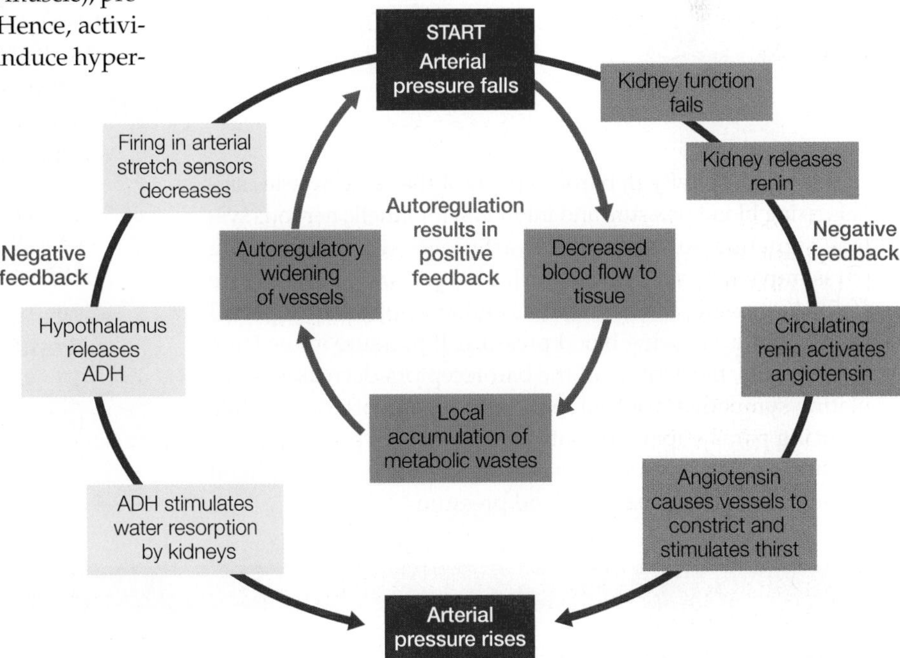
START
Arterial pressure falls
Kidney function fails
Kidney releases renin
Negative feedback
Circulating renin activates angiotensin
Angiotensin causes vessels to constrict and stimulates thirst
Arterial pressure rises
ADH stimulates water resorption by kidneys
Hypothalamus releases ADH
Negative feedback
Firing in arterial stretch sensors decreases
Autoregulation results in positive feedback
Autoregulatory widening of vessels
Decreased blood flow to tissue
Local accumulation of metabolic wastes

**50.17 Control of Blood Pressure through Local and Systemic Mechanisms** A drop in arterial pressure reduces blood flow to tissues, resulting in local accumulation of metabolic wastes. This change in the extracellular environment stimulates autoregulatory opening of the arteries. A fall in central blood pressure is prevented by negative feedback mechanisms that constrict arteries in less essential tissues and stimulate maintenance of blood volume and blood pressure.

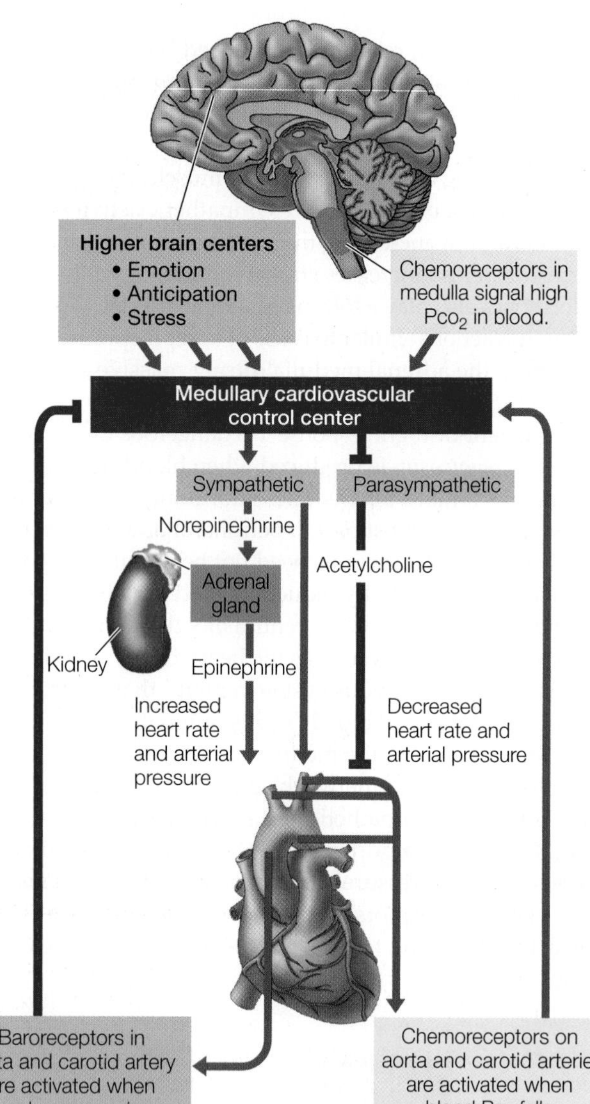


**50.18 Regulating Cardiac Output** The autonomic nervous system controls heart rate in response to information about blood pressure and blood composition originating in baroreceptors and chemosensors shown at the bottom of the figure. Information from these sensors goes to the cardiovascular control center in the medulla, where it is integrated with other information. The medullary center generates responses in the sympathetic and parasympathetic nervous systems that control cardiac output.

Increased activity in baroreceptors of the large arteries signals rising blood pressure and inhibits sympathetic nervous system signaling to arteries and arterioles while increasing parasympathetic signaling to the heart's pacemaker (see Figure 50.5). As a result, the heart slows and arterioles in peripheral tissues dilate, reducing blood pressure. If pressure in the large arteries falls, the activity of the baroreceptors decreases, stimulating sympathetic output to the arteries and arterioles while reducing parasympathetic output to the heart's pacemaker. As a result, the heart beats faster and the arterioles in peripheral tissues constrict, increasing blood pressure.

Another hormone that helps stabilize blood pressure is antidiuretic hormone (ADH, also called vasopressin), which is secreted by the posterior pituitary in response to a fall in the activity of the baroreceptors, signaling a fall in arterial pressure. ADH causes the kidneys to resorb more water and thereby maintain blood volume and increase blood pressure. Increased activity of the baroreceptors inhibits the release of ADH, and as a result the kidneys excrete more water, reducing blood volume and contributing to a fall in arterial pressure (see Figure 50.17).

Other information that causes the cardiovascular control center to increase heart rate and blood pressure comes from chemoreceptors in the medulla, aorta, and the carotid arteries. As we discussed in Section 49.5, the medullary chemosensors are activated by increases in arterial $CO_2$ levels, and the carotid and aortic bodies are activated by falls in arterial $O_2$ levels. Chemosensors send signals to the cardiovascular regulatory center as well as to the respiratory regulatory center.

## 50.5 RECAP

**The delivery of blood to tissues is controlled locally by autoregulatory mechanisms that dilate or constrict arterioles. These local actions are translated into alterations in central blood pressure and composition that are detected by neural and hormonal mechanisms, which then mediate corrective cardiovascular adjustments.**

- How do autoregulatory changes in blood flow to capillary beds result in adjustments to MAP? **See pp. 1062–1063 and Figure 50.16**
- What are the roles of hormones in regulating blood pressure? **See p. 1063 and Figure 50.17**
- Describe the role of baroreceptors and chemoreceptors in regulating blood pressure. **See pp. 1063–1064 and Figure 50.18**

# CHAPTER SUMMARY

## 50.1 Why Do Animals Need a Circulatory System?

- The metabolic needs of the cells of many small animals are met by direct exchange of materials with the external medium. The metabolic needs of the cells of larger animals are met by a **circulatory system** that transports nutrients, respiratory gases, and metabolic wastes throughout the body.
- In **open circulatory systems**, extracellular fluid leaves vessels and percolates through tissues. In **closed circulatory systems**, the blood is contained in a system of vessels. Closed circulatory systems have several advantages, including the ability to selectively direct blood, hormones, and nutrients to specific tissues. **Review Figure 50.1**

## 50.2 How Have Vertebrate Circulatory Systems Evolved?

**SEE WEB ACTIVITY 50.1**

- The circulatory system of vertebrates consists of a heart and a closed system of vessels containing blood that is separate from the interstitial fluid. **Arteries** and **arterioles** carry blood from the heart; **capillaries** are the site of exchange between blood and interstitial fluid; **venules** and **veins** carry blood back to the heart.
- The vertebrate heart evolved from two chambers in fishes to three in amphibians and some reptiles and four in crocodilians, birds, and mammals. This evolutionary progression has led to an increasing separation of blood that flows to the gas exchange organs and blood that flows to the rest of the body.
- The simplest (two-chambered heart) has an **atrium** that receives blood from the body and a **ventricle** that pumps blood out of the heart. An **aorta** distributes blood to arteries.
- In birds and mammals, blood circulates through two completely separate circuits: the **pulmonary**, which transports blood between the heart and lungs, and the **systemic**, which transports oxygen-rich blood between the heart and tissues.

## 50.3 How Does the Mammalian Heart Function?

- The human heart has four chambers. Valves in the heart prevent the backflow of blood. Review Figure 50.2, **WEB ACTIVITY 50.2**
- The **cardiac cycle** has two phases: **systole**, in which the ventricles contract, and **diastole**, in which the ventricles relax. The sequential heart sounds ("lub-dup") are made by the closing of the heart valves. Review Figure 50.3, **ANIMATED TUTORIAL 50.1**
- Blood pressure can be measured using a sphygmomanometer and a stethoscope. Review Figure 50.4
- The autonomic nervous system controls heart rate: sympathetic activity increases heart rate, and parasympathetic activity decreases it. Norepinephrine increases and acetylcholine decreases the rate of depolarization of the plasma membranes of **pacemaker cells**, affecting heart rate. Review Figure 50.5
- The **sinoatrial node** controls the cardiac cycle by initiating a wave of depolarization in the atria, which is conducted to the ventricles through a system consisting of the **atrioventricular node**, **bundle of His**, and **Purkinje fibers**. Review Figure 50.6
- Sustained contraction of ventricular muscle cells is due to long-duration action potentials that are generated by voltage-gated $Na^+$ and $Ca^{2+}$ channels. Review Figure 50.7
- An **electrocardiogram** (**ECG**) records electrical events associated with the contraction and relaxation of the cardiac muscles. Review Figure 50.8

## 50.4 What Are the Properties of Blood and Blood Vessels?

- Blood can be divided into a **plasma** portion (water, salts, and proteins) and a cellular portion (**erythrocytes** or red blood cells, platelets, and white blood cells). All of the cellular components are produced from stem cells in the bone marrow. Review Figure 50.9
- Erythrocytes transport oxygen. Their production in the bone marrow is stimulated by **erythropoietin**, which is produced in response to **hypoxia** (low oxygen levels) in the tissues.
- **Platelets**, along with circulating proteins, are involved in **blood clotting**, which results in a meshwork of **fibrin** threads that help seal vessels. Review Figure 50.10
- Abundant smooth muscle cells allow vessels to change their diameter, altering their resistance and thus blood flow. Arteries and arterioles have many elastic fibers that enable them to withstand high pressures. Review Figure 50.11, **WEB ACTIVITY 50.3**
- Capillary beds are the site of exchange of materials between blood and tissue fluid.
- **Starling's forces** suggest that blood volume is maintained in the capillary beds by an exchange of fluids driven by both blood pressure and osmotic pressure. Review Figure 50.13
- An accumulation of fluid in the extracellular spaces leads to **edema**. Bicarbonate ions in the blood plasma contribute to the osmotic forces that draw water back into capillaries.
- The ability of a specific molecule to cross a capillary wall depends on the architecture of the capillary, the type of substance, and the concentration gradient between the blood and the tissue fluid.
- Veins have a high capacity for storing blood. Aided by gravity, by contractions of skeletal muscle, and by the actions of breathing, they return blood to the heart. Review Figure 50.14
- The **Frank–Starling law** describes forces that increase cardiac output, such as stretch of the cardiac muscles cells caused by increased venous return.
- The **lymphatic system** returns the interstitial fluid to the blood.

## 50.5 How Is the Circulatory System Controlled and Regulated?

- Blood flow through capillary beds is controlled by local **autoregulatory mechanisms**, hormones, and the autonomic nervous system. Review Figure 50.16
- Blood pressure is controlled in part by the hormones ADH and angiotensin, which stimulate contraction of blood vessels. Review Figure 50.17
- Heart rate is controlled by the autonomic nervous system, which responds to information about blood pressure and blood composition that is integrated by regulatory centers in the medulla. Review Figure 50.18

## SELF-QUIZ

1. An open circulatory system is characterized by
   *a.* the absence of a heart.
   *b.* the absence of blood vessels.
   *c.* blood with a composition different from that of tissue fluid.
   *d.* the absence of capillaries.
   *e.* a higher-pressure circuit through gills than to other organs.
2. Which statement about vertebrate circulatory systems is *not* true?
   *a.* In fish, oxygenated blood from the gills returns to the heart through the left atrium.
   *b.* In mammals, deoxygenated blood leaves the heart through the pulmonary artery.

   *c.* In amphibians, deoxygenated blood enters the heart through the right atrium.
   *d.* In reptiles, the blood in the pulmonary artery has a lower oxygen content than the blood in the aorta.
   *e.* In birds, the pressure in the aorta is higher than the pressure in the pulmonary artery.

3. Which statement about the human heart is *true*?
   *a.* The walls of the right ventricle are thicker than the walls of the left ventricle.
   *b.* Blood flowing through atrioventricular valves is always deoxygenated blood.
   *c.* The second heart sound is due to the closing of the aortic valve.
   *d.* Blood returns to the heart from the lungs in the vena cava.
   *e.* During systole, the aortic valve is open and the pulmonary valve is closed.

4. The pacemaker action potentials in the heart
   *a.* are due to opposing actions of norepinephrine and acetylcholine.
   *b.* are generated by the bundle of His.
   *c.* depend on the gap junctions between the cells that make up the atria and those that make up the ventricles.
   *d.* are due to spontaneous depolarization of the plasma membranes of modified cardiac muscle cells.
   *e.* result from hyperpolarization of cells in the sinoatrial node.

5. Blood velocity through capillaries is slow because
   *a.* much blood volume is lost from the capillaries.
   *b.* the pressure in venules is high.
   *c.* the total cross-sectional area of capillaries is larger than that of arterioles.
   *d.* the osmotic pressure in capillaries is very high.
   *e.* erythrocytes must pass through in single file.

6. How are lymphatic vessels like veins?
   *a.* Both have nodes where they join together into larger common vessels.
   *b.* Both carry blood under low pressure.
   *c.* Both are capacitance vessels.
   *d.* Both have valves.
   *e.* Both carry fluids rich in plasma proteins.

7. The production of erythrocytes
   *a.* ceases if the hematocrit falls below normal.
   *b.* is stimulated by erythropoietin.
   *c.* is about equal to the production of white blood cells.
   *d.* is inhibited by prothrombin.
   *e.* occurs in bone marrow before birth and in lymph nodes after birth.

8. Which of the following does *not* increase blood flow through a capillary bed?
   *a.* High concentration of $CO_2$
   *b.* High concentration of lactate and hydrogen ions
   *c.* Histamine
   *d.* ADH
   *e.* Increase in arterial pressure

9. Blood clotting
   *a.* is impaired in patients with hemophilia because they do not produce platelets.
   *b.* is initiated when platelets release fibrinogen.
   *c.* involves a cascade of factors produced in the liver.
   *d.* is initiated by leukocytes forming a meshwork.
   *e.* requires production of angiotensin.

10. Autoregulation of blood flow to a tissue is due to
   *a.* sympathetic innervation.
   *b.* the release of ADH by the hypothalamus.
   *c.* increased activity of baroreceptors.
   *d.* chemoreceptors in the aorta and the carotid arteries.
   *e.* the effect of the local chemical environment on arterioles.

## FOR DISCUSSION

1. At the beginning of a race, cardiac output increases immediately before there is any change in blood $O_2$ or $CO_2$ concentrations. Explain two factors that contribute to this effect. Include the Frank–Starling law in your answer.
2. Explain how the hearts of crocodilians have the advantages of mammalian hearts during exercise but the efficiency of reptilian hearts during rest.
3. A sudden and massive loss of blood results in a decrease in blood pressure. Describe several mechanisms that help return blood pressure to normal.
4. You can describe the cycle of events in a ventricle of the heart by a graph that plots the pressure in the ventricle on the $y$ axis and the volume of blood in the ventricle on the $x$ axis. What would such a graph look like? Where would the heart sounds be on this graph? How would the graph differ for the left and the right ventricles?
5. If the major arteries become clogged with plaque and become less elastic because of atherosclerosis, the left ventricle must work harder and harder to pump an adequate supply of blood to the body. As a result, the left ventricle can become weakened and begin to fail, even though the right ventricle is healthy. A heart attack primarily affecting the left ventricle can have the same effect. This condition is known as congestive heart failure, and it commonly leads to fatal pulmonary edema. Explain how left ventricular failure can result in pulmonary edema, and why is it said that this condition creates a vicious circle that makes itself worse rapidly.

## ADDITIONAL INVESTIGATION

1. There are strains of rats that are spontaneously hypertensive—they naturally have high blood pressure. Can you formulate a hypothesis about the basis for such inherited hypertension and explain experiments you would do to test your hypothesis?
2. If you were in training to compete in the Tour de France, a grueling 3-week-long bicycle race, a trainer might tell you to "train high, compete low." What do you think is the physiological basis for such advice, and how would you test whether or not that reasoning was correct?

# 51 Nutrition, Digestion, and Absorption

## An obesity epidemic

For thousands of years, the Pima of southwestern North America were hunters and gatherers who supplemented their diet with subsistence agriculture. Their environment was arid, so they developed sophisticated irrigation systems; even so, they frequently encountered drought and subsequent starvation. Today most individuals of the ethnic Pima population in North America are clinically obese. In fact, as a population, they are one of the heaviest in the world.

With obesity come related health problems such as diabetes, high blood pressure, and heart disease. The incidence of diabetes among the Pima rose from an extremely high level of 45 percent of adults in 1965 to a staggering 80 percent in 1999. Moreover, diabetes is occurring in younger individuals than ever before. What has caused such a radical health change in an entire population? At least two interacting factors are involved: genetics and lifestyle.

Geneticists hypothesize that recurring episodes of starvation produce strong selective pressure for "thrifty genes"—particular alleles of the genes involved in digestion, absorption, and energy storage that result in greater-than-average efficiency in converting food into energy and into energy reserves, such as fat. Thrifty genes would give individuals a strong selective advantage when food is scarce. An example of a "thrifty" phenotype is seen among the Pima. They have a very low resting metabolic rate and convert food into fat readily. As we will see later in this chapter, the hormone insulin facilitates the conversion of dietary sugar into fat tissue. For many Pima, consuming a standard amount of glucose causes their insulin levels to rise three times higher than it does in Americans of European ancestry.

The other factor in the Pima obesity epidemic is an abrupt change in their traditional lifestyle. When food is plentiful and has high caloric content, thrifty genes con-

**Efficiency Genes** The Pima are an example of a human population that repeatedly experienced periods of severe food deprivation. These historic occurrences may have imposed selection for genes that improve the efficiency of managing the energy obtained from food. With modern diets and lifestyles, these "efficiency genes" can contribute to obesity.

**The Great American Lunch** High-fat fast foods have become prevalent in much of the developed world. A steady diet of such foods will mean weight gain for most people, and is a major reason for obesity in the United States.

tribute to obesity by maximizing fat storage. Today the Pima eat a modern Western diet that includes high-fat, high-calorie fast foods. In general, they also engage in less physical activity than their ancestors did.

A comparative study supports the hypotheses that have been put forward to explain the obesity epidemic in the Pima. Another population of Pima live in the Sierra Madre of northern Mexico. Genetically, they are the same as the Arizona population. However, they live a traditional lifestyle and eat traditional foods. Whereas the Arizona Pima engage in an average of only 2 hours of physical work per week, the Mexico Pima average 23 hours per week. Obesity and diabetes are not prevalent among the Mexico Pima.

A high-calorie diet and sedentary lifestyle affect not just the Pima but contribute to the overall increase in obesity throughout the U.S. population. Researchers are studying the Pima to learn more about the genetics of obesity and related pathologies.

**IN THIS CHAPTER** we will review the nutrients organisms require for energy, for molecular building blocks, and for specific biochemical functions. We will examine diverse adaptations for acquiring, ingesting, and digesting food and absorbing nutrients. Finally, we will learn how the body regulates its traffic in metabolic fuels and return to the question of control of body mass.

CHAPTER OUTLINE



## 51.1 What Do Animals Require from Food?

Animals are **heterotrophs**—they derive their nutrition from eating other organisms. In contrast, **autotrophs** (most plants, some bacteria, some archaea, and some protists) can use solar energy or inorganic chemical energy to synthesize all of their components. Directly and indirectly, heterotrophs take advantage of—indeed, depend on—the organic synthesis carried out by autotrophs, and have evolved an enormous diversity of adaptations to exploit this resource (**Figure 51.1**). In this section we will discover how animals use food, be it plants or other animals, to obtain energy and building blocks of complex molecules. We will also consider the need for special mineral nutrients and organic molecules we call vitamins and the diseases that result when they are lacking in the diet.

### Energy needs and expenditures can be measured

Energy is the capacity to do work, and it comes in different forms—electrical energy, heat energy, chemical energy, nuclear energy. As discussed in Chapter 8, a *calorie* (note the small *c*) is a unit of heat energy; specifically, it is the amount of heat necessary to raise the temperature of 1 gram of water 1°C. Since this is a tiny amount of energy, physiologists commonly use the **kilocalorie** (**kcal**) as a unit of measure (1 kcal = 1,000 calories). Nutritionists also use the kilocalorie as a standard unit of energy, but they traditionally refer to it as the **Calorie** (**Cal**) which is always capitalized to distinguish it from the single calorie.

Just about any food container you pick up in the U.S. carries a label, "Nutrition Facts," that includes the item "Calories." How do Calories (or kcal) relate to the discussion in Chapter 9 about how energy in the chemical bonds of food molecules is transferred to the high-energy phosphate bonds of adenosine triphosphate (ATP) and is then used to do cellular work? And why do we use heat energy as a measure of nutrition?

The reason is found in the laws of thermodynamics, which tell us that energy cannot be created or destroyed, but can be converted from one form to another (see Section 8.1). However, every energy conversion is inefficient and a large portion of the original energy always ends up as heat. Whether we are using the energy in glucose to make ATP or are using that ATP to power muscle contraction or ion transport, most of the available chemical energy is lost as heat. The bottom line is that if an animal is not growing, not doing work on its environment, and not changing its body temperature, the heat it loses to the environment is a measure of its total energy expenditure—its *metabolic rate.*

(A)

(B)

**51.1 Heterotrophs Get Energy from Autotrophs** (A) Herbivores get their energy directly from autotrophs. The large herbivores of the African grasslands must consume huge amounts of plant matter to fulfill their nutritional needs. (B) A carnivore's energy is indirectly obtained from autotrophs, since the energy stored in a prey animal was originally obtained from autotrophs.

The metabolic rate of an animal is a measure of its overall energy needs that must be met by the ingestion, digestion, and assimilation of food (see Section 40.3). The basal metabolic rate of a human is 1,300–1,500 Cal/day for an adult female and 1,600–1,800 Cal/day for an adult male. Physical activity adds to this basal energy requirement. For a person doing sedentary work, about 30 percent of the Calories expended are used for skeletal muscle activity; for a person doing heavy physical labor, more than 95 percent of total caloric expenditure is due to skeletal muscle activity.

The components of food that provide energy are fats, carbohydrates, and proteins. Fats yield 9.5 Cal/gram, carbohydrates 4.2 Cal/gram, and proteins about 4.1 Cal/gram. **Figure 51.2** shows some equivalencies of food, energy, and energy consumption.

Even though the units calorie, kilocalorie, and Calorie remain in popular use, most scientists now use the International System of Units. In this system, the basic unit of energy is the joule: 1 joule = 0.239 calories, and the measure of energy utilization is 1 joule/second = 1 watt. You are familiar with light bulb ratings, so think about that when you convert kcal/day into watts. The metabolic rate of the average man is 1,700 kcal/day or 82 watts (note that a watt includes the time dimension, so it is a rate of energy utilization).

Thus it is possible to quantify the caloric value of any food an animal eats. It is also possible to quantify the caloric expenditure of any activity or behavior an animal performs. By comparing calories consumed with calories expended, we can construct *energy budgets* that allow ecologists and evolutionary biologists to apply a *cost–benefit analysis* to feeding behavior, as we explain in Section 53.4.

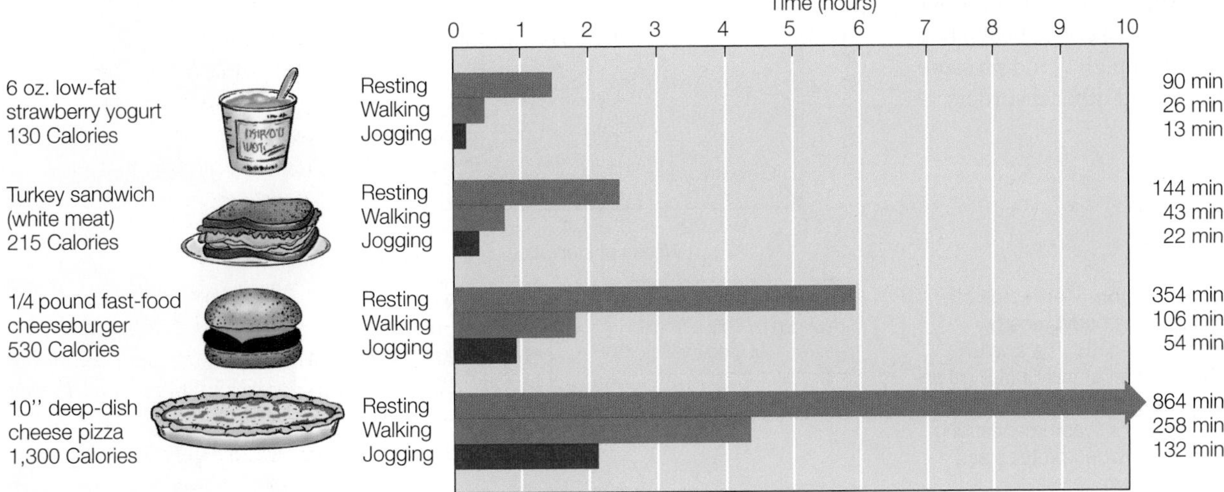


**51.2 Food Energy and How We Use It** The energy contained in several common food items is shown at the left. The graphs indicate about how long it would take a person with a basal metabolic rate of about 1,800 Cal/day to utilize the equivalent amount of energy while resting, walking, or jogging.

### Sources of energy can be stored in the body

Although the cells of the body use energy continuously, most animals do not eat continuously. Therefore, animals must store fuel molecules that can be released as needed between meals. Carbohydrates are stored in liver and muscle cells as glycogen, but the total glycogen store represents only about a day's basal energy requirements (1,500–2,000 Cal). Fat is the most important form of stored energy in the bodies of animals. Not only does fat have more energy per gram than glycogen, but it can be stored with little associated water, making it more compact. Migrating birds store energy as fat to fuel their long flights; if they had to store the same amount of energy as glycogen, they would be too heavy to fly! Proteins are not used as energy storage compounds, although body protein can be metabolized as an energy source of last resort.

If an animal takes in too little food to meet its energy requirements, it is **undernourished** and must start metabolizing some of the molecules of its own body. This "self-consumption" begins with the energy storage compounds glycogen and fat. There is some protein loss due to normal protein turnover not being fully replaced because of a lack of amino acids for new protein synthesis. Once fat reserves are seriously depleted, the body increases its metabolism of proteins for energy (**Figure 51.3A**). The first proteins to be sacrificed are those of the blood plasma. The loss of plasma proteins decreases the osmotic concentration of the plasma, resulting in increased loss of fluid from the blood to the interstitial spaces (*edema*; see Section 50.4). Accumulation of fluid in the extremities and abdomen is the classic sign of *kwashiorkor*, a disease caused by chronic protein deficiency (**Figure 51.3B**). Continued protein loss damages the body's organs, leading eventually to death.

When an animal consistently takes in more food than it needs to meet its energy requirements, it is **overnourished**, and the excess nutrients are stored as increased body mass. First, glycogen reserves build up; then additional dietary carbohydrates, fats, and proteins are converted to body fat. In some species, such as hibernators, seasonal overnutrition is an important adaptation for surviving periods when food is not available. In humans, however, overnutrition can be a serious health hazard, increasing the risk of high blood pressure, heart attack, diabetes, and other disorders.

### Food provides carbon skeletons for biosynthesis

Every animal requires certain basic organic molecules that it cannot synthesize for itself but needs as building blocks for its own complex organic molecules. The acetyl group ($CH_3CO$—) is one such required building block supplying the **carbon skeleton** of larger organic molecules (**Figure 51.4**). Animals cannot make acetyl groups from carbon, oxygen, and hydrogen molecules; they must obtain acetyl groups from food. Acetyl groups can be derived from the metabolism of almost any food, but they originate in plants. Acetyl groups are never in short supply for an adequately nourished animal. However, some groups supplying carbon skeletons can be deficient in an animal's diet even if caloric intake is adequate.

Amino acids, the building blocks of proteins, are a good example of carbon skeletons that can be in short supply. Animals can synthesize some of their own amino acids by utilizing carbon skeletons from acetyl or other groups and transferring to them amino groups (—$NH_2$) derived from other amino acids. Most animals cannot synthesize all the amino acids they need. Each species must obtain certain **essential amino acids** from food. Essential amino acids vary by species. In general, herbivores have fewer essential amino acids than carnivores. If an animal does not take in enough of even one of its essential amino acids, its protein synthesis is impaired and its capacity to maintain enzymatic and transport functions is challenged.

Most researchers agree that adult humans must obtain eight essential amino acids from their food: isoleucine, leucine, lysine, methionine, phenylalanine, threonine, tryptophan, and valine. All eight are available in milk, eggs, meat, and soybean products, but most plant foods do not contain adequate quantities

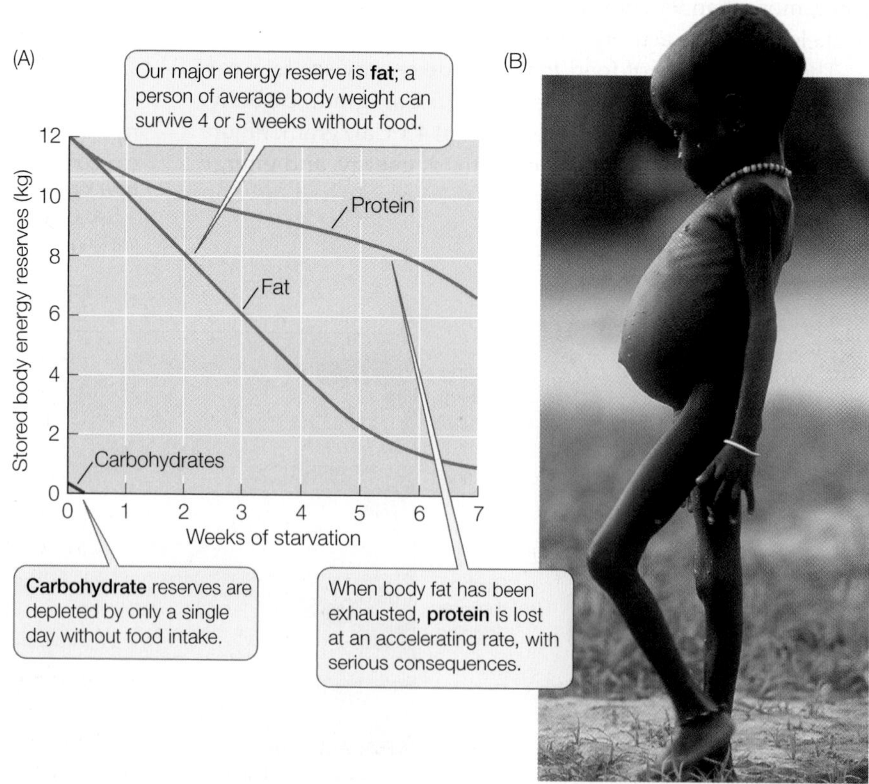

(A)
(B)
Our major energy reserve is **fat**; a person of average body weight can survive 4 or 5 weeks without food.
Stored body energy reserves (kg)
12
10
8
6
4
2
0
Protein
Fat
Carbohydrates
0 1 2 3 4 5 6 7
Weeks of starvation
**Carbohydrate** reserves are depleted by only a single day without food intake.
When body fat has been exhausted, **protein** is lost at an accelerating rate, with serious consequences.


**51.3 The Course of Starvation** (A) In a person subjected to undernutrition, the body's energy reserves are eventually depleted. (B) The swollen abdomen, face, hands, and feet of this Somali girl are due to edema. Along with her spindly limbs, these are symptoms of kwashiorkor, a syndrome resulting from the body breaking down blood proteins and muscle tissue to fuel metabolism.

Animals use acetyl groups obtained from their food to build more complex organic molecules.

The acetyl group is present in virtually all of the foods animals ingest.

Protein, carbohydrate, or fat metabolism

Acetyl group carbon skeleton

Steroid hormones

Oxaloacetate

Citrate

Amino acids, heme, and other compounds

Palmitic acid (and other fatty acids)

**51.4 The Acetyl Group Is an Acquired Carbon Skeleton** Animals cannot synthesize the acetyl group for themselves, but they ingest it in their food and use it to synthesize a wide variety of molecules.

of all eight, so a strict vegetarian diet carries a risk of protein malnutrition. A *complementary* diet of plant foods, however, supplies all eight essential amino acids (**Figure 51.5**). In general, grains (such as rice, wheat, and corn) are complemented by legumes (such as beans and peas). Long before the chemical basis for this complementarity was understood, societies with little access to meat developed complementary diets. Many Central and South American peoples traditionally eat beans with corn, and the native peoples of North America complemented their beans with squash.

Human infants are thought to require four additional amino acids in their diets: histidine, tyrosine, cysteine, and arginine. Also, some amino acids are *conditionally essential* for certain populations that cannot synthesize them in adequate amounts. For example, individuals with the genetic disease phenylketonuria lack the enzyme for converting phenylalanine to tyrosine (see Section 15.3) and thus must obtain tyrosine from their diets. They must keep their dietary intake of phenylalanine low to prevent its accumulation to toxic levels.

Why are dietary proteins completely digested to their constituent amino acids before being used by the body? Wouldn't it be more energy-efficient to reuse some dietary proteins directly? There are several reasons why ingested proteins are not used "as is":

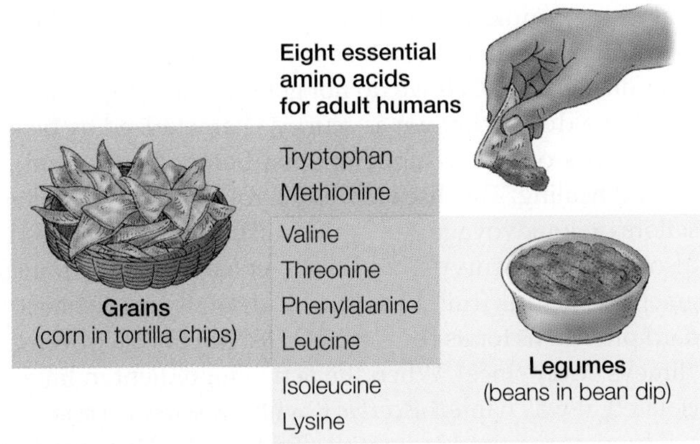


**51.5 A Strategy for Vegetarians** By combining cereal grains with legumes, an adult vegetarian can obtain all eight essential amino acids.

- Macromolecules such as proteins are not readily absorbed by the cells of the gut, but their constituent monomers (such as amino acids) are readily absorbed.
- Protein structure and function are highly species-specific. A protein that functions optimally in one species might not function well in another.
- Foreign proteins entering the body directly from the gut would be recognized as invaders and would be attacked by the immune system.

Humans can synthesize almost all the lipids required by the body using acetyl groups obtained from food (see Figure 51.4), but we must have a dietary source of certain **essential fatty acids**—notably, linoleic acid—that we cannot synthesize. Linoleic acid is an unsaturated fatty acid needed by mammals to synthesize other unsaturated fatty acids, such as arachidonic acid, which is a component of several signaling molecules, including prostaglandins. Essential fatty acids are also necessary components of membrane phospholipids. A deficiency of linoleic acid can lead to problems such as infertility and impaired lactation, but since it is commonly present in vegetable oils, a deficiency is unlikely in an adequately nourished individual.

### Animals need mineral elements for a variety of functions

**Table 51.1** lists the principal mineral elements that animals require. Elements required in large amounts are called **macronutrients**; those required in only tiny amounts (generally less than 100 mg/day) are called **micronutrients**. Some micronutrients are required in such minute amounts that deficiencies are never observed, but they are nevertheless essential elements.

Calcium is an example of a macronutrient. It is the fifth most abundant element in the body; a 70-kg person contains about 1.2 kg of calcium. Calcium phosphate is the principal structural material in bones and teeth. Muscle contraction, neural function, and many other intracellular functions in animals require calcium ions ($Ca^{2+}$). The turnover of calcium in the extracellular fluid is high, as bones are constantly being remodeled and calcium is constantly entering and leaving cells. Calcium is lost from the body

TABLE 51.1

**Mineral Elements Required by Animals**

| ELEMENT | SOURCE IN HUMAN DIET | MAJOR FUNCTIONS |
|---|---|---|
| **MACRONUTRIENTS** | | |
| Calcium (Ca) | Dairy foods, eggs, green leafy vegetables, whole grains, legumes, nuts, meat | Found in bones and teeth; blood clotting; nerve and muscle action; enzyme activation |
| Chlorine (Cl) | Table salt (NaCl), meat, eggs, vegetables, dairy foods | Water balance; digestion (as HCl); principal negative ion in extracellular fluid |
| Magnesium (Mg) | Green vegetables, meat, whole grains, nuts, milk, legumes | Required by many enzymes; found in bones and teeth |
| Phosphorus (P) | Dairy, eggs, meat, whole grains, legumes, nuts | Found in nucleic acids, ATP, and phospholipids; bone formation; buffers; metabolism of sugars |
| Potassium (K) | Meat, whole grains, fruits, vegetables | Nerve and muscle action; protein synthesis; principal positive ion in cells |
| Sodium (Na) | Table salt, dairy foods, meat, eggs | Nerve and muscle action; water balance; principal positive ion in extracellular fluid |
| Sulfur (S) | Meat, eggs, dairy foods, nuts, legumes | Found in proteins and coenzymes; detoxification of harmful substances |
| **MICRONUTRIENTS** | | |
| Chromium (Cr) | Meat, dairy, whole grains, legumes, yeast | Glucose metabolism |
| Cobalt (Co) | Meat, tap water | Found in vitamin $B_{12}$; formation of red blood cells |
| Copper (Cu) | Liver, meat, fish, shellfish, legumes, whole grains, nuts | Found in active site of many redox enzymes and electron carriers; production of hemoglobin; bone formation |
| Fluorine (F) | Most water supplies | Found in teeth; helps prevent decay |
| Iodine (I) | Fish, shellfish, iodized salt | Found in thyroid hormones |
| Iron (Fe) | Liver, meat, green vegetables, eggs, whole grains, legumes, nuts | Found in active sites of many redox enzymes and electron carriers, hemoglobin, and myoglobin |
| Manganese (Mn) | Organ meats, whole grains, legumes, nuts, tea, coffee | Activates many enzymes |
| Molybdenum (Mo) | Organ meats, dairy, whole grains, green vegetables, legumes | Found in some enzymes |
| Selenium (Se) | Meat, seafood, whole grains, eggs, milk, garlic | Fat metabolism |
| Zinc (Zn) | Liver, fish, shellfish, and many other foods | Found in some enzymes and some transcription factors; insulin physiology |

in urine, sweat, and feces, so it must be replaced regularly. Humans require 800–1,000 mg of calcium per day in their diet.

Iron is an example of a micronutrient. It is found throughout the body because it is the oxygen-binding atom in hemoglobin and myoglobin and is a component of enzymes in the electron transport chain. Nevertheless, the total amount of iron in a 70-kg person is only about 4 grams, and since iron is recycled efficiently in the body and is not lost in the urine, we require only about 15 mg per day in our food. Despite the small amount required, insufficient iron is the most common mineral nutrient deficiency in the world today, leading to *anemia*. Anemic individuals are weak and tired all the time.

**yourBioPortal.com**

GO TO Web Activity 51.1 • Mineral Elements Required by Animals

### Animals must obtain vitamins from food

Like essential amino acids and fatty acids, **vitamins** are carbon compounds that an animal requires for growth and metabolism but cannot synthesize for itself. They are required in very small amounts compared with the essential amino acids and fatty acids, which are incorporated into large body structures. Most vitamins function as coenzymes or parts of coenzymes.

The list of vitamins varies from species to species. Most mammals, for example, can make their own ascorbic acid. Primates (including humans) cannot, so for primates, ascorbic acid is a vitamin—*vitamin C*. If we do not get vitamin C in our food, we develop a disease known as scurvy, characterized by bleeding gums, loss of teeth, subcutaneous hemorrhages, and slow wound healing. The disease was a frequently fatal problem for sailors on long voyages until a Scottish physician, James Lind, discovered that scurvy could be prevented if the sailors ate fresh greens and citrus fruit. The British Admiralty made limes standard provisions for its ships (and British sailors have been called "limeys" ever since). When the active ingredient in limes was isolated, it was named *ascorbic* ("without scurvy") *acid*.

Humans require 13 vitamins (**Table 51.2**). They are divided into two groups: water-soluble and fat-soluble. When water-soluble vitamins are ingested in excess of bodily needs, they are

TABLE 51.2

**Vitamins in the Human Diet**

| VITAMIN | SOURCE | FUNCTION | DEFICIENCY SYMPTOMS |
|---|---|---|---|
| **WATER-SOLUBLE** | | | |
| $B_1$ (thiamin) | Liver, legumes, whole grains | Coenzyme in cellular respiration | Beriberi, loss of appetite, fatigue |
| $B_2$ (riboflavin) | Dairy, meat, eggs, green leafy vegetables | Coenzyme in FAD | Lesions in corners of mouth, eye irritation, skin disorders |
| Niacin | Meat, fowl, liver, yeast | Coenzyme in NAD and NADP | Pellagra, skin disorders, diarrhea, mental disorders |
| $B_6$ (pyridoxine) | Liver, whole grains, dairy foods | Coenzyme in amino acid metabolism | Anemia, slow growth, skin problems, convulsions |
| Pantothenic acid | Liver, eggs, yeast | Found in acetyl CoA | Adrenal problems, reproductive problems |
| Biotin | Liver, yeast, bacteria in gut | Found in coenzymes | Skin problems, loss of hair |
| $B_{12}$ (cobalamin) | Liver, meat, dairy foods, eggs | Formation of nucleic acids, proteins, and red blood cells | Pernicious anemia |
| Folic acid | Vegetables, eggs, liver, whole grains | Coenzyme in formation of heme and nucleotides | Anemia |
| C (ascorbic acid) | Citrus fruits, tomatoes, potatoes | Formation of connective tissues; antioxidant | Scurvy, slow healing, poor bone growth |
| **FAT-SOLUBLE** | | | |
| A (retinol) | Fruits, vegetables, liver, dairy | Found in visual pigments | Night blindness |
| D (cholecalciferol) | Fortified milk, fish oils, sunshine | Absorption of calcium and phosphate | Rickets |
| E (tocopherol) | Meat, dairy foods, whole grains | Muscle maintenance, antioxidant | Anemia |
| K (menadione) | Intestinal bacteria, liver | Blood clotting | Blood-clotting problems |

simply eliminated in the urine. (This is the fate of much of the excess vitamin C that people take.) Fat-soluble vitamins, however, can accumulate in body fat and may build up to toxic levels in the liver if taken in excess.

The fat-soluble vitamin D (cholecalciferol), which is essential for absorbing and metabolizing calcium, is a special case because the body can synthesize it. (As noted in Section 41.3, vitamin D is by definition a hormone.) Certain lipids present in the human body can be converted into vitamin D by the action of ultraviolet light on the skin. Thus vitamin D must be obtained in the diet by individuals with inadequate exposure to the sun.

The need for vitamin D may have been an important factor in the evolution of skin color. Human races that are adapted to equatorial and low latitudes have dark skin pigmentation as a protection against the damaging effects of ultraviolet radiation. These peoples generally expose extensive areas of skin to the sun on a regular basis, so their skin synthesizes adequate amounts of vitamin D. Most races that adapted to life in the higher latitudes lost this dark skin pigmentation, probably because lighter skin facilitates vitamin D production in the relatively small areas of skin exposed to sunlight during the short days of winter. The Inuit peoples of the Arctic seem to represent an exception to the correlation between latitude and skin pigmentation, but these dark-skinned people obtain ample vitamin D from the large amounts of animal fat (especially whale blubber) and fish oils in their diet.

**yourBioPortal.com**

GO TO **Web Activity 51.2 • Vitamins in the Human Diet**

### Nutrient deficiencies result in diseases

The lack of any essential nutrient in the diet produces a state of deficiency called **malnutrition**, and chronic malnutrition leads to a characteristic **deficiency disease**. We have already discussed kwashiorkor (protein deficiency) and scurvy (vitamin C deficiency). A shortage of any of the vitamins results in specific deficiency symptoms (see Table 51.2). Another deficiency disease, *beriberi*, was directly involved in the discovery of vitamins.

Beriberi means "extreme weakness." It became prevalent in Asia in the nineteenth century after it became standard practice to mill rice to a high, white polish and discard the hulls present in brown rice. A critical observation was that birds—chickens and pigeons—developed beriberi-like symptoms when fed only polished rice. In 1912, Casimir Funk, a Polish scientist working in England, cured pigeons of beriberi by feeding them discarded rice hulls.

At the time of Funk's discovery, all diseases were thought to be either caused by microorganisms or inherited. Funk suggested the radical idea that beriberi and some other diseases are dietary in origin and result from deficiencies in specific substances. Funk coined the term "vitamines" because he mistakenly thought that all these substances vital for life were compounds with amino groups (vital amines). In 1926, thiamin (vitamin $B_1$)—the substance lost in the rice milling process—was the first vitamin to be isolated in pure form.

Deficiency diseases can also result from an inability to absorb or process an essential nutrient even if it is present in the diet. Vitamin $B_{12}$ (cobalamin), for example, is present in all foods of animal origin. Since plants neither use nor produce vitamin $B_{12}$, a strictly vegetarian diet (not supplemented with dairy products or vitamin pills) can lead to a $B_{12}$ deficiency disease called pernicious anemia, characterized by a failure of red blood cells to mature. The most common cause of pernicious anemia, however, is not a lack of vitamin $B_{12}$ in the diet but an inability to absorb it. Normally, cells in the stomach lining secrete a peptide called *intrinsic factor*, which binds to vitamin $B_{12}$ and makes it possible for it to be absorbed in the small intestine. Conditions that damage the stomach lining, such as alcoholism or gastritis, can thus lead to pernicious anemia.

Inadequate mineral nutrition can also lead to deficiency diseases. Examples are hypothyroidism and goiter resulting from iodine deficiency (see Section 41.3), and anemia resulting from iron deficiency. Iodine deficiency is almost unheard of today in the developed world because of the addition of iodide to salt. However, it is still a major health problem in large segments of the human population. Probably the single least expensive effective action to improve global health would be to provide iodized salt for everyone.

## 51.1 RECAP

**As heterotrophs, animals must obtain the energy and molecular building blocks for biosynthesis from their food. Energy can come from the metabolism of carbohydrates, fats, and proteins. Molecular building blocks include carbon skeletons, vitamins, and minerals.**

- Explain the different roles of food energy from proteins, carbohydrates, and fats in the mammalian body. **See p. 1070 and Figure 51.3**
- Give an example of an essential carbon skeleton, a micronutrient, and a macronutrient. **See pp. 1070–1073, Figure 51.4 and Table 51.1**
- Why should fat-soluble vitamins not be taken in excess? **See p. 1073**

We have surveyed the essential elements of nutrition in animals. Next we will look at various methods and adaptations by which animals obtain the food they need, and mechanisms by which the food is processed in the body to extract nutrients.

# 51.2 How Do Animals Ingest and Digest Food?

Heterotrophic organisms can be classified by how they acquire their nutrition. **Saprobes** (also called *saprotrophs* or *decomposers*) are organisms—mostly protists and fungi—that absorb nutrients from dead organic matter. **Detritivores**, such as earthworms and crabs, actively feed on dead organic material. Animals that feed on living organisms are **predators**: **herbivores** prey on plants, **carnivores** prey on animals, and **omnivores** prey on both. **Filter feeders**, such as clams and blue whales, prey on small organisms by filtering them from the aquatic environment. **Fluid feeders** include mosquitoes, aphids, leeches, and hummingbirds. The anatomical adaptations that enable a species to exploit a particular source of nutrition are usually obvious, but physiological and biochemical adaptations are also important.

### The food of herbivores is often low in energy and hard to digest

Most vegetation is coarse and difficult to break down, but herbivores process large amounts of it because its energy content is low. Therefore, herbivores spend a great deal of time feeding. Many have striking adaptations for feeding, such as the trunk (a flexible, gripping nose) of the elephant or the huge bill of the fruit-eating toucan, which can be half as long as its body. Many types of grinding, rasping, cutting, and shredding mouthparts have evolved in invertebrates for ingesting plant material, and the teeth of herbivorous vertebrates have been shaped by selection to tear, crush, and grind coarse plant matter.

The digestive processes of herbivores can also be quite specialized. An example is the koala, which almost exclusively eats leaves of eucalyptus trees. These leaves are very fibrous, low in usable energy and protein, and high in toxic chemicals. Koalas actually smell like eucalyptol, a common ingredient in cough drops. The koala has strong jaws for grinding the leaves, a very long gut for fermenting them, enzymes in its liver for detoxifying chemicals in the leaves, and a low metabolic rate to compensate for low energy intake.

### Carnivores must detect, capture, and kill prey

The predatory behaviors of many carnivores are legendary—the hunting skills of hawks, wolves, and tigers, for example. Carnivores have evolved stealth, speed, power, large jaws, sharp teeth, and strong gripping appendages. They also have evolved remarkable means of detecting prey. Bats use echolocation, pit vipers sense infrared radiation from the warm bodies of their prey, and certain fishes detect electric fields created in the water by their prey.

Adaptations for killing and ingesting prey are diverse and can be highly specialized. These adaptations are especially important when the prey can inflict damage on the predator. There are many fascinating examples of adaptations for capturing and immobilizing prey, such as the immobilizing venom in the bite of many snakes, the long sticky tongues of chameleons, and the webs of spiders. The long tentacles of certain jellyfish contain some of the most deadly toxins known and can kill humans. Specialized cells called *nematocysts* inject toxin into the victim (see Figure 31.9). Some jellyfish harbor enough nematocysts cells to kill 50 people. Some predators digest their prey externally. For example, a spider injects its insect prey with digestive enzymes and then sucks out the liquefied contents, leaving behind the empty exoskeletons frequently seen in old spider webs.

### Vertebrate species have distinctive teeth

Teeth are adapted for the acquisition and initial processing of specific types of foods. Because they are among the hardest struc-

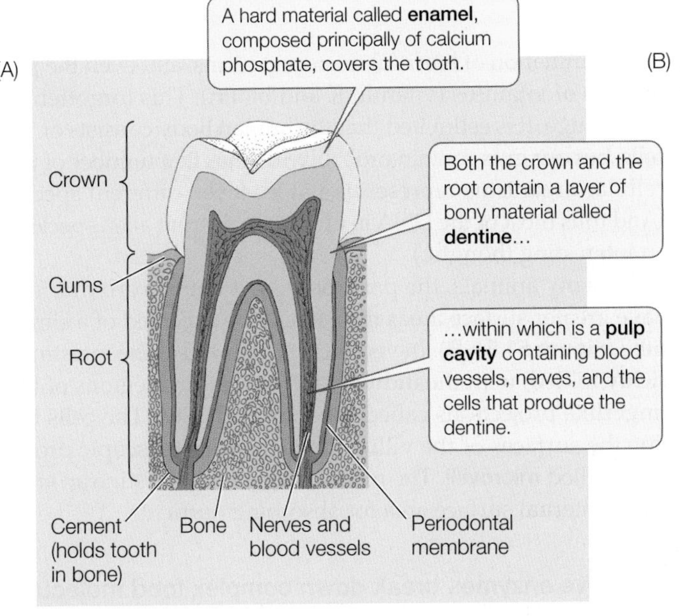


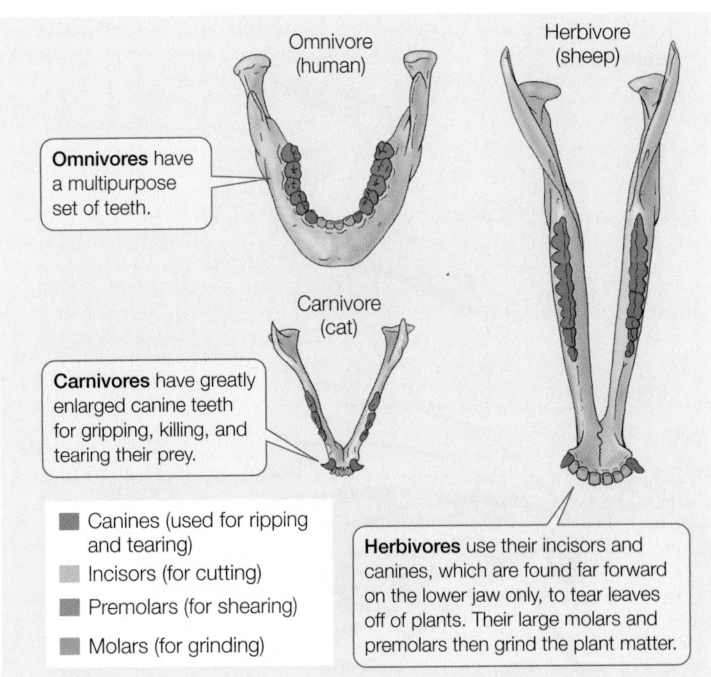


**51.6 Mammalian Teeth** (A) A mammalian tooth has three layers: enamel, dentine, and a pulp cavity. (B) The teeth of different mammalian species are specialized for different diets. This illustration depicts the teeth of the lower jaw, viewed from above.

**yourBioPortal.com**

GO TO **Web Activity 51.3 • Mammalian Teeth**

tures of the body, an animal's teeth remain in the environment long after it dies. Paleontologists use teeth to identify animals that lived in the distant past and to deduce their feeding behavior.

All mammalian teeth have the same general, three-layered structure (**Figure 51.6A**). An extremely hard material called **enamel**, composed principally of calcium phosphate, covers the crown of the tooth. Both the crown and root contain a layer of bony material called **dentine**, inside of which is a **pulp cavity** containing blood vessels, nerves, and the cells that produce the dentine.

There is a great deal of homology in the dentition of mammals, but the shapes and organization of mammalian teeth are adaptations to different diets (**Figure 51.6B**). In general, *incisors* are used for cutting, chopping, and gnawing; *canines* are used for stabbing, gripping, and ripping; and *molars* and *premolars* (the cheek teeth) are used for shearing, crushing, and grinding. The highly varied diet of humans is reflected in our multipurpose set of teeth, as is common among omnivores.

## Digestion usually begins in a body cavity

Animals take food into a body cavity that is continuous with the outside environment. They secrete digestive enzymes into that cavity, and the enzymes break down the food into nutrient molecules that can be absorbed by the cells lining the cavity.

The simplest digestive systems are **gastrovascular cavities**, which connect to the outside world through a single opening. Cnidarians, such the jellyfish mentioned above, capture prey using stinging nematocysts and use their tentacles to cram the prey into their gastrovascular cavities. Enzymes in the gastrovascular cavity partly digest the prey. Cells lining the cavity take in small food particles by endocytosis. The vesicles created by endocytosis then fuse with lysosomes containing digestive enzymes, and intracellular digestion completes the breakdown of the food. Nutrients are released to the cytoplasm as the vesicles break down.

## Tubular guts have an opening at each end

The guts of most animals are tubular: a **mouth** takes in food; molecules are digested and absorbed throughout the length of the gut; and solid digestive wastes are eliminated through an **anus**. Different regions in the tubular gut are specialized for particular functions (**Figure 51.7**). These functions must be coordinated so they occur in the proper sequence and at rates that maximize the efficiency of digestion and absorption of nutrients.

At the anterior end of the gut is the mouth cavity where food can be fragmented, for example by teeth (in many vertebrates), by **radula** (in snails), or by **mandibles** (in many arthropods). In most birds, food is ground by small stones in an early, muscular, portion of the gut called the **gizzard**. Some animals, such as snakes, simply ingest whole prey with little or no fragmentation. **Stomachs** and **crops** are storage chambers that enable animals to ingest relatively large amounts of food when it is available, then digest it gradually. In these storage chambers, food may be further fragmented and mixed, and in most vertebrates it is an important site of digestion. Food delivered into the next section of the gut, the **intestine**, is in small particles, well mixed, and usually partially digested.

Most digestion occurs in the intestine, and nutrients, water, and ions are absorbed across its walls. Glands secrete digestive enzymes into the intestine, and other enzymes are produced by cells lining the intestine. The final segment of the intestine recovers water and ions and stores undigested wastes, or **feces**, so they can be released to the environment at an appropriate

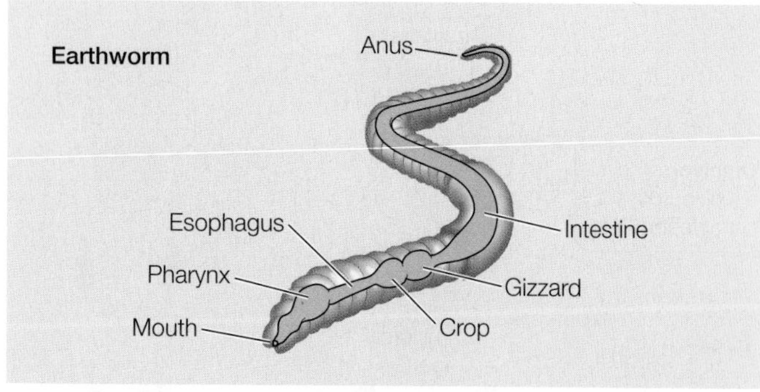


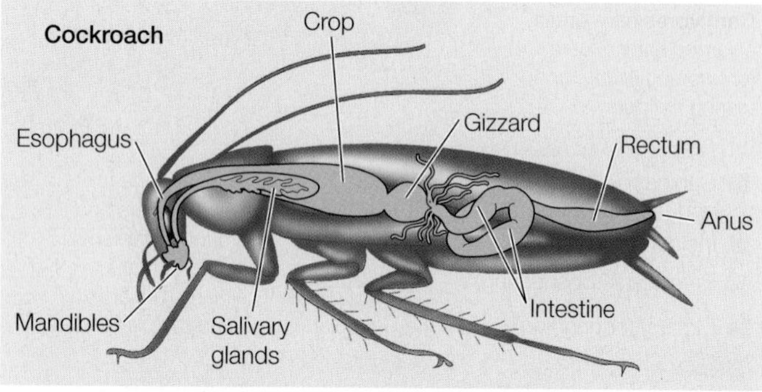


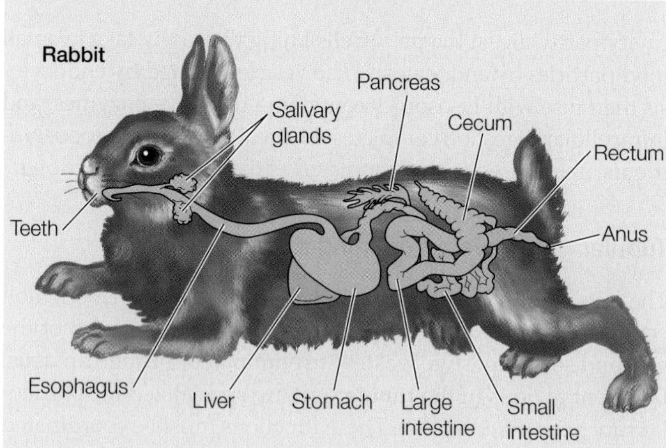


**51.7 Compartments for Digestion and Absorption** Most invertebrates and all vertebrates have a tubular gut that begins with a mouth, which takes in food, and ends in an anus, which eliminates wastes. Between these two structures are specialized regions for digestion and nutrient absorption; the structures in these regions are adapted to different diets and vary from species to species.

time or place. A muscular **rectum** near the anus assists in expelling feces.

Endosymbiotic bacteria colonize the intestines. These bacteria obtain their nutrition from the food passing through the host's gut while contributing to the host's digestive processes. Members of the leech genus *Hirudo*, for example, produce no enzymes that can digest the proteins in the blood they suck from vertebrates; instead, they depend on bacteria to perform this service. The resulting amino acids are subsequently used by both the leech and the bacteria.

The microorganisms in the human gut are called the "forgotten organ" because they provide important services in digestion, elimination of harmful microorganisms, and even the production of vitamins (vitamin K and biotin). This forgotten organ is huge. It is estimated that the human body consists of $10^{13}$ cells, but our guts contain probably 10 times that number of unicellular organisms representing at least 500 different species. (And thus most of the DNA in our bodies is from alien species—an interesting thought.)

In many animals, the parts of the gut that absorb nutrients have greater surface areas than would be expected of a simple tube (**Figure 51.8A,B**). In vertebrates, the wall of the intestine is richly folded, with the individual folds bearing legions of tiny fingerlike projections called **villi** (**Figure 51.8C**). The cells that line the surfaces of the villi, in turn, have microscopic projections called **microvilli**. The microvilli give the intestine an enormous internal surface area for absorbing nutrients.

### Digestive enzymes break down complex food molecules

Protein, carbohydrate, and fat macromolecules are broken down into their simplest monomeric units by hydrolytic enzymes produced at different locations in the digestive tract. Many are secreted into the lumen of the gut, and others remain associated with the membranes of the microvilli. All of these enzymes cleave the chemical bonds of macromolecules through *hydrolysis*, a reaction that adds a water molecule (see Figure 3.4B). Digestive enzymes are classified according to the substances they hydrolyze: *proteases* break the bonds between adjacent amino acids in proteins; *carbohydrases* hydrolyze carbohydrates; *peptidases*, peptides; *lipases*, fats; and *nucleases*, nucleic acids.

How can an organism produce enzymes that hydrolyze biological macromolecules without digesting itself? Many digestive enzymes are produced in an inactive form known as a **zymogen**, so that they cannot act on the cells that produce them. When secreted into the gut, a zymogen is generally activated by another enzyme. The cells lining the gut are not digested because they are protected by a covering of mucus.

## 51.2 RECAP

**Heterotrophs have diverse adaptations for acquiring food. Once captured and/or ingested, food is digested extracellularly by secreted enzymes to release nutrients, which are absorbed into the animal's body, usually via a tubular gut.**

- Why do herbivores typically spend a great deal of their time feeding? See p. 1074
- What is the primary purpose of the intestinal microvilli? See p. 1076 and Figure 51.8
- What is a zymogen, and why is it important? See p. 1076

Once ingested by an animal, food may be fragmented and moved into the gut for digestion by hydrolytic enzymes. The processes of digestion release the nutrients needed by the animal. Let's focus next on how those processes occur in vertebrates.

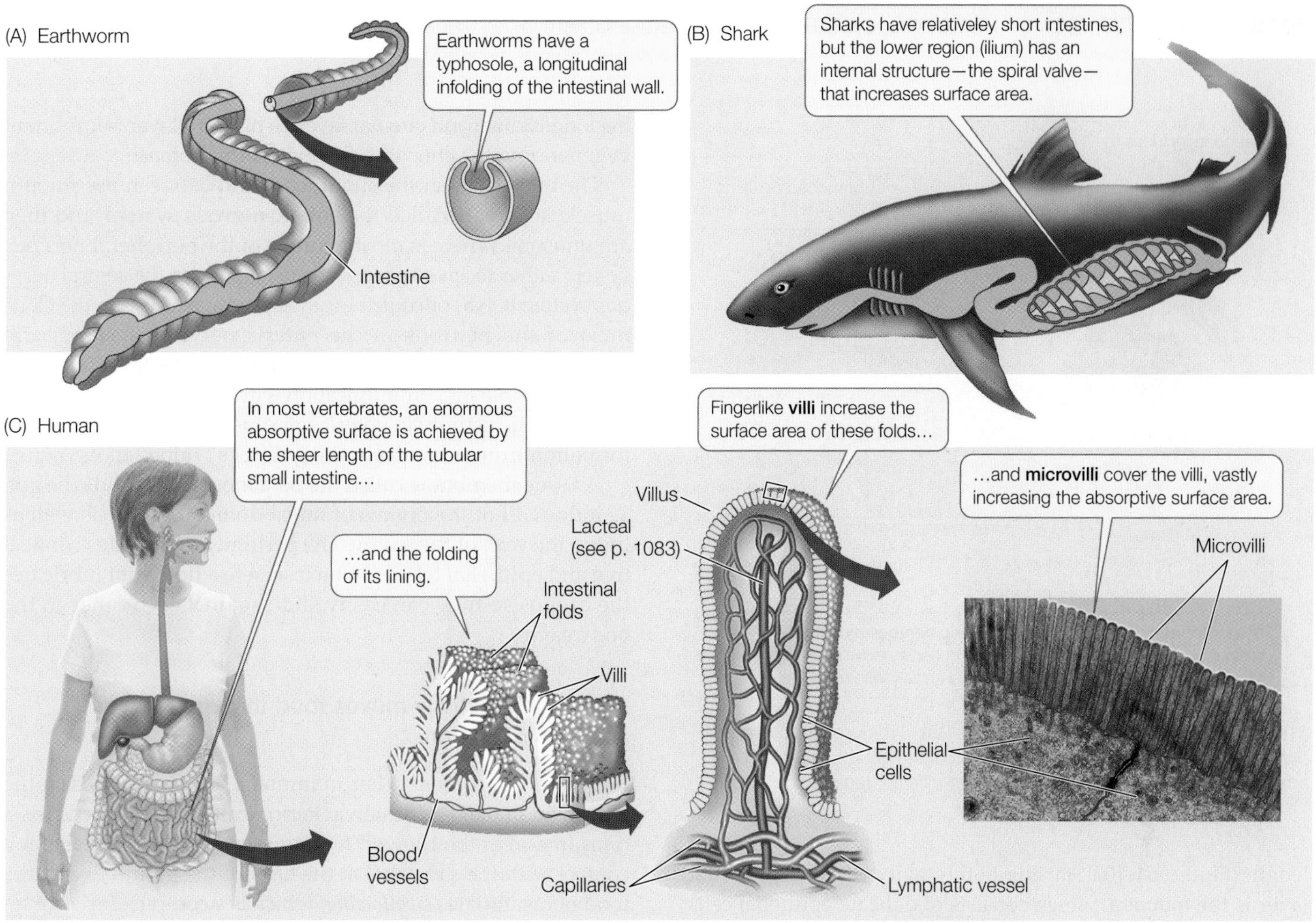

(A) Earthworm

Earthworms have a typhosole, a longitudinal infolding of the intestinal wall.

Intestine

(B) Shark

Sharks have relativeley short intestines, but the lower region (ilium) has an internal structure—the spiral valve—that increases surface area.

(C) Human

In most vertebrates, an enormous absorptive surface is achieved by the sheer length of the tubular small intestine...

...and the folding of its lining.

Fingerlike **villi** increase the surface area of these folds...

...and **microvilli** cover the villi, vastly increasing the absorptive surface area.

Villus

Lacteal (see p. 1083)

Intestinal folds

Villi

Microvilli

Epithelial cells

Blood vessels

Capillaries

Lymphatic vessel

**51.8 Intestinal Surface Area and Nutrient Absorption** Maximizing the surface area of the gut increases an animal's ability to absorb nutrients.

# 51.3 How Does the Vertebrate Gastrointestinal System Function?

Digestion in vertebrates occurs in the gastrointestinal system, which includes a tubular gut running from mouth to anus and several accessory structures that produce secretions that play important roles in digestion (**Figure 51.9**). In this section we will consider three important processes of this system: the movement of food through it, the sequential steps of digestion, and the absorption of nutrients. Our focus will be the human system.

## The vertebrate gut consists of concentric tissue layers

The tissues of the vertebrate gut are arranged in concentric layers that have a similar organization throughout its

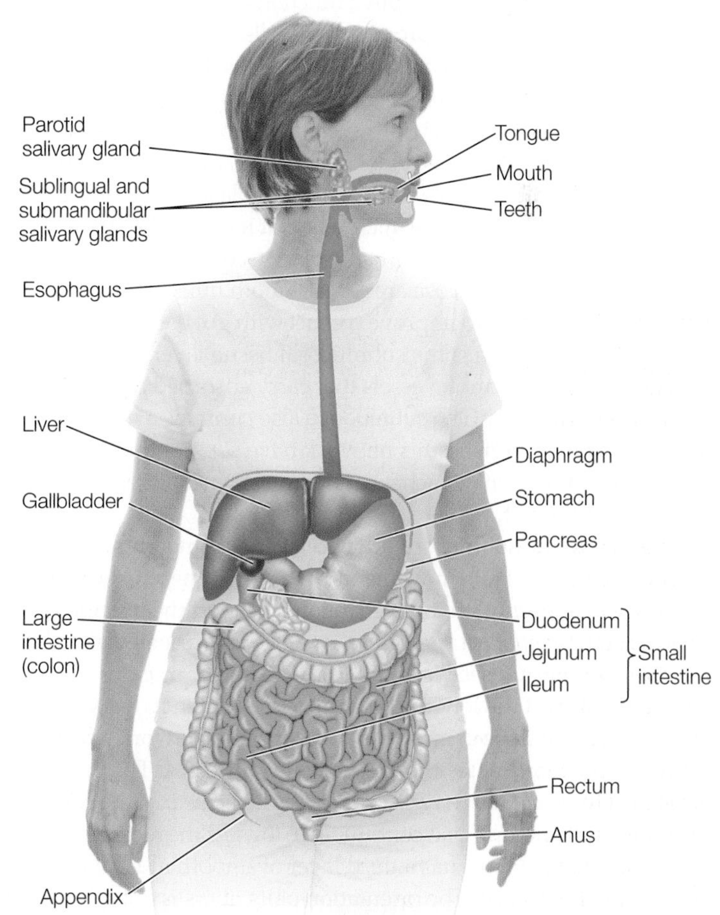


**51.9 The Human Digestive System** Different compartments in the long tubular gut specialize in digesting food, absorbing nutrients, and storing and expelling wastes. Accessory organs contribute secretions containing enzymes and other molecules.

yourBioPortal.com

GO TO Web Activity 51.4 • The Human Digestive System

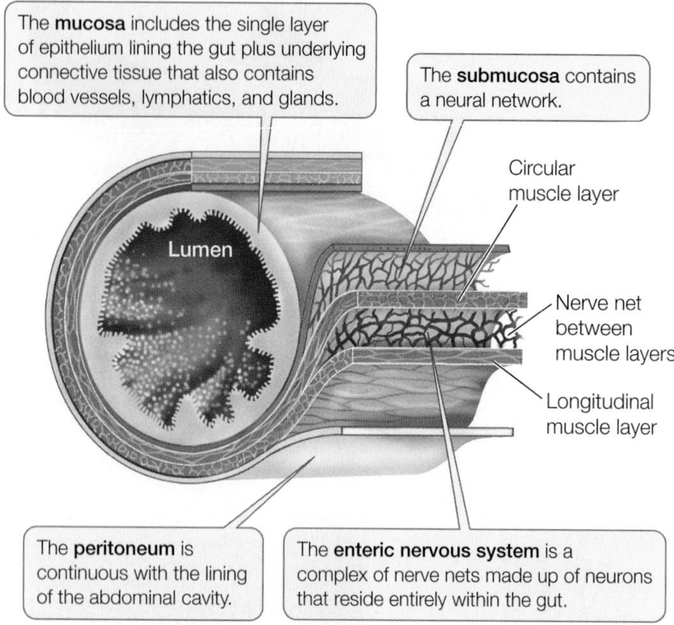

The **mucosa** includes the single layer of epithelium lining the gut plus underlying connective tissue that also contains blood vessels, lymphatics, and glands.
The **submucosa** contains a neural network.
Circular muscle layer
Lumen
Nerve net between muscle layers
Longitudinal muscle layer
The **peritoneum** is continuous with the lining of the abdominal cavity.
The **enteric nervous system** is a complex of nerve nets made up of neurons that reside entirely within the gut.

**51.10 Tissue Layers of the Vertebrate Gut** The organization of tissue layers is the same in all compartments of the gut, but specialized adaptations of specific tissues characterize different regions.

length (**Figure 51.10**). Starting in the cavity, or **lumen**, the first layer is the **mucosa**, which consists of delicate epithelial cells underlain by connective tissue. Some cells of this *mucosal epithelium* have secretory and absorptive functions. Some secrete mucus, which lubricates and protects the walls of the gut; others secrete digestive enzymes or hormones. Mucosal epithelial cells in the stomach secrete hydrochloric acid (HCl) and, as we noted in Section 51.1, some secrete intrinsic factor to aid the absorption of vitamin $B_{12}$. In some regions of the gut, nutrients are absorbed by mucosal epithelial cells. The apical plasma membranes of these absorptive cells have microvilli that increase the surface area over which absorption can take place (see Figure 51.8C).

At the base of the mucosa are some smooth muscle cells that can move the mucosa to improve contact with gut contents, and just under the mucosa is the submucosal tissue layer. Here we find the blood and lymph vessels that carry absorbed nutrients to the rest of the body. The **submucosa** also contains a network of nerves; the neurons in this network have sensory functions (responsible for stomach aches) and control various secretory functions of the gut.

External to the submucosa are two layers of smooth muscle responsible for the large movements of the gut. Innermost is the *circular muscle layer*, with its cells oriented around the gut. Outermost is the *longitudinal muscle layer*, with its cells oriented along the length of the gut (see Figure 51.10). The circular muscles constrict the gut, and the longitudinal muscles shorten it. Between the two layers of smooth muscle is another nerve network which controls and coordinates the movements of the gut. The coordinated activity of the two smooth muscle layers mixes the content of the gut and moves it continuously toward the rectum. Interestingly, the stomach has a third layer of smooth muscle that is closest to the lumen. The orientation of its fibers is oblique to the longitudinal and circular layers. This third layer is important in generating the churning motions of the stomach.

The nerve nets in the submucosa and between the smooth muscle layers are called the **enteric nervous system**, and they are unusual. Whereas most neurons of the peripheral nervous system either receive synapses from neurons in the central nervous system (CNS) or contribute synapses to neurons in the CNS, most of the neurons in the enteric nervous system form synapses only with other neurons in their network. Thus, they are responsible for communication within the gut. The CNS can influence the activity in enteric nervous system and receive information from it, but truly, the gut has a "mind" of its own.

A tissue membrane called the **peritoneum** surrounds the gut, as it does all of the organs of the abdominal cavity as well as lining the wall of the cavity. The peritoneum includes connective and epithelial tissues, which secrete a fluid that lubricates the organs so they can easily slide against each other in the body cavity.

### Mechanical activity moves food through the gut and aids digestion

In humans and most other mammals, food is chewed in the mouth and mixed with saliva. Periodically the *tongue* pushes a *bolus* (mass) of the chewed food toward the throat. By making contact with the *soft palate* at the back of the mouth cavity, the food bolus initiates *swallowing*, which is a complex series of reflexes. Swallowing propels the food through the pharynx (where the mouth cavity and nasal passages join) and into the **esophagus** (food tube). To prevent food from entering the trachea (windpipe), the *larynx* (voice box) closes, and a flap of tissue called the *epiglottis* covers the entrance to the larynx (**Figure 51.11A**).

Once a bolus of food enters the esophagus, it is moved toward the stomach both by the force of gravity and by waves of muscle contraction called **peristalsis** (**Figure 51.11B**). The muscle of the upper region of the esophagus is striated (i.e., skeletal muscle) and is controlled by the central nervous system; the muscles of the rest of the esophagus are smooth muscle controlled by the autonomic and enteric nervous systems.

The smooth muscles of the gut contract in response to being stretched. When a bolus of food reaches the smooth-muscle region of the esophagus and stretches it, the muscle responds by contracting, thus pushing the food toward the stomach. Why doesn't the contraction of the esophageal smooth muscle push the food back toward the mouth? The nerve net between the two smooth muscle layers coordinates the muscles so that contraction is always preceded by an *anticipatory wave of relaxation*. When a region of the gut smooth muscle contracts, the circular smooth muscle just beyond it relaxes while the longitudinal smooth muscle contracts, pushing the food into that area. The resulting stretch causes that circular smooth muscle to contract while the next region relaxes. In this way peristalsis moves food down the gut from the mouth to the anus.

The backward movement of food from the stomach into the esophagus is prevented by the *lower esophageal sphincter*, a thick ring of circular smooth muscle at the junction of the esopha-

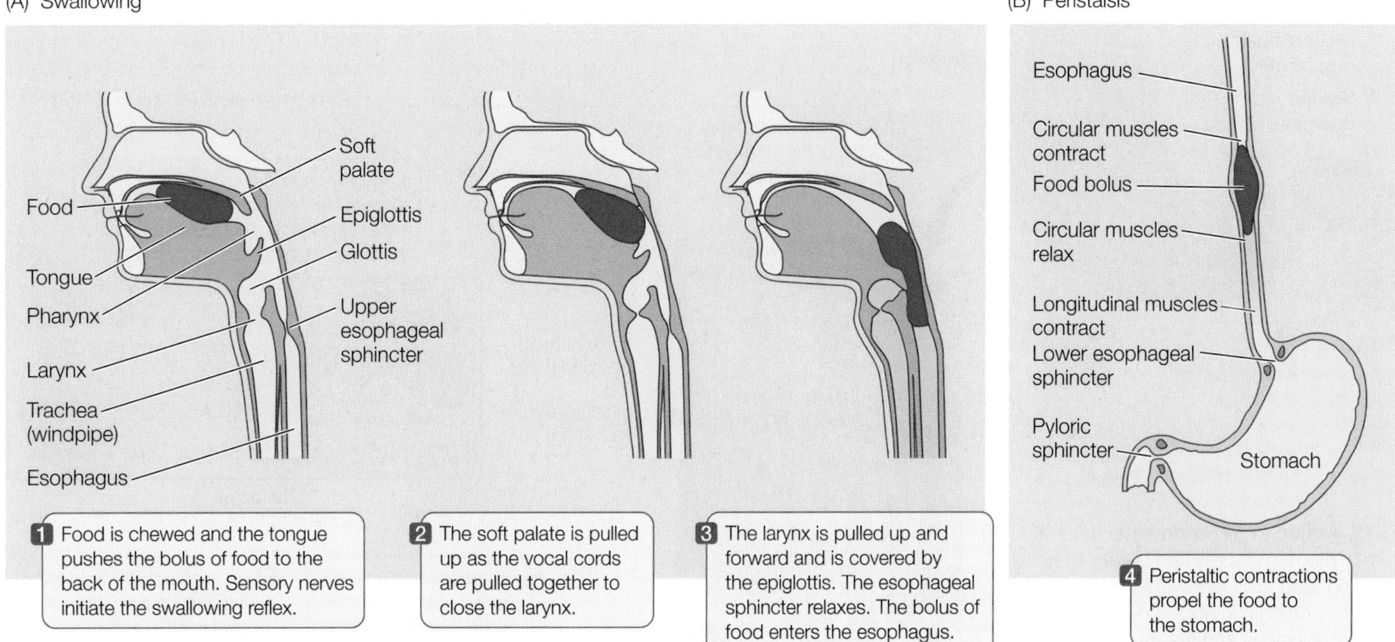
(A) Swallowing
Soft palate
Food
Epiglottis
Glottis
Tongue
Pharynx
Upper esophageal sphincter
Larynx
Trachea (windpipe)
Esophagus
1 Food is chewed and the tongue pushes the bolus of food to the back of the mouth. Sensory nerves initiate the swallowing reflex.
2 The soft palate is pulled up as the vocal cords are pulled together to close the larynx.
3 The larynx is pulled up and forward and is covered by the epiglottis. The esophageal sphincter relaxes. The bolus of food enters the esophagus.
(B) Peristalsis
Esophagus
Circular muscles contract
Food bolus
Circular muscles relax
Longitudinal muscles contract
Lower esophageal sphincter
Pyloric sphincter
Stomach
4 Peristaltic contractions propel the food to the stomach.

**51.11 Swallowing and Peristalsis** (A) Food pushed to the back of the mouth triggers the swallowing reflex. (B) Once a food bolus enters the esophagus, peristalsis propels it from mouth to anus by coordinated actions of the circular and longitudinal muscle layers of the gut.

gus and stomach. This sphincter is normally constricted, but waves of peristalsis cause it to relax enough to let food pass from the esophagus into the stomach. Sphincter muscles are found elsewhere in the digestive tract as well. The *pyloric sphincter* governs the passage of stomach contents into the intestine. Two other important sphincters also control movement of gut contents: the *ileocaecal sphincter* between the small and large intestine, and the internal *anal sphincter* which has to relax to allow defecation.

### Chemical digestion begins in the mouth and the stomach

The main role of the stomach is to store food so that digestion can occur more slowly than ingestion. However, the enzyme **amylase** is secreted by the salivary glands and mixed with food as it is chewed. Amylase hydrolyzes the bonds between the glucose monomers that make up starch molecules. The action of amylase is what makes a chewed piece of bread or cracker taste slightly sweet if you hold it in your mouth long enough.

The stomach also has secretory functions. Deep infoldings in the walls of the stomach called **gastric pits** are lined with three types of secretory cells (**Figure 51.12A**). One type secretes a proteolytic enzyme, **pepsin**, that begins the digestion of protein. Another type secretes hydrochloric acid (HCl), which kills most ingested microorganisms. This nasty mix of substances could damage the stomach walls, but a third cell type secretes mucus that provides a protective coating for the walls of the gastric pits and stomach. This mucus keeps the pH at the surface of the mucosa near neutrality and also prevents pepsin from acting on the stomach cells.

The cells in the gastric pits that secrete pepsin are called *chief cells*, and they secrete it as an inactive zymogen called **pepsinogen**. The extremely low pH of the stomach juices initiates the conversion of pepsinogen to pepsin by cleaving away a sequence of amino acids that masks the active site of the enzyme. Newly activated pepsin can act on other pepsinogen molecules to activate them, creating a positive feedback process called **autocatalysis** (**Figure 51.12B**).

The cells in the gastric pits that secrete HCl are called *parietal cells*. They can secrete so much HCl—about 2 liters per day—that they can bring the pH of the stomach contents below 1, which is the same as battery acid and 10 times more acidic than pure lemon juice. This means that across their plasma membranes, gastric pits can create a $H^+$ concentration difference of 3 million-fold. Such a feat of transport is not seen anywhere else in the body. How do they do it? Enzymes and transporters are involved.

The enzyme carbonic anhydrase in these cells catalyzes the hydration of $CO_2$ to $H_2CO_3$, which dissociates into $H^+$ and bicarbonate ion ($HCO_3^-$). An *antiporter* transport protein (see Figure 6.15) exchanges $HCO_3^-$ for $Cl^-$ on the blood side of the gastric pits, and an antiporter on the gastric pit side exchanges $H^+$ for $K^+$ (**Figure 51.12C**). However, this $K^+$ can leak out again down its concentration gradient. Thus the inward transport of $K^+$ acts like an endless conveyer belt moving $H^+$ out into the stomach lumen. $Cl^-$ also passively leaks out of the gastric lumen side of the parietal cells to maintain electrical neutrality.

### Stomach ulcers can be caused by a bacterium

The secretions of the stomach are highly corrosive to living tissues and can result in *ulcers*—places where the mucosal lining of the stomach is damaged. Stomach ulcers can lead to maladies ranging from indigestion and heartburn to gastric bleeding and

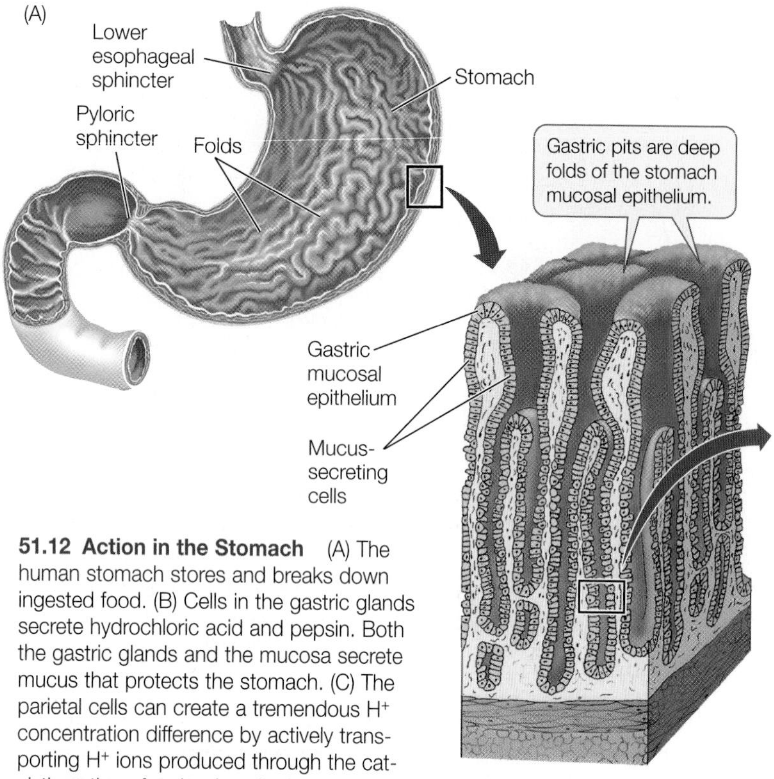

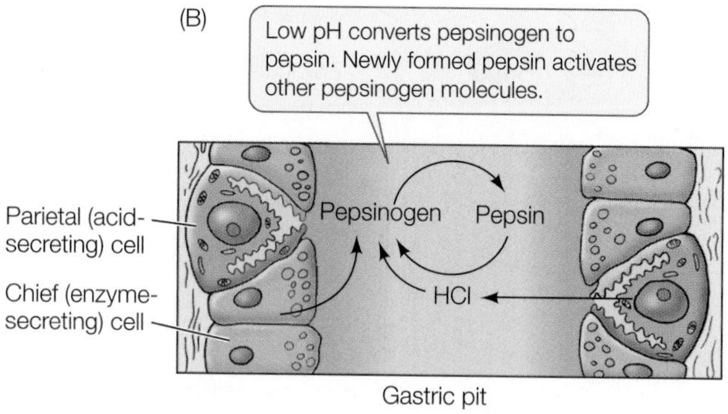

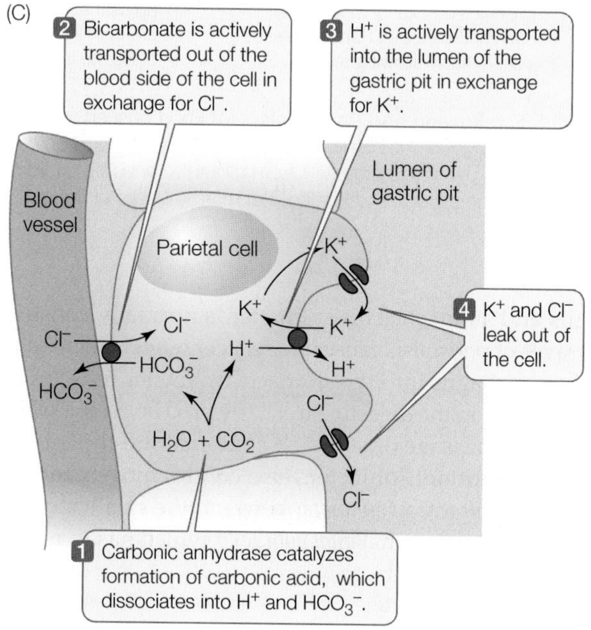

**51.12 Action in the Stomach** (A) The human stomach stores and breaks down ingested food. (B) Cells in the gastric glands secrete hydrochloric acid and pepsin. Both the gastric glands and the mucosa secrete mucus that protects the stomach. (C) The parietal cells can create a tremendous $H^+$ concentration difference by actively transporting $H^+$ ions produced through the catalytic action of carbonic anhydrase into the lumen of the gastric pit.

stomach cancer. It was logical to assume that all ulcers were due to actions of the stomach secretions on the stomach mucosa, and that stress and lifestyle issues leading to excess stomach secretions (especially HCl) were the major causes of ulcers. This view led the pharmaceutical industry to develop a plethora of drugs to decrease stomach acid production, which became "billion dollar drugs" because they were prescribed so widely. What seemed like the simplest fact in gastrointestinal medicine, however, was turned on its head by the work of two Australian researchers. Their work is a perfect example of the application of Koch's postulates (see Section 26.6) for the proof that a microorganism causes a disease (**Figure 51.13**).

In 1982, pathologist Robin Warren observed an unknown bacterium in biopsies from the stomachs of patients with ulcers. In a study of 100 patients, he and Barry Marshall of the University of Western Australia found that the bacterium was always present in the patients with ulcers. They isolated the bacterium, which they named *Helicobacter pylori*, and grew it in culture. Having thus satisfied Koch's first two postulates (microbe shown to be present in all instances of the disease, and isolating the microbe from diseased tissue), they turned to the last two, as described in Figure 51.13. Marshall actually drank a flask of the cultured bacteria (don't try this at home) and developed a pre-ulcerous condition that was subsequently cured with antibiotic. Their research showed not only that *H. pylori* causes stomach ulcers but that ulcer patients can be cured with antibiotics.

The medical profession was so certain that no microorganisms could live in the stomach and that stomach acid was the cause of ulcers that at first Warren and Marshall's findings were resisted and even ridiculed. But in 2005 they received the Nobel Prize in Medicine for their important discovery, and antibiotic therapy is now a primary treatment for stomach ulcers worldwide.

Drugs that decrease stomach acid production are still important ulcer medications for several reasons. Once a person has an ulcer, stomach acid exacerbates it. In addition, in many individuals the lower esophageal sphincter muscle (see Figure 51.12A) is inadequate to prevent acid from entering the esophagus, resulting in irritations and lesions there.

### The stomach gradually releases its contents to the small intestine

Contractions of the smooth muscles in the walls of the stomach churn its contents, thoroughly mixing them with the stomach secretions. The acidic, fluid mixture of gastric juice and partly digested food in the stomach is called **chyme**. A few substances can be absorbed across the stomach wall, including alcohol (hence its

Marshall and Warren set out to satisfy Koch's postulates:

**Test 1**

**The microorganism must be present in every case of the disease.**

**Results:** Biopsies from the stomachs of many patients revealed that the bacterium was always present if the stomach was inflamed or ulcerated.

**Test 2**

**The microorganism must be cultured from a sick host.**

**Results:** The bacterium was isolated from biopsy material and eventually grown in culture media in the laboratory.

**Test 3**

**The isolated and cultured bacteria must be able to induce the disease.**

**Results:** Marshall was examined and found to be free of bacteria and inflammation in his stomach. After drinking a pure culture of the bacterium, he developed stomach inflammation (gastritis).

**Test 4**

**The bacteria must be recoverable from the infected volunteers.**

**Results:** Biopsy of Marshall's stomach 2 weeks after he ingested the bacteria revealed the presence of the bacterium, now christened *Helicobacter pylori*, in the inflamed tissue.

**Conclusion**

Antibiotic treatment eliminated the bacteria and the inflammation in Marshall. The experiment was repeated on healthy volunteers, and many patients with gastric ulcers were cured with antibiotics. Thus Marshall and Warren demonstrated that the stomach inflammation leading to ulcers is caused by *H. pylori* infections in the stomach.

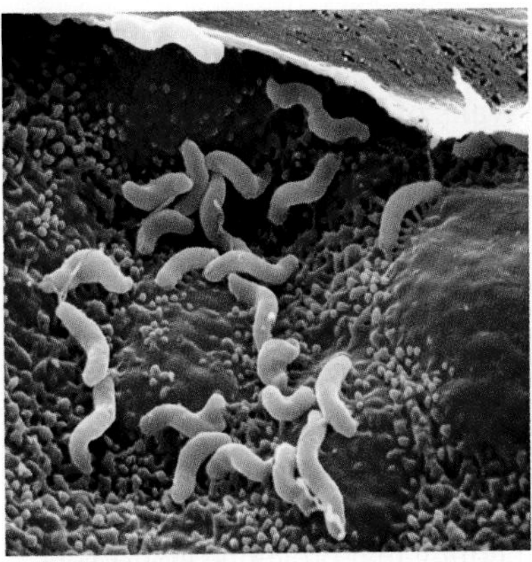

*Helicobacter pylori*

**51.13 Satisfying Koch's Postulates** Marshall and Warren showed that ulcers are caused not by the action of stomach acid but by infection with the bacterium *Helicobacter pylori*.

rapid effects), aspirin, and caffeine, but even these substances are absorbed in rather small quantities from the stomach.

Contractions of the stomach walls push the chyme toward the bottom of the stomach. These waves of contractions cause the pyloric sphincter to relax briefly so that little squirts of the chyme can enter the small intestine. In this manner, the human stomach empties itself gradually over a period of approximately 4 hours. This slow introduction of food into the small intestine enables it to work on a little material at a time.

## Most chemical digestion occurs in the small intestine

In the **small intestine**, the digestion of carbohydrates and proteins continues, and the digestion of fats and absorption of nutrients begin. The small intestine takes its name from its diameter; it is in fact a very large organ, about 6 meters long in an adult human. Given its length, and because of the folds, villi, and microvilli of its lining, its inner surface area is roughly the size of a tennis court. Across this surface, the small intestine absorbs all the nutrient molecules derived from food.

The small intestine of humans has three sections. The initial section (about 25 cm long) is called the **duodenum** and is the site of most digestion; the **jejunum** and the **ileum** (together about 600 cm) carry out 90 percent of the absorption of nutrients (see Figure 51.9).

Digestion in the small intestine requires many specialized enzymes, as well as several other secretions. Two accessory organs that are not part of the digestive tract—the liver and the pancreas—produce many of these secretions and deliver them to the lumen of the intestine through ducts.

**LIVER** The liver synthesizes **bile salts** from cholesterol and secretes them as **bile**. Bile includes bile salts and other substances, such as phospholipids and bilirubin (the breakdown product of hemoglobin). Bile flows from the liver through the *hepatic duct*. A side branch off the hepatic duct called the *cystic duct* goes to the **gallbladder**, where it is stored. Below this junction, the hepatic duct is called the *common bile duct*. Before it reaches the duodenum, it is joined by the *pancreatic duct* (**Figure 51.14**).

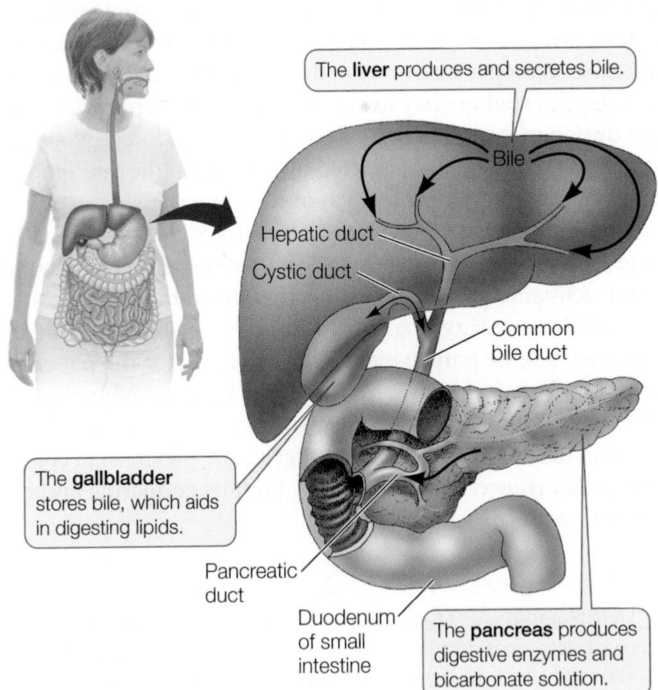

The **liver** produces and secretes bile.
Bile
Hepatic duct
Cystic duct
Common bile duct
The **gallbladder** stores bile, which aids in digesting lipids.
Pancreatic duct
Duodenum of small intestine
The **pancreas** produces digestive enzymes and bicarbonate solution.

**51.14 Ducts of the Gallbladder and Pancreas** Bile produced in the liver leaves the liver via the hepatic duct. Branching off this duct is the gallbladder, which stores bile. Below the gallbladder, the hepatic duct is called the common bile duct and is joined by the pancreatic duct before entering the duodenum.

Fat entering the duodenum stimulates cells of the duodenal epithelium to release the hormone **cholecystokinin** (**CCK**), which in turn stimulates the walls of the gallbladder to contract rhythmically. As a result, bile is squeezed out of the gallbladder and through the cystic duct to the common bile duct. A small sphincter at the junction of the common bile duct with the duodenum, the *sphincter of Oddi*, relaxes in response to waves of peristalsis and allows squirts of bile to enter the duodenal lumen.

To understand the role of bile in fat digestion, think of an oil and vinegar salad dressing. The oil, which is hydrophobic, tends to aggregate in large globules. For that reason, many salad dressings include an *emulsifier*—something that prevents oil droplets from aggregating. Mayonnaise, for example, is oil and vinegar with egg yolk added as an emulsifier. Bile contains salts that emulsify fats in the chyme and thereby greatly enlarge the surface area of the fats exposed to the **lipases**—the enzymes that digest fats. One end of each bile salt molecule is lipophilic (soluble in fat), and the other end is hydrophilic (soluble in water). The lipophilic ends of bile molecules merge with the fat droplets, leaving their hydrophilic ends sticking out. As a result, bile salts prevent the fat droplets from sticking together. The very small fat particles that result are called **micelles** (**Figure 51.15A**).

**PANCREAS** The **pancreas** is a large gland that lies just behind and below the stomach (see Figures 51.9 and 51.14). It is both an endocrine gland (secreting hormones into the interstitial fluid; see Section 41.1) and an exocrine gland (secreting digestive juices through the pancreatic duct to the gut lumen). The exocrine tissues of the pancreas produce a host of digestive enzymes, including lipases, amylases, proteases, and nucleases (**Table 51.3**). As in the stomach, the protease enzymes are released as zymogens; otherwise, they would digest the pancreas and its ducts before they ever reached the duodenum. Once in the duodenum, one of these zymogens is activated by an enzyme called *enterokinase* (secreted by cells lining the duodenum) to produce the active protease **trypsin**. Active trypsin can cleave other zymogens to release more active trypsin, as well as activate other proteases as well. This activation process is similar to the activation of pepsinogen by low pH in the stomach.

The mixture of zymogens produced by the pancreas can be dangerous if the pancreatic duct is blocked or if the pancreas is injured by infection or physical trauma such as a blow to the abdomen. A few activated trypsin molecules can initiate a chain reaction of enzyme activity that digests the pancreas (a condition called *pancreatitis*), destroying both its endocrine and exocrine functions.

The pancreas also produces a secretion rich in bicarbonate ions ($HCO_3^-$). Bicarbonate ions are alkaline (basic) and neutralize the acidic pH of the chyme that enters the duodenum from the stomach. Intestinal enzymes function best at a neutral or slightly alkaline pH.

### Nutrients are absorbed in the small intestine

The final step in digesting proteins and carbohydrates and absorbing their components occurs among the microvilli. Mucosal epithelial cells produce peptidases that cleave small peptides into absorbable amino acids. These epithelial cells also produce the enzymes maltase, lactase, and sucrase, which cleave the common disaccharides into absorbable monosaccharides—glucose, galactose, and fructose. There is also some lipase activity for fat digestion.

Many humans stop producing the enzyme lactase in childhood and thereafter have difficulty digesting lactose (the sugar in milk). Lactose is a disaccharide and cannot be absorbed without being cleaved into its constituents, glucose and galactose. Unabsorbed lactose is metabolized by bacteria in the large intestine, causing gas, diarrhea, and abdominal cramps.

The mechanisms by which cells of the intestinal epithelium absorb nutrients and inorganic ions are diverse and include dif-

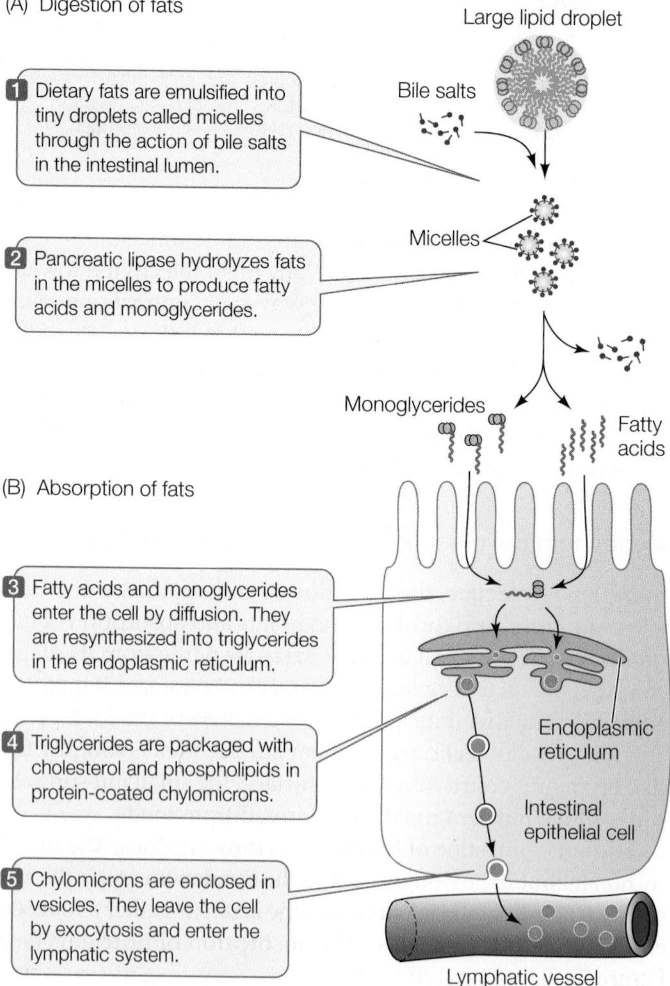

(A) Digestion of fats
Large lipid droplet
Bile salts
1 Dietary fats are emulsified into tiny droplets called micelles through the action of bile salts in the intestinal lumen.
Micelles
2 Pancreatic lipase hydrolyzes fats in the micelles to produce fatty acids and monoglycerides.
Monoglycerides
Fatty acids
(B) Absorption of fats
3 Fatty acids and monoglycerides enter the cell by diffusion. They are resynthesized into triglycerides in the endoplasmic reticulum.
4 Triglycerides are packaged with cholesterol and phospholipids in protein-coated chylomicrons.
Endoplasmic reticulum
Intestinal epithelial cell
5 Chylomicrons are enclosed in vesicles. They leave the cell by exocytosis and enter the lymphatic system.
Lymphatic vessel

**51.15 Digestion and Absorption of Fats** (A) Dietary fats are broken up by bile into small micelles that present a large surface area to lipases. (B) The products of fat digestion are absorbed by intestinal mucosal cells, where they are resynthesized into triglycerides and exported to lymphatic vessels.

**yourBioPortal.com**
GO TO **Animated Tutorial 51.1 • The Digestion and Absorption of Fats**

TABLE 51.3

**Major Digestive Enzymes of Humans**

| SOURCE/ENZYME | ACTION |
|---|---|
| **SALIVARY GLANDS** | |
| Salivary amylase | Starch → Maltose |
| **STOMACH** | |
| Pepsin | Proteins → Peptides; autocatalysis |
| **PANCREAS** | |
| Pancreatic amylase | Starch → Maltose |
| Lipase | Fats → Fatty acids and glycerol |
| Nuclease | Nucleic acids → Nucleotides |
| Trypsin | Proteins → Peptides; zymogen activation |
| Chymotrypsin | Proteins → Peptides |
| Carboxypeptidase | Peptides → Shorter peptides and amino acids |
| **SMALL INTESTINE** | |
| Aminopeptidase | Peptides → Shorter peptides and amino acids |
| Dipeptidase | Dipeptides → Amino acids |
| Enterokinase | Trypsinogen → Trypsin |
| Nuclease | Nucleic acids → Nucleotides |
| Maltase | Maltose → Glucose |
| Lactase | Lactose → Galactose and glucose |
| Sucrase | Sucrose → Fructose and glucose |

fusion, facilitated diffusion, osmosis, active transport, and co-transport. Many inorganic ions such as sodium, calcium, and iron are actively transported by these cells. For example, active $Na^+$ transporters exist on the basal and lateral sides of the epithelial cells. They maintain a low concentration of $Na^+$ in those cells so that $Na^+$ can diffuse in from the chyme in the intestinal lumen. About 30 grams of $Na^+$ are transported this way every day, and $Cl^-$ follows.

The transport of $Na^+$ and other ions is also important for water absorption because it creates an osmotic concentration gradient. At least 7 to 8 liters of water per day move through the spaces between the epithelial cells in response to this osmotic gradient. Because the water moves through *spaces* between the cells and not through the cells themselves, it can carry with it nutrients that are in solution—a transport mechanism called *solvent drag*.

Many different kinds of transport proteins exist in the epithelial cell membranes. Some, such as the transport protein for fructose, only facilitate diffusion, and that requires a concentration gradient. This mechanism works for fructose because once fructose enters the cell it is converted to glucose. Thus the concentration of fructose in the cell is always low and the concentration gradient is maintained. Transport proteins known as *symporters* (see Figure 6.15) exploit the concentration gradient of $Na^+$ between the inside and outside of the cell, which is maintained by the $Na^+$/ $Ka^+$ ATPase common to all cells. Symporters combine the transport of $Na^+$ and another molecule, such as glucose, galactose, or an amino acid. As $Na^+$ is pulled down its concentration gradient into the cell, the "hitchhiking" molecules are carried along with it.

The absorption of the products of fat digestion is relatively simple. Triglycerides are hydrolyzed to diglycerides, monoglycerides, and fatty acids, all of which are lipid-soluble and thus able to pass through the plasma membranes of the microvilli. In the intestinal epithelial cells, these molecules are resynthesized into triglycerides, combined with cholesterol and phospholipids, and coated with protein to form water-soluble **chylomicrons** (**Figure 51.15B**). Rather than enter the blood directly, chylomicrons pass into blind-ended lymph vessels called **lacteals** that are inside each villus (see Figure 51.8). They then flow through the lymphatic system, entering the bloodstream through the thoracic ducts at the base of the neck. After a meal rich in fats, chylomicrons can be so abundant in the blood that they give the plasma a milky appearance. Chylomicrons deliver their triglyceride and cholesterol cargo as they circulate through tissues.

The bile salts that emulsify fats are not absorbed along with the monoglycerides, diglycerides, and the fatty acids, but are shuttled back and forth between the gut contents and the microvilli. In the ileum, bile salts are actively reabsorbed and returned to the liver via the bloodstream.

### Absorbed nutrients go to the liver

Blood leaving the digestive tract flows to the liver in the *hepatic portal vein*. This large vein delivers the blood to small spaces called sinusoids between groups of liver cells. These cells absorb the nutrients coming from the digestive tract and either store them or convert them to molecules the body needs. Glucose, sucrose, and fructose are used to synthesize glycogen. Amino acids are used to build proteins. Lipids from the chylomicrons are either stored as triglycerides or used to make *lipoproteins*, which are released by the liver and deliver the triglycerides and cholesterol to other tissues (see Section 51.4).

### Water and ions are absorbed in the large intestine

The motility of the small intestine gradually pushes its contents into the *large intestine*, or **colon**. Most of the available nutrients have been removed from the chyme that enters the colon, but it contains a lot of water and inorganic ions.

The colon absorbs water and ions, producing semisolid feces from the chyme it receives from the small intestine. Feces are stored in the rectum of the colon until they are eliminated. Absorption of too much water from the colon can cause *constipation*. The opposite condition, *diarrhea*, results if too little water is absorbed; in this case, water in the colon is excreted with the feces. The excessive diarrhea caused by diseases such as cholera can produce such rapid loss of water and electrolytes that death can occur in hours.

### Herbivores rely on microorganisms to digest cellulose

As the primary component of plant cell walls, cellulose is the principal component of the food of herbivores. Most herbivores, however, cannot produce *cellulases*—enzymes that hydrolyze

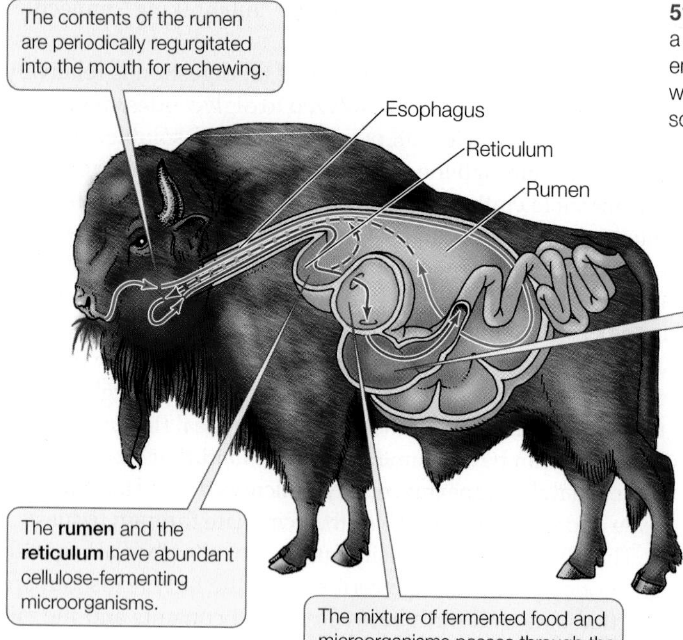
The contents of the rumen are periodically regurgitated into the mouth for rechewing.
Esophagus
Reticulum
Rumen
The **abomasum** is the "true" stomach, secreting HCl and proteases. The microorganisms are killed by the HCl, digested by the proteases, and passed on to the small intestine for further digestion.
The **rumen** and the **reticulum** have abundant cellulose-fermenting microorganisms.
The mixture of fermented food and microorganisms passes through the **omasum**, where it is concentrated by water absorption.

**51.16 A Ruminant's Stomach** Bison, like their relatives domestic cattle, have a specialized stomach with four compartments that enables them to obtain energy from coarse plant material through bacterial fermentation of the otherwise indigestible plant material. The bacteria themselves become an important source of nutrition.

cellulose. Exceptions include silverfish (insects that eat books and stored papers), earthworms, and shipworms. From termites to cattle, herbivores rely on microorganisms in their digestive tracts to digest cellulose.

The stomachs of **ruminants** (cud chewers) such as cattle are large, four-chambered organs that take advantage of their endosymbiotic microorganisms (**Figure 51.16**). The first two chambers, the **rumen** and the **reticulum**, are packed with microorganisms that break down cellulose by fermentation. The ruminant periodically regurgitates the contents of the rumen (the *cud*) into the mouth for rechewing. When swallowed again, the vegetal fibers present more surface area to the microorganisms. The microorganisms metabolize cellulose and other nutrients to simple fatty acids, which become nutrients for their host.

Enormous numbers of microorganisms leave the rumen along with the partially digested food. This mass is concentrated by water absorption in the **omasum** before it enters the true stomach, the **abomasum**, where the microorganisms are killed by secreted hydrochloric acid, digested by proteases, and passed on to the small intestine for further digestion and absorption. A cow derives more than 100 grams of protein per day from digestion of its endosymbiotic microorganisms. The rate of multiplication of microorganisms in the rumen offsets their loss, so a well-balanced, mutually beneficial relationship is maintained.

Some mammalian herbivores have a microbial fermentation chamber called a **cecum** extending from the large intestine. An example is the rabbit (see Figure 51.7). Since the cecum empties into the large intestine, absorption of some nutrients produced by the microorganisms is inefficient, because of the large intestine's limited surface area. Such species frequently produce two kinds of feces—ones that are pure waste and ones that contain cecal material. In a behavior known as **coprophagy**, these species reingest the cecal feces directly from the anus so they can digest and absorb the nutrients that would otherwise be lost. In humans, the cecum has become the vestigial **appendix**, which serves no digestive function.

## 51.3 RECAP

**The vertebrate gastrointestinal system is a tubular gut that is adapted to ingest food, fragment it, digest it, and absorb nutrients. Peristalsis moves food through the gut. Digestion and absorption of nutrients occur mostly in the small intestine; water and ions are absorbed in the large intestine.**

- What digestive functions occur in the mouth and stomach? See p. 1079 and Figure 51.12
- How do bile salts assist in the digestion of fats? See p. 1082 and Figure 51.15
- Describe how symporters drive the absorption of nutrients. See p. 1083

The steps included in ingestion and digestion of food—from fragmentation in the mouth to the digestive processes in the gastrointestinal tract—make the nutrients in food available for absorption and ultimately for metabolism. Let's look at how the processes of digestion are controlled and how nutrients are handled by the body once food has been digested.

## 51.4 How Is the Flow of Nutrients Controlled and Regulated?

The vertebrate gut is an assembly line in reverse—a *dis*assembly line. As with a standard assembly line, the control and coordination of the sequential processes of digestion is critical. Both neuronal and hormonal controls govern these processes. Once the products of digestion are absorbed, their availability to the cells of the body must also be controlled.

You have certainly experienced salivation at the sight or smell of food. That response is an *unconscious reflex*, as is swallowing. Many such autonomic reflexes coordinate activity in different regions of the digestive tract. For example, the introduction of food into the stomach stimulates increased activity in the colon, which can lead to defecation.

As we already mentioned, the digestive tract has an intrinsic nervous system. Neuronal messages can travel from one region of the digestive tract to another without being processed by the CNS. One function of the gut's nervous system is coordinating the movement of food through the gut. Of course, this intrinsic nervous system communicates information to the CNS and receives input from the CNS, but its most important role is to coordinate actions throughout the digestive tract. In spite of this marvelous intrinsic nervous system, however, much of the control and regulation of the digestive system and nutrient management involves hormonal mechanisms.

### Hormones control many digestive functions

Several hormones control the activities of the digestive tract and its accessory organs (**Figure 51.17**). The first hormone ever discovered came from the duodenum; it was called **secretin** because it causes the pancreas to secrete digestive juices. We now know that secretin is only one of several hormones that control pancreatic secretion; specifically, secretin stimulates the pancreas to secrete a solution rich in bicarbonate ions.

The stimulus that causes the duodenum to release secretin is low pH—the arrival of acidic chyme from the stomach. Similarly, the presence of fats and proteins in the chyme stimulates the release of cholecystokinin (CCK), the hormone mentioned earlier that stimulates the gallbladder to release bile. CCK also stimulates the pancreas to release digestive enzymes. Both CCK and secretin slow the movements of the stomach, which slows the delivery of chyme into the small intestine, allowing more complete digestion in the duodenum.

The presence of food in the stomach stimulates cells in the lower region of the stomach to secrete a hormone called **gastrin**. Gastrin returns to the stomach in the blood and stimulates the secretion of digestive juices and also the movements of the stomach. Gastrin release begins to be inhibited when the pH of the stomach contents falls below 3—an example of negative feedback.

Most animals do not eat continuously, so they can be either in an **absorptive state** (food in the gut) or in a **postabsorptive state** (no food in the gut). Nutrient requirements for energy metabolism and biosynthesis are continuous, however. Thus, nutrient traffic must be controlled so that reserves accumulate in the liver, muscle, and adipose (fat) tissue while the animal is in the absorptive state and are used appropriately during the postabsorptive state.

### The liver directs the traffic of the molecules that fuel metabolism

When fuel molecules are abundant in the blood, the liver stores them in the form of glycogen and fats. The liver also synthesizes blood plasma proteins from circulating amino acids. When fuel molecule levels in the blood decline, the liver taps its reserves and delivers nutrients into the blood.

The liver has an enormous capacity to interconvert fuel molecules. Liver cells can convert monosaccharides into either glycogen or fats, and vice versa. The liver can also convert certain amino acids and some other molecules, such as pyruvate and lactate, into glucose—a process termed *gluconeogenesis* (see Section 9.5). The liver is also the major controller of fat metabolism through its production of lipoproteins. A **lipoprotein** is a particle made up of a core of hydrophobic fat and cholesterol with a covering of hydrophilic protein that allows it to be suspended in water.

**LIPOPROTEINS: THE GOOD, THE BAD, AND THE UGLY** Lipoproteins move fats, the most abundant fuel reserve in the body, from sites of storage to sites of use. We saw in the previous section how,

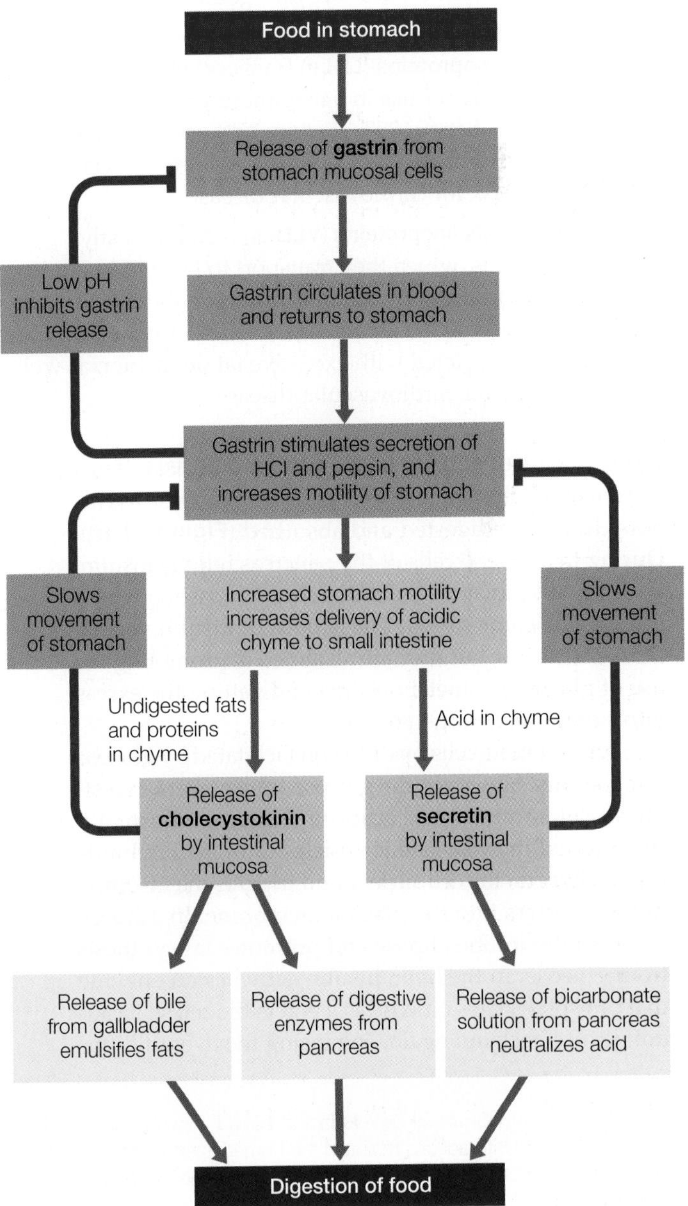


**51.17 Hormones Control Digestion** The hormones gastrin, cholecystokinin, and secretin are involved in feedback loops that control the sequential processing of food in the digestive tract. Red lines indicate inhibitory actions; green lines indicate stimulatory actions.

in the intestine, bile solves the problem of processing hydrophobic fats in an aqueous medium. The transport of fats in the circulatory system presents the same problem, and lipoproteins provide the solution.

The chylomicrons (see p. 1083) produced by the mucosal cells of the intestine are the largest lipoprotein particles in the blood. As the circulation carries chylomicrons through the liver and adipose tissue, lipoprotein lipases begin to break them down, and their triglyceride and cholesterol cargo is absorbed into the liver or fat cells.

Lipoproteins other than chylomicrons are synthesized in the liver. These lipoproteins can be classified according to their density. Fat has a low density (it floats in water) and protein has a high density, so the greater the fat-to-protein ratio in the lipoprotein, the lower its density.

- **High-density lipoproteins (HDLs)** remove cholesterol from tissues and carry it to the liver, where it can be used to synthesize bile. HDL consists of about 50% protein, 35% lipids, and 15% cholesterol. These are the "good" lipoproteins, and their levels are higher in people who exercise and are fit.
- **Low-density lipoproteins (LDLs)** transport cholesterol around the body for use in biosynthesis and for storage. LDL consists of about 25% protein, 25% lipids, and 50% cholesterol. These are the "bad" lipoproteins associated with a high risk for cardiovascular disease.
- **Very low-density lipoproteins (VLDLs)** contain mostly triglyceride fats, which they transport to fat cells in adipose tissues around the body. VLDL consists of about 2% protein, 94% lipids, and 3% cholesterol. These are the "ugly," as they are associated with excessive fat deposition as well as a high risk for cardiovascular disease.

**INSULIN AND GLUCAGON CONTROL FUEL METABOLISM** During the absorptive state, blood glucose levels rise as carbohydrates are digested and absorbed (**Figure 51.18**). During this time, β cells of the pancreas release insulin, which plays a major role in directing glucose to where it will be used or stored. The actions of insulin vary in different tissues, but they are all aimed at promoting the use of glucose for metabolic fuel and getting the excess into storage as glycogen or fat.

Glucose enters cells by diffusion facilitated by glucose transporters. However, the glucose transporters in resting skeletal muscle and adipose tissues are normally sequestered in cytoplasmic vesicles until insulin binds its receptors on the cell surface and triggers the insertion of transporters into the plasma membrane. In adipose cells, insulin inhibits lipase and promotes fat synthesis from glucose. In the liver, insulin activates an enzyme that phosphorylates glucose as it enters the cell so it cannot diffuse back out again, enhancing the overall diffusion of glucose into the cells. In liver cells, insulin also activates the enzymes that catalyze the synthesis of glycogen.

During the postabsorptive state, a fall in blood glucose decreases the release of insulin, and the uptake of glucose by most cells is curtailed (see Figure 51.18). To maintain blood glucose levels, liver cells break down their stored glycogen, which releases glucose into the blood. The liver and adipose tissues supply fatty acids to the blood, and most cells preferentially use fatty acids as their metabolic fuel.

One tissue that does not switch fuel sources when an animal is postabsorptive is the nervous system. The cells of the nervous system require a constant supply of glucose, and they can use other fuels to a very limited extent. Most neurons do not require insulin to absorb glucose from the blood, but they do need an adequate glucose concentration gradient to drive the facilitated diffusion of glucose across their plasma membranes. Therefore it is critical that blood glucose levels are maintained when an animal is postabsorptive. The overall dependence of neural tissues on glucose, and their requirement for constant blood glucose levels, are the reasons it is so important for other cells of the body to shift to fat metabolism during the postabsorptive state.

The metabolism of fuel molecules during the postabsorptive state is mostly controlled by the lack of insulin, but if blood glucose falls below a certain level, another pancreatic hormone, **glucagon**, is called into play. Glucagon's effect is opposite that of insulin: it stimulates liver cells to break down glycogen and to carry out gluconeogenesis. Thus, under the influence of glucagon, the liver produces glucose and releases it into the blood.

yourBioPortal.com
GO TO Animated Tutorial 51.2 • Insulin and Glucose Regulation

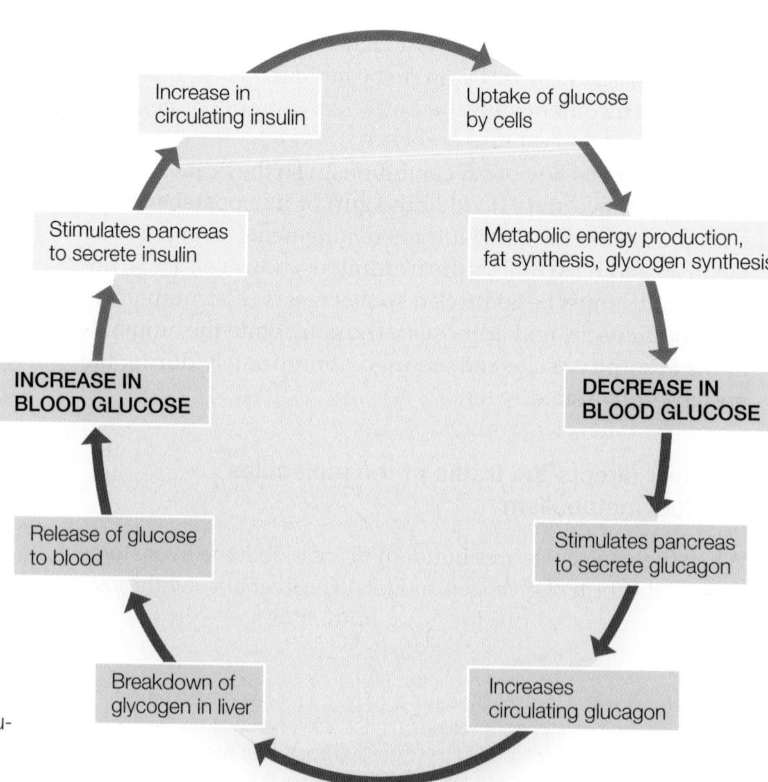


**51.18 Regulating Glucose Levels in the Blood** Insulin (blue) and glucagon (brown) interactions maintain the homeostasis of circulating glucose. It is important for blood glucose to remain stable because it is the essential source fuel for the nervous system.

### Regulating food intake is important

Obesity is a major health issue in the United States. A simple rule—take in fewer calories than your body burns while eating a balanced diet—should solve the problem, but it doesn't. Why? As we noted at the beginning of this chapter, lifestyle plays a major role in obesity, and genetic and regulatory factors "weigh in" as well.

The amount of food an animal eats is governed by its sensations of hunger and satiety. These sensations are influenced by the hypothalamus (see Section 41.2). If a region in the middle of the hypothalamus of rats (the *ventromedial hypothalamus*) is experimentally damaged, the animals will increase their food intake and become obese. If a different region, called the lateral hypothalamus, is damaged, they will decrease their food intake and become thin. In both cases, the rats eventually reach a new equilibrium body weight, which they maintain. Thus, regulation is maintained, but the set point is changed. Other brain regions have also been implicated in control of hunger and satiety.

In Section 40.1 we noted that regulation involves feedback information and a means of comparing that information with a set point. Criteria for a negative feedback signal for food intake include:

- Levels of the signal in the circulation reflect body energy reserves.
- The signal crosses the blood–brain barrier.
- Presence of the signal in the brain decreases food intake and body mass.
- Blocking the signal's actions in the brain results in increased food intake and gain of body mass.

Two molecules—insulin and leptin—satisfy these criteria. We have already discussed the roles of insulin in fuel management (see Figure 51.18); it may also have a role in signaling the hypothalamus about the fuel stores of the body.

The protein **leptin** (Greek *leptos*, "thin") was discovered in a strain of mice that eat enormous amounts of food and become obese. Experiments revealed this trait to be due to a recessive allele of a single gene, which they designate *ob*. It was subsequently learned that the wild-type *Ob* gene codes for leptin. The obese mice carried two copies of the recessive allele for the leptin gene and thus were designated *ob/ob*.

Leptin is a circulating hormone produced by fat cells. Receptors for leptin are found in the regions of the hypothalamus involved in controlling hunger and satiety. It seems that leptin provides feedback information to the brain about the status of the body's fat reserves. When leptin was injected into *ob/ob* mice, they ate less and lost body fat. However, a different strain of obese mice, *db/db*, did not respond to leptin injections. Further experiments revealed that *db/db* mice lacked the receptor for leptin (**Figure 51.19**).

Could leptin be used to treat human obesity? In the few obese people who do not produce leptin, injections curb appetite and facilitate weight loss. Most obese people, however, have higher than normal levels of leptin in their blood, so the likelihood is that it is their leptin receptors that are not completely functional.

## INVESTIGATING LIFE

**51.19 A Single-Gene Mutation Leads to Obesity in Mice** In mice the *Ob* gene codes for the protein leptin, a satiety factor that signals the brain when enough food has been consumed. The recessive *ob* allele is a loss-of-function allele, so *ob/ob* mice do not produce leptin; they do not experience satiety and become obese. The *Db* gene encodes the leptin receptor, so mice homozygous for the recessive loss-of-function allele *db*, even if they produce leptin, cannot use it and so become obese.

**HYPOTHESIS** **Mice who cannot produce the satiety signal protein leptin will not become obese if they are able to obtain leptin from an outside source.**

**METHOD**

1. Create two strains of genetically obese laboratory mice, one of which lacks functional leptin (genotype *ob/ob*) and one which lacks the receptor for leptin (genotype *db/db*).
2. Create parabiotic pairs by surgically joining the circulatory systems of a non-obese (wild-type) mouse with a partner from one of the obese strains.
3. Allow mice to feed at will.

Parabiotic pair

Wild-type mouse (*Ob/– and Db/–*) Genetically obese mouse (either *ob/ob or db/db*)

**RESULTS** Parabiotic *ob/ob* mice obtain leptin from the wild-type partner and lose fat. Parabiotic *db/db* mice remain obese because they lack the leptin receptor and thus the leptin they obtain from their partner has no effect.

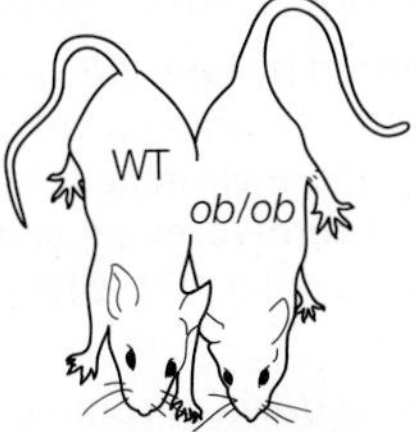


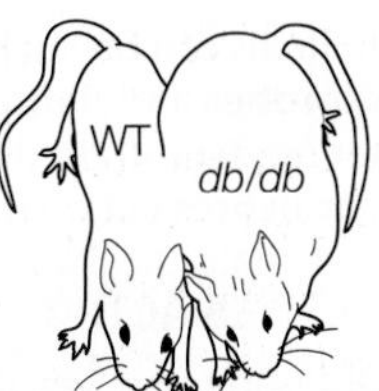


**CONCLUSION** **The protein leptin is a satiety signal that acts to prevent overeating and resultant obesity.**

Go to **yourBioPortal.com** for original citations, discussions, and relevant links for all INVESTIGATING LIFE figures.

Additional feedback signals are involved in regulating food intake. A hormone called *peptide YY* comes from the gut and reduces appetite in both rodents and humans. In contrast, *ghrelin*, a hormone produced and secreted by cells in the stomach, rises before a meal and falls after a meal. Fasting causes an increase in ghrelin levels. Ghrelin binding to its receptors in the hypothalamus stimulates appetite. Ghrelin also stimulates cells in the pituitary gland to release growth hormone.

Most interesting, however, is the discovery of the central role of the enzyme AMP-activated protein kinase (AMPK) as a signal of nutrient insufficiency. When most cells are nutrient deprived, they produce AMPK, which stimulates the oxidation of substrates to replenish ATP. However, in the hypothalamus, increased levels of AMPK stimulate food intake, and lowered levels of AMPK inhibit food intake. In addition, AMPK activity in the hypothalamus is inhibited by leptin and insulin but is stimulated by ghrelin. Thus, AMPK could be a final common pathway for various signals controlling food intake.

## 51.4 RECAP

**The major controlling factors of gut function are an intrinsic nervous system and the hormones gastrin, secretin, and cholecystokinin. Insulin is the major hormonal controller of fuel metabolism. The hypothalamus controls food intake by generating sensations of hunger and satiety influenced by feedback from blood glucose and hormones, including leptin and ghrelin.**

- What are the roles of the three different classes of lipoproteins? See pp. 1085–1086
- By what actions does insulin promote uptake and storage of energy during the absorptive state? See p. 1086 and Figure 51.18
- What evidence supports the hypothesis that leptin influences satiety? See p. 1087 and Figure 51.19

## CHAPTER SUMMARY

### 51.1 What Do Animals Require from Food?

- Animals are **heterotrophs** that derive their energy and molecular building blocks, directly or indirectly, from **autotrophs**.
- Carbohydrates, fats, and proteins in food supply animals with metabolic energy. A measure of the energy content of food is the **kilocalorie** (**kcal**). Excess caloric intake is stored as glycogen and fat. Review Figure 51.2
- For many animals, food provides essential **carbon skeletons** that they cannot synthesize themselves. Review Figure 51.4
- Most researchers consider 8 amino acids to be essential for adult humans; some believe that infants require as many as 12 **essential amino acids** in their diet. **Macronutrients** are mineral elements needed in large quantities, **micronutrients** are needed in small amounts. Review Figure 51.5, Table 51.1, **WEB ACTIVITY 51.1**
- **Vitamins** are organic molecules that must be obtained in food. Review Table 51.2, **WEB ACTIVITY 51.2**
- **Malnutrition** results when any essential nutrient is lacking from the diet. Chronic malnutrition causes **deficiency disease**.

### 51.2 How Do Animals Ingest and Digest Food?

- Animals can be characterized by how they acquire nutrients: **saprobes** and **detritivores** depend on dead organic matter, **filter feeders** strain the aquatic environment for small food items, **herbivores** eat plants, and **carnivores** eat animals. Behavioral and anatomical adaptations reflect these feeding strategies. **SEE WEB ACTIVITY 51.3**
- Digestion involves the breakdown of complex food molecules into monomers that can be absorbed and utilized by cells. In most animals, digestion takes place in a tubular gut. Review Figure 51.7
- Absorptive areas of the gut are characterized by a large surface area produced by extensive folding and numerous **villi** and **microvilli**. Review Figure 51.8
- Hydrolytic enzymes break down proteins, carbohydrates, and fats into their monomeric units. To prevent the organism itself from being digested, many of these enzymes are released as inactive **zymogens**, which become activated when secreted into the gut.

### 51.3 How Does the Vertebrate Gastrointestinal System Function?

- The vertebrate gut can be divided into several compartments with different functions. Review Figure 51.9, **WEB ACTIVITY 51.4**
- The cells and tissues of the vertebrate gut are organized in the same way throughout its length. The innermost tissue layer, the **mucosa**, is the secretory and absorptive surface. The **submucosa** contains blood and lymph vessels and a nerve network. External to the submucosa are two smooth muscle layers. Between the two muscle layers is another nerve network that controls the movements of the gut. Review Figure 51.10
- Swallowing is a reflex that pushes a bolus of food into the **esophagus**. **Peristalsis** and other movements of the gut move the bolus down the esophagus and through the entire length of the gut. Sphincters block the gut at certain locations, but they relax as a wave of peristalsis approaches. Review Figure 51.11
- Digestion begins in the **mouth**, where **amylase** is secreted with the saliva. Digestion of protein begins in the **stomach**, where parietal cells secrete HCl and chief cells secrete **pepsinogen**, a zymogen that becomes **pepsin** when activated by low pH and **autocatalysis**. The mucosa also secretes mucus, which protects the tissues of the gut. Review Figure 51.12
- In the **duodenum**, pancreatic enzymes carry out most of the digestion of food. Bile from the liver and **gallbladder** emulsify fats into micelles. Bicarbonate ions from the **pancreas** neutralize the pH of the **chyme** entering from the stomach to produce an environment conducive to the actions of pancreatic enzymes such as **trypsin**. Review Figure 51.14 and Table 51.3
- Final enzymatic cleavage of polypeptides and disaccharides occurs among the microvilli of the intestinal mucosa. Amino acids, monosaccharides, and inorganic ions are absorbed by the microvilli. Specific transporter proteins are sometimes involved. Symporters often power the active transport of nutrients.

- Fats broken down by **lipases** are absorbed mostly as monoglycerides and fatty acids and are resynthesized into triglycerides within cells. The triglycerides are combined with cholesterol and phospholipids and coated with protein to form **chylomicrons**, which pass out of the mucosal cells and into lymphatic vessels in the submucosa. Review Figure 51.15, **ANIMATED TUTORIAL 51.1**
- Water and ions are absorbed in the large intestine as waste matter and consolidated into **feces**, which are periodically eliminated.
- Microorganisms in some compartments of the gut digest materials that their host cannot. Review Figure 51.16

## 51.4 How Is the Flow of Nutrients Controlled and Regulated?

- Autonomic reflexes coordinate activity of the digestive tract, which has an intrinsic nervous system that can act independently of the CNS.
- The actions of the stomach and small intestine are largely controlled by the hormones **gastrin**, **secretin**, and **cholecystokinin**. Review Figure 51.17
- The liver plays a central role in directing the traffic of fuel molecules. In the **absorptive state**, the liver takes up and stores fats and carbohydrates, converting monosaccharides to glycogen or fats. The liver also takes up amino acids and uses them to produce blood plasma proteins, and can engage in gluconeogenesis.
- Fat and cholesterol are shipped out of the liver as **low-density lipoproteins**. **High-density lipoproteins** act as acceptors of cholesterol and are believed to bring fat and cholesterol back to the liver.
- Insulin largely controls fuel metabolism during the absorptive state and promotes glucose uptake as well as glycogen and fat synthesis. In the **postabsorptive state**, lack of insulin blocks the uptake and utilization of glucose by most cells of the body except neurons. If blood glucose levels fall, **glucagon** secretion increases, stimulating the liver to break down glycogen and release glucose to the blood. Review Figure 51.18, **ANIMATED TUTORIAL 51.2**
- Food intake is governed by sensations of hunger and satiety, which are determined by brain mechanisms responding to feedback signals such as insulin, **leptin**, and **ghrelin**. Review Figure 51.19

## SELF-QUIZ

1. Most of the metabolic energy needed by a bird for a long-distance migratory flight is stored as
   *a.* glycogen.
   *b.* fat.
   *c.* protein.
   *d.* carbohydrates.
   *e.* ATP.
2. Which statement about essential amino acids is true?
   *a.* They are not found in vegetarian diets.
   *b.* They are stored by the body until they are needed.
   *c.* Without them, one is undernourished.
   *d.* All animals require the same ones.
   *e.* Humans can acquire all of theirs by eating milk, eggs, and meat.
3. Which statement about vitamins is true?
   *a.* They are essential inorganic nutrients.
   *b.* They are required in larger amounts than are essential amino acids.
   *c.* Many serve as coenzymes.
   *d.* Vitamin D can be acquired only by eating meat or dairy foods.
   *e.* When vitamin C is eaten in large quantities, the excess is stored in fat for later use.
4. The digestive enzymes of the small intestine
   *a.* do not function best at a low pH.
   *b.* are produced and released in response to circulating secretin.
   *c.* are produced and released under neural control.
   *d.* are all secreted by the pancreas.
   *e.* are all activated by an acidic environment.
5. Which statement about nutrient absorption by the intestinal mucosal cells is true?
   *a.* Carbohydrates are absorbed as disaccharides.
   *b.* Fats are absorbed as fatty acids and monoglycerides.
   *c.* Amino acids move across the plasma membrane only by diffusion.
   *d.* Bile transports fats across the plasma membrane.
   *e.* Most nutrients are absorbed in the duodenum.
6. Chylomicrons are like the tiny micelles of dietary fat in the lumen of the small intestine in that both
   *a.* are coated with bile.
   *b.* are lipid soluble.
   *c.* travel through the lymphatic system.
   *d.* contain triglycerides.
   *e.* are coated with lipoproteins.
7. Microbial fermentation in the gut of a cow
   *a.* produces fatty acids as a major nutrient for the cow.
   *b.* occurs in specialized regions of the small intestine.
   *c.* occurs in the cecum, from which food is regurgitated, chewed again, and swallowed into the true stomach.
   *d.* produces methane as a major nutrient.
   *e.* is possible because the stomach wall does not secrete hydrochloric acid.
8. Which of the following is stimulated by cholecystokinin?
   *a.* Stomach motility
   *b.* Release of bile
   *c.* Secretion of hydrochloric acid
   *d.* Secretion of bicarbonate ions
   *e.* Secretion of mucus
9. During the absorptive state,
   *a.* breakdown of glycogen supplies glucose to the blood.
   *b.* glucagon secretion is high.
   *c.* the number of circulating lipoproteins is low.
   *d.* glucose is the major metabolic fuel.
   *e.* the synthesis of fats and glycogen in muscle is inhibited.
10. During the postabsorptive state,
   *a.* glucose is the major metabolic fuel.
   *b.* glucagon stimulates the liver to produce glycogen.
   *c.* insulin facilitates the uptake of glucose by brain cells.
   *d.* fatty acids constitute the major metabolic fuel.
   *e.* liver functions slow down because of low insulin levels.

## FOR DISCUSSION

1. Several currently popular diet books recommend high fat and protein intake and low carbohydrate intake as a means of losing body mass. What is the rationale for a high-fat and high-protein diet, and what health issues should be considered when someone considers going on such a diet?
2. Carnivores generally have more dietary vitamin requirements than herbivores do. Why?
3. It is said that the most important hormonal control of fuel metabolism in the postabsorptive state is the lack of insulin. Explain.
4. Why is obstruction of the common bile duct so serious? Consider in your answer the multiple functions of the pancreas and the way in which digestive enzymes are processed.
5. Trace the history of a fatty acid molecule from a slice of cheese pizza to a plaque on a coronary artery. Into what possible forms and structures might it have been converted as it passed through the body? Describe a direct and an indirect route it could have taken.

## ADDITIONAL INVESTIGATION

You learned about brown fat in Chapter 40. This adipose tissue has lots of mitochondria and a rich blood supply and is found in many newborn mammals and cold-adapted mammals such as hibernators. The function of brown fat is to produce heat through uncoupling of oxidative phosphorylation and therefore the metabolism of lipids without the production of ATP. It has been thought for a long time that adult humans do not have brown fat, but recent studies have indicated that they do. It has further been hypothesized that the amount of brown fat an individual has is highly variable, and that this might explain why some individuals are more resistant to weight gain than others. Keeping in mind that brown fat is activated by the sympathetic nervous system, how would you investigate the possible role of brown fat in body weight control?

# 52 Salt and Water Balance and Nitrogen Excretion

## Blood, sweat, and tears

Blood, sweat, and tears taste salty because they have ionic concentrations similar to those of the interstitial fluids that bathe the cells of the body. The volume and composition of the interstitial fluids must remain within certain limits and be kept relatively free of wastes. Maintaining homeostasis of the interstitial fluids is the job of the excretory system, and it can be challenging. Sometimes it requires getting rid of excess fluids, and at other times it requires conserving fluids.

The nature of the challenge depends on the environment of the animal and its lifestyle. Some desert animals rarely if ever encounter free water; they must be able to live their entire lives without drinking. All animals derive water from the metabolism of food, but to survive on that amount of water, desert animals must conserve it to an exceptional degree. Accordingly, they excrete wastes that are extremely concentrated. Insects excrete semisolid wastes, and desert rodents excrete urine that is so concentrated it forms crystals of solute. Animals that live in fresh water have the opposite problem: water continuously enters their bodies by osmosis and with the food they eat, so they must constantly bail themselves out by producing copious amounts of dilute urine while at the same time conserving the solutes their bodies need.

The physiological adaptations animals have to maintain salt and water balance through excretion and conservation are remarkably flexible. Consider, for example, the vampire bat, a small tropical mammal that experiences extreme conditions within minutes of each other. The vampire bat feeds on the blood of other mammals, such as goats and cattle. Landing on an unsuspecting (usually sleeping) victim, the bat uses its sharp incisor teeth to make a small incision in a blood vessel and then laps up the blood that wells out.

Blood contains lots of nutritious protein, but it consists mostly of water. Blood meals may be few and far between, so the bat must quickly consume as much as it can—up to half its body mass. To maximize its protein intake and keep its weight low enough to fly, it rapidly eliminates the water from its meal. Within minutes of starting to feed, the bat is producing a lot of very dilute urine. The warm trickle down the neck of the victim is not blood!

**Blood as a Fast Food** The vampire bat *Desmodus rotundus* is able to adjust its excretory physiology from water-excreting to water-conserving, depending on whether it is ingesting or digesting its blood meal.

**Living without Water** Many kangaroo rat species (*Dipodomys*) inhabit arid deserts, and they rarely see free water.

Once feeding ends, this high rate of water loss must stop. Now the vampire bat is digesting and metabolizing protein and must excrete large amounts of nitrogenous wastes while conserving its body water. Within minutes, the bat's excretory system switches from producing abundant, dilute urine to producing a tiny amount of highly concentrated urine. In one feeding cycle, the vampire bat rapidly transitions from an excretory physiology typical of a mammal living in an environment with abundant fresh water to that of a desert mammal that never sees water. What an amazing case of physiological flexibility!

**IN THIS CHAPTER** we will describe adaptations for maintaining salt and water balance and for excreting nitrogenous wastes in environments that present animals with different challenges. Using some invertebrate examples, we will review the basic mechanisms common to all animal excretory systems. We will learn about the basic anatomical unit of the vertebrate excretory system—the nephron—and how it evolved. Finally, we will consider the mechanisms that control and regulate salt and water balance in mammals.

CHAPTER OUTLINE



## 52.1 How Do Excretory Systems Maintain Homeostasis?

Controlling the volume, solute concentration, and composition of extracellular fluids are not huge problems for marine invertebrates because their extracellular fluids are very similar to seawater, and their nitrogenous wastes can simply diffuse into the surrounding seawater. The situation is different for marine vertebrates and for all freshwater and terrestrial animals because their extracellular fluids differ considerably from the external environment. These animals depend on **excretory systems** to maintain the volume, concentration, and composition of their extracellular fluids, and to excrete wastes.

Homeostasis of the extracellular fluids is critical because the concentration of solutes in them determines the water balance of the cells those fluids bathe, and the composition of the extracellular fluid influences the health and functions of all the cells of the body. Consider, for example, the importance of ion concentration gradients between the extracellular fluid and the cytoplasm of nerve and muscle cells (see Sections 45.2 and 48.1).

### Water enters or leaves cells by osmosis

The volume of cells depends on whether they take up water from or lose water to the extracellular fluids. The movement of water across cell plasma membranes depends on differences in solute concentration on the two sides of the membrane and the permeability of the membrane. This is the process of osmosis, which we discuss in Sections 6.3 and 35.1. If the solute concentration of the extracellular fluid is less than that of the cytoplasm, water moves into the cells, causing them to swell and possibly burst. If the solute concentration of the extracellular fluid is greater than that of the cytoplasm, the cells lose water and shrink. Thus the solute concentration of the extracellular fluid affects both the volume and the solute concentration of the cells.

Animal physiologists use the term **osmolarity** in discussing osmosis. The osmolarity of a solution is the number of moles of osmotically active solutes per liter of solvent. Thus, a 1 molar solution of glucose is also a 1 osmolar (1 osmole per liter) solution, but a 1 molar solution of sodium chloride (NaCl) is a 2 osmolar solution, because each NaCl molecule dissociates into two osmotically active ions.

To achieve cellular water balance, animals must maintain the osmolarity of their extracellular fluid within an appropriate range. In addition, they must maintain an appropriate solute composition by saving some substances (reabsorption) and eliminating others (secretion). To accommodate these needs, most animals have excretory systems.

### Excretory systems control extracellular fluid osmolarity and composition

Excretory systems control the osmolarity and composition of the extracellular fluids by excreting solutes that are present in excess (such as NaCl when we eat lots of salty food) and conserving solutes that are valuable or in short supply (such as glucose and amino acids). Excretory systems also eliminate the toxic waste products of protein metabolism. The output of the excretory system is called **urine**.

Three basic processes are common to a wide variety of animal excretory systems: filtration, secretion, and reabsorption. Extracellular fluid is filtered to produce a filtrate that contains no cells or large molecules, such as proteins. In animals with a closed circulatory system, the blood plasma is usually filtered from capillaries into associated tubules. The walls of the capillaries and of the tubules are the filter, and the filtration is driven by blood pressure. As the filtrate flows through the tubules, its composition and concentration are modified through processes of secretion and reabsorption to form the urine that leaves the body.

The processing of the filtrate into urine involves movement of water into and out of the tubules. There are no mechanisms for the active transport of water. The movement of water is due either to a pressure difference (filtration) or to a difference in solute concentration (osmosis). Water always flows down a pressure gradient or up a solute concentration gradient.

### Animals can be osmoconformers or osmoregulators

Animals that live in marine, freshwater, or terrestrial environments face different salt and water balance problems. In the terrestrial environment, salts and water can be scarce and usually must be conserved by excretory systems. In the freshwater environment, water is plentiful but salts are scarce. Freshwater animals have to conserve salts and excrete the water that continuously invades their bodies through osmosis.

In contrast to fresh water, ocean water has a high osmolarity—over 1,000 milliosmoles/liter (mosm/l). Most marine invertebrates equilibrate their extracellular fluid osmolarity with the ocean water and are therefore called **osmoconformers**. Other marine animals maintain extracellular fluid osmolarities much lower than seawater and are therefore called **osmoregulators**. All marine vertebrates except for sharks and rays are osmoregulators. These animals maintain their extracellular fluids at about 300 mosm/l, like that of humans.

There are limits to osmoconformity. Even animals that can osmoconform over a wide range of osmolarities must osmoregulate in extreme environments. No animal could survive if its extracellular fluid had the osmolarity of fresh water. There would be too few solutes, including nutrients and ions essential for cell functions, and water would flow into the cells by osmosis, causing them to swell and burst. Nor could animals survive with internal osmolarities as high as those that may be reached in an evaporating tidal pool. High solute concentrations cause proteins to denature.

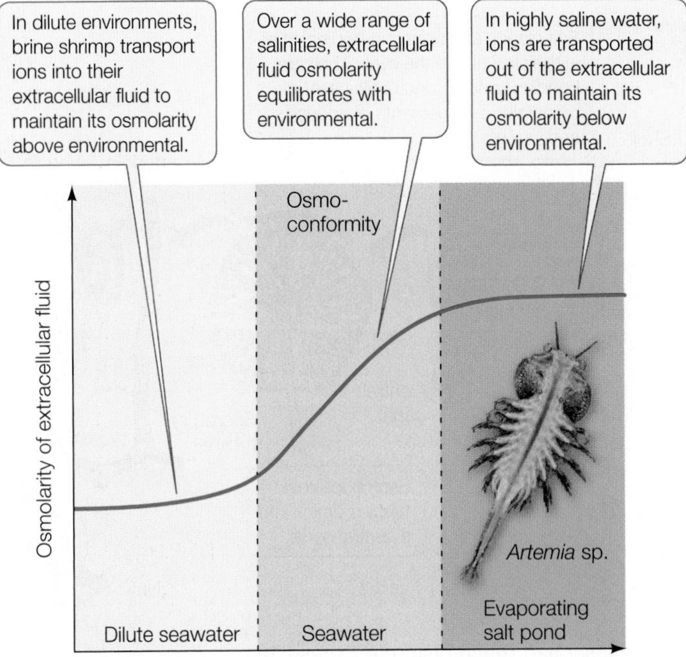
In dilute environments, brine shrimp transport ions into their extracellular fluid to maintain its osmolarity above environmental.
Over a wide range of salinities, extracellular fluid osmolarity equilibrates with environmental.
In highly saline water, ions are transported out of the extracellular fluid to maintain its osmolarity below environmental.
Osmo-conformity
Osmolarity of extracellular fluid
Artemia sp.
Dilute seawater
Seawater
Evaporating salt pond
Osmolarity of environment

**52.1 Osmoconformity Has Limits** Animals that experience an extreme range of salt concentrations in their environment can be osmoconformers over much of that range but must osmoregulate at the extremes. They regulate by actively transporting ions across their gill membranes—in either direction, depending on the challenge.

The brine shrimp *Artemia* is both an osmoconformer and an osmoregulator; this animal can live in environments of almost any osmolarity (**Figure 52.1**). *Artemia* are found in huge numbers in the most salty environments known, such as Utah's Great Salt Lake and in coastal evaporation ponds where salt is concentrated for commercial purposes (see Figure 26.23) and can reach an osmolarity of 2,500 mosm/l. At these high environmental osmolarities, *Artemia* maintains its tissue fluid osmolarity considerably below that of the environment by actively transporting NaCl from its extracellular fluid out across its gill membranes to the environment. *Artemia* cannot survive in fresh water, but it can live in dilute seawater by reversing the direction of transport of NaCl across its gill membranes to maintain its extracellular fluid above that of the environment.

### Animals can be ionic conformers or ionic regulators

Osmoconformers can also be ionic conformers, allowing the ionic composition, as well as the osmolarity, of their extracellular fluid to match that of the environment. Most osmoconformers, however, are ionic regulators to some degree; they employ active transport mechanisms to excrete some ions and to maintain other ions in their extracellular fluid at optimal concentrations.

Terrestrial animals obtain their salts from food and regulate the ionic composition of their extracellular fluids by conserving some ions and excreting others. For example, herbivores have to conserve $Na^+$ because the plants they eat have low concen-

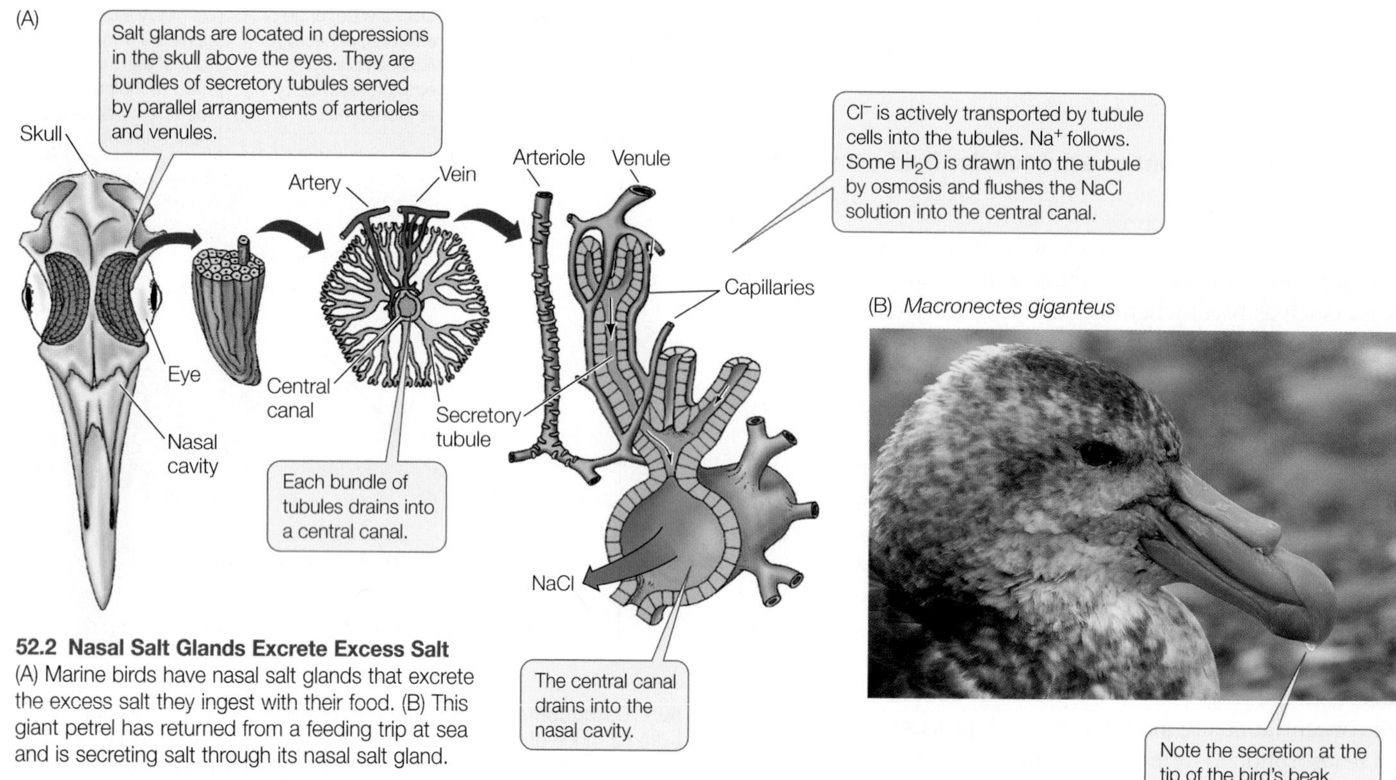
(A)
Salt glands are located in depressions in the skull above the eyes. They are bundles of secretory tubules served by parallel arrangements of arterioles and venules.
Skull
Artery
Vein
Arteriole
Venule
$Cl^-$ is actively transported by tubule cells into the tubules. $Na^+$ follows. Some $H_2O$ is drawn into the tubule by osmosis and flushes the NaCl solution into the central canal.
Capillaries
(B) *Macronectes giganteus*
Eye
Central canal
Secretory tubule
Nasal cavity
Each bundle of tubules drains into a central canal.
NaCl
The central canal drains into the nasal cavity.
Note the secretion at the tip of the bird's beak.

**52.2 Nasal Salt Glands Excrete Excess Salt** (A) Marine birds have nasal salt glands that excrete the excess salt they ingest with their food. (B) This giant petrel has returned from a feeding trip at sea and is secreting salt through its nasal salt gland.

trations of $Na^+$. By contrast, birds that feed on marine animals must excrete the excess of sodium they ingest with their food. Their *nasal salt glands* excrete a concentrated solution of NaCl via a duct that empties into the nasal cavity. Birds, such as penguins and gulls, that have nasal salt glands can be seen frequently sneezing or shaking their heads to get rid of the salty droplets excreted from their nasal salt glands (**Figure 52.2**).

## 52.1 RECAP

**Excretory systems control water and salt balance and the excretion of nitrogenous waste products through three mechanisms: filtration of body fluids to form urine, active secretion of substances into the urine, and active reabsorption of substances from the urine.**

- Describe the two mechanisms used to move water across membranes. See p. 1093
- What different salt and water balance problems might animals encounter in marine, freshwater, and terrestrial environments? What are some of the ways they meet those challenges? See pp. 1093–1094 and Figures 52.1 and 52.2

In addition to maintaining salt and water balance, animals must eliminate the waste products of metabolism from their extracellular fluids. The major problem is nitrogen. When nitrogen-containing molecules are broken down by metabolism, the end product can be toxic.

# 52.2 How Do Animals Excrete Nitrogen?

The end products of the metabolism of carbohydrates and fats are water and carbon dioxide, which are not difficult to eliminate. Proteins and nucleic acids, however, contain nitrogen, so their metabolism produces *nitrogenous wastes* in addition to water and carbon dioxide.

### Animals excrete nitrogen in a number of forms

The most common nitrogenous waste is **ammonia** ($NH_3$). Because it is highly toxic, ammonia is either excreted continuously to prevent its accumulation or is detoxified by conversion into **urea** or **uric acid** (**Figure 52.3**).

**AMMONIA** Ammonia is highly soluble in water and diffuses rapidly, so its continuous excretion is relatively simple for many aquatic animals that continuously lose ammonia from their blood to the environment by diffusion across their gill membranes. Animals that excrete ammonia, such as aquatic invertebrates and bony fishes, are **ammonotelic**.

If ammonia builds up in the extracellular fluids, it becomes toxic at rather low levels and is a dangerous metabolite for terrestrial animals and for those aquatic animals that cannot continuously excrete ammonia. These animals must convert ammonia into urea or uric acid.

**UREA** **Ureotelic** animals, such as mammals, amphibians, and cartilaginous fishes (sharks and rays), excrete urea as their principal nitrogenous waste product. Urea is quite soluble in water,

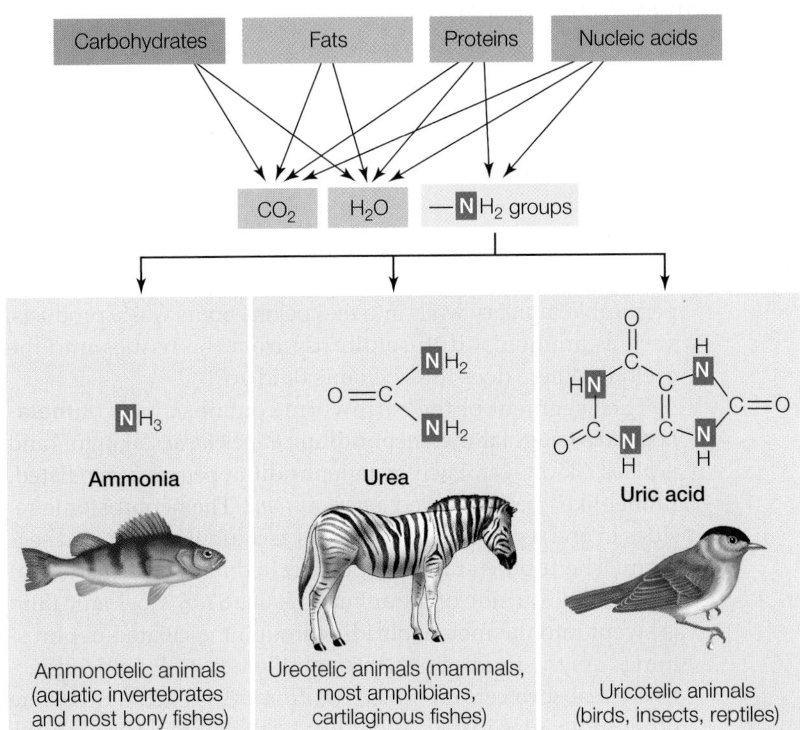


**52.3 Waste Products of Metabolism** The metabolism of proteins and nucleic acids produces nitrogenous wastes. Many aquatic animals, including most fishes, excrete nitrogenous wastes as ammonia, which is highly diffusible and soluble in an aqueous environment. Most terrestrial animals and some aquatic animals excrete either urea or uric acid. Urea is more soluble in water and is the major nitrogenous excretory product for mammals, amphibians, and some fishes. Uric acid is not very soluble in water and is the major nitrogenous excretory product for reptiles, birds, insects, and some amphibians.

but its excretion can result in a large loss of water that many animals can ill afford. As we will see later in this chapter, mammals have evolved excretory systems that conserve water by producing concentrated urea solutions. The sharks and rays are another story. These marine species keep their extracellular fluids almost isosmotic (same osmotic concentration) to the marine environment by retaining high concentrations of urea and another nitrogen-containing compound, trimethylamine oxide, in their body fluids.

**URIC ACID** Animals that conserve water by excreting nitrogenous wastes mostly as uric acid are **uricotelic**. Insects, reptiles (including birds), and some amphibians are uricotelic. Uric acid is not very soluble in water, so it forms a colloidal suspension in the urine and is excreted as a semisolid (for example, the whitish material in bird droppings). A uricotelic animal loses very little water as it disposes of its nitrogenous wastes.

### Most species produce more than one nitrogenous waste

Humans are ureotelic, but we also excrete uric acid. The uric acid in human urine comes largely from the metabolism of nucleic acids and caffeine. If uric acid levels in the extracellular fluids rise too high, uric acid crystals can precipitate in joints and cause the age-old malady called gout. Because solubility goes down with temperature, uric acid crystals usually precipitate first in the extremities, especially the big toe. Pain in the big toe is a telltale symptom of gout.

Humans can also excrete ammonia, which is an important mechanism for regulating the pH of the extracellular fluids. As we will see later in this chapter, excreted ammonia buffers the urine and enables the secretion of more hydrogen ions.

Species that live in different habitats at different developmental stages may use more than one mechanism of nitrogen excretion. The tadpoles of frogs and toads, for example, excrete ammonia across their gill membranes, but adult frogs and toads generally excrete urea. Some adult amphibians that live in arid habitats excrete uric acid.

## 52.2 RECAP

**Ammonia is a common metabolic waste product of nitrogen-containing molecules. Most aquatic animals excrete ammonia by diffusion into the water. Terrestrial animals and some aquatic animals detoxify it by conversion to urea or uric acid.**

- Why might you expect a species from an arid habitat to use uric acid as its primary nitrogenous waste product? See p. 1095 and Figure 52.3

Animals exhibit a variety of adaptations for dealing with the challenges of salt and water balance in different environments. All of these adaptations, however, are based on two basic mechanisms—namely, filtration and tubular processing of the filtrate to conserve some solutes and excrete others.

## 52.3 How Do Invertebrate Excretory Systems Work?

Freshwater and terrestrial invertebrates have a wide variety of adaptations for maintaining salt and water balance and excreting nitrogen. In this section, we will explore three examples of invertebrate excretory systems: *protonephridia*, *metanephridia*, and *Malpighian tubules*. Each of these systems produces an extract of interstitial fluid lacking in large molecules. They then change the solute composition (ions and small molecules) of that fluid to form an excretory product.

### The protonephridia of flatworms excrete water and conserve salts

Many flatworms, such as *Planaria*, live in fresh water. These animals excrete water through an elaborate network of tubules

**52.4 Protonephridia in Flatworms** The protonephridia of the flatworm *Planaria* consist of tubules ending in flame cells. In the region of the flame cells, body fluid is filtered between the tubule cells. The composition of the filtrate is modified as it flows down the tubule.

running throughout their bodies. The tubules end in *flame cells*, so called because each cell has a tuft of cilia projecting into the tubule (**Figure 52.4**). The beating of the cilia gives the appearance of a flickering flame. A flame cell and a tubule together form a **protonephridium** (plural protonephridia; Greek *proto*, "before," and *nephros*, "kidney").

Extracellular fluid enters the tubules by filtration. The beating of the cilia causes a slight negative pressure in the tubule, and movements of the animal create positive pressure in the extracellular fluid. This pressure difference causes extracellular fluid to be filtered through tiny spaces between tubule cells. The filtrate flows toward the animal's excretory pore, and along the way the cells of the tubules modify the composition of the fluid by reabsorption and secretion of specific ions and molecules. Because more ions are reabsorbed than are secreted, the urine that leaves the flatworm's body is less concentrated than the extracellular fluid. Thus, the protonephridium conserves ions and excretes water and wastes.

### The metanephridia of annelids process coelomic fluid

Filtration of body fluids and modification of urine by tubules are highly developed processes in annelid worms such as the earthworm. Annelids are segmented, and in each segment they have a fluid-filled body cavity called a *coelom* (see Figure 32.12). Annelids have a closed circulatory system through which blood is pumped under pressure. The pressure causes the blood to be filtered across the thin, permeable capillary walls into the coelom. Some waste products, such as ammonia, diffuse directly from the tissues into the coelom. Where does this coelomic fluid go?

Each segment of the earthworm contains a pair of **metanephridia** (singular metanephridium; Greek *meta*, "akin to," and *nephros*, "kidney"). Each metanephridium begins as a ciliated, funnel-like opening called a *nephrostome*. The nephrostome resides in one segment and continues as a tubule in the next segment. The tubule ends in a pore, called a *nephridiopore*, that opens to the outside of the animal (**Figure 52.5**). Coelomic fluid is swept into the metanephridia through the ciliated nephrostomes. As the fluid passes through the tubules, their cells actively reabsorb certain molecules from it and actively secrete other molecules into it. What leaves the animal through the nephridiopores is a dilute urine containing nitrogenous wastes and other solutes.

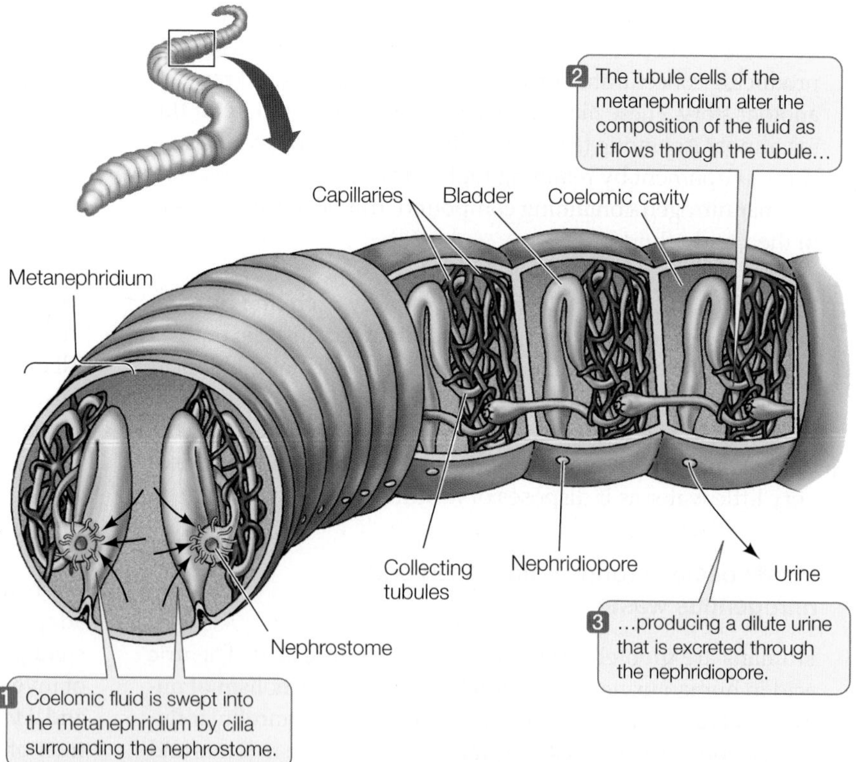

2 The tubule cells of the metanephridium alter the composition of the fluid as it flows through the tubule...
Capillaries
Bladder
Coelomic cavity
Metanephridium
Collecting tubules
Nephridiopore
Urine
3 ...producing a dilute urine that is excreted through the nephridiopore.
Nephrostome
1 Coelomic fluid is swept into the metanephridium by cilia surrounding the nephrostome.

**52.5 Metanephridia in Earthworms** The metanephridia of annelids are arranged segmentally. The cross section at the left end shows a pair of metanephridia. Three longitudinal sections (right) show only one metanephridium of the two in each segment. Coelomic fluid enters the nephrostome and flows through tubules leading to the nephridiopore. A close association of the tubules and blood capillaries facilitates the active exchange of substances between the blood and the tubular fluid.

yourBioPortal.com
GO TO Web Activity 52.1 • Annelid Metanephridia

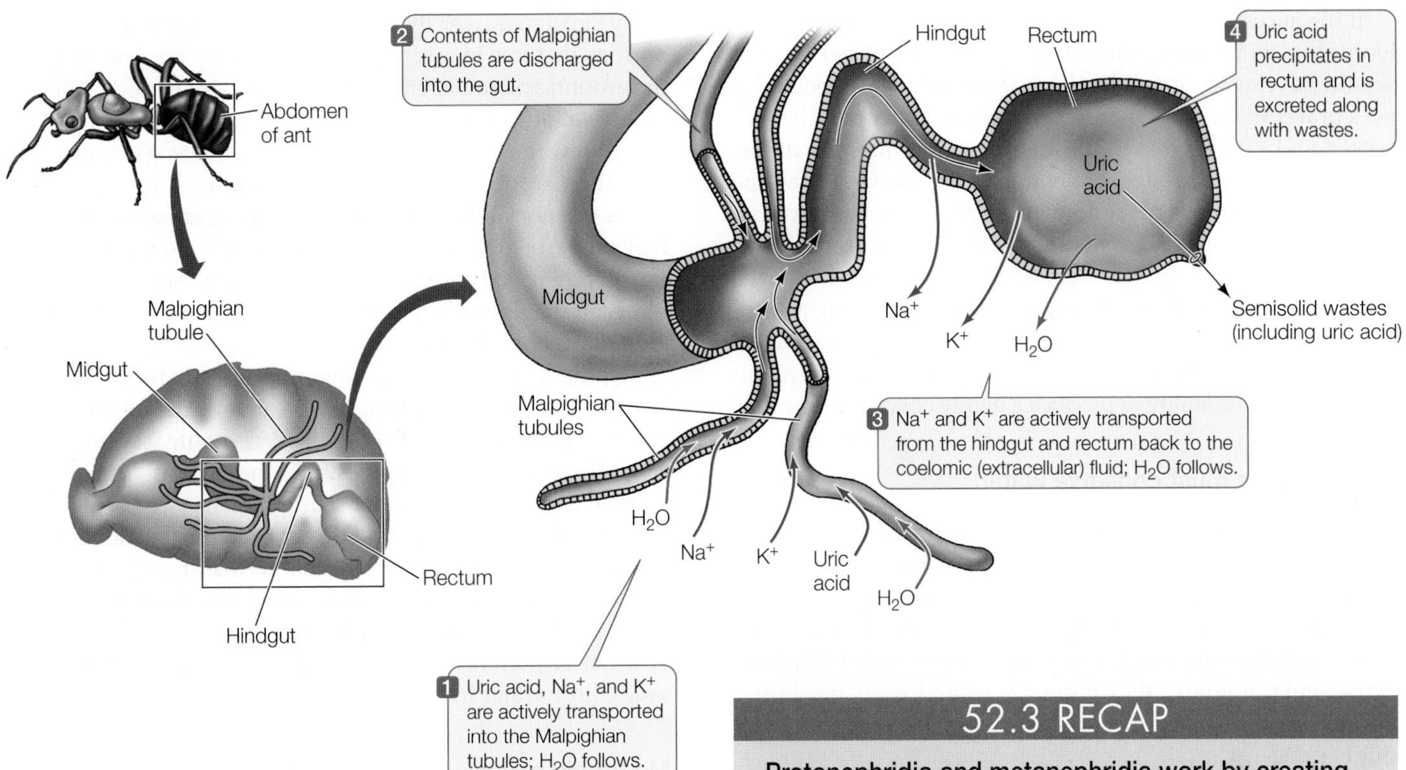

Abdomen of ant
Malpighian tubule
Midgut
Rectum
Hindgut
2 Contents of Malpighian tubules are discharged into the gut.
Midgut
Malpighian tubules
$H_2O$
$Na^+$
$K^+$
Uric acid
$H_2O$
Hindgut
Rectum
Uric acid
4 Uric acid precipitates in rectum and is excreted along with wastes.
$Na^+$
$K^+$
$H_2O$
Semisolid wastes (including uric acid)
3 $Na^+$ and $K^+$ are actively transported from the hindgut and rectum back to the coelomic (extracellular) fluid; $H_2O$ follows.
1 Uric acid, $Na^+$, and $K^+$ are actively transported into the Malpighian tubules; $H_2O$ follows.

**52.6 Malpighian Tubules in Insects** The blind, thin-walled Malpighian tubules are attached to the junction of the insect's midgut and hindgut and project into the spaces containing extracellular fluid. This system makes it possible to excrete wastes with very little loss of water.

### The Malpighian tubules of insects depend on active transport

Insects can excrete nitrogenous wastes with very little loss of water and can therefore live in the driest habitats on Earth. The insect excretory system consists of **Malpighian tubules**. An individual insect has from 2 to more than 100 of these blind-ended tubules that open into the gut between the midgut and hindgut (**Figure 52.6**).

Insects have an open circulatory system and therefore cannot use a pressure difference to filter extracellular fluids into the Malpighian tubules. Instead, the cells of the tubules actively transport uric acid, potassium ions, and sodium ions from the extracellular fluid into the tubules. The high concentration of solutes in the tubules causes water to follow osmotically, which flushes the tubule contents toward the gut.

The epithelial cells of the hindgut and rectum actively transport sodium and potassium ions from the gut contents back into the extracellular fluid. This local transport of salts creates an osmotic gradient that pulls water out of the rectal contents. As its concentration increases, the uric acid forms a colloidal suspension, freeing even more water to be reabsorbed. Remaining in the rectum is the uric acid mixed with other wastes; this semisolid matter is what the insect excretes. The Malpighian tubule system is a highly effective mechanism for excreting nitrogenous wastes and some salts without giving up much water.

### 52.3 RECAP

**Protonephridia and metanephridia work by creating a filtrate of the body fluids that is modified by the secretion and reabsorption of specific substances before being excreted. Insect Malpighian tubules actively secrete uric acid and other solutes into closed tubules.**

- Describe how an earthworm filters its blood and produces urine. **See p. 1096 and Figure 52.5**
- How do Malpighian tubules function, and how does this system make it possible for some insect species to survive in the desert? **See p. 1097 and Figure 52.6**

Having described how several invertebrate groups handle nitrogen excretion, we will next consider the nephron—the basic unit of the vertebrate excretory system—and how it evolved to be able to respond to a variety of salt and water balance challenges and maintain a relatively constant internal environment.

## 52.4 How Do Vertebrates Maintain Salt and Water Balance?

The main excretory organ of vertebrates is the **kidney**, and the functional unit of the kidney is the **nephron**, which has a blood vessel component and a tubule component. The vascular component begins with a knot of capillaries that are highly permeable and filter the blood into the tubule component. The blood vessels also carry substances to the tubules for secretion and carry away substances that the tubule cells reabsorb. Nephrons can filter large volumes of blood and achieve bulk reabsorption of salts and other valuable molecules such as glucose, making the vertebrate kidney well adapted for the excretion of excess water.

If the ancestors of vertebrates evolved in fresh water, as paleontologists propose, the excretory systems of vertebrates would have evolved to excrete excess water. How then have vertebrates adapted to environments where water must be conserved and salts excreted? The answer to this question differs among vertebrate groups. Even among the marine fishes, the excretory adaptations of the bony fishes are different from those of the cartilaginous fishes. Reptiles, birds, and mammals have excretory systems that conserve water. Reptiles and birds achieve this mainly by being uricotelic and producing a semisolid excretory product that contains little water. Mammals in contrast are ammonotelic, excrete a liquid waste product, but have evolved the ability to produce a highly concentrated urine.

### Marine fishes must conserve water

Marine bony fishes osmoregulate their extracellular fluids to maintain them at one-third to one-half the osmolarity of seawater. Their only source of water is the sea around them, so they must conserve water and excrete excess solutes. Marine bony fishes cannot produce urine that is more concentrated than their extracellular fluids, so they minimize water loss by producing very little urine. In contrast, freshwater fishes produce lots of dilute urine.

If marine bony fishes cannot excrete excess solutes in their urine, how do they deal with the large salt loads they ingest with food? Marine bony fishes do not absorb from their guts some of the ions they take in, especially divalent ions such as $Mg^{2+}$ or $SO_4^{2-}$. NaCl, the major salt ingested, is actively excreted across the gill membranes. As we mentioned earlier, bony fishes can lose their nitrogenous waste, ammonia, by diffusion across their gill membranes.

Cartilaginous fishes (sharks and rays) are osmoconformers but not ionic conformers. They raise the osmolarity of their body fluids in a unique way. Unlike boney fishes, they convert nitrogenous wastes to urea and trimethylamine oxide, and they retain large amounts of these compounds in their body fluids. As a result, their body fluids have an osmolarity close to that of seawater, so they do not lose body water to the environment by osmosis, and may actually gain water. These species have adapted to a concentration of urea in the body fluids that would be toxic to other vertebrates.

Sharks and rays still have the problem of excreting the large amounts of salts they take in with their food. They solve this problem by having a gland in the rectum that actively secretes NaCl by a mechanism similar to that of the nasal salt glands of seabirds.

### Terrestrial amphibians and reptiles must avoid desiccation

Most amphibians live in or near fresh water, and they stay in humid habitats when they do venture from the water. Like freshwater fishes, most amphibians produce large amounts of dilute urine and conserve salts. Some amphibians, however, have adapted to habitats that require water conservation.

Amphibians living in dry terrestrial environments have reduced the permeability of their skin to water. Some secrete a waxy substance over the skin to waterproof it. Several species of frogs that live in arid regions of Australia burrow deep into the ground and remain there during long dry periods. They enter **estivation**, a state of very low metabolic activity and therefore low water turnover. When it rains, the frogs come out of estivation, feed, and reproduce. Their most interesting adaptation is an enormous urinary bladder. Before entering estivation, they fill the bladder with dilute urine, which can amount to one-third of their body weight. This dilute urine serves as a water reservoir that is gradually reabsorbed into the blood during the long period of estivation.

Reptiles occupy habitats ranging from aquatic to extremely hot and dry. In fact, snakes, lizards, and birds are among the most prominent members of many desert faunas. Three major adaptations have freed reptiles from the close association with water that is necessary for most amphibians (see Section 33.4). First, reptiles are *amniotes* that do not need fresh water to reproduce because they employ internal fertilization and lay eggs with shells that retard evaporative water loss. Second, they have a dry, epidermis (skin) that retards evaporative water loss. Third, they excrete nitrogenous wastes as uric acid semisolids, losing little water in the process.

### Mammals can produce highly concentrated urine

Mammals occupy diverse habitats, many of which present special excretory system challenges. The most challenging environments are those in which water is severely limited. Mammals have a variety of adaptations to conserve water, but chief among them is the ability to produce a urine that is more concentrated than their extracellular fluids. They are able to concentrate their urine because of adaptations of their kidneys that we will explore in detail in Section 52.5. To understand how these adaptations work, however, we must first describe the structure and function of the vertebrate nephron.

### The nephron is the functional unit of the vertebrate kidney

Urine formation in vertebrate nephrons involves three main processes (**Figure 52.7**):

- *Filtration*. Each nephron has a dense ball of capillaries called a **glomerulus** (plural glomeruli). The glomerulus is highly permeable to water, ions, and small molecules but impermeable to large molecules. Blood pressure drives the movement of water and small molecular weight solutes out of the glomerular capillaries.
- *Tubular reabsorption*. The filtrate from the glomerulus flows into the **renal tubule**. Cells in the renal tubule modify the filtrate by reabsorbing specific ions, nutrients, and water, returning these to the blood, and leaving behind and concentrating excess ions and waste products such as urea.
- *Tubular secretion*. The filtrate in the renal tubule is further modified by tubule cells transporting substances into the tubular contents. These are substances that the body needs to excrete.

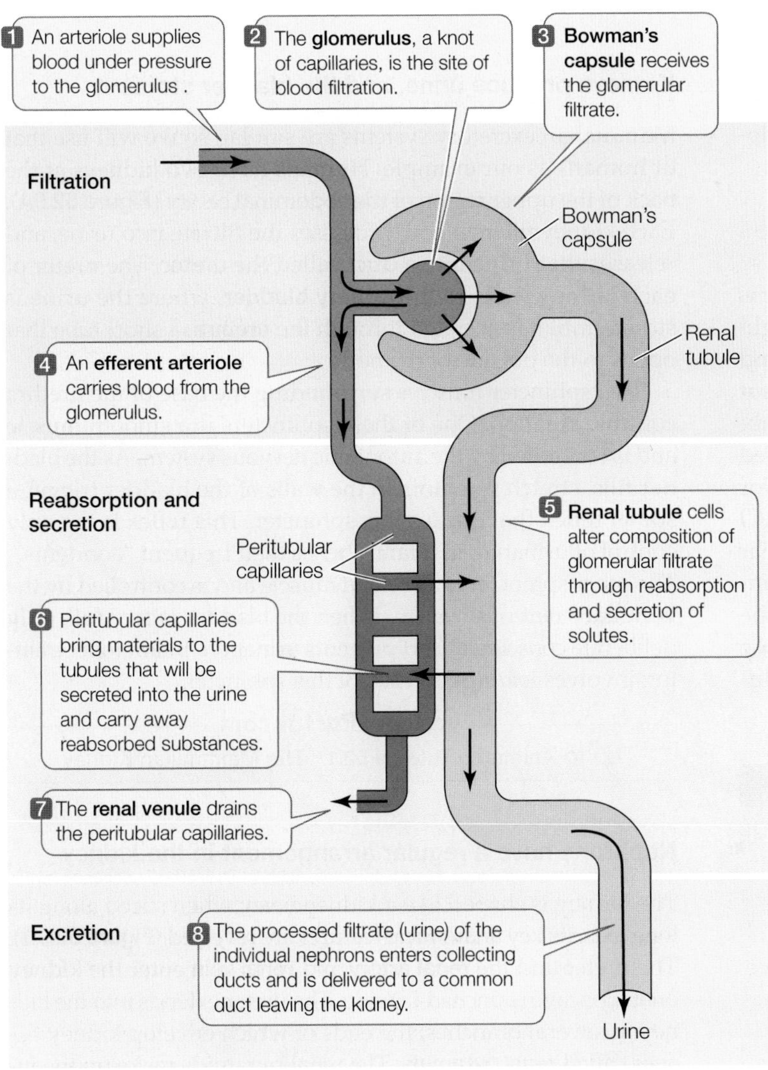

1 An arteriole supplies blood under pressure to the glomerulus.
2 The **glomerulus**, a knot of capillaries, is the site of blood filtration.
3 **Bowman's capsule** receives the glomerular filtrate.
Filtration
Bowman's capsule
Renal tubule
4 An **efferent arteriole** carries blood from the glomerulus.
Reabsorption and secretion
Peritubular capillaries
5 **Renal tubule** cells alter composition of glomerular filtrate through reabsorption and secretion of solutes.
6 Peritubular capillaries bring materials to the tubules that will be secreted into the urine and carry away reabsorbed substances.
7 The **renal venule** drains the peritubular capillaries.
Excretion
8 The processed filtrate (urine) of the individual nephrons enters collecting ducts and is delivered to a common duct leaving the kidney.
Urine


**52.7 The Vertebrate Nephron** The vertebrate nephron consists of a renal tubule closely associated with two capillary beds, the glomerulus and the peritubular capillaries.

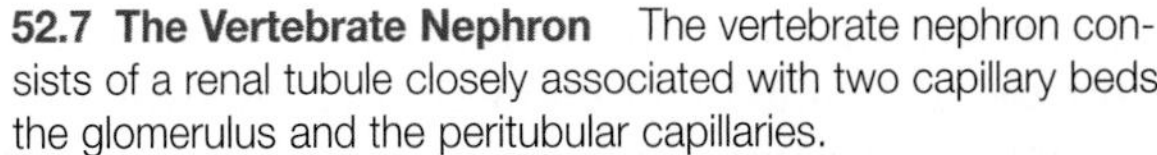

**yourBioPortal.com**

GO TO **Web Activity 52.2 • The Vertebrate Nephron**

Blood enters the glomerular capillaries via an **afferent arteriole** and leaves the glomerulus in an **efferent arteriole**. This short vessel is called an arteriole because it feeds another capillary bed called the **peritubular capillaries** that intimately surround the renal tubules (**Figure 52.8A**). Peritubular capillaries deliver substances to the renal tubule cells that they will secrete into the urine and carry away substances these cells absorb from the urine.

### Blood is filtered into Bowman's capsule

The renal tubule begins with **Bowman's capsule** (see Figure 52.7), which encloses the glomerulus. The glomerulus appears to be pushed into Bowman's capsule much like a fist pushed into an inflated balloon. The cells of the capsule that are in direct contact with the glomerular capillaries are called **podocytes** (**Figure 52.8B**). These highly specialized cells have numerous armlike extensions, each with hundreds of fine, fingerlike processes. The podocytes wrap around the capillaries so that their processes interdigitate and intimately cover the capillaries.

The glomerulus filters the blood to produce a fluid (the renal filtrate) that lacks cells and large molecules. The walls of the capillaries, the basal lamina of the capillary endothelium, and the podocytes of Bowman's capsule all participate in filtration. *Fenestrations* in the walls of the capillaries (see Section 50.4) allow water and many solute molecules, but not red blood cells, to pass through. The meshwork of the basal lamina and the spaces between the processes of the podocytes are even finer and prevent large molecules from leaving the capillaries (**Figure 52.8C**). The arterial pressure of the blood entering the permeable capillaries causes the filtration of water and small molecules in the glomerulus. The glomerular filtration rate is high because blood

**52.8 A Tour of the Nephron** Scanning electron micrographs illustrate the anatomical basis for blood filtration by the kidneys. (A) In a preparation showing only the blood vessels (tubular tissue has been digested away), the glomeruli appear as balls of capillaries served by arterioles. (B) Higher magnification of a glomerulus with the tubule cells intact shows the podocytes that wrap around the glomerular capillaries. (C) This cross section of an intact glomerulus shows the tubule cells that form Bowman's capsule.

(A) Afferent arterioles Glomeruli

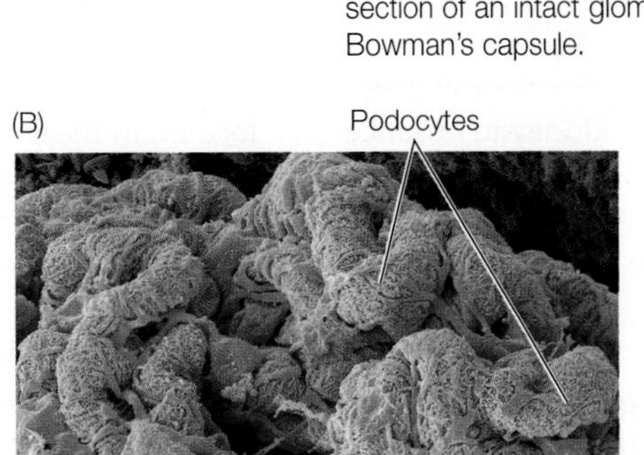


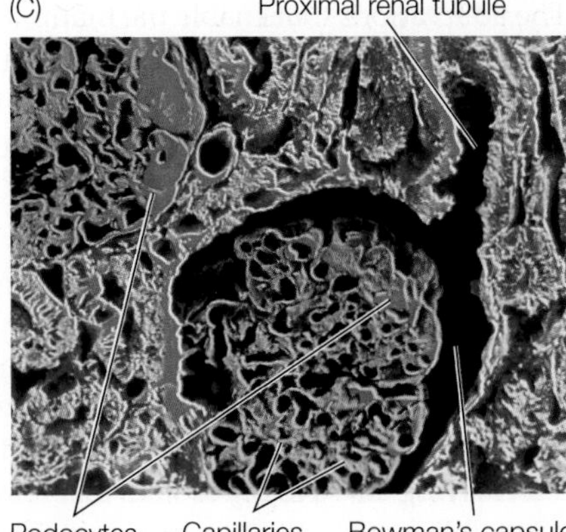

pressure in the glomerular capillaries is unusually high, and because the capillaries of the glomerulus, along with their covering of podocytes, are more permeable to water than other capillary beds in the body.

### The renal tubules convert glomerular filtrate to urine

The composition of the filtrate that enters the renal tubule is similar to that of the blood plasma, with the exception of high-molecular-weight solutes such as proteins. Reabsorption and secretion cause the composition of this fluid to change as it passes down the renal tubule. Cells of the tubule actively reabsorb certain molecules from the tubule fluid (which are returned to the blood flowing through the peritubular capillaries). For example, glucose and amino acids are reabsorbed. Most NaCl is reabsorbed. Other substances in the blood of the peritubular capillaries are actively secreted into the tubule fluid. An example is paraminohippuric acid (PAH), which is produced in the liver from benzoic acid, a common food preservative. Because of the actions of the renal tubules, the excreted urine is very different from the original filtrate.

### 52.4 RECAP

**The kidney is the major excretory organ of vertebrates. Its functional unit is the nephron, which includes a glomerulus that filters blood and a renal tubule that secretes and reabsorbs solutes, modifying the filtrate to produce urine. The nephron is a mechanism for excreting excess water while conserving valuable solutes.**

- Explain the difference in the osmoregulatory adaptations of marine bony and cartilaginous fishes. See p. 1098
- What are the functional relationships between the glomerular and peritubular capillaries? See p. 1099 and Figure 52.7
- Describe how blood is filtered by the glomerulus. See pp. 1099–1100
- How is the composition of the urine made different from the composition of the blood? See p. 1100

The adaptations that enable the mammalian kidney to produce urine more concentrated than extracellular fluids were important steps in vertebrate evolution, and they were largely achieved through changes in the structure and regional functions of the renal tubules.

## 52.5 How Does the Mammalian Kidney Produce Concentrated Urine?

Mammals have high body temperatures and high metabolic rates, and therefore have the potential for a high rate of water loss. Having an excretory system that minimizes water loss made it possible for these highly active species to occupy arid habitats.

### Kidneys produce urine, and the bladder stores it

Mammalian excretory systems are similar, so we will use that of humans as our example. Humans have two kidneys at the back of the upper region of the abdominal cavity (**Figure 52.9A**). Each kidney filters blood, processes the filtrate into urine, and releases that urine into a duct called the **ureter**. The ureter of each kidney leads to the **urinary bladder**, where the urine is stored until it is excreted through the **urethra**, a short tube that opens to the outside of the body.

Two sphincter muscles surrounding the base of the urethra control urination. One of these sphincters is a smooth muscle and is controlled by the autonomic nervous system. As the bladder fills, stretch receptors in the walls of the bladder trigger a spinal reflex that relaxes this sphincter. This reflex is the only control of urination in infants, hence their frequent "accidents." The other sphincter is a skeletal muscle and is controlled by the voluntary nervous system. When the bladder is *very* full, only deliberate conscious effort prevents urination. Infant toilet training involves learning to control this sphincter.

yourBioPortal.com
GO TO Animated Tutorial 52.1 • The Mammalian Kidney

### Nephrons have a regular arrangement in the kidney

The kidney is shaped like a kidney bean; when sliced along its long axis, its key anatomical features are revealed (**Figure 52.9B**). The ureter and the **renal artery** and **renal vein** enter the kidney on its concave (punched-in) side. The ureter extends into the kidney in several branches, the ends of which envelop kidney tissues called **renal pyramids**. The renal pyramids make up the internal core, or **medulla**, of the kidney. The medulla is covered by an outer layer, or **cortex**, that has a granular appearance. Between the cortex and the medulla, the renal artery divides into the many arterioles that serve the nephrons. In this same region, the renal vein collects blood from the many venules that drain the peritubular capillaries.

The organization of the nephrons within the kidney is very regular. All of the glomeruli with their Bowman's capsules are located in the cortex. The initial segment of a renal tubule is called the **proximal convoluted tubule**—"proximal" because it is closest to the glomerulus, and "convoluted" because it is twisted (**Figure 52.9C**). All of the proximal convoluted tubules are also located in the cortex.

At the point at which the renal tubule descends into the medulla, it straightens and descends directly down into the medulla. In the medulla the tubule makes a hairpin turn and ascends back to the cortex, forming what is called the **loop of Henle**. Some nephrons have longer loops of Henle than others. Some 20 to 30 percent of human nephrons that have glomeruli deep in the cortex (i.e., near the border with the medulla) have long loops of Henle that go deep into the medulla. Nephrons that have glomeruli farther up in the cortex generally have short loops of Henle that descend only a short distance into the medulla. As we will see, the long loops are the critical adaptation of the mammalian nephron that enables the kidney to concentrate the urine.

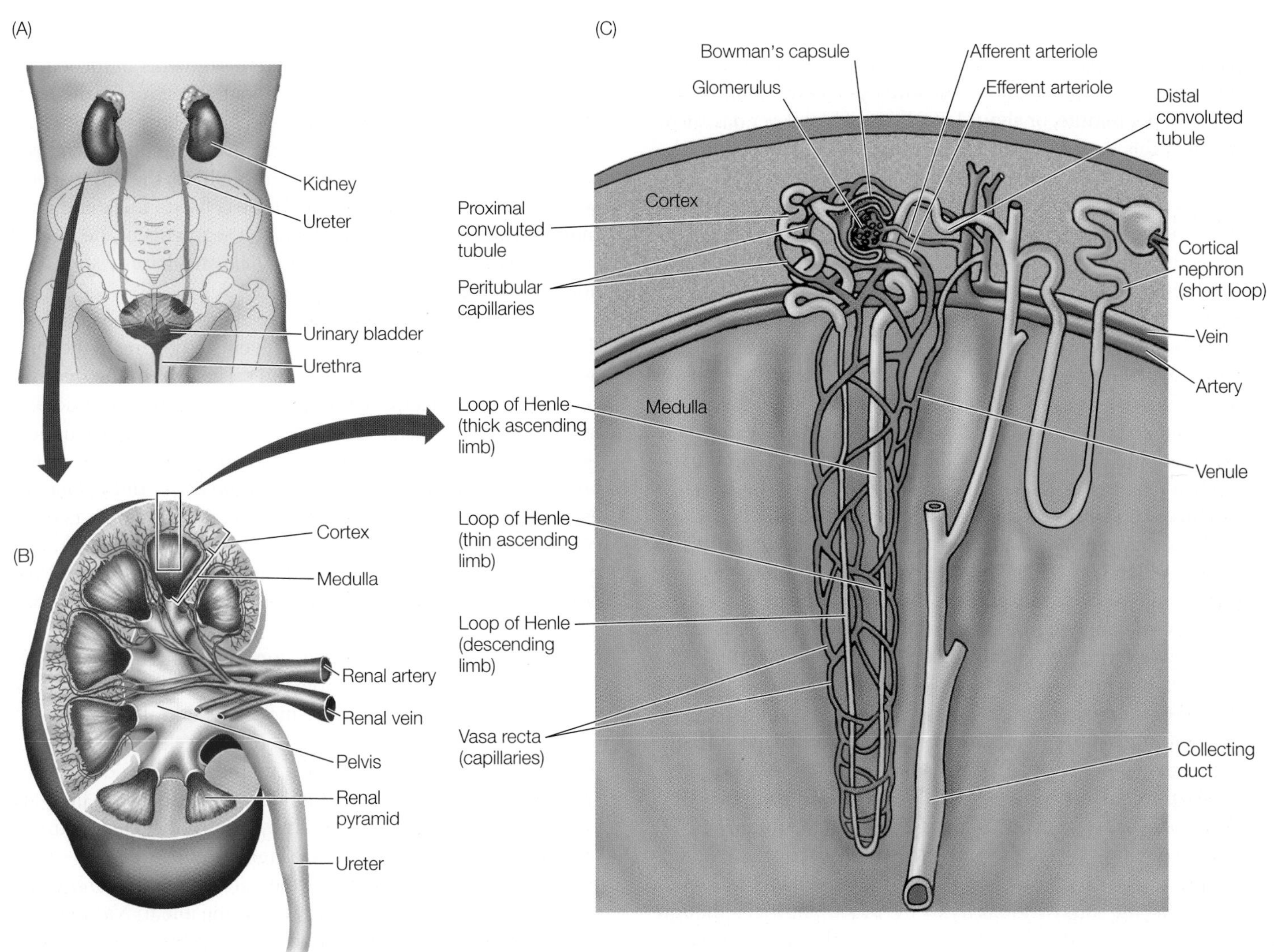


**52.9 The Human Excretory System** (A) The human kidneys lie against the back wall of the abdominal cavity, in the region of the middle back. (B) A highly organized internal structure is the basis for kidney function. Certain parts of the nephrons are in the organ's outer region, called the cortex; other parts are in the internal region, called the medulla. (C) The glomeruli and the proximal and distal convoluted tubules are located in the cortex of the kidney. The loops of Henle run in parallel as straight sections down into the renal medulla and back up to the cortex. Collecting ducts run from the cortex to the inner surface of the medulla, where they open into the ureter. The vasa recta are peritubular capillaries that parallel the loops of Henle.

**yourBioPortal.com**

GO TO **Web Activity 52.3 • The Human Excretory System**

The *ascending limb* of the loop of Henle becomes the **distal convoluted tubule** when it reaches the cortex—"distal" because it is farther from the glomerulus. The distal convoluted tubules of many nephrons join a common **collecting duct** in the cortex. The collecting ducts descend back down through the renal pyramid, parallel to and past the tips of the loops of Henle, and empty into the funnel-shaped *pelvis*. Divisions of the pelvis that surround each renal pyramid join together to leave the kidney as the ureter (see Figure 52.9B).

The organization of the blood vessels of the kidney closely parallels the organization of the nephrons (see Figure 52.9C). Smaller arteries branch from the renal artery and radiate into the cortex, forming the afferent arterioles that carry blood to each glomerulus. Each glomerulus is drained by an efferent arteriole that gives rise to the peritubular capillaries, most of which surround the cortical portions of the tubules. As we have seen, the intimate association between the glomerular and peritubular capillaries permits exchanges between the blood and the specialized regions of the nephron tubule.

Some of the peritubular capillaries run into the medulla in parallel with the loops of Henle and the collecting ducts, forming a vascular network called the **vasa recta**. All of the peritubular capillaries from a nephron join back together into a venule that joins with venules from other nephrons and eventually leads to the renal vein. As we will see, the concentrating ability of the mammalian kidney depends on water reabsorption in the renal medulla, and the vasa recta are the avenue by which that water gets out of the renal medulla and into the circulation.

### Most of the glomerular filtrate is reabsorbed by the proximal convoluted tubule

Most of the water and solutes filtered by the glomerulus are reabsorbed and do not appear in the urine. We can reach this conclusion by comparing the rate of filtration by the glomeruli with the rate of urine production. The kidneys receive about 1 liter of blood per minute, or about 1,500 liters of blood per day. How much of this huge volume is filtered out of the glomeruli? The answer is about 12 percent. This is still a large volume—180 liters per day! We normally urinate 2 to 3 liters per day, so about 98 percent of the fluid volume that is filtered out of the glomerulus is returned to the blood. Where and how is this enormous fluid volume reabsorbed?

The proximal convoluted tubule (PCT) is responsible for most of the reabsorption of water and solutes from the glomerular filtrate. The cells of this section of the renal tubule have many microvilli that increase their apical (facing into the tubule) surface area for reabsorption, and they have many mitochondria—an indication that they are metabolically active. PCT cells actively transport $Na^+$ (with $Cl^-$ following) and other solutes, such as glucose and amino acids, out of the tubule fluid.

Almost all glucose and amino acid molecules that are filtered from the blood are actively reabsorbed by PCT cells and transported into the extracellular fluid. The active transport of solutes from the proximal tubule into the interstitial fluid causes water to follow osmotically. The water and solutes moved into the interstitial fluid are taken up by the peritubular capillaries and returned to the venous blood. These processes accomplish the reabsorption of more than 75 percent of the fluid that initially enters the nephron.

Despite the bulk reabsorption of water and solutes by the PCT, the overall osmolarity of the fluid flowing through the PCT does not change. Thus, the process that is occurring in the PCT is called *isosmotic reabsorption*. The fluid that enters the loop of Henle has the same osmolarity as the blood plasma, although its composition is different. How then does the kidney produce urine that is more concentrated than the blood plasma?

### The loop of Henle creates a concentration gradient in the renal medulla

Humans can produce urine that is four times more concentrated than their blood plasma. The vampire bat we encountered at the beginning of this chapter can produce urine that is twenty times more concentrated than its blood plasma. The concentrating ability of the mammalian kidney arises from a **countercurrent multiplier** mechanism made possible by the anatomical arrangement of the loops of Henle. The term "countercurrent" refers to the opposing directions in which the tubule fluid in the descending and ascending limbs flows. The term "multiplier" refers to the ability of this system to create a solute concentration gradient in the renal medulla.

The loops of Henle do not themselves produce concentrated urine; rather, they increase the osmolarity of the extracellular fluid in the medulla in a graduated way. In humans, for example, the extracellular fluid may be 300 mosm/l (the concentration of blood plasma) at the top of the medulla bordering the cortex, and increase to about 1,200 mosm/l near the hairpin turn of the loop of Henle (see Figure 52.10). How do the loops produce this effect?

The cells that make up the different segments of the loop of Henle differ anatomically and functionally. Cells of the descending limb and the initial cells of the ascending limb are thin, with no microvilli and few mitochondria. They are not specialized for transport. Partway up the ascending limb, the cells become specialized for active transport. These cells are thick and have many mitochondria. Accordingly, the segments of the loop of Henle are named the *thin descending limb*, the *thin ascending limb*, and the *thick ascending limb* (**Figure 52.10**).

The countercurrent multiplier mechanism may be more easily understood by first considering events occurring in the thick ascending limb (Figure 52.10, note 1). The cells of the thick ascending limb reabsorb $Na^+$ and $Cl^-$ from the tubule fluid and move it into the interstitial fluid. (In the following discussion, we will distinguish between the two components of extracellular fluid—the blood plasma and the interstitial fluid.) The thick ascending limb is not permeable to water, so the reabsorption of $Na^+$ and $Cl^-$ from the tubular fluid raises the concentration of those solutes in the surrounding interstitial fluid and decreases the concentration of the tubular fluid entering the distal convoluted tubule.

The thin descending limb, in contrast, is highly permeable to water but not very permeable to $Na^+$ and $Cl^-$. Since the local interstitial fluid has been made more concentrated by the $Na^+$ and $Cl^-$ reabsorbed from the neighboring thick ascending limb, water is withdrawn osmotically from the tubule fluid in the descending limb. Therefore, the fluid in the descending limb becomes more concentrated as it flows toward the hairpin turn at the bottom of the renal medulla (Figure 52.10, note 2).

The thin ascending limb, like the thick ascending limb, is not permeable to water. It is, however, permeable to $Na^+$ and $Cl^-$. As the concentrated tubule fluid flows up the thin ascending limb, it is more concentrated than the surrounding interstitial fluid, so $Na^+$ and $Cl^-$ diffuse out. When the tubule fluid reaches the thick ascending limb, active transport continues to move $Na^+$ and $Cl^-$ from the tubule fluid to the interstitial fluid.

As a result of the processes described above, the tubule fluid reaching the distal convoluted tubule is less concentrated than the blood plasma (Figure 52.10, note 3), and the solutes that have been left behind in the renal medulla have created a concentration gradient in the interstitial fluid of the medulla (indicated by the background color gradient in Figure 52.10).

You may wonder why the blood flow through the medulla does not wash out the concentration gradient established by the loops of Henle. The parallel arrangement of the descending and ascending peritubular capillaries—the vasa recta—in the medulla helps preserve the concentration gradient in the medulla. These capillaries are permeable to both salt and water. Therefore, as blood flows down the descending limb of the vasa recta into the increasingly concentrated interstitial fluid of the medulla, it loses water and gains solutes. As blood flows up from the bottom of the medulla in the ascending limb of the vasa recta, the opposite happens (water is gained and solutes are lost) because now the blood is more concentrated than the surrounding interstitial fluid (Figure 52.10, notes 4–6). The dynamics of this countercurrent exchange of salts and water be-

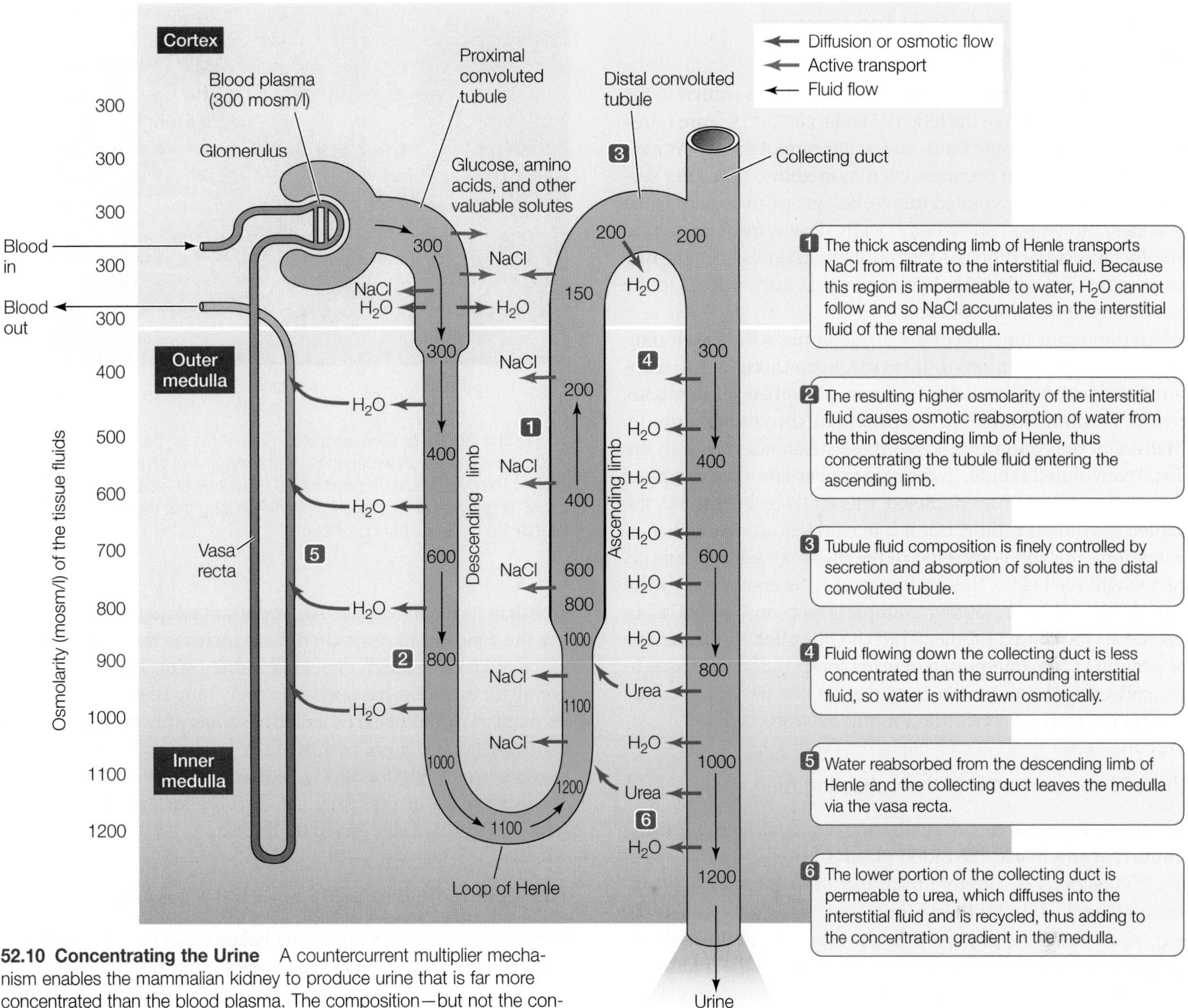


**52.10 Concentrating the Urine** A countercurrent multiplier mechanism enables the mammalian kidney to produce urine that is far more concentrated than the blood plasma. The composition—but not the concentration—of the filtrate is changed by the proximal convoluted tubule, which reabsorbs valuable molecules (including NaCl). Bulk reabsorption of water follows osmotically. The urine concentration process begins in the thick ascending limb of Henle, which reabsorbs NaCl but is impermeable to $H_2O$. Some of the reabsorbed NaCl enters the descending limb and is thereby trapped in the renal medulla, creating a concentration gradient in the interstitial fluid. As urine in the collecting duct passes through this concentration gradient, it can lose water osmotically and become highly concentrated.

tween the blood in the vasa recta and the interstitial fluids result in little net change in the composition of the interstitial fluid in the medulla.

### Water permeability of kidney tubules depends on water channels

We have noted that some tubule regions, such as the PCT, are highly permeable to water while others, such as the thick ascending limb of the loop of Henle, are impermeable to water. How do differences in water permeability in different regions of the nephron arise? Regions of the nephron that are highly permeable to water have greater numbers of *aquaporins*, a class of membrane proteins that form water channels (see Section 6.3). Aquaporins are abundant in kidney PCT cells and in descending limbs of the loops of Henle, but not in the ascending limbs of the loop of Henle. The discovery of aquaporins resulted in Peter Agre of Johns Hopkins University receiving the Nobel Prize in Chemistry in 2003.

As an interesting evolutionary note, aquaporins are also important in maintaining water balance in amphibians. When not in an aqueous environment, many amphibians can gain water from a moist substrate because they have aquaporins in the epithelial cells of their belly skin. Thus, water can cross their skin into the interstitial fluid by osmosis.

### The distal convoluted tubule fine-tunes the composition of the urine

The first portion of the distal convoluted tubule is similar to the thick ascending limb of the loop of Henle. $Na^+$ and $Cl^-$ are transported out of the tubule fluid, and water cannot follow. As a result, the tubule fluid becomes even more dilute. The later sections of the distal convoluted tubule, however, can be permeable to water, and water can be osmotically drawn from the tubule into the interstitial fluid. As the tubule fluid flows from the distal tubule to the collecting duct, it can be below or equal to the osmolarity of the blood plasma.

An important function of the distal tubule is the fine-tuning of the ionic composition of the urine. Even though bulk reabsorption of substances such as calcium, phosphate, bicarbonate, and potassium occurs in the proximal convoluted tubule, changes in the concentrations of these substances occur in the distal convoluted tubule. In the case of potassium, for example, if a person is potassium depleted, this ion is reabsorbed in the distal convoluted tubule, but if a person has an abundance of potassium, this ion is secreted in the distal convoluted tubule. As we will see below, this exchange of $K^+$ is controlled by the hormone *aldosterone*. Another example is reabsorption of $Ca^{2+}$ in the distal convoluted tubule, which is controlled by the actions of vitamin D. The fine-tuning of urine composition continues in the collecting duct. As you can imagine, the list of ion transporters in the distal convoluted tubule is large.

### Urine is concentrated in the collecting duct

The tubule fluid entering the collecting duct is at about the same solute *concentration* as the blood plasma, but its solute *composition* is considerably different from that of the plasma. The major solute in the tubular fluid is now urea, since salts were reabsorbed earlier in the nephron. As the tubule fluid flows down the collecting duct, it loses water osmotically to the interstitial fluid, and that water returns to the circulatory system via the vasa recta (see Figure 52.10).

The concentration gradient established in the renal medulla by the countercurrent multiplier actions of the loops of Henle creates the osmotic potential that withdraws water from the collecting ducts. The collecting ducts begin in the renal cortex and run through the renal medulla before emptying into the ureter at the tips of the renal pyramids. During this journey, the solute concentration of the surrounding interstitial fluid increases, and more and more water can be absorbed from the urine in the collecting duct. By the time it reaches the ureter, the urine can become greatly concentrated, with urea as a major solute.

As water is withdrawn from the collecting duct, some urea also leaks out into the medullary interstitial fluid, adding to its osmotic potential. This urea diffuses back into the loop of Henle and is returned to the collecting duct. The recycling of urea in the renal medulla contributes significantly to the concentration gradient and therefore the ability of the kidney to concentrate the urine in the collecting duct.

Overall, the ability of a mammal to concentrate its urine is determined by the maximum concentration gradient it can establish in its renal medulla. An important adaptation for increasing the concentration gradient is to increase the lengths of the loops of Henle relative to overall kidney size. The small desert gerbil, for example, has such extremely long loops of Henle that its renal pyramid (each of its kidneys has only one, in contrast to ours) extends far out of the concave surface of the kidney and into the ureter (**Figure 52.11**). These animals are so effective in conserving water that they can survive just on the water released by the metabolism of their food.

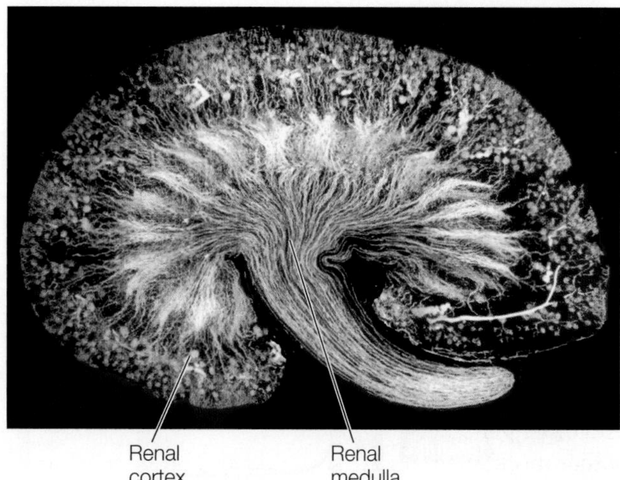


**52.11 The Ability to Concentrate** The ability of the mammalian kidney to concentrate urine depends on the lengths of its loops of Henle relative to the overall size of the kidney. The kidney of a desert gerbil has a single renal pyramid with loops of Henle so long that the pyramid extends far into the ureter (ureter not shown).

### The kidneys help regulate acid–base balance

Besides regulating salt and water balance and excreting nitrogenous wastes, the kidneys have another important role: they regulate the hydrogen ion concentration (the pH) of the extracellular fluids. pH is a critical variable because it influences the structure and function of proteins.

One way to minimize pH changes in a chemical solution is to add a *buffer*—a substance that can either absorb or release hydrogen ions (see Section 2.4). The major buffers in the blood are bicarbonate ions ($HCO_3^-$; see Figure 49.14) that are formed from the dissociation of carbonic acid, which in turn is formed by the hydration of $CO_2$ according to the following equilibrium reaction:

$$CO_2 + H_2O \rightleftharpoons H_2CO_3 \rightleftharpoons H^+ + HCO_3^-$$

From this equation, you can see that if excess hydrogen ions are added to this reaction mixture, the reaction will move to the left and absorb the excess $H^+$. If hydrogen ions are removed from the reaction mixture, however, the reaction will move to the right and supply more $H^+$.

The $HCO_3^-$ buffer system is important for controlling the pH of the blood, and therefore of the interstitial fluids as well, because the reaction can be pushed to the right and pulled to the left physiologically. The lungs control the levels of $CO_2$ in the blood, thus altering the acid portion of the reaction. $CO_2$ is considered the acid portion of the reaction because if you add additional $CO_2$, the reaction shifts to the right, producing more $H^+$ ions. The kidneys control the base portion of the reaction by removing $H^+$ from the blood and returning $HCO_3^-$ to the blood. How are $H^+$ ions removed from the blood?

One mechanism for $H^+$ secretion and $HCO_3^-$ reabsorption involves ammonia ($NH_3$) and ammonium ion ($NH_4^+$). The metabolism of glutamine in tubule cells produces $NH_3$ and $HCO_3^-$. The $HCO_3^-$ is reabsorbed into the interstitial fluid. The $NH_3$ is transported into the tubule fluid by means of an $NH_3$ transporter that has been characterized only recently in an effort to identify novel proteins coming from the sequencing of the human and other genomes (**Figure 52.12**).

Another mechanism for $H^+$ secretion and $HCO_3^-$ reabsorption is shown in **Figure 52.13**. The $H^+$ is transported into the tubule fluid in exchange for $Na^+$. The $H^+$ combines with $HCO_3^-$ that has been filtered in the glomerulus, producing $H_2CO_3$ that disassociates into $H_2O$ and $CO_2$. The $CO_2$ diffuses into the tubule cell where in the presence of the enzyme carbonic anhydrase it produces $HCO_3^-$ that is transported into the interstitial fluid and thence to the blood.

### Kidney failure is treated with dialysis

Loss of kidney function, or *renal failure*, results in the retention of salts and water (hence high blood pressure), retention of urea (uremic poisoning), and a decreasing pH (acidosis). A person who suffers complete renal failure will die within 2 weeks if not treated. A drastic but highly successful treatment is kidney transplant, but it is usually necessary to sustain a patient for considerable time while waiting for a kidney to become available. Therefore, artificial kidneys, or renal dialysis machines, are essential modes of treatment.

In a dialysis machine, the patient's blood flows through many small channels made of semipermeable membranes

## INVESTIGATING LIFE

### 52.12 An Ammonium Transporter in the Renal Tubules?

An important way the kidney excretes hydrogen ions and buffers the blood is to secrete ammonia ($NH_3$) into the renal tubules. It was thought that ammonia simply diffused into the tubules until the function of Rhcg, a protein in the Rh blood antigen family, was discovered. This experiment demonstrated that loss of the *Rhcg* gene in mice impairs their ability to buffer their blood pH by excreting excess $H^+$ (in the form of ammonium).

**HYPOTHESIS** **The protein Rhcg is an ammonia transporter and is critical for the kidney's role in acid-base balance.**

**METHOD**

1. Create a line of mice in which the gene for the protein Rhcg is knocked out (see Section 18.4).
2. Measure starting blood pH, plasma bicarbonate ($HCO_3^-$) levels, and urine ammonium ($NH_4^+$) levels in experimental and control (wild-type) mice.
3. For 6 days, administer drinking water containing $NH_4Cl$ (ammonium chloride, a mild acid) to control mice and to Rhcg knockout mice.
4. Measure the three variables (see item 2 above) at days 2 and 6.

**RESULTS**

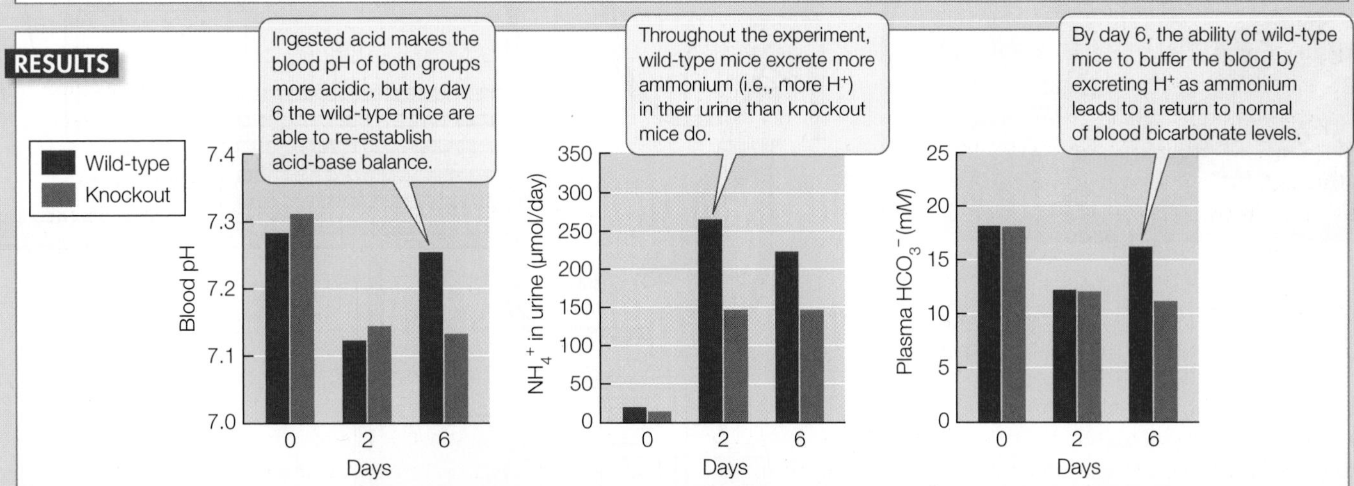
Wild-type
Knockout
Ingested acid makes the blood pH of both groups more acidic, but by day 6 the wild-type mice are able to re-establish acid-base balance.
Blood pH
7.4
7.3
7.2
7.1
7.0
0
2
6
Days
Throughout the experiment, wild-type mice excrete more ammonium (i.e., more H⁺) in their urine than knockout mice do.
NH4+ in urine (µmol/day)
350
300
250
200
150
100
50
0
0
2
6
Days
By day 6, the ability of wild-type mice to buffer the blood by excreting H⁺ as ammonium leads to a return to normal of blood bicarbonate levels.
Plasma HCO3− (mM)
25
20
15
10
5
0
0
2
6
Days

**CONCLUSION** **Lack of functional Rhcg protein impairs a mouse's ability to secrete ammonium ions in its urine, thus limiting the capacity to regulate acid-base balance. This protein is probably an ammonia transporter in the renal tubules.**

Go to **yourBioPortal.com** for original citations, discussions, and relevant links for all INVESTIGATING LIFE figures.

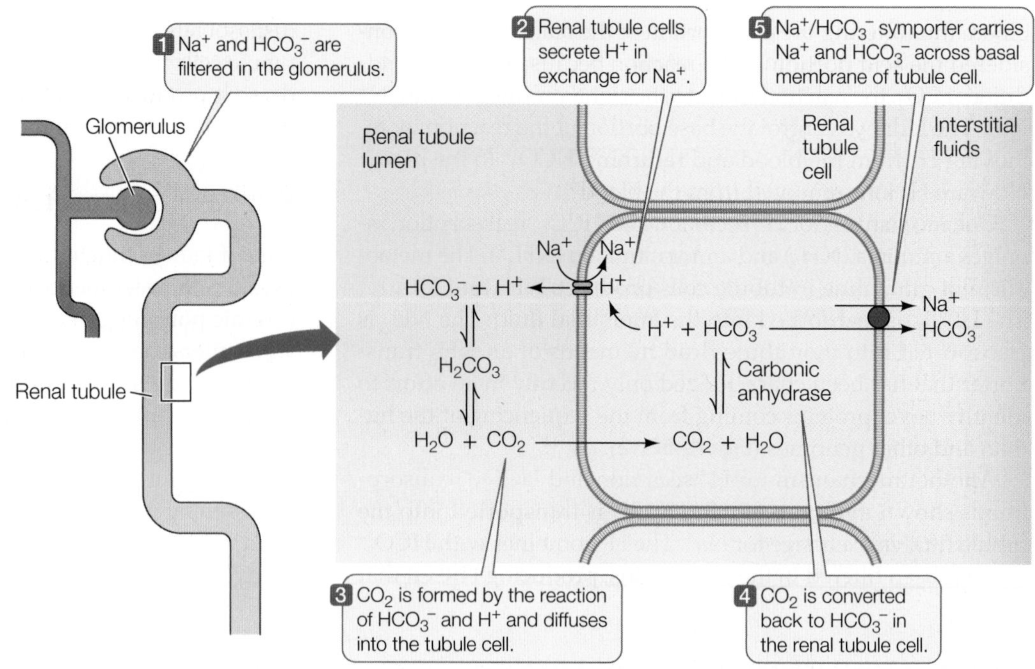
1 $Na^+$ and $HCO_3^-$ are filtered in the glomerulus.
2 Renal tubule cells secrete $H^+$ in exchange for $Na^+$.
5 $Na^+/HCO_3^-$ symporter carries $Na^+$ and $HCO_3^-$ across basal membrane of tubule cell.
Glomerulus
Renal tubule
Renal tubule lumen
Renal tubule cell
Interstitial fluids
$Na^+$
$Na^+$
$HCO_3^- + H^+$
$H^+$
$H^+ + HCO_3^-$
$Na^+$
$HCO_3^-$
$H_2CO_3$
Carbonic anhydrase
$H_2O + CO_2$
$CO_2 + H_2O$
3 $CO_2$ is formed by the reaction of $HCO_3^-$ and $H^+$ and diffuses into the tubule cell.
4 $CO_2$ is converted back to $HCO_3^-$ in the renal tubule cell.

**52.13 The Kidney Excretes Acids and Conserves Bases** Bicarbonate ions are filtered out of the blood at the glomerulus, and renal tubule cells secrete hydrogen ions into the tubule fluid. In the renal tubule, the filtered bicarbonate buffers the secreted hydrogen ions and keeps the urine from becoming too acidic. The $CO_2$ formed by the reaction of bicarbonate and hydrogen ions is converted back to bicarbonate by the renal tubule cells and transported back into the interstitial fluid.

(**Figure 52.14**). A dialysis solution flows on the other side of these membranes, through which small molecules can diffuse. Molecules and ions diffuse from an area of higher concentration to an area of lower concentration, so the composition of the dialysis fluid is crucial. The concentrations of the molecules or ions that need to be conserved must be at the same concentration in the dialysis fluid as they are in the blood. The concentrations of molecules and ions that need to be removed from the blood are zero in the dialysis fluid. The total osmotic potential of the dialysis fluid must equal that of the plasma.

About 500 ml of the patient's blood is in the dialysis machine at any one time, and the unit processes several hundred milli-

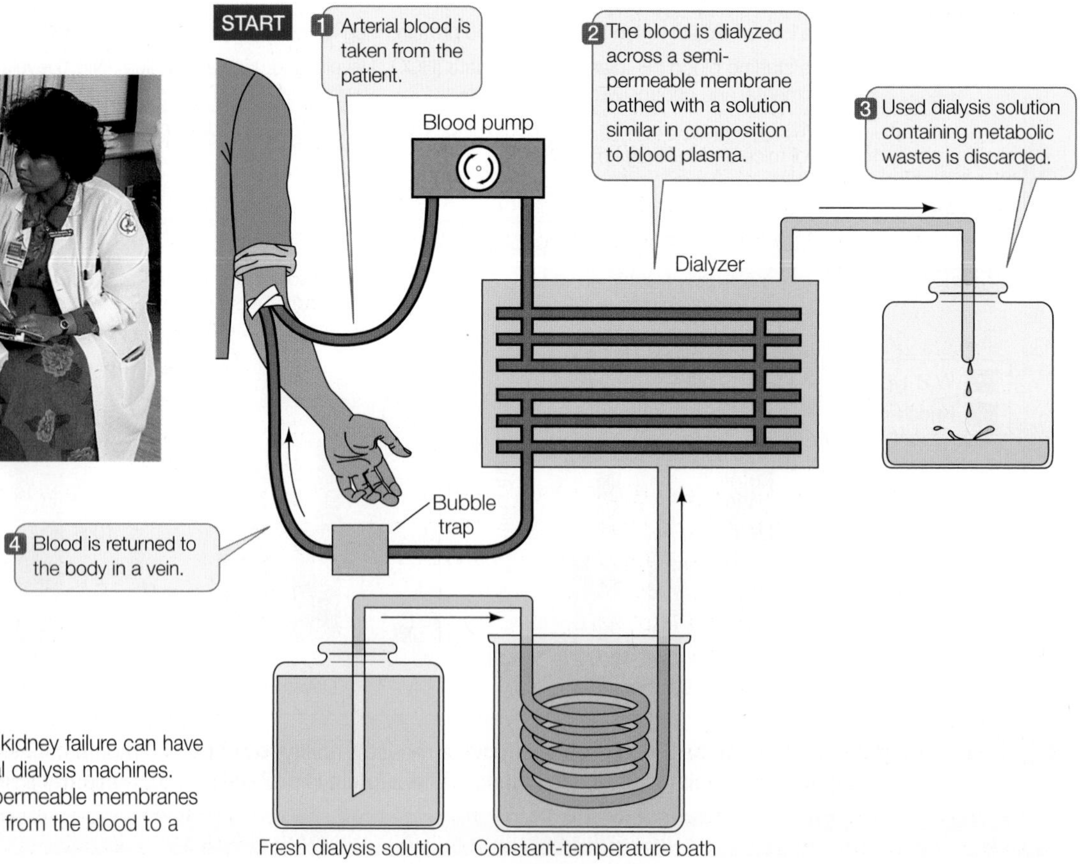

START
1 Arterial blood is taken from the patient.
Blood pump
2 The blood is dialyzed across a semipermeable membrane bathed with a solution similar in composition to blood plasma.
3 Used dialysis solution containing metabolic wastes is discarded.
Dialyzer
Bubble trap
4 Blood is returned to the body in a vein.
Fresh dialysis solution
Constant-temperature bath

**52.14 Renal Dialysis** Patients with kidney failure can have their blood cleansed of wastes by renal dialysis machines. Blood flows through channels of semipermeable membranes that allow diffusion of waste molecules from the blood to a dialysis fluid.

liters of blood per minute. A patient with no kidney function must be on the dialysis machine for 4 to 6 hours three times a week.

## 52.5 RECAP

**The anatomical organization of the nephrons makes it possible for the mammalian kidney to produce a urine more concentrated than the blood, thereby conserving water to maintain extracellular fluid volume. Bulk reabsorption of salts, other valuable solutes, and water takes place in the proximal convoluted tubule. The loops of Henle act as a countercurrent multiplier, creating a concentration gradient of the interstitial fluids in the renal medulla. Collecting ducts run through the renal medulla and lose water osmotically to the surrounding interstitial fluids, concentrating the urine.**

- Explain how the countercurrent multiplier mechanism of the nephron makes it possible for the kidney to form a concentrated urine. **See p. 1102 and Figure 52.10**
- How does the kidney contribute to acid–base balance? **See pp. 1104–1105 and Figure 52.13**

The kidney contributes to homeostasis in several ways, including regulating extracellular fluid volume, maintaining the osmotic concentration and ionic composition of the extracellular fluid, and regulating pH. As we will see next, the kidneys also play a major role in regulating blood pressure.

# 52.6 How Are Kidney Functions Regulated?

Several regulatory mechanisms act on the kidneys to maintain blood pressure, blood osmolarity, and blood composition. We will discuss these mechanisms separately, but keep in mind that they are always working together.

### Glomerular filtration rate is regulated

If the kidneys stop filtering blood, they cannot accomplish any of their functions. The maintenance of a constant **glomerular filtration rate** (**GFR**) depends on an adequate blood supply to the kidneys at an adequate blood pressure. Renal arteries usually deliver blood to the kidneys at high pressure because they are early branches off the aorta. In addition, *autoregulatory mechanisms* ensure adequate blood supply and blood pressure for kidney function regardless of what is happening elsewhere in the body. The kidney's autoregulatory adjustments compensate for decreases in cardiac output or decreases in blood pressure so that the GFR remains constant.

One autoregulatory mechanism is the dilation (expansion) of the afferent renal arterioles when blood pressure falls. This dilation decreases the resistance in the arterioles and helps maintain blood pressure in the glomerulus. If arteriole dilation does not keep the GFR from falling, the kidney releases an enzyme, **renin**, into the blood. Renin converts a circulating protein, angiotensinogen, into angiotensin I, which is then acted on by angiotensin converting enzyme (ACE) to form the active hormone angiotensin II, or simply, **angiotensin** (**Figure 52.15**). Angiotensin has several effects that help restore the GFR to normal:

- It constricts the efferent renal arterioles, raising the resistance for blood leaving the glomerulus. Like putting a finger over the end of a garden hose, this restriction of drainage elevates blood pressure in the glomerular capillaries.
- It constricts peripheral blood vessels all over the body, an action that elevates central blood pressure.
- It stimulates the adrenal cortex to release the hormone **aldosterone**. Aldosterone stimulates sodium reabsorption by the kidney, thereby making its reabsorption of water more effective. Enhanced water reabsorption helps maintain blood volume and therefore central blood pressure.
- It acts on the brain to stimulate thirst. Increased water intake in response to thirst increases blood volume and blood pressure.

Thus the renin-angiotensin-aldosterone system, or RAAS, coordinates many responses to maintain blood pressure and kidney function.

### Blood osmolarity and blood pressure are regulated by ADH

Cells in the hypothalamus can stimulate the release from the posterior pituitary of a hormone called *antidiuretic hormone* (*ADH*, also called *vasopressin*) that can act on cells of the collecting duct to insert aquaporins (water channels) into their plasma membranes. The aquaporins increase the permeability of these membranes to water, and therefore more water is reabsorbed from the collecting duct fluid into the interstitial spaces of the renal medulla. The higher the circulating levels of ADH, the greater the number of aquaporins. Various factors can stimulate or inhibit the release of ADH. Of key importance to kidney function are osmoreceptors that monitor blood osmolarity and stretch receptors that monitor blood pressure (**Figure 52.16**).

Osmoreceptor neurons in the hypothalamus are activated by a rise in blood osmolarity, and they increase the release of ADH. ADH helps regulate blood osmolarity by controlling water reabsorption. The osmoreceptors also stimulate thirst. The resulting water retention and water intake dilute the blood as they expand blood volume.

Stretch receptors in the walls of the aorta and the carotid arteries (see Figure 50.17) that detect an increase in blood pressure will *inhibit* the release of ADH. With less circulating ADH, less water is reabsorbed, which decreases blood volume and hence acts to lower blood pressure.

If blood pressure falls, as when you lose blood volume through hemorrhage or excessive evaporative water loss, activity of the stretch receptors in the aorta and carotid arteries decreases. Input via cranial nerves to the hypothalamus from these receptors inhibits the release of ADH, so when the firing rates

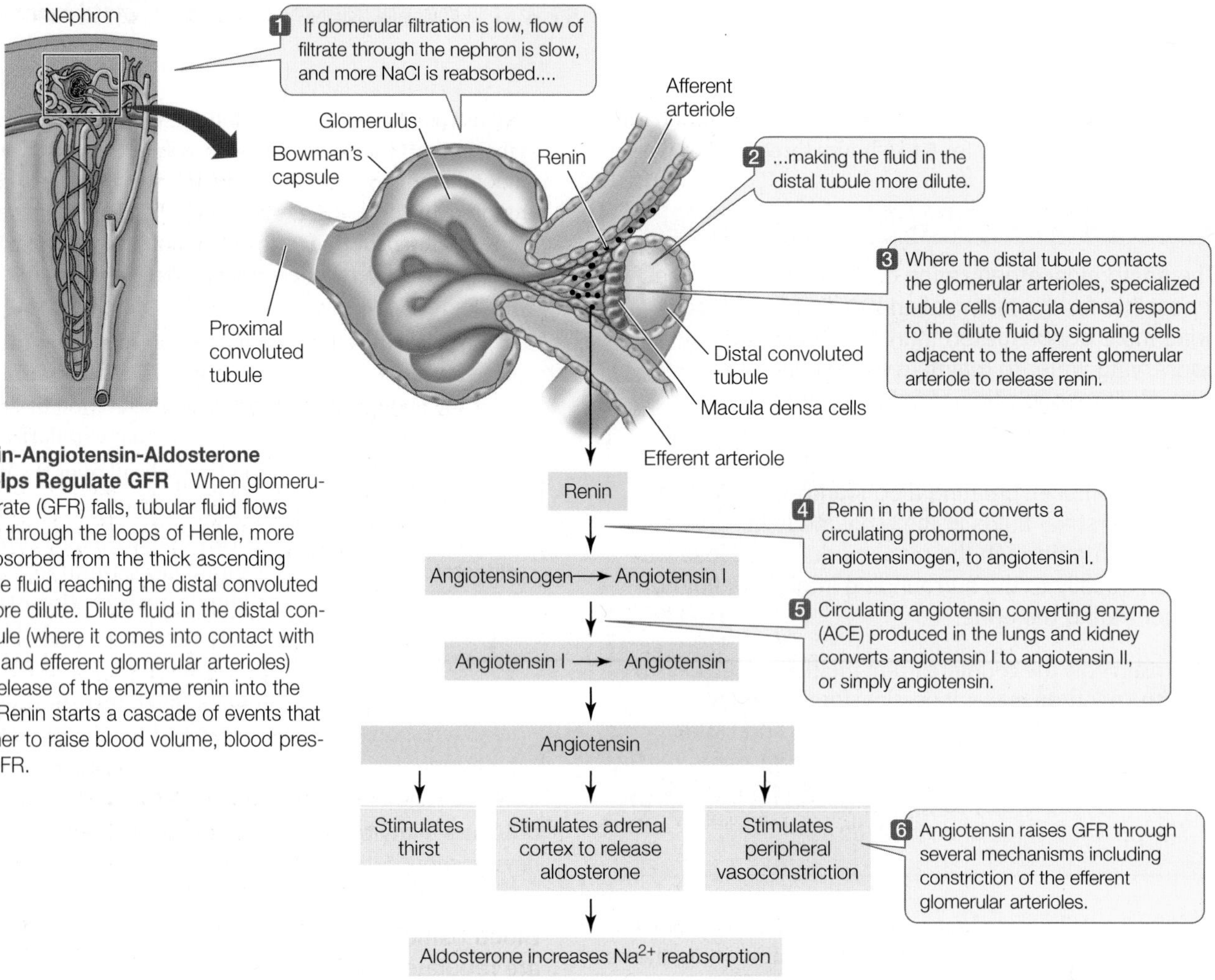

Nephron

**1** If glomerular filtration is low, flow of filtrate through the nephron is slow, and more NaCl is reabsorbed....

Glomerulus

Bowman's capsule

Afferent arteriole

Renin

**2** ...making the fluid in the distal tubule more dilute.

**3** Where the distal tubule contacts the glomerular arterioles, specialized tubule cells (macula densa) respond to the dilute fluid by signaling cells adjacent to the afferent glomerular arteriole to release renin.

Proximal convoluted tubule

Distal convoluted tubule

Macula densa cells

Efferent arteriole

Renin

**4** Renin in the blood converts a circulating prohormone, angiotensinogen, to angiotensin I.

Angiotensinogen ⟶ Angiotensin I

**5** Circulating angiotensin converting enzyme (ACE) produced in the lungs and kidney converts angiotensin I to angiotensin II, or simply angiotensin.

Angiotensin I ⟶ Angiotensin

Angiotensin

Stimulates thirst

Stimulates adrenal cortex to release aldosterone

Stimulates peripheral vasoconstriction

**6** Angiotensin raises GFR through several mechanisms including constriction of the efferent glomerular arterioles.

Aldosterone increases $Na^{2+}$ reabsorption

**52.15 Renin-Angiotensin-Aldosterone System Helps Regulate GFR** When glomerular filtration rate (GFR) falls, tubular fluid flows more slowly through the loops of Henle, more NaCl is reabsorbed from the thick ascending limb, and the fluid reaching the distal convoluted tubule is more dilute. Dilute fluid in the distal convoluted tubule (where it comes into contact with the afferent and efferent glomerular arterioles) stimulates release of the enzyme renin into the circulation. Renin starts a cascade of events that work together to raise blood volume, blood pressure, and GFR.

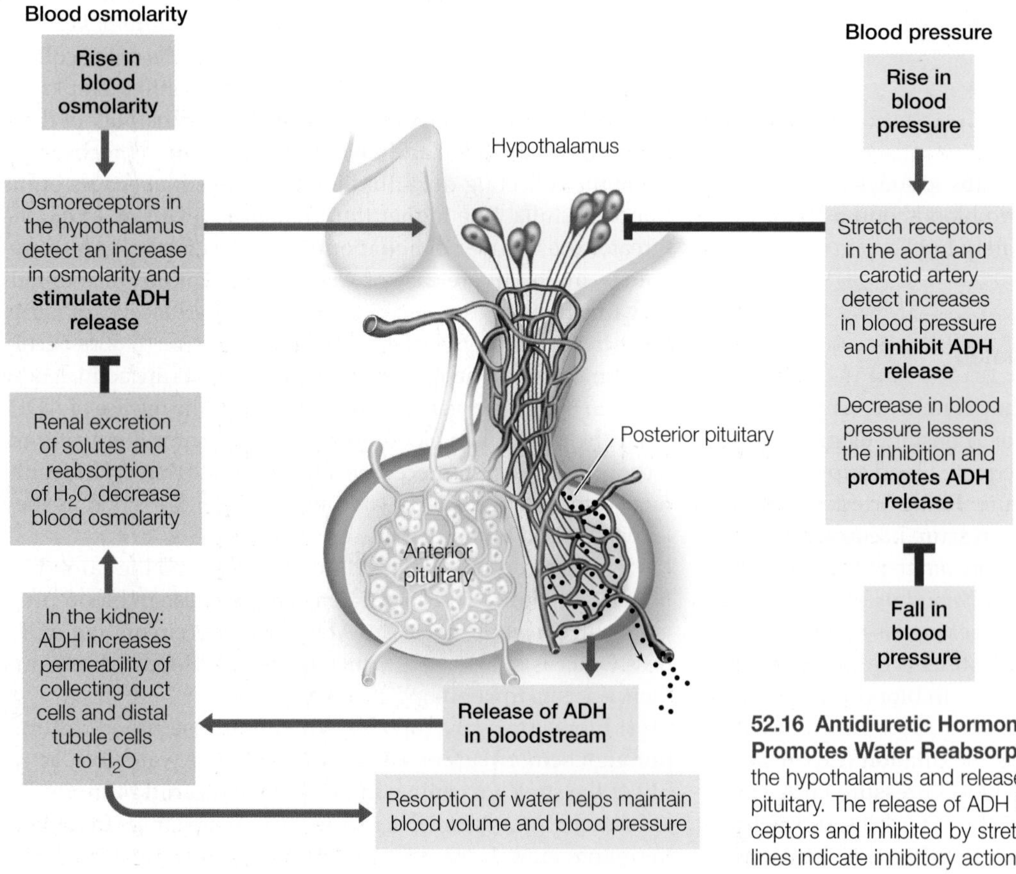

**Blood osmolarity**

**Rise in blood osmolarity**

Osmoreceptors in the hypothalamus detect an increase in osmolarity and **stimulate ADH release**

Renal excretion of solutes and reabsorption of $H_2O$ decrease blood osmolarity

In the kidney: ADH increases permeability of collecting duct cells and distal tubule cells to $H_2O$

Hypothalamus

Posterior pituitary

Anterior pituitary

**Release of ADH in bloodstream**

Resorption of water helps maintain blood volume and blood pressure

**Blood pressure**

**Rise in blood pressure**

Stretch receptors in the aorta and carotid artery detect increases in blood pressure and **inhibit ADH release**

Decrease in blood pressure lessens the inhibition and **promotes ADH release**

**Fall in blood pressure**

**52.16 Antidiuretic Hormone Increases Blood Pressure and Promotes Water Reabsorption** ADH is produced by neurons in the hypothalamus and released from nerve endings in the posterior pituitary. The release of ADH is stimulated by hypothalamic osmoreceptors and inhibited by stretch receptors in the great arteries. Red lines indicate inhibitory actions; green lines show stimulatory actions.

## INVESTIGATING LIFE

### 52.17 ADH Induces Insertion of Aquaporins into Plasma Membranes

Aquaporin proteins make different regions of renal tubules permeable to water. One of these, AQP-2, is responsible for the permeability of the collecting duct cells. How does antidiuretic hormone act on these proteins to control the level of permeability in renal cells?

**HYPOTHESIS** **Antidiuretic hormone (ADH) controls the location of aquaporin proteins.**

**METHOD**

1. Isolate collecting ducts from rat kidney.
2. Use immunochemical staining to localize the AQP-2 aquaporins in collecting duct cells both with and without the presence of ADH. Also localize the aquaporins after ADH is applied and then washed away.

Collecting duct cell

Inside collecting duct

Vesicle

AQP-2

Without ADH, AQP-2 are mostly found in plasma membranes of intracellular vesicles.

With ADH, AQP-2 are mostly found in plasma membranes of collecting duct cells.

After ADH washout, AQP-2 are again sequestered in intracellular vesicles.

3. Perfuse collecting ducts and measure water permeability under the same three conditions.

**RESULTS**

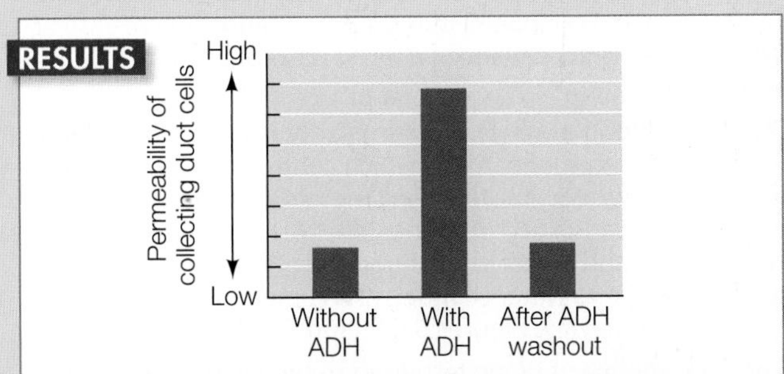


**CONCLUSION** **In the absence of ADH, AQP-2 is sequestered intracellularly. When ADH is present, these water channels are inserted into the plasma membranes, making the cells more permeable to water.**

Go to **yourBioPortal.com** for original citations, discussions, and relevant links for all INVESTIGATING LIFE figures.

of these stretch receptors fall, ADH release increases. More ADH results in more efficient water reabsorption and therefore a protection of blood volume and blood pressure.

Alcohol inhibits ADH release, explaining why excessive beer drinking leads to excessive urination and dehydration, which contributes to the symptoms of a hangover.

As mentioned earlier, the presence of aquaporins in plasma membranes determines their water permeability. Aquaporins play an important and unique role in the collecting duct. Several members of the aquaporin family of water channels are found in the plasma membranes of cells in the distal tubules. At least two aquaporins are localized in the basolateral membranes (facing the blood vessels). A different aquaporin, called AQP-2, is found in the apical plasma membranes (facing into the tubule). But the presence of AQP-2 in these membranes is controlled by ADH, as described in **Figure 52.17**.

Sometimes AQP-2 is sequestered in the membranes of intracellular vesicles, and when it is, the collecting duct permeability is low. ADH causes these vesicles to insert their AQP-2 channels into the apical plasma membrane, making that membrane (and therefore the cell) permeable to water. ADH also stimulates the synthesis of new AQP-2 proteins. Thus, circulating ADH controls the number of AQP-2 water channels in the plasma membranes of the collecting duct cells, and therefore their water permeability.

### The heart produces a hormone that helps lower blood pressure

You may not think of the heart as an endocrine organ, but it is. When blood volume is high, blood pressure is high, and that puts strain on the heart. Under these conditions, the increased venous return stretches the atria of the heart. When the atrial muscle fibers are overly stretched, they release a peptide hormone called **atrial natriuretic peptide** (**ANP**). This peptide hormone enters the circulation, and in the kidney it decreases the reabsorption of sodium. If less sodium is reabsorbed, less water is reabsorbed, and more passes into the urine. Thus, ANP has the effect of lowering blood volume and therefore blood pressure.

## 52.6 RECAP

**Glomerular filtration is essential for kidney function and is sustained by autoregulatory mechanisms. Sensors that monitor blood pressure and blood osmolarity may stimulate or inhibit the release of hormones that regulate kidney function.**

- Explain how falling GFR results in an increase in circulating angiotensin and how angiotensin restores GFR. **See p. 1107**
- Explain how falling blood pressure or increasing blood osmolarity result in changes in permeability of the collecting ducts. **See Figure 52.16**

# CHAPTER SUMMARY

## 52.1 How Do Excretory Systems Maintain Homeostasis?

- **Excretory systems** maintain the osmolarity and volume of the extracellular fluids and eliminate the waste products of nitrogen metabolism through the processes of filtration, reabsorption, and secretion. **Urine** is the output of excretory systems.
- There is no active transport of water, so water must be moved across membranes by a difference in either **osmolarity** or pressure.
- Water enters or leaves cells by osmosis. To achieve cellular water balance, animals must maintain the osmolarity of their extracellular fluids within an acceptable range.
- Marine animals can be **osmoconformers** or **osmoregulators**. Freshwater animals must be osmoregulators and must continually excrete water and conserve salts. Most animals are ionic regulators to some degree.
- Apart from regulating osmolarity of cells and extracellular fluids, most animals must also regulate their ionic composition by conserving some ions and secreting others. Salt glands are adaptations for secretion of NaCl. Review Figure 52.2

## 52.2 How Do Animals Excrete Nitrogen?

- Aquatic animals that breathe water can eliminate nitrogenous wastes such as **ammonia** by diffusion across their gill membranes. Terrestrial animals and some aquatic animals must detoxify ammonia by converting it to **urea** or **uric acid** before excretion. Review Figure 52.3
- Depending on the form in which they excrete their nitrogenous wastes, animals are classified as **ammonotelic**, **ureotelic**, or **uricotelic**.

## 52.3 How Do Invertebrate Excretory Systems Work?

- The **protonephridia** of flatworms consist of flame cells and excretory tubules. Extracellular fluid is filtered into the tubules, which process the filtrate to produce a dilute urine. Review Figure 52.4
- In annelid worms, blood pressure causes filtration of the blood across capillary walls. The filtrate enters the coelomic cavity, where it is taken up by **metanephridia**, which alter the composition of the filtrate by active transport mechanisms. Review Figure 52.5, WEB ACTIVITY 52.1
- The **Malpighian tubules** of insects receive ions and nitrogenous wastes by active transport across the tubule cells. Water follows by osmosis. Ions and water are reabsorbed from the rectum, so the insect excretes semisolid wastes. Review Figure 52.6

## 52.4 How Do Vertebrates Maintain Salt and Water Balance?

- Marine and terrestrial animals conserve water in various ways. Marine bony fishes produce little urine. Cartilaginous fishes retain urea so that the osmolarity of their body fluids remains close to that of seawater. Amphibians remain close to water. Reptiles have skin with low water permeability, lay shelled eggs, and excrete nitrogenous wastes as uric acid.
- Mammals produce urine more concentrated than their extracellular fluids.
- The **nephron**, the functional unit of the vertebrate **kidney**, consists of a **glomerulus**, in which blood is filtered, a **renal tubule**, which processes the filtrate into urine to be excreted, and a system of **peritubular capillaries**, which surround the tubule and serve as a site of secretion and reabsorption. Review Figure 52.7, WEB ACTIVITY 52.2

## 52.5 How Does the Mammalian Kidney Produce Concentrated Urine?

- The concentrating ability of the mammalian kidney is a function of its anatomy, which is ideal for countercurrent exchange. Review Figure 52.9, ANIMATED TUTORIAL 52.1
- The glomeruli and the **proximal** and **distal convoluted tubules** are located in the **cortex** of the kidney. Certain molecules are actively reabsorbed from the glomerular filtrate by the tubule cells, and other molecules are actively secreted. Straight sections of renal tubules called **loops of Henle** and **collecting ducts** are arranged in parallel in the **medulla** of the kidney. SEE WEB ACTIVITY 52.3
- Salts and water are reabsorbed in the proximal convoluted tubule without the renal filtrate becoming more concentrated, although its composition changes.
- The loops of Henle create a concentration gradient in the interstitial fluid of the renal medulla by a **countercurrent multiplier** mechanism. Urine flowing down the collecting ducts to the **ureter** is concentrated by the osmotic reabsorption of water caused by the concentration gradient in the surrounding interstitial fluid. Review Figure 52.10
- Hydrogen ions secreted by the renal tubules are buffered in the urine by bicarbonate and other chemical buffering systems. Review Figures 52.12 and 52.13

## 52.6 How Are Kidney Functions Regulated?

- Kidney function in mammals is controlled by autoregulatory mechanisms that maintain a constant high **glomerular filtration rate** (**GFR**) even if blood pressure varies.
- An important autoregulatory mechanism is the release of **renin** by the kidney when blood pressure falls. Renin activates **angiotensin**, which causes the constriction of efferent glomerular arterioles and peripheral blood vessels, causes the release of **aldosterone** (which enhances water reabsorption), and stimulates thirst.
- Changes in blood pressure and osmolarity influence the release of antidiuretic hormone (ADH), which controls the permeability of the collecting duct to water and therefore the amount of water that is reabsorbed from the urine. ADH stimulates the expression of and controls the intracellular location of aquaporins, which serve as water channels in the membranes of collecting duct cells. Review Figures 52.16 and 52.17
- When the volume of blood returning to the heart increases and stretches the atrial walls, **atrial natriuretic peptide** (**ANP**) is released, which causes increased excretion of salt and water.

**SEE WEB ACTIVITY 52.4 for a review of the major human organ systems.**

## SELF-QUIZ

1. Which statement is *true*?
   *a.* Most marine invertebrates are osmoregulators.
   *b.* All freshwater invertebrates are osmoconformers.
   *c.* Marine bony and cartilaginous fishes have similar interstitial osmolarities.
   *d.* Freshwater fishes are ionic regulators.
   *e.* Marine mammals gain water osmotically.
2. The excretion of nitrogenous wastes
   *a.* by humans can be in the form of urea and uric acid.
   *b.* by mammals is never in the form of uric acid.
   *c.* by marine fishes is mostly in the form of urea.
   *d.* does not contribute to the osmolarity of the urine.
   *e.* requires more water if the waste product is the rather insoluble uric acid.
3. How are earthworm metanephridia like mammalian nephrons?
   *a.* Both process coelomic fluid.
   *b.* Both take in fluid through a ciliated opening.
   *c.* Both produce urine more concentrated than the blood.
   *d.* Both employ tubular secretion and reabsorption to control urine composition.
   *e.* Both involve a countercurrent multiplier effect.
4. What is the role of renal podocytes?
   *a.* They prevent red blood cells and large molecules from entering the renal tubules.
   *b.* They reabsorb most of the glucose that is filtered from the plasma.
   *c.* They control the glomerular filtration rate by changing the resistance of renal arterioles.
   *d.* They provide a large surface area for tubular secretion and reabsorption.
   *e.* They release renin when the glomerular filtration rate falls.
5. Which of the following are *not* found in a renal pyramid?
   *a.* Collecting ducts
   *b.* Vasa recta
   *c.* Peritubular capillaries
   *d.* Convoluted tubules
   *e.* Loops of Henle
6. Which part of the nephron is responsible for most of the difference in mammals between the glomerular filtration rate and the urine production rate?
   *a.* The glomerulus
   *b.* The proximal convoluted tubule
   *c.* The loops of Henle
   *d.* The distal convoluted tubule
   *e.* The collecting duct
7. For mammals of the same size, what feature of their excretory systems would give them the greatest ability to concentrate their urine?
   *a.* Higher glomerular filtration rate
   *b.* Longer convoluted tubules
   *c.* Increased number of nephrons
   *d.* More permeable collecting ducts
   *e.* Longer loops of Henle
8. Which of the following would *not* be a response stimulated by a large drop in blood pressure?
   *a.* Constriction of afferent renal arterioles
   *b.* Increased release of renin
   *c.* Increased release of antidiuretic hormone
   *d.* Increased thirst
   *e.* Constriction of efferent renal arterioles
9. Which statement about angiotensin is *true*?
   *a.* It is secreted by the kidney when the glomerular filtration rate falls.
   *b.* It is released by the posterior pituitary when blood pressure falls.
   *c.* It stimulates thirst.
   *d.* It increases the permeability of the collecting ducts to water.
   *e.* It decreases glomerular filtration rate when blood pressure rises.
10. Birds that feed on marine animals ingest a lot of salt, but they excrete most of it by means of
    *a.* Malpighian tubules.
    *b.* rectal salt glands.
    *c.* gill membranes.
    *d.* concentrated urine.
    *e.* nasal salt glands.

## FOR DISCUSSION

1. Why is it said that the oceans are a physiological desert? For what animals would this apply?
2. Persons with uncontrolled diabetes mellitus can have very high levels of glucose in their blood. Why do such individuals have a high level of urine production?
3. Inulin is a molecule that is filtered out of the glomerulus but is not secreted or reabsorbed by the renal tubules. If you injected inulin into an animal and after a brief time measured the concentration of inulin in the animal's blood and urine, how could you determine the animal's glomerular filtration rate? Assume that the rate of urine production is 1 milliliter per minute.
4. After you did the inulin experiment to measure glomerular filtration rate, how could you use that information to determine whether another substance is secreted or reabsorbed by the renal tubules? Assume you can measure the concentration of that substance in the blood and urine. Urine production is still 1 milliliter per minute.
5. Explain what happens with respect to control and regulation of your salt and water balance when you eat a lot of very salty popcorn.
6. Review Figure 52.2, noting that the venules and the secretory tubules are arranged in parallel with blood flowing in one direction and the contents of the tubule flowing in the opposite direction. What is the advantage of this arrangement?

## ADDITIONAL INVESTIGATION

We mention on p. 1107 that the glomerular filtration rate (GFR) is autoregulated, and this regulation is achieved by control over constriction/dilation of the afferent and efferent renal arterioles. What information could be used in this process? A clue comes from the anatomy of the nephron. When the ascending limb of Henle reaches the cortex, it makes direct contact with the afferent and efferent arterioles of the glomerulus that produced the filtrate flowing in that particular nephron. What aspect of the tubular fluid in the early distal convoluted tubule would reflect changes in the volume of filtrate entering the nephron? How might the cells of the early distal convoluted tubule communicate with the smooth muscle cells of the arterioles?

## WORKING WITH DATA (GO TO yourBioPortal.com)

**What Kidney Characteristic Determines Urine Concentrating Ability?** Since the loops of Henle create the concentration gradient in the renal medulla, it is reasonable to expect that longer loops would make it possible to generate greater concentration gradients and therefore increase the ability to concentrate urine. However, mammals come in many different sizes and so do their kidneys, so would you expect all big mammals to be able to concentrate their urine more than small mammals? In this exercise, we will explore anatomical data and urine concentration data from many species to see what characteristic of their kidneys relates to their urine concentrating abilities.

# 53 Animal Behavior

## Monkey see, monkey do

Many years ago, scientists studying a troop of macaques (*Macaca fuscata*) on the Japanese island of Koshima lured the monkeys into the open by throwing pieces of sweet potato onto the beach from a passing boat. The monkeys tried to brush the sand off the potatoes, but the pieces remained gritty. One day, a young female named Imo took her sweet potatoes to the water and washed off the sand. Soon her siblings and other juveniles in her playgroup were imitating Imo's new behavior. Then their mothers began washing their potatoes. None of the adult males imitated this behavior, but young males learned the behavior from their mothers and their siblings and continued to wash their potatoes as they grew older.

The scientists were fascinated by the way Imo's creative, insightful behavior spread throughout the population, so they presented the monkeys with a new challenge: they threw wheat onto the beach. Picking wheat grains out of the sand was tedious and difficult. Imo, apparently a prodigy in the macaque world, came up with a solution: she carried handfuls of sand and wheat to the water and threw them in. The sand sank, but the wheat floated, enabling her to skim the grain off the surface and eat it. This efficient feeding behavior spread throughout the population just as washing sweet potatoes had—first to other juveniles, then to mothers, and then from mothers to their male and female offspring.

The macaques of that troop now routinely wash their food. They also play in the ocean, which they never did before, and they have added marine food items to their diet. Clearly, this population of monkeys has a *culture*: a set of invented behaviors shared by members of the population and transmitted through social learning.

Animals also display elaborate behaviors they do not learn. Web spinning by spiders, for example, requires no practice or prior experience. In many spider species, females die immediately after they lay their eggs; the hatching spiderlings never see their mother, her web, or webs spun by other spiders. Yet when they first construct their own webs,

**Shared Learned Behaviors Become a Culture** In the space of a single generation, a population of Japanese macaques (*Macaca fuscata*) learned and transmitted a set of behaviors that included washing food, playing in the water, and eating marine food items—a new "culture" of water-related behaviors.

**Spiders Are Born Web Designers** Each spider performs a stereotypic sequence of movements that results in a species-specific web design. Spiders such as this banded garden spider (*Argiope trifasciata*) are born with this ability, with no need to learn from experience or to model their movements after those of a parent.

they do it perfectly, without the benefit of experience or a model to copy.

Web spinning requires thousands of movements performed in just the right sequence. Most of that sequence is *stereotypic*—performed the same way every time. Different spider species spin webs of different designs, using different sequences of movements. Yet every spider knows how to spin its species-specific web at birth; the behavior is part of its genetic inheritance.

Most animal behavior stems from an interaction of genetic inheritance and learning, and modern studies of animal behavior focus on the mechanisms of both. The evolution of behavior is also a topic of great interest, and research on that topic asks how particular behaviors are adaptive in specific environments. In short, behavioral biology is a highly integrative field that incorporates approaches from virtually all of the biological subdisciplines you have studied in this book.

**IN THIS CHAPTER** we will describe how the modern study of animal behavior emerged from earlier research focused on learning and on inherited behavior. We will give examples of how methods from genetics and molecular genetics, physiology, neurobiology, ecology, and evolution have contributed to our understanding of the behavior of animals—including ourselves.

CHAPTER OUTLINE



## 53.1 What Are the Origins of Behavioral Biology?

Humans have studied animal behavior since prehistoric times. Understanding the habits of potential prey, as well as those of their predators, was of great value to hunters. Appreciation of behavioral traits led to the domestication of animal species. Accounts of dramatic animal behaviors such as seasonal appearances and disappearances, mating displays, aggression, prey capture, and even less dramatic behaviors such as parental care, communication, and sleep are found throughout recorded history. Yet the scientific study of animal behavior did not truly get under way until the early 1900s.

### Conditioning was the focus of behaviorists

The discovery of the *conditioned reflex*, a simple form of learning, by the Russian physiologist Ivan Pavlov in the late 1800s was a seminal event. Pavlov was studying the neural control of digestive juice secretion, and he observed that a dog salivated when it smelled food. He also noticed that the dog salivated when the technician who normally fed the dog entered the room even when no food was present. Following up this observation, he substituted a sound stimulus for the technician. When the dog was fed, a metronome ticked. After a number of trials, the dog salivated when it heard the metronome even if no food was offered. Pavlov showed that other sounds worked as well (including a bell, which is commonly associated with his work).

Salivation in response to the sight, smell, or taste of food is a natural reflex response to a stimulus, but salivation in response to a sound was a learned response. The pairing of a sound with the experience of receiving food *conditioned* the dog's nervous system to generate a response, which Pavlov dubbed the **conditioned reflex**. The food was referred to as the *unconditioned stimulus* (US), and the sound as the *conditioned stimulus* (CS) (**Figure 53.1**). Thus Pavlov showed that a simple behavior controlled by the nervous system could be modified through experience.

Pavlov received a Nobel prize in 1904, and his work stimulated much new research because he had developed an experimental model of learning. All sorts of changes in the relationship between a CS and a US were investigated. A particularly interesting variation was developed by the psychologist B. F. Skinner, who showed that any random action of an animal could become a conditioned response to a stimulus if a reward was temporally associated with the action and the stimulus. For example, a rat could be conditioned to press a lever in response to a stimulus if it got a reward when it behaved as the experi-

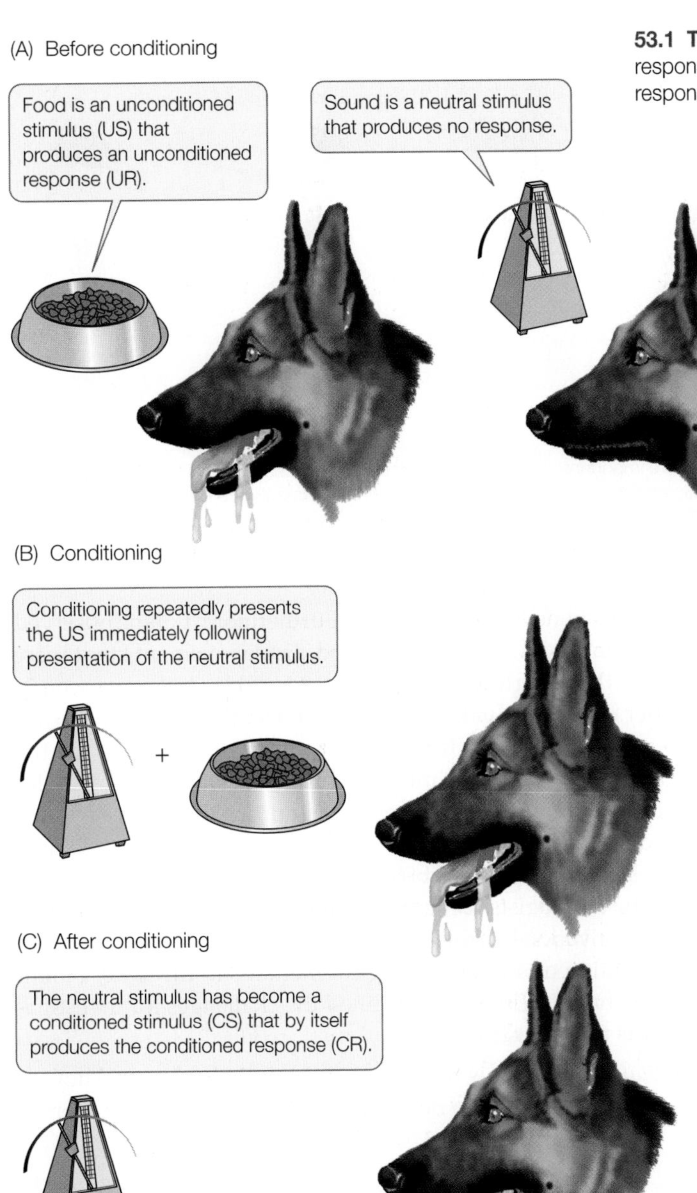

(A) Before conditioning
Food is an unconditioned stimulus (US) that produces an unconditioned response (UR).
Sound is a neutral stimulus that produces no response.
(B) Conditioning
Conditioning repeatedly presents the US immediately following presentation of the neutral stimulus.
+
(C) After conditioning
The neutral stimulus has become a conditioned stimulus (CS) that by itself produces the conditioned response (CR).


**53.1 The Conditioned Reflex** Ivan Pavlov discovered that when a normal response is paired with an artificial stimulus, an animal learns to produce the response even when only the artificial stimulus is presented.

menter desired. Because the animal was conditioned to perform an operation on its environment, this experimental protocol was known as **operant conditioning** and was viewed as another model of learning.

The experimental approach to behavior initiated by Pavlov and Skinner had powerful effects on the nature of research on animal behavior. The focus of scientists using this approach was quite specific:

- They focused on laboratory environments rather than natural environments because in the laboratory, the variables in their conditioning experiments could be precisely controlled.
- They focused on only a few species (predominantly the albino rat) as model systems rather than studying diverse species from nature.
- They focused on questions of learning and memory, largely to the exclusion of other types of behavior (e.g., mating, feeding, communication).

Thus defined, the field of animal behavior research became known as **behaviorism**, and it was largely the domain of psychologists.

## Fixed action patterns were the focus of ethologists

An alternative approach to the study of animal behavior arose at the same time as behaviorism, but largely in Europe. Scientists there focused on describing the characteristics of animals in their natural environment, an approach that became known as **ethology** (Greek *ethos*, "character"; *logos*, "study"). In contrast to the behaviorists, the ethologists were interested in a wide variety of species, their evolutionary relationships, and the ways in which their behaviors were adapted to their environments. The leaders of the ethology movement were Karl von Frisch, who discovered the dance language of bees; Konrad Lorenz, who discovered that the strong bond between parent and offspring develops during a "critical period" following birth; and Niko Tinbergen, who studied inborn patterns of behavior commonly known as instincts. These three scientists shared the Nobel Prize in 1973 for "their discoveries concerning organization and elicitation of individual and social behavior patterns."

The instinctive behaviors that were the main interest of the ethologists were genetically determined patterns. Genetically determined behaviors

- are performed without learning
- are *stereotypic* (i.e., performed the same way each time)
- cannot be modified by learning

The ethologists called such behaviors **fixed action patterns**.

To demonstrate that a behavior was genetically determined, ethologists performed **deprivation experiments** in which an animal was raised in an environment devoid of opportunities to learn its species-specific behavior. An example of a natural deprivation experiment is the web spinning of the spider featured at the opening of this chapter. The parents of the young spider die before it hatches, and in a seasonal environment, it has no model webs to copy when it spins its first web, yet it creates a perfect, species-specific web the first and every time it spins a web.

Fixed action patterns are usually responses to specific stimuli. The ethologists carefully characterized such stimuli, which they called **releasers**. In general, releasers are very simple subsets of the information available in the environment. For example, Tinbergen studied the begging behavior of gull chicks. Adult gulls have a red dot on their bills. When a parent gull returns to the nest to feed its chicks, the chicks peck on the red

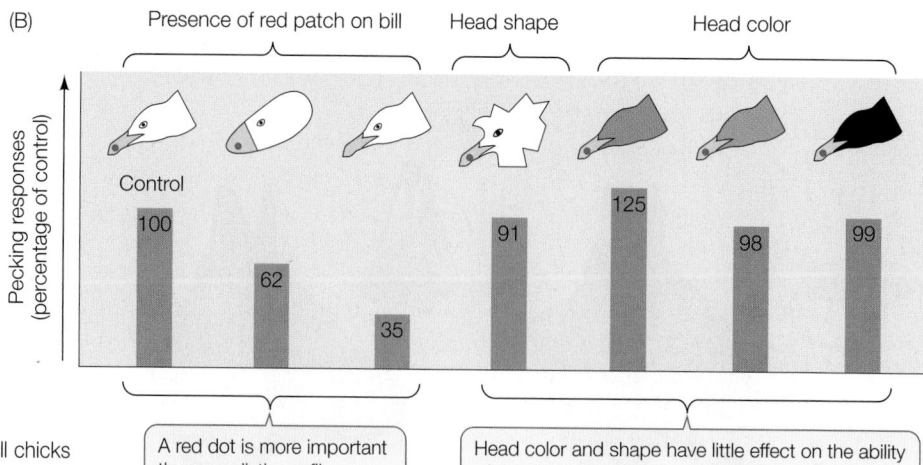


**53.2 Releasing a Fixed Action Pattern** (A) Gull chicks instinctively peck at the red dot on the parent's lower bill, a behavior that his induces the parent to regurgitate food into the chick's mouth. (B) Tinbergen's work showed that the red dot on the parent's beak was the critical component that released the pecking response.

dot, which stimulates the parent to regurgitate food (**Figure 53.2A**). Experimenters investigated what stimulated the chicks to peck their parents' bills. Models of gull heads of different shapes and colors were tested (**Figure 53.2B**), as were models of a beak without a head. The results showed that the red dot was necessary for the release of chick pecking behavior. In fact, a pencil with a red eraser elicited a more robust pecking response than an accurate model of a gull head without a red dot.

### Ethologists probed the causes of behavior

The ethologists demonstrated the genetic basis for fixed action patterns by interbreeding closely related species. Konrad Lorenz studied the courtship behaviors of different species of dabbling ducks. Some of these species, such as mallards, teals, pintails, and gadwalls, are closely related and can interbreed, but they rarely do so in nature. Each male duck performs a courtship display consisting of a precise series of movements that is typical of his species (**Figure 53.3**). A female is not likely to accept him unless the entire display is successfully and correctly completed.

When Lorenz crossbred these duck species, the hybrid offspring expressed some elements of each parent's courtship display, but in novel combinations. Furthermore, Lorenz observed that hybrids sometimes exhibited display elements that were not in the repertoire of either parent species but were seen in other dabbling duck species. Lorenz's interbreeding studies demonstrated that the stereotypic motor patterns of the courtship displays are inherited. The observation that females were not interested in males performing hybrid displays was evidence that *sexual selection* (see Section 21.3) had shaped these genetically determined behaviors.

The ethologists recognized the importance of development and motivation in behavior, and they laid the foundation for the application of modern biological methods to the study of animal behavior. Tinbergen outlined the challenges for investigators as four questions:

- *Causation:* What is the stimulus for the behavior, and how has the relationship between stimulus and behavior been modified by learning?
- *Development:* How does the behavior change with age, and what experiences are necessary for it to be displayed?

**53.3 Courtship Displays Are Highly Specific** The courtship display of the male mallard (*Anas platyrhynchos*) contains the precise 10 elements illustrated here. The displays of closely related dabbling duck species contain some of the same elements, but have other elements not displayed by mallards.

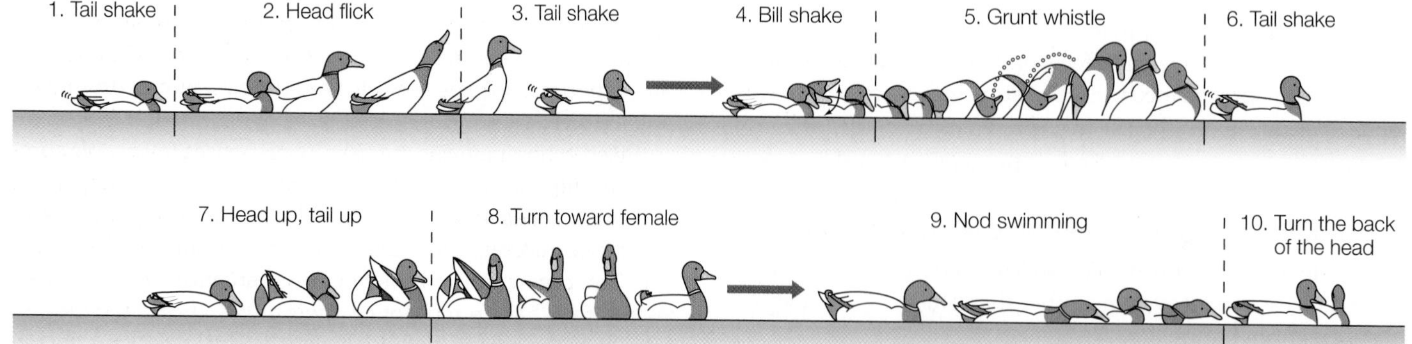

- *Function:* How does the behavior affect the animal's chances for survival and reproduction?
- *Evolution:* How does the behavior compare with similar behaviors in related species, and how might it have evolved?

The first two questions refer to the **proximate causes** of behavior: the immediate genetic, physiological, neurological, and developmental mechanisms that determine how an individual is behaving at a particular time. The third and fourth questions refer to the **ultimate causes** of behavior: the evolutionary processes that produced the animal's capacity and tendency to behave in particular ways. In the sections that follow, we will describe many experiments on animal behavior. For each one, ask yourself which of Tinbergen's four questions it addresses and whether it focuses on proximate or ultimate causes of behavior.

## 53.1 RECAP

**Early scientific studies of animal behavior took two approaches. Behaviorists focused on the study of conditioned behavior in a few species of laboratory animals and asked questions about learning. Ethologists studied genetically determined behavior in many species in their natural environments and asked evolutionary questions.**

- Describe the difference between conditioned reflexes and operant conditioning. **See pp. 1114–1115 and Figure 53.1**
- What is the relationship between a releaser and a fixed action pattern? **See p. 1115 and Figure 53.2**
- Explain the difference between proximate and ultimate causes of behavior and how Tinbergen's questions address these two types of causes. **See pp. 1116–1117**

The work of the ethologists left no doubt that behavior can be genetically determined, but how? Genes code for proteins. Behaviors are highly complex traits involving sensory input and intricate patterns of control over responses to that input. Is it reasonable to think that a single gene can have a specific effect on a behavior?

# 53.2 How Can Genes Influence Behavior?

Behaviors are complex traits that depend on many genes. Nevertheless, evidence from multiple approaches shows that alterations in single genes can result in discrete behavioral phenotypes on which natural selection can operate.

### Breeding experiments can show whether behavioral phenotypes are genetically determined

Behavioral geneticists identify individuals with unusual behavioral phenotypes and conduct breeding experiments to see if those traits are inherited. If the inheritance of a trait follows a Mendelian inheritance pattern (see Section 12.1), then that trait is likely to be controlled by a small number of genes. A classic example is the genetics of hive cleaning by honey bees.

As we saw at the opening of Chapter 43, a honey bee colony consists of a reproductive queen bee and a huge number of nonreproductive workers. The worker bees have many tasks to perform. The queen deposits eggs in cells of the honeycomb structure that the workers construct. The eggs hatch into larvae, which are fed in their cells by workers. When the larvae are ready to pupate, the workers cap and seal their cells. If a pupa dies before emerging as an adult, workers will normally uncap the cell and drag the carcass outside the hive. This *hygienic* behavior increases the resistance of the hive to a bacterium that infects and kills bee larvae. Some hives do not show this behavior; these *nonhygienic* hives are more susceptible to the spread of this disease.

When a nonhygienic female bee was crossed with a hygienic male, the offspring were all nonhygienic, indicating that the genetic determinant of nonhygienic behavior is dominant. When these nonhygienic $F_1$ offspring were backcrossed to hygienic males, however, four phenotypes were produced. Two phenotypes were like those of the parents, either hygienic or nonhygienic, but the other two showed interesting deficits. One phenotype opened the cells of dead pupae, but did not remove them. The other phenotype did not open cells, but if a cell was already open, it would dispose of the dead pupa (**Figure 53.4**). Thus the complex hygienic behavior has at least two components that are controlled by separate genes. But the physical identity of these genes, and how they influence the behavior, could not be revealed by breeding experiments.

### Knockout experiments can reveal the roles of specific genes

Some modern molecular genetic approaches start with identified genes and eliminate or silence them to see what effects their elimination has on a behavioral phenotype (see Section 18.4). As you might expect, knocking out genes involved in sensory pathways can have pronounced effects on behavior. One example is a gene for a specific olfactory receptor in mice.

As we saw in Section 46.2, mice have two olfactory organs: the nasal olfactory epithelium common to all mammals, and a small organ in the nasal passages, called the *vomeronasal organ*, or VNO (**Figure 53.5**). Catherine Dulac at Harvard University discovered that a large number of pheromone receptors were expressed in that organ. (As described in Section 46.2, *pheromones* are signaling molecules released into the environment.) Dulac hypothesized that when the receptors in the male's VNO bound to sex pheromones produced by female mice, they stimulated mating behavior. To test this hypothesis, Dulac created a genetically engineered male mouse in which the gene for an ion channel necessary for VNO receptor signaling was knocked out. Contrary to the prediction of the hypothesis, the knockout males in fact did pursue and mate with females placed in their cages. However, they also pursued and tried to mate with *males* placed in their cages. Normally a male mouse

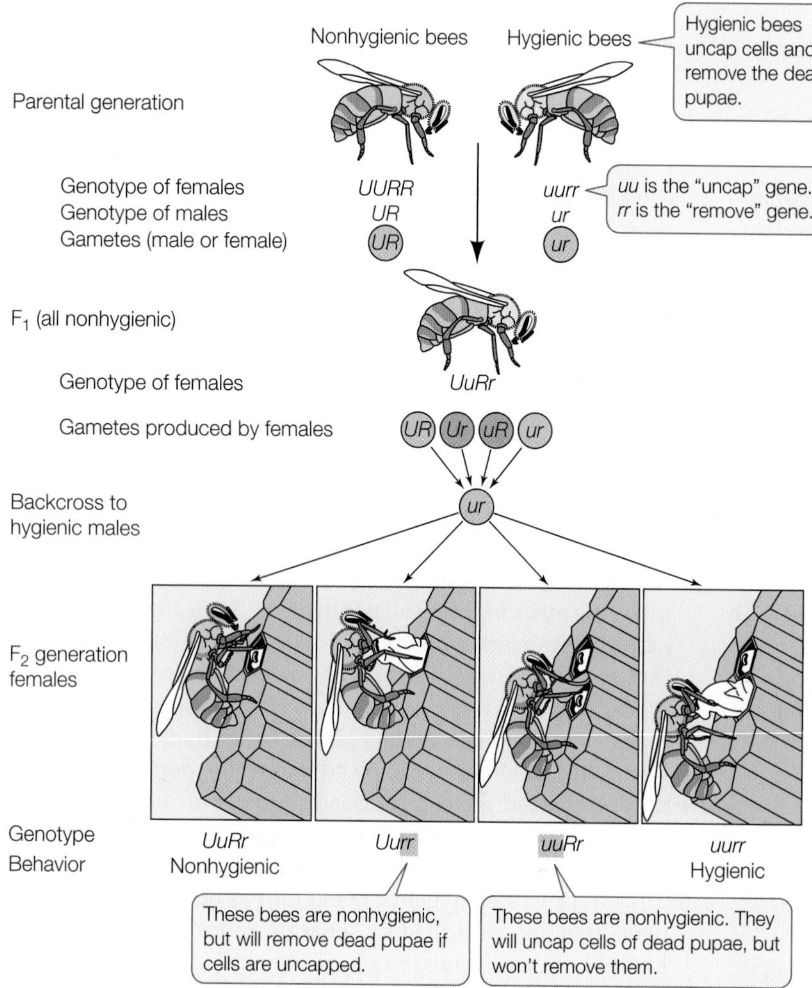


**53.4 Genes and Hygienic Behavior** Some worker honey bees remove carcasses of dead pupae from their hive. Two components of this behavior—cell uncapping and removal of the carcass—are under the control of separate recessive genes, designated *u* and *r*.

reacts aggressively to a strange male, but the knockout male could not discriminate between males and females placed in his cage. Thus properly functioning VNO receptors appear to be essential not for sexual attraction, but for gender identification. It is possible to imagine how selection working on this one gene could modify the intensity of male–male aggression and lead to changes in social behavior.

### Behaviors are controlled by gene cascades

Male courtship behavior in *Drosophila melanogaster*, the laboratory fruit fly, is stereotypic, species-specific, and requires no learning—a classic fixed action pattern. When a male encounters a potential mate, he follows her, taps her body with his foreleg, extends and vibrates one wing, and licks her genitals (**Figure 53.6A**). The development of this complex male behavior is under the control of a single gene, called *fruitless* (*fru*), just as the development of male anatomy is under the control of another gene, called *doublesex* (*dsx*). In both males and females, these two genes are part of a gene expression cascade that results in different *dsx* and *fru* gene products in males and females (**Figure 53.6B**). The female version of the Dsx protein controls the development of female anatomy, and the expression of *fru* in the male nervous system results in the organization of the neural circuitry controlling male sexual behavior.

There are two take-home lessons from this example. First, genes that control aspects of behavior, like other genes, are generally embedded in gene cascades that offer multiple opportunities for simple genetic changes that will alter the phenotype of even complex behaviors. Second, certain genes, such as *dsx* and *fru*, influence a whole range of genes that contribute to complex behaviors. Modifications in any one of those genes or its ex-

**53.5 The Mouse Vomeronasal Organ Identifies Gender** (A) The mouse VNO is located adjacent to the nasal passages. It contains pheromone receptors whose input travels to a specific region of the olfactory bulb (the accessory olfactory bulb). (B) In mice, information from the VNO appears to be crucial in identifying gender and thus a potential sexual partner.

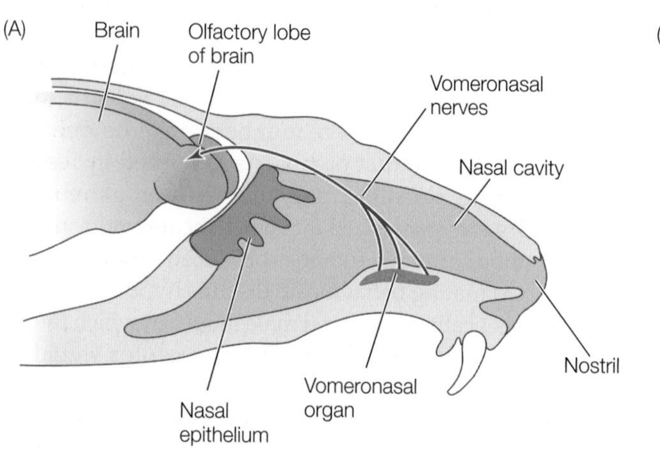

**53.6 The *fruitless* Gene** (A) Male fruit flies display stereotypic, species-specific courtship behavior. (B) Sexual differentiation in *Drosophila* is controlled by a cascade of genes, including the *fru* gene, whose expression results in male sexual behavior.

pression can alter behavior. Thus, even though no behavior is coded for by a single gene, alterations in single genes can influence behavior in ways that affect an animal's fitness.

## 53.2 RECAP

**Breeding experiments show that behavioral phenotypes can be inherited and modified by natural selection. Although most behaviors are controlled by complex cascades of genes, molecular genetic methods have shown that a single, identifiable gene in the cascade can control a complex behavior.**

- How did breeding experiments on honey bees show the genetic basis for hygienic behavior? See p. 1117 and Figure 53.4
- What is the evidence that the vomeronasal organ in mice is responsible for gender identification? See pp. 1117–1118
- Describe some of the different ways a gene can control expression of a complex behavior. See p. 1118 and Figure 53.6

How can the genetic cascades that underlie complex behaviors be programmed to respond selectively to specific sets of stimuli? How can their expression be limited to appropriate times in an animal's life? The answers to these questions can by found by studying how behaviors develop over the life span.

# 53.3 How Does Behavior Develop?

The emergence of behavior as an animal develops and matures depends on the development of the nervous system as well as on the growth and maturation of other body systems. A bird cannot fly until its wings grow and its muscles and flight feathers mature. But even with anatomical and physiological competence, specific behaviors may not be expressed. Behaviors that are adaptive at one stage in an animal's life may not be adaptive at other stages. Behaviors typical of juvenile animals, such as begging for food, may disappear and new behavior patterns of a mature individual, such as courtship displays, appear.

## Hormones can determine behavioral potential and timing

Hormones can determine the development of a behavioral potential at an early age and the expression of that behavior at a later age. An excellent example of this is sexual behavior in rats

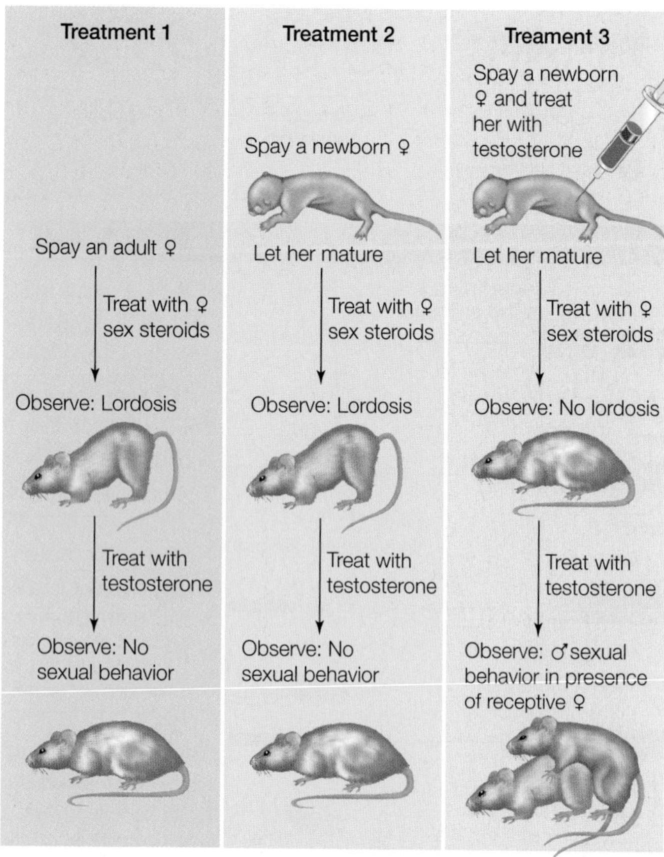


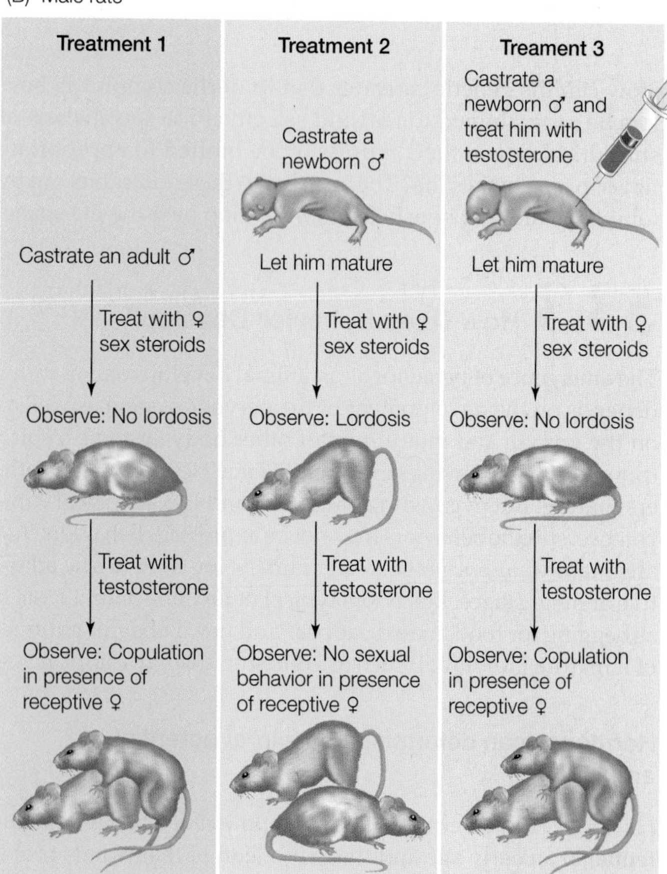


**53.7 Hormonal Control of Sexual Behavior** Experimental hormonal treatments of rats demonstrated that the sex steroids present during early development determine what sexual behavior patterns develop, but the sex steroids present in adulthood control the expression of those patterns.

(**Figure 53.7**). Normally, adult male and female rats exhibit different patterns of sexual behavior: females adopt a sexually receptive posture, called *lordosis*, in the presence of males, and males copulate with receptive females. Neither sex, however, expresses these behaviors until the animals have reached adulthood. Experiments in which newborn and adult rats were neutered (to remove the influence of sex steroids naturally produced by their gonads) and artificially treated with hormones led to the following conclusions:

- Development of male sexual behavior requires that the brain of the newborn rat be exposed to testosterone, but development of female sexual behavior does not require exposure to estrogen.
- Testosterone masculinizes the nervous systems of both genetic males and genetic females.
- Exposure to sex steroids in adulthood is necessary for the expression of sexual behavior, but testosterone produces male sexual behavior only in adult rats whose brains were masculinized when they were newborns, and estrogen produces female sexual behavior only in adult rats whose brains were not masculinized when they were newborns.

Thus, the sex steroids that are present at birth determine which pattern of behavior develops, and the sex steroids that are present in adulthood determine when that pattern is expressed.

### Some behaviors can be acquired only at certain times

Responsiveness to simple releasers is sufficient for certain behaviors such as begging behavior in gull chicks, but more complete information that cannot be genetically programmed is required for other behaviors. An example is parent–offspring recognition. When animals live in close proximity to other individuals, as in a herd or a nesting colony, it is important for a mother and her offspring to learn each other's identity soon after birth so they will be able to find each other in a crowded situation. In many such cases, a parent–offspring bond is formed by **imprinting**. What characterizes imprinting is that an animal learns a specific set of stimuli during a limited time called a **critical period** or **sensitive period**.

Konrad Lorenz demonstrated that young goslings imprint on their parents between 12 and 16 hours after hatching. By positioning himself to be present during this critical period, Lorenz succeeded in imprinting goslings on himself. The imprinted goslings followed him around as if he were their parent (**Figure 53.8A**).

Imprinting requires only a brief exposure, but its effects are strong and long-lasting. Emperor penguins reproduce during the coldest, darkest time of year in Antarctica. The parents walk up to 150 km inland to form a dense colony, where the female lays her egg. She then walks back to the ocean to feed while her

(A)

(B)

**53.8 Imprinting Helps Parents and Offspring Recognize Each Other** (A) Greylag geese that imprinted on Konrad Lorenz as hatchlings followed him everywhere he went. (B) Imprinting allows a male emperor penguin to find his own chick among many others.

mate incubates the egg. By the time she returns, the chick has hatched. She then takes over its care and feeding, and the father walks back to the ocean to feed. Generally, he is away so long that the mother must leave to find food as well to avoid starvation. Thus, after being away for weeks, the father must find his chick in a crowded, milling colony of chicks, all calling for their parents (**Figure 53.8B**). Yet he can unerringly locate his own offspring by recognizing its call, which he learned before he left to feed.

The critical or sensitive period for imprinting may be determined by a brief developmental or hormonal state. For example, if a mother goat does not nuzzle and lick her newborn within 10 minutes after its birth, she will not recognize it as her own offspring later. For goats, the sensitive period is associated with peaking levels of the hormone oxytocin in the mother's circulatory system at the time she gives birth and is sensing the olfactory cues emanating from her newborn kid. A female goat rendered incapable of smelling before giving birth is unable to differentiate between her own kid and other kids after giving birth.

## Bird song learning involves genetics, imprinting, and hormonal timing

Male songbirds use species-specific song to claim and advertise a breeding territory, compete with other males, and declare dominance. They also use song to attract females, who recognize the song of their species even though they do not sing it. For males of many species, such as the white-crowned sparrow, learning is an essential step in the acquisition of song, but *what* they can learn seems to be influenced by genes, and there is a limited *developmental time frame* for learning. A hatchling in nature hears his father and other white-crowned sparrows singing. He also hears the songs of many other bird species. But he does not sing until he approaches sexual maturity almost a year later, and when he does, he sings his father's type of song.

Studies of song learning in this species were initiated in the 1960s by Peter Marler at the University of California at Berkeley. Marler incubated eggs of white-crowned sparrows and hand-reared the hatchlings in the laboratory. He was therefore able to expose them to different tape-recorded songs at different times in their development. He discovered that male birds cannot produce their species-specific song as adults unless they hear it as nestlings in the first 2 months of their lives (**Figure 53.9**). If the nestlings hear tapes of white-crowned sparrow song during those first 2 months, they begin to sing, but poorly, as they approach sexual maturity. Through trial and error, the birds

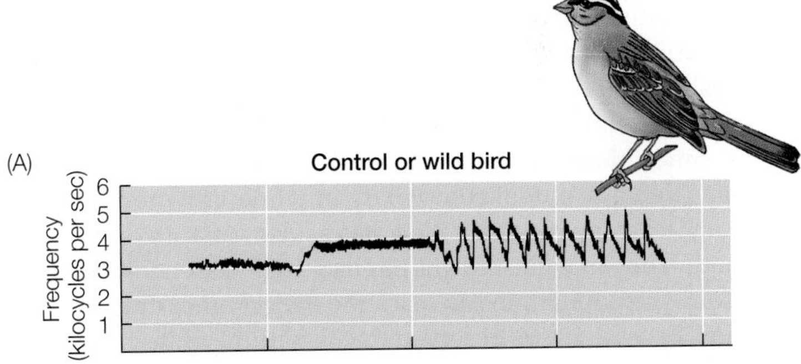


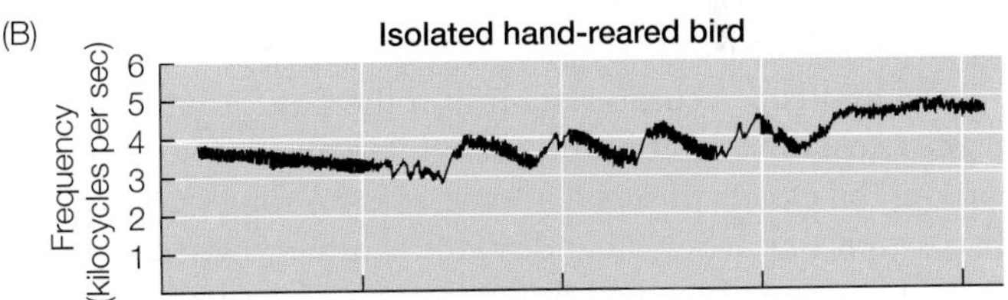


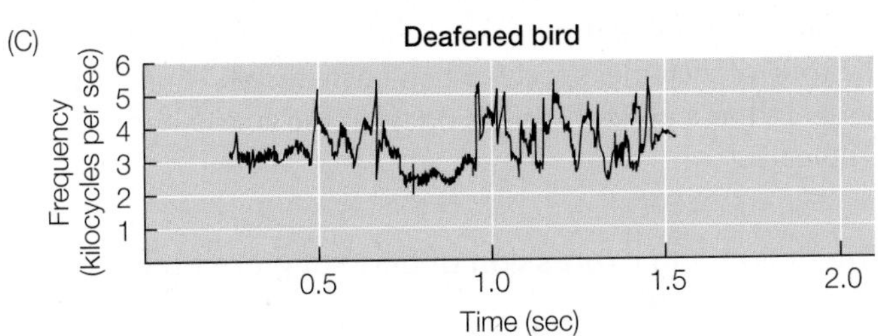


**53.9 Sensitive Periods for Song Learning** (A) Sonogram showing the species-specific song of an adult male white-crowned sparrow (*Zonotrichia leucophrys*). (B) Song of an adult male raised in isolation (never having heard the song as a nestling). (C) Song of an adult male who heard the song as a nestling but was deafened prior to ever singing himself. Marler's experiments showed the bird must first acquire a song memory by hearing the song as a nestling, and must then be able to hear himself as he attempts to match his singing to that song memory.

match their singing to their stored song memory, and from then on they sing their species-specific song. To reach this point, the young bird must be able to hear himself sing. If he is deafened just before he begins to sing, he will not be able to match his stored song memory. If he is deafened *after* he sings his correct species-specific song, however, he will continue to sing like a normal bird. We say that at this point the song behavior pattern is *crystallized*. Thus there are two sensitive periods for song learning: the first in the nestling stage, when a song memory is imprinted; the second as the bird approaches sexual maturity, when he learns to match that song memory.

In nature, nestling male white-crowned sparrows hear the songs of many species, so why do they learn only the song of their own species? Marler investigated this question in his isolation experiments by playing tape recordings of other songs to the hatchlings. The young male sparrows did not learn the songs of other species, even if they heard them many times, but hearing songs of their own species just a few times was sufficient for imprinting. Thus, although male sparrows must learn their song, they seem to have a genetic predisposition to learn their own song and not the songs of other species. Marler called this phenomenon "an instinct to learn." There is an important reason why male white-crowned sparrows should learn only their species-specific song: female white-crowned sparrows also listen to their fathers' songs while they are nestlings, and when they mature, they choose mates who sing like their fathers did.

Since Marler's pioneering work, more investigations have revealed additional complexity in white-crowned sparrow song learning capacity. First, it was demonstrated that each population of white-crowns has its own dialect, and that males from one population can learn the dialect of another population. Second, it was observed that in nature, white-crowns are occasionally heard singing the songs of other species. Could Marler's laboratory experiments have missed a critical natural variable?

Luis Baptista, a curator of birds at the California Academy of Sciences, took up the study of white-crowned sparrows to explore the effects of social interactions on song learning. He discovered that when a hand-reared white-crowned sparrow nestling was exposed to the sight and sound of a related species in an adjacent cage while tapes of his own species' song played in the background, the white-crowned sparrow sang the song of the other species when he matured. Thus social experience has a powerful effect on what the young bird can and will learn.

### The timing and expression of bird song are under hormonal control

As we have seen, both male and female songbirds hear their species-specific song as nestlings, but only the males of most species sing as adults, and they do so only in spring. Hormones underlie both the difference in song expression between male and female songbirds and the timing of song expression.

To determine whether the absence of testosterone is the main reason female songbirds don't sing, investigators injected adult female songbirds with testosterone in spring. In response to these injections, the females sang their species-specific song just as the males did. Apparently females form a memory of their species-specific song when they are nestlings, and have the physical capacity to sing, but under normal circumstances they simply lack the hormonal stimulation.

How does testosterone cause a songbird to sing? A remarkable study revealed that each spring, an increase in circulating testosterone levels causes certain parts of the male's brain necessary for learning and developing song to grow larger. Individual neurons in those regions of the brain increase in size and grow longer extensions, and the number of neurons in those regions increases. Such research on the neurobiology of bird song revealed that, contrary to the once widely held belief that new neurons are not produced in the brains of adult vertebrates, hormones can control behavior by changing brain structure as well as brain function, both developmentally and in response to environmental cues.

What triggers the release of testosterone in response to the onset of spring? Takashi Yoshimura conducted a DNA microarray analysis of brain samples from another bird species, the Japanese quail, which is a model species for avian genetic analyses. Knowing that photoperiod was the environmental cue, he examined the responses of 38,000 genes to determine which genes were activated by changes in day length. Fourteen hours after dawn on the first day with the critical photoperiod that induces singing behavior, genes in the brain that code for thyroid-stimulating hormone were switched on. This hormone acts on the pituitary gland, which in turn regulates the production of other hormones that stimulate the growth of the testes and the production of testosterone (see Figure 43.10).

## 53.3 RECAP

**Hormones can control what behavior patterns are programmed in the brain long before those behaviors are expressed. Learning and expression of behaviors can also be controlled by hormones, and therefore timed for particular stages in the life cycle.**

- What hormonal conditions are necessary for the development of adult sexual behavior in male and female rats? **See pp. 1119–1120 and Figure 53.7**
- What is the adaptive value of imprinting? **See pp. 1120–1121**
- Describe the series of events necessary for a male white-crowned sparrow to sing its species-specific song in spring. **See pp. 1121–1122 and Figure 53.9**

Complex behaviors are the product of interactions of genetic, physiological, and environmental factors. Many genes are involved in shaping behavior, and therefore there are multiple opportunities for selection to favor behavioral modifications. Questions about how changes in behavior adapt animals to environmental conditions are the province of an evolution-based field called *behavioral ecology*.

# 53.4 How Does Behavior Evolve?

Environmental conditions are highly variable over both time and space. Animal behaviors are also variable within and between species. Behavioral ecologists strive to discover the relationships between behavior and environment with the intent of understanding the evolutionary mechanisms underlying behavior.

## Animals must make many behavioral choices

Over an animal's lifetime, its behavior is largely a sequence of choices: where and when to move, where to build a nest, what to eat, when to fight and when to flee, with whom to associate, with whom to mate. Inability to make choices would make it difficult to survive, and making wrong choices reduces its fitness. Behavioral ecologists seek to discover what information animals use to make behavioral choices and how that information relates to aspects of the environment that influence their fitness.

The choice of a place to live, for example, is one that has many consequences for an animal. The environment in which an animal lives is its **habitat**. In most cases the habitat provides not only a protected nest site, but also food and access to mates. In many cases the environmental cues animals use to make their habitat choices are quite simple. For example, seabirds select cliffs or offshore rocks for nesting because those sites offer protection from predators. Animals with very specialized food requirements obviously select habitats where those foods are abundant. The general hypothesis that guides behavioral ecologists is that the cues animals use to select habitats are *reliable predictors of conditions suitable for future survival and reproduction.*

For many species, the presence of **conspecifics**—other members of the same species—is a reliable indicator of the suitability of a particular site for sustaining life. After all, you can't argue with success. Consider the planktonic larvae of marine organisms that drift freely in the water until they initiate a *settling response* that deposits them on the substratum to which they will attach themselves, usually for the rest of their lives. Molecules released into the water by conspecifics can aid larvae in detecting an appropriate substratum, and such molecules induce a settling response in many species.

Observing conspecifics can provide animals with information about the quality of a habitat in other ways. Collared flycatchers (*Ficedula albicollis*) on their breeding grounds in spring are nosy neighbors: they regularly visit the nests of conspecifics. Researchers hypothesized that this behavior allows the flycatchers to assess the quality of the habitat by seeing how well their neighbors are faring. To test this hypothesis, they created some areas with supersized broods—normally an indication of abundant food—by taking young birds from some nests and adding them to nests in another area. The next year, flycatchers preferentially settled in the areas where broods had been artificially enlarged (**Figure 53.10**).

INVESTIGATING LIFE

**53.10 Flycatchers Use Neighbors' Success to Assess Habitat Quality**

Collared flycatchers (*Ficedula albicollis*) regularly "spy" on other flycatchers in their breeding range. Researchers hypothesized that the birds were gathering information on the reproductive success of already established pairs to help them decide where to settle and breed.

**HYPOTHESIS** Collared flycatchers will nest at higher densities in areas where their conspecifics appear to have reproduced successfully.

**METHOD**

1. Transfer nestlings from one territory into nests in another, thereby creating "supersized" broods (an indication of abundant food) in one area and undersized broods in the other. A third research territory is left unaltered to serve as a control.
2. A year later, observe settlement rates (i.e., number of pairs establishing nests) in each territory at the start of the breeding season.

**RESULTS**

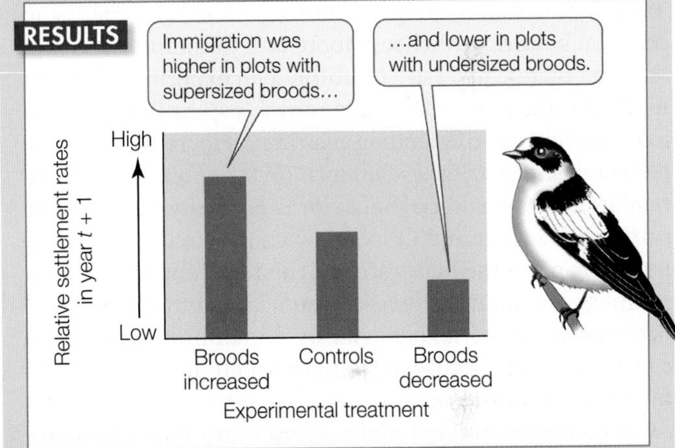


**CONCLUSION** Collared flycatchers use the brood sizes of already settled conspecifics to assess how good a territory is for breeding.

Go to **yourBioPortal.com** for original citations, discussions, and relevant links for all INVESTIGATING LIFE figures.

## Behaviors have costs and benefits

Behavioral ecologists often use a cost–benefit approach to investigate the relationship between behavior, environment, and fitness. A **cost–benefit approach** assumes that an animal has only a limited amount of time and energy, and therefore cannot afford to engage in behaviors that cost more to perform than they bring it in benefits. A cost–benefit approach provides a framework that behavioral ecologists can use to make observations, construct hypotheses, and design experiments to investigate why behavior patterns evolve as they do.

The benefits of a behavior can be measured in terms of the enhancement in fitness an animal accrues by performing the behavior. The cost of a behavior typically has three components:

- **Energetic cost** is the difference between the energy the animal expends performing the behavior and not performing it.

- **Risk cost** is the increased chance of being injured or killed as a result of performing the behavior.
- **Opportunity cost** is the benefit the animal forgoes by not being able to perform other behaviors during the same time interval.

Cost–benefit analysis has been used extensively in the study of **territorial behavior**: aggressive behavior used by an animal to actively deny other animals access to a habitat or resource. Optimal habitats and resources are frequently in short supply, so conspecifics must compete for them. Many animals—usually males—defend all-purpose territories that provide a nest site, food, and access to mates. The territory holder stakes out his boundaries by engaging in aggressive interactions with neighbors, and must then patrol those boundaries constantly and respond to trespassers. These aggressive interactions usually consist of highly stereotypic, species-specific displays such as bird song. Through territorial behavior, the male obtains the resources he needs for reproductive success, but he also pays a price.

Territorial displays require considerable expenditure of energy, they make a male more vulnerable to predation, and they detract from the time he has for feeding or engaging in parental behavior. Michael Moore and Catherine Marler at Arizona State University performed an experiment to estimate the costs incurred by male Yarrow's spiny lizards (*Sceloporus jarrovii*) when defending a territory (**Figure 53.11**). These lizards defend territories that include the habitats of several females. Their territorial behavior is normally most intense during September and October, when the circulating testosterone levels of the males are high and the females are most receptive to mating. The researchers varied the intensity of the lizards' territorial behavior by implanting testosterone capsules in some males in summer, when they are not normally highly territorial.

Testosterone-treated males spent more time patrolling their territories, performed more displays, and expended about one-third more energy than control males (an energetic cost). As a result, they had less time to feed (an opportunity cost), captured fewer insects, stored less energy, and had a higher death rate (a risk cost). In summer, when females are not normally receptive, these high costs of vigorous territorial defense outweigh the reproductive benefits of territoriality. Thus natural selection has favored seasonal variation in the level of the hormone controlling territorial behavior in this species.

**yourBioPortal.com**
GO TO Animated Tutorial 53.1 • The Costs of Defending a Territory

The cost–benefit approach explains the diversity of territorial behaviors seen in different species. Even if a resource is absolutely essential to an animal, if it cannot be defended economically, the animal will not engage in territorial behavior. Food is essential for all animals, but if the food is widely distributed in space or fluctuating in availability, there is no benefit to balance the high costs of trying to defend it. For ex-

## INVESTIGATING LIFE

**53.11 The Costs of Defending a Territory**

By using testosterone implants to increase territorial behavior, Moore and Marler measured the costs to male Yarrow's spiny lizards (*Sceloporus jarrovii*) of defending a territory during the summer, when they do not normally do so.

**HYPOTHESIS** **Yarrow's spiny lizards do not defend a territory during summer because the energetic costs of territorial behavior in that season outweigh the benefits.**

**METHOD**

1. During the summer, when female lizards are not sexually receptive, insert testosterone capsules under the skin of some males; leave other males untreated as controls.
2. Observe the patterns of territorial behavior and the survival rate of the two groups of males.

**RESULTS**

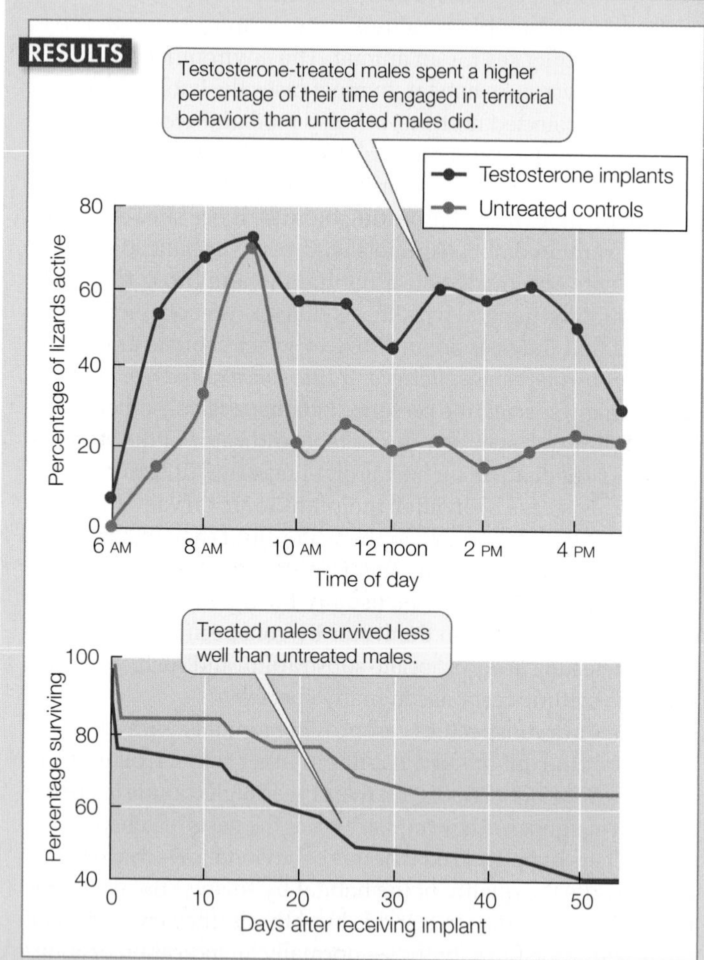


**CONCLUSION** **For these lizards, the cost of defending territories during summer significantly reduces their survival rate without increasing their reproductive success.**

Go to **yourBioPortal.com** for original citations, discussions, and relevant links for all INVESTIGATING LIFE figures.

(A) *Diomedea melanophris*

(B) *Mirounga angustirostris*

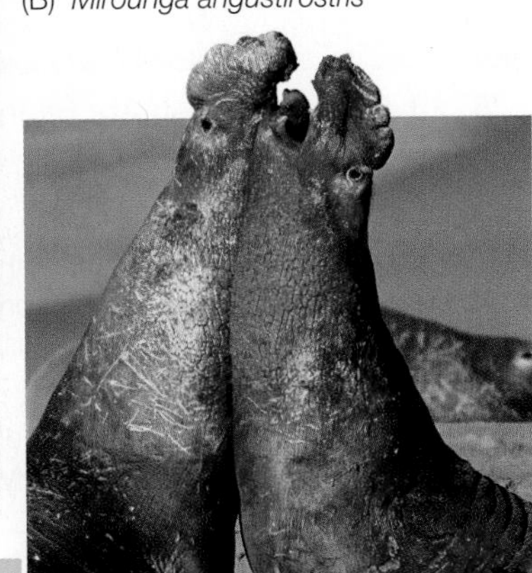

(C) *Centrocercus urophasianus*

**53.12 Animals Defend Territories of Different Sizes** (A) The nesting territories of many seabirds consist of only as much space as they can defend without leaving the nest. (B) Male elephant seals fight vigorously to defend areas of beach where females haul out of the water to give birth to their pups. (C) Male sage grouse gather at a lek in Colorado to perform displays aimed at impressing females and winning the opportunity to mate.

ample, the open ocean where seabirds feed cannot be defended. But safe nest sites on islands or rocky cliffs are in short supply, and they can be defended. Thus the territories of seabirds may be no larger than the distance the birds can reach while sitting on their nests (**Figure 53.12A**).

In some cases the resource that is defended is the female herself. Elephant seals spend most of their lives at sea, but females come to land at traditional beach sites to give birth to their pups. Male elephant seals arrive at these sites ahead of time and stake out territories through vigorous fighting (**Figure 53.12B**). When the females arrive on the beaches, they enter the territories of the males. As long as the male territory holder can fend off challengers, he will be able to mate with all the females using his piece of the beach.

The most unusual form of male territorial behavior arises in situations in which neither food, nest sites, nor females are defended. A **lek** is an area where males gather for the purpose of engaging in communal displays of their territorial prowess aimed at impressing females and winning the opportunity to mate. Even though space is not limited, each male defends a small piece of real estate on which he performs a display (**Figure 53.12C**). Those pieces of territory closest to the center of the lek are the prime sites, and males compete intensely for those locations. The females stroll into the lek, observe the males, and generally mate with the males holding the prime sites. The benefit of this system to the female is that she is inseminated by a successful competitor, and therefore her offspring will carry the genes that contributed to his success. The costs of lekking to males are high, as they engage in continuous, intense territorial behavior that precludes eating, drinking, and sleeping until they are displaced. The benefit is the chance to maximize their fitness by mating with many females.

### Cost–benefit analysis can be applied to foraging behavior

Cost–benefit analyses have also been used to investigate the food choices animals make. When an animal *forages* (searches for food), among the decisions it must make are how much time to spend in each location before giving up and moving on, what resources at each location are actually edible, and which of the different types of potential food should be eaten and which should be left alone. By applying cost–benefit approaches to feeding behavior, scientists have produced a body of knowledge known as **optimal foraging theory**, which helps them to identify the fitness value of feeding choices.

The primary benefit of foraging is the nutritional value of the food obtained: the energy, minerals, and vitamins it contains (see Section 51.1). The costs of foraging are similar to those of other behaviors: energy expended, time lost from other activities that could enhance fitness, and the risk of increased exposure to predators.

Animals frequently have to make choices among food items that may differ not only in terms of energy content, but also abundance or ease of acquisition and processing. Optimal foraging theory predicts that in such situations, animals will make choices that will maximize the rate at which they obtain energy. The more rapidly a foraging animal satisfies its energetic requirements, the lower the opportunity costs and risk costs of foraging.

Earl Werner and Donald Hall of Michigan State University performed laboratory experiments with bluegill sunfish to test this energy maximization hypothesis. In preparation for their experiments, they measured the energy content of water fleas (*Daphnia*) of different sizes (the different food types), how much time bluegill sunfish (the foragers) needed to capture and eat those different food types, the energy they spent pursuing and capturing the different food types, and the rates at which they encountered the different food types under different food densities. Werner and Hall then stocked experimental environments with different densities and proportions of large, medium, and small water fleas. They made two predictions from the energy maximization hypothesis: first, that in an environment with abundant large water fleas, the fish would ignore smaller water fleas; and second, that in an environment stocked with low densities of all three sizes of water fleas, the fish would eat every water flea they encountered. The proportions of large, medium, and small water fleas eaten by the fish under different conditions were close to those predicted by the hypothesis (**Figure 53.13**).

**yourBioPortal.com**
GO TO **Animated Tutorial 53.2 • Foraging Behavior**

The energy maximization hypothesis considers food items in terms of the energy they provide, but animals have nutrient requirements in addition to energy that can play a role in shaping their foraging behavior. Essential minerals, for example, are in short supply in some animals' diets, and those animals may incur large energetic costs and risks to obtain them. Many herbivores and seed eaters, whose food contains low levels of mineral nutrients relative to their needs, obtain those nutrients by eating soil at particular sites where mineral-rich soil is exposed (**Figure 53.14**).

Foods may also be chosen for their medicinal value. Chimpanzees, for example, have been observed eating the pith of a plant called *Vernonia amygdalina*. The pith contains very small quantities of a secondary metabolite called vernonioside B1, which is toxic to chimps at high concentrations, but at low concentrations can kill their intestinal parasites. Chimps that consume this pith have fewer parasites.

## INVESTIGATING LIFE

### 53.13 Bluegill Sunfish Are Energy Maximizers

Based on energy maximization calculations, Werner and Hall predicted (1) that in an environment with abundant large food items (i.e., the water flea *Daphnia*), bluegill sunfish (*Lepomis macrochirus*) would ignore smaller food items and feed preferentially on larger water fleas; and (2) that in an environment where all sizes of *Daphnia* were scarce, the fish would eat every one they encountered. Such a strategy is in keeping with the cost–benefit hypothesis of foraging behavior.

**HYPOTHESIS** **Bluegill food item selection will match the energy maximization predictions of cost–benefit analysis.**

**METHOD**

1. Measure the respective energy content of large, medium, and small water fleas (*Daphnia*).
2. Use cost–benefit energy maximization calculations to predict the rate at which bluegill sunfish will consume the different sized water fleas under different levels of food abundance (i.e., density of water fleas).
3. Provide bluegills with *Daphnia* of different sizes in varying proportions (represented by the different colored bars) and at different densities.

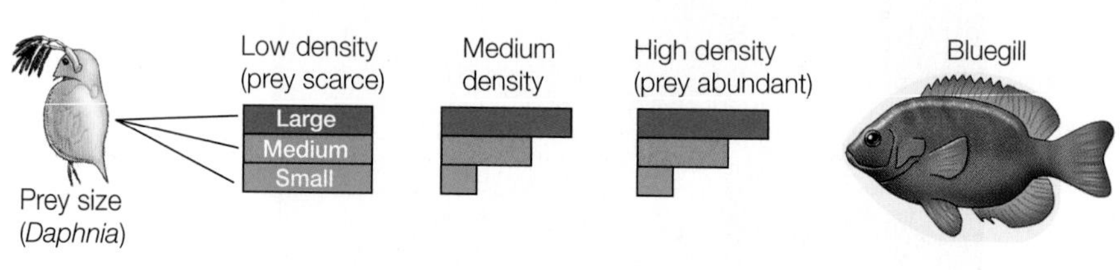


4. Note the proportions of small, medium, and large *Daphnia* actually eaten by the fish under the different conditions and compare these proportions to the predictions of energy maximization.

**RESULTS** The food choices made by bluegills match the predictions of the energy maximization hypothesis.

Low density | Medium density | High density

Proportions in diet predicted

Actual proportions in diet

When food items were scarce, the bluegills ate *Daphnia* of all sizes, as predicted.

When large *Daphnia* were abundant, fish ate the larger prey preferentially.

**CONCLUSION** **Bluegills select food items in accordance with the predictions of energy maximization calculations.**

Go to **yourBioPortal.com** for original citations, discussions, and relevant links for all INVESTIGATING LIFE figures.

(A)

(B)

**53.14 Herbivores Seek Out Unusual Sources of Minerals** (A) Red-and-green macaws of the Amazon jungle obtain essential minerals by eating dried clay. (B) Pierid butterflies obtain needed salts by drinking secretions from the skin and nostrils of a caiman.

## 53.4 RECAP

**Behavioral ecologists seek to explain relationships between variation in behavior and variation in environmental conditions. They seek to discover what information animals use to make behavioral choices and how that information relates to aspects of the environment that influence their fitness. Cost–benefit analysis has been applied to territorial and foraging behavior.**

- What might the presence of conspecifics in a habitat tell an animal about that habitat? **See p. 1123 and Figure 53.10**
- Describe three types of territorial behavior and the costs and benefits of each. **See pp. 1124–1125 and Figure 53.11**
- How can cost–benefit analysis be applied to foraging behavior? **See pp. 1125–1126 and Figure 53.13**

Whereas behavioral ecologists are interested in understanding how the natural environment influences the fitness value of behavioral choices—the ultimate causes of those behaviors, in Tinbergen's terms—other behavioral biologists focus on the physiological mechanisms and principles that underlie behavior—the proximate causes.

# 53.5 What Physiological Mechanisms Underlie Behavior?

Control of behavior involves the nervous and endocrine systems. Execution of behavior involves the musculoskeletal system as well as other *effector mechanisms*, such as those that produce secretions, color changes, electrical impulses, sound, and even light. We have already considered many of the physiological systems that are involved in these processes, including endocrine mechanisms, reproductive systems, nervous systems, sensory systems, and feeding mechanisms. The field of behavioral physiology, which encompasses aspects of all of these systems, is enormous, so here we will dig deeper into just three different phenomena studied by behavioral physiologists: the timing of behavior, navigation, and communication.

## Biological rhythms coordinate behavior with environmental cycles

Earth turns on its axis once every 24 hours, generating daily cycles of light and dark, temperature, humidity, and tides. In addition, Earth is tilted on its axis, so the light–dark cycle changes as it revolves around the sun. These daily and seasonal cycles profoundly influence the physiology and behavior of animals. Animals tend to be either day-active (diurnal) or night-active (nocturnal), and have appropriate sensory capabilities. Therefore it is adaptive to organize behavior on a cycle that corresponds with the environmental cycle of light and dark. Similarly, a behavior that is adaptive at one time of year may not be adaptive at other times. Thus it is important for animals to be able to time their behavior to appropriate times of the day or year and to be able to anticipate those times.

**CIRCADIAN RHYTHMS** Experimental animals kept in constant darkness and at a constant temperature, with food and water available all the time, still demonstrate daily cycles of activities such as sleeping, eating, drinking, and just about anything else that can be measured. The persistence of these daily cycles in the absence of environmental time cues suggests that animals have an internal clock. Because these daily cycles are not exactly 24 hours long, they are known as **circadian rhythms** (*circa*, "about," and *dies*, "day").

As described in Section 37.5, any biological rhythm can be viewed as a series of cycles, and the length of one of those cycles is the *period* of the rhythm. Any point in the cycle is a *phase* of that cycle. Hence, when two rhythms completely match, they are *in phase*, and if a rhythm is shifted (as in the resetting of a clock) it is *phase-advanced* or *phase-delayed*. Because the period of a circadian rhythm is not exactly 24 hours, it must be phase-advanced or phase-delayed each day to remain in phase with

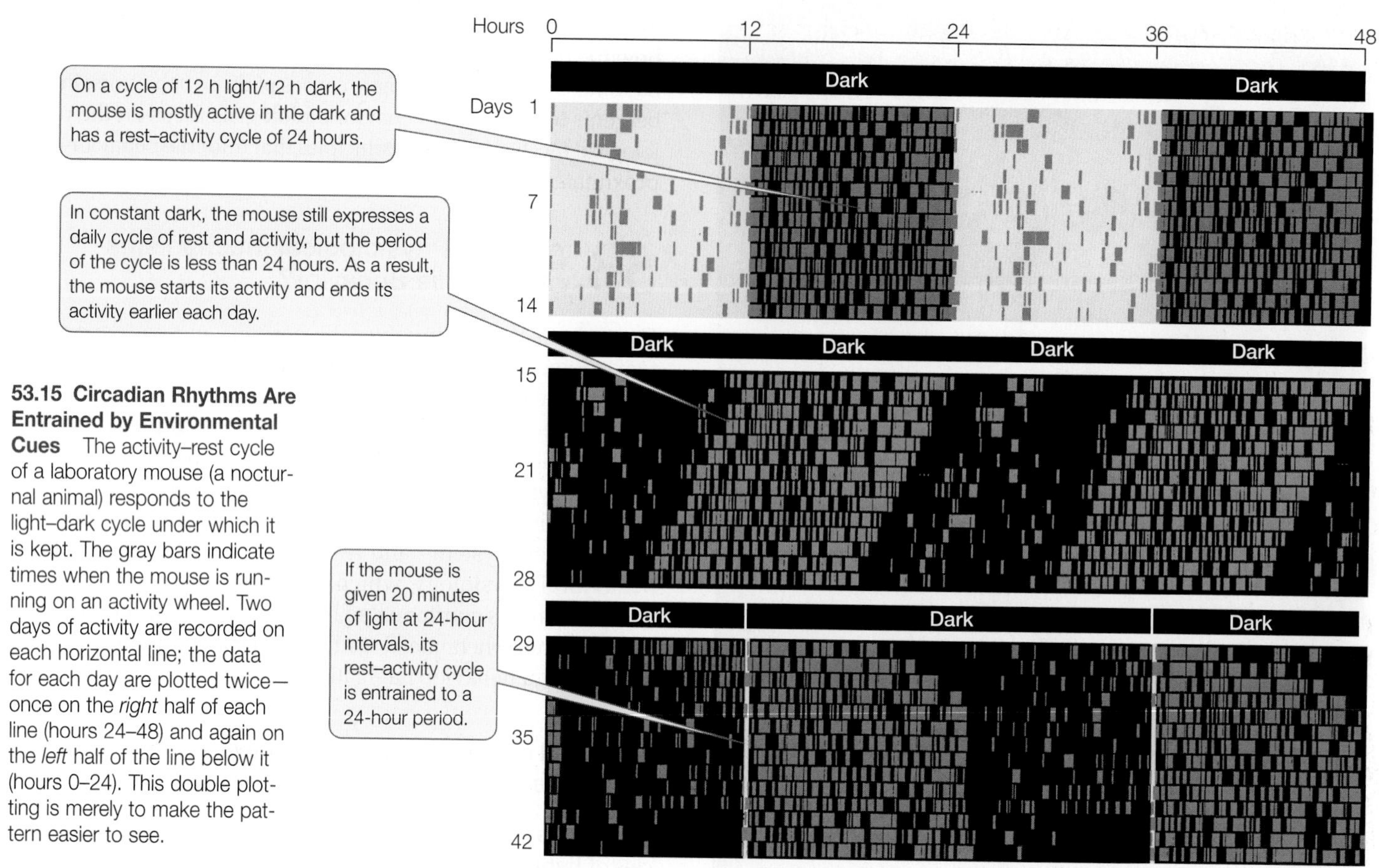

Hours 0
12
24
36
48
Dark
Dark
Days 1
7
14
15
21
28
29
35
42
On a cycle of 12 h light/12 h dark, the mouse is mostly active in the dark and has a rest–activity cycle of 24 hours.
In constant dark, the mouse still expresses a daily cycle of rest and activity, but the period of the cycle is less than 24 hours. As a result, the mouse starts its activity and ends its activity earlier each day.
If the mouse is given 20 minutes of light at 24-hour intervals, its rest–activity cycle is entrained to a 24-hour period.

**53.15 Circadian Rhythms Are Entrained by Environmental Cues** The activity–rest cycle of a laboratory mouse (a nocturnal animal) responds to the light–dark cycle under which it is kept. The gray bars indicate times when the mouse is running on an activity wheel. Two days of activity are recorded on each horizontal line; the data for each day are plotted twice—once on the *right* half of each line (hours 24–48) and again on the *left* half of the line below it (hours 0–24). This double plotting is merely to make the pattern easier to see.

yourBioPortal.com
GO TO Animated Tutorial 53.3 • Circadian Rhythms

the daily cycle of the environment. In other words, the rhythm has to be **entrained** to the cycle of light and dark in the animal's environment.

An animal kept under constant conditions will not be entrained to the light–dark cycle of the environment, and its circadian clock will run according to its natural period—it will be *free-running*. If the period is less than 24 hours, the animal will begin its activity a little earlier each day (**Figure 53.15**). The period of the free-running circadian rhythm is under genetic control. Different species may have different average periods, and within a species, mutations can lead to different period lengths.

Under natural conditions, environmental time cues, such as the onset of light or dark, entrain the free-running rhythm to the light–dark cycle of the environment. In the laboratory, it is possible to entrain the circadian rhythms of free-running animals with short pulses of light or dark administered every 24 hours (see the bottom panel of Figure 53.15).

In mammals, the master circadian "clock" consists of two clusters of neurons just above the optic chiasm (the area of the brain where the optic nerves come together). These structures are called the **suprachiasmatic nuclei** (**SCN**). If they are destroyed, the animal becomes *arrhythmic*: it is just as likely to eat, drink, sleep, or wake at any time of day.

Martin Ralph (now at University of Toronto) and colleagues, who were then at the University of Virginia, used artificial selection to produce two strains of hamsters: one with a short circadian period and one with a long circadian period. When adult hamsters' SCN were destroyed, the animals became arrhythmic. After several weeks, the scientists transplanted SCN tissue from hamster fetuses into arrhythmic hamsters' brains. The long-period adult hamsters received tissue from short-period fetuses, and vice versa. This experiment produced two remarkable results. First, circadian rhythms were restored by the transplanted SCN tissue, demonstrating that the SCN is sufficient to generate circadian rhythms. This is the only case in which a behavior has been restored by a neural transplant. Second, the restored circadian rhythms had the period length of the donor strain, demonstrating that the specific phenotype of the behavior was a property of the donor neural tissue.

The molecular mechanism of the circadian clock involves negative feedback loops. Although there are a number of genes involved, we can generalize the mechanism by saying that when certain *clock genes* are expressed in SCN cells, the mRNA enters the cytoplasm, where it is translated. The resulting proteins combine and the dimer returns to the nucleus as a transcription factor that shuts off the expression of the clock genes. The period of this cycle is about a day. These findings show that it is possible to understand circadian rhythms of behavior at all levels, from the molecular rhythm generators to the environmental stimuli that entrain them to the daily cycle of light and dark.

**CIRCANNUAL RHYTHMS** Seasonal changes in the environment present challenges to many species. Most animals reproduce most successfully if they time their reproductive behavior to coincide with the most favorable time of year for the survival of their offspring. Many species require considerable advance preparation for reproduction. Migratory animals must arrive on their breeding grounds at the right time, and animals that have specialized structures used in mating displays, such as the antlers of deer, moose, and caribou, must grow their equipment before the breeding season arrives.

For many species, a change in day length—the *photoperiod*—is a reliable indicator of seasonal changes to come. For others, however, change in day length is not a reliable seasonal cue. Hibernators, for example, spend long months in dark burrows underground, but must be physiologically prepared to breed almost as soon as they emerge in the spring. A bird overwintering near the equator cannot use changes in photoperiod as a cue to time its migration to its temperate-zone breeding grounds. When held under constant laboratory conditions, such animals show endogenous **circannual rhythms**, built-in neural calendars that keep track of the time of year. Unlike circadian rythms, the neural basis for circannual rhythms is unknown.

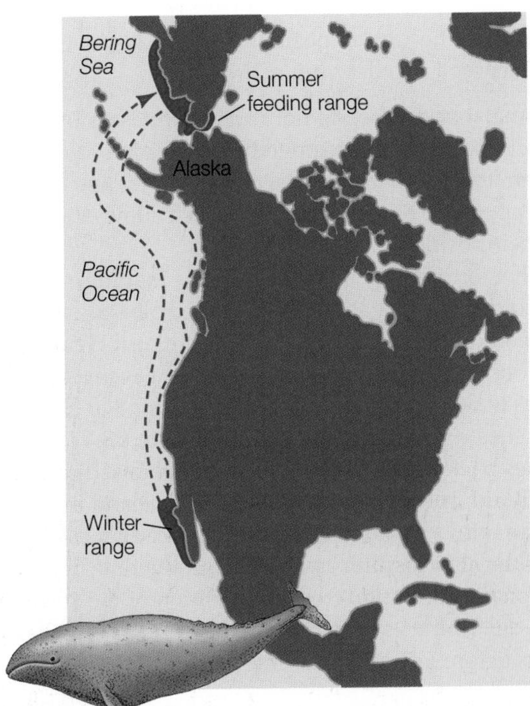


**53.16 Piloting** Gray whales (*Eschrichtius robustus*) migrate south in winter from the Bering Sea to the coast of Baja California by piloting, in part by following the western coast of North America.

## Animals must find their way around their environment

To locate suitable habitats, find food and mates, and escape from predators and bad weather, an animal needs to be able to find its way around its environment. Within its local habitat, an animal can organize its behavior spatially by orienting to landmarks. But what if its destination is a considerable distance away?

**PILOTING: ORIENTATION BY LANDMARKS** Most animals find their way by knowing and remembering the structure of their environment. This form of navigation is called **piloting**. Gray whales, for example, migrate seasonally between the Bering Sea and the coastal lagoons of Mexico (**Figure 53.16**). They find their way in part by following the west coast of North America. Coastlines, mountain chains, rivers, water currents, and wind patterns can all serve as piloting cues for the whales. But some remarkable cases of long-distance orientation and movement cannot be explained by piloting.

**HOMING: RETURN TO A SPECIFIC LOCATION** The ability to return to a nest site, burrow, or other specific location is called **homing**. Homing can be accomplished by piloting in a known environment, but some animals that travel long distances through unfamiliar territory perform much more sophisticated homing. The ability of pigeons to return to their home loft even after being transported to remote sites is well known. How do they find their way home? Experiments have shown that pigeons use the sun as a compass, but they can still find their way home when the sun is not visible. Other experiments have shown that pigeons equipped with frosted contact lenses can find their way home, suggesting that visual cues are not essential. Most amazing has been the demonstration that pigeons can detect Earth's magnetic field and orient to it much as a human orients with a compass. Taken together, the studies of homing by pigeons suggest that they can use multiple, redundant sources of directional information and can switch among those sources depending on the circumstances.

**MIGRATION AND NAVIGATION OVER GREAT DISTANCES** For as long as humans have inhabited temperate latitudes, they have been aware that whole populations of animals, especially birds, disappear and reappear seasonally. Not until the early nineteenth century, however, were patterns of migration traced by marking individual birds with identification bands around their legs. Only when individuals could be unmistakably identified was it possible to show that the same birds and their offspring returned to the same breeding grounds year after year, and that these same birds could be found during the nonbreeding season at locations hundreds or even thousands of kilometers from their breeding grounds.

Many homing and migrating species are able to take direct routes to their destinations through environments they have never experienced because they use mechanisms of navigation other than piloting. Humans use two major forms of navigation:

- *Distance–direction navigation* requires knowing in what direction and how far away the destination is. With a compass to determine direction and a means of measuring distance, humans can navigate.
- *Bicoordinate navigation*, also known as *true navigation*, requires knowing the latitude and longitude (the map coordi-

nates) of both the current position and the destination, as well as a compass to determine direction.

Other animals, of course, do not carry compasses or GPS receivers, but many of them seem to have a *compass sense*, which allows them to use environmental cues to determine direction, and some appear to have a *map sense*, which allows them to determine their position.

The behavior of many animals suggests that they are capable of bicoordinate navigation. Gray-headed albatrosses, for example, breed on oceanic islands in the Southern Hemisphere. When a young gray-headed albatross first leaves its parents' nest, it flies widely over the southern oceans for 8 or 9 years before it reaches reproductive maturity (**Figure 53.17**). At that time, it flies back to the island where it was raised to select a mate and build a nest. How can it find a tiny island in an enormous ocean after years of wandering? A circadian clock probably gives the albatross information about the time of day and the additional information from the position of the sun might allow it to determine its map coordinates—much as sailors did in the days before global positioning satellites.

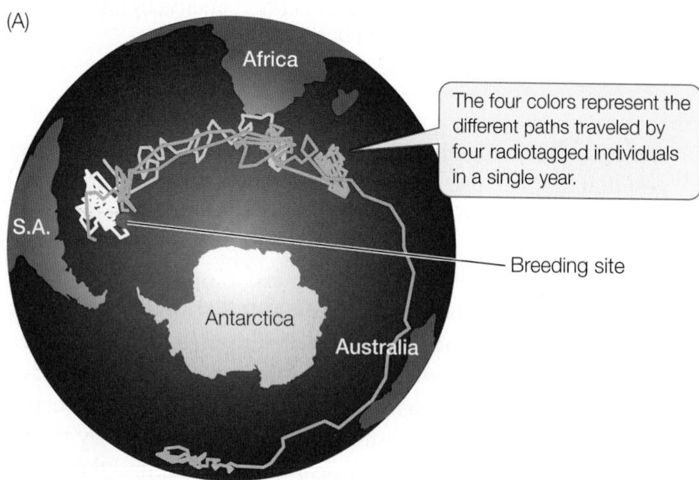
(A)
Africa
The four colors represent the different paths traveled by four radiotagged individuals in a single year.
S.A.
Breeding site
Antarctica
Australia

(B) *Thalassarche chrysostoma*

**53.17 Coming Home** (A) Gray-headed albatrosses are born on islands in the subantarctic oceans. Young birds roam widely over the southern oceans for 8 or 9 years. (B) Once they reach maturity, the birds return to the island where they were born to mate and raise their own young. A courting couple is shown here.

The ability to locate a position by calculating the angles between celestial objects such as the sun and stars and the horizon at specific times of day is called *celestial navigation*. During the day, the sun can serve as a compass for those species that can tell the time of day by means of their circadian clocks. This capacity has been demonstrated by "clock-shifting" experiments.

Researchers placed pigeons in a circular cage that enabled them to see the sun and sky, but no other visual cues. Food bins were arranged around the sides of the cage, and the birds were trained to expect food in the bin at one particular direction—south, for example. After training, no matter what time they were fed, and even if the cage was rotated between feedings, the birds always went to the bin at the southern end of the cage for food, even if that bin contained no food (**Figure 53.18**).

Next, the birds were placed in a room with a controlled light cycle in which the lights were turned on at midnight and off at noon. After 2 weeks of this treatment, the birds' circadian rhythms had been phase-advanced by 6 hours. Then the birds were returned to the circular cage under natural light conditions, with sunrise at 6:00 A.M. At sunrise, if the birds expected to find food in the south bin, then they should have looked for food 90 degrees to the right of the direction of the sun. But because their circadian clocks were telling them it was noon, they looked for food in the direction of the sun—in the east bin. The 6-hour shift in their circadian clocks resulted in a 90-degree error in their orientation, which showed that they were indeed using the sun as a compass. Similar experiments have shown that many animal species can orient by means of a *time-compensated solar compass*.

yourBioPortal.com
GO TO Animated Tutorial 53.4 • Time-Compensated Solar Compass

Many animals are normally nocturnal; in addition, many diurnal bird species migrate at night and thus cannot use the sun to determine direction. The stars offer two sources of information about direction: moving constellations and a fixed point. The positions of constellations (like that of the sun) change because Earth is rotating. With a star map and a clock, direction can be determined using any constellation. But one point that does not change position during the night is the point directly over the axis on which Earth turns. In the Northern Hemisphere, the star Polaris—the "North Star"—lies in that region of the sky and reliably indicates north.

Stephen Emlen at Cornell University showed that birds can learn to use the stars for orientation. As the time of year approaches when young birds would normally migrate to their winter range, young captive birds become more active and orient their activity in the direction they would fly. How do they know that direction? If these birds are raised in a planetarium with a natural star pattern, but one that does not rotate, the birds do not learn to orient, and their premigratory activity is random. However, if the planetarium sky rotates, and even if it rotates around a different point than the North Star, the birds orient their premigratory activity as if the fixed point in the sky was north.

INVESTIGATING LIFE

**53.18 A Time-Compensated Solar Compass**

Experiments show that pigeons use the sun to establish directions for navigation and finding food. "Clock-shifting" experiments demonstrate that the birds' circadian clocks factor into their ability to judge direction correctly based on the sun's position.

**HYPOTHESIS** **Pigeons determine compass direction from the position of the sun with respect to their internal circadian clocks.**

**METHOD**

1. Place a pigeon in a circular cage from which it can see the sun and sky, but not the horizon or any other visual cue.
2. Surround the cage with multiple food bins but place food only in the southernmost bin, thus training the bird to look for food in the south. (Rotating the cage but always placing the food in the southernmost bin confirms that the bird is navigating to find south.)

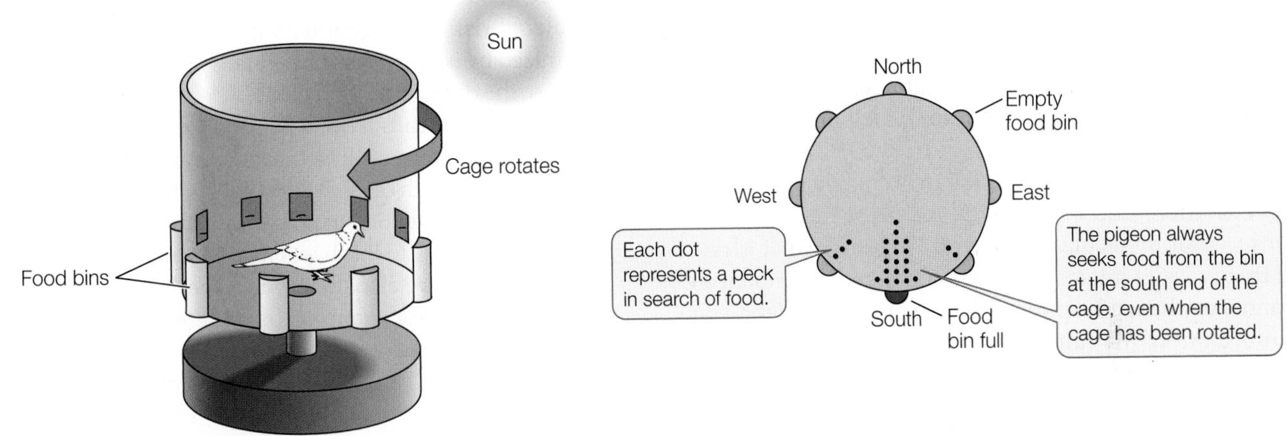


Cage rotates

3. Place the trained pigeon in a room with a controlled light cycle for 2 weeks. Turn the lights on at midnight and off at noon to phase-advance its circadian rhythm by 6 hours (i.e., 6 A.M. feels like noon to the bird).
4. Return the pigeon to the circular cage under natural light and observe its food-seeking behavior.

**RESULTS** A 6-hour shift in the circadian clock results in a 90-degree error in the pigeon's orientation.

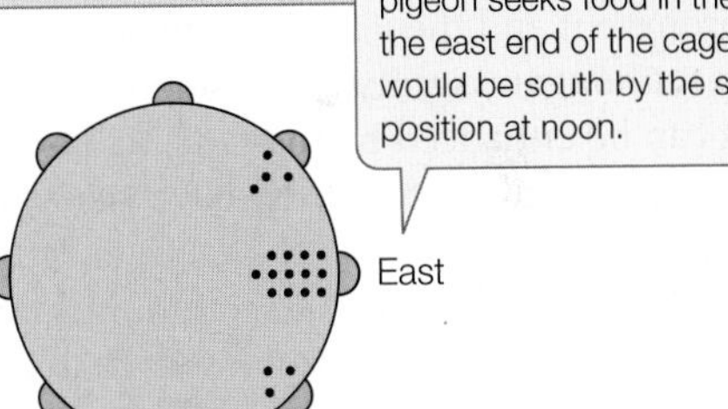

At sunrise, the phase-advanced pigeon seeks food in the bin at the east end of the cage—which would be south by the sun's position at noon.
Empty food bin
East
Each dot represents a peck in search of food.
Food bin full
The pigeon always seeks food from the bin at the south end of the cage, even when the cage has been rotated.

**CONCLUSION** **Pigeons have the ability to determine direction using the position of the sun as a compass.**

Go to **yourBioPortal.com** for original citations, discussions, and relevant links for all INVESTIGATING LIFE figures.

## Animals use multiple modalities to communicate

As individual animals interact, they exchange information; therefore, animal behavior can evolve into systems of information exchange, or **communication**. The behaviors of individuals may become elaborated into communication signals, but only if the transmission of information benefits both the sender and the receiver. To understand why these conditions must be met, consider male courtship displays such as that of the mallard shown in Figure 53.3. This display will be favored if it increases the sender's probability of mating and passing on his genes, and sexual selection will occur if the display conveys information to a female (the receiver) about his qualities as a potential father for her offspring.

Animals communicate using a variety of sensory modalities that vary in the nature of the signal produced, the specificity of the information conveyed, the speed and persistence of the signal, and its suitability in different environments. Behavioral physiologists interested in communication must take into consideration the sensory and motor characteristics of their study animals as well as the physics of the communication modalities they use and the environment in which the communication takes place.

**CHEMICAL SIGNALS** Because of the diversity of their molecular structures, pheromones can communicate very specific, information-rich messages (see Section 46.2). Pheromones are effective day and night, and they can cover a broad range of transmission distances. Pheromones used in different types of communication vary in their volatility (ease of vaporization) and diffusibility; these chemical properties are functions of the nature and size of the pheromone molecule. Pheromones that act as alarm signals, for example, are highly volatile and diffusible, so their message spreads rapidly, but disappears rapidly. Territory-marking and trail-marking pheromones have low volatilities and diffusibilities and stay effective for a long time, so they can convey directional information. Sex pheromones, such as that of the gypsy moth (see Figure 46.3), are intermediate in these properties, so they can spread a long distance, but do not disappear rapidly.

Pheromones are an effective way to exchange species-specific information, and because the recipient must have the proper receptor molecule to detect the pheromone, it is not a signal that is easily intercepted by predators. Pheromonal signals cannot be changed rapidly, but they can convey static, complex information. Mammals that mark their territories with pheromones reveal a great deal of information about themselves: species, individual identity, reproductive status, size (indicated by the height of the marking), and how recently the animal has been in the area (indicated by the strength of the scent).

**VISUAL SIGNALS** Visual signals offer the advantage of rapid delivery of information over considerable distances (depending on the environment and the visual acuity of the receiver); they also convey without ambiguity the position of the signaler. Signal content can be enhanced by movements (as in a courtship display) or by different postures. Effective visual signals, however, require sufficient light, and the receiver must be looking directly at the signaler. Thus, visual communication is not particularly useful at night or in environments that lack light, such as caves and ocean depths. Some species have overcome this constraint with light-emitting mechanisms. Fireflies, for example, use an enzymatic mechanism to create flashes of light. By emitting flashes in species-specific patterns, fireflies advertise for mates at night.

Another drawback of visual signals is that they can be intercepted by other species. There are predatory firefly species, for example, that mimic the flash pattern of females of other species. A male that approaches the mimicking "female," becomes a meal rather than a mate. Thus deception can be part of animal communication systems, just as it is part of human communication.

**ACOUSTIC SIGNALS** Sound cannot convey complex information as rapidly as visual signals can. But acoustic signals, unlike visual signals, can be used at night and in dark environments. They are not hindered by objects that would interfere with visual signals, so they can be transmitted in complex environments such as forests. They are often better than visual signals at getting the attention of a receiver because the receiver does not have to be looking at the signaler for the message to be received. Sounds are also useful for communicating over long distances. Even though the intensity of a sound decreases with distance from the source, loud sounds can transmit information over much longer distances than visual signals can. The complex songs of humpback whales, for example, when produced at ocean depths of about 1,000 meters, can be heard hundreds of kilometers away. In this way, humpback whales can locate one another across vast expanses of ocean.

The information content of acoustic signals can be increased by varying their frequency, as we can see in the sonograms of the species-specific songs of birds shown in Figure 53.9, and as we practice in our own speech. However, acoustic signals place the signaler at risk for detection by predators. This danger can be minimized by adjustments of frequency and signal structure that decrease the directional information the receiver can extract from the signal. Alarm calls tend to be pure tones (a single frequency) without much temporal structure (starts and stops). It is very difficult to localize such calls. On the other hand, territorial calls tend to cover a broad frequency range and have temporal structure. These calls are easy to localize. The frequencies and structures of acoustic signals are also adapted to specific habitats. Different vegetation types, for example, have different sound-absorbing properties: pure tones at lower frequencies carry better in forests, and more complex calls at higher frequencies carry well in open habitats.

**MECHANOSENSORY SIGNALS** Animals in close contact with one another communicate by touch, especially under conditions in which visual communication is difficult. The best-studied use of mechanosensory communication is the dance of honey bees, first described by Karl von Frisch. Honey bees have a spectacular ability to navigate and can reliably communicate the location of food sources as far as 2.5 km away from their hive. When a forager bee finds food, she returns to the hive and communicates her discovery to her hive-mates by dancing in the dark hive on the vertical surface of the honeycomb. Other bees follow the dancer and receive her message.

If the food source is more than about 100 m away from the hive, the forager performs a *waggle dance* (**Figure 53.19**), which conveys information about both the distance and the direction of the food source. She repeatedly traces out a figure-eight pattern as she runs on the honeycomb. She alternates half-circles to the left and right with vigorous wagging of her abdomen in the short, straight run between turns. Bees use the sun as their compass, and the angle of the straight run indicates the direction of the food source relative to the position of the sun projected down to the horizon. Even under cloudy conditions, the forager can provide directions to the food source because she can see polarized light. The honey bee's internal clock allows the dancer to adjust her dance to take into account the sun's movement during her return flight. The speed of the dance indicates the distance to the food source: the farther away it is, the longer the duration of each waggle run.

If the food she has found is less than 80 m from the hive, the forager performs a *round dance*, running rapidly in a circle and

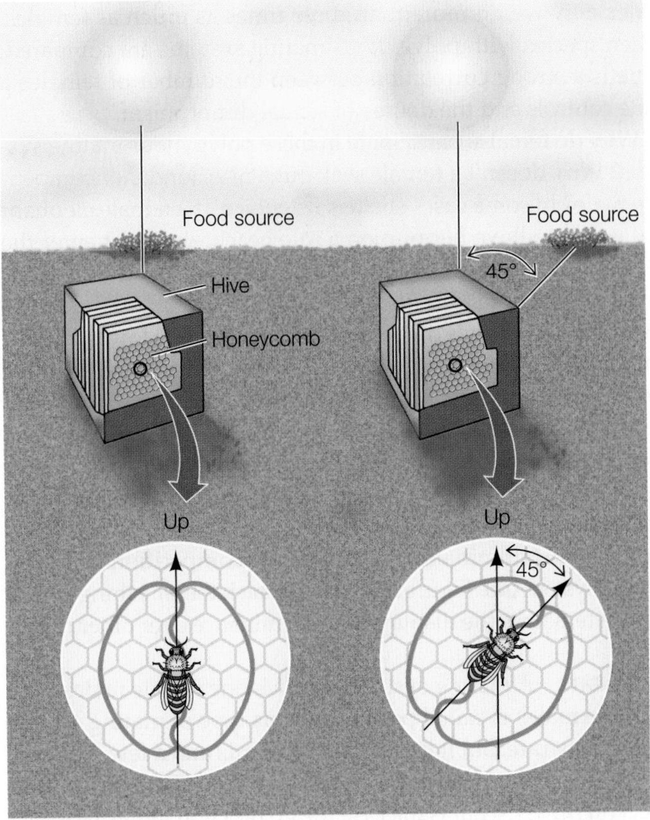


**53.19 The Waggle Dance of the Honey Bee** (A) A honey bee runs straight up on the vertical surface of the honeycomb in the dark hive while wagging her abdomen to tell her hivemates that there is a food source in the direction of the sun. (B) When her waggle runs are at an angle from the vertical, the other bees know that the same angle separates the direction of the food source from the direction of the sun.

**yourBioPortal.com**

GO TO **Web Activity 53.1 • Honey Bee Dance Communication**

reversing her direction after each full circle. The odor on her body and the round dance combine tactile and chemical cues: the odor indicates the flower to be looked for, and the dance communicates the fact that the food source is close to the hive.

**COMMUNICATION IN MULTIPLE SENSORY MODALITIES** Avoiding ambiguity is a high priority in any signaling system. Signal specificity is enhanced if multiple sensory modalities are used. Courtship behavior in fruit flies, for example, involves tactile, chemical, visual, and acoustic signals (see Figure 53.6A). The male fruit fly orients toward the female and taps her body with his foreleg (touch). Upon detection of pheromones in her cuticle (chemical), the male begins to vibrate one wing, producing a species-specific "courtship song" (acoustic). The male then extends his mouthparts to taste the female's genitalia (chemical and tactile); if she is receptive, he initiates copulation. At any point, if any sensory feedback indicates to either the male or the female that their pairing is inappropriate, the courtship abruptly ends.

## 53.5 RECAP

**Biological rhythms allow an animal to anticipate changes in its environment. In mammals, a circadian clock located in the suprachiasmatic nuclei generates a rhythm that is entrainable by environmental information. The navigational abilities of animals range from simple piloting by landmarks to distance-and-direction and bicoordinate navigation. Behaviors may evolve into communication signals if the transmission of information benefits both the sender and the receiver.**

- What is meant by a "free-running" rhythm? Describe how it can be entrained to the 24-hour day. **See pp. 1127–1128 and Figure 53.15**
- Explain the difference between piloting, distance-and-direction navigation, and bicoordinate navigation. **See pp. 1129–1130**
- Give an advantage and a disadvantage for each of three modalities of communication. **See pp. 1131–1133**

We have just seen how a behavioral physiologist might look at the courtship behavior of a male fruit fly. A behavioral ecologist would probably approach this behavior in a different way, asking how it influences the fitness of both the male fly and his mate. Given that interactions between individual animals, especially those involved in mating and the care of offspring, are so directly tied to fitness, it is not surprising that they are an important focus of behavioral studies.

# 53.6 How Does Social Behavior Evolve?

The evolution of social behavior became a field of study in its own right in 1975, with the publication of E. O. Wilson's landmark book, *Sociobiology*. We begin our consideration of this enormous field with simple interactions that involve a single male and a single female, but already we see diversity. Species differ in their mating systems, which vary from monogamous to promiscuous; in the amount of parental care they give their young; and in the degree to which the male contributes to raising the young. Beyond these relatively simple mating systems, there are associations of larger numbers of reproductive individuals in *polygynous* mating systems, in which a male has more than one mate, or *polyandrous* mating systems, in which a female has more than one mate. Even more complex interactions exist in which extended families participate in raising young, and finally there are societies such as honey bee colonies in which large numbers of nonreproductive individuals assist a single reproductive individual. Sociobiology proposes that the evolution of all these variations can be understood by asking how social behavior contributes to the fitness of the *individuals* involved.

### Mating systems maximize the fitness of both partners

At the start of Chapter 7, we learned about the mating behavior of two species of voles. Prairie voles (*Microtus ochrogaster*) are *monogamous*, forming strong pair bonds that can last for life, and both parents participate in rearing the young. In contrast, montane voles (*M. montanus*) are *promiscuous*: males mate with many females, and the young are raised by the females alone. Behavioral physiologists have explained the proximate mechanisms behind these stark behavioral differences in terms of the release of neurohormones and the distribution of the receptors for those hormones in the brains of the two species. The ultimate question—and the one asked here—is why two such different mating systems evolved in two species that are so closely related.

We begin with the premise that there is an asymmetry in the contributions of male and female animals to their offspring at the time of fertilization. As we learned in Section 44.1, females produce a limited number of eggs, and each egg is generously stocked with resources. Males produce an almost infinite number of sperm, which contain next to no resources. So the energetic and opportunity costs of reproduction are greater for the female than for the male. In mammals, this asymmetry increases throughout gestation as the female bears most of the costs. By the time of birth or hatching, the female's investment in the young is much greater than the male's investment, and the main way for the female to maximize her fitness is to make sure her young are healthy and survive to pass on her genes.

The male has different options for maximizing his fitness. He can simply move on after inseminating the female and seek additional mates as a means of maximizing his reproductive success—as in the case of the meadow vole. Or he can stay with the female he inseminated, protect her, and help care for their young—as in the case of the prairie vole. Which strategy maximizes his fitness depends on a number of factors that are influenced by the species' environment, such as the likelihood that a female and her offspring will survive without a male's help, and a male's likelihood of finding another fertile female. Thus sociobiologists seek to quantify these factors in nature as a means of explaining observed differences in mating systems.

**POLYGYNY** **Polygynous** mating systems, in which a male has more than one mate, involve a different asymmetry. In situations in which a male can sequester a group of females from other males, he can increase his fitness by increasing the number of females in his group. As we saw in Section 53.4, male elephant seals accomplish this by protecting an area of beach where females give birth. Male baboons do so by herding females. Male red-winged blackbirds may acquire more than one mate by defending high-quality nesting territories, building more than one nest in their territories, and attracting females to those nests. Since sex ratios in all these species are close to 50-50, a large differential in male fitness is established: some males have high reproductive success while many males have none. Thus selection favors males that are successful in competing with other males to obtain and protect access to many females. In general, bigger, stronger males are the winners, and sexual dimorphism in body size evolves. The elephant seal is an extreme example: males may weigh more than three times as much as females. When species with polygynous mating systems are compared, there is a strong correlation between the number of females a male controls and the degree of sexual dimorphism.

Why do females participate in these polygynous mating systems? Why doesn't a female seek out a nice, kind, noncompetitive male? In some cases she has no choice. If a female elephant seal wants to have her pup on a safe beach, she must enter the territory of a male. If a female red-winged blackbird wants to nest in an optimal territory, she will have to share the attentions of the territory's owner with other females. However, even if the female has a choice of mates, she is likely to maximize her fitness by mating with a male who is strong and dominant enough to control a number of females. Why? If her mate is a dominant male, her male offspring are likely to have their father's traits, become dominant males, and give her more grandchildren. The ultimate result of females selecting males for their prowess and dominance in male–male competition is the lek mating system (see Figure 53.12C), in which the *only* thing a male offers a female is the display of his dominance over other males.

**POLYANDRY** A mating system in which one female mates with multiple males is called **polyandry**. This type of mating system is relatively rare, but it is seen in some birds and a few mammal species in which paternal care for the young can have a large effect on fitness. That is the case the golden lion tamarins (*Leontopithecus rosalia*), primates native to the Brazilian tropical forests (**Figure 53.20**). In comparison to other primates, tamarins are tiny—under 1 kg—so they face high predation pressure. Females usually give birth to twins and thus newborns constitute a higher percentage of maternal weight than those of other primates. They also grow faster, so nursing costs are high. For all these reasons, young tamarins cared for by their mother alone are unlikely to survive.

*Leontopithecus rosalia*

**53.20 Polyandry in a Small Primate** The endangered golden lion tamarins of Brazil are small primates whose unique life history has given rise to polyandry in some groups, with males playing a major role in rearing the young.

What can a male tamarin do to help guarantee his reproductive success? Watching out for predators is one obvious contribution; gathering food for the female and her young is another. Like other primate parents, tamarins carry their young most of the time, but most other primates have single offspring. When tamarin mothers are carrying twins, they spend 92 percent of the time resting, in comparison to only 58 percent of the time when they are not carrying young. Resting is not compatible with foraging and filling the mother's high energy requirements. When a male is present, however, he carries the young about one-third of the time, so the mother has much more time for foraging and feeding.

If one male is helpful in protecting and raising young, then two should be even more helpful. Some females can attract a second mate by being sexually receptive to him. Neither male can be sure that any eventual offspring are his, so it is in the best interest of both to help in their rearing. Of the social groups observed in field studies, only 22 percent had one male and one female, whereas 61 percent had multiple males and one female.

### Fitness can include more than producing offspring

As humans, we readily understand the concept of extended family—brothers, sisters, aunts, uncles, nieces, nephews. Extended families are a common form of social organization in other species as well, and members of these families may cooperate in territory defense, predator avoidance, foraging, and rearing of young. If behavior is favored when it increases the fitness of the individual performing it, then how can we explain the evolution of social behaviors that do not lead to the performer having more offspring and may even appear to be **altruistic**—benefiting another individual at a cost to the performer?

An individual's fitness is increased by having offspring because those offspring carry the parent's genes into the next generation. Fitness gained by producing offspring is referred to as **individual fitness**. However, an individual's genes are carried into the next generation by more than his or her own offspring. In diploid organisms, two offspring of the same parents share, on average, 50 percent of the same alleles, and an individual is likely to share 25 percent of its alleles with its sibling's offspring. Therefore, by helping parents and other relatives raise their offspring, an individual increases the transmission of those shared alleles to the next generation. **Inclusive fitness** is the fitness derived from an individual's own reproductive success plus that derived through the reproductive success of its relatives.

The maximization of inclusive fitness is the mechanism behind **kin selection**: selection for behaviors that increase the reproductive success of relatives even when they come at a cost to the performer. An example is the phenomenon of *helping at the nest*, which was extensively studied in Florida scrub jays by Glen Woolfenden. Scrub jay pairs mate for life and establish large territories, which they defend aggressively. The mating pair may be assisted in rearing their young by 3 to 5 helpers (**Figure 53.21**). The helpers guard against predators, feed the young, clean the nest, and fly with fledglings. Why are these birds helping others rather than rearing their own young? Through a long-term study, Woolfenden was able to establish a number of important facts:

- The helpers are prior offspring of the mating pair and are usually 1 to 3 years old.
- Young birds that attempt to breed have almost zero reproductive success.
- Mating pairs with helpers have approximately 3 times the reproductive success of those without helpers.

These results support the conclusion that helper scrub jays are maximizing their inclusive fitness by helping their parents raise siblings until they are old and mature enough to have a reasonable probability of successfully raising their own offspring.

The concept of kin selection was formalized by W. D. Hamilton in what has become known as **Hamilton's rule**. He argued that, for an apparent altruistic behavior to be adaptive, the fitness benefit of that act to the recipient times the degree of relatedness between the performer and the recipient had to be greater than the cost to the performer. This relationship was clearly stated years before by the eminent geneticist J. B. S. Haldane, who said during an argument about altruism that he would not be willing to risk his life to save his brother—but for 2 brothers or 8 cousins, he would consider it.

**53.21 Helpers at the Nest** Young Florida scrub jays (*Aphelocoma coerulescens*) often forego reproduction in their first few years of adulthood to help their parents raise their siblings. These young birds help their parents feed the nestlings, defend the territory, and protect the nest from predators.

## Eusociality is the extreme result of kin selection

Species such as honey bees whose social groups include nonreproductive workers are called **eusocial** species. Prime examples are wasps, bees, and ants—members of the large insect order *Hymenoptera*. In a honey bee colony, the single reproductive female—the queen—occasionally produces a few male offspring, but most of the thousands of other individuals in the colony—her offspring—are sterile female workers. The key to understanding the evolution of eusociality in hymenopterans is their sex determination mechanism, called **haplodiploidy** because diploid individuals are female and haploid individuals are male. The queen carries a lifetime supply of sperm obtained during her single mating flight, and she controls whether her eggs are fertilized or not. An unfertilized egg develops into a male, a fertilized egg develops into a female.

If a queen copulates with only one male, all the sperm she receives are identical because a haploid male has only one set of chromosomes, all of which are transmitted to every sperm cell. Therefore, the queen's daughters share all of their father's genes and, on average, half of their mother's genes. As a result, the workers in the hive—all sisters—share, on average, 75 percent of their alleles. Were they to reproduce, they would share only 50 percent of their alleles with their own female offspring. Thus they can potentially increase their inclusive fitness more by caring for their sisters than by producing and caring for their own offspring.

Haplodiploidy is not essential for eusociality to evolve. This social system may also arise if it is costly or dangerous to establish new colonies. Nearly all eusocial animals construct elaborate nests or burrow systems within which their offspring are reared. Such a structure represents an enormous investment of resources. Naked mole-rats—the most eusocial mammals—live in elaborate underground tunnel systems that can extend for as much as 5 kilometers in cumulative length (**Figure 53.22**). A colony includes 70 to 80 individuals but only one reproductive female and a few reproductive males. The other colony members are sterile workers that dig and maintain the tunnels, guard against intruders, harvest food (tubers), and use their feces to feed the queen and her offspring. Individuals attempting to found new colonies have a high risk of failing or being captured by predators. When chances of individual reproductive success are practically zero, an individual can best maximize its inclusive fitness by staying with and helping to maintaining the colony.

*Heterocephalus glaber*

**53.22 A Eusocial Mammal** Naked mole-rats live in a large colony with only one reproductive female and a few reproductive males. They live in an elaborate tunnel system excavated by the colony over time.

(A)

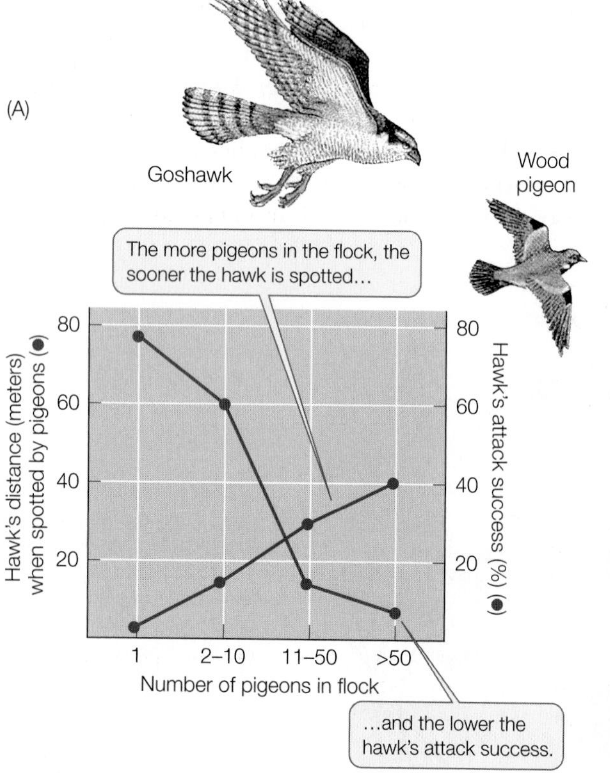


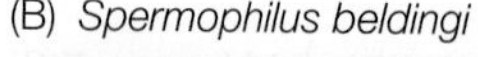

(B) *Spermophilus beldingi*

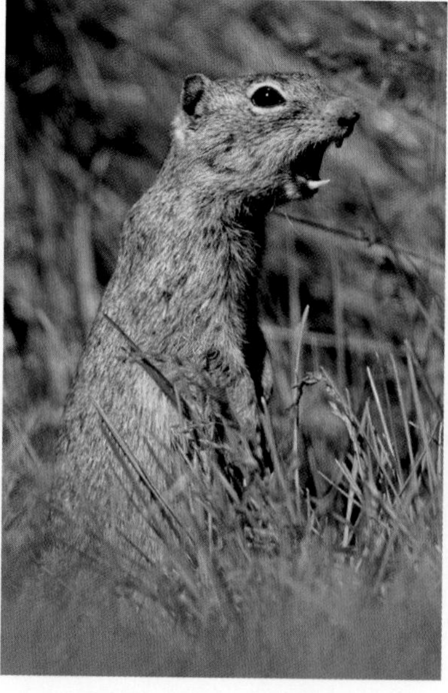

**53.23 Group Living Provides Protection from Predators** Animals that live in groups can spread the cost of looking out for predators. (A) The larger the number of pigeons in a flock, the greater the chances that one of the pigeons will spot a hawk before it attacks, and the lower the chances that the hawk will capture one of the pigeons. (B) A young male Belding's ground squirrel gives an alarm call upon spotting a predator. Although this behavior increases his individual risk of becoming its prey, he increases the survival chances of many of his close relatives.

### Group living has benefits and costs

Apart from their direct influences on reproductive success, social systems can contribute to survival in many ways, but they can also involve costs. Thus the cost–benefit approach of behavioral ecology is relevant to understanding the evolution of social behavior.

An obvious example of a benefit of group living is improved foraging efficiency. By hunting in packs, African wild dogs (see Figure 59.17) employ cooperative strategies that enable them to bring down larger prey than could a single dog. The larger the pack, the greater the hunting success rate. Once the prey is killed, the presence of conspecifics also reduces the risk that the wild dogs will lose their prey to larger scavengers, such as hyenas.

Living in a group can also reduce the risk of the members' becoming prey themselves. Many small birds forage in flocks. To test the hypothesis that flocking provides protection against predators, one investigator released a trained goshawk near wild wood pigeons in England. The hawk was most successful when it attacked solitary pigeons. Its success in capturing a pigeon in a flock decreased as the number of pigeons in the flock increased (**Figure 53.23A**). The larger the flock, the sooner some individual in the flock spotted the hawk and flew away. This escape behavior stimulated other individuals in the flock to take flight as well.

Alarm calling is another means of reducing predation risk, but the caller incurs a risk cost by calling attention to itself. Belding's ground squirrels live in large colonies in open meadows. When one squirrel announces the presence of a predator with loud, sharp barks, all the nearby squirrels dive into their burrows (**Figure 53.23B**). Paul Sherman showed that callers double their risk of being preyed on—so why do they do it? Sherman's and others' research has shown that this altruistic behavior is a product of kin selection. In this polygynous species, males establish large territories in the spring that include the territories of several females, whom they inseminate. The females then drive the males off. Female offspring settle near their mothers, so neighboring females in a colony tend to be sisters, and they defend each other's young. Sherman showed that males are less likely to give alarm calls than females, and that females are more likely to give alarm calls when related individuals are nearby.

Social behavior has many costs as well as benefits. Foraging in a group may reduce the amount of food available to each individual, and the foraging individuals may interfere with one another's foraging activities. Individuals living in groups may face more competition for mates, as well as for food, than solitary individuals would. A large group may actually attract the attention of predators. And living at high population densities can increase the risk of disease transmission. The study of disease transmission in wild animal populations is a relatively new field, but such studies have made it apparent that species living in social groups are more prone to outbreaks of disease than are solitary species.

## 53.6 RECAP

**Social behavior can be understood by asking how it contributes to the fitness of the individuals involved. Asymmetry between the sexes in parental investment is a key factor in the evolution of mating systems. According to the theory of kin selection, an individual can increase its fitness by helping related individuals with whom it shares alleles. In extreme cases, kin selection has given rise to eusociality.**

- What environmental conditions can lead to monogamy, polygamy, or polyandry? See pp. 1134–1135
- Explain how an individual can increase its fitness by helping its relatives. See p. 1135
- Why is eusociality so common among hymenopterans? See p. 1136
- What are some of the costs and benefits of group living? See p. 1137 and Figure 53.23

Knowledge of the behavior of particular species—how they use the environment, how they obtain food, how they organize their activities spatially and temporally—is essential for understanding how species interact in nature. These interactions are one focus of the science of ecology, the subject of Part Ten of this book.

## CHAPTER SUMMARY

### 53.1 What Are the Origins of Behavioral Biology?

- Ivan Pavlov's discovery of **conditioned reflexes** and B. F. Skinner's research on **operant conditioning** as a model for learning led to an approach called **behaviorism** that mainly carried out laboratory experiments on rats and a few other animal models. Review Figure 53.1
- **Ethology** focuses on both the **proximate causes** of behavior—what is the immediate cause of the behavior, and how does the behavior develop?— and on the **ultimate causes**—how does the behavior affect the animal's fitness, and how does it compare with similar behaviors in related species?
- A major focus of the ethologists was **fixed action patterns** and their **releasers**. They performed **deprivation experiments** as well as breeding experiments to demonstrate that certain behaviors are genetically determined.

### 53.2 How Can Genes Influence Behavior?

- Breeding experiments can reveal whether a behavioral phenotype is inherited. Gene knockout experiments can reveal the roles of specific genes underlying a behavioral phenotype. Review Figures 53.4 and 53.5
- Most behaviors are complex traits involving many genes that function in cascades and offer many points for a change in a single gene to influence behavior. Review Figure 53.6

## 53.3 How Does Behavior Develop?

- Hormones can determine the pattern of behavior that develops and the timing of its expression. Review Figure 53.7
- **Imprinting** is a process by which an animal learns a specific set of stimuli during a limited **critical** or **sensitive period**. That critical period may be determined by hormones.
- The development and expression of song in white-crowned sparrows involves a genetic predisposition to learn the species-specific song, a critical period for imprinting of a song memory, and hormonally controlled timing of song expression. Social interactions may also play a role. Review Figure 53.9

## 53.4 How Does Behavior Evolve?

- An animal's behavior is a series of choices that influence its fitness. To make these choices, animals use environmental cues that are reliable predictors of the potential effects of their choices on their fitness. Review Figure 53.10
- The **cost–benefit approach** can be used to investigate the fitness value of specific behaviors. The cost of a behavior typically has three components: **energetic cost**, **risk cost**, and **opportunity cost**. Review Figure 53.11, ANIMATED TUTORIAL 53.1
- According to **optimal foraging theory**, animals should practice feeding behaviors that maximize their energetic gain at the least cost. Review Figure 53.13, ANIMATED TUTORIAL 53.2

## 53.5 What Physiological Mechanisms Underlie Behavior?

- **Circadian rhythms** control the daily cycle of behavior. Without environmental time cues, circadian rhythms free-run with a period that is genetically programmed. They are normally **entrained** to the light–dark cycle by environmental cues. Review Figure 53.15, ANIMATED TUTORIAL 53.3
- **Circannual rhythms** are endogenous yearly cycles whose neural basis is still unknown.
- Three forms of navigation used by animals to find their way in the environment are piloting (orienting to landmarks), distance–direction navigation, and bicoordinate navigation. Navigation mechanisms include celestial navigation and a time-compensated solar compass. Review Figures 53.16–53.18, ANIMATED TUTORIAL 53.4
- The behaviors of individuals may become communication signals if the transmission of information benefits both the sender and the receiver.
- Chemical communication signals (pheromones) can be highly specific and have different time courses depending on their volatility and diffusibility. Visual signals can convey complex messages rapidly, but the recipient must be looking at the sender. Acoustic signals travel well over distances, do not require a focused recipient, and can be modified to reveal or conceal directional information. Tactile signals are used by animals in close proximity and can convey complex messages. Review Figure 53.19, WEB ACTIVITY 53.1

## 53.6 How Does Social Behavior Evolve?

- The evolution of mating systems can be understood by looking at the fitness costs and benefits incurred by each partner in the species' environment. **Polygynous** mating systems, in which one male controls and mates with many females, can result in great variation in male reproductive success. **Polyandry**—a female mating with multiple males—can evolve in circumstances in which a male can make a substantial contribution to the survival of his offspring.
- The fitness an individual gains by producing offspring (**individual fitness**) plus the fitness it gains by increasing the reproductive success of relatives with whom it shares alleles is called **inclusive fitness**. **Kin selection** may favor **altruistic** behavior toward relatives, despite its cost to the performer, if it increases the performer's inclusive fitness.
- As a result of **haplodiploidy**, the sex determination mechanism of the Hymenoptera, nonreproductive workers share more alleles with one another than females share with their own offspring. Haplodiploidy has probably facilitated the evolution of **eusociality** in this group through kin selection. Eusociality has also risen in diploid species in which chances of individual reproductive success are extremely low.
- Group living confers benefits such as greater foraging efficiency, protection from predators, but it also has costs, such as increased competition for food and ease of transmission of diseases.

**SEE WEB ACTIVITY 53.2 for a concept review of this chapter.**

## SELF-QUIZ

1. Which of the following is *not* true of a fixed action pattern?
   *a.* Its expression may depend on hormonal conditions.
   *b.* It is induced by complex, species-specific stimuli.
   *c.* It is highly stereotypic and species-specific.
   *d.* It can be expressed even if the animal has never seen it performed.
   *e.* Its genetic basis can be demonstrated by breeding experiments.
2. Which of the following is *not* a component of the cost of performing a behavior?
   *a.* Its energetic cost
   *b.* The risk of being injured
   *c.* Its opportunity cost
   *d.* The risk of being attacked by a predator
   *e.* Its information cost
3. Birds that migrate at night
   *a.* inherit a star map.
   *b.* determine direction by knowing the time and the position of a constellation in the sky.
   *c.* orient to a particular point in the sky.
   *d.* imprint on one or more key constellations.
   *e.* determine distance, but not direction, from the stars.
4. If a bird is trained to seek food on the western side of a cage open to the sky, and is then placed in a chamber with a controlled light cycle so that its circadian rhythm becomes phase-delayed by 6 hours (i.e., its circadian rhythm is 6 hours behind real time), when it is returned to the open cage at noon in real time, it will seek food in the
   *a.* north.
   *b.* south.
   *c.* east.
   *d.* west.
5. To be able to pilot, an animal must
   *a.* have a time-compensated solar compass.
   *b.* orient to a fixed point in the night sky.
   *c.* know the distance between two points.
   *d.* know landmarks.
   *e.* know its longitude and latitude.

6. Which of the following statements about communication is *true*?
   *a.* Complex information cannot be conveyed by pheromones.
   *b.* Visual signaling is advantageous in complex environments.
   *c.* Acoustic communication always reveals the location of the signaler.
   *d.* An advantage of pheromones is that the message can persist over time.
   *e.* The dance of honey bees is an example of visual signaling.
7. A cost commonly associated with group living is
   *a.* increased risk of predation.
   *b.* interference with foraging.
   *c.* higher exposure to diseases and parasites.
   *d.* poorer access to mates.
   *e.* All of the above
8. The choice of a mating partner may be based on
   *a.* the competitive ability of a potential mate.
   *b.* the territory held by a potential mate.
   *c.* both the competitive ability of a potential mate and the territory it holds.
   *d.* the courtship display of a potential mate.
   *e.* all of the above
9. Altruistic behavior
   *a.* can increase an individual's inclusive fitness.
   *b.* depends on haplodiploid sex determination.
   *c.* is most common between unrelated individuals.
   *d.* always causes a net decrease in the performer's fitness.
   *e.* characterizes a monogamous mating system.
10. A social group is said to be eusocial if
   *a.* group members interact very intensively.
   *b.* some group members produce many more offspring than others do.
   *c.* a dominance hierarchy exists among group members.
   *d.* young individuals remain in the group to help their parents rear other offspring.
   *e.* the group contains nonreproductive helper individuals.

## FOR DISCUSSION

1. Cowbirds are nest parasites. A female cowbird lays an egg in the nest of another bird species, which then incubates the egg and raises the chick. What do you think would characterize the acquisition of song in cowbirds? In a given area, cowbirds tend to parasitize the nests of particular bird species. How do you think female cowbirds learn this behavior? How would you test your hypothesis?
2. Male dogs lift a hind leg when they urinate; female dogs squat. If a male puppy receives an injection of estrogen when it is a newborn, it will never lift its leg to urinate for the rest of its life; it will always squat. How might this result be explained?
3. The short-tailed shearwater is a bird that winters in Antarctica and summers in the Arctic. What problems would this species have in using either the sun or the stars for navigation? What is the most likely means it uses to find its way to its summer and its winter feeding grounds?
4. In most vertebrate species with helpers, the helpers are individuals capable of reproducing, and most of them later breed on their own. Among eusocial insects, sterile castes have evolved repeatedly. What differences between vertebrates and insects might explain the failure of sterile castes to evolve in the former?

## ADDITIONAL INVESTIGATION

Experiments on animal migration show that animals can use a time-compensated solar compass or identification of a fixed point in the night sky as a basis for distance-and-direction navigation. However, the observations on the homing ability of seabirds such as the albatross indicate that animals might be capable of true or bicoordinate navigation. What experiments could you do to prove that animals have that capability, and what experiments could you do to test hypotheses about the mechanisms they might use? (*Hint*: At least two hypotheses might involve using the elevation of the sun or the angles of Earth's magnetic lines of force as a means of determining latitude.)

## WORKING WITH DATA (GO TO yourBioPortal.com)

**The Costs and Benefits of Foraging Behavior** In this exercise based on Figure 53.13, you will analyze the techniques and results of Werner and Hall's experiments with bluegill sunfish and answer questions pertaining to experiments such as this one that seek to quantify behaviors.

# 54 Ecology and the Distribution of Life

## Dung happens

In 1788, eleven ships carrying settlers from Great Britain landed on the shores of Australia (then called New South Wales), laden with all the supplies the settlers anticipated would be needed to found a new European colony on that continent. The arriving cargo included two bulls and five cows, which would provide the foundation of an important livestock industry. The settlers, however, made a grievous error by failing to bring along dung beetles.

Dung beetles are beetle species that during their larval life consume the dung, or excrement, of vertebrates, primarily grazing mammals. Some species lay eggs directly into the dung pat where it was deposited; others carve up the dung pat and form the pieces into balls, which are then rolled away and buried. Australia's native dung beetles, accustomed to processing and burying the dry, fibrous dung pats of kangaroos, wombats, and other indigenous marsupial mammals, were unable to handle the copious quantities of wet manure produced by the rapidly expanding numbers of cattle and sheep. A cow can drop 10 to 12 dung pats every day and, in the absence of dung beetles to break up and bury the dung, thousands of acres of Australian pasture were rendered unusable for grazing livestock.

In addition, the cattle dung proved an exceptionally suitable habitat for the larvae of the bush fly, *Musca vetustissima*, so fly populations exploded. The flies became so pestiferous, flying into people's faces, that eating, drinking, and even breathing were rendered nearly impossible in some parts of Australia at certain times of year.

In 1964, Australia's national science organization tackled the dung problem. Entomologists traveled to Africa and other parts of the world to find dung beetle species that could process and bury cattle dung. The first such species, *Onthophagus gazella,* was imported from South Africa in 1968. Four more species arrived the next year, and by 1984, over 50 dung beetle species from around the world had been brought to Australia to take up the task of cattle dung cleanup.

Multiple species with different habitat and climate requirements were needed to cover the different environments across the Australian continent. The project, deemed a resounding success, not only reduced bush

***Onthophagus gazella* in Australia** Dung beetles from Africa are capable of processing the moist manure of introduced mammals such as cattle.

**Freed from Bush Flies** The importation of dung beetles essentially saved the livestock industries of Australia from plagues of bush flies breeding in the moist dung of cattle and other imported mammals.

fly populations, but had other desirable effects. The beetles reduced the incidence of parasitic infections in cattle by roundworms, hookworms, and other dung-breeding worms, and they increased pastureland productivity by releasing nitrogen from the dung into the soil.

The passengers and crew of the ships that brought cattle to Australia had no idea that they were carrying out a massive ecological experiment. By intent or by accident, people have been moving all kinds of organisms from their native habitats to places they have never lived before. It is now abundantly clear that introducing organisms into existing communities can have unanticipated repercussions and that even small organisms can have large effects on energy flow and nutrient cycling in ecosystems.

**IN THIS CHAPTER** we will introduce and define the field of ecology. We will see how the physical features of the environment, including climate and geological history, influence the distribution and abundance of organisms. Next we will introduce and describe the major biomes, biogeographic regions, and aquatic environments across which Earth's organisms are distributed.

## CHAPTER OUTLINE



# 54.1 What Is Ecology?

**Ecology** is the scientific investigation of interactions among organisms and between organisms and their physical environment. Ecology is a relatively new branch of the biological sciences; in fact, it did not even have a formal name until 1866. Like evolutionary biology, ecology owes its existence to Charles Darwin, who described the focus of *The Origin of Species* as "the coadaptation of organic beings to each other and to their physical conditions of life." Ernst Haeckel, a German biologist who was profoundly influenced by Darwin, constructed a new word for this new enterprise: *ecology*, from the Greek root *oikos*, "house," where "house" embraces all of an organism's environment. As Haeckel explained, "Ecology is the study of all those complex interrelations referred to by Darwin as 'the conditions of the struggle for existence.'"

Ecology provides explanations of the perceptible, palpable world. Although the *consequences* of enzyme–substrate interactions can be visible, the reactions themselves are not readily visible to the casual observer. In contrast, interactions among organisms, and between organisms and their environment, can, with persistence and ingenuity, be observed. Analysis of those observations, however, often requires additional investigation. That butterflies visit flowers has been observed, and admired aesthetically, for centuries; that butterflies perform a useful service to the flowers they visit by transporting pollen is an ecological insight less than 250 years old.

The need for sound science in making decisions about our own interactions with the environment is another important reason for studying ecology. Humans are part of the biotic environment of a tremendous variety of organisms, and our activities have profound effects on energy flow and nutrient cycling through the physical environment. Ecologists can provide objective data on the consequences of human activities. An understanding of ecology greatly improves our ability to grow food sustainably; to manage pests and diseases safely and effectively; and to deal with natural disasters such as floods and fires. The greater our understanding of ecological interactions, the greater the likelihood that we can accomplish these things without causing a cascade of unanticipated consequences for ourselves and other organisms.

### Ecology is not the same as environmentalism

In defining what ecology is, it is important to emphasize what ecology is *not*. "Ecology" is sometimes equated with "environmentalism," but the two terms are not equivalent. Ecology is a science that generates knowledge about interactions in the natural world; as a field of inquiry, it is not inherently focused on human concerns. **Environmentalism** is the use of ecological knowledge, along with economics, ethics, and many other considerations, to inform both personal decisions and public policy relating to stewardship of natural resources and ecosystems.

### Ecologists study biotic and abiotic components of ecosystems

From its beginnings, ecology has encompassed both the living, or **biotic**, components and the **abiotic**, or physical and chemical, components of the environment. But the field has advanced and diversified considerably since Haeckel's time. A continuous influx of new tools—mathematical models, molecular techniques, and satellite imaging, to name just a few—as well as new connections to other fields, particularly the physical sciences—have dramatically changed ecological research. Its ultimate goal, however, remains the same: to identify and understand the biotic and abiotic forces that influence the distribution and abundance of organisms.

## 54.1 RECAP

**Ecology is the scientific investigation of interactions among organisms and between organisms and their physical environment.**

- What are some reasons to consider ecology a useful scientific enterprise? See p. 1141
- How does ecology differ from environmentalism? See pp. 1142–1143

Climate is one of the abiotic factors that determine the kinds of organisms that can survive and reproduce in a particular place. Because climates set limits on the distribution and abundance of organisms, understanding Earth's climates is a good place to begin our study of ecology.

## 54.2 Why Do Climates Vary Geographically?

The terms *weather* and *climate* both refer to atmospheric conditions—temperature, humidity, precipitation, wind direction and velocity—but they refer to different time scales. **Weather** is the short-term state of atmospheric conditions at a particular place and time, whereas **climate** refers to the average atmospheric conditions, and the extent of their variation, at a particular place over a longer time. In other words, climate is what you expect; weather is what you get. The responses of organisms to weather are usually short-term—seeking shelter from a sudden rainstorm, for example, or shivering to keep warm when the temperature drops. Responses to climate, on the other hand, tend to be adaptations that affect physiology, morphology, and behavior. These adaptations are among the forces driving speciation over evolutionary time.

### Global climates are solar powered

Solar energy input, which determines air temperature, is the major determinant of climates on Earth. The intensity of solar radiation varies from place to place due to the shape of Earth and the tilt of its axis, and its orbit around the sun. The amount of solar energy reaching a given unit of Earth's surface depends primarily on the angle of the sun's rays (**Figure 54.1**). At high latitudes (areas close to the poles), sunlight strikes Earth's surface at an angle, so the incoming solar energy is distributed over a larger area (and is thus less intense) than at the Equator, where sunlight strikes Earth's surface perpendicularly. Moreover, when coming in at an angle, the sun's rays must pass through more of Earth's atmosphere, so more of their energy is absorbed and reflected before it reaches the ground. Because of this difference in solar energy input, air at the poles is colder than air at the Equator. The average temperature of the air over the course of the year decreases about 0.4°C for every degree of latitude (about 110 km) at sea level. In addition, because of the tilt of Earth's axis relative to the sun, higher latitudes experience greater variation in both day length and the angle of arriving solar energy over the course of a year, leading to more seasonal variation in temperature.

In the atmosphere closest to Earth's surface, air temperatures also decrease with elevation, so temperatures at sea level are warmer than temperatures on mountaintops at the same latitude.

### Solar energy input determines atmospheric circulation patterns

When a parcel of air is warmed, it expands (that is, its molecules move farther apart) and, because it is less dense than cool air,

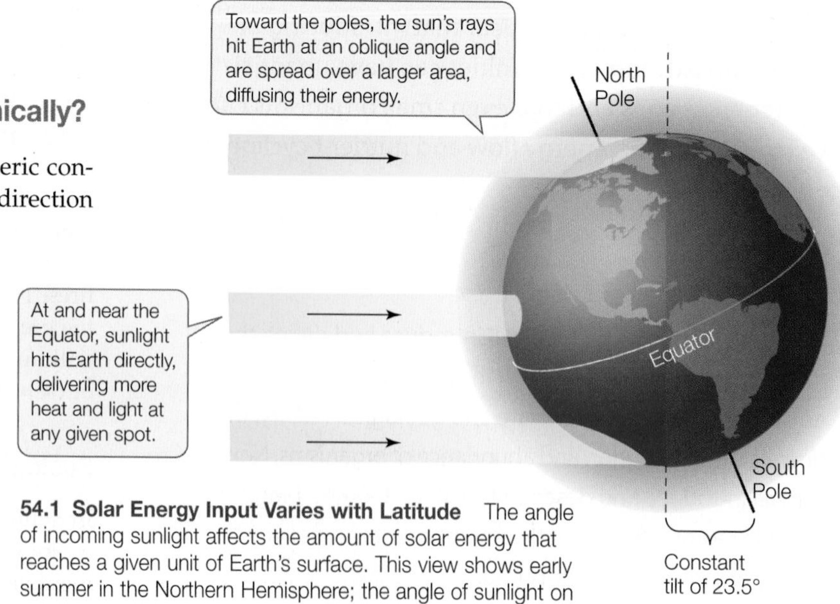

Toward the poles, the sun's rays hit Earth at an oblique angle and are spread over a larger area, diffusing their energy.
North Pole
At and near the Equator, sunlight hits Earth directly, delivering more heat and light at any given spot.
Equator
South Pole
Constant tilt of 23.5°


**54.1 Solar Energy Input Varies with Latitude** The angle of incoming sunlight affects the amount of solar energy that reaches a given unit of Earth's surface. This view shows early summer in the Northern Hemisphere; the angle of sunlight on the hemispheres will shift as the seasons change.

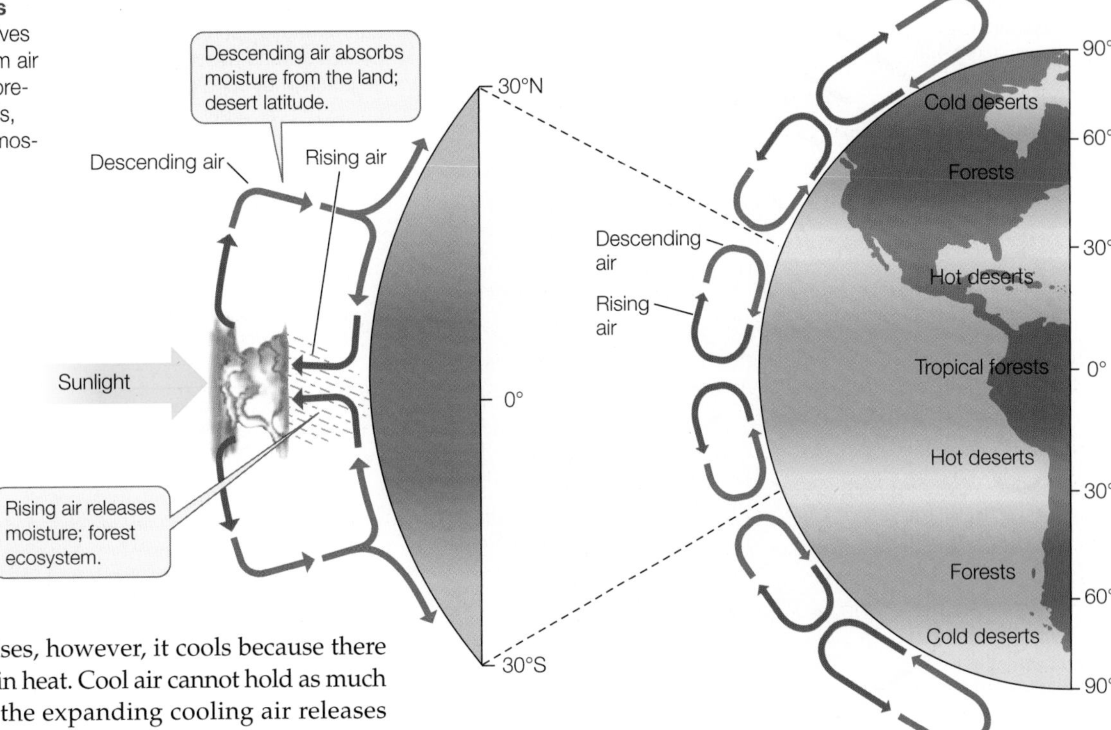


**54.2 Air Circulation in Earth's Atmosphere** Solar energy drives surface air patterns. Moist, warm air rises and releases moisture as precipitation. Cool, dry air descends, absorbing moisture from the atmosphere and surface.

the warm air rises. As it rises, however, it cools because there are fewer molecules to retain heat. Cool air cannot hold as much moisture as warm air, so the expanding cooling air releases moisture in the form of precipitation. Air in the region surrounding the equator receives the greatest input of solar energy. As this tropical air warms, it rises, cools, and releases large amounts of rainfall. And, as this warm air rises into the atmosphere, it is replaced by surface air flowing in from the north and south (**Figure 54.2**).

The air that moves from the north and south into the equatorial tropics is replaced by air from aloft that descends at latitudes of roughly 30°N and 30°S, having traveled away from the equator high in the atmosphere. This air, which cooled and lost its moisture as it rose, now descends, warms, and *takes up* moisture rather than releasing it. Earth's great deserts, such as the Sahara of Africa, the Gobi of China, and the deserts of Australia and the American southwest are located at these latitudes.

At about 60° latitude, air rises again and moves either toward or away from the Equator. At the poles, where there is little solar energy input, air descends again. These cyclic movements of air masses are largely responsible for the prevailing wind patterns at Earth's surface.

### Global oceanic circulation is driven by wind patterns

The rotation of Earth on its axis also influences prevailing winds. The velocity of rotation is fastest at the equator, where Earth's diameter is greatest, and slowest close to the poles. A stationary air mass has the same rotational velocity as does the Earth at the same latitude. As an air mass moves toward the equator, however, its rotational movement becomes slower than that of Earth beneath it. Conversely, the rotational movement of an air mass moving poleward is faster than that of the surface beneath it. Therefore, moving air masses are deflected to the right in the Northern Hemisphere and to the left in the Southern Hemisphere. Air masses moving toward the equator from the north and south veer to become the northeast and southeast *trade winds*, respectively. Air masses moving away from the equator also veer, becoming the *westerly* winds that prevail at mid-latitudes (**Figure 54.3A**).

Incoming solar radiation also affects patterns of water movement in Earth's oceans. Solar energy heats the atmosphere and powers the wind, which in turn, through frictional drag, moves the water it blows over (**Figure 54.3B**). Thus global air circulation patterns drive the circulation patterns of surface ocean waters, known as *currents*. The trade winds, for example, cause currents to converge at the equator and move westward until they encounter a continental land mass. At that point, the currents divide, with some water moving north and some moving south along continental shores. As these currents move toward the poles, the water veers right in the Northern Hemisphere and left in the Southern Hemisphere, just as the winds do.

Ocean currents transport heat, and thus have a large effect on Earth's climates. The poleward movement of water that has warmed while in the tropics transfers large amounts of heat to high latitudes. The Gulf Stream, for example, brings warm water from the tropical Atlantic Ocean (including the Gulf of Mexico) north across the Atlantic to northern Europe, making the climate there considerably milder than that of corresponding latitudes in North America.

In addition to moving water laterally, the trade winds can also cause water to move vertically to influence climate. As trade winds pull surface water away from an area, deeper colder water can rise to replace it. An *upwelling* is an area where water from depths below 50 meters rises to mix with and replace warmer surface waters. The trade winds drive upwellings at the equator; other wind-driven currents create upwellings along coastlines.

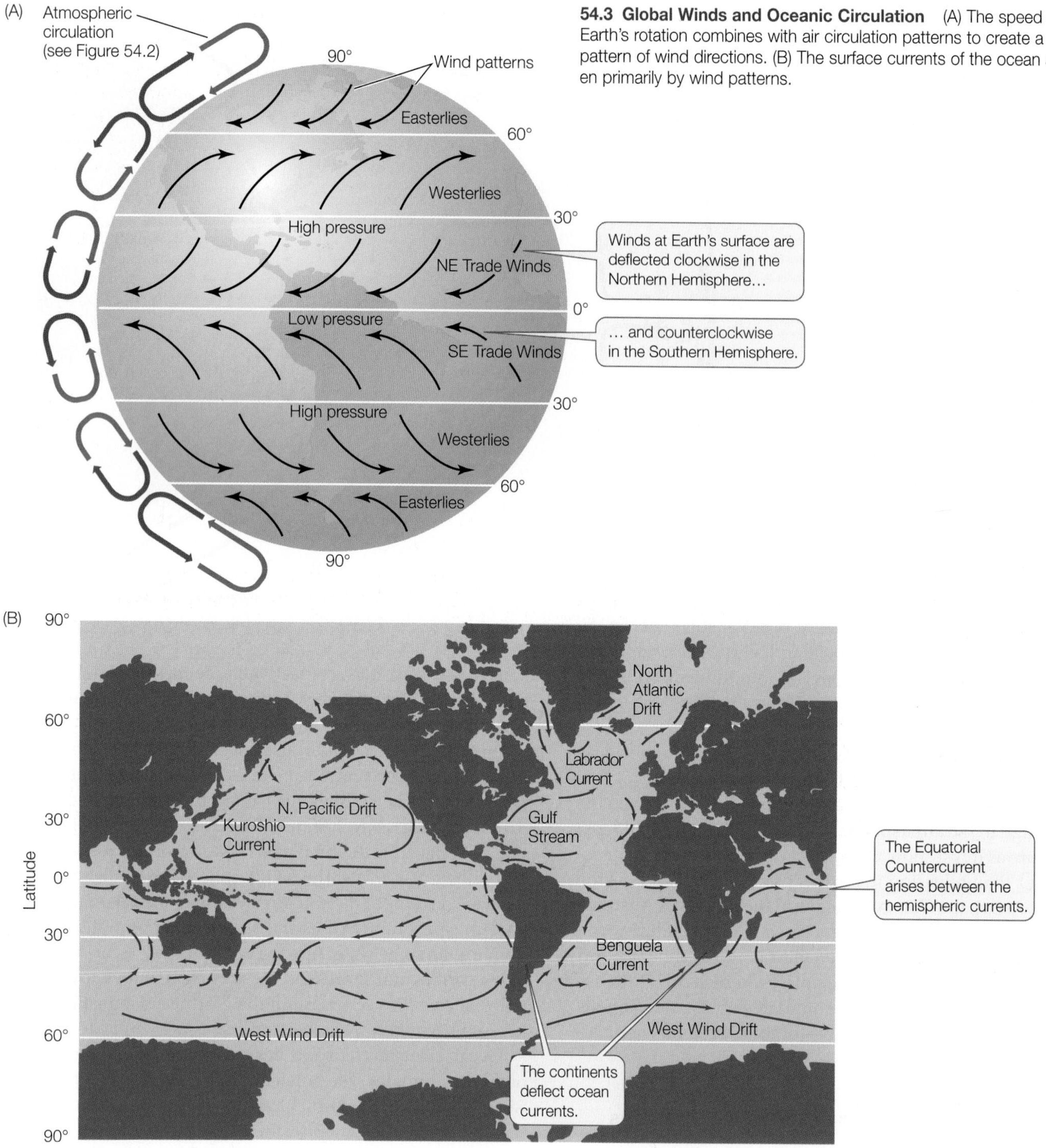

(A) Atmospheric circulation (see Figure 54.2)
90°
Wind patterns
Easterlies
60°
Westerlies
30°
High pressure
NE Trade Winds
Winds at Earth's surface are deflected clockwise in the Northern Hemisphere…
0°
Low pressure
SE Trade Winds
… and counterclockwise in the Southern Hemisphere.
30°
High pressure
Westerlies
60°
Easterlies
90°

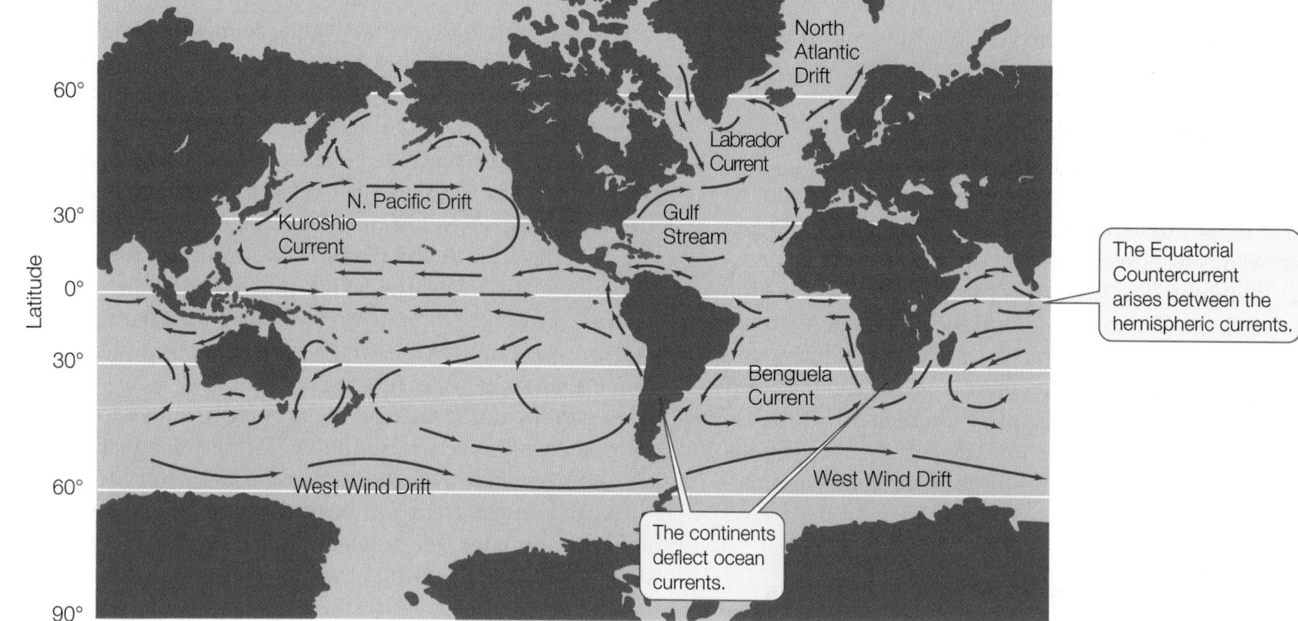


**54.3 Global Winds and Oceanic Circulation** (A) The speed of Earth's rotation combines with air circulation patterns to create a global pattern of wind directions. (B) The surface currents of the ocean are driven primarily by wind patterns.

### Organisms adapt to climatic challenges

The climatic conditions in a region—especially temperature and water availability— act as selective agents on the organisms that live there. As a consequence, many organisms display adaptations to climatic conditions that can involve physiological, morphological, and/or behavioral specializations.

Metabolic specializations help many organisms cope with an environment that periodically becomes too hot, too cold, too dry, or short of food. These specializations are manifested in various forms of resting states that can become a programmed part of the organism's life cycle. *Estivation* (torpor or dormancy during summer), *hibernation* (torpor or dormancy during winter), and *diapause* (the invertebrate equivalent of hibernation) are such states, characterized by dramatically reduced metabolic activity and enhanced physiological resistance to the adverse conditions (see Section 40.4). These states persist until environmental signals indicate that conditions are again conducive to growth.

To deal with extremely low temperatures, cold-hardy organisms are capable of surviving by a variety of strategies, chief among which is producing antifreezes. Microbes, sponges, Arctic fishes, and many temperate-zone insects are among the organisms that produce antifreezes to lower the freezing point of their cell contents or body fluids. One of only two flowering plant species native to Antarctica, the freeze-tolerant hairgrass (*Deschampsia antarctica*) survives by producing proteins that inhibit the formation of damaging large ice crystals. In Alaska, the wood frog (*Rana sylvatica*), the only amphibian found north of the Arctic Circle, survives temperatures of –6°C for over a month with up to 65 percent of its body fluid frozen solid (**Figure 54.4A**). The frog avoids damage to its cells by allowing fluids to freeze in extracellular spaces.

Adaptation to climatic conditions is often reflected in differences in morphology. For example, some endotherms living in cold climates have proportionally rounder shapes and shorter appendages than their relatives adapted to warmer climates, which gives them a smaller surface area and allows them to conserve heat more easily (see Figure 40.14). Pigmentation can also influence rates of heat loss and gain. The larvae of some hornworm moths of the genus *Manduca* are black if they hatch at temperatures below 20°C but are green if the temperature is above that threshold. Black individuals absorb more sunlight and survive at temperatures likely to be lethal to their green conspecifics. When temperature is not an issue, the cryptic green pigmentation is favored because it provides protection from predators (**Figure 54.4B**).

Behavioral mechanisms for temperature regulation often complement physiological and morphological adaptations, particularly among ectotherms. Basking by lizards and evaporative cooling by honeybees are just two examples. Another important behavioral adaptation to climatic conditions is changing one's location to find a more suitable microclimate. A **microclimate** is a subset of climatic conditions in a small specific area that generally differ from those in the environment at large. For example, desert lizards maintain their body temperature by spending time in an underground burrow and moving about the surface only as dictated by environmental conditions (see Figure 40.10). While surface temperatures may fluctuate wildly, the microclimate of the burrow is buffered against such changes.

In addition to such localized movements, long-distance movements can be key adaptations to climatic challenges. Many organisms seek new places to live when local conditions deteriorate. If repeated seasonal changes alter an environment in predictable ways, organisms may evolve life cycles that appear to anticipate those changes. *Migration*, one response to such cyclic environmental changes is discussed in Section 53.5.

Most changes in the abiotic environment, both short-term and long-term, happen independently of anything organisms do. Storms rage, the seasons come and go, and glaciers advance and retreat uninfluenced by the organisms that must deal with their effects. Other environmental changes, however, are influenced by the activities of organisms. As we will see in subsequent chapters, many significant changes in the environment to which organisms must adapt are caused by other organisms, and many characteristics of organisms have been shaped by the history of such interactions.

## 54.2 RECAP

**Latitudinal differences in solar energy input create patterns of atmospheric circulation, and those prevailing winds drive oceanic circulation. Organisms deal with climatic challenges by means of physiological, morphological, and behavioral adaptations.**

- How does latitudinal variation in solar energy input drive global air circulation patterns? See pp. 1142–1143 and Figure 54.2
- How do global air circulation patterns drive ocean currents? See p. 1143 and Figure 54.3

The tremendous variation in Earth's climates has given rise to many different assemblages of organisms. Ecologists have found it useful to classify and name environments based on their ecological similarities.

(A) *Rana sylvatica*

(B) *Manduca quinquemaculata*

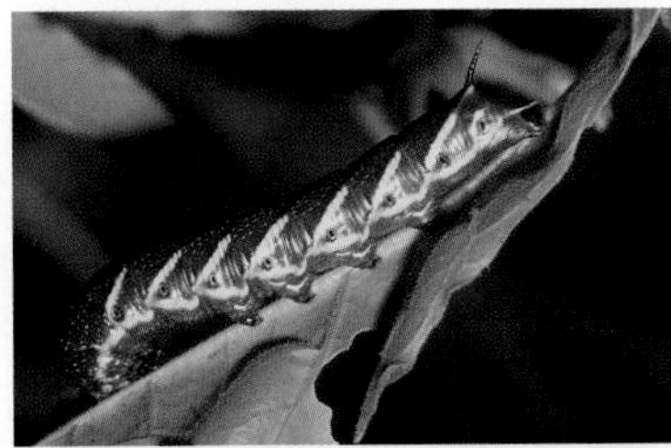

**54.4 Evolutionary Adaptations to Climate** (A) Wood frogs can survive frigid Alaskan winters by allowing up to 65 percent of their body fluids to freeze. (B) The larvae of some hornworm moths (genus *Manduca*) have adaptive pigmentation. The black phenotype allows the caterpillar to absorb sunlight more efficiently and survive at cooler temperatures; the cryptic green phenotype is favored if temperature is not an issue.

## 54.3 What Is a Biome?

A **biome** is an environment that is defined by its climatic and geographic attributes and characterized by ecologically similar organisms, particularly its dominant plants. By providing three-dimensional structure and by modifying microclimates near the ground, the dominant plants of a terrestrial environment strongly influence the lives of the other organisms living there. In contrast, aquatic environments, discussed in Section 54.5, are less dependent on plants for their structure.

The same biome may be present in several widely separated places across the globe, depending in large part on the distribution of suitable climatic conditions (**Figure 54.5**). The identities of the species in geographically separate regions may differ, but through convergent evolution in response to similar selective forces (see Section 22.1), organisms in similar biomes are likely to share many physiological, morphological, and behavioral adaptations.

The distributions of biomes are determined largely by annual patterns of temperature and rainfall. In some biomes, such as temperate deciduous forest, precipitation is relatively constant throughout the year, but temperature varies strikingly between summer and winter. In other biomes, both temperature and precipitation change seasonally. In the tropics, seasonal temperature fluctuations are small and annual cycles are dominated by wet and dry seasons. Tropical biome types are determined primarily by the length of the dry season.

The descriptions of biomes presented here are very general and fall far short of encompassing the variation that can be found within each biome. Furthermore, the boundaries between biomes tend to be arbitrary. Although sometimes an abrupt change is apparent in a landscape, more often one biome gradually merges into another. Despite these uncertainties, recognizing the major biomes of the world is useful inasmuch as these environments share ecological attributes irrespective of their location.

yourBioPortal.com
GO TO Animated Tutorial 54.1 • Biomes

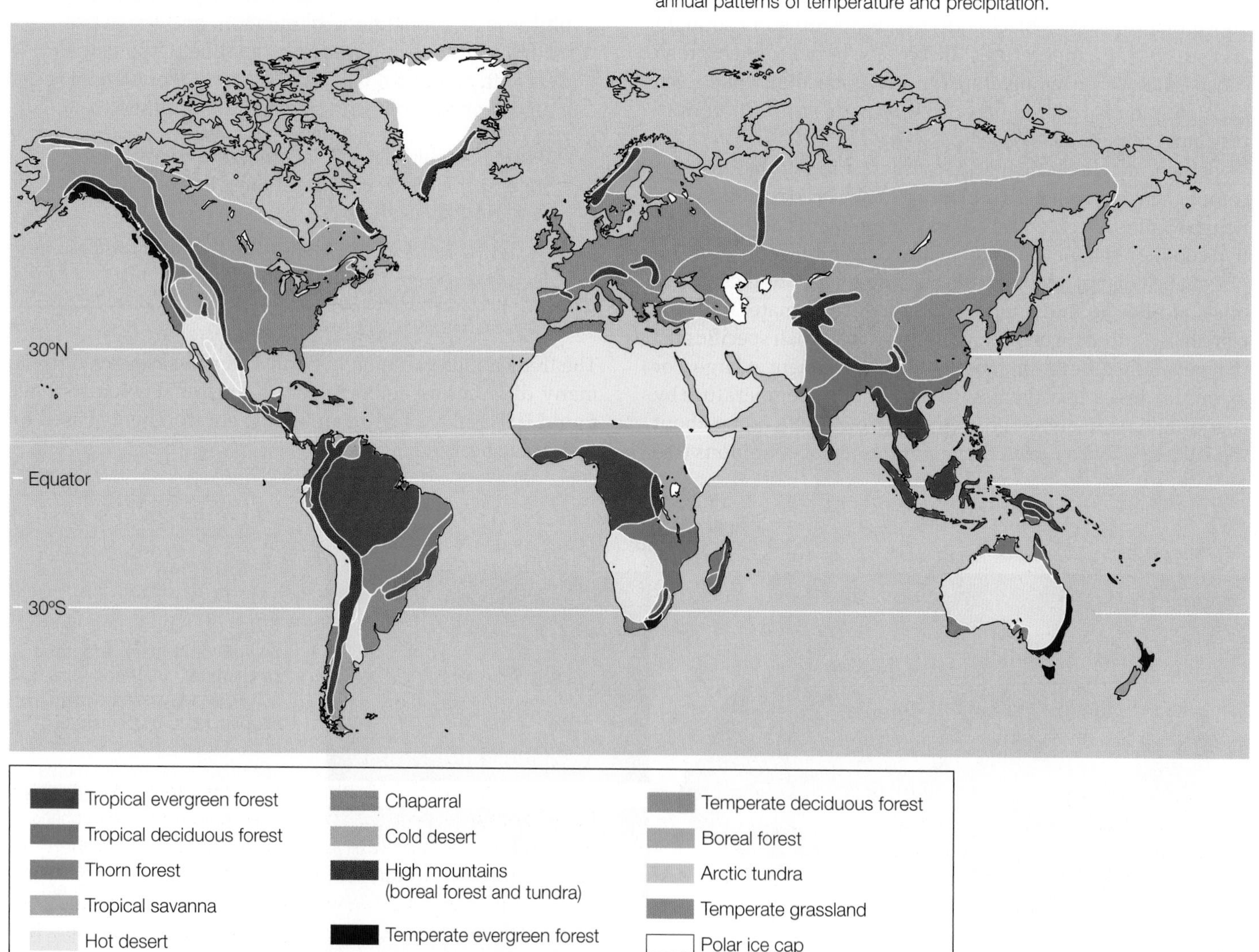


**54.5 The Global Distribution of Biomes** The distribution of biomes is based primarily on vegetation patterns and is strongly influenced by annual patterns of temperature and precipitation.

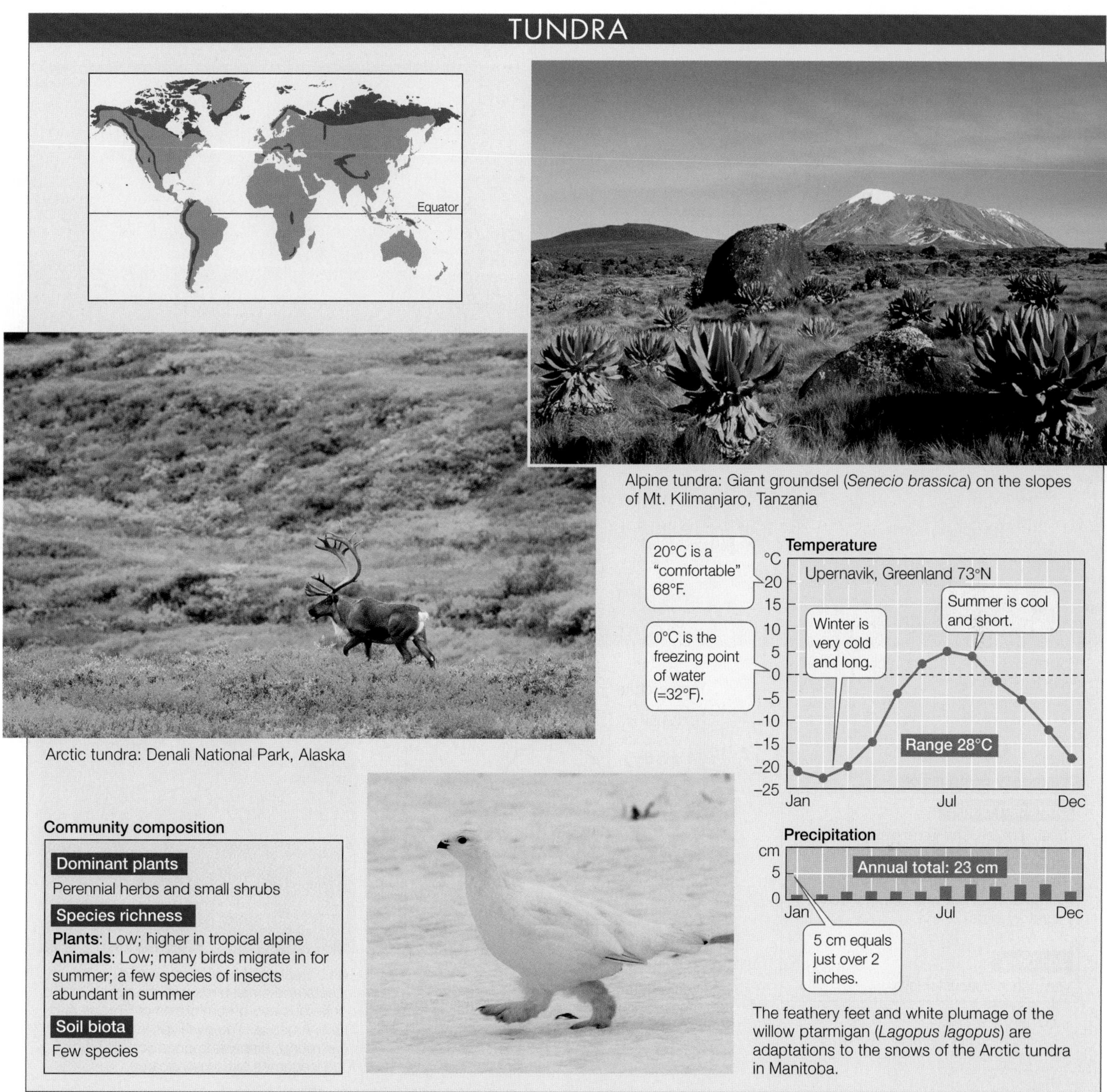


Alpine tundra: Giant groundsel (*Senecio brassica*) on the slopes of Mt. Kilimanjaro, Tanzania

Arctic tundra: Denali National Park, Alaska

The feathery feet and white plumage of the willow ptarmigan (*Lagopus lagopus*) are adaptations to the snows of the Arctic tundra in Manitoba.

### Tundra is found at high latitudes and high elevations

The **tundra** biome is found in the Arctic and at high elevations in mountains at all latitudes. In Arctic tundra, the vegetation consists of low-growing perennial plants and is underlain by *permafrost*—soil permeated with permanently frozen water. The top few centimeters of the soil thaw during the short summers, when the sun may be above the horizon 24 hours a day. Even though there is little precipitation, lowland Arctic tundra is wet because water cannot drain through the permafrost.

Alpine tundra is found at high elevations outside polar regions. *Tropical alpine tundra* is not underlain by permafrost, so photosynthesis and most other biological activities continue (albeit slowly) throughout the year. A variety of plant growth forms are present in tropical alpine tundra, including low-growing shrubs, rosette perennials, and grasses.

Many tundra plants have hairy leaves that trap heat; the flowers of some species move over the course of the day, tracking the sun's warmth. Most animals are either summer migrants or are dormant for much of the year. Resident birds and mammals, such as the ptarmigan and Arctic fox, have thick fur or feathers that may change color from brown in summer to white in winter.

## BOREAL and TEMPERATE EVERGREEN FOREST

Equator

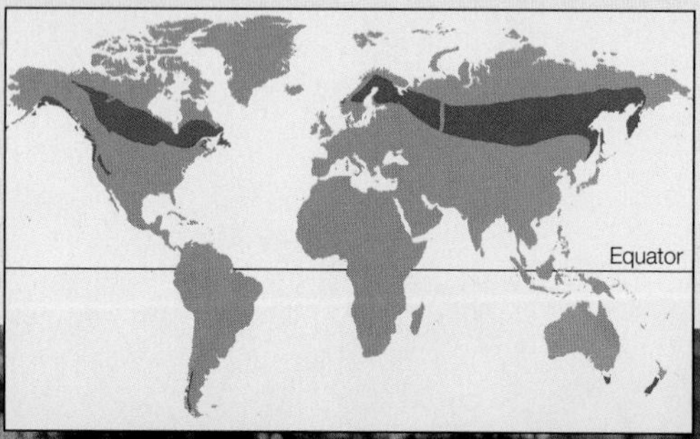

Southern beech (*Nothofagus*) forest, Fiordland National Park, New Zealand

Bull moose (*Alces alces*) in boreal forest, Alberta, Canada

**Community composition**

**Dominant plants**

Trees, shrubs, and perennial herbs

**Species richness**

**Plants**: Low in trees, higher in understory
**Animals**: Low, but with summer peaks in migratory birds

**Soil biota**

Very rich in deep litter layer

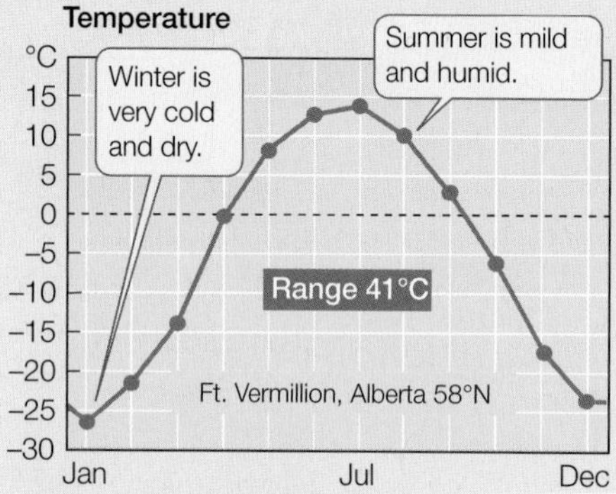

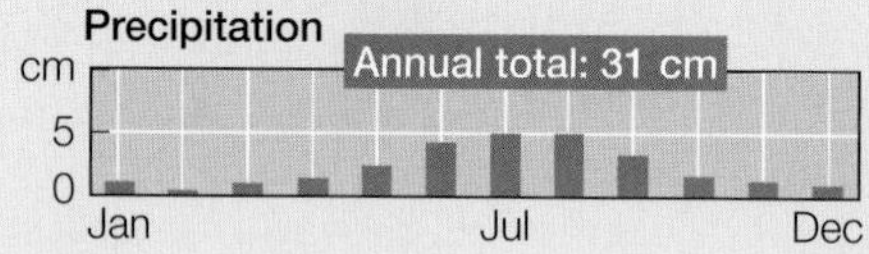

**Temperature**

Winter is very cold and dry.

Summer is mild and humid.

Range 41°C

Ft. Vermillion, Alberta 58°N

**Precipitation**

Annual total: 31 cm

The boreal owl (*Aegolius funereus*) feeds on insects and small birds and mammals of the coniferous evergreen forests of Europe and North America. The owl lives in the forests year-round, dispersing occasionally when food becomes extremely scarce.

### Evergreen trees dominate boreal forests

The **boreal forest** biome (also known as *taiga*) occurs at latitudes below Arctic tundra and at elevations below alpine tundra on temperate-zone mountains. Winters are long and very cold; summers are short, although often relatively warm. The short summer favors trees with evergreen leaves that are ready to photosynthesize as soon as temperatures warm.

The evergreen forests of the Northern Hemisphere are dominated by coniferous gymnosperm species such as spruce and fir. The leaves of conifers are needlelike rather than flat; their reduced surface area cuts down on evaporative water loss. In the Southern Hemisphere forests, the dominant trees are southern beeches (*Nothofagus*), some of which are evergreen.

The dominant mammals of the boreal forest, such as moose and hares, eat leaves, but the seeds in conifer cones support a variety of rodents, birds, and insects. While some small mammals hibernate in winter, voles, lemmings, and mice remain active in snow tunnels, providing food for predators such as foxes and owls.

**Temperate evergreen forests** grow along the coasts of continents in both hemispheres at middle to high latitudes, where winters are mild but wet and summers are cool and dry.

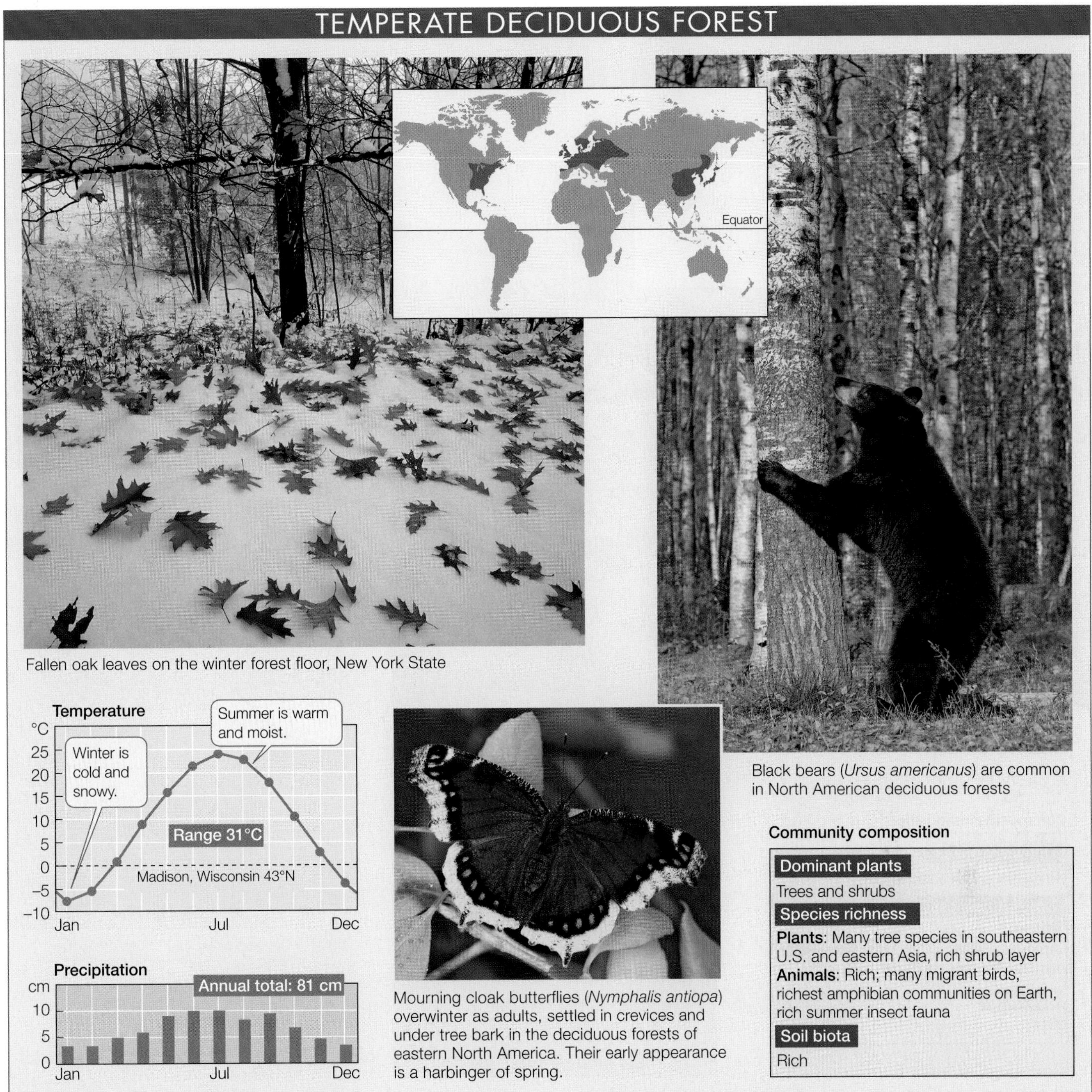


Fallen oak leaves on the winter forest floor, New York State

Black bears (*Ursus americanus*) are common in North American deciduous forests

Mourning cloak butterflies (*Nymphalis antiopa*) overwinter as adults, settled in crevices and under tree bark in the deciduous forests of eastern North America. Their early appearance is a harbinger of spring.

### Temperate deciduous forests change with the seasons

The **temperate deciduous forest** biome is found in eastern North America, eastern Asia, and Europe. Temperatures in these regions fluctuate dramatically between summer and winter, although precipitation is fairly evenly distributed throughout the year. *Deciduous* trees, which dominate these forests, lose their leaves during the cold winters and produce new leaves that photosynthesize rapidly during the warm, moist summers.

Many more tree species live here than in boreal forests. The temperate forests richest in species are those of the southern Appalachian Mountains of the United States and the ones found in eastern China and Japan—areas that were not covered by glaciers during the Pleistocene. Many plant genera are shared among the three geographically separate regions with a deciduous forest biome.

Although many animals are permanent residents of deciduous forests, some (including many birds) migrate to escape the winter cold. Others that remain through the winter acquire massive fat stores in fall and hibernate, often in underground burrows. Many resident insects pass the winter in a state of diapause, the onset of which is triggered by a decreasing number of daylight hours, a reliable predictor of the onset of winter.

## TEMPERATE GRASSLANDS

Przewalski's horse in Khustain Nuruu National Park, Mongolia

American bison herd grazing shortgrass prairie, North Dakota

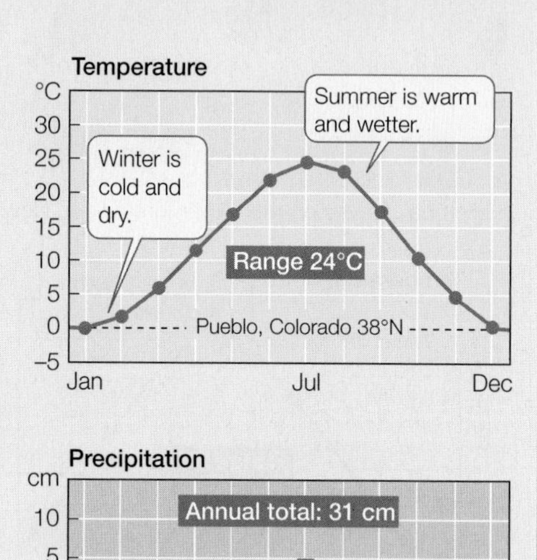


**Community composition**

**Dominant plants**
Perennial grasses and forbs

**Species richness**
**Plants**: Fairly high
**Animals**: Relatively few birds because of simple structure; mammals fairly rich

**Soil biota**
Rich

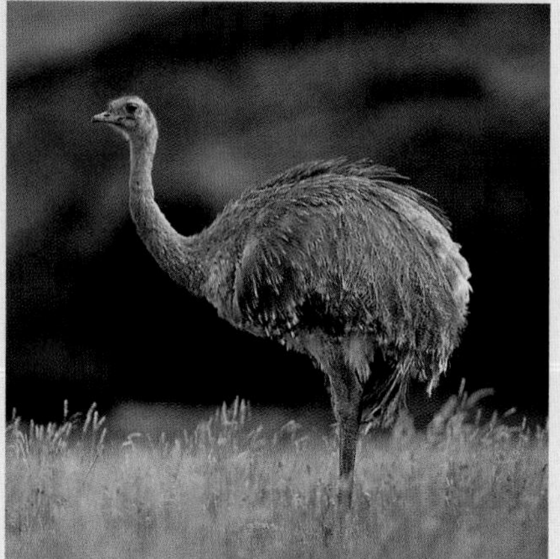

The flightless Darwin's rhea (*Rhea pennata*) is adapted to grazing the grasslands of the Andean plateau (the Altiplano) of Chile and Peru. It is also found on the pampas of Argentina.

### Temperate grasslands are widespread

**Temperate grasslands** are found in many parts of the world, all of which are relatively dry for much of the year. Most grasslands, such as the pampas of Argentina, the veldt of South Africa, and the Great Plains of North America, have hot summers and relatively cold winters. In some grasslands, most of the precipitation falls in winter (as in California grasslands); in others, the majority falls in summer (as in the Great Plains and the Russian steppe).

Grassland vegetation is structurally simple but rich in species of perennial grasses and *forbs* (herbaceous plants other than grasses). This abundance of plant biomass supports herds of large grazing mammals. Grassland plants are adapted to grazing and to fire. They store much of their energy underground and sprout quickly after being burned or grazed. There are comparatively few trees in temperate grasslands because they cannot survive the periodic fires.

The topsoil of grasslands is usually rich and deep, and thus exceptionally well suited to growing crops, particularly grains such as corn and wheat. As a consequence, most of the world's temperate grasslands have been turned over to agriculture and no longer exist in their natural state.

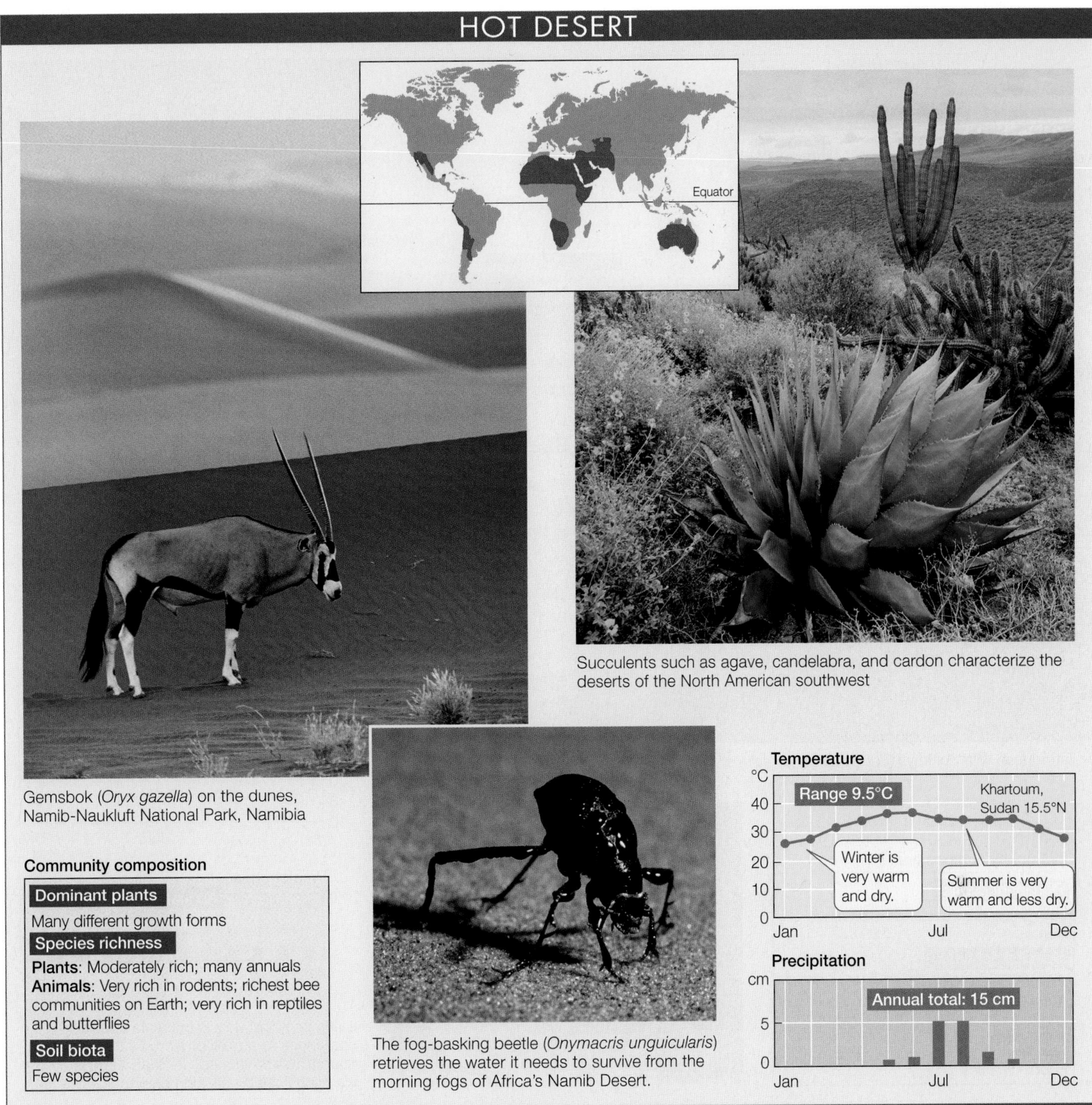


Gemsbok (*Oryx gazella*) on the dunes, Namib-Naukluft National Park, Namibia

Succulents such as agave, candelabra, and cardon characterize the deserts of the North American southwest

The fog-basking beetle (*Onymacris unguicularis*) retrieves the water it needs to survive from the morning fogs of Africa's Namib Desert.

### Hot deserts form around 30° latitude

The **hot desert** biome is found in two belts, centered around latitudes 30°N and 30°S (see Figure 54.2). The driest regions, where rains rarely penetrate, are far from the oceans. Examples of this are in the center of Australia and the middle of the Sahara in Africa.

The structural features of desert plants reflect the need to conserve water. These include taproots that reach water far below the surface; a shallow but extensive root system that absorbs water quickly after rare rain events; and small, wax-covered leaves that open their stomata only at night. Most perennials go dormant during dry seasons, growing quickly as soon as rains return. Their seeds tend to be heat- and drought-resistant and accumulate in a dormant state in the soil.

Small animals are inactive during the hottest part of the day, remaining in underground burrows. When the heat becomes extreme, many small invertebrates estivate. Desert mammals have physiological adaptations for conserving water, including reduced number of sweat glands and kidneys that produce highly concentrated urine. Many desert animals require no water beyond what they can extract from the carbohydrates in their food.

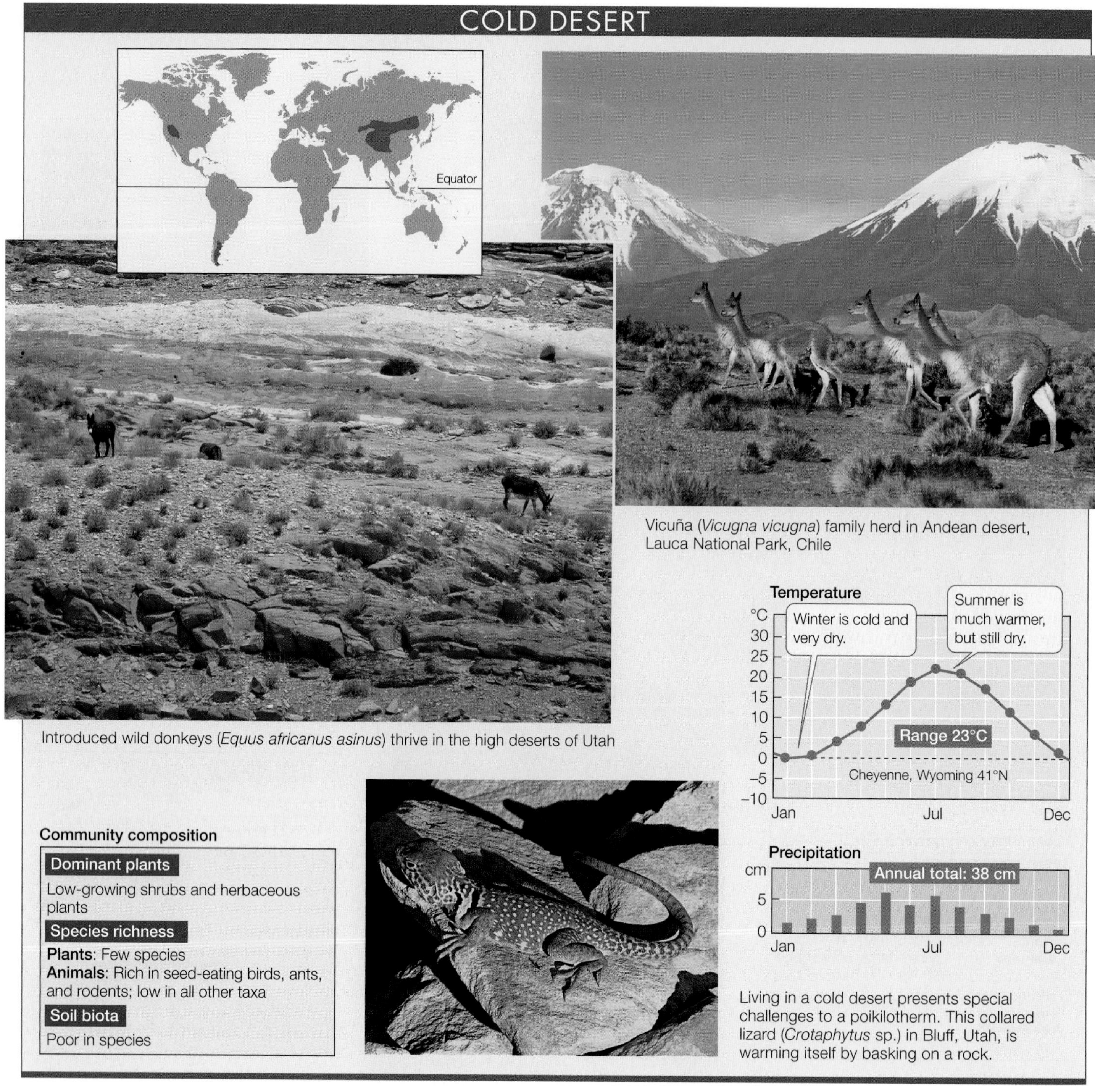


Vicuña (*Vicugna vicugna*) family herd in Andean desert, Lauca National Park, Chile

Introduced wild donkeys (*Equus africanus asinus*) thrive in the high deserts of Utah

Living in a cold desert presents special challenges to a poikilotherm. This collared lizard (*Crotaphytus* sp.) in Bluff, Utah, is warming itself by basking on a rock.

### Cold deserts are high and dry

The **cold desert** biome is found in dry regions at mid- to high latitudes, especially in the interiors of continents in the rain shadows of mountain ranges. Seasonal changes in temperature are great.

Cold deserts are dominated by a few species of low-growing shrubs. The surface layers of the soil are recharged with moisture in winter, and plant growth is concentrated in spring. Cold deserts are relatively poor in species diversity, but the plants that do grow there tend to produce large numbers of seeds, supporting many species of seed-eating birds, ants, and rodents. Burrowing behavior is widespread among cold desert dwellers, but, in contrast with hot desert animals, they burrow to escape cold temperatures, not excessive heat.

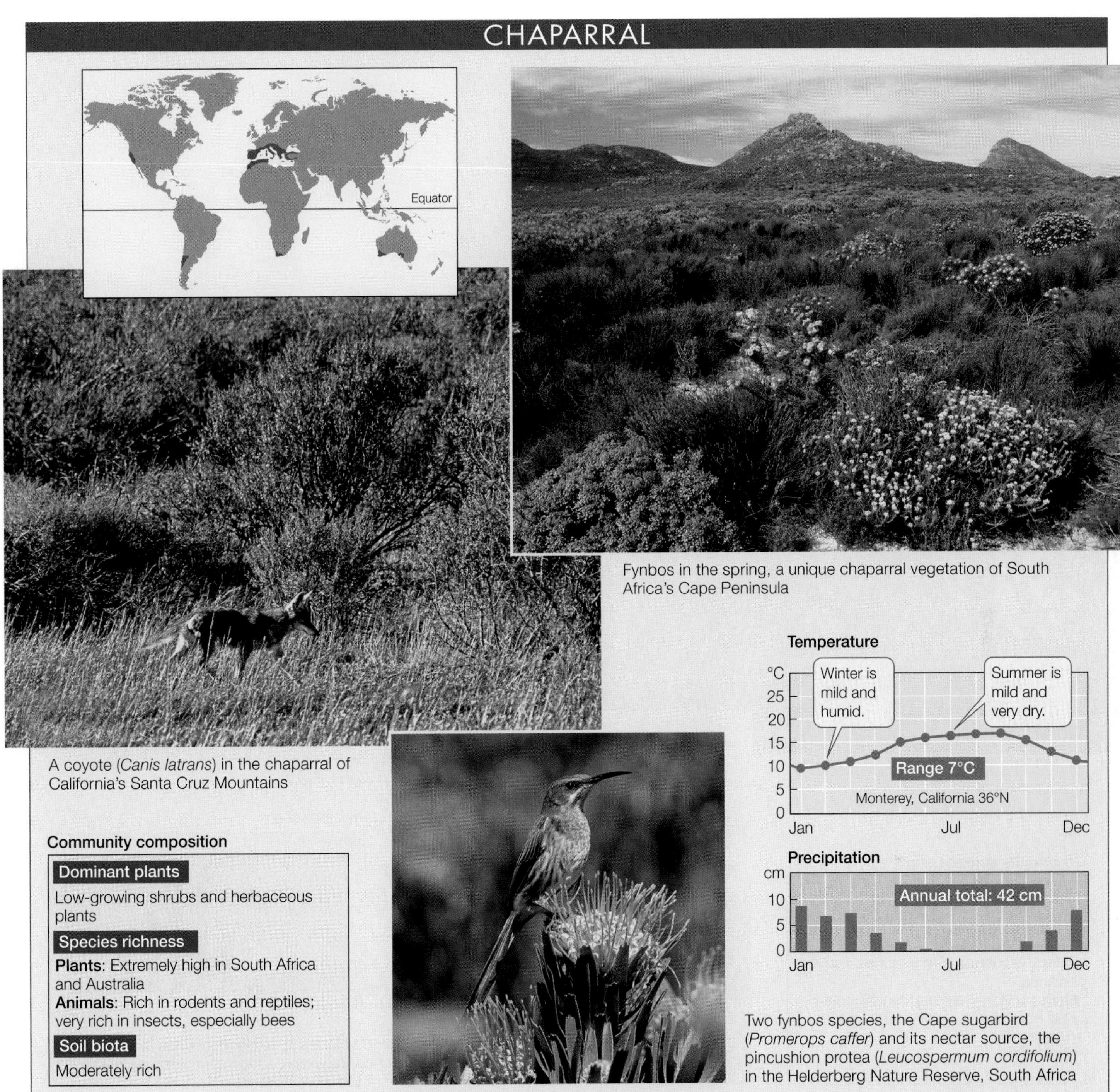


Fynbos in the spring, a unique chaparral vegetation of South Africa's Cape Peninsula

A coyote (*Canis latrans*) in the chaparral of California's Santa Cruz Mountains

Two fynbos species, the Cape sugarbird (*Promerops caffer*) and its nectar source, the pincushion protea (*Leucospermum cordifolium*) in the Helderberg Nature Reserve, South Africa

**Community composition**

**Dominant plants**
Low-growing shrubs and herbaceous plants

**Species richness**
**Plants**: Extremely high in South Africa and Australia
**Animals**: Rich in rodents and reptiles; very rich in insects, especially bees

**Soil biota**
Moderately rich

### Chaparral has hot, dry summers and wet, cool winters

The **chaparral** biome is found on the western sides of continents at mid-latitudes (around 40°) where cool ocean currents flow offshore. Winters in this biome are cool and wet; summers are warm and dry. Such climates are found in the Mediterranean region of Europe, coastal California, central Chile, extreme southern Africa, and southwestern Australia.

The dominant plants of chaparral vegetation are low-growing shrubs and trees with tough evergreen leaves that conserve water. The shrubs carry out most of their growth and photosynthesis in early spring, when insects are active and birds breed. Many chaparral species produce strong-smelling defensive chemicals to reduce losses of hard-to-replace foliage to herbivores. Annual plants are abundant and produce large quantities of seeds that fall onto the soil, supporting many small rodents, most of which store seeds in underground burrows. Burrowing to avoid midday heat and nocturnal foraging are strategies used by many chaparral animals. Chaparral vegetation is adapted to periodic fires; the seeds of some species do not germinate until after they have survived a fire. Many shrubs of Northern Hemisphere chaparral produce bird-dispersed fruits that ripen in late fall, when large numbers of migrant birds arrive from the north.

Plains zebra (*Equus quagga*) in large herd of blue wildebeest (*Connochaetes taurinus*) on the Tanzania savanna, Africa

*Ocotillo* species native to Madagascar dominate this thorn forest

Termite colonies build huge mounds on the African savanna, providing a food source for mammals such as this baboon (*Papio* sp.).

### Thorn forests and tropical savannas have similar climates

The **thorn forest** and **tropical savanna** biomes are found at latitudes below the hot deserts of Africa, South America, and Australia. Little or no rain falls in winter, but rainfall may be heavy during summer. Thorn forests contain many plants similar to those found in hot deserts. The dominant plants are spiny shrubs and small trees, many of which drop their leaves during the long, dry winter. Trees of the genus *Acacia* are common in thorn forests and savannas worldwide. In Africa, *Andansonia* (baobob) trees are also a hallmark of this biome.

Savanna is characterized by expanses of grasses and grasslike plants with scattered individual trees. The largest tropical savannas are found in central and eastern Africa, where they are populated by herds of grazing and browsing mammals and the large carnivores that prey on them. Grazers and browsers maintain the savannas. If savanna vegetation is not grazed, browsed, or burned, it typically reverts to dense thorn forest. The migration of vast herds of African herbivores in search of "greener pastures" during the dry season is another characteristic of this impressive region.

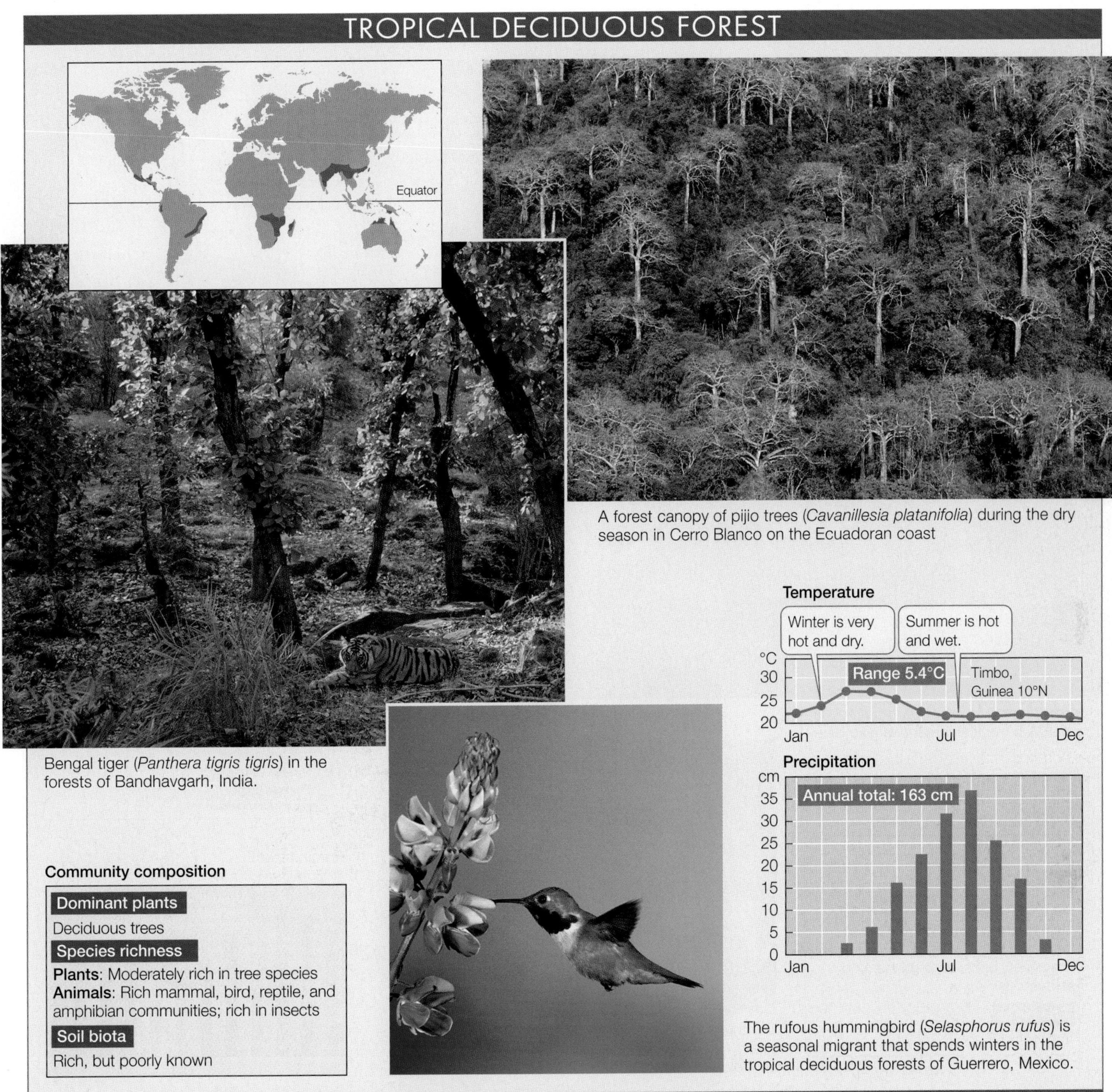


A forest canopy of pijio trees (*Cavanillesia platanifolia*) during the dry season in Cerro Blanco on the Ecuadoran coast

Bengal tiger (*Panthera tigris tigris*) in the forests of Bandhavgarh, India.

The rufous hummingbird (*Selasphorus rufus*) is a seasonal migrant that spends winters in the tropical deciduous forests of Guerrero, Mexico.

### Tropical deciduous forests occur in hot lowlands

As the length of the rainy season increases toward the Equator, the **tropical deciduous forest** biome replaces thorn forest. Tropical deciduous forests have taller trees and fewer succulent plants than thorn forests, and they support a much greater number of plant and animal species. Most of the trees, except for those growing along rivers, lose their leaves during the long, hot dry season. Activity picks up in the rainy season; many plants flower while they are still leafless.

Most plant species in this biome are pollinated by animals. In the Sierra Madre Occidental, a mountain range in the extreme southwestern United States extending into western Mexico, tropical deciduous forests are part of a "nectar corridor": a series of patches of flowering plants that provide refueling stops for long-distance migrants traveling north from overwintering sites to breeding sites in the Rocky Mountains. Among the migratory pollinators that fly this corridor are lesser long-nosed bats, white-winged doves, and rufous hummingbirds.

The soils of this biome are among the best in the tropics for agriculture because they contain more nutrients than the soils of wetter areas. As a result, most tropical deciduous forests worldwide have been cleared for agriculture and grazing.

## TROPICAL EVERGREEN FOREST

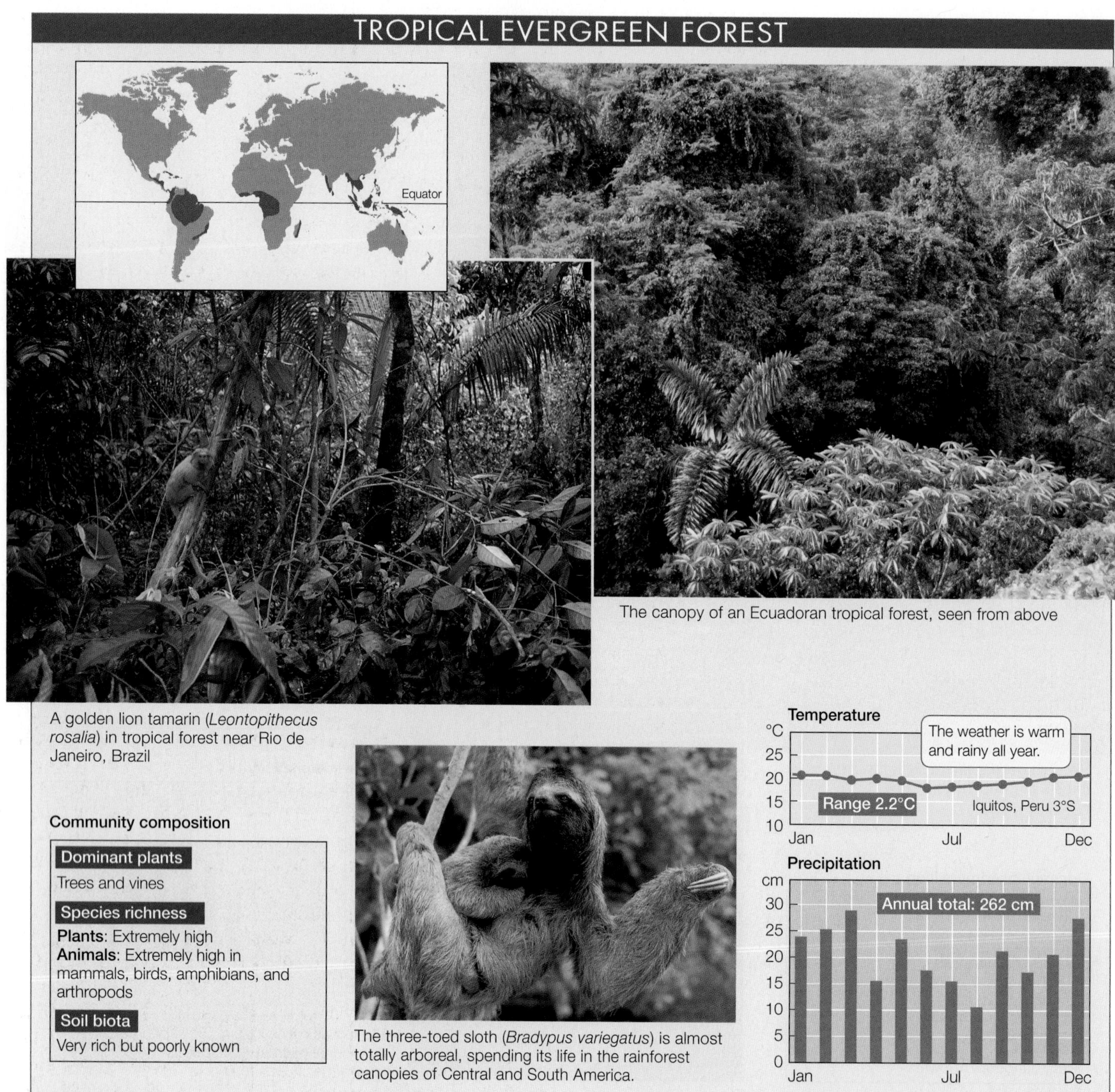


The canopy of an Ecuadoran tropical forest, seen from above

A golden lion tamarin (*Leontopithecus rosalia*) in tropical forest near Rio de Janeiro, Brazil

The three-toed sloth (*Bradypus variegatus*) is almost totally arboreal, spending its life in the rainforest canopies of Central and South America.

**Community composition**

**Dominant plants**
Trees and vines

**Species richness**
**Plants**: Extremely high
**Animals**: Extremely high in mammals, birds, amphibians, and arthropods

**Soil biota**
Very rich but poorly known

### Tropical evergreen forests are rich in species

The **tropical evergreen forest** biome (commonly called the rainforest) is found in Equatorial regions where total rainfall exceeds 250 centimeters annually and the dry season lasts no longer than 2 or 3 months. With no seasons unsuitable for growth, it is the most species-rich of all biomes, with up to 500 species of trees per km$^2$. Although these forests cover less than 2 percent of Earth's surface, they are home to over half of all known species.

Along with the immense number of species they support, tropical evergreen forests have the highest overall productivity of all ecological communities. However, most mineral nutrients are tied up in the vegetation. The soils usually cannot support agriculture without massive applications of fertilizers. These forests are home to many *epiphytes*, plants that grow on other plants, deriving their nutrients and moisture from air and water rather than soil.

Tropical evergreen forests provide humans with a dazzling range of products, including fruits, nuts, medicines, fuel, pulp, and furniture wood. Many more useful species undoubtedly await discovery, as only a small proportion of this biome's species have been inventoried. The rainforests, however, are currently being deforested or converted to agriculture at a rate of almost 20 million hectares per year.

Major topographic features, such as mountains or large lakes, may have regional effects on temperature and precipitation. When prevailing winds bring air masses into contact with a mountain range, for example, the air rises to pass over the mountains, expanding and cooling as it does so. Thus clouds frequently form on the windward side of mountains (the side facing into the winds) and release moisture as rain or snow. On the leeward side (opposite from the direction of the winds), the now-dry air descends, warms, and once again picks up moisture. This pattern often results in a dry area called a **rain shadow** on the leeward side of a mountain range (**Figure 54.6**).

### Biome distribution is not determined solely by temperature

Although warm temperatures and high precipitation support greater numbers of species than do colder, drier conditions, other climatic factors—particularly soil characteristics and the occurrence of wildfires—also influence the structure and life cycles of the dominant vegetation in an area and, consequently, the ecological attributes of the other organisms living there. For example, the plants of Australian deserts grow on extremely nutrient-poor soils and thus have difficulty growing new foliage. Such plants often protect their leaves against consumers by producing large quantities of digestibility-reducing chemicals. Such leaves are generally left uneaten, and when they senesce they drop to the ground, providing an abundance of highly flammable litter to feed the intense fires that periodically sweep across the landscape. As a result, succulent plants, which are easily killed by fires, are not found in Australia, although they are common in deserts on other continents (see p. 1151).

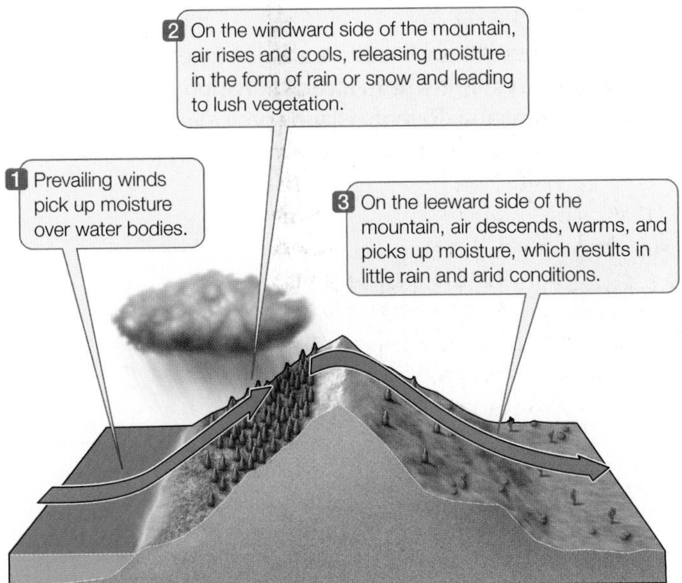

2 On the windward side of the mountain, air rises and cools, releasing moisture in the form of rain or snow and leading to lush vegetation.
1 Prevailing winds pick up moisture over water bodies.
3 On the leeward side of the mountain, air descends, warms, and picks up moisture, which results in little rain and arid conditions.


**54.6 A Rain Shadow** Mean annual precipitation tends to be lower on the leeward side of a mountain range than on the windward side.

**yourBioPortal.com**
GO TO Animated Tutorial 54.2 • Rain Shadow

## 54.3 RECAP

**Ecologists recognize a number of terrestrial environment types called biomes. The geographic distribution of biomes is determined primarily by temperature and precipitation, but is also influenced by soil characteristics and fire.**

- How do temperate grasslands differ from tropical savannas? See pp. 1150 and 1154
- What primary factor distinguishes a tropical biome? See pp. 1154–1156

Climatic conditions explain why chaparral vegetation is confined to the western coasts of continents at mid-latitudes, but it cannot explain why the chaparral in California and the chaparral in South Africa have no species in common. To explain the distribution of species on Earth, another aspect of the abiotic environment—geological history—must be taken into account.

## 54.4 What Is a Biogeographic Region?

Climate interacts with local abiotic features to influence where and how organisms live, but these are not the only factors that determine where organisms can be found. Evolutionary history—where and when groups of organisms originated and diverged—is key to determining the distributions of organisms, which in turn is profoundly influenced by geological history.

### Geological history influenced the distribution of organisms

Until European naturalists traveled the globe, they had no way of knowing how organisms were distributed elsewhere. Alfred Russel Wallace, who along with Charles Darwin advanced the idea that natural selection could account for the evolution of life on Earth (see Section 21.1), was one of those global travelers. Wallace returned to England in 1862 after seven years in the Malay Archipelago, where he noticed some remarkable patterns in the distributions of organisms. For example, he described the dramatically different birds that inhabited the adjacent islands Bali and Lombok:

> In Bali we have barbets, fruit-thrushes and woodpeckers; on passing over to Lombock these are seen no more, but we have an abundance of cockatoos, honeysuckers, and brush-turkeys, which are equally unknown in Bali, or any island further west. The strait here is fifteen miles wide, so that we may pass in two hours from one great division of the earth to another, differing as essentially in their animal life as Europe does from America.

Wallace pointed out that these differences could not be explained by climate or by soil characteristics, because in those respects Bali and Lombok are essentially identical.

Wallace saw that, based on the distributions of plant and animal species, he could draw a line that divided the Malay Archipelago into two distinct halves (**Figure 54.7**). He correctly de-

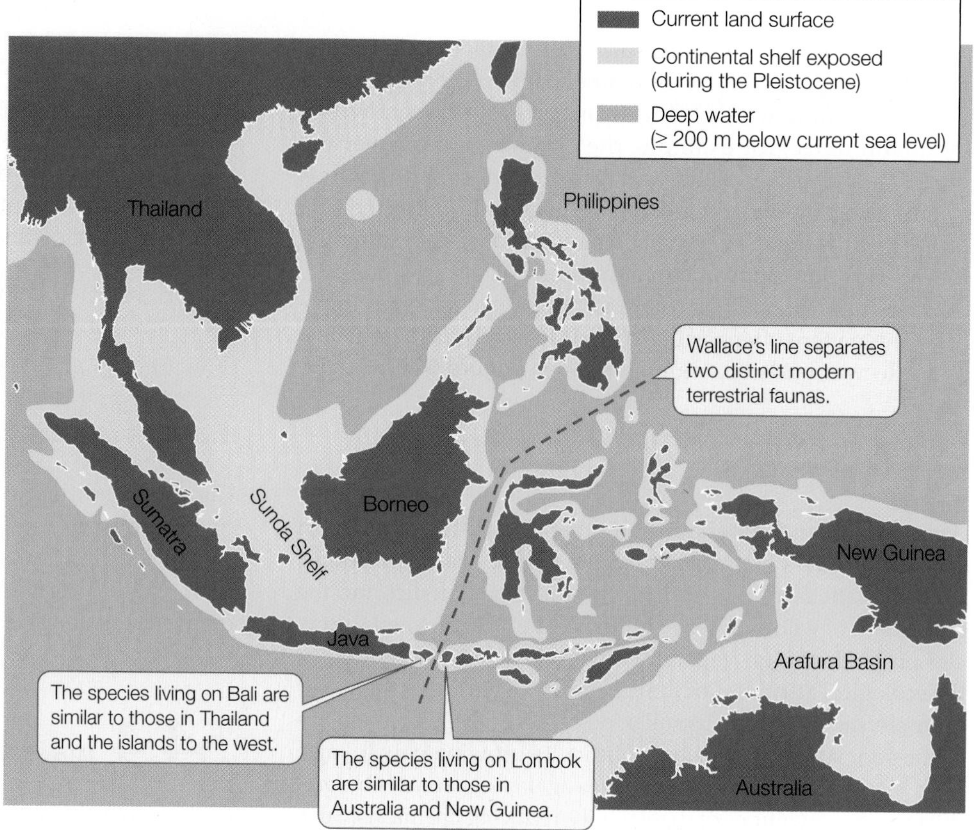


**54.7 Wallace's Line** Wallace's line corresponds to a deep-water channel between the islands of Bali and Lombok. This channel would have blocked the movement of terrestrial organisms even during the Pleistocene glaciations, when sea level was 100 meters lower than it is today.

duced that the dramatic differences in flora and fauna were related to the depth of the channel separating Bali and Lombok. This channel is so deep that it would have remained full of water, and thus would have been a barrier to the movement of terrestrial animals, even during the glaciations of the Pleistocene era, when sea level dropped more than 100 m and Bali and the islands to the west were connected to the Asian mainland.

With these insights, Wallace established the conceptual foundations of **biogeography**, the scientific study of the patterns of distribution of populations, species, and ecological communities across Earth. In *The Geographical Distribution of Animals*, published in 1876, he detailed the factors known at the time that influence the distributions of animals, including past glaciation, land bridges, deep ocean channels, and mountain ranges. He earned some measure of scientific immortality in that the Malay discontinuity that first piqued his curiosity is known to this day as "Wallace's line."

The biotas of different parts of the world differ enough to allow us to divide Earth into several continental-scale areas, called **biogeographic regions** (**Figure 54.8**), each containing characteristic assemblages of species occupying many different biomes. The boundaries of these biogeographic regions are drawn where assemblages of species change dramatically over short distances. Wallace's line, for example, separates the Oriental biogeographic region from the Australasian biogeographic region. The biotas of biogeographic regions differ because oceans, mountains, deserts, and other barriers restrict the dispersal of organisms from one region to another. Although organisms do disperse between adjacent biogeographic regions, such interchanges have not been frequent or massive enough to eliminate the striking differences between them.

**yourBioPortal.com**
GO TO **Web Activity 54.1 • Major Biogeographic Regions**

Most species are confined to a single biogeographic region. A species found only within a certain region is said to be **endemic** to that region. Remote islands typically have distinctive endemic biotas because water barriers limit opportunities for

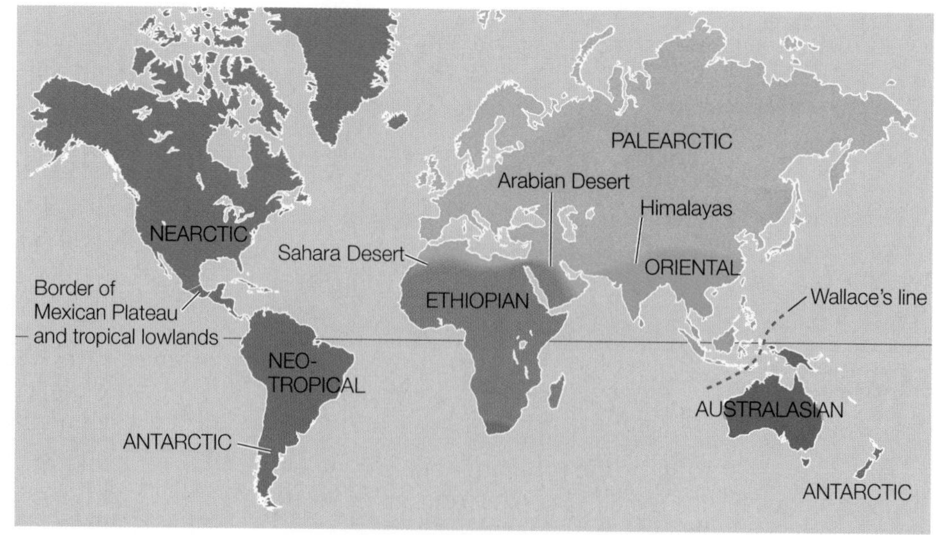


**54.8 Earth's Biogeographic Regions** The major biogeographic regions are separated by climatic, topographic, and/or aquatic barriers to dispersal that cause their biotas to differ strikingly from one another.

**54.9 Madagascar's Endemic Species** The majority of the vascular plant and vertebrate species found on the island of Madagascar are found nowhere else on Earth. Unique reptile and insect species also abound there.

*Adansonia grandidieri* (giant baobab tree)

*Lemur catta* (ring-tailed lemur) climbing *Alluaudia procera* (Madagascar ocotillo)

*Chamaeleo parsonii* (chameleon)

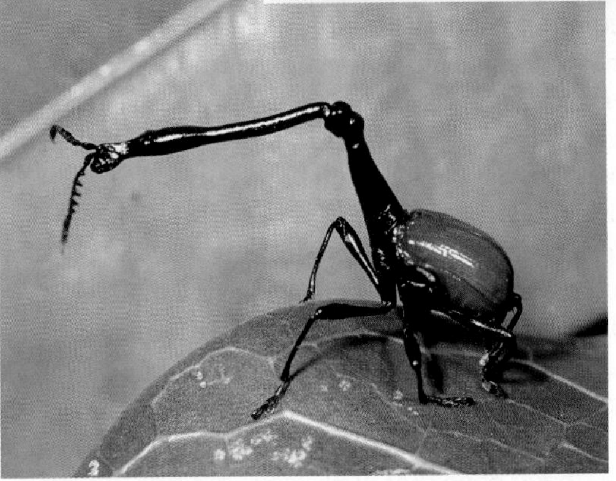

*Trachelophorus giraffa* (giraffe-necked weevil)

most species to colonize them. For example, nearly all the species of vascular plants and vertebrates on Madagascar, a large island 400 km off the eastern coast of Africa, are endemic to that island (**Figure 54.9**). Regions like Madagascar that are home to large numbers of endemics are classified as "biodiversity hotspots" (see Section 59.4).

### Two scientific advances changed the field of biogeography

For many decades after observing that the biotas of the major biogeographic regions are strikingly different, biogeographers speculated about the causes of these differences but the field remained primarily descriptive. It was not until the second half of the twentieth century that two scientific advances transformed biogeography into a dynamic, multidisciplinary field: the acceptance of the theory of continental drift, and the development of phylogenetic taxonomy.

**CONTINENTAL DRIFT** Prior to the mid-nineteenth century, most people believed that Earth was young and had changed little over time. No one at that time imagined that organisms—much less the continents themselves—had moved. But that continents can and do move was confirmed when geological evidence of continental drift was eventually discovered (see Section 25.2).

During the Triassic period, the supercontinent Pangaea broke into two great land masses, Laurasia and Gondwana (see Figure 25.12), which subsequently separated into the continents we know today. Those groups of organisms that are represented on two or more continents are believed to be ancient groups whose ancestors were distributed widely over these great landmasses before they broke apart. After the breakup, however, their de-

(A)

(B)

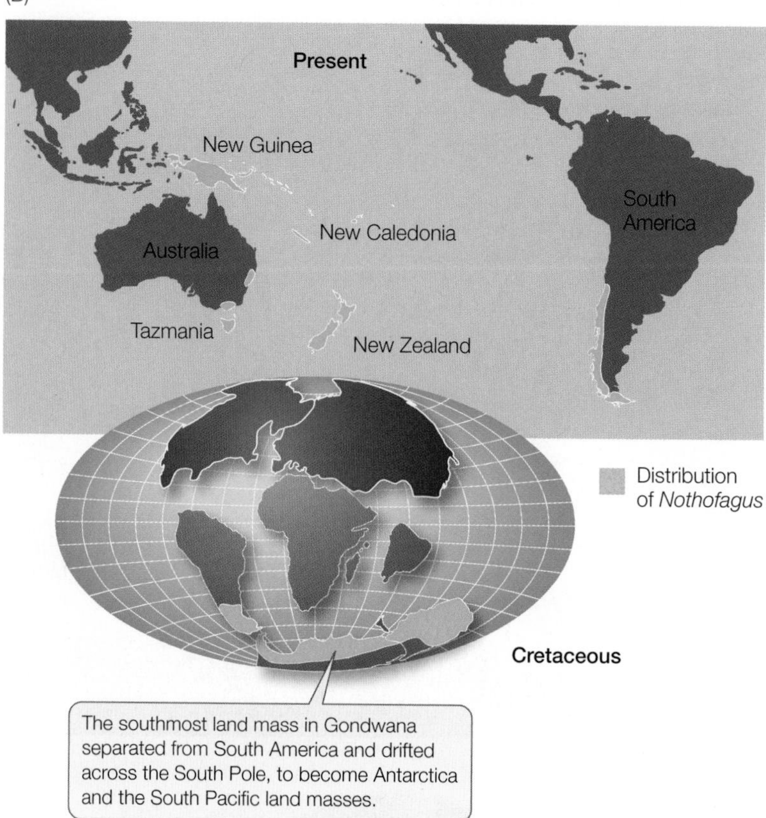


**54.10 *Nothofagus* Has a Gondwanan Distribution** The modern range of southern beeches is best explained by their origin in Gondwana during the Cretaceous. Continental drift resulted in the distribution of *Nothofagus* in South America, New Zealand, and the islands of the South Pacific.

scendants evolved independently, so groups that did not originate until after that time have more discrete distributions. Thus continental drift is at least partly responsible for the existence of the seven regions shown in Figure 54.8.

Continental drift also explains certain biogeographic distributions that would otherwise be difficult to understand. For example, the southern beeches—trees of the genus *Nothofagus*—are found in both the Neotropical and the Australasian biogeographic regions. Their distribution suggests that the genus originated in Gondwana during the Cretaceous period and was geographically separated by the breakup of that land mass (**Figure 54.10**).

What evidence do we have that *Nothofagus* did not simply leapfrog from one biogeographic region to another? Fossilized *Nothofagus* pollen from 55–34 million years ago has been found in Australia, New Zealand, western Antarctica, and South America, suggesting that *Nothophagus* was once continuously distributed across a single land mass. Moreover, the modern distribution of aphid genera that feed exclusively on *Nothofagus* parallels the distribution of the trees. There are no air or water currents between Chile and New Zealand that would be likely to disperse insects, indicating that the aphids arose at a time when their host plants grew on a common land mass.

**PHYLOGENETIC TAXONOMY** As we saw in Chapter 22, taxonomists have developed powerful methods of reconstructing the phylogenetic relationships among organisms. Biogeographers have adapted these methods to help them answer biogeographic questions. Biogeographers can transform phylogenetic trees into **area phylogenies** by replacing the names of the taxa on a tree with the names of the places where those taxa now live or once lived.

Suppose, for example, we wonder why zebras, which are members of the horse family (Equidae), live in Africa when the fossil record indicates that the Equidae evolved in North America. An area phylogeny of living species suggests that the ancestors of today's horses dispersed from North America to Asia, and then from Asia to Africa, and that the subsequent speciation of zebras took place entirely in Africa (**Figure 54.11**).

### Biotic interchange follows fusion of land masses

Geological events not only separate land masses, but may also bring them together. When formerly separated land masses fuse, as may happen through sea level changes or continental drift, species from two different biotas may disperse into a region they had not previously inhabited. This phenomenon is called **biotic interchange**.

A great biotic interchange of mammals occurred when the Central American land bridge formed about 4 million years ago, connecting North and South America for the first time in about 65 million years. While the two continents were separated, their mammalian faunas had diverged greatly. Many of these divergent species dispersed across the newly established land bridge. Only a few South American species—the porcupine, nine-banded armadillo, and Virginia opossum—became established north of the tropical forests of Mexico. However, many North American mammals—rabbits, mice, foxes, bears, raccoons, weasels, cats, tapirs, peccaries, camels, and deer—successfully colonized South America (**Figure 54.12**).

**54.11 Taxonomic Phylogeny to Area Phylogeny** The conversion of a phylogenetic tree to an area phylogeny helps biogeographers explain how the current distribution of a taxon came about.

Onager
African ass
Przewalski's horse
Grevy's zebra
Mountain zebra
Plains zebra

Phylogenetic tree

Ancestral horse
This lineage leads to the modern horse.
Przewalski's horse
Onager
African ass
Mountain zebra
Grevy's zebra
Plains zebra

Area phylogeny

Origin in North America
Central Asia
Middle East and Central Asia
North Africa
South West Africa
East Africa
Eastern and Southern Africa
Horses speciated as they moved from Asia to Africa.
Speciation of zebras has taken place entirely in Africa.
3.9 3 2 1 0
Time (Mya)

### Vicariant events influence distribution patterns

The appearance of a physical barrier to dispersal that splits the range of a species is called a **vicariant event**. A vicariant event divides the species into two or more discontinuous populations, even though no individuals have dispersed to new areas. If, however, members of a species cross an existing barrier and establish a new population, the discontinuous range of the species is considered to be the result of dispersal.

Given that these two major processes—vicariance and dispersal—both influence distribution patterns, how can biogeographers determine the role of each process when reconstructing the evolutionary history of a particular distribution? By

**54.12 Results of a Biotic Interchange** (A) The nine-banded armadillo is among only a handful of mammalian species that colonized North America from South America when the Central American land bridge formed. (B) Some species that exist today only in South America are descended from ancestors that dispersed across the land bridge from North America, including the Chacoan peccary and the guanaco (a member of the camel family).

(A) *Dasypus novemcinctus*

(B) *Catagonus wagneri*

*Lama guanicoe*

studying area phylogenies, a biogeographer may discover evidence suggesting that the distribution of an ancestral species was influenced by a vicariant event, such as continental drift or a change in sea level (see Figure 23.3). If that inference is correct, it is reasonable to assume that ancestral species in other lineages have been affected by the same event and that similar distribution patterns should therefore be seen in other taxonomic groups. Differences in distribution patterns among taxonomic groups may indicate that they responded differently to the same vicariant events, that they diverged at different times, or that they had very different dispersal histories. By analyzing such similarities and differences, biogeographers seek to discover the relative roles of vicariant events and dispersal in determining today's distribution patterns.

The parsimony principle, used to great effect in the reconstruction of phylogenies (see Section 22.2), can also be helpful in biogeographic studies. For example, the New Zealand flightless weevil *Lyperobius huttoni* is found in the mountains of South Island and on sea cliffs at the extreme southwestern corner of North Island (**Figure 54.13**). At first glance, its distribution might suggest that, even though this weevil cannot fly, some individuals in the distant past managed to cross Cook Strait, the 25-km body of water that separates the two islands. However, more than 60 other animal and plant species, including other flightless insects, are found on both sides of Cook Strait. Irrespective of their ability to fly, wade, or swim, it is unlikely that all 60 of these species made the same ocean crossing independently at different times over the course of their evolutionary history. In fact, geological evidence indicates that the present-day southwestern tip of North Island was once united with South Island. Thus a single vicariant event—the separation of the northern tip of South Island from the remainder of the island by the newly formed Cook Strait—could have produced the distribution pattern shared by all 60 species today.

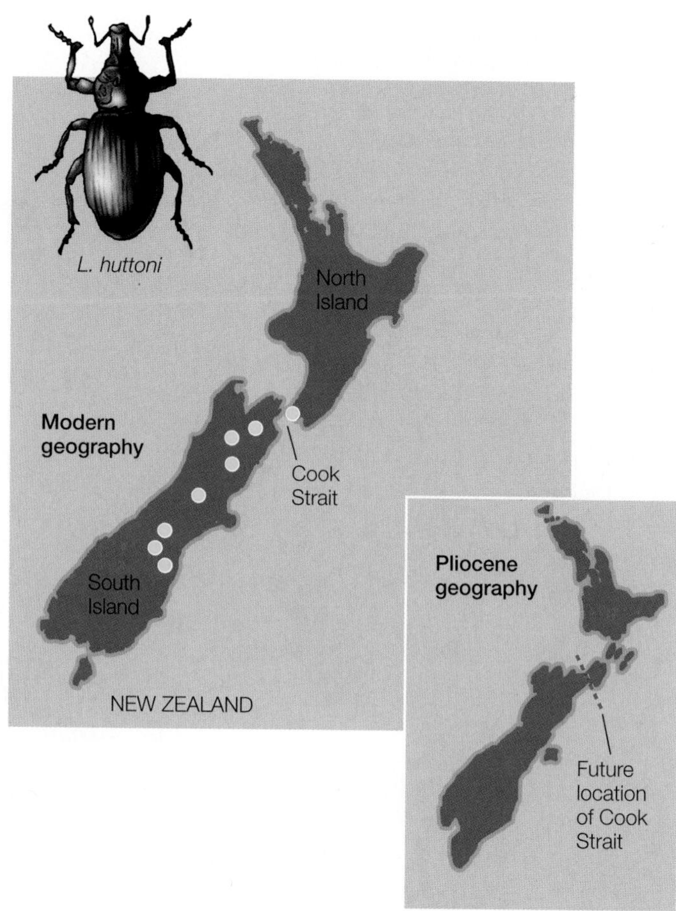


**54.13 A Vicariant Distribution Explained** Yellow circles indicate the current distribution of the flightless weevil *Lyperobius huttoni*. A comparison of New Zealand's present-day geography with its geography in the Pliocene, when the southern part of today's North Island was part of South Island, suggests that a vicariant event—a physical split separating populations—explains this distribution.

While the concept of vicariance is easily applied in cases attributable to continental drift (due to the ease with which continental movements can be correlated with biological patterns), vicariance patterns on a smaller scale are difficult to identify unequivocally because the underlying geologic events, such as mountain uplift or glacial ice sheet movement, may have been less dramatic than continental movements. In other cases, the vicariance approach is conceptually inappropriate simply because no separation of continuous populations ever took place. There are, for example, more than 135 species of long-horned beetles endemic to the Hawaiian Islands. The islands have never been attached to any continent (they arose by volcanism from the deep ocean floor), so this distribution must have been the result of long-distance dispersal. The beetles, which are most closely related to a genus found in North and Central America, probably colonized the islands and subsequently speciated as a consequence of specializing on different host plants on different islands (see Chapter 23). Founder events must have been infrequent, because the beetles are short-lived and vulnerable to desiccation.

In fact, the present distributions of some mammalian groups provide little hint of their evolutionary origins. Canids (members of the dog family), for example, originated in North America but experienced greater evolutionary success after moving into South America; today, only four canid genera are found in North America, but there are six in South America—more than on any other continent.

### Humans exert a powerful influence on biogeographic patterns

One more force capable of generating unusual distribution patterns is human activity. Of the insects found both in Europe and in North America, it has been estimated that over half have been transported between the continents by humans. Some species introductions, such as those of dung beetles to Australia, have been deliberate. That introduction turned out to be beneficial, but many others have not. Species that have been introduced to new regions inadvertently include disease vectors (such as the mosquito that transmits yellow fever), household pests (including at least four cockroach species), and crop pests that can result in economic disaster. The effects of such introductions by humans is an important topic and will be discussed at length in Chapter 59.

## 54.4 RECAP

**Biogeographers divide Earth into seven regions, each containing unique assemblages of species. Vicariant events may generate distribution patterns by splitting the ranges of species; distribution patterns may also change when species disperse across barriers.**

- What determines the boundaries of Earth's major biogeographic regions? How are these boundaries different from those of the biomes described in Section 54.3? **See p. 1146, p. 1158, and Figures 54.5 and 54.8**
- Explain how the concepts of continental drift and phylogenetic taxonomy transformed the field of biogeography. **See pp. 1159–1160**
- How do vicariance and dispersal interact to generate major biogeographic patterns? **See pp. 1161–1162**

As described in Chapter 25, life originated in the oceans. Today the three-quarters of the planet's surface that is covered with water sustains abundant life as well.

# 54.5 How Is Life Distributed in Aquatic Environments?

Most of Earth's aquatic environments are oceans. However, the small percentage of the watery world that consists of freshwater lakes and streams also hosts a significant proportion of Earth's aquatic organisms.

### The oceans can be divided into several life zones

Earth's oceans form one large, interconnected water mass, interrupted in places by continents. Even organisms with limited swimming abilities can travel long distances simply by floating with the great currents depicted in Figure 54.3. Even though the oceans have no physical barriers to dispersal, however, most marine organisms have restricted ranges. Water temperature, salinity, and food supply all vary spatially. Changes in water temperature, for example, can be barriers to dispersal because many marine organisms function well in only a narrow range of temperatures.

The striking physical and biological discontinuities within the oceans divide them into several distinct "life zones." These zones are identified principally by their distance from the surface, because depth determines how much light is available to sustain the photosynthetic organisms that make chemical energy available to consumer organisms. In both marine and freshwater environments, the depth of water reached by enough sunlight to support photosynthesis is called the **photic zone** (**Figure 54.14**). Approximately 90 percent of all aquatic life is found in the photic zone.

The **coastal zone**, extending from the shoreline to the edge of the continental shelf, is characterized by relatively shallow, well-oxygenated water and relatively stable temperatures and salinities. These conditions support high densities of phytoplankton, which in turn support some of the world's most important fisheries. Structure in coastal-zone communities can be provided by a variety of organisms. In warmer coastal waters, corals generate complex reef structures that support ecosystems rivaling the rainforest in diversity, and "forests" of the multicellular algal species (seaweeds and giant kelps) form along many coasts.

Near the shoreline, the area of the coastal zone that is affected by wave action is called the **littoral zone**. The principal autotrophs in this zone—sea grasses and algae—support a diversity of invertebrates as well as small fishes. The portion of the coastal zone lying between high- and low-tide levels is the **intertidal**, where tidal movements create highly variable conditions of temperature and salinity. Intertidal organisms, including clams, mussels, copepods, and burrowing worms, are alternately exposed to air or submerged on a regular basis.

Throughout most of the oceans, the dominant autotrophs are phytoplankton (photosynthetic floating protists; see Chapter 27). In the open ocean, referred to as the **pelagic zone**, the principal consumers of phytoplankton are zooplankton—mainly small crustaceans and larval stages of marine animals—which

The **coastal zone** extends from the shoreline to the edge of the continental shelf. The area affected by wave action is the **littoral zone**.
The open water above the ocean floor is the **pelagic zone**.
Intertidal
Coastal zone
Pelagic zone
Photic zone (limit of sunlight penetration)
Continental shelf
Continental slope
Benthic zone
Trench
Aphotic zone
The ocean floor constitutes the **benthic zone**.
The deepest ocean floor is called the **abyssal zone** or **abyssal plain**.
Abyssal plain
Water depth (m)
200
2,000
4,000
6,000
8,000
10,000

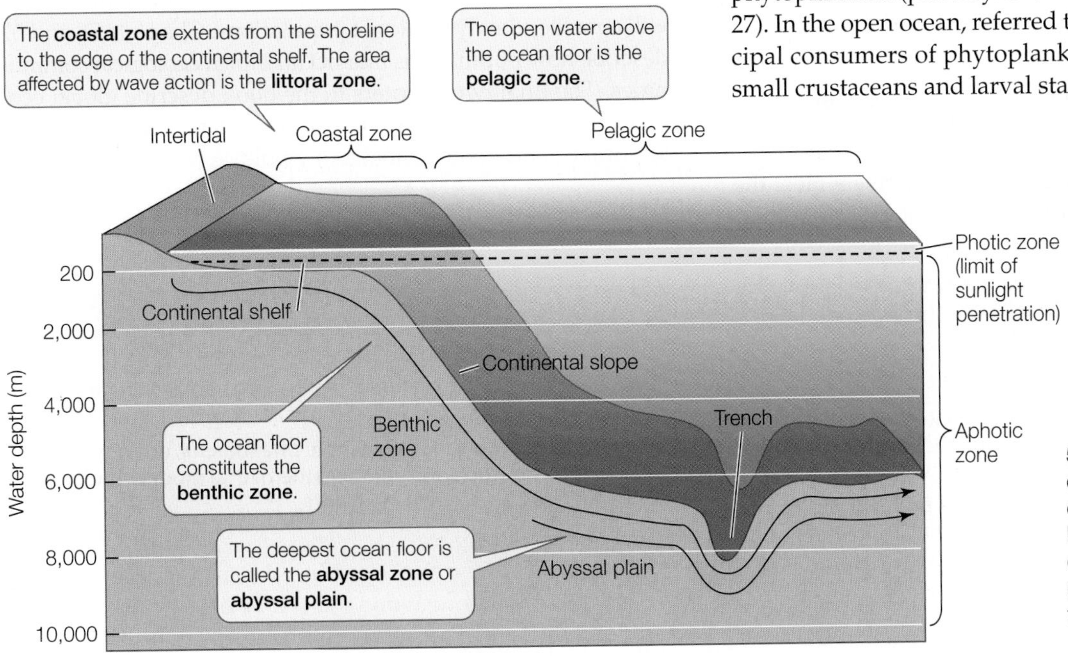

**54.14 Oceanic Life Zones** The ocean's life zones are in large part determined by sunlight penetration. More than 90 percent of ocean-dwelling species live in the sunlit photic zone, which comprises less than 2 percent of the open water.

in turn support many larger free-swimming vertebrate and invertebrate species.

The ocean bottom is referred to as the **benthic zone**. Many benthic organisms are adapted to life on the seafloor substrate. They include sessile animals such as sponges, bryozoans, ribbon worms, and brachiopods (see Chapters 31 and 32), as well as motile bottom feeders such as crabs and sea slugs (see Figure 32.15C).

Where water is too deep for light to penetrate, little photosynthesis can take place, and both plant and animal diversity are low. Depths reached by less than 1 percent of incoming sunlight constitute the **aphotic zone** (including the **abyssal plain** of the deep ocean floor). Many of the organisms inhabiting these regions subsist on decaying organic matter that sinks down from the photic zone. Some produce their own light by way of bioluminescent organs (see Figure 33.13D). Even deep-ocean trenches and rift valleys support ecosystems, often sustained by chemosynthetic microbes that can metabolize the nutrients in seawater without the aid of sunlight (see Section 26.2).

### Freshwater environments may be rich in species

In contrast to the vast oceans, the world's freshwater environments cover less than 3 percent of Earth's surface. However, they are home to about 10 percent of all aquatic species. Freshwater environments may be found in running water (streams and rivers) or in standing water (lakes and ponds). Like the oceans, bodies of standing water can be divided into zones based on depth and light penetration, with a littoral zone close to shore characterized by warm temperatures and high species diversity, a pelagic zone with its upper layers teeming with phytoplankton, and deeper waters below, where little light penetrates and oxygen levels are low, characterized by low diversity.

Discontinuities in the ranges of aquatic organisms are seen because most animals that live in the oceans cannot survive in fresh water, and vice versa. For example, most groups of freshwater fishes cannot tolerate salt water, so they are restricted to a single continent. They can disperse only within the connected rivers and lakes of a drainage basin or in brackish coastal waters.

### Estuaries have characteristics of both freshwater and marine environments

*Estuaries* are bodies of water that form where rivers meet the ocean, in which salt water mixes with fresh water. Having characteristics of both freshwater and marine systems, estuaries are home to many unique species, and they play an important role for other species as a conduit between marine and freshwater environments. Some salmon species that hatch in rivers, for example, spend many months in estuaries adjusting to higher salinities before swimming out to sea to grow into adults. Diversity within estuaries tends to be very high. Estuarine environments have long been an important source of human resources, not the least of which is their role in purifying groundwater. In many places around the world, however, overfishing and pollution threaten estuarine viability.

## 54.5 RECAP

**Despite being connected, the oceans are divided into distinct regions that generate striking physical and biological discontinuities. Oceanic life zones are determined by distance from the surface, which influences how much light is available to sustain photosynthesis. Fresh waters, comprising river basins and thousands of relatively isolated lakes, are also divided into life zones according to the depth and light penetration. Estuaries are bodies of water where salt water mixes with fresh water and support many unique species.**

- How does light penetration affect diversity in different life zones of the oceans? See p. 1163 and Figure 54.14
- How do estuaries link freshwater and marine systems, and why are these ecosystems so important? See p. 1164

The physical environment exerts a tremendous influence on the distribution of organisms on Earth, but to understand how and where life flourishes on Earth, it is also important to understand the influence of organisms on one another. In Chapter 55 we will consider the dynamics of populations. Chapters 56 and 57 describe how populations interact and form ecological communities. The final two chapters in the book describe global ecological patterns and emphasize the importance of ecology both to humans and to our planet as a whole.

## CHAPTER SUMMARY

### 54.1 What Is Ecology?

- **Ecology** is the scientific investigation of interactions among organisms and between organisms and their physical environment.
- An organism's environment encompasses both **abiotic** (physical and chemical) and **biotic** components (other living organisms).
- The science of ecology generates knowledge about interactions in the natural world.
- In addition to observations of visible phenomena, ecology relies on a diversity of tools, including mathematical models, molecular techniques, and satellite imaging

- **Environmentalism** is the use of ecological knowledge to inform our decisions about the stewardship of natural resources and ecosystems.

## 54.2 Why Do Climates Vary Geographically?

- Weather refers to atmospheric conditions at a particular place and time. Climate is the average of atmospheric conditions, and the variation in those conditions, found in a particular place over time.
- The solar energy that reaches a given unit of Earth's surface depends primarily on the angle of the sun's rays, which in turn is a function of latitude. Latitudinal variation in air temperature drives atmospheric circulation patterns of precipitation. See Figures 54.1 and 54.2
- Global atmospheric circulation patterns are also driven by the Earth's rotation; these wind patterns in turn drive ocean surface currents. Review Figure 54.3
- Organisms respond to climatic challenges with physiological, morphological, and behavioral adaptations.

## 54.3 What Is a Biome?

- A **biome** is an environment that is defined by its climatic and geographic attributes and characterized by ecologically similar organisms, particularly its dominant vegetation. Review Figure 54.5, **ANIMATED TUTORIAL 54.1**
- The distribution of biomes is determined primarily by climate, but other factors, such as soil characteristics and fire, also influence vegetation. Terrestrial biomes include Arctic and alpine tundra, boreal forest, temperate evergreen and temperate deciduous forests, temperate grasslands, hot and cold deserts, chaparral, thorn forest and savanna, tropical deciduous forest, and tropical evergreen forest. SEE **ANIMATED TUTORIAL 54.2**

## 54.4 What Is a Biogeographic Region?

- **Biogeography** is the scientific study of the patterns of distribution of populations, species, and ecological communities.
- The boundaries of **biogeographic regions** are drawn where assemblages of species change dramatically over short distances. These boundaries generally correspond to present or past barriers to dispersal. Review Figures 54.7 and 54.8, **WEB ACTIVITY 54.1**
- A species that is found only in a certain region is said to be **endemic** to that region.
- The theory of continental drift explains some discontinuous distributions that include multiple biogeographic regions. Review Figure 54.10
- Biogeographers can transform phylogenetic trees into **area phylogenies** that show where organisms now live or once lived. Review Figure 54.11
- When formerly separated land masses come together, species from both biotas may disperse into the other region, a phenomenon known as **biotic interchange**.
- Both **vicariant events** and dispersal across boundaries generate distinctive patterns of species distributions. Review Figure 54.13

## 54.5 How Is Life Distributed in Aquatic Environments?

- Oceanic **life zones** are determined by the distance from the surface, which in turn influences how much light is available to sustain photosynthetic organisms. Review Figure 54.14
- Fresh waters—river basins and lakes—are also divided into life zones according to depth and light penetration.
- Estuaries are bodies of water where salt and fresh water mix. This globally small but crucial life zone supports many unique species.

# SELF-QUIZ

1. Ecology and environmentalism are
   *a.* synonymous; the terms can be used interchangeably.
   *b.* differentiated by the emphasis placed by ecology on the biotic rather than the abiotic world.
   *c.* differentiated by the lack of utility of ecology in solving world problems.
   *d.* differentiated by the inherent focus of environmentalism on human concerns.
   *e.* both scientific fields of inquiry that generate knowledge about the natural world but use different tools.
2. A biome is
   *a.* a large ecological unit defined by its climatic and geographic attributes and characterized by ecologically similar organisms.
   *b.* a large ecological unit defined by its dominant animals.
   *c.* a large ecological unit defined by climate characteristics.
   *d.* a large ecological unit defined by the number of photosynthetic species.
   *e.* a large ecological unit defined by biogeochemical cycles.
3. The solar energy that reaches a given unit of Earth's surface depends primarily on
   *a.* the angle of the sun's rays.
   *b.* the moisture content of the air.
   *c.* the amount of cloud cover.
   *d.* the strength of the winds.
   *e.* day length.
4. Energy from the sun determines
   *a.* air temperature.
   *b.* air and wind circulation patterns.
   *c.* ocean surface currents.
   *d.* All of the above
   *e.* None of the above
5. Biogeography as a science began when
   *a.* European naturalists first traveled the globe and noted patterns in the distributions of organisms.
   *b.* Europeans traveled to the Middle East during the Crusades.
   *c.* phylogenetic methods were developed.
   *d.* the theory of continental drift was accepted.
   *e.* Charles Darwin proposed the theory of natural selection.
6. Vicariant events
   *a.* result from organisms dispersing across a barrier.
   *b.* were common in the past but are rare today.
   *c.* divide a species into two or more discontinuous populations.
   *d.* were rare in the past but are common today.
   *e.* caused most of today's discontinuous species ranges.

7. The oceans can be divided into life zones even though they are all connected because
   *a.* the rate of photosynthesis is low in the oceans.
   *b.* ocean currents keep organisms close to where they were born.
   *c.* light penetration, and hence photosynthesis, vary with depth and determine where organisms live.
   *d.* trade winds keep warm and cold waters separate.
   *e.* continents provide barriers to movement of planktonic life forms.
8. A parsimonious interpretation of a distribution pattern is one that
   *a.* requires the smallest number of undocumented vicariant events.
   *b.* requires the smallest number of undocumented dispersal events.
   *c.* requires the smallest total number of undocumented vicariant plus dispersal events.
   *d.* accords with the phylogeny of a lineage.
   *e.* accounts for centers of endemism.
9. The only biogeographic region that today is completely isolated from the other regions by water is
   *a.* the Madagascan region.
   *b.* the Ethiopian region.
   *c.* the Neotropical region.
   *d.* the Australasian region.
   *e.* the Nearctic region.
10. Which of the following characteristics is unique to Arctic tundra?
   *a.* Winters are long and cold.
   *b.* The soil is underlain by a layer of permafrost.
   *c.* Many small mammals hibernate in winter.
   *d.* Many of the plants are perennials.
   *e.* None of these characteristics apply to Arctic tundra.

## FOR DISCUSSION

1. Discussions about the likelihood of life on other planets often focus on their proximity to a sunlike star. How different might life on Earth be if the planet's distance from the sun were considerably greater?
2. Today, by far the greatest number of species of fruit flies (genus *Drosophila*) is found in the Hawaiian Islands. Would you conclude that the genus originally evolved in Hawaii and spread to other regions? Under what circumstances do you think it is safe to conclude that a group of organisms evolved in the same region where the greatest number of species live today? (*Hint:* Review the discussion of equid phylogeny and Figure 54.11.)
3. The desert locust (*Schistocerca gregaria*), which has been a major agricultural pest for millennia, is found throughout the Middle East, western Asia, and Africa. Every other species in the genus *Schistocerca* is found in the Western Hemisphere. How could you determine whether the current distribution of species resulted from eastward dispersal (from North America to Africa) or from westward dispersal (from Africa to the Americas)?
4. Processes in nature do not always conform to the parsimony principle. Why, then, do biogeographers often use the parsimony principle to infer the histories of species and lineages?
5. A well-known legend states that Saint Patrick drove the snakes out of Ireland. Give some alternative explanations, based on biogeographic principles, for the absence of indigenous snakes in that country.
6. Most of the world's flightless birds are either nocturnal and secretive (such as the kiwi of New Zealand) or large, swift, and powerful (such as the ostrich of Africa). The exceptions are found primarily on islands, and many of these island species became extinct when humans (and their domesticated animals) arrived. What conditions on islands might permit the survival of flightless birds? Why has human colonization so often resulted in the extinction of such birds? The power of flight has been lost secondarily in representatives of many groups of birds and insects; what are some possible evolutionary advantages of flightlessness that might offset its obvious disadvantages?

## ADDITIONAL INVESTIGATION

Alfred Russel Wallace observed that dramatically different birds inhabited Bali and Lombok, even though the strait that separates them is only 15 miles wide. But most birds can easily fly 15 miles, and can probably see the other island across the water. What kinds of data, in addition to those gathered by Wallace, could be obtained to determine why birds do not fly across the strait or, if indeed they do cross the water, why they fail to colonize the island on the other side?

# 55 Population Ecology

## Reindeer games

In 1944, in the midst of World War II, the U.S. Coast Guard established a LORAN (Long-Range Aids to Navigation) tracking station on the tiny island of St. Matthew in Alaska, an isolated and otherwise unoccupied patch of tundra more than 300 km from the nearest village. As an emergency food supply for the 19 men assigned to the island, the Coast Guard brought in 29 reindeer (*Rangifer tarandus*) by barge and released them.

The reindeer thrived on the thick, lush mat of lichens that covered the 128-square-mile island. Other than the men, the island had no reindeer predators; its only other animal occupants were Arctic foxes, one species of vole, and a few ground-nesting birds. As the war wound down and the men left the island, the reindeer were left behind in an environment of plentiful food and no natural predators.

In 1957, David Klein, at the time a U.S. Fish and Wildlife biologist, visited St. Matthew with a field assistant; together they counted more than 1,350 reindeer, most of which appeared fat and healthy. However, they also noticed areas of seriously overgrazed lichen mats. Klein did not return to the island until 1963, when he and three colleagues hitched a ride with a Coast Guard cutter. By that time, there were more than 6,000 reindeer packed in at a density of 47 per square mile. The island was covered with reindeer tracks and droppings, and the animals were distinctly smaller than the ones sighted 6 years earlier.

The winter of 1963–1964 brought punishing storms, record low temperatures, and tremendous snowfalls to St. Matthew Island. In August 1965, Coast Guard personnel reported massive reindeer deaths. Klein and two colleagues arranged a return visit in the summer of 1966, at which time they found the island littered with reindeer skeletons. The scientists could locate only 42 living reindeer, 41 of which were adult females; the lone male appeared to have deformed antlers. In a remarkably short period, the reindeer population had declined by over 99 percent. Lichens had essentially disappeared from the island, replaced almost entirely by sedges and grasses, on which reindeer cannot subsist. By 1980, the reindeer had entirely disappeared from the island.

Introducing large hoofed mammals to small islands is an inherently risky enterprise, as the experience on St. Matthew graphically illustrates. But

**Reindeer Pause** Part of the St. Matthew reindeer herd is seen In this photograph taken in 1963, shortly before a particularly severe winter destroyed most of this isolated population. The herd had grown exponentially for almost 40 years since being introduced to the island as a food source for soldiers during World War II.

**Predators and Population Cycles** Wolves are the principal non-human predators of reindeer, moose, and other large grazing mammals in northern forests and tundra. There were no wolves on St. Matthew Island, a fact largely responsible for the reindeer population's ability to grow exponentially.

such introductions do not always end in disaster. Reindeer populations introduced to the sub-Antarctic island of South Georgia almost a century ago have persisted and appear to be stable.

Why would populations of a particular species in one place explode and crash, but in another seemingly similar place remain stable over time? Understanding how and why populations change in size is more than an academic pursuit. Ecologists study how populations change over time because that knowledge is critical for understanding why some species become pests in some places and not in others, for managing sustainable harvests of economically important species, and for designing plans for conserving endangered species.

**IN THIS CHAPTER** we will examine how ecologists study populations and investigate how reproductive capacity and environmental resources affect the dynamics of population growth. We will identify factors that limit population densities and determine the effects of environmental variation on population dynamics. Finally, we will show how an understanding of population dynamics is applied to managing populations of importance to humans.

## CHAPTER OUTLINE



# 55.1 How Do Ecologists Study Populations?

Well before ecology became a distinct biological discipline, people engaged in population management. Whenever we grow crops or raise livestock, we are explicitly increasing populations of domesticated plants and animals. Pest control strategies aim to reduce populations of organisms whose presence we consider undesirable. Game wardens, park managers, and conservation biologists aim to maintain stable populations of fish, wildlife, and threatened or endangered species. All of these activities require an understanding of **population dynamics**: the patterns and processes of change in populations. The study of population dynamics also allows us to understand the changes in populations we make inadvertently in the course of other human activities—as when the Coast Guard introduced reindeer to St. Matthew.

A **population** consists of the individuals of a species that interact with one another within a given area at a particular time. Populations are important units for study because groups of individuals that interact in time and space have ecological characteristics that individuals do not. At any given moment, an individual organism occupies only one point in space and is a particular age and size. The members of a population, however, are distributed over space and vary in age and size.

**Population density** is the number of individuals per unit of area or volume. Density is a property of all populations and is a function of the processes that add individuals to the population (births and immigration, or movement of individuals into the population) and the processes that reduce the number of individuals in the population (deaths and emigration, or movement of individuals out of the population).

Populations also have a characteristic *age structure*, or distribution of individuals across age categories, and a characteristic *dispersion* pattern, or spatial distribution of individuals in the environment. These characteristics, which are constantly changing due to births, deaths, and movement, influence the stability of populations and affect the ways in which populations of one species interact with populations of other species. Thus, to study populations, ecologists need to count the individuals in a given area, determine their ages, and calculate the rates at which individuals enter and leave the population. The study of these processes is known as **demography**.

### Ecologists use a variety of approaches to count and track individuals

How individuals are counted depends on the nature of the organism under study. Populations of animals, for example, can

(A)
The pattern of folds and notches on each elephant's ears is as unique as a fingerprint.

(B)
A computer chip on a bee's back logs her movements between the hive and flowers.


**55.1 Identifying Individuals** (A) The pattern of folds and notches on the ears of an elephant is as unique and distinctive as a fingerprint and can be used to recognize individuals in a population. (B) Worker bees in a hive are individually indistinguishable to ecologists, who have come up with ingenious methods of marking. This female honey bee sports a computer chip on her back, which not only identifies her but also logs her movements between the hive and flowers.

be more challenging to count than populations of trees. Most animals can move, so, to avoid double counting, individuals must be identified. Nevertheless, counting every tree in a forest can be logistically difficult, even though the trees are standing still.

In some species, individuals are large and distinct enough, and populations small enough, that investigators can identify all the individuals and count them. Biologists performed this type of count, called a *full census*, on the elephant population of Samburu and Buffalo Springs national reserves in Kenya. By monitoring the population for 21 months, they learned to recognize each of the 760 individual elephants, primarily by their unique and distinctive ear markings (**Figure 55.1A**). Individual recognition is impossible or impractical for most species, however, and to identify such individual organisms they must be marked in some artificial way (**Figure 55.1B**).

In most species, populations are too large and their individual members too small, too similar in appearance, and/or too mobile for a full census to be conducted. Thus population sizes are often estimated from representative samples using statistical methods.

### Population densities can be estimated from samples

Ecologists usually measure the densities of terrestrial organisms as the number of individuals per unit of area; for organisms living in soil, air, or water, the number or mass per unit of volume may be used. Ecologists obtain these measurements from sample units, then extrapolate from these samples to estimate the total population density.

Estimating population densities is easiest for sedentary organisms. Investigators need only count the individuals in a sample of representative locations and extrapolate the counts to the entire geographic range of the population. Individuals may be counted within marked and measured areas called *quadrats*. Plants are often counted along a linear *transect*: a line drawn across the population's range (often designated by a string marked at regular intervals). Any individual that touches the line is counted. By making repeated counts with either of these methods, investigators can make reasonably good estimates of the size of a population.

Counting mobile organisms is more difficult because individuals move into and out of sampling areas. In such cases, investigators may use the **mark–recapture method**. They begin by capturing, marking, and then releasing a number of individuals. Later, after the marked individuals have had time to mix with unmarked individuals in the population, another sample of individuals is captured. The proportion of individuals in the new sample that are marked can be used to estimate the total size of the population in the defined area, as follows:

$$\text{Estimated population size } N = \frac{n_1 \times n_2}{n_{1\cap 2}}$$

where

$n_1$ = the total number of individuals in the first sample (captured, marked, and released)

$n_2$ = the total number of individuals in the second sample

$n_{1\cap 2}$ = the number of marked individuals recaptured in second sample (the "intersection set" of samples 1 and 2)

In other words, we assume that the proportion of individuals in the second sample that were captured and marked in the first sample is about the same as the proportion of individuals in the sampling area that were captured in the first sample.

Quantifying the size and determining the density of populations are important, but these number are only a starting point for understanding population dynamics, because not all individuals contribute equally to population growth.

### Populations have age structures and dispersion patterns

The **age structure** of a population—the distribution of individuals across all age groups—has a profound effect on population growth because reproductive capacity varies with age. Populations with a large proportion of young individuals have a

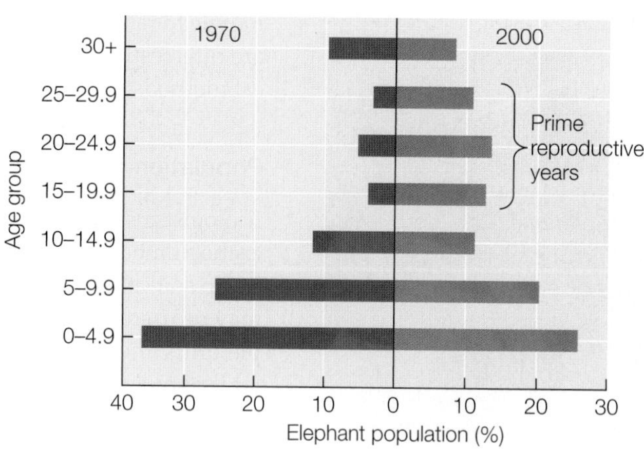

**55.2 Changes in Age Structure Influence Population Growth** The elephant population in Kidepo Valley National Park, Uganda, was monitored between 1970 and 2000. During this time, the proportion of the population achieving prime reproductive age range (15–30 years) grew considerably. Such an age structure in a population is likely to result in an increased rate of growth.

greater potential to grow than populations dominated by individuals that are beyond their peak reproductive years.

For some species, reproduction may be the province of only a tiny fraction of the entire population for only a short interval during the life cycle. Adults of the tiny insect *Clunio maritimus* (the "one-hour midge") mate, lay eggs, and die within about an hour after completing their larval development. In contrast, some vertebrates are capable of reproducing for years. **Figure 55.2** shows the results of a long-term study of the age structure of the elephant population in Kidepo Valley National Park, Uganda. Relative to 1970, the 2000 population skewed significantly toward females, particularly those over 25 years of age. These two disparate age structures were the result of years of differential mortality among young elephants due to drought, and of increased male death rates due to ivory poachers. The age structure as of 2000 portends an increase in the population's growth rate, given that female African elephants become fertile around age 10–15 and can continue producing offspring through their fifties.

**Dispersion** refers to the distribution of individuals in space within a population. Dispersion determines patterns of interaction among individuals and can thus have important effects on population growth. In addition, ecologists must understand the dispersion patterns of a species to choose appropriate sampling areas and methods for estimating population sizes.

Ecologists recognize three basic dispersion patterns (**Figure 55.3**):

- A *clumped* dispersion pattern occurs when the presence of one individual at any point in space increases the probability of others being near that point.
- A *regular* dispersion pattern occurs when the presence of one individual at any point in space reduces the probability of others being near that point.
- A *random* dispersion pattern occurs when there is an equal probability of an individual occupying any point in space.

(A) Clumped dispersion

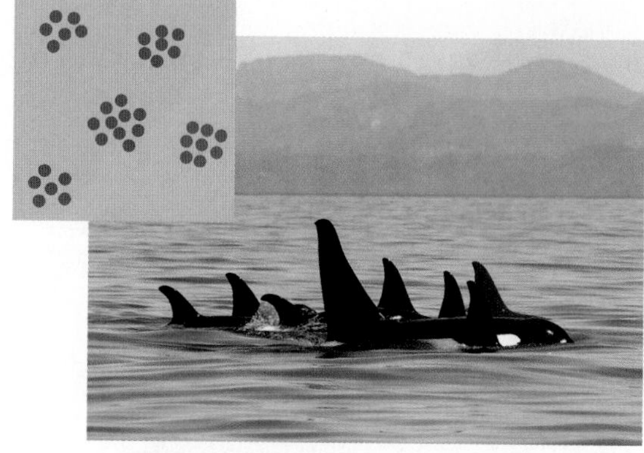

(B) Regular dispersion

(C) Random dispersion

**55.3 Dispersion Patterns** (A) Orcas hunting in pods display a clumped dispersion pattern. (B) Nesting seabirds often stake out territories with a radius defined by their wingspans—an amount of space they can defend without leaving the nest. This results in even dispersion. (C) Dandelion seeds are dispersed by the wind in random fashion.

Spatial variation in environmental conditions strongly influences dispersion patterns. Small-scale differences in temperature, humidity, or wind speed can make particular places more or less suitable for certain organisms. Aphids, for example, cluster along the protruding veins of leaves, where they are sheltered from wind. Interactions among individuals may also bring about characteristic dispersion patterns. Social life, with the cooperation it involves, tends to promote clumped dispersion patterns, as seen in Figure 55.3A. Intraspecific competition for food, space, or mates, on the other hand, tends to space individuals apart in regular dispersion patterns (Figure 55.3B).

### Changes in population size can be estimated from repeated density measurements

Ecologists can use multiple estimates of population densities made over time to estimate the rate at which a population is growing or decreasing. Over any given interval of time, the number of individuals in a population increases by the number of individuals added to the population by birth and immigration and decreases by the number of individuals lost from the population by death and emigration. This relationship is expressed mathematically in the equation

$$N_1 = N_0 + (B - D) + (I - E)$$

where

$N_1$ = the number of individuals at time 1

$N_0$ = the number of individuals at time 0

$B$ = the number of individuals born between time 0 and time 1

$D$ = the number that died between time 0 and time 1

$I$ = the number that immigrated between time 0 and time 1

$E$ = the number that emigrated between time 0 and time 1

Using this equation to estimate $N_1$ over multiple time intervals helps researchers estimate changes in population density.

### Life tables track demographic events

The study of population dynamics requires keeping track of *demographic events* (births, deaths, immigration, and emigration) in populations and determining the *rate* (number per unit of time) at which they occur. A **life table** is a tool that ecologists use for these purposes. Life insurance companies use similar tables (called "actuarial tables") to determine how much to charge people of different ages for insurance policies. Data from life tables can be used to identify the principal *mortality factors*, or causes of death, at particular life stages, to predict future population trends, and to develop strategies for managing populations of species of commercial or ecological value.

**COHORT LIFE TABLES** Life tables can be constructed by a number of methods. To construct a *cohort life table*, investigators start with a **cohort**—a group of individuals born within the same time frame, or *age class*—and record their deaths until no individuals from the cohort remain alive. This type of life table is sometimes called a *horizontal life table* because it is based on data collected across the life span.

TABLE 55.1

**Life Table for the 1978 Cohort of Cactus Ground Finch (*G. scandens*) on Isla Daphne**

| AGE CLASS (YEARS) | NUMBER ALIVE | SURVIVORSHIP[a] ($l_x$) | MORTALITY[b] |
|---|---|---|---|
| 0–1 | 210 | — | 0.57 |
| 1–2 | 91 | 0.43 | 0.14 |
| 2–3 | 78 | 0.37 | 0.10 |
| 3–4 | 70 | 0.33 | 0.07 |
| 4–5 | 65 | 0.31 | 0.05 |
| 5–6 | 62 | 0.30 | 0.32 |
| 6–7 | 42 | 0.20 | 0.45 |
| 7–8 | 23 | 0.11 | 0.35 |
| 8–9 | 15 | 0.07 | 0.07 |
| 9–10 | 14 | 0.07 | 0.21 |
| 10–11 | 11 | 0.05 | 0.09 |
| 11–12 | 10 | 0.05 | 0.60 |
| 12–13 | 4 | 0.02 | 0.25 |
| 13 | 3 | 0.01 | — |

[a] Survivorship = the proportion of the original cohort (here, of 210 individuals) who survive to age $x$.

[b] Mortality ("death rate") = the proportion of individuals of age $x$ who die before reaching age $x + 1$.

The age categories used in a cohort life table depend on the life cycle of the organism of interest. *Age-dependent life tables* track demographic events as a function of calendar age. *Stage-dependent life tables* track demographic events at various stages of the life cycle (e.g., eggs, larvae, pupae, and adults in insects). They are commonly used when survival and reproduction depend more on developmental stage than on calendar age, as is the case, for example, with insects and other animals that undergo metamorphosis.

Using the data in a cohort life table, investigators can calculate **mortality**: the proportion of individuals of each age class that die before reaching the next age class. They can also calculate **survivorship** (represented by the term $l_x$), which is the likelihood of an individual member of the cohort surviving to reach age $x$ (**Table 55.1**).

**ESTIMATING REPRODUCTIVE CAPACITY** A cohort life table can also be used to track the degree to which individuals in different age categories contribute to reproduction (and hence population growth). Because only females produce offspring, life tables generally track the number of offspring produced by each female during each time period—a factor called **fecundity** (indicated in a life table by the term $m_x$). The portion of the life table that tracks fecundity is called a *fecundity schedule* (**Table 55.2**). Such data allows conservation biologists and other scientists to estimate a population's potential for growth (see Chapter 59).

The data shown in Tables 55.1 and 55.2 track the survivorship ($l_x$) and fecundity ($m_x$), respectively, of a cohort of cactus ground finches (*Geospiza scandens*) on Isla Daphne in the Galá-

TABLE 55.2

**Fecundity Schedule for the 90 Females of the 1978 Cohort of Cactus Ground Finch (Table 55.1)**

| AGE CLASS (YEARS) | FECUNDITY[a] ($m_x$) | |
|---|---|---|
| 0–1 | 0.00 | |
| 1–2 | 0.05 | |
| 2–3 | 0.67 | |
| 3–4 | 1.50 | |
| 4–5 | 0.66 | El Niño event; increased rainfall |
| 5–6 | 5.50 | |
| 6–7 | 0.69 | Drought |
| 7–8 | 0.00 | Drought |
| 8–9 | 0.00 | |
| 9–10 | 2.20 | |
| 10–11 | 0.00 | |
| 11–12 | 0.00 | |

[a] Fecundity = number of fledglings per female per breeding season.

pagos archipelago. Peter and Rosemary Grant followed a cohort of 210 birds from 1978, when they hatched, until 1991, when only 3 individuals—all males—remained alive. All of the cactus ground finches on the island were banded so that the Grants could recognize them as individuals.

The *G. scandens* life tables show that mortality was high during the first year of life, then dropped dramatically for several years before increasing in later years. The fecundity data indicate that females of all ages breed, and breeding success does not correlate exclusively with age. Other observations of conditions on Isla Daphne revealed a correlation with rainfall, which in the Galápagos varies dramatically from year to year (see Table 55.2). These life tables and other ecological data, taken together, suggest that the survival of adult birds and the number of offspring they are able to fledge depend on food availability—that is, on cactus flower and fruit production, which are strongly correlated with rainfall. In short, life table data can be useful in separating out the many ecological factors that affect population dynamics.

Fecundity schedules vary greatly among species not only because organisms differ in the number of offspring they can produce, but also because they vary in the timing of reproduction. Female cactus ground finches begin breeding at the age of 1 or 2 years and, in favorable conditions, can fledge multiple offspring each breeding season. In contrast, female African elephants do not produce offspring until they are at least 15 years old and usually produce only one calf about every 5 years.

**VERTICAL LIFE TABLES** Because not all species can be easily followed through time, some life tables are constructed by sampling a population at a single time. These life tables cut across all age categories and thus are known as *vertical life tables*. One way to construct a vertical life table is to record information from a *death assemblage*, a collection of bodies or fossils of individuals that lived together in a particular place at a given time. The birth and death dates on tombstones in a cemetery, for example, can be used to construct a vertical life table for a human population and to estimate its probability of reaching different ages. A 1944 study of Dall mountain sheep (*Ovis dalli dalli*) in Mt. McKinley (now Denali) National Park, Alaska, was based on 608 sheep skulls collected throughout the park. Age at death was estimated by counting growth rings on the horns.

**yourBioPortal.com**

GO TO **Working With Data • Dead Sheep Tell a Tale**

**SURVIVORSHIP CURVES** The construction of life tables has allowed ecologists to observe common patterns, reflecting common solutions to ecological challenges, across a tremendous diversity of organisms. For example, mortality data from a life table can be used to plot a **survivorship curve**. Survivorship curves are correlated with general patterns of life history traits, which in turn suggest the mortality factors the population faces in its environment. Ecologists classify survivorship curves into three types (**Figure 55.4A**):

- *Type I*, or physiological, survivorship curves are typical of organisms that experience high overall survivorship through adulthood (such as humans and many other large mammals). Parental care and low fecundity are typical of species with this type of survivorship curve.
- *Type II*, or ecological, survivorship curves are typical of organisms faced with a constant risk of mortality at all ages (such as most birds).
- *Type III*, or maturational, survivorship curves are typical of organisms that experience low juvenile survivorship (such as most insects and annual plants). Species with this type of survivorship curve tend to produce many offspring, but provide little or no parental care.

Note that survivorship curves can differ even among close relatives within a taxonomic group (**Figure 55.4B**).

55.1 RECAP

**To understand the dynamics of populations, ecologists measure population density, age structure, and dispersion patterns. Life tables can be constructed either by following a cohort of individuals through time or by recording age at death in a vertical life table.**

- What are some of the ways in which population density can be measured? See p. 1169
- What kinds of information do life tables provide about a population? See pp. 1171–1172 and Tables 55.1, 55.2
- Describe the three types of survivorship curves. See p. 1171 and Figure 55.4

Survivorship and fecundity are attributes of all populations, but even within species, environmental variation across a species'

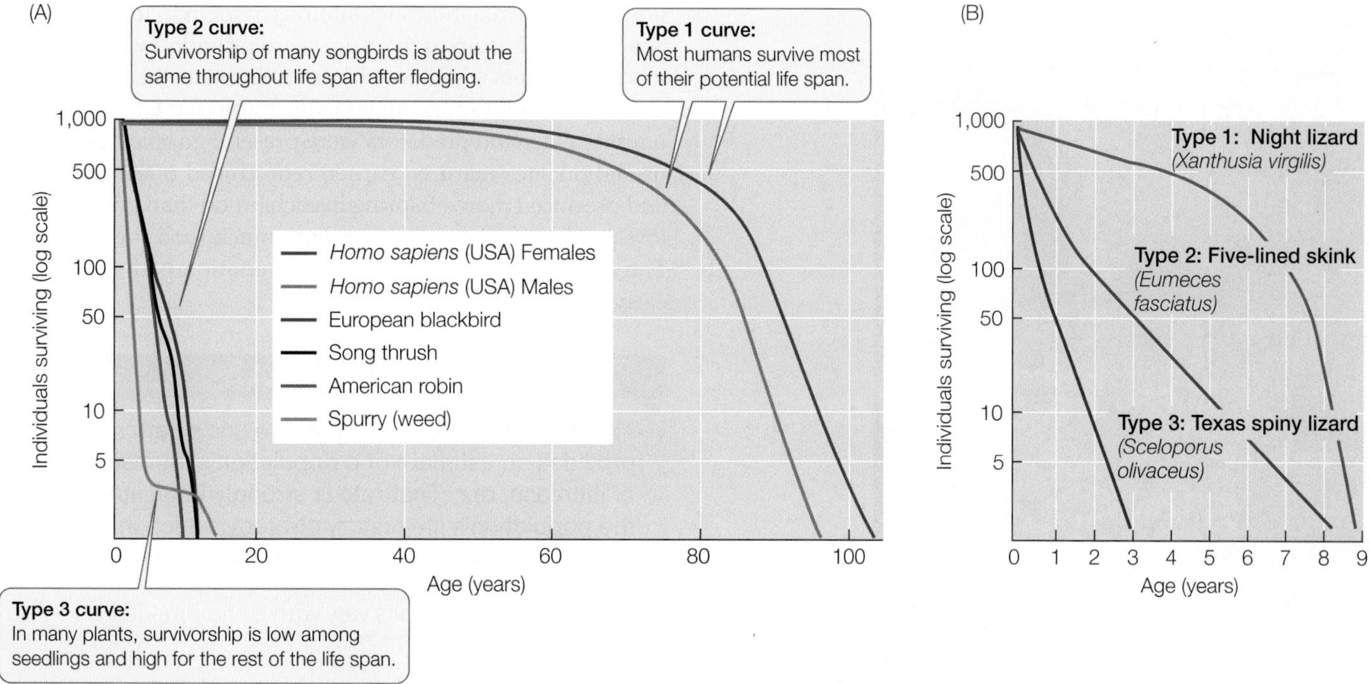


**55.4 Survivorship Curves** (A) Ecologists recognize three types of survivorship curves. (B) Even species that are closely related may have survivorship curves of different types, as indicated here by the survivorship curves of three species of lizards.

range influences these measures. Comparisons across populations and species reveal different patterns in life history traits, which allow organisms to cope with different environmental challenges.

# 55.2 How Do Environmental Conditions Affect Life Histories?

The way in which an organism partitions its time and energy among growth, maintenance, and reproduction is called its **life history strategy**. Because resources and mortality factors vary greatly among environments, life history strategies also vary dramatically. Those variations, in turn, determine how fast populations can grow.

## Survivorship and fecundity determine a population's intrinsic rate of increase

To see how a population is likely to grow, ecologists can use life table data to calculate the population's per capita growth rate, also known as the **intrinsic rate of increase**, symbolized as ***r***. A population's intrinsic rate of increase is the difference between the birth rate ($b$) and the death rate ($d$) per individual (*per capita*) (leaving aside, for the moment, immigration and emigration). It is expressed by the equation

$$r = b - d$$

If the birth rate is greater than the death rate, then $r > 0$, and the population is growing. If the death rate is greater than the birth rate, then $r < 0$, and the population is declining. The equilibrium state $r = 0$ would indicate a stable population that is neither growing nor declining significantly. When applied to the human population, it is sometimes referred to as "zero population growth," with implications we will see in Section 55.5.

## Life history traits vary with environmental conditions

Both survivorship ($l_x$) and fecundity ($m_x$) are highly habitat-dependent in practice, and the intrinsic rate of increase can thus change as the environment changes. The life history traits most influenced by environmental conditions include:

- age at first reproduction (generation time)
- number of broods per female (the number of times a female produces offspring)
- number of offspring per brood (the number of offspring produced each time a female reproduces)

These factors vary not only between species, but also between populations within species.

Opportunities for reproduction for some species are limited to certain locales or certain times of year, whereas other species and populations can breed continuously over their life span. Many desert wildflowers grow and flower only during the spring rainy season, and they may not be able to reproduce at all in years when rainfall is inadequate. In contrast, some tropical vines are able to flower continuously in their warm, moist rainforest environment.

Species that can reproduce multiple times over the course of their adult lives are **iteroparous** (*itero*, to repeat; *pario*, to beget). **Semelparous** species (*semel*, once), on the other hand, reproduce only once in their lives (**Figure 55.5**). Generally, semelparous species produce many more offspring in their single brood than iteroparous species do over their lifetime; semelparity is thus sometimes called "big bang" reproduction.

*Orgyia antiqua*

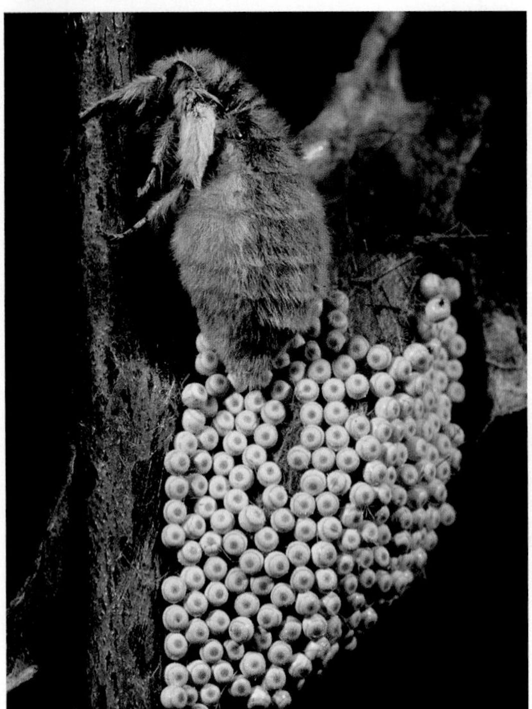

**55.5 Big Bang Reproduction** Semelparous species reproduce only once and invest a great deal of energy in producing the maximum number of offspring. Female rusty tussock moths do not fly but remain with their empty cocoons, which are attached to the plants that are the caterpillar-stage food source. The stationary female lays a large number of eggs and then dies. When the eggs hatch the following spring, the larvae are surrounded by foliage they can eat.

Semelparity is typical of organisms that experience no great survival advantage upon reaching adulthood, including some fishes, many insects, and all annual plants (i.e., organisms with a type III survivorship curve). In contrast, iteroparity is typical of organisms whose survival chances increase once they reach maturity (those with a type I or II survivorship curve). For example, because environmental conditions within the nests of social insects (such as honey bees and ants) are remarkably stable, iteroparity is the rule; some queens may live 10 years or longer and reproduce over their entire adult lives.

### Life history traits are influenced by interspecific interactions

Predation and other interactions among species can influence life history strategies in many ways. Some populations of guppies (*Poecilia reticulata*) in Trinidad, for example, live in streams where they are attacked and eaten by larger fish. But some streams have waterfalls that predatory fishes are unable to negotiate. Guppies that live in the predator-free areas upstream from those waterfalls have lower death rates than guppies below the falls. To see whether the risk of being eaten by a predator influenced the life history strategies of these guppies, David Reznick and his colleagues collected guppies from high-predation and low-predation sites and raised them in the laboratory. Some guppies from each group were provided with plentiful food, and others with limited food, to simulate the variation the fish would typically encounter in their home streams. In the laboratory, where no predators were present, guppies from high-predation sites matured earlier, reproduced more frequently, and produced more offspring in each brood than guppies from low-predation sites, no matter how much food they received. The investigators concluded that predation had selected for early and frequent reproduction.

## 55.2 RECAP

**The difference between birth rate and death rate provides an estimate of a population's intrinsic rate of increase, or *r*. That rate is strongly influenced by the population's life history strategy, which in turn is highly dependent on ecological conditions.**

- What life history traits vary with environmental conditions? See p. 1173
- Explain the difference between iteroparity and semelparity. See p. 1173
- How can predation affect the evolution of life history strategies? See p. 1174

A species may be rare in one location, yet elsewhere within its range it may be superabundant. A species that is rare in one locality one year may be abundant the next year. For any given species, environmental factors limit the growth of its populations in different places and at different times. Population ecologists work to identify and understand these factors.

## 55.3 What Factors Limit Population Densities?

What would happen if all the offspring produced by a population survived to reproduce themselves? The prospects are alarming. In 1911, L. O. Howard, then chief entomologist of the U.S. Department of Agriculture, estimated that a pair of flies beginning to reproduce in Washington, D.C., on April 15 could produce a population of 5,598,720,000,000 adults by September 10. Other entomologists took issue with Howard's calculation—they pegged the number much *higher*. Given such amazing reproductive capacities, it is clear there are forces at work that limit the growth of fly populations (and populations of every other organism).

### All populations have the potential for exponential growth

As the number of individuals in a population increases, the number of new individuals added per unit of time accelerates, even if the intrinsic rate of increase remains constant. If births and deaths occur continuously and at constant rates, a graph of

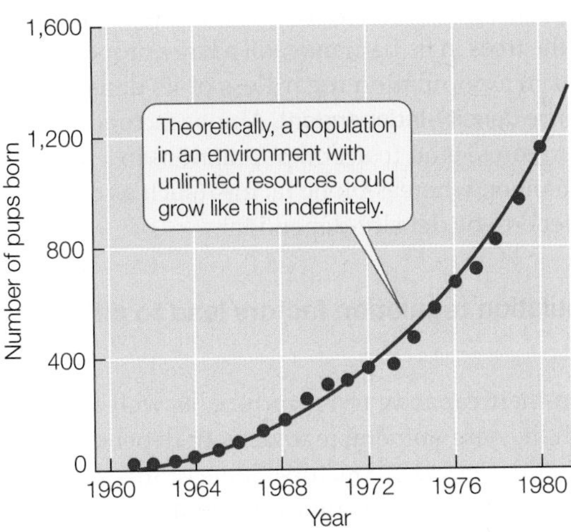


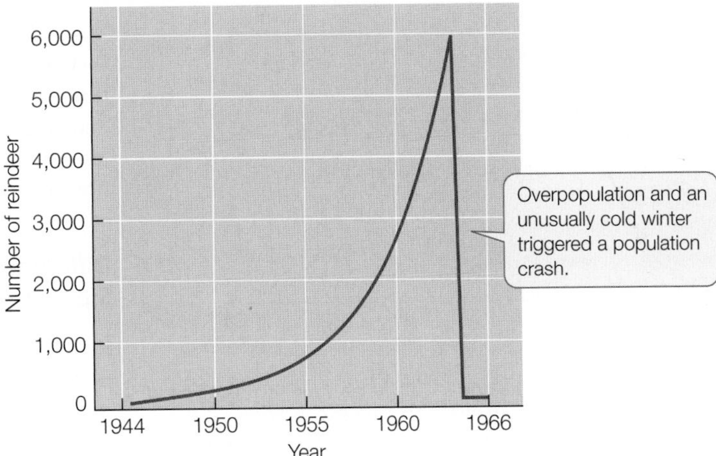


**55.6 Exponential Population Growth Can Lead to a Population Crash** (A) Abundant resources allow a population to grow exponentially, as did the elephant seal population on Año Nuevo Island, California. (B) The reindeer herd introduced on St. Matthew Island grew exponentially for many years despite a decreasing food supply. A single catastrophically cold winter triggered a population crash that eventually resulted in the death of the entire island reindeer population.

**yourBioPortal.com**

GO TO Animated Tutorial 55.1 • Exponential Population Growth

the population size over time forms a continuous upward curve (**Figure 55.6A**). This pattern, known as **exponential growth**, can be expressed as

$$\frac{\text{change in number of individuals}}{\text{change in time}} = b - d$$

$$\frac{\Delta N}{\Delta t} = b - d$$

Using the notation of differential calculus, which is better suited to short time intervals, this equation can be expressed as

$$\frac{dN}{dt} = (b - d)N$$

Because the intrinsic rate of increase, $r$, is equivalent to $b - d$, the **equation for exponential growth** can be simplified to

$$\frac{dN}{dt} = rN$$

The term $dN/dt$ is the rate of change in the size of the population over time, and the expression $rN$ is sometimes called the *biotic potential* of the population.

For very short periods, some populations may grow at rates close to their biotic potential. During the 20 years following their introduction, the reindeer population on St. Matthew, described at the opening of this chapter, grew exponentially. When the reindeer herd was introduced to the island, it had ample habitat, abundant food, and no predators, so there was nothing to limit the population's growth. Favorable climatic conditions also allowed the population to grow exponentially. A rapid change in those conditions—deep snow during one particularly cold winter—was a major factor leading to the population's crash (**Figure 55.6B**).

### Logistic growth occurs as a population approaches its carrying capacity

No real population can maintain exponential growth for very long. As a population increases in density, the resources it requires—such as food, nest sites, or shelter—become depleted. In the absence of adequate resources to sustain more individuals, birth rates drop and death rates rise.

Any given environment has only enough resources to support a finite number of individuals of a species indefinitely. That number of individuals, referred to as the environment's **carrying capacity** (***K***), is a function of its resources. The growth of a population typically slows down as its density approaches the environmental carrying capacity. A population that exhibits decreasing growth as resources become limiting will display a pattern called **logistic growth**, in which a graph of the population size over time forms an S-shaped curve. **Figure 55.7** shows this growth pattern in a laboratory population of beetles that was maintained on a constant food supply.

An S-shaped curve can be generated from the equation for exponential growth (left) by adding a term that slows the population's growth as it approaches carrying capacity:

$$K - \frac{N}{K}$$

The above term represents the *reduction in population growth caused by preemption of available resources* and is referred to as *environmental resistance*. As long as population size is less than carrying capacity (i.e., $N < K$), only a fraction of the available resources are being used. As the population approaches carrying capacity, however, the fraction of resources available for any new

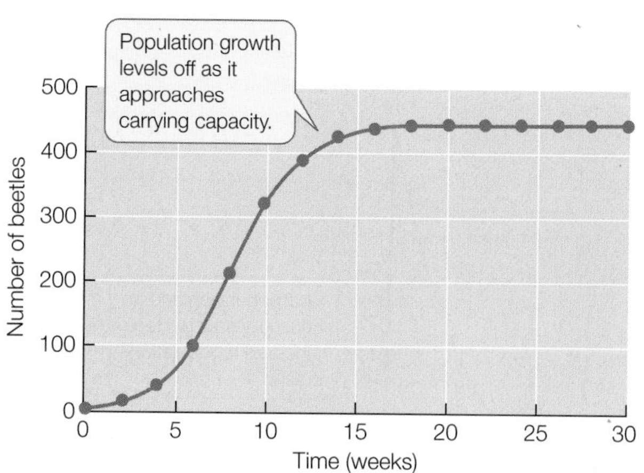


**55.7 Logistic Population Growth Levels Off** In an environment with limited resources, a population typically stops growing exponentially before it reaches the environmental carrying capacity (*K*). The data here was recorded from a laboratory population of sawtoothed grain beetles maintained on a constant food supply; it is a typical logistic growth pattern, which forms an S-shaped curve.

**yourBioPortal.com**

GO TO **Web Activity 55.1 and Animated Tutorial 55.2 • Logistic Population Growth**

individual becomes smaller. The implication is that each individual added to the population depresses population growth by an equal amount. Thus,

$$\text{rate of change in population size over time} = \text{biotic potential} \times \text{environmental resistance}$$

or, in mathematical terms,

$$\frac{dN}{dt} = rN\frac{K-N}{K}$$

Population growth stops when $N = K$ because at that point, $K - N = 0$, so $(K - N)/K = 0$, and thus $dN/dt = 0$.

### Population growth can be limited by density-dependent or density-independent factors

When resources are limited, adding more individuals to a population runs the risk of making things worse for everyone. Factors with an effect on population size that increases in proportion to population density are called **density-dependent** regulation factors. These include:

- *Food supply.* As a population increases, it may deplete its food supply, reducing the amount of food available to each individual. Poor nutrition may then increase the death rate and/or decrease the birth rate.
- *Predators* may be attracted to areas with high densities of their prey. If predators capture a larger proportion of the prey population than they did when that population was small, the death rate of the prey population rises.
- *Pathogens* spread more easily in dense populations than in populations with fewer individuals per unit of area, resulting in a rise in the death rate.

Not all population regulation factors act in a density-dependent manner. A period of extreme cold or a hurricane that blows down most of the trees in its path may kill a large proportion of the individuals in a population regardless of its density, and thus are said to be **density-independent**. Abiotic factors (such as extreme temperatures) tend to act on populations in a density-independent manner, whereas biotic factors (such as competition for food) tend to be density-dependent.

### Different population regulation factors lead to different life histories

Species vary in their capacity to reproduce, as well as in the extent to which they are vulnerable to density-dependent and density-independent mortality factors. Some of this variation in life history traits appears to result from adaptation to different habitat conditions. Generally, unpredictable habitats are associated with high fecundity and correspondingly high *r*, as organisms make the most of rare opportunities to reproduce; conversely, predictable habitats, where organisms have a high probability of reproductive success, are associated with low fecundity and low *r*.

Species whose life history strategies allow for high intrinsic rates of increase are called ***r*-strategists**, and species whose life history strategies allow them to persist at or near the carrying capacity (*K*) of their environment are called ***K*-strategists** (**Figure 55.8**). Many species display elements of both strategies.

For *r*-strategists, life is uncertain. Individuals tend to reproduce only once and to produce large numbers of offspring. They can generally tolerate a wide range of resource conditions. *K*-strategists are adapted to predictable environments, are long-lived, and reproduce several times; their smaller numbers of offspring have a high probability of surviving to adulthood. *K*-strategists tend to be more specialized in their resource use and less tolerant of variation in resource quality. That life history strategies can evolve is suggested by genetic correlations among suites of life history traits. Such genetic correlations imply either simultaneous selection on two life history traits or linkages among the genes that code for those traits. Across *Drosophila melanogaster* strains, for example, intrinsic rate of increase is positively correlated with the ability to reproduce under starvation conditions and with the ability to develop on a variety of media in the laboratory—both of which are consistent with the *r* strategy of tolerating a wide range of conditions.

### Several factors explain why some species achieve higher population densities than others

Density-dependent and density-independent factors can explain how populations grow or decline, but they do not explain why some species are common whereas others are rare—that is, why the characteristic densities of species differ. Many factors explain why typical population densities vary so greatly among species, but four of them are especially influential:

- *Species that use abundant resources generally reach higher population densities than species that use scarce resources.* Thus, on average, the fruit fly *Drosophila melanogaster*, which feeds on

| | *r*-strategists | *K*-strategists |
|---|---|---|
| HABITAT | Can inhabit a broad range of habitats. High tolerance for both environmental instability and low-quality resources. | Specific habitat requirements, including environmental stability. Efficient users of specific and usually high-quality resources. |
| PHYSIOLOGY | Rapid embryonic development, rapid maturation to reproductive age, small body size. | Extended embryonic development, long maturation to reproductive age, large body size. |
| REPRODUCTIVE STRATEGY | Random mating. Reproduce once (semelparity) resulting in a large number of offspring. Little or no parental investment in each offspring. | Mate choice, pair bonds. Reproduce many times (iteroparity), each event producing few offspring. Large parental investment in each offspring. |
| SURVIVORSHIP | Short life span, density-independent mortality, typically a Type 3 survivorship curve (see Figure 55.4). | Long life span, density-dependent mortality, typically a Type 1 or 2 survivorship curve (see Figure 55.4). |
| POPULATION FLUCTUATION | Short periods of exponential population growth (*r*) followed by periodic or seasonal population crashes. | Slowly rising population growth that stabilizes and levels off at carrying capacity (*K*). |
| EXAMPLES | Dandelions, house flies, rabbits | Oak trees, bluebirds, polar bears |

**55.8 Two Life History Strategies** Species whose life histories are geared to achieve the maximum possible rate of population increase are referred to as *r*-strategists; those whose population dynamics are bounded by carrying capacity are *K*-strategists. The life histories of most species combine elements of both types.

yeasts and other microbes found on just about any kind of rotten fruit, reaches substantially higher population densities than do other fruit fly species that feed on the microbes found on specific fruits.

- *Species with small body sizes generally reach higher population densities than species with large body sizes.* In general, population density decreases as body size increases because, on a per capita basis, small individuals require less energy to survive than large individuals.
- *Complex social organization may facilitate high population densities.* Highly social species, including ants, termites, and humans, can achieve remarkably high population densities.
- *Some newly introduced species reach high population densities.* Species that are introduced into a new region, where their normal predators and pathogens are absent, sometimes reach population densities much higher than those in their native ranges. Sometimes these high population densities are only temporary; these densities decline if and when new mortality factors exert an influence. However, in many cases, in the absence of such factors, populations in the newly colonized habitat remain so dense that the introduced species becomes a major problem for native species.

The population of zebra mussels (*Dreissena polymorpha*) in North America demonstrates the speed with which newly introduced populations can grow. Zebra mussels first appeared in Lake St. Clair, between Lake Erie and Lake Huron, in 1988. They were probably transported there from their native Europe in the ballast water of transoceanic cargo ships. They spread rapidly and today occupy most of the Great Lakes and the Mississippi River drainage (**Figure 55.9**), reaching densities as high as 400,000 individuals per square meter in some places. Because they attach to any stable un-

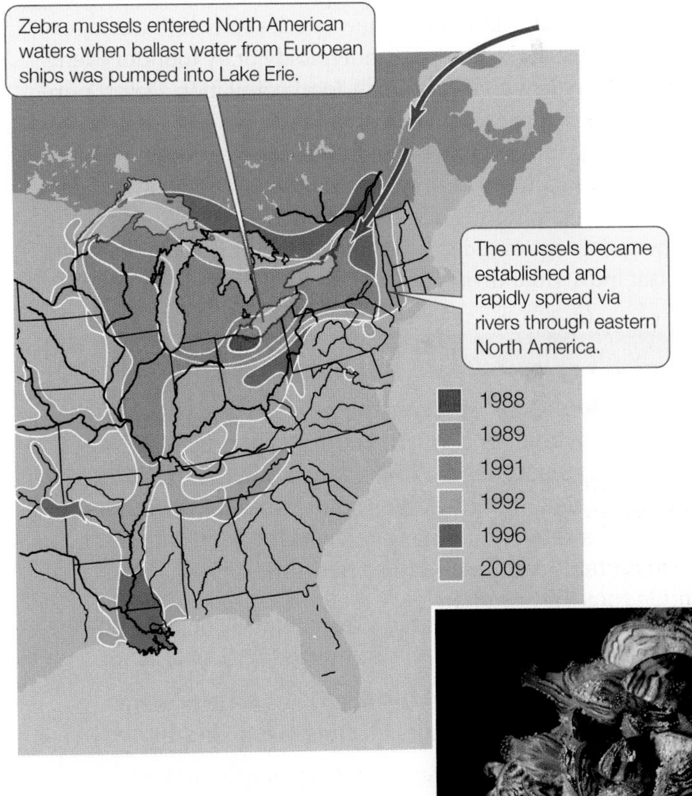

Zebra mussels entered North American waters when ballast water from European ships was pumped into Lake Erie.
The mussels became established and rapidly spread via rivers through eastern North America.
1988
1989
1991
1992
1996
2009

**55.9 Newly Introduced Populations Can Grow Rapidly** Between 1989 and 2005, the range of zebra mussels in eastern North America increased exponentially. Humans can unwittingly transport zebra mussel larvae from one body of water to another in their fishing boats and other watercraft, and in recent years the invasive pest has begun to appear in lakes and streams of the American West.

*Dreissena polymorpha*

derwater substratum, zebra mussels can cover the bottoms of boats and docks and clog municipal water supply intakes and power plant pipelines. They even settle on other aquatic organisms, causing problems for native mussels. Such high densities are never found in their native Europe, where over 100 species of predators and parasites keep their population densities under control.

### Evolutionary history may explain species abundances

The four key factors that explain variation in population densities, as important as they are, cannot explain all differences in species abundances. For example, although Douglas firs and giant sequoias are both large trees that use the same sources of energy (sunlight) and nutrients, Douglas firs are widespread and abundant in western North America, whereas giant sequoias are restricted to a few groves in southern California. Similarly, each of several species of desert pupfish is restricted to a single spring in Death Valley, California, whereas smallmouth bass live in most of the rivers and lakes in eastern North America. To explain these differences, it is important to know not just the contemporary ecology of these organisms, but also their long-term evolutionary history.

As Section 23.2 describes, a new species can originate in several ways. Species that arise by polyploidy or by founder events inevitably begin with a very small, local population. Desert pupfish species appear to have evolved in isolation as increasing aridity in Death Valley over the past 50,000 years cut once continuous populations off from one another. Conversely, when a species is declining toward extinction (as may be happening to the giant sequoia), its range shrinks until it vanishes when the last individual dies.

## 55.3 RECAP

**Population sizes are limited by the carrying capacity of the environment, which is determined by the availability of resources as well as by biotic interactions. Species associated with unpredictable habitats tend to be *r*-strategists, and species associated with predictable habitats tend to be *K*-strategists.**

- Why can populations grow exponentially only for short periods? See p. 1175 and Figures 55.6 and 55.7
- What is the difference between density-dependent and density-independent factors? See p. 1176
- Describe the characteristics of *r*-strategists and *K*-strategists. See p. 1176 and Figure 55.8

All species, no matter how abundant, are found only in those habitats in which they can survive and reproduce well enough to persist over time. Yet a species is rarely found in all of the habitats that seem suitable for it. The evolutionary histories of species supply one explanation for this observation. The next section explores another explanation: spatial variation in habitat suitability.

## 55.4 How Does Habitat Variation Affect Population Dynamics?

Most natural history field guides display maps that show the geographic range over which a species is found. But no species, not even the most abundant, is found everywhere within its mapped range. Every species has particular habitat requirements that determine where within its range it will occur.

### Many populations live in separated habitat patches

Most organisms live in distinct **habitat patches**: areas of a particular habitat type that are surrounded by other, less suitable habitats. Some populations living in separated habitat patches are effectively divided into separate, discrete *subpopulations*, linked together by regular movement of individuals between patches. The larger population to which such subpopulations belong is known as a **metapopulation**.

Each subpopulation has a probability of "birth" (colonization of its habitat patch) and "death" (extinction in that patch). Each subpopulation grows in the ways we have described, but because the subpopulations are much smaller than the metapopulation, local disturbances and random fluctuations in numbers of individuals are more likely to cause the extinction of a subpopulation than of the entire metapopulation. However, if individuals move frequently between subpopulations, immigration may prevent declining subpopulations from becoming extinct, a process called the **rescue effect**.

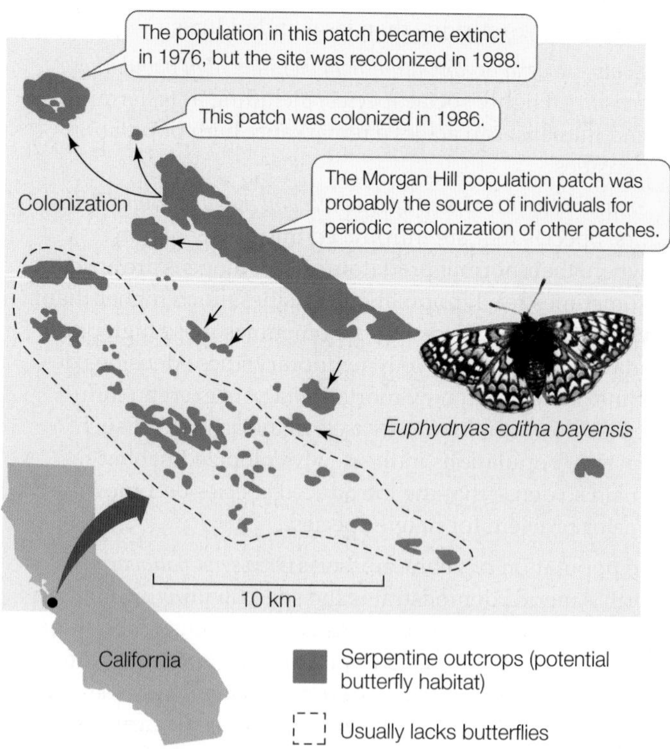


**55.10 A Checkerboard of Checkerspots** The Bay checkerspot butterfly metapopulation is divided into a number of subpopulations confined to patches of habitat (serpentine rock outcrops) that contain its food plants.

The Bay checkerspot butterfly (*Euphydryas editha bayensis*) provides a dramatic illustration of the dynamics of metapopulations. The caterpillars of this butterfly feed on only two species of annual plants, California plantain and purple owl's clover, which are restricted to outcrops of serpentine rock on hills south of San Francisco, California. In 1960, Paul Ehrlich and his colleagues at Stanford University began studying a population of this butterfly in the Jasper Ridge Biological Preserve near the Stanford campus. They determined that the Jasper Ridge population was actually a subpopulation within a large, very fragmented metapopulation. They found that, over the years, three subpopulations within this metapopulation varied enormously and asynchronously in size. Larval survival depends on climatic factors, particularly temperature, the timing of rainfall, and host plant survival. During drought years, most host plants die early in spring, before the caterpillars have developed enough to enter their summer resting stage.

At least three butterfly subpopulations went extinct during a severe drought in 1975–1977. One of the empty patches was repopulated a few years later, most likely by individuals from the largest single subpopulation, near Morgan Hill, which as late as 1989 contained several hundred thousand butterflies (**Figure 55.10**). In 1998, however, the Morgan Hill subpopulation went extinct. Ehrlich and colleagues examined 70 years of climate data for the region and concluded that increasing climatic variation accounted for the extinction. Without a stable source subpopulation to provide emigrants for recolonization, it is likely that none of the other subpopulations will persist without human intervention.

### Corridors allow subpopulations to persist

In any metapopulation, connections between patches, known as **corridors**, play a critical role in facilitating dispersal to maintain subpopulations. Studying corridors is experimentally daunting because long distances may separate patches; moreover, movements of animals, particularly those that fly, can be difficult to monitor. Therefore, to test the importance of corridors, ecologists conducted a small-scale, manipulative experiment using mosses growing on rocks, which provide habitat for a number of small arthropod species, including springtails (minute wingless hexapods) and mites.

yourBioPortal.com
GO TO Animated Tutorial 55.3 • Habitat Fragmentation

In one experiment, the investigators created patches of habitat by clearing away the mosses surrounding the patches (**Figure 55.11**). In small, completely isolated patches, the number of small arthropod species present declined about 40 percent within a year. The investigators also created patches that were connected by narrow corridors of moss. In some cases the corridors were left intact; in others, "pseudocorridors" were disrupted by a barrier 10 mm wide. A 10-mm barrier may seem small, but it presents a daunting obstacle to arthropods only 2 mm wide. Six months later, patches connected by unbroken corridors contained more small arthropod species than did patches connected by the disrupted pseudocorridors.

## INVESTIGATING LIFE

**55.11 Corridors Can Affect Metapopulations**
The manipulative experiments summarized here suggest that corridors between patches of fragmented habitat increase the chances of recolonization and thus subpopulation persistence.

**HYPOTHESIS** **Subpopulations of a fragmented metapopulation are more likely to persist if there is no barrier to recolonization.**

**METHOD**

1. Determine the number of species of small organisms (mostly arthropods) living in an established habitat of mosses growing on a large rock.
2. Fragment the habitat by trimming the mosses into patches of various configurations, leaving bare rock surrounding the patches.

Isolated 20-cm patches
20-cm patches connected by pseudocorridors (with gaps)
20-cm patches connected by corridors

3. After 6 months, determine the number of species present in the various patch configurations.

**RESULTS**

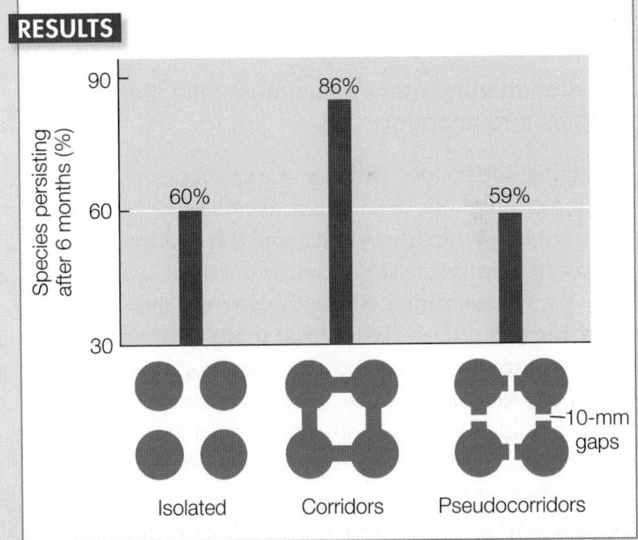


**CONCLUSION** **Even small barriers to recolonization reduce the number of subpopulations persisting in a fragmented habitat.**

**FURTHER INVESTIGATION:** These studies involved the dispersal of very small organisms over very short distances over a very short span of time. How might the longer term effects of imposing barriers to dispersal be investigated?

Go to yourBioPortal.com for original citations, discussions, and relevant links for all INVESTIGATING LIFE figures.

## 55.4 RECAP

**A metapopulation consists of separate subpopulations living in distinct habitat patches, among which individuals move regularly by means of corridors.**

- What effects do patches of unsuitable habitat have on population structure? See p. 1178 and Figure 55.10
- What effects do corridors between habitat patches have on subpopulations? See p. 1179 and Figure 55.11

For many centuries, people have tried to reduce populations of species they consider undesirable and maintain or increase populations of desirable or useful species. Such efforts to manage populations are more likely to be successful if they are based on knowledge of how those populations grow and what determines their densities.

# 55.5 How Can Populations Be Managed Scientifically?

If we wish to increase or decrease populations of other species, we need to understand the life histories and the population dynamics of the organisms we wish to manage. The principles of population dynamics can also help us understand the effects our own population and its activities are having on other species.

### Population management plans must take life history strategies into account

Knowing the life history strategy of a species can be helpful in managing populations of commercial value. The black rockfish (*Sebastes melanops*), an important game fish that lives off the Pacific coast of North America, provides one such example. Rockfish have an indeterminate growth pattern—they continue to grow throughout their lives. As in many other animals, the number of eggs a female produces is proportional to her size, so larger females can produce more eggs than smaller females. In addition, older, larger females are better able to provision the eggs they produce with oil droplets, which provide energy to the newly hatched larvae, giving them a head start in life (**Figure 55.12**). Larvae from eggs with larger oil droplets, produced by larger females, grow faster and survive better than do larvae from eggs with smaller oil droplets.

These life history traits have important implications for the management of rockfish populations. Because fishermen prefer to catch big fish, intensive fishing off the Oregon coast from 1996 to 1999 reduced the average age of female rockfish from 9.5 to 6.5 years. Thus the females reproducing in 1999 were, on average, smaller than the females reproducing in 1996. This change decreased the average number of eggs produced by females and reduced the average growth rate of larvae by about 50 percent. Maintaining productive populations of rockfish may require setting aside no-fishing zones where some females can be protected from fishing and allowed to grow to large sizes.

### Population management plans must be guided by the principles of population dynamics

If we look at a logistic growth curve (see Figure 55.7), we can see that the number of births tends to be highest when a population is well below its carrying capacity. Therefore, if we wish to maximize the number of individuals that can be harvested from a population, we should manage the population so that it is far enough below the carrying capacity to have a high birth rate. Hunting and fishing regulations are established with this objective in mind.

Populations that have high reproductive capacities can persist even if harvest rates are high. In such populations (which include many fish species), each female may produce thousands or millions of eggs. In many of these fast-reproducing populations, the growth rates of individuals are density-dependent. Therefore, if prereproductive individuals are harvested at a high rate, the remaining individuals may grow faster. Some fish populations can be harvested heavily on a sustained basis because a relatively small number of females can produce sufficient numbers of eggs to maintain the population.

(A) *Sebastes melanops*

(B)

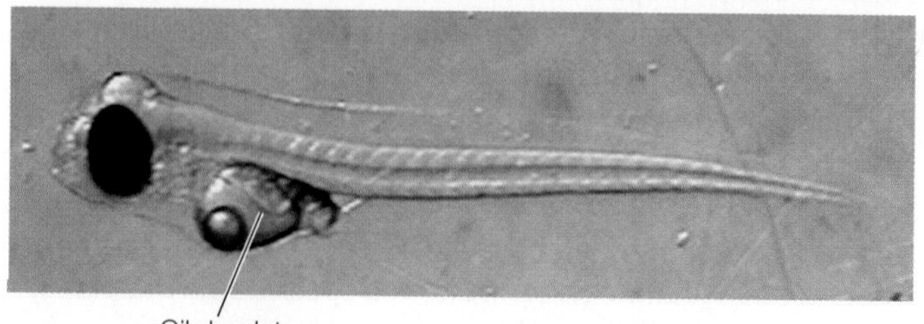


**55.12 Energy Stocks Give Rockfish a Head Start** (A) Among black rockfish, older, larger females are more reproductively successful, producing both more eggs and eggs with larger nutritive oil droplets. (B) The oil droplet attached to this rockfish larva provides it with nutrition to fuel its growth until it can feed on its own.

**55.13 Overharvesting Can Reduce Fish Populations** Populations of cod and haddock on Georges Bank have crashed due to overfishing.

Fish can, of course, be overharvested, as illustrated by the story of the black rockfish. Many fish populations have been greatly reduced because so many individuals were harvested that the few surviving reproductive adults could not maintain the population. Georges Bank, off the coast of New England—a source of cod, haddock, and other prime food fishes—was exploited so heavily during the twentieth century that many fish stocks have been reduced to levels insufficient to support a commercial fishery (**Figure 55.13**). The haddock population has rebounded enough to support a fishery because commercial fishing of that species ceased and was restarted only after the population had recovered. In contrast, managers reduced fishing pressure on cod only slowly, and the cod population has failed to increase.

Many rapidly reproducing species can recover if overharvesting is stopped, but recovery is more difficult for slowly reproducing species. Twentieth-century whalers hunted the blue whale (*Balaenoptera musculus*), Earth's largest animal, nearly to extinction. These whales reproduce very slowly: they live up to 10 years before becoming reproductively mature, produce only one offspring at a time, and have long intervals between births. Not surprisingly, the population has failed to recover.

Whether we want to manage the sizes of populations of desirable species for sustainable harvesting or of undesirable species for control purposes, the same principles apply. If the dynamics of a pest population are influenced primarily by density-dependent factors, then killing part of that population will only reduce it to a density at which it reproduces at a higher rate. A more effective approach to reducing such a population is to remove its resources, thereby lowering the carrying capacity of its environment. For example, we can rid our cities of rats more easily by making garbage unavailable (reducing the carrying capacity of the rats' environment) than by poisoning rats (which only increases their reproductive rate).

**Biological control** is the use of the natural enemies (predators, parasites, or pathogens) to reduce the population density of an economically damaging species (see the opening of Chapter 30). In many cases, the target species is a pest only because it has been introduced to a new area. Natural enemies used for biological control are often obtained from the native region of the pest species. Biological control became popular in the nineteenth century after an outbreak of cottony-cushion scale, an Australian insect that attacks citrus, appeared in the citrus groves in California. A predaceous ladybeetle and a parasitic fly were then introduced from Australia and, within a year of their release, brought the scales under control.

On occasion, introduced predators and parasites not only fail to have any effect on the pest they were imported to control but, freed of their own enemies, they themselves become pests. This fact underlies the horror story of the cane toad (*Bufo marinus*) in Australia. These Central American toads (**Figure 55.14**) were introduced to control cane beetles attacking Australian sugarcane fields. But Australian cane beetles stay high on the upper stalks of the plants; the toads could not reach that high, and thus had no effect on the beetle population. Unfortunately, they had massive effects on other species.

All stages of the *B. marinus* life cycle are poisonous, and Australian reptiles (including snakes and lizards) and mammals that eat them usually die. With no enemies to limit their population growth, cane toads grow fast and outcompete native amphibian species for resources. The toads have spread from northern Australia down the east coast, where they threaten native frog species. The Australian government is forced to spend millions of dollars in attempting to reduce their numbers.

### Human population increase has been exponential

In the nineteenth century, the historian Thomas Carlyle declared economics to be "the dismal science." He was referring in part to a famous essay written a century earlier by Thomas Malthus, in which Malthus pointed out that the human population was growing exponentially but its food supply was not, and at some point famine and death would be the ultimate fate of the

**55.14 Biological Control Gone Awry** The Central American cane toad not only failed to control destructive beetles in Australia's sugarcane fields, but increased dramatically in abundance and now threatens many native Australian species.

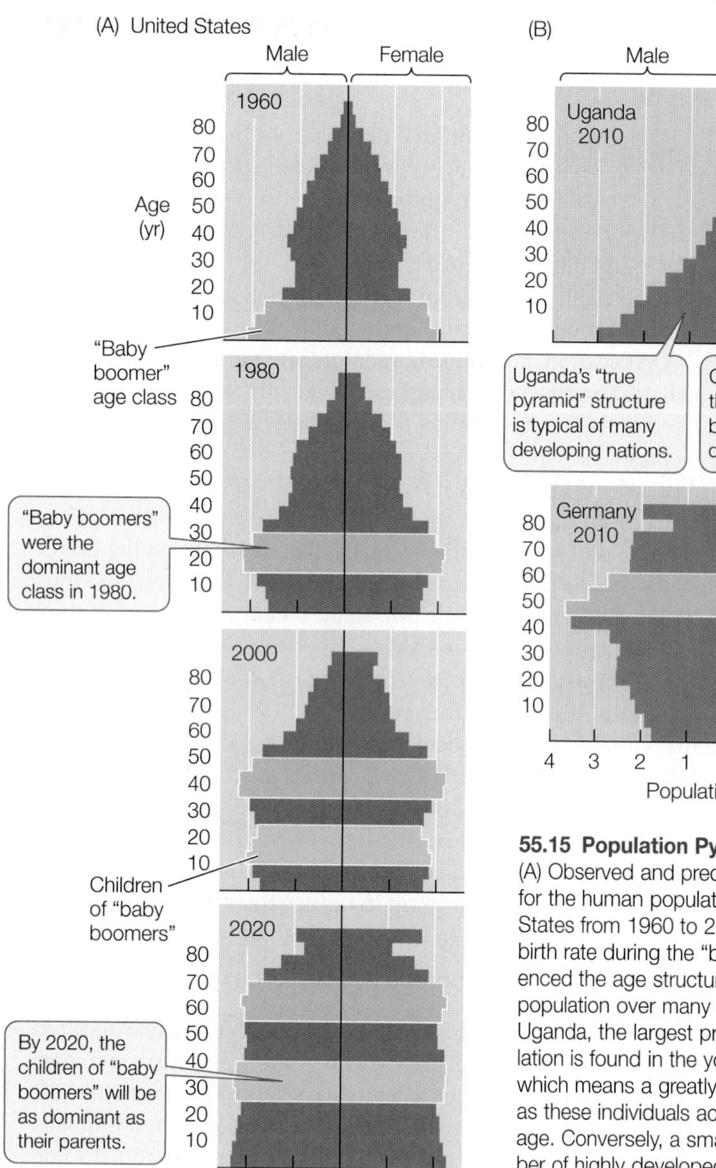


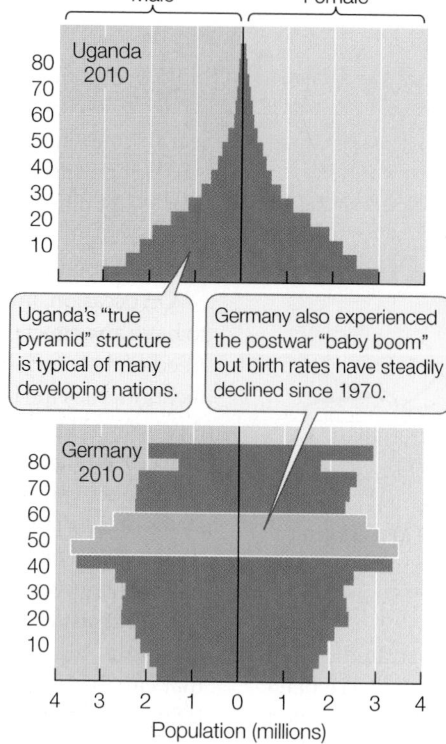


**55.15 Population Pyramids**
(A) Observed and predicted age distributions for the human population of the United States from 1960 to 2020 show how the birth rate during the "baby boom" has influenced the age structure of this country's population over many decades. (B) In Uganda, the largest proportion of the population is found in the youngest age groups, which means a greatly increased birth rate as these individuals achieve reproductive age. Conversely, a small but increasing number of highly developed nations have the population structure seen here for Germany, which presages a population decline.

human race. Neither Malthus nor Carlyle could have anticipated the technological innovations that greatly enhanced the capacity of humans to produce food. Today, however, the size of the human population is once again a serious concern as we confront the effects of our contributions to pollution, widespread habitat destruction, and the extinction of other species.

For thousands of years, Earth's carrying capacity for humans was low due to the relative inefficiency with which humans could obtain food and water. The development of social systems and communication, the domestication of plants and animals, ever-increasing crop and livestock yields through ongoing technological advances, and increasing proficiency at managing diseases have all contributed to unprecedented growth of the human population. It took more than 10,000 years for the population to reach 1 billion, which happened at some point in the late nineteenth century. Today, a mere 125 years later, the planet is home to approximately 7 billion human beings. Population growth has slowed somewhat from its post-World War II highs—the World Resources Institute estimates the current worldwide rate of increase to be about 1.1 percent per year—but with a base of 7 billion, even a minimal growth rate means millions more individuals.

Human populations are not growing at the same pace across the world, however. As we saw in Figure 55.2, in populations of long-lived species such as elephants and humans, the timing of births and deaths affects the age distribution for many years. Between 1946 and 1964, the United States experienced a "baby boom." During those years, almost 75 million babies were born, and the average number of children per family grew from 2.5 to 3.8. Birth rates declined during the 1960s, but in the 1970s and 1980s, the baby boomers became parents, generating another demographic bulge—a "baby boom echo" (**Figure 55.15A**). Today, the echo babies are entering college in record numbers (and reading about themselves in introductory biology textbooks), and are on the threshold of becoming the dominant age class.

The age structure of the U.S. population is typical of many industrialized nations, but a few highly developed nations (particularly in Europe) are experiencing population declines. In the developing world, however, many countries are experiencing exponential growth rates and have populations highly skewed toward younger age classes, portending high rates of population growth in the future (**Figure 55.15B**).

## 55.5 RECAP

**Efforts to manage populations are more likely to be successful if they are based on an understanding of life histories and population dynamics.**

- Describe an effective strategy for reducing a pest population and explain why it is effective. **See p. 1181**
- How have humans changed the carrying capacity of Earth for our own population? **See p. 1182**

Earth's carrying capacity for humans is set in part by the biosphere's ability to absorb the by-products—especially carbon dioxide—of our enormous consumption of fossil fuel energy; by water availability; and by whether we are willing to tolerate the extinction of millions of other species in order to accommodate our prodigious use of Earth's resources. We enumerate some of the effects of human resource use on other species in Chapters 58 and 59, but first Chapters 56 and 57 discuss the nature of species interactions and the formation of ecological communities.

# CHAPTER SUMMARY

## 55.1 How Do Ecologists Study Populations?

- A **population** consists of the individuals of a species that interact with one another within a particular area at a particular time.
- The **density** of a population is the number of individuals per unit of area or volume.
- The **age structure** and **dispersion** pattern of a population influence its dynamics. Review Figures 55.2 and 55.3
- **Life tables** provide summaries of demographic events in a population. A cohort life table tracks a **cohort** of individuals born at the same time and records the **survivorship** or **fecundity** of those individuals over time. Review Tables 55.1 and 55.2
- Life table data can be used to construct a **survivorship curve**. Review Figure 55.4

## 55.2 How Do Environmental Conditions Affect Life Histories?

- The **life history strategy** of an organism describes how it partitions its time and energy among growth, maintenance, and reproduction.
- A population's **intrinsic rate of increase** (*r*) is the difference between the per capita birth rate (*b*) and the per capita death rate (*d*).
- Life history traits within a species may vary with habitat and interactions with other species can influence the evolution of life history traits.

## 55.3 What Factors Limit Population Densities?

- Populations can exhibit **exponential growth** for short periods, but eventually their resources become depleted, causing birth rates to drop and death rates to rise. Review Figure 55.6, **ANIMATED TUTORIAL 55.1**
- **Logistic growth** is the pattern seen when the growth of a population slows as its density approaches the environmental **carrying capacity** (*K*). Review Figure 55.7, **ANIMATED TUTORIAL 55.2 and WEB ACTIVITY 55.1**
- Species that are *r*-strategists have life histories that allow for high intrinsic rates of increase. *K*-strategists persist at or near the carrying capacity (*K*) of their environment. Many species' life history strategies fall along a continuum between these two extremes. Review Figure 55.8
- Population densities are determined by both **density-dependent** and **density-independent** factors. Several factors—including resource abundance, body size, social organization, and the length of time a species has occupied an area—influence population densities.

## 55.4 How Does Habitat Variation Affect Population Dynamics?

- No species is found everywhere within its mapped range. Members of most species live in distinct **habitat patches**.
- A **metapopulation** consists of separate subpopulations among which some individuals move on a regular basis. Review Figure 55.10
- Extinction of a subpopulation may be prevented by immigration of individuals from another subpopulation, a process known as the **rescue effect**. **Corridors** between patches may facilitate such movement. Review Figure 55.11, **ANIMATED TUTORIAL 55.3**

## 55.5 How Can Populations Be Managed Scientifically?

- To manage populations, it is important to understand their life histories and population dynamics. To maximize the number of individuals that can be harvested from a population, the population should be kept well below carrying capacity.
- Reducing the carrying capacity of the environment for a pest species is a more effective way to reduce its population than killing its members.
- Earth's carrying capacity for humans depends on our use of resources and the effects of our activities on the environment. Human populations grow at different rates in different parts of the world. Review Figure 55.15

# SELF-QUIZ

1. The way in which the individuals in a population are spread over the environment describes that population's
   *a.* dynamics.
   *b.* regulation.
   *c.* dispersion.
   *d.* strategy.
   *e.* survivorship.
2. The dispersion pattern in which the presence of one individual increases the likelihood of another individual being nearby is
   *a.* random.
   *b.* regular.
   *c.* clumped.
   *d.* uniform.
   *e.* irregular.
3. Which of the following is *not* a demographic event?
   *a.* Growth
   *b.* Birth
   *c.* Death
   *d.* Immigration
   *e.* Emigration
4. A group of individuals born at the same time is known as a
   *a.* deme.
   *b.* subpopulation.
   *c.* Mendelian population.
   *d.* cohort.
   *e.* taxon.
5. A population whose size levels off at its carrying capacity is exhibiting
   *a.* exponential growth.
   *b.* geometric growth.
   *c.* logistic growth.
   *d.* J-shaped growth.
   *e.* negative growth.

6. The process by which immigrants prevent a subpopulation from becoming extinct is called the
   *a.* colonization effect.
   *b.* rescue effect.
   *c.* metapopulation effect.
   *d.* genetic drift effect.
   *e.* salvage effect.
7. Density-dependent factors are population-regulating factors that
   *a.* affect the same proportion of individuals regardless of population size.
   *b.* are usually weather-related.
   *c.* affect only sessile organisms.
   *d.* influence only fecundity.
   *e.* have effects that vary depending on population size.
8. A metapopulation is
   *a.* an unusually large population.
   *b.* a population that is spread out over a very large area.
   *c.* a group of subpopulations among which some individuals move occasionally.
   *d.* a group of subpopulations that are isolated from one another.
   *e.* a group of subpopulations among which individuals move on a regular basis.
9. The best way to reduce the population of an undesirable species in the long term is to
   *a.* reduce the carrying capacity of the environment for the species.
   *b.* selectively kill reproducing adults.
   *c.* selectively kill prereproductive individuals.
   *d.* attempt to kill individuals of all ages.
   *e.* sterilize individuals.
10. Populations that are most readily overharvested are characterized by having
   *a.* very long-lived adults.
   *b.* short prereproductive periods and many offspring.
   *c.* short prereproductive periods and few offspring.
   *d.* long prereproductive periods and few offspring.
   *e.* long prereproductive periods and many offspring.

## FOR DISCUSSION

1. Most organisms that humans manage for higher densities are long-lived and have low reproductive rates, whereas most organisms that humans want to reduce are short-lived but have high reproductive rates. What is the significance of these differences for management strategies and the effectiveness of management practices?
2. In the mid-nineteenth century, the human population of Ireland was largely dependent on a single food crop, the potato. When a disease caused the potato crop to fail, the Irish population declined drastically for three reasons: (1) a large percentage of the population emigrated to the United States and other countries; (2) the average age of a woman at marriage increased from about 20 to about 30 years; and (3) many families starved to death rather than accept food from Britain. None of these social changes was planned at the national level, yet all contributed to adjusting the population size to the new carrying capacity. Discuss the ecological principles involved, using examples from other species. What would you have done had you been in charge of the national population policy for Ireland at that time?
3. One method of controlling introduced pest species is to introduce a natural enemy (a predator, parasite, or pathogen) from the pest's native habitat to reduce its population density. However, some species introduced to control a pest have become pests themselves. Some scientists argue that biological controls should not be used under any circumstances for pest management. Others argue that, provided they are properly studied and thoroughly vetted, we should continue to use biological control organisms as part of our set of tools for managing pests. Which view do you support, and why?

## ADDITIONAL INVESTIGATION

The arthropods whose metapopulation dynamics were studied in the experiment described in Figure 55.11 were all tiny animals with limited dispersal abilities. How could experiments be designed to test the influence of corridors on the metapopulation dynamics of species such as birds, lizards, and mammals, which readily disperse across large areas? Would you expect the results of such experiments to be similar to those using small arthropods on rocks? Why or why not?

## WORKING WITH DATA (GO TO yourBioPortal.com)

**Dead Sheep Tell a Tale** Vertical life tables are constructed by sampling a population at a single point in time, across all age classes and life stages (thus their designation as "vertical"). One way to construct such a table is to record information from a *death assemblage*—the remains or fossils of the organisms in a given population. In this exercise, you will use data from a 1944 study of the skulls of more than 600 Dall sheep (*Ovis dalli dalli*) in Denali National Park, Alaska to create a vertical life table and survivorship curve for this population.

**Habitat Fragmentation** Figure 55.11 focuses on one of a series of experiments conducted by Andrew Gonzalez and Enrique Chaneton on population density and species diversity among moss-inhabiting mites. This exercise provides further data from these ingenious experiments and asks you to assess their implications for species diversity in fragmented environments.

# 56 Species Interactions and Coevolution

## Anty biotics

Familiar sights in Neotropical forest communities, "leaf-cutter" or "parasol" ants owe their name to their habit of clipping bits of leaf material from trees and carting them off to their nests. The leaves they collect, however, are not for their own consumption. Leaf-cutter ants are fungus farmers: they use the cut leaves as a substrate for growing the fungi on which they feed.

When a new queen ant is ready to leave the nest where she grew up, she packs her cheek pockets with a pellet containing pieces of the fungal mass that dominates her home. After mating, she lands, sheds her wings, and digs into the soil to form a tunnel with an enlarged end. There she ejects the fungal pellet and defecates in it. As the fungus begins to grow, the female lays eggs in it. With time, her eggs are engulfed by fungus, on which the hatching ant grubs feed. These first hatchlings develop into pint-sized workers that leave the nest to collect leaf material to "feed" the growing fungus.

As the colony increases in size, more nest chambers are constructed, and the fungus garden expands; by its third year, the total number of chambers in a colony can exceed several hundred. Colony size is one factor accounting for the reputation of leaf-cutters as pests; it takes more than 2 kg of plant material per day to maintain the enormous fungal garden that nourishes the colony, so these ants can easily strip an area of vegetation.

The ants diligently care for the fungi. In addition to supplying leaves to the garden, they add fertilizer (in the form of their own feces) and regulate the humidity of the chambers. They are even able to distinguish and exclude leaves containing chemicals with fungus-killing properties. The fungal gardens, however, are vulnerable to invasion by undesirable microbes, particularly the virulent green mold *Escovopsis*. To guard against mold invasion, the ants carry around a supply of bacteria in special organs on their exoskeletons, called crypts. These bacteria manufacture powerful antibiotics that suppress the unwelcome mold, but do no harm to the cultivated fungus in the nest.

In the first written description of leaf-cutter ants, the

**Feeding a Partner** Central American leaf-cutter ants (*Atta cephalotes*) transport leaf fragments to their nest. The leaves will serve as food for the fungus they grow.

**A Fungal Garden** Cutaway view of a South American leaf-cutter ant nest chamber filled with fungus. Several winged ants (*Atta colombica*) can be seen in the crevices of the fungal mass.

Spanish priest Bartolome de la Casas decried their assiduous leaf gathering and massive nests, "white as snow," as impediments to the cultivation of citrus and cassava in Hispaniola in 1559. The notion that these ants actually cultivate and eat the white fungus in their nests did not occur to anyone until the naturalist Thomas Belt introduced the idea in 1874. The existence and roles of the intrusive green mold and friendly bacteria were not documented until 1999, after another 125 years. It is not beyond the realm of possibility that even more interactions between species are involved in this system and remain to be revealed.

**IN THIS CHAPTER** we will examine how antagonistic interactions benefit one species at the expense of another, how mutualistic interactions benefit two or more interacting species, and how competition for the same resources affects the species involved. Throughout these discussions, we will see not only how these interactions influence the lives of these organisms, but how they shape species over the long range of evolution.

## CHAPTER OUTLINE



# 56.1 What Types of Interactions Do Ecologists Study?

One of life's certainties is that, at some point between birth and death, every individual organism will encounter and interact with individuals of other species. These interactions have consequences that can affect the individual's fitness. Thus they can influence the densities of populations, the distributions of species, and, over the long term, lead to evolutionary change in one or both of the interacting species.

### Interactions among species can be grouped into several categories

Although the actual number of interactions that take place among living things on Earth is essentially limitless, ecologists group interactions between species into a few basic categories. These categories reflect whether the outcome of the interactions is positive (+), negative (–), or neutral (0) for each of the species involved (**Figure 56.1**). The five broad categories of species interactions that we introduce in this chapter are *antagonistic interactions*, *mutualism*, *competition*, *commensalism*, and *amensalism*.

**Antagonistic interactions** are those in which one species benefits and the other is harmed. Antagonistic interactions include *predation*, in which an individual of one species kills and consumes multiple individuals of other species during its lifetime; *herbivory*, in which an individual of another species consumes part or, more rarely, all of a plant; and *parasitism*, in which one species consumes only certain tissues in one or a few individuals of another species without necessarily killing those individuals—known as *hosts*. Some parasites are *pathogens*, which cause symptoms of disease in their hosts.

**Mutualism** is a type of interaction between species that benefits both species. The interaction between leaf-cutter ants and fungi described at the opening of this chapter is an example of mutualism: the ants feed and cultivate the fungi, and the fungi, in turn, serve as food for the ants. Mutualisms exist between widely varied pairs of partners, including not only animals and fungi, but also fungi and plants, animals and plants, animals and animals, and microbes and all other kinds of organisms.

**Competition** between species refers to interactions in which two or more species use the same resource. The outcomes of these interactions depend on resource availability. Competition can occur along with almost any other kind of interaction: between predators that depend on the same prey species, between herbivores that depend on the same host plant, or between pathogenic microbes attacking the same host. The limiting resource need not be food; species may compete for water, for space, or even, in the case of plants, for sunlight.

(A)

**Major types of species interactions**

| | TYPE OF INTERACTION | EFFECT ON SPECIES 1 | EFFECT ON SPECIES 2 |
|---|---|---|---|
| Antagonistic interactions | Predation (predator–prey) | + | – |
| | Herbivory (plant–herbivore) | – | + |
| | Parasitism (parasite/pathogen–host) | + | – |
| | Mutualism | + | + |
| | Competition | – | – |
| | Commensalism (commensal–host) | + | 0 |
| | Amensalism | 0 | – |

(B)

Parasitism, Predation, Mutualism
The buffalo's hide is infested with parasitic ticks. Oxpecker birds eat the ticks, to the mutual benefit of the birds and the buffalo.
Herbivory
The African buffalo feeds on the grasses of the savanna.
Amensalism, Commensalism
The large mammal unwittingly destroys insects and their nests. The white cattle egrets feed on insects disturbed by the buffalo's passage.
Predation
Carnivores such as timberwolves hunt and kill herbivorous mammals.
Competition
The grizzly bear is attempting to take over the wolves' kill.

**56.1 Types of Interactions** (A) Interactions among species can be grouped into categories based on whether their influence on each of the interacting species is positive (+), negative (–), or neutral (0). These interactions can be broadly grouped as antagonistic or mutualistic. (B) Even a small scene can encompass many different species interactions.

**yourBioPortal.com**

GO TO Web Activity 56.1 • Ecological Interactions

Predation, mutualism, and competition all affect the fitness of both participants, but there are two other types of interactions that affect only one participant. **Commensalism** is a type of interaction in which one participant benefits, but the other is apparently unaffected. Most examples of commensalism (from the Latin, "eating together") involve one species feeding in, on, or around another species. For example, one species may associate with another species that, by virtue of its own feeding behavior, makes food more accessible. The brown-headed cowbird owes its name to its habit of following herds of grazing cattle, foraging on insects flushed from the vegetation by their hooves and teeth (in years past, it was called the buffalo bird because it followed the bison that were once abundant across the North American continent).

Another form of commensalism involves association for the purpose of transport, often to reach food resources that are rare and short-lived. Piles of mammal dung, for example, are a valuable resource for some detritivores, but they can be hard to find and never last long. Many kinds of detritivores that cannot fly—mites, nematodes, and even fungi—attach themselves to the bodies of dung beetles, which not only can fly but are also very good at locating fresh dung (as described in the opening story of Chapter 54). These hitchhikers have no effect on the dung beetle's fitness, nor do cowbirds have any on the cattle that flush insects for them.

**Amensalism** is a type of interaction in which one participant is unaffected while the other is harmed. A herd of elephants moving through a forest crushes insects and plants with each step, but the elephants are unaffected by this carnage. Amensal interactions tend to be more random, and thus less predictable, than other types of interactions.

Although ecologists find it useful to group interactions between species into a few basic categories, the boundaries between categories are not always clear. For example, sea anemones in the Pacific Ocean sting and eat small fish, but a select few fish species (mostly in the genus *Amphiprion*) live inside sea anemones and are unaffected by their stings. These anemonefish move freely among the stinging tentacles to scavenge the cnidarians' leavings and even steal their prey (**Figure 56.2**).

Anemonefish must acclimate to the anemone's venom, and the anemone, in turn, must acclimate to the fish. The acclimation process appears to involve a change in the mucus coat of the fish; wiping off the mucus of an acclimated fish results in immediate stinging, whereas anemones will not attack fish with intact mucus. Although the benefit to the anemonefish is clear—it escapes its own predators by hiding behind the anemone's

*Amphiprion ocellaris*

**56.2 Interactions between Species Are Not Always Clear-Cut** Ecologists long believed that the relationship between sea anemones and anemonefish was a commensalism: that the fish, by living among the anemone's stinging tentacles, gained protection from its predators. But could it also be considered a mutualism—if the fish's feces provide the anemone with beneficial nutrients—or competition—if the fish occasionally steals the anemone's prey?

nematocysts, and it has no need to forage widely for food—the anemones may benefit from the association as well. By defecating while in residence, the anemonefish may provide nitrogen-rich nutrients to the anemones.

The interaction types described in this section are in reality part of a continuum, and their outcomes depend on both ecological and evolutionary circumstances, including the presence and influence of other interacting parties.

### Some types of interactions result in coevolution

All types of interactions have the potential to influence the population densities of the interacting species. By contributing to the differential survival or reproduction of genotypes, they can also alter gene frequencies within the populations over time. Thus these interactions have ecological consequences, as when they affect the distribution and abundance of individuals within a species, as well as evolutionary consequences, as when they lead to adaptations. In some cases, an adaptation in one species may lead to the evolution of an adaptation in a species it interacts with, a process known as **reciprocal adaptation** or **coevolution**.

Darwin saw the fitness of an organism as a measure of whether it gains or loses from interactions with other species. He noted, too, that evolutionary change occurs not only in response to physical conditions, as we saw in Section 54.2, but also in response to interactions among organisms. In his introduction to *The Origin of Species*, Darwin pointed out that woodpeckers have feet, tails, beaks, and tongues "admirably adapted to catch insects under the bark of trees" as a result of their long-standing interactions with their insect prey. Organisms can thus influence the evolution of the organisms with which they interact.

While abiotic factors also act as agents of selection, they differ in a fundamental way from biotic agents of selection in that they do not themselves undergo change as a result of the interaction. Snow and ice cannot increase their killing power as a result of encountering cold-resistant organisms, but predators can, over evolutionary time, become swifter, more powerful, or more efficient at capturing their prey. In response, prey species can, over evolutionary time, become swifter, tougher, less conspicuous, or more poisonous to decrease the likelihood of being consumed. The insect prey of Darwin's woodpeckers, for example, might evolve features that make them more difficult for the woodpeckers to find and capture.

A series of reciprocal adaptations can lead to what has been dubbed a coevolutionary **arms race**. The arms race analogy, first used in the context of interactions between herbivores and plants, can be applied to most antagonistic interactions. The evolution of traits that increase the fitness of a predator or parasite species exerts selection pressure on its prey or host species to counter the consumer's adaptation. The prey or host adaptation, in turn, exerts selection pressure on the consumer to improve its fitness, resulting in an escalating arms race. The types of interactions most likely to lead to coevolution, then, are those that occur predictably with high frequency over time and that have a strong effect on the interacting species; thus most amensal and commensal interactions are less likely to coevolve than are many plant–herbivore, predator–prey, and mutualistic interactions.

## 56.1 RECAP

**Species interactions can be grouped into five categories based on whether they benefit or harm the species involved. Some species interactions can lead to reciprocal adaptations, or coevolution.**

- Describe the five categories of interactions between species. See pp. 1186–1187 and Figure 56.1
- Explain the concept of a coevolutionary arms race. See p. 1188

Now let's take a closer look at antagonistic interactions. In Section 31.3 we looked at a number of feeding strategies from the consumer's point of view. In this section we will see how the interactions of consumers with their resource species influence both species.

## 56.2 How Do Antagonistic Interactions Evolve?

Every species serves as a food resource, in one way or another, for at least one other species. Consumers can increase their fitness by acquiring food, whereas resource species can increase their fitness by avoiding being consumed. Thus the interests of consumer and resource species set up an antagonistic relationship that can lead to a coevolutionary arms race. These consumptive relationships need not, however, be fatal; organisms make meals of one another in many different ways.

(A) *Panthera tigris*

(B) *Cicindela campestris*

(C) *Platythelphusa* sp.

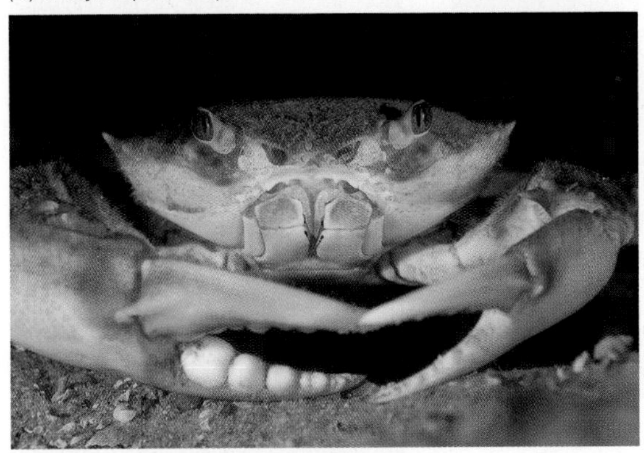

**56.3 Predators Use Many Weapons** (A) Tigers embody most people's image of a predator—a large animal that uses speed, strength, teeth, and claws to capture prey. (B) The 1.3-cm green tiger beetle is also formidable to its prey, including caterpillars. Its huge jaws account for most of its body length, and it is among the swiftest runners among the insects. (C) *Platythelphusa* crabs of Lake Tanganyika have tremendously strong "toothed" claws with which they crush their snail prey. In an evolutionary arms race, thick shells have evolved in the snails.

## Predator–prey interactions result in a range of adaptations

Predator–prey interactions are probably the most familiar, and the most dramatic, type of antagonistic interaction. Predators invariably kill the individuals they consume—referred to as their *prey*—and over its lifetime, a predator kills and consumes many prey individuals. Predators tend to be less specialized than other types of consumers.

The fitness of predators depends on balancing the cost of pursuing, subduing, and handling prey against the energetic return from consuming it. Thus many predators are larger than their prey, and many of them use strength or swiftness to capture prey. This is true of predators of all sizes: tigers pursuing deer and tiger beetles pursuing smaller insects are both swift, powerful, and equipped with strong jaws (**Figure 56.3A,B**). The few predators that are smaller than their prey rely on other strategies that increase their efficiency. Spiders, for example, capture their prey in webs. The short-tailed shrew, among the smallest mammalian predators, produces venomous saliva that paralyzes not only earthworms and snails, but also prey much larger than itself, including mice and small birds.

Prey species have many different kinds of defenses against predators. Many animal species can escape from predators simply by running away. Others have morphological defenses. Tough skin, shells, spines, or hair can foil even a determined predator. A coevolutionary arms race between predators and prey can explain why the lumpy-clawed crabs with the greatest claw strength are found in the vicinity of snail prey with the thickest, most crush-resistant shells (**Figure 56.3C**).

**CHEMICAL DEFENSES** Many animals use chemical defenses to escape or repel their predators. Chemical defenses are generally the province of animal prey that are small, weak, sessile, or otherwise unprotected. Among the mollusks, for example, the weaker a species' shell, the more likely it is to use chemical defenses (see Figure 56.4B). Some vertebrates also rely on chemicals to repel their predators.

Many insects produce sprays, oozes, or froths when attacked. Bombardier beetles, for example, possess a pair of glands near the anal opening. Each gland has two compartments lined with a protective cuticle. The inner compartment contains a mix of relatively nontoxic chemicals, along with hydrogen peroxide. The outer compartment contains enzymes. When the beetle is disturbed, it discharges the contents of the inner compartment into the outer compartment, which leads to an instant, energy-releasing chemical reaction. Oxygen is one of the end products generated by this reaction, and the resulting pressure discharges the mixture with an audible "pop." Due to the energy of the reaction, the temperature of the spray is approximately 100°C. The reaction of predators—including humans—to this hot, explosive secretion is predictable, and bombardier beetles have very few enemies.

But predators may evolve to overcome their prey's chemical defenses, as we saw in the case of the rough-skinned newt and the garter snakes that have become insensitive to its protective toxin (see Figure 21.22). Some predators are not only undeterred by their prey's defensive chemicals, but ingest them and sequester them in their bodies as defenses against their own predators. Some sea slug species, for example, acquire their defensive chemicals from the toxic sponges they can eat with impunity. Many prey species that defend themselves with toxicity advertise that fact, a phenomenon called *aposematism*.

**APOSEMATISM** Some prey species exploit the fact that predators can learn to avoid certain warning signals. Many toxic prey sport bright colors or striking patterns. Such warning coloration, or **aposematism**, increases the probability that a predator will

(A)

*Danaus plexippus* larva

*Tetraopes tetraophthalamus*

*Aphis nerii*

(B) *Chromodoris* sp.

(C) *Dendrobates reticulatus*

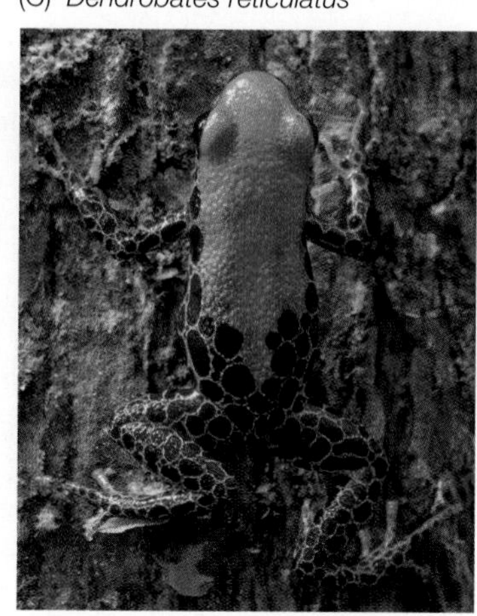

**56.4 Some Prey Come with Warning Labels** (A) Milkweed plants are toxic, and many of the insects that feed on them—including monarch butterfly larvae, milkweed beetles, and aphids—incorporate the plant's toxic chemicals into their systems. (B) Nudibranchs (sea slugs) are mollusks without protective shells, however, they may possess stinging nematocysts (acquired from their hydrozoan prey), and their bright coloration warns predators away. (C) Dart poison frogs of Central and South American sequester highly toxic chemicals in their brightly colored skin.

learn to recognize and avoid a toxic species (**Figure 56.4**). Certain visually hunting predators, particularly among the vertebrates, can learn quickly to associate certain color patterns with an unpleasant dining experience. Thus aposematic species are characteristically tough enough to survive a brief encounter with a predator. Any encounter that results in the death of the aposematic individual is unlikely to result in selection for that aposematic pattern. Aposematic butterflies in the field often sport beak marks, an indication of having survived a taste by an uneducated avian predator. And, as it turns out, even nontoxic species can benefit by mimicking warning coloration.

**MIMICRY SYSTEMS** We have seen that some prey species avoid consumption by mimicking inedible objects. Others do so by mimicking aposematic species. This strategy has led to the evolution of mimicry systems of two types. In **Batesian mimicry**, a nontoxic species (the *mimic*) resembles a toxic species (the *model*) and benefits from the avoidance behavior learned by the predator in response to the toxic model species. For example, venomous bees and wasps sport distinctive yellow and black stripes that are models for harmless hover flies (**Figure 56.5A**). Mimicry may extend beyond physical appearance; many mimics also simulate distinctive behaviors of their models. Some hover flies, for example, vibrate their wings to buzz in beelike fashion.

In **Müllerian mimicry**, a number of aposematic species converge on a common color pattern; all benefit from providing a stronger recognition signal to predators. Many of the Neotropical zebra butterflies, which feed on toxic passionflower plants and incorporate their toxins into their bodies, are Müllerian mimics (**Figure 56.5B**).

**AVOIDING DETECTION** Many prey species escape predators by hiding. One form of hiding is camouflage, or background matching, also called **crypsis** (**Figure 56.6A**). Some animals can even change their coloration to match the substrate they find themselves on. The camouflage of some species not only hides them but allows them to resemble objects their predators consider inedible, a strategy called **homotypy** (**Figure 56.6B**). The dead-leaf butterfly looks very much like a dead leaf, even down to the likeness of a spot of fungal decay, and swallowtail caterpillars look like bird droppings.

Because the vision of many types of predators is adapted to spot moving prey, many prey species simply stop moving if they are being pursued. "Playing possum," a term that is sometimes applied to this strategy, refers to the ability of the opossum (*Didelphus virginianus*) to "play dead."

## Herbivory is a widespread interaction

The most ubiquitous interaction on Earth is that between plants and the herbivores that eat them. Herbivores have a relatively easy time acquiring food: plants are sessile and cannot bite, scratch, or run away. A conservative estimate puts the number of herbivore species (the vast majority of which are insects) at

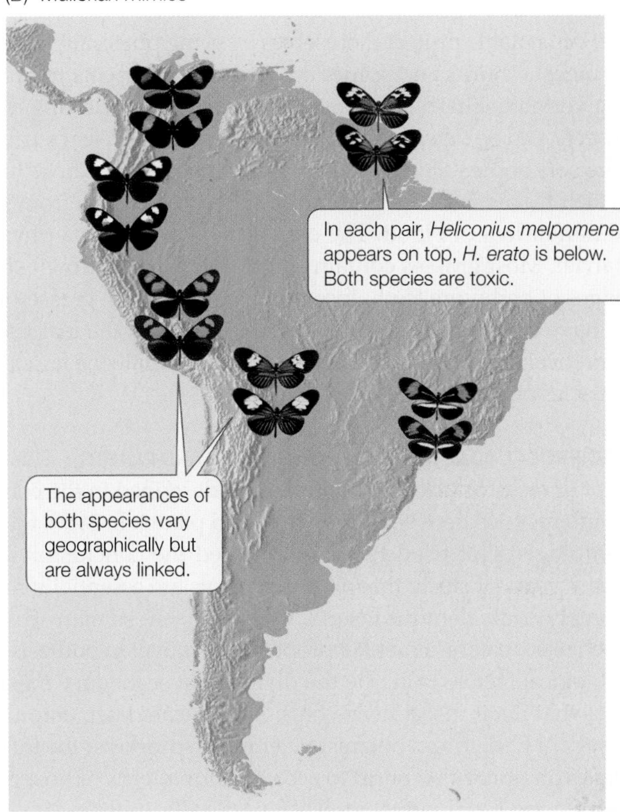


**56.5 What You See Is What You Get—Sometimes**
(A) Batesian mimics are harmless species that gain protection from mimicking dangerous ones. (B) The shared aposematic coloration of Müllerian mimics is an honest advertisement of their toxicity. All the zebra butterflies (genus *Heliconius*) of South America feed on toxic passionflower plants and incorporate the toxins into their bodies. *Heliconius* species living in a particular area have similar warning colors.

400,000 and the number of plant species they consume at over 300,000. Every major class of vertebrates includes at least a few herbivores. In marine systems, organisms that feed on plants and algae include mollusks, crustaceans, echinoderms, and annelids.

In terms of numbers of individuals as well as numbers of species, most of the world's herbivores are insects. Over 90 percent of these insects are *oligophagous*: specialists that dine on just one or a few, often taxonomically related, plant species. *Polyphagous* species, in contrast, feed on as many as dozens of unrelated plant species. Vertebrate herbivores are generally polyphagous; a cow grazing in a pasture, for example, can consume many different plant species in a single afternoon. There are exceptions to this pattern, however; Australian koalas famously feed exclusively on the foliage of eucalyptus trees, and the diet of giant pandas is made up almost entirely of bamboo.

Herbivores, particularly insects, generally consume only parts of their food plants, and usually do not kill them. In most natural ecosystems, insects rarely remove more than 20 percent of plant biomass. For that reason, questions occasionally arise as to the ability of insects to act as selective agents on plant traits. Mortality is not, however, the only form of selection that leads to evolutionary change; herbivores can reduce plant fitness if the plants they attack produce fewer offspring.

**PLANT DEFENSES AGAINST HERBIVORES** The defenses of plants against their diverse consumers are necessarily highly diverse. For most plant species, chemistry is the principal defense mechanism. As we saw at the opening of this chapter, the leaves of some trees contain chemicals that prevent them from being consumed by fungi—and thus, incidentally, from being harvested by leaf-cutter ants. The amazing variety of secondary metabolites produced by plants to defend themselves against herbivores is the topic of Section 39.2. Many plants, however, have additional defenses.

**56.6 Avoiding Consumption by Avoiding Detection** (A) The gray tree frog can change its coloration to blend in with its substrate, an example of crypsis. (B) Homotypy—resemblance to an inedible object—can be an effective defense against visually hunting predators. Birds searching for insect prey are likely to pass by a katydid that looks like a leaf.

Some plants protect themselves by being physically difficult to ingest. Thorns and spines are effective deterrents to browsing vertebrate herbivores. Smaller herbivores, including many insects, can be deterred by small hooked hairs on leaf surfaces. The soft bodies of leafhoppers can be pierced by these hairs, which fix the insect in place until it eventually dies from starvation or loss of blood. The cuticle may also act as a physical barrier. Most grasses contain silica, which wears down sharp edges of herbivore teeth. Insects that feed only on grasses tend to have chisel-like mandibles to slice through the leaf tissue; moreover, their heads are larger to accommodate the larger jaw muscles needed to process their food.

**RECIPROCAL ADAPTATIONS IN HERBIVORES AND PLANTS** The concept of reciprocal adaptation was first described in the context of interactions between herbivores and plants. In 1959, the entomologist Gottfried Fraenkel reached the conclusion after many years of study that all green plants are essentially nutritionally equivalent for insects. Why, then, are so many insects such picky eaters? Fraenkel proposed the novel hypothesis that ecological factors underlie the diversity of secondary metabolites that deter insect herbivores. A few years later, entomologist Paul Ehrlich and botanist Peter Raven proposed the following evolutionary scenario to account for patterns of host plant use among herbivorous insects (specifically, in their case, butterfly families):

- Certain plants, by mutation or recombination, evolve a novel secondary metabolite.
- If the chemical reduces the plant's appeal to herbivores, then plant genotypes producing the chemical are favored.
- Freed from mortality associated with herbivory, plants possessing the novel chemical undergo an adaptive radiation.
- Certain herbivores, by mutation or recombination, evolve resistance to the chemical, and these resistant herbivores undergo their own adaptive radiation.
- With sufficient selection pressure, a resistant herbivore can evolve to use the chemical as a defense against its own predators.

This stepwise coevolutionary process explains not only the biochemical diversity of flowering plants, but also the tremendous diversity of herbivorous insects. The ecological scenario outlined by Ehrlich and Raven is an example of the type of coevolutionary arms race described earlier in the chapter.

A tremendous diversity of adaptations to plant defenses have evolved in herbivores. Many herbivores circumvent plant defenses by behavioral means. For example, the secondary metabolites produced by a plant called Saint-John's-wort (*Hypericum perforatum*) require exposure to sunlight for optimal toxicity, so insects that feed on this plant roll its leaves into a light-impervious cylinder and feed in comfort in the dark. The laticifer-cutting beetles described in Section 39.2 have a different method of detoxifying their food plant. Many large polyphagous herbivores, such as deer, horses and the like, graze on a wide variety of plant species, minimizing their exposure to any particular defensive chemical. Long-lived and with relatively good memories, they can learn to avoid plants with an unpleasant taste.

Unlike large mammalian herbivores, caterpillars and many other insect herbivores may spend their entire lives feeding on a single individual plant. Such oligophagous diets are associated with highly specialized detoxification systems. The diamondback moth caterpillar eats plants in the cabbage family that are rich in toxic mustard oil glycosides. In its gut is an enzyme that breaks down the glycosides into harmless by-products, allowing it to eat these plants with impunity.

Some herbivores take resistance a step further by storing, or sequestering, plant toxins in specialized organs or tissues that are insensitive to those toxins. In this way, they can accumulate large quantities of toxins in their bodies with no ill effects. This strategy also makes the expropriated chemicals available for defense against the herbivores' own enemies. Caterpillars of the monarch butterfly, for example, are insensitive to the neurotoxic glycosides in their milkweed host plants, but most of their enemies, including insect-eating birds, cannot tolerate these compounds.

Yet the plants continue their side of the coevolutionary arms race. As we have seen, the principal consumers of passionflower plants are zebra butterflies. These oligophagous butterflies lay eggs only on passionflower plants, and their larvae sequester host plant toxins in their bodies as they feed on the leaves. Some passionflower species, however, have modified leaf structures that resemble the eggs of butterflies. Female butterflies will not lay eggs on plants already containing eggs, so the egg mimics reduce the plant's probability of being consumed (**Figure 56.7**).

## Microparasite–host interactions may be pathogenic

*Microparasites*, such as viruses, bacteria, and protists, are many orders of magnitude smaller than their hosts and generally live and reproduce inside their hosts. Multiple generations of mi-

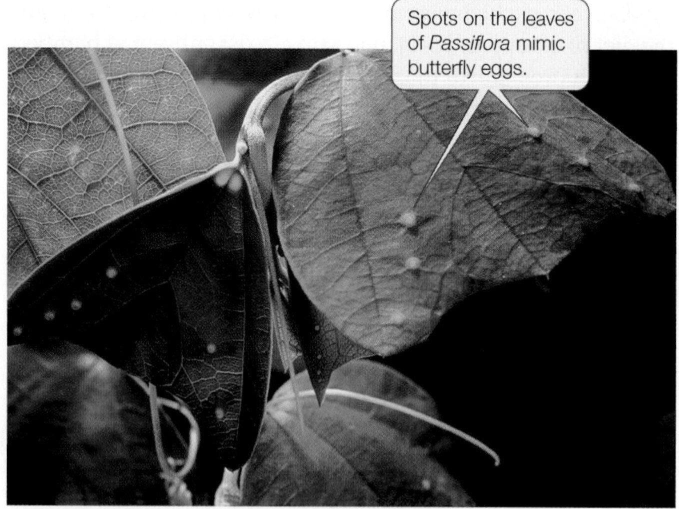

Spots on the leaves of *Passiflora* mimic butterfly eggs.

**56.7 A Plant Uses Mimicry to Avoid Herbivory** Passionflower leaves are modified to resemble the eggs of their principal herbivores, zebra butterflies (*Heliconius* spp.), which will not lay eggs on a plant already containing eggs. The egg mimics deter the female from laying eggs, thus protecting the plant from being eaten by hatchling caterpillars.

croparasites may reside within a single host individual, and a host may harbor thousands or millions of them. Many microparasites, in the process of acquiring nutrients at the expense of their host, cause symptoms of disease—that is, they are *pathogens*. Section 39.1 describes the array of secondary metabolites that plants produce to defend themselves against pathogens, and Chapter 42 describes the defenses of animals.

Infection by pathogens may in some cases result in the death of the host, but death is by no means an inevitable outcome of these interactions. If a pathogen strain is to persist in a host population, the pathogens must continually infect new host individuals. A less deadly strain that kills a smaller proportion of host individuals may succeed in infecting a larger number of new hosts. Thus pathogen and host may reach a state of coexistence as increased host *resistance* (ability to withstand the effects of a pathogen) and decreased pathogen *virulence* (ability to cause disease) evolve. Yet new, virulent strains may also arise, reminding us that the arms race goes on.

Hosts of pathogens fall into three classes: *susceptible* (capable of being infected), *infected*, or *recovered* (and thus, in many cases, *immune*; see Chapter 42). A pathogen can readily invade a host population dominated by susceptible individuals, but as the infection spreads, fewer susceptible individuals remain to be infected. Eventually a point is reached at which most infected individuals no longer transmit the infection to susceptible individuals. Thus rates of infection typically rise, then fall, and do not rise again until a sufficiently large population of susceptible host individuals has reappeared.

### Most ectoparasites have adaptations for holding onto their hosts

While microparasites generally live and reproduce inside their hosts, larger parasites, called *macroparasites*, are associated with their hosts in a slightly less intimate way. Although macroparasites rarely cause the same kinds of disease symptoms that pathogenic microparasites cause, they may nevertheless affect host survival and reproduction and can thereby act as selective agents on their hosts.

Some ectoparasites (external parasites)—leeches, mosquitoes, and the like—are only casually associated with their hosts, interacting with them just long enough to eat their fill and then moving on. Others spend their entire lives in or on their hosts. These ectoparasites have a number of attributes designed to keep them attached to their hosts. Crab lice, ectoparasites that are generally found in the pubic region of their human hosts, have claws on the tips of their legs that clamp around pubic hairs with great precision (**Figure 56.8**). Pulling off a crab louse will often leave the leg behind, still firmly attached to the hair. Ectoparasites have other adaptations designed to reduce the ability of irritated hosts to remove them, such as flattened bodies and a thick, tough cuticle. Many ectoparasites also have adaptations to a sedentary existence, such as loss of sensory organs and wings. Most ectoparasitic insects are highly specialized, sometimes feeding on only a single host species.

Most hosts actively work to rid themselves of their ectoparasites. The Japanese macaque, for example, is prone to infestation by two species of lice, which tend to lay their eggs on the outer surfaces of the host's back, arms, and legs. An individual macaque may harbor more than 500 louse eggs, and would find it difficult to reach and eliminate all of them. To keep louse populations in check, macaques must form and maintain social bonds in order to ensure the consistent presence of grooming partners. Grooming behavior—an important component of the social interactions of many primates—may have evolved in response to ectoparasites. Some biologists believe that human hairlessness and bipedal posture (which freed the hands for manipulating small objects), as well as the opposable thumb, were evolutionary responses to ectoparasites.

*Phthirus pubis*

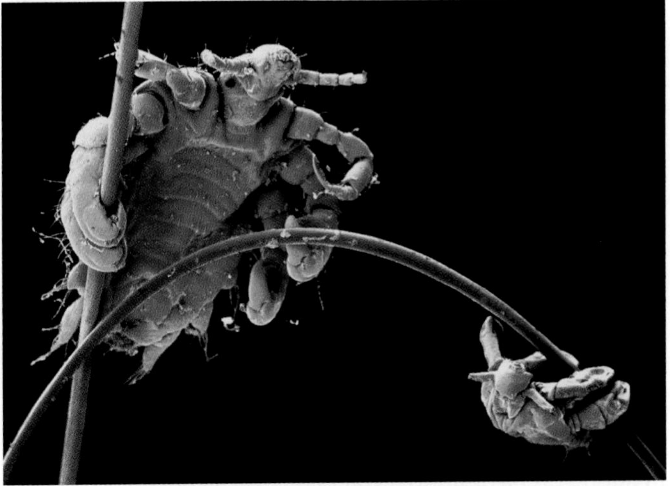


**56.8 Ectoparasites Can Make a Host Crabby** Ectoparasites such as the human pubic (or "crab") lice in this scanning electron micrograph tend to be tiny, wingless, flattened, and equipped with strong claws for gripping their host.

## 56.2 RECAP

**Predator–prey, herbivore–plant, and parasite–host interactions are all antagonistic. Consumers have adaptations for finding and utilizing their resource species with greater efficiency. The resource species in turn have adaptations that reduce their probability of being discovered, captured, and eaten.**

- What are some of the adaptations that help prey species to avoid consumption by predators? See pp. 1189–1190
- How are aposematism and mimicry related? See pp. 1189–1190 and Figures 56.4 and 56.5
- Explain the scenario for coevolution between insect herbivores and their host plants proposed by Ehrlich and Raven. See p. 1192

Like antagonistic interactions, mutually beneficial interactions between species can result in reciprocal adaptation. A mutually beneficial exchange of goods or services can ensure the predictability and frequency of such interactions over evolutionary time; thus many mutualistic interactions are tightly coevolved.

## 56.3 How Do Mutualistic Interactions Evolve?

There are few, if any, taxonomic limits on the formation of mutualisms; animals, for example, can form mutualistic associations with other animals, with plants, and with a wide range of microorganisms. Mutualistic interactions often arise in environments where resources are in short supply. Consequently, many mutualisms involve an exchange of food for housing or defense. In another type of mutualism, many sessile organisms, particularly flowering plants, rely on more mobile species for mating or dispersal.

Many mutualisms are *asymmetrical*—in other words, one party benefits more than the other. One or both partners may evolve adaptations that ensure that the exchange benefits both of them. Reciprocal adaptations are most likely to arise in mutualistic interactions if an increase in dependency on a partner provides an increase in the benefits realized from the interaction. If increased dependence provides no selective advantage, mutualists (particularly those in asymmetrical mutualisms) may evolve into parasites, give up their partners for an independent existence, or even go extinct.

### Plants and pollinators exchange food for pollen transport

For about three-quarters of the planet's 250,000 flowering plant species, reproduction requires the transport of pollen by an animal partner. A mutualistic pollination system requires several features:

- An *attractant* or reward that entices a pollinator to visit the plant
- *Behavior* that ensures that a pollinator visits more than one individual of a plant species
- *Anatomical features* that allow a pollinator to transport the plant's pollen

Flowers entice pollinators in many ways. The most direct reward for pollinators is the pollen itself, which sometimes serves as food. Pollen was probably the original attractant in the evolutionary history of plant–pollinator interactions. Plant reproduction would not be served, however, if pollinators were to eat *all* of a plant's pollen; thus plants have evolved various adaptations to ensure that they benefit from the exchange. For example, some plants have two types of anthers: feeding anthers to produce pollen for pollinators, and fertilization anthers to produce pollen for reproduction. These two types of anthers are shaped and positioned differently, so that as the pollinator dines on pollen from the feeding anthers, the fertilization anthers deposit pollen on a part of its body that will transfer it to the stigma of another flower of the same species.

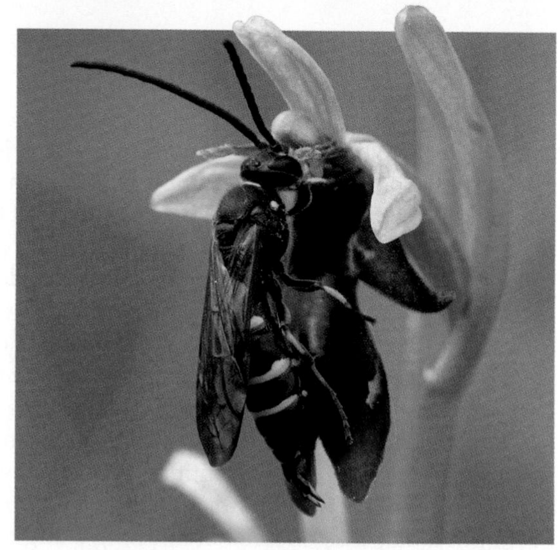

**56.9 Plants Sometimes Take Advantage of Their Pollinators** This orchid (*Ophrys insectifera*) has flowers that look and smell like a female wasp. Male wasps (*Argogorytes mystaceus*) will pollinate the flower while trying to mate with it.

Plants, like pollinators, may take advantage of their partners. Some species have evolved flowers that bear a striking resemblance to females of a particular wasp species (in some cases even producing the same chemical substance that the female wasp uses as a sexual attractant pheromone). Male wasps, in a futile effort to copulate with the flower, get pollen on their bodies, which they deliver to the next flower they visit in their effort to locate a genuine mate (**Figure 56.9**).

Compared with pollen, *nectar*, a sugar-rich solution produced by some angiosperms, is a relatively new evolutionary development. Of all floral rewards, nectar has the greatest appeal and is consumed by the widest range of animal pollinators, including not only insects but also birds (such as hummingbirds) and mammals (such as bats). While nectar serves to attract potential pollinators, it is prone to removal by *thieves*: flower visitors such as ants that can reach and consume the nectar without

**TABLE 56.1**
**Pollination Syndromes Resulting from Diffuse Coevolution**

| | SUITE OF ANATOMICAL TRAITS[a] | | | |
|---|---|---|---|---|
| PREFERRED POLLINATOR | FLOWER SHAPE | FLOWER COLOR | REWARD | ODOR |
| Bees | Irregular | Many | Nectar, pollen | Sweet |
| Flesh flies | Irregular | Purplish | None | Carrion |
| Beetles | Bowl | White or pale | Pollen | Faint |
| Butterflies | Tubular platforms | Many | Nectar | Faint |
| Moths | Often pendant | White or pale | Nectar | Heavy |
| Hummingbirds | Tubular | Red | Nectar | Imperceptible |
| Bats | Cuplike | White or pale | Copious nectar | Musty |

[a] In many plant groups, diffuse coevolution has led to suites of traits (as opposed to a single trait) that are characteristic of the interactions between the plants and their preferred pollinators.

(A)

(B)

*Taraxacum officinale*

**56.10 See Like a Bee** To normal human vision, a dandelion appears solid yellow. Ultraviolet photography reveals the patterns that attract bees to the region of the flower where pollen and nectar will be found. These patterns are invisible to humans and birds.

transporting pollen. Nectar thieves lower plant fitness by depleting nectar that would otherwise attract actual pollinators.

Plants not only need to attract pollinators, but must also ensure that those pollinators carry their pollen to other members of the same species. Repeat visits by a pollinator to different individuals of a particular plant species increase the likelihood that the pollen will end up on the appropriate stigma; thus some plants have adaptations to limit the diversity of their animal visitors. The depth and width of a flower can restrict the size and shape of the pollinator mouthparts that can gain access to their nectar. Timing of flowering can also restrict the number of potential pollinators and encourage pollinator fidelity. Floral characteristics influence the type of pollinator attracted initially. Ultraviolet wavelengths, for example, are highly attractive to bees (**Figure 56.10**) but invisible to most birds.

Most flowers can be successfully pollinated by a number of animal species. The evolution of broad suites of floral characteristics that attract certain groups of pollinators is an example of **diffuse coevolution**: the evolution of similar traits in suites of species experiencing similar selection pressures (**Table 56.1**). Scarlet gilia (*Ipomopsis aggregata*), a common wildflower in the Rocky Mountains, has successfully combined two strategies. Early in its growing season, it produces red flowers that attract hummingbirds; later in the season, the gilia shifts to producing white flowers because by then the most abundant pollinators are hawkmoths, which cannot see red but are attracted to white.

A few plant–pollinator relationships are much more exclusive; these lead to hightly specific, rather than diffuse, coevolution. Yucca plants are pollinated only by a group of moths collectively known as yucca moths, whose larvae feed exclusively on yucca seeds. The stigma of the yucca flower is located fairly deep within the pistil, and fertilization will not occur unless pollen is physically placed there. The specialized mouthparts of female yucca moths have distinctive long tentacles, which the moths use to pack masses of pollen from one yucca flower into transportable balls that they then carry to another flower. The moth pushes the pollen ball deep into the recess in which the flower's stigma is tucked, then turns around and deposits her eggs inside the flower's ovule (**Figure 56.11**). When the eggs hatch, the caterpillars will consume some—but not all—of the

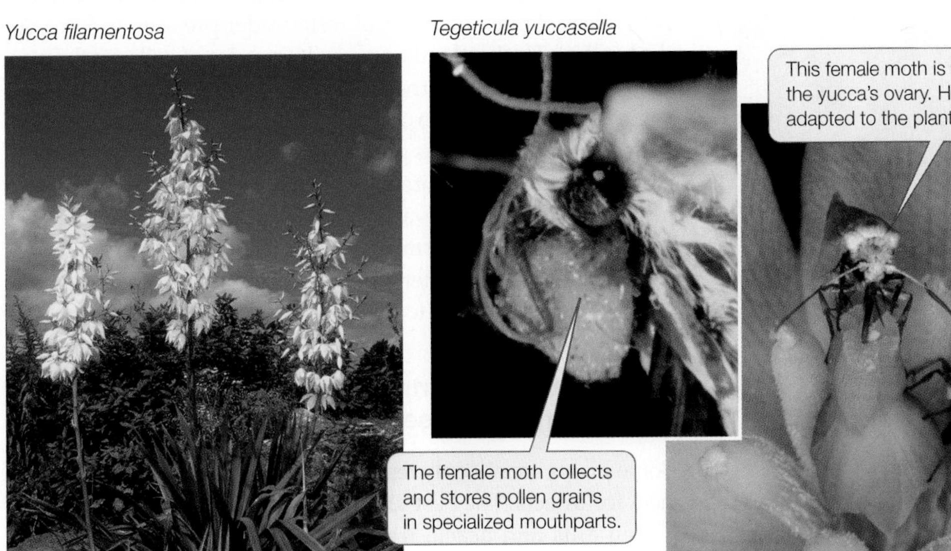

Yucca filamentosa
Tegeticula yuccasella
This female moth is laying eggs in the yucca's ovary. Her anatomy is adapted to the plant's shape.
The female moth collects and stores pollen grains in specialized mouthparts.


**56.11 Pistil-Packing Mama** Yucca flowers are pollinated only by yucca moths, and the larvae of yucca moths feed only on the seeds of yucca plants. The moth *Tegeticula yuccasella* is the exclusive pollinator of the *Yucca filamentosa.*

flower's developing seeds. Neither of these species can reproduce in the absence of the other.

### Plants and frugivores exchange food for seed transport

Many animals that eat fruits provide a valuable service to the plants that produce them by dispersing seeds. Seed dispersal by animals offers plants the advantage of delivery to potential germination sites and comes with the bonus of organic fertilizer for the seeds. Interactions between plants and frugivores, however, are not always reciprocal; in many cases, one party benefits more than the other. Whereas the frugivore is paid "in advance" for transportation services, the seeds may never reach an appropriate destination for germination (your windshield, for example, will not do). From the plant's perspective, a partnership with frugivores requires a delicate balance between discouraging frugivores from eating fruits before the seeds are capable of germinating and attracting them when the seeds are ready. In addition, the plant must protect the seeds from destruction in the frugivore's digestive tract and defend them against inappropriate consumers that damage the seeds or fail to disperse them at all.

The chemical process of fruit ripening ensures that fruits are most attractive to frugivores when the seeds are mature and ready for dispersal. In many fruits, ripening is accompanied by a decrease in organic acids, which make many unripe fruits sour. Color changes, which result from loss of chlorophyll and the accumulation of other pigments (the conversion of peppers from green to red during ripening is an example), have enormous signal value to many frugivores. Green, unripe fruits are generally difficult for vertebrate frugivores to see against green foliage; red and bicolored red and black fruits contrast with foliage. Fruit softens as it ripens to allow for gentle processing by the frugivore and rapid passage through its gut. Another conspicuous change in ripening fruits is an increase in sugar content—the "reward" most sought by frugivores. Seed coats, fruit pulp, and epidermis may all contain secondary chemicals designed to discourage inappropriate frugivores from consuming the fruit.

Because of the often asymmetrical nature of the mutualism between frugivores and plants, relatively few highly specialized frugivores exist. One apparently reciprocal interaction is between mistletoes—parasitic plants that grow on trees—and the mistletoe birds that serve as the plants' primary dispersal agents in Asia and Australia (**Figure 56.12**). These birds dine largely on the fleshy berries of mistletoe. The seeds, covered with a gluelike outer coat, experience little enzymatic or mechanical damage as they pass through the thin-walled guts of the birds that swallow them. When the seeds are voided with the bird's droppings, the sticky outer coat causes the seeds to adhere to the bird's vent feathers, prompting it to wipe its bottom across the tree branch on which it is perched. Once the seed is wiped on the branch, the gluey coat keeps it there—in an ideal location for a mistletoe seed to germinate.

*Dicaeum hirundinaceum*

**56.12 A Frugivore Plants and Fertilizes A Seed at the Same Time** After a mistletoe bird eats the fruit of the parasitic mistletoe plant, the seeds inside the fruit pass through the bird's digestive tract intact. As the seeds are voided, their sticky outer coat makes them stick to the bird's feathers. As the bird wipes itself clean on a branch, the seed sticks to the branch, where it germinates.

### Some mutualistic partners exchange food for care or transport

Some organisms, such as the leaf-cutter ants described at the opening of this chapter, get their food by "farming" fungus. Fungus farming has been documented in a wide variety of species, including bark beetles, termites, and even a snail. In most cases, the farmers provide housing, nutrition, and care for the fungal partner; the fungal species, in turn, provides food for the host. The fungus produces enzymes that can degrade plant proteins and cellulose, converting plant materials that the farmers cannot digest for themselves into an edible form.

Over the past 50 years, one fungus farmer, the southern pine bark beetle (*Dendroctonus frontalis*), has destroyed over a billion dollars' worth of pine forests in the southeastern United States (**Figure 56.13**). The beetle owes much of its efficiency to its mutualistic partners. Masses of adult beetles attack a pine tree at once, overwhelming the tree's ability to defend itself (the tree's defense is to release large quantities of resin under pressure to force out the beetles). The beetles then excavate a series of galleries through the vascular tissue underneath the bark, in which females lay their eggs. Female beetles also carry spores of their partner fungus into the galleries. The fungus grows on and breaks down the gallery walls; the beetles feed directly on the fungus and the partially digested wood. The beetles also transport a bacterium that produces an antibiotic to keep harmful bacteria from attaching the fungus. This insect–fungus–bacteria partnership can overcome the trees' antiherbivore defenses, to their mutual benefit.

### Some mutualistic partners exchange food or housing for defense

Some plants are not only resources, they are also mutualistic partners for insects. The best-known of these interactions is that between ants and acacia trees in Central America. In 1874, in Nicaragua, the naturalist Thomas Belt observed a peculiar inter-

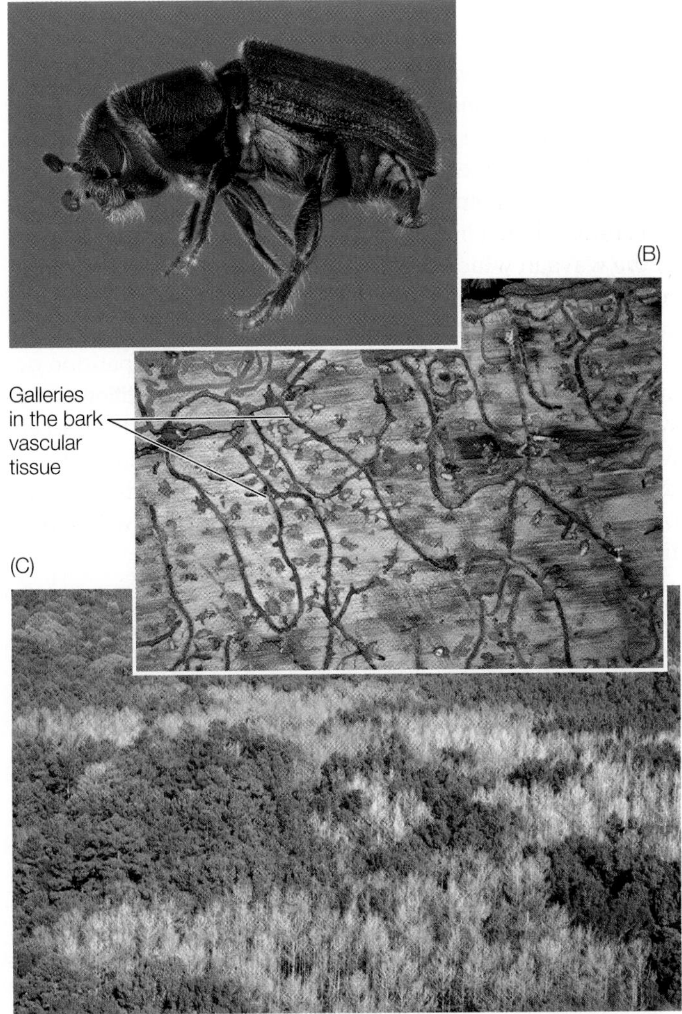


**56.13 A Mutualistic Interaction Brings Death to Pine Trees** (A) The southern pine bark beetle has a mutualistic relationship with a fungus, which it "farms" within the vascular tissue of pine trees. (B) The bark beetle excavates galleries inside the trees' vascular tissue. Here they lay eggs and farm fungus; the fungus overcomes the trees' defenses against the beetle and provides nutrition for the larvae. (C) Masses of bark beetles overwhelm pine forests, resulting in widespread death of pine trees.

action between bull's horn acacia (*Acacia cornigera*) and *Pseudomyrmex* ants, known as acacia ants because they are found only in association with acacias. Bull's horn acacias have enlarged hollow thorns in which the ants build nests. The trees produce nectar in specialized extrafloral structures and modified leaflet tips that are rich in oil and protein, on which the ants feed. These structures have no apparent purpose other than providing food for ants.

Belt suggested that the notoriously aggressive acacia ants defend the plants against their herbivores and competitors in exchange for food and shelter. But his idea was not tested until Daniel Janzen conducted an experiment in 1966. By removing ants from some acacias with insecticide, Janzen demonstrated that trees without ants suffered a reduction in growth and an increase in mortality (**Figure 56.14**). Ants also clipped weeds from around the base of the plants, presumably reducing competition for nutrients.

**yourBioPortal.com**

GO TO Animated Tutorial 56.1 • Mutualism

## INVESTIGATING LIFE

### 56.14 Are Ants and Acacias Mutualists?

Bull's horn acacia trees (*Acacia cornigera*) grow numerous structures that provide food and shelter for ants of the genus *Pseudomyrmex* (acacia ants). Daniel Janzen's experiments demonstrated that the trees benefit greatly from their association with these ants, and that the energy expended in growing ant-attractive structures is repaid with increased growth and survival.

**HYPOTHESIS** ***Acacia cornigera*** **trees deprived of their** ***Pseudomyrmex*** **ant populations will not thrive in comparison with trees populated by ant colonies.**

The "bull's horns" of these *Acacia* trees are enlarged, hollow thorns in which the ants build nests.

**METHOD**

1. Define a population of *A. cornigera* trees; designate some of them as untreated controls and the rest as experiment subjects.
2. Fumigate the experimental *A. cornigera* trees with insecticide to eliminate all *Pseudomyrmex* ants.
3. Apply Tanglefoot® (a sticky material) to the base of the fumigated experimental trees to prevent the ants from recolonizing them.
4. Record the survival and growth rates of the trees in both groups over a 10-month period.

**RESULTS** After 10 months, control trees (with ants) had considerably higher survival rates and greater growth than did trees without ant populations.

Trees surviving for 10 months (%): 0, 20, 40, 60, 80, 100 — Ants present, Ants absent

Growth increments over 10 months (cm): 0, 10, 20, 30, 40, 50, 60, 70, 80 — Ants present, Ants absent

**CONCLUSION** ***Pseudomyrmex*** **ants provide substantial survival benefits to** ***Acacia cornigera*** **trees.**

Go to **yourBioPortal.com** for original citations, discussions, and relevant links for all INVESTIGATING LIFE figures.

## 56.3 RECAP

**Mutualistic interactions involve an exchange of benefits. Most plants rely on mutualisms with mobile animals for fertilization and seed dispersal.**

- Give examples of benefits that are exchanged in at least two mutualisms between plants and animals.
- What three features are required by a plant–pollinator mutualism? See p. 1194
- How are pollinator-plant interactions different from frugivore-fruit interactions? See pp. 1194–1195
- Describe some adaptations in plants that help to maintain the balance in their relationships with frugivores. See p. 1196

From our discussion of interactions between species that benefit both, we move to a type of interaction that benefits neither: competition. This type of interaction is widespread because it can arise wherever two or more species require the same resources.

# 56.4 What Are the Outcomes of Competition?

The antagonistic interactions can be quite attention-getting; the scene of a lion stalking a gazelle is almost emblematic of the African savanna. But at the same time a predator is interacting with its prey, it may also be interacting with other predators that hunt the same prey species. Lions are not the only predators of gazelles; cheetahs, hyenas, and even crocodiles hunt and kill gazelles, potentially reducing the supply available for lions.

Whenever any resource is not sufficiently abundant to meet the needs of all the organisms with an interest in that resource, organisms must compete with one another to gain enough of that resource to survive. Competition not only influences the evolution of species, but also plays an important role in determining the structure and composition of communities, as we will see in the next chapter.

### Competition is widespread because all species share resources

Virtually no species enjoys exclusive access to any given resource; the vast majority must share at least some resources with other species. As we saw in Section 55.3, populations do not grow indefinitely, largely because resources are limited. In the absence of sufficient resources, individuals in the population compete for those resources. Such **intraspecific competition**—competition among individuals of the same species—may result in reduced growth and reproductive rates for some individuals, may exclude some individuals from better habitats, and may cause the deaths of others. **Interspecific competition**—competition among individuals of different species—affects individuals in much the same way. In addition, it can influence the persistence and evolution of species.

The *principle of competitive exclusion* holds that no two species can long coexist sharing the same limiting resource. If one species can prevent all members of another species from utilizing a resource, the inferior competitor may go locally extinct, a result called **competitive exclusion**. In other cases, selection pressures resulting from interspecific competition cause changes in the ways in which the competing species use the limiting resource. If those changes allow them to coexist, the result is called **resource partitioning**.

Whether it is interspecific or intraspecific, competition occurs by two major mechanisms. **Interference competition** occurs when a competitor interferes with another competitor's access to a limiting resource. **Exploitation competition** occurs when a limiting resource is available to all competitors and the outcome of competition depends on the relative efficiency with which the competitors use up the resource.

### Interference competition may restrict habitat use

Interference competition can take many forms. A graphic example involves the desert ant *Conomyrma bicolor* and the honeypot ant *Myrmecocystus mexicanus*. These two ant species occupy the same type of habitat—arid areas containing little vegetation—and feed on similar foods—the sugary excretions of aphids and other sap-feeding insects and occasional arthropods, none of which is in great supply. When *C. bicolor* workers find the entrance of a honeypot ant nest, they pick up small stones in their mandibles, carry them to the rim of the nest opening, and drop them down the hole—up to 200 stones in a 5-minute interval. This activity is enough to stop the honeypot ants from going out foraging. Some honeypot ant colonies, under constant stone-dropping attack for several weeks, may be almost entirely deprived of food.

Even microorganisms interfere with one another's use of resources. In the highly structured environment of the rhizosphere, or "root-world," of the soil, competitive interactions can be locally intense. Many soil bacteria produce substances that subdue their microbial competitors. Actinomycetes, for example, produce chemicals that interfere with essentially every life process in a bacterium. Many of the chemicals these remarkably well-defended microbes produce to defeat their competitors are used as antibiotics by mutualistic partners, such as the bark beetles described in Section 56.3, as well as in human pharmacology.

### Exploitation competition may lead to coexistence

Exploitation competition may lead to coexistence, provided that the species relying on the same resource evolve ways to divide up, or partition, the resource. Resource partitioning can lead to the formation of **guilds**: groups of species that exploit the same resource, but in slightly different ways, making the resource less likely to be preempted. For example, in many Rocky Mountain communities, at least three species of bees consume the nectar of the shindagger agave (*Agave schottii*). The three bee species differ in where and when they collect shindagger nectar. Honey bees tend to forage in places with the greatest numbers of shindagger flowers, bumble bees in places with intermediate numbers of flowers, and carpenter bees where flowers are few and far between. Honey bees also tend to be most active when nectar output is greatest. With their larger nests and greater numbers of offspring to support, honey bees require greater efficiency and greater energy intake. Foraging sites that are not worth their while are left to the other bees.

Sometimes organisms respond to competition by leaving their competitors behind and finding new resources. Robert Denno examined a guild of seven sap-sucking insects called planthoppers living on salt marsh grass in marshes in New Jersey. Six of the seven species partition the resource by feeding on different parts of the plant or by feeding at different times during the season. All but one of the species are wingless and cannot fly. Whenever conditions become crowded in a particular patch of salt marsh grass, the seventh species—the only one with functional wings—flies away to find a new patch in which to feed.

Sometimes individuals within a species display different behaviors or morphologies depending on whether they are competing for resources with another species. Darwin remarked in *The Origin of Species* that "Natural Selection leads to divergence of character; for more living beings can be supported on the same area the more they diverge in structure, habits, and constitutions." This "divergence of character," today called **character displacement**, is most dramatic in cases in which the morphological attributes of a species vary depending on the presence or absence of a competitor. In some of the islands of the Galápagos archipelago, certain cactus species are pollinated exclusively by finches (**Figure 56.15**). On other islands, a carpenter bee (*Xylocopa darwinii*) competes with the finches for cactus nectar; the birds consequently feed more heavily on seeds and insects. On the islands where bees are absent, finches feed on nectar more often, and have significantly smaller wingspans, than finches on islands where they share cacti with bees. On the bee-free islands, a smaller wingspan means a smaller body size, which allows the birds to reach the nectar of the cactus flowers more easily.

## Species may compete indirectly for a resource

Species may compete indirectly for a resource even when they are not present in the same habitat at the same time. Sometimes a species so alters the quality of a resource that it is rendered less usable by other species that may encounter it afterward. For example, feeding by sap-sucking leafhoppers on potato plants early in the growing season can cause leaf curling and chlorosis (loss of chlorophyll); potato beetles that consume these damaged leaves later in the growing season suffer reduced growth and survival rates. Even though these two herbivores do not feed at the same time, one species influences the use of the shared food resource by its competitor.

## Consumers may influence the outcome of competition

Indirect competition can also result when two species share a common predator. For example, the parasitoid wasp *Venturia canescens* is a consumer of two different species of caterpillars that infest stored food products such as flour: the Indianmeal moth caterpillar (*Plodia interpunctella*) and the Mediterranean flour moth caterpillar (*Ephestia kuehniella*). The two species can coexist in a flour bin, but when the wasp is present it preferentially attacks and kills the flour moth caterpillars. Thus, in the presence of the wasp, the competitive balance between the two caterpillar species is altered in the meal moth's favor. This type of competition is indirect because the outcome of competition depends not on how the two competitors utilize the shared resource, but on how the two competitors interact with a shared predator.

## Competition may determine a species' niche

Competition is important in determining where a species can be found. A species' **niche** is the set of physical and biological conditions it requires to survive, grow, and reproduce. Thus a

**56.15 Competition with Carpenter Bees Influences Finch Morphology** On islands in the Galápagos archipelago where finches are the sole pollinators of cactus flowers, a small wingspan increases their ability to negotiate the flowers. On islands where carpenter bees compete with finches for cactus nectar, the birds have a larger wingspan and can also feed on other foods.

*Geospiza fuliginosa*

**Nectar Use and Wingspan of *G. fuliginosa***

| ISLAND | TIME SPENT FEEDING ON FLOWER NECTAR (%) | MEAN WINGSPAN (mm) |
|---|---|---|
| **BEES ABSENT** | | |
| Pinta | 10 | 59.8 |
| Marchena | 28 | 58.2 |
| **BEES PRESENT** | | |
| Fernandina | 1 | 64.8 |
| Santa Cruz | 14 | 64.0 |
| San Salvador | 0 | 63.8 |
| Española | 0 | 64.7 |
| Isabela | 7 | 64.5 |

*Xylocopa darwinii* on *Opuntia* flower

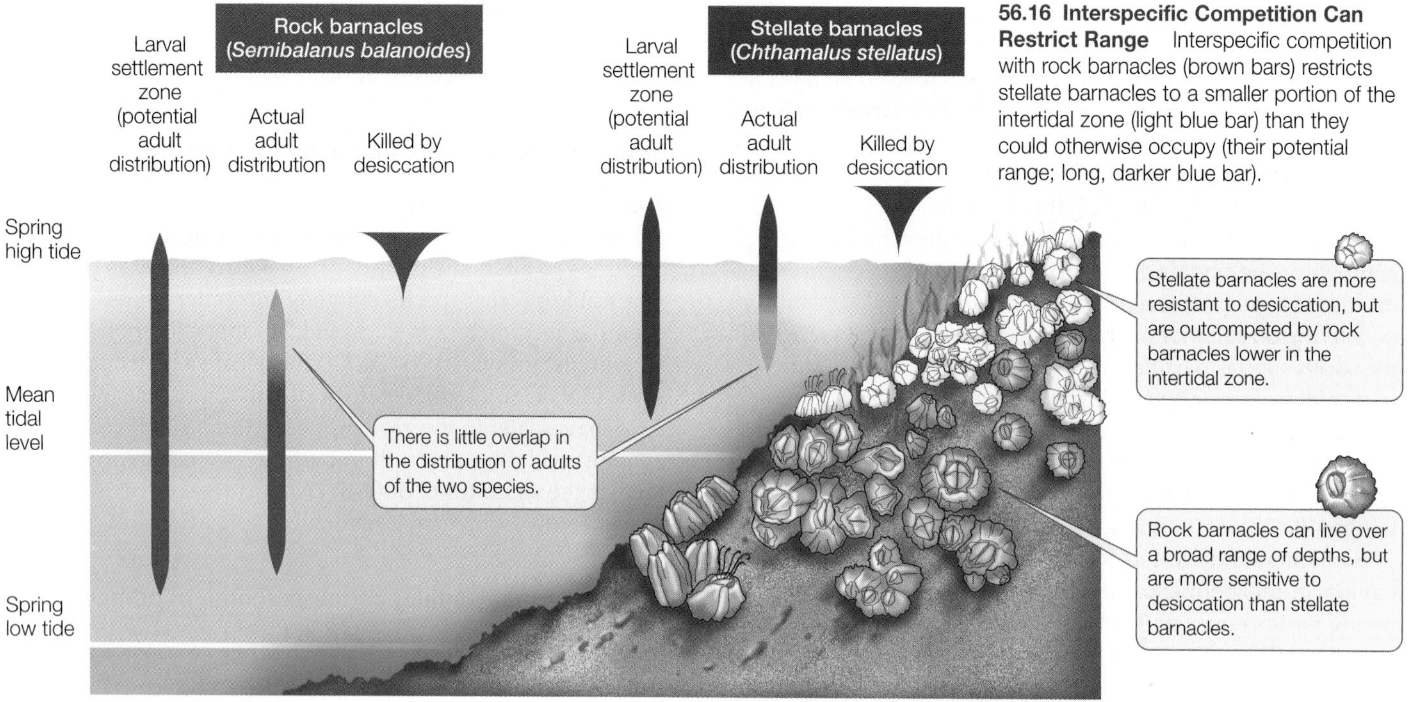

Rock barnacles (Semibalanus balanoides)
Larval settlement zone (potential adult distribution)
Actual adult distribution
Killed by desiccation
Stellate barnacles (Chthamalus stellatus)
Larval settlement zone (potential adult distribution)
Actual adult distribution
Killed by desiccation
Spring high tide
Mean tidal level
Spring low tide
There is little overlap in the distribution of adults of the two species.
Stellate barnacles are more resistant to desiccation, but are outcompeted by rock barnacles lower in the intertidal zone.
Rock barnacles can live over a broad range of depths, but are more sensitive to desiccation than stellate barnacles.

**56.16 Interspecific Competition Can Restrict Range** Interspecific competition with rock barnacles (brown bars) restricts stellate barnacles to a smaller portion of the intertidal zone (light blue bar) than they could otherwise occupy (their potential range; long, darker blue bar).

species' niche is partly defined by the resources available in the environment. Although a species might be physiologically able to live under a wide range of resource conditions, competitors may restrict its use of resources in a particular location. Thus every species has a **fundamental niche**, defined by its physiological capabilities, and a **realized niche**, defined by interactions with other species.

Two species of barnacles, the rock barnacle (*Semibalanus balanoides*) and Poll's stellate barnacle (*Chthamalus stellatus*), compete for space on the rocky shorelines of the North Atlantic Ocean (**Figure 56.16**). The planktonic larvae of both species settle in the intertidal zone and metamorphose into sessile adults. The smaller stellate barnacles generally live higher at higher levels in the intertidal zone, where they face longer periods of exposure and desiccation (drying out) than do rock barnacles. There is little overlap between the areas occupied by adults of the two species. What explains their distinct distributions in the intertidal zone?

In a famous study conducted 50 years ago, Joseph Connell experimentally removed one or the other species from its characteristic zone and observed the response of the other species. Stellate barnacle larvae normally settle in large numbers throughout much of the intertidal zone, including the lower levels where rock barnacles are found, but they thrived at those lower levels only when rock barnacles were not present. The rock barnacles grow so fast that they smother, crush, or undercut the stellate barnacle larvae. In contrast, removing stellate barnacles from their spots higher in the intertidal zone did not lead to their replacement by rock barnacles; the rock barnacles are less tolerant of desiccation and fail to thrive there even when stellate barnacles are not around. The result of the competitive interaction between the two species is a pattern of *intertidal zonation*, with stellate barnacles restricted in their distribution by competition and rock barnacles restricted in their distribution by their physiological limitations.

## 56.4 RECAP

**Competition occurs when two or more species require a resource that is in limited supply. No two species can long coexist sharing a limiting resource. The outcome of competition may be competitive exclusion, in the form of local extinction, or coexistence, in the form of resource partitioning or character displacement.**

- How does exploitation competition differ from interference competition? See pp. 1198–1199
- How can competition lead to character displacement? See p. 1199 and Figure 56.15
- Explain the difference between an organism's fundamental niche and its realized niche. See pp. 1199–1200 and Figure 56.16

The study of interactions among the species in a community is a large part of community ecology—the topic of the next chapter. Every kind of interaction we have studied in this chapter influences the nature and structure of communities. Competition helps determine which species persist and which go extinct, as well as dictating how many different species can be supported by a particular resource. Similarly, antagonistic interactions have important effects on the distribution and abundance of consumer and resource species, and the presence of mutualistic partners may dictate whether a particular species can exist in a particular community.

## CHAPTER SUMMARY

### 56.1 What Types of Interactions Do Ecologists Study?

- Species interactions can be grouped into five categories. **Antagonistic interactions** benefit a consumer while harming the species that is consumed. A **mutualism** benefits both participants, whereas **competition** harms both. **Commensal** interactions benefit one participant with no effect on the other; **amensal** interactions have no effect on one participant but harm the other. Review Figure 56.1, **WEB ACTIVITY 56.1**
- The evolution of an adaptation in one species may lead to the evolution of an adaptation in a species with which it interacts, a process known as **reciprocal adaptation** or **coevolution**. A series of reciprocal adaptations can lead to a coevolutionary **arms race**.

### 56.2 How Do Antagonistic Interactions Evolve?

- Chemically defended animals often advertise their toxicity with **aposematism**, or warning coloration. Review Figure 56.4
- In **Batesian mimicry**, a palatable species mimics an unpalatable species. In **Müllerian mimicry**, two or more unpalatable species converge to resemble one another. Review Figure 56.5
- Some prey species avoid detection by means such as **crypsis** or **homotypy**. Others defend themselves by physical or chemical means. Review Figure 56.6
- Secondary metabolites are the principal defenses used by plants against herbivores, but many plants have physical defenses as well. Many herbivores have evolved resistance to plant secondary metabolites, and some have incorporated them into their own defenses against predators.
- The hosts of pathogenic microparasites fall into three classes: susceptible, infected, and recovered (and thus, in many cases, immune to another infection). Ectoparasites are less intimately associated with their hosts than are microparasites but can nevertheless affect host fitness.

### 56.3 How Do Mutualistic Interactions Evolve?

- Mutualistic interactions involve an exchange of benefits. Many mutualisms arise in environments where resources are in short supply.
- Reciprocal adaptations are most likely to arise when an increase in dependency on a partner provides an increase in the benefits realized from the interaction.
- Many mutualisms between plants and mobile animals involve an exchange of food for transport. In plant–pollinator interactions, animals that collect and transport pollen are rewarded with pollen, nectar, or other food rewards.
- Broad suites of floral characteristics that are attractive to certain types of pollinators exemplify **diffuse coevolution**. Some plant–pollinator mutualisms, however, are much more specific and exclusive. Review Table 56.1 and Figure 56.11
- Plants that depend on frugivores for seed dispersal must balance discouraging frugivores from eating fruits before the seeds are mature, attracting frugivores when the seeds are mature, and protecting the seeds from destruction in an animal's digestive tract.
- Some insects extract energy from plants by "farming" fungal species that possess the enzymes needed to digest plant material. Other mutualisms involve an exchange of food or housing for defense. Review Figure 56.13 and 56.14, **ANIMATED TUTORIAL 56.1**

### 56.4 What Are the Outcomes of Competition?

- Competition occurs whenever a resource is not sufficient to meet the needs of all individuals with an interest in that resource.
- Competition may result in local extinction of the inferior competitor, an outcome called **competitive exclusion**. Alternatively, selection pressures resulting from competition may change the ways in which different species use a limiting resource, an outcome called **resource partitioning**.
- **Interference competition** occurs when an individual interferes with a competitor's access to a limiting resource. **Exploitation competition** occurs when a limiting resource is available to all competitors and the outcome depends on the relative efficiency with which competitors use up the resource.
- Competition may lead to **character displacement**, in which morphological attributes of a species vary geographically depending on whether a competitor is present or absent. Review Figure 56.15
- A species' **niche** is the set of physical and biological conditions it requires to persist. Although a species may be able to persist under a wide range of resource conditions (its fundamental niche), competitors may restrict its use of resources in a particular location (its realized niche). Review Figure 56.16

## SELF-QUIZ

1. Predation, herbivory, and parasitism are all examples of
   *a.* antagonistic interactions.
   *b.* mutualistic interactions.
   *c.* commensal interactions.
   *d.* amensal interactions.
   *e.* competitive interactions.
2. In a coevolutionary arms race, after a plant evolves a novel chemical defense against an herbivore,
   *a.* the herbivore can be expected to go extinct.
   *b.* the herbivore can be expected to disperse to new habitats.
   *c.* the herbivore can be expected to evolve resistance to the plant's defense.
   *d.* the plant can be expected to colonize new habitats.
   *e.* the plant can be expected to outcompete all other plants in the same environment.
3. Two organisms that use the same resources when those resources are in short supply are said to be
   *a.* predators.
   *b.* competitors.
   *c.* mutualists.
   *d.* commensalists.
   *e.* amensalists.
4. Damage caused to shrubs by branches falling from overhead trees is an example of
   *a.* interference competition.
   *b.* partial predation.
   *c.* amensalism.
   *d.* commensalism.
   *e.* diffuse coevolution.

5. A hummingbird sips nectar from the flowers of a plant species and, in the process, pollinates those flowers. This interaction is best classified as
   *a.* parasitism, because the hummingbird consumes the flower's nectar.
   *b.* predation, because the hummingbird eats the plant's seeds.
   *c.* commensalism, because the hummingbird benefits from consuming nectar and the plant is unaffected.
   *d.* mutualistic, because the plant provides nectar for the hummingbird and the hummingbird transports pollen for the plant.
   *e.* Not enough information is provided to classify this interaction.
6. Bright coloration that warns predators of prey toxicity is called
   *a.* aposematism.
   *b.* crypsis.
   *c.* homotypy.
   *d.* amensalism.
   *e.* character displacement.
7. The type of mimicry in which a palatable species resembles a toxic species is called
   *a.* Müllerian.
   *b.* Batesian.
   *c.* diffuse.
   *d.* commensal.
   *e.* homotypy.
8. Interference competition has been demonstrated to occur between
   *a.* species of barnacles in the rocky intertidal zone.
   *b.* species of ants in the deserts of North America.
   *c.* species of soil bacteria that produce antibiotics.
   *d.* all of the above
   *e.* none of the above
9. One factor that can constrain the realized niche occupied by an organism is
   *a.* immigration.
   *b.* aposematism.
   *c.* mimicry.
   *d.* parasitism.
   *e.* competition.
10. The principle of competitive exclusion states that
    *a.* no two species can coexist unless they are sharing the same resource.
    *b.* no two species can long coexist if they are sharing the same limiting resource.
    *c.* if two species are sharing the same resource, one will always go extinct.
    *d.* competition between two species always results in character displacement.
    *e.* none of the above

## FOR DISCUSSION

1. Many species interact, but not all interactions result in coevolution. How could you demonstrate that a particular plant has coevolved with a particular insect? What other explanations might account for adaptations in the plant that reduce the effect of a specific insect herbivore?
2. Aposematism is defined here as warning coloration, but warning colors would only be effective in deterring predators that use vision to find their prey. What other kinds of warning signals do you think might have evolved in interactions involving predators that rely on sensory cues other than vision? How would you investigate whether such signals exist?
3. Many ectoparasites are highly specific and feed on only a narrow range of host species. What do you think happens to host-specific parasites when their hosts go extinct? Should the parasites of an endangered species also be targets for conservation?
4. Even though nectar serves no function in the life of a plant other than to attract and reward pollinators, occasionally pollinators such as honey bees are poisoned by plant nectar. Why do you think some plants produce toxic nectars? Under what circumstances do you think toxic nectars would evolve?

## ADDITIONAL INVESTIGATION

Like the Southern pine beetles (see Figure 56.13), the mountain pine beetle (*Dendroctonus ponderosae*) attacks pine trees with the help of a symbiotic fungus that infects the host tree. In 2009, these beetles infested almost 4 million acres of pine forest across Montana, Wyoming, Colorado, Idaho, Utah, Oregon, and Washington. How would you determine which pine trees are susceptible to mountain pine beetle attack? How could the fact that this beetle has a symbiotic partner affect approaches for managing the outbreak?

# 57 Community Ecology

## Dead reckoning

In March 1996 the body of an apparent suicide victim was discovered under the bushes near the railroad tracks in Cologne, Germany. The corpse was badly decomposed, and most of the abdominal organs had disintegrated completely. When an autopsy was performed, masses of maggots—fly larvae—about 8 mm long were found in what remained of the body cavity, and the leathery patches of dried skin were almost totally covered with a layer of pale yellow eggs. When he examined the body, Mark Benecke, a forensic entomologist, was able to recover a single adult fly, which he identified as *Piophila casei*, also known as the cheese skipper.

The community of species that colonizes a human corpse is diverse, and its composition varies predictably over time as decomposition progresses. Among the many insect species that infest dead bodies, cheese skippers are latecomers—female cheese skippers do not find a corpse attractive until the proteins in the body begin to break down, typically after 1 to 3 months, depending on season and locality. The abundance of *P. casei* eggs suggested to Benecke that the cheese skippers had undergone at least two generations in the decomposing corpse. Knowing that their development time from egg to adult ranged from 11 to 19 days under local weather conditions, he calculated that the first adult cheese skippers probably arrived and laid their eggs about 90 days after death, and that an additional 22 to 38 days would have been required for two generations to complete their development. Thus he calculated that death must have occurred between 112 and 128 days earlier. As it turned out, a 38-year-old woman had been reported missing about 4 months earlier; the estimated postmortem interval helped investigators to identify her as the suicide victim found by the railroad tracks.

Insects are found in decomposing corpses for many reasons. Some species, such as cheese skippers, consume the dead flesh; others, such as hide beetles, eat the hair and nails. Others prey on the insects consuming the corpse or eat their excrement and shed exoskeletons. Thus a dead body can support an entire community of organisms—one with complex connections among its inhabitants. Where and how an individual died can influence the composition of that community; bodies that are submerged, for example, are colonized by a different suite of insects than those that are buried in the ground.

**Dead Reckoning** Attracted by the odor of decaying flesh, bluebottle flies feed on the fur of a dead mole. The insect species on a decomposing corpse change in such a predictable fashion over time that forensic entomologists can estimate the time of death by identifying the species that are present.

**Prairie Potholes** These wetland communities occur in depressions in the prairie landscapes of midwestern North America. The potholes, formed when the glaciers of the last ice age receded, support communities characterized by many aquatic plant species, including sedges, bulrushes, and pondweeds.

What species occur where and when is the concern of community ecologists. The composition of communities can change in a predictable way over time and over space. The spatial scales of these changes can vary from a dead body in a patch of shrubbery to a patch of Amazonian rainforest, and the time scales can range from days to millennia. But the ecological processes affecting a community are similar whatever its scale. The study of the seemingly esoteric patterns of change in the composition of carrion communities has not only allowed ecologists to help find evidence that can identify a missing person or convict a murderer, but has added greatly to our understanding of ecological communities.

**IN THIS CHAPTER** we will investigate how energy flows through communities and how that flow influences their structure, complexity, and composition. We will investigate how communities are assembled over time and how they are reassembled after a disturbance. Finally, we will consider the relationships between species diversity, productivity, and stability in communities.

## CHAPTER OUTLINE



# 57.1 What Are Ecological Communities?

An ecological **community** is a group of species that coexist and interact within a defined area. Although each species has unique interactions with the other species in its community, the community as a whole can be studied on the basis of the distribution of energy and biomass within it.

Communities vary greatly in size and scope. Whereas one ecologist might study the community of species inhabiting a dead body, another might study the community of species that inhabit Lake Superior, the largest of the Great Lakes. The boundaries defining a community are not always easy to recognize. The community of organisms within a pond, for example, is for the most part bounded by the borders of the pond. Pond species interact with one another to a far greater extent than with species outside the pond. The boundaries of other communities are not so clear-cut. For example, while prairie pothole communities (above left) may appear to be self-contained, many of their components originate far away. Seeds arrive at these relatively isolated wetlands via duck excrement; mallards and other ducks may consume, but not digest, seeds in one location and, in the 5–11 hours it takes for the seeds to move from one end of the digestive tract to the other, fly up to 1,400 km before depositing the seeds when they alight to feed again.

Communities may differ conspicuously in **species richness**—that is, in the number of species they contain. Even the same type of community may contain different numbers of species in different places. The number of plant species in prairie pothole communities, for example, may vary more than twofold across the upper Midwest. Although communities vary in size and complexity by orders of magnitude, ecologists have devised methods for quantifying basic properties of community structure and organization irrespective of their scale. These methods have revealed patterns that reflect underlying community assembly rules and general principles.

What determines how many species constitute a community in any particular place? One important factor is the amount of energy available to sustain organisms.

### Energy enters communities through primary producers

Sunlight is the ultimate source of energy for most of Earth's communities. Sunlight makes photosynthesis possible, and photosynthesis, in the vast majority of communities, makes energy

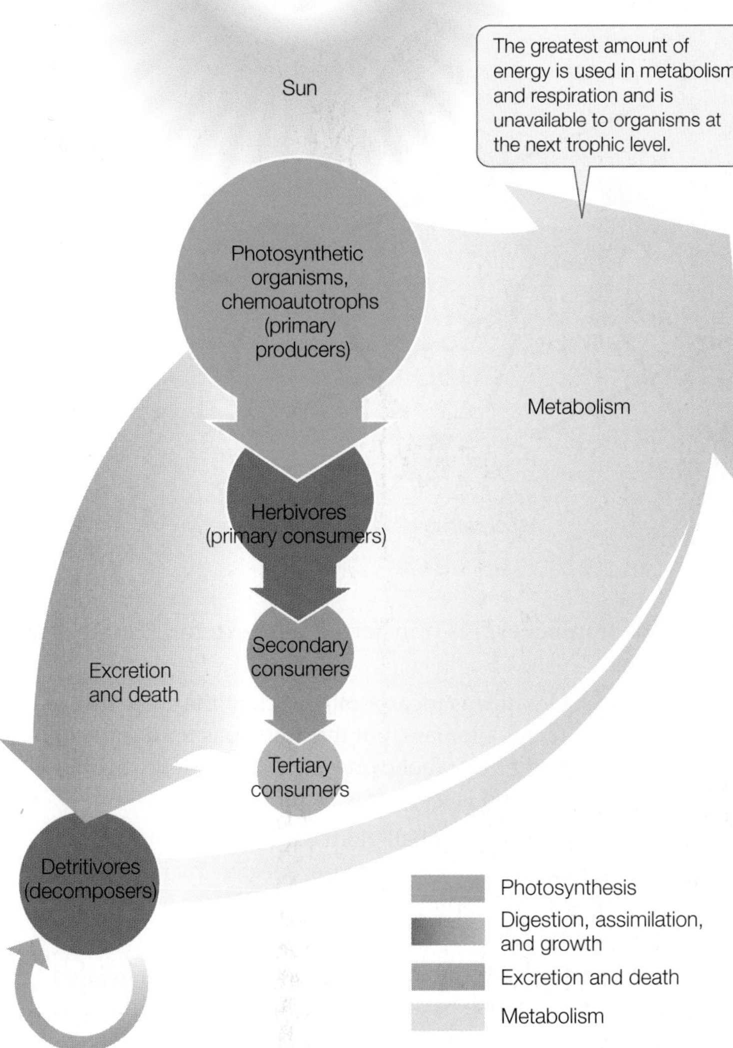


**57.1 Energy Flow through Trophic Levels** Much of the energy of the biomass at each trophic level is lost (often as heat) to metabolism and respiration by the organisms at that level. In this diagram, the width of each arrow is roughly proportional to the amount of energy flowing through that channel. Arrows indicate directions of energy flow.

**yourBioPortal.com**
GO TO **Web Activity 57.1 • Energy Flow through an Ecological Community**

available to other organisms in an edible form. All nonphotosynthetic organisms (*heterotrophs*) consume, either directly or indirectly, the energy-rich organic molecules produced by the plants and other photosynthetic organisms that get their energy directly from sunlight (*autotrophs*). Photosynthetic autotrophs, along with a handful of *chemoautotrophs* (organisms that obtain chemical energy from inorganic molecules in their environment), are known as **primary producers** (**Figure 57.1**).

**Gross primary productivity** (**GPP**) is the rate at which all the primary producers in a particular community turn solar energy into stored chemical energy via photosynthesis. The energy that is accumulated is called **gross primary production**. (The terms "productivity" and "production" are often used interchangeably; *productivity* measures the rate of energy accumulation; *production* measures energy accumulation as a *product*.)

Not all GPP becomes available to heterotrophs, because primary producers use some of that energy for respiration and other metabolic processes. **Net primary productivity** (**NPP**) is the rate at which energy is incorporated into the primary producers' bodies through growth and reproduction. Thus **net primary production** is the amount of primary producer **biomass** (weight of organic matter) available for consumption by heterotrophs. This relationship is described mathematically as

$$NPP = GPP - R$$

where R is the energy lost through respiration.

### Consumers use diverse sources of energy

An organism's **trophic level** indicates where in that sequence it obtains its energy (**Table 57.1**). Primary producers start the chain of trophic levels. At the next level are **primary consumers** — the herbivores that dine on the primary producers. Organisms that eat herbivores, called **secondary consumers**, are the next trophic level. Those that eat secondary consumers are *tertiary consumers*, and so on. The waste products and dead bodies of organisms provide another source of energy, as we saw at the

**TABLE 57.1**
**The Major Trophic Levels**

| TROPHIC LEVEL | SOURCE OF ENERGY | EXAMPLES |
|---|---|---|
| Photosynthesizers (primary producers) | Solar energy | Green plants, photosynthetic bacteria and protists |
| Herbivores (primary consumers) | Tissues of primary producers | Termites, grasshoppers, gypsy moth larvae, anchovies, deer, geese, white-footed mice |
| Primary carnivores (secondary consumers) | Herbivores | Spiders, warblers, wolves, copepods |
| Secondary carnivores (tertiary consumers) | Primary carnivores | Tuna, falcons, killer whales |
| Omnivores | Several trophic levels | Humans, opossums, crabs, robins |
| Detritivores (decomposers) | Dead bodies and waste products of other organisms | Fungi, many bacteria, vultures, earthworms |

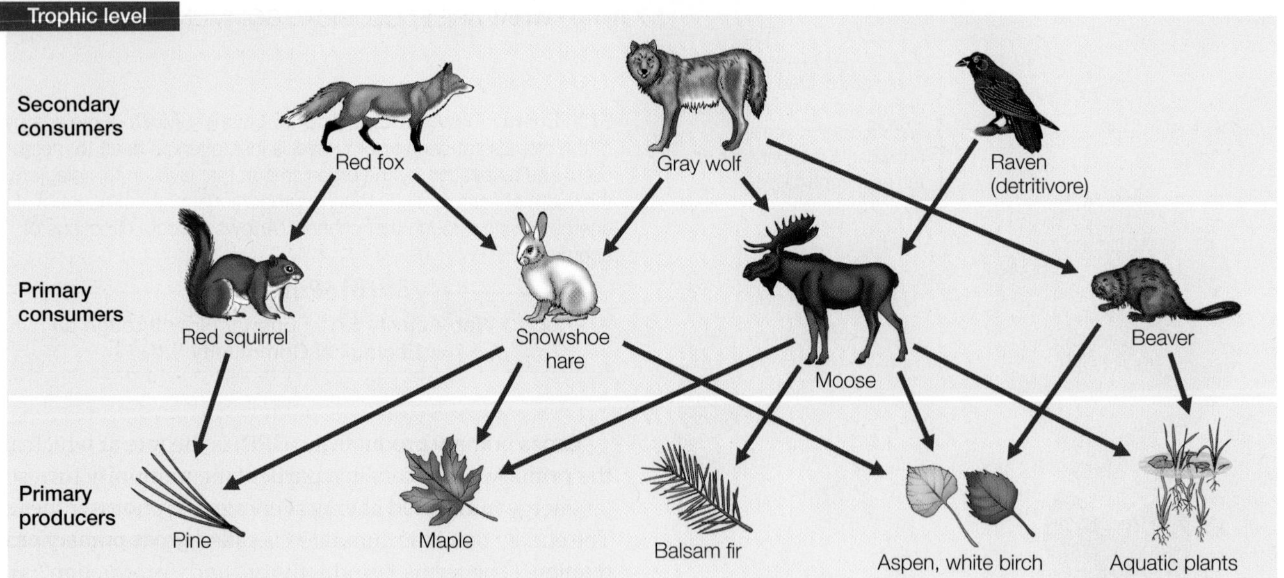


**57.2 Food Webs Show Trophic Interactions in a Community** This food web for Isle Royale National Park, located on a large island in Lake Superior, includes only large vertebrates and the plants on which they depend. Even with these restrictions, the web is complex. The arrows show who eats whom.

opening of this chapter. Organisms that consume such materials are called *detritivores* or *decomposers* (see Figure 57.1).

Some organisms, called *omnivores*, feed on multiple trophic levels. Opossums, for example, are famously omnivorous. Investigators in Portland, Oregon, dissected the stomachs of road-killed opossums and found remains of mammals, birds, insects, earthworms, snails, fruits, bulbs, seeds, leaves, grass, pet food, and garbage, along with some items they couldn't even identify.

Although few organisms are as omnivorous as opossums, most species in a community eat and are eaten by more than one other species. A **food chain** depicts the linear sequence of who eats whom in a given community; food chains are interwoven in a **food web** (**Figure 57.2**). Most communities contain so many species interacting in so many different ways that it is impossible to enumerate (or even identify) all of the links. Nevertheless, simplified diagrams of food webs are useful in envisioning how energy flows through a community.

**yourBioPortal.com**

GO TO Web Activity 57.2 • The Major Trophic Levels

### Fewer individuals and less biomass can be supported at higher trophic levels

One important way to characterize a community is by the distribution of energy and biomass within it. The flow of energy through a food web is governed by the physical laws that regulate energy transformations, foremost among which are the first and second laws of thermodynamics. Recall from Section 8.1 that the amount of energy in the universe is constant, and that when energy is converted from one form to another, part of it becomes unavailable to do work. As Figure 57.1 illustrates, at each trophic level energy is lost to metabolism and respiration. On average, only about 10 percent of the energy of one trophic level is transferred the next, for a number of reasons:

- *Heat loss.* Organisms incorporate much of the energy they accumulate into biomass, but much more is used for respiration and other metabolic processes. That energy is dissipated as heat and is lost to the community.
- *Biomass availability.* Not all biomass can be ingested. Grazers routinely miss blades of grass; effective plant defenses prevent herbivory; prey can escape predators or leave the community.
- *Indigestibility.* Not all biomass ingested can be assimilated by consumers. Tree bark, for example, contains lignin and cellulose, which cannot be digested by most herbivores.

The overall transfer of energy from one trophic level to the next (which can be expressed as the ratio of consumer production to producer production) is called **ecological efficiency**.

*Pyramid diagrams* such as **Figure 57.3A** can be used to illustrate the proportions of energy transferred from each successive trophic levels and to compare those proportions among different communities. Pyramid diagrams can also be used to illustrate the amount of biomass or numbers of individuals found at each level (**Figure 57.3B**). Progressive energy loss through the inefficiencies of energy transfer puts limits on the number of trophic levels in a food chain or food web; largely for this reason, most communities support only three to five trophic levels. Furthermore, each successive trophic level tends to have fewer species, and the species at each level tend to have lower reproductive rates, smaller populations, and larger body sizes.

### Productivity and species richness are linked

Just as the diversity and complexity of trophic levels tend to be positively correlated with the amount of energy available to them, the species richness of a community tends to be positively correlated with its productivity, up to a point. A number of factors that influence productivity, and thus species richness, vary among communities.

The most obvious variation that influences productivity is energy input. The amount of solar energy reaching Earth's surface

(A) Energy flow
(calories/m²/day)

(B) Biomass
(grams/m²)

Forest

In forests, the majority of biomass is tied up in wood, so its energy is not available to most herbivores.

Grassland

Most of the biomass in a grassland is found in the green plants, and most of the energy flows through them.

Open ocean

Trophic level
Secondary consumers
Primary consumers
Primary producers

A marine community produces strikingly different patterns of biomass. The primary producers are unicellular algae, which divide so rapidly that a small biomass can support a much larger biomass of herbivores.

**57.3 Energy and Biomass Distributions** Pyramid diagrams allow ecologists to compare (A) patterns of energy flow through trophic levels in different communities and (B) the amount of biomass present at the different trophic levels.

varies by latitude, as we saw in Figure 54.1. The ability of plants to photosynthesize, however, depends not only on the supply of energy from the sun, but also on the supply of water and nutrients. The species richness of trees across different regions of North America, for example, can be best predicted not by measuring incoming solar radiation, but by measuring an ecological attribute of those regions called *annual evapotranspiration*: the amount of water released from the land surface by evaporation from streams, lakes, and soil and by transpiration from plants. The annual evapotranspiration of a region is measure of the amount of water available to the organisms living there.

The number of species in a community increases with productivity only up to a point, however; if productivity increases beyond that point, species richness may actually decline (**Figure 57.4**). Why should that be? As local productivity increases, so does the number of individuals the local habitat can support (its carrying capacity). Thus populations can grow larger, and larger population sizes should reduce the risk of species extinction. Why, then, should species richness decrease when productivity is very high?

One hypothesis postulates that interspecific competition becomes more intense when productivity is very high, resulting in competitive exclusion of some species (see Section 56.4). This hypothesis is supported by the results of a long-term experiment at the Rothamsted Experiment Station in England, begun in 1857 and still ongoing. Fertilizer has been added regularly to selected plots of land to increase their productivity, and fertilized and unfertilized plots have been monitored continuously. Over 150 years, the number of plant species in the unfertilized plots has remained roughly constant, whereas that number has declined in the fertilized plots, supporting the premise that species richness can decline when productivity rises.

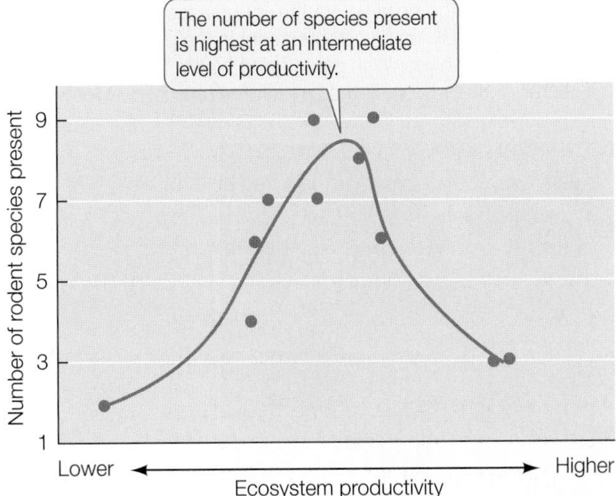


**57.4 Species Richness Peaks at Intermediate Productivity** The number of rodent species living in ecosystems of varying productivity in the Gobi Desert exemplifies a pattern in which species richness increases only up to a certain point. Beyond that point, it can actually decline with productivity.

## 57.1 RECAP

**An ecological community is a group of species that coexist and interact within a defined area. Net primary production—the amount of biomass that primary producers make available to heterotrophs through photosynthesis—is an important factor determining the species richness and number of trophic levels in a community.**

- Explain the difference between gross primary productivity and net primary productivity. See p. 1205
- Describe three trophic levels that might appear in a food web. See pp. 1205–1206, Table 57.1, and Figure Figure 57.2
- What is a typical distribution of energy among the trophic levels of a community? See pp. 1205–1206 and Figures 57.1 and 57.3

Although energy and biomass distributions reveal much about the structure and dynamics of ecological communities, they reflect primarily the abiotic mechanisms that influence the community. In the next section we will see how the species interactions described in Chapter 56 shape community structure.

## 57.2 How Do Interactions among Species Influence Community Structure?

All of the interaction types discussed in Chapter 56, and especially the antagonistic interactions, have a strong influence on an ecological community. Species are not identical bags of biomass through which energy flows. Species in one part of a food web can affect many other species without necessarily eating them.

### Species interactions can cause trophic cascades

The interactions of a single consumer with other species in its community can cause a progression of indirect effects across successive trophic levels, a pattern called a **trophic cascade**. The reintroduction of wolves into Lamar Valley, in Yellowstone National Park, initiated just such a pattern.

The food web in Yellowstone National Park is complex. Wolves in the park feed on moose, pronghorn, mule deer, elk, bison, and bighorn sheep. Although they share these prey with coyotes, mountain lions, and grizzly and black bears, wolves exert particularly strong effects on the park community's structure and dynamics, as demonstrated by the effects of their absence during most of the twentieth century. By 1926, unrestricted hunting had eliminated wolves from the park community.

Annual censuses of elk were initiated by park managers in 1920. To prevent elk from exceeding the park's carrying capacity, the park service culled elk herds (that is, they selectively killed some members of the herds) until 1968, when, in response to public pressure, the culls were stopped. When the culling ended, the elk population rapidly increased (**Figure 57.5A**). The elk browsed aspen trees so intensely that no young trees were added to the population after 1920 (**Figure 57.5B**). The elk also severely browsed streamside willows, with the result that beavers, which depend on willows for food, were nearly exterminated from Lamar Valley. In regions of the park where elk were absent, however, aspen and willow trees flourished. This observation suggested that the decline of the trees elsewhere in the park was indeed due to elk browsing, rather than to climatic conditions or some other factor.

In 1995, after a 70-year absence, park managers reintroduced wolves to Yellowstone, and their population grew rapidly. The wolves preyed primarily on elk. The elk population of Lamar Valley dropped, and elk avoided the aspen groves, where they were especially vulnerable to wolf predation. Young aspen began to grow, willows regrew along streams, and the number of beaver colonies increased from one in 1996 to seven in 2003. Thus the presence or absence of a single predator influenced not only populations of its prey, but also the structure of the vegetation and populations of other species that depend on that vegetation.

(A)
Wolf population
Elk population
Number of elk (thousand)
20
15
10
5
0
1900
1920
1940
1960
1980
2000
2020
Year
High
Number of wolves
Low
Wolves extirpated (1926)
Elk culling suspended (1968)
Wolves restored (1995)

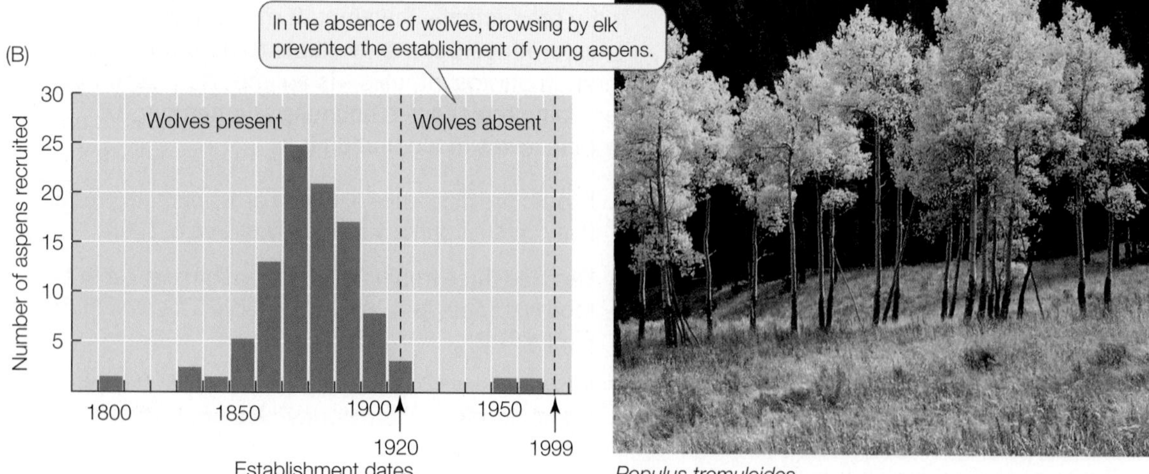


**57.5 Wolves Initiated a Trophic Cascade** (A) Number of elk and relative size of the wolf population in Wyoming's Yellowstone National Park. Once the wolves were gone, the elk population increased to the point where the young aspen trees were seriously overgrazed. (B) Aspen establishment in the presence and in the absence of wolves. With overgrazing by elk, young trees were not recruited; the photo shows a resulting stand of only mature trees.

Herbivores, too, can have indirect effects on other trophic levels, to produce an interaction cascade. The savannas of central Kenya are dominated by large grazing mammals such as zebras, eland, elephants, Grant's gazelles, giraffes, buffaloes, and hartebeests. A team of investigators used *exclosures* (areas protected by a barrier, such as a fence, designed to keep organisms out) to investigate the influence of these grazers on the savanna community. They created six exclosures and compared community structure within the exclosures with that in paired sites where large mammals could graze freely. Over 19 months of monitoring, they found that trees, beetles, and an insectivorous lizard species were all more abundant within the exclosures. The elimination of grazing increased the abundance of trees, which in turn increased the number of beetles by providing more food and habitat, which in turn increased the number of lizards, which feed preferentially on beetles.

Interaction cascades can be initiated by the behaviors of species as well as by their diets. Beavers preferentially cut down some species of trees to build their dams; by so doing, they alter the composition of the vegetation. In addition, the beaver dams create wetlands, meadows, and ponds that provide habitat for species that would otherwise not be able to live in the area. Organisms that build structures that alter existing habitats or create new habitats are called **ecosystem engineers**.

### Keystone species have wide-ranging effects

A species that exerts an influence on a community disproportionate to its abundance is called a **keystone species**. Keystone species influence both the species richness and the number of trophic levels in a community. The ochre sea star (*Pisaster ochraceus*), which lives in rocky intertidal zones on the Pacific coast of North America, is a good example. Its preferred prey is the mussel *Mytilus californianus*. In the absence of sea stars, these mussels crowd out other organisms in a broad belt of the intertidal zone. By consuming mussels, *P. ochraceus* creates bare spaces on the rocky substratum that are taken over by a variety of other species (**Figure 57.6**).

In a classic experiment, Robert Paine of the University of Washington demonstrated the disproportionate influence of ochre sea stars on species richness by removing them from selected sites repeatedly over a 5-year period. Two major changes occurred in the areas where sea stars were absent. First, the lower edge of the mussel bed extended farther down into the intertidal zone, showing that sea stars are able to eliminate mussels completely in areas that are submerged most of the time. Second, and more dramatically, 28 species of animals and algae disappeared from the sea star removal sites. Eventually only *M. californianus*—the dominant competitor for space in the community—occupied the entire substratum. Through its effect on competitive relationships, predation by the sea star determines how many species can thrive in its rocky intertidal community.

Species other than consumers can be keystone species. A plant species that serves as food for many different animals can also be a keystone species. Fig trees in tropical forests produce fruits several times every year, so their fruits are abundant at times when few, if any, other trees are fruiting. Dozens of frugivores depend on figs when no other fruits are present. Fig-eating animals include fruit bats, parrots, toucans, pigeons, flycatchers, trogons, orioles, rodents, howler monkeys, and even fish, which eat figs that fall into nearby streams. All of these animals provide prey for a diverse community of predators. Moreover, the trunks of fig trees provide habitat for several thousand species of insects, reptiles, rodents, and birds. Without fig trees, rainforest communities around the world would be profoundly diminished in species richness.

**57.6 Some Sea Stars Are Keystone Species** Ochre sea stars (*Pisaster ochraceus*) have harvested all the mussels from the lower parts of these rocks on the Olympic Peninsula of Washington. By consuming mussels, the sea star creates bare spaces on the rocks that are occupied by a variety of other intertidal species.

## 57.2 RECAP

**Some interactions between consumers and other species result in a trophic cascade of indirect effects on species at successive trophic levels. Keystone species are species with especially strong effects on the species richness and number of trophic levels in communities.**

- Describe an example of a consumer whose interactions with other species cause indirect effects across trophic levels.
- What are some of the ways in which keystone species can affect other species in their communities? See p. 1209

We have seen how certain keystone species influence the overall species richness of their communities. But species richness is only one measure of community diversity. The next section takes a closer look at other ways in which ecologists measure species diversity.

## 57.3 What Patterns of Species Diversity Have Ecologists Observed?

Communities clearly vary in their diversity, both geographically on scales ranging from local to global, and over time on scales ranging from a day to centuries. Comparing the diversity of two or more communities can be challenging because diversity has many different components depending on the scale at which it is measured:

- **Alpha diversity** is diversity *within* a single community or habitat.
- **Beta diversity** is *between-habitat* diversity, a measure of the change in species composition from one community or habitat to another.
- **Gamma diversity** is the *regional* diversity found over a range of communities or habitats in a geographic region.

### A community's diversity can be measured with a diversity index

The most straightforward way to quantify the alpha diversity of a community is simply to count the number of species present in a sample (that is, the species richness). The larger the area that is sampled, the greater the likelihood that rare species in the community will be found and that the resulting assessment of its species richness will be accurate.

To say that two communities of the same size have the same species richness tells only part of the story, however. Imagine that we take samples of 12 individuals in each of two communities. Our sample from community A contains 3 individuals of each of 4 species (an even distribution of individuals). Our sample from community B, however, contains 9 individuals of one species and only 1 individual each of the other 3 species (an uneven distribution). Even though the species richness of the two communities is the same (4), community B is considered less diverse because the less abundant species are encountered infrequently compared with the single most abundant species (**Figure 57.7**). Thus estimates of diversity should account for both species richness and species evenness.

Ecologists have devised mathematical formulas to quantify diversity that take both species richness and species evenness into account. One widely used measure is the **Shannon diversity index**, based on a mathematical expression of the certainty with which the next item sampled in a series can be predicted. If, for example, we picked individuals from community B in Figure 57.7 at random, we would be fairly certain which species would be picked first, second, third, and so on: most of the time, it would be the most abundant species. The Shannon diversity index ($H$) describes diversity as a measure of that certainty. The higher $H$ is, the lower would be our certainty about the species composition of our sample and, thus, the greater the diversity of the community. In other words, when we know that the abundance of a particular species is high, we can be reasonably certain that we are likely to collect that species in a random sample.

We can calculate $H$ in three steps:

- Determine the proportion of the total number of individuals in the community that are in each species ($p_i$) and multiply that proportion by its logarithm
- Sum the values obtained for all species in the community
- For mathematical reasons, multiply the result by –1

In algebraic notation, this calculation is written as

$$H = -[(p_1 \ln p_1) + (p_2 \ln p_2) + (p_3 \ln p_3) \dots (p_n \ln p_n)]$$

Estimating the alpha diversity of a single community with the Shannon diversity index requires only that we know the total number of individuals collected in a sample and the number of species in the sample. But describing the species diversity of a

Although the two communities have the same species richness, community A has a more even distribution of species, and is thus considered more diverse than community B.
Community A

**Figure 57.7 Species Richness and Species Evenness Contribute to Diversity** These two hypothetical mushroom communities are the same size (12) and have the same species richness (4) but differ in species evenness. The more even distribution results in greater diversity.

geographic region that comprises multiple communities requires a different approach. Recall that measures of beta diversity estimate the change in species composition from one community to another. **Sorenson's index** ($C_S$) is one such measure, calculated

$$C_S = \frac{2j}{a+b}$$

where $2j$ is twice the number of species found in both communities, $a$ is the number of species in the first community, and $b$ is the number of species in the second community.

Gamma diversity—the overall diversity across a geographic region containing many different communities—can be used to make comparisons of diversity among regions. Gamma diversity is equivalent to the alpha diversity of each community combined with the beta diversity between the communities.

Partitioning diversity within a community, between communities, and across an entire region can provide insights into the ecological characteristics of the species making up those communities as well as the processes by which those communities were assembled. High beta diversity, for example, shows that few species are shared between the communities being compared, suggesting that those communities consist mostly of specialists, species with unique habitat requirements, or species with poor dispersal abilities.

A study of diversity patterns in an agricultural area of southern England demonstrates how these three measures of diversity estimates are applied at different scales. Investigators sampled plants and macroinvertebrates in three types of freshwater communities—rivers, ponds, and ditches—in an area of English countryside. Rivers had high species richness (i.e., high alpha diversity), but most rivers in the area contained the same assortment of species, so beta diversity, from river to river, was low. In contrast, the ponds displayed a wide range of variation in species richness (i.e., the ponds, with high beta diversity, were more different from one another in species composition than the rivers were). Thus the ponds contributed more to gamma diversity—the overall diversity of the area—than the rivers did because they contained more unique or rare species. Surprisingly, the ditches, many of which contained water for only a short time, had the lowest alpha diversity (the fewest species overall) of the three community types, but many of the species found in ditches were found nowhere else in the area, including insects that live only in temporary bodies of water (including a very rare water beetle). Ponds and ditches, despite being relatively species-poor, contributed disproportionately to regional diversity (i.e., gamma diversity) because the few species they supported were found nowhere else in the region.

### Latitudinal gradients in diversity are observed in both hemispheres

It has long been known that diversity varies with latitude. About 200 years ago, the German explorer and naturalist Alexander von Humboldt spent 5 years traveling around Latin America. He remarked in the account of his voyages that "the nearer we approach the tropics, the greater the increase in the variety of structure, grace of form, and mixture of colors, as also in perpetual youth and vigour of organic life." Humboldt would not have been surprised to learn that, if he had sailed toward the poles, the diversity he observed would have decreased. These latitudinal gradients in diversity have been observed in a wide variety of taxa, including birds, mammals, and insects (**Figure 57.8**).

Although most ecologists agree that latitudinal gradients in diversity exist, there is less consensus as to why they exist. At least four hypotheses have been advanced to account for latitudinal gradients in diversity:

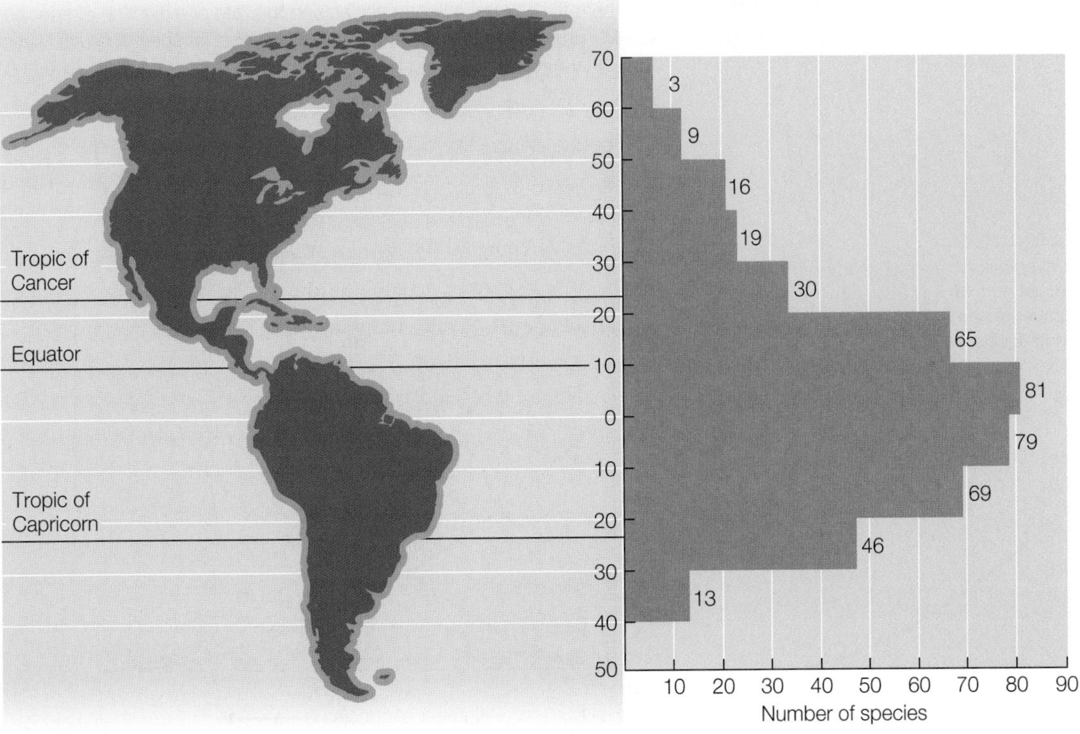


**57.8 Latitudinal Gradients in Diversity** Among swallowtail butterflies (Papilionidae), species diversity decreases with latitude both north and south of the equator. Similar latitudinal gradients of diversity have been observed in many other taxa.

- The *time hypothesis* argues that organisms in tropical regions have had more time to diversify under relatively stable climatic conditions than have more temperate regions.
- The *spatial heterogeneity hypothesis* suggests that tropical regions have high spatial heterogeneity—more different types of microclimates, vegetation, soils, and so forth—and thus contain more different habitats and many more species.
- The *specialization hypothesis* attributes latitudinal gradients in diversity greater of biological competition in the tropics, which lead to narrower niches and more species.
- The *predation hypothesis* proposes that predation intensity is greater in the tropics (and thus conflicts with the specialization hypothesis). Where predation is high, it argues, prey populations are held to levels so low that interspecific competition never comes into play, and rare species can persist.

Why can't ecologists agree on the mechanism underlying latitudinal gradients in species diversity? Corroborative evidence can be found for each of these hypotheses, varying with taxon, locality, and scale. It may be that multiple factors are responsible for this widespread ecological pattern.

### The theory of island biogeography suggests that species richness reaches an equilibrium

While latitudinal gradients prevail on a global scale, other factors influence species diversity within any latitudinal band. In the West Indies, for example, small islands tend to have fewer species of butterflies than large islands, irrespective of latitude. Furthermore, oceanic islands generally have fewer species, irrespective of taxon, than continental areas of comparable size. This **species–area relationship** is a well-established mathematical relationship between the size of an area of habitat and the number of species that area contains. The biologist Edward O. Wilson was struck by this relationship, which he encountered through his exhaustive collection of ant species from all over the world. With Robert MacArthur, Wilson developed the theory of **island biogeography**. They based their theory on just two processes: the immigration of new species to an island and the extinction of species already present on that island (**Figure 57.9A**).

The premise of island biogeography is that the number of species on an island (or in another geographically defined and isolated area) represents a balance, or equilibrium, between the rate at which species immigrate to the island and the rate at which resident species go locally extinct. The rate of immigration is determined in part by the number of species in the area providing the immigrants, known as the **species pool**. In the case of oceanic islands, the species pool comprises all the species on the nearest continent. Not all species that reach the island will persist, however. The more species there are on an island, the higher the likelihood that any one of those species will go extinct.

Rates of immigration and extinction are influenced by two other factors:

- *Distance of the island from the species pool*. The farther the island from the source of immigrants, the fewer species will successfully make the trip and the lower the immigration rate.
- *The size (area) of the island*. The smaller the island, the fewer resources it provides, and the higher the extinction rate. Larger islands provide greater habitat diversity and can sustain larger populations (which in general have lower extinction rates than small populations).

yourBioPortal.com
GO TO Animated Tutorial 57.1 • Biogeography Simulation

**57.9 MacArthur and Wilson's Theory of Island Biogeography** (A) The rate of arrival of new species and the rate of extinction of species already present determine the equilibrium number of species on an island. (B) These rates are affected by the size of the island and its distance from the mainland.



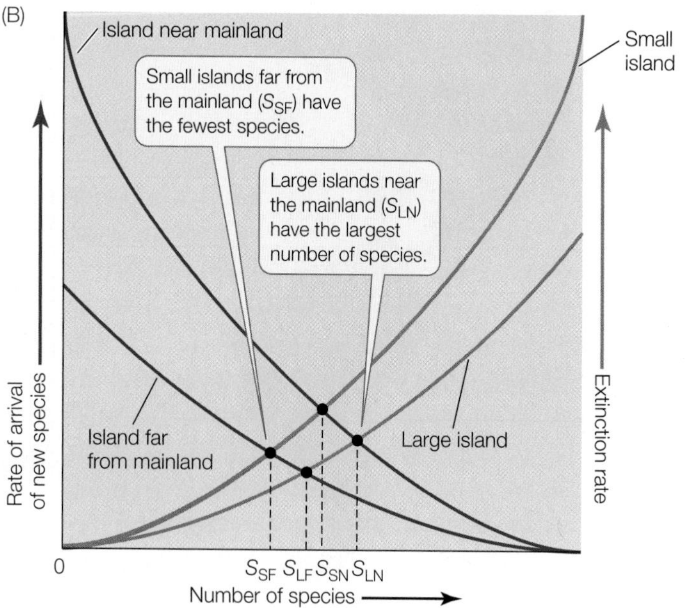

## INVESTIGATING LIFE

### 57.10 Testing the Theory of Island Biogeography

By experimentally removing all the arthropods on four small mangrove islands, Simberloff and Wilson were able to observe the rates at which arthropods recolonized the islands and compare these data with the predictions of island biogeography theory.

**HYPOTHESIS** **Defaunated islands will be rapidly recolonized and will achieve about the same number of species that were present prior to defaunation.**

**METHOD**

1. Census the terrestrial arthropod species on 4 small (11–25 $m^2$) mangrove islets.
2. Erect scaffolding and tent the islets. Fumigate with methyl bromide, a chemical that kills arthropods but does not harm plants.

3. Remove tenting. Monitor recolonization for the following 2 years, periodically censusing terrestrial arthropod species.

**RESULTS** Recolonization was fastest on the closer islands, slowest on the one farthest from the mainland. Two years after defaunation, each island had about the same number of species it had before the experiment.

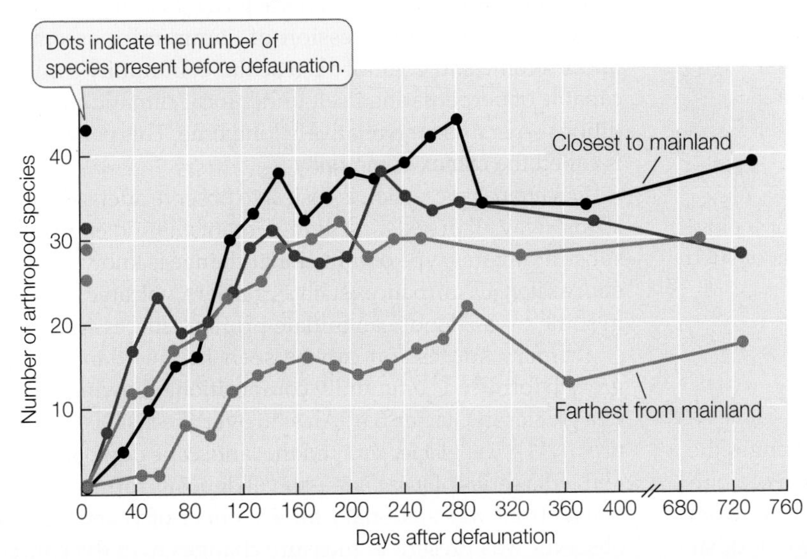


**CONCLUSION** **The data support the premise that an island supports a certain equilibrium number of species.**

**FURTHER INVESTIGATION:** This experiment assessed only arthropod species, and its duration (2 years) was brief in terms of evolutionary time. What experiments might you conduct to assess some of the broader predications of the theory?

Go to **yourBioPortal.com** for original citations, discussions, and relevant links for all INVESTIGATING LIFE figures.

At some point, the number of species arriving from the source and the number of resident species going extinct should balance, and the number should remain stable at this balance point—the *equilibrium number of species* (**Figure 57.9B**).

Wilson and his student Daniel Simberloff devised an ingenious experiment to test the theory of island biogeography. They used four small, isolated clumps of red mangrove (*Rhizophora mangel*), 11–18 m in diameter, in the Florida Keys. These mangrove islands were small enough that Simberloff and Wilson could count the arthropod species on each one, then enclose it in a tent and gas it with methyl bromide to kill all the arthropods. After defaunation, Simberloff and Wilson monitored and tracked recolonization of the islands by arthropods (**Figure 57.10**). Recolonization was generally rapid, but was slowest on the island farthest from the mainland, as predicted by the theory of island biogeography. Furthermore, all of the islands eventually supported roughly the same number of species that they had before defaunation—again, as predicted.

The theory of island biogeography has been widely accepted since Simberloff and Wilson's landmark study. In addition, it has been retrospectively tested by studies of recolonization after major disturbances that serve as "natural experiments." For example, in August 1883, the island of Krakatau was devastated by a series of volcanic eruptions that destroyed all life on its surface. After the lava cooled, plants and animals from Sumatra to the west and Java to the east recolonized Krakatau. By 1933, the island was again covered with a tropical evergreen forest, and 271 species of plants and 27 species of resident land birds were found there (**Table 57.2**). Today the numbers of plant and bird species are not increasing as fast as they did between 1900 and 1930, but colonizations and extinctions continue, as predicted by the theory of island biogeography.

The theory of island biogeography is so robust that it can be applied equally well to *habitat islands*—isolated patches of suitable habitat surrounded by extensive areas of unsuitable habitat. Thus an oasis in the desert or a prairie pothole surrounded by housing subdivisions may acquire an equilibrium number of species in much the same way an oceanic island does.

## TABLE 57.2

**Number of Land Plant and Resident Land Bird Species on Krakatau**

| TIME[a] | NUMBER OF PLANT SPECIES | NUMBER OF BIRD SPECIES |
|---|---|---|
| 1883 ($t$ = 0) | 0 | 0 |
| 1886 ($t$ = 3) | 26 | 0 |
| 1897 ($t$ = 14) | 64 | 8 |
| 1908 ($t$ = 25) | 115 | 13 |
| 1920 ($t$ = 37) | 184 | 27 |
| 1928 ($t$ = 45) | 214 | 27 |
| 1934 ($t$ = 51) | 271 | 29 |

[a] Time is given as the year, followed by the number of years since the volcanic eruption that destroyed all life on the island.

## 57.3 RECAP

**Species diversity encompasses both species richness and species evenness. It is highest in the tropics, decreasing in higher latitudes. Island biogeography theory states that species diversity on an island or other isolated habitat fragment represents a balance between immigration and extinction rates.**

- Explain the concepts of alpha, beta, and gamma diversity. See p. 1210
- What is the difference between species richness and species evenness? See p. 1210 and Figure 57.7
- What factors influence the equilibrium number of species on an island, according to the theory of island biogeography? See p. 1212 and Figure 57.9

Droughts, fires, and volcanic eruptions are examples of ecological disturbances. The next section looks at the effects of such disturbances on ecological communities.

## 57.4 How Do Disturbances Affect Ecological Communities?

An ecological disturbance is any abiotic event that changes the probability of persistence of one or more species in an ecological community. Disturbances may remove some species from a community, but may open up space and resources for other species. The magnitude of the effect of a disturbance varies enormously. Some disturbances are limited to small areas—for example, a log carried by waves may crush algae and animals attached to rocks in an intertidal community. In contrast, hurricanes, forest fires, and volcanic eruptions can affect communities over hundreds or thousands of hectares. Although small-scale disturbances are far more frequent than large ones, a few large events may be responsible for most of the changes in a community. A single hurricane, for example, may fell more trees than years of "normal" storms.

A community's history of disturbance may explain patterns of species diversity that would otherwise be puzzling. The sub-Antarctic Province Islands, located in the South Indian Ocean, provide an example. The climate of these islands is cool, with monthly average temperatures above freezing for only 6 months of the year. Precipitation is high, and gale force winds are not uncommon; the vegetation is primarily tundra. South African entomologist S. L. Chown, an expert on life in the Antarctic, compared the insect faunas on the four largest of these islands and found that the two largest, Marion and Kerguelen, housed fewer arthropod species (16 and 22, respectively) than the significantly smaller Cochons and Possessions (26 and 38).

Why do the species diversity patterns on these islands fail to conform to the theory of island biogeography? One possible explanation is the different geological histories of the islands. The two smaller islands have escaped extensive glaciation, and accordingly have been accumulating species over a longer time than have the larger islands. Thus, the theory of island biogeography notwithstanding, island area may not always be the best predictor of species diversity. Geological history—particularly the history of disturbances—is often a better predictor.

### Succession is the predictable pattern of change in a community after a disturbance

How does a community reassemble itself after a disturbance, particularly one as massive as a glacier? The pattern of change in community composition following a disturbance is known as **succession**. The most common type of succession is *directional* succession, which is characterized by an orderly (or at least predictable) progression of community assemblages. Species come and go until a particular community—one that is capable of perpetuating itself under local climatic and soil conditions—persists for a relatively long time. This persistent stage is called the **climax community**.

Directional succession is easiest to observe after a disturbance strips away all preexisting living organisms and exposes a bare substratum; this type of directional change is known as **primary succession**. Disturbances such as glaciers, volcanic activity, and in some cases, floods can initiate primary succession.

Primary succession can be seen in the change in plant growth form and community composition following the retreat of a glacier in Glacier Bay, Alaska, over the last 200 years (**Figure 57.11**). The glacier scraped the landscape down to bare rock and left a series of *moraines*—gravel deposits formed where the glacial front was stationary for a number of years. No human observer was present to measure changes over the entire 200-year period, but ecologists have inferred the temporal pattern of succession by studying the vegetation on moraines of different ages. The youngest moraines, closest to the current glacial front, are populated with bacteria, fungi, and photosynthetic microorganisms that can support themselves on bare rock. Slightly older moraines farther from the glacial front have lichens, mosses, and a few species of shallow-rooted herbs, such as mountain avens (*Dryas octopetala*). Still farther from the glacial front, successively older moraines have shrubby willows, alders, and spruces.

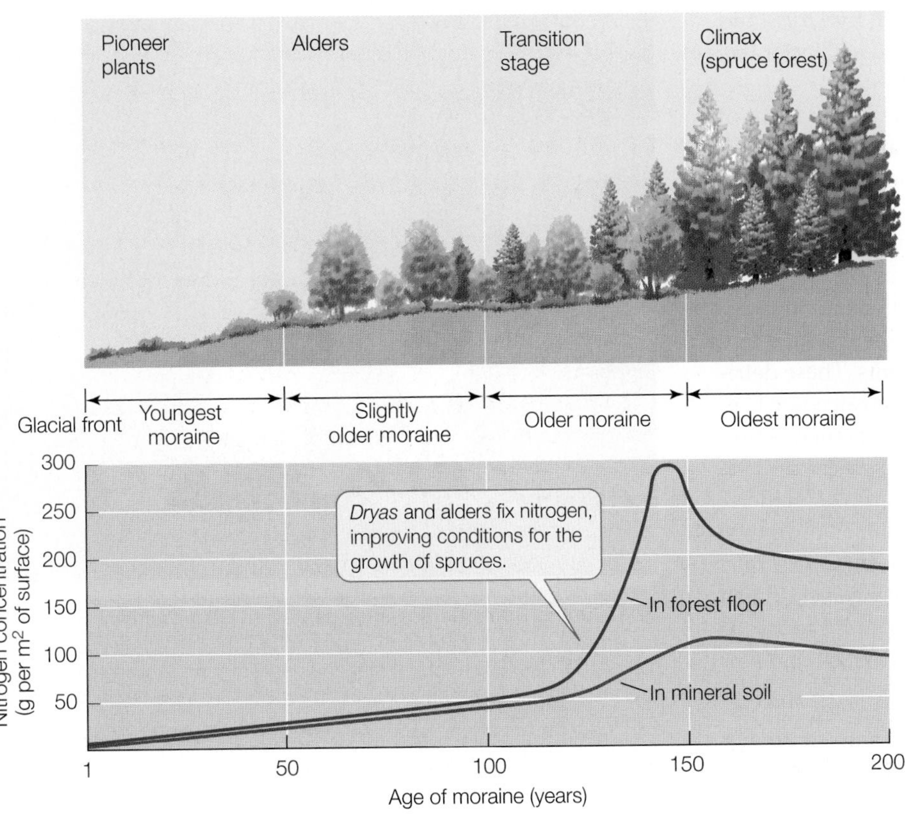


**57.11 Primary Succession** As the community occupying a glacial moraine at Glacier Bay, Alaska, changes from an assemblage of pioneer plants such as *Dryas* to a spruce forest, nitrogen accumulates in the soil.

**yourBioPortal.com**
GO TO **Animated Tutorial 57.2 • Primary Succession on a Glacial Moraine**

Nitrogen is virtually absent from glacial moraines, so the plants that grow best on recently formed moraines at Glacier Bay are *Dryas* and alders (*Alnus*), both of which have nitrogen-fixing bacteria in nodules on their roots (see Figure 36.9). Nitrogen fixation by these plants improves the soil so that spruces can grow. Spruces then outcompete and displace the early colonists. If the local climate does not change dramatically, a climax community dominated by spruce trees is likely to persist for many centuries on old moraines at Glacier Bay.

Directional succession following a disturbance that some organisms, particularly those in the soil, survive is called **secondary succession**. Secondary successions are often initiated by human activities as well as by natural disasters. Generally easier to monitor than primary succession, secondary successions also tend to occur more frequently and progress more rapidly.

A typical sequence of secondary succession in eastern North America begins when forested land that had been cleared for agriculture is abandoned (**Figure 57.12**) The first plants to appear in old agricultural fields are fast-growing annuals such as pigweed, ragweed, and lamb's-quarter. These pioneer species are quickly replaced by more competitive biennials and perennials, such as milkweed, goldenrod, and thistles. Eventually, shrubby plants such as dogwood, eastern red cedar, and sumac become established, followed by tree species such as cottonwood, cherry, and red maple. Ultimately, shade-tolerant trees, including beech and sugar maple, dominate the landscape. The beech–maple forest is the climax community for much of the region.

Directional succession, irrespective of where it takes place, is characterized by certain trends. In general, the colonizing species of early successional stages tend to be good dispersers with high intrinsic rates of reproduction (*r*-strategists). Early stages of succession are characterized by high productivity and simple food webs; most nutrients are present as detritus or in abiotic forms. As succession proceeds, nutrients accumulate in biomass, food webs become more complex, and abiotic sources of nutrients become less important. Species typical of late successional stages tend to be good competitors with relatively low intrinsic rates of reproduction (*K*-strategists).

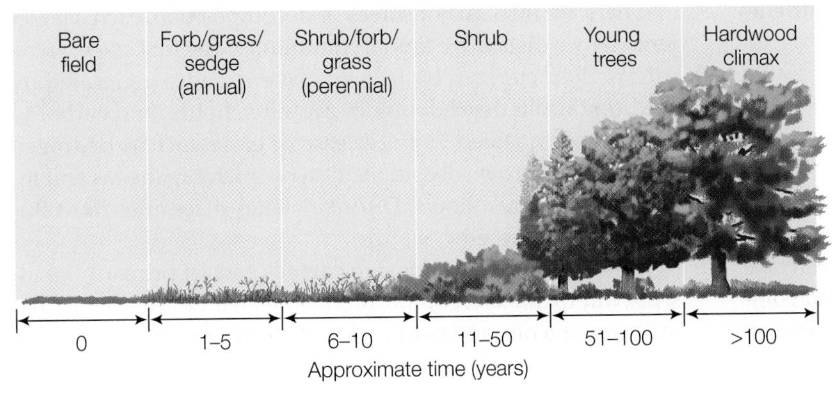


**57.12 Secondary Succession** Secondary succession is the process by which land that once supported an agricultural field can ultimately support a long-lasting climax community characterized by shade-tolerant trees.

### Both facilitation and inhibition influence succession

To some extent, the progress of succession depends on the activity of successive colonists, each of which modifies the environment in such a way as to facilitate colonization by other species. Predators are unlikely to colonize a habitat with no prey species, nor can primary consumers exist before plants are established. The fixation of nitrogen by *Dryas* and alders that allows spruces to become established in Glacier Bay (see Figure 57.11) is an example of such **facilitation**.

Although secondary succession is often described in terms of changes in plant species composition, colonization by plants is actually facilitated by heterotrophs. Often the first organisms to arrive on bare soil after a disturbance are detritivores, which process dead organic matter and release nutrients (especially nitrogen) and thus facilitate the establishment of plants. In a study of intensively burned 20-year-old pine plantations in northern Germany, the first organisms to colonize the burned forests were slime molds, liverworts, mosses, and mushrooms. These detritivores were followed by algae-feeding flies, fungus-feeding beetles, and moss-feeding springtails. Flowering plants such as fireweed moved in, at which point leaf-feeding insects appeared, soon followed by predaceous ground beetles and wolf spiders.

In other cases, the effect of early colonists is **inhibition**, rather than facilitation, of colonization by other species. Old-field species such as goldenrod and thistle produce root exudates that inhibit the germination and growth of potential competitors. Eventually, when these plants grow old and die, other plant species can become established.

**57.13 Some Communities are Adapted to Disturbance** Forest communities dominated by lodgepole pine (*Pinus contorta*) are adapted to periodic fires. Fire removes mature trees weakened by pest infestation, revitalizes the soil, and provides essential conditions for seed germination and new growth.

### Cyclical succession requires adaptation to periodic disturbances

Some patterns of succession are *cyclical*, rather than directional. In such cases, the climax community depends on periodic disturbance in order to persist. The lodgepole pine (*Pinus contorta*) forests that grow in the pumice plateau area of southern Oregon are maintained by periodic forest fires (**Figure 57.13**) In this fire-adapted community, fires return nutrients to the soil in the form of burned organic matter and provide proper conditions for seed germination. The cones of lodgepole pines are sealed shut by resins; only when they are subjected to high temperatures that melt the resins do they open and release their seeds. The lodgepole pine is attacked by *Dendroctonus ponderosae*, the mountain pine beetle, and is also prone to infection by a fungus, *Phaeolus schweinitzii*, which causes the roots and heartwood of the pines to rot. Trees that have lived long enough to experience and be scarred by a fire are much more likely to become infected by the fungus than are trees that have not been scarred. Fungus-infected, weakened trees are preferentially attacked by beetles. After a beetle outbreak, in which many fire-scarred mature trees are killed, the dead trees serve as potential fuel for a fire. The next fire frees up their nutrients for use by the remaining trees as well as new seedlings.

### Heterotrophic succession generates distinctive communities

Plants play a vital role in most patterns of succession because, as autotrophs, they are the source of energy for the other organisms in the community. Successional changes, however, can take place without the participation of plants. Detritus-based communities—found in dung, carrion, or dead plants—undergo a series of changes known as **heterotrophic succession**. These changes differ from those in other types of directional succession. First, succession begins heterotrophically: energy resources are greatest when the habitat first becomes available to colonists and are depleted as succession takes place. There is no mechanism, such as photosynthesis, for generating more energy. Second, total biomass and species diversity decrease over time as the resource base declines. Third, these temporary habitats are not really self-contained, so predators can outnumber primary consumers (detritivores, in these communities) in apparent violation of the laws of thermodynamics.

Because the habitats in which heterotrophic succession occurs are so conveniently small, many patterns of heterotrophic succession have been thoroughly investigated. One such pattern is faunal succession in human corpses, as we saw at the opening of this chapter. M. G. Motter's 1898 study of 150 disinterments, followed by more recent studies on pig decomposition (after disinterments of human bodies for scientific purposes were discouraged), allowed for precise reconstruction of the temporal sequence of species in these communities.

There are three major stages of decomposition, each characterized by a distinctive faunal community. The first stage, *autolysis*, is characterized by fermentative changes generated by the body itself: degradation of proteins, lipids, and carbohydrates, accompanied by the release of gases such as hydrogen sulfide. *Putrefaction*, decomposition by microorganisms and invasive saprobes, follows. *Dry decay* takes place after most fluids have evaporated.

The faunas of decomposing corpses consist primarily of insects, but their exact composition varies with those factors that influence the rate and nature of decomposition: climate, season, and the condition of the body, including whether it is immersed or buried, wrapped or exposed. Typically, immediately after death, as autolysis begins, blow flies, bluebottles, and house flies arrive to lay eggs. As a detectable odor develops, other flies, including greenbottles and flesh flies, arrive. Fat breakdown, with its accompanying release of volatile fatty acids, attracts a range of carrion-feeding beetles. As proteins decompose, cheese skippers move in. Other species have less interest in the corpse than in the corpse-eaters: rove beetles prey on the maggots that develop from the flies' eggs. During dry decay, skin beetles, hide beetles, and clothes moths (which can feed on the keratin in mammalian hair) dominate. In the final stages of decomposition, spider beetles and other scavengers arrive to feed on the excrement and shed exoskeletons of the insects consuming the corpse. This succession varies tremendously with climate and geography, but within any particular region it is sufficiently predictable that it is admissible in court as evidence of the time since death.

## 57.4 RECAP

**Disturbances are abiotic events that change the probability of persistence of one or more species in an ecological community. Ecological succession—a predictable pattern of change in community composition—typically follows a disturbance.**

- How do primary succession and secondary succession differ? See pp. 1214–1215 and Figures 57.11 and 57.12
- Describe some ways in which early colonists facilitate or inhibit colonization by the species that follow them in a pattern of succession. See pp. 1215–1216

Now that we have seen how communities change in composition over time we will look at how certain communities (such as climax communities) can persist over time and withstand disturbance with little change. In this context, we return to productivity as an important arbiter of community stability.

# 57.5 How Does Species Richness Influence Community Stability?

Up to a point, higher productivity favors higher species richness, as we saw in Section 57.1. Does species richness in turn influence productivity? And do both of those properties influence community stability? Species richness might enhance productivity because no two species in a community have the same relationship with the environment, so a mixture of more species might result in a more complete use of the available resources. Moreover, if environmental conditions should change, a species-rich community is more likely to contain some species that can persist under the new conditions. Thus a species-rich community should be more *stable*—that is, it should change less over time in either productivity or species composition—than a species-poor community.

### Species-rich communities use resources more efficiently

To test the hypothesis that species-rich communities are more stable than species-poor communities, David Tilman and his colleagues at the University of Minnesota cleared several outdoor plots, in which they planted grasses in mixtures ranging from a few to 25 species. At the end of each growing season, they measured total plant cover (a measure of grass biomass, and thus of net primary production) and the population densities of all the grasses in each plot. Over a period of 11 years, which included a serious drought, the plots with more species were more productive, and their productivity was less variable from year to year. These findings were consistent with the hypothesis that species richness promotes productivity and stability (**Figure 57.14A**). Moreover, in the plots with greater species richness, soil nitrogen was used more efficiently (**Figure 57.14B**). However, the population densities of *individual* species in the plots were not stable over the years (regardless of a plot's species richness) because different species performed better during drought years and wet years.

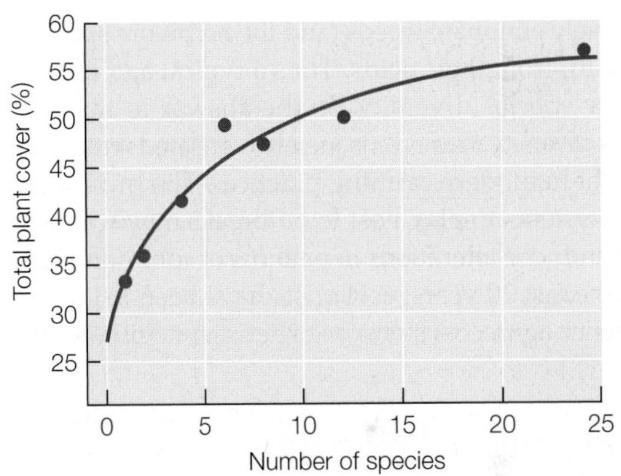


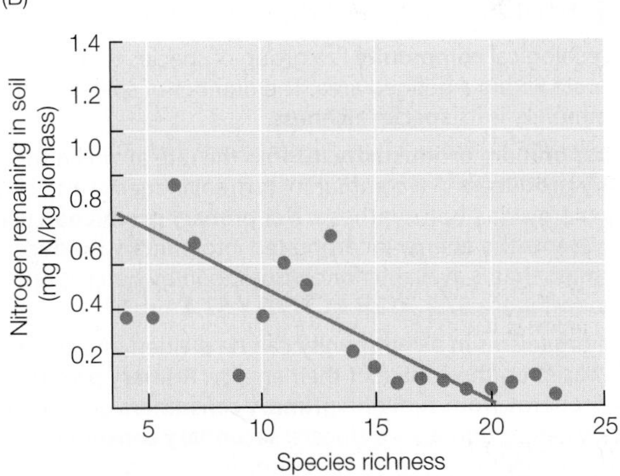


**57.14 Species Richness Enhances Community Productivity** A total of 120 grassland plots, containing from 2 to 22 grass species, were cultivated for 11 years. (A) The plots with greater species richness showed greater stability (with total plant cover being the measure of stability). (B) The plots with greater species richness utilized soil nitrogen more efficiently, as shown by the smaller amount of nitrogen remaining in the soil.

Researchers continue to debate whether species diversity is responsible for maintaining stability or is simply correlated with stability. This question is important because many of the alterations humans have made in the structure of natural communities have reduced their species richness, and many of these human-altered communities—notably, agricultural communities—are notoriously unstable.

### Diversity, productivity, and stability differ between natural and managed communities

Although ecologists have been debating the relationships between community diversity, productivity, and stability for only a few decades, humans have been experimenting with those relationships, albeit inadvertently, for millennia—since plants were domesticated and agriculture was invented. Since the dawn of agriculture, crops have been susceptible to insect **outbreaks**: massive (often sudden) increases in the population sizes of species that destroy or damage crops.

The practice of growing crops as **monocultures**—plantings of only a single species—is one reason why modern agricultural ecosystems are particularly unstable. Most farmers have little tolerance for the presence of any potential competitors for their crops and actively eliminate weeds (and the herbivore species that live with them) from their fields. Thus a typical agroecosystem has very low species diversity. So the answer to the question of whether diversity causes or is merely correlated with stability may be sought in modern farming practices. The predisposition for agroecosystems to play host to outbreaks may well result from human-induced alterations in patterns of community structure.

For the last 20 years, ecologists have been using traditional subsistence agroecosystems as experimental models for testing the relationships between diversity and stability. Throughout the world, many farmers with small land holdings grow multiple crops on the same plot. In Costa Rica, for example, corn is often grown together with sweet potato. Corn–sweet potato dicultures contain fewer sweet potato pests and many more parasitoid wasps than corn monocultures do. Corn pollen provides nutrition for the wasps, and the tall corn plants act as a structural barrier, shade plant, and source of disruptive chemical signals that interfere with the ability of the sweet potato pests to find their host plants.

In recent years, such applications of community ecology have been paying dividends. Although monoculture is overwhelmingly the dominant agricultural practice, polycultures are under development for agricultural production systems as varied as carp and shrimp farming, vermicomposting (raising worms for compost), and biofuel feedstock production.

## 57.5 RECAP

**Communities with higher species diversity tend to be more productive and more stable than less diverse communities because they use resources more efficiently. Modern agricultural monocultures are notoriously unstable.**

- What relationships have ecologists observed between species diversity, community productivity, and community stability? See p. 1217 and Figure 57.14
- Describe some agricultural practices that might result in more stable ecological communities. See p. 1218

## CHAPTER SUMMARY

### 57.1 What Are Ecological Communities?

- An ecological **community** is a group of species that coexist and interact within a defined area. The number of species living in a community is its **species richness**.
- **Gross primary productivity** (**GPP**) is the rate at which the **primary producers** in a community turn solar energy into chemical energy via photosynthesis. **Net primary production** (**NPP**) represents the energy incorporated into primary producer **biomass** that is available for consumption by heterotrophs. Review Figure 57.1, WEB ACTIVITY 57.1
- The organisms in a community can be divided into **trophic levels** based on the source of their energy. **Primary producers** get their energy from sunlight; **primary consumers** get their energy by eating primary producers; **secondary consumers** get their energy by eating primary consumers; and so on. Review Table 57.1, WEB ACTIVITY 57.2
- Organisms that consume the dead bodies of other organisms or their waste products are called detritivores or decomposers. Omnivores are organisms that feed on multiple trophic levels.
- A **food chain** diagrams who eats whom. A **food web** shows how the food chains in a community are interconnected. Review Figure 57.2
- **Ecological efficiency** is the overall transfer of energy from one trophic level to the next. Pyramid diagrams illustrate the proportions of energy or biomass that flow between trophic levels. Review Figure 57.3
- Species richness tends to increase with productivity up to a point; however; if productivity increases beyond that point, species richness may decline. Review Figure 57.4

### 57.2 How Do Interactions among Species Influence Community Structure?

- The interactions of a consumer with other species can result in a a **trophic cascade**: a series of indirect effects across successive trophic levels. Review Figure 57.5
- Organisms that build structures that create habitat for other species are known as **ecosystem engineers**.
- **Keystone species** have an influence on both the species richness and the number of trophic levels in communities out of proportion to their abundance.

## 57.3 What Patterns of Species Diversity Have Ecologists Observed?

- **Alpha diversity** is a measure of diversity within a single community or habitat, whereas **beta diversity** is a measure of the change in species composition from one community or habitat to another. **Gamma diversity** is the diversity found over a range of communities in a geographic region.
- Species diversity encompasses species evenness as well as species richness. Review Figure 57.7
- Diversity indexes are mathematical formulas used to quantify diversity. The **Shannon diversity index** is a measure of alpha diversity that takes species evenness into account. **Sorensen's index** is a measure of beta diversity.
- Latitudinal gradients in diversity, with the greatest diversity at low latitudes, have been observed in many taxa. Review Figure 57.8
- According to the theory of **island biogeography**, the equilibrium number of species on an island represents a balance between the rate at which species immigrate to the island from the mainland **species pool** and the rate at which resident species go extinct. Review Figures 57.9 and 57.10, **ANIMATED TUTORIAL 57.1**

## 57.4 How Do Disturbances Affect Ecological Communities?

- A **disturbance** is an abiotic event that changes the probability of persistence of one or more species in a community.
- **Succession** is a predictable pattern of change in community composition following a disturbance. In directional succession, species come and go in a predictable sequence until a **climax community** persists for an extended time.
- **Primary succession** begins on sites that lack living organisms. **Secondary succession** begins on sites where some organisms have survived a disturbance. Review Figures 57.11 and 57.12, **ANIMATED TUTORIAL 57.2**
- In any pattern of succession, species that have already become established may **facilitate** or **inhibit** colonization by other species.
- In **cyclical succession**, the climax community is maintained by periodic disturbances.
- **Heterotrophic succession** in detritus-based communities does not rely on photosynthesis and therefore differs in a number of ways from other types of succession.

## 57.5 How Does Species Richness Influence Community Stability?

- Species-rich communities use resources more efficiently, and thus tend to vary less in productivity, than do less diverse communities. Review Figure 57.14
- **Monocultures** are subject to pest outbreaks, whereas agroecosystems containing greater species diversity tend to be more stable.

# SELF-QUIZ

1. An ecological community is
   *a.* a group of species that coexist and interact within a defined area.
   *b.* a group of species that coexist and interact in an area together with the abiotic environment.
   *c.* all the species in an area that belong to a particular trophic level.
   *d.* all the species that are members of a local food web.
   *e.* all of the above
2. A trophic level consists of the organisms
   *a.* whose energy has passed through the same number of steps to reach them.
   *b.* that use similar foraging methods to obtain food.
   *c.* that are eaten by a similar set of predators.
   *d.* that eat both plants and other animals.
   *e.* that compete with one another for food.
3. Net primary production is
   *a.* the total amount of photosynthesis in a community.
   *b.* the total amount of primary producer biomass available for consumption by heterotrophs.
   *c.* the total amount of biomass produced by all autotrophs and heterotrophs in a community.
   *d.* the total amount of biomass consumed by heterotrophs.
   *e.* gross primary productivity minus secondary productivity.
4. Pyramid diagrams of energy and biomass distribution for forests and grasslands differ because
   *a.* forests are more productive than grasslands.
   *b.* forests are less productive than grasslands.
   *c.* large mammals avoid living in forests.
   *d.* trees store much of their energy in difficult-to-digest wood, whereas grassland plants produce few difficult-to-digest tissues.
   *e.* grasses grow faster than trees.
5. Keystone species
   *a.* influence the communities in which they live more than expected on the basis of their abundance.
   *b.* may influence the species richness of communities.
   *c.* may influence the number of trophic levels in communities.
   *d.* are not necessarily predators.
   *e.* all of the above
6. Species diversity
   *a.* increases with latitude from the equator to the temperate zone.
   *b.* decreases with latitude from the equator to the temperate zone.
   *c.* shows no consistent pattern from the equator to the temperate zone.
   *d.* increases with latitude from the equator to the temperate zone for terrestrial organisms and decreases with latitude from the equator to the temperate zone for aquatic organisms.
   *e.* increases with latitude from the equator to the temperate zone for animals and decreases with latitude from the equator to the temperate zone for plants.

7. The theory of island biogeography
   *a.* predicts that the equilibrium number of species on an island is a balance between the rate of immigration of new species and the rate of extinction of resident species.
   *b.* predicts that the rate of immigration of new species will decline with island distance from the mainland species pool.
   *c.* predicts that the rate of extinction of resident species will decrease as island size increases.
   *d.* applies to isolated habitat patches as well as to oceanic islands
   *e.* all of the above
8. Ecological succession is
   *a.* the change in species over time.
   *b.* the change in community composition after a disturbance.
   *c.* the change in a forest as the trees grow larger.
   *d.* the process by which a species becomes abundant.
   *e.* the buildup of soil nutrients.
9. Primary succession begins
   *a.* soon after a disturbance ends.
   *b.* at varying times after a disturbance ends.
   *c.* at sites where some organisms have survived a disturbance.
   *d.* at sites where no organisms have survived a disturbance.
   *e.* at sites where only primary producers have survived a disturbance.
10. Early stages of succession are characterized by
   *a.* species that are good dispersers.
   *b.* species with high intrinsic rates of reproduction.
   *c.* simple food webs.
   *d.* nutrients that are available primarily in abiotic sources.
   *e.* all of the above

## FOR DISCUSSION

1. Marek Sammul, Lauri Oksanen, and M. Magi investigated the effect of productivity on species richness by conducting an experiment in 16 different plant tcommunities in western Estonia and northern Norway. They removed one perennial species, the goldenrod *Solidago virgaurea*, from these communities and found that its competitors, particularly the grass species *Anthoxanthum odoratum*, increased in biomass, most noticeably in communities with high productivity (where living plant biomass was greater than 200 $g/m^2$). In less productive communities, such increases could not be detected. How might interspecific competition lead to a decrease in species richness at high levels of productivity? What other hypotheses might explain this puzzling relationship? How would you test them?
2. The increased productivity and stability of species-rich communities could be explained by ecological differences among the species that allow the community as a whole to use resources more efficiently, or by the fact that the more species in a community, the greater the chance that it will contain an unusually productive species. How could you distinguish between these competing hypotheses?
3. If species-rich communities are more productive than species-poor communities, how can modern agriculture, which is based almost entirely on monocultures, be so productive?
4. Figures 57.11 and 57.12 illustrated succession with two examples from forests. How might ecological succession proceed in grasslands? In deserts? In the rocky intertidal zone?

## ADDITIONAL INVESTIGATION

The species diversity of aphids (small sucking insects), parasitoid wasps, and kelps decreases, rather than increases, from the temperate zone to the equator. Can you propose any explanations to account for these unusual patterns?

## WORKING WITH DATA (GO TO yourBioPortal.com)

**Experimental Island Defaunation** This exercise presents the data from Simberloff and Wilson's experiments in the Florida Keys (see Figure 57.10), which were performed in the late 1960s and presented in support of island biogeography theory. You will be asked to analyze different aspects of the experiment and to draw conclusions as to how (and perhaps whether) the data offer significant support for this theory.

# 58 Ecosystems and Global Ecology

## Ecologists swing in the canopy to measure carbon's fate

Photosynthesis by terrestrial plants transforms (*fixes*) the carbon atoms in atmospheric carbon dioxide into the carbohydrate sugars that fuel the fungal and animal worlds. Concentrations of $CO_2$ in the atmosphere have been rising steadily over the last century, largely as a result of human activities, and some scientists have hypothesized that we can expect rates of photosynthesis—and thus $CO_2$ fixation and carbon storage in ecosystems—to increase. But ecologists are discovering that this "silver lining" scenario is probably not going to be the case.

Forest studies are important for determining the responses of vegetation to atmospheric carbon enrichment. Most photosynthesis in any forest takes place high up in tree canopies where sunlight is most intense. Until recently, collecting data from forest canopies required climbing or using balloons. Then, in 1992, Alan Smith at the Smithsonian Tropical Research Institute in Panama realized that he could use a construction crane to gain access to the canopy. He recognized that by swinging the crane's boom in a circle, shuttling the gondola along its length, and lowering the cage to different heights, researchers could reach and measure biological processes over a large volume of the forest canopy. Smith's idea caught on quickly, and today ecologists on four continents use cranes to study forest canopies.

When they monitored growth, reproduction, and carbon storage in large trees, Swiss ecologists found that, as expected, canopy trees increased their growth rates in a carbon-enriched atmosphere; but unlike young, fast-growing saplings, they did not store large amounts of carbon in their trunks and branches. Instead, they transported more carbon first to the soil, and then to the atmosphere (via their roots). The leaf litter they produced was richer in starch and sugar, but poorer in lignin; the litter decomposed faster as soil microbes rapidly respired the carbon back to the atmosphere. These and other experiments suggest that young forests are likely to respond to greater atmospheric $CO_2$ concentrations by growing faster. However, mature forests, along with other types of vegetation such as grasslands, shrublands, and tundra, are not likely to store significantly more carbon in response to $CO_2$ enrichment.

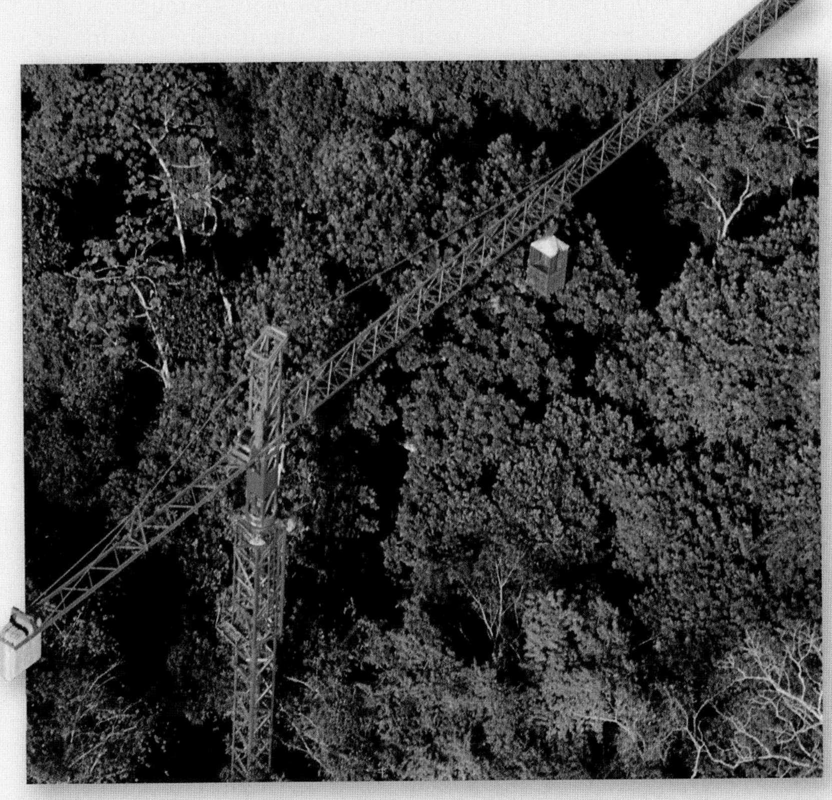

**Studying the Canopy** A crane 42 meters high allows scientists to study the forest canopy trees of Panama's Smithsonian Tropical Research Institute. One aspect of these studies involves taking measurements that reveal the amount of carbon stored in tropical forest trees.

**Enriching the Canopy** Researchers from a team of Swiss ecologists install a mechanism of fine tubes that administer controlled amounts of $CO_2$ to enrich the trees' carbon supply. These experiments simulate what will happen as atmospheric levels of $CO_2$ increase.

As Chapter 54 described, the global ecosystem is composed of many different types of vegetation (biomes), many of which have few woody plants. To understand how ecosystems function at large scales, scientists must make observations and conduct experiments on many vegetation types. Moreover, because the responses of natural ecosystems to carbon enrichment will depend on many other factors—such as temperature, water, wind, and soil fertility—ecologists must study the interactions among these factors to understand how they influence ecosystem processes.

**IN THIS CHAPTER** we introduce the compartments of the physical environment through which energy flows and materials cycle. We will trace these events by following the movements of several chemical elements that are crucial to life. In so doing, we highlight many ways in which human activities have altered these cycles. Finally we will consider the many goods and services that ecosystems provide to human society.

CHAPTER OUTLINE



## 58.1 What Are the Compartments of the Global Ecosystem?

An **ecosystem** is generally defined as any ecological system with defined boundaries. It includes all of the organisms within those boundaries as well as the physical and chemical factors that influence those organisms. In other words, all ecosystems have both biotic and abiotic components. Ecosystems can occupy a wide range of spatial scales, from the global ecosystem to a watershed, a specific forest, a lake or pond, or even a patch of lichen on a rock. In this chapter we focus on the global ecosystem.

### Energy flows and materials cycle through ecosystems

Earth is essentially a closed system with respect to atomic matter, but it is an open system with respect to energy. The sun delivers an essentially constant amount of energy to Earth every day and has done so for billions of years. Energy from the sun, combined with energy from the radioactive decay in Earth's interior, drives the processes that move chemical elements around the planet. The rate at which energy moves through a system is called its *flux*. Elements may accumulate, or *pool*, in some component of an ecosystem; the term *flux* also applies to the movement of elements between pools. Some pools are *sinks*, in which an element is taken out of circulation and locked up for long periods.

Many elements move around the planet in a cyclic fashion (**Figure 58.1**). Nearly all the rocks that make up the continents, for example, have been transformed at least once through a complex cycle of plate tectonic processes (see Figure 25.2). Deep within the Earth, heat, pressure, and melting change the mineral composition of rocks. As the movements of the lithospheric plates push up mountains, the processes of weathering are already wearing them down by breaking rocks into soil and sediment particles (see Section 36.3). Streams carry sediments to the oceans, where they form sedimentary rocks on the seafloor. Eventually, those rocks may be returned to land in another episode of tectonic uplift.

Water is another material that moves cyclically. The water in the oceans today has evaporated, condensed in the atmosphere, precipitated as rain, and returned to the oceans via stream and groundwater flow millions of times. The carbon, nitrogen, oxygen, and sulfur in atmospheric gases have cycled repeatedly through living organisms, which use them to build the proteins that make up their bodies.

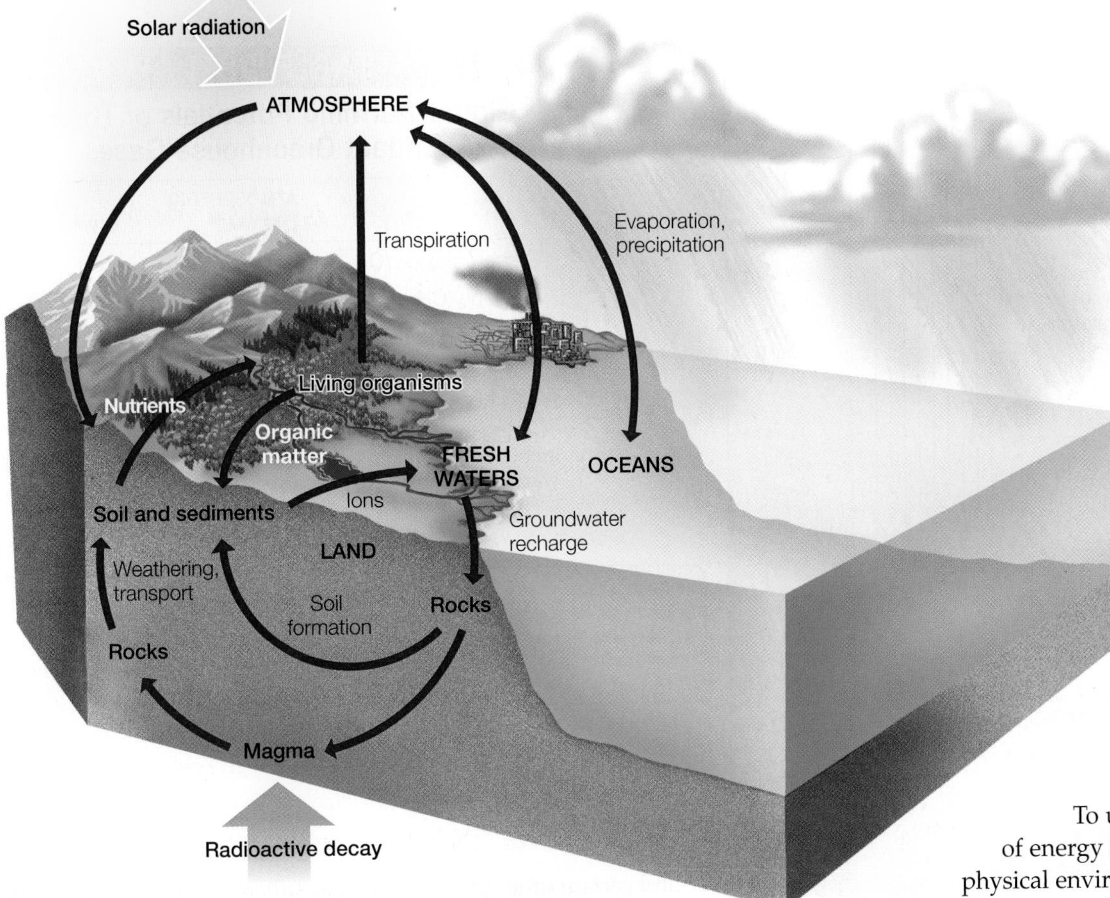


**58.1 Compartments of the Global Ecosystem** The four compartments—atmosphere, oceans, land, and fresh waters—are connected by the flux of materials (arrows). The atmosphere exchanges some components rapidly; exchange of materials within the other compartments (between rocks and soils, for example) is generally much slower.

Earth is a planet with many unusual features: lithospheric plates that move continuously, a moderate surface temperature, a large quantity of liquid water on its surface, and a tremendous diversity of living organisms, to name just a few. The plate tectonic processes that shape Earth's surface also release gases (including carbon dioxide) into the atmosphere through volcanic eruptions. Those atmospheric gases help to moderate planetary temperatures, and those moderate temperatures, in turn, keep water in a liquid state. And living organisms, by virtue of their activities, have an enormous influence on the chemical composition of the atmosphere, the water, and the land.

All organisms depend on inputs of energy (in the form of sunlight or high-energy molecules), liquid water, and nutrients for their metabolism and growth. As described in Section 57.1, energy flows through ecological communities from producers to consumers. At each trophic level, much of this energy is used to power metabolism; that energy is dissipated as heat and lost from the ecosystem (see Figure 57.1). Chemical elements, on the other hand, are not altered when they are transferred among organisms. The availability of the chemical elements of which living organisms are composed is strongly influenced by how organisms get them, how long they retain them, and what they do with them while they have them.

To understand the movements of energy and materials on Earth, the physical environment can be divided into four interacting compartments: the atmosphere, the oceans, fresh waters, and land. These four compartments, and the types of organisms living in them, are very different. Thus the quantities of different elements in each compartment, how long they remain in each compartment (i.e. their *residence time*), and the rates at which they enter and leave the compartments differ strikingly. After describing the four compartments, we will discuss how energy flows through them and elements cycle among them.

## The atmosphere regulates temperatures close to Earth's surface

The outermost compartment of the global ecosystem, the *atmosphere*, is a thin layer of gases surrounding Earth. The atmosphere is 78.08% nitrogen gas ($N_2$), 20.95% oxygen gas ($O_2$), 1% water vapor, 0.93% argon, and 0.03% carbon dioxide ($CO_2$). It also contains traces of hydrogen gas, neon, helium, krypton, xenon, ozone, and methane. The atmosphere contains Earth's biggest pool of $N_2$ as well as a large proportion of its $O_2$. Although $CO_2$ constitutes a very small fraction of the atmosphere, atmospheric $CO_2$ is the source of the carbon used by terrestrial photosynthetic organisms, and atmospheric $CO_2$ that dissolves in ocean water is the source of the carbon used by marine primary producers.

About 80% of the mass of the atmosphere lies in its lowest layer, the **troposphere**. This layer extends upward from Earth's surface about 17 km in the tropics and subtropics, but only about 10 km at high latitudes. Most global air circulation takes place within the troposphere, and virtually all of the atmospheric water vapor is found there (**Figure 58.2**).

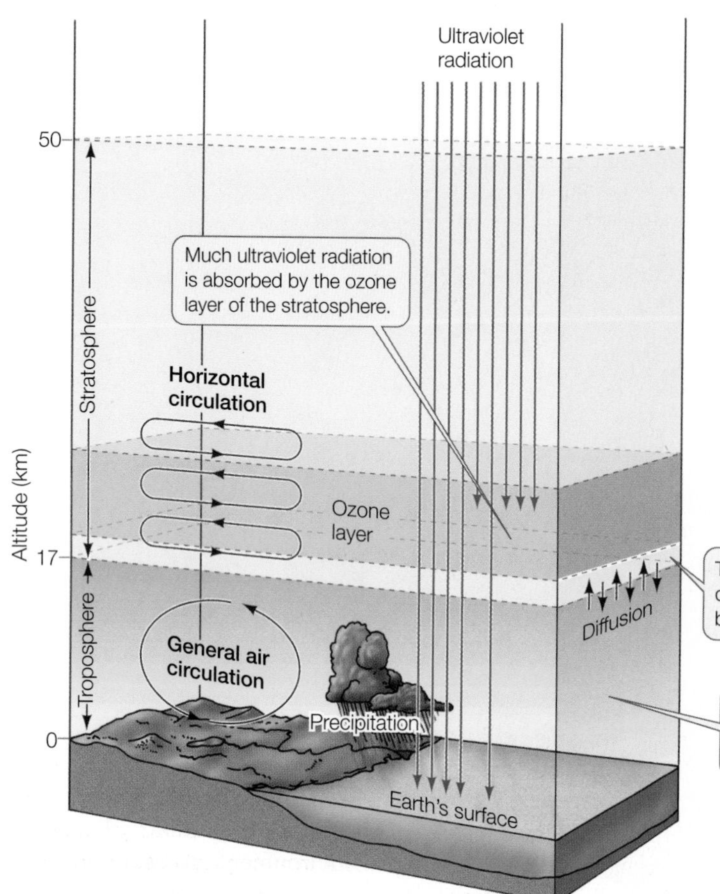


**58.2 The Two Layers of Earth's Atmosphere** The troposphere and the stratosphere differ in their circulation patterns, the amount of water vapor they contain, and the amount of ultraviolet radiation they receive.

The **stratosphere**, which extends from the top of the troposphere up to about 50 kilometers above Earth's surface, contains very little water vapor. Most materials enter the stratosphere from the region of the troposphere that encircles the equator, where air heated by the sun rises to high altitudes (see Section 54.2).

The stratosphere contains a layer of ozone ($O_3$) that absorbs most of the biologically damaging ultraviolet radiation that enters the atmosphere. Over the last several decades, this **ozone layer** has been seriously damaged by human activities. The widespread release of chlorinated fluorocarbons (CFCs), such as the refrigerant freon, caused chemical reactions in the stratosphere that destroyed ozone molecules. This depletion of stratospheric ozone is a source of great concern, as increases in ultraviolet radiation at Earth's surface are associated with increased rates of skin cancer, cataract formation, and crop damage. In 1989, a group of nations signed a global treaty in which they agreed to phase out production of CFCs, so there is hope that the stratospheric ozone layer will one day be restored.

The atmosphere regulates temperatures at and close to Earth's surface by trapping heat. If Earth had no atmosphere, its average surface temperature would be about –18°C, rather than its actual +17°C. Carbon dioxide, methane, water vapor,

**TABLE 58.1**

**Global Warming Potentials of Three Abundant Greenhouse Gases**

| | ATMOSPHERIC LIFETIME[a] | *GWP*[b] |
|---|---|---|
| Carbon dioxide ($CO_2$) | 50–200 | 1 |
| Methane ($CH_4$) | 12 | 21 |
| Nitrous oxide ($N_2O$) | 114 | 289 |

[a] *Atmospheric lifetime* is the number of years these molecules typically remain in the atmosphere before breaking up.
[b] *GWP* (global warming potential) is the ability of a given amount of gas relative to $CO_2$ (GWP = 1) to trap heat and warm the atmosphere over a specific time period (in this case, 100 years).

and certain other gases in the atmosphere are known as **greenhouse gases** because they are transparent to sunlight, but trap heat radiating from Earth's surface back toward space (**Figure 58.3**) The three major greenhouse gases and their potential effects on atmospheric warming are given in **Table 58.1**. Ozone, when it is present in the troposphere rather than the stratosphere, also acts as a greenhouse gas. As we'll see later in this chapter, human activities are increasing the concentrations of greenhouse gases in the atmosphere, and those increases are making the planet warmer.

### The oceans receive materials from the other compartments

Over time scales of hundreds to thousands of years, most materials that cycle through the four compartments of the global ecosystem end up in the oceans. The oceans are enormous, but they exchange materials with the atmosphere only at their surface, so they respond very slowly to inputs from that compartment. They receive materials from land primarily in runoff from rivers.

Except on continental shelves—the shallow ocean waters surrounding large land masses—ocean waters mix slowly. Most materials that enter the oceans from other compartments gradually sink to the seafloor, where they may remain for millions of years until the seafloor sediments are uplifted above sea level by plate tectonic processes. Therefore, concentrations of mineral nutrients are very low in most ocean waters. Some nutrients are brought back to the surface near the coasts of continents, where offshore winds pushing surface waters away from shore cause cold bottom water to rise to the surface. The water in these

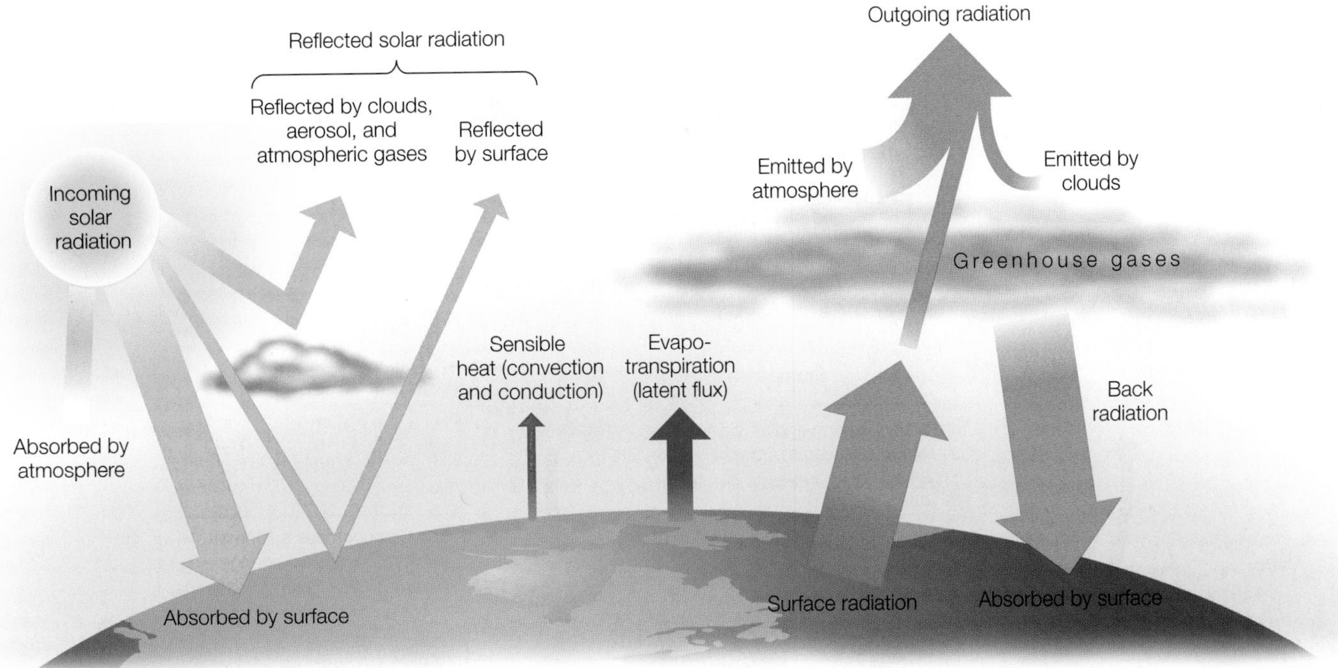


**58.3 Radiant Energy Warms the Planet** Solar energy (yellow arrows) drives the global climate system described in Chapter 54. Much of this energy (in the form of heat) along with the energy generated by Earth's biogeochemical processes (orange arrows) is prevented from radiating back into space by the atmospheric gases. The width of the arrows is roughly proportional to the size of energy flux.

**upwelling zones** is nutrient-rich because when organisms die their bodies sink to the seafloor and return the materials they contain to the bottom sediments. Upwelling zones support high rates of production by phytoplankton, which in turn support dense consumer populations (**Figure 58.4**). Most of the world's great fisheries are concentrated in upwelling zones. The upwelling zone off the coast of Peru is, for example, the source of much of the world's supply of anchovies. In turn, this rich fish population supports vast seabird communities, which in turn produce enormous quantities of guano, or droppings, that has provided Peru with an important raw material to support a major fertilizer industry.

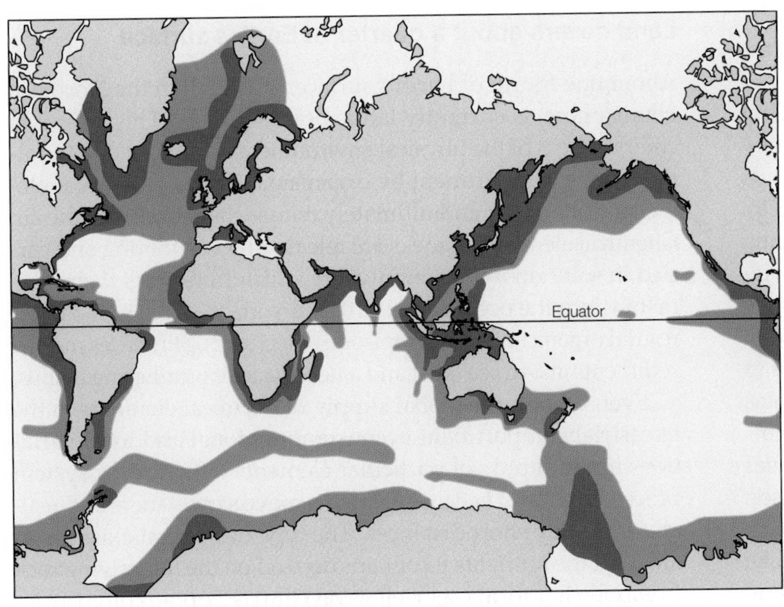


**58.4 Upwelling Zones Are Highly Productive** Biological productivity is greatest near continents where surface waters, driven by prevailing winds, move offshore and are replaced by cool, nutrient-rich water upwelling from below.

### Water moves rapidly through lakes and streams

The freshwater compartment of the global ecosystem consists of streams, lakes, and **groundwater** (water occupying pore spaces in rock, sand, and soil). Only a small fraction of Earth's water resides in lakes and streams at any given time, but water moves rapidly through this compartment, so most of Earth's water spends some time there. Some mineral nutrients enter fresh waters in rainfall, but most are released by the weathering of rocks. They are carried into streams by surface runoff or by movement of groundwater.

After entering streams, mineral nutrients are usually carried rapidly to lakes or to the oceans. In lakes, they are taken up by organisms and incorporated into their bodies. Those organisms eventually die and sink to the lake bottom, taking their nutrients with them. Microbial decomposition of their tissues consumes most or all of the oxygen in the bottom water. The surface waters of lakes thus quickly become depleted of nutrients, while deeper waters become depleted of oxygen. This process is countered, however, by verti-

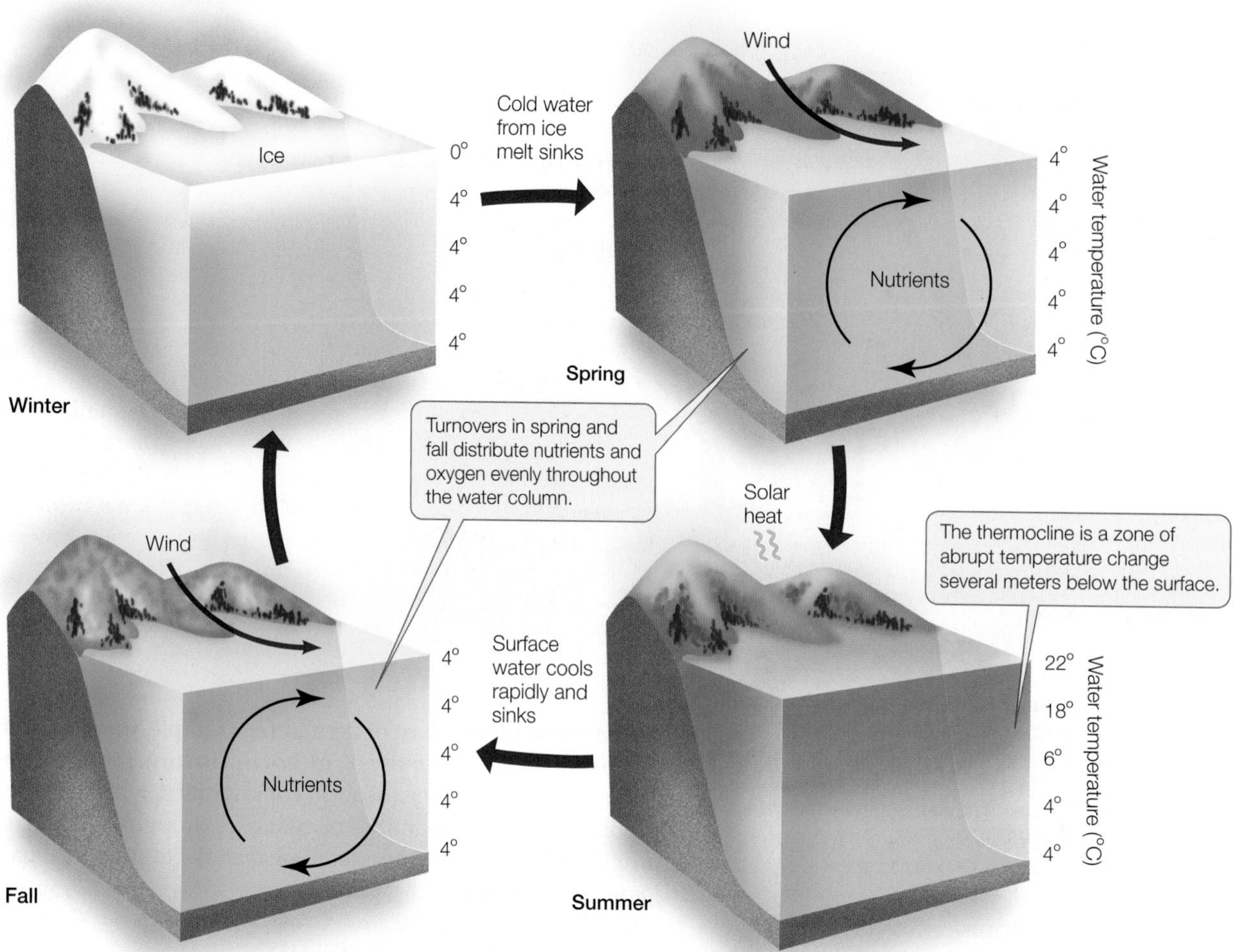


**58.5 The Turnover Cycle in a Temperate Lake** Regular turnovers in spring and fall allow nutrients and oxygen to become evenly distributed in the water column of a lake. The vertical temperature profiles shown here are typical of temperate-zone lakes that freeze in winter.

cal movements of water called **turnover**, which bring nutrients to the surface and oxygen to deeper water.

Wind is an important agent of turnover in shallow lakes, but in deeper lakes it mixes only surface waters. Deep lakes in temperate climates have an annual turnover cycle that depends on certain unique physical properties of water (**Figure 58.5**). Water is densest at 4°C, a few degrees above its freezing point; above and below that temperature, it expands. Thus, in winter, the coldest liquid water in a lake floats at its surface, often covered by a layer of ice. The waters below remain at 4°C. In spring, the sun melts the ice and warms the surface water to 4°C. At this point, water density is uniform throughout the lake, and even modest winds readily mix the entire water column.

As spring and summer progress, the surface water becomes warmer, and the depth of the warm layer gradually increases. However, there is still a well-defined region, called the *thermocline*, where the temperature changes abruptly. Only if the lake is shallow enough to warm right to the bottom does the temperature of the deepest water rise above 4°C. Another turnover occurs in fall as the process reverses itself. The surface of the lake cools until it is denser than the warmer water below it, at which point it sinks and is replaced by warmer water from below, with wind again mixing the entire water column.

### Land covers about a quarter of Earth's surface

About one-fourth of Earth's surface, most of it in the Northern Hemisphere, is currently land above sea level. This *terrestrial* compartment of the physical environment is connected to the atmospheric compartment by organisms that take chemical elements from the air and ultimately release them back into the air. Chemical elements in rocks are released by weathering and carried in solution into groundwater, which transports them ultimately into the oceans, where they are inaccessible to most terrestrial organisms until an episode of tectonic uplift raises marine sediments above sea level and a new cycle of weathering begins.

Even though the global supply of chemical elements in the terrestrial compartment is constant, regional and local variations in the supply of particular elements influence ecosystem processes on land because elements move slowly there and usually over only short distances. The type of soil that exists in an area and the nutrients it contains depend on the underlying rock from which it forms, as well as on climate, topography, the local biota, and the length of time that soil-forming processes have been acting (see Section 36.3).

## 58.1 RECAP

**Earth is a closed system with respect to atomic matter, but an open system with respect to energy. Earth's physical environment can be divided into four compartments—atmosphere, oceans, fresh waters, and land—through which energy flows and elements cycle.**

- How does the atmosphere keep temperatures at and close to Earth's surface warmer than they would be in its absence? See pp. 1223–1224 and Figure 58.3
- Describe the process of turnover in a temperate-zone lake in fall. See p. 1226 and Figure 58.5

Elements cycle through the global ecosystem, but energy flows through it unidirectionally. In the next section we take a brief look at energy flow through ecosystems.

## 58.2 How Does Energy Flow through the Global Ecosystem?

Solar energy drives nearly all ecosystem processes on Earth. In this section we will examine how the geographic distribution of incoming solar radiation influences the amount of energy assimilated by primary producers, and how human activities are modifying energy flow through the global ecosystem.

### The geographic distribution of energy flow is uneven

Nearly all energy utilized by organisms comes (or once came) from the sun. The only exceptions are found in those few ecosystems in which solar energy is not the main energy source (such as some caves and deep-sea hydrothermal vent ecosystems). Even the fossil fuels—coal, oil, and natural gas—on which the economy of modern human civilization is based are reserves of captured solar energy locked in the remains of organisms that lived millions of years ago.

As described in Section 57.1, solar energy enters ecosystems by way of plants and other photosynthetic organisms. These primary producers use some of the energy they assimilate for their own metabolism; the rest—net primary production—is stored in their bodies or used for their growth and reproduction (see Figure 57.1). Because only the energy of net primary production is potentially available to other organisms that consume the producers, we use net primary production as a rough measure of energy flow through ecosystems (**Figure 58.6**).

The geographic distribution of net primary production reflects geographic variation in incoming solar radiation, described in Section 53.2, and the climate patterns that result from it. The distribution of land masses, temperatures, and moisture on Earth makes some ecosystems more productive than others

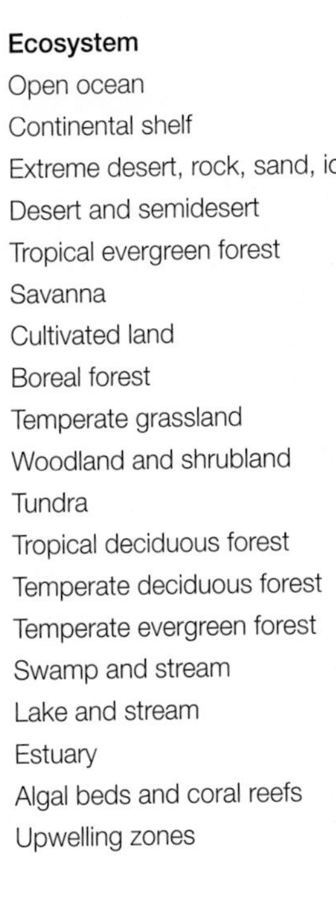


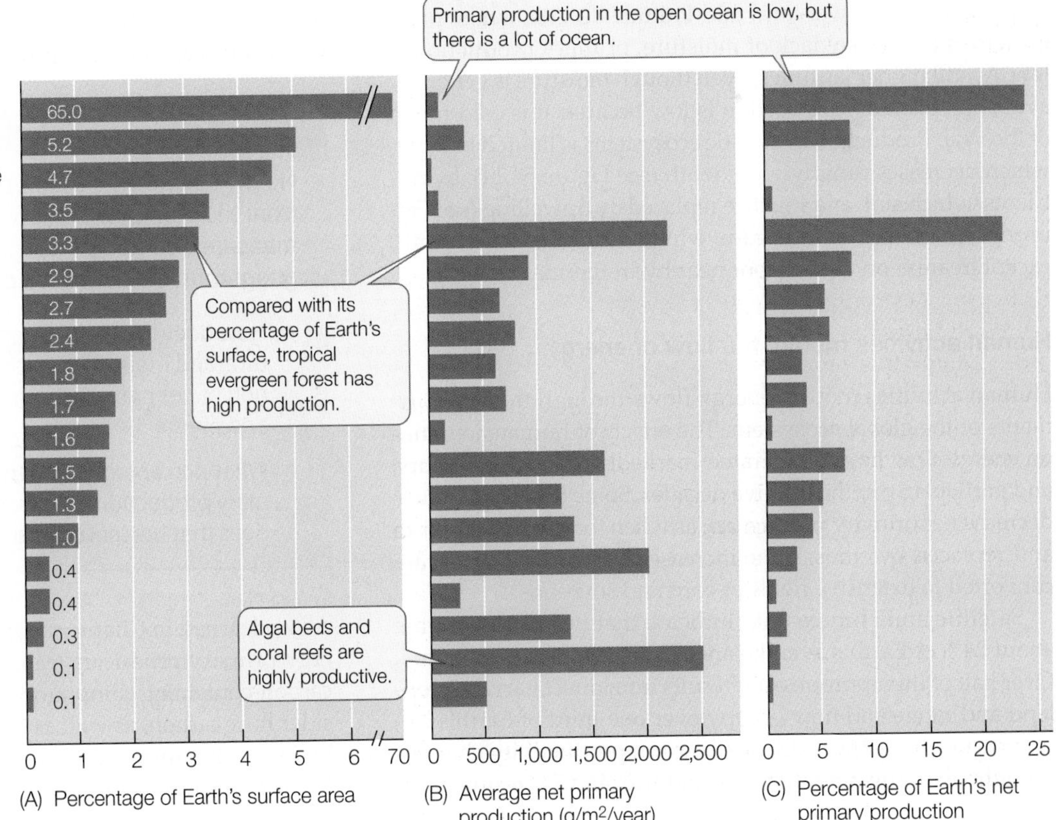


**58.6 Primary Production by Ecosystem Type** The contributions of the different ecosystem types to primary productivity can be measured by (A) their geographic extent; (B) their net primary production; and (C) their percentage of Earth's total primary production.

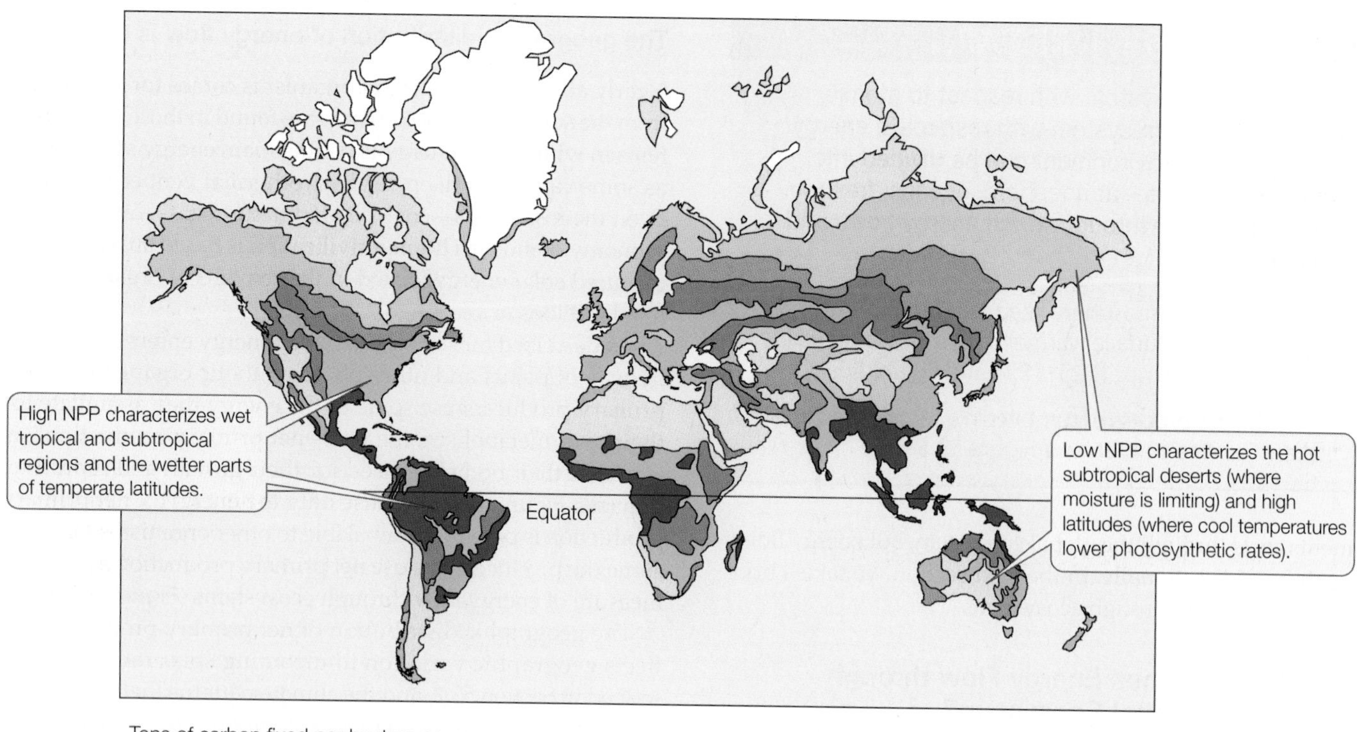
High NPP characterizes wet tropical and subtropical regions and the wetter parts of temperate latitudes.
Equator
Low NPP characterizes the hot subtropical deserts (where moisture is limiting) and high latitudes (where cool temperatures lower photosynthetic rates).
Tons of carbon fixed per hectare per year
Ice caps 0
0–2.5
2.5–6.0
6.0–8.0
8.0–10.0
10.0–30.0
>30.0

**58.7 Geographic Variation in Net Primary Production** Geographic variation in solar energy input results in variation in temperature and moisture over Earth's land surface, and this variation, in turn, influences net primary production in terrestrial ecosystems.

(**Figure 58.7**). Close to the equator at sea level, temperatures are high throughout the year and the water supply is adequate for plant growth much of the time. In these climates, productive forests thrive. In lower- and mid-latitude deserts, where plant growth is limited by lack of moisture, primary production is low. At still higher latitudes, even though moisture is generally available, primary production is low because it is cold much of the year. Production in aquatic ecosystems is limited by light, which decreases rapidly with depth (see Figure 54.14); by nutrients, which sink and must be replaced by upwelling (see Figure 58.4); and by temperature, which decreases with depth, except in areas on the seafloor near hydrothermal vents.

### Human activities modify the flow of energy

Human activities modify energy flows through the compartments of the global ecosystem. The effects of human activities on energy flow have accelerated markedly in the last century, and particularly in the last five decades. Some human activities decrease net primary production, as when forests are cut down and replaced by cities; some increase it, as when prairies are converted to extensive fields of corn and soybeans.

Satellite and climate data indicate that humans consume about 24% of Earth's average annual net primary production. Over half of this appropriation results from direct harvest (cropland and rangeland now occupy over one-third of Earth's ice-free surface); another 40% results from productivity changes brought about by alterations in land use; and 7% results from fires caused by humans. The percentage varies strikingly among regions; urban areas consume as much as 300 times the net primary production they generate, but people in sparsely inhabited parts of the Amazon Basin appropriate vanishingly small amounts of the net primary production there.

## 58.2 RECAP

**Nearly all energy utilized by living organisms comes from the sun. Energy flow through ecosystems, as measured by net primary production, varies with geographic location and climate.**

- How does the distribution of land masses, temperature, and moisture influence the geographic distribution of net primary production? **See pp. 1227–1228 and Figure 58.7**
- What percentage of Earth's average annual net primary production is appropriated by humans, and how does that percentage vary regionally? **See p. 1228**

As we learned in Chapters 3 and 10, most of the biological energy primary producers transform from sunlight is stored in carbon-containing compounds. In the next section we will consider how carbon, as well as other chemical elements required by living organisms, cycle through the four compartments of Earth's physical environment.

## 58.3 How Do Materials Cycle through the Global Ecosystem?

The chemical elements that organisms need in large quantities—carbon, hydrogen, oxygen, nitrogen, phosphorus, and sulfur—cycle through the abiotic and biotic components of ecosystems. The pattern of movement of a chemical element through living organisms and the four compartments of the physical environment is called its **biogeochemical cycle**. Each element has a distinctive biogeochemical cycle whose properties depend on the physical and chemical nature of the element and the ways in which it is used by organisms.

Before we describe the biogeochemical cycles of individual elements, however, we must discuss the influences of water and fire on the rates at which elements cycle through ecosystems.

### Water transfers materials from one compartment to another

The cycling of water through the four compartments of the physical environment is known as the **hydrologic cycle**. Water is an extraordinary molecule by Earth's standard in that it can be found in all three states of matter—gas, liquid, and solid—and can change state fairly readily. The sun powers the hydrologic cycle by causing evaporation from the vast surfaces of the oceans. The hydrologic cycle operates because more water is evaporated from the ocean surface than is returned to them as precipitation. Water also evaporates from soils, from lakes and rivers, and from the leaves of plants (by transpiration), but the total amount evaporated from those surfaces is less than the amount that falls on them as precipitation. The excess terrestrial precipitation eventually returns to the oceans via streams, coastal runoff, and groundwater flows (**Figure 58.8**). Earth's 16 largest rivers account for more than one-third of total runoff from the land. More than half of this total comes from the four largest rivers—the Amazon in South America, the Nile in Africa, the Mississippi in North America, and the Yangtze in Asia.

Despite their relatively small volume, rivers play a disproportionate role in the hydrologic cycle because the average residence time of a water molecule in rivers is only a few years. By comparison, the average residence time of a water molecule in lakes ranges from a few years to centuries. The larger the lake, the longer the residence time; the residence time for Lake Michigan, for example, is about 600 years, and the residence time of the top portion of Lake Superior is 1,500 to 2,000 years (and the brackish water at the bottom never cycles). In the ocean, the average residence time of a water molecule is about 3,000 years. Other accumulations of water include glaciers (with residence times of 20–100 years), seasonal snow cover (a few months), and soil moisture (a month or two). The average residence time of water in organisms is particularly brief—on average, just under a week.

Although large amounts of groundwater are present in underground pools called **aquifers**, this water has a long residence time underground and plays only a small role in the hydrologic

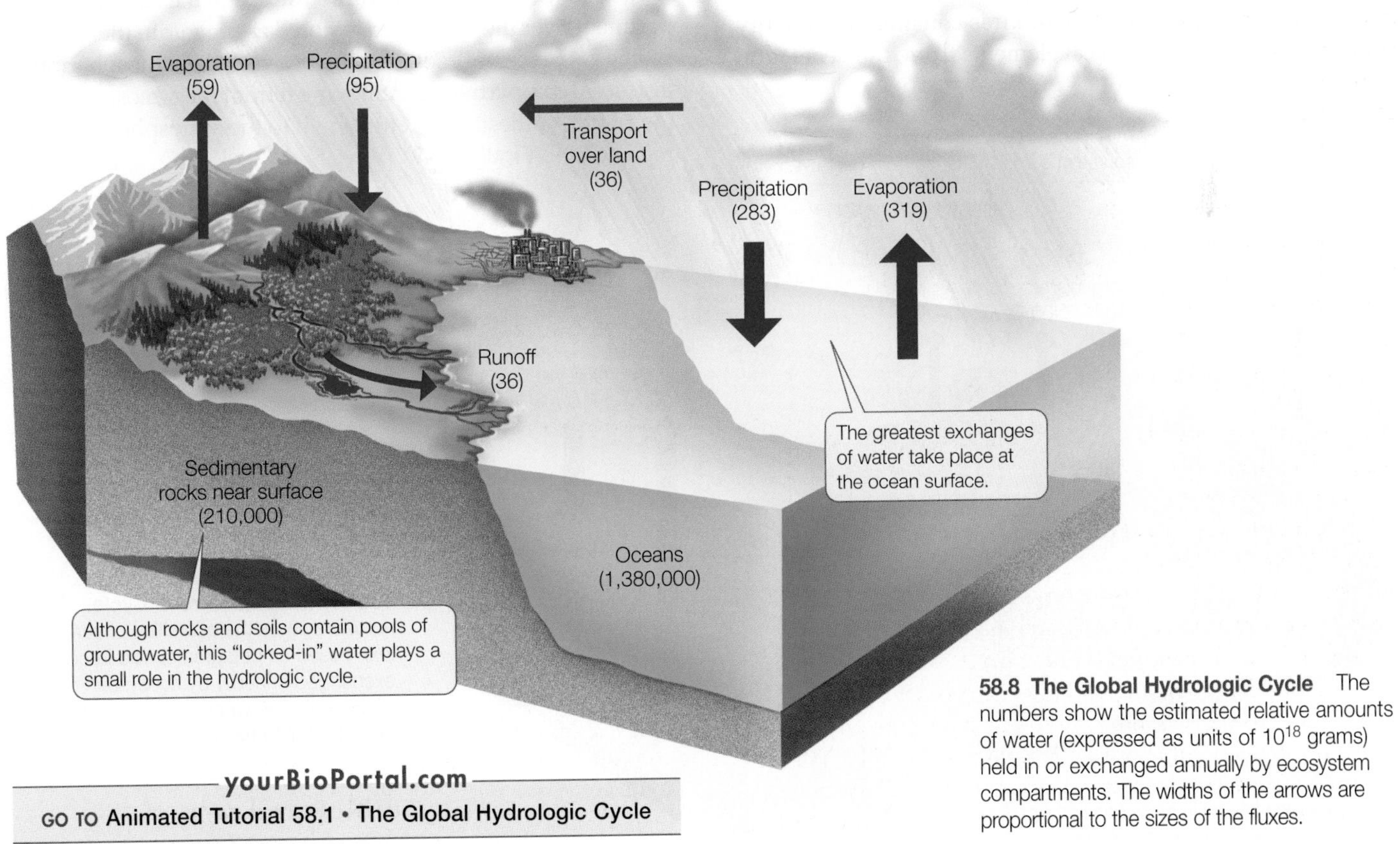
Evaporation (59)
Precipitation (95)
Transport over land (36)
Precipitation (283)
Evaporation (319)
Runoff (36)
The greatest exchanges of water take place at the ocean surface.
Sedimentary rocks near surface (210,000)
Oceans (1,380,000)
Although rocks and soils contain pools of groundwater, this "locked-in" water plays a small role in the hydrologic cycle.

**58.8 The Global Hydrologic Cycle** The numbers show the estimated relative amounts of water (expressed as units of $10^{18}$ grams) held in or exchanged annually by ecosystem compartments. The widths of the arrows are proportional to the sizes of the fluxes.

yourBioPortal.com
GO TO Animated Tutorial 58.1 • The Global Hydrologic Cycle

cycle. In some places, however, groundwater is being seriously depleted because humans are using it more rapidly than it can be replaced, primarily by pumping it for irrigation. Much of the groundwater being used today in the Northern Hemisphere was deposited during the most recent ice age, when regional precipitation was much greater than it is now. Using this groundwater for irrigation and other purposes has increased flows of water to the oceans and has contributed to the sea level rise of the past century.

The effects of groundwater depletion are already being felt. On the North China Plain, for example, depletion of shallow aquifers is forcing people to sink wells more than 1,000 meters deep to reach water. Worldwide, more than 1 billion people have no access to safe drinking water. If current water consumption patterns continue, by 2025 at least 3.5 billion people (48% of the current world population) will live in areas with inadequate water supplies. However, per capita water consumption in the United States and Europe is dropping as a result of increasing use of water-efficient home appliances and the increasing cost of water, as well as new regulations that restrict water use. If such programs are implemented vigorously, global water use in 2025 could be lower than it is today, despite continued population growth.

### Fire is a major mover of elements

Every year, 200–400 million hectares of savannas, 5–15 million hectares of boreal forests, and smaller expanses of other biomes burn. Lightning ignites some fires, but humans start most fires to manage vegetation (as when they cut down and burn forests to clear land for growing crops). Fires consume the energy stored in, and release chemical elements from, the vegetation they burn. Some nutrients, such as nitrogen, are easily vaporized by fire. They are discharged to the atmosphere in smoke or carried into groundwater by rain falling on burned ground.

Fires also release large amounts of carbon into the atmosphere. The global annual flux of carbon to the atmosphere from savanna and forest fires is estimated at 1.7 to 4.1 petagrams (1 pg = $10^{15}$ g or $10^{12}$ kg). Biomass burning (which includes combustion of wood and alcohol, wildfires, and land clearing, but not the burning of fossil fuels) is responsible for about 40% of Earth's annual flux of $CO_2$ into the atmosphere and contributes to the production of other greenhouse gases. Large-scale wildfires in the western and southeastern U.S. can release as much $CO_2$ into the atmosphere as motor vehicles in those regions release over an entire year.

Although biogeochemical cycles interact with one another, it is convenient to discuss the major cycles individually. Our discussion will focus on those elements whose cycles are best characterized and have the greatest effect on living organisms.

### The carbon cycle has been altered by human activities

As described in Part One of this book, all the important macromolecules that compose living organisms contain carbon, and much of the energy that organisms need to fuel their metabolic activities comes from carbon-containing (organic) compounds. Carbon in the atmosphere, in the form of $CO_2$, is incorporated into organic molecules by photosynthesis in the cells of autotrophs. All heterotrophic organisms get their carbon by consuming autotrophs or other heterotrophs, their remains, or their waste products. Most of the carbon stored in the terrestrial compartment over the short term is stored as organic matter in soils. Biological processes move carbon between the atmospheric and terrestrial compartments, removing it from the atmosphere during photosynthesis and returning it to the atmosphere through metabolism (**Figure 58.9**).

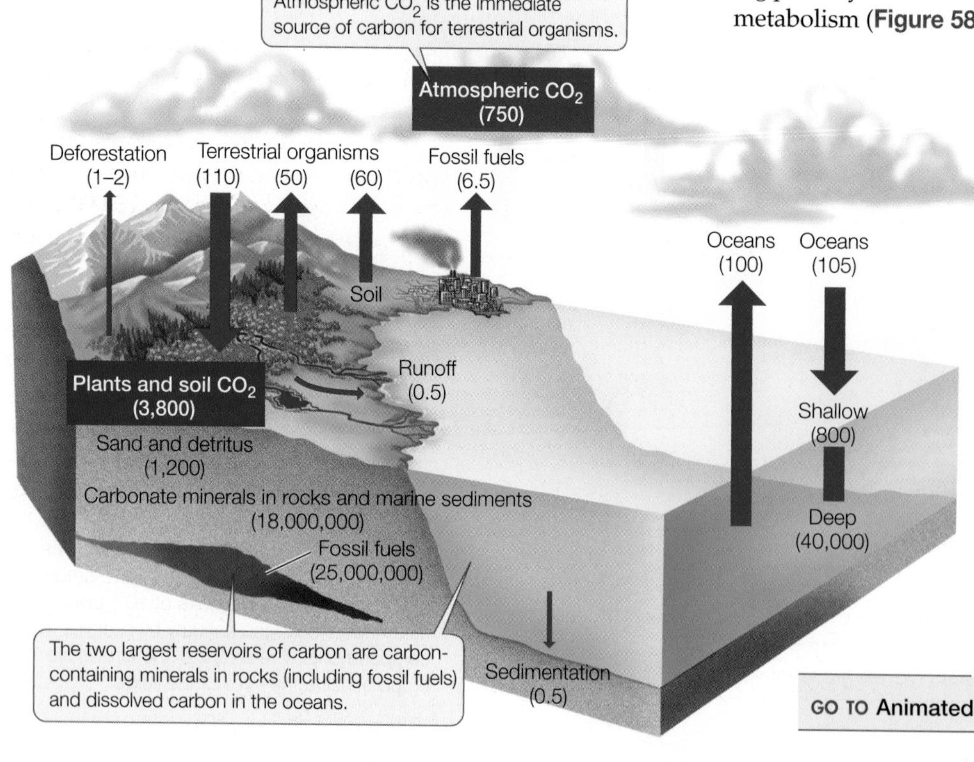


**58.9 The Global Carbon Cycle** The numbers show the quantities of carbon (expressed as petagrams) held in or exchanged annually by ecosystem compartments. The widths of the arrows are proportional to the sizes of the fluxes.

yourBioPortal.com
GO TO Animated Tutorial 58.2 • The Global Carbon Cycle

The amount of carbon in the atmosphere in the form of $CO_2$, and the amount stored in the biosphere in living and dead organisms, are dwarfed by the amount of carbon stored in terrestrial soils and rocks, in marine sediments, and in dissolved form in ocean waters. Carbon dioxide moves into ocean waters from the atmosphere primarily by simple diffusion at the ocean surface; it is stored in seawater in the form of carbonate ($CO_3^{-2}$) or bicarbonate ($HCO_3^-$).

**yourBioPortal.com**
GO TO Animated Tutorial 58.3 • The Ocean Conveyor Belt

Today's oceans absorb 20–25 million tons of $CO_2$ every day—more than at any time during the past 20 million years. As a result, water near the ocean surface is becoming more acidic. An increase in the acidity of ocean surface waters could have negative effects on many marine organisms, particularly corals. The combination of decreasing pH and increasing water temperature kills their symbiotic algae, "bleaching" the corals and killing them as well (see Figure 27.8A). Because so many other reef species depend on corals and the structure they provide, the entire reef community collapses.

At times in the remote past, great quantities of carbon were removed from the global carbon cycle when organisms died in large numbers and were buried in sediments lacking oxygen. In such anaerobic environments, detritivores do not reduce organic carbon to $CO_2$. Instead, organic molecules accumulate and are eventually transformed into deposits of oil, natural gas, coal, or peat. Humans have discovered and used these deposits, known as **fossil fuels**, at ever-increasing rates during the past 150 years. As a result, $CO_2$, one of the final products of the burning of fossil fuels, is being released into the atmosphere faster today than it is dissolving in the oceans or being incorporated into terrestrial biomass. Based on a variety of calculations, atmospheric scientists believe that before the Industrial Revolution, the concentration of atmospheric $CO_2$ was probably about 265 parts per million. Today, it is 387 parts per million (**Figure 58.10**), representing a rate of increase more than 10 times faster than at any other time for millions of years.

**WHERE HAS ALL THE CARBON GONE?** Less than half of the vast amount of $CO_2$ released into the atmosphere by human activities remains in the atmosphere. Where is the rest? Much of it is dissolved in the oceans. Over decades to centuries, the oceans, which contain 50 times more dissolved inorganic carbon than the atmosphere, determine atmospheric $CO_2$ concentrations. The rate at which $CO_2$ moves from the atmosphere to the oceans depends, in part, on photosynthesis by phytoplankton in the surface waters. These organisms remove carbon from the water, thereby increasing its absorption from the atmosphere. In addition, many marine organisms, including clams, oysters, corals, and some algae, combine bicarbonate ions with calcium ($Ca^{2+}$) to form calcium carbonate, which they use to build shells and other structures. When these organisms die, they eventually sink to the ocean floor.

Photosynthesis by terrestrial vegetation, principally in forests and savannas, typically absorbs roughly the same amount of carbon that is released by metabolism—about half of it by plants and half of it by microbes in the soil that break down organic matter produced by plants. Currently, the photosynthetic consumption of $CO_2$ exceeds the metabolic production of $CO_2$, so Earth's terrestrial vegetation is storing carbon that would otherwise be increasing atmospheric $CO_2$ concentrations. Yet, as we saw at the opening of this chapter, we cannot count on terrestrial vegetation to store the extra $CO_2$ that human activities are producing. Furthermore, climate warming—another result of increasing $CO_2$ concentrations, as we'll see next—increases plant metabolism and is thus likely to increase the flux of $CO_2$ from vegetation to the atmosphere.

**ATMOSPHERIC $CO_2$ AND GLOBAL WARMING** What evidence do we have that the buildup of atmospheric $CO_2$ stemming from the burning of fossil fuels is raising Earth's temperature? Measurements of gases in air trapped in the Antarctic and Greenland ice caps show that atmospheric $CO_2$ levels have been higher when Earth has been warmer and lower when it has been cooler (**Figure 58.11**). For example, atmospheric $CO_2$ was very low during the most recent glaciation, between 30,000 and 15,000 years ago, when temperatures were presumably much colder than they are today. In contrast, during a warm interval 5,000 years ago, atmospheric $CO_2$ may have been slightly higher than it is today.

How global climates and ecosystems will change in response to this rapid warming is a subject of intense investigation. Complex computer models of the global ecosystem indicate that a doubling of the atmospheric $CO_2$ concentration from today's level would increase

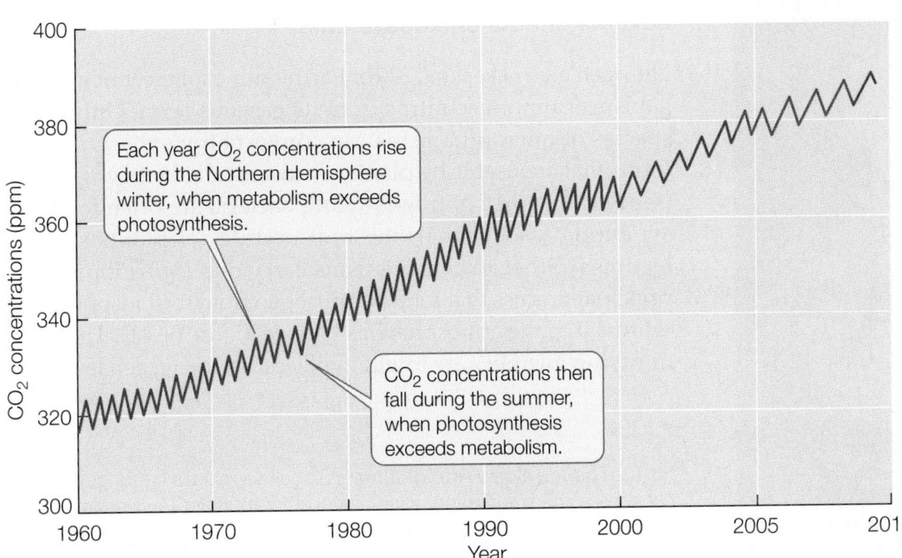
400
380
360
340
320
300
$CO_2$ concentrations (ppm)
Each year $CO_2$ concentrations rise during the Northern Hemisphere winter, when metabolism exceeds photosynthesis.
$CO_2$ concentrations then fall during the summer, when photosynthesis exceeds metabolism.
1960
1970
1980
1990
2000
2005
2010
Year

**58.10 Atmospheric $CO_2$ Concentrations Are Increasing** These carbon dioxide concentrations, expressed as parts per million by volume of dry air, were recorded on top of Mauna Loa, Hawaii, far from most sources of human-generated $CO_2$ emissions.

**58.11 Higher Atmospheric $CO_2$ Concentrations Correlate with Warmer Temperatures** The geological record of variation in atmospheric $CO_2$ concentrations over the last 500 million years shows that $CO_2$ levels have been lowest during times of extensive continental glaciation (red bars), and thus presumably lower temperatures.

mean annual temperatures worldwide and would probably cause droughts in the central regions of continents while increasing precipitation in coastal areas. Global warming has already resulted in the shrinking of Arctic sea ice (**Figure 58.12**). If the polar ice caps continue to melt, sea level will rise (due to both thermal expansion of ocean waters and the addition of glacial meltwater), flooding coastal cities and agricultural lands. Because the rate at which water evaporates from the ocean surface increases with temperature, tropical storms are likely to become more intense. Indeed, during the past two decades, the number of hurricanes generated per year worldwide has not increased, but the number of severe (category 4 and 5) hurricanes has.

Global warming is having a profound effect on the distribution and abundance of interacting species. One clear example is an increase in the severity of insect infestations in certain temperate forest communities. Pine trees in these forests are attacked by pine bark beetles, which carry with them a symbiotic fungus that infects the trees and helps the beetles to overcome the trees' defenses (see Figure 56.13). In Colorado, cold winters have historically limited the ability of these beetles to kill many trees. In 2008, however, Colorado experienced an outbreak of mountain pine beetles larger than any previously recorded; more than a million acres of trees were killed. The lack of an extended period of below-zero temperatures in the previous winter had allowed large numbers of overwintering beetles to survive.

Another result of global warming may be an increase in the frequency of outbreaks of human diseases. Winter cold typically kills many pathogens, sometimes eliminating as much as 99% of a pathogen population. If climate warming reduces this population bottleneck, some diseases will spread to more temperate regions. For example, dengue fever and its mosquito vector are now spreading to higher latitudes where they were formerly absent.

The 2007 Assessment Report of the Intergovernmental Panel on Climate Change (IPCC) forecast these and many other negative effects of global warming on humans, warning, "Human beings are exposed to climate change through changing weather patterns (for example, more intense and frequent extreme events) and indirectly through changes in water, air, food quality and quantity, ecosystems, agriculture, and economy. At this early stage the effects are small but are projected to progressively increase in all countries and regions."

## Recent perturbations of the nitrogen cycle have had adverse effects on ecosystems

Nitrogen gas makes up 78% of Earth's atmosphere, but most organisms cannot use nitrogen in its gaseous form. Only a few species of microorganisms can convert atmospheric $N_2$ into forms that are usable by plants, a process called *nitrogen fixation* (see Section 36.4). Nitrogen-fixing bacteria are often found as symbiotic associates in the root nodules of plants such as legumes. Other microorganisms carry out *denitrification*, the principal process that removes nitrogen from the biosphere and returns it to the atmosphere as $N_2$ (see Figure 36.11). These denitrifying microbes are found in places where oxygen levels are

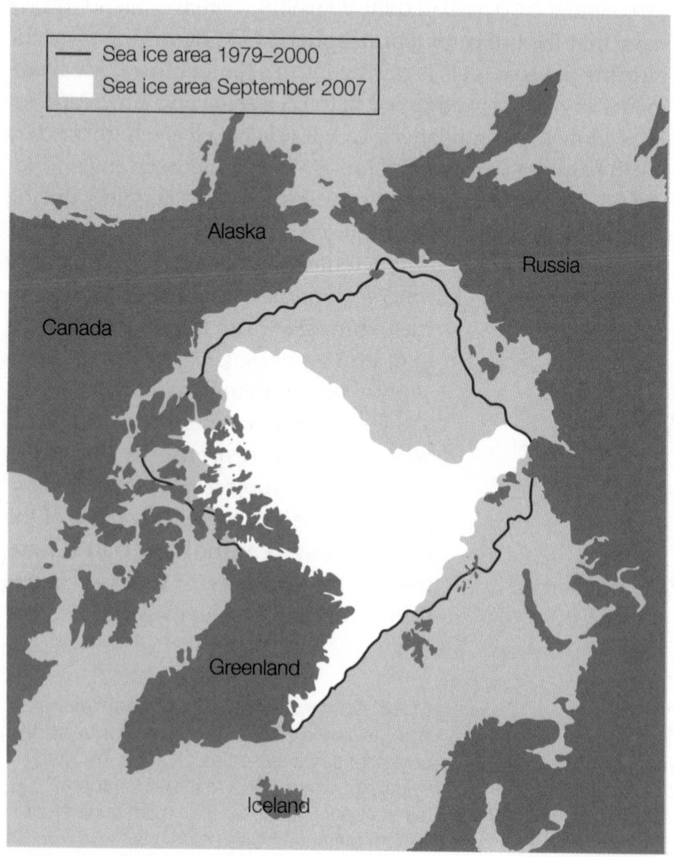


**58.12 The Ice Caps Are Melting** Global warming is causing major changes in the sizes of the polar ice caps. In 2008, for the first time in recorded history, enough of the Arctic Ocean was ice-free that the North Pole could have been circumnavigated by ship.

**58.13 The Global Nitrogen Cycle** The numbers in parentheses show the quantities of nitrogen (expressed as teragrams) held in or exchanged annually by ecosystem compartments. Arrow widths are proportional to the sizes of the fluxes.

yourBioPortal.com

GO TO Animated Tutorial 58.4 • The Global Nitrogen Cycle

low, such as bogs, swamps, and marine sediments. Collectively, this microbial processing of nitrogen accounts for about 95 percent of all natural nitrogen flux on Earth (**Figure 58.13**).

All living organisms require nitrogen, and the inability of the vast majority of them to utilize $N_2$ means that usable nitrogen is often in short supply in ecosystems. Populations of nitrogen-fixing organisms rarely increase in abundance to the extent that nitrogen is no longer limiting because the end products of nitrogen fixation are rapidly lost from ecosystems: ammonia by vaporization and denitrification, and the highly water-soluble nitrates by leaching. Furthermore, fixing nitrogen requires large amounts of energy. Thus nitrogen-fixing organisms often lose out in competition with non-fixers when nitrogen becomes more readily available.

The industrial production of inexpensive artificial fertilizers has had some unanticipated effects on the nitrogen cycle. As a result of extensive use of these fertilizers on agricultural crops, coupled with the burning of fossil fuels (which generates nitric oxide), total industrial nitrogen fixation by humans today is about equal to global natural nitrogen fixation. This human-generated nitrogen flux has been increasing over the past half-century and is expected to continue to increase during the coming decades (**Figure 58.14**).

These large perturbations of the nitrogen cycle have had numerous adverse effects. When more nitrogen is applied to croplands than can be taken up by the crops, the excess nitrogen moves out of the system in surface runoff, or downward into groundwater, and ultimately ends up in rivers, lakes, and oceans. The addition of nutrients to these bodies of water, known as **eutrophication** ("rich food"), can have dramatic effects on aquatic ecosystems. The "dead zone" that has formed

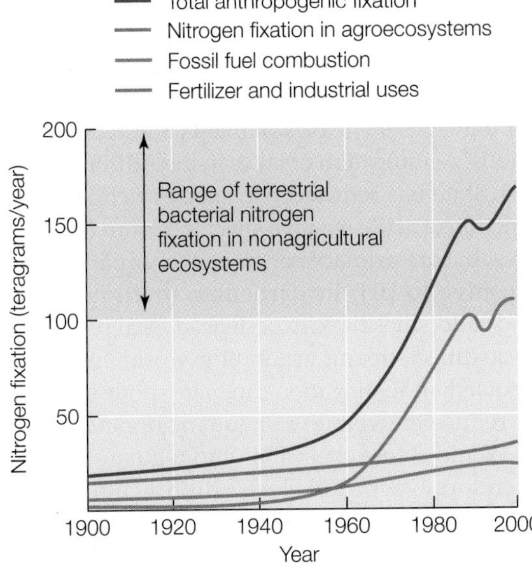


**58.14 Human Activities Have Increased Nitrogen Fixation** Most of the industrially fixed nitrogen (produced by humans by combining, at high pressure and temperatures, atmospheric nitrogen with hydrogen derived from fossil fuels) is manufactured for use in agricultural fertilizers. Some nitrogen fixation is a by-product of fossil fuel combustion.

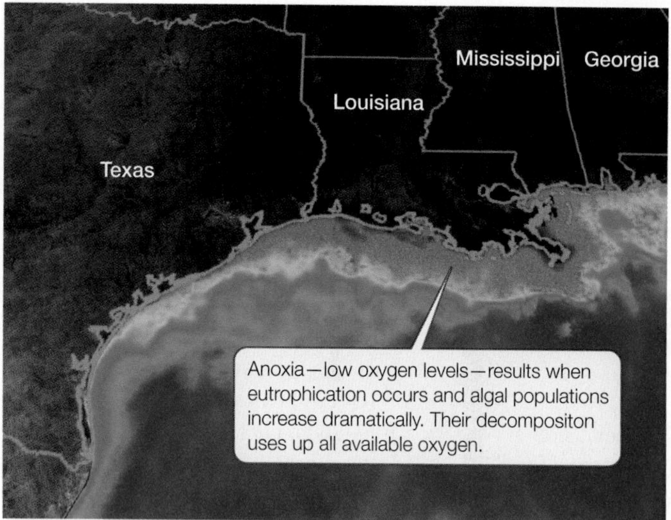

Mississippi
Georgia
Louisiana
Texas
Anoxia—low oxygen levels—results when eutrophication occurs and algal populations increase dramatically. Their decompositon uses up all available oxygen.

**58.15 A Dead Zone** High concentrations of nitrogen and phosphorus in the runoff from agricultural lands in the midwestern United States are carried by the Mississippi River to the Gulf of Mexico. This nutrient enrichment causes excessive algal growth, depleting the oxygen in the water and creating a deoxygenated "dead zone" in which few aquatic organisms can survive.

near the mouth of the Mississippi River in the Gulf of Mexico is a result of flows of water from agricultural fields in the Upper Midwest carrying high concentrations of nitrogen from fertilizer (**Figure 58.15**). The high nutrient levels that result promote luxuriant algal growth. When the algae die, their decomposition by microbes uses up the oxygen in the water, and the resulting anaerobic conditions kill most other marine organisms. Given that the Gulf of Mexico is a key resource for the seafood industry, providing almost three-fourths of the shrimp and two-thirds of the oysters harvested in the United States, an expansion of the dead zone could have a devastating effect on the economies of the Gulf region, over and above its effects on biodiversity and ecosystem function.

The increase in nitrogen fixation has also increased atmospheric concentrations of the greenhouse gas nitrous oxide ($N_2O$; see Table 58.1), resulting in the production of tropospheric $O_3$ —also a greenhouse gas—and smog. Some of the nitrogen that enters the atmosphere falls back to land in precipitation or as dry particles. This deposition of nitrogen affects the composition of terrestrial vegetation by favoring those species that are best adapted to take advantage of high nutrient levels, which then crowd out other species. Atmospheric deposition of nitrogen has increased dramatically during recent decades due to human activity. Spatial variation in nitrogen deposition rates has allowed ecologists to determine that plant species richness in grasslands declines as the rate of nitrogen deposition increases. Rates of nitrogen deposition are high enough over much of Europe and eastern North America to cause substantial reductions in species richness in grasslands on both continents.

### The burning of fossil fuels affects the sulfur cycle

Most of Earth's sulfur supply is locked up in rocks on land and as sulfate salts in deep-sea sediments. Emissions of the gases sulfur dioxide ($SO_2$) and hydrogen sulfide ($H_2S$) from volcanoes account for, on average, 10–20% of the total natural nonbiological flux of sulfur to the atmosphere, but they occur only intermittently. Although volcanic eruptions spew great quantities of sulfur over broad areas, they are rare events. Sulfur in soil is taken up by plants and incorporated into proteins; ultimately, this sulfur is returned to the atmosphere via microbial decomposition. In marine systems, too, microbial decomposition is important in returning sulfur to the atmosphere. In particular, many marine phytoplankton and seaweeds manufacture large quantities of a sulfur-containing compound (DMSP, or dimethylsulfoniopropionate) to maintain their salt and water balance; when broken down, this compound releases dimethyl sulfide ($CH_3SCH_3$), the principal odorant of rotting seaweed stench. Because the quantities of phytoplankton in the vast oceans are enormous, dimethyl sulfide production accounts for about half of the biotic component of the global sulfur cycle.

Atmospheric sulfur plays an important role in global climate. Even if air is moist, clouds do not form readily unless there are small particles in the atmosphere around which water can condense. Dimethyl sulfide is the major component of such particles. Therefore, increases in atmospheric sulfur concentrations can increase cloud cover and reduce the amount of incoming solar radiation that ultimately reaches Earth's surface.

The burning of fossil fuels alters the sulfur cycle as well as the carbon and nitrogen cycles. The combustion of fossil fuels releases sulfur in the form of $SO_2$ and nitrogen in the form of nitrogen dioxide ($NO_2$ ) into the atmosphere, where they react with water molecules to form sulfuric acid ($H_2SO_4$) and nitric acid ($HNO_3$). These acids can travel hundreds of kilometers in the atmosphere before they settle to Earth as dry particles or in precipitation. Rain or snow containing sufficient levels of nitric and sulfuric acids to lower its pH is called **acid precipitation**. Acid precipitation now falls in all major industrialized countries and is particularly widespread in eastern North America and Europe. The normal pH of unpolluted precipitation is about 5.6, but precipitation in New England now averages about pH 4.5 and there have been occasional rainfalls and snowfalls with a pH as low as 3.0. Precipitation with a pH of about 3.5 or lower damages the leaves of plants and reduces rates of photosynthesis. Acidification of lakes in the Adirondack region of New York State has reduced fish species richness by causing the extinction of acid-sensitive species (**Figure 58.16**). Many invertebrates that are primary consumers in aquatic communities are sensitive to pH; in particular, multiple species of mayflies and caddisflies have experienced local population reductions in acidified streams around the world. Even when its effects are not lethal, acidification can have subtle effects on behavior that reduce the viability of aquatic organisms. Diving beetles, for example, lose their ability to regulate their underwater oxygen supply when pH levels drop significantly.

New regulations instituted by the 1990 Clean Air Act have raised the pH of precipitation in much of the eastern United

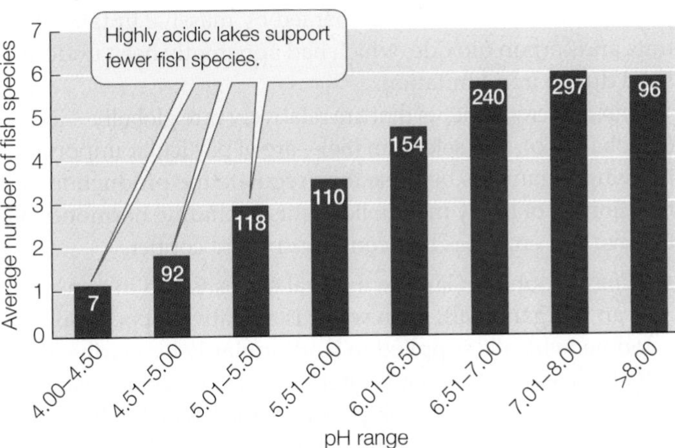


**58.16 Acidification of Lakes Exterminates Fish Species** The number of fish species found in lakes in the Adirondack region of New York State is directly correlated with pH. The numbers in the bars indicate the number of lakes in each pH range.

States, primarily by lowering sulfur emissions. There are indications that, once emissions have been reduced, acidified aquatic systems can recover quickly. David Schindler at the University of Alberta studied the effects of acid precipitation by adding enough $H_2SO_4$ to two small Canadian lakes to reduce their pH from about 6.6 to 5.2. In both lakes, nitrifying bacteria failed to survive under these moderately acidic conditions. As a result, the nitrogen cycle was blocked. When Schindler stopped adding acid to one of the lakes, its pH returned to its original value in about a year, and nitrification resumed. These experiments show that lake ecosystems are very sensitive to acidification but can recover rapidly, presumably because water in lakes is exchanged rapidly.

That said, the larger organisms in such systems may be slower to bounce back. Fourteen rivers in Wales were studied for a quarter-century, to determine the impacts of acid rain reductions. Investigators were disappointed that, even after 25 years, only four insect species—two mayfly species and two caddisfly species—had recolonized the area out of the 29 species expected to come back.

### The global phosphorus cycle lacks a significant atmospheric component

Phosphorus accounts for only about 0.1% of Earth's crust, but it is an essential nutrient for all life forms. It is a key component of DNA, RNA, and ATP. Unlike the elemental cycles discussed thus far, the phosphorus cycle lacks a significant gaseous component. Some phosphorus is transported on dust particles, but very little of the phosphorus cycle takes place in the atmospheric compartment. Most of Earth's phosphorus is present in the form of phosphate salts in rocks and deep-sea sediments. Phosphorus cycling takes millions of years because the processes of sedimentary rock formation, uplift, and weathering all take a long time(**Figure 58.17**). In contrast, phosphorus often cycles rapidly among organisms, and it is often a limiting factor for their growth, particularly for plants. In addition to nitrogen, fertilizers routinely include phosphorus.

Human activity has radically accelerated some parts of the phosphorus cycle. One consequence of the massive use of artificial fertilizers is that between 10.5 and 15.5 teragrams (1 tg = $10^{12}$ g or $10^9$ kg) of phosphorus accumulate each year, primarily in agricultural soils. When the concentration of phosphorus in soils exceeds the capacity of plants to take it up, the excess moves into streams and lakes. Soil erosion due to the clearing of land for purposes such as agriculture and logging also increases the amount of phosphorus and other nutrients in runoff. Phosphorus is a limiting nutrient in many lakes, so when it enters those lakes through runoff, algae previously limited by relatively low levels of phosphorus respond with phenomenal growth. These algal blooms turn the water green and, depending on the metabolic capabilities of the algae, toxic. When the algae die, their decomposition by microbes consumes all of the oxygen in the lake, and anaerobic organisms dominate the bottom sediments. These anaerobic organisms do not break down carbon compounds all the way to $CO_2$, and their metabolic end products, many of which have unpleasant odors, build up.

The capacity of phosphorus-charged runoff to cause extensive eutrophication was graphically illustrated almost 50 years ago. A technological innovation in the formulation of laundry detergents in the early 1960s was the use of sodium tripolyphosphate (STPP) ($Na_5P_3O_{10}$) as a water softener and dirt-breaking agent to enhance cleaning efficiency. Widespread adoption of these new detergents led to massive phosphorus enrichment of lakes and streams and concomitant eutrophication and algal blooms. Water quality declined so much across the United States that phos-

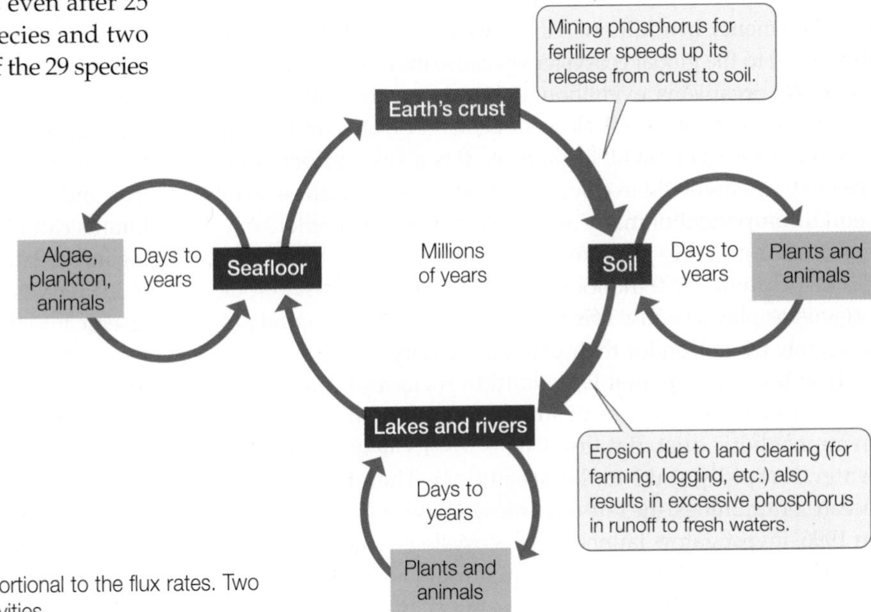


**58.17 The Phosphorus Cycle** Arrow widths are proportional to the flux rates. Two large increases in phosphorus flux are due to human activities.

phate-based detergents were banned in many states, as well as in Europe. In today's detergents, phosphonates—forms of phosphorus that do not appear to promote algal growth—and aluminum silicates perform the functions STPP used to serve.

Detergents, however, are only one agent contributing to eutrophication. Human waste is rich in phosphorus, as are manure from domesticated animals and industrial waste of various types. Two hundred years ago, Lake Erie, one of the Great Lakes on the border between the United States and Canada, had only moderate rates of photosynthesis and clear, oxygenated water. With increasing industrialization in the early part of the twentieth century, nutrient concentrations in the lake increased greatly, and algae proliferated. At the water filtration plant in Cleveland, Ohio, algae increased from 81 per milliliter in 1929 to 2,423 per milliliter in 1962. Populations of bacteria also increased: the numbers of *Escherichia coli* increased enough that many of the lake's beaches were closed and declared health hazards. Since 1972, the United States and Canada have invested more than U.S. $9 billion to improve municipal waste treatment facilities and reduce discharges of pollutants into Lake Erie. As a result, the amount of phosphorus added to Lake Erie has decreased more than 80% from its highest level, and phosphorus concentrations in the lake have declined substantially. The deeper waters of Lake Erie are still oxygen-poor during the summer months, but the rate of oxygen depletion is declining.

We could greatly reduce phosphorus pollution by recovering and recycling phosphorus. The phosphorus contained in sewage and animal wastes could supply much of the needs of the detergent and fertilizer industries. More careful application of fertilizers on agricultural lands could reduce the rate of phosphorus accumulation in soils without reducing crop yields. However, reduction of phosphorus concentrations in soils will take many decades after remedial actions are initiated, and eutrophication of lakes and streams may persist even after these actions are taken.

### Other biogeochemical cycles are also important

Other elements, in addition to those we have just discussed, are important to the global ecosystem because they are essential nutrients for organisms, even though they are needed only in very small amounts. One such element is *iron* (Fe), an essential micronutrient for almost all organisms. It is a key component of the enzymes involved in chlorophyll synthesis, as well as an essential component of many animal enzymes. Iron confers oxygen-binding ability on hemoglobin in vertebrate blood and the cytochrome P450 monooxygenases, which in most aerobic organisms play a central role in detoxifying environmental poisons, rely upon iron for their catalytic activity.

Iron is readily available on land in rocks and minerals. It moves into coastal waters in streams and into the open ocean in atmospheric dust. Because iron is insoluble in oxygenated water, it rapidly sinks to the ocean floor. Therefore, in most ocean communities, the rate of photosynthesis is limited by iron. In 1996, investigators launched an ecosystem-scale experiment in which surface waters of the equatorial Pacific Ocean were seeded with dissolved iron. The response was a tremendous phytoplankton bloom, accompanied by massive uptake of nitrate and carbon dioxide, which had apparently been underutilized due to iron limitation.

Three micronutrients that are relatively rare globally—iodine (I), cobalt (Co), and selenium (Se)—are of particular importance to living organisms because they regulate the production and functioning of many metabolic agents, including hormones, vitamins, and enzymes that contain zinc and copper.

- *Iodine* is found on land in mineral deposits and in seawater as an inorganic salt; fresh water is a relatively poor source. Iodine, which is supplied to land primarily by marine algae that release it to the air, is often deficient in continental interiors. It is an essential component of the hormone thyroxine, which governs many metabolic processes, and a deficiency of iodine in humans leads to the medical condition called goiter (see Figure 41.10).
- *Cobalt* is a rare element found in soil and rocks. The bacteria that fix atmospheric nitrogen require coenzymes derived from cobalt. As an essential component of vitamin $B_{12}$, it is a dietary requirement for almost all animals and is important for the synthesis of proteins.
- *Selenium* is also found in soil and rocks. It is a component of several important mammalian enzymes, such as the antioxidant glutathione peroxidase (which protects tissues from oxidative damage) and several enzymes involved in synthesizing thyroid hormones.

Land plants require all three of these micronutrients in minimal concentrations, but animals, particularly endothermic vertebrates, require them in concentrations that exceed the supply in many environments. Herbivores sometimes obtain these nutrients by drinking water at mineralized springs or by eating enriched earth (see Figure 53.14). Deficiencies of these elements are most common in areas with high rainfall and soils that leach readily.

### Biogeochemical cycles interact

The biogeochemical cycles of different elements interact with one another in complex ways, and perturbations to one cycle can have profound effects on other cycles. In recent years, human-induced perturbations have made these interactions glaringly apparent. For example, nitrate released by human activities can have profound effects on the biogeochemical cycles of iron and arsenic. Nitrate is a powerful oxidant, and nitrate pollutants can increase the movement of arsenic in lakes by causing its oxidation. Some urban lakes have arsenic levels in excess of 2,000 parts per million locked in sediments; oxidation releases arsenic into the water column in a form that is carcinogenic and has negative effects on development. Every year, scientists are discovering interactions of which they were previously unaware, and experiments exploring biogeochemical interactions are increasing in number.

Changes in atmospheric $CO_2$ concentrations have been a particular focus of investigation in recent years because of their potential for interacting with other biogeochemical cycles through photosynthesis. A case in point is a study of whether elevated atmospheric $CO_2$ concentrations affect rates of nitro-

## INVESTIGATING LIFE

**58.18 Effects of Atmospheric $CO_2$ Concentrations on Nitrogen Fixation**

Although the original expectation was that $CO_2$ enrichment would enhance nitrogen fixation, the reverse—decreased fixation—was in fact being observed. In an attempt to explain this result, investigators studied the changes in concentrations of iron and molybdenum, two micronutrient components of enzymes that are essential to nitrogen fixation.

**HYPOTHESIS** **Higher $CO_2$ levels should enhance nitrogen fixation.**

**METHOD**

1. Grow nitrogen-fixing vines (*Galactia elliottii*) in plots under control conditions (normal levels of $CO_2$) and experimental conditions of artificially heightened levels of atmospheric $CO_2$.
2. Measure levels of nitrogen fixation over a period of 7 years.
3. Determine the concentrations of iron and molybdenum in the leaves of *G. elliottii* at year 7.

**RESULTS**

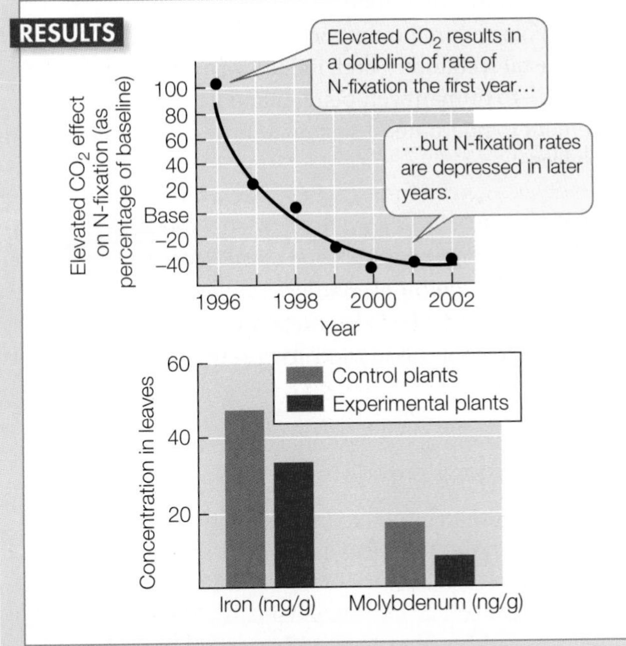


**CONCLUSION** **Decreasing rates of nitrogen fixation associated with increased concentrations of atmospheric $CO_2$ correlate with lower concentrations of essential iron and molybdenum in plant tissues; thus the initial increase in the rate of fixation can result in its eventual depression.**

Go to **yourBioPortal.com** for original citations, discussions, and relevant links for all INVESTIGATING LIFE figures.

gen fixation by microorganisms associated with plant roots. A group of ecologists grew a nitrogen-fixing vine called Elliott's milkpea (*Galactia elliottii*) under artificially increased $CO_2$ concentrations. Higher $CO_2$ concentrations led to an enhancement of nitrogen fixation during the first year of the experiment, but surprisingly, the positive effect disappeared by the third year, and elevated $CO_2$ concentrations actually *reduced* nitrogen fixation below normal levels during the fifth, sixth, and seventh years of the experiment (**Figure 58.18**). The experimenters suspected that depletion of micronutrients such as iron and molybdenum caused the reduction in nitrogen fixation, so they measured concentrations of those elements in the leaves of *G. elliottii*. They found that concentrations of molybdenum were particularly low. Elevated $CO_2$ can increase the acidity of water in soil by enhancing the rate of carbonic acid formation and, by enhancing photosynthesis, increase the accumulation of soil organic matter; both types of change increase the tendency of molybdenum to adsorb onto soil particles, reducing its availability to nitrogen-fixing bacteria and causing a decrease in nitrogen fixation rates.

### 58.3 RECAP

**The pattern of movement of a chemical element through organisms and the physical environment is its biogeochemical cycle. Human activities have affected many biogeochemical cycles, especially those of water, carbon, nitrogen, sulfur, and phosphorus. Increasing concentrations of carbon dioxide and other greenhouse gases in the atmosphere are implicated in global warming.**

- Describe the global hydrologic cycle and explain what drives it. **See pp. 1229–1230 and Figure 58.8**
- How do biological processes move carbon from the atmospheric compartment to the terrestrial compartment and then return it to the atmosphere? **See pp. 1230–1231 and Figure 58.9**
- What are some results of human-induced alterations of the sulfur and nitrogen cycles? **See pp. 1232–1234**
- Name two elements that can cause eutrophication in aquatic ecosystems and describe their effects on those systems. **See pp. 1235–1236**

The biogeochemical cycles of chemical elements are intimately involved in ecosystem function. Just as human alterations of those cycles are having many effects on ecosystems worldwide, the resulting changes in those ecosystems are having profound effects on human lives.

## 58.4 What Services Do Ecosystems Provide?

Although it seems obvious today that humans depend on natural ecosystems for survival, explicit recognition of the value of those ecosystems is rather recent. Environmental writers introduced the idea of "natural capital" in the 1940s; it was in 1970 that ecosystems were first said to provide people with a variety of "goods and services." The goods include food, clean water, clean air, fiber, building materials, and fuel; the services include flood control, soil stabilization, pollination, and climate regulation. Most of these benefits either are irreplaceable, or the technology necessary to replace them is prohibitively expensive. For example, potable fresh water can be provided by desalinating seawater, but only at great cost. The aesthetic, psychological, spiritual, and recreational benefits of ecosystems are less tangible, but no less important, and no more easily replaced.

Humans have increasingly altered Earth's ecosystems to increase their capacity to provide us with goods such as food, fresh water, timber, fiber, and fuel. The benefits have not, however, been equally distributed, and some human populations have actually been harmed by our manipulations of natural ecosystems. Moreover, short-term increases in some ecosystem goods and services have come at the cost of the long-term degradation of others.

The most important driver of change in ecosystems has been changes in land use as natural ecosystems have been converted to other, more intensive uses. Although humans have been altering ecosystems for millennia, the pace and scope of the alterations have increased considerably in recent years. More land was converted to cropland between 1950 and 1980 than in the 150 years between 1700 and 1850. Ecosystem conversions have been particularly rapid in tropical and subtropical biomes. Aquatic ecosystems have suffered losses at an increasing pace as well. Between 1970 and 2000, about 20% of the world's coral reefs were lost, and an additional 20% were degraded. In freshwater ecosystems, so much water is now impounded behind dams that artificial reservoirs hold about six times as much water today as do natural rivers. These freshwater systems are being rapidly depleted: the amount of water withdrawn from rivers—most of it for agriculture—has doubled since 1960. In terms of nutrient cycles, more than half of all the artificial nitrogen fertilizer ever used on Earth has been used since 1985.

Human alteration of ecosystems has had many positive effects on human health and prosperity, but it necessarily involves trade-offs. Agriculture, for example, feeds and employs huge numbers of people. But the spread of agriculture into marginal lands may degrade soils and compromise ecosystem delivery of clean water, as when overuse of artificial fertilizers results in eutrophication. Extensive use of pesticides controls insect pests, but also reduces populations of pollinators and the services they provide to both crops and native plants.

Similarly, the loss of wetlands and other natural buffers has reduced the ability of ecosystems to regulate flooding and other natural hazards. The damage from the tsunami that hit Indonesia and other Southeast Asian countries in December 2004 was greater in many places than it would have been had the mangrove forests that protect the coast not been cut down and converted to cropland. Hurricane Katrina, which struck the U.S. Gulf Coast less than a year later, would not have caused as much flooding in New Orleans had the wetlands surrounding the city been intact. Katrina's devastating effects were due in part to a situation that had been developing for decades.

New Orleans is located on the Mississippi River delta. Much of the city lies below sea level, buffered by dams and levees constructed by the Army Corps of Engineers. The upstream dams that protect New Orleans from flooding also prevent the river from depositing the sediments that have sustained the surrounding delta wetlands for centuries. Oil and natural gas producers have cut thousands of small canals through those wetlands in order to lay pipelines and drilling rigs, and the extraction of oil and gas from beneath the land has caused it to sink. Increased dredging of shipping lanes and rising sea levels due to global warming have contributed to a rise in salinity, killing off many of the great cypress tree swamps. These extensive alterations resulted in the loss of over 80% (more than 1.2 million acres) of the delta wetlands between 1930 and 2005. By the time Katrina made landfall, those wetlands could no longer protect New Orleans from flooding. Storm surges raced along the paths carved by canals and shipping lanes to breach the levees, inundating much of the city.

As it flooded New Orleans, Hurricane Katrina raised awareness and appreciation of an ecosystem hitherto taken for granted by most people. Crucial not only for their flood control services, the delta wetlands provide winter habitat for some 70% of the migrating birds in the huge Mississippi Valley. They are also the spawning grounds for marine organisms, some of which are commercially valuable. The delta's famous shrimping industry contributes about 30% by weight of the total commercial fish harvest in the continental United States. The importance of coastal wetlands—and, indeed, of a wide range of other ecosystems—to human well-being mandates careful ecosystem management to guarantee a sustained flow of ecosystem goods and services.

## 58.4 RECAP

**Ecosystems provide human society with indispensable goods and services. Altering ecosystems can compromise their ability to provide these goods and services.**

- What are some of the essential goods and services that ecosystems provide to humans? See p. 1238
- How do human efforts to increase the provision of some ecosystem services cause degradation of other ecosystem services?

How can we meet the challenge of obtaining goods and services from ecosystems without compromising their ability to provide those goods and services over the long term? Fortunately, options exist for sustainable management of ecosystems.

## 58.5 How Can Ecosystems Be Sustainably Managed?

Practices that allow us to conserve or enhance ecosystems so as to benefit from specific ecosystem goods and services over the long term without compromising others are referred to as **sustainable**. In many cases, the total economic value of a sustainably managed ecosystem is higher than that of an intensively exploited ecosystem (**Figure 58.19**). The long-term economic benefits of preventing overexploitation of marine fish popula-

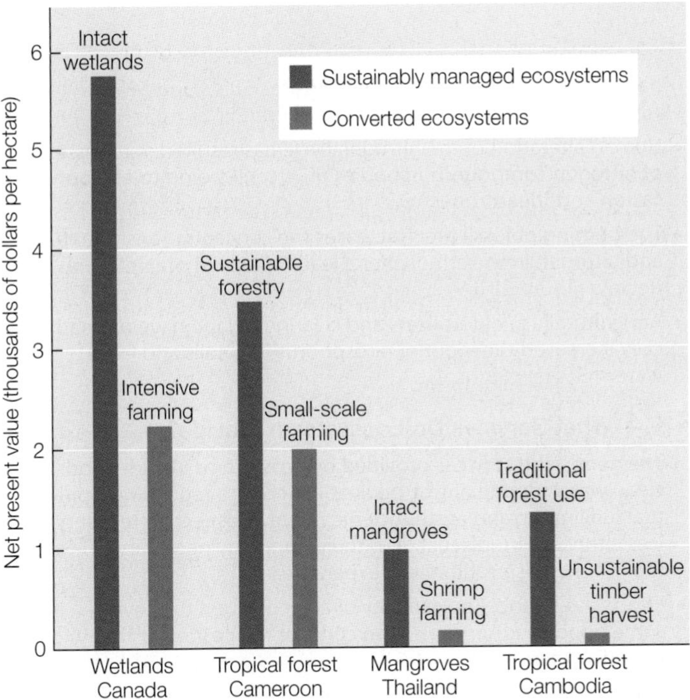


**58.19 Economic Values of Sustainably Managed and Converted Ecosystems** Many types of ecosystems can provide more goods and services when they are sustainably managed than when they are converted and intensively exploited.

tions, for example, are enormous. The collapse of the cod fishery on Georges Bank due to overfishing (see Figure 55.13) resulted in the loss of tens of thousands of jobs and cost at least $2 billion in income support and retraining, and it can take a long time for such fisheries to recover.

Impeding the establishment of policies that encourage sustainable management is the impression that ecosystem services are "public goods" with no market value. People who do not stand to profit from the goods and services provided by natural ecosystems have no incentive to pay for them, whereas individuals who own converted ecosystems can reap great economic benefits from them. As with other "public goods," government action may be needed to create incentives encouraging sustainable ecosystem management. Examples of such action might include:

- Elimination of subsidies that promote damaging exploitation of ecosystems. For example, the billions of dollars paid by governments in developed nations to subsidize domestic agriculture (with the hope of insulating farmers from economic risk) have led to greater food production than the global market warrants, promoted excessive use of fertilizers, and reduced the profitability of agriculture in developing countries.
- More sustainable use of fresh water could be achieved by charging users the full cost of providing the water, by developing methods to use water more efficiently, and by altering the allocation of water rights so that the incentives favor conservation.
- More sustainable use of marine fisheries could be achieved by establishing more protected marine reserves and "no-take" zones where fish can grow to reproductive age. Recent discussions of marine fisheries management center on what is colloquially known as the BOFFF hypothesis—for "Big Old Fat Female Fish"—that the largest and oldest females in a population are the most important to protect because they outproduce younger fish by an enormous margin.

Raising public awareness of endangered ecosystems is essential to the successful implementation of sustainable management programs. Most people do not understand the long-term value of ecosystem goods and services or realize how human activities affect the functioning of ecosystems, so public education programs are needed.

Maintaining and enhancing ecosystem goods and services that have no established market value is especially difficult. Probably the most difficult service to maintain in the face of increasingly intensive human use of the global ecosystem will be biological diversity. The final chapter of this book is devoted to this important topic.

# CHAPTER SUMMARY

## 58.1 What Are the Compartments of the Global Ecosystem?

- Earth is a closed system with respect to atomic matter, but an open system with respect to energy.
- Earth's physical environment can be divided into four compartments: atmosphere, oceans, fresh waters, and land. Review Figure 58.1
- Most global air circulation takes place in the lowest layer of the atmosphere, the **troposphere**. A layer of ozone in the **stratosphere** absorbs ultraviolet radiation. Review Figure 58.2
- Water vapor, carbon dioxide, and other **greenhouse gases** in the atmosphere are transparent to sunlight but trap heat, thus warming Earth's surface. Review Figure 58.3
- Most materials that cycle through the four compartments end up in the oceans, where they eventually sink to the bottom. Cold, nutrient-rich water rises and returns nutrients to the ocean surface in coastal **upwelling zones**. Review Figure 58.4
- Only a small fraction of Earth's water resides in the freshwater compartment, but water cycles rapidly through it. Nutrients reach lakes and streams by surface flow or by the movement of **groundwater**.
- **Turnover** brings nutrients and dissolved carbon dioxide to the surface and oxygen to the deeper waters of freshwater lakes. Turnover occurs regularly in temperate-zone lakes in spring in fall. Review Figure 58.5
- Land covers only one-fourth of Earth's surface but is connected to the atmospheric compartment by living organisms, which

remove chemical elements from the atmosphere and release them back.

## 58.2 How Does Energy Flow through the Global Ecosystem?

- Net primary production varies across the globe, reflecting differences in temperature, moisture, and the distribution of land masses. **Review Figures 58.6 and 58.7**
- Humans appropriate approximately 24% of Earth's average annual net primary production, although the percentage varies regionally.

## 58.3 How Do Materials Cycle through the Global Ecosystem?

- The pattern of movement of a chemical element through organisms and the compartments of the global ecosystem is its **biogeochemical cycle**.
- The **hydrologic cycle** is driven by the sun, which evaporates water from ocean surface than it returns by precipitation. The excess precipitation that falls on land eventually returns to the oceans, primarily in rivers. **Review Figure 58.8, ANIMATED TUTORIAL 58.1**
- Groundwater plays a minor role in the hydrologic cycle, but is being seriously depleted by human activities.
- Carbon is removed from the atmosphere by photosynthesis and returned to the atmosphere by metabolism. The rate at which carbon moves from the atmosphere to the oceans depends in part on photosynthesis by phytoplankton. **Review Figure 58.9, ANIMATED TUTORIALS 58.2 and 58.3**
- The concentration of $CO_2$ in the atmosphere has greatly increased in the last 150 years, largely due to the burning of **fossil fuels**. This buildup of $CO_2$ is warming the global climate. **Review Figures 58.10–58.12**
- As a result of agricultural use of fertilizers and the burning of fossil fuels, total nitrogen fixation by humans is about equal to natural nitrogen fixation. **Review Figures 58.13 and 58.14, ANIMATED TUTORIAL 58.4**
- Human alteration of the nitrogen cycle has resulted in excesses of nitrogen compounds in bodies of water, leading to **eutrophication** and "dead zones."
- The burning of fossil fuels increases the concentrations of sulfur and nitrogen in the atmosphere, leading to **acid precipitation. Review Figure 58.16**
- Agricultural use of fertilizers and clearing of land have dramatically increased the input of phosphorus into soils and fresh waters. **Review Figure 58.17**

## 58.4 What Services Do Ecosystems Provide?

- The goods and services provided by ecosystems include food, clean water, flood control, pollination, climate regulation, spiritual fulfillment, and aesthetic enjoyment. Most ecosystem services either are irreplaceable or the technology necessary to replace them is prohibitively expensive.
- Efforts to enhance the capacity of an ecosystem to provide some goods and services often come at the cost of its ability to provide others.
- The most important driver of change in ecosystems has been changes in land use, as natural ecosystems have been converted to other, more intensive uses.

## 58.5 How Can Ecosystems Be Sustainably Managed?

- In many cases, the total economic value of a sustainably managed ecosystem is higher than that of a converted or intensively exploited ecosystem. **Review Figure 58.19**
- Recognition of the value of ecosystem goods and services that are now perceived as "public goods" may induce government action to protect them. Public education is needed to make people aware of how much they benefit from ecosystem goods and services and how human activities affect the ecosystems that provide them.

**SEE WEB ACTIVITY 58.1 for a concept review of this chapter.**

# SELF-QUIZ

1. Unusual features of Earth as a planet that affect its ecosystem dynamics include
   *a.* lithospheric plates that move continuously.
   *b.* atmospheric gases that moderate surface temperatures.
   *c.* large amounts of water in liquid form.
   *d.* a diversity of living organisms.
   *e.* All of the above
2. Marine upwelling zones are important because
   *a.* they help scientists measure the chemistry of deep ocean water.
   *b.* they bring to the surface organisms that are difficult to observe elsewhere.
   *c.* ships can sail faster in these zones.
   *d.* they increase marine productivity by bringing nutrients back to surface ocean waters.
   *e.* they bring oxygenated water to the surface.
3. Which of the following is *not* true of the troposphere?
   *a.* It contains nearly all of the atmospheric water vapor.
   *b.* It contains a layer of ozone, which filters out ultraviolet light.
   *c.* It is about 17 kilometers thick in the tropics and subtropics.
   *d.* Most global air circulation takes place there.
   *e.* It contains about 80 percent of the mass of the atmosphere.
4. The hydrologic cycle is driven by
   *a.* the flow of water into the oceans via rivers.
   *b.* evaporation (transpiration) of water from the leaves of plants.
   *c.* evaporation of water from the surface of the oceans.
   *d.* precipitation falling on land.
   *e.* the fact that more water falls on the ocean as precipitation than evaporates from its surface.
5. Carbon dioxide is called a greenhouse gas because
   *a.* it is used in greenhouses to increase plant growth.
   *b.* it is transparent to heat, but traps sunlight.
   *c.* it is transparent to sunlight, but traps heat.
   *d.* it is transparent to both sunlight and heat.
   *e.* it traps both sunlight and heat.
6. Iron is an essential micronutrient that
   *a.* is insoluble in oxygenated water.
   *b.* is abundant in rocks and minerals.
   *c.* moves into the ocean via atmospheric dust.
   *d.* is a limiting factor for photosynthesis in oceans.
   *e.* All of the above

7. The biogeochemical cycle of phosphorus differs from the cycles of carbon and nitrogen in that
   *a.* phosphorus lacks an atmospheric component.
   *b.* phosphorus lacks a liquid phase.
   *c.* only phosphorus is cycled through marine organisms.
   *d.* living organisms do not need phosphorus.
   *e.* The phosphorus cycle does not differ importantly from the carbon and nitrogen cycles.
8. The sulfur cycle influences the global climate because
   *a.* sulfur compounds are important greenhouse gases.
   *b.* sulfur compounds help transfer carbon from the atmosphere to the oceans.
   *c.* sulfur compounds in the atmosphere are components of particles around which water condenses to form clouds.
   *d.* sulfur compounds contribute to acid precipitation.
   *e.* Both c and d
9. Acid precipitation results from human modifications of
   *a.* the carbon and nitrogen cycles.
   *b.* the carbon and sulfur cycles.
   *c.* the carbon and phosphorus cycles.
   *d.* the nitrogen and sulfur cycles.
   *e.* the nitrogen and phosphorus cycles.
10. Maintaining the capacity of ecosystems to provide goods and services is important because
   *a.* most ecosystem services cannot be replicated by any other means.
   *b.* replacing them with technological substitutes is possible only in developed nations
   *c.* technological substitutes take up valuable land.
   *d.* governments cannot function without taxing ecosystem services.
   *e.* It is not important. Humans could survive quite well even if ecosystem services declined greatly.

## FOR DISCUSSION

1. A string of powerful hurricanes struck the east coast of the United States over the course of a single year's hurricane season. Some people claim that this disaster was due to warming of the oceans caused by greenhouse gases in the atmosphere. Others assert that global warming is not responsible because hurricanes have occurred for many centuries. How would you evaluate these conflicting claims?
2. The waters of Lake Washington, adjacent to the city of Seattle, rapidly returned to their preindustrial condition when sewage was diverted from the lake to Puget Sound, an arm of the Pacific Ocean. Would all lakes being polluted with sewage clean themselves up as quickly as Lake Washington if sewage inputs were stopped? What characteristics of a lake are most important to its rate of recovery following reduction of nutrient inputs?
3. Tropical forests are being cut down and burned at a rapid rate to clear land for agriculture. Does this necessarily mean that deforestation is a major source of $CO_2$ inputs to the atmosphere? Why or why not?
4. What types of experiments would you conduct to assess the likely consequences of fertilization of the oceans with iron to increase rates of photosynthesis? At what spatial and temporal scales should they be conducted?
5. A government official authorizes the construction of a large coal-burning power plant in a former wilderness area. Its smokestacks discharge great quantities of combustion wastes. List and describe all the likely effects on ecosystems at local, regional, and global levels. If the wastes were thoroughly scrubbed from the stack gases, which of the effects you have just outlined would still happen?
6. Many nations have signed the Kyoto Accord, which commits them to reducing their emissions of $CO_2$ to the atmosphere. This agreement, however, is set to expire in 2012. In 2009, the United Nations drafted a new treaty to control greenhouse gases, proposing a goal of reducing global average annual greenhouse gas emissions to 2 metric tons per person (in 2006, average annual emissions per U.S. citizen were 19.78 tons). One mechanism for achieving this goal is called "cap and trade," whereby a government sets a "cap," or limit, on emissions by polluters, but allows facilities that emit less than their allowance to sell their excess credits to polluters who would otherwise exceed their allowance. What are the relative benefits and drawbacks you can see to such an approach to reducing emissions?

## ADDITIONAL INVESTIGATION

The experiment described in Figure 58.18 showed that interactions between elevated atmospheric $CO_2$ concentrations and nitrogen fixation can reduce the availability of other essential nutrients. The investigators suggested an expansion in the suite of elements that should be studied to determine whether elevated concentrations of $CO_2$ are likely to result in a large carbon sink in terrestrial ecosystems. What other elements should be considered, and how might their influence be investigated?

## WORKING WITH DATA (GO TO yourBioPortal.com)

**Atmospheric $CO_2$ and Nitrogen Fixation** This exercise offers further details and data from the experiments by Bruce Hungate and colleagues illustrated in Figure 58.18. You will be asked to analyze the data and come to your own conclusions about the effects of increasing atmospheric carbon dioxide on global fixation of nitrogen by green plants.

# 59 Conservation Biology

## No laughing matter

By the time the U.S. Congress passed the Endangered Species Act in 1973, Blackburn's sphinx moth (*Manduca blackburni*) was already believed to be extinct. The largest native insect in the Hawaiian Islands, with a wingspan of over 120 mm, this species was once widespread throughout the islands, occurring in dry forests from sea level up to 760 m elevation. As a caterpillar, it fed on native shrubs in the tomato family. Nobody knew exactly what led to its disappearance, but one suspect was a parasitoid wasp imported to the islands for the biological control of the tobacco hornworm (*Manduca sexta*), a close relative of Blackburn's sphinx moth that itself had been accidentally introduced to the islands and had become a pest in commercial tomato fields.

Miraculously, a population of Blackburn's sphinx moths was discovered on Maui in 1984, in an area used by the Hawaii National Guard for military training. Since that time, several other populations have been found on the island of Hawaii and on Kaho'olawe, an island adjacent to Maui. In February 2000, Blackburn's sphinx moth became the first Hawaiian insect species to be protected under the U.S. Endangered Species Act.

To date, no Blackburn's sphinx moths have been rediscovered on the islands of Kauai or Niihau. These islands, however, are home to another endangered species, the Hawaiian alula plant (*Brighamia insignis*). This odd-shaped shrub, one of several species called by the Hawaiian name of "haha," is found in dry forests ranging from sea level to 480 m on sheer cliffs and rocky outcrops. Its bulbous stem, typically 1–2 m long, produces a whorl of leaves at the top, along with clusters of yellow flowers with petals that are fused into a trumpet-shaped tube up to 14 cm in length. Today the species persists only with the assistance of dedicated botanists and conservation biologists who hand-pollinate the flowers, often rappelling down steep cliff faces to do so. Its most important (and possibly only) native pollinator likely went extinct.

The long, tubular flowers are typical of those pollinated by long-tongued moths in the hawk moth family. Short of traveling back in time, there is no sure way to identify the

**Extinction and Back** The large Blackburn's sphinx moth (*Manduca blackburni*) was believed to be extinct, possibly the victim of mortality due to a parasitoid wasp imported to Hawaii to control the tobacco hornworm (*M. sexta*)— another imported insect. A population of *M. blackburni* was rediscovered in 1984.

**One Extinction May Lead to Another** The endangered Hawaiian shrub alula (*Brighamia insignis*), also known as "haha," persists only because humans hand-pollinate the flowers; its natural pollinators may have gone extinct.

native pollinator of *B. insignis*, but evidence points to Blackburn's sphinx moth—whose existence today hangs by a thread on islands many kilometers away, where no alula plants remain. It is no coincidence that both species are endangered.

This contemporary example of the unanticipated effects of species loss is far from unique. Given what we've learned about species interactions, it's not surprising that the loss of one species from a community can result in the endangerment of another. While the effects of biodiversity losses can be difficult to predict, anticipating those effects is a high priority in view of today's escalating rates of species extinctions.

**IN THIS CHAPTER** we will describe the rapidly expanding field of conservation biology, which is devoted to protecting biodiversity. We will show how conservation biologists predict changes in biodiversity and see how some human activities are causing those changes. Finally, we will describe the strategies conservation biologists use to reduce extinction rates, protect species from threats to their persistence, and help populations of endangered species recover.

## CHAPTER OUTLINE



# 59.1 What Is Conservation Biology?

Virtually all natural ecosystems on Earth have been altered by human activities. Many habitats have disappeared completely, and many others have been greatly modified. Even Earth's climate and its great biogeochemical cycles have been altered. One consequence of these changes has been a rapid increase in the rate at which species go extinct. **Conservation biology** is an applied scientific discipline devoted to protecting and managing Earth's biodiversity. The discipline draws heavily upon principles of ecology, ethology, and evolutionary biology, particularly with respect to elucidating the factors that determine whether a given population will persist.

Early conservation efforts were characterized by tensions between those whose principal goal was to conserve natural resources for their economic benefits and those who believed that nature has intrinsic value independent of human economic interests. Today conservation biologists study the full array of goods and services that humans derive from species and ecosystems, including aesthetic and psychological benefits. Understanding the global ecosystem and the effects of human activities on that system is now understood to be essential to the long-term well-being of *Homo sapiens*.

Conservation biology is an *applied* scientific discipline, which is to say that it involves the practical application of knowledge to solve problems. Workers in this field are guided by three basic principles:

- *Evolution is the process that unites all of biology.* To be effective in protecting and managing biodiversity, it is essential to understand the evolutionary processes that generate and maintain it.
- *The ecological world is dynamic.* Because populations and communities change continuously over time, there is no static "balance of nature" that can serve as a goal of conservation activities.
- *Humans are a part of ecosystems.* Human interests and activities must be incorporated into conservation goals and practices.

### Conservation biology aims to protect and manage biodiversity

The term "biodiversity" has multiple meanings. We may speak of biodiversity as the degree of genetic variation within a species. Genetic variation can be measured as the number of alleles at a locus, the number of polymorphic loci in a genome, or the number of individuals in a population that are polymorphic at given loci. As we have seen throughout this book, genetic variation allows organisms to adapt to environmental change. Biodiversity can also be defined in terms of species richness in a particular community. At a larger scale, biodiversity also embraces ecosystem diversity—the complex interactions within and between ecosystems. While we may study these components of biodiversity separately, in life they are intimately interconnected.

The most conspicuous manifestation of biodiversity loss is species extinction. Extinction is a constant theme in the history of life; most of the species that have lived on Earth over the ages are extinct today. Consider, for example, the anaerobic organisms that were lost as early photosynthetic prokaryotes and eukaryotes added oxygen to Earth's atmosphere, as described in Section 25.2. Extinctions have occurred throughout Earth's history at what is referred to as a "background" rate, as changes in environmental conditions favor some species and negatively affect others. But the rate of extinctions taking place today rivals those of the five great *mass extinctions* of life's history (see Table 25.1 and Figure 25.12). Those historical extinction episodes were the result of cataclysmic natural disturbances, whereas the majority of modern extinctions can be attributed to effects of the human population.

Humans have a tremendous capacity to alter ecosystems and, accordingly, to cause extinctions. When humans first arrived in North America from Siberia about 14,000 years ago, they encountered a diverse and spectacular fauna of large mammals, including saber-toothed cats, dire wolves, mammoths, mastodons, giant ground sloths, and giant beavers (**Figure 59.1**). Most of this megafauna went extinct within a few thousand years after humans arrived. Although several hypotheses have been advanced to account for the geologically rapid, simultaneous disappearance of so many large animals, overhunting by humans is widely embraced as the principal cause. Losses of megafauna coinciding with the arrival of humans have also been documented in Australia and Hawaii.

As Chapter 25 described, several mass extinctions have occurred on Earth, each providing ecological opportunities for other groups, which subsequently underwent adaptive radiations. These extinction events occurred relatively far apart over the course of the history of life (see Figure 25.12). Over the past 400 years, increasing industrialization and urbanization have accelerated the rate of species extinctions astronomically. The renowned evolutionary biologist Edward O. Wilson estimates that Earth is losing on the order of 30,000 species per year, putting us in the midst of a sixth mass extinction. Protecting Earth's biodiversity requires maintaining the processes that generate new species as well as providing conditions that bring extinction rates closer to background levels.

**59.1 Extinct North American Megafauna** The extinction of some Pleistocene megafauna in North America may have been partially driven by the arrival of *Homo sapiens*. This reconstruction of California's La Brea Tar Pits shows a dire wolfe (*Canis dirus*, left) and a sabertooth tiger (*Smilodon* sp).

### Biodiversity has great value to human society

Conservation biologists are concerned about the escalating loss of Earth's biodiversity for many reasons:

- *Losing species can threaten the functioning of ecosystems.* The preceding chapters have all described the complex interactions among species. When species are lost, entire communities and ecosystems may change or be lost completely. When that happens, humans may lose the goods and services those ecosystems provide (see Section 58.4).
- *Humans depend on thousands of other species for food, fiber, and medicine.* Over 2,000 species of plants are used for fiber worldwide, and countless more have been domesticated as a source of food. In India alone, over 7,000 species of plants are utilized in traditional medicine, and in the United States more than a quarter of all medical prescriptions contain or are based on plant products. Hundreds of animal species also supply us with food, clothing, and medicine. In addition to goods provided, many organisms provide important services; a wide variety of microbes, for example, provide the fermentation services that render many foods more nutritious and palatable (see Section 51.2).
- *Humans derive enormous psychological benefits, including aesthetic pleasure, from interacting with other organisms.* Aesthetic benefits actually confer economic value on biodiversity. Trees growing on a residential lot, for example, can increase the lot's property value to a point where this added value may be greater than the lumber from the trees could provide.

- *Extinctions deprive the scientific community of opportunities to study and understand ecological relationships among organisms.* The more species that are lost, the more difficult it will be to understand the structure and functioning of ecological communities and ecosystems.
- *Living in ways that cause the extinction of other species raises ethical issues.* Losses of biodiversity increasingly concern philosophers, ethicists, and religious leaders because in such circles species are judged to have intrinsic value.

All of these concerns, to varying degrees, may be integrated by conservation biologists into strategies for protecting biodiversity.

## 59.1 RECAP

**Conservation biology is an applied scientific discipline aimed at protecting and managing biodiversity, which is rapidly decreasing due to extinctions caused by human activities.**

- Explain the distinctions among the multiple meanings of the term "biodiversity." See pp. 1243–1244
- What are some of the ways in which biodiversity is valuable to humans? See pp. 1244–1245

To protect biodiversity, conservation biologists must understand biodiversity as it exists today as well as the ways in which it is changing. Thus one important enterprise for conservation biologists is to predict rates of species extinctions.

# 59.2 How Do Biologists Predict Changes in Biodiversity?

How many, and which, species will go extinct will depend both on human activities and on natural events. Conservation biologists attempt to track the extinctions that are occurring and to predict the number that will occur during the coming century.

### Our knowledge of biodiversity is incomplete

Predicting extinctions is difficult for a number of reasons. First, we do not know how many species are living on Earth today. Many of the species that may go extinct in the near future have not even been named and described. Insects provide a case in point: although over 900,000 species have been described, estimates of the number of species yet to be discovered range from about 2 million to more than 50 million (see Section 32.4). Even in the case of larger organisms, our understanding of biodiversity is far from complete. For example, in an 18-month period in 2005–2006, over 50 species of animals and plants previously unknown to science were discovered in rainforests in Borneo.

Second, we do not know where species live. The ranges of most described species, particularly those that are small, reclusive, and rare to start with, are poorly known. One tiny North American true bug, *Corixidea major* (so rare that it has no common name), had been found in only one location near Clarksville, Tennessee, until entomologists collecting insects attracted to lights at night discovered it in Virginia and Florida, extending its known range by over 1,000 km.

Third, it is difficult to determine whether a species is truly extinct. Rarely is the death of the last surviving member of a species recorded with certainty, as it was in the case of the last passenger pigeon (*Ectopistes migratorius*), a female named Martha who died in the Cincinnati Zoo on September 1, 1914. The status of rare, reclusive species with poorly known life histories is much more difficult to determine, as has been the case with the ivory-billed woodpecker (*Campephilus principalis*) in the United States (**Figure 57.2A**). Pygmy tarsiers (*Tarsius pumilus*, tiny primates weighing less than 60 grams) were thought to have gone extinct from their native cloud forests on the island of Sulewesi in Indonesia. In 2008—85 years after the last live specimen had been sighted—a research team from Texas A&M University discovered individuals of this species living in one of the island's national parks (**Figure 57.2B**).

Fourth, we rarely know all of the connections among species. At the beginning of this chapter, we saw that the extinction of a hawk moth species put at risk the survival of a plant species dependent upon the hawk moth as a pollinator. How many other species have been placed at risk by the extinction of the hawk moth is unknown because the ecological associations of this species were not thoroughly documented before its extinction.

(A) *Campephilus principalis*

(B) *Tarsius pumilus*

**59.2 Is It Really Extinct?** (A) The ivory-billed woodpecker (shown here in a nineteenth-century John James Audubon print) was presumed to be extinct until reports of sightings in 2004. Clear proof of the living bird has so far eluded watchers. (B) Among vertebrates, pygmy tarsiers were recently rediscovered in Indonesia after having been declared extinct.

### We can predict the effects of human activities on biodiversity

Despite these gaps in our understanding of biodiversity, methods exist for estimating probable rates of extinction resulting from human activities. To make such predictions, conservation biologists have applied the principles of the species–area relationship and the theory of island biogeography, which we discussed in Section 57.3. When they have measured the rate at which species richness decreases with decreasing habitat patch size, they have found that, on average, a 90 percent loss of habitat area results in the loss of half of the species that live in and depend on that habitat.

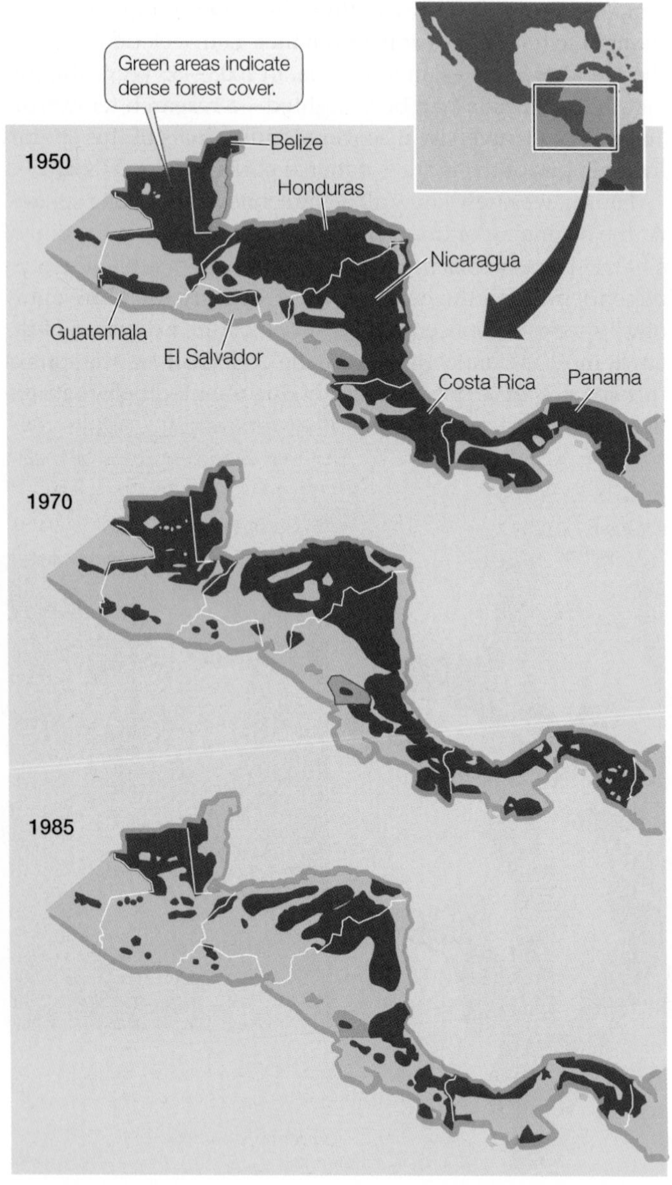


**59.3 Deforestation Rates in Tropical Forests** Less than half the tropical forest that existed in Central America in 1950 remained as of 1985, most of it in small patches. Although the rate of deforestation slowed somewhat after 1985, by 2005 an additional 19 percent had disappeared.

Similar calculations can be made for the total global area of a habitat type. The current rate of loss of tropical evergreen forest—Earth's most species-rich biome—is about 2 percent of the remaining forest each year, due to the increasing demand for forest resources by a rapidly expanding human population (**Figure 59.3**). If this rate of loss continues, at least 1 million species that live in tropical evergreen forests could become extinct this century.

To estimate the risk that a particular population will become extinct, conservation biologists develop statistical models that incorporate information about a population's size, its genetic variation, and the morphology, physiology, and behavior of its members. Species in imminent danger of extinction in all or most of their range are labeled *endangered*. *Threatened* species are those that are likely to become endangered in the near future.

Species with small populations face particular risks. Rarity itself is not always a cause for concern, because some species that live in unusual habitats have probably always been rare and are well adapted to being rare. However, species whose populations suddenly shrink rapidly—the "newly rare"—are usually at high risk, as such reductions in population size can lead to genetic drift and loss of genetic variation. Species with specialized habitat or dietary requirements are more likely to become extinct than species with more generalized requirements.

In addition, populations reduced to a small size or confined to a small range can easily be eliminated by local disturbances. For example, numbers of the Cozumel thrasher (*Toxostoma guttatum*), a member of the mockingbird family known only on the island of Cozumel off the coast of Mexico, had been declining for years due to a combination of factors (including the introduction of bird-eating boa constrictors to the island in 1971). A series of hurricanes, beginning with Hurricane Gilbert in 1988, had a catastrophic effect on the remaining populations. Surveys in 2006 failed to document any surviving individuals, and the species may now be extinct.

## 59.2 RECAP

**Predicting changes in biodiversity is difficult because our knowledge of biodiversity is incomplete. The species–area relationship can be used to predict rates of extinction in areas that are subject to habitat loss.**

- What are some of the gaps in our current knowledge of biodiversity? **See p. 1245**
- Why are species with rapidly shrinking populations especially vulnerable to extinction? **See p. 1246**

Now that we've examined some of the factors that place species at risk of extinction, let's see how human activities contribute to those risks. We begin with habitat loss, the primary threat to species persistence today.

## 59.3 What Factors Threaten Species Persistence?

Human activities that threaten the persistence of species include habitat alteration, introduction of exotic species, overexploitation, and climate alteration. Conservation biologists determine how these activities affect species and use that information to devise strategies to protect species that are endangered or threatened.

### Species are endangered by the degradation, destruction, and fragmentation of their habitats

Loss of suitable habitat by degradation, fragmentation, or destruction is the most important cause of species endangerment in the United States (**Figure 59.4**) and around the world. Many habitats—particularly freshwater habitats—are degraded by pollution. Toxic substances released by human activities have negative effects on the behavior, reproduction, and development of species, reducing both survivorship and competitive ability. Habitat loss can also occur through outright removal. Physical destruction of particular habitat types (such as tropical rainforest) eliminates species that cannot survive anywhere else. Habitat loss also affects nearby habitats that are not destroyed. As portions of a habitat are progressively lost to human activities, the remaining habitat becomes increasingly fragmented: habitat patches become smaller and more isolated.

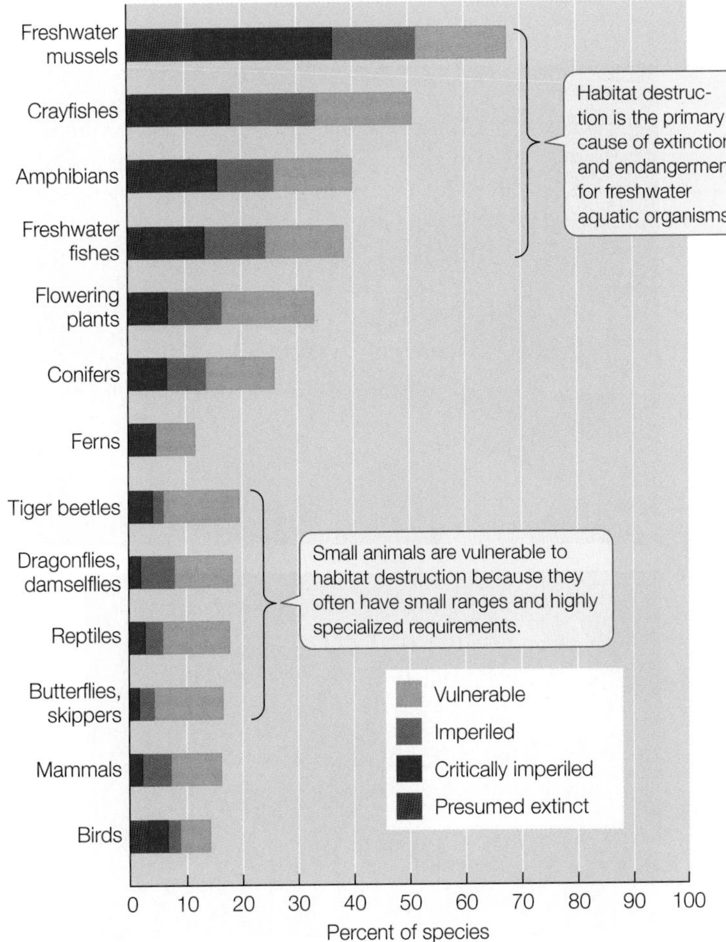
Freshwater mussels
Crayfishes
Amphibians
Freshwater fishes
Flowering plants
Conifers
Ferns
Tiger beetles
Dragonflies, damselflies
Reptiles
Butterflies, skippers
Mammals
Birds
Habitat destruction is the primary cause of extinction and endangerment for freshwater aquatic organisms.
Small animals are vulnerable to habitat destruction because they often have small ranges and highly specialized requirements.
Vulnerable
Imperiled
Critically imperiled
Presumed extinct
0 10 20 30 40 50 60 70 80 90 100
Percent of species

**59.4 Proportions of U.S. Species Extinct or Threatened** The species most threatened with extinction all live in freshwater and wetlands habitats, which are highly vulnerable and have been extensively degraded, polluted, and even destroyed for human development.

Small habitat patches are qualitatively different from larger patches of the same habitat in ways that affect species persistence. Small patches cannot maintain populations of species that require large areas, and they support only small populations of those species that can survive in small patches. In addition, the fraction of a patch influenced by external factors increases rapidly as patch size decreases (**Figure 59.5**). Close to the edges of a forest patch, for example, winds are stronger, temperatures are higher, humidity is lower, and light levels are higher than they are farther inside the forest. Species from surrounding habitats often colonize edges to compete with or prey upon the species living there. The effects of such factors are known as **edge effects**.

One effect of forest fragmentation in midwestern North America has been an increase in the abundance of the brown-headed cowbird (*Molothrus ater*). This bird is a brood parasite—that is, it lays its eggs in the nests of other bird species. When a cowbird egg hatches, the hatchling is raised by the host parents, often to the detriment of their own young. Historically, cowbirds followed bison and other grazing mammals, feeding on insects kicked up by the herds; thus their eggs were laid primarily in nests of grassland host species. Forest fragmentation, however, opened up new opportunities for the cowbirds, which can now lay their eggs in the nests of forest birds in forest edges. Fragmented forests, with relatively more edge than intact forests, thus favor the proliferation of cowbirds at the expense of forest species.

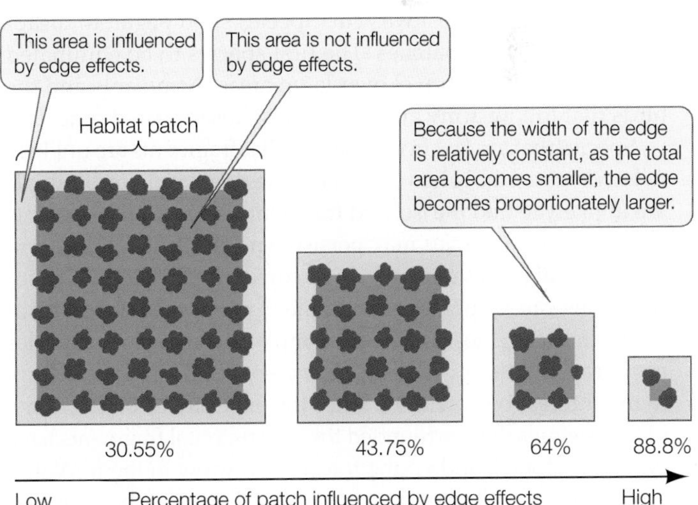
This area is influenced by edge effects.
This area is not influenced by edge effects.
Habitat patch
Because the width of the edge is relatively constant, as the total area becomes smaller, the edge becomes proportionately larger.
30.55%
43.75%
64%
88.8%
Low
Percentage of patch influenced by edge effects
High

**59.5 Edge Effects** The smaller a patch of habitat, the greater the proportion of that patch that is influenced by conditions in the surrounding environment.

yourBioPortal.com
GO TO Animated Tutorial 59.1 • Edge Effects

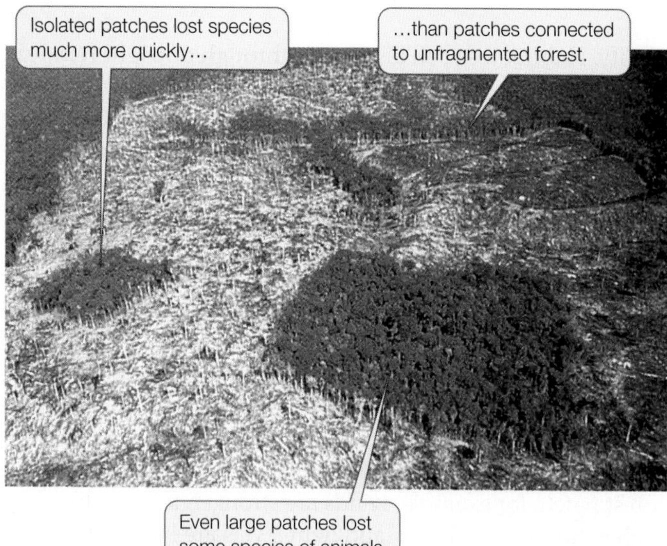
Isolated patches lost species much more quickly...
...than patches connected to unfragmented forest.
Even large patches lost some species of animals.

**59.6 Species Losses in Brazilian Forest Fragments** Biologists studied patches of tropical evergreen forest near Manaus, Brazil, before and after they were isolated by forest clearing. The results of the study demonstrated that small, isolated habitat fragments support fewer species than larger patches of the same habitat.

Because so many habitats have already undergone fragmentation by the time they are first investigated by ecologists, determining the effects of fragmentation on the original communities can be difficult. Timely surveys can provide some of this information. For example, a major research project was launched in 1979 in tropical evergreen forest near Manaus, Brazil, that was slated for conversion to pasture. The landowners agreed to preserve forest patches of certain sizes and configurations laid out by biologists (**Figure 59.6**). The biologists counted the species in the future "patches" while they were still part of the continuous forest, then monitored the patches after the surrounding forest was cut. Species soon began to disappear from isolated patches. The first species to be eliminated were monkeys that travel over large areas. Army ants and the birds that follow army ant swarms also disappeared quickly.

Species that are lost from small habitat fragments are unlikely to become reestablished there because dispersing individuals are unlikely to find the isolated fragments. As Section 55.4 points out, however, a species may persist in a small patch if it is connected to other patches by habitat corridors through which individuals can disperse. Among the experimental forest fragments in Brazil, those that were completely isolated lost species more rapidly than did those that were connected to unfragmented forest by corridors. Since the experiment began, some of the pastures that surrounded the experimental fragments have been abandoned, and young forests now grow in them. Within 7–9 years of abandonment, army ants and some of the birds that follow them recolonized forest fragments connected to larger forest patches by young forests. Other birds that forage in the forest canopy also reestablished themselves. Young forest is not a suitable permanent habitat for most of these species, but they can disperse through it to find more suitable habitat.

Insight into the importance of corridors has led to new conservation initiatives, among the most notable of which is the Yellowstone to Yukon initiative. A joint Canada–United States not-for-profit organization has as its goal sustainable management of the mountain ecosystem extending from Yellowstone National Park to the Yukon Territory. This stretch of land, the largest intact ecosystem of its kind on the planet, contains high-quality habitat for many of North America's most imperiled animals, including grizzly bears, gray wolves, lynx, and native fish. Managing the entire region not only provides corridors to connect habitats, it also provides room for populations to shift in response to global climate change.

### Overexploitation has driven many species to extinction

Although habitat loss presents a greater threat to more species today than does overexploitation, many species are still threatened by overexploitation today. Elephants and rhinoceroses are at risk in much of Africa and Asia because poachers kill them for their tusks and horns, which are used for ornaments and knife handles; in addition, some men believe that powdered rhinoceros horn boosts their sexual potency.

The principal threat to the continued survival of tigers, whose numbers have declined by over 90 percent over the past

(A)

(B)

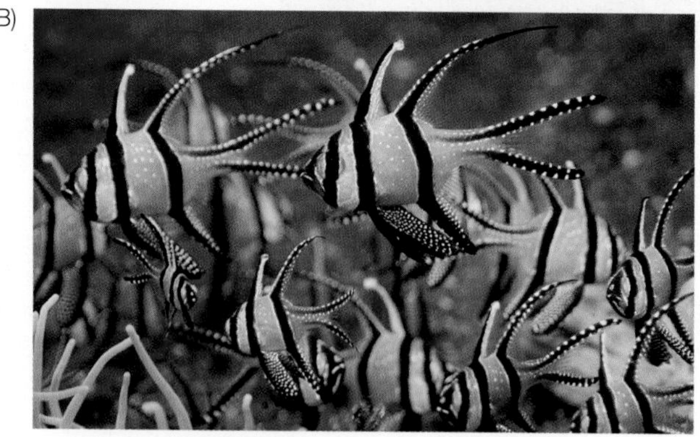

**59.7 Endangered by Exploitation** (A) The principal threat to the continued survival of tigers is the use of their body parts in Asian traditional medicine. (B) The endangered Banggai cardinalfish of Indonesia's coral reefs has been overexploited because of its value to the private aquarium trade.

century, is the use of their body parts in traditional medicine—bones to treat rheumatism, eyes to cure epilepsy, and penises to enhance virility (**Figure 59.7A**). In 2009, a bowl of tiger penis soup could be obtained for $300 in Taiwan. There is some hope that the development of drugs for treating erectile dysfunction will reduce the incidence of poaching of these endangered species, but as yet the drugs are not widely available in Asia, where demand for animal aphrodisiacs is greatest.

Massive international trade in exotic pets and aquarium fishes, ornamental plants, and tropical forest hardwoods has also decimated many species. The Banggai cardinalfish *Pterapogon kauderni* (**Figure 59.7B**) is on the brink of extinction entirely due to the pet trade; almost a million of these critically endangered fish are hauled out of the waters near Sulawesi, Indonesia, to satisfy the demand of saltwater aquarium enthusiasts.

## Invasive predators, competitors, and pathogens threaten many species

As people travel, they deliberately or inadvertently move species to regions outside their original range. Some of these exotic (non-native) species become **invasive**—that is, they reproduce rapidly, spread widely, and have negative effects on the native species of the region. As we saw in Section 55.3, species that are introduced into a region where their natural enemies are absent may reach very high population densities. Moreover, the native species in an invader's new range may not have evolved specific defenses against these new antagonists and competitors.

Invasive species are spread in a number of ways. Marine organisms have been spread throughout the oceans by ballast water from ships, taken on at the port of departure and discharged at the destination port along with its content of surviving animals and plants. The notorious zebra mussel (*Dreissena polymorpha*) is thought to have arrived in North America in this way (see Figure 55.9). The brown tree snake (*Boiga irregularis*; **Figure 59.8**) arrived on Guam in air cargo shortly after World War II. Until then, the only snake on Guam was a tiny, insect-eating species. Although brown tree snakes remained rare for 20 years, in the 1960s they began to multiply and today can be found at densities as high as 5,000 individuals per $km^2$. The snake has exterminated 15 species of land birds, including 3 found only on Guam.

Over the past 400 years, Europeans colonizing new continents deliberately introduced plants and animals to their new homes in an effort to reconstruct their familiar surroundings. Many of these introductions have had disastrous effects on native flora and fauna. In Australia, the introduction of European rabbits and foxes for sport hunting and of dogs and cats as pets led to the extermination of nearly half the small to medium-sized native marsupials over the last 100 years, by a combination of competition with rabbits and predation by foxes, dogs, and cats. Some species deliberately introduced to control other invasive species have themselves become invasive and caused further problems. One example was the case of *Manduca sexta* and its wasp parasite in Hawaii, mentioned at the opening of this chapter; another such case is that of the cane toad (*Bufo marinus*), introduced in Australia to control sugarcane pests (see Figure 55.14).

**59.8 An Agent of Extinction** Since it was accidentally introduced onto the tiny Pacific island of Guam, the brown tree snake (*Boiga irregularis*) has eaten15 species of land birds to extinction.

Invasive plants may also have negative effects on ecosystems. While native plants must devote considerable energy and resources to defending themselves against native herbivores, invasive plants are less prone to attack, in part because their natural enemies have been left behind in their home range. Therefore, invasive plants can devote their resources to growth and reproduction rather than to producing defensive secondary compounds.

Introduced pathogens have also wreaked havoc among native species, as exemplified by the effects of avian malaria in the Hawaiian Islands. Before the arrival of Europeans, no mosquitoes existed anywhere in the islands. The first mosquito species was found there in 1827, and over the next century several others followed. At the start of the twentieth century, the microbial pathogen that causes avian malaria arrived, most likely carried by imported caged birds. Having had no exposure to malaria over the course of their evolutionary history, Hawaii's many endemic bird species were exceptionally vulnerable to infection. Today, nearly all species living below 1,500 m elevation—the current upper limit of the range of the mosquito vectors—have been eliminated by avian malaria. Species living at higher elevations have fared better, but the range of the mosquitoes appears to be expanding upward as the climate warms, placing the surviving endemic species at risk.

## Rapid climate change can cause species extinctions

Human emissions of greenhouse gases are causing global climate warming, as we saw in Section 58.3. Average temperatures in North America, for example, are expected to increase 2°C–5°C by the end of the twenty-first century. If the climate warms to that extent, then the average temperature found at any given location in North America today will be found 500–800 km to the north, and species adapted to particular temperatures will have to shift their ranges by that distance, or evolve new temperature adaptations within a single century. Some habitats, such as alpine tundra, could disappear entirely as temperate forests expand up mountain slopes.

Conservation biologists cannot alter rates of global warming, but they can predict how the resulting climate changes may affect organisms and look for ways to mitigate those effects. Their research activities include analyses of past climate changes and studies of sites currently undergoing rapid climate change. It would be helpful to know, for example, how rapidly species' ranges shifted after the end of the most recent ice age, about 10,000 years ago. Which species did and did not keep pace with climate change? How much, and in what ways, did past ecological communities differ from those of today as a result of differences in the rates at which species ranges shifted?

Species that can disperse easily, such as most birds, may be able to shift their ranges as rapidly as the climate changes, provided that appropriate habitats can be found. However, the ranges of other species are likely to shift more slowly. After the glaciers retreated in North America about 8,000 years ago, for example, the ranges of some coniferous trees such as pine trees, with lightweight seeds that can be carried great distances by wind, expanded northward, so that today they grow as far north as the current climate permits (**Figure 59.9**). Earthworms, on the other hand, fared less well: the glaciers eliminated all earthworm species in Canada and the northern United States, and they have not been replaced by other North American species, which have moved their ranges northward only very slowly. (The earthworms found in many parts of North America today are actually exotic species accidentally introduced from Europe.)

If Earth's surface warms as predicted, entirely new climates will develop, especially at low elevations in the tropics, where a warming of even 2°C would result in conditions warmer than those found anywhere in the humid tropics today. Adaptation to those climates may prove difficult even for many tropical organisms. Since the mid-1980s, the average minimum nightly temperature at the La Selva Biological Station, in the Caribbean lowlands of Costa Rica, has increased from about 20°C to 22°C. On warmer nights, trees use more of their energy reserves just to maintain themselves. As a result, even this small rise in temperature has reduced the average growth rates of six different tree species by about 20 percent.

## 59.3 RECAP

**A number of human activities threaten the persistence of species, including habitat degradation, fragmentation, and destruction; overexploitation; introductions of invasive species; and activities that cause rapid climate change.**

- Describe three ways in which habitat loss is occurring today. See pp. 1247–1248
- Why are extinction rates high in small habitat patches? See pp. 1247–1248 and Figures 59.5 and 59.6
- How can dispersal ability and climate change interact to affect the probability of extinction? See pp. 1249–1250

Demonstrating that species are endangered is an empty exercise if we cannot implement a plan of action to save them. In the next section we consider some of the positive steps that can be taken to protect biodiversity.

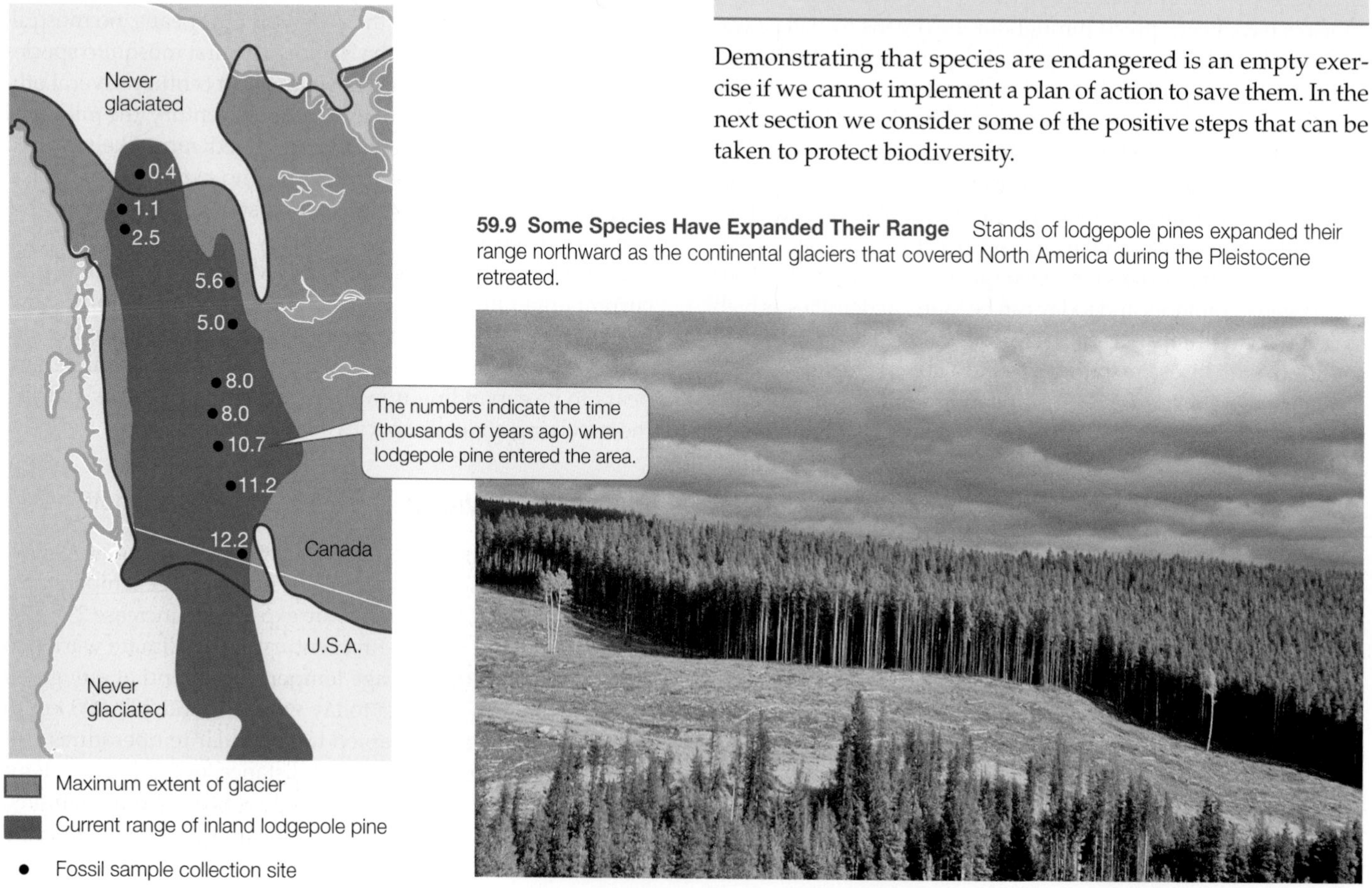

The numbers indicate the time (thousands of years ago) when lodgepole pine entered the area.

Maximum extent of glacier

Current range of inland lodgepole pine

Fossil sample collection site

**59.9 Some Species Have Expanded Their Range** Stands of lodgepole pines expanded their range northward as the continental glaciers that covered North America during the Pleistocene retreated.

## 59.4 What Strategies Do Biologists Use to Protect Biodiversity?

Conservation biologists use scientific theory, empirical data, and tools from a variety of disciplines to help protect endangered and threatened species and ecosystems. They identify the factors that present risks to species and use that information to devise action plans. Implementing those plans, however, often requires the cooperation of many different groups of people, so conservation biologists also work with landowners, politicians, lawyers, environmental activists, and the general public. Let's look at some of the actions that conservation biologists take to protect biodiversity.

### Protected areas preserve habitat and prevent overexploitation

The establishment of *protected areas* in which habitat alteration and exploitation are restricted or prohibited is an important component of efforts to conserve biological diversity. Protected areas preserve habitat and may serve as nurseries from which individuals can disperse into exploited areas, replenishing populations that might otherwise become extinct.

Deciding which areas to protect is a challenging enterprise. Two robust criteria are the total number of species living in an area (its *species richness*; see Chapter 57) and the number of *endemic species* in an area (a measure of its uniqueness). Using these criteria, biologists have identified a number of biodiversity "hotspots" of unusual richness and endemism (**Figure 59.10**). Hotspots occupy only 15.7 percent of Earth's land surface but are home to 77 percent of its terrestrial vertebrate species. Most hotspots are also areas of high human population density, so habitat loss is ongoing. Developing a conservation strategy for any of these regions requires not only a detailed analysis of the distributions of species and the locations of special resources (such as caves, freshwater springs, or migratory

(A) Tropical rainforest hotspots

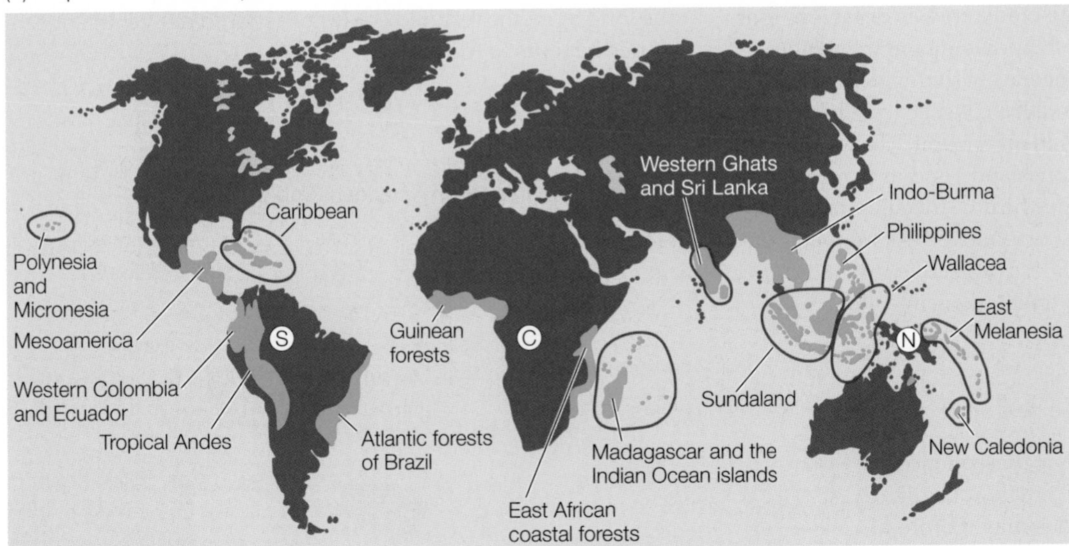


(B) Hotspots in other biomes

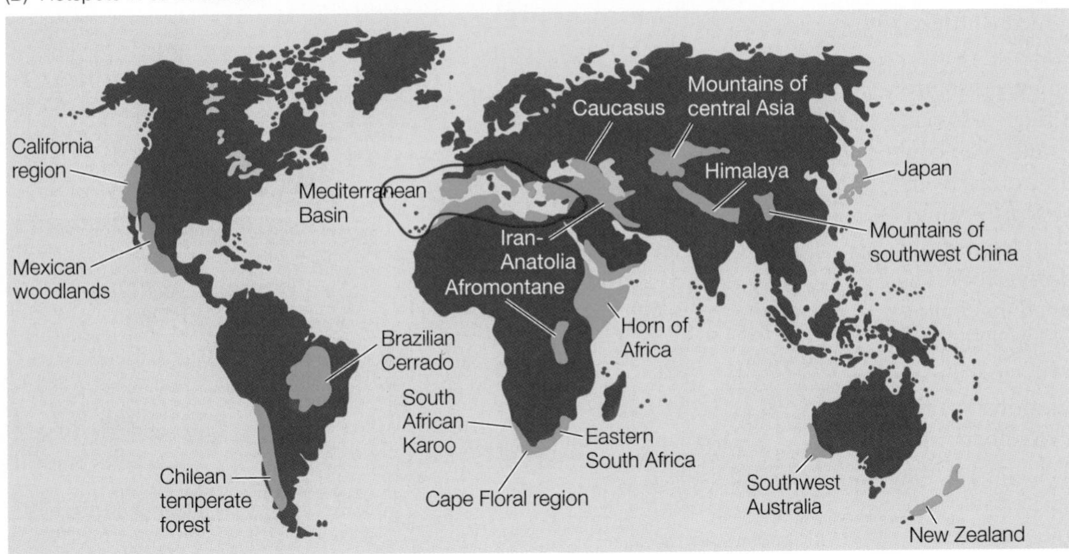


**59.10 Hotspots of Biodiversity** (A) Almost half of the world's highest priority "hotspots" for biodiversity protection are regions of tropical rainforest habitat. Areas circled in red encompass island groups. The circled letters represent the only three remaining areas of extensive unbroken rainforest: S, South American Amazonia; C, the Congo Basin of Africa; and N, the island of New Guinea. (B) Eighteen hotspots representing other (non-rainforest) ecosystems.

"Centers of imminent extinction" are concentrated in tropical forests, islands, and mountainous regions.

**59.11 Centers of Imminent Extinction** The areas shown in yellow include 595 "centers of imminent extinction." These regions are home to 794 species that are at serious risk of extinction.

stopover areas for birds), but also an analysis of factors that threaten and factors that support biodiversity in the region.

In an effort to further pinpoint sites with threatened species that are found nowhere else, in 2005 conservation biologists identified 595 "centers of imminent extinction." These sites are concentrated in tropical forests, on islands, and in mountainous regions (**Figure 59.11**). Only one-third of the sites are legally protected, and most of them are surrounded by land that is undergoing rapid development. Unless protective actions are taken soon, species extinctions are inevitable.

## Degraded ecosystems can be restored

When a species is endangered as a consequence of ecosystem degradation rather than outright habitat loss, protecting the species may require restoring the habitat to a more natural state. Practitioners of **restoration ecology** are developing methods aimed at just such reconstitution, because many degraded ecosystems recover only slowly, if at all, without human assistance.

Because the soil that supports them is so rich, grasslands all over the world have been converted to agriculture. By the middle of the twentieth century, for example, most North American prairies had been converted to cropland or were heavily grazed by domestic livestock. The herds of large mammals that roamed the prairies when European settlers arrived have been reduced to tiny remnants confined to small areas. Most of these remaining populations are too small to maintain their genetic diversity or play their original ecological roles. The species *have* survived, however, so opportunities exist to reintroduce them if their habitat can be restored.

A major prairie restoration project is under way in northeastern Montana. When Lewis and Clark mapped this region 200 years ago, they saw large herds of bison, elk, deer, and pronghorn, as well as abundant populations of their predators. The goal of the restoration project, run by the World Wildlife Fund and the American Prairie Foundation in cooperation with public land managers, is to restore the native prairie and its fauna in a 15,000-km$^2$ area near the Missouri River (**Figure 59.12**).

This ambitious project is feasible for three reasons. First, the private land in the area is owned by a small number of ranchers, each of whom owns extensive grazing leases on public lands administered by either U.S. federal agencies or the State of Montana. Second, most of the land has never been plowed, so native vegetation can recover rapidly when grazing pressures are reduced. Third, the area's human population is decreasing. The ranchers are aging, and their children are abandoning the hard work and uncertain profits of ranching for careers in urban settings. Once a free-ranging herd of several thousand bison and large numbers of elk—along with their predators (wolves)—has been established, nature-minded tourists are expected to flock to the area to view the wildlife spectacle. Over the long term, the restored ecosystem should deliver major economic benefits to the region.

**59.12 Restoring a North American Prairie** (A) A major prairie restoration project is underway north of the Missouri River in the state of Montana (yellow area). (B) Native prairie dogs maintain the vegetation by digging extensive burrows and clipping plants. (C) The first bison were reintroduced to the area in 2005.

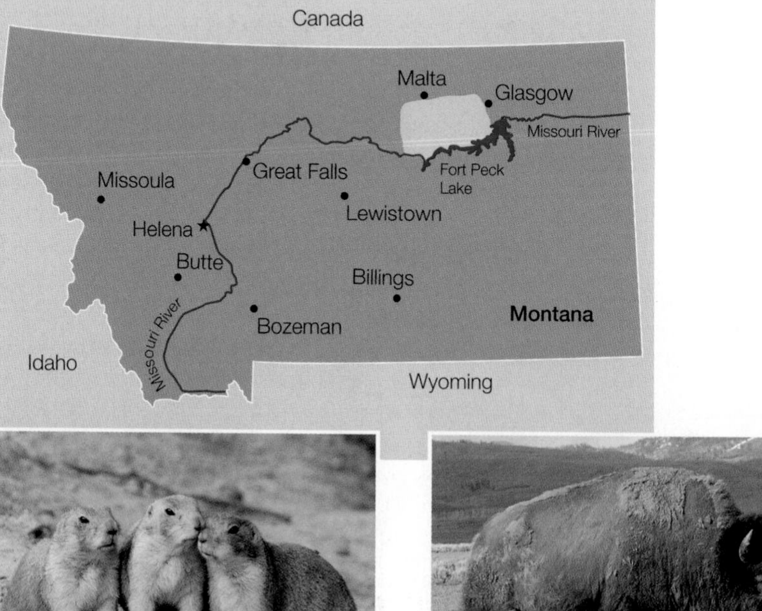


(B) *Cynomys ludovicianus*

(C) *Bison bison*

**59.13 A Wetlands Laboratory** Tijuana Estuary, near San Diego, is a shallow-water wetlands habitat. Experiments at this natural research reserve shed light on ways to restore this valuable ecosystem.

In the United States, the notion that humans are capable of creating functioning replacement ecosystems underlies policies that allow developers to destroy habitats. Destruction of wetlands in particular is often permitted because of developers' assertions that they can be replaced. However, creating new wetlands requires detailed ecological knowledge that almost invariably far surpasses that which is currently available.

In southern California, where 90 percent of the coastal wetlands have been destroyed, wetland restoration is a high priority. Species have been lost from degraded coastal wetlands, so restoration requires species introductions. Early attempts, in which one or two common wetland species were introduced, did not succeed—other wetland-associated species failed to recolonize the "rehabilitated" wetlands. To understand why, conservation biologists established a large field experiment at the Tijuana Estuary (**Figure 59.13**). Here they found that experimental plots planted with species-rich mixtures developed a more complex vegetation structure (which is important to insects and birds) and also accumulated nitrogen (required for plant growth) faster than did species-poor plots (**Figure 59.14**).

### Disturbance patterns sometimes need to be restored

Many species depend on particular patterns of disturbance—such as fires, windstorms, or grazing—to maintain their populations (see Section 57.4). Recognition of the need for periodic disturbance to maintain healthy ecosystems is a rather new dimension of conservation biology. For example, although many

## INVESTIGATING LIFE

**59.14 Species Richness Enhances Wetland Restoration**

When biologists attempted to restore degraded areas of the Tijuana Estuary, they found that several measures of wetland ecosystem health improved more rapidly in species-rich than in species-poor plantings.

**HYPOTHESIS** **When attempting to restore degraded wetlands, planting a mixture of species will approach the ecosystem's original condition more rapidly than planting a single species.**

**METHOD**

1. Delimit experimental plots of degraded wetland habitat, leaving control plots unplanted (i.e., no human intervention).
2. Plant some experimental plots with a single plant species typical of wetlands in the region. Plant other plots with randomly chosen assemblages of 3 and of 6 typical species. Plant the same density of seedlings in all plots.
3. Over the subsequent 18 months, replant and weed the experimental tracts as necessary to compensate for early mortality of seedlings.

**RESULTS** In the plots with higher species richness, more of the ground was covered by plants, the vegetation structure was more complex, and more nitrogen accumulated in the soil.

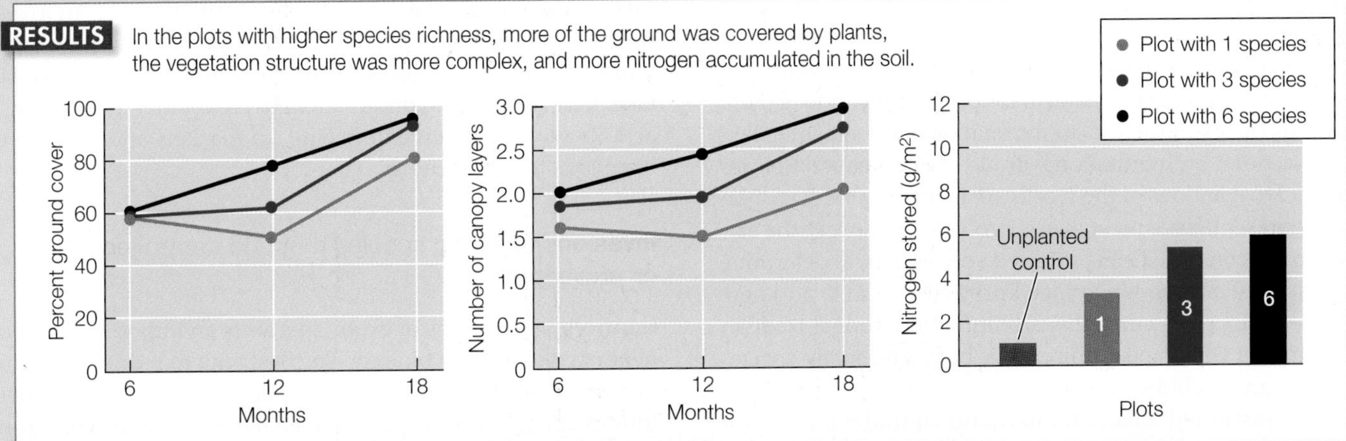


**CONCLUSION** **Planting a rich mixture of species should enhance wetland restoration attempts.**

Go to **yourBioPortal.com** for original citations, discussions, and relevant links for all INVESTIGATING LIFE figures.

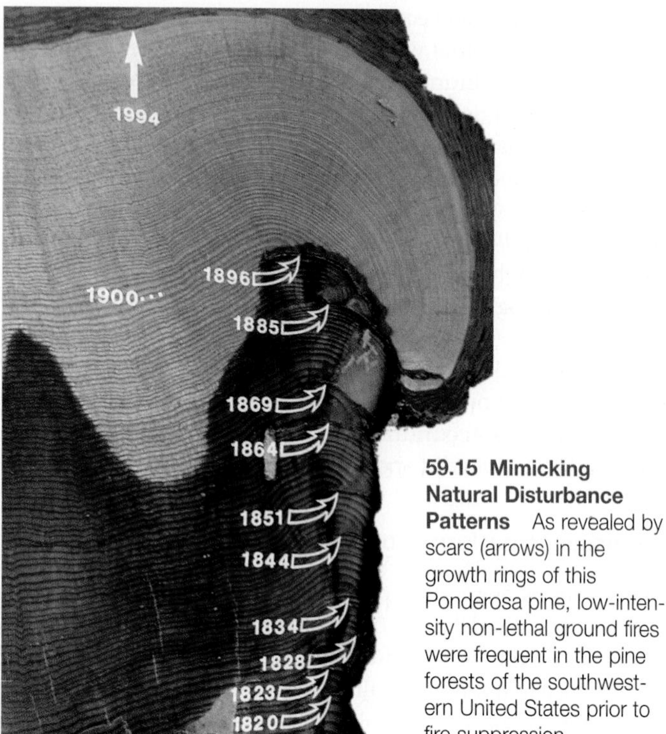


**59.15 Mimicking Natural Disturbance Patterns** As revealed by scars (arrows) in the growth rings of this Ponderosa pine, low-intensity non-lethal ground fires were frequent in the pine forests of the southwestern United States prior to fire suppression.

plant species require periodic fires for successful establishment and survival, for many years the official policy of the U.S. Forest Service, symbolized by the iconic mascot Smokey Bear, was to suppress all forest fires. Today, however, controlled burning is common, particularly in western North America. In order to use fire as an ecosystem management tool, it is important to know the historical pattern of fires in an area, which can be determined in part by studies of annual growth rings of trees (**Figure 59.15**). A schedule of controlled burning that recreates the historical pattern can reduce forest floor litter, avoiding a buildup of fuel that can lead to intense, tree-killing canopy fires.

### Ending trade is crucial to saving some species

Most endangered species cannot survive any further reductions in their breeding populations, so it is important to prevent their exploitation. The legal mechanism for prohibiting trade in these species or their products is an international agreement called the Convention on International Trade in Endangered Species (CITES). Most nations of the world are members of CITES. National representatives meet every two years to review the status of species on the CITES protected species lists, to determine which species may no longer need protection, and to add new species. CITES rules currently prohibit international trade in items such as whale meat, rhinoceros horn, and many species of parrots and orchids.

CITES instituted a ban on international trade in elephant ivory in 1989, but demand for ivory remains strong, especially in Japan and China. As a result, poaching of elephants continues in the forests of central Africa and in East Africa, where the animals are threatened. Southern African countries (Botswana, Namibia, South Africa, Zambia, and Zimbabwe), however, have so many elephants that government officials must kill many of them to control populations in the limited areas where they are allowed to roam to prevent them from dispersing to populated areas and damaging crops. These countries would like to sell the ivory from the culled elephants to fund their conservation efforts. Other countries are worried that if trade restrictions on ivory are relaxed, poaching will escalate over the entire continent.

Control and regulation of ivory trade might be possible if scientists could determine where the ivory comes from. Samuel Wasser and his colleagues at the University of Washington have identified 16 DNA markers that can be extracted from elephant feces. Park rangers in Malawi and Zambia were able to sample the elephant populations in their countries by collecting fresh scat while they were on routine patrols. The source of an elephant tusk can then be determined by matching DNA extracted from the ivory with the geographically based frequencies of the 16 DNA markers found in the dung samples. In June 2002, 6.5 tons of illegal ivory were seized in Singapore. DNA analyses showed that the elephants were killed in Zambia.

Such safeguards were partially responsible for the controversial decision to sanction legal sales of ivory from Namibia, Botswana, Zimbabwe, and South Africa beginning in 2008, the first such sale in close to a decade. Over 100 tons of elephant tusks, the equivalent of over 20,000 dead elephants, were auctioned off over a 2-week period to authorized buyers from China and Japan, for use primarily in folk medicine. The sale of this ivory, all of which was obtained legally from animals that had died of natural causes, generated over $15 million for elephant conservation efforts. Although the sales were monitored by CITES, and although the last sanctioned sale, in 1999, did not appear to increase poaching activities, other African nations and many wildlife conservation groups are concerned that in the absence of oversight and enforcement, the flood of legal ivory will be intermingled with poached ivory in international markets and encourage more illegal elephant slaughter.

One promising development in curbing illegal sales was the decision by eBay, the international online marketplace, to ban sales of ivory on its platform effective January 2009. An investigation by the International Fund for Animal Welfare revealed that two-thirds of online sales of protected wildlife products take place on eBay, so conservationists are optimistic that eBay's actions will be effective in drying up markets and protecting elephants from poaching.

### Invasions of exotic species must be controlled or prevented

Controlling invasions of exotic species is an important component of conservation biology. The best way to reduce the damage caused by invasive species, of course, is to prevent their introduction in the first place. Given the tremendous volume of global trade, it might seem impossible to curtail their spread. Some promising strategies, however, do exist. For example, transoceanic transport of invasive species in ballast water could be largely eliminated by the simple procedure of deoxygenating ballast water before it is pumped out. This practice not only kills most organisms in the water, but also extends the life of ballast tanks, providing an economic benefit to shippers. Although the U.S. House of Representatives passed (by a vote of 395 to 7) a Coast Guard funding bill with language stipulating that freighters operating in the Great Lakes must disinfect their ballast water, the Senate vote has been impeded by legal wrangling. In the absence of federal regulation, some states have already implemented their own ballast water treatment regulations.

Regulating the importation and sale of exotic species can reduce deliberate introductions. In 2002, some members of the American horticulture industry crafted a voluntary code of conduct for their profession, stating that the invasive potential of a plant should be assessed prior to its introduction and marketing. Horticulturists will work with conservation biologists to determine which species are currently invasive, or likely to become so, and to identify suitable alternative species. Stocks of invasive species will be phased out, and gardeners will be encouraged to use noninvasive plants.

Conservation biologists have developed a decision tree, based on the traits that characterize plant species that have become invasive, to help horticulturists and regulators determine whether an exotic plant species should be allowed into North America (**Figure 59.16**). Although the protocols stipulated by this decision tree cannot eliminate all introductions of invasive species, if they are followed conscientiously, they can greatly reduce the risk of such introductions.

## Biodiversity can have market value

Much of the value of ecosystems to human society depends on their biodiversity, but it has been difficult to assess the value of biodiversity in monetary terms. When biodiversity is perceived to have economic value, industries and government agencies have a greater incentive to protect it. The interdisciplinary field of *ecological economics* has provided tools for assessing the economic value of biodiversity.

Many studies have demonstrated the market value of protecting biodiversity. Such cases show that the argument for biodiversity conservation is compelling not only from an ecological or ethical perspective, but also from an economic perspective. Let's take a closer look at two of them.

**WILD DOGS AND ECOTOURISM** *Ecotourism*—environmentally responsible travel to natural areas that contributes to conserving their biodiversity and to the economic well-being of the local communities—is a major source of income for many developing nations. For example, tourists visiting Africa often express interest in seeing wild dogs (*Lycaon pictus*; **Figure 59.17**). However, diseases such as rabies and canine distemper, along with habitat loss, road traffic, deliberate extermination due to a perceived threat to livestock, and many other factors, have decimated wild dog populations throughout the continent, making them the second most endangered carnivore in Africa. Once found in 39 countries, they are today found in only 14 countries. Their endangered status has piqued interest among tourists in catching glimpses of these charismatic animals. South Africa is home to about 400 of Africa's remaining 5,000 dogs, most of which live in Kruger National Park. A survey of visitors to South Africa revealed that nearly three-fourths of them would be willing to pay an extra U.S. $12 for the opportunity to see wild dogs. Conservation biologists are

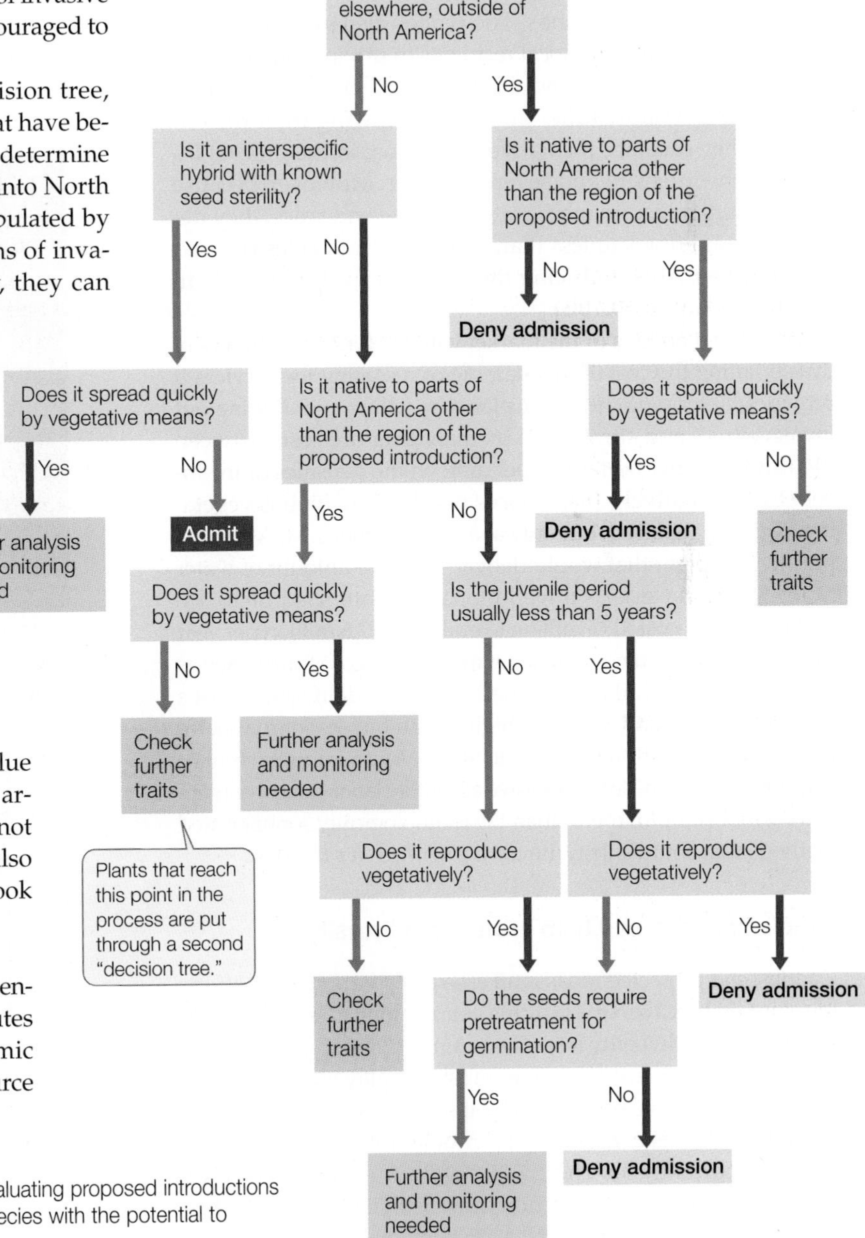


**59.16 A "Decision Tree"** This diagram sets criteria for evaluating proposed introductions of exotic plant species and helps regulators identify those species with the potential to become invasive.

working with lodge owners and ranchers elsewhere in South Africa and in Kenya to encourage them to reestablish wild dogs in areas from which they have disappeared.

**THE FYNBOS** Studies by a group of economists, ecologists, and land managers have attempted to calculate the value of the economic benefits provided by the spectacularly species-rich shrubland community called *fynbos* in the Western Cape Province of South Africa (**Figure 59.18A**). Of the 8,500 plant species in this region, over two-thirds are endemics. Fynbos plants thrive despite the summer droughts, nutrient-poor soils, and periodic fires of the region.

The highlands where fynbos vegetation thrives provide about two-thirds of the Western Cape's water supply. In addition, some of the endemic plants, including proteas, are harvested for cut and dried flowers. An international market has developed for rooibos, a fynbos shrub used for herbal tea. Income also comes from hundreds of thousands of ecotourists who visit the region every year to see the fynbos. The fynbos also provides recreational opportunities for local residents in urban areas nearby.

Recently, several trees and shrubs from other continents have invaded the fynbos. Taller and faster-growing than the endemics, they displace the native vegetation, in the process increasing the intensity and severity of fires. Moreover, because the invaders transpire more water than the endemics, they decrease stream flows to less than half those from areas covered with native plants, reducing the water supply for people in the region (**Figure 59.18B**).

We can get an idea of the market value of fynbos biodiversity by estimating the cost of maintaining or replacing the services it provides. Removing the exotic plants by felling and digging out invasive trees and shrubs and by managing fire costs between \$140 and \$830 per hectare, depending on the densities of the invaders. Alternatively, the services provided by fynbos vegetation could be replaced, but only at a much higher cost. A sewage purification plant that would deliver the same amount of water to the Western Cape as a well-managed watershed of 10,000 hectares would cost \$135 million to build and \$2.6 million per year to operate. Desalination of seawater would cost four times as much. Thus the available alternatives would deliver water at a cost between 1.8 and 6.7 times higher than the cost of maintaining natural vegetation in the watershed. Maintaining the fynbos vegetation would be less expensive and more labor-intensive (generating more employment) than the technologically sophisticated methods that could substitute for the services it provides.

## Simple changes can help protect biodiversity

Establishing protected areas is an essential component of efforts to maintain biodiversity, but this action alone is insufficient to stem global biodiversity loss. The extensive landscapes in which people live and extract resources must also play important roles in biodiversity conservation. The good news is that, carefully used, these lands can contribute much more to conservation than they currently do. The practice of using ecosystems for residences, resources, or recreation in ways that sustain their biodiversity is known as **reconciliation ecology**.

*Lycaon pictus*

**59.17 A Sight for Travelers' Eyes** Tourists visiting Africa to experience its wildlife often express a desire to see the rare African wild dogs. Protecting such species can be in the economic interest of the region.

(A)

(B)

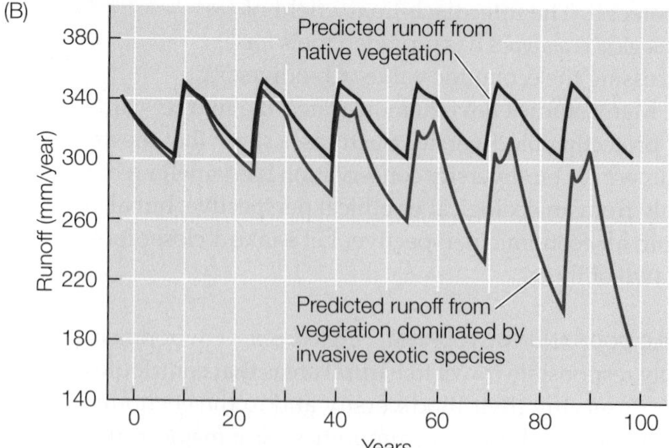


**59.18 Biodiversity Maintains Ecosystem Functioning** (A) The unique fynbos ecosystem of the Western Cape Province of South Africa provides much of the area's water. (B) A computer simulation of stream flows from fynbos watersheds that have and have not been invaded by exotic trees and shrubs.

Reconciliation ecology is based on the principle that most ecosystem services are provided locally, and that people are more motivated to work to protect their local interests than they are to work on national or global issues. The National Wildlife Federation has established a very successful program in which people petition to have their backyards certified as wildlife-friendly. Criteria for certification include planting shrubs that provide food for birds and refraining from applying pesticides on lawns. The city of Tucson, Arizona, has launched a major project to make the city a suitable habitat for many species of birds, not just the typical urban birds that live in most cities in North America.

Even some industrial sites can support biodiversity. The Turkey Point power plant in southern Florida uses large amounts of water to cool its generating units. To cool the heated water before discharging it, the Florida Power & Light Company dug a system of 38 canals that covers 6,000 acres. These cooling canals are separated by low-lying berms that support a variety of native and exotic plants. Red mangroves grow along the edges of the canals. Today, they support a thriving population of American crocodiles, a highly endangered species. Crocodiles living in the canals yield about 10 percent of all young crocodiles in the United States. Having discovered the biodiversity value of its cooling system, the company employs biologists to monitor the crocodiles and works actively to ensure their continued reproductive success.

### Captive breeding programs can maintain a few species

A few of the world's many endangered species can be maintained in captivity while the external threats to their existence are reduced or removed. However, captive propagation is only a temporary measure that buys time to deal with those threats. Zoos, aquariums, and botanical gardens do not have enough space to maintain adequate populations of more than a small fraction of Earth's rare and endangered species. Nonetheless, captive propagation can play an important role by maintaining species during critical periods, providing a source of individuals for reintroduction into the wild, and raising public awareness of threatened and endangered species.

The California condor, North America's largest bird, survives today only because of captive propagation (**Figure 59.19**). Two centuries ago, condors ranged from British Columbia to northern Mexico, but by 1978, the wild population was plunging toward extinction. Many of the birds, which are scavengers, died from ingesting animal carcasses containing lead shot. To save the condor from certain extinction, biologists captured all known remaining condors, which numbered only 22 individuals, and initiated a captive breeding program in 1983.

The first captive-bred birds were released in the mountains north of Los Angeles in 1992. Since that time, captive-bred birds have also been released in northern Arizona and Baja California. These birds are provided with lead-free food in remote areas, and they are trained to avoid power lines prior to release, thereby reducing a major source of mortality. Today, the captive-bred birds use the same roosting sites, bathing pools, and mountain ridges that their wild-born predecessors did. In December 2008, there were 327 known condors in the world, half of which were living in the wild. In 2003, a wild-born chick fledged in the wild for the first time in over two decades.

Most of the major threats to condor survival, including power lines, pesticides, and museum collectors, have been mitigated. Lead poisoning is still a problem, but as of July 1, 2008, under the Ridley–Tree Condor Preservation Act, California hunters are required to use non-lead bullets when hunting within the condor's range. Passage of this legislation marks a change in public attitudes from the days when cattle ranchers, in the mistaken belief that the birds killed livestock, vociferously opposed their reintroduction into the wild.

### The legacy of Samuel Plimsoll

During the nineteenth century, many British merchant ships sailed the world's oceans. Once these ships left harbor, they

(A)

(B) *Gymnogyps californianus*

**59.19 A California Condor Soars** (A) California condors raised in captivity are fed by humans wearing hand puppets so they will not imprint on their captors and will be able to survive in the wild. (B) Numbered wing tags allow conservation biologists to identify and track released adult condors. The survival of North America's largest bird species depends on this captive propagation project.

were out of contact with the rest of the world; in the case of a shipwreck, signaling for rescue was impossible. But ship owners could maximize their profits by overloading their ships, even though this practice caused some of them to become unseaworthy and sink.

Samuel Plimsoll, a member of Parliament, became concerned about the rate of loss of British vessels and sailors. He convinced Parliament to pass legislation requiring that a "load line" be painted on the hull of every large oceangoing vessel. The position of the line was calculated using factors such as the structural strength of the vessel and the shape of its hull. If the load line was under water, the ship was not permitted to leave the harbor. The "Plimsoll line," as it came to be known, dramatically reduced the rate of loss of British ships and sailors. In 1930, the international community agreed on universal adoption of load line regulations.

The increasing loss of Earth's species suggests that the load of human activities has pushed the planet below its Plimsoll line. But where and how should society draw that line? Our decision should be based on scientific information, but, just as in Samuel Plimsoll's time, science cannot determine an "acceptable rate of loss." Ethical considerations will figure prominently in the decisions we make about changing the way we use ecosystems so that other species can survive on Earth with us.

## 59.4 RECAP

**To conserve biodiversity, it is necessary to set aside protected areas, restore ecosystems and natural disturbance patterns, restrict trade in endangered species and transport of invasive species, increase populations of endangered species, and otherwise recognize the benefits of maintaining functioning ecosystems.**

- What are some of the priorities conservation biologists consider when establishing protected areas? **See pp. 1251–1252 and Figures 59.11–59.13**
- Explain the difference between restoration ecology and reconciliation ecology. **See pp. 1252 and 1257**
- Do you think the analogy of the Plimsoll line is a good one to apply to protecting the world's biodiversity? Why or why not?

## CHAPTER SUMMARY

### 59.1 What Is Conservation Biology?

- **Conservation biology** is an applied scientific discipline devoted to protecting and managing biodiversity.
- Conservation biologists recognize that an understanding of the evolutionary processes that generate biodiversity is essential to protecting it. They also understand that ecosystems are dynamic, and that humans are part of those ecosystems.
- Species extinctions have always occurred, but today they are occurring at an alarming rate.
- There are many compelling reasons for protecting biodiversity, including the maintenance of the ecosystems whose many functions provide humans with goods and services, including air and water purification.

### 59.2 How Do Biologists Predict Changes in Biodiversity?

- Our understanding of biodiversity is incomplete. We do not know how many species there are, where they live, or which of the species known to science have gone extinct.
- Biologists can use the species–area relationship and the theory of island biogeography to estimate rates of extinction likely to be caused by habitat loss.
- To estimate a species' risk of extinction, statistical models take into account data on population sizes, genetic variation, physiology, morphology, and behavior.
- Rarity is not always a cause for concern, but species whose populations are shrinking rapidly are usually at risk of extinction.

### 59.3 What Factors Threaten Species Persistence?

- Habitat loss is the most important cause of species endangerment worldwide. As habitats become increasingly fragmented, more species are lost from those habitats. Small habitat patches can support only small populations and are adversely influenced by **edge effects**. **Review Figures 59.5 and 59.6, ANIMATED TUTORIAL 59.1**
- Overexploitation has historically been the most important cause of species extinctions, and it is still a major threat to biodiversity today.
- Some species introduced to regions outside their original range become **invasive**, causing extinctions of native species by competing with them, eating them, or transmitting diseases to them.
- Climate is likely to become an increasingly important cause of extinctions for those species that cannot shift their ranges as rapidly as the climate warms. **Review Figure 59.9**

### 59.4 What Strategies Do Biologists Use to Protect Biodiversity?

- Establishing protected areas is crucial to conserving biodiversity. Protected areas are selected by taking into account species richness, endemism, imminence of threats, and the need to protect representative ecosystems.
- **Restoration ecology** is an important conservation strategy because many degraded ecosystems will not recover, or will do so only very slowly, without human assistance. **Review Figure 59.15**
- International trade in endangered species is controlled by regulations that most countries endorse.
- Conservation biologists work to determine which species are likely to become invasive and prevent their introduction to new areas.
- Even within landscapes where people live and extract resources, steps may be taken to protect biodiversity. This approach is known as **reconciliation ecology**.
- Captive breeding programs can maintain some endangered species for the short term while threats to their persistence are reduced or removed.

**SEE WEB ACTIVITY 59.1 for a concept review of this chapter.**

## SELF-QUIZ

1. Which of the following is *not* currently a major cause of species extinctions?
   *a.* Habitat destruction
   *b.* Rising sea levels
   *c.* Overexploitation
   *d.* Introduction of exotic predators
   *e.* Introduction of exotic pathogens
2. The most important cause of endangerment of species in the United States currently is
   *a.* climate change.
   *b.* invasive species.
   *c.* overexploitation.
   *d.* habitat loss.
   *e.* loss of mutualists.
3. Species extinctions matter to human society because
   *a.* more than a quarter of the medical prescriptions written in the United States contain a plant product.
   *b.* people derive aesthetic pleasure from interacting with other organisms.
   *c.* causing species extinctions raises serious ethical issues.
   *d.* biodiversity helps maintain valuable ecosystem services.
   *e.* All of the above
4. As a habitat patch gets smaller, it
   *a.* cannot support populations of species that require large areas.
   *b.* supports only small populations of many species.
   *c.* is influenced to an increasing degree by edge effects.
   *d.* is invaded by species from surrounding habitats.
   *e.* All of the above
5. A plant species is most likely to become invasive when introduced to a new area if it
   *a.* grows tall.
   *b.* has become invasive in other places where it has been introduced.
   *c.* is closely related to species living in the area where it has been introduced.
   *d.* has specialized disseminators of its seeds.
   *e.* has a long life span.
6. Global warming is a concern because
   *a.* the rate of change in climate is projected to be faster than the rate at which many species can shift their ranges.
   *b.* it is already too hot in the tropics.
   *c.* climates have been so stable for thousands of years that many species lack the ability to tolerate variable temperatures.
   *d.* climate change will be especially harmful to rare species.
   *e.* None of the above
7. Scientists can determine the historical frequency of fires in an area by
   *a.* examining charcoal in sites of ancient villages.
   *b.* measuring carbon in soils.
   *c.* radioactively dating fallen tree trunks.
   *d.* examining fire scars in growth rings of living trees.
   *e.* determining the age structure of forests.
8. Captive propagation is a useful conservation tool, when
   *a.* there is space in zoos, aquariums, and botanical gardens for breeding a few individuals.
   *b.* the areas of origin of the captive individuals are known.
   *c.* the threats that endangered the species are being alleviated so that captive-reared individuals can later be released back into the wild.
   *d.* there are sufficient caretakers.
   *e.* None of the above; captive propagation should never be used because it directs attention away from the need to protect the species in their natural habitats.
9. Restoration ecology is an important field because
   *a.* many areas have been highly degraded.
   *b.* many areas are vulnerable to global climate change.
   *c.* many species suffer from demographic stochasticity.
   *d.* many species are genetically impoverished.
   *e.* fire is a threat to many areas.
10. The field of reconciliation ecology has developed because
   *a.* all other methods of preserving biodiversity have failed.
   *b.* protected areas should be able to maintain biodiversity.
   *c.* protected areas alone are insufficient to maintain biodiversity.
   *d.* scientists are unable to control diseases today.
   *e.* we are not reconciled with other species.

## FOR DISCUSSION

1. Most species driven to extinction by humans in the past were large vertebrates. Do you expect this pattern to persist into the future? If not, why not?
2. Conservation biologists have debated extensively which is better: many small protected areas (which may contain more species) or a few large protected areas (which may be the only ones that can support populations of species that require large areas). What ecological processes should be evaluated in making judgments about the sizes and locations of protected areas?
3. During World War I, doctors adopted a "triage" system for dealing with wounded soldiers. The wounded were divided into three categories: those almost certain to die no matter what was done to help them, those likely to recover even if not assisted, and those whose probability of survival was greatly increased if they were given immediate medical attention. Limited medical resources were directed primarily at the third category. What are some implications of adopting a similar attitude toward species preservation?
4. Utilitarian arguments dominate discussions about the importance of preserving biological richness. In your opinion, what role should ethical and moral arguments play?
5. The desert bighorn sheep of the southwestern United States is endangered. Its major predator, the puma, is also threatened in the region. Under what conditions, if any, would it be appropriate to suppress the population of one rare species to assist another rare species?

## ADDITIONAL INVESTIGATION

Pollination by forest-dwelling bees was found to increase seed production in coffee plants growing within a kilometer of a forest patch in Costa Rica. What forest organisms are likely to provide other benefits (such as control of insect pests) to nearby agricultural crops? Over what distances are those effects likely to be felt? What experiments could be designed to test the importance of those effects and how they vary with distance from forest patches?

# Appendix A: The Tree of Life

Phylogeny is the organizing principle of modern biological taxonomy. A guiding principle of modern phylogeny is *monophyly*. A monophyletic group is considered to be one that contains an ancestral lineage and *all* of its descendants. Any such group can be extracted from a phylogenetic tree with a single cut.

The tree shown here provides a guide to the relationships among the major groups of extant (living) organisms in the tree of life as we have presented them throughout this book. The position of the branching "splits" indicates the relative branching order of the lineages of life, but the time scale is not meant to be uniform. In addition, the groups appearing at the branch tips do not necessarily carry equal phylogenetic "weight." For example, the ginkgo [63] is indeed at the apex of its lineage; this gymnosperm group consists of a single living species. In contrast, a phylogeny of the eudicots [55] could continue on from this point to fill many more trees the size of this one.

The glossary entries that follow are informal descriptions of some major features of the organisms described in Part Seven of this book. Each entry gives the group's common name, followed by the formal scientific name of the group (in parentheses). Numbers in square brackets reference the location of the respective groups on the tree.

It is sometimes convenient to use an informal name to refer to a collection of organisms that are not monophyletic but nonetheless all share (or all lack) some common attribute. We call these "convenience terms"; such groups are indicated in these entries by quotation marks, and we do not give them formal scientific names. Examples include "prokaryotes," "protists," and "algae." Note that these groups cannot be removed with a single cut; they represent a collection of distantly related groups that appear in different parts of the tree. We also use quotation marks here to designate two groups of fungi that are not believed to be monophyletic.

An interactive version of this tree, with links to much greater detail (such as photos, distribution maps, species lists, and identification keys), can be found at yourBioPortal.com.

## – A –

**acorn worms** (*Enteropneusta*) Benthic marine hemichordates [120] with an acorn-shaped proboscis, a short collar (neck), and a long trunk.

**"algae"** A convenience term encompassing various distantly related groups of aquatic, photosynthetic chromalveolates [5] and certain members of the Plantae [8].

**alveolates** (*Alveolata*) [7] Unicellular eukaryotes with a layer of flattened vesicles (alveoli) supporting the plasma membrane. Major alveolate groups include the dinoflagellates [52], apicomplexans [53], and ciliates [54].

**amborella** (*Amborella*) [60] An understory shrub or small tree found in New Caledonia. Thought to be the sister-group of the remaining living angiosperms [14].

**ambulacrarians** (*Ambulacraria*) [30] The echinoderms [119] and hemichordates [120].

**amniotes** (*Amniota*) [37] Mammals, reptiles, and their extinct close relatives. Characterized by many adaptations to terrestrial life, including an amniotic egg (with a unique set of membranes—the amnion, chorion, and allantois), a water-repellant epidermis (with epidermal scales, hair, or feathers), and, in males, a penis that allows internal fertilization.

**amoebozoans** (*Amoebozoa*) [85] A group of eukaryotes [4] that use lobe-shaped pseudopods for locomotion and to engulf food. Major amoebozoan groups include the loboseans, plasmodial slime molds, and cellular slime molds.

**amphibians** (*Amphibia*) [129] Tetrapods [36] with glandular skin that lacks epidermal scales, feathers, or hair. Many amphibian species undergo a complete metamorphosis from an aquatic larval form to a terrestrial adult form, although direct development is also common. Major amphibian groups include frogs and toads (anurans), salamanders, and caecilians.

**amphipods** (*Amphipoda*) Small crustaceans [117] that are abundant in many marine and freshwater habitats. They are important herbivores, scavengers, and micropredators, and are an important food source for many aquatic organisms.

**angiosperms** (*Anthophyta* or *Magnoliophyta*) [14] The flowering plants. Major angiosperm groups include the monocots, eudicots, and magnoliids.

**animals** (*Animalia* or *Metazoa*) [21] Multicellular heterotrophic eukaryotes. The majority of animals are bilaterians [24]. Other groups of animals include the cnidarians [98], ctenophores [97], placozoans [96], and sponges [22]. The closest living relatives of the animals are the choanoflagellates [92].

**annelids** (*Annelida*) [105] Segmented worms, including earthworms, leeches, and polychaetes. One of the major groups of lophotrochozoans [26].

**anthozoans** (*Anthozoa*) One of the major groups of cnidarians [98]. Includes the sea anemones, sea pens, and corals.

**anurans** (*Anura*) Comprising the frogs and toads, this is the largest group of living amphibians [129]. They are tail-less, with a shortened vertebral column and elongate hind legs modified for jumping. Many species have an aquatic larval form known as a tadpole.

**apicomplexans** (*Apicomplexa*) [53] Parasitic alveolates [7] characterized by the possession of an apical complex at some stage in the life cycle.

**arachnids** (*Arachnida*) Chelicerates [115] with a body divided into two parts: a cephalothorax that bears six pairs of appendages (four pairs of which are usually used as legs) and an abdomen that bears the genital opening. Familiar arachnids include spiders, scorpions, mites and ticks, and harvestmen.

**arbuscular mycorrhizal fungi** (*Glomeromycota*) [88] A group of fungi [19] that associate with plant roots in a close symbiotic relationship.

**archaeans** (*Archaea*) [3] Unicellular organisms lacking a nucleus and lacking peptidoglycan in the cell wall. Once grouped with the bacteria, archaeans possess distinctive membrane lipids.

**archosaurs** (*Archosauria*) [39] A group of reptiles [38] that includes dinosaurs and crocodilians [134]. Most dinosaur groups became extinct at the end of the Cretaceous; birds [133] are the only surviving dinosaurs.

**arrow worms** (*Chaetognatha*) [107] Small planktonic or benthic predatory marine worms with fins and a pair of hooked, prey-grasping spines on each side of the head.

**arthropods** (*Arthropoda*) The largest group of ecdysozoans [27]. Arthropods are characterized by a stiff exoskeleton, segmented bodies, and jointed appendages. Includes the chelicerates [115], myriapods [116], crustaceans [117], and hexapods (insects and their relatives) [118].

**ascidians** (*Ascidiacea*) "Sea squirts"; the largest group of urochordates [122]. Also known as tunicates, they are sessile (as adults), marine, saclike filter feeders.

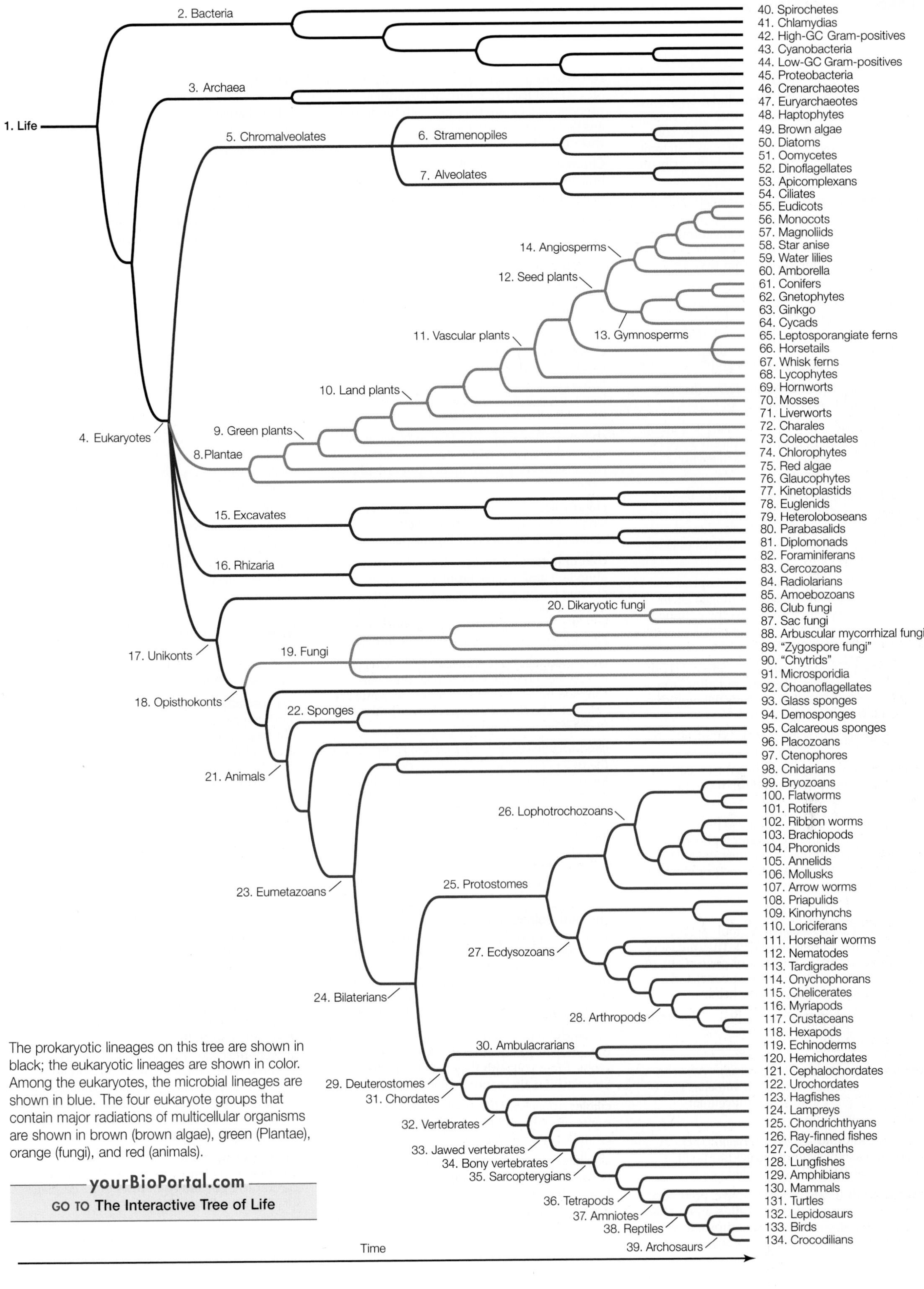


The prokaryotic lineages on this tree are shown in black; the eukaryotic lineages are shown in color. Among the eukaryotes, the microbial lineages are shown in blue. The four eukaryote groups that contain major radiations of multicellular organisms are shown in brown (brown algae), green (Plantae), orange (fungi), and red (animals).

**yourBioPortal.com**

GO TO **The Interactive Tree of Life**

## – B –

**bacteria** (*Eubacteria*) [2] Unicellular organisms lacking a nucleus, possessing distinctive ribosomes and initiator tRNA, and generally containing peptidoglycan in the cell wall. Different bacterial groups are distinguished primarily on nucleotide sequence data.

**barnacles** (*Cirripedia*) Crustaceans [117] that undergo two metamorphoses—first from a feeding planktonic larva to a nonfeeding swimming larva, and then to a sessile adult that forms a "shell" composed of four to eight plates cemented to a hard substrate.

**bilaterians** (*Bilateria*) [24] Those animal groups characterized by bilateral symmetry and three distinct tissue types (endoderm, ectoderm, and mesoderm). Includes the protostomes [25] and deuterostomes [29].

**birds** (*Aves*) [133] Feathered, flying (or secondarily flightless) tetrapods [36].

**bivalves** (*Bivalvia*) Major mollusk [106] group; clams and mussels. Bivalves typically have two similar hinged shells that are each asymmetrical across the midline.

**bony vertebrates** (*Osteichthyes*) [34] Vertebrates [32] in which the skeleton is usually ossified to form bone. Includes the ray-finned fishes [126], coelacanths [127], lungfishes [128], and tetrapods [36].

**brachiopods** (*Brachiopoda*) [103] Lophotrochozoans [26] with two similar hinged shells that are each symmetrical across the midline. Superficially resemble bivalve mollusks, except for the shell symmetry.

**brittle stars** (*Ophiuroidea*) Echinoderms [119] with five long, whip-like arms radiating from a distinct central disk that contains the reproductive and digestive organs.

**brown algae** (*Phaeophyta*) [49] Multicellular, almost exclusively marine stramenopiles [6] generally containing the pigment fucoxanthin as well as chlorophylls *a* and *c* in their chloroplasts.

**bryozoans** (*Ectoprocta* or *Bryozoa*) [99] A group of marine and freshwater lophotrochozoans [26] that live in colonies attached to substrata; also known as ectoprocts or moss animals.

## – C –

**caecilians** (*Gymnophiona*) A group of burrowing or aquatic amphibians [129]. They are elongate, legless, with a short tail (or none at all), reduced eyes covered with skin or bone, and a pair of sensory tentacles on the head.

**calcareous sponges** (*Calcarea*) [95] Filter-feeding marine sponges with spicules composed of calcium carbonate.

**cellular slime molds** (*Dictyostelida*) Amoebozoans [85] in which individual amoebas aggregate under stress to form a multicellular pseudoplasmodium.

**cephalochordates** (*Cephalochordata*) [121] A group of weakly swimming, eel-like benthic marine chordates [31]; also called lancelets.

**cephalopods** (*Cephalopoda*) Active, predatory mollusks [106] in which the molluscan foot has been modified into muscular hydrostatic arms or tentacles. Includes octopuses, squids, and nautiluses.

**cercozoans** (*Cercozoa*) [83] Unicellular eukaryotes [4] that feed by means of threadlike pseudopods. Group together with foraminiferans [82] and radiolarians [84] to comprise the rhizaria [16].

**Charales** [72] Multicellular green algae with branching, apical growth and plasmodesmata between adjacent cells. The closest living relatives of the land plants [10], they retain the egg in the parent organism.

**chelicerates** (*Chelicerata*) [115] A major group of arthropods [28] with pointed appendages (chelicerae) used to grasp food (as opposed to the chewing mandibles of most other arthropods). Includes the arachnids, horseshoe crabs, pycnogonids, and extinct sea scorpions.

**chimaeras** (*Holocephali*) A group of bottom-dwelling, marine, scaleless chondrichthyan fishes [125] with large, permanent, grinding tooth plates (rather than the replaceable teeth found in other chondrichthyans).

**chitons** (*Polyplacophora*) Flattened, slow-moving mollusks [106] with a dorsal protective calcareous covering made up of eight articulating plates.

**chlamydias** (*Chlamydiae*) [41] A group of very small Gram-negative bacteria; they live as intracellular parasites of other organisms.

**chlorophytes** (*Chlorophyta*) [74] The most abundant and diverse group of green algae, including freshwater, marine, and terrestrial forms; some are unicellular, others colonial, and still others multicellular. Chlorophytes use chlorophylls *a* and *c* in their photosynthesis.

**choanoflagellates** (*Choanozoa*) [92] Unicellular eukaryotes [4] with a single flagellum surrounded by a collar. Most are sessile, some are colonial. The closest living relatives of the animals [21].

**chondrichthyans** (*Chondrichthyes*) [125] One of the two main groups of jawed vertebrates [33]; includes sharks, rays, and chimaeras. They have cartilaginous skeletons and paired fins.

**chordates** (*Chordata*) [31] One of the two major groups of deuterostomes [29], characterized by the presence (at some point in development) of a notochord, a hollow dorsal nerve cord, and a post-anal tail. Includes the cephalochordates [121], urochordates [122], and vertebrates [32].

**chromalveolates** (*Chromalveolata*) [5] A contested group, said to have arisen from a common ancestor with chloroplasts derived from a red alga and supported by some molecular evidence. Major chromalveolate groups include the alveolates [7], stramenopiles [6], and haptophytes [48].

**"chytrids"** [90] A convenience term used for a paraphyletic group of mostly aquatic, microscopic fungi [19] with flagellated gametes. Some exhibit alternation of generations.

**ciliates** (*Ciliophora*) [54] Alveolates [7] with numerous cilia and two types of nuclei (micronuclei and macronuclei).

**clitellates** (*Clitellata*) Annelids [105] with gonads contained in a swelling (called a clitellum) toward the head of the animal. Includes earthworms (oligochaetes) and leeches.

**club fungi** (*Basidiomycota*) [86] Fungi [19] that, if multicellular, bear the products of meiosis on club-shaped basidia and possess a long-lasting dikaryotic stage. Some are unicellular.

**club mosses** (*Lycopodiophyta*) [68] Vascular plants [11] characterized by microphylls. See lycophytes.

**cnidarians** (*Cnidaria*) [98] Aquatic, mostly marine eumetazoans [23] with specialized stinging organelles (nematocysts) used for prey capture and defense, and a blind gastrovascular cavity. The closest living relatives of the ctenophores [97].

**coelacanths** (*Actinista*) [127] A group of marine sarcopterygians [35] that was diverse from the Middle Devonian to the Cretaceous, but is now known from just two living species. The pectoral and anal fins are on fleshy stalks supported by skeletal elements, so they are also called lobe-finned fishes.

**Coleochaetales** [73] Multicellular green algae characterized by flattened growth form composed of thin-walled cells. Thought to be the sister-group to the Charales [72] plus land plants [10].

**conifers** (*Pinophyta* or *Coniferophyta*) [61] Cone-bearing, woody seed plants [12].

**copepods** (*Copepoda*) Small, abundant crustaceans [117] found in marine, freshwater, or wet terrestrial habitats. They have a single eye, long antennae, and a body shaped like a teardrop.

**craniates** (*Craniata*) Some biologist exclude the hagfishes [123] from the vertebrates [32], and use the term craniates to refer to the two groups combined.

**crenarchaeotes** (*Crenarchaeota*) [46] A major and diverse group of archaeans [3], defined on the basis of rRNA base sequences. Many are extremophiles (inhabit extreme environments), but the group may also be the most abundant archaeans in the marine environment.

**crinoids** (*Crinoidea*) Echinoderms [119] with a mouth surrounded by feeding arms, and a U-shaped gut with the mouth next to the anus. They attach to the substratum by a stalk or are free-swimming. Crinoids were abundant in the middle and late Paleozoic, but only a few hundred species have survived to the present. Includes the sea lilies and feather stars.

**crocodilians** (*Crocodylia*) [134] A group of large, predatory, aquatic archosaurs [39]. The closest living relatives of birds [133]. Includes alligators, caimans, crocodiles, and gharials.

**crustaceans** (*Crustacea*) [117] Major group of marine, freshwater, and terrestrial arthropods [28] with a head, thorax, and abdomen (although the head and thorax may be fused), covered with a thick exoskeleton, and with two-part appendages. Crustaceans undergo metamorphosis from a nauplius larva. Includes decapods, isopods, krill, barnacles, amphipods, copepods, and ostracods.

**ctenophores** (*Ctenophora*) [97] Radially symmetrical, diploblastic marine animals [21], with a complete gut and eight rows of fused plates of cilia (called ctenes).

**cyanobacteria** (*Cyanobacteria*) [43] A group of unicellular, colonial, or filamentous bacteria that conduct photosynthesis using chlorophyll *a*.

**cycads** (*Cycadophyta*) [64] Palmlike gymnosperms with large, compound leaves.

**cyclostomes** (*Cyclostomata*) This term refers to the possibly monophyletic group of lampreys [124] and hagfishes [123]. Molecular data support this group, but morphological data suggest that lampreys are more closely related to jawed vertebrates [33] than to hagfishes.

## – D –

**decapods** (*Decapoda*) A group of marine, freshwater, and semiterrestrial crustaceans [117] in which five of the eight pairs of thoracic appendages function as legs (the other three pairs, called maxillipeds, function as mouthparts). Includes crabs, lobsters, crayfishes, and shrimps.

**demosponges** (*Demospongiae*) [94] The largest of the three groups of sponges [22], accounting for 90 percent of all sponge species. Demosponges have spicules made of silica, spongin fiber (a protein), or both.

**deuterostomes** (*Deuterostomia*) [29] One of the two major groups of bilaterians [24], in which the mouth forms at the opposite end of the embryo from the blastopore in early development (contrast with protostomes). Includes the ambulacrarians [30] and chordates [31].

**diatoms** (*Bacillariophyta*) [50] Unicellular, photosynthetic stramenopiles [6] with glassy cell walls in two parts.

**dikaryotic fungi** (*Dikarya*) [20] A group of fungi [19] in which two genetically different haploid nuclei coexist and divide within the same hypha; includes club fungi [86] and sac fungi [87].

**dinoflagellates** (*Dinoflagellata*) [52] A group of alveolates [7] usually possessing two flagella, one in an equatorial groove and the other in a longitudinal groove; many are photosynthetic.

**diplomonads** (*Diplomonadida*) [81] A group of eukaryotes [4] lacking mitochondria; most have two nuclei, each with four associated flagella.

## – E –

**ecdysozoans** (*Ecdysozoa*) [27] One of the two major groups of protostomes [25], characterized by periodic molting of their exoskeletons. Nematodes [112] and arthropods [28] are the largest ecdysozoan groups.

**echinoderms** (*Echinodermata*) [119] A major group of marine deuterostomes [29] with fivefold radial symmetry (at some stage of life) and an endoskeleton made of calcified plates and spines. Includes sea stars, crinoids, sea urchins, sea cucumbers, and brittle stars.

**elasmobranchs** (*Elasmobranchii*) The largest group of chondrichthyan fishes [125]. Includes sharks, skates, and rays. In contrast to the other group of living chondrichthyans (the chimaeras), they have replaceable teeth.

**embryophytes** *See* land plants [10].

**eudicots** (*Eudicotyledones*)[55] A group of angiosperms [14] with pollen grains possessing three openings. Typically with two cotyledons, net-veined leaves, taproots, and floral organs typically in multiples of four or five.

**euglenids** (*Euglenida*) [78] Flagellate excavates characterized by a pellicle composed of spiraling strips of protein under the plasma membrane; the mitochondria have disk-shaped cristae. Some are photosynthetic.

**eukaryotes** (*Eukarya*) [4] Organisms made up of one or more complex cells in which the genetic material is contained in nuclei. Contrast with archaeans [3] and bacteria [2].

**eumetazoans** (*Eumetazoa*) [23] Those animals [21] characterized by body symmetry, a gut, a nervous system, specialized types of cell junctions, and well-organized tissues in distinct cell layers (although there have been secondary losses of some of these characteristics in some eumetazoans).

**euphyllophytes** (*Euphyllophyta*) The group of vascular plants [11] that is sister to the lycophytes [68] and which includes all plants with megaphylls.

**euryarchaeotes** (*Euryachaeota*) [47] A major group of archaeans [3], diagnosed on the basis of rRNA sequences. Includes many methanogens, extreme halophiles, and thermophiles.

**eutherians** (*Eutheria*) A group of viviparous mammals [130], eutherians are well developed at birth (contrast to prototherians and marsupials, the other two groups of mammals). Most familiar mammals outside the Australian and South American regions are eutherians (see Table 33.1).

**excavates** (*Excavata*) [15] Diverse group of unicellular, flagellate eukaryotes, many of which possess a feeding groove; some lack mitochondria.

## – F –

**"ferns"** Vascular plants [11] usually possessing large, frondlike leaves that unfold from a "fiddlehead." Not a monophyletic group, although most fern species are encompassed in a monophyletic clade, the leptosporangiate ferns [65].

**flatworms** (*Platyhelminthes*) [100] A group of dorsoventrally flattened and generally elongate soft-bodied lophotrochozoans [26]. May be free-living or parasitic, found in marine, freshwater, or damp terrestrial environments. Major flatworm groups include the tapeworms, flukes, monogeneans, and turbellarians.

**flowering plants** *See* angiosperms [14].

**flukes** (*Trematoda*) A group of wormlike parasitic flatworms [100] with complex life cycles that involve several different host species. May be paraphyletic with respect to tapeworms.

**foraminiferans** (*Foraminifera*) [82] Amoeboid organisms with fine, branched pseudopods that form a food-trapping net. Most produce external shells of calcium carbonate.

**fungi** (*Fungi*) [19] Eukaryotic heterotrophs with absorptive nutrition based on extracellular digestion; cell walls contain chitin. Major fungal groups include the microsporidia [91], "chytrids" [90], "zygospore fungi" [89], arbuscular mycorrhizal fungi [88], sac fungi [87], and club fungi [86].

## – G –

**gastropods** (*Gastropoda*) The largest group of mollusks [106]. Gastropods possess a well-defined head with two or four sensory tentacles (often terminating in eyes) and a ventral foot. Most species have a single coiled or spiraled shell. Common in marine, freshwater, and terrestrial environments.

**ginkgo** (*Ginkgophyta*) [63] A gymnosperm [13] group with only one living species. The ginkgo seed is surrounded by a fleshy tissue not derived from an ovary wall and hence not a fruit.

**glass sponges** (*Hexactinellida*) [93] Sponges [22] with a skeleton composed of four- and/or six-pointed spicules made of silica.

**glaucophytes** (*Glaucophyta*) [76] Unicellular freshwater algae with chloroplasts containing traces of peptidoglycan, the characteristic cell wall material of bacteria.

**gnathostomes** (*Gnathostomata*) *See* jawed vertebrates [33].

**gnetophytes** (*Gnetophyta*) [62] A gymnosperm [13] group with three very different lineages; all have wood with vessels, unlike other gymnosperms.

**green plants** (*Viridiplantae*) [9] Organisms with chlorophylls *a* and *b*, cellulose-containing cell walls, starch as a carbohydrate storage product, and chloroplasts surrounded by two membranes.

**gymnosperms** (*Gymnospermae*) [13] Seed plants [12] with seeds "naked" (i.e., not enclosed in carpels). Probably monophyletic, but status still in doubt. Includes the conifers [61], gnetophytes [62], ginkgo [63], and cycads [64].

## – H –

**hagfishes** (*Myxini*) [123] Elongate, slimy-skinned vertebrates [32] with three small accessory hearts, a partial cranium, and no stomach or paired fins. *See also* craniata; cyclostomes.

**haptophytes** (*Haptophyta*) [48] Unicellular, photosynthetic chromalveolates [5] with two slightly unequal, smooth flagella. Abundant as phytoplankton, some form marine algal blooms.

**hemichordates** (*Hemichordata*) [120] One of the two primary groups of ambulacrarians [30]; marine wormlike organisms with a three-part body plan.

**heteroloboseans** (*Heterolobosea*) [79] Colorless excavates [15] that can transform among amoeboid, flagellate, and encysted stages.

**hexapods** (*Hexapoda*) [118] Major group of arthropods [28] characterized by a reduction (from the ancestral arthropod condition) to six walking appendages, and the consolidation of three body segments to form a thorax. Includes insects and their relatives (see Table 32.2).

**high-GC Gram-positives** (*Actinobacteria*) [42] Gram-positive bacteria with a relatively high (G+C)/(A+T) ratio of their DNA, with a filamentous growth habit.

**hornworts** (*Anthocerophyta*) [69] Nonvascular plants with sporophytes that grow from the base. Cells contain a single large, platelike chloroplast.

**horsehair worms** (*Nematomorpha*) [111] A group of very thin, elongate, wormlike freshwater ecdysozoans [27]. Largely nonfeeding as adults, they are parasites of insects and crayfish as larvae.

**horseshoe crabs** (*Xiphosura*) Marine chelicerates [115] with a large outer shell in three parts: a carapace, an abdomen, and a tail-like telson. There are only five living species, but many additional species are known from fossils.

**horsetails** (*Sphenophyta* or *Equisetophyta*) [66] Vascular plants [11] with reduced megaphylls in whorls.

**hydrozoans** (*Hydrozoa*) A group of cnidarians [98]. Most species go through both polyp and mesuda stages, although one stage or the other is eliminated in some species.

## – I –

**insects** (*Insecta*) The largest group within the hexapods [118]. Insects are characterized by exposed mouthparts and one pair of antennae containing a sensory receptor called a Johnston's organ. Most have two pairs of wings as adults. There are more described species of insects than all other groups of life [1] combined, and many species remain to be discovered. The major insect groups are described in Table 32.2.

**"invertebrates"** A convenience term that encompasses any animal [21] that is not a vertebrate [32].

**isopods** (*Isopoda*) Crustaceans [117] characterized by a compact head, unstalked compound eyes, and mouthparts consisting of four pairs of appendages. Isopods are abundant and widespread in salt, fresh, and brackish water, although some species (the sow bugs) are terrestrial.

## – J –

**jawed vertebrates** (*Gnathostomata*) [33] A major group of vertebrates [32] with jawed mouths. Includes chondrichthyans [125], ray-finned fishes [126], and sarcopterygians [35].

## – K –

**kinetoplastids** (*Kinetoplastida*) [77] Unicellular, flagellate organisms characterized by the presence in their single mitochondrion of a kinetoplast (a structure containing multiple, circular DNA molecules).

**kinorhynchs** (*Kinorhyncha*) [109] Small (< 1 mm) marine ecdysozoans [27] with bodies in 13 segments and a retractable proboscis.

**korarchaeotes** (*Korarchaeota*) A group of archaeans [3] known only by evidence from nucleic acids derived from hot springs. Its phylogenetic relationships within the Archaea are unknown.

**krill** (*Euphausiacea*) A group of shrimplike marine crustaceans [117] that are important components of the zooplankton.

## – L –

**lampreys** (*Petromyzontiformes*) [124] Elongate, eel-like vertebrates [32] that often have rasping and sucking disks for mouths.

**lancelets** (*Cephalochordata*) *See* cephalochordates [121].

**land plants** (*Embryophyta*) [10] Plants with embryos that develop within protective structures; also called embryophytes. Sporophytes and gametophytes are multicellular. Land plants possess a cuticle. Major groups are the liverworts [71], mosses [70], hornworts [69], and vascular plants [11].

**larvaceans** (*Larvacea*) Solitary, planktonic urochordates [122] that retain both notochords and nerve cords throughout their lives.

**lepidosaurs** (*Lepidosauria*) [132] Reptiles [38] with overlapping scales. Includes tuataras and squamates (lizards, snakes, and amphisbaenians).

**leptosporangiate ferns** (*Pteridopsida* or *Polypodiopsida*) [65] Vascular plants [11] usually possessing large, frondlike leaves that unfold from a "fiddlehead," and possessing thin-walled sporangia.

**life** (*Life*) [1] The monophyletic group that includes all known living organisms. Characterized by a nucleic-acid based genetic system (DNA or RNA), metabolism, and cellular structure. Some parasitic forms, such as viruses, have secondarily lost some of these features and rely on the cellular environment of their host.

**liverworts** (*Hepatophyta*) [71] Nonvascular plants lacking stomata; stalk of sporophyte elongates along its entire length.

**loboseans** (*Lobosea*) A group of unicellular amoebozoans [85]; includes the most familiar amoebas (e.g., *Amoeba proteus*).

**"lophophorates"** Not a monophyletic group. A convenience term used to describe several groups of lophotrochozoans [26] that have a feeding structure called a lophophore (a circular or U-shaped ridge around the mouth that bears one or two rows of ciliated, hollow tentacles).

**lophotrochozoans** (*Lophotrochozoa*) [26] One of the two main groups of protostomes [25]. This group is morphologically diverse, and is supported primarily on information from gene sequences. Includes bryozoans [99], flatworms [100], rotifers [101], ribbon worms [102], brachiopods [103], phoronids [104], annelids [105], and mollusks [106].

**loriciferans** (*Loricifera*) [110] Small (< 1 mm) ecdysozoans [27] with bodies in four parts, covered with six plates.

**low-GC Gram-positives** (*Firmicutes*) [44] A diverse group of bacteria [2] with a relatively low (G+C)/(A+T) ratio of their DNA, often but not always Gram-positive, some producing endospores.

**lungfishes** (*Dipnoi*) [128] A group of aquatic sarcopterygians [35] that are the closest living relatives of the tetrapods [36]. They have a modified swim bladder used to absorb oxygen from air, so some species can survive the temporary drying of their habitat.

**lycophytes** (*Lycopodiophyta*) [68] Vascular plants [11] characterized by microphylls; includes club mosses, spike mosses, and quillworts.

## – M –

**magnoliids** (*Magnoliidae*) [57] A major group of angiosperms [14] possessing two cotyledons and pollen grains with a single opening. The group is defined primarily by nucleotide sequence data; it is more closely related to the eudicots and monocots than to three other small angiosperm groups.

**mammals** (*Mammalia*) [130] A group of tetrapods [36] with hair covering all or part of their skin; females produce milk to feed their developing young. Includes the prototherians, marsupials, and eutherians.

**marsupials** (*Marsupialia*) Mammals [130] in which the female typically has a marsupium (a pouch for rearing young, which are born at an extremely early stage in development). Includes such familiar mammals as opossums, koalas, and kangaroos.

**metazoans** (*Metazoa*) *See* animals [21].

**microbial eukaryotes** *See* "protists."

**microsporidia** (*Microsporidia*) [91] A group of parasitic unicellular fungi [19] that lack mitochondria and have walls that contain chitin.

**mollusks** (*Mollusca*) [106] One of the major groups of lophotrochozoans [26], mollusks have bodies composed of a foot, a mantle (which often secretes a hard, calcareous shell), and a visceral mass. Includes monoplacophorans, chitons, bivalves, gastropods, and cephalopods.

**monilophytes** (*Monilophyta*) A group of vascular plants [11], sister to the seed plants [12], characterized by overtopping and possession of megaphylls; includes the ferns [65], horsetails [66], and whisk ferns [67].

**monocots** (*Monocotyledones*) [56] Angiosperms [14] characterized by possession of a single cotyledon, usually parallel leaf veins, a fibrous root system, pollen grains with a single opening, and floral organs usually in multiples of three.

**monogeneans** (*Monogenea*) A group of ectoparasitic flatworms [100].

**monoplacophorans** (*Monoplacophora*) Mollusks [106] with segmented body parts and a single, thin, flat, rounded, bilateral shell.

**mosses** (*Bryophyta*) [70] Nonvascular plants with true stomata and erect, "leafy" gametophytes; sporophytes elongate by apical cell division.

**moss animals** *See* bryozoans [99].

**myriapods** (*Myriapoda*) [116] Arthropods [28] characterized by an elongate, segmented trunk with many legs. Includes centipedes and millipedes.

## – N –

**nanoarchaeotes** (*Nanoarchaeota*) A group of extremely small, thermophilic archaeans [3] with a much-reduced genome. The only described example can survive only when attached to a host organism.

**nematodes** (*Nematoda*) [112] A very large group of elongate, unsegmented ecdysozoans [27] with thick, multilayer cuticles. They are among the most abundant and diverse animals, although most species have not yet been described. Include free-living predators and scavengers, as well as parasites of most species of land plants [10] and animals [21].

**neognaths** (*Neognathae*) The main group of birds [133], including all living species except the ratites (ostrich, emu, rheas, kiwis, cassowaries) and tinamous. *See* palaeognaths.

## – O –

**oligochaetes** (*Oligochaeta*) An annelid [105] group whose members lack parapodia, eyes, and anterior tentacles, and have few setae. Earthworms are the most familiar oligochaetes.

**onychophorans** (*Onychophora*) [114] Elongate, segmented ecdysozoans [27] with many pairs of soft, unjointed, claw-bearing legs. Also known as velvet worms.

**oomycetes** (*Oomycota*) [51] Water molds and relatives; absorptive heterotrophs with nutrient-absorbing, filamentous hyphae.

**opisthokonts** (*Opisthokonta*) [18] A group of unikonts [17] in which the flagellum on motile cells, if present, is posterior. The opisthokonts include the fungi [19], animals [21], and choanoflagellates [92].

**ostracods** (*Ostracoda*) Marine and freshwater crustaceans [117] that are laterally compressed and protected by two clamlike calcareous or chitinous shells.

## – P –

**palaeognaths** (*Palaeognathae*) A group of secondarily flightless or weakly flying birds [133]. Includes the flightless ratites (ostrich, emu, rheas, kiwis, cassowaries) and the weakly flying tinamous.

**parabasalids** (*Parabasalia*) [80] A group of unicellular eukaryotes [4] that lack mitochondria; they possess flagella in clusters near the anterior of the cell.

**phoronids** (*Phoronida*) [104] A small group of sessile, wormlike marine lophotrochozoans [26] that secrete chitinous tubes and feed using a lophophore.

**placoderms** (*Placodermi*) An extinct group of jawed vertebrates [33] that lacked teeth. Placoderms were the dominant predators in Devonian oceans.

**placozoans** (*Placozoa*) [96] A poorly known group of structurally simple, asymmetrical, flattened, transparent animals found in coastal marine tropical and subtropical seas. Most evidence suggests that placozoans are the sister-group of eumetazoans [23].

**Plantae** [8] The most broadly defined plant group. In most parts of this book, we use the word "plant" as synonymous with "land plant" [10], a more restrictive definition.

**plasmodial slime molds** (*Myxogastrida*) Amoebozoans [85] that in their feeding stage consist of a coenocyte called a plasmodium.

**pogonophorans** (*Pogonophora*) Deep-sea annelids [105] that lack a mouth or digestive tract; they feed by taking up dissolved organic matter, facilitated by endosymbiotic bacteria in a specialized organ (the trophosome).

**polychaetes** (*Polychaeta*) A group of mostly marine annelids [105] with one or more pairs of eyes and one or more pairs of feeding tentacles; parapodia and setae extend from most body segments. May be paraphyletic with respect to the clitellates.

**priapulids** (*Priapulida*) [108] A small group of cylindrical, unsegmented, wormlike marine ecdysozoans [27] that takes its name from its phallic appearance.

**"progymnosperms"** Paraphyletic group of extinct vascular plants [11] that flourished from the Devonian through the Mississippian periods. The first truly woody plants, and the first with vascular cambium that produced both secondary xylem and secondary phloem, they reproduced by spores rather than by seeds.

**"prokaryotes"** Not a monophyletic group; as commonly used, includes the bacteria [2] and archaeans [3]. A term of convenience encompassing all cellular organisms that are not eukaryotes.

**proteobacteria** (*Proteobacteria*) [45] A large and extremely diverse group of Gram-negative bacteria that includes many pathogens, nitrogen fixers, and photosynthesizers. Includes the alpha, beta, gamma, delta, and epsilon proteobacteria.

**"protists"** This term of convenience does not describe a monophyletic group but is used to encompass a large number of distinct and distantly related groups of eukaryotes, many but far from all of which are microbial and unicellular. Essentially a "catch-all" term for any eukaryote group not contained within the land plants [10], fungi [19], or animals [21].

**protostomes** (*Protostomia*) [25] One of the two major groups of bilaterians [24]. In protostomes, the mouth typically forms from the blastopore (if present) in early development (contrast with deuterostomes). The major protostome groups are the lophotrochozoans [26] and ecdysozoans [27].

**prototherians** (*Prototheria*) A mostly extinct group of mammals [130], common during the Cretaceous and early Cenozoic. The five living species—the echidnas and the duck-billed platypus—are the only extant egg-laying mammals.

**pterobranchs** (*Pterobranchia*) A small group of sedentary marine hemichordates [120] that live in tubes secreted by the proboscis. They have one to nine pairs of arms, each bearing long tentacles that capture prey and function in gas exchange.

**pycnogonids** (*Pycnogonida*) Treated in this book as a group of chelicerates [115], but sometimes considered an independent group of arthropods [28]. Pycnogonids have reduced bodies and very long, slender legs. Also called sea spiders.

## – R –

**radiolarians** (*Radiolaria*) [84] Amoeboid organisms with needlelike pseudopods supported by microtubules. Most have glassy internal skeletons.

**ray-finned fishes** (*Actinopterygii*) [126] A highly diverse group of freshwater and marine bony vertebrates [34]. They have reduced swim bladders that often function as hydrostatic organs and fins supported by soft rays (lepidotrichia). Includes most familiar fishes.

**red algae** (*Rhodophyta*) [75] Mostly multicellular, marine and freshwater algae characterized by the presence of phycoerythrin in their chloroplasts.

**reptiles** (*Reptilia*) [38] One of the two major groups of extant amniotes [37], supported on the basis of similar skull structure and gene sequences. The term "reptiles" traditionally excluded the birds [133], but the resulting group is then clearly paraphyletic. As used in this book, the reptiles include turtles [131], lepidosaurs [132], birds [133], and crocodilians [134].

**rhizaria** (*Rhizaria*) [16] Mostly amoeboid unicellular eukaryotes with pseudopods, many with external or internal shells. Includes the foraminiferans [82], cercozoans [83], and radiolarians [84].

**rhyniophytes** (*Rhyniophyta*) A group of early vascular plants [11] that appeared in the Silurian and became extinct in the Devonian. Possessed dichotomously branching stems with terminal sporangia but no true leaves or roots.

**ribbon worms** (*Nemertea*) [102] A group of unsegmented lophotrochozoans [26] with an eversible proboscis used to capture prey. Mostly marine, but some species live in fresh water or on land.

**rotifers** (*Rotifera*) [101] Tiny (< 0.5 mm) lophotrochozoans [26] with a pseudocoelomic body cavity that functions as a hydrostatic organ and a ciliated feeding organ called the corona that surrounds the head. They live in freshwater and wet terrestrial habitats.

**roundworms** (*Nematoda*) [112] *See* nematodes.

## – S –

**sac fungi** (*Ascomycota*) [87] Fungi that bear the products of meiosis within sacs (asci) if the organism is multicellular. Some are unicellular.

**salamanders** (*Caudata*) A group of amphibians [129] with distinct tails in both larvae and adults and limbs set at right angles to the body.

**salps** *See* thaliaceans.

**sarcopterygians** (*Sarcopterygii*) [35] One of the two major groups of bony vertebrates [34], characterized by jointed appendages (paired fins or limbs).

**scyphozoans** (*Scyphozoa*) Marine cnidarians [98] in which the medusa stage dominates the life cycle. Commonly known as jellyfish.

**sea cucumbers** (*Holothuroidea*) Echinoderms [119] with an elongate, cucumber-shaped body and leathery skin. They are scavengers on the ocean floor.

**sea spiders** *See* pycnogonids.

**sea squirts** *See* ascidians.

**sea stars** (*Asteroidea*) Echinoderms [119] with five (or more) fleshy "arms" radiating from an indistinct central disk. Also called starfishes.

**sea urchins** (*Echinoidea*) Echinoderms [119] with a test (shell) that is covered in spines. Most are globular in shape, although some groups (such as the sand dollars) are flattened.

**"seed ferns"** A paraphyletic group of loosely related, extinct seed plants that flourished in the Devonian and Carboniferous. Characterized by large, frondlike leaves that bore seeds.

**seed plants** (*Spermatophyta*) [12] Heterosporous vascular plants [11] that produce seeds; most produce wood; branching is axillary (not dichotomous). The major seed plant groups are gymnosperms [13] and angiosperms [14].

**sow bugs** *See* isopods.

**spirochetes** (*Spirochaetes*) [40] Motile, Gram-negative bacteria with a helically coiled structure and characterized by axial filaments.

**sponges** (*Porifera*) [22] A group of relatively asymmetric, filter-feeding animals that lack a gut or nervous system and generally lack differentiated tissues. Includes glass sponges [93], demosponges [94], and calcareous sponges [95].

**springtails** (*Collembola*) Wingless hexapods [118] with springing structures on the third and fourth segments of their bodies. Springtails are extremely abundant in some environments (especially in soil, leaf litter, and vegetation).

**squamates** (*Squamata*) The major group of lepidosaurs [132], characterized by the possession of movable quadrate bones (which allow the upper jaw to move independently of the rest of the skull) and hemipenes (a paired set of eversible penises, or penes) in males. Includes the lizards (a paraphyletic group), snakes, and amphisbaenians.

**star anise** (*Austrobaileyales*) [58] A group of woody angiosperms [14] thought to be the sister-group of the clade of flowering plants that includes eudicots [55], monocots [56], and magnoliids [57].

**starfish** (*Asteroidea*) *See* sea stars.

**stramenopiles** (*Heterokonta* or *Stramenopila*) [6] Organisms having, at some stage in their life cycle, two unequal flagella, the longer possessing rows of tubular hairs. Chloroplasts, when present, surrounded by four membranes. Major stramenopile groups include the brown algae [49], diatoms [50], and oomycetes [51].

## – T –

**tapeworms** (*Cestoda*) Parasitic flatworms [100] that live in the digestive tracts of vertebrates as adults, and usually in various other species of animals as juveniles.

**tardigrades** (*Tardigrada*) [113] Small ($< 0.5$ mm) ecdysozoans [27] with fleshy, unjointed legs and no circulatory or gas exchange organs. They live in marine sands, in temporary freshwater pools, and on the water films of plants. Also called water bears.

**tetrapods** (*Tetrapoda*) [36] The major group of sarcopterygians [35]; includes the amphibians [129] and the amniotes [37]. Named for the presence of four jointed limbs (although limbs have been secondarily reduced or lost completely in several tetrapod groups).

**thaliaceans** (*Thaliacea*) A group of solitary or colonial planktonic marine urochordates [122]. Also called salps.

**therians** (*Theria*) Mammals [130] characterized by viviparity (live birth). Includes eutherians and marsupials.

**theropods** (*Theropoda*) Archosaurs [39] with bipedal stance, hollow bones, a furcula ("wishbone"), elongated metatarsals with three-fingered feet, and a pelvis that points backwards. Includes many well-known extinct dinosaurs (such as *Tyrannosaurus rex*), as well as the living birds [133].

**tracheophytes** *See* vascular plants [11].

**trilobites** (*Trilobita*) An extinct group of arthropods [28] related to the chelicerates [115]. Trilobites flourished from the Cambrian through the Permian.

**tuataras** (*Rhyncocephalia*) A group of lepidosaurs [132] known mostly from fossils; there are just two living tuatara species. The quadrate bone of the upper jaw is fixed firmly to the skull. Sister group of the squamates.

**tunicates** *See* ascidians.

**turbellarians** (*Turbellaria*) A group of free-living, generally carnivorous flatworms [100]. Their monophyly is questionable.

**turtles** (*Testudines*) [131] A group of reptiles [38] with a bony carapace (upper shell) and plastron (lower shell) that encase the body.

## – U –

**unikonts** (*Unikonta*) [17] A group of eukaryotes [4] whose motile cells possess a single flagellum. Major unikont groups include the amoebozoans [85], fungi [19], and animals [21].

**urochordates** (*Urochordata*) [122] A group of chordates [31] that are mostly saclike filter feeders as adults, with motile larvae stages that resemble a tadpole.

## – V –

**vascular plants** (*Tracheophyta*) [11] Plants with xylem and phloem. Major groups include the lycophytes [68] and euphyllophytes.

**vertebrates** (*Vertebrata*) [32] The largest group of chordates [31], characterized by a rigid endoskeleton supported by the vertebral column and an anterior skull encasing a brain. Includes hagfishes [123], lampreys [124], and the jawed vertebrates [33], although some biologists exclude the hagfishes from this group. *See also* craniates.

## – W –

**water bears** *See* tardigrades.

**water lilies** (*Nymphaeaceae*) [59] A group of aquatic, freshwater angiosperms [14] that are rooted in soil in shallow water, with round floating leaves and flowers that extend above the water's surface. They are the sister-group to most of the remaining flowering plants, with the exception of the genus *Amborella* [60].

**whisk ferns** (*Psilotophyta*) [67] Vascular plants [11] lacking leaves and roots.

## – Y –

**"yeasts"** A convenience term for several distantly related groups of unicellular fungi [19].

## – Z –

**"zygospore fungi"** (*Zygomycota*, if monophyletic) [89] A convenience term for a probably paraphyletic group of fungi [19] in which hyphae of differing mating types conjugate to form a zygosporangium.

# Appendix B: Some Measurements Used in Biology

| MEASURES OF | UNIT | EQUIVALENTS | METRIC → ENGLISH CONVERSION |
|---|---|---|---|
| Length | meter (m) | base unit | 1 m = 39.37 inches = 3.28 feet |
| | kilometer (km) | 1 km = 1000 ($10^3$) m | 1 km = 0.62 miles |
| | centimeter (cm) | 1 cm = 0.01 ($10^{-2}$) m | 1 cm = 0.39 inches |
| | millimeter (mm) | 1 mm = 0.1 cm = $10^{-3}$ m | 1 mm = 0.039 inches |
| | micrometer (μm) | 1 μm = 0.001 mm = $10^{-6}$ m | |
| | nanometer (nm) | 1 nm = 0.001 μm = $10^{-9}$ m | |
| Area | square meter ($m^2$) | base unit | 1 $m^2$ = 1.196 square yards |
| | hectare (ha) | 1 ha = 10,000 $m^2$ | 1 ha = 2.47 acres |
| Volume | liter (L) | base unit | 1 L = 1.06 quarts |
| | milliliter (mL) | 1 mL = 0.001 L = $10^{-3}$ L | 1 mL = 0.034 fluid ounces |
| | microliter (μL) | 1 μL = 0.001 mL = $10^{-6}$ L | |
| Mass | gram (g) | base unit | 1 g = 0.035 ounces |
| | kilogram (kg) | 1 kg = 1000 g | 1 kg = 2.20 pounds |
| | metric ton (mt) | 1 mt = 1000 kg | 1 mt = 2,200 pounds = 1.10 ton |
| | milligram (mg) | 1 mg = 0.001 g = $10^{-3}$ g | |
| | microgram (μg) | 1 μg = 0.001 mg = $10^{-6}$ g | |
| Temperature | degree Celsius (°C) | base unit | °C = (°F – 32)/1.8 |
| | | | 0°C = 32°F (water freezes) |
| | | | 100°C = 212°F (water boils) |
| | | | 20°C = 68°F ("room temperature") |
| | | | 37°C = 98.6°F (human internal body temperature) |
| | Kelvin (K)* | K = °C + 273 | 0 K = –460°F |
| Energy | joule (J) | | 1 J ≈ 0.24 calorie = 0.00024 kilocalorie† |

*0 K (–273°C) is "absolute zero," a temperature at which molecular oscillations approach 0—that is, the point at which motion all but stops.

†A *calorie* is the amount of heat necessary to raise the temperature of 1 gram of water 1°C. The *kilocalorie*, or nutritionist's calorie, is what we commonly think of as a calorie in terms of food.

# Answers to Self-Quizzes

## Chapter 2
1. b 6. a
2. d 7. d
3. c 8. a
4. c 9. c
5. d 10. b

## Chapter 3
1. e 6. a
2. e 7. c
3. c 8. e
4. d 9. b
5. b 10. d

## Chapter 4
1. c 6. c
2. c 7. e
3. c 8. b
4. d 9. c
5. e 10. b

## Chapter 5
1. b 6. e
2. d 7. a
3. c 8. d
4. e 9. b
5. a 10. d

## Chapter 6
1. e 6. c
2. c 7. c
3. a 8. b
4. d 9. e
5. c 10. c

## Chapter 7
1. d 6. a
2. d 7. e
3. c 8. d
4. c 9. c
5. d 10. a

## Chapter 8
1. c 6. c
2. e 7. d
3. b 8. b
4. c 9. d
5. c 10. e

## Chapter 9
1. d 6. d
2. d 7. e
3. e 8. d
4. e 9. a
5. c 10. e

## Chapter 10
1. e 6. d
2. b 7. d
3. d 8. d
4. b 9. d
5. e 10. b

## Chapter 11
1. d 6. d
2. c 7. e
3. b 8. d
4. d 9. d
5. c 10. c

## Chapter 12*
1. e 6. d
2. a 7. b
3. d 8. b
4. d 9. b
5. d 10. b

## Chapter 13
1. c 6. b
2. a 7. d
3. c 8. d
4. b 9. c
5. e 10. c

## Chapter 14
1. c 6. d
2. d 7. b
3. e 8. d
4. b 9. d
5. a 10. c

## Chapter 15
1. a 6. b
2. c 7. d
3. b 8. d
4. b 9. d
5. d 10. b

## Chapter 16
1. b 6. c
2. e 7. c
3. a 8. d
4. e 9. b
5. b 10. a

## Chapter 17
1. c 6. e
2. c 7. b
3. a 8. b
4. e 9. b
5. e 10. a

## Chapter 18
1. b 6. b
2. b 7. c
3. a 8. a
4. c 9. e
5. e 10. d

## Chapter 19
1. c 6. c
2. b 7. e
3. a 8. b
4. b 9. a
5. d 10. b

## Chapter 20
1. c 6. b
2. a 7. a
3. c 8. e
4. b 9. d
5. c 10. b

## Chapter 21
1. d 6. d
2. e 7. a
3. d 8. e
4. c 9. b
5. d 10. c

## Chapter 22
1. b 6. b
2. e 7. e
3. a 8. a
4. b 9. e
5. e 10. d

## Chapter 23
1. c 6. c
2. e 7. a
3. d 8. e
4. c 9. c
5. a 10. e

## Chapter 24
1. a 6. e
2. a 7. e
3. d 8. e
4. a 9. b
5. b 10. e

## Chapter 25
1. d 6. a
2. b 7. c
3. e 8. b
4. c 9. c
5. a 10. d

## Chapter 26
1. e 6. b
2. e 7. d
3. a 8. d
4. c 9. b
5. e 10. d

## *Answers to Chapter 12 Genetics Problems

1. *BB* × *bb*; *bb* × *bb*; *Bb* × *bb*; *Bb* × *Bb*
2. 1/32

3*a*. Autosomal dominant

3*b*. 1/4

4*a*. Males (XY) contain only one allele and will show only one color, black ($X^BY$) or yellow ($X^bY$). Females can be heterozygous ($X^BX^b$).

4*b*. $X^bY$, yellow

5. The body color (*G*/*g*) and wing size (*A*/*a*) genes are linked; eye color (*R*/*r*) is unlinked to the other two genes. The map distance is 18.5 units.
6. Yellow, blue, and white in a 1:2:1 ratio
7. $F_1$ all wild-type, *PpSwsw*; $F_2$ will have phenotypes in the ratio 9:3:3:1. See Figure 12.7 (p. 244) for analogous genotypes.

8*a*. Ratio of phenotypes in $F_2$ is 3:1 (double dominant to double recessive).

8*b*. The $F_1$ are *PpByby*; they produce just two kinds of gametes (*Pby* and *pBy*). Combine them carefully and see the 1:2:1 phenotypic ratio fall out in the $F_2$.

8*c*. Pink-blistery

8*d*. See Figures 11.17 and 11.19 (pp. 224–226). Crossing over took place in the $F_1$ generation.

9. *Rraa* and *RrAa*

10*a*. $w^+ > w^e > w$

10*b*. Parents are $w^e w$ and $w^+Y$. Progeny are $w^+w^e$, $w^+w$, $w^eY$, and $wY$.

11*a*. $BX^a$, $BY$, $bX^a$, $bY$

11*b*. Mother $bbX^AX^a$, father $BbX^aY$, son $BbX^aY$, daughter $bbX^aX^a$

12. 75 percent
13. Because the gene is carried on mitochondrial DNA, it is passed through the mother only. Thus if the woman does not have the disease but her husband does, their child will not be affected. On the other hand, if the woman has the disease but her husband does not, their child will have the disease.

**Chapter 27**
1. a 6. b
2. e 7. d
3. c 8. b
4. d 9. a
5. c 10. d

**Chapter 28**
1. d 6. e
2. c 7. c
3. e 8. b
4. b 9. b
5. b 10. d

**Chapter 29**
1. d 6. c
2. c 7. a
3. d 8. e
4. a 9. c
5. d 10. a

**Chapter 30**
1. b 6. a
2. d 7. e
3. e 8. a
4. c 9. e
5. d 10. c

**Chapter 31**
1. c 6. b
2. d 7. c
3. c 8. d
4. d 9. e
5. e 10. d

**Chapter 32**
1. b 6. e
2. e 7. b
3. c 8. d
4. d 9. d
5. a 10. e

**Chapter 33**
1. d 6. e
2. a 7. a
3. c 8. e
4. d 9. c
5. d 10. b

**Chapter 34**
1. c 6. b
2. b 7. b
3. e 8. c
4. e 9. a
5. a 10. d

**Chapter 35**
1. c 6. d
2. d 7. d
3. b 8. e
4. b 9. e
5. b 10. a

**Chapter 36**
1. d 6. c
2. d 7. e
3. c 8. d
4. a 9. d
5. c 10. e

**Chapter 37**
1. a 6. c
2. e 7. b
3. c 8. c
4. d 9. a
5. b 10. b

**Chapter 38**
1. d 6. e
2. b 7. a
3. e 8. c
4. b 9. c
5. d 10. d

**Chapter 39**
1. e 6. a
2. b 7. b
3. c 8. c
4. c 9. e
5. d 10. a

**Chapter 40**
1. c 6. b
2. c 7. e
3. a 8. a
4. d 9. e
5. b 10. b

**Chapter 41**
1. b 6. b
2. a 7. d
3. b 8. e
4. e 9. c
5. e 10. c

**Chapter 42**
1. a 6. a
2. b 7. d
3. e 8. d
4. e 9. a
5. c 10. d

**Chapter 43**
1. c 6. d
2. e 7. d
3. a 8. d
4. d 9. d
5. d 10. a

**Chapter 44**
1. a 6. c
2. c 7. b
3. e 8. b
4. a 9. b
5. d 10. c

**Chapter 45**
1. d 6. e
2. d 7. c
3. c 8. c
4. c 9. d
5. e 10. d

**Chapter 46**
1. d 6. e
2. d 7. b
3. a 8. c
4. e 9. c
5. e 10. d

**Chapter 47**
1. c 6. c
2. a 7. a
3. e 8. c
4. d 9. a
5. d 10. c

**Chapter 48**
1. e 6. d
2. a 7. e
3. b 8. a
4. c 9. a
5. b 10. e

**Chapter 49**
1. e 6. b
2. d 7. c
3. a 8. c
4. b 9. a
5. c 10. d

**Chapter 50**
1. d 6. d
2. a 7. b
3. c 8. d
4. d 9. c
5. c 10. e

**Chapter 51**
1. b 6. d
2. e 7. a
3. c 8. b
4. a 9. d
5. b 10. d

**Chapter 52**
1. d 6. b
2. a 7. e
3. d 8. a
4. a 9. c
5. d 10. e

**Chapter 53**
1. b 6. d
2. e 7. e
3. c 8. e
4. a 9. a
5. d 10. e

**Chapter 54**
1. d 6. c
2. a 7. c
3. a 8. c
4. d 9. d
5. a 10. b

**Chapter 55**
1. c 6. b
2. c 7. e
3. a 8. c
4. d 9. e
5. c 10. d

**Chapter 56**
1. a 6. a
2. c 7. b
3. b 8. d
4. c 9. e
5. d 10. b

**Chapter 57**
1. a 6. b
2. a 7. e
3. b 8. b
4. d 9. d
5. e 10. e

**Chapter 58**
1. e 6. e
2. d 7. a
3. b 8. e
4. c 9. d
5. c 10. b

**Chapter 59**
1. b 6. a
2. d 7. d
3. e 8. c
4. e 9. a
5. b 10. c

# Glossary

## - A -

**abiotic** (a′ bye ah tick) [Gk. *a*: not + *bios*: life] Nonliving. (Contrast with biotic.)

**abscisic acid (ABA)** (ab sighs′ ik) A plant growth substance with growth-inhibiting action. Causes stomata to close; involved in a plant's response to salt and drought stress.

**abscission** (ab sizh′ un) [L. *abscissio*: break off] The process by which leaves, petals, and fruits separate from a plant.

**absorption** (1) Of light: complete retention, without reflection or transmission. (2) Of water or other molecules: soaking up (taking in through pores or by diffusion).

**absorption spectrum** A graph of light absorption versus wavelength of light; shows how much light is absorbed at each wavelength.

**absorptive heterotroph** An organism (usually a fungus) that obtains its food by secreting digestive enzymes into the environment to break down large food molecules, then absorbing the breakdown products.

**abyssal plain** (uh biss′ ul) [Gk. *abyssos*: bottomless] The deep ocean floor.

**accessory pigments** Pigments that absorb light and transfer energy to chlorophylls for photosynthesis.

**acetylcholine (ACh)** A neurotransmitter that carries information across vertebrate neuromuscular junctions and some other synapses. It is then broken down by the enzyme acetylcholinesterase (AChE).

**acetyl coenzyme A (acetyl CoA)** A compound that reacts with oxaloacetate to produce citrate at the beginning of the citric acid cycle; a key metabolic intermediate in the formation of many compounds.

**acid** [L. *acidus*: sharp, sour] A substance that can release a proton in solution. (Contrast with base.)

**acid growth hypothesis** The hypothesis that auxin increases proton pumping, thereby lowering the pH of the cell wall and activating enzymes that loosen polysaccharides. Proposed to explain auxin-induced cell expansion in plants.

**acid precipitation** Precipitation that has a lower pH than normal as a result of acid-forming precursor molecules introduced into the atmosphere by human activities.

**acidic** Having a pH of less than 7.0 (a hydrogen ion concentration greater than $10^{-7}$ molar). (Contrast with basic.)

**acoelomate** An animal that does not have a coelom.

**acrosome** (a′ krow soam) [Gk. *akros*: highest + *soma*: body] The structure at the forward tip of an animal sperm which is the first to fuse with the egg membrane and enter the egg cell.

**ACTH** *See* corticotropin.

**actin** [Gk. *aktis*: ray] A protein that makes up the cytoskeletal microfilaments in eukaryotic cells and is one of the two contractile proteins in muscle.

**action potential** An impulse in a neuron taking the form of a wave of depolarization or hyperpolarization.

**action spectrum** A graph of a biological process versus light wavelength; shows which wavelengths are involved in the process.

**activation energy ($E_a$)** The energy barrier that blocks the tendency for a chemical reaction to occur.

**active site** The region on the surface of an enzyme or ribozyme where the substrate binds, and where catalysis occurs.

**active transport** The energy-dependent transport of a substance across a biological membrane against a concentration gradient—that is, from a region of low concentration (of that substance) to one of high concentration. (*See also* primary active transport, secondary active transport; contrast with facilitated diffusion, passive transport.)

**adaptation** (a dap tay′ shun) (1) In evolutionary biology, a particular structure, physiological process, or behavior that makes an organism better able to survive and reproduce. Also, the evolutionary process that leads to the development or persistence of such a trait. (2) In sensory neurophysiology, a sensory cell's loss of sensitivity as a result of repeated stimulation.

**adaptive radiation** An evolutionary radiation that results in an array of related species that live in a variety of environments and differ in the characteristics they use to exploit those environments.

**adenine (A)** (a′ den een) A nitrogen-containing base found in nucleic acids, ATP, NAD, and other compounds.

**adenosine triphosphate** *See* ATP.

**adrenal gland** (a dree′ nal) [L. *ad*: toward + *renes*: kidneys] An endocrine gland located near the kidneys of vertebrates, consisting of two glandular parts, the cortex and medulla.

**adrenaline** *See* epinephrine.

**adrenocorticotropic hormone** *See* corticotropin.

**adsorption** Binding of a gas or a solute to the surface of a solid.

**adventitious roots** (ad ven ti′ shus) [L. *adventitius*: arriving from outside] Roots originating from the stem at ground level or below; typical of the fibrous root system of monocots.

**aerenchyma** In plants, parenchymal tissue containing air spaces.

**aerobic** (air oh′ bic) [Gk. *aer*: air + *bios:* life] In the presence of oxygen; requiring oxygen. (Contrast with anaerobic.)

**afferent** (af′ ur unt) [L. *ad*: toward + *ferre*: to carry] Carrying to, as in a neuron that carries impulses to the central nervous system (afferent neuron), or a blood vessel that carries blood to a structure. (Contrast with efferent.)

**age structure** The distribution of the individuals in a population across all age groups.

**AIDS** Acquired immune deficiency syndrome, a condition caused by human immunodeficiency virus (HIV) in which the body's T-helper cells are reduced, leaving the victim subject to opportunistic diseases.

**air sacs** Structures in the respiratory system of birds that receive inhaled air; they keep fresh air flowing unidirectionally through the lungs, but are not themselves gas exchange surfaces.

**aldosterone** (al dohs′ ter own) A steroid hormone produced in the adrenal cortex of mammals. Promotes secretion of potassium and reabsorption of sodium in the kidney.

**aleurone layer** In some seeds, a tissue that lies beneath the seed coat and surrounds the endosperm. Secretes digestive enzymes that break down macromolecules stored in the endosperm.

**allantoic membrane** In animal development, an outgrowth of extraembryonic endoderm plus adjacent mesoderm that forms the allantois, a saclike structure that stores metabolic wastes produced by the embryo.

**allantois** (al lun twah′) [Gk. *allant*: sausage] An extraembryonic membrane enclosing a sausage-shaped sac that stores the embryo's nitrogenous wastes.

**allele** (a leel′) [Gk. *allos*: other] The alternate form of a genetic character found at a given locus on a chromosome.

**allele frequency** The relative proportion of a particular allele in a specific population.

**allergic reaction** [Ger. *allergie*: altered] An overreaction of the immune system to amounts of an antigen that do not affect most people; often involves IgE antibodies.

**allometric growth** A pattern of growth in which some parts of the body of an organism grow faster than others, resulting in a change in body proportions as the organism grows.

**allopatric speciation** (al′ lo pat′ rick) [Gk. *allos*: other + *patria*: homeland] The formation of two species from one when reproductive isolation occurs because of the interposition of (or crossing of) a physical geographic barrier such as a river. Also called geographic speciation. (Contrast with sympatric speciation.)

**allopolyploidy** The possession of more than two chromosome sets that are derived from more than one species.

**allosteric regulation** (al lo steer′ ik) [Gk. *allos*: other + *stereos*: structure] Regulation of the activity of a protein (usually an enzyme) by the binding of an effector molecule to a site other than the active site.

**alpha diversity** Species diversity within a single community or habitat. (Contrast with beta diversity, gamma diversity.)

**α (alpha) helix** A prevalent type of secondary protein structure; a right-handed spiral.

**alternation of generations** The succession of multicellular haploid and diploid phases in some sexually reproducing organisms, notably plants.

**alternative splicing** A process for generating different mature mRNAs from a single gene by splicing together different sets of exons during RNA processing.

**altruism** Pertaining to behavior that benefits other individuals at a cost to the individual who performs it.

**alveolus** (al ve′ o lus) (plural: alveoli) [L. *alveus*: cavity] A small, baglike cavity, especially the blind sacs of the lung.

**amensalism** (a men′ sul ism) Interaction in which one animal is harmed and the other is unaffected. (Contrast with commensalism, mutualism.)

**amine** An organic compound containing an amino group ($NH_2$).

**amino acid** An organic compound containing both $NH_2$ and COOH groups. Proteins are polymers of amino acids.

**amino acid replacement** A change in the nucleotide sequence that results in one amino acid being replaced by another.

**ammonotelic** (am moan′ o teel′ ic) [Gk. *telos*: end] Pertaining to an organism in which the final product of breakdown of nitrogen-containing compounds (primarily proteins) is *ammonia*. (Contrast with ureotelic, uricotelic.)

**amnion** (am′ nee on) The fluid-filled sac within which the embryos of reptiles (including birds) and mammals develop.

**amniote egg** A shelled egg surrounding four extraembryonic membranes and embryo-nourishing yolk. This evolutionary adaptation permitted mammals and reptiles to live and reproduce in drier environments than can most amphibians.

**amphipathic** (am′ fi path′ ic) [Gk. *amphi*: both + *pathos*: emotion] Of a molecule, having both hydrophilic and hydrophobic regions.

**amplitude** The magnitude of change over the course of a regular cycle.

**amygdala** A component of the limbic system that is involved in fear and fear memory.

**amylase** (am′ ill ase) An enzyme that catalyzes the hydrolysis of starch, usually to maltose or glucose.

**anabolic reaction** (an uh bah′ lik) [Gk. *ana*: upward + *ballein*: to throw] A synthetic reaction in which simple molecules are linked to form more complex ones; requires an input of energy and captures it in the chemical bonds that are formed. (Contrast with catabolic reaction.)

**anaerobic** (an ur row′ bic) [Gk. *an*: not + *aer*: air + *bios*: life] Occurring without the use of molecular oxygen, $O_2$. (Contrast with aerobic.)

**analogy** (a nal′ o jee) [Gk. *analogia*: resembling] A resemblance between two features that is due to convergent evolution rather than to common ancestry. The structures are said to be *analogous*, and each is an *analog* of the others. (Contrast with homology.)

**anaphase** (an′ a phase) [Gk. *ana*: upward] The stage in cell nuclear division at which the first separation of sister chromatids (or, in the first meiotic division, of paired homologs) occurs.

**ancestral trait** The trait originally present in the ancestor of a given group; may be retained or changed in the descendants of that ancestor.

**androgen** (an′ dro jen) Any of the several male sex steroids (most notably testosterone).

**aneuploidy** (an′ you ploy dee) A condition in which one or more chromosomes or pieces of chromosomes are either lacking or present in excess.

**angiotensin** (an′ jee oh ten′ sin) A peptide hormone that raises blood pressure by causing peripheral vessels to constrict. Also maintains glomerular filtration by constricting efferent vessels and stimulates thirst and the release of aldosterone.

**angular gyrus** A part of the human brain believed to be essential for integrating spoken and written language.

**animal hemisphere** The metabolically active upper portion of some animal eggs, zygotes, and embryos; does not contain the dense nutrient yolk. (Contrast with vegetal hemisphere.)

**anion** (an′ eye on) [Gk. *ana*: upward progress] A negatively charged ion. (Contrast with cation.)

**anisogamous** (an eye sog′ a muss) [Gk. *aniso*: unequal + *gamos*: marriage] Having morphologically dissimilar male and female gametes. (Contrast with isogamous.)

**annual** A plant whose life cycle is completed in one growing season. (Contrast with biennial, perennial.)

**antagonistic interactions** Interactions between two species in which one species benefits and the other is harmed. Includes predation, herbivory, and parasitism.

**antenna system** In photosynthesis, a group of different molecules that cooperate to absorb light energy and transfer it to a reaction center.

**anterior** Toward or pertaining to the tip or headward region of the body axis. (Contrast with posterior.)

**anterior pituitary** The portion of the vertebrate pituitary gland that derives from gut epithelium and produces tropic hormones.

**anther** (an′ thur) [Gk. *anthos*: flower] A pollen-bearing portion of the stamen of a flower.

**antheridium** (an′ thur id′ ee um) [Gk. *antheros*: blooming] The multicellular structure that produces the sperm in nonvascular land plants and ferns.

**antibody** One of the myriad proteins produced by the immune system that specifically binds to a foreign substance in blood or other tissue fluids and initiates its removal from the body.

**anticodon** The three nucleotides in transfer RNA that pair with a complementary triplet (a codon) in messenger RNA.

**antidiuretic hormone (ADH)** A hormone that promotes water reabsorption by the kidney. ADH is produced by neurons in the hypothalamus and released from nerve terminals in the posterior pituitary. Also called vasopressin.

**antigen** (an′ ti jun) Any substance that stimulates the production of an antibody or antibodies in the body of a vertebrate.

**antigenic determinant** The specific region of an antigen that is recognized and bound by a specific antibody. Also called an epitope.

**antiparallel** Pertaining to molecular orientation in which a molecule or parts of a molecule have opposing directions.

**antiporter** A membrane transport protein that moves one substance in one direction and another in the opposite direction. (Contrast with symporter, uniporter.)

**antisense RNA** A single-stranded RNA molecule complementary to, and thus targeted against, an mRNA of interest to block its translation.

**anus** (a′ nus) An opening through which solid digestive wastes are expelled, located at the posterior end of a tubular gut.

**aorta** (a or′ tah) [Gk. *aorte*: aorta] The main trunk of the arteries leading to the systemic (as opposed to the pulmonary) circulation.

**aortic body** A chemosensor in the aorta that senses a decrease in blood supply or a dramatic decrease in partial pressure of oxygen in the blood.

**aortic valve** A one-way valve between the left ventricle of the heart and the aorta that prevents backflow of blood into the ventricle when it relaxes.

**apex** (a′ pecks) The tip or highest point of a structure, as of a growing stem or root.

**aphasia** a deficit in the ability to use or understand words.

**apical** (a′ pi kul) Pertaining to the *apex*, or tip, usually in reference to plants.

**apical dominance** In plants, inhibition by the apical bud of the growth of axillary buds.

**apical hook** A form taken by the stems of many eudicot seedlings that protects the delicate shoot apex while the stem grows through the soil.

**apical meristem** The meristem at the tip of a shoot or root; responsible for a plant's primary growth.

**apomixis** (ap oh mix′ is) [Gk. *apo*: away from + *mixis*: sexual intercourse] The asexual production of seeds.

**apoplast** (ap′ oh plast) In plants, the continuous meshwork of cell walls and extracellular spaces through which material can pass without crossing a plasma membrane. (Contrast with symplast.)

**apoptosis** (ap uh toh′ sis) A series of genetically programmed events leading to cell death.

**aposematism** Warning coloration; bright colors or striking patterns of toxic or mimetic prey species that act as a warning to predators.

**appendix** In the human digestive system, the vestigial equivalent of the cecum, which serves no digestive function.

**aquaporin** A transport protein in plant and animal cell membranes through which water passes in osmosis.

**aquatic** (a kwa′ tic) [L. *aqua*: water] Pertaining to or living in water. (Contrast with marine, terrestrial.)

**aqueous** (a′ kwee us) Pertaining to water or a watery solution.

**aquifer** A large pool of groundwater.

**archegonium** (ar′ ke go′ nee um) The multicellular structure that produces eggs in nonvascular land plants, ferns, and gymnosperms.

**archenteron** (ark en′ ter on) [Gk. *archos*: first + *enteron*: bowel] The earliest primordial animal digestive tract.

**area phylogenies** Phylogenies in which the names of the taxa are replaced with the names of the places where those taxa live or lived.

**arms race** A series of reciprocal adaptations between species involved in antagonistic interactions, in which adaptations that increase the fitness of a consumer species exert selection pressure on its resource species to counter the consumer's adaptation, and vice versa.

**arteriole** A small blood vessel arising from an artery that feeds blood into a capillary bed.

**artery** A muscular blood vessel carrying oxygenated blood away from the heart to other parts of the body. (Contrast with vein.)

**artificial selection** The selection by plant and animal breeders of individuals with certain desirable traits.

**ascus** (ass′ cus) (plural: asci) [Gk. *askos*: bladder] In sac fungi, the club-shaped sporangium within which spores (ascospores) are produced by meiosis.

**asexual reproduction** Reproduction without sex.

**assisted reproductive technologies (ARTs)** Any of several procedures that remove unfertilized eggs from the ovary, combine them with sperm outside the body, and then place fertilized eggs or egg–sperm mixtures in the appropriate location in a female's reproductive tract for development.

**association cortex** In the vertebrate brain, the portion of the cortex involved in higher-order information processing, so named because it integrates, or associates, information from different sensory modalities and from memory.

**associative learning** A form of learning in which two unrelated stimuli become linked to the same response.

**astrocyte** [Gk. *astron*: star] A type of glial cell that contributes to the blood–brain barrier by surrounding the smallest, most permeable blood vessels in the brain.

**atherosclerosis** (ath′ er oh sklair oh′ sis) [Gk. *athero*: gruel, porridge + *skleros*: hard] A disease of the lining of the arteries characterized by fatty, cholesterol-rich deposits in the walls of the arteries. When fibroblasts infiltrate these deposits and calcium precipitates in them, the disease become arteriosclerosis, or "hardening of the arteries."

**atom** [Gk. *atomos*: indivisible] The smallest unit of a chemical element. Consists of a nucleus and one or more electrons.

**atomic mass** *See* atomic weight.

**atomic number** The number of protons in the nucleus of an atom; also equals the number of electrons around the neutral atom. Determines the chemical properties of the atom.

**atomic weight** The average of the mass numbers of a representative sample of atoms of an element, with all the isotopes in their normally occurring proportions. Also called atomic mass.

**ATP (adenosine triphosphate)** An energy-storage compound containing adenine, ribose, and three phosphate groups. When it is formed from ADP, useful energy is stored; when it is broken down (to ADP or AMP), energy is released to drive endergonic reactions.

**ATP synthase** An integral membrane protein that couples the transport of protons with the formation of ATP.

**atrial natriuretic peptide** A hormone released by the atrial muscle fibers of the heart when they are overly stretched, which decreases reabsorption of sodium by the kidney and thus blood volume.

**atrioventricular node** A modified node of cardiac muscle that organizes the action potentials that control contraction of the ventricles.

**atrium** (a′ tree um) [L. *atrium*: central hall] An internal chamber. In the hearts of vertebrates, the thin-walled chamber(s) entered by blood on its way to the ventricle(s). Also, the outer ear.

**auditory system** A sensory system that uses mechanoreceptors to convert pressure waves into receptor potentials; includes structures that gather sound waves, direct them to a sensory organ, and amplify their effect on the mechanoreceptors.

**autocatalysis** [Gk. *autos*: self + *kata*: to break down] A positive feedback process in which an activated enzyme acts on other inactive molecules of the same enzyme to activate them.

**autocrine** A chemical signal that binds to and affects the cell that makes it. (Contrast with paracrine.)

**autoimmunity** An immune response by an organism to its own molecules or cells.

**autonomic nervous system (ANS)** The portion of the peripheral nervous system that controls such involuntary functions as those of guts and glands. Also called the involuntary nervous system.

**autopolyploidy** The possession of more than two entire chromosomes sets that are derived from a single species.

**autosome** Any chromosome (in a eukaryote) other than a sex chromosome.

**autotroph** (au′ tow trowf′) [Gk. *autos*: self + *trophe*: food] An organism that is capable of living exclusively on inorganic materials, water, and some energy source such as sunlight (photoautotrophs) or chemically reduced matter (see chemolithotrophs). (Contrast with heterotroph.)

**auxin** (awk′ sin) [Gk. *auxein*: to grow] In plants, a substance (the most common being indoleacetic acid) that regulates growth and various aspects of development.

**avirulence (*Avr*) genes** Genes in a pathogen that may trigger defenses in plants. *See* gene-for-gene resistance.

**Avogadro's number** The number of atoms or molecules in a mole (weighed out in grams) of a substance, calculated to be $6.022 \times 10^{23}$.

**axillary bud** A bud that forms in the angle (axil) where a leaf meets a stem.

**axon** [Gk. axle] The part of a neuron that conducts action potentials away from the cell body.

**axon hillock** The junction between an axon and its cell body, where action potentials are generated.

**axon terminals** The endings of an axon; they form synapses and release neurotransmitter.

## - B -

**B cell** A type of lymphocyte involved in the humoral immune response of vertebrates. Upon recognizing an antigenic determinant, a B cell develops into a plasma cell, which secretes an antibody. (Contrast with T cell.)

**bacillus** (bah sil′ us) [L: little rod] Any of various rod-shaped bacteria.

**bacterial artificial chromosome (BAC)** A DNA cloning vector used in bacteria that can carry up to 150,000 base pairs of foreign DNA.

**bacteriophage** (bak teer′ ee o fayj) [Gk. *bakterion*: little rod + *phagein*: to eat] Any of a group of viruses that infect bacteria. Also called phage.

**bacteroids** Nitrogen-fixing organelles that develop from endosymbiotic bacteria.

**bark** All tissues external to the vascular cambium of a plant.

**baroreceptor** [Gk. *baros*: weight] A pressure-sensing cell or organ. Sometimes called a stress receptor.

**basal body** A centriole found at the base of a eukaryotic flagellum or cilium.

**basal metabolic rate (BMR)** The minimum rate of energy turnover in an awake (but resting) bird or mammal that is not expending energy for thermoregulation.

**base** (1) A substance that can accept a hydrogen ion in solution. (Contrast with acid.) (2) In nucleic acids, the purine or pyrimidine that is attached to each sugar in the sugar–phosphate backbone.

**base pair (bp)** In double-stranded DNA, a pair of nucleotides formed by the *complementary base pairing* of a purine on one strand and a pyrimidine on the other.

**basic** Having a pH greater than 7.0 (i.e., having a hydrogen ion concentration lower than $10^{-7}$ molar). (Contrast with acidic.)

**basidiocarp** A fruiting structure produced by club fungi.

**basidium** (bass id′ ee yum) In club fungi, the characteristic sporangium in which four spores are formed by meiosis and then borne externally before being shed.

**basilar membrane** A membrane in the human inner ear whose flexion in response to sound waves activates hair cells; flexes at different locations in response to different pitches of sound.

**basophil** A type of phagocytic white blood cell that releases histamine and may promote T cell development.

**Batesian mimicry** The convergence in appearance of an edible species (mimic) with an unpalatable species (model).

**benefit** An improvement in survival and reproductive success resulting from performing a behavior or having a trait. (Contrast with cost.)

**benthic zone** [Gk. *benthos*: bottom] The bottom of the ocean.

**beta diversity** Between-habitat diversity; a measure of the change in species composition from one community or habitat to another. (Contrast with alpha diveristy, gamma diversity.)

**β (beta) pleated sheet** A type of protein secondary structure; results from hydrogen bonding between polypeptide regions running antiparallel to each other.

**biennial** A plant whose life cycle includes vegetative growth in the first year and flowering and senescence in the second year. (Contrast with annual, perennial.)

**bilateral symmetry** The condition in which only the right and left sides of an organism, divided by a single plane through the midline, are mirror images of each other.

**bilayer** A structure that is two layers in thickness. In biology, most often refers to the phospholipid bilayer of membranes. (*See* phospholipid bilayer.)

**bile** A secretion of the liver made up of bile salts synthesized from cholesterol, various phospholipids, and bilirubin (the breakdown product of hemoglobin). Emulsifies fats in the small intestine.

**binary fission** Reproduction of a prokaryote by division of a cell into two comparable progeny cells.

**binocular vision** Overlapping visual fields of an animal's two eyes; allows the animal to see in three dimensions.

**binomial nomenclature** A taxonomic naming system in which each species is given two names (a genus name followed by a species name).

**biofilm** A community of microorganisms embedded in a polysaccharide matrix, forming a highly resistant coating on almost any moist surface.

**biogeochemical cycle** Movement of inorganic elements such as nitrogen, photosphorus, and carbon through living organisms and the physical environment.

**biogeographic region** One of several defined, continental-scale regions of Earth, each of which has a biota distinct from that of the others. (Contrast with biome.)

**biogeography** The scientific study of the patterns of distribution of populations, species, and ecological communities across Earth.

**bioinformatics** The use of computers and/or mathematics to analyze complex biological information, such as DNA sequences.

**biological species concept** The definition of a species as a group of actually or potentially interbreeding natural populations that are reproductively isolated from other such groups. (Contrast with lineage species concept; morphological species concept.)

**biology** [Gk. *bios*: life + *logos*: study] The scientific study of living things.

**bioluminescence** The production of light by biochemical processes in an organism.

**biomass** The total weight of all the organisms, or some designated group of organisms, in a given area.

**biome** (bye' ome) A major division of the ecological communities of Earth, characterized primarily by distinctive vegetation. A given biogeographic region contains many different biomes.

**bioremediation** The use by humans of other organisms to remove contaminants from the environment.

**biosphere** (bye' oh sphere) All regions of Earth (terrestrial and aquatic) and Earth's atmosphere in which organisms can live.

**biota** (bye oh' tah) All of the organisms—animals, plants, fungi, and microorganisms—found in a given area. (Contrast with flora, fauna.)

**biotechnology** The use of cells or living organisms to produce materials useful to humans.

**biotic** (bye ah' tick) [Gk. *bios*: life] Alive. (Contrast with abiotic.)

**biotic interchange** The dispersal of species from two different biotas into the region they had not previously inhabited, as when two formerly separated land masses fuse.

**blade** The thin, flat portion of a leaf.

**blastocoel** (blass' toe seal) [Gk. *blastos*: sprout + *koilos*: hollow] The central, hollow cavity of a blastula.

**blastocyst** (blass' toe cist) An early embryo formed by the first divisions of the fertilized egg (zygote). In mammals, a hollow ball of cells.

**blastodisc** (blass' toe disk) An embryo that forms as a disk of cells on the surface of a large yolk mass; comparable to a blastula, but occurring in animals such as birds and reptiles, in which the massive yolk restricts in incomplete cleavage.

**blastomere** Any of the cells produced by the early divisions of a fertilized animal egg.

**blastula** (blass' chu luh) An early stage of the animal embryo; in many species, a hollow sphere of cells surrounding a central cavity, the blastocoel. (Contrast with blastodisc.)

**block to polyspermy** Any of several responses to entry of a sperm into an egg that prevent more than one sperm from entering the egg.

**blood** A fluid tissue that is pumped around the body; a component of the circulatory system.

**blood–brain barrier** A property of blood vessels in the brain that prevents most chemicals from diffusing from the blood into the brain.

**blue-light receptors** Pigments in plants that absorb blue light (400–500 nm). These pigments mediate many plant responses including phototropism, stomatal movements, and expression of some genes.

**body plan** The general structure of an animal, the arrangement of its organ systems, and the integrated functioning of its parts.

**bond** *See* chemical bond.

**bottleneck** *See* population bottleneck.

**bone** A rigid component of vertebrate skeletal systems that contains an extracellular matrix of insoluble calcium phosphate crystals as well as collagen fibers.

**Bowman's capsule** An elaboration of the renal tubule, composed of podocytes, that surrounds and collects the filtrate from the glomerulus.

**brain** The centralized integrative center of a nervous system.

**brainstem** The portion of the vertebrate brain between the spinal cord and the forebrain, made up of the medulla, pons, and midbrain.

**brassinosteroids** Plant steroid hormones that mediate light effects promoting the elongation of stems and pollen tubes.

**Broca's area** A portion of the human brain essential for speech. Located in the frontal lobe just in front of the primary motor cortex.

**bronchioles** The smallest airways in a vertebrate lung, branching off the bronchi.

**bronchus** (plural: bronchi) The major airway(s) branching off the trachea into the vertebrate lung.

**brown fat** In mammals, fat tissue that is specialized to produce heat. It has many mitochondria and capillaries, and a protein that uncouples oxidative phosphorylation.

**budding** Asexual reproduction in which a more or less complete new organism grows from the body of the parent organism, eventually detaching itself.

**buffer** A substance that can transiently accept or release hydrogen ions and thereby resist changes in pH.

**bulbourethral glands** Secretory structures of the human male reproductive system that produce a small volume of an alkaline, mucoid secretion that helps neutralize acidity in the urethra and lubricate it to facilitate the passage of semen.

**bulk flow** The movement of a solution from a region of higher pressure potential to a region of lower pressure potential.

**bundle of His** Fibers of modified cardiac muscle that conduct action potentials from the atria to the ventricular muscle mass.

**bundle sheath cell** Part of a tissue that surrounds the veins of plants; contains chloroplasts in $C_4$ plants.

## - C -

**C3 plants** Plants that produce 3PG as the first stable product of carbon fixation in photosynthesis and use ribulose bisphosphate as a $CO_2$ receptor.

**C4 plants** Plants that produce oxaloacetate as the first stable product of carbon fixation in photosynthesis and use phosphoenolpyruvate as $CO_2$ acceptor. $C_4$ plants also perform the reactions of $C_3$ photosynthesis.

**calcitonin** Hormone produced by the thyroid gland; lowers blood calcium and promotes bone formation. (Contrast with parathyroid hormone.)

**calorie** [L. *calor*: heat] The amount of heat required to raise the temperature of 1 gram of water by 1°C. Physiologists commonly use the kilocalorie (kcal) as a unit of measure (1 kcal = 1,000 calories). Nutritionists also use the kilocalorie, but refer to it as the *Calorie* (capital C).

**Calvin cycle** The stage of photosynthesis in which $CO_2$ reacts with RuBP to form 3PG, 3PG is reduced to a sugar, and RuBP is regenerated, while other products are released to the rest of the plant. Also known as the Calvin–Benson cycle.

**calyx** (kay' licks) [Gk. *kalyx*: cup] All of the sepals of a flower, collectively.

**CAM** *See* crassulacean acid metabolism.

**Cambrian explosion** The rapid diversification of multicellular life that took place during the Cambrian period.

**cAMP (cyclic AMP)** A compound formed from ATP that acts as a second messenger.

**cancellous bone** A type of bone with numerous internal cavities that make it appear spongy, although it is rigid. (Contrast with compact bone.)

**canopy** The leaf-bearing part of a tree. Collectively, the aggregate of the leaves and branches of the larger woody plants of an ecological community.

**capillaries** [L. *capillaris*: hair] Very small tubes, especially the smallest blood-carrying vessels of animals between the termination of the arteries and the beginnings of the veins. Capillaries are the site of exchange of materials between the blood and the interstitial fluid.

**capsid** The outer shell of a virus that encloses its nucleic acid.

**carbohydrates** Organic compounds containing carbon, hydrogen, and oxygen in the ratio 1:2:1 (i.e., with the general formula $C_nH_{2n}O_n$). Common examples are sugars, starch, and cellulose.

**carbon skeleton** The chains or rings of carbon atoms that form the structural basis of organic molecules. Other atoms or functional groups are attached to the carbon atoms.

**carboxylase** An enzyme that catalyzes the addition of carboxyl groups to a substrate.

**cardiac** (kar' dee ak) [Gk. *kardia*: heart] Pertaining to the heart and its functions.

**cardiac cycle** Contraction of the two atria of the heart, followed by contraction of the two ventricles and then relaxation.

**cardiac muscle** A type of muscle tissue that makes up, and is responsible for the beating of, the heart. Characterized by branching cells with single nuclei and a striated (striped) appearance. (Contrast with smooth muscle, skeletal muscle.)

**cardiovascular system** [Gk. *kardia*: heart + L. *vasculum*: small vessel] The heart, blood, and vessels are of a circulatory system.

**carnivore** [L. *carn*: flesh + *vovare*: to devour] An organism that eats animal tissues. (Contrast with detritivore, herbivore, omnivore.)

**carotenoid** (ka rah' tuh noid) A yellow, orange, or red lipid pigment commonly found as an accessory pigment in photosynthesis; also found in fungi.

**carotid body** A chemosensor in the carotid artery that senses a decrease in blood supply or a dramatic decrease in partial pressure of oxygen in the blood.

**carpel** (kar′ pel) [Gk. *karpos*: fruit] The organ of the flower that contains one or more ovules.

**carrier** (1) In facilitated diffusion, a membrane protein that binds a specific molecule and transports it through the membrane. (2) In respiratory and photosynthetic electron transport, a participating substance such as NAD that exists in both oxidized and reduced forms. (3) In genetics, a person heterozygous for a recessive trait.

**carrying capacity (*K*)** The number of individuals in a population that the resources of its environment can support.

**cartilage** In vertebrates, a tough connective tissue found in joints, the outer ear, and elsewhere. Forms the entire skeleton in some animal groups.

**cartilage bone** A type of bone that begins its development as a cartilaginous structure resembling the future mature bone, then gradually hardens into mature bone. (Contrast with membranous bone.)

**Casparian strip** A band of cell wall containing suberin and lignin, found in the endodermis. Restricts the movement of water across the endodermis.

**caspase** One of a group of proteases that catalyze cleavage of target proteins and are active in apoptosis.

**catabolic reaction** (kat uh bah′ lik) [Gk. *kata*: to break down + *ballein*: to throw] A synthetic reaction in which complex molecules are broken down into simpler ones and energy is released. (Contrast with anabolic reaction.)

**catabolite repression** In the presence of abundant glucose, the diminished synthesis of catabolic enzymes for other energy sources.

**catalyst** (kat′ a list) [Gk. *kata*: to break down] A chemical substance that accelerates a reaction without itself being consumed in the overall course of the reaction. Catalysts lower the activation energy of a reaction. Enzymes are biological catalysts.

**cation** (cat′ eye on) An ion with one or more positive charges. (Contrast with anion.)

**caudal** [L. *cauda*: tail] Pertaining to the tail, or to the posterior part of the body.

**cDNA** *See* complementary DNA.

**cDNA library** A collection of complementary DNAs derived from mRNAs of a particular tissue at a particular time in the life cycle of an organism.

**cecum** (see′ cum) [L. blind] A blind branch off the large intestine. In many nonruminant mammals, the cecum contains a colony of microorganisms that contribute to the digestion of food.

**cell** The simplest structural unit of a living organism. In multicellular organisms, the building blocks of tissues and organs.

**cell adhesion molecules** Molecules on animal cell surfaces that affect the selective association of cells into tissues during development of the embryo.

**cell cycle** The stages through which a cell passes between one division and the next. Includes all stages of interphase and mitosis.

**cell division** The reproduction of a cell to produce two new cells. In eukaryotes, this process involves nuclear division (mitosis) and cytoplasmic division (cytokinesis).

**cell fate** The type of cell that an undifferentiated cell in an embryo will become in the adult.

**cell junctions** Specialized structures associated with the plasma membranes of epithelial cells. Some contribute to cell adhesion, others to intercellular communication.

**cell recognition** Binding of cells to one another mediated by membrane proteins or carbohydrates.

**cell theory** States that cells are the basic structural and physiological units of all living organisms, and that all cells come from preexisting cells.

**cell wall** A relatively rigid structure that encloses cells of plants, fungi, many protists, and most prokaryotes, and which gives these cells their shape and limits their expansion in hypotonic media.

**cellular immune response** Immune system response mediated by T cells and directed against parasites, fungi, intracellular viruses, and foreign tissues (grafts). (Contrast with humoral immune response.)

**cellular respiration** The catabolic pathways by which electrons are removed from various molecules and passed through intermediate electron carriers to $O_2$, generating $H_2O$ and releasing energy.

**cellulose** (sell′ you lowss) A straight-chain polymer of glucose molecules, used by plants as a structural supporting material.

**central dogma** The premise that information flows from DNA to RNA to polypeptide.

**central nervous system (CNS)** That portion of the nervous system that is the site of most information processing, storage, and retrieval; in vertebrates, the brain and spinal cord. (Contrast with peripheral nervous system.)

**central vacuole** In plant cells, a large organelle that stores the waste products of metabolism and maintains turgor.

**centrifuge** [L. *centrum*: center + *fugere*: to flee] A laboratory device in which a sample is spun around a central axis at high speed. Used to separate suspended materials of different densities.

**centriole** (sen′ tree ole) A paired organelle that helps organize the microtubules in animal and protist cells during nuclear division.

**centromere** (sen′ tro meer) [Gk. *centron*: center + *meros*: part] The region where sister chromatids join.

**centrosome** (sen′ tro soam) The major microtubule organizing center of an animal cell.

**cephalization** (sef ah luh zay′ shun) [Gk. *kephale*: head] The evolutionary trend toward increasing concentration of brain and sensory organs at the anterior end of the animal.

**cerebellum** (sair uh bell′ um) [L. diminutive of *cerebrum*, brain] The brain region that controls muscular coordination; located at the anterior end of the hindbrain.

**cerebral cortex** The thin layer of gray matter (neuronal cell bodies) that overlies the cerebrum.

**cerebrum** (su ree′ brum) [L. brain] The dorsal anterior portion of the forebrain, making up the largest part of the brain of mammals; the chief coordination center of the nervous system; consists of two *cerebral hemispheres*.

**cervix** (sir′ vix) [L. neck] The opening of the uterus into the vagina.

**cGMP (cyclic guanosine monophosphate)** An intracellular messenger that is part of signal transmission pathways involving G proteins. (*See* G protein.)

**channel protein** An integral membrane protein that forms an aqueous passageway across the membrane in which it is inserted through which specific solutes may pass.

**chaperone** A protein that guards other proteins by counteracting molecular interactions that threaten their three-dimensional structure.

**character** In genetics, an observable feature, such as eye color. (Contrast with trait.)

**character displacement** An evolutionary phenomenon in which species that compete for the same resources within the same territory tend to diverge in morphology and/or behavior.

**chemical bond** An attractive force stably linking two atoms.

**chemical evolution** The theory that life originated through the chemical transformation of inanimate substances.

**chemical reaction** The change in the composition or distribution of atoms of a substance with consequent alterations in properties.

**chemical synapse** Neural junction at which neurotransmitter molecules released from a presynaptic cell induce changes in a postsynaptic cell. (Contrast with electrical synapse.)

**chemically gated channel** A type of gated channel that opens or closes depending on the presence or absence of a specific molecule, which binds to the channel protein or to a separate receptor that in turn alters the three-dimensional shape of channel protein.

**chemiosmosis** Formation of ATP in mitochondria and chloroplasts, resulting from a pumping of protons across a membrane (against a gradient of electrical charge and of pH), followed by the return of the protons through a protein channel with ATP synthase activity.

**chemoautotroph** *See* chemolithotroph.

**chemoheterotroph** An organism that must obtain both carbon and energy from organic substances. (Contrast with chemolithotroph, photoautotroph, photoheterotroph.)

**chemolithotroph** [Gk. *lithos*: stone, rock] An organism that uses carbon dioxide as a carbon source and obtains energy by oxidizing inorganic substances from its environment; also called chemoautotroph. (Contrast with chemoheterotroph, photoautotroph, photoheterotroph.)

**chemoreceptor** A sensory receptor cell that senses specific molecules (such as odorant molecules or pheromones) in the environment.

**chiasma** (kie az′ muh) (plural: chiasmata) [Gk. cross] An X-shaped connection between paired homologous chromosomes in prophase I of meiosis. A chiasma is the visible manifestation of crossing over between homologous chromosomes.

**chitin** (kye′ tin) [Gk. *kiton*: tunic] The characteristic tough but flexible organic component of the exoskeleton of arthropods, consisting of a complex, nitrogen-containing polysaccharide. Also found in cell walls of fungi.

**chlorophyll** (klor′ o fill) [Gk. *kloros*: green + *phyllon*: leaf] Any of several green pigments associated with chloroplasts or with certain bacterial membranes; responsible for trapping light energy for photosynthesis.

**chloroplast** [Gk. *kloros*: green + *plast*: a particle] An organelle bounded by a double membrane containing the enzymes and pigments that perform photosynthesis. Chloroplasts occur only in eukaryotes.

**choanocyte** (ko' an uh site) The collared, flagellated feeding cells of sponges.

**cholecystokinin** (ko' luh sis tuh kai' nin) A hormone produced and released by the lining of the duodenum when it is stimulated by undigested fats and proteins. It stimulates the gallbladder to release bile and slows stomach activity.

**chorion** (kor' ee on) [Gk. *khorion*: afterbirth] The outermost of the membranes protecting mammal, bird, and reptile embryos; in mammals it forms part of the placenta.

**chromatid** (kro' ma tid) A newly replicated chromosome, from the time molecular duplication occurs until the time the centromeres separate (during anaphase of mitosis or of meiosis II).

**chromatin** The nucleic acid–protein complex that makes up eukaryotic chromosomes.

**chromosomal mutation** Loss of or changes in position/direction of a DNA segment on a chromosome.

**chromosome** (krome' o sowm) [Gk. *kroma*: color + *soma*: body] In bacteria and viruses, the DNA molecule that contains most or all of the genetic information of the cell or virus. In eukaryotes, a structure composed of DNA and proteins that bears part of the genetic information of the cell.

**chylomicron** (ky low my' cron) Particles of lipid coated with protein, produced in the gut from dietary fats and secreted into the extracellular fluids.

**chyme** (kime) [Gk. *kymus*: juice] Created in the stomach; a mixture of ingested food with the digestive juices secreted by the salivary glands and the stomach lining.

**cilium** (sil' ee um) (plural: cilia) [L. eyelash] Hairlike organelle used for locomotion by many unicellular organisms and for moving water and mucus by many multicellular organisms. Generally shorter than a flagellum.

**circadian rhythm** (sir kade' ee an) [L. *circa*: approximately + *dies*: day] A rhythm of growth or activity that recurs about every 24 hours.

**circannual rhythm** [L. *circa*: + *annus*: year) A rhythm of growth or activity that recurs on a yearly basis.

**circulatory system** A system consisting of a muscular pump (heart), a fluid (blood or hemolymph), and a series of conduits (blood vessels) that transports materials around the body.

**citric acid cycle** In cellular respiration, a set of chemical reactions whereby acetyl CoA is oxidized to carbon dioxide and hydrogen atoms are stored as NADH and $FADH_2$. Also called the Krebs cycle.

**clade** [Gk. *klados*: branch] A monophyletic group made up of an ancestor and all of its descendants.

**class I MHC molecules** Cell surface proteins that participate in the cellular immune response directed against virus-infected cells.

**class II MHC molecules** Cell surface proteins that participate in the cell–cell interactions (of T-helper cells, macrophages, and B cells) of the humoral immune response.

**cleavage** The first few cell divisions of an animal zygote. *See also* complete cleavage, incomplete cleavage.

**climate** The long-term average atmospheric conditions (temperature, precipitation, humidity, wind direction and velocity) found in a region.

**climax community** The final stage of succession; a community that is capable of perpetuating itself under local climatic and soil conditions and persists for a relatively long time.

**clinal variation** [Gk. *klinein*: to lean] Gradual change in the phenotype of a species over a geographic gradient.

**clonal deletion** Inactivation or destruction of lymphocyte clones that would produce immune reactions against the animal's own body.

**clonal selection** Mechanism by which exposure to antigen results in the activation of selected T- or B-cell clones, resulting in an immune response.

**clone** [Gk. *klon*: twig, shoot] (1) Genetically identical cells or organisms produced from a common ancestor by asexual means. (2) To produce many identical copies of a DNA sequence by its introduction into, and subsequent asexual reproduction of, a cell or organism.

**closed circulatory system** Circulatory system in which the circulating fluid is contained within a continuous system of vessels. (Contrast with open circulatory system.)

**coastal zone** The marine life zone that extends from the shoreline to the edge of the continental shelf. Characterized by relatively shallow, well-oxygenated water and relatively stable temperatures and salinities.

**coccus** (kock' us) (plural: cocci) [Gk. *kokkos*: berry, pit] Any of various spherical or spheroidal bacteria.

**cochlea** (kock' lee uh) [Gk. *kokhlos*: snail] A spiral tube in the inner ear of vertebrates; it contains the sensory cells involved in hearing.

**codominance** A condition in which two alleles at a locus produce different phenotypic effects and both effects appear in heterozygotes.

**codon** Three nucleotides in messenger RNA that direct the placement of a particular amino acid into a polypeptide chain. (Contrast with anticodon.)

**coelom** (see' loam) [Gk. *koiloma*: cavity] An animal body cavity, enclosed by muscular mesoderm and lined with a mesodermal layer called peritoneum that also surrounds the internal organs.

**coenocytic** (seen' a sit ik) [Gk. *koinos*: common + *kytos*: container] Referring to the condition, found in some fungal hyphae, of "cells" containing many nuclei but enclosed by a single plasma membrane. Results from nuclear division without cytokinesis.

**coenzyme** A nonprotein organic molecule that plays a role in catalysis by an enzyme.

**coevolution** Evolutionary processes in which an adaptation in one species leads to the evolution of an adaptation in a species with which it interacts; also known as reciprocal adaptation.

**cofactor** An inorganic ion that is weakly bound to an enzyme and required for its activity.

**cohesin** A protein involved in binding chromatids together.

**cohesion** The tendency of molecules (or any substances) to stick together.

**cohort** (co' hort) [L. *cohors*: company of soldiers] A group of similar-aged organisms.

**coleoptile** A sheath that surrounds and protects the shoot apical meristem and young primary leaves of a grass seedling as they move through the soil.

**collagen** [Gk. *kolla*: glue] A fibrous protein found extensively in bone and connective tissue.

**collecting duct** In vertebrates, a tubule that receives urine produced in the nephrons of the kidney and delivers that fluid to the ureter for excretion.

**collenchyma** (cull eng' kyma) [Gk. *kolla*: glue + *enchyma*: infusion] A type of plant cell, living at functional maturity, which lends flexible support by virtue of primary cell walls thickened at the corners. (Contrast with parenchyma, sclerenchyma.)

**colon** [Gk. *kolon*] The large intestine.

**commensalism** [L. *com*: together + *mensa*: table] A type of interaction between species in which one participant benefits while the other is unaffected.

**communication** A signal from one organism (or cell) that alters the functioning or behavior of another organism (or cell).

**community** Any ecologically integrated group of species of microorganisms, plants, and animals inhabiting a given area.

**compact bone** A type of bone with a solid, hard structure. (Contrast with cancellous bone.)

**companion cell** In angiosperms, a specialized cell found adjacent to a sieve tube element.

**comparative experiment** Experimental design in which data from various unmanipulated samples or populations are compared, but in which variables are not controlled or even necessarily identified. (Contrast with controlled experiment.)

**comparative genomics** Computer-aided comparison of DNA sequences between different organisms to reveal genes with related functions.

**competition** In ecology, use of the same resource by two or more species when the resource is present in insufficient supply for the combined needs of the species.

**competitive exclusion** A result of competition between species for a limiting resource in which one species completely eliminates the other.

**competitive inhibitor** A nonsubstrate that binds to the active site of an enzyme and thereby inhibits binding of its substrate. (Contrast with noncompetitive inhibitor.)

**complement system** A group of eleven proteins that play a role in some reactions of the immune system. The complement proteins are not immunoglobulins.

**complementary base pairing** The AT (or AU), TA (or UA), CG, and GC pairing of bases in double-stranded DNA, in transcription, and between tRNA and mRNA.

**complementary DNA (cDNA)** DNA formed by reverse transcriptase acting with an RNA template; essential intermediate in the reproduction of retroviruses; used as a tool in recombinant DNA technology; lacks introns.

**complete cleavage** Pattern of cleavage that occurs in eggs that have little yolk. Early cleavage furrows divide the egg completely and the blastomeres are of similar size. (Contrast with incomplete cleavage.)

**complete metamorphosis** A change of state during the life cycle of an organism in which the body is almost completely rebuilt to produce an individual with a very different body form. Characteristic of insects such as butterflies, moths, beetles, ants, wasps, and flies.

**compound** (1) A substance made up of atoms of more than one element. (2) Made up of many units, as in the *compound eyes* of arthropods.

**concerted evolution** The common evolution of a family of repeated genes, such that changes in one copy of the gene family are replicated in other copies of the gene family.

**condensation reaction** A chemical reaction in which two molecules become connected by a covalent bond and a molecule of water is released ($AH + BOH \rightarrow AB + H_2O$.) (Contrast with hydrolysis reaction.)

**conditional mutation** A mutation that results in a characteristic phenotype only under certain environmental conditions.

**conduction** The transfer of heat from one object to another through direct contact.

**cone** (1) In conifers, a reproductive structure consisting of spore-bearing scales extending from a central axis. (Contrast with strobilus.) (2) In the vertebrate retina, a type of photoreceptor cell responsible for color vision.

**conidium** (ko nid' ee um) (plural: conidia) [Gk. *konis*: dust] A type of haploid fungal spore borne at the tips of hyphae, not enclosed in sporangia.

**conjugation** (kon ju gay' shun) [L. *conjugare*: yoke together] (1) A process by which DNA is passed from one cell to another through a *conjugation tube*, as in bacteria. (2) A nonreproductive sexual process by which *Paramecium* and other ciliates exchange genetic material.

**connective tissue** A type of tissue that connects or surrounds other tissues; its cells are embedded in a collagen-containing matrix. One of the four major tissue types in multicellular animals.

**connexon** In a gap junction, a protein channel linking adjacent animal cells.

**consensus sequences** Short stretches of DNA that appear, with little variation, in many different genes.

**conservation biology** An applied science that carries out investigations with the aim of maintaining the diversity of life on Earth.

**conserved** Pertaining to a gene or trait that has evolved very slowly and is similar or even identical in individuals of highly divergent groups.

**conspecifics** Individuals of the same species.

**constant region** The portion of an immunoglobulin molecule whose amino acid composition determines its class and does not vary among immunoglobulins in that class. (Contrast with variable region.)

**constitutive** Always present; produced continually at a constant rate. (Contrast with inducible.)

**consumer** An organism that eats the tissues of some other organism.

**continental drift** The gradual movements of the world's continents that have occurred over billions of years.

**contractile vacuole** (kon trak' tul) A specialized vacuole that collects excess water taken in by osmosis, then contracts to expel the water from the cell.

**controlled experiment** An experiment in which a sample is divided into groups whereby experimental groups are exposed to manipulations of an independent variable while one group serves as an untreated control. The data from the various groups are then compared to see if there are changes in a dependent variable as a result of the experimental manipulation. (Contrast with comparative experiment.)

**convection** The transfer of heat to or from a surface via a moving stream of air or fluid.

**convergent evolution** Independent evolution of similar features from different ancestral traits.

**copulation** Reproductive behavior that results in a male depositing sperm in the reproductive tract of a female.

**cork cambium** [L. *cambiare*: to exchange] In plants, a lateral meristem that produces secondary growth, mainly in the form of waxy-walled protective cells, including some of the cells that become bark.

**cornea** The clear, transparent tissue that covers the eye and allows light to pass through to the retina.

**corolla** (ko role' lah) [L. *corolla*: a small crown] All of the petals of a flower, collectively.

**coronary artery** (kor' oh nair ee) An artery that supplies blood to the heart muscle.

**coronary thrombosis** A fibrous clot that blocks a coronary artery.

**corpus luteum** (kor' pus loo' tee um) (plural: corpora lutea) [L. yellow body] A structure formed from a follicle after ovulation; produces hormones important to the maintenance of pregnancy.

**corridor** A connection between habitat patches through which organisms can disperse; plays a critical role in maintaining subpopulations.

**cortex** [L. *cortex*: covering, rind] (1) In plants, the tissue between the epidermis and the vascular tissue of a stem or root. (2) In animals, the outer tissue of certain organs, such as the adrenal gland (adrenal cortex) and the brain (cerebral cortex).

**corticosteroids** Steroid hormones produced and released by the cortex of the adrenal gland.

**corticotropin** A tropic hormone produced by the anterior pituitary hormone that stimulates cortisol release from the adrenal cortex. Also called adrenocorticotropic hormone (ACTH).

**corticotropin-releasing hormone** A releasing hormone produced by the hypothalamus that controls the release of cortisol from the anterior pituitary.

**cortisol** A corticosteroid that mediates stress responses.

**cost–benefit analysis** An approach to evolutionary studies that assumes an animal has a limited amount of time and energy to devote to each of its activities, and that each activity has fitness costs as well as benefits. (*See also* trade-off.)

**cotyledon** (kot' ul lee' dun) [Gk. *kotyledon*: hollow space] A "seed leaf." An embryonic organ that stores and digests reserve materials; may expand when seed germinates.

**countercurrent flow** An arrangement that promotes the maximum exchange of heat, or of a diffusible substance, between two fluids by having the fluids flow in opposite directions through parallel vessels close together.

**countercurrent multiplier** The mechanism that increases the concentration of the interstitial fluid in the mammalian kidney through countercurrent flow in the loops of Henle and selective permeability and active transport of ions by segments of the loops of Henle.

**covalent bond** Chemical bond based on the sharing of electrons between two atoms.

**CpG islands** DNA regions rich in C resides adjacent to G residues. Especially abundant in promoters, these regions are where methylation of cytosine usually occurs.

**crassulacean acid metabolism (CAM)** A metabolic pathway enabling the plants that possess it to store carbon dioxide at night and then perform photosynthesis during the day with stomata closed.

**critical night length** In the photoperiodic flowering response of short-day plants, the length of night above which flowering occurs and below which the plant remains vegetative. (The reverse applies in the case of long-day plants.)

**critical period** *See* sensitive period.

**cross section** A section taken perpendicular to the longest axis of a structure. Also called a transverse section.

**crossing over** The mechanism by which linked genes undergo recombination. In general, the term refers to the reciprocal exchange of corresponding segments between two homologous chromatids.

**crypsis** [Gk. *kryptos*: hidden] The resemblance of an organism to some part of its environment, which helps it to escape detection by enemies.

**cryptochromes** [Gk. *kryptos*: hidden + *kroma*: color] Photoreceptors mediating some blue-light effects in plants and animals.

**ctene** (teen) [Gk. *cteis*: comb] In ctenophores, a comblike row of cilia-bearing plates. Ctenophores move by beating the cilia on their eight ctenes.

**culture** (1) A laboratory association of organisms under controlled conditions. (2) The collection of knowledge, tools, values, and rules that characterize a human society.

**cuticle** (1) In plants, a waxy layer on the outer body surface that retards water loss. (2) In ecdysozoans, an outer body covering that provides protection and support and is periodically molted.

**cyclic AMP** *See* cAMP.

**cyclic electron transport** In photosynthetic light reactions, the flow of electrons that produces ATP but no NADPH or $O_2$.

**cyclin** A protein that activates a cyclin-dependent kinase, bringing about transitions in the cell cycle.

**cyclin-dependent kinase (Cdk)** A proetin kinase whose target proteins are involved in transitions in the cell cycle and which is active only when complexed with additional protein subunits, called cyclins.

**cytokine** A regulatory protein made by immune system cells that affects other target cells in the immune system.

**cytokinesis** (sy' toe kine ee' sis) [Gk. *kytos*: container + *kinein*: to move] The division of the cytoplasm of a dividing cell. (Contrast with mitosis.)

**cytokinin** (sy' toe kine' in) A member of a class of plant growth substances that plays roles in senescence, cell division, and other phenomena.

**cytoplasm** The contents of the cell, excluding the nucleus.

**cytoplasmic determinants** In animal development, gene products whose spatial distribution may determine such things as embryonic axes.

**cytoplasmic segregation** The asymmetrical distribution of cytoplasmic determinants in a developing animal embryo.

**cytosine (C)** (site′ oh seen) A nitrogen-containing base found in DNA and RNA.

**cytoskeleton** The network of microtubules and microfilaments that gives a eukaryotic cell its shape and its capacity to arrange its organelles and to move.

**cytosol** The fluid portion of the cytoplasm, excluding organelles and other solids.

**cytotoxic T cells ($T_C$)** Cells of the cellular immune system that recognize and directly eliminate virus-infected cells. (Contrast with T-helper cells.)

## - D -

**DAG** *See* diacylglycerol.

**daughter chromosomes** During mitosis, the separated chromatids from the beginning of anaphase onward.

**dead space** The lung volume that fails to be ventilated with fresh air (because the lungs are never completely emptied during exhalation).

**deciduous** [L. *deciduus*: falling off] Pertaining to a woody plant that sheds it leaves but does not die.

**declarative memory** Memory of people, places, events, and things that can be consciously recalled and described. (Contrast with procedural memory.)

**decomposer** An organism that metabolizes organic compounds in debris and dead organisms, releasing inorganic material; found among the bacteria, protists, and fungi. *See also* detritivore, saprobe.

**defensin** A type of protein made by phagocytes that kills bacteria and enveloped viruses by insertion into their plasma membranes.

**degeneracy** The situation in which a single amino acid may be represented by any of two or more different codons in messenger RNA. Most of the amino acids can be represented by more than one codon.

**deletion** A mutation resulting from the loss of a continuous segment of a gene or chromosome. Such mutations almost never revert to wild type. (Contrast with duplication, point mutation.)

**demethylase** An enzyme that catalyzes the removal of the methyl group from cytosine, reversing DNA methylation.

**demography** The study of population structure and of the processes by which it changes.

**denaturation** Loss of activity of an enzyme or nucleic acid molecule as a result of structural changes induced by heat or other means.

**dendrite** [Gk. *dendron*: tree] A fiber of a neuron which often cannot carry action potentials. Usually much branched and relatively short compared with the axon, and commonly carries information to the cell body of the neuron.

**denitrification** Metabolic activity by which nitrate and nitrite ions are reduced to form nitrogen gas; carried by certain soil bacteria.

**denitrifiers** Bacteria that release nitrogen to the atmosphere as nitrogen gas ($N_2$).

**density-dependent** Pertaining to a factor with an effect on population size that increases in proportion to population density.

**density-independent** Pertaining to a factor with an effect on population size that acts independently of population density.

**deoxyribonucleic acid** *See* DNA.

**deoxyribose** A five-carbon sugar found in nucleotides and DNA.

**depolarization** A change in the resting potential across a membrane so that the inside of the cell becomes less negative, or even positive, compared with the outside of the cell. (Contrast with hyperpolarization.)

**derived trait** A trait that differs from the ancestral trait. (Contrast with shared derived trait.)

**dermal tissue system** The outer covering of a plant, consisting of epidermis in the young plant and periderm in a plant with extensive secondary growth. (Contrast with ground tissue system and vascular tissue system.)

**desmosome** (dez′ mo sowm) [Gk. *desmos*: bond + *soma*: body] An adhering junction between animal cells.

**desmotubule** A membrane extension connecting the endoplasmic retitulum of two plant cells that traverses the plasmodesma.

**determination** In development, the process whereby the fate of an embryonic cell or group of cells (e.g., to become epidermal cells or neurons) is set.

**determinate growth** A growth pattern in which the growth of an organism or organ ceases when an adult state is reached; characteristic of most animals and some plant organs. (Contrast with indeterminate growth.)

**detritivore** (di try′ ti vore) [L. *detritus*: worn away + *vorare*: to devour] An organism that obtains its energy from the dead bodies or waste products of other organisms.

**developmental module** A functional entity in the embryo encompassing genes and signaling pathways that determine a physical structure independently of other such modules.

**developmental plasticity** The capacity of an organism to alter its pattern of development in response to environmental conditions.

**diacylglycerol (DAG)** In hormone action, the second messenger produced by hydrolytic removal of the head group of certain phospholipids.

**diapause** A period of developmental or reproductive arrest, entered in response to day length, that enables an organism to better survive.

**diaphragm** (dye′ uh fram) [Gk. *diaphrassein*: barricade] (1) A sheet of muscle that separates the thoracic and abdominal cavities in mammals; responsible for breathing. (2) A method of birth control in which a sheet of rubber is fitted over the woman's cervix, blocking the entry of sperm.

**diastole** (dye ass′ toll ee) [Gk. dilation] The portion of the cardiac cycle when the heart muscle relaxes. (Contrast with systole.)

**diencephalon** The portion of the vertebrate forebrain that develops into the thalamus and hypothalamus.

**differential gene expression** The hypothesis that, given that all cells contain all genes, what makes one cell type different from another is the difference in transcription and translation of those genes.

**differentiation** The process whereby originally similar cells follow different developmental pathways; the actual expression of determination.

**diffuse coevolution** The evolution of similar traits in suites of species experiencing similar selection pressures imposed by other suites of species with which they interact.

**diffusion** Random movement of molecules or other particles, resulting in even distribution of the particles when no barriers are present.

**dihybrid cross** A mating in which the parents differ with respect to the alleles of two loci of interest.

**dikaryon** (di care′ ee ahn) [Gk. *di*: two + *karyon*: kernel] A cell or organism carrying two genetically distinguishable nuclei. Common in fungi.

**dioecious** (die eesh′ us) [Gk. *di*: two + *oikos*: house] Pertaining to organisms in which the two sexes are "housed" in two different individuals, so that eggs and sperm are not produced in the same individuals. Examples: humans, fruit flies, date palms. (Contrast with monoecious.)

**diploblastic** Having two cell layers. (Contrast with triploblastic.)

**diploid** (dip′ loid) [Gk. *diplos*: double] Having a chromosome complement consisting of two copies (homologs) of each chromosome. Designated $2n$.

**diplontic** A type of life cycle in which gametes are the only haploid cells and mitosis occurs only in diploid cells. (Contrast with haplontic.)

**direct transduction** A cell signaling mechanism in which the receptor acts as the effector in the cellular response. (Contrast with indirect transduction.)

**directional selection** Selection in which phenotypes at one extreme of the population distribution are favored. (Contrast with disruptive selection, stabilizing selection.)

**disaccharide** A carbohydrate made up of two monosaccharides (simple sugars).

**dispersal** Movement of organisms away from a parent organism or from an existing population.

**dispersion** The distribution of individuals in space within a population.

**disruptive selection** Selection in which phenotypes at both extremes of the population distribution are favored. (Contrast with directional selection; stabilizing selection.)

**distal** Away from the point of attachment or other reference point. (Contrast with proximal.)

**distal convoluted tubule** The portion of a renal tubule from where it reaches the renal cortex, just past the loop of Henle to where it joins a collecting duct. (Compare with proximal convoluted tubule.)

**disturbance** A short-term event that disrupts populations, communities, or ecosystems by changing the environment.

**disulfide bridge** The covalent bond between two sulfur atoms (–S—S–) linking two molecules or remote parts of the same molecule.

**DNA (deoxyribonucleic acid)** The fundamental hereditary material of all living organisms. In eukaryotes, stored primarily in the cell nucleus. A nucleic acid using deoxyribose rather than ribose.

**DNA fingerprint** An individual's unique pattern of allele sequences, commonly short tandem repeats and single nucleotide polymorphisms.

**DNA helicase** An enzyme that functions to unwind the double helix.

**DNA ligase** Enzyme that unites broken DNA strands during replication and recombination.

**DNA methylation** The addition of methyl groups to to bases in DNA, usually cytosine or guanine.

**DNA methyltransferase** An enzyme that catalyzes the methylation of DNA.

**DNA microarray** A small glass or plastic square onto which thousands of single-stranded DNA sequences are fixed so that hybridization of cell-derived RNA or DNA to the target sequences can be performed.

**DNA polymerase** Any of a group of enzymes that catalyze the formation of DNA strands from a DNA template.

**DNA topoisomerase** An enzyme that unwinds and winds coils of DNA that form during replication and transcription.

**docking protein** A receptor protein that binds (docks) a ribosome to the membrane of the endoplasmic reticulum by binding the signal sequence attached to a new protein being made at the ribosome.

**domain** (1) An independent structural element within a protein. Encoded by recognizable nucleotide sequences, a domain often folds separately from the rest of the protein. Similar domains can appear in a variety of different proteins across phylogenetic groups (e.g., "homeobox domain"; "calcium-binding domain"). (2) In phylogenetics, the three monophyletic branches of life (Bacteria, Archaea, and Eukarya).

**dominance** In genetics, the ability of one allelic form of a gene to determine the phenotype of a heterozygous individual in which the homologous chromosomes carry both it and a different (recessive) allele. (Contrast with recessive.)

**dormancy** A condition in which normal activity is suspended, as in some spores, seeds, and buds.

**dorsal** [L. *dorsum*: back] Toward or pertaining to the back or upper surface. (Contrast with ventral.)

**dorsal lip** In amphibian embryos, the dorsal segment of the blastopore. Also called the "organizer," this region directs the development of nearby embryonic regions.

**double fertilization** In angiosperms, a process in which the nuclei of two sperm fertilize one egg. One sperm's nucleus combines with the egg nucleus to produce a zygote, while the other combines with the same egg's two polar nuclei to produce the first cell of the triploid endosperm (the tissue that will nourish the growing plant embryo).

**double helix** Refers to DNA and the (usually right-handed) coil configuration of two complementary, antiparallel strands.

**downregulation** A negative feedback process in which continuous high concentrations of a hormone can decrease the number of its receptors. (Contrast with upregulation.)

**duodenum** (do' uh dee' num) The beginning portion of the vertebrate small intestine. (Contrast with ileum, jejunum.)

**duplication** A mutation in which a segment of a chromosome is duplicated, often by the attachment of a segment lost from its homolog. (Contrast with deletion.)

## - E -

**ecdysone** (eck die' sone) [Gk. *ek*: out of + *dyo*: to clothe] In insects, a hormone that induces molting.

**ecological efficiency** The overall transfer of energy from one trophic level to the next, expressed as the ratio of consumer production to producer production.

**ecology** [Gk. *oikos*: house] The scientific study of the interaction of organisms with their living (biotic) and nonliving (abiotic) environments.

**ecosystem** (eek' oh sis tum) The organisms of a particular habitat, such as a pond or forest, together with the physical environment in which they live.

**ecosystem engineer** An organism that builds structures that alter existing habitats or create new habitats.

**ecosystem services** Processes by which ecosystems maintain resources that benefit human society.

**ectoderm** [Gk. *ektos*: outside + *derma*: skin] The outermost of the three embryonic germ layers first delineated during gastrulation. Gives rise to the skin, sense organs, and nervous system.

**ectotherm** [Gk. *ektos*: outside + *thermos*: heat] An animal that is dependent on external heat sources for regulating its body temperature (Contrast with endotherm.)

**edema** (i dee' mah) [Gk. *oidema*: swelling] Tissue swelling caused by the accumulation of fluid.

**edge effect** The changes in ecological processes in a community caused by physical and biological factors originating in an adjacent community.

**effector protein** In cell signaling, a protein responsible for the cellular reponse to a signal transduction pathway.

**efferent** (ef' ur unt) [L. *ex*: out + *ferre*: to bear] Carrying outward or away from, as in a neuron that carries impulses outward from the central nervous system (efferent neuron), or a blood vessel that carries blood away from a structure. (Contrast with afferent.)

**egg** In all sexually reproducing organisms, the female gamete; in birds, reptiles, and some other vertebrates, a structure within which early embryonic development occurs. *See also* amniote egg, ovum.

**electrical synapse** A type of synapse at which action potentials spread directly from presynaptic cell to postsynaptic cell. (Contrast with chemical synapse.)

**electrocardiogram (ECG or EKG)** A graphic recording of electrical potentials from the heart.

**electrochemical gradient** The concentration gradient of an ion across a membrane plus the voltage difference across that membrane.

**electroencephalogram (EEG)** A graphic recording of electrical potentials from the brain.

**electromagnetic radiation** A self-propagating wave that travels though space and has both electrical and magnetic properties.

**electron** A subatomic particle outside the nucleus carrying a negative charge and very little mass.

**electron shell** The region surrounding the atomic nucleus at a fixed energy level in which electrons orbit.

**electron transport** The passage of electrons through a series of proteins with a release of energy which may be captured in a concentration gradient or chemical form such as NADH or ATP.

**electronegativity** The tendency of an atom to attract electrons when it occurs as part of a compound.

**electrophoresis** *See* gel electrophoresis.

**element** A substance that cannot be converted to simpler substances by ordinary chemical means.

**elongation** (1) In molecular biology, the addition of monomers to make a longer RNA or protein during transcription or translation. (2) Growth of a plant axis or cell primarily in the longitudinal direction.

**embolus** (em' buh lus) [Gk. *embolos*: stopper] A circulating blood clot. Blockage of a blood vessel by an embolus or by a bubble of gas is referred to as an *embolism*. (Contrast with thrombus.)

**embryo** [Gk. *en*: within + *bryein*: to grow] A young animal, or young plant sporophyte, while it is still contained within a protective structure such as a seed, egg, or uterus.

**embryonic stem cell (ESC)** A pluripotent cell in the blastocyst.

**embryo sac** In angiosperms, the female gametophyte. Found within the ovule, it consists of eight or fewer cells, membrane bounded, but without cellulose walls between them.

**emergent property** A property of a complex system that is not exhibited by its individual component parts.

**emigration** The deliberate and usually oriented departure of an organism from the habitat in which it has been living.

**3′ end** (3 prime) The end of a DNA or RNA strand that has a free hydroxyl group at the 3′ carbon of the sugar (deoxyribose or ribose).

**5′ end** (5 prime) The end of a DNA or RNA strand that has a free phosphate group at the 5′ carbon of the sugar (deoxyribose or ribose).

**endemic** (en dem' ik) [Gk. *endemos*: native] Confined to a particular region, thus often having a comparatively restricted distribution.

**endergonic** A chemical reaction in which the products have higher free energy than the reactants, thereby requiring free energy input to occur. (Contrast with exergonic.)

**endocrine gland** (en' doh krin) [Gk. *endo*: within + *krinein*: to separate] An aggregation of secretory cells that secretes hormones into the blood. The *endocrine system* consists of all *endocrine cells* and endocrine glands in the body that produce and release hormones. (Contrast with exocrine gland.)

**endocytosis** A process by which liquids or solid particles are taken up by a cell through invagination of the plasma membrane. (Contrast with exocytosis.)

**endoderm** [Gk. *endo*: within + *derma*: skin] The innermost of the three embryonic germ layers delineated during gastrulation. Gives rise to the digestive and respiratory tracts and structures associated with them.

**endodermis** In plants, a specialized cell layer marking the inside of the cortex in roots and some stems. Frequently a barrier to free diffusion of solutes.

**endomembrane system** A system of intracellular membranes that exchange material with one another, consisting of the Golgi apparatus, endoplasmic reticulum, and lysosomes when present.

**endometrium** The epithelial lining of the uterus.

**endoplasmic reticulum (ER)** [Gk. *endo*: within + L. *reticulum*: net] A system of membranous tubes and flattened sacs found in the cytoplasm of eukaryotes. Exists in two forms: rough ER, studded with ribosomes; and smooth ER, lacking ribosomes.

**endorphins** Molecules in the mammalian brain act as neurotransmitters in pathways that control pain.

**endoskeleton** [Gk. *endo*: within + *skleros*: hard] An internal skeleton covered by other, soft body tissues. (Contrast with exoskeleton.)

**endosperm** [Gk. *endo*: within + *sperma*: seed] A specialized triploid seed tissue found only in angiosperms; contains stored nutrients for the developing embryo.

**endospore** [Gk. *endo*: within + *spora*: to sow] In some bacteria, a resting structure that can survive harsh environmental conditions.

**endosymbiosis theory** [Gk. *endo*: within + *sym*: together + *bios*: life] The theory that the eukaryotic cell evolved via the engulfing of one prokaryotic cell by another.

**endothelium** The single layer of epithelial cells lining the interior of a blood vessel.

**endotherm** [Gk. *endo*: within + *thermos*: heat] An animal that can control its body temperature by the expenditure of its own metabolic energy. (Contrast with ectotherm.)

**endotoxin** A lipopolysaccharide that forms part of the outer membrane of certain Gram-negative bacteria that is released when the bacteria grow or lyse. (Contrast with exotoxin.)

**energetic cost** The difference between the energy an animal expends in performing a behavior and the energy it would have expended had it rested.

**energy** The capacity to do work or move matter against an opposing force. The capacity to accomplish change in physical and chemical systems.

**energy budget** A quantitative description of all paths of energy exchange between an animal and its environment.

**enkephalins** Molecules in the mammalian brain act as neurotransmitters in pathways that control pain.

**enthalpy (*H*)** The total energy of a system.

**entropy (*S*)** (en' tro pee) [Gk. *tropein*: to change] A measure of the degree of disorder in any system. Spontaneous reactions in a closed system are always accompanied by an increase in entropy.

**enveloped virus** A virus enclosed within a phospholipid membrane derived from its host cell.

**environment** Whatever surrounds and interacts with or otherwise affects a population, organism, or cell. May be external or internal.

**environmentalism** The use of ecological knowledge, along with economics, ethics, and many other considerations, to inform both personal decisions and public policy relating to stewardship of natural resources and ecosystems.

**enzyme** (en' zime) [Gk. *zyme*: to leaven (as in yeast bread)] A catalytic protein that speeds up a biochemical reaction.

**epi-** [Gk. upon, over] A prefix used to designate a structure located on top of another; for example, epidermis, epiphyte.

**epiblast** The upper or overlying portion of the avian blastula which is joined to the hypoblast at the margins of the blastodisc.

**epiboly** The movement of cells over the surface of the blastula toward the forming blastopore.

**epitope** *See* antigenic determinant.

**epidermis** [Gk. *epi*: over + *derma*: skin] In plants and animals, the outermost cell layers. (Only one cell layer thick in plants.)

**epididymis** (epuh did' uh mus) [Gk. *epi*: over + *didymos*: testicle] Coiled tubules in the testes that store sperm and conduct sperm from the seminiferous tubules to the vas deferens.

**epigenetics** The scientific study of changes in the expression of a gene or set of genes that occur without change in the DNA sequence.

**epinephrine** (ep i nef' rin) [Gk. *epi*: over + *nephros*: kidney] The "fight or flight" hormone produced by the medulla of the adrenal gland; it also functions as a neurotransmitter. (Also known as adrenaline.)

**epistasis** Interaction between genes in which the presence of a particular allele of one gene determines whether another gene will be expressed.

**epithelium** A type of animal tissue made up of sheets of cells that lines or covers organs, makes up tubules, and covers the surface of the body; one of the four major tissue types in multicellular animals.

**equilibrium** Any state of balanced opposing forces and no net change.

**ER** *See* endoplasmic reticulum.

**error signal** In regulatory systems, any difference between the set point of the system and its current condition.

**erythrocyte** (ur rith' row site) [Gk. *erythros*: red + *kytos*: container] A red blood cell.

**erythropoietin** A hormone produced by the kidney in response to lack of oxygen that stimulates the production of red blood cells.

**esophagus** (i soff' i gus) [Gk. *oisophagos*: gullet] That part of the gut between the pharynx and the stomach.

**essential acids** Amino acids or fatty acids that an animal cannot synthesize for itself and must obtain from its food.

**essential element** A mineral nutrient required for normal growth and reproduction in plants and animals.

**ester linkage** A condensation (water-releasing) reaction in which the carboxyl group of a fatty acid reacts with the hydroxyl group of an alcohol. Lipids are formed in this way.

**estivation** (ess tuh vay' shun) [L. *aestivalis*: summer] A state of dormancy and hypometabolism that occurs during the summer; usually a means of surviving drought and/or intense heat. (Contrast with hibernation.)

**estrogen** Any of several steroid sex hormones; produced chiefly by the ovaries in mammals.

**estrus** (es' trus) [L. *oestrus*: frenzy] The period of heat, or maximum sexual receptivity, in some female mammals. Ordinarily, the estrus is also the time of release of eggs in the female.

**ethology** [Gk. *ethos*: character + *logos*: study] An approach to the study of animal behavior that focuses on studying many species in natural environments and addresses questions about the evolution of behavior.

**ethylene** One of the plant growth hormones, the gas $H_2C{=}CH_2$. Involved in fruit ripening and other growth and developmental responses.

**eukaryotes** (yew car' ree oats) [Gk. *eu*: true + *karyon*: kernel or nucleus] Organisms whose cells contain their genetic material inside a nucleus. Includes all life other than the viruses, archaea, and bacteria.

**eusocial** Pertaining to a social group that includes nonreproductive individuals, as in honey bees.

**eutrophication** (yoo trofe' ik ay' shun) [Gk. *eu*: truly + *trephein*: to flourish] The addition of nutrient materials to a body of water, resulting in changes in ecological processes and species composition therein.

**evaporation** The transition of water from the liquid to the gaseous phase.

**evolution** Any gradual change. Most often refers to organic or Darwinian evolution, which is the genetic and resulting phenotypic change in populations of organisms from generation to generation. (*See* macroevolution, microevolution; contrast with speciation.)

**evolutionary radiation** The proliferation of many species within a single evolutionary lineage.

**evolutionary reversal** The reappearance of an ancestral trait in a group that had previously acquired a derived trait.

**excision repair** A mechanism that removes damaged DNA and replaces it with the appropriate nucleotide.

**excited state** The state of an atom or molecule when, after absorbing energy, it has more energy than in its normal, ground state. (Contrast with ground state.)

**excretion** Release of metabolic wastes by an organism.

**exergonic** A chemical reaction in which the products of the reaction have lower free energy than the reactants, resulting in a release of free energy. (Contrast with endergonic.)

**exocrine gland** (eks' oh krin) [Gk. *exo*: outside + *krinein*: to separate] Any gland, such as a salivary gland, that secretes to the outside of the body or into the gut. (Contrast with endocrine gland.)

**exocytosis** A process by which a vesicle within a cell fuses with the plasma membrane and releases its contents to the outside. (Contrast with endocytosis.)

**exon** A portion of a DNA molecule, in eukaryotes, that codes for part of a polypeptide. (Contrast with intron.)

**exoskeleton** (eks' oh skel' e ton) [Gk. *exos*: outside + *skleros*: hard] A hard covering on the outside of the body to which muscles are attached. (Contrast with endoskeleton.)

**exotoxin** A highly toxic, usually soluble protein released by living, multiplying bacteria. (Contrast with endotoxin.)

**expanding triplet repeat** A three-base-pair sequence in a human gene that is unstable and can be repeated a few to hundreds of times. Often, the more the repeats, the less the activity of the gene involved. Expanding triplet repeats occur in some human diseases such as Huntington's disease and fragile-X syndrome.

**experiment** A testing process to support or disprove hypotheses and to answer questions. The basis of the scientific method. *See* comparative experiment, controlled experiment.

**expiratory reserve volume** The amount of air that can be forcefully exhaled beyond the normal tidal expiration. (Contrast with inspiratory reserve volume, tidal volume, vital capacity.)

**exploitation competition** Competition in which individuals reduce the quantities of their shared resources. (Contrast with interference competition.)

**exponential growth** Growth, especially in the number of organisms in a population, which is a geometric function of the size of the growing entity: the larger the entity, the faster it grows. (Contrast with logistic growth.)

**expression vector** A DNA vector, such as a plasmid, that carries a DNA sequence that includes the adjacent sequences for its expression into mRNA and protein in a host cell.

**expressivity** The degree to which a genotype is expressed in the phenotype; may be affected by the environment.

**extensor** A muscle that extends an appendage.

**external fertilization** The release of gametes into the environment; typical of aquatic animals. Also called spawning. (Contrast with internal fertilization.)

**external gills** Highly branched and folded extensions of the body surface that provide a large surface area for gas exchange with water; typical of larval amphibians and many larval insects.

**extinction** The termination of a lineage of organisms.

**extracellular matrix** A material of heterogeneous composition surrounding cells and performing many functions including adhesion of cells.

**extraembryonic membranes** Four membranes that support but are not part of the developing embryos of reptiles, birds, and mammals, defining these groups phylogenetically as amniotes. (*See* amnion, allantois, chorion, and yolk sac.)

## - F -

**$F_1$** The first filial generation; the immediate progeny of a parental (P) mating.

**$F_2$** The second filial generation; the immediate progeny of a mating between members of the $F_1$ generation.

**facilitated diffusion** Passive movement through a membrane involving a specific carrier protein; does not proceed against a concentration gradient. (Contrast with active transport, diffusion.)

**facilitation** In succession, modification of the environment by a colonizing species in a way that allows colonization by other species. (Contrast with inhibition.)

**facultative anaerobe** A prokaryote that can shift its metabolism between anaerobic and aerobic operations modes on the presence or absence of $O_2$. (Alternatively, facultative aerobe.)

**fast-twitch fibers** Skeletal muscle fibers that can generate high tension rapidly, but fatigue rapidly ("sprinter" fibers). Characterized by an abundance of enzymes of glycolysis.

**fat** A triglyceride that is solid at room temperature. (Contrast with oil.)

**fate map** A diagram of the blastula showing which cells (blastomeres) are "fated " to contribute to specific tissues and organs in the mature body.

**fatty acid** A molecule made up of a long nonpolar hydrocarbon chain and a polar carboxyl group. Found in many lipids.

**fauna** (faw' nah) All the animals found in a given area. (Contrast with flora.)

**feces** [L. *faeces*: dregs] Waste excreted from the digestive system.

**fecundity ($m_x$)** The average number of offspring produced by each female.

**feedback information** In regulatory systems, information about the relationship between the set point of the system and its current state.

**feedforward information** In regulatory systems, information that changes the set point of the system.

**fermentation** (fur men tay' shun) [L. *fermentum*: yeast] The anaerobic degradation of a substance such as glucose to smaller molecules such as lactic acid or alcohol with the extraction of energy.

**fertilization** Union of gametes. Also known as syngamy.

**fetus** Medical and legal term for the stages of a developing human embryo from about the eighth week of pregnancy (the point at which all major organ systems have formed) to the moment of birth.

**fiber** In angiosperms, an elongated, tapering sclerenchyma cell, usually with a thick cell wall, that serves as a support function in xylem. (*See also* muscle fiber.)

**fibrin** A protein that polymerizes to form long threads that provide structure to a blood clot.

**fibrinogen** A circulating protein that can be stimulated to fall out of solution and provide the structure for a blood clot.

**fibrous root system** A root system typical of monocots composed of numerous thin adventitious roots that are all roughly equal in diameter. (Contrast with taproot system.)

**Fick's law of diffusion** An equation that describes the factors that determine the rate of diffusion of a molecule from an area of higher concentration to an area of lower concentration.

**fight-or-flight response** A rapid physiological response to a sudden threat mediated by the hormone epinephrine.

**filter feeder** An organism that feeds on organisms much smaller than itself that are suspended in water or air by means of a straining device.

**first law of thermodynamics** The principle that energy can be neither created nor destroyed.

**fission** *See* binary fission.

**fitness** The contribution of a genotype or phenotype to the genetic composition of subsequent generations, relative to the contribution of other genotypes or phenotypes. (*See* also inclusive fitness.)

**flagellum** (fla jell' um) (plural: flagella) [L. *flagellum*: whip] Long, whiplike appendage that propels cells. Prokaryotic flagella differ sharply from those found in eukaryotes.

**fixed action pattern** In ethology, a genetically determined behavior that is performed without learning, stereotypic (performed the same way each time), and not modifiable by learning.

**flexor** A muscle that flexes an appendage.

**flora** (flore' ah) All of the plants found in a given area. (Contrast with fauna.)

**floral meristem** In angiosperms, a meristem that forms the floral organs (sepals, petals, stamens, and carpels).

**floral organ identity genes** In angiosperms, genes that determine the fates of floral meristem cells; their expression is triggered by the products of meristem identity genes.

**florigen** A plant hormone involved in the conversion of a vegetative shoot apex to a flower.

**flower** The sexual structure of an angiosperm.

**fluid feeder** An animal that feeds on fluids it extracts from the bodies of other organisms; examples include nectar-feeding birds and blood-sucking insects.

**fluid mosaic model** A molecular model for the structure of biological membranes consisting of a fluid phospholipid bilayer in which suspended proteins are free to move in the plane of the bilayer.

**follicle** [L. *folliculus*: little bag] In female mammals, an immature egg surrounded by nutritive cells.

**follicle-stimulating hormone (FSH)** A gonadotropin produced by the anterior pituitary.

**food chain** A portion of a food web, most commonly a simple sequence of prey species and the predators that consume them.

**food vacuole** Membrane enclosed structure formed by phagocytosis in which engulfed food particles are digested by the action of lysosomal enzymes.

**food web** The complete set of food links between species in a community; a diagram indicating which ones are the eaters and which are eaten.

**forebrain** The region of the vertebrate brain that comprises the cerebrum, thalamus, and hypothalamus.

**fossil** Any recognizable structure originating from an organism, or any impression from such a structure, that has been preserved over geological time.

**fossil fuels** Fuels, including oil, natural gas, coal, and peat, formed over geologic time from organic material buried in anaerobic sediments.

**founder effect** Random changes in allele frequencies resulting from establishment of a population by a very small number of individuals.

**fovea** [L. *fovea*: a small pit] In the vertebrate retina, the area of most distinct vision.

**frame-shift mutation** The addition or deletion of a single or two adjacent nucleotides in a gene's sequence. Results in the misreading of mRNA during translation and the production of a nonfunctional protein. (Contrast with missense mutation, nonsense mutation, silent mutation.)

**Frank–Starling law** The stroke volume of the heart increases with increased return of blood to the heart.

**free energy (*G*)** Energy that is available for doing useful work, after allowance has been made for the increase or decrease of disorder.

**freeze-fracturing** Method of tissue preparation for transmission and scanning electron microscopy in which a tissue is frozen and a knife is then used to crack open the tissue; the fracture often occurs in the path of least resistance, within a membrane.

**frequency-dependent selection** Selection that changes in intensity with the proportion of individuals in a population having the trait.

**fruit** In angiosperms, a ripened and mature ovary (or group of ovaries) containing the seeds. Sometimes applied to reproductive structures of other groups of plants.

**functional genomics** The assignment of functional roles to the proteins encoded by genes identified by sequencing entire genomes.

**functional group** A characteristic combination of atoms that contribute specific properties when attached to larger molecules.

**fundamental niche** A species' niche as defined by its physiological capabilities. (Contrast with realized niche.)

## - G -

**G cap** A chemically modified GTP added to the 5′ end of mRNA; facilitates binding of mRNA to ribosome and prevents mRNA breakdown.

**G1** In the cell cycle, the gap between the end of mitosis and the onset of the S phase.

**G2** In the cell cycle, the gap between the S (synthesis) phase and the onset of mitosis.

**G protein** A membrane protein involved in signal transduction; characterized by binding GDP or GTP.

**gain of function mutation** A mutation that results in a protein with a new function. (Contrast with loss of function mutation.)

**gallbladder** In the human digestive system, an organ in which bile is stored.

**gametangium** (gam uh tan′ gee um) (plural: gametangia) [Gk. *gamos*: marriage + *angeion*: vessel] Any plant or fungal structure within which a gamete is formed.

**gamete** (gam′ eet) [Gk. *gamete/gametes*: wife, husband] The mature sexual reproductive cell: the egg or the sperm.

**gametogenesis** (ga meet′ oh jen′ e sis) The specialized series of cellular divisions that leads to the production of gametes. (*See also* oogenesis, spermatogenesis.)

**gametophyte** (ga meet′ oh fyte) In plants and photosynthetic protists with alternation of generations, the multicellular haploid phase that produces the gametes. (Contrast with sporophyte.)

**gamma diversity** The regional diversity found over a range of communities or habitats in a geographic region. (Contrast with alpha diveristy, beta diversity.)

**ganglion** (gang′ glee un) (plural: ganglia) [Gk. tumor] A cluster of neurons that have similar characteristics or function.

**ganglion cells** Cells at the front of the human retina that transmit information from the bipolar cells to the brain.

**gap junction** A 2.7-nanometer gap between plasma membranes of two animal cells, spanned by protein channels. Gap junctions allow chemical substances or electrical signals to pass from cell to cell.

**gastric pits** Deep infoldings in the walls of the stomach lined with secretory cells.

**gastrin** A hormone secreted by cells in the lower region of the stomach that stimulates the secretion of digestive juices as well as movements of the stomach.

**gastrovascular cavity** Serving for both digestion (gastro) and circulation (vascular); in particular, the central cavity of the body of jellyfish and other cnidarians.

**gastrulation** Development of a blastula into a gastrula. In embryonic development, the process by which a blastula is transformed by massive movements of cells into a *gastrula*, an embryo with three germ layers and distinct body axes.

**gated channel** A membrane protein that changes its three-dimensional shape, and therefore its ion conductance, in response to a stimulus. When open, it allows specific ions to move across the membrane.

**gel electrophoresis** (e lek′ tro fo ree′ sis) [L. *electrum*: amber + Gk. *phorein*: to bear] A technique for separating molecules (such as DNA fragments) from one another on the basis of their electric charges and molecular weights by applying an electric field to a gel.

**gene** [Gk. *genes*: to produce] A unit of heredity. Used here as the unit of genetic function which carries the information for a single polypeptide or RNA.

**gene family** A set of similar genes derived from a single parent gene; need not be on the same chromosomes. The vertebrate globin genes constitute a classic example of a gene family.

**gene flow** Exchange of genes between populations through migration of individuals or movements of gametes.

**gene-for-gene resistance** In plants, a mechanism of resistance to pathogens in which resistance is triggered by the specific interaction of the products of a pathogen's *Avr* genes and a plant's *R* genes.

**gene pool** All of the different alleles of all of the genes existing in all individuals of a population.

**gene therapy** Treatment of a genetic disease by providing patients with cells containing functioning alleles of the genes that are nonfunctional in their bodies.

**gene tree** A graphic representation of the evolutionary relationships of a single gene in different species or of the members of a gene family.

**genetic code** The set of instructions, in the form of nucleotide triplets, that translate a linear sequence of nucleotides in mRNA into a linear sequence of amino acids in a protein.

**genetic drift** Changes in gene frequencies from generation to generation as a result of random (chance) processes.

**genetic map** The positions of genes along a chromosome as revealed by recombination frequencies.

**genetic marker** (1) In gene cloning, a gene of identifiable phenotype that indicates the presence of another gene, DNA segment, or chromosome fragment. (2) In general, a DNA sequence such as a single nucleotide polymorphism whose presence is correlated with the presence of other linked genes on that chromosome.

**genetic structure** The frequencies of different alleles at each locus and the frequencies of different genotypes in a Mendelian population.

**genetic switches** Mechanisms that control how the genetic toolkit is used, such as promoters and the transcription factors that bind them. The signal cascades that converge on and operate these switches determine when and where genes will be turned on and off.

**genetic toolkit** In evolutionary developmental biology, DNA sequences controlling developmental mechanisms that have been conserved over evolutionary time.

**genetics** The scientific study of the structure, functioning, and inheritance of genes, the units of hereditary information.

**genome** (jee′ nome) The complete DNA sequence for a particular organism or individual.

**genomic equivalence** The principle that no information is lost from the nuclei of cells as they pass through the early stages of embryonic development.

**genomic imprinting** The form of a gene's expression is determined by parental source (i.e., whether the gene is inherited from the male or female parent).

**genomic library** All of the cloned DNA fragments generated by the action of a restriction endonuclease on a genome.

**genomics** The scientific study of entire sets of genes and their interactions.

**genotype** (jean′ oh type) [Gk. *gen*: to produce + *typos*: impression] An exact description of the genetic constitution of an individual, either with respect to a single trait or with respect to a larger set of traits. (Contrast with phenotype.)

**genus** (jean′ us) (plural: genera) [Gk. *genos*: stock, kind] A group of related, similar species recognized by taxonomists with a distinct name used in binomial nomenclature.

**germ cell** [L. *germen*: to beget] A reproductive cell or gamete of a multicellular organism. (Contrast with somatic cell.)

**germ layers** The three embryonic layers formed during gastrulation (ectoderm, mesoderm, and endoderm). Also called cell layers or tissue layers.

**germ line mutation** Mutation in a cell that produces gametes (i.e., a germ line cell). (Contrast with somatic mutation.)

**germination** Sprouting of a seed or spore.

**gestation** (jes tay′ shun) [L. *gestare*: to bear] The period during which the embryo of a mammal develops within the uterus. Also known as pregnancy.

**ghrelin** A hormone produced and secreted by cells in the stomach that stimulates appetite.

**gibberellin** (jib er el′ lin) A class of plant growth hormones playing roles in stem elongation, seed germination, flowering of certain plants, etc.

**gill** An organ specialized for gas exchange with water.

**gizzard** (giz′ erd) [L. *gigeria*: cooked chicken parts] A muscular port of the stomach of birds that grinds up food, sometimes with the aid of fragments of stone.

**glia** (glee′ uh) [Gk. *glia*: glue] Cells of the nervous system that do not conduct action potentials.

**glomerular filtration rate (GFR)** The rate at which the blood is filtered in the glomeruli of the kidney.

**glomerulus** (glo mare′ yew lus) [L. *glomus*: ball] Sites in the kidney where blood filtration takes place. Each glomerulus consists of a knot of capillaries served by afferent and efferent arterioles.

**glucagon** Hormone produced by alpha cells of the pancreatic islets of Langerhans. Glucagon stimulates the liver to break down glycogen and release glucose into the circulation.

**gluconeogenesis** The biochemical synthesis of glucose from other substances, such as amino acids, lactate, and glycerol.

**glucose** [Gk. *gleukos*: sugar, sweet] The most common monosaccharide; the monomer of the polysaccharides starch, glycogen, and cellulose.

**glycerol** (gliss′ er ole) A three-carbon alcohol with three hydroxyl groups; a component of phospholipids and triglycerides.

**glycogen** (gly′ ko jen) An energy storage polysaccharide found in animals and fungi; a branched-chain polymer of glucose, similar to starch.

**glycolipid** A lipid to which sugars are attached.

**glycolysis** (gly kol' li sis) [Gk. *gleukos*: sugar + *lysis*: break apart] The enzymatic breakdown of glucose to pyruvic acid.

**glycoprotein** A protein to which sugars are attached.

**glycosidic linkage** Bond between carbohydrate (sugar) molecules through an intervening oxygen atom (–O–).

**glycosylation** The addition of carbohydrates to another type of molecule, such as a protein.

**glyoxysome** (gly ox' ee soam) An organelle found in plants, in which stored lipids are converted to carbohydrates.

**Golgi apparatus** (goal' jee) A system of concentrically folded membranes found in the cytoplasm of eukaryotic cells; functions in secretion from cell by exocytosis.

**gonad** (go' nad) [Gk. *gone*: seed] An organ that produces gametes in animals: either an ovary (female gonad) or testis (male gonad).

**gonadotropin** A type of trophic hormone that stimulates the gonads.

**gonadotropin-releasing hormone (GnRH)** Hormone produced by the hypothalamus that stimulates the anterior pituitary to secrete ("release") gonadotropins.

**Gondwana** The large southern land mass that existed from the Cambrian (540 mya) to the Jurassic (138 mya). Present-day remnants are South America, Africa, India, Australia, and Antarctica.

**grafting** Artificial transplantation of tissue from one organism to another. In horticulture, the transfer of a bud or stem segment from one plant onto the root of another as a form of asexual reproduction.

**Gram stain** A differential purple stain useful in characterizing bacteria. The peptidoglycan-rich cell walls of Gram-positive bacteria stain purple; cell walls of Gram-negative bacteria generally stain orange.

**gravitropism** [Gk. *tropos*: to turn] A directed plant growth response to gravity.

**gray matter** In the nervous system, tissue that is rich in neuronal cell bodies. (Contrast with white matter.)

**greenhouse gases** Gases in the atmosphere, such as carbon dioxide and methane, that are transparent to sunlight, but trap heat radiating from Earth's surface, causing heat to build up at Earth's surface.

**gross primary production** The amount of energy captured by the primary producers in a community.

**gross primary productivity (GPP)** The rate at which the primary producers in a community turn solar energy into stored chemical energy via photosynthesis.

**ground meristem** That part of an apical meristem that gives rise to the ground tissue system of the primary plant body.

**ground tissue system** Those parts of the plant body not included in the dermal or vascular tissue systems. Ground tissues function in storage, photosynthesis, and support.

**growth** An increase in the size of the body and its organs by cell division and cell expansion.

**growth factor** A chemical signal that stimulates cells to divide.

**growth hormone** A peptide hormone released by the anterior pituitary that stimulates many anabolic processes.

**guanine (G)** (gwan' een) A nitrogen-containing base found in DNA, RNA, and GTP.

**guard cells** In plants, specialized, paired epidermal cells that surround and control the opening of a stoma (pore). *See* stoma.

**guild** In ecology, a group of species that exploit the same resource, but in slightly different ways.

**gustation** The sense of taste.

**gut** An animal's digestive tract.

## - H -

**habitat** The particular environment in which an organism lives. A *habitat patch* is an area of a particular habitat surrounded by other habitat types that may be less suitable for that organism.

**hair cell** A type of mechanoreceptor in animals. Detects sound waves and other forms of motion in air or water.

**half-life** The time required for half of a sample of a radioactive isotope to decay to its stable, nonradioactive form, or for a drug or other substance to reach half its initial dosage.

**halophyte** (hal' oh fyte) [Gk. *halos*: salt + *phyton*: plant] A plant that grows in a saline (salty) environment.

**Hamilton's rule** The principle that, for an apparent altruistic behavior to be adaptive, the fitness benefit of that act to the recipient times the degree of relatedness of the performer and the recipient must be greater than the cost to the performer.

**haplodiploidy** A sex determination mechanism in which diploid individuals (which develop from fertilized eggs) are female and haploid individuals (which develop from unfertilized eggs) are male; typical of hymenopterans.

**haploid** (hap' loid) [Gk. *haploeides*: single] Having a chromosome complement consisting of just one copy of each chromosome; designated $1n$ or $n$. (Contrast with diploid.)

**haplontic** A type of life cycle in which the zygote is the only diploid cell and mitosis occurs only in haploid cells. (Contrast with diplontic.)

**haplotype** Linked nucleotide sequences that are usually inherited as a unit (as a "sentence" rather than as individual "words").

**Hardy–Weinberg equilibrium** In a sexually reproducing population, the allele frequency at a given locus that is not being acted on by agents of evolution; the conditions that would result in no evolution in a population.

**haustorium** (haw stor' ee um) (plural: haustoria)[L. *haustus*: draw up] A specialized hypha or other structure by which fungi and some parasitic plants draw nutrients from a host plant.

**Haversian systems** Units of organization in compact bone that reflect the action of intercommunicating osteoblasts.

**heart** In circulatory systems, a muscular pump that moves extracellular fluid around the body.

**heat of vaporization** The energy that must be supplied to convert a molecule from a liquid to a gas at its boiling point.

**heat shock proteins** Chaperone proteins expressed in cells exposed to high or low temperatures or other forms of environmental stress.

**helical** Shaped like a screw or spring; this shape occurs in DNA and proteins.

**helper T cells** *See* T-helper cells.

**hemiparasite** A parasitic plant that can photosynthesize, but derives water and mineral nutrients from the living body of another plant. (Contrast with holoparasite.)

**hemizygous** (hem' ee zie' gus) [Gk. *hemi*: half + *zygotos*: joined] In a diploid organism, having only one allele for a given trait, typically the case for X-linked genes in male mammals and Z-linked genes in female birds. (Contrast with homozygous, heterozygous.)

**hemoglobin** (hee' mo glow bin) [Gk. *heaema*: blood + L. *globus*: globe] Oxygen-transporting protein found in the red blood cells of vertebrates (and found in some invertebrates).

**Hensen's node** In avian embryos, a structure at the anterior end of the primitive groove; determines the fates of cells passing over it during gastrulation.

**hepatic** (heh pat' ik) [Gk. *hepar*: liver] Pertaining to the liver.

**herbivore** (ur' bi vore) [L. *herba*: plant + *vorare*: to devour] An animal that eats plant tissues. (Contrast with carnivore, detritivore, omnivore.)

**heritable trait** A trait that is at least partly determined by genes.

**hermaphroditism** (her maf' row dite ism) The coexistence of both female and male sex organs in the same organism.

**hetero-** [Gk.: *heteros*: other, different] A prefix indicating two or more different conditions, structures, or processes. (Contrast with homo-.)

**heterochrony** Alteration in the timing of developmental events, leading to different results in the adult organism.

**heterocyst** A large, thick-walled cell type in the filaments of certain cyanobacteria that performs nitrogen fixation.

**heteromorphic** (het' er oh more' fik) [Gk. *heteros*: different + *morphe*: form] Having a different form or appearance, as two heteromorphic life stages of a plant. (Contrast with isomorphic.)

**heterosporous** (het' er os' por us) Producing two types of spores, one of which gives rise to a female megaspore and the other to a male microspore. (Contrast with homosporous.)

**heterosis** The superior fitness of heterozygous offspring as compared with that of their dissimilar homozygous parents. Also called hybrid vigor.

**heterotherm** An animal that regulates its body temperature at a constant level at some times but not others, such as a hibernator.

**heterotroph** (het' er oh trof) [Gk. *heteros*: different + *trophe*: feed] An organism that requires preformed organic molecules as food. (Contrast with autotroph.)

**heterotrophic succession** Succession in detritus-based communities, which differs from other types of succession in taking place without the participation of plants.

**heterotypic** Pertaining to adhesion of cells of different types. (Contrast with homotypic.)

**heterozygous** (het' er oh zie' gus) [Gk. *heteros*: different + *zygotos*: joined] In diploid organisms, having different alleles of a given gene on the pair of homologs carrying that gene. (Contrast with homozygous.)

**hexose** [Gk. *hex*: six] A sugar containing six carbon atoms.

**hibernation** [L. *hibernum*: winter] The state of inactivity of some animals during winter; marked by a drop in body temperature and metabolic rate.

**hierarchical sequencing** An approach to DNA sequencing in which genetic markers are mapped and DNA sequences are aligned by matching overlapping sites of known sequence. (Contrast with shotgun sequencing.)

**high-density lipoproteins (HDLs)** Lipoproteins that remove cholesterol from tissues and carry it to the liver; HDLs are the "good" lipoproteins associated with good cardiovascular health.

**high-throughput sequencing** Rapid DNA sequencing on a micro scale in which many fragments of DNA are sequenced in parallel.

**highly repetitive sequences** Short (less than 100 bp), nontranscribed DNA sequences, repeated thousands of times in tandem arrangements.

**hindbrain** The region of the developing vertebrate brain that gives rise to the medulla, pons, and cerebellum.

**hippocampus** [Gr. sea horse] A part of the forebrain that takes part in long-term memory formation.

**histamine** (hiss' tah meen) A substance released by damaged tissue, or by mast cells in response to allergens. Histamine increases vascular permeability, leading to edema (swelling).

**histone** Any one of a group of proteins forming the core of a nucleosome, the structural unit of a eukaryotic chromosome.

**HIV** Human immunodeficiency virus, the retrovirus that causes acquired immune deficiency syndrome (AIDS).

**holoparasite** A fully parasitic plant (i.e., one that does not perform photosynthesis).

**homeobox** 180-base-pair segment of DNA found in certain homeotic genes; regulates the expression of other genes and thus controls large-scale developmental processes.

**homeostasis** (home' ee o sta' sis) [Gk. *homos*: same + *stasis*: position] The maintenance of a steady state, such as a constant temperature or a stable social structure, by means of physiological or behavioral feedback responses.

**homeotic genes** Genes that act during development to determine the formation of an organ from a region of the embryo.

**homeotic mutation** Mutation in a homeotic gene that results in the formation of a different organ than that normally made by a region of the embryo.

**homo-** [Gk. *homos*: same] A prefix indicating two or more similar conditions, structures, or processes. (Contrast with hetero-.)

**homolog** (1) In cytogenetics, one of a pair (or larger set) of chromosomes having the same overall genetic composition and sequence. In diploid organisms, each chromosome inherited from one parent is matched by an identical (except for mutational changes) chromosome—its homolog—from the other parent. (2) In evolutionary biology, one of two or more features in different species that are similar by reason of descent from a common ancestor.

**homology** (ho mol' o jee) [Gk. *homologia*: of one mind; agreement] A similarity between two or more features that is due to inheritance from a common ancestor. The structures are said to be *homologous*, and each is a *homolog* of the others. (Contrast with analogy.)

**homoplasy** (home' uh play zee) [Gk. *homos*: same + *plastikos*: shape, mold] The presence in multiple groups of a trait that is not inherited from the common ancestor of those groups. Can result from convergent evolution, evolutionary reversal, or parallel evolution.

**homosporous** Producing a single type of spore that gives rise to a single type of gametophyte, bearing both female and male reproductive organs. (Contrast with heterosporous.)

**homotypic** Pertaining to adhesion of cells of the same type. (Contrast with heterotypic.)

**homozygous** (home' oh zie' gus) [Gk. *homos*: same + *zygotos*: joined] In diploid organisms, having identical alleles of a given gene on both homologous chromosomes. An individual may be a homozygote with respect to one gene and a heterozygote with respect to another. (Contrast with heterozygous.)

**horizons** The horizontal layers of a soil profile, including the topsoil (A horizon), subsoil (B horizon) and parent rock or bedrom (C horizon)

**hormone** (hore' mone) [Gk. *hormon*: to excite, stimulate] A chemical signal produced in minute amounts at one site in a multicellular organism and transported to another site where it acts on target cells.

**host** An organism that harbors a parasite or symbiont and provides it with nourishment.

**Hox genes** Conserved homeotic genes found in vertebrates, *Drosophila*, and other animal groups. Hox genes contain the homeobox domain and specify pattern and axis formation in these animals.

**human chorionic gonadotropin (hCG)** A hormone secreted by the placenta which sustains the corpus luteum and helps maintain pregnancy.

**humoral immune response** The response of the immune system mediated by B cells that produces circulating antibodies active against extracellular bacterial and viral infections. (Contrast with cellular immune response.)

**humus** (hew' mus) The partly decomposed remains of plants and animals on the surface of a soil.

**hybrid** (high' brid) [L. *hybrida*: mongrel] (1) The offspring of genetically dissimilar parents. (2) In molecular biology, a double helix formed of nucleic acids from different sources.

**hybridize** (1) In genetics, to combine the genetic material of two distinct species or of two distinguishable populations within a species. (2) In molecular biology, to form a double-stranded nucleic acid in which the two strands originate from different sources.

**hybrid vigor** *See* heterosis.

**hybridoma** A cell produced by the fusion of an antibody-producing cell with a myeloma (tumor) cell; it produces monoclonal antibodies.

**hydrocarbon** A compound containing only carbon and hydrogen atoms.

**hydrogen bond** A weak electrostatic bond which arises from the attraction between the slight positive charge on a hydrogen atom and a slight negative charge on a nearby oxygen or nitrogen atom.

**hydrologic cycle** The movement of water from the oceans to the atmosphere, to the soil, and back to the oceans.

**hydrolysis reaction** (high drol' uh sis) [Gk. *hydro*: water + *lysis*: break apart] A chemical reaction that breaks a bond by inserting the components of water ($AB + H_2O \rightarrow AH + BOH$). (Contrast with condensation reaction.)

**hydrophilic** (high dro fill' ik) [Gk. *hydro*: water + *philia*: love] Having an affinity for water. (Contrast with hydrophobic.)

**hydrophobic** (high dro foe' bik) [Gk. *hydro*: water + *phobia*: fear] Having no affinity for water. Uncharged and nonpolar groups of atoms are hydrophobic. (Contrast with hydrophilic.)

**hydroponic** Pertaining to a method of growing plants with their roots suspended in nutrient solutions instead of soil.

**hydrostatic pressure** Pressure generated by compression of liquid in a confined space. Generated in plants, fungi, and some protists with cell walls by the osmotic uptake of water. Generated in animals with closed circulatory systems by the beating of a heart.

**hydrostatic skeleton** A fluid-filled body cavity that transfers forces from one part of the body to another when acted on by surrounding muscles.

**hydroxyl group** The —OH group found on alcohols and sugars.

**hyper-** [Gk. *hyper*: above, over] Prefix indicating above, higher, more. (Contrast with hypo-.)

**hyperpolarization** A change in the resting potential across a membrane so that the inside of a cell becomes more negative compared with the outside of the cell. (Contrast with depolarization.)

**hypersensitive response** A defensive response of plants to microbial infection in which phytoalexins and pathogenesis-related proteins are produced and the infected tissue undergoes apoptosis to isolate the pathogen from the rest of the plant.

**hypertonic** Having a greater solute concentration. Said of one solution compared with another. (Contrast with hypotonic, isotonic.)

**hypha** (high' fuh) (plural: hyphae) [Gk. *hyphe*: web] In the fungi and oomycetes, any single filament.

**hypo-** [Gk. *hypo*: beneath, under] Prefix indicating underneath, below, less. (Contrast with hyper-.)

**hypoblast** The lower tissue portion of the avian blastula which is joined to the epiblast at the margins of the blastodisc.

**hypothalamus** The part of the brain lying below the thalamus; it coordinates water balance, reproduction, temperature regulation, and metabolism.

**hypothermia** Below-normal body temperature.

**hypothesis** A tentative answer to a question, from which testable predictions can be generated. (Contrast with theory.)

**hypotonic** Having a lesser solute concentration. Said of one solution in comparing it to another. (Contrast with hypertonic, isotonic.)

**hypoxia** A deficiency of oxygen.

## - I -

**ileum** The final segment of the small intestine.

**imbibition** Water uptake by a seed; first step in germination.

**immediate hypersensitivity** A rapid, extensive overreaction of the immune system against an allergen, resuting in the release of large amounts of histamine. (Contrast with delayed hypersensitivity.)

**immediate memory** A form of memory for events happening in the present that is almost perfectly photographic, but lasts only seconds. (Contrast with short-term memory, long-term memory.)

**immune system** [L. *immunis*: exempt from] A system in an animal that recognizes and attempts to eliminate or neutralize foreign substances such as bacteria, viruses, and pollutants.

**immunoassay** The use of antibodies to measure the concentration of an antigen in a sample.

**immunoglobulins** A class of proteins containing a tetramer consisting of four polypeptide chains—two identical light chains and two identical heavy chains—held together by disulfide bonds; active as receptors and effectors in the immune system.

**immunological memory** The capacity to more rapidly and massively respond to a second exposure to an antigen than occurred on first exposure.

**imperfect flower** A flower lacking either functional stamens or functional carpels. (Contrast with perfect flower.)

**implantation** The process by which the early mammalian embryo becomes attached to and embedded in the lining of the uterus.

**imprinting** In animal behavior, a rapid form of learning in which an animal learns, during a brief critical period, to make a particular response, which is maintained for life, to some object or other organism. *See also* genomic imprinting.

**in vitro** [L. in glass] A biological process occurring outside of the organism, in the laboratory. (Contrast with in vivo.)

**in vivo** [L. alive] A biological process occurring within a living organism or cell. (Contrast with in vitro.)

**inbreeding** Breeding among close relatives.

**inclusive fitness** The sum of an individual's genetic contribution to subsequent generations both via production of its own offspring and via its influence on the survival of relatives who are not direct descendants.

**incomplete cleavage** A pattern of cleavage that occurs in many eggs that have a lot of yolk, in which the cleavage furrows do not penetrate all of it. (*See also* discoidal cleavage, superficial cleavage; contrast with complete cleavage.)

**incomplete dominance** Condition in which the heterozygous phenotype is intermediate between the two homozygous phenotypes.

**incomplete metamorphosis** Insect development in which changes between instars are gradual.

**independent assortment** During meiosis, the random separation of genes carried on nonhomologous chromosomes into gametes so that inheritance of these genes is random. This principle was articulated by Mendel as his second law.

**indeterminate growth** A open-ended growth pattern in which an organism or organ continues to grow as long as it lives; characteristic of some animals and of plant shoots and roots. (Contrast with determinate growth.)

**indirect transduction** Cell signaling mechanism in which a second messenger mediates the interaction between receptor binding and cellular response. (Contrast with direct transduction.)

**individual fitness** That component of inclusive fitness resulting from an organism producing its own offspring. (Contrast with kin selection.)

**induced fit** A change in the shape of an enzyme caused by binding to its substrate that exposes the active site of the enzyme.

**induced mutation** A mutation resulting from exposure to a mutagen from outside the cell. (Contrast with spontaneous mutation.)

**induced pluripotent stem cells (iPS cells)** Multipotent or pluripotent animal stem cells produced from differentiated cells in vitro by the addition of several genes that are expressed.

**inducer** (1) A compound that stimulates the synthesis of a protein. (2) In embryonic development, a substance that causes a group of target cells to differentiate in a particular way.

**inducible** Produced only in the presence of a particular compound or under particular circumstances. (Contrast with constitutive.)

**induction** In embryonic development, the process by which a factor produced and secreted by certain cells determines the fates other cells.

**inflammation** A nonspecific defense against pathogens; characterized by redness, swelling, pain, and increased temperature.

**inflorescence** A structure composed of several to many flowers.

**inflorescence meristem** A meristem that produces floral meristems as well as other small leafy structures (bracts).

**ingroup** In a phylogenetic study, the group of organisms of primary interest. (Contrast with outgroup.)

**inhibitor** A substance that blocks a biological process.

**initials** Cells that perpetuate plant meristems, comparable to animal stem cells. When an initial divides, one daughter cell develops into another initial, while the other differentiates into a more specialized cell.

**initiation** In molecular biology, the beginning of transcription or translation.

**initiation complex** In protein translation, a combination of a small ribosomal subunit, an mRNA molecule, and the tRNA charged with the first amino acid coded for by the mRNA; formed at the onset of translation.

**initiation site** The part of a promoter where transcription begins.

**inner cell mass** Derived from the mammalian blastula (bastocyst), the inner cell mass will give rise to the yolk sac (via hypoblast) and embryo (via epiblast).

**inositol trisphosphate ($IP_3$)** An intracellular second messenger derived from membrane phospholipids.

**inspiratory reserve volume** The amount of air that can be inhaled above the normal tidal inspiration. (Contrast with expiratory reserve volume, tidal volume, vital capacity.)

**instar** (in' star) An immature stage of an insect between molts.

**insulin** (in' su lin) [L. *insula*: island] A hormone synthesized in islet cells of the pancreas that promotes the conversion of glucose into the storage material, glycogen.

**integrin** In animals, a transmembrane protein that mediates the attachment of epithelial cells to the extracellular matrix.

**integument** [L. *integumentum*: covering] A protective surface structure. In gymnosperms and angiosperms, a layer of tissue around the ovule which will become the seed coat.

**intercostal muscles** Muscles between the ribs that can augment breathing movements by elevating and suppressing the rib cage.

**interference competition** Competition in which individuals actively interfere with one another's access to resources. (Contrast with exploitation competition.)

**interference RNA (RNAi)** *See* RNA interference.

**interferon** A glycoprotein produced by virus-infected animal cells; increases the resistance of neighboring cells to the virus.

**intermediate filaments** Components of the cytoskeleton whose diameters fall between those of the larger microtubules and those of the smaller microfilaments.

**internal environment** In multicelluar organisms, the extracellular fluid surrounding the cells.

**internal fertilization** The release of sperm into the female reproductive tract; typical of most terrestrial animals. (Contrast with external fertilization.)

**internal gills** Gills enclosed in protective body cavities; typical of mollusks, arthropods, and fishes.

**interneuron** A neuron that communicates information between two other neurons.

**internode** The region between two nodes of a plant stem.

**interphase** In the cell cycle, the period between successive nuclear divisions during which the chromosomes are diffuse and the nuclear envelope is intact. During interphase the cell is most active in transcribing and translating genetic information.

**interspecific competition** Competition between members of two or more species. (Contrast with intraspecific competition.)

**interstitial fluid** Extracellular fluid that is not contained in the vessels of a circulatory system.

**intestine** The portion of the gut following the stomach, in which most digestion and absorption occurs.

**intraspecific competition** Competition among members of the same species. (Contrast with interspecific competition.)

**intrinsic rate of increase (*r*)** The rate at which a population can grow when its density is low and environmental conditions are highly favorable.

**intron** Portion of a of a gene within the coding region that is transcribed into pre-mRNA but is spliced out prior to translation. (Contrast with exon.)

**invasive species** An exotic species that reproduces rapidly, spreads widely, and has negative effects on the native species of the region to which it has been introduced.

**invasiveness** The ability of a pathogen to multiply in a host's body. (Contrast with toxigenicity).

**inversion** A rare 180° reversal of the order of genes within a segment of a chromosome.

**ion** (eye' on) [Gk. *ion*: wanderer] An electrically charged particle that forms when an atom gains or loses one or more electrons.

**ion channel** An integral membrane protein that allows ions to diffuse across the membrane in which it is embedded.

**ionic bond** An electrostatic attraction between positively and negatively charged ions.

**ionotropic receptors** A receptor that that directly alters membrane permeability to a type of ion when it combines with its ligand.

**iris** (eye' ris) [Gk. *iris*: rainbow] The round, pigmented membrane that surrounds the pupil of the eye and adjusts its aperture to regulate the amount of light entering the eye.

**island biogeography** A theory proposing that the number of species on an island (or in another geographically defined and isolated area) represents a balance, or equilibrium, between the rate at which species immigrate to the island and the rate at which resident species go extinct.

**islets of Langerhans** Clusters of hormone-producing cells in the pancreas.

**iso-** [Gk. *iso*: equal] Prefix used for two separate entities that share some element of identity.

**isogamous** Having male and female gametes that are morphologically identical. (Contrast with anisogamous.)

**isomers** Molecules consisting of the same numbers and kinds of atoms, but differing in the bonding patterns by which the atoms are held together.

**isomorphic** (eye so more' fik) [Gk. *isos*: equal + *morphe*: form] Having the same form or appearance, as when the haploid and diploid life stages of an organism appear identical. (Contrast with heteromorphic.)

**isotonic** Having the same solute concentration; said of two solutions. (Contrast with hypertonic, hypotonic.)

**isotope** (eye' so tope) [Gk. *isos*: equal + *topos*: place] Isotopes of a given chemical element have the same number of protons in their nuclei (and thus are in the same position on the periodic table), but differ in the number of neutrons.

**isozymes** Enzymes of an organism that have somewhat different amino acid sequences but catalyze the same reaction.

**iteroparous** [L. *itero*, to repeat + *pario*, to beget] Reproducing multiple times in a lifetime. (Contrast with semelparous.)

## - J -

**jejunum** (jih jew' num) The middle division of the small intestine, where most absorption of nutrients occurs. (*See* duodenum, ileum.)

**jelly coat** The outer protective layer of a sea urchin egg, which triggers an acrosomal reaction in sperm.

**joint** In skeletal systems, a junction between two or more bones.

**juvenile hormone** In insects, a hormone maintaining larval growth and preventing maturation or pupation.

## - K -

***K*-strategist** A species whose life history strategy allows it to persist at or near the carrying capacity (*K*) of its environment. (Contrast with *r*-strategist.)

**karyogamy** The fusion of nuclei of two cells. (Contrast with plasmogamy.)

**karyotype** The number, forms, and types of chromosomes in a cell.

**keystone species** Species that have a dominant influence on the composition of a community.

**kidneys** A pair of excretory organs in vertebrates.

**kin selection** That component of inclusive fitness resulting from helping the survival of relatives containing the same alleles by descent from a common ancestor. (Contrast with individual fitness.)

**kinase** *See* protein kinase.

**kinetic energy** (kuh-net' ik) [Gk. *kinetos*: moving] The energy associated with movement. (Contrast with potential energy.)

**kinetochore** (kuh net' oh core) Specialized structure on a centromere to which microtubules attach.

**knockout** A molecular genetic method in which a single gene of an organism is permanently inactivated.

**Koch's postulates** A set of rules for establishing that a particular microorganism causes a particular disease.

**Krebs cycle** *See* citric acid cycle.

## - L -

**lagging strand** In DNA replication, the daughter strand that is synthesized in discontinuous stretches. (*See* Okazaki fragments.)

**larva** (plural: larvae) [L. *lares*: guiding spirits] An immature stage of any animal that differs dramatically in appearance from the adult.

**lateral** [L. *latus*: side] Pertaining to the side.

**lateral gene transfer** The transfer of genes from one species to another, common among bacteria and archaea.

**lateral line** A sensory system in fishes consisting of a canal filled with water and hair cells running down each side under the surface of the skin, which senses disturbances in the surrounding water.

**lateral meristem** Either of the two meristems, the vascular cambium and the cork cambium, that give rise to a plant's secondary growth.

**lateral root** A root extending outward from the taproot in a taproot system; typical of eudicots.

**laticifers** (luh tiss' uh furs) In some plants, elongated cells containing secondary plant products such as latex.

**Laurasia** The northernmost of the two large continents produced by the breakup of Pangaea.

**laws of thermodynamics** [Gk. *thermos*: heat + *dynamis*: power] Laws derived from studies of the physical properties of energy and the ways energy interacts with matter. (*See also* first law of thermodynamics, second law of thermodynamics.)

**leaching** In soils, a process by which mineral nutrients in upper soil horizons are dissolved in water and carried to deeper horizons, where they are unavailable to plant roots.

**leading strand** In DNA replication, the daughter strand that is synthesized continuously. (Contrast with lagging strand.)

**leaf** (plural: leaves) In plants, the chief organ of photosynthesis.

**leaf primordium** (plural: primordia) An outgrowth on the side of the shoot apical meristem that will eventually develop into a leaf.

**leghemoglobin** In nitrogen-fixing plants, an oxygen-carrying protein in the cytoplasm of nodule cells that transports enough oxygen to the nitrogen-fixing bacteria to support their respiration, while keeping free oxygen concentrations low enough to protect nitrogenase.

**lek** A display ground within which male animals compete for and defend small display areas as a means of demonstrating their territorial prowess and winning opportunities to mate.

**lens** In the vertebrate eye, a crystalline protein structure that makes fine adjustments in the focus of images falling on the retina.

**lenticel** (len' ti sill) In plants, a spongy region in the periderm that allows gas exchange.

**leptin** A hormone produced by fat cells that is believed to provide feedback information to the brain about the status of the body's fat reserves.

**leukocyte** *See* white blood cell.

**lichen** (lie' kun) An organism resulting from the symbiotic association of a fungus and either a cyanobacterium or a unicellular alga.

**life cycle** The entire span of the life of an organism from the moment of fertilization (or asexual generation) to the time it reproduces in turn.

**life history strategy** The way in which an organism partitions its time and energy among growth, maintenance, and reproduction.

**ligament** A band of connective tissue linking two bones in a joint.

**ligand** (lig' and) Any molecule that binds to a receptor site of another (usually larger) molecule.

**light reactions** The initial phase of photosynthesis, in which light energy is converted into chemical energy.

**light-independent reactions** The phase of photosynthesis in which chemical energy captured in the light reactions is used to drive the reduction of $CO_2$ to form carbohydrates.

**lignin** A complex, hydrophobic polyphenolic polymer in plant cell walls that crosslinks other wall polymers, strengthening the walls, especially in wood.

**limbic system** A group of evolutionarily primitive structures in the vertebrate telencephalon that are involved in emotions, drives, instinctive behaviors, learning, and memory.

**liming** Application of compounds such as calcium carbonate, calcium hydroxide, or magnesium carbonate—commonly known as *lime*— to soil to reverse its acidification and increase the availability of calcium to plants.

**limiting resource** The required resource whose supply most strongly influences the size of a population.

**lineage species concept** The definition of a species as a branch on the tree of life, which has a history that starts at a speciation event and ends either at extinction or at another speciation event. (Contrast with biological species concept; morphological species concept.)

**linkage** Association between genes on the same chromosome such that they do not show random assortment and seldom recombine; the closer the genes, the lower the frequency of recombination.

**lipase** (lip' ase; lye' pase) An enzyme that digests fats.

**lipid** (lip' id) [Gk. *lipos*: fat] Nonpolar, hydrophobic molecules that include fats, oils, waxes, steroids, and the phospholipids that make up biological membranes.

**lipid bilayer** *See* phospholipid bilayer.

**liver** A large digestive gland. In vertebrates, it secretes bile and is involved in the formation of blood.

**loam** A type of soil consisting of a mixture of sand, silt, clay, and organic matter. One of the best soil types for agriculture.

**locus** (low' kus) (plural: loci, low' sigh) In genetics, a specific location on a chromosome. May be considered synonymous with *gene*.

**logistic growth** Growth, especially in the size of an organism or in the number of organisms in a population, that slows steadily as the entity approaches its maximum size. (Contrast with exponential growth.)

**long-day plant (LDP)** A plant that requires long days (actually, short nights) in order to flower.

**long-term depression (LTD)** A long-lasting decrease in the responsiveness resulting from continuous, repetitive, low-level stimulation. (Contrast with long-term potentiation.)

**long-term potentiation (LTP)** A long-lasting increase in the responsiveness of a neuron resulting from a period of intense stimulation. (Contrast with long-term depression.)

**loop of Henle** (hen' lee) Long, hairpin loop of the mammalian renal tubule that runs from the cortex down into the medulla and back to the cortex; creates a concentration gradient in the interstitial fluids in the medulla.

**lophophore** A U-shaped fold of the body wall with hollow, ciliated tentacles that encircles the mouth of animals in several different groups. Used for filtering prey from the surrounding water.

**loss of function mutation** A mutation that results in the loss of a functional protein. (Contrast with gain of function mutation.)

**low-density lipoproteins (LDLs)** Lipoproteins that transport cholesterol around the body for use in biosynthesis and for storage; LDLs are the "bad" lipoproteins associated with a high risk of cardiovascular disease.

**lumen** (loo' men) [L. *lumen*: light] The open cavity inside any tubular organ or structure, such as the gut or a renal tubule.

**lung** An internal organ specialized for respiratory gas exchange with air.

**luteinizing hormone (LH)** A gonadotropin produced by the anterior pituitary that stimulates the gonads to produce sex hormones.

**lymph** [L. *lympha*: liquid] A fluid derived from blood and other tissues that accumulates in intercellular spaces throughout the body and is returned to the blood by the lymphatic system.

**lymph node** A specialized structure in the vessels of the lymphatic system. Lymph nodes contain lymphocytes,which encounter and respond to foreign cells and molecules in the lymph as it passes through the vessels.

**lymphatic system** A system of vessels that returns interstitial fluid to the blood.

**lymphocyte** One of the two major classes of white blood cells; includes T cells, B cells, and other cell types important in the immune system.

**lymphoid tissues** Tissues of the immune system that are dispersed throughout the body, consisting of the thymus, spleen, bone marrow, and lymph nodes.

**lysis** (lie' sis) [Gk. *lysis*: break apart] Bursting of a cell.

**lysogeny** A form of viral replication in which the virus becomes incorporated into the host chromosome and remains inactive. Also called a lysogenic cycle. (Contrast with lytic cycle.)

**lysosome** (lie' so soam) [Gk. *lysis*: break away + *soma*: body] A membrane-enclosed organelle originating from the Golgi apparatus and containing hydrolytic enzymes. (Contrast with secondary lysosome.)

**lysozyme** (lie' so zyme) An enzyme in saliva, tears, and nasal secretions that hydrolyzes bacterial cell walls.

**lytic cycle** A viral reproductive cycle in which the virus takes over a host cell's synthetic machinery to replicate itself, then bursts (lyses) the host cell, releasing the new viruses. (Contrast with lysogeny.)

## - M -

**M phase** The portion of the cell cycle in which mitosis takes place.

**macroevolution** [Gk. *makros*: large] Evolutionary changes occurring over long time spans and usually involving changes in many traits. (Contrast with microevolution.)

**macromolecule** A giant (molecular weight > 1,000) polymeric molecule. The macromolecules are the proteins, polysaccharides, and nucleic acids.

**macronutrient** In plants, a mineral element required in concentrations of at least 1 milligram per gram of plant dry matter; in animals, a mineral element required in large amounts. (Contrast with micronutrient.)

**macrophage** (mac' roh faj) Phagocyte that engulfs pathogens by endocytosis.

**MADS box** DNA-binding domain in many plant transcription factors that is active in development.

**maintenance methylase** An enzyme that catalyzes the methylation of the new DNA strand when DNA is replicated.

**major histocompatibility complex (MHC)** A complex of linked genes, with multiple alleles, that control a number of cell surface antigens that identify self and can lead to graft rejection.

**malignant** Pertaining to a tumor that can grow indefinitely and/or spread from the original site of growth to other locations in the body. (Contrast with benign.)

**Malpighian tubule** (mal pee' gy un) A type of protonephridium found in insects.

**map unit** The distance between two genes as calculated from genetic crosses; a recombination frequency.

**marine** [L. *mare*: sea, ocean] Pertaining to or living in the ocean. (Contrast with aquatic, terrestrial.)

**mark–recapture method** A method of estimating population sizes of mobile organisms by capturing, marking, and releasing a sample of individuals, then capturing another sample at a later time.

**mass extinction** A period of evolutionary history during which rates of extinction are much higher than during intervening times.

**mass number** The sum of the number of protons and neutrons in an atom's nucleus.

**mast cells** Cells, typically found in connective tissue, that release histamine in response to tissue damage.

**maternal effect genes** Genes coding for morphogens that determine the polarity of the egg and larva in fruit flies.

**mating type** A particular strain of a species that is incapable of sexual reproduction with another member of the same strain but capable of sexual reproduction with members of other strains of the same species.

**maximum likelihood** A statistical method of determining which of two or more hypotheses (such as phylogenetic trees) best fit the observed data, given an explicit model of how the data were generated.

**mechanically gated channel** A molecular channel that opens or closes in response to mechanical force applied to the plasma membrane in which it is inserted.

**mechanoreceptor** A cell that is sensitive to physical movement and generates action potentials in response.

**medulla** (meh dull' luh) (1) The inner, core region of an organ, as in the adrenal medulla (adrenal gland) or the renal medulla (kidneys). (2) The portion of the brainstem that connects to the spinal cord.

**medusa** (plural: medusae) In cnidarians, a free-swimming, sexual life cycle stage shaped like a bell or an umbrella.

**megaphyll** The generally large leaf of a fern, horsetail, or seed plant, with several to many veins. (Contrast with microphyll.)

**megaspore** [Gk. *megas*: large + *spora*: to sow] In plants, a haploid spore that produces a female gametophyte.

**megastrobilus** In conifers, the female (seed-bearing) cone. (Contrast with microstrobilus.)

**meiosis** (my oh' sis) [Gk. *meiosis*: diminution] Division of a diploid nucleus to produce four haploid daughter cells. The process consists of two successive nuclear divisions with only one cycle of chromosome replication. In *meiosis I*, homologous chromosomes separate but retain their chromatids. The second division *meiosis II*, is similar to mitosis, in which chromatids separate.

**melatonin** A hormone released by the pineal gland. Involved in photoperiodicity and circadian rhythms.

**membrane potential** The difference in electrical charge between the inside and the outside of a cell, caused by a difference in the distribution of ions.

**membranous bone** A type of bone that develops by forming on a scaffold of connective tissue. (Contrast with cartilage bone.)

**memory cells** Long-lived lymphocytes produced by exposure to antigen. They persist in the body and are able to mount a rapid response to subsequent exposures to the antigen.

**Mendel's laws** *See* independent assortment; segregation.

**meristem** [Gk. *meristos*: divided] Plant tissue made up of undifferentiated actively dividing cells.

**meristem identity genes** In angiosperms, a group of genes whose expression initiates flower formation, probably by switching meristem cells from a vegetative to a reproductive fate.

**mesenchyme** (mez' en kyme) [Gk. *mesos*: middle + *enchyma*: infusion] Embryonic or unspecialized cells derived from the mesoderm.

**mesoderm** [Gk. *mesos*: middle + *derma*: skin] The middle of the three embryonic germ layers first delineated during gastrulation. Gives rise to the skeleton, circulatory system, muscles, excretory system, and most of the reproductive system.

**mesoglea** (mez' uh glee uh) [Gk. *mesos*: middle + *gloia*, glue] A thick, gelatinous noncellular layer that separates the two cellular tissue layers of ctenophores, cnidarians, and scyphozoans.

**mesophyll** (mez' uh fill) [Gk. *mesos*: middle + *phyllon*: leaf] Chloroplast-containing, photosynthetic cells in the interior of leaves.

**messenger RNA (mRNA)** Transcript of a region of one of the strands of DNA; carries information (as a sequence of codons) for the synthesis of one or more proteins.

**meta-** [Gk.: between, along with, beyond] Prefix denoting a change or a shift to a new form or level; for example, as used in metamorphosis.

**metabolic pathway** A series of enzyme-catalyzed reactions so arranged that the product of one reaction is the substrate of the next.

**metabolism** (meh tab' a lizm) [Gk. *metabole*: change] The sum total of the chemical reactions that occur in an organism, or some subset of that total (as in respiratory metabolism).

**metabolome** The quantitative description of all the small molecules in a cell or organism.

**metabotropic receptor** A receptor that that indirectly alters membrane permeability to a type of ion when it combines with its ligand.

**metagenomics** The practice of analyzing DNA from environmental samples without isolating intact organisms.

**metamorphosis** (met' a mor' fo sis) [Gk. *meta*: between + *morphe*: form, shape] A change occurring between one developmental stage and another, as for example from a tadpole to a frog. (*See* complete metamorphosis, incomplete metamorphosis.)

**metanephridia** The paired excretory organs of annelids.

**metaphase** (met' a phase) The stage in nuclear division at which the centromeres of the highly supercoiled chromosomes are all lying on a plane (the metaphase plane or plate) perpendicular to a line connecting the division poles.

**metapopulation** A population divided into subpopulations, among which there are occasional exchanges of individuals.

**metastasis** (meh tass' tuh sis) The spread of cancer cells from their original site to other parts of the body.

**methylation** The addition of a methyl group (—$CH_3$) to a molecule.

**MHC** *See* major histocompatibility complex.

**micelle** A particle of lipid covered with bile salts that is produced in the duodenum and facilitates digestion and absorption of lipids.

**microclimate** A subset of climatic conditions in a small specific area, which generally differ from those in the environment at large, as in an animal's underground burrow.

**microevolution** Evolutionary changes below the species level, affecting allele frequencies. (Contrast with macroevolution.)

**microfibril** Crosslinked cellulose polymers, forming strong aggregates in the plant cell wall.

**microfilament** In eukaryotic cells, a fibrous structure made up of actin monomers. Microfilaments play roles in the cytoskeleton, in cell movement, and in muscle contraction.

**microglia** Glial cells that act as macrophages and mediators of inflammatory responses in the central nervous system.

**micronutrient** In plants, a mineral element required in concentrations of less than 100 micrograms per gram of plant dry matter; in animals, a mineral element required in concentrations of less than 100 micrograms per day. (Contrast with macronutrient.)

**microphyll** A small leaf with a single vein, found in club mosses and their relatives. (Contrast with megaphyll.)

**micropyle** (mike' roh pile) [Gk. *mikros*: small + *pylon*: gate] Opening in the integument(s) of a seed plant ovule through which pollen grows to reach the female gametophyte within.

**microRNA** A small, noncoding RNA molecule, typically about 21 bases long, that binds to mRNA to inhibit its translation.

**microspore** [Gk. *mikros*: small + *spora*: to sow] In plants, a haploid spore that produces a male gametophyte.

**microstrobilus** In conifers, male pollen-bearing cone. (Contrast with megastrobilus.)

**microtubules** Tubular structures found in centrioles, spindle apparatus, cilia, flagella, and cytoskeleton of eukaryotic cells. These tubules play roles in the motion and maintenance of shape of eukaryotic cells.

**microvilli** (sing.: microvillus) Projections of epithelial cells, such as the cells lining the small intestine, that increase their surface area.

**midbrain** One of the three regions of the vertebrate brain. Part of the brainstem, it serves as a relay station for sensory signals sent to the cerebral hemispheres.

**middle lamella** (la mell' ah) [L. *lamina*: thin sheet] A layer of polysaccharides that separates plant cells; a shared middle lamella lies outside the primary walls of the two cells.

**mineral nutrients** Inorganic ions required by organisms for normal growth and reproduction.

**mismatch repair** A mechanism that scans DNA after it has been replicated and corrects any base-pairing mismatches.

**missense mutation** A change in a gene's sequence that results in a change in the sequence of the amino acid specified by the corresponding codon. (Contrast with frame-shift mutation, nonsense mutation, silent mutation.)

**mitochondrial matrix** The fluid interior of the mitochondrion, enclosed by the inner mitochondrial membrane.

**mitochondrion** (my' toe kon' dree un) (plural: mitochondria) [Gk. *mitos*: thread + *chondros*: grain] An organelle in eukaryotic cells that contains the enzymes of the citric acid cycle, the respiratory chain, and oxidative phosphorylation.

**mitosis** (my toe' sis) [Gk. *mitos*: thread] Nuclear division in eukaryotes leading to the formation of two daughter nuclei, each with a chromosome complement identical to that of the original nucleus.

**model systems** Also known as model organisms, these include the small group of species that are the subject of extensive research. They are organisms that adapt well to laboratory situations and findings from experiments on them can apply across a broad range of species. Classic examples include white rats and the fruit fly *Drosophila.*

**moderately repetitive sequences** DNA sequences repeated 10–1,000 times in the eukaryotic genome. They include the genes that code for rRNAs and tRNAs, as well as the DNA in telomeres.

**Modern Synthesis** An understanding of evolutionary biology that emerged in the early twentieth century as the principles of evolution were integrated with the principles of modern genetics.

**modularity** In evolutionary developmental biology, the principle that the molecular pathways that determine different developmental processes operate independently from one another. *See also* developmental module.

**mole** A quantity of a compound whose weight in grams is numerically equal to its molecular weight expressed in atomic mass units. Avogadro's number of molecules: $6.023 \times 10^{23}$ molecules.

**molecular clock** The approximately constant rate of divergence of macromolecules from one another over evolutionary time; used to date past events in evolutionary history.

**molecular evolution** The scientific study of the mechanisms and consequences of the evolution of macromolecules.

**molecular tool kit** A set of developmental genes and proteins that is common to most animals and is hypothesized to be responsible for the evolution of their differing developmental pathways.

**molecular weight** The sum of the atomic weights of the atoms in a molecule.

**molecule** A chemical substance made up of two or more atoms joined by covalent bonds or ionic attractions.

**molting** The process of shedding part or all of an outer covering, as the shedding of feathers by birds or of the entire exoskeleton by arthropods.

**monoclonal antibody** Antibody produced in the laboratory from a clone of hybridoma cells, each of which produces the same specific antibody.

**monoculture** In agriculture, a large-scale planting of a single species of domesticated crop plant.

**monoecious** (mo nee' shus) [Gk. *mono*: one + *oikos*: house] Pertaining to organisms in which both sexes are "housed" in a single individual that produces both eggs and sperm. (In some plants, these are found in different flowers within the same plant.) Examples include corn, peas, earthworms, hydras. (Contrast with dioecious.)

**monohybrid cross** A mating in which the parents differ with respect to the alleles of only one locus of interest.

**monomer** [Gk. *mono*: one + *meros:* unit] A small molecule, two or more of which can be combined to form oligomers (consisting of a few monomers) or polymers (consisting of many monomers).

**monophyletic** (mon' oh fih leht' ik) [Gk. *mono*: one + *phylon*: tribe] Pertaining to a group that consists of an ancestor and all of its descendants. (Contrast with paraphyletic, polyphyletic.)

**monosaccharide** A simple sugar. Oligosaccharides and polysaccharides are made up of monosaccharides.

**monosomic** Pertaining to an organism with one less than the normal diploid number of chromosomes.

**monosynaptic reflex** A neural reflex that begins in a sensory neuron and makes a single synapse before activating a motor neuron.

**morphogen** A diffusible substance whose concentration gradient determines a developmental pattern in animals and plants.

**morphogenesis** (more′ fo jen′ e sis) [Gk. *morphe*: form + *genesis*: origin] The development of form; the overall consequence of determination, differentiation, and growth.

**morphological species concept** The definition of a species as a group of individuals that look alike. (Contrast with biological species concept; lineage species concept.)

**morphology** (more fol′ o jee) [Gk. *morphe*: form + *logos*: study, discourse] The scientific study of organic form, including both its development and function.

**mosaic development** Pattern of animal embryonic development in which each blastomere contributes a specific part of the adult body. (Contrast with regulative development.)

**motif** *See* structural motif.

**motile** (mo′ tul) Able to move from one place to another. (Contrast with sessile.)

**motor cortex** The region of the cerebral cortex that contains motor neurons that directly stimulate specific muscle fibers to contract.

**motor neuron** A neuron carrying information from the central nervous system to a cell that produces movement.

**motor proteins** Specialized proteins that use energy to change shape and move cells or structures within cells. *See* dynein, kinesin.

**motor unit** A motor neuron and the muscle fibers it controls.

**mouth** An opening through which food is taken in, located at the anterior end of a tubular gut.

**mRNA** *See* messenger RNA.

**mucosal epithelium** An epithelial cell layer containing cells that secrete mucus; found in the digestive and respiratory tracts. Also called mucosa.

**Müllerian mimicry** Convergence in appearance of two or more unpalatable species.

**multipotent** Having the ability to differentiate into a limited number of cell types. (Contrast with pluripotent, totipotent.)

**muscle fiber** A single muscle cell. In the case of skeletal muscle, a syncitial, multinucleate cell.

**muscle tissue** Excitable tissue that can contract through the interactions of actin and myosin; one of the four major tissue types in multicellular animals. There are three types of muscle tissue: skeletal, smooth, and cardiac.

**mutagen** (mute′ ah jen) [L. *mutare*: change + Gk. *genesis*: source] Any agent (e.g., chemicals, radiation) that increases the mutation rate.

**mutation** A change in the genetic material not caused by recombination.

**mutualism** A type of interaction between species that benefits both species.

**mycelium** (my seel′ ee yum) [Gk. *mykes*: fungus] In the fungi, a mass of hyphae.

**mycorrhiza** (my′ ko rye′ za) (plural: mycorrhizae) [Gk. *mykes*: fungus + *rhiza*: root] An association of the root of a plant with the mycelium of a fungus.

**myelin** (my′ a lin) Concentric layers of plasma membrane that form a sheath around some axons; myelin provides the axon with electrical insulation and increases the rate of transmission of action potentials.

**myocardial infarction** Blockage of an artery that carries blood to the heart muscle.

**myofibril** (my′ oh fy′ bril) [Gk. *mys*: muscle + L. *fibrilla*: small fiber] A polymeric unit of actin or myosin in a muscle.

**myoglobin** (my′ oh globe′ in) [Gk. *mys*: muscle + L. *globus*: sphere] An oxygen-binding molecule found in muscle. Consists of a heme unit and a single globin chain; carries less oxygen than hemoglobin.

**myosin** One of the two contractile proteins of muscle.

## - N -

**natural killer cell** A type of lymphocyte that attacks virus-infected cells and some tumor cells as well as antibody-labeled target cells.

**natural selection** The differential contribution of offspring to the next generation by various genetic types belonging to the same population. The mechanism of evolution proposed by Charles Darwin.

**nauplius** (naw′ plee us) [Gk. *nauplios*: shellfish] A bilaterally symmetrical larval form typical of crustaceans.

**necrosis** (nec roh′ sis) [Gk. *nekros*: death] Premature cell death caused by external agents such as toxins.

**negative feedback** In regulatory systems, information that decreases a regulatory response, returning the system to the set point. (Contrast with positive feedback.)

**negative regulation** A type of gene regulation in which a gene is normally transcribed, and the binding of a repressor protein to the promoter prevents transcription. (Contrast with positive regulation.)

**nematocyst** (ne mat′ o sist) [Gk. *nema*: thread + *kystis*: cell] An elaborate, threadlike structure produced by cells of jellyfishes and other cnidarians, used chiefly to paralyze and capture prey.

**nephron** (nef′ ron) [Gk. *nephros*: kidney] The functional unit of the kidney, consisting of a structure for receiving a filtrate of blood and a tubule that reabsorbs selected parts of the filtrate.

**Nernst equation** A mathematical statement that calculates the potential across a membrane permeable to a single type of ion that differs in concentration on the two sides of the membrane.

**nerve** A structure consisting of many neuronal axons and connective tissue.

**nervous tissue** Tissue specialized for processing and communicating information; one of the four major tissue types in multicellular animals.

**net primary productivity (NPP)** The rate at which energy captured by photosynthesis is incorporated into the bodies of primary producers bodies through growth and reproduction.

**net primary production** The amount of primary producer biomass made available for consumption by heterotrophs.

**neural network** An organized group of neurons that contains three functional categories of neurons—afferent neurons, interneurons, and efferent neurons—and is capable of processing information.

**neural tube** An early stage in the development of the vertebrate nervous system consisting of a hollow tube created by two opposing folds of the dorsal ectoderm along the anterior–posterior body axis.

**neurohormone** A chemical signal produced and released by neurons that subsequently acts as a hormone.

**neuromuscular junction** Synapse (point of contact) where a motor neuron axon stimulates a muscle fiber cell.

**neuron** (noor′ on) [Gk. *neuron*: nerve] A nervous system cell that can generate and conduct action potentials along an axon to a synapse with another cell.

**neurotransmitter** A substance produced in and released by a neuron (the presynaptic cell) that diffuses across a synapse and excites or inhibits another cell (the postsynaptic cell).

**neurulation** Stage in vertebrate development during which the nervous system begins to form.

**neutral allele** An allele that does not alter the functioning of the proteins for which it codes.

**neutral theory** A view of molecular evolution that postulates that most mutations do not affect the amino acid being coded for, and that such mutations accumulate in a population at rates driven by genetic drift and mutation rates.

**neutron** (new′ tron) One of the three fundamental particles of matter (along with protons and electrons), with mass approximately 1 amu and no electrical charge.

**niche** (nitch) [L. *nidus*: nest] The set of physical and biological conditions a species requires to survive, grow, and reproduce. (*See also* fundamental niche, realized niche.)

**nitrate reduction** The process by which nitrate ($NO_3^-$) is reduced to ammonia ($NH_3$).

**nitric oxide (NO)** An unstable molecule (a gas) that serves as a second messenger causing smooth muscle to relax. In the nervous system it operates as a neurotransmitter.

**nitrifiers** Chemolithotrophic bacteria that oxidize ammonia to nitrate in soil and in seawater.

**nitrogen fixation** Conversion of atmospheric nitrogen gas ($N_2$) into a more reactive and biologically useful form (ammonia), which makes nitrogen available to living things. Carried out by nitrogen-fixing bacteria, some of them free-living and others living within plant roots.

**nitrogenase** An enzyme complex found in nitrogen-fixing bacteria that mediates the stepwise reduction of atmospheric $N_2$ to ammonia and which is strongly inhibited by oxygen.

**node** [L. *nodus*: knob, knot] In plants, a (sometimes enlarged) point on a stem where a leaf is or was attached.

**node of Ranvier** A gap in the myelin sheath covering an axon; the point where the axonal membrane can fire action potentials.

**nodule** A specialized structure in the roots of nitrogen-fixing plants that houses nitrogen-fixing bacteria, in which oxygen is maintained at a low level by leghemoglobin.

**noncompetitive inhibitor** A nonsubstrate that inhibits the activity of an enzyme by binding to a site other than its active site. (Contrast with competitive inhibitor.)

**noncyclic electron transport** In photosynthesis, the flow of electrons that forms ATP, NADPH, and $O_2$.

**nondisjunction** Failure of sister chromatids to separate in meiosis II or mitosis, or failure of homologous chromosomes to separate in meiosis I. Results in aneuploidy.

**nonpolar** Having electric charges that are evenly balanced from one end to the other. (Contrast with polar.)

**nonrandom mating** Selection of mates on the basis of a particular trait or group of traits.

**nonsense mutation** Change in a gene's sequence that prematurely terminates translation by changing one of its codons to a stop codon.

**nonsynonymous substitution** A change in a gene from one nucleotide to another that changes the amino acid specified by the corresponding codon (i.e., AGC →AGA, or serine → arginine). (Contrast with synonymous substitution.)

**norepinephrine** A neurotransmitter found in the central nervous system and also at the postganglionic nerve endings of the sympathetic nervous system. Also called noradrenaline.

**normal flora** Microorganisms that normally live and reproduce on or in the body without causing disease, and which form a nonspecific defense against pathogens by competing with them for space and nutrients.

**notochord** (no' tow kord) [Gk. *notos*: back + *chorde*: string] A flexible rod of gelatinous material serving as a support in the embryos of all chordates and in the adults of tunicates and lancelets.

**nucleic acid** (new klay' ik) A polymer made up of nucleotides, specialized for the storage, transmission, and expression of genetic information. DNA and RNA are nucleic acids.

**nucleic acid hybridization** A technique in which a single-stranded nucleic acid probe is made that is complementary to, and binds to, a target sequence, either DNA or RNA. The resulting double-stranded molecule is a hybrid.

**nucleoid** (new' klee oid) The region that harbors the chromosomes of a prokaryotic cell. Unlike the eukaryotic nucleus, it is not bounded by a membrane.

**nucleolus** (new klee' oh lus) A small, generally spherical body found within the nucleus of eukaryotic cells. The site of synthesis of ribosomal RNA.

**nucleoside** A nucleotide without the phosphate group; a nitrogenous base attached to a sugar.

**nucleosome** A portion of a eukaryotic chromosome, consisting of part of the DNA molecule wrapped around a group of histone molecules, and held together by another type of histone molecule. The chromosome is made up of many nucleosomes.

**nucleotide** The basic chemical unit in nucleic acids, consisting of a pentose sugar, a phosphate group, and a nitrogen-containing base.

**nucleotide substitution** A change of one base pair to another in a DNA sequence.

**nucleus** (new' klee us) [L. *nux*: kernel or nut] (1) In cells, the centrally located compartment of eukaryotic cells that is bounded by a double membrane and contains the chromosomes. (2) In the brain, an identifiable group of neurons that share common characteristics or functions.

**nutrient** A food substance; or, in the case of mineral nutrients, an inorganic element required for completion of the life cycle of an organism.

## - O -

**obligate anaerobe** An anaerobic prokaryote that cannot survive exposure to $O_2$.

**odorant** A molecule that can bind to an olfactory receptor.

**oil** A triglyceride that is liquid at room temperature. (Contrast with fat.)

**Okazaki fragments** Newly formed DNA making up the lagging strand in DNA replication. DNA ligase links Okazaki fragments together to give a continuous strand.

**olfactory** [L. *olfacere*: to smell] Pertaining to the sense of smell (*olfaction*).

**oligodendrocyte** A type of glial cell that myelinates axons in the central nervous system.

**oligosaccharide** A polymer containing a small number of monosaccharides.

**ommatidia** [Gk. *omma*: eye] The units that make up the compound eye of some arthropods.

**omnivore** [L. *omnis*: everything + *vorare*: to devour] An organism that eats both animal and plant material. (Contrast with carnivore, detritivore, herbivore.)

**oncogene** [Gk. *onkos*: mass, tumor + *genes*: born] A gene that codes for a protein product that stimulates cell proliferation. Mutations in oncogenes that result in excessive cell proliferation can give rise to cancer.

**oocyte** *See* primary oocyte, secondary oocyte.

**oogenesis** (oh' eh jen e sis) [Gk. *oon*: egg + *genesis*: source] Gametogenesis leading to production of an ovum.

**oogonium** (oh' eh go' nee um) (plural: oogonia) (1) In some algae and fungi, a cell in which an egg is produced. (2) In animals, the diploid progeny of a germ cell in females.

**operator** The region of an operon that acts as the binding site for the repressor.

**open circulatory system** Circulatory system in which extracellular fluid leaves the vessels of the circulatory system, percolates between cells and through tissues, and then flows back into the circulatory system to be pumped out again. (Contrast with closed circulatory system.)

**operon** A genetic unit of transcription, typically consisting of several structural genes that are transcribed together; the operon contains at least two control regions: the promoter and the operator.

**opportunity cost** The sum of the benefits an animal forfeits by not being able to perform some other behavior during the time when it is performing a given behavior.

**opsin** (op' sin) [Gk. *opsis*: sight] The protein portion of the visual pigment rhodopsin. (*See* rhodopsin.)

**optic chiasm** [Gk. *chiasma*: cross] Structure on the lower surface of the vertebrate brain where the two optic nerves come together.

**optical isomers** Two isomers that are mirror images of each other.

**optimal foraging theory** The application of a cost–benefit approach to feeding behavior to identify the fitness value of feeding choices.

**orbital** A region in space surrounding the atomic nucleus in which an electron is most likely to be found.

**organ** [Gk. *organon*: tool] A body part, such as the heart, liver, brain, root, or leaf. Organs are composed of different tissues integrated to perform a distinct function. Organs, in turn, are integrated into organ systems.

**organ identity genes** In angiosperms, genes that specify the different organs of the flower. (Compare with homeotic genes.)

**organ of Corti** Structure in the inner ear that transforms mechanical forces produced from pressure waves ("sound waves") into action potentials that are sensed as sound.

**organ system** An interrelated and integrated group of tissues and organs that work together in a physiological function.

**organelle** (or gan el') Any of the membrane-enclosed structures within a eukaryotic cell. Examples include the nucleus, endoplasmic reticulum, and mitochondria.

**organic** (1) Pertaining to any chemical compound that contains carbon. (2) Pertaining to any aspect of living matter, e.g., to its evolution, structure, or chemistry.

**organism** Any living entity.

**organizer** Region of the early amphibian embryo that directs early embryonic development. Also known as the primary embryonic organizer.

**organogenesis** The formation of organs and organ systems during development.

**origin of replication (*ori*)** DNA sequence at which helicase unwinds the DNA double helix and DNA polymerase binds to initiate DNA replication.

**orthology** (or thol' o jee) Type of homology in which the divergence of homologous genes can be traced to speciation events. (Contrast with paralogy.)

**osmoconformer** An aquatic animal that equilibrates the osmolarity of its extracellular fluid that is the same as with that of the external environment.

**osmolarity** The concentration of osmotically active particles in a solution.

**osmoregulation** Regulation of the chemical composition of the body fluids of an organism.

**osmosis** (oz mo' sis) [Gk. *osmos*: to push] Movement of water across a differentially permeable membrane, from one region to another region where the water potential is more negative.

**ossicle** (oss' ick ul) [L. *os*: bone] The calcified construction unit of echinoderm skeletons.

**osteoblast** (oss' tee oh blast) [Gk. *osteon*: bone + *blastos*: sprout] A cell that lay down the protein matrix of bone.

**osteoclast** (oss' tee oh clast) [Gk. *osteon*: bone + *klastos*: broken] A cell that dissolves bone.

**osteocyte** An osteoblast that has become enclosed in lacunae within the bone it has built.

**outgroup** In phylogenetics, a group of organisms used as a point of reference for comparison with the groups of primary interest (the ingroup).

**oval window** The flexible membrane that, when moved by the bones of the middle ear, produces pressure waves in the inner ear.

**ovarian cycle** In human females, the monthly cycle of events by which eggs and hormones are produced. (Contrast with uterine cycle).

**ovary** (oh' var ee) [L. *ovum*: egg] Any female organ, in plants or animals, that produces an egg.

**overtopping** Plant growth pattern in which one branch differentiates from and grows beyond the others.

**oviduct** In mammals, the tube serving to transport eggs to the uterus or to outside of the body.

**oviparity** Reproduction in which eggs are released by the female and development is external to the mother's body. (Contrast with viviparity.)

**ovoviviparity** Pertaining to reproduction in which fertilized eggs develop and hatch within the mother's body but are not attached to the mother by means of a placenta.

**ovulation** Release of an egg from an ovary.

**ovule** (oh' vule) In plants, a structure comprising the megasporangium and the integument, which develops into a seed after fertilization.

**ovum** (oh' vum) (plural: ova) [L. egg] The female gamete.

**oxidation** (ox i day' shun) Relative loss of electrons in a chemical reaction; either outright removal to form an ion, or the sharing of electrons with substances having a greater affinity for them, such as oxygen. Most oxidations, including biological ones, are associated with the liberation of energy. (Contrast with reduction.)

**oxidative phosphorylation** ATP formation in the mitochondrion, associated with flow of electrons through the respiratory chain.

**oxygenase** An enzyme that catalyzes the addition of oxygen to a substrate from $O_2$.

**oxytocin** A hormone released by the posterior pituitary that promotes social bonding.

## - P -

**pancreas** (pan' cree us) A gland located near the stomach of vertebrates that secretes digestive enzymes into the small intestine and releases insulin into the bloodstream.

**Pangaea** (pan jee' uh) [Gk. *pan*: all, every] The single land mass formed when all the continents came together in the Permian period.

**para-** [Gk. *para*: akin to, beside] Prefix indicating association in being along side or accessory to.

**parabronchi** Passages in the lungs of birds through which air flows.

**paracrine** [Gk. *para*: near] Pertaining to a chemical signal, such as a hormone, that acts locally, near the site of its secretion. (Contrast with autocrine.)

**paralogy** (par al' o jee) Type of homology in which the divergence of homologous genes can be traced to gene duplication events. (Contrast with orthology.)

**paraphyletic** (par' a fih leht' ik) [Gk. *para*: beside + *phylon*: tribe] Pertaining to a group that consists of an ancestor and some, but not all, of its descendants. (Contrast with monophyletic, polyphyletic.)

**parasite** An organism that consumes parts of an organism much larger than itself (known as its host). Parasites sometimes, but not always, kill their host.

**parasympathetic nervous system** The division of the autonomic nervous system that works in opposition to the sympathetic nervous system. (Contrast with sympathetic nervous system.)

**parathyroid glands** Four glands on the posterior surface of the thyroid gland that produce and release paraththyroid hormone.

**parathyroid hormone (PTH)** A hormone secreted by the parathyroid glands that stimulates osteoclast activity and raises blood calcium levels. Also called parathormone.

**parenchyma** (pair eng' kyma) A plant tissue composed of relatively unspecialized cells without secondary walls.

**parent rock** The soil horizon consisting of the rock that is breaking down to form the soil. Also called bedrock, or the C horizon.

**parental (P) generation** The individuals that mate in a genetic cross. Their offspring are the first filial ($F_1$) generation.

**parsimony** Preferring the simplest among a set of plausible explanations of any phenomenon.

**parthenocarpy** Formation of fruit from a flower without fertilization.

**parthenogenesis** [Gk. *parthenos*: virgin] Production of an organism from an unfertilized egg.

**particulate theory** In genetics, the theory that genes are physical entities that retain their identities after fertilization.

**passive transport** Diffusion across a membrane; may or may not require a channel or carrier protein. (Contrast with active transport.)

**patch clamping** A technique for isolating a tiny patch of membrane to allow the study of ion movement through a particular channel.

**pathogen** (path' o jen) [Gk. *pathos*: suffering + *genesis:* source] An organism that causes disease.

**pattern formation** In animal embryonic development, the organization of differentiated tissues into specific structures such as wings.

**pedigree** The pattern of transmission of a genetic trait within a family.

**pelagic zone** [Gk. *pelagos*: sea] The open ocean.

**penetrance** The proportion of individuals with a particular genotype that show the expected phenotype.

**penis** An accessory sex organ of male animals that enables the male to deposit sperm in the female's reproductive tract.

**pentose** [Gk. *penta:* five] A sugar containing five carbon atoms.

**PEP carboxylase** The enzyme that combines carbon dioxide with PEP to form a 4-carbon dicarboxylic acid at the start of $C_4$ photosynthesis or of crassulacean acid metabolism (CAM).

**pepsin** [Gk. *pepsis*: digestion] An enzyme in gastric juice that digests protein.

**pepsinogen** Inactive secretory product that is converted into pepsin by low pH or by enzymatic action.

**peptide linkage** The bond between amino acids in a protein; formed between a carboxyl group and amino group ($CO—NH^-$) with the loss of water molecules.

**peptidoglycan** The cell wall material of many bacteria, consisting of a single enormous molecule that surrounds the entire cell.

**perennial** (per ren' ee al) [L. *per*: throughout + *annus*: year] A plant that survives from year to year. (Contrast with annual, biennial.)

**perfect flower** A flower with both stamens and carpels; a hermaphroditic flower. (Contrast with imperfect flower.)

**pericycle** [Gk. *peri*: around + *kyklos*: ring or circle] In plant roots, tissue just within the endodermis, but outside of the root vascular tissue. Meristematic activity of pericycle cells produces lateral root primordia.

**periderm** The outer tissue of the secondary plant body, consisting primarily of cork.

**period** (1) A category in the geological time scale. (2) The duration of a single cycle in a cyclical event, such as a circadian rhythm.

**peripheral nervous system (PNS)** The portion of the nervous system that transmits information to and from the central nervous system, consisting of neurons that extend or reside outside the brain or spinal cord and their supporting cells. (Contrast with central nervous system.)

**peristalsis** (pair' i stall' sis) Wavelike muscular contractions proceeding along a tubular organ, propelling the contents along the tube.

**peritoneum** The mesodermal lining of the body cavity in coelomate animals.

**peroxisome** An organelle that houses reactions in which toxic peroxides are formed and then converted to water.

**petal** [Gk. *petalon*: spread out] In an angiosperm flower, a sterile modified leaf, nonphotosynthetic, frequently brightly colored, and often serving to attract pollinating insects.

**petiole** (pet' ee ole) [L. *petiolus*: small foot] The stalk of a leaf.

**pH** The negative logarithm of the hydrogen ion concentration; a measure of the acidity of a solution. A solution with pH = 7 is said to be neutral; pH values higher than 7 characterize basic solutions, while acidic solutions have pH values less than 7.

**phage** (fayj) *See* bacteriophage.

**phagocyte** [Gk. *phagein*: to eat + *kystos*: sac] One of two major classes of white blood cells; one of the nonspecific defenses of animals; ingests invading microorganisms by phagocytosis.

**phagocytosis** Endocytosis by a cell of another cell or large particle.

**pharming** The use of genetically modified animals to produce medically useful products in their milk.

**pharynx** [Gk. throat] The part of the gut between the mouth and the esophagus.

**phenotype** (fee' no type) [Gk. *phanein*: to show] The observable properties of an individual resulting from both genetic and environmental factors. (Contrast with genotype.)

**phenotypic plasticity** *See* developmental plasticity.

**pheromone** (feer' o mone) [Gk. *pheros*: carry + *hormon*: excite, arouse] A chemical substance used in communication between organisms of the same species.

**phloem** (flo' um) [Gk. *phloos*: bark] In vascular plants, the vascular tissue that transports sugars and other solutes from sources to sinks.

**phosphate group** The functional group $—OPO_3H_2$.

**phosphodiester linkage** The connection in a nucleic acid strand, formed by linking two nucleotides.

**phospholipid** A lipid containing a phosphate group; an important constituent of cellular membranes. (*See* lipid.)

**phospholipid bilayer** The basic structural unit of biological membranes; a sheet of phospholipids two molecules thick in which the phospholipids are lined up with their hydrophobic "tails" packed tightly together and their hydrophilic, phosphate-containing "heads" facing outward. Also called lipid bilayer.

**phosphorylation** Addition of a phosphate group.

**photoautotroph** An organism that obtains energy from light and carbon from carbon dioxide. (Contrast with chemolithotroph, chemoheterotroph, photoheterotroph.)

**photoheterotroph** An organism that obtains energy from light but must obtain its carbon from organic compounds. (Contrast with chemolithotroph, chemoheterotroph, photoautotroph.)

**photon** (foe' ton) [Gk. *photos*: light] A quantum of visible radiation; a "packet" of light energy.

**photoperiodicity** Control of an organism's physiological or behavioral responses by the length of the day or night.

**photoreceptor** (1) In plants, a pigment that triggers a physiological response when it absorbs a photon. (2) In animals, a sensory receptor cell that senses and responds to light energy.

**photorespiration** Light-driven uptake of oxygen and release of carbon dioxide, the carbon being derived from the early reactions of photosynthesis.

**photosynthesis** (foe tow sin' the sis) [literally, "synthesis from light"] Metabolic processes, carried out by green plants, by which visible light is trapped and the energy used to synthesize compounds such as ATP and glucose.

**photosystem** [Gk. *phos:* light + *systema:* assembly] A light-harvesting complex in the chloroplast thylakoid composed of pigments and proteins.

**photosystem I** In photosynthesis, the reactions that absorb light at 700 nm, passing electrons to ferrodoxin and thence to NADPH. Rich in chlorophyll *a*.

**photosystem II** In photosynthesis, the reactions that absorb light at 660 nm, passing electrons to the electron transport chain in the chloroplast. Rich in chlorophyll *b*.

**phototropins** A class of blue light receptors that mediate phototropism and other plant responses.

**phototropism** [Gk. *photos*: light + *trope*: turning] A directed plant growth response to light.

**phycobilin** Photosynthetic pigment that absorbs red, yellow, orange, and green light and is found in cyanobacteria and some red algae.

**phylogenetic tree** A graphic representation of lines of descent among organisms or their genes.

**phylogeny** (fy loj' e nee) [Gk. *phylon*: tribe, race + *genesis*: source] The evolutionary history of a particular group of organisms or their genes.

**physiology** (fiz' ee ol' o jee) [Gk. *physis*: natural form] The scientific study of the functions of living organisms and the individual organs, tissues, and cells of which they are composed.

**phytoalexins** Substances toxic to pathogens, produced by plants in response to fungal or bacterial infection.

**phytochrome** (fy' tow krome) [Gk. *phyton*: plant + *chroma*: color] A plant pigment regulating a large number of developmental and other phenomena in plants.

**phytomers** In plants, the repeating modules that compose a shoot, each consisting of one or more leaves, attached to the stem at a node; an internode; and one or more axillary buds.

**phytoplankton** Photosynthetic plankton.

**phytoremediation** A form of bioremediation that uses plants to clean up environmental pollution.

**pigment** A substance that absorbs visible light.

**pineal gland** Gland located between the cerebral hemispheres that secretes melatonin.

**pinocytosis** Endocytosis by a cell of liquid containing dissolved substances.

**pistil** [L. *pistillum*: pestle] The structure of an angiosperm flower within which the ovules are borne. May consist of a single carpel, or of several carpels fused into a single structure. Usually differentiated into ovary, style, and stigma.

**pith** In plants, relatively unspecialized tissue found within a cylinder of vascular tissue.

**pituitary gland** A small gland attached to the base of the brain in vertebrates. Its hormones control the activities of other glands. Also known as the hypophysis.

**placenta** (pla sen' ta) The organ in female mammals that provides for the nourishment of the fetus and elimination of the fetal waste products.

**plankton** Free-floating small aquatic organisms. Photosynthetic members of the plankton are referred to as phytoplankton.

**planula** (plan' yew la) [L. *planum*: flat] A free-swimming, ciliated larval form typical of the cnidarians.

**plaque** (plack) [Fr.: a metal plate or coin] (1) A circular clearing in a layer (lawn) of bacteria growing on the surface of a nutrient agar gel. (2) An accumulation of prokaryotic organisms on tooth enamel. Acids produced by these microorganisms cause tooth decay. (3) A region of arterial wall invaded by fibroblasts and fatty deposits. (*See* atherosclerosis.)

**plasma** (plaz' muh) The liquid portion of blood, in which blood cells and other particulates are suspended.

**plasma cell** An antibody-secreting cell that develops from a B cell; the effector cell of the humoral immune system.

**plasma membrane** The membrane that surrounds the cell, regulating the entry and exit of molecules and ions. Every cell has a plasma membrane.

**plasmid** A DNA molecule distinct from the chromosome(s); that is, an extrachromosomal element; found in many bacteria. May replicate independently of the chromosome.

**plasmodesma** (plural: plasmodesmata) [Gk. *plassein*: to mold + *desmos*: band] A cytoplasmic strand connecting two adjacent plant cells.

**plasmogamy** The fusion of the cytoplasm of two cells. (Contrast with karyogamy.)

**plastid** Any of the plant cell organelles that house biochemical pathways for photosynthesis.

**plate tectonics** [Gk. *tekton*: builder] The scientific study of the structure and movements of Earth's lithospheric plates, which are the cause of continental drift.

**platelet** A membrane-bounded body without a nucleus, arising as a fragment of a cell in the bone marrow of mammals. Important to blood-clotting action.

**pleiotropy** (plee' a tro pee) [Gk. *pleion*: more] The determination of more than one character by a single gene.

**pleural membrane** [Gk. *pleuras*: rib, side] The membrane lining the outside of the lungs and the walls of the thoracic cavity. Inflammation of these membranes is a condition known as pleurisy.

**pluripotent** [L. *pluri*: many + *potens*: powerful] Having the ability to form all of the cells in the body. (Contrast with multipotent, totipotent.)

**podocytes** Cells of Bowman's capsule of the nephron that cover the capillaries of the glomerulus, forming filtration slits.

**point mutation** A mutation that results from the gain, loss, or substitution of a single nucleotide.

**polar** Having separate and opposite electric charges at two ends, or poles. (Contrast with nonpolar.)

**polar body** A nonfunctional nucleus produced by meiosis during oogenesis.

**polar nuclei** In angiosperms, the two nuclei in the central cell of the megagametophyte; following fertilization they give rise to the endosperm.

**polarity** (1) In chemistry, the property of unequal electron sharing in a covalent bond that defines a polar molecule. (2) In development, the difference between one end of an organism or structure and the other.

**pollen** [L. *pollin*: fine flour] In seed plants, microscopic grains that contain the male gametophyte (microgametophyte) and gamete (microspore).

**pollen tube** A structure that develops from a pollen grain through which sperm are released into the megagametophyte.

**pollination** The process of transferring pollen from an anther to the stigma of a pistil in an angiosperm or from a strobilus to an ovule in a gymnosperm.

**poly-** [Gk. *poly*: many] A prefix denoting multiple entities.

**poly A tail** A long sequence of adenine nucleotides (50–250) added after transcription to the 3′ end of most eukaryotic mRNAs.

**polyandry** Mating system in which one female mates with multiple males.

**polygyny** Mating system in which one male mates with multiple females.

**polymer** [Gk. *poly*: many + *meros*: unit] A large molecule made up of similar or identical subunits called monomers. (Contrast with monomer, oligomer.)

**polymerase chain reaction (PCR)** An enzymatic technique for the rapid production of millions of copies of a particular stretch of DNA where only a small amount of the parent molecule is available.

**polymorphic** (pol' lee mor' fik) [Gk. *poly*: many + *morphe*: form, shape] Coexistence in a population of two or more distinct traits.

**polyp** (pah' lip) [Gk. *poly*: many + *pous:* foot] In cnidarians, a sessile, asexual life cycle stage.

**polypeptide** A large molecule made up of many amino acids joined by peptide linkages. Large polypeptides are called proteins.

**polyphyletic** (pol' lee fih leht' ik) [Gk. *poly*: many + *phylon*: tribe] Pertaining to a group that consists of multiple distantly related organisms, and does not include the common ancestor of the group. (Contrast with monophyletic, paraphyletic.)

**polyploidy** (pol' lee ploid ee) The possession of more than two entire sets of chromosomes.

**polyribosome (polysome)** A complex consisting of a threadlike molecule of messenger RNA and several (or many) ribosomes. The ribosomes move along the mRNA, synthesizing polypeptide chains as they proceed.

**polysaccharide** A macromolecule composed of many monosaccharides (simple sugars). Common examples are cellulose and starch.

**pons** [L. *pons*: bridge] Region of the brainstem anterior to the medulla.

**population** Any group of organisms coexisting at the same time and in the same place and capable of interbreeding with one another.

**population bottleneck** A period during which only a few individuals of a normally large population survive.

**population density** The number of individuals of a population per unit of area or volume.

**population dynamics** The patterns and processes of change in populations.

**population genetics** The study of genetic variation and its causes within populations.

**portal blood vessels** Blood vessels that begin and end in capillary beds.

**positive cooperativity** Occurs when a molecule can bind several ligands and each one that binds alters the conformation of the molecule so that it can bind the next ligand more easily. The binding of four molecules of $O_2$ by hemoglobin is an example of positive cooperativity.

**positive feedback** In regulatory systems, information that amplifies a regulatory response, increasing the deviation of the system from the set point. (Contrast with negative feedback.)

**positive regulation** A form of gene regulation in which a regulatory macromolecule is needed to turn on the transcription of a structural gene; in its absence, transcription will not occur. (Contrast with negative regulation.)

**post-** [L. *postere*: behind, following after] Prefix denoting something that comes after.

**postabsorptive state** State in which no food remains in the gut and thus no nutrients are being absorbed. (Contrast with absorptive state.)

**posterior** Toward or pertaining to the rear. (Contrast with anterior.)

**postsynaptic cell** The cell that receives information from a neuron at a synapse. (Contrast with presynaptic neuron.)

**postzygotic reproductive barriers** Barriers to the reproductive process that occur after the union of the nuclei of two gametes. (Contrast with prezygotic reproductive barriers.)

**potential energy** Energy not doing work, such as the energy stored in chemical bonds. (Contrast with kinetic energy.)

**precapillary sphincter** A cuff of smooth muscle that can shut off the blood flow to a capillary bed.

**pre-mRNA (precursor mRNA)** Initial gene transcript before it is modified to produce functional mRNA. Also known as the primary transcript.

**predator** An organism that kills and eats other organisms.

**prereplication complex** In eukaryotes, a complex of proteins that binds to DNA at the initiation of DNA replication.

**pressure flow model** An effective model for phloem transport in angiosperms. It holds that sieve element transport is driven by an osmotically driven pressure gradient between source and sink.

**pressure potential** The hydrostatic pressure of an enclosed solution in excess of the surrounding atmospheric pressure. (Contrast with solute potential, water potential.)

**presynaptic neuron** The neuron that transmits information to another cell at a synapse. (Contrast with postsynaptic cell.)

**prey** [L. *praeda*: booty] An organism consumed by a predator as an energy source.

**prezygotic reproductive barriers** Barriers to the reproductive process that occur before the union of the nuclei of two gametes (Contrast with postzygotic reproductive barriers.)

**primary active transport** Active transport in which ATP is hydrolyzed, yielding the energy required to transport an ion or molecule against its concentration gradient. (Contrast with secondary active transport.)

**primary cell wall** In plant cells, a structure that forms at the middle lamella after cytokinesis, made up of cellulose microfibrils, hemicelluloses, and pectins. (Contrast with secondary cell wall.)

**primary consumer** An organism (herbivore) that eats plant tissues.

**primary growth** In plants, growth that is characterized by the lengthening of roots and shoots and by the proliferation of new roots and shoots through branching. (Contrast with secondary growth.)

**primary immune response** The first response of the immune system to an antigen, involving recognition by lymphocytes and the production of effector cells and memory cells. (Contrast with secondary immune response.)

**primary lysosome** *See* lysosome.

**primary meristem** Meristem that produces the tissues of the primary plant body.

**primary oocyte** (oh' eh site) [Gk. *oon*: egg + *kytos*: container] The diploid progeny of an oogonium. In many species, a primary oocyte enters prophase of the first meiotic division, then remains in developmental arrest for a long time before resuming meiosis to form a secondary oocyte and a polar body.

**primary plant body** That part of a plant produced by primary growth. Consists of all the *nonwoody* parts of a plant; many herbaceous plants consist entirely of a primary plant body. (Contrast with secondary plant body.)

**primary producer** A photosynthetic or chemosynthetic organism that synthesizes complex organic molecules from simple inorganic ones.

**primary sex determination** Genetic determination of gametic sex, male or female. (Contrast with secondary sex determination.)

**primary spermatocyte** The diploid progeny of a spermatogonium; undergoes the first meiotic division to form secondary spermatocytes.

**primary succession** Succession that begins in an area initially devoid of life, such as on recently exposed glacial till or lava flows. (Contrast with secondary succession.)

**primary structure** The specific sequence of amino acids in a protein.

**primase** An enzyme that catalyzes the synthesis of a primer for DNA replication.

**primer** Strand of nucleic acid, usually RNA, that is the necessary starting material for the synthesis of a new DNA strand, which is synthesized from the 3′ end of the primer.

**primordium** (plural: primordia) [L. origin] The most rudimentary stage of an organ or other part.

**prion** An infectious protein that can proliferate by converting the inactive form of a particular protein into an active protein.

**pro-** [L.: first, before, favoring] A prefix often used in biology to denote a developmental stage that comes first or an evolutionary form that appeared earlier than another. For example, prokaryote, prophase.

**probe** A segment of single stranded nucleic acid used to identify DNA molecules containing the complementary sequence.

**procambium** Primary meristem that produces the vascular tissue.

**procedural memory** Memory of motor tasks. Cannot be consciously recalled and described. (Contrast with declarative memory.)

**processive** Pertaining to an enzyme that catalyzes many reactions each time it binds to a substrate, as DNA polymerase does during DNA replication.

**progesterone** [L. *pro*: favoring + *gestare*: to bear] A female sex hormone that maintains pregnancy.

**prolactin** A hormone released by the anterior pituitary, one of whose functions is the stimulation of milk production in female mammals.

**proliferating cell nuclear antigen (PCNA)** A protein complex that ensures processivity of DNA replication in eukaryotes.

**prometaphase** The phase of nuclear division that begins with the disintegration of the nuclear envelope.

**promoter** A DNA sequence to which RNA polymerase binds to initiate transcription.

**prophage** (pro' fayj) The noninfectious units that are linked with the chromosomes of the host bacteria and multiply with them but do not cause dissolution of the cell. Prophage can later enter into the lytic phase to complete the virus life cycle.

**prophase** (pro' phase) The first stage of nuclear division, during which chromosomes condense from diffuse, threadlike material to discrete, compact bodies.

**prostaglandin** Any one of a group of specialized lipids with hormone-like functions. It is not clear that they act at any considerable distance from the site of their production.

**prostate gland** In male humans, surrounds the urethra at its junction with the vas deferens; supplies an acid-neutralizing fluid to the semen.

**prosthetic group** Any nonprotein portion of an enzyme.

**proteasome** In the eukaryotic cytoplasm, a huge protein structure that binds to and digests cellular proteins that have been tagged by ubiquitin.

**protein** (pro' teen) [Gk. *protos*: first] Long-chain polymer of amino acids with twenty different common side chains. Occurs with its polymer chain extended in fibrous proteins, or coiled into a compact macromolecule in enzymes and other globular proteins.

**protein kinase** (kye' nase) An enzyme that catalyzes the addition of a phosphate group from ATP to a target protein.

**protein kinase cascade** A series of reactions in response to a molecular signal, in which a series of protein kinases activates one another in sequence, amplifying the signal at each step.

**proteoglycan** A glycoprotein containing a protein core with attached long, linear carbohydrate chains.

**proteolysis** [protein + Gk. *lysis*: break apart] An enzymatic digestion of a protein or polypeptide.

**proteome** The set of proteins that can be made by an organism. Because of alternative splicing of pre-mRNA, the number of proteins that can be made is usually much larger than the number of protein-coding genes present in the organism's genome.

**protoderm** Primary meristem that gives rise to the plant epidermis.

**proton** (pro' ton) [Gk. *protos*: first, before] (1) A subatomic particle with a single positive charge.

The number of protons in the nucleus of an atom determine its element. (2) A hydrogen ion, $H^+$.

**proton pump** An active transport system that uses ATP energy to move hydrogen ions across a membrane, generating an electric potential.

**proton-motive force** Force generated across a membrane having two components: a chemical potential (difference in proton concentration) plus an electrical potential due to the electrostatic charge on the proton.

**protonephridium** The excretory organ of flatworms, made up of a tubule and a flame cell.

**protoplast** The living contents of a plant cell; the plasma membrane and everything contained within it.

**provirus** Double-stranded DNA made by a virus that is integrated into the host's chromosome and contains promoters that are recognized by the host cell's transcription apparatus.

**proximal** Near the point of attachment or other reference point. (Contrast with distal.)

**proximal convoluted tubule** The initial segment of a renal tubule, closest to the glomerulus. (Compare with distal convoluted tubule.)

**proximate cause** The immediate genetic, physiological, neurological, and developmental mechanisms responsible for a behavior or morphology. (Contrast with ultimate cause.)

**pseudocoelomate** (soo' do see' low mate) [Gk. *pseudes*: false + *koiloma*: cavity] Having a body cavity, called a pseudocoel, consisting of a fluid-filled space in which many of the internal organs are suspended, but which is enclosed by mesoderm only on its outside.

**pseudogene** [Gk. *pseudes*: false] A DNA segment that is homologous to a functional gene but is not expressed because of changes to its sequence or changes to its location in the genome.

**pseudopod** (soo' do pod) [Gk. *pseudes*: false + *podos*: foot] A temporary, soft extension of the cell body that is used in location, attachment to surfaces, or engulfing particles.

**pulmonary** [L. *pulmo*: lung] Pertaining to the lungs.

**pulmonary circuit** The portion of the circulatory system by which blood is pumped from the heart to the lungs or gills for oxygenation and back to the heart for distribution. (Contrast with systemic circuit.)

**pulmonary valve** A one-way valve between the right ventricle of the heart and the pulmonary artery that prevents backflow of blood into the ventricle when it relaxes.

**Punnett square** Method of predicting the results of a genetic cross by arranging the gametes of each parent at the edges of a square.

**pupa** (pew' pa) [L. *pupa*: doll, puppet] In certain insects (the Holometabola), the encased developmental stage between the larva and the adult.

**pupil** The opening in the vertebrate eye through which light passes.

**purine** (pure' een) One of the two types of nitrogenous bases in nucleic acids. Each of the purines—adenine and guanine—pairs with a specific pyrimidine.

**Purkinje fibers** Specialized heart muscle cells that conduct excitation throughout the ventricular muscle.

**pyrimidine** (per im' a deen) One of the two types of nitrogenous bases in nucleic acids. Each of the pyrimidines—cytosine, thymine, and uracil—pairs with a specific purine.

**pyrogen** Molecule that produces a rise in body temperature (fever); may be produced by an invading pathogen or by cells of the immune system in response to infection.

**pyruvate** A three-carbon acid; the end product of glycolysis and the raw material for the citric acid cycle.

**pyruvate oxidation** Conversion of pyruvate to acetyl CoA and $CO_2$ that occurs in the mitochondrial matrix in the presence of $O_2$.

## - Q -

**$Q_{10}$** A value that compares the rate of a biochemical process or reaction over 10°C temperature ranges. A process that is not temperature-sensitive has a $Q_{10}$ of 1; values of 2 or 3 mean the reaction speeds up as temperature increases.

**quantitative trait loci** A set of genes that determines a complex character that exhibits quantitative variation.

**quaternary structure** The specific three-dimensional arrangement of protein subunits.

## - R -

**R group** The distinguishing group of atoms of a particular amino acid; also known as a side chain.

***r*-strategist** A species whose life history strategy allows for a high intrinsic rate of population increase (*r*). (Contrast with *K*-strategist.)

**radial symmetry** The condition in which any two halves of a body are mirror images of each other, providing the cut passes through the center; a cylinder cut lengthwise down its center displays this form of symmetry.

**radiation** The transfer of heat from warmer objects to cooler ones via the exchange of infrared radiation. *See also* electromagnetic radiation; evolutionary radiation.

**radicle** An embryonic root.

**radioisotope** A radioactive isotope of an element. Examples are carbon-14 ($^{14}C$) and hydrogen-3, or tritium ($^{3}H$).

**reactant** A chemical substance that enters into a chemical reaction with another substance.

**reaction center** A group of electron transfer proteins that receive energy from light-absorbing pigments and convert it to chemical energy by redox reactions.

**realized niche** A species' niche as defined by its interactions with other species. (Contrast with fundamental niche.)

**receptive field** The area of visual space that activates a particular cell in the visual system.

**receptor** *See* receptor protein, sensory receptor cell.

**receptor-mediated endocytosis** Endocytosis initiated by macromolecular binding to a specific membrane receptor.

**receptor potential** The change in the resting potential of a sensory cell when it is stimulated.

**receptor protein** A protein that can bind to a specific molecule, or detect a specific stimulus, within the cell or in the cell's external environment.

**recessive** In genetics, an allele that does not determine phenotype in the presence of a dominant allele. (Contrast with dominance.)

**reciprocal adaptation** *See* coevolution.

**reciprocal crosses** A pair of matings in one of which a female of genotype A mates with a male of genotype B and in the other of which a female of genotype B mates with a male of genotype A.

**recognition sequence** *See* restriction site.

**recombinant** Pertaining to an individual, meiotic product, or chromosome in which genetic materials originally present in two individuals end up in the same haploid complement of genes.

**recombinant DNA** A DNA molecule made in the laboratory that is derived from two or more genetic sources.

**recombinant frequency** The proportion of offspring of a genetic cross that have phenotypes different from the parental phenotypes due to crossing over between linked genes during gamete formation.

**reconciliation ecology** The practice of making exploited lands more biodiversity-friendly.

**rectum** The terminal portion of the gut, ending at the anus.

**redox reaction** A chemical reaction in which one reactant becomes oxidized and the other becomes reduced.

**reduction** Gain of electrons by a chemical reactant; any reduction is accompanied by an oxidation. (Contrast with oxidation.)

**refractory period** The time interval after an action potential during which another action potential cannot be elicited from an excitable membrane.

**regeneration** The development of a complete individual from a fragment of an organism.

**regulative development** A pattern of animal embryonic development in which the fates of the first blastomeres are not absolutely fixed. (Contrast with mosaic development.)

**regulatory sequence** A DNA sequence to which the protein product of a regulatory gene binds.

**regulatory gene** A gene that codes for a protein (or RNA) that in turn controls the expression of another gene.

**regulatory system** A system that uses feedback information to maintain a physiological function or parameter at an optimal level. (Contrast with controlled system.)

**regulatory T cells ($T_{reg}$)** Class of T cells that mediates tolerance to self antigens.

**reinforcement** The evolution of enhanced reproductive isolation between populations due to natural selection for greater isolation.

**releaser** Sensory stimulus that triggers performance of a stereotyped behavior pattern.

**REM (rapid-eye-movement) sleep** A sleep state characterized by vivid dreams, skeletal muscle relaxation, and rapid eye movements. (Contrast with slow-wave sleep.)

**renal** [L. *renes*: kidneys] Relating to the kidneys.

**renal tubule** A structural unit of the kidney that collects filtrate from the blood, reabsorbs specific ions, nutrients, and water and returns them to the blood, and concentrates excess ions and waste products such as urea for excretion from the body.

**replication** The duplication of genetic material.

**replication complex** The close association of several proteins operating in the replication of DNA.

**replication fork** A point at which a DNA molecule is replicating. The fork forms by the unwinding of the parent molecule.

**replicon** A region of DNA replicated from a single origin of replication.

**reporter gene** A genetic marker included in recombinant DNA to indicate the presence of the recombinant DNA in a host cell.

**repressor** A protein encoded by a regulatory gene that can bind to a specific operator and prevent transcription of the operon. (Contrast with activator.)

**reproductive isolation** Condition in which two divergent populations are no longer exchanging genes. Can lead to speciation.

**rescue effect** The process by which individuals moving between subpopulations of a metapopulation may prevent declining subpopulations from becoming extinct.

**resistance (R) genes** Plant genes that confer resistance to specific strains of pathogens.

**resource** Something in the environment required by an organism for its maintenance and growth that is consumed in the process of being used.

**resource partitioning** A situation in which selection pressures resulting from interspecific competition cause changes in the ways in which the competing species use the limiting resource, thereby allowing them to coexist.

**respiration** (res pi ra' shun) [L. *spirare*: to breathe] (1) Cellular respiration. (2) Breathing.

**respiratory chain** The terminal reactions of cellular respiration, in which electrons are passed from NAD or FAD, through a series of intermediate carriers, to molecular oxygen, with the concomitant production of ATP.

**respiratory gases** Oxygen ($O_2$) and carbon dioxide ($CO_2$); the gases that an animal must exchange between its internal body fluids and the outside medium (air or water).

**resting potential** The membrane potential of a living cell at rest. In cells at rest, the interior is negative to the exterior. (Contrast with action potential, electrotonic potential.)

**restoration ecology** The science and practice of restoring damaged or degraded ecosystems.

**restriction enzyme** Any of a type of enzyme that cleaves double-stranded DNA at specific sites; extensively used in recombinant DNA technology. Also called a restriction endonuclease.

**restriction fragment length polymorphism** *See* RFLP.

**restriction point (R)** The specific time during G1 of the cell cycle at which the cell becomes committed to undergo the rest of the cell cycle.

**restriction site** A specific DNA base sequence that is recognized and acted on by a restriction endonuclease.

**reticular system** A central region of the vertebrate brainstem that includes complex fiber tracts conveying neural signals between the forebrain and the spinal cord, with collateral fibers to a variety of nuclei that are involved in autonomic functions, including arousal from sleep.

**retina** (rett' in uh) [L. *rete*: net] The light-sensitive layer of cells in the vertebrate or cephalopod eye.

**retinoblastoma protein** A protein that inhibits an animal cell from passing through the restriction point; inactivation of this protein is necessary for the cell cycle to proceed.

**retrovirus** An RNA virus that contains reverse transcriptase. Its RNA serves as a template for cDNA production, and the cDNA is integrated into a chromosome of the host cell.

**reverse genetics** Method of genetic analysis in which a phenotype is first related to a DNA variation, then the protein involved is identified.

**reverse transcriptase** An enzyme that catalyzes the production of DNA (cDNA), using RNA as a template; essential to the reproduction of retroviruses.

**RFLP** Restriction fragment length polymorphism, the coexistence of two or more patterns of restriction fragments resulting from underlying differences in DNA sequence.

**rhizoids** (rye' zoids) [Gk. root] Hairlike extensions of cells in mosses, liverworts, and a few vascular plants that serve the same function as roots and root hairs in vascular plants. The term is also applied to branched, rootlike extensions of some fungi and algae.

**rhizome** (rye' zome) An underground stem (as opposed to a root) that runs horizontally beneath the ground.

**rhodopsin** A photopigment used in the visual process of transducing photons of light into changes in the membrane potential of photoreceptor cells.

**ribonucleic acid** *See* RNA.

**ribose** A five-carbon sugar in nucleotides and RNA.

**ribosomal RNA (rRNA)** Several species of RNA that are incorporated into the ribosome. Involved in peptide bond formation.

**ribosome** A small particle in the cell that is the site of protein synthesis.

**ribozyme** An RNA molecule with catalytic activity.

**risk cost** The increased chance of being injured or killed as a result of performing a behavior, compared to resting.

**RNA (ribonucleic acid)** An often single stranded nucleic acid whose nucleotides use ribose rather than deoxyribose and in which the base uracil replaces thymine found in DNA. Serves as genome from some viruses. (*See* rRNA, tRNA, mRMA, and ribozyme.)

**RNA interference (RNAi)** A mechanism for reducing mRNA translation whereby a double-stranded RNA, made by the cell or synthetically, is processed into a small, single-stranded RNA, whose binding to a target mRNA results in the latter's breakdown.

**RNA polymerase** An enzyme that catalyzes the formation of RNA from a DNA template.

**RNA splicing** The last stage of RNA processing in eukaryotes, in which the transcripts of introns are excised through the action of small nuclear ribonucleoprotein particles (snRNP).

**rod cells** Light-sensitive cell in the vertebrate retina; these sensory receptor cells are sensitive in extremely dim light and are responsible for dim light, black and white vision.

**root** The organ responsible for anchoring the plant in the soil, absorbing water and minerals, and producing certain hormones. Some roots are storage organs.

**root cap** A thimble-shaped mass of cells, produced by the root apical meristem, that protects the meristem; the organ that perceives the gravitational stimulus in root gravitropism.

**root hair** A long, thin process from a root epidermal cell that absorbs water and minerals from the soil solution.

**root system** The organ system that anchors a plant in place, absorbs water and dissolved minerals, and may store products of photosynthesis from the shoot system.

**rough endoplasmic reticulum (RER)** The portion of the endoplasmic reticulum whose outer surface has attached ribosomes. (Contrast with smooth endoplasmic reticulum.)

**rRNA** *See* ribosomal RNA.

**rubisco** Contraction of ribulose bisphosphate carboxylase/oxygenase, the enzyme that combines carbon dioxide or oxygen with ribulose bisphosphate to catalyze the first step of photosynthetic carbon fixation or photorespiration, respectively.

**ruminant** Herbivorous, cud-chewing mammals such as cows or sheep, characterized by a stomach that consists of four compartments: the rumen, reticulum, omasum, and abomasum.

## - S -

**S phase** In the cell cycle, the stage of interphase during which DNA is replicated. (Contrast with $G_1$ phase, $G_2$ phase, M phase.)

**saltatory conduction** [L. *saltare*: to jump] The rapid conduction of action potentials in myelinated axons; so called because action potentials appear to "jump" between nodes of Ranvier along the axon.

**saprobe** [Gk. *sapros*: rotten] An organism (usually a bacterium or fungus) that obtains its carbon and energy by absorbing nutrients from dead organic matter.

**sarcomere** (sark' o meer) [Gk. *sark*: flesh + *meros*: unit] The contractile unit of a skeletal muscle.

**sarcoplasm** The cytoplasm of a muscle cell.

**sarcoplasmic reticulum** The endoplasmic reticulum of a muscle cell.

**saturated fatty acid** A fatty acid in which all the bonds between carbon atoms in the hydrocarbon chain are single bonds—that is, all the bonds are saturated with hydrogen atoms. (Contrast with unsaturated fatty acid.)

**scientific method** A means of gaining knowledge about the natural world by making observations, posing hypotheses, and conducting experiments to test those hypotheses.

**Schwann cell** A type of glial cell that myelinates axons in the peripheral nervous system.

**scion** In horticulture, the bud or stem from one plant that is grafted to a root or root-bearing stem of another plant (the stock).

**sclerenchyma** (skler eng' kyma) [Gk. *skleros*: hard + *kymus*: juice] A plant tissue composed of cells with heavily thickened cell walls. The cells are dead at functional maturity. The principal types of sclerenchyma cells are fibers and sclereids.

**second law of thermodynamics** The principle that when energy is converted from one form to another, some of that energy becomes unavailable for doing work.

**second messenger** A compound, such as cAMP, that is released within a target cell after a hormone (the first messenger) has bound to a surface receptor on a cell; the second messenger triggers further reactions within the cell.

**secondary active transport** A form of active transport that does not use ATP as an energy source; rather, transport is coupled to ion diffusion down a concentration gradient established by primary active transport.

**secondary cell wall** A thick, cellulosic structure internal to the primary cell wall formed in some plant cells after cell expansion stops (Contrast with primary cell wall.)

**secondary consumer** An organism that eat primary consumers.

**secondary growth** In plants, growth that contributes to an increase in girth. (Contrast with primary growth.)

**secondary immune response** A rapid and intense response to a second or subsequent exposure to an antigen, initiated by memory cells. (Contrast with primary immune response.)

**secondary lysosome** Membrane-enclosed organelle formed by the fusion of a primary lysosome with a phagosome, in which macromolecules taken up by phagocytosis are hydrolyzed into their monomers. (Contrast with lysosome.)

**secondary metabolite** A compound synthesized by a plant that is not needed for basic cellular metabolism. Typically has an antiherbivore or antiparasite function.

**secondary plant body** That part of a plant produced by secondary growth; consists of woody tissues. (Contrast with primary plant body.)

**secondary sex determination** Formation of secondary sexual characteristics (i.e., those other than gonads), such as external sex organs and body hair. (Contrast with primary sex determination.)

**secondary spermatocyte** One of the products of the first meiotic division of a primary spermatocyte.

**secondary structure** Of a protein, localized regularities of structure, such as the α helix and the β pleated sheet.

**secondary succession** Succession after a disturbance that did not eliminate all the organisms originally living on the site. (Contrast with primary succession.)

**secretin** (si kreet' in) A peptide hormone secreted by the upper region of the small intestine when acidic chyme is present. Stimulates the pancreatic duct to secrete bicarbonate ions.

**sedimentary rock** Rock formed by the accumulation of sediment grains on the bottom of a body of water.

**seed** A fertilized, ripened ovule of a gymnosperm or angiosperm. Consists of the embryo, nutritive tissue, and a seed coat.

**segmentation** Division of an animal body into segments.

**segmentation genes** Genes that determine the number and polarity of body segments.

**segregation** In genetics, the separation of alleles, or of homologous chromosomes, from each other during meiosis so that each of the haploid daughter nuclei produced contains one or the other member of the pair found in the diploid parent cell, but never both. This principle was articulated by Mendel as his first law.

**selective permeability** Allowing certain substances to pass through while other substances are excluded; a characteristic of membranes.

**self-incompatability** In plants, the possession of mechanisms that prevent self-fertilization.

**semelparous** [L. *semel*: once + *pario*: to beget] Reproducing only once in a lifetime. (Contrast with iteroparous.)

**semen** (see' men) [L. *semin*: seed] The thick, whitish liquid produced by the male reproductive system in mammals, containing the sperm.

**semiconservative replication** The way in which DNA is synthesized. Each of the two partner strands in a double helix acts as a template for a new partner strand. Hence, after replication, each double helix consists of one old and one new strand.

**seminiferous tubules** The tubules within the testes within which sperm production occurs.

**senescence** [L. *senescere*: to grow old] Aging; deteriorative changes with aging; the increased probability of dying with increasing age.

**sensitive period** The life stage during which some particular type of learning must take place, or during which it occurs much more easily than at other times. Typical of song learning among birds.

**sensor** *See* sensory receptor cell.

**sensory neuron** A specialized neuron that transduces a particular type of sensory stimulus into action potentials.

**sensory receptor cell** Cell that is responsive to a particular type of physical or chemical stimulation.

**sensory transduction** The transformation of environmental stimuli or information into neural signals.

**sepal** (see' pul) [L. *sepalum*: covering] One of the outermost structures of the flower, usually protective in function and enclosing the rest of the flower in the bud stage.

**septum** (plural: septa) [L. wall] (1) A partition or cross-wall appearing in the hyphae of some fungi. (2) The bony structure dividing the nasal passages.

**sequence alignment** A method of identifying homologous positions in DNA or amino acid sequences by pinpointing the locations of deletions and insertions that have occurred since two (or more) organisms diverged from a common ancestor.

**Sertoli cells** Cells in the seminiferous tubules that nuture the developing sperm.

**sessile** (sess' ul) [L. *sedere*: to sit] Permanently attached; not able to move from one place to another. (Contrast with motile.)

**set point** In a regulatory system, the threshold sensitivity to the feedback stimulus.

**sex chromosome** In organisms with a chromosomal mechanism of sex determination, one of the chromosomes involved in sex determination.

**sex linkage** The pattern of inheritance characteristic of genes located on the sex chromosomes of organisms having a chromosomal mechanism for sex determination.

**sexual reproduction** Reproduction involving the union of gametes.

**sexual selection** Selection by one sex of characteristics in individuals of the opposite sex. Also, the favoring of characteristics in one sex as a result of competition among individuals of that sex for mates.

**Shannon diversity index** A formula for quantifying diversity that takes both species richness and species evenness into account; based on a mathematical expression of the certainty with which the next item sampled in a series can be predicted.

**shared derived trait** *See* synapomorphy.

**shoot system** In plants, the organ system consisting of the leaves, stem(s), and flowers.

**short tandem repeat (STR)** A short (1–5 base pairs), moderately repetitive sequence of DNA. The number of copies of an STR at a particular location varies between individuals and is inherited.

**shotgun sequencing** A relatively rapid method of DNA sequencing in which a DNA molecule is broken up into overlapping fragments, each fragment is sequenced, and high-speed computers analyze and realign the fragments. (Contrast with hierarchical sequencing.)

**side chain** *See* R group.

**sieve tube element** The characteristic cell of the phloem in angiosperms, which contains cytoplasm but relatively few organelles, and whose end walls (*sieve plates*) contain pores that form connections with neighboring cells.

**signal sequence** The sequence of a protein that directs the protein protein to a particular organelle.

**signal transduction pathway** The series of biochemical steps whereby a stimulus to a cell (such as a hormone or neurotransmitter binding to a receptor) is translated into a response of the cell.

**silent mutation** A change in a gene's sequence that has no effect on the amino acid sequence of a protein because it occurs in noncoding DNA or because it does not change the amino acid specified by the corresponding codon. (Contrast with frame-shift mutation, missense mutation, nonsense mutation.)

**similarity matrix** A matrix used to compare the degree of divergence among pairs of objects. For molecular sequences, constructed by summing the number or percentage of nucleotides or amino acids that are identical in each pair of sequences.

**single nucleotide polymorphisms (SNPs)** Inherited variations in a single nucleotide base in DNA that differ between individuals.

**single-strand binding protein** In DNA replication, a protein that binds to single strands of DNA after they have been separated from each other, keeping the two strands separate for replication.

**sink** In plants, any organ that imports the products of photosynthesis, such as roots, developing fruits, and immature leaves. (Contrast with source.)

**sinoatrial node** (sigh' no ay' tree al) [L. *sinus*: curve + *atrium*: chamber] The pacemaker of the mammalian heart.

**siRNAs (small interfering RNAs)** Short, double-stranded RNA molecules used in RNA interference.

**sister chromatid** Each of a pair of newly replicated chromatids.

**sister groups** Two phylogenetic groups that are each other's closest relatives.

**skeletal muscle** A type of muscle tissue characterized by multinucleated cells containing highly ordered arrangements of actin and myosin microfilaments. Also called striated muscle. (Contrast with cardiac muscle, smooth muscle.)

**skeletal systems** Organ systems that provide rigid supports against which muscles can pull to create directed movements.

**sliding DNA clamp** Protein complex that keeps DNA polymerase bound to DNA during replication.

**sliding filament theory** Mechanism of muscle contraction based on the formation and breaking of crossbridges between actin and myosin filaments, causing the filaments to slide together.

**slow-wave sleep** A state of deep, restorative sleep characterized by high-amplitude slow waves in the EEG. (Contrast with REM sleep.)

**small intestine** The portion of the gut between the stomach and the colon; consists of the duodenum, the jejunum, and the ileum.

**small nuclear ribonucleoprotein particle (snRNP)** A complex of an enzyme and a small nuclear RNA molecule, functioning in RNA splicing.

**smooth endoplasmic reticulum (SER)** Portion of the endoplasmic reticulum that lacks ribosomes and has a tubular appearance. (Contrast with rough endoplasmic reticulum.)

**smooth muscle** Muscle tissue consisting of sheets of mononucleated cells innervated by the autonomic nervous system. (Contrast with cardiac muscle, skeletal muscle.)

**sodium–potassium ($Na^+$–$K^+$) pump** Antiporter responsible for primary active transport; it pumps sodium ions out of the cell and potassium ions into the cell, both against their concentration gradients. Also called a sodium–potassium ATPase.

**soil horizon** *See* horizon.

**solute** A substance that is dissolved in a liquid (solvent) to form a solution.

**solute potential** A property of any solution, resulting from its solute contents; it may be zero or have a negative value. The more negative the solute potential, the greater the tendency of the solution to take up water through a differentially permeable membrane. (Contrast with pressure potential, water potential.)

**solution** A liquid (the solvent) and its dissolved solutes.

**solvent** Liquid in which a substance (solute) is dissolved to form a solution.

**somatic cell** [Gk. *soma*: body] All the cells of the body that are not specialized for reproduction. (Contrast with germ cell.)

**somatic mutation** Permament genetic change in a somatic cell. These mutations affect the individual only; they are not passed on to offspring. (Contrast with germ line mutation.)

**somatosensory cortex** The region of the cerebral cortex that receives input from mechanosensors distributed throughout the body.

**somatostatin** Peptide hormone made in the hypothalamus that inhibits the release of other hormones from the pituitary and intestine.

**somite** (so' might) One of the segments into which an embryo becomes divided longitudinally, leading to the eventual segmentation of the animal as illustrated by the spinal column, ribs, and associated muscles.

**Sorenson's index** Mathematical formula that measures beta diversity.

**source** In plants, any organ that exports the products of photosynthesis in excess of its own needs, such as a mature leaf or storage organ. (Contrast with sink.)

**spatial summation** In the production or inhibition of action potentials in a postsynaptic cell, the interaction of depolarizations and hyperpolarizations produced at different sites on the postsynaptic cell. (Contrast with temporal summation.)

**spawning** *See* external fertilization.

**speciation** (spee' see ay' shun) The process of splitting one population into two populations that are reproductively isolated from one another.

**species** (spee' sees) [L. kind] The base unit of taxonomic classification, consisting of an ancestor–descendant group of populations of evolutionarily closely related, similar organisms. The more narrowly defined "biological species" consists of individuals capable of interbreeding with each other but not with members of other species.

**species–area relationship** The relationship between the size of an area and the numbers of species it supports.

**species richness** The total number of species living in a region.

**specific defenses** Defensive reactions of the vertebrate immune system that are based on the reaction of an antibody to a specific antigen. (Contrast with nonspecific defenses.)

**specific heat** The amount of energy that must be absorbed by a gram of a substance to raise its temperature by one degree centigrade. By convention, water is assigned a specific heat of one.

**sperm** [Gk. *sperma*: seed] The male gamete.

**spermatid** One of the products of the second meiotic division of a primary spermatocyte; four haploid spermatids, which remain connected by cytoplasmic bridges, are produced for each primary spermatocyte that enters meiosis.

**spermatogenesis** (spur mat' oh jen' e sis) [Gk. *sperma*: seed + *genesis*: source] Gametogenesis leading to the production of sperm.

**spermatogonia** In animals, the diploid progeny of a germ cell in males.

**spherical symmetry** The simplest form of symmetry, in which body parts radiate out from a central point such that an infinite number of planes passing through that central point can divide the organism into similar halves.

**spicule** [L. arrowhead] A hard, calcareous skeletal element typical of sponges.

**spinal reflex** The conversion of afferent to efferent information in the spinal cord without participation of the brain.

**sphincter** (sfink' ter) [Gk. *sphinkter*: something that binds tightly] A ring of muscle that can close an orifice, for example, at the anus.

**spindle** Array of microtubules emanating from both poles of a dividing cell during mitosis and playing a role in the movement of chromosomes at nuclear division. Named for its shape.

**spleen** Organ that serves as a reservoir for venous blood and eliminates old, damaged red blood cells from the circulation.

**spliceosome** RNA–protein complex that splices out introns from eukaryotic pre-mRNAs.

**splicing** *See* RNA splicing.

**spontaneous mutation** A genetic change caused by internal cellular mechanisms, such as an error in DNA replication. (Contrast with induced mutation.)

**sporangiophore** A stalked reproductive structure produced by zygospore fungi that extends from a hypha and bears one or many sporangia.

**sporangium** (spor an' gee um) (plural: sporangia) [Gk. *spora*: seed + *angeion*: vessel or reservoir] In plants and fungi, any specialized stucture within which one or more spores are formed.

**spore** [Gk. *spora*: seed] (1) Any asexual reproductive cell capable of developing into an adult organism without gametic fusion. In plants, haploid spores develop into gametophytes, diploid spores into sporophytes. (2) In prokaryotes, a resistant cell capable of surviving unfavorable periods.

**sporocyte** Specialized cells of the diploid sporophyte that will divide by meiosis to produce four haploid spores. Germination of these spores produces the haploid gametophyte.

**sporophyte** (spor' o fyte) [Gk. *spora*: seed + *phyton*: plant] In plants and protists with alternation of generations, the diploid phase that produces the spores. (Contrast with gametophyte.)

**stabilizing selection** Selection against the extreme phenotypes in a population, so that the intermediate types are favored. (Contrast with disruptive selection.)

**stamen** (stay' men) [L. *stamen*: thread] A male (pollen-producing) unit of a flower, usually composed of an anther, which bears the pollen, and a filament, which is a stalk supporting the anther.

**starch** [O.E. *stearc*: stiff] A polymer of glucose; used by plants to store energy.

**Starling's forces** The two opposing forces responsible for water movement across capillary walls: blood pressure, which squeezes water and small solutes out of the capillaries, and osmotic pressure, which pulls water back into the capillaries.

**start codon** The mRNA triplet (AUG) that acts as a signal for the beginning of translation at the ribosome. (Contrast with stop codon.)

**stele** (steel) [Gk. *stylos*: pillar] The central cylinder of vascular tissue in a plant stem.

**stem** In plants, the organ that holds leaves and/or flowers and transports and distributes materials among the other organs of the plant.

**stem cell** In animals, an undifferentiated cell that is capable of continuous proliferation. A stem cell generates more stem cells and a large clone of differentiated progeny cells. (*See also* embryonic stem cell.)

**steroid** Any of a family of lipids whose multiple rings share carbons. The steroid cholesterol is an important constituent of membranes; other steroids function as hormones.

**sticky ends** On a piece of two-stranded DNA, short, complementary, one-stranded regions produced by the action of a restriction endonuclease. Sticky ends facilitate the joining of segments of DNA from different sources.

**stigma** [L. *stigma*: mark, brand] The part of the pistil at the apex of the style that is receptive to pollen, and on which pollen germinates.

**stimulus** [L. *stimulare*: to goad] Something causing a response; something in the environment detected by a receptor.

**stock** In horticulture, the root or root-bearing stem to which a bud or piece of stem from another plant (the scion) is grafted.

**stoma** (plural: stomata) [Gk. *stoma*: mouth, opening] Small opening in the plant epidermis that permits gas exchange; bounded by a pair of guard cells whose osmotic status regulates the size of the opening.

**stomatal crypt** In plants, a sunken cavity below the leaf surface in which a stoma is sheltered from the drying effects of air currents.

**stop codon** Any of the three mRNA codons that signal the end of protein translation at the ribosome: UAG, UGA, UAA.

**stratosphere** The upper part of Earth's atmosphere, above the troposphere; extends from approximately 18 kilometers upward to approximately 50 kilometers above the surface.

**stratum** (plural strata) [L. *stratos*: layer] A layer of sedimentary rock laid down at a particular time in the past.

**stretch receptor** A modified muscle cell embedded in the connective tissue of a muscle that acts as a mechanoreceptor in response to stretching of that muscle.

**striated muscle** *See* skeletal muscle.

**strobilus** (plural: strobili) One of several cone-like structures in various groups of plants (including club mosses, horsetails, and conifers) associated with the production and dispersal of reproductive products.

**stroma** The fluid contents of an organelle such as a chloroplast or mitochondrion.

**structural gene** A gene that encodes the primary structure of a protein not involved in the regulation of gene expression.

**structural isomers** Molecules made up of the same kinds and numbers of atoms, in which the atoms are bonded differently.

**structural motif** A three-dimensional structural element that is part of a larger molecule. For example, there are four common motifs in DNA-binding proteins: helix-turn-helix, zinc finger, leucine zipper, and helix-loop-helix.

**style** [Gk. *stylos*: pillar or column] In the angiosperm flower, a column of tissue extending from the tip of the ovary, and bearing the stigma or receptive surface for pollen at its apex.

**sub-** [L. under] A prefix used to designate a structure that lies beneath another or is less than another. For example, subcutaneous (beneath the skin); subspecies.

**suberin** A waxlike lipid that is a barrier to water and solute movement across the Casparian strip of the endodermis.

**submucosa** (sub mew koe' sah) The tissue layer just under the epithelial lining of the lumen of the digestive tract.

**subsoil** The soil horizon lying below the topsoil and above the parent rock (bedrock); the zone of infiltration and accumulation of materials leached from the topsoil. Also called the B horizon.

**substrate** (sub' strayte) The molecule or molecules on which an enzyme exerts catalytic action.

**substratum** (plural: substrata) The base material on which a sessile organism lives.

**succession** The gradual, sequential series of changes in the species composition of a community following a disturbance.

**succulence** In plants, possession of fleshy, water-storing leaves or stems; an adaptation to dry environments.

**superficial cleavage** A variation of incomplete cleavage in which cycles of mitosis occur without cell division, producing a syncytium (a single cell with many nuclei).

**suprachiasmatic nuclei (SCN)** In mammals, two clusters of neurons just above the optic chiasm that act as the master circadian clock.

**surface area-to-volume ratio** For any cell, organism, or geometrical solid, the ratio of surface area to volume; this is an important factor in setting an upper limit on the size a cell or organism can attain.

**surface tension** The attractive intermolecular forces at the surface of liquid; especially important in water.

**surfactant** A substance that decreases the surface tension of a liquid. Lung surfactant, secreted by cells of the alveoli, is mostly phospholipid and decreases the amount of work necessary to inflate the lungs.

**survivorship ($l_x$)** In life tables, the proportion of individuals in a cohort that are alive at age $x$. A graph of this data is a survivorship curve.

**suspensor** In the embryos of seed plants, the stalk of cells that pushes the embryo into the endosperm and is a source of nutrient transport to the embryo.

**sustainable** Pertaining to the use and management of ecosystems in such a way that humans benefit over the long term from specific ecosystem goods and services without compromising others.

**symbiosis** (sim' bee oh' sis) [Gk. *sym*: together + *bios*: living] The living together of two or more species in a prolonged and intimate relationship.

**symmetry** Pertaining to an attribute of an animal body in which at least one plane can divide the body into similar, mirror-image halves. (*See* bilateral symmetry, radial symmetry.)

**sympathetic nervous system** The division of the autonomic nervous system that works in opposition to the parasympathetic nervous system. (Contrast with parasympathetic nervous system.)

**sympatric speciation** (sim pat' rik) [Gk. *sym*: same + *patria*: homeland] Speciation due to reproductive isolation without any physical separation of the subpopulation. (Contrast with allopatric speciation.)

**symplast** The continuous meshwork of the interiors of living cells in the plant body, resulting from the presence of plasmodesmata. (Contrast with apoplast.)

**symporter** A membrane transport protein that carries two substances in the same direction. (Contrast with antiporter, uniporter.)

**synapomorphy** A trait that arose in the ancestor of a phylogenetic group and is present (sometimes in modified form) in all of its members, thus helping to delimit and identify that group. Also called a shared derived trait.

**synapse** (sin' aps) [Gk. *syn*: together + *haptein*: to fasten] A specialized type of junction where a neuron meets its target cell (which can be another neuron or some other type of cell) and information in the form of neurotransmitter molecules is exchanged across a synaptic cleft.

**synapsis** (sin ap' sis) The highly specific parallel alignment (pairing) of homologous chromosomes during the first division of meiosis.

**synergids** [Gk. *syn*: together + *ergos*: work] In angiosperms, the two cells accompanying the egg cell at one end of the megagametophyte.

**syngamy** *See* fertilization.

**synonymous (silent) substitution** A change of one nucleotide in a sequence to another when that change does not affect the amino acid specified (i.e., UUA → UUG, both specifying leucine). (Contrast with nonsynonymous substitution, missense substitution, nonsense substitution.)

**systematics** The scientific study of the diversity and relationships among organisms.

**systemic acquired resistance** A general resistance to many plant pathogens following infection by a single agent.

**systemic circuit** Portion of the circulatory system by which oxygenated blood from the lungs or gills is distributed throughout the rest of the body and returned to the heart. (Contrast with pulmonary circuit.)

**systems biology** The scientific study of an organism as an integrated and interacting system of genes, proteins, and biochemical reactions.

**systole** (sis' tuh lee) [Gk. *systole*: contraction] Contraction of a chamber of the heart, driving blood forward in the circulatory system. (Contrast with diastole.)

## - T -

**T cell** A type of lymphocyte involved in the cellular immune response. The final stages of its development occur in the thymus gland. (Contrast with B cell; *see also* cytotoxic T cell, T-helper cell.)

**T cell receptor** A protein on the surface of a T cell that recognizes the antigenic determinant for which the cell is specific.

**T-helper cell ($T_H$)** Type of T cell that stimulates events in both the cellular and humoral immune responses by binding to the antigen on an antigen-presenting cell; target of the HIV-I virus, the agent of AIDS. (Contrast with cytotoxic T cells.)

**T tubules** A system of tubules that runs throughout the cytoplasm of a muscle fiber, through which action potentials spread.

**taproot system** A root system typical of eudicots consisting of a primary root (*taproot*) that extends downward by tip growth and outward by initiating lateral roots. (Contrast with fibrous root system.)

**target cell** A cell with the appropriate receptors to bind and respond to a particular hormone or other chemical mediator.

**taste bud** A structure in the epithelium of the tongue that includes a cluster of chemoreceptors innervated by sensory neurons.

**TATA box** An eight-base-pair sequence, found about 25 base pairs before the starting point for transcription in many eukaryotic promoters, that binds a transcription factor and thus helps initiate transcription.

**taxon** (plural: taxa) [Gk. *taxis*: arrange, put in order] A biological group (typically a species or a clade) that is given a name.

**telencephalon** The outer, surrounding structure of the embryonic vertebrate forebrain, which develops into the cerebrum.

**telomerase** An enzyme that catalyzes the addition of telomeric sequences lost from chromosomes during DNA replication.

**telomeres** (tee' lo merz) [Gk. *telos*: end + *meros*: units, segments] Repeated DNA sequences at the ends of eukaryotic chromosomes.

**telophase** (tee' lo phase) [Gk. *telos*: end] The final phase of mitosis or meiosis during which chromosomes became diffuse, nuclear envelopes reform, and nucleoli begin to reappear in the daughter nuclei.

**template** A molecule or surface on which another molecule is synthesized in complementary fashion, as in the replication of DNA.

**template strand** In double-stranded DNA, the strand that is transcribed to create an RNA transcript that will be processed into a protein. Also refers to a strand of RNA that is used to create a complementary RNA.

**temporal summation** In the production or inhibition of action potentials in a postsynaptic cell, the interaction of depolarizations or hyperpolarizations produced by rapidly repeated stimulation of a single point on the postsynaptic cell. (Contrast with spatial summation.)

**tendon** A collagen-containing band of tissue that connects a muscle with a bone.

**tepal** A sterile, modified, nonphotosynthetic leaf of an angiosperm flower that cannot be distinguished as a petal or a sepal.

**termination** In molecular biology, the end of transcription or translation.

**terminator** A sequence at the 3′ end of mRNA that causes the RNA strand to be released from the transcription complex.

**terrestrial** (ter res′ tree al) [L. *terra*: earth] Pertaining to or living on land. (Contrast with aquatic, marine.)

**tertiary structure** In reference to a protein, the relative locations in three-dimensional space of all the atoms in the molecule. The overall shape of a protein. (Contrast with primary, secondary, and quaternary structures.)

**test cross** Mating of a dominant-phenotype individual (who may be either heterozygous or homozygous) with a homozygous-recessive individual.

**testis** (tes′ tis) (plural: testes) [L. *testis*: witness] The male gonad; the organ that produces the male gametes.

**tetanus** [Gk. *tetanos*: stretched] (1) A state of sustained maximal muscular contraction caused by rapidly repeated stimulation. (2) In medicine, an often fatal disease ("lockjaw") caused by the bacterium *Clostridium tetani.*

**tetrad** [Gk. *tettares*: four] During prophase I of meiosis, the association of a pair of homologous chromosomes or four chromatids.

**thalamus** [Gk. *thalamos*: chamber] A region of the vertebrate forebrain; involved in integration of sensory input.

**theory** [Gk. *theoria*: analysis of facts] A far-reaching explanation of observed facts that is supported by such a wide body of evidence, with no significant contradictory evidence, that it is scientifically accepted as a factual framework. Examples are Newton's theory of gravity and Darwin's theory of evolution. (Contrast with hypothesis.)

**thermoneutral zone** [Gk. *thermos*: temperature] The range of temperatures over which an endotherm does not have to expend extra energy to thermoregulate.

**thermophile** (ther′ muh fyle)[Gk. *thermos*: temperature + *philos*: loving] An organism that lives exclusively in hot environments.

**thoracic cavity** [Gk. *thorax*: breastplate] The portion of the mammalian body cavity bounded by the ribs, shoulders, and diaphragm. Contains the heart and the lungs.

**thoracic duct** The connection between the lymphatic system and the circulatory system.

**threshold** The level of depolarization that causes an electrically excitable membrane to fire an action potential.

**thrombus** (throm′ bus) [Gk. *thrombos*: clot] A blood clot that forms within a blood vessel and remains attached to the wall of the vessel. (Contrast with embolus.)

**thylakoid** (thigh la koid) [Gk. *thylakos*: sack or pouch] A flattened sac within a chloroplast. Thylakoid membranes contain all of the chlorophyll in a plant, in addition to the electron carriers of photophosphorylation. Thylakoids stack to form grana.

**thymine (T)** Nitrogen-containing base found in DNA.

**thymus** [Gk. *thymos*: warty] A ductless, glandular lymphoid tissue, involved in development of the immune system of vertebrates. In humans, the thymus degenerates during puberty.

**thyroid gland** [Gk. *thyreos*: door-shaped] A two-lobed gland in vertebrates. Produces the hormone thyroxin.

**thyrotropin** Hormone produced by the anterior pituitary that stimulates the thyroid gland to produce and release thyroxine. Also called thyroid-stimulating hormone (TSH).

**thyrotropin-releasing hormone (TRH)** Hormone produced by the hypothalamus that stimulates the anterior pituitary to release thyrotropin.

**thyroxine** Hormone produced by the thyroid gland; controls many metabolic processes.

**tidal volume** The amount of air that is exchanged during each breath when a person is at rest.

**tight junction** A junction between epithelial cells in which there is no gap between adjacent cells.

**tissue** A group of similar cells organized into a functional unit; usually integrated with other tissues to form part of an organ.

**tissue system** In plants, any of three organized groups of tissues—dermal tissue, vascular tissue, and ground tissue—that are established during embryogenesis and have distinct functions.

**titin** A protein that holds bundles of myosin filaments in a centered position within the sarcomeres of muscle cells. The largest protein in the human body.

**tonoplast** The membrane of the plant central vacuole.

**topsoil** The uppermost soil horizon; contains most of the organic matter of soil, but may be depleted of most mineral nutrients by leaching. Also called the A horizon.

**totipotent** [L. *toto*: whole, entire + *potens*: powerful] Possessing all the genetic information and other capacities necessary to form an entire individual. (Contrast with multipotent, pluripotent.)

**trachea** (tray′ kee ah) [Gk. *trakhoia*: tube] A tube that carries air to the bronchi of the lungs of vertebrates. When plural (*tracheae*), refers to the major airways of insects.

**tracheary element** Either of two types of xylem cells—tracheids and vessel elements—that undergo apoptosis before assuming their transport function.

**tracheid** (tray′ kee id) A type of tracheary element found in the xylem of nearly all vascular plants, characterized by tapering ends and walls that are pitted but not perforated. (Contrast with vessel element.)

**trade-off** The relationship between the fitness benefits conferred by an adaptation and the fitness costs it imposes. For an adaptation to be favored by natural selection, the benefits must exceed the costs.

**trait** In genetics, a specific form of a character: eye color is a character; brown eyes and blue eyes are traits. (Contrast with character.)

**transcription** The synthesis of RNA using one strand of DNA as a template.

**transcription factors** Proteins that assemble on a eukaryotic chromosome, allowing RNA polymerase II to perform transcription.

**transduction** (1) Transfer of genes from one bacterium to another by a bacteriophage. (2) In sensory cells, the transformation of a stimulus (e.g., light energy, sound pressure waves, chemical or electrical stimulants) into action potentials.

**transfection** Insertion of recombinant DNA into animal cells.

**transfer RNA (tRNA)** A family of double-stranded RNA molecules. Each tRNA carries a specific amino acid and anticodon that will pair with the complementary codon in mRNA during translation.

**transformation** (1) A mechanism for transfer of genetic information in bacteria in which pure DNA from a bacterium of one genotype is taken in through the cell surface of a bacterium of a different genotype and incorporated into the chromosome of the recipient cell. (2) Insertion of recombinant DNA into a host cell.

**transgenic** Containing recombinant DNA incorporated into the genetic material.

**translation** The synthesis of a protein (polypeptide). Takes place on ribosomes, using the information encoded in messenger RNA.

**translocation** (1) In genetics, a rare mutational event that moves a portion of a chromosome to a new location, generally on a nonhomologous chromosome. (2) In vascular plants, movement of solutes in the phloem.

**transmembrane protein** An integral membrane protein that spans the phospholipid bilayer.

**transpiration** [L. *spirare*: to breathe] The evaporation of water from plant leaves and stem, driven by heat from the sun, and providing the motive force to raise water (plus mineral nutrients) from the roots.

**transpiration–cohesion–tension mechanism** Theoretical basis for water movement in plants: evaporation of water from cells within leaves (transpiration) causes an increase ion surface tension, pulling water up through the xylem. Cohesion of water occurs because of hydrogen bonding.

**transposable element** A segment of DNA that can move to, or give rise to copies at, another locus on the same or a different chromosome.

**transposon** Mobile DNA segment that can insert into a chromosome and cause genetic change.

**triglyceride** A simple lipid in which three fatty acids are combined with one molecule of glycerol.

**triploblastic** Having three cell layers.

**trisomic** Containing three rather than two members of a chromosome pair.

**tRNA** *See* transfer RNA.

**trochophore** (troke′ o fore) [Gk. *trochos*: wheel + *phoreus*: bearer] A radially symmetrical larval form typical of annelids and mollusks, distinguished by a wheel-like band of cilia around the middle.

**trophic cascade** The progression over successively lower trophic levels of the indirect effects of a predator.

**trophic level** [Gk *trophes*: nourishment] A group of organisms united by obtaining their energy from the same part of the food web of a biological community.

**trophoblast** [Gk *trophes*: nourishment + *blastos*: sprout] At the 32-cell stage of mammalian development, the outer group of cells that will become part of the placenta and thus nourish the growing embryo. (Contrast with inner cell mass.)

**tropic hormones** Hormones produced by the anterior pituitary that control the secretion of hormones by other endocrine glands.

**tropomyosin** [troe poe my′ oh sin] One of the three protein components of an actin filament; controls the interactions of actin and myosin necessary for muscle contraction.

**troponin** One of the three components of an actin filament; binds to actin, tropomyosin, and $Ca^{2+}$.

**troposphere** The lowest atmospheric zone, reaching upward from the Earth's surface approximately 10–17 km. Zone in which virtually all water vapor is located.

**true-breeding** A genetic cross in which the same result occurs every time with respect to the trait(s) under consideration, due to homozygous parents.

**trypsin** A protein-digesting enzyme. Secreted by the pancreas in its inactive form (trypsinogen), it becomes active in the duodenum of the small intestine.

**tubulin** A protein that polymerizes to form microtubules.

**tumor** [L. *tumor*: a swollen mass] A disorganized mass of cells. Malignant tumors spread to other parts of the body.

**tumor necrosis factor** A family of cytokines (growth factors) that causes cell death and is involved in inflammation.

**tumor suppressor** A gene that codes for a protein product that inhibits cell proliferation; inactive in cancer cells. (Contrast with oncogene.)

**turgor pressure** [L. *turgidus*: swollen] *See* pressure potential.

**turnover** In freshwater ecosystems, vertical movements of water that bring nutrients and dissolved $CO_2$ to the surface and $O_2$ to deeper water.

**tympanic membrane** [Gk. *tympanum*: drum] The eardrum.

## - U -

**ubiquinone** (yoo bic′ kwi known) [L. *ubique*: everywhere] A mobile electron carrier of the mitochondrial respiratory chain. Similar to plastoquinone found in chloroplasts.

**ubiquitin** A small protein that is covalently linked to other cellular proteins identified for breakdown by the proteosome.

**ultimate cause** In ethology, the evolutionary processes that produced an animal's capacity and tendency to behave in particular ways. (Contrast with proximate cause.)

**uniporter** [L. *unus*: one + *portal*: doorway] A membrane transport protein that carries a single substance in one direction. (Contrast with antiporter, symporter.)

**unsaturated fatty acid** A fatty acid whose hydrocarbon chain contains one or more double bonds. (Contrast with saturated fatty acid.)

**upregulation** A process by which the abundance of receptors for a hormone increases when hormone secretion is suppressed. (Contrast with downregulation.)

**upwelling zones** Areas of the ocean where cool, nutrient-rich water from deeper layers rises to the surface.

**uracil (U)** A pyrimidine base found in nucleotides of RNA.

**urea** A compound that is the main excreted form of nitrogen by many animals, including mammals.

**ureotelic** Pertaining to an organism in which the final product of the breakdown of nitrogen-containing compounds (primarily proteins) is urea. (Contrast with ammonotelic, uricotelic.)

**ureter** (your′ uh tur) Long duct leading from the vertebrate kidney to the urinary bladder or the cloaca.

**urethra** (you ree′ thra) In most mammals, the canal through which urine is discharged from the bladder and which serves as the genital duct in males.

**uric acid** A compound that serves as the main excreted form of nitrogen in some animals, particularly those which must conserve water, such as birds, insects, and reptiles.

**uricotelic** Pertaining to an organism in which the final product of the breakdown of nitrogen-containing compounds (primarily proteins) is uric acid. (Contrast with ammonotelic, ureotelic.)

**urinary bladder** A structure in which urine is stored until it can be excreted to the outside of the body.

**urine** (you′ rin) In vertebrates, the fluid waste product containing the toxic nitrogenous by-products of protein and amino acid metabolism.

**uterine cycle** In human females, the monthly cycle of events by which the endometrium is prepared for the arrival of a blastocyst. (Contrast with ovarian cycle).

**uterus** (yoo′ ter us) [L. *utero*: womb] A specialized portion of the female reproductive tract in mammals that receives the fertilized egg and nurtures the embryo in its early development. Also called the womb.

## - V -

**vaccination** Injection of virus or bacteria or their proteins into the body, to induce immunization. The injected material is usually attenuated (weakened) before injection.

**vacuole** (vac′ yew ole) Membrane-enclosed organelle in plant cells that can function for storage, water concentration for turgor, or hydrolysis of stored macromolecules.

**vagina** (vuh jine′ uh) [L. sheath] In female animals, the entry to the reproductive tract.

**van der Waals forces** Weak attractions between atoms resulting from the interaction of the electrons of one atom with the nucleus of another. This type of attraction is about one-fourth as strong as a hydrogen bond.

**variable region** The portion of an immunoglobulin molecule or T-cell receptor that includes the antigen-binding site and is responsible for its specificity. (Contrast with constant region.)

**vas deferens** (plural: vasa deferentia) Duct that transfers sperm from the epididymis to the urethra.

**vasa recta** Blood vessels that parallel the loops of Henle and the collecting ducts in the renal medulla of the kidney.

**vascular** (vas′ kew lar) [L. *vasculum*: a small vessel] Pertaining to organs and tissues that conduct fluid, such as blood vessels in animals and xylem and phloem in plants.

**vascular bundle** In vascular plants, a strand of vascular tissue, including xylem and phloem as well as thick-walled fibers.

**vascular cambium** (kam′ bee um) [L. *cambiare*: to exchange] In plants, a lateral meristem that gives rise to secondary xylem and phloem.

**vascular tissue system** The transport system of a vascular plant, consisting primarily of xylem and phloem.

**vasopressin** *See* antidiuretic hormone.

**vector** (1) An agent, such as an insect, that carries a pathogen affecting another species. (2) A plasmid or virus that carries an inserted piece of DNA into a bacterium for cloning purposes in recombinant DNA technology.

**vegetal hemisphere** The lower portion of some animal eggs, zygotes, and embryos, in which the dense nutrient yolk settles. The *vegetal pole* is to the very bottom of the egg or embyro. (Contrast with animal hemisphere.)

**vegetative** Nonreproductive, nonflowering, or asexual.

**vegetative meristem** An apical meristem that produces leaves.

**vegetative reproduction** Asexual reproduction through the modification of stems, leaves, or roots.

**vein** [L. *vena*: channel] A blood vessel that returns blood to the heart. (Contrast with artery.)

**ventral** [L. *venter*: belly, womb] Toward or pertaining to the belly or lower side. (Contrast with dorsal.)

**ventricle** A muscular heart chamber that pumps blood through the lungs or through the body.

**venule** A small blood vessel draining a capillary bed that joins others of its kind to form a vein. (Contrast with arteriole.)

**vernalization** [L. *vernalis*: spring] Events occurring during a required chilling period, leading eventually to flowering.

**vertebral column** [L. *vertere*: to turn] The jointed, dorsal column that is the primary support structure of vertebrates.

**very low-density lipoproteins (VLDLs)** Lipoproteins that consist mainly of triglyceride fats, which they transport to fat cells in adipose tissues throughout the body; associated with excessive fat deposition and high risk for cardiovascular disease.

**vesicle** Within the cytoplasm, a membrane-enclosed compartment that is associated with other organelles; the Golgi complex is one example.

**vessel element** A type of tracheary element with perforated end walls; found only in angiosperms. (Contrast with tracheid.)

**vestibular system** (ves tib′ yew lar) [L. *vestibulum*: an enclosed passage] Structures within the inner ear that sense changes in position or momentum of the head, affecting balance and motor skills.

**vicariant event** (vye care′ ee unt) [L. *vicus*: change] The splitting of a taxon's range by the imposition of some barrier to dispersal.

**villus** (vil′ lus) (plural: villi) [L. *villus*: shaggy hair or beard] A hairlike projection from a membrane; for example, from many gut walls.

**virion** (veer′ e on) The virus particle, the minimum unit capable of infecting a cell.

**virulence** [L. *virus*: poison, slimy liquid] The ability of a pathogen to cause disease and death.

**virus** Any of a group of ultramicroscopic particles constructed of nucleic acid and protein (and, sometimes, lipid) that require living cells in order to reproduce. Viruses evolved multiple times from different cellular species.

**vital capacity** The maximum capacity for air exchange in one breath; the sum of the tidal volume and the inspiratory and expiratory reserve volumes.

**vitamin** [L. *vita*: life] An organic compound that an organism cannot synthesize, but nevertheless requires in small quantities for normal growth and metabolism.

**vitelline envelope** The inner, proteinaceous protective layer of a sea urchin egg.

**viviparity** (vye vi par′ uh tee) Reproduction in which fertilization of the egg and development of the embryo occur inside the mother's body. (Contrast with oviparity.)

**vivipary** Premature germination in plants.

**voltage-gated channel** A type of gated channel that opens or closes when a certain voltage exists across the membrane in which it is inserted.

**vomeronasal organ (VNO)** Chemosensory structure embedded in the nasal epithelium of amphibians, reptiles, and many mammals. Often specialized for detecting pheromones.

## - W -

**water potential** In osmosis, the tendency for a system (a cell or solution) to take up water from pure water through a differentially permeable membrane. Water flows toward the system with a more negative water potential. (Contrast with solute potential, pressure potential.)

**water vascular system** In echinoderms, a network of water-filled canals that functions in gas exchange, locomotion, and feeding.

**wavelength** The distance between successive peaks of a wave train, such as electromagnetic radiation.

**weathering** The mechanical and chemical processes by which rocks are broken down into soil particles.

**Wernicke's area** A region in the temporal lobe of the human brain that is involved with the sensory aspects of language.

**white blood cells** Cells in the blood plasma that play defensive roles in the immune system. Also called leukocytes.

**white matter** In the central nervous system, tissue that is rich in axons. (Contrast with gray matter.)

**wild-type** Geneticists' term for standard or reference type. Deviants from this standard, even if the deviants are found in the wild, are usually referred to as mutant. (Note that this terminology is not usually applied to human genes.)

**wood** Secondary xylem tissue.

## - X -

**xerophyte** (zee′ row fyte) [Gk. *xerox*: dry + *phyton*: plant] A plant adapted to an environment with limited water supply.

**xylem** (zy′ lum) [Gk. *xylon*: wood] In vascular plants, the tissue that conducts water and minerals; xylem consists, in various plants, of tracheids, vessel elements, fibers, and other highly specialized cells.

## - Y -

**yolk** [M.E. *yolke*: yellow] The stored food material in animal eggs, rich in protein and lipids.

**yolk sac** In reptiles, birds, and mammals, the extraembryonic membrane that forms from the endoderm of the hypoblast; it encloses and digests the yolk.

## - Z -

**zeaxanthin** A blue-light receptor involved in the opening of plant stomata.

**zona pellucida** A jellylike substance that surrounds the mammalian ovum when it is released from the ovary.

**zoospore** (zoe′ o spore) [Gk. *zoon*: animal + *spora*: seed] In algae and fungi, any swimming spore. May be diploid or haploid.

**zygote** (zye′ gote) [Gk. *zygotos*: yoked] The cell created by the union of two gametes, in which the gamete nuclei are also fused. The earliest stage of the diploid generation.

**zymogen** The inactive precursor of a digestive enzyme; secreted into the lumen of the gut, where a protease cleaves it to form the active enzyme.

# Illustration Credits

**Cover**: © Dr. Merlin D. Tuttle/Photo Researchers, Inc.
**Frontispiece**: © Art Wolfe/www.artwolfe.com.

**Photographs appearing behind chapter numbers:**

**Part 1** *HMS Beagle*: Painting by Ronald Dean, reproduced by permission of the artist and Richard Johnson, Esquire.
**Part 2** *Plant cells*: © Ed Reschke/Peter Arnold Inc.
**Part 3** *Stoma*: © Andrew Syred/SPL/Photo Researchers, Inc.
**Part 4** *Anaphase*: © Nasser Rusan.
**Part 5** *Sheep*: © Roddy Field, the Roslin Institute.
**Part 6** *Trilobite*: Courtesy of the Amherst College Museum of Natural History and the Trustees of Amherst College.
**Part 7** *Diatoms*: © Scenics & Science/Alamy.
**Part 8** *Flowers*: © Ed Reschke/Peter Arnold Inc.
**Part 9** *Leopard*: Courtesy of Andrew D. Sinauer.
**Part 10** *Hummingbird*: © Yufeng Zhou/istockphoto.com.

**Table of Contents**
p. xxi: © Corbis RF/AGE Fotostock.
p. xxiii: © Dennis Kunkel Microscopy, Inc.
p. xxiv: © G. Brad Lewis/AGE Fotostock.
p. xxvi: © Dr. Alexey Khodjakov/Photo Researchers, Inc.
p. xxvii: © louise murray/Alamy.
p. xxviii: From B. Palenik et al. 2003. *Nature* 424: 1037.
p. xxx: Courtesy of Andrew D. Sinauer.
p. xxxii: © Ervin Monn/Shutterstock.
p. xxxiv: © Muriel Lasure/Shutterstock.
p. xxxv: © Heather A. Craig/Shutterstock.
p. xxxvii: Courtesy of Jane Sinauer.
p. xxxviii: © Christian Anthony/istockphoto.com.
p. xxxix: © PaulTessier/istockphoto.com.
p. xli: © Marina Cano Trueba/Shutterstock.
p. xlii: © Shutterstock.
p. xliii: © Herbert Hopfensperger/AGE Fotostock.

**Chapter 1** *Frogs*: © Pete Oxford/Minden Pictures. *T. Hayes*: © Pamela S. Turner. 1.1A: © Eye of Science/SPL/Photo Researchers, Inc. 1.1B: Science Photo Library/Photolibrary.com. 1.1C: © Steve Gschmeissner/Photo Researchers, Inc. 1.1D: © Frans Lanting. 1.1E: © Glen Threlfo/Auscape/Minden Pictures. 1.1F: © Piotr Naskrecki/Minden Pictures. 1.1G: © Tui De Roy/Minden Pictures. 1.2A: From R. Hooke, 1664. *Micrographia*. 1.2B: © John Durham/SPL/Photo Researchers, Inc. 1.2C: © Biophoto Associates/Photo Researchers, Inc. 1.3 *Maple*: © Simon Colmer & Abby Rex/Alamy. 1.3 *Spruce*: David McIntyre. 1.3 *Lily pad*: © Pete Oxford/Naturepl.com. 1.3 *Cucumber*: David McIntyre. 1.3 *Pitcher plant*: © Nick Garbutt/Naturepl.com. 1.5A: © Frans Lanting. 1.5B: © Stefan Huwiler/Rolfnp/Alamy. 1.6 *Organism*: © Nico Smit/istockphoto.com. 1.6 *Population*: © blickwinkel/Alamy. 1.6 *Community*: © Georgie Holland/AGE Fotostock. 1.6 *Biosphere*: Courtesy of NASA. 1.7: © P&R Fotos/AGE Fotostock. 1.9: © Michael Abbey/Visuals Unlimited, Inc. 1.11: Courtesy of Wayne Whippen. 1.13: From T. Hayes et al., 2003. *Environ. Health Perspect*. 111: 568.

**Chapter 2** *Hair*: © Steve Gschmeissner/Photo Researchers, Inc. *Barber*: © Digital Vision/Alamy. 2.3: From N. D. Volkow et al., 2001. *Am. J. Psychiatry* 158: 377. 2.14: © Pablo H Caridad/Shutterstock. 2.15: © Denis Miraniuk/Shutterstock.

**Chapter 3** *T. Rex*: © The Natural History Museum, London. *Chicken*: David McIntyre. 3.8: Data from PDB 1IVM. T. Obita, T. Ueda, & T. Imoto, 2003. *Cell. Mol. Life Sci*. 60: 176. 3.10A: Data from PDB 2HHB. G. Fermi et al., 1984. *J. Mol. Biol*. 175: 159. 3.16C *left*: © Biophoto Associates/Photo Researchers, Inc. 3.16C *middle*: © Ken Wagner/Visuals Unlimited. 3.16C *right*: © CNRI/SPL/Photo Researchers, Inc. 3.17 *Cartilage*: © Robert Brons/Biological Photo Service. 3.17 *Beetle*: © Scott Bauer/USDA.

**Chapter 4** *Phoenix*: Courtesy of NASA/JPL/UA/Lockheed Martin. *Ice*: Courtesy of NASA/JPL-Caltech/University of Arizona/Texas A&M University. 4.4: Data from S. Arnott & D. W. Hukins, 1972. *Biochem. Biophys. Res. Commun*. 47(6): 1504. 4.8: Courtesy of the Argonne National Laboratory. 4.13B: Courtesy of Janet Iwasa, Szostak group, MGH/Harvard. 4.14: © Stanley M. Awramik/Biological Photo Service. 4.14 *inset*: © Dennis Kunkel Microscopy, Inc.

**Chapter 5** *Heart cell*: © Roger J. Bick & Brian J. Poindexter/UT-Houston Medical School/Photo Researchers, Inc. *Surgery*: © The Stock Asylum, LLC/Alamy. 5.1: After N. Campbell, 1990. *Biology*, 2nd Ed., Benjamin Cummings. 5.1 *Protein*: Data from PDB 1IVM. T. Obita, T. Ueda, & T. Imoto, 2003. *Cell. Mol. Life Sci*. 60: 176. 5.1 *T4*: © Dept. of Microbiology, Biozentrum/SPL/Photo Researchers, Inc. 5.1 *Bacterium*: © Jim Biddle/Centers for Disease Control. 5.1 *Plant cells*: © Michael Eichelberger/Visuals Unlimited, Inc. 5.1 *Frog egg*: David McIntyre. 5.1 *Bird*: © Steve Byland/Shutterstock. 5.1 *Baby*: Courtesy of Sebastian Grey Miller. 5.3 *Light microscope*: © Radu Razvan/Shutterstock. 5.3 *Bright-field*: Courtesy of the IST Cell Bank, Genoa. 5.3 *Phase-contrast*: © Michael W. Davidson, Florida State U. 5.3 *Stained*: © Richard J. Green/SPL/Photo Researchers, Inc. 5.3 *Confocal*: © Dr. Gopal Murti/SPL/Photo Researchers, Inc. 5.3 *Electron microscope*: © Sinclair Stammers/Photo Researchers, Inc. 5.3 *TEM*: © Dr. Gopal Murti/Visuals Unlimited, Inc. 5.3 *SEM*: © K. R. Porter/SPL/Photo Researchers, Inc. 5.3 *Freeze-fracture*: © D. W. Fawcett/Photo Researchers, Inc. 5.4: © J. J. Cardamone Jr. & B. K. Pugashetti/Biological Photo Service. 5.5A: © Dennis Kunkel Microscopy, Inc. 5.5B: Courtesy of David DeRosier, Brandeis U. 5.7 *Mitochondrion*: © K. Porter, D. Fawcett/Visuals Unlimited. 5.7 *Cytoskeleton*: © Don Fawcett, John Heuser/Photo Researchers, Inc. 5.7 *Nucleolus*: © Richard Rodewald/Biological Photo Service. 5.7 *Peroxisome*: © E. H. Newcomb & S. E. Frederick/Biological Photo Service. 5.7 *Cell wall*: © Biophoto Associates/Photo Researchers, Inc. 5.7 *Ribosome*: From M. Boublik et al., 1990. *The Ribosome*, p. 177. Courtesy of American Society for Microbiology. 5.7 *Centrioles*: © Barry F. King/Biological Photo Service. 5.7 *Plasma membrane*: Courtesy of J. David Robertson, Duke U. Medical Center. 5.7 *Rough ER*: © Don Fawcett/Science Source/Photo Researchers, Inc. 5.7 *Smooth ER*: © Don Fawcett, D. Friend/Science Source/Photo Researchers, Inc. 5.7 *Chloroplast*: © W. P. Wergin, E. H. Newcomb/Biological Photo Service. 5.7 *Golgi apparatus*: Courtesy of L. Andrew Staehelin, U. Colorado. 5.8: © D. W. Fawcett/Photo Researchers, Inc. 5.9A: © Barry King, U. California, Davis/Biological Photo Service. 5.9B: © Biophoto Associates/Science Source/Photo Researchers, Inc. 5.10: © B. Bowers/Photo Researchers, Inc. 5.11: © Sanders/Biological Photo Service. 5.12: © K. Porter, D. Fawcett/Visuals Unlimited. 5.13 *left*: © W. P. Wergin, E. H. Newcomb/Biological Photo Service. 5.13 *right*: © W. P. Wergin/Biological Photo Service. 5.14A: © Michael Eichelberger/Visuals Unlimited, Inc. 5.14B: © Ed Reschke/Peter Arnold Inc. 5.14C: © Gerald & Buff Corsi/Visuals Unlimited. 5.15A: David McIntyre. 5.15A *inset*: © Richard Green/Photo Researchers, Inc. 5.15B: David McIntyre. 5.15B *inset*: Courtesy of R. R. Dute. 5.16: Courtesy of M. C. Ledbetter, Brookhaven National Laboratory. 5.17: Courtesy of Vic Small, Austrian Academy of Sciences, Salzburg, Austria. 5.19: Courtesy of N. Hirokawa. 5.20A *upper*: © SPL/Photo Researchers, Inc. 5.20A *lower*, 5.20B: © W. L. Dentler/Biological Photo Service. 5.22: From N. Pollock et al., 1999. *J. Cell Biol*. 147: 493. Courtesy of R. D. Vale. 5.23: © Michael Abbey/Visuals Unlimited. 5.24: © Biophoto Associates/Photo Researchers, Inc. 5.25 *left*: Courtesy of David Sadava. 5.25 *upper right*: From J. A. Buckwalter & L. Rosenberg, 1983. *Coll. Rel. Res*. 3: 489. Courtesy of L. Rosenberg. 5.25 *lower right*: © J. Gross, Biozentrum/SPL/Photo Researchers, Inc. 5.27: Courtesy of Noriko Okamoto and Isao Inouye.

**Chapter 6** *Patient*: From Alzheimer, A. 1906. Über einen eigenartigen schweren Erkrankungsprozess der Hirnrinde. *Neurologisches Centralblatt* 23: 1129. *Plaques*: © G. W. Willis/Photolibrary.com. 6.2: After L. Stryer, 1981. *Biochemistry*, 2nd Ed., W. H. Freeman. 6.4: © D. W. Fawcett/Photo Researchers, Inc. 6.7A: Courtesy of D. S. Friend, U. California, San Francisco. 6.7B: Courtesy of Darcy E. Kelly, U. Washington. 6.7C: Courtesy of C. Peracchia. 6.10A *left*: © Stanley Flegler/Visuals Unlimited, Inc. 6.10A *center*: © David M. Phillips/Photo Researchers, Inc. 6.10B *left*: © Ed Reschke/Peter Arnold Inc. 6.13: From G. M. Preston et al., 1992. *Science* 256: 385. 6.19: From M. M. Perry, 1979. *J. Cell Sci*. 39: 26.

**Chapter 7** *Voles*: Courtesy of Todd Ahern. *Oxytocin*: Data from PDB 1NPO. J. P. Rose et al., 1996. *Nat. Struct .Biol.* 3: 163. 7.3: © Biophoto Associates/Photo Researchers, Inc. 7.4: Data from PDB 3EML. V. P. Jaakola et al., 2008. *Science* 322: 1211. 7.16: © Stephen A. Stricker, courtesy of Molecular Probes, Inc.

**Chapter 8** *Dairy*: © Bob Randall/istockphoto.com. *Maasai*: ©blickwinkel/Alamy. 8.1: Courtesy of Violet Bedell-McIntyre. 8.5B: © Satoshi Kuribayashi/OSF/Photolibrary.com. 8.11A: Data from PDB 1AL6. B. Schwartz et al., 1997. 8.11B: Data from PDB 1BB6. V. B. Vollan et al., 1999. *Acta Crystallogr. D. Biol. Crystallogr.* 55: 60. 8.11C: Data from PDB 1AB9. N. H. Yennawar, H. P. Yennawar, & G. K. Farber, 1994. *Biochemistry* 33: 7326. 8.13: Data from PDB 1IG8 (P. R. Kuser et al., 2000. *J. Biol. Chem.* 275: 20814) and 1BDG (A. M. Mulichak et al., 1998 *Nat. Struct. Biol.* 5: 555).

**Chapter 9** *Marathoners*: © Chuck Franklin/Alamy. *Mouse*: © Royalty-Free/Corbis. 9.9: From Y. H. Ko et al., 2003. *J. Biol. Chem.* 278: 12305. Courtesy of P. Pedersen.

**Chapter 10** *Rainforest*: © Jon Arnold Images/Photolibrary.com. *FACE*: Courtesy of David F. Karnosky. 10.1: © Andrew Syred/SPL/Photo Researchers, Inc. 10.6: © Martin Shields/Alamy. 10.13: Courtesy of Lawrence Berkeley National Laboratory. 10.17A: © E. H. Newcomb & S. E. Frederick/Biological Photo Service. 10.21: © Aflo Foto Agency/Alamy.

**Chapter 11** *Cells*: © Parviz M. Pour/Photo Researchers, Inc. *Model*: Data from PDB 1GUX. J. Lee et al., 1998. *Nature* 391: 859. 11.1A: © SPL/Photo Researchers, Inc. 11.1B: © Biodisc/Visuals Unlimited/Alamy. 11.1C: © Robert Valentic/Naturepl.com. 11.2B: © John J. Cardamone Jr./Biological Photo Service. 11.8 *Chromosome*: Courtesy of G. F. Bahr. 11.8 *Nucleus*: © D. W. Fawcett/Photo Researchers, Inc. 11.9 *inset*: © Biophoto Associates/Science Source/Photo Researchers, Inc. 11.10B: © Conly L. Rieder/Biological Photo Service. 11.11: © Nasser Rusan. 11.13A: © T. E. Schroeder/Biological Photo Service. 11.13B: © B. A. Palevitz, E. H. Newcomb/Biological Photo Service. 11.14A: © Dr. John Cunningham/Visuals Unlimited. 11.14B: © Steve Gschmeissner/Photo Researchers, Inc. 11.15 *left*: © Andrew Syred/SPL/Photo Researchers, Inc. 11.15 *center*: David McIntyre. 11.15 *right*: © Gerry Ellis, DigitalVision/PictureQuest. 11.16: Courtesy of Dr. Thomas Ried and Dr. Evelin Schröck, NIH. 11.17: © C. A. Hasenkampf/Biological Photo Service. 11.18: Courtesy of J. Kezer. 11.22A: © Gopal Murti/Photo Researchers, Inc. 11.23: © Dennis Kunkel Microscopy, Inc.

**Chapter 12** *Vavilov*: Courtesy of the Library of Congress/New York World-Telegram & Sun Collection. *Harvest*: © Paul Nevin/Photolibrary.com. 12.1: © the Mendelianum. 12.11: Courtesy the American Netherland Dwarf Rabbit Club. 12.14: Courtesy of Madison, Hannah, and Walnut. 12.15: Courtesy of the Plant and Soil Sciences eLibrary (http://plantandsoil.unl.edu); used with permission from the Institute of Agriculture and Natural Resources at the University of Nebraska. 12.16: © Grant Heilman Photography/Alamy. 12.17: © Peter Morenus/U. of Connecticut. 12.26A: Courtesy of L. Caro and R. Curtiss.

**Chapter 13** *Velociraptor*: © Joe Tucciarone/Photo Researchers, Inc. *DNA art*: © Simon Fraser/Karen Rann/SPL/Photo Researchers, Inc. 13.3: © Biozentrum, U. Basel/SPL/Photo Researchers, Inc. 13.6 *X-ray crystallograph*: Courtesy of Prof. M. H. F. Wilkins, Dept. of Biophysics, King's College, U. London. 13.6B: © CSHL Archives/Peter Arnold, Inc. 13.8A: © A. Barrington Brown/Photo Researchers, Inc. 13.8B: Data from S. Arnott & D. W. Hukins, 1972. *Biochem. Biophys. Res. Commun.* 47(6): 1504. 13.14A: Data from PDB 1SKW. Y. Li et al., 2001. *Nat. Struct. Mol. Biol.* 11: 784. 13.20C: © Dr. Peter Lansdorp/Visuals Unlimited.

**Chapter 14** *Mycobacterium*: © Dennis Kunkel Microscopy, Inc. *Ethiopia*: © Jenny Matthews/Alamy. 14.3: Data from PDB 1I3Q. P. Cramer et al., 2001. *Science* 292: 1863. 14.9: From D. C. Tiemeier et al., 1978. *Cell* 14: 237. 14.12: Data from PDB 1EHZ. H. Shi & P. B. Moore, 2000. *RNA* 6: 1091. 14.14: Data from PDB 1GIX and 1G1Y. M. M. Yusupov et al., 2001. *Science* 292: 883. 14.18B: Courtesy of J. E. Edström and *EMBO J.*

**Chapter 15** *Destruction*: © Suzanne Plunkett/AP Images. *Baby* 81: © Gemunu Amarasinghe/AP Images. 15.3: © Stanley Flegler/Visuals Unlimited, Inc. 15.8: © Philippe Plailly/Photo Researchers, Inc. 15.10: Bettmann/CORBIS. 15.11 *Butterfly*: © Bershadsky Yuri/Shutterstock. 15.11 *Bacteria*: Courtesy of Janice Haney Carr/CDC. 15.11 *Fungus*: © Warwick Lister-Kaye/istockphoto.com. 15.17: From C. Harrison et al., 1983. *J. Med. Genet.* 20: 280. 15.19: © Simon Fraser/Photo Researchers, Inc.

**Chapter 16** *Alcoholic*: © LJSphotography/Alamy. *CREB*: Data from PDB 1T2K. D. Panne et al., 2004. *EMBO J.* 23: 4384. 16.2A: © Dennis Kunkel Microscopy, Inc. 16.2B: © Lee D. Simon/Photo Researchers, Inc. 16.6: © NIBSC/Photo Researchers, Inc. 16.21A: Courtesy of the Centers for Disease Control.

**Chapter 17** *Mastiff and chihuahua*: © moodboard RF/Photolibrary.com. *Whippets*: © Stuart Isett/Anzenberger/Eyevine. 17.3: © Sam Ogden/Photo Researchers, Inc. 17.5: Based on an illustration by Anthony R. Kerlavage, Institute for Genomic Research. *Science* 269: 449 (1995). 17.12: Courtesy of O. L. Miller, Jr. 17.17: From P. H. O'Farrell, 1975. *J. Biol. Chem.* 250: 4007. Courtesy of Patrick H. O'Farrell.

**Chapter 18** *Kuwait*: © K. Bry—UNEP/Peter Arnold Inc. *A. Chakrabarty*: © Ted Spiegel/Corbis. 18.5: © Martin Shields/Alamy. 18.15: Courtesy of the Golden Rice Humanitarian Board, www.goldenrice.org. 18.16: Courtesy of Eduardo Blumwald.

**Chapter 19** *Horse*: © Benoit Photo. *Centrifuge*: Courtesy of Cytori Therapeutics. 19.4: © Roddy Field, the Roslin Institute. 19.11A: From J. E. Sulston & H. R. Horvitz, 1977. *Dev. Bio.* 56: 100. 19.14C: Courtesy of J. Bowman. 19.17: Courtesy of W. Driever and C. Nüsslein-Vollhard. 19.18B: Courtesy of C. Rushlow and M. Levine. 19.18C: Courtesy of T. Karr. 19.18D: Courtesy of S. Carroll and S. Paddock. 19.20: Courtesy of F. R. Turner, Indiana U.

**Chapter 20** *Fly head*: © Science Photo Library RF/Photolibrary.com. *Mutant leg*: From G. Halder et al., 1995. *Science* 267: 1788. Courtesy of W. J. Gehring and G. Halder. 20.1 *Mouse*: © orionmystery@flickr/Shutterstock. 20.1 *Fly*: David McIntyre. 20.1 *Shark*: © Kristian Sekulic/Shutterstock. 20.1 *Squid*: © Gergo Orban/Shutterstock. 20.3: © David M. Phillips/Photo Researchers, Inc. 20.5: © Bone Clones, www.boneclones.com. 20.6: Courtesy of J. Hurle and E. Laufer. 20.7: Courtesy of J. Hurle. 20.8 *Cladogram*: After R. Galant & S. Carroll, 2002. *Nature* 415: 910. 20.8 *Insect*: © Stockbyte/PictureQuest. 20.8 *Centipede*: © Burke/Triolo/Brand X Pictures/PictureQuest. 20.9: From M. Kmita and D. Duboule, 2003. *Science* 301: 331. 20.10: © Neil Hardwick/Alamy. 20.11: © Rob Valentic/ANTPhoto.com. 20.12A: © Erick Greene. 20.12B: Courtesy of John Gruber. 20.13: © Nigel Cattlin, Holt Studios International/Photo Researchers, Inc. 20.15: Courtesy of Mike Shapiro and David Kingsley.

**Chapter 21** *Pandemic*: Courtesy of the Naval Historical Foundation. *Researcher*: Courtesy of James Gathany/Centers for Disease Control. 21.1A: Painting by Ronald Dean, reproduced by permission of the artist and Richard Johnson, Esquire. 21.2A: © Luis César Tejo/Shutterstock. 21.2B: © Duncan Usher/Alamy. 21.2C: © PetStockBoys/Alamy. 21.2D: © Arco Images GmbH/Alamy. 21.9A: © Tom Ulrich/OSF/Photolibrary.com. 21.9B: © Kim Karpeles/Alamy. 21.11: David McIntyre. 21.14: Courtesy of David Hillis. 21.16: © Jason Gallier/Alamy. 21.17: David McIntyre. 21.21A: © Reinhard Dirscherl/WaterFrame - Underwater Images/Photolibrary.com. 21.21B: © Marevision/AGE Fotostock. 21.22 *Snake*: © Joseph T. Collins/Photo Researchers, Inc. 21.22 *Newt*: © Robert Clay/Visuals Unlimited.

**Chapter 22** *HIV*: © James Cavallini/Photo Researchers, Inc. *Chimpanzees*: © John Cancalosi/Naturepl.com. 22.6A: Courtesy of William Jeffery. 22.6B: © Jurgen Freund/Naturepl.com. 22.6C: © Michael & Patricia Fogden/Minden Pictures. 22.6D: © Mark Kostich/Shutterstock. 22.8: © Larry Jon Friesen. 22.10: © Alexandra Basolo. 22.13A: © Krieger, C./AGE Fotostock. 22.13B: © Krieger, C./AGE Fotostock. 22.13C: © Ed Reschke/Peter Arnold Inc.

**Chapter 23** *Cuatro Cienegas*: © George Grall/National Geographic. *Fly*: © Dr. David Phillips/Visuals Unlimited, Inc. 23.1A *left*: © R. L. Hambley/Shutterstock. 23.1A *right*: © Frank Leung/istockphoto.com. 23.1B: © Norman Bateman/istockphoto.com. 23.10: © OSF/Photolibrary.com. 23.11 *G. olivacea*: © Charles Melton/Visuals Unlimited, Inc. 23.11 *G. carolinensis*: © Michael Redmer/Visuals Unlimited, Inc. 23.12A: © Yufeng Zhou/istockphoto.com. 23.12B: © P01017 Desmette Frede/AGE Fotostock. 23.12C: © J. S. Sira/Photolibrary.com. 23.12D: © Daniel L. Geiger/SNAP/Alamy. 23.13: Courtesy of Donald A. Levin. 23.14: © Boris I. Timofeev/Pensoft. 23.16A: © Tony Tilford/Photolibrary.com. 23.16B: © W. Peckover/VIREO. 23.17 *Madia*: © Peter K. Ziminsky/Visuals Unlimited. 23.17 *Argyroxiphium*: © Ron Dahlquist/Pacific Stock/Photolibrary.com. 23.17 *Wilkesia*: © Gerald D. Carr. 23.17 *Dubautia*: © Noble Proctor/The National

Audubon Society Collection/Photo Researchers, Inc.

**Chapter 24** *Electric fish*: © Jane Burton/Naturepl.com. *Ray*: © David Fleetham/Alamy. 24.3: *Rice*: data from PDB 1CCR. H. Ochi et al., 1983. *J. Mol. Biol.* 166: 407. *Tuna*: data from PDB 5CYT. T. Takano, 1984. 24.4: From P. B. Rainey & M. Travisano, 1998. *Nature* 394: 69. © Macmillan Publishers Ltd. 24.7A *Langur*: © blickwinkel/Alamy. 24.7A *Longhorn*: Courtesy of David Hillis. 24.7B: © M. Graybill/J. Hodder/Biological Photo Service.

**Chapter 25** *Meganeura*: © Graham Cripps/NHMPL. *Modern dragonfly*: © Natasha Litova/istockphoto.com. *Grand Canyon*: © Tim Fitzharris/Minden Pictures. 25.4A: Photos courtesy of P.F. Hoffman, Geological Survey of Canada. 25.4B: © Robin Smith/Photolibrary.com. 25.8: © Martin Bond/SPL/Photo Researchers, Inc. 25.9: David McIntyre. 25.11 *left*: © Ken Lucas/Visuals Unlimited. 25.11 *center*: © The Natural History Museum, London. 25.11 *right*: Courtesy of Martin Smith. 25.12 *Cambrian*: © John Sibbick/NHMPL. 25.12 *Marella*: Courtesy of the Amherst College Museum of Natural History, The Trustees of Amherst College. 25.12 *Ottoia*: © Alan Sirulnikoff/Photo Researchers, Inc. 25.12 *Anomalocaris*: © Kevin Schafer/Alamy. 25.12 *Devonian*: © Tom McHugh/Field Museum, Chicago/Photo Researchers, Inc. 25.12 *Codiacrinus*: Courtesy of the Amherst College Museum of Natural History, The Trustees of Amherst College. 25.12 *Phacops*: Courtesy of the Amherst College Museum of Natural History, The Trustees of Amherst College. 25.12 *Eridophyllum*: © Mark A. Schneider/Photo Researchers, Inc. 25.12 *Nautiloid*: © The Natural History Museum, London. 25.12 *Permian*: © John Sibbick/NHMPL. 25.12 *Cacops*: © Albert Copley/Visuals Unlimited, Inc. 25.12 *Walchia*: © The Natural History Museum, London. 25.12 *Triassic*: © OSF/Photolibrary.com. 25.12 *Ferns*: © Ken Lucas/Visuals Unlimited. 25.12 *Coelophysis*: © Ken Lucas/Visuals Unlimited. 25.12 *Cretaceous*: © OSF/Photolibrary.com. 25.12 *Gryposaurus*: Courtesy of the Amherst College Museum of Natural History, The Trustees of Amherst College. 25.12 *Magnolia*: © The Natural History Museum, London. 25.12 *Tertiary*: © Publiphoto/Photo Researchers, Inc. 25.12 *Hyracotherium*: Courtesy of the Amherst College Museum of Natural History, The Trustees of Amherst College. 25.12 *Plesiadapis*: © The Natural History Museum, London. 25.13: Courtesy of Conrad C. Labandeira, Department of Paleobiology, National Museum of Natural History, Smithsonian Institution.

**Chapter 26** *Antarctica*: Courtesy of Emily Gercke. *Yellowstone*: Courtesy of Jim Peaco/National Park Service. 26.2: © Dennis Kunkel Microscopy, Inc. 26.3B: © Steve Gschmeissner/Photo Researchers, Inc. 26.4: From F. Balagaddé et al., 2005. *Science* 309: 137. Courtesy of Frederick Balagaddé. 26.5A: © David M. Phillips/Visuals Unlimited. 26.5B: Courtesy of the CDC. 26.6A: © J. A. Breznak & H. S. Pankratz/Biological Photo Service. 26.6B: © J. Robert Waaland/Biological Photo Service. 26.7: © USDA/Visuals Unlimited. 26.8: © Steven Haddock and Steven Miller. 26.12: Courtesy of David Cox/CDC. 26.13: Courtesy of Randall C. Cutlip. 26.14: © David Phillips/Visuals Unlimited. 26.15A: © Paul W. Johnson/Biological Photo Service. 26.15B: © H. S. Pankratz/Biological Photo Service. 26.15C: © Bill Kamin/Visuals Unlimited. 26.16: © Dr. Kari Lounatmaa/Photo Researchers, Inc. 26.17: © Dr. Gary Gaugler/Visuals Unlimited. 26.18: © Michael Gabridge/Visuals Unlimited. 26.20: © Geoff Kidd/Photo Researchers, Inc. 26.21: From K. Kashefi & D. R. Lovley, 2003. *Science* 301: 934. Courtesy of Kazem Kashefi. 26.23: © David Sanger Photography/Alamy. 26.24: From H. Huber et al., 2002. *Nature* 417: 63. © Macmillan Publishers Ltd. Courtesy of Karl O. Stetter. 26.25A: © Science Photo Library RF/Photolibrary.com. 26.25B, C: © Russel Kightley/SPL/Photo Researchers, Inc. 26.25D: © Science Photo Library RF/Photolibrary.com. 26.25E: animate4.com ltd./Photo Researchers, Inc. 26.25F: © Russell Kightley/Photo Researchers, Inc. 26.26: © Nigel Cattlin/Alamy.

**Chapter 27** *Bug*: © Martin Dohrn/Photo Researchers, Inc. *Leishmania*: © Dennis Kunkel Microscopy, Inc. 27.4: © Michael Abbey/Photo Researchers, Inc. 27.7A: © Wim van Egmond/Visuals Unlimited. 27.7B: © Science Photo Library RF/Photolibrary.com. 27.8A: © Georgette Douwma/Naturepl.com. 27.8B: © David Patterson, Linda Amaral Zettler, Mike Peglar, & Tom Nerad/micro*scope. 27.9: © London School of Hygiene/SPL/Photo Researchers, Inc. 27.10A: © Bill Bachman/Photo Researchers, Inc. 27.10B: © Markus Geisen/NHMPL. 27.11: © Science Photo Library RF/Photolibrary.com. 27.16: © Dennis Kunkel Microscopy, Inc. 27.17A: © SPL/Photo Researchers, Inc. 27.17B: © Dennis Kunkel Microscopy, Inc. 27.17C: © Paul W. Johnson/Biological Photo Service. 27.17D: © Steve Gschmeissner/Photo Researchers, Inc. 27.19: © Scenics & Science/Alamy. 27.20A: © Marevision/AGE Fotostock. 27.20B: © Larry Jon Friesen. 27.20C: © J. N. A. Lott/Biological Photo Service. 27.20D: © Gerald & Buff Corsi/Visuals Unlimited, Inc. 27.21: © James W. Richardson/Visuals Unlimited. 27.22A: © Wim van Egmond/Visuals Unlimited, Inc. 27.22B: © Doug Sokell/Visuals Unlimited, Inc. 27.23A: © Carolina Biological/Visuals Unlimited. 27.23B: © Marevision/AGE Fotostock. 27.24A: © J. Paulin/Visuals Unlimited. 27.24B: © Dr. David M. Phillips/Visuals Unlimited. 27.26: © Andrew Syred/SPL/Photo Researchers, Inc. 27.27A: © William Bourland/micro*scope. 27.27B: © David Patterson & Aimlee Laderman/micro*scope. 27.28A: © Eye of Science/Photo Researchers, Inc. 27.29: Courtesy of R. Blanton and M. Grimson.

**Chapter 28** *Silurian*: © Richard Bizley/Photo Researchers, Inc. *Rainforest*: © Photo Resource Hawaii/Alamy. 28.2A: © Bob Gibbons/OSF/Photolibrary.com. 28.2B: © Larry Mellichamp/Visuals Unlimited. 28.4: © J. Robert Waaland/Biological Photo Service. 28.6: © U. Michigan Exhibit Museum. 28.8A: Courtesy of the Biology Department Greenhouses, U. Massachusetts, Amherst. 28.9: After C. P. Osborne et al., 2004. *PNAS* 101: 10360. 28.11A: David McIntyre. 28.11C: © Harold Taylor/Photolibrary.com. 28.12A: © mediacolor's/Alamy. 28.12B: © Ed Reschke/Peter Arnold, Inc. 28.13A: © Dr. John D. Cunningham/Visuals Unlimited, Inc. 28.13B: © Danilo Donadoni/AGE Fotostock. 28.14: © Daniel Vega/AGE Fotostock. 28.15A: David McIntyre. 28.15B: © Carolina Biological/Visuals Unlimited. 28.16A: © Bjorn Svensson/SPL/Photo Researchers, Inc. 28.16B: © J. N. A. Lott/Biological Photo Service. 28.17: Courtesy of the Biology Department Greenhouses, U. Massachusetts, Amherst. 28.18A: © Michael & Patricia Fogden/Minden Pictures. 28.18B: © E.A. Janes/AGE Fotostock. 28.18C: Courtesy of the Talcott Greenhouse, Mount Holyoke College. 28.19 *inset*: David McIntyre.

**Chapter 29** *Masada*: © Eddie Gerald/Alamy. *Coconut*: © Ben Osborne/Naturepl.com. 29.1 *Cycad*: David McIntyre. 29.1 *Ginkgo*: © hypnotype/Shutterstock. 29.1 *Conifer*: © Irina Tischenko/istockphoto.com. 29.1 *Magnolia*: © Dole/Shutterstock. 29.4: © Natural Visions/Alamy. 29.5A: © Susumu Nishinaga/Photo Researchers, Inc. 29.6A: © Victoria Field/Shutterstock. 29.6B: © Topic Photo Agency IN/AGE Fotostock. 29.6C: © Pichugin Dmitry/Shutterstock. 29.6D: © Frans Lanting. 29.7A *left*: David McIntyre. 29.7A *right*: © Stan W. Elems/Visuals Unlimited. 29.7B *left*: David McIntyre. 29.7B *right*: © Dr. John D. Cunningham/Visuals Unlimited. 29.9: Courtesy of Jim Peaco/National Park Service. 29.10A: David McIntyre. 29.10B: © koer/Shutterstock. 29.10C: © Marion Nickig/Picture Press/Photolibrary.com. 29.11A: © Plantography/Alamy. 29.11B: © Thomas Photography LLC/Alamy. 29.16A: © bamby/Shutterstock. 29.16B: © blickwinkel/Alamy. 29.16C: David McIntyre. 29.16D: © Michel de Nijs/istockphoto.com. 29.16E: © Yuri Vainshtein/istockphoto.com. 29.16F: © Denis Pogostin/istockphoto.com. 29.18A: Photo by David McIntyre, courtesy of the University of Massachusetts Biology Department Greenhouses. 29.18B: David McIntyre. 29.18C: © WILDLIFE GmbH/Alamy. 29.18D: © blickwinkel/Alamy. 29.18E: David McIntyre. 29.18F: © Mike Donenfeld/Shutterstock. 29.19A: © amit erez/istockphoto.com. 29.19B: © Tootles/Shutterstock. 29.19C: © Charlotte Erpenbeck/Shutterstock. 29.20A, B: Courtesy of David Hillis. 29.20C: © Susan Law Cain/istockphoto.com.

**Chapter 30** *Fusarium*: © Dr. Gary Gaugler/Visuals Unlimited. *Ant*: © L. E. Gilbert/Biological Photo Service. 30.3: © David M. Phillips/Visuals Unlimited. 30.4B: © Dr. Jeremy Burgess/Photo Researchers, Inc. 30.6: © Richard Packwood/Photolibrary.com. 30.7A: © G. T. Cole/Biological Photo Service. 30.8: © N. Allin & G. L. Barron/Biological Photo Service. 30.9A: Courtesy of David Hillis. 30.9B: David McIntyre. 30.9C: Courtesy of David Hillis. 30.11A: © R. L. Peterson/Biological Photo Service. 30.11B: © Ken Wagner/Visuals Unlimited. 30.12A: © J. Robert Waaland/Biological Photo Service. 30.12B: © M. F. Brown/Visuals Unlimited. 30.12C: © Dr. John D. Cunningham/Visuals Unlimited. 30.12D: © Biophoto Associates/Photo Researchers, Inc. 30.13: © Eye of Science/Photo Researchers, Inc. 30.14: © John Taylor/Visuals Unlimited. 30.15: © G. L. Barron/Biological Photo Service. 30.16A: © Photoshot Holdings Ltd/Alamy. 30.16B: © Biosphoto/Le Moigne Jean-Louis/Peter Arnold Inc. 30.17: © Andrew Syred/SPL/Photo Researchers, Inc. 30.18A: © Jämsen Jorma/AGE Fotostock. 30.18B: © Matt Meadows/Peter Arnold Inc.

**Chapter 31** *Placozoan*: © Ana Yuri Signorovitch. *Sponge*: © Fred Bavendam/Minden Pictures. *Ctenophore*: © Norbert Wu/Minden Pictures. 31.2A: Courtesy of J. B. Morrill. 31.2B: From G. N. Cherr et al., 1992. *Microsc. Res. Tech.* 22: 11. Courtesy of J. B. Morrill. 31.4A: © Ed Robinson/Photolibrary.com. 31.4B: © Steve Gschmeissner/Photo Researchers, Inc. 31.4C: © DEA/Christian Ricci/Photolibrary.com. 31.5A: © Jurgen Freund/Naturepl.com. 31.5B: © John Bell/istockphoto.com. 31.6A: © John A. Anderson/istockphoto.com. 31.6B: © Kevin Schafer/DigitalVision/Photolibrary.com. 31.6B *inset*: © Mike Rogal/Shutterstock. 31.7B: © David Patterson & Aimlee Laderman/micro*scope. 31.8A: © IntraClique LLC/Shutterstock. 31.8B: © Stockbyte/PictureQuest. 31.9: Adapted from F. M. Bayerand & H. B. Owre, 1968. *The Free-Living Lower Invertebrates*, Macmillan Publishing Co. 31.10A: © Cathy Keifer/Shutterstock. 31.10B: David McIntyre. 31.12A: © First Light/Alamy. 31.12B: © Charlie Bishop/istockphoto.com. 31.13A: © Dave Watts/Naturepl.com. 31.13B: © Larry Jon Friesen. 31.14 *inset*: © Scott Camazine/Phototake/Alamy. 31.15: © Larry Jon Friesen. 31.16A: © Jurgen Freund/Naturepl.com. 31.16B: David McIntyre. 31.16C: © David Wrobel/Visuals Unlimited. 31.18B: © Larry Jon Friesen. 31.19: Adapted from F. M. Bayerand & H. B. Owre, 1968. *The Free-Living Lower Invertebrates*, Macmillan Publishing Co. 31.20A: © Larry Jon Friesen. 31.20B: © Georgette Douwma/Naturepl.com. 31.20C: © Larry Jon Friesen. 31.21A: © Jurgen Freund/Naturepl.com. 31.21B: © Stephan Kerkhofs/Shutterstock. 31.22: Adapted from F. M. Bayerand & H. B. Owre, 1968. *The Free-Living Lower Invertebrates*, Macmillan Publishing Co.

**Chapter 32** *Strepsipterans*: Courtesy of Dr. Hans Pohl. *Wasp*: © Sean McCann. 32.2: © blickwinkel/Alamy. 32.3A: From D. C. García-Bellido & D. H. Collins, 2004. *Nature* 429: 40. Courtesy of Diego García-Bellido Capdevila. 32.3B: © Piotr Naskrecki/Minden Pictures. 32.6: © Alexis Rosenfeld/Photo Researchers, Inc. 32.7A: © Larry Jon Friesen. 32.8B: © Robert Brons/Biological Photo Service. 32.9B: © Larry Jon Friesen. 32.10A: © Fred Bavendam/Minden Pictures. 32.11: © David Wrobel/Visuals Unlimited. 32.13A: © Larry Jon Friesen. 32.13B: Courtesy of Cindy Lee Van Dover. 32.13C: © Pakhnyushcha/Shutterstock. 32.13D: © Larry Jon Friesen. 32.15A: © Larry Jon Friesen. 32.15B: © Dave Fleetham/Photolibrary.com. 32.15C: © Larry Jon Friesen. 32.15F: © Reinhard Dirscherl/Alamy. 32.16A: Courtesy of Jen Grenier and Sean Carroll, U.Wisconsin. 32.16B: Courtesy of Graham Budd. 32.16C: Courtesy of Reinhardt Møbjerg Kristensen. 32.17B: © Grave/Photo Researchers, Inc. 32.17C: © Steve Gschmeissner/Photo Researchers, Inc. 32.18: © Pascal Goetgheluck/Photo Researchers, Inc. 32.19A: © Michael Fogden/Photolibrary.com. 32.19B: © Steve Gschmeissner/Photo Researchers, Inc. 32.20: © Kevin Schafer/Alamy. 32.21A: © David M Dennis/OSF/Photolibrary.com. 32.21B: © John R. MacGregor/Peter Arnold, Inc. 32.22A: © David Shale/Naturepl.com. 32.22B: © Frans Lanting. 32.23A: © Kelly Swift, www.swiftinverts.com. 32.23B: © Larry Jon Friesen. 32.23C: © Nigel Cattlin/Alamy. 32.23D: Photo by Eric Erbe; colorization by Chris Pooley/USDA ARS. 32.24A, B: © Larry Jon Friesen. 32.24C: © Solvin Zankl/Naturepl.com. 32.24D: © Larry Jon Friesen. 32.24E: © Norbert Wu/Minden Pictures. 32.27: © Oxford Scientific Films/Photolibrary.com. 32.28: © Mark Moffett/Minden Pictures. 32.29A: © John C. Abbott/Abbott Nature Photography. 32.29B: © Dr. Torsten Heydenreich/imagebroker/Alamy. 32.29C: © Larry Jon Friesen. 32.29D: David McIntyre. 32.29E: © Papilio/Alamy. 32.29F: © CorbisRF/Photolibrary.com. 32.29G: © Larry Jon Friesen. 32.29H: © Juniors Bildarchiv/Alamy.

**Chapter 33** *Gastric-brooding frog*: © Michael Tyler/ANTPhoto.com. *Marsupial frog*: © Michael Fogden/OSF/Photolibrary.com. 33.2: From S. Bengtson, 2000. Teasing fossils out of shales with cameras and computers. *Palaeontologia Electronica* 3(1). 33.4A: © Hal Beral/Visuals Unlimited. 33.4B: © Jose B. Ruiz/Naturepl.com. 33.4C: © WaterFrame/Alamy. 33.4D: © Peter Scoones/Photo Researchers, Inc. 33.4E: © Larry Jon Friesen. 33.5A: © C. R. Wyttenbach/Biological Photo Service. 33.6A: © Stan Elems/Visuals Unlimited, Inc. 33.6B: © Larry Jon Friesen. 33.7A: © Marevision/AGE Fotostock. 33.7B: © David Wrobel/Visuals Unlimited. 33.9A: © Brandon Cole Marine Photography/Alamy. 33.9B *left*: © Marevision/AGE Fotostock. 33.9B *right*: © anne de Haas/istockphoto.com. 33.11B: © Roger Klocek/Visuals Unlimited. 33.12A: © David B. Fleetham/OSF/Photolibrary.com. 33.12B: © Kelvin AitkenAGE Fotostock. 33.12C: © Norbert Wu/Minden Pictures. 33.13A: © Peter Pinnock/ImageState/Alamy. 33.13B, C: © Larry Jon Friesen. 33.13D: © Norbert Wu/Minden Pictures. 33.14A: © Peter Scoones/Photo Researchers, Inc. 33.14B: © Tom McHugh/Photo Researchers, Inc. 33.14C: © Ted Daeschler/Academy of Natural Sciences/VIREO. 33.16A: © Morley Read/Naturepl.com. 33.16B: © Michael & Patricia Fogden/Minden Pictures. 33.16C: © Jack Goldfarb/Design Pics, Inc./Photolibrary.com. 33.16D: Courtesy of David Hillis. 33.19A: © Dave B. Fleetham/OSF/Photolibrary.com. 33.19B: © C. Alan Morgan/Peter Arnold, Inc. 33.19C: © Cathy Keifer/Shutterstock. 33.19D: © Larry Jon Friesen. 33.20A: © Susan Flashman /istockphoto.com. 33.20B: © Gerry Ellis, DigitalVision/PictureQuest. 33.21A: From X. Xu et al., 2003. *Nature* 421: 335. © Macmillan Publishers Ltd. 33.21B: © Tom & Therisa Stack/Painet, Inc. 33.22: © Melinda Fawver/istockphoto.com. 33.23A: © Tim Zurowski/All Canada Photos/Photolibrary.com. 33.23B: © Roger Wilmhurst/Foto Natura/Minden Pictures. 33.23C: Courtesy of Andrew D. Sinauer. 33.23D: © Tom Vezo/Minden Pictures. 33.24A: © imagebroker.net/Photolibrary.com. 33.24B: © Dave Watts/Alamy. 33.25A: © Ingo Arndt/Naturepl.com. 33.25B: © JTB Photo Communications/Photolibrary.com. 33.25C: © Michael & Patricia Fogden/Minden Pictures. 33.26A: © Design Pics Inc/Photolibrary.com. 33.26B: © Barry Mansell/Naturepl.com. 33.26C: © Michael S. Nolan/AGE Fotostock. 33.26D: © John E Marriott/All Canada Photos/AGE Fotostock. 33.28: © Pete Oxford/Minden Pictures. 33.29A: © mike lane/Alamy. 33.29B: © Eric Isselée/Shutterstock. 33.30A: © Steve Bloom Images/Alamy. 33.30B: © Anup Shah/AGE Fotostock. 33.30C: © Lars Christensen/istockphoto.com. 33.30D: © Anup Shah/Minden Pictures.

**Chapter 34** *Svalbard*: Courtesy of the Global Crop Diversity Trust. *Seed*: © Dr. Richard Kessel & Dr. Gene Shih/Visuals Unlimited, Inc. 34.2A: David McIntyre. 34.2C: © Biosphoto/Thiriet Claudius/Peter Arnold Inc. 34.3A: David McIntyre. 34.3B: © Steven Wooster/Garden Picture Library/Photolibrary.com. 34.3C: © Renee Lynn/Photo Researchers, Inc. 34.6: © Biophoto Associates/Photo Researchers, Inc. 34.9A: © Biodisc/Visuals Unlimited. 34.9B: © P. Gates/Biological Photo Service. 34.9C: © Biophoto Associates/Photo Researchers, Inc. 34.9D: © Jack M. Bostrack/Visuals Unlimited. 34.9E: © John D. Cunningham/Visuals Unlimited. 34.9F: © J. Robert Waaland/Biological Photo Service. 34.9G: © Randy Moore/Visuals Unlimited. 34.10 *upper*: © Larry Jon Friesen. 34.10 *lower*: © Biodisc/Visuals Unlimited. 34.11B: © Microfield Scientific LTD/Photo Researchers, Inc. 34.13A: © Larry Jon Friesen. 34.13B: © Ed Reschke/Peter Arnold, Inc. 34.13C: © Dr. James W. Richardson/Visuals Unlimited. 34.14A *left*: © Ed Reschke/Peter Arnold, Inc. 34.14A *right*: © Biodisc/Visuals Unlimited. 34.14B: © Ed Reschke/Peter Arnold, Inc. 34.15B: Courtesy of Thomas Eisner, Cornell U. 34.15C: © Susumu Nishinaga/Photo Researchers, Inc. 34.18: © Biodisc/Visuals Unlimited. 34.19: © David M Dennis/OSF/Photolibrary.com.

**Chapter 35** *Planting*: © Robert Harding Picture Library Ltd/Alamy. *Drought*: Courtesy of the International Rice Research Institute/IRRI file photo. 35.4: © Nigel Cattlin/Alamy. 35.9A: © David M. Phillips/Visuals Unlimited. 35.10: After G. D. Humble & K. Raschke, 1971. *Plant Physiology* 48: 447. 35.12: © R. Kessel & G. Shih/Visuals Unlimited. 35.13: © M. H. Zimmermann.

**Chapter 36** *Family*: Dorothea Lange/Library of Congress, Prints & Photographs Division, FSA/OWI Collection. *Haiti*: Courtesy of the NASA/Goddard Space Flight Center Scientific Visualization Studio. 36.1: David McIntyre. 36.5: © Kathleen Blanchard/Visuals Unlimited. 36.9 *left*: © E. H. Newcomb & S. R. Tandon/Biological Photo Service. 36.9 *right*: © Dr. Jeremy Burgess/SPL/Photo Researchers, Inc. 36.12A: © J. H. Robinson/The National Audubon Society Collection/Photo Researchers, Inc. 36.12B: © Kim Taylor/Naturepl.com. 36.13: Courtesy of Susan and Edwin McGlew.

**Chapter 37** *N. Borlaug*: © Micheline Pelletier/Sygma/Corbis. *Rice*: Courtesy of Drs. Matsuoka and Ashikari. 37.2: © Visions of America/Joe Sohm/Photolibrary.com. 37.3: From J. M. Alonso and J. R. Ecker, 2006. *Nature Reviews Genetics* 7: 524. 37.4: Courtesy of J. A. D. Zeevaart, Michigan State U. 37.5: © Sylvan Wittwer/Visuals Unlimited, Inc. 37.12: © Ed Reschke/Peter Arnold, Inc. 37.16: David McIntyre.

**Chapter 38** *Girl*: © Image Source/ZUMA Press. *Postcards*: © Amoret Tanner/Alamy. 38.1A: David McIntyre. 38.1B1: © Tish1/Shutterstock. 38.1B2: © Pierre BRYE/Alamy. 38.1C: David McIntyre. 38.3A: © Biosphoto/Hazan Muriel/Peter Arnold Inc.

38.3B: © Rolf Nussbaumer/Imagebroker/Photolibrary.com. 38.5: © Biosphoto/Gautier Christian/Peter Arnold Inc. 38.9A: David McIntyre. 38.11: Courtesy of Richard Amasino. 38.17: Courtesy of Richard Amasino and Colleen Bizzell. 38.18A: © Nigel Cattlin, Holt Studios International/Photo Researchers, Inc. 38.18B: © Maurice Nimmo/SPL/Photo Researchers, Inc. 38.19: © Russ Munn/Agstockusa/AGE Fotostock.

**Chapter 39** *Engraving*: © Mary Evans Picture Library/Alamy. *Wormwood*: © Andy Crump, TDR, WHO/SPL/Photo Researchers, Inc. 39.1: Courtesy of Robert Bowden/Kansas State University. 39.1 *inset*: © Nigel Cattlin/Alamy. 39.4: © Holt Studios International Ltd/Alamy. 39.5A: © Kim Taylor/Naturepl.com. 39.5B: © Daniel Borzynski/Alamy. 39.7: After A. Steppuhn et al., 2004. *PLoS Biology* 2: 1074. 39.9: Courtesy of Thomas Eisner, Cornell U. 39.10: © Carr Clifton/Minden Pictures. 39.11: © Dr. Jack Bostrack/Visuals Unlimited, Inc. 39.12: © Martin Harvey/Peter Arnold Inc. 39.13: © Simon Fraser/SPL/Photo Researchers, Inc. 39.14: © imagebroker/Alamy. 39.15: © John N. A. Lott/Biological Photo Service. 39.18: © Jurgen Freund/Naturepl.com. 39.19: Courtesy of Ryan Somma.

**Chapter 40** *P. Radcliffe*: © PCN Black/Alamy. *Soldier*: Courtesy of Major Bryan "Scott" Robison. 40.3A: © Gladden Willis/Visuals Unlimited. 40.3B: From Ross, Pawlina, and Barnash, 2009. *Atlas of Descriptive Histology*. Sinauer Associates: Sunderland, MA. 40.3C: © Ed Reschke/Peter Arnold, Inc. 40.4A: From Ross, Pawlina, and Barnash, 2009. *Atlas of Descriptive Histology*. Sinauer Associates: Sunderland, MA. 40.4B: © Manfred Kage/Peter Arnold Inc. 40.4C: © SPL/Photo Researchers, Inc. 40.5A: © Ed Reschke/Peter Arnold, Inc. 40.5B: From Ross, Pawlina, and Barnash, 2009. *Atlas of Descriptive Histology*. Sinauer Associates: Sunderland, MA. 40.5C: © Dennis Kunkel Microscopy, Inc. 40.6A: © James Cavallini/Photo Researchers, Inc. 40.6B: © Innerspace Imaging/SPL/Photo Researchers, Inc. 40.10B: © Frans Lanting. 40.10C: © Akira Kaede/DigitalVision/Photolibrary.com. 40.11: © Greg Epperson/istockphoto.com. 40.12: © Gerry Ellis/DigitalVision. 40.14: Courtesy of Anton Stabentheiner. 40.17: From Ross, Pawlina, and Barnash, 2009. *Atlas of Descriptive Histology*. Sinauer Associates: Sunderland, MA. 40.18A: © Robert Shantz/Alamy. 40.18B: © Jim Brandenburg/Minden Pictures.

**Chapter 41** *J. Canseco*: © Ezra O. Shaw/Allsport/Getty Images. *Body builder*: © David Reed/Alamy. 41.2 *Prolactin*: Data from PDB 1RW5. K. Teilum et al., 2005. *J. Mol. Biol.* 351: 810. 41.2 *Fish*: © Alaska Stock LLC/Alamy. 41.2 *Amphibian*: © Gustav W. Verderber/Visuals Unlimited, Inc. 41.2 *Birds*: © Peter Bisset/Stock Connection Distribution/Alamy. 41.2 *Mammals*: © Ale Ventura/PhotoAlto/Photolibrary.com. 41.10A: © Ed Reschke/Peter Arnold, Inc. 41.10C: © Scott Camazine/Photo Researchers, Inc. 41.16: Courtesy of Gerhard Heldmaier, Philipps U.

**Chapter 42** *Painting*: © The Bridgeman Art Library/Getty Images. 42.8: © Steve Gschmeissner/Photo Researchers, Inc.

**Chapter 43** *Queen bee*: David McIntyre. *Swarm*: © Stephen Dalton/Minden Pictures. 43.1A: © P&R Photos/AGE Fotostock. 43.1B: © Constantinos Petrinos/Naturepl.com. 43.2A: © Patricia J. Wynne. 43.5: © SIU/Peter Arnold, Inc. 43.6: © ullstein - Ibis Bildagentur/Peter Arnold, Inc. 43.7A: © Morales/AGE Fotostock. 43.7B: © Dave Watts/Naturepl.com. 43.9B: © Ed Reschke/Peter Arnold, Inc. 43.12B: © P. Bagavandoss/Photo Researchers, Inc. 43.15C: © S. I. U. School of Med./Photo Researchers, Inc. 43.17: Courtesy of The Institute for Reproductive Medicine and Science of Saint Barnabas, New Jersey.

**Chapter 44** *Organs*: © Mads Abildgaard/istockphoto.com. *Egg and sperm*: © Dr. David M. Phillips/Visuals Unlimited. 44.1: Courtesy of Richard Elinson, U. Toronto. 44.3A *left*: From H. W. Beams and R. G. Kessel, 1976. *American Scientist* 64: 279. 44.3A *center, right*: © Dr. Lloyd M. Beidler/Photo Researchers, Inc. 44.3B: From H. W. Beams and R. G. Kessel, 1976. *American Scientist* 64: 279. 44.3C: Courtesy of W. Baker and G. Schubiger. 44.4: © Dr. Y. Nikas/Phototake/Alamy. 44.4: From J. G. Mulnard, 1967. *Arch. Biol.* (Liege) 78: 107. Courtesy of J. G. Mulnard. 44.14D: Courtesy of K. W. Tosney and G. Schoenwolf. 44.15B: Courtesy of K. W. Tosney. 44.19A: © CNRI/SPL/Photo Researchers, Inc. 44.19B: © Dr. G. Moscoso/SPL/Photo Researchers, Inc. 44.19C: © Tissuepix/SPL/Photo Researchers, Inc. 44.19D: © Petit Format/Photo Researchers, Inc.

**Chapter 45** *Dentist*: © Image Source/ZUMA Press. *Brain scan*: Courtesy of Dr. Kevin LaBar, Duke U. 45.4B: © C. Raines/Visuals Unlimited. 45.7: From A. L. Hodgkin & R. D. Keynes, 1956. *J. Physiol.* 148: 127.

**Chapter 46** *Snake*: © Casey K. Bishop/Shutterstock. *Bat*: © Stephen Dalton/Photo Researchers, Inc. 46.3A: © Hans Pfletschinger/Peter Arnold, Inc. 46.3B: David McIntyre. 46.10A: © P. Motta/Photo Researchers, Inc. 46.16A: © Dennis Kunkel Microscopy, Inc. 46.19: © Omikron/Science Source/Photo Researchers, Inc.

**Chapter 47** *London*: © Bluesky International Limited. *Rat*: Courtesy of Daoyun Ji and Matthew Wilson. 47.8: Photo from "Brain: The World Inside Your Head," © Evergreen Exhibitions. 47.14A: © David Joel Photography, Inc. 47.16: © Wellcome Dept. of Cognitive Neurology/SPL/Photo Researchers, Inc.

**Chapter 48** *J. Joyner-Kersee*: © AFP/Getty Images. *Frog*: © OSF/Photolibrary.com. 48.1 *Micrograph*: © Frank A. Pepe/Biological Photo Service. 48.2: © Tom Deerinck/Visuals Unlimited. 48.4: © Don W. Fawcett/Photo Researchers, Inc. 48.7: © Manfred Kage/Peter Arnold Inc. 48.8: © SPL/Photo Researchers, Inc. 48.11: Courtesy of Jesper L. Andersen. 48.18: © Robert Brons/Biological Photo Service.

**Chapter 49** *Elephant*: © Steve Bloom, stevebloom.com. *Shark*: © Thomas Aichinger/VW Pics/ZUMA Press. 49.1A: © Ross Armstrong/AGE Fotostock. 49.1B: © WaterFrame/Alamy. 49.1C: © Photoshot Holdings Ltd/Alamy. 49.4B: © Andrew Darrington/Alamy. 49.4C: Courtesy of Thomas Eisner, Cornell U. 49.10 *Bronchi*: © SPL/Photo Researchers, Inc. 49.10 *Alveoli*: © P. Motta/Photo Researchers, Inc. 49.13: © Ross Couper-Johnston/Naturepl.com. 49.17: After C. R. Bainton, 1972. *J. Appl. Physiol.* 33: 775.

**Chapter 50** *R. Lewis*: © NBAE/Getty Images. *Emergency room*: © BananaStock/Alamy. 50.8A: © Brand X Pictures/Alamy. 50.9: After N. Campbell, 1990. *Biology*, 2nd Ed., Benjamin Cummings. 50.10B: © CNRI/Photo Researchers, Inc. 50.12: © Science Source/Photo Researchers, Inc. 50.15A: © Chuck Brown/Science Source/Photo Researchers, Inc. 50.15B: © Biophoto Associates/Science Source/Photo Researchers, Inc.

**Chapter 51** *Pima*: © Marilyn "Angel" Wynn/Nativestock.com. *Burger*: © matt/Shutterstock. 51.1: Courtesy of Andrew D. Sinauer. 51.3: © Hartmut Schwarzbach/Peter Arnold Inc. 51.8C *Microvilli*: © Dennis Kunkel Microscopy, Inc. 51.13: © Eye of Science/SPL/Photo Researchers, Inc. 51.19: © Science VU/Jackson/Visuals Unlimited.

**Chapter 52** *Bat*: © Michael Fogden/OSF/Photolibrary.com. *Kangaroo rat*: © Mary McDonald/Naturepl.com. 52.1 *inset*: © Kim Taylor/Naturepl.com. 52.2B: © Morales/AGE Fotostock. 52.8A: © Susumu Nishinaga/Photo Researchers, Inc. 52.8B: © Science Photo Library/Photolibrary.com. 52.8C: © CNRI/SPL/Photo Researchers, Inc. 52.11: From L. Bankir & C. de Rouffignac, 1985. *Am. J. Physiol.* 249: R643-R666. Courtesy of Lise Bankir, INSERM Unit, Hôpital Necker, Paris. 52.14: © Hank Morgan/Photo Researchers, Inc.

**Chapter 53** *Monkey*: © Kennosuke Tsuda/Minden Pictures. *Spider*: © Don Johnston/All Canada Photos/Photolibrary.com. 53.2A: © Arco Images GmbH/Alamy. 53.5B: © Maximilian Weinzierl/Alamy. 53.8A: © Nina Leen/Time Life Pictures/Getty Images. 53.8B: © Frans Lanting. 53.12A: © Interfoto/Alamy. 53.12B: © Shaun Cunningham/Alamy. 53.12C: © Shattil & Rozinski/Naturepl.com. 53.14A: © Jeremy Woodhouse/Photolibrary.com. 53.14B: © Elvele Images Ltd/Alamy. 53.17B: © Tui De Roy/Minden Pictures. 53.20: © Tui De Roy/Minden Pictures. 53.22: © J. Jarvis/Visuals Unlimited, Inc. 53.23: © GeorgeLepp.com.

**Chapter 54** *Beetle*: David McIntyre. *Cattle*: © Robert Harding Picture Library Ltd/Alamy. 54.4A: Courtesy of Jan M. Storey. 54.4B *upper*: © jack thomas/Alamy. 54.4B *lower*: © Lee & Marleigh Freyenhagen/Shutterstock. *Tundra, left*: © John Schwieder/Alamy. *Tundra, right*: © Mike Harding/Imagestate/Photolibrary.com. *Ptarmigan*: © Rolf Hicker Photography/Alamy. *Boreal, left*: © John E Marriott/All Canada Photos/AGE Photostock. *Boreal, right*: © Robert Harding Picture Library Ltd/Alamy. *Boreal owl*: © John Cancalosi/Alamy. *Temperate deciduous, left*: © Jack Milchanowski/AGE Fotostock. *Temperate deciduous, right*: © Carr Clifton/Minden Pictures. *Butterfly*: © Rick & Nora Bowers/Alamy. *Temperate grasslands, left*: © Tim Fitzharris/Minden Pictures. *Temperate grasslands, right*: © Mikhail Yurenkov/Alamy. *Rhea*: © Biosphoto/Muñoz Juan-Carlos/Peter Arnold Inc. *Hot desert, left*: © Peter Lilja/AGE Fotostock. *Hot desert, right*: © David M. Schrader/Shutterstock. *Beetle*: © Michael

Fogden/OSF/Photolibrary.com. *Cold desert, left*: David McIntyre. *Cold desert, right*: © Tui De Roy/Minden Pictures. *Lizard*: Courtesy of Andrew D. Sinauer. *Chaparral*: © Robert Clay/Alamy. *Fynbos*: © Bob Gibbons/OSF/Photolibrary.com. *Sugarbird*: © Nigel Dennis/AGE Fotostock. *Thorn forest*: © Frans Lanting. *Savanna*: © Eyal Bartov/OSF/Photolibrary.com. *Termite mound*: Courtesy of Andrew D. Sinauer. *Tropical deciduous, left*: © Staffan Widstrand/Naturepl.com. *Tropical deciduous, right*: © Pete Oxford/Naturepl.com. *Hummingbird*: © Claude Steelman/OSF/Photolibrary.com. *Tropical evergreen, left*: © Luciano Candisani/Minden Pictures. *Tropical evergreen, right*: Courtesy of Andrew D. Sinauer. *Sloth*: © Kevin Schafer/AGE Fotostock. 54.9 *Baobab*: © Muriel Lasure/Shutterstock. 54.9 *Lemur*: © Daniel Cox/OSF/Photolibrary.com. 54.9 *Chameleon*: © Pete Oxford/Naturepl.com. 54.9 *Weevil*: © Nick Garbutt/Naturepl.com. 54.10: © Chris Mattison/Alamy. 54.12A: © Norbert Wu/Minden Pictures. 54.12B *left*: © Kenneth W. Fink/Photo Researchers, Inc. 54.12B *right*: © Keren Su/China Span/Alamy.

**Chapter 55** *Reindeer*: Courtesy of Dave Klein. *Wolf*: © Don Johnston/All Canada Photos/Photolibrary.com. 55.1A: Courtesy of Andrew D. Sinauer. 55.1B: © Heidi & Hans-Juergen Koch/Minden Pictures. 55.2 *inset*: © Kitch Bain/Shutterstock. 55.3A: © Rolf Hicker/Photolibrary.com. 55.3B: © Ilmarin/Shutterstock. 55.3C: © Ilya Bushuev/istockphoto.com. 55.5: © M. & C. Photography/Peter Arnold Inc. 55.8: After P. A. Marquet, 2000. *Science* 289: 1487. 55.9: © Wolfgang Poelzer/Photolibrary.com. 55.10 *inset*: © T. W. Davies/California Academy of Sciences. 55.12A: © Michael Durham/Minden Pictures. 55.12B: Courtesy of Colin Chapman. 55.14: © Kathie Atkinson/OSF/Photolibrary.com.

**Chapter 56** *Ants, Fungus*: © Mark Moffett/Minden Pictures. 56.1B *upper*: Courtesy of Andrew D. Sinauer. 56.1B *lower*: © Michael Mauro/Minden Pictures. 56.2: © Jodi Jacobson/Shutterstock. 56.3A: © E. A. Kuttapan/Naturepl.com. 56.3B: © Stephen Dalton/Minden Pictures. 56.3C: © OSF/Photolibrary.com. 56.4A *Caterpillar*: © Cathy Keifer/Shutterstock. 56.4A *Beetle, Aphids*: David McIntyre. 56.4B: © Kelpfish/Shutterstock. 56.4C: © Nicola Vernizzi/Shutterstock. 56.5A *left*: © Juniors Bildarchiv/Alamy. 56.5A *right*: © ranplett/istockphoto.com. 56.5B: Courtesy of Andrew Brower. 56.6A: David McIntyre. 56.6B: © Jason Bazzano/AGE Fotostock. 56.7: © Darlyne A. Murawski/Peter Arnold Inc. 56.8: © Eye of Science/Photo Researchers, Inc. 56.9: © blickwinkel/Alamy. 56.10: © Bjorn Rorslett/Photo Researchers, Inc. 56.11 *left*: © Wildlife/Peter Arnold, Inc. 56.11 *center, right*: Courtesy of Olle Pellmyr. 56.12: © ANT Photo Library/Photo Researchers, Inc. 56.13A, C: Courtesy of Erich G. Vallery, USDA Forest Service/Bugwood.org. 56.13B: Courtesy of James Denny Ward, USDA Forest Service/Bugwood.org. 56.14: © OSF/Photolibrary.com. 56.15 *left*: © Krystyna Szulecka Photography/Alamy. 56.15 *right*: © Pete Oxford/Minden Pictures.

**Chapter 57** *Flies*: © Michael Durham/Minden Pictures. *Potholes*: © Jim Brandenburg/Minden Pictures. 57.5B: Courtesy of William W. Dunmire/National Park Service. 57.6: © Jim Zipp/Photo Researchers, Inc. 57.10: Courtesy of E. O. Wilson. 57.13: Courtesy of the National Park Service.

**Chapter 58** *Crane*: © Mark Moffett/Minden Pictures. *Researchers*: Courtesy of Christian Koerner, U. Basel. 58.1: After M. C. Jacobson et al., 2000. *Introduction to Earth System Science: From Biogeochemical Cycles to Global Change*. Academic Press. 58.15: Courtesy of Goddard SVS/NASA.

**Chapter 59** *Moth*: © Photo Resource Hawaii/Alamy. *Haha*: Courtesy of Forest & Kim Starr. 59.1: © Mark Hallett Paleoart/Photo Researchers, Inc. 59.2B: © John Warburton-Lee Photography/Alamy. 59.6: Richard Bierregaard, Courtesy of the Smithsonian Institution, Office of Environmental Awareness. 59.7A: © Konrad Wothe/Photolibrary.com. 59.7B: © Georgette Douwma/Naturepl.com. 59.8: © Michael & Patricia Fogden/Minden Pictures. 59.9: © John Worrall. 59.12B: © Craig Hosterman/Shutterstock. 59.12C: © Lee Prince/Shutterstock. 59.13: © Richard Wong/www.rwongphoto.com/Alamy. 59.15: Courtesy of Christopher Baisan and the Laboratory of Tree-Ring Research, U. Arizona, Tucson. 59.16: After S. H. Reichard & C. W. Hamilton, 1997. *Conservation Biology* 11: 193. 59.17: © imagebroker.net/Photolibrary.com. 59.18A: © Dennis Cox/Alamy. 59.19A: Courtesy of the U.S. Fish and Wildlife Service. 59.19B: © Tom Vezo/Naturepl.com.

# Index

Numbers in *italic* indicate the information will be found in a figure, caption, or table.

## – E –

## – G –

## – H –

## – N –

## – O –



## – P –

# BIOPORTAL Resources—yourBioPortal.com

## ANIMATED TUTORIALS (AT) and WEB ACTIVITIES (WA)

# BIOPORTAL Resources—yourBioPortal.com

## ANIMATED TUTORIALS (AT) and WEB ACTIVITIES (WA)

| CHAPTER | PAGE | TITLE |
|---|---|---|
| **32** | 668 | Features of the Protostomes **(WA)** |
| **33** | 694 | Deuterostome Phylogeny **(WA)** |
| **33** | 704 | Life Cycle of a Frog **(AT)** |
| **33** | 706 | The Amniote Egg **(WA)** |
| **33** | 717 | Deuterostome Classification **(WA)** |
| **34** | 731 | Eudicot Root **(WA)** |
| **34** | 731 | Monocot Root **(WA)** |
| **34** | 732 | Eudicot Stem **(WA)** |
| **34** | 732 | Monocot Stem **(WA)** |
| **34** | 733 | Eudicot Leaf **(WA)** |
| **34** | 734 | Secondary Growth: The Vascular Cambium **(AT)** |
| **35** | 744 | Apoplast and Symplast of the Root **(WA)** |
| **35** | 746 | Xylem Transport **(AT)** |
| **35** | 752 | The Pressure Flow Model **(AT)** |
| **36** | 758 | Nitrogen and Iron Deficiencies **(AT)** |
| **36** | 766 | The Nitrogen Cycle **(WA)** |
| **37** | 773 | Monocot Shoot Development **(WA)** |
| **37** | 773 | Eudicot Shoot Development **(WA)** |
| **37** | 778 | Events of Seed Germination **(WA)** |
| **37** | 779 | Tropisms **(AT)** |
| **37** | 780 | Went's Experiment **(AT)** |
| **37** | 783 | Auxin Affects Cell Walls **(AT)** |
| **38** | 797 | Double Fertilization **(AT)** |
| **38** | 800 | Early Development of a Eudicot **(WA)** |
| **38** | 806 | The Effect of Interrupted Days and Nights **(AT)** |
| **39** | 816 | Signaling between Plants and Pathogens **(AT)** |
| **39** | 830 | Concept Matching **(WA)** |
| **40** | 845 | Thermoregulation in an Endotherm **(WA)** |
| **40** | 846 | The Hypothalamus **(AT)** |
| **41** | 856 | Complete Metamorphosis **(AT)** |
| **41** | 858 | The Human Endocrine Glands **(WA)** |
| **41** | 859 | The Hypothalamic–Pituitary–Endocrine Axis **(AT)** |
| **41** | 863 | Hormonal Regulation of Calcium **(AT)** |
| **41** | 871 | Concept Matching: Vertebrate Hormones **(WA)** |
| **42** | 875 | The Human Defense System **(WA)** |
| **42** | 876 | Cells of the Immune System **(AT)** |
| **42** | 879 | Inflammation Response **(WA)** |
| **42** | 880 | Pregnancy Test **(AT)** |
| **42** | 885 | Humoral Immune Response **(AT)** |
| **42** | 885 | Immunoglobulin Structure **(WA)** |
| **42** | 887 | Cellular Immune Response **(AT)** |
| **42** | 892 | A B Cell Builds an Antibody **(AT)** |
| **43** | 904 | Fertilization in a Sea Urchin **(AT)** |
| **43** | 909 | The Human Male Reproductive Tract **(WA)** |
| **43** | 910 | Spermatogenesis **(WA)** |
| **43** | 912 | The Human Female Reproductive Tract **(WA)** |
| **43** | 914 | The Ovarian and Uterine Cycles **(AT)** |
| **44** | 929 | Gastrulation **(AT)** |
| **44** | 931 | Tissue Transplants Reveal the Process of Determination **(AT)** |
| **44** | 937 | Extraembryonic Membranes **(WA)** |
| **45** | 948 | The Resting Membrane Potential **(AT)** |
| **45** | 953 | The Action Potential **(AT)** |
| **45** | 957 | Synaptic Transmission **(AT)** |
| **45** | 960 | Neurotransmitters **(WA)** |
| **46** | 972 | Structures of the Human Ear **(WA)** |
| **46** | 973 | Sound Transduction in the Human Ear **(AT)** |
| **46** | 976 | Photosensitivity **(AT)** |
| **46** | 979 | Structure of the Human Eye **(WA)** |
| **46** | 981 | Structure of the Human Retina **(WA)** |
| **47** | 988 | Information Processing in the Spinal Cord **(AT)** |
| **47** | 990 | The Human Cerebrum **(WA)** |
| **47** | 995 | Information Processing in the Retina **(AT)** |
| **47** | 1001 | Language Areas of the Cortex **(WA)** |
| **47** | 1003 | Structures of the Human Brain **(WA)** |
| **48** | 1008 | The Structure of a Sarcomere **(WA)** |
| **48** | 1009 | Molecular Mechanisms of Muscle Contraction **(AT)** |
| **48** | 1010 | The Neuromuscular Junction **(WA)** |
| **48** | 1014 | Smooth Muscle Action **(AT)** |
| **48** | 1021 | Joints **(WA)** |
| **49** | 1032 | Airflow in Birds **(AT)** |
| **49** | 1034 | The Human Respiratory System **(WA)** |
| **49** | 1036 | Airflow in Mammals **(AT)** |
| **49** | 1038 | Oxygen-Binding Curves **(WA)** |
| **49** | 1043 | Concept Matching **(WA)** |
| **50** | 1050 | Vertebrate Circulatory Systems **(WA)** |
| **50** | 1051 | The Human Heart **(WA)** |
| **50** | 1052 | The Cardiac Cycle **(AT)** |
| **50** | 1059 | Structure of a Blood Vessel **(WA)** |
| **51** | 1072 | Mineral Elements Required by Animals **(WA)** |
| **51** | 1073 | Vitamins in the Human Diet **(WA)** |
| **51** | 1075 | Mammalian Teeth **(WA)** |
| **51** | 1077 | The Human Digestive System **(WA)** |
| **51** | 1082 | The Digestion and Absorption of Fats **(AT)** |
| **51** | 1086 | Insulin and Glucose Regulation **(AT)** |
| **52** | 1096 | Annelid Metanephridia **(WA)** |
| **52** | 1099 | The Vertebrate Nephron **(WA)** |
| **52** | 1100 | The Mammalian Kidney **(AT)** |
| **52** | 1101 | The Human Excretory System **(WA)** |
| **52** | 1110 | The Major Organ Systems **(WA)** |
| **53** | 1124 | The Costs of Defending a Territory **(AT)** |
| **53** | 1126 | Foraging Behavior **(AT)** |
| **53** | 1128 | Circadian Rhythms **(AT)** |
| **53** | 1130 | Time-Compensated Solar Compass **(AT)** |
| **53** | 1133 | Honey Bee Dance Communication **(WA)** |
| **53** | 1138 | Concept Matching **(WA)** |
| **54** | 1146 | Biomes **(AT)** |
| **54** | 1157 | Rain Shadow **(AT)** |
| **54** | 1158 | Major Biogeographic Regions **(WA)** |
| **55** | 1175 | Exponential Population Growth **(AT)** |
| **55** | 1176 | Logistic Population Growth **(AT)** |
| **55** | 1176 | Logistic Population Growth **(WA)** |
| **55** | 1179 | Habitat Fragmentation **(AT)** |
| **56** | 1187 | Ecological Interactions **(WA)** |
| **56** | 1197 | Mutualism **(AT)** |
| **57** | 1205 | Energy Flow through an Ecological Community **(WA)** |
| **57** | 1206 | The Major Trophic Levels **(WA)** |
| **57** | 1212 | Biogeography Simulation **(AT)** |
| **57** | 1215 | Primary Succession on a Glacial Moraine **(AT)** |
| **58** | 1229 | The Global Hydrologic Cycle **(AT)** |
| **58** | 1230 | The Global Carbon Cycle **(AT)** |
| **58** | 1231 | The Ocean Conveyor Belt **(AT)** |
| **58** | 1233 | The Global Nitrogen Cycle **(AT)** |
| **58** | 1240 | Concept Matching **(WA)** |
| **59** | 1247 | Edge Effects **(AT)** |
| **59** | 1258 | Concept Matching **(WA)** |